M.I.T. WAVELENGTH TABLES

M.I.T. WAVELENGTH TABLES

Volume 2

Wavelengths by Element

prepared by
Frederick M. Phelps III

The MIT Press
Cambridge, Massachusetts
London, England

This book was printed and bound in the United States of America.

Library of Congress Cataloging in Publication Data

Main entry under title:
Massachusetts Institute of Technology wavelength tables.

Vol. 2 has title: M.I.T. wavelength tables.
Contents: v. 1. Tables of wavelengths—v. 2. Wavelengths by element / by Frederick M. Phelps III.
1. Spectrum analysis—Tables. I. Harrison, George Russell, 1898– . II. Phelps, Frederick M. III. Massachusetts Institute of Technology. Spectroscopy Laboratory. IV. United States. Work Projects Administration. V. Title. VI. Title: Wavelength tables. VII. Title: M.I.T. wavelength tables. VIII. Title: MIT wavelength tables.
QC453.M36 1969 535.8′4′0212 73-95288
ISBN 0-262-16087-0 (v. 2) AACR2

Contents

Introduction

This volume represents the first stage of a project, conceived over twenty years ago, to expand and update the *M.I.T. WAVELENGTH TABLES* compiled by George Harrison and his associates in the 1930s. It presents the same 109,325 atomic emission lines (as corrected in the 1969 edition) rearranged by element, with the addition of wavelengths in vacuum, calculated according to Bengt Edlen's dispersion formula, and corresponding wavenumbers. Intensities are presented as common logarithms with one decimal in the mantissa along with the characteristic. All of these data are now stored on computer-readable magnetic tape.

The title page of the 1939 edition of the tables presented a precise description of its contents and source: "Massachusetts Institute of Technology wavelength tables with intensities in arc, spark, or discharge tube of more than 100,000 spectrum lines most strongly emitted by the atomic elements under normal conditions of excitation between 10,000 A. and 2000 A. arranged in order of decreasing wavelengths, measured and compiled under the direction of George R. Harrison, Professor of Physics, by staff members of the Spectroscopy Laboratory of the Massachusetts Institute of Technology, assisted by the Works Progress Administration." The introductions to the 1939 and 1969 editions have been reprinted in the appendix to this volume.

The larger project to expand the tables had several motivations. First, separation techniques developed during World War II made very pure samples of lanthanide and actinide elements available, allowing a great increase in the accuracy of measurements of these elements. Second, several elements were added to the periodic table after work on the M.I.T. tables began, and it seemed appropriate to include the extensive line lists now available for such elements as neptunium and plutonium in the new tables. Third, Edlen's dispersion formula, published in the *Journal of the Optical Society of America* in 1953, allowed a calculation of wavelengths in vacuum from the original data of wavelengths in air. The wavenumber, which is the reciprocal of the wavelength in vacuum, is important in theoretical work because it is related to the energy levels of the emitting atom. Fourth, the use of modern digital computers would allow the manipulation of huge amounts of data easily and rapidly once the data had been entered on cards or magnetic tape. A computer could sort on any given parameter, compute vacuum wavelengths and wavenumbers, run routine searches, do error checks, and yield direct printouts to avoid typesetting errors. Because of the enormous capacity of computer tape storage, the use of a computer would also allow an extension of the tables into the infrared and ultraviolet and the addition of new data virtually without limit.

The magnitude of the task of expanding the M.I.T. tables to one million spectral lines made it an attractive challenge. The challenge remains to be met, largely because of a lack of funding. In 1965 the Michigan Wavelength Tables Advisory Committee was organized under the chairmanship of Ralph Sawyer and George Harrison. At meetings in Philadelphia in 1965 and in Washington, D.C., in 1966, the committee developed a multiphase program for updating and expanding the tables. This volume is the long-delayed first phase of that program.

Eight students from the Detroit Institute of Technology—W. E. Capers, C. M. Hale, R. A. Kwapisz, D. G. Connors, A. F. Krynicki, E. W. Borosky, W. B. Myers, and T. A. Anastasiou—transferred the original data line by line to computer-readable cards, creating a master file of 109,325 cards. Prudence Riley helped proofread many of these cards. Sandra Flemming worked an entire summer proofreading and correcting errors. My father, Frederick M. Phelps, Jr., proofread about 20,000 cards, and I devoted a sabbatical leave to work on the master card deck. Computer programs were written by Joyce Ablers, Ann Carey, and Sandra Warriner of Central Michigan University.

The spectral code used in the original tables was converted to single symbols for reasons of convenience. Our codes, with those used in the original tables in parentheses, are as follows:

A	(hd)	hazy double line
B	(hl)	hazy, asymmetric toward longer wavelengths
C	(hs)	hazy, asymmetric toward shorter wavelengths
D	(d)	double line
E	(rd)	narrow, self-reversed, double line
F	(rh)	narrow, self-reversed, hazy line
G	(R)	wide, self-reversed line
H	(h)	hazy, diffuse, or nebulous
I	(sd)	asymmetric toward shorter wavelengths, double line
J	(wh or hw)	wide or complex, hazy line
K	(wl)	wide, asymmetric toward longer wavelengths
L	(l)	asymmetric toward longer wavelengths
M	(wr)	complex, self-reversed
N	(ws)	wide, asymmetric toward shorter wavelengths
O	(W)	very wide
P	(hR or Rh)	wide, self-reversed, and hazy line
Q	(whl or hwl)	complex, hazy, asymmetric toward longer wavelengths
R	(r)	narrow, self-reversed
S	(s)	asymmetric toward shorter wavelengths
T	(whs)	complex, hazy, asymmetric toward shorter wavelengths
U	(wR)	complex, wide, self-reversed
V	(Wh)	very wide, hazy
W	(w)	wide or complex
X	(Wl)	very wide, asymmetric toward longer wavelengths
Y	(Ws)	very wide, asymmetric toward shorter wavelengths
Z	(WR or RW)	very wide, self-reversed
&	(Whl)	very wide, hazy, asymmetric toward longer wavelengths
#	(Wd)	very wide, double line
$	(wd)	wide, double line
*	(whd)	wide, hazy, double line
?	(H)	very hazy
@	(Rwh)	self-reversed, wide, hazy

I am grateful to Dewey Barich, President of the Detroit Institute of Technology, for encouraging me to begin work on this project and providing several hundred thousand IBM cards while I was on the faculty there from 1967 to 1969. I am most grateful to my colleagues at Central Michigan University who have assisted me in many ways with this task during the past twelve years and to the CMU Administration for many hours of CPU time in the university computers, for purchasing additional cards, student assistants, and especially for approving the sabbatical leave that allowed me to devote full time to the work for nearly six months.

Frederick M. Phelps III

Tables of Wavelengths and Intensities

Notes:
Wavelengths are all expressed in angstroms (A) $= 10^{-10}$ meters, a unit that is now giving way to nanometers. The wavenumber in kaysers (K) is the reciprocal of the wavelength in centimeters (cm). For example, if the wavelength in air is

$\lambda_{air} = 9057.51$ A,

then the wavelength in vacuum becomes

$\lambda_{vac} = 9060.00$ A $= 0.0000906000$ cm,

and the wavenumber is

$\sigma = 1/0.0000906000$ cm $= 11037.5$ K.

AIR WAVELENGTH (ANGSTRMS)	VACUUM WAVELENGTH (ANGSTRMS)	WAVENUMBER (KAYSERS)	LMENT	LOG ARC	INTENSITY SPK	INTENSITY DIS	REF
9951.88	9954.61	10045.6	A I			1.3	ME
9906.12	9908.84	10092.0	A II			2.0H	BN
9882.18	9884.89	10116.4	A I			0.8	ME
9829.86	9832.56	10170.3	A II			0.6	BN
9800.92	9803.61	10200.3	A I			0.6	ME
9784.501	9787.184	10217.44	A I			3.0	IME
9677.80	9680.45	10330.1	A I			0.9	ME
9673.39	9676.04	10334.8	A I			0.8	ME
9666.86	9669.51	10341.8	A I			1.7	ME
9657.784	9660.433	10351.50	A I			3.2	IME
9595.09	9597.72	10419.1	A I			0.6	ME
9561.60	9564.22	10455.6	A I			0.7	ME
9555.2	9557.8	10463.	A I			0.6	ME
9547.73	9550.35	10470.8	A I			0.3	ME
9486.02	9488.62	10538.9	A I			0.3	ME
9478.39	9480.99	10547.4	A I			1.7	ME
9475.20	9477.80	10551.0	A II			1.5	BN
9459.09	9461.68	10568.9	A I			2.0	ME
9446.57	9449.16	10582.9	A I			0.3	ME
9418.57	9421.15	10614.4	A II			0.3	BN
9408.66	9411.24	10625.6	A I			0.5	ME
9402.69	9405.27	10632.3	A I			1.3	ME
9377.63	9380.20	10660.7	A I			0.7	ME
9374.15	9376.72	10664.7	A II			1.0	BN
9362.50	9365.07	10678.0	A I			0.6	ME
9354.218	9356.785	10687.43	A I			2.3	IME
9340.59	9343.15	10703.0	A I			0.5	ME
9334.80	9337.36	10709.7	A I			0.9	ME
9328.08	9330.64	10717.4	A I			0.3	ME
9322.84	9325.40	10723.4	A			0.3	ME
9291.58	9294.13	10759.5	A I			2.0	ME
9279.72	9282.27	10773.2	A II			1.3	BN
9224.498	9227.029	10837.72	A I			3.0	IME
9221.08	9223.61	10841.7	A I			0.7H	ME
9210.37	9212.90	10854.3	A II			0.3	BN
9198.61	9201.13	10868.2	A			1.7	ME
9194.68	9197.20	10872.9	A I			2.2	ME
9180.17	9182.69	10890.1	A I			0.8	ME
9150.77	9153.28	10925.0	A II			0.3	BN
9138.45	9140.96	10939.8	A II			0.3	BN
9122.966	9125.470	10958.34	A I			2.7	IME
9075.42	9077.91	11015.7	A I			1.8	ME
9073.34	9075.83	11018.3	A I			1.7	ME
9066.77	9069.26	11026.3	A I			1.6	ME
9057.51	9060.00	11037.5	A I			0.3	ME
9057.23	9059.72	11037.9	A I			0.6H	ME
9017.59	9020.07	11086.4	A II			1.7	BN
8994.09	8996.56	11115.4	A I			1.0	ME
8988.20	8990.67	11122.6	A I			0.5	ME
8970.98	8973.44	11144.0	A I			0.3	ME
8967.39	8969.85	11148.5	A I			0.3	ME
8964.48	8966.94	11152.1	A I			1.0	ME
8962.19	8964.65	11154.9	A I			1.6	ME
8926.07	8928.52	11200.1	A II			0.5	BN
8904.53	8906.97	11227.2	A II			0.3	BN
8874.84	8877.28	11264.7	A I			0.6	ME
8849.97	8852.40	11296.4	A I			2.2	ME
8840.82	8843.25	11308.1	A I			1.3	ME
8840.39	8842.82	11308.6	A I			0.5	ME
8805.16	8807.58	11353.9	A			0.5	ME
8799.13	8801.55	11361.6	A I			2.0	ME
8784.59	8787.00	11380.4	A I			1.5	ME
8771.88	8774.29	11396.9	A II			2.0	BN
8761.72	8764.13	11410.2	A I			2.3	ME
8739.51	8741.91	11439.2	A I			0.5	ME
8736.63	8739.03	11442.9	A I			1.3	ME
8736.19	8738.59	11443.5	A I			0.3	ME
8713.79	8716.18	11472.9	A I			0.7	ME
8700.95	8703.34	11489.8	A I			0.5	ME
8690.12	8692.51	11504.2	A I			0.3	ME
8678.43	8680.81	11519.7	A I			1.8	ME
8667.943	8670.324	11533.59	A I			2.6	IME
8620.47	8622.84	11597.1	A I			2.0	ME
8605.78	8608.14	11616.9	A I			2.2	ME
8604.04	8606.40	11619.3	A II			0.9	BN
8579.49	8581.85	11652.5	A I			0.6	ME
8578.06	8580.42	11654.4	A I			0.7	ME
8561.38	8563.73	11677.2	A I			0.5	ME
8521.441	8523.782	11731.88	A I			3.3	IME
8496.64	8498.97	11766.1	A I			0.3	ME
8490.30	8492.63	11774.9	A I			1.6	ME
8443.44	8445.76	11840.3	A I			1.3	ME
8437.71	8440.03	11848.3	A I			0.8	ME
8424.647	8426.962	11866.67	A I			3.3	I
8408.208	8410.519	11889.87	A I			3.3	IME
8399.35	8401.66	11902.4	A I			1.3	ME
8392.28	8394.59	11912.4	A I			1.9	ME
8384.73	8387.03	11923.2	A I			1.8	ME
8367.03	8369.33	11948.4	A I			0.5	ME
8353.50	8355.80	11967.7	A I			0.6	ME
8332.21	8334.50	11998.3	A I			1.3	ME
8291.88	8294.16	12056.7	A I			0.9	ME
8264.521	8266.793	12096.59	A I			3.0	IME
8255.07	8257.34	12110.4	A I			1.7	ME
8249.58	8251.85	12118.5	A I			0.6	ME
8224.72	8226.98	12155.1	A I			0.8	ME
8217.85	8220.11	12165.3	A II			0.5	BN
8178.96	8181.21	12223.1	A I			1.3	ME
8178.84	8181.09	12223.3	A			1.6	ME
8171.95	8174.20	12233.6	A I			1.0	ME
8166.21	8168.46	12242.2	A			0.3	ME
8151.86	8154.10	12263.8	A I			0.5	ME
8143.50	8145.74	12276.4	A I			1.0	ME
8127.30	8129.53	12300.8	A			0.3	ME
8119.18	8121.41	12313.1	A II			1.7	ME
8115.311	8117.542	12319.00	A I			3.7	IME
8103.692	8105.920	12336.66	A I			3.3	IME
8094.06	8096.29	12351.3	A I			1.3	ME
8086.14	8088.36	12363.4	A II			0.3	BN
8082.93	8085.15	12368.3	A I			0.7	ME
8079.68	8081.90	12373.3	A I			1.3	ME
8066.60	8068.82	12393.4	A I			1.3	ME
8053.307	8055.522	12413.84	A I			2.0	ME
8046.13	8048.34	12424.9	A I			1.7	ME
8037.23	8039.44	12438.7	A I			1.3	ME
8021.9	8024.1	12462.	A I			0.3	ME
8017.55	8019.76	12469.2	A II			1.8	BN
8014.786	8016.990	12473.51	A I			2.9	IME
8006.156	8008.358	12486.95	A I			2.8	IME
7965.08	7967.27	12551.3	A I			0.5	ME
7960.84	7963.03	12558.0	A I			0.3	ME
7956.99	7959.18	12564.1	A I			1.0	ME
7948.175	7950.361	12578.05	A I			2.6	IME
7916.45	7918.63	12628.4	A I			1.3	ME
7910.23	7912.41	12638.4	A I			0.6	ME
7891.075	7893.246	12669.06	A I			2.0	IME
7868.20	7870.36	12705.9	A I			1.6	ME
7861.91	7864.07	12716.1	A I			1.2	ME
7860.44	7862.60	12718.4	A I			0.3	ME
7855.73	7857.89	12726.1	A I			0.9	ME
7849.42	7851.58	12736.3	A II			1.2	BN
7814.33	7816.48	12793.5	A I			1.0	ME
7798.55	7800.70	12819.4	A I			1.5	ME
7724.206	7726.332	12942.75	A I			2.3	IME
7723.760	7725.886	12943.50	A I			2.3	IME
7704.81	7706.93	12975.3	A I			1.3	ME
7690.10	7692.22	13000.2	A I			0.3	ME
7689.36	7691.48	13001.4	A			1.0	RT
7683.49	7685.60	13011.3	A II			0.3	BN
7670.04	7672.15	13034.2	A I			1.7	ME

AIR WAVELENGTH (ANGSTRMS)	VACUUM WAVELENGTH (ANGSTRMS)	WAVENUMBER (KAYSERS)	LMENT	LOG ARC	INTENSITY SPK	DIS	REF
7667.03	7669.14	13039.3	A I			0.6	ME
7662.3	7664.4	13047.	A I			0.3	ME
7654.06	7656.17	13061.4	A II			0.8	BN
7635.105	7637.207	13093.79	A I			2.7	IME
7628.86	7630.96	13104.5	A I			1.7	ME
7618.33	7620.43	13122.6	A I			1.5	ME
7618.03	7620.13	13123.1	A II			1.9	BN
7589.33	7591.42	13172.8	A II			2.4	BN
7514.651	7516.720	13303.68	A I			2.3	IME
7510.42	7512.49	13311.2	A I			1.0	MS
7505.13	7507.20	13320.6	A II			2.0	BN
7503.867	7505.933	13322.79	A I			2.8	IME
7491.88	7493.94	13344.1	A I			0.3	RS
7484.24	7486.30	13357.7	A I			1.2	MS
7471.18	7473.24	13381.1	A I			0.6	MS
7443.40	7445.45	13431.0	A I			1.4	MS
7440.46	7442.51	13436.3	A II			2.0	BN
7436.25	7438.30	13443.9	A I			1.0	MS
7435.33	7437.38	13445.6	A I			1.5	MS
7425.24	7427.29	13463.9	A I			1.1	MS
7422.26	7424.30	13469.3	A I			0.8	MS
7412.31	7414.35	13487.4	A I			1.2	MS
7392.97	7395.01	13522.6	A I			1.2	MS
7383.980	7386.014	13539.10	A I			2.6	IME
7380.45	7382.48	13545.6	A II			0.9	BN
7372.118	7374.149	13560.89	A I			2.0	I
7353.316	7355.342	13595.56	A I			2.0	IME
7350.78	7352.81	13600.2	A I			0.8	MS
7348.11	7350.13	13605.2	A II			0.6	RT
7316.00	7318.02	13664.9	A I			1.5	MS
7311.71	7313.72	13672.9	A I			2.0	MS
7285.44	7287.45	13722.2	A I			0.8	MS
7272.936	7274.940	13745.82	A I			2.0	IME
7270.66	7272.66	13750.1	A I			1.0	MS
7267.20	7269.20	13756.7	A I			0.3	MS
7265.23	7267.23	13760.4	A I			0.5	MS
7233.58	7235.57	13820.6	A II			0.9	BN
7229.93	7231.92	13827.6	A I			0.6	MS
7206.986	7208.972	13871.60	A I			2.0	ME
7202.55	7204.54	13880.2	A I			0.3	MS
7176.34	7178.32	13930.8	A I			0.6	MS
7162.57	7164.54	13957.6	A I			0.9	MS
7158.83	7160.80	13964.9	A I			1.5	MS
7147.041	7149.011	13987.95	A I			1.5	IME
7125.80	7127.76	14029.6	A I			1.5	MS
7107.496	7109.455	14065.78	A I			2.3	MS
7086.70	7088.65	14107.1	A I			1.2	MS
7077.03	7078.98	14126.3	A II			0.3	RT
7068.73	7070.68	14142.9	A I			1.5	MS
7067.217	7069.166	14145.94	A I			2.6	IME
7055.01	7056.96	14170.4	A II			0.6	RT
7030.262	7032.201	14220.30	A I			2.0	IME
6992.17	6994.10	14297.8	A I			0.6	MS
6990.16	6992.09	14301.9	A II			0.6	RT
6965.430	6967.351	14352.66	A I			2.6	IME
6960.23	6962.15	14363.4	A I			1.3	MS
6951.46	6953.38	14381.5	A I			1.3	MS
6937.666	6939.580	14410.09	A I			2.0	IME
6888.17	6890.07	14513.6	A I			2.0	MS
6887.10	6889.00	14515.9	A I			1.3	MS
6886.57	6888.47	14517.0	A II			1.2	RT
6879.59	6881.49	14531.7	A I			1.6	MS
6871.290	6873.186	14549.29	A I			2.2	IME
6863.52	6865.41	14565.8	A II			1.2	RT
6861.30	6863.19	14570.5	A II			0.9	RT
6851.86	6853.75	14590.6	A I			0.6	MS
6827.24	6829.12	14643.2	A I			1.5	MS
6818.39	6820.27	14662.2	A II			0.9	RT
6818.26	6820.14	14662.4	A I			0.6	MS
6808.55	6810.43	14683.4	A II			0.8	RT

AIR WAVELENGTH (ANGSTRMS)	VACUUM WAVELENGTH (ANGSTRMS)	WAVENUMBER (KAYSERS)	LMENT	LOG ARC	INTENSITY SPK	DIS	REF
6799.32	6801.20	14703.3	A II			0.6	RT
6779.85	6781.72	14745.5	A I			0.6	MS
6766.56	6768.43	14774.5	A I			2.0	MS
6756.61	6758.48	14796.2	A II			1.0	RT
6756.10	6757.96	14797.3	A I			2.0	MS
6754.30	6756.16	14801.3	A I			0.9	MS
6752.832	6754.696	14804.52	A I			2.3	IME
6722.90	6724.76	14870.4	A I			0.6	MS
6719.20	6721.05	14878.6	A I			2.0	MS
6698.85	6700.70	14923.8	A I			2.0	MS
6698.45	6700.30	14924.7	A I			0.8	MS
6689.91	6691.76	14943.8	A I			0.3	MS
6684.73	6686.58	14955.3	A I			0.8	MS
6684.36	6686.21	14956.2	A II			1.5	RT
6677.282	6679.126	14972.02	A I			1.5	IME
6672.10	6673.94	14983.7	A I			0.3	MS
6668.27	6670.11	14992.2	A I			0.9	RS
6666.36	6668.20	14996.6	A II			1.0	RT
6664.02	6665.86	15001.8	A I			2.0	MS
6660.64	6662.48	15009.4	A I			2.0	MS
6656.88	6658.72	15017.9	A I			0.8	MS
6643.79	6645.62	15047.5	A II			2.0	RT
6639.72	6641.55	15056.7	A II			1.3	RT
6638.24	6640.07	15060.1	A II			1.5	RT
6632.04	6633.87	15074.2	A I			0.9	MS
6623.78	6625.61	15092.9	A I			0.6	RS
6621.01	6622.84	15099.3	A			0.6	RT
6604.853	6606.677	15136.20	A I			1.5	MS
6604.02	6605.84	15138.1	A I			0.3	MS
6598.66	6600.48	15150.4	A I			0.8	MS
6596.10	6597.92	15156.3	A I			0.9	MS
6594.66	6596.48	15159.6	A I			0.3	MS
6581.60	6583.42	15189.7	A I			0.3	MS
6571.37	6573.19	15213.3	A I			0.3	MS
6545.11	6546.92	15274.4	A			0.3	MS
6538.115	6539.921	15290.70	A I			1.5	MS
6513.84	6515.64	15347.7	A I			0.9	MS
6500.25	6502.05	15379.8	A II			0.8	RT
6499.10	6500.90	15382.5	A I			0.8	MS
6493.97	6495.76	15394.7	A I			1.2	MS
6483.10	6484.89	15420.5	A II			1.2	RT
6481.15	6482.94	15425.1	A I			0.9	MS
6472.47	6474.26	15445.8	A			0.6	RT
6468.08	6469.87	15456.3	A			0.6	RT
6466.56	6468.35	15459.9	A I			1.3	MS
6443.89	6445.67	15514.3	A			0.6	RT
6441.95	6443.73	15519.0	A			0.8	RT
6437.63	6439.41	15529.4	A II			0.6	RT
6431.57	6433.35	15544.0	A I			1.2	MS
6422.94	6424.72	15564.9	A			0.7	RT
6418.43	6420.20	15575.8	A			0.6	RT
6416.315	6418.089	15580.96	A I			2.0	IME
6403.10	6404.87	15613.1	A			0.3	RT
6399.23	6401.00	15622.6	A II			0.9	RT
6396.63	6398.40	15628.9	A			0.3	RT
6384.719	6386.484	15658.07	A I			2.0	MS
6369.577	6371.338	15695.29	A I			1.5	MS
6364.89	6366.65	15706.8	A I			1.3	MS
6349.20	6350.96	15745.7	A I			0.3	MS
6348.27	6350.03	15748.0	A			0.3	RT
6333.21	6334.96	15785.4	A			0.6	RT
6332.51	6334.26	15787.2	A			0.6	RT
6324.45	6326.20	15807.3	A			0.8	RT
6309.14	6310.88	15845.6	A I			0.9	MS
6307.662	6309.406	15849.35	A I			1.5	MS
6296.876	6298.617	15876.50	A I			1.3	MS
6278.64	6280.38	15922.6	A I			0.8	MS
6248.40	6250.13	15999.7	A I			1.2	MS
6243.39	6245.12	16012.5	A I			0.8	MS
6243.13	6244.86	16013.2	A II			1.2	RT

AIR WAVELENGTH (ANGSTRMS)	VACUUM WAVELENGTH (ANGSTRMS)	WAVENUMBER (KAYSERS)	LMENT		LOG ARC	INTENSITY SPK	INTENSITY DIS	REF
6239.73	6241.46	16021.9	A	II			0.6	RT
6230.91	6232.63	16044.6	A	I			0.6	MS
6222.61	6224.33	16066.0	A	I			0.6	RS
6215.945	6217.665	16083.21	A	I			1.8	MS
6212.507	6214.226	16092.11	A	I			2.0	MS
6179.41	6181.12	16178.3	A	I			0.6	MS
6173.106	6174.814	16194.82	A	I			2.0	IME
6172.28	6173.99	16197.0	A				1.3	RT
6170.183	6171.890	16202.49	A	I			2.0	IME
6165.11	6166.82	16215.8	A	I			0.9	MS
6155.23	6156.93	16241.8	A	I			1.8	MS
6145.43	6147.13	16267.7	A	I			2.0	MS
6138.67	6140.37	16285.7	A	II			0.8	RT
6128.71	6130.41	16312.1	A	I			0.9	MS
6127.38	6129.08	16315.7	A	I			1.2	MS
6123.38	6125.07	16326.3	A	II			0.8	RT
6119.67	6121.36	16336.2	A	I			0.3	MS
6114.92	6116.61	16348.9	A	II			2.0	RT
6113.47	6115.16	16352.8	A	I			0.9	MS
6105.645	6107.335	16373.75	A	I			1.8	IME
6104.60	6106.29	16376.6	A	I			0.8	MS
6103.56	6105.25	16379.3	A	II			0.9	RT
6101.16	6102.85	16385.8	A	I			0.8	MS
6098.807	6100.495	16392.11	A	I			1.8	MS
6090.76	6092.45	16413.8	A	I			1.0	MS
6085.86	6087.54	16427.0	A	I			0.3	MS
6081.23	6082.91	16439.5	A	I			0.6	MS
6064.75	6066.43	16484.2	A	I			0.8	MS
6059.373	6061.051	16498.79	A	I			2.0	IME
6052.721	6054.397	16516.92	A	I			1.5	IME
6043.230	6044.903	16542.86	A	I			2.0	IME
6032.124	6033.794	16573.32	A	I			1.8	IME
6025.14	6026.81	16592.5	A	I			1.0	MS
6013.68	6015.35	16624.1	A	I			0.8	MS
6005.74	6007.40	16646.1	A	I			0.6	MS
5999.00	6000.66	16664.8	A	I			1.3	MS
5994.66	5996.32	16676.9	A	I			0.3	MS
5988.11	5989.77	16695.1	A	I			0.3	MS
5987.289	5988.947	16697.43	A	I			1.6	MS
5981.90	5983.56	16712.5	A	I			0.7	MS
5971.59	5973.24	16741.3	A	I			0.7	MS
5968.31	5969.96	16750.5	A	I			0.3	MS
5964.46	5966.11	16761.3	A				0.3	MS
5949.26	5950.91	16804.2	A	I			1.0	MS
5943.89	5945.54	16819.3	A	I			0.3	MS
5942.668	5944.314	16822.80	A	I			1.6	MS
5940.86	5942.51	16827.9	A	I			0.3	MS
5928.805	5930.448	16862.13	A	I			2.3	IME
5927.13	5928.77	16866.9	A	I			1.0	MS
5916.58	5918.22	16897.0	A	I			0.7	MS
5912.084	5913.722	16909.82	A	I			2.7	IME
5888.592	5890.224	16977.28	A	I			2.5	IME
5882.625	5884.255	16994.50	A	I			2.0	MS
5870.26	5871.89	17030.3	A	I			0.3	MS
5860.315	5861.939	17059.20	A	I			1.8	IME
5843.74	5845.36	17107.6	A	I			0.3	MS
5834.263	5835.880	17135.37	A	I			1.8	IME
5802.082	5803.691	17230.42	A	I			1.6	MS
5790.39	5792.00	17265.2	A	I			0.7	MS
5789.48	5791.09	17267.9	A	I			1.3	MS
5783.52	5785.12	17285.7	A	I			1.6	MS
5781.55	5783.15	17291.6	A	I			0.3	RS
5774.00	5775.60	17314.2	A	I			1.6	MS
5772.116	5773.717	17319.87	A	II			2.0	MS
5758.84	5760.44	17359.8	A	I			0.7	MS
5747.18	5748.77	17395.0	A	I			0.3	MS
5739.517	5741.109	17418.24	A	I			2.7	IME
5738.40	5739.99	17421.6	A	I			1.3	MS
5737.96	5739.55	17423.0	A	I			0.7	MS
5700.86	5702.44	17536.4	A	I			1.8	MS

AIR WAVELENGTH (ANGSTRMS)	VACUUM WAVELENGTH (ANGSTRMS)	WAVENUMBER (KAYSERS)	LMENT		LOG ARC	INTENSITY SPK	INTENSITY DIS	REF
5691.71	5693.29	17564.5	A	II			0.3	RT
5689.91	5691.49	17570.1	A	I			2.3	MS
5689.64	5691.22	17570.9	A	I			2.3	MS
5687.40	5688.98	17577.9	A	I			1.3	MS
5683.73	5685.31	17589.2	A	I			1.6	MS
5681.900	5683.477	17594.86	A	I			2.7	MS
5673.01	5674.58	17622.4	A				0.3	RT
5665.82	5667.39	17644.8	A	I			0.7	MS
5662.00	5663.57	17656.7	A	I			0.7	MS
5659.130	5660.701	17665.66	A	I			2.7	MS
5654.48	5656.05	17680.2	A				0.7	RT
5650.703	5652.271	17692.00	A	I			3.2	IME
5648.66	5650.23	17698.4	A	I			2.3	MS
5641.34	5642.91	17721.4	A	I			1.8	MS
5639.11	5640.68	17728.4	A	I			2.0	MS
5637.29	5638.85	17734.1	A	I			1.3	MS
5635.91	5637.47	17738.4	A				0.3	RT
5635.54	5637.10	17739.6	A	I			1.8	MS
5630.44	5632.00	17755.7	A	I			1.0	MS
5623.76	5625.32	17776.8	A	I			1.8	MS
5620.89	5622.45	17785.8	A	I			1.8	MS
5620.66	5622.22	17786.6	A	I			0.3	MS
5619.00	5620.56	17791.8	A				0.7	MS
5617.97	5619.53	17795.1	A	I			1.8	MS
5611.35	5612.91	17816.1	A	I			1.3	MS
5608.90	5610.46	17823.9	A	I			1.3	MS
5606.732	5608.289	17830.75	A	I			2.7	IME
5605.25	5606.81	17835.5	A	I			0.7	MS
5604.36	5605.92	17838.3	A	I			1.3	MS
5601.85	5603.41	17846.3	A	I			0.3	MS
5601.08	5602.64	17848.7	A	I			1.8	MS
5600.43	5601.98	17850.8	A	I			1.6	MS
5598.50	5600.05	17857.0	A	I			1.3	MS
5597.46	5599.01	17860.3	A	I			2.7	MS
5591.75	5593.30	17878.5	A				0.7	MS
5588.69	5590.24	17888.3	A	I			2.7	MS
5581.83	5583.38	17910.3	A	I			1.8	MS
5580.95	5582.50	17913.1	A				0.3	MS
5578.56	5580.11	17920.8	A				0.7	RT
5577.70	5579.25	17923.6	A	II			0.7	RT
5574.20	5575.75	17934.8	A	I			0.7	MS
5572.548	5574.095	17940.13	A	I			2.7	IME
5565.96	5567.51	17961.4	A	I			0.7	MS
5560.22	5561.76	17979.9	A	I			1.0	MS
5559.62	5561.16	17981.9	A	I			2.3	MS
5558.702	5560.246	17984.82	A	I			2.7	IME
5554.07	5555.61	17999.8	A				0.7	RT
5553.40	5554.94	18002.0	A	I			0.3	MS
5552.76	5554.30	18004.1	A	I			1.0	MS
5545.08	5546.62	18029.0	A				0.3	RT
5542.73	5544.27	18036.6	A	I			0.3	MS
5541.46	5543.00	18040.8	A	I			0.3	MS
5540.90	5542.44	18042.6	A	I			1.6	MS
5537.39	5538.93	18054.0	A				0.3	RT
5534.45	5535.99	18063.6	A	I			1.8	MS
5528.93	5530.47	18081.7	A	I			1.6	MS
5524.93	5526.46	18094.7	A	I			2.5	MS
5523.70	5525.23	18098.8	A	I			0.7	MS
5518.20	5519.73	18116.8	A	I			0.7	MS
5514.45	5515.98	18129.1	A				0.3	RT
5507.63	5509.16	18151.6	A	I			1.0	MS
5506.112	5507.642	18156.59	A	I			2.7	IME
5505.18	5506.71	18159.7	A	I			1.0	MS
5500.38	5501.91	18175.5	A				0.3	RT
5499.00	5500.53	18180.1	A	I			1.0	MS
5498.24	5499.77	18182.6	A				0.7	RT
5495.872	5497.399	18190.42	A	I			3.0	IME
5493.49	5495.02	18198.3	A	I			1.3	MS
5492.06	5493.59	18203.1	A	I			1.6	MS
5490.13	5491.66	18209.4	A	I			1.8	MS

AIR WAVELENGTH (ANGSTRMS)	VACUUM WAVELENGTH (ANGSTRMS)	WAVENUMBER (KAYSERS)	LMENT		LOG ARC	INTENS ITY SPK	DIS	REF
5488.46	5489.98	18215.0	A	I			0.3	MS
5486.47	5487.99	18221.6	A	I			1.3	MS
5483.32	5484.84	18232.1	A	I			1.0	MS
5473.44	5474.96	18265.0	A	I			2.7	MS
5469.65	5471.17	18277.6	A				1.3	MS
5469.20	5470.72	18279.1	A				0.3	RT
5467.13	5468.65	18286.1	A	I			0.8	MS
5459.61	5461.13	18311.2	A	I			1.3	MS
5457.75	5459.27	18317.5	A	I			1.0	MS
5457.37	5458.89	18318.7	A	I			2.3	MS
5456.01	5457.53	18323.3	A	I			0.7	MS
5454.41	5455.93	18328.7	A				1.0	RT
5451.650	5453.165	18337.97	A	I			2.7	IME
5449.31	5450.82	18345.9	A	I			0.7	RS
5448.61	5450.12	18348.2	A	I			1.0	MS
5443.88	5445.39	18364.1	A	I			1.3	MS
5443.21	5444.72	18366.4	A	I			2.0	MS
5442.22	5443.73	18369.7	A	I			2.7	MS
5439.97	5441.48	18377.4	A	I			2.7	MS
5430.27	5431.78	18410.2	A	I			1.0	MS
5429.69	5431.20	18412.1	A	I			1.3	MS
5422.55	5424.06	18436.4	A	I			0.3	MS
5421.346	5422.853	18440.48	A	I			2.7	IME
5417.22	5418.73	18454.5	A	I			1.0	MS
5413.32	5414.82	18467.8	A	I			1.0	MS
5410.470	5411.974	18477.55	A	I			2.7	MS
5407.44	5408.94	18487.9	A				1.3	RT
5402.69	5404.19	18504.1	A				1.0	RT
5399.01	5400.51	18516.8	A	I			1.3	MS
5397.60	5399.10	18521.6	A				1.0	RT
5393.971	5395.471	18534.06	A	I			2.3	MS
5390.72	5392.22	18545.2	A	I			1.6	MS
5389.10	5390.60	18550.8	A	I			1.6	MS
5387.37	5388.87	18556.8	A	I			1.6	MS
5373.493	5374.987	18604.70	A	I			2.7	MS
5369.97	5371.46	18616.9	A	I			0.7	MS
5356.49	5357.98	18663.7	A				1.0	MS
5353.46	5354.95	18674.3	A				1.3	MS
5350.58	5352.07	18684.4	A	I			1.3	MS
5347.412	5348.899	18695.44	A	I			2.3	MS
5345.81	5347.30	18701.0	A	I			1.3	MS
5344.28	5345.77	18706.4	A	I			0.7	MS
5341.78	5343.27	18715.1	A	I			1.0	MS
5328.02	5329.50	18763.5	A	I			1.3	MS
5324.80	5326.28	18774.8	A	I			0.7	MS
5317.726	5319.205	18799.80	A	I			1.8	MS
5309.517	5310.994	18828.87	A	I			2.3	MS
5305.77	5307.25	18842.2	A	II			1.0	RT
5296.32	5297.79	18875.8	A	I			0.7	MS
5290.00	5291.47	18898.3	A	I			1.3	MS
5286.92	5288.39	18909.3	A				1.3	RT
5286.08	5287.55	18912.4	A	I			1.8	MS
5283.43	5284.90	18921.8	A	I			1.3	MS
5281.66	5283.13	18928.2	A				0.3	RT
5280.40	5281.87	18932.7	A				1.8	MS
5279.05	5280.52	18937.5	A	I			1.3	MS
5267.48	5268.95	18979.1	A	I			0.3	MS
5263.02	5264.48	18995.2	A				0.3	MS
5254.476	5255.939	19026.10	A	I			1.8	MS
5252.786	5254.248	19032.22	A	I			2.5	IME
5249.20	5250.66	19045.2	A	I			1.6	MS
5246.76	5248.22	19054.1	A	I			0.7	MS
5246.24	5247.70	19056.0	A	I			1.6	MS
5242.13	5243.59	19070.9	A	I			0.3	MS
5241.096	5242.555	19074.67	A	I			1.8H	MS
5239.71	5241.17	19079.7	A	I			0.3	MS
5236.21	5237.67	19092.5	A	I			1.3	MS
5234.74	5236.20	19097.8	A	I			0.7	MS
5229.86	5231.32	19115.6	A	I			1.6	MS
5222.90	5224.35	19141.1	A	I			1.3	MS

AIR WAVELENGTH (ANGSTRMS)	VACUUM WAVELENGTH (ANGSTRMS)	WAVENUMBER (KAYSERS)	LMENT		LOG ARC	INTENS ITY SPK	DIS	REF
5221.270	5222.724	19147.10	A	I			2.7	IME
5219.30	5220.75	19154.3	A	I			1.6	MS
5216.84	5218.29	19163.4	A	II			1.0	RT
5216.28	5217.73	19165.4	A	I			1.8	MS
5214.768	5216.220	19170.97	A	I			2.3	MS
5210.488	5211.939	19186.72	A	I			2.3	MS
5208.04	5209.49	19195.7	A	I			1.0	MS
5207.17	5208.62	19198.9	A	I			1.0	MS
5205.79	5207.24	19204.0	A	I			1.0	MS
5198.96	5200.41	19229.3	A	I			0.3	MS
5194.77	5196.22	19244.8	A	I			1.3	MS
5194.02	5195.47	19247.6	A	I			0.7	MS
5192.72	5194.17	19252.4	A	I			1.8	MS
5187.746	5189.191	19270.83	A	I			2.9	IME
5177.535	5178.977	19308.83	A	I			1.6	MS
5176.28	5177.72	19313.5	A	II			1.0	RT
5165.82	5167.26	19352.6	A	II			1.3	RT
5162.80	5164.24	19363.9	A	II			0.3	RT
5162.284	5163.722	19365.88	A	I			2.7	IME
5159.69	5161.13	19375.6	A	I			1.0	MS
5153.11	5154.55	19400.4	A	I			1.3	MS
5151.395	5152.830	19406.81	A	I			2.3	MS
5145.36	5146.79	19429.6	A				2.3	RT
5141.81	5143.24	19443.0	A	II			1.3	MS
5141.81	5143.24	19443.0	A	I			1.3	MS
5134.17	5135.60	19471.9	A	I			0.3	MS
5127.78	5129.21	19496.2	A	I			1.8	MS
5121.88	5123.31	19518.6	A	I			0.7	MS
5118.200	5119.626	19532.68	A	I			1.8	MS
5104.74	5106.16	19584.2	A	I			1.3	MS
5099.64	5101.06	19603.8	A	I			0.7	MS
5098.97	5100.39	19606.3	A				1.3	MS
5093.32	5094.74	19628.1	A	I			1.0	MS
5090.55	5091.97	19638.8	A	II			0.7	RT
5087.087	5088.505	19652.14	A	I			1.8	MS
5082.74	5084.16	19668.9	A	I			1.3	MS
5081.44	5082.86	19674.0	A	I			1.0	MS
5078.03	5079.45	19687.2	A	I			1.6	MS
5073.08	5074.49	19706.4	A	I			2.3	MS
5071.30	5072.71	19713.3	A	I			0.7	MS
5070.99	5072.40	19714.5	A	I			1.6	MS
5069.66	5071.07	19719.7	A	I			0.7	MS
5068.39	5069.80	19724.6	A	I			0.7	MS
5065.48	5066.89	19736.0	A	I			0.7	MS
5063.99	5065.40	19741.8	A	I			0.7	MS
5062.07	5063.48	19749.3	A				2.3	RT
5060.08	5061.49	19757.0	A	I			2.7	MS
5056.53	5057.94	19770.9	A	I		2.3		MS
5054.18	5055.59	19780.1	A	I			2.5	MS
5048.813	5050.221	19801.11	A	I			2.7	MS
5047.30	5048.71	19807.1	A	I			0.3	MS
5044.15	5045.56	19819.4	A	I			0.3	MS
5041.23	5042.64	19830.9	A	I			1.0	MS
5040.51	5041.92	19833.7	A	I			1.0	MS
5035.88	5037.28	19852.0	A	I			0.7	MS
5034.25	5035.65	19858.4	A	I			1.0	MS
5032.025	5033.428	19867.18	A	I			1.8	MS
5029.64	5031.04	19876.6	A				0.7	MS
5017.63	5019.03	19924.2	A	II			0.7	RT
5017.25	5018.65	19925.7	A	I			0.7	MS
5017.16	5018.56	19926.0	A	II			1.8	RT
5009.35	5010.75	19957.1	A				2.3	RT
5007.09	5008.49	19966.1	A	I			0.3	MS
5006.84	5008.24	19967.1	A				0.3	MS
5004.318	5005.714	19977.17	A				1.3	MS
4989.945	4991.337	20034.71	A	I			1.9	MS
4985.09	4986.48	20054.2	A	I			1.0	MS
4975.66	4977.05	20092.2	A	I			0.3	MS
4974.18	4975.57	20098.2	A	I			1.0	MS
4973.53	4974.92	20100.8	A				0.3	MS

AIR WAVELENGTH (ANGSTRMS)	VACUUM WAVELENGTH (ANGSTRMS)	WAVENUMBER (KAYSERS)	LMENT	LOG ARC	INTENSITY SPK	INTENSITY DIS	REF
4972.16	4973.55	20106.4	A			1.3	RT
4965.12	4966.51	20134.9	A			1.6	RT
4956.753	4958.136	20168.87	A I			2.0	MS
4955.21	4956.59	20175.1	A I			0.3	MS
4951.75	4953.13	20189.2	A I			1.0	MS
4949.45	4950.83	20198.6	A II			0.3	RT
4944.80	4946.18	20217.6	A I			0.7	MS
4942.96	4944.34	20225.1	A II			0.7	RT
4937.718	4939.096	20246.62	A I			1.5	MS
4933.24	4934.62	20265.0	A			1.5	RT
4929.16	4930.54	20281.8	A I			0.3	MS
4921.042	4922.416	20315.23	A I			1.9	MS
4917.85	4919.22	20328.4	A I			0.7	MS
4914.32	4915.69	20343.0	A II			0.3	RT
4908.52	4909.89	20367.1	A I			1.0	MS
4904.75	4906.12	20382.7	A II			1.5	RT
4901.26	4902.63	20397.2	A			0.3	MS
4894.692	4896.059	20424.59	A I			2.2	MS
4889.06	4890.43	20448.1	A			0.5	RT
4887.947	4889.312	20452.77	A I			2.3	IME
4886.29	4887.65	20459.7	A I			1.5	MS
4883.86	4885.22	20469.9	A I			0.7	MS
4883.27	4884.63	20472.4	A I			1.5	MS
4882.25	4883.61	20476.6	A			1.0	RT
4879.90	4881.26	20486.5	A			2.5	RT
4878.37	4879.73	20492.9	A			0.3	MS
4876.260	4877.622	20501.79	A I			2.3	IME
4872.73	4874.09	20516.6	A I			1.0	MS
4867.84	4869.20	20537.3	A I			1.0	MS
4867.59	4868.95	20538.3	A II			0.7	RT
4865.91	4867.27	20545.4	A I			1.0	MS
4859.44	4860.80	20572.8	A			0.7	MS
4847.90	4849.25	20621.7	A			1.9	RT
4846.73	4848.08	20626.7	A I			0.7	MS
4836.691	4838.042	20669.52	A I			2.2	MS
4835.97	4837.32	20672.6	A I			1.5	MS
4834.10	4835.45	20680.6	A I			1.5	MS
4832.79	4834.14	20686.2	A I			0.7	MS
4832.38	4833.73	20688.0	A I			0.7	MS
4825.97	4827.32	20715.4	A I			0.3	MS
4806.07	4807.41	20801.2	A			2.7	RS
4804.33	4805.67	20808.7	A I			0.7	MS
4798.742	4800.083	20832.97	A I			1.5	MS
4792.12	4793.46	20861.8	A II			1.3	RT
4791.15	4792.49	20866.0	A I			0.3	MS
4786.19	4787.53	20887.6	A			0.3	RT
4770.34	4771.67	20957.0	A I			0.3	MS
4768.672	4770.005	20964.34	A I			2.2	IME
4764.89	4766.22	20981.0	A			2.2	RT
4752.938	4754.267	21033.74	A I			2.2	IME
4748.23	4749.56	21054.6	A I			0.7	MS
4746.823	4748.151	21060.83	A I			1.9	MS
4735.93	4737.25	21109.3	A			2.6	RT
4732.08	4733.40	21126.4	A II			0.7	RT
4730.66	4731.98	21132.8	A II			0.7	MS
4730.66	4731.98	21132.8	A I			0.7	MS
4727.48	4728.80	21147.0	A			0.7	MS
4726.91	4728.23	21149.6	A			2.3	RS
4724.10	4725.42	21162.1	A			0.7	MS
4721.62	4722.94	21173.2	A			1.0	RT
4719.94	4721.26	21180.8	A I			1.3	MS
4719.22	4720.54	21184.0	A I			0.3	MS
4718.10	4719.42	21189.0	A I			0.3	MS
4709.50	4710.82	21227.7	A I			1.5	MS
4709.08	4710.40	21229.6	A I			1.0	MS
4708.46	4709.78	21232.4	A I			0.3	MS
4703.36	4704.68	21255.4	A II			1.0	RT
4702.316	4703.632	21260.17	A I			3.1	I
4682.29	4683.60	21351.1	A II			1.0	RT
4666.28	4667.59	21424.4	A II			0.3H	RT

AIR WAVELENGTH (ANGSTRMS)	VACUUM WAVELENGTH (ANGSTRMS)	WAVENUMBER (KAYSERS)	LMENT	LOG ARC	INTENSITY SPK	INTENSITY DIS	REF
4657.94	4659.24	21462.7	A			2.2	RT
4651.388	4652.690	21492.94	A I			1.3	MS
4647.493	4648.794	21510.95	A I			1.6	MS
4642.148	4643.448	21535.72	A I			1.9	MS
4637.25	4638.55	21558.5	A II			1.5	RT
4628.441	4629.737	21599.50	A I			3.0	I
4626.78	4628.08	21607.2	A I			1.5	MS
4625.46	4626.76	21613.4	A I			1.0	MS
4611.25	4612.54	21680.0	A II			0.7	RT
4609.60	4610.89	21687.8	A II			2.5	RT
4598.77	4600.06	21738.9	A II			1.3	RT
4596.097	4597.385	21751.50	A I			3.0	I
4593.44	4594.73	21764.1	A II			0.3	RT
4589.93	4591.22	21780.7	A II			2.2	RT
4589.288	4590.574	21783.77	A I			1.9	IME
4587.90	4589.19	21790.4	A			0.3	RT
4587.21	4588.50	21793.6	A I			0.7	MS
4586.610	4587.895	21796.49	A I			1.0	MS
4584.958	4586.243	21804.34	A I			1.0	MS
4579.39	4580.67	21830.9	A			1.9	RT
4569.69	4570.97	21877.2	A I			0.3	MS
4564.82	4566.10	21900.5	A			0.3	MS
4564.43	4565.71	21902.4	A			1.3	RT
4563.78	4565.06	21905.5	A			1.3	RT
4561.03	4562.31	21918.7	A II			1.0	RT
4554.319	4555.596	21951.03	A I			1.2	MS
4547.78	4549.05	21982.6	A			1.3	RT
4545.08	4546.35	21995.6	A			2.3	RT
4544.746	4546.020	21997.26	A			1.5	MS
4541.60	4542.87	22012.5	A I			1.3	MS
4537.67	4538.94	22031.6	A II			1.0	RT
4535.51	4536.78	22042.1	A			1.0	RT
4534.78	4536.05	22045.6	A			1.3	MS
4530.57	4531.84	22066.1	A II			1.0	RT
4522.323	4523.591	22106.33	A I			2.9	I
4510.733	4511.998	22163.13	A I			3.0	I
4509.87	4511.13	22167.4	A			0.3D	MS
4507.45	4508.71	22179.3	A I			0.3	MS
4505.16	4506.42	22190.6	A			0.3	MS
4502.95	4504.21	22201.4	A			1.3	RT
4498.55	4499.81	22223.1	A II			1.0	RT
4490.99	4492.25	22260.6	A II			1.3	RT
4481.83	4483.09	22306.1	A II			1.9	RT
4479.31	4480.57	22318.6	A I			0.7	MS
4474.77	4476.03	22341.2	A II			1.3	RT
4461.46	4462.71	22407.9	A			0.7	MS
4460.53	4461.78	22412.6	A			1.3	MS
4456.61	4457.86	22432.3	A			0.3	MS
4448.88	4450.13	22471.3	A II			1.2	RT
4445.84	4447.09	22486.6	A I			0.7	MS
4440.09	4441.34	22515.7	A II			0.7	RT
4439.48	4440.73	22518.8	A II			0.3	MS
4433.83	4435.07	22547.5	A II			1.3	RT
4431.02	4432.26	22561.8	A			1.9	RT
4430.18	4431.42	22566.1	A			2.0	RT
4426.01	4427.25	22587.4	A			2.5	RT
4423.994	4425.236	22597.66	A I			1.9	IME
4420.90	4422.14	22613.5	A			1.6	RT
4404.91	4406.15	22695.6	A			0.3	RT
4401.02	4402.26	22715.6	A			1.6	RT
4400.09	4401.33	22720.4	A			1.5	RT
4394.65	4395.88	22748.6	A II			0.3H	RT
4385.08	4386.31	22798.2	A I			0.7	RT
4383.79	4385.02	22804.9	A			1.0	RT
4379.74	4380.97	22826.0	A			1.9	RT
4375.96	4377.19	22845.7	A			1.3	RT
4374.87	4376.10	22851.4	A			0.7	RT
4371.36	4372.59	22869.7	A			1.9	RT
4370.76	4371.99	22872.9	A II			1.5	RT
4368.36	4369.59	22885.4	A			0.7	MS

AIR WAVELENGTH (ANGSTRMS)	VACUUM WAVELENGTH (ANGSTRMS)	WAVENUMBER (KAYSERS)	LMENT	LOG ARC	INTENSITY SPK	INTENSITY DIS	REF
4367.87	4369.10	22888.0	A II			1.0	RT
4363.794	4365.020	22909.40	A I			1.9	ME
4362.07	4363.30	22918.4	A II			1.3	RT
4359.67	4360.90	22931.1	A II			0.3	RT
4352.23	4353.45	22970.3	A			1.5	RT
4348.11	4349.33	22992.0	A			2.7H	RT
4345.167	4346.388	23007.61	A I			3.0	I
4337.10	4338.32	23050.4	A II			1.5	RT
4336.51	4337.73	23053.5	A			0.7	RT
4335.337	4336.556	23059.77	A I			2.9	I
4333.560	4334.778	23069.23	A I			3.0	I
4332.06	4333.28	23077.2	A			1.9	RT
4331.25	4332.47	23081.5	A			2.3	RT
4310.47	4311.68	23192.8	A			1.3	MS
4309.25	4310.46	23199.4	A II			0.7	RT
4309.11	4310.32	23200.1	A II			0.3	RT
4300.66	4301.87	23245.7	A II			1.5	RT
4300.100	4301.310	23248.73	A I			3.1	I
4299.24	4300.45	23253.4	A			0.7	MS
4297.99	4299.20	23260.1	A II			1.3	RT
4294.97	4296.18	23276.5	A			1.3	MS
4289.09	4290.30	23308.4	A			0.7	MS
4282.90	4284.11	23342.1	A			1.6	RT
4277.55	4278.75	23371.3	A II			1.9	RT
4275.19	4276.39	23384.2	A			1.0	RT
4272.168	4273.370	23400.73	A I			3.1	I
4271.24	4272.44	23405.8	A			0.3	MS
4266.53	4267.73	23431.7	A			2.3	RT
4266.286	4267.487	23433.00	A I			3.1	I
4265.52	4266.72	23437.2	A			0.3	MS
4259.361	4260.560	23471.09	A I			3.1	I
4258.59	4259.79	23475.3	A			0.7	MS
4255.62	4256.82	23491.7	A			0.7	RT
4254.95	4256.15	23495.4	A			1.0	MS
4251.185	4252.382	23516.23	A I			2.9	I
4250.41	4251.61	23520.5	A			0.3	MS
4249.37	4250.57	23526.3	A			1.3	MS
4243.71	4244.90	23557.6	A II			0.3H	RT
4243.57	4244.76	23558.4	A			1.3	MS
4237.23	4238.42	23593.7	A II			1.6	RT
4229.89	4231.08	23634.6	A			1.0	RT
4228.14	4229.33	23644.4	A			1.6	RT
4227.02	4228.21	23650.7	A II			1.0	RT
4226.65	4227.84	23652.7	A II			0.3	RT
4222.67	4223.86	23675.0	A			1.3	RT
4218.69	4219.88	23697.4	A			1.3	RT
4217.45	4218.64	23704.3	A			1.0	RT
4203.43	4204.61	23783.4	A			1.3	RT
4201.99	4203.17	23791.5	A			1.3	RT
4201.58	4202.76	23793.9	A II			0.3	RT
4200.675	4201.859	23798.99	A I			3.1	I
4199.93	4201.11	23803.2	A II			0.7	RT
4198.317	4199.500	23812.36	A I			3.1	I
4191.028	4192.209	23853.77	A I			3.1	I
4190.712	4191.893	23855.57	A I			2.8	I
4189.67	4190.85	23861.5	A			1.0	RT
4181.883	4183.062	23905.94	A I			3.0	IHU
4179.31	4180.49	23920.6	A			1.3	RT
4178.39	4179.57	23925.9	A			1.3	RT
4176.33	4177.51	23937.7	A			1.3	MS
4175.40	4176.58	23943.1	A			1.0	MS
4168.98	4170.16	23979.9	A			0.7	RT
4168.70	4169.88	23981.5	A			0.3	MS
4168.41	4169.59	23983.2	A			0.3	MS
4164.179	4165.353	24007.57	A I			3.0	I
4158.590	4159.762	24039.83	A I			3.1	I
4156.11	4157.28	24054.2	A			1.3	RT
4154.50	4155.67	24063.5	A			1.9	MS
4152.54	4153.71	24074.9	A			1.3	MS
4131.73	4132.90	24196.1	A II			1.9	RT

AIR WAVELENGTH (ANGSTRMS)	VACUUM WAVELENGTH (ANGSTRMS)	WAVENUMBER (KAYSERS)	LMENT	LOG ARC	INTENSITY SPK	INTENSITY DIS	REF
4129.70	4130.86	24208.0	A			1.0	RT
4128.65	4129.81	24214.2	A			1.3	RT
4116.39	4117.55	24286.3	A II			1.0	RT
4114.52	4115.68	24297.3	A II			0.3	RT
4112.83	4113.99	24307.3	A			1.3	RT
4103.91	4105.07	24360.1	A			2.3	RT
4099.47	4100.63	24386.5	A II			0.7	RT
4097.15	4098.31	24400.3	A II			0.7	RT
4082.40	4083.55	24488.5	A			1.5	RT
4080.67	4081.82	24498.9	A			1.0	RT
4079.60	4080.75	24505.3	A II			1.3	RT
4076.96	4078.11	24521.2	A			1.0	RT
4076.64	4077.79	24523.1	A			1.3	RT
4072.40	4073.55	24548.6	A			1.6	RT
4072.01	4073.16	24551.0	A II			2.2	RT
4054.525	4055.670	24656.84	A I			1.9	I
4052.94	4054.08	24666.5	A II			1.3	RT
4047.51	4048.65	24699.6	A II			0.3	RT
4045.966	4047.109	24709.00	A I			2.2	IHU
4044.418	4045.561	24718.45	A I			3.1	I
4042.91	4044.05	24727.7	A II			1.9	RT
4038.82	4039.96	24752.7	A			1.6	RT
4035.47	4036.61	24773.3	A II			1.5	RT
4033.83	4034.97	24783.3	A			1.5	RT
4032.97	4034.11	24788.6	A I			1.3	MS
4031.41	4032.55	24798.2	A			0.3	RT
4013.87	4015.00	24906.6	A			2.3	RT
4011.23	4012.36	24923.0	A			0.7	RT
3994.81	3995.94	25025.4	A II			1.0	RT
3992.06	3993.19	25042.6	A			1.4	RT
3988.18	3989.31	25067.0	A			0.7	RT
3979.71	3980.84	25120.4	A I			1.0	MS
3979.36	3980.49	25122.6	A			1.4	RT
3974.76	3975.88	25151.6	A			1.2	RT
3974.48	3975.60	25153.4	A			1.0	RT
3968.36	3969.48	25192.2	A			2.3	RT
3958.39	3959.51	25255.6	A			1.0	RT
3952.74	3953.86	25291.7	A			1.2	RT
3948.979	3950.097	25315.84	A I			3.3	I
3947.504	3948.621	25325.29	A I			3.0	IHU
3946.10	3947.22	25334.3	A II			1.4	RT
3944.27	3945.39	25346.1	A			1.7	RT
3932.55	3933.66	25421.6	A			1.4	RT
3931.24	3932.35	25430.1	A			1.2	RT
3928.62	3929.73	25447.0	A			2.1	RT
3925.71	3926.82	25465.9	A II			0.5	RT
3917.77	3918.88	25517.5	A			0.5	RT
3914.76	3915.87	25537.1	A			1.4	RT
3911.58	3912.69	25557.9	A			1.0	RT
3900.63	3901.74	25629.6	A			1.0	RT
3899.86	3900.96	25634.7	A I			2.0	MS
3895.26	3896.36	25665.0	A II			0.5	RT
3894.660	3895.763	25668.91	A I			2.5	IHU
3891.97	3893.07	25686.6	A			1.4	RT
3891.40	3892.50	25690.4	A			1.2	RT
3880.34	3881.44	25763.6	A			0.7	RT
3876.07	3877.17	25792.0	A I			1.0	MS
3875.26	3876.36	25797.4	A			1.4	RT
3872.15	3873.25	25818.1	A			1.0	RT
3868.53	3869.63	25842.3	A		1.7		RT
3866.28	3867.38	25857.3	A I			0.7	MS
3864.26	3865.36	25870.8	A I			1.0	MS
3850.57	3851.66	25962.8	A			2.6	RT
3845.42	3846.51	25997.6	A			1.0	RT
3844.75	3845.84	26002.1	A			0.7	RT
3841.54	3842.63	26023.9	A			0.7	RT
3834.679	3835.767	26070.41	A I			2.9	IHU
3830.43	3831.52	26099.3	A II			1.0	RT
3826.83	3827.92	26123.9	A			1.2	RT
3825.70	3826.79	26131.6	A II			1.0	RT

AIR WAVELENGTH (ANGSTRMS)	VACUUM WAVELENGTH (ANGSTRMS)	WAVENUMBER (KAYSERS)	LMENT	LOG ARC	INTENSITY SPK	INTENSITY DIS	REF
3823.29	3824.37	26148.1	A			0.5	RT
3819.04	3820.12	26177.2	A II			0.7	RT
3809.49	3810.57	26242.8	A			1.4	RT
3808.61	3809.69	26248.9	A			1.0	RT
3803.19	3804.27	26286.2	A II			1.2	RT
3799.39	3800.47	26312.5	A			1.2	RT
3796.60	3797.68	26331.9	A II			0.7	RT
3786.40	3787.48	26402.8	A			1.2	RT
3781.361	3782.435	26438.00	A I			2.5	IHU
3780.84	3781.91	26441.6	A			1.7	RT
3777.55	3778.62	26464.7	A II			0.5	RT
3775.45	3776.52	26479.4	A I			1.0	MS
3774.54	3775.61	26485.8	A			0.5	RT
3770.54	3771.61	26513.9	A			1.2	RT
3770.369	3771.440	26515.07	A I			2.6	IHU
3766.13	3767.20	26544.9	A II			1.0	RT
3765.27	3766.34	26551.0	A			1.2	RT
3763.52	3764.59	26563.3	A			1.0	RT
3756.68	3757.75	26611.7	A			0.5	RT
3754.06	3755.13	26630.3	A II			0.5	RT
3753.53	3754.60	26634.0	A II			0.7	RT
3750.50	3751.57	26655.5	A			0.7	RT
3746.92	3747.99	26681.0	A			0.7	RT
3746.46	3747.52	26684.3	A			0.7	RT
3743.76	3744.82	26703.5	A I			2.0	MS
3737.89	3738.95	26745.5	A II			1.2	RT
3735.49	3736.55	26762.6	A			0.7	RT
3729.29	3730.35	26807.1	A			2.3	RT
3724.51	3725.57	26841.5	A II			0.7	RT
3720.43	3721.49	26871.0	A			1.0	RT
3718.21	3719.27	26887.0	A II			1.2	RT
3717.17	3718.23	26894.5	A			1.0	RT
3714.74	3715.80	26912.1	A			0.5	RT
3713.03	3714.09	26924.5	A II			0.5	RT
3709.90	3710.96	26947.2	A			0.7	RT
3706.94	3707.99	26968.7	A II			0.7	RT
3696.51	3697.56	27044.9	A I			1.3	MS
3690.896	3691.947	27085.98	A I			2.5	IHU
3682.56	3683.61	27147.3	A II			0.7	RT
3680.06	3681.11	27165.7	A II			1.0	RT
3678.27	3679.32	27179.0	A			1.0	RT
3675.22	3676.27	27201.5	A I			2.5	MS
3673.26	3674.31	27216.0	A			0.7	RT
3671.01	3672.06	27232.7	A II			0.5	RT
3670.64	3671.69	27235.4	A I			2.5	MS
3669.62	3670.67	27243.0	A			1.0	RT
3663.76	3664.80	27286.6	A I			0.7	MS
3660.44	3661.48	27311.3	A II			1.2	RT
3659.50	3660.54	27318.4	A I			2.0	MS
3656.05	3657.09	27344.1	A			1.0	RT
3655.29	3656.33	27349.8	A II			1.2	RT
3650.90	3651.94	27382.7	A			0.7	RT
3649.832	3650.872	27390.72	A I			2.9	IHU
3643.09	3644.13	27441.4	A I			2.0	MS
3639.85	3640.89	27465.8	A II			1.4	RT
3637.89	3638.93	27480.6	A			0.5	RT
3637.05	3638.09	27487.0	A			1.0	RT
3635.67	3636.71	27497.4	A			0.5	RT
3634.83	3635.87	27503.8	A II			0.7	RT
3634.461	3635.497	27506.56	A I			2.5	IHU
3632.684	3633.720	27520.01	A I			2.5	IHU
3622.15	3623.18	27600.0	A			1.2	RT
3611.84	3612.87	27678.8	A II			0.7	RT
3606.522	3607.551	27719.64	A I			3.0	IHU
3605.89	3606.92	27724.5	A II			1.2	RT
3603.91	3604.94	27739.7	A			0.5	RT
3603.46	3604.49	27743.2	A			0.5	RT
3601.51	3602.54	27758.2	A II			0.7	RT
3600.22	3601.25	27768.2	A II			0.5	RT
3599.67	3600.70	27772.4	A I			1.3	MS

AIR WAVELENGTH (ANGSTRMS)	VACUUM WAVELENGTH (ANGSTRMS)	WAVENUMBER (KAYSERS)	LMENT	LOG ARC	INTENSITY SPK	INTENSITY DIS	REF
3588.44	3589.46	27859.3	A			2.5	RT
3588.11	3589.13	27861.9	A I			0.5	MS
3582.70	3583.72	27903.9	A I			1.5	MS
3582.35	3583.37	27906.7	A			1.7	RT
3581.62	3582.64	27912.4	A			1.2	RT
3576.62	3577.64	27951.4	A			2.5	RT
3572.29	3573.31	27985.3	A I			2.5	MS
3567.657	3568.676	28021.60	A I			2.5	IHU
3565.02	3566.04	28042.3	A			1.0	RT
3564.27	3565.29	28048.2	A I			2.0	MS
3563.26	3564.28	28056.2	A I			2.0	MS
3562.19	3563.21	28064.6	A II			0.7	RT
3561.04	3562.06	28073.7	A II			1.2	RT
3559.51	3560.53	28085.7	A			1.2	MS
3556.91	3557.93	28106.3	A II			1.0	RT
3555.97	3556.99	28113.7	A I			2.0	MS
3554.306	3555.321	28126.85	A I			2.5	IHU
3553.58	3554.60	28132.6	A I			1.2	MS
3550.03	3551.04	28160.7	A			0.7	RT
3548.51	3549.52	28172.8	A			1.4	RT
3545.84	3546.85	28194.0	A II			2.1	RT
3545.58	3546.59	28196.1	A			2.5	RT
3543.16	3544.17	28215.3	A			1.0	RT
3535.33	3536.34	28277.8	A			1.2	RT
3531.22	3532.23	28310.7	A II			0.5	RT
3521.98	3522.99	28385.0	A II			0.7	RT
3521.27	3522.28	28390.7	A			1.0	RT
3520.00	3521.01	28401.0	A			1.2	RT
3517.90	3518.91	28417.9	A			0.5	RT
3514.39	3515.40	28446.3	A			2.1	RT
3511.15	3512.15	28472.6	A			0.5	RT
3509.78	3510.78	28483.7	A			1.2	RT
3506.46	3507.46	28510.6	A I			1.5	MS
3499.49	3500.49	28567.4	A			1.0	RT
3493.25	3494.25	28618.4	A I			1.3	MS
3491.54	3492.54	28632.5	A			1.7	RT
3491.24	3492.24	28634.9	A			1.2	RT
3490.89	3491.89	28637.8	A II			0.7	RT
3490.50	3491.50	28641.0	A I			0.5	MS
3487.33	3488.33	28667.0	A II			0.5	RT
3483.17	3484.17	28701.3	A I			0.7	MS
3480.52	3481.52	28723.1	A			0.7	RT
3478.24	3479.24	28741.9	A II			0.7	RT
3476.74	3477.74	28754.3	A			1.3	RT
3470.27	3471.26	28807.9	A			0.5	RT
3466.34	3467.33	28840.6	A			1.0	RT
3465.80	3466.79	28845.1	A II			0.5	RT
3464.14	3465.13	28858.9	A II			1.2	RT
3461.078	3462.069	28884.46	A I			2.5	IHU
3457.81	3458.80	28911.8	A I			0.5	MS
3454.94	3455.93	28935.8	A I			1.3	MS
3454.10	3455.09	28942.8	A			1.0	RT
3452.32	3453.31	28957.7	A			0.5	MS
3442.58	3443.57	29039.7	A I			1.0	MS
3430.44	3431.42	29142.4	A			0.7	RT
3429.64	3430.62	29149.2	A II			0.5	RT
3421.64	3422.62	29217.4	A			1.0	RT
3418.51	3419.49	29244.1	A I			0.5	MS
3417.68	3418.66	29251.2	A I			0.5	MS
3416.80	3417.78	29258.8	A I			0.7	MS
3414.46	3415.44	29278.8	A II			0.7	RT
3406.17	3407.15	29350.1	A I			1.5	MS
3397.90	3398.88	29421.5	A I			1.3	MS
3397.90	3398.88	29421.5	A II			1.3	MS
3393.752	3394.726	29457.46	A I			2.4	IHU
3392.81	3393.78	29465.6	A I			2.0	MS
3392.31	3393.28	29470.0	A I			0.5	MS
3390.29	3391.26	29487.5	A I			0.5	MS
3389.85	3390.82	29491.4	A I			1.3	MS
3388.54	3389.51	29502.8	A			1.4	RT

AIR WAVELENGTH (ANGSTRMS)	VACUUM WAVELENGTH (ANGSTRMS)	WAVENUMBER (KAYSERS)	LMENT	LOG ARC	INTENSITY SPK	DIS	REF
3388.35	3389.32	29504.4	A I			1.3	MS
3387.58	3388.55	29511.1	A I			1.3	MS
3381.49	3382.46	29564.3	A I			1.3	MS
3379.58	3380.55	29581.0	A			1.0	RT
3379.48	3380.45	29581.9	A II			0.5H	RT
3376.46	3377.43	29608.3	A II			1.4	RT
3373.481	3374.450	29634.46	A I			2.5	IHU
3372.88	3373.85	29639.7	A I			0.5	MS
3370.97	3371.94	29656.5	A			1.0	RT
3366.59	3367.56	29695.1	A			0.7	RT
3365.54	3366.51	29704.4	A II			0.7	RT
3363.47	3364.44	29722.7	A I			1.3	MS
3361.73	3362.70	29738.0	A II			0.7	RT
3359.48	3360.45	29758.0	A I			1.0	MS
3358.56	3359.53	29766.1	A			0.5	RT
3350.94	3351.90	29833.8	A II			0.7	RT
3344.79	3345.75	29888.6	A			0.5	RT
3336.20	3337.16	29965.6	A			0.7	RT
3325.50	3326.46	30062.0	A I			2.0	MS
3323.82	3324.78	30077.2	A I			1.5	MS
3322.44	3323.40	30089.7	A I			0.7	MS
3321.58	3322.54	30097.5	A I			0.7	MS
3320.06	3321.02	30111.3	A I			0.5	MS
3319.345	3320.300	30117.76	A I			2.5	IHU
3312.97	3313.92	30175.7	A			0.5	RT
3311.26	3312.21	30191.3	A			0.7	RT
3310.47	3311.42	30198.5	A			0.5	MS
3309.39	3310.34	30208.4	A			0.5	RT
3307.24	3308.19	30228.0	A			1.2	RT
3306.50	3307.45	30234.8	A			0.5	RT
3301.87	3302.82	30277.1	A			1.0	RT
3300.39	3301.34	30290.7	A I			1.3	MS
3293.95	3294.90	30349.9	A			0.7	RT
3293.66	3294.61	30352.6	A			1.4	RT
3291.47	3292.42	30372.8	A			0.7	RT
3289.95	3290.90	30386.9	A I			0.5	MS
3289.39	3290.34	30392.0	A I			0.5	MS
3285.87	3286.82	30424.6	A			1.0	RT
3281.72	3282.67	30463.1	A			1.2	RT
3279.25	3280.19	30486.0	A I			0.5	MS
3278.93	3279.87	30489.0	A I			0.5	MS
3273.36	3274.30	30540.9	A			0.7	RT
3271.16	3272.10	30561.4	A I			1.0	MS
3269.05	3269.99	30581.1	A II			0.7	RT
3264.29	3265.23	30625.7	A I			0.5	MS
3263.78	3264.72	30630.5	A I			0.5	MS
3263.60	3264.54	30632.2	A			1.0	RT
3259.71	3260.65	30668.7	A			0.7	RT
3257.58	3258.52	30688.8	A I			2.0	MS
3249.82	3250.76	30762.1	A			1.4	RT
3247.55	3248.49	30783.6	A			0.5	RT
3243.71	3244.65	30820.0	A			1.3	MS
3236.82	3237.75	30885.6	A			0.7	RT
3234.51	3235.44	30907.7	A I			2.0	MS
3229.91	3230.84	30951.7	A I			0.5	MS
3226.00	3226.93	30989.2	A			0.5	RT
3225.58	3226.51	30993.2	A I			1.3	MS
3222.42	3223.35	31023.6	A II			0.5	RT
3221.64	3222.57	31031.1	A II			0.5	RT
3217.70	3218.63	31069.1	A			0.5	RT
3216.75	3217.68	31078.3	A			1.0	RT
3212.54	3213.47	31119.0	A			1.0	RT
3207.61	3208.54	31166.9	A II			1.0	RT
3207.50	3208.43	31167.9	A I			1.0	MS
3205.03	3205.96	31191.9	A II			0.7	RT
3204.34	3205.27	31198.7	A			1.0	RT
3203.66	3204.59	31205.3	A I			1.0	MS
3202.85	3203.78	31213.2	A I			0.7	MS
3201.12	3202.05	31230.0	A I			0.5	MS
3200.39	3201.31	31237.2	A I			2.0	MS

AIR WAVELENGTH (ANGSTRMS)	VACUUM WAVELENGTH (ANGSTRMS)	WAVENUMBER (KAYSERS)	LMENT	LOG ARC	INTENSITY SPK	DIS	REF
3195.12	3196.04	31288.7	A I			0.7	MS
3194.25	3195.17	31297.2	A			1.0	RT
3186.63	3187.55	31372.0	A I			0.7	MS
3186.19	3187.11	31376.4	A II			0.5	RT
3181.05	3181.97	31427.1	A			1.4	RT
3172.96	3173.88	31507.2	A I			2.2	MS
3172.18	3173.10	31514.9	A I			0.7	MS
3169.68	3170.60	31539.8	A			1.7	RT
3165.31	3166.23	31583.3	A			0.7	RT
3161.45	3162.37	31621.9	A			0.5	RT
3161.38	3162.30	31622.6	A II			0.7	RT
3160.06	3160.97	31635.8	A I			0.7	MS
3153.80	3154.71	31698.6	A II			0.7	RT
3152.29	3153.20	31713.8	A I			0.5	MS
3151.52	3152.43	31721.5	A I			0.5	MS
3142.60	3143.51	31811.6	A I			0.5	MS
3139.02	3139.93	31847.9	A			1.4	RT
3137.66	3138.57	31861.6	A II			0.7	RT
3136.55	3137.46	31872.9	A			0.7	RT
3132.87	3133.78	31910.4	A I			0.5	MS
3130.80	3131.71	31931.5	A I			1.3	MS
3120.06	3120.96	32041.4	A I			0.5	MS
3117.85	3118.75	32064.1	A			0.5	MS
3116.63	3117.53	32076.6	A			0.5	MS
3110.66	3111.56	32138.2	A I			0.5	MS
3109.75	3110.65	32147.6	A			0.5	RT
3104.38	3105.28	32203.2	A II			1.0	RT
3102.63	3103.53	32221.4	A			0.5	RT
3100.09	3100.99	32247.8	A I			0.7	MS
3094.98	3095.88	32301.0	A			0.5	RT
3093.41	3094.31	32317.4	A II			1.7	RT
3088.24	3089.14	32371.5	A II			1.0	RT
3085.05	3085.95	32405.0	A			0.7	RT
3082.99	3083.89	32426.6	A II			0.7	RT
3066.89	3067.78	32596.9	A			0.5	RT
3063.44	3064.33	32633.6	A I			0.7	MS
3062.06	3062.95	32648.3	A I			0.5	MS
3060.94	3061.83	32660.2	A			1.0	RT
3056.28	3057.17	32710.0	A I			0.5	MS
3053.20	3054.09	32743.0	A			0.5	RT
3046.10	3046.99	32819.3	A			0.7	RT
3033.52	3034.40	32955.4	A II			0.9	RT
3028.93	3029.81	33005.4	A II			1.2	RT
3026.75	3027.63	33029.1	A II			0.7	RT
3014.49	3015.37	33163.5	A II			1.0	RT
3003.00	3003.88	33290.3	A			0.5	RT
3000.45	3001.32	33318.6	A II			1.0	RT
3000.14	3001.01	33322.1	A			0.7	RT
2979.05	2979.92	33558.0	A II			1.6	RT
2960.27	2961.13	33770.8	A			1.3	RT
2957.56	2958.42	33801.8	A			0.7	RT
2956.55	2957.41	33813.3	A			1.0	RT
2955.39	2956.25	33826.6	A II			1.6	RT
2942.90	2943.76	33970.2	A II			2.0	RT
2935.57	2936.43	34055.0	A II			0.7	RT
2932.60	2933.46	34089.5	A II			1.3	RT
2931.49	2932.35	34102.4	A			1.3	RT
2924.66	2925.52	34182.0	A			1.3	RT
2915.62	2916.47	34288.0	A			0.7	RT
2897.30	2898.15	34504.8	A			1.0	RT
2896.75	2897.60	34511.3	A			1.3	RT
2893.98	2894.83	34544.4	A			0.3	RT
2891.61	2892.46	34572.7	A II			1.6	RT
2884.21	2885.06	34661.4	A			0.7	RT
2879.31	2880.15	34720.4	A			0.7	RT
2878.76	2879.60	34727.0	A			0.3	RT
2874.55	2875.39	34777.9	A			0.7	RT
2865.85	2866.69	34883.4	A II			1.3	RT
2860.73	2861.57	34945.9	A			0.7	RT
2855.33	2856.17	35011.9	A			0.7	RT

AIR WAVELENGTH (ANGSTRMS)	VACUUM WAVELENGTH (ANGSTRMS)	WAVENUMBER (KAYSERS)	LMENT	LOG ARC	INTENSITY SPK	INTENSITY DIS	REF
2847.81	2848.65	35104.4	A II			0.7	RT
2844.12	2844.96	35149.9	A II			0.7	RT
2843.37	2844.21	35159.2	A			0.3	RT
2806.16	2806.99	35625.4	A			1.6	RT
2796.65	2797.47	35746.5	A			0.3	RT
2795.45	2796.27	35761.9	A			0.3	RT
2795.33	2796.15	35763.4	A			0.3	RT
2769.74	2770.56	36093.8	A			1.3	RT
2767.99	2768.81	36116.6	A			0.3	RT
2764.66	2765.48	36160.1	A II			1.0	RT
2762.18	2763.00	36192.6	A			0.7	RT
2757.26	2758.07	36257.2	A			0.7	RT
2754.91	2755.72	36288.1	A II			0.3	RT
2753.92	2754.73	36301.1	A			1.0	RT
2744.82	2745.63	36421.5	A			1.3	RT
2741.09	2741.90	36471.0	A			0.3	RT
2733.06	2733.87	36578.2	A			0.7	RT
2732.53	2733.34	36585.3	A			1.3	RT
2724.84	2725.65	36688.5	A			0.3	RT
2708.28	2709.08	36912.9	A			1.6	RT
2692.62	2693.42	37127.5	A II			1.3	RT
2683.14	2683.94	37258.7	A			0.3	RT
2651.95	2652.74	37696.9	A			0.3	RT
2647.29	2648.08	37763.2	A			1.0	RT
2627.41	2628.19	38049.0	A II			0.3	RT
2624.63	2625.41	38089.2	A			0.7	RT
2621.03	2621.81	38141.6	A			1.0	RT
2617.64	2618.42	38191.0	A			0.3	RT
2570.46	2571.23	38891.9	A			0.7	RT
2570.01	2570.78	38898.7	A II			0.3	RT
2569.25	2570.02	38910.2	A			0.3	RT
2564.45	2565.22	38983.0	A II			1.0	RT
2562.12	2562.89	39018.5	A II			1.3	RT
2559.31	2560.08	39061.3	A			0.7	RT
2556.63	2557.40	39102.3	A			0.7	RT
2549.84	2550.61	39206.4	A II			0.3	RT
2544.72	2545.48	39285.3	A			1.6	RT
2540.08	2540.84	39357.0	A			0.3	RT
2536.04	2536.80	39419.7	A II			1.3	RT
2535.28	2536.04	39431.5	A II			0.3	RT
2534.74	2535.50	39439.9	A			1.6	RT
2528.71	2529.47	39534.0	A II			0.7	RT
2528.33	2529.09	39539.9	A			1.0	RT
2525.51	2526.27	39584.1	A			0.7	RT
2522.53	2523.29	39630.8	A II			0.7	RT
2516.81	2517.57	39720.9	A II			1.3	RT
2515.60	2516.36	39740.0	A II			1.0	RT
2500.42	2501.17	39981.2	A			0.7	RT
2499.55	2500.30	39995.1	A II			0.3	RT
2491.06	2491.81	40131.5	A			0.7	RT
2482.17	2482.92	40275.2	A II			0.7	RT
2481.50	2482.25	40286.0	A			0.7	RT
2480.87	2481.62	40296.3	A			1.0	RT
2479.08	2479.83	40325.4	A II			0.7	RT
2475.48	2476.23	40384.0	A II			0.7	RT
2474.02	2474.77	40407.8	A			0.7	RT
2459.97	2460.71	40638.6	A II			0.7	RT
2454.27	2455.01	40733.0	A			1.0	RT
2440.07	2440.81	40970.0	A II			0.3	RT
2438.78	2439.52	40991.7	A			0.3	RT
2430.06	2430.80	41138.8	A II			0.7	RT
2424.70	2425.44	41229.7	A II			0.3	RT
2424.00	2424.74	41241.6	A			0.3	RT
2422.73	2423.47	41263.2	A			0.3	RT
2420.49	2421.23	41301.4	A II			1.0	RT
2415.80	2416.53	41381.6	A			0.3H	RT
2414.26	2414.99	41408.0	A II			0.7	RT
2412.50	2413.23	41438.2	A			0.3	RT
2410.97	2411.70	41464.5	A II			0.7	RT
2405.27	2406.00	41562.7	A			0.3	RT

AIR WAVELENGTH (ANGSTRMS)	VACUUM WAVELENGTH (ANGSTRMS)	WAVENUMBER (KAYSERS)	LMENT	LOG ARC	INTENSITY SPK	INTENSITY DIS	REF
2404.40	2405.13	41577.8	A II			1.0	RT
2399.26	2399.99	41666.8	A			0.3	RT
2398.39	2399.12	41682.0	A			0.3	RT
2390.94	2391.67	41811.8	A			1.0	RT
2389.11	2389.84	41843.8	A			0.7	RT
2387.96	2388.69	41864.0	A II			1.6	RT
2385.00	2385.73	41915.9	A			1.6	RT
2383.50	2384.23	41942.3	A II			1.6	RT
2381.18	2381.91	41983.2	A			0.3	RT
2379.88	2380.61	42006.1	A			0.3	RT
2377.2	2377.9	42053.	A			0.3	RT
2376.3	2377.0	42069.	A			0.3	RT
2371.74	2372.46	42150.3	A			1.3	RT
2369.97	2370.69	42181.7	A			0.3	RT
2369.28	2370.00	42194.0	A II			0.3	RT
2366.78	2367.50	42238.6	A			1.0	RT
2364.14	2364.86	42285.8	A			1.6	RT
2360.07	2360.79	42358.7	A II			1.6	RT
2357.60	2358.32	42403.0	A			1.8	RT
2354.79	2355.51	42453.6	A II			0.7	RT
2354.14	2354.86	42465.4	A			1.3	RT
2353.42	2354.14	42478.4	A			1.0	RT
2352.73	2353.45	42490.8	A			0.7	RT
2350.50	2351.22	42531.1	A			1.3	RT
2344.22	2344.94	42645.0	A			0.6	RT
2339.80	2340.52	42725.6	A			1.0	RT
2337.79	2338.51	42762.3	A			1.8	RT
2331.45	2332.17	42878.6	A			1.8	RT
2324.40	2325.11	43008.6	A			0.7	RT
2322.11	2322.82	43051.1	A			0.7	RT
2317.77	2318.48	43131.7	A II			0.3	RT
2317.37	2318.08	43139.1	A			1.0	RT
2316.31	2317.02	43158.9	A			2.0	RT
2314.99	2315.70	43183.5	A			1.6	RT
2313.74	2314.45	43206.8	A			1.8	RT
2309.16	2309.87	43292.5	A			1.6	RT
2304.5	2305.2	43380.	A			0.3	RT
2304.0	2304.7	43389.	A			0.3	RT
2300.19	2300.90	43461.3	A II			1.0	RT
2300.07	2300.78	43463.5	A			0.3	RT
2298.17	2298.88	43499.5	A			1.6	RT
2293.27	2293.98	43592.4	A			1.3	RT
2289.78	2290.49	43658.9	A			1.3	RT
2288.4	2289.1	43685.	A			0.3	RT
2287.8	2288.5	43697.	A			1.8	RT
2287.19	2287.90	43708.3	A			1.3	RT
2284.91	2285.62	43751.9	A			0.7	RT
2284.02	2284.72	43769.0	A			1.0	RT
2282.64	2283.34	43795.4	A II			1.3	RT
2282.05	2282.75	43806.7	A			1.6	RT
2281.2	2281.9	43823.	A			1.3	RT
2280.6	2281.3	43835.	A			1.8	RT
2272.6	2273.3	43989.	A			0.3	RT
2261.8	2262.5	44199.	A II			1.0	RT
2252.87	2253.57	44374.1	A			0.3	RT
2252.26	2252.96	44386.1	A II			0.7	RT
2250.3	2251.0	44425.	A II			1.3	RT
2248.0	2248.7	44470.	A			1.3	RT
2245.97	2246.67	44510.4	A			0.7	RT
2244.8	2245.5	44534.	A			0.3	RT
2238.2	2238.9	44665.	A II			1.8	RT
2235.77	2236.46	44713.4	A II			1.6	RT
2229.5	2230.2	44839.	A II			0.3	RT
2227.70	2228.39	44875.4	A			1.8	RT
2224.76	2225.45	44934.7	A			1.6	RT
2222.93	2223.62	44971.7	A			1.8	RT
2221.94	2222.63	44991.7	A			1.6	RT
2218.87	2219.56	45054.0	A			1.8	RT
2215.2	2215.9	45129.	A			0.3	RT
2213.0	2213.7	45173.	A			1.8	RT

AIR WAVELENGTH (ANGSTRMS)	VACUUM WAVELENGTH (ANGSTRMS)	WAVENUMBER (KAYSERS)	LMENT		LOG ARC	INTENSITY SPK	DIS	REF
2212.22	2212.91	45189.4	A				1.3	RT
2208.9	2209.6	45257.	A				1.0	RT
2203.22	2203.91	45374.0	A				0.3	RT
2202.77	2203.46	45383.2	A	II			0.3	RT
2197.56	2198.25	45490.8	A				1.0	RT
2187.34	2188.02	45703.3	A				1.3	RT
2183.7	2184.4	45779.	A				1.3	RT
2183.3	2184.0	45788.	A	II			1.6	RT
2182.74	2183.42	45799.6	A	II			0.7H	RT
2179.25	2179.93	45873.0	A	II			1.6	RT
2177.04	2177.72	45919.5	A				1.3	RT
2171.41	2172.09	46038.6	A	II			1.8	RT
2170.8	2171.5	46051.	A	II			0.3	RT
2164.17	2164.85	46192.6	A				2.0	RT
2162.74	2163.42	46223.1	A				1.8H	RT
2159.0	2159.7	46303.	A	II			1.8H	RT
2152.6	2153.3	46441.	A	II			1.8H	RT
2148.15	2148.83	46537.0	A				0.7	RT
2140.01	2140.68	46714.0	A				1.0	RT
2139.2	2139.9	46732.	A				0.7	RT
2137.2	2137.9	46775.	A				1.6	RT
2131.0	2131.7	46911.	A				0.3	RT
2122.6	2123.3	47097.	A				0.7	RT
4812.25	4813.59	20774.5	AC			1.8		LX
4413.17	4414.41	22653.1	AC			2.0		LX
4386.37	4387.60	22791.5	AC			2.0		LX
4359.09	4360.32	22934.1	AC			1.5		LX
4179.93	4181.11	23917.1	AC			1.8		LX
4168.40	4169.58	23983.3	AC			2.0		LX
4088.37	4089.52	24452.7	AC			2.0		LX
8273.519	8275.793	12083.43	AG	I	0.7			IHZ
7687.779	7689.895	13004.08	AG	I	1.3			IHZ
5666.34	5667.91	17643.2	AG		0.7	0.3		WT
5608.95	5610.51	17823.7	AG	I	0.7H			BX
5545.94	5547.48	18026.2	AG	I	1.5L			BX
5471.551	5473.071	18271.28	AG	I	2.7H	2.0		MIT
5465.487	5467.006	18291.55	AG	I	3.0G	2.7G		MIT
5434.2	5435.7	18397.	AG		0.3	1.0H		KP
5333.73	5335.21	18743.4	AG	I	0.9			BX
5329.73	5331.21	18757.5	AG	I	0.3H			KP
5276.47	5277.94	18946.8	AG	I	0.8			BX
5209.067	5210.517	19191.95	AG	I	3.2G	3.0G		MIT
5123.68	5125.11	19511.8	AG		0.6	1.5		KP
4888.28	4889.65	20451.4	AG		1.0	1.3		KP
4874.18	4875.54	20510.5	AG		1.5			KP
4848.156	4849.510	20620.64	AG			0.3H		MIT
4677.87	4679.18	21371.3	AG		0.3H	0.0H		KP
4668.483	4669.790	21414.24	AG	I	2.3	1.8		MIT
4620.48	4621.77	21636.7	AG			0.6		MIT
4620.02	4621.31	21638.9	AG			0.6H		MIT
4615.86	4617.15	21658.4	AG		0.6H	0.0H		KP
4476.079	4477.335	22334.72	AG	I	1.6	0.9		MIT
4396.32	4397.55	22739.9	AG		2.0			KP
4395.930	4397.165	22741.93	AG		1.0	1.5		MIT
4385.056	4386.288	22798.32	AG			1.1		MIT
4379.25	4380.48	22828.6	AG		0.7	0.0		MIT
4311.074	4312.287	23189.55	AG		0.7	1.4L		MIT
4212.68	4213.87	23731.2	AG	I	2.2H	1.3H		BX
4210.936	4212.122	23741.00	AG	I	2.3H	1.5H		MIT
4151.26	4152.43	24082.3	AG			0.3H		MIT
4096.56	4097.72	24403.8	AG			0.7H		MIT
4085.871	4087.024	24467.68	AG			1.4		MIT
4068.01	4069.16	24575.1	AG			0.3H		FN
4055.264	4056.409	24652.34	AG	I	2.9G	2.7G		MIT
4045.82	4046.96	24709.9	AG		1.0	0.3		MIT
4037.26	4038.40	24762.3	AG			0.3H		MIT
4036.76	4037.90	24765.3	AG			0.5		MIT
4005.32	4006.45	24959.7	AG		1.0	0.3		WX
3985.19	3986.32	25085.8	AG		0.3	1.3		MIT
3981.641	3982.767	25108.17	AG	I	1.5	1.3		FN

AIR WAVELENGTH (ANGSTRMS)	VACUUM WAVELENGTH (ANGSTRMS)	WAVENUMBER (KAYSERS)	LMENT		LOG ARC	INTENSITY SPK	DIS	REF
3961.3	3962.4	25237.	AG		1.2	0.7		CT
3949.44	3950.56	25312.9	AG		0.5	0.9		MIT
3942.95	3944.07	25354.5	AG		0.7H	1.0H		MIT
3907.59	3908.70	25584.0	AG		0.5	0.3H		MIT
3877.47	3878.57	25782.7	AG		0.3	0.0		MIT
3840.80	3841.89	26028.9	AG	I	1.3	0.5H		FN
3811.79	3812.87	26226.9	AG	I	0.7			BX
3810.86	3811.94	26233.4	AG	I	2.0J	1.0H		MIT
3709.248	3710.303	26951.97	AG	I	1.0	0.5		MIT
3694.62	3695.67	27058.7	AG		0.3H	0.0H		MIT
3683.453	3684.502	27140.71	AG		0.6	0.3		MIT
3683.316	3684.365	27141.72	AG			1.0		MIT
3682.47	3683.52	27148.0	AG	I	1.7	0.6		BX
3671.82	3672.87	27226.7	AG			0.8H		FN
3656.64	3657.68	27339.7	AG			0.9H		FN
3644.42	3645.46	27431.4	AG			0.3H		MIT
3624.71	3625.74	27580.6	AG	I	1.4H			BX
3619.77	3620.80	27618.2	AG			0.8H		FN
3617.004	3618.035	27639.31	AG		0.0	0.3H		MIT
3587.424	3588.448	27867.20	AG		0.5H			MIT
3586.98	3588.00	27870.6	AG	I	0.8H			MIT
3569.76	3570.78	28005.1	AG	I	0.6H			BX
3568.435	3569.454	28015.49	AG		0.3	0.0		MIT
3542.612	3543.624	28219.70	AG		1.5	0.7		FN
3538.273	3539.284	28254.30	AG		1.0	0.3		MIT
3521.16	3522.17	28391.6	AG	I	0.7	0.3		BX
3508.09	3509.09	28497.4	AG	I	1.0	0.0		MIT
3501.943	3502.945	28547.41	AG		0.7	0.3		FN
3501.68	3502.68	28549.6	AG		0.9H	1.3H		MIT
3487.76	3488.76	28663.5	AG	I	0.5H	0.5H		BX
3486.97	3487.97	28670.0	AG		0.3H	0.7H		MIT
3480.361	3481.357	28724.43	AG			0.8H		MIT
3475.784	3476.779	28762.25	AG			1.3		MIT
3469.206	3470.199	28816.79	AG		1.0	0.9		FN
3457.09	3458.08	28917.8	AG	I	0.3	0.0H		MIT
3434.646	3435.631	29106.74	AG	I	0.3H			MIT
3419.985	3420.966	29231.51	AG		0.0	0.3H		MIT
3414.55	3415.53	29278.0	AG	I	0.6			BX
3410.78	3411.76	29310.4	AG	I	0.9			BX
3405.03	3406.01	29359.9	AG		0.5	0.3H		MIT
3389.38	3390.35	29495.4	AG			0.3H		FN
3387.106	3388.078	29515.25	AG			0.5H		MIT
3382.891	3383.862	29552.03	AG		3.0G	2.8G		MIT
3372.505	3373.474	29643.03	AG	II	0.0	0.5H		MIT
3369.673	3370.641	29667.95	AG			0.7H		MIT
3352.08	3353.04	29823.6	AG		1.0H			MIT
3350.56	3351.52	29837.2	AG	I	0.3H			BX
3344.66	3345.62	29889.8	AG			0.3H		MIT
3339.22	3340.18	29938.5	AG		0.3H	0.3H		MIT
3333.68	3334.64	29988.3	AG		0.3H	0.6H		FN
3331.85	3332.81	30004.7	AG		0.0	0.6H		MIT
3330.55	3331.51	30016.4	AG			0.3H		FN
3329.83	3330.79	30022.9	AG	II		0.3H		MIT
3327.726	3328.683	30041.91	AG		0.7	0.3H		MIT
3327.150	3328.107	30047.11	AG			0.3H		FN
3316.31	3317.26	30145.3	AG			0.6J		MIT
3312.64	3313.59	30178.7	AG			0.9H		MIT
3310.51	3311.46	30198.1	AG	I	0.3			BX
3305.64	3306.59	30242.6	AG		0.7	0.3H		MIT
3301.55	3302.50	30280.1	AG			0.9H		MIT
3299.453	3300.403	30299.33	AG			0.7H		MIT
3297.64	3298.59	30316.0	AG		0.5	0.6H		FN
3295.467	3296.416	30335.98	AG			0.6H		MIT
3293.01	3293.96	30358.6	AG			0.3H		MIT
3289.170	3290.117	30394.05	AG			0.7		MIT
3282.525	3283.471	30455.58	AG	I	0.5	0.0H		MIT
3280.683	3281.628	30472.68	AG	I	3.3G	3.0G		MIT
3274.410	3275.354	30531.05	AG			0.7H		FN
3269.819	3270.762	30573.92	AG	II		1.0		MIT
3267.35	3268.29	30597.0	AG			1.1		MIT

AIR WAVELENGTH (ANGSTRMS)	VACUUM WAVELENGTH (ANGSTRMS)	WAVENUMBER (KAYSERS)	LMENT	LOG ARC	INTENSITY SPK	INTENSITY DIS	REF
3264.190	3265.131	30626.64	AG	0.0	0.5H		MIT
3256.70	3257.64	30697.1	AG		0.3H		FN
3252.75	3253.69	30734.4	AG		0.9H		MIT
3247.55	3248.49	30783.6	AG	1.2	1.2		MIT
3244.954	3245.890	30808.19	AG		1.0H		MIT
3233.251	3234.184	30919.70	AG	0.7	0.3		MIT
3233.154	3234.087	30920.62	AG		1.0		MIT
3229.99	3230.92	30950.9	AG	0.0	0.6H		MIT
3223.51	3224.44	31013.1	AG II	0.0	1.4H		MIT
3216.708	3217.637	31078.71	AG		0.9H		FN
3215.678	3216.607	31088.66	AG	0.7	0.5		FN
3209.98	3210.91	31143.8	AG	0.0H	0.7H		MIT
3208.227	3209.154	31160.86	AG	0.3	0.6H		MIT
3207.35	3208.28	31169.4	AG		0.6H		MIT
3200.894	3201.819	31232.25	AG	0.0H	0.9H		MIT
3200.030	3200.955	31240.68	AG		0.6H		MIT
3191.848	3192.771	31320.76	AG		0.6H		MIT
3187.83	3188.75	31360.2	AG	0.0	0.9H		MIT
3185.088	3186.009	31387.23	AG		0.6H		MIT
3184.15	3185.07	31396.5	AG II		1.3		MIT
3181.61	3182.53	31421.5	AG	0.3	0.5		MIT
3180.71	3181.63	31430.4	AG	0.3H	1.2		MIT
3179.24	3180.16	31445.0	AG	0.3	1.2H		FN
3179.14	3180.06	31445.9	AG	0.3	0.3		MIT
3173.59	3174.51	31500.9	AG	0.5	0.7H		MIT
3172.24	3173.16	31514.4	AG	0.3H	0.6H		MIT
3170.579	3171.496	31530.86	AG	0.7	0.5		MIT
3158.40	3159.31	31652.4	AG	0.5	0.0H		MIT
3153.11	3154.02	31705.5	AG	0.3	0.7H		MIT
3146.10	3147.01	31776.2	AG II			0.9	BX
3138.60	3139.51	31852.1	AG		0.5H		MIT
3135.023	3135.931	31888.45	AG		0.9H		MIT
3132.775	3133.683	31911.33	AG	0.0H	0.3H		MIT
3130.009	3130.916	31939.53	AG	1.4H	1.2H		MIT
3117.88	3118.78	32063.8	AG	0.3	0.5H		MIT
3115.674	3116.578	32086.48	AG		0.5H		MIT
3112.986	3113.889	32114.18	AG	0.7			MIT
3102.66	3103.56	32221.1	AG	0.5			MIT
3099.122	3100.021	32257.84	AG	1.0	0.9		MIT
3097.985	3098.884	32269.68	AG	0.7			MIT
3096.54	3097.44	32284.7	AG	0.0	0.5H		MIT
3086.389	3087.285	32390.92	AG	0.7	0.0H		MIT
3047.040	3047.926	32809.19	AG		0.7J		MIT
3034.112	3034.995	32948.98	AG	0.6	0.0H		MIT
3012.927	3013.805	33180.65	AG		1.0		MIT
2938.55	2939.41	34020.4	AG II	2.3	2.3J		MIT
2934.23	2935.09	34070.5	AG II	1.0	2.3H		MIT
2929.351	2930.208	34127.27	AG II	1.3	1.6		MIT
2920.04	2920.89	34236.1	AG II		2.0J		MIT
2902.07	2902.92	34448.1	AG II	0.7	2.3J		MIT
2896.49	2897.34	34514.4	AG II	0.3	2.2J		MIT
2892.22	2893.07	34565.4	AG II		0.3H		MIT
2882.20	2883.05	34685.5	AG II		1.0H		MIT
2873.65	2874.49	34788.7	AG II	0.5	2.0J		MIT
2873.53	2874.37	34790.2	AG	0.5	0.3		MIT
2857.26	2858.10	34988.3	AG	0.3H	0.0H		FN
2852.53	2853.37	35046.3	AG	0.0	0.7J		FN
2849.42	2850.26	35084.5	AG		0.3H		MIT
2843.95	2844.79	35152.0	AG	0.7	0.3		MIT
2837.76	2838.59	35228.7	AG II		0.7		MIT
2830.06	2830.89	35324.5	AG	0.3	0.0H		MIT
2829.15	2829.98	35335.9	AG II	0.5	0.5		MIT
2824.370	2825.201	35395.71	AG	2.2J	2.3W		MIT
2823.95	2824.78	35401.0	AG		0.5H		MIT
2819.212	2820.042	35460.46	AG		0.3H		FN
2815.54	2816.37	35506.7	AG II	0.5	1.9J		MIT
2801.93	2802.76	35679.2	AG II		1.0W		BX
2799.66	2800.49	35708.1	AG II	1.3	2.0H		MIT
2786.49	2787.31	35876.9	AG II	1.0	1.0H		MIT
2767.783	2768.600	36119.33	AG	0.7	0.0H		MIT

AIR WAVELENGTH (ANGSTRMS)	VACUUM WAVELENGTH (ANGSTRMS)	WAVENUMBER (KAYSERS)	LMENT	LOG ARC	INTENSITY SPK	INTENSITY DIS	REF
2767.523	2768.340	36122.73	AG II	1.5	2.3		MIT
2756.510	2757.325	36267.04	AG II	0.7	2.3		MIT
2752.19	2753.00	36324.0	AG II	0.3		0.6	BX
2743.899	2744.711	36433.71	AG II		1.7		MIT
2728.73	2729.54	36636.2	AG II			0.7	BX
2721.768	2722.574	36729.94	AG	1.3	1.4		MIT
2716.218	2717.023	36804.99	AG		0.9		MIT
2712.06	2712.86	36861.4	AG II	0.5	2.3H		MIT
2711.21	2712.01	36873.0	AG II	0.0H	2.5J		MIT
2688.35	2689.15	37186.5	AG II		1.0J		MIT
2681.379	2682.176	37283.17	AG II		2.0J		MIT
2660.456	2661.248	37576.36	AG II	1.5	2.2		MIT
2656.92	2657.71	37626.4	AG II		1.3H		FN
2656.66	2657.45	37630.0	AG II	0.0H	1.2H		MIT
2637.87	2638.66	37898.1	AG	0.3	0.5		MIT
2628.58	2629.36	38032.0	AG	0.0	1.4		MIT
2625.668	2626.451	38074.19	AG II	0.8	1.2H		MIT
2617.01	2617.79	38200.1	AG II			1.0H	BX
2614.59	2615.37	38235.5	AG II		2.5J		MIT
2606.16	2606.94	38359.2	AG II	1.0	2.3J		MIT
2602.142	2602.920	38418.40	AG		0.3H		MIT
2598.68	2599.46	38469.6	AG II	1.0	1.0H		M
2595.627	2596.403	38514.82	AG II		1.6J		MIT
2587.26	2588.03	38639.4	AG II	0.0H		0.5	MIT
2584.19	2584.96	38685.3	AG II		0.5J		MIT
2580.742	2581.515	38736.95	AG II	0.0	2.2J		MIT
2580.64	2581.41	38738.5	AG		0.3		MIT
2575.744	2576.515	38812.11	AG	1.0H	0.5H		MIT
2575.354	2576.125	38817.99	AG	0.3H	0.0		MIT
2567.16	2567.93	38941.9	AG II		1.2J		MIT
2564.42	2565.19	38983.5	AG II		1.2J		M
2562.907	2563.675	39006.50	AG		1.0		MIT
2553.407	2554.173	39151.61	AG II	0.3	1.0		MIT
2535.307	2536.069	39431.11	AG II	1.0	1.4		MIT
2535.07	2535.83	39434.8	AG II			1.0	BX
2507.22	2507.98	39872.8	AG II		0.7H		MIT
2506.88	2507.64	39878.2	AG II	0.3		0.3H	BX
2506.60	2507.36	39882.7	AG II	1.2	1.7H		MIT
2504.76	2505.51	39912.0	AG		0.5		MIT
2504.08	2504.83	39922.8	AG II		1.7H		M
2498.81	2499.56	40007.0	AG	0.0	0.5		FR
2486.72	2487.47	40201.5	AG II		0.6H		MIT
2485.775	2486.525	40216.76	AG II	0.0	1.0H		MIT
2480.407	2481.156	40303.79	AG II		1.2H		MIT
2477.284	2478.032	40354.60	AG II		2.2J		MIT
2476.740	2477.488	40363.46	AG II		0.3J		MIT
2476.24	2476.99	40371.6	AG II		0.3J		MIT
2473.800	2474.548	40411.43	AG II	1.3	2.2H		MIT
2472.919	2473.666	40425.82	AG II		1.4H		MIT
2471.34	2472.09	40451.6	AG		0.3H		MIT
2469.562	2470.309	40480.77	AG		0.6		MIT
2462.236	2462.981	40601.21	AG II	0.7	1.9H		MIT
2461.277	2462.022	40617.03	AG II		0.9H		MIT
2460.31	2461.05	40633.0	AG II		1.9J		MIT
2454.22	2454.96	40733.8	AG II			0.3H	BX
2453.328	2454.071	40748.62	AG II		2.1H		MIT
2447.929	2448.671	40838.49	AG II	1.5	2.3J		MIT
2446.32	2447.06	40865.3	AG II		1.4J		MIT
2444.209	2444.950	40900.64	AG II		1.9W		MIT
2437.791	2438.530	41008.31	AG II	1.8	2.7J		FN
2436.57	2437.31	41028.9	AG II		1.3J		MIT
2435.07	2435.81	41054.1	AG II			0.3	BX
2433.57	2434.31	41079.4	AG		0.6H		MIT
2430.352	2431.090	41133.82	AG		0.5		MIT
2429.635	2430.372	41145.96	AG II		2.2J		MIT
2428.196	2428.933	41170.34	AG II	0.0	1.6J		MIT
2422.592	2423.328	41265.57	AG II		1.0J		MIT
2420.067	2420.802	41308.62	AG II		2.0J		MIT
2413.184	2413.918	41426.43	AG II	1.7	2.5H		MIT
2411.59	2412.32	41453.8	AG II			1.3	BX

AIR WAVELENGTH (ANGSTRMS)	VACUUM WAVELENGTH (ANGSTRMS)	WAVENUMBER (KAYSERS)	LMENT	LOG ARC	INTENSITY SPK	INTENSITY DIS	REF
2411.350	2412.083	41457.94	AG II	1.4	2.2H		MIT
2410.090	2410.823	41479.61	AG		0.7		MIT
2409.02	2409.75	41498.0	AG II		1.0H		M
2405.00	2405.73	41567.4	AG II		0.9H		MIT
2402.57	2403.30	41609.4	AG II		1.5J		MIT
2395.627	2396.357	41730.02	AG	0.6	0.7		MIT
2392.96	2393.69	41776.5	AG II		1.4H		MIT
2390.540	2391.268	41818.81	AG II		1.9H		MIT
2386.816	2387.544	41884.05	AG		0.9		MIT
2386.33	2387.06	41892.6	AG II		1.4H		MIT
2383.21	2383.94	41947.4	AG II		1.4		MIT
2375.06	2375.78	42091.4	AG	2.5J	2.5J		MIT
2373.70	2374.42	42115.5	AG II		0.3H		MIT
2369.883	2370.607	42183.29	AG		0.3		MIT
2365.67	2366.39	42258.4	AG II		1.3H		MIT
2364.001	2364.724	42288.24	AG II		2.0V		MIT
2362.179	2362.901	42320.86	AG II		1.9W		MIT
2358.858	2359.579	42380.44	AG II		2.0V		MIT
2357.920	2358.641	42397.29	AG II	1.2	2.0H		MIT
2353.45	2354.17	42477.8	AG II		0.3		EX
2343.746	2344.464	42653.67	AG II		0.5H		MIT
2343.51	2344.23	42658.0	AG	0.6	0.7		MIT
2341.898	2342.616	42687.33	AG		0.6		MIT
2339.178	2339.895	42736.96	AG II		0.6		MIT
2332.274	2332.989	42863.46	AG		0.5		MIT
2331.370	2332.085	42880.08	AG II	1.3	2.2J		MIT
2325.054	2325.768	42996.55	AG II	0.6	1.4J		MIT
2324.677	2325.391	43003.52	AG II	1.2	2.0J		MIT
2321.546	2322.259	43061.52	AG II		1.4W		MIT
2320.246	2320.959	43085.64	AG II	1.2	2.2J		MIT
2318.499	2319.211	43118.10	AG		0.7		MIT
2317.033	2317.745	43145.38	AG II	1.2	2.0		MIT
2315.30	2316.01	43177.7	AG II		0.3		MIT
2312.41	2313.12	43231.6	AG II	1.4J	1.7W		MIT
2309.644	2310.354	43283.40	AG	2.2H	2.3H		MIT
2296.043	2296.750	43539.78	AG II		1.3V		MIT
2290.99	2291.70	43635.8	AG		0.5		MIT
2286.46	2287.17	43722.2	AG		1.1		MIT
2282.62	2283.32	43795.8	AG		0.6		MIT
2282.27	2282.97	43802.5	AG		0.5		MIT
2279.982	2280.686	43846.46	AG II	1.0	2.1H		MIT
2277.405	2278.108	43896.07	AG II		1.4J		MIT
2275.26	2275.96	43937.5	AG II		1.7J		MIT
2274.196	2274.899	43958.00	AG		0.3		MIT
2273.25	2273.95	43976.3	AG		0.5		MIT
2265.88	2266.58	44119.3	AG II		0.3H		MIT
2257.375	2258.074	44285.53	AG II		0.7		MIT
2253.457	2254.155	44362.52	AG II		1.7J		MIT
2250.24	2250.94	44425.9	AG		0.5		MIT
2248.740	2249.437	44455.56	AG II	1.2	2.2J		MIT
2246.412	2247.109	44501.63	AG II	1.4	2.5C		MIT
2246.13	2246.83	44507.2	AG II	0.0	0.6H		MIT
2244.20	2244.90	44545.5	AG	0.3H			MIT
2243.44	2244.14	44560.6	AG II		1.2H		BX
2241.807	2242.503	44593.03	AG II		1.0H		MIT
2241.34	2242.04	44602.3	AG		1.0		MIT
2240.388	2241.083	44621.27	AG II		1.3H		MIT
2238.41	2239.10	44660.7	AG		1.0		MIT
2229.523	2230.216	44838.70	AG II		1.4J		MIT
2226.11	2226.80	44907.4	AG II		1.3H		MIT
2219.70	2220.39	45037.1	AG II		1.2H		MIT
2217.87	2218.56	45074.3	AG II		0.6H		MIT
2211.21	2211.90	45210.0	AG		0.5		MIT
2210.32	2211.01	45228.2	AG II			0.7	BX
2208.496	2209.185	45265.57	AG II		1.4J		MIT
2205.95	2206.64	45317.8	AG II		1.7J		M
2204.40	2205.09	45349.7	AG II		1.0H		MIT
2203.662	2204.350	45364.85	AG II		1.3H		MIT
2202.10	2202.79	45397.0	AG II		1.7J		MIT
2191.86	2192.55	45609.1	AG		1.0		MIT

AIR WAVELENGTH (ANGSTRMS)	VACUUM WAVELENGTH (ANGSTRMS)	WAVENUMBER (KAYSERS)	LMENT	LOG ARC	INTENSITY SPK	INTENSITY DIS	REF
2186.768	2187.452	45715.29	AG II	1.0	2.0		MIT
2171.698	2172.379	46032.48	AG II		1.0		MIT
2170.862	2171.543	46050.21	AG II		1.5H		MIT
2166.506	2167.186	46142.79	AG II	0.7	2.0W		MIT
2166.048	2166.728	46152.54	AG II		0.9		MIT
2161.94	2162.62	46240.2	AG		1.2		MIT
2158.95	2159.63	46304.3	AG II		0.6H		MIT
2149.21	2149.89	46514.1	AG		0.9		MIT
2145.80	2146.48	46588.0	AG II		0.7H		MIT
2145.61	2146.29	46592.1	AG II	0.9	2.2		MIT
2129.12	2129.79	46952.9	AG II		1.0		MIT
2125.52	2126.19	47032.5	AG II		1.4J		MIT
2120.91	2121.58	47134.7	AG II		1.2		MIT
2120.45	2121.12	47144.9	AG II	0.7	1.9		MIT
2113.83	2114.50	47292.5	AG II	1.3	2.2J		MIT
2075.607	2076.269	48163.32	AG II		1.0J		MIT
2069.83	2070.49	48297.7	AG I	1.0	0.5		MIT
2065.91	2066.57	48389.4	AG II	0.6	1.9		MIT
2061.17	2061.83	48500.6	AG I	1.4	1.0H		MIT
2033.931	2034.585	49150.07	AG II		1.4		MIT
2015.90	2016.55	49589.6	AG II		1.4		MIT
2000.684	2001.332	49966.72	AG II		1.5		MIT
4640.46	4641.76	21543.6	AIR		1.0		SQ
4633.97	4635.27	21573.7	AIR		1.0		SQ
4514.8	4516.1	22143.	AIR		1.0		M
4103.4	4104.6	24363.	AIR		1.3		M
3759.8	3760.9	26590.	AIR		1.0		M
3754.6	3755.7	26626.	AIR		1.0		M
3707.3	3708.4	26966.	AIR		0.5		M
3702.9	3704.0	26998.	AIR		0.5		M
3639.6	3640.6	27468.	AIR		1.2		M
3594.6	3595.6	27812.	AIR		0.5		M
3577.2	3578.2	27947.	AIR		0.5		M
3491.98	3492.98	28628.9	AIR		0.8		SQ
3374.1	3375.1	29629.	AIR		1.0		M
3367.3	3368.3	29689.	AIR		1.2J		M
3365.8	3366.8	29702.	AIR		1.0		M
3354.05	3355.01	29806.1	AIR		1.1		SQ
3331.92	3332.88	30004.1	AIR		1.1J		SQ
3329.5	3330.5	30026.	AIR		1.0		M
3320.81	3321.77	30104.5	AIR		0.8		SQ
3318.87	3319.83	30122.1	AIR		0.8J		SQ
3312.4	3313.4	30181.	AIR		0.8		MIT
3265.2	3266.1	30617.	AIR		0.7		M
3261.0	3261.9	30657.	AIR		0.9		M
3238.6	3239.5	30869.	AIR		0.7		MIT
3139.3	3140.2	31845.	AIR		0.8		M
3121.7	3122.6	32024.	AIR		0.7		MIT
3071.6	3072.5	32547.	AIR		0.7		MIT
2794.2	2795.0	35778.	AIR		0.5		MIT
2677.9	2678.7	37332.	AIR		0.5		MIT
2665.8	2666.6	37501.	AIR		0.9		MIT
2631.3	2632.1	37993.	AIR		0.6		MIT
2542.7	2543.5	39316.	AIR		0.7		MIT
2534.1	2534.9	39450.	AIR		0.8		MIT
2524.1	2524.9	39606.	AIR		0.6		MIT
2516.1	2516.9	39732.	AIR		0.8		MIT
2514.3	2515.1	39760.	AIR		0.6		MIT
2507.35	2508.11	39870.7	AIR		0.9		SQ
2486.5	2487.3	40205.	AIR		0.5		MIT
2404.87	2405.60	41569.6	AIR		0.5		SQ
2395.66	2396.39	41729.4	AIR		0.5		SQ
2370.5	2371.2	42172.	AIR		0.5		MIT
2367.4	2368.1	42227.	AIR		0.6		MIT
2070.6	2071.3	48280.	AIR		0.7		MIT
9331.979	9334.540	10712.90	AL II			0.7	PS
9331.546	9334.106	10713.40	AL II			1.0	PS
9290.747	9293.296	10760.44	AL II			1.3	PS
9290.649	9293.198	10760.56	AL II			1.3	PS
9288.550	9291.099	10762.99	AL II			0.7	PS

AIR WAVELENGTH (ANGSTRMS)	VACUUM WAVELENGTH (ANGSTRMS)	WAVENUMBER (KAYSERS)	LMENT	LOG ARC	INTENSITY SPK	DIS	REF
9288.145	9290.694	10763.46	AL II			1.0	PS
9286.794	9289.342	10765.03	AL II			0.7	PS
9286.578	9289.126	10765.28	AL II			0.3	PS
9249.41	9251.95	10808.5	AL II			0.3	PS
8923.56	8926.01	11203.2	AL I			0.7	PS
8912.88	8915.33	11216.6	AL I			0.3	PS
8858.39	8860.82	11285.6	AL II			0.3	PS
8841.26	8843.69	11307.5	AL I			1.0	PS
8828.91	8831.33	11323.3	AL I			0.3	PS
8774.56	8776.97	11393.4	AL I	2.0			MS
8773.91	8776.32	11394.3	AL I			2.2	PS
8772.88	8775.29	11395.6	AL I			1.9	PS
8680.36	8682.74	11517.1	AL II			0.9	PS
8680.27	8682.65	11517.2	AL II			1.0	PS
8675.28	8677.66	11523.8	AL II			0.3	PS
8674.92	8677.30	11524.3	AL II			0.7	PS
8671.28	8673.66	11529.2	AL II			0.3	PS
8640.70	8643.07	11570.0	AL II			1.5	PS
8363.52	8365.82	11953.4	AL II			1.5	PS
8363.30	8365.60	11953.7	AL II			0.3	PS
8359.57	8361.87	11959.1	AL II			1.6	PS
8359.23	8361.53	11959.5	AL II			0.3	PS
8354.35	8356.65	11966.5	AL II			1.7	PS
8160.15	8162.39	12251.3	AL II			1.0	PS
8119.72	8121.95	12312.3	AL II			0.5	PS
8075.37	8077.59	12379.9	AL I			1.1	PS
8065.99	8068.21	12394.3	AL I			1.0	PS
8003.18	8005.38	12491.6	AL I			0.7	PS
7836.85	7839.01	12756.7	AL	1.2L			MS
7836.15	7838.31	12757.9	AL I			1.7	PS
7835.33	7837.49	12759.2	AL I			1.6	PS
7823.72	7825.87	12778.1	AL II			0.7	PS
7815.83	7817.98	12791.0	AL II			0.3	PS
7635.33	7637.43	13093.4	AL II			0.7	PS
7627.85	7629.95	13106.2	AL II			0.3	PS
7471.37	7473.43	13380.7	AL II			0.7	SY
7449.42	7451.47	13420.2	AL II			1.3	PS
7362.31	7364.34	13578.9	AL I	0.7B		1.7	PS
7361.59	7363.62	13580.3	AL I			1.3	PS
7335.72	7337.74	13628.2	AL I			0.3	PS
7138.81	7140.78	14004.1	AL II			0.3	SY
7134.66	7136.63	14012.2	AL II			0.3	SY
7063.62	7065.57	14153.1	AL II			1.2	SY
7056.56	7058.51	14167.3	AL II			1.3	SY
7042.06	7044.00	14196.5	AL II			1.4	SY
6919.96	6921.87	14447.0	AL II			1.4	SY
6917.84	6919.75	14451.4	AL II			1.4	SY
6837.09	6838.98	14622.1	AL II			1.2	SY
6823.38	6825.26	14651.4	AL II			1.0	SY
6816.83	6818.71	14665.5	AL II			0.7	SY
6775.97	6777.84	14754.0	AL II			0.3	SY
6698.78	6700.63	14924.0	AL	1.0			WT
6696.39	6698.24	14929.3	AL II	1.0		0.3	SY
6695.97	6697.82	14930.2	AL I			1.7	PS
6495.45	6497.24	15391.1	AL II			0.3	SY
6335.70	6337.45	15779.2	AL II			1.4	SY
6245.05	6246.78	16008.2	AL		0.7J		GN
6243.36	6245.09	16012.6	AL II			1.9	PS
6231.76	6233.48	16042.4	AL II			1.5	SY
6226.19	6227.91	16056.7	AL II			1.4	SY
6201.49	6203.21	16120.7	AL II			1.2	SY
6183.39	6185.10	16167.9	AL II			1.3	SY
6182.28	6183.99	16170.8	AL II			1.2	SY
6181.57	6183.28	16172.6	AL II			1.0	SY
6176.36	6178.07	16186.3	AL	0.7H			WT
6151.69	6153.39	16251.2	AL	0.7H			WT
6073.17	6074.85	16461.3	AL II			1.2	SY
6068.53	6070.21	16473.9	AL II			1.5	PS
6068.43	6070.11	16474.2	AL II			1.8	PS
6066.40	6068.08	16479.7	AL II			0.3	SY

AIR WAVELENGTH (ANGSTRMS)	VACUUM WAVELENGTH (ANGSTRMS)	WAVENUMBER (KAYSERS)	LMENT	LOG ARC	INTENSITY SPK	DIS	REF
6061.06	6062.74	16494.2	AL II			0.3	SY
6006.42	6008.08	16644.2	AL II			1.5	PS
6001.88	6003.54	16656.8	AL II			1.7	PS
6001.76	6003.42	16657.2	AL II			1.8	PS
6001.18	6002.84	16658.8	AL II			0.7	SY
5999.83	6001.49	16662.5	AL II			1.0	PS
5999.70	6001.36	16662.9	AL II			1.0	PS
5971.94	5973.59	16740.3	AL II			1.5	PS
5867.81	5869.44	17037.4	AL II			1.2	SY
5861.53	5863.15	17055.7	AL II			1.4	SY
5853.62	5855.24	17078.7	AL II			1.5	SY
5722.65	5724.24	17469.6	AL		1.3H		GN
5696.47	5698.05	17549.9	AL		1.2H		GN
5651.69	5653.26	17688.9	AL	1.0H			WT
5635.15	5636.71	17740.8	AL	1.0			WT
5613.19	5614.75	17810.2	AL II			1.2	SY
5593.23	5594.78	17873.8	AL II			2.3	SY
5557.904	5559.447	17987.40	AL I	1.2			MIT
5557.02	5558.56	17990.3	AL I	1.2			WT
5502.88	5504.41	18167.3	AL II			1.2	SY
5388.48	5389.98	18552.9	AL II			0.7	SY
5371.84	5373.33	18610.4	AL II			1.7	SY
5324.61	5326.09	18775.5	AL II			1.4	SY
5316.07	5317.55	18805.7	AL II			1.8	SY
5312.32	5313.80	18818.9	AL II			1.5	SY
5310.76	5312.24	18824.5	AL II			1.0	SY
5285.85	5287.32	18913.2	AL II			1.7	SY
5283.77	5285.24	18920.6	AL II			2.0	SY
5280.21	5281.68	18933.4	AL II			1.7	SY
5278.62	5280.09	18939.1	AL II			1.2	SY
5277.68	5279.15	18942.4	AL II			1.0	SY
5276.81	5278.28	18945.6	AL II			1.0	SY
5276.42	5277.89	18947.0	AL II			1.0	SY
5158.187	5159.624	19381.26	AL II			0.7	SY
5145.654	5147.088	19428.46	AL II			0.9D	SY
5144.998	5146.431	19430.94	AL II			0.7	SY
5144.875	5146.308	19431.41	AL II			0.3	SY
5144.413	5145.846	19433.15	AL II			0.3	SY
5100.34	5101.76	19601.1	AL II			0.7	SY
5093.65	5095.07	19626.8	AL II			1.0	SY
5085.02	5086.44	19660.1	AL II			1.4	SY
5000.97	5002.37	19990.5	AL II			1.2	SY
4962.10	4963.48	20147.1	AL II	1.2	0.5		SY
4918.98	4920.35	20323.7	AL II			1.3	SY
4902.77	4904.14	20390.9	AL II			1.5	SY
4899.64	4901.01	20404.0	AL II			1.2	SY
4898.76	4900.13	20407.6	AL II			1.5	SY
4898.52	4899.89	20408.6	AL II			0.9	SY
4879.0	4880.4	20490.	AL		0.7C		GN
4774.3	4775.6	20940.	AL II			0.3	SY
4701.296	4702.612	21264.78	AL		0.8		MIT
4666.8	4668.1	21422.	AL II			0.7	SY
4655.05	4656.35	21476.0	AL II			0.3	SY
4653.0	4654.3	21485.	AL II			0.3	SY
4650.646	4651.948	21496.37	AL II			0.8	SY
4650.544	4651.846	21496.84	AL II			0.9	SY
4648.62	4649.92	21505.7	AL II			0.6	SY
4640.384	4641.683	21543.91	AL II			1.3	SY
4640.362	4641.661	21544.01	AL II			1.3	SY
4639.833	4641.132	21546.47	AL II			0.8	SY
4639.725	4641.024	21546.97	AL II			0.9	SY
4639.326	4640.625	21548.82	AL II			0.3	SY
4636.384	4637.682	21562.49	AL II			0.6	SY
4635.7	4637.0	21566.	AL II			0.6	SY
4633.2	4634.5	21577.	AL II			0.3	SY
4631.5	4632.8	21585.	AL II			0.3	SY
4629.7	4631.0	21594.	AL II			0.6	SY
4609.7	4611.0	21687.	AL II			0.6	SY
4589.750	4591.036	21781.58	AL II			1.3	SY
4589.689	4590.975	21781.86	AL II			0.6	SY

AIR WAVELENGTH (ANGSTRMS)	VACUUM WAVELENGTH (ANGSTRMS)	WAVENUMBER (KAYSERS)	LMENT	LOG ARC	INTENSITY SPK	DIS	REF
4588.194	4589.480	21788.96	AL II			1.5	SY
4588.082	4589.368	21789.50	AL II			0.3	SY
4585.820	4587.105	21800.24	AL II			1.6	SY
4529.49	4530.76	22071.4	AL		0.7		GN
4510.84	4512.11	22162.6	AL		0.8H		GN
4489.87	4491.13	22266.1	AL II			0.3	SY
4447.8	4449.0	22477.	AL II			1.2	SY
4432.82	4434.06	22552.7	AL II			0.6	SY
4356.807	4358.032	22946.14	AL II			0.8	SY
4356.711	4357.936	22946.65	AL II			0.8J	SY
4347.802	4349.024	22993.66	AL II			1.3	SY
4347.785	4349.007	22993.75	AL II			1.3	SY
4347.316	4348.538	22996.23	AL II			0.8	SY
4347.223	4348.445	22996.73	AL II			0.9	SY
4346.918	4348.140	22998.34	AL II			0.6	SY
4346.866	4348.088	22998.61	AL II			0.3	SY
4331.995	4333.213	23077.56	AL II			0.3	SY
4307.165	4308.376	23210.60	AL II			1.3	SY
4282.97	4284.18	23341.7	AL II			0.3	SY
4240.75	4241.94	23574.1	AL II			1.2	SY
4227.982	4229.173	23645.29	AL II			1.3	SY
4227.923	4229.114	23645.62	AL II			0.8	SY
4227.861	4229.052	23645.96	AL II			0.3	SY
4227.500	4228.691	23647.98	AL II			1.5	SY
4227.406	4228.597	23648.51	AL II			0.9	SY
4226.809	4227.999	23651.85	AL II			1.5	SY
4202.4	4203.6	23789.	AL II			0.9	SY
4168.516	4169.691	23982.59	AL II			0.3H	SY
4168.424	4169.599	23983.12	AL II			0.6	SY
4160.263	4161.436	24030.17	AL II			1.2	SY
4160.239	4161.412	24030.31	AL II			1.1	SY
4159.809	4160.982	24032.79	AL II			0.6	SY
4159.725	4160.898	24033.27	AL II			0.8	SY
4159.450	4160.623	24034.86	AL II			0.6	SY
4159.407	4160.580	24035.11	AL II			0.3	SY
4056.8	4057.9	24643.	AL II			0.3	SY
4039.302	4040.443	24749.76	AL II			0.3	SY
4031.633	4032.772	24796.84	AL II			0.3	SY
4031.135	4032.274	24799.90	AL II			0.3	SY
4030.867	4032.006	24801.55	AL II			0.9	SY
4026.5	4027.6	24828.	AL II			1.5	SY
4009.58	4010.71	24933.2	AL II			0.6	SY
3996.381	3997.511	25015.57	AL II			1.0	SY
3996.323	3997.453	25015.93	AL II			0.3	SY
3996.182	3997.312	25016.81	AL II			0.3	SY
3996.159	3997.289	25016.96	AL II			1.3	SY
3996.075	3997.205	25017.48	AL II			0.3	SY
3995.860	3996.990	25018.83	AL II			1.5	SY
3983.7	3984.8	25095.	AL II			0.3	SY
3961.527	3962.648	25235.65	AL I	3.5	3.3		KS
3946.406	3947.523	25332.34	AL II			0.3	SY
3944.032	3945.148	25347.59	AL I	3.3	3.0		GN
3939.066	3940.181	25379.54	AL II			0.3	SY
3938.621	3939.736	25382.41	AL II			0.3	SY
3900.680	3901.785	25629.29	AL II			2.3	SY
3870.057	3871.154	25832.09	AL II			0.3	SY
3866.160	3867.256	25858.13	AL II			0.7	SY
3859.33	3860.42	25903.9	AL II			1.0	SY
3842.317	3843.407	26018.58	AL II			0.3	SY
3842.213	3843.303	26019.29	AL II			0.7	SY
3842.037	3843.127	26020.48	AL II			1.0	SY
3753.10	3754.17	26637.1	AL II			0.3	SY
3738.003	3739.066	26744.65	AL II			1.0	SY
3734.715	3735.777	26768.19	AL II			0.3	SY
3734.567	3735.629	26769.25	AL II			0.3	SY
3733.910	3734.972	26773.96	AL II			0.7	SY
3731.950	3733.011	26788.03	AL II			0.3	SY
3713.60	3714.66	26920.4	AL		1.0H		GN
3703.217	3704.271	26995.87	AL II			1.3	SY
3702.57	3703.62	27000.6	AL		0.7H		GN

AIR WAVELENGTH (ANGSTRMS)	VACUUM WAVELENGTH (ANGSTRMS)	WAVENUMBER (KAYSERS)	LMENT	LOG ARC	INTENSITY SPK	DIS	REF
3656.319	3657.361	27342.12	AL II			0.3	SY
3655.000	3656.041	27351.99	AL II			2.0	SY
3654.979	3656.020	27352.15	AL II			1.3	SY
3651.090	3652.130	27381.28	AL II			1.3	SY
3651.064	3652.104	27381.48	AL II			1.7	SY
3649.221	3650.261	27395.30	AL II			0.3	SY
3649.182	3650.222	27395.60	AL II			0.7	SY
3612.467	3613.497	27674.02	AL		1.9H		MIT
3603.597	3604.625	27742.14	AL			0.3	SY
3601.74	3602.77	27756.4	AL		1.2		GN
3597.50	3598.53	27789.1	AL II			0.7	SY
3587.441	3588.465	27867.07	AL II			1.9	SY
3587.327	3588.351	27867.96	AL II			0.7H	SY
3587.176	3588.200	27869.13	AL II			0.3	SY
3587.057	3588.081	27870.05	AL II			2.0	SY
3586.908	3587.932	27871.21	AL II			2.7H	SY
3586.802	3587.826	27872.03	AL II			2.3J	SY
3586.692	3587.716	27872.89	AL II			2.3	SY
3586.546	3587.570	27874.02	AL II			2.3	SY
3561.73	3562.75	28068.2	AL		0.3		GN
3552.00	3553.01	28145.1	AL II			0.3	SY
3539.15	3540.16	28247.3	AL		0.9		GN
3534.2	3535.2	28287.	AL		1.2		GN
3503.56	3504.56	28534.2	AL		0.3		GN
3491.80	3492.80	28630.3	AL		0.3		GN
3463.63	3464.62	28863.2	AL II	0.5		0.0	SY
3458.230	3459.221	28908.25	AL			1.0	SY
3444.871	3445.858	29020.35	AL			0.7	SY
3443.651	3444.638	29030.63	AL			1.0	SY
3440.597	3441.583	29056.40	AL			0.3	SY
3439.352	3440.338	29066.91	AL			0.3	SY
3428.916	3429.899	29155.38	AL II			1.7	SY
3351.456	3352.419	29829.20	AL II			1.0	SY
3336.12	3337.08	29966.3	AL		0.3		GN
3318.47	3319.42	30125.7	AL		0.3		GN
3315.614	3316.568	30151.65	AL II			0.3	SY
3314.889	3315.843	30158.24	AL II			0.7	SY
3313.467	3314.421	30171.19	AL II			0.3	SY
3313.351	3314.305	30172.24	AL II			1.0	SY
3301.86	3302.81	30277.2	AL		0.3		GN
3286.834	3287.781	30415.65	AL		0.3		GN
3275.776	3276.720	30518.32	AL II			1.3	SY
3135.875	3136.784	31879.79	AL II			1.0	SY
3092.842	3093.740	32323.34	AL	1.76	1.3		MIT
3092.713	3093.611	32324.69	AL I	3.0	3.0		MIT
3088.523	3089.420	32368.54	AL II			1.0	SY
3082.155	3083.050	32435.41	AL I	2.9	2.9		MIT
3074.665	3075.558	32514.42	AL II			1.7	SY
3066.162	3067.053	32604.59	AL	1.4	1.4		GN
3064.304	3065.195	32624.35	AL	1.3	1.3		GN
3059.933	3060.823	32670.96	AL	0.9	1.0		GN
3057.154	3058.043	32700.65	AL	1.2	1.3		MIT
3054.697	3055.585	32726.95	AL	1.3	1.0		GN
3050.079	3050.966	32776.50	AL	1.3	1.0		GN
3041.278	3042.163	32871.35	AL II			1.7	SY
3026.75	3027.63	33029.1	AL II			0.5	SY
3024.074	3024.955	33058.35	AL II			0.3	SY
3001.82	3002.70	33303.4	AL			1.0	SY
2998.174	2999.048	33343.91	AL II			0.9	SY
2995.524	2996.398	33373.41	AL II			0.8	SY
2994.280	2995.153	33387.27	AL II			0.5	S
2924.52	2925.38	34183.6	AL			1.2	SY
2903.74	2904.59	34428.3	AL II			0.3	SY
2903.19	2904.04	34434.8	AL II			0.5	SY
2902.08	2902.93	34448.0	AL II			0.9	SY
2884.20	2885.05	34661.5	AL			1.5	SY
2881.463	2882.308	34694.42	AL II			1.5	SY
2868.52	2869.36	34851.0	AL II			1.9	SY
2840.05	2840.89	35200.3	AL			1.2	SY
2837.95	2838.78	35226.3	AL			0.9	SY

AIR WAVELENGTH (ANGSTRMS)	VACUUM WAVELENGTH (ANGSTRMS)	WAVENUMBER (KAYSERS)	LMENT	LOG INTENSITY ARC	SPK	DIS	REF
2820.632	2821.462	35442.61	AL II			0.5	SY
2816.179	2817.008	35498.65	AL II			1.2	SY
2805.65	2806.48	35631.9	AL			1.5	SY
2801.173	2801.999	35688.81	AL II			0.5	SY
2769.69	2770.51	36094.5	AL			0.5	SY
2762.460	2763.276	36188.93	AL II			0.9	SY
2760.852	2761.668	36210.00	AL II			0.5	SY
2748.86	2749.67	36368.0	AL			1.5H	SY
2723.091	2723.898	36712.10	AL II			0.9	SY
2720.918	2721.724	36741.42	AL II			0.5	SY
2719.862	2720.668	36755.68	AL II			0.3	SY
2718.96	2719.77	36767.9	AL			0.9	SY
2709.582	2710.385	36895.12	AL II			0.8	SY
2707.444	2708.247	36924.26	AL II			0.3	SY
2688.728	2689.526	37181.27	AL II			1.4	SY
2683.280	2684.077	37256.76	AL II			1.2	SY
2669.166	2669.960	37453.75	AL II	0.5	2.0		SY
2660.393	2661.184	37577.25	AL I	2.2G	1.8		GN
2652.489	2653.279	37689.22	AL I	2.2G	1.8		MIT
2650.10	2650.89	37723.2	AL II			1.5	SY
2640.36	2641.15	37862.3	AL II			1.2	SY
2638.695	2639.481	37886.23	AL II			0.9	SY
2638.625	2639.411	37887.24	AL II			0.3	SY
2638.263	2639.049	37892.44	AL II			1.5	SY
2638.182	2638.968	37893.60	AL II			0.7	SY
2637.696	2638.482	37900.58	AL II			1.6	SY
2636.725	2637.511	37914.54	AL II			0.5	SY
2635.17	2635.96	37936.9	AL II			0.5	SY
2635.03	2635.82	37938.9	AL II			1.2	SY
2631.74	2632.52	37986.4	AL	0.7	0.5H		MIT
2631.553	2632.338	37989.05	AL II			1.8	SY
2627.68	2628.46	38045.0	AL II			1.8	SY
2597.18	2597.96	38491.8	AL II			1.7	SY
2586.95	2587.72	38644.0	AL II			1.7	SY
2575.411	2576.182	38817.13	AL	1.5	1.5		MIT
2575.100	2575.871	38821.82	AL	2.3G	1.9G		MIT
2567.987	2568.757	38929.34	AL	2.3G	1.9G		MIT
2565.68	2566.45	38964.3	AL II			1.5	SY
2559.614	2560.382	39056.68	AL II			1.2	SY
2557.71	2558.48	39085.7	AL II			1.6	SY
2556.78	2557.55	39100.0	AL II			1.2	SY
2556.01	2556.78	39111.7	AL II			1.5	SY
2552.12	2552.89	39171.4	AL II			1.6	SY
2550.23	2551.00	39200.4	AL II			1.2	SY
2549.30	2550.07	39214.7	AL II			0.8	SY
2545.60	2546.36	39271.7	AL II			1.7	SY
2544.79	2545.55	39284.2	AL			0.7	SY
2540.70	2541.46	39347.4	AL II			1.2	SY
2540.12	2540.88	39356.4	AL II			1.5	SY
2533.41	2534.17	39460.6	AL II			0.3	SY
2533.16	2533.92	39464.5	AL II			0.5	SY
2532.655	2533.416	39472.39	AL II			0.9	SY
2532.10	2532.86	39481.0	AL II			1.2	SY
2527.47	2528.23	39553.4	AL II			0.5	SY
2526.477	2527.237	39568.91	AL II			0.5	SY
2520.64	2521.40	39660.5	AL II			0.9	SY
2513.15	2513.91	39778.7	AL II			1.2	SY
2504.25	2505.00	39920.1	AL II			0.5	SY
2497.85	2498.60	40022.4	AL II			0.9	SY
2485.35	2486.10	40223.6	AL II			0.5	SY
2485.16	2485.91	40226.7	AL			0.5	SY
2476.30	2477.05	40370.6	AL II			1.5	SY
2475.260	2476.008	40387.59	AL II			1.5	SY
2472.95	2473.70	40425.3	AL II			0.5	SY
2466.28	2467.03	40534.6	AL II			0.3	SY
2459.82	2460.56	40641.1	AL II			1.5	SY
2458.88	2459.62	40656.6	AL II			0.3	SY
2458.05	2458.79	40670.4	AL II			0.9	SY
2457.20	2457.94	40684.4	AL II			0.5	SY
2455.22	2455.96	40717.2	AL II			0.9	SY

AIR WAVELENGTH (ANGSTRMS)	VACUUM WAVELENGTH (ANGSTRMS)	WAVENUMBER (KAYSERS)	LMENT	LOG INTENSITY ARC	SPK	DIS	REF
2452.59	2453.33	40760.9	AL II			0.3	SY
2452.111	2452.854	40768.84	AL	0.3H			MIT
2427.70	2428.44	41178.7	AL II			1.2	SY
2425.61	2426.35	41214.2	AL II			0.3	SY
2393.835	2394.564	41761.25	AL II			0.9	SY
2392.15	2392.88	41790.7	AL II			1.5	SY
2391.35	2392.08	41804.6	AL II			0.5	SY
2390.755	2391.484	41815.05	AL II			0.9	SY
2389.08	2389.81	41844.4	AL II			0.5	SY
2388.25	2388.98	41858.9	AL II			0.3	SY
2378.408	2379.134	42032.10	AL I	1.6	1.3		GN
2373.362	2374.087	42121.46	AL I	2.3G	2.0G		MIT
2373.132	2373.857	42125.54	AL I	2.0G	1.5		GN
2372.084	2372.808	42144.15	AL I	1.3	1.0		GN
2370.226	2370.950	42177.19	AL	0.7	0.7		GN
2369.309	2370.033	42193.51	AL	1.0	0.6		MIT
2368.114	2368.837	42214.80	AL	0.6	0.0		MIT
2367.616	2368.339	42223.68	AL	0.3		1.0	MIT
2367.062	2367.785	42233.56	AL I	2.2G	1.7G		MIT
2365.49	2366.21	42261.6	AL II			0.8	SY
2350.20	2350.92	42536.5	AL			0.9	SY
2347.54	2348.26	42584.7	AL II			1.1	SY
2345.92	2346.64	42614.1	AL II			0.8	SY
2345.47	2346.19	42622.3	AL II			0.3	SY
2344.69	2345.41	42636.5	AL II			0.5	SY
2340.047	2340.764	42721.09	AL	0.3	0.0		MIT
2328.20	2328.91	42938.5	AL II			0.3	SY
2326.498	2327.212	42969.87	AL II			0.9	SY
2325.497	2326.211	42988.36	AL II			1.2	SY
2325.427	2326.141	42989.66	AL II			0.3	SY
2324.200	2324.914	43012.35	AL II			1.4	SY
2321.567	2322.280	43061.13	AL	0.7	0.8		GN
2319.05	2319.76	43107.9	AL	0.7	0.9		SY
2317.48	2318.19	43137.1	AL	0.3	0.5		SY
2314.002	2314.713	43201.89	AL	0.5	0.5		MIT
2313.77	2314.48	43206.2	AL II			0.5	SY
2313.53	2314.24	43210.7	AL			0.3	SY
2312.47	2313.18	43230.5	AL I	0.3		0.3	SY
2312.225	2312.936	43235.09	AL II			0.3	SY
2304.95	2305.66	43371.5	AL	0.7			MIT
2285.69	2286.40	43737.0	AL II			0.3	SY
2285.52	2286.23	43740.2	AL II			0.9	SY
2285.17	2285.88	43746.9	AL II			1.2	SY
2269.212	2269.914	44054.54	AL I	1.2G			GN
2269.093	2269.795	44056.85	AL I	1.8G	1.4		GN
2263.731	2264.431	44161.20	AL I	0.6H	0.0H		GN
2263.453	2264.153	44166.62	AL I	1.8G	1.4		GN
2257.999	2258.698	44273.29	AL I	0.9			GN
2244.21	2244.91	44545.3	AL II			0.3	SY
2243.05	2243.75	44568.3	AL II			1.5	SY
2216.86	2217.55	45094.8	AL	0.7			HR
2210.046	2210.735	45233.83	AL I	1.1G	1.2		GN
2204.627	2205.315	45345.00	AL	1.0G	1.0		GN
2201.08	2201.77	45418.1	AL	0.7G			HR
2199.569	2200.256	45449.26	AL	0.7			GN
2195.502	2196.188	45533.44	AL II			0.3	SY
2194.251	2194.937	45559.40	AL II			0.5	SY
2192.607	2193.292	45593.56	AL II			0.8	SY
2174.028	2174.709	45983.15	AL I	0.9G	1.0		GN
2168.805	2169.485	46093.88	AL I	0.9G	1.1		GN
2150.59	2151.27	46484.2	AL I	0.7			GN
2145.39	2146.07	46596.9	AL I	0.7			GN
2134.70	2135.37	46830.2	AL I	0.7			GN
2129.44	2130.11	46945.9	AL I	0.7			GN
2099.68	2100.35	47611.2	AL II			1.6	SY
2095.2	2095.9	47713.	AL II			1.6	SY
2094.8	2095.5	47722.	AL II			1.7	SY
2094.3	2095.0	47733.	AL II			1.7	SY
2087.0	2087.7	47900.	AL II			1.6	SY
2081.5	2082.2	48027.	AL II			0.9	SY

AIR WAVELENGTH (ANGSTRMS)	VACUUM WAVELENGTH (ANGSTRMS)	WAVENUMBER (KAYSERS)	LMENT	LOG ARC	INTENSITY SPK	INTENSITY DIS	REF
2073.8	2074.5	48205.	AL II			1.2	SY
2048.39	2049.05	48803.2	AL II			0.5	SY
2039.93	2040.59	49005.5	AL II			1.2	SY
2022.80	2023.45	49420.5	AL II			0.9	SY
2016.91	2017.56	49564.8	AL II			1.9	SY
9923.03	9925.75	10074.8	AS I	0.7			ME
9833.76	9836.46	10166.3	AS I	0.7			ME
9826.69	9829.38	10173.6	AS I	0.9			ME
9597.94	9600.57	10416.1	AS I	1.0			ME
9300.62	9303.17	10749.0	AS I	1.7			ME
9267.29	9269.83	10787.7	AS I	1.4			ME
9134.81	9137.32	10944.1	AS I	1.2			ME
8993.08	8995.55	11116.6	AS I	1.3			ME
8935.58	8938.03	11188.1	AS I	1.7			ME
8869.69	8872.13	11271.3	AS I	2.0			ME
8821.76	8824.18	11332.5	AS I	2.2			ME
8654.16	8656.54	11552.0	AS I	2.0			ME
8564.71	8567.06	11672.6	AS I	2.0			ME
8541.65	8544.00	11704.1	AS I	1.7			ME
8428.94	8431.26	11860.6	AS I	2.0			ME
8355.00	8357.30	11965.6	AS I	1.0			ME
8305.62	8307.90	12036.7	AS I	1.7			ME
8055.72	8057.94	12410.1	AS I	0.7			ME
8042.95	8045.16	12429.8	AS I	1.0			ME
7960.26	7962.45	12558.9	AS I	1.4			ME
7863.45	7865.61	12713.6	AS I	0.6			ME
7410.07	7412.11	13491.4	AS I	0.9			ME
6528.06	6529.86	15314.3	AS		0.7		RO
6512.42	6514.22	15351.0	AS II		0.9		RO
6489.35	6491.14	15405.6	AS II		0.5		RO
6483.86	6485.65	15418.7	AS II		0.5		RO
6443.96	6445.74	15514.1	AS		0.5		RO
6405.95	6407.72	15606.2	AS II		1.0		RO
6254.77	6256.50	15983.4	AS II		1.0		RO
6170.47	6172.18	16201.7	AS II		2.2		RO
6136.01	6137.71	16292.7	AS		1.2		RO
6110.66	6112.35	16360.3	AS II		2.2		RO
6110.30	6111.99	16361.3	AS II		2.2		RO
6081.06	6082.74	16439.9	AS		0.5		RO
6076.82	6078.50	16451.4	AS		1.2		RO
6022.81	6024.48	16598.9	AS II		2.2		RO
5972.34	5973.99	16739.2	AS		0.8		RO
5966.79	5968.44	16754.8	AS		0.6		RO
5838.19	5839.81	17123.9	AS		2.1		RO
5826.15	5827.77	17159.2	AS II		0.8		RO
5783.51	5785.11	17285.7	AS II		1.3		RO
5731.96	5733.55	17441.2	AS II		1.2		RO
5685.74	5687.32	17583.0	AS		1.8		RO
5657.20	5658.77	17671.7	AS II		1.8		RO
5651.53	5653.10	17689.4	AS II		2.3		RO
5630.42	5631.98	17755.7	AS II		0.8		RO
5620.82	5622.38	17786.1	AS II		1.1		RO
5558.31	5559.85	17986.1	AS II		2.3		RO
5497.98	5499.51	18183.4	AS II	2.3			RO
5497.09	5498.62	18186.4	AS		1.1		RO
5491.81	5493.34	18203.9	AS		1.1		RO
5471.92	5473.44	18270.0	AS II		1.2		RO
5451.40	5452.92	18338.8	AS		0.6		RO
5385.52	5387.02	18563.1	AS II		1.2		RO
5331.54	5333.02	18751.1	AS II		2.3		RO
5252.32	5253.78	19033.9	AS II		0.9		RO
5238.69	5240.15	19083.4	AS II		1.1		RO
5231.50	5232.96	19109.7	AS II		1.8		RO
5223.27	5224.72	19139.8	AS II		1.2		RO
5215.56	5217.01	19168.1	AS		0.8		RO
5205.40	5206.85	19205.5	AS		1.1		RO
5182.32	5183.76	19291.0	AS II		1.5		RO
5161.25	5162.69	19369.8	AS		1.5		RO
5125.18	5126.61	19506.1	AS II		0.8		RO
5107.80	5109.22	19572.4	AS II		2.2		RO

AIR WAVELENGTH (ANGSTRMS)	VACUUM WAVELENGTH (ANGSTRMS)	WAVENUMBER (KAYSERS)	LMENT	LOG ARC	INTENSITY SPK	INTENSITY DIS	REF
5105.80	5107.22	19580.1	AS II		2.2		RO
5072.07	5073.48	19710.3	AS II		0.6		RO
4985.60	4986.99	20052.2	AS II		1.7		RO
4932.24	4933.62	20269.1	AS II		0.7		RO
4888.74	4890.11	20449.5	AS II		1.7		RO
4846.13	4847.48	20629.3	AS II		1.0		RO
4812.01	4813.35	20775.5	AS		1.0		RO
4802.25	4803.59	20817.7	AS		1.0		RO
4799.68	4801.02	20828.9	AS II		1.0		RO
4787.29	4788.63	20882.8	AS II		1.2		RO
4730.92	4732.24	21131.6	AS II		2.1		RO
4720.38	4721.70	21178.8	AS II		0.5		RO
4707.82	4709.14	21235.3	AS II		2.3		RO
4672.70	4674.01	21394.9	AS II		1.7		RO
4648.32	4649.62	21507.1	AS		1.5		RO
4632.67	4633.97	21579.8	AS II		1.0		RO
4630.14	4631.44	21591.6	AS II		2.3		RO
4627.80	4629.10	21602.5	AS		2.3		RO
4619.63	4620.92	21640.7	AS		1.0		RO
4617.30	4618.59	21651.6	AS II		1.0		RO
4607.46	4608.75	21697.9	AS		2.3		RO
4602.73	4604.02	21720.1	AS II		2.3		RO
4591.02	4592.31	21775.6	AS		1.5		RO
4582.50	4583.78	21816.0	AS		0.5		RO
4580.92	4582.20	21823.6	AS		1.0		RO
4574.80	4576.08	21852.7	AS		0.5		RO
4574.18	4575.46	21855.7	AS		0.5		RO
4567.95	4569.23	21885.5	AS		0.7		RO
4560.27	4561.55	21922.4	AS		0.5		RO
4552.37	4553.65	21960.4	AS II		1.7		RO
4549.87	4551.15	21972.5	AS II		1.0		RO
4549.23	4550.51	21975.6	AS II		2.1		RO
4543.76	4545.03	22002.0	AS		2.3		RO
4539.97	4541.24	22020.4	AS II		2.3		RO
4532.36	4533.63	22057.4	AS		1.0		RO
4528.61	4529.88	22075.6	AS II		1.0		RO
4516.14	4517.41	22136.6	AS II		1.2		RO
4509.33	4510.59	22170.0	AS		1.0		RO
4507.92	4509.18	22177.0	AS II		1.5		RO
4494.59	4495.85	22242.7	AS II		2.3		RO
4475.69	4476.95	22336.7	AS		1.5		RO
4474.60	4475.86	22342.1	AS		2.3		RO
4471.43	4472.68	22357.9	AS II		1.0		RO
4466.60	4467.85	22382.1	AS II		1.9		RO
4461.92	4463.17	22405.6	AS		1.0		RO
4461.27	4462.52	22408.9	AS		1.5		RO
4460.29	4461.54	22413.8	AS		1.0		RO
4458.74	4459.99	22421.6	AS		1.5		RO
4456.86	4458.11	22431.0	AS		1.2		RO
4447.62	4448.87	22477.6	AS	0.5			RO
4431.73	4432.97	22558.2	AS II		2.3		RO
4427.38	4428.62	22580.4	AS II		2.3		RO
4421.06	4422.30	22612.7	AS		1.0		RO
4413.64	4414.88	22650.7	AS II		1.7		RO
4412.25	4413.49	22657.8	AS		1.0		RO
4404.53	4405.77	22697.5	AS II		1.2		RO
4392.76	4393.99	22758.3	AS II		0.7		RO
4371.38	4372.61	22869.6	AS II		1.7		RO
4367.10	4368.33	22892.1	AS		0.7		RO
4359.90	4361.13	22929.9	AS		1.0		RO
4353.02	4354.24	22966.1	AS II		2.0		RO
4352.25	4353.47	22970.2	AS II		2.3		RO
4336.85	4338.07	23051.7	AS II		2.0		RO
4324.10	4325.32	23119.7	AS II		1.7		RO
4315.86	4317.07	23163.8	AS II		1.7		RO
4302.26	4303.47	23237.1	AS		0.7		RO
4299.51	4300.72	23251.9	AS II		1.0		RO
4297.43	4298.64	23263.2	AS II		1.0		RO
4278.82	4280.02	23364.4	AS II		1.0		RO
4267.24	4268.44	23427.8	AS		0.5		RO

AIR WAVELENGTH (ANGSTRMS)	VACUUM WAVELENGTH (ANGSTRMS)	WAVENUMBER (KAYSERS)	LMENT	LOG ARC	INTENSITY SPK	DIS	REF
4265.13	4266.33	23439.4	AS		1.5		RO
4249.40	4250.60	23526.1	AS II		1.0		RO
4243.26	4244.45	23560.1	AS		2.0		RO
4228.42	4229.61	23642.8	AS		1.0		RO
4225.87	4227.06	23657.1	AS II		1.0		RO
4221.23	4222.42	23683.1	AS		0.7		RO
4209.71	4210.90	23747.9	AS II		1.0		RO
4208.00	4209.19	23757.6	AS II		1.5		RO
4197.61	4198.79	23816.4	AS II		1.5		RO
4190.37	4191.55	23857.5	AS II		1.0		RO
4160.72	4161.89	24027.5	AS II		1.0		RO
4157.64	4158.81	24045.3	AS II		1.5		RO
4123.71	4124.87	24243.2	AS II		0.7		RO
4119.85	4121.01	24265.9	AS		1.7		RO
4089.87	4091.02	24443.8	AS II		0.5		RO
4084.13	4085.28	24478.1	AS II		1.2		RO
4082.57	4083.72	24487.5	AS II		1.2		RO
4077.08	4078.23	24520.4	AS		1.0		RO
4071.64	4072.79	24553.2	AS		1.0		RO
4070.92	4072.07	24557.5	AS		1.0		RO
4065.54	4066.69	24590.0	AS II		1.0		RO
4062.73	4063.88	24607.0	AS II		1.0		RO
4022.32	4023.46	24854.2	AS II		0.7		RO
4010.66	4011.79	24926.5	AS II		1.0		RO
4006.34	4007.47	24953.4	AS II		1.7		RO
3982.45	3983.58	25103.1	AS		1.4		RO
3948.74	3949.86	25317.4	AS II		1.7		RO
3945.93	3947.05	25335.4	AS II		1.0		RO
3943.47	3944.59	25351.2	AS II		0.5		RO
3931.28	3932.39	25429.8	AS II		1.2		RO
3927.96	3929.07	25451.3	AS		0.5		RO
3890.46	3891.56	25696.6	AS II		0.7		RO
3842.82	3843.91	26015.2	AS II		1.7		RO
3841.03	3842.12	26027.3	AS		1.0		RO
3828.74	3829.83	26110.8	AS II		1.0		RO
3799.58	3800.66	26311.2	AS II		1.0		RO
3787.18	3788.26	26397.4	AS II		1.2		RO
3771.10	3772.17	26509.9	AS II		1.0		RO
3749.77	3750.84	26660.7	AS II		2.0		RO
3720.31	3721.37	26871.8	AS II		1.2		RO
3683.58	3684.63	27139.8	AS		1.2		RO
3671.92	3672.97	27226.0	AS II		1.2		RO
3656.78	3657.82	27338.7	AS II		0.7		RO
3551.82	3552.83	28146.5	AS II		1.0		RO
3545.77	3546.78	28194.6	AS		0.7		RO
3513.13	3514.13	28456.5	AS		0.7		RO
3504.30	3505.30	28528.2	AS II		0.7		RO
3126.99	3127.90	31970.4	AS II		1.2		RO
3122.18	3123.09	32019.6	AS II		1.0		RO
3119.60	3120.50	32046.1	AS I	2.0	1.7		ME
3116.63	3117.53	32076.6	AS II		2.2		RO
3075.317	3076.210	32507.53	AS I	1.8	1.5		RO
3058.10	3058.99	32690.5	AS II		1.2		RO
3053.49	3054.38	32739.9	AS II		1.2		RO
3048.36	3049.25	32795.0	AS II		1.0		RO
3032.84	3033.72	32962.8	AS I	2.1	1.8		M
3003.93	3004.81	33280.0	AS		1.7		RO
2990.99	2991.86	33424.0	AS I	1.0	1.3		ME
2959.70	2960.56	33777.3	AS		1.9		RO
2898.71	2899.56	34488.0	AS I	1.4R	1.6		ME
2890.24	2891.09	34589.1	AS II		0.5		RO
2884.51	2885.36	34657.8	AS II		1.4		RO
2860.452	2861.292	34949.25	AS I	1.7R	1.7		RO
2831.26	2832.09	35309.6	AS II		0.9		RO
2830.45	2831.28	35319.7	AS II		1.4		RO
2790.28	2791.10	35828.1	AS		0.5		RO
2780.197	2781.017	35958.06	AS I	1.9G	1.9		MIT
2744.991	2745.803	36419.22	AS I	0.9G	0.9		RO
2741.38	2742.19	36467.2	AS		0.3		RO
2733.30	2734.11	36575.0	AS		0.3		RO

AIR WAVELENGTH (ANGSTRMS)	VACUUM WAVELENGTH (ANGSTRMS)	WAVENUMBER (KAYSERS)	LMENT	LOG ARC	INTENSITY SPK	DIS	REF
2602.96	2603.74	38406.3	AS		1.1		RO
2577.21	2577.98	38790.0	AS		0.5		RO
2573.25	2574.02	38849.7	AS		0.7		RO
2561.18	2561.95	39032.8	AS		0.7		RO
2537.79	2538.55	39392.5	AS		0.7		RO
2498.25	2499.00	40016.0	AS		0.7		RO
2492.91	2493.66	40101.7	AS I	1.4	0.7		RO
2490.44	2491.19	40141.4	AS		0.3		RO
2465.65	2466.40	40545.0	AS		0.5		RO
2463.89	2464.64	40574.0	AS		0.5		RO
2461.76	2462.50	40609.1	AS		0.5		RO
2456.53	2457.27	40695.5	AS I	2.0R	0.9		ME
2445.65	2446.39	40876.5	AS		0.5		RO
2437.23	2437.97	41017.7	AS I	1.4	0.3		ME
2436.00	2436.74	41038.5	AS		0.6		RO
2433.76	2434.50	41076.2	AS		0.9		RO
2381.18	2381.91	41983.2	AS I	1.9	0.6		ME
2370.77	2371.49	42167.5	AS I	1.7R	0.5		ME
2369.67	2370.39	42187.1	AS I	1.6R	1.3		ME
2363.05	2363.77	42305.3	AS I	0.7			ME
2361.42	2362.14	42334.5	AS		0.3		RO
2349.84	2350.56	42543.1	AS I	2.4G	1.3		ME
2344.03	2344.75	42648.5	AS I	1.4			ME
2288.12	2288.83	43690.5	AS I	2.4G	0.7		ME
2271.36	2272.06	44012.9	AS I	1.4	0.0		ME
2266.70	2267.40	44103.4	AS I	1.1			ME
2228.66	2229.35	44856.1	AS I	1.0	0.0		ME
2205.97	2206.66	45317.4	AS I	0.9			ME
2205.16	2205.85	45334.0	AS I	0.7			ME
2198.34	2199.03	45474.7	AS I	0.5			ME
2187.75	2188.43	45694.8	AS I	0.5			ME
2182.94	2183.62	45795.4	AS I	1.0	0.0		ME
2176.26	2176.94	45936.0	AS I	0.9			ME
2166.21	2166.89	46149.1	AS		0.9		RO
2165.52	2166.20	46163.8	AS I	1.7R	0.5		ME
2147.46	2148.14	46552.0	AS		1.1		RO
2144.10	2144.78	46624.9	AS I	1.7R			ME
2138.53	2139.20	46746.4	AS	0.3			RO
2133.81	2134.48	46849.7	AS I	1.3			ME
2113.01	2113.68	47310.9	AS I	1.7	0.7		ME
2095.10	2095.77	47715.3	AS	0.5	0.5H		WT
2089.79	2090.45	47836.5	AS I	0.7	0.0		ME
2085.29	2085.95	47939.7	AS I	1.1	0.5		ME
2079.41	2080.07	48075.2	AS I	0.3	0.0		RO
2074.5	2075.2	48189.	AS		1.3		LG
2069.83	2070.49	48297.7	AS I	1.3	0.5		ME
2067.16	2067.82	48360.1	AS I	1.1	0.0		ME
2065.41	2066.07	48401.1	AS I	1.3	0.5		ME
2047.59	2048.25	48822.2	AS I	1.4	0.5		ME
2040.9	2041.6	48982.	AS		0.5		LG
2031.4	2032.1	49211.	AS		1.9		LG
2028.87	2029.52	49272.7	AS I	0.3			ME
2013.32	2013.97	49653.2	AS I	1.4R	0.9		ME
2012.77	2013.42	49666.7	AS I	1.0			ME
2010.04	2010.69	49734.2	AS I	1.0			ME
2009.18	2009.83	49755.5	AS I	1.7R	0.9		ME
2003.34	2003.99	49900.5	AS I	2.5G	1.0H		ME
2002.54	2003.19	49920.4	AS I	1.3R			ME
7510.75	7512.82	13310.6	AU I	1.3			MIT
7281.8	7283.8	13729.	AU I	0.3			ML
7156.82	7158.79	13968.8	AU	0.3			WT
7023.28	7025.22	14234.4	AU	0.3H			WT
6873.5	6875.4	14545.	AU I	0.7			ML
6781.97	6783.84	14740.9	AU	0.5H			WT
6692.811	6694.659	14937.28	AU	0.3			QI
6652.6	6654.4	15028.	AU I	1.2			ML
6634.22	6636.05	15069.2	AU	0.3			QI
6530.2	6532.0	15309.	AU I	0.7			ML
6456.52	6458.30	15483.9	AU II	1.0	0.7		ED
6436.2	6438.0	15533.	AU I	0.7			ML

AIR WAVELENGTH (ANGSTRMS)	VACUUM WAVELENGTH (ANGSTRMS)	WAVENUMBER (KAYSERS)	LMENT	LOG ARC	INTENSITY SPK	INTENSITY DIS	REF
6377.41	6379.17	15676.0	AU	0.7			WT
6337.4	6339.2	15775.	AU I	0.3			ML
6312.8	6314.5	15836.	AU I	0.9			ML
6302.2	6303.9	15863.	AU I	0.6			ML
6278.179	6279.915	15923.78	AU I	2.8	1.3		QI
6162.228	6163.933	16223.41	AU	1.0	0.7		QI
6160.1	6161.8	16229.	AU I	1.3			ML
6122.28	6123.97	16329.3	AU	0.3			QI
6101.654	6103.343	16384.46	AU	0.7			QI
6022.74	6024.41	16599.1	AU II	1.2	0.7		ED
5968.96	5970.61	16748.7	AU	0.9			WT
5962.72	5964.37	16766.2	AU	1.5	0.5		WT
5956.984	5958.634	16782.37	AU I	1.5	0.9		QI
5933.7	5935.3	16848.	AU I	0.7			ML
5862.943	5864.568	17051.55	AU I	1.5	0.7		QI
5841.44	5843.06	17114.3	AU	0.7			WT
5837.396	5839.014	17126.18	AU I	2.6H	1.0		QI
5726.82	5728.41	17456.9	AU I	1.5	1.0		ML
5721.26	5722.85	17473.8	AU I	1.2			ML
5655.766	5657.336	17676.17	AU I	1.5	0.7		MIT
5651.07	5652.64	17690.9	AU	0.3			WT
5629.27	5630.83	17759.4	AU	0.5			WT
5261.82	5263.28	18999.5	AU I	1.6	0.5		ML
5230.264	5231.720	19114.17	AU I	1.6	1.2		MIT
5147.39	5148.82	19421.9	AU I	1.6	0.7		ML
5064.61	5066.02	19739.4	AU I	1.6	1.0		EX
4902.27	4903.64	20393.0	AU	0.9	0.7		MIT
4811.62	4812.96	20777.2	AU I	1.7	1.2		MIT
4792.60	4793.94	20859.7	AU I	2.30	1.8		MIT
4791.54	4792.88	20864.3	AU		0.3		MIT
4760.20	4761.53	21001.6	AU II		0.6W		MIT
4637.36	4638.66	21558.0	AU		0.3		MIT
4633.09	4634.39	21577.8	AU		1.0		MIT
4620.70	4621.99	21635.7	AU I	0.6			ML
4607.50	4608.79	21697.7	AU		1.2R		MIT
4607.34	4608.63	21698.4	AU I	1.5	1.2		MIT
4587.89	4589.18	21790.4	AU		1.2		MIT
4549.46	4550.74	21974.5	AU		0.7		MIT
4488.25	4489.51	22274.1	AU I	1.6	1.5L		MIT
4437.27	4438.52	22530.1	AU I	1.7	1.0		MIT
4420.63	4421.87	22614.9	AU II		0.6		MIT
4395.29	4396.52	22745.2	AU II	0.7	0.3		MIT
4315.17	4316.38	23167.5	AU II	0.9	1.0H		MIT
4315.09	4316.30	23168.0	AU I	1.6	1.3		MIT
4259.89	4261.09	23468.2	AU II		1.2		MIT
4241.77	4242.96	23568.4	AU I	1.6	1.5		MIT
4084.12	4085.27	24478.2	AU I	1.4	1.1		MIT
4083.24	4084.39	24483.4	AU II	0.7	0.9		MIT
4076.33	4077.48	24524.9	AU II	0.6	1.4W		MIT
4065.08	4066.23	24592.8	AU I	1.7	1.5		MIT
4060.99	4062.14	24617.6	AU II	0.5	0.6		MIT
4052.81	4053.95	24667.3	AU II		1.8		MIT
4040.94	4042.08	24739.7	AU I	1.7	1.6		MIT
4028.48	4029.62	24816.2	AU		1.0		MIT
4016.050	4017.185	24893.05	AU II	1.0	1.2		MIT
3979.59	3980.72	25121.1	AU II	0.7	1.1		MIT
3959.113	3960.233	25251.04	AU	0.9	1.3		MIT
3927.66	3928.77	25453.2	AU I	0.7	0.8		MIT
3914.717	3915.826	25537.40	AU I	1.2	0.5		MIT
3914.19	3915.30	25540.8	AU	1.0	0.6		MIT
3909.376	3910.483	25572.29	AU I	1.0	1.2		MIT
3907.650	3908.757	25583.58	AU	0.7	0.5		MIT
3897.89	3898.99	25647.6	AU I	1.5	1.4		MIT
3889.47	3890.57	25703.2	AU I	0.5	0.9		MIT
3883.350	3884.451	25743.67	AU		1.3		MIT
3880.200	3881.300	25764.56	AU II		1.3		MIT
3875.072	3876.170	25798.66	AU	0.5			MIT
3874.672	3875.770	25801.32	AU	0.7	1.2		MIT
3871.35	3872.45	25823.5	AU	1.4	1.3		MIT
3869.617	3870.714	25835.03	AU II		1.0		MIT

AIR WAVELENGTH (ANGSTRMS)	VACUUM WAVELENGTH (ANGSTRMS)	WAVENUMBER (KAYSERS)	LMENT	LOG ARC	INTENSITY SPK	INTENSITY DIS	REF
3859.37	3860.46	25903.6	AU II		0.5		MIT
3855.60	3856.69	25928.9	AU	0.6	0.6		MIT
3854.855	3855.948	25933.96	AU	0.7	0.7		MIT
3847.423	3848.514	25984.05	AU	0.7	0.7		MIT
3844.89	3845.98	26001.2	AU		0.7		MIT
3836.477	3837.565	26058.19	AU	1.5	1.5		MIT
3831.14	3832.23	26094.5	AU	0.9	0.7		MIT
3829.382	3830.468	26106.47	AU	1.4	1.3		MIT
3825.646	3826.732	26131.96	AU II	1.3	1.5		MIT
3821.91	3822.99	26157.5	AU		0.7J		MIT
3816.130	3817.213	26197.12	AU	1.3	1.6		MIT
3804.00	3805.08	26280.7	AU II	1.4	2.2		MIT
3801.985	3803.064	26294.58	AU I	0.3			MIT
3795.90	3796.98	26336.7	AU I	1.3	0.5H		ML
3773.157	3774.229	26495.48	AU II		1.0		MIT
3769.99	3771.06	26517.7	AU		0.7H		MIT
3764.83	3765.90	26554.1	AU	0.7	0.5		MIT
3752.68	3753.75	26640.1	AU	0.7	1.0		MIT
3708.05	3709.11	26960.7	AU		0.7		MIT
3706.82	3707.87	26969.6	AU		1.2H		MIT
3702.30	3703.35	27002.6	AU		0.7H		MIT
3698.41	3699.46	27030.9	AU		0.3		MIT
3682.86	3683.91	27145.1	AU	1.3	1.0H		MIT
3674.90	3675.95	27203.9	AU	0.6	0.7		MIT
3654.67	3655.71	27354.5	AU	0.5	0.5		MIT
3653.93	3654.97	27360.0	AU	0.7	0.3		MIT
3653.491	3654.532	27363.29	AU II	0.5	0.7		MIT
3650.747	3651.787	27383.85	AU I	0.7	1.0		MIT
3649.09	3650.13	27396.3	AU	0.5	0.7H		MIT
3642.8	3643.8	27444.	AU II	1.0	0.7H		EX
3642.34	3643.38	27447.1	AU	0.7	0.0J		MIT
3637.34	3638.38	27484.8	AU	0.3	0.5H		MIT
3635.133	3636.169	27501.47	AU II	0.5	1.0		MIT
3633.24	3634.28	27515.8	AU II	0.0	1.2		MIT
3632.56	3633.60	27520.9	AU		0.3J		MIT
3623.44	3624.47	27590.2	AU		0.5		MIT
3620.23	3621.26	27614.7	AU		1.0		MIT
3614.00	3615.03	27662.3	AU		1.3		MIT
3607.54	3608.57	27711.8	AU		1.3		MIT
3598.08	3599.11	27784.7	AU I	1.5	1.0		ML
3594.15	3595.18	27815.1	AU	1.4	0.8		MIT
3591.99	3593.02	27831.8	AU	1.2	0.7		MIT
3586.74	3587.76	27872.5	AU I	1.3	1.2		MIT
3580.08	3581.10	27924.4	AU	1.3	1.2		MIT
3565.93	3566.95	28035.2	AU I	1.3	1.2		MIT
3553.572	3554.587	28132.66	AU I	1.3	1.0		MIT
3492.954	3493.954	28620.87	AU II		0.6		MIT
3404.918	3405.895	29360.86	AU		0.5H		MIT
3393.61	3394.58	29458.7	AU		0.5H		MIT
3382.00	3382.97	29559.8	AU II	0.8	0.9		MIT
3361.21	3362.18	29742.6	AU	0.5	0.6		MIT
3358.44	3359.41	29767.2	AU	0.7			MIT
3355.14	3356.10	29796.4	AU I	1.4			MIT
3349.40	3350.36	29847.5	AU	1.2	0.7		MIT
3323.19	3324.15	30082.9	AU I	1.0	0.7		MIT
3320.151	3321.106	30110.45	AU I	1.3	0.7		MIT
3318.65	3319.60	30124.1	AU	0.5	0.6H		MIT
3309.90	3310.85	30203.7	AU II		0.5H		EX
3308.310	3309.262	30218.22	AU I	1.7	1.2		MIT
3273.68	3274.62	30537.9	AU II		0.3		MIT
3267.08	3268.02	30599.6	AU	1.0	0.9		MIT
3243.40	3244.34	30822.9	AU II	0.5	0.9		MIT
3230.636	3231.569	30944.73	AU I	1.2	1.9		MIT
3230.636	3231.569	30944.73	AU II	1.2	1.9		MIT
3228.001	3228.933	30969.98	AU II		1.0		MIT
3204.74	3205.67	31194.8	AU I	1.7	1.5		MIT
3194.71	3195.63	31292.7	AU I	1.4	1.3		MIT
3172.31	3173.23	31513.6	AU II				ME
3172.31	3173.23	31513.6	AU II				ME
3164.9	3165.8	31587.	AU II		0.9H		EX

AIR WAVELENGTH (ANGSTRMS)	VACUUM WAVELENGTH (ANGSTRMS)	WAVENUMBER (KAYSERS)	LMENT	LOG ARC	INTENSITY SPK DIS	REF
3156.588	3157.502	31670.61	AU II		1.0	MIT
3147.58	3148.49	31761.2	AU I	1.0		MIT
3146.37	3147.28	31773.4	AU II	0.5	0.9	EX
3145.521	3146.432	31782.03	AU II	0.5	0.9	MIT
3122.781	3123.686	32013.46	AU I	2.7H	0.7	MIT
3122.50	3123.41	32016.3	AU II		0.3	MIT
3119.67	3120.57	32045.4	AU	0.7		MIT
3106.67	3107.57	32179.5	AU	0.7	0.7	MIT
3103.94	3104.84	32207.8	AU II		0.9	MIT
3093.3	3094.2	32319.	AU II		0.7	EX
3091.30	3092.20	32339.5	AU	0.7		MIT
3033.18	3034.06	32959.1	AU	1.4	1.5H	MIT
3029.205	3030.087	33002.35	AU I	1.4	1.5	MIT
3015.825	3016.704	33148.77	AU II		1.1	MIT
3014.23	3015.11	33166.3	AU	0.7	1.2	MIT
2994.99	2995.86	33379.4	AU II		2.0	MIT
2990.28	2991.15	33431.9	AU II		1.7	MIT
2982.11	2982.98	33523.5	AU II		1.2	MIT
2963.74	2964.61	33731.3	AU		1.0	MIT
2954.39	2955.25	33838.0	AU		1.7	MIT
2932.19	2933.05	34094.2	AU II	0.9	1.6	MIT
2932.19	2933.05	34094.2	AU I	0.9	1.6	MIT
2918.37	2919.22	34255.7	AU II		1.4	MIT
2913.54	2914.39	34312.5	AU II		1.7	MIT
2907.06	2907.91	34388.9	AU		1.4	MIT
2905.90	2906.75	34402.7	AU I	1.0	1.5	MIT
2905.90	2906.75	34402.7	AU II	1.0	1.5	MIT
2893.42	2894.27	34551.0	AU		1.5	MIT
2891.96	2892.81	34568.5	AU I	0.9	1.3	MIT
2885.60	2886.45	34644.7	AU II		1.1	MIT
2883.45	2884.30	34670.5	AU I	1.2	1.3	MIT
2873.42	2874.26	34791.5	AU I	1.3	1.3	ME
2872.38	2873.22	34804.1	AU I			ME
2864.52	2865.36	34899.6	AU II		1.0	MIT
2864.33	2865.17	34901.9	AU I			ME
2856.91	2857.75	34992.6	AU		1.3	MIT
2852.54	2853.38	35046.2	AU II		0.7	MIT
2847.09	2847.93	35113.3	AU II		1.4	MIT
2838.03	2838.86	35225.4	AU		1.9	MIT
2835.89	2836.72	35251.9	AU		1.0	MIT
2835.43	2836.26	35257.6	AU		0.9	MIT
2830.27	2831.10	35321.9	AU II		0.3	MIT
2825.45	2826.28	35382.2	AU	1.0	1.6	MIT
2822.72	2823.55	35416.4	AU		1.9	MIT
2819.95	2820.78	35451.2	AU		2.2	MIT
2805.32	2806.15	35636.1	AU II		1.5	MIT
2802.19	2803.02	35675.9	AU		2.3	MIT
2795.53	2796.35	35760.9	AU II		1.2	MIT
2780.83	2781.65	35949.9	AU		1.3	MIT
2751.02	2751.83	36339.4	AU		0.7H	MIT
2748.85	2749.66	36368.1	AU II		0.7R	MIT
2748.26	2749.07	36375.9	AU II	1.6	1.9	MIT
2748.26	2749.07	36375.9	AU I	1.6	1.9	MIT
2732.01	2732.82	36592.2	AU II		1.5	MIT
2703.36	2704.16	36980.0	AU		0.9	MIT
2700.89	2701.69	37013.9	AU I	1.3	1.4	MIT
2694.37	2695.17	37103.4	AU		0.5	MIT
2688.71	2689.51	37181.5	AU I	0.3	1.3	MIT
2688.16	2688.96	37189.1	AU II		1.0	MIT
2687.63	2688.43	37196.5	AU		1.2	MIT
2675.95	2676.75	37358.8	AU I	2.4G	2.0	MIT
2666.97	2667.76	37484.6	AU II		0.7	MIT
2665.154	2665.947	37510.13	AU		0.7	MIT
2659.43	2660.22	37590.9	AU		0.7	MIT
2641.49	2642.28	37846.1	AU I	0.7	1.3	MIT
2627.03	2627.81	38054.5	AU II		0.8	MIT
2625.50	2626.28	38076.6	AU		1.0	MIT
2617.41	2618.19	38194.3	AU		0.8	MIT
2616.57	2617.35	38206.6	AU		0.9	MIT
2612.74	2613.52	38262.6	AU		0.5	MIT

AIR WAVELENGTH (ANGSTRMS)	VACUUM WAVELENGTH (ANGSTRMS)	WAVENUMBER (KAYSERS)	LMENT	LOG ARC	INTENSITY SPK DIS	REF
2609.48	2610.26	38310.4	AU		0.7	MIT
2602.06	2602.84	38419.6	AU		0.5	MIT
2592.09	2592.87	38567.4	AU		1.0	MIT
2590.04	2590.81	38597.9	AU I	1.5	1.7	MIT
2590.04	2590.81	38597.9	AU II	1.5	1.7	MIT
2565.71	2566.48	38963.9	AU		1.1	MIT
2562.64	2563.41	39010.6	AU II		0.7	MIT
2552.80	2553.57	39160.9	AU		0.7H	MIT
2551.90	2552.67	39174.7	AU		0.7	MIT
2550.24	2551.01	39200.2	AU		0.7H	MIT
2544.25	2545.01	39292.5	AU II		1.0	EX
2544.19	2544.95	39293.4	AU I	1.5	0.9	MIT
2538.00	2538.76	39389.3	AU		0.7H	MIT
2533.69	2534.45	39456.3	AU		1.0H	MIT
2528.11	2528.87	39543.4	AU		0.7H	MIT
2515.16	2515.92	39746.9	AU II		0.7	MIT
2510.49	2511.25	39820.9	AU I	1.4	1.2	MIT
2506.30	2507.06	39887.4	AU		0.7	MIT
2503.32	2504.07	39934.9	AU		1.5	MIT
2498.88	2499.63	40005.9	AU		0.7	MIT
2492.64	2493.39	40106.0	AU II		0.7	MIT
2491.24	2491.99	40128.5	AU I	1.0	0.7H	MIT
2490.38	2491.13	40142.4	AU		0.7H	MIT
2480.28	2481.03	40305.9	AU		1.2H	MIT
2477.71	2478.46	40347.7	AU II		1.2	MIT
2476.04	2476.79	40374.9	AU		0.7H	MIT
2458.12	2458.86	40669.2	AU		0.7	MIT
2446.15	2446.89	40868.2	AU II		0.7	EX
2445.53	2446.27	40878.5	AU II		0.7	MIT
2442.34	2443.08	40931.9	AU		0.7H	MIT
2435.43	2436.17	41048.1	AU		0.7H	MIT
2433.54	2434.28	41079.9	AU		0.7H	MIT
2432.95	2433.69	41089.9	AU		0.7H	MIT
2427.95	2428.69	41174.5	AU I	2.6G	2.0	MIT
2419.34	2420.08	41321.0	AU II		0.7H	MIT
2416.61	2417.34	41367.7	AU II		0.7H	MIT
2405.13	2405.86	41565.1	AU		0.7	MIT
2404.81	2405.54	41570.7	AU I	1.3	0.6	MIT
2402.71	2403.44	41607.0	AU		0.7	MIT
2401.64	2402.37	41625.5	AU		0.7	MIT
2393.52	2394.25	41766.7	AU		0.7	MIT
2388.39	2389.12	41856.5	AU II		0.5	EX
2388.10	2388.83	41861.5	AU II		0.3	MIT
2387.75	2388.48	41867.7	AU I	1.5	1.0	MIT
2384.15	2384.88	41930.9	AU		0.7H	MIT
2382.41	2383.14	41961.5	AU II		0.7	MIT
2376.24	2376.97	42070.5	AU I	1.4	0.5	MIT
2373.14	2373.86	42125.4	AU II		0.7	MIT
2371.60	2372.32	42152.7	AU II		0.7	MIT
2369.36	2370.08	42192.6	AU		0.6	MIT
2367.95	2368.67	42217.7	AU		0.7H	MIT
2364.89	2365.61	42272.4	AU		1.2H	MIT
2364.56	2365.28	42278.2	AU II	1.3	0.9	MIT
2364.56	2365.28	42278.2	AU I	1.3	0.9	MIT
2355.51	2356.23	42440.7	AU		0.7	MIT
2352.65	2353.37	42492.3	AU I	1.4		MIT
2351.58	2352.30	42511.6	AU II		0.7	MIT
2347.11	2347.83	42592.5	AU II		0.7	MIT
2344.28	2345.00	42644.0	AU II		0.7	MIT
2341.64	2342.36	42692.0	AU II		0.7	MIT
2340.19	2340.91	42718.5	AU II		1.0	MIT
2334.06	2334.78	42830.7	AU II		0.7H	MIT
2325.71	2326.42	42984.4	AU II		0.7	MIT
2325.26	2325.97	42992.7	AU		0.7H	MIT
2324.67	2325.38	43003.6	AU		0.7	MIT
2322.28	2322.99	43047.9	AU		0.8	MIT
2318.33	2319.04	43121.2	AU		0.7H	MIT
2315.85	2316.56	43167.4	AU		0.8	MIT
2314.64	2315.35	43190.0	AU		1.2	MIT
2312.32	2313.03	43233.3	AU II		0.7	MIT

AIR WAVELENGTH (ANGSTRMS)	VACUUM WAVELENGTH (ANGSTRMS)	WAVENUMBER (KAYSERS)	LMENT	LOG ARC	INTENSITY SPK	INTENSITY DIS	REF
2310.95	2311.66	43258.9	AU		0.5		MIT
2309.43	2310.14	43287.4	AU		1.1		MIT
2308.21	2308.92	43310.3	AU		0.5		MIT
2304.81	2305.52	43374.2	AU II		1.5		MIT
2301.03	2301.74	43445.4	AU II		0.5		MIT
2298.06	2298.77	43501.6	AU II		0.5		MIT
2296.51	2297.22	43530.9	AU		0.5		MIT
2295.15	2295.86	43556.7	AU		0.5		MIT
2293.98	2294.69	43578.9	AU		0.3H		MIT
2291.52	2292.23	43625.7	AU II		1.2		MIT
2288.63	2289.34	43680.8	AU		0.5		MIT
2288.25	2288.96	43688.0	AU II		0.3H		EX
2283.32	2284.02	43782.4	AU II		1.2		MIT
2282.90	2283.60	43790.4	AU		0.8		MIT
2279.41	2280.11	43857.5	AU		0.5H		MIT
2277.64	2278.34	43891.5	AU II		1.0		MIT
2277.33	2278.03	43897.5	AU		0.9		MIT
2265.99	2266.69	44117.2	AU II		0.5H		MIT
2263.72	2264.42	44161.4	AU II		0.5H		MIT
2262.73	2263.43	44180.7	AU		0.9		MIT
2261.34	2262.04	44207.9	AU II		0.5		MIT
2260.36	2261.06	44227.0	AU		0.7		MIT
2255.91	2256.61	44314.3	AU II		0.7		MIT
2253.46	2254.16	44362.5	AU II		0.7		MIT
2249.07	2249.77	44449.0	AU II		0.9H		MIT
2248.74	2249.44	44455.6	AU		0.8		MIT
2246.68	2247.38	44496.3	AU II		1.3		MIT
2246.43	2247.13	44501.3	AU		1.2		MIT
2245.44	2246.14	44520.9	AU		1.0		MIT
2242.71	2243.41	44575.1	AU		1.5		MIT
2240.28	2240.98	44623.4	AU II		0.7		MIT
2237.49	2238.18	44679.1	AU II		1.0		MIT
2233.70	2234.39	44754.9	AU		0.7		MIT
2231.31	2232.00	44802.8	AU II		1.3		MIT
2229.03	2229.72	44848.6	AU II		1.7		MIT
2222.56	2223.25	44979.2	AU		1.2		MIT
2220.76	2221.45	45015.6	AU II		1.0		MIT
2220.53	2221.22	45020.3	AU		0.8		MIT
2219.22	2219.91	45046.9	AU		0.5H		MIT
2215.76	2216.45	45117.2	AU		1.2		MIT
2213.18	2213.87	45169.8	AU II		1.1		MIT
2210.66	2211.35	45221.3	AU II		0.6		MIT
2205.89	2206.58	45319.0	AU II		1.0		MIT
2201.35	2202.04	45412.5	AU		1.5		MIT
2188.98	2189.66	45669.1	AU II		1.2		MIT
2185.54	2186.22	45741.0	AU II		1.0		MIT
2184.113	2184.797	45770.85	AU II		1.0		MIT
2176.33	2177.01	45934.5	AU I	1.2			ML
2129.48	2130.15	46945.0	AU I	0.8			MIT
2126.644	2127.316	47007.59	AU I	0.8			MIT
2125.34	2126.01	47036.4	AU		1.5		MIT
2110.80	2111.47	47360.4	AU		1.8		MIT
2021.38	2022.03	49455.2	AU I	0.9			ML
2012.00	2012.65	49685.7	AU I	1.2	1.3		MIT
6364.75	6366.51	15707.2	B	0.3			WT
6081.0	6082.7	16440.	B II		0.7		EN
4940.64	4942.02	20234.6	B II		0.3		EN
4937.70	4939.08	20246.7	B		0.3		SY
4901.85	4903.22	20394.8	B		0.3		SV
4784.29	4785.63	20895.9	B II		0.6		EN
4472.82	4474.08	22351.0	B II		0.6		EN
4472.08	4473.33	22354.7	B II		0.6		EN
4402.95	4404.19	22705.7	B		0.3		SY
4382.95	4384.18	22809.3	B		0.6		SY
4375.60	4376.83	22847.6	B		0.3		SY
4371.69	4372.92	22868.0	B		0.3		SY
4355.68	4356.90	22952.1	B		0.8		SY
4353.17	4354.39	22965.3	B		0.3		SY
4344.77	4345.99	23009.7	B		0.6		SY
4231.07	4232.26	23628.0	B	0.5	0.3		SY

AIR WAVELENGTH (ANGSTRMS)	VACUUM WAVELENGTH (ANGSTRMS)	WAVENUMBER (KAYSERS)	LMENT	LOG ARC	INTENSITY SPK	INTENSITY DIS	REF
4226.15	4227.34	23655.5	B	0.6	0.3		SY
4194.82	4196.00	23832.2	B II		0.3		EN
4164.43	4165.60	24006.1	B		0.6		SY
4150.71	4151.88	24085.5	B		0.3		SY
4121.95	4123.11	24253.5	B II		1.3		EN
4082.62	4083.77	24487.2	B		0.3		SY
4050.13	4051.27	24683.6	B		0.3		SY
4039.57	4040.71	24748.1	B		0.6		SY
3901.95	3903.06	25620.9	B	0.5	0.3		EN
3871.39	3872.49	25823.2	B		1.3H		SY
3792.5	3793.6	26360.	B II			2.7	ML
3541.2	3542.2	28231.	B				ME
3452.280	3453.269	28958.07	B II	0.7	1.5		MIT
3451.41	3452.40	28965.4	B II		2.0		EN
3360.09	3361.06	29752.6	B	0.3	1.3		SY
3323.61	3324.57	30079.1	B II		1.0		EN
3323.34	3324.30	30081.6	B II		1.0		SY
3323.20	3324.16	30082.8	B II		0.7		EN
3302.51	3303.46	30271.3	B		1.0		SY
3282.01	3282.96	30460.4	B	0.6	1.1		SY
3260.74	3261.68	30659.0	B II	0.6	1.0		SY
3225.289	3226.220	30996.02	B II	0.7			MIT
3224.25	3225.18	31006.0	B II		0.7		EN
3222.069	3222.999	31027.00	B II		0.7		MIT
3191.84	3192.76	31320.8	B	0.5	1.0		SY
3179.351	3180.271	31443.86	B II	0.7	2.0		MIT
3158.54	3159.45	31651.0	B	0.7	0.3		SY
3136.85	3137.76	31869.9	B II	0.5	0.5H		SY
3135.64	3136.55	31882.2	B II		0.5		SY
3112.50	3113.40	32119.2	B		0.7		SY
3102.09	3102.99	32227.0	B II		0.7		SY
3086.06	3086.96	32394.4	B		0.3		SY
3032.28	3033.16	32968.9	B		1.0		EN
3013.28	3014.16	33176.8	B II		0.5		SY
2981.53	2982.40	33530.0	B		0.3		SY
2918.15	2919.00	34258.3	B II		0.3		SY
2889.72	2890.57	34595.3	B II		0.3		SY
2888.32	2889.17	34612.0	B		0.3		SY
2886.80	2887.65	34630.3	B		0.3		SY
2809.72	2810.55	35580.2	B		0.3		SY
2785.14	2785.96	35894.2	B		1.5		SY
2779.26	2780.08	35970.2	B		2.0		SY
2749.89	2750.70	36354.3	B		0.3		SY
2731.94	2732.75	36593.2	B		0.3		SY
2709.00	2709.80	36903.0	B II		0.3		EN
2696.84	2697.64	37069.4	B		0.3		SY
2695.18	2695.98	37092.3	B		0.8		SY
2652.81	2653.60	37684.7	B II		0.3		SY
2652.58	2653.37	37687.9	B II	0.6	0.3		SY
2610.26	2611.04	38298.9	B		0.3		SY
2566.40	2567.17	38953.4	B II		0.3		SY
2566.26	2567.03	38955.5	B II		1.2		SY
2557.52	2558.29	39088.6	B		0.8		SY
2515.06	2515.82	39748.5	B II		0.8		SY
2514.96	2515.72	39750.1	B		0.8		SY
2514.39	2515.15	39759.1	B		1.7		SY
2508.45	2509.21	39853.2	B		0.3		SY
2497.733	2498.486	40024.24	B I	2.7	2.6		M
2496.778	2497.531	40039.54	B I	2.5	2.5		M
2459.923	2460.667	40639.38	B II		0.3		MIT
2446.10	2446.84	40869.0	B			0.8	SY
2445.11	2445.85	40885.6	B		0.8		SY
2436.95	2437.69	41022.5	B		0.3		SY
2434.95	2435.69	41056.1	B		1.3		SY
2432.29	2433.03	41101.0	B II		1.5		SY
2430.82	2431.56	41125.9	B		0.3		SY
2415.06	2415.79	41394.3	B		0.3		SY
2400.03	2400.76	41653.5	B		0.3		SY
2395.07	2395.80	41739.7	B II		1.2		SY
2393.234	2393.963	41771.74	B II		0.3		MIT

AIR WAVELENGTH (ANGSTRMS)	VACUUM WAVELENGTH (ANGSTRMS)	WAVENUMBER (KAYSERS)	LMENT	LOG ARC	INTENSITY SPK	INTENSITY DIS	REF
2369.96	2370.68	42181.9	B		1.3		SY
2363.88	2364.60	42290.4	B II		1.2		SY
2363.51	2364.23	42297.0	B		0.3		MIT
2357.03	2357.75	42413.3	B		0.8		SY
2355.25	2355.97	42445.4	B		0.8		MIT
2328.67	2329.38	42929.8	B II		0.3		EN
2323.02	2323.73	43034.2	B II		0.3		EN
2266.946	2267.647	44098.57	B II		0.3W		MIT
2266.324	2267.025	44110.67	B II		0.3W		MIT
2220.31	2221.00	45024.7	B II		0.8		EN
2089.59	2090.25	47841.1	B I	2.2	1.3		SY
2088.93	2089.59	47856.2	B I	2.0	1.2		SY
9830.37	9833.07	10169.8	BA I	2.5B			ME
9821.60	9824.29	10178.8	BA I	0.3H			ME
9792.91	9795.60	10208.7	BA I	0.6H			ME
9772.62	9775.30	10229.9	BA I	0.6H			ME
9713.77	9716.43	10291.8	BA I	1.4H			MIT
9704.45	9707.11	10301.7	BA I	1.0H			MIT
9645.76	9648.41	10364.4	BA I	1.4H			MIT
9608.90	9611.54	10404.2	BA I	2.2			MIT
9589.37	9592.00	10425.3	BA I	1.7			MIT
9530.30	9532.91	10490.0	BA I	0.5H			ME
9524.76	9527.37	10496.1	BA I	1.6H			MIT
9455.98	9458.57	10572.4	BA I	2.0			MIT
9450.08	9452.67	10579.0	BA I	1.0H			MIT
9414.6	9417.2	10619.	BA I	0.6H			ME
9403.58	9406.16	10631.3	BA I	1.0			MIT
9398.8	9401.4	10637.	BA I	0.7H			ME
9370.09	9372.66	10669.3	BA I	2.5			MIT
9367.49	9370.06	10672.3	BA I	1.6H			MIT
9324.58	9327.14	10721.4	BA I	1.7H			MIT
9308.16	9310.71	10740.3	BA I	1.7H			MIT
9306.50	9309.05	10742.2	BA I	0.6H			MIT
9253.09	9255.63	10804.2	BA I	1.4			MIT
9219.72	9222.25	10843.3	BA I	2.1			MIT
9215.36	9217.89	10848.5	BA I	1.4H			MIT
9189.58	9192.10	10878.9	BA I	1.8			MIT
9159.66	9162.17	10914.4	BA I	1.0H			ME
9133.29	9135.80	10945.9	BA I	1.2H			ME
9097.8	9100.3	10989.	BA	0.3H			ME
9018.63	9021.11	11085.1	BA I	0.5H			ME
8975.6	8978.1	11138.	BA	0.3H			ME
8937.93	8940.38	11185.2	BA I	1.0H			ME
8927.41	8929.86	11198.4	BA I	0.8B			ME
8914.96	8917.41	11214.0	BA I	2.0			MIT
8909.83	8912.28	11220.5	BA I	0.5H			ME
8860.99	8863.42	11282.3	BA I	1.9			MIT
8799.76	8802.18	11360.8	BA I	2.0H			ME
8793.36	8795.77	11369.1	BA I	0.3H			ME
8737.74	8740.14	11441.5	BA II	0.3H			ME
8710.82	8713.21	11476.8	BA I	0.3H			ME
8654.07	8656.45	11552.1	BA I	1.6			MIT
8593.47	8595.83	11633.6	BA I	0.3			ME
8581.98	8584.34	11649.1	BA I	1.7B			ME
8567.58	8569.93	11668.7	BA I	0.7H			ME
8559.95	8562.30	11679.1	BA I	2.6			MIT
8521.96	8524.30	11731.2	BA I	0.3			ME
8514.28	8516.62	11741.7	BA I	1.6H			MIT
8414.58	8416.89	11880.9	BA I	0.6B			ME
8350.76	8353.06	11971.7	BA I	0.3H			ME
8325.39	8327.68	12008.2	BA I	1.4			MIT
8285.00	8287.28	12066.7	BA I	0.7H			ME
8264.01	8266.28	12097.3	BA I	0.5H			ME
8224.41	8226.67	12155.6	BA	0.8H			ME
8210.24	8212.50	12176.6	BA I	2.3			MIT
8161.64	8163.88	12249.1	BA	1.2W			MIT
8158.12	8160.36	12254.4	BA I	0.5H			ME
8147.75	8149.99	12269.9	BA I	1.3H			MIT
8120.49	8122.72	12311.1	BA I	1.3H			MIT
8018.25	8020.46	12468.1	BA I	1.0H			MIT

AIR WAVELENGTH (ANGSTRMS)	VACUUM WAVELENGTH (ANGSTRMS)	WAVENUMBER (KAYSERS)	LMENT	LOG ARC	INTENSITY SPK	INTENSITY DIS	REF
7982.40	7984.60	12524.1	BA I	1.3H			ME
7961.26	7963.45	12557.4	BA I	1.2H			MIT
7957.33	7959.52	12563.6	BA I	0.3H			ME
7939.42	7941.60	12591.9	BA I	0.5H			ME
7911.30	7913.48	12636.7	BA I	2.3			MIT
7905.72	7907.89	12645.6	BA I	2.5			MIT
7877.98	7880.15	12690.1	BA I	1.3L			MIT
7839.51	7841.67	12752.4	BA I	1.2			MIT
7798.26	7800.41	12819.8	BA I	0.6H			ME
7780.43	7782.57	12849.2	BA I	2.5			MIT
7775.32	7777.46	12857.7	BA I	1.4			MIT
7766.86	7769.00	12871.7	BA I	1.2			MIT
7751.73	7753.86	12896.8	BA I	1.6			MIT
7721.87	7724.00	12946.7	BA I	1.0H			MIT
7706.50	7708.62	12972.5	BA I	1.4			MIT
7672.02	7674.13	13030.8	BA I	2.6			MIT
7642.88	7644.98	13080.5	BA I	2.0			MIT
7636.85	7638.95	13090.8	BA I	1.6			MIT
7610.46	7612.56	13136.2	BA I	1.8			MIT
7583.806	7585.894	13182.36	BA	1.0			BU
7575.22	7577.31	13197.3	BA I	0.3H			ME
7543.41	7545.49	13253.0	BA I	1.2H			MIT
7528.20	7530.27	13279.7	BA I	1.2H			ME
7523.60	7525.67	13287.8	BA I	1.2H			ME
7513.40	7515.47	13305.9	BA I	0.3H			ME
7488.06	7490.12	13350.9	BA I	2.3			MIT
7476.18	7478.24	13372.1	BA I	1.5H			MIT
7459.75	7461.80	13401.6	BA I	2.5B			MIT
7417.52	7419.56	13477.9	BA I	2.0			MIT
7414.26	7416.30	13483.8	BA I	0.6B			ME
7409.96	7412.00	13491.6	BA I	1.5B			MIT
7392.42	7394.46	13523.6	BA I	2.6			MIT
7375.53	7377.56	13554.6	BA I	1.7B			MIT
7359.29	7361.32	13584.5	BA I	1.3			ME
7339.57	7341.59	13621.0	BA I	0.6H			ME
7326.50	7328.52	13645.3	BA I	1.0H			ME
7307.23	7309.24	13681.3	BA I	1.0H			ME
7304.46	7306.47	13686.5	BA	0.7H			ME
7280.27	7282.28	13732.0	BA I	3.0			MIT
7272.33	7274.33	13747.0	BA I	0.3H			ME
7228.81	7230.80	13829.7	BA I	2.3B			MIT
7213.56	7215.55	13859.0	BA I	1.0			ME
7208.12	7210.11	13869.4	BA I	1.3H			MIT
7195.23	7197.21	13894.3	BA I	2.3			MIT
7153.58	7155.55	13975.2	BA I	1.9B			MIT
7148.12	7150.09	13985.8	BA	0.3			MIT
7133.24	7135.21	14015.0	BA	0.6H			MIT
7127.444	7129.409	14026.41	BA	0.3			BU
7126.60	7128.56	14028.1	BA	1.0H			ME
7120.31	7122.27	14040.5	BA I	2.9B			MIT
7090.00	7091.95	14100.5	BA I	2.0B			MIT
7069.43	7071.38	14141.5	BA I	0.5			ME
7059.96	7061.91	14160.5	BA I	3.3			MIT
6986.79	6988.72	14308.8	BA I	0.7			ME
6961.63	6963.55	14360.5	BA I	0.3H			ME
6932.92	6934.83	14420.0	BA I	0.3H			ME
6874.09	6875.99	14543.4	BA II			1.0	RS
6867.87	6869.76	14556.5	BA I	2.0H			MIT
6865.71	6867.60	14561.1	BA I	2.3			MIT
6771.85	6773.72	14762.9	BA I	1.8H			ME
6769.62	6771.49	14767.8	BA II			1.0	RS
6761.86	6763.73	14784.7	BA I	0.3H			ME
6714.03	6715.88	14890.1	BA I	0.3H			ME
6693.88	6695.73	14934.9	BA I	2.8	2.0		MIT
6675.268	6677.111	14976.54	BA I	2.7	2.0		MIT
6654.05	6655.89	15024.3	BA I	1.7			ME
6643.77	6645.60	15047.5	BA	0.6H			LR
6595.32	6597.14	15158.1	BA I	3.0	2.5		MIT
6580.73	6582.55	15191.7	BA	1.3H			ME
6565.91	6567.72	15226.0	BA	0.5			LR

AIR WAVELENGTH (ANGSTRMS)	VACUUM WAVELENGTH (ANGSTRMS)	WAVENUMBER (KAYSERS)	LMENT	LOG ARC	INTENSITY SPK	DIS	REF
6564.33	6566.14	15229.6	BA I	0.5			ME
6544.89	6546.70	15274.9	BA	0.7			LR
6527.314	6529.117	15316.01	BA	2.3	1.3		IKS
6503.433	6505.230	15372.25	BA	0.3	0.3		MIT
6498.762	6500.558	15383.30	BA	1.8	1.3		IKS
6496.901	6498.696	15387.70	BA	2.9R	2.5V		IKS
6490.44	6492.23	15403.0	BA	0.3			BU
6482.912	6484.703	15420.91	BA	2.0V	1.7		IKS
6450.854	6452.637	15497.54	BA	2.0	1.3		IKS
6446.902	6448.684	15507.04	BA	1.3G	1.2		SZ
6411.34	6413.11	15593.1	BA	1.0	0.5		LR
6378.91	6380.67	15672.3	BA II	0.5		0.7	RS
6341.683	6343.436	15764.33	BA	2.0	1.7		IKS
6339.57	6341.32	15769.6	BA	0.7			BU
6300.30	6302.04	15867.9	BA	0.3H			LR
6245.83	6247.56	16006.2	BA	0.3H	0.3		BU
6243.85	6245.58	16011.3	BA	0.8	0.3		BU
6233.586	6235.310	16037.69	BA	1.3W	0.5		SZ
6174.54	6176.25	16191.1	BA	0.8	1.1		LR
6141.716	6143.416	16277.59	BA	3.3J	3.3J		IKS
6140.39	6142.09	16281.1	BA	1.0			LR
6135.83	6137.53	16293.2	BA II			0.6	RS
6129.335	6131.031	16310.47	BA	0.8			SZ
6110.785	6112.477	16359.98	BA I	2.3V	1.8		IKS
6083.441	6085.125	16433.52	BA	1.0	0.5		SZ
6063.118	6064.797	16488.60	BA I	2.3J	1.8		IKS
6035.52	6037.19	16564.0	BA	0.5			LR
6019.474	6021.141	16608.15	BA I	2.2V	1.7		IKS
5999.85	6001.51	16662.5	BA II			1.2	RS
5997.091	5998.752	16670.13	BA I	2.2J	1.7		IKS
5985.54	5987.20	16702.3	BA	0.5			LR
5981.25	5982.91	16714.3	BA II			1.6	RS
5978.496	5980.152	16721.98	BA	0.8	0.5		SZ
5972.78	5974.43	16738.0	BA	2.0			EX
5971.701	5973.355	16741.01	BA I	2.2	1.7		IKS
5964.787	5966.439	16760.42	BA	0.8			SZ
5962.445	5964.097	16767.00	BA	0.7			SZ
5935.421	5937.066	16843.34	BA	0.6			MIT
5927.86	5929.50	16864.8	BA	0.6			LR
5907.650	5909.287	16922.51	BA	1.3			MIT
5853.679	5855.302	17078.54	BA	2.5	2.0H		IKS
5826.295	5827.910	17158.81	BA I	2.2J			MIT
5818.890	5820.503	17180.65	BA I	0.8	0.5		MIT
5805.690	5807.300	17219.71	BA	1.8			MIT
5800.283	5801.891	17235.76	BA I	2.0	1.3		MIT
5784.18	5785.78	17283.7	BA II			1.6	RS
5784.065	5785.669	17284.09	BA	0.8	0.3		MIT
5777.665	5779.267	17303.23	BA I	2.7G	2.0G		MIT
5720.712	5722.299	17475.49	BA	0.9			SZ
5718.364	5719.950	17482.67	BA	0.8			SZ
5715.953	5717.539	17490.04	BA	0.6			SZ
5713.554	5715.139	17497.39	BA	1.0			SZ
5709.546	5711.130	17509.67	BA	0.7W			SZ
5706.042	5707.625	17520.42	BA	0.5			SZ
5704.820	5706.403	17524.18	BA	0.7			SZ
5680.199	5681.775	17600.13	BA	1.8			MIT
5628.938	5630.500	17760.41	BA	0.6			SZ
5625.701	5627.263	17770.63	BA	0.8			SZ
5619.099	5620.659	17791.51	BA	1.0	0.5		SZ
5618.7	5620.3	17793.	BA	1.0			EX
5593.294	5594.847	17873.59	BA	1.1			MIT
5546.148	5547.688	18025.53	BA	0.5J			MIT
5535.551	5537.089	18060.03	BA I	3.0G	2.3G		KN
5519.115	5520.648	18113.81	BA I	2.3J	1.8		SZ
5480.30	5481.82	18242.1	BA II			1.2	RS
5473.689	5475.210	18264.14	BA	1.0			MIT
5437.393	5438.904	18386.06	BA	0.7			SZ
5428.79	5430.30	18415.2	BA II			0.7	RS
5425.55	5427.06	18426.2	BA I	1.0			FL
5424.616	5426.124	18429.36	BA I	2.0G	1.5G		SZ

AIR WAVELENGTH (ANGSTRMS)	VACUUM WAVELENGTH (ANGSTRMS)	WAVENUMBER (KAYSERS)	LMENT	LOG ARC	INTENSITY SPK	DIS	REF
5421.05	5422.56	18441.5	BA II			0.7	RS
5416.344	5417.850	18457.51	BA	0.9			SZ
5404.920	5406.423	18496.52	BA	1.2			SZ
5391.60	5393.10	18542.2	BA II			1.7	RS
5361.35	5362.84	18646.8	BA II			1.6	RS
5349.621	5351.109	18687.72	BA	0.8			SZ
5308.952	5310.429	18830.87	BA	1.0			SZ
5305.758	5307.234	18842.21	BA	0.9			SZ
5302.808	5304.283	18852.69	BA	1.3	0.7		SZ
5294.130	5295.603	18883.59	BA	1.0			SZ
5290.945	5292.417	18894.96	BA	1.0			SZ
5279.619	5281.088	18935.49	BA	0.7			SZ
5277.625	5279.094	18942.65	BA	1.0			SZ
5267.033	5268.499	18980.74	BA I	1.4	1.1		SZ
5253.807	5255.269	19028.52	BA	1.0			SZ
5177.448	5178.890	19309.16	BA	0.6	0.3		SZ
5175.619	5177.061	19315.98	BA	1.0	0.8		SZ
5159.919	5161.356	19374.75	BA	1.7H	1.0		SZ
5054.975	5056.384	19776.98	BA	1.1	0.3		SZ
5013.082	5014.480	19942.25	BA II		1.7		MIT
4997.81	4999.20	20003.2	BA II			1.0	RS
4995.730	4997.124	20011.51	BA	0.3			MIT
4957.162	4958.545	20167.21	BA II			1.7	MIT
4947.333	4948.714	20207.27	BA	0.5	0.3		MIT
4934.086	4935.463	20261.52	BA II	2.6H	2.6H		MIT
4902.898	4904.267	20390.41	BA I	1.2	0.5		SZ
4899.971	4901.339	20402.59	BA II	1.5H	2.3L		SZ
4877.650	4879.012	20495.95	BA I	1.5J	0.9		SZ
4850.84	4852.20	20609.2	BA II			0.7	RS
4847.14	4848.49	20625.0	BA II			1.0	RS
4843.46	4844.81	20640.6	BA II			1.9	RS
4726.450	4727.772	21151.61	BA I	1.9	1.5		MIT
4724.734	4726.056	21159.29	BA	0.7	0.5		MIT
4708.94	4710.26	21230.3	BA II			1.9	RS
4700.436	4701.751	21268.67	BA I	1.4	0.5		MIT
4691.622	4692.935	21308.63	BA	2.0	1.6		MIT
4673.621	4674.929	21390.70	BA	1.6	0.7		MIT
4663.00	4664.31	21439.4	BA I	0.7			FL
4644.10	4645.40	21526.7	BA			1.0	RS
4642.038	4643.338	21536.23	BA	0.5			SZ
4636.333	4637.631	21562.73	BA	1.3			SZ
4628.332	4629.628	21600.01	BA	1.6	0.6		MIT
4619.978	4621.272	21639.06	BA I	1.4			SZ
4604.990	4606.280	21709.49	BA	1.0	0.5		MIT
4599.751	4601.040	21734.22	BA	1.7	1.0		SZ
4591.829	4593.116	21771.71	BA	1.2	0.5		MIT
4589.759	4591.045	21781.53	BA	1.0	0.5		MIT
4579.667	4580.950	21829.53	BA	1.9	1.6		SZ
4573.855	4575.137	21857.27	BA	1.7	1.3		MIT
4554.042	4555.319	21952.36	BA II	3.0G	2.3		MIT
4524.946	4526.215	22093.52	BA II	1.9	1.5		SZ
4523.237	4524.505	22101.86	BA	1.8	1.0		MIT
4509.63	4510.89	22168.6	BA II			0.7	RS
4505.934	4507.198	22186.73	BA	1.8	1.3		MIT
4493.639	4494.900	22247.44	BA I	1.8	0.7		MIT
4488.969	4490.228	22270.58	BA I	1.9	0.5		MIT
4467.110	4468.364	22379.56	BA	1.3	0.5		MIT
4431.898	4433.142	22557.36	BA	1.8	1.5		MIT
4413.683	4414.923	22650.45	BA	1.0	0.5		MIT
4406.845	4408.083	22685.60	BA	1.3		0.5	MIT
4405.23	4406.47	22693.9	BA II			1.3	RS
4402.551	4403.788	22707.73	BA	1.9	1.0		MIT
4359.554	4360.779	22931.68	BA	1.2	0.5		SZ
4350.375	4351.598	22980.06	BA	1.6	1.3		SZ
4332.905	4334.123	23072.72	BA I	1.3	0.5		MIT
4329.62	4330.84	23090.2	BA II			1.0	RS
4326.74	4327.96	23105.6	BA II			0.7	RS
4325.73	4326.95	23111.0	BA II			1.7	RS
4325.163	4326.379	23114.02	BA	1.2	0.5		MIT
4323.612	4324.828	23122.31	BA I	1.0	0.5		MIT

AIR WAVELENGTH (ANGSTRMS)	VACUUM WAVELENGTH (ANGSTRMS)	WAVENUMBER (KAYSERS)	LMENT	LOG ARC	INTENSITY SPK	INTENSITY DIS	REF
4322.999	4324.215	23125.59	BA	1.2	0.7		MIT
4309.32	4310.53	23199.0	BA II			1.9	RS
4297.60	4298.81	23262.3	BA II			0.7	RS
4291.165	4292.372	23297.14	BA	1.0	0.5		SZ
4287.795	4289.001	23315.45	BA II			1.0	MIT
4283.109	4284.314	23340.96	BA	1.4	1.3		MIT
4267.95	4269.15	23423.9	BA II			1.9	RS
4264.386	4265.586	23443.44	BA I	1.2	0.6		SZ
4242.617	4243.812	23563.72	BA	1.0	0.7		MIT
4239.573	4240.767	23580.64	BA I	1.0	0.6		MIT
4223.963	4225.153	23667.78	BA	0.7	0.5		MIT
4216.04	4217.23	23712.3	BA II			1.4	RS
4179.372	4180.550	23920.30	BA	0.8			MIT
4166.010	4167.184	23997.02	BA II	1.1	1.7		MIT
4132.431	4133.597	24192.01	BA	1.0	0.6		MIT
4130.664	4131.829	24202.36	BA II	1.7R	1.8J		MIT
4109.875	4111.035	24324.78	BA	0.3			MIT
4087.377	4088.531	24458.66	BA	0.5			MIT
4084.87	4086.02	24473.7	BA I	0.5			FL
4083.77	4084.92	24480.3	BA II			0.7	RS
4081.333	4082.485	24494.88	BA	0.3			MIT
4080.93	4082.08	24497.3	BA I	0.3	0.5		FL
4036.26	4037.40	24768.4	BA II			1.0	RS
4026.30	4027.44	24829.7	BA I	0.3	0.3H		SD
3997.92	3999.05	25005.9	BA I	0.7			SD
3995.664	3996.794	25020.05	BA I	1.3	0.7		MIT
3993.404	3994.533	25034.21	BA I	2.0G	1.7R		MIT
3982.90	3984.03	25100.2	BA	0.7			MIT
3975.362	3976.487	25147.83	BA I	1.0R			SZ
3949.51	3950.63	25312.4	BA II			1.3	RS
3947.475	3948.592	25325.48	BA I	0.7H			SZ
3945.61	3946.73	25337.4	BA I	0.7H			SD
3939.672	3940.787	25375.64	BA II	0.3		0.7	MIT
3937.876	3938.991	25387.21	BA I	1.1	1.0		MIT
3935.725	3936.839	25401.09	BA I	1.9G	1.5R		MIT
3926.834	3927.946	25458.60	BA	0.7	0.7		MIT
3914.73	3915.84	25537.3	BA II			1.2	RS
3909.917	3911.024	25568.75	BA I	1.7G	1.3R		MIT
3906.010	3907.116	25594.32	BA I	0.6	0.3		MIT
3898.58	3899.68	25643.1	BA I	0.7			SD
3894.352	3895.455	25670.94	BA I	0.7			MIT
3892.662	3893.765	25682.08	BA	0.7H			MIT
3891.785	3892.888	25687.87	BA II	1.3	1.4		MIT
3890.583	3891.685	25695.81	BA I	0.5			MIT
3889.316	3890.418	25704.18	BA	1.0	0.3		M
3861.91	3863.00	25886.6	BA	0.5			M
3854.76	3855.85	25934.6	BA II			0.3	RS
3842.80	3843.89	26015.3	BA II			0.7	RS
3841.15	3842.24	26026.5	BA I	0.7			FL
3828.92	3830.01	26109.6	BA I	0.7			MIT
3816.69	3817.77	26193.3	BA II			0.7	RS
3794.771	3795.849	26344.57	BA	0.9			SZ
3789.72	3790.80	26379.7	BA I	0.5			SD
3788.18	3789.26	26390.4	BA I	0.5			SD
3787.235	3788.311	26396.99	BA I	0.6	0.6		MIT
3779.34	3780.41	26452.1	BA		0.3		PY
3776.0	3777.1	26475.	BA	0.5	0.8		PY
3771.93	3773.00	26504.1	BA	0.5			SD
3769.48	3770.55	26521.3	BA I	0.6			SD
3735.75	3736.81	26760.8	BA II	0.0	0.3		RS
3721.18	3722.24	26865.6	BA I	0.3			MIT
3720.85	3721.91	26867.9	BA I	0.3			SD
3719.93	3720.99	26874.6	BA I	0.3			MIT
3704.22	3705.27	26988.6	BA I	0.3	0.3H		MIT
3701.716	3702.769	27006.81	BA	0.5			SZ
3688.467	3689.517	27103.82	BA I	1.1			MIT
3675.268	3676.315	27201.16	BA	0.3			SZ
3667.93	3668.97	27255.6	BA I	0.3			SD
3667.60	3668.64	27258.0	BA I	0.3			SD
3664.598	3665.642	27280.35	BA	0.3			SZ

AIR WAVELENGTH (ANGSTRMS)	VACUUM WAVELENGTH (ANGSTRMS)	WAVENUMBER (KAYSERS)	LMENT	LOG ARC	INTENSITY SPK	INTENSITY DIS	REF
3662.528	3663.571	27295.77	BA	1.0	0.7L		M
3640.391	3641.429	27461.75	BA	0.9	0.6		MIT
3636.946	3637.983	27487.76	BA	1.3	0.6		SZ
3630.642	3631.677	27535.49	BA	1.2	0.7		MIT
3611.002	3612.032	27685.25	BA	1.0	0.5		SZ
3603.40	3604.43	27743.6	BA I	0.3			MIT
3599.420	3600.447	27774.33	BA I	1.0	0.5		M
3596.57	3597.60	27796.3	BA II	0.0	0.6		RS
3596.33	3597.36	27798.2	BA I	0.3			SD
3593.282	3594.307	27821.77	BA I	0.5			SZ
3588.132	3589.156	27861.70	BA	0.7	0.5		SZ
3586.523	3587.547	27874.20	BA	0.7	0.5		MIT
3579.672	3580.694	27927.55	BA I	1.0	0.9		MIT
3577.615	3578.636	27943.60	BA	0.6			MIT
3576.28	3577.30	27954.0	BA II		0.5		RS
3576.047	3577.068	27955.86	BA	0.5	0.5		SZ
3567.73	3568.75	28021.0	BA II	0.3	0.5		MIT
3566.687	3567.705	28029.22	BA	0.7			MIT
3561.981	3562.998	28066.25	BA	0.5	0.5		SZ
3552.45	3553.46	28141.6	BA II	0.0	0.9		RS
3547.767	3548.781	28178.69	BA I	1.0	0.8		SZ
3544.713	3545.726	28202.97	BA I	1.3	0.7		SZ
3529.495	3530.504	28324.57	BA	0.6			SZ
3524.985	3525.993	28360.81	BA I	1.3	0.7		MIT
3501.116	3502.118	28554.15	BA I	3.0	1.3		MIT
3472.71	3473.70	28787.7	BA II		0.7		SD
3464.233	3465.225	28858.15	BA	0.7			SZ
3430.18	3431.16	29144.6	BA		0.3		SD
3426.454	3427.436	29176.32	BA	0.3			MIT
3421.48	3422.46	29218.7	BA I	0.3	0.3H		SD
3421.01	3421.99	29222.7	BA I	0.5	0.3H		SD
3420.338	3421.319	29228.49	BA I	0.9R	0.3H		MIT
3413.836	3414.815	29284.16	BA I		0.7		MIT
3377.39	3378.36	29600.2	BA I	0.3			SD
3376.993	3377.963	29603.64	BA I	1.2R	0.8		MIT
3356.894	3357.859	29780.88	BA I	1.6R	0.5		SZ
3332.15	3333.11	30002.0	BA	0.0	1.0		MIT
3323.06	3324.02	30084.1	BA I	0.3			SD
3322.874	3323.830	30085.77	BA I	1.5R			MIT
3315.803	3316.757	30149.93	BA	1.0			SZ
3298.20	3299.15	30310.8	BA	0.5	0.6		MIT
3286.755	3287.702	30416.38	BA	0.0	0.8		MIT
3281.735	3282.681	30462.91	BA I	0.9			SZ
3281.50	3282.45	30465.1	BA I	1.4			SD
3272.405	3273.348	30549.76	BA	0.5			KN
3270.108	3271.051	30571.22	BA	0.6			SZ
3262.275	3263.216	30644.62	BA I	0.5	0.5		MIT
3261.96	3262.90	30647.6	BA I	1.6			SD
3253.067	3254.005	30731.36	BA	0.7H			KN
3244.189	3245.125	30815.45	BA I	0.5			MIT
3234.96	3235.89	30903.4	BA		0.5		MIT
3222.44	3223.37	31023.4	BA I	0.5L			MIT
3222.278	3223.208	31024.99	BA I	0.3			MIT
3222.22	3223.15	31025.6	BA I	1.3			MIT
3221.628	3222.558	31031.25	BA I	0.3			MIT
3212.78	3213.71	31116.7	BA		0.7		PY
3203.862	3204.788	31203.31	BA I	0.3H			MIT
3195.21	3196.13	31287.8	BA		0.6		MIT
3193.978	3194.901	31299.87	BA I	0.3			MIT
3193.91	3194.83	31300.5	BA I	0.3			SD
3184.41	3185.33	31393.9	BA	0.5			MIT
3183.948	3184.869	31398.47	BA I	0.3			MIT
3183.16	3184.08	31406.2	BA I	0.7			SD
3173.691	3174.609	31499.94	BA I	0.5			MIT
3165.60	3166.52	31580.4	BA I	0.3	0.3		SD
3163.272	3164.188	31603.69	BA		0.3H		MIT
3158.514	3159.428	31651.29	BA I	0.3			MIT
3155.659	3156.573	31679.93	BA I	0.3			MIT
3155.343	3156.257	31683.10	BA I	0.3			MIT
3146.90	3147.81	31768.1	BA I	0.3	0.3		SD

AIR WAVELENGTH (ANGSTRMS)	VACUUM WAVELENGTH (ANGSTRMS)	WAVENUMBER (KAYSERS)	LMENT	LOG ARC	INTENSITY SPK	DIS	REF
3137.80	3138.71	31860.2	BA I	0.3			SD
3137.73	3138.64	31860.9	BA I	0.3	0.3		MIT
3135.718	3136.627	31881.39	BA I	0.3	0.5		MIT
3130.570	3131.477	31933.81	BA I	0.3	0.5		MIT
3126.48	3127.39	31975.6	BA		0.3		PY
3121.02	3121.92	32031.5	BA I	0.3			SD
3117.94	3118.84	32063.2	BA I	0.3			SD
3109.624	3110.526	32148.90	BA I	0.3			MIT
3108.176	3109.078	32163.88	BA	1.0	0.8		MIT
3079.124	3080.018	32467.34	BA		0.7W		SZ
3071.591	3072.484	32546.96	BA I	2.06	1.7G		MIT
3043.35	3044.24	32849.0	BA		0.6		MIT
2972.22	2973.09	33635.1	BA		0.3		PY
2962.341	2963.206	33747.23	BA		0.5H		SZ
2959.998	2960.863	33773.94	BA		0.5J		MIT
2939.318	2940.178	34011.55	BA		0.3H		SZ
2923.83	2924.69	34191.7	BA		0.5		PY
2831.67	2832.50	35304.5	BA	0.0	1.3		PY
2816.07	2816.90	35500.0	BA		1.5		PY
2785.264	2786.086	35892.65	BA I	1.7			SZ
2771.358	2772.176	36072.74	BA II	1.1	0.9		MIT
2764.98	2765.80	36156.0	BA		0.9		PY
2762.35	2763.17	36190.4	BA		0.3H		PY
2739.236	2740.046	36495.73	BA		1.0		MIT
2709.07	2709.87	36902.1	BA		0.3		PY
2702.639	2703.441	36989.90	BA I	1.7	0.7H		MIT
2647.289	2648.077	37763.25	BA II	1.0	1.6		MIT
2641.375	2642.162	37847.79	BA II	0.7	0.7		MIT
2634.783	2635.568	37942.48	BA II	1.5	1.7		MIT
2596.678	2597.454	38499.23	BA I	1.6			SZ
2567.04	2567.81	38943.7	BA		0.3		PY
2559.587	2560.355	39057.09	BA		0.3H		MIT
2546.95	2547.71	39250.9	BA		0.3		PY
2530.992	2531.753	39498.33	BA		0.3H		SZ
2528.508	2529.268	39537.13	BA II		1.7R		MIT
2524.113	2524.872	39605.96	BA		0.7		SZ
2519.249	2520.007	39682.43	BA		0.3H		SZ
2509.154	2509.910	39842.07	BA		0.3H		SZ
2505.089	2505.844	39906.72	BA		0.3H		SZ
2476.162	2476.910	40372.88	BA		0.7W		SZ
2373.108	2373.833	42125.97	BA I	1.1	0.3W		MIT
2347.577	2348.296	42584.07	BA	1.5	1.6		SZ
2335.269	2335.985	42808.49	BA II	1.8G	2.0G		MIT
2331.136	2331.851	42884.38	BA		0.5H		MIT
2323.578	2324.292	43023.86	BA		0.3H		MIT
2304.235	2304.944	43385.00	BA II	1.8G	1.9G		MIT
2280.757	2281.461	43831.56	BA		0.3W		SZ
2254.732	2255.430	44337.43	BA II	1.0	1.0		SZ
2245.613	2246.310	44517.46	BA II	1.1	1.1		SZ
2216.577	2217.267	45100.56	BA	0.3H			SZ
2214.639	2215.329	45140.02	BA II	0.3	0.8		SZ
2014.18	2014.83	49632.0	BA II			0.6	RS
8254.10	8256.37	12111.9	BE I	2.0			PS
8216.05	8218.31	12167.9	BE I	0.7			PS
7209.134	7211.121	13867.47	BE I	1.0			MIT
6981.0	6982.9	14321.	BE I	1.2			PS
5353.513	5355.002	18674.13	BE I			0.7	PS
5272.7	5274.2	18960.	BE		1.3		SX
5270.843	5272.310	18967.02	BE II			1.1	PS
5270.322	5271.789	18968.89	BE II			1.0	PS
4883.2	4884.6	20473.	BE		0.3		SX
4830.8	4832.1	20695.	BE		0.9		SX
4828.119	4829.468	20706.21	BE II			1.4	PS
4816.71	4818.06	20755.3	BE II			2.5	BL
4673.462	4674.770	21391.43	BE II			2.0	PS
4672.2	4673.5	21397.	BE		2.0		SX
4573.1	4574.4	21861.	BE		1.6		SX
4572.671	4573.953	21862.93	BE I	1.2	1.2		IHZ
4407.91	4409.15	22680.1	BE I	1.3		1.5	PS
4364.0	4365.2	22908.	BE		1.7		SX

AIR WAVELENGTH (ANGSTRMS)	VACUUM WAVELENGTH (ANGSTRMS)	WAVENUMBER (KAYSERS)	LMENT	LOG ARC	INTENSITY SPK	DIS	REF
4361.025	4362.251	22923.95	BE II			1.6	PS
4360.690	4361.916	22925.71	BE II			1.5	P
4254.12	4255.32	23500.0	BE I			0.7	PS
4253.76	4254.96	23502.0	BE I			1.2	PS
4253.05	4254.25	23505.9	BE I			1.3	PS
3866.03	3867.13	25859.0	BE I	1.2			PS
3865.74	3866.84	25860.9	BE I	0.3			PS
3865.521	3866.617	25862.40	BE I	1.0			MIT
3865.43	3866.53	25863.0	BE I	1.5			PS
3865.146	3866.242	25864.91	BE I	0.7			MIT
3813.417	3814.499	26215.76	BE I	1.7			MIT
3736.28	3737.34	26757.0	BE I	1.0			PS
3515.549	3516.554	28436.93	BE I	1.5			MIT
3476.61	3477.61	28755.4	BE I	0.7			PS
3455.20	3456.19	28933.6	BE I	1.3			PS
3367.624	3368.591	29686.00	BE I	1.4			MIT
3345.451	3346.413	29882.75	BE I	0.3			MIT
3321.343	3322.299	30099.64	BE I	3.0R	1.5		MIT
3321.086	3322.042	30101.97	BE I	2.0	1.2		PS
3321.013	3321.969	30102.63	BE I	1.7			PS
3282.890	3283.836	30452.19	BE I	0.9			MIT
3274.640	3275.584	30528.91	BE II			1.7	PS
3241.835	3242.770	30837.83	BE II	0.7		1.7	PS
3241.646	3242.581	30839.63	BE II	0.7		1.2	PS
3233.538	3234.471	30916.95	BE II	0.7		0.7	PS
3229.620	3230.552	30954.46	BE I	1.2			JO
3197.158	3198.082	31268.74	BE II		1.2		MIT
3193.801	3194.724	31301.61	BE I	0.3			MIT
3131.072	3131.979	31928.69	BE II	2.3	2.2		MIT
3130.416	3131.323	31935.38	BE II	2.3	2.3		MIT
3110.965	3111.867	32135.04	BE I	1.2			MIT
3110.842	3111.744	32136.32	BE I	1.3			MIT
3046.676	3047.562	32813.11	BE II	1.0		1.3	PS
3046.520	3047.406	32814.79	BE II			1.2	PS
3019.60	3020.48	33107.3	BE I	0.7			PS
3019.511	3020.390	33108.30	BE I	1.2			MIT
3019.34	3020.22	33110.2	BE I	1.5			PS
2986.62	2987.49	33472.9	BE I	1.0			PS
2986.457	2987.328	33474.73	BE I	1.2			MIT
2986.09	2986.96	33478.8	BE I	0.9			PS
2898.27	2899.12	34493.2	BE I	1.3			PS
2898.19	2899.04	34494.2	BE I	1.2			PS
2738.09	2738.90	36511.0	BE I	1.0			PS
2728.826	2729.634	36634.95	BE II		0.9		MIT
2697.33	2698.13	37062.7	BE II			0.3	PS
2650.781	2651.570	37713.50	BE I	1.4			MIT
2650.760	2651.549	37713.80	BE I	1.3	1.2		ME
2650.702	2651.491	37714.63	BE I	1.0			MIT
2650.550	2651.339	37716.79	BE I	1.5			MIT
2650.470	2651.259	37717.93	BE I	2.0	1.2		PS
2618.10	2618.88	38184.2	BE II			0.6	PS
2507.40	2508.16	39869.9	BE II			1.0	PS
2494.733	2495.485	40072.37	BE I	1.5	1.3		ME
2494.576	2495.328	40074.89	BE I	1.4			ME
2494.559	2495.311	40075.16	BE I	1.5			ME
2453.89	2454.63	40739.3	BE II			0.8	PS
2413.45	2414.18	41421.9	BE II			1.4	PS
2350.835	2351.555	42525.06	BE I	1.1			MIT
2350.685	2351.405	42527.77	BE I	1.4	0.3		MIT
2348.610	2349.329	42565.34	BE I	3.3G	1.7		MIT
2175.069	2175.751	45961.15	BE I	1.4	0.5		PS
2174.942	2175.624	45963.83	BE I	1.0			PS
2161.275	2161.954	46254.46	BE II			0.3	PS
2160.5	2161.2	46271.	BE		0.3		MD
2087.31	2087.97	47893.3	BE	0.3			MIT
2084.0	2084.7	47969.	BE	0.3			MD
2056.52	2057.18	48610.3	BE I	2.0			PS
2056.0	2056.7	48623.	BE		0.8		MD
2032.72	2033.37	49179.4	BE I	1.0			ME
2032.6	2033.3	49182.	BE		0.3		MD

AIR WAVELENGTH (ANGSTRMS)	VACUUM WAVELENGTH (ANGSTRMS)	WAVENUMBER (KAYSERS)	LMENT	LOG ARC	INTENSITY SPK	INTENSITY DIS	REF
5551.5	5553.0	18008.	BH B	0.9			L
5040.1	5041.5	19835.	BH B	1.1			L
4746.9	4748.2	21060.	BH B	1.4			L
4744.0	4745.3	21073.	BH B	1.4			L
4615.4	4616.7	21660.	BH B	1.6			L
4588.8	4590.1	21786.	BH B	1.4			L
4365.9	4367.1	22898.	BH B	1.4			L
4363.4	4364.6	22911.	BH B	1.6			L
4342.0	4343.2	23024.	BH B	1.4			L
4339.6	4340.8	23037.	BH B	1.4			L
4037.4	4038.5	24761.	BH B	1.4			L
3848.7	3849.8	25975.	BH B	2.3			L
3847.0	3848.1	25987.	BH B	2.0			L
3830.2	3831.3	26101.	BH B	1.7			L
3679.1	3680.1	27173.	BH B	2.3			L
3677.8	3678.8	27182.	BH B	1.7			L
3305.4	3306.4	30245.	BH B	1.7			L
3257.0	3257.9	30694.	BH B	2.0			L
3088.6	3089.5	32368.	BH B	2.0			L
3043.6	3044.5	32846.	BH B	2.0			L
2934.9	2935.8	34063.	BH B	2.0			L
2892.2	2893.0	34566.	BH B	2.3			L
2850.6	2851.4	35070.	BH B	1.7			L
2809.9	2810.7	35578.	BH B	1.8			L
2753.4	2754.2	36308.	BH B	2.0			L
2713.8	2714.6	36838.	BH B	2.3			L
2675.3	2676.1	37368.	BH B	1.8			L
2588.0	2588.8	38628.	BH B	1.7			L
2551.4	2552.2	39182.	BH B	2.2			L
2437.1	2437.8	41020.	BH B	2.4			L
2398.5	2399.2	41680.	BH B	2.3			L
2364.5	2365.2	42279.	BH B	1.7			L
9384.	9387.	10650.	BH C				L
9141.	9144.	10940.	BH C				L
8708.	8710.	11480.	BH C				L
8485.	8487.	11780.	BH C				L
8272.	8274.	12090.	BH C				L
8067.	8069.	12390.	BH C				L
7874.	7876.	12700.	BH C				L
7435.	7437.	13450.	BH C				L
7259.	7261.	13770.	BH C				L
7091.	7093.	14100.	BH C				L
6928.	6930.	14430.	BH C				L
6791.9	6793.8	14719.	BH C				L
6677.3	6679.1	14972.	BH C				L
6631.3	6633.1	15076.	BH C				L
6599.3	6601.1	15149.	BH C				L
6533.7	6535.5	15301.	BH C				L
6478.4	6480.2	15432.	BH C				L
6332.2	6334.0	15788.	BH C				L
6278.9	6280.6	15922.	BH C				L
6195.4	6197.1	16136.	BH C				L
6191.2	6192.9	16147.	BH C				L
6132.4	6134.1	16302.	BH C				L
6122.1	6123.8	16330.	BH C				L
6059.7	6061.4	16498.	BH C				L
6004.9	6006.6	16648.	BH C				L
5992.6	5994.3	16683.	BH C	2.0			L
5958.7	5960.4	16777.	BH C				L
5923.4	5925.0	16877.	BH C				L
5858.2	5859.8	17065.	BH C	2.6			L
5730.0	5731.6	17447.	BH C	2.2			L
5635.5	5637.1	17740.	BH C				L
5607.6	5609.2	17828.	BH C	1.2			L
5597.9	5599.5	17859.	BH C	1.7			L
5585.5	5587.1	17898.	BH C				L
5540.7	5542.2	18043.	BH C				L
5501.9	5503.4	18170.	BH C				L
5473.3	5474.8	18265.	BH C	1.8			L
5470.3	5471.8	18275.	BH C				L

AIR WAVELENGTH (ANGSTRMS)	VACUUM WAVELENGTH (ANGSTRMS)	WAVENUMBER (KAYSERS)	LMENT	LOG ARC	INTENSITY SPK	INTENSITY DIS	REF
5354.1	5355.6	18672.	BH C	2.0			L
5239.3	5240.8	19081.	BH C	1.8			L
5165.2	5166.6	19355.	BH C				L
5129.3	5130.7	19490.	BH C				L
5097.7	5099.1	19611.	BH C				L
4935.7	4937.1	20255.	BH C				L
4832.4	4833.8	20688.	BH C				L
4737.1	4738.4	21104.	BH C				L
4733.3	4734.6	21121.	BH C				L
4715.2	4716.5	21202.	BH C				L
4697.6	4698.9	21281.	BH C				L
4684.8	4686.1	21340.	BH C				L
4681.9	4683.2	21353.	BH C				L
4678.6	4679.9	21368.	BH C				L
4612.7	4614.0	21673.	BH C	1.7			L
4606.1	4607.4	21704.	BH C				L
4577.9	4579.2	21838.	BH C				L
4553.1	4554.4	21957.	BH C				L
4531.8	4533.1	22060.	BH C				L
4514.7	4516.0	22144.	BH C				L
4502.1	4503.4	22206.	BH C				L
4495.1	4496.4	22240.	BH C				L
4493.9	4495.2	22246.	BH C				L
4382.5	4383.7	22812.	BH C				L
4371.4	4372.6	22869.	BH C				L
4365.2	4366.4	22902.	BH C				L
4216.0	4217.2	23712.	BH C				L
4197.1	4198.3	23819.	BH C				L
4180.8	4182.0	23912.	BH C				L
4167.6	4168.8	23988.	BH C				L
4166.7	4167.9	23993.	BH C				L
4158.0	4159.2	24043.	BH C				L
3883.4	3884.5	25743.	BH C		0.5		L
3871.4	3872.5	25823.	BH C				L
3861.7	3862.8	25888.	BH C				L
3590.3	3591.3	27845.	BH C				L
3585.8	3586.8	27880.	BH C				L
3583.9	3584.9	27895.	BH C				L
2799.7	2800.5	35708.	BH C	1.7			L
2785.4	2786.2	35891.	BH C	1.5			L
2742.6	2743.4	36451.	BH C	1.3			L
2740.0	2740.8	36486.	BH C	1.1			L
2712.1	2712.9	36861.	BH C	1.1			L
2698.3	2699.1	37049.	BH C	1.3			L
2684.0	2684.8	37247.	BH C	1.0			L
2680.8	2681.6	37291.	BH C	1.2			L
2662.9	2663.7	37542.	BH C	1.1			L
2661.500	2662.292	37561.62	BH C	1.1			L
2659.6	2660.4	37588.	BH C	1.3			L
2630.3	2631.1	38007.	BH C	1.1			L
2600.3	2601.1	38446.	BH C	1.3			L
2594.5	2595.3	38532.	BH C	0.3			L
2569.5	2570.3	38906.	BH C	1.1			L
2557.0	2557.8	39097.	BH C	0.7			L
2539.1	2539.9	39372.	BH C	1.1			L
2525.1	2525.9	39590.	BH C	1.1			L
2511.0	2511.8	39813.	BH C	1.3			L
2493.3	2494.1	40095.	BH C	1.1			L
2484.5	2485.3	40237.	BH C	1.1			L
2463.7	2464.4	40577.	BH C	1.1			L
2459.0	2459.7	40655.	BH C	0.7			L
2435.0	2435.7	41055.	BH C	1.5			L
2425.0	2425.7	41225.	BH C	1.3			L
2408.0	2408.7	41516.	BH C	1.5			L
2394.6	2395.3	41748.	BH C	1.3			L
2382.0	2382.7	41969.	BH C	1.5			L
2365.7	2366.4	42258.	BH C	1.3			L
2356.9	2357.6	42416.	BH C	1.1			L
2338.3	2339.0	42753.	BH C	1.3			L
2333.0	2333.7	42850.	BH C	0.7			L

AIR WAVELENGTH (ANGSTRMS)	VACUUM WAVELENGTH (ANGSTRMS)	WAVENUMBER (KAYSERS)	LMENT	LOG ARC	INTENSITY SPK	INTENSITY DIS	REF
2312.0	2312.7	43239.	BH C	1.5			L
2310.25	2310.96	43272.0	BH C	1.1			L
2302.7	2303.4	43414.	BH C	0.7			L
2286.7	2287.4	43718.	BH C	1.5			L
2274.2	2274.9	43958.	BH C	1.1			L
2272.2	2272.9	43997.	BH C	1.3			L
2262.4	2263.1	44187.	BH C	1.1			L
2247.5	2248.2	44480.	BH C	1.1			L
2238.6	2239.3	44657.	BH C	1.1			L
2221.8	2222.5	44994.	BH C	1.5			L
2216.1	2216.8	45110.	BH C	1.3			L
2209.9	2210.6	45237.	BH C	0.3			L
2196.9	2197.6	45504.	BH C	1.5			L
2195.0	2195.7	45544.	BH C	1.3			L
2189.2	2189.9	45664.	BH C	0.7			L
2173.4	2174.1	45996.	BH C	1.5			L
2162.6	2163.3	46226.	BH C	1.5			L
2150.6	2151.3	46484.	BH C	1.5			L
2137.5	2138.2	46769.	BH C	1.3			L
2128.6	2129.3	46964.	BH C	0.7			L
2113.2	2113.9	47307.	BH C	0.7			L
2107.5	2108.2	47434.	BH C	0.7			L
2090.0	2090.7	47832.	BH C	1.5			L
2068.8	2069.5	48322.	BH C	1.5			L
2047.5	2048.2	48824.	BH C	1.5			L
2035.0	2035.7	49124.	BH C	1.1			L
2031.8	2032.5	49202.	BH C	0.3			L
2026.2	2026.9	49338.	BH C	1.4			L
2012.3	2013.0	49678.	BH C	1.5			L
2006.5	2007.1	49822.	BH C	1.2			L
9834.7	9837.4	10165.	BH CA	1.5			L
9807.3	9810.0	10194.	BH CA	1.3			L
9775.0	9777.7	10227.	BH CA	1.2			L
9741.0	9743.7	10263.	BH CA	0.7			L
9700.0	9702.7	10306.	BH CA	1.0			L
9228.9	9231.4	10833.	BH CA	1.3			L
9215.1	9217.6	10849.	BH CA	0.7			L
9193.4	9195.9	10874.	BH CA	0.3			L
8652.2	8654.6	11555.	BH CA	1.3			L
8642.7	8645.1	11567.	BH CA	0.3			L
8167.2	8169.4	12241.	BH CA	0.7			L
8153.0	8155.2	12262.	BH CA	1.6			L
7721.1	7723.2	12948.	BH CA	0.6			L
7715.6	7717.7	12957.	BH CA	0.7			L
7327.6	7329.6	13643.	BH CA	0.3			L
7318.4	7320.4	13660.	BH CA	0.5			L
7308.3	7310.3	13679.	BH CA	0.3			L
6982.9	6984.8	14317.	BH CA	0.5			L
6968.4	6970.3	14346.	BH CA	0.3			L
6956.2	6958.1	14372.	BH CA	1.3			L
4535.1	4536.4	22044.	BH CA	0.3			L
4519.1	4520.4	22122.	BH CA	0.5			L
4505.0	4506.3	22191.	BH CA	0.6			L
4449.9	4451.1	22466.	BH CA	0.3			L
4425.8	4427.0	22588.	BH CA	0.5			L
4403.9	4405.1	22701.	BH CA	0.8			L
4384.3	4385.5	22802.	BH CA	0.8			L
4366.7	4367.9	22894.	BH CA	0.7			L
4351.2	4352.4	22976.	BH CA	0.7			L
4261.8	4263.0	23458.	BH CA	0.3			L
4240.8	4242.0	23574.	BH CA	0.5			L
4221.9	4223.1	23679.	BH CA	0.7			L
4205.1	4206.3	23774.	BH CA	0.8			L
4149.8	4151.0	24091.	BH CA	0.3			L
4126.0	4127.2	24230.	BH CA	0.5			L
4104.1	4105.3	24359.	BH CA	0.6			L
4084.3	4085.5	24477.	BH CA	0.7			L
3996.5	3997.6	25015.	BH CA	0.6			L
3973.9	3975.0	25157.	BH CA	0.9			L
3959.4	3960.5	25249.	BH CA	0.6			L

AIR WAVELENGTH (ANGSTRMS)	VACUUM WAVELENGTH (ANGSTRMS)	WAVENUMBER (KAYSERS)	LMENT	LOG ARC	INTENSITY SPK	INTENSITY DIS	REF
3894.7	3895.8	25669.	BH CA	0.6			L
3879.2	3880.3	25771.	BH CA	0.6			L
3874.2	3875.3	25804.	BH CA	0.6			L
3872.9	3874.0	25813.	BH CA	0.6			L
3854.1	3855.2	25939.	BH CA	0.6			L
3793.7	3794.8	26352.	BH CA	0.6			L
3773.2	3774.3	26495.	BH CA	0.6			L
3753.2	3754.3	26636.	BH CA	0.9			L
3696.9	3698.0	27042.	BH CA	0.6			L
3676.5	3677.5	27192.	BH CA	0.9			L
3656.6	3657.6	27340.	BH CA	1.1			L
3624.8	3625.8	27580.	BH CA	0.6			L
3604.0	3605.0	27739.	BH CA	0.6			L
3583.7	3584.7	27896.	BH CA	0.9			L
3564.0	3565.0	28050.	BH CA	1.3			L
3514.8	3515.8	28443.	BH CA	0.6			L
3494.7	3495.7	28607.	BH CA	0.9			L
3475.0	3476.0	28769.	BH CA	1.5			L
3429.1	3430.1	29154.	BH CA	0.6			L
3409.1	3410.1	29325.	BH CA	1.1			L
3346.7	3347.7	29872.	BH CA	0.9			L
3308.0	3309.0	30221.	BH CA	0.6			L
3287.4	3288.3	30410.	BH CA	0.9			L
3231.2	3232.1	30939.	BH CA	0.6			L
8360.3	8362.6	11958.	BH CR				L
7908.0	7910.2	12642.	BH CR				L
7842.8	7845.0	12747.	BH CR				L
7778.1	7780.2	12853.	BH CR				L
7249.1	7251.1	13791.	BH CR	0.3			L
7187.1	7189.1	13910.	BH CR	0.3			L
6829.2	6831.1	14639.	BH CR	0.5			L
6771.8	6773.7	14763.	BH CR	0.6			L
6451.5	6453.3	15496.	BH CR	0.3			L
6395.0	6396.8	15633.	BH CR	0.5			L
6168.1	6169.8	16208.	BH CR	0.3			L
6052.3	6054.0	16518.	BH CR	0.8			L
5914.5	5916.1	16903.	BH CR	0.3			L
5795.5	5797.1	17250.	BH CR	1.0			L
5623.1	5624.7	17779.	BH CR	0.3			L
5564.2	5565.7	17967.	BH CR	0.8			L
6847.6	6849.5	14600.	BH F	0.3			L
6842.2	6844.1	14611.	BH F	0.7			L
6836.8	6838.7	14623.	BH F	0.7			L
6831.5	6833.4	14634.	BH F	0.7			L
6821.2	6823.1	14656.	BH F	0.3			L
6816.4	6818.3	14666.	BH F	0.3			L
6735.4	6737.3	14843.	BH F	0.3			L
6729.4	6731.3	14856.	BH F	1.3			L
6724.0	6725.9	14868.	BH F	0.7			L
6719.1	6721.0	14879.	BH F	0.7			L
6718.6	6720.5	14880.	BH F	1.0			L
6655.6	6657.4	15021.	BH F	2.0			L
6652.1	6653.9	15029.	BH F	2.0			L
6648.7	6650.5	15036.	BH F	2.0			L
6645.4	6647.2	15044.	BH F	1.9			L
6642.0	6643.8	15051.	BH F	1.7			L
6638.8	6640.6	15059.	BH F	1.5			L
6635.6	6637.4	15066.	BH F	1.5			L
6632.7	6634.5	15073.	BH F	2.5			L
6629.4	6631.2	15080.	BH F	2.3			L
6626.1	6627.9	15088.	BH F	2.3			L
6622.9	6624.7	15095.	BH F	2.3			L
6619.8	6621.6	15102.	BH F	2.2			L
6616.6	6618.4	15109.	BH F	2.0			L
6613.6	6615.4	15116.	BH F	1.9			L
6610.5	6612.3	15123.	BH F	1.7			L
6607.6	6609.4	15130.	BH F	1.5			L
6604.6	6606.4	15137.	BH F	1.5			L
6601.8	6603.6	15143.	BH F	1.3			L
6527.6	6529.4	15315.	BH F	2.3			L

AIR WAVELENGTH (ANGSTRMS)	VACUUM WAVELENGTH (ANGSTRMS)	WAVENUMBER (KAYSERS)	LMENT	LOG ARC	INTENSITY SPK	DIS	REF
6524.2	6526.0	15323.	BH F	2.2			L
6520.8	6522.6	15331.	BH F	2.0			L
6517.5	6519.3	15339.	BH F	2.0			L
6514.2	6516.0	15347.	BH F	1.9			L
6512.0	6513.8	15352.	BH F	2.5			L
6511.1	6512.9	15354.	BH F	1.3			L
6508.7	6510.5	15360.	BH F	2.3			L
6507.9	6509.7	15362.	BH F	1.3			L
6505.5	6507.3	15367.	BH F	2.0			L
6502.2	6504.0	15375.	BH F	1.9			L
6499.0	6500.8	15383.	BH F	1.9			L
6495.9	6497.7	15390.	BH F	1.5			L
6492.9	6494.7	15397.	BH F	1.5			L
6489.9	6491.7	15404.	BH F	1.3			L
6486.9	6488.7	15411.	BH F	1.3			L
6484.0	6485.8	15418.	BH F	1.0			L
6481.2	6483.0	15425.	BH F	0.7			L
6478.5	6480.3	15431.	BH F	0.3			L
6475.9	6477.7	15438.	BH F	0.3			L
6419.0	6420.8	15574.	BH F	1.7			L
6417.7	6419.5	15578.	BH F	2.0			L
6416.5	6418.3	15580.	BH F	2.0			L
6415.3	6417.1	15583.	BH F	2.0			L
6414.1	6415.9	15586.	BH F	2.0			L
6412.9	6414.7	15589.	BH F	1.9			L
6411.8	6413.6	15592.	BH F	1.7			L
6410.7	6412.5	15595.	BH F	1.5			L
6409.7	6411.5	15597.	BH F	1.3			L
6407.7	6409.5	15602.	BH F	1.0			L
6406.7	6408.5	15604.	BH F	1.3			L
6405.8	6407.6	15606.	BH F	1.0			L
6404.9	6406.7	15609.	BH F	0.7			L
6404.0	6405.8	15611.	BH F	0.7			L
6403.2	6405.0	15613.	BH F	0.7			L
6402.4	6404.2	15615.	BH F	0.7			L
6401.7	6403.5	15616.	BH F	0.3			L
6400.9	6402.7	15618.	BH F	0.3			L
6399.6	6401.4	15622.	BH F	0.3			L
6399.0	6400.8	15623.	BH F	0.3			L
6394.7	6396.5	15634.	BH F	1.0			L
6332.3	6334.1	15788.	BH F	0.3			L
6330.9	6332.7	15791.	BH F	0.3			L
6329.6	6331.4	15794.	BH F	0.7			L
6328.3	6330.0	15798.	BH F	0.7			L
6327.0	6328.7	15801.	BH F	1.0			L
6325.9	6327.6	15804.	BH F	1.0			L
6324.7	6326.4	15807.	BH F	1.0			L
6323.6	6325.3	15809.	BH F	0.7			L
6306.1	6307.8	15853.	BH F	2.0			L
6304.8	6306.5	15856.	BH F	2.2			L
6303.5	6305.2	15860.	BH F	2.2			L
6302.2	6303.9	15863.	BH F	2.2			L
6300.9	6302.6	15866.	BH F	2.2			L
6299.8	6301.5	15869.	BH F	2.0			L
6298.7	6300.4	15872.	BH F	2.0			L
6297.5	6299.2	15875.	BH F	2.0			L
6296.5	6298.2	15877.	BH F	2.0			L
6295.4	6297.1	15880.	BH F	1.9			L
6293.4	6295.1	15885.	BH F	1.9			L
6292.5	6294.2	15887.	BH F	1.7			L
6291.6	6293.3	15890.	BH F	1.7			L
6290.7	6292.4	15892.	BH F	1.7			L
6289.9	6291.6	15894.	BH F	1.5			L
6289.2	6290.9	15896.	BH F	1.5			L
6288.4	6290.1	15898.	BH F	1.3			L
6288.1	6289.8	15899.	BH F	0.3			L
6287.8	6289.5	15899.	BH F	1.0			L
6287.1	6288.8	15901.	BH F	0.7			L
6286.5	6288.2	15903.	BH F	0.7			L
6285.9	6287.6	15904.	BH F	0.3			L

AIR WAVELENGTH (ANGSTRMS)	VACUUM WAVELENGTH (ANGSTRMS)	WAVENUMBER (KAYSERS)	LMENT	LOG ARC	INTENSITY SPK	DIS	REF
6285.5	6287.2	15905.	BH F	0.3			L
6285.4	6287.1	15905.	BH F	1.0			L
6283.1	6284.8	15911.	BH F	1.3			L
6280.9	6282.6	15917.	BH F	1.3			L
6276.5	6278.2	15928.	BH F	1.3			L
6272.4	6274.1	15938.	BH F	1.5			L
6268.4	6270.1	15949.	BH F	1.5			L
6264.5	6266.2	15959.	BH F	1.3			L
6260.9	6262.6	15968.	BH F	1.0			L
6257.4	6259.1	15977.	BH F	1.3			L
6256.6	6258.3	15979.	BH F	1.3			L
6254.1	6255.8	15985.	BH F	1.0			L
6252.2	6253.9	15990.	BH F	1.0			L
6250.3	6252.0	15995.	BH F	1.0			L
6247.8	6249.5	16001.	BH F	1.0			L
6243.7	6245.4	16012.	BH F	0.7			L
6239.8	6241.5	16022.	BH F	1.0			L
6235.9	6237.6	16032.	BH F	1.0			L
6232.3	6234.0	16041.	BH F	0.7			L
6225.7	6227.4	16058.	BH F	0.7			L
6222.3	6224.0	16067.	BH F	1.0			L
6086.9	6088.6	16424.	BH F	1.5			L
6084.9	6086.6	16430.	BH F	1.5			L
6083.2	6084.9	16434.	BH F	1.3			L
6081.6	6083.3	16438.	BH F	1.3			L
6080.2	6081.9	16442.	BH F	0.7			L
6078.8	6080.5	16446.	BH F	0.7			L
6064.4	6066.1	16485.	BH F	2.3			L
6062.3	6064.0	16491.	BH F	2.2			L
6060.4	6062.1	16496.	BH F	2.0			L
6058.6	6060.3	16501.	BH F	1.9			L
6056.9	6058.6	16505.	BH F	1.7			L
6055.5	6057.2	16509.	BH F	1.5			L
6054.0	6055.7	16513.	BH F	1.5			L
6052.8	6054.5	16517.	BH F	1.3			L
6051.6	6053.3	16520.	BH F	1.3			L
6050.8	6052.5	16522.	BH F	1.7			L
6048.8	6050.5	16528.	BH F	1.5			L
6046.9	6048.6	16533.	BH F	1.3			L
6045.3	6047.0	16537.	BH F	1.3			L
6043.6	6045.3	16542.	BH F	1.0			L
6042.2	6043.9	16546.	BH F	1.0			L
6040.9	6042.6	16549.	BH F	1.0			L
6036.9	6038.6	16560.	BH F	1.7			L
6034.8	6036.5	16566.	BH F	1.7			L
6032.9	6034.6	16571.	BH F	1.5			L
6031.1	6032.8	16576.	BH F	1.5			L
6029.3	6031.0	16581.	BH F	1.3			L
6027.8	6029.5	16585.	BH F	1.3			L
6026.0	6027.7	16590.	BH F	1.3			L
6024.3	6026.0	16595.	BH F	1.3			L
5852.9	5854.5	17081.	BH F	0.3			L
5849.4	5851.0	17091.	BH F	0.7			L
5847.9	5849.5	17095.	BH F	1.0			L
5846.0	5847.6	17101.	BH F	0.7			L
5845.8	5847.4	17102.	BH F	1.5			L
5843.8	5845.4	17107.	BH F	1.5			L
5842.6	5844.2	17111.	BH F	0.7			L
5842.0	5843.6	17113.	BH F	1.5			L
5840.4	5842.0	17117.	BH F	1.5			L
5839.3	5840.9	17121.	BH F	1.0			L
5838.9	5840.5	17122.	BH F	1.5			L
5837.5	5839.1	17126.	BH F	1.5			L
5836.3	5837.9	17129.	BH F	1.5			L
5835.9	5837.5	17131.	BH F	1.3			L
5835.2	5836.8	17133.	BH F	1.5			L
5834.2	5835.8	17136.	BH F	1.5			L
5834.0	5835.6	17136.	BH F	0.7			L
5833.4	5835.0	17138.	BH F	1.5			L
5832.7	5834.3	17140.	BH F	1.3			L

AIR WAVELENGTH (ANGSTRMS)	VACUUM WAVELENGTH (ANGSTRMS)	WAVENUMBER (KAYSERS)	LMENT	LOG ARC	INTENSITY SPK	DIS	REF
5832.6	5834.2	17140.	BH F	1.7			L
5832.0	5833.6	17142.	BH F	1.7			L
5831.5	5833.1	17143.	BH F	1.7			L
5831.1	5832.7	17145.	BH F	1.7			L
5830.8	5832.4	17146.	BH F	1.5			L
5830.6	5832.2	17146.	BH F	0.3			L
5829.4	5831.0	17150.	BH F	1.3			L
5827.3	5828.9	17156.	BH F	0.3			L
5826.1	5827.7	17159.	BH F	1.3			L
5824.0	5825.6	17166.	BH F	0.3			L
5823.0	5824.6	17168.	BH F	1.5			L
5820.8	5822.4	17175.	BH F	0.3			L
5819.9	5821.5	17178.	BH F	1.5			L
5817.4	5819.0	17185.	BH F	0.7			L
5816.7	5818.3	17187.	BH F	1.7			L
5814.2	5815.8	17194.	BH F	0.7			L
5813.7	5815.3	17196.	BH F	1.7			L
5810.6	5812.2	17205.	BH F	1.7			L
5807.7	5809.3	17214.	BH F	1.9			L
5804.7	5806.3	17223.	BH F	1.9			L
5801.8	5803.4	17231.	BH F	1.9			L
5801.6	5803.2	17232.	BH F	0.3			L
5798.9	5800.5	17240.	BH F	1.9			L
5798.5	5800.1	17241.	BH F	1.0			L
5795.9	5797.5	17249.	BH F	2.0			L
5795.4	5797.0	17250.	BH F	1.0			L
5793.1	5794.7	17257.	BH F	2.0			L
5792.4	5794.0	17259.	BH F	1.3			L
5790.3	5791.9	17265.	BH F	2.0			L
5789.3	5790.9	17268.	BH F	1.5			L
5787.6	5789.2	17273.	BH F	2.0			L
5786.4	5788.0	17277.	BH F	1.5			L
5784.8	5786.4	17282.	BH F	2.0			L
5783.4	5785.0	17286.	BH F	1.5			L
5782.1	5783.7	17290.	BH F	2.2			L
5780.5	5782.1	17295.	BH F	1.7			L
5779.4	5781.0	17298.	BH F	2.3			L
5777.6	5779.2	17303.	BH F	2.0			L
5774.8	5776.4	17312.	BH F	2.0			L
5771.9	5773.5	17320.	BH F	2.0			L
5705.7	5707.3	17521.	BH F	0.3			L
5698.4	5700.0	17544.	BH F	0.3			L
5697.2	5698.8	17548.	BH F	0.3			L
5680.4	5682.0	17599.	BH F	0.3			MIT
5677.1	5678.7	17610.	BH F	0.7			L
5676.2	5677.8	17612.	BH F	0.7			L
5672.9	5674.5	17623.	BH F	0.7			L
5672.1	5673.7	17625.	BH F	0.7			L
5668.8	5670.4	17635.	BH F	0.7			L
5668.0	5669.6	17638.	BH F	0.7			L
5664.7	5666.3	17648.	BH F	1.0			L
5664.0	5665.6	17650.	BH F	1.0			L
5660.9	5662.5	17660.	BH F	1.0			L
5660.7	5662.3	17661.	BH F	1.0			L
5656.6	5658.2	17674.	BH F	1.0			L
5656.0	5657.6	17675.	BH F	1.0			L
5652.6	5654.2	17686.	BH F	1.3			L
5652.1	5653.7	17688.	BH F	1.3			L
5648.6	5650.2	17699.	BH F	1.3			L
5648.2	5649.8	17700.	BH F	1.3			L
5644.7	5646.3	17711.	BH F	1.3			L
5644.3	5645.9	17712.	BH F	1.3			L
5640.8	5642.4	17723.	BH F	1.3			L
5640.5	5642.1	17724.	BH F	1.3			L
5636.9	5638.5	17735.	BH F	1.3			L
5636.7	5638.3	17736.	BH F	1.3			L
5632.9	5634.5	17748.	BH F	1.0			L
5629.2	5630.8	17760.	BH F	0.7			L
5625.5	5627.1	17771.	BH F	0.3			L
5421.9	5423.4	18439.	BH F	0.7			L
5421.1	5422.6	18441.	BH F	0.7			L
5413.9	5415.4	18466.	BH F	0.3			L
5407.6	5409.1	18487.	BH F	0.7			L
5406.6	5408.1	18491.	BH F	0.3			L
5400.5	5402.0	18512.	BH F	1.0			L
5399.5	5401.0	18515.	BH F	0.7			L
5393.4	5394.9	18536.	BH F	1.0			L
5392.3	5393.8	18540.	BH F	0.7			L
5386.6	5388.1	18559.	BH F	1.0			L
5385.5	5387.0	18563.	BH F	1.0			L
5379.8	5381.3	18583.	BH F	1.0			MIT
5378.7	5380.2	18587.	BH F	1.0			L
5373.1	5374.6	18606.	BH F	1.0			L
5371.9	5373.4	18610.	BH F	1.0			L
5366.4	5367.9	18629.	BH F	1.0			L
5365.2	5366.7	18633.	BH F	1.0			L
5359.8	5361.3	18652.	BH F	1.0			L
5358.6	5360.1	18656.	BH F	1.0			L
5353.4	5354.9	18674.	BH F	1.0			L
5352.1	5353.6	18679.	BH F	1.0			L
5347.0	5348.5	18697.	BH F	1.3			L
5345.6	5347.1	18702.	BH F	1.3			L
5340.7	5342.2	18719.	BH F	1.3			L
5339.3	5340.8	18724.	BH F	1.3			L
5334.5	5336.0	18741.	BH F	1.5			L
5333.1	5334.6	18746.	BH F	1.5			L
5328.3	5329.8	18762.	BH F	1.5			L
5326.9	5328.4	18767.	BH F	1.5			L
5322.2	5323.7	18784.	BH F	1.5			L
5320.7	5322.2	18789.	BH F	1.5			L
5316.2	5317.7	18805.	BH F	1.9			L
5314.7	5316.2	18810.	BH F	1.7			L
5310.3	5311.8	18826.	BH F	1.9			L
5308.7	5310.2	18832.	BH F	1.9			L
5304.4	5305.9	18847.	BH F	2.0			L
5302.7	5304.2	18853.	BH F	2.0			L
5298.6	5300.1	18868.	BH F	2.0			L
5296.8	5298.3	18874.	BH F	2.0			L
5292.9	5294.4	18888.	BH F	2.2			L
5291.0	5292.5	18895.	BH F	2.3			L
5222.6	5224.1	19142.	BH F	0.3			L
5221.8	5223.3	19145.	BH F	0.3			L
5214.3	5215.8	19173.	BH F	0.3			L
5207.8	5209.3	19197.	BH F	0.3			L
5206.9	5208.3	19200.	BH F	0.3			L
5200.5	5201.9	19224.	BH F	0.3			L
5199.7	5201.1	19226.	BH F	0.3			L
5193.4	5194.8	19250.	BH F	0.3			L
5192.6	5194.0	19253.	BH F	0.3			L
5186.4	5187.8	19276.	BH F	0.7			L
5185.6	5187.0	19279.	BH F	0.7			L
5179.5	5180.9	19301.	BH F	0.7			L
5178.6	5180.0	19305.	BH F	0.7			L
5171.8	5173.2	19330.	BH F	0.3			L
5165.9	5167.3	19352.	BH F	0.7			L
5165.0	5166.4	19356.	BH F	0.7			L
5159.3	5160.7	19377.	BH F	0.7			L
5158.4	5159.8	19380.	BH F	0.7			L
5152.8	5154.2	19401.	BH F	0.3			L
5151.9	5153.3	19405.	BH F	0.3			L
5136.9	5138.3	19462.	BH F	0.3			L
5135.5	5136.9	19467.	BH F	0.7			L
5133.9	5135.3	19473.	BH F	0.7			L
5132.4	5133.8	19479.	BH F	0.7			L
5130.8	5132.2	19485.	BH F	0.3			L
5129.0	5130.4	19491.	BH F	0.3			L
5127.2	5128.6	19498.	BH F	0.7			L
5125.1	5126.5	19506.	BH F	0.3			L
5123.3	5124.7	19513.	BH F	0.3			L
5036.4	5037.8	19850.	BH F	0.3			L

AIR WAVELENGTH (ANGSTRMS)	VACUUM WAVELENGTH (ANGSTRMS)	WAVENUMBER (KAYSERS)	LMENT	LOG ARC	INTENSITY SPK	DIS	REF
5034.3	5035.7	19858.	BH F	0.3			L
5032.1	5033.5	19867.	BH F	0.3			L
5029.9	5031.3	19876.	BH F	0.7			L
5027.5	5028.9	19885.	BH F	0.7			L
5025.1	5026.5	19895.	BH F	0.7			L
5022.7	5024.1	19904.	BH F	0.7			L
5020.2	5021.6	19914.	BH F	0.7			L
5017.6	5019.0	19924.	BH F	1.0			L
5014.9	5016.3	19935.	BH F	1.3			L
5012.2	5013.6	19946.	BH F	1.3			L
5009.4	5010.8	19957.	BH F	1.5			L
5008.1	5009.5	19962.	BH F	0.3			L
5006.6	5008.0	19968.	BH F	1.5			L
5003.6	5005.0	19980.	BH F	1.7			L
5003.1	5004.5	19982.	BH F	0.3			L
5000.6	5002.0	19992.	BH F	1.9			L
5000.5	5001.9	19992.	BH F	1.0			L
4997.7	4999.1	20004.	BH F	1.3			L
4994.9	4996.3	20015.	BH F	1.5			L
4992.1	4993.5	20026.	BH F	1.7			L
4967.9	4969.3	20124.	BH F	0.3			L
4967.1	4968.5	20127.	BH F	0.3			L
4965.1	4966.5	20135.	BH F	0.7			L
4964.1	4965.5	20139.	BH F	0.7			L
4962.9	4964.3	20144.	BH F	1.0			L
4961.7	4963.1	20149.	BH F	1.0			L
4960.3	4961.7	20154.	BH F	1.3			L
4958.9	4960.3	20160.	BH F	1.3			L
4955.9	4957.3	20172.	BH F	1.5			L
4954.3	4955.7	20179.	BH F	1.7			L
4952.6	4954.0	20186.	BH F	1.9			L
4950.8	4952.2	20193.	BH F	2.0			L
5789.649	5791.254	17267.42	BH FE0	0.9			ME
6148.7	6150.4	16259.	BH LA	1.3			ME
5920.8	5922.4	16885.	BH LA	1.7			ME
5896.7	5898.3	16954.	BH LA	1.9			ME
5893.6	5895.2	16963.	BH LA	1.8			ME
5869.5	5871.1	17032.	BH LA	1.7			ME
5866.4	5868.0	17041.	BH LA	1.6			ME
5707.9	5709.5	17515.	BH LA	1.0			ME
5705.3	5706.9	17523.	BH LA	0.9			ME
5681.1	5682.7	17597.	BH LA	1.3J			ME
5679.2	5680.8	17603.	BH LA	1.0J			ME
5678.6	5680.2	17605.	BH LA	1.3J			ME
5654.8	5656.4	17679.	BH LA	1.7			ME
5652.3	5653.9	17687.	BH LA	1.7			ME
5628.6	5630.2	17761.	BH LA	2.0			ME
5626.0	5627.6	17770.	BH LA	2.0			ME
5602.5	5604.1	17844.	BH LA	2.3			ME
5602.4	5604.0	17844.	BH LA	2.0			ME
5600.0	5601.6	17852.	BH LA	2.0			ME
5599.9	5601.5	17852.	BH LA	1.7			ME
5588.8	5590.4	17888.	BH LA	0.3			ME
5586.9	5588.5	17894.	BH LA	0.8			ME
5562.6	5564.1	17972.	BH LA	0.7			ME
5560.4	5561.9	17979.	BH LA	0.9			ME
5536.4	5537.9	18057.	BH LA	1.0			ME
5534.2	5535.7	18064.	BH LA	0.9			ME
5510.2	5511.7	18143.	BH LA	0.9			ME
5508.4	5509.9	18149.	BH LA	1.0			ME
5484.3	5485.8	18229.	BH LA	1.0			ME
5458.7	5460.2	18314.	BH LA	1.3			ME
5456.6	5458.1	18321.	BH LA	1.0			ME
5433.0	5434.5	18401.	BH LA	1.5			ME
5431.1	5432.6	18407.	BH LA	1.3			ME
5407.7	5409.2	18487.	BH LA	1.5			ME
5405.7	5407.2	18494.	BH LA	1.3			ME
5382.4	5383.9	18574.	BH LA	1.3			ME
5380.5	5382.0	18580.	BH LA	1.0			ME
4627.3	4628.6	21605.	BH LA	0.5H			ME

AIR WAVELENGTH (ANGSTRMS)	VACUUM WAVELENGTH (ANGSTRMS)	WAVENUMBER (KAYSERS)	LMENT	LOG ARC	INTENSITY SPK	DIS	REF
4622.7	4624.0	21626.	BH LA	0.5			ME
4612.0	4613.3	21676.	BH LA	0.5			ME
4607.3	4608.6	21699.	BH LA	0.5			ME
4602.7	4604.0	21720.	BH LA	0.5			ME
4598.2	4599.5	21741.	BH LA	0.6			ME
4593.9	4595.2	21762.	BH LA	0.7			ME
4589.5	4590.8	21783.	BH LA	0.7			ME
4589.4	4590.7	21783.	BH LA	0.3			ME
4585.1	4586.4	21804.	BH LA	0.6			ME
4580.8	4582.1	21824.	BH LA	0.5			ME
4471.3	4472.6	22359.	BH LA	1.0			ME
4471.2	4472.5	22359.	BH LA	0.5			ME
4464.2	4465.5	22394.	BH LA	1.0			ME
4464.1	4465.4	22395.	BH LA	0.5			ME
4458.5	4459.8	22423.	BH LA	1.2			ME
4458.4	4459.7	22423.	BH LA	0.7			ME
4453.1	4454.3	22450.	BH LA	1.3			ME
4448.0	4449.2	22476.	BH LA	1.3			ME
4447.9	4449.1	22476.	BH LA	1.0			ME
4443.0	4444.2	22501.	BH LA	1.4			ME
4442.9	4444.1	22501.	BH LA	1.1			ME
4438.0	4439.2	22526.	BH LA	1.5			ME
4437.9	4439.1	22527.	BH LA	1.2			ME
4433.0	4434.2	22552.	BH LA	1.2			ME
4428.2	4429.4	22576.	BH LA	1.6			ME
4428.1	4429.3	22577.	BH LA	1.5			ME
4428.0	4429.2	22577.	BH LA	1.2			ME
4423.1	4424.3	22602.	BH LA	1.3			ME
4418.2	4419.4	22627.	BH LA	1.7			ME
4418.1	4419.3	22628.	BH LA	1.4			ME
4414.7	4415.9	22645.	BH LA	0.5			ME
4409.6	4410.8	22671.	BH LA	0.6			ME
4404.9	4406.1	22696.	BH LA	0.5			ME
4400.1	4401.3	22720.	BH LA	0.6			ME
4400.0	4401.2	22721.	BH LA	0.3			ME
4395.7	4396.9	22743.	BH LA	0.8			ME
4395.6	4396.8	22744.	BH LA	0.5			ME
4391.6	4392.8	22764.	BH LA	0.9			ME
4391.5	4392.7	22765.	BH LA	0.6			ME
4387.6	4388.8	22785.	BH LA	1.0			ME
4387.5	4388.7	22786.	BH LA	0.7			ME
4383.5	4384.7	22806.	BH LA	1.2			ME
4383.4	4384.6	22807.	BH LA	0.9			ME
4379.7	4380.9	22826.	BH LA	1.3			ME
4379.6	4380.8	22827.	BH LA	1.0			ME
4375.8	4377.0	22846.	BH LA	1.5			ME
4375.7	4376.9	22847.	BH LA	1.2			ME
4372.0	4373.2	22866.	BH LA	1.6			ME
4371.9	4373.1	22867.	BH LA	1.3			ME
4369.7	4370.9	22878.	BH LA	0.7			ME
3622.6	3623.6	27597.	BH LA	0.8H			ME
3622.0	3623.0	27601.	BH LA	0.6H			ME
3621.2	3622.2	27607.	BH LA	0.6H			ME
3620.8	3621.8	27610.	BH LA	0.5H			ME
3620.6	3621.6	27612.	BH LA	0.3H			ME
3620.1	3621.1	27616.	BH LA	0.8H			ME
3619.4	3620.4	27621.	BH LA	0.6H			ME
3616.3	3617.3	27645.	BH LA	0.5H			ME
3614.8	3615.8	27656.	BH LA	0.7H			ME
3614.4	3615.4	27659.	BH LA	0.6H			ME
3614.2	3615.2	27661.	BH LA	0.5H			ME
3613.8	3614.8	27664.	BH LA	0.5H			ME
3611.4	3612.4	27682.	BH LA	0.6D			ME
3611.1	3612.1	27684.	BH LA	0.8H			ME
3608.1	3609.1	27707.	BH LA	0.5H			ME
3607.8	3608.8	27710.	BH LA	0.7H			ME
3604.5	3605.5	27735.	BH LA	0.7H			ME
3566.1	3567.1	28034.	BH LA	0.7H			ME
3565.8	3566.8	28036.	BH LA	0.6H			ME
6581.1	6582.9	15191.	BH MG	0.5			L

AIR WAVELENGTH (ANGSTRMS)	VACUUM WAVELENGTH (ANGSTRMS)	WAVENUMBER (KAYSERS)	LMENT	LOG ARC	INTENSITY SPK	INTENSITY DIS	REF
6311.8	6313.5	15839.	BH MG	0.6			L
6246.4	6248.1	16005.	BH MG	0.5			L
6060.3	6062.0	16496.	BH MG	0.8			L
5775.5	5777.1	17310.	BH MG	0.7			L
5726.3	5727.9	17458.	BH MG	0.3			L
5518.8	5520.3	18115.	BH MG	0.5			L
5475.9	5477.4	18257.	BH MG	0.3			L
5285.7	5287.2	18914.	BH MG	0.3			L
5206.0	5207.4	19203.	BH MG	0.5			L
5192.0	5193.4	19255.	BH MG	0.5			L
5177.4	5178.8	19309.	BH MG	0.3			L
5162.5	5163.9	19365.	BH MG	0.3			L
5146.9	5148.3	19424.	BH MG	0.3			L
5007.3	5008.7	19965.	BH MG	1.1			L
4996.7	4998.1	20008.	BH MG	1.0			L
4985.9	4987.3	20051.	BH MG	0.9			L
4974.5	4975.9	20097.	BH MG	0.8			L
4962.2	4963.6	20147.	BH MG	0.8			L
4949.4	4950.8	20199.	BH MG	0.7			L
4935.3	4936.7	20256.	BH MG	0.6			L
4923.9	4925.3	20303.	BH MG	0.5			L
6433.6	6435.4	15539.	BH PB	0.5			L
6427.7	6429.5	15553.	BH PB	0.5			L
6342.0	6343.8	15763.	BH PB	0.5			L
6250.8	6252.5	15993.	BH PB	0.7			L
6160.5	6162.2	16228.	BH PB	0.6			L
5910.7	5912.3	16914.	BH PB	0.9			L
5677.8	5679.4	17608.	BH PB	0.9			L
5617.7	5619.3	17796.	BH PB	0.5			L
5459.4	5460.9	18312.	BH PB	0.9			L
5353.8	5355.3	18673.	BH PB	0.5			L
5331.1	5332.6	18753.	BH PB	0.5			L
5162.3	5163.7	19366.	BH PB	0.9			L
5138.2	5139.6	19457.	BH PB	0.5			L
4983.8	4985.2	20059.	BH PB	0.8			L
4816.9	4818.2	20754.	BH PB	0.8			L
4658.0	4659.3	21462.	BH PB	0.7			L
4553.7	4555.0	21954.	BH PB	0.8			L
4410.4	4411.6	22667.	BH PB	0.7			L
4317.1	4318.3	23157.	BH PB	0.6			L
4229.0	4230.2	23640.	BH PB	0.6			L
3485.7	3486.7	28680.	BH PB	1.5			L
3401.9	3402.9	29387.	BH PB	1.3			L
7293.4	7295.4	13707.	BH SC	0.3			ME
7287.9	7289.9	13718.	BH SC	0.3			ME
7275.6	7277.6	13741.	BH SC	1.0			ME
7240.3	7242.3	13808.	BH SC	0.5			ME
7233.2	7235.2	13821.	BH SC	0.5			ME
7221.6	7223.6	13843.	BH SC	0.5			ME
7189.7	7191.7	13905.	BH SC	0.5			ME
7180.7	7182.7	13922.	BH SC	0.6			ME
7141.2	7143.2	13999.	BH SC	0.7			ME
7130.6	7132.6	14020.	BH SC	0.6			ME
7120.5	7122.5	14040.	BH SC	0.6			ME
7120.2	7122.2	14041.	BH SC	0.5			ME
7094.4	7096.4	14092.	BH SC	0.7			ME
7082.4	7084.4	14116.	BH SC	0.7			ME
7072.4	7074.4	14136.	BH SC	0.7			ME
7049.4	7051.3	14182.	BH SC	0.7			ME
7035.8	7037.7	14209.	BH SC	0.8			ME
7025.7	7027.6	14229.	BH SC	0.7			ME
7005.5	7007.4	14271.	BH SC	0.7			ME
6990.7	6992.6	14301.	BH SC	0.8			ME
6985.5	6987.4	14311.	BH SC	0.8			ME
6980.6	6982.5	14321.	BH SC	0.7			ME
6963.1	6965.0	14357.	BH SC	0.6			ME
6961.6	6963.5	14361.	BH SC	0.6			ME
6948.3	6950.2	14388.	BH SC	0.3			ME
6936.7	6938.6	14412.	BH SC	0.6			ME
6914.5	6916.4	14458.	BH SC	0.6			ME

AIR WAVELENGTH (ANGSTRMS)	VACUUM WAVELENGTH (ANGSTRMS)	WAVENUMBER (KAYSERS)	LMENT	LOG ARC	INTENSITY SPK	INTENSITY DIS	REF
6885.3	6887.2	14520.	BH SC	0.9			ME
6881.1	6883.0	14529.	BH SC	0.7			ME
6877.4	6879.3	14536.	BH SC	0.6			ME
6874.3	6876.2	14543.	BH SC	0.6			ME
6864.6	6866.5	14563.	BH SC	0.8			ME
6850.5	6852.4	14593.	BH SC	0.6			ME
6837.6	6839.5	14621.	BH SC	0.6			ME
6800.5	6802.4	14701.	BH SC	0.5			ME
6798.2	6800.1	14706.	BH SC	0.3			ME
6784.0	6785.9	14736.	BH SC	0.5			ME
6752.6	6754.5	14805.	BH SC	0.6			ME
6748.3	6750.2	14814.	BH SC	0.5			ME
6735.4	6737.3	14843.	BH SC	0.3			ME
6705.9	6707.8	14908.	BH SC	0.8			ME
6700.5	6702.3	14920.	BH SC	0.7			ME
6691.2	6693.0	14941.	BH SC	0.6			ME
6661.0	6662.8	15009.	BH SC	0.8			ME
6654.4	6656.2	15023.	BH SC	0.8			ME
6645.1	6646.9	15044.	BH SC	0.8			ME
6617.9	6619.7	15106.	BH SC	0.9			ME
6610.0	6611.8	15124.	BH SC	0.9			ME
6600.8	6602.6	15145.	BH SC	1.0			ME
6575.9	6577.7	15203.	BH SC	1.0			ME
6566.9	6568.7	15224.	BH SC	1.0			ME
6557.8	6559.6	15245.	BH SC	1.0			ME
6535.3	6537.1	15297.	BH SC	1.0			ME
6525.6	6527.4	15320.	BH SC	1.2			ME
6517.7	6519.5	15339.	BH SC	0.8			ME
6503.1	6504.9	15373.	BH SC	0.6			ME
6495.9	6497.7	15390.	BH SC	1.2			ME
6485.4	6487.2	15415.	BH SC	1.3			ME
6476.5	6478.3	15436.	BH SC	1.0			ME
6462.8	6464.6	15469.	BH SC	0.5			ME
6461.1	6462.9	15473.	BH SC	0.5			ME
6457.8	6459.6	15481.	BH SC	1.2			ME
6446.2	6448.0	15509.	BH SC	1.5			ME
6424.0	6425.8	15562.	BH SC	0.3			ME
6420.3	6422.1	15571.	BH SC	0.3			ME
6408.4	6410.2	15600.	BH SC	0.3			ME
6361.7	6363.5	15715.	BH SC	0.3			ME
6317.5	6319.2	15825.	BH SC	0.5			ME
6316.1	6317.8	15828.	BH SC	0.3			ME
6274.6	6276.3	15933.	BH SC	0.6			ME
6271.8	6273.5	15940.	BH SC	0.3			ME
6262.3	6264.0	15964.	BH SC	0.8			ME
6233.1	6234.8	16039.	BH SC	0.8			ME
6229.4	6231.1	16048.	BH SC	0.7			ME
6220.2	6221.9	16072.	BH SC	0.8			ME
6192.9	6194.6	16143.	BH SC	1.0			ME
6188.1	6189.8	16156.	BH SC	1.0			ME
6180.6	6182.3	16175.	BH SC	0.6			ME
6173.1	6174.8	16195.	BH SC	0.3			ME
6153.9	6155.6	16245.	BH SC	1.3			ME
6140.3	6142.0	16281.	BH SC	1.0			ME
6132.0	6133.7	16303.	BH SC	0.5			ME
6116.0	6117.7	16346.	BH SC	1.6			ME
6109.9	6111.6	16362.	BH SC	1.6			ME
6101.9	6103.6	16384.	BH SC	1.5			ME
6092.5	6094.2	16409.	BH SC	0.7			ME
6079.3	6081.0	16445.	BH SC	2.0			ME
6072.7	6074.4	16463.	BH SC	2.0			ME
6064.3	6066.0	16485.	BH SC	1.9			ME
6055.2	6056.9	16510.	BH SC	1.0			ME
6036.2	6037.9	16562.	BH SC	2.3			ME
6017.1	6018.8	16615.	BH SC	1.6			ME
6002.3	6004.0	16656.	BH SC	0.5			ME
5970.5	5972.2	16744.	BH SC	0.7			ME
5968.5	5970.2	16750.	BH SC	1.0			ME
5959.0	5960.7	16777.	BH SC	0.8			ME
5928.1	5929.7	16864.	BH SC	1.2			ME

AIR WAVELENGTH (ANGSTRMS)	VACUUM WAVELENGTH (ANGSTRMS)	WAVENUMBER (KAYSERS)	LMENT	LOG ARC	INTENSITY SPK	INTENSITY DIS	REF
5918.0	5919.6	16893.	BH SC	1.0			ME
5887.4	5889.0	16981.	BH SC	1.3			ME
5877.8	5879.4	17008.	BH SC	1.0			ME
5876.9	5878.5	17011.	BH SC	0.5			ME
5849.1	5850.7	17092.	BH SC	1.3			ME
5847.7	5849.3	17096.	BH SC	1.3			ME
5839.6	5841.2	17120.	BH SC	1.0			ME
5836.5	5838.1	17129.	BH SC	0.8			ME
5811.6	5813.2	17202.	BH SC	1.2			ME
5809.8	5811.4	17207.	BH SC	1.3			ME
5801.3	5802.9	17233.	BH SC	1.0			ME
5797.6	5799.2	17244.	BH SC	0.6			ME
5775.3	5776.9	17310.	BH SC	1.0			ME
5772.7	5774.3	17318.	BH SC	1.2			ME
5764.5	5766.1	17343.	BH SC	0.9			ME
5759.0	5760.6	17359.	BH SC	0.3			ME
5736.9	5738.5	17426.	BH SC	1.0			ME
5396.4	5397.9	18526.	BH SC	0.3			ME
5171.1	5172.5	19333.	BH SC	0.6			ME
5133.7	5135.1	19474.	BH SC	0.7			ME
5133.3	5134.7	19475.	BH SC	0.5			ME
5096.7	5098.1	19615.	BH SC	1.3			ME
4893.3	4894.7	20430.	BH SC	0.6			ME
4893.0	4894.4	20432.	BH SC	0.5			ME
4858.1	4859.5	20578.	BH SC	1.5			ME
4857.8	4859.2	20580.	BH SC	1.3			ME
4778.5	4779.8	20921.	BH SC	0.5			ME
4742.5	4743.8	21080.	BH SC	0.7			ME
4707.3	4708.6	21238.	BH SC	0.7			ME
4707.0	4708.3	21239.	BH SC	1.0			ME
4673.1	4674.4	21393.	BH SC	1.0			ME
4672.6	4673.9	21395.	BH SC	1.0			ME
4606.8	4608.1	21701.	BH SC	1.0			ME
4571.3	4572.6	21869.	BH SC	1.0			ME
4536.6	4537.9	22037.	BH SC	1.6			ME
4502.8	4504.1	22202.	BH SC	0.5			ME
9776.26	9778.94	10226.1	BH SR	1.3L			L
9196.1	9198.6	10871.	BH SR	1.0L			L
8700.05	8702.44	11491.0	BH SR	1.0L			L
8257.78	8260.05	12106.5	BH SR	0.7L			L
7882.36	7884.53	12683.1	BH SR	0.5L			L
4787.2	4788.5	20883.	BH SR	0.3			L
4712.8	4714.1	21213.	BH SR	0.3			L
4692.7	4694.0	21304.	BH SR	0.7			L
4672.6	4673.9	21395.	BH SR	0.6			L
4652.4	4653.7	21488.	BH SR	0.5			L
4606.2	4607.5	21704.	BH SR	0.3			L
4585.5	4586.8	21802.	BH SR	0.5			L
4564.8	4566.1	21901.	BH SR	0.7			L
4544.1	4545.4	22000.	BH SR	0.7			L
4523.3	4524.6	22102.	BH SR	0.6			L
4484.5	4485.8	22293.	BH SR	0.5			L
4463.3	4464.6	22399.	BH SR	0.6			L
4442.1	4443.3	22506.	BH SR	0.5			L
4420.9	4422.1	22613.	BH SR	0.7			L
4399.6	4400.8	22723.	BH SR	0.8			L
4367.6	4368.8	22889.	BH SR	0.3			L
4345.9	4347.1	23004.	BH SR	0.5			L
4324.3	4325.5	23119.	BH SR	0.5			L
4302.7	4303.9	23235.	BH SR	0.6			L
4281.0	4282.2	23352.	BH SR	0.7			L
4233.2	4234.4	23616.	BH SR	0.3			L
4211.2	4212.4	23739.	BH SR	0.5			L
4189.1	4190.3	23865.	BH SR	0.6			L
4167.2	4168.4	23990.	BH SR	0.7			L
4135.0	4136.2	24177.	BH SR	0.6			L
4124.7	4125.9	24237.	BH SR	0.3			L
4113.6	4114.8	24303.	BH SR	0.7			L
4102.3	4103.5	24370.	BH SR	0.3			L
4092.6	4093.8	24427.	BH SR	0.6			L

AIR WAVELENGTH (ANGSTRMS)	VACUUM WAVELENGTH (ANGSTRMS)	WAVENUMBER (KAYSERS)	LMENT	LOG ARC	INTENSITY SPK	INTENSITY DIS	REF
4080.1	4081.3	24502.	BH SR	0.3			L
4058.0	4059.1	24636.	BH SR	0.5			L
4020.1	4021.2	24868.	BH SR	0.3			L
3997.7	3998.8	25007.	BH SR	0.6			L
3975.4	3976.5	25148.	BH SR	0.9			L
3919.6	3920.7	25506.	BH SR	0.6			L
3897.1	3898.2	25653.	BH SR	0.9			L
3822.4	3823.5	26154.	BH SR	0.6			L
3671.5	3672.5	27229.	BH SR	0.6			L
3646.0	3647.0	27419.	BH SR	0.6			L
3607.1	3608.1	27715.	BH SR	0.6			L
3586.9	3587.9	27871.	BH SR	0.9			L
3546.6	3547.6	28188.	BH SR	0.6			L
3525.4	3526.4	28357.	BH SR	0.9			L
3503.8	3504.8	28532.	BH SR	1.5			L
3467.5	3468.5	28831.	BH SR	0.6			L
3445.2	3446.2	29018.	BH SR	1.1			L
3412.8	3413.8	29293.	BH SR	0.6			L
3389.8	3390.8	29492.	BH SR	1.1			L
3337.5	3338.5	29954.	BH SR	0.9			L
3288.1	3289.0	30404.	BH SR	0.6			L
7996.0	7998.2	12503.	BH TI	0.3			L
7988.1	7990.3	12515.	BH TI	0.6			L
7948.6	7950.8	12577.	BH TI	0.9			L
7938.5	7940.7	12593.	BH TI	0.7			L
7907.3	7909.5	12643.	BH TI	0.8			L
7900.9	7903.1	12653.	BH TI	0.6			L
7861.0	7863.2	12717.	BH TI	0.8			L
7828.0	7830.2	12771.	BH TI	1.0			L
7820.1	7822.3	12784.	BH TI	0.9			L
7783.4	7785.5	12844.	BH TI	0.3			L
7755.8	7757.9	12890.	BH TI	0.6			L
7749.8	7751.9	12900.	BH TI	0.6			L
7743.2	7745.3	12911.	BH TI	0.6			L
7713.4	7715.5	12961.	BH TI	0.6			L
7705.2	7707.3	12975.	BH TI	0.9			L
7679.3	7681.4	13018.	BH TI	0.6			L
7672.1	7674.2	13031.	BH TI	1.0			L
7666.4	7668.5	13040.	BH TI	0.8			L
7634.1	7636.2	13095.	BH TI	0.6			L
7628.1	7630.2	13106.	BH TI	0.9			L
7596.0	7598.1	13161.	BH TI	0.5			L
7589.6	7591.7	13172.	BH TI	0.9			L
7303.0	7305.0	13689.	BH TI	0.7			L
7297.0	7299.0	13700.	BH TI	0.5			L
7290.8	7292.8	13712.	BH TI	0.7			L
7273.9	7275.9	13744.	BH TI	0.7			L
7269.1	7271.1	13753.	BH TI	0.8			L
7230.4	7232.4	13827.	BH TI	0.6			L
7219.4	7221.4	13848.	BH TI	0.8			L
7203.6	7205.6	13878.	BH TI	0.6			L
7197.7	7199.7	13889.	BH TI	0.9			L
7183.0	7185.0	13918.	BH TI	1.0			L
7171.2	7173.2	13941.	BH TI	0.7			L
7164.3	7166.3	13954.	BH TI	0.6			L
7159.0	7161.0	13965.	BH TI	0.8			L
7131.3	7133.3	14019.	BH TI	0.7			L
7125.6	7127.6	14030.	BH TI	1.2			L
7124.9	7126.9	14031.	BH TI	0.6			L
7093.2	7095.2	14094.	BH TI	0.6			L
7087.9	7089.9	14105.	BH TI	1.1			L
7060.0	7061.9	14160.	BH TI	0.5			L
7054.5	7056.4	14171.	BH TI	0.9			L
6925.8	6927.7	14435.	BH TI	0.3			L
6919.6	6921.5	14448.	BH TI	0.6			L
6883.8	6885.7	14523.	BH TI	0.5			L
6858.1	6860.0	14577.	BH TI	0.6			L
6852.3	6854.2	14590.	BH TI	0.8			L
6850.2	6852.1	14594.	BH TI	0.7			L
6820.2	6822.1	14658.	BH TI	0.5			L

AIR WAVELENGTH (ANGSTRMS)	VACUUM WAVELENGTH (ANGSTRMS)	WAVENUMBER (KAYSERS)	LMENT	LOG ARC	INTENSITY SPK	INTENSITY DIS	REF
6815.3	6817.2	14669.	BH TI	0.7			L
6785.8	6787.7	14733.	BH TI	0.6			L
6781.9	6783.8	14741.	BH TI	0.7			L
6756.3	6758.2	14797.	BH TI	0.8			L
6751.8	6753.7	14807.	BH TI	0.6			L
6747.8	6749.7	14816.	BH TI	0.7			L
6724.0	6725.9	14868.	BH TI	0.7			L
6719.3	6721.2	14878.	BH TI	0.8			L
6714.4	6716.3	14889.	BH TI	0.8			L
6691.2	6693.0	14941.	BH TI	0.6			L
6681.1	6682.9	14963.	BH TI	0.8			L
6657.7	6659.5	15016.	BH TI	0.3			L
6651.5	6653.3	15030.	BH TI	0.7			L
6634.3	6636.1	15069.	BH TI	0.3			L
6626.1	6627.9	15088.	BH TI	0.3			L
6483.6	6485.4	15419.	BH TI	0.3			L
6268.3	6270.0	15949.	BH TI	0.5			L
6226.3	6228.0	16056.	BH TI	0.5			L
6222.4	6224.1	16066.	BH TI	0.6			L
6215.2	6216.9	16085.	BH TI	0.8			L
6190.1	6191.8	16150.	BH TI	0.5			L
6186.6	6188.3	16159.	BH TI	0.7			L
6183.5	6185.2	16168.	BH TI	0.3			L
6174.6	6176.3	16191.	BH TI	0.5			L
6162.2	6163.9	16223.	BH TI	0.3			L
6159.3	6161.0	16231.	BH TI	0.3			L
6148.8	6150.5	16259.	BH TI	0.3			L
6064.7	6066.4	16484.	BH TI	0.3			L
6057.6	6059.3	16504.	BH TI	0.3			L
6051.0	6052.7	16522.	BH TI	0.5			L
6006.5	6008.2	16644.	BH TI	0.5			L
5954.6	5956.2	16789.	BH TI	0.3			L
5949.9	5951.5	16802.	BH TI	0.7			L
5926.1	5927.7	16870.	BH TI	0.7			L
5922.0	5923.6	16881.	BH TI	1.1			L
5904.9	5906.5	16930.	BH TI	1.1			L
5872.7	5874.3	17023.	BH TI	0.3			L
5870.6	5872.2	17029.	BH TI	0.3			L
5863.0	5864.6	17051.	BH TI	0.9			L
5815.0	5816.6	17192.	BH TI	1.3			L
5811.2	5812.8	17203.	BH TI	1.1			L
5762.0	5763.6	17350.	BH TI	1.3			L
5759.6	5761.2	17357.	BH TI	0.9			L
5733.4	5735.0	17437.	BH TI	0.3			L
5727.4	5729.0	17455.	BH TI	0.7			L
5700.6	5702.2	17537.	BH TI	0.3			L
5694.4	5696.0	17556.	BH TI	0.9			L
5667.6	5669.2	17639.	BH TI	0.3			L
5661.6	5663.2	17658.	BH TI	1.5			L
5635.3	5636.9	17740.	BH TI	0.3			L
5629.3	5630.9	17759.	BH TI	1.5			L
5603.8	5605.4	17840.	BH TI	0.7			L
5597.9	5599.5	17859.	BH TI	1.6			L
8701.6	8704.0	11489.	BH V	0.0			L
8639.9	8642.3	11571.	BH V				L
7533.7	7535.8	13270.	BH V	0.3			L
7474.0	7476.1	13376.	BH V	0.5			L
7034.4	7036.3	14212.	BH V	0.5			L
6974.5	6976.4	14334.	BH V	0.5			L
6916.6	6918.5	14454.	BH V	0.5			L
6477.8	6479.6	15433.	BH V	0.3			L
6087.3	6089.0	16423.	BH V	0.6			L
5837.8	5839.4	17125.	BH V	0.8			L
5737.3	5738.9	17425.	BH V	1.0			L
5516.3	5517.8	18123.	BH V	0.3			L
5469.2	5470.7	18279.	BH V	1.2			L
5323.6	5325.1	18779.	BH V	0.3			L
5275.3	5276.8	18951.	BH V	1.0			L
5228.4	5229.9	19121.	BH V	1.0			L
5104.3	5105.7	19586.	BH V	0.8			L

AIR WAVELENGTH (ANGSTRMS)	VACUUM WAVELENGTH (ANGSTRMS)	WAVENUMBER (KAYSERS)	LMENT	LOG ARC	INTENSITY SPK	INTENSITY DIS	REF
5057.3	5058.7	19768.	BH V	1.0			L
5011.1	5012.5	19950.	BH V	0.8			L
4951.3	4952.7	20191.	BH V	0.3			L
4904.7	4906.1	20383.	BH V	0.3			L
4858.2	4859.6	20578.	BH V	0.3			L
4812.6	4813.9	20773.	BH V	0.3			L
6631.3	6633.1	15076.	BH YT	0.3			ME
6610.5	6612.3	15123.	BH YT	0.6			ME
6591.4	6593.2	15167.	BH YT	0.3			ME
6572.5	6574.3	15211.	BH YT	0.8			ME
6553.8	6555.6	15254.	BH YT	0.9			ME
6535.8	6537.6	15296.	BH YT	1.0			ME
6518.3	6520.1	15337.	BH YT	1.2			ME
6501.2	6503.0	15377.	BH YT	1.2			ME
6484.6	6486.4	15417.	BH YT	1.0			ME
6468.2	6470.0	15456.	BH YT	0.9			ME
6424.1	6425.9	15562.	BH YT	0.3			ME
6405.6	6407.4	15607.	BH YT	0.7			ME
6387.1	6388.9	15652.	BH YT	1.0			ME
6369.9	6371.7	15694.	BH YT	1.0			ME
6359.5	6361.3	15720.	BH YT	0.9			ME
6352.3	6354.1	15738.	BH YT	0.7			ME
6338.1	6339.9	15773.	BH YT	1.0			ME
6335.5	6337.3	15780.	BH YT	0.6			ME
6318.4	6320.1	15822.	BH YT	0.5			ME
6316.2	6317.9	15828.	BH YT	0.9			ME
6302.7	6304.4	15862.	BH YT	0.3			ME
6295.5	6297.2	15880.	BH YT	1.0			ME
6275.0	6276.7	15932.	BH YT	1.2			ME
6255.9	6257.6	15980.	BH YT	0.6			ME
6251.0	6252.7	15993.	BH YT	0.7			ME
6236.7	6238.4	16030.	BH YT	1.5			ME
6230.8	6232.5	16045.	BH YT	0.7			ME
6221.8	6223.5	16068.	BH YT	0.8			ME
6218.0	6219.7	16078.	BH YT	1.6			ME
6209.9	6211.6	16099.	BH YT	0.7			ME
6199.8	6201.5	16125.	BH YT	1.7			ME
6190.7	6192.4	16149.	BH YT	0.7			ME
6182.2	6183.9	16171.	BH YT	1.8			ME
6170.8	6172.5	16201.	BH YT	0.6			ME
6165.1	6166.8	16216.	BH YT	1.9			ME
6151.7	6153.4	16251.	BH YT	0.7			ME
6148.4	6150.1	16260.	BH YT	2.0			ME
6132.1	6133.8	16303.	BH YT	2.3			ME
6127.4	6129.1	16316.	BH YT	0.9			ME
6114.7	6116.4	16349.	BH YT	1.3			ME
6107.8	6109.5	16368.	BH YT	1.2			ME
6096.8	6098.5	16397.	BH YT	1.5			ME
6089.4	6091.1	16417.	BH YT	1.5			ME
6072.8	6074.5	16462.	BH YT	1.2			ME
6070.0	6071.7	16470.	BH YT	0.7			ME
6053.8	6055.5	16514.	BH YT	1.7			ME
6049.2	6050.9	16526.	BH YT	0.5			ME
6030.6	6032.3	16577.	BH YT	2.0			ME
6030.5	6032.2	16578.	BH YT	0.8			ME
6019.9	6021.6	16607.	BH YT	2.2			ME
6011.1	6012.8	16631.	BH YT	0.3			ME
6009.2	6010.9	16636.	BH YT	0.9			ME
6003.6	6005.3	16652.	BH YT	2.3			ME
5992.1	5993.8	16684.	BH YT	1.0			ME
5987.6	5989.3	16697.	BH YT	2.5			ME
5972.1	5973.8	16740.	BH YT	2.8			MIT
5969.6	5971.3	16747.	BH YT	0.6			ME
5956.4	5958.1	16784.	BH YT	1.9			ME
5950.0	5951.6	16802.	BH YT	0.6			ME
5939.1	5940.7	16833.	BH YT	2.0			ME
5933.2	5934.8	16850.	BH YT	0.3			ME
5931.1	5932.7	16856.	BH YT	0.8			ME
5912.7	5914.3	16908.	BH YT	0.5			ME
5912.2	5913.8	16909.	BH YT	1.0			ME

AIR WAVELENGTH (ANGSTRMS)	VACUUM WAVELENGTH (ANGSTRMS)	WAVENUMBER (KAYSERS)	LMENT	LOG ARC	INTENSITY SPK	INTENSITY DIS	REF
5893.9	5895.5	16962.	BH YT	1.0			ME
5876.1	5877.7	17013.	BH YT	1.0			ME
5874.0	5875.6	17019.	BH YT	0.7			ME
5858.8	5860.4	17064.	BH YT	1.0			ME
5855.3	5856.9	17074.	BH YT	0.7			ME
5841.9	5843.5	17113.	BH YT	0.9			ME
5838.1	5839.7	17124.	BH YT	1.0			ME
5837.2	5838.8	17127.	BH YT	0.8			ME
5819.3	5820.9	17179.	BH YT	0.7			ME
5818.6	5820.2	17181.	BH YT	1.2			ME
5800.0	5801.6	17237.	BH YT	1.2			ME
5782.7	5784.3	17288.	BH YT	1.0			ME
5764.2	5765.8	17344.	BH YT	1.3			ME
5746.9	5748.5	17396.	BH YT	1.3			ME
5730.1	5731.7	17447.	BH YT	1.2			ME
5713.8	5715.4	17497.	BH YT	1.0			ME
5697.8	5699.4	17546.	BH YT	0.7			ME
5078.7	5080.1	19685.	BH YT	0.8			ME
5077.9	5079.3	19688.	BH YT	0.6			ME
5050.6	5052.0	19794.	BH YT	1.0			ME
5049.7	5051.1	19798.	BH YT	0.9			ME
5025.3	5026.7	19894.	BH YT	1.3			ME
5024.3	5025.7	19898.	BH YT	1.0			ME
4842.5	4843.9	20645.	BH YT	0.5			ME
4842.0	4843.4	20647.	BH YT	0.7			ME
4818.2	4819.5	20749.	BH YT	1.5			ME
4817.4	4818.7	20752.	BH YT	1.3			ME
4744.9	4746.2	21069.	BH YT	0.8			ME
4744.5	4745.8	21071.	BH YT	0.7			ME
4706.7	4708.0	21240.	BH YT	1.0			ME
4706.3	4707.6	21242.	BH YT	0.6			ME
4676.9	4678.2	21376.	BH YT	0.5			ME
4676.3	4677.6	21378.	BH YT	0.7			ME
4650.2	4651.5	21498.	BH YT	1.2			ME
4649.5	4650.8	21502.	BH YT	1.0			ME
4646.9	4648.2	21514.	BH YT	0.7			ME
4603.7	4605.0	21716.	BH YT	0.7			ME
4561.8	4563.1	21915.	BH YT	0.8			ME
4525.7	4527.0	22090.	BH YT	0.3			ME
4525.3	4526.6	22092.	BH YT	0.7			ME
4496.4	4497.7	22234.	BH YT	0.7			ME
7597.5	7599.6	13159.	BH ZR	0.6			L
7541.5	7543.6	13256.	BH ZR	0.6			L
7484.4	7486.5	13357.	BH ZR	0.5			L
7468.2	7470.3	13386.	BH ZR	0.5			L
7426.3	7428.3	13462.	BH ZR	0.3			L
7335.6	7337.6	13628.	BH ZR	0.3			L
7127.2	7129.2	14027.	BH ZR	0.3			L
7109.8	7111.8	14061.	BH ZR	0.3			L
7079.6	7081.6	14121.	BH ZR	0.5			L
7071.3	7073.2	14138.	BH ZR	0.6			L
7043.1	7045.0	14194.	BH ZR	0.6			L
7033.2	7035.1	14214.	BH ZR	0.7			L
7005.6	7007.5	14270.	BH ZR	0.5			L
6996.3	6998.2	14289.	BH ZR	0.8			L
6968.9	6970.8	14345.	BH ZR	0.3			L
6959.9	6961.8	14364.	BH ZR	0.8			L
6931.8	6933.7	14422.	BH ZR	0.7			L
6923.7	6925.6	14439.	BH ZR	0.7			L
6887.8	6889.7	14514.	BH ZR	0.6			L
6884.8	6886.7	14521.	BH ZR	0.3			L
6848.9	6850.8	14597.	BH ZR	0.5			L
6847.3	6849.2	14600.	BH ZR	0.5			L
6812.5	6814.4	14675.	BH ZR	0.5			L
6777.2	6779.1	14751.	BH ZR	0.6			L
6742.5	6744.4	14827.	BH ZR	0.6			L
6718.0	6719.9	14881.	BH ZR	0.3			L
6710.9	6712.8	14897.	BH ZR	0.3			L
6686.6	6688.4	14951.	BH ZR	0.3			L
6678.0	6679.8	14970.	BH ZR	0.7			L

AIR WAVELENGTH (ANGSTRMS)	VACUUM WAVELENGTH (ANGSTRMS)	WAVENUMBER (KAYSERS)	LMENT	LOG ARC	INTENSITY SPK	INTENSITY DIS	REF
6649.5	6651.3	15035.	BH ZR	0.3			L
6645.3	6647.1	15044.	BH ZR	0.3			L
6613.1	6614.9	15117.	BH ZR	0.5			L
6584.9	6586.7	15182.	BH ZR	0.6			L
6578.2	6580.0	15197.	BH ZR	0.6			L
6549.7	6551.5	15264.	BH ZR	0.3			L
6543.0	6544.8	15279.	BH ZR	1.0			L
6515.2	6517.0	15344.	BH ZR	0.6			L
6508.2	6510.0	15361.	BH ZR	1.3			L
6480.7	6482.5	15426.	BH ZR	0.6			L
6473.7	6475.5	15443.	BH ZR	1.3			L
6452.5	6454.3	15494.	BH ZR	0.3			L
6446.5	6448.3	15508.	BH ZR	0.5			L
6419.3	6421.1	15574.	BH ZR	0.3			L
6412.3	6414.1	15591.	BH ZR	1.1			L
6396.0	6397.8	15630.	BH ZR	0.6			L
6394.1	6395.9	15635.	BH ZR	0.7			L
6387.8	6389.6	15650.	BH ZR	0.6			L
6384.6	6386.4	15658.	BH ZR	0.6			L
6378.3	6380.1	15674.	BH ZR	1.2			L
6362.5	6364.3	15713.	BH ZR	0.7			L
6356.3	6358.1	15728.	BH ZR	0.8			L
6351.2	6353.0	15741.	BH ZR	0.8			L
6344.9	6346.7	15756.	BH ZR	1.3			L
6324.3	6326.0	15808.	BH ZR	0.8			L
6317.6	6319.3	15824.	BH ZR	0.3			L
6308.7	6310.4	15847.	BH ZR	0.3			L
6299.0	6300.7	15871.	BH ZR	0.6			L
6292.8	6294.5	15887.	BH ZR	1.2			L
6273.0	6274.7	15937.	BH ZR	0.6			L
6266.5	6268.2	15953.	BH ZR	0.8			L
6260.9	6262.6	15968.	BH ZR	1.2			L
6238.9	6240.6	16024.	BH ZR	0.8			L
6235.1	6236.8	16034.	BH ZR	0.8			L
6229.4	6231.1	16048.	BH ZR	1.3			L
6222.8	6224.5	16065.	BH ZR	0.3			L
6210.2	6211.9	16098.	BH ZR	0.6			L
6203.9	6205.6	16114.	BH ZR	0.6			L
6188.0	6189.7	16156.	BH ZR	0.6			L
6175.2	6176.9	16189.	BH ZR	0.3			L
6170.2	6171.9	16202.	BH ZR	0.6			L
6153.9	6155.6	16245.	BH ZR	0.9			L
6137.2	6138.9	16290.	BH ZR	0.6			L
6120.1	6121.8	16335.	BH ZR	0.5			L
6099.9	6101.6	16389.	BH ZR	0.5			L
6091.5	6093.2	16412.	BH ZR	0.5			L
6086.9	6088.6	16424.	BH ZR	0.7			L
6070.0	6071.7	16470.	BH ZR	0.8			L
6059.3	6061.0	16499.	BH ZR	0.3			L
6053.8	6055.5	16514.	BH ZR	0.7			L
6039.1	6040.8	16554.	BH ZR	0.6			L
6026.0	6027.7	16590.	BH ZR	0.3			L
6021.3	6023.0	16603.	BH ZR	0.8			L
6014.5	6016.2	16622.	BH ZR	0.3			L
6008.5	6010.2	16638.	BH ZR	0.6			L
5999.4	6001.1	16664.	BH ZR	0.7			L
5983.3	5985.0	16709.	BH ZR	0.9			L
5977.7	5979.4	16724.	BH ZR	1.3			L
5951.9	5953.5	16797.	BH ZR	0.7			L
5947.0	5948.6	16810.	BH ZR	1.3			L
5939.1	5940.7	16833.	BH ZR	0.9			L
5923.3	5924.9	16878.	BH ZR	0.9			L
5917.7	5919.3	16894.	BH ZR	0.9			L
5916.4	5918.0	16897.	BH ZR	0.9			L
5911.5	5913.1	16911.	BH ZR	0.9			L
5908.5	5910.1	16920.	BH ZR	1.8			L
5868.0	5869.6	17037.	BH ZR	1.5			L
5860.1	5861.7	17060.	BH ZR	1.9			L
5839.8	5841.4	17119.	BH ZR	0.3			L
5814.5	5816.1	17194.	BH ZR	0.9			L

AIR WAVELENGTH (ANGSTRMS)	VACUUM WAVELENGTH (ANGSTRMS)	WAVENUMBER (KAYSERS)	LMENT	LOG ARC	INTENSITY SPK	DIS	REF
5809.2	5810.8	17209.	BH ZR	1.5			L
5783.8	5785.4	17285.	BH ZR	0.7			L
5778.5	5780.1	17301.	BH ZR	1.8			L
5753.8	5755.4	17375.	BH ZR	1.3			L
5748.1	5749.7	17392.	BH ZR	2.0			MIT
5724.1	5725.7	17465.	BH ZR	1.8			L
5718.1	5719.7	17483.	BH ZR	2.2			L
5692.3	5693.9	17563.	BH ZR	0.9			L
5687.4	5689.0	17578.	BH ZR	1.3			L
5681.1	5682.7	17597.	BH ZR	0.9			L
5670.7	5672.3	17630.	BH ZR	0.3			L
5666.8	5668.4	17642.	BH ZR	0.9			L
5665.5	5667.1	17646.	BH ZR	0.7			L
5664.2	5665.8	17650.	BH ZR	0.7			L
5663.1	5664.7	17653.	BH ZR	0.9			L
5658.1	5659.7	17669.	BH ZR	1.7			L
5637.0	5638.6	17735.	BH ZR	1.1			L
5634.9	5636.5	17742.	BH ZR	1.1			L
5633.9	5635.5	17745.	BH ZR	1.3			L
5629.5	5631.1	17759.	BH ZR	2.0			L
5629.0	5630.6	17760.	BH ZR	1.9			L
5624.0	5625.6	17776.	BH ZR	0.3			L
5612.6	5614.2	17812.	BH ZR	0.7			L
5610.1	5611.7	17820.	BH ZR	1.8			L
5608.9	5610.5	17824.	BH ZR	0.3			L
5596.5	5598.1	17863.	BH ZR	0.3			L
5592.5	5594.1	17876.	BH ZR	0.3			L
5584.6	5586.2	17901.	BH ZR	0.3			L
5581.7	5583.2	17911.	BH ZR	1.3			L
5580.1	5581.6	17916.	BH ZR	1.1			L
5575.7	5577.2	17930.	BH ZR	1.1			L
5566.9	5568.4	17958.	BH ZR	0.7			L
5562.7	5564.2	17972.	BH ZR	0.3			L
5556.1	5557.6	17993.	BH ZR	0.7			L
5553.1	5554.6	18003.	BH ZR	1.8			L
5551.7	5553.2	18007.	BH ZR	1.8			L
5547.4	5548.9	18021.	BH ZR	0.3			L
5545.2	5546.7	18029.	BH ZR	1.3			L
5539.4	5540.9	18047.	BH ZR	1.3			L
5538.8	5540.3	18049.	BH ZR	1.1			L
5519.6	5521.1	18112.	BH ZR	0.7			L
5515.3	5516.8	18126.	BH ZR	1.3			L
5502.8	5504.3	18167.	BH ZR	0.7			L
5491.7	5493.2	18204.	BH ZR	1.3			L
5490.3	5491.8	18209.	BH ZR	0.7			L
5485.7	5487.2	18224.	BH ZR	1.3			L
5466.0	5467.5	18290.	BH ZR	0.9			L
5465.4	5466.9	18292.	BH ZR	1.3			L
5462.2	5463.7	18303.	BH ZR	1.1			L
5461.6	5463.1	18305.	BH ZR	1.1			L
5456.5	5458.0	18322.	BH ZR	1.6			L
5439.4	5440.9	18379.	BH ZR	1.5			L
5437.0	5438.5	18387.	BH ZR	1.6			L
5433.2	5434.7	18400.	BH ZR	0.7			L
5409.1	5410.6	18482.	BH ZR	0.7			L
5408.2	5409.7	18485.	BH ZR	0.3			L
5407.1	5408.6	18489.	BH ZR	0.7			L
5404.4	5405.9	18498.	BH ZR	0.9			L
5393.0	5394.5	18537.	BH ZR	0.5			L
5390.4	5391.9	18546.	BH ZR	0.5			L
5389.3	5390.8	18550.	BH ZR	0.5			L
5379.5	5381.0	18584.	BH ZR	0.7			L
5377.3	5378.8	18591.	BH ZR	0.3			L
5375.7	5377.2	18597.	BH ZR	0.3			L
5364.3	5365.8	18637.	BH ZR	0.3			L
5360.7	5362.2	18649.	BH ZR	0.7			L
5332.5	5334.0	18748.	BH ZR	0.3			L
5322.0	5323.5	18785.	BH ZR	0.3			L
5308.4	5309.9	18833.	BH ZR	0.3			L
5306.0	5307.5	18841.	BH ZR	0.3			L
5298.3	5299.8	18869.	BH ZR	0.3			L
5212.2	5213.7	19180.	BH ZR	0.9			L
5185.0	5186.4	19281.	BH ZR	1.5			L
5156.3	5157.7	19388.	BH ZR	0.3			L
4902.5	4903.9	20392.	BH ZR	0.9			L
4899.8	4901.2	20403.	BH ZR	0.9			L
4876.1	4877.5	20502.	BH ZR	0.9			L
4863.1	4864.5	20557.	BH ZR	1.1			L
4850.2	4851.6	20612.	BH ZR	1.3			L
4849.7	4851.1	20614.	BH ZR	0.9			L
4847.2	4848.6	20625.	BH ZR	1.3			L
4827.5	4828.8	20709.	BH ZR	1.5			L
4827.4	4828.7	20709.	BH ZR	0.9			L
4771.3	4772.6	20953.	BH ZR	0.6			L
4769.4	4770.7	20961.	BH ZR	0.6			L
4736.9	4738.2	21105.	BH ZR	1.8			L
4716.9	4718.2	21194.	BH ZR	0.6			L
4644.7	4646.0	21524.	BH ZR	1.6			L
4619.8	4621.1	21640.	BH ZR	1.9			L
4610.3	4611.6	21684.	BH ZR	0.6			L
4545.2	4546.5	21995.	BH ZR	1.3			L
4542.6	4543.9	22008.	BH ZR	1.4			L
4534.4	4535.7	22047.	BH ZR	1.5			L
4521.26	4522.53	22111.5	BH ZR	1.3			L
4519.3	4520.6	22121.	BH ZR	1.3			L
4496.2	4497.5	22235.	BH ZR	1.5			L
4493.8	4495.1	22247.	BH ZR	1.4			L
4474.9	4476.2	22341.	BH ZR	0.9			L
4471.5	4472.8	22358.	BH ZR	1.6			L
4469.4	4470.7	22368.	BH ZR	1.5			L
4460.4	4461.7	22413.	BH ZR	0.6			L
4375.1	4376.3	22850.	BH ZR	0.9			L
4368.2	4369.4	22886.	BH ZR	0.9			L
4365.1	4366.3	22902.	BH ZR	1.1			L
4364.4	4365.6	22906.	BH ZR	1.1			L
4363.0	4364.2	22914.	BH ZR	0.6			L
4362.0	4363.2	22919.	BH ZR	0.9			L
4340.2	4341.4	23034.	BH ZR	0.6			L
4313.3	4314.5	23178.	BH ZR	0.9			L
4287.8	4289.0	23315.	BH ZR	0.6			L
4257.1	4258.3	23484.	BH ZR	0.6			L
3726.3	3727.4	26829.	BH ZR	0.7			L
3682.4	3683.4	27148.	BH ZR	1.5			L
3598.9	3599.9	27778.	BH ZR	0.3			L
3593.1	3594.1	27823.	BH ZR	0.3			L
3589.8	3590.8	27849.	BH ZR	0.7			L
3515.8	3516.8	28435.	BH ZR	0.3			L
3511.9	3512.9	28466.	BH ZR	0.3			L
3508.2	3509.2	28496.	BH ZR	1.3			L
3506.2	3507.2	28513.	BH ZR	1.3			L
3492.0	3493.0	28629.	BH ZR	1.5			L
3491.8	3492.8	28630.	BH ZR	1.3			L
3472.4	3473.4	28790.	BH ZR	1.3			L
9828.02	9830.71	10172.2	BI	2.5			ME
9657.2	9659.8	10352.	BI	3.3D			ME
9650.1	9652.7	10360.	BI	1.7H			ME
9417.0	9419.6	10616.	BI	2.0H			ME
9342.55	9345.11	10700.8	BI	2.7H			ME
9149.75	9152.26	10926.3	BI	0.8D			ME
9058.6	9061.1	11036.	BI	1.7X			ME
8982.1	8984.6	11130.	BI	1.2B			ME
8907.9	8910.3	11223.	BI I	2.3J			ME
8863.	8865.	11280.	BI II			1.8	CF
8761.53	8763.94	11410.4	BI	2.0J			ME
8754.92	8757.32	11419.0	BI I	1.6			ME
8682.	8684.	11520.	BI II			1.3	CF
8653.	8655.	11550.	BI II			1.8	CF
8628.0	8630.4	11587.	BI	2.0V			ME
8579.81	8582.17	11652.1	BI I	1.3			ME
8544.52	8546.87	11700.2	BI I	1.6			ME

AIR WAVELENGTH (ANGSTRMS)	VACUUM WAVELENGTH (ANGSTRMS)	WAVENUMBER (KAYSERS)	LMENT	LOG ARC	INTENSITY SPK	DIS	REF
8532.	8534.	11720.	BI II			1.9	CF
8501.8	8504.1	11759.	BI	0.3V			WT
8388.	8390.	11920.	BI II			1.6	CF
8328.	8330.	12000.	BI II			1.6	CF
8210.81	8213.07	12175.7	BI I	0.7			ME
8183.	8185.	12220.	BI II			1.3	CF
8050.	8052.	12420.	BI II			1.5	CF
7975.9	7978.1	12534.	BI	1.5V			ME
7965.	7967.	12550.	BI II			1.7	CF
7908.	7910.	12640.	BI II			1.0	CF
7840.61	7842.77	12750.6	BI I	0.9H			ME
7840.05	7842.21	12751.5	BI	0.8H			ME
7838.70	7840.86	12753.7	BI	2.6B			ME
7825.	7827.	12780.	BI II			1.0	CF
7750.	7752.	12900.	BI II			1.3	CF
7637.	7639.	13090.	BI II			1.3	CF
7502.33	7504.40	13325.5	BI	0.8			M
7388.	7390.	13530.	BI II		0.7W		CF
7381.	7383.	13550.	BI II		1.3W		CF
7262.	7264.	13770.	BI II		0.3		CF
7036.15	7038.09	14208.4	BI I	0.7			MIT
7033.	7035.	14220.	BI II		1.4		CF
7010.	7012.	14260.	BI II		0.5		CF
6991.12	6993.05	14299.9	BI	1.0H			WT
6808.6	6810.5	14683.	BI II		1.8		CF
6600.2	6602.0	15147.	BI II		1.8		CF
6577.2	6579.0	15200.	BI II	0.3H	1.3		CF
6497.65	6499.45	15385.9	BI II	0.3H	1.1		OM
6476.24	6478.03	15436.8	BI I	0.9			WT
6475.73	6477.52	15438.0	BI I	0.9			WT
6184.99	6186.70	16163.7	BI	0.5J			WT
6134.82	6136.52	16295.9	BI I	1.7	1.5		WT
6128.115	6129.811	16313.72	BI II		1.5		OM
6058.963	6060.641	16499.91	BI II			1.6	OM
6052.3	6054.0	16518.	BI II	0.5		1.0	CF
6035.6	6037.3	16564.	BI II		1.0H		CF
5973.01	5974.66	16737.3	BI II		1.6		OM
5894.6	5896.2	16960.	BI II		0.8		ML
5861.155	5862.780	17056.76	BI		0.8		OM
5860.2	5861.8	17059.	BI II		1.2J		ML
5853.9	5855.5	17078.	BI II	0.5	1.0		ML
5819.11	5820.72	17180.0	BI		0.3		OM
5818.3	5819.9	17182.	BI II	0.3	0.8		ML
5812.3	5813.9	17200.	BI II		0.9		ML
5742.55	5744.14	17409.0	BI	1.5	1.0		WT
5719.8	5721.4	17478.	BI II		1.2		ML
5719.21	5720.80	17480.1	BI II		1.3		OM
5718.81	5720.40	17481.3	BI II		1.3		WT
5655.42	5656.99	17677.2	BI II		1.1		OM
5599.41	5600.96	17854.1	BI I	1.0			WT
5552.35	5553.89	18005.4	BI	2.7Q	2.0L		WT
5540.2	5541.7	18045.	BI II		1.3		ML
5525.5	5527.0	18093.	BI II	0.3	0.6		ML
5508.9	5510.4	18147.	BI	0.3			ED
5486.6	5488.1	18221.	BI	0.6			ED
5450.8	5452.3	18341.	BI		0.8		OM
5397.8	5399.3	18521.	BI II	0.7	0.9		ML
5298.36	5299.83	18868.5	BI	1.3L	0.9		OM
5271.07	5272.54	18966.2	BI II	0.3	1.2		ML
5270.34	5271.81	18968.8	BI II	0.8	1.3		ML
5269.71	5271.18	18971.1	BI II	0.9	1.5		ML
5260.8	5262.3	19003.	BI II		1.4		ML
5209.29	5210.74	19191.1	BI II		2.8H		OM
5208.8	5210.3	19193.	BI II	0.7J	1.8		ML
5201.01	5202.46	19221.7	BI II		1.5		OM
5169.9	5171.3	19337.	BI II	0.3	1.1		ML
5144.484	5145.917	19432.88	BI II	0.3	2.5H		OM
5124.3	5125.7	19509.	BI II		2.0J		ML
5118.2	5119.6	19533.	BI II			1.4	ML
5111.0	5112.4	19560.	BI II		1.2		ML

AIR WAVELENGTH (ANGSTRMS)	VACUUM WAVELENGTH (ANGSTRMS)	WAVENUMBER (KAYSERS)	LMENT	LOG ARC	INTENSITY SPK	DIS	REF
5091.293	5092.712	19635.90	BI II	0.3H	1.5J		OM
5079.503	5080.919	19681.48	BI		0.6		OM
5029.5	5030.9	19877.	BI II		1.2		ML
4993.92	4995.31	20018.8	BI II			1.3	OM
4969.7	4971.1	20116.	BI II	0.3H	0.9		ML
4916.6	4918.0	20334.	BI II			1.0	ML
4907.98	4909.35	20369.3	BI II		1.1J		OM
4759.7	4761.0	21004.	BI II		0.3		ML
4749.738	4751.066	21047.91	BI II			1.3	MIT
4733.776	4735.100	21118.88	BI	0.3L			OM
4730.3	4731.6	21134.	BI II			1.4	ML
4729.82	4731.14	21136.5	BI	0.3H			MIT
4722.831	4724.152	21167.82	BI I	1.0	0.7		OM
4722.552	4723.873	21169.07	BI I	3.0	2.0		MIT
4722.190	4723.511	21170.69	BI I	1.0	0.7		OM
4705.35	4706.67	21246.5	BI II		1.7		OM
4696.3	4697.6	21287.	BI II		0.8		ML
4561.152	4562.430	21918.14	BI		1.0		OM
4477.12	4478.38	22329.5	BI II		1.0W		OM
4386.8	4388.0	22789.	BI		0.3H		RR
4379.4	4380.6	22828.	BI II	1.4	1.3		ML
4344.5	4345.7	23011.	BI II		0.3		ML
4340.59	4341.81	23031.9	BI II		1.6H		OM
4339.8	4341.0	23036.	BI II			1.1W	ML
4328.71	4329.93	23095.1	BI		0.6H		OM
4310.3	4311.5	23194.	BI II			0.9	ML
4308.525	4309.737	23203.27	BI I	0.6	0.0		MIT
4308.177	4309.389	23205.15	BI I	1.7	1.1		MIT
4305.4	4306.6	23220.	BI II			0.7	ML
4302.136	4303.346	23237.73	BI II	0.3H	1.7J		OM
4272.49	4273.69	23399.0	BI II		1.0J		OM
4259.62	4260.82	23469.7	BI II		1.8J		OM
4254.152	4255.350	23499.83	BI I	1.0	1.0		OM
4220.83	4222.02	23685.4	BI	1.1	0.3		TO
4192.56	4193.74	23845.1	BI II		0.9		ML
4192.31	4193.49	23846.5	BI II		0.8		ML
4192.10	4193.28	23847.7	BI II		0.3		ML
4127.36	4128.52	24221.7	BI	0.0	0.3		TO
4121.855	4123.018	24254.08	BI I	0.7	0.3		MIT
4121.527	4122.690	24256.01	BI I	2.1J	1.7		MIT
4097.2	4098.4	24400.	BI II	0.6	0.5		ML
4079.207	4080.359	24507.65	BI II	0.3H		1.6W	OM
4048.4	4049.5	24694.	BI II	0.5H	1.0		ML
4018.6	4019.7	24877.	BI II		0.3		CF
3931.9	3933.0	25426.	BI II			1.0	ML
3912.895	3914.003	25549.29	BI	0.3H	0.3H		OM
3888.233	3889.335	25711.34	BI I	1.6	0.3		MIT
3887.935	3889.037	25713.31	BI I	0.7	0.3		OM
3864.2	3865.3	25871.	BI		2.2H		OM
3863.9	3865.0	25873.	BI II			2.0	ML
3849.007	3850.099	25973.36	BI		0.3		MIT
3846.027	3847.118	25993.49	BI		2.0		MIT
3845.8	3846.9	25995.	BI II			2.0	ML
3841.6	3842.7	26023.	BI II			1.4	ML
3818.6	3819.7	26180.	BI II			1.0	ML
3816.173	3817.256	26196.83	BI		1.4H		MIT
3815.8	3816.9	26199.	BI II			2.5	ML
3811.14	3812.22	26231.4	BI II			2.2	OM
3797.0	3798.1	26329.	BI II		0.7		ML
3793.0	3794.1	26357.	BI II			1.4	ML
3792.8	3793.9	26358.	BI		2.7H		WT
3775.745	3776.818	26477.32	BI	0.3			OM
3756.375	3757.443	26613.85	BI		0.7H		MIT
3754.0	3755.1	26631.	BI II		0.5		ML
3724.9	3726.0	26839.	BI II			1.8	CF
3695.55	3696.60	27051.9	BI		1.7		WT
3659.4	3660.4	27319.	BI II		0.3		ML
3654.38	3655.42	27356.6	BI II	0.8	0.7S		MIT
3648.3	3649.3	27402.	BI II		0.5		ML
3619.37	3620.40	27621.2	BI	0.7			TO

AIR WAVELENGTH (ANGSTRMS)	VACUUM WAVELENGTH (ANGSTRMS)	WAVENUMBER (KAYSERS)	LMENT	LOG ARC	INTENSITY SPK	DIS	REF
3615.8	3616.8	27648.	BI II			0.7	ML
3613.817	3614.848	27663.68	BI		1.5		OM
3599.94	3600.97	27770.3	BI	0.3			TO
3596.110	3597.136	27799.89	BI I	2.2J	1.7		MIT
3592.4	3593.4	27829.	BI II			0.7	ML
3588.5	3589.5	27859.	BI II			1.8	ML
3541.36	3542.37	28229.7	BI		0.7H		OM
3527.91	3528.92	28337.3	BI		0.3H		MIT
3519.18	3520.19	28407.6	BI	1.0			TO
3510.853	3511.857	28474.96	BI I	2.3J	1.5		OM
3504.2	3505.2	28529.	BI II		0.5		ML
3473.729	3474.724	28779.27	BI	0.3H	0.5		MIT
3455.275	3456.265	28932.97	BI II		2.0H		MIT
3455.01	3456.00	28935.2	BI II			1.0	ML
3454.82	3455.81	28936.8	BI II			0.7	ML
3454.5	3455.5	28939.	BI II			0.7	ML
3431.23	3432.21	29135.7	BI II			2.2	ML
3430.83	3431.81	29139.1	BI II			2.3	ML
3430.53	3431.51	29141.7	BI II			1.8	ML
3430.30	3431.28	29143.6	BI II			1.5	ML
3430.10	3431.08	29145.3	BI II			1.4D	ML
3411.8	3412.8	29302.	BI II			1.2H	ML
3410.1	3411.1	29316.	BI II			0.7H	ML
3408.6	3409.6	29329.	BI II			1.2	ML
3405.660	3406.637	29354.46	BI I	1.8			MIT
3405.326	3406.303	29357.34	BI	1.6	1.0		MIT
3402.80	3403.78	29379.1	BI I	0.5			TO
3397.213	3398.188	29427.45	BI I	2.0J	1.7		MIT
3393.530	3394.504	29459.38	BI		0.3G		OM
3382.28	3383.25	29557.4	BI	0.3			TO
3361.209	3362.175	29742.65	BI	0.5	0.3		MIT
3355.1	3356.1	29797.	BI II			1.5	ML
3330.5	3331.5	30017.	BI II			1.0	ML
3299.79	3300.74	30296.2	BI II			1.2	ML
3299.57	3300.52	30298.3	BI II			1.0	ML
3299.3	3300.3	30301.	BI II	0.3H	0.5		ML
3295.91	3296.86	30331.9	BI		0.3		OM
3267.97	3268.91	30591.2	BI I	0.3H			TO
3239.73	3240.66	30857.9	BI I	1.0	0.5		TO
3216.8	3217.7	31078.	BI	0.9H	0.3H		TO
3187.6	3188.5	31362.	BI		0.3H		WT
3144.6	3145.5	31791.	BI I	0.9J			TO
3115.423	3116.327	32089.06	BI		2.7		OM
3111.67	3112.57	32127.8	BI II			1.4	ML
3111.41	3112.31	32130.4	BI II			1.2	ML
3111.20	3112.10	32132.6	BI II			1.0	ML
3111.02	3111.92	32134.5	BI II			0.8	ML
3110.88	3111.78	32135.9	BI II			0.7	ML
3093.58	3094.48	32315.6	BI I	1.0W	0.9		TO
3076.662	3077.556	32493.32	BI I	1.3	1.3		MIT
3067.716	3068.608	32588.07	BI I	3.5P	3.3J		MIT
3053.7	3054.6	32738.	BI II			1.8	ML
3039.71	3040.59	32888.3	BI		0.3H		OM
3039.12	3040.00	32894.7	BI		0.3		OM
3035.18	3036.06	32937.4	BI I	1.8H			TO
3034.873	3035.756	32940.72	BI	1.5	1.5		MIT
3033.5	3034.4	32956.	BI II			1.2	ML
3024.635	3025.516	33052.22	BI I	2.4J	1.7		MIT
2993.342	2994.215	33397.74	BI I	2.3J	2.0J		MIT
2989.029	2989.901	33445.92	BI I	2.4J	2.0J		MIT
2968.3	2969.2	33679.	BI II		0.5		CF
2963.4	2964.3	33735.	BI II		0.6		CF
2961.5	2962.4	33757.	BI II			0.3H	ML
2954.7	2955.6	33834.	BI II			0.3	CF
2950.4	2951.3	33884.	BI II			0.5H	CF
2944.287	2945.148	33954.15	BI	0.7	0.6		MIT
2938.298	2939.157	34023.36	BI I	2.5W	2.5W		MIT
2936.7	2937.6	34042.	BI II			0.5	CF
2926.1	2927.0	34165.	BI II			0.3	CF
2897.975	2898.824	34496.74	BI I	2.7Z	2.7Z		MIT

AIR WAVELENGTH (ANGSTRMS)	VACUUM WAVELENGTH (ANGSTRMS)	WAVENUMBER (KAYSERS)	LMENT	LOG ARC	INTENSITY SPK	DIS	REF
2892.905	2893.753	34557.20	BI	1.1V	0.9		OM
2883.81	2884.66	34666.2	BI	0.5	0.5		TO
2863.754	2864.595	34908.95	BI I	1.9W	1.3		OM
2855.674	2856.513	35007.72	BI		1.1		OM
2817.36	2818.19	35483.8	BI II		0.8		CF
2817.17	2818.00	35486.2	BI II		0.5		CF
2809.625	2810.453	35581.46	BI I	2.30	2.0		OM
2803.653	2804.479	35657.25	BI II		1.2		OM
2803.570	2804.396	35658.30	BI II	0.3	1.0		MIT
2803.482	2804.308	35659.42	BI II	0.3H	1.5		MIT
2802.707	2803.533	35669.28	BI II		1.0		MIT
2802.55	2803.38	35671.3	BI II		0.5		CF
2802.44	2803.27	35672.7	BI II		0.3		CF
2798.685	2799.510	35720.54	BI I	2.3	1.4		MIT
2780.521	2781.342	35953.87	BI I	2.30	2.0		OM
2767.88	2768.70	36118.1	BI	0.6			TO
2746.36	2747.17	36401.1	BI II		0.3		CF
2746.2	2747.0	36403.	BI II		0.3		CF
2730.505	2731.313	36612.42	BI I	2.3	2.0		OM
2713.3	2714.1	36845.	BI II		1.0H		CF
2701.8	2702.6	37001.	BI II		0.8		CF
2696.763	2697.563	37070.49	BI I	1.4G	1.2G		MIT
2696.614	2697.414	37072.54	BI	2.0	2.0		MIT
2693.0	2693.8	37122.	BI II		1.2		CF
2679.2	2680.0	37313.	BI II		0.3		CF
2627.906	2628.690	38041.77	BI I	2.3W	2.3		MIT
2613.715	2614.495	38248.30	BI	0.9	0.3		OM
2604.95	2605.73	38377.0	BI II		0.3		CF
2600.61	2601.39	38441.0	BI	1.1H	0.3		TO
2594.039	2594.815	38538.40	BI I	1.1H	0.6		MIT
2582.145	2582.918	38715.90	BI	1.5	0.7		MIT
2545.6	2546.4	39272.	BI II		0.3		ML
2536.56	2537.32	39411.6	BI	0.7H	0.3H		TO
2532.569	2533.330	39473.73	BI	1.4J			MIT
2530.56	2531.32	39505.1	BI II		0.5		ML
2530.41	2531.17	39507.4	BI II		0.5		ML
2530.28	2531.04	39509.4	BI II		0.5		ML
2524.492	2525.251	39600.02	BI I	2.0	1.4		MIT
2515.686	2516.443	39738.63	BI I	2.0	1.4		OM
2501.0	2501.8	39972.	BI II		0.9		ML
2499.51	2500.26	39995.8	BI	1.4	1.1		RK
2489.4	2490.2	40158.	BI	0.9H	0.3		OM
2480.25	2481.00	40306.3	BI II		0.5		ML
2480.174	2480.923	40307.58	BI II		0.3		MIT
2480.03	2480.78	40309.9	BI II		0.7		ML
2448.057	2448.799	40836.35	BI I	1.7	0.9		OM
2435.81	2436.55	41041.7	BI	0.7H			TO
2433.446	2434.184	41081.52	BI	1.5			MIT
2430.454	2431.192	41132.09	BI I	1.5	0.8		OM
2414.80	2415.53	41398.7	BI		0.8		RK
2409.62	2410.35	41487.7	BI	0.9V			MIT
2400.884	2401.615	41638.65	BI I	2.3G	2.0		MIT
2379.73	2380.46	42008.8	BI	0.5			TO
2369.174	2369.898	42195.92	BI	0.7			MIT
2368.551	2369.275	42207.01	BI II	0.3	1.0		MIT
2368.473	2369.197	42208.40	BI II	0.3	1.1		MIT
2368.387	2369.110	42209.94	BI II	0.3	1.2		MIT
2368.253	2368.976	42212.32	BI II	0.5	1.3		OM
2368.195	2368.918	42213.36	BI II	0.0	1.1		MIT
2354.467	2355.187	42459.47	BI	1.2H	0.7		MIT
2349.10	2349.82	42556.5	BI I	1.0			TO
2347.909	2348.628	42578.05	BI	0.7			MIT
2345.91	2346.63	42614.3	BI	1.1	0.7J		TO
2337.514	2338.231	42767.38	BI	0.7			MIT
2333.795	2334.511	42835.53	BI I	1.1			OM
2328.19	2328.90	42938.6	BI	1.2			TO
2317.43	2318.14	43138.0	BI	0.9			TO
2313.80	2314.51	43205.7	BI I	0.3			TO
2309.73	2310.44	43281.8	BI	1.3J			TO
2304.968	2305.677	43371.20	BI	0.3H			MIT

AIR WAVELENGTH (ANGSTRMS)	VACUUM WAVELENGTH (ANGSTRMS)	WAVENUMBER (KAYSERS)	LMENT	LOG INTENSITY ARC	SPK	DIS	REF
2297.58	2298.29	43510.6	BI	0.3H			TO
2293.850	2294.557	43581.40	BI	0.3H			MIT
2289.98	2290.69	43655.0	BI	0.3			TO
2281.345	2282.049	43820.26	BI	1.0			OM
2276.578	2277.281	43912.01	BI I	2.0G	1.6		OM
2249.382	2250.079	44442.88	BI I	1.2			MIT
2246.77	2247.47	44494.5	BI	0.9			TO
2246.418	2247.115	44501.51	BI		0.3		OM
2237.84	2238.53	44672.1	BI	0.3			TO
2230.608	2231.301	44816.90	BI I	2.0G	1.5G		MIT
2228.251	2228.944	44864.30	BI I	2.0G	1.7		OM
2224.205	2224.897	44945.90	BI	0.9			MIT
2214.121	2214.811	45150.58	BI I	0.8H	0.0		MIT
2203.12	2203.81	45376.0	BI	1.3H			OM
2202.83	2203.52	45382.0	BI	1.1	0.9		MIT
2189.586	2190.271	45656.46	BI I	1.4H	0.7H		OM
2186.920	2187.604	45712.11	BI II		1.5		MIT
2177.210	2177.892	45915.96	BI I	1.2			MIT
2176.618	2177.300	45928.44	BI I	1.3	0.6		OM
2164.098	2164.777	46194.12	BI	1.3	0.3		OM
2156.949	2157.627	46347.21	BI	1.9G			OM
2153.532	2154.209	46420.75	BI	1.6J			OM
2152.914	2153.591	46434.07	BI	1.7G			OM
2144.41	2145.09	46618.2	BI		0.7		OM
2143.46	2144.14	46638.9	BI II		1.1		ML
2143.40	2144.08	46640.1	BI II		1.1		ML
2143.35	2144.03	46641.2	BI II		0.9		ML
2134.308	2134.981	46838.82	BI I	2.0G	0.7H		OM
2133.626	2134.299	46853.79	BI I	2.00	1.6		MIT
2110.263	2110.932	47372.45	BI I	2.4G	1.7		OM
2097.63	2098.30	47657.7	BI	0.3H			TO
2077.8	2078.5	48112.	BI II		0.7		ML
2073.2	2073.9	48219.	BI	1.0			RK
2069.70	2070.36	48300.8	BI	0.3			TO
2068.99	2069.65	48317.3	BI II	0.3	1.0		RK
2064.79	2065.45	48415.6	BI	1.7	0.6H		TO
2061.70	2062.36	48488.2	BI I	2.5G	2.0		TO
2057.68	2058.34	48582.9	BI I	1.6	0.6H		TO
2053.52	2054.18	48681.3	BI	0.3H			TO
2049.69	2050.35	48772.2	BI	1.4	1.3		TO
2041.96	2042.62	48956.8	BI	2.0	1.2		TO
2023.99	2024.64	49391.4	BI	1.7			TO
2021.21	2021.86	49459.4	BI I	1.6J	1.2		TO
2020.5	2021.2	49477.	BI	0.3			RK
2011.39	2012.04	49700.8	BI	0.3H			TO
2001.59	2002.24	49944.1	BI	0.3H			TO
9320.83	9323.39	10725.7	BR I			0.6	KS
9265.39	9267.93	10789.9	BR I			0.9	KS
9178.16	9180.68	10892.4	BR I			0.6	KS
9173.59	9176.11	10897.9	BR I			0.6	KS
9166.07	9168.59	10906.8	BR I			0.8	KS
8963.99	8966.45	11152.7	BR I			0.7	KS
8949.31	8951.77	11171.0	BR I			0.3	KS
8932.39	8934.84	11192.1	BR I			0.5	KS
8897.64	8900.08	11235.8	BR I			1.2	KS
8888.83	8891.27	11247.0	BR I			0.3	KS
8825.26	8827.68	11328.0	BR I			1.2	KS
8819.95	8822.37	11334.8	BR I			1.0	KS
8807.52	8809.94	11350.8	BR I			0.3	KS
8793.46	8795.87	11369.0	BR I			0.8	KS
8698.51	8700.90	11493.1	BR I			1.0	KS
8638.66	8641.03	11572.7	BR I			1.4	KS
8625.40	8627.77	11590.5	BR I			0.8	KS
8566.28	8568.63	11670.5	BR			0.8	KS
8557.73	8560.08	11682.1	BR I			0.7	KS
8513.38	8515.72	11743.0	BR I			0.7	KS
8477.47	8479.80	11792.7	BR I			1.3	KS
8471.58	8473.91	11800.9	BR			0.5	KS
8446.55	8448.87	11835.9	BR I			1.7	KS
8384.02	8386.32	11924.2	BR I			0.7	KS

AIR WAVELENGTH (ANGSTRMS)	VACUUM WAVELENGTH (ANGSTRMS)	WAVENUMBER (KAYSERS)	LMENT	LOG INTENSITY ARC	SPK	DIS	REF
8343.70	8345.99	11981.8	BR I			1.3	KS
8334.69	8336.98	11994.7	BR I			1.3	KS
8272.46	8274.73	12085.0	BR I			1.8L	KS
8264.95	8267.22	12096.0	BR I			1.0	KS
8246.87	8249.14	12122.5	BR I			0.7	KS
8153.94	8156.18	12260.6	BR I			1.1	KS
8152.71	8154.95	12262.5	BR			0.3	KS
8131.51	8133.75	12294.5	BR I			1.1	KS
8026.35	8028.56	12455.5	BR I			0.8	KS
8023.93	8026.14	12459.3	BR			0.3	KS
8014.72	8016.92	12473.6	BR I			0.3	KS
8009.98	8012.18	12481.0	BR I			0.6	KS
7989.94	7992.14	12512.3	BR I			1.1	KS
7978.50	7980.69	12530.2	BR I			1.0	KS
7966.95	7969.14	12548.4	BR I			0.3	KS
7950.19	7952.38	12574.9	BR I			0.7	KS
7947.95	7950.14	12578.4	BR I			0.6	KS
7938.64	7940.82	12593.2	BR I			1.1L	KS
7925.88	7928.06	12613.4	BR I			0.6	KS
7881.58	7883.75	12684.3	BR I			0.3	KS
7843.60	7845.76	12745.7	BR I			0.3	KS
7803.03	7805.18	12812.0	BR I			1.2L	KS
7733.64	7735.77	12927.0	BR I			0.3	KS
7715.38	7717.50	12957.6	BR I			0.6	KS
7641.62	7643.72	13082.6	BR I			0.3	KS
7616.46	7618.56	13125.8	BR I			0.8	KS
7607.40	7609.49	13141.5	BR I			0.6	KS
7595.13	7597.22	13162.7	BR			0.5	KS
7591.59	7593.68	13168.8	BR I			0.3	KS
7570.87	7572.95	13204.9	BR I			0.3	KS
7569.16	7571.24	13207.9	BR			0.3	KS
7551.59	7553.67	13238.6	BR I			0.5	KS
7535.81	7537.88	13266.3	BR I			0.5	KS
7513.01	7515.08	13306.6	BR I			1.7L	KS
7425.89	7427.94	13462.7	BR I			2.0	KS
7348.56	7350.58	13604.4	BR I			2.7L	KS
7344.55	7346.57	13611.8	BR I			1.3	KS
7333.70	7335.72	13631.9	BR			0.7	KS
7319.19	7321.21	13658.9	BR			1.0	KS
7311.59	7313.60	13673.2	BR			0.7	KS
7288.49	7290.50	13716.5	BR I			1.0	KS
7260.49	7262.49	13769.4	BR			1.7	KS
7232.47	7234.46	13822.7	BR I			0.7	KS
7184.34	7186.32	13915.3	BR I			1.5	KS
7162.14	7164.11	13958.5	BR			1.7	KS
7142.28	7144.25	13997.3	BR I			1.6	KS
7113.53	7115.49	14053.8	BR I			1.0	KS
7111.63	7113.59	14057.6	BR I			1.5	KS
7058.40	7060.35	14163.6	BR I			1.3	KS
7005.21	7007.14	14271.2	BR I			2.3L	KS
6971.97	6973.89	14339.2	BR I			1.6	KS
6929.79	6931.70	14426.5	BR I			1.0	KS
6904.95	6906.85	14478.4	BR I			1.0	KS
6861.21	6863.10	14570.7	BR I			1.5	KS
6826.09	6827.97	14645.6	BR			0.7	KS
6820.39	6822.27	14657.9	BR I			0.7	KS
6791.50	6793.37	14720.2	BR I			0.7	KS
6790.05	6791.92	14723.4	BR I			1.8	KS
6786.77	6788.64	14730.5	BR I			1.5	KS
6779.48	6781.35	14746.3	BR I			1.0	KS
6760.11	6761.98	14788.6	BR I			1.0	MIT
6728.29	6730.15	14858.5	BR I			1.6	KS
6692.16	6694.01	14938.7	BR I			1.8	KS
6682.29	6684.14	14960.8	BR I			2.0	KS
6631.64	6633.47	15075.1	BR I			2.3L	KS
6620.50	6622.33	15100.4	BR I			1.0	KS
6599.41	6601.23	15148.7	BR			1.2	BL
6582.19	6584.01	15188.3	BR I			2.0	KS
6579.20	6581.02	15195.2	BR I			1.3	KS
6571.35	6573.17	15213.4	BR I			0.7	KS

AIR WAVELENGTH (ANGSTRMS)	VACUUM WAVELENGTH (ANGSTRMS)	WAVENUMBER (KAYSERS)	LMENT	LOG INTENSITY ARC	LOG INTENSITY SPK	LOG INTENSITY DIS	REF
6567.53	6569.34	15222.2	BR			1.4	BL
6559.81	6561.62	15240.1	BR I			2.2	KS
6548.14	6549.95	15267.3	BR I			1.3	KS
6544.61	6546.42	15275.5	BR I			2.0	KS
6541.99	6543.80	15281.7	BR			1.2	BL
6520.02	6521.82	15333.1	BR			1.2	BL
6483.60	6485.39	15419.3	BR I			0.7	KS
6471.66	6473.45	15447.7	BR			0.9	BL
6464.26	6466.05	15465.4	BR			0.5	BL
6446.15	6447.93	15508.8	BR			1.2	BL
6438.08	6439.86	15528.3	BR			0.3	KS
6410.32	6412.09	15595.5	BR I			1.5	KS
6374.88	6376.64	15682.2	BR			0.9	BL
6352.94	6354.70	15736.4	BR			1.4	BL
6350.74	6352.50	15741.8	BR I			2.3L	KS
6335.50	6337.25	15779.7	BR I			0.7	KS
6326.87	6328.62	15801.2	BR			0.9	BL
6296.70	6298.44	15876.9	BR			1.3	KS
6290.13	6291.87	15893.5	BR I			0.7	KS
6285.04	6286.78	15906.4	BR			1.4	BL
6282.50	6284.24	15912.8	BR			0.7	KS
6259.23	6260.96	15972.0	BR			0.9	BL
6203.08	6204.80	16116.6	BR I			1.0	KS
6190.25	6191.96	16150.0	BR			0.9	BL
6177.40	6179.11	16183.6	BR I			1.6	KS
6168.73	6170.44	16206.3	BR			0.5	BL
6161.74	6163.45	16224.7	BR			0.9	BL
6158.20	6159.90	16234.0	BR I			1.0	KS
6148.62	6150.32	16259.3	BR I			2.3	KS
6140.66	6142.36	16280.4	BR			1.2	BL
6132.70	6134.40	16301.5	BR			0.7	KS
6122.12	6123.81	16329.7	BR I			1.7L	KS
6117.62	6119.31	16341.7	BR II			1.2	BL
6095.74	6097.43	16400.4	BR			0.7	KS
6026.50	6028.17	16588.8	BR			1.2	BL
5979.63	5981.29	16718.8	BR			0.7	BL
5979.04	5980.70	16720.5	BR			0.7	BL
5954.00	5955.65	16790.8	BR I			1.3	KS
5950.30	5951.95	16801.2	BR I			1.3	KS
5945.50	5947.15	16814.8	BR			1.0	KS
5940.53	5942.18	16828.9	BR I			1.8L	KS
5926.48	5928.12	16868.7	BR			0.7	BL
5905.45	5907.09	16928.8	BR I			1.3	KS
5871.61	5873.24	17026.4	BR			1.2	BL
5852.10	5853.72	17083.1	BR I			2.2	KS
5833.43	5835.05	17137.8	BR I			1.9	KS
5830.74	5832.36	17145.7	BR			2.0	BL
5783.31	5784.91	17286.3	BR			1.6	KS
5737.13	5738.72	17425.5	BR			1.5	BL
5718.91	5720.50	17481.0	BR			1.5	BL
5710.97	5712.55	17505.3	BR			1.2	BL
5691.43	5693.01	17565.4	BR			1.5	BL
5657.61	5659.18	17670.4	BR			0.7	BL
5600.83	5602.38	17849.5	BR			1.5	BL
5589.93	5591.48	17884.4	BR			2.4	BL
5588.18	5589.73	17889.9	BR			1.2	BL
5536.40	5537.94	18057.3	BR I			1.3	KS
5536.30	5537.84	18057.6	BR			1.7	BL
5528.97	5530.51	18081.5	BR			1.2	BL
5510.90	5512.43	18140.8	BR			0.7	BL
5508.38	5509.91	18149.1	BR			0.7	BL
5506.78	5508.31	18154.4	BR			2.5	BL
5495.06	5496.59	18193.1	BR			2.2	BL
5488.79	5490.32	18213.9	BR			1.8	BL
5482.89	5484.41	18233.5	BR II			1.2	BL
5481.20	5482.72	18239.1	BR			1.2	BL
5479.99	5481.51	18243.1	BR			1.5	BL
5466.23	5467.75	18289.1	BR I			2.2L	KS
5460.70	5462.22	18307.6	BR			1.0L	KS
5450.06	5451.57	18343.3	BR I			1.5	KS

AIR WAVELENGTH (ANGSTRMS)	VACUUM WAVELENGTH (ANGSTRMS)	WAVENUMBER (KAYSERS)	LMENT	LOG INTENSITY ARC	LOG INTENSITY SPK	LOG INTENSITY DIS	REF
5435.11	5436.62	18393.8	BR			1.2	BL
5425.00	5426.51	18428.1	BR			1.2	BL
5422.78	5424.29	18435.6	BR			1.5	BL
5395.52	5397.02	18528.7	BR I			2.2L	KS
5370.27	5371.76	18615.9	BR I			1.3	KS
5364.20	5365.69	18636.9	BR I			1.3	KS
5360.80	5362.29	18648.7	BR			1.2	BL
5345.43	5346.92	18702.4	BR I			1.9L	KS
5335.11	5336.59	18738.5	BR			1.8	BL
5332.04	5333.52	18749.3	BR			2.0	BL
5330.57	5332.05	18754.5	BR			1.2	BL
5304.10	5305.58	18848.1	BR			1.5	BL
5249.05	5250.51	19045.8	BR			1.5	BL
5245.13	5246.59	19060.0	BR I			1.3	KS
5238.23	5239.69	19085.1	BR II			2.0	BL
5199.33	5200.78	19227.9	BR			1.5	BL
5193.89	5195.34	19248.0	BR			1.0	BL
5183.88	5185.32	19285.2	BR			1.0	BL
5182.36	5183.80	19290.9	BR II			2.0	BL
5164.40	5165.84	19357.9	BR II			1.3	BL
5143.46	5144.89	19436.7	BR			1.2	BL
5054.65	5056.06	19778.2	BR			2.3	BL
5038.77	5040.18	19840.6	BR			1.8	BL
5029.35	5030.75	19877.7	BR I			1.0	KS
5020.58	5021.98	19912.5	BR			1.5	BL
5002.70	5004.10	19983.6	BR I			1.6	KS
4986.98	4988.37	20046.6	BR			1.0	BL
4979.76	4981.15	20075.7	BR I			2.1L	KS
4959.37	4960.75	20158.2	BR			1.1	BL
4954.73	4956.11	20177.1	BR I			1.0	KS
4950.72	4952.10	20193.4	BR			0.6	BL
4945.61	4946.99	20214.3	BR			1.2	BL
4942.01	4943.39	20229.0	BR			0.7	BL
4930.66	4932.04	20275.6	BR			1.7	BL
4928.79	4930.17	20283.3	BR			2.2	BL
4926.56	4927.94	20292.5	BR			1.0	BL
4921.27	4922.64	20314.3	BR I			1.3	BL
4921.00	4922.37	20315.4	BR I			0.3	KS
4910.28	4911.65	20359.7	BR			0.3	BL
4867.75	4869.11	20537.6	BR			1.0	BL
4866.70	4868.06	20542.1	BR			1.3	BL
4866.52	4867.88	20542.8	BR			0.6	BL
4860.04	4861.40	20570.2	BR			1.1	KS
4849.37	4850.72	20615.5	BR I			1.4	KS
4848.75	4850.10	20618.1	BR II			2.2	BL
4844.86	4846.21	20634.7	BR			1.3	KS
4838.53	4839.88	20661.7	BR			0.7	BL
4834.46	4835.81	20679.1	BR I			1.4	KS
4818.41	4819.76	20747.9	BR			0.7	BL
4807.62	4808.96	20794.5	BR I			1.0	KS
4802.65	4803.99	20816.0	BR I			1.4	KS
4802.34	4803.68	20817.4	BR			1.3	BL
4799.61	4800.95	20829.2	BR			0.9	BL
4798.24	4799.58	20835.1	BR			0.9	BL
4795.23	4796.57	20848.2	BR			0.7	BL
4791.81	4793.15	20863.1	BR			0.3	BL
4785.50	4786.84	20890.6	BR II			2.6	BL
4785.19	4786.53	20892.0	BR I			1.3	KS
4780.31	4781.65	20913.3	BR I			2.1	KS
4777.14	4778.48	20927.2	BR			0.7	BL
4776.42	4777.76	20930.3	BR			2.3	BL
4775.21	4776.55	20935.6	BR I			1.4	KS
4773.85	4775.18	20941.6	BR			1.1	BL
4772.73	4774.06	20946.5	BR			0.9	BL
4767.10	4768.43	20971.2	BR			2.3	BL
4766.00	4767.33	20976.1	BR II			1.7	BL
4765.62	4766.95	20977.8	BR I			0.7	KS
4752.27	4753.60	21036.7	BR I			2.0	KS
4749.92	4751.25	21047.1	BR			0.6	BL
4744.33	4745.66	21071.9	BR			0.7	BL

AIR WAVELENGTH (ANGSTRMS)	VACUUM WAVELENGTH (ANGSTRMS)	WAVENUMBER (KAYSERS)	LMENT	LOG ARC	INTENSITY SPK	INTENSITY DIS	REF
4742.70	4744.03	21079.1	BR			2.3	BL
4735.42	4736.74	21111.6	BR II			1.3	KS
4728.24	4729.56	21143.6	BR			1.1	BL
4720.37	4721.69	21178.9	BR			1.0	BL
4719.77	4721.09	21181.6	BR			1.9	BL
4717.39	4718.71	21192.2	BR			0.9	BL
4704.86	4706.18	21248.7	BR II			2.4	BL
4701.40	4702.72	21264.3	BR			0.7	BL
4698.56	4699.87	21277.2	BR			0.3	BL
4696.43	4697.74	21286.8	BR			0.3	BL
4693.27	4694.58	21301.1	BR			1.6	BL
4692.33	4693.64	21305.4	BR			0.9	BL
4691.24	4692.55	21310.4	BR			0.3	BL
4678.69	4680.00	21367.5	BR			2.3	BL
4675.64	4676.95	21381.5	BR			0.6	BL
4673.38	4674.69	21391.8	BR			0.6	BL
4672.56	4673.87	21395.6	BR			1.1	BL
4672.11	4673.42	21397.6	BR			0.6	BL
4666.24	4667.55	21424.5	BR			0.6	BL
4651.99	4653.29	21490.2	BR			1.4	BL
4651.38	4652.68	21493.0	BR			0.6	BL
4644.01	4645.31	21527.1	BR			0.6	BL
4643.52	4644.82	21529.4	BR I			1.4L	KS
4642.63	4643.93	21533.5	BR			0.6	BL
4642.03	4643.33	21536.3	BR			1.3	BL
4640.98	4642.28	21541.1	BR			0.6	KS
4629.42	4630.72	21594.9	BR			1.2	BL
4622.75	4624.04	21626.1	BR			2.3	BL
4616.18	4617.47	21656.9	BR			1.3	BL
4614.60	4615.89	21664.3	BR I			2.0L	KS
4605.66	4606.95	21706.3	BR			1.0	BL
4601.36	4602.65	21726.6	BR			1.3	BL
4596.89	4598.18	21747.7	BR			1.0	BL
4593.30	4594.59	21764.7	BR			0.6	BL
4588.77	4590.06	21786.2	BR			0.6	BL
4579.95	4581.23	21828.2	BR			1.4	BL
4575.75	4577.03	21848.2	BR I			2.0L	KS
4573.41	4574.69	21859.4	BR			0.9	BL
4558.03	4559.31	21933.1	BR			1.2	BL
4556.55	4557.83	21940.3	BR			0.6	BL
4542.93	4544.20	22006.1	BR			2.4	BL
4538.77	4540.04	22026.2	BR			1.2	BL
4529.77	4531.04	22070.0	BR I			1.9	KS
4529.60	4530.87	22070.8	BR			1.0	BL
4525.62	4526.89	22090.2	BR I			2.1L	KS
4516.18	4517.45	22136.4	BR			1.3	BL
4514.06	4515.33	22146.8	BR			0.9	BL
4513.44	4514.71	22149.8	BR I			2.0L	KS
4508.07	4509.33	22176.2	BR			0.6	BL
4495.15	4496.41	22240.0	BR			0.6	BL
4490.43	4491.69	22263.3	BR I			1.4	KS
4477.75	4479.01	22326.4	BR I			2.30	KS
4472.62	4473.88	22352.0	BR I			2.1L	KS
4470.01	4471.26	22365.0	BR I			0.3	KS
4466.35	4467.60	22383.4	BR			1.3	BL
4460.13	4461.38	22414.6	BR			0.3	BL
4441.74	4442.99	22507.4	BR I			1.9	KS
4435.76	4437.01	22537.7	BR			0.6	BL
4430.97	4432.21	22562.1	BR			0.3	BL
4429.90	4431.14	22567.5	BR			0.6	BL
4425.14	4426.38	22591.8	BR I			1.1	KS
4423.04	4424.28	22602.5	BR I			0.6	KS
4407.62	4408.86	22681.6	BR			1.1	BL
4399.72	4400.96	22722.3	BR I			1.0	KS
4396.75	4397.99	22737.7	BR			0.6	BL
4396.40	4397.63	22739.5	BR II			1.3	BL
4394.95	4396.18	22747.0	BR			1.1	KS
4393.56	4394.79	22754.2	BR			1.4	BL
4391.61	4392.84	22764.3	BR I			1.4	KS
4386.69	4387.92	22789.8	BR			0.6	BL

AIR WAVELENGTH (ANGSTRMS)	VACUUM WAVELENGTH (ANGSTRMS)	WAVENUMBER (KAYSERS)	LMENT	LOG ARC	INTENSITY SPK	INTENSITY DIS	REF
4384.00	4385.23	22803.8	BR			1.3	BL
4379.78	4381.01	22825.8	PR			0.8	KS
4378.97	4380.20	22830.0	BR			0.6	BL
4377.94	4379.17	22835.4	BR			1.1	BL
4377.23	4378.46	22839.1	BR			0.6	BL
4372.00	4373.23	22866.4	BR			0.9	BL
4369.25	4370.48	22880.8	BR I			0.3	KS
4365.60	4366.83	22899.9	BR			2.3	BL
4365.15	4366.38	22902.3	BR I			1.2	KS
4356.93	4358.15	22945.5	BR II			0.6	BL
4351.22	4352.44	22975.6	BR			1.3	BL
4330.43	4331.65	23085.9	BR			0.9	BL
4307.80	4309.01	23207.2	BR			1.0	BL
4297.12	4298.33	23264.9	BR			1.2	BL
4291.40	4292.61	23295.9	BR			2.2	BL
4260.58	4261.78	23464.4	BR			0.6	BL
4249.89	4251.09	23523.4	BR			1.3	BL
4239.74	4240.93	23579.7	BR			0.6	BL
4236.88	4238.07	23595.6	BR			1.4	BL
4232.31	4233.50	23621.1	BR			0.9	BL
4230.00	4231.19	23634.0	BR II			1.1	BL
4226.15	4227.34	23655.5	BR			0.6	BL
4223.88	4225.07	23668.2	BR			1.9	BL
4222.40	4223.59	23676.5	BR			0.6	BL
4220.81	4222.00	23685.5	BR			0.6	BL
4219.21	4220.40	23694.4	BR			0.6	BL
4206.07	4207.25	23768.5	BR			0.3	BL
4202.88	4204.06	23786.5	BR			0.6	BL
4202.50	4203.68	23788.7	BR			1.4	KS
4201.35	4202.53	23795.2	BR			0.7	BL
4193.46	4194.64	23839.9	BR II			1.4	BL
4193.11	4194.29	23841.9	BR			0.7	BL
4192.35	4193.53	23846.2	BR			0.6	BL
4190.82	4192.00	23855.0	BR			0.9	BL
4181.75	4182.93	23906.7	BR			0.9	BL
4179.64	4180.82	23918.8	BR			1.6	BL
4175.79	4176.97	23940.8	BR I			1.7	KS
4160.00	4161.17	24031.7	BR			0.9	BL
4157.39	4158.56	24046.8	BR I			0.6	KS
4157.14	4158.31	24048.2	BR			1.0	BL
4143.98	4145.15	24124.6	BR I			1.3	KS
4140.21	4141.38	24146.6	BR			1.5	BL
4138.58	4139.75	24156.1	BR			0.7	BL
4138.38	4139.55	24157.2	BR			0.9	BL
4135.66	4136.83	24173.1	BR			1.3	BL
4118.68	4119.84	24272.8	BR			0.3	BL
4117.45	4118.61	24280.0	BR			1.3	BL
4116.65	4117.81	24284.7	BR			0.6	BL
4110.00	4111.16	24324.0	BR			1.0	BL
4106.41	4107.57	24345.3	BR			0.7	BL
4105.47	4106.63	24350.9	BR			0.3	BL
4104.80	4105.96	24354.9	BR			0.3	BL
4102.53	4103.69	24368.3	BR			1.0	BL
4097.88	4099.04	24396.0	BR			0.3	BL
4096.13	4097.29	24406.4	BR			0.7	BL
4090.62	4091.77	24439.3	BR			0.7	BL
4089.73	4090.88	24444.6	BR			0.9	BL
4089.20	4090.35	24447.8	BR			0.7	BL
4075.51	4076.66	24529.9	BR			1.1	BL
4074.49	4075.64	24536.0	BR			0.6	BL
4063.60	4064.75	24601.8	BR			1.0	BL
4054.03	4055.18	24659.9	BR			0.6	BL
4041.61	4042.75	24735.6	BR			0.9	BL
4037.35	4038.49	24761.7	BR			0.7	KS
4036.43	4037.57	24767.4	BR			1.0	BL
4032.85	4033.99	24789.4	BR			1.3	BL
4030.29	4031.43	24805.1	BR			1.1	BL
4024.04	4025.18	24843.6	BR			1.3	BL
4021.89	4023.03	24856.9	BR			0.6	BL
4021.78	4022.92	24857.6	BR I			0.3	KS

AIR WAVELENGTH (ANGSTRMS)	VACUUM WAVELENGTH (ANGSTRMS)	WAVENUMBER (KAYSERS)	LMENT		LOG ARC	INTENSITY SPK	DIS	REF
4018.33	4019.47	24878.9	BR	I			0.6	KS
4014.32	4015.45	24903.8	BR				1.4	BL
4008.76	4009.89	24938.3	BR				1.3	BL
4007.33	4008.46	24947.2	BR				1.0	BL
4005.58	4006.71	24958.1	BR				0.6	BL
4001.45	4002.58	24983.9	BR				0.6	BL
3999.62	4000.75	24995.3	BR				0.9	BL
3999.10	4000.23	24998.6	BR	I			0.5	BL
3997.13	3998.26	25010.9	BR				1.1	BL
3992.39	3993.52	25040.6	BR	I			1.3	KS
3991.39	3992.52	25046.9	BR	I			0.6	KS
3989.23	3990.36	25060.4	BR				0.5	BL
3986.75	3987.88	25076.0	BR				0.5	BL
3986.54	3987.67	25077.3	BR				0.5	BL
3982.80	3983.93	25100.9	BR				0.5	BL
3980.39	3981.52	25116.1	BR	II			1.4	BL
3980.02	3981.15	25118.4	BR	II			1.1	BL
3977.86	3978.99	25132.0	BR				0.5	BL
3970.60	3971.72	25178.0	BR				1.0	BL
3968.66	3969.78	25190.3	BR				0.9	BL
3960.47	3961.59	25242.4	BR				0.8	BL
3955.35	3956.47	25275.1	BR				0.9	BL
3950.61	3951.73	25305.4	BR				1.5	BL
3946.98	3948.10	25328.7	BR				0.5	BL
3946.66	3947.78	25330.7	BR	II			0.3	BL
3946.01	3947.13	25334.9	BR				0.5	BL
3941.05	3942.17	25366.8	BR				0.8	BL
3939.69	3940.81	25375.5	BR				1.2	BL
3938.62	3939.73	25382.4	BR				1.0	BL
3935.15	3936.26	25404.8	BR	II			1.2	BL
3934.11	3935.22	25411.5	BR	I			0.3	KS
3929.56	3930.67	25440.9	BR				1.2	BL
3924.54	3925.65	25473.5	BR				0.5	BL
3924.10	3925.21	25476.3	BR	II			1.5	BL
3923.35	3924.46	25481.2	BR				1.2	BL
3920.65	3921.76	25498.7	BR				1.2	BL
3919.51	3920.62	25506.2	BR				1.2	BL
3917.83	3918.94	25517.1	BR	I			0.6	KS
3915.18	3916.29	25534.4	BR				0.5	BL
3914.28	3915.39	25540.2	BR	II			2.2	BL
3914.10	3915.21	25541.4	BR	II			1.0	BL
3909.37	3910.48	25572.3	BR				0.3	M
3901.25	3902.36	25625.6	BR				0.9	BL
3897.73	3898.83	25648.7	BR				0.5	BL
3891.63	3892.73	25688.9	BR	II			1.4	BL
3888.52	3889.62	25709.4	BR				1.0	BL
3876.89	3877.99	25786.6	BR				0.5	BL
3871.21	3872.31	25824.4	BR	II			1.1	BL
3864.12	3865.22	25871.8	BR				0.3	KS
3863.71	3864.81	25874.5	BR				0.5	BL
3857.21	3858.30	25918.1	BR				0.8	BL
3834.69	3835.78	26070.3	BR				0.3	BL
3829.76	3830.85	26103.9	BR				0.5	KS
3828.55	3829.64	26112.1	BR	I			1.1	KS
3828.49	3829.58	26112.6	BR				0.8	BL
3815.68	3816.76	26200.2	BR				1.2	KS
3811.40	3812.48	26229.6	BR				0.8	BL
3798.28	3799.36	26320.2	BR				0.8	KS
3794.04	3795.12	26349.6	BR	I			0.8	KS
3783.42	3784.49	26423.6	BR				0.5	BL
3773.83	3774.90	26490.7	BR				0.5	BL
3740.51	3741.57	26726.7	BR				0.6	BL
3735.83	3736.89	26760.2	BR				0.8	KS
3714.30	3715.36	26915.3	BR				0.9	BL
3684.64	3685.69	27132.0	BR				0.3	BL
3680.66	3681.71	27161.3	BR				0.8	BL
3659.50	3660.54	27318.4	BR				1.4	BL
3641.50	3642.54	27453.4	BR				0.5	BL
3633.64	3634.68	27512.8	BR				0.8	BL
3620.93	3621.96	27609.3	BR				0.8	BL

AIR WAVELENGTH (ANGSTRMS)	VACUUM WAVELENGTH (ANGSTRMS)	WAVENUMBER (KAYSERS)	LMENT		LOG ARC	INTENSITY SPK	DIS	REF
3606.66	3607.69	27718.6	BR				1.2	BL
3591.31	3592.33	27837.1	BR				0.6	BL
3585.44	3586.46	27882.6	BR				0.8	BL
3525.34	3526.35	28357.9	BR				0.5	BL
3423.82	3424.80	29198.8	BR				0.5	BL
3409.81	3410.79	29318.7	BR				0.5	BL
3346.96	3347.92	29869.3	BR				0.5	BL
3321.03	3321.99	30102.5	BR				0.8	BL
3301.13	3302.08	30283.9	BR				0.6	BL
3291.1	3292.0	30376.	BR				0.5	BL
3239.64	3240.57	30858.7	BR				0.5	BL
3208.33	3209.26	31159.9	BR				0.8	BL
3199.66	3200.58	31244.3	BR				0.9	BL
3121.18	3122.08	32029.9	BR				0.5	BL
3028.86	3029.74	33006.1	BR				0.5	BL
3016.45	3017.33	33141.9	BR				0.6	BL
3015.87	3016.75	33148.3	BR				0.5	BL
2996.72	2997.59	33360.1	BR				0.6	BL
2986.48	2987.35	33474.5	BR				0.6	BL
2985.79	2986.66	33482.2	BR				0.6	BL
2981.80	2982.67	33527.0	BR				0.5	BL
2975.70	2976.57	33595.7	BR				0.5	BL
2973.45	2974.32	33621.1	BR				0.7	BL
2972.22	2973.09	33635.1	BR				1.4	BL
2967.23	2968.10	33691.6	BR				1.3	BL
2961.07	2961.93	33761.7	BR				0.7	BL
2951.92	2952.78	33866.4	BR				0.6	BL
2946.17	2947.03	33932.5	BR				0.3	BL
2936.16	2937.02	34048.1	BR				0.7	BL
2928.93	2929.79	34132.2	BR				0.3	BL
2917.23	2918.08	34269.1	BR				1.2	BL
2910.94	2911.79	34343.1	BR				0.7	BL
2907.70	2908.55	34381.4	BR				0.7	BL
2905.88	2906.73	34402.9	BR				0.7	BL
2904.15	2905.00	34423.4	BR				0.7	BL
2901.92	2902.77	34449.9	BR				0.7	BL
2900.16	2901.01	34470.7	BR				0.5	BL
2893.44	2894.29	34550.8	BR				1.5	BL
2892.93	2893.78	34556.9	BR				0.7	BL
2892.06	2892.91	34567.3	BR				0.7	BL
2887.17	2888.02	34625.8	BR				0.7	BL
2875.42	2876.26	34767.3	BR				0.8	BL
2872.59	2873.43	34801.6	BR				1.4	BL
2867.04	2867.88	34868.9	BR				0.8	BL
2846.15	2846.99	35124.9	BR				0.8	BL
2843.69	2844.53	35155.2	BR				0.3	BL
2836.27	2837.10	35247.2	BR				0.3	BL
2815.41	2816.24	35508.4	BR				0.3	BL
2807.57	2808.40	35607.5	BR				1.2	BL
2802.27	2803.10	35674.8	BR				0.5	BL
2799.00	2799.83	35716.5	BR				1.5	BL
2785.26	2786.08	35892.7	BR				0.5	BL
2782.73	2783.55	35925.3	BR				0.3	BL
2772.62	2773.44	36056.3	BR				0.5	BL
2770.52	2771.34	36083.6	BR				0.7	BL
2769.61	2770.43	36095.5	BR				0.3	BL
2755.31	2756.12	36282.8	BR				0.7	BL
2749.81	2750.62	36355.4	BR				0.3	BL
2747.16	2747.97	36390.5	BR				0.3	BL
2744.79	2745.60	36421.9	BR				0.3	BL
2728.45	2729.26	36640.0	BR				0.7	BL
2727.05	2727.86	36658.8	BR				1.2	BL
2720.44	2721.25	36747.9	BR				1.2	BL
2719.03	2719.84	36766.9	BR				0.7	BL
2718.42	2719.23	36775.2	BR				0.7	BL
2717.51	2718.32	36787.5	BR				0.3	BL
2713.74	2714.54	36838.6	BR				1.4	BL
2709.67	2710.47	36893.9	BR				1.5	BL
2708.66	2709.46	36907.7	BR				0.5	BL
2705.04	2705.84	36957.1	BR				0.3	BL

AIR WAVELENGTH (ANGSTRMS)	VACUUM WAVELENGTH (ANGSTRMS)	WAVENUMBER (KAYSERS)	LMENT	LOG ARC	INTENSITY SPK	INTENSITY DIS	REF
2704.26	2705.06	36967.7	BR			0.5	BL
2702.78	2703.58	36988.0	BR			1.0	BL
2699.84	2700.64	37028.2	BR			0.5	BL
2694.23	2695.03	37105.3	BR			0.7	BL
2690.72	2691.52	37153.7	BR			0.3	BL
2690.13	2690.93	37161.9	BR			1.3	BL
2685.25	2686.05	37229.4	BR			1.0	BL
2681.57	2682.37	37280.5	BR			0.5	BL
2677.34	2678.14	37339.4	BR			0.6	BL
2674.89	2675.68	37373.6	BR			0.3	BL
2672.53	2673.32	37406.6	BR			0.7	BL
2671.53	2672.32	37420.6	BR			0.5	BL
2661.38	2662.17	37563.3	BR			0.3	BL
2660.49	2661.28	37575.9	BR			1.4	BL
2658.65	2659.44	37601.9	BR			0.8	BL
2656.83	2657.62	37627.6	BR			1.4	BL
2653.91	2654.70	37669.0	BR			0.3	BL
2650.76	2651.55	37713.8	BR			0.3	BL
2642.22	2643.01	37835.7	BR			0.7	BL
2639.60	2640.39	37873.2	BR			0.5	BL
2632.87	2633.65	37970.0	BR			0.3	BL
2629.20	2629.98	38023.0	BR			0.5	BL
2626.49	2627.27	38062.3	BR			0.8	BL
2622.31	2623.09	38122.9	BR			0.3	BL
2618.19	2618.97	38182.9	BR			0.5	BL
2616.21	2616.99	38211.8	BR			0.5	BL
2610.93	2611.71	38289.1	BR			0.7	BL
2608.07	2608.85	38331.1	BR			0.7	BL
2607.05	2607.83	38346.1	BR			0.7	BL
2606.18	2606.96	38358.9	BR			0.8	BL
2603.15	2603.93	38403.5	BR			0.7	BL
2601.54	2602.32	38427.3	BR			0.6	BL
2601.16	2601.94	38432.9	BR			1.2	BL
2595.94	2596.72	38510.2	BR			0.7	BL
2594.45	2595.23	38532.3	BR			0.6	BL
2593.76	2594.54	38542.5	BR			1.3	BL
2593.34	2594.12	38548.8	BR			0.3	BL
2592.60	2593.38	38559.8	BR			0.3	BL
2590.34	2591.11	38593.4	BR			0.5	BL
2589.10	2589.87	38611.9	BR			0.8	BL
2588.25	2589.02	38624.6	BR			0.5	BL
2586.93	2587.70	38644.3	BR			1.0	BL
2584.96	2585.73	38673.7	BR			0.6	BL
2584.10	2584.87	38686.6	BR			0.3	BL
2582.20	2582.97	38715.1	BR			0.5	BL
2578.95	2579.72	38763.9	BR			0.6	BL
2578.17	2578.94	38775.6	BR			0.7	BL
2577.84	2578.61	38780.6	BR			0.7	BL
2576.17	2576.94	38805.7	BR			1.2	BL
2573.14	2573.91	38851.4	BR			0.7	BL
2570.80	2571.57	38886.7	BR			0.7	BL
2569.23	2570.00	38910.5	BR			1.2	BL
2567.30	2568.07	38939.8	BR			0.7	BL
2566.52	2567.29	38951.6	BR			0.6	BL
2563.19	2563.96	39002.2	BR			0.3	BL
2557.82	2558.59	39084.1	BR			0.6	BL
2556.93	2557.70	39097.7	BR			1.4	BL
2546.79	2547.55	39253.3	BR			0.3	BL
2546.41	2547.17	39259.2	BR			0.7	BL
2543.06	2543.82	39310.9	BR			0.3	BL
2541.45	2542.21	39335.8	BR			1.6	BL
2538.68	2539.44	39378.7	BR			0.3	BL
2537.02	2537.78	39404.5	BR			0.6	BL
2532.43	2533.19	39475.9	BR			0.7	BL
2524.93	2525.69	39593.1	BR			1.2	BL
2524.33	2525.09	39602.6	BR			0.3	BL
2521.86	2522.62	39641.4	BR			0.7	BL
2521.66	2522.42	39644.5	BR			1.7	BL
2520.27	2521.03	39666.4	BR			0.8	BL
2516.05	2516.81	39732.9	BR			0.5	BL

AIR WAVELENGTH (ANGSTRMS)	VACUUM WAVELENGTH (ANGSTRMS)	WAVENUMBER (KAYSERS)	LMENT	LOG ARC	INTENSITY SPK	INTENSITY DIS	REF
2504.92	2505.67	39909.4	BR			0.5	BL
2499.25	2500.00	40000.0	BR			0.8	BL
2498.40	2499.15	40013.5	BR			0.5	BL
2497.42	2498.17	40029.2	BR			0.3	BL
2495.17	2495.92	40065.4	BR			1.2	BL
2492.15	2492.90	40113.9	BR			0.3	BL
2489.47	2490.22	40157.1	BR			0.5	BL
2488.37	2489.12	40174.8	BR			1.6	BL
2488.29	2489.04	40176.1	BR			0.3	BL
2485.38	2486.13	40223.1	BR			0.5	BL
2482.57	2483.32	40268.7	BR			0.5	BL
2473.18	2473.93	40421.6	BR			0.7	BL
2464.81	2465.56	40558.8	BR			1.2	BL
2462.35	2463.09	40599.3	BR			0.5	BL
2459.46	2460.20	40647.0	BR			0.3	BL
2435.72	2436.46	41043.2	BR			0.5	BL
2416.63	2417.36	41367.4	BR			0.3	BL
2408.20	2408.93	41512.2	BR			0.5	BL
2404.33	2405.06	41579.0	BR			0.6	BL
2397.30	2398.03	41700.9	BR			0.5	BL
2395.34	2396.07	41735.0	BR			1.4	BL
2392.44	2393.17	41785.6	BR			1.3	BL
2392.23	2392.96	41789.3	BR			1.0	BL
2391.88	2392.61	41795.4	BR			0.3	BL
2389.69	2390.42	41833.7	BR			1.8	BL
2388.94	2389.67	41846.8	BR			1.2	BL
2388.63	2389.36	41852.2	BR			0.7	BL
2386.74	2387.47	41885.4	BR			1.4	BL
2386.51	2387.24	41889.4	BR			0.5	BL
2355.69	2356.41	42437.4	BR			1.2	BL
2343.40	2344.12	42660.0	BR			0.7	BL
2337.90	2338.62	42760.3	BR			1.4	BL
2336.92	2337.64	42778.2	BR			0.7	BL
2320.74	2321.45	43076.5	BR			0.3	BL
2317.27	2317.98	43141.0	BR			1.2	BL
2293.41	2294.12	43589.8	BR			0.3	BL
2262.29	2262.99	44189.3	BR			0.3	LC
9658.49	9661.14	10350.7	C I	2.4		1.2	KS
9620.86	9623.50	10391.2	C I	2.1		0.7	KS
9602.8	9605.4	10411.	C I	1.8			IG
9405.77	9408.35	10628.9	C I	2.5		2.3	KS
9111.85	9114.35	10971.7	C I	2.2		2.0	KS
9094.89	9097.39	10992.2	C I	2.7		2.5	KS
9088.57	9091.06	10999.8	C I	2.3		2.0	KS
9078.32	9080.81	11012.2	C I	2.2		1.8	KS
9062.53	9065.02	11031.4	C I			2.2	KS
9061.48	9063.97	11032.7	C I	2.5		2.3	KS
9045.1	9047.6	11053.	C	2.2			IG
8335.19	8337.48	11994.0	C I			2.2	KS
7236.19	7238.18	13815.6	C II		2.2H		FL
7231.12	7233.11	13825.3	C II		2.0		FL
7119.45	7121.41	14042.2	C II		1.2		FL
7118.5	7120.5	14044.	C I	1.3H			EN
7115.13	7117.09	14050.7	C II		1.2		FL
7112.36	7114.32	14056.2	C II		0.7		FL
7063.4	7065.3	14154.	C II		0.7H		FL
7052.9	7054.8	14175.	C II		0.7H		FL
6800.50	6802.38	14700.7	C II		1.5		FL
6798.04	6799.92	14706.1	C II		0.7		FL
6791.30	6793.17	14720.7	C II		1.5		FL
6787.09	6788.96	14729.8	C II		1.2		FL
6783.75	6785.62	14737.0	C II		2.0		FL
6780.27	6782.14	14744.6	C II		1.2		FL
6779.74	6781.61	14745.8	C II		1.7		FL
6750.22	6752.08	14810.2	C II		1.2		FL
6738.36	6740.22	14836.3	C II		0.7		FL
6587.75	6589.57	15175.5	C I			1.7	EN
6582.85	6584.67	15186.8	C II		2.3		FL
6578.03	6579.85	15197.9	C II		2.7		FL
6098.62	6100.31	16392.6	C II		1.5		FL

AIR WAVELENGTH (ANGSTRMS)	VACUUM WAVELENGTH (ANGSTRMS)	WAVENUMBER (KAYSERS)	LMENT	LOG ARC	INTENSITY SPK	INTENSITY DIS	REF
6095.37	6097.06	16401.4	C II		1.2		FL
5907.36	5909.00	16923.4	C II		0.7		FL
5891.65	5893.28	16968.5	C II		1.5		FL
5889.97	5891.60	16973.3	C II		1.8		FL
5856.09	5857.71	17071.5	C II		1.2		FL
5836.31	5837.93	17129.4	C II		0.7		FL
5827.80	5829.42	17154.4	C II		0.7		FL
5805.76	5807.37	17219.5	C I	0.7			RY
5801.17	5802.78	17233.1	C I	1.2			RY
5793.51	5795.12	17255.9	C I	1.5			RY
5662.51	5664.08	17655.1	C II		1.7		FL
5648.08	5649.65	17700.2	C II		1.5		FL
5640.50	5642.07	17724.0	C II		1.2		FL
5536.0	5537.5	18059.	C II		0.7H		EN
5380.242	5381.738	18581.36	C I			2.5	JN
5272.56	5274.03	18960.8	C II		0.7		FL
5259.62	5261.08	19007.5	C II		1.5		FL
5257.36	5258.82	19015.7	C II		1.2		FL
5253.55	5255.01	19029.4	C II		0.7		FL
5151.08	5152.51	19408.0	C II		1.5		FL
5145.16	5146.59	19430.3	C II		1.8		FL
5143.49	5144.92	19436.6	C II		1.2		FL
5139.21	5140.64	19452.8	C II		0.7		FL
5133.29	5134.72	19475.3	C II		1.2		FL
5132.96	5134.39	19476.5	C II		1.5		FL
5121.69	5123.12	19519.4	C II		0.5		FL
5119.55	5120.98	19527.5	C II		1.2		FL
5114.07	5115.50	19548.4	C II		1.2		FL
5052.122	5053.531	19788.15	C I			2.0	JN
5044.8	5046.2	19817.	C II		0.7		EN
5041.66	5043.07	19829.2	C I			1.5	JN
5039.05	5040.46	19839.5	C I			1.5	JN
5033.2	5034.6	19862.	C II		0.7H		EN
5023.79	5025.19	19899.7	C			0.7	JN
4964.90	4966.29	20135.8	C II		1.0		FL
4954.16	4955.54	20179.4	C II		0.7		EN
4932.00	4933.38	20270.1	C I		1.6		JN
4817.33	4818.68	20752.6	C I			0.7	JN
4812.84	4814.19	20771.9	C I		0.7		FL
4775.87	4777.21	20932.7	C I			1.3	JN
4771.72	4773.05	20950.9	C I			1.5	JN
4770.00	4771.33	20958.5	C I			1.0	JN
4766.62	4767.95	20973.4	C I			1.0	JN
4762.41	4763.74	20991.9	C I			1.5	JN
4758.78	4760.11	21007.9	C			1.0	JN
4744.90	4746.23	21069.4	C II		0.7		EN
4734.75	4736.07	21114.5	C II		1.0H		EN
4727.21	4728.53	21148.2	C II		0.7H		EN
4630.52	4631.82	21589.8	C		0.3D		EN
4618.85	4620.14	21644.4	C II		1.4H		FL
4411.52	4412.76	22661.6	C II		1.6		FL
4411.20	4412.44	22663.2	C II		1.6		FL
4410.06	4411.30	22669.1	C		1.5		FL
4376.78	4378.01	22841.4	C		1.0H		EN
4374.28	4375.51	22854.5	C II		1.6		FL
4372.49	4373.72	22863.8	C II		1.5		FL
4371.59	4372.82	22868.6	C II		0.8		FL
4371.33	4372.56	22869.9	C I		1.5		JN
4368.14	4369.37	22886.6	C II			1.5H	EN
4348.07	4349.29	22992.2	C			1.5	JN
4325.88	4327.10	23110.2	C II		1.0		EN
4318.92	4320.13	23147.4	C II		1.0		EN
4317.42	4318.63	23155.5	C II		1.5		EN
4313.50	4314.71	23176.5	C II		1.2		EN
4312.4	4313.6	23182.	C			1.3	JN
4304.0	4305.2	23228.	C			1.0	JN
4296.11	4297.32	23270.3	C II		0.8		EN
4285.96	4287.17	23325.4	C II		0.7		EN
4268.99	4270.19	23418.1	C I			1.0	JN
4267.27	4268.47	23427.6	C II		2.7		FL
4267.02	4268.22	23429.0	C II		2.5		FL
4236.0	4237.2	23600.	C			1.0H	JN
4231.35	4232.54	23626.5	C I			0.7	JN
4228.28	4229.47	23643.6	C			0.7	JN
4212.90	4214.09	23729.9	C			0.7	JN
4212.36	4213.55	23733.0	C			0.7	JN
4076.00	4077.15	24526.9	C II		1.9		EN
4074.89	4076.04	24533.6	C II		1.6		FL
4074.53	4075.68	24535.8	C II		1.7		FL
4065.1	4066.2	24593.	C			1.0	JN
4064.2	4065.3	24598.	C			0.7	JN
4017.27	4018.41	24885.5	C II		0.7		FL
4009.90	4011.03	24931.2	C II		1.0		FL
3980.35	3981.48	25116.3	C II			1.1	FL
3973.84	3974.96	25157.5	C II		0.3H		EN
3952.08	3953.20	25296.0	C II		0.8		FL
3948.15	3949.27	25321.1	C II		0.3		EN
3930.2	3931.3	25437.	C			0.8	JN
3920.677	3921.787	25498.58	C II		2.3		FL
3918.977	3920.087	25509.64	C II		1.9		FL
3904.4	3905.5	25605.	C			0.8H	JN
3880.59	3881.69	25762.0	C II		0.3		EN
3879.60	3880.70	25768.6	C II		0.3		EN
3878.22	3879.32	25777.7	C II		0.3		EN
3876.670	3877.769	25788.02	C II		1.6		FL
3876.409	3877.508	25789.76	C II		1.8		FL
3876.188	3877.287	25791.23	C II		2.1		FL
3876.051	3877.150	25792.14	C II		1.6		FL
3871.62	3872.72	25821.7	C II		0.8		EN
3868.84	3869.94	25840.2	C II		0.3		EN
3836.10	3837.19	26060.7	C II		0.3H		EN
3832.12	3833.21	26087.8	C II		0.8H		EN
3757.1	3758.2	26609.	C			1.1	JN
3729.6	3730.7	26805.	C			0.8	JN
3590.87	3591.89	27840.5	C II		0.8		FL
3589.67	3590.69	27849.8	C II		1.3		FL
3588.92	3589.94	27855.6	C II		0.3		FL
3587.68	3588.70	27865.2	C II		0.8		FL
3585.83	3586.85	27879.6	C II		0.8		FL
3584.98	3586.00	27886.2	C II		0.3		FL
3370.5	3371.5	29661.	C			1.3	JN
3361.75	3362.72	29737.9	C II		0.8		FL
3361.09	3362.06	29743.7	C II		1.1		FL
3167.95	3168.87	31557.0	C II		1.1		FL
3165.99	3166.91	31576.6	C II		0.3		EN
3165.51	3166.43	31581.4	C II		1.3		FL
3158.6	3159.5	31650.	C			0.8	JN
3049.44	3050.33	32783.4	C II		0.3H		EN
2992.63	2993.50	33405.7	C II		1.3H		FL
2967.91	2968.78	33683.9	C II		0.8		FL
2967.31	2968.18	33690.7	C		0.8		FL
2966.88	2967.75	33695.6	C II		0.3		EN
2910.82	2911.67	34344.5	C II		0.3		EN
2885.50	2886.35	34645.9	C II		0.8H		EN
2884.93	2885.78	34652.7	C II		0.3H		EN
2837.602	2838.436	35230.66	C II		1.6		FL
2836.710	2837.544	35241.74	C II		2.3		FL
2801.31	2802.14	35687.1	C II		0.8H		EN
2766.18	2767.00	36140.3	C II		0.3		FL
2747.31	2748.12	36388.5	C II		1.6		FL
2746.50	2747.31	36399.2	C II		1.4		FL
2641.44	2642.23	37846.9	C II		1.3		FL
2640.93	2641.72	37854.2	C II		1.1		FL
2640.58	2641.37	37859.2	C II		1.0		FL
2604.89	2605.67	38377.9	C II		0.3		EN
2591.96	2592.74	38569.3	C II		0.6H		EN
2582.88	2583.65	38704.9	C I			0.6	PS
2574.86	2575.63	38825.4	C II		0.3H		FL
2512.03	2512.79	39796.5	C II		2.6		FL
2511.71	2512.47	39801.5	C II		1.8		FL

AIR WAVELENGTH (ANGSTRMS)	VACUUM WAVELENGTH (ANGSTRMS)	WAVENUMBER (KAYSERS)	LMENT	LOG ARC	INTENSITY SPK	INTENSITY DIS	REF
2509.11	2509.87	39842.8	C II		2.3		FL
2478.573	2479.322	40333.61	C I	2.6		2.6	PS
2402.40	2403.13	41612.4	C II		1.1		FL
2401.77	2402.50	41623.3	C II		0.8		FL
2302.12	2302.83	43424.9	C		1.2W		MIT
2296.89	2297.60	43523.7	C		2.3		MIT
2191.37	2192.06	45619.3	C		0.8		FL
2174.14	2174.82	45980.8	C II		0.3		FL
2173.86	2174.54	45986.7	C II		0.8		FL
2163.60	2164.28	46204.8	C		1.5		FL
2137.93	2138.60	46759.5	C II		0.8		FL
2137.45	2138.12	46770.0	C II		0.3		FL
9701.7	9704.4	10305.	CA I	1.3			ME
9688.6	9691.3	10319.	CA I	1.2			ME
9676.0	9678.7	10332.	CA I	0.7			ME
9664.0	9666.7	10345.	CA I	0.7			ME
9233.4	9235.9	10827.	CA	1.3			ME
8662.140	8664.519	11541.32	CA II	3.0			IWG
8542.089	8544.436	11703.52	CA II	3.0			IWG
8498.018	8500.353	11764.22	CA II	2.5			IWG
8169.8	8172.0	12237.	CA I	0.7			ME
8005.26	8007.46	12488.3	CA	0.5			ME
7326.146	7328.164	13645.98	CA I	2.6			IWG
7202.194	7204.179	13880.83	CA I	1.5			IWG
7148.147	7150.117	13985.78	CA I	2.7			IWG
6798.51	6800.39	14705.1	CA I	0.8H			ME
6784.01	6785.88	14736.5	CA	0.7H			ME
6717.685	6719.540	14881.97	CA I	2.2			IWG
6572.781	6574.597	15210.06	CA I	1.7			IWG
6508.742	6510.540	15359.71	CA	0.3			CW
6499.650	6501.446	15381.19	CA I	1.5	1.2		I
6493.780	6495.574	15395.10	CA I	1.9	1.5		I
6485.38	6487.17	15415.0	CA	0.3H	0.0		AD
6476.74	6478.53	15435.6	CA	0.3H			AD
6471.660	6473.448	15447.72	CA	1.6	1.2		I
6462.566	6464.352	15469.45	CA I	2.1	1.7		I
6455.600	6457.384	15486.15	CA	1.0	0.8		IWG
6449.810	6451.593	15500.05	CA I	1.8	1.1		I
6439.073	6440.853	15525.90	CA I	2.2	1.7		I
6432.07	6433.85	15542.8	CA		0.8		AD
6424.52	6426.30	15561.1	CA	0.3H			AD
6387.59	6389.36	15651.0	CA	0.3H			AD
6370.15	6371.91	15693.9	CA	0.6			AD
6214.10	6215.82	16088.0	CA		0.3		AD
6169.556	6171.263	16204.14	CA I	1.6	1.3		I
6169.052	6170.759	16205.46	CA I	1.4	1.2		I
6166.443	6168.149	16212.32	CA I	1.2	0.7		IWG
6163.758	6165.464	16219.38	CA I	1.0	0.8		IWG
6162.172	6163.877	16223.56	CA I	1.6	1.7		I
6161.289	6162.994	16225.88	CA I	1.0	0.8		IWG
6122.218	6123.913	16329.43	CA I	2.0	2.0		I
6102.721	6104.410	16381.60	CA I	1.9	1.7		I
6069.98	6071.66	16470.0	CA		0.3		AD
6006.15	6007.81	16645.0	CA	0.9			MIT
5867.572	5869.198	17038.10	CA I	0.9			MIT
5857.456	5859.080	17067.53	CA I	1.6	1.5		MIT
5752.64	5754.24	17378.5	CA		0.3		AD
5602.836	5604.391	17843.15	CA I	1.2	1.0		MIT
5601.264	5602.819	17848.16	CA I	1.2	1.0		MI
5598.474	5600.028	17857.05	CA I	1.5	1.3		MIT
5594.454	5596.007	17869.88	CA I	1.5	1.3		MIT
5590.110	5591.662	17883.77	CA I	1.2	1.0		MIT
5588.748	5590.300	17888.13	CA I	1.5	1.4		MIT
5581.969	5583.519	17909.85	CA I	1.3	1.1		MIT
5579.13	5580.68	17919.0	CA		0.3		AD
5570.49	5572.04	17946.8	CA		0.3		AD
5512.955	5514.486	18134.05	CA I	0.6	0.7D		MIT
5495.95	5497.48	18190.2	CA	0.3H	0.0		AD
5349.474	5350.962	18688.23	CA I	1.1	1.1		MIT
5333.78	5335.26	18743.2	CA		0.3		AD

AIR WAVELENGTH (ANGSTRMS)	VACUUM WAVELENGTH (ANGSTRMS)	WAVENUMBER (KAYSERS)	LMENT	LOG ARC	INTENSITY SPK	INTENSITY DIS	REF
5321.30	5322.78	18787.2	CA		0.5		AD
5301.34	5302.82	18857.9	CA		0.3		AD
5294.48	5295.95	18882.3	CA	0.3H	0.0		AD
5271.97	5273.44	18963.0	CA	1.0			AD
5270.276	5271.743	18969.06	CA I	1.3	1.0		MIT
5265.562	5267.027	18986.04	CA I	1.3	1.0		MIT
5264.240	5265.705	18990.81	CA I	1.2	0.9		MIT
5262.248	5263.713	18998.00	CA I	1.3	0.9		MIT
5261.704	5263.168	18999.96	CA I	1.3	0.8		MIT
5260.391	5261.855	19004.70	CA I	0.6	0.6		MIT
5247.42	5248.88	19051.7	CA	0.6	0.5		AD
5231.83	5233.29	19108.4	CA		0.6		AD
5188.850	5190.295	19266.73	CA I	1.7	0.8W		MIT
5137.76	5139.19	19458.3	CA	0.3H	0.3		MIT
5071.89	5073.30	19711.0	CA	0.8	0.3		AD
5050.048	5051.456	19796.27	CA		0.6		MIT
5046.95	5048.36	19808.4	CA	0.8	0.3		AD
5041.625	5043.031	19829.34	CA I	1.5			MIT
5021.152	5022.552	19910.20	CA	1.0			MIT
5019.985	5021.385	19914.82	CA	0.6			MIT
5014.23	5015.63	19937.7	CA	0.7	0.3		AD
5009.02	5010.42	19958.4	CA		0.6		AD
5001.489	5002.884	19988.47	CA	0.3			CW
4929.25	4930.63	20281.4	CA I	0.3			SD
4922.73	4924.10	20308.3	CA		0.3		AD
4919.409	4920.782	20321.97	CA	0.5	0.3L		MIT
4889.96	4891.33	20444.4	CA		0.3L		AD
4878.132	4879.494	20493.93	CA I	2.0	1.0		IWG
4859.31	4860.67	20573.3	CA		0.5H		MIT
4852.16	4853.52	20603.6	CA		0.3H		AD
4847.296	4848.650	20624.30	CA I	0.5			IWG
4807.53	4808.87	20794.9	CA	0.3H			MIT
4794.48	4795.82	20851.5	CA		0.3		MIT
4736.782	4738.107	21105.48	CA	0.7	0.3B		MIT
4735.94	4737.26	21109.2	CA		0.3		AD
4720.25	4721.57	21179.4	CA		0.3		AD
4716.26	4717.58	21197.3	CA		0.6		AD
4708.81	4710.13	21230.9	CA		0.7		AD
4685.265	4686.576	21337.54	CA I	1.4	0.0		IWG
4650.38	4651.68	21497.6	CA		0.5J		MIT
4643.66	4644.96	21528.7	CA		0.6H		AD
4642.88	4644.18	21532.3	CA		0.6H		AD
4634.214	4635.512	21572.59	CA		0.3		MIT
4632.67	4633.97	21579.8	CA		0.6H		AD
4631.39	4632.69	21585.7	CA		0.6H		AD
4629.693	4630.990	21593.66	CA		0.3		MIT
4621.70	4622.99	21631.0	CA		0.7C		AD
4608.54	4609.83	21692.8	CA		0.3		AD
4600.40	4601.69	21731.1	CA		0.5H		AD
4597.4	4598.7	21745.	CA		0.6J		AD
4588.9	4590.2	21786.	CA		0.6J		AD
4586.82	4588.11	21795.5	CA		0.3		AD
4586.03	4587.32	21799.2	CA	0.3	0.5H		AD
4585.871	4587.156	21800.00	CA I	2.1	1.0		IWG
4581.402	4582.686	21821.26	CA I	2.0	1.0		IWG
4578.558	4579.841	21834.82	CA I	1.9	0.7		IWG
4576.57	4577.85	21844.3	CA		0.6H		AD
4572.11	4573.39	21865.6	CA		0.9H		AD
4567.79	4569.07	21886.3	CA		0.3		AD
4566.35	4567.63	21893.2	CA		0.3H		AD
4564.44	4565.72	21902.4	CA		0.6H		AD
4554.590	4555.867	21949.72	CA		0.3		MIT
4553.274	4554.550	21956.06	CA		0.6H		MIT
4552.64	4553.92	21959.1	CA		0.5		AD
4526.935	4528.204	22083.81	CA I	2.0	0.5J		IWG
4516.60	4517.87	22134.3	CA		0.9		AD
4512.88	4514.15	22152.6	CA		0.6		AD
4512.282	4513.548	22155.52	CA I	1.0			IWG
4508.80	4510.06	22172.6	CA		0.6		MIT
4505.00	4506.26	22191.3	CA I	0.3	0.5H		SD

AIR WAVELENGTH (ANGSTRMS)	VACUUM WAVELENGTH (ANGSTRMS)	WAVENUMBER (KAYSERS)	LMENT	LOG ARC	INTENSITY SPK	INTENSITY DIS	REF
4499.90	4501.16	22216.5	CA		1.0H		AD
4484.959	4486.217	22290.49	CA		0.7		MIT
4484.41	4485.67	22293.2	CA		0.5		AD
4479.44	4480.70	22318.0	CA		0.3		AD
4456.620	4457.871	22432.23	CA I	1.3	1.2		MIT
4455.887	4457.138	22435.92	CA I	2.0	1.9		IWG
4454.781	4456.031	22441.49	CA I	2.3	0.7B		IWG
4435.688	4436.933	22538.09	CA I	2.0	1.2		IWG
4434.960	4436.205	22541.79	CA I	2.2	1.4		IWG
4431.35	4432.59	22560.1	CA		0.8		MIT
4425.99	4427.23	22587.5	CA		0.3		AD
4425.441	4426.684	22590.27	CA I	2.0	1.3		IWG
4413.74	4414.98	22650.2	CA		0.5		AD
4412.30	4413.54	22657.6	CA I	0.7			CW
4406.33	4407.57	22688.2	CA		0.9		AD
4399.64	4400.88	22722.7	CA		1.0		AD
4394.00	4395.23	22751.9	CA		0.3		AD
4393.168	4394.402	22756.22	CA	0.7	0.3		MIT
4382.927	4384.158	22809.40	CA	0.7	0.3		MIT
4381.19	4382.42	22818.4	CA		0.5		AD
4374.61	4375.84	22852.8	CA	1.0	0.3H		AD
4371.28	4372.51	22870.2	CA		0.3		AD
4370.73	4371.96	22873.1	CA		0.3		AD
4366.63	4367.86	22894.5	CA		0.3		MIT
4364.12	4365.35	22907.7	CA		0.3		AD
4362.88	4364.11	22914.2	CA		0.3		AD
4358.42	4359.64	22937.6	CA		0.7		AD
4355.096	4356.320	22955.15	CA I	1.7			IWG
4348.06	4349.28	22992.3	CA		0.5		AD
4344.74	4345.96	23009.9	CA		0.3H		MIT
4343.41	4344.63	23016.9	CA		0.3		MIT
4341.110	4342.330	23029.11	CA	0.7	0.0		MIT
4329.22	4330.44	23092.4	CA		0.8H		AD
4325.665	4326.881	23111.33	CA		0.6		MIT
4321.54	4322.76	23133.4	CA	0.5	0.0		AD
4319.54	4320.75	23144.1	CA	0.3H	0.3		AD
4319.13	4320.34	23146.3	CA		0.5		AD
4318.652	4319.867	23148.86	CA I	1.8	1.3		IWG
4307.741	4308.953	23207.50	CA I	1.7	1.3		IWG
4302.527	4303.737	23235.62	CA I	1.7	1.4		IWG
4301.54	4302.75	23240.9	CA		0.7		AD
4300.97	4302.18	23244.0	CA		0.7		AD
4298.986	4300.195	23254.76	CA I	1.5	1.3		IWG
4297.26	4298.47	23264.1	CA	0.3H	0.0		AD
4295.631	4296.839	23272.92	CA	0.3H	0.3		MIT
4290.12	4291.33	23302.8	CA		0.7H		AD
4289.364	4290.571	23306.92	CA I	1.5	1.3		IWG
4284.40	4285.61	23333.9	CA	0.3	0.8H		AD
4283.010	4284.215	23341.50	CA I	1.6	1.3		IWG
4279.76	4280.96	23359.2	CA	0.3	0.6		AD
4278.28	4279.48	23367.3	CA		0.6		AD
4273.907	4275.110	23391.21	CA		0.3		MIT
4271.87	4273.07	23402.4	CA		0.8H		AD
4271.558	4272.760	23404.07	CA	0.3H	0.5		MIT
4258.330	4259.529	23476.78	CA	0.3	0.6		MIT
4256.69	4257.89	23485.8	CA		0.8		AD
4255.509	4256.707	23492.34	CA		0.8		MIT
4254.701	4255.899	23496.80	CA	1.3			MIT
4248.956	4250.152	23528.57	CA	0.3H	1.0		MIT
4248.089	4249.285	23533.37	CA	0.8	0.0		MIT
4246.095	4247.290	23544.42	CA	0.3H	0.7		MIT
4240.74	4241.93	23574.1	CA		0.9H		AD
4240.456	4241.650	23575.73	CA I	1.0	1.0		IWG
4233.738	4234.930	23613.14	CA		0.6		MIT
4228.23	4229.42	23643.9	CA		0.5		AD
4226.728	4227.918	23652.30	CA I	2.76	1.7G		IWG
4213.167	4214.354	23728.43	CA		0.7		MIT
4211.641	4212.827	23737.03	CA		0.3		MIT
4207.276	4208.461	23761.65	CA		1.0		MIT
4204.623	4205.808	23776.65	CA		0.3		MIT

AIR WAVELENGTH (ANGSTRMS)	VACUUM WAVELENGTH (ANGSTRMS)	WAVENUMBER (KAYSERS)	LMENT	LOG ARC	INTENSITY SPK	INTENSITY DIS	REF
4203.22	4204.40	23784.6	CA I	0.3			CW
4201.768	4202.952	23792.80	CA		0.3		MIT
4199.099	4200.282	23807.93	CA	0.3	0.3		MIT
4196.60	4197.78	23822.1	CA	1.1W	0.0H		AD
4194.229	4195.411	23835.57	CA		0.5		MIT
4184.282	4185.461	23892.23	CA	0.3	0.9L		MIT
4180.69	4181.87	23912.8	CA		0.3		AD
4179.74	4180.92	23918.2	CA		0.3		AD
4175.639	4176.816	23941.68	CA		0.7		MIT
4169.60	4170.78	23976.4	CA	0.3H	0.3B		AD
4166.833	4168.008	23992.28	CA	0.3	0.0H		MIT
4165.48	4166.65	24000.1	CA	0.6	0.3H		AD
4164.359	4165.533	24006.53	CA		0.7H		MIT
4161.806	4162.979	24021.26	CA	0.3H	0.0H		MIT
4159.120	4160.293	24036.77	CA	0.3H			MIT
4153.65	4154.82	24068.4	CA		0.8		MIT
4148.403	4149.573	24098.87	CA	0.3	0.3		MIT
4147.673	4148.843	24103.11	CA	0.5	0.3		MIT
4144.147	4145.316	24123.61	CA		0.5		MIT
4141.96	4143.13	24136.4	CA	0.3	0.5		AD
4136.347	4137.514	24169.10	CA	0.3H	0.5		MIT
4132.497	4133.663	24191.62	CA I		0.3		MIT
4130.43	4131.60	24203.7	CA		0.3		AD
4129.12	4130.28	24211.4	CA	0.3H	0.3		AD
4128.080	4129.244	24217.50	CA	0.3	0.6		MIT
4116.095	4117.256	24288.02	CA	0.3	0.5		MIT
4114.19	4115.35	24299.3	CA	0.3	0.3		AD
4109.98	4111.14	24324.1	CA	0.3	0.5		AD
4108.559	4109.718	24332.57	CA I	0.8	0.5		MIT
4099.029	4100.186	24389.14	CA	0.3H	0.0		MIT
4098.533	4099.690	24392.09	CA I	1.2	0.5		IWG
4094.930	4096.086	24413.55	CA I	1.2	0.8		IWG
4092.633	4093.788	24427.25	CA I	1.2	0.3		IWG
4088.82	4089.97	24450.0	CA	0.3	0.5		AD
4084.501	4085.654	24475.89	CA I	0.3	0.3		MIT
4081.74	4082.89	24492.4	CA	0.3	0.6		AD
4071.30	4072.45	24555.2	CA		0.3		AD
4069.01	4070.16	24569.1	CA		0.5		AD
4066.782	4067.930	24582.52	CA	0.3H	0.3		MIT
4064.67	4065.82	24595.3	CA		0.3		AD
4062.49	4063.64	24608.5	CA I	0.3			FL
4059.875	4061.022	24624.35	CA	0.3H	0.5		MIT
4058.930	4060.076	24630.08	CA I	0.5D			MIT
4057.10	4058.25	24641.2	CA	0.3	0.5		AD
4054.64	4055.79	24656.1	CA	0.3	0.3		AD
4051.11	4052.25	24677.6	CA	0.3	0.5		AD
4047.351	4048.494	24700.54	CA	0.3	0.5		MIT
4044.419	4045.562	24718.45	CA	0.7D	0.5		MIT
4040.3	4041.4	24744.	CA	0.3	0.5		AD
4038.519	4039.660	24754.56	CA	0.3	0.3		MIT
4037.18	4038.32	24762.8	CA	0.3	0.3		AD
4030.3	4031.4	24805.	CA	1.0	0.3H		AD
4024.68	4025.82	24839.7	CA	0.5J	0.5		AD
4018.88	4020.02	24875.5	CA		0.6		MIT
4016.15	4017.29	24892.4	CA	0.3J	0.3		AD
4013.88	4015.01	24906.5	CA	0.3	0.3H		AD
4010.26	4011.39	24929.0	CA	0.3	0.3H		AD
4006.301	4007.434	24953.63	CA	0.3	0.3		MIT
4003.35	4004.48	24972.0	CA	0.3	0.3H		AD
3999.13	4000.26	24998.4	CA	0.3	0.3H		AD
3990.69	3991.82	25051.2	CA	0.5	0.9		AD
3984.722	3985.849	25088.76	CA	0.3	0.3		MIT
3980.31	3981.44	25116.6	CA	0.3	0.3		AD
3977.82	3978.95	25132.3	CA		0.3		AD
3973.707	3974.831	25158.30	CA I	2.3	1.2		IWG
3972.570	3973.694	25165.50	CA I	1.1			IWG
3968.468	3969.591	25191.51	CA II	2.76	2.76		I
3962.28	3963.40	25230.9	CA	0.5W	0.3		AD
3960.46	3961.58	25242.4	CA	0.5	0.3H		MIT
3957.053	3958.173	25264.18	CA I	1.9	0.5		IWG

AIR WAVELENGTH (ANGSTRMS)	VACUUM WAVELENGTH (ANGSTRMS)	WAVENUMBER (KAYSERS)	LMENT	LOG ARC	INTENSITY SPK	DIS	REF
3954.64	3955.76	25279.6	CA		0.3		AD
3949.92	3951.04	25309.8	CA		0.3R		HT
3949.57	3950.69	25312.1	CA		0.3		AD
3948.901	3950.019	25316.34	CA I	1.6	1.2		IWG
3946.05	3947.17	25334.6	CA I	0.3			CW
3942.84	3943.96	25355.2	CA	0.3	0.5		MIT
3933.666	3934.780	25414.38	CA II	2.8G	2.8G		I
3933.664	3934.778	25414.40	CA II	2.8G	2.8G		I
3923.75	3924.86	25478.6	CA	0.3	0.3		AD
3918.59	3919.70	25512.2	CA	0.5	0.5		AD
3914.131	3915.240	25541.22	CA		0.3H		MIT
3897.23	3898.33	25652.0	CA	0.3	0.5		MIT
3894.659	3895.762	25668.92	CA		0.5		MIT
3889.141	3890.243	25705.33	CA I	0.9			CW
3885.304	3886.405	25730.72	CA	0.7	0.5		MIT
3875.807	3876.906	25793.77	CA I	1.7			CW
3872.560	3873.658	25815.39	CA I	1.5			MIT
3870.506	3871.603	25829.09	CA I	1.2			CW
3868.410	3869.507	25843.09	CA		0.3		MIT
3859.675	3860.769	25901.57	CA	0.8	0.7H		MIT
3858.81	3859.90	25907.4	CA		0.3H		MIT
3856.024	3857.117	25926.10	CA		0.3H		MIT
3842.57	3843.66	26016.9	CA		0.3		MIT
3838.318	3839.407	26045.69	CA	0.3	0.3		MIT
3836.489	3837.577	26058.11	CA	0.3	0.3H		MIT
3833.963	3835.051	26075.27	CA I	0.7			MIT
3823.760	3824.845	26144.85	CA	0.3	0.3		MIT
3806.49	3807.57	26263.5	CA	0.5	0.3		MIT
3799.93	3801.01	26308.8	CA	0.3	0.3		MIT
3796.112	3797.190	26335.26	CA	0.3	0.3		MIT
3795.63	3796.71	26338.6	CA I	0.5			MIT
3793.358	3794.435	26354.38	CA		0.3		MIT
3789.64	3790.72	26380.2	CA	0.5	0.3		AD
3785.76	3786.84	26407.3	CA	0.3	0.3H		MIT
3783.24	3784.31	26424.9	CA	0.5	0.5		AD
3780.862	3781.936	26441.49	CA	0.3	0.5		MIT
3775.86	3776.93	26476.5	CA	0.3	0.3		MIT
3772.06	3773.13	26503.2	CA	0.3	0.5		AD
3767.43	3768.50	26535.7	CA I	0.3			MIT
3762.64	3763.71	26569.5	CA	0.3	0.5		MIT
3761.72	3762.79	26576.0	CA I	0.3	0.6		CW
3758.718	3759.786	26597.26	CA	0.5	0.5		MIT
3757.26	3758.33	26607.6	CA	0.5	0.5		MIT
3756.58	3757.65	26612.4	CA	0.5	0.6		MIT
3753.367	3754.434	26635.18	CA I	1.5	0.5		CW
3752.49	3753.56	26641.4	CA	0.3H	0.3H		MIT
3750.349	3751.415	26656.61	CA I	1.3	0.5		CW
3748.374	3749.439	26670.65	CA I	1.1			CW
3736.901	3737.963	26752.53	CA II	1.1	1.7		IWG
3734.91	3735.97	26766.8	CA		0.5		AD
3733.92	3734.98	26773.9	CA		0.3		AD
3731.99	3733.05	26787.7	CA	0.3	0.5		MIT
3728.46	3729.52	26813.1	CA	0.3	0.5		AD
3725.764	3726.824	26832.50	CA	0.3	0.6		MIT
3723.87	3724.93	26846.1	CA		0.7		AD
3720.40	3721.46	26871.2	CA		0.5		AD
3717.03	3718.09	26895.6	CA		0.5		AD
3715.87	3716.93	26903.9	CA	0.3	0.6		AD
3712.37	3713.43	26929.3	CA	0.3	0.5		AD
3710.753	3711.809	26941.04	CA		0.5		MIT
3706.026	3707.080	26975.41	CA II	1.2	1.6		IWG
3704.03	3705.08	26989.9	CA	0.5	0.6		MIT
3701.15	3702.20	27010.9	CA		0.3		AD
3690.84	3691.89	27086.4	CA	0.3	0.5		AD
3686.65	3687.70	27117.2	CA		0.7		MIT
3682.03	3683.08	27151.2	CA	0.3	0.6		AD
3681.08	3682.13	27158.2	CA	0.3	0.3H		AD
3678.234	3679.281	27179.22	CA I	1.2	0.3H		MIT
3675.307	3676.354	27200.87	CA I	1.0H	0.3		CW
3673.448	3674.494	27214.63	CA I	0.7	0.5		CW

AIR WAVELENGTH (ANGSTRMS)	VACUUM WAVELENGTH (ANGSTRMS)	WAVENUMBER (KAYSERS)	LMENT	LOG ARC	INTENSITY SPK	DIS	REF
3670.21	3671.26	27238.6	CA	0.3	0.6		MIT
3657.174	3658.216	27335.73	CA	0.3	0.3		MIT
3656.28	3657.32	27342.4	CA	0.3	0.5		MIT
3653.495	3654.536	27363.26	CA		0.5		MIT
3652.56	3653.60	27370.3	CA	0.5	0.3		MIT
3650.62	3651.66	27384.8	CA		0.5		AD
3647.88	3648.92	27405.4	CA		0.5		MIT
3646.053	3647.092	27419.11	CA		0.3H		MIT
3645.080	3646.119	27426.43	CA I	0.5	0.3H		MIT
3644.765	3645.804	27428.79	CA I	1.5			IWG
3644.410	3645.449	27431.47	CA I	2.3	1.2		IWG
3635.95	3636.99	27495.3	CA	0.6	0.3H		AD
3631.51	3632.55	27528.9	CA		0.3		AD
3630.947	3631.982	27533.18	CA I	1.0			IWG
3630.748	3631.783	27534.69	CA I	2.2	1.0		IWG
3628.60	3629.63	27551.0	CA I	0.9			CW
3625.75	3626.78	27572.6	CA I	0.7	0.3		MIT
3624.96	3625.99	27578.6	CA		0.5		MIT
3624.111	3625.144	27585.11	CA I	2.2	1.2		IWG
3612.77	3613.80	27671.7	CA		0.5		AD
3609.30	3610.33	27698.3	CA	0.5	0.3H		MIT
3605.21	3606.24	27729.7	CA	0.3	0.3R		MIT
3602.08	3603.11	27753.8	CA		0.5		MIT
3594.12	3595.15	27815.3	CA I	0.5	0.3		MIT
3591.26	3592.28	27837.4	CA I	0.3	0.3H		CW
3589.49	3590.51	27851.2	CA I	0.3			CW
3588.539	3589.563	27858.54	CA		0.5		MIT
3581.295	3582.317	27914.89	CA		0.3		MIT
3579.450	3580.472	27929.28	CA		0.3		MIT
3568.91	3569.93	28011.8	CA I	0.3			CW
3566.10	3567.12	28033.8	CA I	0.3			MIT
3560.603	3561.620	28077.11	CA		0.3		MIT
3554.85	3555.87	28122.6	CA	0.3	0.6		MIT
3550.03	3551.04	28160.7	CA I	0.5			SD
3549.45	3550.46	28165.3	CA	0.3	0.3		AD
3547.382	3548.396	28181.75	CA I	0.3	0.5		MIT
3545.58	3546.59	28196.1	CA I	0.5	0.6		CW
3542.75	3543.76	28218.6	CA	0.5	0.3		MIT
3541.45	3542.46	28228.9	CA	0.5	0.5		MIT
3537.733	3538.744	28258.61	CA	0.5	0.6		MIT
3535.534	3536.544	28276.19	CA I	0.6			MIT
3533.010	3534.020	28296.39	CA	0.3	0.6		MIT
3531.09	3532.10	28311.8	CA		0.3		AD
3529.967	3530.976	28320.78	CA	0.3	0.5		MIT
3522.700	3523.707	28379.20	CA		0.5		MIT
3518.706	3519.712	28411.41	CA		0.3H		MIT
3510.84	3511.84	28475.1	CA	0.3	0.6		MIT
3506.282	3507.285	28512.08	CA		0.3H		MIT
3502.02	3503.02	28546.8	CA		0.5		AD
3500.26	3501.26	28561.1	CA		0.5		AD
3491.47	3492.47	28633.0	CA		0.6		AD
3487.598	3488.596	28664.82	CA I	2.0	0.3		IWG
3486.40	3487.40	28674.7	CA		0.3		AD
3483.18	3484.18	28701.2	CA	0.3	0.7		AD
3478.623	3479.619	28738.78	CA		0.3		MIT
3477.72	3478.72	28746.2	CA	0.3	0.3		AD
3474.763	3475.758	28770.70	CA I	1.6	0.7		IWG
3468.476	3469.469	28822.85	CA I	1.3	0.5		IWG
3467.76	3468.75	28828.8	CA		0.5		AD
3462.62	3463.61	28871.6	CA	0.3	0.6		AD
3457.13	3458.12	28917.4	CA	0.3	0.5		AD
3451.03	3452.02	28968.6	CA		0.3		AD
3448.31	3449.30	28991.4	CA		0.3		AD
3428.51	3429.49	29158.8	CA	0.3	0.6		AD
3424.92	3425.90	29189.4	CA		0.3		AD
3422.878	3423.860	29206.80	CA		0.3		MIT
3420.82	3421.80	29224.4	CA		0.3H		MIT
3417.72	3418.70	29250.9	CA	0.3	0.6		AD
3415.38	3416.36	29270.9	CA	0.3	0.5		AD
3414.60	3415.58	29277.6	CA		0.3H		AD

AIR WAVELENGTH (ANGSTRMS)	VACUUM WAVELENGTH (ANGSTRMS)	WAVENUMBER (KAYSERS)	LMENT	LOG ARC	INTENSITY SPK	DIS	REF
3408.39	3409.37	29330.9	CA		0.5R		AD
3406.61	3407.59	29346.3	CA		0.5		AD
3392.81	3393.78	29465.6	CA	0.0	0.3		AD
3389.89	3390.86	29491.0	CA	0.0H	0.5		AD
3386.39	3387.36	29521.5	CA	0.0H	0.5		AD
3372.68	3373.65	29641.5	CA	0.0	0.5		AD
3372.00	3372.97	29647.5	CA	0.3			CW
3367.81	3368.78	29684.4	CA	0.0H	0.7R		AD
3362.28	3363.25	29733.2	CA I	0.3	0.0H		CW
3362.131	3363.097	29734.50	CA I	1.2	0.7		CW
3361.918	3362.884	29736.38	CA I	2.1	1.0		IWG
3360.06	3361.03	29752.8	CA		0.6		MIT
3359.34	3360.31	29759.2	CA		0.6		AD
3358.34	3359.31	29768.1	CA		0.3		MIT
3355.36	3356.32	29794.5	CA	0.0H	0.3H		AD
3354.14	3355.10	29805.3	CA	0.0	0.3		AD
3350.36	3351.32	29839.0	CA I	1.2	0.5		CW
3350.209	3351.172	29840.31	CA I	2.0	1.0		IWG
3344.513	3345.475	29891.13	CA I	2.0	0.8		IWG
3343.15	3344.11	29903.3	CA		0.5		MIT
3342.278	3343.239	29911.11	CA		0.3		MIT
3340.65	3341.61	29925.7	CA		0.5		AD
3340.42	3341.38	29927.7	CA		0.3		AD
3324.390	3325.346	30072.05	CA	0.0H	0.8		MIT
3323.933	3324.889	30076.19	CA		0.7		MIT
3322.30	3323.26	30091.0	CA	0.0	0.8		MIT
3320.39	3321.35	30108.3	CA		0.8H		AD
3319.63	3320.59	30115.2	CA		0.9		MIT
3317.53	3318.48	30134.2	CA	0.0H	0.9		AD
3311.46	3312.41	30189.5	CA		0.8		AD
3308.80	3309.75	30213.7	CA	0.0H	0.8		MIT
3302.94	3303.89	30267.3	CA	0.6	0.8		MIT
3301.52	3302.47	30280.4	CA	0.6	0.8		AD
3294.53	3295.48	30344.6	CA		0.3		AD
3292.51	3293.46	30363.2	CA	0.0H	0.5		AD
3290.42	3291.37	30382.5	CA	0.0H	0.5		AD
3286.067	3287.014	30422.75	CA I	1.5	0.3		IWG
3285.81	3286.76	30425.1	CA		0.3		AD
3285.34	3286.29	30429.5	CA	0.5	0.7		MIT
3283.32	3284.27	30448.2	CA		0.3		AD
3281.48	3282.43	30465.3	CA	0.0H	0.6		AD
3274.661	3275.605	30528.71	CA I	1.3			CW
3273.958	3274.902	30535.27	CA	0.3	0.6		MIT
3271.683	3272.626	30556.50	CA	0.0H	0.5		MIT
3270.841	3271.784	30564.37	CA	0.0H	0.3		MIT
3270.502	3271.445	30567.53	CA	0.0	0.3		MIT
3269.101	3270.043	30580.63	CA I	1.0	0.3		MIT
3265.34	3266.28	30615.9	CA	0.0	0.3		MIT
3264.09	3265.03	30627.6	CA	0.0	0.5		MIT
3260.93	3261.87	30657.3	CA	0.0H	0.5		AD
3255.811	3256.750	30705.46	CA	0.0H	0.5		MIT
3254.363	3255.302	30719.12	CA	0.0H	0.3		MIT
3250.26	3251.20	30757.9	CA		0.3		MIT
3248.28	3249.22	30776.6	CA	0.0H	0.5		MIT
3241.76	3242.70	30838.5	CA	0.0	0.3H		MIT
3240.167	3241.102	30853.70	CA		0.3		MIT
3234.990	3235.924	30903.08	CA		0.3		MIT
3233.66	3234.59	30915.8	CA		0.3H		MIT
3233.02	3233.95	30921.9	CA		0.6		AD
3226.13	3227.06	30987.9	CA I	0.9			CW
3225.896	3226.827	30990.19	CA I	1.9	1.0		IWG
3224.33	3225.26	31005.2	CA		0.5		MIT
3220.75	3221.68	31039.7	CA		0.5		AD
3215.327	3216.256	31092.05	CA I	0.7			MIT
3215.129	3216.058	31093.97	CA I	1.3	0.6W		MIT
3212.50	3213.43	31119.4	CA		0.3		AD
3209.930	3210.857	31144.33	CA I	1.5	0.3		CW
3208.71	3209.64	31156.2	CA		0.3		AD
3207.23	3208.16	31170.6	CA	0.0H	0.3		MIT
3206.10	3207.03	31181.5	CA		0.7		AD
3205.08	3206.01	31191.5	CA	0.0H	0.3		AD
3203.22	3204.15	31209.6	CA	0.0	0.5		MIT
3200.84	3201.77	31232.8	CA	0.0	0.3		MIT
3194.75	3195.67	31292.3	CA	0.0	0.3		MIT
3191.28	3192.20	31326.3	CA	0.0	0.3		MIT
3181.275	3182.195	31424.85	CA II	0.9	1.2		IWG
3180.516	3181.436	31432.35	CA I	1.3			MIT
3179.332	3180.252	31444.05	CA II	2.0	2.6W		IWG
3171.75	3172.67	31519.2	CA	0.0H	0.3		AD
3169.854	3170.771	31538.07	CA I	1.0	0.3H		CW
3167.87	3168.79	31557.8	CA	0.0	0.3		AD
3165.78	3166.70	31578.6	CA	0.0H	0.3		AD
3164.618	3165.534	31590.25	CA I	0.7	0.5		CW
3158.869	3159.783	31647.74	CA II	2.0	2.3		ME
3156.822	3157.736	31668.26	CA		0.3		MIT
3154.856	3155.769	31687.99	CA	0.0	0.3		MIT
3151.280	3152.193	31723.95	CA I	0.8	0.3		CW
3150.738	3151.650	31729.41	CA I	1.7	0.3		IWG
3149.504	3150.416	31741.84	CA	0.0	0.5		MIT
3145.76	3146.67	31779.6	CA		0.3		AD
3142.812	3143.722	31809.42	CA		0.3		MIT
3141.164	3142.074	31826.11	CA I	0.5			CW
3140.783	3141.693	31829.97	CA I	1.2	0.5		MIT
3136.003	3136.912	31878.49	CA I	1.2	0.5		CW
3134.56	3135.47	31893.2	CA	0.0	0.3		MIT
3133.342	3134.250	31905.56	CA		0.3		MIT
3129.331	3130.238	31946.45	CA		0.3		MIT
3119.671	3120.576	32045.37	CA		0.9		MIT
3117.653	3118.557	32066.11	CA	1.0	0.3		MIT
3109.51	3110.41	32150.1	CA I	0.6			SD
3108.58	3109.48	32159.7	CA I	1.5	0.5		SD
3107.388	3108.289	32172.03	CA	0.7	0.3		CW
3106.927	3107.828	32176.81	CA	0.0	0.3		MIT
3102.359	3103.259	32224.19	CA	0.3H	0.6		MIT
3100.22	3101.12	32246.4	CA I	0.3			CW
3099.80	3100.70	32250.8	CA	0.0	0.3H		AD
3099.34	3100.24	32255.6	CA I	1.0	0.3		SD
3096.898	3097.797	32281.01	CA	0.0	0.3H		MIT
3096.324	3097.223	32286.99	CA	0.0	0.3		MIT
3095.278	3096.176	32297.90	CA I	1.0	0.3		MIT
3093.460	3094.358	32316.88	CA	0.3	0.5		MIT
3089.015	3089.912	32363.38	CA	0.0	0.5		MIT
3086.229	3087.125	32392.60	CA	0.3	0.6		MIT
3083.740	3084.636	32418.74	CA	0.0	0.3		MIT
3081.547	3082.442	32441.81	CA I	0.3			MIT
3080.826	3081.721	32449.40	CA I	1.3	0.3		MIT
3079.10	3079.99	32467.6	CA	0.0	0.5		AD
3076.99	3077.88	32489.9	CA I	0.8			SD
3073.18	3074.07	32530.1	CA	0.0	0.3		AD
3071.587	3072.480	32547.00	CA I	0.7	0.0		MIT
3068.82	3069.71	32576.4	CA		0.3		MIT
3067.009	3067.900	32595.58	CA I	0.8	0.3		MIT
3063.49	3064.38	32633.0	CA	0.3	0.5		AD
3062.05	3062.94	32648.4	CA I	0.0H	0.3		SD
3060.27	3061.16	32667.4	CA		0.5		AD
3055.325	3056.213	32720.23	CA I	0.7H			MIT
3049.01	3049.90	32788.0	CA I	0.7	0.3H		SD
3045.75	3046.64	32823.1	CA I	0.3H			SD
3043.354	3044.239	32848.93	CA	0.0	0.3		MIT
3041.47	3042.35	32869.3	CA		0.5		AD
3041.05	3041.93	32873.8	CA I	0.3			SD
3039.21	3040.09	32893.7	CA I	0.0H	0.6		SD
3035.808	3036.692	32930.58	CA		0.5		MIT
3034.535	3035.418	32944.39	CA I	0.3			MIT
3030.214	3031.096	32991.37	CA	0.0	0.3		MIT
3028.97	3029.85	33004.9	CA I	0.3	0.3		SD
3028.665	3029.547	33008.24	CA	0.0	0.3		MIT
3024.95	3025.83	33048.8	CA I	0.3			MIT
3024.70	3025.58	33051.5	CA		0.6		AD
3020.15	3021.03	33101.3	CA I	0.3	0.3H		SD

AIR WAVELENGTH (ANGSTRMS)	VACUUM WAVELENGTH (ANGSTRMS)	WAVENUMBER (KAYSERS)	LMENT	LOG ARC	INTENSITY SPK	INTENSITY DIS	REF
3019.37	3020.25	33109.9	CA I	0.0	0.5		SD
3009.205	3010.082	33221.69	CA I	1.3	0.7		IWG
3006.858	3007.734	33247.62	CA I	1.4	0.7		IWG
3000.863	3001.738	33314.04	CA I	1.3	0.8		IWG
2999.641	3000.516	33327.61	CA I	1.3	1.0		IWG
2997.314	2998.188	33353.48	CA	1.4	0.7		MIT
2996.67	2997.54	33360.6	CA I	0.8			SD
2996.47	2997.34	33362.9	CA		0.7		AD
2994.958	2995.831	33379.72	CA I	1.4	0.5		IWG
2989.30	2990.17	33442.9	CA		0.8		AD
2988.98	2989.85	33446.5	CA I	0.7			CW
2988.60	2989.47	33450.7	CA		0.8		MIT
2982.89	2983.76	33514.8	CA I	0.3	0.3		FL
2972.89	2973.76	33627.5	CA		0.7		MIT
2962.8	2963.7	33742.	CA			0.3	BS
2961.77	2962.64	33753.7	CA	0.0H	0.5		AD
2951.24	2952.10	33874.2	CA	0.0H	0.3H		MIT
2949.96	2950.82	33888.9	CA		0.3		MIT
2949.20	2950.06	33897.6	CA	0.0H	0.3H		MIT
2947.25	2948.11	33920.0	CA		0.3H		AD
2936.72	2937.58	34041.6	CA		0.6H		ED
2934.4	2935.3	34069.	CA		0.3C		AD
2933.529	2934.387	34078.67	CA	0.8			MIT
2930.16	2931.02	34117.9	CA		0.3H		AD
2929.71	2930.57	34123.1	CA		0.6H		AD
2929.44	2930.30	34126.2	CA		0.3H		MIT
2928.23	2929.09	34140.3	CA	0.9	0.3		MIT
2924.33	2925.19	34185.9	CA		0.7		AD
2920.47	2921.32	34231.0	CA		0.7		MIT
2919.14	2919.99	34246.6	CA		0.3		AD
2918.42	2919.27	34255.1	CA		0.3S		AD
2914.75	2915.60	34298.2	CA		0.3		AD
2913.68	2914.53	34310.8	CA		0.3		AD
2911.65	2912.50	34334.7	CA		0.3		AD
2910.31	2911.16	34350.5	CA		0.7D		AD
2907.90	2908.75	34379.0	CA		0.7		AD
2899.78	2900.63	34475.3	CA		0.9		AD
2891.96	2892.81	34568.5	CA		0.3		AD
2883.38	2884.23	34671.4	CA		0.5		AD
2881.80	2882.65	34690.4	CA		0.6		AD
2880.87	2881.72	34701.6	CA		0.3		AD
2877.91	2878.75	34737.2	CA	0.0H	0.6		MIT
2870.98	2871.82	34821.1	CA		0.3		AD
2869.95	2870.79	34833.6	CA		0.8		AD
2866.57	2867.41	34874.7	CA		0.8		AD
2859.36	2860.20	34962.6	CA		0.3		AD
2847.48	2848.32	35108.5	CA		0.6		AD
2834.14	2834.97	35273.7	CA		0.5		AD
2825.56	2826.39	35380.8	CA		0.3		MIT
2816.33	2817.16	35496.7	CA		0.5		ED
2813.88	2814.71	35527.7	CA		0.6		AD
2804.855	2805.681	35641.96	CA		0.5J		MIT
2803.72	2804.55	35656.4	CA		0.3		AD
2797.15	2797.97	35740.1	CA		0.6		AD
2791.63	2792.45	35810.8	CA		0.8		AD
2771.27	2772.09	36073.9	CA		0.6		AD
2762.12	2762.94	36193.4	CA		0.3H		AD
2754.18	2754.99	36297.7	CA	0.0	0.5		AD
2736.531	2737.341	36531.80	CA		0.8		MIT
2734.82	2735.63	36554.7	CA	0.9	0.3		SD
2732.56	2733.37	36584.9	CA		0.6		AD
2721.645	2722.451	36731.60	CA I	1.3	0.3		IWG
2718.38	2719.19	36775.7	CA		0.7		AD
2715.08	2715.88	36820.4	CA		0.5		AD
2710.60	2711.40	36881.3	CA		0.6		AD
2704.87	2705.67	36959.4	CA		0.8		AD
2699.63	2700.43	37031.1	CA		0.3		AD
2687.78	2688.58	37194.4	CA		0.9		AD
2686.73	2687.53	37208.9	CA		0.5		AD
2684.38	2685.18	37241.5	CA		0.7		AD
2680.36	2681.16	37297.3	CA I	1.2			SD
2680.12	2680.92	37300.7	CA		0.3		AD
2653.32	2654.11	37677.4	CA		0.3		AD
2642.74	2643.53	37828.2	CA		0.5		AD
2634.17	2634.96	37951.3	CA		0.8		AD
2624.11	2624.89	38096.8	CA		0.6		AD
2620.82	2621.60	38144.6	CA		0.8		AD
2609.57	2610.35	38309.0	CA		0.6		AD
2590.34	2591.11	38593.4	CA		0.3		AD
2587.09	2587.86	38641.9	CA		0.5		AD
2573.09	2573.86	38852.1	CA II	0.5	2.2		MIT
2565.48	2566.25	38967.4	CA		0.6		AD
2556.45	2557.22	39105.0	CA		0.6		AD
2547.88	2548.64	39236.5	CA		0.8		AD
2541.49	2542.25	39335.2	CA		0.8		AD
2530.57	2531.33	39504.9	CA		0.8		AD
2522.06	2522.82	39638.2	CA		0.3		AD
2517.03	2517.79	39717.4	CA		0.8		AD
2497.66	2498.41	40025.4	CA		0.7		MIT
2493.00	2493.75	40100.2	CA		0.8		AD
2447.23	2447.97	40850.1	CA		0.6		AD
2398.559	2399.289	41679.01	CA I	2.0G	1.3		IWG
2393.20	2393.93	41772.3	CA		0.5		AD
2378.59	2379.32	42028.9	CA		0.3		AD
2343.04	2343.76	42666.5	CA		0.3		AD
2334.13	2334.85	42829.4	CA		0.3		AD
2322.91	2323.62	43036.2	CA		0.5		AD
2316.27	2316.98	43159.6	CA		0.3		AD
2312.08	2312.79	43237.8	CA		0.5		AD
2289.41	2290.12	43665.9	CA		0.5		AD
2275.471	2276.174	43933.37	CA I	1.6	0.7		IWG
2267.19	2267.89	44093.8	CA		0.3		AD
2257.40	2258.10	44285.0	CA I	0.3			SD
2252.65	2253.35	44378.4	CA		0.3		AD
2208.606	2209.295	45263.31	CA II	1.3	1.7		CW
2200.720	2201.407	45425.49	CA I	1.3			MIT
2197.791	2198.477	45486.03	CA II	1.3	1.6		CW
2177.8	2178.5	45903.	CA I	0.3			SD
2150.78	2151.46	46480.1	CA I	1.0			CW
2118.68	2119.35	47184.3	CA I	0.7			SD
2112.763	2113.432	47316.40	CA II	1.0	1.4		CW
2103.239	2103.906	47530.63	CA II	1.0	1.4		CW
2097.49	2098.16	47660.9	CA I	0.5			SD
2082.73	2083.39	47998.6	CA I	0.3			SD
2073.04	2073.70	48223.0	CA I	0.3			SD
2064.77	2065.43	48416.1	CA I	0.3			SD
9965.44	9968.17	10031.9	CB	0.6			ME
9957.29	9960.02	10040.1	CB	1.2			ME
9912.26	9914.98	10085.7	CB	1.4			ME
9910.35	9913.07	10087.7	CB	1.3			ME
9896.6	9899.3	10102.	CB	0.7			ME
9890.09	9892.80	10108.4	CB	0.6			ME
9677.5	9680.2	10330.	CB	0.6			ME
9676.75	9679.40	10331.2	CB	1.7			ME
9669.8	9672.5	10339.	CB	0.3			ME
9650.97	9653.62	10358.8	CB	1.1			ME
9631.11	9633.75	10380.2	CB	1.7			ME
9626.88	9629.52	10384.7	CB	2.0			ME
9620.96	9623.60	10391.1	CB	1.0			ME
9598.72	9601.35	10415.2	CB	0.8			ME
9595.06	9597.69	10419.2	CB	1.8			ME
9474.57	9477.17	10551.7	CB	0.7			ME
9472.02	9474.62	10554.5	CB	0.6			ME
9470.93	9473.53	10555.7	CB	0.8			ME
9438.7	9441.3	10592.	CB	0.9H			ME
9435.48	9438.07	10595.4	CB	0.9			ME
9412.39	9414.97	10621.4	CB	0.6			ME
9408.60	9411.18	10625.7	CB	1.3			ME
9393.56	9396.14	10642.7	CB	0.6			ME
9353.17	9355.74	10688.6	CB	1.0			ME

AIR WAVELENGTH (ANGSTRMS)	VACUUM WAVELENGTH (ANGSTRMS)	WAVENUMBER (KAYSERS)	LMENT	LOG ARC	INTENSITY SPK	DIS	REF
9344.4	9347.0	10699.	CB	0.6			ME
9323.54	9326.10	10722.6	CB	1.6			ME
9299.2	9301.8	10751.	CB	0.9			ME
9240.9	9243.4	10818.	CB	1.0			ME
9197.60	9200.12	10869.4	CB	1.2			ME
9186.96	9189.48	10882.0	CB	1.3			ME
9141.31	9143.82	10936.3	CB	1.7			ME
9129.44	9131.95	10950.6	CB	1.0			ME
9125.25	9127.75	10955.6	CB	1.0			ME
9123.60	9126.10	10957.6	CB	0.3			ME
9117.68	9120.18	10964.7	CB	0.8			ME
9084.91	9087.40	11004.2	CB	0.8			ME
9066.54	9069.03	11026.5	CB	0.6			ME
9061.43	9063.92	11032.8	CB	1.3			ME
9039.18	9041.66	11059.9	CB	0.9			ME
9011.74	9014.21	11093.6	CB	0.8			ME
8983.15	8985.62	11128.9	CB	0.6			ME
8967.76	8970.22	11148.0	CB	1.3			ME
8959.75	8962.21	11158.0	CB	1.3			ME
8933.43	8935.88	11190.8	CB	0.8			ME
8930.70	8933.15	11194.3	CB	0.3			ME
8915.76	8918.21	11213.0	CB	0.6			ME
8905.78	8908.23	11225.6	CB	1.5			ME
8897.5	8899.9	11236.	CB	0.8			ME
8896.4	8898.8	11237.	CB	0.7			ME
8815.56	8817.98	11340.5	CB	2.0			ME
8799.76	8802.18	11360.8	CB	0.8			ME
8798.22	8800.64	11362.8	CB	0.3			ME
8769.57	8771.98	11399.9	CB	0.8			ME
8767.97	8770.38	11402.0	CB	1.1			ME
8740.96	8743.36	11437.2	CB	1.3			ME
8717.09	8719.48	11468.6	CB	0.6			ME
8697.55	8699.94	11494.3	CB	1.6			ME
8614.45	8616.82	11605.2	CB	1.3			ME
8575.87	8578.23	11657.4	CB	1.5			ME
8560.54	8562.89	11678.3	CB	1.5			ME
8547.25	8549.60	11696.5	CB	1.3			ME
8526.99	8529.33	11724.2	CB	1.7			ME
8475.98	8478.31	11794.8	CB	2.2			ME
8439.77	8442.09	11845.4	CB	1.4			ME
8434.31	8436.63	11853.1	CB	0.3H			ME
8433.90	8436.22	11853.7	CB	0.6H			ME
8417.08	8419.39	11877.3	CB	0.6			ME
8414.74	8417.05	11880.6	CB	0.3			ME
8406.23	8408.54	11892.7	CB	1.2			ME
8387.88	8390.19	11918.7	CB	0.6H			ME
8371.39	8373.69	11942.2	CB	0.6H			ME
8350.04	8352.33	11972.7	CB	1.0H			ME
8346.08	8348.37	11978.4	CB	1.8			ME
8320.93	8323.22	12014.6	CB	2.7			ME
8262.41	8264.68	12099.7	CB	0.3H			ME
8240.00	8242.27	12132.6	CB	1.7			ME
8228.98	8231.24	12148.8	CB	0.3H			ME
8226.08	8228.34	12153.1	CB	0.3H			ME
8211.23	8213.49	12175.1	CB	0.3H			ME
8190.66	8192.91	12205.7	CB	0.3			ME
8189.40	8191.65	12207.6	CB	0.3H			ME
8176.23	8178.48	12227.2	CB	0.3			ME
8173.03	8175.28	12232.0	CB	0.8			ME
8135.20	8137.44	12288.9	CB	1.9			ME
8118.67	8120.90	12313.9	CB	0.8H			ME
8118.17	8120.40	12314.7	CB	0.8			ME
8017.23	8019.43	12469.7	CB	0.7H			ME
7954.76	7956.95	12567.6	CB	1.0			ME
7938.89	7941.07	12592.8	CB	1.5			ME
7885.31	7887.48	12678.3	CB	1.8			ME
7873.41	7875.58	12697.5	CB	1.4			ME
7868.58	7870.74	12705.3	CB	0.6H			ME
7862.68	7864.84	12714.8	CB	0.3H			ME
7787.11	7789.25	12838.2	CB	0.3			ME

AIR WAVELENGTH (ANGSTRMS)	VACUUM WAVELENGTH (ANGSTRMS)	WAVENUMBER (KAYSERS)	LMENT	LOG ARC	INTENSITY SPK	DIS	REF
7772.11	7774.25	12863.0	CB	0.7H			ME
7757.31	7759.44	12887.5	CB	1.3			ME
7726.68	7728.81	12938.6	CB	1.8			ME
7703.33	7705.45	12977.8	CB	0.8			ME
7647.64	7649.75	13072.3	CB	0.6			ME
7603.40	7605.49	13148.4	CB	0.7			ME
7585.28	7587.37	13179.8	CB	0.7H	0.0H		ME
7583.21	7585.30	13183.4	CB	0.8	0.0		ME
7574.58	7576.67	13198.4	CB	2.0	1.3		ME
7547.71	7549.79	13245.4	CB	0.3	0.0		ME
7532.11	7534.18	13272.8	CB	0.3	0.0		ME
7519.77	7521.84	13294.6	CB	1.3	0.6		ME
7515.93	7518.00	13301.4	CB	1.6	1.0		ME
7478.20	7480.26	13368.5	CB	1.2	0.3		ME
7456.32	7458.37	13407.7	CB	0.8	0.0		ME
7452.88	7454.93	13413.9	CB	0.8			ME
7436.02	7438.07	13444.3	CB	0.8	0.0		ME
7428.50	7430.55	13458.0	CB	0.8	0.0		ME
7419.83	7421.87	13473.7	CB	0.8	0.0		ME
7372.50	7374.53	13560.2	CB	2.2	1.5		ME
7353.16	7355.19	13595.8	CB	1.8	1.0		ME
7332.30	7334.32	13634.5	CB	0.8	0.0		ME
7328.38	7330.40	13641.8	CB	1.3	0.5		ME
7323.92	7325.94	13650.1	CB	1.3	0.3		ME
7317.03	7319.05	13663.0	CB	1.2	0.3		ME
7311.93	7313.94	13672.5	CB	0.8	0.0		ME
7276.76	7278.77	13738.6	CB	0.8	0.0		ME
7274.81	7276.81	13742.3	CB	0.9	0.3		ME
7268.90	7270.90	13753.4	CB	0.8	0.0		ME
7258.90	7260.90	13772.4	CB	1.0	0.3		ME
7256.18	7258.18	13777.6	CB	0.8	0.0		ME
7252.35	7254.35	13784.8	CB	1.6	0.8		ME
7241.80	7243.80	13804.9	CB	0.8	0.0		ME
7208.94	7210.93	13867.8	CB	1.3	0.5		ME
7191.37	7193.35	13901.7	CB	1.2	0.3		ME
7180.01	7181.99	13923.7	CB	0.9	0.0		ME
7178.27	7180.25	13927.1	CB	1.1	0.3		ME
7165.82	7167.80	13951.3	CB	0.8	0.0		ME
7159.43	7161.40	13963.7	CB	2.0	1.2		ME
7130.06	7132.03	14021.3	CB	1.2	0.3		ME
7126.17	7128.13	14028.9	CB	1.5	0.9		ME
7122.95	7124.91	14035.3	CB	0.9	0.3		ME
7119.31	7121.27	14042.4	CB	1.2	0.5		ME
7102.01	7103.97	14076.6	CB	1.5	0.9		ME
7098.94	7100.90	14082.7	CB	1.8	1.0		ME
7091.27	7093.23	14098.0	CB	0.8	0.0		ME
7075.23	7077.18	14129.9	CB	1.0	0.3		ME
7066.41	7068.36	14147.6	CB	0.9	0.3		ME
7051.20	7053.14	14178.1	CB	1.2	0.3		ME
7046.81	7048.75	14186.9	CB	2.3	1.6		ME
7038.04	7039.98	14204.6	CB	1.2	0.3		ME
7026.15	7028.09	14228.6	CB	1.2	0.9		ME
7023.48	7025.42	14234.0	CB	1.5	0.9		ME
7014.90	7016.83	14251.4	CB	0.8	0.0		ME
6996.11	6998.04	14289.7	CB	1.2	0.6		ME
6990.32	6992.25	14301.6	CB	2.0	1.2		ME
6989.38	6991.31	14303.5	CB	0.8	0.0		ME
6986.09	6988.02	14310.2	CB	1.3	0.5		ME
6978.39	6980.31	14326.0	CB	0.9	0.3		ME
6975.05	6976.97	14332.9	CB	1.1	0.3		ME
6972.49	6974.41	14338.1	CB	1.3	0.6		ME
6971.61	6973.53	14339.9	CB	1.2	0.5		ME
6966.89	6968.81	14349.7	CB	1.2	0.5		ME
6946.07	6947.99	14392.7	CB	1.2	0.5		ME
6940.90	6942.81	14403.4	CB	1.0	1.2		ME
6929.05	6930.96	14428.0	CB		0.5		ME
6918.32	6920.23	14450.4	CB	1.9	1.0		ME
6908.07	6909.98	14471.8	CB	1.6	0.9		ME
6906.59	6908.50	14474.9	CB	0.9	0.3		ME
6902.89	6904.79	14482.7	CB	1.8	1.0		ME

AIR WAVELENGTH (ANGSTRMS)	VACUUM WAVELENGTH (ANGSTRMS)	WAVENUMBER (KAYSERS)	LMENT	LOG ARC	INTENSITY SPK	DIS	REF
6888.48	6890.38	14513.0	CB	1.2	0.3		ME
6886.33	6888.23	14517.5	CB	1.5	0.9		ME
6879.90	6881.80	14531.1	CB	1.2	0.3		ME
6876.36	6878.26	14538.6	CB	1.9	1.1		ME
6870.92	6872.82	14550.1	CB	1.3	0.7		ME
6849.35	6851.24	14595.9	CB	1.1	0.8		MIT
6828.11	6829.99	14641.3	CB	2.2	1.5		ME
6795.31	6797.19	14712.0	CB	1.0	0.8		MIT
6739.88	6741.74	14833.0	CB	1.9	1.2		ME
6737.16	6739.02	14838.9	CB	0.5K	0.0K		ME
6723.62	6725.48	14868.8	CB	2.0	1.5		ME
6721.92	6723.78	14872.6	CB	0.3			MIT
6709.88	6711.73	14899.3	CB	1.2	0.3		ME
6703.18	6705.03	14914.2	CB	0.8	0.0		ME
6701.205	6703.055	14918.57	CB	2.0	1.2		MIT
6696.44	6698.29	14929.2	CB	0.8	0.5		ME
6677.33	6679.17	14971.9	CB	2.3	1.7		ME
6660.84	6662.68	15009.0	CB	2.5	1.9		ME
6629.11	6630.94	15080.8	CB	0.3	1.0H		ME
6626.976	6628.806	15085.67	CB	0.8	0.3		MIT
6616.698	6618.525	15109.11	CB	0.8	0.3		MIT
6614.153	6615.980	15114.92	CB	1.4	0.9		MIT
6607.276	6609.101	15130.65	CB	1.2	0.6		MIT
6606.158	6607.983	15133.21	CB	1.2	0.6		MIT
6591.00	6592.82	15168.0	CB	1.3	1.0		ME
6574.73	6576.55	15205.6	CB	1.1	0.3		ME
6556.92	6558.73	15246.8	CB	0.8	0.3		ME
6551.61	6553.42	15259.2	CB	0.8			ME
6544.61	6546.42	15275.5	CB	1.9	1.0		ME
6497.84	6499.64	15385.5	CB	1.1	0.3		ME
6464.32	6466.11	15465.3	CB	0.6	0.3D		ME
6453.11	6454.89	15492.1	CB		0.5H		ME
6449.83	6451.61	15500.0	CB	0.8	0.3		ME
6433.22	6435.00	15540.0	CB	1.5	0.6		ME
6430.46	6432.24	15546.7	CB	1.9	1.0		ME
6422.06	6423.84	15567.0	CB	0.8			ME
6348.98	6350.74	15746.2	CB	0.9	0.3		ME
6346.24	6347.99	15753.0	CB	0.8	0.3		ME
6325.78	6327.53	15804.0	CB	0.8	0.3		ME
6312.122	6313.868	15838.15	CB	0.8	0.0		MIT
6309.24	6310.98	15845.4	CB	0.8	0.3		ME
6286.38	6288.12	15903.0	CB	1.0	0.0		ME
6275.424	6277.160	15930.77	CB	0.8	0.3H		MIT
6260.768	6262.500	15968.06	CB	1.3	0.5H		MIT
6251.76	6253.49	15991.1	CB	1.5	1.0		ME
6221.958	6223.679	16067.67	CB	1.3	0.7		MIT
6219.636	6221.357	16073.66	CB	0.8V	0.3H		MIT
6213.062	6214.781	16090.67	CB	1.0	1.0H		MIT
6183.24	6184.95	16168.3	CB		0.3H		ME
6164.316	6166.022	16217.91	CB	1.0	0.7		MIT
6154.91	6156.61	16242.7	CB	1.0J	1.0H		ME
6148.134	6149.836	16260.60	CB	1.3	0.7		MIT
6142.506	6144.206	16275.50	CB	1.0	0.7H		MIT
6128.623	6130.319	16312.36	CB	1.0	0.6		MIT
6118.10	6119.79	16340.4	CB	0.8	0.3H		ME
6110.947	6112.639	16359.55	CB	0.8	0.5		MIT
6107.709	6109.400	16368.22	CB	1.3	0.5H		MIT
6105.296	6106.986	16374.69	CB	1.0	0.0H		MIT
6103.49	6105.18	16379.5	CB	0.8	0.0H		ME
6083.558	6085.242	16433.20	CB	1.0	0.3		MIT
6056.650	6058.327	16506.21	CB	1.0	0.7		MIT
6048.720	6050.395	16527.85	CB	0.8	0.5H		MIT
6045.499	6047.173	16536.65	CB	1.3	1.0		MIT
6041.96	6043.63	16546.3	CB	1.1H	0.3		ME
6031.840	6033.510	16574.10	CB	1.0	0.7		MIT
6029.746	6031.416	16579.85	CB	1.0	0.5		MIT
6028.695	6030.364	16582.75	CB	0.6	1.0H		MIT
6021.373	6023.041	16602.91	CB	0.6H	1.0H		MIT
6000.25	6001.91	16661.4	CB		0.3H		ME
5997.926	5999.587	16667.81	CB	1.7	0.7		MIT

AIR WAVELENGTH (ANGSTRMS)	VACUUM WAVELENGTH (ANGSTRMS)	WAVENUMBER (KAYSERS)	LMENT	LOG ARC	INTENSITY SPK	DIS	REF
5986.081	5987.739	16700.79	CB	0.7	0.7		MIT
5983.216	5984.873	16708.79	CB	1.3	1.3		MIT
5957.704	5959.354	16780.34	CB	0.9H	0.5H		MIT
5934.163	5935.807	16846.91	CB	0.7	0.3		MIT
5928.205	5929.848	16863.84	CB	0.7	0.3		MIT
5927.406	5929.048	16866.11	CB	0.7	0.7		MIT
5904.480	5906.116	16931.60	CB	0.5H	0.3H		MIT
5903.80	5905.44	16933.6	CB	0.7	0.3		ME
5900.616	5902.251	16942.69	CB	2.3	2.3		MIT
5893.442	5895.075	16963.31	CB	1.2	0.5		MIT
5877.786	5879.415	17008.50	CB	0.7	0.7		MIT
5876.315	5877.944	17012.75	CB	1.0	0.0		MIT
5875.258	5876.886	17015.81	CB	0.7	0.5J		MIT
5874.700	5876.328	17017.43	CB	1.5	0.7		MIT
5868.898	5870.525	17034.25	CB	0.5	0.0		MIT
5866.474	5868.100	17041.29	CB	1.7	1.2		MIT
5864.967	5866.593	17045.67	CB	0.5H	0.0H		MIT
5846.092	5847.713	17100.70	CB	0.7	0.0		MIT
5842.467	5844.087	17111.31	CB	1.2	0.7		MIT
5838.637	5840.256	17122.54	CB	2.3	2.0		MIT
5838.148	5839.766	17123.97	CB	1.2	0.3		MIT
5834.897	5836.515	17133.51	CB	1.2	1.0		MIT
5820.620	5822.234	17175.54	CB	1.0	0.7		MIT
5819.430	5821.043	17179.05	CB	1.3	1.3		MIT
5815.327	5816.939	17191.17	CB	0.7	0.5		MIT
5804.028	5805.637	17224.64	CB	1.2	0.7		MIT
5794.244	5795.851	17253.72	CB	1.2	1.0		MIT
5790.80	5792.41	17264.0	CB		0.7H		MIT
5789.792	5791.398	17266.99	CB	0.5	0.7H		MIT
5787.542	5789.147	17273.70	CB	1.9	1.2		MIT
5780.335	5781.938	17295.24	CB	0.5	0.5		MIT
5776.069	5777.671	17308.01	CB	1.5	0.5		MIT
5771.08	5772.68	17323.0	CB	1.0	0.3		MIT
5764.987	5766.586	17341.28	CB	1.0	1.2		MIT
5760.343	5761.941	17355.26	CB	1.5	1.5		MIT
5754.44	5756.04	17373.1	CB	0.5	0.0		ME
5753.06	5754.66	17377.2	CB		1.5H		ME
5751.440	5753.035	17382.13	CB	1.2	1.0		MIT
5746.90	5748.49	17395.9	CB		1.3H		ME
5738.196	5739.788	17422.25	CB	0.5	0.0		MIT
5737.355	5738.946	17424.80	CB	0.7	0.3		MIT
5729.193	5730.782	17449.62	CB	1.3	1.0		MIT
5725.661	5727.249	17460.39	CB	0.7	0.3H		MIT
5722.707	5724.295	17469.40	CB	0.5H	0.0		MIT
5716.347	5717.933	17488.84	CB	1.0	1.0		MIT
5715.587	5717.173	17491.16	CB	0.5	0.0		MIT
5713.855	5715.440	17496.47	CB	0.3			MIT
5709.330	5710.914	17510.33	CB	1.0H	0.5H		MIT
5708.467	5710.051	17512.98	CB	0.0	0.5H		MIT
5706.485	5708.068	17519.06	CB	1.3	1.0		MIT
5706.157	5707.740	17520.07	CB	0.9	0.5		MIT
5698.03	5699.61	17545.1	CB	0.6	0.5		MIT
5697.897	5699.478	17545.47	CB	0.7	0.5		MIT
5693.090	5694.670	17560.28	CB	0.7	0.3		MIT
5677.470	5679.045	17608.59	CB	0.7	0.5		MIT
5671.907	5673.481	17625.86	CB	1.2	1.0		MIT
5671.021	5672.595	17628.62	CB	2.3	1.0		MIT
5666.864	5668.437	17641.55	CB	0.5	0.3		MIT
5665.631	5667.203	17645.39	CB	2.0	1.5		MIT
5664.71	5666.28	17648.3	CB	2.0R	1.5R		MIT
5654.138	5655.707	17681.25	CB	0.5	0.0		MIT
5645.305	5646.872	17708.92	CB	0.7H	0.3H		MIT
5642.108	5643.674	17718.95	CB	1.9	1.3		MIT
5635.422	5636.986	17739.98	CB	1.0			MIT
5629.168	5630.731	17759.69	CB	1.3	1.0		MIT
5628.257	5629.819	17762.56	CB	0.7	0.0		MIT
5612.30	5613.86	17813.1	CB		0.6J		ME
5603.930	5605.486	17839.67	CB	0.7H	0.0		MIT
5599.591	5601.146	17853.49	CB	0.7	0.0H		MIT
5595.72	5597.27	17865.8	CB		0.6J		ME

AIR WAVELENGTH (ANGSTRMS)	VACUUM WAVELENGTH (ANGSTRMS)	WAVENUMBER (KAYSERS)	LMENT	LOG ARC	INTENSITY SPK	INTENSITY DIS	REF
5594.89	5596.44	17868.5	CB	0.5	0.0		MIT
5590.946	5592.498	17881.10	CB	0.9	0.3H		MIT
5586.97	5588.52	17893.8	CB	1.2	0.7		MIT
5578.29	5579.84	17921.7	CB	1.2	0.7H		MIT
5578.06	5579.61	17922.4	CB	0.5	0.0		MIT
5576.155	5577.703	17928.53	CB	1.9	0.7		MIT
5572.00	5573.55	17941.9	CB		0.3J		ME
5571.436	5572.983	17943.71	CB	0.7	0.5H		MIT
5563.000	5564.545	17970.92	CB	1.0	0.5		MIT
5553.122	5554.664	18002.89	CB	0.5	0.0		MIT
5551.347	5552.889	18008.65	CB	1.5	1.0		MIT
5549.607	5551.148	18014.29	CB	0.7			MIT
5546.419	5547.959	18024.65	CB	0.0	1.0		MIT
5545.623	5547.163	18027.23	CB	0.3H	1.0H		MIT
5541.468	5543.007	18040.75	CB	0.7	0.3		MIT
5532.597	5534.134	18069.68	CB	0.5	0.0		MIT
5523.569	5525.103	18099.21	CB	1.5	1.0		MIT
5521.884	5523.418	18104.73	CB	0.7H	0.0		MIT
5517.388	5518.921	18119.49	CB	0.5	0.0		MIT
5512.816	5514.347	18134.51	CB	1.3	0.5		MIT
5511.21	5512.74	18139.8	CB	0.5	0.0		MIT
5510.68	5512.21	18141.5	CB	0.5H	0.0		MIT
5509.122	5510.652	18146.67	CB	0.7	0.3H		MIT
5504.585	5506.114	18161.63	CB	1.5W	0.5H		MIT
5501.747	5503.275	18171.00	CB	0.5	0.3		MIT
5499.533	5501.061	18178.31	CB	0.7	0.0		MIT
5491.063	5492.589	18206.35	CB	0.5	0.0		MIT
5487.598	5489.123	18217.85	CB	0.0	1.0		MIT
5483.487	5485.011	18231.50	CB	0.7	0.3		MIT
5483.088	5484.611	18232.83	CB	0.7	0.0H		MIT
5481.002	5482.525	18239.77	CB	0.9	0.7		MIT
5476.069	5477.591	18256.20	CB	0.5	0.0		MIT
5460.928	5462.446	18306.82	CB	0.5	0.0		MIT
5458.043	5459.560	18316.50	CB	0.7	0.5		MIT
5457.60	5459.12	18318.0	CB		0.7J		ME
5456.187	5457.703	18322.73	CB	0.5H	0.0		MIT
5455.084	5456.600	18326.43	CB		0.6H		MIT
5448.310	5449.824	18349.22	CB	0.7	0.3H		MIT
5439.78	5441.29	18378.0	CB		1.0J		ME
5437.274	5438.785	18386.46	CB	1.5	1.0		MIT
5433.54	5435.05	18399.1	CB		1.0J		ME
5431.261	5432.771	18406.81	CB	1.0H	0.5		MIT
5422.438	5423.945	18436.76	CB	1.2	0.7		MIT
5416.303	5417.809	18457.65	CB	0.7	0.0H		MIT
5411.245	5412.749	18474.90	CB	0.7	0.5		MIT
5408.916	5410.420	18482.85	CB		0.6J		MIT
5396.33	5397.83	18526.0	CB	1.0	0.0		MIT
5395.859	5397.359	18527.58	CB	0.3H	0.5		MIT
5394.374	5395.874	18532.68	CB	0.5	0.0		MIT
5393.98	5395.48	18534.0	CB		0.3J		ME
5388.302	5389.800	18553.56	CB	0.7	0.5		MIT
5382.849	5384.346	18572.36	CB	0.5	0.0		MIT
5381.336	5382.832	18577.58	CB	0.9	0.5		MIT
5380.709	5382.205	18579.75	CB	0.7	0.0		MIT
5375.918	5377.413	18596.30	CB	0.7	0.0		MIT
5375.266	5376.761	18598.56	CB	1.2	0.5		MIT
5371.117	5372.611	18612.93	CB	0.5	0.0		MIT
5368.399	5369.892	18622.35	CB	0.5H	0.0		MIT
5365.878	5367.370	18631.10	CB	0.0	0.7		MIT
5363.084	5364.575	18640.80	CB	0.3	0.3		MIT
5362.009	5363.500	18644.54	CB	0.5H	0.3H		MIT
5359.194	5360.684	18654.33	CB	0.5	0.5		MIT
5356.84	5358.33	18662.5	CB		0.5H		MIT
5355.702	5357.191	18666.50	CB	0.9	0.5		MIT
5355.310	5356.799	18667.86	CB	0.5	0.3H		MIT
5353.296	5354.785	18674.89	CB	0.7	0.6		MIT
5351.045	5352.533	18682.74	CB	0.5	0.3		MIT
5350.742	5352.230	18683.80	CB	2.2	1.7		MIT
5344.170	5345.656	18706.78	CB	2.6	2.3		MIT
5343.580	5345.066	18708.84	CB	0.7H	0.0H		MIT

AIR WAVELENGTH (ANGSTRMS)	VACUUM WAVELENGTH (ANGSTRMS)	WAVENUMBER (KAYSERS)	LMENT	LOG ARC	INTENSITY SPK	INTENSITY DIS	REF
5340.804	5342.290	18718.57	CB	0.9	0.7		MIT
5336.812	5338.296	18732.57	CB	0.7	0.5		MIT
5334.870	5336.354	18739.39	CB	1.7	1.0		MIT
5331.19	5332.67	18752.3	CB		0.3H		ME
5326.351	5327.833	18769.36	CB	0.5	0.0		MIT
5323.360	5324.841	18779.90	CB	0.6	0.0		MIT
5321.330	5322.810	18787.07	CB	0.7	0.3		MIT
5319.492	5320.972	18793.56	CB	1.2	0.7		MIT
5318.602	5320.082	18796.70	CB	2.0	1.1		MIT
5317.806	5319.285	18799.52	CB	0.7	0.3		MIT
5317.013	5318.492	18802.32	CB	0.5	0.0H		MIT
5315.552	5317.031	18807.49	CB	1.0	0.7		MIT
5306.959	5308.436	18837.94	CB	0.6H	0.0		MIT
5306.351	5307.827	18840.10	CB	0.5	0.0		MIT
5298.981	5300.455	18866.30	CB	0.5	0.3H		MIT
5298.122	5299.596	18869.36	CB	0.5	0.3		MIT
5296.338	5297.812	18875.72	CB	0.7	0.3		MIT
5293.289	5294.762	18886.59	CB	0.6	0.0		MIT
5285.44	5286.91	18914.6	CB	0.5H			MIT
5285.258	5286.729	18915.29	CB	1.3	1.0		MIT
5279.433	5280.902	18936.16	CB	1.0	0.0		MIT
5276.195	5277.663	18947.78	CB	2.3	1.7		MIT
5272.483	5273.950	18961.12	CB	0.7	0.3		MIT
5271.532	5272.999	18964.54	CB	2.3	1.7		MIT
5269.920	5271.387	18970.34	CB	0.7	0.5		MIT
5260.850	5262.314	19003.05	CB	0.5	0.0		MIT
5260.131	5261.595	19005.64	CB	0.7	0.3		MIT
5253.928	5255.390	19028.08	CB	1.3	0.9		MIT
5253.031	5254.493	19031.33	CB	1.0	0.7		MIT
5251.847	5253.309	19035.62	CB	0.7	0.7		MIT
5251.743	5253.205	19036.00	CB	0.7H	0.7		MIT
5251.664	5253.126	19036.28	CB	1.0	1.3		MIT
5247.385	5248.846	19051.81	CB	1.5	0.3		MIT
5241.507	5242.966	19073.17	CB	0.7	0.3H		MIT
5240.391	5241.850	19077.24	CB	0.7	0.3		MIT
5237.48	5238.94	19087.8	CB	0.9	0.7		MIT
5237.370	5238.828	19088.24	CB	0.7	0.7		MIT
5236.836	5238.294	19090.19	CB	0.5	0.6		MIT
5235.101	5236.558	19096.51	CB	0.5	0.0		MIT
5232.809	5234.266	19104.88	CB	1.7	1.0		MIT
5229.374	5230.830	19117.43	CB	0.5	0.0		MIT
5225.155	5226.610	19132.86	CB	1.2	0.7		MIT
5219.095	5220.548	19155.08	CB	2.0	1.0		MIT
5218.465	5219.918	19157.39	CB	0.5	0.0		MIT
5211.234	5212.685	19183.97	CB	0.5	0.0		MIT
5205.130	5206.579	19206.47	CB	0.9	0.7		MIT
5204.033	5205.482	19210.52	CB	0.5	0.0		MIT
5203.224	5204.673	19213.50	CB	1.2	0.9		MIT
5195.837	5197.284	19240.82	CB	1.5	1.0		MIT
5193.448	5194.894	19249.67	CB	0.5	0.3		MIT
5193.075	5194.521	19251.05	CB	2.0	1.3		MIT
5189.197	5190.642	19265.44	CB	1.9	1.1		MIT
5186.984	5188.429	19273.66	CB	1.7	1.0		MIT
5183.82	5185.26	19285.4	CB	0.7H	0.0		MIT
5183.33	5184.77	19287.2	CB	0.7H	0.0		MIT
5180.306	5181.749	19298.50	CB	2.2	1.2		MIT
5178.216	5179.658	19306.29	CB	0.5	0.0		MIT
5174.205	5175.646	19321.26	CB	0.5	0.3		MIT
5164.377	5165.816	19358.03	CB	2.2	1.3		MIT
5160.333	5161.770	19373.20	CB	2.3	1.2		MIT
5158.037	5159.474	19381.82	CB	0.5	0.0		MIT
5157.487	5158.924	19383.89	CB	0.6	0.9		MIT
5153.031	5154.466	19400.65	CB	0.9	0.3		MIT
5152.629	5154.064	19402.16	CB	2.0	1.0		MIT
5150.635	5152.070	19409.68	CB	1.0	0.5		MIT
5147.974	5149.408	19419.71	CB	0.5H	0.0		MIT
5147.538	5148.972	19421.35	CB	1.5	0.7		MIT
5140.689	5142.121	19447.23	CB	0.5	0.3		MIT
5140.575	5142.007	19447.66	CB	1.0	0.7		MIT
5137.396	5138.827	19459.69	CB	0.7	0.0		MIT

AIR WAVELENGTH (ANGSTRMS)	VACUUM WAVELENGTH (ANGSTRMS)	WAVENUMBER (KAYSERS)	LMENT	LOG ARC	INTENSITY SPK	DIS	REF
5135.475	5136.906	19466.97	CB	0.5	0.0		MIT
5134.750	5136.181	19469.72	CB	2.3	1.2		MIT
5133.336	5134.766	19475.08	CB	1.0	0.5		MIT
5129.743	5131.172	19488.72	CB	0.5H	0.0		MIT
5127.660	5129.089	19496.64	CB	1.0	0.5		MIT
5124.695	5126.123	19507.92	CB	0.7	0.0		MIT
5121.800	5123.227	19518.95	CB	1.2	0.7		MIT
5120.298	5121.725	19524.67	CB	1.7	1.0		MIT
5118.064	5119.490	19533.20	CB	0.5	0.0		MIT
5116.740	5118.166	19538.25	CB	0.9	0.0		MIT
5110.912	5112.336	19560.53	CB	1.5	0.3		MIT
5102.38	5103.80	19593.2	CB	1.3	0.3		MIT
5100.162	5101.583	19601.76	CB	2.0	1.2		MIT
5097.763	5099.184	19610.98	CB	0.7	0.0		MIT
5095.598	5097.018	19619.31	CB	1.0	0.0		MIT
5095.299	5096.719	19620.47	CB	1.7	1.5		MIT
5095.124	5096.544	19621.14	CB	0.5H	0.5H		MIT
5094.490	5095.910	19623.58	CB	0.7	0.0		MIT
5094.408	5095.828	19623.90	CB	0.7	0.0		MIT
5094.339	5095.759	19624.16	CB	0.7	0.0		MIT
5088.822	5090.240	19645.44	CB	1.2	0.3		MIT
5086.832	5088.250	19653.12	CB	0.7H	0.0H		MIT
5084.842	5086.259	19660.81	CB	0.7	0.0		MIT
5078.964	5080.380	19683.57	CB	2.5	1.7		MIT
5077.395	5078.810	19689.65	CB	0.9	0.5		MIT
5075.974	5077.389	19695.16	CB	0.7	0.3		MIT
5072.560	5073.974	19708.42	CB	1.0	0.0		MIT
5071.668	5073.082	19711.88	CB	1.0			MIT
5065.254	5066.666	19736.84	CB	1.9	1.0		MIT
5064.455	5065.867	19739.96	CB	0.5	0.0		MIT
5059.352	5060.763	19759.87	CB	1.2	0.7		MIT
5058.007	5059.417	19765.12	CB	1.7	1.0		MIT
5055.626	5057.036	19774.43	CB	0.5	0.0		MIT
5054.667	5056.076	19778.18	CB	0.7	0.5		MIT
5047.957	5049.365	19804.47	CB	1.0	0.7		MIT
5046.755	5048.162	19809.19	CB	0.5H	0.0		MIT
5043.984	5045.390	19820.07	CB	0.5	0.5		MIT
5043.029	5044.435	19823.82	CB	0.6	0.3		MIT
5039.038	5040.443	19839.52	CB	2.3	1.5		MIT
5035.993	5037.397	19851.52	CB	1.0	0.7		MIT
5031.881	5033.284	19867.74	CB	0.7	0.0		MIT
5030.129	5031.532	19874.66	CB	0.7	0.3		MIT
5029.668	5031.071	19876.49	CB	0.5			MIT
5026.361	5027.763	19889.56	CB	1.7	0.9		MIT
5019.512	5020.912	19916.70	CB	1.0	0.5		MIT
5017.747	5019.146	19923.71	CB	1.9	1.0		MIT
5017.355	5018.754	19925.26	CB	0.7			MIT
5013.275	5014.673	19941.48	CB	1.0	0.6		MIT
5011.749	5013.147	19947.55	CB	0.5	0.3		MIT
5008.039	5009.436	19962.33	CB	0.7	0.0		MIT
5002.248	5003.643	19985.44	CB	1.0	0.7		MIT
5000.954	5002.349	19990.61	CB	1.5	0.7		MIT
5000.704	5002.099	19991.61	CB	0.5	0.0		MIT
4997.877	4999.271	20002.92	CB	1.2	0.7		MIT
4994.305	4995.698	20017.22	CB	0.7	0.5		MIT
4992.468	4993.861	20024.59	CB	1.0	0.3		MIT
4988.973	4990.365	20038.61	CB	2.2	1.0		MIT
4976.768	4978.157	20087.76	CB	0.7J			MIT
4975.137	4976.525	20094.34	CB	1.5	0.5		MIT
4973.144	4974.532	20102.40	CB	1.3	0.7		MIT
4971.926	4973.313	20107.32	CB	1.3	0.7		MIT
4967.782	4969.168	20124.09	CB	2.2	1.7		MIT
4965.373	4966.759	20133.86	CB	2.0	1.2		MIT
4963.195	4964.580	20142.69	CB	0.9	0.3		MIT
4962.946	4964.331	20143.70	CB	1.0C	0.0H		MIT
4957.393	4958.776	20166.26	CB	0.9	0.5		MIT
4956.588	4957.971	20169.54	CB	0.5	0.3		MIT
4953.129	4954.511	20183.62	CB	0.7	0.3		MIT
4945.442	4946.822	20215.00	CB	1.2	0.0		MIT
4941.921	4943.300	20229.40	CB	0.5H	0.0H		MIT

AIR WAVELENGTH (ANGSTRMS)	VACUUM WAVELENGTH (ANGSTRMS)	WAVENUMBER (KAYSERS)	LMENT	LOG ARC	INTENSITY SPK	DIS	REF
4941.516	4942.895	20231.06	CB	0.7	0.5		MIT
4936.706	4938.084	20250.77	CB	0.5H			MIT
4933.220	4934.597	20265.08	CB	0.7H	0.8W		MIT
4928.975	4930.351	20282.53	CB	0.7	0.7		MIT
4924.867	4926.242	20299.45	CB	0.5	0.3		MIT
4916.390	4917.763	20334.45	CB	1.0	0.3		MIT
4910.948	4912.319	20356.98	CB	1.2	1.0		MIT
4908.722	4910.093	20366.22	CB	0.5	0.3		MIT
4904.530	4905.899	20383.62	CB	1.0	0.9		MIT
4900.792	4902.160	20399.17	CB	0.7	0.7		MIT
4898.499	4899.867	20408.72	CB	0.5	0.0		MIT
4895.585	4896.952	20420.87	CB	0.7	0.0		MIT
4892.503	4893.869	20433.73	CB	0.7	0.3		MIT
4890.753	4892.119	20441.04	CB	1.2	0.9		MIT
4889.557	4890.922	20446.04	CB	0.5	0.3		MIT
4885.775	4887.139	20461.87	CB	0.5	0.0		MIT
4881.697	4883.060	20478.96	CB	0.3	0.0		MIT
4880.718	4882.081	20483.07	CB	0.7	0.0		MIT
4868.987	4870.347	20532.42	CB	1.0	1.0		MIT
4866.853	4868.212	20541.42	CB	1.2	0.0		MIT
4851.890	4853.245	20604.77	CB	0.5	0.0		MIT
4848.369	4849.724	20619.73	CB	2.2	2.0		MIT
4845.171	4846.525	20633.34	CB	0.9	0.7		MIT
4842.148	4843.501	20646.22	CB	1.0	0.7		MIT
4837.990	4839.342	20663.97	CB	1.2	0.0		MIT
4837.624	4838.976	20665.53	CB	0.7	0.3		MIT
4833.369	4834.720	20683.72	CB	1.0	1.0		MIT
4829.301	4830.650	20701.15	CB	1.0	0.7		MIT
4816.377	4817.723	20756.69	CB	1.7	1.7		MIT
4811.300	4812.645	20778.60	CB	1.2	0.3		MIT
4810.597	4811.941	20781.63	CB	2.0	1.0		MIT
4809.368	4810.712	20786.94	CB	0.7	0.9		MIT
4807.058	4808.402	20796.93	CB	0.7	0.7		MIT
4802.449	4803.791	20816.89	CB	0.5	0.3		MIT
4793.058	4794.398	20857.68	CB	0.5			MIT
4790.915	4792.254	20867.01	CB	0.5	0.3		MIT
4789.961	4791.300	20871.16	CB	0.7	1.2		MIT
4789.803	4791.142	20871.85	CB		0.5H		MIT
4787.735	4789.073	20880.87	CB	0.5H			MIT
4785.702	4787.040	20889.74	CB	0.5	0.3		MIT
4784.278	4785.616	20895.95	CB	0.5	0.0		MIT
4773.247	4774.582	20944.24	CB	1.0	1.0		MIT
4772.813	4774.147	20946.15	CB	0.5	0.0		MIT
4771.854	4773.188	20950.36	CB	0.5H	0.7H		MIT
4766.812	4768.145	20972.52	CB	0.7	1.0		MIT
4766.521	4767.854	20973.80	CB	0.7	0.5		MIT
4758.120	4759.451	21010.83	CB	0.3	0.5		MIT
4755.318	4756.648	21023.21	CB	0.7	0.5		MIT
4754.966	4756.296	21024.76	CB	0.7H	0.3		MIT
4753.481	4754.810	21031.33	CB	0.5H			MIT
4751.423	4752.752	21040.44	CB	0.5	0.7		MIT
4749.705	4751.033	21048.05	CB	2.0	1.7		MIT
4746.986	4748.314	21060.11	CB	0.5	0.0		MIT
4745.028	4746.355	21068.80	CB	0.5	0.3		MIT
4744.623	4745.950	21070.60	CB	1.0	0.9H		MIT
4743.838	4745.165	21074.08	CB	0.5			MIT
4740.614	4741.940	21088.42	CB	0.5	0.5		MIT
4736.491	4737.816	21106.77	CB	0.5	0.7		MIT
4735.334	4736.659	21111.93	CB	0.5H	0.7		MIT
4733.887	4735.211	21118.38	CB	1.5	1.5		MIT
4733.484	4734.808	21120.18	CB	0.9	0.7		MIT
4730.314	4731.637	21134.33	CB	0.7	0.7H		MIT
4727.328	4728.650	21147.68	CB	0.5	0.7		MIT
4726.783	4728.105	21150.12	CB	0.3H	0.5		MIT
4723.795	4725.116	21163.50	CB	0.7	0.7		MIT
4723.149	4724.470	21166.39	CB	0.0	0.5H		MIT
4718.024	4719.344	21189.39	CB	0.3	0.3H		ME
4717.52	4718.84	21191.6	CB	0.0	0.5		ME
4715.826	4717.145	21199.26	CB	0.5	1.7H		MIT
4713.500	4714.819	21209.72	CB	1.2	1.2		MIT

AIR WAVELENGTH (ANGSTRMS)	VACUUM WAVELENGTH (ANGSTRMS)	WAVENUMBER (KAYSERS)	LMENT	LOG ARC	INTENSITY SPK	DIS	REF
4713.046	4714.365	21211.77	CB	0.3	0.5		MIT
4711.876	4713.194	21217.03	CB	0.5H	0.5		MIT
4708.288	4709.605	21233.20	CB	1.7	1.5		MIT
4706.138	4707.455	21242.90	CB	1.7	1.7		MIT
4703.930	4705.246	21252.87	CB	0.5H	0.7H		MIT
4702.053	4703.369	21261.36	CB	0.5H	0.6		MIT
4700.910	4702.225	21266.53	CB	0.5H	0.7		MIT
4699.552	4700.867	21272.67	CB		1.2H		MIT
4697.752	4699.067	21280.82	CB	0.3H	0.5H		ME
4697.470	4698.785	21282.10	CB	0.9	0.7		MIT
4695.467	4696.781	21291.18	CB	0.7	0.9		MIT
4694.513	4695.827	21295.50	CB	0.7	0.9		MIT
4689.160	4690.472	21319.81	CB	0.5	0.3		MIT
4687.781	4689.093	21326.09	CB	0.7	0.7		MIT
4685.927	4687.238	21334.52	CB	0.3	0.3		MIT
4685.527	4686.838	21336.34	CB	0.3	0.0		MIT
4685.135	4686.446	21338.13	CB	1.2	1.3		MIT
4682.985	4684.296	21347.93	CB	0.3	0.3		MIT
4682.664	4683.975	21349.39	CB	0.3	0.5		MIT
4678.523	4679.833	21368.29	CB	0.7	0.7		MIT
4678.426	4679.735	21368.73	CB	0.7	0.7		MIT
4675.370	4676.679	21382.70	CB	1.7W	1.5W		MIT
4673.589	4674.897	21390.84	CB	0.3	0.7		MIT
4672.091	4673.399	21397.70	CB	2.2	2.0		MIT
4670.104	4671.411	21406.81	CB	0.0	0.6H		MIT
4669.871	4671.178	21407.87	CB	0.5	0.3		MIT
4667.223	4668.530	21420.02	CB	1.2	1.0		MIT
4666.240	4667.546	21424.53	CB	1.5	1.2		MIT
4665.330	4666.636	21428.71	CB	0.3H	0.3H		MIT
4663.831	4665.137	21435.60	CB	1.5	1.3		MIT
4660.41	4661.71	21451.3	CB		0.3H		ME
4658.185	4659.489	21461.58	CB	0.3	0.5		MIT
4655.323	4656.626	21474.77	CB		0.3		MIT
4652.182	4653.485	21489.27	CB		0.3H		MIT
4649.269	4650.571	21502.74	CB	0.5J	0.7J		MIT
4648.949	4650.251	21504.22	CB	1.7	1.3		MIT
4646.950	4648.251	21513.47	CB	0.7	1.0		MIT
4643.677	4644.977	21528.63	CB	0.5	0.5H		MIT
4643.311	4644.611	21530.33	CB	0.5	0.5		MIT
4638.172	4639.471	21554.18	CB		0.3		MIT
4638.103	4639.402	21554.50	CB	1.0H	1.0H		MIT
4637.571	4638.870	21556.98	CB		1.3H		MIT
4634.715	4636.013	21570.26	CB	0.3	0.5		ME
4630.112	4631.409	21591.70	CB	1.5	1.3		MIT
4627.475	4628.771	21604.01	CB	0.3	0.5H		MIT
4618.425	4619.719	21646.34	CB		0.5J		MIT
4616.167	4617.460	21656.93	CB	1.0	1.0		MIT
4614.741	4616.034	21663.62	CB	0.0	0.5H		MIT
4612.451	4613.743	21674.38	CB	0.3	0.5H		MIT
4612.122	4613.414	21675.92	CB	0.5	0.5		MIT
4612.041	4613.333	21676.30	CB	0.3H	0.9H		MIT
4610.686	4611.978	21682.67	CB	0.3	0.5		MIT
4608.583	4609.874	21692.57	CB	0.3	0.5H		MIT
4606.766	4608.057	21701.12	CB	1.7	1.7		MIT
4603.801	4605.091	21715.10	CB	0.3	0.5		MIT
4602.860	4604.149	21719.54	CB	0.3	0.7		MIT
4600.206	4601.495	21732.07	CB	1.0	1.2		MIT
4599.476	4600.765	21735.52	CB	0.5H	1.0		MIT
4593.785	4595.072	21762.44	CB		1.2H		MIT
4589.021	4590.307	21785.04	CB		0.9H		MIT
4584.848	4586.133	21804.86	CB	0.5	0.5H		MIT
4584.10	4585.38	21808.4	CB		0.5H		ME
4583.49	4584.77	21811.3	CB	0.3	0.5J		ME
4582.286	4583.570	21817.05	CB	0.7	0.7		MIT
4581.619	4582.903	21820.23	CB	1.5	1.7		MIT
4580.50	4581.78	21825.6	CB	0.0	0.3		ME
4579.453	4580.736	21830.55	CB	0.7H	1.5H		MIT
4575.369	4576.651	21850.04	CB	0.3	0.5		MIT
4574.840	4576.122	21852.56	CB	1.0	1.0		MIT
4574.329	4575.611	21855.00	CB	0.3	0.5		MIT

AIR WAVELENGTH (ANGSTRMS)	VACUUM WAVELENGTH (ANGSTRMS)	WAVENUMBER (KAYSERS)	LMENT	LOG ARC	INTENSITY SPK	DIS	REF
4573.075	4574.357	21861.00	CB	1.5	1.7		MIT
4570.948	4572.229	21871.17	CB	0.3	0.5		MIT
4569.484	4570.765	21878.18	CB	0.3	0.7H		MIT
4569.156	4570.437	21879.75	CB	1.0	0.7		MIT
4564.528	4565.807	21901.93	CB	1.3	1.5		MIT
4564.200	4565.479	21903.50	CB	0.3	0.5		MIT
4561.534	4562.813	21916.31	CB	0.5	0.3H		MIT
4559.420	4560.698	21926.47	CB	0.5H	0.7H		MIT
4556.835	4558.112	21938.91	CB	0.5	0.7H		MIT
4555.561	4556.838	21945.04	CB	0.5H	0.3H		MIT
4553.836	4555.113	21953.35	CB	0.7	0.9		MIT
4551.517	4552.793	21964.54	CB	0.5H	0.3		MIT
4551.035	4552.311	21966.87	CB	0.0	0.5H		MIT
4550.643	4551.919	21968.76	CB	0.0	0.5		MIT
4550.047	4551.323	21971.64	CB	0.5H	1.0H		MIT
4548.768	4550.043	21977.81	CB	0.0	0.5H		MIT
4547.851	4549.126	21982.25	CB	0.3	0.5		MIT
4546.822	4548.097	21987.22	CB	1.2	1.5		MIT
4542.799	4544.073	22006.69	CB	0.7	0.7		MIT
4540.084	4541.357	22019.85	CB	0.7	0.0		MIT
4537.591	4538.863	22031.95	CB	0.7	0.7H		MIT
4535.707	4536.979	22041.10	CB	0.0	0.5		MIT
4533.925	4535.196	22049.76	CB	0.0H	0.7H		MIT
4532.483	4533.754	22056.78	CB	0.0	0.7		MIT
4529.416	4530.686	22071.71	CB	0.3	0.5		MIT
4527.650	4528.920	22080.32	CB	0.0	1.5		MIT
4524.125	4525.394	22097.52	CB	0.7H	0.7H		MIT
4523.410	4524.678	22101.02	CB	1.5	1.5		MIT
4522.544	4523.812	22105.25	CB	0.0	0.5		MIT
4522.192	4523.460	22106.97	CB	0.0H	1.0H		MIT
4520.836	4522.104	22113.60	CB	0.0	0.5		MIT
4516.911	4518.178	22132.82	CB	0.0	0.5		MIT
4513.474	4514.740	22149.67	CB		0.5		MIT
4512.129	4513.395	22156.27	CB	0.3	0.5		MIT
4511.089	4512.354	22161.38	CB	0.7	1.2H		MIT
4508.408	4509.673	22174.56	CB	0.7	0.7		MIT
4503.418	4504.681	22199.13	CB	0.3	0.7		MIT
4503.043	4504.306	22200.98	CB	1.0	1.3		MIT
4499.802	4501.064	22216.97	CB	0.7	1.0		MIT
4495.462	4496.723	22238.42	CB		0.5		MIT
4494.568	4495.829	22242.84	CB	1.0	0.7		MIT
4492.961	4494.221	22250.79	CB	0.0	1.7H		MIT
4481.443	4482.700	22307.98	CB	0.3	0.5		MIT
4475.279	4476.535	22338.71	CB	0.3H	0.7H		MIT
4474.621	4475.877	22341.99	CB	0.7H			MIT
4472.533	4473.788	22352.42	CB	1.0	1.2		MIT
4471.287	4472.542	22358.65	CB	1.3	1.5		MIT
4469.713	4470.967	22366.52	CB	1.1	1.2		MIT
4469.322	4470.576	22368.48	CB	1.0H	0.7H		MIT
4467.910	4469.164	22375.55	CB		1.0H		MIT
4466.423	4467.676	22383.00	CB	0.3	0.3		MIT
4465.924	4467.177	22385.50	CB	0.3	0.3		MIT
4465.230	4466.483	22388.98	CB	0.3	0.3		MIT
4464.148	4465.401	22394.41	CB	0.7	1.0		MIT
4460.423	4461.675	22413.11	CB	0.7	1.0		MIT
4460.204	4461.456	22414.21	CB	0.9			MIT
4460.18	4461.43	22414.3	CB	0.3H	1.0		MIT
4458.118	4459.369	22424.70	CB	0.7	0.7		MIT
4457.424	4458.675	22428.19	CB	1.2	1.2		MIT
4456.800	4458.051	22431.33	CB	0.7	0.7		MIT
4449.908	4451.157	22466.07	CB	0.0H	1.3H		MIT
4448.760	4450.009	22471.87	CB	0.3	0.3		MIT
4447.722	4448.971	22477.11	CB		0.7H		MIT
4447.184	4448.432	22479.83	CB	1.2	1.3H		MIT
4446.168	4447.416	22484.97	CB	0.7	0.9		MIT
4445.851	4447.099	22486.57	CB	0.5	0.9		MIT
4443.297	4444.544	22499.49	CB	0.3	0.0		MIT
4443.014	4444.261	22500.93	CB	0.0	0.7		MIT
4441.808	4443.055	22507.04	CB	0.5	0.7		MIT
4441.591	4442.838	22508.14	CB	0.0	0.3H		MIT

AIR WAVELENGTH (ANGSTRMS)	VACUUM WAVELENGTH (ANGSTRMS)	WAVENUMBER (KAYSERS)	LMENT	LOG ARC	INTENSITY SPK	DIS	REF
4440.433	4441.680	22514.01	CB	0.0	0.7		MIT
4437.901	4439.147	22526.85	CB	0.0	0.7		MIT
4437.217	4438.463	22530.32	CB	1.6	1.7		MIT
4436.709	4437.955	22532.90	CB	0.0	0.3		MIT
4429.445	4430.689	22569.85	CB	1.0	1.0		MIT
4426.685	4427.928	22583.93	CB	0.9	0.9		MIT
4424.657	4425.899	22594.28	CB		0.7H		MIT
4423.871	4425.113	22598.29	CB	0.0	0.7		MIT
4421.656	4422.898	22609.61	CB	0.0H	1.0H		MIT
4420.635	4421.876	22614.83	CB	1.2	1.2		MIT
4420.452	4421.693	22615.77	CB	0.5	0.5		MIT
4419.834	4421.075	22618.93	CB	0.5	0.7		MIT
4419.444	4420.685	22620.93	CB	1.0	1.3		MIT
4416.879	4418.119	22634.06	CB	0.3W	0.3H		ME
4414.881	4416.121	22644.31	CB	0.6W	0.5W		MIT
4411.522	4412.761	22661.55	CB	0.7	0.9		MIT
4410.213	4411.452	22668.27	CB	1.2	1.5		MIT
4406.543	4407.781	22687.15	CB	0.5	0.7		MIT
4402.054	4403.290	22710.29	CB	0.0	0.9		MIT
4401.172	4402.408	22714.84	CB		1.0		MIT
4400.831	4402.067	22716.60	CB	0.3	0.5		MIT
4400.354	4401.590	22719.06	CB	0.7	1.0		MIT
4397.038	4398.273	22736.20	CB	0.3	0.5		MIT
4392.691	4393.925	22758.70	CB	0.7	1.0		MIT
4388.358	4389.591	22781.17	CB	1.0	1.2		MIT
4387.745	4388.978	22784.35	CB	0.5	0.7		MIT
4384.864	4386.096	22799.32	CB	0.3	0.7		MIT
4382.843	4384.074	22809.83	CB	0.5	0.7		MIT
4382.492	4383.723	22811.66	CB	0.5H	0.7H		MIT
4381.126	4382.357	22818.77	CB	0.5	1.3		MIT
4379.525	4380.756	22827.11	CB	0.3	0.5		MIT
4377.956	4379.186	22835.29	CB	1.0	1.5		MIT
4374.783	4376.012	22851.85	CB	0.5	0.7		MIT
4372.645	4373.874	22863.03	CB	0.3	1.3H		MIT
4370.359	4371.587	22874.99	CB	0.5	0.7		MIT
4368.432	4369.660	22885.08	CB	1.2	1.5		MIT
4367.967	4369.195	22887.51	CB	0.3	1.7H		MIT
4367.390	4368.617	22890.54	CB	0.3	1.0H		MIT
4361.653	4362.879	22920.65	CB	0.3	0.3		MIT
4359.854	4361.079	22930.10	CB	1.7	1.7		MIT
4353.272	4354.496	22964.77	CB	0.5H	0.7		MIT
4351.571	4352.794	22973.75	CB	1.0	1.3		MIT
4349.029	4350.252	22987.18	CB	1.0	1.0		MIT
4348.654	4349.876	22989.16	CB	1.2	1.3		MIT
4347.310	4348.532	22996.27	CB	0.6W	0.5		MIT
4346.115	4347.337	23002.59	CB	0.5	0.7		MIT
4345.518	4346.740	23005.75	CB	0.3	0.7		ME
4345.347	4346.569	23006.65	CB	0.7	0.9		MIT
4345.285	4346.507	23006.98	CB	0.7	0.9		MIT
4342.815	4344.036	23020.07	CB	0.7	1.0		MIT
4337.561	4338.780	23047.95	CB	1.0	1.5		ME
4331.374	4332.592	23080.87	CB	1.0	1.0		MIT
4329.732	4330.949	23089.62	CB	0.7	1.0		MIT
4329.476	4330.693	23090.99	CB		0.7		MIT
4328.430	4329.647	23096.57	CB	1.0	1.3		MIT
4327.381	4328.598	23102.17	CB	1.0	1.0		MIT
4326.327	4327.544	23107.80	CB	1.5	0.7		MIT
4321.492	4322.707	23133.65	CB		1.5		MIT
4318.006	4319.220	23152.33	CB	0.5	0.7		MIT
4317.714	4318.928	23153.89	CB		1.3		MIT
4316.481	4317.695	23160.51	CB	0.7	1.2		MIT
4313.877	4315.090	23174.49	CB	0.3	1.0		MIT
4312.453	4313.666	23182.14	CB	0.7	0.7		MIT
4311.700	4312.913	23186.19	CB	0.5	0.5		MIT
4311.391	4312.604	23187.85	CB	0.7	0.7		MIT
4311.265	4312.478	23188.53	CB	1.5	2.0		MIT
4309.561	4310.773	23197.70	CB	0.5	1.0		MIT
4308.691	4309.903	23202.38	CB	0.7	1.0		MIT
4308.120	4309.332	23205.45	CB	0.7	1.0		MIT
4306.284	4307.495	23215.35	CB	0.5	0.7		MIT

AIR WAVELENGTH (ANGSTRMS)	VACUUM WAVELENGTH (ANGSTRMS)	WAVENUMBER (KAYSERS)	LMENT	LOG ARC	INTENSITY SPK	DIS	REF
4304.687	4305.898	23223.96	CB	0.5	0.7		MIT
4303.881	4305.092	23228.31	CB	0.5	1.0		MIT
4302.906	4304.116	23233.57	CB	0.3	1.0H		MIT
4301.210	4302.420	23242.73	CB	0.0	0.3		MIT
4300.993	4302.203	23243.91	CB	1.5	1.5		MIT
4299.599	4300.809	23251.44	CB	1.3	1.5		MIT
4296.158	4297.367	23270.06	CB	1.0	1.5		MIT
4295.618	4296.826	23272.99	CB	0.5	0.7		MIT
4292.479	4293.687	23290.01	CB	1.5	1.7		MIT
4292.035	4293.243	23292.42	CB	1.2	1.3		MIT
4291.192	4292.399	23296.99	CB	1.0	1.2		MIT
4290.35	4291.56	23301.6	CB		0.3H		MIT
4289.444	4290.651	23306.49	CB	1.0	1.3		MIT
4286.990	4288.196	23319.83	CB	1.2	1.5		MIT
4286.215	4287.421	23324.04	CB	0.7	0.7		MIT
4280.598	4281.803	23354.65	CB	1.0	1.2		MIT
4279.499	4280.703	23360.65	CB	0.7	0.7		MIT
4277.496	4278.700	23371.59	CB	0.7	0.0		MIT
4274.892	4276.095	23385.82	CB	0.7	0.7H		MIT
4274.689	4275.892	23386.93	CB	0.7	0.5		MIT
4272.969	4274.172	23396.35	CB	0.3	0.6		MIT
4270.691	4271.893	23408.83	CB	1.5	1.7		MIT
4268.668	4269.869	23419.92	CB	0.7	1.0		MIT
4267.635	4268.836	23425.59	CB		1.0H		MIT
4266.018	4267.219	23434.47	CB	1.2	1.3		MIT
4262.053	4263.253	23456.27	CB	1.3	1.2		MIT
4261.714	4262.914	23458.14	CB	0.7	0.9		MIT
4258.913	4260.112	23473.56	CB	0.7	1.0		MIT
4255.944	4257.142	23489.94	CB	0.5	1.0		MIT
4255.442	4256.640	23492.71	CB	1.5	1.7		MIT
4254.694	4255.892	23496.84	CB	1.0	1.0		MIT
4254.392	4255.590	23498.51	CB	1.0	1.2		ME
4253.695	4254.892	23502.36	CB	1.4	1.6		MIT
4252.972	4254.169	23506.35	CB	1.5	1.7		MIT
4249.457	4250.653	23525.79	CB	0.7	0.9		MIT
4248.658	4249.854	23530.22	CB	0.5H	0.7		MIT
4247.690	4248.886	23535.58	CB	0.3	0.7H		MIT
4246.295	4247.490	23543.31	CB	0.9	1.0		MIT
4242.632	4243.827	23563.64	CB	1.0	1.3		MIT
4241.449	4242.643	23570.21	CB	0.5	0.7		MIT
4237.812	4239.005	23590.44	CB	0.5	0.7H		MIT
4236.996	4238.189	23594.98	CB	0.0	0.7H		MIT
4231.952	4233.144	23623.10	CB	0.9	1.0		MIT
4230.316	4231.507	23632.24	CB	0.5	0.7		MIT
4229.831	4231.022	23634.95	CB	0.7	0.7		MIT
4229.149	4230.340	23638.76	CB	1.7	2.0		MIT
4226.247	4227.437	23654.99	CB	0.6	0.5		MIT
4226.209	4227.399	23655.21	CB	0.5W	0.7		MIT
4220.586	4221.775	23686.72	CB	0.3	0.9		MIT
4218.520	4219.708	23698.32	CB		1.0H		MIT
4217.945	4219.133	23701.55	CB	1.7	1.7		MIT
4216.228	4217.416	23711.20	CB	0.0	1.0		ME
4214.817	4216.004	23719.14	CB	0.7	1.0		MIT
4214.733	4215.920	23719.61	CB	1.6			MIT
4214.672	4215.859	23719.96	CB		2.0		MIT
4212.534	4213.721	23732.00	CB	0.3	0.5		MIT
4212.040	4213.227	23734.78	CB	0.6	0.5		MIT
4208.160	4209.345	23756.66	CB	0.6	0.6		MIT
4206.132	4207.317	23768.12	CB	0.6	0.5		MIT
4205.311	4206.496	23772.76	CB	1.2	1.2		MIT
4204.321	4205.505	23778.35	CB	0.6	0.5		MIT
4203.413	4204.597	23783.49	CB	0.6	0.5		MIT
4201.519	4202.703	23794.21	CB	1.0	1.0		MIT
4198.847	4200.030	23809.35	CB	0.6	0.5		MIT
4198.510	4199.693	23811.26	CB	0.5	0.7		MIT
4198.367	4199.550	23812.08	CB	0.3	0.7H		MIT
4197.613	4198.796	23816.35	CB	0.3	0.7		MIT
4196.950	4198.133	23820.11	CB	0.5	0.7		MIT
4195.657	4196.839	23827.46	CB	0.7	0.5		MIT
4195.094	4196.276	23830.65	CB	1.3	1.3		MIT

AIR WAVELENGTH (ANGSTRMS)	VACUUM WAVELENGTH (ANGSTRMS)	WAVENUMBER (KAYSERS)	LMENT	LOG ARC	INTENSITY SPK	DIS	REF
4193.828	4195.010	23837.85	CB	0.7			ME
4193.800	4194.982	23838.01	CB		1.2		MIT
4192.070	4193.251	23847.84	CB	1.3	1.3		MIT
4190.884	4192.065	23854.59	CB	1.3	1.5		MIT
4190.650	4191.831	23855.92	CB	0.7	1.0		MIT
4189.990	4191.171	23859.68	CB	0.7	1.0		MIT
4189.587	4190.768	23861.98	CB	0.3	0.7H		MIT
4186.105	4187.285	23881.82	CB	0.7	0.9		MIT
4185.67	4186.85	23884.3	CB	0.3	0.3H		MIT
4185.541	4186.721	23885.04	CB	0.0	1.5		MIT
4184.444	4185.623	23891.30	CB	1.3	1.7		MIT
4183.385	4184.564	23897.35	CB		1.0H		MIT
4181.512	4182.690	23908.06	CB		0.3		MIT
4181.340	4182.518	23909.04	CB	0.7	0.7		MIT
4179.755	4180.933	23918.11	CB	1.0	1.3		MIT
4178.406	4179.584	23925.83	CB	0.0	1.0		MIT
4177.867	4179.045	23928.92	CB		0.7		MIT
4176.169	4177.346	23938.64	CB	0.3			MIT
4174.342	4175.519	23949.12	CB	0.5	0.7		MIT
4173.953	4175.130	23951.35	CB	0.7	1.2		MIT
4169.566	4170.741	23976.55	CB	1.0	1.0		MIT
4168.126	4169.301	23984.84	CB	2.0	1.9		MIT
4165.847	4167.021	23997.96	CB	0.5	0.7		MIT
4164.660	4165.834	24004.80	CB	1.5	1.7		MIT
4163.657	4164.831	24010.58	CB	1.8	1.6		MIT
4163.472	4164.646	24011.65	CB	0.5	0.7		MIT
4162.813	4163.987	24015.45	CB	0.5	0.7		MIT
4161.251	4162.424	24024.46	CB	0.7	1.2		MIT
4158.010	4159.182	24043.19	CB	0.5	0.7		MIT
4156.682	4157.854	24050.87	CB	0.7	2.0		MIT
4152.576	4153.747	24074.65	CB	2.0	2.5		MIT
4152.035	4153.206	24077.79	CB	1.0	0.9		MIT
4150.122	4151.292	24088.89	CB	1.2	1.3		MIT
4147.190	4148.359	24105.92	CB	0.9	1.0		MIT
4146.003	4147.172	24112.82	CB	0.6	0.5		MIT
4145.155	4146.324	24117.75	CB	0.6	0.3		ME
4143.931	4145.100	24124.87	CB	0.7H	0.7H		ME
4143.206	4144.374	24129.09	CB	1.0	1.0		MIT
4142.245	4143.413	24134.69	CB	0.6	0.5		MIT
4139.707	4140.875	24149.49	CB	1.7	1.7		MIT
4139.436	4140.603	24151.07	CB	1.0	1.0		MIT
4138.300	4139.467	24157.70	CB	0.6	0.3		MIT
4137.593	4138.760	24161.83	CB	0.5	0.7		MIT
4137.095	4138.262	24164.74	CB I	2.0	1.8		MIT
4135.424	4136.590	24174.50	CB	0.6	0.3		MIT
4134.591	4135.757	24179.37	CB	1.0	1.0		MIT
4133.418	4134.584	24186.23	CB	0.3	0.7H		MIT
4131.527	4132.692	24197.30	CB	0.3	1.0		MIT
4129.927	4131.092	24206.68	CB	1.2	1.5		MIT
4129.426	4130.591	24209.61	CB	1.2	1.3		MIT
4127.454	4128.618	24221.18	CB	0.6	1.0		MIT
4126.902	4128.066	24224.42	CB	0.5	1.0		MIT
4126.185	4127.349	24228.63	CB	0.0	1.0		MIT
4125.575	4126.739	24232.21	CB	0.5	0.7		MIT
4125.248	4126.412	24234.13	CB	1.0	1.3		MIT
4123.810	4124.973	24242.58	CB I	2.3	2.1		MIT
4122.808	4123.971	24248.47	CB	0.7	1.0		MIT
4119.724	4120.886	24266.62	CB	0.0H	1.3		MIT
4119.283	4120.445	24269.22	CB	0.3	2.3		MIT
4116.896	4118.058	24283.29	CB	1.0	1.0		MIT
4115.597	4116.758	24290.96	CB		0.7		MIT
4114.575	4115.736	24296.99	CB		0.7		MIT
4113.944	4115.105	24300.72	CB	1.0	1.0		MIT
4112.481	4113.641	24309.36	CB	0.3			MIT
4112.127	4113.287	24311.46	CB	0.7	1.2		MIT
4110.832	4111.992	24319.11	CB		0.7H		MIT
4110.295	4111.455	24322.29	CB	0.0	1.0		MIT
4108.702	4109.861	24331.72	CB	0.3	0.0H		MIT
4106.175	4107.334	24346.70	CB	0.6	0.5		MIT
4104.169	4105.327	24358.59	CB	0.5	1.5		MIT

AIR WAVELENGTH (ANGSTRMS)	VACUUM WAVELENGTH (ANGSTRMS)	WAVENUMBER (KAYSERS)	LMENT	LOG ARC	INTENSITY SPK	DIS	REF
4100.923	4102.080	24377.87	CB I	2.5W	2.3W		MIT
4100.398	4101.555	24381.00	CB	1.2	1.3		MIT
4099.072	4100.229	24388.88	CB	0.7	0.7		MIT
4098.218	4099.375	24393.97	CB	0.7	0.7		MIT
4097.642	4098.799	24397.39	CB	0.6	0.5H		MIT
4095.936	4097.092	24407.55	CB	0.5	1.0		M
4095.556	4096.712	24409.82	CB	0.6	0.5		MIT
4095.093	4096.249	24412.58	CB	0.6	0.5H		MIT
4090.164	4091.319	24442.00	CB	0.7	0.7		MIT
4089.399	4090.553	24446.57	CB		0.7		MIT
4086.632	4087.786	24463.12	CB	0.5	0.3		MIT
4085.332	4086.485	24470.91	CB	0.0	0.7H		MIT
4084.860	4086.013	24473.73	CB	1.0	1.0		MIT
4084.177	4085.330	24477.83	CB	0.5	0.5		MIT
4079.729	4080.881	24504.51	CB I	2.7W	2.3W		MIT
4079.135	4080.287	24508.08	CB	0.0H	0.5W		MIT
4078.352	4079.503	24512.79	CB	0.6	0.5		MIT
4077.088	4078.239	24520.39	CB	0.5	0.7		MIT
4076.093	4077.244	24526.37	CB	0.6	0.5		MIT
4073.086	4074.236	24544.48	CB	0.3	0.9		MIT
4072.068	4073.218	24550.61	CB		1.2		MIT
4070.964	4072.114	24557.27	CB	0.5	0.7		MIT
4068.256	4069.405	24573.62	CB	1.0R	1.2R		MIT
4067.159	4068.308	24580.25	CB	0.5	0.7		MIT
4066.124	4067.272	24586.50	CB	0.5	0.7		MIT
4064.810	4065.958	24594.45	CB	0.6	0.5		MIT
4063.732	4064.880	24600.97	CB		0.7		MIT
4061.977	4063.124	24611.60	CB	0.3	1.0		MIT
4061.263	4062.410	24615.93	CB	0.6	0.5		MIT
4060.787	4061.934	24618.81	CB	1.0	1.0O		MIT
4060.313	4061.460	24621.69	CB	0.7	0.7		MIT
4059.506	4060.653	24626.58	CB	0.7	0.3		MIT
4058.938	4060.084	24630.03	CB I	3.0W	2.6W		MIT
4056.941	4058.087	24642.15	CB	0.5	0.7		MIT
4051.515	4052.659	24675.16	CB	0.7	0.7		MIT
4049.760	4050.904	24685.85	CB	0.7	0.7		MIT
4048.661	4049.805	24692.55	CB	0.5	0.7		MIT
4046.27	4047.41	24707.1	CB	0.5	0.0		ME
4045.590	4046.733	24711.29	CB	0.0	0.7H		MIT
4044.712	4045.855	24716.66	CB	0.7	0.5		MIT
4044.105	4045.247	24720.37	CB	0.7	1.0		MIT
4043.162	4044.304	24726.13	CB	0.5	1.0		MIT
4042.570	4043.712	24729.75	CB	0.5	0.7		MIT
4041.529	4042.671	24736.12	CB	0.3			MIT
4040.468	4041.610	24742.62	CB	0.3	0.7H		ME
4039.529	4040.670	24748.37	CB	1.5	1.7		MIT
4039.095	4040.236	24751.03	CB	0.7	1.0		MIT
4038.179	4039.320	24756.64	CB	0.5	0.7H		MIT
4037.659	4038.800	24759.83	CB		1.3		MIT
4035.928	4037.068	24770.45	CB	0.5	0.7		MIT
4035.098	4036.238	24775.54	CB	0.6	0.5		MIT
4034.523	4035.663	24779.08	CB	1.0	0.7		M
4033.203	4034.343	24787.19	CB	0.7	0.7		MIT
4032.524	4033.663	24791.36	CB	1.5	1.7		MIT
4031.314	4032.453	24798.80	CB		0.7		MIT
4027.976	4029.114	24819.35	CB	0.7	1.0		MIT
4027.312	4028.450	24823.44	CB	0.7	0.7		MIT
4026.325	4027.463	24829.53	CB		1.0		MIT
4023.141	4024.278	24849.18	CB	0.6	0.5		MIT
4022.388	4023.525	24853.83	CB	0.3	1.0		MIT
4020.237	4021.373	24867.13	CB	0.3	1.0H		MIT
4017.559	4018.695	24883.70	CB	0.5	1.0		MIT
4016.079	4017.214	24892.87	CB	0.7	1.5		MIT
4014.926	4016.061	24900.02	CB	0.7	0.9		MIT
4013.270	4014.404	24910.29	CB	0.5	0.7		MIT
4012.168	4013.302	24917.14	CB		2.0		MIT
4009.714	4010.848	24932.39	CB	0.7	1.0		MIT
4008.280	4009.413	24941.31	CB	0.7	1.0		MIT
4006.995	4008.128	24949.30	CB		0.7		MIT
4005.927	4007.060	24955.96	CB	0.3	0.7H		MIT

AIR WAVELENGTH (ANGSTRMS)	VACUUM WAVELENGTH (ANGSTRMS)	WAVENUMBER (KAYSERS)	LMENT	LOG ARC	INTENSITY SPK	DIS	REF
4002.262	4003.394	24978.81	CB	0.3	0.7H		MIT
4001.130	4002.261	24985.87	CB	1.0	1.2		MIT
4000.605	4001.736	24989.15	CB	0.3	1.7		MIT
3999.709	4000.840	24994.75	CB	0.0	0.7		MIT
3999.182	4000.313	24998.04	CB	0.7	1.0		MIT
3998.444	3999.575	25002.66	CB	0.3	0.7		MIT
3994.461	3995.591	25027.59	CB		0.3H		MIT
3994.426	3995.556	25027.81	CB	0.6			MIT
3991.678	3992.807	25045.04	CB	1.2	1.3		MIT
3990.667	3991.796	25051.38	CB	0.3	0.7		MIT
3988.157	3989.285	25067.15	CB	0.7	1.0		MIT
3986.176	3987.303	25079.61	CB	0.3	0.7		MIT
3984.814	3985.941	25088.18	CB	0.3	0.7		MIT
3983.939	3985.066	25093.69	CB		0.7H		ME
3982.59	3983.72	25102.2	CB	0.0H	0.3		MIT
3982.056	3983.182	25105.55	CB	0.8	0.7		MIT
3980.484	3981.610	25115.47	CB	1.0	1.2		MIT
3979.372	3980.498	25122.49	CB	0.7	0.7		MIT
3978.753	3979.878	25126.40	CB	0.7	0.7		MIT
3977.941	3979.066	25131.52	CB	1.0	1.2		MIT
3976.674	3977.799	25139.53	CB	1.0	1.0		MIT
3976.510	3977.635	25140.57	CB		1.9H		MIT
3973.870	3974.994	25157.27	CB	0.3D	0.0H		MIT
3973.624	3974.748	25158.83	CB	0.7	1.0		MIT
3972.523	3973.647	25165.80	CB	1.3	1.2		MIT
3971.925	3973.049	25169.59	CB	1.2	0.7		MIT
3971.852	3972.976	25170.05	CB	0.7	0.7		MIT
3971.695	3972.819	25171.05	CB		1.2H		MIT
3970.652	3971.775	25177.66	CB	0.5	1.0S		MIT
3969.135	3970.258	25187.28	CB		1.3H		MIT
3968.471	3969.594	25191.50	CB	0.5	1.0		MIT
3967.374	3968.496	25198.46	CB		1.7H		MIT
3966.250	3967.372	25205.60	CB	1.0	1.5		MIT
3966.092	3967.214	25206.60	CB	1.2	1.0		MIT
3965.690	3966.812	25209.16	CB	1.0	1.2		MIT
3964.279	3965.401	25218.13	CB	0.0	1.7		MIT
3963.582	3964.703	25222.57	CB	0.7H			MIT
3962.163	3963.284	25231.60	CB		1.0H		MIT
3962.153	3963.274	25231.66	CB	0.5			MIT
3961.62	3962.74	25235.1	CB		1.0H		ME
3960.984	3962.105	25239.11	CB	0.9	0.7		MIT
3959.359	3960.479	25249.47	CB	1.0	1.0		MIT
3958.135	3959.255	25257.28	CB	0.5	1.3		MIT
3956.624	3957.744	25266.92	CB	0.3	0.7		MIT
3956.40	3957.52	25268.4	CB	0.3W	0.0		ME
3955.881	3957.000	25271.67	CB	0.3	1.3		MIT
3955.681	3956.800	25272.95	CB	0.7	0.5		MIT
3953.509	3954.628	25286.83	CB	0.0H	0.7H		MIT
3953.078	3954.197	25289.59	CB	0.5	0.7		MIT
3953.015	3954.134	25289.99	CB	0.6			ME
3952.370	3953.489	25294.12	CB	0.5	1.7		MIT
3949.935	3951.053	25309.71	CB	0.6	0.5		MIT
3949.455	3950.573	25312.78	CB	0.0	1.7		MIT
3949.328	3950.446	25313.60	CB	0.5	0.3H		MIT
3947.528	3948.645	25325.14	CB	1.0	1.0		MIT
3943.666	3944.782	25349.94	CB	1.3	1.7		MIT
3941.273	3942.389	25365.33	CB	1.0	1.2		MIT
3938.547	3939.662	25382.89	CB		2.0H		ME
3937.965	3939.080	25386.64	CB	0.7	1.0		MIT
3937.438	3938.553	25390.04	CB	1.5	1.5		MIT
3936.447	3937.561	25396.43	CB	0.5	0.7		MIT
3936.023	3937.137	25399.17	CB	0.7	2.3		MIT
3935.448	3936.562	25402.88	CB	0.7	1.0		MIT
3934.813	3935.927	25406.98	CB	0.0H	0.7H		MIT
3934.408	3935.522	25409.59	CB	0.7	0.7		MIT
3934.139	3935.253	25411.33	CB	0.7	1.2		MIT
3933.394	3934.508	25416.14	CB	0.5	0.5		MIT
3933.013	3934.126	25418.60	CB	0.5	0.5		MIT
3931.794	3932.907	25426.48	CB	0.0	1.3H		MIT
3931.460	3932.573	25428.64	CB	0.6	0.5		MIT

AIR WAVELENGTH (ANGSTRMS)	VACUUM WAVELENGTH (ANGSTRMS)	WAVENUMBER (KAYSERS)	LMENT	LOG ARC	INTENSITY SPK	DIS	REF
3930.022	3931.135	25437.95	CB		1.0H		ME
3929.293	3930.406	25442.67	CB	1.2	1.2		MIT
3926.614	3927.726	25460.03	CB	0.7	0.7		MIT
3924.996	3926.107	25470.52	CB	1.0	1.2		MIT
3924.488	3925.599	25473.82	CB	0.7	1.0		MIT
3922.351	3923.462	25487.70	CB	0.7	1.0		MIT
3921.674	3922.785	25492.10	CB	0.5	0.3		MIT
3921.346	3922.456	25494.23	CB		1.0H		MIT
3920.758	3921.868	25498.05	CB	0.0	1.7H		MIT
3920.198	3921.308	25501.69	CB	1.5	2.0		MIT
3919.720	3920.830	25504.80	CB	0.3	2.0		MIT
3919.165	3920.275	25508.42	CB	0.7	0.7		MIT
3919.000	3920.110	25509.49	CB	0.7	0.7		MIT
3914.701	3915.810	25537.50	CB	1.5	2.0		MIT
3913.14	3914.25	25547.7	CB	0.7	0.7		M
3913.013	3914.121	25548.52	CB	0.7	1.0		MIT
3909.595	3910.702	25570.85	CB	0.9L	1.0W		MIT
3908.973	3910.080	25574.92	CB	1.0	0.9		MIT
3908.594	3909.701	25577.40	CB	0.5	0.7		MIT
3906.906	3908.013	25588.45	CB	0.7	0.7		MIT
3904.182	3905.288	25606.31	CB	1.0	1.3		MIT
3900.823	3901.928	25628.35	CB	0.3	0.3		MIT
3900.528	3901.633	25630.29	CB	1.5	1.0		MIT
3899.246	3900.351	25638.72	CB	1.0	1.2		MIT
3898.557	3899.662	25643.25	CB	0.7	0.7		MIT
3898.285	3899.389	25645.04	CB	0.5	2.3		MIT
3895.900	3897.004	25660.74	CB	1.0	1.5		MIT
3894.700	3895.803	25668.65	CB	0.5	0.5		MIT
3894.571	3895.674	25669.50	CB		0.7H		MIT
3894.034	3895.137	25673.04	CB	1.2	1.5		MIT
3893.732	3894.835	25675.03	CB	1.2	1.0		MIT
3891.300	3892.403	25691.07	CB	1.7	2.0		MIT
3890.749	3891.851	25694.71	CB	0.3	0.5		MIT
3889.634	3890.736	25702.08	CB	0.5	0.7		MIT
3887.318	3888.420	25717.39	CB		0.7		MIT
3886.672	3887.773	25721.66	CB	0.3H	0.7J		MIT
3886.072	3887.173	25725.64	CB	1.0	1.0		MIT
3885.682	3886.783	25728.22	CB	1.2	1.5		MIT
3885.441	3886.542	25729.81	CB	1.7	2.0		MIT
3883.137	3884.237	25745.08	CB	1.5	1.5		MIT
3882.679	3883.779	25748.12	CB	0.3	0.5		MIT
3879.642	3880.742	25768.27	CB	0.3	0.3		MIT
3879.346	3880.445	25770.24	CB	0.7	2.5		MIT
3878.969	3880.068	25772.74	CB	0.7	0.5		MIT
3878.822	3879.921	25773.72	CB	0.7	0.7		MIT
3877.558	3878.657	25782.12	CB	1.7	1.3		MIT
3876.965	3878.064	25786.06	CB	0.7	0.7		MIT
3875.763	3876.862	25794.06	CB	1.0	1.7		MIT
3875.697	3876.796	25794.50	CB	0.5H			ME
3875.418	3876.516	25796.36	CB	0.7	1.0		MIT
3873.277	3874.375	25810.61	CB	0.3	0.5		MIT
3871.186	3872.283	25824.56	CB	1.2	1.3		MIT
3870.661	3871.758	25828.06	CB	0.3	0.5		MIT
3868.826	3869.923	25840.31	CB	0.6	0.6		MIT
3868.573	3869.670	25842.00	CB	0.5	0.7		MIT
3867.917	3869.014	25846.38	CB	1.5	1.3		MIT
3865.039	3866.135	25865.63	CB	1.0D			MIT
3865.024	3866.120	25865.73	CB		2.3H		MIT
3864.358	3865.454	25870.19	CB	0.5	0.7		MIT
3863.78	3864.88	25874.1	CB	0.3			ME
3863.384	3864.479	25876.71	CB	1.2	1.0		MIT
3863.052	3864.147	25878.93	CB	0.5	1.3		MIT
3862.931	3864.026	25879.74	CB	1.0	0.7		MIT
3860.857	3861.952	25893.64	CB	0.7	1.0		MIT
3858.950	3860.044	25906.44	CB	1.3	1.7		MIT
3856.678	3857.772	25921.70	CB	1.3	0.7		MIT
3855.500	3856.593	25929.62	CB		1.7H		ME
3855.453	3856.546	25929.94	CB	1.0			MIT
3855.146	3856.239	25932.00	CB	0.7	0.7		MIT
3854.696	3855.789	25935.03	CB	0.5	0.6		MIT

AIR WAVELENGTH (ANGSTRMS)	VACUUM WAVELENGTH (ANGSTRMS)	WAVENUMBER (KAYSERS)	LMENT	LOG INTENSITY ARC	SPK	DIS	REF
3854.123	3855.216	25938.88	CB	0.3	0.5		MIT
3853.59	3854.68	25942.5	CB	0.5W	0.0		ME
3853.382	3854.475	25943.87	CB	1.0	1.3		MIT
3853.095	3854.188	25945.80	CB	0.0	0.5		MIT
3852.615	3853.708	25949.04	CB	0.0	0.9H		MIT
3849.746	3850.838	25968.37	CB	0.3	0.3		MIT
3845.897	3846.988	25994.36	CB	1.0	1.5		MIT
3844.085	3845.175	26006.62	CB	0.6	0.5		MIT
3843.926	3845.016	26007.69	CB	0.7	0.5		MIT
3843.453	3844.543	26010.89	CB	0.3	0.3		MIT
3842.707	3843.797	26015.94	CB	0.7	0.7		MIT
3841.810	3842.900	26022.02	CB	1.0	1.0		MIT
3841.691	3842.781	26022.82	CB		1.0		MIT
3837.080	3838.168	26054.09	CB	0.7	1.0		MIT
3836.742	3837.830	26056.39	CB	0.6W	1.2		MIT
3836.448	3837.536	26058.39	CB	0.7	0.5		MIT
3835.175	3836.263	26067.03	CB	1.3	1.3		MIT
3833.936	3835.024	26075.46	CB	0.5	0.5		MIT
3833.783	3834.871	26076.50	CB	0.3	0.7		MIT
3833.261	3834.349	26080.05	CB	0.7W	1.0		MIT
3831.838	3832.925	26089.74	CB	0.7	2.5		MIT
3831.201	3832.288	26094.07	CB	0.3	0.3		MIT
3830.641	3831.728	26097.89	CB	0.0H	1.3W		MIT
3830.002	3831.089	26102.24	CB	0.7	0.7		MIT
3829.659	3830.746	26104.58	CB	0.5H	0.5H		MIT
3829.211	3830.297	26107.63	CB		1.0H		MIT
3828.238	3829.324	26114.27	CB	0.7	1.2		MIT
3827.012	3828.098	26122.63	CB	0.7	0.7		MIT
3825.409	3826.494	26133.58	CB	0.0	0.3		MIT
3824.876	3825.961	26137.22	CB	1.5	1.7		MIT
3823.995	3825.080	26143.24	CB		1.0J		MIT
3823.135	3824.220	26149.12	CB	0.3	0.5		MIT
3822.954	3824.039	26150.36	CB	0.3	0.5		MIT
3821.191	3822.275	26162.43	CB	1.0O	1.0O		MIT
3820.744	3821.828	26165.49	CB	0.3			ME
3819.147	3820.231	26176.43	CB	1.1	1.0		MIT
3818.858	3819.942	26178.41	CB	0.9	2.5		MIT
3818.204	3819.288	26182.89	CB	0.7W	0.7H		MIT
3816.634	3817.717	26193.66	CB	0.3	0.7J		MIT
3816.341	3817.424	26195.68	CB	0.7	0.5		MIT
3815.511	3816.594	26201.37	CB	1.3	1.5		MIT
3813.472	3814.554	26215.38	CB	0.5	0.7H		MIT
3811.428	3812.510	26229.44	CB		0.7H		MIT
3811.032	3812.114	26232.17	CB	1.0	1.1		MIT
3810.491	3811.573	26235.89	CB	1.5	1.7		MIT
3806.633	3807.714	26262.48	CB	0.5H	1.0H		MIT
3806.196	3807.276	26265.50	CB	0.9	1.0		MIT
3804.736	3805.816	26275.57	CB	1.3	1.7		MIT
3804.204	3805.284	26279.25	CB	0.7	0.5H		ME
3803.882	3804.962	26281.47	CB	1.5	1.3		MIT
3802.923	3804.003	26288.10	CB	1.7	1.7		MIT
3802.636	3803.716	26290.08	CB	0.5	0.3		ME
3802.555	3803.635	26290.64	CB	-0.3	0.0		ME
3801.298	3802.377	26299.34	CB	0.7	0.9		MIT
3801.154	3802.233	26300.33	CB	0.5	1.3		MIT
3800.941	3802.020	26301.81	CB	1.0	0.9		MIT
3800.197	3801.276	26306.96	CB	0.5	0.5		ME
3799.488	3800.567	26311.86	CB	0.7	0.3		MIT
3798.121	3799.199	26321.33	CB	1.7	1.9		MIT
3796.849	3797.927	26330.15	CB	1.0W	1.2W		MIT
3796.591	3797.669	26331.94	CB	0.7	0.7		MIT
3796.440	3797.518	26332.99	CB	1.2	1.2		MIT
3795.542	3796.620	26339.22	CB	1.0	1.3H		MIT
3794.471	3795.548	26346.65	CB	0.7	0.7		MIT
3793.73	3794.81	26351.8	CB	0.6W	0.3		MIT
3792.79	3793.87	26358.3	CB		1.3H		ME
3792.015	3793.092	26363.72	CB	0.7O	0.5O		MIT
3791.447	3792.524	26367.67	CB	0.5	0.5		MIT
3791.207	3792.284	26369.34	CB	1.9	1.9		MIT
3790.152	3791.228	26376.68	CB	2.0R	1.7R		MIT

AIR WAVELENGTH (ANGSTRMS)	VACUUM WAVELENGTH (ANGSTRMS)	WAVENUMBER (KAYSERS)	LMENT	LOG INTENSITY ARC	SPK	DIS	REF
3789.495	3790.571	26381.25	CB	1.0	1.0		MIT
3788.684	3789.760	26386.90	CB	1.0O	0.7H		MIT
3787.483	3788.559	26395.26	CB	0.7	0.7		MIT
3787.280	3788.356	26396.68	CB	0.6H	0.5H		ME
3787.062	3788.138	26398.20	CB	1.5	1.5		MIT
3786.222	3787.297	26404.05	CB	1.0	0.9		MIT
3786.100	3787.175	26404.90	CB	0.5H	0.5J		MIT
3784.881	3785.956	26413.41	CB	0.0	1.0		MIT
3783.839	3784.914	26420.68	CB	1.2	1.3		MIT
3781.379	3782.453	26437.87	CB	0.7	2.3		MIT
3781.014	3782.088	26440.42	CB	1.3	1.3		MIT
3779.580	3780.654	26450.45	CB	0.3	1.3		MIT
3779.227	3780.300	26452.92	CB	0.3	0.3H		MIT
3778.676	3779.749	26456.78	CB	0.5	0.7H		MIT
3778.512	3779.585	26457.93	CB		0.7J		MIT
3777.673	3778.746	26463.80	CB	1.3	1.5		MIT
3776.601	3777.674	26471.32	CB	0.5	0.5		MIT
3776.161	3777.234	26474.40	CB	0.3	0.7		MIT
3775.448	3776.520	26479.40	CB	0.7	1.0		MIT
3774.440	3775.512	26486.47	CB	0.5	0.7		MIT
3774.382	3775.454	26486.88	CB	0.3	0.7		MIT
3773.625	3774.697	26492.19	CB	0.5	0.5		MIT
3773.152	3774.224	26495.51	CB	0.7	0.9		MIT
3771.852	3772.924	26504.65	CB	1.3	1.3		MIT
3770.867	3771.938	26511.57	CB	1.0	1.0H		MIT
3770.71	3771.78	26512.7	CB	0.7W			MIT
3770.653	3771.724	26513.07	CB		1.3H		MIT
3770.332	3771.403	26515.33	CB	0.5	0.0H		ME
3769.984	3771.055	26517.78	CB	0.7	0.7		MIT
3769.146	3770.217	26523.67	CB	1.0	1.2		MIT
3767.427	3768.497	26535.77	CB	0.5	0.6		MIT
3766.134	3767.204	26544.89	CB	1.2	1.3		MIT
3765.787	3766.857	26547.33	CB	0.3	0.5		MIT
3765.076	3766.146	26552.34	CB	1.4W	1.0W		MIT
3764.641	3765.711	26555.41	CB		0.5		MIT
3764.118	3765.188	26559.10	CB	1.0	1.0		MIT
3763.72	3764.79	26561.9	CB	0.7	0.0		MIT
3763.492	3764.561	26563.52	CB	1.0	1.0		MIT
3763.121	3764.190	26566.14	CB	0.0H	1.0H		MIT
3762.982	3764.051	26567.12	CB	0.0	0.7H		MIT
3762.452	3763.521	26570.86	CB	0.7	1.3H		MIT
3761.126	3762.195	26580.23	CB	1.2	1.3		MIT
3760.760	3761.829	26582.82	CB	0.3	0.9H		MIT
3760.644	3761.713	26583.64	CB	0.7	0.6		MIT
3759.553	3760.621	26591.35	CB	1.6	1.7		MIT
3756.252	3757.320	26614.72	CB	0.7	1.0		MIT
3755.948	3757.015	26616.87	CB	0.3O	0.7H		MIT
3755.768	3756.835	26618.15	CB	1.0	1.0		MIT
3755.639	3756.706	26619.06	CB	0.7W	0.3		MIT
3755.285	3756.352	26621.57	CB	0.7	0.7		MIT
3753.180	3754.247	26636.50	CB	1.0	1.0		MIT
3752.723	3753.790	26639.75	CB	0.6	0.5		ME
3752.297	3753.363	26642.77	CB	0.3	0.0		ME
3751.289	3752.355	26649.93	CB	0.0	1.2		MIT
3750.634	3751.700	26654.58	CB	0.6	0.6		MIT
3750.223	3751.289	26657.50	CB	0.5	0.7		MIT
3748.554	3749.620	26669.37	CB	1.0	1.0		MIT
3746.907	3747.972	26681.09	CB	1.3	1.9		MIT
3744.002	3745.066	26701.80	CB	1.5	1.5		MIT
3742.389	3743.453	26713.30	CB	1.5	1.7		MIT
3741.778	3742.842	26717.67	CB	1.3	1.2		MIT
3741.299	3742.363	26721.09	CB	0.3	1.2		MIT
3740.838	3741.902	26724.38	CB	1.0	1.0		MIT
3740.725	3741.788	26725.19	CB	0.5	1.7		MIT
3740.537	3741.600	26726.53	CB	0.5H	0.3H		ME
3739.795	3740.858	26731.83	CB	2.0	2.3		MIT
3738.424	3739.487	26741.64	CB	1.0	1.2		MIT
3736.333	3737.395	26756.60	CB	0.5	0.7H		MIT
3734.736	3735.798	26768.04	CB	0.3	0.7		MIT
3733.622	3734.684	26776.03	CB	1.0	1.0		MIT

AIR WAVELENGTH (ANGSTRMS)	VACUUM WAVELENGTH (ANGSTRMS)	WAVENUMBER (KAYSERS)	LMENT	LOG ARC	INTENSITY SPK	INTENSITY DIS	REF
3733.324	3734.386	26778.17	CB	1.0D	0.9		MIT
3732.034	3733.095	26787.42	CB	0.7	1.0		MIT
3731.539	3732.600	26790.98	CB	0.6	0.9		MIT
3729.616	3730.677	26804.79	CB	0.5	0.7		MIT
3727.230	3728.290	26821.95	CB	0.7	1.0		MIT
3726.236	3727.296	26829.10	CB	1.5	2.0		MIT
3725.219	3726.278	26836.43	CB	1.5W	1.0W		MIT
3723.444	3724.503	26849.22	CB	0.0	1.5		MIT
3723.104	3724.163	26851.67	CB	0.3	0.0		MIT
3722.950	3724.009	26852.78	CB	0.9	0.5		MIT
3722.566	3723.625	26855.55	CB		1.8H		MIT
3722.320	3723.379	26857.33	CB	0.5	0.7		MIT
3721.517	3722.575	26863.12	CB	0.6	0.5		ME
3720.456	3721.514	26870.78	CB	0.7	2.0H		MIT
3719.635	3720.693	26876.71	CB		1.7W		MIT
3718.523	3719.581	26884.75	CB	0.3	0.5		MIT
3717.542	3718.599	26891.84	CB	1.0	0.9H		MIT
3717.066	3718.123	26895.29	CB	0.9	3.0		MIT
3716.991	3718.048	26895.83	CB	1.00			MIT
3716.209	3717.266	26901.49	CB	1.50	1.0W		MIT
3715.972	3717.029	26903.21	CB	0.5	0.7		MIT
3714.852	3715.909	26911.32	CB	0.5H	0.3		ME
3713.817	3714.874	26918.82	CB	0.9	0.7		MIT
3713.72	3714.78	26919.5	CB		0.7H		ME
3713.358	3714.414	26922.14	CB		1.0H		MIT
3713.014	3714.070	26924.64	CB	2.0	1.9H		MIT
3712.554	3713.610	26927.97	CB	0.5	0.5		ME
3711.779	3712.835	26933.60	CB	0.6	0.7		MIT
3711.338	3712.394	26936.80	CB	1.3	1.3		MIT
3710.448	3711.504	26943.26	CB	1.2	1.3		MIT
3709.742	3710.797	26948.39	CB	0.3	0.3		MIT
3709.417	3710.472	26950.75	CB	0.7	1.0		MIT
3709.247	3710.302	26951.98	CB	0.7	1.5		MIT
3708.897	3709.952	26954.52	CB	0.5	0.6		MIT
3707.918	3708.973	26961.64	CB	0.50	2.00		MIT
3707.804	3708.859	26962.47	CB	0.7			MIT
3707.047	3708.102	26967.98	CB	0.8W			M
3705.605	3706.659	26978.47	CB	0.3			MIT
3704.141	3705.195	26989.13	CB	1.5	1.5		MIT
3703.914	3704.968	26990.79	CB	1.2	1.3		MIT
3703.165	3704.219	26996.25	CB	1.3	1.5		MIT
3699.929	3700.982	27019.86	CB	1.2W	1.5W		MIT
3699.578	3700.631	27022.42	CB	0.3	0.5		MIT
3699.077	3700.130	27026.08	CB	0.3	0.5		MIT
3697.846	3698.898	27035.08	CB	1.7	1.7		MIT
3697.392	3698.444	27038.40	CB	1.0	1.3		MIT
3696.677	3697.729	27043.62	CB	0.3	1.7		MIT
3695.898	3696.950	27049.32	CB	0.6	2.3		MIT
3694.795	3695.847	27057.40	CB		0.7		MIT
3694.666	3695.718	27058.34	CB	1.0	1.0		MIT
3693.765	3694.816	27064.95	CB	0.0	0.3		MIT
3693.368	3694.419	27067.85	CB	0.9	1.0		MIT
3692.183	3693.234	27076.54	CB		0.7		MIT
3691.183	3692.234	27083.88	CB	0.7	1.2		MIT
3689.404	3690.454	27096.94	CB	0.3	0.7		MIT
3689.038	3690.088	27099.62	CB	0.7	0.7		MIT
3688.697	3689.747	27102.13	CB	0.7	0.7		MIT
3688.182	3689.232	27105.91	CB	0.7	1.3		MIT
3687.971	3689.021	27107.46	CB	1.3W	2.5W		MIT
3687.445	3688.495	27111.33	CB	0.7	0.5		MIT
3686.557	3687.606	27117.86	CB	0.7	0.7		MIT
3685.128	3686.177	27128.38	CB	1.00	0.30		ME
3684.930	3685.979	27129.83	CB	0.0	0.7		MIT
3684.252	3685.301	27134.83	CB	0.0	0.7		MIT
3683.973	3685.022	27136.88	CB	0.3H	0.5H		ME
3682.953	3684.002	27144.40	CB		1.0H		MIT
3681.683	3682.731	27153.76	CB	0.0	1.0		MIT
3680.857	3681.905	27159.85	CB	0.3	0.3		MIT
3679.608	3680.656	27169.07	CB		1.0H		MIT
3678.717	3679.764	27175.65	CB	0.7	0.5H		MIT

AIR WAVELENGTH (ANGSTRMS)	VACUUM WAVELENGTH (ANGSTRMS)	WAVENUMBER (KAYSERS)	LMENT	LOG ARC	INTENSITY SPK	INTENSITY DIS	REF
3678.078	3679.125	27180.37	CB	0.0	1.0		MIT
3677.778	3678.825	27182.59	CB	0.7	0.7H		MIT
3677.080	3678.127	27187.75	CB	0.7	0.9		MIT
3676.335	3677.382	27193.26	CB		0.7		ME
3676.312	3677.359	27193.43	CB	1.0	1.3		MIT
3675.305	3676.352	27200.88	CB	0.5	0.7		MIT
3675.172	3676.219	27201.86	CB	0.5	0.0		MIT
3674.778	3675.824	27204.78	CB	1.3	0.7		MIT
3674.678	3675.724	27205.52	CB	1.3	0.6		MIT
3673.228	3674.274	27216.26	CB	0.7	1.0		MIT
3672.576	3673.622	27221.09	CB	0.5	0.5		MIT
3672.441	3673.487	27222.09	CB	0.7	0.7		MIT
3671.734	3672.780	27227.33	CB	0.3	0.7		MIT
3671.368	3672.414	27230.05	CB	0.5	0.7		MIT
3670.053	3671.098	27239.80	CB	0.0	1.2		MIT
3669.737	3670.782	27242.15	CB	0.7	0.9		MIT
3669.346	3670.391	27245.05	CB	0.3	0.5H		MIT
3669.007	3670.052	27247.57	CB	1.0	1.0		MIT
3668.622	3669.667	27250.43	CB	0.7	0.7		MIT
3667.757	3668.802	27256.86	CB	0.5	0.7		MIT
3667.662	3668.707	27257.56	CB	0.5	1.0		MIT
3667.001	3668.045	27262.48	CB	0.9	1.2		MIT
3666.533	3667.577	27265.96	CB	0.9	1.0		MIT
3665.159	3666.203	27276.18	CB	0.6H	0.7H		MIT
3664.822	3665.866	27278.69	CB	0.5	0.3		ME
3664.695	3665.739	27279.63	CB	1.5	1.5		MIT
3663.751	3664.795	27286.66	CB		0.9H		ME
3663.439	3664.482	27288.98	CB	0.5	0.5		MIT
3663.178	3664.221	27290.93	CB	0.7	0.5H		MIT
3662.921	3663.964	27292.84	CB	0.3	0.3		MIT
3662.050	3663.093	27299.33	CB	0.7	0.5		MIT
3661.677	3662.720	27302.11	CB	0.7	0.7H		MIT
3660.366	3661.409	27311.89	CB	1.3	1.5		MIT
3659.606	3660.648	27317.56	CB	1.2	2.7		MIT
3658.598	3659.640	27325.09	CB	0.7H	0.7H		MIT
3657.897	3658.939	27330.33	CB	0.6	0.7		MIT
3657.750	3658.792	27331.43	CB	0.5D			ME
3657.110	3658.152	27336.21	CB	0.7	0.7		MIT
3656.495	3657.537	27340.81	CB	0.3	0.3		MIT
3655.975	3657.017	27344.69	CB	0.7	1.0		MIT
3654.423	3655.464	27356.31	CB	1.0	1.0		MIT
3654.227	3655.268	27357.77	CB		0.7H		MIT
3653.615	3654.656	27362.36	CB	1.0	0.7		MIT
3652.275	3653.316	27372.40	CB	0.0	0.5H		ME
3651.305	3652.345	27379.67	CB	0.3			MIT
3651.186	3652.226	27380.56	CB	1.0	2.6		MIT
3650.806	3651.846	27383.41	CB	1.2	1.2		MIT
3650.517	3651.557	27385.58	CB	0.3	0.5		MIT
3649.852	3650.892	27390.57	CB	1.3	1.3		MIT
3647.867	3648.906	27405.47	CB	0.5			ME
3647.724	3648.763	27406.54	CB	0.3	0.3		ME
3647.309	3648.348	27409.66	CB	1.0	1.3		MIT
3645.929	3646.968	27420.04	CB	0.0	0.7		MIT
3645.356	3646.395	27424.35	CB	0.7	0.7		MIT
3644.935	3645.974	27427.52	CB	0.7	1.0		MIT
3643.721	3644.759	27436.65	CB	1.2	1.2		MIT
3643.522	3644.560	27438.15	CB	0.3	0.3		MIT
3643.337	3644.375	27439.54	CB	0.7	1.0		MIT
3641.385	3642.423	27454.25	CB	0.7	0.0		MIT
3641.289	3642.327	27454.98	CB	0.0	0.7		MIT
3640.636	3641.674	27459.90	CB	0.9	1.0		MIT
3639.331	3640.368	27469.75	CB	1.2	1.3		MIT
3639.055	3640.092	27471.83	CB	0.3	1.3		MIT
3638.786	3639.823	27473.86	CB	1.0	1.0		MIT
3637.827	3638.864	27481.10	CB	1.3	1.5		MIT
3637.538	3638.575	27483.29	CB	1.0	1.0		MIT
3636.957	3637.994	27487.68	CB	1.30	1.50		MIT
3635.854	3636.890	27496.02	CB	0.5	0.9		MIT
3635.463	3636.499	27498.98	CB	0.7	0.7		MIT
3635.324	3636.360	27500.03	CB	0.7	0.9		MIT

AIR WAVELENGTH (ANGSTRMS)	VACUUM WAVELENGTH (ANGSTRMS)	WAVENUMBER (KAYSERS)	LMENT	LOG ARC	INTENSITY SPK	DIS	REF
3634.600	3635.636	27505.50	CB	0.3			MIT
3634.443	3635.479	27506.69	CB	0.7	1.2		MIT
3633.714	3634.750	27512.21	CB	0.9	0.5		MIT
3633.311	3634.347	27515.26	CB	0.5	1.5		MIT
3632.999	3634.035	27517.62	CB	0.7	0.5		MIT
3630.695	3631.730	27535.09	CB	0.3	0.0		MIT
3630.616	3631.651	27535.69	CB	0.7	1.2		MIT
3629.467	3630.502	27544.40	CB		1.5		MIT
3628.176	3629.210	27554.20	CB	0.0	1.7		MIT
3627.870	3628.904	27556.53	CB	0.5	0.5		ME
3625.713	3626.747	27572.92	CB	1.0	1.2		MIT
3625.171	3626.205	27577.04	CB	0.9	1.2		MIT
3624.357	3625.390	27583.24	CB	0.5	0.5		ME
3622.613	3623.646	27596.51	CB	0.6	0.0		MIT
3622.493	3623.526	27597.43	CB		0.7H		MIT
3621.027	3622.060	27608.60	CB	1.0	1.2		MIT
3620.590	3621.622	27611.93	CB		0.7H		MIT
3619.727	3620.759	27618.52	CB	0.5	2.5		MIT
3619.513	3620.545	27620.15	CB	0.7	2.3		MIT
3619.200	3620.232	27622.54	CB	0.3	0.0		MIT
3618.901	3619.933	27624.82	CB	0.7	0.7		MIT
3618.442	3619.474	27628.32	CB	1.0	0.7		MIT
3617.713	3618.745	27633.89	CB	0.5	1.2		MIT
3616.498	3617.529	27643.18	CB	0.3	0.3		MIT
3616.209	3617.240	27645.39	CB	0.7	0.7		MIT
3615.498	3616.529	27650.82	CB	1.5	1.5		MIT
3613.452	3614.483	27666.48	CB	0.7	0.5		MIT
3613.239	3614.269	27668.11	CB	0.3	0.7		MIT
3613.013	3614.043	27669.84	CB	0.3	0.3		ME
3612.654	3613.684	27672.59	CB	0.5	0.6		MIT
3611.285	3612.315	27683.08	CB	0.5	0.7H		MIT
3610.764	3611.794	27687.07	CB	0.5	0.7		MIT
3610.003	3611.033	27692.91	CB	0.5	0.7		MIT
3609.360	3610.389	27697.84	CB	0.0	0.7		MIT
3608.313	3609.342	27705.88	CB	0.5	0.5		MIT
3608.014	3609.043	27708.18	CB	0.7	0.7		MIT
3607.327	3608.356	27713.45	CB	0.3	0.7		MIT
3607.015	3608.044	27715.85	CB	0.0	1.0		MIT
3606.806	3607.835	27717.46	CB	0.5	0.0		ME
3606.487	3607.516	27719.91	CB	0.5H	0.7		MIT
3606.35	3607.38	27721.0	CB		0.5H		ME
3606.271	3607.300	27721.57	CB	0.3	0.7J		MIT
3604.641	3605.669	27734.10	CB	0.7W	1.0W		MIT
3604.075	3605.103	27738.46	CB	0.5O	0.3O		MIT
3603.957	3604.985	27739.37	CB	0.3	0.7		MIT
3603.434	3604.462	27743.39	CB	0.6	0.7		MIT
3602.875	3603.903	27747.70	CB	0.3	0.3		MIT
3602.562	3603.590	27750.11	CB	1.5	1.5		MIT
3599.633	3600.660	27772.69	CB	1.0	1.0		MIT
3599.278	3600.305	27775.43	CB	1.2	1.2		MIT
3598.352	3599.379	27782.57	CB	0.7D	0.3H		MIT
3597.514	3598.540	27789.04	CB	0.5	0.5		ME
3597.255	3598.281	27791.04	CB	0.9	1.0		MIT
3593.966	3594.992	27816.48	CB	1.9	1.7		MIT
3593.547	3594.572	27819.72	CB	1.2	0.5		MIT
3591.792	3592.817	27833.31	CB	0.7	0.7		MIT
3591.198	3592.223	27837.92	CB	0.3	1.7		MIT
3590.905	3591.930	27840.19	CB	0.5	0.3		MIT
3590.714	3591.739	27841.67	CB	0.5	0.6		MIT
3589.959	3590.983	27847.52	CB	0.3	0.5		MIT
3589.356	3590.380	27852.20	CB	2.0	2.0		MIT
3589.107	3590.131	27854.14	CB	1.7	1.5		MIT
3588.02	3589.04	27862.6	CB		0.5H		ME
3587.956	3588.980	27863.07	CB	0.3			MIT
3587.406	3588.430	27867.34	CB	0.3	0.6		MIT
3586.869	3587.893	27871.51	CB	0.3			MIT
3586.756	3587.780	27872.39	CB	0.3	1.3		MIT
3586.691	3587.715	27872.90	CB	0.0	1.3		MIT
3584.969	3585.992	27886.28	CB	1.5	1.7		MIT
3582.363	3583.386	27906.57	CB	0.7	1.0		MIT

AIR WAVELENGTH (ANGSTRMS)	VACUUM WAVELENGTH (ANGSTRMS)	WAVENUMBER (KAYSERS)	LMENT	LOG ARC	INTENSITY SPK	DIS	REF
3582.063	3583.085	27908.91	CB	0.5	0.7		MIT
3580.273	3581.295	27922.86	CB	2.0	2.5		MIT
3578.584	3579.606	27936.04	CB	1.2	0.0		MIT
3578.234	3579.255	27938.77	CB	0.0	0.7		MIT
3577.718	3578.739	27942.80	CB	1.0	1.2		MIT
3577.231	3578.252	27946.60	CB	0.6	0.5		MIT
3575.848	3576.869	27957.41	CB	1.7	1.9		MIT
3575.129	3576.150	27963.03	CB	1.0	1.0		MIT
3574.201	3575.221	27970.30	CB	0.0	1.0		MIT
3573.094	3574.114	27978.96	CB	0.3	0.7		MIT
3571.480	3572.500	27991.60	CB	0.3	0.5		MIT
3569.852	3570.871	28004.37	CB	0.5	0.5		M
3569.695	3570.714	28005.60	CB	0.3	0.3		M
3569.466	3570.485	28007.40	CB	1.3	1.2		MIT
3568.725	3569.744	28013.21	CB	0.7	0.7		MIT
3568.506	3569.525	28014.93	CB	1.0	1.7		MIT
3567.995	3569.014	28018.94	CB	0.3	1.7		MIT
3567.103	3568.122	28025.95	CB	0.7	1.5W		MIT
3566.089	3567.107	28033.92	CB	0.7	0.7		MIT
3565.855	3566.873	28035.76	CB	0.5	0.3		ME
3565.691	3566.709	28037.05	CB		0.7H		MIT
3565.056	3566.074	28042.04	CB	1.0	1.2H		MIT
3564.077	3565.095	28049.75	CB	0.3	1.3		MIT
3563.620	3564.638	28053.34	CB	1.0	1.2		MIT
3563.497	3564.515	28054.31	CB	1.5	1.5		MIT
3563.230	3564.248	28056.41	CB	0.3	0.3		MIT
3562.649	3563.666	28060.99	CB	0.5	0.5H		ME
3561.881	3562.898	28067.04	CB		0.9		MIT
3561.691	3562.708	28068.53	CB	0.6	1.0		MIT
3561.143	3562.160	28072.85	CB	0.7	1.0		MIT
3560.482	3561.499	28078.07	CB		1.0		MIT
3560.357	3561.374	28079.05	CB	0.5	0.3		ME
3559.901	3560.918	28082.65	CB		1.0		MIT
3559.596	3560.613	28085.05	CB	0.3	2.0		MIT
3559.123	3560.140	28088.79	CB	0.9	0.9		MIT
3558.012	3559.028	28097.56	CB	0.7	0.7		MIT
3556.017	3557.033	28113.32	CB	0.7	1.0		MIT
3554.664	3555.679	28124.02	CB	1.2	1.3		MIT
3554.519	3555.534	28125.17	CB	1.2	1.3		MIT
3554.125	3555.140	28128.28	CB		1.5		MIT
3553.616	3554.631	28132.31	CB	1.3W	1.2L		MIT
3552.225	3553.240	28143.33	CB	0.3	0.0		MIT
3551.961	3552.976	28145.42	CB	0.3	0.7		MIT
3551.109	3552.123	28152.17	CB	0.7	1.0H		MIT
3550.615	3551.629	28156.09	CB	0.3	0.5		MIT
3550.450	3551.464	28157.40	CB	1.6	1.5		MIT
3550.239	3551.253	28159.07	CB	0.6	0.7		MIT
3549.263	3550.277	28166.82	CB	0.9	0.9		MIT
3548.131	3549.145	28175.80	CB	0.3H	0.7		MIT
3548.089	3549.103	28176.14	CB	0.7	0.7H		MIT
3546.487	3547.500	28188.86	CB	0.7	0.7		MIT
3546.160	3547.173	28191.46	CB	0.3	0.5		ME
3546.031	3547.044	28192.49	CB	0.7	0.9		MIT
3545.396	3546.409	28197.54	CB	0.5W	0.7W		MIT
3544.651	3545.664	28203.46	CB	1.3	1.2		MIT
3544.345	3545.358	28205.90	CB	0.0	0.7		MIT
3544.024	3545.037	28208.45	CB	0.9	1.0		MIT
3543.933	3544.946	28209.18	CB	0.5	0.7		MIT
3542.981	3543.993	28216.76	CB	1.0	1.0		MIT
3542.558	3543.570	28220.13	CB	0.7	1.5D		MIT
3541.897	3542.909	28225.39	CB	1.0	1.2		MIT
3541.252	3542.264	28230.53	CB	1.7	0.7		MIT
3540.962	3541.974	28232.84	CB	1.2	2.7		MIT
3539.648	3540.660	28243.33	CB	1.2	1.2		MIT
3539.116	3540.127	28247.57	CB	0.0	1.2		MIT
3537.624	3538.635	28259.48	CB	0.3	1.5		MIT
3537.475	3538.486	28260.67	CB	1.5	1.5		MIT
3536.213	3537.224	28270.76	CB	0.5	0.7H		MIT
3535.301	3536.311	28278.05	CB	2.5	2.7		MIT
3534.425	3535.435	28285.06	CB	0.5	0.3		MIT

AIR WAVELENGTH (ANGSTRMS)	VACUUM WAVELENGTH (ANGSTRMS)	WAVENUMBER (KAYSERS)	LMENT	LOG ARC	INTENSITY SPK	INTENSITY DIS	REF
3534.224	3535.234	28286.67	CB	0.0	1.2		MIT
3534.117	3535.127	28287.53	CB	1.2	1.2H		MIT
3534.037	3535.047	28288.17	CB	0.0	0.6H		MIT
3533.665	3534.675	28291.14	CB	1.3	1.5		MIT
3532.530	3533.540	28300.23	CB	0.5	0.7H		MIT
3530.824	3531.833	28313.91	CB	0.5	1.2		MIT
3530.089	3531.098	28319.80	CB	0.7	0.7		MIT
3529.391	3530.400	28325.40	CB	0.5	0.7		MIT
3528.896	3529.905	28329.38	CB	0.3	1.6		MIT
3528.476	3529.485	28332.75	CB	0.5	1.7		MIT
3528.307	3529.316	28334.10	CB	1.0	0.0		MIT
3528.105	3529.114	28335.73	CB		0.5		MIT
3527.955	3528.964	28336.93	CB	1.0	0.9		MIT
3527.293	3528.301	28342.25	CB	0.5H	0.3H		MIT
3527.101	3528.109	28343.79	CB	0.7	0.7		MIT
3527.031	3528.039	28344.36	CB		0.5H		MIT
3525.987	3526.995	28352.75	CB		1.2H		MIT
3525.885	3526.893	28353.57	CB	0.5	0.7H		MIT
3525.232	3526.240	28358.82	CB	1.2W	1.5W		MIT
3524.936	3525.944	28361.20	CB	0.3	0.7		MIT
3523.154	3524.161	28375.55	CB	0.7	1.0		MIT
3522.363	3523.370	28381.92	CB		1.70		MIT
3521.602	3522.609	28388.05	CB		1.0H		MIT
3521.140	3522.147	28391.77	CB	0.3	1.0H		MIT
3520.715	3521.722	28395.20	CB	0.5	0.7		MIT
3520.055	3521.061	28400.53	CB	1.3	1.3		MIT
3519.649	3520.655	28403.80	CB	0.7	1.3		MIT
3519.334	3520.340	28406.34	CB	0.3	0.5		MIT
3518.178	3519.184	28415.68	CB	0.5	0.6		MIT
3517.761	3518.767	28419.05	CB	0.7	0.0		MIT
3517.671	3518.677	28419.77	CB	0.3	2.3		MIT
3517.106	3518.112	28424.34	CB	0.0	0.9		MIT
3516.857	3517.863	28426.35	CB	0.3	1.0		MIT
3516.198	3517.204	28431.68	CB	0.5	0.7		MIT
3515.423	3516.428	28437.95	CB	1.3	2.5		MIT
3514.044	3515.049	28449.11	CB		1.7		MIT
3512.756	3513.761	28459.54	CB	0.5	0.3		MIT
3512.309	3513.314	28463.16	CB	0.5	0.3		MIT
3511.191	3512.195	28472.22	CB	0.7	0.7		MIT
3511.157	3512.161	28472.50	CB	1.7	0.0		MIT
3510.257	3511.261	28479.80	CB	1.2	2.3		MIT
3508.537	3509.541	28493.76	CB	0.7	0.7H		MIT
3507.960	3508.963	28498.45	CB	1.3	1.3		MIT
3506.994	3507.997	28506.29	CB	1.0	1.0		MIT
3506.025	3507.028	28514.17	CB	0.5	0.5		MIT
3505.808	3506.811	28515.94	CB	0.7	0.3		MIT
3505.627	3506.630	28517.41	CB	0.5	0.7		MIT
3503.200	3504.202	28537.17	CB	1.0	0.7		MIT
3501.341	3502.343	28552.32	CB	0.5	1.5		MIT
3500.741	3501.743	28557.21	CB		1.5H		MIT
3500.107	3501.108	28562.38	CB	0.7	0.0		MIT
3499.949	3500.950	28563.67	CB	0.7	1.7		MIT
3498.629	3499.630	28574.45	CB	1.5	1.7		MIT
3498.434	3499.435	28576.04	CB	0.3	0.0		MIT
3497.812	3498.813	28581.12	CB	1.5	1.2		MIT
3496.282	3497.282	28593.63	CB	0.5D	0.0W		MIT
3496.026	3497.026	28595.72	CB	1.0	1.0		MIT
3493.473	3494.473	28616.62	CB	0.5	0.5		MIT
3491.892	3492.891	28629.58	CB	0.0	1.2		MIT
3491.477	3492.476	28632.98	CB	1.0	0.7		MIT
3491.031	3492.030	28636.64	CB	1.5	1.7		MIT
3490.418	3491.417	28641.67	CB	0.0	1.0		ME
3489.090	3490.089	28652.57	CB	0.7	1.7		MIT
3488.832	3489.830	28654.69	CB	0.3	1.5		MIT
3488.742	3489.740	28655.43	CB	0.0	1.0H		ME
3487.533	3488.531	28665.36	CB	0.5	0.3		MIT
3486.717	3487.715	28672.07	CB	0.7	0.0		MIT
3485.931	3486.929	28678.53	CB	1.0	0.7		MIT
3485.097	3486.095	28685.40	CB	0.7	0.7H		MIT
3484.627	3485.624	28689.26	CB	0.5	1.2		MIT

AIR WAVELENGTH (ANGSTRMS)	VACUUM WAVELENGTH (ANGSTRMS)	WAVENUMBER (KAYSERS)	LMENT	LOG ARC	INTENSITY SPK	INTENSITY DIS	REF
3484.049	3485.046	28694.02	CB	1.0	2.0		MIT
3482.951	3483.948	28703.07	CB	0.3	2.0		MIT
3482.525	3483.522	28706.58	CB	0.7H	0.0		MIT
3481.051	3482.047	28718.74	CB	0.7	0.3		MIT
3480.213	3481.209	28725.65	CB	0.3	1.3		MIT
3479.564	3480.560	28731.01	CB	0.7	2.3		MIT
3478.778	3479.774	28737.50	CB	0.3	1.7		MIT
3478.692	3479.688	28738.21	CB	1.5	1.2		MIT
3478.006	3479.002	28743.88	CB	0.5	0.3		MIT
3476.28	3477.28	28758.1	CB		1.2		MIT
3475.991	3476.986	28760.54	CB	1.2	0.5H		MIT
3475.585	3476.580	28763.90	CB	1.0	1.2		MIT
3474.668	3475.663	28771.49	CB	2.0	0.6		MIT
3473.994	3474.989	28777.07	CB	0.3	1.3		MIT
3473.126	3474.120	28784.26	CB	0.5	0.3		MIT
3473.024	3474.018	28785.11	CB	1.5	1.5		MIT
3472.774	3473.768	28787.18	CB	0.3	0.3		MIT
3471.523	3472.517	28797.55	CB	0.7	0.7		MIT
3471.191	3472.185	28800.31	CB	0.7	0.6		MIT
3471.039	3472.033	28801.57	CB	0.3	0.0		MIT
3470.254	3471.248	28808.08	CB	0.5H	2.0		MIT
3469.438	3470.431	28814.86	CB	1.0	0.7		MIT
3469.130	3470.123	28817.42	CB	0.3	0.0		MIT
3468.541	3469.534	28822.31	CB	1.0	0.5		MIT
3468.127	3469.120	28825.75	CB		1.7		ME
3467.468	3468.461	28831.23	CB	1.2	1.3		MIT
3465.865	3466.858	28844.56	CB	1.5	1.6		MIT
3463.811	3464.803	28861.67	CB	1.5	1.7		MIT
3463.683	3464.675	28862.74	CB	0.7	0.3		MIT
3463.030	3464.022	28868.18	CB	0.7	1.5		MIT
3462.647	3463.639	28871.37	CB	0.7	0.5		MIT
3461.611	3462.602	28880.01	CB	0.7H	0.3		MIT
3459.703	3460.694	28895.94	CB	1.5	1.3		MIT
3459.54	3460.53	28897.3	CB		1.3		MIT
3458.952	3459.943	28902.21	CB	1.0	1.0		MIT
3458.87	3459.86	28902.9	CB	0.3	0.0		MIT
3458.741	3459.732	28903.98	CB	0.7			MIT
3458.728	3459.719	28904.08	CB		1.0		ME
3457.794	3458.785	28911.89	CB	1.2	1.0		MIT
3457.203	3458.193	28916.83	CB	0.7	0.5		MIT
3457.04	3458.03	28918.2	CB		1.2J		MIT
3456.988	3457.978	28918.63	CB	0.7H			MIT
3456.538	3457.528	28922.40	CB	1.2	1.0		MIT
3454.914	3455.904	28935.99	CB	0.5	1.9		MIT
3454.706	3455.696	28937.73	CB	0.5	1.7		MIT
3453.97	3454.96	28943.9	CB		2.0		MIT
3452.647	3453.636	28954.99	CB	1.2	0.7		MIT
3452.37	3453.36	28957.3	CB	1.2O			MIT
3452.346	3453.335	28957.51	CB	0.7	2.3		MIT
3451.636	3452.625	28963.47	CB	0.0	1.5		MIT
3450.762	3451.751	28970.81	CB	0.3	1.7		MIT
3448.673	3449.661	28988.35	CB	0.0	1.5		MIT
3448.221	3449.209	28992.15	CB	0.5H	1.7		MIT
3446.930	3447.918	29003.01	CB		1.2		MIT
3445.681	3446.668	29013.52	CB	1.7	1.9		MIT
3444.279	3445.266	29025.33	CB	0.0H	1.7		MIT
3443.742	3444.729	29029.86	CB	0.0	1.3		MIT
3442.793	3443.780	29037.86	CB	0.5	0.5		MIT
3442.650	3443.637	29039.07	CB	0.7	0.7		MIT
3441.653	3442.639	29047.48	CB	0.3	1.6		MIT
3440.590	3441.576	29056.46	CB	1.2	1.9		MIT
3439.924	3440.910	29062.08	CB	0.6	1.7		MIT
3439.339	3440.325	29067.02	CB	0.3	0.5		MIT
3438.419	3439.405	29074.80	CB	0.0	1.7J		MIT
3436.962	3437.947	29087.12	CB	1.3R	1.7		MIT
3436.830	3437.815	29088.24	CB	0.3	1.3		MIT
3435.587	3436.572	29098.77	CB		1.0J		MIT
3433.950	3434.934	29112.64	CB		1.0		MIT
3433.747	3434.731	29114.36	CB		1.0		MIT
3433.160	3434.144	29119.34	CB	0.3H	0.3		MIT

AIR WAVELENGTH (ANGSTRMS)	VACUUM WAVELENGTH (ANGSTRMS)	WAVENUMBER (KAYSERS)	LMENT	LOG ARC	INTENSITY SPK	DIS	REF
3433.088	3434.072	29119.95	CB	1.0	0.3		MIT
3432.701	3433.685	29123.23	CB	1.0	2.0		MIT
3432.424	3433.408	29125.58	CB	1.0	0.5		MIT
3431.948	3432.932	29129.62	CB	0.9	0.7		MIT
3431.173	3432.157	29136.20	CB		0.7H		MIT
3431.066	3432.050	29137.11	CB	0.9M			MIT
3429.043	3430.026	29154.30	CB	1.0L	0.7L		MIT
3428.789	3429.772	29156.46	CB	1.0	1.0		MIT
3428.361	3429.344	29160.10	CB	0.5	0.5		MIT
3427.449	3428.432	29167.85	CB	1.5R	1.5		MIT
3426.570	3427.553	29175.34	CB	0.7	2.3		MIT
3425.848	3426.830	29181.49	CB	1.7R	1.5		MIT
3425.424	3426.406	29185.10	CB	1.5R	2.5		MIT
3423.765	3424.747	29199.24	CB	1.2	0.7		MIT
3423.648	3424.630	29200.24	CB		1.0W		MIT
3422.849	3423.831	29207.05	CB	0.0	1.5		MIT
3422.786	3423.768	29207.59	CB	0.5R	0.7		MIT
3421.162	3422.143	29221.46	CB	1.0W	1.7W		MIT
3420.631	3421.612	29225.99	CB	0.7	1.7		MIT
3420.099	3421.080	29230.54	CB	0.5H	0.5J		MIT
3417.860	3418.840	29249.69	CB	0.7	0.7		MIT
3417.262	3418.242	29254.80	CB	0.7B			MIT
3417.18	3418.16	29255.5	CB		1.0		MIT
3415.974	3416.954	29265.83	CB	1.7	1.7		MIT
3414.066	3415.045	29282.19	CB	1.0	0.9		MIT
3413.501	3414.480	29287.03	CB	0.7O	0.3H		MIT
3413.213	3414.192	29289.51	CB	0.0	1.6		MIT
3412.935	3413.914	29291.89	CB	0.7	2.2		MIT
3412.481	3413.460	29295.79	CB	0.3H	1.3H		MIT
3409.908	3410.886	29317.89	CB	0.9	0.5		MIT
3409.188	3410.166	29324.08	CB	1.0	2.0		MIT
3408.678	3409.656	29328.47	CB	0.7	1.7		MIT
3408.376	3409.354	29331.07	CB	1.0	0.7		MIT
3407.976	3408.954	29334.51	CB	0.9	1.0		MIT
3407.303	3408.281	29340.31	CB		1.5		MIT
3406.948	3407.925	29343.36	CB	0.0	1.5		MIT
3406.613	3407.590	29346.25	CB	0.9	0.7		MIT
3406.133	3407.110	29350.39	CB	1.5	1.5		MIT
3405.411	3406.388	29356.61	CB	1.9	1.7		MIT
3403.750	3404.727	29370.93	CB	0.7	0.6		MIT
3403.49	3404.47	29373.2	CB		0.7		MIT
3403.015	3403.991	29377.28	CB	1.3W	1.9W		MIT
3402.017	3402.993	29385.90	CB		1.7		MIT
3401.226	3402.202	29392.73	CB	0.3	1.6		MIT
3399.963	3400.939	29403.65	CB	1.0	0.3		MIT
3399.711	3400.687	29405.83	CB	1.2	1.2		MIT
3399.400	3400.376	29408.52	CB	1.3	1.5		MIT
3398.254	3399.229	29418.43	CB	1.2	1.6		MIT
3397.517	3398.492	29424.81	CB		1.2		MIT
3397.320	3398.295	29426.52	CB		1.7		MIT
3396.372	3397.347	29434.73	CB	0.3	2.2		MIT
3395.929	3396.904	29438.57	CB	1.0	0.7		MIT
3395.725	3396.700	29440.34	CB	0.0	1.5		MIT
3394.975	3395.949	29446.85	CB	0.3	1.7		MIT
3394.083	3395.057	29454.58	CB	0.7R	0.5		MIT
3393.810	3394.784	29456.95	CB	0.0H	1.2		MIT
3392.338	3393.312	29469.74	CB	1.3R	1.5		MIT
3391.591	3392.565	29476.23	CB	0.0	0.7		MIT
3391.330	3392.303	29478.50	CB	0.7	0.5		MIT
3390.800	3391.773	29483.10	CB I	1.3			MIT
3390.630	3391.603	29484.58	CB	1.5	1.7		MIT
3390.412	3391.385	29486.48	CB	1.0	0.5J		MIT
3388.941	3389.914	29499.27	CB	0.0	1.9		MIT
3387.928	3388.901	29508.09	CB		1.3W		MIT
3387.752	3388.725	29509.63	CB	0.9	0.7		MIT
3387.576	3388.549	29511.16	CB	0.5	0.7		MIT
3386.993	3387.965	29516.24	CB	1.0	1.2		MIT
3386.244	3387.216	29522.77	CB	0.7	2.0		MIT
3385.810	3386.782	29526.55	CB	0.5H	0.7H		MIT
3385.668	3386.640	29527.79	CB	0.7	0.9		MIT
3384.658	3385.630	29536.60	CB	0.3	0.9		MIT
3383.800	3384.772	29544.09	CB	1.2	0.7		MIT
3382.407	3383.378	29556.26	CB	0.7	1.6H		MIT
3380.938	3381.909	29569.10	CB		2.3		MIT
3380.861	3381.832	29569.77	CB	0.7			MIT
3380.489	3381.460	29573.03	CB	0.5	0.3H		MIT
3380.410	3381.381	29573.72	CB	1.3	0.5		MIT
3380.053	3381.024	29576.84	CB	0.7	0.7		MIT
3379.300	3380.270	29583.43	CB	0.0	2.0		MIT
3377.739	3378.709	29597.10	CB	0.3	0.3		MIT
3377.374	3378.344	29600.30	CB		1.3W		MIT
3376.730	3377.700	29605.95	CB	0.9	1.2		MIT
3376.335	3377.305	29609.41	CB	0.5	0.3		MIT
3374.925	3375.894	29621.78	CB	1.3	1.3		MIT
3374.247	3375.216	29627.73	CB	0.7	1.2		MIT
3374.090	3375.059	29629.11	CB		1.5H		MIT
3372.562	3373.531	29642.53	CB	1.0H	2.3		MIT
3372.090	3373.059	29646.68	CB	0.7	0.5		MIT
3371.329	3372.297	29653.37	CB	1.0	1.2		MIT
3370.611	3371.579	29659.69	CB	0.5	1.5		MIT
3370.158	3371.126	29663.68	CB		1.9		MIT
3369.831	3370.799	29666.56	CB	1.0	0.5		MIT
3369.157	3370.125	29672.49	CB	0.7	1.7		MIT
3369.075	3370.043	29673.21	CB	1.2			MIT
3368.432	3369.400	29678.88	CB	0.6H	0.9H		MIT
3367.376	3368.343	29688.19	CB	0.7	0.5		MIT
3367.085	3368.052	29690.75	CB	0.3	0.3		MIT
3366.956	3367.923	29691.89	CB	1.3	1.3		MIT
3365.944	3366.911	29700.81	CB	0.0	1.5		MIT
3365.876	3366.843	29701.42	CB	0.3	1.0		MIT
3365.584	3366.551	29703.99	CB	0.7	1.7		MIT
3363.745	3364.711	29720.23	CB	1.0	0.6		MIT
3362.860	3363.826	29728.05	CB	0.5	0.3		MIT
3362.172	3363.138	29734.14	CB		2.0		MIT
3360.902	3361.868	29745.37	CB	0.5	1.7		MIT
3358.417	3359.382	29767.38	CB	2.0	2.0		MIT
3357.043	3358.008	29779.56	CB	1.0	1.0		MIT
3356.46	3357.42	29784.7	CB	1.0W	0.5		MIT
3355.419	3356.383	29793.98	CB	0.7	0.7		MIT
3354.741	3355.705	29800.00	CB	1.3	1.2		MIT
3353.687	3354.651	29809.36	CB	0.7H	0.3		MIT
3353.503	3354.467	29811.00	CB	1.0	1.3		MIT
3353.359	3354.323	29812.28	CB	0.7W	0.5H		MIT
3352.869	3353.833	29816.63	CB	1.0	0.3H		MIT
3352.834	3353.798	29816.95	CB		0.7W		MIT
3352.590	3353.554	29819.11	CB	1.2	1.0		MIT
3352.284	3353.248	29821.84	CB	0.7	0.7		MIT
3350.692	3351.655	29836.00	CB	1.0	0.8H		MIT
3349.523	3350.486	29846.42	CB	1.5	0.7H		MIT
3349.348	3350.311	29847.98	CB	0.7	2.0		MIT
3349.059	3350.022	29850.55	CB	1.9	2.0		MIT
3348.784	3349.747	29853.00	CB	0.0	1.5		MIT
3348.284	3349.247	29857.46	CB	0.0H	1.7		MIT
3347.555	3348.517	29863.96	CB	0.5	0.0		MIT
3346.931	3347.893	29869.53	CB	1.2	1.0		MIT
3346.750	3347.712	29871.15	CB	0.7	1.9		MIT
3346.286	3347.248	29875.29	CB		1.6		ME
3344.245	3345.206	29893.52	CB		1.0		MIT
3343.965	3344.926	29896.02	CB	0.5	1.2		MIT
3343.709	3344.670	29898.31	CB	1.2	1.3		MIT
3343.28	3344.24	29902.1	CB		1.2		MIT
3342.900	3343.861	29905.55	CB	0.3			MIT
3341.974	3342.935	29913.83	CB	2.0R	1.7		MIT
3341.600	3342.561	29917.18	CB	0.5	1.7		MIT
3340.464	3341.425	29927.36	CB	0.3	1.3D		MIT
3339.265	3340.225	29938.10	CB	0.5	0.7H		MIT
3339.158	3340.118	29939.06	CB	1.0W	0.7W		ME
3336.315	3337.274	29964.57	CB	1.0	0.7		MIT
3335.668	3336.627	29970.38	CB		1.5W		MIT
3335.422	3336.381	29972.59	CB	1.5	1.6		MIT

AIR WAVELENGTH (ANGSTRMS)	VACUUM WAVELENGTH (ANGSTRMS)	WAVENUMBER (KAYSERS)	LMENT	LOG ARC	INTENSITY SPK	DIS	REF
3335.244	3336.203	29974.19	CB	0.0W	1.5W		ME
3334.830	3335.789	29977.92	CB	0.0	0.7		MIT
3334.525	3335.484	29980.66	CB		0.7		MIT
3333.974	3334.933	29985.61	CB	0.7	0.3		MIT
3332.697	3333.656	29997.10	CB	0.7W	0.9		MIT
3332.159	3333.117	30001.94	CB	1.0	1.0		MIT
3331.889	3332.847	30004.37	CB	0.7	0.5		MIT
3329.619	3330.577	30024.83	CB	0.5	0.5		MIT
3329.361	3330.319	30027.16	CB	1.0	1.2		MIT
3329.16	3330.12	30029.0	CB		0.9H		ME
3328.157	3329.114	30038.02	CB	0.5	0.5		MIT
3327.920	3328.877	30040.16	CB	0.6	0.7		MIT
3326.619	3327.576	30051.91	CB	1.0	0.7		MIT
3326.54	3327.50	30052.6	CB		0.5H		MIT
3325.436	3326.393	30062.60	CB		0.9H		ME
3325.21	3326.17	30064.6	CB		1.2H		MIT
3324.660	3325.616	30069.61	CB	0.6	1.7		MIT
3324.556	3325.512	30070.55	CB	0.3	1.0		MIT
3323.894	3324.850	30076.54	CB	0.0	1.7		MIT
3322.807	3323.763	30086.38	CB	0.5	0.3		MIT
3320.810	3321.766	30104.47	CB	0.5	2.0		MIT
3319.584	3320.539	30115.59	CB	0.6	1.7		MIT
3318.984	3319.939	30121.03	CB	1.0	0.7		MIT
3318.292	3319.247	30127.32	CB	0.0	0.7		MIT
3316.621	3317.575	30142.50	CB	0.0	1.3		MIT
3315.219	3316.173	30155.24	CB	1.3	1.3		MIT
3313.65	3314.60	30169.5	CB		0.7S		MIT
3313.395	3314.349	30171.84	CB		1.0H		MIT
3313.086	3314.040	30174.66	CB	0.7H	0.5H		MIT
3312.600	3313.553	30179.08	CB	1.6	1.7		MIT
3311.338	3312.291	30190.58	CB	0.7	1.0		MIT
3310.662	3311.615	30196.75	CB		1.2J		MIT
3310.468	3311.421	30198.52	CB	1.0	1.0		MIT
3310.093	3311.046	30201.94	CB	0.5	0.7H		MIT
3309.800	3310.753	30204.61	CB	0.5	0.6		MIT
3309.281	3310.234	30209.35	CB		1.0		MIT
3308.046	3308.998	30220.63	CB	1.0	1.0		MIT
3305.609	3306.561	30242.91	CB	0.0	2.0		MIT
3304.832	3305.783	30250.01	CB	1.0	0.7		MIT
3304.717	3305.668	30251.07	CB	0.0	1.5		MIT
3303.320	3304.271	30263.86	CB	0.0	1.5		MIT
3302.621	3303.572	30270.27	CB	0.0	1.0		MIT
3302.176	3303.127	30274.34	CB	0.7	1.0		MIT
3301.491	3302.442	30280.63	CB	0.0	2.0		MIT
3300.345	3301.295	30291.14	CB		1.0H		MIT
3299.609	3300.559	30297.90	CB	0.7	1.0		MIT
3299.56	3300.51	30298.4	CB		1.0		MIT
3298.405	3299.355	30308.96	CB	0.5	0.7		MIT
3297.664	3298.614	30315.77	CB	0.0	1.5		MIT
3297.290	3298.240	30319.21	CB	0.7R	0.3H		MIT
3297.048	3297.997	30321.43	CB	0.5	1.7		MIT
3296.478	3297.427	30326.67	CB	0.7	0.5		ME
3296.012	3296.961	30330.96	CB	1.3	1.6		MIT
3295.503	3296.452	30335.65	CB	0.0	1.3		MIT
3294.360	3295.309	30346.17	CB	0.3	2.0		MIT
3292.372	3293.320	30364.49	CB		1.3		MIT
3292.020	3292.968	30367.74	CB	0.5	2.0		MIT
3291.918	3292.866	30368.68	CB	0.5	0.3		MIT
3291.059	3292.007	30376.61	CB	1.0	2.0		MIT
3290.008	3290.956	30386.31	CB	1.0	1.0		MIT
3289.548	3290.496	30390.56	CB		1.0		MIT
3289.450	3290.398	30391.47	CB	0.3	0.0		MIT
3287.920	3288.867	30405.61	CB	0.7	0.7		MIT
3287.593	3288.540	30408.63	CB	1.4	0.3		MIT
3286.333	3287.280	30420.29	CB	0.0	1.3		MIT
3285.70	3286.65	30426.1	CB		1.0W		MIT
3285.659	3286.606	30426.53	CB	0.7	0.7		MIT
3283.463	3284.409	30446.88	CB	0.3	2.0		MIT
3279.974	3280.919	30479.26	CB		0.7		MIT
3279.826	3280.771	30480.64	CB	0.6			MIT

AIR WAVELENGTH (ANGSTRMS)	VACUUM WAVELENGTH (ANGSTRMS)	WAVENUMBER (KAYSERS)	LMENT	LOG ARC	INTENSITY SPK	DIS	REF
3279.250	3280.195	30485.99	CB		1.0		MIT
3277.672	3278.617	30500.67	CB	0.5	0.7		MIT
3276.470	3277.414	30511.86	CB		0.7		MIT
3275.954	3276.898	30516.67	CB	0.3	0.3		MIT
3274.788	3275.732	30527.53	CB	0.0	1.0		MIT
3273.886	3274.830	30535.94	CB	1.3R	2.0O		MIT
3273.509	3274.452	30539.46	CB	0.0	1.3		MIT
3272.349	3273.292	30550.28	CB		1.0J		MIT
3272.224	3273.167	30551.45	CB	0.0	1.0		MIT
3272.070	3273.013	30552.89	CB	1.1	0.3		MIT
3271.978	3272.921	30553.75	CB	0.6	0.3		MIT
3271.56	3272.50	30557.6	CB		0.5J		MIT
3270.761	3271.704	30565.11	CB	0.7	0.9		MIT
3270.467	3271.410	30567.86	CB	1.2	1.0		MIT
3269.117	3270.059	30580.49	CB	0.3	1.0		MIT
3267.677	3268.619	30593.96	CB	0.0	1.0		MIT
3267.050	3267.992	30599.83	CB	0.7	0.7		MIT
3266.412	3267.354	30605.81	CB	0.3	0.6H		MIT
3266.004	3266.946	30609.63	CB	0.5W			MIT
3265.310	3266.251	30616.14	CB	0.3	0.3		MIT
3264.592	3265.533	30622.87	CB	1.2	1.0		MIT
3264.260	3265.201	30625.99	CB	0.3H	0.0H		MIT
3263.366	3264.307	30634.37	CB	0.5	2.7		MIT
3262.561	3263.502	30641.93	CB		0.7H		MIT
3261.879	3262.820	30648.34	CB	1.0	0.5		MIT
3261.695	3262.635	30650.07	CB	0.0	1.7		MIT
3260.562	3261.502	30660.72	CB	1.2	2.5		MIT
3260.138	3261.078	30664.71	CB	0.7	0.7		MIT
3259.138	3260.078	30674.11	CB	0.5	0.5		MIT
3257.011	3257.950	30694.15	CB	0.6	0.5H		MIT
3256.74	3257.68	30696.7	CB		0.6		MIT
3256.133	3257.072	30702.42	CB		0.3J		MIT
3255.269	3256.208	30710.57	CB	0.3H	1.3		MIT
3254.880	3255.819	30714.24	CB	0.0	1.7		MIT
3254.067	3255.006	30721.91	CB	1.3	2.5		MIT
3252.764	3253.702	30734.22	CB	0.7	0.5		MIT
3252.429	3253.367	30737.39	CB	0.0	1.0		MIT
3251.625	3252.563	30744.99	CB	0.7	0.5		MIT
3251.487	3252.425	30746.29	CB	0.6	0.3		MIT
3251.260	3252.198	30748.44	CB	0.3	1.2		ME
3250.27	3251.21	30757.8	CB	0.7H	2.0		MIT
3249.515	3250.452	30764.95	CB	1.0	1.0		MIT
3248.935	3249.872	30770.44	CB	0.7	1.7		MIT
3247.474	3248.411	30784.28	CB	1.7W	2.0W		MIT
3246.780	3247.717	30790.86	CB	0.7			MIT
3246.689	3247.626	30791.73	CB		0.7		MIT
3245.07	3246.01	30807.1	CB		0.7		MIT
3244.511	3245.447	30812.40	CB	0.3H	1.5		MIT
3243.828	3244.764	30818.88	CB	0.3	0.7		MIT
3243.329	3244.265	30823.62	CB	0.0H	0.7H		MIT
3242.531	3243.467	30831.21	CB	0.5	1.0		MIT
3242.415	3243.351	30832.31	CB	0.0	0.7		MIT
3241.818	3242.753	30837.99	CB	0.3	1.0		MIT
3238.024	3238.958	30874.12	CB	1.3	2.3		MIT
3237.685	3238.619	30877.35	CB	0.3	1.7		MIT
3237.189	3238.123	30882.09	CB	0.3	0.3		MIT
3236.403	3237.337	30889.59	CB	1.0	2.3		MIT
3232.793	3233.726	30924.08	CB		0.3H		MIT
3230.240	3231.172	30948.52	CB	0.3	1.5		MIT
3229.563	3230.495	30955.00	CB	0.7	1.7		MIT
3228.952	3229.884	30960.86	CB	0.0	0.5		MIT
3228.486	3229.418	30965.33	CB		0.3H		MIT
3227.69	3228.62	30973.0	CB		0.5H		MIT
3225.479	3226.410	30994.20	CB	2.2W	2.9M		MIT
3225.193	3226.124	30996.95	CB	0.5	0.0		MIT
3224.430	3225.361	31004.28	CB	0.5	0.6		MIT
3223.901	3224.832	31009.37	CB	0.7W	0.5H		MIT
3223.324	3224.255	31014.92	CB	1.0	2.0		MIT
3222.952	3223.883	31018.50	CB	0.3	0.3		MIT
3222.751	3223.682	31020.43	CB	0.5	0.3		MIT

AIR WAVELENGTH (ANGSTRMS)	VACUUM WAVELENGTH (ANGSTRMS)	WAVENUMBER (KAYSERS)	LMENT		LOG ARC	INTENSITY SPK	INTENSITY DIS	REF
3222.070	3223.000	31026.99	CB		0.7	1.5		MIT
3221.651	3222.581	31031.02	CB			1.0W		MIT
3221.125	3222.055	31036.09	CB		0.6	0.7		MIT
3220.927	3221.857	31038.00	CB		1.0	1.0		MIT
3220.488	3221.418	31042.23	CB		0.5H	0.7		MIT
3219.671	3220.601	31050.11	CB		0.3	0.3		MIT
3217.864	3218.793	31067.54	CB		0.7	0.7		MIT
3217.797	3218.726	31068.19	CB		0.5	0.0		ME
3217.286	3218.215	31073.12	CB		1.0	0.7		MIT
3217.02	3217.95	31075.7	CB		0.3W	2.3W		MIT
3216.189	3217.118	31083.72	CB		0.3	1.0		MIT
3215.595	3216.524	31089.46	CB		1.7	2.3		MIT
3215.226	3216.155	31093.03	CB		0.7	0.3		MIT
3214.999	3215.928	31095.23	CB			0.7		MIT
3213.882	3214.810	31106.03	CB		0.3	0.3		MIT
3212.143	3213.071	31122.87	CB		0.0	0.5		MIT
3211.812	3212.740	31126.08	CB		0.0	1.0		MIT
3211.40	3212.33	31130.1	CB			0.7		MIT
3210.290	3211.217	31140.84	CB		0.5	0.6		MIT
3210.207	3211.134	31141.64	CB		0.0	0.5		MIT
3208.585	3209.512	31157.39	CB		0.5W	2.0W		MIT
3208.392	3209.319	31159.26	CB			0.3		MIT
3208.09	3209.02	31162.2	CB		0.3H	1.0		MIT
3207.335	3208.262	31169.53	CB		0.7	1.5		MIT
3206.808	3207.735	31174.65	CB		0.5	0.3		MIT
3206.343	3207.269	31179.17	CB		1.0	2.5		MIT
3204.970	3205.896	31192.53	CB		1.0	2.2		MIT
3204.654	3205.580	31195.60	CB		0.7	0.0H		MIT
3203.353	3204.279	31208.27	CB		0.9	1.7		MIT
3203.138	3204.064	31210.37	CB			0.7J		MIT
3200.534	3201.459	31235.76	CB		0.6	0.7		MIT
3198.221	3199.145	31258.35	CB		0.0	1.7		MIT
3197.33	3198.25	31267.1	CB	3		1.3		IS
3196.184	3197.108	31278.27	CB		0.5H			MIT
3196.179	3197.103	31278.32	CB			0.7		MIT
3194.977	3195.901	31290.08	CB	II	1.5	2.5		MIT
3194.275	3195.198	31296.96	CB		0.3W	2.2W		MIT
3192.418	3193.341	31315.17	CB			0.3		MIT
3191.427	3192.350	31324.89	CB		0.5	2.3		MIT
3191.096	3192.019	31328.14	CB		2.0W	2.5W		MIT
3190.429	3191.351	31334.69	CB			0.5		MIT
3189.282	3190.204	31345.96	CB		1.0W	2.5R		MIT
3188.823	3189.745	31350.47	CB			0.7H		MIT
3187.493	3188.415	31363.55	CB		1.1	1.0		MIT
3186.544	3187.465	31372.89	CB		0.7	0.7		MIT
3184.223	3185.144	31395.76	CB		0.7	2.2		MIT
3181.399	3182.319	31423.62	CB		0.7	1.3		MIT
3180.290	3181.210	31434.58	CB		0.7	2.3		MIT
3179.233	3180.153	31445.03	CB		0.3	1.0		MIT
3178.630	3179.549	31451.00	CB		0.3	0.7J		MIT
3177.762	3178.681	31459.59	CB		0.0	0.5		MIT
3175.850	3176.769	31478.53	CB		0.7	1.7		MIT
3175.785	3176.704	31479.17	CB		0.0	1.3		MIT
3175.683	3176.602	31480.18	CB		0.0	1.3		MIT
3174.44	3175.36	31492.5	CB			1.0H		MIT
3173.200	3174.118	31504.81	CB		0.3	2.0		MIT
3172.510	3173.428	31511.67	CB		0.5	0.5		MIT
3172.303	3173.221	31513.72	CB		0.3	0.3		MIT
3172.26	3173.18	31514.1	CB			0.5H		MIT
3171.79	3172.71	31518.8	CB			1.4H		MIT
3171.428	3172.346	31522.42	CB		0.7	1.0W		MIT
3171.168	3172.086	31525.00	CB		0.3	0.5		MIT
3170.160	3171.077	31535.02	CB		0.3	0.5		MIT
3168.598	3169.515	31550.57	CB		0.3	0.3		MIT
3163.402	3164.318	31602.39	CB	II	1.2	0.9		MIT
3163.146	3164.062	31604.95	CB		0.0	0.6		MIT
3159.853	3160.768	31637.88	CB		0.0	1.0		MIT
3159.325	3160.240	31643.17	CB		0.3	0.5		MIT
3158.102	3159.016	31655.42	CB		0.0	0.7		MIT
3155.597	3156.511	31680.55	CB			1.2		MIT
3154.815	3155.728	31688.41	CB		0.5W	2.3W		MIT
3153.987	3154.900	31696.72	CB		0.3			MIT
3153.853	3154.766	31698.07	CB		0.0	1.3		MIT
3153.372	3154.285	31702.91	CB		0.0	1.5H		MIT
3152.782	3153.695	31708.84	CB		0.3	1.5		MIT
3152.159	3153.072	31715.10	CB		0.3	1.7		MIT
3151.870	3152.783	31718.01	CB		0.7	0.3		MIT
3150.407	3151.319	31732.74	CB		0.0	1.7		MIT
3148.713	3149.625	31749.81	CB		0.3	0.3		MIT
3146.926	3147.837	31767.84	CB		0.3	1.0		MIT
3146.868	3147.779	31768.43	CB		0.0	1.2		MIT
3145.400	3146.311	31783.25	CB		1.0	2.0		MIT
3144.343	3145.254	31793.94	CB		0.7H	1.0		MIT
3142.256	3143.166	31815.05	CB			1.2		MIT
3140.499	3141.409	31832.85	CB		0.5	1.0		MIT
3136.970	3137.879	31868.66	CB		0.9	0.5		MIT
3135.916	3136.825	31879.37	CB		0.3	1.0		MIT
3135.404	3136.313	31884.58	CB		0.3	1.0		MIT
3134.338	3135.246	31895.42	CB		0.3	1.2		MIT
3133.083	3133.991	31908.20	CB		0.5	0.6		MIT
3132.764	3133.672	31911.45	CB		0.0	1.2		MIT
3132.137	3133.045	31917.83	CB		0.0	0.3		MIT
3132.011	3132.919	31919.12	CB		0.0	0.7		MIT
3130.786	3131.693	31931.61	CB		2.0	2.0		MIT
3129.644	3130.551	31943.26	CB		0.0	0.7		MIT
3128.899	3129.806	31950.86	CB		0.3H	0.7H		MIT
3128.368	3129.275	31956.29	CB		0.3	1.0		MIT
3127.526	3128.433	31964.89	CB		1.0W	1.7		ME
3125.892	3126.798	31981.60	CB		0.0	1.2		ME
3122.646	3123.551	32014.84	CB		0.9	0.5		MIT
3121.960	3122.865	32021.88	CB		0.3	0.5		MIT
3116.57	3117.47	32077.2	CB		0.0	0.6J		MIT
3116.365	3117.269	32079.36	CB		0.9	0.5		MIT
3115.537	3116.441	32087.89	CB		0.0	0.7		MIT
3115.15	3116.05	32091.9	CB			0.9H		MIT
3113.175	3114.078	32112.23	CB		0.0	0.7		MIT
3112.375	3113.278	32120.49	CB		0.0H	0.6H		MIT
3111.612	3112.515	32128.36	CB			0.6H		MIT
3111.450	3112.353	32130.04	CB		0.6	0.7		MIT
3110.800	3111.702	32136.75	CB		0.0H	0.7		MIT
3109.736	3110.638	32147.75	CB		0.3	0.3		MIT
3106.984	3107.885	32176.22	CB		0.3	1.0		MIT
3106.520	3107.421	32181.02	CB		0.3	0.7		ME
3104.260	3105.161	32204.45	CB		0.0	0.7		MIT
3101.917	3102.817	32228.78	CB		0.3	1.0		MIT
3100.97	3101.87	32238.6	CB		0.3	0.0		MIT
3100.810	3101.710	32240.28	CB		0.0	0.7		MIT
3100.25	3101.15	32246.1	CB			1.3		ME
3100.168	3101.068	32246.96	CB		0.0	0.7		MIT
3099.186	3100.085	32257.18	CB		0.5	1.2		MIT
3098.470	3099.369	32264.63	CB		0.0	0.7		MIT
3097.122	3098.021	32278.67	CB		0.5W	2.0W		MIT
3096.495	3097.394	32285.21	CB		0.6	0.5		MIT
3094.359	3095.257	32307.49	CB		0.3			MIT
3094.183	3095.081	32309.33	CB	II	2.0	3.0		MIT
3092.886	3093.784	32322.88	CB		0.0H	0.7		MIT
3088.267	3089.164	32371.22	CB		0.3			MIT
3088.029	3088.926	32373.72	CB		0.5	0.7		MIT
3087.859	3088.756	32375.50	CB		0.3	1.0		MIT
3086.072	3086.968	32394.24	CB			1.0		MIT
3084.373	3085.269	32412.09	CB		0.0	1.2		MIT
3083.333	3084.228	32423.02	CB		0.3	1.2H		MIT
3082.857	3083.752	32428.03	CB		0.5	0.7		MIT
3081.769	3082.664	32439.47	CB		0.0	1.2		MIT
3081.097	3081.992	32446.55	CB			1.0		MIT
3080.350	3081.245	32454.42	CB		0.9	2.0		MIT
3077.443	3078.337	32485.07	CB		0.0	1.0		MIT
3076.873	3077.767	32491.09	CB		1.0W	1.7		MIT
3075.244	3076.137	32508.30	CB		0.0	1.2		MIT
3074.27	3075.16	32518.6	CB		0.0	0.7		MIT

AIR WAVELENGTH (ANGSTRMS)	VACUUM WAVELENGTH (ANGSTRMS)	WAVENUMBER (KAYSERS)	LMENT	LOG ARC	INTENSITY SPK	DIS	REF
3073.236	3074.129	32529.54	CB	0.3	1.2		MIT
3072.512	3073.405	32537.21	CB	0.3	1.2		MIT
3072.406	3073.299	32538.33	CB	0.7	0.3		MIT
3072.305	3073.198	32539.40	CB	0.3	0.0		MIT
3072.18	3073.07	32540.7	CB	0.0	0.7		MIT
3071.56	3072.45	32547.3	CB	1.0	1.7		MIT
3071.365	3072.257	32549.35	CB	0.3			MIT
3071.178	3072.070	32551.34	CB	0.5	1.2		MIT
3070.901	3071.793	32554.27	CB	0.5	1.2		MIT
3069.680	3070.572	32567.22	CB	0.6	1.2		MIT
3069.518	3070.410	32568.94	CB		0.5		MIT
3069.031	3069.923	32574.11	CB	0.6	0.5		MIT
3068.93	3069.82	32575.2	CB		0.7		MIT
3068.056	3068.948	32584.46	CB		1.0		MIT
3067.533	3068.424	32590.01	CB		1.2		MIT
3066.097	3066.988	32605.28	CB	0.3	1.5		MIT
3065.264	3066.155	32614.14	CB	1.0	2.3		MIT
3064.533	3065.424	32621.92	CB	0.7W	2.3		MIT
3063.790	3064.681	32629.83	CB	0.6	1.0		MIT
3063.130	3064.020	32636.86	CB	0.6D	1.2		MIT
3061.959	3062.849	32649.34	CB		1.0		MIT
3061.404	3062.294	32655.26	CB		0.7H		ME
3061.237	3062.127	32657.04	CB	0.6	0.3		MIT
3061.110	3062.000	32658.39	CB	0.5	0.3		MIT
3060.400	3061.290	32665.97	CB	0.3	0.3		MIT
3059.299	3060.188	32677.73	CB	0.3	1.0		MIT
3057.03	3057.92	32702.0	CB		1.0		MIT
3056.615	3057.504	32706.42	CB	0.5	0.5		MIT
3055.522	3056.410	32718.12	CB	0.3	2.0		MIT
3053.637	3054.525	32738.31	CB	0.0	1.5		MIT
3053.088	3053.976	32744.20	CB	0.5	0.7		MIT
3052.992	3053.880	32745.23	CB	0.3	0.5		MIT
3052.729	3053.617	32748.05	CB	0.3	0.5		MIT
3051.994	3052.882	32755.94	CB	0.3	0.6		MIT
3051.676	3052.564	32759.35	CB	0.3	0.5		MIT
3051.341	3052.228	32762.95	CB	0.0	1.0		MIT
3049.525	3050.412	32782.46	CB		2.2		MIT
3048.634	3049.521	32792.04	CB		0.7		MIT
3048.204	3049.091	32796.66	CB	0.0	1.7		MIT
3048.096	3048.983	32797.83	CB	1.0	0.3		MIT
3046.673	3047.559	32813.14	CB		1.0W		MIT
3045.545	3046.431	32825.30	CB	0.5H	0.3		MIT
3044.763	3045.649	32833.73	CB	0.3	1.5		MIT
3043.277	3044.162	32849.76	CB	0.0	0.7		MIT
3042.790	3043.675	32855.02	CB	0.0	1.0J		MIT
3041.99	3042.88	32863.7	CB		0.7W		MIT
3041.891	3042.776	32864.73	CB	0.0	0.7J		MIT
3041.359	3042.244	32870.47	CB	0.3	0.5		MIT
3040.54	3041.42	32879.3	CB		0.7H		MIT
3039.815	3040.700	32887.17	CB	0.7	2.5		MIT
3039.682	3040.567	32888.61	CB	0.5	1.0H		MIT
3039.406	3040.290	32891.59	CB	0.5	0.7		MIT
3039.187	3040.071	32893.96	CB	0.3	0.5		MIT
3038.16	3039.04	32905.1	CB		1.0J		MIT
3035.025	3035.908	32939.07	CB	0.3	1.5		MIT
3034.95	3035.83	32939.9	CB	0.0	2.0W		ME
3033.393	3034.276	32956.79	CB	0.3	0.3		MIT
3032.768	3033.651	32963.58	CB	0.5	2.5		MIT
3029.86	3030.74	32995.2	CB		0.7J		MIT
3029.745	3030.627	32996.47	CB	0.3	2.0		MIT
3028.75	3029.63	33007.3	CB		0.6		MIT
3028.684	3029.566	33008.03	CB	0.3	0.7H		MIT
3028.443	3029.325	33010.66	CB	1.7	2.3		MIT
3027.888	3028.770	33016.71	CB	0.3	0.3		MIT
3025.372	3026.253	33044.16	CB	0.0	1.0		MIT
3024.738	3025.619	33051.09	CB	1.0W	2.3		MIT
3024.247	3025.128	33056.46	CB	0.3	0.7		MIT
3022.739	3023.619	33072.95	CB	0.7W	2.0		MIT
3022.47	3023.35	33075.9	CB		0.7		MIT
3021.881	3022.761	33082.34	CB		1.20		MIT

AIR WAVELENGTH (ANGSTRMS)	VACUUM WAVELENGTH (ANGSTRMS)	WAVENUMBER (KAYSERS)	LMENT	LOG ARC	INTENSITY SPK	DIS	REF
3020.666	3021.546	33095.64	CB	0.7	0.9		MIT
3019.790	3020.670	33105.24	CB	0.0	0.3H		MIT
3019.567	3020.447	33107.69	CB		0.7		MIT
3019.196	3020.075	33111.76	CB		0.3		MIT
3019.07	3019.95	33113.1	CB	0.0	0.5		MIT
3018.855	3019.734	33115.50	CB	0.0H	1.0		MIT
3018.31	3019.19	33121.5	CB	0.0	0.7H		MIT
3015.823	3016.702	33148.79	CB		1.3		MIT
3015.237	3016.115	33155.23	CB	0.5	0.0		MIT
3015.00	3015.88	33157.8	CB		1.2		MIT
3014.445	3015.323	33163.94	CB	0.5	1.0		MIT
3012.549	3013.427	33184.81	CB	0.3	0.7		MIT
3011.61	3012.49	33195.2	CB		0.6H		MIT
3010.686	3011.563	33205.35	CB	0.0	1.2		MIT
3010.375	3011.252	33208.78	CB	0.0	1.2		MIT
3008.964	3009.841	33224.35	CB		1.0		MIT
3008.415	3009.292	33230.41	CB		1.7		MIT
3005.767	3006.643	33259.69	CB	0.0	1.7		MIT
3005.143	3006.019	33266.59	CB	0.3	0.3		MIT
3004.625	3005.501	33272.33	CB		1.2		MIT
3003.74	3004.62	33282.1	CB		1.0		MIT
3002.212	3003.087	33299.07	CB	0.70	1.5		MIT
3001.85	3002.73	33303.1	CB		1.2J		MIT
3001.125	3002.000	33311.13	CB	0.0	1.2		MIT
3000.119	3000.994	33322.30	CB	0.3	0.7		MIT
2998.226	2999.100	33343.33	CB	0.3	0.3		MIT
2997.486	2998.360	33351.57	CB	0.0	1.0		MIT
2996.797	2997.671	33359.23	CB	0.0	0.7		MIT
2996.513	2997.387	33362.40	CB	0.5	0.5H		MIT
2994.728	2995.601	33382.28	CB	2.0	2.5		MIT
2993.96	2994.83	33390.8	CB		1.7		MIT
2993.807	2994.680	33392.55	CB	0.3	0.7		MIT
2991.950	2992.823	33413.27	CB	0.0	2.0		MIT
2991.435	2992.307	33419.02	CB		1.0		MIT
2990.258	2991.130	33432.18	CB	0.7	2.3		MIT
2990.044	2990.916	33434.57	CB		0.5		MIT
2989.944	2990.816	33435.69	CB		0.7J		MIT
2988.792	2989.664	33448.58	CB	0.3	0.3		MIT
2987.55	2988.42	33462.5	CB	0.0W	1.0W		MIT
2987.289	2988.160	33465.40	CB	0.7	0.6		MIT
2986.840	2987.711	33470.44	CB	0.0	0.7H		MIT
2986.40	2987.27	33475.4	CB		0.7J		MIT
2985.049	2985.920	33490.52	CB	0.3H	1.7		MIT
2983.574	2984.445	33507.07	CB	0.7	1.2J		MIT
2983.141	2984.011	33511.94	CB	0.3	0.3		MIT
2982.91	2983.78	33514.5	CB		0.7H		MIT
2982.106	2982.976	33523.57	CB	1.0W	1.9		MIT
2981.643	2982.513	33528.77	CB	0.5	0.6		MIT
2980.717	2981.587	33539.19	CB	0.5	1.7		MIT
2979.875	2980.745	33548.66	CB	0.3	1.5		MIT
2978.943	2979.812	33559.16	CB	0.0	1.7		MIT
2977.681	2978.550	33573.38	CB	0.0	2.5		MIT
2974.725	2975.593	33606.74	CB		1.0		MIT
2974.555	2975.423	33608.66	CB	0.3	0.0		MIT
2974.098	2974.966	33613.83	CB	0.7	2.3		MIT
2973.31	2974.18	33622.7	CB		0.7		MIT
2972.572	2973.440	33631.08	CB	1.6	2.0		MIT
2970.468	2971.335	33654.90	CB	0.0	0.7		MIT
2970.394	2971.261	33655.74	CB		0.7		MIT
2968.292	2969.159	33679.57	CB	0.0D	0.7		MIT
2965.871	2966.737	33707.06	CB		0.5H		ME
2965.64	2966.51	33709.7	CB		0.7H		MIT
2965.480	2966.346	33711.51	CB	0.7	0.3		MIT
2963.683	2964.549	33731.95	CB	0.3	0.5		MIT
2961.628	2962.493	33755.35	CB	0.0	1.0		MIT
2956.890	2957.754	33809.44	CB	0.0	1.2		MIT
2955.446	2956.310	33825.96	CB	0.3H	0.3		MIT
2954.531	2955.394	33836.43	CB	0.3	0.7		MIT
2954.022	2954.885	33842.26	CB	0.0	0.7		MIT
2950.878	2951.740	33878.32	CB	2.2	2.3		MIT

AIR WAVELENGTH (ANGSTRMS)	VACUUM WAVELENGTH (ANGSTRMS)	WAVENUMBER (KAYSERS)	LMENT	LOG ARC	INTENSITY SPK	DIS	REF
2949.502	2950.364	33894.12	CB		0.7		MIT
2946.898	2947.759	33924.07	CB	0.5	1.5		MIT
2946.116	2946.977	33933.08	CB	0.6	1.3		MIT
2945.883	2946.744	33935.76	CB	0.3	2.0		MIT
2945.748	2946.609	33937.31	CB	0.5			MIT
2944.644	2945.505	33950.04	CB	0.6	0.3H		MIT
2941.543	2942.403	33985.83	CB	1.7	2.5		MIT
2940.85	2941.71	33993.8	CB		0.7		MIT
2938.073	2938.932	34025.96	CB	0.5	0.7		MIT
2937.707	2938.566	34030.20	CB		1.3J		ME
2937.333	2938.192	34034.54	CB	0.3	1.2		MIT
2936.66	2937.52	34042.3	CB		1.5		MIT
2936.457	2937.316	34044.69	CB	0.0	0.3H		MIT
2935.287	2936.146	34058.26	CB	0.3	1.2		MIT
2934.976	2935.834	34061.87	CB	0.3	0.3		MIT
2932.662	2933.520	34088.74	CB	0.0H	1.9		MIT
2932.13	2932.99	34094.9	CB		1.7		MIT
2931.469	2932.327	34102.61	CB	0.5	1.7		MIT
2930.845	2931.702	34109.87	CB		0.7		MIT
2930.643	2931.500	34112.23	CB		1.0		MIT
2930.263	2931.120	34116.65	CB		1.3J		MIT
2927.810	2928.667	34145.23	CB	2.3	2.9R		MIT
2924.827	2925.683	34180.05	CB	0.6	0.5		MIT
2923.028	2923.884	34201.09	CB	0.5	0.3		MIT
2922.47	2923.33	34207.6	CB		0.7W		MIT
2919.32	2920.17	34244.5	CB		0.7H		MIT
2918.917	2919.771	34249.25	CB	0.0	1.0H		MIT
2918.56	2919.41	34253.4	CB		1.0		MIT
2917.052	2917.906	34271.15	CB	1.0W	2.0R		MIT
2916.504	2917.358	34277.59	CB	0.5	0.5		MIT
2916.094	2916.948	34282.41	CB		1.0		MIT
2915.408	2916.262	34290.48	CB	0.0	1.0		MIT
2911.745	2912.598	34333.61	CB	0.9	2.0		MIT
2910.686	2911.538	34346.10	CB	0.3	0.7		MIT
2910.587	2911.439	34347.27	CB	1.0	2.0		MIT
2908.979	2909.831	34366.26	CB	0.0	0.7		MIT
2908.881	2909.733	34367.41	CB	0.3	1.3		MIT
2908.243	2909.095	34374.95	CB	1.3R	2.3		MIT
2907.520	2908.372	34383.50	CB		1.2H		MIT
2906.795	2907.647	34392.08	CB		0.8J		MIT
2904.86	2905.71	34415.0	CB		1.0		MIT
2904.160	2905.011	34423.28	CB	0.0	0.6		MIT
2903.654	2904.505	34429.28	CB	0.7	0.5		MIT
2900.675	2901.525	34464.63	CB		2.0J		MIT
2899.239	2900.089	34481.71	CB	1.3	2.7		MIT
2897.812	2898.661	34498.68	CB	1.2	2.2		MIT
2897.351	2898.200	34504.17	CB	0.5	0.3		MIT
2897.002	2897.851	34508.33	CB	0.3	0.5		MIT
2896.248	2897.097	34517.31	CB	0.3	0.0		MIT
2895.364	2896.213	34527.85	CB	0.0	1.0		MIT
2894.911	2895.760	34533.25	CB	0.3	0.7J		MIT
2894.425	2895.273	34539.05	CB	0.0	1.3		MIT
2893.069	2893.917	34555.24	CB		2.0O		MIT
2891.41	2892.26	34575.1	CB		1.0J		MIT
2890.558	2891.405	34585.26	CB	0.0	0.7		MIT
2890.351	2891.198	34587.73	CB	0.0H	1.0		MIT
2890.181	2891.028	34589.77	CB	0.5	0.0		MIT
2889.900	2890.747	34593.13	CB	0.6	0.3		MIT
2888.833	2889.680	34605.91	CB	1.0	2.0		MIT
2887.692	2888.539	34619.58	CB	0.3	1.2		MIT
2887.298	2888.145	34624.30	CB	0.3	0.0		MIT
2887.093	2887.940	34626.76	CB	0.0H	1.0		MIT
2886.62	2887.47	34632.4	CB		0.9		MIT
2886.322	2887.168	34636.01	CB	0.3	0.7		MIT
2884.972	2885.818	34652.22	CB	0.5	0.6		MIT
2883.178	2884.024	34673.78	CB	2.0	2.9G		MIT
2882.474	2883.320	34682.25	CB	0.0	0.7		MIT
2880.715	2881.560	34703.42	CB	0.6W	1.7		MIT
2879.495	2880.340	34718.13	CB	1.4	0.3		MIT
2879.364	2880.209	34719.71	CB	0.3D	1.0		MIT

AIR WAVELENGTH (ANGSTRMS)	VACUUM WAVELENGTH (ANGSTRMS)	WAVENUMBER (KAYSERS)	LMENT	LOG ARC	INTENSITY SPK	DIS	REF
2879.22	2880.06	34721.4	CB		0.7H		MIT
2878.739	2879.584	34727.24	CB	0.5	1.0		MIT
2878.165	2879.009	34734.17	CB	0.3			MIT
2877.852	2878.696	34737.95	CB	0.3	0.6		MIT
2877.62	2878.46	34740.7	CB		0.7O		MIT
2877.032	2877.876	34747.85	CB	0.5	1.0W		MIT
2876.947	2877.791	34748.87	CB	1.6O	2.7O		MIT
2875.392	2876.236	34767.67	CB	1.7R	2.5		MIT
2874.566	2875.410	34777.65	CB	0.7	0.5		MIT
2872.799	2873.642	34799.04	CB	0.0	1.0		MIT
2870.654	2871.497	34825.05	CB	0.3			MIT
2870.35	2871.19	34828.7	CB		1.0J		MIT
2869.828	2870.670	34835.07	CB	0.0	1.0		MIT
2868.525	2869.367	34850.89	CB	1.2	2.5		MIT
2866.840	2867.682	34871.37	CB		0.3		MIT
2866.669	2867.511	34873.45	CB	0.5	0.5		MIT
2865.609	2866.450	34886.35	CB	0.7	1.5		MIT
2864.321	2865.162	34902.04	CB	0.7	1.0		MIT
2861.647	2862.487	34934.65	CB		1.0W		MIT
2861.093	2861.933	34941.42	CB	1.0	2.0		MIT
2859.963	2860.803	34955.22	CB	0.7	0.5		MIT
2859.435	2860.275	34961.67	CB		0.5		MIT
2859.038	2859.878	34966.53	CB	0.0	1.7		MIT
2857.362	2858.201	34987.04	CB		0.7		MIT
2857.293	2858.132	34987.88	CB	0.6	0.5		MIT
2855.538	2856.377	35009.39	CB	0.0	1.0		MIT
2855.082	2855.921	35014.98	CB		0.6		MIT
2854.168	2855.007	35026.19	CB	0.7	0.6		MIT
2853.519	2854.357	35034.15	CB	0.0	1.0		MIT
2851.977	2852.815	35053.10	CB	0.6	0.7		MIT
2851.447	2852.285	35059.61	CB	0.7	0.5		MIT
2851.295	2852.133	35061.48	CB		0.7		MIT
2850.381	2851.219	35072.72	CB	0.0	0.7		MIT
2849.557	2850.394	35082.86	CB	0.3W	2.0W		MIT
2848.295	2849.132	35098.41	CB	0.3	1.0		MIT
2848.017	2848.854	35101.83	CB	0.0	0.9		MIT
2847.237	2848.074	35111.45	CB	0.0	1.0		MIT
2846.285	2847.122	35123.19	CB	1.0	1.7		MIT
2845.802	2846.638	35129.15	CB	0.5	1.0		MIT
2844.435	2845.271	35146.04	CB	0.3	1.0		MIT
2843.637	2844.473	35155.90	CB	0.5	1.0		MIT
2842.648	2843.484	35168.13	CB	1.0	2.0		MIT
2842.021	2842.857	35175.89	CB	0.5	0.3		MIT
2841.146	2841.981	35186.72	CB	1.0	2.0		MIT
2840.936	2841.771	35189.32	CB	0.7	0.3		MIT
2839.80	2840.64	35203.4	CB	0.3	0.7		MIT
2839.598	2840.433	35205.90	CB		1.0H		MIT
2836.241	2837.075	35247.57	CB	0.5	0.7		MIT
2835.116	2835.950	35261.55	CB	0.7D	2.0		MIT
2833.304	2834.137	35284.10	CB	0.0	1.0		MIT
2832.790	2833.623	35290.51	CB		0.7H		MIT
2830.836	2831.669	35314.86	CB		0.7		MIT
2830.571	2831.404	35318.17	CB	0.0	1.5		MIT
2829.752	2830.585	35328.39	CB	0.5	1.7		MIT
2827.116	2827.948	35361.33	CB		1.3		ME
2827.077	2827.909	35361.82	CB	0.9	1.7		MIT
2826.99	2827.82	35362.9	CB		0.6H		MIT
2826.475	2827.307	35369.35	CB	0.5	0.7		MIT
2825.855	2826.687	35377.11	CB	0.3	0.7		MIT
2825.183	2826.014	35385.52	CB	0.7	0.5		MIT
2823.881	2824.712	35401.84	CB	0.0	1.0		MIT
2823.327	2824.158	35408.78	CB	0.0	0.7		MIT
2821.921	2822.752	35426.42	CB	0.6	0.5		MIT
2820.804	2821.634	35440.45	CB	0.5	1.3		MIT
2819.894	2820.724	35451.89	CB		1.0J		MIT
2819.214	2820.044	35460.44	CB	0.9	0.5		MIT
2818.740	2819.570	35466.40	CB	0.3	1.0		MIT
2818.199	2819.029	35473.21	CB		1.0O		ME
2817.310	2818.140	35484.40	CB	0.0	0.7		MIT
2816.677	2817.506	35492.38	CB	0.3	1.7		MIT

AIR WAVELENGTH (ANGSTRMS)	VACUUM WAVELENGTH (ANGSTRMS)	WAVENUMBER (KAYSERS)	LMENT	LOG ARC	INTENSITY SPK	INTENSITY DIS	REF
2815.399	2816.228	35508.49	CB		1.1		ME
2811.627	2812.455	35556.12	CB	0.3	1.2		MIT
2810.812	2811.640	35566.43	CB	0.5	2.0		MIT
2809.657	2810.485	35581.05	CB		1.0		MIT
2809.172	2810.000	35587.19	CB	0.3	1.0		MIT
2808.750	2809.577	35592.54	CB	0.0H	1.2		MIT
2808.054	2808.881	35601.36	CB	0.7	0.3		MIT
2806.913	2807.740	35615.83	CB		0.9H		ME
2805.95	2806.78	35628.1	CB		1.0H		MIT
2805.808	2806.635	35629.86	CB	0.0	1.0H		MIT
2803.808	2804.634	35655.27	CB	0.5	1.2		MIT
2803.54	2804.37	35658.7	CB		0.5V		ME
2802.715	2803.541	35669.18	CB	0.3	0.7		MIT
2801.551	2802.377	35684.00	CB		1.0W		ME
2800.318	2801.143	35699.71	CB	0.5	0.6		MIT
2799.357	2800.182	35711.96	CB	0.5	0.3		MIT
2799.175	2800.000	35714.29	CB	0.3	0.9		MIT
2798.909	2799.734	35717.68	CB	0.3	1.2		MIT
2797.693	2798.518	35733.20	CB	1.0	2.3		MIT
2795.865	2796.689	35756.56	CB	0.6	0.5		MIT
2795.138	2795.962	35765.86	CB	0.0	1.2		MIT
2793.885	2794.709	35781.90	CB		1.0		MIT
2793.695	2794.519	35784.34	CB		0.7		MIT
2793.048	2793.872	35792.63	CB	1.0W	2.0		MIT
2791.740	2792.563	35809.40	CB	0.5	2.0		MIT
2791.372	2792.195	35814.12	CB	0.0	0.7		MIT
2790.573	2791.396	35824.37	CB	0.3	1.0		MIT
2789.742	2790.565	35835.04	CB		0.5H		MIT
2789.206	2790.029	35841.93	CB	0.3	0.7H		MIT
2788.686	2789.509	35848.61	CB		0.7		MIT
2785.068	2785.890	35895.18	CB	0.0	0.7		MIT
2784.445	2785.266	35903.21	CB		1.0		MIT
2782.805	2783.626	35924.37	CB		0.7		MIT
2782.363	2783.184	35930.07	CB	1.0	1.3		MIT
2780.983	2781.804	35947.90	CB	0.0	0.6		MIT
2780.245	2781.065	35957.44	CB	1.5	2.3R		MIT
2779.719	2780.539	35964.25	CB	0.7	0.6		MIT
2779.361	2780.181	35968.88	CB	0.7	0.7		MIT
2778.016	2778.836	35986.29	CB		0.7		MIT
2774.486	2775.305	36032.08	CB		1.0		MIT
2774.060	2774.879	36037.61	CB	0.3	0.7D		MIT
2773.203	2774.022	36048.74	CB	1.2	1.0		MIT
2771.654	2772.472	36068.89	CB	0.3	1.3		MIT
2771.404	2772.222	36072.14	CB	0.3	1.2		MIT
2769.567	2770.385	36096.07	CB	0.3	1.5		MIT
2769.300	2770.118	36099.55	CB		1.2		MIT
2768.128	2768.946	36114.83	CB	1.0	2.0		MIT
2767.208	2768.025	36126.84	CB	0.0	0.7		MIT
2766.184	2767.001	36140.21	CB	0.5	0.3		MIT
2765.928	2766.745	36143.56	CB	0.7	1.0		MIT
2765.279	2766.096	36152.04	CB	0.5	1.0		MIT
2764.562	2765.379	36161.41	CB	0.5	1.0		MIT
2763.597	2764.413	36174.04	CB	0.3	1.0		MIT
2763.384	2764.200	36176.83	CB	0.7	0.7		MIT
2762.489	2763.305	36188.55	CB	0.3	0.7		MIT
2762.325	2763.141	36190.70	CB	0.3	1.0		MIT
2760.998	2761.814	36208.09	CB	0.7	0.7		MIT
2759.970	2760.786	36221.58	CB	0.3	1.0H		MIT
2759.161	2759.976	36232.19	CB		1.0		MIT
2758.78	2759.60	36237.2	CB	0.0	2.0W		MIT
2758.612	2759.427	36239.40	CB	1.0	1.2		MIT
2757.506	2758.321	36253.94	C*	0.3	1.5		MIT
2757.260	2758.075	36257.17	CB	0.5	1.7		MIT
2755.637	2756.451	36278.53	CB	0.7	0.3		MIT
2755.565	2756.379	36279.48	CB	0.0	0.6		MIT
2755.289	2756.103	36283.11	CB	0.7	1.0		MIT
2754.522	2755.336	36293.21	CB	1.0W	2.0		MIT
2754.073	2754.887	36299.13	CB	0.5	0.5		MIT
2753.742	2754.556	36303.49	CB	0.3	0.7		MIT
2753.553	2754.367	36305.98	CB	0.3	0.5		MIT
2753.138	2753.952	36311.46	CB	0.3	1.9		MIT
2753.006	2753.820	36313.20	CB	0.7	1.7H		MIT
2752.018	2752.832	36326.23	CB	0.3	0.7		MIT
2750.578	2751.391	36345.25	CB	0.3	1.5		MIT
2749.824	2750.637	36355.21	CB	0.3	0.9		MIT
2749.685	2750.498	36357.05	CB	0.0	0.7		MIT
2748.849	2749.662	36368.11	CB	1.0	1.0		MIT
2748.072	2748.885	36378.39	CB	0.0	1.2		MIT
2747.600	2748.413	36384.64	CB		1.0		MIT
2747.376	2748.188	36387.61	CB	0.0	0.7		MIT
2746.914	2747.726	36393.73	CB	0.7	1.3		MIT
2746.104	2746.916	36404.46	CB		1.7H		MIT
2745.729	2746.541	36409.43	CB	0.5	1.7		MIT
2745.302	2746.114	36415.10	CB	0.7W	1.5W		MIT
2744.965	2745.777	36419.56	CB	0.0	0.7W		MIT
2744.447	2745.259	36426.44	CB	0.0	1.0		MIT
2743.478	2744.290	36439.30	CB	0.0	1.3		MIT
2742.596	2743.407	36451.02	CB	0.0	1.3		MIT
2741.147	2741.958	36470.29	CB	0.6	1.0		MIT
2740.184	2740.995	36483.11	CB	0.5	1.7		MIT
2739.238	2740.048	36495.71	CB		1.7		MIT
2737.088	2737.898	36524.37	CB	0.7	2.0		MIT
2736.521	2737.331	36531.94	CB	0.0	0.5W		ME
2735.952	2736.762	36539.54	CB		1.6		MIT
2734.732	2735.541	36555.83	CB		1.0		MIT
2734.350	2735.159	36560.94	CB	0.7	1.9		MIT
2733.737	2734.546	36569.14	CB	0.3	1.7		MIT
2733.463	2734.272	36572.81	CB	0.5	1.6		MIT
2733.259	2734.068	36575.54	CB	1.5	1.7		MIT
2730.318	2731.126	36614.93	CB	0.3	2.3		MIT
2729.828	2730.636	36621.50	CB	0.5			MIT
2729.374	2730.182	36627.59	CB	0.3D	0.7		MIT
2728.078	2728.886	36644.99	CB	0.5	0.0		MIT
2727.433	2728.241	36653.66	CB	0.3	1.0		MIT
2726.080	2726.887	36671.85	CB	0.5	0.3		MIT
2724.955	2725.762	36686.99	CB	0.0	0.6		MIT
2723.984	2724.791	36700.06	CB	1.0	0.3		MIT
2723.662	2724.469	36704.40	CB	0.3	2.3		MIT
2722.683	2723.489	36717.60	CB	0.0	1.0		MIT
2722.306	2723.112	36722.69	CB	0.5	0.0		MIT
2721.983	2722.789	36727.04	CB	1.0	2.3		MIT
2721.627	2722.433	36731.85	CB	0.6	0.9		MIT
2721.156	2721.962	36738.20	CB	0.3	0.7		MIT
2720.932	2721.738	36741.23	CB	0.5	0.7		MIT
2720.261	2721.067	36750.29	CB	0.3	0.7		MIT
2720.022	2720.828	36753.52	CB	0.5			MIT
2717.629	2718.434	36785.88	CB	0.3	1.3		MIT
2717.329	2718.134	36789.94	CB	0.3	1.3		MIT
2716.624	2717.429	36799.49	CB	1.0	2.3		MIT
2716.308	2717.113	36803.77	CB	0.5	1.5		MIT
2716.095	2716.900	36806.66	CB	0.7	0.0		MIT
2715.879	2716.684	36809.58	CB	0.3	2.0		MIT
2715.687	2716.492	36812.19	CB	0.6	0.0		MIT
2715.344	2716.149	36816.83	CB	0.3	2.0		MIT
2714.198	2715.002	36832.38	CB	0.5	0.7		MIT
2713.734	2714.538	36838.68	CB		0.7		MIT
2711.368	2712.172	36870.82	CB	0.0	0.7		MIT
2709.598	2710.401	36894.90	CB	0.0	0.7		MIT
2707.833	2708.636	36918.95	CB	0.5	1.9		MIT
2707.211	2708.014	36927.43	CB	0.0	0.7		MIT
2706.395	2707.198	36938.57	CB	0.3	1.2		MIT
2705.326	2706.128	36953.16	CB	0.0	0.7		MIT
2704.899	2705.701	36959.00	CB		0.7		MIT
2704.694	2705.496	36961.80	CB	0.3	1.0		MIT
2704.427	2705.229	36965.45	CB	0.3	1.0H		MIT
2704.257	2705.059	36967.77	CB	0.3W	1.7		MIT
2702.519	2703.321	36991.54	CB	0.9	1.9		MIT
2702.196	2702.998	36995.96	CB	1.0	2.0		MIT
2700.876	2701.677	37014.04	CB	0.3	1.3		MIT
2700.555	2701.356	37018.44	CB	0.7D	1.0		MIT

AIR WAVELENGTH (ANGSTRMS)	VACUUM WAVELENGTH (ANGSTRMS)	WAVENUMBER (KAYSERS)	LMENT	LOG ARC	INTENSITY SPK	INTENSITY DIS	REF
2700.314	2701.115	37021.75	CB	0.0	0.7		MIT
2700.153	2700.954	37023.96	CB	0.5	1.0		MIT
2698.862	2699.663	37041.66	CB	0.7	2.3		MIT
2697.063	2697.863	37066.37	CB	1.0	2.7		MIT
2696.050	2696.850	37080.30	CB	0.7	0.3		MIT
2695.597	2696.397	37086.53	CB		0.7		MIT
2695.040	2695.840	37094.19	CB	1.0	0.5		MIT
2694.755	2695.555	37098.12	CB	0.0	0.7		MIT
2694.315	2695.115	37104.17	CB	0.0H	0.7		MIT
2692.653	2693.452	37127.07	CB		0.6H		MIT
2692.000	2692.799	37136.08	CB	0.0	0.9		MIT
2691.773	2692.572	37139.21	CB	1.0	2.0		MIT
2689.878	2690.677	37165.37	CB		0.6H		MIT
2689.072	2689.870	37176.51	CB	0.3	0.7H		MIT
2687.151	2687.949	37203.09	CB	1.0	0.7		MIT
2686.391	2687.189	37213.61	CB	0.3	2.5		MIT
2683.219	2684.016	37257.60	CB	0.3	0.7		MIT
2682.467	2683.264	37268.05	CB	0.0	1.0		MIT
2682.130	2682.927	37272.73	CB	0.7	0.3		MIT
2680.057	2680.853	37301.56	CB	0.6	1.9		MIT
2679.881	2680.677	37304.01	CB	0.6	0.0		MIT
2679.008	2679.804	37316.16	CB	1.0	0.3		MIT
2678.656	2679.452	37321.07	CB	0.5	1.0		MIT
2678.098	2678.894	37328.84	CB	0.0	1.0		MIT
2677.660	2678.456	37334.95	CB	0.3	1.7		MIT
2676.125	2676.920	37356.36	CB	0.3	1.0H		MIT
2675.944	2676.739	37358.89	CB	1.0	2.0		MIT
2675.101	2675.896	37370.66	CB		0.6H		MIT
2674.838	2675.633	37374.33	CB	0.0	0.7		MIT
2674.700	2675.495	37376.26	CB	0.3			MIT
2673.567	2674.362	37392.10	CB	1.0	2.7		MIT
2671.931	2672.725	37414.99	CB	1.3	2.3		MIT
2671.256	2672.050	37424.45	CB	0.3	1.2		MIT
2670.152	2670.946	37439.92	CB	0.0	0.7		MIT
2668.499	2669.292	37463.11	CB	0.0	0.7		MIT
2668.291	2669.084	37466.03	CB	1.0	0.5		MIT
2667.765	2668.558	37473.42	CB	0.5	1.5		MIT
2667.297	2668.090	37479.99	CB	0.7	1.0		MIT
2667.149	2667.942	37482.07	CB	0.3	0.7		MIT
2667.082	2667.875	37483.01	CB	0.0	0.7		MIT
2666.594	2667.387	37489.87	CB	0.7	1.7		MIT
2665.251	2666.044	37508.76	CB	0.5	2.5		MIT
2663.559	2664.351	37532.59	CB	0.3	1.0		MIT
2662.721	2663.513	37544.40	CB	0.0	0.5		MIT
2661.856	2662.648	37556.60	CB	0.5	0.0		MIT
2660.035	2660.826	37582.31	CB	0.3	1.5		MIT
2659.052	2659.843	37596.20	CB	0.5H	1.5		MIT
2658.876	2659.667	37598.69	CB	0.5	0.7		MIT
2658.027	2658.818	37610.70	CB		2.3		MIT
2657.616	2658.407	37616.52	CB	1.2	0.5		MIT
2656.984	2657.775	37625.46	CB	0.6	0.3		MIT
2656.078	2656.868	37638.29	CB	0.9	2.3		MIT
2655.865	2656.655	37641.31	CB	0.3	0.7		MIT
2655.698	2656.488	37643.68	CB	0.7	0.0		MIT
2654.448	2655.238	37661.41	CB	1.0	0.7		MIT
2653.475	2654.265	37675.22	CB	0.5	0.3		MIT
2653.381	2654.171	37676.55	CB	0.7	0.5		MIT
2651.807	2652.596	37698.91	CB	0.3	1.2		MIT
2651.122	2651.911	37708.65	CB	0.5	2.3		MIT
2649.712	2650.501	37728.72	CB	0.0	0.6		MIT
2649.517	2650.306	37731.49	CB	1.0	0.7		MIT
2648.033	2648.822	37752.64	CB		0.6		MIT
2647.505	2648.293	37760.17	CB	1.2H	0.7		MIT
2646.258	2647.046	37777.96	CB	0.9	2.3J		MIT
2645.273	2646.061	37792.02	CB		1.0J		MIT
2644.927	2645.715	37796.97	CB	0.0	0.7		MIT
2642.572	2643.359	37830.65	CB	0.0	0.7		MIT
2642.238	2643.025	37835.43	CB	0.7	2.5		MIT
2641.058	2641.845	37852.34	CB	0.3	1.5		MIT
2640.918	2641.705	37854.34	CB	0.7	0.3		MIT

AIR WAVELENGTH (ANGSTRMS)	VACUUM WAVELENGTH (ANGSTRMS)	WAVENUMBER (KAYSERS)	LMENT	LOG ARC	INTENSITY SPK	INTENSITY DIS	REF
2639.886	2640.673	37869.14	CB	0.3	1.0		MIT
2638.880	2639.666	37883.58	CB	0.5	0.9		MIT
2638.597	2639.383	37887.64	CB	0.3	1.0		MIT
2638.130	2638.916	37894.35	CB		1.3H		MIT
2637.982	2638.768	37896.47	CB	0.3	1.2		MIT
2635.837	2636.623	37927.31	CB		0.7H		MIT
2634.712	2635.497	37943.50	CB	0.7	0.5		MIT
2634.157	2634.942	37951.50	CB		1.2		MIT
2633.161	2633.946	37965.85	CB		2.3		MIT
2632.516	2633.301	37975.15	CB	0.6W	2.5H		MIT
2631.449	2632.234	37990.55	CB	0.0	0.7J		MIT
2630.977	2631.761	37997.37	CB	0.3	1.0		MIT
2629.208	2629.992	38022.93	CB	0.0	0.7		MIT
2628.673	2629.457	38030.67	CB		1.5J		MIT
2628.493	2629.277	38033.27	CB	1.0	0.0		MIT
2628.408	2629.192	38034.50	CB	0.0	0.5		MIT
2627.440	2628.224	38048.51	CB	1.3	0.7		MIT
2627.055	2627.839	38054.09	CB		0.7W		MIT
2626.633	2627.416	38060.20	CB	0.0	0.7		MIT
2626.398	2627.181	38063.61	CB	0.0	0.7		MIT
2623.509	2624.292	38105.52	CB	1.0	0.5		MIT
2623.313	2624.096	38108.37	CB	0.0	0.9		MIT
2623.172	2623.955	38110.42	CB	0.3W	0.7		MIT
2622.954	2623.737	38113.58	CB	0.3	1.3		MIT
2622.003	2622.785	38127.41	CB	0.5	0.3		MIT
2620.588	2621.370	38147.99	CB	0.5	0.0		MIT
2620.448	2621.230	38150.03	CB	0.5	2.3		MIT
2618.446	2619.227	38179.20	CB	0.0	0.7		MIT
2617.430	2618.211	38194.02	CB		1.0		MIT
2617.094	2617.875	38198.92	CB		0.7H		MIT
2616.479	2617.260	38207.90	CB	1.2	0.5		MIT
2616.218	2616.999	38211.71	CB	0.3	1.5		MIT
2614.763	2615.544	38232.97	CB		1.5		MIT
2614.31	2615.09	38239.6	CB	0.3	1.3		MIT
2613.931	2614.711	38245.14	CB	0.0	1.5		MIT
2613.854	2614.634	38246.27	CB	0.5	1.5H		MIT
2612.382	2613.162	38267.81	CB	0.9	0.3		MIT
2612.26	2613.04	38269.6	CB		0.7J		MIT
2610.275	2611.055	38298.70	CB	1.0	0.3		MIT
2609.435	2610.214	38311.03	CB		0.6		MIT
2608.962	2609.741	38317.98	CB	0.3	1.0		MIT
2608.841	2609.620	38319.75	CB	0.5	0.0		MIT
2608.685	2609.464	38322.04	CB	0.0	0.6		MIT
2606.204	2606.983	38358.52	CB		0.7J		ME
2605.068	2605.846	38375.25	CB	0.0	1.6		MIT
2605.016	2605.794	38376.02	CB	0.0	1.6		MIT
2604.754	2605.532	38379.87	CB	0.3	1.6		MIT
2603.734	2604.512	38394.91	CB	0.3	1.0		MIT
2603.309	2604.087	38401.18	CB	0.5	0.0		MIT
2602.488	2603.266	38413.29	CB	0.0	0.7		MIT
2602.007	2602.785	38420.39	CB	0.6	0.3L		MIT
2601.836	2602.614	38422.92	CB	0.6	0.3		MIT
2601.291	2602.068	38430.96	CB	0.6	2.3		MIT
2600.152	2600.929	38447.80	CB	1.1	1.0		MIT
2599.521	2600.298	38457.13	CB		0.7J		MIT
2598.883	2599.660	38466.57	CB		2.2		MIT
2598.041	2598.818	38479.04	CB	0.3	1.0		MIT
2597.752	2598.529	38483.32	CB		1.2J		MIT
2597.139	2597.915	38492.40	CB	0.9	0.5		MIT
2596.968	2597.744	38494.94	CB		0.7H		MIT
2594.967	2595.743	38524.62	CB	0.0	0.7		MIT
2594.740	2595.516	38527.99	CB	0.5	1.9		MIT
2594.338	2595.114	38533.96	CB	0.3	1.0		MIT
2593.764	2594.540	38542.48	CB	0.3H	2.0		MIT
2592.198	2592.973	38565.77	CB	1.3	0.5		MIT
2592.032	2592.807	38568.24	CB	0.3	0.6		MIT
2590.944	2591.719	38584.43	CB	1.2	2.9		MIT
2589.267	2590.042	38609.42	CB	0.5	0.3		MIT
2588.965	2589.739	38613.92	CB	0.0	1.3		MIT
2588.362	2589.136	38622.92	CB	0.3			MIT

AIR WAVELENGTH (ANGSTRMS)	VACUUM WAVELENGTH (ANGSTRMS)	WAVENUMBER (KAYSERS)	LMENT	LOG ARC	INTENSITY SPK	DIS	REF
2587.957	2588.731	38628.96	CB	0.5	1.3		MIT
2587.404	2588.178	38637.22	CB		1.5		MIT
2586.910	2587.684	38644.60	CB	0.0	1.0		MIT
2586.096	2586.870	38656.76	CB		1.6J		MIT
2583.986	2584.759	38688.32	CB	1.0	2.9J		MIT
2583.220	2583.993	38699.79	CB	0.7	0.3		MIT
2583.108	2583.881	38701.47	CB	0.9	0.3		MIT
2581.195	2581.968	38730.15	CB	0.6	0.3		MIT
2580.285	2581.057	38743.81	CB	0.3	2.0		MIT
2578.738	2579.510	38767.05	CB	1.3	1.0		MIT
2578.197	2578.969	38775.19	CB	0.6	0.3		MIT
2576.597	2577.369	38799.26	CB	0.5	0.3		MIT
2575.963	2576.734	38808.81	CB	0.0	1.0		MIT
2574.843	2575.614	38825.69	CB	0.3	2.0		MIT
2574.073	2574.844	38837.31	CB	0.3	1.0		MIT
2573.910	2574.681	38839.76	CB		0.6H		MIT
2573.133	2573.904	38851.49	CB	0.0	1.3		MIT
2573.017	2573.788	38853.24	CB	0.3	1.0		MIT
2572.102	2572.872	38867.06	CB	0.8	0.5		MIT
2571.326	2572.096	38878.79	CB	0.6	2.0		MIT
2571.054	2571.824	38882.91	CB	0.5	0.3		MIT
2570.784	2571.554	38886.99	CB	0.5	0.3		MIT
2569.185	2569.955	38911.19	CB	0.0	0.7		MIT
2569.032	2569.802	38913.51	CB	1.0	0.5		MIT
2568.676	2569.446	38918.90	CB	0.0	0.6		MIT
2568.407	2569.177	38922.98	CB	0.3	1.3		MIT
2567.511	2568.280	38936.56	CB	1.0	0.7		MIT
2567.444	2568.213	38937.57	CB		0.7H		MIT
2566.072	2566.841	38958.39	CB	0.3	1.5		MIT
2565.676	2566.445	38964.40	CB	0.0	1.0		MIT
2565.502	2566.271	38967.05	CB	0.3	1.1		MIT
2565.408	2566.177	38968.48	CB	1.2	1.0		MIT
2564.843	2565.612	38977.06	CB	0.0	1.2		MIT
2564.732	2565.501	38978.75	CB	0.3	0.7H		MIT
2564.069	2564.838	38988.82	CB	0.0	1.5		MIT
2563.918	2564.687	38991.12	CB	0.3	1.0		MIT
2562.406	2563.174	39014.12	CB	0.6	2.0		MIT
2561.703	2562.471	39024.83	CB	0.3	1.0		MIT
2560.741	2561.509	39039.49	CB	0.0	0.7H		MIT
2560.626	2561.394	39041.24	CB	0.3	1.2		MIT
2560.109	2560.877	39049.13	CB	0.0	1.0		MIT
2559.429	2560.197	39059.50	CB		0.9		MIT
2558.940	2559.707	39066.96	CB	1.1	0.3		MIT
2558.622	2559.389	39071.82	CB	0.0	0.7		MIT
2557.944	2558.711	39082.18	CB		2.0H		MIT
2556.936	2557.703	39097.58	CB	0.7	2.3		MIT
2555.633	2556.400	39117.52	CB	0.3	1.9		MIT
2555.319	2556.086	39122.32	CB	0.0	1.5		MIT
2554.795	2555.561	39130.35	CB	0.0	1.3		MIT
2554.107	2554.873	39140.88	CB	0.7	0.3		MIT
2553.492	2554.258	39150.31	CB	0.0	1.5		MIT
2552.014	2552.780	39172.98	CB		0.5H		ME
2551.382	2552.148	39182.69	CB	0.7	2.0		MIT
2550.035	2550.800	39203.38	CB	0.0	0.7		MIT
2549.904	2550.669	39205.40	CB	0.6	0.0		MIT
2549.433	2550.198	39212.64	CB		0.5H		MIT
2548.633	2549.398	39224.95	CB	0.3	1.9		MIT
2545.637	2546.401	39271.11	CB	0.0	2.2		MIT
2544.803	2545.567	39283.98	CB	0.7	2.5		MIT
2543.981	2544.745	39296.67	CB	0.6	0.0		MIT
2541.972	2542.735	39327.73	CB	0.5	0.0		MIT
2541.424	2542.187	39336.21	CB	0.5	1.9		MIT
2541.089	2541.852	39341.39	CB		0.5H		MIT
2540.811	2541.574	39345.69	CB	0.0	0.5H		MIT
2540.615	2541.378	39348.73	CB	0.3	2.2		MIT
2539.226	2539.989	39370.25	CB	0.0	1.0		MIT
2538.058	2538.820	39388.37	CB	0.5	0.6H		MIT
2534.441	2535.203	39444.58	CB	0.0	1.6		MIT
2533.920	2534.682	39452.69	CB	0.0	1.1B		MIT
2533.188	2533.949	39464.09	CB	0.3	1.9		MIT
2531.254	2532.015	39494.24	CB	0.0	1.9		MIT
2530.970	2531.731	39498.67	CB	0.3	2.0		MIT
2530.165	2530.926	39511.24	CB		0.5		MIT
2530.105	2530.866	39512.17	CB		0.7		MIT
2529.546	2530.306	39520.90	CB		0.7		MIT
2527.920	2528.680	39546.32	CB	0.3	1.7		MIT
2525.808	2526.568	39579.39	CB	0.5	1.9		MIT
2524.987	2525.746	39592.26	CB	0.7	0.3		MIT
2523.755	2524.514	39611.58	CB	0.0W	0.7W		MIT
2522.882	2523.641	39625.29	CB		1.0		MIT
2522.340	2523.099	39633.80	CB	0.0	1.2		MIT
2521.404	2522.163	39648.51	CB	0.5	2.0		MIT
2520.659	2521.417	39660.23	CB		0.5H		MIT
2520.511	2521.269	39662.56	CB	0.8	0.3		MIT
2519.841	2520.599	39673.11	CB	0.0	0.5		MIT
2519.689	2520.447	39675.50	CB	0.3W	1.2W		MIT
2519.025	2519.783	39685.96	CB	0.3			MIT
2517.659	2518.417	39707.49	CB	0.0	0.7H		MIT
2517.488	2518.246	39710.18	CB	0.0	1.0		MIT
2514.353	2515.110	39759.69	CB	0.5	0.9		MIT
2511.972	2512.728	39797.38	CB		1.7H		MIT
2511.760	2512.516	39800.74	CB		1.0		MIT
2511.005	2511.761	39812.70	CB	0.5	2.0		MIT
2508.538	2509.294	39851.85	CB		1.4		MIT
2507.137	2507.892	39874.12	CB	0.5	0.3		MIT
2505.907	2506.662	39893.69	CB	0.0	1.0		MIT
2504.944	2505.699	39909.03	CB		0.6		MIT
2504.651	2505.406	39913.70	CB	1.0	0.6		MIT
2504.249	2505.004	39920.10	CB		0.6J		MIT
2502.491	2503.245	39948.14	CB	0.5	1.7		MIT
2501.405	2502.159	39965.49	CB		2.2		MIT
2500.420	2501.174	39981.23	CB	0.3	0.9		MIT
2499.750	2500.504	39991.94	CB		2.3		MIT
2498.253	2499.006	40015.91	CB	0.0	1.4		MIT
2496.973	2497.726	40036.42	CB	0.0	0.5		MIT
2493.022	2493.774	40099.87	CB		1.3J		MIT
2490.988	2491.740	40132.61	CB		1.0V		ME
2490.848	2491.599	40134.86	CB		1.3J		MIT
2490.213	2490.964	40145.10	CB	0.0	1.2		MIT
2490.107	2490.858	40146.80	CB		1.0		MIT
2489.463	2490.214	40157.19	CB	0.0	0.5		MIT
2488.747	2489.498	40168.74	CB		1.6		MIT
2488.236	2488.987	40176.99	CB	0.3	1.0		MIT
2486.028	2486.778	40212.67	CB		1.3J		MIT
2485.418	2486.168	40222.54	CB	0.0	1.1		MIT
2484.932	2485.682	40230.41	CB	0.3	1.4		MIT
2483.882	2484.632	40247.41	CB	0.5	1.9		MIT
2483.717	2484.467	40250.08	CB	0.3	1.0		MIT
2479.938	2480.687	40311.42	CB	0.5	1.7		MIT
2478.566	2479.315	40333.73	CB		0.6J		MIT
2478.290	2479.039	40338.22	CB	0.6	0.9		MIT
2477.945	2478.693	40343.83	CB	0.3	1.0		MIT
2477.384	2478.132	40352.97	CB	0.7	2.3		MIT
2476.479	2477.227	40367.71	CB	0.6			MIT
2475.887	2476.635	40377.37	CB		1.4H		MIT
2474.812	2475.560	40394.90	CB	0.3	1.5H		MIT
2474.663	2475.411	40397.34	CB	0.7H	0.3		MIT
2474.186	2474.934	40405.12	CB		0.5H		MIT
2472.982	2473.729	40424.79	CB		0.5H		MIT
2472.378	2473.125	40434.67	CB	0.6D	1.4		MIT
2471.317	2472.064	40452.03	CB	0.3W	1.1W		MIT
2469.412	2470.159	40483.23	CB	0.0	1.9		MIT
2469.080	2469.826	40488.67	CB	1.0H	0.3		MIT
2468.740	2469.486	40494.25	CB		1.7		MIT
2468.031	2468.777	40505.88	CB	0.3	0.5H		MIT
2466.727	2467.473	40527.29	CB	1.0	0.3		MIT
2466.562	2467.308	40530.00	CB	0.0	0.7		MIT
2466.313	2467.059	40534.10	CB	0.6	0.0		MIT
2465.203	2465.949	40552.35	CB	0.5	0.7		MIT
2464.648	2465.393	40561.48	CB	0.0	1.2		MIT

AIR WAVELENGTH (ANGSTRMS)	VACUUM WAVELENGTH (ANGSTRMS)	WAVENUMBER (KAYSERS)	LMENT	LOG ARC	INTENSITY SPK	DIS	REF
2464.432	2465.177	40565.03	CB	0.7	0.0		MIT
2463.728	2464.473	40576.62	CB	0.0	0.7H		MIT
2463.637	2464.382	40578.12	CB	0.5	0.3		MIT
2462.892	2463.637	40590.40	CB	0.9	0.5		MIT
2462.496	2463.241	40596.92	CB	0.0W	0.7W		MIT
2462.050	2462.795	40604.28	CB	0.0	2.0		MIT
2461.757	2462.502	40609.11	CB	0.7			MIT
2461.175	2461.920	40618.71	CB	0.3W	0.9W		MIT
2460.402	2461.146	40631.47	CB		2.30		MIT
2459.564	2460.308	40645.31	CB	0.0	0.7		MIT
2458.317	2459.061	40665.93	CB		0.7J		MIT
2458.087	2458.831	40669.73	CB	0.3	1.7		MIT
2457.245	2457.989	40683.67	CB		1.2J		MIT
2456.996	2457.740	40687.79	CB	0.0	2.3		MIT
2456.675	2457.419	40693.11	CB	0.0W	0.7W		MIT
2455.533	2456.276	40712.03	CB	0.0	0.5H		MIT
2454.378	2455.121	40731.19	CB		0.5		MIT
2453.946	2454.689	40738.36	CB	0.3	1.3		MIT
2453.859	2454.602	40739.80	CB	0.3	1.0		MIT
2453.373	2454.116	40747.87	CB	0.6	0.3		MIT
2453.089	2453.832	40752.59	CB	0.9	0.3		MIT
2452.467	2453.210	40762.92	CB		0.3		MIT
2451.874	2452.616	40772.78	CB	0.5	2.0		MIT
2450.433	2451.175	40796.76	CB	0.3W	0.9W		MIT
2450.250	2450.992	40799.81	CB	0.0	0.7		MIT
2450.065	2450.807	40802.89	CB	0.0	0.5		MIT
2448.261	2449.003	40832.95	CB	0.0	1.30		MIT
2447.970	2448.712	40837.80	CB	0.3	1.5		MIT
2446.440	2447.181	40863.34	CB		1.0		MIT
2446.133	2446.874	40868.47	CB	0.7			MIT
2446.085	2446.826	40869.27	CB		1.0H		MIT
2445.830	2446.571	40873.53	CB		2.0		MIT
2445.072	2445.813	40886.20	CB	1.0	0.3		MIT
2444.852	2445.593	40889.88	CB		0.5H		MIT
2444.481	2445.222	40896.09	CB	0.3	1.2		MIT
2443.529	2444.270	40912.02	CB	0.7	0.3		MIT
2442.677	2443.417	40926.29	CB	0.0	1.7W		MIT
2442.142	2442.882	40935.25	CB	0.5	1.9		MIT
2441.859	2442.599	40940.00	CB	0.5W	1.4W		MIT
2440.973	2441.713	40954.85	CB	0.0	0.7		MIT
2437.721	2438.460	41009.49	CB		0.7J		ME
2437.415	2438.154	41014.63	CB	0.6	2.0		MIT
2437.166	2437.905	41018.82	CB	0.7	0.0		MIT
2436.330	2437.069	41032.90	CB	0.7	0.5		MIT
2435.952	2436.691	41039.27	CB	0.5	1.7		MIT
2435.376	2436.115	41048.97	CB	0.0	0.5		MIT
2435.071	2435.810	41054.11	CB	0.0	1.0W		MIT
2434.962	2435.701	41055.95	CB	0.5	0.0		MIT
2434.665	2435.404	41060.96	CB		0.5		MIT
2434.30	2435.04	41067.1	CB II			0.3	TK
2433.795	2434.533	41075.63	CB	0.5	2.0		MIT
2433.686	2434.424	41077.47	CB	0.7	0.5		MIT
2433.557	2434.295	41079.65	CB		0.6J		MIT
2432.832	2433.570	41091.89	CB		0.5W		MIT
2432.322	2433.060	41100.51	CB	0.3W	1.6W		MIT
2431.676	2432.414	41111.42	CB	0.0W	1.6W		MIT
2430.311	2431.049	41134.51	CB	0.0	0.5H		MIT
2428.882	2429.619	41158.71	CB		1.1J		MIT
2428.603	2429.340	41163.44	CB	0.0	1.2		MIT
2427.539	2428.276	41181.48	CB	0.7	0.3H		MIT
2426.956	2427.693	41191.37	CB	0.3			MIT
2426.791	2427.528	41194.17	CB		1.6		MIT
2426.636	2427.373	41196.81	CB	0.5	0.3		MIT
2426.240	2426.977	41203.53	CB	0.0	0.6		MIT
2426.130	2426.867	41205.40	CB		0.5H		MIT
2425.114	2425.850	41222.66	CB	0.0	0.5		MIT
2424.002	2424.738	41241.57	CB	0.6	0.0		MIT
2421.909	2422.645	41277.21	CB		1.9		MIT
2420.144	2420.879	41307.31	CB	0.6	0.0		MIT
2419.966	2420.701	41310.35	CB	0.5			MIT

AIR WAVELENGTH (ANGSTRMS)	VACUUM WAVELENGTH (ANGSTRMS)	WAVENUMBER (KAYSERS)	LMENT	LOG ARC	INTENSITY SPK	DIS	REF
2419.803	2420.538	41313.13	CB	0.6			MIT
2419.468	2420.203	41318.85	CB	0.0	1.0		MIT
2418.692	2419.427	41332.10	CB	0.7	2.7		MIT
2417.326	2418.061	41355.46	CB	0.3	1.2		MIT
2417.157	2417.892	41358.35	CB		0.7H		MIT
2416.993	2417.727	41361.15	CB	0.9	2.3		MIT
2416.245	2416.979	41373.96	CB	0.3	0.5H		MIT
2416.170	2416.904	41375.24	CB	0.7	1.4		MIT
2415.951	2416.685	41378.99	CB	0.3W	1.2W		MIT
2414.871	2415.605	41397.50	CB		0.7J		MIT
2414.485	2415.219	41404.11	CB		2.0		MIT
2414.211	2414.945	41408.81	CB	0.5			MIT
2413.939	2414.673	41413.48	CB		2.5		MIT
2412.804	2413.538	41432.96	CB		1.6H		MIT
2412.645	2413.379	41435.69	CB	0.3			MIT
2412.464	2413.197	41438.80	CB	0.7	2.0		MIT
2412.273	2413.006	41442.08	CB		0.7J		MIT
2411.233	2411.966	41459.95	CB		0.5H		MIT
2410.275	2411.008	41476.43	CB	0.6W	1.4W		MIT
2410.083	2410.816	41479.73	CB	0.0	0.7H		MIT
2407.687	2408.419	41521.01	CB	0.5	1.2		MIT
2406.939	2407.671	41533.91	CB	0.7W	0.0		MIT
2405.850	2406.582	41552.71	CB	0.7W	1.7W		MIT
2405.340	2406.072	41561.52	CB	0.7W	1.7W		MIT
2404.886	2405.618	41569.36	CB	0.7	1.7		MIT
2404.278	2405.010	41579.88	CB	0.3	0.7H		ME
2404.211	2404.943	41581.03	CB	0.0	1.0		MIT
2402.657	2403.388	41607.93	CB		0.6H		MIT
2402.344	2403.075	41613.35	CB	0.7	0.0		MIT
2401.873	2402.604	41621.51	CB	0.3			ME
2401.034	2401.765	41636.05	CB	0.0	0.5		MIT
2400.914	2401.645	41638.13	CB	0.5	0.0		MIT
2399.711	2400.442	41659.00	CB	0.7			MIT
2398.482	2399.212	41680.35	CB	1.0	1.5		MIT
2397.962	2398.692	41689.38	CB	0.3H	0.6		MIT
2397.667	2398.397	41694.51	CB	0.3	0.3		MIT
2397.18	2397.91	41703.0	CB		0.3		MIT
2396.770	2397.500	41710.12	CB		0.5H		MIT
2396.31	2397.04	41718.1	CB		0.3W		MIT
2395.84	2396.57	41726.3	CB	0.3H	0.9		MIT
2395.323	2396.053	41735.31	CB	1.2	0.3		MIT
2394.675	2395.404	41746.60	CB		0.3J		MIT
2394.404	2395.133	41751.33	CB	0.6			MIT
2394.059	2394.788	41757.35	CB	0.7			MIT
2391.910	2392.639	41794.86	CB		1.3		MIT
2390.745	2391.474	41815.22	CB		0.3W		MIT
2390.075	2390.803	41826.94	CB	0.3H			MIT
2388.267	2388.995	41858.61	CB	0.7	1.4W		MIT
2387.524	2388.252	41871.63	CB	1.0	1.4J		MIT
2387.088	2387.816	41879.28	CB	0.3	0.9		MIT
2386.404	2387.132	41891.28	CB	0.7			MIT
2385.244	2385.971	41911.65	CB		0.8		MIT
2384.847	2385.574	41918.63	CB	0.3	0.6		MIT
2384.223	2384.950	41929.60	CB		0.5W		MIT
2383.836	2384.563	41936.41	CB	0.3H	0.0H		MIT
2382.246	2382.973	41964.39	CB	0.3			MIT
2381.128	2381.854	41984.10	CB		0.7		MIT
2380.143	2380.869	42001.47	CB		1.0J		MIT
2379.112	2379.838	42019.67	CB		0.6H		MIT
2378.937	2379.663	42022.76	CB	0.3H			MIT
2377.977	2378.703	42039.72	CB	0.5	1.0		MIT
2377.63	2378.36	42045.9	CB II		1.2		MIT
2376.400	2377.125	42067.62	CB	0.9	1.8		MIT
2374.163	2374.888	42107.25	CB		0.7		MIT
2373.960	2374.685	42110.85	CB	0.3W	0.7W		MIT
2373.075	2373.800	42126.56	CB	0.7			MIT
2372.725	2373.449	42132.77	CB	0.3	1.7		MIT
2372.230	2372.954	42141.56	CB		0.9W		MIT
2370.732	2371.456	42168.19	CB		0.6J		ME
2369.952	2370.676	42182.06	CB	0.7	1.4		MIT

AIR WAVELENGTH (ANGSTRMS)	VACUUM WAVELENGTH (ANGSTRMS)	WAVENUMBER (KAYSERS)	LMENT	LOG ARC	INTENSITY SPK	DIS	REF
2368.945	2369.669	42199.99	CB		0.7		MIT
2368.866	2369.590	42201.40	CB	0.6			MIT
2367.360	2368.083	42228.25	CB		0.5		MIT
2366.218	2366.941	42248.62	CB		0.9		MIT
2365.957	2366.680	42253.28	CB	0.3			MIT
2365.740	2366.463	42257.16	CB		0.7		MIT
2365.624	2366.347	42259.23	CB		1.3W		MIT
2365.299	2366.022	42265.04	CB		0.3		MIT
2365.216	2365.939	42266.52	CB	0.7	1.3L		MIT
2364.325	2365.048	42282.45	CB	0.7H			ME
2363.594	2364.316	42295.52	CB	0.3	0.3		MIT
2362.486	2363.208	42315.36	CB		1.3		MIT
2362.049	2362.771	42323.19	CB		1.4W		MIT
2361.052	2361.774	42341.06	CB		0.5J		MIT
2360.304	2361.026	42354.47	CB	0.3	0.8		MIT
2357.445	2358.166	42405.83	CB		0.8W		MIT
2356.301	2357.022	42426.42	CB	0.5H	1.0		MIT
2356.031	2356.752	42431.28	CB	0.3	1.0		MIT
2355.685	2356.406	42437.52	CB	0.3	0.6		MIT
2355.546	2356.267	42440.02	CB		1.2L		MIT
2354.949	2355.669	42450.78	CB		0.3W		MIT
2354.038	2354.758	42467.20	CB	0.6	1.1		MIT
2353.82	2354.54	42471.1	CB	0.6			MIT
2353.783	2354.503	42471.80	CB	0.8W	0.3		MIT
2353.507	2354.227	42476.78	CB	0.6			MIT
2352.838	2353.558	42488.86	CB	1.0W	1.3		MIT
2352.342	2353.062	42497.82	CB	0.7	1.3		MIT
2352.124	2352.844	42501.76	CB	0.5			MIT
2351.671	2352.391	42509.94	CB	0.3H			MIT
2351.634	2352.354	42510.61	CB		0.3H		MIT
2350.712	2351.432	42527.29	CB	0.3H			MIT
2350.488	2351.207	42531.34	CB	0.0	0.6		ME
2350.028	2350.747	42539.66	CB	0.3			MIT
2349.857	2350.576	42542.76	CB	0.3			MIT
2349.421	2350.140	42550.65	CB		0.3H		MIT
2349.213	2349.932	42554.42	CB		1.2		MIT
2348.748	2349.467	42562.84	CB	0.5W			MIT
2348.660	2349.379	42564.44	CB		0.7J		MIT
2346.949	2347.668	42595.47	CB		0.3H		MIT
2346.686	2347.405	42600.24	CB	0.5			MIT
2346.529	2347.248	42603.09	CB	0.5	1.0W		MIT
2345.346	2346.064	42624.58	CB	0.3	0.7		MIT
2344.641	2345.359	42637.39	CB	0.3			MIT
2344.513	2345.231	42639.72	CB	0.3			MIT
2344.14	2344.86	42646.5	CB		0.6		MIT
2343.68	2344.40	42654.9	CB		0.5H		MIT
2343.273	2343.991	42662.28	CB	0.3	0.5		MIT
2342.843	2343.561	42670.11	CB		0.5		MIT
2340.276	2340.993	42716.91	CB	0.6			MIT
2340.154	2340.871	42719.14	CB	0.3H			MIT
2340.040	2340.757	42721.22	CB	0.9	1.5		MIT
2339.59	2340.31	42729.4	CB		0.5W		MIT
2338.740	2339.457	42744.96	CB	0.5			MIT
2338.41	2339.13	42751.0	CB	0.3H			MIT
2338.093	2338.810	42756.79	CB		0.9W		MIT
2337.756	2338.473	42762.96	CB	0.5			MIT
2336.413	2337.129	42787.53	CB	0.5	0.6W		MIT
2335.617	2336.333	42802.11	CB	0.3H	1.2		MIT
2335.314	2336.030	42807.67	CB	0.5H	1.0W		MIT
2334.81	2335.53	42816.9	CB	0.5	1.4W		MIT
2333.658	2334.374	42838.04	CB	0.5			MIT
2332.883	2333.599	42852.27	CB		0.3H		MIT
2330.186	2330.901	42901.87	CB		0.9W		MIT
2328.222	2328.937	42938.05	CB	0.5			ME
2328.026	2328.740	42941.67	CB	0.3	0.5		MIT
2327.522	2328.236	42950.96	CB		0.5		MIT
2327.129	2327.843	42958.22	CB	0.3	0.7		MIT
2326.221	2326.935	42974.98	CB	0.7	1.2		ME
2325.511	2326.225	42988.10	CB	0.0	0.7		MIT
2324.584	2325.298	43005.25	CB	0.3			MIT
2324.406	2325.120	43008.54	CB		0.5H		MIT
2324.237	2324.951	43011.67	CB	1.2	1.0		MIT
2324.063	2324.777	43014.88	CB	0.3	1.0		MIT
2323.515	2324.228	43025.03	CB		1.0		MIT
2322.982	2323.695	43034.90	CB	0.3			MIT
2321.989	2322.702	43053.30	CB	0.9	1.3		MIT
2320.656	2321.369	43078.03	CB	0.3	0.5		MIT
2320.238	2320.951	43085.79	CB	0.6	0.7		MIT
2319.855	2320.568	43092.90	CB	0.5	0.3		MIT
2319.598	2320.311	43097.68	CB	0.6	1.0		MIT
2318.432	2319.144	43119.35	CB	0.5J			ME
2318.179	2318.891	43124.06	CB		1.2J		MIT
2317.783	2318.495	43131.42	CB	0.3	0.5		MIT
2317.24	2317.95	43141.5	CB		1.1		MIT
2316.929	2317.641	43147.32	CB	0.3H	1.0		MIT
2315.179	2315.891	43179.93	CB	0.7	0.7W		MIT
2314.854	2315.566	43185.99	CB	0.8	1.3		MIT
2314.356	2315.067	43195.29	CB	0.5			MIT
2313.525	2314.236	43210.80	CB	0.5	1.0		MIT
2313.32	2314.03	43214.6	CB		1.5W		MIT
2311.679	2312.390	43245.30	CB	0.3			MIT
2311.455	2312.166	43249.49	CB	0.3	1.2		MIT
2310.318	2311.029	43270.78	CB	0.3	0.9		MIT
2309.94	2310.65	43277.9	CB		1.2W		MIT
2309.234	2309.944	43291.09	CB	1.0L	1.5L		MIT
2308.801	2309.511	43299.20	CB	0.0H	0.3W		MIT
2305.432	2306.142	43362.47	CB		0.3H		MIT
2304.77	2305.48	43374.9	CB		1.3W		MIT
2304.362	2305.071	43382.61	CB		0.6J		MIT
2303.153	2303.862	43405.38	CB	0.3	0.3		MIT
2302.698	2303.407	43413.95	CB	0.9	0.9		MIT
2302.085	2302.794	43425.51	CB	1.2	1.5		MIT
2300.784	2301.492	43450.07	CB	1.0	1.2		MIT
2300.515	2301.223	43455.15	CB		0.3		MIT
2300.339	2301.047	43458.47	CB	0.6	1.0		MIT
2299.221	2299.929	43479.60	CB		0.7		MIT
2298.655	2299.363	43490.31	CB		0.3H		MIT
2298.380	2299.088	43495.51	CB	0.3H	0.6		MIT
2297.856	2298.564	43505.43	CB		0.7W		MIT
2297.615	2298.323	43509.99	CB	0.5	0.7		MIT
2295.979	2296.686	43540.99	CB	0.3	0.3		MIT
2295.676	2296.383	43546.74	CB	1.2	1.5		MIT
2294.982	2295.689	43559.90	CB	0.3	0.9		MIT
2294.706	2295.413	43565.14	CB		0.3J		MIT
2293.922	2294.629	43580.03	CB	0.3	1.2		MIT
2293.271	2293.978	43592.40	CB	0.3			MIT
2293.148	2293.855	43594.74	CB		0.3		MIT
2292.324	2293.031	43610.41	CB		0.3		MIT
2291.645	2292.351	43623.33	CB		0.7		MIT
2291.384	2292.090	43628.30	CB	0.3	0.3		MIT
2290.38	2291.09	43647.4	CB		1.4W		MIT
2289.837	2290.543	43657.77	CB	0.3			MIT
2288.866	2289.572	43676.29	CB	0.3	0.7		MIT
2286.888	2287.593	43714.06	CB	0.3	0.5H		MIT
2286.750	2287.455	43716.70	CB		0.6		MIT
2286.352	2287.057	43724.31	CB		0.6W		MIT
2285.673	2286.378	43737.30	CB		0.3W		ME
2285.221	2285.926	43745.95	CB	0.9	1.2		MIT
2284.348	2285.053	43762.66	CB		1.2W		MIT
2283.377	2284.082	43781.27	CB	0.3			ME
2283.005	2283.710	43788.40	CB	1.0	1.4L		MIT
2281.825	2282.529	43811.05	CB	0.3	1.0		MIT
2281.506	2282.210	43817.17	CB		1.4		MIT
2281.131	2281.835	43824.37	CB	0.3	0.9		MIT
2280.445	2281.149	43837.56	CB	0.7	0.9		MIT
2279.388	2280.092	43857.88	CB		1.3W		MIT
2277.423	2278.126	43895.72	CB	0.5			MIT
2276.214	2276.917	43919.03	CB	0.3			MIT
2276.170	2276.873	43919.88	CB	0.3	0.5		MIT
2275.219	2275.922	43938.24	CB		1.3		MIT

AIR WAVELENGTH (ANGSTRMS)	VACUUM WAVELENGTH (ANGSTRMS)	WAVENUMBER (KAYSERS)	LMENT	LOG ARC	INTENSITY SPK	INTENSITY DIS	REF
2274.763	2275.466	43947.05	CB	0.5			MIT
2274.201	2274.904	43957.90	CB	0.3	1.9		MIT
2274.131	2274.834	43959.26	CB	1.1	2.5		MIT
2273.92	2274.62	43963.3	CB		0.9		MIT
2273.565	2274.268	43970.20	CB	1.2	1.4		MIT
2273.120	2273.822	43978.81	CB		0.5		ME
2272.723	2273.425	43986.49	CB	0.7R	1.0		MIT
2270.179	2270.881	44035.78	CB	0.8	1.3L		MIT
2269.860	2270.562	44041.96	CB	0.5	1.2		MIT
2269.647	2270.349	44046.10	CB	0.3			MIT
2269.535	2270.237	44048.27	CB	0.6	0.3		ME
2269.202	2269.904	44054.73	CB	0.0	0.5W		ME
2268.523	2269.224	44067.92	CB	0.9	1.2		MIT
2266.726	2267.427	44102.85	CB	0.5	1.0		MIT
2265.671	2266.372	44123.39	CB	0.7	1.3W		MIT
2265.592	2266.293	44124.92	CB		1.7V		ME
2264.553	2265.254	44145.17	CB	0.3	0.9		MIT
2264.074	2264.775	44154.51	CB	0.3	0.5H		ME
2263.309	2264.009	44169.43	CB	0.3	0.5W		MIT
2263.220	2263.920	44171.17	CB	0.0	0.7		MIT
2262.29	2262.99	44189.3	CB		0.6		MIT
2262.131	2262.831	44192.43	CB	0.5	0.7		MIT
2261.727	2262.427	44200.32	CB	0.3	0.3		MIT
2261.530	2262.230	44204.17	CB	0.3	0.5		MIT
2260.854	2261.554	44217.39	CB	0.5			MIT
2257.881	2258.580	44275.60	CB	1.2			MIT
2257.537	2258.236	44282.35	CB	0.5	1.0		MIT
2256.070	2256.769	44311.14	CB	0.3			MIT
2255.593	2256.292	44320.51	CB	0.9	1.3		MIT
2255.172	2255.871	44328.78	CB		0.3		MIT
2254.949	2255.648	44333.17	CB	0.7	1.0		MIT
2254.557	2255.255	44340.87	CB	1.0	0.3		MIT
2253.802	2254.500	44355.73	CB	0.5			ME
2253.31	2254.01	44365.4	CB	0.3H			MIT
2252.620	2253.318	44379.00	CB		0.6		MIT
2252.206	2252.904	44387.16	CB	0.7	1.5		MIT
2250.465	2251.163	44421.49	CB	0.7	1.2		MIT
2250.303	2251.001	44424.69	CB	1.0	0.3		MIT
2250.004	2250.701	44430.59	CB	0.3	0.7W		MIT
2249.53	2250.23	44440.0	CB		1.3W		MIT
2248.84	2249.54	44453.6	CB		0.7		ME
2248.275	2248.972	44464.76	CB		1.0		MIT
2247.993	2248.690	44470.33	CB	1.0			MIT
2246.98	2247.68	44490.4	CB		0.3		MIT
2246.761	2247.458	44494.72	CB		1.0		MIT
2246.495	2247.192	44499.99	CB		0.6		MIT
2246.423	2247.120	44501.41	CB	0.5H			MIT
2246.175	2246.872	44506.33	CB	1.0			MIT
2244.93	2245.63	44531.0	CB		0.3		MIT
2244.29	2244.99	44543.7	CB		0.9		MIT
2242.962	2243.658	44570.07	CB	0.6			MIT
2242.579	2243.275	44577.69	CB	0.7	1.7B		ME
2242.288	2242.984	44583.47	CB	0.7	0.5		MIT
2241.017	2241.713	44608.75	CB		0.8		MIT
2240.646	2241.341	44616.14	CB	1.0	1.2		MIT
2238.516	2239.211	44658.59	CB	0.9			MIT
2237.496	2238.191	44678.94	CB	1.0	1.3L		MIT
2237.30	2237.99	44682.9	CB		0.3H		MIT
2236.96	2237.65	44689.6	CB		0.3H		MIT
2236.725	2237.420	44694.34	CB	0.7	1.0		MIT
2236.42	2237.11	44700.4	CB		0.3		MIT
2236.22	2236.91	44704.4	CB	0.6			MIT
2233.56	2234.25	44757.7	CB		0.6		MIT
2233.172	2233.866	44765.44	CB	0.5			MIT
2232.547	2233.241	44777.98	CB	1.0	0.0		MIT
2231.434	2232.128	44800.31	CB	0.5			MIT
2230.853	2231.546	44811.98	CB		0.6J		MIT
2229.720	2230.413	44834.74	CB	1.5	1.3		MIT
2229.66	2230.35	44836.0	CB	1.1			MIT
2228.028	2228.721	44868.79	CB	0.9W	0.0		MIT
2227.697	2228.390	44875.45	CB	0.7	0.5		MIT
2227.274	2227.967	44883.98	CB	0.6			MIT
2226.927	2227.620	44890.97	CB	0.7			MIT
2225.550	2226.242	44918.74	CB	0.5	0.3W		MIT
2225.339	2226.031	44923.00	CB	0.9			MIT
2225.096	2225.788	44927.91	CB		0.7J		ME
2224.668	2225.360	44936.55	CB	0.7	1.5B		MIT
2223.678	2224.370	44956.55	CB	1.0			MIT
2221.71	2222.40	44996.4	CB		0.3W		MIT
2221.474	2222.165	45001.15	CB	0.3			ME
2221.42	2222.11	45002.2	CB		0.6		MIT
2220.281	2220.972	45025.33	CB		0.5W		MIT
2220.184	2220.875	45027.29	CB	1.0	0.3		ME
2219.336	2220.027	45044.50	CB	0.5	0.6		MIT
2219.169	2219.860	45047.89	CB	0.5	0.3		MIT
2217.862	2218.553	45074.43	CB	0.7			MIT
2217.23	2217.92	45087.3	CB	0.3J	1.3W		MIT
2215.53	2216.22	45121.9	CB	0.6			MIT
2215.29	2215.98	45126.8	CB	0.3	0.6		MIT
2214.033	2214.723	45152.38	CB	0.9W			MIT
2212.931	2213.621	45174.86	CB		0.6W		MIT
2211.46	2212.15	45204.9	CB	1.1	0.3		MIT
2211.267	2211.956	45208.85	CB		0.5		MIT
2210.923	2211.612	45215.88	CB	0.7	1.0		MIT
2210.627	2211.316	45221.94	CB	0.3			MIT
2210.537	2211.226	45223.78	CB	0.6	1.0		MIT
2210.442	2211.131	45225.72	CB	0.5			MIT
2210.31	2211.00	45228.4	CB		0.3		ME
2206.632	2207.320	45303.80	CB		0.7		MIT
2206.010	2206.698	45316.57	CB		1.4W		MIT
2204.613	2205.301	45345.29	CB	0.7			MIT
2203.630	2204.318	45365.51	CB	1.2	1.6		MIT
2203.168	2203.856	45375.02	CB	0.6	0.7		MIT
2202.007	2202.694	45398.95	CB	0.5	0.7		ME
2201.370	2202.057	45412.08	CB		0.9		MIT
2199.961	2200.648	45441.16	CB	0.6	1.2		MIT
2199.594	2200.281	45448.75	CB	0.5	0.6		MIT
2199.05	2199.74	45460.0	CB		0.5		MIT
2198.31	2199.00	45475.3	CB		0.5W		MIT
2196.84	2197.53	45505.7	CB	0.3			MIT
2196.40	2197.09	45514.8	CB		1.3		MIT
2195.82	2196.51	45526.9	CB	0.5			MIT
2195.78	2196.47	45527.7	CB		0.9		MIT
2195.72	2196.41	45528.9	CB		0.3		MIT
2195.653	2196.339	45530.31	CB		0.7W		MIT
2193.803	2194.489	45568.70	CB	0.7	0.3		MIT
2193.009	2193.694	45585.20	CB	0.7			MIT
2192.417	2193.102	45597.51	CB	0.3	0.5		MIT
2188.942	2189.627	45669.89	CB	0.7			MIT
2188.141	2188.825	45686.60	CB		0.9W		MIT
2187.033	2187.717	45709.75	CB		0.6		MIT
2185.872	2186.556	45734.02	CB	0.0	0.7		ME
2185.398	2186.082	45743.94	CB	0.3	0.5L		ME
2180.667	2181.350	45843.17	CB		1.0W		MIT
2178.227	2178.909	45894.52	CB	0.5			MIT
2178.08	2178.76	45897.6	CB	0.7			MIT
2177.25	2177.93	45915.1	CB	0.0	0.3O		MIT
2176.762	2177.444	45925.40	CB	0.9	1.5		MIT
2175.843	2176.525	45944.80	CB	0.7	1.4		MIT
2175.551	2176.233	45950.97	CB	0.7			MIT
2170.69	2171.37	46053.9	CB	0.6	0.6		MIT
2169.895	2170.576	46070.73	CB		0.7		MIT
2167.238	2167.918	46127.20	CB	0.6	1.3		MIT
2164.27	2164.95	46190.5	CB		0.3W		MIT
2163.072	2163.751	46216.03	CB	0.5	0.7		MIT
2161.53	2162.21	46249.0	CB	0.6			MIT
2160.272	2160.951	46275.93	CB	0.7	1.7		MIT
2158.133	2158.811	46321.79	CB	0.7	1.0		MIT
2157.51	2158.19	46335.2	CB		0.3		MIT
2157.27	2157.95	46340.3	CB		0.7		MIT

AIR WAVELENGTH (ANGSTRMS)	VACUUM WAVELENGTH (ANGSTRMS)	WAVENUMBER (KAYSERS)	LMENT	LOG ARC	INTENSITY SPK	INTENSITY DIS	REF
2156.730	2157.408	46351.92	CB	0.9	1.3		MIT
2156.26	2156.94	46362.0	CB		0.6W		MIT
2155.623	2156.301	46375.72	CB	0.7	1.4		MIT
2155.14	2155.82	46386.1	CB		0.3W		MIT
2154.73	2155.41	46394.9	CB	0.5			MIT
2154.203	2154.880	46406.29	CB	0.7	1.1		MIT
2153.65	2154.33	46418.2	CB		0.3		MIT
2153.547	2154.224	46420.42	CB	0.5	0.0		MIT
2153.30	2153.98	46425.7	CB		0.5W		MIT
2152.55	2153.23	46441.9	CB		0.3W		MIT
2152.06	2152.74	46452.5	CB		0.7		MIT
2149.535	2150.211	46507.05	CB	1.0	1.3		MIT
2149.03	2149.71	46518.0	CB		0.3		MIT
2148.720	2149.396	46524.69	CB	0.7			MIT
2148.646	2149.322	46526.29	CB	0.9	1.2		MIT
2147.194	2147.870	46557.75	CB	1.0	1.3		MIT
2146.368	2147.044	46575.67	CB		1.2W		MIT
2146.14	2146.82	46580.6	CB		0.5J		ME
2144.80	2145.48	46609.7	CB	0.0J	0.6		MIT
2144.492	2145.167	46616.41	CB	0.6	0.3		MIT
2144.20	2144.88	46622.8	CB		1.0W		MIT
2143.21	2143.89	46644.3	CB	0.6	0.8		MIT
2142.91	2143.59	46650.8	CB	0.0H	0.8		MIT
2142.02	2142.69	46670.2	CB	0.3	0.8		MIT
2140.39	2141.06	46705.7	CB	0.7	1.0		MIT
2138.88	2139.55	46738.7	CB	0.3	0.9		MIT
2138.552	2139.226	46745.87	CB	0.3			ME
2137.545	2138.219	46767.89	CB	0.7	1.6		MIT
2137.054	2137.728	46778.64	CB	0.7	1.5		MIT
2134.950	2135.624	46824.73	CB	0.3	1.3		MIT
2134.709	2135.382	46830.02	CB	1.1	1.3		MIT
2134.492	2135.165	46834.78	CB	0.8	1.1		MIT
2132.83	2133.50	46871.3	CB		1.1		MIT
2131.180	2131.853	46907.56	CB	1.1	1.6		MIT
2130.25	2130.92	46928.0	CB		0.9		MIT
2129.70	2130.37	46940.1	CB	0.0	0.6		MIT
2129.02	2129.69	46955.1	CB	0.0	0.5		MIT
2128.23	2128.90	46972.6	CB		0.3		MIT
2128.14	2128.81	46974.5	CB		0.5		MIT
2126.543	2127.215	47009.83	CB	1.2	1.7		MIT
2125.206	2125.878	47039.40	CB	1.2	1.6		MIT
2124.34	2125.01	47058.6	CB	0.3	0.9		MIT
2122.74	2123.41	47094.0	CB		1.2S		MIT
2121.66	2122.33	47118.0	CB		1.1		MIT
2120.52	2121.19	47143.3	CB	0.3	0.9		MIT
2120.01	2120.68	47154.7	CB	0.3	0.5		MIT
2119.63	2120.30	47163.1	CB	0.5	0.0		ME
2119.09	2119.76	47175.1	CB		0.9W		MIT
2118.873	2119.543	47179.98	CB	1.0	1.3		MIT
2116.390	2117.060	47235.32	CB	0.7	1.2		MIT
2114.78	2115.45	47271.3	CB	0.3	0.3		MIT
2113.085	2113.754	47309.19	CB	1.0	1.3L		MIT
2112.32	2112.99	47326.3	CB		0.9		MIT
2111.21	2111.88	47351.2	CB		0.7		MIT
2110.98	2111.65	47356.4	CB		0.5		MIT
2110.07	2110.74	47376.8	CB	0.5	0.7		MIT
2109.424	2110.092	47391.29	CB	1.2	1.7		MIT
2108.35	2109.02	47415.4	CB	0.3	0.3H		MIT
2107.265	2107.933	47439.84	CB	1.0	1.4		MIT
2103.59	2104.26	47522.7	CB	1.0	1.2		MIT
2093.28	2093.95	47756.7	CB		0.9		MIT
2093.08	2093.75	47761.3	CB		0.7		MIT
2091.37	2092.03	47800.4	CB		0.7H		MIT
2087.09	2087.75	47898.4	CB		0.3		MIT
2082.89	2083.55	47994.9	CB	0.8	1.4		A
2080.05	2080.71	48060.5	CB	0.5	1.2		A
2065.71	2066.37	48394.0	CB	0.7	1.3		A
2064.21	2064.87	48429.2	CB	0.3	1.5		A
2061.45	2062.11	48494.0	CB	0.3	1.1		A
2060.27	2060.93	48521.8	CB		1.4		MIT

AIR WAVELENGTH (ANGSTRMS)	VACUUM WAVELENGTH (ANGSTRMS)	WAVENUMBER (KAYSERS)	LMENT	LOG ARC	INTENSITY SPK	INTENSITY DIS	REF
2057.05	2057.71	48597.7	CB	0.6	1.5		A
2049.87	2050.53	48768.0	CB	0.3	1.6		MIT
2047.50	2048.16	48824.4	CB	0.3	0.3		A
2044.41	2045.07	48898.2	CB	0.3	0.6		A
2044.22	2044.88	48902.7	CB	0.5	1.3		A
2044.12	2044.78	48905.1	CB	0.5	0.6		A
2043.52	2044.18	48919.5	CB	0.3	1.4		A
2043.16	2043.82	48928.1	CB	0.5	1.4		A
2037.94	2038.59	49053.4	CB	0.5	1.4		A
2033.56	2034.21	49159.0	CB	0.3	0.6		A
2032.46	2033.11	49185.6	CB		1.4W		MIT
2029.32	2029.97	49261.7	CB	1.0	1.7		A
2025.31	2025.96	49359.2	CB	1.0	1.5		A
2019.80	2020.45	49493.9	CB	0.6	1.2		A
2018.67	2019.32	49521.6	CB	0.3	1.2		A
2017.28	2017.93	49555.7	CB	0.5	1.4		A
2016.05	2016.70	49585.9	CB	0.7	1.2		A
2011.99	2012.64	49686.0	CB	0.6	1.4		A
2010.00	2010.65	49735.2	CB	0.7	1.4		A
2008.75	2009.40	49766.1	CB		0.3		MIT
2008.45	2009.10	49773.5	CB	0.6	1.3		A
2005.03	2005.68	49858.4	CB	0.7	1.4		A
2004.76	2005.41	49865.1	CB	0.6	1.3		MIT
2002.09	2002.74	49931.6	CB		0.3		MIT
8200.2	8202.5	12191.	CD I	0.7			PS
8066.99	8069.21	12392.8	CD II		1.2		VS
7664.74	7666.85	13043.2	CD		0.5		VS
7399.2	7401.2	13511.	CD I	1.8			PS
7393.0	7395.0	13523.	CD	1.8			PS
7385.3	7387.3	13537.	CD I	2.9			PS
7383.9	7385.9	13539.	CD	3.0			PS
7346.2	7348.2	13609.	CD I	3.0			PS
7284.38	7286.39	13724.2	CD II		1.4		VS
7237.01	7239.00	13814.1	CD		1.2		VS
7132.27	7134.24	14016.9	CD I	1.5			PS
6948.720	6950.637	14387.17	CD	0.5	0.5H		MIT
6778.10	6779.97	14749.3	CD I	1.5			PS
6759.26	6761.13	14790.4	CD II		1.5		VS
6725.83	6727.69	14863.9	CD II		2.0		VS
6567.73	6569.54	15221.8	CD II	0.8	0.5		VS
6464.98	6466.77	15463.7	CD II	0.7	1.7		VS
6438.4696	6440.2491	15527.350	CD I	3.3	3.0		IS
6359.93	6361.69	15719.1	CD II	1.0	1.7		VS
6354.72	6356.48	15732.0	CD II	0.7	1.6		VS
6329.97	6331.72	15793.5	CD I	1.5			PS
6325.19	6326.94	15805.4	CD I	2.0			WD
6198.22	6199.93	16129.2	CD I	1.2			PS
6128.66	6130.36	16312.3	CD I	1.2			PS
6116.19	6117.88	16345.5	CD I	1.7			PS
6111.52	6113.21	16358.0	CD I	2.0			PS
6099.18	6100.87	16391.1	CD I	2.5			PS
6031.38	6033.05	16575.4	CD I	1.5			PS
5880.19	5881.82	17001.5	CD II	0.6	0.5		VS
5843.175	5844.795	17109.24	CD II	0.5		1.6	MIT
5783.93	5785.53	17284.5	CD	0.7			PS
5637.26	5638.82	17734.2	CD	1.0			PS
5629.830	5631.393	17757.60	CD	0.5	0.0		MIT
5606.85	5608.41	17830.4	CD I	0.7			PS
5604.683	5606.239	17837.27	CD I	1.0			PS
5603.518	5605.074	17840.98	CD	1.2	0.7		MIT
5598.769	5600.323	17856.11	CD I	1.2			PS
5381.82	5383.32	18575.9	CD II	0.3	1.3		VS
5378.037	5379.532	18588.98	CD II	0.7	1.7		MIT
5337.492	5338.977	18730.18	CD II	0.7	1.4		MIT
5297.64	5299.11	18871.1	CD I	0.5			PS
5271.57	5273.04	18964.4	CD II	0.3H	1.1		VS
5267.96	5269.43	18977.4	CD II	0.3H	1.0		VS
5154.68	5156.12	19394.4	CD I	0.8R			PS
5085.824	5087.242	19657.02	CD I	3.0J	2.7		HZ
5025.95	5027.35	19891.2	CD II	0.5	0.3H		TK

AIR WAVELENGTH (ANGSTRMS)	VACUUM WAVELENGTH (ANGSTRMS)	WAVENUMBER (KAYSERS)	LMENT	LOG ARC	INTENSITY SPK	DIS	REF
4882.04	4883.40	20477.5	CD II		1.0		TK
4799.918	4801.260	20827.87	CD I	2.5W	2.5W		HZ
4744.72	4746.05	21070.2	CD II		0.6		VS
4741.78	4743.11	21083.2	CD II		0.5		VS
4678.156	4679.465	21369.96	CD I	2.30	2.30		MIT
4662.352	4663.657	21442.40	CD I	0.9R			I
4615.75	4617.04	21658.9	CD	0.3			PS
4615.39	4616.68	21660.6	CD I	0.5			PS
4614.17	4615.46	21666.3	CD I	0.6			PS
4511.34	4512.61	22160.1	CD	0.7			PS
4415.700	4416.940	22640.11	CD II	0.0	1.3		MIT
4414.63	4415.87	22645.6	CD		2.3		TK
4413.042	4414.281	22653.74	CD I	0.5	0.3		MIT
4412.31	4413.55	22657.5	CD II		1.0		VS
4306.82	4308.03	23212.5	CD I	0.9	0.5		PS
4285.070	4286.276	23330.28	CD II		0.9		VS
4245.869	4247.064	23545.68	CD		0.3		MIT
4243.39	4244.58	23559.4	CD		0.3		TK
4141.58	4142.75	24138.6	CD II		1.0		MIT
4140.5	4141.7	24145.	CD	0.7			SD
4134.78	4135.95	24178.3	CD II		1.2		VS
4029.08	4030.22	24812.6	CD II		0.7		VS
4006.68	4007.81	24951.3	CD		0.7		TK
3981.77	3982.90	25107.4	CD I	1.0R			PS
3957.40	3958.52	25262.0	CD II		0.9		TK
3905.1	3906.2	25600.	CD I	0.9			SD
3827.41	3828.50	26119.9	CD II		0.7		M
3818.5	3819.6	26181.	CD I	0.7			SD
3776.32	3777.39	26473.3	CD		0.5		TK
3768.10	3769.17	26531.0	CD II		0.3		MIT
3729.06	3730.12	26808.8	CD I	1.2R			FL
3697.50	3698.55	27037.6	CD		0.3		TK
3688.282	3689.332	27105.18	CD II		0.3		MIT
3667.32	3668.36	27260.1	CD II		1.0		M
3649.597	3650.637	27392.48	CD I	1.3	1.2		MIT
3614.450	3615.481	27658.84	CD I	1.8	2.0		MIT
3612.875	3613.905	27670.90	CD I	2.9	2.7		IME
3610.510	3611.540	27689.02	CD I	3.0	2.7		IME
3576.51	3577.53	27952.2	CD		0.5		MIT
3535.687	3536.698	28274.97	CD II	0.7	1.2		MIT
3524.072	3525.080	28368.15	CD II		0.9		MIT
3500.00	3501.00	28563.3	CD I	1.4	1.2		MIT
3495.34	3496.34	28601.3	CD			2.0	MIT
3483.04	3484.04	28702.3	CD II		0.8		VS
3467.656	3468.649	28829.67	CD I	2.9	2.6		IME
3466.201	3467.194	28841.77	CD I	3.0	2.7		IME
3464.37	3465.36	28857.0	CD II		1.0		MIT
3422.964	3423.946	29206.07	CD II		0.5		MIT
3420.16	3421.14	29230.0	CD II		0.7		MIT
3417.396	3418.376	29253.66	CD II	1.0	1.2		MIT
3403.653	3404.630	29371.77	CD I	2.9	2.7H		IME
3402.16	3403.14	29384.7	CD II		0.7		MIT
3399.67	3400.65	29406.2	CD II		0.7		MIT
3388.85	3389.82	29500.1	CD II		0.8		VS
3385.40	3386.37	29530.1	CD II			1.6	TK
3385.346	3386.318	29530.60	CD	0.9			MIT
3376.778	3377.748	29605.53	CD		0.5		MIT
3370.910	3371.878	29657.06	CD		0.3		MIT
3355.30	3356.26	29795.0	CD II			0.5	TK
3343.15	3344.11	29903.3	CD II		1.2		VS
3298.97	3299.92	30303.8	CD	1.2			PS
3292.11	3293.06	30366.9	CD II			0.3	TK
3285.98	3286.93	30423.6	CD			1.1	ES
3283.82	3284.77	30443.6	CD			1.1	ES
3283.543	3284.489	30446.14	CD II		0.3		MIT
3276.8	3277.7	30509.	CD			1.1	ES
3264.4	3265.3	30625.	CD			1.0	ES
3261.057	3261.997	30656.06	CD I	2.5	2.5		MIT
3252.525	3253.463	30736.48	CD I	2.5	2.5		IME
3250.301	3251.239	30757.51	CD		1.4		MIT

AIR WAVELENGTH (ANGSTRMS)	VACUUM WAVELENGTH (ANGSTRMS)	WAVENUMBER (KAYSERS)	LMENT	LOG ARC	INTENSITY SPK	DIS	REF
3250.17	3251.11	30758.7	CD II		2.0		MIT
3243.15	3244.09	30825.3	CD		1.2		VS
3238.81	3239.74	30866.6	CD II			0.7	TK
3236.67	3237.60	30887.0	CD			1.1	ES
3232.356	3233.289	30928.26	CD II		0.3		MIT
3224.21	3225.14	31006.4	CD			1.0	ES
3222.565	3223.496	31022.22	CD II		0.3		MIT
3221.63	3222.56	31031.2	CD			1.1	ES
3217.8	3218.7	31068.	CD			1.2	ES
3210.1	3211.0	31143.	CD			1.1	ES
3201.8	3202.7	31223.	CD			0.9	ES
3197.8	3198.7	31262.	CD			0.9	ES
3194.36	3195.28	31296.1	CD		0.3		MIT
3185.55	3186.47	31382.7	CD II			1.2	TK
3182.91	3183.83	31408.7	CD			0.9	ES
3180.012	3180.932	31437.33	CD II		0.3		MIT
3174.489	3175.407	31492.02	CD II		0.3		MIT
3173.613	3174.531	31500.71	CD		0.3		MIT
3164.42	3165.34	31592.2	CD II			0.3	TK
3161.85	3162.77	31617.9	CD II	0.3	0.5		MIT
3160.814	3161.729	31628.26	CD	0.3	0.0		MIT
3157.111	3158.025	31665.36	CD II		0.3		MIT
3154.673	3155.586	31689.83	CD	0.5			MIT
3149.920	3150.832	31737.65	CD II	0.7	0.5		MIT
3146.78	3147.69	31769.3	CD II			1.0	VS
3141.6	3142.5	31822.	CD			1.1	ES
3133.167	3134.075	31907.34	CD I	2.3	2.5		IME
3129.206	3130.113	31947.73	CD		1.0		MIT
3124.4	3125.3	31997.	CD			1.0	ES
3121.8	3122.7	32023.	CD			1.0	ES
3121.770	3122.675	32023.82	CD		0.3		MIT
3118.915	3119.819	32053.14	CD	0.5	0.3		MIT
3112.964	3113.867	32114.41	CD II		0.3		MIT
3112.206	3113.109	32122.23	CD	0.5	0.3		MIT
3107.81	3108.71	32167.7	CD		0.3		MIT
3106.71	3107.61	32179.1	CD	0.3	0.6		MIT
3104.59	3105.49	32201.0	CD		0.6		MIT
3095.45	3096.35	32296.1	CD			1.0	M
3093.740	3094.638	32313.96	CD II	0.5	0.3		MIT
3092.393	3093.291	32328.03	CD II	1.0	1.2		MIT
3089.856	3090.753	32354.57	CD II		0.5		MIT
3084.866	3085.762	32406.91	CD	1.0	1.6H		MIT
3082.68	3083.58	32429.9	CD I	1.5			PS
3081.58	3082.48	32441.5	CD II		0.3		VS
3080.827	3081.722	32449.39	CD I	2.2	2.0L		IME
3077.2	3078.1	32488.	CD			1.1	ES
3074.62	3075.51	32514.9	CD	0.0H	0.5		MIT
3073.9	3074.8	32522.	CD			1.0	ES
3071.648	3072.541	32546.36	CD II		0.3		MIT
3068.790	3069.682	32576.67	CD II		0.3H		MIT
3064.955	3065.846	32617.43	CD		1.2H		MIT
3063.725	3064.616	32630.52	CD	0.3	0.3		MIT
3060.28	3061.17	32667.2	CD	0.6	0.7		MIT
3059.22	3060.11	32678.6	CD II		0.3		MIT
3057.51	3058.40	32696.9	CD		0.3		MIT
3056.41	3057.30	32708.6	CD	0.0	0.3		VS
3053.1	3054.0	32744.	CD			1.0	ES
3048.82	3049.71	32790.0	CD			0.3H	TK
3039.572	3040.457	32889.80	CD	0.6			MIT
3035.761	3036.645	32931.08	CD		0.3		MIT
3030.67	3031.55	32986.4	CD II		0.8		MIT
3017.32	3018.20	33132.3	CD			1.1	ES
3011.3	3012.2	33199.	CD			1.0	ES
3008.02	3008.90	33234.8	CD II	1.0	0.7		MIT
3005.41	3006.29	33263.6	CD I	1.4	0.6		FL
3001.51	3002.39	33306.9	CD II			1.0	TK
2996.5	2997.4	33362.	CD			1.0	ES
2996.03	2996.90	33367.8	CD		1.4		MIT
2992.3	2993.2	33409.	CD	1.0			ES
2987.2	2988.1	33466.	CD			1.4	ES

AIR WAVELENGTH (ANGSTRMS)	VACUUM WAVELENGTH (ANGSTRMS)	WAVENUMBER (KAYSERS)	LMENT	LOG ARC	INTENSITY SPK	DIS	REF
2981.89	2982.76	33526.0	CD I	1.7		1.0	FL
2981.34	2982.21	33532.2	CD I	2.3G		1.6	FL
2980.628	2981.498	33540.19	CD I	3.0G	2.7		MIT
2971.2	2972.1	33647.	CD			1.0	ES
2965.1	2966.0	33716.	CD			0.7	ES
2964.7	2965.6	33720.	CD			0.7	ES
2964.3	2965.2	33725.	CD			1.0	ES
2961.47	2962.34	33757.1	CD I	1.3	1.2		M
2960.83	2961.69	33764.5	CD II			0.7	VS
2953.2	2954.1	33852.	CD			0.7	ES
2951.82	2952.68	33867.5	CD		1.4		MIT
2948.16	2949.02	33909.5	CD		1.5		MIT
2943.89	2944.75	33958.7	CD II			0.7	VS
2934.15	2935.01	34071.5	CD			0.3	VS
2931.14	2932.00	34106.4	CD			0.7D	VS
2929.285	2930.142	34128.04	CD II		1.7		MIT
2927.91	2928.77	34144.1	CD II		0.9		MIT
2926.93	2927.79	34155.5	CD			0.5	VS
2919.13	2919.98	34246.8	CD			0.5	VS
2914.69	2915.54	34298.9	CD II			1.7	VS
2912.66	2913.51	34322.8	CD II		0.3		MIT
2911.64	2912.49	34334.9	CD II			1.2	VS
2910.8	2911.7	34345.	CD			1.5	ES
2908.74	2909.59	34369.1	CD I	0.7L			FL
2903.13	2903.98	34435.5	CD I	0.7H			FL
2893.74	2894.59	34547.2	CD II			0.5	TK
2893.28	2894.13	34552.7	CD			0.3	VS
2886.60	2887.45	34632.7	CD II			0.5	VS
2881.23	2882.08	34697.2	CD I	1.7G		1.5	FL
2880.77	2881.62	34702.8	CD I	2.3G	2.1		MIT
2868.26	2869.10	34854.1	CD I	2.0	1.9		M
2862.31	2863.15	34926.6	CD I	1.2	1.0		MIT
2856.45	2857.29	34998.2	CD II			0.9	VS
2841.60	2842.44	35181.1	CD			0.7	ES
2836.907	2837.741	35239.29	CD I	2.3	1.9		MIT
2834.19	2835.02	35273.1	CD II			2.0	TK
2823.19	2824.02	35410.5	CD II			1.3	TK
2819.89	2820.72	35451.9	CD II			0.5	VS
2818.73	2819.56	35466.5	CD I	1.0	1.0		MIT
2813.41	2814.24	35533.6	CD II			1.0	TK
2810.93	2811.76	35564.9	CD II			0.5	TK
2809.01	2809.84	35589.2	CD			0.5	VS
2805.59	2806.42	35632.6	CD		1.5		MIT
2798.98	2799.81	35716.8	CD II		0.5		MIT
2798.14	2798.96	35727.5	CD		0.5		MIT
2780.28	2781.10	35957.0	CD			1.4	ES
2776.08	2776.90	36011.4	CD		0.5		MIT
2775.047	2775.866	36024.79	CD I	1.7	1.3		HZ
2772.16	2772.98	36062.3	CD			0.7	ES
2771.96	2772.78	36064.9	CD II		0.7		MIT
2768.47	2769.29	36110.4	CD			0.7	ES
2767.49	2768.31	36123.2	CD II			0.3	VS
2767.15	2767.97	36127.6	CD			1.1	ES
2764.11	2764.93	36167.3	CD	1.7H	1.4		ES
2763.89	2764.71	36170.2	CD I	2.0H	1.7		MIT
2757.83	2758.64	36249.7	CD			0.7	ES
2756.79	2757.60	36263.4	CD I	1.7H			MIT
2753.80	2754.61	36302.7	CD II		0.3		MIT
2751.88	2752.69	36328.0	CD			0.7	ES
2748.58	2749.39	36371.7	CD II	0.7	2.3		MIT
2733.86	2734.67	36567.5	CD I	1.7	1.4		MIT
2728.26	2729.07	36642.5	CD			0.7	ES
2726.93	2727.74	36660.4	CD			1.2	ES
2720.2	2721.0	36751.	CD			0.3	ES
2716.34	2717.14	36803.3	CD			0.7	ES
2716.00	2716.80	36807.9	CD			0.7	ES
2712.57	2713.37	36854.5	CD I	1.9	1.3		MIT
2707.93	2708.73	36917.6	CD II			0.5	TK
2707.14	2707.94	36928.4	CD II			1.5	TK
2702.7	2703.5	36989.	CD			0.7	ES

AIR WAVELENGTH (ANGSTRMS)	VACUUM WAVELENGTH (ANGSTRMS)	WAVENUMBER (KAYSERS)	LMENT	LOG ARC	INTENSITY SPK	DIS	REF
2687.69	2688.49	37195.6	CD			0.5	VS
2685.08	2685.88	37231.8	CD			0.5	VS
2680.08	2680.88	37301.2	CD II			1.2	TK
2677.64	2678.44	37335.2	CD I	2.0	1.4		FL
2675.36	2676.16	37367.0	CD			0.3D	TK
2674.82	2675.61	37374.6	CD II			0.3	TK
2672.68	2673.47	37404.5	CD II			1.0	VS
2670.66	2671.45	37432.8	CD I	1.0W			FL
2670.21	2671.00	37439.1	CD II			0.3	VS
2668.33	2669.12	37465.5	CD II			1.0	TK
2660.40	2661.19	37577.1	CD I	1.7H		0.7	FL
2659.29	2660.08	37592.8	CD II			0.7	VS
2657.00	2657.79	37625.2	CD	1.0		0.0	FL
2654.27	2655.06	37663.9	CD II			0.3	VS
2650.44	2651.23	37718.4	CD II			0.5	VS
2649.52	2650.31	37731.5	CD			0.7	ES
2646.84	2647.63	37769.6	CD			0.3	VS
2645.86	2646.65	37783.6	CD			0.7	VS
2640.69	2641.48	37857.6	CD			0.7	VS
2639.50	2640.29	37874.7	CD I	1.9	1.2		FL
2638.49	2639.28	37889.2	CD			0.5	TK
2636.29	2637.08	37920.8	CD			0.3	VS
2633.63	2634.42	37959.1	CD		0.5		MIT
2633.20	2633.98	37965.3	CD			0.3	VS
2632.244	2633.029	37979.08	CD I	1.6		0.5	MIT
2630.63	2631.41	38002.4	CD			0.5	VS
2630.22	2631.00	38008.3	CD			0.3	VS
2629.05	2629.83	38025.2	CD I	1.7	1.0		MIT
2626.11	2626.89	38067.8	CD		0.5		MIT
2618.97	2619.75	38171.6	CD II			1.5	TK
2617.87	2618.65	38187.6	CD			0.3	VS
2617.13	2617.91	38198.4	CD			0.3	VS
2616.02	2616.80	38214.6	CD			0.3	VS
2614.96	2615.74	38230.1	CD			0.3	VS
2613.12	2613.90	38257.0	CD			0.3	VS
2612.19	2612.97	38270.6	CD			0.5	VS
2611.81	2612.59	38276.2	CD			0.3	VS
2602.18	2602.96	38417.8	CD I	1.4H		0.7	FL
2601.48	2602.26	38428.2	CD			0.3	VS
2600.79	2601.57	38438.4	CD			0.3	VS
2600.32	2601.10	38445.3	CD			0.3	VS
2596.13	2596.91	38507.4	CD			0.5	VS
2592.14	2592.92	38566.6	CD I	1.5			FL
2588.51	2589.28	38620.7	CD			0.6	VS
2586.43	2587.20	38651.8	CD			0.7	VS
2585.07	2585.84	38672.1	CD I	0.5		0.0	M
2580.30	2581.07	38743.6	CD I	1.7		0.7	MIT
2573.02	2573.79	38853.2	CD II				MIT
2565.88	2566.65	38961.3	CD I	0.5		1.0	FL
2564.02	2564.79	38989.6	CD			0.3	VS
2554.51	2555.28	39134.7	CD I	0.5		0.0	FL
2553.56	2554.33	39149.3	CD I	1.4		0.3	M
2552.99	2553.76	39158.0	CD II			1.0	TK
2552.91	2553.68	39159.2	CD II			1.0	TK
2552.18	2552.95	39170.4	CD II	0.7H		2.0	MIT
2544.710	2545.474	39285.41	CD I	1.7		0.7	MIT
2541.64	2542.40	39332.9	CD I	0.3		0.0	FL
2533.91	2534.67	39452.8	CD I	0.3		0.0	FL
2526.200	2526.960	39573.25	CD II		0.3		MIT
2525.389	2526.149	39585.95	CD	1.4H			MIT
2518.79	2519.55	39689.7	CD I	1.2H		0.0	MIT
2517.25	2518.01	39713.9	CD II			0.5	TK
2516.34	2517.10	39728.3	CD II		1.4		MIT
2510.15	2510.91	39826.3	CD II			0.3	TK
2509.25	2510.01	39840.5	CD II			1.5	TK
2509.043	2509.799	39843.83	CD I	1.0		0.5	MIT
2500.86	2501.61	39974.2	CD II			0.3	TK
2499.849	2500.603	39990.36	CD II		1.2		MIT
2496.533	2497.286	40043.47	CD II		0.6		MIT
2495.73	2496.48	40056.4	CD II			1.5	TK

AIR WAVELENGTH (ANGSTRMS)	VACUUM WAVELENGTH (ANGSTRMS)	WAVENUMBER (KAYSERS)	LMENT	LOG ARC	INTENSITY SPK	DIS	REF
2491.16	2491.91	40129.8	CD I	0.5		0.3	FL
2488.69	2489.44	40169.7	CD II			0.3	TK
2487.937	2488.688	40181.82	CD II		1.2		MIT
2475.23	2475.98	40388.1	CD I			0.3	M
2470.61	2471.36	40463.6	CD II			1.7	TK
2469.84	2470.59	40476.2	CD II			2.7	TK
2468.25	2469.00	40502.3	CD I	0.3		0.0	FL
2457.87	2458.61	40673.3	CD I	0.3		0.0	FL
2443.715	2444.456	40908.90	CD II		0.6		MIT
2433.444	2434.182	41081.56	CD II		0.9		MIT
2427.076	2427.813	41189.34	CD II		0.5		MIT
2426.385	2427.122	41201.07	CD II		0.7		MIT
2419.49	2420.23	41318.5	CD II			1.4	TK
2418.70	2419.43	41332.0	CD II		1.3		MIT
2418.26	2418.99	41339.5	CD II		0.7		MIT
2376.839	2377.564	42059.85	CD II		0.7		MIT
2367.933	2368.656	42218.03	CD II		0.3		MIT
2367.195	2367.918	42231.19	CD II		1.0		MIT
2366.19	2366.91	42249.1	CD II			0.5	TK
2365.39	2366.11	42263.4	CD II		1.0		MIT
2360.643	2361.365	42348.39	CD	0.3H			MIT
2350.30	2351.02	42534.7	CD			1.0	ES
2347.65	2348.37	42582.7	CD	0.3			HR
2346.625	2347.344	42601.35	CD	0.3			MIT
2345.51	2346.23	42621.6	CD	1.0H			HR
2337.54	2338.26	42766.9	CD	0.7H			HU
2332.98	2333.70	42850.5	CD II			0.3	TK
2330.11	2330.82	42903.3	CD II			0.3	TK
2329.282	2329.997	42918.51	CD	1.7	1.8		MIT
2322.56	2323.27	43042.7	CD II			0.7	TK
2321.94	2322.65	43054.2	CD			1.3	BL
2321.15	2321.86	43068.9	CD II		2.0H		M
2317.46	2318.17	43137.4	CD		0.5		MIT
2316.51	2317.22	43155.1	CD			0.3	TK
2313.49	2314.20	43211.5	CD			1.5	BL
2312.84	2313.55	43223.6	CD II	0.0	2.3		MIT
2312.57	2313.28	43228.6	CD	1.4			MIT
2307.78	2308.49	43318.4	CD II			0.5	TK
2306.609	2307.319	43340.35	CD	1.3	1.5		MIT
2304.46	2305.17	43380.8	CD			0.5	BL
2301.18	2301.89	43442.6	CD			0.5	BL
2295.83	2296.54	43543.8	CD			0.3	BL
2295.32	2296.03	43553.5	CD II			0.5	TK
2294.28	2294.99	43573.2	CD			0.3	TK
2292.02	2292.73	43616.2	CD			0.6	TK
2290.86	2291.57	43638.3	CD II			0.6	TK
2288.74	2289.45	43678.7	CD		1.3		MIT
2288.018	2288.724	43692.47	CD I	3.26	2.5G		HZ
2284.67	2285.37	43756.5	CD			0.3	BL
2282.52	2283.22	43797.7	CD II			0.5	TK
2272.38	2273.08	43993.1	CD II			0.3	TK
2271.63	2272.33	44007.6	CD			0.3	BL
2269.82	2270.52	44042.7	CD		0.5		MIT
2267.47	2268.17	44088.4	CD	1.3	1.5		MIT
2265.81	2266.51	44120.7	CD II		1.5		MIT
2265.017	2265.718	44136.12	CD II	1.4D	2.5		HZ
2262.29	2262.99	44189.3	CD	0.5			FL
2256.11	2256.81	44310.4	CD			0.3	BL
2255.77	2256.47	44317.0	CD			0.3	BL
2246.929	2247.626	44491.39	CD II		0.3		MIT
2246.80	2247.50	44494.0	CD			0.3D	BL
2243.80	2244.50	44553.4	CD			0.3	BL
2242.43	2243.13	44580.6	CD			0.3	BL
2240.99	2241.69	44609.3	CD			0.3	BL
2240.45	2241.15	44620.0	CD			0.5	BL
2239.86	2240.56	44631.8	CD	1.9	1.5		M
2238.57	2239.27	44657.5	CD		0.5		MIT
2236.32	2237.01	44702.4	CD		0.7		MIT
2233.96	2234.65	44749.7	CD			0.3	BL
2230.37	2231.06	44821.7	CD	0.3			MIT
2225.20	2225.89	44925.8	CD			1.0	BL
2224.39	2225.08	44942.2	CD		0.5		MIT
2223.57	2224.26	44958.7	CD			0.3	BL
2222.67	2223.36	44976.9	CD			0.5	BL
2222.08	2222.77	44988.9	CD II			0.3	TK
2213.70	2214.39	45159.2	CD		0.3		MIT
2211.15	2211.84	45211.2	CD			0.3	TK
2210.37	2211.06	45227.2	CD II			1.0	TK
2209.70	2210.39	45240.9	CD		0.5H		MIT
2204.18	2204.87	45354.2	CD II			0.3	TK
2201.51	2202.20	45409.2	CD			0.3	TK
2199.81	2200.50	45444.3	CD			0.3	BL
2195.35	2196.04	45536.6	CD		1.2		MIT
2194.63	2195.32	45551.5	CD	0.7	2.0H		MIT
2190.63	2191.31	45634.7	CD			0.5	TK
2188.84	2189.52	45672.0	CD			0.3	BL
2188.55	2189.23	45678.1	CD II			1.7	TK
2187.80	2188.48	45693.7	CD		0.7		M
2186.95	2187.63	45711.5	CD			0.7	TK
2182.64	2183.32	45801.7	CD II			0.5	TK
2181.39	2182.07	45828.0	CD			0.3	BL
2176.88	2177.56	45922.9	CD II			0.5	TK
2162.94	2163.62	46218.9	CD II			0.5	TK
2155.70	2156.38	46374.1	CD II			1.5	TK
2151.06	2151.74	46474.1	CD			0.5	TK
2148.96	2149.64	46519.5	CD			0.3	TK
2147.76	2148.44	46545.5	CD			0.3	TK
2145.04	2145.72	46604.5	CD			1.3	BL
2144.382	2145.057	46618.80	CD II	1.7	2.36		HZ
2129.12	2129.79	46952.9	CD II			1.0	TK
2127.13	2127.80	46996.9	CD			0.3	TK
2125.66	2126.33	47029.4	CD II			0.5	TK
2120.42	2121.09	47145.6	CD			0.3	BL
2112.17	2112.84	47329.7	CD			1.0	BL
2105.25	2105.92	47485.2	CD			0.3	BL
2101.10	2101.77	47579.0	CD			0.7	BL
2096.63	2097.30	47680.4	CD II			2.0	TK
2088.52	2089.18	47865.6	CD		1.0		MIT
2088.39	2089.05	47868.5	CD			0.3	BL
2074.00	2074.66	48200.6	CD			0.7	BL
2072.50	2073.16	48235.5	CD			0.3	BL
2046.22	2046.88	48854.9	CD			0.5	TK
2040.44	2041.10	48993.3	CD II			0.5	TK
2036.79	2037.44	49081.1	CD II			1.5	TK
2033.03	2033.68	49171.9	CD			1.3	TK
2028.86	2029.51	49272.9	CD II			0.7	TK
2020.33	2020.98	49480.9	CD			0.3	BL
2008.06	2008.71	49783.2	CD II			1.0	TK
2004.65	2005.30	49867.9	CD			1.0	BL
2001.28	2001.93	49951.8	CD			0.5	TK
8772.08	8774.49	11396.7	CE I	1.0			KS
8647.59	8649.97	11560.7	CE	0.7			KS
8612.62	8614.99	11607.7	CE	0.7			KS
8495.64	8497.97	11767.5	CE	0.5			KS
8467.38	8469.71	11806.8	CE	0.3W			KS
8371.90	8374.20	11941.4	CE	0.9S			KS
8363.82	8366.12	11953.0	CE	0.3H			KS
8352.39	8354.69	11969.3	CE	0.3			KS
8310.26	8312.54	12030.0	CE	0.6			MIT
8300.70	8302.98	12043.9	CE	0.6			MIT
8261.11	8263.38	12101.6	CE	0.9			MIT
8245.21	8247.48	12124.9	CE	0.6			MIT
8234.21	8236.47	12141.1	CE	0.6J			MIT
8223.61	8225.87	12156.8	CE	0.6			MIT
8120.37	8122.60	12311.3	CE	0.7			MIT
8025.59	8027.80	12456.7	CE	0.8			MIT
8002.64	8004.84	12492.4	CE	0.5			MIT
7975.44	7977.63	12535.1	CE	1.0			MIT
7972.34	7974.53	12539.9	CE	0.6			KS
7927.53	7929.71	12610.8	CE	0.7			KS

AIR WAVELENGTH (ANGSTRMS)	VACUUM WAVELENGTH (ANGSTRMS)	WAVENUMBER (KAYSERS)	LMENT	LOG ARC	INTENSITY SPK	INTENSITY DIS	REF
7879.60	7881.77	12687.5	CE	0.5Q			MIT
7860.54	7862.70	12718.3	CE	0.5			KS
7859.05	7861.21	12720.7	CE	1.0			MIT
7851.20	7853.36	12733.4	CE	0.3			MIT
7844.97	7847.13	12743.5	CE	0.3			MIT
7842.61	7844.77	12747.3	CE	0.5			MIT
7835.85	7838.01	12758.3	CE	0.3			MIT
7797.75	7799.90	12820.7	CE	0.5			MIT
7732.36	7734.49	12929.1	CE	0.5			MIT
7689.18	7691.30	13001.7	CE	0.5			MIT
7682.46	7684.57	13013.1	CE	0.3			MIT
7670.93	7673.04	13032.6	CE	0.8X			MIT
7397.76	7399.80	13513.9	CE	0.9			MIT
7329.90	7331.92	13639.0	CE	0.7			MIT
7252.74	7254.74	13784.1	CE	0.9			MIT
7241.70	7243.70	13805.1	CE	0.7			MIT
7238.37	7240.36	13811.5	CE	0.5			MIT
7217.36	7219.35	13851.7	CE	0.7			MIT
7201.87	7203.85	13881.5	CE	0.6			MIT
7156.99	7158.96	13968.5	CE	0.7			MIT
7151.69	7153.66	13978.9	CE	0.5			MIT
7150.21	7152.18	13981.7	CE	0.5			MIT
7141.47	7143.44	13998.9	CE	0.5			MIT
7086.38	7088.33	14107.7	CE	0.7			MIT
7061.74	7063.69	14156.9	CE	0.9			MIT
7049.70	7051.64	14181.1	CE	0.5			MIT
7031.04	7032.98	14218.7	CE	0.7			MIT
7018.88	7020.82	14243.4	CE	0.8D			MIT
7017.27	7019.21	14246.6	CE	0.6			MIT
6999.96	7001.89	14281.9	CE	1.0			MIT
6986.07	6988.00	14310.2	CE	1.2			MIT
6973.54	6975.46	14336.0	CE	0.5			MIT
6960.77	6962.69	14362.3	CE	1.0			MIT
6939.54	6941.45	14406.2	CE	0.9			MIT
6934.13	6936.04	14417.4	CE	0.3			MIT
6932.12	6934.03	14421.6	CE	0.6			KS
6924.83	6926.74	14436.8	CE	1.3			MIT
6899.10	6901.00	14490.7	CE	0.6			MIT
6898.48	6900.38	14491.9	CE	0.5			KS
6894.55	6896.45	14500.2	CE	0.8			KS
6893.69	6895.59	14502.0	CE	0.7			MIT
6856.53	6858.42	14580.6	CE	0.6			KS
6853.57	6855.46	14586.9	CE	0.6			MIT
6847.25	6849.14	14600.4	CE	0.7			MIT
6846.78	6848.67	14601.4	CE	0.5			KS
6844.43	6846.32	14606.4	CE	0.3			KS
6826.43	6828.31	14644.9	CE	0.5			KS
6818.25	6820.13	14662.5	CE	0.7			MIT
6808.89	6810.77	14682.6	CE	0.5			KS
6807.83	6809.71	14684.9	CE	1.0			MIT
6803.20	6805.08	14694.9	CE	0.3			KS
6780.74	6782.61	14743.6	CE	0.7			MIT
6778.28	6780.15	14748.9	CE	1.0			MIT
6775.59	6777.46	14754.8	CE	1.0			MIT
6774.29	6776.16	14757.6	CE	0.5			MIT
6767.65	6769.52	14772.1	CE	0.5			KS
6764.45	6766.32	14779.1	CE	0.3			MIT
6749.51	6751.37	14811.8	CE	0.3			MIT
6729.54	6731.40	14855.8	CE	1.0			MIT
6728.710	6730.568	14857.59	CE	0.9			MIT
6720.32	6722.18	14876.1	CE	0.3			KS
6713.48	6715.33	14891.3	CE	0.6			MIT
6710.16	6712.01	14898.7	CE	0.3			KS
6705.92	6707.77	14908.1	CE	0.3			KN
6704.32	6706.17	14911.6	CE	1.4			MIT
6700.70	6702.55	14919.7	CE	1.3			MIT
6689.70	6691.55	14944.2	CE	0.5			MIT
6686.59	6688.44	14951.2	CE	0.9			MIT
6679.809	6681.653	14966.36	CE	0.7			MIT
6675.544	6677.387	14975.92	CE	0.5			MIT

AIR WAVELENGTH (ANGSTRMS)	VACUUM WAVELENGTH (ANGSTRMS)	WAVENUMBER (KAYSERS)	LMENT	LOG ARC	INTENSITY SPK	INTENSITY DIS	REF
6665.68	6667.52	14998.1	CE	1.0			MIT
6662.55	6664.39	15005.1	CE	0.6			MIT
6661.41	6663.25	15007.7	CE	0.9			MIT
6655.47	6657.31	15021.1	CE	0.5			KN
6654.809	6656.647	15022.58	CE	0.3			MIT
6654.30	6656.14	15023.7	CE	0.5			MIT
6652.77	6654.61	15027.2	CE	0.7			MIT
6651.423	6653.260	15030.23	CE	1.0			MIT
6650.888	6652.725	15031.44	CE	0.9			MIT
6647.38	6649.22	15039.4	CE	0.3			KS
6641.15	6642.98	15053.5	CE	0.5			MIT
6628.932	6630.763	15081.22	CE	1.0			MIT
6623.004	6624.833	15094.72	CE	0.7			MIT
6612.06	6613.89	15119.7	CE	0.9			MIT
6606.856	6608.681	15131.61	CE	0.6			MIT
6606.33	6608.15	15132.8	CE	1.0			KN
6605.386	6607.210	15134.98	CE	0.5			MIT
6599.61	6601.43	15148.2	CE	0.8			KN
6579.110	6580.927	15195.43	CE	0.9			MIT
6577.470	6579.287	15199.22	CE	0.5			MIT
6559.336	6561.148	15241.23	CE	0.5			MIT
6555.670	6557.481	15249.76	CE	1.2			MIT
6551.722	6553.532	15258.95	CE	1.2			MIT
6543.609	6545.417	15277.87	CE	0.5			MIT
6537.476	6539.282	15292.20	CE	0.3			MIT
6534.501	6536.306	15299.16	CE	0.7			MIT
6530.677	6532.481	15308.12	CE	0.8			MIT
6527.025	6528.828	15316.68	CE	0.5W			MIT
6525.326	6527.129	15320.67	CE	0.7			MIT
6522.55	6524.35	15327.2	CE	0.5			KN
6517.295	6519.096	15339.55	CE	1.0			MIT
6513.60	6515.40	15348.2	CE	0.9W			MIT
6509.011	6510.809	15359.07	CE	0.9			MIT
6507.16	6508.96	15363.4	CE	0.5			MIT
6504.13	6505.93	15370.6	CE	0.7			MIT
6503.266	6505.063	15372.64	CE	0.5			MIT
6490.994	6492.788	15401.70	CE	0.9			MIT
6485.97	6487.76	15413.6	CE	0.7			KN
6482.522	6484.313	15421.83	CE	0.3			MIT
6481.977	6483.768	15423.13	CE	0.8			MIT
6473.707	6475.496	15442.83	CE	0.8	0.0		MIT
6468.971	6470.759	15454.14	CE	0.5			MIT
6467.418	6469.205	15457.85	CE	1.3			MIT
6466.898	6468.685	15459.09	CE	0.7	0.0		MIT
6461.890	6463.676	15471.07	CE	0.7			MIT
6458.052	6459.837	15480.27	CE	1.4			MIT
6446.154	6447.936	15508.84	CE	0.8			MIT
6441.045	6442.825	15521.14	CE	0.5			MIT
6439.970	6441.750	15523.73	CE	0.8			MIT
6436.405	6438.184	15532.33	CE	1.1			MIT
6434.396	6436.174	15537.18	CE	1.1			MIT
6430.068	6431.845	15547.64	CE	1.0			MIT
6425.296	6427.072	15559.19	CE	0.5	0.0		MIT
6424.502	6426.278	15561.11	CE	0.6D			MIT
6399.907	6401.676	15620.91	CE	0.6			MIT
6396.244	6398.012	15629.85	CE	0.8			MIT
6393.023	6394.790	15637.73	CE	0.7			MIT
6390.321	6392.088	15644.34	CE	0.9			MIT
6386.864	6388.630	15652.81	CE	1.1W			MIT
6386.102	6387.867	15654.68	CE	0.7			MIT
6379.75	6381.51	15670.3	CE	0.3			KS
6372.987	6374.749	15686.89	CE	0.8			MIT
6372.486	6374.248	15688.13	CE	0.7			MIT
6371.105	6372.866	15691.53	CE	1.0			MIT
6360.216	6361.974	15718.39	CE	0.6			MIT
6343.963	6345.717	15758.66	CE	1.2			MIT
6340.686	6342.439	15766.81	CE	0.6			MIT
6337.212	6338.964	15775.45	CE	0.9			MIT
6335.369	6337.121	15780.04	CE	1.2			MIT
6331.979	6333.730	15788.48	CE	0.7H			MIT

AIR WAVELENGTH (ANGSTRMS)	VACUUM WAVELENGTH (ANGSTRMS)	WAVENUMBER (KAYSERS)	LMENT	LOG ARC	INTENSITY SPK	DIS	REF
6329.428	6331.178	15794.85	CE	0.5			MIT
6321.247	6322.995	15815.29	CE	0.3			MIT
6318.567	6320.314	15822.00	CE	0.3			MIT
6310.017	6311.762	15843.44	CE	1.3			MIT
6308.026	6309.770	15848.44	CE	0.7			MIT
6306.628	6308.372	15851.95	CE	1.0			MIT
6300.210	6301.952	15868.10	CE	1.3			MIT
6299.514	6301.256	15869.85	CE	0.6			MIT
6295.557	6297.298	15879.83	CE	1.3			MIT
6286.404	6288.143	15902.95	CE	0.5			MIT
6277.109	6278.845	15926.50	CE	0.5			MIT
6276.456	6278.192	15928.15	CE	1.0			MIT
6273.743	6275.478	15935.04	CE	0.7			MIT
6272.051	6273.786	15939.34	CE	1.3			MIT
6270.287	6272.021	15943.82	CE	0.9			MIT
6269.817	6271.551	15945.02	CE	0.6	O		MIT
6264.256	6265.989	15959.17	CE	0.8			MIT
6257.990	6259.721	15975.15	CE	0.9			MIT
6256.349	6258.080	15979.34	CE	0.7			MIT
6253.622	6255.352	15986.31	CE	0.9			MIT
6242.914	6244.641	16013.73	CE	0.9			MIT
6241.907	6243.634	16016.31	CE	0.8			MIT
6238.702	6240.428	16024.54	CE	1.0			MIT
6237.458	6239.183	16027.74	CE	1.0			MIT
6232.452	6234.176	16040.61	CE	0.9			MIT
6228.945	6230.668	16049.64	CE	1.0			MIT
6228.234	6229.957	16051.47	CE	0.9			MIT
6223.681	6225.403	16063.22	CE	0.5			MIT
6223.249	6224.971	16064.33	CE	0.8			MIT
6216.836	6218.556	16080.90	CE	1.0			MIT
6216.078	6217.798	16082.86	CE	0.3			MIT
6212.500	6214.219	16092.13	CE	0.7			MIT
6211.059	6212.777	16095.86	CE	0.9			MIT
6209.563	6211.281	16099.74	CE	0.7			MIT
6208.989	6210.707	16101.23	CE	1.1			MIT
6202.090	6203.806	16119.14	CE	0.3			MIT
6198.047	6199.762	16129.65	CE	0.9			MIT
6196.280	6197.994	16134.25	CE	0.6			MIT
6195.546	6197.260	16136.16	CE	1.0			MIT
6195.246	6196.960	16136.94	CE	0.9			MIT
6192.282	6193.995	16144.67	CE	0.6			MIT
6186.917	6188.629	16158.67	CE	1.0			MIT
6186.157	6187.869	16160.65	CE	1.1			MIT
6178.386	6180.096	16180.98	CE	0.6			MIT
6177.958	6179.668	16182.10	CE	0.5			MIT
6175.285	6176.994	16189.10	CE	0.9			MIT
6172.866	6174.574	16195.45	CE	0.3			MIT
6165.469	6167.175	16214.88	CE	0.6			MIT
6164.697	6166.403	16216.91	CE	0.5			MIT
6163.188	6164.894	16220.88	CE	0.3			MIT
6162.168	6163.873	16223.56	CE	0.9	K		MIT
6159.817	6161.522	16229.76	CE	0.9			MIT
6158.940	6160.644	16232.07	CE	0.5			MIT
6156.741	6158.445	16237.87	CE	0.8			MIT
6155.503	6157.207	16241.13	CE	0.5			MIT
6151.730	6153.432	16251.09	CE	1.0			MIT
6149.585	6151.287	16256.76	CE	0.7			MIT
6147.852	6149.553	16261.34	CE	1.0			MIT
6146.421	6148.122	16265.13	CE	1.1			MIT
6143.362	6145.062	16273.23	CE	0.5			MIT
6142.914	6144.614	16274.42	CE	0.8			MIT
6139.031	6140.730	16284.71	CE	0.5			MIT
6137.234	6138.933	16289.48	CE	0.6			MIT
6135.522	6137.220	16294.02	CE	0.5			MIT
6133.610	6135.308	16299.10	CE	0.6			MIT
6132.004	6133.701	16303.37	CE	0.6			MIT
6130.133	6131.830	16308.35	CE	0.7			MIT
6126.088	6127.784	16319.11	CE	0.3			MIT
6123.673	6125.368	16325.55	CE	1.2			MIT
6119.794	6121.488	16335.90	CE	0.6			MIT

AIR WAVELENGTH (ANGSTRMS)	VACUUM WAVELENGTH (ANGSTRMS)	WAVENUMBER (KAYSERS)	LMENT	LOG ARC	INTENSITY SPK	DIS	REF
6118.892	6120.586	16338.31	CE	0.8			MIT
6118.548	6120.242	16339.22	CE	0.5			MIT
6117.708	6119.401	16341.47	CE	0.3			MIT
6115.437	6117.130	16347.54	CE	0.6			MIT
6114.68	6116.37	16349.6	CE	0.5			MIT
6111.948	6113.640	16356.87	CE	0.7			MIT
6108.738	6110.429	16365.46	CE	0.6			MIT
6102.751	6104.440	16381.52	CE	0.5			MIT
6099.792	6101.481	16389.47	CE	1.0			MIT
6098.336	6100.024	16393.38	CE	1.0			MIT
6097.601	6099.289	16395.35	CE	0.5			MIT
6093.198	6094.885	16407.20	CE	1.0			MIT
6089.657	6091.343	16416.74	CE	0.5			MIT
6088.919	6090.605	16418.73	CE	0.9	W		MIT
6086.617	6088.302	16424.94	CE	0.8			MIT
6081.276	6082.960	16439.37	CE	0.8			MIT
6080.367	6082.050	16441.82	CE	0.8	W		MIT
6077.551	6079.234	16449.44	CE	0.3			MIT
6077.144	6078.826	16450.54	CE	0.9			MIT
6076.598	6078.280	16452.02	CE	0.7	D		MIT
6075.917	6077.599	16453.87	CE	0.6			MIT
6075.566	6077.248	16454.82	CE	0.6	L		MIT
6071.995	6073.676	16464.49	CE	1.2			MIT
6069.465	6071.145	16471.36	CE	1.3			MIT
6068.629	6070.309	16473.62	CE	0.7			MIT
6066.712	6068.392	16478.83	CE	0.9	W		MIT
6060.646	6062.324	16495.32	CE	0.5			MIT
6059.317	6060.995	16498.94	CE	0.5			MIT
6057.987	6059.664	16502.56	CE	1.2			MIT
6057.493	6059.170	16503.91	CE	0.8			MIT
6051.804	6053.480	16519.42	CE	0.7			MIT
6047.387	6049.062	16531.49	CE	1.2			MIT
6045.882	6047.556	16535.60	CE	0.3			MIT
6045.430	6047.104	16536.84	CE	0.8			MIT
6043.386	6045.059	16542.44	CE	1.5			MIT
6040.629	6042.302	16549.99	CE	0.6			MIT
6035.487	6037.158	16564.08	CE	0.6			MIT
6034.204	6035.875	16567.61	CE	0.6			MIT
6033.576	6035.247	16569.33	CE	0.5			MIT
6031.247	6032.917	16575.73	CE	0.7			MIT
6024.189	6025.857	16595.15	CE	1.7			MIT
6021.683	6023.351	16602.05	CE	0.3			MIT
6020.596	6022.263	16605.05	CE	0.8			MIT
6018.794	6020.461	16610.02	CE	0.7			MIT
6016.571	6018.237	16616.16	CE	1.0			MIT
6014.061	6015.727	16623.10	CE	0.5			MIT
6013.418	6015.083	16624.87	CE	1.4			MIT
6011.551	6013.216	16630.04	CE	0.6			MIT
6010.449	6012.114	16633.09	CE	0.7			MIT
6007.358	6009.022	16641.64	CE	0.7			MIT
6006.808	6008.472	16643.17	CE	1.2			MIT
6006.211	6007.874	16644.82	CE	0.8			MIT
6005.855	6007.518	16645.81	CE	1.3			MIT
6003.664	6005.327	16651.88	CE	0.5			MIT
6001.892	6003.554	16656.80	CE	1.0			MIT
6000.184	6001.846	16661.54	CE	0.8			MIT
5997.048	5998.709	16670.25	CE	0.3			MIT
5995.260	5996.921	16675.23	CE	0.5			MIT
5992.657	5994.317	16682.47	CE	1.0			MIT
5991.624	5993.284	16685.34	CE	0.3			MIT
5990.851	5992.510	16687.50	CE	0.5			MIT
5989.380	5991.039	16691.60	CE	1.0			MIT
5988.792	5990.451	16693.24	CE	0.7			MIT
5987.385	5989.043	16697.16	CE	0.6			MIT
5986.246	5987.904	16700.33	CE	0.5			MIT
5981.838	5983.495	16712.64	CE	0.6			MIT
5981.195	5982.852	16714.44	CE	0.5			MIT
5979.397	5981.053	16719.46	CE	0.7	W		MIT
5975.985	5977.640	16729.01	CE	1.2			MIT
5975.830	5977.485	16729.44	CE	1.2			MIT

AIR WAVELENGTH (ANGSTRMS)	VACUUM WAVELENGTH (ANGSTRMS)	WAVENUMBER (KAYSERS)	LMENT	LOG ARC	INTENSITY SPK	DIS	REF
5975.233	5976.888	16731.11	CE	0.5			MIT
5972.096	5973.750	16739.90	CE	0.8			MIT
5966.344	5967.997	16756.04	CE	0.9			MIT
5964.645	5966.297	16760.81	CE	0.7			MIT
5963.359	5965.011	16764.43	CE	0.5			MIT
5960.883	5962.534	16771.39	CE	0.3			MIT
5960.185	5961.836	16773.36	CE	0.6			MIT
5959.698	5961.349	16774.73	CE	0.8			MIT
5956.835	5958.485	16782.79	CE	0.5			MIT
5952.872	5954.521	16793.96	CE	0.5			MIT
5950.607	5952.256	16800.35	CE	0.9			MIT
5947.638	5949.286	16808.74	CE	0.8			MIT
5944.884	5946.531	16816.53	CE	0.9			MIT
5943.519	5945.166	16820.39	CE	0.5			MIT
5942.660	5944.306	16822.82	CE	1.0			MIT
5941.895	5943.541	16824.99	CE	0.5			MIT
5941.537	5943.183	16826.00	CE	0.5			MIT
5940.849	5942.495	16827.95	CE	1.6			MIT
5938.435	5940.080	16834.79	CE	0.6			MIT
5937.715	5939.360	16836.83	CE	1.3			MIT
5934.443	5936.087	16846.11	CE	0.7			MIT
5933.581	5935.225	16848.56	CE	0.5			MIT
5932.163	5933.807	16852.59	CE	0.8			MIT
5931.717	5933.361	16853.85	CE	0.6			MIT
5929.835	5931.478	16859.20	CE	0.9			MIT
5929.498	5931.141	16860.16	CE	0.7			MIT
5928.728	5930.371	16862.35	CE	0.3			MIT
5928.339	5929.982	16863.46	CE	1.4			MIT
5926.303	5927.945	16869.25	CE	1.4			MIT
5924.051	5925.692	16875.67	CE	0.9			MIT
5922.952	5924.593	16878.80	CE	0.5			MIT
5922.125	5923.766	16881.15	CE	0.5			MIT
5920.438	5922.078	16885.96	CE	1.2			MIT
5914.835	5916.474	16901.96	CE	0.9			MIT
5912.912	5914.550	16907.46	CE	1.3			MIT
5910.134	5911.772	16915.40	CE	1.2			MIT
5909.875	5911.513	16916.14	CE	1.2			MIT
5907.486	5909.123	16922.99	CE	0.5			MIT
5906.009	5907.646	16927.22	CE	1.0			MIT
5902.660	5904.296	16936.82	CE	0.3			MIT
5901.321	5902.956	16940.66	CE	0.7	D		MIT
5900.674	5902.309	16942.52	CE	0.5	W		MIT
5893.190	5894.823	16964.04	CE	0.3			MIT
5878.896	5880.525	17005.28	CE	0.5			MIT
5878.078	5879.707	17007.65	CE	0.3			MIT
5873.882	5875.510	17019.80	CE	0.5			MIT
5871.605	5873.232	17026.40	CE	1.3			MIT
5862.511	5864.136	17052.81	CE	1.4			MIT
5859.386	5861.010	17061.91	CE	0.9			MIT
5858.150	5859.774	17065.50	CE	0.6			MIT
5857.133	5858.757	17068.47	CE	0.7			MIT
5853.668	5855.291	17078.57	CE	0.7			MIT
5853.357	5854.980	17079.48	CE	0.6			MIT
5853.071	5854.693	17080.31	CE	0.6			MIT
5851.061	5852.683	17086.18	CE	0.7	D		MIT
5848.340	5849.961	17094.13	CE	1.0			MIT
5843.746	5845.366	17107.57	CE	0.8			MIT
5843.110	5844.730	17109.43	CE	0.5			MIT
5839.379	5840.998	17120.36	CE	1.0			MIT
5838.156	5839.774	17123.95	CE	1.2			MIT
5835.839	5837.457	17130.75	CE	1.4			MIT
5834.244	5835.861	17135.43	CE	0.7			MIT
5831.925	5833.542	17142.25	CE	1.4			MIT
5831.387	5833.004	17143.83	CE	0.5			MIT
5830.130	5831.746	17147.52	CE	0.8			MIT
5827.255	5828.871	17155.98	CE	0.5			MIT
5823.458	5825.073	17167.17	CE	0.3			MIT
5822.993	5824.607	17168.54	CE	1.1			MIT
5820.795	5822.409	17175.02	CE	0.5			MIT
5820.397	5822.011	17176.20	CE	0.9			MIT

AIR WAVELENGTH (ANGSTRMS)	VACUUM WAVELENGTH (ANGSTRMS)	WAVENUMBER (KAYSERS)	LMENT	LOG ARC	INTENSITY SPK	DIS	REF
5817.777	5819.390	17183.93	CE	0.5			MIT
5812.932	5814.544	17198.25	CE	1.6			MIT
5810.720	5812.331	17204.80	CE	1.2			MIT
5804.415	5806.024	17223.49	CE	1.4			MIT
5799.798	5801.406	17237.20	CE	0.5			MIT
5796.063	5797.670	17248.31	CE	0.8			MIT
5794.780	5796.387	17252.13	CE	0.6			MIT
5791.677	5793.283	17261.37	CE	0.8			MIT
5791.323	5792.929	17262.43	CE	0.7			MIT
5789.833	5791.439	17266.87	CE	0.3			MIT
5788.592	5790.197	17270.57	CE	0.5			MIT
5788.133	5789.738	17271.94	CE	1.4			MIT
5787.217	5788.822	17274.67	CE	0.6			MIT
5786.858	5788.463	17275.74	CE	0.5			MIT
5784.848	5786.452	17281.75	CE	1.1			MIT
5783.988	5785.592	17284.32	CE	0.5			MIT
5782.806	5784.410	17287.85	CE	0.7			MIT
5782.436	5784.040	17288.96	CE	0.9			MIT
5780.207	5781.810	17295.62	CE	0.3			MIT
5778.412	5780.014	17301.00	CE	0.6			MIT
5775.803	5777.405	17308.81	CE	0.5			MIT
5774.995	5776.597	17311.23	CE	0.6	D		MIT
5773.585	5775.186	17315.46	CE	0.7			MIT
5773.118	5774.719	17316.86	CE	1.5			MIT
5772.885	5774.486	17317.56	CE	1.2			MIT
5772.220	5773.821	17319.55	CE	1.0			MIT
5770.435	5772.035	17324.91	CE	0.9			MIT
5769.949	5771.549	17326.37	CE	0.8			MIT
5768.895	5770.495	17329.54	CE	1.2			MIT
5766.434	5768.033	17336.93	CE	0.5			MIT
5765.340	5766.939	17340.22	CE	1.0			MIT
5764.774	5766.373	17341.92	CE	0.7			MIT
5760.978	5762.576	17353.35	CE	0.3			MIT
5760.585	5762.183	17354.53	CE	0.7			MIT
5758.301	5759.898	17361.42	CE	0.9			MIT
5752.524	5754.120	17378.85	CE	0.9			MIT
5748.947	5750.542	17389.67	CE	0.9			MIT
5746.487	5748.081	17397.11	CE	0.5			MIT
5744.693	5746.286	17402.54	CE	0.6			MIT
5743.533	5745.126	17406.06	CE	1.4			MIT
5739.628	5741.220	17417.90	CE	0.5			MIT
5735.686	5737.277	17429.87	CE	0.8			MIT
5733.938	5735.529	17435.19	CE	0.8			MIT
5729.379	5730.968	17449.06	CE	0.3			MIT
5726.135	5727.723	17458.94	CE	0.7			MIT
5725.854	5727.442	17459.80	CE	1.3			MIT
5721.957	5723.544	17471.69	CE	1.1			MIT
5719.560	5721.147	17479.01	CE	0.5			MIT
5719.035	5720.622	17480.62	CE	1.6			MIT
5718.365	5719.951	17482.67	CE	0.6			MIT
5716.495	5718.081	17488.39	CE	0.8			MIT
5715.286	5716.872	17492.08	CE	0.5			MIT
5712.291	5713.876	17501.26	CE	0.9			MIT
5711.447	5713.032	17503.84	CE	0.9			MIT
5707.430	5709.013	17516.16	CE	0.5			MIT
5703.226	5704.808	17529.07	CE	0.5			MIT
5702.390	5703.972	17531.64	CE	0.5			MIT
5699.228	5700.809	17541.37	CE	1.6			MIT
5696.995	5698.576	17548.25	CE	1.6			MIT
5695.840	5697.420	17551.80	CE	1.0			MIT
5692.940	5694.520	17560.74	CE	1.4			MIT
5692.128	5693.707	17563.25	CE	0.8			MIT
5688.485	5690.063	17574.50	CE	0.5			MIT
5685.859	5687.437	17582.61	CE	0.7			MIT
5683.773	5685.350	17589.07	CE	0.3			MIT
5682.783	5684.360	17592.13	CE	0.5			MIT
5680.270	5681.846	17599.91	CE	0.3			MIT
5677.756	5679.332	17607.71	CE	1.4			MIT
5677.248	5678.823	17609.28	CE	0.3			MIT
5676.880	5678.455	17610.42	CE	1.0			MIT

AIR WAVELENGTH (ANGSTRMS)	VACUUM WAVELENGTH (ANGSTRMS)	WAVENUMBER (KAYSERS)	LMENT	LOG ARC	INTENSITY SPK	DIS	REF
5675.103	5676.678	17615.94	CE	0.9			MIT
5671.870	5673.444	17625.98	CE	1.0			MIT
5671.412	5672.986	17627.40	CE	0.5			MIT
5669.968	5671.541	17631.89	CE	1.7			MIT
5668.901	5670.474	17635.21	CE	1.3			MIT
5664.683	5666.255	17648.34	CE	0.9			MIT
5663.986	5665.558	17650.51	CE	1.1			MIT
5663.484	5665.056	17652.08	CE	0.7			MIT
5663.183	5664.755	17653.01	CE	0.6			MIT
5659.779	5661.350	17663.63	CE	0.8			MIT
5656.213	5657.783	17674.77	CE	0.5			MIT
5655.128	5656.697	17678.16	CE	1.6			MIT
5652.958	5654.527	17684.95	CE	0.5			MIT
5650.602	5652.170	17692.32	CE	1.2			MIT
5647.098	5648.665	17703.30	CE	0.5			MIT
5646.575	5648.142	17704.94	CE	1.0			MIT
5640.787	5642.353	17723.10	CE	0.8			MIT
5640.113	5641.678	17725.22	CE	0.6			MIT
5638.628	5640.193	17729.89	CE	0.5			MIT
5638.194	5639.759	17731.25	CE	1.0			MIT
5637.386	5638.951	17733.80	CE	0.9			MIT
5634.486	5636.050	17742.92	CE	0.8D			MIT
5633.092	5634.656	17747.31	CE	1.2W			MIT
5632.484	5634.047	17749.23	CE	1.1			MIT
5628.208	5629.770	17762.71	CE	0.9			MIT
5625.229	5626.790	17772.12	CE	0.7W			MIT
5623.754	5625.315	17776.78	CE	0.7			MIT
5623.000	5624.561	17779.17	CE	0.9			MIT
5622.674	5624.235	17780.20	CE	0.6			MIT
5620.390	5621.950	17787.42	CE	0.9			MIT
5615.977	5617.536	17801.40	CE	0.8			MIT
5614.717	5616.276	17805.39	CE	1.3			MIT
5613.698	5615.256	17808.63	CE	0.7			MIT
5610.920	5612.478	17817.44	CE	1.2			MIT
5610.257	5611.814	17819.55	CE	1.2			MIT
5609.449	5611.006	17822.11	CE	0.8			MIT
5606.464	5608.020	17831.60	CE	0.7			MIT
5604.425	5605.981	17838.09	CE	0.6			MIT
5601.303	5602.858	17848.03	CE	1.7			MIT
5598.955	5600.509	17855.52	CE	1.0			MIT
5597.947	5599.501	17858.73	CE	1.1			MIT
5595.866	5597.420	17865.37	CE	1.4			MIT
5594.944	5596.497	17868.32	CE	1.2			MIT
5593.721	5595.274	17872.23	CE	0.6W			MIT
5590.525	5592.077	17882.44	CE	0.8			MIT
5588.330	5589.882	17889.47	CE	1.0			MIT
5584.817	5586.368	17900.72	CE	1.0			MIT
5582.737	5584.287	17907.39	CE	1.3			MIT
5582.573	5584.123	17907.92	CE	0.5			MIT
5578.888	5580.437	17919.74	CE	1.0			MIT
5578.270	5579.819	17921.73	CE	1.1			MIT
5577.283	5578.832	17924.90	CE	0.7			MIT
5572.189	5573.736	17941.29	CE	0.9			MIT
5570.922	5572.469	17945.37	CE	0.6			MIT
5569.294	5570.841	17950.61	CE	0.5			MIT
5567.812	5569.358	17955.39	CE	1.1			MIT
5566.499	5568.045	17959.63	CE	0.6			MIT
5565.970	5567.516	17961.33	CE	1.5			MIT
5564.960	5566.505	17964.59	CE	1.6			MIT
5564.236	5565.781	17966.93	CE	1.0			MIT
5563.028	5564.573	17970.83	CE	1.0			MIT
5559.223	5560.767	17983.13	CE	1.2			MIT
5558.658	5560.202	17984.96	CE	0.8			MIT
5556.945	5558.488	17990.50	CE	0.9			MIT
5556.254	5557.797	17992.74	CE	1.5			MIT
5552.296	5553.838	18005.57	CE	0.8			MIT
5551.409	5552.951	18008.44	CE	0.9			MIT
5550.038	5551.579	18012.89	CE	0.8W			MIT
5548.817	5550.358	18016.86	CE	1.4			MIT
5544.653	5546.193	18030.39	CE	1.0			MIT

AIR WAVELENGTH (ANGSTRMS)	VACUUM WAVELENGTH (ANGSTRMS)	WAVENUMBER (KAYSERS)	LMENT	LOG ARC	INTENSITY SPK	DIS	REF
5542.713	5544.252	18036.70	CE	0.7			MIT
5540.578	5542.117	18043.65	CE	0.9			MIT
5537.535	5539.073	18053.56	CE	1.0			MIT
5535.240	5536.777	18061.05	CE	1.2			MIT
5532.037	5533.574	18071.50	CE	0.5			MIT
5527.895	5529.430	18085.04	CE	0.7			MIT
5527.176	5528.711	18087.40	CE	1.0			MIT
5526.850	5528.385	18088.47	CE	0.7			MIT
5522.994	5524.528	18101.09	CE	2.0			MIT
5522.459	5523.993	18102.85	CE	1.2			MIT
5518.491	5520.024	18115.86	CE	1.1			MIT
5517.395	5518.928	18119.46	CE	1.0			MIT
5516.081	5517.613	18123.78	CE	0.8			MIT
5513.113	5514.645	18133.53	CE	0.6			MIT
5512.085	5513.616	18136.92	CE	1.7S			MIT
5511.689	5513.220	18138.22	CE	0.6			MIT
5510.682	5512.213	18141.53	CE	1.0			MIT
5506.447	5507.977	18155.49	CE	0.8			MIT
5506.090	5507.620	18156.66	CE	0.8			MIT
5498.194	5499.722	18182.74	CE	1.2			MIT
5491.154	5492.680	18206.05	CE	1.0			MIT
5478.598	5480.120	18247.77	CE	1.1			MIT
5473.534	5475.055	18264.66	CE	1.1			MIT
5472.297	5473.818	18268.79	CE	1.4			MIT
5468.371	5469.891	18281.90	CE	1.2			MIT
5465.344	5466.863	18292.03	CE	1.3			MIT
5464.20	5465.72	18295.9	CE	0.7			MIT
5460.086	5461.603	18309.64	CE	1.1			MIT
5459.21	5460.73	18312.6	CE	0.6			MIT
5458.808	5460.325	18313.93	CE	0.9			MIT
5457.211	5458.728	18319.29	CE	1.1			MIT
5456.408	5457.924	18321.98	CE	1.2			MIT
5453.950	5455.466	18330.24	CE	1.1			MIT
5451.723	5453.238	18337.73	CE	1.0			MIT
5450.038	5451.553	18343.40	CE	1.0			MIT
5449.224	5450.738	18346.14	CE	1.4			MIT
5446.196	5447.710	18356.34	CE	1.0			MIT
5445.425	5446.938	18358.94	CE	1.0			MIT
5438.426	5439.938	18382.56	CE	1.0			MIT
5433.340	5434.850	18399.77	CE	0.9			MIT
5430.243	5431.752	18410.26	CE	1.0			MIT
5427.250	5428.759	18420.42	CE	0.9			MIT
5426.366	5427.874	18423.42	CE	0.8			MIT
5423.420	5424.928	18433.43	CE	1.1			MIT
5420.385	5421.892	18443.75	CE	1.3			MIT
5418.701	5420.207	18449.48	CE	1.1			MIT
5414.092	5415.597	18465.19	CE	1.1			MIT
5411.752	5413.256	18473.17	CE	0.9			MIT
5409.224	5410.728	18481.80	CE	1.7			MIT
5401.208	5402.710	18509.23	CE	0.8			MIT
5399.574	5401.075	18514.83	CE	1.1			MIT
5399.035	5400.536	18516.68	CE	0.3			MIT
5397.644	5399.145	18521.45	CE	1.4			MIT
5395.696	5397.196	18528.14	CE	1.0			MIT
5395.242	5396.742	18529.70	CE	1.0			MIT
5393.391	5394.891	18536.06	CE	1.5			MIT
5391.882	5393.381	18541.25	CE	1.0W			MIT
5386.763	5388.261	18558.86	CE	0.9W			MIT
5386.354	5387.852	18560.27	CE	1.2			MIT
5382.613	5384.110	18573.17	CE	1.0			MIT
5379.890	5381.386	18582.57	CE	1.0			MIT
5371.600	5373.094	18611.25	CE	0.8			MIT
5369.119	5370.612	18619.85	CE	0.9			MIT
5363.325	5364.817	18639.97	CE	1.2			MIT
5359.954	5361.445	18651.69	CE	1.0			MIT
5359.298	5360.788	18653.97	CE	0.9			MIT
5357.203	5358.693	18661.27	CE	1.0			MIT
5355.617	5357.106	18666.79	CE	1.0			MIT
5355.181	5356.670	18668.31	CE	1.0			MIT
5353.534	5355.023	18674.06	CE	1.7	1.5		MIT

AIR WAVELENGTH (ANGSTRMS)	VACUUM WAVELENGTH (ANGSTRMS)	WAVENUMBER (KAYSERS)	LMENT	LOG ARC	INTENSITY SPK	DIS	REF
5347.806	5349.293	18694.06	CE	0.9			MIT
5346.529	5348.016	18698.52	CE	0.3			MIT
5345.097	5346.584	18703.53	CE	0.7			MIT
5336.177	5337.661	18734.80	CE	1.2			MIT
5335.708	5337.192	18736.44	CE	1.2			MIT
5334.674	5336.158	18740.07	CE	0.7			MIT
5333.825	5335.309	18743.06	CE	0.8			MIT
5330.582	5332.065	18754.46	CE	1.4			MIT
5329.495	5330.978	18758.29	CE	1.0			MIT
5315.072	5316.551	18809.19	CE	0.8			MIT
5314.895	5316.374	18809.81	CE	0.8			MIT
5314.405	5315.883	18811.55	CE	0.9			MIT
5313.932	5315.410	18813.22	CE	0.9			MIT
5308.554	5310.031	18832.28	CE	0.9			MIT
5308.315	5309.792	18833.13	CE	0.7			MIT
5303.349	5304.825	18850.76	CE	0.9			MIT
5298.29	5299.76	18868.8	CE	1.0			MIT
5296.598	5298.072	18874.79	CE	1.2			MIT
5294.068	5295.541	18883.81	CE	1.1			MIT
5290.941	5292.413	18894.97	CE	1.1			MIT
5286.837	5288.308	18909.64	CE	1.1			MIT
5281.375	5282.845	18929.20	CE	1.0			MIT
5277.543	5279.012	18942.94	CE	0.9			MIT
5276.968	5278.437	18945.00	CE	0.7			MIT
5274.244	5275.712	18954.79	CE	1.7	0.5		MIT
5271.880	5273.347	18963.29	CE	1.2 W			MIT
5271.063	5272.530	18966.23	CE	1.0			MIT
5265.710	5267.176	18985.51	CE	1.2			MIT
5264.213	5265.678	18990.91	CE	1.0			MIT
5261.711	5263.175	18999.94	CE	1.1			MIT
5259.933	5261.397	19006.36	CE	0.9			MIT
5254.84	5256.30	19024.8	CE	1.1			MIT
5250.487	5251.948	19040.55	CE	0.8			MIT
5250.140	5251.601	19041.81	CE	0.8			MIT
5249.618	5251.079	19043.70	CE	1.1			MIT
5249.183	5250.644	19045.28	CE	1.0			MIT
5247.424	5248.885	19051.67	CE	0.7			MIT
5245.920	5247.380	19057.13	CE	1.5			MIT
5245.277	5246.737	19059.47	CE	1.0			MIT
5244.506	5245.966	19062.27	CE	1.3			MIT
5243.079	5244.538	19067.46	CE	0.9			MIT
5241.59	5243.05	19072.9	CE	0.7			MIT
5238.487	5239.945	19084.17	CE	1.1			MIT
5237.050	5238.508	19089.41	CE	0.8			AB
5234.014	5235.471	19100.48	CE	0.9			AB
5232.864	5234.321	19104.68	CE	1.2			MIT
5230.142	5231.598	19114.62	CE	0.9			AB
5229.749	5231.205	19116.05	CE	1.3			MIT
5223.472	5224.926	19139.03	CE	1.3			MIT
5222.948	5224.402	19140.95	CE	0.3			MIT
5221.920	5223.374	19144.71	CE	0.3			MIT
5216.384	5217.836	19165.03	CE	0.9			MIT
5211.921	5213.372	19181.44	CE	1.7 S			MIT
5211.043	5212.494	19184.68	CE	0.9			MIT
5208.901	5210.351	19192.56	CE	0.3			MIT
5207.344	5208.794	19198.30	CE	0.9			MIT
5205.521	5206.970	19205.02	CE	0.9			MIT
5205.142	5206.591	19206.42	CE	1.0			MIT
5204.726	5206.175	19207.96	CE	0.9 W			MIT
5204.274	5205.723	19209.63	CE	1.0			MIT
5203.279	5204.728	19213.30	CE	0.9			MIT
5202.594	5204.043	19215.83	CE	0.7			MIT
5202.464	5203.913	19216.31	CE	0.8			MIT
5201.389	5202.837	19220.28	CE	1.2			MIT
5200.418	5201.866	19223.87	CE	1.1 W			MIT
5200.122	5201.570	19224.97	CE	0.8			MIT
5194.980	5196.427	19243.99	CE	0.7			MIT
5194.750	5196.197	19244.84	CE	0.9			MIT
5191.675	5193.121	19256.24	CE	1.5 J			MIT
5188.652	5190.097	19267.46	CE	1.0			MIT

AIR WAVELENGTH (ANGSTRMS)	VACUUM WAVELENGTH (ANGSTRMS)	WAVENUMBER (KAYSERS)	LMENT	LOG ARC	INTENSITY SPK	DIS	REF
5187.452	5188.897	19271.92	CE	1.7			MIT
5183.197	5184.641	19287.74	CE	1.0			MIT
5181.935	5183.378	19292.44	CE	1.0			MIT
5181.750	5183.193	19293.13	CE	0.7			MIT
5180.890	5182.333	19296.33	CE	1.2			MIT
5179.458	5180.901	19301.66	CE	0.5			MIT
5178.688	5180.130	19304.53	CE	1.0			MIT
5177.730	5179.172	19308.10	CE	0.9			MIT
5174.540	5175.981	19320.01	CE	1.4			MIT
5173.701	5175.142	19323.14	CE	0.5			MIT
5169.718	5171.158	19338.03	CE	1.2			MIT
5164.386	5165.825	19357.99	CE	1.1			MIT
5161.484	5162.922	19368.88	CE	1.5			MIT
5159.689	5161.126	19375.62	CE	1.5			MIT
5154.388	5155.824	19395.54	CE	1.0			AB
5150.405	5151.840	19410.54	CE	1.0			MIT
5149.990	5151.425	19412.11	CE	1.1			MIT
5149.645	5151.080	19413.41	CE	1.0			MIT
5147.549	5148.983	19421.31	CE	1.2			MIT
5145.138	5146.571	19430.41	CE	0.6			MIT
5140.503	5141.935	19447.93	CE	1.0			MIT
5139.760	5141.192	19450.74	CE	1.0			MIT
5138.018	5139.449	19457.34	CE	1.1			MIT
5137.762	5139.193	19458.31	CE	1.0			MIT
5137.117	5138.548	19460.75	CE	1.1			MIT
5135.318	5136.749	19467.57	CE	0.9 W			MIT
5134.465	5135.896	19470.80	CE	1.1			MIT
5129.578	5131.007	19489.35	CE	1.5			MIT
5125.009	5126.437	19506.73	CE	1.0			MIT
5122.674	5124.101	19515.62	CE	1.1			AB
5122.393	5123.820	19516.69	CE	1.1 W			MIT
5120.770	5122.197	19522.87	CE	1.2			MIT
5117.175	5118.601	19536.59	CE	1.3			MIT
5115.634	5117.060	19542.47	CE	1.0			MIT
5115.224	5116.649	19544.04	CE	1.1			MIT
5115.028	5116.453	19544.79	CE	0.6			MIT
5112.688	5114.113	19553.73	CE	1.3			MIT
5111.601	5113.025	19557.89	CE	1.0			MIT
5107.470	5108.893	19573.71	CE	1.0			MIT
5107.203	5108.626	19574.73	CE	1.0			MIT
5105.208	5106.631	19582.38	CE	0.3			MIT
5099.395	5100.816	19604.71	CE	1.1			MIT
5097.242	5098.663	19612.99	CE	0.9			MIT
5091.751	5093.170	19634.14	CE	1.2			MIT
5090.864	5092.283	19637.56	CE	0.5			MIT
5086.567	5087.985	19654.15	CE	0.8			MIT
5084.790	5086.207	19661.01	CE	0.7			MIT
5084.168	5085.585	19663.42	CE	1.1			MIT
5083.538	5084.955	19665.86	CE	1.1			MIT
5080.478	5081.894	19677.70	CE	1.2			MIT
5079.681	5081.097	19680.79	CE	1.5			MIT
5078.338	5079.754	19685.99	CE	0.7			MIT
5076.474	5077.889	19693.22	CE	1.0	0.0		MIT
5075.304	5076.719	19697.76	CE	1.0	0.0		MIT
5075.226	5076.641	19698.06	CE	0.8			MIT
5074.710	5076.125	19700.07	CE	1.0			MIT
5072.912	5074.326	19707.05	CE	0.3			MIT
5071.773	5073.187	19711.48	CE	1.3			MIT
5067.148	5068.561	19729.47	CE	1.0	0.0		MIT
5065.734	5067.146	19734.97	CE	0.9			MIT
5063.919	5065.331	19742.05	CE	1.3			MIT
5056.005	5057.415	19772.95	CE	0.9			MIT
5055.782	5057.192	19773.82	CE	1.1			MIT
5054.174	5055.583	19780.11	CE	1.0			MIT
5053.529	5054.938	19782.64	CE	1.0			MIT
5053.268	5054.677	19783.66	CE	1.0			MIT
5050.986	5052.394	19792.60	CE	1.3			MIT
5048.818	5050.226	19801.09	CE	1.5			MIT
5048.541	5049.949	19802.18	CE	0.7			MIT
5044.008	5045.414	19819.98	CE	1.4	0.0		MIT

AIR WAVELENGTH (ANGSTRMS)	VACUUM WAVELENGTH (ANGSTRMS)	WAVENUMBER (KAYSERS)	LMENT	LOG ARC	INTENSITY SPK	DIS	REF
5042.086	5043.492	19827.53	CE	1.0			MIT
5040.855	5042.261	19832.37	CE	1.5			MIT
5039.933	5041.338	19836.00	CE	0.9			MIT
5039.749	5041.154	19836.73	CE	0.9			MIT
5037.982	5039.387	19843.68	CE	0.8			MIT
5037.766	5039.171	19844.53	CE	1.3			MIT
5036.623	5038.028	19849.04	CE	1.0			MIT
5033.813	5035.217	19860.12	CE	1.3			MIT
5031.742	5033.145	19868.29	CE	1.1			MIT
5028.304	5029.706	19881.88	CE	1.3			MIT
5025.408	5026.810	19893.33	CE	0.5			MIT
5022.871	5024.272	19903.38	CE	1.3	0.0		MIT
5021.444	5022.844	19909.04	CE	1.3			MIT
5016.507	5017.906	19928.63	CE	1.0			MIT
5013.759	5015.157	19939.55	CE	1.3			MIT
5012.510	5013.908	19944.52	CE	1.1W			MIT
5011.765	5013.163	19947.49	CE	1.1	0.0		MIT
5009.444	5010.841	19956.73	CE	0.8			MIT
5009.094	5010.491	19958.12	CE	1.5			MIT
5005.702	5007.098	19971.65	CE	1.0			MIT
5003.271	5004.667	19981.35	CE	0.9			MIT
5001.495	5002.890	19988.45	CE	0.8			MIT
4998.131	4999.525	20001.90	CE	1.3			MIT
4994.613	4996.006	20015.99	CE	1.0			MIT
4992.402	4993.795	20024.85	CE	1.1			MIT
4990.667	4992.059	20031.81	CE	1.0			MIT
4988.688	4990.080	20039.76	CE	1.1			MIT
4987.541	4988.932	20044.37	CE	0.9			MIT
4986.423	4987.814	20048.86	CE	1.3O			MIT
4982.132	4983.522	20066.13	CE	0.9			MIT
4974.104	4975.492	20098.51	CE	1.0			MIT
4972.239	4973.626	20106.05	CE	0.9			MIT
4971.936	4973.323	20107.28	CE	0.9			MIT
4971.660	4973.047	20108.40	CE	0.3			MIT
4971.475	4972.862	20109.14	CE	1.1			MIT
4970.667	4972.054	20112.41	CE	1.1			MIT
4970.122	4971.509	20114.62	CE	0.8			MIT
4966.387	4967.773	20129.75	CE	1.0			MIT
4965.232	4966.618	20134.43	CE	0.5			MIT
4965.177	4966.563	20134.65	CE	0.9W			MIT
4958.225	4959.609	20162.88	CE	0.5			MIT
4956.974	4958.357	20167.97	CE	0.7			MIT
4955.965	4957.348	20172.08	CE	0.8			MIT
4955.388	4956.771	20174.42	CE	0.8			MIT
4954.039	4955.422	20179.92	CE	0.8			MIT
4953.716	4955.098	20181.23	CE	0.9			MIT
4948.674	4950.055	20201.79	CE	1.3J			MIT
4944.617	4945.997	20218.37	CE	1.3			MIT
4943.451	4944.831	20223.14	CE	0.6			MIT
4939.125	4940.504	20240.85	CE	1.3			MIT
4938.820	4940.199	20242.10	CE	0.3			MIT
4930.721	4932.097	20275.35	CE	1.0			MIT
4930.539	4931.915	20276.10	CE	1.0			MIT
4924.258	4925.633	20301.96	CE	1.3			MIT
4924.04	4925.41	20302.9	CE	0.3			KN
4921.918	4923.292	20311.61	CE	0.9			MIT
4920.783	4922.157	20316.30	CE	1.2			MIT
4919.888	4921.261	20319.99	CE	1.1S			MIT
4915.668	4917.040	20337.44	CE	1.2			MIT
4915.321	4916.693	20338.87	CE	1.3			MIT
4914.940	4916.312	20340.45	CE	0.9			MIT
4914.086	4915.458	20343.98	CE	0.7			MIT
4913.423	4914.795	20346.73	CE	0.6			MIT
4908.121	4909.491	20368.71	CE	0.8			MIT
4904.880	4906.250	20382.17	CE	1.1			MIT
4901.676	4903.045	20395.49	CE	1.1			MIT
4899.901	4901.269	20402.88	CE	1.5			MIT
4898.207	4899.575	20409.94	CE	0.6			MIT
4897.078	4898.445	20414.64	CE	1.0			MIT
4894.774	4896.141	20424.25	CE	0.7			MIT

AIR WAVELENGTH (ANGSTRMS)	VACUUM WAVELENGTH (ANGSTRMS)	WAVENUMBER (KAYSERS)	LMENT	LOG ARC	INTENSITY SPK	DIS	REF
4893.968	4895.335	20427.61	CE	1.0			MIT
4893.225	4894.591	20430.72	CE	0.3			MIT
4892.858	4894.224	20432.25	CE	0.9			MIT
4889.587	4890.952	20445.92	CE	1.3			MIT
4882.462	4883.826	20475.75	CE	1.5			MIT
4881.540	4882.903	20479.62	CE	0.8			MIT
4874.350	4875.711	20509.83	CE	0.9			MIT
4872.858	4874.219	20516.11	CE	0.6			MIT
4871.961	4873.322	20519.89	CE	0.6			MIT
4870.158	4871.518	20527.48	CE	0.3			MIT
4868.635	4869.995	20533.90	CE	0.9			MIT
4863.259	4864.617	20556.60	CE	1.0			MIT
4861.732	4863.090	20563.06	CE	1.0			MIT
4859.480	4860.837	20572.59	CE	1.2			MIT
4858.716	4860.073	20575.82	CE	0.8			MIT
4855.005	4856.361	20591.55	CE	0.7			MIT
4853.611	4854.967	20597.46	CE	0.8			MIT
4852.705	4854.061	20601.31	CE	0.7			MIT
4852.619	4853.975	20601.67	CE	0.7			MIT
4849.914	4851.269	20613.16	CE	1.0			MIT
4849.554	4850.909	20614.69	CE	0.6			MIT
4847.754	4849.108	20622.35	CE	1.1			MIT
4846.574	4847.928	20627.37	CE	0.9			MIT
4845.517	4846.871	20631.87	CE	1.3			MIT
4843.033	4844.386	20642.45	CE	0.9			MIT
4842.296	4843.649	20645.59	CE	0.8			MIT
4838.952	4840.304	20659.86	CE	0.8			MIT
4838.449	4839.801	20662.01	CE	0.3			MIT
4837.487	4838.839	20666.12	CE	0.9			MIT
4836.669	4838.020	20669.61	CE	1.2			MIT
4835.447	4836.798	20674.83	CE	0.5			MIT
4834.044	4835.395	20680.83	CE	1.0			MIT
4829.838	4831.188	20698.84	CE	0.7			MIT
4829.037	4830.386	20702.28	CE	0.7			MIT
4822.541	4823.889	20730.16	CE	1.4			MIT
4820.612	4821.959	20738.46	CE	0.9			MIT
4820.032	4821.379	20740.95	CE	1.1			MIT
4817.480	4818.826	20751.94	CE	0.7			MIT
4810.392	4811.736	20782.52	CE	0.8			MIT
4808.502	4809.846	20790.69	CE	1.0			MIT
4807.677	4809.021	20794.25	CE	0.7			MIT
4805.931	4807.274	20801.81	CE	1.3			MIT
4805.532	4806.875	20803.54	CE	0.3			MIT
4803.062	4804.404	20814.23	CE	0.6			MIT
4801.357	4802.699	20821.62	CE	0.3			MIT
4788.868	4790.207	20875.93	CE	0.6			MIT
4788.425	4789.764	20877.86	CE	1.0			MIT
4786.540	4787.878	20886.08	CE	1.0			MIT
4784.777	4786.115	20893.77	CE	1.0			MIT
4782.216	4783.553	20904.96	CE	1.0			MIT
4781.716	4783.053	20907.15	CE	0.9			MIT
4780.733	4782.070	20911.45	CE	0.8			MIT
4776.478	4777.813	20930.08	CE	0.8			MIT
4776.234	4777.569	20931.15	CE	0.8			MIT
4775.077	4776.412	20936.22	CE	1.0			MIT
4773.942	4775.277	20941.20	CE	1.3			MIT
4768.770	4770.103	20963.91	CE	1.0			MIT
4764.722	4766.054	20981.72	CE	0.9			MIT
4758.542	4759.873	21008.97	CE	0.6			MIT
4757.842	4759.173	21012.06	CE	1.2			MIT
4752.577	4753.906	21035.33	CE	1.0			MIT
4751.409	4752.738	21040.50	CE	0.7			MIT
4750.837	4752.166	21043.04	CE	0.9			MIT
4750.227	4751.555	21045.74	CE	0.8			MIT
4747.143	4748.471	21059.41	CE	1.5			MIT
4744.818	4746.145	21069.73	CE	1.0			MIT
4743.256	4744.583	21076.67	CE	1.0			MIT
4739.534	4740.860	21093.22	CE	1.0			MIT
4737.282	4738.607	21103.25	CE	1.3			MIT
4734.697	4736.021	21114.77	CE	0.9			MIT

AIR WAVELENGTH (ANGSTRMS)	VACUUM WAVELENGTH (ANGSTRMS)	WAVENUMBER (KAYSERS)	LMENT	LOG ARC	INTENSITY SPK	DIS	REF
4733.945	4735.269	21118.12	CE	1.0			MIT
4727.588	4728.910	21146.52	CE	0.7			MIT
4725.090	4726.412	21157.70	CE	1.0			MIT
4724.845	4726.167	21158.80	CE	1.0			MIT
4724.315	4725.637	21161.17	CE	0.9			MIT
4719.735	4721.055	21181.70	CE	0.8			MIT
4715.072	4716.391	21202.65	CE	0.8			MIT
4713.996	4715.315	21207.49	CE	1.0	0.5		MIT
4710.223	4711.541	21224.48	CE	0.3			MIT
4710.004	4711.322	21225.47	CE	0.6			MIT
4707.940	4709.257	21234.77	CE	0.9			MIT
4707.277	4708.594	21237.76	CE	0.3H			MIT
4707.005	4708.322	21238.99	CE	0.3	0.0		MIT
4706.461	4707.778	21241.44	CE	0.3H			MIT
4705.579	4706.896	21245.43	CE	0.5H			MIT
4704.012	4705.328	21252.50	CE	0.9			MIT
4703.779	4705.095	21253.55	CE	0.3			MIT
4702.737	4704.053	21258.26	CE	0.5			MIT
4702.008	4703.324	21261.56	CE	0.8			MIT
4701.445	4702.761	21264.11	CE	0.9			MIT
4700.624	4701.939	21267.82	CE	0.3			MIT
4699.128	4700.443	21274.59	CE	0.3			MIT
4696.789	4698.103	21285.19	CE	0.3			MIT
4696.523	4697.837	21286.39	CE	0.5			MIT
4694.880	4696.194	21293.84	CE	0.6			MIT
4694.348	4695.662	21296.25	CE	0.5L			MIT
4693.580	4694.893	21299.74	CE	0.3			MIT
4692.055	4693.368	21306.66	CE	0.6	0.0		MIT
4690.834	4692.147	21312.21	CE	0.3			M
4690.712	4692.025	21312.76	CE	0.3			MIT
4690.497	4691.810	21313.74	CE	0.3			MIT
4690.170	4691.483	21315.22	CE	0.3			MIT
4689.503	4690.815	21318.25	CE	0.3			MIT
4688.889	4690.201	21321.05	CE	0.8			MIT
4688.229	4689.541	21324.05	CE	0.3			MIT
4687.637	4688.949	21326.74	CE	0.3			MIT
4686.806	4688.118	21330.52	CE	0.3			MIT
4685.232	4686.543	21337.69	CE	0.5			MIT
4684.605	4685.916	21340.54	CE	0.9S			MIT
4683.072	4684.383	21347.53	CE	0.3			MIT
4680.993	4682.303	21357.01	CE	0.5			MIT
4680.458	4681.768	21359.45	CE	0.3			MIT
4680.127	4681.437	21360.96	CE	0.8			MIT
4678.621	4679.931	21367.84	CE	0.3			MIT
4678.513	4679.823	21368.33	CE	0.3			MIT
4676.338	4677.647	21378.27	CE	0.3			MIT
4675.312	4676.621	21382.96	CE	0.3			MIT
4674.878	4676.187	21384.95	CE	0.3H			MIT
4674.492	4675.800	21386.71	CE	0.5			MIT
4671.711	4673.019	21399.44	CE	0.3			MIT
4670.913	4672.220	21403.10	CE	0.6			MIT
4670.737	4672.044	21403.91	CE	0.6			MIT
4670.445	4671.752	21405.24	CE	0.3H			MIT
4670.097	4671.404	21406.84	CE	0.3			MIT
4669.638	4670.945	21408.94	CE	0.5			MIT
4669.502	4670.809	21409.57	CE	0.6			MIT
4667.364	4668.671	21419.37	CE	0.3			MIT
4666.707	4668.013	21422.39	CE	0.3S			MIT
4665.276	4666.582	21428.96	CE	0.3			MIT
4664.070	4665.376	21434.50	CE	0.5			MIT
4663.237	4664.542	21438.33	CE	0.5			MIT
4662.258	4663.563	21442.83	CE	0.3			MIT
4661.621	4662.926	21445.76	CE	0.6			MIT
4659.937	4661.242	21453.51	CE	0.5			MIT
4659.400	4660.704	21455.98	CE	0.8	0.8		MIT
4658.042	4659.346	21462.24	CE	0.3			MIT
4657.822	4659.126	21463.25	CE	0.3			MIT
4657.210	4658.514	21466.07	CE	0.5L			MIT
4657.017	4658.321	21466.96	CE	0.3			MIT
4656.603	4657.907	21468.87	CE	0.5			MIT

AIR WAVELENGTH (ANGSTRMS)	VACUUM WAVELENGTH (ANGSTRMS)	WAVENUMBER (KAYSERS)	LMENT	LOG ARC	INTENSITY SPK	DIS	REF
4656.187	4657.491	21470.79	CE	0.3			MIT
4656.005	4657.309	21471.63	CE	0.5			MIT
4655.227	4656.530	21475.22	CE	0.3			MIT
4654.286	4655.589	21479.56	CE	0.8			MIT
4653.866	4655.169	21481.50	CE	0.3			MIT
4653.381	4654.684	21483.74	CE	0.3			MIT
4651.691	4652.993	21491.54	CE	0.3H			MIT
4651.067	4652.369	21494.43	CE	0.3H			MIT
4650.516	4651.818	21496.97	CE	0.8			MIT
4649.889	4651.191	21499.87	CE	0.8			MIT
4649.477	4650.779	21501.77	CE	0.3			MIT
4648.826	4650.128	21504.79	CE	0.6			MIT
4647.280	4648.581	21511.94	CE	0.5			MIT
4646.943	4648.244	21513.50	CE	0.3			MIT
4644.204	4645.504	21526.19	CE	0.8	0.8		MIT
4643.174	4644.474	21530.96	CE	0.7			MIT
4643.079	4644.379	21531.40	CE	0.3			MIT
4642.278	4643.578	21535.12	CE	0.5H			MIT
4641.058	4642.358	21540.78	CE	0.8			MIT
4640.875	4642.175	21541.63	CE	0.5			MIT
4640.093	4641.392	21545.26	CE	0.3			MIT
4638.753	4640.052	21551.48	CE	0.3			MIT
4638.345	4639.644	21553.38	CE	0.3H			MIT
4636.744	4638.042	21560.82	CE	0.3			MIT
4633.599	4634.897	21575.45	CE	0.3			MIT
4633.087	4634.384	21577.84	CE	0.3			MIT
4632.322	4633.619	21581.40	CE	0.9			MIT
4630.816	4632.113	21588.42	CE	0.5H			MIT
4628.161	4629.457	21600.80	CE	1.3	1.3		MIT
4625.295	4626.590	21614.19	CE	0.3H			MIT
4624.898	4626.193	21616.04	CE	0.9	1.0		MIT
4624.367	4625.662	21618.53	CE	0.3			MIT
4624.201	4625.496	21619.30	CE	0.3			MIT
4623.463	4624.758	21622.75	CE	0.3			MIT
4620.048	4621.342	21638.74	CE	0.5			MIT
4618.933	4620.227	21643.96	CE	0.3			MIT
4618.610	4619.904	21645.47	CE	0.5			MIT
4618.067	4619.361	21648.02	CE	0.3			MIT
4616.969	4618.262	21653.17	CE	0.3			MIT
4616.084	4617.377	21657.32	CE	0.5			MIT
4615.325	4616.618	21660.88	CE	0.3			MIT
4615.195	4616.488	21661.49	CE	0.8	0.0		M
4613.023	4614.315	21671.69	CE	0.8			MIT
4611.976	4613.268	21676.61	CE	0.5			MIT
4611.557	4612.849	21678.58	CE	0.6			MIT
4610.474	4611.766	21683.67	CE	0.8			MIT
4609.656	4610.947	21687.52	CE	0.3			MIT
4609.238	4610.529	21689.48	CE	0.3			MIT
4608.750	4610.041	21691.78	CE	0.3			MIT
4608.492	4609.783	21693.00	CE	0.8			MIT
4607.290	4608.581	21698.66	CE	0.5			MIT
4607.087	4608.378	21699.61	CE	0.3			MIT
4606.401	4607.691	21702.84	CE	1.1	1.2		MIT
4605.480	4606.770	21707.18	CE	0.6			MIT
4604.209	4605.499	21713.17	CE	0.5			MIT
4602.752	4604.041	21720.05	CE	0.6			MIT
4601.568	4602.857	21725.64	CE	0.3			MIT
4601.371	4602.660	21726.57	CE	0.6			MIT
4601.019	4602.308	21728.23	CE	0.3			MIT
4600.236	4601.525	21731.93	CE	0.5			MIT
4599.304	4600.593	21736.33	CE	0.3			MIT
4599.023	4600.311	21737.66	CE	0.5			MIT
4598.590	4599.878	21739.71	CE	0.3			MIT
4597.792	4599.080	21743.48	CE	0.5			MIT
4597.170	4598.458	21746.42	CE	0.6			MIT
4596.933	4598.221	21747.54	CE	0.3			MIT
4596.167	4597.455	21751.17	CE	0.3			MIT
4595.627	4596.915	21753.72	CE	0.3			MIT
4594.129	4595.416	21760.81	CE	0.6			MIT
4593.932	4595.219	21761.75	CE	1.5	1.5		MIT

AIR WAVELENGTH (ANGSTRMS)	VACUUM WAVELENGTH (ANGSTRMS)	WAVENUMBER (KAYSERS)	LMENT	LOG ARC	INTENSITY SPK	INTENSITY DIS	REF
4593.716	4595.003	21762.77	CE	0.3			MIT
4593.103	4594.390	21765.68	CE	0.5			MIT
4592.271	4593.558	21769.62	CE	0.5			MIT
4591.577	4592.864	21772.91	CE	0.3			MIT
4591.459	4592.745	21773.47	CE	0.5			MIT
4591.119	4592.405	21775.08	CE	0.9			MIT
4590.739	4592.025	21776.88	CE	0.3			MIT
4589.924	4591.210	21780.75	CE	0.3			MIT
4589.388	4590.674	21783.29	CE	0.3			MIT
4589.171	4590.457	21784.32	CE	0.5			MIT
4588.418	4589.704	21787.90	CE	0.8			MIT
4586.051	4587.336	21799.14	CE	0.3			MIT
4584.183	4585.468	21808.03	CE	0.3H			MIT
4583.096	4584.380	21813.20	CE	0.7			MIT
4582.502	4583.786	21816.03	CE	1.0	0.9		MIT
4582.385	4583.669	21816.58	CE	0.7			MIT
4581.088	4582.372	21822.76	CE	0.6	0.0		MIT
4579.928	4581.211	21828.29	CE	0.3			MIT
4579.277	4580.560	21831.39	CE	0.8			MIT
4578.778	4580.061	21833.77	CE	0.8			MIT
4578.153	4579.436	21836.75	CE	0.3			MIT
4576.481	4577.764	21844.73	CE	0.8			MIT
4575.767	4577.049	21848.14	CE	0.3			MIT
4575.076	4576.358	21851.44	CE	0.3			MIT
4574.746	4576.028	21853.01	CE	0.5			MIT
4574.119	4575.401	21856.01	CE	0.5			MIT
4572.789	4574.071	21862.36	CE	0.8			MIT
4572.277	4573.558	21864.81	CE	1.5S	1.5		MIT
4572.013	4573.294	21866.08	CE	0.5			MIT
4571.481	4572.762	21868.62	CE	0.3			MIT
4570.638	4571.919	21872.65	CE	0.8	0.0		MIT
4570.093	4571.374	21875.26	CE	0.5			MIT
4569.660	4570.941	21877.33	CE	0.6			MIT
4569.140	4570.421	21879.82	CE	0.3			MIT
4568.587	4569.867	21882.47	CE	0.5			MIT
4567.408	4568.688	21888.12	CE	0.5			MIT
4567.158	4568.438	21889.32	CE	0.3			MIT
4566.592	4567.872	21892.03	CE	0.3H			MIT
4565.841	4567.121	21895.63	CE	1.1S	1.1		MIT
4565.235	4566.515	21898.54	CE	0.8	0.0		MIT
4564.774	4566.053	21900.75	CE	0.5			MIT
4563.375	4564.654	21907.47	CE	0.6			MIT
4563.051	4564.330	21909.02	CE	0.3			MIT
4562.361	4563.640	21912.33	CE	1.6	1.6		MIT
4561.566	4562.845	21916.15	CE	0.3			MIT
4561.283	4562.561	21917.51	CE	0.3H			MIT
4560.959	4562.237	21919.07	CE	1.3			MIT
4560.557	4561.835	21921.00	CE	0.3			MIT
4560.280	4561.558	21922.33	CE	1.4	1.4		MIT
4559.622	4560.900	21925.50	CE	0.5			MIT
4559.182	4560.460	21927.61	CE	0.3			MIT
4558.908	4560.186	21928.93	CE	0.3			MIT
4558.604	4559.882	21930.39	CE	0.9			MIT
4558.080	4559.358	21932.91	CE	0.3H			MIT
4557.980	4559.258	21933.40	CE	0.5H			MIT
4557.406	4558.683	21936.16	CE	0.5			MIT
4557.038	4558.315	21937.93	CE	0.5			MIT
4556.224	4557.501	21941.85	CE	0.5			MIT
4555.620	4556.897	21944.76	CE	0.3			MIT
4555.425	4556.702	21945.70	CE	0.7			MIT
4554.557	4555.834	21949.88	CE	0.8			MIT
4554.333	4555.610	21950.96	CE	0.3			MIT
4554.035	4555.312	21952.40	CE	1.5S			MIT
4553.755	4555.032	21953.75	CE	0.3			MIT
4553.419	4554.695	21955.36	CE	0.3			MIT
4553.062	4554.338	21957.09	CE	0.9			MIT
4552.308	4553.584	21960.72	CE	0.3			MIT
4552.068	4553.344	21961.88	CE	0.7			MIT
4551.297	4552.573	21965.60	CE	1.3			MIT
4550.922	4552.198	21967.41	CE	0.3			MIT

AIR WAVELENGTH (ANGSTRMS)	VACUUM WAVELENGTH (ANGSTRMS)	WAVENUMBER (KAYSERS)	LMENT	LOG ARC	INTENSITY SPK	INTENSITY DIS	REF
4550.565	4551.841	21969.14	CE	0.5			MIT
4550.298	4551.574	21970.42	CE	0.8			MIT
4549.967	4551.243	21972.02	CE	0.3			MIT
4549.644	4550.919	21973.58	CE	0.8			MIT
4548.884	4550.159	21977.25	CE	1.0			MIT
4548.387	4549.662	21979.65	CE	0.3			MIT
4547.710	4548.985	21982.93	CE	0.3			MIT
4547.290	4548.565	21984.96	CE	0.3H			MIT
4546.837	4548.112	21987.15	CE	0.3			MIT
4546.063	4547.337	21990.89	CE	0.9			MIT
4545.873	4547.147	21991.81	CE	0.7			MIT
4545.452	4546.726	21993.85	CE	0.6			MIT
4545.197	4546.471	21995.08	CE	0.5			MIT
4544.955	4546.229	21996.25	CE	1.0			MIT
4543.688	4544.962	22002.39	CE	0.3			MIT
4543.575	4544.849	22002.93	CE	0.5			MIT
4542.955	4544.229	22005.94	CE	0.8R			MIT
4542.075	4543.348	22010.20	CE	0.5			MIT
4541.775	4543.048	22011.65	CE	0.5			MIT
4541.565	4542.838	22012.67	CE	0.5			MIT
4541.203	4542.476	22014.43	CE	0.3			MIT
4540.626	4541.899	22017.22	CE	0.6			MIT
4539.746	4541.019	22021.49	CE	1.3	1.0		MIT
4539.588	4540.861	22022.26	CE	0.5			MIT
4539.069	4540.342	22024.77	CE	0.9			MIT
4538.422	4539.694	22027.91	CE	0.3			MIT
4537.884	4539.156	22030.53	CE	0.8			M
4536.886	4538.158	22035.37	CE	0.8			MIT
4536.647	4537.919	22036.53	CE	0.5			MIT
4536.201	4537.473	22038.70	CE	0.6			MIT
4535.528	4536.800	22041.97	CE	0.3			MIT
4535.355	4536.627	22042.81	CE	0.3H			MIT
4535.155	4536.427	22043.78	CE	0.3			MIT
4534.221	4535.492	22048.32	CE	0.6			MIT
4533.561	4534.832	22051.53	CE	0.5H			MIT
4533.225	4534.496	22053.17	CE	0.3H	0.0		MIT
4532.488	4533.759	22056.75	CE	0.8			MIT
4532.010	4533.281	22059.08	CE	0.7			MIT
4531.633	4532.904	22060.91	CE	0.6			MIT
4531.312	4532.583	22062.48	CE	0.9			MIT
4531.067	4532.338	22063.67	CE	0.5			MIT
4530.945	4532.215	22064.26	CE	0.5			MIT
4530.815	4532.085	22064.90	CE	0.5			MIT
4529.914	4531.184	22069.29	CE	0.7			MIT
4529.277	4530.547	22072.39	CE	0.5			MIT
4528.705	4529.975	22075.18	CE	0.3			MIT
4528.472	4529.742	22076.31	CE	1.5	1.2		MIT
4527.957	4529.227	22078.82	CE	0.5			MIT
4527.348	4528.618	22081.79	CE	1.7	1.4		MIT
4526.372	4527.641	22086.55	CE	0.3			MIT
4526.350	4527.619	22086.66	CE	0.3			MIT
4524.591	4525.860	22095.25	CE	0.3			MIT
4524.029	4525.298	22097.99	CE	0.6			MIT
4523.077	4524.345	22102.65	CE	1.5	1.4		MIT
4522.590	4523.858	22105.02	CE	0.5			MIT
4522.468	4523.736	22105.62	CE	0.3			MIT
4522.077	4523.345	22107.53	CE	1.4	1.2		MIT
4521.959	4523.227	22108.11	CE	0.8			MIT
4521.135	4522.403	22112.14	CE	0.5			MIT
4520.405	4521.673	22115.71	CE	0.5			MIT
4519.592	4520.859	22119.69	CE	0.9			MIT
4518.280	4519.547	22126.11	CE	0.6			MIT
4518.017	4519.284	22127.40	CE	1.1			MIT
4516.529	4517.796	22134.69	CE	0.3			MIT
4516.290	4517.557	22135.86	CE	0.6			MIT
4515.857	4517.124	22137.98	CE	1.3			MIT
4514.457	4515.723	22144.85	CE	0.5			MIT
4514.060	4515.326	22146.79	CE	0.6			MIT
4513.138	4514.404	22151.32	CE	0.3			MIT
4511.635	4512.900	22158.70	CE	1.0			MIT

AIR WAVELENGTH (ANGSTRMS)	VACUUM WAVELENGTH (ANGSTRMS)	WAVENUMBER (KAYSERS)	LMENT	LOG ARC	INTENSITY SPK	DIS	REF
4510.921	4512.186	22162.21	CE	0.8			MIT
4510.761	4512.026	22162.99	CE	0.5			MIT
4510.166	4511.431	22165.92	CE	0.6			MIT
4510.082	4511.347	22166.33	CE	0.3H			MIT
4509.497	4510.762	22169.20	CE	0.5			MIT
4509.262	4510.527	22170.36	CE	0.6			MIT
4509.143	4510.408	22170.95	CE	1.1			MIT
4508.729	4509.994	22172.98	CE	0.7			MIT
4508.083	4509.347	22176.16	CE	0.9			MIT
4507.756	4509.020	22177.77	CE	0.5			MIT
4507.343	4508.607	22179.80	CE	0.3			MIT
4507.052	4508.316	22181.23	CE	0.5			MIT
4506.963	4508.227	22181.67	CE	0.5			MIT
4506.632	4507.896	22183.30	CE	0.5			MIT
4506.418	4507.682	22184.35	CE	1.2			MIT
4505.596	4506.860	22188.40	CE	0.5			MIT
4505.125	4506.389	22190.72	CE	0.8			MIT
4504.063	4505.326	22195.95	CE	0.3			MIT
4503.847	4505.110	22197.01	CE	0.5			MIT
4503.353	4504.616	22199.45	CE	0.5H			MIT
4502.575	4503.838	22203.28	CE	0.3			MIT
4502.144	4503.407	22205.41	CE	0.3			MIT
4501.835	4503.098	22206.94	CE	0.3			MIT
4501.697	4502.960	22207.62	CE	0.5			MIT
4501.095	4502.358	22210.59	CE	0.8			MIT
4500.341	4501.603	22214.31	CE	0.9			MIT
4499.752	4501.014	22217.22	CE	0.8			MIT
4499.511	4500.773	22218.41	CE	0.6			MIT
4499.229	4500.491	22219.80	CE	0.6			MIT
4498.415	4499.677	22223.82	CE	0.5H			MIT
4497.849	4499.111	22226.61	CE	1.5	0.6		MIT
4497.613	4498.875	22227.78	CE	0.8S			MIT
4496.231	4497.492	22234.61	CE	1.0			MIT
4495.387	4496.648	22238.79	CE	1.3	0.3		MIT
4494.223	4495.484	22244.55	CE	1.3	0.5		MIT
4493.693	4494.954	22247.17	CE	0.5			MIT
4493.418	4494.679	22248.53	CE	0.6			MIT
4493.321	4494.582	22249.01	CE	0.5			MIT
4492.951	4494.211	22250.84	CE	0.8			MIT
4492.764	4494.024	22251.77	CE	0.5			MIT
4492.343	4493.603	22253.86	CE	0.5H			MIT
4491.290	4492.550	22259.07	CE	0.6			MIT
4490.465	4491.725	22263.16	CE	0.6			MIT
4489.821	4491.081	22266.36	CE	0.3			MIT
4489.530	4490.790	22267.80	CE	0.7			MIT
4488.809	4490.068	22271.38	CE	0.8			MIT
4488.138	4489.397	22274.71	CE	0.3			MIT
4487.871	4489.130	22276.03	CE	0.8			MIT
4487.489	4488.748	22277.93	CE	0.5			MIT
4487.169	4488.428	22279.52	CE	0.5			MIT
4486.909	4488.168	22280.81	CE	1.6	1.2		MIT
4486.404	4487.663	22283.31	CE	0.5H			MIT
4485.522	4486.780	22287.70	CE	1.2	0.0		MIT
4484.819	4486.077	22291.19	CE	1.5	0.5		MIT
4483.897	4485.155	22295.77	CE	1.6	1.0		MIT
4483.347	4484.605	22298.51	CE	0.9	0.0		MIT
4483.053	4484.311	22299.97	CE	0.5			MIT
4482.787	4484.045	22301.29	CE	0.6H			MIT
4482.591	4483.849	22302.27	CE	0.5H			MIT
4481.904	4483.162	22305.69	CE	0.7			MIT
4481.637	4482.894	22307.02	CE	0.5			MIT
4480.849	4482.106	22310.94	CE	0.3			MIT
4480.464	4481.721	22312.86	CE	0.5			MIT
4480.328	4481.585	22313.53	CE	0.5			MIT
4479.979	4481.236	22315.27	CE	0.8			MIT
4479.432	4480.689	22318.00	CE	0.6			MIT
4479.359	4480.616	22318.36	CE	1.6	1.3		MIT
4479.203	4480.460	22319.14	CE	0.6	0.0		MIT
4477.989	4479.246	22325.19	CE	0.9			MIT
4477.630	4478.886	22326.98	CE	0.6			MIT

AIR WAVELENGTH (ANGSTRMS)	VACUUM WAVELENGTH (ANGSTRMS)	WAVENUMBER (KAYSERS)	LMENT	LOG ARC	INTENSITY SPK	DIS	REF
4477.354	4478.610	22328.35	CE	0.3			MIT
4477.218	4478.474	22329.03	CE	0.6			MIT
4476.995	4478.251	22330.15	CE	0.5H			MIT
4476.510	4477.766	22332.56	CE	0.5			MIT
4475.974	4477.230	22335.24	CE	0.5			MIT
4475.650	4476.906	22336.86	CE	0.3			MIT
4475.382	4476.638	22338.19	CE	0.3			MIT
4475.296	4476.552	22338.62	CE	0.6			MIT
4474.865	4476.121	22340.77	CE	0.6			MIT
4474.693	4475.949	22341.63	CE	1.1	0.0		MIT
4473.661	4474.916	22346.79	CE	0.3			MIT
4473.138	4474.393	22349.40	CE	0.5			MIT
4472.718	4473.973	22351.50	CE	1.2	0.5		MIT
4472.082	4473.337	22354.68	CE	0.5			MIT
4471.633	4472.888	22356.92	CE	0.8			MIT
4471.240	4472.495	22358.89	CE	1.5	0.9		MIT
4470.540	4471.795	22362.39	CE	0.6			MIT
4469.850	4471.104	22365.84	CE	0.6			MIT
4469.518	4470.772	22367.50	CE	0.3			MIT
4469.238	4470.492	22368.90	CE	0.3			MIT
4468.626	4469.880	22371.97	CE	0.6			MIT
4468.024	4469.278	22374.98	CE	0.5			MIT
4467.539	4468.793	22377.41	CE	1.5	0.6		MIT
4467.308	4468.562	22378.56	CE	0.8			MIT
4467.080	4468.334	22379.71	CE	0.6			MIT
4466.794	4468.048	22381.14	CE	0.5			MIT
4466.729	4467.983	22381.47	CE	0.5			MIT
4466.007	4467.260	22385.08	CE	0.5			MIT
4465.438	4466.691	22387.94	CE	0.8			MIT
4464.691	4465.944	22391.68	CE	0.9S	0.5		MIT
4464.173	4465.426	22394.28	CE	1.0	0.0		MIT
4463.856	4465.109	22395.87	CE	0.7			MIT
4463.410	4464.663	22398.11	CE	1.5	0.8		MIT
4462.302	4463.554	22403.67	CE	0.6			MIT
4462.037	4463.289	22405.00	CE	0.5			MIT
4461.373	4462.625	22408.33	CE	0.5			MIT
4461.136	4462.388	22409.53	CE	1.5	0.8		MIT
4460.213	4461.465	22414.16	CE	1.8	1.3		MIT
4459.086	4460.338	22419.83	CE	0.6			MIT
4458.848	4460.099	22421.02	CE	0.5			MIT
4458.602	4459.853	22422.26	CE	0.6			MIT
4458.294	4459.545	22423.81	CE	0.3			MIT
4457.776	4459.027	22426.42	CE	0.8			MIT
4457.463	4458.714	22427.99	CE	0.3H			MIT
4457.145	4458.396	22429.59	CE	0.3			MIT
4456.561	4457.812	22432.53	CE	0.7			MIT
4455.902	4457.153	22435.85	CE	0.5			MIT
4455.656	4456.907	22437.09	CE	1.3	0.0		MIT
4455.461	4456.712	22438.07	CE	0.5			MIT
4454.985	4456.235	22440.47	CE	1.1	0.0		MIT
4454.762	4456.012	22441.59	CE	0.7			MIT
4454.531	4455.781	22442.75	CE	0.5			MIT
4454.050	4455.300	22445.18	CE	0.3			MIT
4453.773	4455.023	22446.57	CE	0.5			MIT
4453.160	4454.410	22449.66	CE	1.2			MIT
4452.553	4453.803	22452.72	CE	1.4			MIT
4451.534	4452.784	22457.86	CE	0.5			MIT
4451.124	4452.373	22459.93	CE	0.5			MIT
4450.732	4451.981	22461.91	CE	1.5	0.7		MIT
4450.258	4451.507	22464.30	CE	0.3			MIT
4450.134	4451.383	22464.93	CE	0.5			MIT
4449.635	4450.884	22467.45	CE	1.1	0.0		MIT
4449.336	4450.585	22468.96	CE	1.7	0.9		MIT
4448.924	4450.173	22471.04	CE	0.3			MIT
4448.228	4449.477	22474.55	CE	0.3			MIT
4447.686	4448.935	22477.29	CE	0.8			MIT
4447.280	4448.528	22479.34	CE	0.3			MIT
4446.639	4447.887	22482.58	CE	0.5			MIT
4446.154	4447.402	22485.04	CE	0.9			MIT
4445.265	4446.513	22489.53	CE	0.5			MIT

AIR WAVELENGTH (ANGSTRMS)	VACUUM WAVELENGTH (ANGSTRMS)	WAVENUMBER (KAYSERS)	LMENT	LOG ARC	INTENSITY SPK	INTENSITY DIS	REF
4444.704	4445.952	22492.37	CE	1.5	0.8		MIT
4444.393	4445.641	22493.95	CE	1.5	0.6		MIT
4443.743	4444.990	22497.24	CE	1.3	0.3		MIT
4443.296	4444.543	22499.50	CE	0.5			MIT
4442.463	4443.710	22503.72	CE	0.8			MIT
4442.031	4443.278	22505.91	CE	0.3			MIT
4441.860	4443.107	22506.77	CE	0.5			MIT
4441.604	4442.851	22508.07	CE	0.7			MIT
4440.883	4442.130	22511.72	CE	1.3	0.3		MIT
4440.126	4441.373	22515.56	CE	0.6			MIT
4439.518	4440.764	22518.65	CE	1.1			MIT
4439.244	4440.490	22520.04	CE	1.1	0.0		MIT
4438.184	4439.430	22525.41	CE	0.3			MIT
4438.082	4439.328	22525.93	CE	0.5			MIT
4437.613	4438.859	22528.31	CE	1.3	0.3		MIT
4437.288	4438.534	22529.96	CE	0.3			MIT
4437.075	4438.321	22531.04	CE	0.3			MIT
4436.592	4437.838	22533.50	CE	0.3			MIT
4436.211	4437.456	22535.43	CE	0.8			MIT
4435.620	4436.865	22538.44	CE	0.7			MIT
4435.475	4436.720	22539.17	CE	0.3			MIT
4434.952	4436.197	22541.83	CE	0.9			MIT
4434.372	4435.617	22544.78	CE	0.5			MIT
4433.725	4434.970	22548.07	CE	1.1	0.0		MIT
4433.235	4434.480	22550.56	CE	0.5H			MIT
4432.917	4434.162	22552.18	CE	0.7			MIT
4432.720	4433.965	22553.18	CE	0.9	0.0		MIT
4432.279	4433.523	22555.42	CE	0.6			MIT
4430.147	4431.391	22566.28	CE	0.3			MIT
4429.999	4431.243	22567.03	CE	0.9	0.0		MIT
4429.270	4430.514	22570.75	CE	1.5	0.7		MIT
4428.438	4429.681	22574.99	CE	1.3	0.5		MIT
4428.216	4429.459	22576.12	CE	0.3			MIT
4427.918	4429.161	22577.64	CE	1.4	0.6		MIT
4427.071	4428.314	22581.96	CE	1.3	0.5		MIT
4426.301	4427.544	22585.89	CE	0.6			MIT
4426.079	4427.322	22587.02	CE	0.6			MIT
4425.921	4427.164	22587.82	CE	0.6			MIT
4425.607	4426.850	22589.43	CE	0.8			MIT
4425.326	4426.569	22590.86	CE	0.5S			MIT
4425.118	4426.361	22591.92	CE	0.3			MIT
4424.540	4425.782	22594.87	CE	0.5			MIT
4424.314	4425.556	22596.03	CE	0.8			MIT
4424.140	4425.382	22596.92	CE	0.5			MIT
4423.678	4424.920	22599.28	CE	1.4	0.5		MIT
4423.444	4424.686	22600.47	CE	1.1			MIT
4422.420	4423.662	22605.71	CE	0.5			MIT
4422.140	4423.382	22607.14	CE	0.8			MIT
4421.788	4423.030	22608.94	CE	0.5			MIT
4421.473	4422.715	22610.55	CE	0.5			MIT
4421.321	4422.563	22611.32	CE	0.6			MIT
4421.130	4422.372	22612.30	CE	0.9			MIT
4420.417	4421.658	22615.95	CE	0.5			MIT
4419.927	4421.168	22618.46	CE	0.5			MIT
4419.675	4420.916	22619.75	CE	0.5			MIT
4419.298	4420.539	22621.68	CE	1.3	0.0		MIT
4418.784	4420.025	22624.31	CE	1.6	1.0		MIT
4418.295	4419.536	22626.81	CE	0.5			MIT
4418.070	4419.311	22627.96	CE	0.5			MIT
4417.915	4419.156	22628.76	CE	0.6			MIT
4417.374	4418.615	22631.53	CE	0.6			MIT
4416.902	4418.142	22633.95	CE	1.4S	0.6H		MIT
4416.618	4417.858	22635.40	CE	0.7			MIT
4416.400	4417.640	22636.52	CE	0.3			MIT
4415.770	4417.010	22639.75	CE	0.7			MIT
4415.621	4416.861	22640.51	CE	0.7			MIT
4415.220	4416.460	22642.57	CE	0.7			MIT
4414.508	4415.748	22646.22	CE	0.5L			MIT
4413.804	4415.044	22649.83	CE	1.3	0.0		MIT
4413.190	4414.429	22652.98	CE	1.5	0.3		MIT

AIR WAVELENGTH (ANGSTRMS)	VACUUM WAVELENGTH (ANGSTRMS)	WAVENUMBER (KAYSERS)	LMENT	LOG ARC	INTENSITY SPK	INTENSITY DIS	REF
4412.740	4413.979	22655.29	CE	0.5			MIT
4412.329	4413.568	22657.41	CE	0.5			MIT
4412.017	4413.256	22659.01	CE	1.3	0.3		MIT
4411.691	4412.930	22660.68	CE	0.9			MIT
4410.760	4411.999	22665.46	CE	1.0L	0.3		MIT
4410.641	4411.880	22666.08	CE	1.1S	0.5		MIT
4410.066	4411.305	22669.03	CE	0.5			MIT
4409.301	4410.539	22672.96	CE	0.3			MIT
4409.047	4410.285	22674.27	CE	0.6			MIT
4408.869	4410.107	22675.19	CE	1.1	0.0		MIT
4408.768	4410.006	22675.71	CE	0.6			MIT
4408.341	4409.579	22677.90	CE	0.5			MIT
4407.822	4409.060	22680.57	CE	0.5			MIT
4407.278	4408.516	22683.37	CE	1.6	0.5		MIT
4406.004	4407.242	22689.93	CE	0.3H			MIT
4405.469	4406.706	22692.69	CE	1.3	0.3		MIT
4405.303	4406.540	22693.54	CE	0.6			MIT
4405.153	4406.390	22694.31	CE	0.3			MIT
4404.771	4406.008	22696.28	CE	0.3			MIT
4404.574	4405.811	22697.30	CE	0.5			MIT
4404.382	4405.619	22698.28	CE	0.3			MIT
4404.250	4405.487	22698.97	CE	0.6			MIT
4403.555	4404.792	22702.55	CE	0.8			MIT
4403.295	4404.532	22703.89	CE	0.9	0.0		MIT
4403.053	4404.290	22705.14	CE	0.8			MIT
4402.412	4403.649	22708.44	CE	0.3			MIT
4402.005	4403.241	22710.54	CE	0.6			MIT
4401.515	4402.751	22713.07	CE	0.7			MIT
4401.151	4402.387	22714.95	CE	0.3			MIT
4400.873	4402.109	22716.38	CE	1.0	0.3		MIT
4400.543	4401.779	22718.09	CE	1.0	0.3		MIT
4400.148	4401.384	22720.13	CE	0.8			MIT
4399.541	4400.777	22723.26	CE	0.6			MIT
4399.203	4400.439	22725.01	CE	1.5	0.8		MIT
4398.787	4400.023	22727.16	CE	1.3	0.5		MIT
4398.542	4399.778	22728.42	CE	0.6			MIT
4398.252	4399.487	22729.92	CE	0.3			MIT
4397.975	4399.210	22731.35	CE	0.3			MIT
4397.857	4399.092	22731.96	CE	0.8			MIT
4397.679	4398.914	22732.88	CE	0.6			MIT
4397.277	4398.512	22734.96	CE	0.7			MIT
4397.191	4398.426	22735.41	CE	0.5			MIT
4396.582	4397.817	22738.55	CE	1.3	0.0H		MIT
4396.437	4397.672	22739.30	CE	0.3			MIT
4396.191	4397.426	22740.58	CE	0.7			MIT
4396.028	4397.263	22741.42	CE	0.9	0.3		MIT
4395.725	4396.960	22742.99	CE	0.8			MIT
4395.051	4396.286	22746.48	CE	0.7			M
4394.779	4396.014	22747.88	CE	1.5	0.5		MIT
4394.286	4395.520	22750.44	CE	0.6			MIT
4393.904	4395.138	22752.41	CE	0.5			MIT
4393.552	4394.786	22754.24	CE	0.7			MIT
4393.192	4394.426	22756.10	CE	1.5	0.5		MIT
4392.676	4393.910	22758.77	CE	0.6			MIT
4392.432	4393.666	22760.04	CE	0.5			MIT
4392.167	4393.401	22761.41	CE	0.3			MIT
4392.028	4393.262	22762.13	CE	0.3			MIT
4391.893	4393.127	22762.83	CE	0.3			MIT
4391.661	4392.895	22764.03	CE	1.6	1.2		MIT
4391.337	4392.571	22765.71	CE	0.5			MIT
4390.809	4392.043	22768.45	CE	0.6			MIT
4390.496	4391.729	22770.07	CE	0.7			MIT
4390.279	4391.512	22771.20	CE	1.0			MIT
4390.153	4391.386	22771.85	CE	0.3			MIT
4389.807	4391.040	22773.65	CE	0.6			MIT
4389.112	4390.345	22777.25	CE	0.7			MIT
4388.007	4389.240	22782.99	CE	0.9	0.5		MIT
4387.501	4388.734	22785.62	CE	0.5			MIT
4387.058	4388.291	22787.92	CE	0.7	0.0		MIT
4386.835	4388.067	22789.07	CE	1.2	0.8		MIT

AIR WAVELENGTH (ANGSTRMS)	VACUUM WAVELENGTH (ANGSTRMS)	WAVENUMBER (KAYSERS)	LMENT	LOG ARC	INTENSITY SPK	DIS	REF
4386.700	4387.932	22789.78	CE	0.9	0.5		MIT
4386.346	4387.578	22791.62	CE	1.0S			MIT
4385.503	4386.735	22796.00	CE	0.5			MIT
4385.324	4386.556	22796.93	CE	0.6			MIT
4384.633	4385.865	22800.52	CE	0.5			MIT
4384.454	4385.686	22801.45	CE	0.6			MIT
4383.880	4385.112	22804.44	CE	0.3			MIT
4383.740	4384.972	22805.17	CE	0.6			MIT
4383.555	4384.787	22806.13	CE	0.9	0.0		MIT
4382.963	4384.194	22809.21	CE	0.6			MIT
4382.167	4383.398	22813.35	CE	1.6	1.1		MIT
4381.775	4383.006	22815.39	CE	0.8	0.3		MIT
4381.091	4382.322	22818.95	CE	0.6			MIT
4380.705	4381.936	22820.96	CE	0.5			MIT
4380.333	4381.564	22822.90	CE	0.6			MIT
4380.060	4381.291	22824.32	CE	1.5	0.3		MIT
4379.841	4381.072	22825.47	CE	0.3			MIT
4379.081	4380.311	22829.43	CE	0.5			MIT
4378.818	4380.048	22830.80	CE	0.6			MIT
4378.576	4379.806	22832.06	CE	0.7			MIT
4378.407	4379.637	22832.94	CE	0.5			MIT
4377.046	4378.276	22840.04	CE	0.3			MIT
4376.880	4378.110	22840.91	CE	0.6			MIT
4375.918	4377.148	22845.93	CE	1.6	0.7		MIT
4375.687	4376.917	22847.13	CE	0.3			MIT
4375.174	4376.403	22849.81	CE	1.1			MIT
4374.760	4375.989	22851.98	CE	0.5			MIT
4374.085	4375.314	22855.50	CE	0.5			MIT
4373.818	4375.047	22856.90	CE	1.6	0.6		MIT
4373.219	4374.448	22860.03	CE	1.4			MIT
4372.706	4373.935	22862.71	CE	0.5			MIT
4372.401	4373.630	22864.30	CE	1.5	0.0		MIT
4371.856	4373.085	22867.16	CE	0.3			MIT
4371.710	4372.938	22867.92	CE	0.3			MIT
4371.571	4372.799	22868.65	CE	0.5			MIT
4371.006	4372.234	22871.60	CE	0.3			MIT
4370.651	4371.879	22873.46	CE	1.1R			MIT
4369.243	4370.471	22880.83	CE	1.0	0.0		MIT
4368.886	4370.114	22882.70	CE	0.5			MIT
4368.234	4369.462	22886.11	CE	0.9	0.0		MIT
4367.558	4368.785	22889.66	CE	0.8	0.0		MIT
4367.313	4368.540	22890.94	CE	0.7S			MIT
4367.001	4368.228	22892.58	CE	1.4	0.6S		MIT
4366.535	4367.762	22895.02	CE	0.3			MIT
4366.114	4367.341	22897.23	CE	0.3			MIT
4365.630	4366.857	22899.77	CE	0.3			MIT
4365.520	4366.747	22900.34	CE	0.6			MIT
4365.349	4366.576	22901.24	CE	0.3			MIT
4364.658	4365.885	22904.86	CE	1.5	0.8		MIT
4364.441	4365.668	22906.00	CE	0.5L			MIT
4364.115	4365.341	22907.72	CE	0.7			MIT
4363.487	4364.713	22911.01	CE	0.5			MIT
4363.385	4364.611	22911.55	CE	0.6			MIT
4363.100	4364.326	22913.04	CE	0.5			MIT
4362.444	4363.670	22916.49	CE	0.6			MIT
4361.661	4362.887	22920.60	CE	1.3	0.3		MIT
4361.352	4362.578	22922.23	CE	1.0			MIT
4360.443	4361.669	22927.01	CE	1.2	0.0		MIT
4360.178	4361.403	22928.40	CE	1.0	0.0		MIT
4359.945	4361.170	22929.62	CE	0.3			MIT
4359.629	4360.854	22931.29	CE	0.3			MIT
4359.368	4360.593	22932.66	CE	0.3			MIT
4359.068	4360.293	22934.24	CE	1.2	0.3		MIT
4357.907	4359.132	22940.35	CE	1.1	0.0		MIT
4357.125	4358.350	22944.47	CE	0.5S			MIT
4356.962	4358.187	22945.32	CE	0.6			MIT
4356.748	4357.973	22946.45	CE	1.0			MIT
4356.157	4357.381	22949.56	CE	0.3			MIT
4355.945	4357.169	22950.68	CE	0.3			MIT
4355.813	4357.037	22951.38	CE	0.3			MIT

AIR WAVELENGTH (ANGSTRMS)	VACUUM WAVELENGTH (ANGSTRMS)	WAVENUMBER (KAYSERS)	LMENT	LOG ARC	INTENSITY SPK	DIS	REF
4355.431	4356.655	22953.39	CE	0.8			MIT
4355.157	4356.381	22954.83	CE	0.3S			MIT
4354.852	4356.076	22956.44	CE	0.9L			MIT
4354.419	4355.643	22958.72	CE	0.8			MIT
4353.866	4355.090	22961.64	CE	0.7			MIT
4353.513	4354.737	22963.50	CE	0.5			MIT
4353.370	4354.594	22964.25	CE	0.8	0.3		MIT
4353.132	4354.356	22965.51	CE	0.5			MIT
4352.971	4354.195	22966.36	CE	0.3			MIT
4352.706	4353.929	22967.76	CE I	1.6	0.7		MIT
4352.706	4353.929	22967.76	CE II	1.6	0.7		MIT
4351.982	4353.205	22971.58	CE	0.3H			MIT
4351.812	4353.035	22972.48	CE	0.9			MIT
4351.388	4352.611	22974.72	CE	0.3	0.5J		MIT
4350.500	4351.723	22979.40	CE	0.5			MIT
4350.324	4351.547	22980.33	CE	0.3			MIT
4349.789	4351.012	22983.16	CE II	1.6	0.7		MIT
4349.390	4350.613	22985.27	CE	0.5O	0.8J		MIT
4349.107	4350.330	22986.76	CE	0.3			MIT
4348.600	4349.822	22989.44	CE	0.6			MIT
4348.342	4349.564	22990.81	CE	0.5			MIT
4348.194	4349.416	22991.59	CE II	0.6			MIT
4347.710	4348.932	22994.15	CE	0.6			MIT
4347.598	4348.820	22994.74	CE	0.6			MIT
4347.389	4348.611	22995.85	CE	0.6			MIT
4347.076	4348.298	22997.50	CE	0.5			MIT
4346.833	4348.055	22998.79	CE	2.3	1.6		MIT
4345.961	4347.183	23003.40	CE II	0.8	0.3		MIT
4345.832	4347.054	23004.09	CE	0.7			MIT
4345.457	4346.679	23006.07	CE	0.9	0.5J		MIT
4345.214	4346.436	23007.36	CE	0.3			MIT
4344.924	4346.145	23008.89	CE II	0.8			MIT
4344.750	4345.971	23009.82	CE	0.5			MIT
4344.290	4345.511	23012.25	CE	0.8			MIT
4343.871	4345.092	23014.47	CE	0.6			MIT
4343.558	4344.779	23016.13	CE	0.8			MIT
4342.487	4343.708	23021.81	CE	1.1	0.0		MIT
4342.140	4343.361	23023.65	CE II	1.0			MIT
4340.558	4341.778	23032.04	CE	1.1	0.0		MIT
4339.317	4340.537	23038.62	CE	1.4	0.7		MIT
4339.059	4340.279	23039.99	CE	0.5			MIT
4338.146	4339.366	23044.84	CE	0.3			MIT
4337.777	4338.997	23046.80	CE I	1.4	1.0		MIT
4337.777	4338.997	23046.80	CE II	1.4	1.0		MIT
4337.586	4338.806	23047.82	CE	0.5			MIT
4337.293	4338.512	23049.37	CE	0.5			MIT
4336.255	4337.474	23054.89	CE	1.4	0.8		MIT
4336.088	4337.307	23055.78	CE	0.3			MIT
4335.487	4336.706	23058.98	CE	0.8			MIT
4335.408	4336.627	23059.40	CE	0.3			MIT
4334.868	4336.087	23062.27	CE	1.3			MIT
4334.634	4335.853	23063.51	CE	0.5			MIT
4334.331	4335.550	23065.13	CE	0.3			MIT
4334.227	4335.446	23065.68	CE	0.5			MIT
4333.412	4334.630	23070.02	CE	0.6			MIT
4332.708	4333.926	23073.77	CE	1.5S	0.6		MIT
4332.471	4333.689	23075.03	CE	0.5			MIT
4332.319	4333.537	23075.84	CE	0.3			MIT
4331.941	4333.159	23077.85	CE	0.8			MIT
4331.758	4332.976	23078.83	CE	1.3	0.3		MIT
4330.937	4332.155	23083.20	CE	0.8			MIT
4330.445	4331.663	23085.82	CE II	1.7	0.7		MIT
4329.938	4331.155	23088.53	CE	0.8			MIT
4327.762	4328.979	23100.14	CE	0.5L			MIT
4326.826	4328.043	23105.13	CE	1.2			MIT
4325.314	4326.530	23113.21	CE	0.6			MIT
4325.126	4326.342	23114.21	CE	0.3			MIT
4324.789	4326.005	23116.01	CE	1.3	0.5		MIT
4324.598	4325.814	23117.04	CE	1.0			MIT
4324.091	4325.307	23119.75	CE	0.6H			MIT

AIR WAVELENGTH (ANGSTRMS)	VACUUM WAVELENGTH (ANGSTRMS)	WAVENUMBER (KAYSERS)	LMENT	LOG ARC	INTENSITY SPK	INTENSITY DIS	REF
4322.789	4324.005	23126.71	CE	0.9			MIT
4322.116	4323.331	23130.31	CE	0.6			MIT
4321.519	4322.734	23133.51	CE	0.6			MIT
4321.255	4322.470	23134.92	CE	0.9			MIT
4320.723	4321.938	23137.77	CE II	1.7	0.9		MIT
4319.682	4320.897	23143.34	CE	0.8	0.5J		MIT
4319.087	4320.302	23146.53	CE	0.9			MIT
4318.636	4319.851	23148.95	CE	1.1	0.0		MIT
4318.049	4319.263	23152.10	CE	0.5			MIT
4317.985	4319.199	23152.44	CE	1.1L			MIT
4317.333	4318.547	23155.94	CE	1.4S	0.0		MIT
4317.109	4318.323	23157.14	CE	0.9	0.5H		MIT
4316.420	4317.634	23160.83	CE	0.8			MIT
4316.008	4317.222	23163.04	CE	0.7			MIT
4315.686	4316.900	23164.77	CE	1.0			MIT
4315.406	4316.620	23166.27	CE II	0.8	0.3		MIT
4314.934	4316.148	23168.81	CE II	0.8			MIT
4314.480	4315.693	23171.25	CE	1.0	0.0		MIT
4313.594	4314.807	23176.01	CE	1.0			MIT
4313.103	4314.316	23178.65	CE II	1.1	0.0		MIT
4312.859	4314.072	23179.96	CE	0.8			MIT
4312.560	4313.773	23181.56	CE	1.0			MIT
4311.590	4312.803	23186.78	CE	1.4	0.5		MIT
4311.038	4312.251	23189.75	CE	0.3			MIT
4310.699	4311.911	23191.57	CE II	1.5S	0.5		MIT
4310.388	4311.600	23193.24	CE	1.0			MIT
4309.739	4310.951	23196.74	CE II	1.4	0.6		MIT
4309.583	4310.795	23197.58	CE	0.9			MIT
4306.724	4307.935	23212.98	CE	1.5	1.2		MIT
4305.609	4306.820	23218.99	CE II	0.6			MIT
4305.415	4306.626	23220.03	CE	0.6			MIT
4305.141	4306.352	23221.51	CE	1.3	0.7		MIT
4304.721	4305.932	23223.78	CE II	1.0	0.3H		MIT
4304.278	4305.489	23226.17	CE	1.2	0.3H		MIT
4302.653	4303.863	23234.94	CE	1.0			MIT
4301.533	4302.743	23240.99	CE	1.0			MIT
4300.862	4302.072	23244.61	CE	1.2			MIT
4300.331	4301.541	23247.48	CE II	1.6	1.2		MIT
4299.361	4300.570	23252.73	CE II	1.2	0.3		MIT
4299.092	4300.301	23254.18	CE II	0.7			MIT
4298.160	4299.369	23259.23	CE	0.3			MIT
4297.811	4299.020	23261.11	CE	0.3			MIT
4296.786	4297.995	23266.66	CE	0.7			MIT
4296.680	4297.889	23267.24	CE	1.6	1.4		MIT
4296.371	4297.580	23268.91	CE	1.0			MIT
4296.069	4297.278	23270.55	CE II	1.2	0.3		MIT
4294.756	4295.964	23277.66	CE II	1.0			MIT
4293.131	4294.339	23286.47	CE	1.0			MIT
4292.767	4293.975	23288.45	CE II	1.0	0.3		MIT
4292.581	4293.789	23289.46	CE	1.0	0.3		MIT
4290.593	4291.800	23300.25	CE	0.6			MIT
4290.431	4291.638	23301.13	CE II	0.3			MIT
4290.295	4291.502	23301.86	CE	0.3			MIT
4289.938	4291.145	23303.80	CE	1.7	1.4		MIT
4289.454	4290.661	23306.43	CE II	1.4	0.6		MIT
4288.666	4289.873	23310.72	CE II	1.5			MIT
4287.002	4288.208	23319.76	CE	0.9			MIT
4286.922	4288.128	23320.20	CE	0.9J			MIT
4285.842	4287.048	23326.07	CE	0.3			MIT
4285.366	4286.572	23328.67	CE	1.4	0.9		MIT
4283.550	4284.755	23338.56	CE II	0.7H			MIT
4283.316	4284.521	23339.83	CE	0.9			MIT
4282.813	4284.018	23342.57	CE	0.5			MIT
4281.917	4283.122	23347.46	CE II	0.7			MIT
4281.158	4282.363	23351.59	CE II	1.2			MIT
4280.996	4282.201	23352.48	CE II	1.2	0.6		MIT
4280.141	4281.345	23357.14	CE	1.2	0.3		MIT
4279.324	4280.528	23361.60	CE	0.6			MIT
4279.167	4280.371	23362.46	CE	0.3			MIT
4278.866	4280.070	23364.10	CE II	1.3	0.7		MIT

AIR WAVELENGTH (ANGSTRMS)	VACUUM WAVELENGTH (ANGSTRMS)	WAVENUMBER (KAYSERS)	LMENT	LOG ARC	INTENSITY SPK	INTENSITY DIS	REF
4278.248	4279.452	23367.48	CE	1.0	0.0		MIT
4276.333	4277.536	23377.94	CE	0.6			MIT
4275.561	4276.764	23382.16	CE	1.4	0.6		MIT
4273.788	4274.991	23391.86	CE	0.3			MIT
4273.444	4274.647	23393.75	CE	1.3	0.3		MIT
4273.011	4274.214	23396.12	CE	0.3			MIT
4272.855	4274.057	23396.97	CE	0.3			MIT
4272.339	4273.541	23399.80	CE	0.3			MIT
4272.034	4273.236	23401.47	CE	0.3			MIT
4271.479	4272.681	23404.51	CE	0.7			MIT
4271.070	4272.272	23406.75	CE	0.3			MIT
4270.716	4271.918	23408.69	CE	1.4			MIT
4270.189	4271.391	23411.58	CE II	1.4			MIT
4269.251	4270.453	23416.72	CE	0.7	0.3		MIT
4268.301	4269.502	23421.93	CE	1.0			MIT
4267.852	4269.053	23424.40	CE	0.3			MIT
4267.668	4268.869	23425.41	CE	0.3			MIT
4267.221	4268.422	23427.86	CE	1.0			MIT
4266.698	4267.899	23430.73	CE	0.3			MIT
4266.081	4267.282	23434.12	CE	0.3			MIT
4265.925	4267.126	23434.98	CE	0.3			MIT
4265.710	4266.911	23436.16	CE	0.3			MIT
4265.186	4266.386	23439.04	CE	0.6			MIT
4264.688	4265.888	23441.78	CE	1.0			MIT
4264.370	4265.570	23443.52	CE II	1.0			MIT
4263.947	4265.147	23445.85	CE	1.0			MIT
4263.427	4264.627	23448.71	CE	1.6			MIT
4263.133	4264.333	23450.33	CE	0.5			MIT
4263.066	4264.266	23450.70	CE	0.5			MIT
4262.795	4263.995	23452.19	CE	0.8			MIT
4262.367	4263.567	23454.54	CE	0.3			MIT
4261.480	4262.679	23459.42	CE	0.6			MIT
4261.164	4262.363	23461.16	CE	1.3	0.0		MIT
4261.069	4262.268	23461.69	CE	0.5			MIT
4260.723	4261.922	23463.59	CE	0.3			MIT
4260.345	4261.544	23465.67	CE	0.3			MIT
4259.748	4260.947	23468.96	CE	1.2	0.0		MIT
4259.600	4260.799	23469.78	CE	0.5			MIT
4259.227	4260.426	23471.83	CE	0.3			MIT
4259.064	4260.263	23472.73	CE	1.0			MIT
4258.883	4260.082	23473.73	CE	0.5			MIT
4258.401	4259.600	23476.39	CE	0.9			MIT
4258.321	4259.520	23476.83	CE	0.6			MIT
4258.218	4259.417	23477.39	CE	0.7			MIT
4257.813	4259.012	23479.63	CE	0.3			MIT
4257.476	4258.674	23481.49	CE	0.5			MIT
4257.121	4258.319	23483.44	CE II	1.3	0.0		MIT
4256.714	4257.912	23485.69	CE	0.3			MIT
4256.354	4257.552	23487.68	CE	0.5			MIT
4256.155	4257.353	23488.77	CE II	1.1	0.0		MIT
4255.992	4257.190	23489.67	CE	0.7			MIT
4255.784	4256.982	23490.82	CE	1.6	0.8		MIT
4255.361	4256.559	23493.16	CE II	1.0			MIT
4255.178	4256.376	23494.17	CE	0.3			MIT
4254.905	4256.103	23495.67	CE	0.9			MIT
4254.370	4255.568	23498.63	CE	0.9			MIT
4254.004	4255.202	23500.65	CE	0.5			MIT
4253.831	4255.028	23501.60	CE	0.5			MIT
4253.356	4254.553	23504.23	CE	1.6S	0.5		MIT
4252.455	4253.652	23509.21	CE	0.5			MIT
4251.858	4253.055	23512.51	CE	1.0	0.0		MIT
4251.602	4252.799	23513.93	CE	0.9	0.0		MIT
4251.364	4252.561	23515.24	CE	0.7			MIT
4250.816	4252.013	23518.27	CE	0.9			MIT
4250.664	4251.861	23519.11	CE	1.3			MIT
4249.677	4250.873	23524.58	CE	0.6			MIT
4249.099	4250.295	23527.78	CE	0.3			MIT
4249.005	4250.201	23528.30	CE	0.7			MIT
4248.676	4249.872	23530.12	CE	1.8	0.9		MIT
4247.964	4249.160	23534.06	CE	0.3			MIT

AIR WAVELENGTH (ANGSTRMS)	VACUUM WAVELENGTH (ANGSTRMS)	WAVENUMBER (KAYSERS)	LMENT		LOG ARC	INTENSITY SPK	DIS	REF
4247.695	4248.891	23535.55	CE		0.3			MIT
4247.653	4248.849	23535.79	CE		0.3			MIT
4246.938	4248.134	23539.75	CE		0.5			MIT
4246.711	4247.907	23541.01	CE		1.5	0.6		MIT
4246.398	4247.594	23542.74	CE		1.0			MIT
4245.978	4247.173	23545.07	CE	II	0.9	0.3		MIT
4245.880	4247.075	23545.61	CE		0.8	0.3		MIT
4245.529	4246.724	23547.56	CE		0.6			MIT
4245.222	4246.417	23549.26	CE		0.5			MIT
4244.916	4246.111	23550.96	CE		0.8			MIT
4244.552	4245.747	23552.98	CE		0.7			MIT
4244.470	4245.665	23553.44	CE		0.5			MIT
4243.789	4244.984	23557.22	CE		1.1			MIT
4243.298	4244.493	23559.94	CE		0.8			MIT
4242.852	4244.047	23562.42	CE		0.7			MIT
4242.723	4243.918	23563.13	CE	II	1.2	0.5		MIT
4242.259	4243.453	23565.71	CE		0.7			MIT
4242.009	4243.203	23567.10	CE		1.0			MIT
4241.743	4242.937	23568.58	CE		0.5			MIT
4241.644	4242.838	23569.13	CE		0.3			MIT
4241.405	4242.599	23570.46	CE	II	0.7			MIT
4241.244	4242.438	23571.35	CE		0.5			MIT
4240.669	4241.863	23574.55	CE		0.3			MIT
4240.584	4241.778	23575.02	CE		0.7			MIT
4239.912	4241.106	23578.76	CE		1.5	0.6		MIT
4239.641	4240.835	23580.26	CE		0.7			MIT
4239.021	4240.215	23583.71	CE		0.5			MIT
4238.557	4239.750	23586.29	CE		0.6			MIT
4237.798	4238.991	23590.52	CE		0.9			MIT
4237.205	4238.398	23593.82	CE		0.9S			MIT
4237.139	4238.332	23594.19	CE		0.6			MIT
4236.731	4237.924	23596.46	CE		0.3			MIT
4236.355	4237.548	23598.55	CE		1.2			MIT
4236.023	4237.216	23600.40	CE		1.5	0.0		MIT
4234.727	4235.919	23607.62	CE	II	1.1			MIT
4234.211	4235.403	23610.50	CE		1.5	0.3		MIT
4233.952	4235.144	23611.95	CE	II	0.9			MIT
4233.197	4234.389	23616.16	CE		1.2	0.0		MIT
4232.569	4233.761	23619.66	CE	II	1.3	0.3		MIT
4232.050	4233.242	23622.56	CE		1.3	0.3		MIT
4231.745	4232.937	23624.26	CE		1.5	0.7		MIT
4231.329	4232.521	23626.58	CE		0.3			MIT
4231.174	4232.366	23627.45	CE		0.5			MIT
4230.556	4231.747	23630.90	CE		0.8			MIT
4230.120	4231.311	23633.34	CE		1.1L			MIT
4229.979	4231.170	23634.12	CE		0.5			MIT
4229.632	4230.823	23636.06	CE		0.3			MIT
4228.947	4230.138	23639.89	CE		0.3			MIT
4228.827	4230.018	23640.56	CE		0.7			MIT
4228.297	4229.488	23643.52	CE		1.2			MIT
4227.746	4228.937	23646.61	CE		1.6	0.7		MIT
4227.412	4228.603	23648.48	CE		1.0			MIT
4226.734	4227.924	23652.27	CE		1.7	1.5		MIT
4226.335	4227.525	23654.50	CE		0.3			MIT
4226.196	4227.386	23655.28	CE		0.5			MIT
4225.746	4226.936	23657.80	CE		0.8	0.9		MIT
4224.552	4225.742	23664.48	CE		0.9	0.0		MIT
4224.221	4225.411	23666.34	CE		0.3			MIT
4223.884	4225.074	23668.23	CE	II	1.3	0.0		MIT
4223.153	4224.342	23672.32	CE		0.5			MIT
4222.884	4224.073	23673.83	CE		0.5			MIT
4222.599	4223.788	23675.43	CE		1.9	1.3		MIT
4221.731	4222.920	23680.30	CE		0.3			MIT
4221.629	4222.818	23680.87	CE		0.6			MIT
4221.486	4222.675	23681.67	CE		0.3			MIT
4221.171	4222.360	23683.44	CE		1.1	0.0		MIT
4220.769	4221.958	23685.69	CE		0.8			MIT
4220.549	4221.738	23686.93	CE		0.5			MIT
4220.185	4221.374	23688.97	CE		0.5			MIT
4219.704	4220.893	23691.67	CE		0.8			MIT

AIR WAVELENGTH (ANGSTRMS)	VACUUM WAVELENGTH (ANGSTRMS)	WAVENUMBER (KAYSERS)	LMENT		LOG ARC	INTENSITY SPK	DIS	REF
4219.395	4220.583	23693.41	CE		0.8			MIT
4218.002	4219.190	23701.23	CE		0.5			MIT
4217.591	4218.779	23703.54	CE		1.4	0.5		MIT
4217.226	4218.414	23705.59	CE		0.7			MIT
4214.698	4215.885	23719.81	CE		0.6			MIT
4214.039	4215.226	23723.52	CE	I	1.5	0.6		MIT
4214.039	4215.226	23723.52	CE	II	1.5	0.6		MIT
4213.036	4214.223	23729.17	CE	II	1.2	0.3		MIT
4212.391	4213.578	23732.80	CE		0.0	0.3		MIT
4211.906	4213.092	23735.53	CE		0.0	0.3		MIT
4211.582	4212.768	23737.36	CE		0.5			MIT
4210.240	4211.426	23744.93	CE		0.5			MIT
4209.997	4211.183	23746.30	CE		0.8	0.3H		MIT
4209.409	4210.595	23749.61	CE		1.4	0.5L		MIT
4209.198	4210.384	23750.80	CE		0.6			MIT
4208.699	4209.885	23753.62	CE		0.3			MIT
4208.439	4209.625	23755.09	CE		0.8			MIT
4208.245	4209.431	23756.18	CE		0.6			MIT
4207.577	4208.762	23759.95	CE		0.3			MIT
4206.832	4208.017	23764.16	CE		0.9			MIT
4206.324	4207.509	23767.03	CE		0.5			MIT
4205.892	4207.077	23769.47	CE		0.6	0.0H		MIT
4205.792	4206.977	23770.04	CE		0.6			MIT
4205.161	4206.346	23773.60	CE		0.8	0.3H		MIT
4204.739	4205.924	23775.99	CE		1.2			MIT
4204.296	4205.480	23778.50	CE		0.0	0.5		MIT
4203.505	4204.689	23782.97	CE		0.8	0.3H		MIT
4203.330	4204.514	23783.96	CE		0.3			MIT
4203.174	4204.358	23784.84	CE		0.3			MIT
4202.944	4204.128	23786.15	CE	II	1.6	1.3		MIT
4202.709	4203.893	23787.48	CE		0.3			MIT
4201.300	4202.484	23795.45	CE		0.6	1.3		MIT
4201.239	4202.423	23795.80	CE		0.9L	0.5		MIT
4200.694	4201.878	23798.89	CE			0.3H		MIT
4200.139	4201.322	23802.03	CE		0.5	0.0		MIT
4199.667	4200.850	23804.71	CE		0.5			MIT
4198.724	4199.907	23810.05	CE		0.9	0.5		MIT
4198.432	4199.615	23811.71	CE	II	0.7	0.0		MIT
4197.998	4199.181	23814.17	CE	I	1.1	0.3		MIT
4197.998	4199.181	23814.17	CE	II	1.1	0.3		MIT
4197.669	4198.852	23816.03	CE	II	0.8	0.5		MIT
4197.511	4198.694	23816.93	CE	II	0.8	0.0		MIT
4196.573	4197.755	23822.25	CE		0.6	0.6		MIT
4196.335	4197.517	23823.61	CE	II	1.3	0.5		MIT
4195.817	4196.999	23826.55	CE	II	1.0			MIT
4195.281	4196.463	23829.59	CE		0.8	0.3		MIT
4194.906	4196.088	23831.72	CE		0.9S	0.3		MIT
4194.832	4196.014	23832.14	CE		0.5			MIT
4194.108	4195.290	23836.26	CE		0.6			MIT
4193.874	4195.056	23837.59	CE	II	1.5	0.7		MIT
4193.283	4194.465	23840.95	CE		1.3	0.5		MIT
4193.094	4194.276	23842.02	CE		1.4	0.6		MIT
4192.756	4193.937	23843.94	CE	II	0.8	0.0		MIT
4191.345	4192.526	23851.97	CE		0.8	0.6H		MIT
4191.031	4192.212	23853.76	CE		0.9	0.3		MIT
4190.626	4191.807	23856.06	CE		1.5	0.5		MIT
4190.331	4191.512	23857.74	CE		0.6			MIT
4189.640	4190.821	23861.68	CE		0.9	0.0		MIT
4189.179	4190.360	23864.30	CE	II	0.9			MIT
4188.661	4189.841	23867.25	CE		0.5			MIT
4188.384	4189.564	23868.83	CE		0.7L	0.0H		MIT
4187.323	4188.503	23874.88	CE	II	1.5	1.2		MIT
4186.860	4188.040	23877.52	CE		0.3			MIT
4186.599	4187.779	23879.01	CE	II	1.9	1.4		MIT
4185.528	4186.708	23885.12	CE		0.5			MIT
4185.333	4186.512	23886.23	CE	II	1.5	0.6		MIT
4184.628	4185.807	23890.25	CE		0.3			MIT
4184.379	4185.558	23891.68	CE		0.5	0.0		MIT
4184.066	4185.245	23893.46	CE		0.5	0.6		MIT
4183.801	4184.980	23894.98	CE		0.5			MIT

AIR WAVELENGTH (ANGSTRMS)	VACUUM WAVELENGTH (ANGSTRMS)	WAVENUMBER (KAYSERS)	LMENT		LOG ARC	INTENSITY SPK	DIS	REF
4183.163	4184.342	23898.62	CE		0.8	0.5L		MIT
4181.579	4182.758	23907.67	CE		0.3			MIT
4181.345	4182.523	23909.01	CE			0.3		MIT
4181.081	4182.259	23910.52	CE		1.3	0.5		MIT
4179.806	4180.984	23917.81	CE		0.0	0.3		MIT
4179.289	4180.467	23920.77	CE	II	1.0	0.0		MIT
4179.079	4180.257	23921.98	CE		0.9			MIT
4178.390	4179.568	23925.92	CE		0.0	0.3		MIT
4178.283	4179.461	23926.53	CE		0.0	0.3		MIT
4178.153	4179.331	23927.28	CE		0.5			MIT
4178.084	4179.262	23927.67	CE		0.5			MIT
4177.349	4178.526	23931.88	CE		0.3	0.5L		MIT
4176.703	4177.880	23935.58	CE		1.3	0.3		MIT
4176.080	4177.257	23939.15	CE	II	1.1			MIT
4175.945	4177.122	23939.93	CE		0.8	0.0		MIT
4175.498	4176.675	23942.49	CE		0.5	0.0		MIT
4175.236	4176.413	23943.99	CE		1.2	0.3H		MIT
4174.475	4175.652	23948.36	CE		0.9L			MIT
4174.389	4175.566	23948.85	CE	II	0.9			MIT
4173.952	4175.129	23951.36	CE		0.5			MIT
4173.660	4174.836	23953.03	CE			0.3H		MIT
4173.143	4174.319	23956.00	CE		0.3			MIT
4172.885	4174.061	23957.48	CE		0.3	0.3		MIT
4172.621	4173.797	23959.00	CE		0.3			MIT
4172.161	4173.337	23961.64	CE	II	1.3	0.0		MIT
4171.964	4173.140	23962.77	CE		0.3			MIT
4171.769	4172.945	23963.89	CE		0.3	0.5		MIT
4171.386	4172.562	23966.09	CE	II	1.3			MIT
4171.040	4172.216	23968.08	CE		0.8	0.0H		MIT
4170.656	4171.832	23970.29	CE		0.3			MIT
4169.880	4171.055	23974.75	CE	II	1.1	0.5		MIT
4169.773	4170.948	23975.36	CE		1.1	0.3		MIT
4167.804	4168.979	23986.69	CE		1.1	0.5		MIT
4167.585	4168.760	23987.95	CE		0.6			MIT
4166.884	4168.059	23991.99	CE		1.3	0.8		MIT
4166.654	4167.829	23993.31	CE		1.0	0.0		MIT
4166.202	4167.376	23995.91	CE		0.8			MIT
4165.855	4167.029	23997.91	CE		0.7			MIT
4165.606	4166.780	23999.34	CE	II	1.6	0.8		MIT
4164.514	4165.688	24005.64	CE		0.5			MIT
4163.984	4165.158	24008.69	CE		0.7			MIT
4163.516	4164.690	24011.39	CE		1.3	0.9S		MIT
4162.876	4164.050	24015.08	CE	II	0.6	0.5H		MIT
4162.632	4163.806	24016.49	CE		1.1	0.5		MIT
4162.468	4163.641	24017.44	CE		0.3	0.0H		MIT
4162.296	4163.469	24018.43	CE		0.8	0.3		MIT
4161.947	4163.120	24020.44	CE		0.8	0.5H		MIT
4161.793	4162.966	24021.33	CE		0.8			MIT
4161.175	4162.348	24024.90	CE		1.3	0.0		MIT
4160.418	4161.591	24029.27	CE		0.6			MIT
4160.182	4161.355	24030.64	CE		0.9			MIT
4160.108	4161.281	24031.06	CE	II	0.9			MIT
4159.869	4161.042	24032.44	CE		0.5	0.9H		MIT
4159.727	4160.900	24033.26	CE		0.3			MIT
4159.032	4160.205	24037.28	CE	II	1.5	0.7		MIT
4158.059	4159.231	24042.90	CE		0.3H	0.9H		MIT
4157.572	4158.744	24045.72	CE		0.5H	0.0		MIT
4156.953	4158.125	24049.30	CE		0.3H	0.5H		MIT
4156.388	4157.560	24052.57	CE		0.3	0.0		MIT
4155.978	4157.150	24054.94	CE		0.6			MIT
4155.532	4156.704	24057.52	CE		0.8	0.0		MIT
4154.466	4155.637	24063.70	CE		0.3	0.3		MIT
4153.925	4155.096	24066.83	CE		1.1	0.5		MIT
4153.396	4154.567	24069.90	CE		0.6			MIT
4153.129	4154.300	24071.44	CE	II	1.1	0.3		MIT
4152.928	4154.099	24072.61	CE		0.6	0.0		MIT
4151.970	4153.141	24078.16	CE		1.5	0.9		MIT
4151.721	4152.892	24079.61	CE		0.3			MIT
4150.908	4152.078	24084.32	CE		1.3	0.5		MIT
4150.432	4151.602	24087.08	CE	I	0.5	0.3		MIT
4150.401	4151.571	24087.26	CE		0.3	0.0		MIT
4149.936	4151.106	24089.96	CE	II	0.8L	0.6L		MIT
4149.786	4150.956	24090.83	CE		0.6	0.0		MIT
4149.150	4150.320	24094.53	CE		0.7	0.3L		MIT
4148.901	4150.071	24095.97	CE		1.4	0.5		MIT
4148.478	4149.648	24098.43	CE		0.3	0.5		MIT
4148.161	4149.331	24100.27	CE		1.2	0.3		MIT
4147.525	4148.695	24103.97	CE		1.0	0.3		MIT
4147.314	4148.484	24105.19	CE		0.3			MIT
4147.196	4148.365	24105.88	CE		0.0	0.5		MIT
4146.485	4147.654	24110.01	CE		0.6			MIT
4146.234	4147.403	24111.47	CE	II	1.4	0.6		MIT
4145.856	4147.025	24113.67	CE			0.3		MIT
4145.491	4146.660	24115.79	CE		0.5			MIT
4144.995	4146.164	24118.68	CE		1.0L	0.6		MIT
4144.852	4146.021	24119.51	CE		0.3	0.0		MIT
4144.492	4145.661	24121.61	CE		1.0	0.5		MIT
4144.370	4145.539	24122.32	CE		0.3			MIT
4144.205	4145.374	24123.28	CE			0.5		MIT
4142.826	4143.994	24131.31	CE		0.8	0.3		MIT
4142.717	4143.885	24131.94	CE		0.3			MIT
4142.398	4143.566	24133.80	CE		1.5	1.5		MIT
4142.343	4143.511	24134.12	CE		1.5	0.8		MIT
4142.036	4143.204	24135.91	CE		0.6			MIT
4140.945	4142.113	24142.27	CE		0.7			MIT
4140.751	4141.919	24143.40	CE		0.9	0.8L		MIT
4140.512	4141.680	24144.79	CE		0.7			MIT
4140.200	4141.368	24146.61	CE		0.5	0.8		MIT
4139.819	4140.987	24148.83	CE		0.7	0.0		MIT
4139.426	4140.593	24151.13	CE		0.9	0.3		MIT
4139.031	4140.198	24153.43	CE		0.3	0.3L		MIT
4138.743	4139.910	24155.11	CE		0.3	0.0		MIT
4138.352	4139.519	24157.40	CE		1.1	0.6		MIT
4138.102	4139.269	24158.85	CE		1.1	0.5		MIT
4137.646	4138.813	24161.52	CE		1.4	1.1		MIT
4137.473	4138.640	24162.53	CE	II	0.8	0.3		MIT
4136.897	4138.064	24165.89	CE		0.7	0.3		MIT
4136.769	4137.936	24166.64	CE		0.7			MIT
4135.887	4137.054	24171.79	CE		1.0	0.3		MIT
4135.443	4136.609	24174.39	CE		1.3L	0.6L		MIT
4135.105	4136.271	24176.36	CE		0.7			MIT
4134.565	4135.731	24179.52	CE			0.3S		MIT
4133.800	4134.966	24184.00	CE		1.5	0.9		MIT
4133.715	4134.881	24184.49	CE		0.5	0.0		MIT
4133.048	4134.214	24188.40	CE		0.5			MIT
4132.637	4133.803	24190.80	CE	II	0.9	0.3H		MIT
4132.447	4133.613	24191.91	CE		0.6			KN
4132.313	4133.479	24192.70	CE		0.9	0.3S		MIT
4131.855	4133.020	24195.38	CE		0.7	0.3		MIT
4131.347	4132.512	24198.35	CE		0.3			MIT
4131.100	4132.265	24199.80	CE	II	1.5	0.9		MIT
4130.706	4131.871	24202.11	CE	II	1.4	0.9		MIT
4130.234	4131.399	24204.88	CE			0.6		MIT
4130.123	4131.288	24205.53	CE			0.3		MIT
4129.921	4131.086	24206.71	CE		0.5			MIT
4129.176	4130.341	24211.08	CE		0.7	0.0		MIT
4129.103	4130.268	24211.50	CE		0.5			MIT
4128.905	4130.070	24212.67	CE		0.5	0.0H		MIT
4128.479	4129.644	24215.17	CE		0.5			MIT
4128.363	4129.528	24215.84	CE		1.0S	0.6		MIT
4128.068	4129.232	24217.58	CE	II	1.0	0.5		MIT
4127.743	4128.907	24219.48	CE		0.9	0.6		MIT
4127.367	4128.531	24221.69	CE		1.5	1.1		MIT
4127.104	4128.268	24223.23	CE		0.6	0.0		MIT
4126.995	4128.159	24223.87	CE		0.5	0.0		MIT
4126.883	4128.047	24224.53	CE		0.7	0.0		MIT
4126.660	4127.824	24225.84	CE		0.9	0.5S		MIT
4126.364	4127.528	24227.58	CE		0.8			MIT
4126.022	4127.186	24229.58	CE		0.3			MIT
4125.777	4126.941	24231.02	CE	II	0.8S	0.0		MIT

AIR WAVELENGTH (ANGSTRMS)	VACUUM WAVELENGTH (ANGSTRMS)	WAVENUMBER (KAYSERS)	LMENT		LOG ARC	INTENSITY SPK	DIS	REF
4125.709	4126.873	24231.42	CE		0.3			MIT
4125.429	4126.593	24233.07	CE		0.8	0.3		MIT
4124.785	4125.949	24236.85	CE		1.4	0.7		MIT
4123.872	4125.035	24242.22	CE		1.4	0.8		MIT
4123.488	4124.651	24244.47	CE		1.3	0.7		MIT
4122.857	4124.020	24248.18	CE		0.5			MIT
4122.237	4123.400	24251.83	CE		0.5	0.5L		MIT
4122.145	4123.308	24252.37	CE			0.3		MIT
4121.908	4123.071	24253.77	CE		0.5			MIT
4121.602	4122.765	24255.57	CE	II	0.9	0.0		MIT
4121.278	4122.441	24257.47	CE		0.3			MIT
4121.069	4122.232	24258.71	CE		0.3	0.3		MIT
4120.829	4121.992	24260.12	CE		1.4	0.8		MIT
4120.798	4121.961	24260.30	CE		0.6			MIT
4120.507	4121.669	24262.01	CE		0.3	0.3		MIT
4119.877	4121.039	24265.72	CE	II	1.3D	0.5		MIT
4119.343	4120.505	24268.87	CE		0.3H			MIT
4119.266	4120.428	24269.32	CE			0.3H		MIT
4119.015	4120.177	24270.80	CE		1.4	0.5		MIT
4118.192	4119.354	24275.65	CE		0.5			MIT
4118.144	4119.306	24275.94	CE		1.4	0.9		MIT
4117.987	4119.149	24276.86	CE		0.5	0.0		MIT
4117.826	4118.988	24277.81	CE		0.6	0.0H		MIT
4117.586	4118.748	24279.22	CE		1.5	0.7		MIT
4117.288	4118.450	24280.98	CE		1.3	0.6		MIT
4117.013	4118.175	24282.60	CE		1.5	0.8		MIT
4116.713	4117.875	24284.37	CE		0.8	0.0		MIT
4115.879	4117.040	24289.29	CE			0.8H		MIT
4115.657	4116.818	24290.60	CE		0.5			MIT
4115.374	4116.535	24292.27	CE		1.6	0.8		MIT
4114.916	4116.077	24294.98	CE		1.1			MIT
4114.149	4115.310	24299.51	CE	II	1.1	0.3		MIT
4113.997	4115.158	24300.41	CE		0.3	0.5H		MIT
4113.771	4114.932	24301.74	CE		0.3L			MIT
4113.726	4114.887	24302.01	CE	II	1.5	0.5		MIT
4113.553	4114.714	24303.03	CE		0.8	0.0		MIT
4113.332	4114.493	24304.33	CE		0.5			MIT
4113.170	4114.331	24305.29	CE		0.5	0.0H		MIT
4112.484	4113.644	24309.34	CE		0.8			MIT
4112.257	4113.417	24310.69	CE			0.3		MIT
4112.173	4113.333	24311.18	CE		0.3			MIT
4112.085	4113.245	24311.70	CE		0.5L			MIT
4111.931	4113.091	24312.61	CE	II	0.9	0.0H		MIT
4111.394	4112.554	24315.79	CE		1.5	0.7		MIT
4110.840	4112.000	24319.07	CE	II	1.3			MIT
4110.607	4111.767	24320.45	CE		0.3			MIT
4110.381	4111.541	24321.78	CE		1.5	1.0H		MIT
4110.027	4111.187	24323.88	CE		0.5			MIT
4109.905	4111.065	24324.60	CE			0.5H		MIT
4109.559	4110.719	24326.65	CE		1.1			MIT
4108.729	4109.888	24331.56	CE		0.9			MIT
4108.543	4109.702	24332.66	CE		0.3			MIT
4108.255	4109.414	24334.37	CE		0.9	0.0		MIT
4108.087	4109.246	24335.36	CE			0.3		MIT
4107.797	4108.956	24337.08	CE		1.2	0.3H		MIT
4107.421	4108.580	24339.31	CE	I	1.5	0.9		MIT
4107.195	4108.354	24340.65	CE		0.3			MIT
4106.915	4108.074	24342.31	CE		1.0			MIT
4106.852	4108.011	24342.68	CE	II	1.1	0.3		MIT
4106.652	4107.811	24343.87	CE			0.3		MIT
4106.134	4107.293	24346.94	CE		1.5	0.3		MIT
4105.625	4106.784	24349.96	CE			0.3		MIT
4105.495	4106.654	24350.73	CE		0.3			MIT
4105.343	4106.502	24351.63	CE		0.3			MIT
4104.996	4106.154	24353.69	CE		1.6	0.5		MIT
4104.425	4105.583	24357.07	CE		1.5	0.0		MIT
4103.730	4104.888	24361.20	CE		0.6			MIT
4103.448	4104.606	24362.87	CE		0.6			MIT
4102.724	4103.882	24367.17	CE		0.3	0.0		MIT
4102.364	4103.522	24369.31	CE		1.3	0.3		MIT

AIR WAVELENGTH (ANGSTRMS)	VACUUM WAVELENGTH (ANGSTRMS)	WAVENUMBER (KAYSERS)	LMENT		LOG ARC	INTENSITY SPK	DIS	REF
4102.073	4103.231	24371.04	CE		0.3			MIT
4101.772	4102.930	24372.83	CE		1.5	0.8		MIT
4101.550	4102.708	24374.15	CE		0.3	0.5		MIT
4101.357	4102.514	24375.29	CE		0.0	0.3		MIT
4100.889	4102.046	24378.08	CE		0.9	0.0		MIT
4099.749	4100.906	24384.85	CE		1.1	0.3		MIT
4099.387	4100.544	24387.01	CE		0.8			MIT
4098.981	4100.138	24389.42	CE		1.2	0.0		MIT
4098.466	4099.623	24392.49	CE		0.6			MIT
4098.153	4099.310	24394.35	CE		0.9			MIT
4097.617	4098.773	24397.54	CE		0.3			MIT
4097.338	4098.494	24399.20	CE		0.6			MIT
4096.926	4098.082	24401.66	CE			0.3		MIT
4096.837	4097.993	24402.19	CE		0.3	0.3		MIT
4096.099	4097.255	24406.58	CE		0.6			M
4095.819	4096.975	24408.25	CE		0.8			MIT
4095.448	4096.604	24410.46	CE		0.9			MIT
4095.113	4096.269	24412.46	CE		0.9			MIT
4094.591	4095.747	24415.57	CE		0.3			MIT
4094.312	4095.468	24417.24	CE			0.3		MIT
4094.199	4095.355	24417.91	CE		0.5			MIT
4093.955	4095.111	24419.36	CE		1.3	0.5		MIT
4093.285	4094.440	24423.36	CE		1.2			MIT
4092.715	4093.870	24426.76	CE	II	1.3	0.3		MIT
4092.085	4093.240	24430.52	CE		1.0	0.0		MIT
4091.852	4093.007	24431.92	CE		0.8			MIT
4091.591	4092.746	24433.47	CE		0.5			MIT
4091.058	4092.213	24436.66	CE		0.5			MIT
4090.947	4092.102	24437.32	CE	II	0.8	0.3		MIT
4090.775	4091.930	24438.35	CE			0.3		MIT
4090.466	4091.621	24440.19	CE		1.2	0.0		MIT
4089.855	4091.009	24443.84	CE		0.8			MIT
4089.742	4090.896	24444.52	CE		0.8			MIT
4089.157	4090.311	24448.02	CE		0.6			MIT
4089.009	4090.163	24448.90	CE	II	0.8			MIT
4088.851	4090.005	24449.85	CE		1.2	0.3		MIT
4088.583	4089.737	24451.45	CE		0.9	0.0		MIT
4088.123	4089.277	24454.20	CE		0.5			MIT
4087.566	4088.720	24457.53	CE		0.9	0.3		MIT
4087.360	4088.514	24458.76	CE	II	0.8			MIT
4087.303	4088.457	24459.11	CE	II	0.9			MIT
4087.103	4088.257	24460.30	CE		0.5			MIT
4086.768	4087.922	24462.31	CE		0.5			MIT
4086.424	4087.578	24464.37	CE		0.9	0.3		MIT
4086.072	4087.225	24466.48	CE		0.5			MIT
4085.998	4087.151	24466.92	CE		0.5			MIT
4085.748	4086.901	24468.42	CE		0.9			MIT
4085.232	4086.385	24471.51	CE	II	1.3	0.7		MIT
4084.783	4085.936	24474.20	CE		0.5	0.0H		MIT
4084.646	4085.799	24475.02	CE		0.5			MIT
4084.197	4085.350	24477.71	CE		0.5			MIT
4083.637	4084.790	24481.06	CE		0.8	0.3		MIT
4083.482	4084.635	24481.99	CE		0.8	0.0		MIT
4083.233	4084.386	24483.49	CE		1.5D	0.8		MIT
4083.162	4084.315	24483.91	CE		0.9			MIT
4082.962	4084.115	24485.11	CE		0.6			MIT
4082.100	4083.252	24490.28	CE		0.8			MIT
4081.853	4083.005	24491.76	CE		0.3			MIT
4081.716	4082.868	24492.58	CE		0.3			MIT
4081.615	4082.767	24493.19	CE		0.8			MIT
4081.222	4082.374	24495.55	CE		1.6	0.9		MIT
4080.553	4081.705	24499.57	CE		0.8	0.0		MIT
4080.436	4081.588	24500.27	CE	II	0.9	0.3		MIT
4080.025	4081.177	24502.74	CE		0.7			MIT
4079.667	4080.819	24504.89	CE		1.2			MIT
4079.277	4080.429	24507.23	CE		0.8			MIT
4079.016	4080.168	24508.80	CE		1.2	0.0H		MIT
4078.606	4079.758	24511.26	CE		0.7			MIT
4078.515	4079.666	24511.81	CE		0.7	0.0		MIT
4078.321	4079.472	24512.97	CE		1.2	0.6		MIT

AIR WAVELENGTH (ANGSTRMS)	VACUUM WAVELENGTH (ANGSTRMS)	WAVENUMBER (KAYSERS)	LMENT	LOG ARC	INTENSITY SPK	DIS	REF
4077.470	4078.621	24518.09	CE II	1.3	0.6		MIT
4076.237	4077.388	24525.51	CE	1.1	0.0		MIT
4075.850	4077.001	24527.83	CE II	0.6S			MIT
4075.786	4076.937	24528.22	CE	1.0	0.3		MIT
4075.714	4076.865	24528.65	CE	1.2	0.3		MIT
4075.538	4076.689	24529.71	CE	0.3			MIT
4075.173	4076.324	24531.91	CE	0.3			MIT
4074.651	4075.801	24535.05	CE II	0.8			MIT
4074.558	4075.708	24535.61	CE	0.3			MIT
4074.127	4075.277	24538.21	CE	0.8			MIT
4073.735	4074.885	24540.57	CE	1.5	0.5		MIT
4073.477	4074.627	24542.12	CE	1.7	0.9		MIT
4072.917	4074.067	24545.50	CE	1.3	0.3		MIT
4072.667	4073.817	24547.00	CE	0.3			MIT
4072.549	4073.699	24547.72	CE	0.3			MIT
4071.814	4072.964	24552.15	CE	1.5	0.7		MIT
4071.473	4072.623	24554.20	CE	0.5			MIT
4071.346	4072.496	24554.97	CE	0.6			MIT
4071.080	4072.230	24556.57	CE	1.3	0.3		MIT
4070.836	4071.985	24558.04	CE	1.0			MIT
4070.094	4071.243	24562.52	CE	1.1	0.0		MIT
4069.921	4071.070	24563.56	CE	0.3	0.5		MIT
4069.704	4070.853	24564.87	CE	0.6			MIT
4068.991	4070.140	24569.18	CE	0.5			MIT
4068.836	4069.985	24570.11	CE	1.4	0.6		MIT
4068.444	4069.593	24572.48	CE	1.0H	0.0H		MIT
4068.057	4069.206	24574.82	CE	0.5			MIT
4067.766	4068.915	24576.58	CE	0.5			MIT
4067.279	4068.428	24579.52	CE	1.4			MIT
4066.906	4068.054	24581.77	CE II	0.8			MIT
4066.566	4067.714	24583.83	CE	0.3H			MIT
4066.497	4067.645	24584.25	CE	0.9	0.3		MIT
4066.155	4067.303	24586.31	CE	0.7	0.0		MIT
4065.553	4066.701	24589.96	CE	0.3L			MIT
4065.164	4066.312	24592.31	CE II	1.1	0.5		MIT
4064.907	4066.055	24593.86	CE II	0.9			MIT
4064.793	4065.941	24594.55	CE	0.5	0.0		MIT
4063.919	4065.067	24599.84	CE	0.9L	0.3		MIT
4062.941	4064.088	24605.76	CE II	1.4	0.7		MIT
4062.557	4063.704	24608.09	CE	0.5			MIT
4062.223	4063.370	24610.11	CE	1.6	0.9		MIT
4061.811	4062.958	24612.61	CE	0.7			MIT
4061.705	4062.852	24613.25	CE	0.3			MIT
4061.423	4062.570	24614.96	CE II	0.8			MIT
4060.716	4061.863	24619.25	CE	0.9	0.0		MIT
4060.471	4061.618	24620.73	CE	1.0	0.3		MIT
4060.169	4061.316	24622.56	CE	0.8	0.0		MIT
4059.367	4060.513	24627.43	CE	0.5	0.0H		MIT
4059.322	4060.468	24627.70	CE II	0.9H			MIT
4058.244	4059.390	24634.24	CE II	1.3			MIT
4057.556	4058.702	24638.42	CE	0.3			MIT
4057.303	4058.449	24639.96	CE	0.3			MIT
4056.900	4058.046	24642.40	CE	1.2	0.3		MIT
4056.338	4057.484	24645.82	CE	0.6			MIT
4056.248	4057.394	24646.36	CE	0.5			MIT
4055.838	4056.984	24648.85	CE	1.1			MIT
4055.158	4056.303	24652.99	CE	0.9	0.0		MIT
4054.991	4056.136	24654.00	CE II	1.1	0.8		MIT
4054.658	4055.803	24656.03	CE	0.7			MIT
4054.099	4055.244	24659.43	CE	0.3			MIT
4053.506	4054.651	24663.03	CE II	1.6	0.9		MIT
4053.070	4054.215	24665.69	CE	0.7			MIT
4052.628	4053.773	24668.38	CE	0.3			MIT
4052.058	4053.203	24671.85	CE	0.8			MIT
4051.985	4053.130	24672.29	CE	1.3L	0.5H		MIT
4051.614	4052.758	24674.55	CE	0.6			MIT
4051.427	4052.571	24675.69	CE	1.3	0.5		MIT
4051.233	4052.377	24676.87	CE	0.5			MIT
4050.947	4052.091	24678.61	CE	0.3			MIT
4050.812	4051.956	24679.44	CE	1.1	0.0		MIT

AIR WAVELENGTH (ANGSTRMS)	VACUUM WAVELENGTH (ANGSTRMS)	WAVENUMBER (KAYSERS)	LMENT	LOG ARC	INTENSITY SPK	DIS	REF
4049.790	4050.934	24685.67	CE II	0.9	0.0		MIT
4049.555	4050.699	24687.10	CE	0.3			MIT
4049.362	4050.506	24688.27	CE	0.8			MIT
4049.194	4050.338	24689.30	CE	0.8			MIT
4049.030	4050.174	24690.30	CE	1.3	0.0		MIT
4048.366	4049.510	24694.35	CE II	0.9	0.0		MIT
4047.875	4049.018	24697.34	CE	0.7			MIT
4047.699	4048.842	24698.42	CE	0.5			MIT
4047.620	4048.763	24698.90	CE	0.5			MIT
4047.392	4048.535	24700.29	CE	0.6			MIT
4047.275	4048.418	24701.00	CE	1.3	0.3		MIT
4046.853	4047.996	24703.58	CE	0.5			MIT
4046.734	4047.877	24704.31	CE	0.3			MIT
4046.340	4047.483	24706.71	CE II	1.5	1.0		MIT
4046.340	4047.483	24706.71	CE I	1.5	1.0		MIT
4045.973	4047.116	24708.95	CE II	0.7			MIT
4045.316	4046.459	24712.97	CE	0.5			MIT
4045.209	4046.352	24713.62	CE	0.9	0.5		MIT
4044.330	4045.473	24718.99	CE	0.6H	0.0		MIT
4044.062	4045.204	24720.63	CE	0.5			MIT
4043.955	4045.097	24721.28	CE	0.5			MIT
4043.747	4044.889	24722.55	CE	0.5			MIT
4043.473	4044.615	24724.23	CE	0.6S			MIT
4043.409	4044.551	24724.62	CE	0.3			MIT
4043.078	4044.220	24726.65	CE	0.3			MIT
4042.584	4043.726	24729.67	CE II	1.7	0.5		MIT
4042.584	4043.726	24729.67	CE I	1.7	0.5		MIT
4042.135	4043.277	24732.41	CE	0.9			MIT
4041.270	4042.412	24737.71	CE	0.8			MIT
4040.762	4041.904	24740.82	CE II	1.8	0.7		MIT
4039.890	4041.031	24746.16	CE	1.1L			MIT
4039.298	4040.439	24749.78	CE	0.3			MIT
4038.344	4039.485	24755.63	CE	1.0	0.7		KN
4037.960	4039.101	24757.99	CE	1.3	0.3		MIT
4037.665	4038.806	24759.79	CE II	1.4	0.5		MIT
4037.390	4038.531	24761.48	CE	0.8			MIT
4037.203	4038.344	24762.63	CE I	0.3			M
4036.573	4037.714	24766.49	CE	0.3			MIT
4036.088	4037.228	24769.47	CE	0.7			MIT
4035.990	4037.130	24770.07	CE	0.3			MIT
4034.570	4035.710	24778.79	CE	0.3			MIT
4034.259	4035.399	24780.70	CE	0.3			MIT
4033.786	4034.926	24783.60	CE	0.8			MIT
4033.378	4034.518	24786.11	CE	0.3H			MIT
4032.748	4033.888	24789.98	CE	0.3			MIT
4032.554	4033.693	24791.18	CE	0.5			MIT
4031.669	4032.808	24796.62	CE	1.0D			MIT
4031.336	4032.475	24798.67	CE II	1.6	0.9		MIT
4030.853	4031.992	24801.64	CE	0.3			MIT
4030.344	4031.483	24804.77	CE II	1.3	0.6		MIT
4030.155	4031.294	24805.93	CE	0.7			MIT
4029.753	4030.892	24808.41	CE II	0.6			MIT
4029.260	4030.399	24811.44	CE	0.7	0.0H		MIT
4028.411	4029.549	24816.67	CE II	1.5	0.9		MIT
4028.198	4029.336	24817.98	CE II	0.3			MIT
4027.990	4029.128	24819.26	CE	0.6			MIT
4027.878	4029.016	24819.95	CE	0.6			MIT
4027.693	4028.831	24821.09	CE II	1.3S	0.5		MIT
4027.048	4028.186	24825.07	CE	0.7			MIT
4026.936	4028.074	24825.76	CE	0.3			MIT
4026.39	4027.53	24829.1	CE	0.3			MIT
4026.253	4027.391	24829.97	CE	0.3			MIT
4026.143	4027.281	24830.65	CE	0.3			MIT
4025.650	4026.788	24833.69	CE	0.3			MIT
4025.294	4026.432	24835.89	CE	0.3			MIT
4025.150	4026.288	24836.78	CE	1.1	0.3		MIT
4024.491	4025.628	24840.84	CE	1.2	0.7		MIT
4024.348	4025.485	24841.73	CE	0.7			MIT
4023.640	4024.777	24846.10	CE	0.6			MIT
4023.370	4024.507	24847.76	CE	0.9	0.0		MIT

AIR WAVELENGTH (ANGSTRMS)	VACUUM WAVELENGTH (ANGSTRMS)	WAVENUMBER (KAYSERS)	LMENT	LOG ARC	INTENSITY SPK	DIS	REF
4022.933	4024.070	24850.46	CE	0.5L			MIT
4022.748	4023.885	24851.60	CE	0.6			MIT
4022.453	4023.590	24853.43	CE	0.7			MIT
4022.272	4023.409	24854.55	CE	1.2	0.6		MIT
4021.242	4022.379	24860.91	CE	0.7			MIT
4020.540	4021.676	24865.25	CE II	0.7			MIT
4019.897	4021.033	24869.23	CE	0.9	0.5		MIT
4019.480	4020.616	24871.81	CE	0.8	0.0		MIT
4019.280	4020.416	24873.05	CE II	0.6			MIT
4019.193	4020.329	24873.59	CE	0.3			MIT
4019.044	4020.180	24874.51	CE	1.2	0.6		MIT
4018.919	4020.055	24875.28	CE	0.3			MIT
4018.511	4019.647	24877.81	CE	0.3			MIT
4018.400	4019.536	24878.50	CE	0.3			MIT
4018.227	4019.363	24879.57	CE II	0.3			MIT
4018.066	4019.202	24880.56	CE	0.3			MIT
4017.596	4018.732	24883.47	CE II	1.0S			MIT
4016.419	4017.554	24890.76	CE	0.3			MIT
4016.103	4017.238	24892.72	CE	0.5			MIT
4015.877	4017.012	24894.12	CE	1.3	0.6		MIT
4015.715	4016.850	24895.13	CE	0.3			MIT
4014.899	4016.034	24900.19	CE II	1.8	1.1		MIT
4014.346	4015.481	24903.62	CE	0.3			MIT
4014.150	4015.285	24904.83	CE	0.5			MIT
4013.941	4015.076	24906.13	CE	0.5			MIT
4013.647	4014.782	24907.96	CE	0.5			MIT
4013.265	4014.399	24910.33	CE	0.3			MIT
4012.388	4013.522	24915.77	CE I	1.8	1.3		MIT
4012.388	4013.522	24915.77	CE II	1.8	1.3		MIT
4012.139	4013.273	24917.32	CE	0.6			MIT
4011.773	4012.907	24919.59	CE	0.3			MIT
4011.560	4012.694	24920.91	CE II	1.2	0.5		MIT
4011.296	4012.430	24922.55	CE	0.6			MIT
4010.795	4011.929	24925.67	CE	0.3			MIT
4010.136	4011.270	24929.76	CE	1.2	0.5		MIT
4009.55	4010.68	24933.4	CE	0.3	0.0H		MIT
4009.063	4010.196	24936.44	CE	1.1L			MIT
4008.664	4009.797	24938.92	CE	0.9			MIT
4008.446	4009.579	24940.27	CE	0.8			MIT
4007.588	4008.721	24945.61	CE	1.2	0.6		MIT
4007.451	4008.584	24946.47	CE	0.6H			MIT
4007.022	4008.155	24949.14	CE	0.3			MIT
4005.639	4006.771	24957.75	CE	1.3	0.8		MIT
4005.244	4006.376	24960.21	CE	1.3			MIT
4004.582	4005.714	24964.34	CE	1.1	0.5		MIT
4004.047	4005.179	24967.67	CE	0.8H			MIT
4003.771	4004.903	24969.39	CE	1.6	1.3		MIT
4003.601	4004.733	24970.45	CE	0.5			MIT
4003.320	4004.452	24972.21	CE	0.3			MIT
4003.170	4004.302	24973.14	CE II	1.0	0.0		MIT
4002.974	4004.106	24974.36	CE II	0.9	0.0		MIT
4002.813	4003.945	24975.37	CE	1.3	0.6		MIT
4002.257	4003.389	24978.84	CE	0.6			MIT
4001.732	4002.863	24982.12	CE	0.7	0.3		MIT
4001.555	4002.686	24983.22	CE	1.3S	0.7S		MIT
4001.055	4002.186	24986.34	CE II	1.3	0.3		MIT
4000.799	4001.930	24987.94	CE	0.6	0.0		MIT
4000.675	4001.806	24988.72	CE	0.7	0.0		MIT
3999.240	4000.371	24997.68	CE II	1.9	1.3		MIT
3998.750	3999.881	25000.75	CE	0.3			MIT
3998.235	3999.366	25003.97	CE	0.3			MIT
3997.716	3998.846	25007.21	CE	1.3	0.3		MIT
3997.483	3998.613	25008.67	CE	0.7			MIT
3997.268	3998.398	25010.01	CE	0.5			MIT
3996.768	3997.898	25013.14	CE	0.6			MIT
3996.487	3997.617	25014.90	CE II	1.0			MIT
3996.359	3997.489	25015.70	CE	0.3H			MIT
3995.423	3996.553	25021.56	CE II	0.6			MIT
3994.807	3995.937	25025.42	CE	0.5			MIT
3994.573	3995.703	25026.89	CE	0.8			MIT

AIR WAVELENGTH (ANGSTRMS)	VACUUM WAVELENGTH (ANGSTRMS)	WAVENUMBER (KAYSERS)	LMENT	LOG ARC	INTENSITY SPK	DIS	REF
3993.822	3994.951	25031.59	CE	1.7	0.8		MIT
3993.402	3994.531	25034.23	CE	0.5			MIT
3993.190	3994.319	25035.56	CE	0.7			MIT
3992.913	3994.042	25037.29	CE	1.2	0.5		MIT
3992.711	3993.840	25038.56	CE	0.5			MIT
3992.386	3993.515	25040.60	CE II	1.7	0.9		MIT
3992.126	3993.255	25042.23	CE	0.6			MIT
3991.326	3992.455	25047.25	CE II	0.7			MIT
3991.223	3992.352	25047.89	CE	0.3			MIT
3990.687	3991.816	25051.26	CE	0.8	0.0		MIT
3990.415	3991.543	25052.97	CE	0.5			MIT
3990.106	3991.234	25054.91	CE II	1.3	0.3		MIT
3989.756	3990.884	25057.10	CE	0.6			MIT
3989.441	3990.569	25059.08	CE	1.3	0.8		MIT
3987.699	3988.827	25070.03	CE	0.3L			MIT
3987.508	3988.636	25071.23	CE	0.3S			MIT
3987.058	3988.186	25074.06	CE	0.3			MIT
3986.395	3987.522	25078.23	CE	1.2	0.0		MIT
3986.134	3987.261	25079.87	CE	0.3			MIT
3985.921	3987.048	25081.21	CE	0.3			MIT
3985.238	3986.365	25085.51	CE	0.6			MIT
3984.675	3985.802	25089.05	CE	1.6	0.9		MIT
3984.308	3985.435	25091.36	CE	0.3			MIT
3983.292	3984.419	25097.76	CE	1.0	0.5		MIT
3983.195	3984.322	25098.38	CE	0.3			MIT
3982.903	3984.030	25100.22	CE II	1.5	0.8		MIT
3982.89	3984.02	25100.3	CE	0.8			MIT
3982.238	3983.364	25104.41	CE	0.9			MIT
3982.169	3983.295	25104.84	CE	0.8			MIT
3981.896	3983.022	25106.56	CE	0.7			MIT
3981.234	3982.360	25110.74	CE	0.3			MIT
3980.883	3982.009	25112.95	CE II	1.5	0.9		MIT
3980.453	3981.579	25115.67	CE	0.3			MIT
3980.254	3981.380	25116.92	CE	0.8			MIT
3979.933	3981.059	25118.95	CE	1.1			MIT
3978.895	3980.020	25125.50	CE	1.7	1.7		MIT
3978.650	3979.775	25127.05	CE	1.5	0.8		MIT
3978.336	3979.461	25129.03	CE	0.5			MIT
3977.771	3978.896	25132.60	CE	1.2			MIT
3977.533	3978.658	25134.10	CE	0.8			MIT
3977.093	3978.218	25136.88	CE	0.6			MIT
3976.778	3977.903	25138.87	CE	0.6			MIT
3976.259	3977.384	25142.16	CE	0.5S	0.3		MIT
3976.049	3977.174	25143.48	CE	0.6			MIT
3975.066	3976.190	25149.70	CE	1.3	0.5		MIT
3974.816	3975.940	25151.28	CE	0.5			MIT
3974.489	3975.613	25153.35	CE	0.9			MIT
3974.336	3975.460	25154.32	CE	0.3L			MIT
3974.192	3975.316	25155.23	CE II	0.8			MIT
3974.004	3975.128	25156.42	CE	0.7	0.7		MIT
3973.034	3974.158	25162.56	CE	0.9			MIT
3972.071	3973.195	25168.66	CE II	1.4	0.6		MIT
3971.876	3973.000	25169.90	CE II	0.8			MIT
3971.684	3972.808	25171.12	CE II	1.5	0.8		MIT
3971.014	3972.137	25175.36	CE	0.5			MIT
3970.788	3971.911	25176.80	CE	0.5			MIT
3970.636	3971.759	25177.76	CE II	1.2	0.5		MIT
3970.424	3971.547	25179.10	CE	0.8	0.0		MIT
3970.041	3971.164	25181.53	CE	1.1	0.5		MIT
3969.156	3970.279	25187.15	CE	0.3			MIT
3968.469	3969.592	25191.51	CE	1.5	1.5W		MIT
3967.915	3969.038	25195.02	CE	0.5			MIT
3967.644	3968.767	25196.75	CE	0.5			MIT
3967.53	3968.65	25197.5	CE	0.7	0.0		MIT
3967.178	3968.300	25199.71	CE II	0.8	0.3		MIT
3967.048	3968.170	25200.53	CE II	1.5	0.8		MIT
3965.930	3967.052	25207.64	CE	0.5			MIT
3965.212	3966.334	25212.20	CE	0.3			MIT
3965.077	3966.199	25213.06	CE	0.3			MIT
3964.92	3966.04	25214.1	CE	0.3	0.0H		MIT

AIR WAVELENGTH (ANGSTRMS)	VACUUM WAVELENGTH (ANGSTRMS)	WAVENUMBER (KAYSERS)	LMENT		LOG ARC	INTENSITY SPK	DIS	REF
3964.503	3965.625	25216.71	CE	II	1.4	0.8		MIT
3964.178	3965.300	25218.77	CE		1.2	0.7		MIT
3963.781	3964.903	25221.30	CE		0.5			MIT
3963.374	3964.495	25223.89	CE		0.8	0.3		MIT
3962.899	3964.020	25226.91	CE		0.5			MIT
3962.086	3963.207	25232.09	CE		1.2	0.6		MIT
3961.661	3962.782	25234.80	CE		0.8	0.3		MIT
3961.386	3962.507	25236.55	CE		0.3			MIT
3960.914	3962.035	25239.56	CE	II	1.6	0.9		MIT
3960.376	3961.497	25242.99	CE		1.0	0.3		MIT
3959.799	3960.919	25246.66	CE		1.3	0.5		MIT
3959.616	3960.736	25247.83	CE		1.0	0.3		MIT
3958.868	3959.988	25252.60	CE		1.3	0.8		MIT
3958.266	3959.386	25256.44	CE	II	1.3	0.5		MIT
3957.969	3959.089	25258.34	CE		1.0	0.3		MIT
3957.860	3958.980	25259.03	CE		0.3			MIT
3957.209	3958.329	25263.19	CE		0.7S			MIT
3957.146	3958.266	25263.59	CE		0.5			MIT
3956.898	3958.018	25265.17	CE	II	0.9	0.5		MIT
3956.772	3957.892	25265.98	CE		0.3			MIT
3956.284	3957.404	25269.09	CE		1.5	0.9		MIT
3956.063	3957.182	25270.50	CE		0.7	0.3		MIT
3955.917	3957.036	25271.44	CE		1.3	0.5		MIT
3955.363	3956.482	25274.98	CE		0.7	0.7		MIT
3954.451	3955.570	25280.81	CE		0.3	0.9J		MIT
3953.951	3955.070	25284.00	CE	II	1.3	0.3		MIT
3953.776	3954.895	25285.12	CE		0.3			MIT
3953.660	3954.779	25285.86	CE	II	1.1	0.6		MIT
3952.541	3953.660	25293.02	CE	II	1.8	1.5		MIT
3952.106	3953.224	25295.81	CE		1.0	0.0		MIT
3951.624	3952.742	25298.89	CE		0.9	0.0		MIT
3951.420	3952.538	25300.20	CE		0.6			MIT
3950.802	3951.920	25304.16	CE		0.8	0.0		MIT
3950.424	3951.542	25306.58	CE	II	1.0	0.5		MIT
3949.816	3950.934	25310.47	CE		1.0			MIT
3949.385	3950.503	25313.23	CE	II	1.3	0.5		MIT
3948.949	3950.067	25316.03	CE		0.6			MIT
3948.454	3949.572	25319.20	CE		0.3			MIT
3947.973	3949.090	25322.29	CE		1.3	0.5		MIT
3947.708	3948.825	25323.99	CE		0.3			MIT
3947.593	3948.710	25324.72	CE		0.5			MIT
3947.258	3948.375	25326.87	CE		0.6			MIT
3947.122	3948.239	25327.75	CE		0.6			MIT
3946.681	3947.798	25330.58	CE		1.3	0.0		MIT
3946.166	3947.283	25333.88	CE		0.8			MIT
3946.048	3947.165	25334.64	CE		0.3			MIT
3945.925	3947.042	25335.43	CE		0.3			MIT
3945.82	3946.94	25336.1	CE		0.3			MIT
3945.507	3946.624	25338.11	CE		0.5	0.0		MIT
3945.364	3946.481	25339.03	CE		0.3			MIT
3944.924	3946.041	25341.86	CE		0.8			MIT
3944.835	3945.952	25342.43	CE		0.7			MIT
3944.093	3945.209	25347.20	CE	II	0.9			MIT
3943.888	3945.004	25348.51	CE	II	1.6	1.2		MIT
3943.497	3944.613	25351.03	CE		0.8	0.0		MIT
3943.138	3944.254	25353.34	CE	II	1.1	0.5		MIT
3943.000	3944.116	25354.22	CE		0.5			MIT
3942.746	3943.862	25355.86	CE		1.7	1.3		MIT
3942.151	3943.267	25359.68	CE	II	1.5	0.9		MIT
3941.585	3942.701	25363.32	CE		0.7			MIT
3940.972	3942.088	25367.27	CE		1.0	0.3		MIT
3940.64	3941.76	25369.4	CE		0.9S			MIT
3940.338	3941.453	25371.35	CE	II	1.5	0.8		MIT
3939.77	3940.89	25375.0	CE		0.3H			MIT
3939.656	3940.771	25375.74	CE	II	1.0			MIT
3939.518	3940.633	25376.63	CE		1.3	0.0		MIT
3939.127	3940.242	25379.15	CE		0.6			MIT
3938.774	3939.889	25381.43	CE		0.5			MIT
3938.589	3939.704	25382.62	CE		1.0			MIT
3938.159	3939.274	25385.39	CE		0.9			MIT

AIR WAVELENGTH (ANGSTRMS)	VACUUM WAVELENGTH (ANGSTRMS)	WAVENUMBER (KAYSERS)	LMENT		LOG ARC	INTENSITY SPK	DIS	REF
3938.086	3939.201	25385.86	CE	II	1.5	0.8		MIT
3937.807	3938.922	25387.66	CE		1.1			MIT
3937.634	3938.749	25388.77	CE	II	1.2	0.0		MIT
3937.147	3938.262	25391.91	CE		1.2L	0.3		MIT
3936.962	3938.077	25393.11	CE		0.3			MIT
3935.927	3937.041	25399.78	CE		0.8			MIT
3935.501	3936.615	25402.53	CE		0.8			MIT
3934.752	3935.866	25407.37	CE		1.0			MIT
3934.509	3935.623	25408.94	CE		0.3			MIT
3934.076	3935.190	25411.74	CE		0.8			MIT
3933.731	3934.845	25413.96	CE		1.8	1.8		MIT
3932.979	3934.092	25418.82	CE		0.8S	0.0		MIT
3932.794	3933.907	25420.02	CE		0.5			MIT
3932.393	3933.506	25422.61	CE		0.5			MIT
3932.145	3933.258	25424.21	CE		1.1	0.3		MIT
3931.827	3932.940	25426.27	CE		1.3	0.5		MIT
3931.369	3932.482	25429.23	CE	II	1.2	0.6		MIT
3931.369	3932.482	25429.23	CE	I	1.2	0.6		MIT
3931.088	3932.201	25431.05	CE	I	1.5	0.9		MIT
3931.088	3932.201	25431.05	CE	II	1.5	0.9		MIT
3930.806	3931.919	25432.87	CE		1.1	0.5		MIT
3930.592	3931.705	25434.26	CE		0.5			MIT
3929.961	3931.074	25438.34	CE		0.9	0.0		MIT
3929.671	3930.784	25440.22	CE		0.8			MIT
3929.209	3930.321	25443.21	CE		0.3H			MIT
3929.125	3930.237	25443.75	CE		0.3H			MIT
3928.956	3930.068	25444.85	CE		0.3			MIT
3928.848	3929.960	25445.55	CE		0.8			MIT
3928.553	3929.665	25447.46	CE		0.3			MIT
3928.318	3929.430	25448.98	CE	II	1.3	0.3		MIT
3927.574	3928.686	25453.80	CE		0.3H			MIT
3927.450	3928.562	25454.61	CE	II	1.2L	0.3		MIT
3927.450	3928.562	25454.61	CE	I	1.2L	0.3		MIT
3927.164	3928.276	25456.46	CE		0.3			MIT
3927.001	3928.113	25457.52	CE		0.9	0.0		MIT
3926.904	3928.016	25458.15	CE		0.5			MIT
3926.282	3927.394	25462.18	CE		0.8			MIT
3926.167	3927.279	25462.92	CE	II	0.5			MIT
3926.083	3927.195	25463.47	CE		0.5			MIT
3925.874	3926.986	25464.82	CE		0.5			MIT
3925.100	3926.211	25469.85	CE		0.3			MIT
3925.00	3926.11	25470.5	CE		0.3			MIT
3924.803	3925.914	25471.77	CE		1.0	0.0		MIT
3924.644	3925.755	25472.80	CE	I	0.5			MIT
3924.644	3925.755	25472.80	CE	II	0.5			MIT
3924.543	3925.654	25473.46	CE		1.3	0.7		MIT
3924.248	3925.359	25475.38	CE		0.6			MIT
3923.512	3924.623	25480.15	CE		0.5			MIT
3923.109	3924.220	25482.77	CE		1.2	0.6		MIT
3922.863	3923.974	25484.37	CE		0.9			MIT
3922.005	3923.116	25489.95	CE		0.3S			MIT
3921.731	3922.842	25491.73	CE	I	1.4L	0.0		MIT
3921.082	3922.192	25495.95	CE		0.6			MIT
3920.783	3921.893	25497.89	CE		0.5			MIT
3920.33	3921.44	25500.8	CE		0.3H			MIT
3919.813	3920.923	25504.20	CE		1.7	0.3		MIT
3918.946	3920.056	25509.84	CE		0.3			MIT
3918.276	3919.386	25514.20	CE		1.8	0.8		MIT
3917.644	3918.753	25518.32	CE		1.3	0.5		MIT
3917.252	3918.361	25520.87	CE		0.8			MIT
3916.895	3918.004	25523.20	CE		1.1	0.3H		MIT
3916.683	3917.792	25524.58	CE		0.8			MIT
3916.530	3917.639	25525.58	CE		0.3			MIT
3916.141	3917.250	25528.11	CE		1.3	0.6		MIT
3915.633	3916.742	25531.42	CE		0.5			MIT
3915.522	3916.631	25532.15	CE		1.0	0.5		MIT
3915.225	3916.334	25534.08	CE		0.5			MIT
3914.949	3916.058	25535.89	CE		1.3	0.3		MIT
3914.417	3915.526	25539.35	CE		0.3			MIT
3914.175	3915.284	25540.93	CE		0.8	0.0		MIT

AIR WAVELENGTH (ANGSTRMS)	VACUUM WAVELENGTH (ANGSTRMS)	WAVENUMBER (KAYSERS)	LMENT	LOG ARC	INTENSITY SPK	DIS	REF
3913.994	3915.103	25542.11	CE II	1.0	0.3		MIT
3913.357	3914.465	25546.27	CE	0.5			MIT
3913.145	3914.253	25547.66	CE	0.5			MIT
3912.436	3913.544	25552.29	CE I	1.7	0.7		MIT
3912.436	3913.544	25552.29	CE II	1.7	0.7		MIT
3912.189	3913.297	25553.90	CE II	1.4	0.5		MIT
3911.726	3912.834	25556.92	CE	0.5			MIT
3911.303	3912.411	25559.69	CE	1.2	0.0		MIT
3910.703	3911.811	25563.61	CE	1.1	0.3H		MIT
3909.934	3911.041	25568.64	CE	1.1	0.0		MIT
3909.752	3910.859	25569.83	CE	0.8	0.0		MIT
3909.313	3910.420	25572.70	CE II	1.5	0.5		MIT
3909.045	3910.152	25574.45	CE	0.8	0.0		MIT
3908.765	3909.872	25576.28	CE	1.1	0.3		MIT
3908.543	3909.650	25577.74	CE II	1.3	0.5		MIT
3908.543	3909.650	25577.74	CE I	1.3	0.5		MIT
3908.408	3909.515	25578.62	CE	1.5	0.8		MIT
3908.094	3909.201	25580.68	CE II	0.9	0.0		MIT
3907.964	3909.071	25581.53	CE	0.3			MIT
3907.639	3908.746	25583.65	CE	0.3	0.0		MIT
3907.445	3908.552	25584.92	CE	0.8	0.3		MIT
3907.289	3908.396	25585.95	CE	1.5	0.8		MIT
3906.924	3908.031	25588.33	CE	0.9	0.5		MIT
3906.452	3907.559	25591.43	CE	0.9	0.3		MIT
3906.104	3907.210	25593.71	CE	0.3			MIT
3905.920	3907.026	25594.91	CE	0.5			MIT
3905.85	3906.96	25595.4	CE	0.5			MIT
3905.298	3906.404	25598.99	CE	0.8			MIT
3904.922	3906.028	25601.45	CE	0.3	0.0		MIT
3904.582	3905.688	25603.68	CE II	0.8	0.0		MIT
3904.340	3905.446	25605.27	CE I	1.1	0.5		MIT
3904.340	3905.446	25605.27	CE II	1.1	0.5		MIT
3904.160	3905.266	25606.45	CE	0.6			MIT
3903.932	3905.038	25607.95	CE	1.0	0.5		MIT
3903.534	3904.640	25610.56	CE	0.5			MIT
3903.342	3904.448	25611.82	CE	1.2	0.5		MIT
3903.120	3904.226	25613.27	CE	0.3	0.3		MIT
3902.89	3904.00	25614.8	CE	0.9	0.0		MIT
3902.745	3903.851	25615.73	CE	0.3			MIT
3902.509	3903.615	25617.28	CE	0.5			MIT
3902.44	3903.55	25617.7	CE	0.3	0.0		MIT
3902.272	3903.377	25618.84	CE	0.3			MIT
3902.13	3903.24	25619.8	CE	0.3			MIT
3901.304	3902.409	25625.20	CE	1.0	0.3		MIT
3901.132	3902.237	25626.33	CE	0.5			MIT
3900.883	3901.988	25627.96	CE	0.5	0.0		MIT
3900.204	3901.309	25632.42	CE	1.0	0.3H		MIT
3899.859	3900.964	25634.69	CE	0.3	0.3H		MIT
3899.385	3900.490	25637.81	CE	0.8	0.3		MIT
3899.270	3900.375	25638.56	CE	0.3			MIT
3898.944	3900.049	25640.71	CE I	1.2	0.5		MIT
3898.600	3899.705	25642.97	CE II	0.6			MIT
3898.433	3899.537	25644.07	CE	0.3			MIT
3898.273	3899.377	25645.12	CE	1.9	0.8		MIT
3898.014	3899.118	25646.82	CE	0.7			MIT
3897.432	3898.536	25650.65	CE	0.8			MIT
3896.804	3897.908	25654.79	CE II	1.5	0.8		MIT
3896.804	3897.908	25654.79	CE I	1.5	0.8		MIT
3895.828	3896.932	25661.21	CE	0.3			MIT
3895.119	3896.223	25665.89	CE I	1.6	0.8		MIT
3895.119	3896.223	25665.89	CE II	1.6	0.8		MIT
3893.233	3894.336	25678.32	CE	1.3	0.0		MIT
3892.32	3893.42	25684.3	CE	0.3			MIT
3892.063	3893.166	25686.04	CE	0.5			MIT
3891.774	3892.877	25687.94	CE	0.8			MIT
3890.986	3892.089	25693.15	CE II	1.1	0.5		MIT
3890.761	3891.863	25694.63	CE	0.9	0.3		MIT
3890.527	3891.629	25696.18	CE II	0.6			MIT
3890.441	3891.543	25696.75	CE	0.3			MIT
3889.990	3891.092	25699.73	CE	1.7	0.9		MIT

AIR WAVELENGTH (ANGSTRMS)	VACUUM WAVELENGTH (ANGSTRMS)	WAVENUMBER (KAYSERS)	LMENT	LOG ARC	INTENSITY SPK	DIS	REF
3889.549	3890.651	25702.64	CE	0.5			MIT
3889.476	3890.578	25703.12	CE II	0.6			MIT
3889.304	3890.406	25704.26	CE	0.8	0.0		MIT
3888.997	3890.099	25706.29	CE	1.1	0.5		MIT
3888.388	3889.490	25710.31	CE II	1.2	0.6		MIT
3888.110	3889.212	25712.15	CE	0.3			MIT
3886.496	3887.597	25722.83	CE II	0.8L	0.5H		MIT
3885.769	3886.870	25727.64	CE	0.6	0.0		MIT
3885.236	3886.337	25731.17	CE	0.7			MIT
3884.992	3886.093	25732.79	CE	0.5			MIT
3884.765	3885.866	25734.29	CE	0.9			MIT
3884.745	3885.846	25734.42	CE	0.5	0.0		MIT
3884.562	3885.663	25735.64	CE	0.5			MIT
3884.203	3885.304	25738.01	CE	0.9	0.5		MIT
3883.978	3885.079	25739.50	CE II	0.8			MIT
3883.566	3884.667	25742.24	CE II	0.6S	0.0		MIT
3882.565	3883.665	25748.87	CE	0.3	0.3		MIT
3882.446	3883.546	25749.66	CE	0.7			MIT
3881.874	3882.974	25753.46	CE II	0.8	0.6		MIT
3881.665	3882.765	25754.84	CE II	0.7	0.5		MIT
3881.497	3882.597	25755.96	CE	0.3			MIT
3880.406	3881.506	25763.20	CE	0.3	0.5		MIT
3879.968	3881.068	25766.11	CE	0.6	0.5		MIT
3879.605	3880.705	25768.52	CE	0.6	0.5		MIT
3879.311	3880.410	25770.47	CE II	0.7	0.5		AB
3879.074	3880.173	25772.04	CE	0.7	0.0		MIT
3878.766	3879.865	25774.09	CE	0.5	0.3		MIT
3878.372	3879.471	25776.71	CE II	1.2	1.1		MIT
3878.372	3879.471	25776.71	CE I	1.2	1.1		MIT
3877.984	3879.083	25779.29	CE	0.9	0.9		MIT
3877.559	3878.658	25782.11	CE	0.5S			MIT
3877.122	3878.221	25785.02	CE	0.6			MIT
3876.974	3878.073	25786.00	CE II	1.2	0.5S		MIT
3876.370	3877.469	25790.02	CE	0.3			MIT
3876.140	3877.239	25791.55	CE	0.8	0.0		MIT
3875.902	3877.001	25793.14	CE	0.7	0.3S		MIT
3875.261	3876.359	25797.40	CE	0.6	0.3		MIT
3875.036	3876.134	25798.90	CE II	0.8			MIT
3874.677	3875.775	25801.29	CE	0.8	0.3		MIT
3874.333	3875.431	25803.58	CE	0.6			MIT
3874.134	3875.232	25804.91	CE	0.6	0.6		MIT
3873.255	3874.353	25810.76	CE	0.6			MIT
3873.032	3874.130	25812.25	CE	0.3			MIT
3872.729	3873.827	25814.27	CE	0.9	1.2		MIT
3872.510	3873.608	25815.73	CE	0.3H			MIT
3872.139	3873.237	25818.20	CE II	0.5	0.5		MIT
3871.809	3872.907	25820.40	CE	0.9			MIT
3871.643	3872.740	25821.51	CE		0.5		MIT
3871.400	3872.497	25823.13	CE	0.9	1.2		MIT
3870.866	3871.963	25826.69	CE	0.8	0.3		MIT
3869.566	3870.663	25835.37	CE	0.5	0.0		MIT
3869.323	3870.420	25836.99	CE	0.3	0.5		MIT
3868.644	3869.741	25841.52	CE	0.5	0.3		MIT
3868.502	3869.599	25842.47	CE II	0.5			MIT
3868.138	3869.235	25844.91	CE II	0.8	0.3		MIT
3868.058	3869.155	25845.44	CE	0.3			MIT
3867.604	3868.700	25848.47	CE	0.7	0.8		MIT
3867.110	3868.206	25851.77	CE	0.3			MIT
3866.819	3867.915	25853.72	CE	1.1	0.8		MIT
3865.392	3866.488	25863.26	CE	0.7	0.3		MIT
3865.105	3866.201	25865.19	CE	0.3			MIT
3864.869	3865.965	25866.76	CE	0.3			MIT
3864.464	3865.560	25869.48	CE	0.3	0.5		MIT
3863.746	3864.841	25874.28	CE II	0.5			MIT
3863.595	3864.690	25875.29	CE	0.3	0.0		MIT
3862.946	3864.041	25879.64	CE	0.6	0.3		MIT
3862.791	3863.886	25880.68	CE	0.6	0.3		MIT
3862.465	3863.560	25882.86	CE	1.2	1.0		MIT
3862.134	3863.229	25885.08	CE	0.3			MIT
3861.928	3863.023	25886.46	CE	0.5	0.0		MIT

AIR WAVELENGTH (ANGSTRMS)	VACUUM WAVELENGTH (ANGSTRMS)	WAVENUMBER (KAYSERS)	LMENT	LOG ARC	INTENSITY SPK	INTENSITY DIS	REF
3861.580	3862.675	25888.80	CE	0.5H	0.6H		MIT
3860.993	3862.088	25892.73	CE	0.3	0.3		MIT
3860.626	3861.721	25895.19	CE	0.9	1.2		MIT
3860.401	3861.496	25896.70	CE	0.8	0.5		MIT
3860.178	3861.273	25898.20	CE	0.3	0.6		MIT
3859.942	3861.036	25899.78	CE	0.5	0.7		MIT
3859.792	3860.886	25900.79	CE	0.5			MIT
3857.944	3859.038	25913.19	CE II	0.5	0.3		MIT
3857.821	3858.915	25914.02	CE II	0.5	0.0		MIT
3857.644	3858.738	25915.21	CE II	0.9	0.6L		MIT
3857.240	3858.334	25917.92	CE II	0.7	0.5		MIT
3857.018	3858.112	25919.42	CE	1.0	0.3		MIT
3856.293	3857.386	25924.29	CE	0.3H			MIT
3855.872	3856.965	25927.12	CE	0.5	0.3		MIT
3855.703	3856.796	25928.25	CE	0.3	0.3		MIT
3855.302	3856.395	25930.95	CE	0.9	0.5		MIT
3855.192	3856.285	25931.69	CE	0.3			MIT
3854.965	3856.058	25933.22	CE	0.5	0.5		MIT
3854.322	3855.415	25937.54	CE II	0.8	0.0		MIT
3854.187	3855.280	25938.45	CE II	0.8	0.0		MIT
3854.187	3855.280	25938.45	CE I	0.8	0.0		MIT
3853.164	3854.257	25945.34	CE II	1.4	0.5		MIT
3853.164	3854.257	25945.34	CE I	1.4	0.5		MIT
3852.930	3854.023	25946.92	CE	0.3	0.3		MIT
3852.387	3853.479	25950.57	CE	0.9	1.4		MIT
3852.106	3853.198	25952.47	CE II	0.8	0.5		MIT
3851.352	3852.444	25957.55	CE	0.5	0.6		MIT
3850.779	3851.871	25961.41	CE	0.3			MIT
3850.713	3851.805	25961.85	CE	0.3H			MIT
3850.122	3851.214	25965.84	CE	0.5			MIT
3849.681	3850.773	25968.81	CE	0.3			MIT
3849.569	3850.661	25969.57	CE II	0.5			MIT
3849.068	3850.160	25972.95	CE	0.6L	0.6L		MIT
3848.597	3849.688	25976.13	CE	1.3	0.3		MIT
3848.105	3849.196	25979.45	CE II	0.9	0.0		MIT
3847.812	3848.903	25981.43	CE	0.8L	0.8L		MIT
3847.120	3848.211	25986.10	CE	0.3			MIT
3846.989	3848.080	25986.99	CE	0.3	0.5		MIT
3846.803	3847.894	25988.24	CE	0.6	0.0		MIT
3846.520	3847.611	25990.15	CE	0.6	0.0		MIT
3846.140	3847.231	25992.72	CE	0.3			MIT
3845.934	3847.025	25994.11	CE	0.3	0.0		MIT
3845.483	3846.574	25997.16	CE	0.7S	0.7S		MIT
3845.275	3846.366	25998.57	CE	0.5	0.3		MIT
3844.844	3845.935	26001.48	CE	0.6			MIT
3844.248	3845.338	26005.51	CE	0.7S	0.6S		MIT
3843.768	3844.858	26008.76	CE	1.1	0.0H		MIT
3843.469	3844.559	26010.78	CE	0.5	0.7		MIT
3842.988	3844.078	26014.04	CE	1.3	1.4		MIT
3842.051	3843.141	26020.38	CE	0.5			MIT
3841.716	3842.806	26022.65	CE	0.7L	0.8L		MIT
3841.603	3842.693	26023.42	CE	0.3			MIT
3841.435	3842.525	26024.56	CE	0.5H	0.3		MIT
3841.037	3842.127	26027.25	CE	0.5	0.3		MIT
3840.747	3841.836	26029.22	CE	0.3			MIT
3840.453	3841.542	26031.21	CE	1.5	1.5		MIT
3839.748	3840.837	26035.99	CE	0.3			MIT
3839.497	3840.586	26037.69	CE	0.8	0.0		MIT
3839.155	3840.244	26040.01	CE	0.6	0.5L		MIT
3838.542	3839.631	26044.17	CE	1.5	0.5		MIT
3837.534	3838.623	26051.01	CE	0.3			MIT
3837.212	3838.301	26053.20	CE II	0.9			MIT
3836.112	3837.200	26060.67	CE II	1.2	0.8		MIT
3835.904	3836.992	26062.08	CE	0.5			MIT
3835.748	3836.836	26063.14	CE	0.6	0.3		MIT
3834.782	3835.870	26069.71	CE II	0.8			MIT
3834.556	3835.644	26071.24	CE II	1.0	0.6L		MIT
3834.220	3835.308	26073.53	CE	0.8	1.1		MIT
3833.770	3834.858	26076.59	CE	0.5			MIT
3833.629	3834.717	26077.55	CE	0.6	0.6		MIT

AIR WAVELENGTH (ANGSTRMS)	VACUUM WAVELENGTH (ANGSTRMS)	WAVENUMBER (KAYSERS)	LMENT	LOG ARC	INTENSITY SPK	INTENSITY DIS	REF
3832.884	3833.971	26082.61	CE	0.5	0.3		MIT
3832.753	3833.840	26083.51	CE II	0.7			MIT
3832.659	3833.746	26084.15	CE	0.3	0.6		MIT
3832.336	3833.423	26086.34	CE	0.3			MIT
3832.231	3833.318	26087.06	CE	0.7			MIT
3832.055	3833.142	26088.26	CE	0.5	0.3		MIT
3831.928	3833.015	26089.12	CE	0.5	0.3		MIT
3831.785	3832.872	26090.10	CE	0.5	0.6H		MIT
3831.672	3832.759	26090.86	CE	0.5	0.0		MIT
3831.550	3832.637	26091.70	CE	0.6			MIT
3831.083	3832.170	26094.88	CE	1.0	0.3		MIT
3830.910	3831.997	26096.05	CE	0.5			MIT
3830.555	3831.642	26098.47	CE	0.6			MIT
3830.030	3831.117	26102.05	CE	0.6	0.3		MIT
3829.942	3831.029	26102.65	CE II	0.5			MIT
3829.816	3830.903	26103.51	CE	0.3			MIT
3829.694	3830.781	26104.34	CE II	0.6	0.3		MIT
3829.408	3830.495	26106.29	CE	0.5	0.3		MIT
3829.283	3830.369	26107.14	CE	0.3			MIT
3828.599	3829.685	26111.81	CE	0.6			MIT
3828.389	3829.475	26113.24	CE	0.5	0.3		MIT
3828.206	3829.292	26114.49	CE				MIT
3827.985	3829.071	26115.99	CE	0.5			MIT
3827.855	3828.941	26116.88	CE	0.9	0.0		MIT
3827.605	3828.691	26118.59	CE	0.5	0.5		MIT
3827.375	3828.461	26120.16	CE	0.9	1.0		MIT
3827.214	3828.300	26121.25	CE II	0.8	0.0		MIT
3826.708	3827.794	26124.71	CE	0.6L	0.8L		MIT
3826.530	3827.616	26125.92	CE	0.6			MIT
3825.855	3826.941	26130.53	CE	0.5			MIT
3824.860	3825.945	26137.33	CE	0.7			MIT
3824.419	3825.504	26140.34	CE	0.5	0.5		MIT
3824.116	3825.201	26142.42	CE	0.3	0.0		MIT
3823.903	3824.988	26143.87	CE II	0.9			MIT
3823.903	3824.988	26143.87	CE I	0.9			MIT
3823.695	3824.780	26145.29	CE	0.7			MIT
3823.486	3824.571	26146.72	CE	0.3	0.0		MIT
3823.087	3824.172	26149.45	CE	0.6S	0.6S		MIT
3822.874	3823.959	26150.91	CE	0.3	0.0		MIT
3822.170	3823.255	26155.73	CE	0.3			MIT
3821.700	3822.784	26158.94	CE II	0.8	1.0		MIT
3821.574	3822.658	26159.80	CE	0.3			MIT
3821.270	3822.354	26161.89	CE II	0.9	0.0		MIT
3820.871	3821.955	26164.62	CE II	0.7	0.5		MIT
3820.545	3821.629	26166.85	CE	0.5	0.3		MIT
3820.422	3821.506	26167.69	CE	0.5	0.6		MIT
3819.999	3821.083	26170.59	CE	0.6	0.5		MIT
3819.534	3820.618	26173.78	CE	0.3	0.5		MIT
3819.204	3820.288	26176.04	CE	0.5	0.5		MIT
3819.024	3820.108	26177.27	CE II	1.3	0.8		MIT
3818.810	3819.894	26178.74	CE	0.3			MIT
3818.688	3819.772	26179.57	CE II	0.7	0.3		MIT
3818.240	3819.324	26182.65	CE	0.5	0.7		MIT
3817.847	3818.930	26185.34	CE	0.3	0.5		MIT
3817.455	3818.538	26188.03	CE	1.4	0.3		MIT
3816.793	3817.876	26192.57	CE	0.5			MIT
3816.310	3817.393	26195.89	CE	0.5	0.5		MIT
3816.134	3817.217	26197.10	CE	0.7	0.8		MIT
3815.831	3816.914	26199.18	CE	1.7	0.7		MIT
3815.371	3816.454	26202.33	CE	0.3			MIT
3815.010	3816.093	26204.81	CE	0.5			MIT
3814.932	3816.015	26205.35	CE II	0.3	0.5		MIT
3814.622	3815.705	26207.48	CE	0.3	0.3		MIT
3814.517	3815.600	26208.20	CE	0.6	0.5		MIT
3813.575	3814.657	26214.67	CE	0.7			MIT
3813.059	3814.141	26218.22	CE	0.3			MIT
3812.958	3814.040	26218.92	CE	0.6	0.6		MIT
3812.589	3813.671	26221.45	CE	0.5			MIT
3812.206	3813.288	26224.09	CE	0.9	0.3		MIT
3811.610	3812.692	26228.19	CE	0.7			MIT

AIR WAVELENGTH (ANGSTRMS)	VACUUM WAVELENGTH (ANGSTRMS)	WAVENUMBER (KAYSERS)	LMENT		LOG ARC	INTENSITY SPK	DIS	REF
3811.492	3812.574	26229.00	CE		0.3			MIT
3811.324	3812.406	26230.16	CE		0.6	0.8		MIT
3810.902	3811.984	26233.06	CE		0.7			MIT
3810.245	3811.327	26237.58	CE		0.5	0.0		MIT
3810.099	3811.180	26238.59	CE		0.8			MIT
3809.497	3810.578	26242.74	CE		0.6	0.3		MIT
3809.224	3810.305	26244.62	CE		1.4	0.0H		MIT
3808.659	3809.740	26248.51	CE		0.5	0.5		MIT
3808.392	3809.473	26250.35	CE	II	0.5	0.3H		MIT
3808.121	3809.202	26252.22	CE	I	1.5	1.5		MIT
3808.121	3809.202	26252.22	CE	II	1.5	1.5		MIT
3807.691	3808.772	26255.18	CE		0.7	0.3		MIT
3807.267	3808.348	26258.11	CE		0.5			MIT
3806.710	3807.791	26261.95	CE		0.3			MIT
3806.636	3807.717	26262.46	CE		0.5			MIT
3806.394	3807.475	26264.13	CE		1.1	1.1		MIT
3805.829	3806.909	26268.03	CE		0.3	0.3		MIT
3805.537	3806.617	26270.04	CE		0.5	0.5		MIT
3805.339	3806.419	26271.41	CE		0.3			MIT
3804.959	3806.039	26274.03	CE		0.3			MIT
3804.697	3805.777	26275.84	CE		0.5	0.5		MIT
3804.156	3805.236	26279.58	CE		0.6	0.3		MIT
3803.841	3804.921	26281.76	CE		0.8			MIT
3803.097	3804.177	26286.90	CE	II	1.5	0.7		MIT
3802.802	3803.882	26288.94	CE		0.5	0.6		MIT
3802.592	3803.672	26290.39	CE		0.5	0.3		MIT
3802.340	3803.419	26292.13	CE		0.5			MIT
3801.924	3803.003	26295.01	CE		0.6			MIT
3801.529	3802.608	26297.74	CE		1.4	0.5		MIT
3800.324	3801.403	26306.08	CE	II	1.2	0.5		MIT
3799.903	3800.982	26308.99	CE		0.5	0.5		MIT
3799.713	3800.792	26310.31	CE		0.6	0.3		MIT
3799.547	3800.626	26311.46	CE		0.5	0.6		MIT
3799.097	3800.176	26314.57	CE		0.5			MIT
3799.038	3800.117	26314.98	CE	II	0.6			MIT
3798.624	3799.703	26317.85	CE		0.5	0.5		MIT
3798.51	3799.59	26318.6	CE		0.5	0.5		MIT
3798.238	3799.316	26320.52	CE		0.3			MIT
3798.08	3799.16	26321.6	CE		0.3	0.3		MIT
3797.322	3798.400	26326.87	CE		0.3			MIT
3796.674	3797.752	26331.37	CE		0.8			MIT
3796.432	3797.510	26333.04	CE		0.3			MIT
3795.570	3796.648	26339.02	CE		0.6			MIT
3795.256	3796.334	26341.20	CE		0.7	0.5		MIT
3795.013	3796.091	26342.89	CE		0.8	0.3		MIT
3794.680	3795.757	26345.20	CE		0.7			MIT
3794.208	3795.285	26348.48	CE		0.6			MIT
3793.857	3794.934	26350.92	CE		0.8			MIT
3793.522	3794.599	26353.25	CE		0.7			MIT
3793.425	3794.502	26353.92	CE		0.5	0.0		MIT
3792.809	3793.886	26358.20	CE		0.6	0.5		MIT
3792.745	3793.822	26358.64	CE		0.3			MIT
3792.326	3793.403	26361.55	CE	II	1.3	0.3H		MIT
3791.964	3793.041	26364.07	CE		0.6			MIT
3791.689	3792.766	26365.98	CE		1.2			MIT
3791.223	3792.300	26369.23	CE		0.5	0.3		MIT
3790.999	3792.076	26370.78	CE		0.3			MIT
3790.878	3791.954	26371.62	CE		0.6	0.5		MIT
3790.808	3791.884	26372.11	CE		0.8	0.6		MIT
3790.335	3791.411	26375.40	CE	II	0.6			MIT
3790.086	3791.162	26377.14	CE		0.3H			MIT
3789.837	3790.913	26378.87	CE		0.5	0.5		MIT
3789.473	3790.549	26381.40	CE		0.7			MIT
3788.753	3789.829	26386.42	CE		1.2	0.5H		MIT
3788.430	3789.506	26388.67	CE		0.5	0.5		MIT
3788.206	3789.282	26390.23	CE		0.7	0.3		MIT
3787.906	3788.982	26392.31	CE		1.2	0.3		MIT
3787.572	3788.648	26394.64	CE		0.6			MIT
3787.462	3788.538	26395.41	CE		0.5	0.0		MIT
3787.159	3788.235	26397.52	CE		0.5	0.7		MIT

AIR WAVELENGTH (ANGSTRMS)	VACUUM WAVELENGTH (ANGSTRMS)	WAVENUMBER (KAYSERS)	LMENT		LOG ARC	INTENSITY SPK	DIS	REF
3787.011	3788.086	26398.55	CE		0.5			MIT
3786.632	3787.707	26401.20	CE	II	1.0S	0.5		MIT
3786.532	3787.607	26401.89	CE		0.5			MIT
3785.564	3786.639	26408.64	CE		0.3	0.3		MIT
3785.249	3786.324	26410.84	CE		0.3	0.3		MIT
3785.021	3786.096	26412.43	CE		1.1	0.3		MIT
3784.352	3785.427	26417.10	CE		0.6	0.5		MIT
3783.815	3784.890	26420.85	CE		0.3	0.6		MIT
3783.579	3784.654	26422.50	CE	II	0.9	0.5		MIT
3783.429	3784.504	26423.54	CE		0.5	0.6		MIT
3783.035	3784.109	26426.30	CE		1.1	0.5		MIT
3782.525	3783.599	26429.86	CE	I	1.3	0.3		MIT
3782.525	3783.599	26429.86	CE	II	1.3	0.3		MIT
3782.034	3783.108	26433.29	CE		0.3			MIT
3781.981	3783.055	26433.66	CE		0.3			MIT
3781.620	3782.694	26436.19	CE		1.4	0.6		MIT
3781.102	3782.176	26439.81	CE	II	0.7			MIT
3780.561	3781.635	26443.59	CE		0.3			MIT
3779.869	3780.943	26448.43	CE		0.5			MIT
3779.611	3780.685	26450.24	CE		0.8			MIT
3778.977	3780.050	26454.67	CE		0.3	0.3		MIT
3777.672	3778.745	26463.81	CE	II	1.1	0.5		MIT
3777.091	3778.164	26467.88	CE		0.8	1.2		MIT
3776.711	3777.784	26470.55	CE	II	0.3			MIT
3776.608	3777.681	26471.27	CE	II	0.8	0.5		MIT
3776.146	3777.219	26474.51	CE		0.7	0.7		MIT
3775.986	3777.059	26475.63	CE		1.1	1.2		MIT
3775.564	3776.637	26478.59	CE		0.6			MIT
3775.363	3776.435	26480.00	CE		0.3	0.6		MIT
3774.399	3775.471	26486.76	CE		0.3			MIT
3773.895	3774.967	26490.30	CE		0.3			MIT
3773.437	3774.509	26493.51	CE		0.7	0.5		MIT
3773.214	3774.286	26495.08	CE		0.9	0.3		MIT
3772.863	3773.935	26497.54	CE		0.7	0.5		MIT
3772.650	3773.722	26499.04	CE	II	0.9	0.0		MIT
3772.450	3773.522	26500.44	CE		0.6	0.0		MIT
3772.029	3773.101	26503.40	CE		0.5			MIT
3771.606	3772.677	26506.37	CE	II	1.2	0.3		MIT
3771.381	3772.452	26507.96	CE		0.5	0.3		MIT
3770.766	3771.837	26512.28	CE		1.3	0.5		MIT
3770.388	3771.459	26514.94	CE		0.5			MIT
3769.939	3771.010	26518.09	CE		0.9	0.5		MIT
3769.408	3770.479	26521.83	CE		0.5			MIT
3769.356	3770.427	26522.20	CE		0.3			MIT
3769.250	3770.321	26522.94	CE		0.5			MIT
3769.041	3770.112	26524.41	CE	II	0.8	0.0		MIT
3768.762	3769.833	26526.38	CE		0.9	0.3		MIT
3768.658	3769.729	26527.11	CE		0.3	0.0		MIT
3768.305	3769.376	26529.59	CE		0.6	0.9		MIT
3767.996	3769.067	26531.77	CE		0.5			MIT
3767.903	3768.974	26532.42	CE		0.5			MIT
3767.566	3768.636	26534.80	CE		0.3	0.6		MIT
3767.193	3768.263	26537.42	CE		0.3			MIT
3766.846	3767.916	26539.87	CE		0.5	0.0		MIT
3766.696	3767.766	26540.93	CE		0.6			MIT
3766.509	3767.579	26542.24	CE	II	1.1	0.5		MIT
3766.293	3767.363	26543.76	CE		0.3	0.5		MIT
3766.064	3767.134	26545.38	CE		0.8			MIT
3765.889	3766.959	26546.61	CE		0.7	0.0		MIT
3765.727	3766.797	26547.75	CE		0.3			MIT
3765.527	3766.597	26549.16	CE		0.3			MIT
3765.044	3766.114	26552.57	CE	II	1.1	0.5		MIT
3764.608	3765.678	26555.65	CE		0.6	0.5		MIT
3764.117	3765.187	26559.11	CE	II	1.3	0.7		MIT
3763.724	3764.793	26561.88	CE		0.5			MIT
3763.605	3764.674	26562.72	CE	II	0.8	0.3		MIT
3762.975	3764.044	26567.17	CE		1.2	0.6		MIT
3762.668	3763.737	26569.34	CE		0.3	0.5		MIT
3762.280	3763.349	26572.08	CE		0.6	0.7		MIT
3762.220	3763.289	26572.50	CE		0.5			MIT

AIR WAVELENGTH (ANGSTRMS)	VACUUM WAVELENGTH (ANGSTRMS)	WAVENUMBER (KAYSERS)	LMENT		LOG ARC	INTENSITY SPK	INTENSITY DIS	REF
3761.953	3763.022	26574.39	CE		0.3			MIT
3761.450	3762.519	26577.94	CE		0.3			MIT
3761.185	3762.254	26579.81	CE		0.5			MIT
3760.824	3761.893	26582.36	CE		0.3			MIT
3760.694	3761.763	26583.28	CE		0.8			MIT
3760.399	3761.468	26585.37	CE	II	0.7	0.5		MIT
3760.177	3761.246	26586.94	CE		0.3			MIT
3759.754	3760.822	26589.93	CE		1.0	0.3		MIT
3759.132	3760.200	26594.33	CE		0.6	0.6		MIT
3759.078	3760.146	26594.71	CE		0.6			MIT
3758.699	3759.767	26597.39	CE		0.5			MIT
3758.517	3759.585	26598.68	CE		0.9	0.6		MIT
3758.221	3759.289	26600.77	CE		0.7			MIT
3758.049	3759.117	26601.99	CE		0.5			MIT
3757.862	3758.930	26603.32	CE	II	1.2	0.3		MIT
3757.501	3758.569	26605.87	CE		0.5	0.3		MIT
3757.222	3758.290	26607.85	CE		1.3	0.5		MIT
3756.255	3757.323	26614.70	CE		0.8	0.7		MIT
3756.114	3757.181	26615.70	CE		0.3			MIT
3755.786	3756.853	26618.02	CE		0.5			MIT
3755.722	3756.789	26618.47	CE		0.9	0.3		MIT
3755.425	3756.492	26620.58	CE	II	1.3	0.5		MIT
3754.824	3755.891	26624.84	CE		0.6			MIT
3754.770	3755.837	26625.22	CE		0.3			MIT
3754.479	3755.546	26627.29	CE		0.9	0.0		MIT
3753.958	3755.025	26630.98	CE		0.3	0.3		MIT
3753.767	3754.834	26632.34	CE		0.8	0.0		MIT
3753.059	3754.126	26637.36	CE		0.8	0.3		MIT
3752.714	3753.781	26639.81	CE		0.3			MIT
3752.453	3753.520	26641.66	CE		0.5			MIT
3752.336	3753.402	26642.49	CE		0.7	0.3		MIT
3752.190	3753.256	26643.53	CE		0.5	0.5		MIT
3751.757	3752.823	26646.60	CE		0.8	0.6		MIT
3751.449	3752.515	26648.79	CE		1.2S	0.8		MIT
3751.002	3752.068	26651.97	CE	II	1.2	0.7		MIT
3750.731	3751.797	26653.89	CE		0.3	0.3		MIT
3750.078	3751.144	26658.53	CE		1.1	0.5		MIT
3749.493	3750.559	26662.69	CE		0.3	0.3H		MIT
3749.365	3750.431	26663.60	CE		0.7			MIT
3748.527	3749.593	26669.56	CE		0.3			MIT
3748.056	3749.121	26672.92	CE		1.0	0.5H		MIT
3747.546	3748.611	26676.55	CE		0.8			MIT
3747.087	3748.152	26679.81	CE		0.3	0.0		MIT
3746.568	3747.633	26683.51	CE		0.3	0.0		MIT
3746.374	3747.439	26684.89	CE	II	0.9	0.3		MIT
3746.246	3747.311	26685.80	CE	II	0.5			MIT
3744.723	3745.788	26696.66	CE		0.5	0.3		MIT
3744.075	3745.139	26701.28	CE	II	0.5			MIT
3743.986	3745.050	26701.91	CE		0.8H	0.3H		MIT
3742.309	3743.373	26713.88	CE		0.3	0.5		MIT
3742.223	3743.287	26714.49	CE		0.8	0.5		MIT
3741.727	3742.791	26718.03	CE	II	1.0	0.0		MIT
3741.399	3742.463	26720.37	CE		0.9			MIT
3741.006	3742.070	26723.18	CE		0.9			MIT
3740.574	3741.637	26726.27	CE		0.3			MIT
3740.427	3741.490	26727.32	CE		0.3			MIT
3740.133	3741.196	26729.42	CE		0.8			MIT
3740.036	3741.099	26730.11	CE		0.5			MIT
3739.947	3741.010	26730.75	CE		0.9			MIT
3739.690	3740.753	26732.58	CE	II	0.8	0.3		MIT
3739.028	3740.091	26737.32	CE		0.5			MIT
3738.253	3739.316	26742.86	CE		0.7			MIT
3737.958	3739.021	26744.97	CE		0.6			MIT
3737.736	3738.799	26746.56	CE		1.0	0.3		MIT
3737.516	3738.579	26748.13	CE	II	0.7			MIT
3737.020	3738.083	26751.68	CE		0.5			MIT
3736.913	3737.975	26752.45	CE		0.3	0.0		MIT
3736.469	3737.531	26755.63	CE			0.5		MIT
3736.402	3737.464	26756.11	CE		0.5	0.5		MIT
3736.060	3737.122	26758.56	CE		0.5	0.3		MIT
3735.912	3736.974	26759.62	CE		0.3			MIT
3735.773	3736.835	26760.61	CE		0.3			MIT
3735.599	3736.661	26761.86	CE		0.5	0.6		MIT
3734.856	3735.918	26767.18	CE		0.3	0.3		MIT
3734.211	3735.273	26771.81	CE		0.3			MIT
3734.059	3735.121	26772.90	CE		0.3			MIT
3733.773	3734.835	26774.95	CE		0.3			MIT
3733.517	3734.579	26776.78	CE		0.3			MIT
3733.108	3734.170	26779.72	CE		0.3			MIT
3732.582	3733.643	26783.49	CE		0.3	0.3		MIT
3732.461	3733.522	26784.36	CE		0.3			MIT
3732.186	3733.247	26786.33	CE		0.3			MIT
3731.876	3732.937	26788.56	CE		0.6	0.0		MIT
3731.679	3732.740	26789.97	CE		0.3			MIT
3731.258	3732.319	26792.99	CE		0.3	0.0		MIT
3731.17	3732.23	26793.6	CE		0.3	0.0		MIT
3730.679	3731.740	26797.15	CE		0.3	0.3		MIT
3730.334	3731.395	26799.63	CE		1.0	0.5		MIT
3729.922	3730.983	26802.59	CE	II	0.9	0.0		MIT
3729.001	3730.061	26809.21	CE		0.7			MIT
3728.423	3729.483	26813.37	CE		1.7	1.0		MIT
3728.182	3729.242	26815.10	CE		0.7	0.0		MIT
3728.023	3729.083	26816.24	CE		0.9	0.7		MIT
3727.633	3728.693	26819.05	CE		0.5			MIT
3727.331	3728.391	26821.22	CE		0.5	0.7J		MIT
3726.955	3728.015	26823.93	CE		1.2	0.7		MIT
3726.456	3727.516	26827.52	CE	II	0.8			MIT
3726.019	3727.079	26830.67	CE		0.7			MIT
3725.675	3726.735	26833.14	CE		1.6	1.0		MIT
3725.195	3726.254	26836.60	CE		0.3			MIT
3724.636	3725.695	26840.63	CE	II	1.1	0.3		MIT
3724.221	3725.280	26843.62	CE		0.3			MIT
3723.836	3724.895	26846.39	CE		0.0	0.3		MIT
3723.656	3724.715	26847.69	CE		1.1			MIT
3722.759	3723.818	26854.16	CE		1.1	0.3		MIT
3722.554	3723.613	26855.64	CE		0.3H			MIT
3722.291	3723.350	26857.54	CE	II	1.2	0.0		MIT
3722.098	3723.157	26858.93	CE		1.0	0.0		MIT
3721.949	3723.008	26860.00	CE		0.3			MIT
3721.812	3722.871	26860.99	CE		0.7			MIT
3721.648	3722.707	26862.18	CE		0.6	0.0		MIT
3720.595	3721.653	26869.78	CE		0.3			MIT
3720.380	3721.438	26871.33	CE		0.3	0.3		MIT
3719.946	3721.004	26874.47	CE		0.3W			MIT
3719.797	3720.855	26875.54	CE	II	1.2S	0.7		MIT
3719.430	3720.488	26878.20	CE		0.9			MIT
3719.081	3720.139	26880.72	CE	II	0.5			MIT
3718.380	3719.438	26885.78	CE		1.2	0.7		MIT
3718.190	3719.248	26887.16	CE	II	1.2	0.7		MIT
3718.190	3719.248	26887.16	CE	I	1.2	0.7		MIT
3717.767	3718.825	26890.22	CE		0.5			MIT
3717.484	3718.541	26892.26	CE		0.9	0.0		MIT
3716.930	3717.987	26896.27	CE	II	1.0	0.3		MIT
3716.714	3717.771	26897.84	CE		0.3			MIT
3716.365	3717.422	26900.36	CE	I	1.2	1.0		MIT
3716.365	3717.422	26900.36	CE	II	1.2	1.0		MIT
3715.912	3716.969	26903.64	CE		0.6			MIT
3715.466	3716.523	26906.87	CE		1.0	0.3		MIT
3715.143	3716.200	26909.21	CE		0.9	0.0		MIT
3714.772	3715.829	26911.90	CE		1.0	0.3		MIT
3714.526	3715.583	26913.68	CE		0.5			MIT
3713.988	3715.045	26917.58	CE		0.9			MIT
3713.662	3714.718	26919.94	CE	II	0.7			MIT
3713.456	3714.512	26921.43	CE		0.7W	0.3W		MIT
3713.039	3714.095	26924.46	CE		0.5			MIT
3712.724	3713.780	26926.74	CE		0.3	0.0		MIT
3712.098	3713.154	26931.28	CE		0.3L			MIT
3711.990	3713.046	26932.07	CE		0.3			MIT
3711.687	3712.743	26934.26	CE	II	0.5			MIT
3711.628	3712.684	26934.69	CE		0.3W	0.3		MIT

AIR WAVELENGTH (ANGSTRMS)	VACUUM WAVELENGTH (ANGSTRMS)	WAVENUMBER (KAYSERS)	LMENT	LOG ARC	INTENSITY SPK	DIS	REF
3711.001	3712.057	26939.24	CE	0.3			MIT
3710.253	3711.309	26944.67	CE	0.3	0.0		MIT
3709.933	3710.988	26947.00	CE	1.4	1.0		MIT
3709.588	3710.643	26949.50	CE	0.7			MIT
3709.286	3710.341	26951.70	CE	1.4			M
3708.709	3709.764	26955.89	CE	0.3			MIT
3707.886	3708.941	26961.87	CE	0.5			M
3707.681	3708.736	26963.36	CE	0.5			MIT
3707.602	3708.657	26963.94	CE	0.5			MIT
3707.387	3708.442	26965.50	CE	1.0			MIT
3706.936	3707.991	26968.78	CE	0.6	0.3		MIT
3706.863	3707.918	26969.31	CE	0.3			KN
3705.568	3706.622	26978.74	CE	0.5			MIT
3705.365	3706.419	26980.22	CE	0.3			MIT
3705.048	3706.102	26982.53	CE	0.7	0.3		MIT
3704.977	3706.031	26983.04	CE II	0.9	0.7		MIT
3704.674	3705.728	26985.25	CE	0.8			MIT
3703.912	3704.966	26990.80	CE	0.5			MIT
3703.366	3704.420	26994.78	CE	0.3			MIT
3702.790	3703.844	26998.98	CE II	1.0	0.7W		MIT
3702.588	3703.642	27000.45	CE	0.6			MIT
3702.246	3703.300	27002.95	CE	0.3			KN
3702.192	3703.245	27003.34	CE	0.5			MIT
3701.866	3702.919	27005.72	CE	0.6W			MIT
3701.724	3702.777	27006.75	CE II	0.6			MIT
3701.197	3702.250	27010.60	CE	0.5			MIT
3701.015	3702.068	27011.93	CE	0.5W	0.0		MIT
3700.742	3701.795	27013.92	CE	0.5			MIT
3699.920	3700.973	27019.92	CE I	1.3S	0.3		MIT
3699.920	3700.973	27019.92	CE II	1.3S	0.3		MIT
3699.305	3700.358	27024.41	CE	0.6			MIT
3699.176	3700.229	27025.36	CE	0.9			MIT
3699.135	3700.188	27025.66	CE	0.8	0.0		MIT
3698.658	3699.711	27029.14	CE II	1.3	0.0		MIT
3698.364	3699.417	27031.29	CE	1.0	0.3		MIT
3698.125	3699.177	27033.04	CE	1.1	0.0		MIT
3697.660	3698.712	27036.44	CE	1.0	0.0		MIT
3697.293	3698.345	27039.12	CE	0.3			MIT
3696.672	3697.724	27043.66	CE II	0.5			MIT
3696.485	3697.537	27045.03	CE	0.5			MIT
3695.963	3697.015	27048.85	CE	0.8	0.0		MIT
3695.856	3696.908	27049.63	CE	0.3			MIT
3695.419	3696.471	27052.83	CE	0.3			MIT
3695.237	3696.289	27054.16	CE	0.3			MIT
3694.911	3695.963	27056.55	CE II	1.2S	0.3		MIT
3694.911	3695.963	27056.55	CE I	1.2S	0.3		MIT
3694.173	3695.224	27061.96	CE	0.6W			MIT
3693.708	3694.759	27065.36	CE I	0.9			MIT
3693.708	3694.759	27065.36	CE II	0.9			MIT
3693.422	3694.473	27067.46	CE	1.0	0.0		MIT
3692.552	3693.603	27073.84	CE	0.3			MIT
3692.222	3693.273	27076.25	CE	0.5			MIT
3690.935	3691.986	27085.70	CE	0.5W			MIT
3690.122	3691.172	27091.66	CE	0.9			MIT
3689.685	3690.735	27094.87	CE	0.9			MIT
3689.494	3690.544	27096.27	CE	0.7			MIT
3689.161	3690.211	27098.72	CE I	0.9			MIT
3689.161	3690.211	27098.72	CE II	0.9			MIT
3688.659	3689.709	27102.41	CE	1.0	0.0		MIT
3688.420	3689.470	27104.17	CE	0.3			MIT
3687.802	3688.852	27108.71	CE II	1.0	0.3		MIT
3687.802	3688.852	27108.71	CE I	1.0	0.3		MIT
3687.455	3688.505	27111.26	CE	0.3H			MIT
3686.970	3688.020	27114.82	CE	0.3			MIT
3686.596	3687.645	27117.57	CE	0.5			MIT
3686.263	3687.312	27120.02	CE	0.7			MIT
3686.036	3687.085	27121.69	CE	0.9			MIT
3685.522	3686.571	27125.48	CE II	0.8			MIT
3685.398	3686.447	27126.39	CE	0.3			MIT
3685.234	3686.283	27127.60	CE	0.5	0.0		MIT

AIR WAVELENGTH (ANGSTRMS)	VACUUM WAVELENGTH (ANGSTRMS)	WAVENUMBER (KAYSERS)	LMENT	LOG ARC	INTENSITY SPK	DIS	REF
3685.005	3686.054	27129.28	CE	0.3			MIT
3684.740	3685.789	27131.23	CE	0.3			MIT
3684.243	3685.292	27134.89	CE	0.8			MIT
3683.393	3684.442	27141.16	CE	0.3			MIT
3682.647	3683.695	27146.65	CE II	0.9	0.0		MIT
3682.077	3683.125	27150.85	CE	1.0S	0.3		MIT
3681.791	3682.839	27152.96	CE	0.3			MIT
3681.376	3682.424	27156.02	CE	1.0	0.5		MIT
3681.113	3682.161	27157.97	CE	0.3	0.0		MIT
3680.848	3681.896	27159.92	CE	0.8			MIT
3680.633	3681.681	27161.51	CE	0.5			MIT
3680.425	3681.473	27163.04	CE	0.6			MIT
3680.079	3681.127	27165.59	CE II	1.0	0.3		MIT
3679.879	3680.927	27167.07	CE II	0.3			MIT
3679.424	3680.472	27170.43	CE	1.1	0.6		MIT
3679.157	3680.205	27172.40	CE	0.8	0.0		MIT
3679.067	3680.115	27173.07	CE	0.5			MIT
3678.848	3679.895	27174.69	CE	0.9			MIT
3678.223	3679.270	27179.30	CE	0.9			MIT
3677.927	3678.974	27181.49	CE	0.3			MIT
3677.859	3678.906	27181.99	CE	0.3			MIT
3677.170	3678.217	27187.08	CE II	0.9			MIT
3676.984	3678.031	27188.46	CE	0.5			MIT
3676.156	3677.203	27194.58	CE	1.1	0.0		MIT
3675.730	3676.777	27197.74	CE	0.3			MIT
3675.524	3676.571	27199.26	CE	0.7			MIT
3675.365	3676.412	27200.44	CE	0.9	0.0		MIT
3674.994	3676.040	27203.18	CE	0.5			MIT
3674.648	3675.694	27205.74	CE	0.3			MIT
3674.473	3675.519	27207.04	CE	0.6			MIT
3674.150	3675.196	27209.43	CE	0.8	0.0		MIT
3674.047	3675.093	27210.19	CE	0.7			MIT
3673.747	3674.793	27212.42	CE	0.7			MIT
3673.638	3674.684	27213.22	CE	1.0S	0.0		MIT
3673.265	3674.311	27215.99	CE	0.5			MIT
3672.789	3673.835	27219.51	CE	1.2	0.7		MIT
3672.257	3673.303	27223.46	CE	0.3			MIT
3672.166	3673.212	27224.13	CE I	1.0	0.0		MIT
3671.937	3672.983	27225.83	CE II	1.0	0.0		MIT
3671.515	3672.561	27228.96	CE	0.3			MIT
3671.316	3672.362	27230.44	CE	0.7	0.0		MIT
3670.672	3671.717	27235.21	CE II	0.8			MIT
3670.490	3671.535	27236.56	CE	1.0	0.0		MIT
3670.072	3671.117	27239.66	CE	0.3W			MIT
3669.962	3671.007	27240.48	CE	0.5			MIT
3669.719	3670.764	27242.28	CE	0.3			MIT
3669.313	3670.358	27245.30	CE	0.5			MIT
3669.084	3670.129	27247.00	CE	0.3	0.0		MIT
3668.719	3669.764	27249.71	CE II	1.1S	0.3		MIT
3667.981	3669.026	27255.19	CE	1.9S	1.2		MIT
3667.550	3668.595	27258.40	CE	0.5			MIT
3667.277	3668.321	27260.42	CE	0.6			MIT
3666.348	3667.392	27267.33	CE II	0.7			MIT
3666.024	3667.068	27269.74	CE	1.2			MIT
3665.574	3666.618	27273.09	CE	0.5			MIT
3665.496	3666.540	27273.67	CE	0.6			MIT
3665.050	3666.094	27276.99	CE	0.9			MIT
3664.949	3665.993	27277.74	CE II	0.9			MIT
3664.731	3665.775	27279.36	CE	0.9	0.0		MIT
3664.437	3665.481	27281.55	CE	0.3			MIT
3663.701	3664.745	27287.03	CE	1.0S	0.0		MIT
3662.990	3664.033	27292.33	CE	0.9S	0.0		MIT
3662.731	3663.774	27294.26	CE	0.3			MIT
3662.487	3663.530	27296.08	CE	0.9			MIT
3662.265	3663.308	27297.73	CE	0.3			MIT
3661.911	3662.954	27300.37	CE	0.8			MIT
3661.732	3662.775	27301.70	CE	1.0	0.3		MIT
3660.641	3661.684	27309.84	CE	1.6	1.0		MIT
3660.397	3661.440	27311.66	CE	0.3			MIT
3660.156	3661.199	27313.46	CE	1.1	0.6		MIT

AIR WAVELENGTH (ANGSTRMS)	VACUUM WAVELENGTH (ANGSTRMS)	WAVENUMBER (KAYSERS)	LMENT	LOG ARC	INTENSITY SPK	DIS	REF
3659.972	3661.015	27314.83	CE II	1.3	0.7		MIT
3659.502	3660.544	27318.34	CE	0.3			MIT
3659.227	3660.269	27320.39	CE	1.3	0.8		MIT
3658.774	3659.816	27323.78	CE	0.8			MIT
3658.424	3659.466	27326.39	CE	0.3			MIT
3658.258	3659.300	27327.63	CE II	1.0			MIT
3658.089	3659.131	27328.89	CE	0.7			MIT
3657.681	3658.723	27331.94	CE	0.7			MIT
3657.384	3658.426	27334.16	CE	0.3			MIT
3656.755	3657.797	27338.86	CE II	1.0S			MIT
3656.576	3657.618	27340.20	CE	0.5			MIT
3656.301	3657.343	27342.26	CE	0.3			MIT
3655.851	3656.893	27345.62	CE I	1.4	1.1		MIT
3655.851	3656.893	27345.62	CE II	1.4	1.1		MIT
3655.359	3656.400	27349.30	CE II	0.5			MIT
3655.021	3656.062	27351.83	CE	0.5			KN
3654.974	3656.015	27352.18	CE	1.2	0.5		MIT
3654.592	3655.633	27355.04	CE	0.3			MIT
3654.094	3655.135	27358.77	CE	0.3			MIT
3653.670	3654.711	27361.95	CE	1.3	0.9		MIT
3653.108	3654.149	27366.15	CE	1.2	0.7		MIT
3652.325	3653.366	27372.02	CE	0.3			MIT
3652.263	3653.304	27372.49	CE	0.5			MIT
3651.651	3652.691	27377.07	CE	0.7			MIT
3651.306	3652.346	27379.66	CE	0.3	0.0		MIT
3650.881	3651.921	27382.85	CE	1.1	0.5		MIT
3650.121	3651.161	27388.55	CE	1.0	0.0		MIT
3649.726	3650.766	27391.51	CE II	1.0	0.0		MIT
3648.820	3649.860	27398.31	CE	0.6			MIT
3648.528	3649.568	27400.51	CE	0.3			MIT
3648.096	3649.136	27403.75	CE	0.5			MIT
3647.954	3648.993	27404.82	CE	1.0	0.7		MIT
3647.751	3648.790	27406.34	CE	1.0	0.5		MIT
3647.546	3648.585	27407.88	CE	0.5			M
3646.965	3648.004	27412.25	CE	1.2	0.7		MIT
3646.652	3647.691	27414.60	CE	1.0	0.5		MIT
3645.711	3646.750	27421.68	CE	0.3			MIT
3645.452	3646.491	27423.63	CE II	1.0W	0.3		MIT
3645.228	3646.267	27425.31	CE II	0.7	0.0		MIT
3644.967	3646.006	27427.27	CE	0.3			MIT
3644.548	3645.587	27430.43	CE II	0.7	0.0		MIT
3644.297	3645.336	27432.32	CE	0.9	0.0		MIT
3643.45	3644.49	27438.7	CE	0.7			MIT
3643.014	3644.052	27441.98	CE	0.3			MIT
3642.833	3643.871	27443.34	CE	0.9	0.0		MIT
3642.620	3643.658	27444.95	CE	0.8			MIT
3642.250	3643.288	27447.73	CE	0.6			MIT
3642.174	3643.212	27448.31	CE	0.3			MIT
3641.734	3642.772	27451.62	CE	0.9	0.0		MIT
3641.537	3642.575	27453.11	CE	1.0	0.0		MIT
3640.687	3641.725	27459.52	CE	1.0			MIT
3639.867	3640.904	27465.70	CE II	0.5			MIT
3639.570	3640.607	27467.95	CE	0.7			MIT
3639.312	3640.349	27469.89	CE	0.7			MIT
3638.275	3639.312	27477.72	CE	1.0	0.3		MIT
3638.046	3639.083	27479.45	CE	0.8			MIT
3637.749	3638.786	27481.69	CE	0.9	0.0		MIT
3637.574	3638.611	27483.02	CE	0.3			MIT
3637.460	3638.497	27483.88	CE II	0.7			MIT
3636.736	3637.773	27489.35	CE	0.3			MIT
3636.558	3637.595	27490.70	CE	0.3			MIT
3636.371	3637.407	27492.11	CE	0.5	0.0		MIT
3636.070	3637.106	27494.38	CE	0.7	0.0		MIT
3635.883	3636.919	27495.80	CE	0.3			MIT
3635.784	3636.820	27496.55	CE	0.3			MIT
3634.432	3635.468	27506.77	CE	0.3			MIT
3633.865	3634.901	27511.07	CE	0.3			MIT
3633.400	3634.436	27514.59	CE II	0.9			MIT
3633.075	3634.111	27517.05	CE	0.3			MIT
3632.782	3633.818	27519.27	CE	0.5			MIT

AIR WAVELENGTH (ANGSTRMS)	VACUUM WAVELENGTH (ANGSTRMS)	WAVENUMBER (KAYSERS)	LMENT	LOG ARC	INTENSITY SPK	DIS	REF
3632.303	3633.338	27522.90	CE	0.8			MIT
3632.106	3633.141	27524.39	CE	1.0	0.3		MIT
3631.194	3632.229	27531.30	CE I	1.7	0.5		MIT
3631.194	3632.229	27531.30	CE II	1.7	0.5		MIT
3630.787	3631.822	27534.39	CE	0.5W			MIT
3630.421	3631.456	27537.17	CE	0.8	0.0		MIT
3630.147	3631.182	27539.24	CE	0.7			MIT
3629.806	3630.841	27541.83	CE	0.7			MIT
3629.305	3630.340	27545.63	CE	0.3			MIT
3628.813	3629.848	27549.37	CE	0.9			MIT
3628.622	3629.656	27550.82	CE	1.0S	0.0		MIT
3628.247	3629.281	27553.66	CE	1.0			MIT
3626.986	3628.020	27563.24	CE	0.8W			MIT
3625.768	3626.802	27572.50	CE	0.3			MIT
3625.637	3626.671	27573.50	CE	0.6			MIT
3625.375	3626.409	27575.49	CE I	0.8			MIT
3625.161	3626.195	27577.12	CE	0.6			MIT
3624.179	3625.212	27584.59	CE	1.0	0.0		MIT
3623.843	3624.876	27587.15	CE II	1.8	0.7		MIT
3623.843	3624.876	27587.15	CE I	1.8	0.7		MIT
3623.758	3624.791	27587.80	CE	0.9	0.0		MIT
3622.439	3623.472	27597.84	CE	0.9	0.0		MIT
3622.145	3623.178	27600.08	CE	1.2	0.7		MIT
3621.435	3622.468	27605.49	CE	0.3			MIT
3621.146	3622.179	27607.70	CE	0.9			MIT
3620.328	3621.360	27613.93	CE	0.7			MIT
3619.924	3620.956	27617.01	CE	0.9	0.0		MIT
3619.391	3620.423	27621.08	CE	0.9			MIT
3618.968	3620.000	27624.31	CE	0.3			MIT
3618.780	3619.812	27625.75	CE	0.3			MIT
3618.582	3619.614	27627.26	CE II	1.0	0.5		MIT
3618.423	3619.455	27628.47	CE	0.5			MIT
3618.129	3619.161	27630.72	CE	0.3			MIT
3617.811	3618.843	27633.14	CE	0.3			MIT
3617.101	3618.132	27638.57	CE	0.6			MIT
3617.016	3618.047	27639.22	CE	0.3			MIT
3616.878	3617.909	27640.27	CE	0.5			MIT
3616.642	3617.673	27642.07	CE	0.3			MIT
3616.458	3617.489	27643.48	CE	0.6			MIT
3616.307	3617.338	27644.64	CE	0.3			MIT
3616.204	3617.235	27645.42	CE	0.7	0.0		MIT
3616.112	3617.143	27646.13	CE	0.3	0.0		MIT
3615.625	3616.656	27649.85	CE	1.0			MIT
3615.039	3616.070	27654.33	CE	0.3			MIT
3614.362	3615.393	27659.51	CE	0.5			MIT
3614.233	3615.264	27660.50	CE	0.5			MIT
3614.029	3615.060	27662.06	CE	0.7	0.0		MIT
3613.701	3614.732	27664.57	CE II	1.3	0.7		MIT
3613.701	3614.732	27664.57	CE I	1.3	0.7		MIT
3612.838	3613.868	27671.18	CE	0.5			MIT
3612.609	3613.639	27672.93	CE	1.5	1.4		MIT
3612.466	3613.496	27674.03	CE	0.8			MIT
3612.322	3613.352	27675.13	CE	0.9S	0.0		MIT
3611.733	3612.763	27679.65	CE	0.7			MIT
3611.650	3612.680	27680.28	CE	1.3W	0.3W		MIT
3611.338	3612.368	27682.67	CE II	1.0			MIT
3610.914	3611.944	27685.92	CE	1.0			MIT
3610.446	3611.476	27689.51	CE	0.3			MIT
3610.257	3611.287	27690.96	CE	0.7			MIT
3609.894	3610.924	27693.75	CE	0.3			MIT
3609.687	3610.717	27695.33	CE	1.6	1.0		MIT
3609.212	3610.241	27698.98	CE	0.3			MIT
3608.847	3609.876	27701.78	CE	0.3			MIT
3608.309	3609.338	27705.91	CE	0.3			MIT
3607.874	3608.903	27709.25	CE	0.5W			MIT
3607.625	3608.654	27711.16	CE	1.2	0.9		MIT
3606.130	3607.159	27722.65	CE	0.7			MIT
3605.783	3606.812	27725.32	CE	0.7			MIT
3604.928	3605.956	27731.89	CE	0.8	0.0		MIT
3604.697	3605.725	27733.67	CE	0.5	0.0		MIT

AIR WAVELENGTH (ANGSTRMS)	VACUUM WAVELENGTH (ANGSTRMS)	WAVENUMBER (KAYSERS)	LMENT	LOG ARC	INTENSITY SPK	DIS	REF
3604.655	3605.683	27733.99	CE	0.5			MIT
3604.521	3605.549	27735.02	CE	0.3	0.0		MIT
3604.197	3605.225	27737.52	CE	0.9	0.0		MIT
3604.101	3605.129	27738.26	CE	0.5	0.0		MIT
3603.733	3604.761	27741.09	CE	0.7			MIT
3603.476	3604.504	27743.07	CE	0.7			MIT
3603.358	3604.386	27743.98	CE II	0.8			MIT
3602.803	3603.831	27748.25	CE	0.3	0.0		MIT
3600.583	3601.610	27765.36	CE II	1.2	0.3		MIT
3600.583	3601.610	27765.36	CE I	1.2	0.3		MIT
3599.974	3601.001	27770.06	CE	1.0	0.0		MIT
3598.196	3599.223	27783.78	CE II	1.3	0.0		MIT
3598.196	3599.223	27783.78	CE I	1.3	0.0		MIT
3597.232	3598.258	27791.22	CE	0.7			MIT
3596.725	3597.751	27795.14	CE II	1.0	0.0		MIT
3596.115	3597.141	27799.85	CE	1.1	0.3		MIT
3594.096	3595.122	27815.47	CE	0.7	0.3		MIT
3594.032	3595.058	27815.97	CE	0.7			MIT
3593.134	3594.159	27822.92	CE II	0.9			MIT
3591.745	3592.770	27833.68	CE	0.6	0.0		MIT
3590.598	3591.623	27842.57	CE	1.7	0.0		MIT
3590.351	3591.376	27844.48	CE	0.7J	0.7J		MIT
3588.564	3589.588	27858.35	CE	0.3	0.0		MIT
3588.496	3589.520	27858.88	CE	0.5			MIT
3588.430	3589.454	27859.39	CE	0.7			MIT
3588.130	3589.154	27861.72	CE	0.8	0.0		MIT
3587.680	3588.704	27865.21	CE	0.3			MIT
3587.639	3588.663	27865.53	CE II	1.0	0.3		MIT
3587.223	3588.247	27868.76	CE	0.7	0.0		MIT
3587.079	3588.103	27869.88	CE	0.5	0.3		MIT
3586.753	3587.777	27872.42	CE II	1.0	0.6		MIT
3584.805	3585.828	27887.56	CE	0.3	0.0		MIT
3584.447	3585.470	27890.35	CE	0.3	0.3		MIT
3584.337	3585.360	27891.20	CE	0.8	0.3		MIT
3583.968	3584.991	27894.07	CE	0.3	0.0		MIT
3583.664	3584.687	27896.44	CE	0.5			MIT
3583.107	3584.130	27900.78	CE	0.5			MIT
3582.602	3583.625	27904.71	CE	0.5	0.0		MIT
3582.124	3583.146	27908.43	CE	0.3			MIT
3580.779	3581.801	27918.92	CE	1.0	0.3		MIT
3580.562	3581.584	27920.61	CE	0.3			MIT
3578.950	3579.972	27933.18	CE	0.5			MIT
3578.822	3579.844	27934.18	CE	0.5			MIT
3577.798	3578.819	27942.18	CE	0.3	0.3		MIT
3577.458	3578.479	27944.83	CE	2.5	1.1		MIT
3576.232	3577.253	27954.41	CE	1.3	0.0		MIT
3575.662	3576.683	27958.87	CE	0.5			MIT
3575.288	3576.309	27961.79	CE	0.6	0.0		MIT
3575.177	3576.198	27962.66	CE	0.3	0.0		MIT
3573.701	3574.721	27974.21	CE	1.0	0.5		MIT
3572.754	3573.774	27981.62	CE	0.7	0.0		MIT
3572.428	3573.448	27984.18	CE	1.1	0.7		MIT
3571.454	3572.474	27991.81	CE	0.5			MIT
3571.362	3572.382	27992.53	CE	0.6	0.0		MIT
3570.983	3572.003	27995.50	CE II	0.8	0.0		MIT
3570.088	3571.107	28002.52	CE	0.5			MIT
3569.827	3570.846	28004.57	CE	0.7	0.0		MIT
3569.316	3570.335	28008.57	CE	0.9	0.3		MIT
3568.513	3569.532	28014.88	CE	0.3			MIT
3568.126	3569.145	28017.92	CE	1.0	0.0		MIT
3567.166	3568.185	28025.46	CE	0.3	0.3		MIT
3566.033	3567.051	28034.36	CE	1.0W	0.8W		MIT
3565.424	3566.442	28039.15	CE	0.8S	0.3		MIT
3563.823	3564.841	28051.74	CE II	0.3	0.0		MIT
3563.591	3564.609	28053.57	CE	0.3			MIT
3562.091	3563.108	28065.38	CE	0.8			MIT
3561.538	3562.555	28069.74	CE	0.8			MIT
3560.798	3561.815	28075.57	CE	2.5	0.3		MIT
3559.328	3560.345	28087.17	CE II	0.8			MIT
3558.877	3559.893	28090.73	CE	0.5	0.3		MIT
3558.706	3559.722	28092.08	CE II	0.9	0.0		MIT
3557.487	3558.503	28101.70	CE	0.8	0.0		MIT
3556.893	3557.909	28106.40	CE	0.8			MIT
3556.365	3557.381	28110.57	CE	0.7	0.0		MIT
3556.098	3557.114	28112.68	CE	0.9			MIT
3555.788	3556.804	28115.13	CE II	0.6	0.0		MIT
3555.219	3556.235	28119.63	CE	0.5	0.0		MIT
3555.160	3556.176	28120.10	CE	0.7	0.3		MIT
3554.993	3556.008	28121.42	CE II	1.4			MIT
3554.627	3555.642	28124.31	CE	1.0	0.3		MIT
3552.929	3553.944	28137.75	CE	0.5			MIT
3552.727	3553.742	28139.35	CE II	1.3S	1.0		MIT
3552.367	3553.382	28142.21	CE	0.3	0.3		MIT
3552.241	3553.256	28143.20	CE	0.3			MIT
3552.067	3553.082	28144.58	CE II	0.7	0.3		MIT
3551.775	3552.790	28146.90	CE	1.0	0.0		MIT
3551.664	3552.679	28147.77	CE II	1.0	0.0		MIT
3551.434	3552.449	28149.60	CE	1.0	0.0		MIT
3549.733	3550.747	28163.09	CE	0.3	0.0		MIT
3549.326	3550.340	28166.32	CE	0.5	0.3		MIT
3549.120	3550.134	28167.95	CE	0.8	0.0		MIT
3548.835	3549.849	28170.21	CE	0.9	0.0		MIT
3548.169	3549.183	28175.50	CE	1.0H	0.7		MIT
3547.807	3548.821	28178.37	CE	0.5			MIT
3547.001	3548.014	28184.78	CE	1.2	0.5		MIT
3546.657	3547.670	28187.51	CE II	0.9			MIT
3546.190	3547.203	28191.22	CE II	1.3	0.3		MIT
3545.914	3546.927	28193.42	CE	0.7	0.0		MIT
3545.781	3546.794	28194.48	CE II	0.9	0.0		MIT
3545.603	3546.616	28195.89	CE II	1.0	0.0		MIT
3544.773	3545.786	28202.49	CE	0.5			MIT
3544.389	3545.402	28205.55	CE	0.5			MIT
3544.273	3545.286	28206.47	CE	0.5	0.3		MIT
3543.832	3544.845	28209.98	CE	0.5			MIT
3543.686	3544.699	28211.14	CE	0.5			MIT
3543.523	3544.536	28212.44	CE II	1.0			MIT
3543.283	3544.295	28214.35	CE	1.0	0.3		MIT
3542.279	3543.291	28222.35	CE	0.3	0.0		MIT
3542.182	3543.194	28223.12	CE	0.6	0.0		MIT
3541.911	3542.923	28225.28	CE	0.6	0.0		MIT
3541.662	3542.674	28227.26	CE	1.0	0.3		MIT
3540.792	3541.804	28234.20	CE	0.7	0.0		MIT
3540.380	3541.392	28237.49	CE	0.3	0.3		MIT
3540.312	3541.324	28238.03	CE	0.3	0.5		MIT
3540.117	3541.129	28239.58	CE	0.5			MIT
3539.836	3540.848	28241.83	CE	0.7			MIT
3539.086	3540.097	28247.81	CE II	2.0	1.0		MIT
3538.960	3539.971	28248.82	CE	0.5			MIT
3538.792	3539.803	28250.16	CE	0.5	0.3		MIT
3538.759	3539.770	28250.42	CE	0.7	0.3		MIT
3538.462	3539.473	28252.79	CE	0.3	0.0		MIT
3538.415	3539.426	28253.17	CE	0.3	0.0		MIT
3538.304	3539.315	28254.05	CE	0.3			MIT
3537.849	3538.860	28257.69	CE	0.5			MIT
3537.437	3538.448	28260.98	CE II	1.0	0.5		MIT
3537.134	3538.145	28263.40	CE	1.0	0.0		MIT
3536.696	3537.707	28266.90	CE	1.0	0.3		MIT
3536.482	3537.493	28268.61	CE	0.9	0.3		MIT
3536.006	3537.017	28272.42	CE	0.3			MIT
3535.840	3536.851	28273.74	CE	0.3	0.0		MIT
3535.689	3536.700	28274.95	CE	0.5	0.0		MIT
3535.566	3536.576	28275.93	CE	1.0	0.0		MIT
3535.161	3536.171	28279.17	CE	0.5			MIT
3535.046	3536.056	28280.09	CE	0.9			MIT
3534.742	3535.752	28282.52	CE	0.7			MIT
3534.436	3535.446	28284.97	CE	1.0			MIT
3534.051	3535.061	28288.05	CE	1.5	1.0		MIT
3533.716	3534.726	28290.74	CE	0.7W	0.5		MIT
3533.679	3534.689	28291.03	CE	0.6			MIT
3533.564	3534.574	28291.95	CE	1.1	0.5		MIT

AIR WAVELENGTH (ANGSTRMS)	VACUUM WAVELENGTH (ANGSTRMS)	WAVENUMBER (KAYSERS)	LMENT	LOG ARC	INTENSITY SPK	DIS	REF
3533.441	3534.451	28292.94	CE		0.3H		MIT
3533.114	3534.124	28295.56	CE	0.9			MIT
3532.879	3533.889	28297.44	CE II	1.0	0.0		MIT
3532.608	3533.618	28299.61	CE II	0.8	0.0		MIT
3532.390	3533.400	28301.35	CE	0.6	0.0		MIT
3531.593	3532.602	28307.74	CE	1.3	0.3		MIT
3530.946	3531.955	28312.93	CE	1.0	0.0		MIT
3530.635	3531.644	28315.42	CE	1.0			MIT
3530.403	3531.412	28317.28	CE	0.5			MIT
3530.137	3531.146	28319.42	CE	0.6	0.3		MIT
3530.022	3531.031	28320.34	CE II	1.3	0.5		MIT
3529.732	3530.741	28322.67	CE	0.5			MIT
3529.537	3530.546	28324.23	CE	0.7	0.3		MIT
3529.271	3530.280	28326.37	CE	0.9	0.3		MIT
3529.041	3530.050	28328.21	CE II	1.1	0.3		MIT
3528.644	3529.653	28331.40	CE	0.9	0.0		MIT
3528.053	3529.062	28336.15	CE II	1.0	0.0		MIT
3527.848	3528.856	28337.79	CE II	1.2S	0.3		MIT
3527.610	3528.618	28339.70	CE II	0.9	0.3		MIT
3527.310	3528.318	28342.11	CE	0.3	0.3		MIT
3527.209	3528.217	28342.93	CE	0.3			MIT
3526.682	3527.690	28347.16	CE	1.4	0.5		MIT
3526.006	3527.014	28352.59	CE	0.5			MIT
3525.936	3526.944	28353.16	CE	0.3	0.0		MIT
3524.073	3525.081	28368.15	CE	0.9			MIT
3524.008	3525.016	28368.67	CE	0.9			MIT
3523.735	3524.742	28370.87	CE	0.7			MIT
3523.613	3524.620	28371.85	CE	0.3	0.3		MIT
3523.331	3524.338	28374.12	CE	0.3	0.3		MIT
3523.104	3524.111	28375.95	CE	0.9			MIT
3522.963	3523.970	28377.08	CE	0.3			MIT
3522.797	3523.804	28378.42	CE	0.3W	0.0		MIT
3522.695	3523.702	28379.24	CE	0.0	0.3		MIT
3522.605	3523.612	28379.97	CE	0.0H	0.3		MIT
3522.451	3523.458	28381.21	CE	0.3			MIT
3522.211	3523.218	28383.14	CE	0.3			MIT
3521.880	3522.887	28385.81	CE	1.5	0.7		MIT
3521.535	3522.542	28388.59	CE	0.3			MIT
3521.331	3522.338	28390.24	CE	0.3			MIT
3521.179	3522.186	28391.46	CE	0.3			MIT
3521.128	3522.135	28391.87	CE	0.3	0.0		MIT
3520.976	3521.983	28393.10	CE	0.7	0.0		MIT
3520.522	3521.529	28396.76	CE II	1.5	0.3		MIT
3520.522	3521.529	28396.76	CE I	1.5	0.3		MIT
3520.247	3521.254	28398.98	CE	0.6			MIT
3520.100	3521.107	28400.16	CE	0.3	0.0		MIT
3519.928	3520.934	28401.55	CE	0.5			MIT
3519.736	3520.742	28403.10	CE	1.3			MIT
3519.077	3520.083	28408.42	CE	1.4	0.6		MIT
3518.734	3519.740	28411.19	CE	0.5			MIT
3518.495	3519.501	28413.12	CE	0.5			MIT
3518.371	3519.377	28414.12	CE	1.1	0.0		MIT
3518.040	3519.046	28416.79	CE	0.9	0.0		MIT
3517.906	3518.912	28417.87	CE	0.7	0.0		MIT
3517.380	3518.386	28422.12	CE	1.6	0.8		MIT
3516.973	3517.979	28425.41	CE	0.0	0.5H		MIT
3516.206	3517.212	28431.61	CE	0.3			MIT
3515.936	3516.941	28433.80	CE	0.7			MIT
3515.779	3516.784	28435.07	CE II	0.9	0.0		MIT
3515.637	3516.642	28436.22	CE	0.9			MIT
3515.546	3516.551	28436.95	CE	0.3			MIT
3515.390	3516.395	28438.21	CE	0.3			MIT
3515.285	3516.290	28439.06	CE	0.6			MIT
3514.966	3515.971	28441.64	CE	0.3			MIT
3514.802	3515.807	28442.97	CE	0.3	0.0		MIT
3514.328	3515.333	28446.81	CE	0.8			MIT
3513.856	3514.861	28450.63	CE	0.9			MIT
3513.790	3514.795	28451.16	CE	0.8			MIT
3513.475	3514.480	28453.71	CE	0.6			MIT
3513.283	3514.288	28455.27	CE	0.8			MIT

AIR WAVELENGTH (ANGSTRMS)	VACUUM WAVELENGTH (ANGSTRMS)	WAVENUMBER (KAYSERS)	LMENT	LOG ARC	INTENSITY SPK	DIS	REF
3513.048	3514.053	28457.17	CE	0.0	0.3		MIT
3512.566	3513.571	28461.08	CE	0.7			MIT
3512.272	3513.276	28463.46	CE	0.8			MIT
3511.589	3512.593	28469.00	CE	0.9	0.0		MIT
3511.459	3512.463	28470.05	CE	0.3	0.3		MIT
3511.413	3512.417	28470.42	CE	0.3			MIT
3511.136	3512.140	28472.67	CE	0.5	0.0		MIT
3510.779	3511.783	28475.56	CE	0.5			MIT
3510.688	3511.692	28476.30	CE	1.1	0.0		MIT
3510.426	3511.430	28478.43	CE	0.5			MIT
3510.293	3511.297	28479.50	CE	0.7			MIT
3510.220	3511.224	28480.10	CE	0.9	0.0		MIT
3509.934	3510.938	28482.42	CE	0.9			MIT
3509.729	3510.733	28484.08	CE	1.1			MIT
3509.531	3510.535	28485.69	CE	0.5			MIT
3509.313	3510.317	28487.46	CE	0.5	0.3		MIT
3509.254	3510.258	28487.94	CE II	1.0	0.3		MIT
3509.061	3510.065	28489.50	CE	0.3			MIT
3508.865	3509.869	28491.09	CE	0.3			MIT
3508.711	3509.715	28492.34	CE	1.1	0.3		MIT
3508.467	3509.471	28494.33	CE II	1.0			MIT
3508.335	3509.338	28495.40	CE	0.6			MIT
3508.031	3509.034	28497.87	CE	0.5			MIT
3507.945	3508.948	28498.57	CE II	1.1	0.5		MIT
3507.811	3508.814	28499.66	CE	0.7			MIT
3507.528	3508.531	28501.96	CE	0.3			MIT
3507.345	3508.348	28503.44	CE	0.9			MIT
3507.221	3508.224	28504.45	CE	0.7			MIT
3507.015	3508.018	28506.12	CE	0.3			MIT
3506.726	3507.729	28508.47	CE	0.9			MIT
3506.464	3507.467	28510.60	CE	0.3			MIT
3506.252	3507.255	28512.33	CE	1.2	0.0		MIT
3505.955	3506.958	28514.74	CE	0.6			MIT
3505.688	3506.691	28516.91	CE	0.5			MIT
3505.507	3506.510	28518.39	CE	0.3			MIT
3505.457	3506.460	28518.79	CE	0.3			MIT
3505.181	3506.184	28521.04	CE	0.9			MIT
3504.606	3505.609	28525.72	CE	0.7	1.0J		MIT
3504.521	3505.524	28526.41	CE	0.8			MIT
3504.149	3505.151	28529.44	CE	0.7			MIT
3504.090	3505.092	28529.92	CE	0.7	0.5		MIT
3503.981	3504.983	28530.81	CE II	0.7			MIT
3503.955	3504.957	28531.02	CE	0.6			MIT
3503.634	3504.636	28533.63	CE	0.6			MIT
3503.534	3504.536	28534.45	CE	0.6			MIT
3503.085	3504.087	28538.10	CE	1.0			MIT
3502.978	3503.980	28538.98	CE	0.7			MIT
3502.889	3503.891	28539.70	CE II	0.9S			MIT
3502.646	3503.648	28541.68	CE II	0.9			MIT
3501.815	3502.817	28548.45	CE	0.9			MIT
3501.594	3502.596	28550.25	CE	1.1			MIT
3501.453	3502.455	28551.40	CE I	1.3W	0.5		MIT
3501.453	3502.455	28551.40	CE II	1.3W	0.5		MIT
3501.339	3502.341	28552.33	CE	0.6			MIT
3501.031	3502.033	28554.85	CE	0.7			MIT
3500.834	3501.836	28556.45	CE	0.6			MIT
3500.680	3501.682	28557.71	CE	1.3			MIT
3499.994	3500.995	28563.31	CE	1.2	0.0		MIT
3499.777	3500.778	28565.08	CE	0.5			MIT
3499.622	3500.623	28566.34	CE	0.7			MIT
3499.466	3500.467	28567.61	CE	0.3			MIT
3499.388	3500.389	28568.25	CE	0.6			MIT
3499.300	3500.301	28568.97	CE	0.6			MIT
3498.924	3499.925	28572.04	CE	0.3			MIT
3498.826	3499.827	28572.84	CE	0.3			MIT
3498.679	3499.680	28574.04	CE	1.2			MIT
3498.562	3499.563	28575.00	CE	0.9			MIT
3497.444	3498.445	28584.13	CE	0.3			MIT
3497.309	3498.310	28585.23	CE	1.2			MIT
3496.444	3497.444	28592.31	CE	0.7			MIT

AIR WAVELENGTH (ANGSTRMS)	VACUUM WAVELENGTH (ANGSTRMS)	WAVENUMBER (KAYSERS)	LMENT	LOG ARC	INTENSITY SPK	INTENSITY DIS	REF
3496.326	3497.326	28593.27	CE	1.1			MIT
3496.210	3497.210	28594.22	CE	0.3			MIT
3495.941	3496.941	28596.42	CE II	1.2	0.0		MIT
3495.729	3496.729	28598.15	CE	1.0			MIT
3495.478	3496.478	28600.21	CE	1.0			MIT
3495.007	3496.007	28604.06	CE	1.2			MIT
3494.863	3495.863	28605.24	CE	0.5			MIT
3494.648	3495.648	28607.00	CE	1.0			MIT
3494.397	3495.397	28609.05	CE	0.3			MIT
3493.936	3494.936	28612.83	CE	0.9			MIT
3493.724	3494.724	28614.56	CE II	1.3			MIT
3493.110	3494.110	28619.59	CE	1.1	0.0		MIT
3492.983	3493.983	28620.64	CE	0.5			MIT
3492.559	3493.558	28624.11	CE	0.5			MIT
3492.486	3493.485	28624.71	CE	0.7			MIT
3492.249	3493.248	28626.65	CE	1.1			MIT
3491.696	3492.695	28631.18	CE	1.1			MIT
3491.576	3492.575	28632.17	CE	0.9			MIT
3491.395	3492.394	28633.65	CE	0.5			MIT
3490.713	3491.712	28639.25	CE	0.8			MIT
3490.259	3491.258	28642.97	CE	0.6			MIT
3490.125	3491.124	28644.07	CE	1.4	0.5		MIT
3489.555	3490.554	28648.75	CE	0.7			MIT
3489.455	3490.454	28649.57	CE	0.7			MIT
3489.379	3490.378	28650.20	CE	0.5			MIT
3489.054	3490.053	28652.86	CE	0.5			MIT
3488.811	3489.809	28654.86	CE	0.9			MIT
3488.553	3489.551	28656.98	CE	1.5	0.7H		MIT
3488.372	3489.370	28658.47	CE	0.6			MIT
3488.164	3489.162	28660.17	CE	0.7			MIT
3487.863	3488.861	28662.65	CE	0.6			MIT
3487.566	3488.564	28665.09	CE I	1.5	0.7		MIT
3487.381	3488.379	28666.61	CE	0.6			MIT
3487.288	3488.286	28667.37	CE	0.5			MIT
3487.162	3488.160	28668.41	CE	1.0			MIT
3486.855	3487.853	28670.93	CE	0.9			MIT
3486.664	3487.662	28672.50	CE	0.3			MIT
3486.269	3487.267	28675.75	CE	1.0			MIT
3485.054	3486.052	28685.75	CE	1.5	1.0		MIT
3484.743	3485.740	28688.31	CE	1.0	0.0		MIT
3484.698	3485.695	28688.68	CE	0.7			MIT
3484.559	3485.556	28689.82	CE	0.6			MIT
3484.120	3485.117	28693.44	CE	0.5			MIT
3484.054	3485.051	28693.98	CE	0.5			MIT
3483.985	3484.982	28694.55	CE	0.7			MIT
3483.512	3484.509	28698.45	CE	0.8			MIT
3483.323	3484.320	28700.00	CE	0.5			MIT
3482.807	3483.804	28704.26	CE	1.0			MIT
3482.420	3483.417	28707.45	CE	0.7			MIT
3482.345	3483.342	28708.06	CE II	1.3	0.3		MIT
3482.140	3483.137	28709.75	CE II	1.3			MIT
3481.752	3482.749	28712.95	CE	0.3H			MIT
3481.673	3482.670	28713.60	CE	0.5			MIT
3481.159	3482.156	28717.84	CE II	1.1	0.0		MIT
3480.979	3481.975	28719.33	CE	1.3	0.0		MIT
3480.757	3481.753	28721.16	CE	0.3			MIT
3480.614	3481.610	28722.34	CE	0.6			MIT
3480.379	3481.375	28724.28	CE II	1.2			MIT
3480.273	3481.269	28725.16	CE II	1.2			MIT
3479.982	3480.978	28727.56	CE	0.5			MIT
3479.744	3480.740	28729.52	CE	0.6			MIT
3479.607	3480.603	28730.65	CE	1.2	0.0		MIT
3479.402	3480.398	28732.35	CE	0.3			MIT
3479.028	3480.024	28735.43	CE	1.3			MIT
3478.573	3479.569	28739.19	CE	0.5			MIT
3478.324	3479.320	28741.25	CE	0.3			MIT
3477.989	3478.985	28744.02	CE	1.0			MIT
3477.452	3478.448	28748.46	CE	1.2	0.0		MIT
3477.392	3478.388	28748.95	CE	0.9			MIT
3477.269	3478.265	28749.97	CE	0.3			MIT

AIR WAVELENGTH (ANGSTRMS)	VACUUM WAVELENGTH (ANGSTRMS)	WAVENUMBER (KAYSERS)	LMENT	LOG ARC	INTENSITY SPK	INTENSITY DIS	REF
3476.842	3477.837	28753.50	CE	1.5	1.0		MIT
3476.575	3477.570	28755.71	CE	0.7			MIT
3476.355	3477.350	28757.53	CE	1.2			MIT
3475.675	3476.670	28763.16	CE II	1.2			MIT
3474.780	3475.775	28770.56	CE	1.1			MIT
3474.401	3475.396	28773.70	CE	0.3			MIT
3474.216	3475.211	28775.23	CE II	1.2	0.0		MIT
3473.812	3474.807	28778.58	CE	0.9			MIT
3473.725	3474.720	28779.30	CE	0.7			MIT
3473.478	3474.473	28781.35	CE	0.3			MIT
3473.131	3474.125	28784.22	CE	1.0			MIT
3473.035	3474.029	28785.02	CE	0.3			MIT
3472.027	3473.021	28793.37	CE	1.0			MIT
3471.650	3472.644	28796.50	CE	0.5			MIT
3471.546	3472.540	28797.36	CE II	0.5			MIT
3471.286	3472.280	28799.52	CE	0.3H			MIT
3471.171	3472.165	28800.48	CE	0.5			MIT
3471.008	3472.002	28801.83	CE	0.5	1.0		MIT
3470.407	3471.401	28806.81	CE	1.0			MIT
3470.035	3471.029	28809.90	CE	0.8			MIT
3469.400	3470.393	28815.18	CE	1.0			MIT
3469.007	3470.000	28818.44	CE	0.9			MIT
3468.885	3469.878	28819.45	CE	0.9			MIT
3468.708	3469.701	28820.93	CE	0.5			MIT
3468.383	3469.376	28823.62	CE	0.8			MIT
3468.113	3469.106	28825.87	CE	1.3	0.0		MIT
3467.776	3468.769	28828.67	CE II	1.2			MIT
3467.497	3468.490	28830.99	CE	0.3			MIT
3466.946	3467.939	28835.57	CE II	1.0			MIT
3466.803	3467.796	28836.76	CE	1.0			MIT
3466.646	3467.639	28838.07	CE	0.3			MIT
3466.077	3467.070	28842.80	CE	0.7			MIT
3466.029	3467.022	28843.20	CE	0.5			MIT
3465.503	3466.495	28847.58	CE	0.7			MIT
3465.416	3466.408	28848.30	CE	0.9			MIT
3465.302	3466.294	28849.25	CE	0.9			MIT
3464.988	3465.980	28851.87	CE	1.1			MIT
3464.862	3465.854	28852.92	CE	1.1			MIT
3464.722	3465.714	28854.08	CE	0.3			MIT
3464.214	3465.206	28858.31	CE	0.9	0.0W		MIT
3464.160	3465.152	28858.76	CE II	1.0			MIT
3463.762	3464.754	28862.08	CE	1.2S	0.0		MIT
3463.349	3464.341	28865.52	CE	0.5			MIT
3463.218	3464.210	28866.61	CE	1.1	0.0		MIT
3463.128	3464.120	28867.36	CE	0.5			MIT
3462.767	3463.759	28870.37	CE	1.0			MIT
3462.433	3463.425	28873.16	CE	0.9			MIT
3462.234	3463.226	28874.81	CE	0.5			MIT
3461.793	3462.785	28878.49	CE	1.0			MIT
3461.417	3462.408	28881.63	CE	0.3			MIT
3461.344	3462.335	28882.24	CE	1.2S			MIT
3461.217	3462.208	28883.30	CE	0.6			MIT
3460.999	3461.990	28885.12	CE	1.1			MIT
3460.783	3461.774	28886.92	CE	0.3			MIT
3460.581	3461.572	28888.61	CE	0.3			MIT
3460.163	3461.154	28892.10	CE	0.8			MIT
3459.833	3460.824	28894.85	CE	1.1	0.0		MIT
3459.435	3460.426	28898.18	CE	0.3			MIT
3459.385	3460.376	28898.59	CE	0.3H	1.2J		MIT
3459.144	3460.135	28900.61	CE	0.7			MIT
3458.862	3459.853	28902.96	CE	0.9			MIT
3458.218	3459.209	28908.35	CE	0.6			MIT
3457.560	3458.550	28913.85	CE	1.3			MIT
3457.173	3458.163	28917.08	CE II	0.8			MIT
3456.772	3457.762	28920.44	CE II	0.9	0.0		MIT
3456.674	3457.664	28921.26	CE	1.3	0.0		MIT
3456.340	3457.330	28924.05	CE II	1.0			MIT
3456.031	3457.021	28926.64	CE	0.3			MIT
3455.471	3456.461	28931.33	CE	0.3			MIT
3455.023	3456.013	28935.08	CE	0.7			MIT

AIR WAVELENGTH (ANGSTRMS)	VACUUM WAVELENGTH (ANGSTRMS)	WAVENUMBER (KAYSERS)	LMENT		LOG ARC	INTENSITY SPK	DIS	REF
3454.911	3455.901	28936.02	CE		0.7			MIT
3454.807	3455.797	28936.89	CE		0.9			MIT
3454.475	3455.465	28939.67	CE		1.0	1.6		MIT
3454.256	3455.246	28941.50	CE		0.3			MIT
3454.022	3455.012	28943.46	CE		0.5			MIT
3453.760	3454.749	28945.66	CE		0.5			MIT
3453.639	3454.628	28946.67	CE		0.3			MIT
3453.241	3454.230	28950.01	CE		0.9			MIT
3452.806	3453.795	28953.66	CE		0.7			MIT
3452.623	3453.612	28955.19	CE	II	0.7			MIT
3452.538	3453.527	28955.90	CE		0.7			MIT
3452.279	3453.268	28958.08	CE		0.3			MIT
3452.214	3453.203	28958.62	CE		0.6			MIT
3452.057	3453.046	28959.94	CE		0.3			MIT
3451.903	3452.892	28961.23	CE		0.3			MIT
3451.629	3452.618	28963.53	CE	II	0.5			MIT
3451.564	3452.553	28964.07	CE		0.7			MIT
3451.329	3452.318	28966.05	CE		0.5			MIT
3450.981	3451.970	28968.97	CE		0.3			MIT
3450.917	3451.906	28969.50	CE		0.5			MIT
3450.801	3451.790	28970.48	CE		0.3			MIT
3450.722	3451.711	28971.14	CE		0.3			MIT
3450.361	3451.350	28974.17	CE		0.5			MIT
3450.232	3451.221	28975.26	CE		0.9			MIT
3449.984	3450.973	28977.34	CE		1.0			MIT
3449.910	3450.898	28977.96	CE	II	0.7			MIT
3448.865	3449.853	28986.74	CE		0.7			MIT
3448.645	3449.633	28988.59	CE		0.8			MIT
3448.288	3449.276	28991.59	CE	II	1.1			MIT
3448.023	3449.011	28993.82	CE		0.3			MIT
3447.968	3448.956	28994.28	CE		0.3			MIT
3447.531	3448.519	28997.96	CE		0.3			MIT
3447.273	3448.261	29000.13	CE		0.5			MIT
3446.721	3447.709	29004.77	CE		1.2	0.0		MIT
3446.206	3447.194	29009.10	CE		1.2			MIT
3444.790	3445.777	29021.03	CE		0.7			MIT
3444.185	3445.172	29026.13	CE		0.7			MIT
3443.683	3444.670	29030.36	CE		0.9	1.2		MIT
3443.525	3444.512	29031.69	CE		0.9			MIT
3443.350	3444.337	29033.17	CE		0.3			MIT
3442.955	3443.942	29036.50	CE	II	1.3			MIT
3442.557	3443.544	29039.85	CE		0.6			MIT
3442.251	3443.238	29042.43	CE		0.9			MIT
3442.183	3443.170	29043.01	CE		1.0			MIT
3441.894	3442.880	29045.45	CE		0.3			MIT
3441.401	3442.387	29049.61	CE		1.0			MIT
3441.210	3442.196	29051.22	CE		1.5	0.7		MIT
3440.577	3441.563	29056.56	CE		0.5			MIT
3440.472	3441.458	29057.45	CE		1.2W			MIT
3440.003	3440.989	29061.41	CE		0.3			MIT
3439.831	3440.817	29062.87	CE		1.4	0.3		MIT
3439.440	3440.426	29066.17	CE		0.5			MIT
3438.881	3439.867	29070.89	CE		0.5			MIT
3438.235	3439.220	29076.36	CE		0.5			MIT
3438.066	3439.051	29077.78	CE		1.2			MIT
3437.813	3438.798	29079.93	CE		1.0			MIT
3437.324	3438.309	29084.06	CE		1.1			MIT
3436.959	3437.944	29087.15	CE		0.7			MIT
3436.727	3437.712	29089.11	CE		0.5			MIT
3436.304	3437.289	29092.70	CE		1.2			MIT
3436.199	3437.184	29093.58	CE		0.7			MIT
3435.682	3436.667	29097.96	CE		0.5			MIT
3435.207	3436.192	29101.99	CE		1.3			MIT
3434.290	3435.274	29109.75	CE		0.5			MIT
3434.134	3435.118	29111.08	CE		0.7			MIT
3433.689	3434.673	29114.85	CE		1.0			MIT
3433.569	3434.553	29115.87	CE		1.0			MIT
3433.091	3434.075	29119.92	CE		1.4	0.7		MIT
3432.775	3433.759	29122.60	CE		0.6			MIT
3432.419	3433.403	29125.62	CE		0.5			MIT

AIR WAVELENGTH (ANGSTRMS)	VACUUM WAVELENGTH (ANGSTRMS)	WAVENUMBER (KAYSERS)	LMENT		LOG ARC	INTENSITY SPK	DIS	REF
3431.739	3432.723	29131.39	CE		0.9			MIT
3431.498	3432.482	29133.44	CE		1.1	0.0		MIT
3431.196	3432.180	29136.00	CE		0.9			MIT
3431.019	3432.003	29137.51	CE		1.0			MIT
3430.848	3431.832	29138.96	CE		1.1			MIT
3430.665	3431.649	29140.51	CE		0.6			MIT
3430.600	3431.584	29141.06	CE		0.6			MIT
3430.316	3431.299	29143.48	CE		1.0	0.3		MIT
3430.252	3431.235	29144.02	CE		0.7			MIT
3429.856	3430.839	29147.39	CE		1.1			MIT
3429.638	3430.621	29149.24	CE		0.5			MIT
3429.176	3430.159	29153.17	CE		0.7			MIT
3428.944	3429.927	29155.14	CE		0.3			MIT
3428.870	3429.853	29155.77	CE		0.6			MIT
3428.697	3429.680	29157.24	CE	II	1.0			MIT
3428.502	3429.485	29158.90	CE		0.7			MIT
3428.055	3429.038	29162.70	CE		0.5			MIT
3427.834	3428.817	29164.58	CE		0.5			MIT
3427.605	3428.588	29166.53	CE	II	0.7			MIT
3427.283	3428.266	29169.27	CE		1.1	1.0H		MIT
3427.121	3428.104	29170.65	CE		0.9			MIT
3427.087	3428.070	29170.94	CE		0.7			MIT
3426.879	3427.862	29172.71	CE		0.3			MIT
3426.583	3427.566	29175.23	CE	II	1.3	0.0		MIT
3426.446	3427.428	29176.39	CE		0.3			MIT
3426.208	3427.190	29178.42	CE		1.5	0.8		MIT
3425.944	3426.926	29180.67	CE		1.3			MIT
3425.679	3426.661	29182.93	CE		0.3			MIT
3425.344	3426.326	29185.78	CE		1.0			MIT
3424.366	3425.348	29194.11	CE		0.5			MIT
3424.020	3425.002	29197.06	CE		0.9W			MIT
3423.853	3424.835	29198.49	CE		1.3W	0.3		MIT
3423.616	3424.598	29200.51	CE		0.5			MIT
3423.477	3424.459	29201.70	CE		0.6			MIT
3422.708	3423.690	29208.26	CE		1.5	1.0		MIT
3422.507	3423.488	29209.97	CE		1.3			MIT
3422.420	3423.401	29210.71	CE		0.9			MIT
3422.215	3423.196	29212.46	CE		0.3			MIT
3422.000	3422.981	29214.30	CE		0.7			MIT
3421.714	3422.695	29216.74	CE		0.5			MIT
3421.542	3422.523	29218.21	CE	II	0.7			MIT
3421.071	3422.052	29222.23	CE		0.8			MIT
3420.955	3421.936	29223.22	CE		0.7			MIT
3420.534	3421.515	29226.82	CE	II	0.9			MIT
3420.176	3421.157	29229.88	CE	II	1.5	0.3		MIT
3419.328	3420.309	29237.13	CE		0.3			MIT
3419.246	3420.227	29237.83	CE		0.3			MIT
3418.930	3419.911	29240.53	CE		1.3	0.3		MIT
3417.897	3418.877	29249.37	CE		1.3			MIT
3417.450	3418.430	29253.19	CE		1.5	0.7		MIT
3416.861	3417.841	29258.24	CE		1.0	0.0		MIT
3416.700	3417.680	29259.61	CE		1.1			MIT
3416.556	3417.536	29260.85	CE		1.3			MIT
3415.614	3416.594	29268.92	CE		1.0			MIT
3415.307	3416.287	29271.55	CE		0.5			MIT
3415.069	3416.049	29273.59	CE		0.9			MIT
3414.766	3415.745	29276.19	CE		0.7			MIT
3414.605	3415.584	29277.57	CE		0.7			MIT
3414.313	3415.292	29280.07	CE		0.9			MIT
3414.168	3415.147	29281.31	CE	II	0.9S			MIT
3413.614	3414.593	29286.07	CE		0.5			MIT
3413.452	3414.431	29287.46	CE		0.6			MIT
3413.329	3414.308	29288.51	CE		0.9			MIT
3413.231	3414.210	29289.35	CE		0.6			MIT
3412.334	3413.313	29297.05	CE	II	1.0			MIT
3411.835	3412.814	29301.33	CE		1.1S			MIT
3411.569	3412.548	29303.62	CE		1.1			MIT
3411.434	3412.413	29304.78	CE		1.0			MIT
3411.112	3412.091	29307.55	CE		0.7			MIT
3410.226	3411.204	29315.16	CE		1.1			MIT

AIR WAVELENGTH (ANGSTRMS)	VACUUM WAVELENGTH (ANGSTRMS)	WAVENUMBER (KAYSERS)	LMENT	LOG ARC	INTENSITY SPK	INTENSITY DIS	REF
3409.863	3410.841	29318.28	CE	0.3			MIT
3409.502	3410.480	29321.39	CE	0.9			MIT
3409.405	3410.383	29322.22	CE II	1.0			MIT
3408.953	3409.931	29326.11	CE	0.7			MIT
3408.804	3409.782	29327.39	CE	1.0			MIT
3408.468	3409.446	29330.28	CE	0.3			MIT
3408.394	3409.372	29330.92	CE	0.9			MIT
3408.188	3409.166	29332.69	CE	0.3			MIT
3408.044	3409.022	29333.93	CE	0.3			MIT
3407.695	3408.673	29336.93	CE	0.9H			MIT
3407.599	3408.577	29337.76	CE	0.3			MIT
3407.243	3408.221	29340.82	CE	1.4	0.3		MIT
3406.942	3407.919	29343.42	CE	0.3			MIT
3406.441	3407.418	29347.73	CE	0.3			MIT
3406.426	3407.403	29347.86	CE	0.3			MIT
3406.364	3407.341	29348.40	CE II	0.9			MIT
3406.215	3407.192	29349.68	CE	1.0			MIT
3406.095	3407.072	29350.71	CE	0.7			MIT
3405.977	3406.954	29351.73	CE	1.4	0.5		MIT
3405.807	3406.784	29353.20	CE	1.0			MIT
3405.631	3406.608	29354.71	CE	0.6			MIT
3405.444	3406.421	29356.32	CE	0.8			MIT
3404.910	3405.887	29360.93	CE	1.3	0.3		MIT
3404.428	3405.405	29365.08	CE	1.1			MIT
3404.130	3405.107	29367.66	CE	1.3			MIT
3403.846	3404.823	29370.10	CE	0.9			MIT
3403.728	3404.705	29371.12	CE	0.5			MIT
3403.603	3404.580	29372.20	CE	1.2			MIT
3403.183	3404.160	29375.83	CE	1.0			MIT
3402.542	3403.518	29381.36	CE	0.5			MIT
3402.184	3403.160	29384.45	CE	1.0			MIT
3402.020	3402.996	29385.87	CE	0.3			MIT
3400.254	3401.230	29401.13	CE	0.9O			MIT
3399.074	3400.049	29411.34	CE	0.3			MIT
3398.925	3399.900	29412.63	CE	1.2	0.0		MIT
3398.722	3399.697	29414.38	CE	1.0			MIT
3398.639	3399.614	29415.10	CE	0.6			MIT
3398.339	3399.314	29417.70	CE	1.0			MIT
3397.083	3398.058	29428.57	CE	1.1			MIT
3396.724	3397.699	29431.68	CE	1.2			MIT
3395.414	3396.389	29443.04	CE	1.0			MIT
3395.036	3396.010	29446.32	CE	1.2			MIT
3394.138	3395.112	29454.11	CE II	1.2	0.0		MIT
3393.920	3394.894	29456.00	CE	1.5	0.0		MIT
3393.574	3394.548	29459.00	CE	1.0			MIT
3392.784	3393.758	29465.86	CE II	1.0			MIT
3392.256	3393.230	29470.45	CE	0.5			MIT
3391.884	3392.858	29473.68	CE	0.5			MIT
3391.593	3392.567	29476.21	CE	1.0			MIT
3390.812	3391.785	29483.00	CE	0.9			MIT
3390.515	3391.488	29485.58	CE II	1.3	0.0		MIT
3389.837	3390.810	29491.48	CE	0.9			MIT
3389.643	3390.616	29493.17	CE	0.9			MIT
3389.419	3390.392	29495.11	CE	0.7			MIT
3389.202	3390.175	29497.00	CE II	0.7			MIT
3388.485	3389.458	29503.24	CE	0.7			MIT
3388.407	3389.380	29503.92	CE II	0.9			MIT
3387.780	3388.753	29509.38	CE	1.3	0.3H		MIT
3387.553	3388.526	29511.36	CE	0.3			MIT
3387.342	3388.314	29513.20	CE	0.6			MIT
3386.859	3387.831	29517.41	CE	0.9O			MIT
3386.405	3387.377	29521.36	CE	0.3			MIT
3386.089	3387.061	29524.12	CE	0.9			MIT
3385.749	3386.721	29527.08	CE	0.6			MIT
3385.072	3386.044	29532.99	CE	0.7			MIT
3384.069	3385.041	29541.74	CE	0.9			MIT
3383.925	3384.897	29543.00	CE II	0.9			MIT
3383.687	3384.659	29545.08	CE	1.3	0.3		MIT
3383.394	3384.365	29547.64	CE	1.1	0.0		MIT
3383.279	3384.250	29548.64	CE	0.9			MIT

AIR WAVELENGTH (ANGSTRMS)	VACUUM WAVELENGTH (ANGSTRMS)	WAVENUMBER (KAYSERS)	LMENT	LOG ARC	INTENSITY SPK	INTENSITY DIS	REF
3382.888	3383.859	29552.06	CE	0.9			MIT
3382.698	3383.669	29553.72	CE	0.9			MIT
3382.512	3383.483	29555.34	CE	0.9			MIT
3382.308	3383.279	29557.12	CE	0.9			MIT
3381.491	3382.462	29564.26	CE	1.3			MIT
3381.121	3382.092	29567.50	CE	1.0			MIT
3380.916	3381.887	29569.29	CE	1.0			MIT
3379.172	3380.142	29584.55	CE	1.5	0.3		MIT
3378.815	3379.785	29587.68	CE	0.7			MIT
3378.188	3379.158	29593.17	CE	0.7			MIT
3378.127	3379.097	29593.70	CE	0.7			MIT
3377.846	3378.816	29596.17	CE	0.7			MIT
3377.771	3378.741	29596.82	CE	0.9			MIT
3377.307	3378.277	29600.89	CE	0.3			MIT
3377.127	3378.097	29602.47	CE	1.7	0.7S		MIT
3376.963	3377.933	29603.90	CE	0.3			MIT
3376.681	3377.651	29606.38	CE	0.6			MIT
3376.332	3377.302	29609.44	CE	0.3			MIT
3376.003	3376.973	29612.32	CE	0.3			MIT
3375.775	3376.745	29614.32	CE	1.3	0.3		MIT
3375.588	3376.557	29615.96	CE	0.7			MIT
3375.510	3376.479	29616.65	CE	0.7			MIT
3375.135	3376.104	29619.94	CE	1.0			MIT
3374.826	3375.795	29622.65	CE	0.7			MIT
3374.499	3375.468	29625.52	CE	0.3			MIT
3374.157	3375.126	29628.52	CE	1.0	0.0		MIT
3373.982	3374.951	29630.06	CE	0.6			MIT
3373.729	3374.698	29632.28	CE II	1.4	0.5		MIT
3373.455	3374.424	29634.69	CE	1.4	0.5H		MIT
3372.648	3373.617	29641.78	CE	0.5			MIT
3372.539	3373.508	29642.74	CE	0.9			MIT
3372.391	3373.360	29644.04	CE	0.5			MIT
3371.845	3372.814	29648.84	CE	0.9			MIT
3371.172	3372.140	29654.76	CE	1.4	0.5		MIT
3371.063	3372.031	29655.72	CE	0.9			MIT
3370.863	3371.831	29657.48	CE	0.8			MIT
3370.396	3371.364	29661.58	CE	0.9			MIT
3370.045	3371.013	29664.67	CE	1.1			MIT
3369.994	3370.962	29665.12	CE	0.5			MIT
3369.749	3370.717	29667.28	CE	0.3			MIT
3369.672	3370.640	29667.96	CE	0.3			MIT
3369.387	3370.355	29670.47	CE	1.0			MIT
3369.102	3370.070	29672.98	CE	0.9			MIT
3369.044	3370.012	29673.49	CE	0.7			MIT
3368.937	3369.905	29674.43	CE	0.7			MIT
3368.794	3369.762	29675.69	CE	1.1	0.0		MIT
3368.690	3369.658	29676.60	CE II	1.3	0.3		MIT
3368.375	3369.343	29679.38	CE	1.0			MIT
3368.120	3369.088	29681.63	CE	0.7			MIT
3367.778	3368.745	29684.64	CE	0.7W			MIT
3367.525	3368.492	29686.87	CE	0.7			MIT
3367.316	3368.283	29688.71	CE	0.9			MIT
3367.158	3368.125	29690.11	CE	0.3			MIT
3366.675	3367.642	29694.37	CE	0.3			MIT
3366.554	3367.521	29695.43	CE II	1.5	0.5		MIT
3365.837	3366.804	29701.76	CE	1.2	0.3		MIT
3365.708	3366.675	29702.90	CE	0.7			MIT
3365.329	3366.296	29706.24	CE	0.9			MIT
3364.821	3365.788	29710.73	CE II	1.2	0.6		MIT
3364.623	3365.590	29712.48	CE	1.1	0.6		MIT
3364.345	3365.312	29714.93	CE	1.3	0.0		MIT
3364.019	3364.986	29717.81	CE	1.0			MIT
3363.537	3364.503	29722.07	CE	0.9			MIT
3363.369	3364.335	29723.55	CE	0.3			MIT
3363.207	3364.173	29724.99	CE	0.3			MIT
3362.930	3363.896	29727.43	CE	0.5			MIT
3361.853	3362.819	29736.96	CE II	0.9			MIT
3361.762	3362.728	29737.76	CE	1.4	0.0		MIT
3361.555	3362.521	29739.59	CE II	1.0			MIT
3361.232	3362.198	29742.45	CE	0.9	0.7		MIT

AIR WAVELENGTH (ANGSTRMS)	VACUUM WAVELENGTH (ANGSTRMS)	WAVENUMBER (KAYSERS)	LMENT	LOG ARC	INTENSITY SPK	DIS	REF
3360.993	3361.959	29744.56	CE	0.5			MIT
3360.714	3361.680	29747.03	CE	0.7			MIT
3360.541	3361.507	29748.56	CE	1.5	0.6		MIT
3360.405	3361.371	29749.77	CE	0.3			MIT
3360.366	3361.332	29750.11	CE	0.3			MIT
3360.291	3361.257	29750.78	CE	0.5			MIT
3359.445	3360.410	29758.27	CE	0.9			MIT
3359.305	3360.270	29759.51	CE	1.0			MIT
3358.494	3359.459	29766.70	CE	0.9			MIT
3358.426	3359.391	29767.30	CE	0.9			MIT
3357.920	3358.885	29771.78	CE	0.9			MIT
3357.853	3358.818	29772.38	CE	0.9			MIT
3357.726	3358.691	29773.50	CE	0.9			MIT
3357.215	3358.180	29778.04	CE	1.5	0.5		MIT
3356.765	3357.730	29782.03	CE	0.9			MIT
3356.410	3357.375	29785.18	CE	1.4	0.3		MIT
3356.074	3357.038	29788.16	CE	0.9			MIT
3355.361	3356.325	29794.49	CE	0.6			MIT
3355.021	3355.985	29797.51	CE II	1.3S	0.5		MIT
3354.758	3355.722	29799.84	CE	0.3			MIT
3354.520	3355.484	29801.96	CE II	1.0			MIT
3353.944	3354.908	29807.08	CE	1.0			MIT
3353.330	3354.294	29812.53	CE	1.0	1.5		MIT
3352.986	3353.950	29815.59	CE II	1.0	0.3		MIT
3352.938	3353.902	29816.02	CE	1.3			MIT
3352.283	3353.247	29821.85	CE II	1.4	0.0		MIT
3351.750	3352.713	29826.59	CE	0.5			MIT
3351.520	3352.483	29828.64	CE	1.1			MIT
3351.077	3352.040	29832.58	CE II	0.5			MIT
3350.935	3351.898	29833.84	CE	0.5			MIT
3350.747	3351.710	29835.52	CE	1.0			MIT
3350.683	3351.646	29836.08	CE	1.3			MIT
3349.967	3350.930	29842.46	CE II	1.5	0.0		MIT
3349.405	3350.368	29847.47	CE	0.6	0.7		MIT
3348.775	3349.738	29853.08	CE	0.8			MIT
3348.190	3349.152	29858.30	CE	0.9			MIT
3347.472	3348.434	29864.70	CE	0.9			MIT
3346.517	3347.479	29873.23	CE II	1.2	0.3		MIT
3346.167	3347.129	29876.35	CE	0.6			MIT
3345.436	3346.398	29882.88	CE	1.3			MIT
3345.230	3346.192	29884.72	CE	0.5			MIT
3344.761	3345.723	29888.91	CE II	1.7	0.9		MIT
3344.335	3345.296	29892.72	CE	0.9			MIT
3343.861	3344.822	29896.95	CE	1.7	0.8		MIT
3343.233	3344.194	29902.57	CE	0.5			MIT
3342.845	3343.806	29906.04	CE	0.3			MIT
3342.531	3343.492	29908.85	CE II	0.9			MIT
3342.312	3343.273	29910.81	CE	0.5			MIT
3342.195	3343.156	29911.86	CE	0.5			MIT
3341.868	3342.829	29914.78	CE II	1.6	0.7		MIT
3341.868	3342.829	29914.78	CE I	1.6	0.7		MIT
3341.640	3342.601	29916.82	CE	0.3			MIT
3341.293	3342.254	29919.93	CE	1.0			MIT
3341.010	3341.971	29922.47	CE	0.8			MIT
3340.886	3341.847	29923.58	CE II	1.3			MIT
3340.307	3341.267	29928.76	CE	1.1			MIT
3339.797	3340.757	29933.33	CE	1.0			MIT
3339.508	3340.468	29935.92	CE II	1.1			MIT
3338.534	3339.494	29944.66	CE	0.3			MIT
3337.872	3338.832	29950.59	CE	1.0			MIT
3337.502	3338.462	29953.92	CE	1.3			MIT
3336.742	3337.702	29960.74	CE	1.0			MIT
3336.551	3337.511	29962.45	CE	1.0			MIT
3336.372	3337.331	29964.06	CE	1.3S			MIT
3336.148	3337.107	29966.07	CE	0.3			MIT
3336.073	3337.032	29966.75	CE	0.3			MIT
3335.883	3336.842	29968.45	CE	0.3			MIT
3335.692	3336.651	29970.17	CE	0.9			MIT
3335.552	3336.511	29971.43	CE	0.3			MIT
3334.882	3335.841	29977.45	CE II	0.6			MIT

AIR WAVELENGTH (ANGSTRMS)	VACUUM WAVELENGTH (ANGSTRMS)	WAVENUMBER (KAYSERS)	LMENT	LOG ARC	INTENSITY SPK	DIS	REF
3334.603	3335.562	29979.96	CE	0.6			MIT
3334.455	3335.414	29981.29	CE	1.3S	0.5		MIT
3334.279	3335.238	29982.87	CE	1.0			MIT
3333.901	3334.860	29986.27	CE	0.9			MIT
3333.660	3334.619	29988.44	CE	1.3			MIT
3333.311	3334.270	29991.57	CE	0.7			MIT
3333.037	3333.996	29994.04	CE	1.2			MIT
3332.881	3333.840	29995.45	CE	1.0			MIT
3332.474	3333.432	29999.11	CE	1.0			MIT
3332.204	3333.162	30001.54	CE	0.7			MIT
3331.793	3332.751	30005.24	CE	1.0			MIT
3331.413	3332.371	30008.66	CE	0.5			MIT
3331.224	3332.182	30010.36	CE II	1.0			MIT
3330.638	3331.596	30015.64	CE	0.5			MIT
3330.484	3331.442	30017.03	CE	0.9			MIT
3330.379	3331.337	30017.98	CE	0.6			MIT
3329.002	3329.960	30030.39	CE	1.0			MIT
3328.920	3329.878	30031.13	CE	0.7			MIT
3327.991	3328.948	30039.52	CE	0.7			MIT
3327.899	3328.856	30040.35	CE	1.1			MIT
3327.661	3328.618	30042.50	CE	1.1			MIT
3327.225	3328.182	30046.43	CE	1.0			MIT
3326.947	3327.904	30048.94	CE	1.2			MIT
3326.077	3327.034	30056.80	CE	1.0			MIT
3325.329	3326.286	30063.56	CE	1.4S	0.0		MIT
3325.144	3326.101	30065.24	CE	0.9			MIT
3325.063	3326.020	30065.97	CE	0.6			MIT
3324.985	3325.942	30066.67	CE II	1.2			MIT
3324.764	3325.721	30068.67	CE	1.1			MIT
3324.594	3325.550	30070.21	CE	0.5			MIT
3323.976	3324.932	30075.80	CE	0.3			MIT
3323.452	3324.408	30080.54	CE	0.3			MIT
3323.294	3324.250	30081.97	CE	1.0			MIT
3323.100	3324.056	30083.73	CE	0.5			MIT
3322.916	3323.872	30085.39	CE	0.3	0.0		MIT
3322.634	3323.590	30087.95	CE	1.0			MIT
3322.175	3323.131	30092.10	CE	0.3			MIT
3321.275	3322.231	30100.26	CE	0.5			MIT
3320.940	3321.896	30103.29	CE II	1.0			MIT
3320.787	3321.742	30104.68	CE II	0.9			MIT
3320.553	3321.508	30106.80	CE	0.7			MIT
3320.423	3321.378	30107.98	CE	1.1			MIT
3318.964	3319.919	30121.22	CE II	1.2	0.0		MIT
3318.403	3319.358	30126.31	CE	0.9			MIT
3317.901	3318.856	30130.87	CE	0.6			MIT
3317.797	3318.752	30131.81	CE	1.2	0.3		MIT
3317.224	3318.179	30137.02	CE	0.3			MIT
3316.983	3317.938	30139.21	CE	0.7			MIT
3316.540	3317.494	30143.23	CE	0.7			MIT
3316.113	3317.067	30147.11	CE	0.3			MIT
3315.112	3316.066	30156.22	CE	0.9			MIT
3314.721	3315.675	30159.77	CE II	1.4	0.5		MIT
3314.035	3314.989	30166.01	CE	1.3			MIT
3313.680	3314.634	30169.25	CE	0.3H			MIT
3313.531	3314.485	30170.60	CE	0.9			MIT
3313.300	3314.254	30172.71	CE	1.3	0.0		MIT
3312.215	3313.168	30182.59	CE	1.5	0.7		MIT
3311.732	3312.685	30186.99	CE	0.3H			MIT
3311.497	3312.450	30189.13	CE	1.2	0.0		MIT
3311.388	3312.341	30190.13	CE	0.5			MIT
3310.877	3311.830	30194.79	CE	1.0			MIT
3310.629	3311.582	30197.05	CE	0.7			MIT
3309.834	3310.787	30204.30	CE	0.7			MIT
3309.500	3310.453	30207.35	CE	0.3			MIT
3309.273	3310.226	30209.42	CE	1.0	0.0		MIT
3309.132	3310.085	30210.71	CE	0.9			MIT
3308.691	3309.643	30214.74	CE	0.3			MIT
3308.271	3309.223	30218.57	CE	0.3			MIT
3308.088	3309.040	30220.24	CE	1.0			MIT
3308.017	3308.969	30220.89	CE	1.2			MIT

AIR WAVELENGTH (ANGSTRMS)	VACUUM WAVELENGTH (ANGSTRMS)	WAVENUMBER (KAYSERS)	LMENT	LOG ARC	INTENSITY SPK	DIS	REF
3307.233	3308.185	30228.05	CE II	1.4	0.0		MIT
3306.632	3307.584	30233.55	CE	1.3			MIT
3306.182	3307.134	30237.66	CE	0.7			MIT
3306.048	3307.000	30238.89	CE	0.9			MIT
3305.051	3306.002	30248.01	CE	1.0			MIT
3304.836	3305.787	30249.98	CE	1.5	0.5		MIT
3304.022	3304.973	30257.43	CE	0.5			MIT
3303.771	3304.722	30259.73	CE	0.9			MIT
3303.670	3304.621	30260.66	CE	0.7			MIT
3303.225	3304.176	30264.73	CE II	1.0			MIT
3302.913	3303.864	30267.59	CE	1.0O			MIT
3301.905	3302.856	30276.83	CE	1.0			MIT
3301.652	3302.603	30279.15	CE	0.5			MIT
3301.230	3302.181	30283.02	CE	0.5			MIT
3300.955	3301.905	30285.54	CE	0.9			MIT
3300.152	3301.102	30292.91	CE	1.5	0.5		MIT
3299.985	3300.935	30294.45	CE	1.2	0.0		MIT
3298.346	3299.296	30309.50	CE	1.2			MIT
3298.186	3299.136	30310.97	CE	0.6			MIT
3297.954	3298.904	30313.10	CE	0.6			MIT
3296.883	3297.832	30322.95	CE	1.4	0.3		MIT
3296.763	3297.712	30324.05	CE	0.3			MIT
3296.185	3297.134	30329.37	CE	1.3	0.3		MIT
3295.289	3296.238	30337.61	CE I	1.5	0.5		MIT
3295.289	3296.238	30337.61	CE II	1.5	0.5		MIT
3294.948	3295.897	30340.75	CE	1.3			MIT
3294.606	3295.555	30343.90	CE	0.5			MIT
3293.945	3294.894	30349.99	CE	0.5			MIT
3293.589	3294.538	30353.27	CE	1.3			MIT
3292.929	3293.877	30359.36	CE	1.1			MIT
3292.099	3293.047	30367.01	CE	0.3			MIT
3290.577	3291.525	30381.06	CE	1.0	0.0		MIT
3290.341	3291.289	30383.24	CE II	1.3	0.0		MIT
3289.944	3290.892	30386.90	CE	0.6			MIT
3289.790	3290.738	30388.32	CE	0.3			MIT
3289.523	3290.471	30390.79	CE	0.3			MIT
3289.280	3290.227	30393.03	CE	1.0			MIT
3288.941	3289.888	30396.17	CE	0.3			MIT
3288.770	3289.717	30397.75	CE	1.2			MIT
3288.561	3289.508	30399.68	CE	0.3			MIT
3288.343	3289.290	30401.70	CE	0.3			MIT
3288.199	3289.146	30403.03	CE	0.5			MIT
3288.147	3289.094	30403.51	CE	1.1			MIT
3287.884	3288.831	30405.94	CE	1.0			MIT
3287.790	3288.737	30406.81	CE	0.8			MIT
3287.399	3288.346	30410.43	CE	1.0			MIT
3286.828	3287.775	30415.71	CE	0.6			MIT
3286.676	3287.623	30417.11	CE	0.6			MIT
3286.375	3287.322	30419.90	CE	0.9			MIT
3286.216	3287.163	30421.37	CE	0.7			MIT
3286.029	3286.976	30423.10	CE II	1.3	0.0		MIT
3286.029	3286.976	30423.10	CE I	1.3	0.0		MIT
3285.786	3286.733	30425.35	CE	0.5W			MIT
3285.224	3286.170	30430.56	CE II	1.5	0.7		MIT
3284.606	3285.552	30436.28	CE	0.9			MIT
3284.421	3285.367	30438.00	CE	0.9			MIT
3284.218	3285.164	30439.88	CE II	1.3			MIT
3283.888	3284.834	30442.94	CE	0.5			MIT
3283.680	3284.626	30444.87	CE II	1.4	0.0		MIT
3283.354	3284.300	30447.89	CE	1.3S	0.0		MIT
3283.175	3284.121	30449.55	CE	0.5			MIT
3282.319	3283.265	30457.49	CE	0.3			MIT
3281.996	3282.942	30460.49	CE	0.5			MIT
3281.951	3282.897	30460.90	CE	0.5			MIT
3281.587	3282.533	30464.28	CE	0.5			MIT
3281.095	3282.040	30468.85	CE	1.3			MIT
3280.668	3281.613	30472.82	CE	0.8			MIT
3280.485	3281.430	30474.52	CE II	1.2	0.0		MIT
3279.842	3280.787	30480.49	CE II	1.5	0.7		MIT
3279.842	3280.787	30480.49	CE I	1.5	0.7		MIT
3279.205	3280.150	30486.41	CE	0.3			MIT
3279.009	3279.954	30488.23	CE	1.3	0.5		MIT
3277.989	3278.934	30497.72	CE	0.3			MIT
3277.934	3278.879	30498.23	CE	0.9			MIT
3277.143	3278.087	30505.59	CE	0.7			MIT
3276.966	3277.910	30507.24	CE	0.3			MIT
3276.360	3277.304	30512.88	CE	0.8			MIT
3276.251	3277.195	30513.90	CE	1.3	0.0		MIT
3275.567	3276.511	30520.27	CE	0.3W			MIT
3274.864	3275.808	30526.82	CE	1.5	0.9		MIT
3274.616	3275.560	30529.13	CE	0.9			MIT
3274.112	3275.056	30533.83	CE	0.9			MIT
3274.064	3275.008	30534.28	CE	1.0			MIT
3273.964	3274.908	30535.21	CE	0.7			MIT
3273.926	3274.870	30535.57	CE	0.7			MIT
3273.516	3274.459	30539.39	CE	0.6			MIT
3272.947	3273.890	30544.70	CE	0.5			MIT
3272.725	3273.668	30546.77	CE	1.0			MIT
3272.606	3273.549	30547.88	CE	0.9			MIT
3272.253	3273.196	30551.18	CE	1.6	1.2		MIT
3272.064	3273.007	30552.94	CE	0.9			MIT
3271.961	3272.904	30553.91	CE	0.7			MIT
3271.550	3272.493	30557.74	CE	1.4	0.0		MIT
3271.151	3272.094	30561.47	CE II	1.3	0.3		MIT
3270.686	3271.629	30565.81	CE	0.7			MIT
3270.133	3271.076	30570.98	CE	1.1			MIT
3269.129	3270.071	30580.37	CE	1.0			MIT
3268.022	3268.964	30590.73	CE	0.3			MIT
3267.237	3268.179	30598.08	CE II	1.1			MIT
3266.865	3267.807	30601.56	CE	0.5			MIT
3266.089	3267.031	30608.83	CE	0.5			MIT
3265.828	3266.770	30611.28	CE	1.0			MIT
3265.679	3266.620	30612.68	CE	0.7			MIT
3265.568	3266.509	30613.72	CE	0.9			MIT
3265.423	3266.364	30615.08	CE	1.2	0.0		MIT
3265.145	3266.086	30617.68	CE	0.3			MIT
3264.711	3265.652	30621.75	CE	0.3			MIT
3263.884	3264.825	30629.51	CE II	1.4	0.5		MIT
3263.445	3264.386	30633.63	CE	1.3	0.5		MIT
3263.339	3264.280	30634.63	CE	0.9			MIT
3263.071	3264.012	30637.14	CE II	1.0			MIT
3262.138	3263.079	30645.91	CE	0.7			MIT
3261.633	3262.573	30650.65	CE	0.3	0.3H		MIT
3261.524	3262.464	30651.68	CE	0.3			MIT
3261.242	3262.182	30654.33	CE II	0.9			MIT
3261.116	3262.056	30655.51	CE	0.3			MIT
3260.975	3261.915	30656.83	CE II	1.4	0.5		MIT
3260.320	3261.260	30662.99	CE	0.5			MIT
3259.784	3260.724	30668.04	CE II	1.3			MIT
3259.557	3260.497	30670.17	CE	0.5			MIT
3258.874	3259.814	30676.60	CE	1.4	0.0		MIT
3257.813	3258.752	30686.59	CE	0.8			MIT
3257.132	3258.071	30693.00	CE	0.6			MIT
3256.682	3257.621	30697.25	CE	1.3			MIT
3256.251	3257.190	30701.31	CE	1.1			MIT
3256.232	3257.171	30701.49	CE	0.8			MIT
3255.215	3256.154	30711.08	CE	0.9			MIT
3254.866	3255.805	30714.37	CE	1.1			MIT
3254.667	3255.606	30716.25	CE	0.5			MIT
3254.287	3255.226	30719.84	CE	0.5			MIT
3254.013	3254.952	30722.42	CE II	1.5	0.6		MIT
3253.342	3254.280	30728.76	CE	0.9			MIT
3252.982	3253.920	30732.16	CE	0.3			MIT
3252.483	3253.421	30736.87	CE	1.5	0.5		MIT
3251.885	3252.823	30742.53	CE	1.2			MIT
3251.370	3252.308	30747.40	CE	0.6			MIT
3249.831	3250.768	30761.96	CE	0.6			MIT
3249.498	3250.435	30765.11	CE	0.5			MIT
3249.429	3250.366	30765.76	CE	1.0			MIT
3249.170	3250.107	30768.22	CE	1.4			MIT

AIR WAVELENGTH (ANGSTRMS)	VACUUM WAVELENGTH (ANGSTRMS)	WAVENUMBER (KAYSERS)	LMENT	LOG ARC	INTENSITY SPK	INTENSITY DIS	REF
3248.528	3249.465	30774.29	CE	0.6			MIT
3248.428	3249.365	30775.24	CE	1.1			MIT
3248.065	3249.002	30778.68	CE	0.7			MIT
3247.898	3248.835	30780.26	CE	0.5			MIT
3247.552	3248.489	30783.54	CE	1.2			MIT
3247.250	3248.187	30786.41	CE	0.5			MIT
3247.118	3248.055	30787.66	CE	0.5			MIT
3246.674	3247.611	30791.87	CE II	1.5	0.5		MIT
3246.674	3247.611	30791.87	CE I	1.5	0.5		MIT
3245.684	3246.620	30801.26	CE	0.3			MIT
3245.546	3246.482	30802.57	CE	1.2			MIT
3245.166	3246.102	30806.18	CE	1.4	0.0		MIT
3245.124	3246.060	30806.57	CE	0.9			MIT
3244.950	3245.886	30808.23	CE	1.0			MIT
3243.949	3244.885	30817.73	CE	0.5			MIT
3243.573	3244.509	30821.31	CE	0.7			MIT
3243.370	3244.306	30823.24	CE II		1.0		MIT
3243.214	3244.150	30824.72	CE	0.3			MIT
3242.788	3243.724	30828.77	CE	0.5			MIT
3242.539	3243.475	30831.13	CE	0.3			MIT
3242.135	3243.071	30834.98	CE II	1.1			MIT
3241.358	3242.293	30842.37	CE	0.5			MIT
3241.214	3242.149	30843.74	CE	1.0			MIT
3240.672	3241.607	30848.90	CE	0.5			MIT
3240.398	3241.333	30851.50	CE	0.6			MIT
3238.902	3239.837	30865.75	CE	0.6			MIT
3236.856	3237.790	30885.26	CE	0.9			MIT
3236.735	3237.669	30886.42	CE II	1.5	0.9		MIT
3235.670	3236.604	30896.58	CE	1.2	0.0		MIT
3235.011	3235.945	30902.88	CE	1.0	0.0		MIT
3234.892	3235.826	30904.01	CE	1.3S	0.7		MIT
3234.495	3235.429	30907.81	CE	1.0	0.5		MIT
3234.274	3235.208	30909.92	CE	0.7			MIT
3234.161	3235.094	30911.00	CE II	1.6	0.9		MIT
3234.161	3235.094	30911.00	CE I	1.6	0.9		MIT
3233.773	3234.706	30914.71	CE	1.2			MIT
3233.441	3234.374	30917.88	CE II	1.5	0.3		MIT
3233.208	3234.141	30920.11	CE	0.3			MIT
3232.875	3233.808	30923.29	CE	0.5			MIT
3232.665	3233.598	30925.30	CE	0.5			MIT
3232.290	3233.223	30928.89	CE	1.2			MIT
3231.980	3232.913	30931.86	CE	0.5			MIT
3231.807	3232.740	30933.51	CE	0.3			MIT
3231.236	3232.169	30938.98	CE II	1.5	1.0		MIT
3230.288	3231.221	30948.06	CE	0.9			MIT
3230.079	3231.011	30950.06	CE	1.0	0.0		MIT
3230.023	3230.955	30950.60	CE	0.5			MIT
3229.599	3230.531	30954.66	CE	1.0			MIT
3229.363	3230.295	30956.92	CE II	1.4	0.5		MIT
3229.122	3230.054	30959.23	CE	1.4	0.0		MIT
3228.048	3228.980	30969.53	CE	1.4			MIT
3227.114	3228.046	30978.50	CE	1.4			MIT
3226.036	3226.967	30988.85	CE II	0.9			MIT
3225.672	3226.603	30992.34	CE	1.4	0.5		MIT
3225.485	3226.416	30994.14	CE	0.3			MIT
3225.408	3226.339	30994.88	CE	0.3			MIT
3224.834	3225.765	31000.40	CE	1.0			MIT
3224.295	3225.226	31005.58	CE	0.5			MIT
3223.905	3224.836	31009.33	CE	1.2	0.0		MIT
3223.519	3224.450	31013.04	CE	0.7			MIT
3223.365	3224.296	31014.52	CE	1.2	0.0		MIT
3222.853	3223.784	31019.45	CE	0.8L	0.0		MIT
3222.611	3223.542	31021.78	CE	0.8			MIT
3222.411	3223.342	31023.71	CE	1.3	0.0		MIT
3222.003	3222.933	31027.64	CE	0.3			MIT
3221.470	3222.400	31032.77	CE	0.3			MIT
3221.171	3222.101	31035.65	CE II	1.7	0.9		MIT
3221.061	3221.991	31036.71	CE	0.3			MIT
3220.871	3221.801	31038.54	CE	1.5			MIT
3220.402	3221.332	31043.06	CE	1.1S			MIT

AIR WAVELENGTH (ANGSTRMS)	VACUUM WAVELENGTH (ANGSTRMS)	WAVENUMBER (KAYSERS)	LMENT	LOG ARC	INTENSITY SPK	INTENSITY DIS	REF
3218.944	3219.874	31057.12	CE	1.7	0.9		MIT
3218.380	3219.309	31062.56	CE II	1.5	0.3		MIT
3218.380	3219.309	31062.56	CE I	1.5	0.3		MIT
3217.866	3218.795	31067.52	CE	0.5			MIT
3217.522	3218.451	31070.84	CE	1.3			MIT
3217.132	3218.061	31074.61	CE	1.1			MIT
3216.717	3217.646	31078.62	CE	1.1			MIT
3216.587	3217.516	31079.88	CE	0.5			MIT
3216.395	3217.324	31081.73	CE	0.5			MIT
3216.225	3217.154	31083.37	CE	0.5			MIT
3216.181	3217.110	31083.80	CE	0.5			MIT
3215.896	3216.825	31086.55	CE	0.9			MIT
3215.602	3216.531	31089.40	CE	0.6			MIT
3215.093	3216.022	31094.32	CE	1.0			MIT
3214.883	3215.812	31096.35	CE	0.3			MIT
3214.630	3215.559	31098.80	CE	0.3			MIT
3214.253	3215.181	31102.44	CE	1.1			MIT
3214.024	3214.952	31104.66	CE	1.2			MIT
3213.530	3214.458	31109.44	CE	0.3			MIT
3213.060	3213.988	31113.99	CE	1.3			MIT
3212.899	3213.827	31115.55	CE	0.5			MIT
3212.591	3213.519	31118.53	CE	0.9			MIT
3212.480	3213.408	31119.61	CE	1.3			MIT
3212.320	3213.248	31121.16	CE	0.5			MIT
3211.612	3212.540	31128.02	CE	1.0			MIT
3211.336	3212.264	31130.69	CE	1.0			MIT
3210.951	3211.879	31134.43	CE	1.3	0.3		MIT
3210.094	3211.021	31142.74	CE	0.9			MIT
3209.510	3210.437	31148.41	CE	0.5			MIT
3209.359	3210.286	31149.87	CE	0.5			MIT
3209.048	3209.975	31152.89	CE	0.3			MIT
3207.624	3208.551	31166.72	CE II	1.1			MIT
3207.286	3208.213	31170.00	CE	0.9			MIT
3207.168	3208.095	31171.15	CE	0.5			MIT
3206.921	3207.848	31173.55	CE II	1.1			MIT
3206.509	3207.435	31177.56	CE	1.1			MIT
3206.035	3206.961	31182.17	CE	0.3			MIT
3205.960	3206.886	31182.90	CE	1.0	0.0		MIT
3205.406	3206.332	31188.28	CE	0.7			MIT
3204.920	3205.846	31193.01	CE	1.0			MIT
3204.356	3205.282	31198.50	CE	1.2			MIT
3204.160	3205.086	31200.41	CE	1.5			MIT
3203.144	3204.070	31210.31	CE	0.6			MIT
3202.906	3203.832	31212.63	CE	1.2			MIT
3202.653	3203.579	31215.09	CE	0.3			MIT
3202.443	3203.368	31217.14	CE	0.3			MIT
3202.356	3203.281	31217.99	CE	1.0			MIT
3202.243	3203.168	31219.09	CE	1.2			MIT
3201.714	3202.639	31224.25	CE	1.7	1.0		MIT
3201.421	3202.346	31227.10	CE	1.0			MIT
3201.105	3202.030	31230.19	CE	1.2			MIT
3200.867	3201.792	31232.51	CE	0.9			MIT
3200.625	3201.550	31234.87	CE	0.8			MIT
3200.515	3201.440	31235.94	CE	1.0	0.0		MIT
3200.126	3201.051	31239.74	CE	0.3			MIT
3199.279	3200.204	31248.01	CE	1.4	0.3		MIT
3198.729	3199.654	31253.38	CE	0.6			MIT
3198.240	3199.164	31258.16	CE	0.7			MIT
3197.036	3197.960	31269.93	CE	0.3			MIT
3195.936	3196.860	31280.70	CE	1.3			MIT
3195.713	3196.637	31282.88	CE	1.1			MIT
3195.550	3196.474	31284.47	CE	1.1			MIT
3195.323	3196.247	31286.70	CE	0.7			MIT
3194.825	3195.749	31291.57	CE	1.4			MIT
3194.683	3195.606	31292.96	CE	0.7			MIT
3194.102	3195.025	31298.66	CE	1.2			MIT
3193.981	3194.904	31299.84	CE	0.3			MIT
3193.333	3194.256	31306.19	CE	1.2			MIT
3193.126	3194.049	31308.22	CE	1.0			MIT
3191.395	3192.318	31325.20	CE	0.3			MIT

AIR WAVELENGTH (ANGSTRMS)	VACUUM WAVELENGTH (ANGSTRMS)	WAVENUMBER (KAYSERS)	LMENT		LOG ARC	INTENSITY SPK	DIS	REF
3191.185	3192.108	31327.26	CE		1.2			MIT
3191.081	3192.004	31328.29	CE		0.7			MIT
3190.833	3191.756	31330.72	CE		0.7W			MIT
3190.395	3191.317	31335.02	CE		1.2			MIT
3190.341	3191.263	31335.55	CE	II	1.4			MIT
3190.163	3191.085	31337.30	CE		0.3			MIT
3189.638	3190.560	31342.46	CE	II	1.5			MIT
3189.265	3190.187	31346.12	CE		1.2			MIT
3189.105	3190.027	31347.70	CE		0.3			MIT
3188.787	3189.709	31350.82	CE	II	1.4			MIT
3188.307	3189.229	31355.54	CE		0.7			MIT
3187.862	3188.784	31359.92	CE		1.2			MIT
3187.669	3188.591	31361.82	CE		1.3			MIT
3186.126	3187.047	31377.00	CE		1.6			MIT
3185.506	3186.427	31383.11	CE		0.3			MIT
3185.413	3186.334	31384.03	CE		0.6			MIT
3184.721	3185.642	31390.85	CE		1.0			MIT
3184.622	3185.543	31391.82	CE		1.1			MIT
3184.330	3185.251	31394.70	CE		0.3			MIT
3184.212	3185.133	31395.86	CE	II	1.3			MIT
3183.994	3184.915	31398.01	CE		0.6			MIT
3183.844	3184.765	31399.49	CE		0.3			MIT
3183.523	3184.444	31402.66	CE		1.6			MIT
3183.093	3184.014	31406.90	CE		0.7			MIT
3182.662	3183.582	31411.15	CE		1.3			MIT
3182.200	3183.120	31415.72	CE		1.2			MIT
3181.948	3182.868	31418.20	CE		1.0			MIT
3181.840	3182.760	31419.27	CE		0.6			MIT
3181.592	3182.512	31421.72	CE		1.2			MIT
3181.378	3182.298	31423.83	CE		0.5			MIT
3181.192	3182.112	31425.67	CE		0.6			MIT
3180.824	3181.744	31429.30	CE		1.3			MIT
3179.341	3180.261	31443.96	CE		0.7			MIT
3178.754	3179.673	31449.77	CE		1.2			MIT
3178.485	3179.404	31452.43	CE	II	1.1			MIT
3178.243	3179.162	31454.83	CE		0.5			MIT
3178.107	3179.026	31456.17	CE		1.2			MIT
3177.620	3178.539	31460.99	CE		0.3			MIT
3177.137	3178.056	31465.78	CE		1.3			MIT
3176.799	3177.718	31469.12	CE		1.5			MIT
3175.716	3176.635	31479.85	CE		0.9			MIT
3175.581	3176.500	31481.19	CE		0.3			MIT
3175.059	3175.978	31486.37	CE		1.0			MIT
3174.077	3174.995	31496.11	CE		1.2			MIT
3173.622	3174.540	31500.62	CE		1.2			MIT
3173.325	3174.243	31503.57	CE		0.6			MIT
3172.541	3173.459	31511.36	CE		0.5			MIT
3172.299	3173.217	31513.76	CE	II	1.3			MIT
3171.613	3172.531	31520.58	CE	II	1.3			MIT
3171.613	3172.531	31520.58	CE	I	1.3			MIT
3171.465	3172.383	31522.05	CE		0.3			MIT
3170.528	3171.445	31531.36	CE		0.3			MIT
3170.069	3170.986	31535.93	CE		1.1			MIT
3169.396	3170.313	31542.63	CE		0.5			MIT
3169.183	3170.100	31544.75	CE		1.5			MIT
3168.621	3169.538	31550.34	CE		1.4			MIT
3167.918	3168.835	31557.34	CE	II	1.1			MIT
3167.790	3168.707	31558.62	CE		0.6			MIT
3167.324	3168.241	31563.26	CE	II	1.2			MIT
3167.226	3168.143	31564.24	CE		1.2			MIT
3166.608	3167.524	31570.40	CE		1.3			MIT
3166.557	3167.473	31570.90	CE		0.7			MIT
3166.243	3167.159	31574.03	CE	II	1.3			MIT
3164.154	3165.070	31594.88	CE	II	1.6			MIT
3163.345	3164.261	31602.96	CE		1.2			MIT
3162.996	3163.911	31606.45	CE		0.9H			MIT
3162.835	3163.750	31608.05	CE		0.6			MIT
3161.026	3161.941	31626.14	CE		1.2			MIT
3160.306	3161.221	31633.35	CE		0.7			MIT
3160.280	3161.195	31633.61	CE		0.6			MIT

AIR WAVELENGTH (ANGSTRMS)	VACUUM WAVELENGTH (ANGSTRMS)	WAVENUMBER (KAYSERS)	LMENT		LOG ARC	INTENSITY SPK	DIS	REF
3159.714	3160.629	31639.27	CE		0.6			MIT
3159.378	3160.293	31642.64	CE		0.5			MIT
3159.072	3159.987	31645.71	CE		0.9			MIT
3158.882	3159.796	31647.61	CE		0.7			MIT
3158.812	3159.726	31648.31	CE		1.2			MIT
3158.434	3159.348	31652.10	CE		1.3			MIT
3158.262	3159.176	31653.82	CE		0.3			MIT
3157.718	3158.632	31659.27	CE		0.9			MIT
3157.399	3158.313	31662.47	CE		0.7W			MIT
3156.775	3157.689	31668.73	CE		1.3			MIT
3156.422	3157.336	31672.27	CE		0.7			MIT
3155.793	3156.707	31678.58	CE	II	1.3			MIT
3155.704	3156.618	31679.48	CE		1.3			MIT
3154.506	3155.419	31691.51	CE	II	1.2			MIT
3154.005	3154.918	31696.54	CE		0.7			MIT
3153.474	3154.387	31701.88	CE		0.9			MIT
3153.345	3154.258	31703.18	CE		0.5H			MIT
3152.489	3153.402	31711.78	CE		0.3			MIT
3152.250	3153.163	31714.19	CE		0.9			MIT
3151.127	3152.040	31725.49	CE		1.4			MIT
3150.573	3151.485	31731.07	CE		0.3			MIT
3150.461	3151.373	31732.20	CE		0.7			MIT
3149.937	3150.849	31737.48	CE	II	1.1			MIT
3149.426	3150.338	31742.62	CE		1.4			MIT
3149.253	3150.165	31744.37	CE		1.1			MIT
3149.077	3149.989	31746.14	CE		0.3			MIT
3148.832	3149.744	31748.61	CE		0.6			MIT
3148.651	3149.563	31750.44	CE		1.3			MIT
3148.463	3149.375	31752.33	CE	II	1.3			MIT
3147.837	3148.749	31758.65	CE		1.2			MIT
3147.555	3148.467	31761.49	CE		0.3			MIT
3146.780	3147.691	31769.31	CE		0.7			MIT
3146.410	3147.321	31773.05	CE	II	1.5			MIT
3146.225	3147.136	31774.92	CE		0.6			MIT
3146.165	3147.076	31775.52	CE		1.3			MIT
3145.283	3146.194	31784.44	CE	II	1.5			MIT
3144.817	3145.728	31789.15	CE		0.9			MIT
3144.596	3145.507	31791.38	CE	II	1.4			MIT
3144.154	3145.065	31795.85	CE		0.7			MIT
3143.593	3144.504	31801.52	CE		0.7			MIT
3142.890	3143.800	31808.64	CE		0.3			MIT
3142.547	3143.457	31812.11	CE		1.1			MIT
3142.312	3143.222	31814.49	CE		1.4			MIT
3140.676	3141.586	31831.06	CE		1.1H			MIT
3140.432	3141.342	31833.53	CE		0.3			MIT
3140.346	3141.256	31834.40	CE		0.3			MIT
3139.241	3140.151	31845.61	CE		0.6			MIT
3139.182	3140.091	31846.21	CE		0.6			MIT
3138.296	3139.205	31855.20	CE	II	1.3			MIT
3137.809	3138.718	31860.14	CE		0.6			MIT
3137.601	3138.510	31862.25	CE		1.4			MIT
3137.406	3138.315	31864.23	CE		0.3			MIT
3136.903	3137.812	31869.34	CE		0.7			MIT
3136.722	3137.631	31871.18	CE		1.4			MIT
3136.243	3137.152	31876.05	CE		0.9			MIT
3135.833	3136.742	31880.22	CE		0.6			MIT
3135.569	3136.478	31882.90	CE		1.2			MIT
3135.179	3136.087	31886.87	CE		1.2			MIT
3134.604	3135.512	31892.72	CE		0.6			MIT
3133.533	3134.441	31903.61	CE		0.3			MIT
3133.407	3134.315	31904.90	CE		0.7			MIT
3133.327	3134.235	31905.71	CE		1.3			MIT
3132.858	3133.766	31910.49	CE		0.5			MIT
3132.640	3133.548	31912.71	CE		1.2			MIT
3132.590	3133.498	31913.22	CE		1.4			MIT
3132.043	3132.951	31918.79	CE		1.2			MIT
3131.679	3132.587	31922.50	CE		0.6			MIT
3131.505	3132.413	31924.27	CE		0.3D			MIT
3130.919	3131.826	31930.25	CE		0.7			MIT
3130.872	3131.779	31930.73	CE		1.5	0.3		MIT

AIR WAVELENGTH (ANGSTRMS)	VACUUM WAVELENGTH (ANGSTRMS)	WAVENUMBER (KAYSERS)	LMENT	LOG ARC	INTENSITY SPK	INTENSITY DIS	REF
3130.516	3131.423	31934.36	CE	0.7			MIT
3130.334	3131.241	31936.22	CE	1.5	0.0		MIT
3130.197	3131.104	31937.61	CE	1.2			MIT
3129.316	3130.223	31946.61	CE	1.1			MIT
3129.041	3129.948	31949.41	CE	0.6			MIT
3128.961	3129.868	31950.23	CE	0.3			MIT
3128.002	3128.909	31960.02	CE	1.0			MIT
3127.750	3128.657	31962.60	CE	1.2			MIT
3127.530	3128.437	31964.85	CE II	1.6			MIT
3127.182	3128.088	31968.41	CE	0.7			MIT
3127.100	3128.006	31969.24	CE	1.3			MIT
3125.762	3126.668	31982.93	CE II	1.3			MIT
3125.358	3126.264	31987.06	CE	0.3			MIT
3124.859	3125.765	31992.17	CE	0.5			MIT
3124.101	3125.007	31999.93	CE	1.3			MIT
3123.949	3124.855	32001.49	CE	1.2			MIT
3123.566	3124.472	32005.41	CE	1.1			MIT
3123.349	3124.255	32007.64	CE	0.9			MIT
3122.854	3123.759	32012.71	CE	0.5			MIT
3122.620	3123.525	32015.11	CE	1.0			MIT
3122.213	3123.118	32019.28	CE	0.3			MIT
3121.479	3122.384	32026.81	CE	0.5			MIT
3121.281	3122.186	32028.84	CE	0.3			MIT
3121.085	3121.990	32030.85	CE	0.3			MIT
3120.473	3121.378	32037.14	CE	0.3			MIT
3119.577	3120.482	32046.34	CE	1.1			MIT
3119.351	3120.256	32048.66	CE	0.3			MIT
3118.839	3119.743	32053.92	CE	0.3			MIT
3117.773	3118.677	32064.88	CE	0.5			MIT
3117.471	3118.375	32067.98	CE	0.3			MIT
3117.006	3117.910	32072.77	CE	0.6			MIT
3115.979	3116.883	32083.34	CE	0.7			MIT
3115.647	3116.551	32086.76	CE	0.3H			MIT
3115.116	3116.019	32092.23	CE	0.6			MIT
3113.693	3114.596	32106.89	CE	0.3			MIT
3112.973	3113.876	32114.32	CE	0.9			MIT
3112.755	3113.658	32116.57	CE	0.8			MIT
3112.202	3113.105	32122.27	CE	1.2			MIT
3111.826	3112.729	32126.15	CE	0.6W			MIT
3111.234	3112.136	32132.27	CE	1.0			MIT
3111.165	3112.067	32132.98	CE	1.3			MIT
3110.998	3111.900	32134.71	CE	0.3			MIT
3110.278	3111.180	32142.14	CE	1.5	0.0		MIT
3109.383	3110.285	32151.39	CE	1.2			MIT
3108.964	3109.866	32155.73	CE	1.0			MIT
3108.924	3109.826	32156.14	CE	0.9			MIT
3108.552	3109.454	32159.99	CE	0.3			MIT
3107.636	3108.538	32169.47	CE	0.3			MIT
3107.538	3108.440	32170.48	CE	0.3	0.0		MIT
3107.468	3108.370	32171.21	CE	1.4			MIT
3105.864	3106.765	32187.82	CE	0.3			MIT
3105.501	3106.402	32191.58	CE	0.5			MIT
3105.266	3106.167	32194.02	CE	1.2			MIT
3105.148	3106.049	32195.24	CE	0.6			MIT
3104.928	3105.829	32197.52	CE	0.6W			MIT
3104.593	3105.494	32201.00	CE	0.5			MIT
3104.012	3104.913	32207.02	CE	1.0H			MIT
3103.377	3104.277	32213.61	CE	1.2H	0.3		MIT
3103.269	3104.169	32214.74	CE	0.5			MIT
3103.008	3103.908	32217.45	CE	0.3			MIT
3102.563	3103.463	32222.07	CE	1.2H			MIT
3102.431	3103.331	32223.44	CE	1.0			MIT
3102.358	3103.258	32224.20	CE	1.1			MIT
3101.791	3102.691	32230.09	CE	1.3			MIT
3101.392	3102.292	32234.23	CE	1.0			MIT
3099.426	3100.325	32254.68	CE	0.7	0.0		MIT
3099.142	3100.041	32257.63	CE	0.6			MIT
3098.958	3099.857	32259.55	CE	0.3			MIT
3098.148	3099.047	32267.98	CE	0.7			MIT
3097.079	3097.978	32279.12	CE II	1.3			MIT

AIR WAVELENGTH (ANGSTRMS)	VACUUM WAVELENGTH (ANGSTRMS)	WAVENUMBER (KAYSERS)	LMENT	LOG ARC	INTENSITY SPK	INTENSITY DIS	REF
3096.883	3097.782	32281.16	CE II	1.3			MIT
3096.726	3097.625	32282.80	CE	0.3			MIT
3096.501	3097.400	32285.15	CE	1.4			MIT
3096.402	3097.301	32286.18	CE	0.7			MIT
3095.859	3096.758	32291.84	CE	0.7			MIT
3095.762	3096.661	32292.85	CE	0.9			MIT
3095.581	3096.480	32294.74	CE	1.3			MIT
3095.262	3096.160	32298.07	CE	0.5	0.0		MIT
3095.099	3095.997	32299.77	CE II	1.2			MIT
3095.099	3095.997	32299.77	CE II	1.2			MIT
3093.950	3094.848	32311.76	CE	0.6L			MIT
3093.614	3094.512	32315.27	CE	1.3			MIT
3093.339	3094.237	32318.15	CE II	1.1			MIT
3093.243	3094.141	32319.15	CE	0.8			MIT
3092.818	3093.716	32323.59	CE	0.6	0.0		MIT
3092.724	3093.622	32324.57	CE	0.6			MIT
3092.197	3093.095	32330.08	CE	0.5			MIT
3092.035	3092.933	32331.77	CE	0.3			MIT
3091.918	3092.816	32333.00	CE	1.3			MIT
3091.706	3092.604	32335.22	CE	0.7	0.0		MIT
3091.366	3092.263	32338.77	CE	0.3	0.0		MIT
3091.294	3092.191	32339.52	CE II	1.3			MIT
3091.294	3092.191	32339.52	CE I	1.3			MIT
3091.108	3092.005	32341.47	CE	0.5			MIT
3090.880	3091.777	32343.86	CE	1.3			MIT
3090.701	3091.598	32345.73	CE	1.3			MIT
3090.516	3091.413	32347.67	CE	1.2	0.0		MIT
3090.372	3091.269	32349.17	CE	1.3	0.5		MIT
3089.787	3090.684	32355.30	CE	0.6			MIT
3089.584	3090.481	32357.42	CE	0.5			MIT
3088.312	3089.209	32370.75	CE	0.7			MIT
3087.399	3088.295	32380.32	CE	0.9	0.0		MIT
3087.172	3088.068	32382.70	CE	1.0			MIT
3087.011	3087.907	32384.39	CE	0.3			MIT
3086.678	3087.574	32387.89	CE	0.6			MIT
3085.758	3086.654	32397.54	CE	0.9			MIT
3085.577	3086.473	32399.44	CE	0.3			MIT
3085.453	3086.349	32400.74	CE	0.6			MIT
3085.415	3086.311	32401.14	CE	0.6			MIT
3084.723	3085.619	32408.41	CE	0.6			MIT
3084.466	3085.362	32411.11	CE	1.6S	0.5		MIT
3084.257	3085.153	32413.31	CE	0.7			MIT
3083.959	3084.855	32416.44	CE	0.5			MIT
3083.670	3084.566	32419.48	CE	1.3S	0.7		MIT
3083.408	3084.303	32422.23	CE	0.5			MIT
3083.067	3083.962	32425.82	CE	0.7			MIT
3082.304	3083.199	32433.84	CE	1.3	0.3		MIT
3081.984	3082.879	32437.21	CE	0.7			MIT
3081.245	3082.140	32444.99	CE	0.5			MIT
3080.636	3081.531	32451.40	CE	1.3	0.3		MIT
3079.958	3080.853	32458.55	CE	1.0	0.3		MIT
3079.906	3080.801	32459.09	CE II	0.9			MIT
3079.643	3080.538	32461.87	CE	1.3S	0.3		MIT
3079.453	3080.347	32463.87	CE	1.0			MIT
3078.908	3079.802	32469.62	CE	0.9			MIT
3077.644	3078.538	32482.95	CE	1.2			MIT
3077.334	3078.228	32486.22	CE	1.2			MIT
3076.893	3077.787	32490.88	CE	0.8			MIT
3076.250	3077.144	32497.67	CE	1.2	0.0		MIT
3076.132	3077.026	32498.92	CE	0.5			MIT
3075.534	3076.428	32505.24	CE	0.3			MIT
3075.206	3076.099	32508.70	CE	0.5	0.3		MIT
3074.324	3075.217	32518.03	CE	1.1			MIT
3074.164	3075.057	32519.72	CE	0.8			MIT
3073.989	3074.882	32521.57	CE	0.9			MIT
3073.538	3074.431	32526.34	CE	0.3			MIT
3073.336	3074.229	32528.48	CE	1.0			MIT
3072.886	3073.779	32533.25	CE	1.3	0.0		MIT
3072.391	3073.284	32538.49	CE II	1.3			MIT
3071.617	3072.510	32546.69	CE	1.3S	0.0		MIT

AIR WAVELENGTH (ANGSTRMS)	VACUUM WAVELENGTH (ANGSTRMS)	WAVENUMBER (KAYSERS)	LMENT	LOG ARC	INTENSITY SPK	INTENSITY DIS	REF
3071.110	3072.002	32552.06	CE II	1.3S	0.0		MIT
3069.721	3070.613	32566.79	CE	0.8			MIT
3069.644	3070.536	32567.60	CE	1.3S			MIT
3069.320	3070.212	32571.04	CE	0.6			MIT
3069.147	3070.039	32572.88	CE	0.5			MIT
3068.817	3069.709	32576.38	CE	0.3			MIT
3068.678	3069.570	32577.85	CE	1.3			MIT
3067.895	3068.787	32586.17	CE	0.8			MIT
3067.443	3068.334	32590.97	CE	0.6			MIT
3066.380	3067.271	32602.27	CE	1.3			MIT
3066.133	3067.024	32604.90	CE	0.5			MIT
3065.780	3066.671	32608.65	CE	0.8			MIT
3065.395	3066.286	32612.74	CE	0.7			MIT
3064.024	3064.915	32627.34	CE	1.2			MIT
3063.776	3064.667	32629.98	CE	0.3			MIT
3063.389	3064.279	32634.10	CE	0.6			MIT
3063.010	3063.900	32638.14	CE	1.6	1.0		MIT
3061.218	3062.108	32657.24	CE	1.0			MIT
3059.739	3060.629	32673.03	CE	1.1			MIT
3058.551	3059.440	32685.72	CE	1.1			MIT
3058.107	3058.996	32690.46	CE	0.7			MIT
3057.383	3058.272	32698.20	CE	0.7			MIT
3056.777	3057.666	32704.69	CE II	1.6	0.5		MIT
3056.130	3057.019	32711.61	CE	0.6			MIT
3055.243	3056.131	32721.11	CE	1.3	0.5		MIT
3055.098	3055.986	32722.66	CE	0.6			MIT
3054.613	3055.501	32727.85	CE	0.3			MIT
3054.443	3055.331	32729.68	CE	0.5			MIT
3053.701	3054.589	32737.63	CE	0.3			MIT
3053.289	3054.177	32742.04	CE	0.9			MIT
3053.015	3053.903	32744.98	CE	0.6			MIT
3052.684	3053.572	32748.53	CE	0.6			MIT
3051.975	3052.863	32756.14	CE	1.0	0.3		MIT
3051.924	3052.812	32756.69	CE	1.0			MIT
3051.287	3052.174	32763.53	CE	0.3			MIT
3051.162	3052.049	32764.87	CE II	0.8			MIT
3050.587	3051.474	32771.04	CE	1.1			MIT
3050.300	3051.187	32774.13	CE	0.9			MIT
3048.932	3049.819	32788.83	CE	0.3			MIT
3048.517	3049.404	32793.30	CE	0.5			MIT
3048.437	3049.324	32794.16	CE	0.5			MIT
3048.308	3049.195	32795.54	CE	0.5			MIT
3047.784	3048.671	32801.18	CE	0.3			MIT
3047.503	3048.389	32804.21	CE	0.3			MIT
3047.207	3048.093	32807.39	CE	0.3			MIT
3046.711	3047.597	32812.73	CE	1.2			MIT
3046.124	3047.010	32819.06	CE	0.8			MIT
3045.792	3046.678	32822.63	CE	0.6			MIT
3045.672	3046.558	32823.93	CE	0.8			MIT
3044.395	3045.281	32837.69	CE	1.2			MIT
3043.713	3044.599	32845.05	CE	0.5			MIT
3043.269	3044.154	32849.85	CE	0.5			MIT
3043.094	3043.979	32851.73	CE	1.0W			MIT
3042.077	3042.962	32862.72	CE	0.5			MIT
3041.763	3042.648	32866.11	CE	0.7			MIT
3041.610	3042.495	32867.76	CE	0.9			MIT
3040.252	3041.137	32882.44	CE	0.6			MIT
3039.572	3040.457	32889.80	CE	0.9			MIT
3039.512	3040.396	32890.45	CE	0.9			MIT
3039.253	3040.137	32893.25	CE	0.6			MIT
3038.993	3039.877	32896.06	CE	0.6			MIT
3037.731	3038.615	32909.73	CE	1.4	0.5		MIT
3037.274	3038.158	32914.68	CE	0.8			MIT
3037.049	3037.933	32917.12	CE II	0.9			MIT
3035.863	3036.747	32929.98	CE	1.1			MIT
3035.008	3035.891	32939.25	CE	0.8			MIT
3033.124	3034.007	32959.71	CE	0.9			MIT
3033.050	3033.933	32960.52	CE	1.0			MIT
3032.727	3033.610	32964.03	CE II	1.0			MIT
3032.334	3033.217	32968.30	CE	0.9			MIT

AIR WAVELENGTH (ANGSTRMS)	VACUUM WAVELENGTH (ANGSTRMS)	WAVENUMBER (KAYSERS)	LMENT	LOG ARC	INTENSITY SPK	INTENSITY DIS	REF
3031.621	3032.504	32976.05	CE	1.1	1.0		MIT
3031.497	3032.379	32977.40	CE	0.3			MIT
3030.858	3031.740	32984.35	CE	0.5			MIT
3030.619	3031.501	32986.96	CE	0.7			MIT
3030.471	3031.353	32988.57	CE	0.7			MIT
3030.309	3031.191	32990.33	CE	1.4			MIT
3028.958	3029.840	33005.04	CE	1.1			MIT
3028.753	3029.635	33007.28	CE	0.6			MIT
3028.664	3029.546	33008.25	CE	0.6			MIT
3028.171	3029.053	33013.62	CE	0.7			MIT
3027.626	3028.508	33019.56	CE	1.0			MIT
3026.620	3027.501	33030.54	CE	1.3W			MIT
3025.298	3026.179	33044.97	CE	0.3			MIT
3025.126	3026.007	33046.85	CE II	0.9			MIT
3024.571	3025.452	33052.92	CE	1.1			MIT
3023.966	3024.847	33059.53	CE	0.6			MIT
3023.880	3024.761	33060.47	CE	0.7			MIT
3023.485	3024.365	33064.79	CE	1.0			MIT
3023.433	3024.313	33065.35	CE	0.7			MIT
3022.791	3023.671	33072.38	CE	1.1			MIT
3022.469	3023.349	33075.90	CE	0.7			MIT
3022.353	3023.233	33077.17	CE	0.3			MIT
3021.038	3021.918	33091.57	CE	1.2			MIT
3020.883	3021.763	33093.27	CE	1.2			MIT
3019.419	3020.298	33109.31	CE	0.5			MIT
3018.771	3019.650	33116.42	CE	0.5			MIT
3017.769	3018.648	33127.41	CE	0.8			MIT
3017.195	3018.074	33133.71	CE	1.3	0.7		MIT
3017.096	3017.975	33134.80	CE	1.0			MIT
3016.556	3017.435	33140.73	CE	0.3			MIT
3015.414	3016.292	33153.28	CE	0.9			MIT
3014.554	3015.432	33162.74	CE	0.5			MIT
3014.470	3015.348	33163.67	CE	0.5			MIT
3014.261	3015.139	33165.96	CE	0.9			MIT
3014.105	3014.983	33167.68	CE	0.7			MIT
3013.170	3014.048	33177.97	CE	0.9			MIT
3012.484	3013.362	33185.53	CE	0.7			MIT
3011.881	3012.759	33192.17	CE	1.2			MIT
3011.471	3012.348	33196.69	CE	0.3			MIT
3011.017	3011.894	33201.70	CE	0.5			MIT
3010.956	3011.833	33202.37	CE	0.3			MIT
3010.724	3011.601	33204.93	CE	0.3			MIT
3010.549	3011.426	33206.86	CE	0.7			MIT
3010.462	3011.339	33207.82	CE	0.8			MIT
3009.770	3010.647	33215.45	CE	0.7			MIT
3008.789	3009.666	33226.28	CE	1.6	0.5		MIT
3008.131	3009.008	33233.55	CE	1.1			MIT
3007.797	3008.674	33237.24	CE	0.3			MIT
3007.032	3007.908	33245.69	CE	0.3			MIT
3006.614	3007.490	33250.31	CE	0.6			MIT
3005.213	3006.089	33265.82	CE	0.3			MIT
3003.680	3004.556	33282.79	CE	0.3			MIT
3003.562	3004.438	33284.10	CE	1.1S	0.0		MIT
3002.748	3003.623	33293.12	CE	1.3	0.0		MIT
3002.376	3003.251	33297.25	CE	1.1S			MIT
3002.136	3003.011	33299.91	CE	1.3	0.0		MIT
3000.330	3001.205	33319.95	CE	0.7			MIT
3000.132	3001.007	33322.15	CE	0.3			MIT
3000.068	3000.943	33322.86	CE	1.0			MIT
2999.480	3000.354	33329.40	CE	0.8			MIT
2999.432	3000.306	33329.93	CE	0.8			MIT
2999.202	3000.076	33332.48	CE	0.5			MIT
2999.066	2999.940	33334.00	CE	1.2			MIT
2998.769	2999.643	33337.30	CE	1.2			MIT
2998.014	2998.888	33345.69	CE	0.5			MIT
2997.715	2998.589	33349.02	CE	0.3			MIT
2997.468	2998.342	33351.77	CE	0.8			MIT
2997.143	2998.017	33355.38	CE	0.3			MIT
2995.644	2996.518	33372.07	CE	1.5	0.0		MIT
2995.157	2996.030	33377.50	CE	0.5			MIT

AIR WAVELENGTH (ANGSTRMS)	VACUUM WAVELENGTH (ANGSTRMS)	WAVENUMBER (KAYSERS)	LMENT	LOG ARC	INTENSITY SPK	INTENSITY DIS	REF
2994.418	2995.291	33385.73	CE	1.1			MIT
2994.137	2995.010	33388.87	CE	0.6W			MIT
2993.629	2994.502	33394.53	CE	0.5			MIT
2992.366	2993.239	33408.63	CE	0.8			MIT
2992.229	2993.102	33410.16	CE	1.0			MIT
2991.898	2992.771	33413.85	CE	1.0			MIT
2991.715	2992.588	33415.90	CE	1.0			MIT
2991.400	2992.272	33419.42	CE	0.3			MIT
2990.873	2991.745	33425.31	CE	1.6	0.0		MIT
2989.575	2990.447	33439.82	CE	0.8			MIT
2989.313	2990.185	33442.75	CE	0.6			MIT
2989.019	2989.891	33446.04	CE	0.3H			MIT
2988.878	2989.750	33447.61	CE	0.5			MIT
2987.543	2988.415	33462.56	CE	0.3			MIT
2987.351	2988.222	33464.71	CE	0.9			MIT
2986.668	2987.539	33472.36	CE	1.0			MIT
2986.208	2987.079	33477.52	CE	0.3H			MIT
2985.907	2986.778	33480.89	CE	1.0			MIT
2985.818	2986.689	33481.89	CE	1.1			MIT
2985.032	2985.903	33490.71	CE	0.5			MIT
2984.866	2985.737	33492.57	CE	0.5			MIT
2984.713	2985.584	33494.29	CE	0.6			MIT
2984.560	2985.431	33496.00	CE	1.3			MIT
2984.106	2984.977	33501.10	CE	0.3			MIT
2983.918	2984.789	33503.21	CE	0.8			MIT
2982.966	2983.836	33513.90	CE	0.6			MIT
2982.541	2983.411	33518.68	CE	0.5			MIT
2981.905	2982.775	33525.83	CE	1.2			MIT
2981.395	2982.265	33531.56	CE	0.3			MIT
2980.409	2981.279	33542.65	CE	1.0			MIT
2979.356	2980.225	33554.51	CE	0.3			MIT
2978.810	2979.679	33560.66	CE	0.8			MIT
2977.941	2978.810	33570.45	CE	0.7			MIT
2977.769	2978.638	33572.39	CE	1.0			MIT
2977.548	2978.417	33574.88	CE	0.9			MIT
2977.461	2978.330	33575.86	CE	1.0			MIT
2976.905	2977.774	33582.13	CE	0.9			MIT
2975.939	2976.808	33593.03	CE	1.3			MIT
2974.605	2975.473	33608.10	CE	0.3			MIT
2974.482	2975.350	33609.49	CE	0.5			MIT
2974.018	2974.886	33614.73	CE	0.5			MIT
2973.436	2974.304	33621.31	CE	0.9			MIT
2973.186	2974.054	33624.14	CE	0.3			MIT
2972.582	2973.450	33630.97	CE	0.7H			MIT
2971.840	2972.708	33639.37	CE	0.3H			MIT
2971.196	2972.063	33646.66	CE	0.5			MIT
2970.316	2971.183	33656.62	CE	1.0			MIT
2969.922	2970.789	33661.09	CE	0.5			MIT
2968.691	2969.558	33675.05	CE	0.9			MIT
2968.362	2969.229	33678.78	CE	0.5			MIT
2967.866	2968.733	33684.41	CE	0.3			MIT
2967.315	2968.181	33690.66	CE	0.3			MIT
2967.108	2967.974	33693.01	CE	0.9			MIT
2966.798	2967.664	33696.53	CE	0.5			MIT
2966.123	2966.989	33704.20	CE	0.5			MIT
2965.837	2966.703	33707.45	CE	0.8			MIT
2965.269	2966.135	33713.91	CE	1.2			MIT
2965.135	2966.001	33715.43	CE	0.9			MIT
2964.801	2965.667	33719.23	CE	1.4S			MIT
2964.526	2965.392	33722.36	CE	0.3			MIT
2963.996	2964.862	33728.39	CE	0.3			MIT
2963.876	2964.742	33729.75	CE	0.5			MIT
2961.367	2962.232	33758.33	CE	0.5			MIT
2960.640	2961.505	33766.62	CE	0.8			MIT
2960.438	2961.303	33768.92	CE	1.0			MIT
2959.105	2959.969	33784.13	CE	1.2			MIT
2958.038	2958.902	33796.32	CE	0.5			MIT
2957.841	2958.705	33798.57	CE	0.3			MIT
2957.792	2958.656	33799.13	CE	0.5			MIT
2957.165	2958.029	33806.29	CE	0.5			MIT

AIR WAVELENGTH (ANGSTRMS)	VACUUM WAVELENGTH (ANGSTRMS)	WAVENUMBER (KAYSERS)	LMENT	LOG ARC	INTENSITY SPK	INTENSITY DIS	REF
2956.707	2957.571	33811.53	CE	1.2			MIT
2955.941	2956.805	33820.29	CE	1.4			MIT
2955.807	2956.671	33821.83	CE	0.7			MIT
2955.603	2956.467	33824.16	CE	1.3			MIT
2955.412	2956.276	33826.35	CE	0.6			MIT
2955.100	2955.963	33829.92	CE	0.3			MIT
2954.409	2955.272	33837.83	CE	0.3			MIT
2953.076	2953.939	33853.10	CE	0.6			MIT
2951.777	2952.640	33868.00	CE	0.6			MIT
2951.687	2952.550	33869.03	CE	0.6			MIT
2951.489	2952.352	33871.31	CE	0.6			MIT
2951.291	2952.154	33873.58	CE	1.0			MIT
2950.498	2951.360	33882.68	CE	0.3			MIT
2950.304	2951.166	33884.91	CE	1.1			MIT
2948.857	2949.719	33901.54	CE	0.6			MIT
2948.735	2949.597	33902.94	CE	0.6			MIT
2948.385	2949.247	33906.96	CE	1.0			MIT
2947.742	2948.604	33914.36	CE	0.5			MIT
2947.149	2948.010	33921.18	CE	0.6			MIT
2946.859	2947.720	33924.52	CE	0.6			MIT
2946.382	2947.243	33930.01	CE	0.9			MIT
2946.101	2946.962	33933.25	CE	0.3			MIT
2946.056	2946.917	33933.77	CE	1.0			MIT
2945.856	2946.717	33936.07	CE	0.3			MIT
2945.582	2946.443	33939.23	CE	0.5			MIT
2945.385	2946.246	33941.50	CE	0.6			MIT
2944.773	2945.634	33948.55	CE	0.6			MIT
2944.346	2945.207	33953.47	CE	1.3			MIT
2943.987	2944.848	33957.61	CE	0.8			MIT
2943.673	2944.534	33961.24	CE	0.9			MIT
2943.215	2944.076	33966.52	CE	0.8			MIT
2941.197	2942.057	33989.82	CE	0.7			MIT
2940.877	2941.737	33993.52	CE	0.7			MIT
2940.785	2941.645	33994.59	CE	1.2			MIT
2940.663	2941.523	33996.00	CE	0.6			MIT
2939.770	2940.630	34006.32	CE	0.3			MIT
2939.538	2940.398	34009.01	CE	1.1			MIT
2938.682	2939.541	34018.91	CE	0.5			MIT
2938.317	2939.176	34023.14	CE	0.3			MIT
2938.220	2939.079	34024.26	CE	0.7			MIT
2938.051	2938.910	34026.22	CE	0.7			MIT
2937.707	2938.566	34030.20	CE	0.8			MIT
2935.757	2936.616	34052.81	CE	0.5			MIT
2934.345	2935.203	34069.19	CE	1.0			MIT
2932.274	2933.132	34093.25	CE	1.0			MIT
2932.090	2932.948	34095.39	CE	0.6			MIT
2929.109	2929.966	34130.09	CE	1.0			MIT
2927.864	2928.721	34144.60	CE	0.5			MIT
2927.548	2928.405	34148.29	CE	0.7			MIT
2927.249	2928.106	34151.77	CE	0.6			MIT
2926.928	2927.784	34155.52	CE	0.5			MIT
2925.925	2926.781	34167.23	CE	0.9			MIT
2925.641	2926.497	34170.54	CE	0.8			MIT
2925.185	2926.041	34175.87	CE	1.0			MIT
2923.834	2924.690	34191.66	CE	0.5	0.9H		MIT
2923.455	2924.311	34196.09	CE	0.8			MIT
2922.880	2923.735	34202.82	CE	0.5			MIT
2922.581	2923.436	34206.32	CE	1.0			MIT
2922.370	2923.225	34208.79	CE	0.9			MIT
2922.291	2923.146	34209.71	CE	0.3			MIT
2922.080	2922.935	34212.18	CE	0.9			MIT
2921.421	2922.276	34219.90	CE	0.8			MIT
2921.371	2922.226	34220.49	CE	0.8			MIT
2920.482	2921.337	34230.90	CE	0.8			MIT
2920.292	2921.147	34233.13	CE	0.3			MIT
2919.858	2920.713	34238.22	CE	1.1			MIT
2918.781	2919.635	34250.85	CE	1.0			MIT
2918.665	2919.519	34252.21	CE	1.5S			MIT
2917.886	2918.740	34261.36	CE	0.3			MIT
2917.686	2918.540	34263.71	CE	0.3			MIT

AIR WAVELENGTH (ANGSTRMS)	VACUUM WAVELENGTH (ANGSTRMS)	WAVENUMBER (KAYSERS)	LMENT	LOG ARC	INTENSITY SPK	DIS	REF
2917.491	2918.345	34266.00	CE	0.5			MIT
2917.408	2918.262	34266.97	CE	0.8			MIT
2917.127	2917.981	34270.27	CE	0.5			MIT
2917.027	2917.881	34271.44	CE	0.3			MIT
2916.682	2917.536	34275.50	CE	1.0			MIT
2915.747	2916.601	34286.49	CE	0.3			MIT
2915.555	2916.409	34288.75	CE	1.0			MIT
2915.252	2916.106	34292.31	CE	0.3			MIT
2914.933	2915.787	34296.06	CE	0.3			MIT
2914.747	2915.600	34298.25	CE	0.3			MIT
2914.347	2915.200	34302.96	CE	0.8			MIT
2914.105	2914.958	34305.81	CE	0.3			MIT
2913.738	2914.591	34310.13	CE	0.7			MIT
2913.528	2914.381	34312.60	CE	0.9			MIT
2913.362	2914.215	34314.56	CE	0.7			MIT
2913.195	2914.048	34316.52	CE	0.6			MIT
2912.906	2913.759	34319.93	CE	1.1			MIT
2912.769	2913.622	34321.54	CE	1.1			MIT
2911.514	2912.367	34336.34	CE	0.9			MIT
2911.120	2911.973	34340.98	CE	0.3			MIT
2910.852	2911.705	34344.14	CE	0.3			MIT
2910.385	2911.237	34349.65	CE	0.5			MIT
2910.213	2911.065	34351.69	CE	0.7			MIT
2909.614	2910.466	34358.76	CE	0.7			MIT
2908.421	2909.273	34372.85	CE	1.5	S		MIT
2908.097	2908.949	34376.68	CE	0.5			MIT
2907.819	2908.671	34379.96	CE	0.6			MIT
2907.462	2908.314	34384.19	CE	0.3			MIT
2907.093	2907.945	34388.55	CE	0.6			MIT
2906.088	2906.939	34400.44	CE	0.5			MIT
2906.032	2906.883	34401.11	CE	0.3			MIT
2903.519	2904.370	34430.88	CE	0.5			MIT
2902.715	2903.565	34440.42	CE	0.5			MIT
2902.664	2903.514	34441.02	CE	0.7			MIT
2902.385	2903.235	34444.33	CE	0.3			MIT
2902.323	2903.173	34445.07	CE	0.5			MIT
2902.147	2902.997	34447.15	CE	0.9			MIT
2901.708	2902.558	34452.37	CE	0.3			MIT
2901.545	2902.395	34454.30	CE	0.5			MIT
2900.606	2901.456	34465.46	CE	0.8			MIT
2900.368	2901.218	34468.28	CE	0.8			MIT
2899.620	2900.470	34477.17	CE	0.3			MIT
2899.446	2900.296	34479.24	CE	0.5			MIT
2899.073	2899.923	34483.68	CE	0.7	L		MIT
2898.944	2899.794	34485.21	CE	0.8			MIT
2898.336	2899.185	34492.45	CE	0.8			MIT
2897.172	2898.021	34506.31	CE	0.6			MIT
2896.862	2897.711	34510.00	CE	0.7			MIT
2896.733	2897.582	34511.53	CE	1.3	S		MIT
2896.404	2897.253	34515.45	CE	0.9			MIT
2895.949	2896.798	34520.88	CE	0.6			MIT
2895.639	2896.488	34524.57	CE	0.9			MIT
2894.881	2895.730	34533.61	CE	0.6			MIT
2894.732	2895.581	34535.39	CE	0.3			MIT
2894.296	2895.144	34540.59	CE	0.6			MIT
2894.216	2895.064	34541.55	CE	1.0			MIT
2894.086	2894.934	34543.10	CE	0.9			MIT
2894.023	2894.871	34543.85	CE	0.9			MIT
2893.056	2893.904	34555.40	CE	0.3			MIT
2892.145	2892.993	34566.28	CE	1.2			MIT
2891.728	2892.576	34571.26	CE	0.8			MIT
2891.636	2892.484	34572.36	CE	0.5			MIT
2890.687	2891.535	34583.71	CE	0.3	O		MIT
2890.409	2891.256	34587.04	CE	0.3			MIT
2890.173	2891.020	34589.86	CE	0.9			MIT
2889.460	2890.307	34598.40	CE	0.3			MIT
2888.700	2889.547	34607.50	CE	1.0			MIT
2888.456	2889.303	34610.42	CE	0.8			MIT
2888.074	2888.921	34615.00	CE	0.3			MIT
2886.332	2887.178	34635.89	CE	0.5			MIT
2885.932	2886.778	34640.69	CE	0.3			MIT
2885.291	2886.137	34648.39	CE	1.0			MIT
2884.729	2885.575	34655.14	CE	0.5			MIT
2884.455	2885.301	34658.43	CE	0.3			MIT
2884.052	2884.898	34663.27	CE	0.3			MIT
2883.607	2884.453	34668.62	CE	0.3			MIT
2882.609	2883.455	34680.62	CE	1.2			MIT
2881.772	2882.617	34690.70	CE	0.3			MIT
2881.420	2882.265	34694.93	CE	0.6	S		MIT
2881.131	2881.976	34698.41	CE	1.1			MIT
2880.928	2881.773	34700.86	CE	0.3			MIT
2880.849	2881.694	34701.81	CE	0.3			MIT
2880.637	2881.482	34704.36	CE	1.3			MIT
2880.356	2881.201	34707.75	CE	0.8			MIT
2879.366	2880.211	34719.68	CE	0.3			MIT
2878.786	2879.631	34726.68	CE	0.3			MIT
2878.635	2879.480	34728.50	CE	0.9			MIT
2878.018	2878.862	34735.94	CE	0.3			MIT
2876.183	2877.027	34758.10	CE	0.3			MIT
2874.917	2875.761	34773.41	CE	0.3			MIT
2874.821	2875.665	34774.57	CE	0.3			MIT
2874.551	2875.395	34777.84	CE	0.3			MIT
2874.135	2874.978	34782.87	CE	1.5	W		MIT
2873.354	2874.197	34792.32	CE	0.3			MIT
2872.592	2873.435	34801.55	CE	0.5			MIT
2872.481	2873.324	34802.90	CE	0.5			MIT
2872.435	2873.278	34803.45	CE	0.3			MIT
2871.633	2872.476	34813.17	CE	0.8			MIT
2871.576	2872.419	34813.87	CE	0.8			MIT
2871.075	2871.918	34819.94	CE	1.2			MIT
2870.624	2871.467	34825.41	CE	0.8			MIT
2870.500	2871.343	34826.91	CE	0.3			MIT
2870.045	2870.887	34832.44	CE	0.3			MIT
2869.647	2870.489	34837.27	CE	0.5			MIT
2869.559	2870.401	34838.33	CE	0.3			MIT
2868.960	2869.802	34845.61	CE	0.8			MIT
2868.230	2869.072	34854.48	CE	0.9			MIT
2867.273	2868.115	34866.11	CE	0.7			MIT
2866.809	2867.651	34871.75	CE	1.1			MIT
2866.057	2866.898	34880.90	CE	0.5			MIT
2865.563	2866.404	34886.91	CE	0.3			MIT
2865.513	2866.354	34887.52	CE	0.9			MIT
2865.378	2866.219	34889.17	CE	0.9			MIT
2864.819	2865.660	34895.97	CE	0.6			MIT
2864.484	2865.325	34900.05	CE	0.7			MIT
2863.341	2864.182	34913.98	CE	1.1			MIT
2863.225	2864.066	34915.40	CE	0.3			MIT
2862.785	2863.626	34920.77	CE II	1.2			MIT
2861.621	2862.461	34934.97	CE	1.3			MIT
2861.345	2862.185	34938.34	CE	1.0			MIT
2861.169	2862.009	34940.49	CE	0.5			MIT
2860.864	2861.704	34944.21	CE	0.3			MIT
2860.643	2861.483	34946.91	CE	0.3			MIT
2860.555	2861.395	34947.99	CE	0.5			MIT
2859.521	2860.361	34960.62	CE	1.0			MIT
2859.062	2859.902	34966.24	CE	0.5			MIT
2858.716	2859.556	34970.47	CE	0.3			MIT
2858.472	2859.312	34973.45	CE	0.3			MIT
2858.009	2858.848	34979.12	CE	1.2	S		MIT
2856.976	2857.815	34991.77	CE	0.5			MIT
2856.937	2857.776	34992.24	CE	0.8			MIT
2856.831	2857.670	34993.54	CE	0.5			MIT
2856.674	2857.513	34995.46	CE	0.3			MIT
2856.473	2857.312	34997.93	CE	0.6			MIT
2856.394	2857.233	34998.90	CE	0.8			MIT
2856.046	2856.885	35003.16	CE	0.5			MIT
2855.718	2856.557	35007.18	CE	1.1			MIT
2855.448	2856.287	35010.49	CE	1.2			MIT
2855.321	2856.160	35012.05	CE	1.0			MIT
2854.883	2855.722	35017.42	CE	1.4	S		MIT

AIR WAVELENGTH (ANGSTRMS)	VACUUM WAVELENGTH (ANGSTRMS)	WAVENUMBER (KAYSERS)	LMENT	LOG ARC	INTENSITY SPK	DIS	REF
2854.667	2855.506	35020.07	CE	1.5 S			MIT
2854.491	2855.330	35022.23	CE	0.7			MIT
2852.237	2853.075	35049.90	CE	0.5			MIT
2851.598	2852.436	35057.75	CE	0.5			MIT
2849.198	2850.035	35087.29	CE	0.3			MIT
2849.033	2849.870	35089.32	CE	1.2			MIT
2848.819	2849.656	35091.95	CE	0.3			MIT
2848.079	2848.916	35101.07	CE	0.3			MIT
2847.689	2848.526	35105.88	CE	0.6			MIT
2847.236	2848.073	35111.46	CE	0.5			MIT
2847.005	2847.842	35114.31	CE	0.5			MIT
2846.363	2847.200	35122.23	CE	0.3			MIT
2846.265	2847.102	35123.44	CE	0.6			MIT
2845.805	2846.641	35129.12	CE	0.3			MIT
2845.751	2846.587	35129.78	CE	0.6			MIT
2845.452	2846.288	35133.47	CE	1.0			MIT
2845.357	2846.193	35134.65	CE	0.6 O			MIT
2845.180	2846.016	35136.83	CE	0.3			MIT
2844.882	2845.718	35140.51	CE	0.3			MIT
2844.760	2845.596	35142.02	CE	0.7			MIT
2844.727	2845.563	35142.43	CE	0.7			MIT
2844.311	2845.147	35147.57	CE	0.7			MIT
2842.924	2843.760	35164.71	CE	0.5			MIT
2842.830	2843.666	35165.88	CE	1.4 D			MIT
2842.519	2843.355	35169.72	CE	1.0			MIT
2842.195	2843.031	35173.73	CE	0.3			MIT
2842.075	2842.911	35175.22	CE	0.3			MIT
2841.715	2842.550	35179.67	CE	1.0			MIT
2841.486	2842.321	35182.51	CE	0.9			MIT
2841.153	2841.988	35186.63	CE	0.5			MIT
2840.686	2841.521	35192.42	CE	0.7			MIT
2840.156	2840.991	35198.98	CE	0.5			MIT
2839.561	2840.396	35206.36	CE	0.7			MIT
2839.523	2840.358	35206.83	CE	0.3			MIT
2839.364	2840.199	35208.80	CE	1.3			MIT
2839.235	2840.070	35210.40	CE	0.5			MIT
2839.030	2839.865	35212.94	CE	0.3			MIT
2838.945	2839.780	35214.00	CE	0.3			MIT
2838.857	2839.692	35215.09	CE	0.3			MIT
2838.810	2839.645	35215.67	CE	0.3			MIT
2837.994	2838.829	35225.80	CE	0.9			MIT
2837.896	2838.731	35227.01	CE	1.0			MIT
2837.598	2838.432	35230.71	CE	0.9			MIT
2837.289	2838.123	35234.55	CE	1.7 S			MIT
2836.039	2836.873	35250.08	CE	0.8			MIT
2835.800	2836.634	35253.05	CE	0.3			MIT
2835.604	2836.438	35255.49	CE	1.0			MIT
2835.393	2836.227	35258.11	CE	0.3			MIT
2834.902	2835.736	35264.22	CE	0.3			MIT
2834.744	2835.578	35266.18	CE	0.8			MIT
2834.474	2835.308	35269.54	CE	0.7			MIT
2834.099	2834.933	35274.21	CE	0.3			MIT
2833.580	2834.413	35280.67	CE	0.3			MIT
2833.309	2834.142	35284.04	CE	1.7 D			MIT
2833.044	2833.877	35287.34	CE	0.5			MIT
2832.928	2833.761	35288.79	CE	0.3			MIT
2832.753	2833.586	35290.97	CE	0.3			MIT
2832.568	2833.401	35293.27	CE	0.3			MIT
2832.307	2833.140	35296.52	CE	1.3			MIT
2831.647	2832.480	35304.75	CE	0.5			MIT
2830.899	2831.732	35314.08	CE	1.5			MIT
2830.431	2831.264	35319.92	CE	0.5			MIT
2830.341	2831.174	35321.04	CE	0.3			MIT
2829.618	2830.451	35330.06	CE	0.3			MIT
2829.556	2830.388	35330.84	CE	0.6			MIT
2829.247	2830.079	35334.70	CE	0.3			MIT
2828.703	2829.535	35341.49	CE	0.5			MIT
2828.039	2828.871	35349.79	CE	0.9			MIT
2827.654	2828.486	35354.60	CE	0.7			MIT
2826.647	2827.479	35367.20	CE	0.3			MIT
2826.499	2827.331	35369.05	CE	0.7			MIT
2825.715	2826.547	35378.86	CE	0.3			MIT
2825.296	2826.127	35384.11	CE	0.3			MIT
2824.880	2825.711	35389.32	CE	1.0			MIT
2824.629	2825.460	35392.46	CE	0.3			MIT
2824.034	2824.865	35399.92	CE	1.2			MIT
2823.421	2824.252	35407.60	CE	0.9			MIT
2822.366	2823.197	35420.84	CE	0.8			MIT
2822.017	2822.848	35425.22	CE	0.7			MIT
2821.682	2822.513	35429.43	CE	0.8			MIT
2821.473	2822.304	35432.05	CE	0.9			MIT
2820.744	2821.574	35441.21	CE	0.3			MIT
2820.323	2821.153	35446.50	CE	0.3			MIT
2819.309	2820.139	35459.25	CE	0.5			MIT
2819.253	2820.083	35459.95	CE	0.3			MIT
2818.376	2819.206	35470.98	CE	1.0			MIT
2818.173	2819.003	35473.54	CE	0.3			MIT
2818.029	2818.859	35475.35	CE	0.3 H			MIT
2817.499	2818.329	35482.02	CE	1.2			MIT
2817.364	2818.194	35483.72	CE	0.3			MIT
2816.052	2816.881	35500.25	CE	0.3			MIT
2814.997	2815.826	35513.56	CE	0.5			MIT
2814.955	2815.784	35514.09	CE	1.3			MIT
2814.811	2815.640	35515.90	CE	1.6 D			MIT
2814.568	2815.397	35518.97	CE	0.3			MIT
2814.470	2815.299	35520.21	CE	0.6			MIT
2814.315	2815.144	35522.17	CE	0.5			MIT
2812.903	2813.731	35540.00	CE	1.0			MIT
2811.866	2812.694	35553.10	CE	1.1			MIT
2811.419	2812.247	35558.75	CE	0.8			MIT
2810.177	2811.005	35574.47	CE	1.0			MIT
2809.560	2810.388	35582.28	CE	0.9			MIT
2808.98	2809.81	35589.6	CE	0.7			MIT
2808.823	2809.650	35591.62	CE	0.3			MIT
2808.360	2809.187	35597.48	CE	0.3			MIT
2807.820	2808.647	35604.33	CE	0.6			MIT
2807.674	2808.501	35606.18	CE	0.6			MIT
2807.038	2807.865	35614.25	CE	1.0			MIT
2806.711	2807.538	35618.40	CE	0.3			MIT
2805.624	2806.451	35632.20	CE	0.3			MIT
2805.457	2806.284	35634.32	CE	0.3			MIT
2805.113	2805.940	35638.69	CE	0.5			MIT
2804.814	2805.640	35642.49	CE	0.3			MIT
2804.389	2805.215	35647.89	CE	0.6			MIT
2804.088	2804.914	35651.71	CE	0.3			MIT
2803.943	2804.769	35653.56	CE	0.6			MIT
2803.121	2803.947	35664.01	CE	0.3			MIT
2803.042	2803.868	35665.02	CE	0.9			MIT
2802.699	2803.525	35669.38	CE	1.3	0.7H		MIT
2802.282	2803.108	35674.69	CE	0.3			MIT
2801.747	2802.573	35681.50	CE	0.3			MIT
2801.196	2802.022	35688.52	CE	0.3			MIT
2800.556	2801.381	35696.67	CE	0.3			MIT
2799.155	2799.980	35714.54	CE	0.5			MIT
2799.112	2799.937	35715.09	CE	0.5			MIT
2798.894	2799.719	35717.87	CE	0.3			MIT
2797.724	2798.549	35732.81	CE	0.6			MIT
2797.627	2798.452	35734.04	CE	0.5			MIT
2797.174	2797.999	35739.83	CE	0.6			MIT
2797.009	2797.834	35741.94	CE	0.5			MIT
2796.206	2797.030	35752.20	CE	0.3			MIT
2795.945	2796.769	35755.54	CE	0.3			MIT
2795.384	2796.208	35762.72	CE	0.3			MIT
2794.247	2795.071	35777.27	CE	0.5			MIT
2793.568	2794.392	35785.96	CE	0.6			MIT
2792.884	2793.708	35794.73	CE	0.3			MIT
2792.439	2793.262	35800.43	CE	0.5			MIT
2792.380	2793.203	35801.19	CE	0.3			MIT
2791.787	2792.610	35808.79	CE	0.3			MIT
2791.418	2792.241	35813.52	CE	1.3			MIT

AIR WAVELENGTH (ANGSTRMS)	VACUUM WAVELENGTH (ANGSTRMS)	WAVENUMBER (KAYSERS)	LMENT	LOG ARC	INTENSITY SPK	INTENSITY DIS	REF
2791.008	2791.831	35818.79	CE	0.5			MIT
2790.529	2791.352	35824.93	CE	1.0			MIT
2789.328	2790.151	35840.36	CE	0.3			MIT
2788.689	2789.512	35848.57	CE	0.9			MIT
2788.022	2788.844	35857.15	CE	0.3			MIT
2787.650	2788.472	35861.93	CE	0.3			MIT
2787.128	2787.950	35868.65	CE	0.8			MIT
2786.909	2787.731	35871.47	CE	0.3			MIT
2786.688	2787.510	35874.31	CE	0.3			MIT
2786.520	2787.342	35876.47	CE	0.3			MIT
2785.347	2786.169	35891.58	CE	1.1			MIT
2785.125	2785.947	35894.44	CE	0.6			MIT
2784.971	2785.793	35896.43	CE	0.3			MIT
2784.664	2785.486	35900.38	CE	0.6			MIT
2784.271	2785.092	35905.45	CE	1.2			MIT
2784.056	2784.877	35908.22	CE	0.3			MIT
2783.786	2784.607	35911.71	CE	0.5			MIT
2783.490	2784.311	35915.52	CE	0.3			MIT
2783.043	2783.864	35921.29	CE	0.7			MIT
2782.699	2783.520	35925.73	CE	0.3			MIT
2781.986	2782.807	35934.94	CE	1.0			MIT
2781.894	2782.715	35936.13	CE	0.8			MIT
2781.431	2782.252	35942.11	CE	0.8			MIT
2780.005	2780.825	35960.55	CE	1.2			MIT
2779.825	2780.645	35962.87	CE	0.3			MIT
2778.699	2779.519	35977.45	CE	0.3			MIT
2778.395	2779.215	35981.38	CE	0.3			MIT
2778.056	2778.876	35985.77	CE	0.3			MIT
2777.798	2778.618	35989.12	CE	0.5			MIT
2776.961	2777.781	35999.96	CE	0.3			MIT
2775.158	2775.977	36023.35	CE	1.0			MIT
2774.059	2774.878	36037.62	CE	0.3			MIT
2773.947	2774.766	36039.08	CE	0.3			MIT
2773.022	2773.841	36051.10	CE	0.3			MIT
2772.927	2773.746	36052.33	CE	0.3			MIT
2772.326	2773.145	36060.15	CE	0.5			MIT
2771.508	2772.326	36070.79	CE	0.9			MIT
2771.465	2772.283	36071.35	CE	0.5			MIT
2771.148	2771.966	36075.48	CE	0.9			MIT
2770.811	2771.629	36079.86	CE	1.0			MIT
2769.275	2770.093	36099.87	CE	0.3			MIT
2768.837	2769.655	36105.58	CE	1.1			MIT
2768.522	2769.340	36109.69	CE	0.5			MIT
2768.337	2769.155	36112.10	CE	0.7	0.5W		MIT
2767.007	2767.824	36129.46	CE	1.3			MIT
2766.588	2767.405	36134.93	CE	0.3			MIT
2766.381	2767.198	36137.64	CE	0.3			MIT
2765.534	2766.351	36148.70	CE	0.8			MIT
2765.385	2766.202	36150.65	CE	0.9			MIT
2765.117	2765.934	36154.16	CE	0.9			MIT
2764.696	2765.513	36159.66	CE	0.7			MIT
2764.633	2765.450	36160.48	CE	0.7			MIT
2764.397	2765.214	36163.57	CE	0.5			MIT
2764.142	2764.959	36166.91	CE	0.3			MIT
2763.722	2764.538	36172.40	CE	0.3			MIT
2763.598	2764.414	36174.03	CE	0.7			MIT
2762.898	2763.714	36183.19	CE	1.1			MIT
2762.215	2763.031	36192.14	CE	1.3			MIT
2761.757	2762.573	36198.14	CE	0.3			MIT
2761.672	2762.488	36199.25	CE	0.3			MIT
2761.501	2762.317	36201.50	CE	0.9			MIT
2761.415	2762.231	36202.62	CE	1.3			MIT
2760.384	2761.200	36216.14	CE	0.7			MIT
2760.271	2761.087	36217.62	CE	0.3			MIT
2759.880	2760.695	36222.76	CE	0.3			MIT
2759.054	2759.869	36233.60	CE	0.3			MIT
2757.467	2758.282	36254.45	CE	0.3			MIT
2757.186	2758.001	36258.15	CE	0.3			MIT
2756.798	2757.613	36263.25	CE	1.0			MIT
2756.086	2756.901	36272.62	CE	0.3			MIT
2755.962	2756.777	36274.25	CE	0.3			MIT
2755.413	2756.227	36281.48	CE	0.9			MIT
2754.232	2755.046	36297.03	CE	0.7			MIT
2753.541	2754.355	36306.14	CE	0.3			MIT
2753.302	2754.116	36309.29	CE	0.3			MIT
2753.210	2754.024	36310.51	CE	0.3			MIT
2753.073	2753.887	36312.31	CE	0.8			MIT
2752.945	2753.759	36314.00	CE	0.3			MIT
2752.166	2752.980	36324.28	CE	1.2			MIT
2750.894	2751.707	36341.07	CE	1.1			MIT
2750.451	2751.264	36346.93	CE	0.9			MIT
2749.954	2750.767	36353.50	CE	0.7			MIT
2749.705	2750.518	36356.79	CE	0.3			MIT
2749.526	2750.339	36359.15	CE	0.9			MIT
2748.798	2749.611	36368.78	CE	0.3			MIT
2748.021	2748.834	36379.07	CE	0.3			MIT
2747.833	2748.646	36381.56	CE	0.3			MIT
2747.155	2747.967	36390.53	CE	1.1			MIT
2746.654	2747.466	36397.17	CE	0.5			MIT
2746.562	2747.374	36398.39	CE	0.3			MIT
2745.723	2746.535	36409.51	CE	1.0			MIT
2745.007	2745.819	36419.01	CE	0.7			MIT
2744.540	2745.352	36425.21	CE	0.6			MIT
2743.328	2744.139	36441.30	CE	0.3			MIT
2743.062	2743.873	36444.83	CE	0.9			MIT
2741.957	2742.768	36459.52	CE	1.1			MIT
2741.575	2742.386	36464.60	CE	0.5			MIT
2741.521	2742.332	36465.31	CE	0.5			MIT
2740.577	2741.388	36477.87	CE	0.5			MIT
2738.805	2739.615	36501.47	CE	0.3			MIT
2738.638	2739.448	36503.70	CE	0.6			MIT
2738.316	2739.126	36507.99	CE	0.3			MIT
2737.414	2738.224	36520.02	CE	0.3			MIT
2737.057	2737.867	36524.78	CE	0.3			MIT
2736.326	2737.136	36534.54	CE	1.0			MIT
2734.836	2735.645	36554.44	CE	0.3			MIT
2734.389	2735.198	36560.42	CE	0.9			MIT
2734.077	2734.886	36564.59	CE	0.5			MIT
2733.363	2734.172	36574.14	CE	0.3			MIT
2732.827	2733.636	36581.32	CE	1.3			MIT
2732.689	2733.498	36583.16	CE	0.3			MIT
2732.404	2733.213	36586.98	CE	0.3			MIT
2732.166	2732.975	36590.17	CE	1.0			MIT
2732.035	2732.844	36591.92	CE	0.9			MIT
2731.912	2732.721	36593.57	CE	0.3			MIT
2731.707	2732.516	36596.31	CE	0.5			MIT
2731.407	2732.216	36600.33	CE	0.3			MIT
2731.358	2732.167	36600.99	CE	0.3			MIT
2730.800	2731.608	36608.47	CE	0.8			MIT
2730.699	2731.507	36609.82	CE	0.6			MIT
2730.521	2731.329	36612.21	CE	0.3W			MIT
2730.258	2731.066	36615.73	CE	0.3			MIT
2730.076	2730.884	36618.18	CE	0.3	0.5		MIT
2729.683	2730.491	36623.45	CE	0.8			MIT
2729.538	2730.346	36625.39	CE	0.3			MIT
2729.322	2730.130	36628.29	CE	0.9			MIT
2729.162	2729.970	36630.44	CE	0.9			MIT
2728.902	2729.710	36633.93	CE	0.3			MIT
2728.485	2729.293	36639.53	CE	0.3			MIT
2727.960	2728.768	36646.58	CE	0.5			MIT
2727.685	2728.493	36650.27	CE	0.5			MIT
2726.482	2727.289	36666.44	CE	0.3			MIT
2726.371	2727.178	36667.94	CE	0.3			MIT
2726.144	2726.951	36670.99	CE	1.0			MIT
2725.325	2726.132	36682.01	CE	0.5			MIT
2725.170	2725.977	36684.09	CE	0.8			MIT
2724.948	2725.755	36687.08	CE	1.0			MIT
2724.071	2724.878	36698.89	CE	0.3			MIT
2723.888	2724.695	36701.36	CE	0.3			MIT
2723.381	2724.188	36708.19	CE	0.9			MIT

AIR WAVELENGTH (ANGSTRMS)	VACUUM WAVELENGTH (ANGSTRMS)	WAVENUMBER (KAYSERS)	LMENT	LOG ARC	INTENSITY SPK	DIS	REF
2723.309	2724.116	36709.16	CE	1.0			MIT
2722.594	2723.400	36718.80	CE	0.3			MIT
2722.374	2723.180	36721.77	CE	0.8			MIT
2721.990	2722.796	36726.95	CE	0.5			MIT
2721.685	2722.491	36731.06	CE	1.0			MIT
2721.186	2721.992	36737.80	CE	0.6			MIT
2721.132	2721.938	36738.53	CE	0.6			MIT
2720.124	2720.930	36752.14	CE	0.3			MIT
2719.979	2720.785	36754.10	CE	1.1			MIT
2719.345	2720.151	36762.67	CE	0.0	0.3		MIT
2719.237	2720.043	36764.13	CE	0.8			MIT
2719.117	2719.923	36765.75	CE	0.3			MIT
2718.087	2718.892	36779.68	CE	0.6			MIT
2717.275	2718.080	36790.67	CE	1.0			MIT
2716.291	2717.096	36804.00	CE	0.5			MIT
2716.146	2716.951	36805.96	CE	0.3			MIT
2715.746	2716.551	36811.39	CE	0.7			MIT
2715.426	2716.231	36815.72	CE	0.7			MIT
2715.314	2716.119	36817.24	CE	0.5			MIT
2715.243	2716.048	36818.21	CE	0.8			MIT
2715.170	2715.975	36819.19	CE	1.0			MIT
2714.975	2715.780	36821.84	CE	0.7			MIT
2714.723	2715.528	36825.26	CE	0.7			MIT
2713.295	2714.099	36844.64	CE	0.3			MIT
2712.976	2713.780	36848.97	CE	0.3			MIT
2712.372	2713.176	36857.17	CE	0.3			MIT
2711.935	2712.739	36863.11	CE	0.3			MIT
2711.432	2712.236	36869.95	CE	0.6			MIT
2711.244	2712.048	36872.51	CE	0.6			MIT
2711.011	2711.815	36875.68	CE	0.7			MIT
2709.406	2710.209	36897.52	CE	0.9			MIT
2708.953	2709.756	36903.69	CE	0.9			MIT
2708.680	2709.483	36907.41	CE	0.3			MIT
2708.435	2709.238	36910.75	CE	0.6			MIT
2708.173	2708.976	36914.32	CE	0.8			MIT
2708.131	2708.934	36914.89	CE	0.9			MIT
2707.470	2708.273	36923.90	CE	0.3			MIT
2707.024	2707.827	36929.98	CE	0.3			MIT
2706.884	2707.687	36931.90	CE	1.2			MIT
2706.336	2707.139	36939.37	CE	0.3			MIT
2706.141	2706.943	36942.04	CE	0.5			MIT
2703.944	2704.746	36972.05	CE	1.0			MIT
2703.495	2704.297	36978.19	CE	0.6			MIT
2703.138	2703.940	36983.07	CE	0.8			MIT
2702.879	2703.681	36986.62	CE	0.6W			MIT
2702.539	2703.341	36991.27	CE	0.3			MIT
2702.119	2702.921	36997.02	CE	0.5W			MIT
2701.924	2702.725	36999.69	CE	0.3W			MIT
2701.703	2702.504	37002.71	CE	0.5			MIT
2700.471	2701.272	37019.60	CE	0.5			MIT
2700.231	2701.032	37022.88	CE	0.7			MIT
2699.528	2700.329	37032.53	CE	0.3			MIT
2699.152	2699.953	37037.68	CE	0.5			MIT
2698.731	2699.532	37043.46	CE	0.5			MIT
2698.359	2699.160	37048.57	CE	0.5			MIT
2698.033	2698.834	37053.04	CE	0.8			MIT
2697.535	2698.335	37059.88	CE	0.3			MIT
2697.031	2697.831	37066.81	CE	0.7			MIT
2696.824	2697.624	37069.65	CE	0.7			MIT
2696.072	2696.872	37079.99	CE	1.3			MIT
2695.961	2696.761	37081.52	CE	1.0			MIT
2695.538	2696.338	37087.34	CE	0.7			MIT
2695.215	2696.015	37091.78	CE	0.9			MIT
2693.447	2694.246	37116.13	CE	0.8			MIT
2693.027	2693.826	37121.92	CE	0.5			MIT
2692.407	2693.206	37130.47	CE	1.1			MIT
2692.013	2692.812	37135.90	CE	0.3			MIT
2691.687	2692.486	37140.40	CE	0.8			MIT
2691.219	2692.018	37146.86	CE	0.5			MIT
2691.175	2691.974	37147.46	CE	0.3			MIT
2690.979	2691.778	37150.17	CE	0.5			MIT
2689.483	2690.281	37170.83	CE	0.3			MIT
2689.176	2689.974	37175.08	CE	0.3			MIT
2688.160	2688.958	37189.12	CE	0.3			MIT
2687.988	2688.786	37191.50	CE	0.9			MIT
2687.621	2688.419	37196.58	CE	0.3			MIT
2687.126	2687.924	37203.43	CE	0.7			MIT
2686.766	2687.564	37208.42	CE	0.6			MIT
2686.333	2687.131	37214.42	CE	0.3			MIT
2685.774	2686.572	37222.16	CE	0.3			MIT
2685.428	2686.225	37226.96	CE	0.3			MIT
2685.038	2685.835	37232.36	CE	0.7			MIT
2684.697	2685.494	37237.09	CE	0.3			MIT
2684.418	2685.215	37240.96	CE	0.6			MIT
2684.284	2685.081	37242.82	CE	1.0			MIT
2684.244	2685.041	37243.38	CE	0.5			MIT
2683.963	2684.760	37247.27	CE	0.3			MIT
2683.057	2683.854	37259.85	CE	0.3			MIT
2682.727	2683.524	37264.44	CE	1.0			MIT
2682.351	2683.148	37269.66	CE	0.3			MIT
2681.659	2682.456	37279.27	CE	0.5			MIT
2681.288	2682.085	37284.43	CE	0.3W			MIT
2680.425	2681.221	37296.44	CE	0.6			MIT
2679.827	2680.623	37304.76	CE	0.3			MIT
2678.946	2679.742	37317.03	CE	0.3			MIT
2678.618	2679.414	37321.60	CE	0.3			MIT
2677.592	2678.388	37335.90	CE	0.7			MIT
2677.293	2678.089	37340.06	CE	0.3			MIT
2676.755	2677.550	37347.57	CE	0.6			MIT
2676.623	2677.418	37349.41	CE	0.6			MIT
2676.467	2677.262	37351.59	CE	0.5			MIT
2676.355	2677.150	37353.15	CE	0.3			MIT
2676.121	2676.916	37356.42	CE	0.3			MIT
2675.733	2676.528	37361.83	CE	0.3			MIT
2674.830	2675.625	37374.45	CE	0.5			MIT
2674.587	2675.382	37377.84	CE	0.3			MIT
2674.447	2675.242	37379.80	CE	0.3			MIT
2673.545	2674.340	37392.41	CE	0.3			MIT
2673.073	2673.868	37399.01	CE	1.0			MIT
2672.777	2673.571	37403.15	CE	0.3	0.3		MIT
2672.403	2673.197	37408.39	CE	0.3			MIT
2672.280	2673.074	37410.11	CE	0.3			MIT
2671.908	2672.702	37415.32	CE	0.3			MIT
2671.185	2671.979	37425.44	CE	0.3			MIT
2669.585	2670.379	37447.87	CE	0.3			MIT
2668.219	2669.012	37467.04	CE	0.3			MIT
2666.496	2667.289	37491.25	CE	1.0			MIT
2666.135	2666.928	37496.33	CE	0.3			MIT
2665.903	2666.696	37499.59	CE	0.3			MIT
2665.271	2666.064	37508.48	CE	0.7			MIT
2665.002	2665.795	37512.27	CE	0.3			MIT
2664.737	2665.530	37516.00	CE	0.3			MIT
2663.638	2664.430	37531.48	CE	0.6			MIT
2663.297	2664.089	37536.28	CE	0.3			MIT
2662.861	2663.653	37542.43	CE	0.0	0.6		MIT
2662.587	2663.379	37546.29	CE	0.3			MIT
2662.170	2662.962	37552.17	CE	0.3			MIT
2662.024	2662.816	37554.23	CE	0.6			MIT
2661.617	2662.409	37559.97	CE	0.6			MIT
2661.352	2662.144	37563.71	CE	0.3			MIT
2660.638	2661.430	37573.79	CE	0.6			MIT
2659.716	2660.507	37586.82	CE	0.3			MIT
2658.668	2659.459	37601.63	CE	0.6			MIT
2658.383	2659.174	37605.66	CE	0.5			MIT
2657.789	2658.580	37614.07	CE	0.3			MIT
2656.844	2657.635	37627.44	CE	0.9			MIT
2656.440	2657.231	37633.17	CE	0.6			MIT
2652.531	2653.321	37688.62	CE	0.3			MIT
2652.152	2652.942	37694.01	CE	0.3			MIT
2652.006	2652.795	37696.08	CE	1.0			MIT

AIR WAVELENGTH (ANGSTRMS)	VACUUM WAVELENGTH (ANGSTRMS)	WAVENUMBER (KAYSERS)	LMENT	LOG ARC	INTENSITY SPK	INTENSITY DIS	REF
2651.418	2652.207	37704.44	CE	0.3			MIT
2651.006	2651.795	37710.30	CE	1.4	0.3		MIT
2650.584	2651.373	37716.31	CE	0.6			MIT
2649.330	2650.119	37734.16	CE	1.0			MIT
2648.299	2649.088	37748.85	CE	0.6			MIT
2648.247	2649.036	37749.59	CE	0.3			MIT
2647.105	2647.893	37765.87	CE	0.9			MIT
2647.007	2647.795	37767.27	CE	0.5			MIT
2646.640	2647.428	37772.51	CE	0.3			MIT
2644.027	2644.815	37809.83	CE	0.3			MIT
2642.483	2643.270	37831.92	CE	0.5			MIT
2641.493	2642.280	37846.10	CE	0.7			MIT
2641.456	2642.243	37846.63	CE	0.9			MIT
2640.750	2641.537	37856.75	CE	0.3			MIT
2640.646	2641.433	37858.24	CE	0.3			MIT
2639.770	2640.557	37870.80	CE	0.3			MIT
2639.367	2640.153	37876.59	CE	0.5			MIT
2638.807	2639.593	37884.62	CE	0.7			MIT
2637.958	2638.744	37896.82	CE	0.3			MIT
2637.055	2637.841	37909.79	CE	0.6			MIT
2636.658	2637.444	37915.50	CE	0.3			MIT
2636.355	2637.141	37919.86	CE	0.3			MIT
2635.148	2635.933	37937.23	CE	1.3	0.0		MIT
2634.547	2635.332	37945.88	CE	0.6			MIT
2632.106	2632.891	37981.07	CE	0.3			MIT
2630.616	2631.400	38002.58	CE	0.5			MIT
2629.977	2630.761	38011.81	CE	0.3			MIT
2628.813	2629.597	38028.64	CE	0.3			MIT
2626.265	2627.048	38065.54	CE	0.3			MIT
2625.736	2626.519	38073.20	CE	1.0			MIT
2624.418	2625.201	38092.32	CE	0.3			MIT
2623.442	2624.225	38106.49	CE	0.8			MIT
2623.315	2624.098	38108.34	CE	0.8			MIT
2622.972	2623.755	38113.32	CE	0.8			MIT
2622.784	2623.567	38116.05	CE	0.7			MIT
2620.931	2621.713	38143.00	CE	0.3			MIT
2619.804	2620.586	38159.41	CE	0.7			MIT
2619.715	2620.497	38160.70	CE	0.3			MIT
2618.920	2619.702	38172.29	CE	0.6			MIT
2617.723	2618.504	38189.74	CE	0.5			MIT
2617.108	2617.889	38198.71	CE	0.6			MIT
2615.877	2616.658	38216.69	CE	0.6			MIT
2615.453	2616.234	38222.88	CE	0.7			MIT
2614.566	2615.347	38235.85	CE	0.7			MIT
2614.292	2615.072	38239.86	CE	0.3			MIT
2613.897	2614.677	38245.64	CE	1.2	0.0		MIT
2610.985	2611.765	38288.29	CE	0.3			MIT
2610.775	2611.555	38291.37	CE	0.3			MIT
2610.531	2611.311	38294.95	CE	0.3			MIT
2610.334	2611.114	38297.84	CE	0.3			MIT
2609.900	2610.679	38304.21	CE	0.9			MIT
2609.502	2610.281	38310.05	CE	1.1			MIT
2609.116	2609.895	38315.71	CE	0.3			MIT
2607.344	2608.123	38341.75	CE	0.7			MIT
2606.088	2606.867	38360.23	CE	0.6			MIT
2605.378	2606.156	38370.68	CE	0.6			MIT
2603.590	2604.368	38397.03	CE	0.6			MIT
2603.444	2604.222	38399.19	CE	0.3			MIT
2602.772	2603.550	38409.10	CE	0.3			MIT
2602.088	2602.866	38419.19	CE	0.3			MIT
2600.878	2601.655	38437.07	CE	0.5			MIT
2600.788	2601.565	38438.40	CE	0.3			MIT
2600.434	2601.211	38443.63	CE	0.5			MIT
2598.250	2599.027	38475.94	CE	0.7			MIT
2597.047	2597.823	38493.76	CE	0.9			MIT
2596.485	2597.261	38502.10	CE	0.5			MIT
2596.358	2597.134	38503.98	CE	0.3			MIT
2595.575	2596.351	38515.59	CE	0.6			MIT
2594.787	2595.563	38527.29	CE	0.3			MIT
2593.802	2594.578	38541.92	CE	0.3			MIT

AIR WAVELENGTH (ANGSTRMS)	VACUUM WAVELENGTH (ANGSTRMS)	WAVENUMBER (KAYSERS)	LMENT	LOG ARC	INTENSITY SPK	INTENSITY DIS	REF
2592.801	2593.576	38556.80	CE	0.3			MIT
2592.338	2593.113	38563.68	CE	1.0			MIT
2590.805	2591.580	38586.50	CE	0.3			MIT
2589.559	2590.334	38605.07	CE	0.3			MIT
2589.057	2589.832	38612.55	CE	0.7			MIT
2588.885	2589.659	38615.12	CE	0.5			MIT
2588.426	2589.200	38621.96	CE	0.5			MIT
2587.883	2588.657	38630.07	CE	0.3			MIT
2586.562	2587.336	38649.79	CE	0.3			MIT
2585.187	2585.961	38670.35	CE	0.3			MIT
2584.203	2584.976	38685.07	CE	0.7			MIT
2583.128	2583.901	38701.17	CE	0.7			MIT
2582.793	2583.566	38706.19	CE	0.5			MIT
2582.590	2583.363	38709.23	CE	0.3			MIT
2579.428	2580.200	38756.68	CE	0.3			MIT
2575.933	2576.704	38809.26	CE	0.8			MIT
2573.141	2573.912	38851.37	CE	1.2			MIT
2571.930	2572.700	38869.66	CE	0.7			MIT
2571.361	2572.131	38878.26	CE	0.5			MIT
2569.879	2570.649	38900.68	CE	0.8			MIT
2569.168	2569.938	38911.45	CE	0.9			MIT
2568.705	2569.475	38918.46	CE	0.3			MIT
2567.679	2568.448	38934.01	CE	0.3			MIT
2566.589	2567.358	38950.54	CE	0.7			MIT
2566.374	2567.143	38953.81	CE	0.3			MIT
2565.590	2566.359	38965.71	CE	0.9			MIT
2562.424	2563.192	39013.85	CE	0.7			MIT
2561.511	2562.279	39027.76	CE	0.5			MIT
2561.393	2562.161	39029.55	CE	0.3			MIT
2558.354	2559.121	39075.91	CE	0.3			MIT
2555.949	2556.716	39112.68	CE	0.3			MIT
2553.588	2554.354	39148.84	CE	0.3			MIT
2553.456	2554.222	39150.86	CE	0.3			MIT
2553.344	2554.110	39152.58	CE	0.5			MIT
2552.261	2553.027	39169.19	CE	0.3			MIT
2551.771	2552.537	39176.71	CE	0.7	0.5H		MIT
2551.206	2551.972	39185.39	CE	0.3			MIT
2550.095	2550.860	39202.46	CE	0.3			MIT
2549.260	2550.025	39215.30	CE	0.3			MIT
2548.799	2549.564	39222.39	CE	1.0	1.0H		MIT
2548.683	2549.448	39224.18	CE	1.3	1.3H		MIT
2547.893	2548.658	39236.34	CE	0.6			MIT
2547.187	2547.952	39247.21	CE	0.3			MIT
2545.746	2546.510	39269.43	CE	0.3			MIT
2544.376	2545.140	39290.57	CE	0.3			MIT
2544.269	2545.033	39292.22	CE	0.5			MIT
2543.536	2544.300	39303.54	CE	0.6			MIT
2543.086	2543.850	39310.50	CE	1.1	0.3		MIT
2542.361	2543.124	39321.71	CE	0.5			MIT
2534.866	2535.628	39437.97	CE	0.7			MIT
2534.180	2534.942	39448.64	CE	0.3			MIT
2532.415	2533.176	39476.13	CE	0.6			MIT
2531.147	2531.908	39495.91	CE	0.3			MIT
2528.287	2529.047	39540.58	CE	1.2	0.0		MIT
2524.000	2524.759	39607.74	CE	1.0			MIT
2520.540	2521.298	39662.10	CE	0.3			MIT
2518.505	2519.263	39694.15	CE	1.0	1.3		MIT
2515.360	2516.117	39743.78	CE	0.3			MIT
2514.862	2515.619	39751.65	CE	0.3	0.3		MIT
2513.868	2514.625	39767.36	CE	0.3			MIT
2513.304	2514.061	39776.29	CE	1.1			MIT
2512.814	2513.571	39784.04	CE	0.5			MIT
2496.914	2497.667	40037.37	CE		0.3		MIT
2454.36	2455.10	40731.5	CE		1.1		A
2439.85	2440.59	40973.7	CE		1.3		A
2380.16	2380.89	42001.2	CE		1.7		MIT
2377.52	2378.25	42047.8	CE		0.90		MIT
2377.11	2377.84	42055.0	CE		1.00		MIT
2372.37	2373.09	42139.1	CE		1.7		MIT
2367.80	2368.52	42220.4	CE		1.0		MIT

AIR WAVELENGTH (ANGSTRMS)	VACUUM WAVELENGTH (ANGSTRMS)	WAVENUMBER (KAYSERS)	LMENT	LOG ARC	INTENSITY SPK	INTENSITY DIS	REF
2362.57	2363.29	42313.9	CE		1.0		MIT
2350.15	2350.87	42537.5	CE		1.7		MIT
2337.71	2338.43	42763.8	CE		0.70		MIT
2324.37	2325.08	43009.2	CE		1.10		MIT
2318.69	2319.40	43114.5	CE		1.6		MIT
2317.37	2318.08	43139.1	CE		1.2		MIT
2300.68	2301.39	43452.0	CE		1.00		MIT
2287.85	2288.56	43695.7	CE		1.0		MIT
2264.89	2265.59	44138.6	CE		1.3		MIT
2242.33	2243.03	44582.6	CE		1.8		MIT
2228.09	2228.78	44867.5	CE		1.4		MIT
2227.87	2228.56	44872.0	CE		1.3		MIT
2225.10	2225.79	44927.8	CE		2.0		MIT
2222.04	2222.73	44989.7	CE		2.0		MIT
2203.18	2203.87	45374.8	CE		1.3		A
2180.67	2181.35	45843.1	CE		2.0		MIT
2125.04	2125.71	47043.1	CE		1.0W		A
2083.35	2084.01	47984.3	CE		1.4		A
9876.08	9878.79	10122.7	CL I			0.3	KS
9875.95	9878.66	10122.8	CL I			1.7	KS
9808.46	9811.15	10192.5	CL I			0.7	KS
9744.33	9747.00	10259.6	CL I			1.5	KS
9702.30	9704.96	10304.0	CL I			0.3	KS
9592.25	9594.88	10422.2	CL I			0.6	KS
9452.06	9454.65	10576.8	CL I			0.8	KS
9393.85	9396.43	10642.3	CL I			0.3	KS
9288.84	9291.39	10762.7	CL I			0.9	KS
9197.47	9199.99	10869.6	CL I			0.3	KS
9191.71	9194.23	10876.4	CL I			1.0	KS
9121.12	9123.62	10960.6	CL I			1.2	KS
9073.17	9075.66	11018.5	CL I			1.1	KS
9069.68	9072.17	11022.7	CL			0.9	KS
9045.43	9047.91	11052.3	CL I			1.2	KS
9038.98	9041.46	11060.2	CL I			1.0	KS
8948.01	8950.47	11172.6	CL I			1.3	KS
8912.90	8915.35	11216.6	CL I			1.2	KS
8686.30	8688.69	11509.2	CL I			1.2	KS
8676.7	8679.1	11522.	CL I			0.7	MJ
8585.99	8588.35	11643.7	CL I			1.5	KS
8577.95	8580.31	11654.6	CL I			0.3	KS
8575.27	8577.63	11658.2	CL I			1.4	KS
8550.50	8552.85	11692.0	CL I			0.9	KS
8467.32	8469.65	11806.9	CL I			0.9	KS
8428.27	8430.59	11861.6	CL I			1.5	KS
8406.14	8408.45	11892.8	CL I			0.3	KS
8391.96	8394.27	11912.9	CL II			0.5	KS
8382.76	8385.06	11926.0	CL II			0.7	KS
8376.3	8378.6	11935.	CL I			1.0	MJ
8375.97	8378.27	11935.6	CL I			1.6	KS
8361.81	8364.11	11955.8	CL II			0.9	KS
8360.63	8362.93	11957.5	CL II			1.2	KS
8353.00	8355.30	11968.5	CL II			0.3	KS
8333.31	8335.60	11996.7	CL I			1.5	KS
8221.76	8224.02	12159.5	CL I			1.4	KS
8220.45	8222.71	12161.4	CL I			1.2	KS
8212.03	8214.29	12173.9	CL I			1.3	KS
8211.3	8213.6	12175.	CL I			1.1	MJ
8200.23	8202.48	12191.4	CL I			1.0	KS
8199.06	8201.31	12193.2	CL I			1.0	KS
8194.39	8196.64	12200.1	CL I			1.2	KS
8087.66	8089.88	12361.1	CL I			0.6	KS
8086.71	8088.93	12362.6	CL			1.2	KS
8085.57	8087.79	12364.3	CL I			1.1	KS
8084.51	8086.73	12365.9	CL I			0.9	KS
8051.03	8053.24	12417.4	CL			0.6	KS
8023.31	8025.52	12460.3	CL			0.6	KS
8017.54	8019.75	12469.2	CL I			0.3	KS
8015.58	8017.78	12472.3	CL I			0.9	KS
7997.84	8000.04	12499.9	CL I			1.1	KS
7980.61	7982.81	12526.9	CL I			0.3	KS

AIR WAVELENGTH (ANGSTRMS)	VACUUM WAVELENGTH (ANGSTRMS)	WAVENUMBER (KAYSERS)	LMENT	LOG ARC	INTENSITY SPK	INTENSITY DIS	REF
7976.97	7979.16	12532.6	CL I			0.8	KS
7974.74	7976.93	12536.2	CL			0.6	KS
7935.01	7937.19	12598.9	CL I			0.6	KS
7933.88	7936.06	12600.7	CL I			0.9	KS
7924.67	7926.85	12615.3	CL I			1.2	KS
7915.10	7917.28	12630.6	CL I			0.8	KS
7899.36	7901.53	12655.8	CL I			1.0	KS
7878.24	7880.41	12689.7	CL I			1.2	KS
7830.79	7832.94	12766.6	CL I			0.9	KS
7821.39	7823.54	12781.9	CL I			1.2	KS
7771.13	7773.27	12864.6	CL I			0.3	KS
7769.18	7771.32	12867.8	CL I			0.9	KS
7744.98	7747.11	12908.0	CL I			1.3	KS
7717.60	7719.72	12953.8	CL I			1.3	KS
7702.87	7704.99	12978.6	CL I			0.3	KS
7672.46	7674.57	13030.0	CL I			1.0	KS
7644.80	7646.90	13077.2	CL II			0.6	KS
7620.51	7622.61	13118.9	CL II			0.6	KS
7578.07	7580.16	13192.3	CL II			1.0	KS
7565.53	7567.61	13214.2	CL II			1.3	KS
7547.09	7549.17	13246.5	CL I			1.4	KS
7492.10	7494.16	13343.7	CL I			1.0	KS
7489.46	7491.52	13348.4	CL I			0.7	KS
7462.40	7464.46	13396.8	CL I			0.7	KS
7459.46	7461.51	13402.1	CL I			0.3	KS
7444.35	7446.40	13429.3	CL I			0.3	KS
7435.68	7437.73	13445.0	CL I			0.7	KS
7414.12	7416.16	13484.1	CL I			2.2	KS
7389.28	7391.32	13529.4	CL II			0.8	KS
7342.83	7344.85	13615.0	CL I			0.3	KS
7256.65	7258.65	13776.7	CL I			2.3	KS
7086.83	7088.78	14106.8	CL I			1.0	KS
6981.90	6983.83	14318.8	CL I			0.7	KS
6966.95	6968.87	14349.5	CL I			0.3	KS
6952.13	6954.05	14380.1	CL II			1.4	KS
6932.94	6934.85	14419.9	CL I			1.0	KS
6930.45	6932.36	14425.1	CL II			0.6	KS
6850.21	6852.10	14594.1	CL II			1.6	KS
6841.86	6843.75	14611.9	CL II			1.0	KS
6840.26	6842.15	14615.3	CL I			0.3	KS
6831.62	6833.51	14633.8	CL II			1.5	KS
6759.42	6761.29	14790.1	CL II			1.5	KS
6713.43	6715.28	14891.4	CL II			1.6	KS
6686.04	6687.89	14952.4	CL II			1.7	KS
6681.03	6682.87	14963.6	CL II			1.2	KS
6678.45	6680.29	14969.4	CL I			0.3	KS
6661.68	6663.52	15007.1	CL II			1.9	KS
6653.75	6655.59	15025.0	CL II			1.4	KS
6614.5	6616.3	15114.	CL			1.0	MJ
6609.30	6611.13	15126.0	CL			0.7	KS
6604.61	6606.43	15136.8	CL I			0.7	KS
6542.42	6544.23	15280.6	CL			0.7	KS
6531.43	6533.23	15306.3	CL I			1.2	KS
6522.38	6524.18	15327.6	CL II			1.0	KS
6502.28	6504.08	15375.0	CL I			0.3	KS
6487.1	6488.9	15411.	CL			1.0	MJ
6450.36	6452.14	15498.7	CL I			1.2	KS
6434.80	6436.58	15536.2	CL			1.4	KS
6425.64	6427.42	15558.3	CL			0.7	KS
6419.25	6421.02	15573.8	CL II			0.9	KS
6399.41	6401.18	15622.1	CL II			1.0	KS
6398.64	6400.41	15624.0	CL			1.6	KS
6384.13	6385.89	15659.5	CL II			0.7	KS
6341.70	6343.45	15764.3	CL			1.0	KS
6252.34	6254.07	15989.6	CL I			0.7	KS
6242.66	6244.39	16014.4	CL I			0.3	KS
6231.57	6233.29	16042.9	CL I			0.7	KS
6227.18	6228.90	16054.2	CL II			0.8	KS
6211.61	6213.33	16094.4	CL I			0.7	KS
6194.75	6196.46	16138.2	CL I			1.2	KS

AIR WAVELENGTH (ANGSTRMS)	VACUUM WAVELENGTH (ANGSTRMS)	WAVENUMBER (KAYSERS)	LMENT	LOG INTENSITY ARC	SPK	DIS	REF
6162.14	6163.85	16223.6	CL I			1.0	KS
6140.25	6141.95	16281.5	CL I			1.5	KS
6114.41	6116.10	16350.3	CL I			1.2	KS
6094.65	6096.34	16403.3	CL II			2.0	KS
5930.42	5932.06	16857.5	CL I			0.3	KS
5856.74	5858.36	17069.6	CL I			0.6	KS
5847.74	5849.36	17095.9	CL I			0.6	KS
5846.71	5848.33	17098.9	CL			0.6	KS
5844.27	5845.89	17106.0	CL			0.3	KS
5802.91	5804.52	17228.0	CL I			0.3	KS
5799.94	5801.55	17236.8	CL			0.8	KS
5796.34	5797.95	17247.5	CL			0.8	KS
5790.50	5792.11	17264.9	CL II			1.4	KS
5634.84	5636.40	17741.8	CL II			1.3	KS
5578.22	5579.77	17921.9	CL I			0.3	KS
5569.17	5570.72	17951.0	CL			0.6	KS
5568.81	5570.36	17952.2	CL II			1.2	KS
5532.16	5533.70	18071.1	CL I			0.8	KS
5514.71	5516.24	18128.3	CL			0.6	KS
5513.00	5514.53	18133.9	CL I			0.6	KS
5457.47	5458.99	18318.4	CL II			1.5	KS
5457.02	5458.54	18319.9	CL II			1.9	KS
5456.27	5457.79	18322.4	CL II			1.7	KS
5444.99	5446.50	18360.4	CL II			1.0	KS
5444.25	5445.76	18362.9	CL II			1.8	KS
5443.42	5444.93	18365.7	CL II			2.0	KS
5424.36	5425.87	18430.2	CL II			1.4	KS
5423.52	5425.03	18433.1	CL II			2.0	KS
5423.25	5424.76	18434.0	CL II			2.2	KS
5392.21	5393.71	18540.1	CL			1.0	BL
5341.04	5342.53	18717.7	CL			0.3	KS
5333.70	5335.18	18743.5	CL II			1.2	KS
5285.45	5286.92	18914.6	CL			0.3	BL
5221.340	5222.794	19146.84	CL II			1.8	MU
5218.16	5219.61	19158.5	CL II			0.6	MU
5217.92	5219.37	19159.4	CL II			2.0	MU
5217.35	5218.80	19161.5	CL II			0.6	MU
5189.70	5191.15	19263.6	CL II			1.4	KS
5175.85	5177.29	19315.1	CL II			1.3	KS
5173.15	5174.59	19325.2	CL II			1.4	KS
5162.34	5163.78	19365.7	CL II			1.0	KS
5158.79	5160.23	19379.0	CL II			0.9	KS
5140.38	5141.81	19448.4	CL I			0.3	KS
5113.36	5114.78	19551.2	CL II			1.6	KS
5113.12	5114.54	19552.1	CL II			0.9	MU
5104.08	5105.50	19586.7	CL II			1.4	KS
5103.85	5105.27	19587.6	CL II			0.8	MU
5103.04	5104.46	19590.7	CL II			2.1	KS
5102.86	5104.28	19591.4	CL II			1.5	MU
5098.34	5099.76	19608.8	CL II			1.3	KS
5078.25	5079.67	19686.3	CL II			2.2	KS
5068.10	5069.51	19725.8	CL II			0.6	KS
4995.52	4996.91	20012.4	CL II			1.8	KS
4976.65	4978.04	20088.2	CL I			0.9	KS
4970.12	4971.51	20114.6	CL II			1.7	KS
4967.33	4968.72	20125.9	CL			0.8	BL
4943.24	4944.62	20224.0	CL II			1.2	KS
4936.99	4938.37	20249.6	CL II			1.4	KS
4925.17	4926.54	20298.2	CL II			1.2	KS
4924.83	4926.20	20299.6	CL II			1.0	KS
4924.28	4925.65	20301.9	CL II			1.3	KS
4922.14	4923.51	20310.7	CL II			1.3	KS
4917.72	4919.09	20328.9	CL II			2.1	KS
4914.32	4915.69	20343.0	CL II			1.1	KS
4907.17	4908.54	20372.7	CL II			1.2	KS
4904.76	4906.13	20382.7	CL II			2.1	KS
4896.77	4898.14	20415.9	CL II			2.3	KS
4857.04	4858.40	20582.9	CL II			1.0	KS
4852.72	4854.08	20601.2	CL I			0.9	KS
4842.44	4843.79	20645.0	CL II			0.9	KS

AIR WAVELENGTH (ANGSTRMS)	VACUUM WAVELENGTH (ANGSTRMS)	WAVENUMBER (KAYSERS)	LMENT	LOG INTENSITY ARC	SPK	DIS	REF
4836.79	4838.14	20669.1	CL II			1.3	KS
4819.79	4821.14	20742.0	CL II			1.4	KS
4819.46	4820.81	20743.4	CL II			2.3	KS
4818.55	4819.90	20747.3	CL I			0.6	KS
4811.57	4812.91	20777.4	CL II			1.1	KS
4810.06	4811.40	20783.9	CL II			2.3	KS
4809.05	4810.39	20788.3	CL II			1.0	KS
4798.40	4799.74	20834.5	CL II			1.2	KS
4794.54	4795.88	20851.2	CL II			2.4	KS
4792.04	4793.38	20862.1	CL II			1.1	KS
4785.44	4786.78	20890.9	CL II			1.7	KS
4781.82	4783.16	20906.7	CL II			1.7	KS
4781.32	4782.66	20908.9	CL II			1.9	KS
4778.93	4780.27	20919.3	CL II			1.7	KS
4771.66	4772.99	20951.2	CL II			1.3	KS
4771.09	4772.42	20953.7	CL II			1.6	KS
4768.68	4770.01	20964.3	CL II			2.2	KS
4765.30	4766.63	20979.2	CL II			1.0	KS
4755.64	4756.97	21021.8	CL II			1.7	KS
4748.67	4750.00	21052.6	CL II			1.3	KS
4740.68	4742.01	21088.1	CL I			1.0	KS
4740.40	4741.73	21089.4	CL II			2.2	KS
4739.42	4740.75	21093.7	CL II			1.0	KS
4738.41	4739.74	21098.2	CL II			1.0	KS
4721.43	4722.75	21174.1	CL II			1.4	KS
4721.28	4722.60	21174.8	CL I			0.8	KS
4691.54	4692.85	21309.0	CL I			0.9	KS
4677.76	4679.07	21371.8	CL I			0.6	KS
4661.22	4662.52	21447.6	CL I			1.2	KS
4654.06	4655.36	21480.6	CL I			0.9	KS
4623.96	4625.26	21620.4	CL I			0.8	KS
4601.00	4602.29	21728.3	CL I			1.3	KS
4585.03	4586.31	21804.0	CL II			1.2	KS
4584.28	4585.56	21807.6	CL II			1.3	KS
4582.39	4583.67	21816.6	CL II			0.3	MU
4578.18	4579.46	21836.6	CL I			0.3	KS
4572.35	4573.63	21864.5	CL II			0.3	MU
4572.13	4573.41	21865.5	CL II			2.0	KS
4569.42	4570.70	21878.5	CL II			1.7	KS
4544.48	4545.75	21998.6	CL II			1.0	KS
4536.78	4538.05	22035.9	CL II			1.3	KS
4526.21	4527.48	22087.4	CL I			1.4	KS
4519.19	4520.46	22121.7	CL II			1.3	KS
4504.27	4505.53	22194.9	CL II			1.3	KS
4497.30	4498.56	22229.3	CL II			1.3	KS
4491.08	4492.34	22260.1	CL I			0.9	KS
4490.00	4491.26	22265.5	CL II			1.7	KS
4482.02	4483.28	22305.1	CL II			1.0	KS
4475.28	4476.54	22338.7	CL II			1.1	KS
4469.37	4470.62	22368.2	CL I			1.1	KS
4446.12	4447.37	22485.2	CL I			0.6	KS
4445.82	4447.07	22486.7	CL I			0.6	KS
4445.09	4446.34	22490.4	CL			1.0	L
4438.81	4440.06	22522.2	CL			0.9	BL
4438.48	4439.73	22523.9	CL I			1.2	KS
4403.03	4404.27	22705.3	CL I			1.1	KS
4402.52	4403.76	22707.9	CL I			0.3	KS
4399.14	4400.38	22725.3	CL II			1.2	KS
4390.38	4391.61	22770.7	CL I			0.9	KS
4389.76	4390.99	22773.9	CL I			1.4	KS
4389.32	4390.55	22776.2	CL			0.8	BL
4387.53	4388.76	22785.5	CL I			0.8	KS
4379.91	4381.14	22825.1	CL I			1.2	KS
4372.91	4374.14	22861.6	CL II			1.9	KS
4371.55	4372.78	22868.8	CL I			0.3	KS
4369.52	4370.75	22879.4	CL I			1.1	KS
4363.30	4364.53	22912.0	CL I			1.1	KS
4343.62	4344.84	23015.8	CL II			2.0	KS
4336.26	4337.48	23054.9	CL II			1.7	KS
4332.80	4334.02	23073.3	CL II			1.0	KS

AIR WAVELENGTH (ANGSTRMS)	VACUUM WAVELENGTH (ANGSTRMS)	WAVENUMBER (KAYSERS)	LMENT	LOG ARC	INTENSITY SPK	INTENSITY DIS	REF
4323.34	4324.56	23123.8	CL I			1.1	KS
4309.06	4310.27	23200.4	CL II			1.7	KS
4307.42	4308.63	23209.2	CL II			1.9	KS
4304.07	4305.28	23227.3	CL II			1.6	KS
4291.76	4292.97	23293.9	CL II			1.7	KS
4283.45	4284.66	23339.1	CL I			0.8	MJ
4276.51	4277.71	23377.0	CL II			1.5	KS
4270.61	4271.81	23409.3	CL II			1.4	KS
4264.58	4265.78	23442.4	CL I			0.8	KS
4261.22	4262.42	23460.9	CL II			1.3	KS
4259.52	4260.72	23470.2	CL II			1.5	KS
4253.51	4254.71	23503.4	CL II			1.9	KS
4241.38	4242.57	23570.6	CL II			1.8	KS
4235.49	4236.68	23603.4	CL II			1.4	KS
4234.09	4235.28	23611.2	CL II			1.7	KS
4226.43	4227.62	23654.0	CL I			1.0	KS
4224.92	4226.11	23662.4	CL II			1.2	KS
4216.85	4218.04	23707.7	CL			0.9	BL
4215.64	4216.83	23714.5	CL			0.8	BL
4209.67	4210.86	23748.1	CL I			0.9	KS
4208.03	4209.22	23757.4	CL II			1.5	KS
4205.07	4206.25	23774.1	CL II			1.0	KS
4204.54	4205.72	23777.1	CL II			1.3	KS
4195.11	4196.29	23830.6	CL II			1.3	KS
4191.59	4192.77	23850.6	CL II			1.2	KS
4188.82	4190.00	23866.4	CL II			1.2	KS
4185.61	4186.79	23884.6	CL II			1.3	KS
4157.82	4158.99	24044.3	CL II			1.4	KS
4147.09	4148.26	24106.5	CL II			1.5	KS
4133.66	4134.83	24184.8	CL II			1.3	KS
4132.48	4133.65	24191.7	CL II			2.3	KS
4130.86	4132.03	24201.2	CL II			1.3	KS
4130.21	4131.38	24205.0	CL			0.6	BL
4124.00	4125.16	24241.5	CL II			1.1	KS
4098.74	4099.90	24390.9	CL			0.9	BL
4098.40	4099.56	24392.9	CL			1.1	BL
4096.98	4098.14	24401.3	CL			0.8	BL
4092.53	4093.69	24427.9	CL			0.9	BL
4088.15	4089.30	24454.0	CL			0.8	BL
4079.88	4081.03	24503.6	CL II			1.2	KS
4057.08	4058.23	24641.3	CL			1.0	BL
4054.18	4055.33	24658.9	CL II			1.0	KS
4052.22	4053.36	24670.9	CL II			1.1	KS
4051.64	4052.78	24674.4	CL			0.8	BL
4050.46	4051.60	24681.6	CL			0.9	BL
4047.87	4049.01	24697.4	CL			0.8	BL
4044.58	4045.72	24717.5	CL			1.0	BL
4044.09	4045.23	24720.5	CL II			0.6	KS
4040.64	4041.78	24741.6	CL II			1.0	KS
4038.43	4039.57	24755.1	CL			0.8	BL
4036.53	4037.67	24766.8	CL II			1.0	KS
4032.19	4033.33	24793.4	CL			0.6	KS
4029.20	4030.34	24811.8	CL			0.9	BL
4026.67	4027.81	24827.4	CL			0.6	BL
4025.68	4026.82	24833.5	CL II			0.8	KS
4020.06	4021.20	24868.2	CL II			1.2	KS
4015.06	4016.19	24899.2	CL			1.0	BL
3995.18	3996.31	25023.1	CL II			0.3	MU
3990.19	3991.32	25054.4	CL II			1.3	KS
3981.94	3983.07	25106.3	CL II			1.2	KS
3971.18	3972.30	25174.3	CL II			0.8	KS
3954.21	3955.33	25282.4	CL II			1.3	KS
3949.96	3951.08	25309.6	CL II			1.0	KS
3928.63	3929.74	25447.0	CL II			0.7	KS
3927.88	3928.99	25451.8	CL II			0.8	KS
3917.57	3918.68	25518.8	CL II			1.3	KS
3916.70	3917.81	25524.5	CL II			1.3	KS
3913.92	3915.03	25542.6	CL II			1.5	KS
3902.84	3903.95	25615.1	CL II			1.0	KS
3883.80	3884.90	25740.7	CL II			1.1	KS

AIR WAVELENGTH (ANGSTRMS)	VACUUM WAVELENGTH (ANGSTRMS)	WAVENUMBER (KAYSERS)	LMENT	LOG ARC	INTENSITY SPK	INTENSITY DIS	REF
3868.62	3869.72	25841.7	CL II			1.6	KS
3864.60	3865.70	25868.6	CL II			1.2	KS
3861.88	3862.97	25886.8	CL II			1.3	KS
3861.34	3862.43	25890.4	CL II			1.7	KS
3860.99	3862.08	25892.7	CL II			2.0	KS
3860.83	3861.92	25893.8	CL II			2.2	KS
3854.75	3855.84	25934.7	CL II			1.5	KS
3851.69	3852.78	25955.3	CL II			1.5	KS
3851.42	3852.51	25957.1	CL II			1.9	KS
3851.02	3852.11	25959.8	CL II			2.0	KS
3850.58	3851.67	25962.7	CL II			1.1	KS
3845.82	3846.91	25994.9	CL II			1.5	KS
3845.68	3846.77	25995.8	CL II			1.9	KS
3845.42	3846.51	25997.6	CL II			1.7	KS
3843.26	3844.35	26012.2	CL II			2.0	KS
3838.37	3839.46	26045.3	CL II			1.3	KS
3830.80	3831.89	26096.8	CL II			1.2	KS
3829.27	3830.36	26107.2	CL II			1.2	KS
3827.62	3828.71	26118.5	CL II			2.2	KS
3820.25	3821.33	26168.9	CL II			2.0	KS
3818.40	3819.48	26181.6	CL II			1.5	KS
3810.10	3811.18	26238.6	CL II			1.5	KS
3809.51	3810.59	26242.6	CL II			1.6	KS
3805.24	3806.32	26272.1	CL II			1.9	KS
3798.80	3799.88	26316.6	CL II			1.7	KS
3793.75	3794.83	26351.7	CL II			1.4	KS
3791.46	3792.54	26367.6	CL			0.5	BL
3781.23	3782.30	26438.9	CL II			1.5	KS
3774.25	3775.32	26487.8	CL II			1.4	KS
3773.68	3774.75	26491.8	CL II			1.3	KS
3769.13	3770.20	26523.8	CL II			1.3	KS
3768.13	3769.20	26530.8	CL II			1.3	KS
3767.57	3768.64	26534.8	CL II			1.5	KS
3750.81	3751.88	26653.3	CL			0.9	BL
3750.00	3751.07	26659.1	CL II			1.5	KS
3748.46	3749.53	26670.0	CL II			1.2	KS
3733.73	3734.79	26775.3	CL II			1.0	KS
3725.71	3726.77	26832.9	CL			0.6	BL
3717.94	3719.00	26889.0	CL II			1.2	KS
3705.47	3706.52	26979.4	CL			0.5	BL
3688.44	3689.49	27104.0	CL II			1.2	KS
3682.05	3683.10	27151.1	CL			0.8	BL
3673.83	3674.88	27211.8	CL II			1.3	KS
3670.29	3671.34	27238.1	CL			0.9	BL
3668.03	3669.07	27254.8	CL II			1.3	KS
3659.767	3660.810	27316.36	CL II			0.3	MU
3658.33	3659.37	27327.1	CL II			0.6	MU
3656.95	3657.99	27337.4	CL			1.2	BL
3650.10	3651.14	27388.7	CL II			0.6	MU
3648.01	3649.05	27404.4	CL II			0.3	MU
3639.15	3640.19	27471.1	CL II			0.3	MU
3622.69	3623.72	27595.9	CL			0.9	BL
3618.88	3619.91	27625.0	CL II			0.5	MU
3615.07	3616.10	27654.1	CL II			0.3	MU
3612.86	3613.89	27671.0	CL			1.0	BL
3610.02	3611.05	27692.8	CL			0.6	BL
3609.74	3610.77	27694.9	CL II			0.3	MU
3602.10	3603.13	27753.7	CL			1.1	BL
3584.07	3585.09	27893.3	CL			0.6	BL
3567.98	3569.00	28019.1	CL			0.8	BL
3526.13	3527.14	28351.6	CL II			1.5	KS
3522.14	3523.15	28383.7	CL II			1.6	KS
3513.69	3514.69	28452.0	CL II			1.1	KS
3513.22	3514.22	28455.8	CL II			1.5	KS
3509.39	3510.39	28486.8	CL II			1.6	KS
3508.94	3509.94	28490.5	CL II			1.1	KS
3505.530	3506.533	28518.20	CL			0.6	MU
3505.44	3506.44	28518.9	CL II			1.1	KS
3479.82	3480.82	28728.9	CL II			1.5	KS
3415.75	3416.73	29267.7	CL			0.6H	BL

AIR WAVELENGTH (ANGSTRMS)	VACUUM WAVELENGTH (ANGSTRMS)	WAVENUMBER (KAYSERS)	LMENT	LOG ARC	INTENSITY SPK	DIS	REF
3409.92	3410.90	29317.8	CL II			0.7	KS
3393.46	3394.43	29460.0	CL			1.2	BL
3392.87	3393.84	29465.1	CL			1.2	BL
3387.59	3388.56	29511.0	CL			1.1	BL
3353.39	3354.35	29812.0	CL II			2.1	KS
3340.36	3341.32	29928.3	CL			1.3	JV
3337.5	3338.5	29954.	CL II			0.3	MU
3336.09	3337.05	29966.6	CL			1.0	JV
3333.64	3334.60	29988.6	CL II			1.6	KS
3332.42	3333.38	29999.6	CL II			1.2	KS
3329.12	3330.08	30029.3	CL II			2.2	KS
3320.51	3321.47	30107.2	CL			1.2	JV
3320.14	3321.10	30110.6	CL II			1.5	KS
3316.86	3317.81	30140.3	CL II			1.7	KS
3315.44	3316.39	30153.2	CL II			2.0	KS
3312.78	3313.73	30177.4	CL II			1.2	KS
3307.90	3308.85	30222.0	CL II			1.7	KS
3306.45	3307.40	30235.2	CL II			1.6	KS
3300.86	3301.81	30286.4	CL			0.8	JV
3289.72	3290.67	30389.0	CL			1.1	JV
3283.32	3284.27	30448.2	CL			1.0	JV
3280.60	3281.55	30473.4	CL			0.9	JV
3276.81	3277.75	30508.7	CL II			1.6	KS
3259.18	3260.12	30673.7	CL			0.9	BL
3251.07	3252.01	30750.2	CL			0.7	BL
3244.36	3245.30	30813.8	CL			0.7	JV
3232.58	3233.51	30926.1	CL			0.6	JV
3231.70	3232.63	30934.5	CL			0.7	BL
3230.69	3231.62	30944.2	CL			0.8	BL
3227.75	3228.68	30972.4	CL			0.5	AN
3215.70	3216.63	31088.4	CL II			0.5	MU
3203.05	3203.98	31211.2	CL II			1.3	KS
3202.12	3203.05	31220.3	CL II			0.8	KS
3191.40	3192.32	31325.1	CL			1.2	JV
3189.04	3189.96	31348.3	CL II			1.3	KS
3187.42	3188.34	31364.3	CL II			0.7	KS
3181.70	3182.62	31420.6	CL II			0.8	KS
3181.26	3182.18	31425.0	CL II			0.7	KS
3180.43	3181.35	31433.2	CL II			0.8	KS
3175.30	3176.22	31484.0	CL II			0.8	KS
3173.66	3174.58	31500.2	CL II			1.3	KS
3172.56	3173.48	31511.2	CL II			0.8	KS
3170.23	3171.15	31534.3	CL II			1.2	KS
3169.45	3170.37	31542.1	CL II			0.8	KS
3161.44	3162.36	31622.0	CL II			1.3	KS
3160.52	3161.43	31631.2	CL II			1.0	KS
3147.86	3148.77	31758.4	CL II			1.3	KS
3140.84	3141.75	31829.4	CL			0.6	BL
3139.16	3140.07	31846.4	CL			1.2	BL
3125.96	3126.87	31980.9	CL II			0.7	KS
3125.44	3126.35	31986.2	CL II			0.8	KS
3123.72	3124.63	32003.8	CL II			1.2	KS
3121.62	3122.53	32025.4	CL II			1.0	KS
3119.82	3120.72	32043.8	CL II			1.1	KS
3106.34	3107.24	32182.9	CL			0.6	BL
3104.47	3105.37	32202.3	CL			0.9	BL
3098.25	3099.15	32266.9	CL			0.5	AN
3096.72	3097.62	32282.9	CL II			1.4	KS
3093.00	3093.90	32321.7	CL			0.6	BL
3092.22	3093.12	32329.8	CL II			1.7	KS
3086.37	3087.27	32391.1	CL			0.8	BL
3076.64	3077.53	32493.6	CL			1.2	JV
3071.35	3072.24	32549.5	CL II			1.6	KS
3063.09	3063.98	32637.3	CL			1.0	JV
3059.69	3060.58	32673.6	CL			0.7	JV
3058.00	3058.89	32691.6	CL II			1.6	KS
3057.75	3058.64	32694.3	CL			0.9	AN
3053.74	3054.63	32737.2	CL II			1.0	KS
3045.00	3045.89	32831.2	CL II			1.0	KS
3037.98	3038.86	32907.0	CL II			1.5	KS

AIR WAVELENGTH (ANGSTRMS)	VACUUM WAVELENGTH (ANGSTRMS)	WAVENUMBER (KAYSERS)	LMENT	LOG ARC	INTENSITY SPK	DIS	REF
3022.93	3023.81	33070.9	CL II			1.5	KS
3018.82	3019.70	33115.9	CL II			1.1	KS
3006.98	3007.86	33246.3	CL II			1.3	KS
3006.05	3006.93	33256.5	CL II			1.3	KS
3004.39	3005.27	33274.9	CL II			1.0	KS
3002.45	3003.33	33296.4	CL			0.9	JV
2996.63	2997.50	33361.1	CL II			1.6	KS
2993.09	2993.96	33400.5	CL II			0.9	KS
2990.00	2990.87	33435.1	CL			0.7	AN
2982.78	2983.65	33516.0	CL II			1.3	KS
2978.56	2979.43	33563.5	CL			0.8	AN
2972.60	2973.47	33630.8	CL			0.3	BL
2970.73	2971.60	33651.9	CL			0.7	AN
2965.50	2966.37	33711.3	CL			1.0	JV
2944.20	2945.06	33955.2	CL			0.6	AN
2943.20	2944.06	33966.7	CL			0.6	AN
2934.60	2935.46	34066.2	CL II			0.7	KS
2931.87	2932.73	34098.0	CL			0.9	JV
2915.30	2916.15	34291.7	CL			0.7	JV
2912.06	2912.91	34329.9	CL II			1.2	KS
2906.25	2907.10	34398.5	CL II			1.3	KS
2898.45	2899.30	34491.1	CL			0.9	AN
2897.00	2897.85	34508.4	CL			0.8	AN
2891.72	2892.57	34571.4	CL			0.8	AN
2886.63	2887.48	34632.3	CL II			0.5	KS
2868.41	2869.25	34852.3	CL II			1.0	KS
2868.28	2869.12	34853.9	CL			0.5	AN
2863.55	2864.39	34911.4	CL II			0.8	KS
2863.20	2864.04	34915.7	CL			0.5	AN
2862.06	2862.90	34929.6	CL II			0.7	KS
2860.71	2861.55	34946.1	CL II			0.7	KS
2847.59	2848.43	35107.1	CL			0.7	AN
2844.37	2845.21	35146.8	CL			0.5	AN
2833.03	2833.86	35287.5	CL			0.6	BL
2822.30	2823.13	35421.7	CL			1.0	AN
2800.22	2801.05	35701.0	CL			0.6	BL
2799.43	2800.26	35711.0	CL			0.6	AN
2788.62	2789.44	35849.5	CL			0.6	BL
2783.14	2783.96	35920.0	CL			0.8	JV
2781.01	2781.83	35947.5	CL			0.9	JV
2769.41	2770.23	36098.1	CL			0.6	AN
2767.74	2768.56	36119.9	CL			0.6	AN
2764.95	2765.77	36156.3	CL			0.6	BL
2763.88	2764.70	36170.3	CL II			1.0	KS
2758.69	2759.51	36238.4	CL II			0.7	KS
2758.58	2759.40	36239.8	CL			0.3	AN
2754.10	2754.91	36298.8	CL II			1.4	KS
2754.04	2754.85	36299.6	CL			0.8	BL
2753.09	2753.90	36312.1	CL			0.8	AN
2751.52	2752.33	36332.8	CL II			0.7	KS
2751.23	2752.04	36336.6	CL			0.7	JV
2741.75	2742.56	36462.3	CL			0.7	AN
2724.03	2724.84	36699.5	CL			0.7	JV
2718.30	2719.11	36776.8	CL			0.6	AN
2714.38	2715.18	36829.9	CL II			0.9	KS
2710.38	2711.18	36884.3	CL			1.0	JV
2709.03	2709.83	36902.6	CL II			1.0	KS
2701.36	2702.16	37007.4	CL			0.6	JV
2698.54	2699.34	37046.1	CL			0.5	AN
2688.04	2688.84	37190.8	CL II			2.2	KS
2676.95	2677.75	37344.9	CL II			2.2	KS
2672.19	2672.98	37411.4	CL II			1.7	KS
2672.08	2672.87	37412.9	CL			1.0	AN
2671.43	2672.22	37422.0	CL II			0.8	KS
2667.36	2668.15	37479.1	CL II			1.6	KS
2666.46	2667.25	37491.8	CL II			1.3	KS
2658.74	2659.53	37600.6	CL II			2.0	KS
2651.82	2652.61	37698.7	CL			0.7	AN
2648.19	2648.98	37750.4	CL II			1.0	KS
2646.88	2647.67	37769.1	CL II			1.4	KS

AIR WAVELENGTH (ANGSTRMS)	VACUUM WAVELENGTH (ANGSTRMS)	WAVENUMBER (KAYSERS)	LMENT	LOG INTENSITY ARC	SPK	DIS	REF
2646.82	2647.61	37769.9	CL			0.5	AN
2634.95	2635.74	37940.1	CL II			1.1	KS
2615.13	2615.91	38227.6	CL II			1.0	KS
2614.64	2615.42	38234.8	CL			0.5	AN
2605.70	2606.48	38365.9	CL II			0.3	MU
2605.67	2606.45	38366.4	CL II			0.7	KS
2604.18	2604.96	38388.3	CL II			0.9	KS
2603.36	2604.14	38400.4	CL II			1.0	KS
2580.69	2581.46	38737.7	CL			1.2	JV
2571.10	2571.87	38882.2	CL II			0.9	KS
2565.29	2566.06	38970.3	CL II			1.2	KS
2564.84	2565.61	38977.1	CL II			1.3	KS
2564.13	2564.90	38987.9	CL II			0.8	KS
2549.85	2550.62	39206.2	CL II			1.7	KS
2547.76	2548.52	39238.4	CL II			1.1	KS
2546.94	2547.70	39251.0	CL II			1.3	KS
2544.84	2545.60	39283.4	CL II			1.2	KS
2543.98	2544.74	39296.7	CL II			1.0	KS
2541.95	2542.71	39328.1	CL			0.9	AN
2529.93	2530.69	39514.9	CL			0.7H	JV
2528.88	2529.64	39531.3	CL			0.9	AN
2516.42	2517.18	39727.0	CL			0.3	AN
2502.75	2503.50	39944.0	CL II			1.6	KS
2498.53	2499.28	40011.5	CL II			1.5	KS
2496.04	2496.79	40051.4	CL II			1.3	KS
2492.75	2493.50	40104.2	CL			0.5	AN
2488.74	2489.49	40168.9	CL			0.7	AN
2472.66	2473.41	40430.1	CL			0.6	BL
2459.86	2460.60	40640.4	CL II			1.0	KS
2459.75	2460.49	40642.2	CL			0.8	AN
2452.30	2453.04	40765.7	CL II			1.0	KS
2445.34	2446.08	40881.7	CL II			1.3	KS
2444.12	2444.86	40902.1	CL II			0.8	KS
2440.38	2441.12	40964.8	CL			0.8	AN
2437.74	2438.48	41009.2	CL			0.5	JV
2434.10	2434.84	41070.5	CL II			1.7	KS
2430.16	2430.90	41137.1	CL II			1.5	KS
2430.04	2430.78	41139.1	CL			1.2	AN
2428.02	2428.76	41173.3	CL II			1.0	KS
2427.79	2428.53	41177.2	CL II			1.3	KS
2424.01	2424.75	41241.4	CL II			1.0	KS
2412.48	2413.21	41438.5	CL II			1.0	KS
2407.10	2407.83	41531.1	CL II			0.7	KS
2407.05	2407.78	41532.0	CL			0.3	AN
2401.84	2402.57	41622.1	CL			0.5	JV
2387.18	2387.91	41877.7	CL			0.5	AN
2386.64	2387.37	41887.1	CL			0.6	AN
2327.03	2327.74	42960.0	CL			0.8	BL
2308.93	2309.64	43296.8	CL			0.5	BL
2304.57	2305.28	43378.7	CL			0.5	BL
2294.58	2295.29	43567.5	CL			0.5	JV
2288.16	2288.87	43689.8	CL			0.6	BL
2255.64	2256.34	44319.6	CL			0.6	BL
2253.16	2253.86	44368.4	CL			1.5	KS
2251.50	2252.20	44401.1	CL			1.6	KS
2250.96	2251.66	44411.7	CL			1.3	KS
2231.22	2231.91	44804.6	CL			0.6	JV
2202.25	2202.94	45393.9	CL			1.1	AN
2161.93	2162.61	46240.4	CL			1.0	JV
2093.64	2094.31	47748.5	CL			1.0	JV
2091.40	2092.06	47799.7	CL			0.9	AN
2088.52	2089.18	47865.6	CL			0.9	AN
2087.00	2087.66	47900.4	CL			1.0	AN
9999.7	10002.	9997.6	CO I	0.3H			ME
9952.2	9954.9	10045.	CO I	0.5H			ME
9940.69	9943.42	10056.9	CO I	0.3			ME
9914.12	9916.84	10083.9	CO	0.5			ME
9912.73	9915.45	10085.3	CO I	1.0			ME
9890.92	9893.63	10107.5	CO I	1.5			ME
9869.23	9871.94	10129.7	CO	0.3			ME

AIR WAVELENGTH (ANGSTRMS)	VACUUM WAVELENGTH (ANGSTRMS)	WAVENUMBER (KAYSERS)	LMENT	LOG INTENSITY ARC	SPK	DIS	REF
9847.7	9850.4	10152.	CO I	0.3			ME
9823.52	9826.21	10176.9	CO I	0.6H			ME
9818.39	9821.08	10182.2	CO	0.3			ME
9798.37	9801.06	10203.0	CO I	0.3H			ME
9785.39	9788.07	10216.5	CO	1.6			ME
9764.53	9767.21	10238.3	CO I	1.4			ME
9746.05	9748.72	10257.7	CO I	1.7			MIT
9696.60	9699.26	10310.1	CO I	0.3			ME
9678.21	9680.86	10329.7	CO I	0.5H			ME
9613.46	9616.10	10399.2	CO	0.5H			ME
9606.52	9609.15	10406.7	CO I	0.3			ME
9597.89	9600.52	10416.1	CO I	2.5			MIT
9569.00	9571.62	10447.6	CO I	0.3H			ME
9548.66	9551.28	10469.8	CO	0.3			ME
9544.53	9547.15	10474.3	CO I	2.5			MIT
9527.17	9529.78	10493.4	CO	0.5			ME
9442.34	9444.93	10587.7	CO I	0.3H			ME
9356.98	9359.55	10684.3	CO I	2.3			ME
9344.92	9347.48	10698.1	CO I	1.5H			MIT
9280.42	9282.97	10772.4	CO I	0.5			ME
9258.18	9260.72	10798.3	CO I	0.3			ME
9204.11	9206.64	10861.7	CO I	0.5			ME
9181.75	9184.27	10888.2	CO I	1.2			ME
9177.94	9180.46	10892.7	CO I	1.0			MIT
9165.52	9168.04	10907.5	CO I	0.7H			ME
9133.24	9135.75	10946.0	CO I	0.5H			ME
9095.37	9097.87	10991.6	CO I	1.7			ME
9037.91	9040.39	11061.5	CO I	1.8			MIT
8972.89	8975.35	11141.6	CO	0.3			ME
8958.37	8960.83	11159.7	CO I	0.6			ME
8939.13	8941.58	11183.7	CO	0.3			MIT
8926.28	8928.73	11199.8	CO I	1.3			MIT
8904.73	8907.17	11226.9	CO I	0.7V			MIT
8888.70	8891.14	11247.2	CO I	0.5H			ME
8870.70	8873.14	11270.0	CO I	0.5			ME
8850.70	8853.13	11295.4	CO I	1.0H			ME
8835.21	8837.64	11315.2	CO I	1.0			ME
8819.15	8821.57	11335.8	CO I	1.3			MIT
8766.55	8768.96	11403.9	CO I	0.3H			ME
8750.113	8752.516	11425.29	CO I	1.2H			ME
8745.574	8747.976	11431.22	CO I	0.6H			ME
8744.37	8746.77	11432.8	CO I	1.5			ME
8733.27	8735.67	11447.3	CO I	1.0H			ME
8675.12	8677.50	11524.1	CO I	1.0H			MIT
8661.09	8663.47	11542.7	CO I	1.8			MIT
8596.09	8598.45	11630.0	CO I	0.5H			ME
8589.73	8592.09	11638.6	CO I	1.7			MIT
8586.74	8589.10	11642.7	CO I	1.5			MIT
8575.35	8577.71	11658.1	CO I	1.7			MIT
8574.57	8576.93	11659.2	CO I	1.7			MIT
8569.72	8572.07	11665.8	CO I	1.3			ME
8559.07	8561.42	11680.3	CO I	1.0			MIT
8513.52	8515.86	11742.8	CO I	0.7H			MIT
8489.50	8491.83	11776.0	CO I	1.5			MIT
8478.45	8480.78	11791.4	CO I	0.9			ME
8409.03	8411.34	11888.7	CO I	1.0V			ME
8379.47	8381.77	11930.7	CO I	1.5			MIT
8378.39	8380.69	11932.2	CO I	1.7			MIT
8372.84	8375.14	11940.1	CO I	1.9H			MIT
8345.55	8347.84	11979.1	CO I	1.3H			MIT
8342.63	8344.92	11983.3	CO I	1.7V			MIT
8331.69	8333.98	11999.1	CO I	1.3			MIT
8318.55	8320.84	12018.0	CO I	0.7V			ME
8315.35	8317.64	12022.7	CO I	1.3			MIT
8301.45	8303.73	12042.8	CO I	0.7H			MIT
8299.95	8302.23	12044.9	CO I	1.8H			MIT
8296.85	8299.13	12049.5	CO I	1.7H			MIT
8283.48	8285.76	12068.9	CO I	1.7H			MIT
8275.55	8277.82	12080.5	CO I	0.9H			MIT
8272.38	8274.65	12085.1	CO I	0.7H			MIT

AIR WAVELENGTH (ANGSTRMS)	VACUUM WAVELENGTH (ANGSTRMS)	WAVENUMBER (KAYSERS)	LMENT	LOG ARC	INTENSITY SPK	DIS	REF
8269.38	8271.65	12089.5	CO I	1.9H			MIT
8208.66	8210.92	12178.9	CO I	1.9			MIT
8193.03	8195.28	12202.1	CO I	2.1			MIT
8189.32	8191.57	12207.7	CO I	1.0H			MIT
8167.88	8170.13	12239.7	CO I	1.3V			MIT
8160.65	8162.89	12250.6	CO I	1.6			ME
8154.31	8156.55	12260.1	CO I	1.0H			ME
8152.11	8154.35	12263.4	CO I	1.8			MIT
8150.19	8152.43	12266.3	CO	1.7			MIT
8140.43	8142.67	12281.0	CO I	1.6			MIT
8137.08	8139.32	12286.0	CO I	1.9			MIT
8116.41	8118.64	12317.3	CO I	1.9			MIT
8114.04	8116.27	12320.9	CO I	1.0H			MIT
8112.13	8114.36	12323.8	CO I	0.7V			ME
8093.96	8096.19	12351.5	CO I	1.9H			MIT
8085.54	8087.76	12364.4	CO I	0.7H			MIT
8080.23	8082.45	12372.5	CO I	1.8			MIT
8066.49	8068.71	12393.6	CO I	1.8			MIT
8062.98	8065.20	12398.9	CO I	0.7V			ME
8056.06	8058.28	12409.6	CO I	1.9H			MIT
8053.50	8055.71	12413.6	CO I	0.3H			ME
8050.76	8052.97	12417.8	CO I	0.7V			MIT
8043.33	8045.54	12429.2	CO I	1.9H			MIT
8041.37	8043.58	12432.3	CO I	1.3			MIT
8037.64	8039.85	12438.0	CO I	1.0H			MIT
8032.41	8034.62	12446.1	CO I	0.7H			ME
8029.26	8031.47	12451.0	CO I	1.9			MIT
8024.73	8026.94	12458.1	CO I	1.6			MIT
8022.131	8024.337	12462.09	CO I	1.6			MIT
8017.86	8020.07	12468.7	CO I	1.2H			MIT
8016.56	8018.76	12470.7	CO I	1.2H			MIT
8012.99	8015.19	12476.3	CO I	1.3V			MIT
8007.27	8009.47	12485.2	CO I	1.9H			MIT
7998.09	8000.29	12499.6	CO	1.1H			MIT
7996.80	7999.00	12501.6	CO I	1.0H			MIT
7987.38	7989.58	12516.3	CO I	1.9			MIT
7980.57	7982.77	12527.0	CO I	0.5H			MIT
7966.08	7968.27	12549.8	CO I	1.6			MIT
7962.40	7964.59	12555.6	CO I	0.5V			ME
7960.55	7962.74	12558.5	CO I	1.4H			MIT
7957.76	7959.95	12562.9	CO I	1.6H			MIT
7926.55	7928.73	12612.4	CO I	1.9H			MIT
7919.48	7921.66	12623.6	CO	1.2H			MIT
7908.71	7910.89	12640.8	CO I	1.9H			MIT
7885.25	7887.42	12678.4	CO I	1.0V			MIT
7877.49	7879.66	12690.9	CO	1.3			MIT
7873.32	7875.49	12697.6	CO I	1.0H			MIT
7871.39	7873.56	12700.7	CO I	1.9			MIT
7869.90	7872.07	12703.2	CO I	1.9			MIT
7859.39	7861.55	12720.1	CO I	1.0V			MIT
7855.85	7858.01	12725.9	CO I	1.6H			MIT
7843.64	7845.80	12745.7	CO I	1.1			MIT
7840.05	7842.21	12751.5	CO I	1.6			MIT
7838.17	7840.33	12754.6	CO I	1.9			MIT
7822.12	7824.27	12780.7	CO	0.3V			ME
7818.25	7820.40	12787.1	CO	0.7H			ME
7810.30	7812.45	12800.1	CO I	1.0H			MIT
7809.24	7811.39	12801.8	CO I	1.3			MIT
7794.13	7796.27	12826.6	CO I	0.7H			MIT
7764.02	7766.16	12876.4	CO I	1.1			MIT
7743.27	7745.40	12910.9	CO I	0.6			MIT
7735.45	7737.58	12923.9	CO I	1.0H			MIT
7734.23	7736.36	12926.0	CO I	1.6			MIT
7725.95	7728.08	12939.8	CO I	1.1			MIT
7712.68	7714.80	12962.1	CO I	1.9			MIT
7704.92	7707.04	12975.2	CO I	1.1			MIT
7701.90	7704.02	12980.2	CO I	1.1			MIT
7695.94	7698.06	12990.3	CO	1.0H			MIT
7648.08	7650.19	13071.6	CO I	1.1H			MIT
7637.63	7639.73	13089.5	CO I	0.6			ME

AIR WAVELENGTH (ANGSTRMS)	VACUUM WAVELENGTH (ANGSTRMS)	WAVENUMBER (KAYSERS)	LMENT	LOG ARC	INTENSITY SPK	DIS	REF
7634.50	7636.60	13094.8	CO I	0.5			MIT
7618.64	7620.74	13122.1	CO	0.3			MIT
7610.24	7612.34	13136.6	CO I	1.5			MIT
7590.57	7592.66	13170.6	CO I	1.5			MIT
7586.72	7588.81	13177.3	CO I	1.3			MIT
7564.96	7567.04	13215.2	CO I	1.3			MIT
7561.06	7563.14	13222.0	CO I	1.2			MIT
7559.65	7561.73	13224.5	CO I	1.2			MIT
7553.99	7556.07	13234.4	CO I	2.3			MIT
7533.48	7535.55	13270.4	CO I	1.6			MIT
7526.29	7528.36	13283.1	CO I	0.3			MIT
7502.72	7504.79	13324.8	CO I	0.5H			MIT
7489.37	7491.43	13348.6	CO I	1.0			MIT
7478.77	7480.83	13367.5	CO I	0.5			MIT
7474.35	7476.41	13375.4	CO I	0.6			ME
7457.36	7459.41	13405.9	CO I	2.3O			MIT
7443.43	7445.48	13431.0	CO I	1.0H			MIT
7437.16	7439.21	13442.3	CO I	1.5			MIT
7429.00	7431.05	13457.1	CO I	1.0			MIT
7417.38	7419.42	13478.1	CO I	2.5O			MIT
7414.54	7416.58	13483.3	CO I	0.7			MIT
7409.43	7411.47	13492.6	CO	1.0			SL
7388.70	7390.74	13530.4	CO I	2.3			MIT
7354.59	7356.62	13593.2	CO I	2.2			MIT
7353.47	7355.50	13595.3	CO I	1.4			MIT
7315.73	7317.75	13665.4	CO I	1.4			MIT
7285.28	7287.29	13722.5	CO I	2.3			MIT
7263.58	7265.58	13763.5	CO I	0.8			MIT
7250.12	7252.12	13789.1	CO I	1.9			MIT
7217.34	7219.33	13851.7	CO I	0.9			MIT
7202.17	7204.16	13880.9	CO	0.6			MIT
7193.60	7195.58	13897.4	CO I	2.3W			MIT
7159.18	7161.15	13964.2	CO I	2.4			MIT
7154.71	7156.68	13973.0	CO I	2.3			MIT
7148.139	7150.109	13985.80	CO	0.3			MIT
7134.32	7136.29	14012.9	CO I	2.3			MIT
7124.47	7126.43	14032.3	CO I	1.7			MIT
7122.25	7124.21	14036.6	CO	1.7			MIT
7117.91	7119.87	14045.2	CO I	0.3			ME
7113.56	7115.52	14053.8	CO I	2.2L			MIT
7102.55	7104.51	14075.6	CO	1.4			MIT
7094.53	7096.49	14091.5	CO I	1.6			MIT
7084.99	7086.94	14110.5	CO I	2.7O			MIT
7079.20	7081.15	14122.0	CO I	1.0			MIT
7070.41	7072.36	14139.6	CO I	1.3			MIT
7055.88	7057.83	14168.7	CO I	1.4			MIT
7054.04	7055.99	14172.4	CO I	2.3O			MIT
7052.89	7054.83	14174.7	CO I	2.5O			MIT
7042.58	7044.52	14195.4	CO I	0.7H			MIT
7032.52	7034.46	14215.7	CO I	1.4			MIT
7027.81	7029.75	14225.3	CO I	2.3O			MIT
7016.61	7018.55	14248.0	CO I	2.5O			MIT
7015.18	7017.11	14250.9	CO I	0.7			MIT
7004.81	7006.74	14272.0	CO I	2.2			MIT
6997.22	6999.15	14287.4	CO I	2.3W			MIT
6978.50	6980.42	14325.8	CO I	0.3			ME
6946.31	6948.23	14392.2	CO I	1.0			MIT
6937.81	6939.72	14409.8	CO I	2.2			MIT
6922.21	6924.12	14442.3	CO I	0.7			MIT
6910.84	6912.75	14466.0	CO I	1.2			MIT
6908.08	6909.99	14471.8	CO	1.5			MIT
6901.52	6903.42	14485.6	CO	0.7			MIT
6878.50	6880.40	14534.0	CO	0.3H			ME
6872.40	6874.30	14546.9	CO I	2.3W			MIT
6864.91	6866.80	14562.8	CO	1.0			MIT
6858.38	6860.27	14576.7	CO	1.4			MIT
6846.97	6848.86	14601.0	CO I	1.4			MIT
6845.66	6847.55	14603.8	CO I	1.0			ME
6838.11	6840.00	14619.9	CO	1.2			MIT
6826.96	6828.84	14643.8	CO I	0.5			MIT

AIR WAVELENGTH (ANGSTRMS)	VACUUM WAVELENGTH (ANGSTRMS)	WAVENUMBER (KAYSERS)	LMENT	LOG INTENSITY ARC	SPK	DIS	REF
6819.53	6821.41	14659.7	CO	1.3			MIT
6814.94	6816.82	14669.6	CO I	2.2G			MIT
6808.94	6810.82	14682.5	CO I	1.4H			MIT
6799.40	6801.28	14703.1	CO I	0.8H			MIT
6789.26	6791.13	14725.1	CO I	1.2			MIT
6784.85	6786.72	14734.7	CO I	1.4			MIT
6771.06	6772.93	14764.7	CO	2.3H			MIT
6767.60	6769.47	14772.2	CO	1.5H			ME
6767.394	6769.262	14772.66	CO I	1.2			MIT
6758.10	6759.97	14793.0	CO I	1.4			MIT
6756.57	6758.44	14796.3	CO I	1.4			MIT
6742.17	6744.03	14827.9	CO I	0.7			MIT
6722.71	6724.57	14870.8	CO I	0.6			MIT
6720.95	6722.81	14874.7	CO I	0.6			MIT
6717.64	6719.49	14882.1	CO I	0.7H			MIT
6712.71	6714.56	14893.0	CO I	0.9H			MIT
6703.92	6705.77	14912.5	CO I	1.2			MIT
6692.87	6694.72	14937.2	CO I	0.6			MIT
6684.87	6686.72	14955.0	CO I	1.5			MIT
6684.08	6685.93	14956.8	CO	1.5			MIT
6678.81	6680.65	14968.6	CO I	2.1			MIT
6667.60	6669.44	14993.8	CO	0.7R			ME
6665.29	6667.13	14999.0	CO I	0.6			MIT
6663.72	6665.56	15002.5	CO I	0.5H			MIT
6652.29	6654.13	15028.3	CO I	0.5H			MIT
6649.97	6651.81	15033.5	CO I	0.7			ME
6645.33	6647.17	15044.0	CO I	0.6H			MIT
6643.65	6645.48	15047.8	CO	0.3H			MIT
6635.12	6636.95	15067.2	CO I	1.4H			MIT
6632.445	6634.277	15073.23	CO I	2.2			MIT
6623.791	6625.620	15092.93	CO	1.80			MIT
6617.526	6619.354	15107.21	CO I	1.5			SL
6617.118	6618.945	15108.15	CO I	1.5			SL
6595.905	6597.727	15156.74	CO I	2.2			MIT
6595.38	6597.20	15157.9	CO	0.6			SL
6591.808	6593.629	15166.16	CO I	1.2			SL
6579.37	6581.19	15194.8	CO I	1.2			MIT
6567.079	6568.893	15223.27	CO	0.3H			MIT
6563.421	6565.234	15231.75	CO I	2.3W	0.7		MIT
6551.44	6553.25	15259.6	CO I	1.9W			MIT
6538.40	6540.21	15290.0	CO	0.3H			ME
6535.13	6536.94	15297.7	CO I	0.3H			MIT
6517.00	6518.80	15340.2	CO I	0.5H			MIT
6508.735	6510.533	15359.72	CO I	0.7H			MIT
6504.242	6506.039	15370.33	CO I	1.2			MIT
6499.628	6501.424	15381.25	CO	0.3H			SL
6496.893	6498.688	15387.72	CO	0.5H			DN
6493.738	6495.532	15395.20	CO	1.4			MIT
6490.336	6492.129	15403.27	CO I	1.8			MIT
6482.800	6484.591	15421.17	CO I	0.3H			SL
6477.878	6479.668	15432.89	CO I	1.9			MIT
6474.566	6476.355	15440.78	CO	1.4W			MIT
6471.66	6473.45	15447.7	CO	0.5			MIT
6470.16	6471.95	15451.3	CO I	0.5			MIT
6463.010	6464.796	15468.39	CO I	1.4H			MIT
6462.584	6464.370	15469.41	CO	1.8			MIT
6458.8	6460.6	15478.	CO	0.3H			SL
6454.996	6456.780	15487.60	CO I	2.3W			MIT
6451.14	6452.92	15496.8	CO I	1.8W			MIT
6450.239	6452.022	15499.02	CO I	3.0			MIT
6449.769	6451.552	15500.15	CO	0.5			MIT
6447.055	6448.837	15506.67	CO	0.3H			MIT
6444.697	6446.478	15512.35	CO I	1.4			MIT
6439.83	6441.61	15524.1	CO	0.3H			ME
6439.171	6440.951	15525.66	CO	1.9			MIT
6431.09	6432.87	15545.2	CO I	0.7H			MIT
6430.337	6432.114	15546.99	CO I	1.5H			SL
6429.907	6431.684	15548.03	CO I	1.7			MIT
6425.115	6426.891	15559.62	CO I	0.7			M
6421.743	6423.518	15567.79	CO I	1.3			M

AIR WAVELENGTH (ANGSTRMS)	VACUUM WAVELENGTH (ANGSTRMS)	WAVENUMBER (KAYSERS)	LMENT	LOG INTENSITY ARC	SPK	DIS	REF
6417.824	6419.598	15577.30	CO I	2.3			MIT
6408.458	6410.229	15600.07	CO I	0.5H			MIT
6396.524	6398.292	15629.17	CO I	1.0H			MIT
6395.195	6396.963	15632.42	CO I	2.1			MIT
6386.69	6388.46	15653.2	CO I	0.7H			MIT
6384.487	6386.252	15658.64	CO	0.3H			MIT
6352.750	6354.506	15736.86	CO I	0.3H			M
6351.425	6353.181	15740.15	CO I	1.4			MIT
6347.827	6349.582	15749.07	CO	2.1			MIT
6340.803	6342.556	15766.51	CO	1.0			MIT
6337.952	6339.704	15773.61	CO I	0.7			MIT
6322.94	6324.69	15811.1	CO I	0.3H			ME
6320.411	6322.159	15817.38	CO I	1.9			MIT
6315.779	6317.526	15828.98	CO I	0.3H			MIT
6314.98	6316.73	15831.0	CO	0.3H			DN
6314.529	6316.275	15832.12	CO I	1.7			MIT
6313.047	6314.793	15835.83	CO I	1.7			MIT
6311.28	6313.03	15840.3	CO	1.5H			SL
6296.967	6298.708	15876.27	CO I	0.3H			SL
6291.859	6293.599	15889.16	CO I	0.7			MIT
6282.634	6284.372	15912.49	CO I	2.50			MIT
6276.627	6278.363	15927.72	CO I	1.6			MIT
6275.133	6276.869	15931.51	CO I	1.4			MIT
6273.026	6274.761	15936.86	CO I	1.8W			MIT
6271.476	6273.211	15940.80	CO I	0.7			MIT
6262.825	6264.557	15962.82	CO I	0.7H			MIT
6261.088	6262.820	15967.25	CO	0.3H			M
6257.577	6259.308	15976.21	CO I	1.8			MIT
6257.053	6258.784	15977.55	CO I	0.5			MIT
6253.940	6255.670	15985.50	CO I	0.3H			M
6249.506	6251.235	15996.84	CO I	2.1			MIT
6247.285	6249.013	16002.53	CO I	0.9			MIT
6246.412	6248.140	16004.76	CO I	0.7H			M
6242.48	6244.21	16014.8	CO	0.3			ME
6237.122	6238.847	16028.60	CO	0.3H			MIT
6232.437	6234.161	16040.65	CO I	1.4			MIT
6230.968	6232.692	16044.43	CO I	2.3W			MIT
6223.372	6225.094	16064.02	CO I	1.0			MIT
6222.329	6224.050	16066.71	CO I	0.5			MIT
6211.189	6212.907	16095.52	CO I	1.4			MIT
6205.503	6207.220	16110.27	CO I	0.5H			MIT
6203.701	6205.417	16114.95	CO	0.3H			MIT
6203.50	6205.22	16115.5	CO	0.3H			ME
6203.301	6205.017	16115.99	CO	0.3H			ME
6197.837	6199.552	16130.20	CO I	0.7H			MIT
6193.553	6195.267	16141.35	CO I	1.2			MIT
6188.995	6190.707	16153.24	CO I	2.3W			MIT
6181.030	6182.740	16174.06	CO I	1.0H			MIT
6175.029	6176.738	16189.78	CO I	0.3			MIT
6169.556	6171.263	16204.14	CO	0.3H			MIT
6168.85	6170.56	16206.0	CO I	0.3H			SL
6162.170	6163.875	16223.56	CO	1.8			SL
6158.47	6160.17	16233.3	CO I	0.3H			M
6146.38	6148.08	16265.2	CO I	0.5H			M
6143.743	6145.443	16272.22	CO I	0.7H			SL
6141.722	6143.422	16277.57	CO	0.3			M
6132.410	6134.107	16302.29	CO I	1.0H			MIT
6129.106	6130.802	16311.08	CO I	1.3W			MIT
6128.256	6129.952	16313.34	CO I	0.7			M
6122.653	6124.348	16328.27	CO I	2.1			MIT
6122.218	6123.913	16329.43	CO	1.6			MIT
6116.984	6118.677	16343.40	CO I	1.9			MIT
6107.927	6109.618	16367.64	CO I	1.4			MIT
6105.519	6107.209	16374.09	CO	0.6			SL
6105.470	6107.160	16374.22	CO I	1.0H			MIT
6102.739	6104.428	16381.55	CO	1.0			M
6100.779	6102.468	16386.81	CO I	0.6H			MIT
6093.127	6094.814	16407.39	CO I	2.3			MIT
6086.648	6088.333	16424.86	CO I	1.9			MIT
6083.283	6084.967	16433.94	CO I	0.3H			MIT

AIR WAVELENGTH (ANGSTRMS)	VACUUM WAVELENGTH (ANGSTRMS)	WAVENUMBER (KAYSERS)	LMENT	LOG INTENSITY ARC	SPK	DIS	REF
6082.435	6084.119	16436.23	CO I	2.50			MIT
6070.656	6072.337	16468.12	CO I	1.3			MIT
6058.225	6059.902	16501.92	CO I	0.5			MIT
6049.095	6050.770	16526.82	CO I	1.7H			MIT
6029.89	6031.56	16579.5	CO	0.3H			ME
6021.794	6023.462	16601.75	CO	1.7			MIT
6016.642	6018.308	16615.97	CO	1.6			MIT
6015.380	6017.046	16619.45	CO I	0.3			MIT
6013.582	6015.247	16624.42	CO I	1.5			MIT
6011.419	6013.084	16630.40	CO I	0.3H			MIT
6007.671	6009.335	16640.78	CO I	1.7H			MIT
6006.355	6008.019	16644.42	CO I	1.7H			MIT
6005.017	6006.680	16648.13	CO I	0.9			MIT
6002.457	6004.119	16655.23	CO I	0.3			MIT
6000.668	6002.330	16660.20	CO I	1.9			MIT
5996.902	5998.563	16670.66	CO I	0.3			MIT
5991.878	5993.538	16684.64	CO I	3.0R			MIT
5984.265	5985.923	16705.86	CO I	0.3			MIT
5984.085	5985.743	16706.36	CO	1.0			MIT
5983.267	5984.924	16708.65	CO I	0.3			MIT
5981.994	5983.651	16712.21	CO I	0.5H			MIT
5965.661	5967.314	16757.96	CO I	0.3H			MIT
5965.02	5966.67	16759.8	CO I	0.3H			ME
5946.491	5948.138	16811.98	CO I	1.8			MIT
5935.386	5937.030	16843.44	CO I	2.2			MIT
5922.365	5924.006	16880.47	CO I	0.5			MIT
5916.88	5918.52	16896.1	CO I	0.3H			ME
5915.539	5917.178	16899.95	CO I	2.3W			MIT
5905.587	5907.224	16928.43	CO I	0.3H			MIT
5890.484	5892.116	16971.83	CO I	0.8			MIT
5883.410	5885.041	16992.24	CO I	0.5			MIT
5881.077	5882.707	16998.98	CO I	0.6			MIT
5877.425	5879.054	17009.54	CO I	0.6H			MIT
5876.103	5877.732	17013.37	CO I	0.6H			MIT
5857.464	5859.088	17067.50	CO	1.3			MIT
5846.586	5848.207	17099.26	CO I	0.7			MIT
5830.078	5831.694	17147.68	CO I	1.2			MIT
5826.304	5827.919	17158.78	CO I	0.5			MIT
5790.078	5791.684	17266.14	CO I	0.3			MIT
5774.364	5775.965	17313.12	CO I	0.3			MIT
5770.436	5772.036	17324.91	CO I	0.5			MIT
5752.883	5754.479	17377.77	CO I	0.3H			MIT
5750.952	5752.547	17383.60	CO I	0.3			MIT
5740.986	5742.578	17413.78	CO I	0.3H			MIT
5737.981	5739.573	17422.90	CO	0.3H			MIT
5703.031	5704.613	17529.67	CO I	0.3H			MIT
5688.593	5690.171	17574.16	CO I	1.0			MIT
5675.421	5676.996	17614.95	CO I	0.5H			MIT
5659.109	5660.680	17665.72	CO I	1.4			MIT
5651.734	5653.303	17688.78	CO I	0.5H			MIT
5647.224	5648.791	17702.90	CO I	2.8W			MIT
5643.087	5644.653	17715.88	CO	0.7H			MIT
5642.542	5644.108	17717.59	CO I	0.3H			MIT
5639.991	5641.556	17725.60	CO I	0.7H			MIT
5637.720	5639.285	17732.75	CO I	1.3			MIT
5636.119	5637.683	17737.78	CO I	1.3			MIT
5627.726	5629.288	17764.24	CO I	0.3			MIT
5616.077	5617.636	17801.08	CO I	0.7			MIT
5606.795	5608.352	17830.55	CO	0.3			MIT
5598.478	5600.032	17857.04	CO	1.7			MIT
5594.768	5596.321	17868.88	CO I	0.3			MIT
5594.458	5596.011	17869.87	CO	1.6			MIT
5592.185	5593.738	17877.13	CO I	0.3			MIT
5590.732	5592.284	17881.78	CO I	2.7			MIT
5558.825	5560.369	17984.42	CO	0.7			MIT
5546.968	5548.509	18022.86	CO I	0.6			MIT
5545.930	5547.470	18026.23	CO I	1.4H			MIT
5530.769	5532.305	18075.65	CO I	2.7			MIT
5524.979	5526.514	18094.59	CO I	1.4W			MIT
5523.292	5524.826	18100.12	CO I	2.5W			MIT

AIR WAVELENGTH (ANGSTRMS)	VACUUM WAVELENGTH (ANGSTRMS)	WAVENUMBER (KAYSERS)	LMENT	LOG INTENSITY ARC	SPK	DIS	REF
5515.982	5517.514	18124.10	CO I	0.5H			MIT
5495.668	5497.195	18191.10	CO I	1.2			MIT
5489.649	5491.174	18211.04	CO I	2.2W			MIT
5483.962	5485.486	18229.93	CO I	2.2W			MIT
5483.343	5484.867	18231.98	CO I	2.7W			MIT
5477.077	5478.599	18252.84	CO I	1.6			MIT
5470.457	5471.977	18274.93	CO I	1.7			MIT
5469.305	5470.825	18278.78	CO I	2.1			MIT
5454.555	5456.071	18328.21	CO I	2.5W			MIT
5452.305	5453.820	18335.77	CO I	1.4			MIT
5444.572	5446.085	18361.81	CO I	2.6W			MIT
5436.986	5438.497	18387.43	CO I	1.4			MIT
5434.529	5436.040	18395.75	CO I	1.0			MIT
5425.619	5427.127	18425.96	CO I	0.6			MIT
5408.119	5409.622	18485.58	CO I	1.5			MIT
5407.511	5409.014	18487.66	CO I	2.0			MIT
5401.983	5403.485	18506.57	CO I	2.0W			MIT
5399.758	5401.259	18514.20	CO I	1.0			MIT
5393.729	5395.229	18534.90	CO I	0.5H			MIT
5390.462	5391.961	18546.13	CO I	1.4			MIT
5381.748	5383.244	18576.16	CO I	2.2			MIT
5381.105	5382.601	18578.38	CO I	2.2			MIT
5369.576	5371.069	18618.27	CO I	2.7W			MIT
5368.888	5370.381	18620.65	CO	1.5			MIT
5366.726	5368.218	18628.15	CO I	0.7			MIT
5364.824	5366.316	18634.76	CO I	0.6			MIT
5362.766	5364.257	18641.91	CO I	2.7W			MIT
5359.184	5360.674	18654.37	CO I	2.5W			MIT
5358.923	5360.413	18655.28	CO I	1.6W			MIT
5353.485	5354.974	18674.23	CO I	2.7W			MIT
5352.049	5353.538	18679.24	CO I	2.7W			MIT
5349.087	5350.575	18689.58	CO I	1.9			MIT
5347.490	5348.977	18695.16	CO I	1.9			MIT
5344.566	5346.053	18705.39	CO I	1.0			MIT
5343.388	5344.874	18709.51	CO I	2.8W			MIT
5342.708	5344.194	18711.90	CO I	2.9W			MIT
5341.333	5342.819	18716.71	CO I	2.5W			MIT
5339.528	5341.013	18723.04	CO I	2.0W			MIT
5336.168	5337.652	18734.83	CO I	1.7			MIT
5334.838	5336.322	18739.50	CO I	1.8			MIT
5333.654	5335.138	18743.66	CO I	2.0			MIT
5332.674	5334.157	18747.10	CO I	2.3W			MIT
5331.466	5332.949	18751.35	CO I	2.7W	1.9		MIT
5326.247	5327.729	18769.73	CO I	1.0			MIT
5325.954	5327.436	18770.76	CO I	1.4			MIT
5325.278	5326.759	18773.14	CO I	2.5W			MIT
5316.780	5318.259	18803.15	CO I	2.5W			MIT
5312.658	5314.136	18817.73	CO I	2.6W			MIT
5310.205	5311.682	18826.43	CO I	1.3			MIT
5301.057	5302.532	18858.92	CO I	2.8W			MIT
5287.806	5289.277	18906.17	CO I	1.2			MIT
5287.574	5289.045	18907.00	CO I	1.0			MIT
5283.492	5284.962	18921.61	CO I	2.1W			MIT
5280.649	5282.118	18931.80	CO I	2.7W			MIT
5276.192	5277.660	18947.79	CO I	2.6W			MIT
5268.515	5269.981	18975.40	CO I	2.7W			MIT
5266.487	5267.953	18982.71	CO I	2.7W			MIT
5266.301	5267.767	18983.38	CO I	2.0			MIT
5265.825	5267.291	18985.09	CO I	1.4			MIT
5264.246	5265.711	18990.79	CO I	1.2			MIT
5257.624	5259.087	19014.71	CO I	2.6W			MIT
5254.650	5256.113	19025.47	CO I	2.3W			MIT
5249.998	5251.459	19042.33	CO I	2.3W			MIT
5247.932	5249.393	19049.82	CO I	2.7W			MIT
5237.091	5238.549	19089.26	CO	0.7			MIT
5235.206	5236.663	19096.13	CO	2.0W			MIT
5230.217	5231.673	19114.34	CO I	2.5G			MIT
5222.483	5223.937	19142.65	CO I	1.7			MIT
5219.027	5220.480	19155.33	CO I	1.0			MIT
5212.711	5214.162	19178.54	CO I	2.5W			MIT

AIR WAVELENGTH (ANGSTRMS)	VACUUM WAVELENGTH (ANGSTRMS)	WAVENUMBER (KAYSERS)	LMENT	LOG ARC	INTENSITY SPK	DIS	REF
5211.821	5213.272	19181.81	CO I	2.0			MIT
5210.842	5212.293	19185.42	CO I	1.7			MIT
5210.057	5211.508	19188.30	CO	2.0			MIT
5192.351	5193.797	19253.74	CO I	2.0W			MIT
5183.610	5185.054	19286.20	CO	1.5			MIT
5176.078	5177.520	19314.27	CO I	2.7R			MIT
5172.688	5174.129	19326.93	CO	1.0			MIT
5166.065	5167.504	19351.70	CO I	1.0			MIT
5165.159	5166.598	19355.10	CO I	1.5			MIT
5158.842	5160.279	19378.80	CO I	1.6			MIT
5158.431	5159.868	19380.34	CO I	1.6			MIT
5156.344	5157.780	19388.19	CO I	2.5W			MIT
5154.053	5155.489	19396.80	CO I	2.3O			MIT
5149.789	5151.224	19412.86	CO I	2.0			MIT
5146.744	5148.178	19424.35	CO I	2.6W			MIT
5145.512	5146.945	19429.00	CO I	1.9W			MIT
5133.449	5134.879	19474.66	CO I	1.7W			MIT
5126.195	5127.623	19502.21	CO I	2.3			MIT
5125.692	5127.120	19504.13	CO I	2.0W			MIT
5124.767	5126.195	19507.65	CO I	1.4H			MIT
5122.768	5124.195	19515.26	CO I	2.2			MIT
5113.232	5114.657	19551.65	CO I	2.0			MIT
5108.888	5110.312	19568.28	CO I	2.3W			MIT
5094.945	5096.365	19621.83	CO I	2.0			MIT
5087.855	5089.273	19649.17	CO I	1.2			MIT
5022.141	5023.542	19906.27	CO I	0.3			MIT
5007.301	5008.698	19965.27	CO I	0.7			MIT
4993.049	4994.442	20022.26	CO I	0.7			MIT
4988.040	4989.432	20042.36	CO I	2.7G			MIT
4986.449	4987.840	20048.76	CO I	1.0			MIT
4979.932	4981.321	20074.99	CO I	1.8			MIT
4971.959	4973.346	20107.19	CO I	2.2			MIT
4971.048	4972.435	20110.87	CO I	0.8			MIT
4967.897	4969.283	20123.63	CO I	1.0W			MIT
4967.528	4968.914	20125.12	CO I	0.5			MIT
4966.589	4967.975	20128.93	CO I	2.0			MIT
4959.688	4961.072	20156.93	CO I	0.7			MIT
4953.190	4954.572	20183.38	CO I	1.7			MIT
4951.828	4953.210	20188.93	CO I	0.3			MIT
4948.587	4949.968	20202.15	CO I	0.6			MIT
4945.784	4947.164	20213.60	CO	0.6			MIT
4943.282	4944.662	20223.83	CO I	0.3			MIT
4942.350	4943.729	20227.64	CO I	0.3			MIT
4941.350	4942.729	20231.74	CO I	0.5			MIT
4936.411	4937.789	20251.98	CO I	0.8			MIT
4935.222	4936.600	20256.86	CO I	0.3			MIT
4934.065	4935.442	20261.61	CO	1.4			MIT
4932.879	4934.256	20266.48	CO I	0.7			MIT
4928.283	4929.659	20285.38	CO I	2.3O			MIT
4924.980	4926.355	20298.98	CO I	0.6			MIT
4920.261	4921.635	20318.45	CO I	1.0			MIT
4912.391	4913.763	20351.00	CO I	0.7			MIT
4908.484	4909.854	20367.20	CO	0.5			MIT
4907.125	4908.495	20372.84	CO I	0.3			MIT
4904.173	4905.542	20385.11	CO I	1.9			MIT
4899.522	4900.890	20404.46	CO I	2.6O			MIT
4897.186	4898.553	20414.19	CO	1.2			MIT
4886.990	4888.355	20456.78	CO I	0.7			MIT
4882.718	4884.082	20474.68	CO I	1.0			MIT
4869.385	4870.745	20530.74	CO I	1.0			MIT
4867.878	4869.238	20537.10	CO I	2.9O	2.0		MIT
4863.461	4864.820	20555.75	CO I	0.7			MIT
4843.461	4844.814	20640.63	CO I	2.5			MIT
4840.266	4841.618	20654.25	CO I	2.8W	2.2		MIT
4815.894	4817.240	20758.77	CO I	0.6	0.0		MIT
4813.983	4815.328	20767.02	CO I	2.0	0.3		MIT
4813.484	4814.829	20769.17	CO I	3.0O	0.8		MIT
4796.371	4797.712	20843.27	CO I	2.0			MIT
4795.852	4797.193	20845.52	CO I	2.0			MIT
4792.863	4794.203	20858.52	CO I	2.8O	0.7		MIT

AIR WAVELENGTH (ANGSTRMS)	VACUUM WAVELENGTH (ANGSTRMS)	WAVENUMBER (KAYSERS)	LMENT	LOG ARC	INTENSITY SPK	DIS	REF
4785.069	4786.407	20892.50	CO I	1.7			MIT
4781.433	4782.770	20908.39	CO I	2.6	0.3H		MIT
4780.007	4781.343	20914.62	CO I	2.7W			MIT
4778.255	4779.591	20922.29	CO I	2.0			MIT
4776.317	4777.652	20930.78	CO I	2.5			MIT
4771.108	4772.442	20953.63	CO I	2.7W			MIT
4768.081	4769.414	20966.94	CO I	2.5	1.0		MIT
4767.142	4768.475	20971.06	CO I	2.0			MIT
4756.725	4758.055	21016.99	CO I	2.0			MIT
4754.36	4755.69	21027.4	CO I	2.3	0.3		MIT
4749.683	4751.011	21048.15	CO I	2.7	2.0H		MIT
4746.107	4747.434	21064.01	CO I	2.0			MIT
4737.768	4739.093	21101.08	CO	2.2	0.0		MIT
4734.833	4736.157	21114.16	CO I	2.2			MIT
4732.056	4733.380	21126.55	CO I	1.6			MIT
4729.056	4730.379	21139.96	CO	0.3			M
4727.940	4729.263	21144.95	CO I	2.5			MIT
4721.408	4722.729	21174.20	CO	0.9	2.0H		MIT
4718.478	4719.798	21187.35	CO I	1.7			MIT
4704.386	4705.702	21250.81	CO I	0.5			MIT
4699.187	4700.502	21274.32	CO I	1.4			MIT
4698.379	4699.694	21277.98	CO I	2.5	0.9		MIT
4693.212	4694.525	21301.41	CO I	2.7	1.4		MIT
4688.476	4689.788	21322.93	CO I	1.3			MIT
4685.856	4687.167	21334.85	CO I	1.5			MIT
4682.376	4683.687	21350.70	CO I	2.7			MIT
4677.251	4678.560	21374.10	CO I	0.6			MIT
4667.866	4669.173	21417.07	CO	1.0			MIT
4663.410	4664.716	21437.53	CO I	2.8O			MIT
4657.387	4658.691	21465.26	CO I	2.0	1.5		MIT
4654.849	4656.152	21476.96	CO I	1.4			MIT
4648.652	4649.954	21505.59	CO I	0.7			MIT
4644.324	4645.624	21525.63	CO I	1.8			MIT
4643.722	4645.022	21528.42	CO	1.2			MIT
4640.803	4642.103	21541.96	CO I	1.0			MIT
4629.378	4630.675	21595.13	CO I	2.8O	0.7		MIT
4628.936	4630.232	21597.19	CO I	2.1			MIT
4625.780	4627.076	21611.92	CO I	2.3			MIT
4624.580	4625.875	21617.53	CO I	1.0			MIT
4623.042	4624.337	21624.72	CO I	2.2			MIT
4622.697	4623.992	21626.34	CO I	1.5			MIT
4620.817	4622.111	21635.14	CO I	1.4			MIT
4614.009	4615.301	21667.06	CO I	1.8			MIT
4608.911	4610.202	21691.02	CO I	0.9			MIT
4607.33	4608.62	21698.5	CO	0.3H			DN
4601.165	4602.454	21727.54	CO I	1.5			MIT
4596.905	4598.193	21747.67	CO I	2.6			MIT
4594.633	4595.920	21758.43	CO I	2.6			MIT
4588.699	4589.985	21786.56	CO I	2.0	0.0		MIT
4586.939	4588.224	21794.92	CO I	1.2			MIT
4583.881	4585.165	21809.46	CO	0.3			MIT
4581.597	4582.881	21820.34	CO I	3.0W	1.0		MIT
4580.140	4581.423	21827.28	CO I	2.5	0.5		MIT
4579.364	4580.647	21830.98	CO I	1.4			MIT
4574.940	4576.222	21852.09	CO I	1.3			MIT
4570.025	4571.306	21875.59	CO I	2.5			MIT
4567.863	4569.143	21885.94	CO	0.3H			MIT
4567.533	4568.813	21887.52	CO	0.3			MIT
4566.611	4567.891	21891.94	CO I	2.0	0.3H		MIT
4565.586	4566.866	21896.86	CO I	2.9O	1.1		MIT
4564.843	4566.122	21900.42	CO I	1.0			MIT
4564.170	4565.449	21903.65	CO	1.5	0.0		MIT
4563.989	4565.268	21904.52	CO I	1.0	0.0		MIT
4561.942	4563.221	21914.35	CO I	1.4			MIT
4559.123	4560.401	21927.90	CO I	1.0			MIT
4553.327	4554.603	21955.81	CO I	1.4			MIT
4552.436	4553.712	21960.11	CO	1.4			MIT
4549.656	4550.931	21973.52	CO I	2.8			MIT
4545.988	4547.262	21991.25	CO I	1.0			MIT
4545.243	4546.517	21994.86	CO I	1.7			MIT

AIR WAVELENGTH (ANGSTRMS)	VACUUM WAVELENGTH (ANGSTRMS)	WAVENUMBER (KAYSERS)	LMENT	LOG ARC	INTENSITY SPK	INTENSITY DIS	REF
4543.812	4545.086	22001.78	CO I	2.70			MIT
4540.785	4542.058	22016.45	CO I	1.5			MIT
4533.994	4535.265	22049.43	CO I	2.7	0.9		MIT
4530.963	4532.233	22064.18	CO I	3.0W	0.9		MIT
4527.933	4529.203	22078.94	CO I	2.0	0.3H		MIT
4526.781	4528.050	22084.56	CO I	1.0			MIT
4525.787	4527.056	22089.41	CO I	0.7			MIT
4524.933	4526.202	22093.58	CO	0.5			MIT
4519.290	4520.557	22121.17	CO I	1.6			MIT
4517.109	4518.376	22131.85	CO I	2.5	0.8		MIT
4514.189	4515.455	22146.16	CO I	1.8	0.3		MIT
4500.552	4501.814	22213.27	CO I	0.7			MIT
4499.261	4500.523	22219.64	CO I	0.5			MIT
4494.755	4496.016	22241.91	CO I	2.0			MIT
4492.730	4493.990	22251.94	CO	0.5			M
4492.076	4493.336	22255.18	CO I	0.7			MIT
4490.309	4491.569	22263.94	CO I	0.3			MIT
4486.710	4487.969	22281.79	CO I	1.7	0.0		MIT
4484.515	4485.773	22292.70	CO I	1.8			MIT
4483.927	4485.185	22295.62	CO I	2.0			MIT
4483.586	4484.844	22297.32	CO I	1.3H			MIT
4478.657	4479.914	22321.86	CO I	0.7			MIT
4478.321	4479.578	22323.53	CO I	2.0	0.5		MIT
4477.238	4478.494	22328.93	CO I	1.5J			MIT
4471.818	4473.073	22356.00	CO	0.7			MIT
4471.550	4472.805	22357.34	CO I	2.0	0.6		MIT
4469.555	4470.809	22367.31	CO I	2.5	0.7		MIT
4466.886	4468.140	22380.68	CO I	2.5	0.7		MIT
4465.809	4467.062	22386.08	CO I	0.7			KB
4458.595	4459.846	22422.30	CO I	1.0			MIT
4445.715	4446.963	22487.26	CO I	2.1	0.3		MIT
4445.036	4446.284	22490.69	CO I	1.6	0.3		M
4441.948	4443.195	22506.33	CO I	0.7			MIT
4437.872	4439.118	22527.00	CO I	0.3			KB
4436.204	4437.449	22535.47	CO I	1.2			MIT
4431.619	4432.863	22558.78	CO I	0.7			MIT
4421.345	4422.587	22611.20	CO I	1.0			MIT
4417.403	4418.644	22631.38	CO I	1.0	0.5		MIT
4416.483	4417.723	22636.09	CO	0.5			M
4404.948	4406.185	22695.37	CO I	0.7			MIT
4402.676	4403.913	22707.08	CO I	0.7H			MIT
4395.879	4397.114	22742.19	CO I	0.6J	0.3		MIT
4391.884	4393.118	22762.88	CO I	1.0	0.5		MIT
4391.571	4392.805	22764.50	CO I	1.0J	0.6		MIT
4387.929	4389.162	22783.39	CO I	0.5	0.0		MIT
4380.071	4381.302	22824.27	CO I	0.7J	0.5		MIT
4375.543	4376.772	22847.89	CO	0.7			MIT
4374.925	4376.154	22851.11	CO I	1.0	0.5		MIT
4374.429	4375.658	22853.71	CO I	0.3J			KB
4373.633	4374.862	22857.86	CO I	1.2J			MIT
4371.130	4372.358	22870.95	CO I	1.4J			MIT
4366.217	4367.444	22896.69	CO I	0.3	0.7H		MIT
4361.919	4363.145	22919.25	CO I	0.3H			MIT
4361.031	4362.257	22923.92	CO I	0.3			MIT
4360.829	4362.055	22924.98	CO I	1.0	0.3		MIT
4359.434	4360.659	22932.31	CO I	1.2	0.0		MIT
4357.173	4358.398	22944.21	CO I	1.0			MIT
4356.904	4358.129	22945.63	CO I	0.5			MIT
4353.821	4355.045	22961.88	CO I	0.6	0.3		MIT
4339.625	4340.845	23036.99	CO I	1.7			MIT
4331.235	4332.453	23081.61	CO	0.7H	0.3		MIT
4320.387	4321.602	23139.57	CO I	0.3	0.0		MIT
4310.074	4311.286	23194.93	CO I	0.3			MIT
4309.43	4310.64	23198.4	CO I	0.3			M
4308.738	4309.950	23202.13	CO	0.3			MIT
4307.422	4308.634	23209.21	CO I	0.5			MIT
4303.236	4304.446	23231.79	CO I	1.2	0.3		MIT
4301.031	4302.241	23243.70	CO I	0.5			MIT
4297.934	4299.143	23260.45	CO I	0.3	0.3H		MIT
4292.239	4293.447	23291.31	CO	0.5H			MIT

AIR WAVELENGTH (ANGSTRMS)	VACUUM WAVELENGTH (ANGSTRMS)	WAVENUMBER (KAYSERS)	LMENT	LOG ARC	INTENSITY SPK	INTENSITY DIS	REF
4287.382	4288.588	23317.70	CO I	0.3	0.3		MIT
4285.787	4286.993	23326.37	CO I	2.1			MIT
4276.104	4277.307	23379.19	CO I	0.3	0.0		MIT
4275.069	4276.272	23384.85	CO I	0.5			DN
4270.431	4271.633	23410.25	CO I	0.3			MIT
4268.438	4269.639	23421.18	CO I	0.3	0.3		MIT
4268.032	4269.233	23423.41	CO I	0.5H			MIT
4263.744	4264.944	23446.97	CO I	0.3			MIT
4255.729	4256.927	23491.12	CO	0.5H			M
4252.308	4253.505	23510.02	CO I	2.2			MIT
4248.193	4249.389	23532.79	CO I	0.3			M
4245.565	4246.760	23547.36	CO I	0.3			M
4241.886	4243.080	23567.78	CO I	0.5H			MIT
4241.514	4242.708	23569.85	CO I	0.5H			MIT
4238.440	4239.633	23586.95	CO I	0.3			MIT
4233.996	4235.188	23611.70	CO I	2.0W			DN
4229.955	4231.146	23634.26	CO I	0.5H			MIT
4225.109	4226.299	23661.36	CO	0.7			MIT
4220.27	4221.46	23688.5	CO	0.3	0.0		M
4214.874	4216.061	23718.82	CO	0.3	0.0		MIT
4207.615	4208.800	23759.74	CO I	0.3H			MIT
4198.425	4199.608	23811.75	CO I	0.5H			MIT
4195.626	4196.808	23827.63	CO	0.5			MIT
4194.344	4195.526	23834.91	CO	0.7			MIT
4192.836	4194.017	23843.49	CO I	0.5H			MIT
4190.706	4191.887	23855.60	CO I	1.00	0.7		MIT
4187.250	4188.430	23875.29	CO I	1.7	0.5		MIT
4179.229	4180.407	23921.12	CO I	1.2	0.3		MIT
4170.905	4172.081	23968.86	CO I	0.6	0.3		MIT
4162.169	4163.342	24019.16	CO I	0.8	0.5		MIT
4158.425	4159.597	24040.79	CO	1.4H	0.7		MIT
4139.450	4140.617	24150.99	CO I	1.0			MIT
4132.155	4133.321	24193.62	CO I	1.2H			MIT
4122.271	4123.434	24251.63	CO I	1.0H			MIT
4121.319	4122.482	24257.23	CO I	3.0G	1.4		MIT
4118.773	4119.935	24272.23	CO I	3.0G			MIT
4110.535	4111.695	24320.87	CO I	2.8			MIT
4110.073	4111.233	24323.60	CO I	0.7H			MIT
4104.750	4105.908	24355.15	CO I	1.7			MIT
4104.430	4105.588	24357.04	CO I	1.5	0.3		MIT
4097.203	4098.359	24400.01	CO I	0.3			MIT
4095.945	4097.101	24407.50	CO I	0.5			MIT
4093.053	4094.208	24424.75	CO I	1.0			MIT
4092.848	4094.003	24425.97	CO I	1.4	1.0		MIT
4092.391	4093.546	24428.70	CO I	2.80	1.2		MIT
4090.354	4091.509	24440.86	CO I	1.3H			MIT
4088.299	4089.453	24453.15	CO I	1.7			MIT
4086.307	4087.461	24465.07	CO I	2.6	1.2		MIT
4085.579	4086.732	24469.43	CO I	0.3H	0.0		MIT
4084.113	4085.266	24478.21	CO I	0.3			M
4082.602	4083.755	24487.27	CO I	1.7			MIT
4077.406	4078.557	24518.47	CO I	2.0J	0.3H		MIT
4076.573	4077.724	24523.48	CO I	0.5H	0.0		MIT
4076.132	4077.283	24526.14	CO I	1.8			MIT
4069.548	4070.697	24565.82	CO I	0.3H			MIT
4068.544	4069.693	24571.88	CO I	2.2	2.0		MIT
4066.369	4067.517	24585.02	CO I	0.7	0.7		MIT
4063.177	4064.324	24604.33	CO I	0.5H			MIT
4058.600	4059.746	24632.08	CO I	2.0			MIT
4058.190	4059.336	24634.57	CO I	2.0			MIT
4057.199	4058.345	24640.59	CO I	2.0			MIT
4056.979	4058.125	24641.92	CO I	1.3H	0.3		MIT
4054.622	4055.767	24656.25	CO I	0.3			MIT
4053.922	4055.067	24660.50	CO I	0.8	0.3		MIT
4052.922	4054.067	24666.59	CO I	1.6			MIT
4049.290	4050.434	24688.71	CO I	0.6	0.3		MI
4045.390	4046.533	24712.51	CO I	2.6			MIT
4042.045	4043.187	24732.97	CO	0.5			MIT
4040.802	4041.944	24740.57	CO I	1.2	0.0		M
4040.647	4041.789	24741.52	CO I	0.3			MIT

AIR WAVELENGTH (ANGSTRMS)	VACUUM WAVELENGTH (ANGSTRMS)	WAVENUMBER (KAYSERS)	LMENT	LOG ARC	INTENSITY SPK	DIS	REF
4035.554	4036.694	24772.75	CO I	2.2	0.5		MIT
4034.858	4035.998	24777.02	CO	0.3			MIT
4027.037	4028.175	24825.14	CO I	2.3	0.6		MIT
4023.403	4024.540	24847.56	CO I	2.3			MIT
4020.905	4022.041	24863.00	CO I	2.7W			MIT
4019.302	4020.438	24872.91	CO I	1.9			MIT
4019.140	4020.276	24873.91	CO I	0.7			MIT
4016.883	4018.018	24887.89	CO I	1.0	0.7H		MIT
4016.828	4017.963	24888.23	CO	0.7			M
4015.224	4016.359	24898.17	CO	0.3	0.3		M
4013.945	4015.080	24906.11	CO I	2.5			MIT
4012.160	4013.294	24917.19	CO I	0.3			MIT
4010.941	4012.075	24924.76	CO I	0.5H	0.3		DN
4007.943	4009.076	24943.40	CO I	0.5			MIT
4003.605	4004.737	24970.43	CO I	1.2			MIT
3997.906	3999.036	25006.02	CO I	2.3	1.3		MIT
3995.310	3996.440	25022.27	CO I	3.0G	1.3		MIT
3994.537	3995.667	25027.11	CO I	1.8			MIT
3991.831	3992.960	25044.08	CO I	1.2			MIT
3991.686	3992.815	25044.99	CO I	1.8	0.8		MIT
3991.543	3992.672	25045.89	CO I	1.5			MIT
3990.299	3991.427	25053.69	CO I	1.9	1.0		MIT
3988.878	3990.006	25062.62	CO I	0.3			MIT
3987.115	3988.243	25073.70	CO I	1.9			MIT
3985.448	3986.575	25084.19	CO I	0.3H	0.0		M
3979.519	3980.645	25121.56	CO I	2.2W	1.1		MIT
3978.864	3979.989	25125.70	CO I	1.3	0.5		MIT
3978.655	3979.780	25127.01	CO I	2.0			MIT
3977.183	3978.308	25136.31	CO I	1.3			MIT
3975.322	3976.447	25148.08	CO I	1.5	0.7		M
3974.734	3975.858	25151.80	CO I	2.0	1.0		MIT
3973.561	3974.685	25159.23	CO I	1.2	0.7		MIT
3973.149	3974.273	25161.83	CO I	2.2W	0.8		MIT
3972.527	3973.651	25165.77	CO I	2.0	0.8		M
3969.117	3970.240	25187.40	CO I	2.0W	0.8		MIT
3965.235	3966.357	25212.05	CO I	0.9			MIT
3960.995	3962.116	25239.04	CO I	1.8	1.0		MIT
3957.935	3959.055	25258.55	CO I	2.0G			DN
3957.629	3958.749	25260.51	CO I	1.0	0.3		DN
3956.280	3957.400	25269.12	CO I	1.2	0.3		MIT
3952.917	3954.036	25290.62	CO I	2.0	1.9		MIT
3952.326	3953.445	25294.40	CO I	1.6	0.6		MIT
3951.725	3952.843	25298.25	CO I	0.6H	0.3		MIT
3947.127	3948.244	25327.71	CO I	1.3	0.3		MIT
3946.635	3947.752	25330.87	CO	1.3			M
3945.328	3946.445	25339.26	CO I	2.3	1.2		MIT
3944.950	3946.067	25341.69	CO I	0.7H	0.3		MIT
3942.692	3943.808	25356.20	CO I	0.9			MIT
3941.732	3942.848	25362.38	CO I	2.3J			MIT
3940.887	3942.003	25367.82	CO I	2.0			MIT
3938.903	3940.018	25380.59	CO I	0.7			M
3937.949	3939.064	25386.74	CO I	0.8H			MIT
3935.970	3937.084	25399.51	CO I	2.6G	1.2		MIT
3935.287	3936.401	25403.92	CO I	0.8			MIT
3934.707	3935.821	25407.66	CO	0.7	0.3		MIT
3933.914	3935.028	25412.78	CO I	1.8			MIT
3933.654	3934.768	25414.46	CO	1.9			MIT
3929.254	3930.367	25442.92	CO I	1.2	0.3		MIT
3925.163	3926.274	25469.44	CO I	1.3			MIT
3922.752	3923.863	25485.09	CO I	2.0			MIT
3920.733	3921.843	25498.21	CO I	1.4			MIT
3920.581	3921.691	25499.20	CO I	0.9			MIT
3920.143	3921.253	25502.05	CO I	1.3			MIT
3919.635	3920.745	25505.36	CO I	0.6	0.0		MIT
3917.113	3918.222	25521.78	CO I	1.9	1.0		MIT
3915.510	3916.619	25532.23	CO I	0.3	0.0		MIT
3909.934	3911.041	25568.64	CO I	2.3O			MIT
3906.294	3907.401	25592.46	CO I	2.2			MIT
3904.052	3905.158	25607.16	CO I	0.9			MIT
3898.494	3899.598	25643.67	CO I	1.9R	0.8		MIT
3894.982	3896.086	25666.79	CO I	2.5G	0.5		MIT
3894.081	3895.184	25672.73	CO I	3.0G	2.0		MIT
3893.303	3894.406	25677.86	CO I	1.2			MIT
3893.070	3894.173	25679.39	CO I	1.0	0.3		M
3892.125	3893.228	25685.63	CO I	1.3			MIT
3891.680	3892.783	25688.56	CO I	0.7	0.3		MIT
3885.288	3886.389	25730.83	CO I	1.8	0.6		MIT
3884.615	3885.716	25735.28	CO I	2.0			MIT
3881.874	3882.974	25753.46	CO I	2.5G	1.5		MIT
3881.006	3882.106	25759.22	CO I	0.5			MIT
3880.839	3881.939	25760.32	CO	0.9			MIT
3879.925	3881.025	25766.39	CO	0.3			MIT
3878.737	3879.836	25774.28	CO I	1.8R			MIT
3876.835	3877.934	25786.93	CO I	2.5W	1.6		MIT
3873.955	3875.053	25806.10	CO I	2.6G	1.9		MIT
3873.115	3874.213	25811.70	CO I	2.7G	1.9		MIT
3870.530	3871.627	25828.93	CO I	1.8	0.9		MIT
3866.830	3867.926	25853.65	CO I	0.3	0.3		MIT
3863.607	3864.702	25875.21	CO I	1.5			MIT
3861.165	3862.260	25891.58	CO I	2.5G	1.2		MIT
3858.299	3859.393	25910.81	CO	1.3	0.6		MIT
3856.800	3857.894	25920.88	CO	1.9	0.3		MIT
3850.953	3852.045	25960.24	CO I	2.0			MIT
3850.103	3851.195	25965.97	CO I	0.7			MIT
3845.471	3846.562	25997.24	CO I	2.7G	2.0		MIT
3843.692	3844.782	26009.28	CO I	1.8			MIT
3842.050	3843.140	26020.39	CO I	2.6G	1.3		MIT
3841.457	3842.547	26024.41	CO I	1.8			MIT
3835.900	3836.988	26062.11	CO I	0.9			MIT
3835.689	3836.777	26063.54	CO	1.0	0.6		MIT
3835.497	3836.585	26064.85	CO I	0.5	0.3		MIT
3832.899	3833.986	26082.51	CO I	0.7			MIT
3819.910	3820.994	26171.20	CO I	1.3	0.5		MIT
3817.941	3819.025	26184.70	CO I	1.3	0.6		MIT
3816.874	3817.957	26192.02	CO I	1.8	0.7		MIT
3816.473	3817.556	26194.77	CO I	1.8			MIT
3816.326	3817.409	26195.78	CO I	1.8	1.7R		MIT
3814.459	3815.542	26208.60	CO I	1.5			MIT
3813.925	3815.007	26212.27	CO I	1.5R			MIT
3813.293	3814.375	26216.61	CO	0.3			MIT
3812.470	3813.552	26222.27	CO I	2.0W			MIT
3811.067	3812.149	26231.93	CO I	1.5			MIT
3808.106	3809.187	26252.32	CO I	2.3W	0.8		MIT
3805.769	3806.849	26268.44	CO I	1.3			MIT
3783.728	3784.803	26421.46	CO I	0.7H	0.5		MIT
3777.542	3778.615	26464.72	CO I	2.3			MIT
3777.078	3778.151	26467.97	CO I	0.3	0.3		MIT
3774.605	3775.677	26485.31	CO I	2.3O			MIT
3760.393	3761.462	26585.41	CO I	1.5			MIT
3759.690	3760.758	26590.38	CO I	1.5			MIT
3755.449	3756.516	26620.41	CO I	2.0			MIT
3754.348	3755.415	26628.22	CO I	1.5	0.5		M
3752.787	3753.854	26639.29	CO	1.0	0.5		MIT
3751.629	3752.695	26647.51	CO I	2.0	1.8		MIT
3749.935	3751.001	26659.55	CO I	1.8	0.7		MIT
3745.504	3746.569	26691.09	CO I	2.5G			MIT
3740.195	3741.258	26728.98	CO I	1.8			MIT
3735.928	3736.990	26759.50	CO I	2.3G			MIT
3734.867	3735.929	26767.10	CO I	1.8			MIT
3734.141	3735.203	26772.31	CO I	1.8			MIT
3733.491	3734.553	26776.97	CO I	2.2			MIT
3732.399	3733.460	26784.80	CO I	2.3R			MIT
3731.278	3732.339	26792.85	CO I	1.3	0.5		MIT
3730.485	3731.546	26798.54	CO I	2.3R			MIT
3728.841	3729.901	26810.36	CO I	1.3			MIT
3726.657	3727.717	26826.07	CO I	1.5			MIT
3712.180	3713.236	26930.69	CO I	1.6	0.9		MIT
3711.652	3712.708	26934.52	CO I	1.5			M
3708.823	3709.878	26955.06	CO I	2.0			MIT
3707.468	3708.523	26964.91	CO I	1.5			MIT

AIR WAVELENGTH (ANGSTRMS)	VACUUM WAVELENGTH (ANGSTRMS)	WAVENUMBER (KAYSERS)	LMENT	LOG ARC	INTENSITY SPK	DIS	REF
3704.060	3705.114	26989.72	CO I	2.5R	1.5		MIT
3702.243	3703.297	27002.97	CO	2.3			MIT
3699.017	3700.070	27026.52	CO	0.9H			MIT
3693.478	3694.529	27067.05	CO I	1.5	1.0		MIT
3693.364	3694.415	27067.88	CO I	1.3			MIT
3693.112	3694.163	27069.73	CO I	1.9	1.2		MIT
3690.723	3691.774	27087.25	CO I	1.8	1.0		MIT
3686.484	3687.533	27118.40	CO I	0.9			M
3684.958	3686.007	27129.63	CO I	0.9			M
3684.479	3685.528	27133.16	CO I	2.30			MIT
3683.050	3684.099	27143.68	CO I	2.3G			MIT
3676.554	3677.601	27191.64	CO	2.0	1.5		MIT
3670.058	3671.103	27239.77	CO I	1.3			MIT
3662.161	3663.204	27298.51	CO I	2.0	1.4		MIT
3660.693	3661.736	27309.45	CO	0.7	0.3		MIT
3657.915	3658.957	27330.19	CO I	1.3	0.7		MIT
3656.966	3658.008	27337.28	CO I	1.8	0.8		MIT
3654.446	3655.487	27356.14	CO I	1.5			MIT
3652.543	3653.584	27370.39	CO I	2.3R			MIT
3651.258	3652.298	27380.02	CO I	1.3			MIT
3649.351	3650.391	27394.33	CO	2.3	0.6		M
3648.145	3649.185	27403.38	CO	1.3	0.3		MIT
3647.659	3648.698	27407.03	CO I	2.0	0.9		MIT
3647.388	3648.427	27409.07	CO	0.9H			MIT
3647.089	3648.128	27411.32	CO I	1.5	0.6		MIT
3645.195	3646.234	27425.56	CO I	1.8	0.5		MIT
3643.184	3644.222	27440.70	CO I	1.9	1.2		MIT
3641.788	3642.826	27451.22	CO I	1.8	0.9		MIT
3639.444	3640.481	27468.90	CO I	2.3	1.3		MIT
3637.323	3638.360	27484.91	CO I	1.5	0.7		MIT
3636.721	3637.758	27489.46	CO I	1.6	0.8		MIT
3634.712	3635.748	27504.66	CO	1.8	1.0		MIT
3633.333	3634.369	27515.09	CO I	0.9			MIT
3632.842	3633.878	27518.81	CO	1.8			MIT
3631.948	3632.983	27525.59	CO I	1.3			MIT
3631.391	3632.426	27529.81	CO I	1.70	1.4		M
3627.808	3628.842	27557.00	CO I	2.3			MIT
3626.010	3627.044	27570.66	CO I	0.7			MIT
3624.958	3625.992	27578.66	CO I	1.3			MIT
3624.329	3625.362	27583.45	CO I	0.9	0.6		MIT
3621.179	3622.212	27607.44	CO II	1.2	1.7H		MIT
3620.429	3621.461	27613.16	CO I	0.7			MIT
3618.006	3619.038	27631.65	CO I	0.3			MIT
3615.392	3616.423	27651.63	CO I	0.9			MIT
3611.701	3612.731	27679.89	CO I	1.4			MIT
3609.758	3610.788	27694.79	CO	0.7	0.5		MIT
3608.307	3609.336	27705.93	CO I	0.3H			MIT
3605.356	3606.384	27728.60	CO I	1.8			MIT
3605.015	3606.043	27731.23	CO I	1.2			MIT
3604.470	3605.498	27735.42	CO	0.5			MIT
3602.084	3603.112	27753.79	CO I	2.3	1.5		MIT
3600.809	3601.836	27763.62	CO I	0.6			MIT
3596.514	3597.540	27796.77	CO I	0.7			MIT
3594.872	3595.898	27809.47	CO I	2.30			MIT
3591.749	3592.774	27833.65	CO I	0.5			MIT
3587.190	3588.214	27869.02	CO I	2.3R	1.7H		MIT
3585.808	3586.831	27879.76	CO I	0.3H			MIT
3585.160	3586.183	27884.80	CO I	1.8			MIT
3584.801	3585.824	27887.59	CO I	1.4			MIT
3581.874	3582.896	27910.38	CO I	0.5			MIT
3579.029	3580.051	27932.56	CO I	0.7			MIT
3578.903	3579.925	27933.55	CO I	0.9	0.3		MIT
3578.075	3579.096	27940.01	CO I	1.3			MIT
3578.03	3579.05	27940.4	CO II		1.5		ME
3577.688	3578.709	27943.03	CO I	0.3H			MIT
3577.262	3578.283	27946.36	CO I	0.5			MIT
3575.361	3576.382	27961.22	CO I	2.3R	1.4		MIT
3574.964	3575.985	27964.33	CO I	2.3	1.4		MIT
3569.379	3570.398	28008.08	CO I	2.6G	2.0		MIT
3568.426	3569.445	28015.56	CO I	0.3			MIT

AIR WAVELENGTH (ANGSTRMS)	VACUUM WAVELENGTH (ANGSTRMS)	WAVENUMBER (KAYSERS)	LMENT	LOG ARC	INTENSITY SPK	DIS	REF
3564.951	3565.969	28042.87	CO I	2.2W			MIT
3564.643	3565.661	28045.29	CO I	0.3			MIT
3564.125	3565.143	28049.37	CO I	0.6	0.7		MIT
3562.918	3563.936	28058.87	CO I	1.0			MIT
3562.097	3563.114	28065.34	CO I	1.2			MIT
3560.893	3561.910	28074.82	CO I	2.3	1.4		MIT
3560.306	3561.323	28079.45	CO I	1.3H	0.5		MIT
3558.781	3559.797	28091.49	CO I	1.6W			MIT
3555.940	3556.956	28113.93	CO II		0.8		MIT
3553.161	3554.176	28135.92	CO I	0.7H			MIT
3552.990	3554.005	28137.27	CO I	1.3			MIT
3552.721	3553.736	28139.40	CO I	1.3			MIT
3551.666	3552.681	28147.76	CO I	0.3			MIT
3550.595	3551.609	28156.25	CO I	2.3			MIT
3548.444	3549.458	28173.32	CO I	1.3	0.8		MIT
3546.710	3547.723	28187.09	CO I	0.9			MIT
3545.036	3546.049	28200.40	CO II	0.3	1.5		MIT
3543.259	3544.271	28214.54	CO I	1.5			MIT
3542.508	3543.520	28220.52	CO I	0.3H			MIT
3537.726	3538.737	28258.67	CO I	0.3			MIT
3534.773	3535.783	28282.28	CO I	0.7			MIT
3533.358	3534.368	28293.60	CO I	2.3W			MIT
3530.554	3531.563	28316.07	CO I	0.3			MIT
3529.813	3530.822	28322.02	CO I	3.0G	1.5		MIT
3529.033	3530.042	28328.28	CO I	2.3G			MIT
3527.947	3528.956	28337.00	CO I	0.5			MIT
3526.849	3527.857	28345.82	CO I	2.5G	1.4		MIT
3525.872	3526.880	28353.67	CO I	0.9H			MIT
3525.089	3526.097	28359.97	CO I	0.3H			MIT
3523.701	3524.708	28371.14	CO I	1.2	0.5		MIT
3523.434	3524.441	28373.29	CO I	2.5R	1.4		MIT
3522.847	3523.854	28378.02	CO I	0.7	0.0		MIT
3521.731	3522.738	28387.01	CO I	0.7			MIT
3521.567	3522.574	28388.33	CO I	2.3R	1.4		MIT
3520.081	3521.088	28400.32	CO I	2.00			MIT
3519.818	3520.824	28402.44	CO	0.3H	1.4H		MIT
3518.349	3519.355	28414.30	CO I	2.30	2.0		MIT
3517.446	3518.452	28421.59	CO II	0.3	1.0		MIT
3516.675	3517.681	28427.82	CO I	0.3			MIT
3516.418	3517.424	28429.90	CO I	0.3			MIT
3514.213	3515.218	28447.74	CO II		1.3		MIT
3513.480	3514.485	28453.67	CO I	2.5G	1.4		MIT
3512.641	3513.646	28460.47	CO I	2.6G	2.0		MIT
3510.426	3511.430	28478.43	CO I				ME
3509.843	3510.847	28483.16	CO I	2.6G	1.6		MIT
3506.315	3507.318	28511.81	CO I	2.6G	1.2		MIT
3505.136	3506.139	28521.41	CO	0.9			MIT
3504.732	3505.735	28524.69	CO I	1.3	0.3		MIT
3503.717	3504.719	28532.96	CO	0.7			MIT
3502.624	3503.626	28541.86	CO I	1.8	0.7		MIT
3502.279	3503.281	28544.67	CO I	3.3G	1.3		MIT
3501.725	3502.727	28549.19	CO II	0.7	2.0		M
3496.794	3497.795	28589.44	CO I	1.5			MIT
3496.681	3497.681	28590.37	CO I	2.2G	0.6		MIT
3496.070	3497.070	28595.36	CO	1.0			MIT
3495.687	3496.687	28598.50	CO I	3.0G	1.4		MIT
3491.987	3492.986	28628.80	CO I	1.0			MIT
3491.321	3492.320	28634.26	CO I	2.3G	0.9		MIT
3490.741	3491.740	28639.02	CO I	1.8			MIT
3489.402	3490.401	28650.01	CO I	2.0G	1.4		MIT
3487.713	3488.711	28663.88	CO I	1.3	0.0		MIT
3485.700	3486.698	28680.43	CO I	1.2			MIT
3485.366	3486.364	28683.18	CO	2.0J	0.5		MIT
3483.410	3484.407	28699.29	CO I	2.5G	1.0		MIT
3483.141	3484.138	28701.50	CO I	1.0			MIT
3480.020	3481.016	28727.24	CO I	1.90	0.3H		MIT
3478.744	3479.740	28737.78	CO I	1.8	0.3		MIT
3478.555	3479.551	28739.34	CO I	1.6	0.3		MIT
3477.848	3478.844	28745.18	CO I	1.3			MIT
3476.360	3477.355	28757.49	CO I	2.0G			M

AIR WAVELENGTH (ANGSTRMS)	VACUUM WAVELENGTH (ANGSTRMS)	WAVENUMBER (KAYSERS)	LMENT	LOG ARC	INTENSITY SPK	DIS	REF
3474.530	3475.525	28772.63	CO I	1.5	0.3		MIT
3474.022	3475.017	28776.84	CO I	3.5G	2.0		M
3472.707	3473.701	28787.74	CO I	0.8			MIT
3471.378	3472.372	28798.76	CO I	1.9	1.4J		MIT
3469.705	3470.699	28812.64	CO I	0.9			MIT
3468.974	3469.967	28818.72	CO I	1.3			MIT
3468.592	3469.585	28821.89	CO I	0.7			MIT
3465.800	3466.793	28845.11	CO I	3.3G	1.4		MIT
3463.499	3464.491	28864.27	CO I	0.9			MIT
3462.803	3463.795	28870.07	CO I	3.0G	1.9		MIT
3461.176	3462.167	28883.64	CO I	2.0J	0.5		MIT
3460.719	3461.710	28887.46	CO I	1.3			MIT
3458.028	3459.019	28909.93	CO I	1.8W			MIT
3456.926	3457.916	28919.15	CO I	1.5	0.3		MIT
3456.437	3457.427	28923.24	CO I	0.7			MIT
3455.234	3456.224	28933.31	CO I	3.3G	1.0		MIT
3453.505	3454.494	28947.80	CO I	3.5G	2.3		MIT
3452.315	3453.304	28957.77	CO	0.9	0.7		MIT
3449.706	3450.694	28979.67	CO I	0.7			MIT
3449.441	3450.429	28981.90	CO I	2.7G	2.1		MIT
3449.170	3450.158	28984.18	CO I	2.7G	2.1		MIT
3448.358	3449.346	28991.00	CO I	1.0	0.3		MIT
3447.802	3448.790	28995.68	CO	0.7H			DN
3447.281	3448.269	29000.06	CO I	1.0			MIT
3446.856	3447.844	29003.64	CO	0.3			MIT
3446.388	3447.376	29007.57	CO II	0.5	1.8		MIT
3446.088	3447.076	29010.10	CO I	1.8H			MIT
3443.641	3444.628	29030.71	CO I	2.7G	2.0		MIT
3443.203	3444.190	29034.41	CO	1.4	0.3		MIT
3442.926	3443.913	29036.74	CO I	2.6G	1.2		MIT
3442.380	3443.367	29041.35	CO II	1.4	0.5		MIT
3442.361	3443.348	29041.51	CO II		0.9		MIT
3441.140	3442.126	29051.81	CO	0.9			MIT
3438.909	3439.895	29070.66	CO I	1.5			MIT
3438.713	3439.699	29072.31	CO I	1.9O			MIT
3437.692	3438.677	29080.95	CO I	2.2V			MIT
3436.960	3437.945	29087.14	CO I	1.0			MIT
3435.753	3436.738	29097.36	CO I	0.9			MIT
3433.040	3434.024	29120.35	CO I	3.0G	2.2		MIT
3432.845	3433.829	29122.01	CO	0.5			MIT
3432.318	3433.302	29126.48	CO I	1.8O			MIT
3431.575	3432.559	29132.79	CO I	2.7G	1.6		MIT
3429.681	3430.664	29148.87	CO	1.5			MIT
3428.762	3429.745	29156.69	CO I	0.9			MIT
3428.233	3429.216	29161.19	CO	2.0O	0.3		MIT
3427.768	3428.751	29165.14	CO I	0.8			MIT
3426.453	3427.435	29176.33	CO I	1.2	0.3		MIT
3424.506	3425.488	29192.92	CO I	1.9	0.3		MIT
3423.833	3424.815	29198.66	CO II	1.2	1.3		MIT
3422.896	3423.878	29206.65	CO I	1.3	0.6		MIT
3422.784	3423.766	29207.61	CO	0.8			MIT
3421.626	3422.607	29217.49	CO I	1.3	0.3		MIT
3421.348	3422.329	29219.87	CO I	0.5			MIT
3420.792	3421.773	29224.61	CO I	1.9	0.3		MIT
3420.483	3421.464	29227.25	CO I	0.5	0.3		MIT
3417.795	3418.775	29250.24	CO I	1.5	0.3		MIT
3417.673	3418.653	29251.28	CO I	1.4			MIT
3417.160	3418.140	29255.68	CO I	2.6G			MIT
3415.783	3416.763	29267.47	CO II	2.0G	1.3		MIT
3415.530	3416.510	29269.64	CO I	1.3			MIT
3414.736	3415.715	29276.44	CO I	2.3O			MIT
3412.633	3413.612	29294.48	CO I	3.0G	1.6		MIT
3412.339	3413.318	29297.01	CO I	3.0G	2.0		MIT
3409.646	3410.624	29320.15	CO I	1.2			MIT
3409.177	3410.155	29324.18	CO I	3.0G	2.1		MIT
3408.891	3409.869	29326.64	CO I	0.9			MIT
3405.816	3406.793	29353.12	CO	1.5G			MIT
3405.120	3406.097	29359.12	CO I	3.3G	2.2		MIT
3402.064	3403.040	29385.49	CO I	1.0	0.3		MIT
3401.913	3402.889	29386.79	CO I	1.3			MIT

AIR WAVELENGTH (ANGSTRMS)	VACUUM WAVELENGTH (ANGSTRMS)	WAVENUMBER (KAYSERS)	LMENT	LOG ARC	INTENSITY SPK	DIS	REF
3401.617	3402.593	29389.35	CO I	1.3			MIT
3400.469	3401.445	29399.27	CO I	0.7			MIT
3398.811	3399.786	29413.61	CO	1.3			MIT
3396.457	3397.432	29434.00	CO I	0.5			MIT
3395.375	3396.350	29443.38	CO I	2.6G	1.7		MIT
3394.916	3395.890	29447.36	CO I	0.8			MIT
3390.403	3391.376	29486.55	CO I	1.5	0.3		MIT
3388.494	3389.467	29503.17	CO I	0.5			MIT
3388.173	3389.146	29505.96	CO I	2.4G	1.1		MIT
3388.173	3389.146	29505.96	CO II	2.4G	1.1		MIT
3387.693	3388.666	29510.14	CO II	0.3	1.2		MIT
3387.061	3388.033	29515.65	CO I	0.3			MIT
3385.559	3386.531	29528.74	CO I	0.8			MIT
3385.224	3386.196	29531.66	CO I	2.4G	1.2		MIT
3383.920	3384.892	29543.04	CO I	1.8			MIT
3381.498	3382.469	29564.20	CO I	2.0O			MIT
3378.744	3379.714	29588.30	CO	1.6	0.3		MIT
3378.360	3379.330	29591.66	CO I	1.5	0.3		MIT
3377.063	3378.033	29603.03	CO I	2.0			MIT
3376.934	3377.904	29604.16	CO I	0.3			MIT
3376.206	3377.176	29610.54	CO I	1.3	0.5		MIT
3374.303	3375.272	29627.24	CO I	1.8			MIT
3373.972	3374.941	29630.15	CO I	1.8			MIT
3373.230	3374.199	29636.67	CO I	1.8			MIT
3371.015	3371.983	29656.14	CO I	0.3			MIT
3370.94	3371.91	29656.8	CO II		1.7		ME
3370.326	3371.294	29662.20	CO I	1.9	0.3		MIT
3368.570	3369.538	29677.66	CO I	0.9			MIT
3367.109	3368.076	29690.54	CO I	2.5G	1.5		MIT
3365.014	3365.981	29709.02	CO I	0.9			MIT
3364.255	3365.222	29715.73	CO I	1.5	0.3		MIT
3363.760	3364.726	29720.10	CO I	1.9R			MIT
3363.274	3364.240	29724.39	CO I	1.5	0.3		MIT
3362.802	3363.768	29728.56	CO I	1.9			MIT
3361.558	3362.524	29739.57	CO I	1.9	0.3		MIT
3361.267	3362.233	29742.14	CO I	1.3			MIT
3361.093	3362.059	29743.68	CO I	0.7			MIT
3359.286	3360.251	29759.68	CO I	2.0	0.3		MIT
3359.085	3360.050	29761.46	CO I	1.5			DN
3358.60	3359.57	29765.8	CO II		1.3		MIT
3358.007	3358.972	29771.01	CO I	1.2			MIT
3356.838	3357.803	29781.38	CO I	1.4			MIT
3356.472	3357.437	29784.63	CO I	2.2O	0.3		MIT
3355.944	3356.908	29789.31	CO I	1.0			MIT
3355.118	3356.082	29796.65	CO I	1.0			MIT
3354.377	3355.341	29803.23	CO I	2.3G			MIT
3354.213	3355.177	29804.69	CO	1.5			MIT
3352.80	3353.76	29817.2	CO II		1.5		ME
3351.536	3352.499	29828.49	CO I	1.5	0.3		MIT
3351.148	3352.111	29831.95	CO I	0.6			MIT
3350.376	3351.339	29838.82	CO I	0.6H			MIT
3349.526	3350.489	29846.39	CO	0.8			MIT
3349.222	3350.185	29849.10	CO I	0.5H			MIT
3348.113	3349.075	29858.99	CO I	1.9			MIT
3347.574	3348.536	29863.79	CO I	0.8			MIT
3346.935	3347.897	29869.50	CO I	2.0	0.3		MIT
3346.310	3347.272	29875.07	CO I	0.7			MIT
3344.245	3345.206	29893.52	CO I	0.6			MIT
3342.734	3343.695	29907.03	CO I	2.2O			MIT
3342.564	3343.525	29908.55	CO I	0.6			MIT
3341.947	3342.908	29914.08	CO I	1.4			MIT
3341.344	3342.305	29919.47	CO	1.8	0.3		MIT
3339.783	3340.743	29933.46	CO I	2.2W			MIT
3338.519	3339.479	29944.79	CO I	0.7			MIT
3337.172	3338.132	29956.88	CO I	1.8	0.3		MIT
3334.137	3335.096	29984.15	CO I	2.4G			MIT
3333.388	3334.347	29990.88	CO I	2.0			MIT
3331.678	3332.636	30006.27	CO I	0.3			MIT
3329.471	3330.429	30026.16	CO I	1.9			MIT
3329.025	3329.983	30030.19	CO I	1.3			MIT

AIR WAVELENGTH (ANGSTRMS)	VACUUM WAVELENGTH (ANGSTRMS)	WAVENUMBER (KAYSERS)	LMENT	LOG ARC	INTENSITY SPK	DIS	REF
3328.209	3329.166	30037.55	CO I	1.6			MIT
3326.991	3327.948	30048.55	CO	2.0			MIT
3326.564	3327.521	30052.40	CO I	1.8			MIT
3325.243	3326.200	30064.34	CO I	1.9	0.5		MIT
3322.199	3323.155	30091.89	CO I	2.0O			MIT
3321.912	3322.868	30094.49	CO I	1.4			MIT
3319.825	3320.780	30113.41	CO I	1.5	0.3		MIT
3319.561	3320.516	30115.80	CO I	0.9			MIT
3319.478	3320.433	30116.55	CO	1.9			MIT
3319.156	3320.111	30119.47	CO I	1.8			MIT
3318.400	3319.355	30126.34	CO I	1.5			MIT
3315.035	3315.989	30156.92	CO I	1.0			MIT
3314.345	3315.299	30163.19	CO I	0.9			MIT
3314.076	3315.030	30165.64	CO I	2.0			MIT
3313.120	3314.074	30174.34	CO	1.0			MIT
3312.829	3313.782	30177.00	CO I	1.3			MIT
3312.277	3313.230	30182.02	CO I	0.5			MIT
3312.148	3313.101	30183.20	CO I	1.8	0.3		MIT
3309.017	3309.969	30211.76	CO	0.3			MIT
3308.809	3309.761	30213.66	CO I	1.6			MIT
3308.488	3309.440	30216.59	CO I	1.5			MIT
3307.473	3308.425	30225.86	CO I	0.6			MIT
3307.154	3308.106	30228.78	CO I	1.9			MIT
3305.731	3306.683	30241.79	CO I	1.3			MIT
3305.110	3306.062	30247.47	CO I	0.9			MIT
3304.790	3305.741	30250.40	CO I	0.7			MIT
3304.119	3305.070	30256.54	CO I	1.2			MIT
3303.878	3304.829	30258.75	CO I	1.8R			MIT
3298.676	3299.626	30306.47	CO I	1.8	0.3		MIT
3294.536	3295.485	30344.55	CO I	0.9			MIT
3294.098	3295.047	30348.58	CO I	0.5			MIT
3293.861	3294.810	30350.77	CO I	1.6G			MIT
3293.212	3294.160	30356.75	CO I	1.2			MIT
3292.087	3293.035	30367.12	CO I	1.3			MIT
3287.827	3288.774	30406.47	CO I	0.7			MIT
3287.571	3288.518	30408.83	CO I	1.3	0.3		MIT
3287.194	3288.141	30412.32	CO I	1.8			MIT
3286.545	3287.492	30418.33	CO I	0.5			MIT
3285.874	3286.821	30424.54	CO	0.7			MIT
3285.104	3286.050	30431.67	CO	0.8H			MIT
3283.777	3284.723	30443.97	CO I	1.8O			MIT
3283.462	3284.408	30446.89	CO I	1.9			MIT
3283.329	3284.275	30448.12	CO I	0.8			MIT
3282.232	3283.178	30458.30	CO I	0.5			MIT
3282.041	3282.987	30460.07	CO I	0.6			DN
3281.588	3282.534	30464.27	CO I	0.8			MIT
3280.681	3281.626	30472.70	CO	0.3			MIT
3279.256	3280.201	30485.94	CO I	1.8	0.3		MIT
3278.844	3279.789	30489.77	CO I	1.8	0.3		MIT
3278.107	3279.052	30496.62	CO I	0.9			MIT
3277.664	3278.609	30500.74	CO I	1.3	0.3		MIT
3277.318	3278.262	30503.96	CO I	1.8O			MIT
3276.480	3277.424	30511.77	CO I	1.5	0.3		MIT
3273.931	3274.875	30535.52	CO	1.0			MIT
3272.405	3273.348	30549.76	CO I	0.5			MIT
3271.783	3272.726	30555.57	CO I	1.8			MIT
3270.198	3271.141	30570.38	CO I	1.0			MIT
3268.888	3269.830	30582.63	CO	0.6			MIT
3265.352	3266.293	30615.74	CO I	1.5	0.3		MIT
3264.843	3265.784	30620.52	CO I	1.5	0.3		MIT
3264.718	3265.659	30621.69	CO I	0.8H			DN
3263.213	3264.154	30635.81	CO I	1.5			MIT
3262.435	3263.376	30643.12	CO	0.3			MIT
3260.819	3261.759	30658.30	CO I	1.8	0.6		MIT
3260.286	3261.226	30663.31	CO I	0.7			MIT
3259.847	3260.787	30667.44	CO	0.5			MIT
3258.025	3258.965	30684.59	CO I	1.8			MIT
3254.206	3255.145	30720.60	CO I	2.5G			MIT
3250.335	3251.273	30757.19	CO I	0.6			MIT
3250.002	3250.940	30760.34	CO I	1.8			MIT

AIR WAVELENGTH (ANGSTRMS)	VACUUM WAVELENGTH (ANGSTRMS)	WAVENUMBER (KAYSERS)	LMENT	LOG ARC	INTENSITY SPK	DIS	REF
3247.179	3248.116	30787.08	CO I	1.9			MIT
3246.997	3247.934	30788.80	CO I	1.5			MIT
3245.744	3246.680	30800.69	CO I	0.5			MIT
3243.842	3244.778	30818.75	CO I	2.0			MIT
3243.579	3244.515	30821.25	CO I	0.9			MIT
3241.546	3242.481	30840.58	CO I	0.7			MIT
3237.028	3237.962	30883.62	CO I	2.0			MIT
3235.783	3236.717	30895.50	CO I	0.7H			MIT
3235.538	3236.472	30897.84	CO I	1.8			MIT
3234.119	3235.052	30911.40	CO I	0.3			MIT
3232.874	3233.807	30923.30	CO I	1.8	1.4		MIT
3227.763	3228.695	30972.27	CO I	0.7			MIT
3226.985	3227.917	30979.73	CO	1.9R			MIT
3225.023	3225.954	30998.58	CO	0.5			MIT
3224.642	3225.573	31002.24	CO I	1.8			MIT
3223.147	3224.078	31016.62	CO I	0.5			MIT
3219.153	3220.083	31055.10	CO I	1.8			MIT
3216.996	3217.925	31075.93	CO I	0.3			MIT
3215.336	3216.265	31091.97	CO I	0.7			MIT
3210.838	3211.766	31135.52	CO	1.3J			MIT
3210.226	3211.153	31141.46	CO I	1.9	0.3		MIT
3205.886	3206.812	31183.61	CO I	0.3			MIT
3203.031	3203.957	31211.41	CO I	1.6			MIT
3199.324	3200.249	31247.57	CO I	1.5			MIT
3198.663	3199.588	31254.03	CO I	1.8	0.3		MIT
3193.160	3194.083	31307.89	CO I	1.7			MIT
3192.224	3193.147	31317.07	CO I	1.5			MIT
3191.298	3192.221	31326.16	CO I	1.5			MIT
3189.756	3190.678	31341.30	CO I	1.8	0.5		MIT
3188.373	3189.295	31354.89	CO I	2.0	0.3H		MIT
3186.348	3187.269	31374.82	CO I	1.8			MIT
3185.950	3186.871	31378.74	CO I	1.6			MIT
3182.123	3183.043	31416.48	CO I	1.9	0.3H		MIT
3180.283	3181.203	31434.65	CO I	1.0			MIT
3177.273	3178.192	31464.43	CO	2.0			MIT
3174.905	3175.824	31487.90	CO I	1.9			MIT
3174.140	3175.058	31495.49	CO I	1.3			MIT
3169.767	3170.684	31538.93	CO	2.0			MIT
3168.060	3168.977	31555.93	CO I	2.0			MIT
3161.654	3162.569	31619.86	CO I	1.8			MIT
3159.665	3160.580	31639.77	CO I	2.0	0.3H		MIT
3158.775	3159.689	31648.68	CO I	2.2R			MIT
3154.794	3155.707	31688.62	CO I	2.0	0.6H		MIT
3154.678	3155.591	31689.78	CO	2.0			MIT
3152.707	3153.620	31709.59	CO I	2.0			MIT
3150.819	3151.731	31728.59	CO	0.8			MIT
3150.655	3151.567	31730.24	CO	0.9			MIT
3149.313	3150.225	31743.76	CO I	1.8			MIT
3147.064	3147.975	31766.45	CO I	2.2G			MIT
3145.021	3145.932	31787.08	CO I	1.5			MIT
3144.380	3145.291	31793.56	CO	0.8			MIT
3140.721	3141.631	31830.60	CO I	1.3	0.5		MIT
3139.943	3140.853	31838.49	CO I	2.2R	1.0		MIT
3137.751	3138.660	31860.73	CO I	1.8	0.3		MIT
3137.454	3138.363	31863.75	CO I	1.7			DN
3137.327	3138.236	31865.03	CO I	2.2R			MIT
3136.999	3137.908	31868.37	CO I	1.0			MIT
3136.726	3137.635	31871.14	CO I	1.8			MIT
3132.216	3133.124	31917.03	CO I	1.6			MIT
3131.830	3132.738	31920.96	CO I	0.9			MIT
3129.482	3130.389	31944.91	CO I	1.6	0.3		MIT
3128.997	3129.904	31949.86	CO I	1.4			DN
3127.249	3128.155	31967.72	CO I	2.0			MIT
3126.725	3127.631	31973.08	CO I	1.8			MIT
3126.488	3127.394	31975.50	CO I	1.3			MIT
3121.566	3122.471	32025.92	CO I	1.8R	0.5		MIT
3121.415	3122.320	32027.47	CO I	2.2R	0.8		MIT
3118.246	3119.150	32060.01	CO I	1.8	0.3		MIT
3113.479	3114.382	32109.10	CO I	2.0	0.6		MIT
3111.339	3112.241	32131.18	CO I	1.3	0.5		MIT

AIR WAVELENGTH (ANGSTRMS)	VACUUM WAVELENGTH (ANGSTRMS)	WAVENUMBER (KAYSERS)	LMENT	LOG ARC	INTENSITY SPK	DIS	REF
3110.825	3111.727	32136.49	CO I	1.8	0.3		MIT
3110.021	3110.923	32144.80	CO I	1.8	0.3		MIT
3109.510	3110.412	32150.08	CO I	1.8	0.0H		MIT
3107.542	3108.444	32170.44	CO I	1.2	0.5		MIT
3107.044	3107.945	32175.60	CO I	1.8	0.5		MIT
3105.924	3106.825	32187.20	CO I	1.5			MIT
3103.983	3104.884	32207.33	CO I	1.8R			MIT
3103.742	3104.643	32209.83	CO I	1.9	0.3		MIT
3102.407	3103.307	32223.69	CO I	1.8	0.6		MIT
3099.670	3100.570	32252.14	CO I	1.7			MIT
3098.196	3099.095	32267.48	CO I	2.0R	0.7		MIT
3096.700	3097.599	32283.07	CO I	1.8	0.5		MIT
3096.405	3097.304	32286.15	CO I	1.8	0.5		MIT
3095.718	3096.617	32293.31	CO I	1.8	0.3		MIT
3090.254	3091.151	32350.41	CO I	1.9	0.0		MIT
3089.595	3090.492	32357.31	CO I	2.0R			MIT
3088.680	3089.577	32366.89	CO I	1.0			MIT
3087.806	3088.703	32376.05	CO I	1.8			MIT
3086.777	3087.673	32386.85	CO I	2.3G			MIT
3086.396	3087.292	32390.84	CO I	1.9	0.3		MIT
3082.844	3083.739	32428.16	CO I	1.5			MIT
3082.618	3083.513	32430.54	CO	2.2G	1.7		MIT
3079.398	3080.292	32464.45	CO I	1.9	0.3		MIT
3073.521	3074.414	32526.52	CO I	1.8	0.3		MIT
3072.664	3073.557	32535.59	CO I	1.3			MIT
3072.344	3073.237	32538.98	CO I	2.3G	2.0		MIT
3071.962	3072.855	32543.03	CO	1.9	0.3		MIT
3070.857	3071.749	32554.74	CO I	0.7			MIT
3070.752	3071.644	32555.85	CO I	0.7			MIT
3064.370	3065.261	32623.65	CO I	2.0			MIT
3062.201	3063.091	32646.76	CO I	1.8	0.0		MIT
3061.819	3062.709	32650.83	CO I	2.3G	2.1		MIT
3061.021	3061.911	32659.34	CO I	1.2			MIT
3060.052	3060.942	32669.69	CO I	2.2	0.0		MIT
3054.724	3055.612	32726.67	CO I	1.8			MIT
3054.132	3055.020	32733.01	CO I	1.3			MIT
3050.932	3051.819	32767.34	CO I	1.8			MIT
3050.500	3051.387	32771.98	CO I	1.8			MIT
3048.888	3049.775	32789.31	CO I	2.2R			MIT
3048.113	3049.000	32797.64	CO I	1.4			MIT
3044.005	3044.891	32841.90	CO I	2.6G			MIT
3042.484	3043.369	32858.32	CO I	1.9R	0.9		MIT
3040.814	3041.699	32876.37	CO I	1.0			MIT
3039.567	3040.451	32889.85	CO I	1.8			MIT
3038.306	3039.190	32903.50	CO I	1.4			MIT
3034.433	3035.316	32945.50	CO I	1.9	0.3		MIT
3026.371	3027.252	33033.26	CO I	2.0	1.6		MIT
3022.362	3023.242	33077.07	CO I	1.8	0.3		MIT
3020.639	3021.519	33095.94	CO	1.8			MIT
3017.546	3018.425	33129.86	CO I	2.0R	0.7		MIT
3017.262	3018.141	33132.98	CO I	1.8			MIT
3015.684	3016.563	33150.32	CO I	1.8	0.3		MIT
3013.596	3014.474	33173.28	CO I	1.8R	0.6		MIT
3005.764	3006.640	33259.72	CO I	2.0	0.3		MIT
3000.546	3001.421	33317.56	CO I	1.9	0.0		MIT
2996.552	2997.426	33361.96	CO I	1.0			MIT
2995.150	2996.023	33377.58	CO I	1.7	0.0		MIT
2989.588	2990.460	33439.67	CO I	1.9G	1.5		MIT
2987.585	2988.457	33462.09	CO	0.3			MIT
2987.162	2988.033	33466.83	CO I	1.9G	1.7		MIT
2984.133	2985.004	33500.80	CO	0.6			MIT
2982.263	2983.133	33521.80	CO I	1.0			MIT
2978.014	2978.883	33569.63	CO I	1.5			MIT
2975.464	2976.333	33598.40	CO I	0.6			MIT
2966.900	2967.766	33695.37	CO	1.5			MIT
2957.675	2958.539	33800.47	CO	1.7	0.0		MIT
2955.386	2956.250	33826.64	CO	1.5			MIT
2954.74	2955.60	33834.0	CO	0.3	2.0		MIT
2943.484	2944.345	33963.42	CO I	1.5			MIT
2943.15	2944.01	33967.3	CO II		2.0J		MIT

AIR WAVELENGTH (ANGSTRMS)	VACUUM WAVELENGTH (ANGSTRMS)	WAVENUMBER (KAYSERS)	LMENT	LOG ARC	INTENSITY SPK	DIS	REF
2930.43	2931.29	34114.7	CO II		2.2J		MIT
2929.509	2930.366	34125.43	CO I	1.9			MIT
2928.814	2929.671	34133.53	CO I	1.7	0.0H		MIT
2927.670	2928.527	34146.86	CO I	1.7	0.0		MIT
2919.554	2920.409	34241.78	CO I	1.5			MIT
2907.460	2908.312	34384.21	CO	0.6			MIT
2904.288	2905.139	34421.76	CO I	0.3			MIT
2903.195	2904.046	34434.72	CO	1.4	0.0		MIT
2899.817	2900.667	34474.83	CO I	1.4	0.0		MIT
2895.891	2896.740	34521.57	CO I	0.5			MIT
2895.485	2896.334	34526.41	CO I	1.3			MIT
2895.335	2896.184	34528.20	CO	0.6			MIT
2892.247	2893.095	34565.06	CO	1.4			MIT
2890.48	2891.33	34586.2	CO		1.0		EX
2886.445	2887.291	34634.54	CO I	1.7	0.3		MIT
2885.305	2886.151	34648.22	CO I	0.5			MIT
2883.603	2884.449	34668.67	CO I	1.2			MIT
2882.220	2883.065	34685.30	CO I	1.5			MIT
2881.580	2882.425	34693.01	CO	0.6	0.0		MIT
2880.29	2881.13	34708.5	CO II		1.7J		MIT
2879.621	2880.466	34716.61	CO	1.4			MIT
2878.558	2879.403	34729.43	CO I	1.1			MIT
2871.24	2872.08	34817.9	CO		2.0		EX
2870.03	2870.87	34832.6	CO II		1.7J		MIT
2867.473	2868.315	34863.68	CO	0.6			MIT
2865.501	2866.342	34887.67	CO I	0.3	0.0H		MIT
2863.538	2864.379	34911.58	CO	0.5			MIT
2862.766	2863.607	34921.00	CO I	1.0H			MIT
2862.606	2863.447	34922.95	CO I	1.7			MIT
2861.364	2862.204	34938.11	CO I	1.2R			MIT
2859.659	2860.499	34958.94	CO I	1.6			MIT
2850.951	2851.789	35065.71	CO I	1.5			MIT
2850.043	2850.881	35076.88	CO I	1.9			MIT
2848.37	2849.21	35097.5	CO II		1.8H		MIT
2845.63	2846.47	35131.3	CO II		1.7		MIT
2842.383	2843.219	35171.41	CO I	1.5			MIT
2837.151	2837.985	35236.26	CO I	1.9R			MIT
2834.939	2835.773	35263.76	CO II	0.3	1.9		MIT
2834.428	2835.262	35270.11	CO I	1.7			MIT
2833.923	2834.757	35276.40	CO I	1.6			MIT
2828.478	2829.310	35344.30	CO	1.2			MIT
2826.805	2827.637	35365.22	CO I	1.7O			MIT
2825.242	2826.073	35384.79	CO II	0.7	2.3		MIT
2825.151	2825.982	35385.92	CO I	1.9W			MIT
2824.364	2825.195	35395.78	CO I	0.3			MIT
2823.647	2824.478	35404.77	CO	0.7H			MIT
2821.745	2822.576	35428.63	CO II	1.5H	1.0H		MIT
2820.009	2820.839	35450.44	CO I	1.7			MIT
2819.174	2820.004	35460.94	CO I	1.0			MIT
2818.597	2819.427	35468.20	CO I	1.5			MIT
2815.559	2816.388	35506.47	CO I	1.7R			MIT
2814.981	2815.810	35513.76	CO I	1.4			MIT
2812.449	2813.277	35545.73	CO I	0.5			MIT
2811.518	2812.346	35557.50	CO I	1.7W			MIT
2811.127	2811.955	35562.45	CO I	1.7			MIT
2810.860	2811.688	35565.83	CO II	0.7	1.9H		MIT
2807.174	2808.001	35612.52	CO II	0.0	1.4		MIT
2804.099	2804.925	35651.57	CO I	0.7			MIT
2803.772	2804.598	35655.73	CO I	2.0	1.1		MIT
2798.653	2799.478	35720.95	CO	0.6			MIT
2797.081	2797.906	35741.02	CO I	1.7	0.3		MIT
2796.83	2797.65	35744.2	CO II		0.5		MIT
2796.231	2797.055	35751.88	CO I	1.7	0.7		MIT
2795.819	2796.643	35757.15	CO I	1.2			MIT
2795.60	2796.42	35760.0	CO		1.0J		EX
2793.92	2794.74	35781.5	CO II		0.7H		MIT
2792.439	2793.262	35800.43	CO I	1.6	0.5		MIT
2791.011	2791.834	35818.75	CO I	1.7			MIT
2790.283	2791.106	35828.09	CO	1.5H			MIT
2787.020	2787.842	35870.04	CO I	0.7			MIT

AIR WAVELENGTH (ANGSTRMS)	VACUUM WAVELENGTH (ANGSTRMS)	WAVENUMBER (KAYSERS)	LMENT	LOG ARC	INTENSITY SPK	DIS	REF
2785.902	2786.724	35884.43	CO I	1.7			MIT
2782.261	2783.082	35931.39	CO I	0.5			MIT
2781.032	2781.853	35947.27	CO	0.9			MIT
2779.841	2780.661	35962.67	CO II		0.5		MIT
2778.822	2779.642	35975.85	CO I	1.9	0.9		MIT
2776.21	2777.03	36009.7	CO		1.0		MIT
2775.580	2776.399	36017.87	CO I	1.7			MIT
2775.181	2776.000	36023.05	CO II	1.5H	2.0H		MIT
2774.959	2775.778	36025.93	CO I	1.7			MIT
2773.67	2774.49	36042.7	CO I	0.5			MIT
2772.696	2773.515	36055.34	CO	1.5H			MIT
2772.552	2773.371	36057.21	CO I	1.2H			MIT
2771.708	2772.526	36068.19	CO	1.0H			MIT
2769.670	2770.488	36094.73	CO I	1.0			MIT
2769.085	2769.903	36102.35	CO		1.5H		MIT
2768.689	2769.507	36107.51	CO	1.3			MIT
2768.294	2769.112	36112.67	CO I	1.0			MIT
2766.85	2767.67	36131.5	CO II		0.5H		MIT
2766.387	2767.204	36137.56	CO I	1.7	0.5		MIT
2766.221	2767.038	36139.73	CO	1.7	1.7		MIT
2764.188	2765.005	36166.31	CO I	2.0R			MIT
2761.372	2762.188	36203.19	CO I	1.9	0.7		MIT
2758.538	2759.353	36240.38	CO I	1.5			MIT
2753.34	2754.15	36308.8	CO II		0.6H		MIT
2752.073	2752.887	36325.51	CO I	1.6	0.0		MIT
2750.141	2750.954	36351.02	CO	1.2			MIT
2746.032	2746.844	36405.42	CO I	1.7			MIT
2745.100	2745.912	36417.77	CO I	1.7	1.8		MIT
2740.459	2741.270	36479.44	CO I	1.7	0.6		MIT
2738.96	2739.77	36499.4	CO		1.3		MIT
2734.68	2735.49	36556.5	CO II		0.5H		MIT
2731.115	2731.924	36604.25	CO	1.70	1.2		MIT
2727.92	2728.73	36647.1	CO II		1.2J		MIT
2722.114	2722.920	36725.28	CO I	1.7W			MIT
2719.582	2720.388	36759.47	CO	1.4			MIT
2715.987	2716.792	36808.12	CO I	1.9W	1.9		MIT
2714.418	2715.222	36829.39	CO II	1.1	2.30		MIT
2709.054	2709.857	36902.31	CO II		1.5J		MIT
2708.817	2709.620	36905.54	CO	1.5	0.3		MIT
2707.502	2708.305	36923.47	CO II		2.0J		MIT
2706.739	2707.542	36933.87	CO II		2.0J		MIT
2705.848	2706.650	36946.04	CO I	1.2W	2.0W		MIT
2702.45	2703.25	36992.5	CO		0.5		MIT
2702.114	2702.916	36997.09	CO II		1.4J		MIT
2697.041	2697.841	37066.67	CO II		1.8		MIT
2695.849	2696.649	37083.06	CO I	1.7W			MIT
2694.680	2695.480	37099.15	CO II	1.4	2.3W		MIT
2694.398	2695.198	37103.03	CO I	1.4			MIT
2693.12	2693.92	37120.6	CO		1.4		EX
2693.01	2693.81	37122.1	CO		1.4		EX
2689.82	2690.62	37166.2	CO		1.6		MIT
2685.341	2686.138	37228.16	CO I	1.9W			MIT
2684.55	2685.35	37239.1	CO II		1.3J		MIT
2680.44	2681.24	37296.2	CO		0.8		MIT
2680.108	2680.904	37300.85	CO	1.4	0.5		MIT
2679.756	2680.552	37305.75	CO	1.90			MIT
2678.04	2678.84	37329.6	CO		0.5		MIT
2676.01	2676.81	37358.0	CO II		1.0H		MIT
2675.982	2676.777	37358.36	CO I	1.00	0.7		MIT
2673.930	2674.725	37387.02	CO I	1.4			MIT
2669.910	2670.704	37443.31	CO II		2.0J		MIT
2666.83	2667.62	37486.6	CO II		0.5H		MIT
2663.529	2664.321	37533.01	CO II	1.2W	1.8W		MIT
2662.651	2663.443	37545.39	CO II		0.7		MIT
2661.715	2662.507	37558.59	CO	0.3	0.7		MIT
2653.703	2654.493	37671.98	CO II	0.7	1.6		MIT
2652.83	2653.62	37684.4	CO		0.9		MIT
2652.34	2653.13	37691.3	CO II	0.5	0.6		MIT
2650.270	2651.059	37720.77	CO I	1.7W	1.4		MIT
2649.940	2650.729	37725.47	CO I	1.7W	0.7		MIT

AIR WAVELENGTH (ANGSTRMS)	VACUUM WAVELENGTH (ANGSTRMS)	WAVENUMBER (KAYSERS)	LMENT	LOG ARC	INTENSITY SPK	DIS	REF
2648.635	2649.424	37744.06	CO I	0.7	1.6W		MIT
2646.417	2647.205	37775.69	CO I	1.0W	0.9		MIT
2644.775	2645.563	37799.14	CO I	1.0W	0.7		MIT
2642.874	2643.661	37826.33	CO I	1.0R			MIT
2636.365	2637.151	37919.71	CO I	0.7	1.0		MIT
2636.02	2636.81	37924.7	CO		1.6		EX
2634.91	2635.70	37940.6	CO		1.3		EX
2632.239	2633.024	37979.15	CO II	1.0W	1.6W		MIT
2629.974	2630.758	38011.85	CO I	1.5	0.0		MIT
2628.761	2629.545	38029.39	CO	0.5			MIT
2627.640	2628.424	38045.62	CO I	1.70			MIT
2626.892	2627.675	38056.45	CO II		1.0V		MIT
2623.755	2624.538	38101.95	CO I	1.6	0.5		MIT
2622.429	2623.211	38121.21	CO I	1.5R	1.2		MIT
2622.062	2622.844	38126.55	CO I	1.6W	1.3		MIT
2622.062	2622.844	38126.55	CO II	1.6W	1.3		MIT
2619.802	2620.584	38159.44	CO		1.6		MIT
2619.276	2620.058	38167.10	CO I	1.7W	0.6		MIT
2618.91	2619.69	38172.4	CO		1.5		EX
2617.861	2618.642	38187.73	CO I	1.7W			MIT
2616.261	2617.042	38211.08	CO I	1.6W	0.3		MIT
2615.335	2616.116	38224.61	CO I	1.0	0.3H		MIT
2614.361	2615.142	38238.85	CO II	0.8	1.8W		MIT
2614.128	2614.908	38242.26	CO I	1.5			MIT
2613.492	2614.272	38251.56	CO	1.4			MIT
2612.630	2613.410	38264.18	CO II		1.3J		MIT
2610.759	2611.539	38291.60	CO I	1.6W	0.0H		MIT
2606.122	2606.901	38359.73	CO I	1.6			MIT
2605.677	2606.455	38366.28	CO	1.5	2.3		MIT
2604.408	2605.186	38384.97	CO		1.5		MIT
2600.980	2601.757	38435.56	CO I	1.0W	0.5		MIT
2599.207	2599.984	38461.78	CO I	0.7			MIT
2594.158	2594.934	38536.63	CO I	1.0W			MIT
2592.882	2593.657	38555.59	CO II		0.3		MIT
2591.686	2592.461	38573.38	CO I	1.0R	0.5		MIT
2590.594	2591.369	38589.64	CO I	1.90			MIT
2587.221	2587.995	38639.95	CO II	1.0W	2.0H		MIT
2585.336	2586.110	38668.12	CO I	1.70			MIT
2583.176	2583.949	38700.45	CO		1.6W		MIT
2582.37	2583.14	38712.5	CO II		0.5		FI
2582.241	2583.014	38714.46	CO II	1.7W	2.7J		MIT
2580.955	2581.728	38733.75	CO II		0.6		MIT
2580.837	2581.610	38735.52	CO I	1.70	0.6		MIT
2580.326	2581.098	38743.19	CO II	1.2W	2.0J		MIT
2578.926	2579.698	38764.23	CO I	1.5			MIT
2576.104	2576.875	38806.69	CO	1.5			MIT
2575.56	2576.33	38814.9	CO II		1.2		MIT
2574.862	2575.633	38825.40	CO	0.5	1.6		MIT
2574.350	2575.121	38833.13	CO I	0.8R			MIT
2573.537	2574.308	38845.39	CO	1.5R	1.1		MIT
2573.397	2574.168	38847.51	CO	1.6	1.1		MIT
2572.237	2573.008	38865.02	CO I	1.7W	1.1		MIT
2569.740	2570.510	38902.79	CO		1.5		MIT
2567.346	2568.115	38939.06	CO I	1.7R			MIT
2565.37	2566.14	38969.0	CO		1.3		MIT
2564.036	2564.805	38989.33	CO II	1.2W	2.0J		MIT
2562.146	2562.914	39018.08	CO I	1.0R			MIT
2561.280	2562.048	39031.28	CO	1.4	1.1H		MIT
2560.09	2560.86	39049.4	CO	0.0D	1.8[illegible]		MIT
2559.407	2560.175	39059.84	CO II	1.0	1.8J		MIT
2557.346	2558.113	39091.31	CO	0.3	1.5		MIT
2556.758	2557.525	39100.30	CO I	1.7W	2.2		MIT
2553.374	2554.140	39152.12	CO I	1.0R			MIT
2553.003	2553.769	39157.81	CO I	1.6R			MIT
2552.38	2553.15	39167.4	CO		1.3		MIT
2550.674	2551.439	39193.56	CO II		0.7J		MIT
2550.02	2550.79	39203.6	CO II		1.8		MIT
2549.884	2550.649	39205.70	CO II		1.0H		MIT
2548.336	2549.101	39229.52	CO I	1.3	1.9		MIT
2546.737	2547.502	39254.15	CO	0.0	1.7		MIT

AIR WAVELENGTH (ANGSTRMS)	VACUUM WAVELENGTH (ANGSTRMS)	WAVENUMBER (KAYSERS)	LMENT	LOG ARC	INTENSITY SPK	INTENSITY DIS	REF
2546.165	2546.929	39262.96	CO		1.3		MIT
2545.04	2545.80	39280.3	CO	0.5	1.5		MIT
2544.253	2545.017	39292.47	CO I	1.7R	2.0		MIT
2541.938	2542.701	39328.25	CO II	1.6	2.5H		MIT
2540.65	2541.41	39348.2	CO	0.8	1.6		MIT
2537.46	2538.22	39397.6	CO		1.0		MIT
2536.802	2537.564	39407.87	CO		0.3		MIT
2536.493	2537.255	39412.67	CO I	0.0	0.3		MIT
2535.964	2536.726	39420.89	CO I	1.0R	1.6		MIT
2533.81	2534.57	39454.4	CO	0.7R	1.8		EX
2532.175	2532.936	39479.87	CO I		1.9		MIT
2530.133	2530.894	39511.74	CO I	1.6W	2.5		MIT
2528.967	2529.727	39529.95	CO I	1.7G			MIT
2528.615	2529.375	39535.45	CO II	0.6	2.3		MIT
2528.184	2528.944	39542.19	CO		1.6		MIT
2524.965	2525.724	39592.60	CO II	1.7W	2.8		MIT
2524.601	2525.360	39598.31	CO		1.5		MIT
2522.94	2523.70	39624.4	CO		1.5		EX
2521.363	2522.122	39649.16	CO	1.9G	2.2		MIT
2519.822	2520.580	39673.40	CO II	1.6	2.3		MIT
2517.869	2518.627	39704.18	CO I	1.0W			MIT
2517.800	2518.558	39705.26	CO I	0.8	1.0		MIT
2517.41	2518.17	39711.4	CO		1.6O		MIT
2511.15	2511.91	39810.4	CO	0.5	1.2		MIT
2511.016	2511.772	39812.53	CO I	1.0R	1.3		MIT
2510.12	2510.88	39826.7	CO		1.3		MIT
2507.962	2508.717	39861.00	CO	0.3	1.5		MIT
2507.676	2508.431	39865.55	CO I	1.6W	0.9		MIT
2506.877	2507.632	39878.26	CO I	1.0W	0.5		MIT
2506.462	2507.217	39884.86	CO II	1.7W	2.3H		MIT
2504.517	2505.272	39915.83	CO I	0.6	1.0		MIT
2500.501	2501.255	39979.94	CO	1.0H			MIT
2498.83	2499.58	40006.7	CO	0.3	1.9		MIT
2497.501	2498.254	40027.96	CO	0.0	1.6		MIT
2493.933	2494.685	40085.22	CO	1.5W	0.0		MIT
2490.394	2491.145	40142.18	CO	0.7	1.9		MIT
2487.42	2488.17	40190.2	CO II	1.0	1.3		MIT
2486.44	2487.19	40206.0	CO II	0.0	1.6		MIT
2485.356	2486.106	40223.54	CO II	1.4	1.9		MIT
2483.612	2484.362	40251.79	CO I		1.4		MIT
2478.207	2478.956	40339.57	CO	0.6	1.3		MIT
2477.477	2478.225	40351.46	CO	0.0	1.3		MIT
2476.641	2477.389	40365.08	CO I	1.6W	1.4		MIT
2470.276	2471.023	40469.07	CO	1.3W	0.9		MIT
2467.691	2468.437	40511.46	CO I	1.3W	0.5		MIT
2467.06	2467.81	40521.8	CO	0.3	1.9		MIT
2464.195	2464.940	40568.93	CO II	1.6	2.2		MIT
2462.12	2462.86	40603.1	CO	1.3	0.0		MIT
2460.806	2461.551	40624.80	CO I	1.3	0.5		MIT
2460.206	2460.950	40634.71	CO	1.3			MIT
2459.49	2460.23	40646.5	CO	0.0	1.5		MIT
2456.237	2456.980	40700.36	CO I	1.3W	0.8		MIT
2450.004	2450.746	40803.90	CO	1.6	2.3		MIT
2449.163	2449.905	40817.91	CO II	1.1	1.5		MIT
2447.76	2448.50	40841.3	CO	0.6	2.0G		MIT
2446.020	2446.761	40870.36	CO	0.3	1.6		MIT
2443.781	2444.522	40907.80	CO II	0.9	1.3		MIT
2442.66	2443.40	40926.6	CO	0.3	1.3		MIT
2441.72	2442.46	40942.3	CO		1.2		MIT
2441.050	2441.790	40953.56	CO	1.3	0.8		MIT
2440.12	2440.86	40969.2	CO	0.6W	0.0J		MIT
2439.503	2440.243	40979.53	CO	0.9H			MIT
2439.046	2439.786	40987.21	CO I	1.3G	1.3		MIT
2438.411	2439.150	40997.88	CO I	0.3	0.5H		MIT
2436.979	2437.718	41021.97	CO II	0.3	1.3		MIT
2436.657	2437.396	41027.39	CO	1.7G	1.4		MIT
2436.442	2437.181	41031.01	CO	0.0	0.7		MIT
2435.830	2436.569	41041.32	CO	1.0	0.0		MIT
2435.092	2435.831	41053.76	CO	1.3W	0.3H		IBU
2432.52	2433.26	41097.2	CO		1.9		MIT

AIR WAVELENGTH (ANGSTRMS)	VACUUM WAVELENGTH (ANGSTRMS)	WAVENUMBER (KAYSERS)	LMENT	LOG ARC	INTENSITY SPK	INTENSITY DIS	REF
2432.214	2432.952	41102.33	CO I	1.6G			MIT
2430.175	2430.913	41136.82	CO	1.0			MIT
2429.228	2429.965	41152.85	CO I	1.4	0.5		MIT
2428.596	2429.333	41163.56	CO	1.0	0.0		MIT
2428.293	2429.030	41168.70	CO II	1.0	1.3		MIT
2427.001	2427.738	41190.61	CO	1.1	0.0		MIT
2426.12	2426.86	41205.6	CO		1.3		EX
2425.594	2426.330	41214.50	CO I	0.9	0.0		MIT
2424.930	2425.666	41225.79	CO I	2.4G			MIT
2423.621	2424.357	41248.05	CO II	1.1	1.3		MIT
2422.564	2423.300	41266.05	CO	1.5W	0.5		MIT
2421.688	2422.424	41280.97	CO I	0.9			MIT
2420.726	2421.461	41297.38	CO	0.9	1.9		MIT
2419.832	2420.567	41312.63	CO I	0.8			MIT
2419.336	2420.071	41321.10	CO I	1.0D			MIT
2419.121	2419.856	41324.77	CO	1.3W	0.0		MIT
2417.654	2418.389	41349.85	CO II	1.3W	1.3		MIT
2417.329	2418.064	41355.41	CO	1.4			MIT
2417.049	2417.784	41360.20	CO	1.0			MIT
2416.896	2417.630	41362.81	CO II	0.9	1.3		MIT
2416.213	2416.947	41374.50	CO	0.0	1.2		MIT
2415.99	2416.72	41378.3	CO	1.0R	1.3		MIT
2415.299	2416.033	41390.16	CO I	1.6G	1.3		MIT
2414.459	2415.193	41404.56	CO I	1.6G	1.2		MIT
2414.063	2414.797	41411.35	CO II	0.9	1.5		MIT
2413.580	2414.314	41419.64	CO	1.2	0.3		MIT
2413.191	2413.925	41426.31	CO I	1.2			MIT
2412.895	2413.629	41431.40	CO	0.8	0.0		MIT
2412.757	2413.491	41433.77	CO I	1.1R			MIT
2411.622	2412.355	41453.26	CO I	2.4G	1.7		MIT
2410.513	2411.246	41472.33	CO	1.6W			MIT
2409.665	2410.398	41486.93	CO I	0.9			MIT
2409.126	2409.859	41496.21	CO I	1.3			MIT
2408.750	2409.483	41502.69	CO II	1.4W	1.4		MIT
2408.415	2409.148	41508.46	CO	0.0	1.0		MIT
2407.668	2408.400	41521.34	CO II	0.0	1.2		MIT
2407.254	2407.986	41528.48	CO I	2.0	0.3		MIT
2406.267	2406.999	41545.51	CO I	1.4	0.0		MIT
2404.882	2405.614	41569.43	CO I	1.0			MIT
2404.54	2405.27	41575.4	CO	0.0	1.3		MIT
2404.172	2404.904	41581.71	CO II	1.5W	1.7		MIT
2403.643	2404.374	41590.86	CO	1.2	1.1		MIT
2403.343	2404.074	41596.05	CO I	1.2			MIT
2402.559	2403.290	41609.62	CO I	1.2			MIT
2402.170	2402.901	41616.36	CO I	1.5R			MIT
2402.065	2402.796	41618.18	CO I	1.0R	0.5		MIT
2401.600	2402.331	41626.24	CO I	1.5	0.3		MIT
2401.108	2401.839	41634.77	CO	1.5	0.3		MIT
2400.837	2401.568	41639.46	CO I	1.5	0.3		MIT
2400.561	2401.292	41644.25	CO I	1.5			MIT
2398.55	2399.28	41679.2	CO	0.6			MIT
2398.37	2399.10	41682.3	CO	0.3	1.3		MIT
2397.388	2398.118	41699.37	CO II	0.6	1.4		MIT
2397.27	2398.00	41701.4	CO	0.6			MIT
2397.03	2397.76	41705.6	CO	0.8			MIT
2396.774	2397.504	41710.05	CO	2.0			MIT
2396.58	2397.31	41713.4	CO	0.7			MIT
2396.23	2396.96	41719.5	CO	1.0			MIT
2395.52	2396.25	41731.9	CO		0.8		MIT
2395.416	2396.146	41733.69	CO	0.8	0.3		MIT
2394.227	2394.956	41754.42	CO	0.6			MIT
2393.90	2394.63	41760.1	CO II	0.9	1.2		MIT
2392.60	2393.33	41782.8	CO	1.0	1.2		MIT
2392.03	2392.76	41792.8	CO I	0.3	0.6		MIT
2391.887	2392.616	41795.26	CO	0.8W			MIT
2391.369	2392.098	41804.31	CO	1.0			MIT
2390.420	2391.148	41820.91	CO	0.6			MIT
2389.982	2390.710	41828.57	CO I	0.9			MIT
2389.565	2390.293	41835.87	CO II	1.1	1.3		ME
2388.915	2389.643	41847.25	CO II	1.0	1.5		IBU

AIR WAVELENGTH (ANGSTRMS)	VACUUM WAVELENGTH (ANGSTRMS)	WAVENUMBER (KAYSERS)	LMENT	LOG ARC	INTENSITY SPK	DIS	REF
2388.376	2389.104	41856.70	CO	0.5			MIT
2388.177	2388.905	41860.18	CO I	0.7			MIT
2387.462	2388.190	41872.72	CO I	1.0	1.0		MIT
2386.727	2387.455	41885.61	CO	0.6	1.2		MIT
2386.509	2387.237	41889.44	CO I	0.5			MIT
2386.363	2387.091	41892.00	CO II	1.0	1.4		MIT
2385.816	2386.543	41901.60	CO	1.0			MIT
2384.860	2385.587	41918.40	CO I	1.0G	0.6		MIT
2384.10	2384.83	41931.8	CO		0.3		MIT
2383.46	2384.19	41943.0	CO II	1.2	1.5		MIT
2382.335	2383.062	41962.83	CO	0.3	0.6		MIT
2381.752	2382.479	41973.10	CO	0.6	1.1		MIT
2381.251	2381.977	41981.93	CO I	0.6			MIT
2380.695	2381.421	41991.73	CO I	0.6			MIT
2380.485	2381.211	41995.44	CO I	1.3D	1.0		MIT
2379.358	2380.084	42015.32	CO I	0.6			MIT
2379.154	2379.880	42018.93	CO I	0.6			MIT
2378.908	2379.634	42023.27	CO	0.7			MIT
2378.622	2379.348	42028.32	CO II	1.4	1.7W		MIT
2377.215	2377.940	42053.20	CO I	1.1	0.3		MIT
2376.970	2377.695	42057.53	CO I	0.8			MIT
2376.446	2377.171	42066.80	CO	0.6			MIT
2375.18	2375.91	42089.2	CO II	1.0	1.3		MIT
2374.749	2375.474	42096.86	CO	0.5			MIT
2374.456	2375.181	42102.06	CO I	0.6			MIT
2373.862	2374.587	42112.59	CO I	1.0			MIT
2373.385	2374.110	42121.05	CO	1.3	0.3		MIT
2373.09	2373.81	42126.3	CO	0.6	0.8		MIT
2372.831	2373.555	42130.89	CO I	1.2	0.3		MIT
2372.51	2373.23	42136.6	CO		0.3W		MIT
2371.860	2372.584	42148.13	CO I	0.8	1.1		MIT
2371.60	2372.32	42152.7	CO		0.5		MIT
2371.440	2372.164	42155.60	CO	1.2			MIT
2370.74	2371.46	42168.0	CO		0.9		MIT
2370.509	2371.233	42172.15	CO	1.0			MIT
2369.927	2370.651	42182.51	CO I	1.0			MIT
2369.676	2370.400	42186.98	CO I	1.2	0.7		MIT
2367.20	2367.92	42231.1	CO		1.0		MIT
2366.054	2366.777	42251.55	CO I	0.7			MIT
2365.067	2365.790	42269.18	CO I	1.3D	0.6		MIT
2364.251	2364.974	42283.77	CO	0.5			MIT
2363.787	2364.509	42292.07	CO II	1.4	1.7		MIT
2362.772	2363.494	42310.24	CO	0.3			MIT
2362.333	2363.055	42318.10	CO I	0.9			MIT
2361.53	2362.25	42332.5	CO II	1.0	1.2		MIT
2361.14	2361.86	42339.5	CO		1.0		MIT
2360.509	2361.231	42350.80	CO	0.6	1.2		MIT
2359.08	2359.80	42376.5	CO		0.3		MIT
2358.671	2359.392	42383.79	CO I	1.0			MIT
2358.176	2358.897	42392.69	CO I	1.3	1.5		MIT
2357.502	2358.223	42404.81	CO	1.0			MIT
2356.78	2357.50	42417.8	CO		0.3		MIT
2356.268	2356.989	42427.02	CO I	1.0			MIT
2355.620	2356.341	42438.69	CO I	0.8	0.9		MIT
2355.05	2355.77	42449.0	CO	0.3			MIT
2354.41	2355.13	42460.5	CO	0.7			MIT
2354.179	2354.899	42464.66	CO	0.3			MIT
2353.73	2354.45	42472.8	CO	0.3			MIT
2353.42	2354.14	42478.4	CO II	1.0	1.5		MIT
2352.854	2353.574	42488.57	CO I	1.2D	0.6		MIT
2352.21	2352.93	42500.2	CO		0.6		MIT
2351.978	2352.698	42504.40	CO I	0.6			MIT
2351.848	2352.568	42506.75	CO		0.8		MIT
2351.391	2352.111	42515.01	CO I	1.0			MIT
2351.17	2351.89	42519.0	CO		1.0		MIT
2350.594	2351.314	42529.42	CO I	0.8			MIT
2350.280	2350.999	42535.10	CO I	1.1	0.3		MIT
2349.16	2349.88	42555.4	CO		0.3		MIT
2348.46	2349.18	42568.1	CO		0.30		MIT
2347.826	2348.545	42579.56	CO		0.9		MIT

AIR WAVELENGTH (ANGSTRMS)	VACUUM WAVELENGTH (ANGSTRMS)	WAVENUMBER (KAYSERS)	LMENT	LOG ARC	INTENSITY SPK	DIS	REF
2347.660	2348.379	42582.57	CO I	0.6			MIT
2347.39	2348.11	42587.5	CO II	1.0	1.4		MIT
2347.121	2347.840	42592.35	CO		0.6		MIT
2346.74	2347.46	42599.3	CO	0.3			MIT
2346.60	2347.32	42601.8	CO	0.5	1.3		MIT
2346.162	2346.881	42609.75	CO I	0.8	0.3		MIT
2345.50	2346.22	42621.8	CO	0.8	0.8		MIT
2344.64	2345.36	42637.4	CO	0.3	0.9		MIT
2344.260	2344.978	42644.32	CO II	1.2	1.2		MIT
2342.793	2343.511	42671.02	CO I	0.3			MIT
2342.418	2343.136	42677.85	CO	0.3			MIT
2341.788	2342.506	42689.33	CO	0.6			MIT
2341.12	2341.84	42701.5	CO		1.3		MIT
2340.992	2341.709	42703.85	CO I	0.7			MIT
2339.552	2340.269	42730.13	CO I	0.7			MIT
2339.053	2339.770	42739.25	CO I	0.6	1.0		MIT
2338.717	2339.434	42745.38	CO		0.7		MIT
2338.671	2339.388	42746.23	CO I	1.0D			MIT
2337.936	2338.653	42759.66	CO	0.5W	1.0		MIT
2336.99	2337.71	42777.0	CO		1.0		MIT
2336.807	2337.523	42780.32	CO	0.3			MIT
2336.492	2337.208	42786.09	CO	0.3			MIT
2336.241	2336.957	42790.68	CO II	0.8	1.2		MIT
2335.992	2336.708	42795.24	CO I	1.0D	0.5		MIT
2335.107	2335.823	42811.46	CO I	1.2			MIT
2334.88	2335.60	42815.6	CO		0.6		MIT
2334.128	2334.844	42829.42	CO I	0.7	1.1		MIT
2333.962	2334.678	42832.46	CO	0.5			MIT
2333.07	2333.79	42848.8	CO I	0.8			MIT
2332.095	2332.810	42866.75	CO	1.0			MIT
2331.695	2332.410	42874.10	CO	0.3			MIT
2330.350	2331.065	42898.85	CO II	1.0	1.3		MIT
2329.095	2329.810	42921.96	CO II	0.8	1.0		ME
2328.855	2329.570	42926.38	CO	1.0			MIT
2328.305	2329.020	42936.52	CO	0.8			MIT
2327.68	2328.39	42948.0	CO		1.0		MIT
2327.535	2328.249	42950.73	CO I	0.7			MIT
2326.48	2327.19	42970.2	CO II	1.3	1.5		MIT
2326.136	2326.850	42976.56	CO II	1.2	1.3		MIT
2325.80	2326.51	42982.8	CO	0.8	0.5		MIT
2325.615	2326.329	42986.18	CO I	1.7W			MIT
2325.549	2326.263	42987.40	CO I	1.1	0.3		MIT
2324.32	2325.03	43010.1	CO II	1.3	1.7		MIT
2323.144	2323.857	43031.90	CO I	1.2D	0.7		MIT
2322.681	2323.394	43040.48	CO	0.5			MIT
2322.247	2322.960	43048.52	CO I	0.6			MIT
2322.01	2322.72	43052.9	CO		0.7		MIT
2321.96	2322.67	43053.8	CO	0.5			MIT
2321.38	2322.09	43064.6	CO	1.0	0.9W		MIT
2320.911	2321.624	43073.30	CO I	0.6			MIT
2319.86	2320.57	43092.8	CO		0.7		MIT
2319.27	2319.98	43103.8	CO		0.7		MIT
2319.159	2319.872	43105.83	CO I	0.6			MIT
2318.42	2319.13	43119.6	CO II	0.3	1.0		MIT
2317.523	2318.235	43136.26	CO	0.7			MIT
2317.055	2317.767	43144.97	CO	0.3	1.3		MIT
2316.862	2317.574	43148.57	CO I	0.7			MIT
2316.735	2317.447	43150.93	CO I	0.7			MIT
2316.162	2316.874	43161.61	CO	1.0W			MIT
2315.760	2316.472	43169.10	CO	0.5	0.7		MIT
2314.98	2315.69	43183.6	CO II	1.4	1.5		MIT
2314.651	2315.363	43189.78	CO		0.6		MIT
2314.054	2314.765	43200.92	CO II	1.4	1.5		MIT
2313.64	2314.35	43208.6	CO II	0.8	1.0		MIT
2312.55	2313.26	43229.0	CO II	0.5	1.1		MIT
2311.604	2312.315	43246.71	CO II	1.2	1.7		MIT
2311.358	2312.069	43251.31	CO I	1.0			MIT
2310.962	2311.673	43258.72	CO I	1.1			MIT
2310.83	2311.54	43261.2	CO		0.6		MIT
2309.020	2309.730	43295.10	CO I	1.0G	1.0		MIT

AIR WAVELENGTH (ANGSTRMS)	VACUUM WAVELENGTH (ANGSTRMS)	WAVENUMBER (KAYSERS)	LMENT	LOG ARC	INTENSITY SPK	DIS	REF
2307.857	2308.567	43316.91	CO II	1.4	1.7W		MIT
2307.49	2308.20	43323.8	CO		0.8		MIT
2307.010	2307.720	43332.82	CO		0.8		MIT
2306.78	2307.49	43337.1	CO		0.7		MIT
2306.10	2306.81	43349.9	CO		0.6W		MIT
2305.18	2305.89	43367.2	CO I	1.2	0.3		MIT
2304.185	2304.894	43385.94	CO I	1.0	0.8		MIT
2303.974	2304.683	43389.91	CO I	1.1	0.3		MIT
2303.510	2304.219	43398.65	CO I	1.0			MIT
2301.40	2302.11	43438.4	CO II	1.0	1.4		MIT
2300.78	2301.49	43450.1	CO	0.7	1.2		MIT
2300.48	2301.19	43455.8	CO		0.3		MIT
2299.755	2300.463	43469.50	CO	0.6	1.5		MIT
2299.42	2300.13	43475.8	CO II		1.4		ME
2298.738	2299.446	43488.73	CO II		1.1		MIT
2298.375	2299.083	43495.60	CO I	1.2	0.3		MIT
2297.36	2298.07	43514.8	CO		0.9		MIT
2296.710	2297.418	43527.13	CO I	1.3	0.3		MIT
2296.051	2296.758	43539.62	CO I	1.3	0.3		MIT
2295.92	2296.63	43542.1	CO		0.9		MIT
2295.230	2295.937	43555.20	CO I	1.2	0.5		MIT
2294.006	2294.713	43578.43	CO I	1.0			MIT
2293.39	2294.10	43590.1	CO II	1.0	1.2L		MIT
2292.99	2293.70	43597.7	CO II	1.0	1.5		MIT
2291.455	2292.161	43626.94	CO I	1.1	0.7W		MIT
2290.544	2291.250	43644.29	CO I	1.0			MIT
2290.33	2291.04	43648.4	CO		0.7		MIT
2289.502	2290.208	43664.16	CO	1.0			MIT
2289.06	2289.77	43672.6	CO		0.3		MIT
2288.786	2289.492	43677.81	CO I	1.2			MIT
2288.56	2289.27	43682.1	CO		0.3		MIT
2287.809	2288.515	43696.46	CO	1.1D	0.6		MIT
2286.156	2286.861	43728.06	CO II	1.6	2.5L		MIT
2285.81	2286.52	43734.7	CO		0.3		MIT
2285.412	2286.117	43742.29	CO I	1.1			MIT
2284.846	2285.551	43753.12	CO I	1.5			MIT
2284.381	2285.086	43762.03	CO I	0.9			MIT
2283.519	2284.224	43778.55	CO II	1.0	1.2		MIT
2282.366	2283.070	43800.66	CO		0.7		MIT
2281.88	2282.58	43810.0	CO	0.5	1.4		MIT
2281.34	2282.04	43820.4	CO	0.7			MIT
2280.958	2281.662	43827.70	CO II	0.6	0.7		MIT
2280.46	2281.16	43837.3	CO		0.8W		MIT
2280.15	2280.85	43843.2	CO	0.3			MIT
2279.490	2280.194	43855.92	CO I	1.2			MIT
2279.02	2279.72	43865.0	CO		0.5		MIT
2278.46	2279.16	43875.7	CO		0.9		MIT
2278.298	2279.002	43878.86	CO I	1.0			MIT
2276.91	2277.61	43905.6	CO		0.3		MIT
2276.532	2277.235	43912.90	CO	1.3R	0.7		MIT
2275.876	2276.579	43925.56	CO I	1.0			MIT
2275.41	2276.11	43934.5	CO		0.8		MIT
2274.633	2275.336	43949.56	CO I	0.9			MIT
2274.490	2275.193	43952.32	CO	1.0	0.3		MIT
2273.67	2274.37	43968.2	CO		0.8W		MIT
2272.91	2273.61	43982.9	CO		0.3H		MIT
2272.252	2272.954	43995.61	CO II	0.6	0.9		MIT
2271.21	2271.91	44015.8	CO		0.7W		MIT
2269.98	2270.68	44039.6	CO		0.5		MIT
2268.746	2269.448	44063.59	CO I	1.1			MIT
2268.168	2268.869	44074.82	CO I	1.2	0.3		MIT
2267.101	2267.802	44095.56	CO I	1.1			MIT
2266.80	2267.50	44101.4	CO		0.7		MIT
2266.53	2267.23	44106.7	CO		1.5		MIT
2265.742	2266.443	44122.00	CO II	0.5	0.7		MIT
2264.874	2265.575	44138.91	CO	1.2			MIT
2264.418	2265.119	44147.80	CO	1.0	0.3		MIT
2264.15	2264.85	44153.0	CO		0.3		MIT
2263.55	2264.25	44164.7	CO		0.7W		MIT
2262.598	2263.298	44183.31	CO I	1.0			MIT

AIR WAVELENGTH (ANGSTRMS)	VACUUM WAVELENGTH (ANGSTRMS)	WAVENUMBER (KAYSERS)	LMENT	LOG ARC	INTENSITY SPK	DIS	REF
2261.55	2262.25	44203.8	CO		0.9		MIT
2261.29	2261.99	44208.9	CO		0.3		MIT
2260.01	2260.71	44233.9	CO	0.3	1.4		MIT
2258.59	2259.29	44261.7	CO		0.3		MIT
2258.328	2259.027	44266.84	CO	1.0			MIT
2257.86	2258.56	44276.0	CO		0.5O		MIT
2257.587	2258.286	44281.37	CO	1.0			MIT
2256.74	2257.44	44298.0	CO		1.5		MIT
2256.573	2257.272	44301.27	CO I	1.0			MIT
2255.65	2256.35	44319.4	CO		0.6		MIT
2254.832	2255.531	44335.47	CO I		1.0		MIT
2253.776	2254.474	44356.24	CO	1.0			MIT
2253.50	2254.20	44361.7	CO		1.1		MIT
2252.82	2253.52	44375.1	CO		0.3W		MIT
2252.731	2253.429	44376.81	CO I	1.0			MIT
2252.38	2253.08	44383.7	CO		0.5W		MIT
2251.842	2252.540	44394.33	CO I	0.9			MIT
2251.116	2251.814	44408.65	CO		0.9		MIT
2250.510	2251.208	44420.60	CO	1.0	0.3		MIT
2250.39	2251.09	44423.0	CO		0.5		MIT
2248.988	2249.685	44450.66	CO I	0.7			MIT
2248.66	2249.36	44457.1	CO II	0.3	0.7		MIT
2248.18	2248.88	44466.6	CO		0.5W		MIT
2247.869	2248.566	44472.79	CO I	0.6			MIT
2246.15	2246.85	44506.8	CO		0.5W		MIT
2245.600	2246.297	44517.72	CO I	1.0			MIT
2245.464	2246.160	44520.42	CO I	0.7			MIT
2245.127	2245.823	44527.10	CO II	1.2	1.5		MIT
2244.45	2245.15	44540.5	CO	0.3	0.3		MIT
2243.90	2244.60	44551.4	CO		0.6W		MIT
2243.257	2243.953	44564.21	CO I	1.0			MIT
2242.90	2243.60	44571.3	CO		0.6W		MIT
2241.654	2242.350	44596.08	CO I	1.0			MIT
2241.27	2241.97	44603.7	CO		0.6W		MIT
2240.734	2241.429	44614.39	CO	0.8			MIT
2240.13	2240.83	44626.4	CO		0.3		MIT
2239.80	2240.50	44633.0	CO		0.6		MIT
2237.130	2237.825	44686.25	CO	1.0			MIT
2236.796	2237.491	44692.92	CO I	1.2	0.6		MIT
2235.11	2235.80	44726.6	CO		0.7		MIT
2234.714	2235.408	44734.56	CO I	1.1			MIT
2233.754	2234.448	44753.78	CO I	1.0			MIT
2233.099	2233.793	44766.91	CO	1.0			MIT
2232.877	2233.571	44771.36	CO	0.6			MIT
2232.468	2233.162	44779.56	CO I	0.9	1.2W		MIT
2232.066	2232.760	44787.62	CO II	0.8	1.3		MIT
2231.749	2232.443	44793.98	CO I	0.8			MIT
2230.52	2231.21	44818.7	CO		1.0O		MIT
2229.732	2230.425	44834.50	CO I	1.0	0.3		MIT
2228.806	2229.499	44853.13	CO I	1.1			MIT
2228.334	2229.027	44862.63	CO I	0.6			MIT
2227.868	2228.561	44872.01	CO I	1.0			MIT
2227.657	2228.350	44876.26	CO I	1.1	0.6		MIT
2225.848	2226.540	44912.73	CO	0.7			MIT
2225.350	2226.042	44922.78	CO I	1.1	0.3		MIT
2224.87	2225.56	44932.5	CO		0.7		MIT
2224.09	2224.78	44948.2	CO		1.0		MIT
2222.96	2223.65	44971.1	CO		1.0		MIT
2222.25	2222.94	44985.4	CO		0.6		MIT
2221.54	2222.23	44999.8	CO		0.5		MIT
2221.28	2221.97	45005.1	CO		0.5		MIT
2220.09	2220.78	45029.2	CO II	1.0	1.3		MIT
2219.154	2219.845	45048.19	CO I	1.0	0.6		MIT
2218.813	2219.504	45055.12	CO	1.0			MIT
2217.80	2218.49	45075.7	CO		0.5W		MIT
2217.28	2217.97	45086.3	CO II	0.6	1.0L		MIT
2216.480	2217.170	45102.53	CO	0.6	1.4		MIT
2214.79	2215.48	45137.0	CO II	1.0	1.1		MIT
2213.86	2214.55	45155.9	CO	0.6	0.6		MIT
2213.819	2214.509	45156.74	CO	0.8			MIT

AIR WAVELENGTH (ANGSTRMS)	VACUUM WAVELENGTH (ANGSTRMS)	WAVENUMBER (KAYSERS)	LMENT	LOG ARC	INTENSITY SPK	INTENSITY DIS	REF
2213.18	2213.87	45169.8	CO		1.0		MIT
2212.354	2213.043	45186.64	CO I	1.0			MIT
2211.430	2212.119	45205.52	CO II	1.0	1.3		MIT
2211.07	2211.76	45212.9	CO		0.7		MIT
2209.51	2210.20	45244.8	CO		1.0		MIT
2209.05	2209.74	45254.2	CO		0.3		MIT
2208.522	2209.211	45265.04	CO I	1.1			MIT
2207.92	2208.61	45277.4	CO II		1.5		MIT
2207.87	2208.56	45278.4	CO I	1.0			MIT
2207.697	2208.385	45281.95	CO I	1.0			MIT
2206.19	2206.88	45312.9	CO II	0.6			MIT
2205.874	2206.562	45319.37	CO I	0.8	0.9		MIT
2205.53	2206.22	45326.4	CO II		1.2		MIT
2205.07	2205.76	45335.9	CO II		1.2		MIT
2204.796	2205.484	45341.52	CO I	1.3	0.3		MIT
2203.43	2204.12	45369.6	CO II		0.5		MIT
2202.957	2203.644	45379.37	CO II	0.5	1.5		MIT
2201.235	2201.922	45414.87	CO I	0.6			MIT
2200.693	2201.380	45426.05	CO	1.9			MIT
2200.420	2201.107	45431.69	CO II	0.7	1.2		MIT
2199.740	2200.427	45445.73	CO	1.3			M
2198.764	2199.451	45465.90	CO I	0.3			MIT
2198.300	2198.987	45475.50	CO II	0.5	1.0		MIT
2197.633	2198.319	45489.30	CO	0.7			MIT
2196.904	2197.590	45504.39	CO	0.5			MIT
2196.462	2197.148	45513.54	CO I	1.2	0.6		MIT
2194.612	2195.298	45551.91	CO	0.8			MIT
2193.60	2194.29	45572.9	CO II	0.6	1.5		MIT
2192.49	2193.18	45596.0	CO II	0.3	1.4		MIT
2190.665	2191.350	45633.97	CO II	0.3	1.3		MIT
2190.226	2190.911	45643.12	CO	0.5			MIT
2189.350	2190.035	45661.38	CO I	0.5			MIT
2189.00	2189.68	45668.7	CO II	0.5	1.0		MIT
2187.284	2187.968	45704.50	CO	0.7			MIT
2187.04	2187.72	45709.6	CO II	0.3	0.8		MIT
2186.786	2187.470	45714.91	CO	1.1	0.7		MIT
2186.030	2186.714	45730.72	CO	0.5			MIT
2184.950	2185.634	45753.32	CO I	1.0			MIT
2184.314	2184.998	45766.64	CO	0.9			MIT
2182.587	2183.270	45802.85	CO I	1.2			MIT
2182.381	2183.064	45807.17	CO	0.3			MIT
2182.00	2182.68	45815.2	CO		1.0		MIT
2181.729	2182.412	45820.86	CO II	0.9	1.9		MIT
2181.121	2181.804	45833.63	CO I	1.1	0.3		MIT
2180.60	2181.28	45844.6	CO II		0.8		MIT
2180.065	2180.748	45855.83	CO I	1.0			MIT
2178.952	2179.634	45879.25	CO	1.4			MIT
2176.968	2177.650	45921.06	CO	0.3			MIT
2176.494	2177.176	45931.06	CO	0.6			MIT
2174.93	2175.61	45964.1	CO II		0.3		MIT
2174.605	2175.287	45970.95	CO I	1.5R	1.1		MIT
2174.54	2175.22	45972.3	CO II	1.2	1.4		MIT
2174.03	2174.71	45983.1	CO		1.0		MIT
2173.845	2174.526	45987.02	CO I	1.0			MIT
2173.34	2174.02	45997.7	CO II	0.3	1.3		MIT
2173.173	2173.854	46001.24	CO	1.0			MIT
2172.89	2173.57	46007.2	CO II		0.5		MIT
2172.175	2172.856	46022.38	CO I	0.6			MIT
2170.565	2171.246	46056.51	CO I	1.0			MIT
2168.711	2169.391	46095.88	CO I	1.3	0.3		MIT
2166.764	2167.444	46137.29	CO I	0.7			MIT
2165.54	2166.22	46163.4	CO	0.3	1.0		A
2163.574	2164.253	46205.31	CO I	1.1	0.5		MIT
2163.034	2163.713	46216.85	CO I	1.2			MIT
2162.196	2162.875	46234.76	CO I	0.8			MIT
2158.542	2159.220	46313.01	CO I	1.0			MIT
2158.18	2158.86	46320.8	CO II		0.3		MIT
2156.94	2157.62	46347.4	CO II	0.6	1.0		MIT
2156.69	2157.37	46352.8	CO II		0.5		MIT
2154.074	2154.751	46409.07	CO I	1.0			MIT

AIR WAVELENGTH (ANGSTRMS)	VACUUM WAVELENGTH (ANGSTRMS)	WAVENUMBER (KAYSERS)	LMENT	LOG ARC	INTENSITY SPK	INTENSITY DIS	REF
2152.148	2152.825	46450.59	CO I	1.0			MIT
2148.708	2149.384	46524.95	CO I	0.8			MIT
2147.39	2148.07	46553.5	CO II		0.3		MIT
2146.97	2147.65	46562.6	CO II		1.0		MIT
2146.264	2146.940	46577.92	CO I	1.1	0.8		MIT
2145.45	2146.13	46595.6	CO	1.1			MIT
2143.679	2144.354	46634.08	CO I	0.5			MIT
2138.971	2139.645	46736.72	CO I	1.2			MIT
2137.780	2138.454	46762.75	CO I	1.2	0.3		MIT
2136.49	2137.16	46791.0	CO II		0.3		MIT
2135.798	2136.472	46806.14	CO	0.6			MIT
2135.59	2136.26	46810.7	CO I	0.5			MIT
2133.46	2134.13	46857.4	CO II	0.3	0.6		MIT
2132.767	2133.440	46872.65	CO	1.0	0.5		MIT
2131.052	2131.725	46910.37	CO I	0.5			MIT
2130.276	2130.949	46927.46	CO I	0.9			MIT
2129.508	2130.180	46944.38	CO I	0.7			MIT
2127.147	2127.819	46996.48	CO I	1.0			MIT
2126.199	2126.871	47017.43	CO I	0.7			MIT
2125.949	2126.621	47022.96	CO	0.7			MIT
2125.86	2126.53	47024.9	CO II		0.3		MIT
2125.322	2125.994	47036.83	CO I	0.7			MIT
2125.116	2125.788	47041.39	CO	1.0			MIT
2121.391	2122.062	47123.98	CO I	0.5			MIT
2120.705	2121.376	47139.22	CO I	1.0	0.3		MIT
2119.904	2120.575	47157.03	CO I	1.0			MIT
2119.192	2119.862	47172.87	CO	0.7			MIT
2118.505	2119.175	47188.17	CO	0.8			MIT
2117.95	2118.62	47200.5	CO II	0.3	0.7		MIT
2116.842	2117.512	47225.24	CO I	1.0	0.3		MIT
2115.338	2116.008	47258.81	CO I	1.1			MIT
2114.41	2115.08	47279.5	CO II	0.6	0.3		MIT
2113.536	2114.205	47299.10	CO	1.1	0.3		MIT
2111.46	2112.13	47345.6	CO II		1.2		MIT
2111.416	2112.085	47346.58	CO I	1.0			MIT
2109.206	2109.874	47396.19	CO	0.7			MIT
2108.980	2109.648	47401.27	CO	1.2			MIT
2106.798	2107.466	47450.35	CO	1.4	0.3		MIT
2105.49	2106.16	47479.8	CO II		0.3		MIT
2105.34	2106.01	47483.2	CO II		0.5		MIT
2104.730	2105.398	47496.97	CO	1.4	0.3		MIT
2099.35	2100.02	47618.7	CO I	1.0			MIT
2098.942	2099.608	47627.93	CO I	1.1	0.5		MIT
2097.511	2098.177	47660.42	CO	1.3			MIT
2097.12	2097.79	47669.3	CO II		0.6		MIT
2095.77	2096.44	47700.0	CO	1.2	0.3		MIT
2094.24	2094.91	47734.9	CO II		0.5		MIT
2093.40	2094.07	47754.0	CO	1.2			MIT
2092.80	2093.47	47767.7	CO		0.7		A
2091.05	2091.71	47807.7	CO II	0.3	0.8		MIT
2090.47	2091.13	47820.9	CO		0.3		A
2082.68	2083.34	47999.8	CO	0.5	1.4		MIT
2073.27	2073.93	48217.6	CO	1.0	0.3		MIT
2068.99	2069.65	48317.3	CO I	1.0	0.5		MIT
2065.53	2066.19	48398.3	CO II	1.0	1.5		MIT
2064.86	2065.52	48414.0	CO	0.6			A
2063.76	2064.42	48439.8	CO II	1.1	1.5		MIT
2058.81	2059.47	48556.2	CO II	1.0	1.6		MIT
2054.07	2054.73	48668.2	CO	1.0			A
2050.74	2051.40	48747.3	CO II		0.6		MIT
2049.17	2049.83	48784.6	CO II	0.5	1.0		MIT
2043.00	2043.66	48931.9	CO	0.9			A
2038.678	2039.333	49035.64	CO		1.0		MIT
2036.59	2037.24	49085.9	CO II	0.9	1.5		MIT
2032.728	2033.382	49179.15	CO	0.3	1.3		MIT
2031.948	2032.602	49198.03	CO		1.0		MIT
2027.02	2027.67	49317.6	CO II	1.2	1.7		MIT
2025.75	2026.40	49348.5	CO II	1.0L	1.5		MIT
2024.43	2025.08	49380.7	CO	0.5			A
2022.34	2022.99	49431.7	CO II	1.3	1.9		MIT

AIR WAVELENGTH (ANGSTRMS)	VACUUM WAVELENGTH (ANGSTRMS)	WAVENUMBER (KAYSERS)	LMENT	LOG ARC	INTENSITY SPK	DIS	REF
2017.26	2017.91	49556.2	CO	0.6			A
2016.912	2017.563	49564.75	CO	0.3			MIT
2015.99	2016.64	49587.4	CO	0.6			A
2014.58	2015.23	49622.1	CO	1.3W			A
2013.86	2014.51	49639.9	CO		1.0		MIT
2011.50	2012.15	49698.1	CO	1.2	1.7		MIT
2011.07	2011.72	49708.7	CO	0.7			A
2010.10	2010.75	49732.7	CO	0.9			A
2008.28	2008.93	49777.7	CO	0.7	0.7		MIT
2000.78	2001.43	49964.3	CO II	0.6	1.1		MIT
9949.06	9951.79	10048.4	CR I	0.7			KS
9946.30	9949.03	10051.2	CR I	0.3			KS
9904.47	9907.19	10093.7	CR I	0.3			KS
9900.87	9903.58	10097.3	CR I	0.7			MIT
9773.27	9775.95	10229.2	CR I	0.8			MIT
9752.84	9755.51	10250.6	CR I	0.3			KS
9734.51	9737.18	10269.9	CR I	1.7			MIT
9730.27	9732.94	10274.4	CR I	1.4			MIT
9670.49	9673.14	10337.9	CR I	1.7			MIT
9667.22	9669.87	10341.4	CR I	1.4			MIT
9626.30	9628.94	10385.4	CR I	0.6			KS
9574.24	9576.87	10441.8	CR I	1.7			MIT
9571.74	9574.37	10444.6	CR I	1.4			MIT
9568.58	9571.20	10448.0	CR I	0.5			KS
9520.13	9522.74	10501.2	CR	0.3			KS
9446.95	9449.54	10582.5	CR I	1.9			MIT
9444.36	9446.95	10585.4	CR I	0.6			KS
9362.06	9364.63	10678.5	CR I	1.1			KS
9313.54	9316.10	10734.1	CR I	0.9			MIT
9294.14	9296.69	10756.5	CR I	1.3			MIT
9290.38	9292.93	10760.9	CR I	1.9			MIT
9263.96	9266.50	10791.6	CR I	1.4			MIT
9208.27	9210.80	10856.8	CR I	1.4			MIT
9148.45	9150.96	10927.8	CR I	0.7			KS
9142.60	9145.11	10934.8	CR I	0.9			KS
9141.12	9143.63	10936.6	CR I	0.3			KS
9140.51	9143.02	10937.3	CR I	0.3			KS
9059.74	9062.23	11034.8	CR I	0.7			KS
9035.85	9038.33	11064.0	CR I	1.7			MIT
9021.65	9024.13	11081.4	CR I	2.0			MIT
9017.03	9019.51	11087.1	CR I	2.0			MIT
9009.91	9012.38	11095.8	CR I	2.0			MIT
8976.83	8979.29	11136.7	CR I	1.5			MIT
8955.77	8958.23	11162.9	CR I	0.5			MIT
8947.15	8949.61	11173.7	CR I	1.7			MIT
8939.20	8941.65	11183.6	CR I	0.9			KS
8925.78	8928.23	11200.4	CR I	1.2			MIT
8917.10	8919.55	11211.3	CR I	1.0			MIT
8916.24	8918.69	11212.4	CR I	1.1			MIT
8835.65	8838.08	11314.7	CR I	1.0			MIT
8773.54	8775.95	11394.8	CR I	0.5H			MIT
8767.12	8769.53	11403.1	CR I	0.3			KS
8753.45	8755.85	11420.9	CR I	0.3			MIT
8732.15	8734.55	11448.8	CR I	0.3			MIT
8718.66	8721.05	11466.5	CR I	0.9H			MIT
8707.97	8710.36	11480.6	CR I	1.0			MIT
8707.37	8709.76	11481.4	CR I	0.7			MIT
8687.46	8689.85	11507.7	CR I	0.9			MIT
8643.07	8645.44	11566.8	CR I	1.2			MIT
8636.28	8638.65	11575.9	CR I	1.0			MIT
8583.01	8585.37	11647.7	CR I	1.0			KS
8562.56	8564.91	11675.5	CR I	0.9			KS
8555.50	8557.85	11685.2	CR I	0.9			MIT
8548.86	8551.21	11694.2	CR I	1.2			MIT
8543.72	8546.07	11701.3	CR I	0.3			KS
8537.80	8540.15	11709.4	CR I	0.8H			KS
8510.97	8513.31	11746.3	CR I	0.9			MIT
8483.36	8485.69	11784.5	CR I	0.5H			KS
8455.24	8457.56	11823.7	CR I	0.3			KS
8450.26	8452.58	11830.7	CR I	0.5			KS

AIR WAVELENGTH (ANGSTRMS)	VACUUM WAVELENGTH (ANGSTRMS)	WAVENUMBER (KAYSERS)	LMENT	LOG ARC	INTENSITY SPK	DIS	REF
8397.04	8399.35	11905.7	CR I	0.8			KS
8378.54	8380.84	11932.0	CR I	0.9			MIT
8348.28	8350.57	11975.2	CR I	1.3			MIT
8338.83	8341.12	11988.8	CR I	0.7			KS
8336.81	8339.10	11991.7	CR I	1.0			KS
8323.44	8325.73	12011.0	CR I	0.7			KS
8322.99	8325.28	12011.6	CR I	0.5			KS
8322.06	8324.35	12012.9	CR I	1.3J			MIT
8318.27	8320.56	12018.4	CR I	0.7V			MIT
8303.17	8305.45	12040.3	CR I	0.7			MIT
8297.58	8299.86	12048.4	CR	0.5			KS
8296.90	8299.18	12049.4	CR I	0.6			KS
8290.62	8292.90	12058.5	CR I	1.0			KS
8287.38	8289.66	12063.2	CR I	1.4			KS
8286.33	8288.61	12064.7	CR	0.3H			KS
8285.72	8288.00	12065.6	CR	0.3			KS
8273.80	8276.07	12083.0	CR I	0.5			KS
8273.20	8275.47	12083.9	CR	0.5			KS
8264.27	8266.54	12097.0	CR I	0.3			KS
8262.01	8264.28	12100.3	CR I	1.0			MIT
8240.67	8242.94	12131.6	CR I	0.3			KS
8238.34	8240.60	12135.0	CR I	0.9			MIT
8235.89	8238.15	12138.6	CR I	1.5			MIT
8225.67	8227.93	12153.7	CR I	0.5			KS
8224.07	8226.33	12156.1	CR I	0.9			MIT
8218.07	8220.33	12165.0	CR	0.5			KS
8216.32	8218.58	12167.6	CR I	0.7			MIT
8194.87	8197.12	12199.4	CR I	0.6			KS
8188.77	8191.02	12208.5	CR	0.5H			KS
8185.67	8187.92	12213.1	CR I	0.7			MIT
8169.75	8172.00	12236.9	CR I	0.7			MIT
8167.97	8170.22	12239.6	CR	0.5			MIT
8166.72	8168.97	12241.4	CR I	0.7			MIT
8163.18	8165.42	12246.8	CR I	1.5			MIT
8155.03	8157.27	12259.0	CR I	0.8			MIT
8128.29	8130.52	12299.3	CR I	0.7H			MIT
8119.13	8121.36	12313.2	CR I	0.6H			KS
8098.23	8100.46	12345.0	CR I	0.5H			MIT
8085.00	8087.22	12365.2	CR I	1.1H			MIT
8061.37	8063.59	12401.4	CR I	1.1H			MIT
8045.38	8047.59	12426.1	CR	0.5			MIT
8034.98	8037.19	12442.2	CR	0.7H			SL
8018.03	8020.24	12468.5	CR I	0.3H			MIT
8014.90	8017.10	12473.3	CR I	0.3H			MIT
7990.53	7992.73	12511.4	CR	1.2			MIT
7989.38	7991.58	12513.2	CR I	1.2			MIT
7971.29	7973.48	12541.6	CR I	0.5			MIT
7942.04	7944.22	12587.8	CR I	1.3			MIT
7924.09	7926.27	12616.3	CR I	0.3			MIT
7917.84	7920.02	12626.2	CR I	1.2			MIT
7910.53	7912.71	12637.9	CR II	1.1			MIT
7908.27	7910.45	12641.5	CR I	1.2			MIT
7884.95	7887.12	12678.9	CR I	0.6			MIT
7771.65	7773.79	12863.7	CR	1.1			MIT
7763.00	7765.14	12878.1	CR	0.3H			MIT
7733.99	7736.12	12926.4	CR I	0.7			MIT
7725.99	7728.12	12939.8	CR I	0.6			MIT
7722.89	7725.02	12945.0	CR I	0.7			KS
7513.13	7515.20	13306.4	CR	0.7			MIT
7462.31	7464.37	13397.0	CR I	1.9			MIT
7452.48	7454.53	13414.7	CR	0.5			MIT
7419.59	7421.63	13474.1	CR	0.5			MIT
7413.36	7415.40	13485.4	CR	0.3			MIT
7400.210	7402.248	13509.41	CR I	1.9			MIT
7355.90	7357.93	13590.8	CR I	1.9			MIT
7268.65	7270.65	13753.9	CR	0.7			MIT
7236.20	7238.19	13815.6	CR	0.5			MIT
7218.62	7220.61	13849.2	CR	0.6			MIT
7207.87	7209.86	13869.9	CR	1.2			MIT
7188.06	7190.04	13908.1	CR	0.5			MIT

AIR WAVELENGTH (ANGSTRMS)	VACUUM WAVELENGTH (ANGSTRMS)	WAVENUMBER (KAYSERS)	LMENT	LOG ARC	INTENSITY SPK	INTENSITY DIS	REF
7185.52	7187.50	13913.0	CR	0.7			MIT
7170.63	7172.61	13941.9	CR	0.7			ME
6981.04	6982.97	14320.6	CR	0.9			ME
6980.81	6982.74	14321.0	CR I	0.7			ME
6979.82	6981.75	14323.1	CR I	1.3H			MIT
6978.48	6980.40	14325.8	CR I	2.1J			MIT
6925.20	6927.11	14436.0	CR I	1.7H			ME
6924.13	6926.04	14438.3	CR I	1.8H			MIT
6883.03	6884.93	14524.5	CR I	1.8H			ME
6882.38	6884.28	14525.8	CR I	1.3H			ME
6881.62	6883.52	14527.4	CR I	1.2J			ME
6789.17	6791.04	14725.3	CR	1.6			MIT
6762.41	6764.28	14783.6	CR	1.7			MIT
6757.77	6759.64	14793.7	CR	0.9			MIT
6752.58	6754.44	14805.1	CR	0.3			SL
6751.27	6753.13	14807.9	CR	0.3			MIT
6746.61	6748.47	14818.2	CR	0.8			MIT
6744.64	6746.50	14822.5	CR	0.8			MIT
6737.697	6739.557	14837.77	CR	0.7			MIT
6734.188	6736.047	14845.50	CR I	1.1			MIT
6729.732	6731.590	14855.33	CR	1.0			MIT
6715.36	6717.21	14887.1	CR I	0.9			ME
6701.633	6703.483	14917.62	CR	0.5			SS
6677.20	6679.04	14972.2	CR	0.3			ME
6669.257	6671.099	14990.04	CR I	1.9			MIT
6666.24	6668.08	14996.8	CR	0.9			MIT
6661.076	6662.915	15008.44	CR I	2.0			MIT
6651.886	6653.723	15029.18	CR	0.9			SS
6643.023	6644.857	15049.23	CR	0.7			MIT
6630.015	6631.846	15078.76	CR I	1.7			MIT
6612.180	6614.006	15119.43	CR I	1.1L			MIT
6608.894	6610.719	15126.95	CR	1.4			MIT
6597.556	6599.378	15152.94	CR	1.6			MIT
6594.675	6596.496	15159.56	CR	1.8J			MIT
6580.913	6582.731	15191.26	CR I	1.5			MIT
6572.900	6574.716	15209.78	CR I	1.4			MIT
6537.95	6539.76	15291.1	CR I	1.5			ME
6529.197	6531.001	15311.59	CR	1.6H			HL
6516.026	6517.826	15342.54	CR	0.7H			HL
6501.212	6503.008	15377.50	CR	1.5			HL
6428.125	6429.902	15552.34	CR	1.1			MIT
6362.874	6364.633	15711.82	CR I	2.2	0.9		MIT
6330.101	6331.851	15793.17	CR I	2.3	0.9		MIT
6327.434	6329.184	15799.83	CR	0.9S			MIT
6313.218	6314.964	15835.40	CR	0.5			MIT
6261.285	6263.017	15966.75	CR	0.5			MIT
6190.657	6192.370	16148.91	CR	0.7			MIT
6135.835	6137.533	16293.19	CR	1.0			SS
6135.759	6137.457	16293.39	CR	0.7			MIT
6102.709	6104.398	16381.63	CR	1.0			HL
6047.665	6049.340	16530.73	CR	0.8			HL
6045.403	6047.077	16536.92	CR	0.7			HL
5902.182	5903.818	16938.19	CR	0.6			MIT
5889.989	5891.621	16973.26	CR	1.1			MIT
5888.78	5890.41	16976.7	CR	0.5L			MIT
5888.008	5889.640	16978.97	CR	1.3			MIT
5884.452	5886.083	16989.23	CR I	1.3			MIT
5876.564	5878.193	17012.03	CR	0.5			MIT
5866.93	5868.56	17040.0	CR	0.6			MIT
5844.606	5846.226	17105.05	CR I	1.3			MIT
5843.235	5844.855	17109.06	CR I	0.3			MIT
5838.68	5840.30	17122.4	CR I	0.6			MIT
5801.193	5802.802	17233.05	CR	0.5			MIT
5798.473	5800.081	17241.14	CR I	0.3W			MIT
5797.898	5799.506	17242.85	CR I	0.6			MIT
5796.757	5798.364	17246.24	CR	0.8			MIT
5791.781	5793.387	17261.06	CR	0.7			MIT
5791.005	5792.611	17263.37	CR I	1.6J			MIT
5788.389	5789.994	17271.18	CR I	0.9			MIT
5787.99	5789.60	17272.4	CR	1.7J			MIT

AIR WAVELENGTH (ANGSTRMS)	VACUUM WAVELENGTH (ANGSTRMS)	WAVENUMBER (KAYSERS)	LMENT	LOG ARC	INTENSITY SPK	INTENSITY DIS	REF
5785.820	5787.424	17278.84	CR I	1.2			MIT
5785.002	5786.606	17281.29	CR	1.3			MIT
5783.934	5785.538	17284.48	CR	1.5H			MIT
5783.112	5784.716	17286.93	CR	1.5H			MIT
5781.806	5783.409	17290.84	CR	1.3			MIT
5781.195	5782.798	17292.67	CR I	1.3			MIT
5774.774	5776.375	17311.89	CR	0.3			MIT
5772.676	5774.277	17318.19	CR	0.8			MIT
5753.692	5755.288	17375.33	CR	1.2			MIT
5746.432	5748.026	17397.28	CR	1.1S			MIT
5738.554	5740.146	17421.16	CR	0.6			MIT
5736.632	5738.223	17427.00	CR	0.5			MIT
5729.203	5730.792	17449.59	CR	0.9L			MIT
5719.821	5721.408	17478.22	CR I	1.0			MIT
5712.778	5714.363	17499.76	CR I	1.2			MIT
5712.635	5714.220	17500.20	CR	0.5			MIT
5702.307	5703.889	17531.90	CR	1.3			MIT
5700.514	5702.096	17537.41	CR	0.9			MIT
5698.330	5699.911	17544.13	CR I	1.5	0.3		MIT
5694.730	5696.310	17555.22	CR I	1.5			MIT
5683.514	5685.091	17589.87	CR	0.5			MIT
5682.483	5684.060	17593.06	CR I	0.8J			MIT
5681.198	5682.774	17597.04	CR	0.5J			MIT
5674.180	5675.755	17618.80	CR	0.6			MIT
5664.040	5665.612	17650.34	CR	1.3			MIT
5658.633	5660.203	17667.21	CR	0.7S			MIT
5649.371	5650.939	17696.17	CR I	1.0H			MIT
5642.362	5643.928	17718.16	CR I	0.6			MIT
5638.182	5639.747	17731.29	CR	0.8			MIT
5628.645	5630.207	17761.33	CR I	1.4	0.3		MIT
5620.66	5622.22	17786.6	CR	0.5			MIT
5574.426	5575.974	17934.09	CR	0.9			MIT
5568.290	5569.836	17953.85	CR	0.5			MIT
5554.281	5555.824	17999.13	CR	0.5S			MIT
5509.927	5511.458	18144.02	CR	0.8			MIT
5508.212	5509.742	18149.67	CR	0.8			MIT
5480.502	5482.025	18241.44	CR	1.3			MIT
5463.974	5465.492	18296.61	CR	1.2S			MIT
5442.413	5443.926	18369.10	CR	1.3			MIT
5432.347	5433.857	18403.13	CR	1.0			MIT
5409.791	5411.295	18479.86	CR I	2.5G	1.5		MIT
5405.004	5406.507	18496.23	CR	1.3			MIT
5400.608	5402.109	18511.29	CR	1.5	0.0		MIT
5391.350	5392.849	18543.07	CR	1.2			MIT
5390.394	5391.893	18546.36	CR	1.3			MIT
5387.573	5389.071	18556.07	CR	1.3			MIT
5386.978	5388.476	18558.12	CR	1.3			MIT
5373.715	5375.209	18603.93	CR	0.9			MIT
5370.356	5371.849	18615.56	CR	1.0			MIT
5368.546	5370.039	18621.84	CR	1.3			MIT
5348.319	5349.807	18692.26	CR I	2.2G	1.2		MIT
5345.807	5347.294	18701.05	CR I	2.5G	1.4		MIT
5344.761	5346.248	18704.71	CR I	1.2			MIT
5340.437	5341.922	18719.85	CR I	1.7			MIT
5329.719	5331.202	18757.50	CR I	0.7	0.3		MIT
5328.339	5329.821	18762.35	CR	0.7H	0.9		MIT
5318.775	5320.255	18796.09	CR I	1.5			MIT
5312.878	5314.356	18816.96	CR I	1.6			MIT
5307.281	5308.758	18836.80	CR I	1.1			MIT
5304.211	5305.687	18847.70	CR I	1.3			MIT
5300.749	5302.224	18860.01	CR I	1.4	0.6		MIT
5298.269	5299.743	18868.84	CR I	1.2G	1.4		MIT
5297.976	5299.450	18869.88	CR I	0.7H	0.0H		MIT
5297.360	5298.834	18872.08	CR I	0.7H	0.3H		MIT
5296.686	5298.160	18874.48	CR I	1.2F	1.2		MIT
5293.383	5294.856	18886.25	CR	0.7			MIT
5292.865	5294.338	18888.10	CR	0.5			MIT
5287.188	5288.659	18908.38	CR I	1.6			MIT
5280.289	5281.758	18933.09	CR	1.2	1.3		MIT
5278.262	5279.731	18940.36	CR	0.8			MIT

AIR WAVELENGTH (ANGSTRMS)	VACUUM WAVELENGTH (ANGSTRMS)	WAVENUMBER (KAYSERS)	LMENT	LOG ARC	INTENSITY SPK	DIS	REF
5276.03	5277.50	18948.4	CR I	0.7H	0.5H		HL
5275.689	5277.157	18949.60	CR I	0.6H	0.0J		MIT
5275.171	5276.639	18951.46	CR I	0.5	0.5		MIT
5273.439	5274.907	18957.68	CR I	1.2			MIT
5272.010	5273.477	18962.82	CR I	1.4	0.0		MIT
5265.722	5267.188	18985.46	CR I	1.5	1.0		MIT
5265.160	5266.625	18987.49	CR I	1.2			MIT
5264.152	5265.617	18991.13	CR I	2.0R	1.3		MIT
5263.750	5265.215	18992.58	CR	0.6			HL
5261.754	5263.218	18999.78	CR I	1.4			MIT
5255.132	5256.595	19023.72	CR I	1.3	0.5		MIT
5254.918	5256.381	19024.50	CR I	1.3			MIT
5247.564	5249.025	19051.16	CR I	1.8	1.2		MIT
5243.395	5244.855	19066.31	CR I	1.7			MIT
5242.355	5243.814	19070.09	CR	0.3			MIT
5241.458	5242.917	19073.35	CR	1.1			MIT
5240.468	5241.927	19076.96	CR I	1.3H			MIT
5238.971	5240.429	19082.41	CR I	1.8			MIT
5237.34	5238.80	19088.4	CR II	0.3	0.9		MIT
5230.228	5231.684	19114.30	CR I	1.3			MIT
5228.082	5229.537	19122.15	CR	1.3			MIT
5226.891	5228.346	19126.51	CR I	1.2			MIT
5225.821	5227.276	19130.42	CR	1.2			MIT
5225.032	5226.487	19133.31	CR I	0.8	0.0		MIT
5224.941	5226.396	19133.65	CR I	1.3	0.6		MIT
5224.541	5225.996	19135.11	CR I	1.1			MIT
5224.082	5225.536	19136.79	CR I	1.1			MIT
5222.676	5224.130	19141.94	CR I	1.3			MIT
5221.753	5223.207	19145.33	CR	1.4			MIT
5220.912	5222.366	19148.41	CR	1.0			MIT
5214.127	5215.579	19173.33	CR	1.3			MIT
5212.239	5213.690	19180.27	CR I	0.6H			MIT
5208.436	5209.886	19194.28	CR I	2.7G	2.0		MIT
5206.039	5207.489	19203.11	CR I	2.7G	2.3		MIT
5204.518	5205.967	19208.73	CR I	2.6G	2.0		MIT
5200.188	5201.636	19224.72	CR I	1.5			MIT
5196.443	5197.890	19238.57	CR I	1.7	0.5		MIT
5193.488	5194.934	19249.52	CR	1.2			MIT
5192.000	5193.446	19255.04	CR I	1.7			MIT
5184.590	5186.034	19282.56	CR I	1.8	0.0		MIT
5177.430	5178.872	19309.22	CR I	1.7L			MIT
5166.227	5167.666	19351.10	CR	1.9	0.3		MIT
5161.765	5163.203	19367.82	CR	1.3			MIT
5144.672	5146.105	19432.17	CR I	1.5			MIT
5142.263	5143.696	19441.27	CR I	1.1			MIT
5139.654	5141.086	19451.14	CR	1.7	0.0		MIT
5123.465	5124.893	19512.60	CR I	0.9			MIT
5122.121	5123.548	19517.72	CR I	1.3			MIT
5113.130	5114.555	19552.04	CR I	1.4			MIT
5112.490	5113.915	19554.49	CR I	1.2			MIT
5110.751	5112.175	19561.15	CR I	1.6			MIT
5091.890	5093.309	19633.60	CR I	1.4			MIT
5078.711	5080.127	19684.55	CR	1.3			MIT
5072.920	5074.334	19707.02	CR I	1.5			MIT
5068.290	5069.703	19725.02	CR	1.3			MIT
5067.714	5069.127	19727.26	CR I	1.7			MIT
5065.910	5067.322	19734.29	CR I	1.5			MIT
5051.900	5053.309	19789.01	CR I	1.7			MIT
5048.752	5050.160	19801.35	CR I	1.5			MIT
5021.903	5023.304	19907.22	CR I	1.3			MIT
5019.32	5020.72	19917.5	CR	1.3			GS
5013.316	5014.714	19941.31	CR I	1.8	0.3		MIT
5005.24	5006.64	19973.5	CR	0.0	0.5V		GS
5004.38	5005.78	19976.9	CR	0.8H			MIT
4985.959	4987.350	20050.73	CR	1.4			MIT
4964.928	4966.313	20135.66	CR I	0.9	0.5		MIT
4954.811	4956.194	20176.77	CR	2.0	0.9		MIT
4949.617	4950.998	20197.95	CR	0.9			MIT
4944.576	4945.956	20218.54	CR	1.5			MIT
4942.495	4943.875	20227.05	CR I	2.1	0.5		MIT

AIR WAVELENGTH (ANGSTRMS)	VACUUM WAVELENGTH (ANGSTRMS)	WAVENUMBER (KAYSERS)	LMENT	LOG ARC	INTENSITY SPK	DIS	REF
4936.334	4937.712	20252.30	CR	2.3	0.7		MIT
4931.109	4932.485	20273.75	CR	1.2			MIT
4930.183	4931.559	20277.56	CR	1.5			MIT
4923.284	4924.658	20305.98	CR	1.0			MIT
4922.267	4923.641	20310.17	CR	2.3	1.6		MIT
4920.945	4922.319	20315.63	CR	1.7			MIT
4919.460	4920.833	20321.76	CR	0.8			MIT
4905.050	4906.420	20381.46	CR	1.5			MIT
4903.239	4904.608	20388.99	CR I	2.1			MIT
4894.359	4895.726	20425.98	CR	1.5			MIT
4888.530	4889.895	20450.34	CR I	2.0			MIT
4887.715	4889.080	20453.75	CR	1.2			MIT
4887.013	4888.378	20456.68	CR	2.1	1.5		MIT
4885.957	4887.321	20461.10	CR	1.2			MIT
4885.776	4887.140	20461.86	CR I	1.8			MIT
4884.949	4886.313	20465.33	CR I	1.4			MIT
4880.036	4881.399	20485.93	CR	1.4			MIT
4876.366	4877.728	20501.35	CR II	0.3	1.2		MIT
4875.525	4876.887	20504.89	CR	1.1			MIT
4874.651	4876.012	20508.56	CR	1.3			MIT
4870.796	4872.156	20524.79	CR	2.2	1.4		MIT
4864.312	4865.671	20552.15	CR II	0.5	1.1		MIT
4861.842	4863.200	20562.59	CR	2.1	0.9		MIT
4861.205	4862.563	20565.29	CR I	1.9			MIT
4857.291	4858.648	20581.86	CR I	1.4			MIT
4855.146	4856.502	20590.95	CR I	1.3			MIT
4851.465	4852.820	20606.57	CR	1.5			MIT
4848.273	4849.627	20620.14	CR II	0.3	1.2		MIT
4847.177	4848.531	20624.80	CR	1.0			MIT
4846.301	4847.655	20628.53	CR	1.3			MIT
4843.153	4844.506	20641.94	CR	0.3			MIT
4841.796	4843.149	20647.72	CR	0.5			MIT
4840.440	4841.792	20653.51	CR	0.6			MIT
4838.442	4839.794	20662.04	CR	1.0			MIT
4836.857	4838.208	20668.81	CR I	1.9			MIT
4836.364	4837.715	20670.92	CR	0.8			MIT
4836.229	4837.580	20671.49	CR II		0.3H		MIT
4835.693	4837.044	20673.78	CR	0.8			MIT
4831.627	4832.977	20691.18	CR	1.4			MIT
4829.376	4830.725	20700.82	CR I	2.3	1.6		MIT
4824.124	4825.472	20723.36	CR II	0.6	1.5		MIT
4823.922	4825.270	20724.23	CR	1.4			MIT
4816.142	4817.488	20757.71	CR	1.5			MIT
4814.265	4815.610	20765.80	CR	2.0			MIT
4812.37	4813.71	20774.0	CR II		0.6		MIT
4810.733	4812.078	20781.04	CR	1.5			MIT
4809.289	4810.633	20787.28	CR	1.1			MIT
4806.255	4807.598	20800.41	CR I	1.9			MIT
4804.70	4806.04	20807.1	CR I	1.5			MIT
4801.030	4802.372	20823.04	CR	2.3	1.8		MIT
4797.715	4799.056	20837.43	CR	1.4			MIT
4796.889	4798.230	20841.02	CR I	0.7H			MIT
4796.169	4797.510	20844.15	CR	2.1	0.0H		MIT
4792.513	4793.853	20860.05	CR	2.3	1.6		MIT
4790.337	4791.676	20869.52	CR I	2.0	0.0		MIT
4789.32	4790.66	20874.0	CR I	2.5	2.0		ME
4787.752	4789.090	20880.79	CR	1.1			MIT
4783.079	4784.416	20901.19	CR	1.4H			MIT
4779.939	4781.275	20914.92	CR I	1.0			MIT
4777.579	4778.915	20925.25	CR I	1.0			MIT
4775.520	4776.855	20934.27	CR	0.9H			MIT
4775.141	4776.476	20935.94	CR	1.5			MIT
4774.557	4775.892	20938.50	CR I	1.3			MIT
4771.611	4772.945	20951.42	CR I	1.3			MIT
4770.670	4772.004	20955.56	CR I	1.5			MIT
4769.813	4771.147	20959.32	CR	0.8H			MIT
4767.860	4769.193	20967.91	CR	2.0	0.9		MIT
4766.633	4767.966	20973.30	CR I	1.9	0.8		MIT
4764.643	4765.975	20982.06	CR I	1.5			MIT
4764.294	4765.626	20983.60	CR	2.3	1.5		MIT

AIR WAVELENGTH (ANGSTRMS)	VACUUM WAVELENGTH (ANGSTRMS)	WAVENUMBER (KAYSERS)	LMENT	LOG ARC	INTENSITY SPK	DIS	REF
4761.242	4762.573	20997.05	CR	1.4			MIT
4759.907	4761.238	21002.94	CR I	1.2			MIT
4757.591	4758.921	21013.17	CR	1.5	0.0		MIT
4757.326	4758.656	21014.33	CR	1.4			MIT
4756.113	4757.443	21019.70	CR	2.5	2.0		MIT
4755.137	4756.467	21024.01	CR I	1.8			MIT
4754.743	4756.073	21025.75	CR	1.9			MIT
4752.897	4754.226	21033.92	CR	0.9			MIT
4752.084	4753.413	21037.52	CR	2.0	1.6		MIT
4745.308	4746.635	21067.55	CR I	1.9	0.3		MIT
4743.112	4744.439	21077.31	CR	1.6			MIT
4737.350	4738.675	21102.95	CR	2.3	1.9		MIT
4736.151	4737.476	21108.29	CR	0.8			MIT
4730.711	4732.034	21132.56	CR	2.0	1.7		MIT
4729.723	4731.046	21136.98	CR	1.5	0.8		MIT
4727.153	4728.475	21148.47	CR	1.9	1.3		MIT
4724.416	4725.738	21160.72	CR	2.1	1.0		MIT
4723.102	4724.423	21166.60	CR	2.1	0.9		MIT
4718.429	4719.749	21187.57	CR I	2.3	2.2		MIT
4717.688	4719.008	21190.90	CR	1.3	0.0		MIT
4713.993	4715.312	21207.50	CR	1.2	0.0		MIT
4708.040	4709.357	21234.32	CR I	2.3	2.2		MIT
4707.754	4709.071	21235.61	CR	0.9	0.0		MIT
4706.102	4707.419	21243.06	CR	1.5	0.0		MIT
4700.608	4701.923	21267.89	CR I	1.7	0.6		MIT
4699.589	4700.904	21272.50	CR	1.5	0.0		MIT
4698.947	4700.262	21275.41	CR	1.2			MIT
4698.615	4699.930	21276.91	CR I	1.6	0.9		MIT
4698.456	4699.771	21277.63	CR I	1.8	1.1		MIT
4697.395	4698.710	21282.44	CR	1.2	0.0		MIT
4697.062	4698.376	21283.95	CR I	1.7	1.1		MIT
4695.153	4696.467	21292.60	CR	1.7	0.6		MIT
4693.949	4695.263	21298.06	CR	1.7	1.3		MIT
4689.374	4690.686	21318.84	CR I	1.9	1.5		MIT
4686.193	4687.505	21333.31	CR	1.3	0.0		MIT
4684.605	4685.916	21340.54	CR	1.3			MIT
4680.870	4682.180	21357.57	CR	1.8	0.9		MIT
4680.544	4681.854	21359.06	CR I	1.7	1.4		MIT
4669.336	4670.643	21410.33	CR I	1.7	1.3		MIT
4666.512	4667.818	21423.28	CR I	1.7	1.4		MIT
4666.215	4667.521	21424.65	CR	1.5	0.9		HL
4665.902	4667.208	21426.08	CR	1.3	1.0		MIT
4664.798	4666.104	21431.16	CR I	1.8	1.3		MIT
4663.832	4665.138	21435.59	CR I	1.7	1.2		MIT
4663.328	4664.633	21437.91	CR I	1.6	1.4		MIT
4662.41	4663.72	21442.1	CR	0.7			MIT
4656.189	4657.493	21470.78	CR	1.7	0.6		MIT
4654.736	4656.039	21477.48	CR I	1.8	0.9		MIT
4652.158	4653.461	21489.38	CR I	2.3G	2.2		MIT
4651.285	4652.587	21493.42	CR I	2.0	2.0		MIT
4649.461	4650.763	21501.85	CR	1.8	0.5		MIT
4648.868	4650.170	21504.59	CR	1.7	0.5		MIT
4648.126	4649.427	21508.02	CR	1.6	0.8		MIT
4646.808	4648.109	21514.12	CR I	1.5	0.5		MIT
4646.495	4647.796	21515.57	CR	0.7	0.0		MIT
4646.174	4647.475	21517.06	CR I	2.0	2.2		MIT
4642.011	4643.311	21536.36	CR	1.2	0.0		MIT
4639.715	4641.014	21547.01	CR	1.0	0.3		MIT
4637.772	4639.071	21556.04	CR	1.3	0.8		MIT
4637.182	4638.481	21558.78	CR	1.3	0.9		MIT
4635.444	4636.742	21566.87	CR	0.7			MIT
4634.089	4635.387	21573.17	CR II	0.7	1.9H		MIT
4633.286	4634.584	21576.91	CR I	1.3	0.8		MIT
4632.180	4633.477	21582.06	CR	1.4	0.9		MIT
4628.473	4629.769	21599.35	CR I	1.2			MIT
4627.372	4628.668	21604.49	CR	0.7			MIT
4626.855	4628.151	21606.90	CR	1.0			MIT
4626.188	4627.484	21610.02	CR I	2.0	2.1		MIT
4625.925	4627.221	21611.25	CR	0.9	0.8		MIT
4625.297	4626.592	21614.18	CR	0.8			MIT

AIR WAVELENGTH (ANGSTRMS)	VACUUM WAVELENGTH (ANGSTRMS)	WAVENUMBER (KAYSERS)	LMENT	LOG ARC	INTENSITY SPK	DIS	REF
4624.575	4625.870	21617.55	CR	1.2	0.8		MIT
4622.761	4624.056	21626.04	CR	1.3	1.0		MIT
4622.491	4623.786	21627.30	CR	1.5	1.5		MIT
4621.963	4623.258	21629.77	CR	1.7	1.6		MIT
4621.893	4623.188	21630.10	CR	1.3			MIT
4621.077	4622.371	21633.92	CR	1.0D			MIT
4619.551	4620.845	21641.06	CR	1.7	1.5		MIT
4618.829	4620.123	21644.45	CR II	0.8	1.9H		MIT
4616.665	4617.958	21654.59	CR II		1.7		MIT
4616.137	4617.430	21657.07	CR I	2.5R	2.3		MIT
4614.751	4616.044	21663.57	CR	1.1	0.0		MIT
4614.523	4615.816	21664.64	CR	1.2	0.6		MIT
4614.154	4615.446	21666.38	CR	1.2	0.0		MIT
4613.373	4614.665	21670.04	CR I	2.2	1.8		MIT
4611.968	4613.260	21676.65	CR	1.2	0.6		MIT
4611.044	4612.336	21680.99	CR	1.0			MIT
4609.894	4611.185	21686.40	CR	1.2	0.5		MIT
4606.375	4607.665	21702.97	CR	1.2	0.5		MIT
4605.816	4607.106	21705.60	CR	1.5			MIT
4604.620	4605.910	21711.24	CR	1.2			MIT
4601.021	4602.310	21728.22	CR	1.0	0.6		MIT
4600.752	4602.041	21729.49	CR I	2.2	2.2		MIT
4600.104	4601.393	21732.55	CR	1.3	1.7		MIT
4599.003	4600.291	21737.75	CR	1.0			MIT
4598.441	4599.729	21740.41	CR	1.3	0.6		MIT
4596.927	4598.215	21747.57	CR	0.8			MIT
4596.380	4597.668	21750.16	CR	1.0			MIT
4595.590	4596.878	21753.90	CR	1.7	1.8		MIT
4595.04	4596.33	21756.5	CR	1.2			MIT
4594.403	4595.690	21759.52	CR	1.0			MIT
4593.831	4595.118	21762.23	CR	1.1			MIT
4592.537	4593.824	21768.36	CR	1.4	0.0		MIT
4592.058	4593.345	21770.63	CR II	0.5	1.5H		MIT
4591.394	4592.680	21773.78	CR I	2.3	2.1		ME
4589.935	4591.221	21780.70	CR II		0.6H		MIT
4589.008	4590.294	21785.10	CR	1.0			MIT
4588.217	4589.503	21788.85	CR II	1.0	2.8H		MIT
4587.866	4589.152	21790.52	CR I	1.5			MIT
4586.138	4587.423	21798.73	CR	1.4	0.8		MIT
4585.088	4586.373	21803.72	CR	1.0	0.3		MIT
4584.934	4586.219	21804.46	CR I	1.0	0.0		MIT
4584.095	4585.380	21808.45	CR	1.3	0.0		MIT
4583.897	4585.181	21809.39	CR I	1.0			MIT
4582.451	4583.735	21816.27	CR	0.9			MIT
4582.065	4583.349	21818.11	CR	0.7			MIT
4581.063	4582.347	21822.88	CR	1.2	0.0		MIT
4580.056	4581.339	21827.68	CR I	2.5	2.1		MIT
4579.619	4580.902	21829.76	CR	0.7			MIT
4578.334	4579.617	21835.89	CR	1.4	0.3		MIT
4576.770	4578.053	21843.35	CR	1.0			MIT
4575.121	4576.403	21851.22	CR	1.4	0.7		MIT
4574.212	4575.494	21855.56	CR	1.0			MIT
4571.676	4572.957	21867.69	CR	1.7	1.6		MIT
4571.105	4572.386	21870.42	CR I	1.0	0.0		MIT
4570.534	4571.815	21873.15	CR	0.8			MIT
4570.335	4571.616	21874.10	CR I	1.2J			MIT
4569.644	4570.925	21877.41	CR	1.7	1.0		MIT
4569.530	4570.811	21877.96	CR	1.2	0.6		MIT
4566.602	4567.882	21891.99	CR I	1.0			MIT
4565.512	4566.792	21897.21	CR I	1.3	1.5		MIT
4564.166	4565.445	21903.67	CR	1.2	1.0		MIT
4563.657	4564.936	21906.11	CR	1.2	0.5		MIT
4563.245	4564.524	21908.09	CR	1.1	0.3		MIT
4561.485	4562.764	21916.54	CR	1.2H			MIT
4558.659	4559.937	21930.13	CR II	1.3	2.8J		MIT
4558.251	4559.529	21932.09	CR	1.2			MIT
4556.169	4557.446	21942.11	CR	1.6	1.1		MIT
4555.296	4556.573	21946.32	CR	1.2			MIT
4555.092	4556.369	21947.30	CR	1.2	1.7		MIT
4555.029	4556.306	21947.60	CR II		1.6H		MIT

AIR WAVELENGTH (ANGSTRMS)	VACUUM WAVELENGTH (ANGSTRMS)	WAVENUMBER (KAYSERS)	LMENT	LOG ARC	INTENSITY SPK	DIS	REF
4554.830	4556.107	21948.56	CR	1.4	0.3		MIT
4553.949	4555.226	21952.81	CR	1.3	0.5		MIT
4546.572	4547.847	21988.43	CR	0.5			MIT
4545.956	4547.230	21991.41	CR I	2.3	2.1		MIT
4545.335	4546.609	21994.41	CR I	1.4	1.1		MIT
4544.619	4545.893	21997.88	CR I	2.0	1.8		MIT
4543.74	4545.01	22002.1	CR	1.3	0.3		MIT
4542.621	4543.895	22007.55	CR	1.5	0.9		MIT
4541.513	4542.786	22012.92	CR	1.5	0.8		MIT
4541.071	4542.344	22015.06	CR I	1.5	0.9		MIT
4540.719	4541.992	22016.77	CR	1.6D	1.6		MIT
4540.502	4541.775	22017.82	CR I	1.6D	1.6		MIT
4539.788	4541.061	22021.29	CR I	1.6	1.4		MIT
4535.721	4536.993	22041.03	CR I	2.1	2.0		MIT
4535.146	4536.418	22043.83	CR I	1.7	1.5		MIT
4532.750	4534.021	22055.48	CR	1.2	0.3		MIT
4530.739	4532.009	22065.27	CR I	2.2	2.1		MIT
4529.851	4531.121	22069.59	CR I	1.4	0.9		MIT
4527.471	4528.741	22081.19	CR	1.2D	0.8		MIT
4527.339	4528.609	22081.84	CR I	1.2D	0.9		MIT
4526.466	4527.735	22086.10	CR I	1.7	1.5		MIT
4526.108	4527.377	22087.84	CR	1.3	0.8		MIT
4524.841	4526.110	22094.03	CR	1.3	0.5		MIT
4522.018	4523.286	22107.82	CR	1.1	0.0		MIT
4521.141	4522.409	22112.11	CR	1.4	1.0		MIT
4519.828	4521.096	22118.53	CR	1.0			MIT
4518.628	4519.895	22124.41	CR	0.8			MIT
4515.440	4516.706	22140.03	CR	1.4	1.0		MIT
4514.531	4515.797	22144.48	CR I	1.5	0.9		MIT
4514.373	4515.639	22145.26	CR	1.0	0.9		MIT
4513.216	4514.482	22150.94	CR	1.0			MIT
4512.608	4513.874	22153.92	CR	0.9J			MIT
4511.903	4513.168	22157.38	CR	1.9	2.0		MIT
4510.005	4511.270	22166.71	CR	1.2	0.0		MIT
4508.744	4510.009	22172.91	CR	0.7			MIT
4506.853	4508.117	22182.21	CR	1.5	1.5		MIT
4503.908	4505.171	22196.71	CR	0.3			MIT
4503.045	4504.308	22200.97	CR	1.0	0.0		MIT
4501.788	4503.051	22207.17	CR	1.2	0.8		MIT
4501.112	4502.375	22210.50	CR	1.6	1.5		MIT
4500.295	4501.557	22214.53	CR	1.7	1.5		MIT
4499.269	4500.531	22219.60	CR	0.6			MIT
4498.730	4499.992	22222.26	CR	1.5	1.2		MIT
4496.862	4498.123	22231.49	CR I	2.3	2.3		MIT
4495.275	4496.536	22239.34	CR	1.2	0.3		MIT
4492.312	4493.572	22254.01	CR	1.5	1.2		MIT
4491.858	4493.118	22256.26	CR	1.7	0.6		MIT
4491.678	4492.938	22257.15	CR I	1.4D	0.0		MIT
4490.555	4491.815	22262.72	CR I	1.2H			MIT
4489.471	4490.731	22268.09	CR	1.4	1.2		MIT
4488.051	4489.310	22275.14	CR	1.4	1.2		MIT
4484.683	4485.941	22291.87	CR	0.8			MIT
4482.878	4484.136	22300.84	CR	1.4	1.1		MIT
4481.431	4482.688	22308.04	CR	1.0			MIT
4480.263	4481.520	22313.86	CR	0.5	0.0		MIT
4478.872	4480.129	22320.79	CR	0.5			MIT
4477.982	4479.239	22325.22	CR	0.7			MIT
4477.510	4478.766	22327.58	CR	0.7			MIT
4477.053	4478.309	22329.86	CR I	1.3			MIT
4475.345	4476.601	22338.38	CR I	1.6	0.8H		MIT
4473.782	4475.037	22346.18	CR I	1.4	0.0		MIT
4469.837	4471.091	22365.90	CR	0.5			MIT
4468.649	4469.903	22371.85	CR	0.5			MIT
4468.365	4469.619	22373.27	CR I	0.7			MIT
4467.561	4468.815	22377.30	CR I	1.3	0.3		MIT
4466.165	4467.418	22384.29	CR I	1.2	0.0		MIT
4465.357	4466.610	22388.34	CR I	1.3	1.0		MIT
4464.907	4466.160	22390.60	CR I	1.2	0.9		MIT
4464.669	4465.922	22391.79	CR I	1.2	0.0		MIT
4462.774	4464.026	22401.30	CR I	1.3	0.3		MIT

AIR WAVELENGTH (ANGSTRMS)	VACUUM WAVELENGTH (ANGSTRMS)	WAVENUMBER (KAYSERS)	LMENT	LOG ARC	INTENSITY SPK	DIS	REF
4461.302	4462.554	22408.69	CR	0.7			MIT
4460.769	4462.021	22411.37	CR I	1.4			MIT
4459.738	4460.990	22416.55	CR I	1.4	1.4		MIT
4459.391	4460.643	22418.29	CR I	1.3			MIT
4458.538	4459.789	22422.58	CR I	1.7	2.1		MIT
4455.458	4456.709	22438.08	CR I	0.9			MIT
4443.707	4444.954	22497.42	CR	1.2	0.3		MIT
4442.268	4443.515	22504.71	CR	1.2	0.5		MIT
4433.968	4435.213	22546.83	CR	1.0	0.0		MIT
4432.175	4433.419	22555.95	CR	1.5	1.2		MIT
4430.486	4431.730	22564.55	CR	1.2	0.9		MIT
4429.938	4431.182	22567.34	CR	1.2	0.7		MIT
4428.501	4429.744	22574.67	CR I	1.4	0.8		MIT
4427.725	4428.968	22578.62	CR I	0.9			MIT
4425.129	4426.372	22591.87	CR	1.2	0.0		MIT
4424.281	4425.523	22596.20	CR I	1.4	1.5		MIT
4424.075	4425.317	22597.25	CR	1.0	0.3		MIT
4423.318	4424.560	22601.12	CR	1.2	0.8		MIT
4422.697	4423.939	22604.29	CR	1.0	0.8		MIT
4420.942	4422.183	22613.26	CR I	0.9			MIT
4419.103	4420.344	22622.67	CR	0.9			MIT
4414.355	4415.595	22647.01	CR I	1.2H			MIT
4413.866	4415.106	22649.51	CR	1.4	1.2		MIT
4412.993	4414.232	22654.00	CR	0.7			MIT
4412.250	4413.489	22657.81	CR I	1.5	1.0		MIT
4411.093	4412.332	22663.75	CR I	1.2D	1.1		MIT
4410.967	4412.206	22664.40	CR	1.0D	0.3		MIT
4410.304	4411.543	22667.81	CR I	1.4	0.9		MIT
4409.553	4410.791	22671.67	CR	0.7			MIT
4406.666	4407.904	22686.52	CR	0.7			MIT
4406.273	4407.511	22688.54	CR I	1.0			MIT
4403.498	4404.735	22702.84	CR	1.2D	1.4		MIT
4403.372	4404.609	22703.49	CR	1.2D	0.8		MIT
4399.823	4401.059	22721.80	CR I	1.3	0.5		MIT
4397.251	4398.486	22735.09	CR I	1.4	0.7		MIT
4395.417	4396.652	22744.58	CR	1.2	0.0		MIT
4393.534	4394.768	22754.33	CR	0.7			MIT
4391.753	4392.987	22763.56	CR I	1.7	1.5		MIT
4387.496	4388.729	22785.64	CR	1.2D	1.2		MIT
4387.380	4388.613	22786.24	CR	1.0D	0.3		MIT
4384.977	4386.209	22798.73	CR I	2.2	2.3		MIT
4382.853	4384.084	22809.78	CR I	1.1	0.3		MIT
4381.112	4382.343	22818.84	CR I	1.5	1.4		MIT
4379.782	4381.013	22825.77	CR I	1.2	0.3		MIT
4378.322	4379.552	22833.38	CR	0.7	0.0		MIT
4377.549	4378.779	22837.42	CR	1.4	1.0		MIT
4376.798	4378.028	22841.33	CR	1.3	1.3		MIT
4375.333	4376.562	22848.98	CR	1.4	1.5		MIT
4374.158	4375.387	22855.12	CR	1.7	1.8		MIT
4373.656	4374.885	22857.74	CR	1.2	1.1		MIT
4373.254	4374.483	22859.84	CR I	1.7	1.7		MIT
4371.279	4372.507	22870.17	CR I	2.3	2.2		MIT
4368.910	4370.138	22882.57	CR	0.7			MIT
4368.252	4369.480	22886.02	CR I	1.2	0.8		MIT
4364.142	4365.368	22907.57	CR I	0.9			MIT
4363.134	4364.360	22912.87	CR	1.4	1.5		MIT
4359.992	4361.217	22929.38	CR	1.2	0.3		MIT
4359.631	4360.856	22931.28	CR I	2.3	2.2		MIT
4357.525	4358.750	22942.36	CR	1.1	0.6		MIT
4356.760	4357.985	22946.39	CR I	1.2	0.9		MIT
4356.287	4357.511	22948.88	CR	0.9	0.0		MIT
4353.983	4355.207	22961.02	CR	1.3	0.5		HL
4351.770	4352.993	22972.70	CR I	2.5	2.5		MIT
4351.051	4352.274	22976.49	CR I	2.0	2.2		MIT
4348.835	4350.057	22988.20	CR	0.6J			MIT
4347.504	4348.726	22995.24	CR	1.5	0.0		MIT
4345.809	4347.031	23004.21	CR	1.2			MIT
4345.085	4346.306	23008.04	CR	1.5	0.5		MIT
4344.507	4345.728	23011.10	CR I	2.6R	2.5		MIT
4343.163	4344.384	23018.22	CR I	1.8	1.1		MIT

AIR WAVELENGTH (ANGSTRMS)	VACUUM WAVELENGTH (ANGSTRMS)	WAVENUMBER (KAYSERS)	LMENT	LOG ARC	INTENSITY SPK	INTENSITY DIS	REF
4342.009	4343.230	23024.34	CR	0.9			MIT
4341.450	4342.671	23027.30	CR I	1.1			MIT
4341.185	4342.405	23028.71	CR	0.9			MIT
4340.130	4341.350	23034.31	CR I	1.9	1.5		MIT
4339.718	4340.938	23036.50	CR I	2.2	2.2		MIT
4339.450	4340.670	23037.92	CR I	2.5R	2.5		MIT
4338.799	4340.019	23041.37	CR	1.7	0.9		MIT
4338.412	4339.632	23043.43	CR	1.5	0.6		MIT
4337.566	4338.785	23047.92	CR I	2.7	2.5		MIT
4337.267	4338.486	23049.51	CR	0.9	0.6		MIT
4332.569	4333.787	23074.51	CR	2.1	0.3		MIT
4325.075	4326.291	23114.49	CR	2.1	2.1		MIT
4323.523	4324.739	23122.78	CR	2.0	1.2		MIT
4321.617	4322.832	23132.98	CR	1.8	0.5		MIT
4321.238	4322.453	23135.01	CR	1.8	0.3		MIT
4320.592	4321.807	23138.47	CR I	2.1	0.8		MIT
4319.641	4320.856	23143.56	CR I	2.0	1.3		MIT
4319.321	4320.536	23145.28	CR	1.2			MIT
4312.469	4313.682	23182.05	CR	1.5	0.0		MIT
4309.712	4310.924	23196.88	CR	1.1			MIT
4307.486	4308.698	23208.87	CR	1.5	0.0		MIT
4305.453	4306.664	23219.83	CR I	2.2	1.3		MIT
4302.774	4303.984	23234.28	CR	1.6	0.3		MIT
4301.178	4302.388	23242.91	CR	2.0	1.4		MIT
4300.511	4301.721	23246.51	CR	2.0	1.3		MIT
4299.718	4300.928	23250.80	CR I	2.0	1.7		MIT
4297.738	4298.947	23261.51	CR	2.1	1.5		MIT
4297.050	4298.259	23265.24	CR I	2.0	1.2		MIT
4296.631	4297.840	23267.50	CR	1.2			MIT
4296.275	4297.484	23269.43	CR	1.2			MIT
4295.757	4296.965	23272.24	CR I	2.1	1.6		MIT
4293.565	4294.773	23284.12	CR I	1.7	0.9		MIT
4291.964	4293.172	23292.80	CR	1.5	1.3		MIT
4289.721	4290.928	23304.98	CR I	3.5G	2.9R		MIT
4288.381	4289.588	23312.26	CR	1.2	0.5		MIT
4284.904	4286.110	23331.18	CR I	1.2	1.0		MIT
4284.725	4285.931	23332.16	CR I	1.6	0.9		MIT
4284.217	4285.422	23334.92	CR I		1.5		MIT
4282.988	4284.193	23341.62	CR	1.1	0.0		MIT
4280.405	4281.609	23355.70	CR	1.9	1.7		MIT
4277.792	4278.996	23369.97	CR	1.4J			MIT
4275.973	4277.176	23379.91	CR	1.5	1.2		MIT
4274.803	4276.006	23386.31	CR I	3.6G	2.9R		MIT
4272.910	4274.112	23396.67	CR I	1.6	1.5		MIT
4271.061	4272.263	23406.80	CR	1.5	1.1		MIT
4269.951	4271.153	23412.88	CR	1.6	0.8		MIT
4269.296	4270.498	23416.48	CR		0.5		MIT
4268.788	4269.989	23419.26	CR	1.5	0.5		MIT
4266.821	4268.022	23430.06	CR	1.5	0.0		MIT
4263.141	4264.341	23450.28	CR	2.1	1.9		MIT
4262.356	4263.556	23454.60	CR	1.5	0.3		MIT
4262.133	4263.333	23455.83	CR	1.6	0.9		MIT
4261.907	4263.107	23457.07	CR II		1.5		MIT
4261.615	4262.815	23458.68	CR	1.5	0.9		MIT
4261.354	4262.553	23460.12	CR I	2.1	1.7		MIT
4259.158	4260.357	23472.21	CR I	1.5	0.0		MIT
4257.368	4258.566	23482.08	CR I	1.5	0.0		MIT
4256.620	4257.818	23486.21	CR I	1.1	0.0		MIT
4255.502	4256.700	23492.38	CR	1.5	1.5		MIT
4254.346	4255.544	23498.76	CR I	3.7G	3.0G		MIT
4252.661	4253.858	23508.07	CR II		1.0		MIT
4252.243	4253.440	23510.38	CR I	1.5	1.0		MIT
4248.709	4249.905	23529.94	CR	1.5	0.3		MIT
4248.344	4249.540	23531.96	CR I	1.5	0.3		MIT
4242.853	4244.048	23562.41	CR I	1.2			MIT
4242.381	4243.575	23565.03	CR II	0.6	1.7		MIT
4241.194	4242.388	23571.63	CR	1.0			MIT
4240.705	4241.899	23574.35	CR	2.3	1.5		MIT
4238.957	4240.151	23584.07	CR I	2.0	1.2		MIT
4237.710	4238.903	23591.01	CR I	1.8			MIT

AIR WAVELENGTH (ANGSTRMS)	VACUUM WAVELENGTH (ANGSTRMS)	WAVENUMBER (KAYSERS)	LMENT	LOG ARC	INTENSITY SPK	INTENSITY DIS	REF
4234.515	4235.707	23608.81	CR	1.8	0.3		MIT
4233.266	4234.458	23615.77	CR II		0.8		MIT
4232.866	4234.058	23618.00	CR I	1.8			MIT
4232.222	4233.414	23621.60	CR	1.8	0.7		MIT
4231.664	4232.856	23624.71	CR	0.5			MIT
4230.481	4231.672	23631.32	CR I	1.8	0.9		MIT
4226.758	4227.948	23652.13	CR	2.1	1.5		MIT
4224.514	4225.704	23664.70	CR	1.8	1.1		MIT
4223.470	4224.660	23670.55	CR I	1.2			MIT
4222.732	4223.921	23674.68	CR I	2.0	1.2		MIT
4221.572	4222.761	23681.19	CR	1.9	1.5		MIT
4217.626	4218.814	23703.34	CR I	2.2	1.8		MIT
4216.365	4217.553	23710.43	CR I	1.8	1.4		MIT
4213.179	4214.366	23728.36	CR	1.8	0.9		MIT
4212.660	4213.847	23731.29	CR I	1.9	0.9		MIT
4211.349	4212.535	23738.67	CR	2.0	1.5		MIT
4210.751	4211.937	23742.04	CR	1.1			MIT
4209.756	4210.942	23747.66	CR	1.9	1.3		MIT
4209.368	4210.554	23749.84	CR	2.0	1.6		MIT
4208.357	4209.543	23755.55	CR	2.0	1.4		MIT
4206.899	4208.084	23763.78	CR	1.9	1.4		MIT
4204.471	4205.656	23777.51	CR	1.9	1.5		MIT
4204.198	4205.382	23779.05	CR I	1.7	0.8		MIT
4203.590	4204.774	23782.49	CR I	2.0	1.3		MIT
4200.103	4201.286	23802.23	CR	1.9	0.9		MIT
4198.525	4199.708	23811.18	CR	2.0	1.5		MIT
4197.234	4198.417	23818.50	CR	1.8	1.4		MIT
4194.951	4196.133	23831.47	CR	1.8	1.4		MIT
4193.662	4194.844	23838.79	CR	2.0	1.4		MIT
4192.549	4193.730	23845.12	CR	0.8			MIT
4192.103	4193.284	23847.66	CR	1.6	1.2		MIT
4191.750	4192.931	23849.66	CR	1.7	0.8		MIT
4191.271	4192.452	23852.39	CR I	1.8	1.2		MIT
4190.660	4191.841	23855.87	CR I	1.1			MIT
4190.131	4191.312	23858.88	CR	1.6	1.2		MIT
4186.359	4187.539	23880.38	CR	1.7	1.0		MIT
4185.345	4186.524	23886.16	CR	1.5	0.5		MIT
4184.895	4186.074	23888.73	CR	1.5	1.0		MIT
4179.959	4181.137	23916.94	CR	1.4	0.0		MIT
4179.459	4180.637	23919.80	CR		0.9		MIT
4179.257	4180.435	23920.96	CR	2.0	1.6		MIT
4177.899	4179.077	23928.73	CR	1.6H	0.0H		MIT
4176.686	4177.863	23935.68	CR	1.2	0.3		MIT
4175.945	4177.122	23939.93	CR	1.6	1.0		MIT
4175.227	4176.404	23944.04	CR	1.5	0.9		MIT
4174.941	4176.118	23945.69	CR I	1.0	0.5		MIT
4174.795	4175.972	23946.52	CR I	2.0	1.6		MIT
4174.326	4175.503	23949.21	CR I	0.5			MIT
4172.769	4173.945	23958.15	CR	1.5	1.2		MIT
4171.675	4172.851	23964.43	CR	1.8	0.9		MIT
4170.202	4171.378	23972.90	CR	1.8	1.2		MIT
4169.838	4171.013	23974.99	CR	1.9	1.4		MIT
4165.519	4166.693	23999.85	CR	1.9	1.5		MIT
4163.620	4164.794	24010.79	CR	2.0	1.7		M
4161.415	4162.588	24023.51	CR	1.7	1.5		MIT
4153.816	4154.987	24067.46	CR I	1.7	1.5		MIT
4153.067	4154.238	24071.80	CR I	1.6	0.8		MIT
4152.775	4153.946	24073.50	CR	1.7	1.1		MIT
4146.695	4147.864	24108.79	CR	1.4	0.3		MIT
4146.454	4147.623	24110.19	CR II	1.3	0.0		MIT
4146.235	4147.404	24111.47	CR	1.3	0.3		MIT
4145.804	4146.973	24113.97	CR		1.3		MIT
4142.50	4143.67	24133.2	CR	1.0	0.0		EX
4142.193	4143.361	24135.00	CR	1.5	0.9		MIT
4134.389	4135.555	24180.55	CR	1.4	0.5		MIT
4132.441	4133.607	24191.95	CR II		0.3		MIT
4131.360	4132.525	24198.28	CR	1.5	1.3		MIT
4129.368	4130.533	24209.95	CR I	1.0			HL
4128.405	4129.570	24215.60	CR	1.0	0.8		MIT
4127.643	4128.807	24220.07	CR I	1.5	1.0		MIT

AIR WAVELENGTH (ANGSTRMS)	VACUUM WAVELENGTH (ANGSTRMS)	WAVENUMBER (KAYSERS)	LMENT	LOG ARC	INTENSITY SPK	DIS	REF
4127.302	4128.466	24222.07	CR I	1.3	1.0		MIT
4126.925	4128.089	24224.28	CR	1.5	0.6		MIT
4126.521	4127.685	24226.65	CR I	2.0	1.7		MIT
4126.099	4127.263	24229.13	CR	1.2	0.3		MIT
4123.387	4124.550	24245.07	CR	1.5	1.2		MIT
4122.162	4123.325	24252.27	CR I	1.5	0.7		MIT
4121.817	4122.980	24254.30	CR	1.6	1.0		MIT
4121.473	4122.636	24256.33	CR	1.0	0.0		MIT
4121.260	4122.423	24257.58	CR	1.5	0.9		MIT
4120.613	4121.776	24261.39	CR I	1.6	1.0		MIT
4111.67	4112.83	24314.2	CR I	1.3H			CT
4111.36	4112.52	24316.0	CR I	1.3H			CT
4110.911	4112.071	24318.65	CR I	1.0V			MIT
4109.584	4110.744	24326.50	CR I	1.6	1.0		MIT
4108.400	4109.559	24333.51	CR I	1.5	0.6		MIT
4106.022	4107.181	24347.60	CR	1.1	0.0		MIT
4104.867	4106.025	24354.45	CR	1.5	0.9		MIT
4101.157	4102.314	24376.48	CR	1.5	0.3		M
4099.016	4100.173	24389.22	CR	1.5	0.9		MIT
4098.18	4099.34	24394.2	CR I	1.3H			CT
4097.96	4099.12	24395.5	CR I	1.3			CT
4097.65	4098.81	24397.4	CR I	1.3H			CT
4092.174	4093.329	24429.99	CR	1.4	0.3		MIT
4090.305	4091.460	24441.16	CR	1.5	1.2		MIT
4085.987	4087.140	24466.98	CR	0.9			MIT
4085.021	4086.174	24472.77	CR	1.2	0.0		MIT
4081.737	4082.889	24492.46	CR	1.4	0.3		MIT
4080.221	4081.373	24501.56	CR	1.2	0.0		MIT
4077.677	4078.828	24516.84	CR	1.5	1.0		MIT
4077.089	4078.240	24520.38	CR	1.5	1.0		MIT
4076.061	4077.212	24526.56	CR	1.5	1.2		MIT
4074.863	4076.014	24533.77	CR	1.4	1.0		MIT
4072.908	4074.058	24545.55	CR	0.9			MIT
4072.684	4073.834	24546.90	CR	1.1	0.0		MIT
4071.000	4072.150	24557.05	CR	1.2	0.6		MIT
4067.839	4068.988	24576.14	CR	1.5	0.6		MIT
4066.938	4068.086	24581.58	CR	1.4	1.5		MIT
4065.716	4066.864	24588.97	CR	1.9	1.5		MIT
4060.765	4061.912	24618.95	CR	1.1	0.8		MIT
4058.772	4059.918	24631.04	CR	1.9	1.7		MIT
4056.793	4057.939	24643.05	CR	1.2	0.5		MIT
4056.052	4057.198	24647.55	CR	1.5	0.9		MIT
4054.183	4055.328	24658.92	CR II	0.0	0.5		MIT
4051.329	4052.473	24676.29	CR	1.5	0.9		MIT
4050.035	4051.179	24684.17	CR I	1.5	0.0		MIT
4049.783	4050.927	24685.71	CR	1.5	0.8		MIT
4049.158	4050.302	24689.52	CR	0.6	0.5		MIT
4049.006	4050.150	24690.45	CR	0.6	0.5		MIT
4048.780	4049.924	24691.82	CR	1.9	1.7		MIT
4046.760	4047.903	24704.15	CR I	1.5	0.5		MIT
4043.696	4044.838	24722.87	CR	1.5	0.3		MIT
4042.246	4043.388	24731.74	CR I	1.5	0.0		MIT
4041.79	4042.93	24734.5	CR I	1.3			M
4039.22	4040.36	24750.3	CR	1.2	0.5		MIT
4039.100	4040.241	24751.00	CR	2.0	1.6		MIT
4037.993	4039.134	24757.78	CR	1.2H			MIT
4037.722	4038.863	24759.45	CR	1.2			MIT
4037.294	4038.435	24762.07	CR I	1.9	1.1		MIT
4035.24	4036.38	24774.7	CR	0.9			MIT
4034.998	4036.138	24776.16	CR	0.9			MIT
4034.048	4035.188	24781.99	CR I	1.3			MIT
4033.263	4034.403	24786.82	CR I	1.5	0.9		MIT
4033.072	4034.212	24787.99	CR	1.2	0.3		MIT
4031.130	4032.269	24799.93	CR	1.5	0.8		HL
4030.681	4031.820	24802.70	CR	1.6	1.5		MIT
4028.02	4029.16	24819.1	CR	1.5			MIT
4027.103	4028.241	24824.73	CR	0.9	1.5		MIT
4026.166	4027.304	24830.51	CR	2.0	1.5		MIT
4025.012	4026.150	24837.63	CR	2.0	1.4		MIT
4024.567	4025.704	24840.37	CR I	1.3	0.0		MIT

AIR WAVELENGTH (ANGSTRMS)	VACUUM WAVELENGTH (ANGSTRMS)	WAVENUMBER (KAYSERS)	LMENT	LOG ARC	INTENSITY SPK	DIS	REF
4023.739	4024.876	24845.49	CR	1.6	1.2		MIT
4023.43	4024.57	24847.4	CR I	1.2			MIT
4022.263	4023.400	24854.60	CR	1.9	1.6		MIT
4018.202	4019.338	24879.72	CR I	1.5	0.9		MIT
4014.668	4015.803	24901.62	CR	1.6	0.9		MIT
4012.469	4013.603	24915.27	CR	1.8	1.8		MIT
4003.921	4005.053	24968.46	CR	1.5	1.1		HL
4003.33	4004.46	24972.1	CR		1.3		EX
4001.444	4002.575	24983.91	CR	2.3	1.9		MIT
4000.63	4001.76	24989.0	CR	0.7			EX
3999.945	4001.076	24993.28	CR	0.5			MIT
3999.679	4000.810	24994.94	CR	1.6	1.0		MIT
3998.860	3999.991	25000.06	CR	1.4	0.3		MIT
3996.274	3997.404	25016.24	CR	0.3			MIT
3993.968	3995.097	25030.68	CR I	1.8	1.3		MIT
3992.845	3993.974	25037.72	CR I	2.2	1.8		MIT
3992.109	3993.238	25042.33	CR	1.3	0.0		MIT
3991.673	3992.802	25045.07	CR I	2.0	1.7		MIT
3991.123	3992.252	25048.52	CR I	2.3	1.8		MIT
3989.986	3991.114	25055.66	CR	1.9	1.6		MIT
3988.656	3989.784	25064.01	CR	0.7			MIT
3987.89	3989.02	25068.8	CR	0.5			MIT
3984.338	3985.465	25091.18	CR I	1.9	1.8		MIT
3983.907	3985.034	25093.89	CR I	2.3	1.8		MIT
3983.237	3984.364	25098.11	CR	1.3	0.8		MIT
3981.233	3982.359	25110.74	CR I	2.0	1.7		MIT
3979.798	3980.924	25119.80	CR I	1.9	1.3		MIT
3979.52	3980.65	25121.6	CR		0.7		MIT
3979.324	3980.450	25122.79	CR	1.0	0.3		MIT
3978.677	3979.802	25126.88	CR I	1.9	1.6		MIT
3976.665	3977.790	25139.59	CR I	2.5	2.5		MIT
3976.310	3977.435	25141.83	CR	1.1	0.3		MIT
3972.688	3973.812	25164.75	CR I	1.8	1.1		MIT
3971.255	3972.378	25173.83	CR I	1.9	1.7		MIT
3969.748	3970.871	25183.39	CR I	2.3	2.0		MIT
3969.061	3970.184	25187.75	CR I	1.9	1.7		MIT
3963.690	3964.811	25221.88	CR I	2.5	2.5		MIT
3962.187	3963.308	25231.45	CR	0.9	0.0		MIT
3960.763	3961.884	25240.52	CR	1.6	0.9		MIT
3958.07	3959.19	25257.7	CR	0.7	0.3		MIT
3953.163	3954.282	25289.04	CR I	1.8	1.1		MIT
3952.399	3953.518	25293.93	CR I	1.8	1.3		MIT
3951.765	3952.883	25297.99	CR I	1.6	0.7		MIT
3951.097	3952.215	25302.27	CR I	1.7	0.9		MIT
3949.617	3950.735	25311.75	CR	0.8			MIT
3949.585	3950.703	25311.95	CR I	0.7	0.0		MIT
3948.853	3949.971	25316.64	CR	1.4	0.3		MIT
3945.968	3947.085	25335.15	CR	1.7	0.8		MIT
3945.495	3946.612	25338.19	CR	1.7	1.2		MIT
3943.614	3944.730	25350.28	CR I	1.3	0.6		MIT
3941.490	3942.606	25363.94	CR I	2.3R	1.8		MIT
3941.154	3942.270	25366.10	CR	1.2	0.9		MIT
3939.298	3940.413	25378.05	CR	0.9			MIT
3938.341	3939.456	25384.22	CR	1.6	0.5		MIT
3929.697	3930.810	25440.05	CR	1.5J			MIT
3928.636	3929.748	25446.92	CR I	2.2	1.6		MIT
3926.649	3927.761	25459.80	CR	1.5	1.4		MIT
3921.022	3922.132	25496.33	CR I	2.2	1.6		MIT
3919.159	3920.269	25508.45	CR I	2.5R	2.1		MIT
3917.596	3918.705	25518.63	CR I	1.3	1.1		MIT
3916.980	3918.089	25522.64	CR I	1.5	0.9		MIT
3916.243	3917.352	25527.45	CR I	2.0	1.8		MIT
3915.843	3916.952	25530.05	CR I	2.1	1.9		MIT
3915.507	3916.616	25532.25	CR	1.2	1.0		MIT
3914.335	3915.444	25539.89	CR	0.5	0.3		MIT
3911.999	3913.107	25555.14	CR	1.6J			MIT
3911.82	3912.93	25556.3	CR I	1.0			CT
3908.755	3909.862	25576.35	CR I	2.3	2.2		MIT
3907.778	3908.885	25582.74	CR	1.5	1.0		MIT
3905.659	3906.765	25596.62	CR	0.3	0.6		MIT

AIR WAVELENGTH (ANGSTRMS)	VACUUM WAVELENGTH (ANGSTRMS)	WAVENUMBER (KAYSERS)	LMENT	LOG ARC	INTENSITY SPK	DIS	REF
3903.164	3904.270	25612.99	CR I	1.5	1.5		MIT
3902.915	3904.021	25614.62	CR I	2.0	2.0		MIT
3902.108	3903.213	25619.92	CR	1.6	1.5		MIT
3897.650	3898.754	25649.22	CR I	1.6	1.4		MIT
3894.568	3895.671	25669.52	CR	1.0			MIT
3894.035	3895.138	25673.03	CR I	1.8	1.6		MIT
3891.934	3893.037	25686.89	CR I	1.6	1.4		MIT
3886.789	3887.890	25720.89	CR I	2.1	2.1		MIT
3885.218	3886.319	25731.29	CR I	1.6	1.7		MIT
3885.084	3886.185	25732.18	CR I	1.2	1.0		MIT
3883.660	3884.761	25741.61	CR I	1.5	1.3		MIT
3883.292	3884.393	25744.05	CR I	1.8	1.9		MIT
3881.856	3882.956	25753.57	CR I	1.7	0.8		MIT
3881.214	3882.314	25757.83	CR I	1.8	1.3		MIT
3880.399	3881.499	25763.24	CR	1.1			MIT
3879.222	3880.321	25771.06	CR I	1.8	1.2		MIT
3875.209	3876.307	25797.75	CR I	0.3H	0.0H		MIT
3874.533	3875.631	25802.25	CR I	1.8	1.1		MIT
3870.267	3871.364	25830.69	CR I	1.9W	0.5W		MIT
3868.266	3869.363	25844.05	CR	0.9	0.7		MIT
3866.546	3867.642	25855.55	CR		0.5		MIT
3865.601	3866.697	25861.87	CR	0.7	0.5		MIT
3862.548	3863.643	25882.31	CR	1.4	1.3		MIT
3858.891	3859.985	25906.83	CR I	1.5J	1.3J		MIT
3857.631	3858.725	25915.30	CR I	1.7	1.4		MIT
3856.281	3857.374	25924.37	CR I	1.3	1.2		MIT
3855.571	3856.664	25929.14	CR I	1.5	1.5		MIT
3855.286	3856.379	25931.06	CR I	1.5	1.5		MIT
3854.788	3855.881	25934.41	CR	1.3D	1.2D		MIT
3854.220	3855.313	25938.23	CR I	1.6	1.2		MIT
3853.176	3854.269	25945.26	CR I	1.3	1.0		MIT
3852.218	3853.310	25951.71	CR I	1.8	1.1		MIT
3850.042	3851.134	25966.38	CR I	1.6R	1.6R		MIT
3849.534	3850.626	25969.80	CR I	1.3	1.3		MIT
3849.365	3850.457	25970.95	CR I	1.6H	1.5H		MIT
3848.983	3850.075	25973.52	CR I	1.9D	1.7D		MIT
3847.73	3848.82	25982.0	CR	0.7H			MIT
3847.40	3848.49	25984.2	CR	0.7J			MIT
3842.02	3843.11	26020.6	CR I	0.5	0.3		MIT
3841.277	3842.367	26025.63	CR I	2.2	1.9		MIT
3836.070	3837.158	26060.95	CR I	1.4	0.9		MIT
3834.735	3835.823	26070.02	CR I	1.4	1.1		MIT
3832.349	3833.436	26086.26	CR I	1.3	0.3		MIT
3831.032	3832.119	26095.22	CR I	1.6	1.4		MIT
3830.032	3831.119	26102.04	CR I	2.2W	1.7		MIT
3826.425	3827.511	26126.64	CR I	1.6	1.3		MIT
3825.390	3826.475	26133.71	CR I	1.6	1.2		MIT
3823.522	3824.607	26146.48	CR I	1.6	1.5		MIT
3822.092	3823.177	26156.26	CR I	1.0	0.8		MIT
3821.582	3822.666	26159.75	CR I	1.1	0.9		MIT
3820.874	3821.958	26164.60	CR I	1.0	0.8		MIT
3819.981	3821.065	26170.71	CR I	1.1	1.0		MIT
3819.564	3820.648	26173.57	CR I	1.8	1.6		MIT
3818.481	3819.565	26180.99	CR I	1.7	1.3		MIT
3817.844	3818.927	26185.36	CR I	1.5	1.3		MIT
3816.173	3817.256	26196.83	CR I	1.5	1.0		MIT
3815.433	3816.516	26201.91	CR	1.5	1.1		MIT
3814.622	3815.705	26207.48	CR	1.5	1.5		MIT
3814.007	3815.090	26211.71	CR	0.3	0.6		MIT
3812.250	3813.332	26223.78	CR	1.6	1.2		MIT
3809.496	3810.577	26242.74	CR	1.1	0.8		MIT
3807.926	3809.007	26253.56	CR I	1.4	1.1		MIT
3806.829	3807.910	26261.13	CR	1.5	1.5		MIT
3804.798	3805.878	26275.15	CR I	2.0	1.5		MIT
3802.986	3804.066	26287.66	CR	0.3H			MIT
3801.200	3802.279	26300.01	CR	1.5H	0.5H		MIT
3797.716	3798.794	26324.14	CR I	2.0	1.3		MIT
3797.126	3798.204	26328.23	CR I	1.7	1.5		MIT
3794.608	3795.685	26345.70	CR I	1.7	1.5		MIT
3793.879	3794.956	26350.76	CR I	1.7	1.5		MIT

AIR WAVELENGTH (ANGSTRMS)	VACUUM WAVELENGTH (ANGSTRMS)	WAVENUMBER (KAYSERS)	LMENT	LOG ARC	INTENSITY SPK	DIS	REF
3793.289	3794.366	26354.86	CR I	1.7	1.5		MIT
3792.426	3793.503	26360.86	CR	0.3	0.0		MIT
3792.137	3793.214	26362.87	CR I	1.8	1.6		MIT
3791.376	3792.453	26368.16	CR I	1.9	1.6		MIT
3790.454	3791.530	26374.57	CR I	1.7	1.2		MIT
3790.228	3791.304	26376.15	CR I	1.5	1.0		MIT
3789.723	3790.799	26379.66	CR I	1.7	1.0		MIT
3788.864	3789.940	26385.64	CR I	1.8	1.0		MIT
3786.212	3787.287	26404.12	CR	0.7	0.3		MIT
3780.729	3781.803	26442.42	CR	0.5			MIT
3780.079	3781.153	26446.96	CR	0.3			MIT
3777.919	3778.992	26462.08	CR	0.5			MIT
3768.994	3770.065	26524.74	CR	0.3	0.3		MIT
3768.734	3769.805	26526.57	CR I	1.5	1.4		MIT
3768.240	3769.311	26530.05	CR I	1.8	1.8		MIT
3767.431	3768.501	26535.75	CR	1.3	1.0		MIT
3765.612	3766.682	26548.56	CR II		1.2		MIT
3764.603	3765.673	26555.68	CR	0.3			MIT
3761.868	3762.937	26574.99	CR	0.7H	1.0H		MIT
3761.701	3762.770	26576.17	CR	0.6	0.9		MIT
3758.044	3759.112	26602.03	CR I	1.7	1.5		MIT
3757.662	3758.730	26604.73	CR I	1.7	1.7		MIT
3757.174	3758.242	26608.19	CR I	1.7	1.5		MIT
3755.828	3756.895	26617.72	CR	0.3	0.3		MIT
3754.575	3755.642	26626.61	CR II	0.9	1.5		MIT
3750.562	3751.628	26655.09	CR		1.1H		MIT
3748.998	3750.064	26666.21	CR I	2.1G	2.1G		MIT
3748.614	3749.680	26668.95	CR I	1.6	1.5		MIT
3747.264	3748.329	26678.55	CR	1.1	0.8		MIT
3744.490	3745.554	26698.32	CR I	1.5	1.1		MIT
3743.884	3744.948	26702.64	CR I	1.6	1.6		MIT
3743.578	3744.642	26704.82	CR I	1.6	1.6		MIT
3742.968	3744.032	26709.17	CR I	1.4	1.0		MIT
3738.376	3739.439	26741.98	CR II	0.8	1.6		MIT
3737.554	3738.617	26747.86	CR		1.3		MIT
3732.032	3733.093	26787.44	CR I	1.7	1.2		MIT
3730.807	3731.868	26796.23	CR I	1.8	1.1		MIT
3727.359	3728.419	26821.02	CR		0.7H		MIT
3725.976	3727.036	26830.98	CR	0.3	0.3H		MIT
3723.395	3724.454	26849.57	CR		1.2		MIT
3716.531	3717.588	26899.16	CR	1.3	0.8		MIT
3715.428	3716.485	26907.15	CR		1.5		MIT
3715.186	3716.243	26908.90	CR II	0.8	1.3		MIT
3712.949	3714.005	26925.11	CR II	1.1S	2.1		MIT
3711.282	3712.338	26937.20	CR		1.1		MIT
3710.60	3711.66	26942.1	CR	0.3	0.3		MIT
3710.09	3711.15	26945.9	CR	0.6	0.3		MIT
3698.178	3699.230	27032.65	CR	0.3			MIT
3697.997	3699.049	27033.97	CR		1.6		MIT
3695.858	3696.910	27049.62	CR	1.1	0.5		MIT
3694.974	3696.026	27056.09	CR		0.8H		MIT
3693.089	3694.140	27069.90	CR	1.0	0.7		MIT
3689.628	3690.678	27095.29	CR	0.9	0.7		MIT
3689.302	3690.352	27097.69	CR	0.8	0.5		MIT
3688.457	3689.507	27103.89	CR	1.3	0.8		MIT
3687.545	3688.595	27110.60	CR I	1.1H	0.7		MIT
3687.526	3688.576	27110.74	CR	0.6	0.0		MIT
3687.252	3688.302	27112.75	CR I	0.8H	0.6L		MIT
3686.803	3687.853	27116.05	CR I	1.3H	0.7H		MIT
3685.548	3686.597	27125.28	CR I	0.6	0.5		MIT
3685.255	3686.304	27127.44	CR	0.5H	0.0H		MIT
3684.993	3686.042	27129.37	CR	0.3	0.3H		MIT
3684.247	3685.296	27134.86	CR		0.5		MIT
3681.691	3682.739	27153.70	CR	1.3	0.9		MIT
3679.819	3680.867	27167.51	CR	1.6	0.9		MIT
3679.070	3680.118	27173.04	CR	1.2	0.7		MIT
3677.888	3678.935	27181.78	CR II	0.5L	1.8		MIT
3677.678	3678.725	27183.33	CR II	0.8	1.5		MIT
3676.322	3677.369	27193.36	CR	1.6	1.2		MIT
3668.029	3669.074	27254.84	CR I	1.2	0.6		MIT

AIR WAVELENGTH (ANGSTRMS)	VACUUM WAVELENGTH (ANGSTRMS)	WAVENUMBER (KAYSERS)	LMENT	LOG ARC	INTENSITY SPK	INTENSITY DIS	REF
3666.642	3667.686	27265.15	CR I	1.3	1.2		MIT
3666.171	3667.215	27268.65	CR I	1.1	0.3		MIT
3665.980	3667.024	27270.07	CR	1.3	1.2		MIT
3664.942	3665.986	27277.79	CR	0.0	1.6		MIT
3663.206	3664.249	27290.72	CR I	1.5	1.3		MIT
3662.840	3663.883	27293.45	CR I	1.4	0.9		ME
3662.382	3663.425	27296.86	CR	0.8	0.6H		MIT
3661.239	3662.282	27305.38	CR	0.3	0.0H		MIT
3659.975	3661.018	27314.81	CR	0.3	0.0		MIT
3658.168	3659.210	27328.30	CR		1.3		MIT
3656.261	3657.303	27342.56	CR I	1.9	1.4		MIT
3653.912	3654.953	27360.13	CR I	2.0	1.4		MIT
3652.587	3653.628	27370.06	CR	0.5			MIT
3651.660	3652.700	27377.01	CR II	0.0	1.3		MIT
3650.344	3651.384	27386.87	CR		1.6		MIT
3649.862	3650.902	27390.49	CR	1.2	0.6		MIT
3648.997	3650.037	27396.99	CR I	1.6	1.3		MIT
3648.534	3649.574	27400.46	CR I	1.0	0.9		MIT
3647.394	3648.433	27409.02	CR II	0.3	1.3		MIT
3646.161	3647.200	27418.29	CR	1.3	0.9		MIT
3645.583	3646.622	27422.64	CR	0.6	0.5		MIT
3644.693	3645.732	27429.34	CR II	0.7	1.3		MIT
3643.199	3644.237	27440.58	CR II	0.7	1.5		MIT
3641.830	3642.868	27450.90	CR I	1.6	1.2		MIT
3641.470	3642.508	27453.61	CR I	1.5	1.4		MIT
3640.388	3641.426	27461.77	CR I	1.5	0.7		MIT
3639.802	3640.839	27466.19	CR I	1.8	1.4		MIT
3636.590	3637.627	27490.45	CR I	1.8	1.5		MIT
3635.281	3636.317	27500.35	CR I	1.4	0.9		MIT
3634.994	3636.030	27502.52	CR	1.4	1.1		MIT
3634.014	3635.050	27509.94	CR		0.8		MIT
3632.839	3633.875	27518.84	CR	1.9	1.5		MIT
3631.687	3632.722	27527.56	CR II	1.0	1.8		MIT
3631.449	3632.484	27529.37	CR I				ME
3626.337	3627.371	27568.18	CR	1.3	0.5		MIT
3619.460	3620.492	27620.55	CR	1.5	0.9		MIT
3617.307	3618.339	27636.99	CR		0.8		MIT
3615.645	3616.676	27649.70	CR I	1.5	1.0		MIT
3613.669	3614.700	27664.82	CR	1.0	0.9		MIT
3613.183	3614.213	27668.54	CR	0.7	0.9		MIT
3610.052	3611.082	27692.53	CR	1.3	0.9		MIT
3609.479	3610.509	27696.93	CR	1.3	1.1		MIT
3608.401	3609.430	27705.20	CR	1.1	0.9		MIT
3605.333	3606.361	27728.78	CR I	2.7G	2.6G		MIT
3603.745	3604.773	27741.00	CR I	1.7	1.2		ME
3602.574	3603.602	27750.01	CR I	1.2	1.0		MIT
3601.666	3602.694	27757.01	CR I	1.7	1.5		MIT
3599.395	3600.422	27774.52	CR	1.5	1.3		MIT
3594.31	3595.34	27813.8	CR I	0.9			ME
3593.488	3594.513	27820.18	CR I	2.7G	2.6G		MIT
3585.506	3586.529	27882.11	CR II	0.8	1.5		MIT
3585.295	3586.318	27883.75	CR	0.8	0.7		MIT
3584.33	3585.35	27891.3	CR	1.0J	1.0J		MIT
3582.624	3583.647	27904.54	CR	1.5	1.1		MIT
3578.687	3579.709	27935.24	CR I	2.7G	2.6R		MIT
3574.935	3575.956	27964.55	CR I	0.9	0.8		MIT
3574.805	3575.826	27965.57	CR I	0.9	1.2		MIT
3574.039	3575.059	27971.56	CR I	1.7	1.2		MIT
3573.643	3574.663	27974.66	CR I	1.8	1.2		MIT
3572.748	3573.768	27981.67	CR I	1.4	0.7		MIT
3569.147	3570.166	28009.90	CR	0.6	0.6		MIT
3568.418	3569.437	28015.62	CR	0.8J			MIT
3566.163	3567.181	28033.34	CR	1.9J	1.1J		MIT
3565.161	3566.179	28041.22	CR	0.8	0.0		MIT
3564.953	3565.971	28042.85	CR	0.9	0.0		MIT
3564.710	3565.728	28044.76	CR	1.3	0.9		MIT
3564.293	3565.311	28048.04	CR	1.2	0.8		MIT
3563.925	3564.943	28050.94	CR		0.9		MIT
3562.88	3563.90	28059.2	CR	1.0			MIT
3562.475	3563.492	28062.36	CR	1.0	0.7		MIT
3562.28	3563.30	28063.9	CR	1.3	1.1		MIT
3560.901	3561.918	28074.76	CR	0.3H	0.0H		MIT
3559.781	3560.798	28083.59	CR	1.2	1.0		MIT
3559.21	3560.23	28088.1	CR	0.3H			MIT
3558.58	3559.60	28093.1	CR		0.9H		MIT
3558.519	3559.535	28093.55	CR	1.3			MIT
3556.130	3557.146	28112.43	CR II	1.3	0.6		MIT
3555.788	3556.804	28115.13	CR	1.3	0.8		MIT
3553.968	3554.983	28129.53	CR	0.8	0.8		MIT
3552.953	3553.968	28137.56	CR	0.5J			MIT
3552.846	3553.861	28138.41	CR	0.3H	0.7J		MIT
3550.635	3551.649	28155.93	CR I	1.8	1.8		MIT
3549.283	3550.297	28166.66	CR	0.7J			MIT
3548.731	3549.745	28171.04	CR	0.3J	0.6J		MIT
3547.982	3548.996	28176.99	CR	0.7J			MIT
3537.233	3538.244	28262.61	CR	1.2	0.0		MIT
3532.888	3533.898	28297.37	CR	0.9	0.3		MIT
3531.083	3532.092	28311.83	CR	1.2			MIT
3527.091	3528.099	28343.87	CR	1.5	0.7		MIT
3523.59	3524.60	28372.0	CR	1.0	0.0		MIT
3522.950	3523.957	28377.19	CR	0.9	0.0		MIT
3522.78	3523.79	28378.6	CR I	0.9J			MIT
3522.145	3523.152	28383.67	CR		0.8		MIT
3519.45	3520.46	28405.4	CR	0.8H			MIT
3518.40	3519.41	28413.9	CR	0.9	0.0		MIT
3513.044	3514.049	28457.20	CR		0.9		MIT
3512.65	3513.65	28460.4	CR	0.9H			MIT
3511.836	3512.840	28466.99	CR II	1.3	1.7		MIT
3510.538	3511.542	28477.52	CR	1.6	0.9		MIT
3508.836	3509.840	28491.33	CR	1.1H	0.0H		MIT
3508.098	3509.101	28497.32	CR	1.2	0.3		MIT
3507.209	3508.212	28504.55	CR	1.1H			MIT
3503.876	3504.878	28531.66	CR	1.3J			MIT
3503.390	3504.392	28535.62	CR	0.9			MIT
3502.307	3503.309	28544.44	CR	1.5	0.8		MIT
3495.559	3496.559	28599.54	CR		1.2		MIT
3495.375	3496.375	28601.05	CR II	1.1	1.5		MIT
3494.967	3495.967	28604.39	CR	1.5	1.4		MIT
3494.522	3495.522	28608.03	CR II	0.0	0.3		MIT
3488.453	3489.451	28657.80	CR	1.5	1.0		MIT
3486.48	3487.48	28674.0	CR	1.3	0.6		MIT
3484.15	3485.15	28693.2	CR II	0.9	1.5		MIT
3483.512	3484.509	28698.45	CR	1.2	1.5		KS
3482.591	3483.588	28706.04	CR	0.0	0.8		MIT
3481.536	3482.533	28714.74	CR	1.5	1.5		MIT
3481.303	3482.300	28716.66	CR I	1.2	1.5		MIT
3480.29	3481.29	28725.0	CR I	1.5	0.0		MIT
3479.308	3480.304	28733.12	CR I	1.5R			MIT
3479.124	3480.120	28734.64	CR I	1.5	0.0		MIT
3478.771	3479.767	28737.56	CR I	1.5	0.5		MIT
3477.25	3478.25	28750.1	CR	0.8	0.0		MIT
3477.161	3478.156	28750.86	CR I	1.4	0.7		MIT
3475.129	3476.124	28767.67	CR II	0.8	1.6		MIT
3475.129	3476.124	28767.67	CR I	0.8	1.6		MIT
3474.865	3475.860	28769.86	CR I	1.5	0.5		MIT
3474.379	3475.374	28773.88	CR	1.5	0.9		MIT
3473.612	3474.607	28780.24	CR I	1.5	1.0		MIT
3472.906	3473.900	28786.09	CR	1.5	0.9		MIT
3472.764	3473.758	28787.26	CR I	1.5	0.9		MIT
3472.07	3473.06	28793.0	CR		1.9		MIT
3471.497	3472.491	28797.77	CR I	1.5	0.5		MIT
3470.529	3471.523	28805.80	CR I	1.4	0.8		MIT
3470.401	3471.395	28806.86	CR I	1.5	0.9		MIT
3469.590	3470.584	28813.60	CR I	1.7	1.2		MIT
3468.75	3469.74	28820.6	CR	1.5	0.9		MIT
3467.715	3468.708	28829.18	CR	1.7	1.5		MIT
3467.022	3468.015	28834.94	CR I	1.7	1.3		MIT
3465.583	3466.576	28846.91	CR	1.5	0.9		MIT
3465.250	3466.242	28849.68	CR	1.5	1.5		M
3464.836	3465.828	28853.13	CR	1.5	0.5		MIT

AIR WAVELENGTH (ANGSTRMS)	VACUUM WAVELENGTH (ANGSTRMS)	WAVENUMBER (KAYSERS)	LMENT	LOG ARC	INTENSITY SPK	INTENSITY DIS	REF
3464.00	3464.99	28860.1	CR II	0.0	0.3		MIT
3463.610	3464.602	28863.34	CR	1.2	0.0		MIT
3462.73	3463.72	28870.7	CR II	0.3	0.9		MIT
3460.430	3461.421	28889.87	CR I	1.6	1.5		MIT
3459.29	3460.28	28899.4	CR		1.5		MIT
3458.090	3459.081	28909.42	CR	1.5	1.2		MIT
3457.63	3458.62	28913.3	CR	0.6	2.1		MIT
3455.602	3456.592	28930.23	CR	1.7	1.5		MIT
3455.281	3456.271	28932.92	CR	1.5	1.0		MIT
3454.99	3455.98	28935.4	CR		2.0		MIT
3453.743	3454.732	28945.80	CR I	1.5	1.4		MIT
3453.328	3454.317	28949.28	CR I	1.5	1.5		MIT
3450.83	3451.82	28970.2	CR	1.4	0.6		MIT
3448.190	3449.178	28992.42	CR I	1.1	0.0		MIT
3447.760	3448.748	28996.03	CR I	1.5	1.5		MIT
3447.430	3448.418	28998.81	CR I	1.5	1.5		MIT
3447.015	3448.003	29002.30	CR I	1.5	1.4		MIT
3445.618	3446.605	29014.06	CR	2.0	1.9		MIT
3443.790	3444.777	29029.46	CR	1.5	1.4		MIT
3441.439	3442.425	29049.29	CR I	1.9	2.0		MIT
3441.115	3442.101	29052.02	CR I	1.5	1.4		MIT
3439.37	3440.36	29066.8	CR	1.5	0.3		MIT
3436.187	3437.172	29093.69	CR I	1.7	1.7		MIT
3435.819	3436.804	29096.80	CR	1.5	1.1		MIT
3435.679	3436.664	29097.99	CR I	1.3	1.1		MIT
3435.488	3436.473	29099.60	CR	0.9	0.0		MIT
3434.112	3435.096	29111.26	CR I	1.5	1.4		MIT
3433.598	3434.582	29115.62	CR I	1.7	1.5		MIT
3433.311	3434.295	29118.06	CR II	1.5	2.2		MIT
3432.851	3433.835	29121.96	CR	1.0	0.0		MIT
3432.324	3433.308	29126.43	CR	1.4	0.8		MIT
3431.998	3432.982	29129.20	CR	1.4	0.6		M
3431.69	3432.67	29131.8	CR	1.1	0.3		MIT
3431.59	3432.57	29132.7	CR	0.8			MIT
3431.284	3432.268	29135.26	CR	1.5	0.9		MIT
3427.66	3428.64	29166.1	CR	1.6	0.0		MIT
3425.98	3426.96	29180.4	CR	1.4	0.3		MIT
3422.739	3423.721	29207.99	CR II	1.5	2.1		MIT
3421.72	3422.70	29216.7	CR	1.1			MIT
3421.64	3422.62	29217.4	CR	0.3H	0.8		MIT
3421.212	3422.193	29221.03	CR II	1.7	2.3		MIT
3419.90	3420.88	29232.2	CR	1.1H			MIT
3419.67	3420.65	29234.2	CR	0.9H			MIT
3418.01	3418.99	29248.4	CR	0.9			MIT
3415.57	3416.55	29269.3	CR	0.9J			MIT
3414.305	3415.284	29280.14	CR	0.9J			MIT
3411.051	3412.030	29308.07	CR	1.9J	0.5J		MIT
3409.397	3410.375	29322.29	CR	1.8J	0.0J		MIT
3408.765	3409.743	29327.72	CR II	1.5	2.0		MIT
3408.04	3409.02	29334.0	CR	1.6J	0.0J		MIT
3407.27	3408.25	29340.6	CR	1.7J	0.0J		MIT
3406.93	3407.91	29343.5	CR	0.6	0.0		MIT
3405.22	3406.20	29358.2	CR	1.1	0.0		MIT
3403.595	3404.572	29372.27	CR	1.5	0.5		MIT
3403.322	3404.299	29374.63	CR II	1.5	2.3		MIT
3402.399	3403.375	29382.59	CR II	1.4	1.9		MIT
3399.53	3400.51	29407.4	CR	0.0	1.8		MIT
3395.61	3396.58	29441.3	CR	0.3	2.0		MIT
3394.295	3395.269	29452.75	CR	1.2	2.2		MIT
3393.839	3394.813	29456.70	CR II	1.2	2.1		MIT
3392.987	3393.961	29464.10	CR II	1.0	2.0		MIT
3391.434	3392.408	29477.59	CR II	0.6	2.2		MIT
3391.372	3392.346	29478.13	CR	1.1			MIT
3390.773	3391.746	29483.34	CR	1.5	0.5		MIT
3388.710	3389.683	29501.28	CR I	1.6	0.6		MIT
3386.521	3387.493	29520.35	CR	1.5	0.3		MIT
3385.327	3386.299	29530.76	CR	1.5	0.3		MIT
3384.644	3385.616	29536.72	CR I	1.6	0.0		MIT
3384.241	3385.213	29540.24	CR I	1.5			MIT
3382.683	3383.654	29553.85	CR II	1.5	2.3		MIT

AIR WAVELENGTH (ANGSTRMS)	VACUUM WAVELENGTH (ANGSTRMS)	WAVENUMBER (KAYSERS)	LMENT	LOG ARC	INTENSITY SPK	INTENSITY DIS	REF
3382.08	3383.05	29559.1	CR	1.5	0.0		MIT
3379.825	3380.796	29578.84	CR II	1.2	2.0		MIT
3379.825	3380.796	29578.84	CR I	1.2	2.0		MIT
3379.564	3380.534	29581.12	CR I	1.1H			MIT
3379.371	3380.341	29582.81	CR II	0.8	2.0		MIT
3379.171	3380.141	29584.56	CR I	1.6	0.8		MIT
3378.337	3379.307	29591.86	CR II	1.4	2.2		MIT
3376.397	3377.367	29608.87	CR	1.5	1.3		MIT
3376.15	3377.12	29611.0	CR	1.4H			KN
3375.63	3376.60	29615.6	CR	0.8			MIT
3374.935	3375.904	29621.69	CR	1.4	0.6		MIT
3374.60	3375.57	29624.6	CR	1.4	0.3		MIT
3372.13	3373.10	29646.3	CR		1.5		MIT
3371.43	3372.40	29652.5	CR		0.3H		MIT
3370.231	3371.199	29663.04	CR	1.5	0.0H		MIT
3368.054	3369.022	29682.21	CR II	1.5	2.1		MIT
3367.53	3368.50	29686.8	CR I	1.7J			MIT
3365.523	3366.490	29704.53	CR	1.4	0.3		MIT
3364.691	3365.658	29711.87	CR		0.3		MIT
3363.73	3364.70	29720.4	CR II	0.9	1.5		MIT
3362.711	3363.677	29729.37	CR I	1.6			MIT
3362.213	3363.179	29733.77	CR I	1.9	0.9		MIT
3361.770	3362.736	29737.69	CR II	1.0	2.0		MIT
3360.295	3361.261	29750.74	CR II	1.7	2.3		MIT
3360.13	3361.10	29752.2	CR	1.3			MIT
3359.18	3360.15	29760.6	CR	1.4			MIT
3358.501	3359.466	29766.64	CR II	1.6	2.3		MIT
3357.41	3358.37	29776.3	CR	0.8	2.1		MIT
3356.72	3357.68	29782.4	CR	1.5J			MIT
3356.400	3357.365	29785.27	CR	1.5J			MIT
3353.61	3354.57	29810.1	CR	1.5			MIT
3353.126	3354.090	29814.35	CR	1.2	1.7		MIT
3353.026	3353.990	29815.24	CR	1.3	1.5		MIT
3351.966	3352.929	29824.67	CR I	1.7	1.7		MIT
3351.596	3352.559	29827.96	CR	1.5	0.9		MIT
3349.322	3350.285	29848.21	CR	1.5	1.7		MIT
3349.072	3350.035	29850.44	CR	2.1	1.6		MIT
3347.837	3348.799	29861.45	CR II	1.5	2.1		MIT
3347.48	3348.44	29864.6	CR	1.1			MIT
3346.742	3347.704	29871.22	CR	2.2G	1.9R		MIT
3346.018	3346.980	29877.68	CR	1.5	1.5		MIT
3345.37	3346.33	29883.5	CR	1.3	0.0		MIT
3345.15	3346.11	29885.4	CR	1.2	0.3		MIT
3344.51	3345.47	29891.1	CR	1.3	0.3		MIT
3343.74	3344.70	29898.0	CR	1.5H			MIT
3343.342	3344.303	29901.59	CR I	1.5	1.0		MIT
3343.227	3344.188	29902.62	CR	1.1	0.8		MIT
3342.586	3343.547	29908.36	CR II	1.5	2.1		MIT
3342.021	3342.982	29913.41	CR	0.8	0.5H		MIT
3341.43	3342.39	29918.7	CR	1.7	0.0		MIT
3339.804	3340.764	29933.27	CR II	1.4	2.2		MIT
3337.22	3338.18	29956.4	CR I	1.0			MIT
3336.98	3337.94	29958.6	CR	1.3	0.9		MIT
3336.330	3337.289	29964.44	CR II	1.3	1.9		MIT
3334.925	3335.884	29977.06	CR	1.0	0.8		MIT
3334.690	3335.649	29979.17	CR	2.2J			MIT
3333.605	3334.564	29988.93	CR	2.1	0.5H		MIT
3332.879	3333.838	29995.46	CR	1.5	1.0		MIT
3330.60	3331.56	30016.0	CR	1.9			MIT
3329.053	3330.011	30029.93	CR I	1.5	0.8		MIT
3328.351	3329.308	30036.27	CR II	1.3	1.6		MIT
3327.24	3328.20	30046.3	CR I	0.9H			MIT
3326.590	3327.547	30052.17	CR	1.6	1.3		MIT
3324.346	3325.302	30072.45	CR	1.0	1.8		MIT
3324.060	3325.016	30075.04	CR II	1.3	1.3		MIT
3323.53	3324.49	30079.8	CR		1.1		MIT
3323.25	3324.21	30082.4	CR	1.4			MIT
3322.70	3323.66	30087.4	CR	0.0	1.3		MIT
3321.19	3322.15	30101.0	CR	1.3	0.0		MIT
3318.08	3319.03	30129.2	CR	1.9V			MIT

AIR WAVELENGTH (ANGSTRMS)	VACUUM WAVELENGTH (ANGSTRMS)	WAVENUMBER (KAYSERS)	LMENT	LOG ARC	INTENSITY SPK	DIS	REF
3316.503	3317.457	30143.57	CR I	1.3	1.0		MIT
3316.23	3317.18	30146.1	CR	1.0			MIT
3315.29	3316.24	30154.6	CR		0.9		MIT
3315.20	3316.15	30155.4	CR I	1.0			MIT
3314.563	3315.517	30161.21	CR	1.0	2.0		MIT
3314.19	3315.14	30164.6	CR I	0.9			MIT
3314.053	3315.007	30165.85	CR		1.5		MIT
3313.721	3314.675	30168.87	CR	1.5	0.0		MIT
3312.70	3313.65	30178.2	CR	0.5			MIT
3312.179	3313.132	30182.92	CR II	0.7	2.1		HL
3312.081	3313.034	30183.81	CR I	1.0			MIT
3311.929	3312.882	30185.20	CR II	0.8	2.1		HL
3311.30	3312.25	30190.9	CR I	0.9			MIT
3310.647	3311.600	30196.88	CR	0.3	2.3		MIT
3309.836	3310.789	30204.28	CR	1.5	0.3		MIT
3307.755	3308.707	30223.28	CR I	1.6	0.9		MIT
3305.23	3306.18	30246.4	CR	1.4			MIT
3304.741	3305.692	30250.85	CR		0.3		MIT
3304.326	3305.277	30254.65	CR I	1.5H			MIT
3302.876	3303.827	30267.93	CR	1.5	0.3		MIT
3302.19	3303.14	30274.2	CR	1.7H	0.0H		MIT
3301.22	3302.17	30283.1	CR		1.2		MIT
3300.80	3301.75	30287.0	CR	1.0			MIT
3298.318	3299.268	30309.76	CR	1.5	0.9		MIT
3295.427	3296.376	30336.34	CR II	1.0	2.3		MIT
3293.83	3294.78	30351.1	CR	1.5	0.0		MIT
3291.762	3292.710	30370.12	CR	1.0	2.3		MIT
3290.99	3291.94	30377.2	CR	1.5V			MIT
3286.36	3287.31	30420.0	CR	1.3	0.0		MIT
3283.06	3284.01	30450.6	CR		1.5		MIT
3272.89	3273.83	30545.2	CR		0.6		MIT
3271.96	3272.90	30553.9	CR	1.8J			MIT
3270.72	3271.66	30565.5	CR	1.7	0.0		MIT
3269.765	3270.708	30574.42	CR		1.5		MIT
3266.634	3267.576	30603.73	CR I	1.5	0.6		MIT
3264.264	3265.205	30625.95	CR	0.8	1.7		MIT
3259.975	3260.915	30666.24	CR	1.7	1.5		MIT
3258.77	3259.71	30677.6	CR		1.7		MIT
3257.822	3258.761	30686.50	CR	1.6	1.5		MIT
3254.94	3255.88	30713.7	CR	1.7J			MIT
3253.27	3254.21	30729.4	CR	1.3	0.0		MIT
3252.490	3253.428	30736.81	CR		1.4H		MIT
3251.836	3252.774	30742.99	CR	1.5			MIT
3251.58	3252.52	30745.4	CR	1.5			MIT
3247.274	3248.211	30786.18	CR I	1.3	0.0		MIT
3245.542	3246.478	30802.61	CR	1.3	1.2		MIT
3245.485	3246.421	30803.15	CR I	1.0	0.3		MIT
3244.115	3245.051	30816.16	CR I	1.5	0.6		MIT
3240.951	3241.886	30846.24	CR I	1.5	0.3		MIT
3238.76	3239.69	30867.1	CR	0.8	2.3		MIT
3238.51	3239.44	30869.5	CR	1.4	0.9		MIT
3238.087	3239.021	30873.52	CR	1.5	1.3		MIT
3237.729	3238.663	30876.94	CR	1.6	1.5		MIT
3234.06	3234.99	30912.0	CR	1.0	2.2		MIT
3233.234	3234.167	30919.86	CR I	1.5	0.6		MIT
3229.866	3230.798	30952.10	CR		0.9		MIT
3229.204	3230.136	30958.45	CR	1.5			MIT
3226.555	3227.487	30983.86	CR I	1.5	0.0		MIT
3225.356	3226.287	30995.38	CR		1.3		MIT
3219.133	3220.063	31055.30	CR		1.5		MIT
3218.69	3219.62	31059.6	CR	1.9J	0.3J		MIT
3217.400	3218.329	31072.02	CR	1.5	1.3		MIT
3216.56	3217.49	31080.1	CR	0.5	2.1		MIT
3212.895	3213.823	31115.59	CR		1.5		MIT
3211.309	3212.237	31130.96	CR	1.5	1.1		MIT
3209.183	3210.110	31151.58	CR II	1.6	2.1		MIT
3208.590	3209.517	31157.34	CR II	1.3	1.6		MIT
3207.995	3208.922	31163.11	CR		1.0		MIT
3205.098	3206.024	31191.28	CR	0.0	1.6		MIT
3203.515	3204.441	31206.69	CR		1.1		MIT

AIR WAVELENGTH (ANGSTRMS)	VACUUM WAVELENGTH (ANGSTRMS)	WAVENUMBER (KAYSERS)	LMENT	LOG ARC	INTENSITY SPK	DIS	REF
3202.506	3203.431	31216.52	CR		1.2		MIT
3201.260	3202.185	31228.68	CR	0.0	1.7		MIT
3198.111	3199.035	31259.42	CR	1.6	1.0		MIT
3197.981	3198.905	31260.69	CR		0.9		MIT
3197.079	3198.003	31269.51	CR II	1.5	1.5H		MIT
3194.62	3195.54	31293.6	CR		1.0		ME
3192.114	3193.037	31318.15	CR I	1.5	0.3		MIT
3189.838	3190.760	31340.49	CR		1.0		MIT
3188.011	3188.933	31358.45	CR	2.2H	1.8H		MIT
3186.744	3187.665	31370.92	CR		1.5		MIT
3184.341	3185.262	31394.59	CR		1.5		MIT
3183.325	3184.246	31404.61	CR	0.8	2.2		MIT
3181.428	3182.348	31423.34	CR II	1.2	1.6		MIT
3180.701	3181.621	31430.52	CR II	1.5	2.2		MIT
3179.45	3180.37	31442.9	CR		1.0		MIT
3179.283	3180.203	31444.54	CR	2.0	1.0H		MIT
3178.778	3179.697	31449.53	CR		1.0		MIT
3173.565	3174.483	31501.19	CR		1.4		MIT
3172.079	3172.997	31515.95	CR	0.3	2.3		MIT
3169.584	3170.501	31540.75	CR I	1.4	0.3		MIT
3169.192	3170.109	31544.66	CR	0.3	1.7		MIT
3163.923	3164.839	31597.19	CR		1.0		MIT
3163.756	3164.672	31598.85	CR	1.6	1.4		MIT
3162.436	3163.351	31612.04	CR	1.4			MIT
3160.097	3161.012	31635.44	CR		1.0		MIT
3159.845	3160.760	31637.96	CR		0.8		MIT
3159.59	3160.50	31640.5	CR	1.8J	1.3H		MIT
3159.105	3160.020	31645.37	CR II	0.6	1.4		MIT
3158.833	3159.747	31648.10	CR	0.8	0.0		MIT
3158.024	3158.938	31656.21	CR	0.0	1.5		MIT
3155.149	3156.063	31685.05	CR	1.5	1.4		MIT
3154.107	3155.020	31695.52	CR		0.7		MIT
3150.109	3151.021	31735.74	CR	0.8	1.7		MIT
3149.822	3150.734	31738.63	CR	0.8	1.6		MIT
3148.445	3149.357	31752.51	CR	1.5	1.2		MIT
3147.227	3148.139	31764.80	CR II	1.4	2.2		MIT
3145.766	3146.677	31779.55	CR		1.4		MIT
3145.62	3146.53	31781.0	CR	1.3	0.0		MIT
3145.103	3146.014	31786.25	CR II	1.0	1.5		MIT
3144.409	3145.320	31793.27	CR	1.7	1.1		MIT
3143.68	3144.59	31800.6	CR	0.3	0.8		MIT
3142.742	3143.652	31810.13	CR	0.0	1.3		MIT
3141.82	3142.73	31819.5	CR	0.6	0.3		MIT
3140.21	3141.12	31835.8	CR		0.8		MIT
3138.203	3139.112	31856.14	CR	1.2	0.8		MIT
3137.116	3138.025	31867.18	CR		0.6		MIT
3136.680	3137.589	31871.61	CR II	1.3	1.7		MIT
3135.87	3136.78	31879.8	CR	0.9			EX
3135.341	3136.250	31885.22	CR	0.0	1.4		MIT
3134.92	3135.83	31889.5	CR	1.0			EX
3134.307	3135.215	31895.74	CR	0.5	1.7		MIT
3132.820	3133.728	31910.87	CR	1.3	0.3		MIT
3132.058	3132.966	31918.64	CR II	1.4	2.1		MIT
3131.211	3132.118	31927.27	CR	1.3	0.8		MIT
3130.565	3131.472	31933.86	CR	0.0	1.1		MIT
3128.699	3129.606	31952.91	CR II	1.5	2.2		MIT
3125.470	3126.376	31985.92	CR	0.9	0.6		MIT
3124.978	3125.884	31990.95	CR II	1.3	2.1		MIT
3122.596	3123.501	32015.35	CR II	1.0	1.9		MIT
3121.796	3122.701	32023.56	CR		0.9		MIT
3121.64	3122.55	32025.2	CR	1.0			MIT
3121.20	3122.10	32029.7	CR		0.8		MIT
3121.037	3121.942	32031.34	CR		0.7		MIT
3120.64	3121.54	32035.4	CR	1.0			MIT
3120.371	3121.276	32038.18	CR II	1.6	2.2		MIT
3119.706	3120.611	32045.01	CR	1.5	0.8		MIT
3119.246	3120.150	32049.74	CR	0.6			MIT
3118.652	3119.556	32055.84	CR II	1.5	2.3		MIT
3117.258	3118.162	32070.18	CR	0.0	1.5		MIT
3116.744	3117.648	32075.46	CR		1.5		MIT

AIR WAVELENGTH (ANGSTRMS)	VACUUM WAVELENGTH (ANGSTRMS)	WAVENUMBER (KAYSERS)	LMENT	LOG ARC	INTENSITY SPK	DIS	REF
3115.645	3116.549	32086.78	CR		1.6		MIT
3115.274	3116.177	32090.60	CR II		1.5		MIT
3114.86	3115.76	32094.9	CR	1.2			MIT
3114.45	3115.35	32099.1	CR	1.0			MIT
3114.08	3114.98	32102.9	CR	0.3J			MIT
3112.98	3113.88	32114.2	CR	1.1			MIT
3111.941	3112.844	32124.97	CR II		1.6		MIT
3111.34	3112.24	32131.2	CR	0.8			MIT
3111.00	3111.90	32134.7	CR	0.9			MIT
3110.860	3111.762	32136.13	CR	1.4	0.9		MIT
3109.79	3110.69	32147.2	CR	0.3			MIT
3109.336	3110.238	32151.88	CR	1.5	1.1		MIT
3108.649	3109.551	32158.99	CR		1.5		MIT
3107.572	3108.474	32170.13	CR	0.3	2.1		MIT
3107.24	3108.14	32173.6	CR	0.9H			MIT
3105.56	3106.46	32191.0	CR	1.0			MIT
3104.70	3105.60	32199.9	CR	1.0	1.3		MIT
3103.474	3104.375	32212.61	CR		1.7		MIT
3098.160	3099.059	32267.86	CR		1.5		MIT
3096.70	3097.60	32283.1	CR	1.2	0.3		MIT
3096.531	3097.430	32284.83	CR	1.5			MIT
3096.13	3097.03	32289.0	CR	0.0	2.0		MIT
3095.859	3096.758	32291.84	CR	2.1	0.5		MIT
3095.488	3096.387	32295.71	CR		1.1		MIT
3095.382	3096.280	32296.82	CR	1.2	0.3		MIT
3094.927	3095.825	32301.56	CR		1.2		MIT
3093.947	3094.845	32311.79	CR		1.4		MIT
3093.488	3094.386	32316.59	CR	0.0	2.0		MIT
3087.884	3088.781	32375.24	CR		1.6		MIT
3087.540	3088.437	32378.84	CR	1.0	0.3		MIT
3086.77	3087.67	32386.9	CR	0.9	0.0		MIT
3085.352	3086.248	32401.80	CR		1.2		MIT
3084.56	3085.46	32410.1	CR	1.0	0.0		MIT
3084.453	3085.349	32411.25	CR		1.5		MIT
3083.601	3084.497	32420.20	CR		1.5		MIT
3080.72	3081.61	32450.5	CR	1.1	0.0		MIT
3079.332	3080.226	32465.15	CR		1.4		MIT
3077.831	3078.725	32480.98	CR I	1.4	2.1R		MIT
3077.79	3078.68	32481.4	CR II				ME
3077.247	3078.141	32487.14	CR		1.6		MIT
3076.54	3077.43	32494.6	CR I	1.0			MIT
3076.15	3077.04	32498.7	CR	1.0			MIT
3073.679	3074.572	32524.85	CR	1.5	1.4		MIT
3073.228	3074.121	32529.62	CR II		1.2		MIT
3072.462	3073.355	32537.73	CR		1.2		MIT
3071.572	3072.465	32547.16	CR II		1.1		MIT
3071.30	3072.19	32550.0	CR I	1.2	0.0		MIT
3067.16	3068.05	32594.0	CR II	1.4	1.6		MIT
3065.067	3065.958	32616.23	CR I	1.7	1.3		ME
3063.839	3064.730	32629.31	CR	0.7	1.1		MIT
3063.258	3064.148	32635.50	CR		0.9		MIT
3061.81	3062.70	32650.9	CR	1.0	0.3		MIT
3061.652	3062.542	32652.61	CR I	1.3	0.5		MIT
3059.521	3060.410	32675.35	CR II	0.9	1.8		MIT
3058.345	3059.234	32687.92	CR II		1.5		MIT
3057.861	3058.750	32693.09	CR		1.5		MIT
3056.619	3057.508	32706.38	CR II		0.9		MIT
3055.472	3056.360	32718.65	CR		1.3		MIT
3053.880	3054.768	32735.71	CR I	2.3	2.2		MIT
3053.673	3054.561	32737.93	CR		0.7		MIT
3052.229	3053.117	32753.42	CR	1.3	1.0		MIT
3050.137	3051.024	32775.88	CR	1.0	2.2		MIT
3049.883	3050.770	32778.61	CR I	1.3R			MIT
3047.768	3048.655	32801.35	CR II	0.5	1.2		MIT
3047.455	3048.341	32804.72	CR	1.4	0.8		MIT
3044.223	3045.109	32839.55	CR		1.1		MIT
3043.883	3044.769	32843.22	CR		1.7		MIT
3042.789	3043.674	32855.03	CR II	0.0	2.0		MIT
3041.74	3042.63	32866.4	CR	0.3	2.1		MIT
3040.846	3041.731	32876.02	CR I	2.7G	2.3		MIT

AIR WAVELENGTH (ANGSTRMS)	VACUUM WAVELENGTH (ANGSTRMS)	WAVENUMBER (KAYSERS)	LMENT	LOG ARC	INTENSITY SPK	DIS	REF
3039.780	3040.665	32887.55	CR I	1.9	1.5		MIT
3038.043	3038.927	32906.35	CR		1.0		MIT
3037.044	3037.928	32917.17	CR I	2.3R	2.0		MIT
3034.542	3035.425	32944.31	CR		1.5		MIT
3034.190	3035.073	32948.13	CR I	2.3R	1.8		MIT
3032.927	3033.810	32961.85	CR II	1.0	2.0		MIT
3031.486	3032.368	32977.52	CR	1.2	0.9		MIT
3031.353	3032.235	32978.97	CR I	1.6	1.5		MIT
3030.245	3031.127	32991.03	CR I	2.3R	2.2		MIT
3029.164	3030.046	33002.80	CR I	1.8	1.7		MIT
3028.125	3029.007	33014.12	CR	0.3	2.1		MIT
3026.647	3027.528	33030.25	CR	0.9	2.1		MIT
3024.681	3025.562	33051.71	CR	1.0	0.9		MIT
3024.350	3025.231	33055.33	CR I	2.5R	2.1		MIT
3021.558	3022.438	33085.87	CR I	2.5R	2.3R		MIT
3020.673	3021.553	33095.57	CR I	2.3R	2.0		MIT
3018.821	3019.700	33115.87	CR I	2.3R	1.8		MIT
3018.496	3019.375	33119.43	CR I	2.3R	2.1		MIT
3017.569	3018.448	33129.61	CR I	2.5R	2.3		MIT
3015.510	3016.388	33152.23	CR	0.0	2.2		MIT
3015.194	3016.072	33155.70	CR I	2.3R	1.9		MIT
3014.915	3015.793	33158.77	CR I	2.5R	2.0		MIT
3014.760	3015.638	33160.48	CR I	2.5R	2.0		MIT
3013.713	3014.591	33172.00	CR I	2.3R	2.2		MIT
3013.030	3013.908	33179.51	CR I	1.9	1.6		MIT
3011.09	3011.97	33200.9	CR	1.2	0.3		MIT
3010.642	3011.519	33205.83	CR		1.6		MIT
3005.057	3005.933	33267.54	CR I	2.5R	2.1		MIT
3003.924	3004.800	33280.09	CR	0.0	2.2		MIT
3000.890	3001.765	33313.73	CR I	2.2R	2.1		MIT
2998.787	2999.661	33337.10	CR I	2.3R	1.8		MIT
2998.121	2998.995	33344.50	CR	1.1	0.3		MIT
2996.580	2997.454	33361.65	CR I	2.5R	2.1		MIT
2995.103	2995.976	33378.10	CR I	2.3R	1.9		MIT
2994.737	2995.610	33382.18	CR	0.5	0.3		MIT
2994.069	2994.942	33389.63	CR I	2.2	1.7		MIT
2992.45	2993.32	33407.7	CR		1.3		MIT
2991.886	2992.759	33413.99	CR I	2.1R	1.8		MIT
2989.194	2990.066	33444.08	CR	1.0	2.0		MIT
2988.649	2989.521	33450.18	CR I	2.3R	2.2		MIT
2986.473	2987.344	33474.55	CR I	2.1R	2.1		MIT
2986.137	2987.008	33478.31	CR I	0.8	0.9		MIT
2985.995	2986.866	33479.91	CR I	1.4R	1.2		MIT
2985.849	2986.720	33481.54	CR I	1.4R	1.2		MIT
2985.325	2986.196	33487.42	CR	1.0	1.8		MIT
2984.826	2985.697	33493.02	CR	1.1			MIT
2984.026	2984.897	33502.00	CR	1.2	0.6		MIT
2980.791	2981.661	33538.35	CR I	1.9R	1.7		MIT
2979.741	2980.611	33550.17	CR	1.0	1.8		MIT
2976.718	2977.587	33584.24	CR	0.0	1.1		MIT
2975.483	2976.352	33598.18	CR I	2.0R	1.7		MIT
2971.906	2972.774	33638.62	CR II	0.9	1.5V		ME
2971.112	2971.979	33647.61	CR I	1.9R	1.2		MIT
2969.67	2970.54	33664.0	CR II				ME
2968.997	2969.864	33671.58	CR I	0.5			MIT
2968.177	2969.044	33680.88	CR I	0.7H			MIT
2967.642	2968.509	33686.95	CR I	1.8R	1.5		MIT
2966.051	2966.917	33705.02	CR	0.3	1.8		MIT
2963.469	2964.335	33734.38	CR		1.3		MIT
2962.395	2963.260	33746.61	CR	0.9			MIT
2961.732	2962.597	33754.17	CR	1.0	1.8		MIT
2959.550	2960.415	33779.05	CR II		1.3		MIT
2957.56	2958.42	33801.8	CR I	1.0H	0.0H		MIT
2957.28	2958.14	33805.0	CR I				ME
2956.603	2957.467	33812.72	CR		1.3		MIT
2956.330	2957.194	33815.84	CR	1.3	0.0		MIT
2955.12	2955.98	33829.7	CR II				ME
2954.67	2955.53	33834.8	CR II				ME
2953.706	2954.569	33845.88	CR II	0.8	1.4		ME
2953.358	2954.221	33849.87	CR II	0.6	1.7		ME

AIR WAVELENGTH (ANGSTRMS)	VACUUM WAVELENGTH (ANGSTRMS)	WAVENUMBER (KAYSERS)	LMENT	LOG ARC	INTENSITY SPK	DIS	REF
2952.459	2953.322	33860.18	CR		1.3		MIT
2951.95	2952.81	33866.0	CR II				ME
2951.40	2952.26	33872.3	CR II				ME
2950.687	2951.549	33880.51	CR		0.7		MIT
2949.44	2950.30	33894.8	CR		1.3		MIT
2948.89	2949.75	33901.2	CR I	0.9J			MIT
2947.497	2948.359	33917.18	CR		1.4		MIT
2946.842	2947.703	33924.72	CR	0.7	1.5		MIT
2941.880	2942.740	33981.93	CR I	1.1	1.4		MIT
2940.978	2941.838	33992.35	CR		1.0		MIT
2940.340	2941.200	33999.73	CR	0.9C			MIT
2940.225	2941.085	34001.06	CR		1.5		MIT
2939.45	2940.31	34010.0	CR	0.8	1.3		MIT
2938.85	2939.71	34017.0	CR	0.8J			MIT
2936.93	2937.79	34039.2	CR	0.6	1.3		MIT
2935.538	2936.397	34055.35	CR	0.9			MIT
2935.139	2935.998	34059.97	CR	0.9	1.6		MIT
2934.493	2935.351	34067.47	CR	1.4H			MIT
2934.32	2935.18	34069.5	CR		0.8		MIT
2933.970	2934.828	34073.54	CR	0.3H	1.6		MIT
2933.43	2934.29	34079.8	CR	0.7H			KN
2932.705	2933.563	34088.24	CR	0.3	1.4		MIT
2930.853	2931.710	34109.78	CR		1.7		MIT
2929.445	2930.302	34126.17	CR		1.6		MIT
2928.301	2929.158	34139.51	CR	0.8	1.2		MIT
2928.149	2929.006	34141.28	CR	0.8	1.3		MIT
2927.081	2927.938	34153.73	CR	0.3	1.9		MIT
2926.159	2927.015	34164.50	CR II	0.3	1.6		MIT
2925.60	2926.46	34171.0	CR	0.6H			MIT
2923.684	2924.540	34193.42	CR	0.0	1.4		MIT
2922.451	2923.306	34207.84	CR		0.8		MIT
2921.817	2922.672	34215.26	CR II	0.6	1.8		MIT
2921.354	2922.209	34220.69	CR	0.8			MIT
2921.245	2922.100	34221.96	CR	0.0	1.4		MIT
2916.165	2917.019	34281.58	CR I	1.3	1.2		MIT
2915.232	2916.086	34292.55	CR	0.0	1.4		MIT
2913.727	2914.580	34310.26	CR	1.8R	0.7		MIT
2911.682	2912.535	34334.35	CR		1.6		MIT
2911.145	2911.998	34340.69	CR I	1.6	0.9		MIT
2910.902	2911.755	34343.56	CR I	1.8R	0.9		MIT
2910.653	2911.505	34346.49	CR		1.7		MIT
2909.052	2909.904	34365.39	CR I	1.8R	1.1		MIT
2905.487	2906.338	34407.56	CR I	1.8R	0.9		MIT
2904.681	2905.532	34417.10	CR	1.3	0.3		MIT
2900.265	2901.115	34469.51	CR I	1.3			MIT
2899.483	2900.333	34478.80	CR	0.3	1.6		MIT
2899.208	2900.058	34482.07	CR I	1.7	1.4		MIT
2898.536	2899.385	34490.07	CR	1.1	1.6		MIT
2897.704	2898.553	34499.97	CR	0.5	1.4		MIT
2896.753	2897.602	34511.30	CR I	1.8R	1.5		MIT
2896.461	2897.310	34514.77	CR	0.8	1.5		MIT
2894.25	2895.10	34541.1	CR II		1.4		MIT
2894.171	2895.019	34542.08	CR	1.6	0.3		MIT
2893.486	2894.334	34550.26	CR		0.7H		MIT
2893.254	2894.102	34553.03	CR I	1.9R	1.0		MIT
2892.953	2893.801	34556.62	CR	0.3	1.3		MIT
2891.88	2892.73	34569.5	CR		1.5		MIT
2891.415	2892.263	34575.01	CR	1.5	1.2		MIT
2891.104	2891.952	34578.73	CR	1.0	1.6		MIT
2890.726	2891.574	34583.25	CR	1.3			MIT
2890.16	2891.01	34590.0	CR	1.3			MIT
2889.484	2890.331	34598.11	CR	0.3	1.4		MIT
2889.265	2890.112	34600.73	CR I	1.8R	1.5		MIT
2889.205	2890.052	34601.45	CR	1.2	1.3		MIT
2888.74	2889.59	34607.0	CR	0.5	1.6		MIT
2888.382	2889.229	34611.31	CR I	1.2			MIT
2887.77	2888.62	34618.6	CR		1.5		MIT
2886.997	2887.844	34627.91	CR I	2.0	1.3		MIT
2881.931	2882.776	34688.78	CR	0.0	1.5		HL
2881.141	2881.986	34698.29	CR	1.4	0.3		MIT

AIR WAVELENGTH (ANGSTRMS)	VACUUM WAVELENGTH (ANGSTRMS)	WAVENUMBER (KAYSERS)	LMENT	LOG ARC	INTENSITY SPK	DIS	REF
2880.869	2881.714	34701.57	CR II	1.3	1.4		MIT
2879.274	2880.119	34720.79	CR I	1.8	1.1		MIT
2878.449	2879.294	34730.74	CR	1.3	1.9		MIT
2877.978	2878.822	34736.43	CR II	1.5	2.0		MIT
2876.661	2877.505	34752.33	CR		1.3		MIT
2876.245	2877.089	34757.35	CR II	1.4	1.9V		MIT
2875.993	2876.837	34760.40	CR II	1.5	1.9J		MIT
2873.819	2874.662	34786.69	CR II	1.3	1.6		MIT
2873.485	2874.328	34790.74	CR II	1.5	2.1		MIT
2873.186	2874.029	34794.36	CR	1.4			MIT
2871.632	2872.475	34813.19	CR I	1.7	0.3		MIT
2871.454	2872.297	34815.34	CR		1.9		MIT
2870.436	2871.279	34827.69	CR II	1.4	2.5C		MIT
2870.178	2871.020	34830.82	CR	1.1			MIT
2870.04	2870.88	34832.5	CR	1.1H			MIT
2867.648	2868.490	34861.55	CR II	1.9	2.0G		MIT
2867.096	2867.938	34868.26	CR	1.3	1.5		MIT
2866.742	2867.584	34872.57	CR II	1.9	2.1G		MIT
2865.892	2866.733	34882.91	CR	0.0	1.2		MIT
2865.676	2866.517	34885.54	CR		1.3		MIT
2865.334	2866.175	34889.70	CR	1.2	1.2		MIT
2865.107	2865.948	34892.47	CR II	1.8	2.3G		MIT
2862.571	2863.412	34923.38	CR II	1.9	2.5G		MIT
2860.934	2861.774	34943.36	CR II	1.8	2.0		MIT
2858.911	2859.751	34968.08	CR II	1.7	1.9V		MIT
2858.654	2859.494	34971.23	CR II	1.3	1.5		MIT
2857.97	2858.81	34979.6	CR	0.3	1.6		MIT
2857.402	2858.241	34986.55	CR II	1.3	1.9		MIT
2856.766	2857.605	34994.34	CR II	1.3	1.8		MIT
2855.676	2856.515	35007.69	CR II	1.8	2.3V		MIT
2855.073	2855.912	35015.09	CR	0.6	2.0		MIT
2853.218	2854.056	35037.85	CR II	0.7	2.0G		ME
2851.356	2852.194	35060.73	CR	1.3	1.9		MIT
2850.709	2851.547	35068.69	CR		0.8		MIT
2850.293	2851.131	35073.81	CR		0.7		MIT
2849.838	2850.675	35079.40	CR II	1.9	2.2R		MIT
2849.287	2850.124	35086.19	CR	1.5	1.5		MIT
2848.40	2849.24	35097.1	CR	0.8	1.5		MIT
2846.70	2847.54	35118.1	CR	0.0	1.1		MIT
2846.437	2847.274	35121.32	CR	0.3	1.4		MIT
2846.023	2846.860	35126.43	CR	1.4	0.6		MIT
2843.252	2844.088	35160.66	CR II	2.1	2.6R		MIT
2842.767	2843.603	35166.66	CR		1.1		MIT
2842.428	2843.264	35170.85	CR		0.9		MIT
2840.438	2841.273	35195.49	CR		0.8		MIT
2840.021	2840.856	35200.66	CR	1.4	2.1		MIT
2839.235	2840.070	35210.40	CR		1.3		MIT
2838.786	2839.621	35215.97	CR	0.5			MIT
2837.877	2838.712	35227.25	CR	0.3	1.5		MIT
2836.479	2837.313	35244.61	CR	0.5	1.3		MIT
2835.633	2836.467	35255.13	CR II	2.0	2.6R		MIT
2834.262	2835.096	35272.18	CR		2.1		MIT
2833.393	2834.226	35283.00	CR		0.5		MIT
2832.46	2833.29	35294.6	CR	0.3	2.1		MIT
2831.036	2831.869	35312.37	CR	1.2	0.0		MIT
2830.468	2831.301	35319.46	CR	1.2	1.9H		MIT
2828.786	2829.618	35340.46	CR		1.1		MIT
2828.167	2828.999	35348.19	CR	1.2	0.3		MIT
2827.958	2828.790	35350.80	CR		0.9		MIT
2826.746	2827.578	35365.96	CR	1.8	0.5		MIT
2826.407	2827.239	35370.20	CR		0.8		MIT
2826.151	2826.983	35373.40	CR		1.1		MIT
2825.490	2826.322	35381.68	CR	0.0	1.3		MIT
2824.537	2825.368	35393.62	CR		1.0		MIT
2822.371	2823.202	35420.78	CR	1.3	2.0		MIT
2822.012	2822.843	35425.28	CR	1.0	1.9		MIT
2820.824	2821.654	35440.20	CR	1.3	0.0		MIT
2818.473	2819.303	35469.76	CR	1.2			MIT
2818.359	2819.189	35471.20	CR	0.9	1.9		MIT
2817.949	2818.779	35476.36	CR		1.3		MIT

AIR WAVELENGTH (ANGSTRMS)	VACUUM WAVELENGTH (ANGSTRMS)	WAVENUMBER (KAYSERS)	LMENT	LOG ARC	INTENSITY SPK	DIS	REF
2817.570	2818.400	35481.13	CR		1.2		MIT
2816.842	2817.671	35490.30	CR	0.5	1.5		MIT
2816.699	2817.528	35492.10	CR	1.1			MIT
2814.536	2815.365	35519.37	CR	1.2	0.3		MIT
2814.220	2815.049	35523.36	CR		0.7		MIT
2812.006	2812.834	35551.33	CR	0.9	1.9		MIT
2811.459	2812.287	35558.25	CR		0.8		MIT
2811.171	2811.999	35561.89	CR	1.0			MIT
2811.049	2811.877	35563.43	CR		1.2		MIT
2810.895	2811.723	35565.38	CR		0.5		MIT
2809.940	2810.768	35577.47	CR	1.0			MIT
2809.612	2810.440	35581.62	CR		0.6H		MIT
2809.279	2810.107	35585.84	CR		0.8		MIT
2808.018	2808.845	35601.82	CR		1.5		MIT
2803.357	2804.183	35661.01	CR	0.0	1.5		MIT
2800.771	2801.596	35693.93	CR	1.1	2.2		MIT
2800.174	2800.999	35701.54	CR		1.6		MIT
2798.672	2799.497	35720.70	CR	1.0	1.3		MIT
2795.818	2796.642	35757.17	CR	1.5	0.5		MIT
2792.164	2792.987	35803.96	CR	0.7	1.9		MIT
2789.396	2790.219	35839.48	CR		1.3		MIT
2787.895	2788.717	35858.78	CR		1.3		MIT
2787.628	2788.450	35862.21	CR	0.7	1.5		MIT
2786.486	2787.308	35876.91	CR	0.0	1.5		MIT
2785.700	2786.522	35887.03	CR	0.7	1.9		MIT
2785.104	2785.926	35894.71	CR		1.0		MIT
2784.317	2785.138	35904.86	CR		0.5		MIT
2783.843	2784.664	35910.97	CR		1.5		MIT
2782.592	2783.413	35927.12	CR		1.3		MIT
2782.354	2783.175	35930.19	CR		1.5		MIT
2781.155	2781.976	35945.68	CR	1.0	1.1		MIT
2780.888	2781.709	35949.13	CR		1.3		MIT
2780.703	2781.524	35951.52	CR I	2.8G	1.2		MIT
2780.299	2781.119	35956.74	CR		2.0		MIT
2779.135	2779.955	35971.80	CR	1.3	0.7		MIT
2778.938	2779.758	35974.35	CR		1.1		MIT
2778.060	2778.880	35985.72	CR	1.1	1.8		MIT
2777.667	2778.487	35990.81	CR	1.6	0.0		MIT
2776.652	2777.472	36003.97	CR		1.4		MIT
2775.668	2776.487	36016.73	CR	1.5			MIT
2774.436	2775.255	36032.72	CR		2.0		MIT
2773.312	2774.131	36047.33	CR	0.0	1.6		MIT
2772.341	2773.160	36059.95	CR		0.9		MIT
2771.93	2772.75	36065.3	CR		1.1H		MIT
2771.448	2772.266	36071.57	CR	1.3	0.3		MIT
2771.289	2772.107	36073.64	CR		1.1		MIT
2769.915	2770.733	36091.53	CR I	2.6R	1.6		MIT
2768.591	2769.409	36108.79	CR	0.0	1.8		MIT
2768.154	2768.972	36114.49	CR		1.0		MIT
2767.54	2768.36	36122.5	CR	1.5	0.9D		EX
2767.277	2768.094	36125.94	CR		0.6		MIT
2766.540	2767.357	36135.56	CR II	1.6	2.5R		MIT
2765.863	2766.680	36144.40	CR		1.5		MIT
2765.474	2766.291	36149.49	CR	0.3	1.3		HL
2764.350	2765.167	36164.19	CR I	2.3R	0.8		MIT
2763.972	2764.788	36169.13	CR		1.5		MIT
2763.59	2764.41	36174.1	CR	0.0	1.5		MIT
2763.063	2763.879	36181.03	CR	1.5	0.6		MIT
2762.76	2763.58	36185.0	CR	0.0	0.7		MIT
2762.593	2763.409	36187.19	CR II	1.6	2.0		MIT
2761.755	2762.571	36198.17	CR I	2.5R	1.5		MIT
2760.837	2761.653	36210.20	CR		1.2		MIT
2760.523	2761.339	36214.32	CR		1.3		MIT
2760.357	2761.173	36216.50	CR	0.0	1.3		MIT
2760.054	2760.870	36220.47	CR	0.0	1.0		MIT
2759.729	2760.544	36224.74	CR	0.0	1.2		MIT
2759.391	2760.206	36229.17	CR	1.0	1.5		MIT
2758.977	2759.792	36234.61	CR	0.0	1.6		MIT
2758.616	2759.431	36239.35	CR		1.5		MIT
2757.723	2758.538	36251.09	CR II	1.5	2.2		MIT

AIR WAVELENGTH (ANGSTRMS)	VACUUM WAVELENGTH (ANGSTRMS)	WAVENUMBER (KAYSERS)	LMENT	LOG ARC	INTENSITY SPK	DIS	REF
2757.099	2757.914	36259.29	CR I	2.5R	1.0		MIT
2756.96	2757.77	36261.1	CR II		1.5		ME
2756.89	2757.70	36262.0	CR II		1.5		ME
2756.753	2757.568	36263.84	CR	1.0	0.0		MIT
2756.30	2757.11	36269.8	CR	0.0	2.0		MIT
2755.27	2756.08	36283.4	CR	1.7D	0.3		MIT
2754.90	2755.71	36288.2	CR	1.0			EX
2754.28	2755.09	36296.4	CR	0.5	1.7		MIT
2752.877	2753.691	36314.90	CR I	2.5R	1.6		MIT
2752.387	2753.201	36321.36	CR		1.0		MIT
2751.871	2752.685	36328.17	CR II	1.3	2.1		MIT
2751.599	2752.412	36331.76	CR I	1.5			MIT
2750.728	2751.541	36343.27	CR II	1.5	2.2		MIT
2749.82	2750.63	36355.3	CR		1.3		MIT
2748.985	2749.798	36366.31	CR II	1.5	2.3		MIT
2748.286	2749.099	36375.56	CR I	2.5	0.7		MIT
2747.926	2748.739	36380.32	CR		0.8		MIT
2746.180	2746.992	36403.45	CR		1.4		MIT
2745.427	2746.239	36413.44	CR		1.1		MIT
2744.977	2745.789	36419.41	CR		1.3		MIT
2744.591	2745.403	36424.53	CR		1.6		MIT
2743.643	2744.455	36437.11	CR II	1.5	2.1		MIT
2742.171	2742.982	36456.67	CR	1.5	0.5		MIT
2742.030	2742.841	36458.55	CR II	1.2	1.7		MIT
2741.068	2741.879	36471.34	CR	1.5	1.5		MIT
2740.095	2740.906	36484.29	CR II	1.3	1.8		MIT
2739.760	2740.571	36488.75	CR		1.1		MIT
2739.383	2740.194	36493.77	CR	1.5	1.4		MIT
2737.31	2738.12	36521.4	CR	1.0			MIT
2737.094	2737.904	36524.29	CR		1.0		MIT
2736.473	2737.283	36532.58	CR I	2.5R	1.7		MIT
2735.756	2736.566	36542.15	CR		0.9		MIT
2734.564	2735.373	36558.08	CR		1.0		MIT
2731.908	2732.717	36593.62	CR I	2.5R	1.5		MIT
2729.737	2730.545	36622.72	CR		1.1		MIT
2728.162	2728.970	36643.87	CR		1.4		MIT
2727.257	2728.065	36656.02	CR	0.6	1.8		MIT
2726.511	2727.318	36666.05	CR I	2.5R	1.6		MIT
2726.241	2727.048	36669.68	CR		1.2		MIT
2724.04	2724.85	36699.3	CR	0.0	1.6		MIT
2723.63	2724.44	36704.8	CR		1.5		MIT
2722.749	2723.555	36716.71	CR II	1.5	1.9		MIT
2720.686	2721.492	36744.55	CR		1.3		MIT
2720.26	2721.07	36750.3	CR	0.0	1.5		MIT
2720.07	2720.88	36752.9	CR	0.7	1.5		MIT
2718.070	2718.875	36779.91	CR	0.9	0.8		MIT
2717.509	2718.314	36787.51	CR II	1.1	1.6		MIT
2717.053	2717.858	36793.68	CR		0.3		MIT
2716.181	2716.986	36805.49	CR I	1.3	0.6		MIT
2712.307	2713.111	36858.06	CR II	1.5	1.8		MIT
2710.92	2711.72	36876.9	CR	0.0	1.8		MIT
2710.23	2711.03	36886.3	CR	1.7H			MIT
2709.31	2710.11	36898.8	CR	0.3	1.8		MIT
2708.79	2709.59	36905.9	CR	0.5	1.6		MIT
2705.726	2706.528	36947.70	CR	1.1			MIT
2705.430	2706.232	36951.74	CR	1.1	0.3		MIT
2704.747	2705.549	36961.07	CR	1.2R	0.3		MIT
2703.856	2704.658	36973.25	CR II	0.9	1.5		MIT
2703.554	2704.356	36977.38	CR		1.7		MIT
2703.480	2704.282	36978.39	CR I	1.2			MIT
2702.533	2703.335	36991.35	CR	1.0	0.3		MIT
2701.989	2702.790	36998.80	CR I	1.5	0.9		MIT
2701.656	2702.457	37003.36	CR		0.7		MIT
2701.250	2702.051	37008.92	CR		0.7		MIT
2701.09	2701.89	37011.1	CR	0.0	0.9		MIT
2700.597	2701.398	37017.87	CR I	1.5	0.3		MIT
2699.349	2700.150	37034.98	CR		0.9		MIT
2698.850	2699.651	37041.83	CR		1.5		MIT
2698.686	2699.487	37044.08	CR II	1.1	1.5		MIT
2698.409	2699.210	37047.88	CR II	1.1	1.5		MIT

AIR WAVELENGTH (ANGSTRMS)	VACUUM WAVELENGTH (ANGSTRMS)	WAVENUMBER (KAYSERS)	LMENT	LOG ARC	INTENSITY SPK	INTENSITY DIS	REF
2697.907	2698.708	37054.77	CR	0.5	1.5		MIT
2697.50	2698.30	37060.4	CR	0.0	1.5		MIT
2696.750	2697.550	37070.67	CR		1.3		MIT
2696.543	2697.343	37073.52	CR I	1.5	0.0		MIT
2694.694	2695.494	37098.96	CR		0.3		MIT
2693.52	2694.32	37115.1	CR	0.0	1.6		MIT
2692.12	2692.92	37134.4	CR	0.0	1.1		MIT
2691.041	2691.840	37149.31	CR II	1.5	2.1		MIT
2690.257	2691.056	37160.14	CR I	1.5	0.3		MIT
2689.19	2689.99	37174.9	CR		1.0		MIT
2688.293	2689.091	37187.29	CR	0.0	1.2		MIT
2688.042	2688.840	37190.76	CR I	1.5			MIT
2687.090	2687.888	37203.93	CR II	1.5	1.8		MIT
2685.182	2685.979	37230.37	CR		0.6		MIT
2684.095	2684.892	37245.44	CR		0.3		MIT
2683.441	2684.238	37254.52	CR		1.1		MIT
2681.468	2682.265	37281.93	CR I	0.9	0.5		MIT
2680.341	2681.137	37297.60	CR I	1.0	0.6		MIT
2679.890	2680.686	37303.88	CR		0.5		MIT
2678.792	2679.588	37319.17	CR II	1.0	1.9		MIT
2678.165	2678.961	37327.91	CR I	0.9	0.5		MIT
2677.159	2677.955	37341.93	CR II	1.5	2.5R		MIT
2676.542	2677.337	37350.54	CR		0.5		MIT
2675.682	2676.477	37362.55	CR	0.0	1.2		MIT
2673.656	2674.451	37390.86	CR I	0.8	0.0		MIT
2672.831	2673.625	37402.40	CR II	1.1	1.2		MIT
2672.374	2673.168	37408.79	CR	0.0	0.7		MIT
2671.985	2672.779	37414.24	CR I	0.9			MIT
2671.809	2672.603	37416.70	CR II	1.5	1.2		MIT
2670.240	2671.034	37438.69	CR	0.0	1.0		MIT
2670.07	2670.86	37441.1	CR	0.3	1.1		MIT
2669.371	2670.165	37450.87	CR I	0.8	0.0		MIT
2668.712	2669.505	37460.12	CR II	1.0	1.1		MIT
2667.879	2668.672	37471.82	CR		0.5		MIT
2666.021	2666.814	37497.93	CR II	1.0	0.9		MIT
2663.679	2664.471	37530.90	CR II	0.8	0.7		MIT
2663.423	2664.215	37534.50	CR II	1.1	0.8		MIT
2661.728	2662.520	37558.41	CR II	0.9	0.8		MIT
2661.401	2662.193	37563.02	CR		0.3		MIT
2659.746	2660.537	37586.39	CR		0.3		MIT
2658.92	2659.71	37598.1	CR	0.0	0.7		MIT
2658.592	2659.383	37602.71	CR II	1.3	1.5		MIT
2657.552	2658.343	37617.42	CR		0.5		MIT
2655.778	2656.568	37642.55	CR		0.3		MIT
2653.586	2654.376	37673.64	CR I	1.1	1.5		MIT
2652.097	2652.887	37694.79	CR		0.3		MIT
2648.105	2648.894	37751.61	CR		0.3		MIT
2643.532	2644.319	37816.91	CR		0.8		MIT
2642.116	2642.903	37837.18	CR	1.5	0.5		MIT
2641.796	2642.583	37841.76	CR		0.9		MIT
2641.368	2642.155	37847.89	CR		0.3		MIT
2639.430	2640.216	37875.68	CR	1.5			MIT
2638.895	2639.681	37883.36	CR	1.2	0.0		MIT
2637.477	2638.263	37903.73	CR		0.6		MIT
2637.194	2637.980	37907.79	CR		0.3		MIT
2636.450	2637.236	37918.49	CR		0.3		MIT
2634.311	2635.096	37949.28	CR		0.3		MIT
2630.916	2631.700	37998.25	CR	0.5	1.0		MIT
2629.824	2630.608	38014.02	CR I	0.9	0.0		MIT
2628.005	2628.789	38040.33	CR		0.5		MIT
2626.782	2627.565	38058.04	CR		0.5		MIT
2626.597	2627.380	38060.72	CR	1.1	0.0		MIT
2625.323	2626.106	38079.19	CR I	1.1	0.0		MIT
2623.386	2624.169	38107.31	CR		0.3		MIT
2622.863	2623.646	38114.90	CR	1.4	0.3		MIT
2620.845	2621.627	38144.25	CR	0.8			MIT
2620.510	2621.292	38149.13	CR		0.7		MIT
2620.476	2621.258	38149.62	CR I	0.9	0.5		MIT
2619.978	2620.760	38156.87	CR	0.8	0.0		MIT
2619.650	2620.432	38161.65	CR		1.2		MIT

AIR WAVELENGTH (ANGSTRMS)	VACUUM WAVELENGTH (ANGSTRMS)	WAVENUMBER (KAYSERS)	LMENT	LOG ARC	INTENSITY SPK	INTENSITY DIS	REF
2619.490	2620.272	38163.98	CR	0.9			MIT
2618.269	2619.050	38181.78	CR I	1.2			MIT
2616.230	2617.011	38211.53	CR		0.8		MIT
2614.63	2615.41	38234.9	CR		0.8		MIT
2613.503	2614.283	38251.40	CR		0.3		MIT
2613.311	2614.091	38254.21	CR	1.0	0.0		MIT
2612.555	2613.335	38265.28	CR		0.3		MIT
2612.204	2612.984	38270.42	CR	0.9	0.0		MIT
2612.024	2612.804	38273.06	CR	0.9	0.0		MIT
2611.628	2612.408	38278.86	CR		0.5		MIT
2611.045	2611.825	38287.41	CR		0.3		MIT
2610.790	2611.570	38291.15	CR		0.8		MIT
2610.294	2611.074	38298.42	CR I	0.9			MIT
2608.396	2609.175	38326.29	CR	0.9			MIT
2608.162	2608.941	38329.73	CR		0.6		MIT
2607.906	2608.685	38333.49	CR	0.0	1.5		MIT
2607.633	2608.412	38337.50	CR		0.5		MIT
2606.523	2607.302	38353.83	CR		0.7		MIT
2606.066	2606.845	38360.55	CR		0.3		MIT
2605.611	2606.389	38367.25	CR		0.6		MIT
2604.152	2604.930	38388.75	CR		0.6		MIT
2603.725	2604.503	38395.04	CR		0.3		MIT
2603.570	2604.348	38397.33	CR I	1.5	0.0		MIT
2601.845	2602.623	38422.78	CR		0.5		MIT
2596.173	2596.949	38506.72	CR		0.8		MIT
2595.558	2596.334	38515.85	CR		0.8		MIT
2591.848	2592.623	38570.97	CR I	2.0R	1.1		MIT
2590.729	2591.504	38587.63	CR	0.0	1.1		MIT
2589.695	2590.470	38603.04	CR		0.9		MIT
2589.053	2589.828	38612.61	CR		0.3		MIT
2588.203	2588.977	38625.29	CR I	0.9	0.0		MIT
2587.405	2588.179	38637.20	CR		0.3		MIT
2585.603	2586.377	38664.13	CR		0.3		MIT
2584.904	2585.678	38674.58	CR		0.3		MIT
2584.656	2585.429	38678.29	CR	1.0	0.0		MIT
2584.098	2584.871	38686.65	CR	0.0	1.4		MIT
2583.626	2584.399	38693.71	CR		0.3		MIT
2583.018	2583.791	38702.82	CR	1.0			MIT
2582.118	2582.891	38716.31	CR		0.5		MIT
2579.155	2579.927	38760.78	CR I	1.1	0.5		MIT
2578.264	2579.036	38774.18	CR	1.0	0.7		MIT
2577.651	2578.423	38783.40	CR I	1.5	0.5		MIT
2575.803	2576.574	38811.22	CR		0.6		MIT
2573.540	2574.311	38845.35	CR		0.8		MIT
2572.146	2572.917	38866.40	CR I	0.7	0.5		MIT
2571.741	2572.511	38872.52	CR I	1.7R	1.5		MIT
2568.526	2569.296	38921.17	CR I	0.7	0.3		MIT
2568.100	2568.870	38927.63	CR I	0.9			MIT
2567.599	2568.368	38935.22	CR		0.3		MIT
2566.871	2567.640	38946.27	CR		0.3		MIT
2566.553	2567.322	38951.09	CR I	1.0	0.5		MIT
2563.581	2564.349	38996.25	CR	0.0	1.2		MIT
2563.430	2564.198	38998.54	CR II		0.7		HL
2563.36	2564.13	38999.6	CR		1.0D		MIT
2560.688	2561.456	39040.30	CR I	1.5	1.2		MIT
2559.80	2560.57	39053.8	CR		0.9		MIT
2557.454	2558.221	39089.66	CR		0.3		MIT
2557.153	2557.920	39094.27	CR I	1.5	0.3		MIT
2555.55	2556.32	39118.8	CR	0.8	0.5		MIT
2553.062	2553.828	39156.90	CR I	1.3	0.0		MIT
2551.590	2552.356	39179.49	CR	0.0	1.3		MIT
2549.535	2550.300	39211.07	CR I	1.4	0.3		MIT
2548.593	2549.358	39225.56	CR		0.9		MIT
2548.056	2548.821	39233.83	CR		0.7		MIT
2547.554	2548.319	39241.56	CR		0.5		MIT
2546.456	2547.220	39258.48	CR		0.3		MIT
2545.644	2546.408	39271.00	CR I	1.2	0.0		MIT
2545.215	2545.979	39277.62	CR	0.9			MIT
2544.32	2545.08	39291.4	CR II		0.3		MIT
2543.140	2543.904	39309.66	CR		1.0		MIT

AIR WAVELENGTH (ANGSTRMS)	VACUUM WAVELENGTH (ANGSTRMS)	WAVENUMBER (KAYSERS)	LMENT	LOG ARC	INTENSITY SPK	INTENSITY DIS	REF
2541.669	2542.432	39332.41	CR I	1.1			MIT
2541.353	2542.116	39337.30	CR I	1.8	0.5		MIT
2538.953	2539.716	39374.49	CR		0.3		MIT
2538.29	2539.05	39384.8	CR		1.4		MIT
2534.336	2535.098	39446.21	CR II	0.9	1.0		MIT
2531.840	2532.601	39485.10	CR II	0.6	0.7		MIT
2531.00	2531.76	39498.2	CR		0.9		MIT
2530.447	2531.208	39506.83	CR I	1.5	0.0		MIT
2530.22	2530.98	39510.4	CR		0.8		MIT
2529.95	2530.71	39514.6	CR		1.3		MIT
2529.14	2529.90	39527.2	CR	0.3W			MIT
2528.244	2529.004	39541.25	CR I	1.5			MIT
2528.013	2528.773	39544.87	CR I	1.5	0.0		MIT
2527.115	2527.875	39558.92	CR I	0.9	0.3		MIT
2523.32	2524.08	39618.4	CR		1.3H		MIT
2520.656	2521.414	39660.28	CR		1.2		MIT
2519.515	2520.273	39678.24	CR I	2.2R	0.8		MIT
2518.711	2519.469	39690.90	CR I	1.3	0.0		MIT
2518.35	2519.11	39696.6	CR		1.2H		MIT
2518.085	2518.843	39700.77	CR	0.3			MIT
2517.868	2518.626	39704.19	CR	0.9			MIT
2517.566	2518.324	39708.95	CR I	1.2			MIT
2516.916	2517.674	39719.21	CR I	1.5	0.3		MIT
2516.65	2517.41	39723.4	CR		0.8H		MIT
2515.12	2515.88	39747.6	CR		1.1J		MIT
2513.71	2514.47	39769.9	CR		1.0H		MIT
2513.620	2514.377	39771.29	CR I	1.3	0.3		MIT
2512.40	2513.16	39790.6	CR		0.8		MIT
2511.956	2512.712	39797.63	CR	1.4	0.0		MIT
2510.619	2511.375	39818.82	CR I	0.9			MIT
2510.25	2511.01	39824.7	CR	0.0	0.3		MIT
2508.980	2509.736	39844.83	CR I	1.4	0.0		MIT
2508.110	2508.865	39858.65	CR I	1.5	0.0		MIT
2507.326	2508.081	39871.12	CR	1.2			MIT
2504.306	2505.061	39919.19	CR I	2.2R	0.5		MIT
2502.531	2503.285	39947.51	CR	2.0R	0.5		MIT
2500.671	2501.425	39977.22	CR I	0.9			MIT
2499.844	2500.598	39990.44	CR I	1.3			MIT
2498.822	2499.575	40006.80	CR		0.9		MIT
2496.309	2497.062	40047.07	CR I	2.1R	0.3		MIT
2495.077	2495.829	40066.84	CR I	1.5			MIT
2493.300	2494.052	40095.39	CR		0.3		MIT
2492.889	2493.641	40102.00	CR		0.6		MIT
2492.646	2493.398	40105.91	CR		0.8		MIT
2492.568	2493.320	40107.17	CR I	1.7			MIT
2491.348	2492.100	40126.81	CR I	1.5	0.0		MIT
2490.800	2491.551	40135.63	CR		0.3J		MIT
2490.088	2490.839	40147.11	CR		0.6		MIT
2489.293	2490.044	40159.93	CR		1.0		MIT
2486.681	2487.432	40202.11	CR		0.6		MIT
2486.318	2487.068	40207.98	CR		0.9		MIT
2485.49	2486.24	40221.4	CR		0.3H		MIT
2483.073	2483.823	40260.52	CR		0.8		MIT
2479.825	2480.574	40313.25	CR		0.3		MIT
2479.134	2479.883	40324.49	CR	1.3			MIT
2478.59	2479.34	40333.3	CR		0.5W		MIT
2476.922	2477.670	40360.50	CR		0.6		MIT
2475.695	2476.443	40380.50	CR		1.1		MIT
2474.948	2475.696	40392.69	CR		0.3		MIT
2474.072	2474.820	40406.99	CR	1.5			MIT
2472.86	2473.61	40426.8	CR		0.7		MIT
2469.139	2469.885	40487.71	CR		1.3		MIT
2466.63	2467.38	40528.9	CR		0.7J		MIT
2465.776	2466.522	40542.92	CR		0.5		MIT
2462.366	2463.111	40599.06	CR		0.5		MIT
2460.440	2461.184	40630.84	CR		0.8		MIT
2454.468	2455.211	40729.70	CR		0.9		MIT
2454.063	2454.806	40736.42	CR		0.3		MIT
2450.378	2451.120	40797.67	CR		0.5		MIT
2449.965	2450.707	40804.55	CR		1.0		MIT

AIR WAVELENGTH (ANGSTRMS)	VACUUM WAVELENGTH (ANGSTRMS)	WAVENUMBER (KAYSERS)	LMENT	LOG ARC	INTENSITY SPK	INTENSITY DIS	REF
2449.652	2450.394	40809.76	CR		1.0		MIT
2446.927	2447.668	40855.21	CR		0.6		MIT
2438.475	2439.214	40996.81	CR		1.5		MIT
2433.224	2433.962	41085.27	CR		1.0		MIT
2416.411	2417.145	41371.12	CR		0.7		MIT
2408.747	2409.480	41502.74	CR I	1.6	0.0		MIT
2408.622	2409.355	41504.89	CR I	2.2R	0.3R		MIT
2400.251	2400.982	41649.63	CR		0.3		MIT
2399.58	2400.31	41661.3	CR I	1.6			MIT
2399.06	2399.79	41670.3	CR I	1.7	0.3		MIT
2398.52	2399.25	41679.7	CR		1.1		MIT
2397.77	2398.50	41692.7	CR	0.7	1.5		MIT
2396.37	2397.10	41717.1	CR I	1.5	0.5		MIT
2395.79	2396.52	41727.2	CR I	1.4	0.3H		MIT
2394.01	2394.74	41758.2	CR		1.7		MIT
2392.89	2393.62	41777.7	CR I	1.6	0.3H		MIT
2392.37	2393.10	41786.8	CR	1.4	0.3H		MIT
2389.76	2390.49	41832.5	CR		1.4		MIT
2389.46	2390.19	41837.7	CR I	1.4			MIT
2386.78	2387.51	41884.7	CR I	1.3			MIT
2386.19	2386.92	41895.0	CR I	1.3			MIT
2385.74	2386.47	41902.9	CR I	1.4			MIT
2383.33	2384.06	41945.3	CR I	1.3			MIT
2381.48	2382.21	41977.9	CR	0.5	1.4		MIT
2379.95	2380.68	42004.9	CR	0.9			MIT
2373.73	2374.45	42114.9	CR I	1.8			A
2372.89	2373.61	42129.8	CR	1.6	0.5		MIT
2370.39	2371.11	42174.3	CR I	1.6	0.5		MIT
2368.49	2369.21	42208.1	CR I	1.5	0.3		MIT
2367.88	2368.60	42219.0	CR I	1.3			MIT
2366.85	2367.57	42237.3	CR I	1.8	1.8		MIT
2366.34	2367.06	42246.5	CR I	1.2	0.5		MIT
2366.15	2366.87	42249.8	CR I	1.0	0.5		MIT
2365.97	2366.69	42253.0	CR I	1.4	0.9		MIT
2365.15	2365.87	42267.7	CR	1.0	0.5		MIT
2364.71	2365.43	42275.6	CR I	1.3	0.9		MIT
2362.23	2362.95	42319.9	CR I	1.4	0.3H		MIT
2354.32	2355.04	42462.1	CR I	1.1	0.3H		MIT
2347.90	2348.62	42578.2	CR	0.3			MIT
2345.33	2346.05	42624.9	CR	0.9	1.8		MIT
2344.56	2345.28	42638.9	CR		1.2		MIT
2340.49	2341.21	42713.0	CR		1.0		MIT
2337.74	2338.46	42763.2	CR	0.5	1.1		MIT
2334.39	2335.11	42824.6	CR	0.7	1.0		MIT
2333.48	2334.20	42841.3	CR		1.4		MIT
2330.05	2330.76	42904.4	CR		1.0		MIT
2324.89	2325.60	42999.6	CR		1.7		MIT
2320.08	2320.79	43088.7	CR	1.0	1.5		MIT
2319.38	2320.09	43101.7	CR	1.0	1.3		MIT
2319.07	2319.78	43107.5	CR		0.3		MIT
2314.74	2315.45	43188.1	CR	0.9	1.7		MIT
2314.63	2315.34	43190.2	CR		0.9		MIT
2310.04	2310.75	43276.0	CR		0.8		MIT
2307.208	2307.918	43329.10	CR	1.0	1.3		MIT
2306.84	2307.55	43336.0	CR	0.3H	1.0		MIT
2300.52	2301.23	43455.0	CR		1.2		MIT
2297.188	2297.896	43518.08	CR	1.0	1.7		MIT
2295.57	2296.28	43548.7	CR 3		1.0		MIT
2290.68	2291.39	43641.7	CR 3		1.1		MIT
2289.26	2289.97	43668.8	CR 3		0.9		MIT
2287.16	2287.87	43708.9	CR 3		0.3H		ME
2284.68	2285.38	43756.3	CR	1.3			A
2284.53	2285.23	43759.2	CR	0.3H	1.0		MIT
2284.44	2285.14	43760.9	CR 3	0.3H	1.0		MIT
2277.49	2278.19	43894.4	CR		1.1		MIT
2276.42	2277.12	43915.1	CR		1.2		MIT
2275.48	2276.18	43933.2	CR		0.9		MIT
2273.36	2274.06	43974.2	CR		1.3		MIT
2268.15	2268.85	44075.2	CR	1.0			A
2266.67	2267.37	44103.9	CR	1.0			A

AIR WAVELENGTH (ANGSTRMS)	VACUUM WAVELENGTH (ANGSTRMS)	WAVENUMBER (KAYSERS)	LMENT	LOG ARC	INTENSITY SPK	DIS	REF
2264.92	2265.62	44138.0	CR		1.0		MIT
2261.67	2262.37	44201.4	CR	0.9	1.2		A
2257.81	2258.51	44277.0	CR		1.0		MIT
2257.48	2258.18	44283.5	CR		0.6		MIT
2256.67	2257.37	44299.4	CR		0.7		MIT
2256.05	2256.75	44311.5	CR		1.7		MIT
2251.52	2252.22	44400.7	CR		1.3		MIT
2249.77	2250.47	44435.2	CR	0.5	1.3		A
2248.56	2249.26	44459.1	CR	0.5	1.3		MIT
2248.33	2249.03	44463.7	CR	0.5	1.3		MIT
2247.92	2248.62	44471.8	CR		1.1		MIT
2247.70	2248.40	44476.1	CR		0.9		MIT
2243.62	2244.32	44557.0	CR		1.5		MIT
2243.31	2244.01	44563.2	CR		1.4		MIT
2241.84	2242.54	44592.4	CR		1.5		MIT
2241.36	2242.06	44601.9	CR		1.2		MIT
2228.25	2228.94	44864.3	CR	0.9	0.7		A
2226.70	2227.39	44895.5	CR		1.5		MIT
2213.70	2214.39	45159.2	CR		1.5		MIT
2211.83	2212.52	45197.3	CR II	0.5	1.3		MIT
2203.91	2204.60	45359.7	CR		0.7		MIT
2194.93	2195.62	45545.3	CR	0.9			A
2162.49	2163.17	46228.5	CR	1.3			MIT
2153.97	2154.65	46411.3	CR		0.3H		MIT
2139.15	2139.82	46732.8	CR		1.3		A
2117.53	2118.20	47209.9	CR 3		1.7		ME
2089.16	2089.82	47850.9	CR		1.2		MIT
2079.31	2079.97	48077.6	CR		0.3H		MIT
2073.94	2074.60	48202.0	CR		0.5		MIT
2065.42	2066.08	48400.8	CR II	1.7	2.2		CT
2062.34	2063.00	48473.1	CR		1.3		MIT
2061.49	2062.15	48493.1	CR II	2.0	2.3		CT
2055.52	2056.18	48633.9	CR II	2.0	2.5		CT
2046.93	2047.59	48838.0	CR		0.3H		MIT
2043.10	2043.76	48929.5	CR	0.9			A
2041.86	2042.52	48959.2	CR		1.1		MIT
2039.31	2039.97	49020.5	CR	1.4H			MIT
2038.00	2038.65	49052.0	CR		0.5		MIT
2028.90	2029.55	49271.9	CR		0.3H		MIT
2022.14	2022.79	49436.6	CR		0.5		MIT
2012.45	2013.10	49674.6	CR		0.9		MIT
2011.16	2011.81	49706.5	CR	0.9	1.1		A
2004.96	2005.61	49860.2	CR	1.1			A
9208.46	9210.99	10856.6	CS I	2.3			ME
9172.24	9174.76	10899.5	CS I	3.0			ME
8943.50	8945.96	11178.2	CS I	3.3G			ME
8761.38	8763.79	11410.6	CS I	2.7			ME
8521.10	8523.44	11732.3	CS I	3.7G			ME
8079.021	8081.243	12374.33	CS I	3.0			IMS
8078.923	8081.145	12374.48	CS I	2.0			IMS
8053.35	8055.56	12413.8	CS	2.0S			MS
8015.710	8017.915	12472.07	CS I	2.3			IMS
7990.68	7992.88	12511.1	CS	2.0S			MS
7944.11	7946.30	12584.5	CS	2.9			ME
7609.01	7611.10	13138.7	CS	2.7L			MIT
7279.949	7281.955	13732.58	CS I	1.5L			MS
7270.70	7272.70	13750.0	CS	1.2S			MS
7248.99	7250.99	13791.2	CS		0.3		MIT
7229.01	7231.00	13829.3	CS I	1.5L			MIT
7228.526	7230.518	13830.27	CS	2.7		0.3	MS
7219.70	7221.69	13847.2	CS	1.2S			MS
7205.99	7207.98	13873.5	CS			0.3	SV
7188.32	7190.30	13907.6	CS			0.3	SV
7160.88	7162.85	13960.9	CS			0.3	SV
7149.554	7151.525	13983.03	CS			1.0	SV
7130.532	7132.498	14020.33	CS			0.7	SV
7121.18	7123.14	14038.7	CS			0.3	SV
7085.72	7087.67	14109.0	CS			0.3	SV
6983.488	6985.414	14315.54	CS I	1.4			MS
6979.681	6981.606	14323.35	CS			1.2	SV

AIR WAVELENGTH (ANGSTRMS)	VACUUM WAVELENGTH (ANGSTRMS)	WAVENUMBER (KAYSERS)	LMENT	LOG ARC	INTENSITY SPK	DIS	REF
6973.29	6975.21	14336.5	CS	2.7			MIT
6955.519	6957.438	14373.11	CS II			1.3	SV
6892.42	6894.32	14504.7	CS			0.3	SV
6870.450	6872.346	14551.07	CS I	2.3		0.7	MS
6825.22	6827.10	14647.5	CS I	1.2H			MIT
6724.646	6726.502	14866.57	CS		1.0		MS
6723.279	6725.135	14869.59	CS	2.7	0.8		MS
6646.564	6648.399	15041.21	CS			1.2	SV
6628.654	6630.485	15081.85	CS I	1.5	1.1		MS
6586.506	6588.325	15178.36	CS I	2.7		0.7	MS
6586.019	6587.838	15179.49	CS	1.5			MS
6536.440	6538.246	15294.62	CS II			1.2	LP
6495.528	6497.323	15390.95	CS II			1.2	LP
6472.617	6474.406	15445.43	CS I	1.2			MS
6431.966	6433.744	15543.05	CS I		1.2		MS
6419.541	6421.315	15573.13	CS II			1.0	LP
6386.94	6388.71	15652.6	CS	1.4L			ME
6365.518	6367.278	15705.30	CS I	0.3			MS
6354.98	6356.74	15731.3	CS I	1.2H			ME
6326.204	6327.953	15802.90	CS I		0.7		MS
6217.27	6218.99	16079.8	CS I	1.2W			ME
6212.87	6214.59	16091.2	CS I	2.0		1.0	ME
6128.619	6130.315	16312.37	CS II			1.3	SV
6076.738	6078.420	16451.64	CS II			0.3	SV
6034.089	6035.760	16567.92	CS I	1.5		0.3	MS
6010.33	6011.99	16633.4	CS I	1.7		1.0	ME
5984.393	5986.051	16705.50	CS			1.2	SV
5925.651	5927.293	16871.11	CS			1.8	SV
5844.7	5846.3	17105.	CS I	1.5W			FL
5832.6	5834.2	17140.	CS			1.4	DR
5831.159	5832.776	17144.50	CS II			1.8	SV
5814.181	5815.793	17194.56	CS II			1.4	SV
5663.8	5665.4	17651.	CS I	1.2W			FL
5635.22	5636.78	17740.6	CS I	1.3W			FL
5566.7	5568.2	17959.	CS			1.6	DR
5563.019	5564.564	17970.86	CS II			2.1	SV
5507.174	5508.704	18153.09	CS			1.2	SV
5465.9	5467.4	18290.	CS I	0.7J			FL
5419.687	5421.194	18446.12	CS II			1.8	SV
5402.793	5404.295	18503.80	CS			1.6	SV
5370.979	5372.473	18613.40	CS II			1.9	SV
5358.53	5360.02	18656.6	CS II			2.7	SV
5349.31	5350.80	18688.8	CS			1.2	SV
5349.16	5350.65	18689.3	CS II			1.4	SV
5348.95	5350.44	18690.1	CS			1.4	SV
5306.609	5308.085	18839.18	CS II			1.4	SV
5274.044	5275.512	18955.51	CS II			1.6	SV
5249.373	5250.834	19044.59	CS II			1.9	SV
5227.002	5228.457	19126.10	CS II			2.3	SV
5211.5	5213.0	19183.	CS			1.2	DR
5209.62	5211.07	19189.9	CS			1.2	SV
5209.44	5210.89	19190.6	CS			1.2	SV
5096.604	5098.024	19615.44	CS II			1.6	SV
5081.773	5083.190	19672.69	CS			1.2	SV
5059.866	5061.277	19757.86	CS II			1.4	SV
5052.696	5054.105	19785.90	CS II			1.4	SV
5043.800	5045.206	19820.79	CS II			1.9	SV
5001.641	5003.036	19987.86	CS			0.3	SV
4972.593	4973.980	20104.62	CS II			1.4	SV
4952.835	4954.217	20184.82	CS II			1.5	SV
4870.024	4871.384	20528.05	CS II			1.5	SV
4864.24	4865.60	20552.5	CS			1.0	BS
4851.583	4852.938	20606.07	CS			0.9	SV
4835.03	4836.38	20676.6	CS			1.2	BS
4830.161	4831.511	20697.46	CS II			1.5	SV
4825.42	4826.77	20717.8	CS			1.0	BS
4804.61	4805.95	20807.5	CS			1.0	BS
4786.363	4787.701	20886.85	CS II			1.2	SV
4768.41	4769.74	20965.5	CS			1.0	BS
4763.616	4764.948	20986.59	CS			1.4	SV

AIR WAVELENGTH (ANGSTRMS)	VACUUM WAVELENGTH (ANGSTRMS)	WAVENUMBER (KAYSERS)	LMENT	LOG INTENSITY ARC	SPK	DIS	REF
4758.92	4760.25	21007.3	CS			1.0	BS
4757.87	4759.20	21011.9	CS			1.0	BS
4749.132	4750.460	21050.59	CS II			1.0	OT
4739.665	4740.991	21092.64	CS II			1.3	OT
4733.06	4734.38	21122.1	CS			1.3	BS
4732.975	4734.299	21122.45	CS II			1.3	SV
4728.18	4729.50	21143.9	CS			1.0	BS
4716.19	4717.51	21197.6	CS	1.0			BS
4701.793	4703.109	21262.53	CS II			1.4	OT
4695.610	4696.924	21290.53	CS			1.0	SV
4674.89	4676.20	21384.9	CS			1.0	BS
4670.280	4671.587	21406.00	CS II			1.3	SV
4656.538	4657.842	21469.17	CS			1.1	SV
4651.1	4652.4	21494.	CS			1.0	BS
4646.508	4647.809	21515.51	CS II			1.4	SV
4623.091	4624.386	21624.49	CS II			1.3	SV
4616.28	4617.57	21656.4	CS			1.2	SV
4616.13	4617.42	21657.1	CS II			1.2	SV
4616.01	4617.30	21657.7	CS			1.0	SV
4610.505	4611.797	21683.52	CS			1.0	SV
4603.755	4605.045	21715.32	CS II			1.8	OT
4599.22	4600.51	21736.7	CS			1.2	BS
4597.673	4598.961	21744.04	CS II			1.0	OT
4593.177	4594.464	21765.32	CS I	3.06	1.76		SV
4572.611	4573.892	21863.22	CS			1.0	SV
4571.786	4573.067	21867.16	CS II			1.2	OT
4566.983	4568.263	21890.16	CS II			1.2	OT
4555.355	4556.632	21946.03	CS I	3.36	2.0		SV
4543.71	4544.98	22002.3	CS			1.0	BS
4538.942	4540.215	22025.39	CS II			1.5	SV
4534.64	4535.91	22046.3	CS			1.0	BS
4532.500	4533.771	22056.70	CS			1.0	SV
4531.45	4532.72	22061.8	CS			1.2	BS
4526.725	4527.994	22084.83	CS II			1.5	SV
4522.846	4524.114	22103.77	CS			1.2	SV
4522.36	4523.63	22106.1	CS			1.2	BS
4515.495	4516.761	22139.76	CS II			1.0	OT
4506.834	4508.098	22182.30	CS II			1.0	SV
4506.705	4507.969	22182.94	CS			1.2	SV
4501.525	4502.788	22208.46	CS II			1.5	SV
4496.758	4498.019	22232.01	CS			1.2	SV
4493.660	4494.921	22247.33	CS			1.0	SV
4469.09	4470.34	22369.6	CS			0.3	SV
4457.680	4458.931	22426.90	CS II			1.2	SV
4447.649	4448.898	22477.48	CS			1.0	SV
4444.004	4445.252	22495.92	CS II			1.0	SV
4440.26	4441.51	22514.9	CS			1.2	BS
4436.06	4437.31	22536.2	CS II			0.3	SV
4435.708	4436.953	22537.99	CS I			1.3	SV
4425.663	4426.906	22589.14	CS			1.3	SV
4424.046	4425.288	22597.40	CS II			1.0	SV
4410.208	4411.447	22668.30	CS			1.3	SV
4405.253	4406.490	22693.80	CS II			1.5	SV
4403.854	4405.091	22701.01	CS			1.3	SV
4399.495	4400.731	22723.50	CS			1.3	SV
4397.994	4399.229	22731.25	CS			1.0	SV
4396.909	4398.144	22736.86	CS II			1.2	SV
4388.764	4389.997	22779.06	CS II			1.0	SV
4384.428	4385.660	22801.59	CS II			1.4	SV
4373.018	4374.247	22861.08	CS II			1.5	SV
4368.77	4370.00	22883.3	CS			1.0	BS
4367.66	4368.89	22889.1	CS			1.0	BS
4363.275	4364.501	22912.12	CS II			1.7	SV
4359.02	4360.25	22934.5	CS			1.0	BS
4335.411	4336.630	23059.38	CS			0.9	SV
4330.239	4331.457	23086.92	CS II			1.3	SV
4327.580	4328.797	23101.11	CS II			1.0	OT
4326.315	4327.532	23107.86	CS			1.0	SV
4316.992	4318.206	23157.76	CS II			0.3	OT
4312.778	4313.991	23180.39	CS			1.0	SV

AIR WAVELENGTH (ANGSTRMS)	VACUUM WAVELENGTH (ANGSTRMS)	WAVENUMBER (KAYSERS)	LMENT	LOG INTENSITY ARC	SPK	DIS	REF
4307.942	4309.154	23206.41	CS II			0.9	OT
4306.54	4307.75	23214.0	CS			1.0	BS
4300.636	4301.846	23245.84	CS II			1.5	SV
4297.514	4298.723	23262.72	CS			1.0	SV
4292.008	4293.216	23292.56	CS			1.1	SV
4288.350	4289.557	23312.43	CS II			1.5	SV
4284.13	4285.34	23335.4	CS			1.0	BS
4282.59	4283.80	23343.8	CS			1.0	BS
4281.31	4282.51	23350.8	CS			1.0	BS
4277.100	4278.304	23373.75	CS II			1.7	SV
4272.87	4274.07	23396.9	CS			1.0	BS
4271.744	4272.946	23403.06	CS			1.0	SV
4268.89	4270.09	23418.7	CS			1.0	BS
4264.675	4265.875	23441.85	CS			1.7	SV
4241.973	4243.167	23567.30	CS			1.0	SV
4234.408	4235.600	23609.40	CS			1.3	SV
4232.188	4233.380	23621.79	CS			1.4	SV
4228.350	4229.541	23643.23	CS II			1.5	OT
4227.100	4228.290	23650.22	CS II			1.7	OT
4221.119	4222.308	23683.73	CS II			1.2	SV
4219.516	4220.704	23692.73	CS			0.7	SV
4213.129	4214.316	23728.64	CS II			1.5	SV
4193.198	4194.380	23841.43	CS			0.9	SV
4173.533	4174.709	23953.76	CS			1.2	SV
4163.243	4164.417	24012.97	CS			1.2	SV
4158.610	4159.782	24039.72	CS			1.3	SV
4151.267	4152.438	24082.24	CS II			1.3	SV
4132.003	4133.169	24194.51	CS II			1.0	OT
4121.210	4122.373	24257.87	CS II			1.2	OT
4119.288	4120.450	24269.19	CS			0.9	SV
4108.232	4109.391	24334.50	CS			0.7	SV
4081.471	4082.623	24494.05	CS			1.0	SV
4068.773	4069.922	24570.50	CS II			1.5	SV
4067.958	4069.107	24575.42	CS II			1.5	SV
4064.69	4065.84	24595.2	CS			1.0	BS
4063.93	4065.08	24599.8	CS			1.2	BS
4053.956	4055.101	24660.30	CS			1.2	SV
4047.184	4048.327	24701.56	CS II			1.3	SV
4043.422	4044.564	24724.54	CS			1.3	SV
4039.841	4040.982	24746.46	CS			1.7	SV
4035.83	4036.97	24771.1	CS			1.2	BS
4031.10	4032.24	24800.1	CS			1.0	BS
4025.67	4026.81	24833.6	CS			1.0	SV
4023.582	4024.719	24846.45	CS			1.0	SV
4014.99	4016.12	24899.6	CS			1.0	BS
4010.54	4011.67	24927.2	CS			1.0	BS
4006.772	4007.905	24950.69	CS			1.0	SV
4006.537	4007.670	24952.16	CS			1.5	SV
4001.682	4002.813	24982.43	CS			1.3	SV
3993.863	3994.992	25031.34	CS			0.6	SV
3978.000	3979.125	25131.15	CS			1.0	SV
3974.239	3975.363	25154.93	CS			0.8	SV
3967.30	3968.42	25198.9	CS			0.6	BS
3965.187	3966.309	25212.36	CS II			1.4	SV
3959.495	3960.615	25248.60	CS II			1.3	SV
3955.923	3957.042	25271.40	CS			1.0	SV
3929.46	3930.57	25441.6	CS			0.3	BS
3925.583	3926.695	25466.71	CS II			1.4	SV
3921.69	3922.80	25492.0	CS			0.6	BS
3913.37	3914.48	25546.2	CS			0.3	BS
3906.933	3908.040	25588.28	CS II			1.3	SV
3904.806	3905.912	25602.21	CS			0.6	SV
3900.82	3901.93	25628.4	CS			0.6	BS
3900.09	3901.19	25633.2	CS II			0.6	SV
3896.978	3898.082	25653.64	CS II			0.8	SV
3893.09	3894.19	25679.3	CS			0.6	BS
3892.206	3893.309	25685.09	CS			0.6	SV
3888.65	3889.75	25708.6	CS I	2.2	1.0		FL
3876.39	3877.49	25789.9	CS I	2.5			FL
3870.164	3871.261	25831.38	CS			0.6	SV

AIR WAVELENGTH (ANGSTRMS)	VACUUM WAVELENGTH (ANGSTRMS)	WAVENUMBER (KAYSERS)	LMENT	LOG INTENSITY ARC	LOG INTENSITY SPK	LOG INTENSITY DIS	REF
3864.367	3865.463	25870.12	CS			0.6	SV
3864.249	3865.345	25870.92	CS			0.8	SV
3861.489	3862.584	25889.41	CS			0.6	SV
3837.449	3838.538	26051.59	CS			0.6	SV
3819.61	3820.69	26173.3	CS			0.6	BS
3805.412	3806.492	26270.91	CS II			0.3	MIT
3805.096	3806.176	26273.09	CS II			1.4	SV
3797.908	3798.986	26322.81	CS			0.6	SV
3785.424	3786.499	26409.62	CS II			1.3	SV
3751.402	3752.468	26649.13	CS			0.6	SV
3734.337	3735.399	26770.90	CS			1.0	SV
3732.535	3733.596	26783.83	CS II			0.6	MIT
3729.980	3731.041	26802.17	CS			0.6	SV
3724.9	3726.0	26839.	CS			0.6	BS
3710.774	3711.830	26940.89	CS			0.6	SV
3699.475	3700.528	27023.17	CS			1.0	SV
3680.454	3681.502	27162.83	CS			0.6	SV
3680.101	3681.149	27165.43	CS			0.6	SV
3661.391	3662.434	27304.25	CS			0.8	SV
3655.73	3656.77	27346.5	CS			0.6	BS
3651.073	3652.113	27381.41	CS			0.6	SV
3641.40	3642.44	27454.1	CS			0.6	BS
3641.332	3642.370	27454.65	CS			0.7	SV
3634.75	3635.79	27504.4	CS			0.8	BS
3624.56	3625.59	27581.7	CS			0.6	BS
3622.691	3623.724	27595.92	CS			0.8	SV
3618.161	3619.193	27630.47	CS			0.8	SV
3617.41	3618.44	27636.2	CS I	1.8			BV
3614.989	3616.020	27654.72	CS			0.6	SV
3611.52	3612.55	27681.3	CS I	2.3			BV
3608.285	3609.314	27706.09	CS			1.0	SV
3605.535	3606.564	27727.23	CS			0.6	SV
3602.852	3603.880	27747.87	CS			0.9	SV
3600.73	3601.76	27764.2	CS			1.0	SV
3598.97	3600.00	27777.8	CS			0.6	BS
3597.73	3598.76	27787.4	CS			0.8	SV
3597.430	3598.456	27789.69	CS			1.0	SV
3592.48	3593.51	27828.0	CS			0.6	BS
3581.3	3582.3	27915.	CS			0.6	BS
3573.24	3574.26	27977.8	CS			0.6	BS
3569.28	3570.30	28008.9	CS			0.6	BS
3565.111	3566.129	28041.61	CS			1.0	SV
3559.798	3560.815	28083.46	CS			0.7	SV
3541.45	3542.46	28228.9	CS			0.6	BS
3533.364	3534.374	28293.55	CS			0.8	SV
3531.376	3532.385	28309.48	CS			0.6	SV
3518.15	3519.16	28415.9	CS			0.8	BS
3516.03	3517.04	28433.0	CS			0.6	BS
3514.022	3515.027	28449.28	CS			0.8	SV
3504.85	3505.85	28523.7	CS			0.6	BS
3503.67	3504.67	28533.3	CS			0.6	SV
3480.13	3481.13	28726.3	CS I	1.7			BV
3479.25	3480.25	28733.6	CS			0.6	BS
3476.88	3477.88	28753.2	CS I	2.0			BV
3470.92	3471.91	28802.6	CS			0.6	BS
3469.81	3470.80	28811.8	CS			0.6	BS
3465.20	3466.19	28850.1	CS			0.6	SV
3463.425	3464.417	28864.89	CS			0.8	SV
3459.185	3460.176	28900.26	CS II			1.2	OT
3457.18	3458.17	28917.0	CS			0.6	BS
3455.48	3456.47	28931.2	CS			0.6	BS
3450.36	3451.35	28974.2	CS			0.8	SV
3443.88	3444.87	29028.7	CS			0.6	MIT
3430.4	3431.4	29143.	CS			0.6	BS
3429.47	3430.45	29150.7	CS			0.5	BS
3418.11	3419.09	29247.6	CS			0.8	SV
3411.313	3412.292	29305.82	CS			1.0	SV
3406.626	3407.603	29346.14	CS			1.0	SV
3400.00	3400.98	29403.3	CS I	1.5			BV
3398.14	3399.12	29419.4	CS I	1.8			BV
3397.187	3398.162	29427.67	CS			0.8	SV
3393.25	3394.22	29461.8	CS			0.5	BS
3389.15	3390.12	29497.5	CS			0.8	BS
3379.0	3380.0	29586.	CS			0.8	BS
3368.555	3369.523	29677.79	CS II			1.5	SV
3364.52	3365.49	29713.4	CS			0.6	BS
3358.8	3359.8	29764.	CS			0.6	BS
3357.687	3358.652	29773.85	CS			0.8	SV
3353.88	3354.84	29807.6	CS			0.6	BS
3349.445	3350.408	29847.11	CS			1.0	SV
3348.72	3349.68	29853.6	CS I	1.2			BV
3347.44	3348.40	29865.0	CS I	1.5			BV
3344.004	3344.965	29895.68	CS			1.0	SV
3340.574	3341.535	29926.37	CS			1.0	SV
3330.33	3331.29	30018.4	CS			0.6	BS
3329.428	3330.386	30026.55	CS II			1.0	OT
3324.5	3325.5	30071.	CS			0.6	BS
3322.8	3323.8	30086.	CS			0.6	BS
3315.498	3316.452	30152.70	CS			1.0	SV
3314.04	3314.99	30166.0	CS I	0.7			BV
3313.16	3314.11	30174.0	CS I	1.0			BV
3311.52	3312.47	30188.9	CS			0.6	BS
3303.72	3304.67	30260.2	CS			0.6	BS
3299.86	3300.81	30295.6	CS			0.8	SV
3289.13	3290.08	30394.4	CS I	0.3			BV
3288.56	3289.51	30399.7	CS I	0.6			BV
3282.1	3283.0	30459.	CS			0.5	BS
3278.26	3279.20	30495.2	CS			0.6	BS
3275.68	3276.62	30519.2	CS			0.6	BS
3271.626	3272.569	30557.03	CS II			1.3	OT
3270.44	3271.38	30568.1	CS I	0.3			BV
3268.314	3269.256	30588.00	CS			1.0	SV
3267.135	3268.077	30599.04	CS II			1.5	OT
3265.924	3266.866	30610.38	CS II			1.5	OT
3263.06	3264.00	30637.2	CS			0.6	BS
3262.29	3263.23	30644.5	CS			0.8	BS
3255.35	3256.29	30709.8	CS			1.0	BS
3250.58	3251.52	30754.9	CS			0.8	BS
3247.5	3248.4	30784.	CS			0.6	BS
3242.28	3243.22	30833.6	CS			1.0	BS
3234.16	3235.09	30911.0	CS			0.8	BS
3227.2	3228.1	30978.	CS			0.6	BS
3219.1	3220.0	31056.	CS			0.6	BS
3213.7	3214.6	31108.	CS			0.8	BS
3209.65	3210.58	31147.1	CS			1.0	BS
3207.07	3208.00	31172.1	CS			0.6	BS
3204.27	3205.20	31199.3	CS			0.6	BS
3201.09	3202.02	31230.3	CS			0.6	BS
3198.7	3199.6	31254.	CS			0.6	BS
3195.5	3196.4	31285.	CS			0.6	BS
3193.6	3194.5	31304.	CS			0.6	BS
3192.1	3193.0	31318.	CS			0.6	BS
3189.20	3190.12	31346.8	CS			0.6	BS
3178.61	3179.53	31451.2	CS			1.0	BS
3172.56	3173.48	31511.2	CS			1.0	BS
3169.73	3170.65	31539.3	CS			0.6	BS
3154.55	3155.46	31691.1	CS			0.6	BS
3153.88	3154.79	31697.8	CS			0.8	BS
3152.30	3153.21	31713.7	CS			0.8	BS
3151.14	3152.05	31725.4	CS			0.8	BS
3149.36	3150.27	31743.3	CS			1.0	BS
3145.2	3146.1	31785.	CS			0.6	BS
3141.46	3142.37	31823.1	CS			0.6	BS
3134.8	3135.7	31891.	CS			0.6	BS
3130.7	3131.6	31932.	CS			0.6	BS
3129.1	3130.0	31949.	CS			0.6	BS
3125.3	3126.2	31988.	CS			0.6	BS
3118.35	3119.25	32058.9	CS			0.6	BS
3112.18	3113.08	32122.5	CS			0.6	BS
3109.3	3110.2	32152.	CS			0.6	BS

AIR WAVELENGTH (ANGSTRMS)	VACUUM WAVELENGTH (ANGSTRMS)	WAVENUMBER (KAYSERS)	LMENT	LOG INTENSITY ARC	SPK	DIS	REF
3097.38	3098.28	32276.0	CS			1.0	BS
3095.77	3096.67	32292.8	CS			0.8	BS
3094.82	3095.72	32302.7	CS			0.6	BS
3091.6	3092.5	32336.	CS			0.6	BS
3088.9	3089.8	32365.	CS			0.6	BS
3084.4	3085.3	32412.	CS			0.6	BS
3080.874	3081.769	32448.90	CS II			0.8	OT
3078.09	3078.98	32478.2	CS			0.8	BS
3072.7	3073.6	32535.	CS			0.6	BS
3069.73	3070.62	32566.7	CS			0.6	BS
3067.8	3068.7	32587.	CS			0.6	BS
3067.2	3068.1	32594.	CS			0.6	BS
3066.62	3067.51	32599.7	CS			1.0	BS
3063.7	3064.6	32631.	CS			0.6	BS
3062.7	3063.6	32641.	CS			0.6	BS
3061.24	3062.13	32657.0	CS			0.8	BS
3060.12	3061.01	32669.0	CS			0.8	BS
3058.6	3059.5	32685.	CS			0.8	BS
3054.56	3055.45	32728.4	CS			0.6	BS
3054.20	3055.09	32732.3	CS			0.6	BS
3053.5	3054.4	32740.	CS			0.6	BS
3050.8	3051.7	32769.	CS			0.8	BS
3045.9	3046.8	32821.	CS			0.6	BS
3042.3	3043.2	32860.	CS			0.6	BS
3039.31	3040.19	32892.6	CS			0.6	BS
3032.41	3033.29	32967.5	CS			0.6	BS
3031.5	3032.4	32977.	CS			0.6	BS
3030.35	3031.23	32989.9	CS			0.6	BS
3029.15	3030.03	33003.0	CS			0.6	BS
3028.25	3029.13	33012.8	CS			0.6	BS
3020.9	3021.8	33093.	CS			0.6	BS
3020.3	3021.2	33100.	CS			0.6	BS
3015.8	3016.7	33149.	CS			0.6	BS
3012.041	3012.919	33190.41	CS II			0.9	OT
3002.88	3003.76	33291.7	CS			0.8	BS
3001.271	3002.146	33309.51	CS II			1.0	OT
2999.513	3000.388	33329.03	CS		0.9		MIT
2998.20	2999.07	33343.6	CS			0.3	BS
2997.2	2998.1	33355.	CS			0.3	BS
2996.15	2997.02	33366.4	CS			0.3	BS
2995.34	2996.21	33375.5	CS			1.3	BS
2990.78	2991.65	33426.3	CS			0.3	BS
2986.89	2987.76	33469.9	CS			0.3	BS
2985.3	2986.2	33488.	CS			0.3	BS
2983.91	2984.78	33503.3	CS			0.3	BS
2982.5	2983.4	33519.	CS			0.3	BS
2982.03	2982.90	33524.4	CS			0.3	BS
2977.258	2978.127	33578.15	CS II			0.3	OT
2976.81	2977.68	33583.2	CS			0.3	BS
2975.65	2976.52	33596.3	CS			0.3	BS
2975.13	2976.00	33602.2	CS			0.3	BS
2972.8	2973.7	33628.	CS			0.3	BS
2970.851	2971.718	33650.56	CS II			0.3	OT
2969.0	2969.9	33671.	CS			0.9	BS
2968.383	2969.250	33678.54	CS II			0.3	OT
2965.4	2966.3	33712.	CS			0.3	BS
2965.0	2965.9	33717.	CS			0.9	BS
2962.4	2963.3	33747.	CS			0.3	BS
2951.59	2952.45	33870.1	CS			0.3	BS
2949.800	2950.662	33890.70	CS II			0.3	OT
2947.85	2948.71	33913.1	CS			0.3	BS
2944.1	2945.0	33956.	CS			0.3	BS
2942.21	2943.07	33978.1	CS			0.9	BS
2940.953	2941.813	33992.64	CS II			0.9	OT
2938.5	2939.4	34021.	CS			1.3	BS
2931.11	2931.97	34106.8	CS			1.3	BS
2924.48	2925.34	34184.1	CS			0.3	BS
2922.21	2923.07	34210.7	CS			0.3	BS
2921.83	2922.69	34215.1	CS			0.3	BS
2921.03	2921.89	34224.5	CS			1.3	BS

AIR WAVELENGTH (ANGSTRMS)	VACUUM WAVELENGTH (ANGSTRMS)	WAVENUMBER (KAYSERS)	LMENT	LOG INTENSITY ARC	SPK	DIS	REF
2915.24	2916.09	34292.5	CS			0.3	BS
2914.652	2915.505	34299.37	CS II			0.9	OT
2910.82	2911.67	34344.5	CS			0.3	BS
2906.17	2907.02	34399.5	CS			0.3	BS
2901.1	2902.0	34460.	CS			0.3	BS
2899.7	2900.5	34476.	CS			0.3	BS
2895.32	2896.17	34528.4	CS			0.3	BS
2894.85	2895.70	34534.0	CS			0.3	BS
2893.81	2894.66	34546.4	CS			0.3	BS
2891.75	2892.60	34571.0	CS			0.3	BS
2886.67	2887.52	34631.8	CS			1.3	BS
2884.42	2885.27	34658.9	CS			0.9	BS
2883.745	2884.591	34666.96	CS II			0.3	OT
2881.16	2882.01	34698.1	CS			1.3	BS
2879.25	2880.09	34721.1	CS			0.9	BS
2877.29	2878.13	34744.7	CS			0.9	BS
2875.30	2876.14	34768.8	CS			0.9	BS
2872.35	2873.19	34804.5	CS			0.9	BS
2871.32	2872.16	34817.0	CS			0.3	BS
2868.33	2869.17	34853.3	CS			0.9	BS
2866.90	2867.74	34870.6	CS			0.3	BS
2866.32	2867.16	34877.7	CS			0.9	BS
2865.45	2866.29	34888.3	CS			0.3	BS
2862.40	2863.24	34925.5	CS			0.9	BS
2860.85	2861.69	34944.4	CS			0.9	BS
2859.32	2860.16	34963.1	CS			1.3	BS
2857.83	2858.67	34981.3	CS			0.3	BS
2854.45	2855.29	35022.7	CS			0.9	BS
2852.415	2853.253	35047.71	CS II			0.3	OT
2851.23	2852.07	35062.3	CS			1.3	BS
2850.4	2851.2	35072.	CS			0.3	BS
2847.19	2848.03	35112.0	CS			0.3	BS
2846.193	2847.030	35124.33	CS II			0.3	OT
2845.67	2846.51	35130.8	CS			1.3	BS
2844.48	2845.32	35145.5	CS			0.3	BS
2841.5	2842.3	35182.	CS			0.3	BS
2838.09	2838.92	35224.6	CS			1.3	BS
2835.01	2835.84	35262.9	CS			0.9	BS
2829.423	2830.255	35332.50	CS II			0.3	OT
2829.045	2829.877	35337.22	CS II			0.3	OT
2826.802	2827.634	35365.26	CS II			0.3	OT
2824.12	2824.95	35398.8	CS			0.9	BS
2823.03	2823.86	35412.5	CS			0.9	BS
2820.268	2821.098	35447.19	CS II			0.3	OT
2819.240	2820.070	35460.11	CS			0.3	BS
2817.98	2818.81	35476.0	CS			1.3	BS
2816.943	2817.772	35489.03	CS II			0.3	OT
2815.33	2816.16	35509.4	CS			0.3	BS
2810.81	2811.64	35566.5	CS			1.3	BS
2809.91	2810.74	35577.9	CS			0.9	BS
2799.47	2800.30	35710.5	CS			0.3	BS
2794.50	2795.32	35774.0	CS II			0.9	BS
2793.316	2794.140	35789.19	CS II			0.3	OT
2792.16	2792.98	35804.0	CS			0.9	BS
2789.797	2790.620	35834.33	CS II			0.9	OT
2788.81	2789.63	35847.0	CS			0.3	BS
2788.22	2789.04	35854.6	CS			1.3	SV
2787.02	2787.84	35870.0	CS			0.9	BS
2784.666	2785.488	35900.36	CS II			0.3	OT
2784.10	2784.92	35907.7	CS			0.9	BS
2780.81	2781.63	35950.1	CS			0.3	BS
2779.9	2780.7	35962.	CS			0.9	BS
2779.1	2779.9	35972.	CS			0.9	BS
2776.42	2777.24	36007.0	CS			1.3	BS
2774.46	2775.28	36032.4	CS			0.9	BS
2769.5	2770.3	36097.	CS			0.3	BS
2766.095	2766.912	36141.37	CS II			0.3	OT
2764.42	2765.24	36163.3	CS			1.3	BS
2761.9	2762.7	36196.	CS			0.3	BS
2757.4	2758.2	36255.	CS			0.3	BS

AIR WAVELENGTH (ANGSTRMS)	VACUUM WAVELENGTH (ANGSTRMS)	WAVENUMBER (KAYSERS)	LMENT	LOG ARC	INTENSITY SPK	INTENSITY DIS	REF
2755.20	2756.01	36284.3	CS			1.3	BS
2751.11	2751.92	36338.2	CS			0.3	PS
2748.18	2748.99	36377.0	CS			1.3	BS
2740.67	2741.48	36476.6	CS II			0.9	BS
2734.85	2735.66	36554.3	CS			0.3	OT
2733.879	2734.688	36567.24	CS II			0.3	OT
2731.8	2732.6	36595.	CS			0.3	BS
2727.80	2728.61	36648.7	CS			0.3	BS
2726.802	2727.609	36662.14	CS II			0.3	OT
2723.95	2724.76	36700.5	CS			0.9	BS
2721.6	2722.4	36732.	CS			0.3	BS
2719.0	2719.8	36767.	CS			0.3	BS
2715.8	2716.6	36811.	CS			0.3	BS
2714.0	2714.8	36835.	CS			0.3	BS
2711.6	2712.4	36868.	CS			0.3	BS
2710.5	2711.3	36883.	CS			0.3	BS
2709.0	2709.8	36903.	CS			0.3	BS
2706.79	2707.59	36933.2	CS			1.3	BS
2705.3	2706.1	36953.	CS			0.3	BS
2704.1	2704.9	36970.	CS			0.3	BS
2701.18	2701.98	37009.9	CS			1.3	BS
2700.30	2701.10	37021.9	CS			0.9	BS
2699.16	2699.96	37037.6	CS			0.9	BS
2691.83	2692.63	37138.4	CS			0.3	BS
2689.412	2690.210	37171.81	CS II			0.3	OT
2686.64	2687.44	37210.2	CS			0.9	BS
2681.99	2682.79	37274.7	CS			0.9	BS
2681.34	2682.14	37283.7	CS			0.9	BS
2678.92	2679.72	37317.4	CS			1.3	BS
2677.01	2677.81	37344.0	CS			0.3	BS
2674.62	2675.41	37377.4	CS			0.3	BS
2674.0	2674.8	37386.	CS			0.3	BS
2673.208	2674.003	37397.12	CS		0.3		MIT
2671.70	2672.49	37418.2	CS			0.3	BS
2669.792	2670.586	37444.97	CS II			0.9	OT
2668.76	2669.55	37459.5	CS			0.9	BS
2658.71	2659.50	37601.0	CS			0.3	BS
2656.83	2657.62	37627.6	CS			0.3	BS
2650.7	2651.5	37715.	CS			1.3	BS
2646.20	2646.99	37778.8	CS			0.9	BS
2644.7	2645.5	37800.	CS			1.3	BS
2642.63	2643.42	37829.8	CS			1.3	BS
2641.0	2641.8	37853.	CS			0.3	BS
2640.27	2641.06	37863.6	CS			0.9	BS
2637.6	2638.4	37902.	CS			0.3	BS
2635.871	2636.657	37926.82	CS II		0.3		MIT
2634.17	2634.96	37951.3	CS			0.3	BS
2627.952	2628.736	38041.10	CS II			0.3	OT
2627.84	2628.62	38042.7	CS			0.9	BS
2624.1	2624.9	38097.	CS			0.3	BS
2619.22	2620.00	38167.9	CS			0.3	BS
2616.27	2617.05	38211.0	CS			0.9	BS
2614.62	2615.40	38235.1	CS			0.9	BS
2613.6	2614.4	38250.	CS			0.3	BS
2612.1	2612.9	38272.	CS			0.3	BS
2610.140	2610.920	38300.68	CS II			0.8	OT
2609.43	2610.21	38311.1	CS			1.3	BS
2605.40	2606.18	38370.4	CS			1.3	BS
2603.72	2604.50	38395.1	CS			0.3	BS
2600.36	2601.14	38444.7	CS			1.3	BS
2598.7	2599.5	38469.	CS			0.3	BS
2596.95	2597.73	38495.2	CS			1.3	BS
2591.17	2591.95	38581.1	CS			1.3	BS
2590.2	2591.0	38595.	CS			0.3	BS
2589.3	2590.1	38609.	CS			0.3	BS
2582.5	2583.3	38711.	CS			0.3	BS
2574.59	2575.36	38829.5	CS			0.3	BS
2573.055	2573.826	38852.67	CS		1.3		MIT
2571.41	2572.18	38877.5	CS			0.3	BS
2568.71	2569.48	38918.4	CS			0.9	BS

AIR WAVELENGTH (ANGSTRMS)	VACUUM WAVELENGTH (ANGSTRMS)	WAVENUMBER (KAYSERS)	LMENT	LOG ARC	INTENSITY SPK	INTENSITY DIS	REF
2560.37	2561.14	39045.1	CS			0.9	BS
2554.8	2555.6	39130.	CS			0.3	BS
2551.17	2551.94	39185.9	CS			0.3	BS
2546.1	2546.9	39264.	CS			0.3	BS
2543.92	2544.68	39297.6	CS			1.3	BS
2540.52	2541.28	39350.2	CS			0.3	BS
2539.174	2539.937	39371.06	CS II			0.3	OT
2539.08	2539.84	39372.5	CS			0.3	BS
2538.67	2539.43	39378.9	CS			0.3	BS
2533.44	2534.20	39460.2	CS			1.3	BS
2528.8	2529.6	39533.	CS			0.9	BS
2525.68	2526.44	39581.4	CS			1.3	BS
2520.8	2521.6	39658.	CS			0.3	BS
2512.1	2512.9	39795.	CS			0.3	BS
2511.51	2512.27	39804.7	CS			0.9	BS
2502.2	2503.0	39953.	CS			0.3	BS
2496.9	2497.7	40038.	CS			0.3	BS
2495.04	2495.79	40067.4	CS			1.3	BS
2489.5	2490.3	40157.	CS			0.3	BS
2485.42	2486.17	40222.5	CS			1.3	BS
2483.0	2483.7	40262.	CS			0.3	BS
2480.7	2481.4	40299.	CS			0.3	BS
2477.58	2478.33	40349.8	CS			1.3	BS
2466.8	2467.5	40526.	CS			0.3	BS
2466.3	2467.0	40534.	CS			0.3	BS
2462.0	2462.7	40605.	CS			0.3	BS
2459.23	2459.97	40650.8	CS			0.3	BS
2455.80	2456.54	40707.6	CS			0.9	BS
2443.2	2443.9	40917.	CS			0.3	BS
2439.8	2440.5	40974.	CS			0.3	BS
2437.1	2437.8	41020.	CS			0.3	BS
2432.6	2433.3	41096.	CS			0.9	BS
2427.65	2428.39	41179.6	CS			1.3	BS
2426.41	2427.15	41200.6	CS			0.9	BS
2425.15	2425.89	41222.0	CS			1.3	BS
2422.9	2423.6	41260.	CS			0.3	BS
2421.4	2422.1	41286.	CS			0.3	BS
2420.06	2420.80	41308.7	CS		0.3		MIT
2415.0	2415.7	41395.	CS			0.3	BS
2414.77	2415.50	41399.2	CS		0.3		MIT
2411.98	2412.71	41447.1	CS		0.3		MIT
2406.9	2407.6	41535.	CS			0.3	BS
2401.7	2402.4	41624.	CS			0.3	BS
2394.92	2395.65	41742.3	CS		0.3		MIT
2393.6	2394.3	41765.	CS			0.3	BS
2392.92	2393.65	41777.2	CS		0.9		MIT
2380.2	2380.9	42000.	CS			0.3	BS
2379.54	2380.27	42012.1	CS			0.3	BS
2375.82	2376.55	42077.9	CS			0.3	BS
2373.4	2374.1	42121.	CS			0.3	BS
2364.827	2365.550	42273.47	CS		0.3		MIT
2357.9	2358.6	42398.	CS			0.3	BS
2354.42	2355.14	42460.3	CS			0.3	BS
2351.911	2352.631	42505.61	CS		0.3		MIT
2340.47	2341.19	42713.4	CS			0.3	BS
2332.42	2333.14	42860.8	CS			0.9	BS
2273.88	2274.58	43964.1	CS II			1.3	RF
2267.66	2268.36	44084.7	CS II			1.3	RF
2254.94	2255.64	44333.3	CS II			1.3	RF
2228.94	2229.63	44850.4	CS II			1.2	RF
2179.60	2180.28	45865.6	CS II			1.2	RF
2177.61	2178.29	45907.5	CS II			0.3	RF
2146.75	2147.43	46567.4	CS II			1.2	RF
9960.46	9963.19	10036.9	CU II		1.2		SH
9960.07	9962.80	10037.3	CU II		1.0		SH
9939.05	9941.78	10058.6	CU II		1.3D		SH
9926.10	9928.82	10071.7	CU II		1.0		SH
9925.67	9928.39	10072.1	CU II		1.3		SH
9918.05	9920.77	10079.9	CU II		1.2		SH
9916.52	9919.24	10081.4	CU II		1.5		SH

AIR WAVELENGTH (ANGSTRMS)	VACUUM WAVELENGTH (ANGSTRMS)	WAVENUMBER (KAYSERS)	LMENT	LOG ARC	INTENSITY SPK	INTENSITY DIS	REF
9905.44	9908.16	10092.7	CU II		0.3		SH
9894.44	9897.15	10103.9	CU II		0.7		SH
9893.04	9895.75	10105.3	CU II		0.7H		SH
9884.09	9886.80	10114.5	CU II		1.0H		SH
9881.57	9884.28	10117.1	CU II		1.2		SH
9868.20	9870.91	10130.8	CU II		1.2		SH
9864.26	9866.96	10134.8	CU II		1.6		SH
9861.41	9864.11	10137.8	CU II		1.7		SH
9858.87	9861.57	10140.4	CU II		0.5H		SH
9850.58	9853.28	10148.9	CU II		0.5		SH
9837.94	9840.64	10161.9	CU II		1.4		SH
9830.90	9833.60	10169.2	CU II		0.7		SH
9829.06	9831.76	10171.1	CU II		0.5		SH
9828.06	9830.75	10172.2	CU II		0.7		SH
9813.35	9816.04	10187.4	CU II		1.3		SH
9739.6	9742.3	10264.	CU I	0.3V			KS
9735.94	9738.61	10268.4	CU II		1.2		SH
9732.28	9734.95	10272.3	CU II		0.5		SH
9688.71	9691.37	10318.5	CU II		1.0		SH
9530.3	9532.9	10490.	CU I	0.5V			KS
9512.43	9515.04	10509.7	CU II		0.3		SH
9463.71	9466.31	10563.8	CU II		0.5		SH
9451.59	9454.18	10577.3	CU II		0.3		SH
9332.04	9334.60	10712.8	CU II		0.7		SH
9205.40	9207.93	10860.2	CU II		1.3		SH
9103.33	9105.83	10982.0	CU II		1.0		SH
8996.2	8998.7	11113.	CU I	0.5H			KS
8609.26	8611.62	11612.2	CU II		0.5		SH
8584.0	8586.4	11646.	CU I	0.3W			KS
8511.04	8513.38	11746.2	CU II		1.6		SH
8503.46	8505.80	11756.7	CU II		1.2		SH
8477.26	8479.59	11793.0	CU II		1.0		SH
8408.15	8410.46	11890.0	CU I	0.5W			KS
8283.21	8285.49	12069.3	CU II		1.8		SH
8277.60	8279.88	12077.5	CU II		1.7		SH
8256.90	8259.17	12107.7	CU II		0.7		SH
8235.30	8237.56	12139.5	CU II			1.0	SH
8192.28	8194.53	12203.3	CU II		1.5		SH
8095.55	8097.78	12349.1	CU II		1.6		SH
8092.634	8094.859	12353.52	CU I	2.60			IBU
8088.58	8090.80	12359.7	CU II		1.3		SH
8075.46	8077.68	12379.8	CU II		0.3		SH
8026.45	8028.66	12455.4	CU II		1.0		SH
8006.21	8008.41	12486.9	CU I	0.3			KS
7996.72	7998.92	12501.7	CU II		1.0		SH
7988.17	7990.37	12515.1	CU II		1.8		SH
7972.01	7974.20	12540.4	CU II		0.9		SH
7944.42	7946.61	12584.0	CU II	0.0	1.4		SH
7933.130	7935.312	12601.90	CU I	2.2			IBU
7902.57	7904.74	12650.6	CU II		1.4		SH
7895.83	7898.00	12661.4	CU II		1.3		SH
7890.56	7892.73	12669.9	CU II	0.3	0.5		SH
7860.58	7862.74	12718.2	CU II		0.7		SH
7845.03	7847.19	12743.4	CU II		1.4		SH
7825.66	7827.81	12775.0	CU II		1.7		SH
7820.57	7822.72	12783.3	CU II		0.7		SH
7812.33	7814.48	12796.8	CU II		1.0		SH
7807.66	7809.81	12804.4	CU II		1.8		SH
7805.19	7807.34	12808.5	CU II		1.4		SH
7778.74	7780.88	12852.0	CU II		1.5		SH
7754.37	7756.50	12892.4	CU II		1.0		SH
7744.09	7746.22	12909.5	CU II		0.7		SH
7738.68	7740.81	12918.6	CU II		1.5		SH
7726.64	7728.77	12938.7	CU II		0.7		SH
7664.70	7666.81	13043.2	CU II	0.7	1.8		SH
7652.36	7654.47	13064.3	CU II		1.5		SH
7579.87	7581.96	13189.2	CU II		1.0		SH
7579.02	7581.11	13190.7	CU II		1.5		SH
7570.09	7572.17	13206.2	CU I	1.7			ME
7562.01	7564.09	13220.4	CU II		1.4		SH

AIR WAVELENGTH (ANGSTRMS)	VACUUM WAVELENGTH (ANGSTRMS)	WAVENUMBER (KAYSERS)	LMENT	LOG ARC	INTENSITY SPK	INTENSITY DIS	REF
7438.15	7440.20	13440.5	CU II		1.2		SH
7433.85	7435.90	13448.3	CU II		0.7		SH
7427.26	7429.31	13460.2	CU	0.5			ME
7420.70	7422.74	13472.1	CU II		0.9		SH
7404.34	7406.38	13501.9	CU II		2.0		SH
7399.89	7401.93	13510.0	CU II		1.3		SH
7382.18	7384.21	13542.4	CU II		1.0		SH
7331.74	7333.76	13635.6	CU II		1.2		SH
7326.02	7328.04	13646.2	CU II		1.2		SH
7306.60	7308.61	13682.5	CU II		1.1		SH
7255.83	7257.83	13778.2	CU II		1.3		SH
7194.92	7196.90	13894.9	CU II		1.2		SH
7193.56	7195.54	13897.5	CU	1.0			ME
7154.29	7156.26	13973.8	CU I	0.5H			ME
7124.66	7126.62	14031.9	CU I	0.5H			ME
7039.37	7041.31	14201.9	CU I	1.2			MIT
7022.75	7024.69	14235.5	CU II		0.3		SH
7000.05	7001.98	14281.7	CU I	0.5H			MIT
6968.34	6970.26	14346.7	CU I	0.5			KS
6935.82	6937.73	14413.9	CU I	0.9			MIT
6920.06	6921.97	14446.8	CU I	2.0			MIT
6905.94	6907.85	14476.3	CU I	1.6			MIT
6890.90	6892.80	14507.9	CU I	0.9			ME
6889.92	6891.82	14509.9	CU I	0.9			ME
6881.94	6883.84	14526.8	CU I	0.9			ME
6872.43	6874.33	14546.9	CU II		0.5		SH
6840.99	6842.88	14613.7	CU I	0.5			ME
6835.46	6837.35	14625.6	CU I	0.5			ME
6823.40	6825.28	14651.4	CU II		0.5		SH
6821.86	6823.74	14654.7	CU I	0.5H			ME
6809.90	6811.78	14680.4	CU II		0.6		SH
6806.60	6808.48	14687.6	CU II		0.6		SH
6780.40	6782.27	14744.3	CU II		0.5		SH
6775.64	6777.51	14754.7	CU I	0.9			ME
6770.70	6772.57	14765.4	CU II		0.9		SH
6758.55	6760.42	14792.0	CU II		0.9		SH
6749.97	6751.83	14810.8	CU II		0.5		SH
6749.29	6751.15	14812.3	CU I	0.9			ME
6741.42	6743.28	14829.6	CU I	1.7			ME
6737.64	6739.50	14837.9	CU II		0.7		SH
6672.23	6674.07	14983.4	CU I	1.2			ME
6660.99	6662.83	15008.6	CU II		0.9		SH
6649.22	6651.06	15035.2	CU II		0.3		SH
6641.41	6643.24	15052.9	CU II		1.0		SH
6634.7	6636.5	15068.	CU I	0.5H			KS
6631.85	6633.68	15074.6	CU II		0.3		SH
6629.67	6631.50	15079.5	CU I	0.5			ME
6624.29	6626.12	15091.8	CU II		0.9		SH
6621.61	6623.44	15097.9	CU I	1.3			ME
6599.68	6601.50	15148.1	CU I	1.4			AZ
6583.54	6585.36	15185.2	CU I	0.9			AZ
6565.54	6567.35	15226.8	CU I	1.2			ME
6564.50	6566.31	15229.2	CU II		1.0		SH
6555.05	6556.86	15251.2	CU II		0.7		SH
6551.58	6553.39	15259.3	CU II		0.3		SH
6550.98	6552.79	15260.7	CU I	0.5			ME
6544.51	6546.32	15275.8	CU I	0.5			ME
6541.93	6543.74	15281.8	CU II		0.3		SH
6530.30	6532.10	15309.0	CU II		0.9		SH
6524.12	6525.92	15323.5	CU II		0.3		SH
6506.14	6507.94	15365.8	CU I	0.9			AZ
6494.04	6495.83	15394.5	CU II		1.5		SH
6485.18	6486.97	15415.5	CU I	0.9			ME
6484.46	6486.25	15417.2	CU II		1.3		SH
6481.46	6483.25	15424.4	CU II		1.2		SH
6479.64	6481.43	15428.7	CU	0.3			SH
6474.20	6475.99	15441.7	CU I	1.2			ME
6470.152	6471.940	15451.32	CU II		1.7		SH
6466.60	6468.39	15459.8	CU II		0.5		SH
6457.54	6459.32	15481.5	CU II		0.5		SH

AIR WAVELENGTH (ANGSTRMS)	VACUUM WAVELENGTH (ANGSTRMS)	WAVENUMBER (KAYSERS)	LMENT	LOG ARC	INTENSITY SPK	DIS	REF
6448.49	6450.27	15503.2	CU II		1.0		SH
6443.47	6445.25	15515.3	CU II		0.7		SH
6441.698	6443.478	15519.57	CU II		1.6		SH
6432.78	6434.56	15541.1	CU II		0.5		SH
6427.57	6429.35	15553.7	CU I	0.5			ME
6423.90	6425.68	15562.6	CU II		1.5		SH
6415.18	6416.95	15583.7	CU I	0.5			AZ
6414.62	6416.39	15585.1	CU II		1.3		SH
6411.18	6412.95	15593.4	CU II		1.0		SH
6403.70	6405.47	15611.7	CU II		0.7		SH
6400.59	6402.36	15619.2	CU	0.5			AZ
6377.84	6379.60	15675.0	CU II		1.3		SH
6373.27	6375.03	15686.2	CU II		0.7		SH
6358.09	6359.85	15723.7	CU I	0.9			AZ
6357.45	6359.21	15725.2	CU II		1.2		SH
6326.10	6327.85	15803.2	CU		0.5		SH
6325.45	6327.20	15804.8	CU I	1.3			AZ
6318.00	6319.75	15823.4	CU II		0.5		SH
6312.83	6314.58	15836.4	CU II		1.3		SH
6311.292	6313.037	15840.24	CU II		1.5		SH
6305.956	6307.700	15853.64	CU II		1.2		SH
6300.988	6302.731	15866.14	CU II		1.6		SH
6292.86	6294.60	15886.6	CU I	0.9			AZ
6288.72	6290.46	15897.1	CU II		0.7		SH
6276.708	6278.444	15927.51	CU II		1.0		SH
6276.624	6278.360	15927.73	CU II		1.0		SH
6273.330	6275.065	15936.09	CU II		1.8		SH
6268.30	6270.03	15948.9	CU I	1.6			AZ
6261.826	6263.558	15965.37	CU II		1.6		SH
6257.86	6259.59	15975.5	CU II		0.7		SH
6253.37	6255.10	15987.0	CU I	0.5			AZ
6233.79	6235.51	16037.2	CU I	0.5			AZ
6223.66	6225.38	16063.3	CU I	1.2			AZ
6220.94	6222.66	16070.3	CU I	0.9			AZ
6219.818	6221.539	16073.19	CU II		1.5		SH
6216.910	6218.630	16080.71	CU II		1.8		SH
6216.38	6218.10	16082.1	CU I	0.9			AZ
6208.46	6210.18	16102.6	CU II		1.2		SH
6204.27	6205.99	16113.5	CU II		1.2		SH
6198.11	6199.82	16129.5	CU II		0.7		SH
6188.69	6190.40	16154.0	CU II		1.3		SH
6186.860	6188.572	16158.82	CU II		1.3		SH
6172.020	6173.728	16197.67	CU II		1.3		SH
6165.7	6167.4	16214.	CU I	0.5H			KS
6158.00	6159.70	16234.6	CU II		0.7		SH
6154.24	6155.94	16244.5	CU II		1.5		SH
6150.42	6152.12	16254.6	CU II		1.3		SH
6147.31	6149.01	16262.8	CU I	1.3			AZ
6127.73	6129.43	16314.7	CU I	1.9			AZ
6119.55	6121.24	16336.6	CU	1.4			AZ
6114.468	6116.160	16350.13	CU II		1.3		SH
6110.90	6112.59	16359.7	CU II		0.7		SH
6107.45	6109.14	16368.9	CU II		1.0		SH
6105.97	6107.66	16372.9	CU II		0.7		SH
6100.01	6101.70	16388.9	CU II		0.7		SH
6097.33	6099.02	16396.1	CU II		1.0		SH
6080.320	6082.003	16441.95	CU II		1.5		SH
6072.25	6073.93	16463.8	CU II	0.0H	0.7		SH
6064.69	6066.37	16484.3	CU I	0.5			AZ
6062.73	6064.41	16489.6	CU	0.5			AZ
6032.33	6034.00	16572.7	CU I	0.9			AZ
6023.25	6024.92	16597.7	CU II		1.0		SH
6013.40	6015.07	16624.9	CU II		0.9		SH
6000.104	6001.766	16661.76	CU II		1.6		SH
5995.59	5997.25	16674.3	CU II		1.0		SH
5993.27	5994.93	16680.8	CU II		0.9		SH
5988.30	5989.96	16694.6	CU II		1.4		SH
5979.20	5980.86	16720.0	CU II		0.7		SH
5979.20	5980.86	16720.0	CU II		0.5		SH
5941.168	5942.814	16827.04	CU II		1.7		SH

AIR WAVELENGTH (ANGSTRMS)	VACUUM WAVELENGTH (ANGSTRMS)	WAVENUMBER (KAYSERS)	LMENT	LOG ARC	INTENSITY SPK	DIS	REF
5937.59	5939.24	16837.2	CU II		0.7		SH
5926.90	5928.54	16867.6	CU II		0.5		SH
5901.21	5902.85	16941.0	CU II		0.7		SH
5897.986	5899.620	16950.24	CU II		1.4		SH
5858.63	5860.25	17064.1	CU II		0.7		SH
5851.93	5853.55	17083.6	CU II		0.3		SH
5842.67	5844.29	17110.7	CU II		0.6		SH
5833.68	5835.30	17137.1	CU II		0.7		SH
5826.02	5827.64	17159.6	CU II		1.0		SH
5806.00	5807.61	17218.8	CU II		1.4		SH
5782.132	5783.735	17289.86	CU I	3.0			IBU
5761.37	5762.97	17352.2	CU II		0.3H		SH
5759.43	5761.03	17358.0	CU II		0.7H		SH
5732.325	5733.915	17440.09	CU I	1.9			SH
5721.78	5723.37	17472.2	CU II		1.3		SH
5700.240	5701.822	17538.25	CU I	2.5			IBU
5692.41	5693.99	17562.4	CU II		0.3		SH
5689.86	5691.44	17570.2	CU II		0.7		SH
5682.42	5684.00	17593.2	CU II		1.3		SH
5664.47	5666.04	17649.0	CU II		0.5		SH
5641.30	5642.87	17721.5	CU II		1.3		SH
5635.57	5637.13	17739.5	CU II		0.3		SH
5633.14	5634.70	17747.2	CU II		0.5		SH
5615.20	5616.76	17803.9	CU II		0.7		SH
5593.73	5595.28	17872.2	CU II		0.7		SH
5554.94	5556.48	17997.0	CU I	0.7			SH
5543.48	5545.02	18034.2	CU II		0.3		SH
5534.98	5536.52	18061.9	CU II		0.5		SH
5482.65	5484.17	18234.3	CU II		0.5		SH
5469.63	5471.15	18277.7	CU II		0.5		SH
5463.138	5464.656	18299.41	CU I	2.2			MIT
5437.36	5438.87	18386.2	CU II		0.3H		SH
5393.96	5395.46	18534.1	CU II		0.5		SH
5390.45	5391.95	18546.2	CU II		0.7		SH
5376.85	5378.35	18593.1	CU II		0.5		SH
5368.42	5369.91	18622.3	CU II		1.0		SH
5365.62	5367.11	18632.0	CU II		0.7		SH
5360.030	5361.521	18651.43	CU I	2.3			SH
5352.666	5354.155	18677.08	CU I	2.5			SH
5328.81	5330.29	18760.7	CU		0.5		SH
5292.517	5293.990	18889.35	CU I	1.7			IBU
5276.522	5277.990	18946.60	CU II		1.2		SH
5269.988	5271.455	18970.10	CU II		1.5		SH
5245.36	5246.82	19059.2	CU II		1.0		SH
5229.58	5231.04	19116.7	CU II		0.5		SH
5220.070	5221.523	19151.50	CU I	2.0			IBU
5218.202	5219.655	19158.35	CU I	2.8			IBU
5212.780	5214.231	19178.28	CU I	0.6			HZ
5207.128	5208.578	19199.10	CU II		1.3		SH
5200.87	5202.32	19222.2	CU I	0.7J			SH
5183.364	5184.808	19287.12	CU II		1.3		SH
5175.89	5177.33	19315.0	CU II		0.3		SH
5158.090	5159.527	19381.62	CU II		1.0		SH
5153.235	5154.671	19399.88	CU I	2.8			IBU
5144.12	5145.55	19434.3	CU I	0.7J			SH
5124.461	5125.889	19508.81	CU II		1.3		SH
5120.745	5122.172	19522.97	CU II		1.3		SH
5111.913	5113.338	19556.70	CU I	1.2			IBU
5108.331	5109.755	19570.41	CU II		0.5		SH
5105.541	5106.964	19581.11	CU I	2.7			IBU
5100.08	5101.50	19602.1	CU II		1.0		SH
5093.792	5095.212	19626.27	CU II		1.3		SH
5088.932	5090.350	19645.01	CU II		1.0		SH
5088.487	5089.905	19646.73	CU II		1.0		SH
5088.260	5089.678	19647.61	CU II		1.5		SH
5083.991	5085.408	19664.10	CU II		1.2		SH
5077.805	5079.220	19688.06	CU II		0.7		SH
5076.173	5077.588	19694.39	CU I	0.3H			IBU
5072.293	5073.707	19709.46	CU II		1.3		SH
5067.082	5068.495	19729.72	CU II		1.5		SH

AIR WAVELENGTH (ANGSTRMS)	VACUUM WAVELENGTH (ANGSTRMS)	WAVENUMBER (KAYSERS)	LMENT	LOG ARC	INTENSITY SPK	INTENSITY DIS	REF
5065.448	5066.860	19736.09	CU II		1.6		SH
5060.635	5062.046	19754.86	CU II		1.5		SH
5058.897	5060.307	19761.65	CU II		1.5		SH
5051.778	5053.187	19789.49	CU II		1.8		SH
5047.343	5048.750	19806.88	CU II		1.0		SH
5041.322	5042.728	19830.54	CU II		1.0		SH
5039.002	5040.407	19839.67	CU II		1.0		SH
5030.778	5032.181	19872.10	CU II		0.3		SH
5024.027	5025.428	19898.80	CU II		0.7		SH
5021.285	5022.685	19909.67	CU II		1.3		SH
5020.139	5021.539	19914.21	CU II	0.7			SH
5016.611	5018.010	19928.22	CU I	1.2			IBU
5015.207	5016.606	19933.80	CU II		1.0		SH
5012.611	5014.009	19944.12	CU II		1.3		SH
5009.833	5011.230	19955.18	CU II		1.3		SH
5006.787	5008.184	19967.32	CU II		1.5		SH
4985.503	4986.894	20052.56	CU II		1.0		SH
4980.006	4981.395	20074.70	CU II		0.6		SH
4974.151	4975.539	20098.32	CU II		0.6		SH
4973.689	4975.077	20100.19	CU II		0.6		SH
4955.964	4957.347	20172.08	CU II		0.3		SH
4953.729	4955.111	20181.18	CU II		1.2		MIT
4951.627	4953.009	20189.75	CU II		0.7		SH
4943.020	4944.400	20224.90	CU II		0.8		SH
4940.060	4941.439	20237.02	CU II		0.3H		SH
4937.967	4939.345	20245.60	CU II		0.7		SH
4937.196	4938.574	20248.76	CU II		0.8		SH
4931.653	4933.030	20271.52	CU II		1.4		SH
4931.483	4932.860	20272.22	CU II		0.8		SH
4926.390	4927.765	20293.18	CU II		0.8		SH
4920.031	4921.405	20319.40	CU II		0.5		SH
4918.373	4919.746	20326.25	CU II		1.0		SH
4915.821	4917.193	20336.80	CU II		0.7		SH
4912.909	4914.281	20348.86	CU II		0.8		SH
4912.369	4913.741	20351.09	CU II		0.7		MIT
4909.726	4911.097	20362.05	CU II		1.4		SH
4909.032	4910.403	20364.93	CU II		0.3		SH
4906.548	4907.918	20375.24	CU II		0.8		SH
4901.412	4902.781	20396.59	CU II		0.8		SH
4889.690	4891.055	20445.49	CU II		1.0		SH
4873.291	4874.652	20514.28	CU II		0.7		SH
4854.966	4856.322	20591.71	CU II		1.0		SH
4851.248	4852.603	20607.50	CU II		0.7		SH
4832.236	4833.586	20688.57	CU II		1.0		SH
4812.940	4814.285	20771.52	CU II		1.2		SH
4807.039	4808.383	20797.01	CU II		0.6		SH
4797.042	4798.383	20840.35	CU I	1.1	0.0H		HS
4793.800	4795.140	20854.45	CU	0.7J	0.3J		MX
4766.729	4768.062	20972.88	CU II		0.3		SH
4758.421	4759.752	21009.50	CU II		1.0		SH
4704.596	4705.912	21249.86	CU I	2.3	1.7		MIT
4697.490	4698.805	21282.01	CU I	1.8W	0.7W		MIT
4687.770	4689.082	21326.14	CU II		0.3		SH
4681.990	4683.300	21352.46	CU II		1.3		SH
4674.76	4676.07	21385.5	CU II	2.3	1.50		HS
4673.555	4674.863	21391.00	CU II		0.8		SH
4671.693	4673.001	21399.53	CU II		1.0		MIT
4667.297	4668.604	21419.68	CU II		0.7		SH
4662.634	4663.939	21441.10	CU II		0.7		MIT
4661.350	4662.655	21447.01	CU II		0.5		SH
4660.294	4661.599	21451.87	CU II		0.3		SH
4651.134	4652.436	21494.11	CU I	2.4	1.6		MIT
4649.266	4650.568	21502.75	CU II		1.8		SH
4608.457	4609.748	21693.16	CU II		0.3		SH
4597.936	4599.224	21742.80	CU II		0.3		MIT
4596.903	4598.191	21747.68	CU II		0.6J		SL
4586.954	4588.239	21794.85	CU I	2.4W	1.9W		HS
4555.922	4557.199	21943.30	CU II	0.3	1.8		SH
4541.032	4542.305	22015.25	CU II		0.7		SH
4540.335	4541.608	22018.63	CU		0.5		SH

AIR WAVELENGTH (ANGSTRMS)	VACUUM WAVELENGTH (ANGSTRMS)	WAVENUMBER (KAYSERS)	LMENT	LOG ARC	INTENSITY SPK	INTENSITY DIS	REF
4540.207	4541.480	22019.25	CU		0.3		SH
4539.695	4540.968	22021.74	CU I	2.00	1.90		MIT
4530.819	4532.089	22064.88	CU I	2.3	1.7		MIT
4516.050	4517.317	22137.04	CU II		0.3		SH
4513.20	4514.47	22151.0	CU I	0.5	0.0H		HS
4509.388	4510.653	22169.74	CU I	2.2	1.5		MIT
4507.5	4508.8	22179.	CU I	1.7W	1.5J		HS
4505.997	4507.261	22186.42	CU II	0.0	1.7		SH
4480.359	4481.616	22313.38	CU I	2.3	1.3		MIT
4444.823	4446.071	22491.77	CU II		0.3J		SH
4415.60	4416.84	22640.6	CU I	1.6W			HS
4378.430	4379.660	22832.82	CU II		0.3		SH
4378.20	4379.43	22834.0	CU I	2.3W	1.5W		MIT
4365.362	4366.589	22901.17	CU II		0.8		SH
4336.00	4337.22	23056.2	CU I	0.3			MIT
4328.68	4329.90	23095.2	CU I	0.3			MIT
4292.469	4293.677	23290.06	CU II		0.8		SH
4285.239	4286.445	23329.36	CU II	0.3	0.6		SH
4279.959	4281.163	23358.14	CU II		0.7		SH
4276.039	4277.242	23379.55	CU II		0.8		MIT
4275.131	4276.334	23384.51	CU I	1.9	1.5		MIT
4259.43	4260.63	23470.7	CU I	1.4J	0.3J		HS
4253.34	4254.54	23504.3	CU I	0.8J			SH
4248.957	4250.153	23528.56	CU I	1.9	1.2		MIT
4242.26	4243.45	23565.7	CU I	1.3			MIT
4239.448	4240.642	23581.34	CU II		0.8		SH
4231.364	4232.556	23626.39	CU II		0.9		MIT
4230.444	4231.635	23631.53	CU II		0.5		SH
4228.8	4230.0	23641.	CU	0.5			HS
4227.936	4229.127	23645.54	CU II		0.8		SH
4211.861	4213.047	23735.79	CU II	0.0H	0.8		SH
4179.508	4180.686	23919.52	CU II	0.0H	0.8		MIT
4177.758	4178.936	23929.54	CU I	1.8W	0.0		HS
4171.854	4173.030	23963.40	CU II		0.7		MIT
4164.283	4165.457	24006.97	CU II		0.5H		MIT
4162.298	4163.471	24018.42	CU II		0.3		MIT
4161.155	4162.328	24025.02	CU II		0.8		SH
4153.623	4154.794	24068.58	CU II		0.3H		SH
4143.039	4144.207	24130.07	CU II		0.3J		MIT
4141.296	4142.464	24140.22	CU II		0.5		SH
4123.287	4124.450	24245.66	CU I	1.5W	0.0H		MIT
4121.74	4122.90	24254.8	CU I	1.3			MIT
4104.226	4105.384	24358.26	CU I	1.5	0.0H		MIT
4080.553	4081.705	24499.57	CU I	1.5W			MIT
4077.716	4078.867	24516.61	CU	0.7			MIT
4075.588	4076.739	24529.41	CU I	1.6			MIT
4073.27	4074.42	24543.4	CU I	1.2			EX
4063.293	4064.441	24603.63	CU I	1.5W	0.8		MIT
4062.698	4063.845	24607.24	CU I	2.7W	1.3		MIT
4056.7	4057.8	24644.	CU I	0.9J			HS
4053.651	4054.796	24662.15	CU II		0.3H		MIT
4050.656	4051.800	24680.39	CU I	1.5			HS
4043.751	4044.893	24722.53	CU II		1.0		SH
4043.502	4044.644	24724.05	CU II		1.4		SH
4022.657	4023.794	24852.17	CU I	2.6	1.4		MIT
4015.815	4016.950	24894.51	CU I	0.8J			HS
4010.85	4011.98	24925.3	CU I	0.8			EX
4003.038	4004.170	24973.97	CU I	1.6	0.0H		HS
3993.291	3994.420	25034.92	CU II		0.6H		MIT
3987.012	3988.140	25074.35	CU II		0.5		MIT
3945.570	3946.687	25337.71	CU II		0.3		SH
3932.917	3934.030	25419.22	CU I	1.0			MIT
3925.274	3926.385	25468.72	CU I	1.7	0.3H		MIT
3923.439	3924.550	25480.63	CU II	0.3	0.0H		MIT
3921.267	3922.377	25494.74	CU II	1.6	0.0H		MIT
3921.267	3922.377	25494.74	CU I	1.6	0.0H		MIT
3920.641	3921.751	25498.81	CU II		0.3		SH
3903.163	3904.269	25612.99	CU II		0.7		SH
3899.106	3900.211	25639.64	CU I	0.3J			MIT
3896.675	3897.779	25655.64	CU II		0.3		MIT

AIR WAVELENGTH (ANGSTRMS)	VACUUM WAVELENGTH (ANGSTRMS)	WAVENUMBER (KAYSERS)	LMENT	LOG ARC	INTENSITY SPK	INTENSITY DIS	REF
3892.913	3894.016	25680.43	CU II	0.0	0.3		SH
3890.073	3891.175	25699.18	CU II		0.5		SH
3884.523	3885.624	25735.89	CU II		0.3		SH
3884.129	3885.230	25738.50	CU II		0.7		MIT
3881.707	3882.807	25754.56	CU I	0.7			MIT
3879.387	3880.487	25769.97	CU II	0.0H	0.3H		SH
3864.121	3865.217	25871.77	CU II		0.3J		SH
3862.762	3863.857	25880.87	CU I	1.0			MIT
3861.746	3862.841	25887.68	CU I	1.7	0.3		MIT
3860.472	3861.567	25896.23	CU I	1.5	0.8		MIT
3842.568	3843.658	26016.88	CU II		0.3		MIT
3836.150	3837.238	26060.41	CU II		0.3J		SH
3826.908	3827.994	26123.34	CU II		0.3H		SH
3825.047	3826.132	26136.05	CU I	1.6	0.5		IBU
3820.884	3821.968	26164.53	CU I	1.0	0.5		MIT
3817.509	3818.592	26187.66	CU I	0.8			MIT
3813.542	3814.624	26214.90	CU I	0.9			MIT
3805.30	3806.38	26271.7	CU I	1.3	0.3H		M
3800.502	3801.581	26304.84	CU I	1.3	0.3H		MIT
3797.832	3798.910	26323.34	CU II		0.3H		SH
3786.261	3787.336	26403.78	CU II		0.3H		SH
3771.904	3772.976	26504.28	CU I	1.5	0.7		MIT
3764.837	3765.907	26554.03	CU I	1.0			MIT
3759.492	3760.560	26591.78	CU I	1.3	0.3H		MIT
3748.207	3749.272	26671.84	CU II		1.0		SH
3743.365	3744.429	26706.34	CU I	1.6	1.6		MIT
3741.242	3742.306	26721.50	CU I	1.7	0.3		MIT
3720.771	3721.829	26868.51	CU I	1.0	0.0H		MIT
3712.009	3713.065	26931.93	CU I	1.3	0.6H		MIT
3700.536	3701.589	27015.43	CU I	1.3	0.8		MIT
3695.358	3696.410	27053.28	CU I	0.8			MIT
3686.555	3687.604	27117.88	CU II		1.4		SH
3684.930	3685.979	27129.83	CU I	0.8	0.5		MIT
3684.672	3685.721	27131.73	CU I	1.1	0.6		MIT
3682.428	3683.476	27148.27	CU II		0.6		SH
3676.878	3677.925	27189.24	CU	1.4J	0.0H		MIT
3671.953	3672.999	27225.71	CU I	1.3	0.5		MIT
3665.735	3666.779	27271.89	CU I	1.3	0.7		MIT
3659.353	3660.395	27319.45	CU I	0.9	0.6		MIT
3656.788	3657.830	27338.61	CU I	0.6	0.5		MIT
3655.859	3656.901	27345.56	CU I	1.3	0.8		MIT
3654.30	3655.34	27357.2	CU I	1.0J	0.3H		MIT
3652.43	3653.47	27371.2	CU I	1.0J	0.0H		MIT
3650.855	3651.895	27383.04	CU I	0.6	0.0H		MIT
3648.383	3649.423	27401.59	CU I	1.0	0.8		MIT
3645.232	3646.271	27425.28	CU I	1.3	0.7		MIT
3643.631	3644.669	27437.33	CU I	0.7			MIT
3641.693	3642.731	27451.93	CU I	1.8	0.7		MIT
3635.916	3636.952	27495.55	CU I	1.7	0.8		MIT
3632.558	3633.593	27520.97	CU I	1.4	0.5		MIT
3629.794	3630.829	27541.92	CU I	1.2	0.3		HS
3627.33	3628.36	27560.6	CU I	1.0J	0.7J		HS
3624.236	3625.269	27584.16	CU I	1.50	0.5J		HS
3621.245	3622.278	27606.94	CU I	1.3	0.7		MIT
3620.352	3621.384	27613.75	CU I	1.5	0.7		MIT
3614.218	3615.249	27660.61	CU I	1.7	0.8		MIT
3613.761	3614.792	27664.11	CU I	1.8	0.8		MIT
3610.809	3611.839	27686.73	CU I	1.4	0.8		MIT
3609.307	3610.336	27698.25	CU I	1.4	0.7		MIT
3602.032	3603.060	27754.19	CU II	1.7	1.50		MIT
3602.032	3603.060	27754.19	CU I	1.7	1.40		MIT
3599.140	3600.167	27776.49	CU I	1.8	1.5		MIT
3598.014	3599.041	27785.18	CU I	1.6J			MIT
3594.023	3595.049	27816.04	CU I	1.2	0.3		MIT
3566.14	3567.16	28033.5	CU I	0.5			HS
3548.742	3549.756	28170.95	CU II		0.7		SH
3546.45	3547.46	28189.2	CU I	0.3	0.0H		HS
3544.963	3545.976	28200.98	CU I	1.5	0.8		MIT
3533.746	3534.756	28290.50	CU I	1.7	1.2		MIT
3530.386	3531.395	28317.42	CU I	1.7	1.3		MIT

AIR WAVELENGTH (ANGSTRMS)	VACUUM WAVELENGTH (ANGSTRMS)	WAVENUMBER (KAYSERS)	LMENT	LOG ARC	INTENSITY SPK	INTENSITY DIS	REF
3527.482	3528.490	28340.73	CU I	1.7	1.0		MIT
3524.239	3525.247	28366.81	CU I	1.6	1.0		MIT
3520.031	3521.037	28400.72	CU I	1.5	1.0		MIT
3517.039	3518.045	28424.88	CU I	1.3	0.5H		MIT
3512.121	3513.125	28464.68	CU I	1.7	1.5		MIT
3507.38	3508.38	28503.2	CU I	0.5	0.3H		EX
3501.529	3502.531	28550.78	CU I	0.3H	0.0H		MIT
3501.338	3502.340	28552.34	CU I	0.7	0.0H		MIT
3500.324	3501.325	28560.61	CU I	1.3	0.3H		MIT
3498.063	3499.064	28579.07	CU I	1.3	0.7		MIT
3488.858	3489.856	28654.47	CU I	1.5	0.7		M
3483.761	3484.758	28696.40	CU I	1.8	1.4		MIT
3481.909	3482.906	28711.66	CU I	0.5	0.5		SH
3475.999	3476.994	28760.47	CU I	1.4	1.0		MIT
3474.578	3475.573	28772.24	CU I	0.7			M
3472.141	3473.135	28792.43	CU I	1.3	0.7		MIT
3465.401	3466.393	28848.43	CU I	1.0	0.3H		MIT
3463.499	3464.491	28864.27	CU I	0.8H			MIT
3459.428	3460.419	28898.24	CU I	1.4	0.3H		MIT
3457.851	3458.842	28911.41	CU I	1.7	1.2		MIT
3454.686	3455.676	28937.90	CU	1.6	1.0H		MIT
3450.332	3451.321	28974.42	CU I	2.2	1.5		MIT
3440.507	3441.493	29057.16	CU I	1.0	0.6		IBU
3436.543	3437.528	29090.67	CU I	0.8			MIT
3433.972	3434.956	29112.45	CU I	0.7H			MIT
3420.166	3421.147	29229.96	CU I	1.2	0.5H		MIT
3415.78	3416.76	29267.5	CU	1.0V	0.5V		MIT
3413.343	3414.322	29288.39	CU I	1.3	0.8		MIT
3402.244	3403.220	29383.93	CU I	1.7	1.0		MIT
3396.324	3397.299	29435.15	CU I	1.5	0.5		HS
3395.476	3396.451	29442.50	CU I	1.5	0.8		MIT
3392.988	3393.962	29464.09	CU I	0.7			MIT
3392.016	3392.990	29472.53	CU I	0.8			MIT
3385.4	3386.4	29530.	CU I	0.3			SH
3384.948	3385.920	29534.07	CU II		0.5		SH
3384.815	3385.787	29535.23	CU I	1.0	0.5		HS
3381.421	3382.392	29564.88	CU I	1.3	0.8		MIT
3381.124	3382.095	29567.47	CU I	1.2	0.7H		MIT
3380.717	3381.688	29571.03	CU II		0.7		SH
3379.961	3380.932	29577.65	CU II	0.3H	0.5H		SH
3379.69	3380.66	29580.0	CU I	0.5H			MIT
3378.512	3379.482	29590.33	CU II		0.3		SH
3377.706	3378.676	29597.39	CU II		0.5		SH
3375.672	3376.641	29615.23	CU I	1.4	0.7		MIT
3374.953	3375.922	29621.53	CU II		1.0		SH
3373.594	3374.563	29633.47	CU II		0.9		SH
3371.412	3372.380	29652.65	CU II		0.6		SH
3370.454	3371.422	29661.07	CU		1.2		MIT
3366.560	3367.527	29695.38	CU II		0.3		SH
3366.269	3367.236	29697.95	CU II		0.3H		SH
3365.65	3366.62	29703.4	CU II		0.9		SH
3365.349	3366.316	29706.07	CU I	1.8	1.5		MIT
3354.474	3355.438	29802.37	CU I	1.5	1.0H		MIT
3352.044	3353.007	29823.97	CU II		0.6		SH
3349.463	3350.426	29846.95	CU II		0.3H		SH
3349.292	3350.255	29848.48	CU I	1.8	1.6		MIT
3343.743	3344.704	29898.01	CU II		1.0		SH
3341.182	3342.143	29920.93	CU I	0.7			MIT
3338.640	3339.600	29943.71	CU II	0.0H	0.7		MIT
3337.844	3338.804	29950.85	CU I	1.8	1.7		MIT
3335.215	3336.174	29974.45	CU	1.8	1.2		IBU
3329.636	3330.594	30024.68	CU I	1.8	1.0		MIT
3325.812	3326.769	30059.20	CU II		0.6		MIT
3319.684	3320.639	30114.68	CU I	1.8	1.3		MIT
3317.223	3318.178	30137.02	CU I	1.8	1.3		MIT
3316.277	3317.231	30145.62	CU II		1.0		MIT
3311.00	3311.95	30193.7	CU I	0.5			HS
3307.948	3308.900	30221.52	CU I	1.8	1.5		IBU
3303.516	3304.467	30262.07	CU II		0.3		SH
3301.228	3302.179	30283.04	CU		1.2		SH

AIR WAVELENGTH (ANGSTRMS)	VACUUM WAVELENGTH (ANGSTRMS)	WAVENUMBER (KAYSERS)	LMENT	LOG ARC	INTENSITY SPK	DIS	REF
3300.885	3301.835	30286.19	CU II		0.8		SH
3300.644	3301.594	30288.40	CU II		0.5		SH
3300.444	3301.394	30290.23	CU II		0.3		SH
3297.199	3298.148	30320.04	CU II		0.7		SH
3295.103	3296.052	30339.33	CU II		0.9		SH
3292.903	3293.851	30359.60	CU I	0.9	0.3		HS
3292.830	3293.778	30360.27	CU I	1.0	0.3		MIT
3292.393	3293.341	30364.30	CU I	0.8	0.0H		MIT
3292.124	3293.072	30366.78	CU II		0.7		SH
3290.544	3291.492	30381.36	CU I	1.4	1.4		MIT
3282.716	3283.662	30453.81	CU I	1.4	1.20		MIT
3281.696	3282.642	30463.27	CU II		0.7		SH
3279.816	3280.761	30480.73	CU I	1.4	1.5		MIT
3277.310	3278.254	30504.04	CU I	0.8	0.3		MIT
3273.962	3274.906	30535.23	CU I	2.5G	3.2G		MIT
3268.278	3269.220	30588.33	CU I	1.2	1.0		HS
3266.023	3266.965	30609.45	CU I	1.3	1.2		MIT
3252.220	3253.158	30739.36	CU I	0.6	0.5		MIT
3250.469	3251.407	30755.92	CU II		0.7		SH
3247.540	3248.477	30783.66	CU I	3.7G	3.3G		IBU
3243.164	3244.100	30825.19	CU I	1.2	1.2		MIT
3238.836	3239.771	30866.38	CU II		0.5		MIT
3235.713	3236.647	30896.17	CU I	1.2	0.8		MIT
3233.90	3234.83	30913.5	CU I	0.3H	0.3H		MIT
3231.175	3232.108	30939.56	CU I	1.2	1.0		MIT
3226.602	3227.534	30983.41	CU I	1.1	0.8		MIT
3224.664	3225.595	31002.03	CU I	1.4	1.0		MIT
3223.435	3224.366	31013.85	CU I	1.3	1.0		MIT
3208.234	3209.161	31160.79	CU I	1.4	1.2		MIT
3194.095	3195.018	31298.73	CU I	1.8	1.8		MIT
3193.981	3194.904	31299.84	CU	0.3			MIT
3186.017	3186.938	31378.08	CU II		0.5		SH
3184.843	3185.764	31389.65	CU II		0.9		SH
3182.175	3183.095	31415.96	CU II		0.7		SH
3179.793	3180.713	31439.49	CU II		0.5		SH
3171.663	3172.581	31520.08	CU I	0.8			MIT
3169.681	3170.598	31539.79	CU I	1.7	1.3		MIT
3166.574	3167.490	31570.74	CU II		0.3H		MIT
3160.047	3160.962	31635.94	CU I	1.1	0.3H		HS
3158.62	3159.53	31650.2	CU II		0.3		MIT
3156.627	3157.541	31670.22	CU I	1.7	1.2		MIT
3151.62	3152.53	31720.5	CU I	0.8			MIT
3151.049	3151.961	31726.28	CU II	0.3	0.7		SH
3149.508	3150.420	31741.80	CU I	1.7	0.3		MIT
3146.821	3147.732	31768.90	CU I	2.0	1.3		MIT
3142.444	3143.354	31813.15	CU I	1.8	1.2		MIT
3140.312	3141.222	31834.75	CU I	1.7	1.1		MIT
3128.701	3129.608	31952.89	CU I	1.8	1.2		MIT
3126.108	3127.014	31979.39	CU I	1.9	1.3		MIT
3122.432	3123.337	32017.03	CU	0.8	0.0		MIT
3120.435	3121.340	32037.52	CU I	1.4	0.5		MIT
3118.355	3119.259	32058.89	CU I	0.7	0.0H		MIT
3116.348	3117.252	32079.54	CU I	1.7	1.1		MIT
3113.482	3114.385	32109.07	CU I	1.1	0.3		MIT
3108.605	3109.507	32159.44	CU I	1.3	0.7		MIT
3108.452	3109.354	32161.02	CU	1.2	0.0H		MIT
3099.934	3100.834	32249.39	CU I	1.8	1.0		MIT
3093.992	3094.890	32311.32	CU I	2.2	1.7		MIT
3088.132	3089.029	32372.64	CU I	1.5	0.8		MIT
3085.206	3086.102	32403.34	CU	0.6	0.0H		MIT
3083.371	3084.266	32422.62	CU II	0.3	0.3H		MIT
3073.801	3074.694	32523.56	CU I	1.8	1.3		MIT
3068.906	3069.798	32575.44	CU I	1.2			MIT
3063.415	3064.305	32633.82	CU I	2.5	1.7		MIT
3044.028	3044.914	32841.65	CU I	1.3	0.0H		MIT
3036.104	3036.988	32927.37	CU I	2.3	1.7		MIT
3030.258	3031.140	32990.89	CU I	1.0	0.0H		MIT
3024.994	3025.875	33048.29	CU I	1.5	0.5		MIT
3022.613	3023.493	33074.33	CU I	1.5	0.5		MIT
3021.557	3022.437	33085.88	CU I	1.4	0.7		MIT

AIR WAVELENGTH (ANGSTRMS)	VACUUM WAVELENGTH (ANGSTRMS)	WAVENUMBER (KAYSERS)	LMENT	LOG ARC	INTENSITY SPK	DIS	REF
3012.005	3012.883	33190.81	CU I	1.7	0.8		MIT
3010.839	3011.716	33203.66	CU I	2.4	1.5		MIT
2998.384	2999.258	33341.58	CU I	1.3	0.3H		IBU
2997.364	2998.238	33352.92	CU I	2.5	1.5		IBU
2991.754	2992.627	33415.46	CU	1.2			MIT
2985.954	2986.825	33480.37	CU	0.5			MIT
2982.765	2983.635	33516.16	CU I	0.8			MIT
2979.367	2980.236	33554.38	CU I	1.0			IBU
2978.274	2979.143	33566.70	CU I	1.0	0.0H		IBU
2961.165	2962.030	33760.63	CU I	2.5	2.5		IBU
2951.26	2952.12	33873.9	CU I	0.5H			MIT
2931.699	2932.557	34099.94	CU				ME
2926.830	2927.686	34156.66	CU				ME
2926.057	2926.913	34165.69	CU				ME
2925.439	2926.295	34172.90	CU I	1.0			MIT
2924.882	2925.738	34179.41	CU I	0.3			MIT
2923.704	2924.560	34193.18	CU	0.8			MIT
2920.296	2921.151	34233.08	CU				ME
2911.215	2912.068	34339.86	CU I	0.3H			MIT
2884.383	2885.229	34659.29	CU II		1.5		MIT
2882.934	2883.780	34676.71	CU I	2.5			ME
2877.689	2878.533	34739.91	CU II	0.7	1.3		IBU
2874.56	2875.40	34777.7	CU	0.5H			MIT
2858.734	2859.574	34970.25	CU I	1.5	0.3H		IBU
2858.225	2859.065	34976.48	CU I	1.5	0.3H		IBU
2857.746	2858.585	34982.34	CU II		0.6H		SH
2848.715	2849.552	35093.23	CU II		0.3		MIT
2848.525	2849.362	35095.57	CU		0.3		MIT
2846.478	2847.315	35120.81	CU I	0.9			MIT
2837.547	2838.381	35231.35	CU II		2.4		MIT
2824.369	2825.200	35395.72	CU I	3.0	2.5		MIT
2813.02	2813.85	35538.5	CU		0.5H		MIT
2810.803	2811.631	35566.55	CU II		0.3H		MIT
2802.683	2803.509	35669.58	CU I	1.0	0.3		MIT
2799.536	2800.361	35709.68	CU II		0.5		MIT
2795.331	2796.155	35763.40	CU II		0.3		MIT
2791.96	2792.78	35806.6	CU	0.8			MIT
2791.785	2792.608	35808.82	CU II	0.0H	0.7		MIT
2786.496	2787.318	35876.78	CU I	0.9			MIT
2783.551	2784.372	35914.74	CU I	1.3			MIT
2782.592	2783.413	35927.12	CU I	1.3	0.0H		MIT
2769.666	2770.484	36094.78	CU II	0.7H	2.6		IBU
2768.878	2769.696	36105.05	CU I	1.4	0.0H		MIT
2766.371	2767.188	36137.77	CU I	2.7	1.4		IBU
2763.69	2764.51	36172.8	CU	0.5H	0.3H		MIT
2751.810	2752.624	36328.98	CU I	0.7			MIT
2751.29	2752.10	36335.8	CU	1.0W			MIT
2745.452	2746.264	36413.10	CU I	1.3H			SH
2745.276	2746.088	36415.44	CU II	1.5	1.8		SH
2739.764	2740.575	36488.70	CU II		1.1		MIT
2737.331	2738.141	36521.13	CU II		0.6		MIT
2723.953	2724.760	36700.48	CU I	1.7	0.0H		MIT
2721.675	2722.481	36731.20	CU II		2.2		IBU
2720.199	2721.005	36751.13	CU I	1.3			MIT
2718.775	2719.581	36770.38	CU II	1.6	2.5W		IBU
2715.52	2716.32	36814.5	CU	1.3W			MIT
2713.505	2714.309	36841.79	CU II	1.7	2.5W		IBU
2711.88	2712.68	36863.9	CU II		0.5		SH
2703.184	2703.986	36982.44	CU II	1.0	2.3		IBU
2702.61	2703.41	36990.3	CU I	0.9D			MIT
2700.963	2701.764	37012.85	CU II	1.3	2.6		IBU
2689.299	2690.097	37173.37	CU II		2.5		IBU
2666.288	2667.081	37494.18	CU II		1.3		IBU
2630.004	2630.788	38011.42	CU I	1.1	0.3H		MIT
2627.365	2628.149	38049.60	CU I	0.8			MIT
2626.63	2627.41	38060.2	CU I	0.6			MIT
2620.675	2621.457	38146.73	CU II	0.0H	0.9H		IBU
2618.366	2619.147	38180.36	CU I	2.7W	2.0		IBU
2614.41	2615.19	38238.1	CU		0.7W		MIT
2609.312	2610.091	38312.84	CU		1.5		MIT

AIR WAVELENGTH (ANGSTRMS)	VACUUM WAVELENGTH (ANGSTRMS)	WAVENUMBER (KAYSERS)	LMENT	LOG ARC	INTENSITY SPK	INTENSITY DIS	REF
2600.266	2601.043	38446.11	CU II	0.0H	2.3		IBU
2598.813	2599.590	38467.61	CU II		2.3		IBU
2590.526	2591.301	38590.66	CU II	0.0H	2.4		IBU
2571.742	2572.512	38872.50	CU II		2.2		IBU
2544.802	2545.566	39283.99	CU II		2.8G		IBU
2529.302	2530.062	39524.72	CU II		2.8		IBU
2526.73	2527.49	39565.0	CU II		1.7		MIT
2526.589	2527.349	39567.15	CU II		2.3		IBU
2518.95	2519.71	39687.1	CU II		0.8H		MIT
2506.270	2507.025	39887.91	CU II		2.7R		IBU
2494.89	2495.64	40069.8	CU I	0.3J			MIT
2492.146	2492.898	40113.96	CU I	2.3R	1.7		MIT
2489.652	2490.403	40154.14	CU II	0.5	1.7		IBU
2485.787	2486.537	40216.57	CU II	0.7	1.7		MIT
2482.341	2483.091	40272.39	CU		1.3		MIT
2477.13	2477.88	40357.1	CU		0.3W		MIT
2473.332	2474.079	40419.07	CU II	0.7	1.3		IBU
2468.584	2469.330	40496.81	CU II	0.7	1.8		MIT
2457.68	2458.42	40676.5	CU		0.3O		MIT
2448.217	2448.959	40833.68	CU II	0.0H	0.3H		MIT
2444.437	2445.178	40896.82	CU		1.3		MIT
2443.320	2444.061	40915.52	CU II	0.0	1.0H		MIT
2442.656	2443.396	40926.64	CU II	0.3	0.5H		MIT
2441.637	2442.377	40943.72	CU I	2.3	2.0		IBU
2441.010	2441.750	40954.23	CU	0.3H			MIT
2439.907	2440.647	40972.75	CU I	0.8			MIT
2428.911	2429.648	41158.22	CU II	0.0H	0.7H		MIT
2424.436	2425.172	41234.19	CU II	0.6	2.3		IBU
2414.860	2415.594	41397.69	CU II		0.3		MIT
2414.735	2415.469	41399.83	CU	0.7H			MIT
2414.20	2414.93	41409.0	CU II	0.0H	0.3H		MIT
2412.326	2413.059	41441.17	CU		0.5		MIT
2406.665	2407.397	41538.64	CU I	2.2	1.7		MIT
2405.494	2406.226	41558.86	CU		0.9		MIT
2403.335	2404.066	41596.19	CU II	2.0	2.5		IBU
2400.112	2400.843	41652.04	CU II	0.7	2.0		IBU
2392.627	2393.356	41782.33	CU I	3.0	1.7		IBU
2391.72	2392.45	41798.2	CU		1.3		MIT
2385.07	2385.80	41914.7	CU II		1.0H		MIT
2384.84	2385.57	41918.7	CU II		1.2		MIT
2376.38	2377.11	42068.0	CU II	0.5	1.5		MIT
2376.27	2377.00	42069.9	CU		1.4		MIT
2370.885	2371.609	42165.46	CU II		0.7H		IBU
2369.887	2370.611	42183.22	CU II	1.3	1.5		IBU
2368.17	2368.89	42213.8	CU	0.3	1.2		MIT
2367.51	2368.23	42225.6	CU		0.6H		MIT
2364.14	2364.86	42285.8	CU II		1.0		MIT
2363.95	2364.67	42289.1	CU	0.9	0.3H		MIT
2363.21	2363.93	42302.4	CU I	1.0	0.8		MIT
2361.57	2362.29	42331.8	CU		0.9		MIT
2361.43	2362.15	42334.3	CU	1.1			MIT
2361.23	2361.95	42337.9	CU II	0.3		0.5	MIT
2359.49	2360.21	42369.1	CU	0.7			MIT
2358.53	2359.25	42386.3	CU	1.1			MIT
2356.81	2357.53	42417.3	CU	0.8			MIT
2356.651	2357.372	42420.12	CU II	0.7	1.4		MIT
2355.91	2356.63	42433.5	CU	0.6			MIT
2355.155	2355.876	42447.06	CU II		1.4		IBU
2354.04	2354.76	42467.2	CU	1.0			MIT
2353.96	2354.68	42468.6	CU II			0.3	SH
2353.30	2354.02	42480.5	CU	0.8			MIT
2350.96	2351.68	42522.8	CU	0.5			MIT
2349.85	2350.57	42542.9	CU	0.7			MIT
2348.82	2349.54	42561.5	CU II	1.2D	1.3		MIT
2347.90	2348.62	42578.2	CU	0.6			MIT
2346.13	2346.85	42610.3	CU	0.3	1.4		MIT
2343.63	2344.35	42655.8	CU	0.6			MIT
2341.38	2342.10	42696.8	CU	0.3			I
2340.04	2340.76	42721.2	CU	0.8			MIT
2339.74	2340.46	42726.7	CU II	0.3	0.5J		MIT
2338.67	2339.39	42746.2	CU	0.9			MIT
2337.39	2338.11	42769.6	CU	0.3			MIT
2336.20	2336.92	42791.4	CU II	0.5	1.3		MIT
2325.68	2326.39	42985.0	CU	0.3			MIT
2323.92	2324.63	43017.5	CU II	0.3		0.0	MIT
2323.01	2323.72	43034.4	CU II		0.6H		MIT
2319.561	2320.274	43098.37	CU I	1.3			IBU
2312.38	2313.09	43232.2	CU	0.5	0.3		MIT
2309.61	2310.32	43284.0	CU II		1.3		MIT
2303.116	2303.825	43406.08	CU I	1.5	1.3		IBU
2300.87	2301.58	43448.4	CU	0.9			MIT
2299.44	2300.15	43475.5	CU II		1.2		MIT
2298.84	2299.55	43486.8	CU	0.5W			A
2294.364	2295.071	43571.63	CU II	0.5	1.4		IBU
2293.842	2294.549	43581.55	CU I	1.6J	1.0		IBU
2292.681	2293.388	43603.62	CU II		0.7		IBU
2290.998	2291.704	43635.65	CU II		1.4		SH
2286.733	2287.438	43717.02	CU II		1.4		MIT
2278.441	2279.145	43876.11	CU II		1.3		MIT
2276.253	2276.956	43918.28	CU II	1.0	1.5		IBU
2274.870	2275.573	43944.98	CU II		0.5H		IBU
2265.458	2266.159	44127.53	CU		1.2		MIT
2265.353	2266.054	44129.58	CU II		0.7		MIT
2263.780	2264.480	44160.24	CU II		1.4		SH
2263.212	2263.912	44171.32	CU II		1.0		SH
2263.079	2263.779	44173.92	CU I	1.5	1.0		IBU
2260.528	2261.228	44223.76	CU I	1.4	0.8		IBU
2255.08	2255.78	44330.6	CU	0.0	1.5		MIT
2254.971	2255.670	44332.73	CU II	0.5	1.0		MIT
2251.88	2252.58	44393.6	CU II	0.0	0.3H		MIT
2249.06	2249.76	44449.2	CU II		1.4		MIT
2246.995	2247.692	44490.08	CU II	1.5	2.7		IBU
2244.265	2244.961	44544.20	CU I	1.4	1.0		IBU
2243.10	2243.80	44567.3	CU II			0.8	SH
2242.613	2243.309	44577.01	CU II	1.4	1.7H		IBU
2242.14	2242.84	44586.4	CU II			0.8	SH
2238.454	2239.149	44659.82	CU I	1.6			IBU
2236.278	2236.973	44703.28	CU I	1.5			IBU
2231.571	2232.265	44797.56	CU II		1.2		SH
2230.948	2231.641	44810.07	CU II		1.2		SH
2230.40	2231.09	44821.1	CU II			1.0	SH
2230.084	2230.777	44827.43	CU I	1.5R	1.3		IBU
2230.084	2230.777	44827.43	CU II	1.5R	1.3		IBU
2229.850	2230.543	44832.13	CU II		0.6H		SH
2228.863	2229.556	44851.98	CU II	1.0	1.6		IBU
2227.775	2228.468	44873.88	CU I	1.6R	1.4P		IBU
2226.864	2227.557	44892.24	CU II		1.5		MIT
2225.697	2226.389	44915.77	CU I	2.2	1.3		IBU
2224.779	2225.471	44934.31	CU II		1.4		MIT
2218.62	2219.31	45059.0	CU	0.3R	1.2		MIT
2218.504	2219.195	45061.39	CU II			1.4	SH
2218.100	2218.791	45069.60	CU II	1.4	1.6		IBU
2215.654	2216.344	45119.35	CU I	1.5			IBU
2215.100	2215.790	45130.63	CU II		1.2		SH
2214.581	2215.271	45141.21	CU I	1.7R	0.7H		IBU
2212.832	2213.522	45176.88	CU		1.3		MIT
2212.741	2213.431	45178.74	CU II			1.0	SH
2210.259	2210.948	45229.47	CU II	1.3	1.6		IBU
2209.795	2210.484	45238.96	CU II		1.0		SH
2200.57	2201.26	45428.6	CU II		1.0		MIT
2199.752	2200.439	45445.48	CU I	1.3G	0.7H		IBU
2199.583	2200.270	45448.97	CU I	1.7G	1.3		IBU
2195.77	2196.46	45527.9	CU II		1.5		MIT
2192.260	2192.945	45600.77	CU II	1.4	2.7H		IBU
2190.50	2191.18	45637.4	CU II		0.3H		MIT
2189.621	2190.306	45655.73	CU II	1.1	1.6		IBU
2189.36	2190.04	45661.2	CU II		0.3		SH
2182.90	2183.58	45796.3	CU II		0.9		MIT
2181.72	2182.40	45821.0	CU I	1.7R	1.2		MIT
2181.41	2182.09	45827.6	CU II			0.6	SH

AIR WAVELENGTH (ANGSTRMS)	VACUUM WAVELENGTH (ANGSTRMS)	WAVENUMBER (KAYSERS)	LMENT	LOG ARC	INTENSITY SPK	DIS	REF
2180.75	2181.43	45841.4	CU II	0.5		1.0	MIT
2179.399	2180.082	45869.84	CU II	1.1	1.5		IBU
2178.944	2179.626	45879.42	CU I	1.5R	1.1		IBU
2174.968	2175.650	45963.28	CU II	0.5	1.6		SH
2172.51	2173.19	46015.3	CU		0.7		MIT
2171.76	2172.44	46031.2	CU I	1.5			MIT
2169.53	2170.21	46078.5	CU I	1.5			MIT
2166.87	2167.55	46135.0	CU II		0.3		MIT
2165.09	2165.77	46173.0	CU I	1.8G	1.4		MIT
2161.314	2161.993	46253.62	CU II		1.6		SH
2158.41	2159.09	46315.9	CU II		0.5		MIT
2151.801	2152.478	46458.08	CU II		1.5		SH
2148.974	2149.650	46519.19	CU II	1.2	1.4		IBU
2146.98	2147.66	46562.4	CU II		1.2		MIT
2145.50	2146.18	46594.5	CU II		1.1		MIT
2144.73	2145.41	46611.2	CU II	0.0H	0.5		MIT
2140.67	2141.34	46699.6	CU I	0.6			MIT
2138.507	2139.181	46746.86	CU I	1.4J			MIT
2135.976	2136.650	46802.24	CU II	1.4	2.7W		IBU
2134.355	2135.028	46837.78	CU II	1.5	1.6		IBU
2134.25	2134.92	46840.1	CU	0.5H			MIT
2131.23	2131.90	46906.5	CU II		0.3		SH
2130.761	2131.434	46916.78	CU I	1.4	0.5		MIT
2126.028	2126.700	47021.21	CU II	1.2	1.5		IBU
2125.24	2125.91	47038.6	CU II		0.6		MIT
2125.098	2125.770	47041.79	CU II		1.3		SH
2124.10	2124.77	47063.9	CU I	0.3			MIT
2122.966	2123.637	47089.02	CU II	1.2	1.7		IBU
2118.38	2119.05	47191.0	CU II		0.6		SH
2117.300	2117.970	47215.02	CU II	0.3	1.3		SH
2112.090	2112.759	47331.48	CU II	1.2	1.6		IBU
2112.090	2112.759	47331.48	CU I	1.2	1.6		IBU
2111.30	2111.97	47349.2	CU II			0.8	SH
2110.31	2110.98	47371.4	CU II			0.6	SH
2109.78	2110.45	47383.3	CU	0.3			MIT
2106.39	2107.06	47459.5	CU II		0.3		MIT
2105.112	2105.780	47488.35	CU I	1.2			IBU
2104.782	2105.450	47495.79	CU II	0.9	1.4		IBU
2100.41	2101.08	47594.6	CU		0.5J		MIT
2098.41	2099.08	47640.0	CU II		1.5		MIT
2094.77	2095.44	47722.8	CU II		0.9H		SH
2093.606	2094.271	47749.30	CU II	0.3	1.3		SH
2087.92	2088.58	47879.3	CU II		1.5		MIT
2087.74	2088.40	47883.5	CU	0.3			MIT
2085.303	2085.967	47939.40	CU II	0.3	1.4		MIT
2085.22	2085.88	47941.3	CU I		0.8H		MIT
2084.32	2084.98	47962.0	CU II		0.6		MIT
2082.92	2083.58	47994.2	CU II		0.8		MIT
2080.91	2081.57	48040.6	CU	0.6H			MIT
2080.06	2080.72	48060.2	CU II		0.9H		MIT
2079.46	2080.12	48074.1	CU I	1.2H			MIT
2078.652	2079.314	48092.77	CU II	0.0	1.6		MIT
2069.92	2070.58	48295.6	CU II			0.3	SH
2066.32	2066.98	48379.8	CU II		1.3		MIT
2066.04	2066.70	48386.3	CU	0.3H			MIT
2062.49	2063.15	48469.6	CU II		1.4		MIT
2054.966	2055.624	48647.03	CU II	1.2	1.5		MIT
2054.41	2055.07	48660.2	CU II	0.5R	1.3R		MIT
2054.30	2054.96	48662.8	CU II		0.9J		MIT
2047.65	2048.31	48820.8	CU II		1.0R		SH
2043.787	2044.443	48913.08	CU II	1.2	1.5		MIT
2037.117	2037.772	49073.21	CU II	1.1	1.5		MIT
2036.97	2037.62	49076.7	CU II			0.6	SH
2035.843	2036.497	49103.92	CU II	1.1	1.5		MIT
2031.023	2031.677	49220.43	CU II	0.3	1.5		SH
2029.94	2030.59	49246.7	CU II	0.5	1.2J		MIT
2027.18	2027.83	49313.7	CU II		1.2		MIT
2025.475	2026.128	49355.23	CU II	0.9	1.5		MIT
2025.15	2025.80	49363.1	CU	0.3			MIT
2024.33	2024.98	49383.1	CU I	1.7R	0.8		MIT
2022.24	2022.89	49434.2	CU II		1.0		MIT
2017.61	2018.26	49547.6	CU II	0.6	0.3		MIT
2016.883	2017.534	49565.46	CU II	0.3	1.3		MIT
2015.575	2016.226	49597.62	CU II		1.7		MIT
2012.96	2013.61	49662.0	CU II		1.3		SH
2002.27	2002.92	49927.1	CU		0.5		SH
2000.348	2000.996	49975.11	CU II	0.8	0.8		SH
4987.59	4988.98	20044.2	DX	0.3			M
8438.57	8440.89	11847.1	DY	0.3			KS
8416.61	8418.92	11878.0	DY	0.3			KS
8391.96	8394.27	11912.9	DY	0.7			KS
8326.04	8328.33	12007.2	DY	0.7			MIT
8265.50	8267.77	12095.2	DY	0.7			MIT
8243.95	8246.22	12126.8	DY	0.3			KS
8233.55	8235.81	12142.1	DY	0.3			KS
8218.58	8220.84	12164.2	DY	0.3			KS
8216.98	8219.24	12166.6	DY	0.30			KS
8201.56	8203.81	12189.4	DY	1.2			MIT
8198.75	8201.00	12193.6	DY	0.7			KS
8147.25	8149.49	12270.7	DY	0.3			KS
8108.40	8110.63	12329.5	DY	0.3			KS
8040.01	8042.22	12434.4	DY	0.3			KS
8027.21	8029.42	12454.2	DY	0.3			KS
8025.33	8027.54	12457.1	DY	0.3			KS
8008.68	8010.88	12483.0	DY	0.3			KS
7982.80	7985.00	12523.5	DY	0.7			KS
7973.14	7975.33	12538.7	DY	0.3			KS
7968.63	7970.82	12545.8	DY	0.3			KS
7909.36	7911.54	12639.8	DY	0.7			KS
7865.00	7867.16	12711.1	DY	0.3			KS
7864.32	7866.48	12712.2	DY	0.3			KS
7835.55	7837.71	12758.8	DY	0.3			KS
7832.79	7834.95	12763.3	DY	0.3			KS
7814.69	7816.84	12792.9	DY	0.3			KS
7812.08	7814.23	12797.2	DY	1.2			KS
7790.05	7792.19	12833.4	DY	1.0			MIT
7780.92	7783.06	12848.4	DY	0.7			KS
7760.07	7762.21	12882.9	DY	0.3			KS
7757.34	7759.47	12887.5	DY	0.3			KS
7751.61	7753.74	12897.0	DY	0.7			KS
7750.14	7752.27	12899.4	DY	0.3			KS
7739.38	7741.51	12917.4	DY	0.3			KS
7732.55	7734.68	12928.8	DY	0.3			KS
7729.78	7731.91	12933.4	DY	1.2			KS
7721.08	7723.20	12948.0	DY	0.3			KS
7715.35	7717.47	12957.6	DY	1.2			MIT
7711.93	7714.05	12963.3	DY	0.3			KS
7696.50	7698.62	12989.3	DY	0.3			KS
7693.85	7695.97	12993.8	DY	0.3			KS
7676.74	7678.85	13022.8	DY	0.7			KS
7666.78	7668.89	13039.7	DY	0.3			KS
7662.35	7664.46	13047.2	DY	1.4			MIT
7648.12	7650.23	13071.5	DY	0.3			KS
7646.66	7648.76	13074.0	DY	1.0			KS
7645.87	7647.97	13075.4	DY	1.0			KS
7641.15	7643.25	13083.4	DY	1.2			KS
7639.25	7641.35	13086.7	DY	0.3			KS
7635.29	7637.39	13093.5	DY	0.3			KS
7631.32	7633.42	13100.3	DY	0.3			KS
7617.73	7619.83	13123.7	DY	0.7			KS
7616.21	7618.31	13126.3	DY	0.3			KS
7611.54	7613.64	13134.3	DY	0.7			KS
7609.25	7611.34	13138.3	DY	0.7			KS
7594.95	7597.04	13163.0	DY	0.3			KS
7591.36	7593.45	13169.2	DY	1.0			KS
7587.76	7589.85	13175.5	DY	0.3			KS
7577.47	7579.56	13193.4	DY	0.7			KS
7562.96	7565.04	13218.7	DY	1.0			KS
7559.81	7561.89	13224.2	DY	1.0			MIT
7557.81	7559.89	13227.7	DY	0.3			KS

AIR WAVELENGTH (ANGSTRMS)	VACUUM WAVELENGTH (ANGSTRMS)	WAVENUMBER (KAYSERS)	LMENT	LOG ARC	INTENSITY SPK	DIS	REF
7553.03	7555.11	13236.1	DY	1.0			KS
7543.77	7545.85	13252.3	DY	1.2			MIT
7533.14	7535.21	13271.0	DY	0.7			KS
7521.52	7523.59	13291.5	DY	0.3			KS
7516.59	7518.66	13300.2	DY	1.0			KS
7503.90	7505.97	13322.7	DY	0.3			MIT
7490.70	7492.76	13346.2	DY	0.3			KS
7483.01	7485.07	13359.9	DY	0.5			KS
7459.97	7462.02	13401.2	DY	0.3			KS
7457.05	7459.10	13406.4	DY	0.5			MIT
7451.08	7453.13	13417.2	DY	0.3			KS
7426.97	7429.02	13460.7	DY	0.8			MIT
7412.44	7414.48	13487.1	DY	0.7			MIT
7407.62	7409.66	13495.9	DY	0.3			KS
7403.17	7405.21	13504.0	DY	0.3			KS
7381.60	7383.63	13543.5	DY	0.3			KS
7376.06	7378.09	13553.6	DY	0.3			KS
7345.18	7347.20	13610.6	DY	0.5			KS
7288.26	7290.27	13716.9	DY	0.3			KS
7273.54	7275.54	13744.7	DY	0.3			MIT
7250.04	7252.04	13789.2	DY	0.5			KS
7230.11	7232.10	13827.2	DY	0.5			KS
7213.32	7215.31	13859.4	DY	0.3			KS
7175.16	7177.14	13933.1	DY	0.5			KS
7156.51	7158.48	13969.4	DY	0.3			KS
7120.93	7122.89	14039.2	DY	0.5			KS
7109.24	7111.20	14062.3	DY	0.7			MIT
7101.70	7103.66	14077.2	DY	0.3			KS
7075.15	7077.10	14130.1	DY	0.7			KS
7062.33	7064.28	14155.7	DY	0.3			KS
7055.95	7057.90	14168.5	DY	0.7			KS
7017.43	7019.37	14246.3	DY	0.7			KS
6998.10	7000.03	14285.7	DY	0.5			KS
6991.31	6993.24	14299.5	DY	0.3			KS
6982.47	6984.40	14317.6	DY	0.3			KS
6950.29	6952.21	14383.9	DY	0.5			KS
6932.62	6934.53	14420.6	DY	0.3			KS
6929.54	6931.45	14427.0	DY	0.3			KS
6912.27	6914.18	14463.0	DY	0.3			KS
6906.57	6908.48	14475.0	DY	0.5			KS
6899.34	6901.24	14490.1	DY	0.7			KS
6897.98	6899.88	14493.0	DY	0.3			KS
6888.90	6890.80	14512.1	DY	0.5			KS
6856.50	6858.39	14580.7	DY	0.5			KS
6853.00	6854.89	14588.1	DY	0.8			KS
6835.44	6837.33	14625.6	DY	0.9			KS
6827.16	6829.04	14643.3	DY	0.3			KS
6818.24	6820.12	14662.5	DY	0.3			KS
6803.20	6805.08	14694.9	DY	0.3			KS
6765.96	6767.83	14775.8	DY	0.8			KS
6747.98	6749.84	14815.2	DY	0.6			KS
6713.18	6715.03	14892.0	DY	0.3			KS
6700.66	6702.51	14919.8	DY	0.5			KS
6667.90	6669.74	14993.1	DY	0.9			KS
6661.69	6663.53	15007.1	DY	0.7			KS
6658.40	6660.24	15014.5	DY	0.6			KS
6654.27	6656.11	15023.8	DY	0.5			KS
6643.41	6645.24	15048.4	DY	0.6			KS
6614.91	6616.74	15113.2	DY	0.3			KS
6594.16	6595.98	15160.7	DY	0.5			KS
6579.38	6581.20	15194.8	DY	0.9			KS
6565.17	6566.98	15227.7	DY	0.5			KS
6558.02	6559.83	15244.3	DY	0.5			KS
6548.28	6550.09	15267.0	DY	0.5			KS
6519.18	6520.98	15335.1	DY	0.5			KS
6498.53	6500.33	15383.8	DY	0.3			ED
6486.64	6488.43	15412.0	DY	0.8			KS
6483.62	6485.41	15419.2	DY	0.7			KS
6481.03	6482.82	15425.4	DY	0.3			KS
6474.97	6476.76	15439.8	DY	0.3			KS

AIR WAVELENGTH (ANGSTRMS)	VACUUM WAVELENGTH (ANGSTRMS)	WAVENUMBER (KAYSERS)	LMENT	LOG ARC	INTENSITY SPK	DIS	REF
6472.04	6473.83	15446.8	DY	0.3			KS
6468.60	6470.39	15455.0	DY	0.5			KS
6460.86	6462.65	15473.5	DY	0.3			KS
6450.54	6452.32	15498.3	DY	0.3			KS
6444.17	6445.95	15513.6	DY	0.3			ED
6443.71	6445.49	15514.7	DY	0.5			KS
6441.85	6443.63	15519.2	DY	0.3			KS
6436.57	6438.35	15531.9	DY	0.3			KS
6435.66	6437.44	15534.1	DY	0.3			KS
6434.96	6436.74	15535.8	DY	0.3			ED
6432.96	6434.74	15540.7	DY	0.3			KS
6427.79	6429.57	15553.2	DY	0.3			KS
6427.40	6429.18	15554.1	DY	0.3			KS
6421.93	6423.71	15567.3	DY	0.7			KS
6418.41	6420.18	15575.9	DY	0.3			KS
6402.31	6404.08	15615.1	DY	0.5			KS
6396.61	6398.38	15629.0	DY	0.6			KS
6390.661	6392.428	15643.51	DY	0.3			MIT
6388.38	6390.15	15649.1	DY	0.3			KS
6387.56	6389.33	15651.1	DY	0.3			KS
6386.81	6388.58	15652.9	DY	0.6			KS
6377.72	6379.48	15675.2	DY	0.5			KS
6375.04	6376.80	15681.8	DY	0.3			KS
6369.63	6371.39	15695.2	DY	0.3			KS
6359.78	6361.54	15719.5	DY	0.3			KS
6355.105	6356.862	15731.03	DY	0.3			MIT
6353.66	6355.42	15734.6	DY	0.3			KS
6353.12	6354.88	15735.9	DY	0.3			KS
6348.358	6350.113	15747.75	DY	0.3			MIT
6343.34	6345.09	15760.2	DY	0.6			KS
6342.55	6344.30	15762.2	DY	0.3			KS
6341.38	6343.13	15765.1	DY	0.3			KS
6338.12	6339.87	15773.2	DY	0.3			KS
6331.10	6332.85	15790.7	DY	0.3			KS
6329.23	6330.98	15795.3	DY	0.3			KS
6323.23	6324.98	15810.3	DY	0.3			KS
6317.24	6318.99	15825.3	DY	0.3			KS
6300.39	6302.13	15867.7	DY	0.3			KS
6291.66	6293.40	15889.7	DY	0.5			KS
6273.86	6275.60	15934.7	DY	0.3 H			KS
6273.00	6274.74	15936.9	DY	0.3 H			KS
6271.09	6272.82	15941.8	DY	0.3			KS
6270.76	6272.49	15942.6	DY	0.3			KS
6269.80	6271.53	15945.1	DY	0.3			KS
6265.54	6267.27	15955.9	DY	0.3 H			ED
6260.38	6262.11	15969.1	DY	0.5			KS
6259.095	6260.826	15972.33	DY	1.2			MIT
6255.46	6257.19	15981.6	DY	0.3			KS
6250.65	6252.38	15993.9	DY	0.3			KS
6246.86	6248.59	16003.6	DY	0.3			KS
6239.28	6241.01	16023.1	DY	0.3			KS
6230.62	6232.34	16045.3	DY	0.3			KS
6229.77	6231.49	16047.5	DY	0.5			KS
6216.59	6218.31	16081.5	DY	0.3			KS
6213.432	6215.151	16089.71	DY	0.3			MIT
6212.67	6214.39	16091.7	DY	0.3			KS
6207.978	6209.696	16103.85	DY	0.3			MIT
6199.25	6200.97	16126.5	DY	0.3			KS
6196.23	6197.94	16134.4	DY	0.5			KS
6189.67	6191.38	16151.5	DY	0.3			KS
6184.73	6186.44	16164.4	DY	0.3			KS
6173.38	6175.09	16194.1	DY	0.3			KS
6170.516	6172.224	16201.62	DY	0.3			MIT
6168.433	6170.140	16207.09	DY	0.8			MIT
6165.549	6167.255	16214.67	DY	0.6			MIT
6158.32	6160.02	16233.7	DY	0.5			KS
6151.48	6153.18	16251.7	DY	0.3			KS
6150.67	6152.37	16253.9	DY	0.3			KS
6144.93	6146.63	16269.1	DY	0.3			KS
6138.30	6140.00	16286.7	DY	0.3			KS

AIR WAVELENGTH (ANGSTRMS)	VACUUM WAVELENGTH (ANGSTRMS)	WAVENUMBER (KAYSERS)	LMENT	LOG ARC	INTENSITY SPK	DIS	REF
6133.635	6135.333	16299.04	DY	0.5			MIT
6127.146	6128.842	16316.30	DY	0.5			MIT
6126.475	6128.171	16318.08	DY	0.5			MIT
6121.64	6123.33	16331.0	DY	0.3			KS
6119.629	6121.323	16336.34	DY	0.3			MIT
6115.668	6117.361	16346.92	DY	0.3			MIT
6115.284	6116.977	16347.94	DY	0.3			MIT
6113.97	6115.66	16351.5	DY	0.3			KS
6107.605	6109.296	16368.50	DY	0.3			MIT
6106.23	6107.92	16372.2	DY	0.3			KS
6103.67	6105.36	16379.1	DY	0.3			KS
6103.370	6105.060	16379.86	DY	0.3			MIT
6099.62	6101.31	16389.9	DY	0.3			KS
6090.881	6092.567	16413.44	DY	0.3			MIT
6089.292	6090.978	16417.73	DY	0.3			MIT
6088.265	6089.950	16420.50	DY	0.9			MIT
6085.053	6086.738	16429.16	DY	0.6			MIT
6077.02	6078.70	16450.9	DY	0.3H			KS
6074.61	6076.29	16457.4	DY	0.3			KS
6073.774	6075.456	16459.67	DY	0.3			MIT
6067.896	6069.576	16475.61	DY	0.3			MIT
6063.561	6065.240	16487.39	DY	0.3			MIT
6058.183	6059.860	16502.03	DY	0.5			MIT
6050.03	6051.71	16524.3	DY	0.3			KS
6044.491	6046.165	16539.41	DY	0.3			MIT
6042.86	6044.53	16543.9	DY	0.3			ED
6038.682	6040.354	16555.32	DY	0.3			MIT
6030.979	6032.649	16576.47	DY	0.5			MIT
6024.99	6026.66	16592.9	DY	0.3			KS
6023.59	6025.26	16596.8	DY	0.3			KS
6021.577	6023.245	16602.35	DY	0.3			MIT
6018.50	6020.17	16610.8	DY	0.3			KS
6017.27	6018.94	16614.2	DY	0.5			KS
6013.66	6015.33	16624.2	DY	0.3H			KS
6010.814	6012.479	16632.08	DY	0.7			MIT
6009.30	6010.96	16636.3	DY	0.3			KS
6008.940	6010.604	16637.26	DY	0.5			MIT
6003.244	6004.907	16653.05	DY	0.3			MIT
6000.86	6002.52	16659.7	DY	0.3			KS
5988.575	5990.234	16693.84	DY	0.9			MIT
5984.863	5986.521	16704.19	DY	0.5			MIT
5974.503	5976.158	16733.16	DY	0.9			MIT
5966.483	5968.136	16755.65	DY	0.5			MIT
5964.478	5966.130	16761.28	DY	0.6			MIT
5945.811	5947.458	16813.91	DY	0.7			MIT
5931.88	5933.52	16853.4	DY	0.3			ED
5915.168	5916.807	16901.01	DY	0.7			MIT
5909.155	5910.792	16918.21	DY	0.5			MIT
5894.47	5896.10	16960.4	DY	0.3			MIT
5868.091	5869.717	17036.59	DY	0.5			MIT
5856.040	5857.663	17071.65	DY	0.5H			MIT
5848.089	5849.710	17094.86	DY	0.7H			MIT
5833.858	5835.475	17136.56	DY	0.5H			MIT
5832.027	5833.644	17141.95	DY	1.0H			MIT
5766.99	5768.59	17335.3	DY	0.5			KS
5758.82	5760.42	17359.9	DY	0.5			KS
5745.560	5747.154	17399.92	DY	0.5			MIT
5740.199	5741.791	17416.17	DY	0.5			MIT
5725.84	5727.43	17459.8	DY	0.5H			KS
5718.458	5720.044	17482.38	DY	0.6			MIT
5707.84	5709.42	17514.9	DY	0.5			KS
5702.92	5704.50	17530.0	DY	0.5			KS
5698.721	5700.302	17542.93	DY	0.5			MIT
5694.523	5696.103	17555.86	DY	0.9J			MIT
5693.620	5695.200	17558.65	DY	0.5H			MIT
5692.52	5694.10	17562.0	DY	0.7H			KS
5685.585	5687.163	17583.46	DY	0.5			MIT
5678.339	5679.915	17605.90	DY	0.5			MIT
5677.686	5679.262	17607.92	DY	0.5			MIT
5671.248	5672.822	17627.91	DY	0.5			MIT

AIR WAVELENGTH (ANGSTRMS)	VACUUM WAVELENGTH (ANGSTRMS)	WAVENUMBER (KAYSERS)	LMENT	LOG ARC	INTENSITY SPK	DIS	REF
5663.872	5665.444	17650.87	DY	0.5H			MIT
5660.282	5661.853	17662.06	DY	0.5			MIT
5652.014	5653.583	17687.90	DY	1.0			MIT
5645.990	5647.557	17706.77	DY	0.7			MIT
5641.524	5643.090	17720.79	DY	0.6			MIT
5639.499	5641.064	17727.15	DY	1.1			MIT
5627.485	5629.047	17765.00	DY	0.5			MIT
5613.230	5614.788	17810.11	DY	0.6			MIT
5605.629	5607.185	17834.26	DY	0.5			MIT
5597.346	5598.900	17860.65	DY	0.5			MIT
5592.306	5593.859	17876.75	DY	0.5			MIT
5583.19	5584.74	17905.9	DY	0.5H			KS
5562.462	5564.007	17972.66	DY	0.5			MIT
5547.275	5548.816	18021.86	DY	0.9			MIT
5542.187	5543.726	18038.41	DY	0.5			MIT
5530.556	5532.092	18076.34	DY	0.5			MIT
5528.03	5529.57	18084.6	DY	0.5			KS
5515.402	5516.934	18126.01	DY	0.5			MIT
5506.504	5508.034	18155.30	DY	0.5			MIT
5502.787	5504.316	18167.56	DY	0.5			MIT
5496.837	5498.364	18187.23	DY	0.5			MIT
5477.24	5478.76	18252.3	DY	0.5			ED
5469.096	5470.616	18279.48	DY	0.5			MIT
5455.448	5456.964	18325.21	DY	0.6			MIT
5443.339	5444.852	18365.97	DY	0.6			MIT
5426.706	5428.214	18422.26	DY	0.7			MIT
5424.250	5425.758	18430.60	DY	0.5			MIT
5423.316	5424.824	18433.78	DY	0.7			MIT
5420.754	5422.261	18442.49	DY	0.5			MIT
5419.135	5420.641	18448.00	DY	0.7			MIT
5399.928	5401.429	18513.62	DY	0.5			MIT
5395.576	5397.076	18528.55	DY	0.7			MIT
5392.04	5393.54	18540.7	DY	0.5			MIT
5391.24	5392.74	18543.4	DY	0.3			ED
5389.574	5391.073	18549.19	DY	0.9			MIT
5385.628	5387.125	18562.77	DY	0.5			MIT
5376.103	5377.598	18595.66	DY	0.6			MIT
5369.240	5370.733	18619.43	DY	0.5			MIT
5368.199	5369.692	18623.04	DY	0.5			MIT
5352.123	5353.612	18678.98	DY	0.6			MIT
5340.314	5341.799	18720.28	DY	0.8			MIT
5337.432	5338.917	18730.39	DY	0.5			MIT
5324.697	5326.178	18775.19	DY	0.7			MIT
5312.624	5314.102	18817.85	DY	0.5			MIT
5311.88	5313.36	18820.5	DY	0.5H			ED
5309.010	5310.487	18830.67	DY	0.8			MIT
5301.585	5303.060	18857.04	DY	1.0			MIT
5297.827	5299.301	18870.41	DY	0.5			MIT
5284.987	5286.458	18916.26	DY	0.5			MIT
5279.708	5281.177	18935.17	DY	0.5			MIT
5277.397	5278.866	18943.46	DY	0.5			MIT
5275.314	5276.782	18950.94	DY	0.7			MIT
5274.077	5275.545	18955.39	DY	0.5			MIT
5272.261	5273.728	18961.92	DY	0.6			MIT
5267.122	5268.588	18980.42	DY	0.5			MIT
5260.555	5262.019	19004.11	DY	0.8			MIT
5258.392	5259.856	19011.93	DY	0.5			MIT
5249.63	5251.09	19043.7	DY	0.5H			ED
5236.246	5237.704	19092.34	DY	0.5			MIT
5214.304	5215.756	19172.68	DY	0.8			MIT
5197.663	5199.110	19234.06	DY	1.0			MIT
5192.88	5194.33	19251.8	DY	1.2			MIT
5188.477	5189.922	19268.11	DY	0.5			MIT
5185.154	5186.598	19280.46	DY	0.5			MIT
5169.699	5171.139	19338.10	DY	0.9			MIT
5166.944	5168.283	19348.78	DY	0.6			MIT
5165.373	5166.812	19354.29	DY	0.5			MIT
5160.998	5162.436	19370.70	DY	0.5			MIT
5155.312	5156.748	19392.07	DY	0.7			MIT
5145.203	5146.636	19430.17	DY	0.5			MIT

AIR WAVELENGTH (ANGSTRMS)	VACUUM WAVELENGTH (ANGSTRMS)	WAVENUMBER (KAYSERS)	LMENT	LOG ARC	INTENSITY SPK	DIS	REF
5139.594	5141.026	19451.37	DY	1.0			MIT
5135.016	5136.447	19468.71	DY	0.5			MIT
5129.830	5131.259	19488.39	DY	0.5			MIT
5120.01	5121.44	19525.8	DY	0.9			ED
5110.318	5111.742	19562.80	DY	0.5			MIT
5107.979	5109.402	19571.76	DY	0.5			MIT
5092.206	5093.625	19632.38	DY	0.5H			MIT
5090.83	5092.25	19637.7	DY	0.7			ED
5090.378	5091.797	19639.43	DY	0.7			MIT
5077.66	5079.08	19688.6	DY	0.7			MIT
5070.65	5072.06	19715.8	DY	0.7			ED
5055.42	5056.83	19775.2	DY	0.5			ED
5042.62	5044.03	19825.4	DY	1.0			ED
5022.12	5023.52	19906.4	DY	0.5			ED
5007.37	5008.77	19965.0	DY	0.7			ED
5003.86	5005.26	19979.0	DY	0.5H			ED
4999.00	5000.39	19998.4	DY	0.3H			ED
4998.47	4999.86	20000.5	DY	0.3H			ED
4995.35	4996.74	20013.0	DY	0.3H			ED
4994.83	4996.22	20015.1	DY	0.3H			ED
4993.53	4994.92	20020.3	DY	0.3H			ED
4985.531	4986.922	20052.45	DY	0.3			MIT
4981.96	4983.35	20066.8	DY	0.3H			ED
4980.17	4981.56	20074.0	DY	0.3H			ED
4974.998	4976.386	20094.90	DY	0.5			MIT
4973.575	4974.963	20100.65	DY	0.3			MIT
4971.786	4973.173	20107.89	DY	0.3H			MIT
4969.87	4971.26	20115.6	DY	0.3H			ED
4963.104	4964.489	20143.06	DY	0.3H			MIT
4961.80	4963.18	20148.4	DY	0.7			ED
4959.595	4960.979	20157.31	DY	0.5			MIT
4957.357	4958.740	20166.41	DY	1.3	0.5		MIT
4954.364	4955.747	20178.59	DY	0.5			MIT
4953.38	4954.76	20182.6	DY	0.5			ED
4952.26	4953.64	20187.2	DY	0.3H			ED
4951.450	4952.832	20190.47	DY	0.5			MIT
4951.00	4952.38	20192.3	DY	0.3H			ED
4949.40	4950.78	20198.8	DY	0.3			ED
4948.216	4949.597	20203.67	DY	0.3			MIT
4946.284	4947.665	20211.56	DY	0.3			MIT
4945.510	4946.890	20214.72	DY	0.3			MIT
4942.854	4944.234	20225.58	DY	0.3			MIT
4941.156	4942.535	20232.53	DY	0.3			MIT
4936.835	4938.213	20250.24	DY	0.5			MIT
4936.112	4937.490	20253.21	DY	0.3			MIT
4933.845	4935.222	20262.51	DY	0.3			MIT
4931.00	4932.38	20274.2	DY	0.3H			ED
4929.35	4930.73	20281.0	DY	0.5			ED
4927.330	4928.705	20289.30	DY	0.3			MIT
4923.161	4924.535	20306.48	DY	0.8	0.3		MIT
4922.227	4923.601	20310.34	DY	0.6	0.3		MIT
4921.514	4922.888	20313.28	DY	0.6			MIT
4918.220	4919.593	20326.89	DY	0.6			MIT
4917.180	4918.553	20331.18	DY	0.5			MIT
4916.420	4917.793	20334.33	DY	0.3			MIT
4914.735	4916.107	20341.30	DY	0.5			MIT
4912.52	4913.89	20350.5	DY	0.6			ED
4909.80	4911.17	20361.7	DY	0.3			ED
4908.99	4910.36	20365.1	DY	0.3			ED
4907.433	4908.803	20371.56	DY	0.3			MIT
4906.233	4907.603	20376.55	DY	0.7			MIT
4903.65	4905.02	20387.3	DY	0.3			ED
4901.906	4903.275	20394.53	DY	0.3			MIT
4900.10	4901.47	20402.1	DY	0.5			ED
4899.248	4900.616	20405.60	DY	0.6			MIT
4897.109	4898.476	20414.51	DY	0.5			MIT
4895.858	4897.225	20419.73	DY	0.7			MIT
4893.684	4895.051	20428.80	DY	0.6			MIT
4892.60	4893.97	20433.3	DY	0.5			ED
4890.102	4891.468	20443.76	DY	0.9	0.3		MIT

AIR WAVELENGTH (ANGSTRMS)	VACUUM WAVELENGTH (ANGSTRMS)	WAVENUMBER (KAYSERS)	LMENT	LOG ARC	INTENSITY SPK	DIS	REF
4889.325	4890.690	20447.01	DY	0.7	0.5		MIT
4888.087	4889.452	20452.19	DY	0.7	0.3		MIT
4886.170	4887.535	20460.21	DY	0.3			MIT
4884.546	4885.910	20467.02	DY	0.5			MIT
4884.143	4885.507	20468.71	DY	0.3			MIT
4881.979	4883.342	20477.78	DY	0.3			MIT
4880.167	4881.530	20485.38	DY	0.7			MIT
4875.924	4877.286	20503.21	DY	0.5			MIT
4875.459	4876.821	20505.16	DY	0.5			MIT
4873.166	4874.527	20514.81	DY	0.5			MIT
4869.62	4870.98	20529.7	DY	0.5			ED
4868.058	4869.418	20536.34	DY	0.7	0.3		MIT
4865.68	4867.04	20546.4	DY	0.3H			ED
4862.00	4863.36	20561.9	DY	0.3			ED
4860.67	4862.03	20567.6	DY	0.5			ED
4860.05	4861.41	20570.2	DY	0.3H			ED
4859.06	4860.42	20574.4	DY	0.3H			ED
4857.91	4859.27	20579.2	DY	0.3H			ED
4857.31	4858.67	20581.8	DY	0.3H			ED
4856.238	4857.595	20586.32	DY	0.7	0.3		MIT
4855.581	4856.937	20589.11	DY	0.3			MIT
4852.527	4853.883	20602.06	DY	0.3			MIT
4851.476	4852.831	20606.53	DY	0.6H			MIT
4845.72	4847.07	20631.0	DY	0.3			ED
4844.87	4846.22	20634.6	DY	0.3	0.3		ED
4843.73	4845.08	20639.5	DY	0.3H			ED
4843.414	4844.767	20640.83	DY	0.5H	0.3H		MIT
4841.772	4843.125	20647.83	DY	0.5	0.3H		MIT
4840.467	4841.819	20653.39	DY	0.7	0.3H		MIT
4838.40	4839.75	20662.2	DY	0.3			ED
4837.75	4839.10	20665.0	DY	0.3			ED
4833.768	4835.119	20682.02	DY	0.8	0.3		MIT
4832.388	4833.738	20687.92	DY	0.8	0.3H		MIT
4830.88	4832.23	20694.4	DY	0.3			ED
4829.692	4831.042	20699.47	DY	0.7	0.5H		MIT
4828.880	4830.229	20702.95	DY	0.7			MIT
4827.12	4828.47	20710.5	DY	0.3	0.3		ED
4826.559	4827.908	20712.91	DY	0.6	0.3		MIT
4824.966	4826.314	20719.75	DY	0.7	0.3H		MIT
4823.732	4825.080	20725.04	DY	0.3			MIT
4822.00	4823.35	20732.5	DY	0.3			ED
4821.298	4822.645	20735.51	DY	0.6			MIT
4821.03	4822.38	20736.7	DY	0.3H			ED
4819.041	4820.388	20745.22	DY	0.8	0.3		MIT
4818.231	4819.578	20748.71	DY	0.3			MIT
4817.502	4818.848	20751.85	DY	0.3			MIT
4812.812	4814.157	20772.07	DY	0.3	0.3		MIT
4810.241	4811.585	20783.17	DY	0.3			MIT
4808.75	4810.09	20789.6	DY	0.5			ED
4807.945	4809.289	20793.09	DY	0.6			MIT
4805.77	4807.11	20802.5	DY	0.3			ED
4804.509	4805.852	20807.97	DY	0.5	0.3		MIT
4803.515	4804.858	20812.27	DY	0.3			MIT
4802.08	4803.42	20818.5	DY	0.5			ED
4800.68	4802.02	20824.6	DY	0.6			ED
4798.436	4799.777	20834.30	DY	0.3			MIT
4797.598	4798.939	20837.94	DY	0.3	0.3		MIT
4794.850	4796.190	20849.88	DY	0.5			MIT
4792.89	4794.23	20858.4	DY	0.6H			MIT
4791.89	4793.23	20862.8	DY	0.3			ED
4791.299	4792.638	20865.33	DY	0.9	0.3		MIT
4788.45	4789.79	20877.7	DY	0.3			ED
4786.935	4788.273	20884.35	DY	0.9	0.6		MIT
4786.240	4787.578	20887.39	DY	0.3			MIT
4785.315	4786.653	20891.43	DY	0.5	0.3		MIT
4784.918	4786.256	20893.16	DY	0.6	0.3		MIT
4781.88	4783.22	20906.4	DY	0.5			ED
4781.042	4782.379	20910.10	DY	0.5	0.3		MIT
4780.19	4781.53	20913.8	DY	0.3			ED
4779.44	4780.78	20917.1	DY	0.3H			MIT

AIR WAVELENGTH (ANGSTRMS)	VACUUM WAVELENGTH (ANGSTRMS)	WAVENUMBER (KAYSERS)	LMENT	LOG ARC	INTENSITY SPK	INTENSITY DIS	REF
4776.89	4778.23	20928.3	DY	0.3	0.3		M
4775.804	4777.139	20933.03	DY	0.8	0.3		MIT
4774.804	4776.139	20937.41	DY	0.7	0.3		MIT
4773.69	4775.02	20942.3	DY	0.3	0.3		MIT
4771.930	4773.264	20950.02	DY	0.9	0.3		MIT
4770.81	4772.14	20954.9	DY	0.5	0.3		MIT
4769.630	4770.964	20960.13	DY	0.6	0.3		MIT
4767.162	4768.495	20970.98	DY	0.3			MIT
4765.57	4766.90	20978.0	DY	0.3			ED
4764.73	4766.06	20981.7	DY	0.6			MIT
4763.82	4765.15	20985.7	DY	0.3			MIT
4761.34	4762.67	20996.6	DY	0.3			MIT
4760.010	4761.341	21002.49	DY	1.2	0.6		MIT
4757.415	4758.745	21013.94	DY	0.5			MIT
4756.674	4758.004	21017.22	DY	0.6	0.3		MIT
4755.01	4756.34	21024.6	DY	0.7	0.5		MIT
4752.28	4753.61	21036.6	DY	0.3	0.3		ED
4749.45	4750.78	21049.2	DY	0.3			ED
4748.208	4749.536	21054.69	DY	0.3H			MIT
4745.806	4747.133	21065.34	DY	0.6	0.5		MIT
4743.60	4744.93	21075.1	DY	0.3H			ED
4743.03	4744.36	21077.7	DY	0.5	0.3		MIT
4741.539	4742.865	21084.30	DY	0.5	0.3		MIT
4740.928	4742.254	21087.02	DY	0.5	0.3		MIT
4738.50	4739.83	21097.8	DY	0.3	0.3		ED
4736.79	4738.11	21105.4	DY	0.3			MIT
4735.49	4736.81	21111.2	DY	0.3			ED
4734.39	4735.71	21116.1	DY	0.3			MIT
4731.851	4733.175	21127.47	DY	1.5	1.0		MIT
4728.876	4730.199	21140.76	DY	0.5			MIT
4728.36	4729.68	21143.1	DY	0.3	0.3		MIT
4727.133	4728.455	21148.55	DY	1.0	0.9		MIT
4726.371	4727.693	21151.97	DY	0.5	0.3		MIT
4724.223	4725.545	21161.58	DY	0.5			MIT
4723.921	4725.243	21162.94	DY	0.5			MIT
4723.167	4724.488	21166.31	DY	0.3	0.3		MIT
4721.234	4722.555	21174.98	DY	1.1	0.7		MIT
4709.226	4710.544	21228.97	DY	0.6	0.3		MIT
4706.782	4708.099	21240.00	DY	0.6			MIT
4705.73	4707.05	21244.7	DY	0.5H			ED
4703.471	4704.787	21254.95	DY	0.7	0.3		MIT
4698.687	4700.002	21276.59	DY	1.0	0.9		MIT
4697.08	4698.39	21283.9	DY	0.5			MIT
4695.241	4696.555	21292.20	DY	0.9	0.3		MIT
4694.354	4695.668	21296.23	DY	0.7	0.3		MIT
4693.99	4695.30	21297.9	DY	0.3			MIT
4693.672	4694.986	21299.32	DY	0.5	0.3		MIT
4692.737	4694.050	21303.56	DY	0.5	0.3		MIT
4690.23	4691.54	21314.9	DY	0.5			MIT
4689.768	4691.080	21317.05	DY	0.9	0.9		MIT
4683.792	4685.103	21344.25	DY	0.6	0.3		MIT
4682.03	4683.34	21352.3	DY	0.5	0.3		MIT
4681.00	4682.31	21357.0	DY	0.6			M
4679.12	4680.43	21365.6	DY	0.5	0.3		M
4675.807	4677.116	21380.70	DY	0.8	0.3		MIT
4674.61	4675.92	21386.2	DY	0.5			MIT
4673.615	4674.923	21390.73	DY	1.0	0.9		MIT
4673.21	4674.52	21392.6	DY	0.3			ED
4672.482	4673.790	21395.91	DY	0.7H			MIT
4671.10	4672.41	21402.2	DY	0.6	0.6		M
4669.398	4670.705	21410.04	DY	0.5	0.3		MIT
4668.20	4669.51	21415.5	DY	0.3	0.3		M
4667.787	4669.094	21417.43	DY	0.3	0.3		MIT
4666.548	4667.854	21423.12	DY	0.6	0.3		MIT
4665.92	4667.23	21426.0	DY	0.3			M
4664.675	4665.981	21431.72	DY	1.0	0.9		MIT
4663.10	4664.41	21439.0	DY	0.5	0.3		MIT
4662.746	4664.051	21440.59	DY	0.6	0.3		MIT
4662.20	4663.51	21443.1	DY	0.5			M
4660.802	4662.107	21449.53	DY	0.6	0.6		MIT

AIR WAVELENGTH (ANGSTRMS)	VACUUM WAVELENGTH (ANGSTRMS)	WAVENUMBER (KAYSERS)	LMENT	LOG ARC	INTENSITY SPK	INTENSITY DIS	REF
4659.497	4660.801	21455.54	DY	0.3	0.3		MIT
4657.48	4658.78	21464.8	DY	0.3	0.3		M
4654.74	4656.04	21477.5	DY	0.6	0.3		M
4651.541	4652.843	21492.23	DY	0.6	0.3		MIT
4650.16	4651.46	21498.6	DY	0.6	0.3		MIT
4649.463	4650.765	21501.84	DY	0.5	0.6		MIT
4647.298	4648.599	21511.86	DY	0.5	0.3		MIT
4646.73	4648.03	21514.5	DY	0.3			MIT
4643.461	4644.761	21529.63	DY	0.5			MIT
4642.779	4644.079	21532.79	DY	0.3	0.5H		MIT
4642.212	4643.512	21535.43	DY	0.3			MIT
4638.812	4640.111	21551.21	DY	0.7	0.6		MIT
4637.08	4638.38	21559.3	DY	0.5	0.3		MIT
4635.316	4636.614	21567.46	DY	0.6	0.3		MIT
4634.787	4636.085	21569.92	DY	0.5			MIT
4631.506	4632.803	21585.20	DY	0.6	0.3		MIT
4629.13	4630.43	21596.3	DY	0.6H	0.3		M
4628.081	4629.377	21601.18	DY	0.7	0.6		MIT
4627.20	4628.50	21605.3	DY	0.3			ED
4626.01	4627.31	21610.9	DY	0.5			MIT
4624.42	4625.72	21618.3	DY	0.5	0.3		M
4624.103	4625.398	21619.76	DY	0.5	0.3		MIT
4622.38	4623.67	21627.8	DY	0.5	0.3		MIT
4621.427	4622.721	21632.28	DY	0.5	0.3		MIT
4620.04	4621.33	21638.8	DY	0.9	0.6		MIT
4617.27	4618.56	21651.7	DY	0.7	0.6		MIT
4615.57	4616.86	21659.7	DY	0.5	0.6		MIT
4614.83	4616.12	21663.2	DY	0.5	0.3		MIT
4613.83	4615.12	21667.9	DY	0.5			MIT
4612.275	4613.567	21675.20	DY	1.7	1.0		MIT
4611.71	4613.00	21677.9	DY	0.3	0.3		MIT
4609.06	4610.35	21690.3	DY	0.3	0.3H		M
4606.048	4607.338	21704.50	DY	0.6			MIT
4604.32	4605.61	21712.6	DY	0.3			MIT
4600.68	4601.97	21729.8	DY	0.3	0.3		MIT
4599.86	4601.15	21733.7	DY	0.5	0.3		MIT
4597.50	4598.79	21744.9	DY	0.3	0.3		M
4595.14	4596.43	21756.0	DY	0.6	0.3		MIT
4591.78	4593.07	21771.9	DY	0.5	0.5		MIT
4590.571	4591.857	21777.68	DY	0.5			MIT
4589.376	4590.662	21783.35	DY	1.8	1.2		MIT
4587.931	4589.217	21790.21	DY	0.8	0.5		MIT
4586.642	4587.927	21796.33	DY	0.6	0.3		MIT
4586.215	4587.500	21798.36	DY	0.3	0.3		MIT
4585.72	4587.00	21800.7	DY	0.5			MIT
4584.78	4586.06	21805.2	DY	0.5	0.3		M
4583.644	4584.928	21810.59	DY	0.3			MIT
4583.054	4584.338	21813.40	DY	0.5	0.3		MIT
4581.46	4582.74	21821.0	DY	0.3	0.3		M
4579.63	4580.91	21829.7	DY	0.3			ED
4577.796	4579.079	21838.45	DY	1.2	0.9		MIT
4576.605	4577.888	21844.14	DY	0.7	0.3		MIT
4574.32	4575.60	21855.1	DY	0.3			MIT
4573.87	4575.15	21857.2	DY	0.7	0.3		MIT
4568.42	4569.70	21883.3	DY	0.3			M
4567.07	4568.35	21889.7	DY	0.5	0.3		MIT
4566.21	4567.49	21893.9	DY	0.5			MIT
4565.116	4566.396	21899.11	DY	0.9	0.6		MIT
4560.11	4561.39	21923.1	DY	0.3			MIT
4558.16	4559.44	21932.5	DY	0.5	0.3		MIT
4556.459	4557.736	21940.72	DY	0.5	0.6		MIT
4555.24	4556.52	21946.6	DY	0.6			MIT
4554.234	4555.511	21951.44	DY	0.3			MIT
4551.368	4552.644	21965.26	DY	0.5			MIT
4550.89	4552.17	21967.6	DY	0.7	0.3		MIT
4550.00	4551.28	21971.9	DY	0.5H			M
4547.43	4548.70	21984.3	DY	0.3			MIT
4546.475	4547.750	21988.90	DY	0.3			MIT
4545.353	4546.627	21994.32	DY	0.5	0.6		MIT
4544.12	4545.39	22000.3	DY	0.3			M

AIR WAVELENGTH (ANGSTRMS)	VACUUM WAVELENGTH (ANGSTRMS)	WAVENUMBER (KAYSERS)	LMENT	LOG ARC	INTENSITY SPK	DIS	REF
4541.698	4542.971	22012.02	DY	0.8	0.9		MIT
4539.157	4540.430	22024.35	DY	0.5			MIT
4538.76	4540.03	22026.3	DY	0.7	0.3		MIT
4536.73	4538.00	22036.1	DY	0.3H			MIT
4536.15	4537.42	22038.9	DY	0.3			MIT
4532.54	4533.81	22056.5	DY	0.3			ED
4529.780	4531.050	22069.94	DY	0.3	0.3		MIT
4527.78	4529.05	22079.7	DY	0.9	0.6		MIT
4526.096	4527.365	22087.90	DY	0.5			MIT
4523.50	4524.77	22100.6	DY	0.3			M
4522.68	4523.95	22104.6	DY	0.5H			M
4519.83	4521.10	22118.5	DY	0.7	0.6		MIT
4518.968	4520.235	22122.74	DY	0.3	0.0		MIT
4518.538	4519.805	22124.85	DY	0.8	0.6		MIT
4516.96	4518.23	22132.6	DY	0.8	0.6		M
4515.938	4517.205	22137.58	DY	0.3			MIT
4513.60	4514.87	22149.1	DY	0.5			MIT
4509.97	4511.23	22166.9	DY	0.3			MIT
4508.11	4509.37	22176.0	DY	0.3			M
4506.957	4508.221	22181.70	DY	0.6	0.6		MIT
4506.083	4507.347	22186.00	DY	0.3	0.3		MIT
4503.252	4504.515	22199.95	DY	0.6	0.6		MIT
4502.600	4503.863	22203.16	DY	0.3	0.3		MIT
4499.25	4500.51	22219.7	DY	0.3			MIT
4498.48	4499.74	22223.5	DY	0.5	0.3		ED
4496.41	4497.67	22233.7	DY	0.5			KN
4495.82	4497.08	22236.6	DY	0.3			MIT
4491.32	4492.58	22258.9	DY	0.3			MIT
4491.032	4492.292	22260.35	DY	0.5H			MIT
4488.557	4489.816	22272.63	DY	0.6	0.3		MIT
4486.242	4487.501	22284.12	DY	0.7	0.3		MIT
4484.37	4485.63	22293.4	DY	0.7	0.3		KN
4482.33	4483.59	22303.6	DY	0.6	0.3		KN
4481.99	4483.25	22305.3	DY	0.5			ED
4480.695	4481.952	22311.71	DY	0.6	0.3		MIT
4479.50	4480.76	22317.7	DY	0.3			ED
4477.497	4478.753	22327.64	DY	0.3	0.3		MIT
4476.631	4477.887	22331.96	DY	0.3	0.3		MIT
4473.546	4474.801	22347.36	DY	0.3			MIT
4472.77	4474.03	22351.2	DY	0.3			ED
4471.543	4472.798	22357.37	DY	0.3H			MIT
4468.165	4469.419	22374.27	DY	1.0	0.6		MIT
4467.89	4469.14	22375.6	DY	0.8	0.3		MIT
4464.75	4466.00	22391.4	DY	0.3			ED
4461.19	4462.44	22409.2	DY	0.6	0.3		MIT
4460.31	4461.56	22413.7	DY	0.3H			MIT
4455.49	4456.74	22437.9	DY	1.0			KN
4453.61	4454.86	22447.4	DY	0.3H			ED
4452.81	4454.06	22451.4	DY	0.3			ED
4449.702	4450.951	22467.11	DY	1.3	1.1		KN
4449.16	4450.41	22469.9	DY	0.6	0.3		KN
4448.23	4449.48	22474.5	DY	0.7	0.3		KN
4444.61	4445.86	22492.9	DY	1.0	0.3		KN
4440.13	4441.38	22515.5	DY	0.5			ED
4438.40	4439.65	22524.3	DY	0.3			ED
4436.65	4437.90	22533.2	DY	0.6	0.3		KN
4435.78	4437.03	22537.6	DY	0.6			KN
4435.02	4436.27	22541.5	DY	0.3			ED
4432.78	4434.02	22552.9	DY	0.3			ED
4431.00	4432.24	22561.9	DY	0.8	0.6		KN
4429.23	4430.47	22570.9	DY	0.3			ED
4426.87	4428.11	22583.0	DY	0.6	0.3		KN
4425.82	4427.06	22588.3	DY	0.5			ED
4421.69	4422.93	22609.4	DY	0.6	0.6		KN
4418.85	4420.09	22624.0	DY	0.3	0.3		ED
4418.09	4419.33	22627.9	DY	0.5	0.3		ED
4416.44	4417.68	22636.3	DY	0.3			ED
4412.80	4414.04	22655.0	DY	0.5H			ED
4411.37	4412.61	22662.3	DY	0.5	0.3		KN
4411.00	4412.24	22664.2	DY	0.3			ED

AIR WAVELENGTH (ANGSTRMS)	VACUUM WAVELENGTH (ANGSTRMS)	WAVENUMBER (KAYSERS)	LMENT	LOG ARC	INTENSITY SPK	DIS	REF
4409.384	4410.622	22672.54	DY	1.5	0.9		KN
4408.05	4409.29	22679.4	DY	0.5	0.6		KN
4407.54	4408.78	22682.0	DY	0.3			ED
4405.58	4406.82	22692.1	DY	0.5	0.3		ED
4403.56	4404.80	22702.5	DY	0.5			ED
4400.10	4401.34	22720.4	DY	0.6	0.3		KN
4399.73	4400.97	22722.3	DY	0.5			KN
4396.05	4397.28	22741.3	DY	0.6	0.3		ED
4394.98	4396.21	22746.8	DY	1.4	0.6		KN
4393.28	4394.51	22755.6	DY	0.3H			KN
4389.79	4391.02	22773.7	DY	0.5	0.3		KN
4386.82	4388.05	22789.1	DY	0.3			ED
4385.29	4386.52	22797.1	DY	0.6	0.6		KN
4384.30	4385.53	22802.2	DY	0.6	0.3		KN
4383.18	4384.41	22808.1	DY	0.5			ED
4380.23	4381.46	22823.4	DY	0.7	0.3		KN
4379.41	4380.64	22827.7	DY	0.3			ED
4375.33	4376.56	22849.0	DY	1.0	0.6		KN
4374.80	4376.03	22851.8	DY	1.1	0.6		KN
4374.24	4375.47	22854.7	DY	1.1	0.6		KN
4371.33	4372.56	22869.9	DY	0.3			ED
4366.73	4367.96	22894.0	DY	0.9	0.3		KN
4366.11	4367.34	22897.2	DY	0.5	0.6		ED
4365.62	4366.85	22899.8	DY	0.3			KN
4364.28	4365.51	22906.9	DY	0.8	0.9		KN
4364.06	4365.29	22908.0	DY	0.6			ED
4362.92	4364.15	22914.0	DY	0.5	0.3		ED
4362.29	4363.52	22917.3	DY	0.5	0.3		ED
4361.39	4362.62	22922.0	DY	0.8	0.6		KN
4360.21	4361.44	22928.2	DY	0.7	0.6		MIT
4358.461	4359.686	22937.43	DY	1.4	0.6		MIT
4356.13	4357.35	22949.7	DY	0.5H			KN
4349.09	4350.31	22986.9	DY	0.3	0.3		ED
4347.72	4348.94	22994.1	DY	0.7	0.6		KN
4346.33	4347.55	23001.4	DY	0.7	0.6		ED
4344.62	4345.84	23010.5	DY	0.3			ED
4344.26	4345.48	23012.4	DY	0.6			ED
4341.09	4342.31	23029.2	DY	0.3			ED
4340.47	4341.69	23032.5	DY	0.5			ED
4339.68	4340.90	23036.7	DY	1.2	0.9		KN
4338.45	4339.67	23043.2	DY	0.5			MIT
4337.49	4338.71	23048.3	DY	0.7			ED
4336.01	4337.23	23056.2	DY	0.5			KN
4334.61	4335.83	23063.6	DY	0.7			ED
4333.72	4334.94	23068.4	DY	0.3			ED
4332.92	4334.14	23072.6	DY	0.5			ED
4332.11	4333.33	23076.9	DY	0.3			ED
4331.14	4332.36	23082.1	DY	0.3			ED
4329.89	4331.11	23088.8	DY	0.6	0.3		KN
4328.90	4330.12	23094.1	DY	1.0	0.3		KN
4327.98	4329.20	23099.0	DY	0.6	0.3		ED
4326.39	4327.61	23107.5	DY	0.3			KN
4325.14	4326.36	23114.1	DY	1.0	0.6		KN
4323.794	4325.010	23121.33	DY	0.7			KN
4322.55	4323.77	23128.0	DY	0.7	0.3		KN
4318.985	4320.200	23147.08	DY	0.5			KN
4317.93	4319.14	23152.7	DY	0.3			ED
4313.886	4315.099	23174.44	DY	1.1	0.6		KN
4312.429	4313.642	23182.27	DY	0.3			KN
4311.93	4313.14	23184.9	DY	0.6			KN
4308.623	4309.835	23202.75	DY	2.0	1.1		KN
4308.344	4309.556	23204.25	DY	0.8			KN
4306.77	4307.98	23212.7	DY	0.3			KN
4306.22	4307.43	23215.7	DY	0.5			ED
4304.58	4305.79	23224.5	DY	0.5			ED
4303.60	4304.81	23229.8	DY	0.5			ED
4303.03	4304.24	23232.9	DY	0.9			ED
4302.72	4303.93	23234.6	DY	1.0			KN
4302.57	4303.78	23235.4	DY	0.3			KN
4300.76	4301.97	23245.2	DY	0.7			KN

AIR WAVELENGTH (ANGSTRMS)	VACUUM WAVELENGTH (ANGSTRMS)	WAVENUMBER (KAYSERS)	LMENT	LOG ARC	INTENSITY SPK	INTENSITY DIS	REF
4300.25	4301.46	23247.9	DY	0.3			ED
4298.90	4300.11	23255.2	DY	0.6			KN
4298.44	4299.65	23257.7	DY	0.3			ED
4296.34	4297.55	23269.1	DY	0.3			KN
4295.58	4296.79	23273.2	DY	0.9	0.3		ED
4295.038	4296.246	23276.13	DY	1.3			KN
4294.936	4296.144	23276.69	DY	1.4	1.2		KN
4292.94	4294.15	23287.5	DY	0.7			ED
4291.93	4293.14	23293.0	DY	1.4			KN
4290.44	4291.65	23301.1	DY	0.7			KN
4289.35	4290.56	23307.0	DY	0.3			ED
4288.01	4289.22	23314.3	DY	0.3			KN
4286.89	4288.10	23320.4	DY	0.3			ED
4283.02	4284.23	23341.4	DY	0.3			ED
4279.78	4280.98	23359.1	DY	0.9			KN
4278.61	4279.81	23365.5	DY	0.5			ED
4277.76	4278.96	23370.1	DY	0.3			KN
4276.74	4277.94	23375.7	DY	1.2			KN
4275.96	4277.16	23380.0	DY	0.3			KN
4275.45	4276.65	23382.8	DY	0.7			KN
4275.00	4276.20	23385.2	DY	0.7			KN
4274.04	4275.24	23390.5	DY	0.7	0.3		ED
4273.14	4274.34	23395.4	DY	0.6	0.6		KN
4269.60	4270.80	23414.8	DY	1.0	0.3		KN
4268.31	4269.51	23421.9	DY	0.9			KN
4267.92	4269.12	23424.0	DY	0.8			KN
4266.52	4267.72	23431.7	DY	0.3	0.3H		M
4265.83	4267.03	23435.5	DY	0.8	0.3		KN
4258.56	4259.76	23475.5	DY	1.2	0.3		KN
4258.14	4259.34	23477.8	DY	0.9			KN
4257.72	4258.92	23480.1	DY	0.8	0.3		KN
4256.323	4257.521	23487.85	DY	1.4	0.9		KN
4256.204	4257.402	23488.50	DY	0.9			KN
4253.65	4254.85	23502.6	DY	0.5			ED
4252.79	4253.99	23507.4	DY	0.8	0.3		ED
4252.37	4253.57	23509.7	DY	0.8			ED
4251.73	4252.93	23513.2	DY	0.8			KN
4250.84	4252.04	23518.1	DY	0.7			ED
4250.35	4251.55	23520.9	DY	0.9			KN
4248.44	4249.64	23531.4	DY	0.8	0.3		KN
4247.36	4248.56	23537.4	DY	1.5	0.6		KN
4245.923	4247.118	23545.38	DY	1.4	0.6		MIT
4244.79	4245.99	23551.7	DY	0.7	0.3		KN
4243.45	4244.64	23559.1	DY	1.0	0.3		KN
4243.00	4244.19	23561.6	DY	0.6			KN
4241.83	4243.02	23568.1	DY	0.3			KN
4240.81	4242.00	23573.8	DY	0.3			ED
4240.75	4241.94	23574.1	DY	0.5			ED
4239.868	4241.062	23579.00	DY	1.7	0.6		MIT
4238.44	4239.63	23586.9	DY	0.8	0.3		KN
4237.54	4238.73	23591.9	DY	0.9	0.3		KN
4234.83	4236.02	23607.1	DY	1.0			KN
4233.43	4234.62	23614.9	DY	0.7			ED
4232.033	4233.225	23622.65	DY	1.3	0.3		MIT
4229.63	4230.82	23636.1	DY	0.7H			ED
4225.153	4226.343	23661.12	DY	1.6	0.9		KN
4224.68	4225.87	23663.8	DY	0.8			KN
4223.38	4224.57	23671.1	DY	0.3			ED
4222.221	4223.410	23677.55	DY	1.0			MIT
4221.104	4222.293	23683.81	DY	1.8	0.9		KN
4218.58	4219.77	23698.0	DY	0.8			KN
4218.091	4219.279	23700.73	DY	1.7	0.9		KN
4216.96	4218.15	23707.1	DY	0.5			KN
4215.169	4216.356	23717.16	DY	1.7	0.9		MIT
4214.38	4215.57	23721.6	DY	0.5			KN
4213.182	4214.369	23728.35	DY	0.7	0.9		KN
4211.719	4212.905	23736.59	DY	2.3	1.2		KN
4211.248	4212.434	23739.24	DY	0.6	0.3		MIT
4207.68	4208.87	23759.4	DY	0.7			KN
4206.544	4207.729	23765.79	DY	1.1	0.6		KN

AIR WAVELENGTH (ANGSTRMS)	VACUUM WAVELENGTH (ANGSTRMS)	WAVENUMBER (KAYSERS)	LMENT	LOG ARC	INTENSITY SPK	INTENSITY DIS	REF
4205.64	4206.82	23770.9	DY	0.7	0.3		ED
4205.03	4206.21	23774.4	DY	0.8	0.3		KN
4202.250	4203.434	23790.07	DY	1.3	0.6		MIT
4201.372	4202.556	23795.04	DY	0.9	0.6		KN
4201.318	4202.502	23795.35	DY	1.5W			MIT
4201.01	4202.19	23797.1	DY	0.7			KN
4199.00	4200.18	23808.5	DY	0.3			ED
4198.026	4199.209	23814.01	DY	1.3	0.6		MIT
4197.93	4199.11	23814.6	DY	0.6			MIT
4195.22	4196.40	23829.9	DY	1.1			KN
4194.827	4196.009	23832.17	DY	1.7	1.1		KN
4193.84	4195.02	23837.8	DY	0.5			ED
4191.627	4192.808	23850.36	DY	1.6	0.3		KN
4190.90	4192.08	23854.5	DY	1.0	0.3		KN
4190.27	4191.45	23858.1	DY	0.5			ED
4189.4	4190.6	23863.	DY	0.8			KN
4186.810	4187.990	23877.80	DY	2.0W	1.1		KN
4183.732	4184.911	23895.37	DY	1.2	0.9		MIT
4183.611	4184.790	23896.06	DY	0.9			KN
4182.42	4183.60	23902.9	DY	0.9	0.6		KN
4181.25	4182.43	23909.6	DY	0.7			MIT
4180.91	4182.09	23911.5	DY	0.8	0.6		ED
4180.33	4181.51	23914.8	DY	0.3	0.3		ED
4179.46	4180.64	23919.8	DY	0.5			ED
4178.59	4179.77	23924.8	DY	0.9	0.3		ED
4178.072	4179.250	23927.74	DY	1.1			KN
4177.754	4178.932	23929.56	DY	0.8			KN
4176.64	4177.82	23935.9	DY	0.5			KN
4171.992	4173.168	23962.61	DY	1.2			KN
4171.925	4173.101	23963.00	DY	0.6	0.3		KN
4170.55	4171.73	23970.9	DY	0.8			KN
4169.241	4170.416	23978.42	DY	0.7	0.3		KN
4167.966	4169.141	23985.76	DY	1.7	1.1		KN
4167.40	4168.57	23989.0	DY	0.7			KN
4165.84	4167.01	23998.0	DY	0.3			ED
4164.74	4165.91	24004.3	DY	0.5			KN
4163.83	4165.00	24009.6	DY	0.8			ED
4162.25	4163.42	24018.7	DY	0.9			KN
4161.64	4162.81	24022.2	DY	0.5			ED
4160.24	4161.41	24030.3	DY	1.0	0.3		KN
4159.34	4160.51	24035.5	DY	1.1	0.3		KN
4157.86	4159.03	24044.1	DY	0.7	0.3		KN
4154.27	4155.44	24064.8	DY	1.0	0.3		KN
4153.36	4154.53	24070.1	DY	0.7			KN
4153.11	4154.28	24071.6	DY	0.8			KN
4152.43	4153.60	24075.5	DY	1.2R	0.6		KN
4148.88	4150.05	24096.1	DY	0.5			ED
4147.10	4148.27	24106.4	DY	0.6	0.3		ED
4146.071	4147.240	24112.42	DY	1.6	0.6		MIT
4145.59	4146.76	24115.2	DY	0.7			KN
4143.100	4144.268	24129.71	DY	1.6	0.9		KN
4141.516	4142.684	24138.94	DY	1.2	0.6		MIT
4140.12	4141.29	24147.1	DY	0.5			ED
4139.56	4140.73	24150.4	DY	0.9	0.3		KN
4138.54	4139.71	24156.3	DY	1.0			KN
4138.14	4139.31	24158.6	DY	0.3			ED
4134.75	4135.92	24178.4	DY	1.0			KN
4134.14	4135.31	24182.0	DY	1.1	0.3		KN
4133.863	4135.029	24183.63	DY	1.2	0.3		MIT
4133.370	4134.536	24186.51	DY	1.0	0.6		MIT
4132.85	4134.02	24189.6	DY	1.0	0.3		KN
4131.038	4132.203	24200.17	DY	1.1	0.6		MIT
4130.42	4131.59	24203.8	DY	1.1	0.3		KN
4129.425	4130.590	24209.62	DY	1.3	0.9		KN
4129.13	4130.29	24211.4	DY	1.0	0.3		KN
4128.241	4129.406	24216.56	DY	1.3	0.6		KN
4126.14	4127.30	24228.9	DY	0.9	0.3		KN
4124.626	4125.790	24237.78	DY	1.2	0.9		KN
4120.66	4121.82	24261.1	DY	0.7			KN
4119.325	4120.487	24268.98	DY	1.2	0.6		MIT

AIR WAVELENGTH (ANGSTRMS)	VACUUM WAVELENGTH (ANGSTRMS)	WAVENUMBER (KAYSERS)	LMENT	LOG ARC	INTENSITY SPK	INTENSITY DIS	REF
4118.01	4119.17	24276.7	DY	0.6			ED
4117.05	4118.21	24282.4	DY	0.3H			ED
4114.09	4115.25	24299.9	DY	1.1	0.6		KN
4113.05	4114.21	24306.0	DY	1.1	0.3		KN
4111.346	4112.506	24316.07	DY	1.5	1.1		KN
4109.20	4110.36	24328.8	DY	0.3			ED
4108.28	4109.44	24334.2	DY	0.3			ED
4107.45	4108.61	24339.1	DY	0.6			KN
4107.16	4108.32	24340.9	DY	0.8	0.3		ED
4106.70	4107.86	24343.6	DY	0.9	0.3		KN
4106.39	4107.55	24345.4	DY	0.9			KN
4105.97	4107.13	24347.9	DY	0.7			ED
4105.051	4106.209	24353.36	DY	1.0	0.6		MIT
4103.878	4105.036	24360.32	DY	1.5	0.6		KN
4103.312	4104.470	24363.68	DY	1.7	1.7		KN
4101.95	4103.11	24371.8	DY	0.9	0.3		KN
4101.43	4102.59	24374.9	DY	0.7	0.3		KN
4099.886	4101.043	24384.04	DY	0.8	0.3		KN
4099.21	4100.37	24388.1	DY	0.6			ED
4097.24	4098.40	24399.8	DY	0.3			ED
4096.78	4097.94	24402.5	DY	0.8			ED
4096.109	4097.265	24406.52	DY	1.2	0.3		KN
4093.647	4094.802	24421.20	DY	0.7			KN
4092.43	4093.59	24428.5	DY	0.6			ED
4091.77	4092.92	24432.4	DY	0.9	0.6		KN
4091.533	4092.688	24433.82	DY	1.0	0.6		MIT
4090.399	4091.554	24440.59	DY	0.6			KN
4089.511	4090.665	24445.90	DY	0.8			KN
4087.388	4088.542	24458.60	DY	0.8			KN
4087.213	4088.367	24459.65	DY	1.2	0.6		MIT
4085.344	4086.497	24470.83	DY	1.0	0.3		KN
4085.140	4086.293	24472.06	DY	0.9	0.3		KN
4083.74	4084.89	24480.4	DY	0.6			KN
4083.109	4084.262	24484.23	DY	0.6			KN
4081.83	4082.98	24491.9	DY	0.7	0.3		ED
4080.37	4081.52	24500.7	DY	0.5			ED
4079.595	4080.747	24505.32	DY	0.9			KN
4079.27	4080.42	24507.3	DY	0.8			KN
4077.974	4079.125	24515.06	DY	2.2R	2.0		KN
4077.35	4078.50	24518.8	DY	0.6			KN
4076.01	4077.16	24526.9	DY	0.3			ED
4074.02	4075.17	24538.9	DY	0.8	0.3		KN
4073.110	4074.260	24544.33	DY	1.9	1.2		KN
4072.65	4073.80	24547.1	DY	0.8	0.3		KN
4071.032	4072.182	24556.86	DY	0.7			KN
4070.24	4071.39	24561.6	DY	0.6	0.3H		ED
4066.69	4067.84	24583.1	DY	0.8			MIT
4066.30	4067.45	24585.4	DY	0.7	0.3		ED
4065.400	4066.548	24590.88	DY	0.8			KN
4061.574	4062.721	24614.04	DY	0.5			MIT
4061.054	4062.201	24617.20	DY	0.5	0.3		KN
4060.58	4061.73	24620.1	DY	0.9	0.3		KN
4058.25	4059.40	24634.2	DY	0.3H			ED
4057.40	4058.55	24639.4	DY	0.6	0.6		KN
4055.159	4056.304	24652.98	DY	0.7	0.9		MIT
4055.013	4056.158	24653.87	DY	0.3H			KN
4053.86	4055.01	24660.9	DY	0.5			ED
4053.34	4054.48	24664.1	DY	0.3	0.3		ED
4052.40	4053.54	24669.8	DY	0.3			KN
4050.579	4051.723	24680.86	DY	1.5	1.2		MIT
4049.374	4050.518	24688.20	DY	0.8			KN
4048.942	4050.086	24690.83	DY	0.7			KN
4048.35	4049.49	24694.4	DY	1.2	0.3		ED
4047.740	4048.883	24698.17	DY	0.6	0.3		KN
4045.983	4047.126	24708.89	DY	2.2	1.1		MIT
4045.281	4046.424	24713.18	DY	0.5			KN
4041.989	4043.131	24733.31	DY	1.1	0.6		KN
4038.834	4039.975	24752.63	DY	0.7	0.3		KN
4038.528	4039.669	24754.50	DY	1.2	0.6		KN
4037.628	4038.769	24760.02	DY	0.8			KN

AIR WAVELENGTH (ANGSTRMS)	VACUUM WAVELENGTH (ANGSTRMS)	WAVENUMBER (KAYSERS)	LMENT	LOG ARC	INTENSITY SPK	INTENSITY DIS	REF
4036.335	4037.475	24767.95	DY	1.2	0.6		KN
4033.666	4034.806	24784.34	DY	1.2	0.6		KN
4032.847	4033.987	24789.37	DY	0.9			KN
4032.480	4033.619	24791.63	DY	1.3	1.1		KN
4031.081	4032.220	24800.23	DY	0.8			KN
4029.41	4030.55	24810.5	DY	0.7	0.3		ED
4028.418	4029.556	24816.63	DY	0.7	0.9		KN
4028.325	4029.463	24817.20	DY	1.0			KN
4027.787	4028.925	24820.51	DY	1.2	0.6		KN
4025.75	4026.89	24833.1	DY	0.3			KN
4025.605	4026.743	24833.97	DY	1.0	0.3		KN
4024.906	4026.043	24838.28	DY	1.0			KN
4024.437	4025.574	24841.18	DY	1.3			KN
4023.722	4024.859	24845.59	DY	1.0	0.3		KN
4020.897	4022.033	24863.04	DY	1.0	0.6		KN
4019.48	4020.62	24871.8	DY	0.7	0.3		ED
4017.76	4018.90	24882.5	DY	0.8	0.3		ED
4015.18	4016.31	24898.4	DY	1.0	0.3		KN
4014.713	4015.848	24901.34	DY	1.3	0.6		KN
4013.826	4014.961	24906.84	DY	1.1	0.3		KN
4012.82	4013.95	24913.1	DY	0.7	0.3		ED
4012.523	4013.657	24914.93	DY	0.6			KN
4011.295	4012.429	24922.56	DY	1.1	0.9		KN
4010.08	4011.21	24930.1	DY	1.0	0.6		KN
4008.49	4009.62	24940.0	DY	0.7			ED
4007.77	4008.90	24944.5	DY	1.1	0.3		KN
4007.137	4008.270	24948.42	DY	0.8			KN
4006.071	4007.204	24955.06	DY	1.0	0.3		KN
4005.840	4006.972	24956.50	DY	1.1	0.3		KN
4004.48	4005.61	24965.0	DY	0.7			KN
4004.33	4005.46	24965.9	DY	0.9	0.3		ED
4000.454	4001.585	24990.10	DY	2.6	2.5		KN
3999.82	4000.95	24994.1	DY	0.5	0.3		ED
3998.940	4000.071	24999.56	DY	0.7H			KN
3998.10	3999.23	25004.8	DY	0.5			KN
3996.699	3997.829	25013.57	DY	2.3	1.9		MIT
3994.535	3995.665	25027.13	DY	0.7			KN
3993.575	3994.704	25033.14	DY	0.9			KN
3992.45	3993.58	25040.2	DY	0.5			ED
3991.89	3993.02	25043.7	DY	0.7			KN
3991.331	3992.460	25047.22	DY	1.6			MIT
3990.35	3991.48	25053.4	DY	0.8			KN
3988.89	3990.02	25062.5	DY	0.7			M
3988.21	3989.34	25066.8	DY	0.6			M
3987.06	3988.19	25074.1	DY	0.8			KN
3986.05	3987.18	25080.4	DY	0.7			KN
3984.70	3985.83	25088.9	DY	0.9			KN
3984.226	3985.353	25091.88	DY	1.9			MIT
3983.664	3984.791	25095.42	DY	2.2	0.9		MIT
3981.938	3983.064	25106.30	DY	2.2	2.0		MIT
3981.372	3982.498	25109.87	DY	0.6			KN
3979.95	3981.08	25118.8	DY	0.7			ED
3979.481	3980.607	25121.80	DY	1.5			MIT
3978.573	3979.698	25127.53	DY	2.3	1.2		MIT
3975.04	3976.16	25149.9	DY	0.5	0.0		ED
3973.879	3975.003	25157.21	DY	1.0			KN
3972.414	3973.538	25166.49	DY	0.8			KN
3971.214	3972.337	25174.09	DY	1.1			KN
3969.233	3970.356	25186.66	DY	0.8			KN
3968.395	3969.518	25191.98	DY	2.5			MIT
3967.517	3968.639	25197.55	DY	0.9			KN
3966.362	3967.484	25204.89	DY	0.7			KN
3964.70	3965.82	25215.5	DY	0.7			KN
3963.80	3964.92	25221.2	DY	1.0			KN
3963.16	3964.28	25225.2	DY	0.5			M
3962.601	3963.722	25228.81	DY	0.9			KN
3959.69	3960.81	25247.4	DY	0.5	0.0		M
3959.35	3960.47	25249.5	DY	1.0			KN
3958.004	3959.124	25258.11	DY	0.3			KN
3957.802	3958.922	25259.40	DY	1.8			MIT

AIR WAVELENGTH (ANGSTRMS)	VACUUM WAVELENGTH (ANGSTRMS)	WAVENUMBER (KAYSERS)	LMENT	LOG ARC	INTENSITY SPK	INTENSITY DIS	REF
3957.21	3958.33	25263.2	DY	0.5	0.0		M
3956.81	3957.93	25265.7	DY	0.5	0.0		M
3956.24	3957.36	25269.4	DY	0.6			M
3955.12	3956.24	25276.5	DY	0.6			ED
3954.565	3955.684	25280.08	DY	1.6			MIT
3953.515	3954.634	25286.79	DY	0.7			KN
3953.12	3954.24	25289.3	DY	0.5	0.0		M
3950.399	3951.517	25306.74	DY	1.7			MIT
3946.939	3948.056	25328.92	DY	1.5			KN
3946.35	3947.47	25332.7	DY	0.5			M
3945.94	3947.06	25335.3	DY	0.5			MIT
3944.692	3945.809	25343.35	DY	2.5	2.2		MIT
3942.536	3943.652	25357.21	DY	1.5	0.7		MIT
3942.050	3943.166	25360.33	DY	0.6			KN
3939.70	3940.82	25375.5	DY	0.5	0.3		ED
3939.23	3940.35	25378.5	DY	0.5			KN
3938.209	3939.324	25385.07	DY	0.7			KN
3938.06	3939.17	25386.0	DY	1.3			MIT
3937.990	3939.105	25386.48	DY	0.7			KN
3937.165	3938.280	25391.80	DY	0.6			KN
3936.715	3937.829	25394.70	DY	1.0			MIT
3936.29	3937.40	25397.4	DY	0.6			KN
3936.03	3937.14	25399.1	DY	1.2	0.3		M
3934.17	3935.28	25411.1	DY	1.1			M
3932.98	3934.09	25418.8	DY	1.0			M
3932.228	3933.341	25423.68	DY	1.5			MIT
3931.537	3932.650	25428.15	DY	2.3			MIT
3931.290	3932.403	25429.74	DY	1.3			MIT
3930.153	3931.266	25437.10	DY	1.0			KN
3929.33	3930.44	25442.4	DY	1.0			KN
3927.866	3928.978	25451.91	DY	0.7			KN
3923.394	3924.505	25480.92	DY	1.5			MIT
3923.300	3924.411	25481.53	DY	1.0			KN
3921.39	3922.50	25493.9	DY	0.7H			ED
3919.145	3920.255	25508.55	DY	0.6			KN
3918.54	3919.65	25512.5	DY	0.9			M
3917.372	3918.481	25520.09	DY	0.7			KN
3917.298	3918.407	25520.57	DY	0.7			MIT
3915.605	3916.714	25531.61	DY	1.9			MIT
3914.877	3915.986	25536.35	DY	1.7			MIT
3913.95	3915.06	25542.4	DY	0.9			M
3913.72	3914.83	25543.9	DY	0.3			M
3913.628	3914.736	25544.50	DY	0.8			KN
3912.85	3913.96	25549.6	DY	0.7	0.0		M
3912.544	3913.652	25551.58	DY	0.7			KN
3911.677	3912.785	25557.24	DY	0.7	0.0		KN
3905.95	3907.06	25594.7	DY	0.8	0.0		ED
3905.56	3906.67	25597.3	DY	0.8			M
3904.14	3905.25	25606.6	DY	1.3			KN
3903.332	3904.438	25611.88	DY	0.9			KN
3902.39	3903.50	25618.1	DY	0.7			M
3901.338	3902.443	25624.97	DY	1.4			MIT
3901.09	3902.20	25626.6	DY	0.6			M
3900.39	3901.49	25631.2	DY	0.5			M
3899.15	3900.25	25639.4	DY	1.3			M
3898.544	3899.648	25643.34	DY	2.0			MIT
3896.656	3897.760	25655.76	DY	0.9			KN
3895.35	3896.45	25664.4	DY	1.2	0.5		M
3894.533	3895.636	25669.75	DY	0.6			KN
3892.87	3893.97	25680.7	DY	0.9	0.3		M
3891.982	3893.085	25686.57	DY	1.1			KN
3891.85	3892.95	25687.4	DY	0.8	0.6		KN
3888.99	3890.09	25706.3	DY	1.3			M
3888.40	3889.50	25710.2	DY	0.8			M
3887.54	3888.64	25715.9	DY	1.1	0.3		MIT
3883.06	3884.16	25745.6	DY	0.7			MIT
3882.001	3883.101	25752.61	DY	1.4			MIT
3879.05	3880.15	25772.2	DY	1.4	0.3		M
3877.94	3879.04	25779.6	DY	0.6	0.3		M
3875.15	3876.25	25798.1	DY	0.5			M

AIR WAVELENGTH (ANGSTRMS)	VACUUM WAVELENGTH (ANGSTRMS)	WAVENUMBER (KAYSERS)	LMENT	LOG ARC	INTENSITY SPK	INTENSITY DIS	REF
3873.995	3875.093	25805.83	DY	2.0			MIT
3872.117	3873.215	25818.35	DY	2.5	2.2		MIT
3871.635	3872.732	25821.56	DY	1.3			MIT
3869.870	3870.967	25833.34	DY	2.0			MIT
3869.430	3870.527	25836.28	DY	1.4			MIT
3869.10	3870.20	25838.5	DY	0.7	0.3		M
3868.814	3869.911	25840.39	DY	1.8			MIT
3868.461	3869.558	25842.75	DY	1.7			MIT
3867.84	3868.94	25846.9	DY	0.8			M
3866.592	3867.688	25855.24	DY	1.4			MIT
3865.45	3866.55	25862.9	DY	1.0	0.3		M
3863.20	3864.30	25877.9	DY	0.7			M
3862.66	3863.76	25881.6	DY	0.8	0.0		M
3858.40	3859.49	25910.1	DY	0.7			M
3858.09	3859.18	25912.2	DY	0.8			M
3855.60	3856.69	25928.9	DY	0.9	0.3		M
3854.90	3855.99	25933.7	DY	0.7	0.3		M
3853.041	3854.134	25946.17	DY	2.0			MIT
3850.527	3851.619	25963.11	DY	0.6			KN
3850.449	3851.541	25963.63	DY	0.6	0.0		KN
3850.049	3851.141	25966.33	DY	0.5			KN
3849.400	3850.492	25970.71	DY	1.3			MIT
3848.36	3849.45	25977.7	DY	0.6			M
3846.99	3848.08	25987.0	DY	1.0	0.0		KN
3846.358	3847.449	25991.25	DY	1.4			MIT
3844.30	3845.39	26005.2	DY	0.8	0.5		M
3842.98	3844.07	26014.1	DY	0.7	0.3		ED
3842.016	3843.106	26020.62	DY	1.3			MIT
3841.316	3842.406	26025.36	DY	2.0			MIT
3840.91	3842.00	26028.1	DY	0.7			MIT
3840.44	3841.53	26031.3	DY	0.6H			ED
3839.80	3840.89	26035.6	DY	0.6H	0.0		M
3838.67	3839.76	26043.3	DY	1.0			MIT
3837.86	3838.95	26048.8	DY	0.3	0.0		M
3836.51	3837.60	26058.0	DY	2.0	1.6		MIT
3834.55	3835.64	26071.3	DY	0.5	0.0		M
3832.90	3833.99	26082.5	DY	1.0			MIT
3831.643	3832.730	26091.06	DY	1.0			MIT
3831.04	3832.13	26095.2	DY	0.7	0.5		M
3828.19	3829.28	26114.6	DY	1.0	0.5		M
3824.00	3825.09	26143.2	DY	0.7			M
3822.591	3823.676	26152.84	DY	1.0			MIT
3821.47	3822.55	26160.5	DY	0.6			M
3818.76	3819.84	26179.1	DY	0.7			MIT
3817.54	3818.62	26187.4	DY	0.3			ED
3816.770	3817.853	26192.73	DY	2.0	1.7		MIT
3816.16	3817.24	26196.9	DY	0.6	0.3		M
3814.57	3815.65	26207.8	DY	0.6H	0.5		M
3813.684	3814.766	26213.93	DY	1.3			MIT
3813.12	3814.20	26217.8	DY	0.7			MIT
3812.29	3813.37	26223.5	DY	0.5			MIT
3809.84	3810.92	26240.4	DY	0.5	0.3		ED
3809.035	3810.116	26245.92	DY	0.7			MIT
3807.45	3808.53	26256.8	DY	0.3H	0.0		ED
3806.280	3807.360	26264.92	DY	1.4	1.9		MIT
3804.152	3805.232	26279.61	DY	1.3			MIT
3803.59	3804.67	26283.5	DY	0.3			M
3802.68	3803.76	26289.8	DY	0.6	0.3		M
3801.88	3802.96	26295.3	DY	0.9	0.3		ED
3799.93	3801.01	26308.8	DY	0.3			M
3798.65	3799.73	26317.7	DY	0.3			M
3797.76	3798.84	26323.8	DY	0.6			M
3796.84	3797.92	26330.2	DY	0.3H	0.0H		M
3793.565	3794.642	26352.95	DY	0.6	0.3		MIT
3791.82	3792.90	26365.1	DY	0.8	0.6		ED
3788.449	3789.525	26388.53	DY	2.0	1.6		MIT
3787.270	3788.346	26396.75	DY	0.8	0.7		MIT
3786.20	3787.28	26404.2	DY	1.3	1.2		M
3785.424	3786.499	26409.62	DY	0.8	0.5		MIT
3784.65	3785.72	26415.0	DY	0.8	0.3		M

AIR WAVELENGTH (ANGSTRMS)	VACUUM WAVELENGTH (ANGSTRMS)	WAVENUMBER (KAYSERS)	LMENT	LOG ARC	INTENSITY SPK	DIS	REF
3783.96	3785.03	26419.8	DY	0.8	0.6		ED
3783.608	3784.683	26422.29	DY	0.9	0.5		MIT
3782.888	3783.962	26427.32	DY	0.9	0.3		MIT
3781.48	3782.55	26437.2	DY	0.9	0.5		M
3780.93	3782.00	26441.0	DY	0.5			M
3780.33	3781.40	26445.2	DY	0.6	0.0		MIT
3779.63	3780.70	26450.1	DY	0.5	0.0		MIT
3779.247	3780.320	26452.78	DY	0.8	0.5		MIT
3777.45	3778.52	26465.4	DY	0.8	0.3		ED
3776.01	3777.08	26475.5	DY	0.3			ED
3774.836	3775.908	26483.69	DY	0.6	0.0H		ED
3773.319	3774.391	26494.34	DY	0.9	0.5		MIT
3773.062	3774.134	26496.15	DY	0.9	0.5		MIT
3771.18	3772.25	26509.4	DY	0.5			ED
3771.08	3772.15	26510.1	DY	1.0	0.7		ED
3767.642	3768.712	26534.26	DY	1.2	0.7		MIT
3764.86	3765.93	26553.9	DY	0.8	0.6H		M
3762.74	3763.81	26568.8	DY	0.8	0.3		M
3762.28	3763.35	26572.1	DY	0.5	0.0H		ED
3761.40	3762.47	26578.3	DY	0.3H			ED
3757.372	3758.440	26606.78	DY	2.3	1.7		MIT
3757.063	3758.131	26608.97	DY	1.0	0.7		MIT
3756.10	3757.17	26615.8	DY	0.6			M
3755.37	3756.44	26621.0	DY	0.8	0.3H		M
3754.77	3755.84	26625.2	DY	0.8	0.3		M
3753.762	3754.829	26632.37	DY	1.9	1.0		MIT
3753.514	3754.581	26634.13	DY	1.7	1.3		MIT
3751.814	3752.880	26646.20	DY	0.8	0.5		MIT
3750.334	3751.400	26656.72	DY	0.9	0.6		MIT
3747.827	3748.892	26674.55	DY	1.8	1.3		MIT
3745.64	3746.70	26690.1	DY	0.6			ED
3741.18	3742.24	26721.9	DY	0.8	0.3		ED
3740.074	3741.137	26729.84	DY	0.8	0.3		MIT
3739.355	3740.418	26734.98	DY	1.1	0.7		MIT
3734.21	3735.27	26771.8	DY	0.6	0.3		ED
3731.35	3732.41	26792.3	DY	0.7	0.3		M
3729.20	3730.26	26807.8	DY	0.3	0.3		M
3728.74	3729.80	26811.1	DY	0.6	0.3		M
3727.99	3729.05	26816.5	DY	1.2	0.5		M
3725.90	3726.96	26831.5	DY	0.7	0.3		M
3725.44	3726.50	26834.8	DY	0.8	0.6		M
3724.91	3725.97	26838.6	DY	0.6			M
3724.42	3725.48	26842.2	DY	1.2	1.2		ED
3717.29	3718.35	26893.7	DY	0.5			M
3716.93	3717.99	26896.3	DY	0.7	0.7		M
3715.54	3716.60	26906.3	DY	0.6	0.6		M
3715.28	3716.34	26908.2	DY	0.6	0.6		M
3713.84	3714.90	26918.6	DY	0.8	0.6		ED
3712.86	3713.92	26925.8	DY	0.3	0.3		M
3711.65	3712.71	26934.5	DY	0.8	0.8		ED
3710.73	3711.79	26941.2	DY	0.6	0.3		M
3710.080	3711.136	26945.93	DY	1.3	0.7		MIT
3708.227	3709.282	26959.40	DY	1.3	1.0		ED
3707.416	3708.471	26965.29	DY	0.7	0.3		MIT
3703.21	3704.26	26995.9	DY	0.5			ED
3701.62	3702.67	27007.5	DY	0.8	1.0		ED
3700.59	3701.64	27015.0	DY	0.7	0.3		M
3698.94	3699.99	27027.1	DY	0.8	0.3		M
3698.28	3699.33	27031.9	DY	0.8			ED
3698.17	3699.22	27032.7	DY	1.0	1.5		M
3697.90	3698.95	27034.7	DY	0.5			ED
3697.25	3698.30	27039.4	DY	1.0	0.6		ED
3696.87	3697.92	27042.2	DY	1.0H	0.3		M
3696.08	3697.13	27048.0	DY	0.3H			ED
3695.61	3696.66	27051.4	DY	0.6	0.3		ED
3694.75	3695.80	27057.7	DY	1.3	1.5		M
3694.36	3695.41	27060.6	DY	0.8	0.6		ED
3693.84	3694.89	27064.4	DY	0.6	0.0		ED
3691.16	3692.21	27084.1	DY	0.5	0.0		M
3690.57	3691.62	27088.4	DY	0.5	0.0H		M

AIR WAVELENGTH (ANGSTRMS)	VACUUM WAVELENGTH (ANGSTRMS)	WAVENUMBER (KAYSERS)	LMENT	LOG ARC	INTENSITY SPK	DIS	REF
3688.90	3689.95	27100.6	DY	0.5			ED
3688.32	3689.37	27104.9	DY	0.8	0.5		M
3685.74	3686.79	27123.9	DY	0.8	0.8		ED
3684.83	3685.88	27130.6	DY	0.5H	0.3H		M
3682.52	3683.57	27147.6	DY	0.7H	0.3J		ED
3681.93	3682.98	27151.9	DY	0.3	0.0		ED
3678.519	3679.566	27177.11	DY	0.7	0.0		M
3678.00	3679.05	27180.9	DY	0.5	0.0		M
3677.26	3678.31	27186.4	DY	0.5	0.0		ED
3676.56	3677.61	27191.6	DY	0.8	1.5		ED
3676.01	3677.06	27195.7	DY	0.8	0.3		M
3674.454	3675.500	27207.18	DY	0.8	0.0		MIT
3674.093	3675.139	27209.85	DY	2.0	1.7		MIT
3673.152	3674.198	27216.82	DY	1.4	0.7		MIT
3672.67	3673.72	27220.4	DY	1.0	0.6		M
3672.312	3673.358	27223.05	DY	2.0	2.0		ED
3671.697	3672.743	27227.61	DY	0.8	0.0H		MIT
3669.39	3670.44	27244.7	DY	0.3			EX
3668.911	3669.956	27248.28	DY	0.8	0.5		MIT
3668.50	3669.54	27251.3	DY	0.5	0.0		ED
3666.850	3667.894	27263.60	DY	0.8	0.5		MIT
3666.31	3667.35	27267.6	DY	0.7	0.0		ED
3665.40	3666.44	27274.4	DY	0.6			M
3665.20	3666.24	27275.9	DY	0.8	0.7		ED
3664.96	3666.00	27277.7	DY	0.5			ED
3664.63	3665.67	27280.1	DY	1.0	0.8		M
3662.24	3663.28	27297.9	DY	0.8			ED
3661.75	3662.79	27301.6	DY	0.8	0.6		ED
3661.06	3662.10	27306.7	DY	0.5	0.0		M
3659.90	3660.94	27315.4	DY	0.5	0.0		M
3658.64	3659.68	27324.8	DY	0.5	0.0H		ED
3656.85	3657.89	27338.1	DY	0.6	0.3		M
3656.38	3657.42	27341.7	DY	0.5			ED
3655.90	3656.94	27345.3	DY	0.5	0.0		ED
3655.60	3656.64	27347.5	DY	1.0	0.3		ED
3654.878	3655.919	27352.90	DY	0.7	0.0		MIT
3654.17	3655.21	27358.2	DY	0.6	0.0		M
3650.00	3651.04	27389.5	DY	0.7	0.3		M
3648.807	3649.847	27398.41	DY	1.7	1.5		MIT
3648.40	3649.44	27401.5	DY	0.5	0.0		ED
3646.85	3647.89	27413.1	DY	0.6			ED
3646.60	3647.64	27415.0	DY	0.7	0.3		M
3645.86	3646.90	27420.6	DY	0.8	0.6		ED
3645.416	3646.455	27423.90	DY	2.5	2.0		MIT
3643.89	3644.93	27435.4	DY	1.2	0.8		ED
3643.50	3644.54	27438.3	DY	0.5			ED
3641.07	3642.11	27456.6	DY	0.5	0.0		M
3640.80	3641.84	27458.7	DY	0.6	0.0		ED
3640.24	3641.28	27462.9	DY	1.2	0.8		ED
3639.86	3640.90	27465.8	DY	0.6	0.3		ED
3638.89	3639.93	27473.1	DY	0.3	0.0H		M
3637.27	3638.31	27485.3	DY	1.0	0.5		ED
3635.26	3636.30	27500.5	DY	1.0	0.8		ED
3633.77	3634.81	27511.8	DY	0.6	0.0		M
3633.26	3634.30	27515.6	DY	1.4			M
3633.00	3634.04	27517.6	DY	1.0	0.5		ED
3632.73	3633.77	27519.7	DY	1.0	0.6		ED
3630.46	3631.49	27536.9	DY	1.0			M
3630.18	3631.21	27539.0	DY	1.2	1.2		ED
3629.440	3630.475	27544.61	DY	2.0	1.7		MIT
3626.76	3627.79	27565.0	DY	0.6H	0.0H		ED
3624.25	3625.28	27584.1	DY	1.0	1.0		ED
3620.56	3621.59	27612.2	DY	0.5	0.0		M
3620.176	3621.208	27615.09	DY	1.9	1.3		MIT
3619.96	3620.99	27616.7	DY	1.4	0.7		ED
3619.47	3620.50	27620.5	DY	0.5	0.0		ED
3618.522	3619.554	27627.71	DY	1.9	1.0		MIT
3618.07	3619.10	27631.2	DY	0.8			M
3617.24	3618.27	27637.5	DY	0.8	0.3		ED
3616.35	3617.38	27644.3	DY	0.5	0.0		MIT

AIR WAVELENGTH (ANGSTRMS)	VACUUM WAVELENGTH (ANGSTRMS)	WAVENUMBER (KAYSERS)	LMENT	LOG ARC	INTENSITY SPK	INTENSITY DIS	REF
3616.085	3617.116	27646.33	DY	0.5	0.0		MIT
3614.707	3615.738	27656.87	DY	1.2	0.7		MIT
3614.083	3615.114	27661.65	DY	1.5	1.0		MIT
3613.06	3614.09	27669.5	DY	0.8	0.6		ED
3612.788	3613.818	27671.56	DY	1.4	1.0		MIT
3611.90	3612.93	27678.4	DY	0.3	0.0		M
3611.16	3612.19	27684.0	DY	0.5	0.0		M
3610.77	3611.80	27687.0	DY	0.6	0.3		ED
3609.250	3610.279	27698.69	DY	0.5	0.0		MIT
3608.06	3609.09	27707.8	DY	0.6	0.3		M
3606.126	3607.155	27722.68	DY	2.3	2.0		MIT
3605.09	3606.12	27730.6	DY	0.6	0.3		M
3604.85	3605.88	27732.5	DY	0.6W	0.3		M
3604.36	3605.39	27736.3	DY	0.5	0.0		ED
3603.15	3604.18	27745.6	DY	0.5	0.0		ED
3602.82	3603.85	27748.1	DY	1.0	0.6		ED
3601.39	3602.42	27759.1	DY	0.5	0.0H		M
3600.34	3601.37	27767.2	DY	1.3	1.5		ED
3598.265	3599.292	27783.25	DY	1.0	0.7H		ED
3597.943	3598.970	27785.73	DY	0.5	0.3		MIT
3596.491	3597.517	27796.95	DY	0.5	0.0		MIT
3596.067	3597.093	27800.23	DY	1.7	1.3		MIT
3595.046	3596.072	27808.12	DY	2.3	2.0		MIT
3594.57	3595.60	27811.8	DY	0.6	0.3		M
3593.15	3594.18	27822.8	DY	0.8	0.6		MIT
3592.117	3593.142	27830.79	DY	1.9	1.5		MIT
3591.821	3592.846	27833.09	DY	1.9	1.3		MIT
3591.423	3592.448	27836.17	DY	2.3	2.0		MIT
3590.667	3591.692	27842.03	DY	1.5	1.3		MIT
3586.116	3587.140	27877.37	DY	0.6	0.3H		MIT
3585.776	3586.799	27880.01	DY	2.2	2.0		MIT
3585.066	3586.089	27885.53	DY	2.5	2.0		MIT
3584.426	3585.449	27890.51	DY	1.7	1.3		MIT
3582.030	3583.052	27909.16	DY	1.4	1.0		MIT
3580.043	3581.065	27924.65	DY	1.9	1.5		MIT
3579.42	3580.44	27929.5	DY	0.3	0.0		M
3579.125	3580.147	27931.82	DY	0.5	0.0		MIT
3578.80	3579.82	27934.4	DY	0.3	0.0		M
3578.47	3579.49	27936.9	DY	0.6	0.3		ED
3577.989	3579.010	27940.68	DY	2.2	1.7		MIT
3576.873	3577.894	27949.40	DY	2.3	1.7		MIT
3576.250	3577.271	27954.27	DY	2.5			MIT
3574.63	3575.65	27966.9	DY	0.3	0.0		M
3574.160	3575.180	27970.62	DY	2.3	2.0		MIT
3573.838	3574.858	27973.14	DY	1.9	1.9		MIT
3573.03	3574.05	27979.5	DY	0.5	0.0		M
3571.67	3572.69	27990.1	DY	0.5	0.3		M
3571.02	3572.04	27995.2	DY	0.5			ED
3569.669	3570.688	28005.80	DY	1.6	1.0		MIT
3569.25	3570.27	28009.1	DY	0.3			M
3568.994	3570.013	28011.10	DY	2.0			MIT
3568.65	3569.67	28013.8	DY	0.6			ED
3568.33	3569.35	28016.3	DY	0.6	0.3		M
3567.30	3568.32	28024.4	DY	0.6			ED
3566.80	3567.82	28028.3	DY	0.6	0.3H		M
3566.06	3567.08	28034.1	DY	0.6	0.3		M
3565.65	3566.67	28037.4	DY	0.8	0.3		ED
3565.34	3566.36	28039.8	DY	0.8			ED
3564.244	3565.262	28048.43	DY	1.4	1.0		MIT
3563.699	3564.717	28052.72	DY	1.7	1.5		MIT
3563.154	3564.172	28057.01	DY	2.3	2.0		MIT
3562.67	3563.69	28060.8	DY	0.7	0.0		M
3560.149	3561.166	28080.69	DY	1.6	1.3		ED
3559.27	3560.29	28087.6	DY	1.0	0.6		M
3558.20	3559.22	28096.1	DY	1.2	0.6		M
3557.632	3558.648	28100.56	DY	0.5	0.3		MIT
3556.36	3557.38	28110.6	DY	0.3			ED
3555.96	3556.98	28113.8	DY	0.8	0.6		ED
3555.69	3556.71	28115.9	DY	0.3			ED
3555.27	3556.29	28119.2	DY	0.8			M

AIR WAVELENGTH (ANGSTRMS)	VACUUM WAVELENGTH (ANGSTRMS)	WAVENUMBER (KAYSERS)	LMENT	LOG ARC	INTENSITY SPK	INTENSITY DIS	REF
3554.83	3555.85	28122.7	DY	0.6	0.6		M
3553.55	3554.57	28132.8	DY	0.6			ED
3553.19	3554.21	28135.7	DY	0.5	0.3		ED
3551.99	3553.00	28145.2	DY	0.6			ED
3551.59	3552.60	28148.4	DY	1.0	0.8		ED
3551.156	3552.170	28151.80	DY	1.4	0.0		MIT
3550.228	3551.242	28159.16	DY	2.3	2.0		MIT
3549.80	3550.81	28162.6	DY	0.8			ED
3549.255	3550.269	28166.88	DY	0.6	0.3		MIT
3548.726	3549.740	28171.08	DY	0.5	0.3		MIT
3548.202	3549.216	28175.24	DY	0.9	0.3		MIT
3547.89	3548.90	28177.7	DY	0.6			M
3547.53	3548.54	28180.6	DY	0.8	0.3		ED
3546.841	3547.854	28186.05	DY	1.7	1.0		MIT
3545.74	3546.75	28194.8	DY	0.8	0.3		ED
3544.211	3545.224	28206.97	DY	1.4	1.0H		MIT
3542.86	3543.87	28217.7	DY	0.8	0.3		M
3542.333	3543.345	28221.92	DY	2.0	1.3		ED
3541.86	3542.87	28225.7	DY	0.5			ED
3541.29	3542.30	28230.2	DY	0.8			ED
3540.67	3541.68	28235.2	DY	1.0	0.3		ED
3540.34	3541.35	28237.8	DY	0.6			ED
3539.88	3540.89	28241.5	DY	0.6			ED
3539.62	3540.63	28243.6	DY	0.8	0.6		ED
3539.376	3540.387	28245.50	DY	1.3	0.3H		MIT
3538.523	3539.534	28252.30	DY	2.2	1.6		MIT
3538.50	3539.51	28252.5	DY	0.7	0.3		ED
3537.673	3538.684	28259.09	DY	1.0			MIT
3536.56	3537.57	28268.0	DY	1.0			ED
3536.024	3537.035	28272.27	DY	2.1	1.0		MIT
3535.60	3536.61	28275.7	DY	0.6			M
3534.963	3535.973	28280.76	DY	2.1			MIT
3534.443	3535.453	28284.92	DY	0.8	0.3		ED
3534.05	3535.06	28288.1	DY	0.6			ED
3533.70	3534.71	28290.9	DY	0.8	0.3		ED
3532.44	3533.45	28300.9	DY	0.8	0.3		M
3531.712	3532.721	28306.79	DY	2.0	2.0		MIT
3530.88	3531.89	28313.5	DY	0.6			ED
3530.54	3531.55	28316.2	DY	0.8	0.6		ED
3529.52	3530.53	28324.4	DY	0.6			ED
3529.036	3530.045	28328.25	DY	0.7			MIT
3528.926	3529.935	28329.14	DY	0.3	1.0		MIT
3528.51	3529.52	28332.5	DY	0.8	0.3		ED
3527.98	3528.99	28336.7	DY	0.8	0.3		M
3526.909	3527.917	28345.34	DY	0.6	0.0		MIT
3526.61	3527.62	28347.7	DY	0.6	0.3		M
3525.751	3526.759	28354.65	DY	1.1			MIT
3524.92	3525.93	28361.3	DY	0.8			M
3524.61	3525.62	28363.8	DY	0.8	0.6		M
3524.03	3525.04	28368.5	DY	1.2	1.3		ED
3523.22	3524.23	28375.0	DY	0.6			ED
3522.868	3523.875	28377.85	DY	1.0	0.3		MIT
3522.275	3523.282	28382.63	DY	0.5			MIT
3521.85	3522.86	28386.1	DY	0.6	0.3		ED
3521.13	3522.14	28391.9	DY	1.2	0.8		M
3519.770	3520.776	28402.83	DY	0.3	0.0H		MIT
3518.71	3519.72	28411.4	DY	0.6	0.3		ED
3517.58	3518.59	28420.5	DY	0.6			M
3517.270	3518.276	28423.01	DY	1.8	0.6		MIT
3516.96	3517.97	28425.5	DY	0.6			ED
3516.48	3517.49	28429.4	DY	0.6			ED
3516.15	3517.16	28432.1	DY	0.8			ED
3515.64	3516.65	28436.2	DY	0.8			ED
3515.053	3516.058	28440.94	DY	0.5	0.0		MIT
3514.80	3515.81	28443.0	DY	0.3			M
3514.43	3515.44	28446.0	DY	0.8			ED
3514.10	3515.10	28448.6	DY	0.6	0.3		M
3513.84	3514.84	28450.8	DY	0.6			M
3513.10	3514.10	28456.7	DY	0.6			ED
3512.707	3513.712	28459.93	DY	1.1			MIT

AIR WAVELENGTH (ANGSTRMS)	VACUUM WAVELENGTH (ANGSTRMS)	WAVENUMBER (KAYSERS)	LMENT	LOG ARC	INTENSITY SPK	DIS	REF
3512.563	3513.568	28461.10	DY	1.0			MIT
3511.98	3512.98	28465.8	DY	0.6			ED
3511.694	3512.698	28468.14	DY	1.0	0.3		MIT
3511.40	3512.40	28470.5	DY	0.6	0.3		M
3511.00	3512.00	28473.8	DY	0.8	0.3		ED
3510.088	3511.092	28481.17	DY	0.3	0.0		MIT
3509.41	3510.41	28486.7	DY	0.8	0.3		ED
3509.00	3510.00	28490.0	DY	0.8	0.6		ED
3507.53	3508.53	28501.9	DY	0.8			ED
3506.820	3507.823	28507.71	DY	1.9			MIT
3505.83	3506.83	28515.8	DY	1.3	0.3		M
3505.457	3506.460	28518.79	DY	1.8	0.3		MIT
3504.522	3505.525	28526.40	DY	2.0	0.5		MIT
3503.66	3504.66	28533.4	DY	0.6	0.3		M
3503.179	3504.181	28537.34	DY	0.9	0.5H		MIT
3502.09	3503.09	28546.2	DY	0.8	0.6		M
3501.869	3502.871	28548.01	DY	0.6	0.0		MIT
3501.436	3502.438	28551.54	DY	1.0	0.5		MIT
3500.50	3501.50	28559.2	DY	0.3	0.6		M
3499.82	3500.82	28564.7	DY	0.7			M
3499.61	3500.61	28566.4	DY	0.6			MIT
3498.939	3499.940	28571.92	DY	1.2			MIT
3498.67	3499.67	28574.1	DY	1.7	1.7		M
3497.841	3498.842	28580.89	DY	1.5	0.5		MIT
3497.111	3498.112	28586.85	DY	0.5	0.5		MIT
3496.72	3497.72	28590.1	DY	0.5	0.3		MIT
3496.27	3497.27	28593.7	DY	0.6			ED
3494.496	3495.496	28608.24	DY	2.0	0.7		MIT
3494.135	3495.135	28611.20	DY	1.3	0.3H		MIT
3493.26	3494.26	28618.4	DY	0.5			M
3492.52	3493.52	28624.4	DY	0.3	0.0		MIT
3490.63	3491.63	28639.9	DY	0.7			M
3488.97	3489.97	28653.6	DY	0.7	0.0H		M
3487.58	3488.58	28665.0	DY	0.8	0.3		MIT
3487.202	3488.200	28668.08	DY	0.8			MIT
3486.142	3487.140	28676.80	DY	0.3			MIT
3485.90	3486.90	28678.8	DY	0.7	0.0		MIT
3484.683	3485.680	28688.80	DY	1.0	0.5		MIT
3483.29	3484.29	28700.3	DY	0.3	0.0H		MIT
3482.77	3483.77	28704.6	DY	0.8	0.0		MIT
3482.09	3483.09	28710.2	DY	0.3	0.3		MIT
3481.62	3482.62	28714.0	DY	0.3	0.0H		MIT
3480.817	3481.813	28720.67	DY	1.1			MIT
3480.42	3481.42	28723.9	DY	0.7	0.0		MIT
3479.76	3480.76	28729.4	DY	0.3	0.0		MIT
3478.48	3479.48	28740.0	DY	0.8			MIT
3478.23	3479.23	28742.0	DY	0.3	0.0		M
3477.94	3478.94	28744.4	DY	0.6	0.0		MIT
3477.074	3478.069	28751.58	DY	2.0	1.5		MIT
3476.74	3477.74	28754.3	DY	0.3			MIT
3476.36	3477.36	28757.5	DY	0.6	0.0		MIT
3475.65	3476.65	28763.4	DY	0.3	0.0		MIT
3475.34	3476.34	28765.9	DY	0.3	0.3H		MIT
3474.71	3475.70	28771.1	DY	0.3	0.0		MIT
3474.29	3475.28	28774.6	DY	1.1	0.3		MIT
3473.702	3474.697	28779.49	DY	1.9	1.6		MIT
3473.30	3474.29	28782.8	DY	0.5			MIT
3472.94	3473.93	28785.8	DY	0.5	0.0		MIT
3472.25	3473.24	28791.5	DY	0.6	0.0		MIT
3471.95	3472.94	28794.0	DY	0.5			MIT
3471.531	3472.525	28797.49	DY	1.0	0.7H		MIT
3471.145	3472.139	28800.69	DY	1.0			MIT
3470.17	3471.16	28808.8	DY	0.8	0.0		MIT
3469.87	3470.86	28811.3	DY	0.5	0.0		MIT
3468.77	3469.76	28820.4	DY	0.5	0.0		MIT
3468.435	3469.428	28823.19	DY	1.7	0.5		MIT
3467.86	3468.85	28828.0	DY	0.6	0.0		MIT
3467.49	3468.48	28831.1	DY	0.3	0.0		MIT
3467.07	3468.06	28834.5	DY	0.5	0.0		MIT
3466.12	3467.11	28842.4	DY	0.3	0.0		MIT

AIR WAVELENGTH (ANGSTRMS)	VACUUM WAVELENGTH (ANGSTRMS)	WAVENUMBER (KAYSERS)	LMENT	LOG ARC	INTENSITY SPK	DIS	REF
3465.30	3466.29	28849.3	DY	0.3			MIT
3464.46	3465.45	28856.3	DY	0.3	0.0		MIT
3463.876	3464.868	28861.13	DY	1.1	0.3		MIT
3463.36	3464.35	28865.4	DY	0.7	0.3		MIT
3461.31	3462.30	28882.5	DY	0.5	0.0		MIT
3460.971	3461.962	28885.35	DY	2.0	0.5		MIT
3460.64	3461.63	28888.1	DY	0.6	0.0		MIT
3460.40	3461.39	28890.1	DY	1.3	0.3		MIT
3460.05	3461.04	28893.0	DY	0.7	0.0		M
3459.30	3460.29	28899.3	DY	0.6	0.0		MIT
3458.99	3459.98	28901.9	DY	1.0	0.0		MIT
3458.57	3459.56	28905.4	DY	0.3			MIT
3456.566	3457.556	28922.16	DY	1.7	1.5		MIT
3456.01	3457.00	28926.8	DY	1.6O			MIT
3455.08	3456.07	28934.6	DY	0.5	0.0H		MIT
3454.516	3455.506	28939.32	DY	1.3	0.3		MIT
3454.326	3455.316	28940.92	DY	2.0	1.0		MIT
3453.12	3454.11	28951.0	DY	0.7J	0.3		MIT
3451.70	3452.69	28962.9	DY	0.5	0.3		MIT
3450.22	3451.21	28975.4	DY	0.6	0.3		M
3449.898	3450.886	28978.06	DY	1.6	0.7		MIT
3448.50	3449.49	28989.8	DY	0.3	0.3		MIT
3448.08	3449.07	28993.3	DY	0.3	0.3		MIT
3447.78	3448.77	28995.9	DY	1.0	0.6		M
3447.28	3448.27	29000.1	DY	1.0	0.3		MIT
3447.001	3447.989	29002.42	DY	1.7	0.9		MIT
3445.582	3446.569	29014.36	DY	1.9	0.9		MIT
3444.26	3445.25	29025.5	DY	1.0	0.3		MIT
3443.46	3444.45	29032.2	DY	0.5	0.3		M
3442.53	3443.52	29040.1	DY	0.3	0.6		M
3441.453	3442.439	29049.17	DY	1.7	0.7		MIT
3440.94	3441.93	29053.5	DY	1.4	0.7		MIT
3440.46	3441.45	29057.6	DY	0.6			MIT
3439.33	3440.32	29067.1	DY	1.0	0.7		MIT
3438.952	3439.938	29070.29	DY	1.4	1.0		MIT
3436.95	3437.94	29087.2	DY	0.5			MIT
3435.91	3436.89	29096.0	DY	0.5	0.6		MIT
3435.27	3436.25	29101.4	DY	0.6	0.3		MIT
3434.373	3435.358	29109.05	DY	1.9	1.5		MIT
3432.87	3433.85	29121.8	DY	0.7	0.6		MIT
3432.58	3433.56	29124.3	DY	1.4	0.7		MIT
3431.79	3432.77	29131.0	DY	1.3	0.7		MIT
3431.03	3432.01	29137.4	DY	0.3			MIT
3430.21	3431.19	29144.4	DY	0.3			MIT
3429.442	3430.425	29150.91	DY	1.7	0.7		MIT
3429.00	3429.98	29154.7	DY	0.6	0.6		MIT
3428.47	3429.45	29159.2	DY	0.5	0.3		MIT
3426.07	3427.05	29179.6	DY	0.3			MIT
3425.60	3426.58	29183.6	DY	0.3			MIT
3425.06	3426.04	29188.2	DY	1.6	0.7		MIT
3423.82	3424.80	29198.8	DY	0.5	0.6		MIT
3423.24	3424.22	29203.7	DY	0.6	0.3		MIT
3422.878	3423.860	29206.80	DY	1.2	0.7		MIT
3422.57	3423.55	29209.4	DY	1.0D			MIT
3422.07	3423.05	29213.7	DY	0.5	0.3		MIT
3421.32	3422.30	29220.1	DY	0.8	0.7		MIT
3420.81	3421.79	29224.5	DY	0.7	0.3		MIT
3419.99	3420.97	29231.5	DY	0.3	0.3		MIT
3419.643	3420.624	29234.43	DY	1.3	0.3		MIT
3418.14	3419.12	29247.3	DY	1.2	0.7		MIT
3417.140	3418.120	29255.85	DY	1.2	0.7		MIT
3415.35	3416.33	29271.2	DY	0.6			M
3414.830	3415.810	29275.64	DY	1.5	0.7		MIT
3414.30	3415.28	29280.2	DY	0.3	0.3		MIT
3413.794	3414.773	29284.52	DY	1.6	1.0		MIT
3412.47	3413.45	29295.9	DY	0.3	0.3		MIT
3411.52	3412.50	29304.0	DY	0.5	0.3		MIT
3411.21	3412.19	29306.7	DY	0.3	0.3		MIT
3410.72	3411.70	29310.9	DY	0.5	0.3		MIT
3409.45	3410.43	29321.8	DY	0.5	0.3		MIT

AIR WAVELENGTH (ANGSTRMS)	VACUUM WAVELENGTH (ANGSTRMS)	WAVENUMBER (KAYSERS)	LMENT	LOG ARC	INTENSITY SPK	INTENSITY DIS	REF
3408.157	3409.135	29332.96	DY	1.1	0.7		MIT
3407.80	3408.78	29336.0	DY	2.2	1.0		MIT
3407.168	3408.146	29341.47	DY	1.3	0.7		MIT
3406.75	3407.73	29345.1	DY	0.5			MIT
3405.663	3406.640	29354.44	DY	1.0	0.7		MIT
3404.99	3405.97	29360.2	DY	0.6	0.6		MIT
3403.45	3404.43	29373.5	DY	0.6	0.6		M
3403.27	3404.25	29375.1	DY	0.8	0.7		MIT
3401.98	3402.96	29386.2	DY	0.7	0.3		MIT
3401.28	3402.26	29392.3	DY	0.3	0.3		MIT
3399.34	3400.32	29409.0	DY	0.5			MIT
3398.34	3399.32	29417.7	DY	0.6	0.3		MIT
3397.99	3398.97	29420.7	DY	0.5	0.3		MIT
3397.39	3398.37	29425.9	DY	0.3	0.3		MIT
3396.169	3397.144	29436.49	DY	1.3	1.0		MIT
3394.84	3395.81	29448.0	DY	0.3	0.3		MIT
3393.99	3394.96	29455.4	DY	0.6	0.5		MIT
3393.583	3394.557	29458.92	DY	2.0	1.0		MIT
3392.99	3393.96	29464.1	DY	0.3	0.3		MIT
3391.99	3392.96	29472.8	DY	0.6			MIT
3391.13	3392.10	29480.2	DY	0.5D	0.5D		MIT
3390.81	3391.78	29483.0	DY	0.3	0.3		MIT
3389.46	3390.43	29494.8	DY	0.3	0.3		MIT
3388.863	3389.836	29499.95	DY	1.3	0.7		MIT
3386.58	3387.55	29519.8	DY	0.3	0.3		MIT
3385.027	3385.999	29533.38	DY	1.4	0.9		MIT
3378.899	3379.869	29586.94	DY	0.3	0.3		MIT
3378.43	3379.40	29591.1	DY	0.6	0.3		MIT
3378.20	3379.17	29593.1	DY	0.3D	0.3D		MIT
3377.11	3378.08	29602.6	DY	0.6	0.6		MIT
3376.63	3377.60	29606.8	DY	0.6	0.6		MIT
3376.38	3377.35	29609.0	DY	0.5	0.5		MIT
3376.01	3376.98	29612.3	DY	0.6	0.3		MIT
3375.75	3376.72	29614.5	DY	0.7	0.6		MIT
3374.28	3375.25	29627.4	DY	0.6	0.3		MIT
3371.76	3372.73	29649.6	DY	0.9D	0.7D		MIT
3370.86	3371.83	29657.5	DY	0.8	0.6		MIT
3369.64	3370.61	29668.2	DY	0.3	0.3		MIT
3369.27	3370.24	29671.5	DY	0.3	0.3		MIT
3368.116	3369.084	29681.66	DY	1.8	0.7		MIT
3367.53	3368.50	29686.8	DY	0.6	0.3		MIT
3365.80	3366.77	29702.1	DY	1.0	0.7		MIT
3362.17	3363.14	29734.1	DY	0.5	0.3		MIT
3360.65	3361.62	29747.6	DY	0.3	0.3		MIT
3359.478	3360.443	29757.98	DY	1.1	0.7		MIT
3358.96	3359.93	29762.6	DY	0.3	0.3		MIT
3358.61	3359.58	29765.7	DY	1.2	0.3		MIT
3358.27	3359.24	29768.7	DY	0.5	0.3		MIT
3357.61	3358.57	29774.5	DY	0.3	0.3		MIT
3357.31	3358.27	29777.2	DY	0.3	0.3		MIT
3356.220	3357.185	29786.86	DY	1.0	0.7		MIT
3355.60	3356.56	29792.4	DY	0.3	0.3		MIT
3355.07	3356.03	29797.1	DY	0.6	0.6		MIT
3353.595	3354.559	29810.18	DY	1.5	0.7		MIT
3352.696	3353.660	29818.17	DY	1.3	0.7		MIT
3350.95	3351.91	29833.7	DY	0.3	0.3		MIT
3350.66	3351.62	29836.3	DY	0.5	0.3		MIT
3348.88	3349.84	29852.1	DY	0.3			MIT
3347.82	3348.78	29861.6	DY	1.1	1.0		MIT
3346.64	3347.60	29872.1	DY	0.3	0.3		MIT
3346.14	3347.10	29876.6	DY	0.5			MIT
3345.79	3346.75	29879.7	DY	0.5	0.3		MIT
3345.37	3346.33	29883.5	DY	0.5	0.3		MIT
3344.49	3345.45	29891.3	DY	0.5	0.3		MIT
3343.48	3344.44	29900.4	DY	0.5			MIT
3342.65	3343.61	29907.8	DY	0.7	0.3		MIT
3341.88	3342.84	29914.7	DY	1.5			MIT
3341.433	3342.394	29918.68	DY	1.0	0.3		MIT
3340.986	3341.947	29922.68	DY	0.5			MIT
3340.60	3341.56	29926.1	DY	0.7			MIT

AIR WAVELENGTH (ANGSTRMS)	VACUUM WAVELENGTH (ANGSTRMS)	WAVENUMBER (KAYSERS)	LMENT	LOG ARC	INTENSITY SPK	INTENSITY DIS	REF
3340.010	3340.970	29931.42	DY	1.9	0.6		MIT
3339.51	3340.47	29935.9	DY	1.4	0.6		MIT
3338.64	3339.60	29943.7	DY	0.6			MIT
3337.12	3338.08	29957.3	DY	0.3	0.3		MIT
3336.83	3337.79	29959.9	DY	0.3	0.3		MIT
3335.83	3336.79	29968.9	DY	0.6	0.3		MIT
3334.86	3335.82	29977.6	DY	0.3	0.3		MIT
3334.45	3335.41	29981.3	DY	0.6	0.3		MIT
3334.14	3335.10	29984.1	DY	1.1	0.3		MIT
3333.54	3334.50	29989.5	DY	0.5	0.3		MIT
3332.04	3333.00	30003.0	DY	0.3			MIT
3331.28	3332.24	30009.9	DY	0.6	0.3		MIT
3330.60	3331.56	30016.0	DY	0.6	0.3		MIT
3328.79	3329.75	30032.3	DY	0.7			MIT
3327.88	3328.84	30040.5	DY	0.3			MIT
3327.30	3328.26	30045.8	DY	0.7	0.3		MIT
3327.08	3328.04	30047.7	DY	1.0	0.3		MIT
3326.42	3327.38	30053.7	DY	0.7	0.6		MIT
3326.19	3327.15	30055.8	DY	1.3	1.0		MIT
3325.28	3326.24	30064.0	DY	0.3			MIT
3324.20	3325.16	30073.8	DY	0.5			MIT
3323.95	3324.91	30076.0	DY	0.3			MIT
3323.27	3324.23	30082.2	DY	0.6			MIT
3319.887	3320.842	30112.84	DY	2.2	1.0		MIT
3319.88	3320.84	30112.9	DY II	0.8	0.8		ME
3319.421	3320.376	30117.07	DY	0.6			MIT
3318.42	3319.37	30126.1	DY	0.3	0.3		MIT
3318.13	3319.08	30128.8	DY	0.6	0.3		MIT
3317.119	3318.074	30137.97	DY	1.5	0.6		MIT
3316.325	3317.279	30145.19	DY	1.7	0.6		MIT
3314.95	3315.90	30157.7	DY	0.9	0.3		MIT
3313.31	3314.26	30172.6	DY	1.0	0.6		MIT
3313.04	3313.99	30175.1	DY	0.5			MIT
3312.729	3313.682	30177.91	DY	1.7	0.7		MIT
3312.33	3313.28	30181.5	DY	0.6	0.3		MIT
3311.51	3312.46	30189.0	DY	0.7			MIT
3310.96	3311.91	30194.0	DY	0.7	0.5		MIT
3308.89	3309.84	30212.9	DY	1.4			MIT
3308.05	3309.00	30220.6	DY	0.3			MIT
3306.79	3307.74	30232.1	DY	0.3	0.3		MIT
3306.19	3307.14	30237.6	DY	1.1	0.8		MIT
3305.47	3306.42	30244.2	DY	1.0D	0.7D		MIT
3304.71	3305.66	30251.1	DY	0.5			MIT
3304.30	3305.25	30254.9	DY	0.6			MIT
3303.60	3304.55	30261.3	DY	0.3			MIT
3302.472	3303.423	30271.63	DY	0.6	0.3		MIT
3302.02	3302.97	30275.8	DY	0.3			MIT
3301.72	3302.67	30278.5	DY	0.3			MIT
3300.93	3301.88	30285.8	DY	0.6	0.3		MIT
3299.15	3300.10	30302.1	DY	0.3			MIT
3297.607	3298.557	30316.29	DY	0.8	0.3		MIT
3296.31	3297.26	30328.2	DY	1.0	0.3		MIT
3295.92	3296.87	30331.8	DY	0.3	0.0		MIT
3295.205	3296.154	30338.39	DY	0.5	0.0H		MIT
3294.94	3295.89	30340.8	DY	0.5	0.0		MIT
3294.661	3295.610	30343.40	DY	0.3			MIT
3293.82	3294.77	30351.1	DY	1.0	0.0		MIT
3291.119	3292.067	30376.05	DY	0.6			MIT
3289.34	3290.29	30392.5	DY	0.3	0.3H		MIT
3288.635	3289.582	30399.00	DY	0.7			MIT
3287.953	3288.900	30405.30	DY	0.7	0.3		MIT
3286.573	3287.520	30418.07	DY	0.3			MIT
3284.37	3285.32	30438.5	DY	0.3	0.0		MIT
3282.79	3283.74	30453.1	DY	0.3	0.0		MIT
3280.10	3281.05	30478.1	DY	1.8	0.7		MIT
3272.73	3273.67	30546.7	DY	0.8			MIT
3272.073	3273.016	30552.86	DY	0.3	0.0		MIT
3269.527	3270.469	30576.65	DY	0.3	0.0		MIT
3269.12	3270.06	30580.5	DY	1.3	0.5		MIT
3267.764	3268.706	30593.15	DY	0.3			MIT

AIR WAVELENGTH (ANGSTRMS)	VACUUM WAVELENGTH (ANGSTRMS)	WAVENUMBER (KAYSERS)	LMENT	LOG ARC	INTENSITY SPK	DIS	REF
3266.207	3267.149	30607.73	DY	1.1	0.7		MIT
3266.00	3266.94	30609.7	DY	1.0	0.7		MIT
3264.290	3265.231	30625.70	DY	0.5			MIT
3262.27	3263.21	30644.7	DY	0.3			MIT
3262.02	3262.96	30647.0	DY	0.3	0.0		MIT
3261.22	3262.16	30654.5	DY	0.7	0.3		MIT
3260.69	3261.63	30659.5	DY	1.0	0.3		MIT
3260.00	3260.94	30666.0	DY	1.0	0.3		MIT
3257.361	3258.300	30690.85	DY	0.6	0.3H		MIT
3256.25	3257.19	30701.3	DY	1.4	0.7		MIT
3254.483	3255.422	30717.99	DY	0.7	0.3		MIT
3253.91	3254.85	30723.4	DY	0.6	0.0		MIT
3252.19	3253.13	30739.6	DY	1.2	0.3		MIT
3251.90	3252.84	30742.4	DY	1.0	0.3		MIT
3251.260	3252.198	30748.44	DY	2.0	2.0		MIT
3248.364	3249.301	30775.85	DY	1.1	0.6		MIT
3245.14	3246.08	30806.4	DY	0.3	0.0		MIT
3243.78	3244.72	30819.3	DY	1.1	0.5		MIT
3242.51	3243.45	30831.4	DY	0.6			MIT
3242.285	3243.221	30833.55	DY	0.6			MIT
3240.878	3241.813	30846.94	DY	1.0	0.7H		MIT
3240.043	3240.978	30854.88	DY	0.8			MIT
3239.59	3240.52	30859.2	DY	0.7			MIT
3238.70	3239.63	30867.7	DY	0.6			MIT
3238.178	3239.113	30872.65	DY	0.5			MIT
3237.098	3238.032	30882.95	DY	0.5			MIT
3236.63	3237.56	30887.4	DY	0.7	0.5		MIT
3235.90	3236.83	30894.4	DY	1.5S	0.9		MIT
3233.422	3234.355	30918.06	DY	0.3			MIT
3232.652	3233.585	30925.43	DY	1.2	0.6		MIT
3230.32	3231.25	30947.7	DY	0.3			MIT
3229.94	3230.87	30951.4	DY	0.9D	0.3		MIT
3229.37	3230.30	30956.9	DY	1.0	0.3		MIT
3228.97	3229.90	30960.7	DY	1.0	0.0		MIT
3227.71	3228.64	30972.8	DY	0.5	0.0		MIT
3227.43	3228.36	30975.5	DY	0.3			MIT
3226.38	3227.31	30985.5	DY	0.8	0.0		MIT
3225.96	3226.89	30989.6	DY	1.3	0.5		MIT
3225.08	3226.01	30998.0	DY	1.0	0.0		MIT
3223.29	3224.22	31015.2	DY	1.3	0.0		MIT
3221.50	3222.43	31032.5	DY	1.1	0.3		MIT
3220.46	3221.39	31042.5	DY	1.0	0.3		MIT
3216.63	3217.56	31079.5	DY	1.4	1.2		MIT
3215.189	3216.118	31093.39	DY	1.6	1.0		MIT
3214.636	3215.565	31098.74	DY	1.2	0.7		MIT
3212.684	3213.612	31117.63	DY	0.8	0.3		MIT
3212.447	3213.375	31119.93	DY	0.6	0.3		MIT
3212.03	3212.96	31124.0	DY	0.7	0.0		MIT
3208.81	3209.74	31155.2	DY	1.0	0.3		MIT
3207.10	3208.03	31171.8	DY	1.3	0.5		MIT
3206.40	3207.33	31178.6	DY	1.6	0.7		MIT
3205.46	3206.39	31187.8	DY	0.7	0.3		MIT
3202.56	3203.49	31216.0	DY	0.6	0.0		MIT
3202.239	3203.164	31219.13	DY	0.3			MIT
3201.63	3202.56	31225.1	DY	0.6	0.0		MIT
3201.29	3202.22	31228.4	DY	0.6	0.0		MIT
3197.592	3198.516	31264.50	DY	0.6	0.3		MIT
3196.429	3197.353	31275.87	DY	0.3	0.0		MIT
3193.66	3194.58	31303.0	DY	0.3			MIT
3193.308	3194.231	31306.44	DY	1.4	0.7		MIT
3192.031	3192.954	31318.96	DY	0.5			MIT
3190.648	3191.570	31332.54	DY	0.5			MIT
3189.76	3190.68	31341.3	DY	0.6	0.0		MIT
3189.064	3189.986	31348.10	DY	0.5	0.0		MIT
3188.667	3189.589	31352.00	DY	0.7	0.3		MIT
3187.676	3188.598	31361.75	DY	1.4	1.0		MIT
3186.375	3187.296	31374.55	DY	1.4	1.0		MIT
3184.777	3185.698	31390.29	DY	1.3	0.7		MIT
3184.205	3185.126	31395.93	DY	0.3	0.0		MIT
3183.196	3184.117	31405.89	DY	0.8	0.5		MIT

AIR WAVELENGTH (ANGSTRMS)	VACUUM WAVELENGTH (ANGSTRMS)	WAVENUMBER (KAYSERS)	LMENT	LOG ARC	INTENSITY SPK	DIS	REF
3181.940	3182.860	31418.28	DY	1.0	0.7H		MIT
3179.026	3179.946	31447.08	DY	0.5			MIT
3178.373	3179.292	31453.54	DY	1.3	0.7		MIT
3177.88	3178.80	31458.4	DY	1.5	0.7		MIT
3177.531	3178.450	31461.87	DY	0.8	0.5		MIT
3174.883	3175.801	31488.11	DY	1.1	0.6		MIT
3171.36	3172.28	31523.1	DY	0.7			MIT
3170.92	3171.84	31527.5	DY	0.7			MIT
3170.746	3171.663	31529.20	DY	1.0	0.5		MIT
3169.978	3170.895	31536.83	DY	2.0	1.7		MIT
3169.553	3170.470	31541.06	DY	0.7	0.0H		MIT
3168.95	3169.87	31547.1	DY	0.8			MIT
3168.58	3169.50	31550.7	DY	0.3	0.3H		ED
3168.12	3169.04	31555.3	DY	0.8	0.3H		MIT
3167.771	3168.688	31558.81	DY	0.3	0.0H		MIT
3167.42	3168.34	31562.3	DY	1.0D	0.3		MIT
3164.041	3164.957	31596.01	DY	0.5	0.0		MIT
3162.824	3163.739	31608.17	DY	1.9	1.8		MIT
3161.028	3161.943	31626.12	DY	1.0	0.7		MIT
3160.499	3161.414	31631.42	DY	1.0	0.5		MIT
3159.309	3160.224	31643.33	DY	0.6			MIT
3157.549	3158.463	31660.97	DY	0.9	0.0H		MIT
3157.203	3158.117	31664.44	DY	0.6	0.0H		MIT
3156.52	3157.43	31671.3	DY	1.7	1.3		MIT
3153.307	3154.220	31703.56	DY	1.0	0.5		MIT
3152.379	3153.292	31712.89	DY	1.1			MIT
3151.893	3152.806	31717.78	DY	1.3	0.7		MIT
3150.12	3151.03	31735.6	DY	1.0	0.7		M
3147.533	3148.445	31761.72	DY	1.0	0.3		MIT
3146.882	3147.793	31768.28	DY	0.3	0.0H		MIT
3146.165	3147.076	31775.52	DY	1.3	0.7		MIT
3145.222	3146.133	31785.05	DY	1.3	0.7		MIT
3143.831	3144.742	31799.11	DY	1.3	0.7		MIT
3143.181	3144.092	31805.69	DY	0.8	0.3		MIT
3142.303	3143.213	31814.58	DY	1.0	0.5		MIT
3141.13	3142.04	31826.5	DY	1.7	1.3		MIT
3140.645	3141.555	31831.37	DY	1.6	1.3		MIT
3139.503	3140.413	31842.95	DY	0.6	0.0		MIT
3136.692	3137.601	31871.49	DY	0.7			MIT
3135.68	3136.59	31881.8	DY	0.7			MIT
3135.37	3136.28	31884.9	DY	2.0	1.7		MIT
3132.98	3133.89	31909.2	DY	0.8	0.3		M
3132.122	3133.030	31917.99	DY	0.7	0.3		MIT
3130.159	3131.066	31938.00	DY	0.6	0.3H		MIT
3129.24	3130.15	31947.4	DY	0.5			M
3128.409	3129.316	31955.87	DY	1.6	1.0		MIT
3126.802	3127.708	31972.29	DY	0.5	0.0		MIT
3126.18	3127.09	31978.6	DY	1.2	0.0H		MIT
3122.946	3123.851	32011.77	DY	0.5	0.0H		MIT
3122.03	3122.94	32021.2	DY	0.7	0.7H		M
3120.184	3121.089	32040.10	DY	1.4	1.0		MIT
3117.898	3118.802	32063.59	DY	0.3	0.0H		MIT
3117.510	3118.414	32067.58	DY	0.8	0.3		MIT
3116.869	3117.773	32074.18	DY	0.6	0.3		MIT
3115.91	3116.81	32084.1	DY	0.5			MIT
3113.104	3114.007	32112.97	DY	0.7	0.3H		MIT
3112.879	3113.782	32115.29	DY	0.5			MIT
3111.548	3112.451	32129.02	DY	0.6			MIT
3110.757	3111.659	32137.19	DY	1.0	0.7H		MIT
3109.768	3110.670	32147.41	DY	1.6	1.3		MIT
3106.93	3107.83	32176.8	DY	0.8			MIT
3105.001	3105.902	32196.77	DY	1.3	0.7		MIT
3104.121	3105.022	32205.89	DY	0.7			MIT
3103.839	3104.740	32208.82	DY	1.5	1.0		MIT
3103.246	3104.146	32214.98	DY	1.5	0.0		MIT
3102.19	3103.09	32225.9	DY	0.6	0.0H		MIT
3101.91	3102.81	32228.9	DY	0.6	0.3		MIT
3098.050	3098.949	32269.00	DY	0.6			MIT
3095.75	3096.65	32293.0	DY	1.2	0.0H		MIT
3093.82	3094.72	32313.1	DY	1.0	0.0		MIT

AIR WAVELENGTH (ANGSTRMS)	VACUUM WAVELENGTH (ANGSTRMS)	WAVENUMBER (KAYSERS)	LMENT	LOG ARC	INTENSITY SPK	INTENSITY DIS	REF
3093.108	3094.006	32320.56	DY	1.2	0.7		MIT
3090.190	3091.087	32351.08	DY	0.7			MIT
3089.604	3090.501	32357.21	DY	0.7			MIT
3088.42	3089.32	32369.6	DY	0.3			MIT
3087.52	3088.42	32379.1	DY	0.3	0.0H		MIT
3082.515	3083.410	32431.62	DY	1.2	0.7		MIT
3080.927	3081.822	32448.34	DY	1.0S	0.7		MIT
3079.341	3080.235	32465.05	DY	1.2	0.7		MIT
3078.686	3079.580	32471.96	DY	1.4			MIT
3078.35	3079.24	32475.5	DY	1.1	0.0		MIT
3076.90	3077.79	32490.8	DY	1.0S			MIT
3075.20	3076.09	32508.8	DY	0.5	0.6		MIT
3074.00	3074.89	32521.5	DY	0.5			M
3073.542	3074.435	32526.30	DY	1.5	0.9		MIT
3071.920	3072.813	32543.48	DY	1.4	0.6		MIT
3066.994	3067.885	32595.74	DY	1.3	0.6		MIT
3065.14	3066.03	32615.5	DY	0.6	0.0H		MIT
3064.04	3064.93	32627.2	DY	0.9			MIT
3062.620	3063.510	32642.29	DY	1.4	1.0		MIT
3062.190	3063.080	32646.88	DY	1.1	0.8		MIT
3061.506	3062.396	32654.17	DY	1.0			MIT
3060.653	3061.543	32663.27	DY	1.5	1.0		MIT
3060.30	3061.19	32667.0	DY	1.0	0.0		M
3060.00	3060.89	32670.2	DY	1.0	0.7		M
3059.48	3060.37	32675.8	DY	0.8	0.6		MIT
3056.97	3057.86	32702.6	DY	1.1	0.6		MIT
3052.324	3053.212	32752.40	DY	1.3	0.6		MIT
3049.133	3050.020	32786.67	DY	1.4	0.6		MIT
3048.41	3049.30	32794.5	DY	0.8			MIT
3047.610	3048.497	32803.06	DY	1.0			MIT
3046.38	3047.27	32816.3	DY	0.6			MIT
3044.547	3045.433	32836.06	DY	0.8	0.6		MIT
3043.44	3044.33	32848.0	DY II	1.3	0.3		MIT
3043.144	3044.029	32851.19	DY	1.5	1.0		MIT
3041.648	3042.533	32867.35	DY	0.8			MIT
3040.28	3041.16	32882.1	DY	0.6			MIT
3038.291	3039.175	32903.66	DY	1.3	1.0		MIT
3036.709	3037.593	32920.81	DY	1.3	0.6		MIT
3033.18	3034.06	32959.1	DY	0.9	0.0		MIT
3031.18	3032.06	32980.9	DY	0.6	0.0H		MIT
3030.41	3031.29	32989.2	DY	0.9S	0.6		MIT
3029.826	3030.708	32995.59	DY	1.5	0.6		MIT
3027.57	3028.45	33020.2	DY	1.0	0.6		MIT
3026.163	3027.044	33035.53	DY	1.5	0.6		MIT
3025.61	3026.49	33041.6	DY	0.9			MIT
3021.79	3022.67	33083.3	DY	0.7			MIT
3020.65	3021.53	33095.8	DY	1.0			MIT
3017.73	3018.61	33127.8	DY	0.3			MIT
3016.96	3017.84	33136.3	DY	0.9	0.5		MIT
3015.694	3016.573	33150.21	DY	1.0	0.5		MIT
3015.074	3015.952	33157.02	DY	1.3	0.7		MIT
3012.30	3013.18	33187.5	DY	0.5			MIT
3008.821	3009.698	33225.93	DY	0.7			MIT
3003.762	3004.638	33281.88	DY	1.2	0.7		MIT
3002.38	3003.26	33297.2	DY	1.0	0.0H		MIT
3001.03	3001.90	33312.2	DY	0.5			M
2996.76	2997.63	33359.6	DY	0.3			MIT
2995.451	2996.324	33374.22	DY	0.3			MIT
2993.117	2993.990	33400.25	DY	0.3			MIT
2992.420	2993.293	33408.02	DY	0.5			MIT
2991.618	2992.491	33416.98	DY	0.3	0.0H		MIT
2991.364	2992.236	33419.82	DY	0.3	0.0		MIT
2988.715	2989.587	33449.44	DY	0.3			MIT
2987.892	2988.764	33458.65	DY	0.3			MIT
2985.93	2986.80	33480.6	DY	0.6	0.3		ED
2979.658	2980.528	33551.11	DY	0.3			MIT
2977.426	2978.295	33576.26	DY	0.3			MIT
2975.82	2976.69	33594.4	DY	0.3			MIT
2973.18	2974.05	33624.2	DY	0.6			MIT
2969.334	2970.201	33667.75	DY	0.3			MIT

AIR WAVELENGTH (ANGSTRMS)	VACUUM WAVELENGTH (ANGSTRMS)	WAVENUMBER (KAYSERS)	LMENT	LOG ARC	INTENSITY SPK	INTENSITY DIS	REF
2964.62	2965.49	33721.3	DY	0.3			MIT
2962.372	2963.237	33746.88	DY	0.3	0.0		MIT
2961.58	2962.45	33755.9	DY	0.3			MIT
2957.762	2958.626	33799.47	DY	0.5			MIT
2953.709	2954.572	33845.85	DY	0.3	0.0		MIT
2953.039	2953.902	33853.53	DY	0.3			MIT
2952.128	2952.991	33863.97	DY	0.3			MIT
2950.339	2951.201	33884.51	DY	0.3			MIT
2948.323	2949.185	33907.67	DY	0.3	0.0		MIT
2947.221	2948.083	33920.35	DY	0.3	0.0H		MIT
2947.062	2947.923	33922.18	DY	0.3			MIT
2946.783	2947.644	33925.40	DY	0.3			MIT
2946.316	2947.177	33930.77	DY	0.3			MIT
2944.564	2945.425	33950.96	DY	0.3	0.0H		MIT
2941.064	2941.924	33991.36	DY	0.3	0.0		MIT
2934.529	2935.387	34067.05	DY	0.3	0.0		MIT
2927.067	2927.924	34153.90	DY	0.3			MIT
2918.670	2919.524	34252.15	DY	0.3	0.0		MIT
2917.923	2918.777	34260.92	DY	0.3			MIT
2913.952	2914.805	34307.61	DY	0.3	0.0		MIT
2909.325	2910.177	34362.17	DY	0.3			MIT
2906.394	2907.245	34396.82	DY	0.3	0.0		MIT
2904.68	2905.53	34417.1	DY	0.3			MIT
2900.825	2901.675	34462.85	DY	0.3			MIT
2891.026	2891.874	34579.66	DY	0.5	0.0		MIT
2890.746	2891.594	34583.01	DY	0.3	0.0		MIT
2890.444	2891.291	34586.62	DY	0.3			MIT
2888.54	2889.39	34609.4	DY	0.3			EX
2885.50	2886.35	34645.9	DY	0.3	0.0		MIT
2884.82	2885.67	34654.0	DY	0.3	0.0H		MIT
2884.28	2885.13	34660.5	DY	0.3H	0.0H		MIT
2881.59	2882.44	34692.9	DY	0.3H			MIT
2881.058	2881.903	34699.29	DY	0.3	0.0H		MIT
2878.703	2879.548	34727.68	DY	0.3	0.0H		MIT
2877.885	2878.729	34737.55	DY	0.3	0.0H		MIT
2876.392	2877.236	34755.58	DY	0.3			MIT
2875.190	2876.034	34770.11	DY	0.3			MIT
2867.62	2868.46	34861.9	DY	0.3			MIT
2866.271	2867.113	34878.30	DY	0.3			MIT
2862.695	2863.536	34921.86	DY	0.3			MIT
2860.70	2861.54	34946.2	DY	0.3			MIT
2860.172	2861.012	34952.67	DY	0.3			MIT
2859.727	2860.567	34958.11	DY	0.3			MIT
2856.42	2857.26	34998.6	DY	0.3			MIT
2852.129	2852.967	35051.23	DY	0.7			MIT
2837.613	2838.447	35230.53	DY	0.3			MIT
2836.994	2837.828	35238.21	DY	0.3H			MIT
2835.136	2835.970	35261.31	DY	0.3			MIT
2828.376	2829.208	35345.58	DY	0.3			MIT
2825.431	2826.262	35382.42	DY	0.3			MIT
2816.395	2817.224	35495.93	DY	0.7	0.3		MIT
2815.231	2816.060	35510.61	DY	0.3			MIT
2811.427	2812.255	35558.65	DY	0.5	0.0H		MIT
2810.855	2811.683	35565.89	DY	0.5			MIT
2802.70	2803.53	35669.4	DY	0.3			MIT
2801.410	2802.236	35685.79	DY	0.5	0.0H		MIT
2800.33	2801.16	35699.6	DY	0.5	0.0H		MIT
2799.72	2800.55	35707.3	DY	0.3H			MIT
2798.86	2799.68	35718.3	DY	0.3			MIT
2781.573	2782.394	35940.28	DY	0.3			MIT
2772.61	2773.43	36056.5	DY	0.5			MIT
2771.225	2772.043	36074.47	DY	0.3			MIT
2766.507	2767.324	36135.99	DY	0.3	0.0		MIT
2757.089	2757.904	36259.42	DY	0.3			MIT
2755.76	2756.57	36276.9	DY	0.8			MIT
2740.71	2741.52	36476.1	DY	0.5			MIT
2711.09	2711.89	36874.6	DY	0.3			ED
2692.85	2693.65	37124.4	DY	0.6			MIT
2679.87	2680.67	37304.2	DY	0.3			MIT
2676.84	2677.64	37346.4	DY	0.5			MIT

AIR WAVELENGTH (ANGSTRMS)	VACUUM WAVELENGTH (ANGSTRMS)	WAVENUMBER (KAYSERS)	LMENT	LOG ARC	INTENSITY SPK	INTENSITY DIS	REF
2668.07	2668.86	37469.1	DY	0.3			MIT
2645.32	2646.11	37791.4	DY	0.5			ED
2634.811	2635.596	37942.08	DY	1.6	1.3		MIT
2626.86	2627.64	38056.9	DY	0.3			MIT
2623.71	2624.49	38102.6	DY	0.7			MIT
2608.67	2609.45	38322.3	DY	0.5			ED
2600.77	2601.55	38438.7	DY	0.6			MIT
2600.17	2600.95	38447.5	DY	0.7			MIT
2598.97	2599.75	38465.3	DY	0.5			MIT
2592.56	2593.34	38560.4	DY	0.5			MIT
2591.51	2592.29	38576.0	DY	0.7			ED
2585.32	2586.09	38668.4	DY	0.6			MIT
2579.97	2580.74	38748.5	DY	0.5			MIT
2573.35	2574.12	38848.2	DY	0.5			MIT
2572.52	2573.29	38860.7	DY	0.5			MIT
2571.25	2572.02	38879.9	DY	0.5			MIT
2566.27	2567.04	38955.4	DY	0.8			MIT
2560.20	2560.97	39047.7	DY	0.8			MIT
2557.94	2558.71	39082.2	DY	0.9			ED
2552.29	2553.06	39168.7	DY	0.9			MIT
2545.13	2545.89	39278.9	DY	0.7			MIT
2543.82	2544.58	39299.2	DY	1.0			MIT
2532.33	2533.09	39477.5	DY	0.5			MIT
2528.49	2529.25	39537.4	DY	0.6			MIT
2524.12	2524.88	39605.9	DY	0.3			MIT
2517.61	2518.37	39708.3	DY	1.0			MIT
2516.11	2516.87	39731.9	DY	0.7			MIT
2514.34	2515.10	39759.9	DY	0.5			MIT
2513.57	2514.33	39772.1	DY	0.7			MIT
2510.32	2511.08	39823.6	DY	0.3			MIT
2490.62	2491.37	40138.5	DY	0.5			MIT
2485.12	2485.87	40227.4	DY	0.3			MIT
2480.92	2481.67	40295.5	DY	0.7			MIT
2471.41	2472.16	40450.5	DY	0.5			MIT
2460.00	2460.74	40638.1	DY	0.3			MIT
2455.15	2455.89	40718.4	DY	0.5			MIT
2444.151	2444.892	40901.61	DY	0.3			MIT
2438.86	2439.60	40990.3	DY	0.3			ED
2436.93	2437.67	41022.8	DY	0.5			MIT
2436.12	2436.86	41036.4	DY	0.6			MIT
2422.75	2423.49	41262.9	DY	0.5			MIT
2410.01	2410.74	41481.0	DY	0.7			MIT
2387.36	2388.09	41874.5	DY	0.9			MIT
2382.85	2383.58	41953.8	DY	0.3			MIT
2381.97	2382.70	41969.3	DY	0.9			MIT
2373.83	2374.55	42113.2	DY	0.3			MIT
2370.27	2370.99	42176.4	DY	0.3			MIT
2362.62	2363.34	42313.0	DY	0.5			MIT
2357.40	2358.12	42406.6	DY	0.8			MIT
2356.92	2357.64	42415.3	DY	0.9			MIT
2352.60	2353.32	42493.2	DY	0.6			MIT
2352.08	2352.80	42502.5	DY	0.5			MIT
2349.63	2350.35	42546.9	DY	0.5			MIT
2325.31	2326.02	42991.8	DY	0.6			ED
2319.73	2320.44	43095.2	DY	0.5			MIT
7937.73	7939.91	12594.6	ER	0.5			ED
7921.78	7923.96	12619.9	ER	0.7			ED
7797.42	7799.57	12821.2	ER	0.5			ED
7680.00	7682.11	13017.2	ER	0.3			ED
7654.43	7656.54	13060.7	ER	0.6			ED
7490.18	7492.24	13347.1	ER	0.9			ED
7469.46	7471.52	13384.2	ER	1.0			ED
7459.53	7461.58	13402.0	ER	0.6			ED
7362.59	7364.62	13578.4	ER	0.6			ED
7355.34	7357.37	13591.8	ER	0.6			ED
7316.29	7318.31	13664.4	ER	0.8			ED
7196.99	7198.97	13890.9	ER	0.6			MIT
7145.18	7147.15	13991.6	ER	0.6			ED
7135.69	7137.66	14010.2	ER	0.8			ED
7101.23	7103.19	14078.2	ER	0.6			ED

AIR WAVELENGTH (ANGSTRMS)	VACUUM WAVELENGTH (ANGSTRMS)	WAVENUMBER (KAYSERS)	LMENT	LOG ARC	INTENSITY SPK	INTENSITY DIS	REF
7001.44	7003.37	14278.8	ER	0.8			ED
6994.39	6996.32	14293.2	ER	0.8			MIT
6973.03	6974.95	14337.0	ER	0.6			MIT
6951.87	6953.79	14380.7	ER	0.8			MIT
6944.95	6946.87	14395.0	ER	0.8			MIT
6938.36	6940.27	14408.7	ER	0.8			MIT
6926.08	6927.99	14434.2	ER	0.8			MIT
6908.26	6910.17	14471.4	ER	0.6			MIT
6897.53	6899.43	14493.9	ER	0.8			MIT
6895.70	6897.60	14497.8	ER	0.8			MIT
6884.13	6886.03	14522.2	ER	0.6			MIT
6880.01	6881.91	14530.8	ER	0.8			ED
6865.20	6867.09	14562.2	ER	0.6			ED
6825.99	6827.87	14645.8	ER	0.6			ED
6825.46	6827.34	14647.0	ER	0.8			ED
6796.93	6798.81	14708.5	ER	0.8			ED
6779.86	6781.73	14745.5	ER	0.6			ED
6776.14	6778.01	14753.6	ER	0.6			ED
6774.08	6775.95	14758.1	ER	0.8			MIT
6773.40	6775.27	14759.6	ER	0.6			ED
6768.94	6770.81	14769.3	ER	0.6			ED
6762.96	6764.83	14782.3	ER	0.6			ED
6761.69	6763.56	14785.1	ER	0.6			ED
6759.87	6761.74	14789.1	ER	0.9			ED
6722.76	6724.62	14870.7	ER	0.8			ED
6721.93	6723.79	14872.6	ER	0.9			ED
6687.12	6688.97	14950.0	ER	0.6			ED
6637.62	6639.45	15061.5	ER	0.6			ED
6616.75	6618.58	15109.0	ER	0.9			ED
6612.17	6614.00	15119.4	ER	0.6			MIT
6604.97	6606.79	15135.9	ER	0.6			ED
6601.10	6602.92	15144.8	ER	1.1			ED
6593.52	6595.34	15162.2	ER	0.6			ED
6583.46	6585.28	15185.4	ER	1.0			ED
6557.78	6559.59	15244.8	ER	0.8			ED
6556.31	6558.12	15248.3	ER	0.8			ED
6541.53	6543.34	15282.7	ER	0.8			ED
6522.90	6524.70	15326.4	ER	0.6			ED
6520.51	6522.31	15332.0	ER	0.8			ED
6519.71	6521.51	15333.9	ER	0.6			ED
6518.92	6520.72	15335.7	ER	0.6			ED
6503.63	6505.43	15371.8	ER	0.6			ED
6492.352	6494.146	15398.48	ER	1.1			MIT
6489.15	6490.94	15406.1	ER	0.9			ED
6485.87	6487.66	15413.9	ER	0.6			ED
6481.77	6483.56	15423.6	ER	0.6			ED
6460.27	6462.06	15474.9	ER	0.8			ED
6454.02	6455.80	15489.9	ER	0.6			ED
6441.31	6443.09	15520.5	ER	1.0			ED
6432.50	6434.28	15541.8	ER	0.6			ED
6425.574	6427.350	15558.51	ER	0.6			MIT
6423.10	6424.88	15564.5	ER	0.8			ED
6418.88	6420.65	15574.7	ER	0.6			ED
6415.51	6417.28	15582.9	ER	0.6			ED
6413.59	6415.36	15587.6	ER	0.8			ED
6410.327	6412.099	15595.52	ER	0.6			MIT
6408.423	6410.194	15600.15	ER	0.6			MIT
6405.54	6407.31	15607.2	ER	0.6			ED
6399.79	6401.56	15621.2	ER	0.6			ED
6398.13	6399.90	15625.2	ER	0.8			ED
6388.194	6389.960	15649.55	ER	1.1			MIT
6384.04	6385.80	15659.7	ER	0.6			ED
6375.57	6377.33	15680.5	ER	0.6			ED
6374.49	6376.25	15683.2	ER	0.6			ED
6372.69	6374.45	15687.6	ER	0.6			ED
6371.632	6373.394	15690.23	ER	0.6			MIT
6366.572	6368.332	15702.70	ER	0.6			MIT
6361.80	6363.56	15714.5	ER	0.6			ED
6351.57	6353.33	15739.8	ER	0.8			ED
6347.17	6348.92	15750.7	ER	0.9			ED

AIR WAVELENGTH (ANGSTRMS)	VACUUM WAVELENGTH (ANGSTRMS)	WAVENUMBER (KAYSERS)	LMENT	LOG ARC	INTENSITY SPK	DIS	REF
6340.66	6342.41	15766.9	ER	0.6			ED
6332.24	6333.99	15787.8	ER	0.6			ED
6326.11	6327.86	15803.1	ER	1.0			ED
6320.63	6322.38	15816.8	ER	0.6			ED
6319.17	6320.92	15820.5	ER	0.6			ED
6308.79	6310.53	15846.5	ER	1.2			ED
6299.41	6301.15	15870.1	ER	1.0			ED
6288.61	6290.35	15897.4	ER	0.6			ED
6286.86	6288.60	15901.8	ER	0.9			ED
6281.40	6283.14	15915.6	ER	0.6			ED
6274.96	6276.70	15931.9	ER	0.9			ED
6271.63	6273.36	15940.4	ER	0.8			ED
6267.94	6269.67	15949.8	ER	0.8			ED
6262.56	6264.29	15963.5	ER	0.9			ED
6221.01	6222.73	16070.1	ER	1.1			ED
6213.67	6215.39	16089.1	ER	0.6			ED
6209.29	6211.01	16100.4	ER	0.6			ED
6204.63	6206.35	16112.5	ER	0.6			ED
6183.20	6184.91	16168.4	ER	0.8			ED
6170.05	6171.76	16202.8	ER	0.9			ED
6140.33	6142.03	16281.3	ER	0.6			ED
6132.23	6133.93	16302.8	ER	0.6			ED
6126.90	6128.60	16316.9	ER	0.6			ED
6125.32	6127.02	16321.2	ER	0.8			ED
6116.004	6117.697	16346.02	ER	0.8			MIT
6115.38	6117.07	16347.7	ER	0.6			ED
6105.20	6106.89	16374.9	ER	0.8			ED
6097.929	6099.617	16394.47	ER	0.6			MIT
6079.64	6081.32	16443.8	ER	0.6			ED
6076.44	6078.12	16452.4	ER	1.0			ED
6067.50	6069.18	16476.7	ER	0.6			ED
6061.26	6062.94	16493.6	ER	0.8			ED
6054.84	6056.52	16511.1	ER	0.8			ED
6052.79	6054.47	16516.7	ER	0.6			ED
6048.13	6049.80	16529.5	ER	0.8			ED
6045.65	6047.32	16536.2	ER	0.8			ED
6032.14	6033.81	16573.3	ER	0.8			ED
6022.551	6024.219	16599.66	ER	0.9			MIT
6015.76	6017.43	16618.4	ER	0.6			ED
6014.837	6016.503	16620.95	ER	0.9			MIT
6008.755	6010.419	16637.77	ER	0.6			MIT
6007.954	6009.618	16639.99	ER	0.6			MIT
6006.81	6008.47	16643.2	ER	1.0			M
5989.88	5991.54	16690.2	ER	0.9			ED
5975.50	5977.16	16730.4	ER	0.9			ED
5970.903	5972.557	16743.25	ER	0.9			MIT
5968.70	5970.35	16749.4	ER	1.1			ED
5966.76	5968.41	16754.9	ER	0.9			ED
5958.942	5960.593	16776.85	ER	1.3			MIT
5956.07	5957.72	16784.9	ER	0.9			ED
5955.49	5957.14	16786.6	ER	0.9			ED
5948.70	5950.35	16805.7	ER	1.1			ED
5946.36	5948.01	16812.4	ER	0.9			ED
5937.20	5938.84	16838.3	ER	1.1			ED
5934.320	5935.964	16846.46	ER	0.9			MIT
5933.495	5935.139	16848.80	ER	0.9			MIT
5916.458	5918.097	16897.32	ER	0.9			MIT
5914.12	5915.76	16904.0	ER	0.9			ED
5909.25	5910.89	16917.9	ER	1.3			ED
5906.07	5907.71	16927.0	ER	1.3			ED
5902.09	5903.73	16938.5	ER	1.5			ED
5886.458	5888.089	16983.44	ER	1.5			MIT
5886.30	5887.93	16983.9	ER	1.1			ED
5881.14	5882.77	16998.8	ER	1.5			ED
5873.52	5875.15	17020.9	ER	0.9			ED
5872.338	5873.966	17024.27	ER	1.3			ED
5856.949	5858.572	17069.00	ER	0.9			MIT
5855.294	5856.917	17073.83	ER	1.5			ED
5850.064	5851.686	17089.09	ER	1.3			MIT
5842.67	5844.29	17110.7	ER	0.9			ED

AIR WAVELENGTH (ANGSTRMS)	VACUUM WAVELENGTH (ANGSTRMS)	WAVENUMBER (KAYSERS)	LMENT	LOG ARC	INTENSITY SPK	DIS	REF
5841.105	5842.724	17115.30	ER	0.9			MIT
5835.826	5837.444	17130.79	ER	0.9			MIT
5833.945	5835.562	17136.31	ER	1.1			MIT
5826.788	5828.403	17157.36	ER	1.7			MIT
5814.28	5815.89	17194.3	ER	0.9			ED
5812.07	5813.68	17200.8	ER	0.9			ED
5806.13	5807.74	17218.4	ER	0.9			ED
5800.770	5802.378	17234.31	ER	1.1			MIT
5791.117	5792.723	17263.04	ER	1.3			MIT
5788.91	5790.52	17269.6	ER	0.9			ED
5788.63	5790.24	17270.5	ER	0.9			ED
5784.631	5786.235	17282.40	ER	1.1			MIT
5782.824	5784.428	17287.79	ER	1.1			MIT
5769.930	5771.530	17326.43	ER	1.3			MIT
5762.790	5764.388	17347.90	ER	1.5			MIT
5757.617	5759.214	17363.48	ER	1.5			MIT
5752.548	5754.144	17378.78	ER	1.1			MIT
5752.023	5753.618	17380.37	ER	0.6			MIT
5748.645	5750.239	17390.58	ER	1.1			MIT
5740.625	5742.217	17414.87	ER	0.9			MIT
5739.188	5740.780	17419.24	ER	1.5			MIT
5733.42	5735.01	17436.8	ER	1.1			ED
5727.020	5728.609	17456.25	ER	0.9			MIT
5719.531	5721.118	17479.10	ER	1.1			MIT
5717.82	5719.41	17484.3	ER	0.9			ED
5717.494	5719.080	17485.33	ER	1.1			MIT
5710.879	5712.463	17505.58	ER	1.3			MIT
5698.95	5700.53	17542.2	ER	0.9			ED
5695.55	5697.13	17552.7	ER	0.9			ED
5694.87	5696.45	17554.8	ER	0.9			ED
5687.35	5688.93	17578.0	ER	0.9			ED
5684.756	5686.333	17586.02	ER	0.9			MIT
5682.53	5684.11	17592.9	ER	0.9			ED
5675.821	5677.396	17613.71	ER	1.1			MIT
5675.491	5677.066	17614.73	ER	0.9			MIT
5667.66	5669.23	17639.1	ER	0.9			ED
5666.64	5668.21	17642.2	ER	0.9			ED
5665.427	5666.999	17646.02	ER	1.3			MIT
5664.946	5666.518	17647.52	ER	1.1			MIT
5660.52	5662.09	17661.3	ER	0.9			ED
5658.48	5660.05	17667.7	ER	1.1			ED
5657.18	5658.75	17671.7	ER	0.9			ED
5651.49	5653.06	17689.5	ER	0.9			ED
5642.62	5644.19	17717.4	ER	0.9			ED
5640.344	5641.910	17724.50	ER	0.9			MIT
5639.80	5641.37	17726.2	ER	1.1			ED
5632.57	5634.13	17749.0	ER	1.1			ED
5631.413	5632.976	17752.60	ER	1.1			MIT
5626.522	5628.084	17768.04	ER	1.3			MIT
5622.02	5623.58	17782.3	ER	0.5			ED
5620.74	5622.30	17786.3	ER	0.9			ED
5617.62	5619.18	17796.2	ER	0.9			ED
5611.807	5613.365	17814.63	ER	1.1			MIT
5604.59	5606.15	17837.6	ER	0.9			ED
5601.19	5602.75	17848.4	ER	1.3			ED
5594.79	5596.34	17868.8	ER	0.9			ED
5593.450	5595.003	17873.09	ER	1.3			MIT
5590.84	5592.39	17881.4	ER	0.9			ED
5585.16	5586.71	17899.6	ER	0.9			ED
5583.621	5585.171	17904.55	ER	0.9			MIT
5581.858	5583.408	17910.21	ER	0.9			MIT
5576.33	5577.88	17928.0	ER	0.9			ED
5572.46	5574.01	17940.4	ER	0.9			ED
5563.11	5564.65	17970.6	ER	0.9			ED
5556.59	5558.13	17991.6	ER	0.9			ED
5555.85	5557.39	17994.1	ER	0.9			ED
5553.16	5554.70	18002.8	ER	1.1			ED
5551.50	5553.04	18008.1	ER	0.9			ED
5540.73	5542.27	18043.1	ER	0.9			ED
5536.26	5537.80	18057.7	ER	0.9			ED

AIR WAVELENGTH (ANGSTRMS)	VACUUM WAVELENGTH (ANGSTRMS)	WAVENUMBER (KAYSERS)	LMENT	LOG ARC	INTENSITY SPK	INTENSITY DIS	REF
5533.64	5535.18	18066.3	ER	0.9			ED
5532.35	5533.89	18070.5	ER	0.9			ED
5529.80	5531.34	18078.8	ER	0.9			ED
5523.16	5524.69	18100.6	ER	0.9			ED
5518.75	5520.28	18115.0	ER	1.1			ED
5518.12	5519.65	18117.1	ER	1.1			ED
5516.82	5518.35	18121.4	ER	0.9			ED
5516.005	5517.537	18124.03	ER	0.9			MIT
5514.73	5516.26	18128.2	ER	1.1			ED
5513.40	5514.93	18132.6	ER	0.9			ED
5505.67	5507.20	18158.1	ER	1.1			ED
5503.78	5505.31	18164.3	ER	0.9			ED
5502.802	5504.331	18167.51	ER	0.9			ED
5501.61	5503.14	18171.4	ER	0.9			ED
5497.407	5498.934	18185.34	ER	0.9			MIT
5491.72	5493.25	18204.2	ER	1.1			ED
5485.93	5487.45	18223.4	ER	1.5			ED
5477.45	5478.97	18251.6	ER	1.3			ED
5475.15	5476.67	18259.3	ER	0.9			ED
5469.63	5471.15	18277.7	ER	1.1			ED
5468.320	5469.840	18282.07	ER	0.9			MIT
5466.459	5467.978	18288.30	ER	1.1			MIT
5462.451	5463.969	18301.71	ER	1.3			MIT
5458.406	5459.923	18315.28	ER	0.9			MIT
5456.618	5458.134	18321.28	ER	1.5			MIT
5455.619	5457.135	18324.63	ER	0.9			MIT
5454.268	5455.784	18329.17	ER	1.5			MIT
5451.30	5452.82	18339.1	ER	0.9			ED
5448.23	5449.74	18349.5	ER	1.1			ED
5445.660	5447.174	18358.15	ER	1.1			MIT
5444.62	5446.13	18361.6	ER	0.9			ED
5442.74	5444.25	18368.0	ER	0.9			ED
5439.83	5441.34	18377.8	ER	0.9			ED
5438.74	5440.25	18381.5	ER	1.1			ED
5436.882	5438.393	18387.78	ER	0.9			MIT
5435.41	5436.92	18392.8	ER	0.9			ED
5434.156	5435.666	18397.01	ER	1.1			MIT
5433.81	5435.32	18398.2	ER	1.1			ED
5433.36	5434.87	18399.7	ER	0.9			ED
5432.47	5433.98	18402.7	ER	1.1			ED
5429.54	5431.05	18412.6	ER	1.1			ED
5427.50	5429.01	18419.6	ER	1.1			ED
5422.80	5424.31	18435.5	ER	1.5			ED
5419.887	5421.394	18445.44	ER	1.3			MIT
5415.690	5417.196	18459.74	ER	0.9			MIT
5414.636	5416.141	18463.33	ER	1.7			MIT
5414.18	5415.69	18464.9	ER	1.1			ED
5410.54	5412.04	18477.3	ER	0.9			ED
5407.09	5408.59	18489.1	ER	0.9			ED
5400.77	5402.27	18510.7	ER	0.9			ED
5395.87	5397.37	18527.5	ER	1.7			ED
5388.72	5390.22	18552.1	ER	1.1			ED
5386.18	5387.68	18560.9	ER	1.1			ED
5385.623	5387.120	18562.79	ER	1.1			MIT
5384.93	5386.43	18565.2	ER	0.9			ED
5384.13	5385.63	18567.9	ER	0.9			ED
5383.41	5384.91	18570.4	ER	0.9			ED
5382.13	5383.63	18574.8	ER	1.1			ED
5380.35	5381.85	18581.0	ER	0.9			ED
5377.79	5379.29	18589.8	ER	0.9			ED
5376.69	5378.19	18593.6	ER	1.1			ED
5368.85	5370.34	18620.8	ER	1.3			ED
5367.762	5369.255	18624.56	ER	1.1			MIT
5360.917	5362.408	18648.34	ER	0.9			MIT
5352.21	5353.70	18678.7	ER	0.9			ED
5350.44	5351.93	18684.9	ER	1.1			ED
5348.047	5349.534	18693.22	ER	1.3			MIT
5346.93	5348.42	18697.1	ER	0.9			ED
5346.481	5347.968	18698.69	ER	0.9			MIT
5346.02	5347.51	18700.3	ER	0.9			ED

AIR WAVELENGTH (ANGSTRMS)	VACUUM WAVELENGTH (ANGSTRMS)	WAVENUMBER (KAYSERS)	LMENT	LOG ARC	INTENSITY SPK	INTENSITY DIS	REF
5344.935	5346.422	18704.10	ER	0.9			MIT
5344.507	5345.994	18705.60	ER	1.5			MIT
5343.933	5345.419	18707.61	ER	1.5			MIT
5337.133	5338.618	18731.44	ER	0.9			MIT
5336.36	5337.84	18734.1	ER	0.9			ED
5335.228	5336.712	18738.13	ER	0.9			MIT
5334.20	5335.68	18741.7	ER	1.3			ED
5333.36	5334.84	18744.7	ER	0.9			ED
5333.017	5334.500	18745.90	ER	1.1			MIT
5330.69	5332.17	18754.1	ER	1.1			ED
5318.92	5320.40	18795.6	ER	1.1			ED
5317.61	5319.09	18800.2	ER	1.1			ED
5316.45	5317.93	18804.3	ER	0.9			ED
5315.27	5316.75	18808.5	ER	0.9			ED
5313.92	5315.40	18813.3	ER	0.9			ED
5311.89	5313.37	18820.5	ER	0.9			ED
5310.03	5311.51	18827.1	ER	1.1			ED
5307.115	5308.592	18837.39	ER	1.1			MIT
5305.88	5307.36	18841.8	ER	0.9			MIT
5304.51	5305.99	18846.6	ER	1.1			ED
5303.15	5304.63	18851.5	ER	1.1			ED
5302.30	5303.78	18854.5	ER	1.5			ED
5301.26	5302.73	18858.2	ER	1.1			ED
5300.59	5302.06	18860.6	ER	0.9			ED
5298.64	5300.11	18867.5	ER	1.1			ED
5292.37	5293.84	18889.9	ER	0.9			ED
5288.60	5290.07	18903.3	ER	1.1			ED
5286.04	5287.51	18912.5	ER	1.1			ED
5285.61	5287.08	18914.0	ER	0.9			ED
5284.02	5285.49	18919.7	ER	0.9			ED
5282.65	5284.12	18924.6	ER	1.1			ED
5279.33	5280.80	18936.5	ER	1.5			ED
5278.91	5280.38	18938.0	ER	0.9			ED
5277.70	5279.17	18942.4	ER	1.1			ED
5276.90	5278.37	18945.2	ER	0.9			ED
5276.03	5277.50	18948.4	ER	1.3			ED
5273.86	5275.33	18956.2	ER	1.1			ED
5272.90	5274.37	18959.6	ER	1.3			MIT
5270.54	5272.01	18968.1	ER	0.9			ED
5265.01	5266.48	18988.0	ER	1.1			ED
5264.47	5265.94	18990.0	ER	1.1			ED
5261.40	5262.86	19001.1	ER	0.9			ED
5260.26	5261.72	19005.2	ER	0.9			ED
5257.480	5258.943	19015.23	ER	1.1			MIT
5257.02	5258.48	19016.9	ER	1.3			MIT
5255.955	5257.418	19020.74	ER	1.7			MIT
5253.38	5254.84	19030.1	ER	0.9			ED
5250.38	5251.84	19040.9	ER	0.9			ED
5248.677	5250.138	19047.12	ER	1.1			MIT
5247.71	5249.17	19050.6	ER	0.9			ED
5246.12	5247.58	19056.4	ER	1.3			MIT
5243.52	5244.98	19065.9	ER	1.1			ED
5239.68	5241.14	19079.8	ER	0.9			ED
5239.20	5240.66	19081.6	ER	0.9			ED
5237.75	5239.21	19086.9	ER	0.9			ED
5233.42	5234.88	19102.6	ER	0.9			ED
5229.320	5230.776	19117.62	ER	1.5			MIT
5226.06	5227.51	19129.6	ER	1.1			ED
5218.25	5219.70	19158.2	ER	1.5			MIT
5215.58	5217.03	19168.0	ER	0.9			ED
5215.13	5216.58	19169.6	ER	1.3			MIT
5214.36	5215.81	19172.5	ER	0.9			ED
5212.922	5214.373	19177.76	ER	1.3			MIT
5211.63	5213.08	19182.5	ER	0.6			MIT
5211.16	5212.61	19184.2	ER	1.1			ED
5210.28	5211.73	19187.5	ER	1.1			ED
5206.519	5207.969	19201.34	ER	1.3			MIT
5191.657	5193.103	19256.31	ER	0.9			MIT
5188.888	5190.333	19266.59	ER	1.5			MIT
5180.961	5182.404	19296.06	ER	0.9			MIT

AIR WAVELENGTH (ANGSTRMS)	VACUUM WAVELENGTH (ANGSTRMS)	WAVENUMBER (KAYSERS)	LMENT	LOG ARC	INTENSITY SPK	DIS	REF
5179.48	5180.92	19301.6	ER	0.9			ED
5172.761	5174.202	19326.65	ER	1.1			MIT
5164.761	5166.200	19356.59	ER	1.5			MIT
5163.845	5165.283	19360.02	ER	1.1			MIT
5160.322	5161.759	19373.24	ER	1.1			MIT
5149.37	5150.80	19414.4	ER	1.1			ED
5144.900	5146.333	19431.31	ER	0.9			MIT
5144.132	5145.565	19434.21	ER	1.1			MIT
5143.576	5145.009	19436.31	ER	1.1			MIT
5133.827	5135.257	19473.22	ER	1.6			MIT
5133.02	5134.45	19476.3	ER	0.9			MIT
5131.524	5132.954	19481.96	ER	0.9			MIT
5127.397	5128.826	19497.64	ER	1.4			MIT
5124.568	5125.996	19508.40	ER	0.9			MIT
5119.638	5121.065	19527.19	ER	0.9			MIT
5115.47	5116.90	19543.1	ER	0.9			ED
5114.037	5115.462	19548.58	ER	0.3W			MIT
5101.133	5102.555	19598.03	ER	0.9			MIT
5093.755	5095.175	19626.41	ER	0.9			MIT
5092.224	5093.643	19632.31	ER	0.3			MIT
5090.272	5091.691	19639.84	ER	0.9			MIT
5080.51	5081.93	19677.6	ER	0.9			ED
5077.657	5079.072	19688.63	ER	1.1			MIT
5072.84	5074.25	19707.3	ER	0.9			MIT
5070.33	5071.74	19717.1	ER	1.1			MIT
5066.701	5068.114	19731.21	ER	0.6			MIT
5063.396	5064.808	19744.09	ER	0.9			MIT
5060.884	5062.295	19753.89	ER	0.8			MIT
5060.423	5061.834	19755.69	ER	0.6			MIT
5058.623	5060.033	19762.72	ER	0.9			MIT
5045.99	5047.40	19812.2	ER	1.1			ED
5044.882	5046.289	19816.54	ER	0.9			MIT
5043.844	5045.250	19820.62	ER	1.1			MIT
5042.038	5043.444	19827.72	ER	1.8			MIT
5035.934	5037.338	19851.75	ER	1.1			MIT
5034.246	5035.650	19858.41	ER	0.8			MIT
5028.903	5030.305	19879.51	ER	1.3			MIT
5028.309	5029.711	19881.86	ER	1.1			MIT
5025.787	5027.189	19891.83	ER	0.6			MIT
5024.266	5025.667	19897.85	ER	1.1			MIT
5017.65	5019.05	19924.1	ER	0.9			ED
5013.76	5015.16	19939.6	ER	0.9			MIT
5009.761	5011.158	19955.47	ER	0.8			MIT
5008.969	5010.366	19958.62	ER	1.1			MIT
5007.238	5008.635	19965.52	ER	1.2			MIT
5000.379	5001.774	19992.91	ER	1.2			MIT
4992.346	4993.739	20025.08	ER	0.3			MIT
4980.71	4982.10	20071.9	ER	0.3			MIT
4976.385	4977.774	20089.30	ER	0.3			MIT
4974.52	4975.91	20096.8	ER	0.5			MIT
4966.903	4968.289	20127.65	ER	1.2			MIT
4966.635	4968.021	20128.74	ER	0.3			MIT
4953.578	4954.960	20181.80	ER	0.5			MIT
4951.739	4953.121	20189.29	ER	1.2			MIT
4950.657	4952.039	20193.70	ER	0.7			MIT
4944.358	4945.738	20219.43	ER	0.5			MIT
4941.846	4943.225	20229.71	ER	0.3			MIT
4935.498	4936.876	20255.73	ER	1.5			MIT
4934.074	4935.451	20261.57	ER	1.3			MIT
4931.808	4933.185	20270.88	ER	0.5			MIT
4928.849	4930.225	20283.05	ER	0.3			MIT
4927.359	4928.734	20289.18	ER	0.3			MIT
4926.966	4928.341	20290.80	ER	0.3			MIT
4925.435	4926.810	20297.11	ER	0.3			MIT
4917.053	4918.426	20331.71	ER	0.5			MIT
4912.034	4913.405	20352.48	ER	0.3			MIT
4906.023	4907.393	20377.42	ER	1.3			MIT
4905.215	4906.585	20380.78	ER	0.3			MIT
4904.432	4905.801	20384.03	ER	0.3			MIT
4903.619	4904.988	20387.41	ER	0.7			MIT

AIR WAVELENGTH (ANGSTRMS)	VACUUM WAVELENGTH (ANGSTRMS)	WAVENUMBER (KAYSERS)	LMENT	LOG ARC	INTENSITY SPK	DIS	REF
4901.620	4902.989	20395.72	ER	0.3			MIT
4900.108	4901.476	20402.02	ER	1.5	1.3		MIT
4898.161	4899.529	20410.13	ER	0.5S			MIT
4896.958	4898.325	20415.14	ER	0.3			MIT
4891.67	4893.04	20437.2	ER	0.5			MIT
4888.874	4890.239	20448.90	ER	0.5			MIT
4886.299	4887.664	20459.67	ER	1.2			MIT
4881.054	4882.417	20481.66	ER	0.3W			MIT
4879.896	4881.259	20486.52	ER	0.6	0.0		MIT
4878.33	4879.69	20493.1	ER	0.6	0.0		MIT
4872.479	4873.840	20517.70	ER	0.8	0.0		MIT
4872.099	4873.460	20519.30	ER	1.4	0.5		MIT
4870.44	4871.80	20526.3	ER	0.3			ED
4864.534	4865.893	20551.21	ER	0.5	0.0H		MIT
4863.61	4864.97	20555.1	ER	0.3			M
4861.589	4862.947	20563.66	ER	0.7	0.0		MIT
4858.468	4859.825	20576.87	ER	0.9	0.5		MIT
4857.425	4858.782	20581.29	ER	1.0			MIT
4856.709	4858.066	20584.32	ER	1.1			MIT
4854.857	4856.213	20592.18	ER	1.6S			MIT
4854.425	4855.781	20594.01	ER	0.3	0.0		MIT
4853.117	4854.473	20599.56	ER	0.8			MIT
4852.679	4854.035	20601.42	ER	1.7			MIT
4851.635	4852.990	20605.85	ER	0.8			MIT
4848.822	4850.177	20617.81	ER	1.0			MIT
4847.634	4848.988	20622.86	ER	0.3			MIT
4846.643	4847.997	20627.07	ER	0.3			MIT
4845.655	4847.009	20631.28	ER	1.7			MIT
4844.045	4845.398	20638.14	ER	0.3			MIT
4842.98	4844.33	20642.7	ER	0.3			ED
4842.030	4843.383	20646.73	ER	0.6			MIT
4840.462	4841.814	20653.41	ER	0.3			MIT
4839.531	4840.883	20657.39	ER	0.5			MIT
4838.961	4840.313	20659.82	ER	0.5			MIT
4837.835	4839.187	20664.63	ER	0.5			MIT
4837.460	4838.812	20666.23	ER	0.6			MIT
4836.208	4837.559	20671.58	ER	0.7			MIT
4834.737	4836.088	20677.87	ER	0.5			MIT
4831.163	4832.513	20693.17	ER	1.4			MIT
4830.723	4832.073	20695.05	ER	0.7			MIT
4828.562	4829.911	20704.31	ER	0.3			MIT
4828.062	4829.411	20706.46	ER	0.5			MIT
4827.243	4828.592	20709.97	ER	0.5			MIT
4825.207	4826.555	20718.71	ER	0.7			MIT
4824.561	4825.909	20721.48	ER	0.5			MIT
4823.297	4824.645	20726.91	ER	1.2	1.2		MIT
4822.119	4823.467	20731.98	ER	1.0	0.3		MIT
4820.747	4822.094	20737.88	ER	0.5			MIT
4820.344	4821.691	20739.61	ER	1.4	0.3		MIT
4818.171	4819.518	20748.97	ER	1.0			MIT
4817.769	4819.115	20750.70	ER	0.3			MIT
4817.369	4818.715	20752.42	ER	1.1			MIT
4816.620	4817.966	20755.65	ER	0.3			MIT
4811.473	4812.818	20777.85	ER	0.3			MIT
4804.792	4806.135	20806.74	ER	1.2	0.0		MIT
4804.218	4805.561	20809.23	ER	1.0			MIT
4802.760	4804.102	20815.54	ER	0.3			MIT
4795.504	4796.844	20847.04	ER	1.0	0.0		MIT
4786.603	4787.941	20885.80	ER	0.3	0.3		MIT
4784.676	4786.014	20894.22	ER	0.3			MIT
4781.849	4783.186	20906.57	ER	0.5			MIT
4781.019	4782.356	20910.20	ER	1.5	0.3		MIT
4779.743	4781.079	20915.78	ER	0.3			MIT
4778.500	4779.836	20921.22	ER	0.5			MIT
4775.740	4777.075	20933.31	ER	0.6			MIT
4775.215	4776.550	20935.61	ER	0.3			MIT
4773.429	4774.764	20943.45	ER	0.3			MIT
4766.065	4767.398	20975.80	ER	0.6			MIT
4764.638	4765.970	20982.09	ER	0.3			MIT
4762.643	4763.975	20990.87	ER	0.7	0.0		MIT

AIR WAVELENGTH (ANGSTRMS)	VACUUM WAVELENGTH (ANGSTRMS)	WAVENUMBER (KAYSERS)	LMENT	LOG ARC	INTENSITY SPK	DIS	REF
4761.02	4762.35	20998.0	ER	0.7	0.7		M
4759.651	4760.982	21004.07	ER	1.2	0.3		MIT
4757.116	4758.446	21015.26	ER	0.3			MIT
4754.659	4755.989	21026.12	ER	0.3W			MIT
4752.772	4754.101	21034.47	ER	1.3	0.3		MIT
4751.526	4752.855	21039.99	ER	1.1	0.0		MIT
4747.078	4748.406	21059.70	ER	0.5			MIT
4745.283	4746.610	21067.67	ER	0.3			MIT
4743.104	4744.431	21077.34	ER	0.3			MIT
4742.549	4743.875	21079.81	ER	0.5W			MIT
4741.997	4743.323	21082.26	ER	0.5W			MIT
4741.398	4742.724	21084.93	ER	1.3			MIT
4736.945	4738.270	21104.75	ER	0.5			MIT
4734.032	4735.356	21117.74	ER	0.3			MIT
4733.337	4734.661	21120.84	ER	0.8			MIT
4732.242	4733.566	21125.72	ER	0.5			MIT
4731.586	4732.910	21128.65	ER	0.9	0.0		MIT
4730.800	4732.123	21132.16	ER	0.3			MIT
4729.028	4730.351	21140.08	ER	0.9			MIT
4726.076	4727.398	21153.28	ER	1.1	0.7		MIT
4724.517	4725.839	21160.26	ER	0.9	0.3		MIT
4723.245	4724.566	21165.96	ER	0.5L			MIT
4722.703	4724.024	21168.39	ER	1.1			MIT
4722.020	4723.341	21171.46	ER	0.6			MIT
4721.065	4722.386	21175.74	ER	0.7			MIT
4719.766	4721.086	21181.56	ER	0.3			MIT
4719.31	4720.63	21183.6	ER	0.6			ED
4718.715	4720.035	21186.28	ER	0.3			MIT
4715.951	4717.270	21198.70	ER	0.5			MIT
4714.995	4716.314	21203.00	ER	0.6	0.0		MIT
4713.769	4715.088	21208.51	ER	0.6			MIT
4713.060	4714.379	21211.70	ER	0.3			MIT
4709.78	4711.10	21226.5	ER	0.5	0.0		M
4706.730	4708.047	21240.23	ER	0.3			MIT
4705.808	4707.125	21244.39	ER	0.8D			MIT
4705.305	4706.622	21246.66	ER	0.6			MIT
4704.493	4705.809	21250.33	ER	0.3			MIT
4703.768	4705.084	21253.60	ER	0.5			MIT
4703.054	4704.370	21256.83	ER	0.3			MIT
4702.167	4703.483	21260.84	ER	0.5	0.3		MIT
4700.783	4702.098	21267.10	ER	0.6			MIT
4700.272	4701.587	21269.41	ER	0.6			MIT
4699.710	4701.025	21271.96	ER	0.3			MIT
4699.227	4700.542	21274.14	ER	0.7L			MIT
4698.22	4699.53	21278.7	ER	0.5			M
4697.173	4698.487	21283.45	ER	0.6	0.3		MIT
4691.097	4692.410	21311.01	ER	0.3			MIT
4688.63	4689.94	21322.2	ER	0.7	0.3		M
4687.835	4689.147	21325.84	ER	0.5			MIT
4684.592	4685.903	21340.60	ER	0.3			MIT
4681.944	4683.254	21352.67	ER	0.6D			MIT
4681.558	4682.868	21354.43	ER	0.5			MIT
4679.070	4680.380	21365.79	ER	0.8	0.3		MIT
4675.622	4676.931	21381.54	ER	1.2	0.6		MIT
4674.849	4676.158	21385.08	ER	1.7	1.2		MIT
4671.582	4672.890	21400.03	ER	0.5	0.0		MIT
4671.092	4672.400	21402.28	ER	0.3	0.0		MIT
4670.095	4671.402	21406.85	ER	0.3			MIT
4669.518	4670.825	21409.49	ER	0.5			MIT
4667.59	4668.90	21418.3	ER	0.5	0.3		M
4666.14	4667.45	21425.0	ER	0.3			ED
4665.42	4666.73	21428.3	ER	0.8	0.3		MIT
4664.235	4665.541	21433.74	ER	0.3			MIT
4663.235	4664.540	21438.34	ER	0.3			MIT
4661.390	4662.695	21446.82	ER	0.3			MIT
4661.225	4662.530	21447.58	ER	0.5			MIT
4659.96	4661.26	21453.4	ER	0.5V			M
4658.47	4659.77	21460.3	ER	0.3			ED
4657.044	4658.348	21466.84	ER	0.3	0.0		MIT
4656.677	4657.981	21468.53	ER	0.5	0.3		MIT

AIR WAVELENGTH (ANGSTRMS)	VACUUM WAVELENGTH (ANGSTRMS)	WAVENUMBER (KAYSERS)	LMENT	LOG ARC	INTENSITY SPK	DIS	REF
4655.905	4657.209	21472.09	ER	0.5B			MIT
4655.119	4656.422	21475.72	ER	0.6			MIT
4653.309	4654.612	21484.07	ER	0.5			MIT
4652.015	4653.317	21490.04	ER	0.5	0.3		MIT
4651.40	4652.70	21492.9	ER	0.3			MIT
4650.81	4652.11	21495.6	ER	0.3			M
4650.345	4651.647	21497.76	ER	0.3			MIT
4649.08	4650.38	21503.6	ER	0.3			ED
4646.825	4648.126	21514.05	ER	0.5			MIT
4645.86	4647.16	21518.5	ER	0.3			MIT
4644.37	4645.67	21525.4	ER	0.5			ED
4643.691	4644.991	21528.57	ER	1.7	1.2		MIT
4643.19	4644.49	21530.9	ER	0.3			M
4640.606	4641.905	21542.88	ER	0.6	0.3		MIT
4637.85	4639.15	21555.7	ER	0.3	0.0		M
4634.27	4635.57	21572.3	ER	0.3	0.0		M
4632.36	4633.66	21581.2	ER	0.3			M
4631.58	4632.88	21584.9	ER	0.3	0.0		M
4630.885	4632.182	21588.10	ER	1.2	0.5		MIT
4630.58	4631.88	21589.5	ER	0.3			ED
4628.47	4629.77	21599.4	ER	0.3			ED
4627.26	4628.56	21605.0	ER	0.3			ED
4626.60	4627.90	21608.1	ER	0.3			EX
4626.28	4627.58	21609.6	ER	0.3	0.0W		M
4625.56	4626.86	21612.9	ER	0.5	0.3		M
4625.14	4626.44	21614.9	ER	0.3			ED
4624.78	4626.08	21616.6	ER	0.5	0.0		M
4623.02	4624.31	21624.8	ER	0.3			ED
4622.53	4623.82	21627.1	ER	0.5			ED
4621.58	4622.87	21631.6	ER	0.3			M
4620.40	4621.69	21637.1	ER	0.3			ED
4619.03	4620.32	21643.5	ER	0.3			ED
4618.33	4619.62	21646.8	ER	0.3			M
4617.64	4618.93	21650.0	ER	0.3			ED
4615.91	4617.20	21658.1	ER	0.6	0.5		M
4613.95	4615.24	21667.3	ER	0.3			ED
4611.80	4613.09	21677.4	ER	0.5			ED
4611.27	4612.56	21679.9	ER	0.6	0.3		M
4610.26	4611.55	21684.7	ER	0.3			ED
4609.43	4610.72	21688.6	ER	0.5			ED
4608.784	4610.075	21691.62	ER	0.3			MIT
4607.89	4609.18	21695.8	ER	0.3			ED
4606.613	4607.903	21701.84	ER	1.3	0.0		MIT
4604.890	4606.180	21709.96	ER	0.6			MIT
4603.73	4605.02	21715.4	ER	0.5	0.0		M
4602.06	4603.35	21723.3	ER	0.3	0.0		MIT
4601.10	4602.39	21727.9	ER	0.6			ED
4598.130	4599.418	21741.88	ER	0.3	0.0		MIT
4596.73	4598.02	21748.5	ER	0.6			ED
4594.01	4595.30	21761.4	ER	0.3			M
4592.93	4594.22	21766.5	ER	0.5			ED
4590.77	4592.06	21776.7	ER	0.6	0.0		M
4589.30	4590.59	21783.7	ER	0.5	0.3		M
4588.14	4589.43	21789.2	ER	0.5			M
4586.848	4588.133	21795.36	ER	0.5			MIT
4584.855	4586.140	21804.83	ER	0.9			MIT
4581.70	4582.98	21819.9	ER	0.8			M
4581.070	4582.354	21822.85	ER	0.3			MIT
4577.804	4579.087	21838.42	ER	0.9			MIT
4575.08	4576.36	21851.4	ER	0.3			ED
4572.98	4574.26	21861.4	ER	0.5W	0.0		M
4572.51	4573.79	21863.7	ER	0.3			M
4571.88	4573.16	21866.7	ER	0.3			M
4569.63	4570.91	21877.5	ER	0.6O			M
4569.330	4570.611	21878.92	ER	0.7			MIT
4568.827	4570.107	21881.32	ER	0.6			MIT
4567.740	4569.020	21886.53	ER	0.5			MIT
4566.370	4567.650	21893.10	ER	0.7	0.0		MIT
4564.830	4566.109	21900.48	ER	0.5			MIT
4563.970	4565.249	21904.61	ER	0.7			MIT

AIR WAVELENGTH (ANGSTRMS)	VACUUM WAVELENGTH (ANGSTRMS)	WAVENUMBER (KAYSERS)	LMENT	LOG ARC	INTENSITY SPK	DIS	REF
4563.270	4564.549	21907.97	ER	1.0	0.5		MIT
4561.160	4562.438	21918.10	ER	0.3			MIT
4557.30	4558.58	21936.7	ER	0.5			ED
4555.693	4556.970	21944.41	ER	0.5			MIT
4553.129	4554.405	21956.76	ER	0.7			MIT
4552.144	4553.420	21961.51	ER	0.9	0.0		MIT
4546.966	4548.241	21986.52	ER	0.3			MIT
4544.834	4546.108	21996.84	ER	0.6			MIT
4542.229	4543.502	22009.45	ER	0.3			MIT
4541.673	4542.946	22012.15	ER	0.7	0.0		MIT
4541.276	4542.549	22014.07	ER	0.7	0.0		MIT
4540.180	4541.453	22019.39	ER	0.7			MIT
4539.416	4540.689	22023.09	ER	0.3			MIT
4538.155	4539.427	22029.21	ER	0.3			MIT
4535.872	4537.144	22040.30	ER	0.6	0.0		MIT
4533.628	4534.899	22051.21	ER	0.5			MIT
4533.228	4534.499	22053.15	ER	0.8			MIT
4532.717	4533.988	22055.64	ER	0.3			MIT
4532.156	4533.427	22058.37	ER	0.5			MIT
4531.849	4533.120	22059.86	ER	0.3			MIT
4531.11	4532.38	22063.5	ER	0.7			ED
4530.309	4531.579	22067.36	ER	0.3	0.3		MIT
4529.378	4530.648	22071.90	ER	0.6			MIT
4527.783	4529.053	22079.67	ER	1.6	1.0		MIT
4527.236	4528.506	22082.34	ER	1.7	1.0		MIT
4526.918	4528.187	22083.89	ER	0.3	0.3		MIT
4522.737	4524.005	22104.31	ER	0.6L			MIT
4522.037	4523.305	22107.73	ER	1.0			MIT
4519.440	4520.707	22120.43	ER	0.7			MIT
4518.640	4519.907	22124.35	ER	1.6W	0.0H		MIT
4517.942	4519.209	22127.77	ER	0.3W			MIT
4516.814	4518.081	22133.29	ER	0.3			MIT
4516.534	4517.801	22134.66	ER	0.3			MIT
4514.751	4516.017	22143.41	ER	0.6W			MIT
4512.564	4513.830	22154.14	ER	0.6			MIT
4512.205	4513.471	22155.90	ER	0.5			MIT
4511.715	4512.980	22158.31	ER	0.6W	0.0H		MIT
4510.917	4512.182	22162.23	ER	0.3W			MIT
4509.968	4511.233	22166.89	ER	0.3W			MIT
4509.447	4510.712	22169.45	ER	0.3			MIT
4509.161	4510.426	22170.86	ER	0.3			MIT
4508.78	4510.04	22172.7	ER	0.3			MIT
4508.01	4509.27	22176.5	ER	0.3			MIT
4507.402	4508.666	22179.51	ER	0.5			MIT
4505.943	4507.207	22186.69	ER	1.6	0.7		MIT
4503.772	4505.035	22197.38	ER	0.3			MIT
4501.804	4503.067	22207.09	ER	0.6			MIT
4500.752	4502.015	22212.28	ER	1.3	1.3		MIT
4499.404	4500.666	22218.93	ER	0.8			MIT
4498.426	4499.688	22223.76	ER	0.7			MIT
4496.383	4497.644	22233.86	ER	0.8			MIT
4495.529	4496.790	22238.08	ER	0.6			MIT
4493.828	4495.089	22246.50	ER	0.7$			MIT
4493.070	4494.330	22250.25	ER	0.5			MIT
4492.412	4493.672	22253.51	ER	0.8			MIT
4491.738	4492.998	22256.85	ER	1.0			MIT
4490.590	4491.850	22262.54	ER	0.3			MIT
4490.171	4491.431	22264.62	ER	0.7			MIT
4488.96	4490.22	22270.6	ER	0.7K	0.0		M
4487.273	4488.532	22279.00	ER	1.3	0.0		MIT
4486.682	4487.941	22281.93	ER	0.8			M
4484.452	4485.710	22293.01	ER	0.9			MIT
4483.662	4484.920	22296.94	ER	0.5			MIT
4482.601	4483.859	22302.22	ER	0.3			MIT
4482.03	4483.29	22305.1	ER	0.3	0.3H		ED
4481.273	4482.530	22308.83	ER	1.3			MIT
4480.868	4482.125	22310.84	ER	0.3			MIT
4480.24	4481.50	22314.0	ER	0.5	0.0		MIT
4479.358	4480.615	22318.36	ER	0.3W			MIT
4477.69	4478.95	22326.7	ER	0.6	0.0		M

AIR WAVELENGTH (ANGSTRMS)	VACUUM WAVELENGTH (ANGSTRMS)	WAVENUMBER (KAYSERS)	LMENT	LOG ARC	INTENSITY SPK	DIS	REF
4475.544	4476.800	22337.39	ER	1.3	0.0		MIT
4474.498	4475.754	22342.61	ER	0.6			MIT
4473.500	4474.755	22347.59	ER	1.0	0.0		MIT
4471.932	4473.187	22355.43	ER	0.3W			MIT
4468.140	4469.394	22374.40	ER	1.1			MIT
4464.65	4465.90	22391.9	ER	0.9D	0.3		MIT
4459.242	4460.494	22419.04	ER	0.8			MIT
4451.581	4452.831	22457.62	ER	1.0	0.0		MIT
4449.703	4450.952	22467.10	ER	1.5	0.0		MIT
4449.326	4450.575	22469.01	ER	0.3			MIT
4448.611	4449.860	22472.62	ER	0.8			MIT
4448.070	4449.319	22475.35	ER	0.7			MIT
4446.258	4447.506	22484.51	ER	0.3			MIT
4444.571	4445.819	22493.04	ER	0.9D			MIT
4444.258	4445.506	22494.63	ER	0.8			MIT
4441.213	4442.460	22510.05	ER	0.7			MIT
4439.636	4440.882	22518.05	ER	0.3			MIT
4438.958	4440.204	22521.49	ER	0.3			MIT
4436.348	4437.594	22534.74	ER	0.3			MIT
4432.224	4433.468	22555.70	ER	0.6L			MIT
4429.896	4431.140	22567.56	ER	0.6W			MIT
4426.771	4428.014	22583.49	ER	1.1			MIT
4424.570	4425.812	22594.72	ER	1.0			MIT
4422.47	4423.71	22605.4	ER	1.5	1.3		M
4421.709	4422.951	22609.34	ER	0.7			MIT
4420.534	4421.775	22615.35	ER	1.2			MIT
4419.611	4420.852	22620.07	ER	1.4	0.6		MIT
4418.702	4419.943	22624.73	ER	0.7			MIT
4409.359	4410.597	22672.67	ER	1.5	0.3		MIT
4401.848	4403.084	22711.35	ER	1.1			MIT
4394.413	4395.647	22749.78	ER	0.5			MIT
4391.657	4392.891	22764.05	ER	0.7			MIT
4390.19	4391.42	22771.7	ER	0.8			M
4388.36	4389.59	22781.2	ER	0.9	0.0		MIT
4384.698	4385.930	22800.18	ER	1.5	0.7		MIT
4382.168	4383.399	22813.34	ER	1.0	0.0		MIT
4380.648	4381.879	22821.26	ER	0.8			MIT
4378.342	4379.572	22833.28	ER	1.0			MIT
4374.925	4376.154	22851.11	ER	1.6J	1.4J		MIT
4374.241	4375.470	22854.69	ER	0.8			MIT
4373.829	4375.058	22856.84	ER	0.7			MIT
4373.320	4374.549	22859.50	ER	0.3			MIT
4371.435	4372.663	22869.36	ER	0.7			MIT
4369.384	4370.612	22880.09	ER	1.1			MIT
4366.725	4367.952	22894.02	ER	0.9D			MIT
4362.260	4363.486	22917.46	ER	0.6	0.0		MIT
4361.659	4362.885	22920.61	ER	0.3	0.0		MIT
4360.337	4361.562	22927.56	ER	0.3			MIT
4358.172	4359.397	22938.95	ER	0.6			MIT
4357.39	4358.61	22943.1	ER	0.3			MIT
4356.695	4357.920	22946.73	ER	1.1			MIT
4353.639	4354.863	22962.84	ER	0.7			MIT
4351.645	4352.868	22973.36	ER	0.6			MIT
4351.179	4352.402	22975.82	ER	0.3			MIT
4350.733	4351.956	22978.17	ER	1.2			MIT
4349.773	4350.996	22983.25	ER	0.5			MIT
4348.341	4349.563	22990.81	ER	1.0			MIT
4347.308	4348.530	22996.28	ER	0.8	0.0H		MIT
4345.076	4346.297	23008.09	ER	0.3			MIT
4340.918	4342.138	23030.13	ER	0.9D			MIT
4339.637	4340.857	23036.93	ER	1.2			MIT
4338.997	4340.217	23040.32	ER	1.0			MIT
4335.027	4336.246	23061.42	ER	0.3			MIT
4333.748	4334.966	23068.23	ER	1.0			MIT
4331.368	4332.586	23080.90	ER	0.7			MIT
4330.273	4331.491	23086.74	ER	0.7			MIT
4328.81	4330.03	23094.5	ER	0.8	0.0		M
4322.637	4323.853	23127.52	ER	0.8			MIT
4319.938	4321.153	23141.97	ER	1.1			MIT
4316.41	4317.62	23160.9	ER	0.6	0.0		M

AIR WAVELENGTH (ANGSTRMS)	VACUUM WAVELENGTH (ANGSTRMS)	WAVENUMBER (KAYSERS)	LMENT	LOG ARC	INTENSITY SPK	DIS	REF
4315.772	4316.986	23164.31	ER	0.6			MIT
4313.689	4314.902	23175.50	ER	0.3			MIT
4313.416	4314.629	23176.96	ER	0.3			MIT
4310.435	4311.647	23192.99	ER	0.3			MIT
4308.630	4309.842	23202.71	ER	1.5	0.5		MIT
4306.343	4307.554	23215.03	ER	0.8			MIT
4303.813	4305.024	23228.68	ER	1.1			MIT
4301.604	4302.814	23240.60	ER	1.4	0.3		MIT
4301.260	4302.470	23242.46	ER	0.6			MIT
4298.913	4300.122	23255.15	ER	1.0			MIT
4297.30	4298.51	23263.9	ER	0.3			ED
4296.752	4297.961	23266.85	ER	1.0D			MIT
4295.039	4296.247	23276.13	ER	1.2			MIT
4292.17	4293.38	23291.7	ER	0.8	0.0		MIT
4286.560	4287.766	23322.17	ER	1.3			MIT
4285.59	4286.80	23327.4	ER	0.5	0.0		MIT
4281.34	4282.54	23350.6	ER	0.3	0.0		M
4280.85	4282.05	23353.3	ER	1.1	0.3		M
4276.686	4277.889	23376.01	ER	0.8			MIT
4276.54	4277.74	23376.8	ER	0.8	0.5		ED
4272.44	4273.64	23399.2	ER	0.6	0.0H		MIT
4271.970	4273.172	23401.82	ER	0.6	0.0		MIT
4269.930	4271.132	23413.00	ER	0.6	0.0		MIT
4264.86	4266.06	23440.8	ER	0.6	0.3		MIT
4254.32	4255.52	23498.9	ER	0.8	0.0		MIT
4253.582	4254.779	23502.98	ER	1.0	0.3		MIT
4251.938	4253.135	23512.07	ER	1.3	0.0		MIT
4239.860	4241.054	23579.04	ER	1.1			MIT
4237.04	4238.23	23594.7	ER	0.3			MIT
4235.18	4236.37	23605.1	ER	0.3			MIT
4234.78	4235.97	23607.3	ER	1.1	0.3		MIT
4232.026	4233.218	23622.69	ER	1.0			MIT
4230.202	4231.393	23632.88	ER	1.4			MIT
4223.72	4224.91	23669.1	ER	1.0	0.3		MIT
4220.993	4222.182	23684.44	ER	1.3	0.0		MIT
4218.426	4219.614	23698.85	ER	1.1			MIT
4218.094	4219.282	23700.71	ER	1.2	0.0		MIT
4215.964	4217.152	23712.69	ER	1.0			MIT
4211.718	4212.904	23736.59	ER	1.5	0.7		MIT
4205.92	4207.10	23769.3	ER	0.8			MIT
4201.60	4202.78	23793.7	ER	0.9D			MIT
4200.64	4201.82	23799.2	ER	1.0D	0.0		MIT
4198.69	4199.87	23810.2	ER	1.0D	0.3		MIT
4197.075	4198.258	23819.41	ER	1.1W	0.0		MIT
4194.845	4196.027	23832.07	ER	1.3	0.0		MIT
4192.563	4193.744	23845.04	ER	0.8			MIT
4192.410	4193.591	23845.91	ER	0.9			MIT
4191.32	4192.50	23852.1	ER	0.8W			MIT
4190.698	4191.879	23855.65	ER	1.0L			MIT
4189.984	4191.165	23859.72	ER	1.0			MIT
4189.503	4190.684	23862.46	ER	1.0			MIT
4188.93	4190.11	23865.7	ER	1.0D			MIT
4187.983	4189.163	23871.12	ER	0.8$			MIT
4186.71	4187.89	23878.4	ER	0.9	0.3		MIT
4185.724	4186.904	23884.00	ER	1.2W			MIT
4184.995	4186.174	23888.16	ER	0.8			MIT
4184.063	4185.242	23893.48	ER	0.6			MIT
4182.270	4183.449	23903.72	ER	1.0			MIT
4180.875	4182.053	23911.70	ER	1.4W			MIT
4180.28	4181.46	23915.1	ER	0.8D	0.3		MIT
4179.401	4180.579	23920.13	ER	0.7			MIT
4178.602	4179.780	23924.71	ER	0.9S			MIT
4176.32	4177.50	23937.8	ER	1.0D			MIT
4175.180	4176.357	23944.31	ER	0.6			MIT
4174.905	4176.082	23945.89	ER	1.0D			MIT
4171.708	4172.884	23964.24	ER	1.2			MIT
4170.492	4171.668	23971.23	ER	0.8			MIT
4168.490	4169.665	23982.74	ER	0.9			MIT
4167.98	4169.15	23985.7	ER	0.3	0.0		MIT
4165.395	4166.569	24000.56	ER	1.2			MIT

AIR WAVELENGTH (ANGSTRMS)	VACUUM WAVELENGTH (ANGSTRMS)	WAVENUMBER (KAYSERS)	LMENT	LOG ARC	INTENSITY SPK	DIS	REF
4164.834	4166.008	24003.79	ER	0.7			MIT
4164.247	4165.421	24007.18	ER	0.8			MIT
4163.453	4164.627	24011.76	ER	0.9			MIT
4163.031	4164.205	24014.19	ER	1.1			MIT
4161.642	4162.815	24022.21	ER	1.1			MIT
4160.310	4161.483	24029.90	ER	0.9			MIT
4159.881	4161.054	24032.37	ER	1.3W			MIT
4158.428	4159.600	24040.77	ER	1.0			MIT
4155.38	4156.55	24058.4	ER	0.6D			ED
4154.501	4155.672	24063.49	ER	0.8			MIT
4152.771	4153.942	24073.52	ER	0.8			MIT
4152.308	4153.479	24076.20	ER	0.8W			MIT
4151.109	4152.280	24083.16	ER	1.3	0.5		MIT
4150.78	4151.95	24085.1	ER	0.8W	0.3		MIT
4145.884	4147.053	24113.51	ER	1.0			MIT
4144.766	4145.935	24120.01	ER	1.1D			MIT
4144.232	4145.401	24123.12	ER	0.7			MIT
4142.96	4144.13	24130.5	ER	1.5	1.2		M
4140.769	4141.937	24143.29	ER	1.1			MIT
4140.192	4141.360	24146.66	ER	1.1			MIT
4138.335	4139.502	24157.49	ER	1.0			MIT
4137.272	4138.439	24163.70	ER	0.9			MIT
4131.502	4132.667	24197.45	ER	1.0			MIT
4129.433	4130.598	24209.57	ER	1.0	0.3		MIT
4127.113	4128.277	24223.18	ER	1.0			MIT
4126.227	4127.391	24228.38	ER	1.0S	0.3		MIT
4125.635	4126.799	24231.86	ER	0.7			MIT
4124.796	4125.960	24236.79	ER	0.3			MIT
4123.237	4124.400	24245.95	ER	0.7			MIT
4123.053	4124.216	24247.03	ER	0.9			MIT
4120.842	4122.005	24260.04	ER	0.9			MIT
4120.188	4121.350	24263.89	ER	0.9			MIT
4119.329	4120.491	24268.95	ER	1.3K	0.3J		MIT
4116.362	4117.523	24286.44	ER	0.6D			MIT
4114.074	4115.235	24299.95	ER	1.1W			MIT
4112.622	4113.782	24308.53	ER	0.8	0.0		MIT
4111.351	4112.511	24316.04	ER	1.2	0.0		MIT
4109.900	4111.060	24324.63	ER	0.8W			MIT
4109.087	4110.247	24329.44	ER	0.7			MIT
4108.551	4109.710	24332.61	ER	1.0			MIT
4106.661	4107.820	24343.81	ER	1.1D			MIT
4104.34	4105.50	24357.6	ER	0.5			ED
4103.806	4104.964	24360.75	ER	1.1			MIT
4103.311	4104.469	24363.69	ER	1.3	0.0		MIT
4101.093	4102.250	24376.86	ER	0.5			MIT
4100.563	4101.720	24380.01	ER	1.2	0.0		MIT
4098.103	4099.260	24394.65	ER	0.9			MIT
4096.926	4098.082	24401.66	ER	0.9			MIT
4096.839	4097.995	24402.18	ER	0.9			MIT
4094.656	4095.812	24415.19	ER	0.8			MIT
4092.901	4094.056	24425.65	ER	0.8			MIT
4091.770	4092.925	24432.41	ER	0.8			MIT
4090.771	4091.926	24438.37	ER	1.0			MIT
4090.135	4091.290	24442.17	ER	0.8D			MIT
4087.638	4088.792	24457.10	ER	1.3	0.0		MIT
4081.237	4082.389	24495.46	ER	1.0S	0.0		MIT
4077.88	4079.03	24515.6	ER I	0.9			ME
4076.01	4077.16	24526.9	ER	0.3			M
4074.003	4075.153	24538.95	ER	0.8			MIT
4073.650	4074.800	24541.08	ER	0.5H			MIT
4073.120	4074.270	24544.27	ER	1.3	0.0		MIT
4072.377	4073.527	24548.75	ER	1.0D	0.5J		MIT
4069.562	4070.711	24565.73	ER	0.8			MIT
4069.248	4070.397	24567.63	ER	0.8			MIT
4068.332	4069.481	24573.16	ER	0.7S			MIT
4065.065	4066.213	24592.91	ER	0.9			MIT
4062.948	4064.095	24605.72	ER	0.8			MIT
4059.785	4060.932	24624.89	ER	1.3	0.3		MIT
4059.509	4060.656	24626.57	ER	0.8			MIT
4057.819	4058.965	24636.82	ER	1.5			MIT

AIR WAVELENGTH (ANGSTRMS)	VACUUM WAVELENGTH (ANGSTRMS)	WAVENUMBER (KAYSERS)	LMENT	LOG ARC	INTENSITY SPK	DIS	REF
4055.990	4057.136	24647.93	ER	0.7			MIT
4055.848	4056.994	24648.79	ER	0.8			MIT
4055.471	4056.616	24651.09	ER	1.1	0.3		MIT
4055.03	4056.18	24653.8	ER	0.5			M
4053.885	4055.030	24660.73	ER	1.3	0.0		MIT
4050.570	4051.714	24680.91	ER	1.4	0.0		MIT
4049.48	4050.62	24687.6	ER	1.0D	0.3		MIT
4048.346	4049.490	24694.47	ER	1.2	0.0		MIT
4047.850	4048.993	24697.50	ER	0.7			MIT
4046.960	4048.103	24702.93	ER	0.9			MIT
4045.89	4047.03	24709.5	ER	0.3			ED
4045.43	4046.57	24712.3	ER	0.9$	0.0$		M
4043.015	4044.157	24727.03	ER	1.0			MIT
4040.777	4041.919	24740.73	ER	1.3			MIT
4037.680	4038.821	24759.70	ER	0.6			MIT
4036.116	4037.256	24769.30	ER	0.5			MIT
4032.477	4033.616	24791.65	ER	1.0			MIT
4027.066	4028.204	24824.96	ER	0.3			MIT
4025.537	4026.675	24834.39	ER	0.5			MIT
4021.96	4023.10	24856.5	ER	0.7			ED
4021.558	4022.695	24858.96	ER	0.7			MIT
4020.519	4021.655	24865.38	ER	1.3	0.0		MIT
4017.714	4018.850	24882.74	ER	0.5D			MIT
4015.579	4016.714	24895.97	ER	1.3	0.0		MIT
4012.58	4013.71	24914.6	ER	0.6	0.3		M
4010.54	4011.67	24927.2	ER	0.8D	0.3		M
4009.785	4010.919	24931.95	ER	0.5			MIT
4009.165	4010.298	24935.80	ER	1.2	0.0		MIT
4008.185	4009.318	24941.90	ER	0.9	0.0		MIT
4007.967	4009.100	24943.25	ER	1.5	0.8		MIT
4004.579	4005.711	24964.36	ER	0.6D			MIT
4004.046	4005.178	24967.68	ER	0.7			MIT
3999.17	4000.30	24998.1	ER	0.7	0.3		MIT
3996.695	3997.825	25013.60	ER	1.4			MIT
3995.754	3996.884	25019.49	ER	0.6			MIT
3995.268	3996.398	25022.53	ER	0.8			MIT
3994.858	3995.988	25025.10	ER	0.7			MIT
3991.158	3992.287	25048.30	ER	1.2			MIT
3987.663	3988.791	25070.25	ER	0.6			MIT
3986.687	3987.814	25076.39	ER	0.8D			MIT
3984.214	3985.341	25091.96	ER	1.1	0.0		MIT
3983.655	3984.782	25095.48	ER	1.3	0.5		MIT
3983.143	3984.270	25098.70	ER	0.5			MIT
3982.333	3983.459	25103.81	ER	0.6			MIT
3981.930	3983.056	25106.35	ER	1.0	0.0		MIT
3981.225	3982.351	25110.79	ER	0.6			MIT
3980.609	3981.735	25114.68	ER	0.6			MIT
3980.149	3981.275	25117.58	ER	1.0			MIT
3979.13	3980.26	25124.0	ER	0.8D	0.3		MIT
3978.565	3979.690	25127.58	ER	1.3S	0.5		MIT
3977.024	3978.149	25137.32	ER	1.0			MIT
3976.743	3977.868	25139.09	ER	0.5$			MIT
3975.296	3976.421	25148.25	ER	0.3			MIT
3974.723	3975.847	25151.87	ER	1.2	0.5		MIT
3973.577	3974.701	25159.12	ER	1.3	0.3		MIT
3973.263	3974.387	25161.11	ER	0.5	0.0J		MIT
3973.038	3974.162	25162.54	ER	1.4			MIT
3969.434	3970.557	25185.38	ER	0.8	0.0		MIT
3966.355	3967.477	25204.93	ER	0.5			MIT
3964.507	3965.629	25216.68	ER	1.1			MIT
3963.362	3964.483	25223.97	ER	0.9			MIT
3961.210	3962.331	25237.67	ER	0.8			MIT
3959.908	3961.028	25245.97	ER	0.5	0.0		MIT
3957.45	3958.57	25261.6	ER	0.7			MIT
3956.432	3957.552	25268.15	ER	1.0			MIT
3951.49	3952.61	25299.7	ER	0.3	0.0		M
3950.351	3951.469	25307.04	ER	1.5	1.0H		MIT
3948.067	3949.184	25321.68	ER	0.9L			MIT
3944.423	3945.539	25345.08	ER	1.1			MIT
3944.014	3945.130	25347.71	ER	0.3			MIT

AIR WAVELENGTH (ANGSTRMS)	VACUUM WAVELENGTH (ANGSTRMS)	WAVENUMBER (KAYSERS)	LMENT	LOG ARC	INTENSITY SPK	DIS	REF
3943.191	3944.307	25353.00	ER	1.0			MIT
3942.533	3943.649	25357.23	ER	0.5			MIT
3939.806	3940.921	25374.78	ER	0.3			MIT
3939.355	3940.470	25377.68	ER	0.6			MIT
3938.923	3940.038	25380.47	ER	0.6			MIT
3938.631	3939.746	25382.35	ER	1.3	0.5		MIT
3937.018	3938.133	25392.75	ER	1.3			MIT
3934.812	3935.926	25406.98	ER	0.7			MIT
3932.253	3933.366	25423.52	ER	1.3	0.7		MIT
3931.528	3932.641	25428.20	ER	1.3	0.6		MIT
3929.225	3930.337	25443.11	ER	0.5			MIT
3923.80	3924.91	25478.3	ER	0.5			MIT
3921.886	3922.997	25490.72	ER	1.3			MIT
3918.556	3919.666	25512.38	ER	0.6			MIT
3918.270	3919.380	25514.24	ER	0.6			MIT
3918.052	3919.162	25515.66	ER	0.7S			MIT
3915.69	3916.80	25531.1	ER	0.5	0.0		MIT
3914.867	3915.976	25536.42	ER	1.2	0.0		MIT
3912.429	3913.537	25552.33	ER	1.1			MIT
3911.911	3913.019	25555.72	ER	0.7			MIT
3911.558	3912.666	25558.02	ER	0.9			MIT
3910.512	3911.620	25564.86	ER	0.5			MIT
3909.57	3910.68	25571.0	ER	0.6			MIT
3908.418	3909.525	25578.55	ER	1.0			MIT
3906.316	3907.423	25592.32	ER	1.4	1.1		MIT
3905.415	3906.521	25598.22	ER	1.3	0.0		MIT
3904.587	3905.693	25603.65	ER	1.0	0.0		MIT
3903.993	3905.099	25607.55	ER	0.7			MIT
3903.906	3905.012	25608.12	ER	0.7	0.0H		MIT
3902.766	3903.872	25615.60	ER	1.0	0.0		MIT
3901.66	3902.77	25622.9	ER	0.3			MIT
3900.863	3901.968	25628.09	ER	1.3	0.0		MIT
3899.036	3900.141	25640.10	ER	0.3	0.0		MIT
3898.530	3899.634	25643.43	ER	1.5	1.0		MIT
3896.234	3897.338	25658.54	ER	1.5	1.2		MIT
3895.805	3896.909	25661.36	ER	0.8	0.0		MIT
3892.692	3893.795	25681.89	ER	1.4	0.3		MIT
3892.422	3893.525	25683.67	ER	0.3			MIT
3890.619	3891.721	25695.57	ER	1.0	0.3		MIT
3889.799	3890.901	25700.99	ER	1.2	0.0		MIT
3889.024	3890.126	25706.11	ER	1.0	0.0		MIT
3888.093	3889.195	25712.26	ER	1.3	0.0		MIT
3887.160	3888.262	25718.44	ER	0.3	0.0		MIT
3885.291	3886.392	25730.81	ER	1.0			MIT
3882.868	3883.968	25746.86	ER	0.9	0.3		MIT
3880.66	3881.76	25761.5	ER	1.4$	0.8$		MIT
3880.043	3881.143	25765.61	ER	1.3$	0.3O		MIT
3879.66	3880.76	25768.1	ER	1.3D	0.3W		MIT
3879.27	3880.37	25770.7	ER	1.2$	0.0W		MIT
3878.84	3879.94	25773.6	ER	1.0	0.0		MIT
3877.93	3879.03	25779.6	ER	0.8	0.0		MIT
3877.52	3878.62	25782.4	ER	1.2D	0.0		MIT
3876.95	3878.05	25786.2	ER	1.2D	0.0W		MIT
3876.45	3877.55	25789.5	ER	1.2D	0.0		MIT
3874.125	3875.223	25804.97	ER	1.3D	0.0W		MIT
3873.53	3874.63	25808.9	ER	0.7D	0.0W		MIT
3872.15	3873.25	25818.1	ER	0.5	0.3		M
3870.355	3871.452	25830.10	ER	1.0	0.0		MIT
3864.81	3865.91	25867.2	ER	0.8	0.0		MIT
3863.46	3864.56	25876.2	ER	0.7	0.5		MIT
3863.08	3864.18	25878.7	ER	0.5	0.0		MIT
3862.411	3863.506	25883.23	ER	1.2	0.5		MIT
3858.395	3859.489	25910.17	ER	0.6	0.3		MIT
3857.79	3858.88	25914.2	ER	0.8D			MIT
3854.565	3855.658	25935.91	ER	0.6	0.0		MIT
3853.215	3854.308	25945.00	ER	0.7	0.0		MIT
3849.002	3850.094	25973.40	ER	1.1	0.3		MIT
3847.89	3848.98	25980.9	ER	1.3D	0.0		MIT
3847.262	3848.353	25985.14	ER	0.6			MIT
3846.974	3848.065	25987.09	ER	0.8			MIT

AIR WAVELENGTH (ANGSTRMS)	VACUUM WAVELENGTH (ANGSTRMS)	WAVENUMBER (KAYSERS)	LMENT	LOG ARC	INTENSITY SPK	DIS	REF
3844.231	3845.321	26005.63	ER	1.3	0.3		MIT
3842.98	3844.07	26014.1	ER	1.2	0.3		MIT
3841.844	3842.934	26021.79	ER	1.1D	0.0D		MIT
3840.921	3842.010	26028.04	ER	0.8			MIT
3838.339	3839.428	26045.55	ER	1.0			MIT
3837.631	3838.720	26050.35	ER	1.2	0.0		MIT
3837.418	3838.507	26051.80	ER	1.0	0.0		MIT
3836.505	3837.593	26058.00	ER	1.6	1.0		MIT
3835.65	3836.74	26063.8	ER	1.0W			MIT
3835.26	3836.35	26066.5	ER	1.4D	0.3D		MIT
3833.041	3834.128	26081.55	ER	0.6			MIT
3832.436	3833.523	26085.66	ER	1.1	0.0		MIT
3830.51	3831.60	26098.8	ER	0.9W	0.0H		MIT
3830.067	3831.154	26101.80	ER	0.8			MIT
3829.52	3830.61	26105.5	ER	1.1	0.3		MIT
3828.18	3829.27	26114.7	ER	1.2W	0.5W		MIT
3826.819	3827.905	26123.95	ER	1.1D	0.3D		MIT
3825.247	3826.332	26134.69	ER	1.0D	0.3W		MIT
3824.764	3825.849	26137.99	ER	0.9W	0.0		MIT
3823.04	3824.12	26149.8	ER	1.0	0.0		MIT
3822.312	3823.397	26154.75	ER	1.3D	0.3W		MIT
3821.735	3822.820	26158.70	ER	1.0H	0.0		MIT
3820.801	3821.885	26165.10	ER	1.2	0.7		MIT
3819.271	3820.355	26175.58	ER	1.1D	0.3W		MIT
3818.704	3819.788	26179.47	ER	1.0D	0.0D		MIT
3817.69	3818.77	26186.4	ER	1.2W	0.3W		MIT
3815.49	3816.57	26201.5	ER	0.8D			MIT
3813.323	3814.405	26216.41	ER	1.0D	0.0D		MIT
3812.96	3814.04	26218.9	ER	1.2	0.3		MIT
3812.061	3813.143	26225.08	ER	1.0			MIT
3810.787	3811.869	26233.85	ER	1.0	0.0		MIT
3810.330	3811.412	26237.00	ER	1.0	0.0		MIT
3808.149	3809.230	26252.02	ER	1.4	0.6		MIT
3807.943	3809.024	26253.45	ER	1.3	0.5		MIT
3807.082	3808.163	26259.38	ER	0.6			MIT
3804.778	3805.858	26275.28	ER	1.0	0.0		MIT
3803.123	3804.203	26286.72	ER	0.9D	0.0D		MIT
3798.65	3799.73	26317.7	ER	0.6			MIT
3798.25	3799.33	26320.4	ER	0.7			MIT
3797.065	3798.143	26328.66	ER	0.3			MIT
3795.379	3796.457	26340.35	ER	1.0			MIT
3795.29	3796.37	26341.0	ER	1.3D	0.0D		MIT
3794.779	3795.857	26344.51	ER	0.3			MIT
3794.38	3795.46	26347.3	ER	0.7	0.0		MIT
3793.786	3794.863	26351.41	ER	1.0	0.0		MIT
3792.915	3793.992	26357.46	ER	0.9			MIT
3792.812	3793.889	26358.18	ER	0.9	0.0		MIT
3791.845	3792.922	26364.90	ER	1.3	0.8		MIT
3790.836	3791.912	26371.92	ER	1.0	0.0		MIT
3790.537	3791.613	26374.00	ER	1.1	0.7J		MIT
3789.919	3790.995	26378.30	ER	0.9			MIT
3787.910	3788.986	26392.29	ER	1.2	0.3		MIT
3787.42	3788.50	26395.7	ER	0.5D	0.0D		MIT
3787.27	3788.35	26396.7	ER	1.4D	0.7D		MIT
3786.839	3787.914	26399.75	ER	1.3	0.8		MIT
3786.177	3787.252	26404.37	ER	1.3	1.1		MIT
3785.356	3786.431	26410.09	ER	1.2D	0.0D		MIT
3783.857	3784.932	26420.56	ER	1.0D	0.0D		MIT
3783.544	3784.619	26422.74	ER	1.0	0.0		MIT
3782.26	3783.33	26431.7	ER	0.5	0.0		MIT
3781.018	3782.092	26440.39	ER	1.0	0.5		MIT
3780.430	3781.504	26444.51	ER	1.0D	0.0		MIT
3779.803	3780.877	26448.89	ER	1.0			MIT
3778.915	3779.988	26455.11	ER	0.5			MIT
3778.62	3779.69	26457.2	ER	1.0	0.9		MIT
3778.23	3779.30	26459.9	ER	0.9	0.3		MIT
3777.63	3778.70	26464.1	ER	0.8	0.3		MIT
3777.089	3778.162	26467.90	ER	1.4	1.0H		MIT
3775.665	3776.738	26477.88	ER	0.9	0.3		MIT
3772.469	3773.541	26500.31	ER	0.3			MIT

AIR WAVELENGTH (ANGSTRMS)	VACUUM WAVELENGTH (ANGSTRMS)	WAVENUMBER (KAYSERS)	LMENT	LOG ARC	INTENSITY SPK	DIS	REF
3771.110	3772.181	26509.86	ER	1.1	0.3		MIT
3770.244	3771.315	26515.95	ER	1.0D	0.0D		MIT
3769.948	3771.019	26518.03	ER	0.9	0.3		MIT
3768.791	3769.862	26526.17	ER	0.7			MIT
3768.218	3769.289	26530.21	ER	1.0D			MIT
3767.761	3768.831	26533.42	ER	0.3			MIT
3766.249	3767.319	26544.07	ER	0.7			MIT
3766.162	3767.232	26544.69	ER	1.0	0.5		MIT
3764.31	3765.38	26557.7	ER	1.3D			MIT
3763.24	3764.31	26565.3	ER	1.2D	0.0		MIT
3760.36	3761.43	26585.6	ER	1.2D	0.0D		MIT
3759.94	3761.01	26588.6	ER	0.6D			MIT
3759.54	3760.61	26591.4	ER	0.3			MIT
3759.093	3760.161	26594.60	ER	0.7			MIT
3757.369	3758.437	26606.81	ER	1.3	0.8		MIT
3757.056	3758.124	26609.02	ER	0.7			MIT
3756.340	3757.408	26614.09	ER	1.3$			MIT
3756.053	3757.120	26616.13	ER	1.1			MIT
3754.266	3755.333	26628.80	ER	0.3			MIT
3752.191	3753.257	26643.52	ER	1.3			MIT
3751.97	3753.04	26645.1	ER	0.3			MIT
3750.544	3751.610	26655.22	ER	0.8			MIT
3750.167	3751.233	26657.90	ER	1.1D			MIT
3747.554	3748.619	26676.49	ER	1.3	1.0		MIT
3747.437	3748.502	26677.32	ER	0.3			MIT
3746.62	3747.69	26683.1	ER	0.3			MIT
3746.06	3747.12	26687.1	ER	1.1O	0.0O		MIT
3745.111	3746.176	26693.89	ER	0.6	0.0		MIT
3744.987	3746.052	26694.77	ER	1.0	0.3		MIT
3742.643	3743.707	26711.49	ER	1.3	1.1		MIT
3741.851	3742.915	26717.15	ER	1.3D			MIT
3741.101	3742.165	26722.50	ER	1.2	0.3		MIT
3740.279	3741.342	26728.37	ER	1.0	0.5		MIT
3738.168	3739.231	26743.47	ER	0.9	0.3		MIT
3734.589	3735.651	26769.10	ER	0.6			MIT
3734.459	3735.521	26770.03	ER	0.6			MIT
3733.76	3734.82	26775.0	ER	0.3			MIT
3732.16	3733.22	26786.5	ER	0.3			MIT
3731.794	3732.855	26789.15	ER	0.3			MIT
3731.273	3732.334	26792.89	ER	1.2S	0.6		MIT
3730.599	3731.660	26797.73	ER	0.9D			MIT
3729.526	3730.587	26805.44	ER	1.4	0.8		MIT
3727.999	3729.059	26816.42	ER	0.7			MIT
3724.916	3725.975	26838.61	ER	0.8	0.0		MIT
3724.449	3725.508	26841.98	ER	1.2	0.6		MIT
3721.465	3722.523	26863.50	ER	0.9			MIT
3719.295	3720.353	26879.17	ER	0.6D			MIT
3718.60	3719.66	26884.2	ER	0.3			MIT
3717.262	3718.319	26893.87	ER	1.0			MIT
3715.956	3717.013	26903.32	ER	0.7D			MIT
3715.15	3716.21	26909.2	ER	0.3D			MIT
3714.03	3715.09	26917.3	ER	0.5H			MIT
3713.851	3714.908	26918.57	ER	1.0			MIT
3712.392	3713.448	26929.15	ER	1.3	0.3		MIT
3711.825	3712.881	26933.26	ER	0.8			MIT
3707.644	3708.699	26963.63	ER	1.2D	0.3D		MIT
3706.524	3707.579	26971.78	ER	1.1D			MIT
3705.776	3706.830	26977.23	ER	0.7W			MIT
3702.506	3703.560	27001.05	ER	0.9			MIT
3701.57	3702.62	27007.9	ER	1.1D	0.3O		MIT
3700.728	3701.781	27014.02	ER	1.0	0.3		MIT
3699.744	3700.797	27021.21	ER	0.5			MIT
3698.969	3700.022	27026.87	ER	1.0			MIT
3698.209	3699.261	27032.42	ER	0.5			MIT
3696.922	3697.974	27041.83	ER	1.1	0.0		MIT
3696.255	3697.307	27046.71	ER	1.3	0.3		MIT
3695.091	3696.143	27055.23	ER	0.3			MIT
3694.811	3695.863	27057.28	ER	1.2	0.8		MIT
3694.193	3695.244	27061.81	ER	1.4D	1.2		MIT
3692.652	3693.703	27073.10	ER	1.3	1.1		MIT

AIR WAVELENGTH (ANGSTRMS)	VACUUM WAVELENGTH (ANGSTRMS)	WAVENUMBER (KAYSERS)	LMENT	LOG ARC	INTENSITY SPK	INTENSITY DIS	REF
3689.12	3690.17	27099.0	ER	0.3			MIT
3687.758	3688.808	27109.03	ER	1.0	0.5		MIT
3687.100	3688.150	27113.87	ER	1.0			MIT
3685.17	3686.22	27128.1	ER	0.8W			MIT
3684.861	3685.910	27130.34	ER	1.00	0.0		MIT
3684.284	3685.333	27134.59	ER	1.2	0.5		MIT
3684.014	3685.063	27136.58	ER	0.8			MIT
3683.472	3684.521	27140.57	ER	1.4			MIT
3682.707	3683.755	27146.21	ER	1.3	0.6		MIT
3680.10	3681.15	27165.4	ER	0.8			MIT
3678.958	3680.005	27173.87	ER	1.1			MIT
3676.508	3677.555	27191.98	ER	1.3	0.3		MIT
3675.85	3676.90	27196.9	ER	0.30	0.0		MIT
3674.087	3675.133	27209.90	ER	1.3	0.3		MIT
3673.142	3674.188	27216.90	ER	1.1	0.0		MIT
3672.303	3673.349	27223.12	ER	1.3	0.3		MIT
3671.306	3672.352	27230.51	ER	1.1	0.0		MIT
3669.023	3670.068	27247.45	ER	0.6			MIT
3668.732	3669.777	27249.61	ER	0.0	0.6J		MIT
3668.491	3669.536	27251.40	ER	1.4	0.5		MIT
3668.04	3669.08	27254.7	ER	0.8D	0.0		MIT
3665.21	3666.25	27275.8	ER	0.9			MIT
3664.440	3665.484	27281.53	ER	1.6	1.3H		MIT
3662.867	3663.910	27293.24	ER	1.0			MIT
3662.275	3663.318	27297.66	ER	1.2	0.0		MIT
3662.04	3663.08	27299.4	ER	1.0	0.0		MIT
3660.783	3661.826	27308.78	ER	1.0	0.0		MIT
3659.55	3660.59	27318.0	ER	1.1W	0.0		MIT
3658.84	3659.88	27323.3	ER	1.00	0.00		MIT
3655.732	3656.773	27346.51	ER	0.7			MIT
3654.363	3655.404	27356.76	ER	0.6			MIT
3652.879	3653.920	27367.87	ER	1.3	0.6		MIT
3652.583	3653.624	27370.09	ER	1.3	0.5		MIT
3650.414	3651.454	27386.35	ER	1.2	0.3		MIT
3648.978	3650.018	27397.13	ER	0.6	0.0		MIT
3647.23	3648.27	27410.3	ER	0.6			MIT
3646.782	3647.821	27413.62	ER	0.7			MIT
3645.938	3646.977	27419.97	ER	1.2	0.3		MIT
3645.399	3646.438	27424.02	ER	1.4S	1.1		MIT
3644.73	3645.77	27429.1	ER	0.5	0.3		MIT
3641.88	3642.92	27450.5	ER	1.1	0.0		MIT
3641.269	3642.307	27455.13	ER	1.3	0.3		MIT
3640.252	3641.289	27462.80	ER	1.3	0.0		MIT
3639.906	3640.943	27465.41	ER	0.9			MIT
3639.014	3640.051	27472.14	ER	0.8L			MIT
3638.678	3639.715	27474.68	ER	1.2	0.0		MIT
3637.168	3638.205	27486.08	ER	1.1	0.0		MIT
3634.679	3635.715	27504.91	ER	0.9			MIT
3633.541	3634.577	27513.52	ER	1.2	0.3		MIT
3632.785	3633.821	27519.25	ER	0.7	0.0		MIT
3630.242	3631.277	27538.52	ER	1.4	0.8		MIT
3630.061	3631.096	27539.90	ER	0.9	0.0		MIT
3629.853	3630.888	27541.47	ER	0.6	0.0		MIT
3629.39	3630.42	27545.0	ER	1.0	0.0		MIT
3628.94	3629.97	27548.4	ER	0.3			MIT
3628.705	3629.739	27550.19	ER	1.4	1.1H		MIT
3628.041	3629.075	27555.23	ER	1.2	0.0		MIT
3627.843	3628.877	27556.73	ER	0.7	0.0		MIT
3626.873	3627.907	27564.10	ER	0.5			MIT
3626.42	3627.45	27567.6	ER	0.3D	0.3		MIT
3625.282	3626.316	27576.20	ER	0.9	0.0		MIT
3624.601	3625.634	27581.38	ER	1.0	0.0		MIT
3624.278	3625.311	27583.84	ER	1.0	0.0		MIT
3622.157	3623.190	27599.99	ER	0.6			MIT
3620.945	3621.977	27609.23	ER	1.4	1.2		MIT
3620.177	3621.209	27615.08	ER	1.7	0.5		MIT
3619.981	3621.013	27616.58	ER	1.0	0.0		MIT
3618.92	3619.95	27624.7	ER	1.0W	0.8		MIT
3617.82	3618.85	27633.1	ER	1.1	0.7		MIT
3617.183	3618.215	27637.94	ER	0.8			MIT

AIR WAVELENGTH (ANGSTRMS)	VACUUM WAVELENGTH (ANGSTRMS)	WAVENUMBER (KAYSERS)	LMENT	LOG ARC	INTENSITY SPK	INTENSITY DIS	REF
3616.573	3617.604	27642.60	ER	1.5	1.3		MIT
3616.071	3617.102	27646.44	ER	1.0			MIT
3614.637	3615.668	27657.41	ER	1.1L	0.0		MIT
3613.096	3614.126	27669.20	ER	0.8			MIT
3612.87	3613.90	27670.9	ER	1.0W	0.3H		MIT
3612.562	3613.592	27673.29	ER	0.6H			MIT
3612.398	3613.428	27674.55	ER	0.9			MIT
3609.44	3610.47	27697.2	ER	0.8D	0.0H		MIT
3608.22	3609.25	27706.6	ER	1.0			MIT
3607.431	3608.460	27712.65	ER	0.9			MIT
3606.08	3607.11	27723.0	ER	1.2D	0.6		MIT
3605.695	3606.724	27726.00	ER	0.5			MIT
3604.901	3605.929	27732.10	ER	1.1	0.6		MIT
3604.715	3605.743	27733.53	ER	1.0	0.3		MIT
3600.742	3601.769	27764.13	ER	1.5W	1.3W		MIT
3599.829	3600.856	27771.17	ER	1.5W	1.3S		MIT
3599.508	3600.535	27773.65	ER	1.3	0.7		MIT
3598.85	3599.88	27778.7	ER	0.9C	0.0		MIT
3597.773	3598.800	27787.04	ER	0.8			MIT
3596.23	3597.26	27799.0	ER	0.3D			MIT
3595.838	3596.864	27802.00	ER	0.8			MIT
3595.47	3596.50	27804.8	ER	0.8D	0.0		MIT
3595.042	3596.068	27808.15	ER	1.3	0.9		MIT
3594.128	3595.154	27815.22	ER	1.0			MIT
3593.952	3594.978	27816.59	ER	0.9			MIT
3591.158	3592.183	27838.23	ER	0.8H			MIT
3590.72	3591.74	27841.6	ER	0.8D	0.0		MIT
3590.35	3591.37	27844.5	ER	1.4			MIT
3588.948	3589.972	27855.37	ER	1.0	0.5		MIT
3588.764	3589.788	27856.80	ER	0.8	0.0		MIT
3588.343	3589.367	27860.06	ER	1.3			MIT
3587.748	3588.772	27864.69	ER	1.0	0.5		MIT
3587.26	3588.28	27868.5	ER	0.5			MIT
3586.64	3587.66	27873.3	ER	0.9D			MIT
3585.79	3586.81	27879.9	ER	1.00	0.3H		MIT
3584.518	3585.541	27889.79	ER	1.4	1.2H		MIT
3581.842	3582.864	27910.63	ER	1.0H	0.3		MIT
3580.48	3581.50	27921.2	ER	1.1D	0.7		MIT
3579.44	3580.46	27929.4	ER	1.2D			MIT
3578.33	3579.35	27938.0	ER	0.5W			MIT
3575.68	3576.70	27958.7	ER	0.5			MIT
3574.743	3575.764	27966.05	ER	1.0	0.0		MIT
3573.843	3574.863	27973.10	ER	1.0	0.6		MIT
3573.077	3574.097	27979.09	ER	1.0			MIT
3572.117	3573.137	27986.61	ER	0.7			MIT
3571.73	3572.75	27989.6	ER	1.2D	0.0		MIT
3570.756	3571.776	27997.28	ER	1.0	0.0		MIT
3569.252	3570.271	28009.08	ER	0.9			MIT
3567.576	3568.595	28022.24	ER	0.9			MIT
3566.090	3567.108	28033.91	ER	1.1			MIT
3565.177	3566.195	28041.09	ER	0.8			MIT
3564.40	3565.42	28047.2	ER	0.9D	0.0H		MIT
3563.533	3564.551	28054.03	ER	0.7			MIT
3563.402	3564.420	28055.06	ER	1.1	0.0		MIT
3563.16	3564.18	28057.0	ER	1.2W	0.8		MIT
3561.891	3562.908	28066.96	ER	0.9			MIT
3561.273	3562.290	28071.83	ER	1.1S			MIT
3559.902	3560.919	28082.64	ER	1.2	0.8		MIT
3558.736	3559.752	28091.84	ER	1.0	0.0		MIT
3558.019	3559.035	28097.50	ER	1.0	0.0		MIT
3557.81	3558.83	28099.1	ER	1.0$	0.3W		MIT
3557.11	3558.13	28104.7	ER	0.6	0.0		MIT
3554.67	3555.69	28124.0	ER	0.8D	0.0		MIT
3554.293	3555.308	28126.96	ER	1.0			MIT
3553.204	3554.219	28135.58	ER	1.4	1.2		MIT
3552.317	3553.332	28142.60	ER	1.2			MIT
3551.795	3552.810	28146.74	ER	0.7S	0.0		MIT
3551.29	3552.30	28150.7	ER	1.3	0.3		MIT
3550.222	3551.236	28159.21	ER	1.1	0.9		MIT
3549.848	3550.862	28162.18	ER	1.4	1.3S		MIT

AIR WAVELENGTH (ANGSTRMS)	VACUUM WAVELENGTH (ANGSTRMS)	WAVENUMBER (KAYSERS)	LMENT	LOG ARC	INTENSITY SPK	DIS	REF
3549.553	3550.567	28164.51	ER	0.8	0.0		MIT
3548.23	3549.24	28175.0	ER	0.9D	0.0		MIT
3547.514	3548.528	28180.70	ER	1.4L	0.0H		MIT
3546.58	3547.59	28188.1	ER	0.9W			MIT
3545.16	3546.17	28199.4	ER	0.3			MIT
3544.350	3545.363	28205.86	ER	1.0	0.0		MIT
3542.98	3543.99	28216.8	ER	0.8D	0.5		MIT
3542.36	3543.37	28221.7	ER	1.0D	0.6		MIT
3539.598	3540.610	28243.73	ER	1.0			MIT
3539.289	3540.300	28246.19	ER	0.5			MIT
3538.516	3539.527	28252.36	ER	1.3	1.0		MIT
3537.282	3538.293	28262.22	ER	0.8			MIT
3536.764	3537.775	28266.36	ER	0.3			MIT
3536.023	3537.034	28272.28	ER	1.3S	1.1		MIT
3534.55	3535.56	28284.1	ER	1.2D	0.3		MIT
3534.346	3535.356	28285.69	ER	0.3			MIT
3531.264	3532.273	28310.38	ER	1.1L	0.0		MIT
3530.36	3531.37	28317.6	ER	0.8W	0.0		MIT
3529.34	3530.35	28325.8	ER	0.8W			MIT
3527.133	3528.141	28343.54	ER	1.2D			MIT
3526.813	3527.821	28346.11	ER	1.0W			MIT
3524.920	3525.928	28361.33	ER	1.3	0.9		MIT
3523.983	3524.991	28368.87	ER	1.4	0.9		MIT
3522.524	3523.531	28380.62	ER	0.5			MIT
3520.732	3521.739	28395.06	ER	1.3R			MIT
3520.034	3521.040	28400.70	ER	1.1	0.6		MIT
3519.094	3520.100	28408.28	ER	1.3F	0.3F		MIT
3518.176	3519.182	28415.69	ER	1.3	0.9		MIT
3517.894	3518.900	28417.97	ER	0.9	0.0		MIT
3516.99	3518.00	28425.3	ER	1.3W	0.3		MIT
3516.71	3517.72	28427.5	ER	0.7			MIT
3516.520	3517.526	28429.07	ER	0.9			MIT
3516.003	3517.008	28433.26	ER	1.0	0.0		MIT
3514.894	3515.899	28442.23	ER	1.2	0.7		MIT
3512.772	3513.777	28459.41	ER	0.3			MIT
3512.695	3513.700	28460.03	ER	1.0	0.0		MIT
3511.689	3512.693	28468.18	ER	1.0			MIT
3508.938	3509.942	28490.50	ER	0.7			MIT
3508.811	3509.815	28491.53	ER	0.9			MIT
3508.395	3509.398	28494.91	ER	1.2	0.8		MIT
3507.644	3508.647	28501.01	ER	0.8			MIT
3507.480	3508.483	28502.34	ER	1.3	0.6		MIT
3506.815	3507.818	28507.75	ER	1.2	0.8		MIT
3505.686	3506.689	28516.93	ER	0.9			MIT
3505.074	3506.077	28521.91	ER	1.1	0.9		MIT
3504.518	3505.521	28526.43	ER	1.4	0.6H		MIT
3504.06	3505.06	28530.2	ER	1.0D	0.0		MIT
3503.373	3504.375	28535.76	ER	1.0	0.0		MIT
3502.783	3503.785	28540.56	ER	1.0	0.0		MIT
3501.144	3502.146	28553.92	ER	1.0			MIT
3499.965	3500.966	28563.54	ER	1.0			MIT
3499.104	3500.105	28570.57	ER	1.3	1.2		MIT
3498.711	3499.712	28573.78	ER	1.0	0.5		MIT
3497.03	3498.03	28587.5	ER	0.5	0.0H		MIT
3496.860	3497.861	28588.90	ER	1.4	1.3		MIT
3494.815	3495.815	28605.63	ER	1.3	0.5		MIT
3494.136	3495.136	28611.19	ER	1.1	0.5		MIT
3493.721	3494.721	28614.59	ER	0.7			MIT
3492.543	3493.542	28624.24	ER	1.4D	0.3D		MIT
3490.236	3491.235	28643.16	ER	0.9			MIT
3490.057	3491.056	28644.63	ER	1.0	0.5		MIT
3489.353	3490.352	28650.41	ER	1.0			MIT
3488.529	3489.527	28657.18	ER	0.8			MIT
3486.828	3487.826	28671.16	ER	1.3	0.8		MIT
3485.860	3486.858	28679.12	ER	1.4	0.8		MIT
3485.169	3486.167	28684.80	ER	1.0	0.3		MIT
3484.830	3485.827	28687.59	ER	1.5D	0.8D		MIT
3484.557	3485.554	28689.84	ER	1.0	0.0		MIT
3482.605	3483.602	28705.92	ER	1.0	0.0		MIT
3481.807	3482.804	28712.50	ER	0.9			MIT

AIR WAVELENGTH (ANGSTRMS)	VACUUM WAVELENGTH (ANGSTRMS)	WAVENUMBER (KAYSERS)	LMENT	LOG ARC	INTENSITY SPK	DIS	REF
3480.722	3481.718	28721.45	ER	0.6			MIT
3480.441	3481.437	28723.77	ER	1.4	0.9		MIT
3479.416	3480.412	28732.23	ER	1.4	1.0		MIT
3477.938	3478.934	28744.44	ER	1.0	0.0		MIT
3477.068	3478.063	28751.63	ER	1.2	0.8		MIT
3476.301	3477.296	28757.98	ER	1.5	0.8		MIT
3474.707	3475.702	28771.17	ER	0.8			MIT
3474.275	3475.270	28774.75	ER	0.9	0.8W		MIT
3474.093	3475.088	28776.25	ER	0.9	0.8W		MIT
3473.95	3474.94	28777.4	ER	0.3			MIT
3473.86	3474.85	28778.2	ER	1.1	0.0		MIT
3472.82	3473.81	28786.8	ER	0.9H			MIT
3471.714	3472.708	28795.97	ER	1.4	0.9		MIT
3471.137	3472.131	28800.76	ER	1.0	0.0		MIT
3470.914	3471.908	28802.61	ER	1.0			MIT
3469.722	3470.716	28812.50	ER	1.1	0.5		MIT
3469.478	3470.472	28814.53	ER	0.6D			MIT
3465.12	3466.11	28850.8	ER	0.9D			MIT
3464.536	3465.528	28855.63	ER	1.0	0.5		MIT
3462.58	3463.57	28871.9	ER	1.0			MIT
3461.40	3462.39	28881.8	ER	1.0	0.0		MIT
3460.968	3461.959	28885.38	ER	1.3L	0.8		MIT
3458.792	3459.783	28903.55	ER	0.7			MIT
3457.26	3458.25	28916.4	ER	1.0	0.0		MIT
3456.561	3457.551	28922.20	ER	1.1	0.8		MIT
3454.318	3455.308	28940.98	ER	1.3	0.9		MIT
3453.10	3454.09	28951.2	ER	1.1	0.0		MIT
3453.04	3454.03	28951.7	ER	1.1	0.0		ED
3452.26	3453.25	28958.2	ER	0.9	0.0		MIT
3451.239	3452.228	28966.80	ER	1.0	0.0		MIT
3450.946	3451.935	28969.26	ER	1.3			MIT
3450.47	3451.46	28973.3	ER	0.7			MIT
3448.069	3449.057	28993.43	ER	1.3	0.3		MIT
3447.520	3448.508	28998.05	ER	1.0D			MIT
3446.87	3447.86	29003.5	ER	1.2	0.0		MIT
3446.36	3447.35	29007.8	ER	0.9	0.0		MIT
3445.575	3446.562	29014.42	ER	1.3	0.9		MIT
3445.46	3446.45	29015.4	ER	0.8			MIT
3444.29	3445.28	29025.2	ER	0.3			MIT
3443.70	3444.69	29030.2	ER	1.4	0.3		MIT
3443.23	3444.22	29034.2	ER	0.7			MIT
3442.638	3443.625	29039.17	ER	1.0	0.0		MIT
3442.37	3443.36	29041.4	ER	1.0	0.0		MIT
3441.135	3442.121	29051.85	ER	1.1	0.7		MIT
3438.473	3439.459	29074.34	ER	1.0	0.3H		MIT
3438.32	3439.31	29075.6	ER	1.1	0.0		MIT
3437.64	3438.63	29081.4	ER	1.0	0.0		MIT
3436.336	3437.321	29092.42	ER	1.0	0.0		MIT
3434.634	3435.619	29106.84	ER	1.1	0.3		MIT
3434.367	3435.352	29109.10	ER	1.4	0.9		MIT
3433.131	3434.115	29119.58	ER	1.4	0.3		MIT
3431.06	3432.04	29137.2	ER	0.8D	0.0H		MIT
3429.215	3430.198	29152.83	ER	0.7			MIT
3428.394	3429.377	29159.82	ER	0.6	0.0		MIT
3424.42	3425.40	29193.6	ER	1.0	0.0		MIT
3422.87	3423.85	29206.9	ER	0.5D	0.0		MIT
3422.62	3423.60	29209.0	ER	0.6			MIT
3422.471	3423.452	29210.28	ER	0.8			MIT
3421.065	3422.046	29222.28	ER	1.0	0.0		MIT
3420.183	3421.164	29229.82	ER	0.6H	0.3H		MIT
3419.962	3420.943	29231.71	ER	1.0	0.5		MIT
3419.22	3420.20	29238.1	ER	0.3			ED
3418.731	3419.712	29242.23	ER	1.0	0.3		MIT
3417.638	3418.618	29251.58	ER	1.0	0.6		MIT
3417.29	3418.27	29254.6	ER	1.0	0.5		MIT
3416.45	3417.43	29261.8	ER	1.3D	0.8D		MIT
3416.132	3417.112	29264.48	ER	0.9	0.0		MIT
3414.79	3415.77	29276.0	ER	0.9	0.0		MIT
3413.69	3414.67	29285.4	ER	0.7	0.0		MIT
3413.36	3414.34	29288.2	ER	0.9	0.0		MIT

AIR WAVELENGTH (ANGSTRMS)	VACUUM WAVELENGTH (ANGSTRMS)	WAVENUMBER (KAYSERS)	LMENT	LOG ARC	INTENSITY SPK	DIS	REF
3411.89	3412.87	29300.9	ER	0.8	0.0H		MIT
3409.872	3410.850	29318.20	ER	1.0	0.0		MIT
3409.24	3410.22	29323.6	ER	0.9	0.0		MIT
3408.678	3409.656	29328.47	ER	1.0	0.5		MIT
3407.791	3408.769	29336.11	ER	1.2	0.9		MIT
3407.303	3408.281	29340.31	ER	0.6			MIT
3406.96	3407.94	29343.3	ER	1.0	0.0		MIT
3404.13	3405.11	29367.6	ER	0.9			MIT
3403.678	3404.655	29371.55	ER	1.0	0.0		MIT
3402.87	3403.85	29378.5	ER	0.9	0.0		MIT
3401.830	3402.806	29387.51	ER	1.1	0.8		MIT
3401.201	3402.177	29392.94	ER	1.0	0.0		MIT
3399.601	3400.577	29406.78	ER	0.7			MIT
3398.94	3399.92	29412.5	ER	1.6$	0.8$		MIT
3398.276	3399.251	29418.24	ER	1.1L	0.3		MIT
3396.83	3397.80	29430.8	ER	1.1S	0.3		MIT
3396.05	3397.02	29437.5	ER	1.1D	0.5		MIT
3395.79	3396.76	29439.8	ER	0.6			MIT
3395.280	3396.255	29444.20	ER	0.9	0.0		MIT
3394.860	3395.834	29447.84	ER	1.2	0.5		MIT
3394.394	3395.368	29451.89	ER	1.0	0.0		MIT
3394.099	3395.073	29454.45	ER	1.0	0.0		MIT
3393.572	3394.546	29459.02	ER	1.4	1.0		MIT
3391.989	3392.963	29472.77	ER	1.5	1.1		MIT
3389.744	3390.717	29492.29	ER	1.3	0.5		MIT
3389.599	3390.572	29493.55	ER	1.0	0.3		MIT
3387.73	3388.70	29509.8	ER	0.8D			MIT
3385.087	3386.059	29532.86	ER	1.4D	1.2D		MIT
3384.25	3385.22	29540.2	ER	0.8	0.0		MIT
3384.09	3385.06	29541.6	ER	0.8	0.0		MIT
3383.78	3384.75	29544.3	ER	1.1	0.0		MIT
3382.892	3383.863	29552.02	ER	1.3	0.0		MIT
3382.067	3383.038	29559.23	ER	1.2	0.0		MIT
3381.32	3382.29	29565.8	ER	1.3	0.6		MIT
3381.079	3382.050	29567.87	ER	1.2	0.3		MIT
3378.75	3379.72	29588.2	ER	0.8			MIT
3377.00	3377.97	29603.6	ER	1.0			MIT
3376.09	3377.06	29611.6	ER	1.1	0.0		MIT
3374.170	3375.139	29628.41	ER	1.4	0.7		MIT
3373.67	3374.64	29632.8	ER	0.7	0.0		MIT
3372.750	3373.719	29640.88	ER	1.5	1.3		MIT
3370.59	3371.56	29659.9	ER	1.2L	0.5		MIT
3369.620	3370.588	29668.41	ER	1.1	0.0		MIT
3368.953	3369.921	29674.29	ER	1.0	0.0		MIT
3368.07	3369.04	29682.1	ER	1.3	0.7		M
3367.634	3368.601	29685.91	ER	1.3			MIT
3366.69	3367.66	29694.2	ER	1.2	0.0		MIT
3364.43	3365.40	29714.2	ER	0.7			MIT
3364.09	3365.06	29717.2	ER	1.4	0.8		MIT
3361.67	3362.64	29738.6	ER	0.8	0.0		MIT
3361.02	3361.99	29744.3	ER	1.1	0.0		MIT
3358.15	3359.12	29769.7	ER	1.0	0.0		MIT
3356.21	3357.17	29786.9	ER	1.3	0.3		MIT
3354.58	3355.54	29801.4	ER	1.0			MIT
3351.30	3352.26	29830.6	ER	1.0W	0.3		MIT
3350.255	3351.218	29839.90	ER	1.0	0.3		MIT
3350.063	3351.026	29841.61	ER	1.3	0.8		MIT
3348.74	3349.70	29853.4	ER	1.0	0.0		MIT
3348.140	3349.102	29858.75	ER	1.2	0.3		MIT
3347.86	3348.82	29861.2	ER	0.8	0.0		MIT
3347.734	3348.696	29862.37	ER	1.0	0.0		MIT
3347.64	3348.60	29863.2	ER	0.7	0.3H		MIT
3346.685	3347.647	29871.73	ER	0.5			MIT
3346.350	3347.312	29874.72	ER	1.0	0.3		MIT
3346.03	3346.99	29877.6	ER	1.5	1.0		MIT
3345.46	3346.42	29882.7	ER	0.8	0.0		MIT
3344.75	3345.71	29889.0	ER	0.9	0.0		MIT
3344.36	3345.32	29892.5	ER	1.1	0.0		MIT
3344.17	3345.13	29894.2	ER	0.6			MIT
3343.70	3344.66	29898.4	ER	1.0W	0.3		MIT
3343.43	3344.39	29900.8	ER	0.6J	0.0H		MIT
3343.12	3344.08	29903.6	ER	0.5			MIT
3342.896	3343.857	29905.58	ER	1.0			MIT
3342.11	3343.07	29912.6	ER	0.8	0.0		MIT
3341.836	3342.797	29915.07	ER	1.3	0.5		MIT
3341.61	3342.57	29917.1	ER	1.1	0.3		MIT
3340.48	3341.44	29927.2	ER	0.5	0.3		MIT
3340.02	3340.98	29931.3	ER	1.3	0.5		MIT
3339.12	3340.08	29939.4	ER	0.6	0.0		MIT
3338.018	3338.978	29949.28	ER	1.1	0.0H		MIT
3337.79	3338.75	29951.3	ER	1.3	0.5		MIT
3337.25	3338.21	29956.2	ER	1.2	0.5		MIT
3336.76	3337.72	29960.6	ER	1.2	0.0		MIT
3335.43	3336.39	29972.5	ER	0.9			MIT
3335.32	3336.28	29973.5	ER	0.8	0.0		MIT
3334.88	3335.84	29977.5	ER	0.7	0.0		MIT
3332.703	3333.662	29997.05	ER	1.4	0.3		MIT
3331.558	3332.516	30007.36	ER	0.8			MIT
3331.05	3332.01	30011.9	ER	0.6	0.0		MIT
3329.659	3330.617	30024.47	ER	1.4	0.5		MIT
3326.88	3327.84	30049.6	ER	1.0	0.0		MIT
3323.20	3324.16	30082.8	ER	1.4	0.6		MIT
3319.872	3320.827	30112.98	ER	1.3	0.0		MIT
3319.780	3320.735	30113.81	ER	0.6	0.0		MIT
3319.54	3320.50	30116.0	ER	0.3	0.0		MIT
3318.771	3319.726	30122.97	ER	1.2	0.0		MIT
3318.23	3319.18	30127.9	ER	1.0	0.0		MIT
3317.47	3318.42	30134.8	ER	1.2	0.3		MIT
3316.385	3317.339	30144.64	ER	1.1	0.7		MIT
3316.06	3317.01	30147.6	ER	0.9			MIT
3314.941	3315.895	30157.77	ER	1.1	0.0		MIT
3314.427	3315.381	30162.45	ER	1.2	0.3		MIT
3313.653	3314.607	30169.49	ER	1.0	0.3		MIT
3313.49	3314.44	30171.0	ER	0.8	0.0		MIT
3312.424	3313.377	30180.69	ER	1.4	1.2		MIT
3311.09	3312.04	30192.8	ER	0.5			MIT
3308.88	3309.83	30213.0	ER	1.2	0.3		MIT
3308.69	3309.64	30214.7	ER	0.5			MIT
3307.46	3308.41	30226.0	ER	1.1	0.3		MIT
3305.568	3306.520	30243.28	ER	1.2	0.5		MIT
3305.120	3306.072	30247.38	ER	0.7J			MIT
3304.074	3305.025	30256.96	ER	1.0			MIT
3303.95	3304.90	30258.1	ER	1.0	0.3		MIT
3301.93	3302.88	30276.6	ER	1.2	0.3		MIT
3301.68	3302.63	30278.9	ER	0.5			MIT
3301.08	3302.03	30284.4	ER	1.0			MIT
3300.577	3301.527	30289.01	ER	1.1			MIT
3299.42	3300.37	30299.6	ER	1.0			MIT
3291.27	3292.22	30374.7	ER	0.7	0.0		MIT
3287.96	3288.91	30405.2	ER	1.0S	0.3		MIT
3286.754	3287.701	30416.39	ER	1.1	0.3		MIT
3286.18	3287.13	30421.7	ER	1.0	0.3		MIT
3285.60	3286.55	30427.1	ER	1.2	0.3		MIT
3280.22	3281.17	30477.0	ER	1.1	0.5		MIT
3279.32	3280.26	30485.3	ER	1.3	0.7		MIT
3278.222	3279.167	30495.55	ER	1.2	0.5		MIT
3277.698	3278.643	30500.43	ER	1.1	0.3		MIT
3275.442	3276.386	30521.44	ER	1.2	0.3		MIT
3274.747	3275.691	30527.91	ER	1.0	0.0		MIT
3273.325	3274.268	30541.17	ER	0.8			MIT
3273.08	3274.02	30543.5	ER	1.4	1.2		MIT
3272.36	3273.30	30550.2	ER	0.7			MIT
3272.10	3273.04	30552.6	ER	1.1	0.0		MIT
3269.411	3270.353	30577.74	ER	1.3	0.6		MIT
3268.80	3269.74	30583.4	ER	1.1	0.0		MIT
3267.15	3268.09	30598.9	ER	0.6	0.6		MIT
3264.781	3265.722	30621.10	ER	1.4	0.7		MIT
3262.80	3263.74	30639.7	ER	1.2	0.6		MIT
3259.048	3259.988	30674.96	ER	1.3D	0.8D		MIT
3258.484	3259.424	30680.27	ER	1.0	0.3		MIT

AIR WAVELENGTH (ANGSTRMS)	VACUUM WAVELENGTH (ANGSTRMS)	WAVENUMBER (KAYSERS)	LMENT	LOG ARC	INTENSITY SPK	DIS	REF
3257.69	3258.63	30687.7	ER	0.3			ED
3256.35	3257.29	30700.4	ER	1.0	0.3		MIT
3255.787	3256.726	30705.68	ER	1.0	0.0		MIT
3249.342	3250.279	30766.59	ER	1.4	0.9		MIT
3247.52	3248.46	30783.9	ER	1.3D	0.3		MIT
3243.469	3244.405	30822.29	ER	0.6	0.0		MIT
3243.252	3244.188	30824.36	ER	0.5	0.0		MIT
3240.484	3241.419	30850.69	ER	1.1	0.3		MIT
3237.979	3238.913	30874.55	ER	1.3	0.5		MIT
3234.798	3235.732	30904.91	ER	0.6			MIT
3232.026	3232.959	30931.42	ER	1.3	0.5		MIT
3230.585	3231.518	30945.21	ER	1.4	1.2		MIT
3230.12	3231.05	30949.7	ER	0.6	0.0		MIT
3229.93	3230.86	30951.5	ER	1.0	0.3		MIT
3227.16	3228.09	30978.1	ER	1.3	0.3		MIT
3223.308	3224.239	31015.07	ER	1.4	0.7		MIT
3220.730	3221.660	31039.90	ER	1.4	0.7		MIT
3219.72	3220.65	31049.6	ER	1.0			MIT
3217.790	3218.719	31068.26	ER	0.6			MIT
3214.446	3215.374	31100.58	ER	1.2	0.6		MIT
3208.049	3208.976	31162.59	ER	1.1	0.3		MIT
3205.148	3206.074	31190.79	ER	1.2	0.5		MIT
3203.950	3204.876	31202.46	ER	0.8	0.0		MIT
3203.36	3204.29	31208.2	ER	1.3	1.1		MIT
3200.59	3201.51	31235.2	ER	0.8	0.0		MIT
3194.09	3195.01	31298.8	ER	1.0	0.0		MIT
3192.634	3193.557	31313.05	ER	1.3	1.1		MIT
3187.785	3188.707	31360.68	ER	1.4	1.1		MIT
3185.247	3186.168	31385.66	ER	1.2	0.3		MIT
3183.41	3184.33	31403.8	ER	1.2	0.6		MIT
3181.923	3182.843	31418.45	ER	1.3	0.7		MIT
3181.68	3182.60	31420.9	ER	0.7D	0.3H		MIT
3179.61	3180.53	31441.3	ER	0.6	0.6		MIT
3175.52	3176.44	31481.8	ER	1.0	0.3		MIT
3172.64	3173.56	31510.4	ER	0.8	0.0		MIT
3171.518	3172.436	31521.52	ER	0.8	0.3		MIT
3169.29	3170.21	31543.7	ER	0.8	0.0		MIT
3167.103	3168.020	31565.46	ER	1.2	0.5		MIT
3164.53	3165.45	31591.1	ER	0.8	0.3		MIT
3161.37	3162.29	31622.7	ER	1.3	0.5		MIT
3160.35	3161.26	31632.9	ER	1.3	0.3		MIT
3154.286	3155.199	31693.72	ER	1.2	0.7		MIT
3152.377	3153.290	31712.91	ER	1.1	0.5		MIT
3150.55	3151.46	31731.3	ER	1.1	0.0		MIT
3144.504	3145.415	31792.31	ER	1.0	0.3		MIT
3144.32	3145.23	31794.2	ER	1.0S	0.3		MIT
3143.63	3144.54	31801.1	ER	1.2	0.0		MIT
3143.23	3144.14	31805.2	ER	0.6	0.0		MIT
3142.79	3143.70	31809.6	ER	1.0	0.3		MIT
3141.80	3142.71	31819.7	ER	0.7	0.3D		MIT
3141.12	3142.03	31826.6	ER	1.3	0.5		MIT
3138.49	3139.40	31853.2	ER	1.0	0.0		MIT
3137.85	3138.76	31859.7	ER	1.0	0.3		MIT
3135.61	3136.52	31882.5	ER	0.7	0.0		MIT
3132.775	3133.683	31911.33	ER	1.2	0.7		MIT
3132.51	3133.42	31914.0	ER	1.1	0.5		MIT
3132.03	3132.94	31918.9	ER	0.8	0.3		MIT
3131.07	3131.98	31928.7	ER	0.5			MIT
3127.32	3128.23	31967.0	ER	1.0	0.0		EX
3126.18	3127.09	31978.6	ER	0.6			MIT
3125.65	3126.56	31984.1	ER	1.0	0.3		MIT
3125.17	3126.08	31989.0	ER	1.2	0.3		MIT
3123.087	3123.992	32010.32	ER	1.1	0.0		MIT
3122.65	3123.56	32014.8	ER	1.1	0.9		MIT
3121.89	3122.80	32022.6	ER	1.1	0.0		MIT
3119.05	3119.95	32051.7	ER	0.9	0.3		MIT
3118.833	3119.737	32053.98	ER	1.1	0.5		MIT
3118.440	3119.344	32058.02	ER	0.8			MIT
3116.947	3117.851	32073.37	ER	1.3	0.3		MIT
3115.534	3116.438	32087.92	ER	1.2	0.0		MIT

AIR WAVELENGTH (ANGSTRMS)	VACUUM WAVELENGTH (ANGSTRMS)	WAVENUMBER (KAYSERS)	LMENT	LOG ARC	INTENSITY SPK	DIS	REF
3115.09	3115.99	32092.5	ER	1.1	0.0		MIT
3113.536	3114.439	32108.51	ER	1.1	0.5		MIT
3112.03	3112.93	32124.1	ER	1.3	0.8		MIT
3110.87	3111.77	32136.0	ER	1.1	0.5		MIT
3106.787	3107.688	32178.26	ER	1.2	0.3		MIT
3102.686	3103.586	32220.79	ER	1.2	0.3		MIT
3099.19	3100.09	32257.1	ER	1.3	0.8		MIT
3095.88	3096.78	32291.6	ER	1.4	0.9		MIT
3093.75	3094.65	32313.9	ER	0.8	0.0		MIT
3091.93	3092.83	32332.9	ER	0.7			MIT
3089.77	3090.67	32355.5	ER	0.7	0.0		MIT
3088.76	3089.66	32366.1	ER	0.9	0.3		MIT
3087.79	3088.69	32376.2	ER	1.0	0.0		MIT
3087.12	3088.02	32383.2	ER	0.7			MIT
3084.36	3085.26	32412.2	ER	1.0	0.3		MIT
3084.02	3084.92	32415.8	ER	1.2	0.6		MIT
3082.08	3082.98	32436.2	ER	1.1	0.5		MIT
3081.38	3082.27	32443.6	ER	0.5	0.0		MIT
3078.873	3079.767	32469.99	ER	0.9S	0.3		MIT
3078.233	3079.127	32476.74	ER	0.7	0.0		MIT
3073.347	3074.240	32528.36	ER	1.1	0.6		MIT
3072.52	3073.41	32537.1	ER	1.2	0.8		MIT
3072.13	3073.02	32541.2	ER	0.8	0.0		MIT
3070.743	3071.635	32555.95	ER	1.2	0.6		MIT
3069.229	3070.121	32572.01	ER	1.2	0.0		MIT
3066.225	3067.116	32603.92	ER	1.2	0.3		MIT
3064.97	3065.86	32617.3	ER	0.5	0.0		MIT
3064.84	3065.73	32618.6	ER	0.8			MIT
3061.683	3062.573	32652.28	ER	1.2	0.3		MIT
3061.28	3062.17	32656.6	ER	0.7	0.0H		MIT
3057.52	3058.41	32696.7	ER	0.7			MIT
3054.42	3055.31	32729.9	ER	1.0	0.3		MIT
3053.78	3054.67	32736.8	ER	0.8	0.0		MIT
3050.83	3051.72	32768.4	ER	0.7	0.0H		MIT
3049.99	3050.88	32777.5	ER	1.0	0.0		MIT
3049.29	3050.18	32785.0	ER	0.6	0.0		MIT
3048.40	3049.29	32794.5	ER	1.0	0.0		MIT
3036.222	3037.106	32926.08	ER	1.3	0.6		MIT
3031.308	3032.190	32979.46	ER	0.9	0.5		MIT
3028.278	3029.160	33012.46	ER	1.3	0.3		MIT
3025.92	3026.80	33038.2	ER	0.7	0.3		MIT
3019.769	3020.649	33105.47	ER	1.1	0.0		MIT
3017.740	3018.619	33127.73	ER	1.0	0.0		MIT
3016.838	3017.717	33137.63	ER	1.0	0.3		MIT
3012.468	3013.346	33185.70	ER	1.0	0.3		MIT
3011.16	3012.04	33200.1	ER	0.9	0.0		MIT
3010.29	3011.17	33209.7	ER	0.8			MIT
3003.831	3004.707	33281.12	ER	1.0			MIT
3002.66	3003.54	33294.1	ER	1.2	0.5		MIT
2998.058	2998.932	33345.20	ER	1.0	0.0		MIT
2996.379	2997.253	33363.89	ER	1.1	0.0H		MIT
2994.48	2995.35	33385.0	ER	0.8D	0.0		MIT
2994.09	2994.96	33389.4	ER	0.5			MIT
2989.55	2990.42	33440.1	ER	0.7D	0.0		MIT
2989.287	2990.159	33443.04	ER	1.3	0.0		MIT
2983.793	2984.664	33504.61	ER	1.0	0.0		MIT
2983.223	2984.093	33511.02	ER	1.1	0.0		MIT
2983.058	2983.928	33512.87	ER	1.0	0.0		MIT
2979.950	2980.820	33547.82	ER	1.0			MIT
2975.679	2976.548	33595.97	ER	0.7	0.3		MIT
2974.48	2975.35	33609.5	ER	0.6	0.0		MIT
2973.743	2974.611	33617.84	ER	0.5			MIT
2972.279	2973.147	33634.40	ER	0.7L			MIT
2971.260	2972.127	33645.93	ER	0.6D			MIT
2970.07	2970.94	33659.4	ER	0.3			MIT
2968.761	2969.628	33674.25	ER	1.1	0.3		MIT
2967.938	2968.805	33683.59	ER	0.7	0.0		MIT
2966.185	2967.051	33703.50	ER	0.6	0.0		MIT
2965.22	2966.09	33714.5	ER	0.6			MIT
2964.518	2965.384	33722.45	ER	1.3	0.6		MIT

AIR WAVELENGTH (ANGSTRMS)	VACUUM WAVELENGTH (ANGSTRMS)	WAVENUMBER (KAYSERS)	LMENT	LOG ARC	INTENSITY SPK	DIS	REF
2964.254	2965.120	33725.45	ER	0.5	0.0		MIT
2963.901	2964.767	33729.47	ER	0.7	0.0		MIT
2963.706	2964.572	33731.69	ER	0.8	0.0		MIT
2962.502	2963.367	33745.40	ER	0.9	0.3		MIT
2962.334	2963.199	33747.31	ER	0.5			MIT
2960.120	2960.985	33772.55	ER	0.7	0.0		MIT
2958.89	2959.75	33786.6	ER	0.8	0.0H		MIT
2956.955	2957.819	33808.70	ER	0.8	0.0		MIT
2956.448	2957.312	33814.49	ER	0.7	0.0H		MIT
2953.706	2954.569	33845.88	ER	0.6			MIT
2950.070	2950.932	33887.60	ER	0.7			MIT
2948.802	2949.664	33902.17	ER	1.0	0.0		MIT
2946.615	2947.476	33927.33	ER	0.9	0.7		MIT
2945.280	2946.141	33942.71	ER	1.2			MIT
2944.071	2944.932	33956.65	ER	1.1			MIT
2942.211	2943.071	33978.11	ER	0.9	0.0		MIT
2941.715	2942.575	33983.84	ER	0.8	0.0		MIT
2941.179	2942.039	33990.03	ER	0.8	0.0		MIT
2939.310	2940.170	34011.64	ER	0.7			MIT
2934.638	2935.496	34065.79	ER	0.6	0.0		MIT
2934.517	2935.375	34067.19	ER	1.2	0.0		MIT
2930.647	2931.504	34112.18	ER	1.1			MIT
2929.734	2930.591	34122.81	ER	1.1			MIT
2928.40	2929.26	34138.4	ER	0.6	0.0		MIT
2928.284	2929.141	34139.70	ER	1.2	0.3		MIT
2927.717	2928.574	34146.31	ER	0.8	0.0		MIT
2926.349	2927.205	34162.28	ER	0.5L			MIT
2920.243	2921.098	34233.71	ER	1.0			MIT
2919.27	2920.12	34245.1	ER	1.1D	0.8		MIT
2915.626	2916.480	34287.91	ER	1.3	0.3		MIT
2914.667	2915.520	34299.19	ER	0.6			MIT
2911.417	2912.270	34337.48	ER	1.5	1.2		MIT
2911.069	2911.922	34341.58	ER	0.6S			MIT
2910.357	2911.209	34349.98	ER	1.3	0.8		MIT
2909.577	2910.429	34359.19	ER	0.8	0.0		MIT
2908.535	2909.387	34371.50	ER	0.8			MIT
2906.500	2907.351	34395.57	ER	0.8			MIT
2904.467	2905.318	34419.64	ER	1.4	0.7		MIT
2903.443	2904.294	34431.78	ER	0.6			MIT
2897.518	2898.367	34502.18	ER	1.1	0.3		MIT
2896.960	2897.809	34508.83	ER	1.2	0.5		MIT
2896.574	2897.423	34513.43	ER	0.7			MIT
2895.927	2896.776	34521.14	ER	0.6			MIT
2893.907	2894.755	34545.23	ER	1.1	0.0		MIT
2893.496	2894.344	34550.14	ER	0.6			MIT
2891.387	2892.235	34575.34	ER	1.3	1.1		MIT
2888.160	2889.007	34613.97	ER	0.8	0.0		MIT
2887.106	2887.953	34626.61	ER	1.0	0.3H		MIT
2886.825	2887.672	34629.98	ER	0.3			MIT
2882.603	2883.449	34680.69	ER	0.8			MIT
2878.913	2879.758	34725.15	ER	0.8	0.3H		MIT
2874.83	2875.67	34774.5	ER	1.1	0.0		MIT
2873.808	2874.651	34786.83	ER	1.0	0.3		MIT
2872.837	2873.680	34798.58	ER	0.8	0.0		MIT
2871.681	2872.524	34812.59	ER	1.0	0.0		MIT
2870.527	2871.370	34826.59	ER	0.5			MIT
2866.355	2867.197	34877.27	ER	0.5			MIT
2862.605	2863.446	34922.96	ER	0.7			MIT
2860.257	2861.097	34951.63	ER	0.6			MIT
2859.814	2860.654	34957.04	ER	1.4	1.0		MIT
2859.385	2860.225	34962.29	ER	0.8	0.8		MIT
2858.558	2859.398	34972.40	ER	1.0	0.3		MIT
2855.414	2856.253	35010.91	ER	1.2	0.5		MIT
2848.368	2849.205	35097.51	ER	1.0	0.6		MIT
2847.018	2847.855	35114.15	ER	0.5			MIT
2846.265	2847.102	35123.44	ER	0.7	0.3H		MIT
2845.866	2846.702	35128.36	ER	0.9	0.0		MIT
2841.895	2842.731	35177.45	ER	0.7			MIT
2841.343	2842.178	35184.28	ER	0.7			MIT
2840.830	2841.665	35190.63	ER	1.0	0.3		MIT

AIR WAVELENGTH (ANGSTRMS)	VACUUM WAVELENGTH (ANGSTRMS)	WAVENUMBER (KAYSERS)	LMENT	LOG ARC	INTENSITY SPK	DIS	REF
2838.714	2839.549	35216.86	ER	1.2	0.5		MIT
2837.106	2837.940	35236.82	ER	0.7			MIT
2833.910	2834.744	35276.56	ER	0.7	0.3H		MIT
2833.061	2833.894	35287.13	ER	1.4			MIT
2831.794	2832.627	35302.92	ER	0.5			MIT
2830.416	2831.249	35320.10	ER	0.8	0.3		MIT
2829.373	2830.205	35333.12	ER	0.7			MIT
2827.900	2828.732	35351.53	ER	1.1	0.3		MIT
2824.322	2825.153	35396.31	ER	0.5			MIT
2820.186	2821.016	35448.22	ER	1.3	0.5		MIT
2819.815	2820.645	35452.88	ER	0.8	0.3		MIT
2818.852	2819.682	35464.99	ER	1.0	0.9		MIT
2807.370	2808.197	35610.04	ER	0.6			MIT
2805.898	2806.725	35628.72	ER		0.5		MIT
2804.358	2805.184	35648.28	ER	1.1	0.3		MIT
2803.541	2804.367	35658.67	ER	0.8	0.9		MIT
2802.868	2803.694	35667.23	ER	0.8	0.0		MIT
2802.528	2803.354	35671.56	ER	1.1	0.0		MIT
2799.728	2800.553	35707.23	ER	0.6			MIT
2793.846	2794.670	35782.40	ER	0.6	0.0		MIT
2793.168	2793.992	35791.09	ER	0.5			MIT
2792.524	2793.347	35799.34	ER		0.8		MIT
2788.462	2789.284	35851.49	ER	0.6			MIT
2787.711	2788.533	35861.15	ER	1.1	0.5		MIT
2787.406	2788.228	35865.07	ER	0.9	0.0		MIT
2786.109	2786.931	35881.77	ER	1.0	0.0		MIT
2784.954	2785.776	35896.65	ER	1.0	0.0		MIT
2783.691	2784.512	35912.93	ER	0.3	0.3		MIT
2782.118	2782.939	35933.24	ER	0.7			MIT
2781.187	2782.008	35945.26	ER	0.6S	0.0		MIT
2774.620	2775.439	36030.33	ER	0.8	0.3		MIT
2768.741	2769.559	36106.84	ER		0.3H		MIT
2766.387	2767.204	36137.56	ER	1.1	0.3		MIT
2765.617	2766.434	36147.62	ER	1.0	0.0		MIT
2764.701	2765.518	36159.60	ER	0.5			MIT
2761.905	2762.721	36196.20	ER		0.5		MIT
2759.204	2760.019	36231.63	ER		0.8		MIT
2755.643	2756.457	36278.45	ER	1.3	0.5		MIT
2754.944	2755.758	36287.65	ER	1.2	0.0		MIT
2750.17	2750.98	36350.6	ER	1.2	0.5		MIT
2746.059	2746.871	36405.06	ER		0.3		MIT
2744.13	2744.94	36430.6	ER	0.3	0.3H		MIT
2739.306	2740.117	36494.80	ER	1.3	0.6		MIT
2738.50	2739.31	36505.5	ER		0.3		MIT
2730.093	2730.901	36617.95	ER	0.9	0.0		MIT
2726.230	2727.037	36669.83	ER	1.2	0.0		MIT
2724.660	2725.467	36690.96	ER	0.5			MIT
2723.23	2724.04	36710.2	ER		0.7		MIT
2721.592	2722.398	36732.32	ER	0.7	0.0		MIT
2720.74	2721.55	36743.8	ER	0.6			MIT
2712.662	2713.466	36853.23	ER	0.6	0.3		MIT
2712.120	2712.924	36860.60	ER	0.8	0.0		MIT
2704.379	2705.181	36966.10	ER	0.5			MIT
2698.392	2699.193	37048.12	ER		0.7		MIT
2692.832	2693.631	37124.61	ER	0.5	0.3H		MIT
2688.403	2689.201	37185.76	ER	0.5			MIT
2684.747	2685.544	37236.40	ER	0.6	0.5		MIT
2683.14	2683.94	37258.7	ER		0.3		MIT
2677.589	2678.385	37335.94	ER	0.3			MIT
2675.347	2676.142	37367.22	ER	0.7	0.0		MIT
2672.246	2673.040	37410.58	ER	1.1	0.3		MIT
2670.255	2671.049	37438.48	ER	1.4	0.5		MIT
2666.296	2667.089	37494.06	ER	0.7			MIT
2665.044	2665.837	37511.68	ER	1.3	0.7		MIT
2663.457	2664.249	37534.03	ER	0.3			MIT
2655.246	2656.036	37650.09	ER	0.6			MIT
2654.096	2654.886	37666.40	ER	0.3			MIT
2653.743	2654.533	37671.41	ER	1.2	1.0		MIT
2637.806	2638.592	37899.00	ER		0.8		MIT
2627.767	2628.551	38043.78	ER	0.6			MIT

AIR WAVELENGTH (ANGSTRMS)	VACUUM WAVELENGTH (ANGSTRMS)	WAVENUMBER (KAYSERS)	LMENT	LOG ARC	INTENSITY SPK	DIS	REF
2624.181	2624.964	38095.76	ER	1.0			MIT
2616.873	2617.654	38202.14	ER	0.6	0.0		MIT
2615.420	2616.201	38223.37	ER	1.2L	1.0		MIT
2614.555	2615.336	38236.01	ER	0.5	0.3		MIT
2612.379	2613.159	38267.86	ER	0.6			MIT
2604.88	2605.66	38378.0	ER	0.6	0.0		MIT
2602.672	2603.450	38410.58	ER	0.5			MIT
2595.036	2595.812	38523.59	ER	1.0	0.5		MIT
2592.581	2593.356	38560.07	ER	1.0	0.0		MIT
2591.905	2592.680	38570.13	ER	0.0	0.5		MIT
2590.81	2591.58	38586.4	ER		0.3		MIT
2587.358	2588.132	38637.90	ER	0.6			MIT
2587.045	2587.819	38642.58	ER	1.0	0.0		MIT
2581.582	2582.355	38724.35	ER	0.9	0.3		MIT
2579.611	2580.383	38753.93	ER	0.8	0.7H		MIT
2578.794	2579.566	38766.21	ER		0.7		MIT
2570.803	2571.573	38886.70	ER	0.3	0.0H		MIT
2557.27	2558.04	39092.5	ER		0.3		MIT
2547.270	2548.035	39245.93	ER	0.9	0.0		MIT
2534.979	2535.741	39436.21	ER	0.6	0.0		MIT
2522.185	2522.944	39636.24	ER	0.7	0.0		MIT
2513.926	2514.683	39766.45	ER	0.6			MIT
2508.63	2509.39	39850.4	ER		0.5		MIT
2507.620	2508.375	39866.44	ER	0.6			MIT
2503.488	2504.242	39932.24	ER	0.6	0.0		MIT
2493.26	2494.01	40096.0	ER	0.3	0.0		MIT
2460.70	2461.44	40626.5	ER	0.5	0.6W		MIT
2446.37	2447.11	40864.5	ER	1.2	0.3		MIT
2445.46	2446.20	40879.7	ER		0.7		MIT
2422.53	2423.27	41266.6	ER		0.5		EX
2420.28	2421.02	41305.0	ER	0.7	0.3		MIT
2419.79	2420.53	41313.4	ER		0.7		MIT
2414.39	2415.12	41405.7	ER	0.0H	0.3H		MIT
2407.94	2408.67	41516.6	ER		0.5		EX
2404.61	2405.34	41574.1	ER		0.5		MIT
2403.752	2404.483	41588.97	ER		0.5		MIT
2403.35	2404.08	41595.9	ER		0.8W		MIT
2399.57	2400.30	41661.5	ER	0.7			A
2398.87	2399.60	41673.6	ER		0.9		MIT
2398.15	2398.88	41686.1	ER		1.0		MIT
2397.25	2397.98	41701.8	ER	0.8	0.5		MIT
2396.38	2397.11	41716.9	ER		1.5		MIT
2395.81	2396.54	41726.8	ER		0.6		MIT
2395.09	2395.82	41739.4	ER		0.5		MIT
2393.01	2393.74	41775.6	ER	0.3H	0.9		MIT
2391.94	2392.67	41794.3	ER	0.5	0.3H		MIT
2391.63	2392.36	41799.7	ER		0.8		MIT
2390.15	2390.88	41825.6	ER	0.8	0.5		MIT
2389.79	2390.52	41831.9	ER		0.5		MIT
2387.56	2388.29	41871.0	ER		0.3H		MIT
2387.16	2387.89	41878.0	ER	0.9	0.5		MIT
2386.58	2387.31	41888.2	ER	0.7	0.6		MIT
2384.09	2384.82	41931.9	ER		0.6		MIT
2383.84	2384.57	41936.3	ER		0.6		MIT
2381.72	2382.45	41973.7	ER		0.7		MIT
2381.41	2382.14	41979.1	ER		0.5		MIT
2378.85	2379.58	42024.3	ER		0.8		MIT
2377.83	2378.56	42042.3	ER	0.8	0.3H		MIT
2377.14	2377.87	42054.5	ER		0.8		MIT
2373.82	2374.54	42113.3	ER	1.0			A
2372.06	2372.78	42144.6	ER		0.5H		MIT
2371.31	2372.03	42157.9	ER		0.3H		MIT
2367.68	2368.40	42222.5	ER	0.8	0.6		MIT
2365.47	2366.19	42262.0	ER		0.9		MIT
2363.94	2364.66	42289.3	ER	0.8	0.3H		MIT
2359.33	2360.05	42372.0	ER		0.5		MIT
2358.71	2359.43	42383.1	ER	0.7	0.7		MIT
2358.51	2359.23	42386.7	ER	1.1			MIT
2355.55	2356.27	42440.0	ER		0.6		MIT
2354.18	2354.90	42464.6	ER	1.2	0.3H		MIT

AIR WAVELENGTH (ANGSTRMS)	VACUUM WAVELENGTH (ANGSTRMS)	WAVENUMBER (KAYSERS)	LMENT	LOG ARC	INTENSITY SPK	DIS	REF
2349.26	2349.98	42553.6	ER	0.5			MIT
2345.10	2345.82	42629.0	ER	0.3H			MIT
2341.84	2342.56	42688.4	ER	1.0			MIT
2338.29	2339.01	42753.2	ER	0.5	0.8		A
2329.33	2330.04	42917.6	ER		0.8		MIT
2328.043	2328.758	42941.35	ER	0.3H			MIT
2327.36	2328.07	42954.0	ER		2.2		A
2323.64	2324.35	43022.7	ER	0.5			MIT
2314.43	2315.14	43193.9	ER	0.5	1.6		A
2283.41	2284.11	43780.6	ER	0.5	0.7		A
2275.30	2276.00	43936.7	ER	0.8	0.3		A
2191.24	2191.93	45622.0	ER		2.3C		A
2185.71	2186.39	45737.4	ER	1.5			A
2170.00	2170.68	46068.5	ER	0.9			A
2131.39	2132.06	46902.9	ER	0.8	0.3H		A
2128.07	2128.74	46976.1	ER		2.4W		A
2089.96	2090.62	47832.6	ER	0.9			A
2085.50	2086.16	47934.9	ER	0.8			A
2080.94	2081.60	48039.9	ER	0.9			A
2068.65	2069.31	48325.3	ER	0.7			A
2058.28	2058.94	48568.7	ER	0.8			A
2025.33	2025.98	49358.8	ER	0.7			A
2001.77	2002.42	49939.6	ER	0.7			A
9988.65	9991.39	10008.6	EU	1.2			KN
9898.30	9901.01	10100.0	EU	1.6			KN
9883.16	9885.87	10115.4	EU	1.0			KN
9138.8	9141.3	10939.	EU I	0.3			KN
9112.0	9114.5	10971.	EU	0.3			KN
9094.8	9097.3	10992.	EU	0.5			KN
9085.3	9087.8	11004.	EU	1.7			KN
9083.2	9085.7	11006.	EU	1.5			KN
9058.0	9060.5	11037.	EU	1.9			KN
9045.0	9047.5	11053.	EU	0.5			KN
9024.5	9027.0	11078.	EU	1.8	0.0		KN
9018.0	9020.5	11086.	EU	1.7			KN
8982.57	8985.04	11129.6	EU	1.2			KN
8965.5	8968.0	11151.	EU	0.6			KN
8961.9	8964.4	11155.	EU	0.3			KN
8935.7	8938.2	11188.	EU	0.6			KN
8934.41	8936.86	11189.6	EU	1.7			KN
8917.65	8920.10	11210.6	EU	1.8			KN
8899.8	8902.2	11233.	EU	0.9			KN
8886.8	8889.2	11250.	EU	0.3			KN
8884.2	8886.6	11253.	EU	0.3			KN
8870.34	8872.78	11270.4	EU	2.0			KN
8817.0	8819.4	11339.	EU	0.3			KN
8790.89	8793.30	11372.3	EU	1.6			KN
8785.06	8787.47	11379.8	EU	0.7			KN
8782.46	8784.87	11383.2	EU	1.2			KN
8751.6	8754.0	11423.	EU	0.6			KN
8749.5	8751.9	11426.	EU	0.3			KN
8743.9	8746.3	11433.	EU	0.8			KN
8727.78	8730.18	11454.5	EU	1.6			KN
8704.47	8706.86	11485.2	EU	1.0			KN
8665.20	8667.58	11537.2	EU	0.9			KN
8642.69	8645.06	11567.3	EU	2.5			KN
8641.69	8644.06	11568.6	EU	1.0			KN
8631.7	8634.1	11582.	EU	0.3			KN
8597.19	8599.55	11628.5	EU	1.0			KN
8542.12	8544.47	11703.5	EU	2.0			KN
8514.63	8516.97	11741.3	EU	0.9			KN
8464.75	8467.08	11810.4	EU	1.7			KN
8450.1	8452.4	11831.	EU	0.3			KN
8391.6	8393.9	11913.	EU	0.3			KN
8335.7	8338.0	11993.	EU	0.6			KN
8331.6	8333.9	11999.	EU	0.3			KN
8298.6	8300.9	12047.	EU	0.3			KN
8274.6	8276.9	12082.	EU	0.8			KN
8253.8	8256.1	12112.	EU	0.8			KN
8247.2	8249.5	12122.	EU	0.3			KN

AIR WAVELENGTH (ANGSTRMS)	VACUUM WAVELENGTH (ANGSTRMS)	WAVENUMBER (KAYSERS)	LMENT	LOG INTENSITY ARC	SPK	DIS	REF
8243.7	8246.0	12127.	EU	0.5			KN
8237.5	8239.8	12136.	EU	0.6			KN
8226.82	8229.08	12152.0	EU	2.0			KN
8211.0	8213.3	12175.	EU	0.3			KN
8209.84	8212.10	12177.2	EU	2.3			KN
8188.1	8190.4	12209.	EU	0.5			KN
8184.7	8187.0	12215.	EU	0.3			KN
8168.13	8170.38	12239.3	EU	1.0			KN
8155.5	8157.7	12258.	EU	0.6			KN
8150.2	8152.4	12266.	EU	0.5			KN
8133.7	8135.9	12291.	EU	0.7			KN
8080.5	8082.7	12372.	EU	0.5			KN
8050.3	8052.5	12418.	EU	0.8			KN
8015.47	8017.67	12472.4	EU	1.3			KN
7998.5	8000.7	12499.	EU	0.3			KN
7908.69	7910.87	12640.8	EU	0.5			KN
7901.86	7904.03	12651.8	EU	0.7			KN
7890.56	7892.73	12669.9	EU	0.7			KN
7887.98	7890.15	12674.0	EU	2.5			KN
7882.35	7884.52	12683.1	EU	1.0			KN
7874.8	7877.0	12695.	EU	0.5			KN
7850.2	7852.4	12735.	EU	0.3			KN
7848.72	7850.88	12737.4	EU	0.7			KN
7818.22	7820.37	12787.1	EU	1.3			KN
7803.32	7805.47	12811.5	EU	1.5			KN
7759.41	7761.55	12884.0	EU	0.6			KN
7746.20	7748.33	12906.0	EU	2.8			KN
7742.58	7744.71	12912.0	EU	2.6			KN
7627.67	7629.77	13106.6	EU	0.9			KN
7603.88	7605.97	13147.6	EU	0.6			KN
7583.91	7586.00	13182.2	EU	2.6			KN
7552.01	7554.09	13237.9	EU	0.6			KN
7547.28	7549.36	13246.2	EU	1.2			KN
7533.00	7535.07	13271.3	EU	1.2			KN
7532.20	7534.27	13272.7	EU	0.5			KN
7528.70	7530.77	13278.8	EU	2.3			KN
7521.24	7523.31	13292.0	EU	0.6			KN
7508.47	7510.54	13314.6	EU	1.1			KN
7507.05	7509.12	13317.1	EU	0.5			KN
7491.00	7493.06	13345.7	EU	1.20			KN
7481.6	7483.7	13362.	EU	0.30			KN
7470.53	7472.59	13382.2	EU	1.50			KN
7466.2	7468.3	13390.	EU	0.3			KN
7450.36	7452.41	13418.5	EU	0.8			KN
7444.59	7446.64	13428.9	EU	1.0			KN
7436.59	7438.64	13443.3	EU	2.00			KN
7426.57	7428.62	13461.5	EU	2.7			KN
7409.51	7411.55	13492.4	EU	0.5			KN
7404.68	7406.72	13501.2	EU	1.00			KN
7404.41	7406.45	13501.7	EU	1.50			KN
7402.04	7404.08	13506.1	EU	0.60			KN
7400.0	7402.0	13510.	EU	0.60			KN
7394.86	7396.90	13519.2	EU	1.10			KN
7392.71	7394.75	13523.1	EU	0.70			KN
7389.15	7391.19	13529.6	EU	1.60			KN
7387.34	7389.37	13532.9	EU	1.20			KN
7374.1	7376.1	13557.	EU	0.30			KN
7370.27	7372.30	13564.3	EU I	2.8			MIT
7369.69	7371.72	13565.3	EU	2.30			KN
7367.70	7369.73	13569.0	EU	1.00			KN
7365.73	7367.76	13572.7	EU	0.6			KN
7362.25	7364.28	13579.1	EU	1.70			KN
7356.65	7358.68	13589.4	EU	1.50			KN
7346.24	7348.26	13608.7	EU	1.30			MIT
7340.11	7342.13	13620.0	EU	0.7			KN
7336.28	7338.30	13627.1	EU	2.50			KN
7331.0	7333.0	13637.	EU	0.5			KN
7326.1	7328.1	13646.	EU	0.5			KN
7313.64	7315.66	13669.3	EU	1.80			MIT
7310.33	7312.34	13675.5	EU	1.50			KN
7301.16	7303.17	13692.7	EU II	2.8			KN
7297.57	7299.58	13699.4	EU	1.50			MIT
7281.57	7283.58	13729.5	EU	1.70			MIT
7262.80	7264.80	13765.0	EU	2.30			MIT
7258.74	7260.74	13772.7	EU	2.00			MIT
7248.15	7250.15	13792.8	EU	1.30			MIT
7224.72	7226.71	13837.6	EU	2.20			MIT
7217.60	7219.59	13851.2	EU I	2.8			MIT
7204.48	7206.47	13876.4	EU	1.0			KN
7194.85	7196.83	13895.0	EU I	2.8			MIT
7188.0	7190.0	13908.	EU	0.30			KN
7187.0	7189.0	13910.	EU	0.30			KN
7181.7	7183.7	13920.	EU	0.60			KN
7180.0	7182.0	13924.	EU	0.3			KN
7175.50	7177.48	13932.5	EU	2.30			KN
7174.9	7176.9	13934.	EU	1.00			KN
7164.67	7166.64	13953.5	EU	1.70			KN
7155.31	7157.28	13971.8	EU	0.3			MIT
7148.76	7150.73	13984.6	EU	0.5			KN
7139.6	7141.6	14002.	EU	0.7			KN
7129.40	7131.37	14022.6	EU	0.6			KN
7120.0	7122.0	14041.	EU	0.5			KN
7114.9	7116.9	14051.	EU	0.6			KN
7114.47	7116.43	14052.0	EU	0.6			KN
7109.01	7110.97	14062.8	EU	0.5			KN
7106.48	7108.44	14067.8	EU	2.80			MIT
7103.43	7105.39	14073.8	EU	0.90			KN
7077.09	7079.04	14126.2	EU	2.9			MIT
7074.56	7076.51	14131.3	EU	1.8			MIT
7058.4	7060.3	14164.	EU	0.50			KN
7052.5	7054.4	14175.	EU	0.3			KN
7040.18	7042.12	14200.3	EU	2.9			MIT
7036.0	7037.9	14209.	EU	0.50			KN
7015.07	7017.00	14251.1	EU	0.6			KN
6985.6	6987.5	14311.	EU	0.30			KN
6983.3	6985.2	14316.	EU	0.3			KN
6982.4	6984.3	14318.	EU	0.3			KN
6978.19	6980.11	14326.4	EU	1.0			KN
6973.35	6975.27	14336.4	EU	0.90			KN
6965.9	6967.8	14352.	EU	0.60			KN
6965.4	6967.3	14353.	EU	0.3			KN
6950.69	6952.61	14383.1	EU	0.3			KN
6947.49	6949.41	14389.7	EU	0.6			KN
6914.82	6916.73	14457.7	EU	1.60			MIT
6910.20	6912.11	14467.4	EU	1.00			MIT
6908.74	6910.65	14470.4	EU	0.6			KN
6903.71	6905.61	14481.0	EU	2.30			KN
6898.27	6900.17	14492.4	EU	1.50			KN
6890.76	6892.66	14508.2	EU	0.5			KN
6885.4	6887.3	14519.	EU	0.30			KN
6864.55	6866.44	14563.6	EU	3.00			MIT
6847.21	6849.10	14600.5	EU	1.00			KN
6844.98	6846.87	14605.2	EU	1.60			MIT
6840.93	6842.82	14613.9	EU I	2.0			KN
6834.41	6836.30	14627.8	EU	1.50			KN
6822.65	6824.53	14653.0	EU	1.20			KN
6821.2	6823.1	14656.	EU	0.30			KN
6816.11	6817.99	14667.1	EU	2.20			MIT
6802.79	6804.67	14695.8	EU	2.7			KN
6791.6	6793.5	14720.	EU	0.30			KN
6787.52	6789.39	14728.9	EU	1.50			KN
6784.9	6786.8	14734.	EU	0.60			KN
6782.7	6784.6	14739.	EU	1.3			KN
6782.59	6784.46	14739.6	EU	2.0			KN
6772.42	6774.29	14761.7	EU	0.60			KN
6766.6	6768.5	14774.	EU	0.30			KN
6758.53	6760.40	14792.0	EU	0.60			KN
6750.30	6752.16	14810.1	EU	1.0			KN
6744.96	6746.82	14821.8	EU	2.0			KN
6741.9	6743.8	14828.	EU	0.3			KN

AIR WAVELENGTH (ANGSTRMS)	VACUUM WAVELENGTH (ANGSTRMS)	WAVENUMBER (KAYSERS)	LMENT	LOG ARC	INTENSITY SPK	DIS	REF
6735.0	6736.9	14844.	EU	0.50			KN
6733.8	6735.7	14846.	EU	0.30			KN
6732.3	6734.2	14850.	EU	0.60			KN
6717.9	6719.8	14881.	EU	0.50			KN
6710.50	6712.35	14897.9	EU	1.40			MIT
6701.11	6702.96	14918.8	EU	1.00			KN
6695.86	6697.71	14930.5	EU	0.60			MIT
6693.97	6695.82	14934.7	EU	2.7			MIT
6692.98	6694.83	14936.9	EU	1.6			KN
6685.27	6687.12	14954.1	EU	1.9			KN
6682.04	6683.88	14961.4	EU	0.90			KN
6671.90	6673.74	14984.1	EU	0.60			KN
6648.3	6650.1	15037.	EU	0.3			KN
6645.15	6646.99	15044.4	EU	3.0			MIT
6640.8	6642.6	15054.	EU	0.3			KN
6626.1	6627.9	15088.	EU	0.3			KN
6623.7	6625.5	15093.	EU	0.3			KN
6621.3	6623.1	15099.	EU	0.3			KN
6619.28	6621.11	15103.2	EU	0.70			KN
6603.7	6605.5	15139.	EU	0.70			KN
6603.60	6605.42	15139.1	EU	1.9			KN
6601.2	6603.0	15145.	EU	0.30			KN
6595.4	6597.2	15158.	EU	0.3			KN
6593.82	6595.64	15161.5	EU	2.6			MIT
6587.0	6588.8	15177.	EU	0.30			KN
6572.9	6574.7	15210.	EU	0.5			KN
6570.760	6572.575	15214.74	EU	1.2			KN
6567.87	6569.68	15221.4	EU	2.8			MIT
6561.18	6562.99	15236.9	EU	0.80			KN
6549.14	6550.95	15265.0	EU	1.30			KN
6543.8	6545.6	15277.	EU	0.50			KN
6532.96	6534.76	15302.8	EU	0.60			KN
6530.606	6532.410	15308.29	EU	0.8			KN
6522.76	6524.56	15326.7	EU	1.4			MIT
6519.60	6521.40	15334.1	EU	2.7			MIT
6501.535	6503.331	15376.73	EU	2.5			MIT
6493.90	6495.69	15394.8	EU	1.6			KN
6483.07	6484.86	15420.5	EU	1.5			KN
6476.54	6478.33	15436.1	EU	0.80			MIT
6473.26	6475.05	15443.9	EU	0.5			KN
6470.75	6472.54	15449.9	EU	1.50			KN
6467.47	6469.26	15457.7	EU	0.3			KN
6465.3	6467.1	15463.	EU	0.30			KN
6457.995	6459.780	15480.40	EU	2.7			MIT
6449.8	6451.6	15500.	EU	0.3			KN
6439.97	6441.75	15523.7	EU	1.6			KN
6437.69	6439.47	15529.2	EU II	2.8			MIT
6428.28	6430.06	15552.0	EU	2.3			KN
6411.344	6413.116	15593.04	EU	2.7			MIT
6410.103	6411.875	15596.06	EU	2.70			MIT
6406.110	6407.881	15605.78	EU	2.0			MIT
6403.01	6404.78	15613.3	EU	0.5			KN
6400.932	6402.701	15618.41	EU	2.80			MIT
6389.595	6391.361	15646.12	EU	1.0			MIT
6383.861	6385.626	15660.17	EU	2.5			MIT
6382.741	6384.506	15662.92	EU	2.3			MIT
6373.276	6375.038	15686.18	EU	0.70			MIT
6369.243	6371.004	15696.11	EU	2.5			MIT
6366.74	6368.50	15702.3	EU	0.70			KN
6360.466	6362.225	15717.77	EU	0.6			MIT
6355.885	6357.642	15729.10	EU	2.3			MIT
6350.044	6351.800	15743.57	EU	2.80			MIT
6346.0	6347.8	15754.	EU	1.0			KN
6335.783	6337.535	15779.01	EU	2.3			MIT
6324.436	6326.185	15807.32	EU	1.0			MIT
6318.561	6320.308	15822.01	EU	1.1			MIT
6313.789	6315.535	15833.97	EU	1.40			MIT
6303.395	6305.138	15860.08	EU II	2.8			MIT
6300.50	6302.24	15867.4	EU	0.60			KN
6299.764	6301.506	15869.22	EU	2.70			MIT

AIR WAVELENGTH (ANGSTRMS)	VACUUM WAVELENGTH (ANGSTRMS)	WAVENUMBER (KAYSERS)	LMENT	LOG ARC	INTENSITY SPK	DIS	REF
6296.98	6298.72	15876.2	EU	0.5			KN
6291.341	6293.081	15890.47	EU	2.4			MIT
6288.34	6290.08	15898.1	EU	2.5			KN
6285.941	6287.680	15904.12	EU	1.30			MIT
6266.948	6268.681	15952.32	EU	2.1			MIT
6264.594	6266.327	15958.31	EU	1.0			MIT
6263.437	6265.169	15961.26	EU	1.0			MIT
6262.285	6264.017	15964.20	EU	2.8			MIT
6250.457	6252.186	15994.41	EU	1.8			MIT
6245.924	6247.652	16006.01	EU	0.7			MIT
6233.743	6235.467	16037.29	EU	2.0			MIT
6230.511	6232.235	16045.61	EU	0.8			MIT
6209.352	6211.070	16100.29	EU	0.7			MIT
6207.603	6209.320	16104.82	EU	0.8			MIT
6195.063	6196.777	16137.42	EU	2.5			MIT
6191.564	6193.277	16146.54	EU	0.7			MIT
6188.098	6189.810	16155.58	EU	2.70			MIT
6178.746	6180.456	16180.04	EU	2.1			MIT
6173.046	6174.754	16194.98	EU II	2.8			MIT
6164.44	6166.15	16217.6	EU	0.9			KN
6158.708	6160.412	16232.68	EU	0.50			MIT
6155.37	6157.07	16241.5	EU	1.00			KN
6153.225	6154.928	16247.14	EU	0.90			MIT
6150.11	6151.81	16255.4	EU	0.70			KN
6149.27	6150.97	16257.6	EU	0.7			KN
6140.11	6141.81	16281.8	EU	0.50			KN
6139.15	6140.85	16284.4	EU	0.80			KN
6124.679	6126.374	16322.87	EU	2.2			MIT
6118.779	6120.473	16338.61	EU	2.60			MIT
6118.10	6119.79	16340.4	EU	0.6			KN
6111.882	6113.574	16357.05	EU	0.6W			MIT
6111.019	6112.711	16359.35	EU	0.7			MIT
6108.134	6109.825	16367.08	EU	2.2			MIT
6107.477	6109.168	16368.84	EU	1.5			MIT
6100.033	6101.722	16388.82	EU	1.1			MIT
6099.385	6101.073	16390.56	EU	2.8			MIT
6083.873	6085.557	16432.35	EU	2.7			MIT
6079.997	6081.680	16442.82	EU	1.30			MIT
6077.366	6079.049	16449.94	EU	2.00			MIT
6075.590	6077.272	16454.75	EU	2.5			MIT
6074.268	6075.950	16458.33	EU	1.0			MIT
6072.738	6074.419	16462.48	EU	0.30			MIT
6061.642	6063.320	16492.61	EU	1.20			MIT
6057.356	6059.033	16504.28	EU	2.8			MIT
6052.933	6054.609	16516.34	EU	0.90			MIT
6052.183	6053.859	16518.39	EU	0.7			MIT
6051.280	6052.956	16520.85	EU	0.8			MIT
6049.503	6051.178	16525.71	EU II	3.0			MIT
6044.668	6046.342	16538.93	EU	2.3			MIT
6043.109	6044.782	16543.19	EU	0.30			MIT
6040.815	6042.488	16549.48	EU	0.8			MIT
6032.383	6034.053	16572.61	EU	0.9			MIT
6029.009	6030.679	16581.88	EU	2.7			MIT
6025.059	6026.728	16592.75	EU	0.3			MIT
6023.157	6024.825	16597.99	EU	2.3			MIT
6018.195	6019.862	16611.68	EU	3.0			MIT
6016.358	6018.024	16616.75	EU	0.50			MIT
6016.027	6017.693	16617.66	EU	0.9			MIT
6015.580	6017.246	16618.90	EU	2.2			MIT
6012.558	6014.223	16627.25	EU	2.6			MIT
6012.210	6013.875	16628.21	EU	2.50			MIT
6005.68	6007.34	16646.3	EU	1.8			KN
6004.41	6006.07	16649.8	EU	2.5			KN
6003.08	6004.74	16653.5	EU	1.3W			KN
5992.860	5994.520	16681.90	EU	3.0			MIT
5986.814	5988.472	16698.75	EU	0.3H			MIT
5983.855	5985.512	16707.01	EU	1.3			MIT
5983.26	5984.92	16708.7	EU	1.3			KN
5980.481	5982.138	16716.43	EU	1.0			MIT
5979.975	5981.631	16717.85	EU	0.6			MIT

AIR WAVELENGTH (ANGSTRMS)	VACUUM WAVELENGTH (ANGSTRMS)	WAVENUMBER (KAYSERS)	LMENT	LOG INTENSITY ARC	SPK	DIS	REF
5977.403	5979.059	16725.04	EU	0.3H			MIT
5973.672	5975.327	16735.49	EU	0.3			MIT
5972.776	5974.431	16738.00	EU	2.5			MIT
5971.687	5973.341	16741.05	EU	1.6			MIT
5970.93	5972.58	16743.2	EU	0.7			KN
5968.439	5970.092	16750.16	EU	1.0			MIT
5967.159	5968.812	16753.75	EU	3.3S			MIT
5966.066	5967.719	16756.82	EU	3.0			MIT
5963.765	5965.417	16763.29	EU	1.8			MIT
5954.269	5955.919	16790.02	EU	1.5			MIT
5953.828	5955.477	16791.26	EU	1.6			MIT
5953.496	5955.145	16792.20	EU	1.5			MIT
5942.729	5944.375	16822.62	EU	1.5			MIT
5937.779	5939.424	16836.65	EU	1.0			MIT
5926.527	5928.169	16868.61	EU	1.9			MIT
5925.307	5926.949	16872.09	EU	1.3			MIT
5924.920	5926.562	16873.19	EU	1.0			MIT
5915.764	5917.403	16899.30	EU	2.3L			MIT
5914.68	5916.32	16902.4	EU	0.7H			KN
5909.955	5911.593	16915.92	EU	1.5			MIT
5909.375	5911.013	16917.57	EU	0.3H			MIT
5908.755	5910.392	16919.35	EU	0.8			MIT
5906.26	5907.90	16926.5	EU	0.6			KN
5902.758	5904.394	16936.54	EU	1.5			MIT
5895.288	5896.922	16958.00	EU	1.4			MIT
5891.269	5892.902	16969.57	EU	2.3			MIT
5885.164	5886.795	16987.17	EU	0.8			MIT
5884.815	5886.446	16988.18	EU	0.8			MIT
5880.784	5882.414	16999.82	EU	0.5			MIT
5872.975	5874.603	17022.43	EU II	2.5			MIT
5867.783	5869.409	17037.49	EU	0.5			MIT
5866.638	5868.264	17040.81	EU	1.0L			MIT
5864.765	5866.391	17046.26	EU	1.3			MIT
5862.369	5863.994	17053.22	EU	0.3			MIT
5860.960	5862.585	17057.32	EU	1.5			MIT
5856.928	5858.551	17069.07	EU	1.0			MIT
5852.43	5854.05	17082.2	EU	0.6			KN
5845.762	5847.382	17101.67	EU	1.5			MIT
5843.528	5845.148	17108.21	EU	0.6			MIT
5838.031	5839.649	17124.32	EU	1.0			MIT
5831.047	5832.664	17144.83	EU	3.3O			MIT
5829.46	5831.08	17149.5	EU	0.6W			MIT
5820.907	5822.521	17174.69	EU	1.5			MIT
5820.002	5821.616	17177.36	EU	1.5			MIT
5818.738	5820.351	17181.09	EU II	3.0			MIT
5805.678	5807.288	17219.74	EU	1.3			MIT
5800.275	5801.883	17235.78	EU	2.3			MIT
5794.578	5796.185	17252.73	EU	0.6J			MIT
5792.719	5794.325	17258.26	EU	1.5			MIT
5783.711	5785.315	17285.14	EU	2.2S			MIT
5781.363	5782.966	17292.16	EU	0.6			MIT
5769.557	5771.157	17327.55	EU	0.3H			MIT
5767.620	5769.220	17333.37	EU	1.0			MIT
5766.774	5768.373	17335.91	EU	0.5			MIT
5765.200	5766.799	17340.64	EU	3.3			MIT
5744.364	5745.957	17403.54	EU	0.6			MIT
5738.998	5740.590	17419.81	EU	2.5			MIT
5736.615	5738.206	17427.05	EU	1.1			MIT
5730.888	5732.478	17444.46	EU	2.5			MIT
5728.224	5729.813	17452.58	EU	0.5			MIT
5718.79	5720.38	17481.4	EU	0.6J			KN
5707.92	5709.50	17514.7	EU	0.6			KN
5688.345	5689.923	17574.93	EU	0.3H			MIT
5684.27	5685.85	17587.5	EU	2.1			KN
5681.053	5682.629	17597.49	EU	1.6			MIT
5674.964	5676.539	17616.37	EU	0.9			MIT
5673.840	5675.414	17619.86	EU	2.3			MIT
5668.29	5669.86	17637.1	EU	0.8H			KN
5665.368	5666.940	17646.21	EU	0.6H			MIT
5654.645	5656.214	17679.67	EU	0.9H			MIT

AIR WAVELENGTH (ANGSTRMS)	VACUUM WAVELENGTH (ANGSTRMS)	WAVENUMBER (KAYSERS)	LMENT	LOG INTENSITY ARC	SPK	DIS	REF
5651.114	5652.682	17690.72	EU	0.6			MIT
5650.284	5651.852	17693.31	EU	0.6			MIT
5649.867	5651.435	17694.62	EU	0.8			MIT
5645.801	5647.368	17707.36	EU	3.0			MIT
5640.21	5641.78	17724.9	EU	0.6H			KN
5632.561	5634.124	17748.99	EU	2.3S			MIT
5627.068	5628.630	17766.31	EU	0.8			MIT
5622.448	5624.009	17780.91	EU	2.3S			MIT
5618.811	5620.371	17792.42	EU	2.1			MIT
5617.047	5618.606	17798.01	EU	0.6			MIT
5615.648	5617.207	17802.44	EU	0.3			MIT
5609.724	5611.281	17821.24	EU	0.6			MIT
5607.368	5608.925	17828.73	EU	0.6			MIT
5605.853	5607.409	17833.55	EU	1.6			MIT
5599.811	5601.366	17852.79	EU	1.6			MIT
5599.101	5600.655	17855.05	EU	1.0			MIT
5598.486	5600.040	17857.01	EU	0.3			MIT
5594.652	5596.205	17869.25	EU	0.6			MIT
5593.093	5594.646	17874.23	EU	0.8			MIT
5592.258	5593.811	17876.90	EU	1.4			MIT
5591.598	5593.150	17879.01	EU	0.6			MIT
5586.770	5588.321	17894.46	EU	2.3			MIT
5586.249	5587.800	17896.13	EU	2.5			MIT
5580.039	5581.588	17916.05	EU	2.5			MIT
5579.647	5581.196	17917.31	EU	2.3S			MIT
5577.126	5578.675	17925.41	EU	3.0			MIT
5572.649	5574.196	17939.81	EU	0.6			MIT
5570.359	5571.906	17947.18	EU	3.0			MIT
5566.894	5568.440	17958.35	EU	0.6			MIT
5566.410	5567.956	17959.91	EU	0.6			MIT
5547.446	5548.987	18021.31	EU	3.0			MIT
5542.541	5544.080	18037.26	EU	1.9			MIT
5541.596	5543.135	18040.33	EU	0.8			MIT
5536.83	5538.37	18055.9	EU	0.5			KN
5536.13	5537.67	18058.1	EU	1.5			KN
5533.244	5534.781	18067.56	EU	1.5			MIT
5526.620	5528.155	18089.22	EU	1.8			MIT
5519.636	5521.169	18112.10	EU	0.5			KN
5511.787	5513.318	18137.90	EU	0.6			MIT
5511.091	5512.622	18140.19	EU	0.8			MIT
5510.547	5512.078	18141.98	EU	2.5S			MIT
5504.914	5506.443	18160.54	EU	0.3H			MIT
5502.299	5503.828	18169.17	EU	0.3			MIT
5500.824	5502.352	18174.05	EU	1.3			MIT
5500.469	5501.997	18175.22	EU	0.8			MIT
5497.909	5499.436	18183.68	EU	0.6			MIT
5495.805	5497.332	18190.64	EU	0.6			MIT
5495.170	5496.697	18192.75	EU	2.1			MIT
5490.03	5491.56	18209.8	EU	0.5			KN
5488.648	5490.173	18214.36	EU	2.7			MIT
5485.478	5487.002	18224.89	EU	0.8			MIT
5484.40	5485.92	18228.5	EU				KN
5481.843	5483.366	18236.97	EU	0.3H			MIT
5472.328	5473.849	18268.68	EU	3.0S			MIT
5467.05	5468.57	18286.3	EU	1.0O			KN
5457.617	5459.134	18317.93	EU	1.3			MIT
5452.965	5454.480	18333.55	EU	3.0S			MIT
5451.530	5453.045	18338.38	EU	3.0S			MIT
5447.157	5448.671	18353.10	EU	0.7			MIT
5446.547	5448.061	18355.16	EU	0.3			MIT
5443.556	5445.069	18365.24	EU	2.1			MIT
5426.932	5428.441	18421.50	EU	2.3			MIT
5426.736	5428.244	18422.16	EU	0.3H			MIT
5421.080	5422.587	18441.38	EU	2.1			MIT
5413.760	5415.265	18466.32	EU	0.6J			MIT
5411.842	5413.346	18472.86	EU	1.9			MIT
5407.33	5408.83	18488.3	EU	0.6H			KN
5405.311	5406.814	18495.18	EU	1.6			MIT
5402.791	5404.293	18503.81	EU	3.0			MIT
5399.69	5401.19	18514.4	EU	0.6H			KN

AIR WAVELENGTH (ANGSTRMS)	VACUUM WAVELENGTH (ANGSTRMS)	WAVENUMBER (KAYSERS)	LMENT	LOG ARC	INTENSITY SPK	DIS	REF
5397.383	5398.884	18522.35	EU	1.0			MIT
5393.454	5394.954	18535.84	EU	0.3			MIT
5392.912	5394.411	18537.70	EU	2.2			MIT
5391.00	5392.50	18544.3	EU	0.6			KN
5376.911	5378.406	18592.87	EU	2.3			MIT
5375.047	5376.542	18599.32	EU	0.6H			MIT
5368.148	5369.641	18623.22	EU	0.3			MIT
5364.588	5366.080	18635.58	EU	0.6			MIT
5361.592	5363.083	18645.99	EU	2.5			MIT
5360.809	5362.300	18648.72	EU	2.2			MIT
5357.608	5359.098	18659.86	EU	3.0			MIT
5356.721	5358.211	18662.95	EU	1.6			MIT
5355.707	5357.197	18666.48	EU	0.6H			MIT
5355.081	5356.570	18668.66	EU	2.3			MIT
5352.824	5354.313	18676.53	EU	1.9			MIT
5351.671	5353.159	18680.56	EU	2.2			MIT
5350.399	5351.887	18685.00	EU	1.8H			MIT
5349.474	5350.962	18688.23	EU	0.3			MIT
5343.76	5345.25	18708.2	EU	0.3			MIT
5341.877	5343.363	18714.81	EU	1.3			MIT
5338.267	5339.752	18727.46	EU	0.6H			MIT
5329.279	5330.761	18759.05	EU	0.3			MIT
5323.017	5324.498	18781.11	EU	0.3H			MIT
5316.942	5318.421	18802.57	EU	1.2			MIT
5312.172	5313.650	18819.46	EU	0.6H			MIT
5310.023	5311.500	18827.07	EU	0.6H			MIT
5305.501	5306.977	18843.12	EU	0.6			MIT
5303.873	5305.349	18848.90	EU	2.5			MIT
5302.724	5304.199	18852.99	EU	1.5			MIT
5298.156	5299.630	18869.24	EU	1.3			MIT
5295.56	5297.03	18878.5	EU	0.3			KN
5294.596	5296.069	18881.93	EU	2.5			MIT
5293.68	5295.15	18885.2	EU	1.7W			KN
5291.251	5292.723	18893.86	EU	2.3			MIT
5289.247	5290.719	18901.02	EU	2.1			MIT
5287.231	5288.702	18908.23	EU	2.1			MIT
5285.727	5287.198	18913.61	EU	1.6			MIT
5285.459	5286.930	18914.57	EU	1.6			MIT
5282.823	5284.293	18924.01	EU	3.0			MIT
5280.65	5282.12	18931.8	EU	1.3H			KN
5279.4	5280.9	18936.	EU	0.3			KN
5278.16	5279.63	18940.7	EU	1.2			KN
5277.10	5278.57	18944.5	EU	0.3			KN
5275.66	5277.13	18949.7	EU	1.5			KN
5275.05	5276.52	18951.9	EU	1.0			KN
5274.42	5275.89	18954.2	EU	0.6H			KN
5272.48	5273.95	18961.1	EU	2.6			KN
5271.95	5273.42	18963.0	EU	3.3			KN
5271.7	5273.2	18964.	EU	0.7			KN
5268.77	5270.24	18974.5	EU	0.3			KN
5266.40	5267.87	18983.0	EU	3.0			KN
5263.04	5264.50	18995.1	EU	1.5			KN
5260.90	5262.36	19002.9	EU	0.3			KN
5259.86	5261.32	19006.6	EU	0.3			KN
5256.075	5257.538	19020.31	EU	1.3			MIT
5253.38	5254.84	19030.1	EU	0.3			KN
5252.86	5254.32	19031.9	EU	0.3			KN
5250.38	5251.84	19040.9	EU	0.3			KN
5249.112	5250.573	19045.54	EU	1.8			MIT
5248.615	5250.076	19047.34	EU	1.9			MIT
5245.55	5247.01	19058.5	EU	0.9H			KN
5242.682	5244.141	19068.90	EU	1.5			MIT
5240.41	5241.87	19077.2	EU	0.3H			KN
5239.78	5241.24	19079.5	EU	0.3H			KN
5239.218	5240.676	19081.51	EU	1.9			MIT
5236.119	5237.577	19092.80	EU	1.4			MIT
5233.888	5235.345	19100.94	EU	1.5			MIT
5224.72	5226.17	19134.4	EU	0.6H			KN
5223.478	5224.932	19139.00	EU	2.8			MIT
5223.053	5224.507	19140.56	EU	0.9			MIT

AIR WAVELENGTH (ANGSTRMS)	VACUUM WAVELENGTH (ANGSTRMS)	WAVENUMBER (KAYSERS)	LMENT	LOG ARC	INTENSITY SPK	DIS	REF
5222.28	5223.73	19143.4	EU	0.3H			KN
5219.421	5220.874	19153.88	EU	0.6			MIT
5219.209	5220.662	19154.66	EU	1.1			MIT
5217.021	5218.474	19162.69	EU	2.1			MIT
5215.088	5216.540	19169.79	EU	3.0			MIT
5213.368	5214.820	19176.12	EU	2.2			MIT
5207.893	5209.343	19196.28	EU	1.3H			MIT
5206.430	5207.880	19201.67	EU	1.6			MIT
5200.925	5202.373	19222.00	EU	2.5			MIT
5199.846	5201.294	19225.99	EU	2.7			MIT
5199.28	5200.73	19228.1	EU	0.6			KN
5193.731	5195.177	19248.62	EU	1.3			MIT
5189.839	5191.284	19263.06	EU	1.0			MIT
5188.576	5190.021	19267.75	EU	0.6			MIT
5184.274	5185.718	19283.73	EU	0.3			MIT
5183.606	5185.050	19286.22	EU	0.6			MIT
5181.014	5182.457	19295.87	EU	0.6			MIT
5178.701	5180.143	19304.49	EU	0.9			MIT
5177.977	5179.419	19307.18	EU	0.8			MIT
5176.411	5177.853	19313.02	EU	0.8			MIT
5174.889	5176.330	19318.71	EU	0.9			MIT
5173.991	5175.432	19322.06	EU	0.6			MIT
5170.492	5171.932	19335.13	EU	0.6			MIT
5169.756	5171.196	19337.89	EU	0.6			MIT
5169.320	5170.760	19339.52	EU	0.8			MIT
5168.226	5169.666	19343.61	EU	0.6			MIT
5167.187	5168.626	19347.50	EU	0.6			MIT
5166.716	5168.155	19349.26	EU	2.1			MIT
5163.779	5165.217	19360.27	EU	0.5			MIT
5162.39	5163.83	19365.5	EU	0.5			KN
5160.409	5161.846	19372.91	EU	0.8			MIT
5160.070	5161.507	19374.19	EU	2.3			MIT
5156.410	5157.846	19387.94	EU	0.9			MIT
5155.407	5156.843	19391.71	EU	1.2			MIT
5150.821	5152.256	19408.97	EU	1.1			MIT
5149.30	5150.73	19414.7	EU	1.0			KN
5148.41	5149.84	19418.1	EU	0.9			KN
5147.782	5149.216	19420.43	EU	0.8			MIT
5146.72	5148.15	19424.4	EU	0.5			KN
5146.408	5147.842	19425.62	EU	0.9			MIT
5146.024	5147.458	19427.07	EU	0.6			MIT
5145.70	5147.13	19428.3	EU	0.3			KN
5141.059	5142.491	19445.83	EU	0.8			MIT
5138.5	5139.9	19455.	EU	0.5			KN
5137.52	5138.95	19459.2	EU	0.6			KN
5135.433	5136.864	19467.13	EU	0.9			MIT
5133.485	5134.915	19474.52	EU	2.2			MIT
5132.365	5133.795	19478.77	EU	1.0			MIT
5131.397	5132.827	19482.44	EU	0.3			MIT
5130.820	5132.250	19484.63	EU	0.5			MIT
5130.451	5131.880	19486.03	EU	0.5			MIT
5130.088	5131.517	19487.41	EU	1.3			MIT
5129.082	5130.511	19491.24	EU	2.3			MIT
5127.895	5129.324	19495.75	EU	0.6			MIT
5124.772	5126.200	19507.63	EU	1.8			MIT
5119.453	5120.880	19527.90	EU	0.8			MIT
5115.740	5117.166	19542.07	EU	0.6			MIT
5114.363	5115.788	19547.33	EU	2.2			MIT
5112.824	5114.249	19553.21	EU	0.6			MIT
5110.966	5112.390	19560.32	EU	0.8			MIT
5102.417	5103.839	19593.09	EU	0.6			MIT
5101.252	5102.674	19597.57	EU	0.8			MIT
5098.712	5100.133	19607.33	EU	1.2			MIT
5098.536	5099.957	19608.01	EU	0.6			MIT
5096.415	5097.835	19616.17	EU	1.5			MIT
5094.44	5095.86	19623.8	EU	0.5			KN
5092.664	5094.083	19630.62	EU	1.3W			MIT
5092.27	5093.69	19632.1	EU	0.5			KN
5091.40	5092.82	19635.5	EU	0.5			KN
5089.079	5090.497	19644.45	EU	1.3			MIT

AIR WAVELENGTH (ANGSTRMS)	VACUUM WAVELENGTH (ANGSTRMS)	WAVENUMBER (KAYSERS)	LMENT	LOG ARC	INTENSITY SPK	INTENSITY DIS	REF
5087.18	5088.60	19651.8	EU	0.6			KN
5085.603	5087.021	19657.87	EU	0.9			MIT
5079.955	5081.371	19679.73	EU	0.7			MIT
5078.005	5079.421	19687.28	EU	1.2			MIT
5077.396	5078.811	19689.65	EU	0.8			MIT
5074.751	5076.166	19699.91	EU	0.3			MIT
5070.272	5071.685	19717.31	EU	0.9			MIT
5069.26	5070.67	19721.2	EU	0.3			KN
5067.948	5069.361	19726.35	EU	1.5			MIT
5065.685	5067.097	19735.17	EU	0.5			MIT
5063.733	5065.145	19742.77	EU	1.3			MIT
5059.965	5061.376	19757.47	EU	0.7			MIT
5057.482	5058.892	19767.17	EU	0.6			MIT
5056.019	5057.429	19772.89	EU	0.6			MIT
5052.592	5054.001	19786.30	EU	0.6			MIT
5045.335	5046.742	19814.76	EU	1.2W			MIT
5035.426	5036.830	19853.76	EU	1.3W			MIT
5033.541	5034.945	19861.19	EU	1.8	0.3		MIT
5032.42	5033.82	19865.6	EU	0.8			KN
5029.549	5030.952	19876.96	EU	1.5	0.0		MIT
5029.48	5030.88	19877.2	EU	2.7W			KN
5022.905	5024.306	19903.25	EU	2.1			MIT
5018.599	5019.999	19920.32	EU	0.6W			MIT
5015.64	5017.04	19932.1	EU	1.0W			KN
5013.136	5014.534	19942.03	EU	2.1	0.0		MIT
4995.554	4996.948	20012.22	EU	0.5	0.0		MIT
4986.754	4988.145	20047.53	EU	0.7			MIT
4983.984	4985.375	20058.67	EU	0.3			MIT
4976.423	4977.812	20089.15	EU	0.5			MIT
4975.757	4977.145	20091.84	EU	1.0			MIT
4968.708	4970.094	20120.34	EU	0.7			MIT
4962.525	4963.910	20145.41	EU	1.3			MIT
4961.388	4962.773	20150.03	EU	0.3			MIT
4960.183	4961.567	20154.92	EU	1.3			MIT
4953.495	4954.877	20182.13	EU	1.1			MIT
4950.108	4951.490	20195.94	EU	0.3			MIT
4947.356	4948.737	20207.18	EU	0.7			MIT
4938.31	4939.69	20244.2	EU	2.40			KN
4932.818	4934.195	20266.73	EU	0.9			MIT
4928.005	4929.381	20286.52	EU	0.7			MIT
4924.734	4926.109	20300.00	EU	1.0			MIT
4922.235	4923.609	20310.30	EU	0.3			MIT
4917.491	4918.864	20329.90	EU	0.5			MIT
4911.406	4912.777	20355.08	EU	1.5	0.8		MIT
4907.173	4908.543	20372.64	EU	1.3	0.8		MIT
4900.838	4902.206	20398.98	EU	1.4	0.3		MIT
4894.68	4896.05	20424.6	EU	1.00	0.3		KN
4891.02	4892.39	20439.9	EU	0.3			KN
4884.055	4885.419	20469.07	EU	1.2	0.5		MIT
4879.484	4880.847	20488.25	EU	0.50			MIT
4879.167	4880.530	20489.58	EU	0.5	0.5		MIT
4867.602	4868.962	20538.26	EU	1.5	0.5		MIT
4866.398	4867.757	20543.34	EU	0.3			MIT
4860.86	4862.22	20566.7	EU	1.10			KN
4859.515	4860.872	20572.44	EU	0.3			MIT
4852.03	4853.39	20604.2	EU	0.7			KN
4851.246	4852.601	20607.50	EU	0.3			MIT
4849.647	4851.002	20614.30	EU	1.2W	0.3		MIT
4845.62	4846.97	20631.4	EU	0.0W	0.3		KN
4844.30	4845.65	20637.1	EU	0.3			KN
4843.372	4844.725	20641.01	EU	0.3			MIT
4840.480	4841.832	20653.34	EU	1.2	0.5		MIT
4839.067	4840.419	20659.37	EU	0.50			MIT
4838.99	4840.34	20659.7	EU	1.30			KN
4838.141	4839.493	20663.32	EU	0.5	0.3		MIT
4836.734	4838.085	20669.33	EU	0.3			MIT
4830.338	4831.688	20696.70	EU	1.3	0.3		MIT
4829.865	4831.215	20698.73	EU	0.5			MIT
4829.298	4830.647	20701.16	EU	0.8	0.3		MIT
4825.646	4826.994	20716.82	EU	0.7	0.3		MIT

AIR WAVELENGTH (ANGSTRMS)	VACUUM WAVELENGTH (ANGSTRMS)	WAVENUMBER (KAYSERS)	LMENT	LOG ARC	INTENSITY SPK	INTENSITY DIS	REF
4825.390	4826.738	20717.92	EU	0.30			MIT
4824.26	4825.61	20722.8	EU	0.3	0.3		KN
4820.81	4822.16	20737.6	EU	0.60			KN
4820.494	4821.841	20738.97	EU	0.5	0.3		MIT
4819.872	4821.219	20741.64	EU	0.5			MIT
4819.56	4820.91	20743.0	EU	0.3W			KN
4814.521	4815.867	20764.70	EU	0.3	0.3		MIT
4813.55	4814.90	20768.9	EU	0.9			KN
4809.289	4810.633	20787.28	EU	1.0	0.3		MIT
4808.621	4809.965	20790.17	EU	0.0	0.3		MIT
4808.29	4809.63	20791.6	EU	0.7			ED
4806.94	4808.28	20797.4	EU	0.3W			KN
4805.196	4806.539	20804.99	EU	0.0	0.3		MIT
4804.086	4805.429	20809.80	EU	0.9			MIT
4800.82	4802.16	20823.9	EU	0.3W			KN
4799.383	4800.725	20830.19	EU	0.7			MIT
4798.925	4800.266	20832.18	EU	0.7			MIT
4798.065	4799.406	20835.91	EU	1.0	0.0		MIT
4795.888	4797.229	20845.37	EU	0.3			MIT
4792.566	4793.906	20859.82	EU	1.3	0.3		MIT
4791.84	4793.18	20863.0	EU	1.00			KN
4791.158	4792.497	20865.95	EU	0.5			MIT
4789.62	4790.96	20872.6	EU	0.30			KN
4787.992	4789.331	20879.75	EU	0.3			MIT
4787.624	4788.962	20881.35	EU	0.3			MIT
4784.016	4785.353	20897.10	EU	1.1			MIT
4781.318	4782.655	20908.89	EU	1.2			MIT
4779.689	4781.025	20916.02	EU	0.3			MIT
4778.647	4779.983	20920.58	EU	1.3			MIT
4777.700	4779.036	20924.72	EU	1.6			MIT
4777.151	4778.487	20927.13	EU	0.5			MIT
4773.01	4774.34	20945.3	EU	0.30			KN
4772.66	4773.99	20946.8	EU	0.7W			MIT
4771.72	4773.05	20950.9	EU	0.5W	0.0		M
4770.776	4772.110	20955.09	EU	1.2			MIT
4769.615	4770.949	20960.19	EU	0.7			MIT
4768.293	4769.626	20966.00	EU	0.3			MIT
4766.656	4767.989	20973.20	EU	0.3	0.0		MIT
4765.60	4766.93	20977.9	EU	0.5W			KN
4765.16	4766.49	20979.8	EU	0.7	0.0		KN
4763.971	4765.303	20985.02	EU	0.6	0.3		MIT
4763.242	4764.574	20988.24	EU	1.20			MIT
4762.912	4764.244	20989.69	EU	1.8	0.3		MIT
4762.383	4763.715	20992.02	EU	0.5	0.0		MIT
4758.734	4760.065	21008.12	EU	1.80	0.3		MIT
4755.937	4757.267	21020.47	EU	1.8W			MIT
4753.716	4755.045	21030.29	EU	0.5	0.0		MIT
4752.37	4753.70	21036.2	EU	0.50			KN
4751.365	4752.694	21040.70	EU	0.8			MIT
4750.08	4751.41	21046.4	EU	0.3W			KN
4749.634	4750.962	21048.37	EU	0.8			MIT
4741.775	4743.101	21083.25	EU	1.00			MIT
4740.524	4741.850	21088.82	EU	2.7	0.3		MIT
4739.962	4741.288	21091.32	EU	0.3	0.3		MIT
4739.17	4740.50	21094.8	EU	1.9			M
4736.608	4737.933	21106.25	EU	1.8			MIT
4735.847	4737.172	21109.64	EU	0.5	0.0		MIT
4734.199	4735.523	21116.99	EU	0.50			MIT
4731.818	4733.142	21127.62	EU	0.6W			MIT
4730.741	4732.064	21132.43	EU	0.60			MIT
4728.148	4729.471	21144.01	EU	1.6			MIT
4727.47	4728.79	21147.1	EU	0.90			KN
4725.703	4727.025	21154.96	EU	0.9	0.3		MIT
4724.936	4726.258	21158.39	EU	0.6W			MIT
4724.057	4725.379	21162.33	EU	1.5			MIT
4723.913	4725.235	21162.97	EU	1.3			MIT
4720.550	4721.871	21178.05	EU	1.00			MIT
4720.223	4721.544	21179.51	EU	0.90			MIT
4718.614	4719.934	21186.74	EU	1.80	0.3		MIT
4717.225	4718.545	21192.98	EU	1.80	0.3		MIT

AIR WAVELENGTH (ANGSTRMS)	VACUUM WAVELENGTH (ANGSTRMS)	WAVENUMBER (KAYSERS)	LMENT	LOG ARC	INTENSITY SPK	DIS	REF
4716.26	4717.58	21197.3	EU	0.50			MIT
4713.606	4714.925	21209.25	EU	2.6	0.3		MIT
4712.421	4713.739	21214.58	EU	0.6			MIT
4712.105	4713.423	21216.00	EU	0.3			MIT
4709.82	4711.14	21226.3	EU	1.50	0.3		MIT
4705.98	4707.30	21243.6	EU	0.60			KN
4704.601	4705.917	21249.84	EU	1.5W			MIT
4703.887	4705.203	21253.07	EU	0.90			MIT
4703.358	4704.674	21255.46	EU	0.6			MIT
4702.621	4703.937	21258.79	EU	1.2			MIT
4700.419	4701.734	21268.75	EU	0.80			KN
4699.395	4700.710	21273.38	EU	0.80			MIT
4698.136	4699.451	21279.08	EU	2.5	0.3		MIT
4697.589	4698.904	21281.56	EU	1.0			MIT
4695.339	4696.653	21291.76	EU	1.0W	0.0		MIT
4692.635	4693.948	21304.03	EU	1.8W	0.3		MIT
4690.27	4691.58	21314.8	EU	0.50			MIT
4689.73	4691.04	21317.2	EU	0.60			MIT
4688.50	4689.81	21322.8	EU	0.5W	0.0		KN
4688.23	4689.54	21324.0	EU	2.0	0.3		MIT
4686.85	4688.16	21330.3	EU	0.50			KN
4686.66	4687.97	21331.2	EU	0.30			KN
4685.70	4687.01	21335.6	EU	0.5W			KN
4685.25	4686.56	21337.6	EU	1.80			KN
4684.82	4686.13	21339.6	EU	0.6W			MIT
4681.51	4682.82	21354.6	EU	1.0	0.0		KN
4681.06	4682.37	21356.7	EU	1.2	0.0		KN
4679.48	4680.79	21363.9	EU	0.6W			KN
4675.48	4676.79	21382.2	EU	1.30			KN
4672.85	4674.16	21394.2	EU	0.8			MIT
4671.38	4672.69	21401.0	EU	1.0			MIT
4671.18	4672.49	21401.9	EU	1.5	0.0		M
4668.14	4669.45	21415.8	EU	1.2W			MIT
4667.41	4668.72	21419.2	EU	0.5			KN
4666.59	4667.90	21422.9	EU	0.7W			MIT
4665.07	4666.38	21429.9	EU	1.50			KN
4661.872	4663.177	21444.61	EU	1.96	1.3		MIT
4661.05	4662.35	21448.4	EU	1.20			MIT
4660.37	4661.67	21451.5	EU	1.7	0.0		MIT
4658.61	4659.91	21459.6	EU	0.9			MIT
4656.73	4658.03	21468.3	EU	1.7	0.0		MIT
4654.0	4655.3	21481.	EU	0.8W			KN
4653.48	4654.78	21483.3	EU	1.4W			M
4651.55	4652.85	21492.2	EU	0.7			KN
4650.48	4651.78	21497.1	EU	1.0			MIT
4649.04	4650.34	21503.8	EU	0.9W			MIT
4647.49	4648.79	21511.0	EU	1.2W			KN
4646.30	4647.60	21516.5	EU	0.5W			MIT
4645.73	4647.03	21519.1	EU	0.9W			MIT
4644.24	4645.54	21526.0	EU	1.7			KN
4642.77	4644.07	21532.8	EU	1.2W			MIT
4641.415	4642.715	21539.12	EU	1.3			KN
4635.39	4636.69	21567.1	EU	0.8			KN
4634.72	4636.02	21570.2	EU	0.60			MIT
4633.1	4634.4	21578.	EU	0.9W			KN
4632.58	4633.88	21580.2	EU	0.5W			MIT
4629.82	4631.12	21593.1	EU	1.20			KN
4627.225	4628.521	21605.17	EU	1.7	1.2		MIT
4626.21	4627.51	21609.9	EU	0.80			MIT
4625.30	4626.60	21614.2	EU	1.7W			M
4624.9	4626.2	21616.	EU	0.6W			KN
4623.39	4624.68	21623.1	EU	0.90			KN
4621.93	4623.22	21629.9	EU	0.3			MIT
4621.7	4623.0	21631.	EU	0.3			KN
4621.34	4622.63	21632.7	EU	1.20			KN
4620.32	4621.61	21637.5	EU	0.6			KN
4616.49	4617.78	21655.4	EU	1.5W			MIT
4614.63	4615.92	21664.1	EU	0.8	0.7		MIT
4612.35	4613.64	21674.9	EU	0.3	0.3		MIT
4611.52	4612.81	21678.7	EU	1.7W			KN

AIR WAVELENGTH (ANGSTRMS)	VACUUM WAVELENGTH (ANGSTRMS)	WAVENUMBER (KAYSERS)	LMENT	LOG ARC	INTENSITY SPK	DIS	REF
4609.20	4610.49	21689.7	EU	0.3W			MIT
4608.15	4609.44	21694.6	EU	0.3W			M
4605.84	4607.13	21705.5	EU	0.9			M
4604.83	4606.12	21710.2	EU	0.8			MIT
4602.63	4603.92	21720.6	EU	1.2W			KN
4601.19	4602.48	21727.4	EU	1.10			M
4599.19	4600.48	21736.9	EU	0.80			KN
4597.33	4598.62	21745.7	EU	1.60			MIT
4594.02	4595.31	21761.3	EU	2.76	2.3		MIT
4592.39	4593.68	21769.1	EU	0.6W			MIT
4591.07	4592.36	21775.3	EU	0.6			M
4588.96	4590.25	21785.3	EU	0.9W			KN
4586.44	4587.73	21797.3	EU	1.5W			MIT
4585.89	4587.18	21799.9	EU	0.3W			KN
4585.67	4586.95	21801.0	EU	0.9W	0.7		MIT
4580.76	4582.04	21824.3	EU	0.8W			KN
4580.20	4581.48	21827.0	EU	0.7W	0.7		MIT
4579.81	4581.09	21828.9	EU	0.5			MIT
4578.60	4579.88	21834.6	EU	0.7W			MIT
4578.07	4579.35	21837.1	EU	1.00			MIT
4576.92	4578.20	21842.6	EU	1.00			MIT
4576.36	4577.64	21845.3	EU	1.10	1.0		M
4575.783	4577.065	21848.06	EU	1.3			MIT
4575.22	4576.50	21850.7	EU	1.20			KN
4574.88	4576.16	21852.4	EU	0.3W			M
4574.20	4575.48	21855.6	EU	1.0	0.7		M
4573.7	4575.0	21858.	EU	0.6W	0.6		KN
4572.5	4573.8	21864.	EU	0.50			KN
4571.28	4572.56	21869.6	EU	0.30			MIT
4570.50	4571.78	21873.3	EU	0.50			M
4569.95	4571.23	21875.9	EU	0.3			MIT
4569.05	4570.33	21880.3	EU	0.8W			MIT
4568.90	4570.18	21881.0	EU	0.8W			KN
4565.44	4566.72	21897.6	EU	0.6W			MIT
4564.53	4565.81	21901.9	EU	1.2W			KN
4562.68	4563.96	21910.8	EU	1.00			KN
4556.96	4558.24	21938.3	EU	0.6			M
4555.71	4556.99	21944.3	EU	1.10			M
4555.38	4556.66	21945.9	EU	0.6	0.6		MIT
4553.3	4554.6	21956.	EU	0.5W			KN
4552.19	4553.47	21961.3	EU	0.70	0.7		KN
4551.7	4553.0	21964.	EU	0.3W			MIT
4551.08	4552.36	21966.6	EU	0.7W			MIT
4550.67	4551.95	21968.6	EU	1.3W			MIT
4549.74	4551.02	21973.1	EU	0.8W			MIT
4549.52	4550.80	21974.2	EU	0.8	0.7		MIT
4547.75	4549.02	21982.7	EU	0.60	0.6		MIT
4547.221	4548.496	21985.29	EU	0.5			MIT
4545.62	4546.89	21993.0	EU	0.50			MIT
4543.793	4545.067	22001.88	EU	0.8			MIT
4543.18	4544.45	22004.9	EU	1.00			KN
4541.96	4543.23	22010.8	EU	0.60			MIT
4540.66	4541.93	22017.1	EU	0.60	0.6		MIT
4539.69	4540.96	22021.8	EU	0.5W	0.5		M
4539.29	4540.56	22023.7	EU I	1.1	0.3		KN
4538.07	4539.34	22029.6	EU		1.30		KN
4535.59	4536.86	22041.7	EU	1.2			KN
4535.578	4536.850	22041.73	EU	0.6			MIT
4528.49	4529.76	22076.2	EU	0.9W			KN
4527.74	4529.01	22079.9	EU	0.3			KN
4527.34	4528.61	22081.8	EU	0.80	0.7		MIT
4527.035	4528.304	22083.32	EU	0.5			MIT
4526.686	4527.955	22085.02	EU	2.0			MIT
4526.67	4527.94	22085.1	EU	0.7	0.3		ED
4526.083	4527.352	22087.97	EU	0.8			MIT
4524.495	4525.764	22095.72	EU	1.1			MIT
4522.602	4523.870	22104.97	EU I	2.36	2.3		KN
4522.58	4523.85	22105.1	EU	2.7			KN
4520.986	4522.254	22112.87	EU	0.7			MIT
4519.556	4520.823	22119.86	EU	0.5			MIT

AIR WAVELENGTH (ANGSTRMS)	VACUUM WAVELENGTH (ANGSTRMS)	WAVENUMBER (KAYSERS)	LMENT	LOG ARC	INTENSITY SPK	DIS	REF
4518.68	4519.95	22124.1	EU	0.9W	0.7		KN
4517.76	4519.03	22128.7	EU	1.3W			KN
4517.356	4518.623	22130.64	EU	0.8			MIT
4516.938	4518.205	22132.68	EU	0.6			MIT
4516.55	4517.82	22134.6	EU	0.3	0.3		ED
4514.047	4515.313	22146.86	EU	1.2			MIT
4513.20	4514.47	22151.0	EU	1.3W			M
4512.616	4513.882	22153.88	EU	0.3			MIT
4512.14	4513.41	22156.2	EU	0.3			MIT
4511.53	4512.80	22159.2	EU	0.5W	0.5		MIT
4509.048	4510.313	22171.41	EU	0.6			MIT
4508.695	4509.960	22173.15	EU	1.0W	0.7H		MIT
4507.424	4508.688	22179.40	EU	0.8			MIT
4505.26	4506.52	22190.1	EU	0.9W			KN
4504.539	4505.803	22193.60	EU	0.6			MIT
4503.43	4504.69	22199.1	EU	0.50			KN
4502.116	4503.379	22205.55	EU	0.3			MIT
4497.53	4498.79	22228.2	EU	0.60			KN
4494.34	4495.60	22244.0	EU	0.6			KN
4492.40	4493.66	22253.6	EU	0.6W			MIT
4488.283	4489.542	22273.99	EU	1.1W	0.3		MIT
4485.59	4486.85	22287.4	EU	0.5W	0.0		MIT
4485.148	4486.406	22289.55	EU II	1.6	0.3		MIT
4484.67	4485.93	22291.9	EU	0.9W			KN
4480.11	4481.37	22314.6	EU	1.20	0.0		KN
4478.4	4479.7	22323.	EU	0.5W			KN
4477.179	4478.435	22329.23	EU	1.10			MIT
4475.78	4477.04	22336.2	EU	0.80			MIT
4474.605	4475.861	22342.07	EU	0.30			MIT
4474.102	4475.357	22344.58	EU	0.70			MIT
4471.992	4473.247	22355.13	EU	1.5W	0.3		MIT
4471.64	4472.89	22356.9	EU	0.6	0.3		MIT
4469.66	4470.91	22366.8	EU	0.80			KN
4468.03	4469.28	22374.9	EU	1.2W			KN
4466.34	4467.59	22383.4	EU	0.60			MIT
4466.02	4467.27	22385.0	EU	0.30			MIT
4464.968	4466.221	22390.29	EU	1.9	1.0		MIT
4464.565	4465.818	22392.31	EU	1.5			MIT
4463.829	4465.082	22396.01	EU	0.3W	0.3		MIT
4462.14	4463.39	22404.5	EU	0.7W			KN
4461.548	4462.800	22407.46	EU	0.3	0.3		MIT
4456.97	4458.22	22430.5	EU	0.7W			MIT
4456.07	4457.32	22435.0	EU	0.6W			KN
4451.97	4453.22	22455.7	EU	0.80			KN
4449.13	4450.38	22470.0	EU	1.0W			KN
4446.714	4447.962	22482.21	EU	0.80			MIT
4441.487	4442.734	22508.66	EU	1.1W			MIT
4440.041	4441.287	22515.99	EU	0.8W			MIT
4437.952	4439.198	22526.59	EU	0.6	0.0		MIT
4437.03	4438.28	22531.3	EU	0.5W			MIT
4436.67	4437.92	22533.1	EU	0.3W			KN
4435.602	4436.847	22538.53	EU	2.66	2.0		KN
4435.53	4436.78	22538.9	EU	3.3			KN
4434.803	4436.048	22542.59	EU II	1.3W	0.3		MIT
4433.22	4434.46	22550.6	EU	0.8W	0.0		MIT
4430.80	4432.04	22562.9	EU	1.00			KN
4429.756	4431.000	22568.27	EU	0.5	0.3		MIT
4426.80	4428.04	22583.3	EU	1.00			MIT
4426.422	4427.665	22585.27	EU	0.60	0.3		MIT
4423.35	4424.59	22600.9	EU	0.60			KN
4422.96	4424.20	22602.9	EU	0.5W			KN
4419.657	4420.898	22619.84	EU	0.50	0.3		MIT
4417.25	4418.49	22632.2	EU	1.8W			M
4416.71	4417.95	22634.9	EU	0.5W			MIT
4414.647	4415.887	22645.51	EU	0.8	0.3		MIT
4413.51	4414.75	22651.3	EU	1.0W			MIT
4412.03	4413.27	22658.9	EU	0.8W			KN
4411.11	4412.35	22663.7	EU	0.3W			MIT
4410.65	4411.89	22666.0	EU	0.6W	0.0		KN
4409.64	4410.88	22671.2	EU	0.80			MIT

AIR WAVELENGTH (ANGSTRMS)	VACUUM WAVELENGTH (ANGSTRMS)	WAVENUMBER (KAYSERS)	LMENT	LOG ARC	INTENSITY SPK	DIS	REF
4407.083	4408.321	22684.37	EU	1.10			MIT
4406.782	4408.020	22685.92	EU	1.2W			MIT
4405.271	4406.508	22693.71	EU	1.20			MIT
4403.165	4404.402	22704.56	EU	0.90	0.3		MIT
4402.89	4404.13	22706.0	EU	0.3	0.0		MIT
4399.60	4400.84	22723.0	EU	0.3W			KN
4399.313	4400.549	22724.44	EU	1.1W	0.0		MIT
4397.95	4399.19	22731.5	EU	0.8W			MIT
4397.71	4398.95	22732.7	EU	0.8	0.3		MIT
4391.369	4392.603	22765.55	EU	0.9W			MIT
4391.19	4392.42	22766.5	EU	1.0W			MIT
4390.363	4391.596	22770.76	EU	0.6W	0.3		MIT
4389.22	4390.45	22776.7	EU	0.3W			MIT
4389.071	4390.304	22777.47	EU	0.7	0.3		MIT
4387.884	4389.117	22783.63	EU	2.3			MIT
4384.94	4386.17	22798.9	EU	0.6W			MIT
4383.158	4384.390	22808.19	EU I	2.00	1.3		MIT
4382.052	4383.283	22813.95	EU	0.3	0.0		MIT
4380.977	4382.208	22819.55	EU	0.3	0.0		MIT
4380.167	4381.398	22823.77	EU	0.5W			MIT
4379.813	4381.044	22825.61	EU	0.7W	0.3		MIT
4377.33	4378.56	22838.6	EU	0.3W			MIT
4376.412	4377.642	22843.35	EU	0.50	0.3		MIT
4376.16	4377.39	22844.7	EU	0.3			KN
4375.12	4376.35	22850.1	EU	0.6	0.3		MIT
4372.207	4373.436	22865.32	EU	1.00	0.6		MIT
4371.546	4372.774	22868.78	EU	0.6W			MIT
4370.464	4371.692	22874.44	EU	1.80	0.3		MIT
4369.467	4370.695	22879.66	EU	1.3W	0.3		MIT
4368.42	4369.65	22885.1	EU	0.9W	0.3		MIT
4366.52	4367.75	22895.1	EU	1.00			MIT
4361.576	4362.802	22921.05	EU	1.0W	0.3		MIT
4357.761	4358.986	22941.12	EU	0.8	0.3		MIT
4355.097	4356.321	22955.15	EU II	2.2	1.3		ME
4354.787	4356.011	22956.78	EU	2.0W	0.3		MIT
4353.517	4354.741	22963.48	EU	0.3	0.0		MIT
4352.241	4353.464	22970.21	EU	0.6	0.3		MIT
4351.265	4352.488	22975.36	EU	0.5W			MIT
4350.119	4351.342	22981.42	EU	1.0W			MIT
4349.715	4350.938	22983.55	EU	1.3W			MIT
4349.485	4350.708	22984.77	EU	0.3	0.3		MIT
4349.25	4350.47	22986.0	EU	0.3W			KN
4348.126	4349.348	22991.95	EU	0.6W			MIT
4347.269	4348.491	22996.48	EU	0.7W			MIT
4345.904	4347.126	23003.71	EU	1.90	0.0		MIT
4343.252	4344.473	23017.75	EU	1.30			MIT
4342.49	4343.71	23021.8	EU	0.6W			MIT
4340.59	4341.81	23031.9	EU	0.9W	0.0		KN
4339.938	4341.158	23035.33	EU	0.5W	0.0		MIT
4338.46	4339.68	23043.2	EU	0.5W			KN
4337.684	4338.904	23047.30	EU	2.00	0.3		MIT
4336.708	4337.927	23052.48	EU	0.9W			MIT
4335.459	4336.678	23059.12	EU	0.3	0.0		MIT
4334.752	4335.971	23062.89	EU	0.3	0.3		MIT
4333.749	4334.967	23068.22	EU	0.8W			MIT
4332.85	4334.07	23073.0	EU	0.5W	0.0		KN
4332.403	4333.621	23075.39	EU	0.3W	0.0		MIT
4331.176	4332.394	23081.93	EU	1.9			MIT
4330.611	4331.829	23084.94	EU	1.3	0.6		MIT
4329.97	4331.19	23088.4	EU	2.0W	0.0		MIT
4329.361	4330.578	23091.60	EU	2.00	0.3		MIT
4326.44	4327.66	23107.2	EU	0.9W			KN
4326.13	4327.35	23108.9	EU	0.6W			KN
4325.53	4326.75	23112.1	EU	1.5W			KN
4322.575	4323.791	23127.85	EU	1.8			MIT
4321.207	4322.422	23135.18	EU	0.30			MIT
4320.156	4321.371	23140.80	EU	1.1	0.0		MIT
4317.668	4318.882	23154.14	EU	1.3			MIT
4317.4	4318.6	23156.	EU	1.0W			KN
4315.07	4316.28	23168.1	EU	0.9W			KN

AIR WAVELENGTH (ANGSTRMS)	VACUUM WAVELENGTH (ANGSTRMS)	WAVENUMBER (KAYSERS)	LMENT	LOG ARC	INTENSITY SPK	DIS	REF
4314.297	4315.510	23172.23	EU	0.6W			MIT
4313.851	4315.064	23174.63	EU	1.0	0.3		MIT
4313.451	4314.664	23176.77	EU	0.7			MIT
4311.305	4312.518	23188.31	EU	0.60	0.0		MIT
4311.028	4312.241	23189.80	EU	1.0	0.3		MIT
4310.585	4311.797	23192.18	EU	0.8	0.3		MIT
4310.19	4311.40	23194.3	EU	0.6W			KN
4306.861	4308.072	23212.24	EU	0.3W			MIT
4301.56	4302.77	23240.8	EU	0.7	0.0		MIT
4300.838	4302.048	23244.74	EU	1.0	0.3		MIT
4298.724	4299.933	23256.18	EU	1.30	0.3		MIT
4298.07	4299.28	23259.7	EU	0.8W			KN
4297.86	4299.07	23260.9	EU	0.3W			MIT
4297.428	4298.637	23263.19	EU	1.00			MIT
4295.441	4296.649	23273.95	EU	1.3	0.7		MIT
4294.67	4295.88	23278.1	EU	0.6	0.3		MIT
4293.87	4295.08	23282.5	EU	0.8W			MIT
4292.991	4294.199	23287.23	EU	1.00			MIT
4292.424	4293.632	23290.31	EU	0.60			MIT
4288.605	4289.812	23311.05	EU	1.0	0.0		MIT
4287.8	4289.0	23315.	EU	0.7			KN
4287.431	4288.637	23317.43	EU	1.0	0.0		MIT
4287.27	4288.48	23318.3	EU	0.3			KN
4285.717	4286.923	23326.75	EU	0.5L			MIT
4284.652	4285.858	23332.55	EU	0.3W			MIT
4283.869	4285.074	23336.82	EU	1.1	0.3		MIT
4282.6	4283.8	23344.	EU	0.8			KN
4281.922	4283.127	23347.43	EU	1.4	0.3		MIT
4281.09	4282.29	23352.0	EU	0.3			MIT
4279.69	4280.89	23359.6	EU	1.0			KN
4279.25	4280.45	23362.0	EU	0.9			KN
4277.64	4278.84	23370.8	EU	0.6			M
4276.204	4277.407	23378.65	EU	1.0	0.3		MIT
4275.912	4277.115	23380.24	EU	0.6			MIT
4273.61	4274.81	23392.8	EU	0.3			MIT
4272.757	4273.959	23397.51	EU	0.7	0.0		MIT
4272.34	4273.54	23399.8	EU	0.6			MIT
4270.508	4271.710	23409.83	EU	1.2			MIT
4270.45	4271.65	23410.1	EU	1.2	0.3		KN
4270.245	4271.447	23411.27	EU	1.3	0.3		MIT
4269.495	4270.697	23415.38	EU	0.7	0.0H		MIT
4267.85	4269.05	23424.4	EU	1.0			KN
4266.76	4267.96	23430.4	EU	0.3			MIT
4266.37	4267.57	23432.5	EU	0.5			MIT
4265.811	4267.012	23435.60	EU	0.8			MIT
4264.909	4266.109	23440.56	EU	1.1	0.3H		MIT
4264.69	4265.89	23441.8	EU	0.6			MIT
4263.812	4265.012	23446.59	EU	0.5	0.0		MIT
4262.177	4263.377	23455.59	EU	0.3			MIT
4261.794	4262.994	23457.69	EU	0.6			MIT
4261.17	4262.37	23461.1	EU	0.6	0.3		MIT
4260.980	4262.179	23462.18	EU	1.1			MIT
4258.12	4259.32	23477.9	EU	0.8			MIT
4257.850	4259.049	23479.42	EU	1.0	0.0		MIT
4255.941	4257.139	23489.95	EU	0.6	0.0H		MIT
4255.27	4256.47	23493.7	EU	1.2			MIT
4253.814	4255.011	23501.70	EU	1.2	0.5		MIT
4249.43	4250.63	23525.9	EU	1.3			KN
4247.831	4249.027	23534.80	EU	0.7H	0.3		MIT
4247.73	4248.93	23535.4	EU	0.9			KN
4247.069	4248.265	23539.02	EU	1.2	0.5		MIT
4246.13	4247.33	23544.2	EU	1.1			KN
4245.86	4247.06	23545.7	EU	0.7	0.0		MIT
4245.39	4246.59	23548.3	EU	1.2D	0.0		M
4244.751	4245.946	23551.88	EU	1.0D			MIT
4242.47	4243.66	23564.5	EU	0.8	0.0H		KN
4240.9	4242.1	23573.	EU	0.3	0.0		KN
4240.120	4241.314	23577.60	EU	0.5			MIT
4238.705	4239.899	23585.47	EU	1.2	0.3		MIT
4237.519	4238.712	23592.07	EU	1.2	0.3		MIT

AIR WAVELENGTH (ANGSTRMS)	VACUUM WAVELENGTH (ANGSTRMS)	WAVENUMBER (KAYSERS)	LMENT	LOG ARC	INTENSITY SPK	DIS	REF
4236.23	4237.42	23599.2	EU	0.5			MIT
4235.14	4236.33	23605.3	EU	0.3			MIT
4235.02	4236.21	23606.0	EU	0.8			KN
4234.29	4235.48	23610.1	EU	0.5			KN
4234.099	4235.291	23611.13	EU	0.3			MIT
4233.609	4234.801	23613.86	EU	0.9			MIT
4232.462	4233.654	23620.26	EU	1.1	0.5		MIT
4231.23	4232.42	23627.1	EU	0.3			KN
4230.63	4231.82	23630.5	EU	0.7	0.0		MIT
4229.333	4230.524	23637.73	EU	1.0	0.5		MIT
4228.04	4229.23	23645.0	EU	0.5W			MIT
4227.58	4228.77	23647.5	EU	0.9			KN
4224.88	4226.07	23662.6	EU	0.6			KN
4224.30	4225.49	23665.9	EU	0.5			KN
4222.31	4223.50	23677.1	EU	0.7			MIT
4221.080	4222.269	23683.95	EU	1.0	0.3		MIT
4220.675	4221.864	23686.22	EU	0.3			MIT
4217.779	4218.967	23702.48	EU	0.8	0.0H		MIT
4217.27	4218.46	23705.4	EU	0.6			MIT
4213.912	4215.099	23724.24	EU	0.3L			MIT
4213.54	4214.73	23726.3	EU	0.5	0.3		MIT
4210.67	4211.86	23742.5	EU	0.3	0.0		MIT
4209.11	4210.30	23751.3	EU	0.3			KN
4208.93	4210.12	23752.3	EU	0.7			KN
4208.16	4209.35	23756.7	EU	1.0	0.0		MIT
4205.046	4206.231	23774.25	EU II	2.3G	1.7		KN
4204.909	4206.094	23775.03	EU	0.7			KN
4202.68	4203.86	23787.6	EU	1.2			MIT
4202.03	4203.21	23791.3	EU	0.7			MIT
4200.81	4201.99	23798.2	EU	0.3			KN
4198.66	4199.84	23810.4	EU	0.5			MIT
4196.188	4197.370	23824.44	EU	0.8	0.5		MIT
4195.40	4196.58	23828.9	EU	0.3			ED
4195.347	4196.529	23829.22	EU	1.0	0.3		MIT
4194.474	4195.656	23834.18	EU	1.0			MIT
4192.564	4193.745	23845.03	EU	0.9			MIT
4189.755	4190.936	23861.02	EU	0.7	0.0H		MIT
4186.42	4187.60	23880.0	EU	0.8			KN
4184.985	4186.164	23888.22	EU	0.8			MIT
4183.19	4184.37	23898.5	EU	0.6			MIT
4182.26	4183.44	23903.8	EU	0.9	0.0		M
4180.88	4182.06	23911.7	EU	1.0	0.0		MIT
4179.40	4180.58	23920.1	EU	0.3	0.0H		MIT
4177.57	4178.75	23930.6	EU	0.9	0.3		KN
4176.725	4177.902	23935.46	EU	0.7			MIT
4176.62	4177.80	23936.1	EU	0.9	0.3		KN
4175.17	4176.35	23944.4	EU	0.9	0.3		MIT
4173.75	4174.93	23952.5	EU	0.9			MIT
4172.80	4173.98	23958.0	EU	1.1	0.5		KN
4169.46	4170.64	23977.2	EU	1.0	0.0H		MIT
4166.83	4168.00	23992.3	EU	0.3	0.0H		MIT
4164.23	4165.40	24007.3	EU	0.3			MIT
4163.455	4164.629	24011.74	EU	0.6			MIT
4162.17	4163.34	24019.2	EU	0.7	0.0		M
4161.36	4162.53	24023.8	EU	1.0			KN
4160.476	4161.649	24028.94	EU	1.0	0.0		MIT
4159.23	4160.40	24036.1	EU	1.0	0.0		MIT
4158.78	4159.95	24038.7	EU	0.9			KN
4157.80	4158.97	24044.4	EU	1.3			MIT
4153.44	4154.61	24069.6	EU	0.7	0.3		MIT
4152.25	4153.42	24076.5	EU	0.3	0.0H		M
4151.66	4152.83	24080.0	EU	0.6			MIT
4151.55	4152.72	24080.6	EU	0.6	0.0H		M
4149.23	4150.40	24094.1	EU	1.1			KN
4148.61	4149.78	24097.7	EU	0.3			MIT
4147.78	4148.95	24102.5	EU	0.5			MIT
4147.19	4148.36	24105.9	EU	0.9	0.0		MIT
4146.35	4147.52	24110.8	EU	0.9	0.0H		KN
4145.97	4147.14	24113.0	EU	0.3			ED
4145.244	4146.413	24117.23	EU	0.9	0.0		MIT

AIR WAVELENGTH (ANGSTRMS)	VACUUM WAVELENGTH (ANGSTRMS)	WAVENUMBER (KAYSERS)	LMENT	LOG ARC	INTENSITY SPK	DIS	REF
4144.51	4145.68	24121.5	EU	0.9	0.0		KN
4144.22	4145.39	24123.2	EU	0.3			MIT
4142.97	4144.14	24130.5	EU	0.3			MIT
4141.746	4142.914	24137.60	EU	1.0			MIT
4141.034	4142.202	24141.75	EU	1.0	0.3		MIT
4140.02	4141.19	24147.7	EU	0.7	0.3H		MIT
4139.693	4140.861	24149.57	EU	0.8			MIT
4139.47	4140.64	24150.9	EU	0.5			MIT
4138.30	4139.47	24157.7	EU	0.3			MIT
4137.076	4138.243	24164.85	EU	1.0			MIT
4136.612	4137.779	24167.56	EU	1.1	0.3		MIT
4136.19	4137.36	24170.0	EU	0.8			KN
4134.955	4136.121	24177.24	EU	0.8	0.0H		MIT
4133.115	4134.281	24188.00	EU	0.6	0.0H		MIT
4132.24	4133.41	24193.1	EU	0.6	0.0H		MIT
4129.737	4130.902	24207.79	EU II	2.2R	1.7R		MIT
4129.621	4130.786	24208.47	EU	1.2			MIT
4128.10	4129.26	24217.4	EU	0.3			KN
4127.288	4128.452	24222.15	EU	0.5			MIT
4125.53	4126.69	24232.5	EU	1.0			KN
4124.910	4126.074	24236.12	EU	0.9	0.3		MIT
4124.553	4125.717	24238.21	EU	0.9	0.3		MIT
4124.02	4125.18	24241.4	EU	0.3			KN
4122.942	4124.105	24247.68	EU	0.5	0.0		MIT
4122.786	4123.949	24248.60	EU	0.5			MIT
4121.81	4122.97	24254.3	EU	0.3W			KN
4119.33	4120.49	24268.9	EU	1.0	0.5		M
4117.73	4118.89	24278.4	EU	0.3D			MIT
4116.978	4118.140	24282.81	EU	0.3S			MIT
4115.806	4116.967	24289.72	EU	0.8			MIT
4115.54	4116.70	24291.3	EU	0.3			MIT
4114.58	4115.74	24297.0	EU	0.3			KN
4113.55	4114.71	24303.1	EU	0.3			MIT
4113.03	4114.19	24306.1	EU	0.3			KN
4112.174	4113.334	24311.18	EU	1.2W	0.3W		MIT
4112.055	4113.215	24311.88	EU	1.0	0.3		MIT
4111.069	4112.229	24317.71	EU	1.0	0.0		MIT
4109.89	4111.05	24324.7	EU	0.3			MIT
4108.57	4109.73	24332.5	EU	0.6			KN
4107.87	4109.03	24336.6	EU	0.3	0.8		MIT
4106.88	4108.04	24342.5	EU	1.0	0.9		MIT
4106.345	4107.504	24345.69	EU	0.5			KN
4105.83	4106.99	24348.7	EU	0.3	0.3		M
4103.88	4105.04	24360.3	EU	0.7			MIT
4103.41	4104.57	24363.1	EU	0.30	0.5		KN
4102.72	4103.88	24367.2	EU	1.0	0.5		MIT
4101.44	4102.60	24374.8	EU	0.3			MIT
4100.71	4101.87	24379.1	EU	0.3			KN
4100.25	4101.41	24381.9	EU	0.3			MIT
4099.72	4100.88	24385.0	EU	1.1D	0.7		MIT
4099.05	4100.21	24389.0	EU	0.9D			MIT
4096.815	4097.971	24402.32	EU	1.2	0.3		MIT
4095.93	4097.09	24407.6	EU	0.3			MIT
4094.90	4096.06	24413.7	EU	0.8D			MIT
4094.309	4095.465	24417.25	EU	0.6	0.0H		MIT
4092.95	4094.11	24425.4	EU	0.7			KN
4091.75	4092.90	24432.5	EU	0.3			ED
4090.770	4091.925	24438.38	EU	0.5			MIT
4090.20	4091.35	24441.8	EU	0.8D			KN
4089.745	4090.899	24444.50	EU	1.0	0.0H		MIT
4088.793	4089.947	24450.19	EU	0.5			MIT
4087.86	4089.01	24455.8	EU	0.9			KN
4087.609	4088.763	24457.27	EU	0.3			MIT
4087.106	4088.260	24460.28	EU	0.6	0.0H		MIT
4086.422	4087.576	24464.38	EU	0.6			MIT
4085.51	4086.66	24469.8	EU	0.5			MIT
4085.388	4086.541	24470.57	EU	1.0L	0.3		MIT
4085.03	4086.18	24472.7	EU	0.9	0.0		MIT
4084.89	4086.04	24473.6	EU	0.5			MIT
4081.042	4082.194	24496.63	EU	0.7	0.0H		MIT

AIR WAVELENGTH (ANGSTRMS)	VACUUM WAVELENGTH (ANGSTRMS)	WAVENUMBER (KAYSERS)	LMENT	LOG ARC	INTENSITY SPK	DIS	REF
4080.772	4081.924	24498.25	EU	0.8			MIT
4080.51	4081.66	24499.8	EU	0.3			KN
4078.231	4079.382	24513.51	EU	0.5			MIT
4076.952	4078.103	24521.21	EU	1.0			MIT
4075.96	4077.11	24527.2	EU	0.8			MIT
4075.616	4076.767	24529.24	EU	0.3			MIT
4074.514	4075.664	24535.88	EU	0.6D			MIT
4073.76	4074.91	24540.4	EU	0.5			KN
4073.52	4074.67	24541.9	EU	0.5L			KN
4071.387	4072.537	24554.72	EU	0.7			MIT
4071.201	4072.351	24555.84	EU	0.8	0.3H		MIT
4070.53	4071.68	24559.9	EU	0.7			KN
4069.02	4070.17	24569.0	EU	0.6			ED
4068.97	4070.12	24569.3	EU	0.6			KN
4067.110	4068.259	24580.54	EU	0.9	0.0		MIT
4066.065	4067.213	24586.86	EU	1.0	0.0		MIT
4065.406	4066.554	24590.84	EU	0.7			MIT
4062.65	4063.80	24607.5	EU	1.00			KN
4062.20	4063.35	24610.2	EU	0.5			M
4062.152	4063.299	24610.54	EU	1.0	0.3		MIT
4061.552	4062.699	24614.18	EU	1.1	0.3		MIT
4060.301	4061.448	24621.76	EU	0.3			MIT
4060.02	4061.17	24623.5	EU	0.5			KN
4059.376	4060.522	24627.37	EU	1.4			MIT
4059.035	4060.181	24629.44	EU	0.6			KN
4058.449	4059.595	24633.00	EU	0.3			MIT
4055.255	4056.400	24652.40	EU	0.3			MIT
4053.094	4054.239	24665.54	EU	0.3			MIT
4052.084	4053.229	24671.69	EU	0.3			MIT
4050.43	4051.57	24681.8	EU	0.8			KN
4048.654	4049.798	24692.59	EU	0.3			MIT
4048.266	4049.410	24694.96	EU	0.6			MIT
4047.74	4048.88	24698.2	EU	0.3			KN
4043.97	4045.11	24721.2	EU	1.3			KN
4042.06	4043.20	24732.9	EU	1.0W	0.3		MIT
4040.47	4041.61	24742.6	EU	0.7			MIT
4039.192	4040.333	24750.43	EU	1.2			MIT
4038.37	4039.51	24755.5	EU	0.3			MIT
4037.665	4038.806	24759.79	EU	0.90			MIT
4037.135	4038.276	24763.04	EU	0.8	0.3		MIT
4036.56	4037.70	24766.6	EU	0.5			MIT
4036.11	4037.25	24769.3	EU	1.70			KN
4034.11	4035.25	24781.6	EU	0.3	0.5		MIT
4033.69	4034.83	24784.2	EU	0.9W			MIT
4031.38	4032.52	24798.4	EU	0.8	0.5		MIT
4030.66	4031.80	24802.8	EU	0.7W			KN
4030.203	4031.342	24805.64	EU	1.0W			MIT
4029.996	4031.135	24806.91	EU	0.7			MIT
4029.58	4030.72	24809.5	EU	0.8			MIT
4028.63	4029.77	24815.3	EU	1.2W			MIT
4028.52	4029.66	24816.0	EU	1.00			KN
4027.28	4028.42	24823.6	EU	0.3W			KN
4026.625	4027.763	24827.68	EU	0.7W			MIT
4025.95	4027.09	24831.8	EU	0.7			MIT
4025.443	4026.581	24834.97	EU	0.3			MIT
4024.34	4025.48	24841.8	EU	0.50			KN
4023.84	4024.98	24844.9	EU	0.3			MIT
4022.91	4024.05	24850.6	EU	0.80			KN
4021.01	4022.15	24862.4	EU	0.7W			MIT
4019.712	4020.848	24870.37	EU	0.5			MIT
4018.400	4019.536	24878.50	EU	1.0	0.6		MIT
4017.731	4018.867	24882.64	EU	0.5			MIT
4017.582	4018.718	24883.56	EU	1.40	1.4		MIT
4016.703	4017.838	24889.00	EU	0.9			MIT
4014.901	4016.036	24900.18	EU	0.5W			MIT
4014.655	4015.790	24901.70	EU	0.5			MIT
4014.39	4015.52	24903.4	EU	0.3			KN
4013.55	4014.68	24908.6	EU	0.30			MIT
4012.816	4013.950	24913.11	EU	1.3	1.4		MIT
4011.683	4012.817	24920.15	EU II	1.4			MIT

AIR WAVELENGTH (ANGSTRMS)	VACUUM WAVELENGTH (ANGSTRMS)	WAVENUMBER (KAYSERS)	LMENT		LOG ARC	INTENSITY SPK	DIS	REF
4010.805	4011.939	24925.60	EU		0.3			MIT
4010.422	4011.556	24927.99	EU		1.30			MIT
4010.18	4011.31	24929.5	EU		0.9	0.6		MIT
4008.872	4010.005	24937.62	EU		0.90	0.3		MIT
4007.98	4009.11	24943.2	EU		0.8			M
4007.687	4008.820	24945.00	EU		0.7D			MIT
4006.56	4007.69	24952.0	EU		0.8			MIT
4006.105	4007.238	24954.85	EU		0.3			MIT
4004.592	4005.724	24964.27	EU		0.7			MIT
4003.715	4004.847	24969.74	EU		1.3	1.4		MIT
4002.91	4004.04	24974.8	EU		1.2			MIT
4002.624	4003.756	24976.55	EU		0.7			MIT
4002.56	4003.69	24976.9	EU		1.5	0.3		KN
4002.020	4003.151	24980.32	EU		0.6			MIT
4001.31	4002.44	24984.7	EU		0.8	0.3		MIT
4000.807	4001.938	24987.89	EU		0.5			MIT
4000.698	4001.829	24988.57	EU		0.5			MIT
3998.818	3999.949	25000.32	EU		0.50			MIT
3995.994	3997.124	25017.99	EU		0.5	0.3		MIT
3995.755	3996.885	25019.49	EU		1.1			MIT
3995.07	3996.20	25023.8	EU		0.6W			KN
3994.68	3995.81	25026.2	EU		0.7			MIT
3993.928	3995.057	25030.93	EU		0.9	0.3		MIT
3992.367	3993.496	25040.72	EU		0.6			MIT
3991.15	3992.28	25048.4	EU		0.50			KN
3988.583	3989.711	25064.47	EU		0.7			MIT
3988.253	3989.381	25066.55	EU		1.00	0.3		MIT
3987.83	3988.96	25069.2	EU		1.3W			MIT
3987.26	3988.39	25072.8	EU		0.50			MIT
3987.098	3988.226	25073.81	EU		0.3			MIT
3986.923	3988.051	25074.91	EU		0.8			MIT
3986.618	3987.745	25076.83	EU		1.0			MIT
3986.08	3987.21	25080.2	EU		0.60			MIT
3985.65	3986.78	25082.9	EU		0.60			MIT
3985.39	3986.52	25084.6	EU		1.0			MIT
3985.222	3986.349	25085.61	EU		0.8			MIT
3984.923	3986.050	25087.49	EU		0.9			MIT
3982.991	3984.118	25099.66	EU		0.3H			MIT
3981.452	3982.578	25109.36	EU		0.5			MIT
3981.313	3982.439	25110.24	EU		0.6			MIT
3981.108	3982.234	25111.53	EU		0.6			MIT
3980.042	3981.168	25118.26	EU		0.5			MIT
3979.735	3980.861	25120.20	EU		0.5	0.3		MIT
3979.628	3980.754	25120.87	EU		1.0	0.3		MIT
3979.018	3980.143	25124.72	EU		0.6			MIT
3978.446	3979.571	25128.33	EU		1.50			MIT
3977.628	3978.753	25133.50	EU		0.7			MIT
3976.859	3977.984	25138.36	EU		1.1			MIT
3975.943	3977.068	25144.15	EU		0.9J			MIT
3975.562	3976.687	25146.56	EU		0.9			MIT
3975.200	3976.324	25148.85	EU		0.80			MIT
3973.837	3974.961	25157.48	EU		0.80			MIT
3973.38	3974.50	25160.4	EU		0.5			MIT
3972.047	3973.171	25168.82	EU		0.7	0.7		MIT
3971.99	3973.11	25169.2	EU	II		3.3		ME
3971.90	3973.02	25169.7	EU		2.0@			MIT
3971.088	3972.211	25174.89	EU		0.6			MIT
3970.612	3971.735	25177.91	EU		0.6			MIT
3969.90	3971.02	25182.4	EU		0.90			MIT
3969.23	3970.35	25186.7	EU		1.30			MIT
3968.870	3969.993	25188.96	EU		0.5J			MIT
3967.13	3968.25	25200.0	EU		1.40			M
3966.615	3967.737	25203.28	EU		0.70			MIT
3965.449	3966.571	25210.69	EU		0.5	0.3		MIT
3965.01	3966.13	25213.5	EU		1.2W			KN
3964.89	3966.01	25214.2	EU		1.6$			MIT
3964.49	3965.61	25216.8	EU		1.30			MIT
3963.62	3964.74	25222.3	EU		0.90			M
3961.99	3963.11	25232.7	EU		1.00			KN
3961.14	3962.26	25238.1	EU		1.70			MIT

AIR WAVELENGTH (ANGSTRMS)	VACUUM WAVELENGTH (ANGSTRMS)	WAVENUMBER (KAYSERS)	LMENT		LOG ARC	INTENSITY SPK	DIS	REF
3960.733	3961.854	25240.71	EU		0.5			MIT
3959.25	3960.37	25250.2	EU		0.50			MIT
3958.88	3960.00	25252.5	EU		0.50			MIT
3957.915	3959.035	25258.68	EU		1.0			MIT
3957.040	3958.160	25264.27	EU		1.1			MIT
3956.68	3957.80	25266.6	EU		0.60			MIT
3955.754	3956.873	25272.48	EU		1.70			MIT
3954.255	3955.374	25282.06	EU		0.8$			MIT
3954.11	3955.23	25283.0	EU		0.30			MIT
3953.47	3954.59	25287.1	EU		0.80	0.0		MIT
3953.154	3954.273	25289.10	EU		0.6			MIT
3952.25	3953.37	25294.9	EU		0.50			MIT
3951.313	3952.431	25300.88	EU		0.5			MIT
3950.854	3951.972	25303.82	EU		0.8			MIT
3949.837	3950.955	25310.34	EU		0.7			MIT
3949.586	3950.704	25311.95	EU		1.70			MIT
3949.123	3950.241	25314.91	EU		1.4W	0.0		MIT
3948.779	3949.897	25317.12	EU		0.6			MIT
3948.606	3949.724	25318.23	EU		0.5			MIT
3947.58	3948.70	25324.8	EU		0.6W			MIT
3946.552	3947.669	25331.40	EU		0.8			MIT
3946.18	3947.30	25333.8	EU		0.6W			KN
3945.67	3946.79	25337.1	EU		0.9W	0.0		MIT
3945.160	3946.277	25340.34	EU		0.7			MIT
3944.592	3945.709	25343.99	EU		0.70			MIT
3944.01	3945.13	25347.7	EU		0.9H			MIT
3943.602	3944.718	25350.35	EU		0.5			MIT
3943.085	3944.201	25353.68	EU		1.7	1.2		MIT
3942.938	3944.054	25354.62	EU		1.20			MIT
3942.69	3943.81	25356.2	EU		0.3W			KN
3942.352	3943.468	25358.39	EU		1.30			MIT
3942.203	3943.319	25359.35	EU		1.2	0.3		MIT
3941.563	3942.679	25363.47	EU		0.70	0.6		MIT
3940.357	3941.472	25371.23	EU		0.50			MIT
3939.985	3941.100	25373.62	EU		0.30	0.0		MIT
3939.455	3940.570	25377.04	EU		0.80			MIT
3939.195	3940.310	25378.71	EU		0.7			MIT
3938.38	3939.49	25384.0	EU		0.6			MIT
3938.26	3939.37	25384.7	EU		0.5			MIT
3937.619	3938.734	25388.87	EU		0.30			MIT
3937.468	3938.583	25389.84	EU		0.3W			MIT
3936.632	3937.746	25395.24	EU		1.3			MIT
3935.985	3937.099	25399.41	EU		1.0			MIT
3934.387	3935.501	25409.73	EU		0.7			MIT
3933.677	3934.791	25414.31	EU		1.0			MIT
3932.28	3933.39	25423.3	EU		0.50			MIT
3930.503	3931.616	25434.83	EU	II	3.0G	2.66		MIT
3930.44	3931.55	25435.2	EU		2.00			MIT
3929.892	3931.005	25438.79	EU	II	0.8W			MIT
3929.819	3930.932	25439.26	EU		1.0W			MIT
3928.98	3930.09	25444.7	EU		0.9W			KN
3928.921	3930.033	25445.08	EU		1.0W	0.3		MIT
3927.446	3928.558	25454.63	EU		0.3H			MIT
3926.97	3928.08	25457.7	EU		0.30			KN
3922.533	3923.644	25486.51	EU		0.5			MIT
3919.2	3920.3	25508.	EU		0.5W			KN
3919.106	3920.216	25508.80	EU		1.3			MIT
3918.518	3919.628	25512.63	EU		1.3W			MIT
3918.158	3919.268	25514.97	EU		0.8	0.3		MIT
3917.818	3918.928	25517.19	EU		0.7W			MIT
3917.681	3918.790	25518.08	EU		0.7W			MIT
3917.282	3918.391	25520.68	EU		1.3W			MIT
3916.836	3917.945	25523.58	EU		1.00			MIT
3915.995	3917.104	25529.06	EU		1.3			MIT
3915.242	3916.351	25533.97	EU		0.7W			MIT
3914.84	3915.95	25536.6	EU		0.6W			MIT
3914.155	3915.264	25541.06	EU		0.7W			MIT
3913.745	3914.853	25543.74	EU		0.70	0.0		MIT
3913.245	3914.353	25547.00	EU		0.6	0.0		MIT
3912.43	3913.54	25552.3	EU		0.7	0.0		MIT

AIR WAVELENGTH (ANGSTRMS)	VACUUM WAVELENGTH (ANGSTRMS)	WAVENUMBER (KAYSERS)	LMENT	LOG ARC	INTENSITY SPK	DIS	REF
3911.97	3913.08	25555.3	EU	0.6W			KN
3911.606	3912.714	25557.71	EU	0.7			MIT
3910.14	3911.25	25567.3	EU	0.7W	0.0		M
3909.908	3911.015	25568.81	EU	1.0	0.0		MIT
3907.110	3908.217	25587.12	EU II	3.0Z	2.7G		MIT
3905.657	3906.763	25596.64	EU	0.5			MIT
3904.914	3906.020	25601.51	EU	0.9			MIT
3903.918	3905.024	25608.04	EU	1.00			MIT
3903.634	3904.740	25609.90	EU	1.0			MIT
3903.240	3904.346	25612.49	EU	1.00			MIT
3901.681	3902.786	25622.72	EU	0.7W			MIT
3900.918	3902.023	25627.73	EU	0.7W			MIT
3900.519	3901.624	25630.35	EU	1.1W			MIT
3900.423	3901.528	25630.98	EU	0.9W	0.3		MIT
3900.169	3901.274	25632.65	EU	1.1W			MIT
3899.708	3900.813	25635.68	EU	0.6			MIT
3899.455	3900.560	25637.35	EU	0.7W	0.5		MIT
3898.78	3899.88	25641.8	EU	1.30			MIT
3898.50	3899.60	25643.6	EU	0.30			MIT
3898.26	3899.36	25645.2	EU	0.70	0.9V		MIT
3897.73	3898.83	25648.7	EU	1.30			MIT
3897.23	3898.33	25652.0	EU	1.00	1.00		MIT
3896.82	3897.92	25654.7	EU	0.60			MIT
3896.45	3897.55	25657.1	EU	0.30			KN
3895.46	3896.56	25663.6	EU	0.30			MIT
3895.12	3896.22	25665.9	EU	0.30			MIT
3894.97	3896.07	25666.9	EU	0.60	0.6V		MIT
3894.72	3895.82	25668.5	EU	0.50			MIT
3894.34	3895.44	25671.0	EU	0.3			MIT
3894.17	3895.27	25672.1	EU	1.0W	0.7J		MIT
3893.38	3894.48	25677.4	EU	0.3W			MIT
3893.11	3894.21	25679.1	EU	0.30			MIT
3892.81	3893.91	25681.1	EU	0.8W			MIT
3891.48	3892.58	25689.9	EU	0.5W			MIT
3891.34	3892.44	25690.8	EU	0.6W			MIT
3889.51	3890.61	25702.9	EU	1.00	0.70		MIT
3888.68	3889.78	25708.4	EU	0.9W	0.7W		MIT
3887.81	3888.91	25714.1	EU	0.7W			MIT
3885.98	3887.08	25726.2	EU	0.5W			MIT
3885.30	3886.40	25730.7	EU	0.5W	0.6J		MIT
3884.76	3885.86	25734.3	EU	1.70	0.7V		MIT
3884.44	3885.54	25736.4	EU	0.90	0.70		MIT
3883.64	3884.74	25741.7	EU	1.00	1.00		MIT
3881.31	3882.41	25757.2	EU	0.60			MIT
3880.21	3881.31	25764.5	EU	0.30			MIT
3877.88	3878.98	25780.0	EU	0.6K			MIT
3877.27	3878.37	25784.0	EU	0.7W			KN
3875.95	3877.05	25792.8	EU	0.50			MIT
3875.35	3876.45	25796.8	EU	1.00			MIT
3873.54	3874.64	25808.9	EU	0.9W	0.7W		MIT
3872.73	3873.83	25814.3	EU	0.8W			MIT
3870.87	3871.97	25826.7	EU	0.3W			MIT
3869.73	3870.83	25834.3	EU	1.3J			MIT
3869.16	3870.26	25838.1	EU	0.80			MIT
3866.21	3867.31	25857.8	EU	0.70	0.80		MIT
3865.56	3866.66	25862.1	EU	1.50			MIT
3865.29	3866.39	25863.9	EU	0.3	0.3		KN
3864.78	3865.88	25867.4	EU	0.30			MIT
3864.10	3865.20	25871.9	EU	0.8W	1.0W		MIT
3864.04	3865.14	25872.3	EU	0.3W			KN
3863.655	3864.750	25874.89	EU	0.5			MIT
3862.94	3864.04	25879.7	EU	0.7J			MIT
3861.959	3863.054	25886.25	EU	0.3	0.0H		MIT
3861.18	3862.27	25891.5	EU	1.5W	1.5W		MIT
3860.72	3861.81	25894.6	EU	0.7D	0.8		MIT
3859.52	3860.61	25902.6	EU	0.6D			KN
3859.22	3860.31	25904.6	EU	0.3J			MIT
3857.68	3858.77	25915.0	EU	0.6W			MIT
3856.523	3857.617	25922.74	EU	0.6W			MIT
3854.82	3855.91	25934.2	EU	0.8			KN

AIR WAVELENGTH (ANGSTRMS)	VACUUM WAVELENGTH (ANGSTRMS)	WAVENUMBER (KAYSERS)	LMENT	LOG ARC	INTENSITY SPK	DIS	REF
3854.64	3855.73	25935.4	EU	0.5			KN
3853.90	3854.99	25940.4	EU	0.5			MIT
3851.59	3852.68	25955.9	EU	0.9W	0.7J		MIT
3850.53	3851.62	25963.1	EU	0.90			MIT
3849.63	3850.72	25969.2	EU	0.3			MIT
3849.40	3850.49	25970.7	EU	0.6W	0.5W		MIT
3848.40	3849.49	25977.5	EU	0.3	0.0H		MIT
3848.20	3849.29	25978.8	EU	0.3W	0.0		MIT
3847.84	3848.93	25981.2	EU	1.1W			MIT
3847.19	3848.28	25985.6	EU	0.3			MIT
3846.39	3847.48	25991.0	EU	0.6	0.3		KN
3844.23	3845.32	26005.6	EU	1.00			MIT
3843.14	3844.23	26013.0	EU	0.5W	0.3		MIT
3842.54	3843.63	26017.1	EU	1.00			KN
3842.36	3843.45	26018.3	EU	0.50	0.3		MIT
3840.96	3842.05	26027.8	EU	0.8D	0.7		MIT
3838.36	3839.45	26045.4	EU	0.3W			MIT
3838.24	3839.33	26046.2	EU	0.6			KN
3837.88	3838.97	26048.7	EU	0.80	0.80		MIT
3837.43	3838.52	26051.7	EU	0.50	0.5V		MIT
3836.512	3837.600	26057.95	EU	1.0			MIT
3836.13	3837.22	26060.6	EU	0.80			MIT
3834.91	3836.00	26068.8	EU	0.60	0.60		MIT
3832.43	3833.52	26085.7	EU	0.80	0.60		MIT
3831.17	3832.26	26094.3	EU	0.60	0.8V		MIT
3829.435	3830.522	26106.11	EU	0.7	1.0		MIT
3828.93	3830.02	26109.6	EU	0.8J	0.9J		MIT
3827.29	3828.38	26120.7	EU	0.60	0.70		MIT
3826.68	3827.77	26124.9	EU	1.20	1.00		MIT
3826.41	3827.50	26126.7	EU	0.50			MIT
3825.66	3826.75	26131.9	EU	0.7	0.7		MIT
3825.134	3826.219	26135.46	EU	0.5	0.7H		MIT
3823.349	3824.434	26147.66	EU	0.5	0.7		MIT
3823.06	3824.14	26149.6	EU	0.60	0.60		MIT
3821.43	3822.51	26160.8	EU	0.5W	0.5W		MIT
3819.66	3820.74	26172.9	EU II	2.7$	2.7$		MIT
3819.068	3820.152	26176.97	EU	0.3	0.0H		MIT
3817.69	3818.77	26186.4	EU	0.6J	1.0J		MIT
3817.01	3818.09	26191.1	EU	0.7D	0.6D		MIT
3815.81	3816.89	26199.3	EU	0.5J	0.5J		MIT
3815.50	3816.58	26201.4	EU	1.3J	1.2J		MIT
3814.87	3815.95	26205.8	EU	0.7J	0.7J		MIT
3813.07	3814.15	26218.1	EU	0.3			KN
3812.03	3813.11	26225.3	EU	0.90	0.7J		MIT
3811.34	3812.42	26230.1	EU	1.00			MIT
3810.76	3811.84	26234.0	EU	0.5	0.5		MIT
3809.74	3810.82	26241.1	EU	0.80	0.70		MIT
3809.099	3810.180	26245.48	EU	1.0	0.7		MIT
3808.161	3809.242	26251.94	EU	0.8	0.7		MIT
3807.89	3808.97	26253.8	EU	0.7	0.6		KN
3807.59	3808.67	26255.9	EU	1.30	1.20		MIT
3806.76	3807.84	26261.6	EU	0.6			KN
3806.16	3807.24	26265.7	EU	0.80	0.70		MIT
3805.81	3806.89	26268.2	EU	0.80	0.70		MIT
3804.26	3805.34	26278.9	EU	1.00	1.00		MIT
3803.099	3804.179	26286.88	EU	1.0W	1.0W		MIT
3802.749	3803.829	26289.30	EU	0.7W	0.7W		MIT
3802.228	3803.307	26292.91	EU	0.6	0.5		MIT
3801.61	3802.69	26297.2	EU	0.80	1.00		MIT
3801.36	3802.44	26298.9	EU	1.00	1.00		MIT
3800.55	3801.63	26304.5	EU	1.00	1.00		MIT
3799.50	3800.58	26311.8	EU	1.3$			KN
3799.01	3800.09	26315.2	EU II	2.0J			MIT
3798.71	3799.79	26317.2	EU	0.3J			MIT
3796.69	3797.77	26331.3	EU	0.30			MIT
3796.313	3797.391	26333.87	EU	0.7W	1.0W		MIT
3796.16	3797.24	26334.9	EU	0.6W	0.9W		MIT
3795.95	3797.03	26336.4	EU	0.6			KN
3795.34	3796.42	26340.6	EU	1.00	0.70		MIT
3795.03	3796.11	26342.8	EU	1.00			MIT

AIR WAVELENGTH (ANGSTRMS)	VACUUM WAVELENGTH (ANGSTRMS)	WAVENUMBER (KAYSERS)	LMENT	LOG ARC	INTENSITY SPK	DIS	REF
3794.794	3795.872	26344.41	EU	0.6	0.5		MIT
3794.370	3795.447	26347.35	EU	0.9W	0.7W		MIT
3794.158	3795.235	26348.83	EU	0.3			MIT
3793.85	3794.93	26351.0	EU	1.3W	0.7		KN
3793.02	3794.10	26356.7	EU	0.70			MIT
3791.63	3792.71	26366.4	EU	0.30			MIT
3791.49	3792.57	26367.4	EU II	0.70			MIT
3790.71	3791.79	26372.8	EU	1.0	0.3		KN
3790.50	3791.58	26374.2	EU	0.90	0.70		MIT
3789.170	3790.246	26383.51	EU	0.8	0.6		MIT
3788.76	3789.84	26386.4	EU	1.2W	1.0W		MIT
3787.19	3788.27	26397.3	EU	1.00	0.70		MIT
3786.975	3788.050	26398.80	EU	0.3			MIT
3786.83	3787.91	26399.8	EU	0.70	1.00		MIT
3786.32	3787.40	26403.4	EU	0.6	0.3		KN
3785.80	3786.88	26407.0	EU	1.0J	1.0J		MIT
3785.47	3786.55	26409.3	EU	1.40	1.0J		MIT
3784.23	3785.30	26417.9	EU	1.00	1.00		MIT
3783.87	3784.94	26420.5	EU	1.00	0.70		MIT
3783.52	3784.59	26422.9	EU	1.0D	0.7		MIT
3781.39	3782.46	26437.8	EU	1.40	1.50		MIT
3781.09	3782.16	26439.9	EU	0.60			MIT
3780.51	3781.58	26443.9	EU	1.00	1.00		MIT
3779.86	3780.93	26448.5	EU	1.3	1.0		KN
3779.32	3780.39	26452.3	EU	0.5W	0.7W		MIT
3778.814	3779.887	26455.81	EU	1.3W	0.7W		MIT
3778.40	3779.47	26458.7	EU	0.7X	0.7V		MIT
3777.60	3778.67	26464.3	EU	1.20	0.70		MIT
3777.16	3778.23	26467.4	EU	1.0	0.7		KN
3776.52	3777.59	26471.9	EU	0.60	0.60		MIT
3776.20	3777.27	26474.1	EU	1.00			MIT
3775.69	3776.76	26477.7	EU	0.60	0.60		MIT
3775.41	3776.48	26479.7	EU	0.60			MIT
3774.08	3775.15	26489.0	EU	1.70	1.00		MIT
3772.72	3773.79	26498.6	EU	0.30			MIT
3772.16	3773.23	26502.5	EU	1.3D			MIT
3771.76	3772.83	26505.3	EU	1.0D			MIT
3771.42	3772.49	26507.7	EU	0.6D			MIT
3771.16	3772.23	26509.5	EU	1.00	1.30		MIT
3770.23	3771.30	26516.1	EU	1.20	1.40		MIT
3769.32	3770.39	26522.4	EU	1.1J	0.7J		MIT
3769.0	3770.1	26525.	EU	1.0D			KN
3768.51	3769.58	26528.1	EU	0.9J			MIT
3768.23	3769.30	26530.1	EU	1.3W			MIT
3767.60	3768.67	26534.6	EU	1.2J	0.7J		MIT
3766.30	3767.37	26543.7	EU	0.8D			MIT
3766.08	3767.15	26545.3	EU	0.30			MIT
3765.940	3767.010	26546.25	EU II	1.0			MIT
3764.76	3765.83	26554.6	EU	0.70	0.0V		MIT
3764.33	3765.40	26557.6	EU	0.9			MIT
3764.02	3765.09	26559.8	EU	0.30			MIT
3763.018	3764.087	26566.87	EU	0.6	0.3		MIT
3762.33	3763.40	26571.7	EU	1.0W			MIT
3761.134	3762.203	26580.17	EU II	1.0	1.0		MIT
3760.76	3761.83	26582.8	EU	1.00			MIT
3760.34	3761.41	26585.8	EU	1.00	1.00		MIT
3759.43	3760.50	26592.2	EU	0.30			MIT
3758.56	3759.63	26598.4	EU	0.70	0.0V		MIT
3757.648	3758.716	26604.83	EU	0.8	0.3		MIT
3757.42	3758.49	26606.4	EU	0.50	0.30		MIT
3756.81	3757.88	26610.8	EU	0.50	0.0V		MIT
3756.00	3757.07	26616.5	EU	0.50	0.00		MIT
3755.45	3756.52	26620.4	EU	0.8D	0.0H		MIT
3754.90	3755.97	26624.3	EU	0.6			MIT
3753.763	3754.830	26632.36	EU	0.7	0.0		MIT
3753.061	3754.128	26637.35	EU	0.7			MIT
3752.860	3753.927	26638.77	EU	0.7	0.0		MIT
3752.54	3753.61	26641.0	EU	0.80			MIT
3751.72	3752.79	26646.9	EU	0.80			MIT
3751.53	3752.60	26648.2	EU	0.7W	0.0J		KN

AIR WAVELENGTH (ANGSTRMS)	VACUUM WAVELENGTH (ANGSTRMS)	WAVENUMBER (KAYSERS)	LMENT	LOG ARC	INTENSITY SPK	DIS	REF
3751.05	3752.12	26651.6	EU	0.8W	0.0J		MIT
3750.705	3751.771	26654.08	EU	0.3			MIT
3750.13	3751.20	26658.2	EU	0.9J	0.3J		KN
3749.75	3750.82	26660.9	EU	0.50			MIT
3747.24	3748.31	26678.7	EU	0.70	0.0J		MIT
3746.716	3747.781	26682.46	EU	0.6	0.5		MIT
3746.03	3747.09	26687.3	EU	1.0			KN
3744.55	3745.61	26697.9	EU	1.0W	0.7J		MIT
3744.20	3745.26	26700.4	EU II	1.30			MIT
3743.561	3744.625	26704.94	EU	1.0			MIT
3742.34	3743.40	26713.6	EU	0.50			KN
3742.23	3743.29	26714.4	EU	1.40			MIT
3741.62	3742.68	26718.8	EU	0.5			MIT
3741.32	3742.38	26720.9	EU	1.0	0.7		MIT
3740.27	3741.33	26728.4	EU	0.9	0.3		MIT
3739.50	3740.56	26733.9	EU	1.3W			MIT
3738.76	3739.82	26739.2	EU	0.6W			MIT
3738.068	3739.131	26744.18	EU	1.0	1.0		MIT
3737.37	3738.43	26749.2	EU	0.3			KN
3736.26	3737.32	26757.1	EU	0.6W			MIT
3736.038	3737.100	26758.72	EU	0.8	0.0		MIT
3735.95	3737.01	26759.4	EU	0.7			MIT
3735.62	3736.68	26761.7	EU	0.50			MIT
3734.66	3735.72	26768.6	EU	0.30			MIT
3734.23	3735.29	26771.7	EU	0.70			MIT
3733.66	3734.72	26775.8	EU	0.9W	0.3W		MIT
3733.23	3734.29	26778.8	EU	0.3W			KN
3732.73	3733.79	26782.4	EU	0.70	0.0V		MIT
3732.21	3733.27	26786.2	EU	1.50			MIT
3731.82	3732.88	26789.0	EU	1.00	0.30		MIT
3731.26	3732.32	26793.0	EU	1.0W	0.0W		MIT
3729.752	3730.813	26803.81	EU	1.0			MIT
3729.697	3730.758	26804.21	EU	1.0	0.0H		MIT
3729.43	3730.49	26806.1	EU	0.5	0.3H		MIT
3729.10	3730.16	26808.5	EU	0.80	0.00		MIT
3728.65	3729.71	26811.7	EU	0.3W			MIT
3728.23	3729.29	26814.7	EU	0.7W			MIT
3726.92	3727.98	26824.2	EU	0.6W	0.0W		MIT
3726.62	3727.68	26826.3	EU	0.5W			MIT
3726.37	3727.43	26828.1	EU	0.5W	0.0		MIT
3725.74	3726.80	26832.7	EU	0.3W			MIT
3724.99	3726.05	26838.1	EU	2.4	1.7		KN
3724.20	3725.26	26843.8	EU	0.3			MIT
3723.84	3724.90	26846.4	EU	0.7W	0.0W		MIT
3722.61	3723.67	26855.2	EU	1.60	1.0		MIT
3721.82	3722.88	26860.9	EU	0.3			KN
3720.695	3721.753	26869.06	EU	0.6W			MIT
3719.167	3720.225	26880.09	EU	1.5			KN
3717.69	3718.75	26890.8	EU	1.3W	0.3		MIT
3716.91	3717.97	26896.4	EU	1.0W	0.3		MIT
3715.65	3716.71	26905.5	EU	0.9W			MIT
3714.911	3715.968	26910.89	EU I	1.6	0.3		MIT
3713.459	3714.515	26921.41	EU	1.7	0.3		MIT
3712.50	3713.56	26928.4	EU	0.50			MIT
3712.40	3713.46	26929.1	EU	0.8			MIT
3712.18	3713.24	26930.7	EU	0.6	0.0H		MIT
3711.66	3712.72	26934.5	EU	1.0W	0.0H		MIT
3710.881	3711.937	26940.11	EU	1.5L	1.5H		MIT
3710.29	3711.35	26944.4	EU	1.3W	0.0J		MIT
3710.01	3711.07	26946.4	EU	0.7			KN
3708.37	3709.43	26958.4	EU	1.0			KN
3707.40	3708.45	26965.4	EU	0.9W	0.7		KN
3706.57	3707.62	26971.4	EU	1.10			MIT
3706.3	3707.4	26973.	EU	0.5			KN
3705.05	3706.10	26982.5	EU	1.1			MIT
3704.98	3706.03	26983.0	EU	1.1W	0.0J		MIT
3703.9	3705.0	26991.	EU	0.7	0.5		KN
3703.55	3704.60	26993.4	EU	0.8			KN
3702.3	3703.4	27003.	EU	0.6			KN
3702.02	3703.07	27004.6	EU	0.5			KN

AIR WAVELENGTH (ANGSTRMS)	VACUUM WAVELENGTH (ANGSTRMS)	WAVENUMBER (KAYSERS)	LMENT	LOG ARC	INTENSITY SPK	DIS	REF
3701.16	3702.21	27010.9	EU	0.7W	0.0H		MIT
3699.24	3700.29	27024.9	EU	0.3W			MIT
3697.95	3699.00	27034.3	EU	0.6W	0.0H		MIT
3696.44	3697.49	27045.4	EU	1.1D	0.0		MIT
3695.84	3696.89	27049.7	EU	0.7W	0.0		MIT
3693.82	3694.87	27064.5	EU	1.4W	0.0		MIT
3692.66	3693.71	27073.0	EU	0.7W			MIT
3692.22	3693.27	27076.3	EU	0.6J			MIT
3691.98	3693.03	27078.0	EU	0.50	0.0		MIT
3691.55	3692.60	27081.2	EU	1.3D			MIT
3689.42	3690.47	27096.8	EU	1.0W			MIT
3688.44	3689.49	27104.0	EU II	3.00	2.70		MIT
3687.79	3688.84	27108.8	EU	1.6W	1.7W		MIT
3685.67	3686.72	27124.4	EU	0.5W	0.0		MIT
3683.85	3684.90	27137.8	EU	0.90	0.0H		MIT
3683.27	3684.32	27142.1	EU	1.3W	0.0H		MIT
3682.44	3683.49	27148.2	EU	1.1	0.0		MIT
3681.54	3682.59	27154.8	EU	0.5	0.0H		MIT
3680.76	3681.81	27160.6	EU	0.6	0.5		KN
3679.49	3680.54	27169.9	EU	1.2	0.0		MIT
3678.27	3679.32	27179.0	EU	1.7			MIT
3678.12	3679.17	27180.1	EU	0.9W	0.3		MIT
3676.87	3677.92	27189.3	EU	1.2	0.0		MIT
3676.62	3677.67	27191.1	EU	1.30	0.3		MIT
3676.14	3677.19	27194.7	EU	0.3W			MIT
3674.67	3675.72	27205.6	EU	1.4			MIT
3674.18	3675.23	27209.2	EU	1.2W			KN
3673.196	3674.242	27216.50	EU	1.3	0.3		MIT
3672.21	3673.26	27223.8	EU	0.8W			MIT
3670.830	3671.875	27234.04	EU	0.6	0.0H		MIT
3669.81	3670.86	27241.6	EU	0.90	0.0J		MIT
3669.13	3670.17	27246.7	EU	0.8W	0.0H		MIT
3668.51	3669.55	27251.3	EU	0.3W	0.3W		MIT
3666.84	3667.88	27263.7	EU	0.7W	0.6W		MIT
3665.878	3666.922	27270.83	EU	0.3W			MIT
3665.15	3666.19	27276.2	EU	0.6W			MIT
3663.44	3664.48	27289.0	EU	0.6W	0.6W		MIT
3662.94	3663.98	27292.7	EU	1.2W	0.7W		MIT
3662.52	3663.56	27295.8	EU	0.5W			MIT
3662.33	3663.37	27297.2	EU	0.9			KN
3661.91	3662.95	27300.4	EU	0.5W			MIT
3660.60	3661.64	27310.1	EU	1.5D			MIT
3659.91	3660.95	27315.3	EU	0.5J			KN
3659.13	3660.17	27321.1	EU	0.3H	0.3H		MIT
3658.80	3659.84	27323.6	EU	0.3W			MIT
3657.605	3658.647	27332.51	EU	0.3W			MIT
3657.408	3658.450	27333.98	EU	0.3H	0.3H		MIT
3656.27	3657.31	27342.5	EU	0.6			KN
3655.25	3656.29	27350.1	EU	0.80			MIT
3652.65	3653.69	27369.6	EU	0.7W	0.6J		MIT
3649.818	3650.858	27390.82	EU	0.7	0.3		MIT
3648.27	3649.31	27402.4	EU	0.8W	0.7		KN
3647.03	3648.07	27411.8	EU	0.3H			MIT
3646.750	3647.789	27413.86	EU	1.3	0.7H		KN
3646.660	3647.699	27414.54	EU	1.00			MIT
3645.43	3646.47	27423.8	EU	0.7			MIT
3645.17	3646.21	27425.7	EU	1.0W	0.7J		MIT
3644.99	3646.03	27427.1	EU	0.7W			MIT
3644.50	3645.54	27430.8	EU	0.8W	0.0J		MIT
3644.27	3645.31	27432.5	EU	0.70			KN
3644.08	3645.12	27433.9	EU	0.3	0.0		MIT
3641.24	3642.28	27455.4	EU	0.3J	0.0H		MIT
3638.32	3639.36	27477.4	EU	0.7	0.3		KN
3638.083	3639.120	27479.17	EU	0.7	0.7		MIT
3637.676	3638.713	27482.25	EU	1.1	1.0		MIT
3636.73	3637.77	27489.4	EU	0.7	0.6		MIT
3635.87	3636.91	27495.9	EU		0.6		MIT
3635.65	3636.69	27497.6	EU	0.80			KN
3632.175	3633.210	27523.87	EU	1.5	1.0		MIT
3631.78	3632.82	27526.9	EU	0.5			M
3630.50	3631.53	27536.6	EU	0.5D			MIT
3629.797	3630.832	27541.90	EU	1.6	1.7H		MIT
3629.058	3630.093	27547.51	EU	0.6			MIT
3628.46	3629.49	27552.1	EU		0.3		MIT
3627.785	3628.819	27557.17	EU	0.3			MIT
3627.424	3628.458	27559.92	EU	0.5	0.7		MIT
3627.048	3628.082	27562.77	EU	0.3			MIT
3626.33	3627.36	27568.2	EU		0.3		MIT
3623.72	3624.75	27588.1	EU	0.6D	0.6D		MIT
3623.44	3624.47	27590.2	EU	0.6			KN
3623.13	3624.16	27592.6	EU	0.5D	0.3		MIT
3622.558	3623.591	27596.93	EU	1.3H	1.7		MIT
3621.903	3622.936	27601.93	EU	0.9	1.0		MIT
3620.890	3621.922	27609.65	EU	0.8	0.7		MIT
3618.191	3619.223	27630.24	EU	0.9H			MIT
3616.51	3617.54	27643.1	EU	1.0D			MIT
3616.154	3617.185	27645.81	EU	1.6	1.0		MIT
3614.86	3615.89	27655.7	EU		0.3		MIT
3614.70	3615.73	27656.9	EU	0.5			MIT
3614.084	3615.115	27661.64	EU	0.3D	0.3D		MIT
3613.76	3614.79	27664.1	EU	0.3H	0.0J		MIT
3612.19	3613.22	27676.1	EU	0.8W	0.3		MIT
3611.580	3612.610	27680.82	EU	1.6	0.7		MIT
3611.332	3612.362	27682.72	EU	1.0H			MIT
3611.02	3612.05	27685.1	EU	1.0W	0.0		KN
3610.572	3611.602	27688.54	EU		0.3H		MIT
3609.94	3610.97	27693.4	EU	0.5W			KN
3608.72	3609.75	27702.7	EU	1.0	0.0H		KN
3606.71	3607.74	27718.2	EU	1.7			KN
3606.382	3607.411	27720.71	EU	0.7			MIT
3606.21	3607.24	27722.0	EU	0.3H	0.0H		MIT
3605.61	3606.64	27726.6	EU	0.6W			KN
3605.338	3606.366	27728.74	EU	0.5	0.0H		MIT
3604.684	3605.712	27733.77	EU	0.8	0.0H		MIT
3603.20	3604.23	27745.2	EU	2.0W	1.7		KN
3602.50	3603.53	27750.6	EU	0.8W	0.0		MIT
3601.73	3602.76	27756.5	EU	0.3W			MIT
3598.57	3599.60	27780.9	EU	0.3	0.0		MIT
3598.114	3599.141	27784.41	EU	0.3H			MIT
3596.86	3597.89	27794.1	EU	0.5W	0.3H		KN
3596.23	3597.26	27799.0	EU	0.50	0.0H		KN
3593.13	3594.16	27822.9	EU	0.3W			MIT
3592.927	3593.952	27824.52	EU	0.8	0.0H		MIT
3591.79	3592.81	27833.3	EU	0.5	0.0		MIT
3591.34	3592.36	27836.8	EU	1.0	0.0		MIT
3590.62	3591.64	27842.4	EU	0.3H	0.0H		MIT
3590.45	3591.47	27843.7	EU	1.0J			KN
3589.96	3590.98	27847.5	EU	0.3H	0.0H		MIT
3589.267	3590.291	27852.89	EU	0.5			KN
3588.55	3589.57	27858.5	EU	0.5	0.0		KN
3588.108	3589.132	27861.89	EU	0.3	0.0		MIT
3587.15	3588.17	27869.3	EU	0.5	0.0H		KN
3580.57	3581.59	27920.5	EU	0.5	0.0H		MIT
3580.232	3581.254	27923.18	EU	0.6	0.0H		MIT
3579.36	3580.38	27930.0	EU	0.6	0.0H		MIT
3579.08	3580.10	27932.2	EU	0.8W	0.0H		MIT
3578.49	3579.51	27936.8	EU	0.3	0.0		MIT
3578.110	3579.131	27939.74	EU	0.8	0.0		MIT
3576.01	3577.03	27956.1	EU	0.7	0.0H		KN
3575.42	3576.44	27960.8	EU	0.3J	0.0H		MIT
3573.904	3574.924	27972.62	EU	0.5	0.0		MIT
3573.45	3574.47	27976.2	EU	0.50			MIT
3572.60	3573.62	27982.8	EU	0.5	0.0H		MIT
3572.11	3573.13	27986.7	EU	0.5W	0.0		MIT
3571.66	3572.68	27990.2	EU	0.3W			MIT
3571.32	3572.34	27992.9	EU	0.30			MIT
3571.099	3572.119	27994.59	EU	0.5			MIT
3569.82	3570.84	28004.6	EU	0.6			MIT
3569.17	3570.19	28009.7	EU	0.60	0.0H		MIT
3567.723	3568.742	28021.08	EU	0.5			MIT

AIR WAVELENGTH (ANGSTRMS)	VACUUM WAVELENGTH (ANGSTRMS)	WAVENUMBER (KAYSERS)	LMENT	LOG ARC	INTENSITY SPK DIS	REF
3566.79	3567.81	28028.4	EU	0.6	0.0H	MIT
3565.80	3566.82	28036.2	EU	0.3	0.0	MIT
3565.23	3566.25	28040.7	EU	0.7		KN
3563.72	3564.74	28052.6	EU	0.8W	0.0H	MIT
3562.18	3563.20	28064.7	EU	0.6	0.0	MIT
3561.28	3562.30	28071.8	EU	0.5	0.0H	MIT
3560.57	3561.59	28077.4	EU	0.8W	0.0H	MIT
3557.75	3558.77	28099.6	EU	0.5	0.0	MIT
3556.85	3557.87	28106.7	EU	1.0W		KN
3556.487	3557.503	28109.60	EU	0.3	0.0H	MIT
3555.37	3556.39	28118.4	EU	0.6D	0.0H	MIT
3552.52	3553.53	28141.0	EU	0.8	0.3	MIT
3551.28	3552.29	28150.8	EU	0.5D	0.0H	MIT
3549.68	3550.69	28163.5	EU	0.5	0.3H	MIT
3548.927	3549.941	28169.48	EU	0.6		MIT
3548.51	3549.52	28172.8	EU	0.6W	0.0H	MIT
3548.03	3549.04	28176.6	EU	0.6	0.0H	MIT
3547.37	3548.38	28181.9	EU	0.5W		MIT
3547.11	3548.12	28183.9	EU	1.2W		KN
3544.77	3545.78	28202.5	EU	1.0W	0.0	MIT
3544.15	3545.16	28207.4	EU	1.0W	0.0	MIT
3543.88	3544.89	28209.6	EU	1.20	0.6	MIT
3542.16	3543.17	28223.3	EU	1.3W	0.7W	MIT
3541.34	3542.35	28229.8	EU	1.1	0.3	KN
3540.77	3541.78	28234.4	EU	0.5W		MIT
3540.44	3541.45	28237.0	EU	0.8		KN
3540.23	3541.24	28238.7	EU	0.50		MIT
3539.76	3540.77	28242.4	EU	0.5		MIT
3539.24	3540.25	28246.6	EU	0.3		MIT
3538.09	3539.10	28255.8	EU	1.3W	1.0H	MIT
3537.75	3538.76	28258.5	EU	1.0	1.0H	MIT
3537.56	3538.57	28260.0	EU	0.6		KN
3537.276	3538.287	28262.26	EU	0.6		MIT
3534.14	3535.15	28287.3	EU	0.8W	0.3	MIT
3532.23	3533.24	28302.6	EU	0.7	0.0	MIT
3531.832	3532.842	28305.83	EU	1.1	0.3	MIT
3531.14	3532.15	28311.4	EU	1.0	1.0	MIT
3530.36	3531.37	28317.6	EU	1.1W	0.0	MIT
3528.60	3529.61	28331.7	EU	0.9W		MIT
3527.87	3528.88	28337.6	EU	1.3W	0.7W	MIT
3526.70	3527.71	28347.0	EU	1.1	0.0H	MIT
3526.01	3527.02	28352.6	EU	0.6		MIT
3525.784	3526.792	28354.38	EU	1.0	0.7H	MIT
3524.34	3525.35	28366.0	EU	0.5W	0.0H	MIT
3523.67	3524.68	28371.4	EU	0.3W	0.0H	MIT
3523.50	3524.51	28372.8	EU	1.2W		MIT
3523.18	3524.19	28375.3	EU	1.2W		KN
3522.367	3523.374	28381.89	EU	1.3		MIT
3521.09	3522.10	28392.2	EU	1.7	0.6H	MIT
3520.14	3521.15	28399.8	EU	0.6	0.6	MIT
3518.904	3519.910	28409.82	EU	0.7		KN
3518.48	3519.49	28413.2	EU	1.3	1.0	MIT
3514.49	3515.50	28445.5	EU	1.3	0.7	MIT
3514.22	3515.22	28447.7	EU	1.2		MIT
3513.34	3514.34	28454.8	EU	1.0W		MIT
3512.27	3513.27	28463.5	EU	1.00		MIT
3511.85	3512.85	28466.9	EU	1.4W	0.0	MIT
3511.18	3512.18	28472.3	EU	0.6W	0.0	MIT
3511.07	3512.07	28473.2	EU	1.5W	0.3	MIT
3510.07	3511.07	28481.3	EU	0.6		KN
3508.859	3509.863	28491.14	EU	1.0	0.0	MIT
3508.740	3509.744	28492.11	EU	0.9		MIT
3507.55	3508.55	28501.8	EU	0.9W	0.7J	MIT
3505.97	3506.97	28514.6	EU	1.0W	0.0	MIT
3505.305	3506.308	28520.03	EU	1.4W	0.7W	MIT
3505.06	3506.06	28522.0	EU	0.3		MIT
3504.91	3505.91	28523.2	EU	0.60		MIT
3503.75	3504.75	28532.7	EU	0.9D		MIT
3503.232	3504.234	28536.91	EU	0.7		MIT
3502.81	3503.81	28540.3	EU	1.2D	0.7	MIT

AIR WAVELENGTH (ANGSTRMS)	VACUUM WAVELENGTH (ANGSTRMS)	WAVENUMBER (KAYSERS)	LMENT	LOG ARC	INTENSITY SPK DIS	REF
3502.46	3503.46	28543.2	EU	1.0W	0.3	MIT
3497.65	3498.65	28582.4	EU	1.0W		MIT
3495.155	3496.155	28602.85	EU	1.2N	0.3	MIT
3493.407	3494.407	28617.16	EU	0.8		MIT
3492.969	3493.969	28620.75	EU	0.3		MIT
3491.09	3492.09	28636.1	EU	1.0	0.0	MIT
3489.25	3490.25	28651.2	EU	1.3W	0.3	MIT
3488.306	3489.304	28659.01	EU	1.3	0.0	MIT
3485.865	3486.863	28679.08	EU	1.3	0.3	MIT
3485.435	3486.433	28682.61	EU	1.3		MIT
3485.16	3486.16	28684.9	EU	1.3S	0.3H	MIT
3484.979	3485.976	28686.37	EU	0.6		MIT
3484.070	3485.067	28693.85	EU	0.3		MIT
3482.69	3483.69	28705.2	EU	0.5		MIT
3482.54	3483.54	28706.5	EU	0.8W	0.3	MIT
3481.61	3482.61	28714.1	EU	1.70	0.3	MIT
3480.834	3481.830	28720.53	EU	1.0		MIT
3480.44	3481.44	28723.8	EU	0.9	0.0	MIT
3478.20	3479.20	28742.3	EU	1.0		MIT
3477.56	3478.56	28747.6	EU	0.9		MIT
3477.19	3478.19	28750.6	EU	0.6		MIT
3477.07	3478.07	28751.6	EU	0.7W		MIT
3476.98	3477.98	28752.4	EU	1.2	0.3	MIT
3476.61	3477.61	28755.4	EU	1.3	0.3	MIT
3474.52	3475.51	28772.7	EU	1.2	0.3	MIT
3471.725	3472.719	28795.88	EU	0.8	0.0H	MIT
3469.30	3470.29	28816.0	EU	0.7	0.3	MIT
3467.887	3468.880	28827.75	EU	0.7		MIT
3466.88	3467.87	28836.1	EU	1.3	0.0	MIT
3466.416	3467.409	28839.98	EU	1.5	0.3	MIT
3464.36	3465.35	28857.1	EU	1.2D		MIT
3463.28	3464.27	28866.1	EU	1.30	0.0H	MIT
3463.03	3464.02	28868.2	EU	0.9		MIT
3461.665	3462.657	28879.56	EU	0.5		MIT
3461.380	3462.371	28881.94	EU	1.4	0.3	MIT
3460.290	3461.281	28891.04	EU	1.2	0.0H	MIT
3457.570	3458.560	28913.76	EU	1.4	0.0	MIT
3457.054	3458.044	28918.08	EU	1.4	0.3	MIT
3454.76	3455.75	28937.3	EU	1.60	0.3	MIT
3454.15	3455.14	28942.4	EU	0.5	0.0	MIT
3453.88	3454.87	28944.6	EU	0.3W		MIT
3453.47	3454.46	28948.1	EU	0.5	0.3	MIT
3452.245	3453.234	28958.36	EU	1.0		KN
3450.18	3451.17	28975.7	EU	0.5		MIT
3448.43	3449.42	28990.4	EU	0.5	0.3	MIT
3448.163	3449.151	28992.64	EU	0.6	0.3	MIT
3447.874	3448.862	28995.07	EU	0.3		MIT
3446.372	3447.360	29007.71	EU	0.7	0.5	MIT
3445.84	3446.83	29012.2	EU	0.5	0.3	MIT
3445.17	3446.16	29017.8	EU	1.6	0.3	M
3442.365	3443.352	29041.47	EU	0.3H		MIT
3440.996	3441.982	29053.03	EU	1.6	1.4	MIT
3440.819	3441.805	29054.52	EU	0.7		MIT
3439.587	3440.573	29064.93	EU	0.5		MIT
3438.31	3439.30	29075.7	EU	0.7		MIT
3436.07	3437.05	29094.7	EU	0.7		MIT
3435.734	3436.719	29097.52	EU	1.1	1.0	MIT
3435.65	3436.63	29098.2	EU	0.6H		KN
3435.205	3436.190	29102.00	EU	1.5	0.3	MIT
3435.064	3436.049	29103.20	EU	1.1	0.5	MIT
3434.74	3435.72	29105.9	EU	0.5		KN
3432.523	3433.507	29124.74	EU	1.0		MIT
3430.90	3431.88	29138.5	EU	0.7	0.3	MIT
3430.371	3431.354	29143.01	EU	0.7	0.5	MIT
3429.29	3430.27	29152.2	EU	0.3A	0.3D	MIT
3428.77	3429.75	29156.6	EU	1.1S	0.3	MIT
3428.33	3429.31	29160.4	EU	0.5		MIT
3427.756	3428.739	29165.24	EU	0.7		MIT
3427.05	3428.03	29171.2	EU	0.6		KN
3426.44	3427.42	29176.4	EU	0.9	0.3	MIT

AIR WAVELENGTH (ANGSTRMS)	VACUUM WAVELENGTH (ANGSTRMS)	WAVENUMBER (KAYSERS)	LMENT	LOG ARC	INTENSITY SPK	DIS	REF
3425.02	3426.00	29188.5	EU	1.7	1.6		MIT
3424.596	3425.578	29192.15	EU	0.5	0.3		MIT
3423.110	3424.092	29204.83	EU	1.0	1.0		MIT
3421.667	3422.648	29217.14	EU	0.7	0.7		MIT
3419.852	3420.833	29232.65	EU	1.3	0.5		MIT
3418.809	3419.790	29241.57	EU	0.9	0.3		MIT
3417.42	3418.40	29253.4	EU	0.6	0.3		MIT
3416.879	3417.859	29258.08	EU	0.7			MIT
3416.735	3417.715	29259.31	EU	0.8	0.8		MIT
3414.773	3415.752	29276.13	EU	1.6			MIT
3414.02	3415.00	29282.6	EU	0.6H	0.3		MIT
3412.731	3413.710	29293.64	EU	1.3	0.9		MIT
3412.261	3413.240	29297.68	EU	0.3H	0.6		MIT
3410.620	3411.598	29311.77	EU	0.6	0.3		MIT
3410.10	3411.08	29316.2	EU	0.5D	0.3D		MIT
3409.953	3410.931	29317.51	EU	0.6			MIT
3409.647	3410.625	29320.14	EU	0.8	0.3		MIT
3406.12	3407.10	29350.5	EU	1.2W	0.6		MIT
3405.67	3406.65	29354.4	EU	0.8W			MIT
3403.163	3404.140	29376.00	EU	0.7	0.3		MIT
3403.03	3404.01	29377.1	EU	0.7			MIT
3402.457	3403.433	29382.09	EU	0.6	0.6		MIT
3398.47	3399.45	29416.6	EU		0.3		MIT
3396.58	3397.55	29432.9	EU	2.0	1.0		MIT
3395.35	3396.32	29443.6	EU	1.2	0.5		MIT
3395.175	3396.149	29445.11	EU	0.3H	0.3		MIT
3394.069	3395.043	29454.71	EU	1.0			MIT
3393.769	3394.743	29457.31	EU	0.5	0.3		MIT
3393.25	3394.22	29461.8	EU	0.9	0.6		MIT
3391.992	3392.966	29472.74	EU	1.6	0.7		MIT
3390.787	3391.760	29483.22	EU	1.1	0.9		MIT
3387.418	3388.390	29512.54	EU	0.6	0.3		MIT
3382.412	3383.383	29556.21	EU	0.3H			MIT
3381.73	3382.70	29562.2	EU	0.3	0.5		MIT
3380.25	3381.22	29575.1	EU	1.2	1.0		MIT
3379.263	3380.233	29583.76	EU	0.5	0.3		MIT
3377.703	3378.673	29597.42	EU	1.0	0.3		MIT
3373.228	3374.197	29636.68	EU	0.5	0.3		MIT
3371.698	3372.666	29650.13	EU	0.6	0.3		MIT
3370.52	3371.49	29660.5	EU	0.5D	0.5D		MIT
3369.052	3370.020	29673.42	EU	1.6	0.7		MIT
3366.13	3367.10	29699.2	EU	0.3D	0.3D		MIT
3365.10	3366.07	29708.3	EU	0.3D			MIT
3363.351	3364.317	29723.71	EU	0.6	0.3		MIT
3362.783	3363.749	29728.73	EU	0.7	0.3		MIT
3361.615	3362.581	29739.06	EU	0.3H	0.5		MIT
3361.214	3362.180	29742.61	EU	1.3	0.3		MIT
3357.106	3358.071	29779.00	EU	0.3	0.3		MIT
3354.375	3355.339	29803.25	EU	1.1			MIT
3353.703	3354.667	29809.22	EU	1.0			MIT
3352.592	3353.556	29819.10	EU	0.3H	0.3		MIT
3351.556	3352.519	29828.31	EU	1.1	0.3		MIT
3351.19	3352.15	29831.6	EU	1.2O			MIT
3351.036	3351.999	29832.94	EU	0.3	0.3		MIT
3350.43	3351.39	29838.3	EU	1.5			MIT
3346.96	3347.92	29869.3	EU	0.3	0.3		MIT
3346.629	3347.591	29872.23	EU	0.6	0.3		MIT
3346.40	3347.36	29874.3	EU	0.6$			MIT
3346.10	3347.06	29876.9	EU	0.3	0.3		MIT
3345.311	3346.273	29884.00	EU	0.5	0.3		MIT
3342.675	3343.636	29907.56	EU	0.3H	0.3		MIT
3341.892	3342.853	29914.57	EU	0.7	0.3		MIT
3341.076	3342.037	29921.87	EU	0.5D	0.3D		MIT
3338.742	3339.702	29942.79	EU	1.2	0.7		MIT
3338.51	3339.47	29944.9	EU	0.6	0.5		MIT
3337.584	3338.544	29953.18	EU	0.3H	0.3		MIT
3336.47	3337.43	29963.2	EU	1.2	0.6		KN
3334.326	3335.285	29982.45	EU	1.7	0.7		MIT
3334.144	3335.103	29984.08	EU	0.5	0.3		MIT
3333.654	3334.613	29988.49	EU	0.5	0.6		MIT

AIR WAVELENGTH (ANGSTRMS)	VACUUM WAVELENGTH (ANGSTRMS)	WAVENUMBER (KAYSERS)	LMENT	LOG ARC	INTENSITY SPK	DIS	REF
3331.17	3332.13	30010.9	EU	0.3H	0.6H		MIT
3329.804	3330.762	30023.16	EU	0.5			MIT
3329.533	3330.491	30025.60	EU	1.0	0.7		MIT
3328.640	3329.597	30033.66	EU	0.5	0.6		MIT
3328.057	3329.014	30038.92	EU	0.5	0.5		MIT
3327.087	3328.044	30047.68	EU	0.5	0.3		MIT
3325.976	3326.933	30057.72	EU	1.3	0.3		MIT
3325.76	3326.72	30059.7	EU	0.3H			MIT
3322.743	3323.699	30086.96	EU	0.5			MIT
3322.261	3323.217	30091.32	EU	1.3	0.3		MIT
3321.857	3322.813	30094.99	EU	1.5	0.7		MIT
3320.14	3321.10	30110.6	EU	0.5	0.3H		MIT
3319.887	3320.842	30112.84	EU	1.3	0.7		MIT
3317.268	3318.223	30136.62	EU	0.5	0.3		MIT
3316.476	3317.430	30143.81	EU	0.3			MIT
3313.323	3314.277	30172.50	EU	1.5	1.6		MIT
3312.17	3313.12	30183.0	EU	0.5W	0.0		MIT
3310.80	3311.75	30195.5	EU	0.5J	0.0		MIT
3310.34	3311.29	30199.7	EU	0.3W	0.3		MIT
3308.977	3309.929	30212.12	EU	0.3H	0.3		MIT
3308.018	3308.970	30220.88	EU	1.3	1.5		MIT
3307.30	3308.25	30227.4	EU	0.3H	0.3		MIT
3304.495	3305.446	30253.10	EU	1.3			MIT
3304.200	3305.151	30255.80	EU	0.8	0.3		MIT
3301.95	3302.90	30276.4	EU	1.4	0.3		MIT
3301.57	3302.52	30279.9	EU	1.2W	0.0		MIT
3301.24	3302.19	30282.9	EU	0.6W	0.0		MIT
3300.75	3301.70	30287.4	EU	0.6	0.0H		KN
3300.38	3301.33	30290.8	EU	0.5	0.3		MIT
3299.14	3300.09	30302.2	EU	0.3	0.0H		MIT
3298.89	3299.84	30304.5	EU	0.3	0.0H		MIT
3298.805	3299.755	30305.28	EU	0.5			MIT
3298.32	3299.27	30309.7	EU	0.7W	0.3W		MIT
3297.48	3298.43	30317.5	EU	0.5			MIT
3295.45	3296.40	30336.1	EU	0.3W			MIT
3291.64	3292.59	30371.2	EU	0.5			MIT
3291.41	3292.36	30373.4	EU	0.3H			MIT
3289.38	3290.33	30392.1	EU	0.3			MIT
3288.53	3289.48	30400.0	EU	0.7			MIT
3288.43	3289.38	30400.9	EU	0.3	0.0H		MIT
3287.51	3288.46	30409.4	EU	0.6	0.6		MIT
3285.902	3286.849	30424.28	EU	1.3			MIT
3284.395	3285.341	30438.24	EU	0.6			MIT
3283.706	3284.652	30444.62	EU	0.3	0.7		MIT
3282.522	3283.468	30455.61	EU	0.5O	0.3		MIT
3279.52	3280.46	30483.5	EU	1.0			KN
3277.78	3278.72	30499.7	EU	1.3	0.5		MIT
3274.29	3275.23	30532.2	EU	0.5	0.0H		MIT
3272.76	3273.70	30546.4	EU	1.3	0.5		MIT
3269.66	3270.60	30575.4	EU	0.6J			MIT
3269.414	3270.356	30577.71	EU	0.3	0.0H		MIT
3268.66	3269.60	30584.8	EU	0.8O			KN
3267.01	3267.95	30600.2	EU	0.6J	0.3		KN
3266.39	3267.33	30606.0	EU	1.3	1.3		MIT
3265.78	3266.72	30611.7	EU	1.3			KN
3263.70	3264.64	30631.2	EU	0.5			MIT
3263.50	3264.44	30633.1	EU	0.8O	0.6		MIT
3263.02	3263.96	30637.6	EU	0.3O			MIT
3262.61	3263.55	30641.5	EU	1.2			MIT
3262.495	3263.436	30642.55	EU	0.7	0.3H		MIT
3261.59	3262.53	30651.1	EU	0.3H			MIT
3260.883	3261.823	30657.70	EU	0.7			MIT
3258.668	3259.608	30678.54	EU	0.3H	0.3H		MIT
3254.23	3255.17	30720.4	EU	0.7	0.6H		MIT
3253.36	3254.30	30728.6	EU	0.5			MIT
3251.45	3252.39	30746.6	EU	0.7	0.5		MIT
3247.530	3248.467	30783.75	EU	1.7O	0.7		MIT
3247.30	3248.24	30785.9	EU	0.6	0.0		MIT
3246.780	3247.717	30790.86	EU	0.6J	0.3		MIT
3246.41	3247.35	30794.4	EU	1.1	0.7		KN

AIR WAVELENGTH (ANGSTRMS)	VACUUM WAVELENGTH (ANGSTRMS)	WAVENUMBER (KAYSERS)	LMENT	LOG ARC	INTENSITY SPK	INTENSITY DIS	REF
3246.02	3246.96	30798.1	EU	1.0			MIT
3245.469	3246.405	30803.30	EU	0.3H			MIT
3245.13	3246.07	30806.5	EU	0.5W			MIT
3244.645	3245.581	30811.12	EU	0.6			MIT
3244.47	3245.41	30812.8	EU	0.30	0.0		MIT
3244.20	3245.14	30815.4	EU	1.0	0.3		KN
3241.397	3242.332	30842.00	EU	1.4	0.0H		MIT
3240.67	3241.61	30848.9	EU	0.60	0.0		MIT
3240.11	3241.04	30854.2	EU	0.5	0.0J		MIT
3239.93	3240.86	30856.0	EU	0.8			MIT
3238.89	3239.82	30865.9	EU	0.3H	0.5J		KN
3237.39	3238.32	30880.2	EU	1.1W	0.7		MIT
3235.81	3236.74	30895.2	EU	0.9	0.7		MIT
3235.13	3236.06	30901.7	EU	1.0			MIT
3234.30	3235.23	30909.7	EU	0.5W			MIT
3232.31	3233.24	30928.7	EU	0.6W	0.0		MIT
3231.86	3232.79	30933.0	EU	0.5W			MIT
3230.58	3231.51	30945.3	EU	0.6			MIT
3221.69	3222.62	31030.6	EU	0.9W	0.0		MIT
3218.63	3219.56	31060.1	EU	0.6W			MIT
3213.747	3214.675	31107.34	EU	2.0	1.3H		MIT
3212.808	3213.736	31116.43	EU	2.3	1.3		MIT
3210.566	3211.494	31138.16	EU	1.9			MIT
3207.08	3208.01	31172.0	EU	0.3H			MIT
3198.76	3199.68	31253.1	EU	1.2W	0.0H		MIT
3196.545	3197.469	31274.74	EU	0.9			MIT
3195.08	3196.00	31289.1	EU	0.6W			MIT
3194.37	3195.29	31296.0	EU	0.3	0.7H		MIT
3191.39	3192.31	31325.2	EU	0.7	0.7H		MIT
3187.00	3187.92	31368.4	EU	0.8	0.3		KN
3185.56	3186.48	31382.6	EU	1.0J	1.0J		MIT
3184.22	3185.14	31395.8	EU	1.00			MIT
3183.81	3184.73	31399.8	EU	0.3H	0.9		MIT
3181.28	3182.20	31424.8	EU	0.6	0.7H		MIT
3173.601	3174.519	31500.83	EU	1.5	0.0		MIT
3171.939	3172.857	31517.34	EU	1.0	0.0		MIT
3170.97	3171.89	31527.0	EU	0.8	1.0		MIT
3170.38	3171.30	31532.8	EU	1.2	0.0		MIT
3168.282	3169.199	31553.72	EU	1.0			MIT
3166.50	3167.42	31571.5	EU	1.3W	0.0H		MIT
3164.81	3165.73	31588.3	EU	0.7W	0.0H		MIT
3164.643	3165.559	31590.00	EU	0.7	0.5		MIT
3157.29	3158.20	31663.6	EU	1.2			KN
3151.88	3152.79	31717.9	EU	0.8	0.7		KN
3150.46	3151.37	31732.2	EU	0.6H	0.0H		MIT
3149.883	3150.795	31738.02	EU	1.6W	1.3W		MIT
3147.42	3148.33	31762.9	EU	1.3	0.7H		KN
3146.715	3147.626	31769.97	EU	0.7	0.3		MIT
3144.217	3145.128	31795.21	EU	0.5			MIT
3140.37	3141.28	31834.2	EU	0.3H	0.0H		MIT
3139.330	3140.240	31844.71	EU	0.8	0.7		MIT
3136.94	3137.85	31869.0	EU	0.5W			MIT
3136.36	3137.27	31874.9	EU	0.5			MIT
3134.70	3135.61	31891.7	EU	0.50	0.0		MIT
3133.24	3134.15	31906.6	EU	1.0			MIT
3132.16	3133.07	31917.6	EU	1.2W			MIT
3131.75	3132.66	31921.8	EU	1.30	0.7		KN
3130.74	3131.65	31932.1	EU	2.00	2.0		MIT
3125.105	3126.011	31989.65	EU	0.7	0.0H		MIT
3124.87	3125.78	31992.1	EU	0.8W	0.7		MIT
3124.21	3125.12	31998.8	EU	0.3H	0.0H		MIT
3123.957	3124.863	32001.41	EU	1.3			MIT
3122.49	3123.40	32016.4	EU	0.7D			MIT
3121.78	3122.69	32023.7	EU	0.3H	0.0H		MIT
3117.629	3118.533	32066.36	EU	0.3H			MIT
3115.74	3116.64	32085.8	EU	0.8			KN
3113.02	3113.92	32113.8	EU	1.00	0.7		MIT
3111.432	3112.335	32130.22	EU	2.3			MIT
3110.24	3111.14	32142.5	EU	1.3			MIT
3108.203	3109.105	32163.60	EU	1.0	0.0H		MIT

AIR WAVELENGTH (ANGSTRMS)	VACUUM WAVELENGTH (ANGSTRMS)	WAVENUMBER (KAYSERS)	LMENT	LOG ARC	INTENSITY SPK	INTENSITY DIS	REF
3106.172	3107.073	32184.63	EU	1.3	0.0		MIT
3099.547	3100.447	32253.42	EU	0.8W	0.5H		MIT
3098.17	3099.07	32267.7	EU	1.4D	0.7		MIT
3097.456	3098.355	32275.19	EU	2.3W	0.7		MIT
3094.629	3095.527	32304.67	EU	0.5	0.0H		MIT
3094.184	3095.082	32309.32	EU	1.0W			MIT
3091.200	3092.097	32340.51	EU	1.0			MIT
3089.345	3090.242	32359.93	EU	1.3			MIT
3085.68	3086.58	32398.4	EU	0.6W			MIT
3084.475	3085.371	32411.02	EU	0.3			MIT
3079.056	3079.950	32468.05	EU	0.5			MIT
3077.353	3078.247	32486.02	EU	1.5	1.3		MIT
3076.065	3076.959	32499.62	EU	0.9	0.0		MIT
3069.099	3069.991	32573.39	EU	0.8			MIT
3067.784	3068.676	32587.35	EU	0.8			MIT
3067.672	3068.564	32588.54	EU	0.9			MIT
3066.95	3067.84	32596.2	EU	1.0			MIT
3059.00	3059.89	32680.9	EU	1.3			MIT
3054.93	3055.82	32724.5	EU	2.6W	0.5		MIT
3049.31	3050.20	32784.8	EU	0.5D			MIT
3049.03	3049.92	32787.8	EU	1.20			MIT
3040.77	3041.65	32876.8	EU	1.4W			MIT
3040.57	3041.45	32879.0	EU	1.0			MIT
3039.89	3040.77	32886.4	EU	1.2	0.7		MIT
3039.70	3040.58	32888.4	EU	1.0			MIT
3035.38	3036.26	32935.2	EU	1.0W			MIT
3031.17	3032.05	32981.0	EU		1.0H		MIT
3029.82	3030.70	32995.7	EU	0.6	1.0		MIT
3029.41	3030.29	33000.1	EU	0.3			MIT
3028.67	3029.55	33008.2	EU	0.5			MIT
3026.75	3027.63	33029.1	EU		1.2		MIT
3023.34	3024.22	33066.4	EU		1.0H		MIT
3022.154	3023.034	33079.35	EU	1.4	1.0		MIT
3017.47	3018.35	33130.7	EU	1.2D			MIT
3012.38	3013.26	33186.7	EU	1.3W	0.0		MIT
3009.53	3010.41	33218.1	EU	1.00			KN
3006.273	3007.149	33254.09	EU	1.5			MIT
3004.80	3005.68	33270.4	EU	1.3	0.0H		MIT
3001.35	3002.22	33308.6	EU	0.60			MIT
2998.14	2999.01	33344.3	EU	0.6			MIT
2995.62	2996.49	33372.3	EU	0.3			MIT
2995.24	2996.11	33376.6	EU	1.2			MIT
2991.34	2992.21	33420.1	EU	1.9	1.6		MIT
2990.258	2991.130	33432.18	EU	1.1			MIT
2986.920	2987.791	33469.54	EU	1.4			MIT
2979.680	2980.550	33550.86	EU	0.5			MIT
2979.08	2979.95	33557.6	EU	0.3W			MIT
2975.59	2976.46	33597.0	EU	0.3			MIT
2974.79	2975.66	33606.0	EU	0.5			MIT
2970.36	2971.23	33656.1	EU	0.3			MIT
2966.52	2967.39	33699.7	EU	0.3H			MIT
2964.93	2965.80	33717.8	EU	0.3D			MIT
2960.23	2961.09	33771.3	EU	2.20			MIT
2959.479	2960.344	33779.86	EU	1.5			MIT
2958.97	2959.83	33785.7	EU	0.8			MIT
2958.89	2959.75	33786.6	EU	0.7L			MIT
2952.68	2953.54	33857.6	EU	2.3W	2.2		MIT
2951.19	2952.05	33874.7	EU	0.3H			MIT
2950.847	2951.709	33878.67	EU	0.6			MIT
2949.16	2950.02	33898.0	EU	0.7D	0.7		MIT
2947.300	2948.162	33919.44	EU	1.5			MIT
2944.72	2945.58	33949.2	EU	0.5			MIT
2940.820	2941.680	33994.18	EU	1.0			MIT
2940.46	2941.32	33998.3	EU	1.2			MIT
2939.769	2940.629	34006.33	EU	0.5			MIT
2932.84	2933.70	34086.7	EU	0.6D			MIT
2932.523	2933.381	34090.36	EU	0.8	0.5		MIT
2926.19	2927.05	34164.1	EU	0.6W			MIT
2925.035	2925.891	34177.62	EU	2.2L	2.0		MIT
2917.46	2918.31	34266.4	EU	1.2			MIT

AIR WAVELENGTH (ANGSTRMS)	VACUUM WAVELENGTH (ANGSTRMS)	WAVENUMBER (KAYSERS)	LMENT	LOG ARC	INTENSITY SPK	DIS	REF
2917.252	2918.106	34268.80	EU	0.3			MIT
2914.43	2915.28	34302.0	EU	0.5			MIT
2912.34	2913.19	34326.6	EU	0.5W	0.7W		MIT
2912.12	2912.97	34329.2	EU	0.7			KN
2909.01	2909.86	34365.9	EU	1.6			MIT
2906.676	2907.527	34393.48	EU	2.50	2.5		MIT
2904.212	2905.063	34422.66	EU	1.0			MIT
2893.85	2894.70	34545.9	EU	2.2	2.0		MIT
2893.03	2893.88	34555.7	EU	1.6			MIT
2892.54	2893.39	34561.6	EU	1.6W	1.5H		MIT
2887.88	2888.73	34617.3	EU	1.5	1.5		MIT
2886.92	2887.77	34628.8	EU	0.7			MIT
2878.86	2879.70	34725.8	EU	1.3			MIT
2877.890	2878.734	34737.49	EU	0.3	0.0		MIT
2877.76	2878.60	34739.1	EU	0.7			MIT
2876.06	2876.90	34759.6	EU	1.3			MIT
2871.57	2872.41	34813.9	EU	0.6	0.0H		MIT
2867.30	2868.14	34865.8	EU	0.3			MIT
2864.42	2865.26	34900.8	EU	1.0	1.0H		MIT
2862.57	2863.41	34923.4	EU	2.00	1.8		MIT
2859.667	2860.507	34958.84	EU II	1.6			MIT
2852.56	2853.40	35045.9	EU	0.70			MIT
2844.54	2845.38	35144.7	EU	0.3			MIT
2843.97	2844.81	35151.8	EU	1.3W			MIT
2842.82	2843.66	35166.0	EU	0.3	0.0H		MIT
2833.25	2834.08	35284.8	EU	1.0W	0.7J		MIT
2833.056	2833.889	35287.19	EU	0.5			MIT
2829.300	2830.132	35334.04	EU	1.3	1.0		MIT
2828.691	2829.523	35341.64	EU	2.30	2.2		MIT
2827.26	2828.09	35359.5	EU	1.1	0.8		MIT
2825.18	2826.01	35385.6	EU	0.5			MIT
2824.54	2825.37	35393.6	EU	0.3W			MIT
2820.77	2821.60	35440.9	EU	2.30	2.30		MIT
2816.18	2817.01	35498.6	EU	1.7W	1.7		MIT
2813.95	2814.78	35526.8	EU	2.5W	2.5J		MIT
2813.08	2813.91	35537.8	EU	1.2	0.7		MIT
2811.747	2812.575	35554.61	EU II	1.7			MIT
2810.70	2811.53	35567.9	EU	0.7W			MIT
2807.18	2808.01	35612.5	EU	0.3			MIT
2806.41	2807.24	35622.2	EU	0.3W			MIT
2802.86	2803.69	35667.3	EU	2.2W			MIT
2792.51	2793.33	35799.5	EU		0.3		MIT
2787.22	2788.04	35867.5	EU	0.6W			MIT
2781.89	2782.71	35936.2	EU	2.0W	2.0W		MIT
2780.526	2781.347	35953.81	EU	1.3			MIT
2776.52	2777.34	36005.7	EU	0.6			MIT
2772.90	2773.72	36052.7	EU	0.5			MIT
2752.16	2752.97	36324.4	EU	1.3W			MIT
2747.84	2748.65	36381.5	EU	1.10			MIT
2747.28	2748.09	36388.9	EU	1.3W			MIT
2745.60	2746.41	36411.1	EU	1.0			MIT
2744.255	2745.067	36428.99	EU	1.3			MIT
2743.25	2744.06	36442.3	EU	0.5D			MIT
2740.628	2741.439	36477.19	EU	1.3	1.0		MIT
2738.56	2739.37	36504.7	EU	0.8	0.0H		MIT
2735.246	2736.056	36548.97	EU	1.2			MIT
2732.60	2733.41	36584.4	EU	1.1			MIT
2731.32	2732.13	36601.5	EU	1.1			MIT
2729.39	2730.20	36627.4	EU II	2.0D	1.8		MIT
2727.780	2728.588	36649.00	EU	2.5L	2.7		MIT
2723.92	2724.73	36700.9	EU	1.5			MIT
2720.521	2721.327	36746.78	EU	1.0	0.6		MIT
2716.97	2717.78	36794.8	EU	2.5	2.5		MIT
2709.990	2710.793	36889.57	EU	1.3			MIT
2709.73	2710.53	36893.1	EU	0.9W			MIT
2708.23	2709.03	36913.5	EU		0.7		MIT
2705.263	2706.065	36954.02	EU	1.7L			MIT
2701.894	2702.695	37000.10	EU	2.50	2.3		MIT
2701.125	2701.926	37010.63	EU II	1.8H	1.8H		MIT
2692.019	2692.818	37135.82	EU	2.3	2.3		MIT

AIR WAVELENGTH (ANGSTRMS)	VACUUM WAVELENGTH (ANGSTRMS)	WAVENUMBER (KAYSERS)	LMENT	LOG ARC	INTENSITY SPK	DIS	REF
2685.65	2686.45	37223.9	EU II	2.2	2.0		MIT
2682.592	2683.389	37266.31	EU	0.8			MIT
2678.28	2679.08	37326.3	EU	2.2	2.0		MIT
2676.05	2676.85	37357.4	EU		0.7		EX
2673.409	2674.204	37394.31	EU	2.1	1.7		MIT
2670.84	2671.63	37430.3	EU	0.5			MIT
2668.33	2669.12	37465.5	EU	2.5W	2.6		MIT
2659.42	2660.21	37591.0	EU	0.6			MIT
2658.40	2659.19	37605.4	EU	1.2W			MIT
2653.598	2654.388	37673.47	EU	1.2	1.2H		MIT
2645.29	2646.08	37791.8	EU		0.7		MIT
2643.83	2644.62	37812.6	EU	0.6			MIT
2641.26	2642.05	37849.4	EU	2.0W	1.8		MIT
2638.764	2639.550	37885.24	EU	2.5	2.3		MIT
2635.464	2636.250	37932.68	EU	1.5			MIT
2626.780	2627.563	38058.07	EU	0.9			MIT
2616.36	2617.14	38209.6	EU		0.3H		MIT
2604.61	2605.39	38382.0	EU	1.3W			MIT
2596.44	2597.22	38502.8	EU		0.7		MIT
2581.86	2582.63	38720.2	EU	0.5	0.0		MIT
2580.60	2581.37	38739.1	EU	1.4W			MIT
2577.56	2578.33	38784.8	EU	0.70			MIT
2577.153	2577.925	38790.89	EU	1.3	1.3		MIT
2574.764	2575.535	38826.88	EU	0.3	0.0		MIT
2568.548	2569.318	38920.84	EU	0.8	0.0		MIT
2568.174	2568.944	38926.51	EU	1.2	1.2		MIT
2564.182	2564.951	38987.10	EU	1.3	1.8		MIT
2559.177	2559.944	39063.35	EU	1.0	1.3		MIT
2554.785	2555.551	39130.50	EU	1.0	0.0		MIT
2522.17	2522.93	39636.5	EU		0.6W		MIT
2513.79	2514.55	39768.6	EU		0.8W		MIT
2471.12	2471.87	40455.2	EU	0.6H			MIT
2460.47	2461.21	40630.4	EU	0.3H			MIT
2454.95	2455.69	40721.7	EU	0.7			MIT
2448.73	2449.47	40825.1	EU		0.3W		MIT
2446.04	2446.78	40870.0	EU	0.0	0.8W		MIT
2444.39	2445.13	40897.6	EU		0.7W		MIT
2436.76	2437.50	41025.7	EU		0.3J		MIT
2435.19	2435.93	41052.1	EU		0.3W		MIT
2421.53	2422.27	41283.7	EU	0.5W			MIT
2412.10	2412.83	41445.0	EU		0.3		MIT
2398.909	2399.639	41672.93	EU	0.6	0.7H		MIT
2392.00	2392.73	41793.3	EU	0.7W	0.7		MIT
2387.37	2388.10	41874.3	EU	0.6	0.5		MIT
2379.65	2380.38	42010.2	EU	0.3			MIT
2375.46	2376.19	42084.3	EU	0.5	1.3		MIT
2374.17	2374.89	42107.1	EU	0.5	0.8		MIT
2372.91	2373.63	42129.5	EU	0.3			MIT
2350.53	2351.25	42530.6	EU	0.5	0.8		MIT
2340.62	2341.34	42710.6	EU	0.7			MIT
9433.63	9436.22	10597.5	F I			0.5	EN
9178.68	9181.20	10891.8	F I			0.5	EN
9042.11	9044.59	11056.3	F I			1.0	EN
9025.49	9027.97	11076.7	F I			0.7	EN
8912.78	8915.23	11216.8	F I			1.4	EN
8910.24	8912.69	11220.0	F I			1.2	EN
8900.92	8903.36	11231.7	F I			1.5	EN
8849.07	8851.50	11297.5	F I			0.5	EN
8844.50	8846.93	11303.4	F I			0.7	EN
8831.22	8833.65	11320.4	F I			1.0	EN
8807.59	8810.01	11350.7	F			1.4	EN
8777.75	8780.16	11389.3	F I			1.0	EN
8737.28	8739.68	11442.1	F I			0.7	EN
8345.57	8347.86	11979.1	F I			0.7	EN
8302.40	8304.68	12041.4	F I			1.0	EN
8298.59	8300.87	12046.9	F I			1.3	EN
8274.62	8276.89	12081.8	F I			1.2	EN
8232.18	8234.44	12144.1	F I			1.0	EN
8230.77	8233.03	12146.2	F I			1.5	EN
8214.72	8216.98	12169.9	F I			1.3	EN

AIR WAVELENGTH (ANGSTRMS)	VACUUM WAVELENGTH (ANGSTRMS)	WAVENUMBER (KAYSERS)	LMENT	LOG ARC	INTENSITY SPK	DIS	REF
8208.63	8210.89	12178.9	F I			0.7	EN
8197.74	8199.99	12195.1	F I			0.5	EN
8194.6	8196.9	12200.	F I			1.1	DI
8191.24	8193.49	12204.8	F I			0.7	EN
8183.5	8185.7	12216.	F			0.7	DI
8179.34	8181.59	12222.6	F I			1.0	EN
8159.52	8161.76	12252.2	F I			1.2	EN
8129.27	8131.51	12297.8	F I			1.0	EN
8126.56	8128.79	12301.9	F I			0.8	EN
8102.9	8105.1	12338.	F			0.6	DI
8077.53	8079.75	12376.6	F			1.0	EN
8075.52	8077.74	12379.7	F I			1.2	EN
8040.94	8043.15	12432.9	F I			1.5	EN
8016.6	8018.8	12471.	F			1.0H	DI
8007.77	8009.97	12484.4	F I			0.5	EN
7956.31	7958.50	12565.2	F I			1.0	EN
7954.08	7956.27	12568.7	F I			0.5	EN
7948.52	7950.71	12577.5	F I			1.0	EN
7936.32	7938.50	12596.8	F I			1.1	EN
7930.94	7933.12	12605.4	F I			0.7	EN
7898.59	7900.76	12657.0	F I			1.2	EN
7879.19	7881.36	12688.2	F I			1.0	EN
7822.61	7824.76	12779.9	F I			0.7	EN
7800.22	7802.37	12816.6	F I			1.7	EN
7754.70	7756.83	12891.9	F I			1.8	EN
7607.17	7609.26	13141.9	F I			1.7	EN
7573.41	7575.50	13200.5	F I			1.6	EN
7552.24	7554.32	13237.5	F I			1.6	EN
7514.93	7517.00	13303.2	F I			1.5	EN
7489.14	7491.20	13349.0	F I			1.7	EN
7482.72	7484.78	13360.4	F I			1.9	EN
7476.54	7478.60	13371.5	F I			1.2	EN
7425.64	7427.69	13463.1	F I			2.2	EN
7398.68	7400.72	13512.2	F I			2.6	EN
7331.95	7333.97	13635.2	F I			2.3	EN
7314.31	7316.33	13668.1	F I			1.6	EN
7313.74	7315.76	13669.1	F I			1.0	EN
7311.02	7313.03	13674.2	F I			2.1	EN
7309.03	7311.04	13677.9	F			1.7	EN
7299.00	7301.01	13696.7	F I			1.4	EN
7202.37	7204.36	13880.5	F I			2.1	EN
7127.989	7129.954	14025.34	F I			2.2	DI
7037.45	7039.39	14205.8	F I			2.3	EN
6966.35	6968.27	14350.8	F I			1.8	EN
6909.82	6911.73	14468.2	F I			2.2	EN
6902.46	6904.36	14483.6	F I			2.7	EN
6870.22	6872.12	14551.6	F I			2.2	EN
6856.02	6857.91	14581.7	F I			3.0	EN
6834.26	6836.15	14628.1	F I			2.5	EN
6795.52	6797.40	14711.5	F I			1.8	EN
6773.97	6775.84	14758.3	F			2.0	EN
6766.97	6768.84	14773.6	F			0.7	EN
6762.89	6764.76	14782.5	F			1.0	EN
6762.13	6764.00	14784.2	F			0.7	GL
6708.284	6710.136	14902.83	F I			1.6	DI
6690.47	6692.32	14942.5	F I			1.8	EN
6650.39	6652.23	15032.6	F I			1.6	EN
6580.38	6582.20	15192.5	F I			1.5	EN
6569.69	6571.50	15217.2	F I			1.7	EN
6522.67	6524.47	15326.9	F			0.3	GL
6502.04	6503.84	15375.5	F I			0.3	GL
6463.52	6465.31	15467.2	F I			1.2	EN
6416.30	6418.07	15581.0	F			0.6	GL
6413.66	6415.43	15587.4	F I			2.2	EN
6405.15	6406.92	15608.1	F			0.6	GL
6356.87	6358.63	15726.7	F I			0.3	GL
6348.50	6350.26	15747.4	F I			2.3	EN
6254.68	6256.41	15983.6	F I			0.5	GL
6239.64	6241.37	16022.1	F I			2.5	EN
6210.83	6212.55	16096.4	F I			0.7	EN

AIR WAVELENGTH (ANGSTRMS)	VACUUM WAVELENGTH (ANGSTRMS)	WAVENUMBER (KAYSERS)	LMENT	LOG ARC	INTENSITY SPK	DIS	REF
6149.73	6151.43	16256.4	F I			1.0	EN
6080.13	6081.81	16442.5	F I			0.3	GL
6047.53	6049.20	16531.1	F I			1.2	EN
6038.04	6039.71	16557.1	F I			0.3	GL
6017.90	6019.57	16612.5	F I			0.5	GL
6015.90	6017.57	16618.0	F			0.3	GL
5996.76	5998.42	16671.1	F			0.3	GL
5707.23	5708.81	17516.8	F I			0.6	GL
5700.72	5702.30	17536.8	F I			0.5	GL
5671.62	5673.19	17626.8	F I			0.3	GL
5173.16	5174.60	19325.2	F II			1.2	DI
5110.85	5112.27	19560.8	F			0.7	DI
5001.98	5003.38	19986.5	F II			1.5	DI
4933.25	4934.63	20265.0	F II			1.5	DI
4859.37	4860.73	20573.1	F II			1.7	DI
4734.37	4735.69	21116.2	F II			0.5	DI
4527.3	4528.6	22082.	F			0.5H	DI
4522.1	4523.4	22107.	F			0.3H	DI
4505.9	4507.2	22187.	F			0.3D	DI
4447.18	4448.43	22479.9	F II			2.3H	DI
4446.71	4447.96	22482.2	F II			2.2H	DI
4446.51	4447.76	22483.2	F II			1.6H	DI
4317.3	4318.5	23156.	F II			0.8	DI
4299.177	4300.386	23253.72	F II			2.2	DI
4278.89	4280.09	23364.0	F II			1.3H	DI
4277.51	4278.71	23371.5	F II			1.6H	DI
4275.21	4276.41	23384.1	F II			2.0H	DI
4246.16	4247.36	23544.1	F II			2.5H	DI
4225.12	4226.31	23661.3	F			1.3H	DI
4217.9	4219.1	23702.	F			0.3H	DI
4207.87	4209.06	23758.3	F II			0.5	DI
4207.442	4208.627	23760.72	F II			1.5	DI
4207.162	4208.347	23762.30	F II			1.7	DI
4192.62	4193.80	23844.7	F II			0.5	DI
4126.96	4128.12	24224.1	F II			0.5	DI
4119.219	4120.381	24269.60	F II			1.7	DI
4118.756	4119.918	24272.33	F II			0.8	DI
4117.008	4118.170	24282.63	F II			1.5	DI
4116.547	4117.708	24285.35	F II			1.7	DI
4112.975	4114.136	24306.44	F II			1.5	DI
4112.734	4113.894	24307.87	F II			1.3	DI
4109.173	4110.333	24328.93	F II			2.0	DI
4103.871	4105.029	24360.36	F II			1.7	DI
4103.724	4104.882	24361.24	F II			1.7	DI
4103.525	4104.683	24362.42	F II			2.5	DI
4103.217	4104.375	24364.25	F II			1.5	DI
4103.085	4104.243	24365.03	F II			2.2	DI
4083.919	4085.072	24479.37	F II			1.6	DI
4025.495	4026.633	24834.65	F II			2.5	DI
4025.010	4026.148	24837.64	F II			2.2	DI
4024.727	4025.864	24839.39	F II			2.7	DI
3974.791	3975.915	25151.44	F II			1.3	DI
3972.670	3973.794	25164.87	F II			1.0	DI
3972.411	3973.535	25166.51	F II			0.3	DI
3972.047	3973.171	25168.82	F II			0.8	DI
3971.626	3972.750	25171.48	F II			0.5	DI
3952.26	3953.38	25294.8	F II			0.5	DI
3945.65	3946.77	25337.2	F II			1.0	DI
3944.33	3945.45	25345.7	F II			1.3	DI
3941.52	3942.64	25363.7	F II			0.8	DI
3939.03	3940.15	25379.8	F II			1.5	DI
3935.00	3936.11	25405.8	F II			0.8	DI
3903.819	3904.925	25608.69	F II			1.0	DI
3901.955	3903.060	25620.92	F II			1.2	DI
3901.852	3902.957	25621.60	F II			0.5	DI
3898.833	3899.938	25641.44	F II			1.3	DI
3898.725	3899.830	25642.15	F II			0.5	DI
3896.66	3897.76	25655.7	F II			0.8	DI
3893.04	3894.14	25679.6	F II			0.5	DI
3851.667	3852.759	25955.42	F II			2.3	DI

AIR WAVELENGTH (ANGSTRMS)	VACUUM WAVELENGTH (ANGSTRMS)	WAVENUMBER (KAYSERS)	LMENT	LOG INTENSITY ARC	SPK	DIS	REF
3849.987	3851.079	25966.75	F II				DI
3847.086	3848.177	25986.33	F II			2.9	DI
3821.5	3822.6	26160.	F			0.5H	DI
3818.52	3819.60	26180.7	F II			0.5	DI
3814.65	3815.73	26207.3	F II			1.0	DI
3805.90	3806.98	26267.5	F II			1.2	DI
3801.09	3802.17	26300.8	F II			0.8	DI
3798.46	3799.54	26319.0	F II			0.8	DI
3794.60	3795.68	26345.8	F II			0.5	DI
3792.40	3793.48	26361.0	F II			1.0	DI
3785.97	3787.05	26405.8	F II			0.5	DI
3739.60	3740.66	26733.2	F II			1.0	DI
3710.365	3711.421	26943.86	F			1.0	DI
3704.51	3705.56	26986.4	F II			1.8	DI
3679.67	3680.72	27168.6	F II			1.2	DI
3642.798	3643.836	27443.60	F II			1.5	DI
3641.985	3643.023	27449.73	F II			1.8	DI
3641.011	3642.049	27457.07	F II			0.8	DI
3640.891	3641.929	27457.98	F II			2.0	DI
3608.89	3609.92	27701.4	F II			0.8	DI
3607.32	3608.35	27713.5	F II			0.8	DI
3606.80	3607.83	27717.5	F II			1.0	DI
3603.72	3604.75	27741.2	F II			1.3	DI
3602.85	3603.88	27747.9	F			1.8D	DI
3601.403	3602.430	27759.04	F			1.5	DI
3598.704	3599.731	27779.85	F			1.5	DI
3595.917	3596.943	27801.39	F			1.2	DI
3589.345	3590.369	27852.29	F			1.3	DI
3587.980	3589.004	27862.88	F			1.2	DI
3587.42	3588.44	27867.2	F			0.8	DI
3587.13	3588.15	27869.5	F			0.8	DI
3577.23	3578.25	27946.6	F			1.8	DI
3574.92	3575.94	27964.7	F II			0.8	DI
3571.68	3572.70	27990.0	F II			0.8	DI
3569.47	3570.49	28007.4	F II			0.5	DI
3544.392	3545.405	28205.52	F II			0.8	DI
3541.937	3542.949	28225.07	F II			1.8	DI
3541.765	3542.777	28226.44	F II			2.0	DI
3538.474	3539.485	28252.70	F II			0.8	DI
3536.838	3537.849	28265.76	F II			1.5	DI
3535.162	3536.172	28279.16	F II			1.0	DI
3522.883	3523.890	28377.73	F II			1.3	DI
3516.32	3517.33	28430.7	F			0.5L	DI
3505.763	3506.766	28516.30	F			1.0	DI
3505.614	3506.617	28517.52	F II			2.8	DI
3505.508	3506.511	28518.38	F II			1.3	DI
3503.095	3504.097	28538.02	F II			2.6	DI
3502.954	3503.956	28539.17	F II			1.8	DI
3502.859	3503.861	28539.94	F II			1.0	DI
3501.562	3502.564	28550.51	F II			1.2	DI
3501.487	3502.489	28551.13	F II			0.8	DI
3501.416	3502.418	28551.71	F II			2.3	DI
3493.215	3494.215	28618.74	F II			0.5	DI
3475.68	3476.68	28763.1	F II			0.5	DI
3474.800	3475.795	28770.40	F II			1.5	DI
3473.621	3474.616	28780.16	F II			0.5	DI
3473.314	3474.308	28782.71	F II			1.2	DI
3472.964	3473.958	28785.61	F II			1.3	DI
3417.21	3418.19	29255.2	F			1.0	DI
3417.02	3418.00	29256.9	F II			1.3	DI
3416.58	3417.56	29260.6	F			1.0	DI
3416.45	3417.43	29261.8	F II			1.0	DI
3414.663	3415.642	29277.07	F II			1.2	DI
3412.04	3413.02	29299.6	F II			0.7	DI
3411.66	3412.64	29302.8	F II			0.8	DI
3410.82	3411.80	29310.1	F			0.3	DI
3409.02	3410.00	29325.5	F II			0.8	DI
3408.68	3409.66	29328.5	F II			0.3	DI
3406.56	3407.54	29346.7	F II			0.5	DI
3405.980	3406.957	29351.70	F			1.0	DI

AIR WAVELENGTH (ANGSTRMS)	VACUUM WAVELENGTH (ANGSTRMS)	WAVENUMBER (KAYSERS)	LMENT	LOG INTENSITY ARC	SPK	DIS	REF
3399.29	3400.27	29409.5	F II			0.8H	DI
3394.22	3395.19	29453.4	F II			0.5A	DI
3379.29	3380.26	29583.5	F II			0.5	DI
3377.44	3378.41	29599.7	F II			1.0	DI
3373.49	3374.46	29634.4	F II			1.2	DI
3358.32	3359.29	29768.2	F			1.0	DI
3355.98	3356.94	29789.0	F			0.8	DI
3354.34	3355.30	29803.6	F			0.5	DI
3321.30	3322.26	30100.0	F			0.5	DI
3303.89	3304.84	30258.6	F II			1.3	DI
3301.41	3302.36	30281.4	F II			0.8	DI
3300.09	3301.04	30293.5	F			0.5A	DI
3296.56	3297.51	30325.9	F II			1.2	DI
3296.19	3297.14	30329.3	F II			0.5	DI
3294.37	3295.32	30346.1	F II			1.0	DI
3264.16	3265.10	30626.9	F II			1.5	DI
3230.89	3231.82	30942.3	F			0.5H	DI
3229.00	3229.93	30960.4	F			0.3H	DI
3202.740	3203.666	31214.25	F II			2.3	DI
3153.492	3154.405	31701.70	F II			1.3	DI
3147.965	3148.877	31757.36	F II			1.2	DI
3124.19	3125.10	31999.0	F II			0.5H	DI
3106.16	3107.06	32184.7	F II			1.0	DI
3059.960	3060.850	32670.67	F II			1.8	DI
3058.141	3059.030	32690.10	F II			1.5	DI
3057.083	3057.972	32701.41	F II			1.3	DI
2988.45	2989.32	33452.4	F			1.0R	DI
2947.27	2948.13	33919.8	F			0.3H	DI
2884.13	2884.98	34662.3	F			0.3	DI
2876.49	2877.33	34754.4	F II			1.0	DI
2875.88	2876.72	34761.8	F II			0.7	DI
2874.80	2875.64	34774.8	F II			1.2	DI
2874.22	2875.06	34781.8	F II			0.3	DI
2873.13	2873.97	34795.0	F II			0.7	DI
2871.40	2872.24	34816.0	F II			1.4	DI
2867.30	2868.14	34865.8	F II			1.0	DI
2759.86	2760.68	36223.0	F			0.3	DI
2556.10	2556.87	39110.4	F II			1.2	DI
9980.55	9983.29	10016.7	FE I	0.3H			ME
9944.13	9946.86	10053.4	FE	0.5H			ME
9889.082	9891.793	10109.39	FE I	1.6			ME
9868.09	9870.80	10130.9	FE I	0.5			ME
9861.793	9864.497	10137.36	FE I	1.5			ME
9834.04	9836.74	10166.0	FE I	0.5H			ME
9811.36	9814.05	10189.5	FE	0.3			ME
9800.335	9803.022	10200.94	FE I	1.3			ME
9786.62	9789.30	10215.2	FE I	0.3			ME
9783.96	9786.64	10218.0	FE I	0.5			ME
9763.913	9766.591	10238.99	FE I	1.2			ME
9763.450	9766.127	10239.47	FE I	1.2			ME
9753.129	9755.804	10250.31	FE I	1.0			ME
9747.24	9749.91	10256.5	FE	0.3			ME
9738.624	9741.295	10265.58	FE I	2.3			ME
9699.70	9702.36	10306.8	FE I	0.8H			ME
9666.59	9669.24	10342.1	FE	0.3			ME
9658.94	9661.59	10350.3	FE I	0.3			ME
9657.30	9659.95	10352.0	FE I	0.5			ME
9653.143	9655.791	10356.48	FE I	1.3			ME
9637.55	9640.19	10373.2	FE	0.3			ME
9634.22	9636.86	10376.8	FE I	0.7			ME
9626.562	9629.202	10385.08	FE I	1.5H			ME
9602.07	9604.70	10411.6	FE	0.3			ME
9569.960	9572.585	10446.50	FE I	1.6H			ME
9550.90	9553.52	10467.3	FE I	0.3			ME
9529.31	9531.92	10491.1	FE I	0.6H			ME
9513.24	9515.85	10508.8	FE I	1.0H			MIT
9462.98	9465.58	10564.6	FE I	0.3			ME
9454.24	9456.83	10574.4	FE I	0.6H			ME
9452.45	9455.04	10576.4	FE I	0.3			ME
9443.98	9446.57	10585.8	FE I	1.0H			ME

AIR WAVELENGTH (ANGSTRMS)	VACUUM WAVELENGTH (ANGSTRMS)	WAVENUMBER (KAYSERS)	LMENT	LOG ARC	INTENSITY SPK	INTENSITY DIS	REF
9437.91	9440.50	10592.7	FE I	0.3			ME
9430.08	9432.67	10601.5	FE	0.6			ME
9414.14	9416.72	10619.4	FE I	1.3H			ME
9401.14	9403.72	10634.1	FE I	1.0H			MIT
9394.71	9397.29	10641.4	FE I	0.5H			ME
9388.28	9390.86	10648.7	FE I	0.5H			ME
9382.81	9385.38	10654.9	FE	0.5H			MIT
9372.900	9375.472	10666.13	FE I	0.8			ME
9362.370	9364.939	10678.13	FE I	0.6			ME
9359.420	9361.988	10681.49	FE I	0.5			ME
9350.44	9353.01	10691.7	FE I	1.0			MIT
9343.44	9346.00	10699.8	FE I	0.5			MIT
9333.94	9336.50	10710.7	FE I	0.3			ME
9318.15	9320.71	10728.8	FE I	0.5			MIT
9307.94	9310.49	10740.6	FE I	0.3			ME
9294.66	9297.21	10755.9	FE I	0.3			ME
9259.06	9261.60	10797.3	FE I	1.2			MIT
9258.49	9261.03	10797.9	FE	0.5			ME
9258.31	9260.85	10798.1	FE I	1.3			MIT
9246.54	9249.08	10811.9	FE I	0.5			ME
9242.30	9244.84	10816.8	FE I	0.3			MIT
9217.54	9220.07	10845.9	FE I	0.7H			ME
9214.41	9216.94	10849.6	FE I	0.8			MIT
9210.030	9212.558	10854.75	FE I	0.8			ME
9199.52	9202.04	10867.2	FE	0.3H			ME
9173.46	9175.98	10898.0	FE I	0.6D			ME
9166.44	9168.96	10906.4	FE I	0.5H			ME
9148.08	9150.59	10928.3	FE	0.5			DI
9147.800	9150.311	10928.59	FE I	0.7H			ME
9146.11	9148.62	10930.6	FE I	0.5			MIT
9118.888	9121.391	10963.24	FE I	1.3			ME
9116.14	9118.64	10966.6	FE	0.3			ME
9100.47	9102.97	10985.4	FE I	0.7H			MIT
9089.413	9091.908	10998.79	FE I	1.5			ME
9088.326	9090.821	11000.11	FE I	1.6			ME
9080.48	9082.97	11009.6	FE I	0.5H			ME
9079.599	9082.091	11010.68	FE I	0.9			ME
9070.40	9072.89	11021.8	FE I	0.3			MIT
9062.24	9064.73	11031.8	FE I	0.3			ME
9024.47	9026.95	11077.9	FE I	1.2			ME
9019.84	9022.32	11083.6	FE I	0.3			ME
9012.098	9014.572	11093.15	FE I	1.5			ME
9010.55	9013.02	11095.1	FE I	0.3			ME
9008.37	9010.84	11097.7	FE	0.3			ME
8999.561	9002.032	11108.60	FE I	2.0			ME
8984.87	8987.34	11126.8	FE I	0.5			ME
8975.408	8977.872	11138.50	FE I	1.2			ME
8945.204	8947.660	11176.11	FE I	1.3			ME
8943.00	8945.46	11178.9	FE I	0.5			ME
8929.04	8931.49	11196.3	FE I	0.7			ME
8919.95	8922.40	11207.7	FE I	1.0			ME
8876.13	8878.57	11263.1	FE I	0.3			ME
8868.40	8870.84	11272.9	FE I	0.5			MIT
8866.961	8869.396	11274.72	FE I	2.2			ME
8846.82	8849.25	11300.4	FE I	0.7			ME
8838.433	8840.860	11311.12	FE I	1.5			ME
8824.227	8826.650	11329.33	FE I	2.3			ME
8809.1	8811.5	11349.	FE	0.3			ME
8808.3	8810.7	11350.	FE I	0.6H			ME
8804.624	8807.042	11354.55	FE I	1.0			ME
8796.42	8798.84	11365.1	FE I	0.3			ME
8793.376	8795.791	11369.07	FE I	2.1			ME
8790.62	8793.03	11372.6	FE I	1.0H			ME
8784.44	8786.85	11380.6	FE I	0.7			ME
8764.000	8766.407	11407.18	FE I	2.0			ME
8757.192	8759.597	11416.05	FE I	1.7			ME
8747.32	8749.72	11428.9	FE I	0.3			ME
8729.1	8731.5	11453.	FE I	0.3			ME
8713.19	8715.58	11473.7	FE I	1.0			ME
8710.29	8712.68	11477.5	FE I	1.3H			ME

AIR WAVELENGTH (ANGSTRMS)	VACUUM WAVELENGTH (ANGSTRMS)	WAVENUMBER (KAYSERS)	LMENT	LOG ARC	INTENSITY SPK	INTENSITY DIS	REF
8688.633	8691.019	11506.13	FE I	2.2			ME
8674.751	8677.134	11524.54	FE I	1.7			ME
8661.908	8664.287	11541.63	FE I	2.0			ME
8621.612	8623.980	11595.57	FE I	0.7			ME
8611.807	8614.173	11608.78	FE I	1.0			ME
8598.21	8600.57	11627.1	FE	0.5			BU
8582.267	8584.625	11648.73	FE I	0.9			ME
8566.21	8568.56	11670.6	FE	0.3			BU
8526.685	8529.028	11724.67	FE I	0.5			ME
8515.11	8517.45	11740.6	FE I	0.3			MIT
8514.075	8516.414	11742.03	FE I	1.0			ME
8497.00	8499.33	11765.6	FE I	0.3			ME
8468.413	8470.740	11805.34	FE I	1.3			ME
8439.603	8441.922	11845.64	FE I	0.7			ME
8387.781	8390.086	11918.83	FE I	1.5	0.0		ME
8365.642	8367.941	11950.37	FE I	1.2			ME
8360.822	8363.120	11957.26	FE I	0.9			ME
8342.95	8345.24	11982.9	FE I	0.3			KN
8339.431	8341.723	11987.93	FE I	1.3			ME
8331.941	8334.231	11998.71	FE I	1.3			ME
8327.063	8329.352	12005.74	FE I	1.6	0.3		ME
8293.527	8295.807	12054.28	FE I	0.9			ME
8275.90	8278.17	12080.0	FE I	0.8			MIT
8274.31	8276.58	12082.3	FE	0.8			MIT
8264.27	8266.54	12097.0	FE	1.3			KN
8248.151	8250.418	12120.60	FE I	1.0			ME
8247.45	8249.72	12121.6	FE	0.3			BU
8239.130	8241.395	12133.87	FE I	0.7			ME
8232.94	8235.20	12143.0	FE	0.5			BU
8232.347	8234.610	12143.87	FE I	1.0			ME
8220.406	8222.666	12161.51	FE I	0.7	0.5		ME
8207.767	8210.023	12180.23	FE I	2.0			ME
8198.951	8201.205	12193.33	FE I	1.3			ME
8186.81	8189.06	12211.4	FE I	0.6			MIT
8149.58	8151.82	12267.2	FE	0.6			MIT
8096.875	8099.101	12347.05	FE	0.6			ME
8085.200	8087.423	12364.88	FE I	1.3	0.0H		ME
8080.668	8082.890	12371.81	FE I	0.6			ME
8075.13	8077.35	12380.3	FE I	0.7			ME
8046.073	8048.286	12425.01	FE I	1.4	0.0		ME
8028.341	8030.549	12452.45	FE I	2.0			ME
7998.972	8001.172	12498.17	FE I	1.5	0.0		ME
7994.473	7996.672	12505.20	FE	0.8			ME
7959.23	7961.42	12560.6	FE I	0.3			MIT
7955.81	7958.00	12566.0	FE	0.3			ME
7945.878	7948.064	12581.68	FE I	1.5	0.3		ME
7941.09	7943.27	12589.3	FE I	0.7			ME
7937.166	7939.349	12595.49	FE I	1.6	0.0		ME
7912.866	7915.043	12634.17	FE I	0.7			ME
7869.65	7871.82	12703.6	FE I	0.5			MIT
7855.44	7857.60	12726.5	FE I	0.5			MIT
7844.63	7846.79	12744.1	FE I	0.3			MIT
7832.224	7834.379	12764.25	FE I	1.5	0.0		ME
7808.03	7810.18	12803.8	FE	0.6			MIT
7780.586	7782.727	12848.97	FE I	1.4			ME
7771.93	7774.07	12863.3	FE	0.5			MIT
7751.13	7753.26	12897.8	FE I	0.7			MIT
7748.281	7750.413	12902.54	FE I	1.4			ME
7742.68	7744.81	12911.9	FE I	0.5			MIT
7723.20	7725.33	12944.4	FE I	0.6			ME
7711.73	7713.85	12963.7	FE II	1.4	1.2		MIT
7710.390	7712.512	12965.94	FE I	1.0			ME
7664.302	7666.412	13043.91	FE I	1.2			ME
7661.223	7663.332	13049.15	FE I	1.0			ME
7658.97	7661.08	13053.0	FE	0.3			ME
7653.76	7655.87	13061.9	FE I	1.9			MIT
7620.538	7622.636	13118.82	FE I	0.7			ME
7586.044	7588.133	13178.47	FE I	1.0			ME
7583.796	7585.884	13182.38	FE I	0.7			ME
7568.925	7571.009	13208.28	FE I	0.3			ME

AIR WAVELENGTH (ANGSTRMS)	VACUUM WAVELENGTH (ANGSTRMS)	WAVENUMBER (KAYSERS)	LMENT	LOG ARC	INTENSITY SPK	DIS	REF
7533.42	7535.49	13270.5	FE II		0.3		KN
7531.171	7533.245	13274.49	FE I	1.9			ME
7515.88	7517.95	13301.5	FE II		0.9		KN
7511.045	7513.113	13310.06	FE I	1.3	0.0		ME
7507.28	7509.35	13316.7	FE I	1.6			MIT
7495.088	7497.152	13338.40	FE I	2.3			ME
7491.678	7493.741	13344.47	FE I	1.3			BU
7476.30	7478.36	13371.9	FE I	1.1H			ME
7462.38	7464.44	13396.9	FE II	0.5	1.3		KN
7461.534	7463.589	13398.38	FE I	0.5			BU
7449.34	7451.39	13420.3	FE II		0.8		KN
7447.39	7449.44	13423.8	FE I	0.5H			MIT
7445.776	7447.827	13426.74	FE I	2.2			ME
7443.022	7445.072	13431.70	FE I	0.9			BU
7440.98	7443.03	13435.4	FE I	1.3H			BU
7430.87	7432.92	13453.7	FE I	0.7J			MIT
7430.58	7432.63	13454.2	FE I	0.6			KN
7418.674	7420.717	13475.79	FE	1.0			ME
7411.178	7413.219	13489.42	FE I	2.0			ME
7401.689	7403.728	13506.71	FE	1.0			ME
7389.425	7391.460	13529.13	FE I	2.0	1.9H		ME
7386.402	7388.437	13534.66	FE	1.6	1.4H		BU
7382.92	7384.95	13541.1	FE I	0.8H			MIT
7376.46	7378.49	13552.9	FE		1.0J		KN
7376.434	7378.466	13552.95	FE	0.9H			KN
7363.95	7365.98	13575.9	FE I	0.9H			MIT
7351.56	7353.59	13598.8	FE I	1.3	0.8H		MIT
7351.160	7353.185	13599.55	FE I	0.9H	0.8H		BU
7334.66	7336.68	13630.1	FE		0.9		KN
7333.62	7335.64	13632.1	FE I	0.9H			BU
7320.70	7322.72	13656.1	FE II	1.4	1.3H		KN
7311.101	7313.115	13674.06	FE I	1.8	1.4H		BB
7310.24	7312.25	13675.7	FE II		0.8		KN
7307.97	7309.98	13679.9	FE II	1.4H			KN
7307.957	7309.971	13679.95	FE	1.5			BU
7306.61	7308.62	13682.5	FE I	1.4	1.1H		BU
7300.47	7302.48	13694.0	FE	0.9H			ME
7294.98	7296.99	13704.3	FE	0.7H			MIT
7293.068	7295.078	13707.87	FE I	2.0	1.7H		BB
7288.760	7290.768	13715.98	FE I	1.5	1.3H		BB
7287.36	7289.37	13718.6	FE		0.8		KN
7285.286	7287.293	13722.52	FE I	0.3H			BU
7284.853	7286.860	13723.33	FE	1.2	1.2H		BU
7282.39	7284.40	13728.0	FE	0.7H			BU
7264.99	7266.99	13760.8	FE		1.0		KN
7261.54	7263.54	13767.4	FE I	1.3H			BU
7254.649	7256.648	13780.47	FE	1.0H	0.9J		BU
7244.86	7246.86	13799.1	FE I	1.3			BU
7239.885	7241.880	13808.57	FE I	1.4	1.3H		BB
7228.69	7230.68	13829.9	FE I	0.9H			ME
7228.03	7230.02	13831.2	FE I	1.6			ME
7224.51	7226.50	13838.0	FE II		1.1		KN
7223.668	7225.659	13839.57	FE I	1.3			BB
7222.39	7224.38	13842.0	FE II		1.2H		KN
7221.23	7223.22	13844.2	FE I	1.0	0.9H		MIT
7219.686	7221.676	13847.20	FE	1.1	1.3H		BB
7212.48	7214.47	13861.0	FE I	0.9H			MIT
7207.406	7209.392	13870.79	FE I	2.5	2.5		ME
7194.92	7196.90	13894.9	FE	0.9H			ME
7193.23	7195.21	13898.1	FE		0.9		KN
7189.17	7191.15	13906.0	FE I	0.8	0.9H		BU
7187.341	7189.322	13909.52	FE I	2.7	2.5		ME
7181.93	7183.91	13920.0	FE I	0.9H			BU
7176.886	7178.864	13929.78	FE I	1.0H			BU
7175.937	7177.915	13931.62	FE I	0.9H			BU
7164.469	7166.444	13953.92	FE I	2.3	2.0H		MIT
7155.64	7157.61	13971.1	FE I	1.0H			BU
7151.495	7153.466	13979.24	FE I	0.9H			BU
7145.317	7147.287	13991.32	FE I	1.2H	1.0H		BU
7142.522	7144.491	13996.80	FE I	1.1H	0.9H		BU

AIR WAVELENGTH (ANGSTRMS)	VACUUM WAVELENGTH (ANGSTRMS)	WAVENUMBER (KAYSERS)	LMENT	LOG ARC	INTENSITY SPK	DIS	REF
7134.99	7136.96	14011.6	FE		0.7		KN
7132.989	7134.955	14015.51	FE	1.2	1.0H		BB
7130.942	7132.908	14019.53	FE I	2.0	1.9H		BB
7107.461	7109.420	14065.84	FE I	0.7	0.9J		BB
7095.425	7097.381	14089.70	FE I	0.9H			BB
7090.404	7092.359	14099.68	FE I	1.7	1.4H		BB
7086.76	7088.71	14106.9	FE I	0.8H			BU
7083.396	7085.349	14113.63	FE I	0.7H			BU
7071.88	7073.83	14136.6	FE I	0.6			BU
7068.415	7070.364	14143.54	FE I	1.6	1.5		BB
7067.44	7069.39	14145.5	FE		1.0H		KN
7038.251	7040.192	14204.16	FE I	1.6			KN
7024.649	7026.586	14231.66	FE I	1.2	0.9J		BU
7024.084	7026.021	14232.81	FE	0.7			BU
7023.003	7024.940	14235.00	FE I	1.6	1.5H		BU
7016.436	7018.371	14248.32	FE I	2.0	1.4H		BU
7011.364	7013.298	14258.63	FE	0.7H			BU
7010.362	7012.295	14260.67	FE I	0.5H			BU
7008.014	7009.947	14265.44	FE I	1.0			BU
7000.633	7002.564	14280.48	FE	0.5H			BU
6999.902	7001.833	14281.97	FE I	1.4			BB
6988.530	6990.457	14305.21	FE I	0.9			BB
6978.855	6980.780	14325.05	FE I	1.8	1.1H		BB
6977.445	6979.369	14327.94	FE	0.6			BU
6976.934	6978.858	14328.99	FE	0.5			BB
6951.261	6953.178	14381.91	FE I	1.0H			BB
6945.208	6947.124	14394.45	FE I	1.8	1.3H		BB
6933.632	6935.545	14418.48	FE I	0.7H			BU
6916.702	6918.610	14453.77	FE I	1.5	0.7J		BB
6902.80	6904.70	14482.9	FE		0.7H		BU
6885.77	6887.67	14518.7	FE I	1.0H	0.9H		BU
6862.481	6864.375	14567.97	FE I	0.8			BU
6858.164	6860.056	14577.14	FE I	1.2	1.2H		BU
6857.25	6859.14	14579.1	FE	0.7J	0.7J		BU
6855.179	6857.071	14583.49	FE I	1.8	1.9H		BU
6843.671	6845.559	14608.01	FE I	1.5	1.5H		BB
6842.668	6844.556	14610.15	FE I	0.9	0.7H		BU
6841.349	6843.237	14612.97	FE I	1.7	1.7H		BB
6839.828	6841.715	14616.22	FE I	0.5H			BU
6838.86	6840.75	14618.3	FE	0.7H	0.7J		BU
6837.04	6838.93	14622.2	FE	0.7H	0.7H		MIT
6828.610	6830.494	14640.23	FE I	1.3	1.4H		BB
6820.38	6822.26	14657.9	FE	1.0	0.8H		MIT
6810.28	6812.16	14679.6	FE I	1.2			MIT
6806.859	6808.738	14687.01	FE I	0.9H	0.8J		BU
6804.020	6805.898	14693.14	FE I	0.8H			BB
6786.88	6788.75	14730.2	FE I	0.8H			BU
6752.734	6754.598	14804.73	FE I	1.0	1.1J		BB
6750.157	6752.020	14810.38	FE I	1.7	1.3H		MIT
6733.171	6735.030	14847.74	FE I	0.7			BU
6726.78	6728.64	14861.8	FE	1.0			MIT
6726.668	6728.525	14862.10	FE I	1.3			KN
6717.556	6719.411	14882.26	FE I	0.7H			BU
6715.410	6717.264	14887.01	FE I	0.7H			BU
6705.117	6706.968	14909.87	FE I	1.1H	1.1J		BB
6703.575	6705.426	14913.30	FE I	0.8	0.7J		MIT
6677.993	6679.837	14970.43	FE I	2.4	2.2		S
6663.446	6665.286	15003.11	FE I	1.8	1.4H		S
6633.772	6635.604	15070.22	FE I	1.8H	1.4V		MIT
6627.566	6629.396	15084.33	FE I	0.6H	0.9J		MIT
6609.116	6610.941	15126.44	FE I	1.4	1.1H		MIT
6597.607	6599.429	15152.83	FE I	1.0			BU
6593.878	6595.699	15161.40	FE I	1.5	1.3		MIT
6592.919	6594.740	15163.60	FE I	2.2	1.9		S
6575.022	6576.838	15204.87	FE I	1.1H	1.2H		MIT
6569.224	6571.039	15218.29	FE I	1.7	1.4H		MIT
6546.245	6548.053	15271.71	FE I	2.2	1.7		S
6533.97	6535.78	15300.4	FE	0.5	0.5H		BU
6518.374	6520.175	15337.01	FE I	1.0H	0.8H		MIT
6516.053	6517.853	15342.47	FE II		1.3		KN

AIR WAVELENGTH (ANGSTRMS)	VACUUM WAVELENGTH (ANGSTRMS)	WAVENUMBER (KAYSERS)	LMENT	LOG ARC	INTENSITY SPK	DIS	REF
6501.681	6503.477	15376.39	FE	0.3H			BU
6498.951	6500.747	15382.85	FE I	0.5H			MIT
6496.456	6498.251	15388.76	FE	1.0			BU
6494.985	6496.780	15392.24	FE I	2.6	2.2		S
6493.05	6494.84	15396.8	FE		0.9		KN
6481.874	6483.665	15423.37	FE I	1.1H	1.9J		MIT
6475.625	6477.414	15438.26	FE I	0.9H	0.9H		MIT
6469.214	6471.002	15453.56	FE I	0.9H	0.7H		BB
6462.729	6464.515	15469.06	FE I	1.3	0.8H		MIT
6456.376	6458.160	15484.29	FE II		0.9H		DO
6446.432	6448.214	15508.17	FE II		1.3		DO
6432.654	6434.432	15541.39	FE II		0.3		KN
6430.851	6432.628	15545.74	FE I	2.0	1.9		S
6421.368	6423.143	15568.70	FE I	1.5			MIT
6421.355	6423.130	15568.73	FE I	1.8	1.6H		S
6419.977	6421.752	15572.08	FE I	1.3H	1.2H		MIT
6416.942	6418.716	15579.44	FE II		0.3		KN
6411.664	6413.436	15592.27	FE I	2.0	1.9H		MIT
6408.029	6409.800	15601.11	FE I	1.7	1.5H		MIT
6400.318	6402.087	15619.91	FE I	0.3			MIT
6400.018	6401.787	15620.64	FE I	2.3	2.2H		MIT
6393.605	6395.372	15636.31	FE I	2.0	1.9H		S
6380.747	6382.511	15667.81	FE	1.4H	0.9H		MIT
6358.690	6360.448	15722.16	FE I	0.9H	0.8H		MIT
6355.039	6356.796	15731.19	FE I	1.2H	0.9H		MIT
6344.153	6345.907	15758.19	FE I	0.7H	0.3H		MIT
6336.839	6338.591	15776.38	FE I	1.8	1.5H		MIT
6335.335	6337.087	15780.12	FE I	1.7	1.3H		S
6322.691	6324.439	15811.68	FE I	0.9H	0.9H		MIT
6318.022	6319.769	15823.36	FE I	1.6	1.4H		S
6315.814	6317.561	15828.90	FE	0.3			BU
6315.310	6317.056	15830.16	FE I	0.7H			MIT
6302.511	6304.254	15862.31	FE I	1.2H	1.2H		MIT
6301.517	6303.260	15864.81	FE I	1.7	1.7H		MIT
6297.797	6299.539	15874.18	FE I	1.0H	1.2H		MIT
6290.977	6292.717	15891.39	FE I	0.7H			MIT
6270.220	6271.954	15943.99	FE I	0.3			MIT
6265.140	6266.873	15956.92	FE I	1.1H	0.7H		S
6256.369	6258.100	15979.29	FE I	0.9H			MIT
6254.261	6255.991	15984.68	FE I	1.0			BU
6252.561	6254.291	15989.02	FE I	1.8	1.4H		S
6247.56	6249.29	16001.8	FE II	0.0	0.9		KN
6246.335	6248.063	16004.96	FE I	1.3	1.3H		MIT
6240.647	6242.373	16019.55	FE I	0.3			MIT
6238.612	6240.338	16024.77	FE	0.3			MIT
6238.411	6240.137	16025.29	FE II		0.3		KN
6232.657	6234.381	16040.08	FE I	0.7H	0.7V		MIT
6230.728	6232.452	16045.05	FE I	1.8	1.7		S
6229.236	6230.959	16048.89	FE I	0.3			MIT
6227.241	6228.964	16054.04	FE	0.7			MIT
6221.36	6223.08	16069.2	FE I	0.5H			MIT
6219.284	6221.005	16074.57	FE I	1.6	0.7		MIT
6217.289	6219.009	16079.73	FE	0.3			MIT
6215.153	6216.873	16085.26	FE	0.5			MIT
6213.443	6215.162	16089.69	FE I	1.3			MIT
6212.055	6213.774	16093.28	FE I	0.3			MIT
6210.811	6212.529	16096.50	FE	0.3			MIT
6209.592	6211.310	16099.66	FE	0.3			MIT
6208.428	6210.146	16102.68	FE	0.3			MIT
6207.259	6208.976	16105.71	FE	0.3			MIT
6200.317	6202.033	16123.75	FE I	1.2			MIT
6197.75	6199.46	16130.4	FE	0.3			MIT
6191.562	6193.275	16146.55	FE I	2.0	1.3H		S
6188.748	6190.460	16153.89	FE	0.5			MIT
6188.023	6189.735	16155.78	FE I	0.3H			MIT
6185.578	6187.290	16162.17	FE	0.3H			MIT
6184.536	6186.247	16164.89	FE	0.5			MIT
6184.125	6185.836	16165.96	FE	0.5			MIT
6183.727	6185.438	16167.00	FE	0.3			MIT
6180.216	6181.926	16176.19	FE I	0.8			MIT

AIR WAVELENGTH (ANGSTRMS)	VACUUM WAVELENGTH (ANGSTRMS)	WAVENUMBER (KAYSERS)	LMENT	LOG ARC	INTENSITY SPK	DIS	REF
6173.339	6175.047	16194.21	FE I	1.3			MIT
6170.502	6172.210	16201.65	FE I	1.2			MIT
6165.362	6167.068	16215.16	FE	0.6			MIT
6163.553	6165.259	16219.92	FE I	0.3			MIT
6157.734	6159.438	16235.25	FE	1.2			MIT
6151.632	6153.334	16251.35	FE I	0.9			MIT
6149.265	6150.967	16257.61	FE II		0.6		KN
6147.849	6149.550	16261.35	FE II	0.7	0.8		MIT
6144.40	6146.10	16270.5	FE	0.3			BU
6141.759	6143.459	16277.47	FE I	1.0			MIT
6137.696	6139.395	16288.25	FE I	2.0			S
6137.001	6138.700	16290.09	FE I	1.0			MIT
6136.620	6138.318	16291.11	FE I	2.0			S
6134.06	6135.76	16297.9	FE	0.3			BU
6127.915	6129.611	16314.25	FE I	0.9			MIT
6127.225	6128.921	16316.09	FE	0.3			MIT
6109.318	6111.009	16363.91	FE	0.6			BU
6103.185	6104.874	16380.35	FE I	0.9H			MIT
6102.182	6103.871	16383.05	FE I	1.2	1.3H		MIT
6098.501	6100.189	16392.93	FE	0.3			MIT
6096.683	6098.371	16397.82	FE I	0.3			MIT
6095.53	6097.22	16400.9	FE	0.5H			BU
6093.637	6095.324	16406.02	FE I	0.5			MIT
6089.564	6091.250	16416.99	FE	0.6			MIT
6085.26	6086.94	16428.6	FE	0.3			MIT
6082.714	6084.398	16435.48	FE I	0.3			MIT
6079.022	6080.705	16445.46	FE I	0.3			MIT
6078.481	6080.164	16446.93	FE I	0.9	0.8H		MIT
6065.487	6067.166	16482.16	FE I	1.7	1.5		S
6062.89	6064.57	16489.2	FE I	0.3H			BU
6056.001	6057.678	16507.98	FE I	1.0	1.0H		MIT
6042.089	6043.762	16545.99	FE I	0.5	0.6		MIT
6027.057	6028.726	16587.25	FE	0.8	1.1		S
6024.063	6025.731	16595.50	FE I	1.3	1.3H		MIT
6021.829	6023.497	16601.65	FE I	2.5			MIT
6020.179	6021.846	16606.20	FE I	0.9	1.0H		MIT
6020.047	6021.714	16606.57	FE	0.3			MIT
6018.040	6019.707	16612.10	FE	0.3			MIT
6016.655	6018.321	16615.93	FE I	2.0			MIT
6013.505	6015.170	16624.63	FE	2.0			MIT
6008.576	6010.240	16638.27	FE I	1.3	1.0		MIT
6007.966	6009.630	16639.96	FE I	1.0H	1.0H		MIT
6006.638	6008.302	16643.64	FE	0.3			MIT
6005.146	6006.809	16647.77	FE	0.3H			MIT
6003.034	6004.697	16653.63	FE I	1.5	1.2		MIT
6001.198	6002.860	16658.73	FE	0.5H			MIT
5999.950	6001.612	16662.19	FE	1.0H			BU
5997.805	5999.466	16668.15	FE I	0.6			MIT
5991.383	5993.042	16686.02	FE II	0.3H	0.3H		MIT
5988.239	5989.898	16694.78	FE	0.5			MIT
5987.055	5988.713	16698.08	FE I	1.4H	1.1H		MIT
5984.804	5986.462	16704.36	FE I	1.7	1.3H		MIT
5983.705	5985.362	16707.43	FE I	1.5	1.1H		MIT
5976.799	5978.455	16726.73	FE	0.3			MIT
5975.358	5977.013	16730.76	FE	1.0	1.0		MIT
5969.554	5971.208	16747.03	FE I	0.7			MIT
5956.708	5958.358	16783.15	FE I	1.1			MIT
5952.740	5954.389	16794.33	FE I	0.9H			MIT
5952.362	5954.011	16795.40	FE	0.5			MIT
5949.357	5951.005	16803.88	FE I	0.6H			MIT
5942.119	5943.765	16824.35	FE	0.3			MIT
5940.968	5942.614	16827.61	FE I	0.8			MIT
5939.218	5940.864	16832.57	FE	0.6H			MIT
5938.735	5940.380	16833.94	FE	0.6			MIT
5937.115	5938.760	16838.53	FE	0.5			MIT
5934.680	5936.324	16845.44	FE I	1.2	1.1H		MIT
5933.085	5934.729	16849.97	FE	0.6			MIT
5931.715	5933.359	16853.86	FE	0.5			MIT
5930.186	5931.829	16858.21	FE I	1.5	1.0H		MIT
5927.806	5929.448	16864.98	FE	1.0			MIT

AIR WAVELENGTH (ANGSTRMS)	VACUUM WAVELENGTH (ANGSTRMS)	WAVENUMBER (KAYSERS)	LMENT	LOG ARC	INTENSITY SPK	DIS	REF
5926.189	5927.831	16869.58	FE	0.6			MIT
5924.73	5926.37	16873.7	FE	0.3			MIT
5923.45	5925.09	16877.4	FE	0.6H			MIT
5923.058	5924.699	16878.49	FE	0.6			MIT
5919.997	5921.637	16887.22	FE	0.6	0.6		MIT
5918.05	5919.69	16892.8	FE	0.7			MIT
5916.258	5917.897	16897.89	FE I	1.4	0.6		MIT
5914.162	5915.801	16903.88	FE I	1.7	1.4H		MIT
5913.062	5914.701	16907.03	FE	0.5			MIT
5912.834	5914.472	16907.68	FE	1.0			MIT
5910.597	5912.235	16914.08	FE	0.7			MIT
5909.986	5911.624	16915.83	FE I	0.3			MIT
5908.409	5910.046	16920.34	FE	0.7			MIT
5908.252	5909.889	16920.79	FE	0.9			MIT
5906.015	5907.652	16927.20	FE	0.8			MIT
5905.683	5907.320	16928.15	FE	1.1	0.9H		MIT
5902.527	5904.163	16937.20	FE	0.8			MIT
5901.68	5903.32	16939.6	FE	0.5			MIT
5895.497	5897.131	16957.40	FE	0.6			MIT
5883.848	5885.479	16990.97	FE I	1.2	1.0		MIT
5879.994	5881.624	17002.11	FE I	0.8			MIT
5879.782	5881.412	17002.72	FE	0.9			MIT
5878.002	5879.631	17007.87	FE	0.7H			MIT
5875.372	5877.000	17015.48	FE	1.2H			MIT
5873.219	5874.847	17021.72	FE I	0.9	0.3		MIT
5872.912	5874.540	17022.61	FE	0.9			MIT
5871.286	5872.913	17027.32	FE	0.5			MIT
5871.039	5872.666	17028.04	FE	0.6			MIT
5869.756	5871.383	17031.76	FE	1.0			MIT
5862.363	5863.988	17053.24	FE I	1.5	1.5		MIT
5859.608	5861.232	17061.26	FE I	1.2	1.1		MIT
5859.197	5860.821	17062.46	FE	0.8	0.3		MIT
5857.127	5858.751	17068.49	FE	0.3	0.5		MIT
5856.081	5857.704	17071.53	FE	0.9H	0.5		MIT
5855.130	5856.753	17074.31	FE	0.7			MIT
5854.462	5856.085	17076.25	FE	0.5			MIT
5853.187	5854.809	17079.98	FE I	0.5H			MIT
5845.868	5847.489	17101.36	FE	0.9			MIT
5844.145	5845.765	17106.40	FE	0.6			MIT
5842.484	5844.104	17111.26	FE	0.8			MIT
5837.703	5839.321	17125.28	FE	0.5			MIT
5836.558	5838.176	17128.64	FE	0.3			MIT
5835.281	5836.899	17132.39	FE	0.5			MIT
5834.779	5836.397	17133.86	FE	0.8			MIT
5833.64	5835.26	17137.2	FE	0.3	1.0		MIT
5832.848	5834.465	17139.53	FE	0.5			MIT
5832.073	5833.690	17141.81	FE	0.6			MIT
5831.693	5833.310	17142.93	FE	0.5			MIT
5830.799	5832.415	17145.55	FE	0.5H			MIT
5830.587	5832.203	17146.18	FE	0.7H			MIT
5828.487	5830.103	17152.36	FE	0.6			MIT
5826.585	5828.200	17157.96	FE	0.3			MIT
5825.691	5827.306	17160.59	FE	0.8			MIT
5824.851	5826.466	17163.06	FE	0.5			MIT
5816.376	5817.989	17188.07	FE	1.2D	1.0H		MIT
5815.158	5816.770	17191.67	FE	0.5H			MIT
5811.936	5813.547	17201.20	FE	0.5			MIT
5809.245	5810.856	17209.17	FE I	0.5	0.7H		MIT
5806.730	5808.340	17216.62	FE I	1.0	0.7H		MIT
5798.198	5799.806	17241.96	FE				ME
5791.041	5792.647	17263.26	FE I	0.8H	0.3H		MIT
5780.62	5782.22	17294.4	FE				ME
5775.091	5776.693	17310.94	FE I	1.1	0.3		MIT
5763.011	5764.609	17347.23	FE I	1.9	1.5		MIT
5753.136	5754.732	17377.00	FE I	1.6	1.3		MIT
5752.056	5753.651	17380.27	FE I	0.9H	0.3H		MIT
5741.861	5743.454	17411.13	FE				ME
5731.770	5733.360	17441.78	FE I	1.0	0.5		MIT
5717.840	5719.426	17484.27	FE I	1.0	0.3		MIT
5715.107	5716.693	17492.63	FE I	0.6H			MIT

AIR WAVELENGTH (ANGSTRMS)	VACUUM WAVELENGTH (ANGSTRMS)	WAVENUMBER (KAYSERS)	LMENT	LOG ARC	INTENSITY SPK	DIS	REF
5712.145	5713.730	17501.70	FE I	0.6	0.3		MIT
5709.379	5710.963	17510.18	FE I	2.0H			MIT
5705.978	5707.561	17520.62	FE I	1.2	1.0		MIT
5701.556	5703.138	17534.21	FE I	1.7	1.4		MIT
5693.634	5695.214	17558.60	FE	0.5	0.5		MIT
5686.522	5688.100	17580.56	FE	1.0	0.9		MIT
5679.022	5680.598	17603.78	FE I	0.7	0.6		MIT
5667.525	5669.098	17639.49	FE I	0.7H	0.5H		MIT
5662.525	5664.096	17655.07	FE I	1.7	1.7		S
5658.826	5660.396	17666.61	FE I	2.0	1.9		S
5658.537	5660.107	17667.51	FE I	1.5	0.3		MIT
5655.498	5657.068	17677.00	FE I	1.0	0.7		MIT
5655.179	5656.748	17678.00	FE	0.6	0.3		BU
5641.464	5643.030	17720.98	FE I	1.2	0.9		BU
5638.272	5639.837	17731.01	FE I	1.6	1.3		MIT
5636.698	5638.263	17735.96	FE	0.3	0.0		MIT
5635.83	5637.39	17738.7	FE I	0.6H	0.3H		MIT
5633.962	5635.526	17744.57	FE	1.3	1.0		MIT
5630.35	5631.91	17756.0	FE	0.7	0.3		MIT
5624.549	5626.110	17774.27	FE I	2.2	2.1		S
5624.069	5625.630	17775.79	FE I	0.6H	0.3H		MIT
5621.28	5622.84	17784.6	FE	0.6	0.3		MIT
5620.527	5622.087	17786.99	FE I	0.5H	0.3H		BU
5620.040	5621.600	17788.53	FE	0.3	0.3		MIT
5618.640	5620.200	17792.96	FE I	1.0	0.9		MIT
5615.652	5617.211	17802.43	FE I	2.6	2.5		S
5615.301	5616.860	17803.54	FE I	0.6	1.1		MIT
5602.956	5604.512	17842.77	FE I	1.7	1.5		MIT
5602.779	5604.334	17843.33	FE I	0.9	0.9		MIT
5600.234	5601.789	17851.44	FE I	0.3			MIT
5598.289	5599.843	17857.64	FE I	1.3			MIT
5594.659	5596.212	17869.23	FE	1.0			MIT
5587.576	5589.127	17891.88	FE	0.8			MIT
5586.763	5588.314	17894.48	FE I	2.6	1.7		S
5584.764	5586.315	17900.89	FE	1.4	0.3		MIT
5576.106	5577.654	17928.68	FE I	2.2			MIT
5573.105	5574.653	17938.34	FE I	0.9			MIT
5572.849	5574.396	17939.16	FE I	2.5	1.4		S
5569.625	5571.172	17949.55	FE I	2.5	1.2		S
5567.399	5568.945	17956.72	FE I	1.5			MIT
5565.702	5567.248	17962.20	FE I	1.8			MIT
5563.604	5565.149	17968.97	FE I	2.0	0.7		BB
5562.709	5564.254	17971.86	FE I	1.2H			MIT
5560.223	5561.767	17979.90	FE I	0.7H			MIT
5554.887	5556.430	17997.17	FE I	2.0			MIT
5553.585	5555.127	18001.39	FE I	0.8			MIT
5549.95	5551.49	18013.2	FE	0.9H			MIT
5546.490	5548.030	18024.41	FE I	1.6			MIT
5543.928	5545.468	18032.74	FE I	1.0			MIT
5543.183	5544.723	18035.17	FE	1.4			MIT
5539.28	5540.82	18047.9	FE	1.5			MIT
5538.57	5540.11	18050.2	FE	1.7			MIT
5535.411	5536.948	18060.49	FE	1.7			MIT
5534.860	5536.397	18062.29	FE II		1.0		DO
5534.66	5536.20	18062.9	FE I	1.3			MIT
5532.749	5534.286	18069.18	FE I	0.6			MIT
5525.553	5527.088	18092.71	FE I	1.6	0.3		MIT
5522.461	5523.995	18102.84	FE I	0.9			MIT
5514.631	5516.163	18128.54	FE	1.7	1.0		MIT
5506.782	5508.312	18154.38	FE I	2.2	1.0		S
5505.885	5507.415	18157.34	FE I	1.0H			MIT
5501.469	5502.997	18171.92	FE I	2.2			S
5497.519	5499.046	18184.97	FE I	2.2	0.7		S
5494.462	5495.989	18195.09	FE	0.3			MIT
5493.511	5495.037	18198.24	FE I	0.6			MIT
5487.775	5489.300	18217.26	FE I	1.0H			MIT
5487.138	5488.663	18219.37	FE I	1.7	0.7		MIT
5483.116	5484.640	18232.74	FE I	1.2			MIT
5481.459	5482.982	18238.25	FE I	0.7			MIT
5481.252	5482.775	18238.94	FE I	0.7			MIT

AIR WAVELENGTH (ANGSTRMS)	VACUUM WAVELENGTH (ANGSTRMS)	WAVENUMBER (KAYSERS)	LMENT	LOG ARC	INTENSITY SPK	DIS	REF
5480.873	5482.396	18240.20	FE I	1.0			MIT
5478.473	5479.995	18248.19	FE I	0.3			MIT
5476.578	5478.100	18254.50	FE I	1.9			MIT
5476.295	5477.817	18255.45	FE	1.1			MIT
5474.917	5476.438	18260.04	FE	2.0			RL
5473.920	5475.441	18263.37	FE I	2.0			MIT
5472.729	5474.250	18267.34	FE I	0.8			MIT
5466.947	5468.466	18286.66	FE		0.7		DO
5466.407	5467.926	18288.47	FE I	1.4			MIT
5464.279	5465.797	18295.59	FE	0.8			MIT
5463.283	5464.801	18298.93	FE I	2.0			MIT
5462.969	5464.487	18299.98	FE I	1.7			MIT
5457.651	5459.168	18317.81	FE	1.3	0.3		MIT
5455.613	5457.129	18324.65	FE I	2.5	1.5		S
5455.433	5456.949	18325.26	FE I	1.7			BU
5449.779	5451.294	18344.27	FE	1.0			MIT
5448.374	5449.888	18349.00	FE	0.3H			MIT
5446.920	5448.434	18353.90	FE I	2.5	1.5		S
5445.037	5446.550	18360.25	FE I	2.2			MIT
5436.595	5438.106	18388.75	FE I	0.7			MIT
5434.527	5436.038	18395.75	FE I	2.5	1.5		S
5432.950	5434.460	18401.09	FE I	0.5			MIT
5429.699	5431.208	18412.11	FE I	2.7	1.6		S
5427.832	5429.341	18418.44	FE		0.9		KN
5425.262	5426.770	18427.17	FE II		0.3		KN
5424.076	5425.584	18431.20	FE	2.6	1.3		MIT
5415.207	5416.712	18461.38	FE I	2.7	1.3		MIT
5410.909	5412.413	18476.05	FE I	2.3	1.0		MIT
5409.132	5410.636	18482.12	FE I	1.0H			MIT
5405.778	5407.281	18493.58	FE I	2.6	1.8		S
5404.148	5405.650	18499.16	FE I	2.5	1.5		MIT
5403.819	5405.321	18500.29	FE	1.5			MIT
5402.550	5404.052	18504.63	FE	0.6			MIT
5400.503	5402.004	18511.65	FE I	2.1			MIT
5398.285	5399.786	18519.25	FE I	1.8			BU
5397.616	5399.117	18521.55	FE	0.5			MIT
5397.131	5398.632	18523.21	FE I	2.6	1.7		S
5393.182	5394.681	18536.78	FE I	2.2	1.0		MIT
5391.475	5392.974	18542.64	FE I	1.4			MIT
5389.461	5390.960	18549.57	FE I	1.8			BU
5387.50	5389.00	18556.3	FE	0.3			MIT
5386.342	5387.840	18560.31	FE	0.5			MIT
5383.371	5384.868	18570.56	FE	2.6H	1.6		MIT
5379.577	5381.073	18583.66	FE	1.5			MIT
5378.858	5380.354	18586.14	FE	0.7			MIT
5376.849	5378.344	18593.08	FE I	0.7H			MIT
5373.712	5375.206	18603.94	FE I	1.2			MIT
5371.493	5372.987	18611.62	FE I	2.8			S
5369.957	5371.450	18616.95	FE	2.2H	1.3H		MIT
5367.460	5368.953	18625.61	FE	2.3H	1.2H		MIT
5365.405	5366.897	18632.74	FE	1.6			MIT
5365.400	5366.892	18632.76	FE				ME
5364.883	5366.375	18634.55	FE	2.3H	1.0H		MIT
5362.864	5364.355	18641.57	FE II		1.2		DO
5362.751	5364.242	18641.96	FE	0.8			MIT
5353.389	5354.878	18674.56	FE I	1.8	0.3		MIT
5349.736	5351.224	18687.31	FE I	0.6H			MIT
5343.470	5344.956	18709.23	FE I	1.1H			MIT
5341.026	5342.512	18717.79	FE I	2.3	1.2		S
5339.938	5341.423	18721.60	FE I	2.3	1.5		MIT
5333.30	5334.78	18744.9	FE	0.5			RL
5332.901	5334.384	18746.31	FE				ME
5329.999	5331.482	18756.51	FE	1.2			MIT
5328.534	5330.016	18761.67	FE I	2.2	1.5		S
5328.050	5329.532	18763.37	FE I	2.6	2.0		MIT
5326.157	5327.639	18770.04	FE	0.8			MIT
5325.563	5327.044	18772.14	FE II		0.6		KN
5324.182	5325.663	18777.00	FE I	2.6	1.8		MIT
5322.054	5323.535	18784.51	FE I	1.5			MIT
5321.113	5322.593	18787.83	FE	0.9H			MIT

AIR WAVELENGTH (ANGSTRMS)	VACUUM WAVELENGTH (ANGSTRMS)	WAVENUMBER (KAYSERS)	LMENT	LOG ARC	INTENSITY SPK	DIS	REF
5320.048	5321.528	18791.59	FE	0.8			BU
5316.609	5318.088	18803.75	FE II		2.2		DO
5315.07	5316.55	18809.2	FE I	0.7H			RL
5307.365	5308.842	18836.50	FE I	2.1			S
5303.419	5304.895	18850.52	FE II	1.4	1.4		DO
5302.314	5303.789	18854.44	FE I	2.5			MIT
5298.780	5300.254	18867.02	FE	1.1			MIT
5293.973	5295.446	18884.15	FE	0.9			BU
5290.850	5292.322	18895.30	FE I	1.2			MIT
5288.530	5290.002	18903.59	FE	1.5			MIT
5287.922	5289.393	18905.76	FE	2.0	1.3		BU
5284.416	5285.886	18918.30	FE	0.3			RL
5284.092	5285.562	18919.46	FE II		1.8H		DO
5283.626	5285.096	18921.13	FE I	2.6	1.6		MIT
5281.799	5283.269	18927.68	FE I	2.5	1.3		MIT
5280.360	5281.829	18932.83	FE	1.2			MIT
5275.994	5277.462	18948.50	FE II	0.3	1.2		DO
5273.378	5274.846	18957.90	FE I	1.7	0.6		MIT
5273.170	5274.637	18958.65	FE I	1.9	0.6		MIT
5270.360	5271.827	18968.76	FE I	2.6	1.9		S
5269.541	5271.008	18971.71	FE I	2.9	2.3		MIT
5266.579	5268.045	18982.37	FE I	2.7	1.6		MIT
5266.03	5267.50	18984.4	FE I	0.8H			BB
5264.796	5266.261	18988.80	FE II		0.6		KN
5263.874	5265.339	18992.13	FE	0.8			MIT
5263.330	5264.795	18994.09	FE I	2.5			MIT
5254.961	5256.424	19024.34	FE I	1.7			MIT
5253.480	5254.942	19029.71	FE I	1.8			MIT
5251.971	5253.433	19035.17	FE	1.1			MIT
5250.650	5252.111	19039.96	FE I	2.2			S
5250.211	5251.672	19041.55	FE I	1.5			MIT
5249.09	5250.55	19045.6	FE I	0.6H			RL
5247.063	5248.524	19052.98	FE I	1.7	1.0		MIT
5243.789	5245.249	19064.87	FE I	1.3			MIT
5242.495	5243.954	19069.58	FE	2.1	0.7		S
5238.09	5239.55	19085.6	FE	0.6H			MIT
5236.197	5237.655	19092.51	FE	0.8			MIT
5235.390	5236.847	19095.46	FE I	1.5			MIT
5234.627	5236.084	19098.24	FE II		0.7		MIT
5232.940	5234.397	19104.40	FE I	2.9	2.2		MIT
5229.867	5231.323	19115.62	FE I	2.3	1.2H		MIT
5228.413	5229.869	19120.94	FE	1.2H	0.8H		MIT
5227.192	5228.647	19125.41	FE I	2.6	1.8		S
5226.875	5228.330	19126.57	FE I	2.3	1.2		MIT
5225.531	5226.986	19131.49	FE I	1.8			MIT
5223.193	5224.647	19140.05	FE	0.8			MIT
5217.924	5219.377	19159.38	FE	0.8			MIT
5217.398	5218.851	19161.31	FE I	2.2	0.5		MIT
5216.278	5217.730	19165.42	FE I	2.5	1.0		S
5215.185	5216.637	19169.44	FE I	2.3	0.7		MIT
5208.601	5210.051	19193.67	FE I	2.3	0.9		MIT
5204.583	5206.032	19208.49	FE I	2.1			MIT
5202.339	5203.788	19216.77	FE I	2.5	1.0		S
5198.837	5200.285	19229.72	FE	1.0			MIT
5198.714	5200.162	19230.17	FE I	1.9			S
5197.590	5199.037	19234.33	FE II		1.0		KN
5196.098	5197.545	19239.85	FE	1.4H			MIT
5195.478	5196.925	19242.15	FE I	2.0H			MIT
5194.948	5196.395	19244.11	FE I	2.3	1.2		MIT
5192.357	5193.803	19253.72	FE I	2.6	1.7		MIT
5191.467	5192.913	19257.01	FE I	2.6	1.5		MIT
5187.922	5189.367	19270.17	FE				ME
5184.292	5185.736	19283.67	FE I	1.3			MIT
5180.058	5181.501	19299.43	FE	1.0			MIT
5178.78	5180.22	19304.2	FE	0.3			MIT
5177.239	5178.681	19309.94	FE	0.6			MIT
5171.599	5173.039	19331.00	FE I	2.5	1.8H		S
5170.75	5172.19	19334.2	FE	0.6H			MIT
5169.027	5170.467	19340.61	FE II	0.3	2.3H		MIT
5168.901	5170.341	19341.08	FE I	1.9			S

AIR WAVELENGTH (ANGSTRMS)	VACUUM WAVELENGTH (ANGSTRMS)	WAVENUMBER (KAYSERS)	LMENT	LOG ARC	INTENSITY SPK	INTENSITY DIS	REF
5167.491	5168.930	19346.36	FE I	2.8	2.2		S
5166.296	5167.735	19350.84	FE I	2.1			MIT
5165.424	5166.863	19354.10	FE I	1.7			MIT
5164.614	5166.053	19357.14	FE	1.8H			MIT
5162.288	5163.726	19365.86	FE I	2.5H			MIT
5159.053	5160.490	19378.00	FE	1.5H			MIT
5151.914	5153.349	19404.86	FE I	1.8			MIT
5150.843	5152.278	19408.89	FE I	2.2			S
5148.258	5149.692	19418.64	FE I	1.5			MIT
5148.052	5149.486	19419.41	FE I	1.3			MIT
5145.098	5146.531	19430.56	FE I	1.0			MIT
5142.940	5144.373	19438.72	FE I	2.1			MIT
5142.540	5143.973	19440.23	FE I	2.0H			MIT
5141.747	5143.179	19443.23	FE I	2.0	2.0H		MIT
5139.480	5140.912	19451.80	FE I	2.3	1.6		MIT
5139.260	5140.692	19452.64	FE I	2.1			MIT
5137.388	5138.819	19459.72	FE I	2.3H			MIT
5136.795	5138.226	19461.97	FE II	0.5	2.0		MIT
5133.680	5135.110	19473.78	FE I	2.3H	0.0H		MIT
5131.474	5132.904	19482.15	FE I	2.1			MIT
5129.658	5131.087	19489.05	FE I	0.6			MIT
5127.363	5128.792	19497.77	FE I	2.0			S
5126.213	5127.641	19502.14	FE I	0.7H			MIT
5125.130	5126.558	19506.26	FE I	2.0H			MIT
5123.723	5125.151	19511.62	FE I	2.3			S
5121.646	5123.073	19519.53	FE				ME
5110.414	5111.838	19562.44	FE I	2.5			S
5107.645	5109.068	19573.04	FE I	0.3			MIT
5107.452	5108.875	19573.78	FE I	2.0			MIT
5106.441	5107.864	19577.66	FE	1.4			BU
5102.200	5103.622	19593.93	FE I	1.9			MIT
5098.710	5100.131	19607.34	FE I	2.3			MIT
5096.995	5098.416	19613.94	FE I	1.5H			MIT
5090.789	5092.208	19637.85	FE I	1.6H			MIT
5086.765	5088.183	19653.38	FE	0.3	2.0		MIT
5083.342	5084.759	19666.62	FE I	2.3			S
5079.753	5081.169	19680.51	FE I	2.0			MIT
5079.236	5080.652	19682.51	FE I	2.0			MIT
5078.993	5080.409	19683.46	FE I	1.3H			MIT
5076.280	5077.695	19693.98	FE	0.5H			MIT
5074.760	5076.175	19699.87	FE I	1.9			MIT
5073.560	5074.974	19704.53	FE		1.3H		MIT
5070.957	5072.371	19714.65	FE	0.0	1.8		MIT
5068.786	5070.199	19723.09	FE I	2.6	2.3		MIT
5065.201	5066.613	19737.05	FE	1.2			MIT
5065.016	5066.428	19737.77	FE I	1.4H			MIT
5060.079	5061.490	19757.03	FE I	0.3			MIT
5054.647	5056.056	19778.26	FE	0.5			MIT
5051.636	5053.045	19790.05	FE I	2.3			S
5049.825	5051.233	19797.15	FE I	2.6	0.0		S
5048.454	5049.862	19802.52	FE I	1.7			MIT
5044.221	5045.628	19819.14	FE I	1.4			MIT
5041.759	5043.165	19828.82	FE I	2.5			S
5041.081	5042.487	19831.49	FE I	2.1			MIT
5039.261	5040.666	19838.65	FE I	2.0	0.3H		MIT
5030.784	5032.187	19872.08	FE II	0.0	2.1		MIT
5029.623	5031.026	19876.66	FE				ME
5028.137	5029.539	19882.54	FE	2.0			MIT
5027.212	5028.614	19886.20	FE	1.8			MIT
5027.136	5028.538	19886.50	FE I	1.8			MIT
5023.476	5024.877	19900.99	FE	1.0	2.5		RL
5022.250	5023.651	19905.84	FE I	2.2			MIT
5018.440	5019.840	19920.96	FE II	1.9	1.7		MIT
5014.959	5016.358	19934.78	FE I	2.7			MIT
5012.071	5013.469	19946.27	FE I	2.5			S
5011.31	5012.71	19949.3	FE	0.7			MIT
5010.852	5012.250	19951.12	FE	1.7			MIT
5010.029	5011.426	19954.40	FE	0.8			MIT
5007.288	5008.685	19965.32	FE I	1.4H			MIT
5006.133	5007.529	19969.93	FE I	2.5	0.7		MIT
5005.725	5007.121	19971.55	FE	2.3			MIT
5004.99	5006.39	19974.5	FE	0.8			MIT
5004.794	5006.190	19975.27	FE	0.5	2.0		MIT
5002.809	5004.205	19983.20	FE I	1.3			MIT
5001.871	5003.266	19986.94	FE I	2.5	1.6		S
4999.117	5000.512	19997.95	FE	0.6			MIT
4997.801	4999.195	20003.22	FE	1.3	2.5		MIT
4995.627	4997.021	20011.93	FE	0.5	1.8		MIT
4994.133	4995.526	20017.91	FE I	2.3			S
4991.277	4992.669	20029.36	FE I	1.9			MIT
4990.462	4991.854	20032.64	FE	0.7			MIT
4988.963	4990.355	20038.66	FE I	2.0H			MIT
4985.559	4986.950	20052.34	FE I	2.0			MIT
4985.260	4986.651	20053.54	FE I	2.0			MIT
4983.855	4985.245	20059.19	FE I	2.3H			MIT
4983.258	4984.648	20061.60	FE I	2.0H			MIT
4982.507	4983.897	20064.62	FE	2.3			MIT
4978.606	4979.995	20080.34	FE I	1.9			MIT
4973.108	4974.496	20102.54	FE I	2.0			MIT
4970.496	4971.883	20113.10	FE	1.3			MIT
4969.927	4971.314	20115.41	FE	1.7			MIT
4968.709	4970.095	20120.34	FE	0.5			MIT
4966.096	4967.482	20130.92	FE I	2.5	0.0		S
4962.564	4963.949	20145.25	FE	1.0H			MIT
4961.91	4963.29	20147.9	FE	0.3			MIT
4957.609	4958.993	20165.39	FE I	2.5	2.2		MIT
4957.307	4958.690	20166.61	FE I	2.0	1.3		MIT
4952.646	4954.028	20185.59	FE I	0.7H			MIT
4950.105	4951.487	20195.96	FE I		1.0		MIT
4946.639	4948.020	20210.11	FE	0.0	1.7		MIT
4946.400	4947.781	20211.08	FE I	0.7	1.6		MIT
4939.690	4941.069	20238.54	FE I	2.2	0.3		S
4939.244	4940.623	20240.36	FE I	1.0H			MIT
4938.817	4940.196	20242.11	FE I	2.5	0.0		MIT
4938.183	4939.561	20244.71	FE I	2.0			MIT
4934.023	4935.400	20261.78	FE I	1.6			MIT
4933.627	4935.004	20263.41	FE	0.3	1.8		MIT
4933.348	4934.725	20264.55	FE	1.7	1.5		MIT
4930.331	4931.707	20276.95	FE I	1.4			MIT
4927.875	4929.251	20287.06	FE	1.3			MIT
4927.447	4928.823	20288.82	FE	1.7	0.8		MIT
4924.776	4926.151	20299.82	FE I	2.0			S
4923.916	4925.291	20303.37	FE II	1.5	1.7		MIT
4923.720	4925.095	20304.18	FE	0.7	2.0		MIT
4920.505	4921.879	20317.45	FE I	2.7	2.1		MIT
4918.999	4920.372	20323.67	FE I	2.5	1.7		S
4917.250	4918.623	20330.89	FE	0.5H			MIT
4915.605	4916.977	20337.70	FE	0.3H			MIT
4911.800	4913.171	20353.45	FE I	1.0			MIT
4910.570	4911.941	20358.55	FE I	1.2			MIT
4910.328	4911.699	20359.55	FE I	1.2			MIT
4910.026	4911.397	20360.81	FE I	2.0			MIT
4909.390	4910.761	20363.44	FE I	1.7			MIT
4907.745	4909.115	20370.27	FE I	1.4			MIT
4905.182	4906.552	20380.91	FE I	1.0			MIT
4903.317	4904.686	20388.67	FE I	2.7	0.3		S
4891.498	4892.864	20437.93	FE I	1.8	1.2		MIT
4890.769	4892.135	20440.97	FE I	2.0	1.2		MIT
4888.651	4890.016	20449.83	FE I	0.3	0.0		MIT
4886.346	4887.711	20459.48	FE I	0.7			MIT
4885.436	4886.800	20463.29	FE I	0.3			MIT
4882.168	4883.531	20476.99	FE				ME
4881.724	4883.087	20478.85	FE I	0.5			MIT
4878.218	4879.580	20493.56	FE I	1.9	0.6		S
4872.146	4873.507	20519.10	FE I	2.0	1.5		MIT
4871.325	4872.686	20522.56	FE I	2.3	2.0		MIT
4863.655	4865.014	20554.93	FE I	0.3			MIT
4859.748	4861.106	20571.45	FE I	2.2	1.6		S
4855.686	4857.042	20588.66	FE I	0.9	0.0		MIT
4845.656	4847.010	20631.28	FE I	0.5			MIT

AIR WAVELENGTH (ANGSTRMS)	VACUUM WAVELENGTH (ANGSTRMS)	WAVENUMBER (KAYSERS)	LMENT	LOG ARC	INTENSITY SPK	DIS	REF
4844.016	4845.369	20638.26	FE	0.3			MIT
4843.150	4844.503	20641.95	FE I	0.6			MIT
4839.545	4840.897	20657.33	FE I	0.7	0.0		MIT
4835.867	4837.218	20673.04	FE I	0.3			MIT
4832.730	4834.080	20686.46	FE I	0.7			MIT
4821.047	4822.394	20736.59	FE	2.3H	2.3H		MIT
4802.886	4804.228	20815.00	FE	0.3			MIT
4800.656	4801.998	20824.67	FE	1.2			MIT
4791.248	4792.587	20865.56	FE I	2.3	2.3G		MIT
4789.654	4790.993	20872.50	FE I	2.0			S
4788.753	4790.092	20876.43	FE I	1.6			MIT
4786.810	4788.148	20884.90	FE I	2.2			S
4772.817	4774.151	20946.13	FE I	1.0	0.6		S
4768.397	4769.730	20965.55	FE I	0.5			MIT
4768.340	4769.673	20965.80	FE	0.8			MIT
4757.582	4758.912	21013.20	FE I	0.5			MIT
4747.480	4748.808	21057.92	FE	1.5	1.4H		MIT
4745.806	4747.133	21065.34	FE I	0.9	0.0		S
4741.533	4742.859	21084.33	FE I	1.1	0.0		S
4736.780	4738.105	21105.48	FE I	2.1	1.7		MIT
4735.848	4737.173	21109.64	FE	1.0	0.0		MIT
4733.596	4734.920	21119.68	FE I	1.2	0.0		S
4731.492	4732.816	21129.07	FE II	0.7	0.0		MIT
4729.699	4731.022	21137.08	FE I	1.4	1.4		MIT
4728.560	4729.883	21142.17	FE I	1.3	0.0		MIT
4727.407	4728.729	21147.33	FE I	1.0			MIT
4714.074	4715.393	21207.14	FE	1.7	1.7		MIT
4710.286	4711.604	21224.20	FE I	1.3	0.3		S
4709.098	4710.416	21229.55	FE I	1.3	0.3		MIT
4708.960	4710.278	21230.17	FE	1.7	1.7		MIT
4707.489	4708.806	21236.80	FE I	0.5			MIT
4707.281	4708.598	21237.74	FE I	2.0	1.1		S
4705.464	4706.781	21245.94	FE	0.3			MIT
4704.963	4706.280	21248.21	FE	1.0	0.0		MIT
4701.052	4702.367	21265.88	FE I	0.3			MIT
4691.414	4692.727	21309.57	FE I	1.9	1.0		S
4690.144	4691.457	21315.34	FE I	0.8			MIT
4683.566	4684.877	21345.28	FE I	0.8			MIT
4680.305	4681.615	21360.15	FE I	1.0			MIT
4678.852	4680.162	21366.78	FE I	2.2	2.0		S
4673.172	4674.480	21392.75	FE I	1.3	0.3		MIT
4669.184	4670.491	21411.02	FE I	1.2	0.3		MIT
4668.140	4669.447	21415.81	FE I	2.1	1.0		MIT
4667.459	4668.766	21418.94	FE I	2.2	1.3		S
4661.976	4663.281	21444.13	FE I	1.0			MIT
4661.537	4662.842	21446.15	FE	0.3H			MIT
4654.624	4655.927	21478.00	FE I	1.0	0.3		MIT
4654.503	4655.806	21478.56	FE I	1.3	0.5		MIT
4647.437	4648.738	21511.21	FE I	2.1	1.6		S
4643.475	4644.775	21529.57	FE I	1.5	0.3		MIT
4638.018	4639.317	21554.90	FE I	1.9	1.0		MIT
4637.518	4638.817	21557.22	FE I	2.0	1.0		MIT
4635.852	4637.150	21564.97	FE I	1.1	0.0		MIT
4635.328	4636.626	21567.41	FE II	0.3			MIT
4632.918	4634.215	21578.63	FE I	1.8	0.6		MIT
4630.128	4631.425	21591.63	FE I	1.0	0.3		MIT
4629.328	4630.624	21595.36	FE II	0.8	0.9		MIT
4625.055	4626.350	21615.31	FE I	2.0	1.1		MIT
4619.298	4620.592	21642.25	FE I	2.0	0.9		MIT
4618.765	4620.059	21644.75	FE I	1.0	0.0		MIT
4613.219	4614.511	21670.77	FE I	1.5	0.3		MIT
4611.289	4612.581	21679.84	FE I	2.3	1.4		MIT
4607.654	4608.945	21696.94	FE I	1.7	0.7		MIT
4602.944	4604.234	21719.14	FE I	2.5	2.0		S
4602.010	4603.299	21723.55	FE I	1.3	0.3		MIT
4598.135	4599.423	21741.86	FE I	1.7	0.6		MIT
4596.062	4597.350	21751.66	FE I	1.0	0.3		MIT
4595.365	4596.653	21754.96	FE	1.2	0.3		MIT
4592.655	4593.942	21767.80	FE I	2.3	1.7		S
4587.136	4588.421	21793.99	FE	1.1	0.3		MIT

AIR WAVELENGTH (ANGSTRMS)	VACUUM WAVELENGTH (ANGSTRMS)	WAVENUMBER (KAYSERS)	LMENT	LOG ARC	INTENSITY SPK	DIS	REF
4584.824	4586.109	21804.98	FE I	0.9	0.0		MIT
4583.848	4585.132	21809.62	FE II	2.2	2.2		MIT
4582.835	4584.119	21814.44	FE II	0.0	0.0		MIT
4581.522	4582.806	21820.69	FE I	1.8	0.3		MIT
4580.600	4581.884	21825.08	FE I	0.8			MIT
4576.331	4577.613	21845.44	FE II	0.5	0.3		MIT
4574.722	4576.004	21853.13	FE I	1.1	0.0		MIT
4574.503	4575.785	21854.17	FE	0.8	0.5		MIT
4568.777	4570.057	21881.56	FE I	1.0	0.0		MIT
4566.520	4567.800	21892.38	FE	0.7	0.0		BB
4565.667	4566.947	21896.47	FE I	0.9	0.0		MIT
4565.329	4566.609	21898.09	FE	0.8			MIT
4560.096	4561.374	21923.22	FE I	1.3			MIT
4556.125	4557.402	21942.32	FE I	2.2	1.5		MIT
4555.895	4557.172	21943.43	FE II	1.1	1.1		MIT
4552.549	4553.825	21959.56	FE	1.0	0.0		MIT
4550.795	4552.071	21968.02	FE	1.7			MIT
4549.470	4550.745	21974.42	FE II	2.0	2.0		MIT
4547.851	4549.126	21982.25	FE	2.3	2.0		S
4547.026	4548.301	21986.23	FE I	0.8	0.0		MIT
4542.424	4543.698	22008.51	FE	0.5	0.0		MIT
4541.523	4542.796	22012.87	FE II	0.3	0.3		MIT
4534.166	4535.437	22048.59	FE II	0.5	0.0		DO
4531.652	4532.923	22060.82	FE I	0.9	0.0		MIT
4531.152	4532.423	22063.26	FE I	2.1			S
4529.676	4530.946	22070.45	FE	1.0	0.3		MIT
4529.563	4530.833	22071.00	FE I	0.8			MIT
4528.619	4529.889	22075.60	FE I	2.8	2.3		S
4526.419	4527.688	22086.33	FE	1.0	0.0		MIT
4525.146	4526.415	22092.54	FE I	2.0	1.7		MIT
4522.634	4523.902	22104.81	FE II	1.8	1.7		MIT
4520.236	4521.504	22116.54	FE II	1.6	1.5		MIT
4517.530	4518.797	22129.78	FE I	1.5	0.5		S
4515.345	4516.611	22140.49	FE II	1.0	1.0		MIT
4514.191	4515.457	22146.15	FE				ME
4508.285	4509.550	22175.16	FE II	1.6	1.5		MIT
4504.849	4506.113	22192.08	FE I	0.8			MIT
4502.598	4503.861	22203.17	FE	0.5			MIT
4495.966	4497.227	22235.92	FE I	0.8			MIT
4494.568	4495.829	22242.84	FE I	2.6	2.2		S
4491.406	4492.666	22258.50	FE II	0.3	0.3		MIT
4490.765	4492.025	22261.68	FE I	1.6	0.0		MIT
4490.087	4491.347	22265.04	FE I	1.6	1.0		MIT
4489.741	4491.001	22266.75	FE I	2.0	1.1		S
4489.185	4490.444	22269.51	FE II	0.3	0.3		MIT
4488.918	4490.177	22270.83	FE I	0.5	0.0		MIT
4488.132	4489.391	22274.74	FE I	0.8			MIT
4485.680	4486.939	22286.91	FE I	1.7	0.3		MIT
4484.225	4485.483	22294.14	FE I	2.1	1.6		MIT
4482.752	4484.010	22301.47	FE I	1.3	0.3		MIT
4482.258	4483.516	22303.93	FE I	2.2	1.8		MIT
4482.172	4483.430	22304.35	FE I	2.2	1.8		MIT
4480.144	4481.401	22314.45	FE I	1.0	0.3		MIT
4479.622	4480.879	22317.05	FE I	1.2	0.3		MIT
4476.022	4477.278	22335.00	FE I	2.7	2.5		MIT
4472.721	4473.976	22351.48	FE	1.0	0.0		MIT
4469.380	4470.634	22368.19	FE I	2.3	2.0		MIT
4466.940	4468.194	22380.41	FE	0.8			MIT
4466.554	4467.807	22382.34	FE I	2.7	2.5		S
4464.776	4466.029	22391.26	FE I	1.5	0.5		MIT
4461.654	4462.906	22406.92	FE I	2.5	2.1		S
4461.224	4462.476	22409.08	FE I	0.7	0.0		MIT
4459.121	4460.373	22419.65	FE I	2.6	2.3		S
4458.107	4459.358	22424.75	FE I	1.5	0.0		MIT
4455.258	4456.509	22439.09	FE	0.3	0.5		DO
4455.035	4456.285	22440.21	FE I	1.3	0.0		MIT
4454.383	4455.633	22443.50	FE I	2.3	1.9		S
4451.545	4452.795	22457.81	FE	0.3	0.6		DO
4450.311	4451.560	22464.03	FE I	1.1	0.3		MIT
4449.214	4450.463	22469.57	FE II	0.3	0.6		DO

AIR WAVELENGTH (ANGSTRMS)	VACUUM WAVELENGTH (ANGSTRMS)	WAVENUMBER (KAYSERS)	LMENT	LOG ARC	INTENSITY SPK	DIS	REF
4447.722	4448.971	22477.11	FE I	2.3	2.0		S
4446.842	4448.090	22481.56	FE I	1.0	1.0		MIT
4443.197	4444.444	22500.00	FE I	2.3	2.0		S
4442.343	4443.590	22504.33	FE I	2.6	2.3		S
4439.885	4441.131	22516.78	FE I	0.8	0.0		MIT
4438.353	4439.599	22524.56	FE I	1.0	0.0		MIT
4436.930	4438.176	22531.78	FE I	1.2	0.3		MIT
4435.152	4436.397	22540.81	FE I	1.8	0.5		MIT
4433.786	4435.031	22547.76	FE I	1.5	0.3		MIT
4433.221	4434.466	22550.63	FE I	2.2	1.3		MIT
4432.573	4433.818	22553.93	FE	0.7	0.0		MIT
4430.618	4431.862	22563.88	FE I	2.3	0.9		S
4430.207	4431.451	22565.97	FE I	1.0	0.0		MIT
4427.312	4428.555	22580.73	FE I	2.7	2.3		S
4422.570	4423.812	22604.94	FE I	2.5	2.1		S
4419.543	4420.784	22620.42	FE	0.9	0.0		MIT
4416.817	4418.057	22634.38	FE II		0.8		DO
4415.125	4416.365	22643.06	FE I	2.8	2.6		S
4410.714	4411.953	22665.70	FE	1.3			MIT
4408.419	4409.657	22677.50	FE I	2.1	1.8		S
4407.716	4408.954	22681.12	FE I	2.0	1.7		MIT
4404.752	4405.989	22696.38	FE I	3.0	2.8		S
4401.450	4402.686	22713.41	FE I	0.8	0.3		MIT
4401.300	4402.536	22714.18	FE I	1.8	1.2		MIT
4400.351	4401.587	22719.08	FE	1.3	0.0		MIT
4395.286	4396.521	22745.26	FE I	1.9			MIT
4390.954	4392.188	22767.70	FE I	2.0	1.5		S
4389.247	4390.480	22776.55	FE I	1.5	0.3		MIT
4388.411	4389.644	22780.89	FE I	2.1	1.7		MIT
4387.897	4389.130	22783.56	FE I	2.2	1.5		MIT
4385.381	4386.613	22796.63	FE II	0.6	1.0		MIT
4384.699	4385.931	22800.18	FE I	0.7	0.3		MIT
4383.547	4384.779	22806.17	FE I	3.0	2.9		S
4382.773	4384.004	22810.20	FE	1.0	1.0		MIT
4376.783	4378.013	22841.41	FE I	0.8	0.0		MIT
4375.932	4377.162	22845.85	FE I	2.7	2.3		S
4373.566	4374.795	22858.21	FE I	1.7	0.5		MIT
4369.774	4371.002	22878.05	FE I	2.3	2.0		S
4367.906	4369.133	22887.83	FE I	1.8	1.8		MIT
4367.582	4368.809	22889.53	FE I	2.0	1.7		MIT
4358.505	4359.730	22937.20	FE I	1.8	1.3		S
4357.574	4358.799	22942.10	FE	0.3	0.5		DO
4352.737	4353.960	22967.59	FE I	2.5	2.2		S
4351.762	4352.985	22972.74	FE II	1.5	1.5		MIT
4351.552	4352.775	22973.85	FE I	1.5	0.7		MIT
4348.939	4350.161	22987.65	FE I	0.9	0.3		MIT
4347.851	4349.073	22993.41	FE I	0.7	0.3		MIT
4346.556	4347.778	23000.25	FE	1.7	1.0		MIT
4343.705	4344.926	23015.35	FE	1.1	0.3		MIT
4343.257	4344.478	23017.73	FE	1.3	0.5		MIT
4337.049	4338.268	23050.67	FE I	2.6	2.2		S
4330.959	4332.177	23083.08	FE	0.7			MIT
4330.155	4331.373	23087.37	FE	1.0	0.3		MIT
4327.098	4328.315	23103.68	FE	2.0	1.7		MIT
4326.760	4327.977	23105.49	FE I	1.0	0.6		MIT
4325.765	4326.981	23110.80	FE I	3.0	2.8		S
4321.800	4323.015	23132.00	FE	1.3	0.6		MIT
4320.84	4322.06	23137.1	FE	0.5H	0.0H		MIT
4319.717	4320.932	23143.16	FE II	0.0	0.0		DO
4315.087	4316.301	23167.99	FE I	2.7	2.5		S
4314.289	4315.502	23172.27	FE II	0.5	0.5		DO
4309.380	4310.592	23198.67	FE I	2.1	1.8		MIT
4309.039	4310.251	23200.50	FE	1.6	1.3		MIT
4307.906	4309.118	23206.61	FE I	3.0G	2.9G		S
4305.455	4306.666	23219.82	FE I	2.0	1.7		S
4303.587	4304.798	23229.90	FE	1.4			ME
4303.168	4304.378	23232.16	FE II	1.1	1.2		MIT
4302.192	4303.402	23237.43	FE I	1.7	1.0		MIT
4299.241	4300.450	23253.38	FE I	2.7	2.6		MIT
4298.040	4299.249	23259.88	FE I	2.0	2.6		S

AIR WAVELENGTH (ANGSTRMS)	VACUUM WAVELENGTH (ANGSTRMS)	WAVENUMBER (KAYSERS)	LMENT	LOG ARC	INTENSITY SPK	DIS	REF
4296.585	4297.794	23267.75	FE II	0.3	0.3		MIT
4294.128	4295.336	23281.06	FE I	2.8	2.6		S
4292.290	4293.498	23291.03	FE I	1.2			MIT
4291.466	4292.673	23295.51	FE I	2.1	1.3		MIT
4290.868	4292.075	23298.75	FE I	1.3	0.3		MIT
4290.381	4291.588	23301.40	FE	1.5	0.7		MIT
4288.965	4290.172	23309.09	FE I	0.7	0.0		MIT
4288.154	4289.360	23313.50	FE I	1.7	0.8		MIT
4286.992	4288.198	23319.82	FE I	1.0	0.0		MIT
4286.889	4288.095	23320.38	FE	0.3	0.0		MIT
4286.437	4287.643	23322.84	FE I	0.5	0.0		MIT
4285.445	4286.651	23328.24	FE	2.1	1.7		S
4282.406	4283.611	23344.79	FE I	2.8	2.5		S
4279.480	4280.684	23360.75	FE I	0.7	0.0		MIT
4276.678	4277.881	23376.06	FE I	1.0	0.0		MIT
4273.87	4275.07	23391.4	FE I	1.0	0.3		MIT
4273.317	4274.520	23394.44	FE II	0.5H	0.3H		DO
4271.764	4272.966	23402.95	FE I	3.0	2.8		S
4271.161	4272.363	23406.25	FE I	2.6	2.5		MIT
4268.758	4269.959	23419.43	FE	1.5	1.0		MIT
4267.830	4269.031	23424.52	FE	2.1	1.8		S
4266.968	4268.169	23429.25	FE I	1.8	1.5		MIT
4264.74	4265.94	23441.5	FE I	1.1	0.3		MIT
4264.215	4265.415	23444.38	FE I	1.5	0.6		MIT
4260.479	4261.678	23464.93	FE I	2.6	2.5		S
4260.134	4261.333	23466.83	FE I	0.8	0.5		MIT
4260.003	4261.202	23467.56	FE I	1.2	0.7		MIT
4258.958	4260.157	23473.31	FE I	0.9	0.3		MIT
4258.614	4259.813	23475.21	FE I	1.1	0.3		MIT
4258.324	4259.523	23476.81	FE I	1.8	0.6		MIT
4258.155	4259.354	23477.74	FE II	0.3	0.3		DO
4256.223	4257.421	23488.40	FE I	0.6	0.0		MIT
4255.852	4257.050	23490.45	FE	0.3	0.0		MIT
4255.499	4256.697	23492.39	FE	0.7	0.3		MIT
4254.938	4256.136	23495.49	FE	0.3	0.0		MIT
4253.93	4255.13	23501.1	FE	0.3	0.0		MIT
4250.790	4251.987	23518.42	FE I	2.6	2.4		S
4250.130	4251.327	23522.07	FE I	2.4	2.2		MIT
4248.226	4249.422	23532.61	FE I	2.2	1.6		MIT
4247.433	4248.629	23537.01	FE I	2.3	2.0		MIT
4246.090	4247.285	23544.45	FE	1.9	1.5		MIT
4245.260	4246.455	23549.05	FE I	1.9	1.6		MIT
4243.368	4244.563	23559.55	FE	1.0	0.5		MIT
4242.588	4243.783	23563.88	FE I	0.5			MIT
4241.112	4242.306	23572.08	FE I	0.0	0.6		MIT
4240.370	4241.564	23576.21	FE	1.5	0.7		MIT
4239.849	4241.043	23579.11	FE I	1.6	1.2		MIT
4239.733	4240.927	23579.75	FE	1.5	1.0		MIT
4238.821	4240.015	23584.82	FE I	2.3	2.0		ME
4238.039	4239.232	23589.18	FE I	1.9	1.2		MIT
4237.167	4238.360	23594.03	FE	0.7	0.3		MIT
4235.942	4237.135	23600.85	FE I	2.5	2.3		MIT
4233.612	4234.804	23613.84	FE I	2.4	2.2		MIT
4233.168	4234.360	23616.32	FE II	2.0	2.0		MIT
4232.732	4233.924	23618.75	FE I	1.0	0.0		MIT
4231.703	4232.895	23624.50	FE	0.5	0.0		MIT
4229.761	4230.952	23635.34	FE I	1.3	0.3		MIT
4227.432	4228.623	23648.36	FE I	2.5	2.4		MIT
4226.430	4227.620	23653.97	FE I	1.9	1.4		MIT
4225.957	4227.147	23656.62	FE I	1.9	1.5		MIT
4225.465	4226.655	23659.37	FE I	1.9	1.3		MIT
4224.517	4225.707	23664.68	FE I	1.8	1.2		MIT
4224.176	4225.366	23666.59	FE I	2.3	1.9		MIT
4222.221	4223.410	23677.55	FE I	2.3	2.3		MIT
4220.348	4221.537	23688.06	FE	1.9	1.6		MIT
4219.364	4220.552	23693.58	FE I	2.4	2.3		S
4217.555	4218.743	23703.74	FE I	2.3	2.0		MIT
4216.186	4217.374	23711.44	FE I	2.3	2.0		S
4215.970	4217.158	23712.66	FE I	0.3	0.0		MIT
4215.425	4216.612	23715.72	FE I	1.8	1.2		MIT

AIR WAVELENGTH (ANGSTRMS)	VACUUM WAVELENGTH (ANGSTRMS)	WAVENUMBER (KAYSERS)	LMENT	LOG ARC	INTENSITY SPK	DIS	REF
4213.650	4214.837	23725.71	FE	2.0	1.8		S
4210.352	4211.538	23744.29	FE I	2.5	2.3		MIT
4208.615	4209.801	23754.09	FE	2.0	1.7		MIT
4207.131	4208.316	23762.47	FE I	1.9	1.6		MIT
4206.702	4207.887	23764.90	FE I	2.1	1.4		MIT
4205.546	4206.731	23771.43	FE I	1.7	0.8		MIT
4203.987	4205.171	23780.24	FE I	2.3	2.1		S
4203.572	4204.756	23782.59	FE I	1.0	0.0		MIT
4202.758	4203.942	23787.20	FE	1.0	0.6		MIT
4202.031	4203.215	23791.31	FE I	2.6	2.5		S
4200.927	4202.111	23797.56	FE I	1.9	1.3		MIT
4199.099	4200.282	23807.93	FE I	2.5	2.3		MIT
4198.643	4199.826	23810.51	FE I	1.0	0.3		MIT
4198.312	4199.495	23812.39	FE I	2.4	2.2		MIT
4196.533	4197.715	23822.48	FE	0.5			MIT
4196.214	4197.396	23824.29	FE I	2.0	1.7		MIT
4195.623	4196.805	23827.65	FE I	1.4	0.5		MIT
4195.337	4196.519	23829.27	FE I	2.2	2.0		MIT
4192.51	4193.69	23845.3	FE	0.3			MIT
4191.680	4192.861	23850.06	FE I	1.3	0.8		MIT
4191.436	4192.617	23851.45	FE I	2.3	2.0		MIT
4189.564	4190.745	23862.11	FE	0.5			MIT
4187.801	4188.981	23872.15	FE I	2.3	2.2		MIT
4187.589	4188.769	23873.36	FE	0.5	0.0		MIT
4187.044	4188.224	23876.47	FE I	2.4	2.3		MIT
4184.895	4186.074	23888.73	FE I	2.0	1.9		S
4182.386	4183.565	23903.06	FE I	1.9	1.5		MIT
4181.757	4182.936	23906.66	FE I	2.3	2.2		MIT
4178.868	4180.046	23923.18	FE II	1.0	1.0		RL
4178.051	4179.229	23927.86	FE	0.5	0.0		MIT
4177.596	4178.773	23930.47	FE I	2.0	1.4		MIT
4176.572	4177.749	23936.33	FE I	2.0	1.7		MIT
4175.640	4176.817	23941.68	FE I	2.0	1.9		S
4174.917	4176.094	23945.82	FE I	2.0	1.4		MIT
4173.926	4175.102	23951.51	FE I	1.7	0.7		MIT
4173.475	4174.651	23954.10	FE II	0.9	0.9		RL
4173.323	4174.499	23954.97	FE I	1.4	0.7		MIT
4172.750	4173.926	23958.26	FE I	1.8	1.0		MIT
4172.652	4173.828	23958.82	FE I	0.5	0.0		MIT
4172.127	4173.303	23961.83	FE	1.9	1.7		MIT
4171.700	4172.876	23964.29	FE	0.9	0.3		MIT
4170.906	4172.082	23968.85	FE I	1.9	1.6		S
4169.773	4170.948	23975.36	FE I	0.5			MIT
4168.946	4170.121	23980.12	FE	1.0	0.0		MIT
4167.960	4169.135	23985.79	FE	1.0	0.3		MIT
4167.862	4169.037	23986.35	FE	0.9	0.3		MIT
4165.420	4166.594	24000.42	FE	1.1	0.3		MIT
4163.676	4164.850	24010.47	FE I	1.1	0.0		MIT
4161.492	4162.665	24023.07	FE	1.2			MIT
4161.080	4162.253	24025.45	FE I	1.0			MIT
4158.798	4159.971	24038.63	FE I	2.0	1.4		MIT
4157.791	4158.963	24044.45	FE I	2.2	1.9		MIT
4156.803	4157.975	24050.17	FE I	2.0	1.9		S
4156.681	4157.853	24050.87	FE I	0.3			MIT
4154.812	4155.983	24061.69	FE I	2.0	0.9		MIT
4154.503	4155.674	24063.48	FE I	2.0	1.9		MIT
4154.108	4155.279	24065.77	FE	0.5			MIT
4153.910	4155.081	24066.92	FE I	2.1	2.0		MIT
4153.405	4154.576	24069.84	FE	0.0	0.0H		MIT
4152.170	4153.341	24077.00	FE I	1.8	0.7		MIT
4151.955	4153.126	24078.25	FE	0.6	0.0		MIT
4150.264	4151.434	24088.06	FE I	1.7	0.3		MIT
4149.370	4150.540	24093.25	FE I	2.0	1.5		MIT
4147.673	4148.843	24103.11	FE I	2.3	2.0		S
4146.070	4147.239	24112.43	FE I	1.2	0.5		MIT
4143.871	4145.040	24125.22	FE I	2.6	2.4		S
4143.420	4144.588	24127.85	FE I	2.3	2.0		MIT
4141.862	4143.030	24136.92	FE I	1.2	0.7		BU
4139.923	4141.091	24148.23	FE I	1.6	1.5		MIT
4137.004	4138.171	24165.27	FE	2.0	1.9		MIT

AIR WAVELENGTH (ANGSTRMS)	VACUUM WAVELENGTH (ANGSTRMS)	WAVENUMBER (KAYSERS)	LMENT	LOG ARC	INTENSITY SPK	DIS	REF
4134.681	4135.847	24178.84	FE I	2.2	2.0		S
4134.425	4135.591	24180.34	FE I	1.0	0.5		MIT
4134.340	4135.506	24180.84	FE I	0.3			MIT
4133.862	4135.028	24183.63	FE I	1.7	0.8		MIT
4132.903	4134.069	24189.25	FE I	0.9			ME
4132.060	4133.226	24194.18	FE I	2.5	2.3		S
4130.04	4131.20	24206.0	FE I	1.3	0.5		MIT
4129.22	4130.38	24210.8	FE	0.7	0.0		MIT
4128.735	4129.900	24213.66	FE II	0.3	0.3H		MIT
4127.803	4128.967	24219.13	FE	1.4	1.2		MIT
4127.612	4128.776	24220.25	FE I	2.0	1.9		S
4126.88	4128.04	24224.6	FE I	0.9	0.0		MIT
4126.190	4127.354	24228.60	FE I	1.9	1.8		MIT
4125.883	4127.047	24230.40	FE I	1.4	1.2		MIT
4125.621	4126.785	24231.94	FE	1.9	1.5		MIT
4123.745	4124.908	24242.96	FE	1.9	1.3		MIT
4122.510	4123.673	24250.23	FE I	1.8	1.5		MIT
4121.806	4122.969	24254.37	FE I	2.0	1.6		S
4120.211	4121.373	24263.76	FE	1.9	1.5		MIT
4119.394	4120.556	24268.57	FE	0.3	0.0		MIT
4118.903	4120.065	24271.46	FE I	0.3	0.0		MIT
4118.549	4119.711	24273.55	FE	2.3	2.0		S
4117.863	4119.025	24277.59	FE I	0.8	0.0		MIT
4114.957	4116.118	24294.74	FE	1.0	0.3		MIT
4114.449	4115.610	24297.74	FE I	1.9	1.7		S
4112.966	4114.127	24306.50	FE	1.8	1.0		MIT
4112.347	4113.507	24310.16	FE I	0.8	0.0		MIT
4109.808	4110.968	24325.17	FE I	2.1	2.0		MIT
4109.072	4110.231	24329.53	FE I	1.1	0.3		MIT
4107.492	4108.651	24338.89	FE I	2.1	2.0		S
4106.439	4107.598	24345.13	FE I	1.0	0.3		MIT
4106.265	4107.424	24346.16	FE I	0.8	0.0		MIT
4104.95	4106.11	24354.0	FE	0.6H	0.6H		MIT
4104.128	4105.286	24358.84	FE I	2.0	1.4		MIT
4101.679	4102.837	24373.38	FE I	0.7	0.3		MIT
4101.274	4102.431	24375.79	FE I	1.6	1.0		MIT
4100.743	4101.900	24378.94	FE I	1.9	1.5		MIT
4100.166	4101.323	24382.37	FE	1.0	0.0		MIT
4098.187	4099.344	24394.15	FE I	2.0	1.6		MIT
4097.112	4098.268	24400.55	FE I	0.6	0.0		MIT
4096.97	4098.13	24401.4	FE	0.3			MIT
4095.976	4097.132	24407.32	FE I	1.9	1.6		MIT
4092.287	4093.442	24429.32	FE	0.8	0.0		MIT
4091.562	4092.717	24433.65	FE I	0.9	0.5		MIT
4090.077	4091.232	24442.52	FE I	0.3			MIT
4089.222	4090.376	24447.63	FE I	1.0	0.3		MIT
4088.577	4089.731	24451.49	FE	0.8	0.3		MIT
4087.103	4088.257	24460.30	FE I	1.7	0.7		MIT
4086.000	4087.153	24466.91	FE	0.3			MIT
4085.324	4086.477	24470.96	FE I	2.0	1.8		MIT
4085.008	4086.161	24472.85	FE I	1.9	1.5		MIT
4084.499	4085.652	24475.90	FE I	2.1	1.9		MIT
4083.780	4084.933	24480.21	FE I	1.2	0.3		MIT
4083.554	4084.707	24481.56	FE I	1.0	0.3		MIT
4082.117	4083.269	24490.18	FE I	1.0	0.3		MIT
4080.887	4082.039	24497.56	FE	0.7	0.0		MIT
4080.221	4081.373	24501.56	FE I	1.8	1.0		MIT
4079.845	4080.997	24503.82	FE I	1.9	1.6		MIT
4078.358	4079.509	24512.75	FE I	1.9	1.6		MIT
4076.803	4077.954	24522.10	FE	0.9	0.0		MIT
4076.637	4077.788	24523.10	FE I	1.9	1.7		MIT
4076.502	4077.653	24523.91	FE I	0.3	0.0		MIT
4076.222	4077.373	24525.60	FE I	0.3	0.0		MIT
4075.937	4077.088	24527.31	FE	0.7	0.7		MIT
4074.791	4075.942	24534.21	FE I	1.9	1.6		MIT
4073.775	4074.925	24540.33	FE I	1.9	1.3		MIT
4073.450	4074.600	24542.28	FE II	0.9	0.9		DO
4072.509	4073.659	24547.96	FE I	0.3	0.0		MIT
4071.740	4072.890	24552.59	FE I	2.5	2.3		S
4070.779	4071.928	24558.39	FE I	1.7	1.3		MIT

AIR WAVELENGTH (ANGSTRMS)	VACUUM WAVELENGTH (ANGSTRMS)	WAVENUMBER (KAYSERS)	LMENT	LOG ARC	INTENSITY SPK	DIS	REF
4070.277	4071.426	24561.42	FE	0.3			MIT
4069.883	4071.032	24563.79	FE II	0.0	0.0		DO
4067.982	4069.131	24575.27	FE I	2.2	2.0		MIT
4067.275	4068.424	24579.54	FE I	1.9	1.8		S
4066.979	4068.127	24581.33	FE I	2.0	1.9		S
4066.590	4067.738	24583.69	FE	1.6	1.3		MIT
4065.392	4066.540	24590.93	FE I	1.2	0.8		MIT
4064.450	4065.598	24596.63	FE I	0.3	0.0		MIT
4063.597	4064.745	24601.79	FE I	2.6	2.5		S
4063.284	4064.432	24603.69	FE I	1.0	1.0		MIT
4062.444	4063.591	24608.77	FE I	2.1	2.0		MIT
4061.955	4063.102	24611.74	FE	0.5	0.0		MIT
4061.110	4062.257	24616.86	FE	0.5	0.0		MIT
4059.721	4060.868	24625.28	FE	1.2	0.9		MIT
4058.760	4059.906	24631.11	FE I	1.6	1.0		MIT
4058.229	4059.375	24634.33	FE	1.9	1.4		MIT
4057.346	4058.492	24639.69	FE I	1.3	0.5		MIT
4055.039	4056.184	24653.71	FE I	1.6	1.0		MIT
4054.883	4056.028	24654.66	FE I	1.4	0.7		MIT
4054.833	4055.978	24654.96	FE I	1.4	0.7		MIT
4054.183	4055.328	24658.92	FE I	0.3			MIT
4053.273	4054.418	24664.45	FE	0.8	0.3		MIT
4052.664	4053.809	24668.16	FE I	0.5	0.0		MIT
4052.307	4053.452	24670.33	FE I	0.3			MIT
4051.908	4053.053	24672.76	FE	1.0	0.3		MIT
4050.688	4051.832	24680.19	FE	0.7			MIT
4049.87	4051.01	24685.2	FE	1.5	0.5		MIT
4049.331	4050.475	24688.46	FE I	1.0	0.3		MIT
4047.310	4048.453	24700.79	FE I	0.5			MIT
4045.815	4046.958	24709.92	FE I	2.6	2.5		S
4044.611	4045.754	24717.27	FE I	1.8	1.5		MIT
4043.905	4045.047	24721.59	FE I	1.4	0.8		MIT
4041.285	4042.427	24737.62	FE	0.7	0.3		MIT
4040.644	4041.786	24741.54	FE	1.3	0.8		MIT
4040.096	4041.237	24744.90	FE	0.0	0.3		MIT
4038.805	4039.946	24752.80	FE	0.3			MIT
4032.630	4033.769	24790.71	FE I	1.9	1.2		MIT
4032.469	4033.608	24791.70	FE I	0.6	0.0		MIT
4031.965	4033.104	24794.80	FE	1.9	1.7		MIT
4031.243	4032.382	24799.24	FE I	0.3			MIT
4030.492	4031.631	24803.86	FE I	2.1	1.8		MIT
4030.194	4031.333	24805.69	FE I	1.3	0.6		MIT
4029.639	4030.778	24809.11	FE I	1.9	1.4		MIT
4028.777	4029.915	24814.42	FE	0.3			MIT
4024.739	4025.876	24839.31	FE I	2.1	1.5		MIT
4024.104	4025.241	24843.23	FE I	0.9	0.3		MIT
4022.744	4023.881	24851.63	FE I	0.5			MIT
4021.870	4023.007	24857.03	FE I	2.3	2.0		MIT
4021.618	4022.755	24858.59	FE I	0.3			MIT
4020.487	4021.623	24865.58	FE	0.3			MIT
4018.275	4019.411	24879.27	FE I	1.7	0.8		MIT
4017.153	4018.288	24886.22	FE I	1.9	1.7		MIT
4017.093	4018.228	24886.59	FE I	0.7			MIT
4016.429	4017.564	24890.70	FE I	1.2	0.6		MIT
4014.534	4015.669	24902.45	FE	2.3	2.0		S
4014.271	4015.406	24904.08	FE I	0.5	0.0		MIT
4013.824	4014.959	24906.86	FE	2.3			MIT
4013.795	4014.930	24907.04	FE I	1.9	1.6		MIT
4013.647	4014.782	24907.96	FE I	0.9	0.0		MIT
4011.728	4012.862	24919.87	FE I	0.3			MIT
4011.412	4012.546	24921.83	FE I	0.7	0.0		MIT
4010.947	4012.081	24924.72	FE I	0.3	0.0		MIT
4009.717	4010.851	24932.37	FE I	2.1	2.0		MIT
4008.873	4010.006	24937.62	FE	0.7	0.0		MIT
4007.273	4008.406	24947.57	FE I	1.9	1.7		MIT
4006.761	4007.894	24950.76	FE I	0.8	0.3		MIT
4006.631	4007.764	24951.57	FE	1.3	1.2		MIT
4006.315	4007.448	24953.54	FE	1.8	1.5		MIT
4005.246	4006.378	24960.20	FE I	2.4	2.3		S
4004.843	4005.975	24962.71	FE	1.0	0.8		MIT
4003.767	4004.899	24969.42	FE	1.5	1.9		MIT
4002.663	4003.795	24976.31	FE I	0.3			MIT
4001.667	4002.798	24982.52	FE I	1.9	1.7		MIT
4000.452	4001.583	24990.11	FE	1.5	1.0		MIT
4000.266	4001.397	24991.27	FE I	0.9	0.0		MIT
3998.055	3999.185	25005.09	FE I	2.2	2.0		MIT
3997.396	3998.526	25009.21	FE I	2.5	2.2		MIT
3996.972	3998.102	25011.87	FE	1.6	1.3		MIT
3995.989	3997.119	25018.02	FE I	1.8	1.3		MIT
3995.200	3996.330	25022.96	FE	1.0			MIT
3994.121	3995.250	25029.72	FE I	1.4	1.0		MIT
3993.106	3994.235	25036.08	FE	0.3			MIT
3990.379	3991.507	25053.19	FE	1.8	1.4		MIT
3989.860	3990.988	25056.45	FE	1.5	0.7		MIT
3989.010	3990.138	25061.79	FE	1.2J	0.0J		MIT
3986.172	3987.299	25079.63	FE I	2.1	0.9		MIT
3985.388	3986.515	25084.56	FE	2.1	1.6		MIT
3984.935	3986.062	25087.42	FE	0.3			MIT
3983.962	3985.089	25093.54	FE I	2.3	2.1		MIT
3981.774	3982.900	25107.33	FE I	2.2	2.0		MIT
3981.104	3982.230	25111.56	FE I	0.3			MIT
3979.641	3980.767	25120.79	FE	0.8			MIT
3978.464	3979.589	25128.22	FE I	0.8	0.0		MIT
3977.744	3978.869	25132.77	FE I	2.5	2.2		MIT
3976.864	3977.989	25138.33	FE I	1.5	1.0		MIT
3976.614	3977.739	25139.91	FE	0.9	1.5		MIT
3976.562	3977.687	25140.24	FE	0.3	0.0		MIT
3976.390	3977.515	25141.33	FE	0.5	0.0		MIT
3975.842	3976.967	25144.79	FE	0.9	0.0		MIT
3975.210	3976.334	25148.79	FE I	0.6			MIT
3974.764	3975.888	25151.61	FE I	0.9			MIT
3974.398	3975.522	25153.93	FE I	1.0	0.0		MIT
3973.654	3974.778	25158.64	FE	1.6	1.0		MIT
3973.264	3974.388	25161.11	FE	0.3			MIT
3971.811	3972.935	25170.31	FE I	0.3			MIT
3971.329	3972.452	25173.37	FE I	2.3	2.1		MIT
3970.394	3971.517	25179.29	FE	1.7	1.5		MIT
3970.263	3971.386	25180.12	FE	0.7	0.3		MIT
3969.633	3970.756	25184.12	FE	0.7	0.7		MIT
3969.261	3970.384	25186.48	FE I	2.8	2.6		S
3968.370	3969.493	25192.14	FE	0.3			MIT
3967.969	3969.092	25194.68	FE I	1.8	1.2		MIT
3967.423	3968.545	25198.15	FE	2.1	2.0		S
3966.629	3967.751	25203.19	FE I	1.9	1.6		MIT
3966.518	3967.640	25203.90	FE I	0.6H	0.0		MIT
3966.066	3967.188	25206.77	FE I	2.0	1.8		S
3965.513	3966.635	25210.29	FE I	1.0	0.5		MIT
3964.520	3965.642	25216.60	FE I	1.9	1.4		MIT
3963.109	3964.230	25225.58	FE I	2.1	1.7		MIT
3962.722	3963.843	25228.04	FE	0.3H			MIT
3962.353	3963.474	25230.39	FE I	0.8	0.0		MIT
3961.145	3962.266	25238.08	FE I	1.4	0.8		MIT
3960.895	3962.016	25239.68	FE	0.5	0.5H		DO
3960.285	3961.406	25243.56	FE	1.5	0.8		MIT
3958.740	3959.860	25253.42	FE	0.3H			MIT
3958.400	3959.520	25255.59	FE	0.3			MIT
3957.030	3958.150	25264.33	FE	1.7	1.2		MIT
3956.681	3957.801	25266.56	FE I	2.2	2.2		S
3956.460	3957.580	25267.97	FE	2.0	2.0		MIT
3955.960	3957.079	25271.16	FE	1.0	0.7		MIT
3955.354	3956.473	25275.03	FE I	1.4	0.7		MIT
3954.715	3955.834	25279.12	FE	0.3			MIT
3953.861	3954.980	25284.58	FE	0.9	0.3		MIT
3953.155	3954.274	25289.09	FE I	1.9	1.6		MIT
3952.702	3953.821	25291.99	FE I	0.9	0.0		MIT
3952.606	3953.725	25292.61	FE I	1.9	1.7		MIT
3951.168	3952.286	25301.81	FE	2.2	2.1		MIT
3949.957	3951.075	25309.57	FE I	2.2	2.0		MIT
3949.15	3950.27	25314.7	FE	0.6	0.0		MIT
3948.779	3949.897	25317.12	FE	2.2	2.0		S

AIR WAVELENGTH (ANGSTRMS)	VACUUM WAVELENGTH (ANGSTRMS)	WAVENUMBER (KAYSERS)	LMENT	LOG ARC	INTENSITY SPK	INTENSITY DIS	REF
3948.107	3949.224	25321.43	FE I	2.1	1.7		MIT
3947.532	3948.649	25325.12	FE I	1.8	1.3		MIT
3947.391	3948.508	25326.02	FE I	0.0			MIT
3947.005	3948.122	25328.50	FE I	1.7	1.3		MIT
3945.126	3946.243	25340.56	FE	1.5	1.0		MIT
3944.896	3946.013	25342.04	FE I	1.2	0.9		MIT
3944.752	3945.869	25342.96	FE I	0.6	0.0		MIT
3943.594	3944.710	25350.40	FE	0.3H			MIT
3943.348	3944.464	25351.99	FE I	1.6	0.9		MIT
3942.443	3943.559	25357.80	FE I	2.0	1.8		S
3941.280	3942.396	25365.29	FE I	1.8	1.0		MIT
3940.882	3941.998	25367.85	FE I	2.2	1.9		S
3940.044	3941.159	25373.25	FE	0.3			MIT
3938.969	3940.084	25380.17	FE II	0.6	0.6		DO
3938.020	3939.135	25386.28	FE	0.5			MIT
3937.332	3938.447	25390.72	FE I	1.9	1.5		MIT
3935.953	3937.067	25399.62	FE II	0.3	0.0		MIT
3935.815	3936.929	25400.51	FE I	2.0	0.9		S
3935.306	3936.420	25403.79	FE I	1.6	0.9		MIT
3934.230	3935.344	25410.74	FE	0.3			MIT
3933.605	3934.719	25414.78	FE I	2.3	2.3		MIT
3932.921	3934.034	25419.20	FE	0.9	0.6		MIT
3932.630	3933.743	25421.08	FE I	1.9	1.6		MIT
3932.27	3933.38	25423.4	FE	0.5			MIT
3931.123	3932.236	25430.82	FE I	1.5	1.2		MIT
3930.299	3931.412	25436.16	FE I	2.8	2.6		S
3929.212	3930.324	25443.19	FE	1.2	0.9		MIT
3929.124	3930.236	25443.76	FE	1.0	0.7		MIT
3928.083	3929.195	25450.50	FE I	1.2	1.2		MIT
3927.922	3929.034	25451.55	FE I	2.7	2.5		S
3926.836	3927.948	25458.59	FE	0.7	0.0		MIT
3925.947	3927.059	25464.35	FE I	1.7	1.5		MIT
3925.646	3926.758	25466.30	FE I	1.9	1.7		MIT
3925.200	3926.311	25469.20	FE I	1.2	0.5		MIT
3924.179	3925.290	25475.82	FE	0.3			MIT
3923.970	3925.081	25477.18	FE	0.3			MIT
3922.914	3924.025	25484.04	FE I	2.8	2.6		S
3921.260	3922.370	25494.79	FE	0.3			MIT
3921.180	3922.290	25495.31	FE	0.3H			MIT
3920.840	3921.950	25497.52	FE I	0.8	0.3		MIT
3920.260	3921.370	25501.29	FE I	2.7	2.5		S
3919.070	3920.180	25509.03	FE I	1.2	0.8		MIT
3918.646	3919.756	25511.79	FE I	1.8	1.6		MIT
3918.420	3919.530	25513.26	FE	1.2	1.0		MIT
3918.320	3919.430	25513.92	FE I	1.3	1.0		MIT
3917.185	3918.294	25521.31	FE I	2.2	1.8		S
3916.733	3917.842	25524.25	FE	2.0	1.9		MIT
3914.281	3915.390	25540.24	FE I	1.2	0.5		MIT
3913.635	3914.743	25544.46	FE I	2.0	1.4		MIT
3913.210	3914.318	25547.23	FE	0.3H	0.0H		MIT
3912.448	3913.556	25552.21	FE	0.3H	0.0H		MIT
3912.05	3913.16	25554.8	FE	0.7J	0.7J		MIT
3911.699	3912.807	25557.10	FE	0.0			MIT
3911.001	3912.109	25561.66	FE	0.8	0.3		MIT
3910.844	3911.952	25562.69	FE I	1.5	1.0		MIT
3909.835	3910.942	25569.28	FE I	1.6	1.1		MIT
3909.669	3910.776	25570.37	FE I	1.3	0.7		MIT
3907.937	3909.044	25581.70	FE	2.0	1.8		S
3907.675	3908.782	25583.42	FE	0.3	0.0		MIT
3907.470	3908.577	25584.76	FE	1.2	0.8		MIT
3906.751	3907.858	25589.47	FE	1.0	1.0		MIT
3906.482	3907.589	25591.23	FE I	2.5	2.3		S
3906.037	3907.143	25594.15	FE II	0.3W	0.3W		MIT
3904.63	3905.74	25603.4	FE	0.3			MIT
3903.900	3905.006	25608.16	FE I	2.0	1.9		MIT
3902.948	3904.054	25614.40	FE I	2.7	2.6		S
3900.519	3901.624	25630.35	FE I	1.8	0.5D		MIT
3899.709	3900.814	25635.68	FE I	2.7	2.5		S
3899.038	3900.143	25640.09	FE I	1.3	0.9		MIT
3899.01	3900.11	25640.3	FE	0.8			MIT

AIR WAVELENGTH (ANGSTRMS)	VACUUM WAVELENGTH (ANGSTRMS)	WAVENUMBER (KAYSERS)	LMENT	LOG ARC	INTENSITY SPK	INTENSITY DIS	REF
3898.013	3899.117	25646.83	FE I	1.9	1.7		MIT
3897.895	3898.999	25647.61	FE	2.0	1.8		MIT
3897.447	3898.551	25650.55	FE I	1.0	0.7		MIT
3895.658	3896.762	25662.33	FE I	2.6	2.5		S
3895.230	3896.334	25665.15	FE	0.3H			MIT
3895.146	3896.250	25665.71	FE	0.3			MIT
3894.012	3895.115	25673.18	FE	0.9	0.9		MIT
3893.915	3895.018	25673.82	FE I	1.0	0.6		MIT
3893.394	3894.497	25677.26	FE I	2.0	0.9		MIT
3893.318	3894.421	25677.76	FE I	0.7	0.6		MIT
3892.894	3893.997	25680.55	FE	0.7	0.3		MIT
3892.410	3893.513	25683.75	FE I	0.3			MIT
3891.929	3893.032	25686.92	FE	2.0	1.8		MIT
3890.844	3891.946	25694.08	FE	1.8	1.5		MIT
3890.390	3891.492	25697.08	FE I	0.6	0.3		MIT
3890.237	3891.339	25698.09	FE	1.2			MIT
3889.923	3891.025	25700.17	FE	0.3			MIT
3888.823	3889.925	25707.44	FE	1.6	1.2		MIT
3888.517	3889.619	25709.46	FE I	0.6W	0.5W		S
3887.051	3888.153	25719.16	FE I	0.5W	0.3W		S
3886.284	3887.385	25724.23	FE I	2.8	2.6		S
3885.514	3886.615	25729.33	FE I	2.0	1.8		MIT
3885.153	3886.254	25731.72	FE I	0.8	0.5		MIT
3884.667	3885.768	25734.94	FE I	0.0			MIT
3884.362	3885.463	25736.96	FE	1.9	1.5		MIT
3883.289	3884.390	25744.07	FE	1.8	1.6		MIT
3882.480	3883.580	25749.44	FE	0.3			MIT
3881.980	3883.080	25752.75	FE	0.3			MIT
3881.015	3882.115	25759.16	FE	0.3H	0.0H		MIT
3880.780	3881.880	25760.72	FE	0.5H	0.3H		MIT
3880.225	3881.325	25764.40	FE	0.3			MIT
3879.650	3880.750	25768.22	FE	0.3J			MIT
3879.274	3880.373	25770.72	FE	0.3O	0.3O		MIT
3878.736	3879.835	25774.29	FE	1.0	1.0		MIT
3878.676	3879.775	25774.69	FE I	1.5			MIT
3878.575	3879.674	25775.36	FE I	2.5G	2.5		S
3878.021	3879.120	25779.04	FE I	2.6	2.5		S
3877.510	3878.609	25782.44	FE	0.3			MIT
3876.671	3877.770	25788.02	FE I	0.0			MIT
3876.378	3877.477	25789.97	FE	0.3			MIT
3876.045	3877.144	25792.18	FE I	1.6	1.2		MIT
3875.383	3876.481	25796.59	FE	0.3H	0.0H		MIT
3874.692	3875.790	25801.19	FE	0.5H			MIT
3874.051	3875.149	25805.46	FE I	0.0			MIT
3873.948	3875.046	25806.15	FE	0.0			MIT
3873.763	3874.861	25807.38	FE I	2.1	1.9		S
3872.924	3874.022	25812.97	FE I	0.5	0.0		MIT
3872.504	3873.602	25815.77	FE I	2.5	2.5		S
3871.750	3872.848	25820.79	FE I	2.0	1.8		MIT
3870.810	3871.907	25827.06	FE	0.3H			MIT
3869.564	3870.661	25835.38	FE I	2.0	1.9		MIT
3869.563	3870.660	25835.39	FE	0.6			MIT
3868.620	3869.717	25841.69	FE	0.5			MIT
3868.241	3869.338	25844.22	FE I	0.5	0.0		MIT
3867.923	3869.020	25846.34	FE I	1.5	0.9		MIT
3867.219	3868.315	25851.05	FE I	2.2	2.0		S
3865.526	3866.622	25862.37	FE I	2.8	2.6		S
3864.767	3865.863	25867.45	FE	0.3H			MIT
3864.306	3865.402	25870.53	FE	0.3			MIT
3863.742	3864.837	25874.31	FE I	1.8	1.5		MIT
3863.413	3864.508	25876.51	FE II	0.0H	0.0H		MIT
3861.60	3862.69	25888.7	FE	0.5	0.3		MIT
3861.342	3862.437	25890.39	FE	1.9	1.7		MIT
3860.915	3862.010	25893.25	FE	0.0	0.3		DO
3859.913	3861.007	25899.98	FE I	3.0R	2.8		S
3859.216	3860.310	25904.65	FE I	2.0	2.0		MIT
3856.373	3857.467	25923.75	FE I	2.7	2.5		S
3855.862	3856.955	25927.19	FE	0.6	0.3		MIT
3855.328	3856.421	25930.78	FE	0.3	0.0		MIT
3853.824	3854.917	25940.90	FE	0.5			MIT

AIR WAVELENGTH (ANGSTRMS)	VACUUM WAVELENGTH (ANGSTRMS)	WAVENUMBER (KAYSERS)	LMENT	LOG ARC	INTENSITY SPK	DIS	REF
3853.463	3854.556	25943.33	FE I	0.8	0.5		MIT
3852.575	3853.668	25949.31	FE I	2.2	2.0		MIT
3851.58	3852.67	25956.0	FE	0.6			MIT
3850.978	3852.070	25960.07	FE	0.6			MIT
3850.820	3851.912	25961.13	FE I	2.3	1.9		S
3849.969	3851.061	25966.87	FE I	2.7	2.6		S
3848.298	3849.389	25978.15	FE	0.7	0.3		MIT
3846.803	3847.894	25988.24	FE I	2.1	2.0		S
3846.415	3847.506	25990.86	FE	1.7	1.9		MIT
3846.001	3847.092	25993.66	FE	1.0	0.0		MIT
3845.704	3846.795	25995.67	FE	1.0	0.8		MIT
3845.174	3846.265	25999.25	FE I	2.0	1.8		MIT
3844.283	3845.373	26005.28	FE	1.0	0.0		MIT
3843.259	3844.349	26012.21	FE	2.1	2.0		S
3842.975	3844.065	26014.13	FE I	0.5			MIT
3842.885	3843.975	26014.74	FE	0.8	0.3		MIT
3842.65	3843.74	26016.3	FE	0.0			MIT
3841.051	3842.141	26027.16	FE I	2.7	2.6		S
3840.439	3841.528	26031.31	FE I	2.6	2.5		S
3839.630	3840.719	26036.79	FE	0.5	0.6		MIT
3839.259	3840.348	26039.31	FE	2.0	1.9		S
3838.036	3839.125	26047.60	FE	0.0	0.0		DO
3837.142	3838.231	26053.67	FE I	1.4	0.8		MIT
3836.333	3837.421	26059.17	FE	2.0	1.8		MIT
3834.225	3835.313	26073.49	FE I	2.6	2.6		S
3833.310	3834.398	26079.72	FE I	2.0	1.8		MIT
3830.860	3831.947	26096.40	FE I	1.1	0.6		MIT
3830.761	3831.848	26097.07	FE I	1.0	0.5		MIT
3829.77	3830.86	26103.8	FE	0.9	0.3		MIT
3829.459	3830.546	26105.94	FE	1.2	0.9		MIT
3829.126	3830.212	26108.21	FE	0.6	0.3		MIT
3828.500	3829.586	26112.48	FE	0.3	0.0		MIT
3827.825	3828.911	26117.09	FE I	2.3	2.3		S
3827.572	3828.658	26118.81	FE I	0.6	0.3		MIT
3826.846	3827.932	26123.77	FE	0.8	0.3		MIT
3825.884	3826.970	26130.34	FE I	2.7	2.6		S
3825.406	3826.491	26133.60	FE	0.3	0.0		MIT
3824.444	3825.529	26140.17	FE I	2.2	2.0		S
3824.302	3825.387	26141.15	FE	0.8	0.8		MIT
3824.078	3825.163	26142.68	FE I	0.7	0.5		MIT
3821.838	3822.923	26158.00	FE I	1.7	1.5		MIT
3821.178	3822.262	26162.52	FE	2.0	2.0		MIT
3820.429	3821.513	26167.65	FE I	2.9	2.8		MIT
3819.50	3820.58	26174.0	FE	0.3	0.0		MIT
3818.62	3819.70	26180.0	FE	0.3	0.0		MIT
3817.650	3818.733	26186.69	FE	1.7	1.2		MIT
3816.340	3817.423	26195.68	FE I	1.4	1.3		MIT
3815.842	3816.925	26199.10	FE I	2.8	2.8		S
3814.785	3815.868	26206.36	FE	1.0	0.7		MIT
3814.523	3815.606	26208.16	FE I	1.9	1.6		MIT
3813.890	3814.972	26212.51	FE	1.7	1.4		MIT
3813.632	3814.714	26214.28	FE	1.5	1.2		MIT
3813.059	3814.141	26218.22	FE I	0.3	0.3		MIT
3812.964	3814.046	26218.87	FE I	2.6	2.5		MIT
3811.895	3812.977	26226.23	FE I	1.2	1.0		MIT
3810.761	3811.843	26234.03	FE	1.8	1.4		MIT
3809.574	3810.655	26242.21	FE	1.2	0.7		MIT
3809.151	3810.232	26245.12	FE	0.0			MIT
3809.044	3810.125	26245.86	FE I	0.3			MIT
3808.730	3809.811	26248.02	FE I	2.0	1.8		MIT
3808.287	3809.368	26251.07	FE	0.6	0.0		MIT
3807.537	3808.618	26256.24	FE I	2.2	2.0		MIT
3806.699	3807.780	26262.02	FE I	2.3	2.2		MIT
3806.219	3807.299	26265.34	FE	1.6	1.3		MIT
3805.345	3806.425	26271.37	FE	2.6	2.5		S
3804.013	3805.093	26280.57	FE I	1.6	1.0		MIT
3802.285	3803.364	26292.51	FE	1.4	1.0		MIT
3801.985	3803.064	26294.58	FE I	1.4	0.7		MIT
3801.807	3802.886	26295.82	FE	0.8	0.5		MIT
3801.682	3802.761	26296.68	FE	1.7	1.4		MIT

AIR WAVELENGTH (ANGSTRMS)	VACUUM WAVELENGTH (ANGSTRMS)	WAVENUMBER (KAYSERS)	LMENT	LOG ARC	INTENSITY SPK	DIS	REF
3801.33	3802.41	26299.1	FE	0.0			MIT
3799.549	3800.628	26311.44	FE I	2.6	2.5		S
3798.513	3799.591	26318.62	FE I	2.6	2.5		S
3797.949	3799.027	26322.53	FE I	0.6	0.3		MIT
3797.517	3798.595	26325.52	FE	2.5	2.3		S
3796.00	3797.08	26336.0	FE I	0.0			MIT
3795.532	3796.610	26339.29	FE	0.3	0.0		MIT
3795.004	3796.082	26342.95	FE I	2.7	2.6		S
3794.340	3795.417	26347.56	FE	1.9	1.7		MIT
3793.877	3794.954	26350.78	FE I	1.4	1.0		MIT
3793.484	3794.561	26353.51	FE I	1.0	0.7		MIT
3793.360	3794.437	26354.37	FE I	0.7	0.3		MIT
3792.833	3793.910	26358.03	FE I	1.0	0.5		MIT
3792.160	3793.237	26362.71	FE I	1.6	1.3		MIT
3791.73	3792.81	26365.7	FE	0.7H	0.3H		MIT
3791.507	3792.584	26367.25	FE I	0.8	0.3		MIT
3791.20	3792.28	26369.4	FE	0.3			MIT
3790.760	3791.836	26372.45	FE I	1.0	0.5		MIT
3790.664	3791.740	26373.11	FE I	0.3	0.0		MIT
3790.095	3791.171	26377.07	FE I	2.3	2.0		S
3789.810	3790.886	26379.06	FE	0.6	0.0H		MIT
3789.571	3790.647	26380.72	FE I	0.0			MIT
3789.438	3790.514	26381.65	FE	0.3	0.0		MIT
3789.177	3790.253	26383.46	FE	1.9	1.7		MIT
3787.883	3788.959	26392.48	FE I	2.7	2.5		S
3787.425	3788.501	26395.67	FE	0.3			MIT
3787.167	3788.243	26397.47	FE	1.4	1.2		MIT
3786.679	3787.754	26400.87	FE I	2.1	1.7		MIT
3786.176	3787.251	26404.37	FE I	2.0	1.8		MIT
3785.951	3787.026	26405.94	FE	2.1	1.9		MIT
3785.713	3786.788	26407.60	FE	1.0	0.6		MIT
3783.347	3784.422	26424.12	FE II	0.0H	0.0H		MIT
3782.612	3783.686	26429.25	FE	0.8	0.3		MIT
3782.456	3783.530	26430.34	FE I	1.0	0.3		MIT
3782.130	3783.204	26432.62	FE	0.9	0.3		MIT
3781.935	3783.009	26433.98	FE	1.1	0.7		MIT
3781.190	3782.264	26439.19	FE I	1.6	1.1		MIT
3779.483	3780.557	26451.13	FE	0.3			MIT
3779.446	3780.520	26451.39	FE I	2.0	1.8		MIT
3778.701	3779.774	26456.61	FE I	1.0	0.6		MIT
3778.513	3779.586	26457.92	FE	1.8	1.4		MIT
3778.320	3779.393	26459.27	FE I	0.9	0.6		MIT
3777.447	3778.520	26465.39	FE I	1.3	1.0		MIT
3777.062	3778.135	26468.09	FE	1.1	0.8		MIT
3776.459	3777.532	26472.31	FE	2.1	1.8		MIT
3775.868	3776.941	26476.46	FE I	0.3			MIT
3774.827	3775.899	26483.76	FE I	2.0	1.6		MIT
3773.699	3774.771	26491.67	FE I	1.6	1.0		MIT
3773.366	3774.438	26494.01	FE	0.3			MIT
3771.493	3772.564	26507.17	FE I	0.3			MIT
3770.412	3771.483	26514.77	FE	0.6	0.0		MIT
3770.303	3771.374	26515.53	FE I	1.5	1.1		MIT
3769.990	3771.061	26517.74	FE I	1.9	1.5		MIT
3768.029	3769.100	26531.54	FE I	1.2	0.9		MIT
3767.194	3768.264	26537.42	FE I	2.7	2.6		S
3766.666	3767.736	26541.14	FE I	0.5	0.3		MIT
3766.087	3767.157	26545.22	FE I	0.3	0.0		MIT
3765.706	3766.776	26547.90	FE	0.0	0.0		MIT
3765.542	3766.612	26549.06	FE	2.3	2.2		S
3765.033	3766.103	26552.65	FE	0.5			MIT
3763.790	3764.859	26561.42	FE I	2.7	2.6		S
3762.913	3763.982	26567.61	FE II	0.3	0.6		MIT
3762.209	3763.278	26572.58	FE I	0.8	0.5		MIT
3761.408	3762.477	26578.24	FE I	1.3	0.9		MIT
3760.534	3761.603	26584.41	FE	2.0	1.8		MIT
3760.052	3761.120	26587.82	FE I	2.2	2.0		S
3759.460	3760.528	26592.01	FE II		0.3		MIT
3759.158	3760.226	26594.14	FE	0.5	0.3		MIT
3758.235	3759.303	26600.68	FE I	2.8	2.8		S
3757.455	3758.523	26606.20	FE	1.2	1.0		MIT

AIR WAVELENGTH (ANGSTRMS)	VACUUM WAVELENGTH (ANGSTRMS)	WAVENUMBER (KAYSERS)	LMENT	LOG ARC	INTENSITY SPK	DIS	REF
3756.941	3758.009	26609.84	FE	1.9	1.8		MIT
3756.071	3757.138	26616.00	FE	1.2	0.9		MIT
3754.504	3755.571	26627.11	FE I	0.3H	1.0		MIT
3753.614	3754.681	26633.42	FE I	2.2	2.0		MIT
3753.155	3754.222	26636.68	FE	0.5	0.0		MIT
3752.420	3753.487	26641.90	FE I	1.1	0.6		MIT
3752.058	3753.124	26644.47	FE	0.3H			MIT
3751.820	3752.886	26646.16	FE I	0.7	0.3		MIT
3751.060	3752.126	26651.56	FE	0.0			MIT
3749.487	3750.553	26662.74	FE I	3.0R	2.8		S
3748.969	3750.035	26666.42	FE I	1.5	1.3		MIT
3748.490	3749.555	26669.83	FE II	0.3H			M
3748.264	3749.329	26671.44	FE I	2.7	2.3		S
3746.929	3747.994	26680.94	FE	1.6	1.4		MIT
3746.485	3747.550	26684.10	FE I	0.5	0.0		MIT
3745.903	3746.968	26688.25	FE I	2.2	2.0		MIT
3745.564	3746.629	26690.66	FE I	2.7	2.7		MIT
3744.105	3745.169	26701.06	FE I	1.6	1.3		MIT
3743.787	3744.851	26703.33	FE I	0.0			MIT
3743.482	3744.546	26705.51	FE	0.6	1.0		MIT
3743.363	3744.427	26706.35	FE I	2.3	2.2		MIT
3742.624	3743.688	26711.63	FE I	1.7	1.4		MIT
3741.791	3742.855	26717.57	FE	0.5H			MIT
3741.484	3742.548	26719.77	FE I	0.5			MIT
3740.251	3741.314	26728.57	FE	1.8	1.5		MIT
3740.061	3741.124	26729.93	FE	0.9	0.6		MIT
3739.531	3740.594	26733.72	FE	1.9	1.5		MIT
3739.312	3740.375	26735.29	FE I	0.8	0.3		MIT
3739.129	3740.192	26736.59	FE I	1.0	0.7		MIT
3738.308	3739.371	26742.47	FE	2.0	2.0		S
3737.133	3738.196	26750.87	FE I	3.0R	2.8		S
3735.330	3736.392	26763.79	FE I	1.5	1.3		MIT
3734.867	3735.929	26767.10	FE I	3.0P	2.8		MIT
3733.319	3734.381	26778.20	FE I	2.6	2.5		S
3732.399	3733.460	26784.80	FE	2.3	2.2		S
3731.91	3732.97	26788.3	FE	0.3			MIT
3731.378	3732.439	26792.13	FE	1.6	1.3		MIT
3730.947	3732.008	26795.23	FE	1.7	1.5		MIT
3730.390	3731.451	26799.23	FE	1.8	1.6		MIT
3728.667	3729.727	26811.61	FE I	1.3	1.0		MIT
3727.815	3728.875	26817.74	FE I	0.7	0.5		MIT
3727.621	3728.681	26819.14	FE I	2.3	2.2		S
3727.095	3728.155	26822.92	FE I	1.5	1.0		MIT
3726.925	3727.985	26824.14	FE I	2.0	1.8		MIT
3725.490	3726.550	26834.48	FE	1.2	0.9		MIT
3724.380	3725.439	26842.47	FE I	2.3	2.2		S
3723.88	3724.94	26846.1	FE	0.3			MIT
3722.564	3723.623	26855.57	FE I	2.7	2.6		S
3721.921	3722.980	26860.21	FE	1.2	1.0		MIT
3721.611	3722.670	26862.44	FE	0.8	0.6		MIT
3721.510	3722.568	26863.17	FE I	1.0	0.6		MIT
3721.395	3722.453	26864.00	FE I	0.5	0.3		MIT
3721.271	3722.329	26864.90	FE I	0.8	0.6		MIT
3721.192	3722.250	26865.47	FE	0.0	0.0		MIT
3719.935	3720.993	26874.55	FE I	3.0G	2.8		S
3718.408	3719.466	26885.58	FE I	1.9	1.7		MIT
3716.448	3717.505	26899.76	FE I	2.2	2.0		MIT
3715.915	3716.972	26903.62	FE I	1.9	1.7		MIT
3712.964	3714.020	26925.00	FE	0.7	0.0		MIT
3711.410	3712.466	26936.27	FE	1.7	1.4		MIT
3711.225	3712.281	26937.62	FE I	1.9	1.7		MIT
3709.665	3710.720	26948.95	FE	0.6	0.6		MIT
3709.533	3710.588	26949.90	FE	0.8	0.8		MIT
3709.249	3710.304	26951.97	FE I	2.8	2.6		MIT
3708.602	3709.657	26956.67	FE I	0.7	0.3		MIT
3707.925	3708.980	26961.59	FE I	1.9	1.8		MIT
3707.824	3708.879	26962.32	FE I	1.9	1.7		MIT
3707.561	3708.616	26964.24	FE	0.3	0.0		MIT
3707.459	3708.514	26964.98	FE	0.5	0.0		MIT
3707.049	3708.104	26967.96	FE I	2.2	2.0		MIT

AIR WAVELENGTH (ANGSTRMS)	VACUUM WAVELENGTH (ANGSTRMS)	WAVENUMBER (KAYSERS)	LMENT	LOG ARC	INTENSITY SPK	DIS	REF
3705.567	3706.621	26978.75	FE I	2.8	2.7		S
3704.463	3705.517	26986.79	FE I	2.1	2.0		S
3704.021	3705.075	26990.01	FE	0.3	0.0		MIT
3703.823	3704.877	26991.45	FE	1.2	1.0		MIT
3703.696	3704.750	26992.37	FE I	1.1	0.8		MIT
3703.555	3704.609	26993.40	FE I	1.5	1.4		MIT
3702.495	3703.549	27001.13	FE I	1.3	0.8		MIT
3702.031	3703.084	27004.51	FE	1.7	1.5		MIT
3701.090	3702.143	27011.38	FE I	2.5	2.3		MIT
3699.141	3700.194	27025.61	FE I	1.2	0.5		MIT
3698.605	3699.658	27029.53	FE	1.6	1.3		MIT
3698.15	3699.20	27032.9	FE	0.5			MIT
3697.536	3698.588	27037.34	FE	0.6			MIT
3697.432	3698.484	27038.10	FE I	2.0	1.8		MIT
3696.027	3697.079	27048.38	FE I	0.0			MIT
3695.650	3696.702	27051.14	FE	0.3			MIT
3695.515	3696.567	27052.13	FE I	0.0			MIT
3695.054	3696.106	27055.50	FE	2.3	2.2		S
3694.010	3695.061	27063.15	FE I	2.6	2.5		MIT
3693.032	3694.083	27070.32	FE	1.2	0.8		MIT
3692.652	3693.703	27073.10	FE	0.7	0.0		MIT
3691.335	3692.386	27082.76	FE	0.3			MIT
3691.152	3692.203	27084.10	FE	0.6			MIT
3690.730	3691.781	27087.20	FE	1.9	1.8		MIT
3690.459	3691.509	27089.19	FE	1.2	0.8		MIT
3689.897	3690.947	27093.31	FE	0.7	0.0		MIT
3689.463	3690.513	27096.50	FE I	2.3	2.2		MIT
3688.476	3689.526	27103.75	FE	1.6	0.9		MIT
3687.662	3688.712	27109.74	FE I	1.2	1.2		MIT
3687.458	3688.508	27111.24	FE I	2.6	2.5		S
3687.101	3688.151	27113.86	FE I	1.2	0.8		MIT
3686.258	3687.307	27120.06	FE	1.0	0.6		MIT
3686.003	3687.052	27121.94	FE I	2.2	2.1		MIT
3684.112	3685.161	27135.86	FE I	2.5	2.3		MIT
3683.616	3684.665	27139.51	FE I	0.5	0.0		MIT
3683.058	3684.107	27143.62	FE I	2.3	2.0		MIT
3682.209	3683.257	27149.88	FE I	2.6	2.5		MIT
3681.651	3682.699	27154.00	FE I	0.8	0.3		MIT
3681.227	3682.275	27157.12	FE	0.8	0.3		MIT
3680.798	3681.846	27160.29	FE	1.1	0.8		MIT
3680.675	3681.723	27161.20	FE	0.5			MIT
3680.385	3681.433	27163.34	FE	0.3			MIT
3679.915	3680.963	27166.81	FE I	2.7	2.5		S
3679.002	3680.049	27173.55	FE	0.3	0.3		MIT
3678.98	3680.03	27173.7	FE I	0.7	0.0		MIT
3678.864	3679.911	27174.57	FE I	2.0	1.7		MIT
3677.630	3678.677	27183.69	FE	1.9	1.8		S
3677.477	3678.524	27184.82	FE I	0.3	0.0		MIT
3677.308	3678.355	27186.06	FE	1.6	1.5		MIT
3676.879	3677.926	27189.24	FE I	1.0	0.3		MIT
3676.314	3677.361	27193.42	FE I	2.3	2.0		S
3675.730	3676.777	27197.74	FE	0.3H			MIT
3674.768	3675.814	27204.86	FE	1.6	1.4		MIT
3674.415	3675.461	27207.47	FE	1.1	0.7		MIT
3674.043	3675.089	27210.22	FE	0.8	0.0		MIT
3673.90	3674.95	27211.3	FE	0.8	0.3		MIT
3673.087	3674.133	27217.31	FE	1.0	0.6		MIT
3672.712	3673.758	27220.08	FE I	0.6	0.3		MIT
3671.700	3672.746	27227.59	FE	0.5	0.0		MIT
3671.525	3672.571	27228.89	FE I	0.3			MIT
3670.812	3671.857	27234.17	FE I	1.3	0.8		MIT
3670.071	3671.116	27239.67	FE	2.3	2.3		MIT
3670.028	3671.073	27239.99	FE I	2.0			MIT
3669.757	3670.802	27242.00	FE	0.3			MIT
3669.523	3670.568	27243.74	FE I	2.3	2.2		S
3669.157	3670.202	27246.46	FE I	1.7	1.5		MIT
3668.893	3669.938	27248.42	FE	0.7	0.0		MIT
3668.214	3669.259	27253.46	FE I	1.2	0.6		MIT
3667.999	3669.044	27255.06	FE I	1.8	1.0		MIT
3667.262	3668.306	27260.54	FE	1.9	1.4		MIT

AIR WAVELENGTH (ANGSTRMS)	VACUUM WAVELENGTH (ANGSTRMS)	WAVENUMBER (KAYSERS)	LMENT	LOG ARC	INTENSITY SPK	INTENSITY DIS	REF
3666.948	3667.992	27262.87	FE I	0.5			MIT
3666.781	3667.825	27264.11	FE	0.5			MIT
3666.249	3667.293	27268.07	FE I	1.3	0.8		MIT
3664.694	3665.738	27279.64	FE	1.1	0.5		MIT
3664.540	3665.584	27280.78	FE I	1.5	0.9		MIT
3663.954	3664.998	27285.15	FE	0.6	0.3		MIT
3663.458	3664.501	27288.84	FE I	1.4	0.8		MIT
3663.267	3664.310	27290.26	FE	0.9	0.5		MIT
3662.849	3663.892	27293.38	FE	1.5	0.7		MIT
3661.374	3662.417	27304.37	FE I	1.0	0.0		MIT
3660.800	3661.843	27308.66	FE	0.3			MIT
3660.335	3661.378	27312.12	FE I	0.7	0.0		MIT
3659.75	3660.79	27316.5	FE	0.5			MIT
3659.521	3660.563	27318.20	FE I	2.1	1.9		MIT
3658.552	3659.594	27325.43	FE I	0.7	0.0		MIT
3658.020	3659.062	27329.41	FE	0.5	0.0		MIT
3657.904	3658.946	27330.27	FE I	1.3	0.6		MIT
3657.43	3658.47	27333.8	FE	0.3			MIT
3657.139	3658.181	27335.99	FE I	1.3	0.8		MIT
3656.227	3657.269	27342.81	FE	1.2	0.7		MIT
3655.675	3656.716	27346.94	FE	1.2	0.3H		MIT
3655.467	3656.508	27348.49	FE I	1.4	1.4		MIT
3654.99	3656.03	27352.1	FE	0.5	0.0		MIT
3654.66	3655.70	27354.5	FE I	0.6			MIT
3653.976	3655.017	27359.65	FE	0.6	0.3H		MIT
3653.763	3654.804	27361.25	FE I	1.4	1.0		MIT
3652.73	3653.77	27369.0	FE	0.8			MIT
3652.47	3653.51	27370.9	FE	0.6			MIT
3651.469	3652.509	27378.44	FE I	2.5	2.3		S
3651.10	3652.14	27381.2	FE I	1.0	0.5		MIT
3650.53	3651.57	27385.5	FE	0.5			MIT
3650.280	3651.320	27387.35	FE I	1.8	1.7		MIT
3650.032	3651.072	27389.22	FE I	1.8	1.5		MIT
3649.508	3650.548	27393.15	FE I	2.0	2.0		S
3649.304	3650.344	27394.68	FE I	1.8	1.4		MIT
3647.844	3648.883	27405.64	FE I	2.7	2.6		S
3647.426	3648.465	27408.78	FE I	1.3	1.2		MIT
3645.825	3646.864	27420.82	FE	1.9	1.8		MIT
3645.494	3646.533	27423.31	FE I	1.2	0.8		MIT
3645.222	3646.261	27425.36	FE	1.0			MIT
3645.080	3646.119	27426.43	FE I	1.3	0.9		MIT
3644.798	3645.837	27428.55	FE	1.3	0.8		MIT
3644.587	3645.626	27430.13	FE	0.3	0.0		MIT
3644.185	3645.223	27433.16	FE	0.6			MIT
3643.815	3644.853	27435.95	FE	0.8	0.0		MIT
3643.713	3644.751	27436.71	FE I	0.8	0.6		MIT
3643.626	3644.664	27437.37	FE I	1.3	0.9		MIT
3643.11	3644.15	27441.2	FE	1.5	0.7		MIT
3642.81	3643.85	27443.5	FE	1.3			MIT
3640.390	3641.428	27461.76	FE I	2.5	2.3		MIT
3638.298	3639.335	27477.55	FE I	2.0	1.9		MIT
3637.866	3638.903	27480.81	FE I	1.3	0.8		MIT
3637.250	3638.287	27485.47	FE I	1.1	0.7		MIT
3636.995	3638.032	27487.39	FE I	1.3	1.0		MIT
3636.750	3637.787	27489.24	FE	0.5			MIT
3636.650	3637.687	27490.00	FE	0.7	0.5		MIT
3636.490	3637.527	27491.21	FE	0.5	0.0		MIT
3636.234	3637.270	27493.14	FE	1.2	1.0		MIT
3636.186	3637.222	27493.51	FE I	1.6	1.0		MIT
3635.196	3636.232	27500.99	FE I	1.1	0.3		MIT
3634.687	3635.723	27504.84	FE	1.3	0.3		MIT
3634.334	3635.370	27507.52	FE I	1.2	0.7		MIT
3633.833	3634.869	27511.31	FE	0.8	0.5		MIT
3633.078	3634.114	27517.03	FE	1.0	0.5		MIT
3632.979	3634.015	27517.78	FE I	1.1	0.9		MIT
3632.556	3633.591	27520.98	FE	1.5	1.4		MIT
3632.040	3633.075	27524.89	FE	1.7	1.7		MIT
3631.464	3632.499	27529.26	FE I	2.7	2.5		S
3631.096	3632.131	27532.05	FE I	1.4	1.0		MIT
3630.349	3631.384	27537.71	FE I	1.6	1.2		MIT

AIR WAVELENGTH (ANGSTRMS)	VACUUM WAVELENGTH (ANGSTRMS)	WAVENUMBER (KAYSERS)	LMENT	LOG ARC	INTENSITY SPK	INTENSITY DIS	REF
3629.61	3630.64	27543.3	FE	0.6			MIT
3628.810	3629.845	27549.39	FE	0.6	0.0		MIT
3628.094	3629.128	27554.83	FE I	1.0	0.5		MIT
3627.790	3628.824	27557.14	FE	0.9	0.0		MIT
3627.55	3628.58	27559.0	FE	0.5			MIT
3627.04	3628.07	27562.8	FE	1.0	0.5		MIT
3626.715	3627.749	27565.30	FE	0.3H			MIT
3626.171	3627.205	27569.44	FE	0.3			MIT
3625.148	3626.182	27577.22	FE I	1.8	1.5		MIT
3624.890	3625.924	27579.18	FE II		0.3		MIT
3624.814	3625.847	27579.76	FE	1.1	0.3		MIT
3624.307	3625.340	27583.62	FE I	1.0	0.3		MIT
3624.055	3625.088	27585.53	FE	0.3			MIT
3623.772	3624.805	27587.69	FE I	1.5	0.8		MIT
3623.446	3624.479	27590.17	FE I	1.2	0.7		MIT
3623.186	3624.219	27592.15	FE I	2.0	1.9		MIT
3622.005	3623.038	27601.15	FE I	2.1	2.0		MIT
3621.715	3622.748	27603.36	FE	0.8	0.7		MIT
3621.463	3622.496	27605.28	FE I	2.1	2.0		S
3621.273	3622.306	27606.73	FE II		0.0		MIT
3621.108	3622.141	27607.99	FE	0.3			MIT
3620.78	3621.81	27610.5	FE	0.5			MIT
3620.471	3621.503	27612.84	FE	1.2	0.3		MIT
3620.228	3621.260	27614.70	FE I	0.6	0.0		MIT
3619.772	3620.804	27618.17	FE I	0.5	0.0		MIT
3619.390	3620.422	27621.09	FE	1.1	0.0		MIT
3618.769	3619.801	27625.83	FE I	2.6	2.6		S
3618.389	3619.421	27628.73	FE I	0.9	0.6		MIT
3618.301	3619.333	27629.40	FE I	0.6	0.0		MIT
3617.788	3618.820	27633.32	FE	2.1	1.9		S
3617.318	3618.350	27636.91	FE	1.4	1.2		MIT
3617.095	3618.126	27638.61	FE	0.0			MIT
3616.572	3617.603	27642.61	FE I	1.5	0.8		MIT
3616.324	3617.355	27644.51	FE I	0.6	0.0		MIT
3616.145	3617.176	27645.87	FE	0.8			MIT
3615.665	3616.696	27649.54	FE I	1.0	0.3		MIT
3615.35	3616.38	27651.9	FE	1.0			MIT
3615.201	3616.232	27653.09	FE	1.0	0.0		MIT
3614.873	3615.904	27655.60	FE II		0.3		MIT
3614.715	3615.746	27656.81	FE	0.8	0.0H		MIT
3614.561	3615.592	27657.99	FE	1.2	0.8		MIT
3614.116	3615.147	27661.40	FE	1.0	0.5		MIT
3613.609	3614.640	27665.27	FE	0.5			MIT
3613.452	3614.483	27666.48	FE	1.0	0.3		MIT
3613.148	3614.178	27668.80	FE I	0.8	0.3		MIT
3612.945	3613.975	27670.36	FE I	1.3	0.6		MIT
3612.503	3613.533	27673.75	FE	0.5	0.0		MIT
3612.074	3613.104	27677.03	FE I	1.9	1.7		MIT
3610.702	3611.732	27687.55	FE I	1.0	0.5		MIT
3610.162	3611.192	27691.69	FE I	2.0	2.0		MIT
3608.861	3609.890	27701.67	FE I	2.7	2.6		S
3608.150	3609.179	27707.13	FE I	1.2	1.4		MIT
3606.682	3607.711	27718.41	FE I	2.3	2.2		MIT
3605.908	3606.937	27724.36	FE	0.5	0.0		MIT
3605.458	3606.486	27727.82	FE I	2.5	2.2		MIT
3605.211	3606.239	27729.72	FE I	1.1	0.3		MIT
3604.380	3605.408	27736.11	FE	1.0	0.0		MIT
3604.27	3605.30	27737.0	FE	0.7			MIT
3603.949	3604.977	27739.43	FE	0.3			MIT
3603.823	3604.851	27740.40	FE I	1.3	1.1		MIT
3603.572	3604.600	27742.33	FE	0.0			MIT
3603.208	3604.236	27745.13	FE I	2.2	1.9		MIT
3602.529	3603.557	27750.36	FE I	1.7	1.5		MIT
3602.466	3603.494	27750.85	FE	1.0	0.7		MIT
3602.085	3603.113	27753.78	FE I	1.3	0.7		MIT
3599.974	3601.001	27770.06	FE	0.7			MIT
3599.623	3600.650	27772.76	FE	1.6	1.5		MIT
3599.146	3600.173	27776.44	FE	1.0	0.7		MIT
3598.984	3600.011	27777.69	FE	0.5	0.0		MIT
3598.721	3599.748	27779.72	FE	0.8	0.3		MIT

AIR WAVELENGTH (ANGSTRMS)	VACUUM WAVELENGTH (ANGSTRMS)	WAVENUMBER (KAYSERS)	LMENT		LOG ARC	INTENSITY SPK	INTENSITY DIS	REF
3597.061	3598.087	27792.54	FE		1.6	1.0		MIT
3596.198	3597.224	27799.21	FE		1.2	0.7		MIT
3595.857	3596.883	27801.85	FE		0.6	0.0		MIT
3595.308	3596.334	27806.09	FE	I	1.3	0.8		MIT
3594.636	3595.662	27811.29	FE	I	2.1	2.0		MIT
3593.329	3594.354	27821.41	FE	I	0.8	0.3		MIT
3592.890	3593.915	27824.81	FE	I	0.5			MIT
3592.692	3593.717	27826.34	FE	I	1.1	0.3		MIT
3592.486	3593.511	27827.94	FE		0.5	0.0		MIT
3592.209	3593.234	27830.08	FE		0.8			MIT
3591.487	3592.512	27835.68	FE		0.8			MIT
3591.345	3592.370	27836.78	FE	I	1.1	0.5		MIT
3591.001	3592.026	27839.44	FE	I	0.6	0.0		MIT
3590.592	3591.617	27842.61	FE		0.5			MIT
3590.085	3591.110	27846.55	FE		0.8	0.5		MIT
3589.609	3590.633	27850.24	FE		0.3			MIT
3589.456	3590.480	27851.43	FE	I	1.7	1.5		MIT
3589.107	3590.131	27854.14	FE	I	1.8	1.5		S
3588.914	3589.938	27855.63	FE	I	1.0	0.6		MIT
3588.616	3589.640	27857.95	FE	I	1.5	1.0		MIT
3588.09	3589.11	27862.0	FE		0.6			MIT
3587.758	3588.782	27864.61	FE	I	1.7	1.4		MIT
3587.422	3588.446	27867.22	FE	I	1.0	0.7		MIT
3587.240	3588.264	27868.63	FE		0.7	0.3		MIT
3586.986	3588.010	27870.60	FE	I	2.3	2.2		MIT
3586.747	3587.771	27872.46	FE		0.6	0.3		MIT
3586.114	3587.138	27877.38	FE		1.9	1.9		S
3585.707	3586.730	27880.54	FE	I	2.1	1.9		MIT
3585.320	3586.343	27883.55	FE	I	2.2	2.0		S
3584.960	3585.983	27886.35	FE		1.5	1.4		MIT
3584.790	3585.813	27887.68	FE	I	0.5			MIT
3584.663	3585.686	27888.67	FE	I	2.0	1.8		S
3583.692	3584.715	27896.22	FE		0.8	0.0		MIT
3583.341	3584.364	27898.95	FE		1.7	1.2		MIT
3582.688	3583.711	27904.04	FE		0.5			MIT
3582.201	3583.223	27907.83	FE		1.5	1.5		MIT
3581.810	3582.832	27910.88	FE		0.5	0.3		MIT
3581.649	3582.671	27912.13	FE	I	0.6	0.5		MIT
3581.195	3582.217	27915.67	FE	I	3.0G	2.8R		S
3579.828	3580.850	27926.33	FE	I	0.0			MIT
3579.558	3580.580	27928.44	FE		0.3	0.0		MIT
3578.380	3579.402	27937.63	FE	I	1.6	0.7		MIT
3577.75	3578.77	27942.6	FE		0.3			MIT
3576.760	3577.781	27950.28	FE		1.9	1.6		S
3575.980	3577.001	27956.38	FE	I	1.9	1.4		MIT
3575.375	3576.396	27961.11	FE		1.6	1.5		MIT
3575.250	3576.271	27962.09	FE	I	1.0	0.7		MIT
3575.118	3576.139	27963.12	FE	I	1.2	0.7		MIT
3573.888	3574.908	27972.75	FE		1.6	1.5		MIT
3573.836	3574.856	27973.15	FE		1.3	1.2		MIT
3573.400	3574.420	27976.56	FE		1.7	1.3		MIT
3571.996	3573.016	27987.56	FE	I	2.0	1.9		MIT
3571.68	3572.70	27990.0	FE		0.3			MIT
3571.227	3572.247	27993.59	FE	I	1.6	0.8		MIT
3570.258	3571.277	28001.19	FE		1.7	1.2		MIT
3570.097	3571.116	28002.45	FE	I	2.5	2.5		MIT
3568.979	3569.998	28011.22	FE	I	1.7	1.5		MIT
3568.822	3569.841	28012.45	FE		1.2	0.8		MIT
3568.423	3569.442	28015.58	FE	I	1.3	0.6		MIT
3567.720	3568.739	28021.10	FE		0.3			MIT
3567.383	3568.402	28023.75	FE	I	1.0	0.3		MIT
3567.038	3568.057	28026.46	FE	I	1.7	1.2		MIT
3566.582	3567.600	28030.04	FE		0.3			MIT
3566.311	3567.329	28032.17	FE	I	0.5	0.0		MIT
3566.148	3567.166	28033.46	FE	II	0.3			DO
3565.588	3566.606	28037.86	FE	I	1.0	0.6		MIT
3565.381	3566.399	28039.49	FE	I	2.6	2.5		S
3564.530	3565.548	28046.18	FE		1.5	1.2		MIT
3564.116	3565.134	28049.44	FE	I	1.2	0.3		MIT
3561.812	3562.829	28067.58	FE		0.6			MIT
3560.887	3561.904	28074.87	FE		0.3			MIT
3560.697	3561.714	28076.37	FE		1.7	1.2		MIT
3559.509	3560.526	28085.74	FE		1.7	1.4		MIT
3559.083	3560.100	28089.10	FE		0.6	0.3		MIT
3558.518	3559.534	28093.56	FE	I	2.6	2.5		S
3558.100	3559.116	28096.86	FE		0.3			MIT
3557.298	3558.314	28103.20	FE		1.2			MIT
3556.883	3557.899	28106.48	FE	I	2.5	2.2		MIT
3556.696	3557.712	28107.95	FE	I	0.6	0.0		MIT
3555.475	3556.491	28117.61	FE		0.3			MIT
3554.929	3555.944	28121.92	FE	I	2.6	2.5		MIT
3554.513	3555.528	28125.22	FE	I	0.6	0.3		MIT
3554.120	3555.135	28128.32	FE	I	1.7	1.3		MIT
3553.741	3554.756	28131.32	FE		2.0	2.0		MIT
3552.831	3553.846	28138.53	FE	I	1.9	1.7		MIT
3552.118	3553.133	28144.18	FE		1.0	0.8		MIT
3551.120	3552.134	28152.09	FE		0.3			MIT
3549.872	3550.886	28161.98	FE	I	1.2	0.7		MIT
3548.089	3549.103	28176.14	FE		1.0			MIT
3548.024	3549.038	28176.65	FE	I	1.0	0.8		MIT
3547.198	3548.211	28183.21	FE	I	1.4	0.9		MIT
3546.22	3547.23	28191.0	FE	I	0.3			MIT
3545.834	3546.847	28194.05	FE		0.3	0.0		MIT
3545.640	3546.653	28195.60	FE	I	2.0	1.8		MIT
3544.633	3545.646	28203.61	FE		1.7	0.8		MIT
3543.674	3544.687	28211.24	FE		1.8	1.5		MIT
3543.392	3544.404	28213.48	FE		1.0	0.3		MIT
3542.235	3543.247	28222.70	FE	I	0.7	0.0		MIT
3542.078	3543.090	28223.95	FE	I	2.2	2.0		MIT
3541.086	3542.098	28231.86	FE	I	2.3	2.3		MIT
3540.713	3541.725	28234.83	FE	I	1.0	0.6		MIT
3540.118	3541.130	28239.58	FE	I	2.0	1.8		MIT
3539.200	3540.211	28246.90	FE		0.0H	0.0H		MIT
3538.795	3539.806	28250.13	FE		0.0H			MIT
3538.550	3539.561	28252.09	FE	I	0.0H			MIT
3538.290	3539.301	28254.17	FE		0.3H	0.0H		MIT
3537.897	3538.908	28257.30	FE	I	1.7	1.4		MIT
3537.731	3538.742	28258.63	FE	I	1.4	1.2		MIT
3537.493	3538.504	28260.53	FE	I	1.4	0.9		MIT
3536.557	3537.568	28268.01	FE	I	2.5	2.3		MIT
3536.186	3537.197	28270.98	FE	I	1.6	1.0		MIT
3534.914	3535.924	28281.15	FE	I	0.7	0.3		MIT
3534.529	3535.539	28284.23	FE		0.6	0.3		MIT
3533.202	3534.212	28294.85	FE	I	1.7	1.7		MIT
3533.010	3534.020	28296.39	FE		1.7	1.9		MIT
3532.576	3533.586	28299.86	FE		0.7	0.3		MIT
3531.446	3532.455	28308.92	FE	I	0.5	0.0		MIT
3530.389	3531.398	28317.40	FE	I	1.7	1.4		MIT
3529.820	3530.829	28321.96	FE		2.1	1.9		MIT
3529.532	3530.541	28324.27	FE		0.3			MIT
3527.797	3528.805	28338.20	FE	I	2.0	1.9		MIT
3526.676	3527.684	28347.21	FE	I	1.9	1.7		MIT
3526.464	3527.472	28348.91	FE	I	1.3	1.0		MIT
3526.383	3527.391	28349.56	FE		1.3	1.0		MIT
3526.169	3527.177	28351.28	FE	I	1.7	1.4		MIT
3526.042	3527.050	28352.30	FE	I	1.9	1.7		MIT
3524.241	3525.249	28366.79	FE	I	1.8	1.7		MIT
3524.071	3525.079	28368.16	FE		1.7	1.6		MIT
3523.312	3524.319	28374.27	FE		1.0	0.6		MIT
3522.893	3523.900	28377.65	FE	I	1.0	0.5		MIT
3522.276	3523.283	28382.62	FE	I	1.7	1.5		MIT
3521.841	3522.848	28386.12	FE	I	1.7	1.3		MIT
3521.264	3522.271	28390.78	FE	I	2.5	2.3		S
3520.855	3521.862	28394.07	FE		1.0	0.6		MIT
3518.882	3519.888	28409.99	FE	I	1.0	0.3		MIT
3518.685	3519.691	28411.58	FE		0.8	0.0		MIT
3516.558	3517.564	28428.77	FE		1.5	0.6		MIT
3516.403	3517.409	28430.02	FE	I	1.6	1.2		BU
3514.626	3515.631	28444.40	FE		0.8	0.3		MIT
3513.820	3514.825	28450.92	FE	I	2.6	2.5		S

AIR WAVELENGTH (ANGSTRMS)	VACUUM WAVELENGTH (ANGSTRMS)	WAVENUMBER (KAYSERS)	LMENT	LOG ARC	INTENSITY SPK	DIS	REF
3513.065	3514.070	28457.03	FE I	1.0	0.3		MIT
3512.958	3513.963	28457.90	FE	0.3	0.0		MIT
3512.227	3513.231	28463.82	FE	0.7	0.0		MIT
3512.08	3513.08	28465.0	FE	0.0			MIT
3511.741	3512.745	28467.76	FE	0.3	0.0		MIT
3510.446	3511.450	28478.26	FE I	1.2	0.9		MIT
3509.868	3510.872	28482.95	FE I	1.2	0.6		MIT
3509.132	3510.136	28488.93	FE I	0.3			MIT
3508.535	3509.539	28493.77	FE I	1.3	1.0		MIT
3508.485	3509.489	28494.18	FE	1.6	1.3		MIT
3508.213	3509.216	28496.39	FE II	0.0H	0.0H		MIT
3507.397	3508.400	28503.02	FE II	0.3	0.0		MIT
3507.170	3508.173	28504.86	FE	0.5H	0.0H		MIT
3506.498	3507.501	28510.33	FE I	1.7	1.5		MIT
3505.061	3506.064	28522.01	FE	1.0	1.0		MIT
3504.859	3505.862	28523.66	FE I	1.0	0.7		MIT
3503.474	3504.476	28534.94	FE II	0.0H	0.0H		MIT
3502.635	3503.637	28541.77	FE	0.3			MIT
3500.864	3501.866	28556.21	FE	0.3			MIT
3500.567	3501.568	28558.63	FE	1.7	1.3		MIT
3499.877	3500.878	28564.26	FE II	0.3H	0.3H		MIT
3498.758	3499.759	28573.40	FE	0.0H			MIT
3497.843	3498.844	28580.87	FE I	2.3	2.3		S
3497.108	3498.109	28586.88	FE I	2.3	2.0		MIT
3495.896	3496.896	28596.79	FE	0.8	0.0		MIT
3495.289	3496.289	28601.75	FE I	2.0	1.8		MIT
3494.672	3495.672	28606.80	FE II		0.5		DO
3494.17	3495.17	28610.9	FE I	0.3			MIT
3493.697	3494.697	28614.79	FE	0.5	0.0		MIT
3493.474	3494.474	28616.61	FE II	1.6	1.9		MIT
3493.290	3494.290	28618.12	FE I	0.0			MIT
3490.575	3491.574	28640.38	FE I	2.6	2.5		S
3489.670	3490.669	28647.81	FE	1.3	1.2		MIT
3487.996	3488.994	28661.55	FE II	0.0H	0.0H		MIT
3486.557	3487.555	28673.38	FE I	0.3			MIT
3485.342	3486.340	28683.38	FE I	2.0	1.7		S
3484.972	3485.969	28686.42	FE I	0.6	0.0		MIT
3484.855	3485.852	28687.39	FE I	0.0			MIT
3483.010	3484.007	28702.58	FE I	1.7	1.0		MIT
3481.863	3482.860	28712.04	FE	0.0H			MIT
3481.565	3482.562	28714.50	FE I	0.0			MIT
3480.347	3481.343	28724.54	FE	0.3			MIT
3479.686	3480.682	28730.00	FE	0.3			MIT
3478.787	3479.783	28737.43	FE I	0.0			MIT
3478.629	3479.625	28738.73	FE	1.3	0.8		MIT
3478.377	3479.373	28740.81	FE I	0.0			MIT
3477.856	3478.852	28745.12	FE	1.3	0.6		MIT
3477.007	3478.002	28752.14	FE I	0.5	0.0		MIT
3476.858	3477.853	28753.37	FE I	0.5	0.6		MIT
3476.704	3477.699	28754.64	FE I	2.5	2.3		S
3476.344	3477.339	28757.62	FE I	0.7	0.3		MIT
3475.866	3476.861	28761.57	FE	0.3	0.0		MIT
3475.651	3476.646	28763.35	FE I	0.8	0.8		MIT
3475.454	3476.449	28764.98	FE I	2.6	2.5		MIT
3474.440	3475.435	28773.38	FE	1.0	0.8		MIT
3473.682	3474.677	28779.66	FE	0.5H			M
3473.502	3474.497	28781.15	FE I	0.5			MIT
3473.309	3474.303	28782.75	FE	1.0	0.3		MIT
3472.36	3473.35	28790.6	FE	0.0J			MIT
3472.03	3473.02	28793.4	FE	0.0			MIT
3471.915	3472.909	28794.30	FE	0.0			MIT
3471.34	3472.33	28799.1	FE I	1.6	1.2		MIT
3471.270	3472.264	28799.65	FE I	1.6	1.2		MIT
3469.834	3470.828	28811.57	FE I	1.5	1.0		MIT
3469.600	3470.594	28813.51	FE	0.0			MIT
3469.395	3470.388	28815.22	FE	0.3H	0.0		MIT
3469.011	3470.004	28818.41	FE	1.3	0.0		MIT
3468.848	3469.841	28819.76	FE I	1.5	1.1		MIT
3468.679	3469.672	28821.17	FE II	1.0	1.3		MIT
3466.895	3467.888	28836.00	FE I	1.0	0.6		MIT

AIR WAVELENGTH (ANGSTRMS)	VACUUM WAVELENGTH (ANGSTRMS)	WAVENUMBER (KAYSERS)	LMENT	LOG ARC	INTENSITY SPK	DIS	REF
3466.500	3467.493	28839.28	FE I	1.5	1.8		MIT
3465.863	3466.856	28844.58	FE I	2.7	2.6		S
3464.917	3465.909	28852.46	FE I	0.3			MIT
3464.493	3465.485	28855.99	FE II	0.0H	0.0H		MIT
3463.302	3464.294	28865.91	FE I	0.8	0.0		MIT
3462.361	3463.353	28873.76	FE I	1.0	0.5		MIT
3459.918	3460.909	28894.14	FE I	1.9	1.7		MIT
3459.740	3460.731	28895.63	FE	0.3	0.0		MIT
3459.429	3460.420	28898.23	FE	1.0	0.5		MIT
3458.309	3459.300	28907.58	FE	1.8	1.4		MIT
3457.515	3458.505	28914.22	FE I	0.0			MIT
3457.086	3458.076	28917.81	FE	0.7			MIT
3456.934	3457.924	28919.08	FE II	0.5	0.6		MIT
3455.234	3456.224	28933.31	FE	0.3			MIT
3453.023	3454.012	28951.84	FE	1.5	1.2		MIT
3452.277	3453.266	28958.09	FE I	2.2	0.9		MIT
3451.918	3452.907	28961.10	FE I	2.0	1.8		MIT
3451.619	3452.608	28963.61	FE I	1.2	0.6		MIT
3451.228	3452.217	28966.89	FE	0.0	0.0		DO
3450.330	3451.319	28974.43	FE I	2.2	1.9		MIT
3447.281	3448.269	29000.06	FE I	2.0	1.8		MIT
3446.792	3447.780	29004.17	FE	0.0			MIT
3445.771	3446.758	29012.77	FE	1.0	0.5		MIT
3445.151	3446.138	29017.99	FE I	2.5	2.2		S
3443.878	3444.865	29028.71	FE I	2.7	2.3		S
3443.650	3444.637	29030.64	FE	0.6	0.0		MIT
3442.677	3443.664	29038.84	FE I	1.5	0.7		MIT
3442.361	3443.348	29041.51	FE I	1.7	1.2		MIT
3442.239	3443.226	29042.54	FE II	0.3			MIT
3440.991	3441.977	29053.07	FE I	2.5	2.3		MIT
3440.610	3441.596	29056.29	FE I	2.7	2.5		MIT
3439.872	3440.858	29062.52	FE	1.2	0.8		MIT
3438.306	3439.292	29075.76	FE	1.0	0.5H		MIT
3437.952	3438.937	29078.75	FE	1.2	0.8		MIT
3437.051	3438.036	29086.37	FE	1.9	1.2		MIT
3436.112	3437.097	29094.32	FE II	0.7	1.2		DO
3434.025	3435.009	29112.00	FE I	0.3H			MIT
3433.045	3434.029	29120.31	FE	1.7	0.0H		MIT
3431.814	3432.798	29130.76	FE	1.7	1.3		MIT
3428.753	3429.736	29156.76	FE	0.7	0.3		MIT
3428.197	3429.180	29161.49	FE I	1.7	1.7		MIT
3427.121	3428.104	29170.65	FE I	1.7	1.7		S
3427.014	3427.997	29171.56	FE I	0.0			MIT
3426.636	3427.619	29174.77	FE I	1.9	1.8		MIT
3426.388	3427.370	29176.89	FE I	1.9	1.3		MIT
3426.32	3427.30	29177.5	FE I	0.7	0.3		MIT
3425.582	3426.564	29183.75	FE II	0.3	0.6		DO
3425.015	3425.997	29188.58	FE	1.8	1.6		MIT
3424.288	3425.270	29194.78	FE I	2.3	2.2		MIT
3422.660	3423.642	29208.67	FE I	2.0	1.7		MIT
3422.493	3423.474	29210.09	FE	1.6	1.0		MIT
3422.14	3423.12	29213.1	FE	0.3			MIT
3419.699	3420.680	29233.96	FE	0.8	0.3		MIT
3419.154	3420.135	29238.61	FE	0.3			MIT
3418.88	3419.86	29241.0	FE	0.0			MIT
3418.512	3419.492	29244.11	FE I	2.2	2.0		MIT
3418.176	3419.156	29246.98	FE	0.6	0.0H		MIT
3417.843	3418.823	29249.83	FE I	2.2	2.0		MIT
3417.262	3418.242	29254.80	FE I	0.8			MIT
3417.164	3418.144	29255.64	FE	0.7			MIT
3416.678	3417.658	29259.80	FE I	0.0			MIT
3416.291	3417.271	29263.12	FE	0.3			MIT
3416.028	3417.008	29265.37	FE II	0.3H	0.3H		MIT
3415.534	3416.514	29269.60	FE I	1.8	1.3		MIT
3413.48	3414.46	29287.2	FE	0.3			MIT
3413.135	3414.114	29290.18	FE I	2.6	2.5		S
3412.640	3413.619	29294.42	FE	0.9			MIT
3411.356	3412.335	29305.45	FE	1.9	1.5		MIT
3411.133	3412.112	29307.36	FE	0.3			MIT
3410.900	3411.878	29309.37	FE I	1.0	0.3		MIT

AIR WAVELENGTH (ANGSTRMS)	VACUUM WAVELENGTH (ANGSTRMS)	WAVENUMBER (KAYSERS)	LMENT	LOG ARC	INTENSITY SPK	INTENSITY DIS	REF
3410.176	3411.154	29315.59	FE	1.5	1.3		MIT
3409.20	3410.18	29324.0	FE	1.6	0.6		MIT
3407.461	3408.439	29338.95	FE I	2.6	2.6		S
3406.805	3407.782	29344.60	FE I	2.0	1.8		MIT
3406.442	3407.419	29347.72	FE	1.5	1.0		MIT
3405.833	3406.810	29352.97	FE	0.5	0.0		MIT
3405.578	3406.555	29355.17	FE	0.0			MIT
3404.754	3405.731	29362.27	FE I	0.3			MIT
3404.359	3405.336	29365.68	FE I	2.0	1.7		MIT
3404.304	3405.281	29366.15	FE I	1.4	1.4		MIT
3403.32	3404.30	29374.6	FE	0.8	0.5		MIT
3402.262	3403.238	29383.78	FE	2.2	2.2		MIT
3401.521	3402.497	29390.18	FE I	2.2	2.0		S
3399.336	3400.312	29409.07	FE I	2.3	2.3		S
3399.234	3400.210	29409.95	FE I	0.7			MIT
3398.220	3399.195	29418.73	FE	0.3	0.0		MIT
3397.642	3398.617	29423.73	FE I	1.0	0.3		MIT
3397.564	3398.539	29424.41	FE	0.0	0.3		MIT
3397.213	3398.188	29427.45	FE	0.0			MIT
3396.978	3397.953	29429.48	FE I	2.1	1.4		S
3396.385	3397.360	29434.62	FE I	0.0			MIT
3395.942	3396.917	29438.46	FE	0.5			MIT
3395.329	3396.304	29443.78	FE II		0.3		MIT
3394.589	3395.563	29450.19	FE I	2.2	1.9		MIT
3394.087	3395.061	29454.55	FE I	0.8	0.3		MIT
3393.917	3394.891	29456.02	FE	0.8	0.0		MIT
3393.381	3394.355	29460.68	FE I	0.8	0.5		MIT
3392.657	3393.631	29466.97	FE I	2.5	2.3		MIT
3392.313	3393.287	29469.95	FE I	2.1	1.9		MIT
3392.014	3392.988	29472.55	FE	1.3	0.8		MIT
3390.080	3391.053	29489.36	FE	0.3			MIT
3389.749	3390.722	29492.24	FE I	1.2	0.6		MIT
3388.964	3389.937	29499.07	FE I	0.5			MIT
3388.626	3389.599	29502.02	FE	0.7	0.0		MIT
3387.630	3388.603	29510.69	FE	0.7			MIT
3387.411	3388.383	29512.60	FE	1.5	0.9		MIT
3385.55	3386.52	29528.8	FE	0.3			MIT
3385.443	3386.415	29529.75	FE	0.5			MIT
3383.980	3384.952	29542.52	FE I	2.3	2.0		I
3383.698	3384.670	29544.98	FE I	2.0	1.8		MIT
3382.409	3383.380	29556.24	FE I	1.7	1.0		MIT
3381.337	3382.308	29565.61	FE I	0.7	0.3		MIT
3380.998	3381.969	29568.57	FE II		0.3		MIT
3380.111	3381.082	29576.33	FE	2.3	1.4		I
3379.022	3379.992	29585.87	FE I	1.9	1.7		MIT
3378.685	3379.655	29588.82	FE	2.2	1.9		MIT
3377.20	3378.17	29601.8	FE	0.7J	0.7J		MIT
3376.50	3377.47	29608.0	FE	0.7	0.0H		MIT
3374.458	3375.427	29625.88	FE	0.6			MIT
3374.226	3375.195	29627.92	FE I	0.3			MIT
3373.875	3374.844	29631.00	FE	0.5	0.0		MIT
3372.861	3373.830	29639.91	FE	0.3			MIT
3372.351	3373.320	29644.39	FE	0.0			MIT
3372.080	3373.049	29646.77	FE I	1.6	0.8		MIT
3370.786	3371.754	29658.15	FE	2.5	2.3		S
3369.549	3370.517	29669.04	FE	2.5	2.3		MIT
3369.349	3370.317	29670.80	FE	0.3			MIT
3369.147	3370.115	29672.58	FE	0.3			MIT
3368.168	3369.136	29681.20	FE	0.7	0.0		MIT
3367.161	3368.128	29690.08	FE	1.0	0.5		MIT
3366.960	3367.927	29691.85	FE II	0.3	0.6		DO
3366.867	3367.834	29692.67	FE I	1.7	1.2		MIT
3366.789	3367.756	29693.36	FE	1.7	1.4		MIT
3364.40	3365.37	29714.4	FE	0.3			MIT
3364.278	3365.245	29715.52	FE	0.8	0.0		MIT
3363.811	3364.777	29719.65	FE	0.8	0.0		MIT
3362.279	3363.245	29733.19	FE	0.8			MIT
3361.952	3362.918	29736.08	FE	0.8	0.0		MIT
3360.928	3361.894	29745.14	FE I	0.8	0.0		MIT
3360.313	3361.279	29750.58	FE	0.5	0.0		MIT

AIR WAVELENGTH (ANGSTRMS)	VACUUM WAVELENGTH (ANGSTRMS)	WAVENUMBER (KAYSERS)	LMENT	LOG ARC	INTENSITY SPK	INTENSITY DIS	REF
3360.103	3361.069	29752.44	FE II	0.6	0.7		DO
3359.815	3360.780	29754.99	FE	1.0	0.3		MIT
3359.491	3360.456	29757.86	FE I	1.2	0.5		MIT
3358.909	3359.874	29763.02	FE	0.7			MIT
3358.252	3359.217	29768.84	FE II	0.5H	0.8		DO
3357.56	3358.52	29775.0	FE	0.3			MIT
3356.693	3357.658	29782.67	FE	1.2	0.0		MIT
3356.402	3357.367	29785.25	FE I	1.5	0.9		MIT
3356.266	3357.231	29786.46	FE II	0.0	0.3		MIT
3355.229	3356.193	29795.66	FE	2.0	2.0		I
3354.064	3355.028	29806.01	FE	1.6	1.6		MIT
3353.267	3354.231	29813.09	FE I	1.0	0.7		MIT
3352.930	3353.894	29816.09	FE I	0.7	0.5		MIT
3351.745	3352.708	29826.63	FE	1.9	1.8		MIT
3351.521	3352.484	29828.63	FE I	1.8	1.8		MIT
3350.285	3351.248	29839.63	FE I	0.9	0.6		MIT
3349.738	3350.701	29844.50	FE	0.0			MIT
3347.927	3348.889	29860.65	FE I	2.2	2.0		S
3347.508	3348.470	29864.38	FE	0.3	0.0		MIT
3346.936	3347.898	29869.49	FE I	1.5	1.2		MIT
3345.682	3346.644	29880.68	FE	0.0			MIT
3343.676	3344.637	29898.61	FE	0.3	0.0		MIT
3343.242	3344.203	29902.49	FE I	0.7	0.3		MIT
3342.292	3343.253	29910.99	FE	1.3	1.3		MIT
3342.216	3343.177	29911.67	FE I	1.6	1.6		MIT
3341.905	3342.866	29914.45	FE	2.0	1.9		MIT
3340.566	3341.527	29926.44	FE I	2.1	2.0		S
3339.692	3340.652	29934.27	FE	0.3	0.0		MIT
3339.582	3340.542	29935.26	FE I	1.0	0.7		MIT
3339.195	3340.155	29938.73	FE I	1.9	1.7		MIT
3338.636	3339.596	29943.74	FE	1.8	1.4		MIT
3338.518	3339.478	29944.80	FE II		0.0		MIT
3337.666	3338.626	29952.44	FE	2.1	2.0		I
3336.254	3337.213	29965.12	FE	1.6	1.5		MIT
3335.768	3336.727	29969.49	FE	2.1	2.0		MIT
3335.514	3336.473	29971.77	FE I	0.3	0.3		MIT
3334.217	3335.176	29983.43	FE I	2.2H	2.0H		MIT
3331.776	3332.734	30005.39	FE	1.6	1.0		MIT
3331.612	3332.570	30006.87	FE I	2.1	1.8		MIT
3330.313	3331.271	30018.57	FE	0.7	0.3		MIT
3329.532	3330.490	30025.61	FE	1.5	0.8		MIT
3329.054	3330.012	30029.93	FE	0.9	0.5		MIT
3328.867	3329.825	30031.61	FE	2.2	2.0		I
3327.958	3328.915	30039.81	FE	0.3			MIT
3327.491	3328.448	30044.03	FE I	1.2	0.8		MIT
3325.462	3326.419	30062.36	FE I	2.0	1.9		MIT
3324.789	3325.746	30068.45	FE	0.3			MIT
3324.537	3325.493	30070.73	FE I	2.0	1.9		MIT
3324.366	3325.322	30072.27	FE	0.7	0.5		MIT
3323.737	3324.693	30077.96	FE	2.2	2.2		I
3323.068	3324.024	30084.02	FE II		2.0		MIT
3322.477	3323.433	30089.37	FE	2.2	2.0		MIT
3320.779	3321.734	30104.75	FE	1.5	1.1		MIT
3320.650	3321.605	30105.92	FE I	1.3	1.0		MIT
3319.249	3320.204	30118.63	FE	1.8	1.7		MIT
3317.121	3318.076	30137.95	FE I	2.0	1.9		MIT
3314.742	3315.696	30159.58	FE	2.3	2.3		I
3314.447	3315.401	30162.26	FE	1.2	0.8		MIT
3314.072	3315.026	30165.68	FE I	0.7	0.3		MIT
3313.728	3314.682	30168.81	FE I	0.5	0.0		MIT
3312.703	3313.656	30178.14	FE II	0.7	0.5		MIT
3312.226	3313.179	30182.49	FE	0.7	0.3		MIT
3311.453	3312.406	30189.53	FE I	0.0	0.0		MIT
3310.489	3311.442	30198.33	FE	1.7	1.6		MIT
3310.342	3311.295	30199.67	FE	2.0	1.9		MIT
3308.749	3309.701	30214.21	FE	0.7	0.5		MIT
3307.232	3308.184	30228.06	FE	1.9	1.8		MIT
3307.146	3308.098	30228.85	FE	0.5	0.3		MIT
3307.012	3307.964	30230.07	FE	0.7	0.6		MIT
3306.354	3307.306	30236.09	FE I	2.3	2.2		I

AIR WAVELENGTH (ANGSTRMS)	VACUUM WAVELENGTH (ANGSTRMS)	WAVENUMBER (KAYSERS)	LMENT	LOG ARC	INTENSITY SPK	DIS	REF
3305.971	3306.923	30239.59	FE I	2.6	2.5		I
3305.135	3306.087	30247.24	FE	0.7	0.3		MIT
3303.567	3304.518	30261.60	FE	1.8	1.0		MIT
3303.471	3304.422	30262.48	FE II	0.7	0.7		MIT
3302.859	3303.810	30268.08	FE II	0.0	0.7		MIT
3301.426	3302.377	30281.22	FE	0.0			MIT
3301.222	3302.173	30283.09	FE	1.2	0.8		MIT
3299.506	3300.456	30298.84	FE I	0.6	0.3		MIT
3299.075	3300.025	30302.80	FE	0.7	0.3		MIT
3298.133	3299.083	30311.46	FE I	2.3	2.2		S
3297.891	3298.841	30313.68	FE II	0.6	1.2		MIT
3296.803	3297.752	30323.68	FE	0.5	0.3H		MIT
3296.467	3297.416	30326.77	FE	1.1	0.8		MIT
3295.816	3296.765	30332.76	FE II	0.6	1.5		MIT
3295.420	3296.369	30336.41	FE II	0.0	0.6		DO
3293.142	3294.090	30357.39	FE	1.0	0.7		MIT
3292.590	3293.538	30362.48	FE I	2.5	2.2		MIT
3292.023	3292.971	30367.71	FE	2.2	2.1		MIT
3290.989	3291.937	30377.25	FE I	2.1	1.9		MIT
3290.719	3291.667	30379.75	FE	1.2	0.8		MIT
3290.041	3290.989	30386.01	FE	0.5	0.3		MIT
3289.436	3290.384	30391.59	FE	1.0	0.6		MIT
3289.346	3290.293	30392.43	FE II		1.6		MIT
3288.967	3289.914	30395.93	FE	1.5	1.2		MIT
3288.654	3289.601	30398.82	FE	1.2	0.8		MIT
3287.676	3288.623	30407.86	FE	0.0			MIT
3287.113	3288.060	30413.07	FE	0.7	0.5		MIT
3286.755	3287.702	30416.38	FE I	2.7	2.6		S
3286.446	3287.393	30419.24	FE	0.7H	0.5H		MIT
3286.022	3286.969	30423.17	FE I	1.5	1.2		MIT
3285.415	3286.361	30428.79	FE II	1.8	1.6		MIT
3285.20	3286.15	30430.8	FE	0.6	0.3		MIT
3284.588	3285.534	30436.45	FE I	2.3	2.1		S
3283.543	3284.489	30446.14	FE	0.8	0.5		MIT
3283.427	3284.373	30447.21	FE I	0.6	0.3		MIT
3282.892	3283.838	30452.17	FE	1.9	1.9		MIT
3282.44	3283.39	30456.4	FE	0.3	0.0		MIT
3281.300	3282.245	30466.95	FE II	1.2	2.0		MIT
3280.261	3281.206	30476.60	FE	2.2	2.2		I
3279.739	3280.684	30481.45	FE	0.8	0.5		MIT
3279.653	3280.598	30482.25	FE		0.3		MIT
3279.149	3280.094	30486.93	FE	0.3	0.0		MIT
3278.734	3279.679	30490.79	FE	2.0	1.8		MIT
3277.346	3278.290	30503.70	FE II	1.6	2.3		MIT
3276.606	3277.550	30510.59	FE II		1.0		MIT
3276.468	3277.412	30511.88	FE I	2.0	1.7		MIT
3274.453	3275.397	30530.65	FE	1.9	1.8		MIT
3273.499	3274.442	30539.55	FE II		0.8		MIT
3272.70	3273.64	30547.0	FE	0.3			MIT
3272.595	3273.538	30547.99	FE	0.8	0.5		MIT
3271.683	3272.626	30556.50	FE I	1.4	1.2		MIT
3271.486	3272.429	30558.34	FE	1.2	0.8		MIT
3271.002	3271.945	30562.86	FE I	2.5	2.5		S
3269.959	3270.902	30572.61	FE	0.7	0.5		MIT
3269.772	3270.715	30574.36	FE II		0.6		MIT
3269.235	3270.177	30579.38	FE	1.3	0.8		MIT
3268.509	3269.451	30586.17	FE II		0.7		MIT
3268.236	3269.178	30588.73	FE I	2.1	2.0		MIT
3267.036	3267.978	30599.96	FE II		0.6		MIT
3266.938	3267.880	30600.88	FE II		1.2		MIT
3265.619	3266.560	30613.24	FE I	2.5	2.5		MIT
3265.048	3265.989	30618.59	FE I	2.3	2.2		MIT
3264.710	3265.651	30621.76	FE I	0.6	0.3		MIT
3264.513	3265.454	30623.61	FE I	1.9	1.8		MIT
3263.369	3264.310	30634.35	FE I	1.5	1.2		MIT
3263.07	3264.01	30637.1	FE	0.0			MIT
3262.280	3263.221	30644.57	FE	1.7	1.4		MIT
3262.013	3262.954	30647.08	FE	1.5	1.2		MIT
3261.333	3262.273	30653.47	FE	1.4	0.8		MIT
3260.261	3261.201	30663.55	FE	1.3	1.2		MIT

AIR WAVELENGTH (ANGSTRMS)	VACUUM WAVELENGTH (ANGSTRMS)	WAVENUMBER (KAYSERS)	LMENT	LOG ARC	INTENSITY SPK	DIS	REF
3259.994	3260.934	30666.06	FE I	2.2	2.0		MIT
3259.048	3259.988	30674.96	FE II	0.0	2.3		MIT
3258.773	3259.713	30677.55	FE II		2.2		MIT
3257.894	3258.834	30685.83	FE II		0.5		MIT
3257.594	3258.533	30688.65	FE I	2.0	2.0		S
3257.240	3258.179	30691.99	FE I	1.4	1.1		MIT
3257.116	3258.055	30693.16	FE	0.0			MIT
3256.698	3257.637	30697.09	FE	1.3	0.8		MIT
3255.890	3256.829	30704.71	FE II	1.3	2.0		MIT
3254.734	3255.673	30715.62	FE	1.2	0.8		MIT
3254.363	3255.302	30719.12	FE	2.3	2.2		I
3253.949	3254.888	30723.03	FE	1.3	0.9		MIT
3253.835	3254.773	30724.10	FE	0.3	0.0		MIT
3253.602	3254.540	30726.30	FE	2.0	1.9		MIT
3252.926	3253.864	30732.69	FE	1.9	1.7		MIT
3252.437	3253.375	30737.31	FE	2.0	1.6		MIT
3251.235	3252.173	30748.67	FE I	2.5	2.2		MIT
3250.627	3251.565	30754.43	FE I	1.8	1.6		MIT
3250.394	3251.332	30756.63	FE	1.3	0.8		MIT
3249.85	3250.79	30761.8	FE	0.0			MIT
3249.657	3250.594	30763.60	FE II		1.0		MIT
3249.192	3250.129	30768.01	FE	1.8	1.5		MIT
3249.033	3249.970	30769.51	FE	0.0			MIT
3248.206	3249.143	30777.35	FE I	2.3	2.2		MIT
3247.393	3248.330	30785.05	FE II		0.3		MIT
3247.278	3248.215	30786.14	FE I	1.3	1.0		MIT
3247.213	3248.150	30786.76	FE	1.0	1.0		MIT
3247.171	3248.108	30787.16	FE II		1.0		MIT
3246.962	3247.899	30789.14	FE I	2.0	1.8		MIT
3246.482	3247.419	30793.69	FE	1.6	1.4		MIT
3245.984	3246.920	30798.41	FE I	2.3	2.2		BU
3244.190	3245.126	30815.44	FE I	2.5	2.3		S
3243.724	3244.660	30819.87	FE II		1.8		MIT
3243.403	3244.339	30822.92	FE	1.8	1.3		MIT
3243.109	3244.045	30825.72	FE	1.7	1.3		MIT
3242.272	3243.208	30833.67	FE	0.5	0.0		MIT
3241.685	3242.620	30839.26	FE II		0.3		MIT
3240.015	3240.950	30855.15	FE	0.0	0.0		MIT
3239.436	3240.371	30860.67	FE I	2.6	2.5		S
3239.034	3239.969	30864.50	FE I	0.7	0.5		MIT
3237.819	3238.753	30876.08	FE II	0.0	2.0		MIT
3237.402	3238.336	30880.05	FE II		1.0		MIT
3237.222	3238.156	30881.77	FE	0.5	0.0		MIT
3236.223	3237.157	30891.30	FE I	2.5	2.3		S
3235.836	3236.770	30895.00	FE	0.3	0.0		MIT
3235.59	3236.52	30897.4	FE	0.7	0.3		MIT
3234.611	3235.545	30906.70	FE I	2.3	2.1		MIT
3233.971	3234.904	30912.81	FE I	2.5	2.2		MIT
3233.054	3233.987	30921.58	FE	2.0	1.8		MIT
3232.791	3233.724	30924.10	FE II		1.7		MIT
3231.712	3232.645	30934.42	FE II		1.5		MIT
3231.582	3232.515	30935.67	FE I	0.5	0.0		MIT
3230.967	3231.900	30941.55	FE I	2.5	2.3		MIT
3230.211	3231.143	30948.80	FE I	2.0	1.9		MIT
3229.994	3230.926	30950.87	FE	1.3	1.3		MIT
3229.873	3230.805	30952.03	FE	1.0	1.0		MIT
3229.791	3230.723	30952.82	FE I	0.0	0.0		MIT
3229.593	3230.525	30954.72	FE	1.0	0.7		MIT
3229.123	3230.055	30959.22	FE I	1.9	1.7		MIT
3228.905	3229.837	30961.31	FE	1.9	1.6		MIT
3228.597	3229.529	30964.27	FE	0.7	0.3		MIT
3228.254	3229.186	30967.56	FE I	2.0	1.9		MIT
3227.996	3228.928	30970.03	FE	0.5	0.0		MIT
3227.802	3228.734	30971.89	FE I	0.6			MIT
3227.747	3228.679	30972.42	FE II	2.3	2.5		MIT
3227.063	3227.995	30978.99	FE I	1.5	1.0		MIT
3226.720	3227.652	30982.28	FE I	0.9	0.5		MIT
3226.383	3227.315	30985.51	FE II		0.0		MIT
3226.021	3226.952	30988.99	FE	0.3	0.0		MIT
3225.789	3226.720	30991.22	FE I	2.5	2.2		S

AIR WAVELENGTH (ANGSTRMS)	VACUUM WAVELENGTH (ANGSTRMS)	WAVENUMBER (KAYSERS)	LMENT	LOG ARC	INTENSITY SPK	INTENSITY DIS	REF
3224.931	3225.862	30999.47	FE	0.5	0.0		MIT
3223.844	3224.775	31009.92	FE I	0.3	0.3		MIT
3223.452	3224.383	31013.69	FE	0.0	0.0		MIT
3223.268	3224.199	31015.46	FE	0.6	0.3		MIT
3222.069	3222.999	31027.00	FE I	2.3	2.0		S
3221.931	3222.861	31028.33	FE I	0.3	0.0		MIT
3219.810	3220.740	31048.77	FE I	2.0	1.9		MIT
3219.581	3220.511	31050.98	FE I	2.3	2.1		MIT
3217.380	3218.309	31072.22	FE I	2.3	2.1		S
3215.940	3216.869	31086.13	FE I	2.5	2.2		S
3215.416	3216.345	31091.20	FE	1.0	0.6		MIT
3215.228	3216.157	31093.01	FE	0.6	0.3		MIT
3214.399	3215.327	31101.03	FE I	2.0	1.7		MIT
3214.040	3214.968	31104.50	FE I	2.6	2.3		MIT
3213.314	3214.242	31111.53	FE II	1.7	2.5		MIT
3212.172	3213.100	31122.59	FE	0.6	0.3		MIT
3211.992	3212.920	31124.34	FE I	1.8	1.7		MIT
3211.881	3212.809	31125.41	FE	1.0	1.0		MIT
3211.683	3212.611	31127.33	FE I	1.9	1.7		MIT
3211.487	3212.415	31129.23	FE	1.9	1.6		MIT
3210.834	3211.762	31135.56	FE I	2.2	2.0		MIT
3210.656	3211.584	31137.29	FE	0.0			MIT
3210.451	3211.378	31139.28	FE II	0.7	1.7		MIT
3210.236	3211.163	31141.36	FE I	2.2	2.0		MIT
3209.297	3210.224	31150.47	FE	2.3	2.1		MIT
3208.475	3209.402	31158.45	FE	2.0	1.9		MIT
3207.654	3208.581	31166.43	FE	0.3H	0.0H		MIT
3207.089	3208.016	31171.92	FE I	1.9	1.7		MIT
3205.400	3206.326	31188.34	FE I	2.5	2.3		S
3202.951	3203.877	31212.19	FE	0.5	0.3		MIT
3202.654	3203.580	31215.08	FE I	0.3	0.0		MIT
3202.560	3203.485	31216.00	FE	1.6	1.3		MIT
3200.784	3201.709	31233.32	FE I	1.4	1.2		MIT
3200.475	3201.400	31236.33	FE I	2.2	2.2		S
3199.525	3200.450	31245.61	FE I	2.5	2.3		MIT
3198.485	3199.409	31255.77	FE	0.3	0.0		MIT
3198.266	3199.190	31257.91	FE	0.9	0.8		MIT
3197.517	3198.441	31265.23	FE	1.0	0.9		MIT
3196.997	3197.921	31270.31	FE I	2.2			MIT
3196.930	3197.854	31270.97	FE I	2.7	2.5		S
3196.129	3197.053	31278.81	FE	2.0			MIT
3196.076	3197.000	31279.33	FE II	1.0	2.2		MIT
3195.231	3196.155	31287.60	FE	0.8	0.5		MIT
3194.597	3195.520	31293.81	FE	1.8	1.3		MIT
3194.423	3195.346	31295.51	FE I	2.0	1.8		MIT
3193.978	3194.901	31299.87	FE	0.5	0.0		MIT
3193.802	3194.725	31301.60	FE II	1.0	1.7		MIT
3193.802	3194.725	31301.60	FE I	1.0	1.7		MIT
3193.303	3194.226	31306.49	FE I	1.3	1.2		MIT
3193.228	3194.151	31307.22	FE I	2.0	1.8		MIT
3192.926	3193.849	31310.18	FE II	0.7			MIT
3192.802	3193.725	31311.40	FE I	2.2	0.9		MIT
3192.503	3193.426	31314.33	FE	0.3	0.0		MIT
3192.408	3193.331	31315.26	FE	1.1	0.8		MIT
3192.064	3192.987	31318.64	FE II		1.0		MIT
3191.659	3192.582	31322.61	FE I	2.3	2.2		MIT
3191.112	3192.035	31327.98	FE	1.3	0.9		S
3190.816	3191.739	31330.89	FE	1.6	1.3		MIT
3190.651	3191.573	31332.51	FE I	1.7	1.4		MIT
3190.016	3190.938	31338.74	FE	1.5	1.0		MIT
3188.821	3189.743	31350.49	FE I	2.2	2.0		MIT
3188.571	3189.493	31352.95	FE I	2.2	2.0		MIT
3188.026	3188.948	31358.31	FE	0.6	1.0		MIT
3187.680	3188.602	31361.71	FE	0.8	0.3		MIT
3187.293	3188.215	31365.52	FE II		1.8		MIT
3187.163	3188.085	31366.80	FE	0.7	0.0		MIT
3186.741	3187.662	31370.95	FE II	1.3	2.5		MIT
3185.316	3186.237	31384.98	FE II		1.4		MIT
3184.896	3185.817	31389.12	FE I	2.3	2.2		S
3184.622	3185.543	31391.82	FE I	1.8	1.6		MIT

AIR WAVELENGTH (ANGSTRMS)	VACUUM WAVELENGTH (ANGSTRMS)	WAVENUMBER (KAYSERS)	LMENT	LOG ARC	INTENSITY SPK	INTENSITY DIS	REF
3184.107	3185.028	31396.90	FE	0.7	0.3		MIT
3183.573	3184.494	31402.17	FE	0.7	0.5		MIT
3183.108	3184.029	31406.75	FE II	0.8	1.7		MIT
3182.975	3183.896	31408.07	FE	2.1	1.8		MIT
3182.061	3182.981	31417.09	FE	1.9	1.9		MIT
3181.908	3182.828	31418.60	FE	1.2	1.2		MIT
3181.855	3182.775	31419.12	FE	1.2	1.2		MIT
3181.520	3182.440	31422.43	FE	1.9	1.8		MIT
3180.755	3181.675	31429.99	FE I	2.0	2.0		MIT
3180.226	3181.146	31435.21	FE I	2.5	2.5		MIT
3180.164	3181.084	31435.83	FE II	1.0	1.2		DO
3179.507	3180.427	31442.32	FE II		0.8		MIT
3178.967	3179.887	31447.66	FE	1.5	1.2		MIT
3178.546	3179.465	31451.83	FE	1.0	0.8		MIT
3178.015	3178.934	31457.08	FE I	2.5	2.2		S
3177.677	3178.596	31460.43	FE	0.3	0.0		MIT
3177.535	3178.454	31461.83	FE II	0.7	2.5		MIT
3176.363	3177.282	31473.44	FE I	1.3	1.0		MIT
3175.989	3176.908	31477.15	FE	1.1	0.7		MIT
3175.447	3176.366	31482.52	FE I	2.3	2.3		S
3175.080	3175.999	31486.16	FE II		0.3		MIT
3175.035	3175.954	31486.61	FE	0.0			MIT
3174.96	3175.88	31487.4	FE	0.7	0.6		MIT
3173.687	3174.605	31499.98	FE	1.3	1.3		MIT
3173.614	3174.532	31500.70	FE	1.3	1.3		MIT
3173.412	3174.330	31502.71	FE	1.3	1.0		MIT
3172.067	3172.985	31516.07	FE I	2.0	2.0		MIT
3171.663	3172.581	31520.08	FE I	1.5	1.0		MIT
3171.353	3172.271	31523.16	FE I	2.0	1.9		MIT
3170.346	3171.263	31533.17	FE II	1.0	1.7		MIT
3169.615	3170.532	31540.45	FE	0.3	0.3		MIT
3168.857	3169.774	31547.99	FE	1.5	1.2		MIT
3168.151	3169.068	31555.02	FE	1.0	0.5		MIT
3167.922	3168.839	31557.30	FE	2.0	1.5		MIT
3167.859	3168.776	31557.93	FE II	0.3	2.0		MIT
3166.699	3167.615	31569.49	FE II		0.5		MIT
3166.438	3167.354	31572.09	FE	2.0	1.9		MIT
3166.249	3167.165	31573.98	FE	0.7	0.5		MIT
3165.861	3166.777	31577.84	FE I	2.0	1.9		MIT
3165.006	3165.922	31586.37	FE I	2.0	1.8		MIT
3164.299	3165.215	31593.43	FE I	1.3	1.0		MIT
3163.874	3164.790	31597.68	FE	1.6	1.4		MIT
3163.098	3164.014	31605.43	FE II	0.0	1.0		MIT
3162.800	3163.715	31608.41	FE II	0.0	2.0		MIT
3162.335	3163.250	31613.05	FE I	1.8	1.7		MIT
3161.949	3162.864	31616.91	FE I	2.3	2.2		MIT
3161.556	3162.471	31620.84	FE	0.3	0.0		MIT
3161.371	3162.286	31622.69	FE I	1.9	1.8		MIT
3160.917	3161.832	31627.23	FE	0.7	0.6		MIT
3160.658	3161.573	31629.83	FE I	2.2	2.1		S
3160.344	3161.259	31632.97	FE	1.6	1.3		MIT
3160.200	3161.115	31634.41	FE	1.8	1.7		MIT
3159.023	3159.937	31646.20	FE	0.8	0.5		MIT
3158.511	3159.425	31651.32	FE	0.0			MIT
3158.400	3159.314	31652.44	FE	0.8	0.5		MIT
3158.163	3159.077	31654.81	FE		0.5		MIT
3157.98	3158.89	31656.6	FE I	1.0	0.8		MIT
3157.887	3158.801	31657.58	FE I	2.0	2.0		MIT
3157.451	3158.365	31661.95	FE	1.0	0.7		MIT
3157.151	3158.065	31664.96	FE	0.6	0.5		MIT
3157.040	3157.954	31666.07	FE I	2.2	2.0		S
3156.458	3157.372	31671.91	FE	0.5	0.3		MIT
3156.274	3157.188	31673.76	FE	2.1	2.0		MIT
3155.953	3156.867	31676.98	FE II		0.3		MIT
3155.794	3156.708	31678.57	FE	0.7	0.0		MIT
3155.299	3156.213	31683.54	FE	1.7	1.5		MIT
3155.121	3156.035	31685.33	FE	0.6	0.0		MIT
3154.504	3155.417	31691.53	FE	1.3	0.7		MIT
3154.206	3155.119	31694.52	FE II		2.6		MIT
3153.748	3154.661	31699.12	FE	1.6	1.2		MIT

AIR WAVELENGTH (ANGSTRMS)	VACUUM WAVELENGTH (ANGSTRMS)	WAVENUMBER (KAYSERS)	LMENT	LOG ARC	INTENSITY SPK	DIS	REF
3153.367	3154.280	31702.96	FE	1.2	1.2		MIT
3153.206	3154.119	31704.57	FE I	2.0	1.9		MIT
3152.026	3152.939	31716.44	FE	0.3	0.0		MIT
3151.869	3152.782	31718.02	FE I	1.6	1.2		MIT
3151.351	3152.264	31723.24	FE	2.5	2.2		MIT
3151.005	3151.917	31726.72	FE	0.3	0.3		MIT
3150.304	3151.216	31733.78	FE	1.8	1.5		MIT
3149.848	3150.760	31738.37	FE	0.0	0.0		MIT
3148.915	3149.827	31747.78	FE	0.3H	0.3H		MIT
3148.414	3149.326	31752.83	FE	2.0	1.6		MIT
3147.793	3148.705	31759.09	FE	1.6	1.2		MIT
3147.605	3148.517	31760.99	FE	1.0	0.7		MIT
3147.290	3148.202	31764.17	FE	1.3	1.0		MIT
3146.748	3147.659	31769.64	FE II		0.6		DO
3146.471	3147.382	31772.44	FE I	0.7	0.3		MIT
3145.519	3146.430	31782.05	FE	0.3	0.3		MIT
3145.057	3145.968	31786.72	FE	1.6	1.4		MIT
3144.758	3145.669	31789.74	FE II		1.7		MIT
3144.495	3145.406	31792.40	FE I	2.2	2.0		MIT
3143.9896	3144.9003	31797.510	FE	2.3	2.2		IME
3143.245	3144.156	31805.04	FE	1.8	1.5		MIT
3142.883	3143.793	31808.71	FE	1.9	1.8		MIT
3142.453	3143.363	31813.06	FE I	2.1	2.0		MIT
3141.887	3142.797	31818.79	FE	0.3	0.3		MIT
3140.734	3141.644	31830.47	FE	0.3	0.3		MIT
3140.391	3141.301	31833.95	FE	2.0	1.9		MIT
3139.908	3140.818	31838.84	FE	1.8	1.6		MIT
3139.658	3140.568	31841.38	FE I	1.2	0.9		MIT
3139.166	3140.075	31846.37	FE	0.3	0.3		MIT
3138.518	3139.427	31852.94	FE	1.0	0.7		MIT
3137.785	3138.694	31860.38	FE	0.3	0.0		MIT
3136.893	3137.802	31869.44	FE	0.3	0.3		MIT
3136.499	3137.408	31873.45	FE	1.8	1.6		MIT
3135.861	3136.770	31879.93	FE I	0.8	0.6		MIT
3135.680	3136.589	31881.77	FE	0.5	0.3		MIT
3135.590	3136.499	31882.69	FE	0.5	0.3		MIT
3135.453	3136.362	31884.08	FE	1.0	0.5		MIT
3135.364	3136.273	31884.98	FE II	0.0	2.0		MIT
3134.405	3135.313	31894.74	FE	0.5H	0.0H		MIT
3134.111	3135.019	31897.73	FE I	2.3	2.1		S
3133.226	3134.134	31906.74	FE	0.7	0.6		MIT
3133.053	3133.961	31908.50	FE II		1.5		MIT
3132.680	3133.588	31912.30	FE	0.7	0.5		MIT
3132.514	3133.422	31913.99	FE	1.8	1.6		MIT
3131.718	3132.626	31922.10	FE II		1.5		MIT
3131.455	3132.363	31924.78	FE	0.3H			MIT
3130.567	3131.474	31933.84	FE II	0.6	0.6		MIT
3130.278	3131.185	31936.79	FE	0.7	0.6		MIT
3129.330	3130.237	31946.46	FE I	2.0	1.8		MIT
3129.105	3130.012	31948.76	FE	1.2	0.9		MIT
3128.896	3129.803	31950.89	FE	1.0	0.7		MIT
3128.285	3129.192	31957.13	FE	0.7	0.6		MIT
3127.687	3128.594	31963.24	FE	0.5	0.3		MIT
3126.758	3127.664	31972.74	FE	0.8	0.5		MIT
3126.173	3127.079	31978.72	FE	2.2	1.8		MIT
3125.654	3126.560	31984.03	FE I	2.6	2.5		I
3125.268	3126.174	31987.98	FE	0.7	0.3		MIT
3124.889	3125.795	31991.86	FE	1.2	0.8		MIT
3124.09	3125.00	32000.0	FE I	0.0			MIT
3123.951	3124.857	32001.47	FE	0.7	0.7		MIT
3123.548	3124.454	32005.60	FE	0.5	0.0		MIT
3123.346	3124.252	32007.67	FE	1.0	0.6		MIT
3122.547	3123.452	32015.86	FE	0.6	0.6		MIT
3122.301	3123.206	32018.38	FE	1.8	1.3		MIT
3121.773	3122.678	32023.79	FE	0.8	0.7		MIT
3120.874	3121.779	32033.02	FE	1.9	1.7		MIT
3120.434	3121.339	32037.53	FE	2.0	1.9		MIT
3120.231	3121.136	32039.62	FE	0.7	0.3		MIT
3120.022	3120.927	32041.76	FE II		0.3		MIT
3119.667	3120.572	32045.41	FE	0.0	0.0		MIT

AIR WAVELENGTH (ANGSTRMS)	VACUUM WAVELENGTH (ANGSTRMS)	WAVENUMBER (KAYSERS)	LMENT	LOG ARC	INTENSITY SPK	DIS	REF
3119.494	3120.399	32047.19	FE	2.0	1.9		MIT
3117.888	3118.792	32063.70	FE	0.5	0.5		MIT
3117.762	3118.666	32064.99	FE	0.3	0.3		MIT
3117.640	3118.544	32066.25	FE	1.3	1.0		MIT
3117.191	3118.095	32070.86	FE	0.3	0.3		MIT
3116.982	3117.886	32073.01	FE	0.3			MIT
3116.633	3117.537	32076.61	FE I	2.2			S
3116.590	3117.494	32077.05	FE II		2.2		DO
3116.250	3117.154	32080.55	FE I	0.7	0.6		MIT
3115.658	3116.562	32086.64	FE	0.5	0.0		MIT
3115.359	3116.263	32089.72	FE		0.0		MIT
3114.775	3115.678	32095.74	FE	0.6	0.3		MIT
3114.682	3115.585	32096.70	FE II		1.0		MIT
3114.293	3115.196	32100.71	FE II		1.9		MIT
3113.592	3114.495	32107.93	FE	1.4	1.0		MIT
3113.367	3114.270	32110.25	FE	0.3	0.3		MIT
3112.081	3112.984	32123.52	FE	1.5	1.3		MIT
3111.821	3112.724	32126.21	FE	1.0	0.8		MIT
3111.684	3112.587	32127.62	FE	0.9	0.5		MIT
3110.841	3111.743	32136.33	FE	1.3	1.0		MIT
3110.71	3111.61	32137.7	FE	0.3			MIT
3110.281	3111.183	32142.11	FE	1.6	1.5		MIT
3110.188	3111.090	32143.07	FE	0.8	0.8		MIT
3109.620	3110.522	32148.94	FE	0.3	0.0		MIT
3109.337	3110.239	32151.87	FE	0.5	0.5		MIT
3109.070	3109.972	32154.63	FE	0.3	0.3		MIT
3108.954	3109.856	32155.83	FE	0.3	0.3		MIT
3107.978	3108.880	32165.93	FE	1.3	1.0		MIT
3107.327	3108.228	32172.67	FE	0.3	0.0		MIT
3106.559	3107.460	32180.62	FE II	0.0	1.5		DO
3106.018	3106.919	32186.23	FE	0.6	0.6		MIT
3105.666	3106.567	32189.87	FE	0.6	0.3		MIT
3105.548	3106.449	32191.10	FE II	0.0	1.5		DO
3105.168	3106.069	32195.04	FE II		1.8		MIT
3104.168	3105.069	32205.41	FE	0.3	0.3		MIT
3103.849	3104.750	32208.72	FE	0.3	0.3		MIT
3103.767	3104.668	32209.57	FE	0.3	0.3		MIT
3102.872	3103.772	32218.86	FE	1.5	1.3		MIT
3102.638	3103.538	32221.29	FE	0.8	0.7		MIT
3102.361	3103.261	32224.16	FE	0.3	0.3		MIT
3102.144	3103.044	32226.42	FE	0.3	0.3		MIT
3101.004	3101.904	32238.26	FE	0.9	0.9		MIT
3100.839	3101.739	32239.98	FE	0.5	0.5		MIT
3100.666	3101.566	32241.78	FE I	2.0	2.0		MIT
3100.304	3101.204	32245.54	FE I	2.0	2.0		MIT
3099.417	3100.316	32254.77	FE	0.3	0.3		MIT
3098.588	3099.487	32263.40	FE	0.7	0.7		MIT
3098.192	3099.091	32267.52	FE I	1.8	1.8		MIT
3097.815	3098.714	32271.45	FE	1.3H	1.0H		MIT
3097.506	3098.405	32274.67	FE	0.0H	0.0H		MIT
3096.836	3097.735	32281.65	FE	1.5	1.3		MIT
3096.296	3097.195	32287.28	FE II	0.3	1.5		DO
3096.044	3096.943	32289.91	FE	0.5	0.0		MIT
3095.269	3096.167	32298.00	FE	1.0	0.8		MIT
3094.900	3095.798	32301.84	FE I	1.5	1.2		MIT
3094.622	3095.520	32304.75	FE	0.5	0.6		MIT
3094.328	3095.226	32307.82	FE	0.3H	0.3H		MIT
3093.883	3094.781	32312.46	FE	1.6	1.5		MIT
3093.806	3094.704	32313.27	FE I	1.7	1.6		MIT
3093.357	3094.255	32317.96	FE	1.8	1.6		MIT
3092.778	3093.676	32324.01	FE	1.7	1.5		MIT
3092.399	3093.297	32327.97	FE	0.6	0.7		MIT
3091.578	3092.476	32336.55	FE I	2.5	2.3		S
3091.097	3091.994	32341.58	FE	0.3	0.0		MIT
3090.209	3091.106	32350.88	FE	1.5	1.2		MIT
3089.741	3090.638	32355.78	FE	0.3	0.3		MIT
3089.512	3090.409	32358.18	FE	0.3			MIT
3089.388	3090.285	32359.48	FE II		1.0		MIT
3088.799	3089.696	32365.65	FE	0.3H	0.3H		MIT
3088.350	3089.247	32370.35	FE	0.3	0.0		MIT

AIR WAVELENGTH (ANGSTRMS)	VACUUM WAVELENGTH (ANGSTRMS)	WAVENUMBER (KAYSERS)	LMENT	LOG ARC	INTENSITY SPK	DIS	REF
3087.348	3088.244	32380.86	FE	0.3	0.3		MIT
3086.69	3087.59	32387.8	FE	0.3H	0.3H		MIT
3086.560	3087.456	32389.12	FE	0.3H	0.3H		MIT
3085.202	3086.098	32403.38	FE	0.5	0.5		MIT
3083.742	3084.638	32418.72	FE I	2.7	2.7		S
3083.15	3084.05	32424.9	FE	0.0	0.0		MIT
3083.027	3083.922	32426.24	FE II	0.0	0.3		MIT
3081.84	3082.74	32438.7	FE	0.3	0.3		MIT
3081.656	3082.551	32440.66	FE	0.3H	0.3H		MIT
3081.550	3082.445	32441.78	FE	0.3	0.3		MIT
3081.225	3082.120	32445.20	FE	0.3H	0.3H		MIT
3080.985	3081.880	32447.73	FE	0.6	0.6		MIT
3080.754	3081.649	32450.16	FE	0.3	0.3		MIT
3080.403	3081.298	32453.86	FE II		0.3		MIT
3080.112	3081.007	32456.92	FE	1.5	1.2		MIT
3079.985	3080.880	32458.26	FE	1.5	1.3		MIT
3078.698	3079.592	32471.83	FE II	0.6	1.2H		DO
3078.434	3079.328	32474.61	FE I	1.9	1.7		MIT
3078.018	3078.912	32479.00	FE	2.0	1.9		MIT
3077.643	3078.537	32482.96	FE	1.8	1.4		MIT
3077.168	3078.062	32487.98	FE II	0.0	2.5		MIT
3076.455	3077.349	32495.50	FE II		0.7		DO
3075.721	3076.615	32503.26	FE I	2.6	2.6		S
3074.438	3075.331	32516.82	FE	1.6	1.4		MIT
3074.151	3075.044	32519.86	FE	1.6	1.4		MIT
3073.983	3074.876	32521.64	FE	1.6	1.4		MIT
3073.244	3074.137	32529.46	FE	0.3			MIT
3072.335	3073.228	32539.08	FE	0.9	0.8		MIT
3072.05	3072.94	32542.1	FE	0.8	0.5		MIT
3071.267	3072.159	32550.39	FE II	0.0	0.3		MIT
3071.146	3072.038	32551.68	FE II	0.3	0.3		MIT
3070.694	3071.586	32556.47	FE II		0.8		MIT
3069.453	3070.345	32569.63	FE	0.8	0.6		MIT
3069.34	3070.23	32570.8	FE	0.3	0.0		MIT
3068.176	3069.068	32583.19	FE I	2.2	2.2		MIT
3067.944	3068.836	32585.65	FE	1.2	1.0		MIT
3067.244	3068.135	32593.08	FE I	2.5	2.5		S
3067.120	3068.011	32594.40	FE I	0.8	0.8		MIT
3067.004	3067.895	32595.64	FE	0.3	0.3		MIT
3066.483	3067.374	32601.17	FE	1.8	1.6		MIT
3065.315	3066.206	32613.59	FE II		1.8		DO
3064.217	3065.108	32625.28	FE	0.6	0.6		MIT
3063.933	3064.824	32628.30	FE	1.6	1.5		MIT
3063.728	3064.619	32630.49	FE	0.5	0.5		MIT
3063.563	3064.454	32632.25	FE	0.3	0.3		MIT
3063.152	3064.042	32636.62	FE	0.3	0.0		MIT
3062.871	3063.761	32639.62	FE	0.6	0.3		MIT
3062.233	3063.123	32646.42	FE II	0.3	2.6		MIT
3061.824	3062.714	32650.78	FE	0.7	0.5		MIT
3060.988	3061.878	32659.70	FE I	1.7	1.5		MIT
3060.792	3061.682	32661.79	FE	0.6	0.6		MIT
3060.645	3061.535	32663.36	FE	0.3	0.3		MIT
3060.538	3061.428	32664.50	FE	0.8	0.8		MIT
3059.086	3059.975	32680.00	FE I	2.8R	2.6		S
3058.493	3059.382	32686.34	FE	0.5	0.5		MIT
3057.790	3058.679	32693.85	FE	0.3	0.3		MIT
3057.65	3058.54	32695.4	FE	0.3	0.3		MIT
3057.446	3058.335	32697.53	FE I	2.6	2.6		S
3056.802	3057.691	32704.42	FE II	0.6	1.4		DO
3056.25	3057.14	32710.3	FE	0.9	0.6		MIT
3055.71	3056.60	32716.1	FE	1.0	0.8		MIT
3055.368	3056.256	32719.77	FE II		0.3		DO
3055.263	3056.151	32720.89	FE I	2.3	2.2		I
3053.44	3054.33	32740.4	FE I	1.9	1.7		MIT
3053.070	3053.958	32744.39	FE	2.0	1.9		MIT
3049.36	3050.25	32784.2	FE	1.4	0.9		MIT
3049.011	3049.898	32787.98	FE II	0.6	0.5		DO
3048.455	3049.342	32793.96	FE	2.0	0.9		MIT
3047.605	3048.492	32803.11	FE I	2.9R	2.7R		S
3047.050	3047.936	32809.08	FE	0.3	0.5		MIT

AIR WAVELENGTH (ANGSTRMS)	VACUUM WAVELENGTH (ANGSTRMS)	WAVENUMBER (KAYSERS)	LMENT	LOG ARC	INTENSITY SPK	DIS	REF
3046.930	3047.816	32810.38	FE	0.6	0.6		MIT
3045.594	3046.480	32824.77	FE	1.0	0.8		MIT
3045.078	3045.964	32830.33	FE	2.2	2.0		MIT
3044.843	3045.729	32832.86	FE II		1.1		DO
3042.665	3043.550	32856.37	FE I	2.5	2.3		MIT
3042.022	3042.907	32863.31	FE I	2.1	2.0		MIT
3041.738	3042.623	32866.38	FE I	2.0	1.9		MIT
3041.639	3042.524	32867.45	FE I	1.9	1.9		MIT
3040.96	3041.84	32874.8	FE	0.9	0.6		MIT
3040.428	3041.313	32880.54	FE I	2.6	2.6		I
3039.316	3040.200	32892.57	FE	1.3	1.2		MIT
3038.779	3039.663	32898.38	FE II		0.5		MIT
3037.782	3038.666	32909.18	FE I	0.6	0.6		MIT
3037.388	3038.272	32913.45	FE I	2.8G	2.6R		S
3036.986	3037.870	32917.80	FE II		0.3		DO
3035.74	3036.62	32931.3	FE	2.0	1.8		MIT
3034.538	3035.421	32944.36	FE I	1.8	1.6		MIT
3034.12	3035.00	32948.9	FE	0.3			MIT
3033.445	3034.328	32956.23	FE II	0.6	0.3		DO
3033.101	3033.984	32959.96	FE	1.6	1.3		MIT
3031.639	3032.522	32975.86	FE I	2.3	2.3		MIT
3031.215	3032.097	32980.47	FE	2.2	2.2		MIT
3030.61	3031.49	32987.0	FE	0.7	0.5		MIT
3030.149	3031.031	32992.07	FE	2.5	2.5		I
3029.233	3030.115	33002.05	FE I	1.9	1.8		MIT
3026.461	3027.342	33032.27	FE I	2.3	2.3		MIT
3025.842	3026.723	33039.03	FE I	2.6R	2.5R		MIT
3025.638	3026.519	33041.26	FE	2.0	2.0		MIT
3025.282	3026.163	33045.15	FE	1.7	1.5		MIT
3024.033	3024.914	33058.79	FE I	2.5	2.3		I
3021.407	3022.287	33087.53	FE II		0.0		ME
3021.073	3021.953	33091.18	FE I	2.8G	2.5R		MIT
3020.640	3021.520	33095.93	FE I	3.0G	2.8R		MIT
3020.489	3021.369	33097.58	FE I	2.5R	2.5R		MIT
3019.290	3020.169	33110.73	FE	0.3	0.3		MIT
3018.982	3019.861	33114.10	FE I	2.2	2.2		MIT
3018.13	3019.01	33123.5	FE	0.6	0.6		MIT
3017.632	3018.511	33128.92	FE I	2.2	2.2		MIT
3017.43	3018.31	33131.1	FE	0.5	0.5		MIT
3016.185	3017.064	33144.81	FE I	2.3	2.2		MIT
3015.9129	3016.7915	33147.800	FE	1.8	1.7		IME
3014.175	3015.053	33166.91	FE I	1.8	1.5		MIT
3012.450	3013.328	33185.90	FE	1.7	1.5		MIT
3011.94	3012.82	33191.5	FE	0.5	0.3		MIT
3011.482	3012.359	33196.57	FE	2.1	2.1		MIT
3009.570	3010.447	33217.66	FE I	2.7	2.6		I
3009.092	3009.969	33222.94	FE	1.9	1.8		MIT
3008.139	3009.016	33233.46	FE I	2.8R	2.6R		MIT
3007.281	3008.157	33242.94	FE I	1.9	1.8		MIT
3007.145	3008.021	33244.44	FE I	2.0	1.9		MIT
3005.306	3006.182	33264.79	FE	1.8	1.6		MIT
3004.630	3005.506	33272.27	FE I	0.7	0.5		MIT
3004.263	3005.139	33276.33	FE II		0.3		MIT
3004.123	3004.999	33277.88	FE	1.3	1.0		MIT
3003.031	3003.906	33289.99	FE I	2.3	2.0		I
3002.649	3003.524	33294.22	FE II	1.3	2.2		MIT
3002.334	3003.209	33297.71	FE II		0.7		MIT
3000.951	3001.826	33313.06	FE I	2.9G	2.5R		MIT
3000.452	3001.327	33318.60	FE I	2.0	1.9		MIT
3000.065	3000.940	33322.90	FE II		1.0		MIT
2999.512	3000.387	33329.04	FE I	2.7	2.5		S
2999.20	3000.07	33332.5	FE	0.7	0.5		MIT
2998.855	2999.729	33336.34	FE II		0.5		DO
2997.301	2998.175	33353.62	FE II		1.8		MIT
2996.391	2997.265	33363.75	FE	2.0	1.7		MIT
2995.840	2996.714	33369.89	FE	0.5	0.0		MIT
2995.27	2996.14	33376.2	FE	0.6	0.3		MIT
2994.507	2995.380	33384.74	FE I	0.3			MIT
2994.429	2995.302	33385.61	FE I	3.0G	2.8R		MIT
2992.61	2993.48	33405.9	FE	0.3	0.3		MIT

AIR WAVELENGTH (ANGSTRMS)	VACUUM WAVELENGTH (ANGSTRMS)	WAVENUMBER (KAYSERS)	LMENT	LOG ARC	INTENSITY SPK	DIS	REF
2992.41	2993.28	33408.1	FE	0.3H	0.0H		MIT
2991.637	2992.510	33416.77	FE	2.0	1.9		MIT
2991.095	2991.967	33422.82	FE	0.0			MIT
2990.392	2991.264	33430.68	FE	2.2	2.0		I
2988.940	2989.812	33446.92	FE	0.3	0.0		MIT
2988.474	2989.346	33452.14	FE I	1.8	1.5		MIT
2987.292	2988.163	33465.37	FE I	2.5	2.3		S
2986.653	2987.524	33472.53	FE	1.2			MIT
2986.614	2987.485	33472.97	FE II		1.5		MIT
2986.460	2987.331	33474.69	FE I	2.0	1.8		MIT
2986.311	2987.182	33476.36	FE	0.5			MIT
2985.74	2986.61	33482.8	FE	0.5	0.3		MIT
2985.550	2986.421	33484.90	FE II	1.9	2.5		MIT
2984.830	2985.701	33492.97	FE II	2.3R	2.6		MIT
2984.830	2985.701	33492.97	FE I	2.3R	2.6		MIT
2984.575	2985.446	33495.83	FE	0.3			MIT
2983.572	2984.443	33507.10	FE I	3.0G	2.6R		MIT
2982.232	2983.102	33522.15	FE II	0.7	1.0		MIT
2982.062	2982.932	33524.06	FE II		2.0		MIT
2981.854	2982.724	33526.40	FE I	2.0	1.7		MIT
2981.446	2982.316	33530.99	FE I	2.5	2.3		S
2980.958	2981.828	33536.48	FE II		1.0		MIT
2980.539	2981.409	33541.19	FE	2.0	1.8		MIT
2979.87	2980.74	33548.7	FE	0.6	0.3		MIT
2979.352	2980.221	33554.55	FE II	1.3	2.0		MIT
2979.095	2979.964	33557.45	FE II		1.5		MIT
2978.848	2979.717	33560.23	FE II		0.8		MIT
2978.05	2978.92	33569.2	FE	0.9	0.5		MIT
2976.55	2977.42	33586.1	FE I	1.2	1.0		MIT
2976.131	2977.000	33590.87	FE	2.0	1.8		MIT
2975.936	2976.805	33593.07	FE II	0.5	1.6		MIT
2974.778	2975.646	33606.14	FE	1.0	0.8		MIT
2973.237	2974.105	33623.56	FE I	2.7G	2.6G		MIT
2973.134	2974.002	33624.73	FE I	2.7G	2.6G		MIT
2972.279	2973.147	33634.40	FE I	2.0	1.6		MIT
2971.76	2972.63	33640.3	FE	0.7	0.3		MIT
2970.684	2971.551	33652.46	FE II		1.0		MIT
2970.513	2971.380	33654.39	FE II	1.5	2.0		MIT
2970.105	2970.972	33659.02	FE I	2.6	2.3		MIT
2969.934	2970.801	33660.95	FE II		1.2		MIT
2969.477	2970.344	33666.13	FE I	1.8	1.8		MIT
2969.362	2970.229	33667.44	FE I	1.9	1.9		MIT
2968.732	2969.599	33674.58	FE II		0.3		MIT
2968.481	2969.348	33677.43	FE	1.5	1.3		MIT
2966.900	2967.766	33695.37	FE I	3.0G	2.8R		MIT
2966.24	2967.11	33702.9	FE I	0.6	0.6		MIT
2965.811	2966.677	33707.75	FE	1.4	1.2		MIT
2965.255	2966.121	33714.07	FE I	2.6	2.2		S
2965.037	2965.903	33716.54	FE II	0.6	1.7		MIT
2964.630	2965.496	33721.17	FE II	0.9	2.2		MIT
2964.22	2965.09	33725.8	FE	0.5			MIT
2964.133	2964.999	33726.83	FE II		0.8		MIT
2963.71	2964.58	33731.6	FE	0.3H	0.0H		MIT
2963.56	2964.43	33733.4	FE	0.3	0.0		MIT
2962.59	2963.46	33744.4	FE	0.3	0.0		MIT
2962.12	2962.99	33749.7	FE I	0.9	0.6		MIT
2961.281	2962.146	33759.31	FE II	0.5	1.6		MIT
2960.66	2961.52	33766.4	FE	0.8	0.5		MIT
2960.555	2961.420	33767.59	FE	1.0	0.7		MIT
2960.299	2961.164	33770.51	FE	1.8	1.5		MIT
2959.992	2960.857	33774.01	FE	2.2	1.9		MIT
2959.835	2960.700	33775.80	FE		0.3		MIT
2959.682	2960.547	33777.55	FE	2.2	1.0		MIT
2959.599	2960.464	33778.49	FE II		1.8		MIT
2959.34	2960.20	33781.5	FE	1.8	1.4		MIT
2958.44	2959.30	33791.7	FE	0.5	0.3		MIT
2957.633	2958.497	33800.95	FE	0.3	0.3		MIT
2957.491	2958.355	33802.57	FE	0.6	0.5		MIT
2957.365	2958.229	33804.01	FE I	2.5	2.5		S
2956.859	2957.723	33809.79	FE	0.7	0.6		MIT

AIR WAVELENGTH (ANGSTRMS)	VACUUM WAVELENGTH (ANGSTRMS)	WAVENUMBER (KAYSERS)	LMENT	LOG ARC	INTENSITY SPK	DIS	REF
2956.70	2957.56	33811.6	FE I	1.4H	0.9H		MIT
2954.655	2955.518	33835.01	FE	2.0	1.8		MIT
2954.346	2955.209	33838.55	FE	0.3	0.3		MIT
2954.050	2954.913	33841.94	FE II		0.5		DO
2953.940	2954.803	33843.20	FE I	2.6R	2.2		S
2953.778	2954.641	33845.06	FE II	0.7	1.9		MIT
2953.486	2954.349	33848.40	FE	2.0	1.7		MIT
2951.56	2952.42	33870.5	FE	1.0	0.6		MIT
2951.379	2952.242	33872.57	FE	0.6	0.3		MIT
2951.098	2951.960	33875.79	FE II		0.5		MIT
2950.93	2951.79	33877.7	FE	0.3			MIT
2950.243	2951.105	33885.61	FE	2.8	2.5		MIT
2949.70	2950.56	33891.9	FE	1.0	0.7		MIT
2948.95	2949.81	33900.5	FE I	1.0	0.6		MIT
2948.726	2949.588	33903.04	FE I	1.0	0.8		MIT
2948.433	2949.295	33906.41	FE	1.9	1.8		MIT
2947.876	2948.738	33912.82	FE I	2.8R	2.3		MIT
2947.658	2948.520	33915.33	FE II	1.0	2.0		MIT
2947.363	2948.225	33918.72	FE	1.5	1.3		MIT
2947.13	2947.99	33921.4	FE	0.6	0.3		MIT
2946.10	2946.96	33933.3	FE	0.5	0.0		MIT
2945.88	2946.74	33935.8	FE	0.3			MIT
2945.70	2946.56	33937.9	FE	1.0	0.7		MIT
2945.262	2946.123	33942.91	FE II	0.0	0.5		MIT
2945.055	2945.916	33945.30	FE	2.0	1.5		MIT
2944.51	2945.37	33951.6	FE	0.3			MIT
2944.398	2945.259	33952.87	FE II	1.8	2.8		MIT
2943.57	2944.43	33962.4	FE	1.1	0.8		MIT
2942.63	2943.49	33973.3	FE	1.0	0.7		MIT
2941.75	2942.61	33983.4	FE	0.3			MIT
2941.343	2942.203	33988.14	FE I	2.8	2.5		S
2940.591	2941.451	33996.83	FE	2.3	1.9		MIT
2939.508	2940.368	34009.35	FE II	0.5	1.5		MIT
2939.080	2939.939	34014.31	FE I	1.9	1.3		MIT
2938.73	2939.59	34018.4	FE	0.3			MIT
2938.05	2938.91	34026.2	FE	0.3	0.0		MIT
2937.811	2938.670	34029.00	FE	2.5	2.2		MIT
2936.905	2937.764	34039.50	FE I	2.8R	2.7R		MIT
2936.449	2937.308	34044.78	FE	0.7	0.7		MIT
2936.024	2936.883	34049.71	FE II	0.7	1.0		MIT
2934.493	2935.351	34067.47	FE II		0.9		MIT
2934.37	2935.23	34068.9	FE I	0.8	0.5		MIT
2933.81	2934.67	34075.4	FE	0.5	0.0		MIT
2931.98	2932.84	34096.7	FE	0.7	0.5		MIT
2931.81	2932.67	34098.6	FE	1.0	0.8		MIT
2931.600	2932.458	34101.09	FE		1.2		MIT
2931.43	2932.29	34103.1	FE	1.0	0.5		MIT
2930.66	2931.52	34112.0	FE	0.3			MIT
2929.624	2930.481	34124.09	FE I	1.7	1.0		MIT
2929.24	2930.10	34128.6	FE	0.5	0.0		MIT
2929.121	2929.978	34129.95	FE	1.0	1.0		MIT
2929.008	2929.865	34131.27	FE I	2.2	2.0		S
2928.753	2929.610	34134.24	FE I	0.7			MIT
2928.105	2928.962	34141.79	FE I	1.0	0.6		MIT
2927.88	2928.74	34144.4	FE	0.5	0.3		MIT
2927.552	2928.409	34148.24	FE	1.3	1.1		MIT
2927.05	2927.91	34154.1	FE	0.0			MIT
2926.587	2927.443	34159.50	FE II	2.2	2.6		MIT
2926.01	2926.87	34166.2	FE	0.3	0.0		MIT
2925.900	2926.756	34167.52	FE I	1.2	1.0		MIT
2925.791	2926.647	34168.79	FE	1.2	0.8		MIT
2925.363	2926.219	34173.79	FE	1.8	1.7		MIT
2924.35	2925.21	34185.6	FE	0.8	0.5		MIT
2923.99	2924.85	34189.8	FE	0.7	0.3		MIT
2923.852	2924.708	34191.45	FE	2.0	1.8		MIT
2923.439	2924.295	34196.28	FE	1.5	1.1		MIT
2923.290	2924.146	34198.02	FE	1.7	1.5		MIT
2923.17	2924.03	34199.4	FE	0.5	0.0		MIT
2922.620	2923.475	34205.86	FE	1.7	1.4		MIT
2922.38	2923.24	34208.7	FE I	1.0	0.6		MIT

AIR WAVELENGTH (ANGSTRMS)	VACUUM WAVELENGTH (ANGSTRMS)	WAVENUMBER (KAYSERS)	LMENT	LOG ARC	INTENSITY SPK	DIS	REF
2922.22	2923.08	34210.5	FE	0.5	0.0		MIT
2922.024	2922.879	34212.84	FE II		1.7		MIT
2921.71	2922.57	34216.5	FE	0.6	0.3		MIT
2921.60	2922.46	34217.8	FE	0.5	0.0		MIT
2920.99	2921.85	34225.0	FE	1.1	0.8		MIT
2920.690	2921.545	34228.46	FE I	2.2	1.9		I
2920.29	2921.14	34233.1	FE I	0.3	0.0		MIT
2920.15	2921.00	34234.8	FE	0.3	0.0		MIT
2919.847	2920.702	34238.35	FE	1.9	1.5		MIT
2919.21	2920.06	34245.8	FE	1.2	1.0		MIT
2919.04	2919.89	34247.8	FE	0.0			MIT
2918.824	2919.678	34250.35	FE	1.2	0.9		MIT
2918.357	2919.211	34255.83	FE	1.6	1.4		MIT
2918.16	2919.01	34258.1	FE	0.5	0.3		MIT
2918.027	2918.881	34259.70	FE	2.1	2.0		MIT
2917.874	2918.728	34261.50	FE	0.3			MIT
2917.470	2918.324	34266.24	FE II	0.3	1.4		MIT
2917.094	2917.948	34270.66	FE II		1.3		MIT
2916.920	2917.774	34272.70	FE II		0.5		MIT
2916.153	2917.007	34281.72	FE II	0.3H	0.8		MIT
2914.306	2915.159	34303.44	FE I	1.7	1.4		MIT
2914.20	2915.05	34304.7	FE	1.0	0.7		MIT
2912.257	2913.110	34327.58	FE I	0.9	0.6		MIT
2912.158	2913.011	34328.74	FE I	2.2	2.2		S
2911.08	2911.93	34341.5	FE	0.5			MIT
2911.01	2911.86	34342.3	FE	0.5			MIT
2910.92	2911.77	34343.3	FE	1.0	0.6		MIT
2910.759	2911.611	34345.24	FE II		0.3		MIT
2910.67	2911.52	34346.3	FE	0.6			MIT
2909.86	2910.71	34355.9	FE	0.6	0.0		MIT
2909.59	2910.44	34359.0	FE	0.3			MIT
2909.503	2910.355	34360.07	FE	1.8	1.5		MIT
2909.31	2910.16	34362.4	FE	0.6	0.3		MIT
2908.859	2909.711	34367.67	FE	1.9	1.6		MIT
2907.857	2908.709	34379.52	FE II	0.3	1.3		MIT
2907.520	2908.372	34383.50	FE I	2.0	1.9		MIT
2906.97	2907.82	34390.0	FE	0.3			MIT
2906.61	2907.46	34394.3	FE	0.3H	0.0		MIT
2906.421	2907.272	34396.50	FE	1.8	1.4		MIT
2906.122	2906.973	34400.04	FE II		1.6		MIT
2905.38	2906.23	34408.8	FE	0.8	0.6		MIT
2904.60	2905.45	34418.1	FE	0.0			MIT
2904.163	2905.014	34423.24	FE	1.2	0.9		MIT
2904.08	2904.93	34424.2	FE	0.8	0.5		MIT
2903.76	2904.61	34428.0	FE	0.3			MIT
2903.36	2904.21	34432.8	FE	0.3			MIT
2902.86	2903.71	34438.7	FE	0.3			MIT
2902.469	2903.319	34443.33	FE II	0.0	1.5		MIT
2902.317	2903.167	34445.14	FE II		0.3		MIT
2901.915	2902.765	34449.91	FE	2.1	1.6		MIT
2901.81	2902.66	34451.2	FE	0.5	0.3		MIT
2901.385	2902.235	34456.20	FE I	2.0	1.9		MIT
2899.415	2900.265	34479.61	FE	2.1	2.0		I
2898.93	2899.78	34485.4	FE	0.3			MIT
2898.86	2899.71	34486.2	FE	0.8	0.6		MIT
2898.735	2899.585	34487.70	FE II		0.5		MIT
2898.47	2899.32	34490.9	FE	0.3			MIT
2898.355	2899.204	34492.22	FE	2.0	1.5		MIT
2898.06	2898.91	34495.7	FE	0.3			MIT
2897.85	2898.70	34498.2	FE	0.3			MIT
2897.64	2898.49	34500.7	FE	0.6	0.0		MIT
2897.262	2898.111	34505.23	FE II		2.3		MIT
2896.96	2897.81	34508.8	FE	0.5			MIT
2896.69	2897.54	34512.0	FE	0.5			MIT
2896.40	2897.25	34515.5	FE	0.3			MIT
2895.215	2896.064	34529.63	FE II		1.9		MIT
2895.035	2895.884	34531.77	FE I	2.1	1.8		I
2894.778	2895.627	34534.84	FE II		1.9		MIT
2894.505	2895.353	34538.10	FE	2.2	2.2		I
2893.884	2894.732	34545.51	FE I	1.4	1.3		MIT
2893.764	2894.612	34546.94	FE I	1.2	0.9		MIT
2893.43	2894.28	34550.9	FE	0.0H			MIT
2892.828	2893.676	34558.12	FE II	0.3	1.3		MIT
2892.483	2893.331	34562.24	FE	2.0	1.6		MIT
2892.215	2893.063	34565.44	FE II	0.3	0.0		MIT
2891.906	2892.754	34569.14	FE	1.4	1.0		MIT
2891.71	2892.56	34571.5	FE	1.2	1.0		MIT
2891.40	2892.25	34575.2	FE I	0.6	0.3		MIT
2890.42	2891.27	34586.9	FE	0.6	0.0		MIT
2889.992	2890.839	34592.03	FE	1.2	0.9		MIT
2889.88	2890.73	34593.4	FE	1.3	1.2		MIT
2888.093	2888.940	34614.77	FE II		1.9		MIT
2887.96	2888.81	34616.4	FE	0.6	0.0		MIT
2887.807	2888.654	34618.20	FE	1.9	1.8		MIT
2887.312	2888.159	34624.14	FE II	0.5	1.3		DO
2887.154	2888.001	34626.03	FE	0.8	0.3		MIT
2886.318	2887.164	34636.06	FE I	1.7	1.2		MIT
2886.228	2887.074	34637.14	FE II		0.5		MIT
2885.928	2886.774	34640.74	FE II		1.8		MIT
2885.350	2886.196	34647.68	FE	0.8	0.3		MIT
2884.779	2885.625	34654.54	FE II		1.4H		DO
2883.735	2884.581	34667.08	FE	1.5			MIT
2883.702	2884.548	34667.48	FE II		2.5		MIT
2880.827	2881.672	34702.08	FE II	0.0	1.4		MIT
2880.756	2881.601	34702.93	FE II	1.2	1.7		MIT
2880.582	2881.427	34705.03	FE I	1.2	0.7		MIT
2879.43	2880.27	34718.9	FE	0.8	0.3		MIT
2879.242	2880.087	34721.18	FE II		1.4		MIT
2879.12	2879.96	34722.6	FE	0.3			MIT
2878.761	2879.606	34726.98	FE	0.9	0.7		MIT
2878.64	2879.48	34728.4	FE	0.5	0.3		MIT
2877.301	2878.145	34744.60	FE I	2.3	2.1		I
2876.802	2877.646	34750.62	FE II		2.0		MIT
2876.71	2877.55	34751.7	FE	0.7	0.3		MIT
2876.56	2877.40	34753.5	FE	0.5			MIT
2876.01	2876.85	34760.2	FE	1.2			MIT
2875.346	2876.190	34768.22	FE II		1.8		MIT
2875.304	2876.148	34768.73	FE I	2.1	1.7		MIT
2874.882	2875.726	34773.83	FE	1.8	1.3		MIT
2874.30	2875.14	34780.9	FE	0.3			MIT
2874.172	2875.015	34782.42	FE I	2.5	2.3		I
2873.65	2874.49	34788.7	FE	0.9	0.3		MIT
2873.528	2874.371	34790.22	FE	1.2	0.0		MIT
2873.401	2874.244	34791.75	FE II		2.5		MIT
2872.50	2873.34	34802.7	FE	0.6			MIT
2872.382	2873.225	34804.10	FE II	0.3	1.3		DO
2872.336	2873.179	34804.65	FE I	2.2	1.7		MIT
2871.73	2872.57	34812.0	FE	0.5	0.0		MIT
2871.129	2871.972	34819.29	FE II		1.3		MIT
2871.060	2871.903	34820.12	FE II		1.6		MIT
2870.602	2871.445	34825.68	FE II		1.2		MIT
2869.83	2870.67	34835.0	FE	1.0	0.7		MIT
2869.690	2870.532	34836.74	FE II		0.5		MIT
2869.308	2870.150	34841.38	FE I	2.5	1.8		S
2869.163	2870.005	34843.14	FE II		0.5		MIT
2868.871	2869.713	34846.69	FE II				MIT
2868.455	2869.297	34851.74	FE II	1.9	1.6		MIT
2868.214	2869.056	34854.67	FE	1.2	0.9		MIT
2867.880	2868.722	34858.73	FE I	0.7	0.3		MIT
2867.563	2868.405	34862.58	FE	1.8	1.5		MIT
2867.314	2868.156	34865.61	FE	1.8	1.5		MIT
2866.629	2867.471	34873.94	FE I	2.1	1.9		MIT
2865.87	2866.71	34883.2	FE	0.3			MIT
2865.189	2866.030	34891.47	FE	0.3			MIT
2864.973	2865.814	34894.10	FE II		1.7		MIT
2864.367	2865.208	34901.48	FE II		1.0		DO
2863.864	2864.705	34907.61	FE I	2.1	2.0		I
2863.435	2864.276	34912.84	FE I	2.0	1.9		MIT
2863.33	2864.17	34914.1	FE	0.7			MIT
2863.08	2863.92	34917.2	FE	0.5	0.3		MIT

AIR WAVELENGTH (ANGSTRMS)	VACUUM WAVELENGTH (ANGSTRMS)	WAVENUMBER (KAYSERS)	LMENT	LOG ARC	INTENSITY SPK	DIS	REF
2862.498	2863.339	34924.27	FE I	2.0	1.7		MIT
2861.189	2862.029	34940.24	FE II	0.0	1.5		MIT
2860.21	2861.05	34952.2	FE	0.3			MIT
2859.48	2860.32	34961.1	FE	0.3			MIT
2858.897	2859.737	34968.25	FE I	2.0	1.5		MIT
2858.343	2859.183	34975.03	FE II	0.5	2.3		MIT
2858.08	2858.92	34978.2	FE	0.6	0.0		MIT
2857.996	2858.835	34979.28	FE	1.1	0.3		MIT
2857.807	2858.646	34981.59	FE	1.1	0.3		MIT
2857.415	2858.254	34986.39	FE II		1.0		MIT
2857.172	2858.011	34989.37	FE II		1.5		MIT
2856.928	2857.767	34992.35	FE II		1.2H		DO
2856.390	2857.229	34998.94	FE II		0.7H		MIT
2856.136	2856.975	35002.06	FE II	0.0	1.6		MIT
2856.05	2856.89	35003.1	FE	0.3			MIT
2855.91	2856.75	35004.8	FE	0.3			MIT
2855.670	2856.509	35007.77	FE II	0.3	2.3		MIT
2853.774	2854.612	35031.02	FE	1.2	0.8		MIT
2853.688	2854.526	35032.08	FE I	1.2	0.8		MIT
2853.204	2854.042	35038.02	FE II		1.0		MIT
2852.958	2853.796	35041.04	FE I	0.8	0.3		M
2852.35	2853.19	35048.5	FE	0.3			MIT
2851.798	2852.636	35055.30	FE I	2.3	2.2		S
2851.50	2852.34	35059.0	FE	0.7	0.3		MIT
2850.11	2850.95	35076.1	FE	0.3			MIT
2849.606	2850.443	35082.26	FE II		1.7		MIT
2849.19	2850.03	35087.4	FE	0.3			MIT
2848.909	2849.746	35090.84	FE II	0.0	0.9		MIT
2848.718	2849.555	35093.20	FE I	1.8	1.5		MIT
2848.332	2849.169	35097.95	FE II		0.7H		DO
2848.122	2848.959	35100.54	FE II		1.0		DO
2848.046	2848.883	35101.48	FE II		1.8		MIT
2847.221	2848.058	35111.65	FE II		0.5		MIT
2846.825	2847.662	35116.53	FE I	1.3	1.1		MIT
2845.95	2846.79	35127.3	FE	0.6			MIT
2845.714	2846.550	35130.24	FE I	0.9	0.5		MIT
2845.595	2846.431	35131.71	FE I	2.1	0.8		MIT
2845.544	2846.380	35132.34	FE	2.1	0.8		MIT
2845.450	2846.286	35133.50	FE II		0.8		DO
2845.385	2846.221	35134.30	FE II		0.5		MIT
2844.973	2845.809	35139.39	FE II		0.7		DO
2843.979	2844.815	35151.67	FE I	2.5	2.5		MIT
2843.923	2844.759	35152.36	FE I	0.3			MIT
2843.632	2844.468	35155.96	FE I	2.1	2.0		MIT
2843.484	2844.320	35157.79	FE II		0.5		MIT
2843.319	2844.155	35159.83	FE II		0.5		MIT
2843.24	2844.08	35160.8	FE	0.8	0.3		MIT
2842.93	2843.77	35164.6	FE	0.7	0.3		MIT
2842.673	2843.509	35167.82	FE II		0.0H		MIT
2842.076	2842.912	35175.21	FE II		0.7		MIT
2840.932	2841.767	35189.37	FE	0.8	0.5		MIT
2840.762	2841.597	35191.48	FE II		1.5		MIT
2840.647	2841.482	35192.90	FE II		1.8		MIT
2840.423	2841.258	35195.67	FE I	2.1	1.3		MIT
2840.342	2841.177	35196.68	FE II		0.9		MIT
2839.819	2840.654	35203.16	FE II		1.0H		DO
2839.529	2840.364	35206.76	FE II	0.6	1.4		MIT
2838.45	2839.28	35220.1	FE I	0.9	0.7		MIT
2838.235	2839.070	35222.81	FE II		0.9		DO
2838.120	2838.955	35224.23	FE I	2.2	2.2		S
2837.299	2838.133	35234.43	FE II		1.4		MIT
2837.03	2837.86	35237.8	FE	0.0	0.3		MIT
2836.716	2837.550	35241.67	FE II		1.3		DO
2836.509	2837.343	35244.24	FE II	0.5	1.1		DO
2836.322	2837.156	35246.56	FE	0.9	0.7		MIT
2836.188	2837.022	35248.23	FE II	0.5	1.0		MIT
2835.955	2836.789	35251.12	FE I	1.2	1.0		MIT
2835.459	2836.293	35257.29	FE I	2.0	2.0		MIT
2834.755	2835.589	35266.04	FE	1.2	1.0		MIT
2834.41	2835.24	35270.3	FE I	0.5	0.3		MIT

AIR WAVELENGTH (ANGSTRMS)	VACUUM WAVELENGTH (ANGSTRMS)	WAVENUMBER (KAYSERS)	LMENT	LOG ARC	INTENSITY SPK	DIS	REF
2834.18	2835.01	35273.2	FE	0.7	0.5		MIT
2833.40	2834.23	35282.9	FE	1.0	0.9		MIT
2833.100	2833.933	35286.65	FE II		0.7H		DO
2832.436	2833.269	35294.92	FE I	2.5	2.3		S
2831.562	2832.395	35305.81	FE II	0.0	2.7		MIT
2830.964	2831.797	35313.27	FE	1.0			MIT
2828.813	2829.645	35340.12	FE	2.0	1.8		MIT
2828.681	2829.513	35341.77	FE II		0.9		DO
2828.634	2829.466	35342.35	FE II		1.9		MIT
2827.895	2828.727	35351.59	FE I	1.8	1.7		MIT
2827.60	2828.43	35355.3	FE	1.2	1.1		MIT
2827.434	2828.266	35357.35	FE II		1.4		MIT
2827.08	2827.91	35361.8	FE	0.0			MIT
2826.50	2827.33	35369.0	FE	1.0	0.9		MIT
2826.026	2826.858	35374.97	FE II		1.4		MIT
2826.00	2826.83	35375.3	FE I	0.8			MIT
2825.689	2826.521	35379.19	FE I	1.8	1.8		MIT
2825.560	2826.392	35380.80	FE	2.2	2.2		MIT
2824.67	2825.50	35392.0	FE	0.3D	0.3D		MIT
2823.276	2824.107	35409.42	FE I	2.3	2.5		S
2821.88	2822.71	35426.9	FE	0.3	0.0		MIT
2821.63	2822.46	35430.1	FE	0.3	0.0		MIT
2821.01	2821.84	35437.9	FE	0.7	0.5		MIT
2820.809	2821.639	35440.39	FE I	1.3	1.2		MIT
2820.22	2821.05	35447.8	FE	0.6	0.5		MIT
2819.905	2820.735	35451.75	FE		0.5H		MIT
2819.47	2820.30	35457.2	FE	0.6	0.3		MIT
2819.333	2820.163	35458.94	FE II		0.5		MIT
2819.294	2820.124	35459.43	FE	1.0			MIT
2818.04	2818.87	35475.2	FE	0.6	0.5		MIT
2817.938	2818.768	35476.50	FE I	0.6	0.5		MIT
2817.508	2818.338	35481.91	FE I	2.0	1.8		MIT
2817.107	2817.936	35486.96	FE II		1.3H		DO
2816.66	2817.49	35492.6	FE	0.7	0.6		MIT
2815.507	2816.336	35507.13	FE I	1.6	1.4		MIT
2815.016	2815.845	35513.32	FE	1.2	0.9		MIT
2813.613	2814.442	35531.03	FE II	0.7	1.8		MIT
2813.288	2814.117	35535.13	FE I	2.6	2.6		S
2812.493	2813.321	35545.17	FE II	0.3	1.4		MIT
2812.31	2813.14	35547.5	FE I	0.3	0.0		MIT
2812.049	2812.877	35550.79	FE	1.2	1.0		MIT
2811.268	2812.096	35560.66	FE II		1.6		MIT
2810.265	2811.093	35573.35	FE	1.6	1.2		MIT
2809.806	2810.634	35579.17	FE II	0.0	2.0H		MIT
2808.79	2809.62	35592.0	FE	0.0			MIT
2808.62	2809.45	35594.2	FE	0.8	0.3		MIT
2808.320	2809.147	35597.99	FE	2.0	1.6		MIT
2807.85	2808.68	35604.0	FE	0.3	0.0		MIT
2807.245	2808.072	35611.62	FE I	1.2	0.9		MIT
2806.984	2807.811	35614.93	FE I	2.3	2.3		I
2806.070	2806.897	35626.53	FE	1.2	0.7		MIT
2805.791	2806.618	35630.08	FE II		1.7		MIT
2805.315	2806.142	35636.12	FE II		1.2		MIT
2804.865	2805.691	35641.84	FE	1.3	1.2		MIT
2804.521	2805.347	35646.21	FE I	2.5	2.3		S
2804.021	2804.847	35652.56	FE II		1.2		MIT
2803.620	2804.446	35657.66	FE	1.7	1.3		MIT
2803.172	2803.998	35663.36	FE I	1.2			MIT
2803.12	2803.95	35664.0	FE	1.5	1.2		MIT
2802.25	2803.08	35675.1	FE	0.5			MIT
2800.464	2801.289	35697.85	FE	1.7	1.0		MIT
2800.28	2801.11	35700.2	FE	0.8	0.3		MIT
2799.721	2800.546	35707.32	FE II		1.0		MIT
2799.286	2800.111	35712.87	FE II	0.0	2.0		MIT
2799.153	2799.978	35714.57	FE	1.7	1.0		MIT
2797.914	2798.739	35730.38	FE II		1.3		MIT
2797.775	2798.600	35732.15	FE I	2.2	1.9		I
2796.871	2797.696	35743.70	FE I	1.2	0.5		MIT
2796.649	2797.473	35746.54	FE II		1.3		MIT
2795.85	2796.67	35756.8	FE	1.20	1.0		MIT

AIR WAVELENGTH (ANGSTRMS)	VACUUM WAVELENGTH (ANGSTRMS)	WAVENUMBER (KAYSERS)	LMENT	LOG ARC	INTENSITY SPK	INTENSITY DIS	REF
2795.767	2796.591	35757.82	FE II		0.5H		MIT
2795.544	2796.368	35760.67	FE I	2.0	1.8		MIT
2795.007	2795.831	35767.54	FE I	1.7	1.5		MIT
2794.704	2795.528	35771.42	FE I	1.7	1.5		MIT
2794.16	2794.98	35778.4	FE	1.0			MIT
2793.888	2794.712	35781.87	FE II	0.9	2.2		MIT
2792.402	2793.225	35800.91	FE	1.7	1.4		MIT
2792.053	2792.876	35805.38	FE II		0.3		MIT
2791.792	2792.615	35808.73	FE	1.8	1.6		MIT
2791.462	2792.285	35812.96	FE	1.6	1.3		MIT
2791.008	2791.831	35818.79	FE		1.0		MIT
2790.93	2791.75	35819.8	FE	0.3			MIT
2790.559	2791.382	35824.55	FE II		1.5		MIT
2789.802	2790.625	35834.27	FE	1.7	1.4		MIT
2789.685	2790.508	35835.77	FE	0.5	0.0		MIT
2789.483	2790.306	35838.37	FE I	1.8	1.5		MIT
2788.105	2788.927	35856.08	FE I	2.2	2.2		MIT
2787.935	2788.757	35858.27	FE I	1.4	1.2		MIT
2787.260	2788.082	35866.95	FE II		1.0H		DO
2787.11	2787.93	35868.9	FE	0.0			MIT
2786.780	2787.602	35873.13	FE	1.2	0.8		MIT
2786.19	2787.01	35880.7	FE	0.9	0.3		MIT
2785.213	2786.035	35893.31	FE II		1.6H		DO
2784.346	2785.167	35904.48	FE	1.2	0.8		MIT
2784.282	2785.103	35905.31	FE II		0.7		MIT
2784.011	2784.832	35908.80	FE	1.1	0.5		MIT
2783.696	2784.517	35912.87	FE II	1.3	2.6		MIT
2782.05	2782.87	35934.1	FE	1.1	0.8		MIT
2781.835	2782.656	35936.89	FE I	2.0	1.8		I
2781.01	2781.83	35947.5	FE	0.9	0.3		MIT
2780.890	2781.711	35949.10	FE	1.0	0.5		MIT
2780.700	2781.521	35951.56	FE	1.5	1.2		MIT
2780.54	2781.36	35953.6	FE	1.0	0.3		MIT
2780.045	2780.865	35960.03	FE II		1.3		MIT
2779.906	2780.726	35961.83	FE II		1.6		DO
2779.70	2780.52	35964.5	FE	0.6			MIT
2779.299	2780.119	35969.68	FE II	1.4	2.5		MIT
2778.845	2779.665	35975.56	FE	1.8	1.6		MIT
2778.59	2779.41	35978.9	FE	0.8	0.0		MIT
2778.221	2779.041	35983.64	FE I	2.0	1.9		S
2778.071	2778.891	35985.58	FE	1.5	1.0		MIT
2776.923	2777.743	36000.46	FE II		1.4H		DO
2776.399	2777.219	36007.25	FE	2.0	1.5		MIT
2776.175	2776.994	36010.15	FE II		1.6		MIT
2774.734	2775.553	36028.85	FE I	1.9	1.0		MIT
2774.691	2775.510	36029.41	FE II		1.7		MIT
2774.14	2774.96	36036.6	FE	0.0			MIT
2773.90	2774.72	36039.7	FE	1.0	0.7		MIT
2773.680	2774.499	36042.54	FE II		0.0		MIT
2773.236	2774.055	36048.31	FE	2.0	1.6		MIT
2772.827	2773.646	36053.63	FE	1.2	0.6		MIT
2772.511	2773.330	36057.74	FE	1.5	1.3		MIT
2772.33	2773.15	36060.1	FE	1.7	1.0		MIT
2772.109	2772.927	36062.97	FE I	2.5	2.5		IJA
2771.89	2772.71	36065.8	FE	1.1	0.8		MIT
2771.560	2772.378	36070.11	FE II	0.0	1.0		MIT
2771.184	2772.002	36075.01	FE II		1.7		DO
2770.701	2771.519	36081.29	FE	1.3	0.8		MIT
2770.508	2771.326	36083.81	FE II		1.7		MIT
2769.671	2770.489	36094.71	FE I	1.8	1.3		MIT
2769.351	2770.169	36098.88	FE II	0.0	1.3		MIT
2769.302	2770.120	36099.52	FE	2.0	1.0		MIT
2769.145	2769.963	36101.57	FE II	0.0	1.2		MIT
2768.934	2769.752	36104.32	FE II	1.6	2.0		MIT
2768.439	2769.257	36110.77	FE	1.4	0.7		MIT
2768.337	2769.155	36112.10	FE II		0.0		MIT
2768.112	2768.929	36115.04	FE	1.5	0.9		MIT
2767.523	2768.340	36122.73	FE I	2.5			S
2767.503	2768.320	36122.99	FE II	1.0	2.6J		MIT
2766.912	2767.729	36130.70	FE I	2.0	1.6		MIT

AIR WAVELENGTH (ANGSTRMS)	VACUUM WAVELENGTH (ANGSTRMS)	WAVENUMBER (KAYSERS)	LMENT	LOG ARC	INTENSITY SPK	INTENSITY DIS	REF
2766.662	2767.479	36133.97	FE	1.2	0.8		MIT
2766.03	2766.85	36142.2	FE	0.0			MIT
2765.70	2766.52	36146.5	FE	0.8	0.0		MIT
2765.495	2766.312	36149.21	FE II		0.7		MIT
2764.784	2765.601	36158.51	FE II		1.3		MIT
2764.330	2765.147	36164.45	FE	1.8	1.6		MIT
2763.907	2764.723	36169.98	FE II	0.0	1.4		MIT
2763.41	2764.23	36176.5	FE	0.3			MIT
2763.108	2763.924	36180.44	FE I	2.0	1.8		I
2762.778	2763.594	36184.76	FE I	1.4	1.0		MIT
2762.68	2763.50	36186.0	FE	1.7	0.0		MIT
2762.441	2763.257	36189.18	FE II		1.0		MIT
2762.032	2762.848	36194.54	FE I	2.0	1.8		MIT
2761.813	2762.629	36197.40	FE II	1.7	2.3		BU
2761.785	2762.601	36197.77	FE I	2.3			MIT
2761.51	2762.33	36201.4	FE	0.6			MIT
2760.897	2761.713	36209.41	FE	1.2	0.9		MIT
2759.817	2760.632	36223.58	FE I	2.0	1.8		MIT
2759.335	2760.150	36229.91	FE II	0.3	1.1		MIT
2758.51	2759.33	36240.7	FE	0.3	1.4H		MIT
2757.865	2758.680	36249.22	FE	1.4	1.0		MIT
2757.320	2758.135	36256.38	FE I	2.0	1.8		MIT
2757.025	2757.840	36260.26	FE II	1.0	1.5		MIT
2756.512	2757.327	36267.01	FE II	1.0	0.8		MIT
2756.334	2757.149	36269.35	FE I	2.5	2.0		MIT
2756.264	2757.079	36270.27	FE I	2.5	2.0		MIT
2755.737	2756.551	36277.21	FE II	2.5	2.0		I
2755.18	2755.99	36284.5	FE	1.2			MIT
2754.95	2755.76	36287.6	FE	1.4			MIT
2754.907	2755.721	36288.14	FE II		1.3		DO
2754.426	2755.240	36294.48	FE	1.8	1.3		MIT
2754.037	2754.851	36299.60	FE I	2.0	1.5		MIT
2753.81	2754.62	36302.6	FE	0.7			MIT
2753.692	2754.506	36304.15	FE I	1.8	1.4		MIT
2753.287	2754.101	36309.49	FE II	1.4	2.2H		MIT
2753.102	2753.916	36311.93	FE	1.4			MIT
2753.029	2753.843	36312.89	FE	0.3			MIT
2752.159	2752.973	36324.37	FE II		1.0		DO
2752.095	2752.909	36325.22	FE	0.0	1.3		MIT
2751.811	2752.625	36328.96	FE	1.2	0.7		MIT
2751.37	2752.18	36334.8	FE	1.2			MIT
2751.123	2751.936	36338.05	FE II		1.8		MIT
2750.878	2751.691	36341.29	FE	1.8	1.3		MIT
2750.722	2751.535	36343.35	FE I	1.2			MIT
2750.144	2750.957	36350.98	FE I	2.5H	2.0		MIT
2749.68	2750.49	36357.1	FE	0.5			MIT
2749.484	2750.297	36359.71	FE II	1.2	1.3		MIT
2749.324	2750.137	36361.83	FE II	1.5	1.5		I
2749.184	2749.997	36363.68	FE II	1.6	1.6		MIT
2747.68	2748.49	36383.6	FE	0.3			MIT
2747.557	2748.370	36385.21	FE I	1.5	0.7		MIT
2746.982	2747.794	36392.83	FE II	2.3	2.5J		I
2746.483	2747.295	36399.44	FE II	2.2	2.5J		I
2746.157	2746.969	36403.76	FE II		1.0		DO
2745.97	2746.78	36406.2	FE	0.5			MIT
2745.82	2746.63	36408.2	FE	0.9			MIT
2745.082	2745.894	36418.01	FE	1.0	0.0		MIT
2744.894	2745.706	36420.51	FE II		0.3		MIT
2744.527	2745.339	36425.38	FE I	1.8	1.7		MIT
2744.072	2744.884	36431.42	FE I	2.2	0.9		MIT
2743.565	2744.377	36438.15	FE	1.7	1.2		MIT
2743.196	2744.007	36443.05	FE II	1.9	2.2		MIT
2742.408	2743.219	36453.52	FE I	1.7	1.7		MIT
2742.256	2743.067	36455.54	FE I	1.4	1.4		MIT
2742.017	2742.828	36458.72	FE I	1.5	1.2		MIT
2741.828	2742.639	36461.23	FE	0.0			MIT
2741.58	2742.39	36464.5	FE	0.6	0.0		MIT
2741.397	2742.208	36466.96	FE II		1.8		MIT
2741.11	2741.92	36470.8	FE	1.0	0.5		MIT
2739.546	2740.357	36491.60	FE II	2.3	2.5H		I

AIR WAVELENGTH (ANGSTRMS)	VACUUM WAVELENGTH (ANGSTRMS)	WAVENUMBER (KAYSERS)	LMENT	LOG ARC	INTENSITY SPK	DIS	REF
2738.45	2739.26	36506.2	FE	0.9	0.0		MIT
2738.213	2739.023	36509.37	FE I	1.0	0.3		MIT
2737.834	2738.644	36514.42	FE	1.4	1.0		MIT
2737.642	2738.452	36516.98	FE	1.0			MIT
2737.634	2738.444	36517.09	FE II		1.0		MIT
2737.314	2738.124	36521.35	FE I	2.5R	2.2		MIT
2736.968	2737.778	36525.97	FE II		1.1		ME
2736.78	2737.59	36528.5	FE	1.0H	0.3H		MIT
2736.528	2737.338	36531.84	FE II	0.3	0.5		MIT
2735.614	2736.424	36544.05	FE I	1.3	1.0		MIT
2735.475	2736.285	36545.91	FE I	2.1	2.0		S
2734.97	2735.78	36552.6	FE	0.6			MIT
2734.888	2735.697	36553.75	FE	0.7			MIT
2734.803	2735.612	36554.89	FE		0.7		DO
2734.619	2735.428	36557.35	FE I	1.3	0.9		MIT
2734.273	2735.082	36561.97	FE I	1.6	1.4		MIT
2734.004	2734.813	36565.57	FE I	1.6	1.4		MIT
2733.584	2734.393	36571.19	FE I	2.5	2.3		MIT
2733.363	2734.172	36574.14	FE	0.3	0.0		MIT
2732.936	2733.745	36579.86	FE		1.6		DO
2732.448	2733.257	36586.39	FE II	0.9	1.3		MIT
2732.005	2732.814	36592.32	FE II		1.6		MIT
2731.349	2732.158	36601.11	FE	0.6	0.0		MIT
2731.28	2732.09	36602.0	FE	0.6	0.0		MIT
2730.984	2731.792	36606.00	FE I	1.8	1.2		MIT
2730.738	2731.546	36609.30	FE II	1.9	2.2		MIT
2730.16	2730.97	36617.0	FE	0.0			MIT
2730.03	2730.84	36618.8	FE	0.9	0.0		MIT
2728.972	2729.780	36632.99	FE I	1.2	0.7		MIT
2728.905	2729.713	36633.89	FE	0.5	0.7		MIT
2728.825	2729.633	36634.96	FE	1.7	1.2		MIT
2728.021	2728.829	36645.76	FE	2.0	1.6		MIT
2727.539	2728.347	36652.23	FE II	2.2	2.2		I
2727.384	2728.192	36654.32	FE II	0.0H	1.6		MIT
2726.78	2727.59	36662.4	FE	0.3			MIT
2726.509	2727.316	36666.08	FE II		1.2		DO
2726.254	2727.061	36669.51	FE		0.9		DO
2726.240	2727.047	36669.70	FE	1.4	1.1		MIT
2726.054	2726.861	36672.20	FE	2.0	1.9		MIT
2725.81	2726.62	36675.5	FE	0.6	0.3		MIT
2725.606	2726.413	36678.23	FE I	1.2	0.7		MIT
2725.33	2726.14	36681.9	FE	0.9	0.6		MIT
2725.29	2726.10	36682.5	FE I	0.9	0.6		MIT
2724.959	2725.766	36686.94	FE I	1.4	1.2		MIT
2724.885	2725.692	36687.93	FE II	1.2	1.4		MIT
2724.68	2725.49	36690.7	FE	1.0	0.3		MIT
2723.79	2724.60	36702.7	FE	0.5	0.0		MIT
2723.577	2724.384	36705.55	FE I	2.5	2.3		S
2722.737	2723.543	36716.87	FE		1.9		MIT
2722.040	2722.846	36726.27	FE II	1.3	1.8		MIT
2721.819	2722.625	36729.26	FE II		1.5		MIT
2721.12	2721.93	36738.7	FE	0.5	0.3		MIT
2720.905	2721.711	36741.59	FE I	2.8R			MIT
2720.52	2721.33	36746.8	FE	0.9	0.7		MIT
2720.199	2721.005	36751.13	FE	1.5	1.4		MIT
2719.423	2720.229	36761.62	FE	1.3	1.1		MIT
2719.301	2720.107	36763.26	FE II		0.6		MIT
2719.025	2719.831	36767.00	FE I	2.7R	2.5R		MIT
2718.635	2719.441	36772.27	FE II		0.8		MIT
2718.435	2719.240	36774.98	FE I	1.9	1.8		I
2717.790	2718.595	36783.70	FE	1.7	1.4		MIT
2717.367	2718.172	36789.43	FE	1.2	0.7		MIT
2717.20	2718.01	36791.7	FE	0.3H	0.0H		MIT
2716.258	2717.063	36804.45	FE	0.7			MIT
2716.218	2717.023	36804.99	FE II	1.3	2.2		MIT
2715.323	2716.128	36817.12	FE I	1.1	0.7		MIT
2715.125	2715.930	36819.81	FE	0.7	0.3		MIT
2714.868	2715.673	36823.29	FE I	1.6	1.2		MIT
2714.412	2715.216	36829.48	FE II	2.3	2.6		I
2714.062	2714.866	36834.23	FE	1.3	0.5		MIT

AIR WAVELENGTH (ANGSTRMS)	VACUUM WAVELENGTH (ANGSTRMS)	WAVENUMBER (KAYSERS)	LMENT	LOG ARC	INTENSITY SPK	DIS	REF
2712.388	2713.192	36856.96	FE II	0.3	2.0		MIT
2711.845	2712.649	36864.34	FE II	0.6	2.0		MIT
2711.654	2712.458	36866.93	FE I	2.0	1.7		I
2711.463	2712.267	36869.53	FE	1.1	0.5		MIT
2710.547	2711.351	36881.99	FE I	1.9	1.5		MIT
2709.993	2710.796	36889.53	FE	1.6	1.0		MIT
2709.374	2710.177	36897.96	FE II		0.6		MIT
2709.056	2709.859	36902.29	FE II	0.5	2.0		MIT
2708.570	2709.373	36908.91	FE	1.9	1.7		MIT
2707.94	2708.74	36917.5	FE	0.3			MIT
2707.451	2708.254	36924.16	FE	1.3	0.8		MIT
2707.132	2707.935	36928.51	FE II		1.8		MIT
2707.05	2707.85	36929.6	FE	0.6			MIT
2706.581	2707.384	36936.03	FE I	2.2	2.2		I
2706.072	2706.874	36942.98	FE	0.9	0.5		MIT
2706.015	2706.817	36943.75	FE	1.8	1.6		MIT
2705.23	2706.03	36954.5	FE	0.0H			MIT
2704.582	2705.384	36963.33	FE II	0.7	1.0		MIT
2703.989	2704.791	36971.43	FE II	1.5	2.6		MIT
2702.55	2703.35	36991.1	FE	0.8	0.5		MIT
2702.453	2703.255	36992.45	FE	1.2	1.0		MIT
2701.911	2702.712	36999.87	FE	1.3	0.7		MIT
2701.160	2701.961	37010.15	FE		0.3		DO
2699.77	2700.57	37029.2	FE	0.8	0.6		MIT
2699.185	2699.986	37037.23	FE		0.7		DO
2699.107	2699.908	37038.30	FE I	2.0	1.8		S
2698.165	2698.966	37051.23	FE	1.5	0.8		MIT
2697.734	2698.534	37057.15	FE	0.0	0.5		MIT
2697.462	2698.262	37060.89	FE II	0.5	1.7		MIT
2697.308	2698.108	37063.00	FE II	0.3	1.2		MIT
2697.021	2697.821	37066.95	FE I	1.7	1.4		MIT
2696.284	2697.084	37077.08	FE	2.0	1.7		MIT
2695.996	2696.796	37081.04	FE	1.9	1.7		MIT
2695.662	2696.462	37085.63	FE	1.3	1.1		MIT
2695.534	2696.334	37087.39	FE	1.6	1.5		MIT
2695.039	2695.839	37094.21	FE	1.5	1.3		MIT
2694.538	2695.338	37101.10	FE	2.0	1.5		MIT
2693.857	2694.657	37110.48	FE II		1.5		MIT
2692.836	2693.635	37124.55	FE II	1.2	1.3		MIT
2692.655	2693.454	37127.05	FE I	1.3			MIT
2692.597	2693.396	37127.85	FE II		2.5		MIT
2692.254	2693.053	37132.58	FE I	1.3	0.6		MIT
2691.732	2692.531	37139.78	FE II		1.5		MIT
2691.49	2692.29	37143.1	FE	0.7	0.0		MIT
2690.071	2690.870	37162.71	FE I	1.5	1.5		MIT
2689.88	2690.68	37165.4	FE	1.0			MIT
2689.829	2690.628	37166.05	FE I	1.6	1.6		MIT
2689.50	2690.30	37170.6	FE	0.6	0.3		MIT
2689.423	2690.221	37171.66	FE	0.7	0.5		MIT
2689.212	2690.010	37174.58	FE I	2.2	2.2		S
2687.802	2688.600	37194.08	FE	1.3	1.0		MIT
2686.748	2687.546	37208.67	FE	1.7	0.8		MIT
2686.488	2687.286	37212.27	FE	0.0	0.6		MIT
2686.394	2687.192	37213.57	FE II		1.0		MIT
2686.100	2686.898	37217.64	FE II		0.8		MIT
2685.76	2686.56	37222.4	FE	0.6			MIT
2684.857	2685.654	37234.87	FE I	0.9			MIT
2684.751	2685.548	37236.34	FE II	0.5	2.6		MIT
2684.071	2684.868	37245.78	FE	1.5	1.2		MIT
2683.949	2684.746	37247.47	FE	1.2	0.9		MIT
2682.986	2683.783	37260.84	FE		1.6		MIT
2682.518	2683.315	37267.34	FE		1.6		MIT
2682.214	2683.011	37271.56	FE	1.5	1.2		MIT
2681.589	2682.386	37280.25	FE	1.7	1.4		MIT
2681.46	2682.26	37282.0	FE	0.3			MIT
2681.029	2681.825	37288.04	FE II		1.3		MIT
2680.91	2681.71	37289.7	FE I	0.6			MIT
2680.793	2681.589	37291.32	FE II		1.1		MIT
2680.454	2681.250	37296.03	FE I	1.8	1.5		MIT
2680.16	2680.96	37300.1	FE	1.3	0.9		MIT

AIR WAVELENGTH (ANGSTRMS)	VACUUM WAVELENGTH (ANGSTRMS)	WAVENUMBER (KAYSERS)	LMENT	LOG ARC	INTENSITY SPK	DIS	REF
2679.062	2679.858	37315.41	FE I	2.3	2.3		S
2678.05	2678.85	37329.5	FE	1.3	1.0		MIT
2676.877	2677.672	37345.87	FE	0.3	1.4		MIT
2676.11	2676.91	37356.6	FE	1.2W	0.8W		MIT
2675.283	2676.078	37368.12	FE	1.5	1.2		MIT
2674.72	2675.51	37376.0	FE	0.3			MIT
2674.18	2674.97	37383.5	FE	0.8H	0.5H		MIT
2673.213	2674.008	37397.05	FE I	1.5	1.2		I
2672.77	2673.56	37403.2	FE	0.3			MIT
2672.506	2673.300	37406.94	FE II		1.2J		DO
2672.148	2672.942	37411.96	FE II	0.0V	1.4J		MIT
2671.404	2672.198	37422.37	FE II		1.4J		DO
2670.80	2671.59	37430.8	FE	1.0	0.6		MIT
2670.384	2671.178	37436.67	FE II		1.0		DO
2669.934	2670.728	37442.98	FE		1.4		MIT
2669.496	2670.290	37449.12	FE	1.7	1.4		MIT
2667.919	2668.712	37471.25	FE I	1.7	1.3		MIT
2666.970	2667.763	37484.59	FE I	1.5	1.0		MIT
2666.818	2667.611	37486.73	FE I	1.9	1.2		MIT
2666.635	2667.428	37489.30	FE II	0.7	1.9		MIT
2666.405	2667.198	37492.53	FE I	1.8	1.0		MIT
2664.664	2665.457	37517.02	FE II	1.3	2.5		MIT
2664.259	2665.051	37522.73	FE		1.0		DO
2664.209	2665.001	37523.43	FE II		0.5		DO
2664.040	2664.832	37525.81	FE	1.2	0.7		MIT
2663.78	2664.57	37529.5	FE	1.0H	0.5H		MIT
2662.563	2663.355	37546.63	FE II	0.3H	1.2H		DO
2662.315	2663.107	37550.12	FE	1.4	1.0		MIT
2662.056	2662.848	37553.78	FE I	1.8	1.6		I
2661.312	2662.104	37564.28	FE	1.1	1.0		MIT
2661.196	2661.988	37565.91	FE I	1.0	0.8		MIT
2660.402	2661.193	37577.12	FE I	1.6	1.2		MIT
2659.24	2660.03	37593.5	FE	0.9	0.3		MIT
2658.93	2659.72	37597.9	FE	1.0	0.3		MIT
2658.47	2659.26	37604.4	FE	1.3	0.3		MIT
2658.251	2659.042	37607.53	FE II		1.9		DO
2657.921	2658.712	37612.20	FE II		1.3		MIT
2656.796	2657.587	37628.12	FE I	1.7	1.4		MIT
2656.149	2656.939	37637.29	FE	1.8	1.6		MIT
2655.13	2655.92	37651.7	FE	0.8	0.5		MIT
2654.27	2655.06	37663.9	FE	0.6	0.3		MIT
2652.571	2653.361	37688.06	FE II	0.0	1.6		MIT
2651.706	2652.495	37700.35	FE I	1.8	1.8		I
2651.291	2652.080	37706.25	FE II		0.3		MIT
2650.492	2651.281	37717.62	FE II		2.2H		DO
2649.464	2650.253	37732.25	FE II		1.8		MIT
2648.54	2649.33	37745.4	FE	0.8	0.3		MIT
2648.43	2649.22	37747.0	FE	0.8	0.3		MIT
2648.162	2648.951	37750.80	FE II	0.3	0.3		MIT
2647.557	2648.345	37759.42	FE I	2.0	1.8		I
2646.32	2647.11	37777.1	FE	0.7			MIT
2646.217	2647.005	37778.54	FE II		1.0		MIT
2645.427	2646.215	37789.83	FE I	1.7	1.0		MIT
2645.330	2646.118	37791.21	FE		0.9		MIT
2645.191	2645.979	37793.20	FE		0.9		DO
2645.084	2645.872	37794.73	FE II		1.3		MIT
2644.000	2644.788	37810.22	FE I	2.2	2.2		I
2642.977	2643.764	37824.85	FE		0.3		MIT
2642.015	2642.802	37838.63	FE II		1.3		DO
2641.649	2642.436	37843.87	FE I	2.0	1.8		MIT
2641.126	2641.913	37851.36	FE II		1.3		MIT
2640.22	2641.01	37864.4	FE	0.50	0.30		MIT
2639.553	2640.340	37873.92	FE II	0.0	2.0		MIT
2638.243	2639.029	37892.72	FE		0.3		MIT
2637.644	2638.430	37901.33	FE II	0.3	2.3		MIT
2637.10	2637.89	37909.1	FE	0.5J	0.0H		MIT
2636.482	2637.268	37918.03	FE I	1.7	1.3		MIT
2635.808	2636.594	37927.73	FE I	2.5	2.3		S
2635.393	2636.179	37933.70	FE II		1.3		MIT
2634.772	2635.557	37942.64	FE	0.0	0.0		MIT

AIR WAVELENGTH (ANGSTRMS)	VACUUM WAVELENGTH (ANGSTRMS)	WAVENUMBER (KAYSERS)	LMENT	LOG ARC	INTENSITY SPK	DIS	REF
2633.64	2634.43	37959.0	FE	0.7	0.0		MIT
2633.194	2633.979	37965.37	FE		1.9		MIT
2632.99	2633.77	37968.3	FE	1.0			MIT
2632.598	2633.383	37973.97	FE I	1.9	1.6		MIT
2632.236	2633.021	37979.19	FE I	2.0	1.8		MIT
2631.609	2632.394	37988.24	FE II	1.0	1.7		MIT
2631.323	2632.108	37992.37	FE II	2.2	1.8		MIT
2631.051	2631.835	37996.30	FE II	2.3	2.1		MIT
2630.066	2630.850	38010.52	FE II	1.0	2.0		MIT
2629.587	2630.371	38017.45	FE I	1.8	2.2		MIT
2629.587	2630.371	38017.45	FE II	1.8	2.2		MIT
2629.19	2629.97	38023.2	FE	0.3			MIT
2628.570	2629.354	38032.16	FE II	0.0	0.3		MIT
2628.292	2629.076	38036.18	FE II	2.6	2.6		I
2627.24	2628.02	38051.4	FE	0.9			MIT
2627.14	2627.92	38052.9	FE	0.0	1.0		MIT
2626.500	2627.283	38062.13	FE II	0.5	1.9		MIT
2625.666	2626.449	38074.22	FE II	1.2	1.1		ME
2625.490	2626.273	38076.77	FE	0.0	1.8		MIT
2624.37	2625.15	38093.0	FE	0.7			MIT
2624.14	2624.92	38096.4	FE	1.0	0.0		MIT
2623.727	2624.510	38102.35	FE II		1.3		MIT
2623.63	2624.41	38103.8	FE	0.3			MIT
2623.532	2624.315	38105.19	FE I	2.0	1.9		MIT
2623.366	2624.149	38107.60	FE I	1.4	1.0		MIT
2623.123	2623.906	38111.13	FE		1.5		MIT
2622.83	2623.61	38115.4	FE	0.3J	0.3J		MIT
2621.94	2622.72	38128.3	FE	0.5			MIT
2621.668	2622.450	38132.28	FE II	2.3	2.6		I
2620.695	2621.477	38146.43	FE II	0.5	1.9		MIT
2620.407	2621.189	38150.63	FE II	1.8	1.6		MIT
2620.173	2620.955	38154.03	FE II		1.6		MIT
2619.076	2619.858	38170.01	FE II	0.7	2.2		MIT
2618.706	2619.488	38175.41	FE I	1.8	1.4		MIT
2618.017	2618.798	38185.45	FE I	2.2	1.8		MIT
2617.616	2618.397	38191.30	FE II	2.5	2.6		I
2617.14	2617.92	38198.2	FE	0.9	0.0		MIT
2616.44	2617.22	38208.5	FE	0.7	0.5		MIT
2616.32	2617.10	38210.2	FE	0.5			MIT
2615.87	2616.65	38216.8	FE	0.5			MIT
2615.422	2616.203	38223.34	FE	1.4	1.0		MIT
2614.872	2615.653	38231.38	FE II		1.0		MIT
2614.490	2615.271	38236.96	FE I	1.6	0.7		MIT
2614.180	2614.960	38241.50	FE II		0.5		MIT
2613.823	2614.603	38246.72	FE II	2.6	2.6		I
2613.576	2614.356	38250.33	FE II	0.3	0.7		DO
2613.24	2614.02	38255.2	FE	1.0	0.3		MIT
2613.19	2613.97	38256.0	FE	1.2			MIT
2612.771	2613.551	38262.12	FE I	1.7	1.0		MIT
2611.872	2612.652	38275.29	FE II	2.7	2.7		I
2611.332	2612.112	38283.20	FE II		0.0		MIT
2611.072	2611.852	38287.01	FE II	1.3	1.9		MIT
2610.749	2611.529	38291.75	FE I	1.6	0.6		MIT
2610.515	2611.295	38295.18	FE		0.0H		MIT
2609.866	2610.645	38304.70	FE II		1.6		MIT
2609.437	2610.216	38311.00	FE II		1.0		MIT
2609.220	2609.999	38314.19	FE	1.0	0.0		MIT
2609.126	2609.905	38315.57	FE II		1.3		MIT
2608.851	2609.630	38319.61	FE II		1.3		MIT
2608.578	2609.357	38323.62	FE	2.0	1.0		MIT
2607.81	2608.59	38334.9	FE	0.3W			MIT
2607.087	2607.866	38345.53	FE II	2.5	2.6		MIT
2606.825	2607.604	38349.38	FE I	2.3	1.5		MIT
2606.651	2607.430	38351.94	FE	0.6			MIT
2606.504	2607.283	38354.11	FE II		1.9		MIT
2606.31	2607.09	38357.0	FE	1.0			MIT
2605.905	2606.683	38362.92	FE		1.3		MIT
2605.653	2606.431	38366.63	FE I	1.9	1.0		MIT
2605.415	2606.193	38370.14	FE II		1.6		MIT
2605.303	2606.081	38371.79	FE II		1.7		MIT

AIR WAVELENGTH (ANGSTRMS)	VACUUM WAVELENGTH (ANGSTRMS)	WAVENUMBER (KAYSERS)	LMENT	LOG ARC	INTENSITY SPK	DIS	REF
2605.040	2605.818	38375.66	FE		1.9		MIT
2604.867	2605.645	38378.21	FE	1.3			MIT
2604.758	2605.536	38379.82	FE	1.3	0.0		MIT
2604.049	2604.827	38390.27	FE		0.6		MIT
2604.01	2604.79	38390.8	FE	0.5			MIT
2603.565	2604.343	38397.40	FE	1.3	0.5		MIT
2600.202	2600.979	38447.06	FE	1.0			BU
2600.10	2600.88	38448.6	FE	1.0			MIT
2599.661	2600.438	38455.06	FE	0.8			MIT
2599.570	2600.347	38456.41	FE I	3.0			MIT
2599.396	2600.173	38458.98	FE II	3.0	3.0H		MIT
2599.131	2599.908	38462.90	FE	0.0			MIT
2598.854	2599.631	38467.00	FE	1.3			MIT
2598.47	2599.25	38472.7	FE	0.0			MIT
2598.369	2599.146	38474.18	FE II	2.8	3.0H		I
2598.028	2598.805	38479.23	FE II	0.6	0.6		DO
2597.943	2598.720	38480.49	FE II		0.5		DO
2597.828	2598.605	38482.19	FE	0.8			MIT
2596.434	2597.210	38502.85	FE	0.5			MIT
2595.428	2596.204	38517.77	FE	0.5			MIT
2595.294	2596.070	38519.76	FE II		0.5		MIT
2594.150	2594.926	38536.75	FE I	1.3	0.3		MIT
2594.038	2594.814	38538.41	FE	1.3	0.3		MIT
2593.726	2594.502	38543.05	FE II	1.2	1.8		MIT
2593.518	2594.294	38546.14	FE	1.4			MIT
2592.779	2593.554	38557.12	FE	1.3	2.0		MIT
2592.292	2593.067	38564.37	FE	1.1			MIT
2591.543	2592.318	38575.51	FE II	1.7	2.0		MIT
2591.256	2592.031	38579.79	FE	1.3			MIT
2590.545	2591.320	38590.37	FE II	0.3	1.5		MIT
2590.03	2590.80	38598.0	FE	0.3			MIT
2588.795	2589.569	38616.46	FE II	0.0	1.3		MIT
2588.191	2588.965	38625.47	FE		0.3		MIT
2587.999	2588.773	38628.33	FE	1.6			MIT
2587.948	2588.722	38629.10	FE II		1.7		MIT
2587.160	2587.934	38640.86	FE	0.8			MIT
2585.876	2586.650	38660.05	FE II	1.8	2.0		I
2585.620	2586.394	38663.87	FE II		0.5		MIT
2585.00	2585.77	38673.1	FE	0.5			MIT
2584.72	2585.49	38677.3	FE	0.5			MIT
2584.536	2585.309	38680.09	FE I	2.0	1.5		S
2584.303	2585.076	38683.58	FE	0.3			MIT
2583.048	2583.821	38702.37	FE II		0.5		MIT
2582.92	2583.69	38704.3	FE	0.5			MIT
2582.816	2583.589	38705.85	FE	0.7			MIT
2582.585	2583.358	38709.31	FE II	1.4	1.9		MIT
2582.299	2583.072	38713.60	FE	1.7			MIT
2581.463	2582.236	38726.13	FE	1.1			MIT
2581.113	2581.886	38731.38	FE	0.0	1.4		MIT
2580.96	2581.73	38733.7	FE	0.7			MIT
2580.714	2581.487	38737.37	FE II		1.0		MIT
2580.457	2581.229	38741.23	FE I	0.9			MIT
2580.30	2581.07	38743.6	FE	0.8	0.0		MIT
2580.064	2580.836	38747.13	FE I	1.0			MIT
2579.845	2580.617	38750.42	FE	1.0	0.0		MIT
2579.414	2580.186	38756.89	FE II	0.0	1.0		MIT
2579.272	2580.044	38759.03	FE I	1.1	0.8H		MIT
2578.821	2579.593	38765.80	FE	0.5H	0.7		MIT
2578.02	2578.79	38777.9	FE	0.3			MIT
2577.923	2578.695	38779.31	FE II	1.5	2.0		MIT
2577.427	2578.199	38786.77	FE II		0.3		MIT
2577.293	2578.065	38788.79	FE	0.5			MIT
2576.865	2577.637	38795.23	FE II	0.3	1.8		MIT
2576.691	2577.463	38797.85	FE I	1.6	0.7		MIT
2575.744	2576.515	38812.11	FE	1.9	1.0		I
2575.344	2576.115	38818.14	FE	0.5			MIT
2574.89	2575.66	38825.0	FE	0.5			MIT
2574.368	2575.139	38832.85	FE II	1.7	2.2		MIT
2573.77	2574.54	38841.9	FE	0.3	0.3H		MIT
2573.742	2574.513	38842.30	FE II	0.3	0.5		MIT
2573.49	2574.26	38846.1	FE	0.5			MIT
2573.206	2573.977	38850.39	FE II		1.6		MIT
2572.966	2573.737	38854.01	FE	0.0	1.2		MIT
2572.14	2572.91	38866.5	FE	0.3			MIT
2571.548	2572.318	38875.44	FE II		1.2		MIT
2570.841	2571.611	38886.13	FE II	1.8	2.0		MIT
2570.525	2571.295	38890.91	FE	1.0	0.7		MIT
2569.764	2570.534	38902.42	FE II		1.8		MIT
2569.743	2570.513	38902.74	FE I	1.0	1.1		MIT
2569.601	2570.371	38904.89	FE I	1.3	0.0		MIT
2569.361	2570.131	38908.52	FE	0.5			MIT
2569.293	2570.063	38909.56	FE	0.5			MIT
2568.882	2569.652	38915.78	FE II		1.4		MIT
2568.865	2569.635	38916.04	FE I	1.3	1.0		MIT
2568.400	2569.170	38923.08	FE II		1.9		MIT
2567.87	2568.64	38931.1	FE	1.1			MIT
2567.805	2568.574	38932.10	FE	0.8			MIT
2567.326	2568.095	38939.36	FE	0.3			MIT
2566.912	2567.681	38945.64	FE II	1.8	2.2		MIT
2566.619	2567.388	38950.09	FE II	0.5	1.2		MIT
2566.406	2567.175	38953.32	FE	0.0	1.2		MIT
2566.29	2567.06	38955.1	FE	0.5	1.0		MIT
2566.214	2566.983	38956.24	FE	0.3	1.6		MIT
2565.708	2566.477	38963.92	FE	0.6			MIT
2565.46	2566.23	38967.7	FE	0.3			MIT
2564.706	2565.475	38979.14	FE	0.5			MIT
2564.548	2565.317	38981.54	FE I	1.2			MIT
2563.834	2564.603	38992.40	FE I	0.3	1.0		MIT
2563.834	2564.603	38992.40	FE II	0.3	1.0		MIT
2563.474	2564.242	38997.87	FE II	1.8	2.1		MIT
2563.402	2564.170	38998.97	FE	0.7	1.4H		MIT
2563.095	2563.863	39003.64	FE	0.8			MIT
2562.63	2563.40	39010.7	FE	0.3			MIT
2562.534	2563.302	39012.18	FE II	1.7	2.2		I
2562.229	2562.997	39016.82	FE I	1.2			MIT
2562.097	2562.865	39018.83	FE II	0.3	1.4		MIT
2561.855	2562.623	39022.52	FE I	1.2			MIT
2561.71	2562.48	39024.7	FE	0.5			MIT
2561.46	2562.23	39028.5	FE	0.5			MIT
2561.27	2562.04	39031.4	FE I	0.6			MIT
2560.559	2561.327	39042.27	FE I	1.2			MIT
2560.272	2561.040	39046.64	FE II	1.0	1.9		MIT
2559.928	2560.696	39051.89	FE II	0.3	1.2		MIT
2559.768	2560.536	39054.33	FE II	0.5	1.5		MIT
2559.243	2560.010	39062.34	FE II		1.3		MIT
2558.483	2559.250	39073.94	FE	1.0	1.3		MIT
2557.502	2558.269	39088.93	FE II	0.0	1.7		MIT
2557.265	2558.032	39092.55	FE	0.0			MIT
2557.081	2557.848	39095.37	FE II		1.0		MIT
2556.864	2557.631	39098.68	FE	1.3	0.0		MIT
2556.57	2557.34	39103.2	FE	0.3			MIT
2556.30	2557.07	39107.3	FE	1.2	0.0		MIT
2555.73	2556.50	39116.0	FE	0.7			MIT
2555.648	2556.415	39117.29	FE I	0.5			MIT
2555.442	2556.209	39120.44	FE II	0.3	1.5		MIT
2555.220	2555.987	39123.84	FE	1.0			MIT
2555.066	2555.832	39126.19	FE II	1.3	1.3		MIT
2554.520	2555.286	39134.56	FE	0.3			MIT
2553.826	2554.592	39145.19	FE	1.0	0.8		MIT
2553.734	2554.500	39146.60	FE II		1.2		MIT
2553.50	2554.27	39150.2	FE	0.3			MIT
2553.185	2553.951	39155.02	FE	1.0	1.3		MIT
2552.832	2553.598	39160.43	FE I	0.7	0.0		MIT
2552.773	2553.539	39161.34	FE	1.3			MIT
2552.607	2553.373	39163.88	FE	1.2	0.3		MIT
2551.44	2552.21	39181.8	FE	0.3			MIT
2551.212	2551.978	39185.30	FE II		0.8		MIT
2551.092	2551.858	39187.14	FE	1.4	0.0		I
2550.811	2551.576	39191.46	FE	0.3			MIT
2550.681	2551.446	39193.45	FE II	0.5	1.5		MIT

AIR WAVELENGTH (ANGSTRMS)	VACUUM WAVELENGTH (ANGSTRMS)	WAVENUMBER (KAYSERS)	LMENT		LOG ARC	INTENSITY SPK	DIS	REF
2550.508	2551.273	39196.11	FE		1.4			MIT
2550.020	2550.785	39203.61	FE	II	1.2	1.6		MIT
2549.774	2550.539	39207.40	FE	II	0.0	0.3		MIT
2549.613	2550.378	39209.87	FE	I	1.8G	0.3		MIT
2549.461	2550.226	39212.21	FE	II	0.0	1.2		MIT
2549.395	2550.160	39213.22	FE	II	0.0	1.0		MIT
2549.081	2549.846	39218.05	FE	II	0.5	1.5		MIT
2548.919	2549.684	39220.55	FE	II	0.0	0.5		MIT
2548.739	2549.504	39223.32	FE			1.1		MIT
2548.586	2549.351	39225.67	FE	II	0.3	0.9		MIT
2548.326	2549.091	39229.67	FE	II		1.0		MIT
2548.08	2548.84	39233.5	FE		1.7			MIT
2547.330	2548.095	39245.01	FE	II	0.0	1.5		MIT
2546.874	2547.639	39252.04	FE		1.7	0.0		MIT
2546.661	2547.426	39255.32	FE	II	0.0	1.5		MIT
2545.979	2546.743	39265.83	FE	I	2.0G	1.5		MIT
2545.520	2546.284	39272.91	FE	II		0.0		MIT
2545.433	2546.197	39274.25	FE	II		0.6		MIT
2545.217	2545.981	39277.59	FE	II	0.6	1.4		MIT
2544.971	2545.735	39281.38	FE	II	0.3	1.0		MIT
2544.708	2545.472	39285.44	FE		2.0	0.7		MIT
2543.920	2544.684	39297.61	FE		1.6	1.3		MIT
2543.647	2544.411	39301.83	FE		2.8			MIT
2543.431	2544.195	39305.17	FE	II		0.7		MIT
2543.384	2544.148	39305.89	FE	II	0.7	1.7		MIT
2542.732	2543.496	39315.97	FE	II	0.0	1.6		MIT
2542.101	2542.864	39325.73	FE		1.6	0.9		I
2541.835	2542.598	39329.85	FE	II	0.3	0.7		MIT
2541.097	2541.860	39341.27	FE	II	0.0	1.2		MIT
2540.976	2541.739	39343.14	FE	I	2.0G	1.0		MIT
2540.666	2541.429	39347.94	FE	II	0.8	1.5		MIT
2540.531	2541.294	39350.03	FE			0.3		MIT
2539.979	2540.742	39358.58	FE		0.5	1.3H		MIT
2539.357	2540.120	39368.22	FE	I	1.2			MIT
2538.997	2539.760	39373.80	FE		1.0	1.3		MIT
2538.904	2539.667	39375.25	FE	II	0.0	0.5		MIT
2538.809	2539.572	39376.72	FE	II	1.2	1.5		M
2538.699	2539.462	39378.42	FE	II		0.7		MIT
2538.497	2539.260	39381.56	FE	II	0.3	1.0		MIT
2538.201	2538.964	39386.15	FE	II	0.3	1.5		MIT
2537.950	2538.712	39390.05	FE		0.8			MIT
2537.458	2538.220	39397.68	FE		0.5			MIT
2537.173	2537.935	39402.11	FE	I	0.5			MIT
2537.141	2537.903	39402.60	FE	II		0.3		MIT
2536.817	2537.579	39407.64	FE	II	1.0	0.6		MIT
2536.673	2537.435	39409.87	FE	II	0.0	0.7		DO
2536.224	2536.986	39416.85	FE		0.5			MIT
2535.604	2536.366	39426.49	FE	I	3.0			MIT
2535.477	2536.239	39428.46	FE	II		1.3		MIT
2535.364	2536.126	39430.22	FE		0.0	0.5		MIT
2535.124	2535.886	39433.95	FE	I	0.3			MIT
2534.416	2535.178	39444.97	FE	II	0.8	1.7		MIT
2533.803	2534.564	39454.51	FE		1.1			MIT
2533.627	2534.388	39457.25	FE	II	0.9	1.7		MIT
2533.143	2533.904	39464.79	FE		0.3	0.0		MIT
2532.877	2533.638	39468.93	FE	I	0.5	0.0		MIT
2532.53	2533.29	39474.3	FE		1.0			MIT
2531.745	2532.506	39486.58	FE		0.6			MIT
2531.44	2532.20	39491.3	FE		0.6			MIT
2531.084	2531.845	39496.89	FE	II		0.6		MIT
2530.692	2531.453	39503.01	FE		1.4	1.8		I
2530.106	2530.867	39512.16	FE	II	0.3	1.5		MIT
2529.833	2530.594	39516.42	FE	I	1.7G	0.3		IJA
2529.547	2530.307	39520.89	FE	II	1.2	2.0		MIT
2529.306	2530.066	39524.65	FE		1.0			MIT
2529.224	2529.984	39525.94	FE	II		0.3		MIT
2529.133	2529.893	39527.36	FE	I	1.9G	0.7		MIT
2529.080	2529.840	39528.19	FE			1.8		MIT
2528.879	2529.639	39531.33	FE		0.3	0.3		MIT
2528.174	2528.934	39542.35	FE		0.8			MIT

AIR WAVELENGTH (ANGSTRMS)	VACUUM WAVELENGTH (ANGSTRMS)	WAVENUMBER (KAYSERS)	LMENT		LOG ARC	INTENSITY SPK	DIS	REF
2527.927	2528.687	39546.21	FE		0.3			MIT
2527.694	2528.454	39549.86	FE	II	0.0	1.5		MIT
2527.433	2528.193	39553.94	FE	I	2.3G	1.7		BU
2527.102	2527.862	39559.12	FE	II	0.0	1.8		MIT
2526.834	2527.594	39563.32	FE	II		0.0		MIT
2526.295	2527.055	39571.76	FE	II	1.0	1.8		MIT
2526.075	2526.835	39575.21	FE	II	0.5	1.3		MIT
2525.865	2526.625	39578.50	FE	II		0.7		MIT
2525.387	2526.147	39585.99	FE	II	1.3	1.8		MIT
2525.107	2525.866	39590.37	FE			0.3		MIT
2525.022	2525.781	39591.71	FE		1.3	0.0		MIT
2524.289	2525.048	39603.20	FE	I	2.0G	1.7		MIT
2523.660	2524.419	39613.07	FE		1.2	1.0		MIT
2523.14	2523.90	39621.2	FE		0.0			MIT
2522.848	2523.607	39625.82	FE	I	2.5G	1.7		MIT
2522.495	2523.254	39631.37	FE	I	0.5			MIT
2522.196	2522.955	39636.06	FE	II	0.0	1.0		MIT
2521.917	2522.676	39640.45	FE	I	0.8			MIT
2521.814	2522.573	39642.07	FE		0.0	1.8		MIT
2521.485	2522.244	39647.24	FE			0.7		MIT
2521.208	2521.967	39651.60	FE		0.0	0.3		MIT
2521.094	2521.853	39653.39	FE	II	0.3	1.3		MIT
2520.968	2521.726	39655.37	FE		0.5			MIT
2520.883	2521.641	39656.71	FE		1.9			MIT
2520.677	2521.435	39659.95	FE	II		0.8		MIT
2519.627	2520.385	39676.48	FE		1.5	1.3		I
2519.405	2520.163	39679.97	FE	II		0.6		MIT
2519.047	2519.805	39685.61	FE	II		1.8		MIT
2518.824	2519.582	39689.12	FE		0.8			MIT
2518.101	2518.859	39700.52	FE	I	2.3G	1.7		MIT
2517.659	2518.417	39707.49	FE	I	1.3	1.1		MIT
2517.211	2517.969	39714.55	FE	II	0.0	0.5		MIT
2517.120	2517.878	39715.99	FE	II	1.0	1.8		MIT
2516.570	2517.327	39724.67	FE	I	1.0	0.0		MIT
2516.250	2517.007	39729.72	FE	I	0.3			MIT
2515.105	2515.862	39747.81	FE			1.5		MIT
2514.914	2515.671	39750.83	FE	II	0.0	1.4		MIT
2514.711	2515.468	39754.03	FE		0.3			MIT
2514.559	2515.316	39756.44	FE		0.8			MIT
2514.378	2515.135	39759.30	FE	II	0.3	0.9		MIT
2513.856	2514.613	39767.55	FE		0.7			MIT
2513.330	2514.087	39775.88	FE		0.8	0.6J		MIT
2512.848	2513.605	39783.50	FE		0.5			MIT
2512.524	2513.281	39788.63	FE	II		1.6		MIT
2512.363	2513.119	39791.18	FE		1.1	0.0		MIT
2512.105	2512.861	39795.27	FE			1.5		MIT
2511.759	2512.515	39800.75	FE	II	1.4	2.0		MIT
2511.370	2512.126	39806.92	FE	II	0.5	0.5		MIT
2510.834	2511.590	39815.41	FE	I	2.5G	1.7		MIT
2509.122	2509.878	39842.58	FE	II	0.0	1.7		MIT
2508.949	2509.705	39845.33	FE		0.0			MIT
2508.749	2509.505	39848.50	FE		1.3	0.0		MIT
2508.495	2509.251	39852.54	FE		0.3			MIT
2508.333	2509.089	39855.11	FE			1.3H		MIT
2507.899	2508.654	39862.01	FE		1.6	0.8		I
2507.730	2508.485	39864.69	FE		0.5			MIT
2507.430	2508.185	39869.46	FE		0.3			MIT
2507.014	2507.769	39876.08	FE	II	0.0	1.0		DO
2506.569	2507.324	39883.16	FE		1.0			MIT
2506.429	2507.184	39885.38	FE	II	0.0	0.5		MIT
2506.093	2506.848	39890.73	FE	II	0.3	1.8		MIT
2505.628	2506.383	39898.13	FE		1.1			MIT
2505.486	2506.241	39900.40	FE		1.0	0.0		MIT
2505.218	2505.973	39904.66	FE	II	0.3	1.4		MIT
2505.006	2505.761	39908.04	FE		1.0	0.0		MIT
2504.098	2504.853	39922.51	FE		0.3			MIT
2503.870	2504.625	39926.15	FE	II	0.9	1.8		MIT
2503.655	2504.409	39929.57	FE		1.9			MIT
2503.556	2504.310	39931.15	FE	II	0.3	1.3		MIT
2503.491	2504.245	39932.19	FE		0.6			MIT

AIR WAVELENGTH (ANGSTRMS)	VACUUM WAVELENGTH (ANGSTRMS)	WAVENUMBER (KAYSERS)	LMENT	LOG ARC	INTENSITY SPK	DIS	REF
2503.325	2504.079	39934.84	FE II	0.7	1.7		MIT
2502.390	2503.144	39949.76	FE II	0.5	1.8		M
2501.695	2502.449	39960.85	FE I	1.3	0.6		MIT
2501.127	2501.881	39969.93	FE I	2.0G	1.4		MIT
2500.926	2501.680	39973.14	FE	1.1	1.6		MIT
2498.894	2499.647	40005.64	FE I	1.3	1.8		MIT
2498.894	2499.647	40005.64	FE II	1.3	1.8		MIT
2498.208	2498.961	40016.63	FE	0.5			MIT
2497.820	2498.573	40022.84	FE II	1.2	1.7		MIT
2497.717	2498.470	40024.49	FE II	0.0	0.5		MIT
2497.300	2498.053	40031.18	FE II		1.2		MIT
2496.991	2497.744	40036.13	FE	1.3	0.0		MIT
2496.533	2497.286	40043.47	FE	1.6	1.2		I
2496.071	2496.824	40050.89	FE	0.6			MIT
2495.863	2496.616	40054.22	FE I	1.4	1.5		MIT
2494.250	2495.002	40080.12	FE I	1.0			MIT
2494.109	2494.861	40082.39	FE II		0.3		MIT
2493.999	2494.751	40084.16	FE I	1.3	0.0		MIT
2493.261	2494.013	40096.02	FE II	1.0	2.0		MIT
2493.180	2493.932	40097.32	FE II	1.0	2.0		MIT
2492.644	2493.396	40105.95	FE I	0.5			MIT
2492.337	2493.089	40110.88	FE II		1.5		MIT
2491.984	2492.736	40116.57	FE	1.0			MIT
2491.390	2492.142	40126.13	FE II		1.5		MIT
2491.155	2491.907	40129.92	FE I	2.2G	1.0		MIT
2490.861	2491.612	40134.65	FE II		1.2		MIT
2490.728	2491.479	40136.79	FE II	0.0	1.0		MIT
2490.644	2491.395	40138.15	FE I	2.3G	1.0		MIT
2489.822	2490.573	40151.40	FE II	0.0	1.7		MIT
2489.751	2490.502	40152.54	FE I	2.3G	0.3		MIT
2489.486	2490.237	40156.82	FE II	0.0	1.6		MIT
2488.950	2489.701	40165.46	FE	1.0	0.0		MIT
2488.148	2488.899	40178.41	FE I	2.8G	2.0R		MIT
2487.367	2488.118	40191.02	FE I	0.8	0.3		MIT
2487.064	2487.815	40195.92	FE I	1.4	1.0		I
2486.693	2487.444	40201.92	FE	1.5	0.5		MIT
2486.371	2487.121	40207.12	FE	1.6			MIT
2486.345	2487.095	40207.54	FE II		1.9		MIT
2485.985	2486.735	40213.37	FE	1.0	0.0		MIT
2485.075	2485.825	40228.09	FE II		0.3		MIT
2484.556	2485.306	40236.49	FE II		1.0H		MIT
2484.236	2484.986	40241.68	FE II		1.3		MIT
2484.187	2484.937	40242.47	FE I	2.0G	0.7		MIT
2483.723	2484.473	40249.99	FE II		1.4		MIT
2483.535	2484.285	40253.03	FE I	0.0			MIT
2483.270	2484.020	40257.33	FE I	2.7F	1.7		MIT
2482.654	2483.404	40267.32	FE II	0.0	1.7		MIT
2482.326	2483.076	40272.64	FE		0.8		M
2482.115	2482.864	40276.06	FE II	0.5	1.7		MIT
2481.574	2482.323	40284.84	FE II		1.2		MIT
2481.049	2481.798	40293.37	FE II		1.6		MIT
2480.158	2480.907	40307.84	FE II	1.0	1.9		MIT
2479.776	2480.525	40314.05	FE I	2.3G	1.5		MIT
2479.624	2480.373	40316.52	FE	0.8			MIT
2479.476	2480.225	40318.92	FE I	0.9	0.5		MIT
2478.569	2479.318	40333.68	FE II	0.6	1.6		MIT
2478.448	2479.197	40335.65	FE II		0.7H		MIT
2478.116	2478.865	40341.05	FE II		1.3		MIT
2477.480	2478.228	40351.41	FE II		0.0		MIT
2477.338	2478.086	40353.72	FE II	0.0	1.6		MIT
2476.861	2477.609	40361.49	FE I	0.5			MIT
2476.654	2477.402	40364.86	FE I	1.2	1.4		MIT
2476.267	2477.015	40371.17	FE II		1.5		MIT
2476.029	2476.777	40375.05	FE	1.3H			MIT
2475.548	2476.296	40382.90	FE II		1.0H		MIT
2474.814	2475.562	40394.87	FE	1.6	1.7		I
2474.765	2475.513	40395.67	FE II		1.8		MIT
2473.321	2474.068	40419.25	FE II	0.0	0.6		MIT
2473.156	2473.903	40421.95	FE	0.7			MIT
2472.909	2473.656	40425.99	FE I	3.0			MIT
2472.880	2473.627	40426.46	FE I	2.5G	1.4		MIT
2472.430	2473.177	40433.82	FE II		1.2		MIT
2472.344	2473.091	40435.23	FE I	1.5	0.7		MIT
2472.106	2472.853	40439.12	FE	1.0			MIT
2472.072	2472.819	40439.67	FE II		1.4		MIT
2470.966	2471.713	40457.77	FE I	1.4	0.0		MIT
2470.752	2471.499	40461.28	FE II		0.5		MIT
2470.658	2471.405	40462.82	FE II	0.9	1.7		MIT
2470.406	2471.153	40466.94	FE II	0.0	1.6		MIT
2470.06	2470.81	40472.6	FE		0.8		MIT
2469.834	2470.581	40476.31	FE		0.9		MIT
2469.512	2470.259	40481.59	FE	0.3	1.6		MIT
2468.878	2469.624	40491.99	FE	1.6	1.2		I
2468.295	2469.041	40501.55	FE II	0.0	1.5		MIT
2468.106	2468.852	40504.65	FE	0.3			MIT
2467.730	2468.476	40510.82	FE I	1.0	0.3		MIT
2466.818	2467.564	40525.80	FE II	0.0	1.5		MIT
2466.680	2467.426	40528.07	FE II	0.0	1.0		MIT
2465.914	2466.660	40540.65	FE II	0.8	2.0		MIT
2465.455	2466.201	40548.20	FE	1.2			MIT
2465.320	2466.066	40550.42	FE	0.5			MIT
2465.200	2465.946	40552.40	FE II		1.7		MIT
2465.148	2465.894	40553.25	FE I	1.8			I
2464.903	2465.648	40557.28	FE II	0.6	1.7		MIT
2464.007	2464.752	40572.03	FE II	0.7	1.9		MIT
2463.728	2464.473	40576.62	FE I	1.2	0.8		MIT
2463.507	2464.252	40580.26	FE	0.7			MIT
2463.278	2464.023	40584.04	FE II	0.8	1.8		MIT
2462.644	2463.389	40594.48	FE I	2.3G	1.7		MIT
2462.180	2462.925	40602.13	FE I	1.7R	0.5		MIT
2461.857	2462.602	40607.46	FE II	1.2	1.8		MIT
2461.671	2462.416	40610.53	FE II		0.3		MIT
2461.279	2462.024	40616.99	FE II	0.7	1.7		MIT
2461.056	2461.801	40620.67	FE	1.3			BU
2460.653	2461.397	40627.33	FE		0.7H		MIT
2460.596	2461.340	40628.27	FE	1.2			MIT
2460.453	2461.197	40630.63	FE II	0.0	1.2H		MIT
2460.31	2461.05	40633.0	FE	0.8			MIT
2459.172	2459.916	40651.79	FE	0.8			MIT
2459.097	2459.841	40653.03	FE II	0.0	0.5		MIT
2458.964	2459.708	40655.23	FE		1.0		DO
2458.782	2459.526	40658.24	FE II	0.8	1.8		MIT
2457.595	2458.339	40677.88	FE	1.8	1.5		I
2456.819	2457.563	40690.72	FE II		1.3		MIT
2456.641	2457.385	40693.67	FE II		1.2		MIT
2455.901	2456.644	40705.93	FE II		1.4H		MIT
2455.56	2456.30	40711.6	FE	1.2H			MIT
2454.576	2455.319	40727.90	FE II		1.9		MIT
2454.163	2454.906	40734.76	FE II		0.8		MIT
2453.975	2454.718	40737.88	FE II		0.5H		MIT
2453.796	2454.539	40740.85	FE II	0.0	1.0		MIT
2453.474	2454.217	40746.20	FE	1.3	0.5		I
2452.918	2453.661	40755.43	FE II		0.8		MIT
2452.590	2453.333	40760.88	FE	0.8	0.0		MIT
2451.388	2452.130	40780.87	FE	0.7			MIT
2451.214	2451.956	40783.76	FE II		1.2		MIT
2451.103	2451.845	40785.61	FE II	0.0	0.6		MIT
2450.438	2451.180	40796.67	FE	1.2			MIT
2450.202	2450.944	40800.60	FE II	0.0	1.2		MIT
2449.960	2450.702	40804.63	FE II	0.0	1.5		MIT
2449.736	2450.478	40808.37	FE II	0.0	0.3		MIT
2449.272	2450.014	40816.10	FE II		0.0		MIT
2449.187	2449.929	40817.51	FE II		0.6		MIT
2448.732	2449.474	40825.10	FE II		0.3		MIT
2448.388	2449.130	40830.83	FE	0.5			MIT
2447.747	2448.489	40841.52	FE II		1.9		MIT
2447.708	2448.450	40842.17	FE I	1.8	2.0		S
2447.441	2448.182	40846.63	FE	0.7			MIT
2447.320	2448.061	40848.65	FE II	0.0	1.2		MIT
2447.200	2447.941	40850.65	FE II	0.0	1.5		MIT

AIR WAVELENGTH (ANGSTRMS)	VACUUM WAVELENGTH (ANGSTRMS)	WAVENUMBER (KAYSERS)	LMENT	LOG ARC	INTENSITY SPK	INTENSITY DIS	REF
2446.465	2447.206	40862.92	FE II	0.5	1.6		MIT
2446.104	2446.845	40868.95	FE II	0.5	1.5		MIT
2445.782	2446.523	40874.33	FE II	0.3	1.2		MIT
2445.558	2446.299	40878.08	FE II	1.2	1.6		MIT
2445.211	2445.952	40883.88	FE I	0.8	0.0		MIT
2444.74	2445.48	40891.7	FE	0.5			MIT
2444.512	2445.253	40895.57	FE II	1.4	1.8		MIT
2444.206	2444.947	40900.69	FE	1.0			MIT
2443.870	2444.611	40906.31	FE I	1.6	0.6		I
2443.287	2444.028	40916.07	FE	0.5			MIT
2443.166	2443.906	40918.10	FE	0.8			MIT
2442.567	2443.307	40928.13	FE	1.8	1.0		I
2441.546	2442.286	40945.24	FE II	0.0	0.7		MIT
2440.863	2441.603	40956.70	FE	0.5			MIT
2440.425	2441.165	40964.05	FE II	0.0	1.6		MIT
2440.107	2440.847	40969.39	FE	1.4	0.9		MIT
2439.744	2440.484	40975.48	FE	1.3	1.1		MIT
2439.300	2440.040	40982.94	FE II	1.2	2.0		MIT
2438.181	2438.920	41001.75	FE I	1.5	0.6		I
2437.666	2438.405	41010.41	FE II	0.0	0.5H		MIT
2437.267	2438.006	41017.12	FE II	0.0	0.0		MIT
2437.154	2437.893	41019.03	FE II	0.3	0.3		MIT
2436.626	2437.365	41027.91	FE II	0.6	1.4H		MIT
2436.344	2437.083	41032.66	FE	1.4			MIT
2436.221	2436.960	41034.73	FE II		1.0		MIT
2435.855	2436.594	41040.90	FE	0.3H	0.3		MIT
2435.809	2436.548	41041.67	FE II		0.5		MIT
2434.944	2435.683	41056.25	FE II	0.8	1.3		MIT
2434.728	2435.467	41059.90	FE	0.8	1.5		MIT
2434.645	2435.384	41061.29	FE II	0.0	0.7H		MIT
2434.245	2434.983	41068.04	FE II	0.5	1.0H		MIT
2433.908	2434.646	41073.73	FE	0.3			MIT
2433.496	2434.234	41080.68	FE II	0.3	1.2		MIT
2432.868	2433.606	41091.28	FE	0.7	1.6		MIT
2432.267	2433.005	41101.44	FE II	1.4	1.8		MIT
2431.963	2432.701	41106.57	FE	0.5			MIT
2431.30	2432.04	41117.8	FE	0.8	0.0		MIT
2431.024	2431.762	41122.45	FE	1.3	0.5H		MIT
2430.839	2431.577	41125.58	FE		0.3		I
2430.186	2430.924	41136.63	FE II	0.3	0.3		MIT
2430.072	2430.809	41138.56	FE II	1.2	1.8		MIT
2429.813	2430.550	41142.94	FE I	0.8			MIT
2429.495	2430.232	41148.33	FE II		0.3		MIT
2429.385	2430.122	41150.19	FE II		1.2		MIT
2429.036	2429.773	41156.10	FE II	0.0	0.9		MIT
2428.795	2429.532	41160.19	FE II	0.0	1.2		MIT
2428.361	2429.098	41167.54	FE II	0.5	1.2		MIT
2428.286	2429.023	41168.81	FE II	0.0	1.0		MIT
2428.20	2428.94	41170.3	FE	1.0			MIT
2427.197	2427.934	41187.28	FE II		0.7		MIT
2426.309	2427.046	41202.36	FE	0.6			MIT
2425.930	2426.667	41208.79	FE II		1.0		MIT
2425.684	2426.420	41212.97	FE II	0.0	1.6		MIT
2425.361	2426.097	41218.46	FE II		1.3		MIT
2424.585	2425.321	41231.65	FE II	0.0	1.3		MIT
2424.390	2425.126	41234.97	FE II	0.3	0.6		MIT
2424.143	2424.879	41239.17	FE II	1.4	1.8		MIT
2423.210	2423.946	41255.05	FE II	0.3	1.6		MIT
2423.094	2423.830	41257.02	FE I	0.7			MIT
2422.941	2423.677	41259.63	FE II		0.3		MIT
2422.679	2423.415	41264.09	FE II	0.5	1.8		MIT
2420.383	2421.118	41303.23	FE I	0.3			MIT
2419.876	2420.611	41311.88	FE II	0.3	0.5		MIT
2419.876	2420.611	41311.88	FE I	0.3	0.5		MIT
2419.78	2420.52	41313.5	FE	0.5			MIT
2419.41	2420.15	41319.8	FE	0.5			MIT
2419.058	2419.793	41325.85	FE I	0.3			MIT
2418.440	2419.175	41336.41	FE II	0.0	1.0H		MIT
2417.866	2418.601	41346.22	FE II	1.0	2.0		MIT
2417.490	2418.225	41352.65	FE I	0.8			MIT
2416.714	2417.448	41365.93	FE II		0.7		MIT
2416.457	2417.191	41370.33	FE II	0.0	1.6H		MIT
2415.063	2415.797	41394.21	FE II	0.0	1.7		MIT
2414.080	2414.814	41411.06	FE II		0.6		MIT
2413.309	2414.043	41424.29	FE II	1.8	2.0H		I
2411.066	2411.799	41462.82	FE II	1.5	1.8H		I
2410.517	2411.250	41472.27	FE II	1.7	1.8H		I
2409.706	2410.439	41486.22	FE II		0.8		MIT
2409.376	2410.109	41491.90	FE II		0.3		MIT
2408.653	2409.386	41504.36	FE II		0.7H		MIT
2408.045	2408.777	41514.84	FE I	1.0			MIT
2407.949	2408.681	41516.49	FE II		1.3		MIT
2407.58	2408.31	41522.9	FE	0.8			MIT
2407.229	2407.961	41528.91	FE	1.3	0.3		MIT
2406.993	2407.725	41532.98	FE II	0.0	0.7		MIT
2406.658	2407.390	41538.76	FE II	0.7	1.0		DB
2404.882	2405.614	41569.43	FE II	1.7	2.0J		MIT
2404.430	2405.162	41577.25	FE II	1.4	1.6		I
2402.600	2403.331	41608.91	FE II	1.0	1.3		MIT
2401.52	2402.25	41627.6	FE	0.5			MIT
2400.338	2401.069	41648.12	FE II	0.6	1.4		MIT
2400.275	2401.006	41649.21	FE II		0.5		MIT
2399.239	2399.969	41667.20	FE II	1.3	1.5		I
2398.664	2399.394	41677.18	FE II		0.7H		DO
2398.21	2398.94	41685.1	FE	0.5			MIT
2396.712	2397.442	41711.13	FE II		0.9		MIT
2395.625	2396.355	41730.05	FE II	1.7	2.0V		MIT
2395.408	2396.138	41733.83	FE II	0.8	0.7H		MIT
2394.898	2395.627	41742.72	FE II	0.5	0.7		MIT
2391.82	2392.55	41796.4	FE	0.5			MIT
2391.474	2392.203	41802.48	FE II	0.9	1.3		MIT
2390.223	2390.951	41824.35	FE II				MIT
2389.971	2390.699	41828.76	FE I	1.2			I
2389.402	2390.130	41838.72	FE	0.3	0.3		MIT
2388.627	2389.355	41852.30	FE II	1.4	1.5		I
2388.387	2389.115	41856.50	FE II		0.3		MIT
2388.202	2388.930	41859.75	FE II		0.5		MIT
2387.427	2388.155	41873.33	FE II	0.3	0.9		MIT
2387.279	2388.007	41875.93	FE	0.8			MIT
2386.396	2387.124	41891.42	FE II	0.3	0.7		MIT
2385.95	2386.68	41899.2	FE I	0.6			MIT
2385.579	2386.306	41905.77	FE	0.3			MIT
2385.007	2385.734	41915.82	FE II	0.9	1.2		MIT
2384.386	2385.113	41926.73	FE	1.3	0.7		I
2383.241	2383.968	41946.87	FE II	0.9	1.1		MIT
2383.055	2383.782	41950.15	FE II	0.8	0.3		MIT
2382.893	2383.620	41953.00	FE II		0.6		MIT
2382.355	2383.082	41962.47	FE II	0.5			MIT
2382.039	2382.766	41968.04	FE II	1.6R	2.0G		MIT
2381.83	2382.56	41971.7	FE I	0.5			MIT
2380.759	2381.485	41990.60	FE II	1.1	1.2		I
2379.275	2380.001	42016.79	FE II	1.1	1.2		I
2378.980	2379.706	42022.00	FE II	0.3	0.3		MIT
2378.527	2379.253	42030.00	FE II		0.5		MIT
2377.87	2378.60	42041.6	FE	0.3			MIT
2377.521	2378.247	42047.79	FE	0.0	0.7H		MIT
2377.23	2377.96	42052.9	FE	0.7			MIT
2376.435	2377.160	42067.00	FE II	0.5	1.7H		MIT
2375.192	2375.917	42089.01	FE II		1.2		I
2374.873	2375.598	42094.67	FE	0.3			MIT
2374.517	2375.242	42100.98	FE I	0.8			I
2373.730	2374.455	42114.93	FE II	0.8	1.2		MIT
2373.624	2374.349	42116.81	FE I	0.3			MIT
2372.627	2373.351	42134.51	FE II		0.5		DB
2371.428	2372.152	42155.81	FE I	0.9			I
2370.496	2371.220	42172.38	FE II	0.9	0.8		I
2369.960	2370.684	42181.92	FE II	0.5	1.2H		MIT
2369.445	2370.169	42191.09	FE I	0.9	0.0		MIT
2368.595	2369.319	42206.23	FE II	1.2	1.4		I
2366.595	2367.318	42241.89	FE II	1.0	1.3		I

AIR WAVELENGTH (ANGSTRMS)	VACUUM WAVELENGTH (ANGSTRMS)	WAVENUMBER (KAYSERS)	LMENT	LOG ARC	INTENSITY SPK	DIS	REF
2365.851	2366.574	42255.18	FE	0.8H			MIT
2365.771	2366.494	42256.61	FE	0.5	0.7		DO
2364.826	2365.549	42273.49	FE II	1.2	1.5		I
2363.94	2364.66	42289.3	FE	1.0W			MIT
2363.635	2364.357	42294.79	FE II	0.3	0.3		MIT
2362.019	2362.741	42323.72	FE II	0.9	1.2		I
2361.724	2362.446	42329.01	FE II		0.6		MIT
2360.293	2361.015	42354.67	FE II	1.0	0.9		I
2359.997	2360.719	42359.98	FE II	1.0	0.9		I
2359.589	2360.311	42367.31	FE II		0.5		MIT
2359.104	2359.825	42376.02	FE II	1.2	1.3		I
2357.008	2357.729	42413.70	FE II		1.0		MIT
2355.915	2356.636	42433.37	FE I		0.3		MIT
2355.328	2356.049	42443.95	FE I	0.5			MIT
2355.328	2356.049	42443.95	FE II	0.5			MIT
2355.223	2355.944	42445.84	FE II		0.5		MIT
2354.889	2355.609	42451.86	FE II	1.0	1.0		I
2354.466	2355.186	42459.48	FE II	0.7	1.2		MIT
2352.44	2353.16	42496.0	FE		0.9H		MIT
2351.191	2351.911	42518.62	FE II	0.3	0.8		MIT
2350.399	2351.118	42532.95	FE I	0.5			MIT
2348.303	2349.022	42570.91	FE II	0.7	1.3		MIT
2348.099	2348.818	42574.61	FE II	0.7	1.3		MIT
2346.293	2347.012	42607.37	FE II	0.0	0.3		MIT
2345.340	2346.058	42624.69	FE II		1.3		MIT
2345.070	2345.788	42629.59	FE	1.3H			MIT
2344.280	2344.998	42643.96	FE II	0.8	1.3		I
2343.959	2344.677	42649.80	FE II	0.7	1.0		MIT
2343.492	2344.210	42658.29	FE II	1.0	1.7		MIT
2342.252	2342.970	42680.88	FE II		0.5		MIT
2341.963	2342.681	42686.14	FE II		0.3		MIT
2340.932	2341.649	42704.94	FE II		0.3		MIT
2340.449	2341.166	42713.75	FE II		0.3		MIT
2340.26	2340.98	42717.2	FE	0.5H			MIT
2339.58	2340.30	42729.6	FE	1.0J			MIT
2339.418	2340.135	42732.58	FE II		0.9		MIT
2338.005	2338.722	42758.40	FE II	1.2	1.5		I
2337.526	2338.243	42767.16	FE		0.3H		MIT
2336.861	2337.577	42779.33	FE II		1.3H		MIT
2334.528	2335.244	42822.08	FE	0.5			MIT
2332.796	2333.512	42853.87	FE II	1.2	1.6		I
2332.028	2332.743	42867.98	FE		0.6H		MIT
2331.306	2332.021	42881.26	FE II	1.0	0.8		I
2329.630	2330.345	42912.10	FE I	0.7			MIT
2329.357	2330.072	42917.13	FE		0.8		MIT
2327.954	2328.668	42943.00	FE II		0.8H		MIT
2327.394	2328.108	42953.33	FE II	1.0	1.4		I
2326.352	2327.066	42972.56	FE II		0.9		MIT
2325.574	2326.288	42986.94	FE II		0.7		MIT
2325.299	2326.013	42992.02	FE II		0.7		MIT
2324.476	2325.190	43007.24	FE		0.7J		MIT
2322.949	2323.662	43035.51	FE		1.0J		DO
2322.326	2323.039	43047.06	FE II		0.7		DO
2320.356	2321.069	43083.60	FE I	1.4	0.7		I
2319.771	2320.484	43094.46	FE	0.0	0.0H		DO
2318.541	2319.253	43117.32	FE II		0.9		MIT
2318.347	2319.059	43120.93	FE II		0.6		MIT
2318.17	2318.88	43124.2	FE	1.0W			MIT
2317.889	2318.601	43129.45	FE	0.9			MIT
2317.377	2318.089	43138.98	FE II		1.0H		DO
2314.848	2315.560	43186.10	FE		0.7J		MIT
2313.962	2314.673	43202.64	FE II	0.3	0.8		DO
2313.300	2314.011	43215.00	FE II		0.8		MIT
2313.102	2313.813	43218.70	FE I	1.4	0.5		I
2312.034	2312.745	43238.66	FE II		1.2		MIT
2311.288	2311.999	43252.62	FE		1.5		DO
2311.224	2311.935	43253.81	FE II		1.3		DO
2310.097	2310.808	43274.92	FE		0.7		DO
2308.997	2309.707	43295.53	FE I	1.3	0.7		I
2308.765	2309.475	43299.88	FE		1.2		MIT
2307.314	2308.024	43327.11	FE		1.4		MIT
2306.380	2307.090	43344.65	FE	1.0	0.5		MIT
2306.171	2306.881	43348.58	FE I	1.0			MIT
2304.729	2305.438	43375.70	FE I	1.0	1.3		MIT
2303.833	2304.542	43392.57	FE		0.3		MIT
2303.578	2304.287	43397.37	FE I	1.0	0.5		I
2303.422	2304.131	43400.31	FE	1.0			I
2303.351	2304.060	43401.65	FE II		1.2		MIT
2301.681	2302.390	43433.13	FE I	1.2			I
2301.424	2302.133	43437.98	FE II		0.6		MIT
2301.173	2301.882	43442.72	FE	1.0			MIT
2300.59	2301.30	43453.7	FE	0.9			MIT
2300.139	2300.847	43462.25	FE I	1.2	0.5		I
2299.455	2300.163	43475.18	FE I	0.9			MIT
2299.218	2299.926	43479.66	FE I	1.2	0.6		I
2298.662	2299.370	43490.17	FE	0.9	0.3		MIT
2298.231	2298.939	43498.33	FE II	0.7	1.4		MIT
2298.17	2298.88	43499.5	FE I	0.7			MIT
2297.786	2298.494	43506.75	FE I	1.5G	0.8		I
2296.925	2297.633	43523.06	FE I	1.2	0.3		I
2296.666	2297.374	43527.97	FE II	0.7			MIT
2296.17	2296.88	43537.4	FE	0.3	0.7J		MIT
2296.108	2296.815	43538.54	FE		0.3		DO
2295.040	2295.747	43558.80	FE	0.3			MIT
2294.612	2295.319	43566.93	FE II	1.2	0.7		MIT
2294.406	2295.113	43570.84	FE I	1.2	0.3		I
2293.846	2294.553	43581.47	FE	1.0	0.8		I
2293.765	2294.472	43583.01	FE II		1.0		DO
2293.134	2293.841	43595.00	FE II		0.3		MIT
2292.770	2293.477	43601.92	FE II		0.6		DO
2292.523	2293.230	43606.62	FE I	1.2	0.3		I
2291.624	2292.330	43623.73	FE I	1.0	0.3		MIT
2291.120	2291.826	43633.32	FE I	1.2			I
2290.768	2291.474	43640.03	FE I	0.8			MIT
2290.545	2291.251	43644.27	FE I	1.2			MIT
2290.065	2290.771	43653.42	FE I	1.0			MIT
2289.031	2289.737	43673.14	FE I	1.4			MIT
2288.678	2289.384	43679.87	FE		0.5		DO
2287.630	2288.336	43699.88	FE I	0.7			I
2287.248	2287.954	43707.18	FE I	1.3	0.8		I
2286.428	2287.133	43722.85	FE	0.3			MIT
2286.152	2286.857	43728.13	FE		0.3		MIT
2284.085	2284.790	43767.70	FE I	1.4	1.3		I
2283.653	2284.358	43775.98	FE I	1.0			I
2283.305	2284.010	43782.65	FE I	0.9			MIT
2283.082	2283.787	43786.93	FE I	0.9			MIT
2282.863	2283.568	43791.13	FE I	0.9			MIT
2281.99	2282.69	43807.9	FE	0.8W			MIT
2280.220	2280.924	43841.88	FE I	1.2			MIT
2279.924	2280.628	43847.57	FE I	1.2	1.6		I
2277.672	2278.375	43890.92	FE I	1.4			MIT
2277.096	2277.799	43902.02	FE I	1.2			I
2276.024	2276.727	43922.70	FE I	1.3	1.0		I
2275.595	2276.298	43930.98	FE I	0.9			MIT
2275.187	2275.890	43938.86	FE	1.0			MIT
2274.088	2274.791	43960.09	FE I	1.2	1.2		I
2272.816	2273.518	43984.69	FE I	1.3			MIT
2272.61	2273.31	43988.7	FE	0.5			MIT
2272.067	2272.769	43999.19	FE I	1.2	0.5		I
2271.780	2272.482	44004.75	FE I	1.5			I
2270.860	2271.562	44022.57	FE I	1.0			I
2270.348	2271.050	44032.50	FE II	0.3	0.6		M
2269.097	2269.799	44056.77	FE I	1.1			MIT
2268.847	2269.549	44061.63	FE II		1.0		MIT
2268.562	2269.263	44067.16	FE II		0.9		MIT
2268.139	2268.840	44075.38	FE II		1.1		MIT
2267.588	2268.289	44086.09	FE II	0.3	1.5		MIT
2267.466	2268.167	44088.46	FE I	1.2			MIT
2267.078	2267.779	44096.00	FE I	0.9	0.3		MIT
2266.899	2267.600	44099.49	FE I	1.2			MIT

AIR WAVELENGTH (ANGSTRMS)	VACUUM WAVELENGTH (ANGSTRMS)	WAVENUMBER (KAYSERS)	LMENT	LOG ARC	INTENSITY SPK	INTENSITY DIS	REF
2266.699	2267.400	44103.38	FE II		0.7		DO
2266.249	2266.950	44112.13	FE II		0.9		MIT
2265.994	2266.695	44117.10	FE II	0.6	1.0		MIT
2265.59	2266.29	44125.0	FE I	0.6			MIT
2265.050	2265.751	44135.48	FE I	1.0			I
2264.591	2265.292	44144.43	FE II		1.3		MIT
2264.389	2265.090	44148.37	FE I	1.5	0.7		I
2263.474	2264.174	44166.21	FE I	0.8			MIT
2263.228	2263.928	44171.01	FE II		1.5		MIT
2262.680	2263.380	44181.71	FE II	0.8	1.1		MIT
2262.261	2262.961	44189.89	FE II		0.6		DO
2260.853	2261.553	44217.41	FE II	1.1	1.1		MIT
2260.600	2261.300	44222.35	FE	1.0			MIT
2260.228	2260.928	44229.63	FE II		0.7		DO
2260.079	2260.779	44232.55	FE II	1.0	1.2		I
2259.589	2260.289	44242.14	FE II		1.3		DO
2259.511	2260.211	44243.67	FE I	1.4			I
2259.279	2259.978	44248.21	FE I	0.8			MIT
2258.372	2259.071	44265.98	FE II		0.3		DO
2257.67	2258.37	44279.7	FE	0.5			MIT
2256.897	2257.596	44294.91	FE II		1.2H		DO
2256.433	2257.132	44304.01	FE II		1.0		MIT
2255.980	2256.679	44312.91	FE II	0.0	0.5		MIT
2255.860	2256.559	44315.27	FE I	1.3			I
2255.763	2256.462	44317.17	FE II		1.3		MIT
2255.150	2255.849	44329.22	FE II		1.1		MIT
2254.897	2255.596	44334.19	FE II		0.5H		MIT
2254.390	2255.088	44344.16	FE II		0.3		MIT
2254.078	2254.776	44350.30	FE II		1.0H		DO
2253.124	2253.822	44369.07	FE II	1.1	1.5		I
2251.872	2252.570	44393.74	FE I	1.1	1.8		MIT
2251.559	2252.257	44399.91	FE II	0.6	0.6		MIT
2250.930	2251.628	44412.32	FE II	0.7	1.3		MIT
2250.776	2251.474	44415.35	FE I	1.0			MIT
2250.169	2250.867	44427.33	FE II	0.7	1.2		MIT
2249.175	2249.872	44446.97	FE II	1.0	1.7		I
2248.857	2249.554	44453.25	FE I	1.5			I
2247.692	2248.389	44476.29	FE II		1.7		DO
2247.46	2248.16	44480.9	FE	0.5W			MIT
2246.907	2247.604	44491.83	FE II		1.3		MIT
2245.654	2246.351	44516.65	FE I	1.1			MIT
2245.505	2246.202	44519.60	FE II		1.7		DO
2244.611	2245.307	44537.33	FE II		0.5		MIT
2244.388	2245.084	44541.76	FE II		1.0		MIT
2243.91	2244.61	44551.2	FE I	0.5			MIT
2243.587	2244.283	44557.66	FE II		0.3		MIT
2243.150	2243.846	44566.34	FE		1.0H		MIT
2242.338	2243.034	44582.48	FE II		0.5		MIT
2241.840	2242.536	44592.38	FE	0.8	0.3		MIT
2240.627	2241.322	44616.52	FE	1.3	0.5H		IME
2240.339	2241.034	44622.25	FE II	0.3	0.5		MIT
2239.643	2240.338	44636.12	FE		0.6		MIT
2238.629	2239.324	44656.33	FE		1.0		DO
2238.259	2238.954	44663.71	FE I	0.7			MIT
2237.897	2238.592	44670.94	FE		1.1		MIT
2237.814	2238.509	44672.60	FE	0.3			MIT
2236.685	2237.380	44695.14	FE II		0.5		MIT
2236.313	2237.008	44702.58	FE	1.0			MIT
2235.520	2236.214	44718.43	FE II		0.9		MIT
2234.86	2235.55	44731.6	FE	0.3			MIT
2234.429	2235.123	44740.26	FE	0.6			MIT
2233.914	2234.608	44750.58	FE II		1.5		MIT
2232.078	2232.772	44787.38	FE II	0.3	1.1		MIT
2231.211	2231.904	44804.79	FE I	0.9			I
2229.07	2229.76	44847.8	FE	0.8			MIT
2228.168	2228.861	44865.97	FE I	1.0	0.8		I
2227.616	2228.309	44877.09	FE II		0.3		MIT
2227.407	2228.100	44881.30	FE II		0.3		DO
2227.180	2227.873	44885.87	FE II		0.9		MIT
2224.464	2225.156	44940.67	FE II	0.0	1.1		DO

AIR WAVELENGTH (ANGSTRMS)	VACUUM WAVELENGTH (ANGSTRMS)	WAVENUMBER (KAYSERS)	LMENT	LOG ARC	INTENSITY SPK	INTENSITY DIS	REF
2223.486	2224.178	44960.43	FE II		1.4		MIT
2222.763	2223.455	44975.06	FE	0.7	0.7H		MIT
2222.450	2223.142	44981.39	FE II		1.0		MIT
2221.83	2222.52	44993.9	FE	1.1			MIT
2221.159	2221.850	45007.53	FE II		1.4		MIT
2220.91	2221.60	45012.6	FE I	0.5			MIT
2220.374	2221.065	45023.44	FE II	0.3	1.5		MIT
2219.893	2220.584	45033.20	FE II	0.3	1.3		DO
2219.712	2220.403	45036.87	FE II		0.3		DO
2217.74	2218.43	45076.9	FE I	0.5			MIT
2217.58	2218.27	45080.2	FE	0.5			MIT
2217.056	2217.746	45090.82	FE II		1.0		MIT
2214.407	2215.097	45144.75	FE II		0.6		MIT
2213.654	2214.344	45160.11	FE II	0.3	1.5		DB
2211.234	2211.923	45209.53	FE I	0.8			IME
2211.101	2211.790	45212.25	FE II		1.1		MIT
2210.686	2211.375	45220.73	FE I	0.9			IME
2208.81	2209.50	45259.1	FE	0.8W			MIT
2208.010	2208.699	45275.53	FE	0.7			MIT
2207.068	2207.756	45294.85	FE I	0.8			IME
2206.572	2207.260	45305.03	FE II		0.8		MIT
2204.08	2204.77	45356.2	FE	0.3			MIT
2203.463	2204.151	45368.95	FE	0.6			MIT
2201.408	2202.095	45411.30	FE		0.3		DO
2201.117	2201.804	45417.30	FE I	1.0			IME
2200.722	2201.409	45425.45	FE I	1.5	0.9		I
2200.39	2201.08	45432.3	FE I	1.7G			ME
2199.75	2200.44	45445.5	FE	0.6			MIT
2199.573	2200.260	45449.18	FE	1.0			MIT
2197.230	2197.916	45497.64	FE I	0.3			MIT
2196.039	2196.725	45522.31	FE I	1.9G	1.5		I
2193.564	2194.250	45573.67	FE	0.3			MIT
2193.411	2194.096	45576.85	FE	0.9			MIT
2192.82	2193.51	45589.1	FE	0.7			MIT
2191.838	2192.523	45609.55	FE I	2.0G	1.1		MIT
2191.202	2191.887	45622.79	FE I	1.0	0.3		I
2190.772	2191.457	45631.74	FE	1.3	0.3		MIT
2189.38	2190.06	45660.7	FE I	0.6			MIT
2189.18	2189.86	45664.9	FE	0.8			MIT
2189.032	2189.717	45668.01	FE II	0.0	1.3		MIT
2187.876	2188.560	45692.14	FE II		0.8		MIT
2187.688	2188.372	45696.06	FE II		1.1		MIT
2187.436	2188.120	45701.33	FE II		0.9		MIT
2187.191	2187.875	45706.45	FE I	1.7G	1.0		I
2186.890	2187.574	45712.74	FE I	0.8	1.1		I
2186.485	2187.169	45721.20	FE	1.7G	1.9		MIT
2186.241	2186.925	45726.31	FE I	0.7			MIT
2185.21	2185.89	45747.9	FE	0.5			MIT
2184.46	2185.14	45763.6	FE	0.6			MIT
2183.979	2184.663	45773.66	FE	1.3			IME
2183.835	2184.518	45776.68	FE		1.1		MIT
2183.47	2184.15	45784.3	FE	0.8			MIT
2182.38	2183.06	45807.2	FE		0.7H		MIT
2181.14	2181.82	45833.2	FE I	0.3			MIT
2180.866	2181.549	45838.99	FE I	0.9			IME
2180.26	2180.94	45851.7	FE	0.3			MIT
2178.96	2179.64	45879.1	FE	0.9			MIT
2178.090	2178.772	45897.41	FE I	2.0G	1.3		MIT
2177.037	2177.719	45919.60	FE II		0.9		MIT
2176.837	2177.519	45923.82	FE I	0.9			IME
2176.516	2177.198	45930.60	FE		0.9		MIT
2176.40	2177.08	45933.0	FE	0.3			MIT
2176.03	2176.71	45940.9	FE	0.5			MIT
2175.447	2176.129	45953.16	FE I	0.9	1.4		DO
2174.862	2175.544	45965.52	FE II		0.9		MIT
2174.686	2175.368	45969.24	FE		0.5		MIT
2174.14	2174.82	45980.8	FE	0.6			MIT
2173.71	2174.39	45989.9	FE	0.3	0.3		MIT
2173.211	2173.892	46000.44	FE I	0.9	0.3		I
2172.975	2173.656	46005.43	FE II	0.5	1.0		MIT

AIR WAVELENGTH (ANGSTRMS)	VACUUM WAVELENGTH (ANGSTRMS)	WAVENUMBER (KAYSERS)	LMENT	LOG ARC	INTENSITY SPK	DIS	REF
2172.581	2173.262	46013.78	FE I	0.6			IME
2172.143	2172.824	46023.05	FE I	0.5			MIT
2171.292	2171.973	46041.09	FE I	1.0	0.5		BU
2170.54	2171.22	46057.0	FE	1.0			MIT
2169.946	2170.627	46069.65	FE	0.9			MIT
2168.38	2169.06	46102.9	FE	0.9	0.5		MIT
2168.05	2168.73	46109.9	FE	0.9			MIT
2167.879	2168.559	46113.57	FE II		1.0		MIT
2167.400	2168.080	46123.76	FE	0.5	0.8		MIT
2166.773	2167.453	46137.10	FE I	2.0G	1.5		MIT
2166.22	2166.90	46148.9	FE	0.6			MIT
2165.860	2166.540	46156.55	FE	1.3H	0.0H		I
2164.547	2165.227	46184.54	FE I	0.3H	0.3H		IME
2164.39	2165.07	46187.9	FE	0.9			MIT
2164.325	2165.004	46189.28	FE II	0.3	1.1		DO
2163.860	2164.539	46199.21	FE I	0.9	0.0		IME
2163.366	2164.045	46209.75	FE	1.1			I
2162.24	2162.92	46233.8	FE	1.0H			MIT
2162.022	2162.701	46238.48	FE I	0.7	1.4		MIT
2162.022	2162.701	46238.48	FE II	0.7	1.4		MIT
2161.577	2162.256	46248.00	FE I	0.8			IME
2161.158	2161.837	46256.96	FE II		1.1		MIT
2160.24	2160.92	46276.6	FE	0.3			MIT
2159.893	2160.572	46284.05	FE I	1.1	0.3		MIT
2159.650	2160.329	46289.26	FE I	0.9	0.5		MIT
2159.420	2160.098	46294.19	FE I	0.5			MIT
2159.15	2159.83	46300.0	FE	0.6			MIT
2158.920	2159.598	46304.91	FE I	0.8			MIT
2158.48	2159.16	46314.3	FE I	1.1	0.9		MIT
2157.792	2158.470	46329.11	FE I	1.0	0.6		I
2156.47	2157.15	46357.5	FE	1.3			MIT
2155.81	2156.49	46371.7	FE	0.7			MIT
2155.64	2156.32	46375.4	FE	0.9			MIT
2155.242	2155.920	46383.92	FE I	0.5			MIT
2155.020	2155.698	46388.70	FE I	0.6			MIT
2154.458	2155.135	46400.80	FE	1.2			IME
2154.188	2154.865	46406.61	FE I	0.5			IME
2153.004	2153.681	46432.13	FE I	1.0	0.0		IME
2152.24	2152.92	46448.6	FE	0.9			MIT
2151.897	2152.574	46456.01	FE	0.7			MIT
2151.76	2152.44	46459.0	FE		0.7H		MIT
2151.70	2152.38	46460.3	FE	1.3			MIT
2151.099	2151.776	46473.24	FE I	0.9	1.0		IME
2150.625	2151.302	46483.48	FE II		1.2		MIT
2150.182	2150.859	46493.06	FE I	0.9	0.3H		IME
2149.62	2150.30	46505.2	FE	0.3			MIT
2149.17	2149.85	46515.0	FE	1.0			MIT
2148.50	2149.18	46529.5	FE	0.7			MIT
2148.394	2149.070	46531.75	FE I	0.5			MIT
2147.787	2148.463	46544.90	FE	0.9			IME
2147.04	2147.72	46561.1	FE	0.9			MIT
2146.71	2147.39	46568.2	FE	0.3			MIT
2146.04	2146.72	46582.8	FE	0.0	1.0		MIT
2145.188	2145.864	46601.29	FE I	1.1	0.3		IME
2144.45	2145.13	46617.3	FE	1.4			MIT
2143.90	2144.58	46629.3	FE	0.9			MIT
2141.715	2142.390	46676.85	FE I	0.9			IME
2141.471	2142.146	46682.16	FE	1.0			MIT
2141.08	2141.75	46690.7	FE I	0.3	0.9		MIT
2139.93	2140.60	46715.8	FE I	0.6			MIT
2139.695	2140.369	46720.91	FE I	0.9	0.9		IME
2138.589	2139.263	46745.06	FE I	0.9			IME
2138.01	2138.68	46757.7	FE	0.3	0.5		MIT
2136.19	2136.86	46797.6	FE	0.3			MIT
2135.957	2136.631	46802.66	FE	1.0	0.7		IME
2135.10	2135.77	46821.4	FE	0.3			MIT
2133.31	2133.98	46860.7	FE	0.6			MIT
2132.015	2132.688	46889.19	FE I	1.1	0.5		IME
2130.962	2131.635	46912.35	FE	1.3			IME
2130.42	2131.09	46924.3	FE	0.7			MIT
2127.87	2128.54	46980.5	FE	0.7			MIT
2125.01	2125.68	47043.7	FE	1.5			MIT
2121.94	2122.61	47111.8	FE	0.8			MIT
2115.168	2115.838	47262.61	FE I	0.9			IME
2114.60	2115.27	47275.3	FE	0.9	0.0		MIT
2113.09	2113.76	47309.1	FE	0.6			MIT
2112.966	2113.635	47311.85	FE I	0.9	0.3		IME
2110.233	2110.902	47373.12	FE I	0.9			IME
2109.11	2109.78	47398.3	FE I		1.0		MIT
2108.955	2109.623	47401.83	FE I	1.0	0.3		IME
2108.202	2108.870	47418.75	FE I	0.7	0.6		BU
2107.13	2107.80	47442.9	FE	0.9H			A
2102.902	2103.569	47538.25	FE I	0.6			MIT
2102.349	2103.016	47550.75	FE I	1.2	0.5		IME
2100.795	2101.462	47585.92	FE I	1.2	0.3		IME
2044.93	2045.59	48885.7	FE	0.0H	0.7		A
2020.52	2021.17	49476.2	FE	0.3	2.0		A
2017.07	2017.72	49560.9	FE	0.3	1.0		A
7793.0	7795.1	12828.	GA II			1.0	SY
7198.7	7200.7	13888.	GA II			1.8	SY
7000.0	7001.9	14282.	GA II			0.3	SY
6456.3	6458.1	15484.	GA II			1.2	SY
6419.4	6421.2	15573.	GA II			1.4	SY
6414.01	6415.78	15586.6	GA		1.2		KL
6396.61	6398.38	15629.0	GA		1.3		KL
6334.2	6336.0	15783.	GA II			2.0	SY
5992.34	5994.00	16683.4	GA		1.0		KL
5844.22	5845.84	17106.2	GA		1.1		KL
5421.6	5423.1	18440.	GA II			0.3	SY
5416.8	5418.3	18456.	GA II			1.0	SY
5360.6	5362.1	18649.	GA II			0.7	SY
5360.09	5361.58	18651.2	GA	0.3	1.1		KL
5353.81	5355.30	18673.1	GA	0.3			FL
5338.88	5340.37	18725.3	GA		0.7		KL
5338.3	5339.8	18727.	GA II			0.3	SY
5218.21	5219.66	19158.3	GA II			1.0	SY
4995.61	4997.00	20012.0	GA		0.5		KL
4864.92	4866.28	20549.6	GA		0.6		KL
4820.59	4821.94	20738.6	GA		0.3		KL
4616.43	4617.72	21655.7	GA		0.3		KL
4417.69	4418.93	22629.9	GA		0.7		KL
4261.93	4263.13	23456.9	GA		0.6		KL
4261.78	4262.98	23457.8	GA II		0.3		SY
4256.17	4257.37	23488.7	GA		0.3		KL
4255.52	4256.72	23492.3	GA II			0.5	SY
4254.13	4255.33	23499.9	GA		0.3		KL
4250.91	4252.11	23517.7	GA II			0.3	SY
4250.67	4251.87	23519.1	GA		0.6		KL
4172.056	4173.232	23962.24	GA I	3.3G	3.0G		MIT
4078.90	4080.05	24509.5	GA		0.7		MIT
4069.52	4070.67	24566.0	GA		0.3		KL
4032.982	4034.122	24788.54	GA I	3.0G	2.7G		MIT
3944.29	3945.41	25345.9	GA		0.3		KL
3924.39	3925.50	25474.4	GA II		1.40		SY
3806.87	3807.95	26260.8	GA		0.6		KL
3734.87	3735.93	26767.1	GA II			0.6	MIT
3730.04	3731.10	26801.7	GA		1.3		MIT
3705.85	3706.90	26976.7	GA II			0.3	SY
3701.60	3702.65	27007.7	GA		0.3		MIT
3583.60	3584.62	27896.9	GA II			0.3	SY
3576.98	3578.00	27948.6	GA		0.6		MIT
3517.21	3518.22	28423.5	GA		0.5		KL
3479.46	3480.46	28731.9	GA		0.8		KL
3473.39	3474.38	28782.1	GA		0.6		KL
3470.34	3471.33	28807.4	GA II			0.7	SY
3447.280	3448.268	29000.07	GA II			0.3	MIT
3446.46	3447.45	29007.0	GA II			0.5	SY
3436.66	3437.65	29089.7	GA II			0.3	SY
3417.12	3418.10	29256.0	GA		0.7		KL
3375.30	3376.27	29618.5	GA		1.1		MIT

AIR WAVELENGTH (ANGSTRMS)	VACUUM WAVELENGTH (ANGSTRMS)	WAVENUMBER (KAYSERS)	LMENT	LOG ARC	INTENSITY SPK	INTENSITY DIS	REF
3374.94	3375.91	29621.6	GA II			0.6	SY
3181.44	3182.36	31423.2	GA		0.3		KL
3158.18	3159.09	31654.6	GA II		0.3		MIT
3089.957	3090.854	32353.52	GA	0.9	0.9		MIT
3075.87	3076.76	32501.7	GA		0.3		KL
3072.14	3073.03	32541.1	GA		0.5		KL
3064.45	3065.34	32622.8	GA		0.3		KL
3058.15	3059.04	32690.0	GA		0.6W		KL
3024.33	3025.21	33055.5	GA		0.3W		KL
3015.60	3016.48	33151.2	GA		0.3		KL
3011.90	3012.78	33192.0	GA II		0.3		SY
3004.06	3004.94	33278.6	GA		1.2		MIT
2997.93	2998.80	33346.6	GA			0.7	KL
2992.84	2993.71	33403.3	GA II			0.3	SY
2987.99	2988.86	33457.5	GA		0.8		KL
2974.77	2975.64	33606.2	GA II			0.9	SY
2971.60	2972.47	33642.1	GA II			0.7	SY
2971.01	2971.88	33648.8	GA II			0.3	SY
2969.41	2970.28	33666.9	GA II			0.7	SY
2944.175	2945.036	33955.44	GA	1.0	1.2R		MIT
2943.637	2944.498	33961.65	GA I	1.0	1.3R		MIT
2910.77	2911.62	34345.1	GA II			0.5	SY
2874.244	2875.087	34781.55	GA I	1.0	1.2R		MIT
2836.87	2837.70	35239.7	GA		0.3		KL
2830.19	2831.02	35322.9	GA		0.3		KL
2780.15	2780.97	35958.7	GA II			1.6	SY
2719.653	2720.459	36758.51	GA	0.7	1.2		MIT
2700.47	2701.27	37019.6	GA II			1.8	SY
2691.29	2692.09	37145.9	GA I			1.4	SY
2665.05	2665.84	37511.6	GA I			1.6	SY
2659.866	2660.657	37584.70	GA	0.7	1.1		MIT
2632.66	2633.44	37973.1	GA I			1.6	SY
2624.82	2625.60	38086.5	GA I			1.4	SY
2607.47	2608.25	38339.9	GA I			1.2	SY
2602.04	2602.82	38419.9	GA		0.5		KL
2588.92	2589.69	38614.6	GA		0.7		KL
2555.28	2556.05	39122.9	GA II			0.8	SY
2552.87	2553.64	39159.9	GA II			0.7	SY
2551.26	2552.03	39184.6	GA II			0.3	SY
2534.83	2535.59	39438.5	GA I			1.3	SY
2514.15	2514.91	39762.9	GA II			0.7	SY
2513.55	2514.31	39772.4	GA II			0.8	SY
2500.705	2501.459	39976.67	GA	0.7	0.7R		MIT
2500.173	2500.927	39985.18	GA	1.1	1.0R		MIT
2482.76	2483.51	40265.6	GA I			1.2	SY
2450.071	2450.813	40802.79	GA	1.0	1.0		M
2438.88	2439.62	40990.0	GA II			0.9	SY
2424.29	2425.03	41236.7	GA		0.7		KL
2423.13	2423.87	41256.4	GA		0.5		MIT
2422.63	2423.37	41264.9	GA		0.6		M
2418.699	2419.434	41331.98	GA	0.5	0.7		UH
2377.53	2378.26	42047.6	GA II			0.5	SY
2371.325	2372.049	42157.64	GA	0.5	0.7		UH
2363.98	2364.70	42288.6	GA		0.3		MIT
2360.95	2361.67	42342.9	GA		0.5		MIT
2359.47	2360.19	42369.4	GA		0.3		MIT
2358.84	2359.56	42380.8	GA		0.5		MIT
2347.82	2348.54	42579.7	GA		0.3		MIT
2343.27	2343.99	42662.3	GA		0.8		MIT
2342.4	2343.1	42678.	GA		0.5		WB
2341.55	2342.27	42693.7	GA		0.3		MIT
2338.596	2339.313	42747.60	GA	0.3			UH
2338.283	2339.000	42753.32	GA	0.5	0.6		MIT
2337.26	2337.98	42772.0	GA		0.3		MIT
2318.76	2319.47	43113.2	GA		0.5H		MIT
2316.90	2317.61	43147.9	GA		0.3H		MIT
2297.869	2298.577	43505.18	GA	0.3	0.5		UH
2294.202	2294.909	43574.71	GA	0.3	0.5		UH
2285.19	2285.90	43746.5	GA		0.5		KL
2266.84	2267.54	44100.6	GA		0.5		MIT

AIR WAVELENGTH (ANGSTRMS)	VACUUM WAVELENGTH (ANGSTRMS)	WAVENUMBER (KAYSERS)	LMENT	LOG ARC	INTENSITY SPK	INTENSITY DIS	REF
2259.224	2259.923	44249.29	GA	0.3	0.3		MIT
2255.29	2255.99	44326.5	GA		0.5		MIT
2255.034	2255.733	44331.50	GA	0.3			UH
2236.103	2236.797	44706.77	GA	0.3	0.7		UH
2218.039	2218.730	45070.84	GA	0.3			UH
2211.753	2212.442	45198.92	GA	0.3			UH
2205.91	2206.60	45318.6	GA		0.5		KL
2195.665	2196.351	45530.06	GA	0.3	0.5		UH
2176.9	2177.6	45922.	GA		0.7		KL
2171.9	2172.6	46028.	GA	0.3			UH
2091.34	2092.00	47801.0	GA II			1.6	SY
2090.5	2091.2	47820.	GA		0.7		WB
2073.8	2074.5	48205.	GA		0.5H		WB
2064.3	2065.0	48427.	GA		0.3H		WB
2061.9	2062.6	48483.	GA		0.5H		WB
2040.9	2041.6	48982.	GA		0.5H		WB
2040.2	2040.9	48999.	GA		0.7H		WB
2039.0	2039.7	49028.	GA		0.7H		WB
2021.6	2022.3	49450.	GA		0.5H		WB
2019.6	2020.3	49499.	GA		0.7H		WB
2007.7	2008.3	49792.	GA		0.6H		WB
8668.61	8670.99	11532.7	GD	0.6			KS
8561.70	8564.05	11676.7	GD	0.6			KS
8527.79	8530.13	11723.2	GD	0.6			KS
8445.45	8447.77	11837.4	GD	0.6			KS
8442.58	8444.90	11841.5	GD	0.6			KS
8398.24	8400.55	11904.0	GD	0.6			KS
8349.77	8352.06	11973.1	GD	0.6			KS
8316.38	8318.67	12021.2	GD	0.6			KS
8315.04	8317.33	12023.1	GD	0.6W			KS
8275.42	8277.69	12080.7	GD	0.6			KS
8218.08	8220.34	12164.9	GD	0.6			KS
8146.10	8148.34	12272.4	GD	0.9			KS
8077.58	8079.80	12376.5	GD	0.6			KS
8048.56	8050.77	12421.2	GD	0.6			KS
8019.76	8021.97	12465.8	GD	0.6			KS
8010.40	8012.60	12480.3	GD	0.6			KS
7993.75	7995.95	12506.3	GD	0.6			KS
7978.15	7980.34	12530.8	GD	0.6			KS
7963.22	7965.41	12554.3	GD	0.6			KS
7930.23	7932.41	12606.5	GD	0.8			KS
7910.08	7912.26	12638.6	GD	0.8			KS
7908.11	7910.29	12641.8	GD	0.8			KS
7902.33	7904.50	12651.0	GD	0.8			KS
7884.38	7886.55	12679.8	GD	0.6			KS
7869.73	7871.90	12703.4	GD	0.8			KS
7867.72	7869.88	12706.7	GD	0.6			KS
7856.94	7859.10	12724.1	GD	1.0			KS
7846.35	7848.51	12741.3	GD	1.1			KS
7844.84	7847.00	12743.7	GD	0.8			KS
7838.82	7840.98	12753.5	GD	0.6			KS
7834.50	7836.66	12760.6	GD	0.6L			KS
7787.22	7789.36	12838.0	GD	0.6			KS
7769.26	7771.40	12867.7	GD	0.6			KS
7766.50	7768.64	12872.3	GD	0.6			KS
7760.98	7763.12	12881.4	GD	0.6			KS
7755.97	7758.10	12889.7	GD	0.9			KS
7749.26	7751.39	12900.9	GD	1.1			KS
7746.20	7748.33	12906.0	GD	0.6			KS
7738.09	7740.22	12919.5	GD	0.6			KS
7733.50	7735.63	12927.2	GD	1.1			KS
7717.70	7719.82	12953.7	GD	0.6			KS
7694.50	7696.62	12992.7	GD	0.9			KS
7683.36	7685.47	13011.6	GD	0.8			KS
7677.16	7679.27	13022.1	GD	0.8			KS
7676.06	7678.17	13023.9	GD	0.8			KS
7672.60	7674.71	13029.8	GD	1.0			KS
7666.95	7669.06	13039.4	GD	0.6			KS
7650.32	7652.43	13067.7	GD	1.0			KS
7644.88	7646.98	13077.1	GD	0.6			KS

AIR WAVELENGTH (ANGSTRMS)	VACUUM WAVELENGTH (ANGSTRMS)	WAVENUMBER (KAYSERS)	LMENT	LOG ARC	INTENSITY SPK	DIS	REF
7638.60	7640.70	13087.8	GD	0.6S			KS
7621.96	7624.06	13116.4	GD	1.0			KS
7611.78	7613.88	13133.9	GD	0.8			KS
7594.56	7596.65	13163.7	GD	0.6			KS
7588.22	7590.31	13174.7	GD	0.8			MIT
7583.94	7586.03	13182.1	GD	0.8			MIT
7566.08	7568.16	13213.2	GD	0.8			MIT
7563.03	7565.11	13218.6	GD	1.1			MIT
7528.72	7530.79	13278.8	GD	0.6			KS
7526.57	7528.64	13282.6	GD	0.6			KS
7505.34	7507.41	13320.2	GD	0.8			MIT
7489.47	7491.53	13348.4	GD	1.6			MIT
7477.47	7479.53	13369.8	GD	1.3			KS
7470.45	7472.51	13382.4	GD	1.3			KS
7464.38	7466.44	13393.3	GD	2.3			MIT
7442.17	7444.22	13433.2	GD	0.7			ED
7441.89	7443.94	13433.7	GD	2.3			MIT
7434.50	7436.55	13447.1	GD	1.3			MIT
7432.59	7434.64	13450.6	GD	1.3			MIT
7430.20	7432.25	13454.9	GD	1.6			MIT
7426.56	7428.61	13461.5	GD	1.9			MIT
7417.99	7420.03	13477.0	GD	1.7			MIT
7405.76	7407.80	13499.3	GD	1.6			KS
7397.96	7400.00	13513.5	GD	0.7			KS
7394.91	7396.95	13519.1	GD	1.6			MIT
7385.91	7387.94	13535.6	GD	1.6			KS
7384.52	7386.55	13538.1	GD	0.7			KS
7380.30	7382.33	13545.8	GD	1.6			MIT
7377.59	7379.62	13550.8	GD	1.6			KS
7377.27	7379.30	13551.4	GD	0.9			MIT
7376.42	7378.45	13553.0	GD	1.6			MIT
7373.83	7375.86	13557.7	GD	1.6			MIT
7370.28	7372.31	13564.3	GD	1.9			MIT
7362.62	7364.65	13578.4	GD	0.7			KS
7330.23	7332.25	13638.4	GD	0.7			KS
7327.09	7329.11	13644.2	GD	1.6			MIT
7324.89	7326.91	13648.3	GD	1.9			MIT
7323.35	7325.37	13651.2	GD	0.7			KS
7318.90	7320.92	13659.5	GD	1.3			KS
7313.28	7315.29	13670.0	GD	2.2			MIT
7312.66	7314.67	13671.2	GD	0.9			ED
7301.23	7303.24	13692.6	GD	2.2			MIT
7298.64	7300.65	13697.4	GD	0.9			ED
7294.06	7296.07	13706.0	GD	0.9			ED
7292.23	7294.24	13709.4	GD	0.7			KS
7291.35	7293.36	13711.1	GD	1.8			MIT
7282.54	7284.55	13727.7	GD	0.7			KS
7279.40	7281.41	13733.6	GD	0.8			KS
7274.08	7276.08	13743.7	GD	0.8			KS
7272.96	7274.96	13745.8	GD	0.7			ED
7270.97	7272.97	13749.5	GD	0.7			ED
7270.42	7272.42	13750.6	GD	0.7			ED
7264.70	7266.70	13761.4	GD	0.7			ED
7263.65	7265.65	13763.4	GD	0.8			KS
7262.67	7264.67	13765.2	GD	1.9			KS
7254.75	7256.75	13780.3	GD	0.7			MIT
7252.72	7254.72	13784.1	GD	1.9			MIT
7246.49	7248.49	13796.0	GD	0.7			KS
7242.27	7244.27	13804.0	GD	1.3			KS
7233.44	7235.43	13820.9	GD	1.9			MIT
7229.18	7231.17	13829.0	GD	0.7			KS
7228.05	7230.04	13831.2	GD	1.6			MIT
7222.56	7224.55	13841.7	GD	0.7			ED
7208.75	7210.74	13868.2	GD	0.7			ED
7207.71	7209.70	13870.2	GD	0.7			KS
7205.12	7207.11	13875.2	GD	0.9			ED
7201.42	7203.40	13882.3	GD	1.6			MIT
7197.09	7199.07	13890.7	GD	1.2			MIT
7194.87	7196.85	13895.0	GD	1.7			MIT
7191.93	7193.91	13900.6	GD	0.7			KS

AIR WAVELENGTH (ANGSTRMS)	VACUUM WAVELENGTH (ANGSTRMS)	WAVENUMBER (KAYSERS)	LMENT	LOG ARC	INTENSITY SPK	DIS	REF
7189.59	7191.57	13905.2	GD	1.9			MIT
7187.12	7189.10	13909.9	GD	0.7			KS
7180.88	7182.86	13922.0	GD	0.9			ED
7180.35	7182.33	13923.1	GD	0.9			ED
7178.80	7180.78	13926.1	GD	0.8			KS
7176.03	7178.01	13931.4	GD	0.9			ED
7173.45	7175.43	13936.4	GD	0.7			MIT
7172.29	7174.27	13938.7	GD	1.7			MIT
7168.41	7170.39	13946.2	GD	2.7			MIT
7166.61	7168.59	13949.7	GD	0.7			KS
7164.33	7166.30	13954.2	GD	0.8			KS
7163.52	7165.49	13955.8	GD	0.9			ED
7160.25	7162.22	13962.1	GD	0.7			KS
7158.28	7160.25	13966.0	GD	1.3			MIT
7153.16	7155.13	13976.0	GD	1.3			KS
7150.84	7152.81	13980.5	GD	0.7			MIT
7149.84	7151.81	13982.5	GD	0.9			ED
7148.20	7150.17	13985.7	GD	0.9			MIT
7147.36	7149.33	13987.3	GD	1.7			MIT
7146.16	7148.13	13989.7	GD	0.8			MIT
7141.17	7143.14	13999.4	GD	0.8			MIT
7139.76	7141.73	14002.2	GD	0.7			KS
7135.72	7137.69	14010.1	GD	1.2			MIT
7133.16	7135.13	14015.2	GD	0.8			MIT
7129.80	7131.77	14021.8	GD	0.8			KS
7128.13	7130.10	14025.1	GD	0.9			ED
7126.80	7128.76	14027.7	GD	0.9			ED
7122.59	7124.55	14036.0	GD	1.7			MIT
7118.91	7120.87	14043.2	GD	1.6			MIT
7116.76	7118.72	14047.5	GD	0.9			MIT
7115.49	7117.45	14050.0	GD	0.9			ED
7115.01	7116.97	14050.9	GD	0.9			ED
7113.15	7115.11	14054.6	GD	1.2			ED
7112.02	7113.98	14056.8	GD	0.9			ED
7105.66	7107.62	14069.4	GD	1.2			ED
7100.71	7102.67	14079.2	GD	0.7			MIT
7098.75	7100.71	14083.1	GD	0.7			MIT
7098.10	7100.06	14084.4	GD	0.9			MIT
7085.52	7087.47	14109.4	GD	1.3			KS
7078.77	7080.72	14122.8	GD	0.9			ED
7075.94	7077.89	14128.5	GD	0.8			KS
7074.60	7076.55	14131.2	GD	0.7			KS
7073.63	7075.58	14133.1	GD	1.3			MIT
7071.00	7072.95	14138.4	GD	1.3			MIT
7069.93	7071.88	14140.5	GD	1.0			MIT
7068.08	7070.03	14144.2	GD	1.4			MIT
7065.86	7067.81	14148.7	GD	0.7			KS
7058.00	7059.95	14164.4	GD	1.2			MIT
7055.67	7057.62	14169.1	GD	0.9			ED
7054.61	7056.56	14171.2	GD	1.4			MIT
7050.95	7052.89	14178.6	GD	1.4			MIT
7045.02	7046.96	14190.5	GD	0.9			KS
7043.75	7045.69	14193.1	GD	0.9			ED
7043.27	7045.21	14194.0	GD	0.7			KS
7038.73	7040.67	14203.2	GD	0.7			KS
7037.86	7039.80	14204.9	GD	0.8			KS
7037.25	7039.19	14206.2	GD	1.7			MIT
7033.39	7035.33	14214.0	GD	0.9			ED
7024.48	7026.42	14232.0	GD	0.9			ED
7022.02	7023.96	14237.0	GD	0.9			ED
7019.22	7021.16	14242.7	GD	0.9			ED
7017.68	7019.62	14245.8	GD	0.8			KS
7007.46	7009.39	14266.6	GD	0.7			KS
7006.16	7008.09	14269.2	GD	1.9			MIT
7000.74	7002.67	14280.3	GD	1.3			MIT
6996.78	6998.71	14288.3	GD	2.3			MIT
6993.14	6995.07	14295.8	GD	1.3			KS
6991.93	6993.86	14298.3	GD	2.2			MIT
6988.70	6990.63	14304.9	GD	1.0			KS
6985.88	6987.81	14310.6	GD	2.3			MIT

AIR WAVELENGTH (ANGSTRMS)	VACUUM WAVELENGTH (ANGSTRMS)	WAVENUMBER (KAYSERS)	LMENT	LOG ARC	INTENSITY SPK	DIS	REF
6983.51	6985.44	14315.5	GD	0.7			MIT
6980.85	6982.78	14320.9	GD	1.4			KS
6978.26	6980.18	14326.3	GD	1.3			KS
6976.35	6978.27	14330.2	GD	1.3			KS
6971.64	6973.56	14339.9	GD	1.3			KS
6969.09	6971.01	14345.1	GD	0.9			ED
6964.32	6966.24	14354.9	GD	0.7			MIT
6959.24	6961.16	14365.4	GD	1.3			KS
6957.72	6959.64	14368.6	GD	1.3			KS
6955.20	6957.12	14373.8	GD	1.2			ED
6954.43	6956.35	14375.4	GD	0.5			ED
6951.77	6953.69	14380.9	GD	0.9			ED
6947.24	6949.16	14390.2	GD	0.8			KS
6945.96	6947.88	14392.9	GD	1.3			KS
6944.99	6946.91	14394.9	GD	0.7			KS
6943.07	6944.99	14398.9	GD	0.7			KS
6937.49	6939.40	14410.5	GD	0.9			ED
6935.51	6937.42	14414.6	GD	0.7			KS
6932.36	6934.27	14421.1	GD	0.7			KS
6929.13	6931.04	14427.8	GD	0.7			KS
6926.48	6928.39	14433.4	GD	1.0			MIT
6924.95	6926.86	14436.6	GD	1.2			KS
6923.20	6925.11	14440.2	GD	0.5			ED
6920.60	6922.51	14445.6	GD	1.5			KS
6919.61	6921.52	14447.7	GD	0.7			KS
6916.57	6918.48	14454.1	GD	2.3			MIT
6906.90	6908.81	14474.3	GD	0.5			KS
6906.34	6908.25	14475.5	GD	0.3			KS
6905.50	6907.41	14477.2	GD	0.5			KS
6900.68	6902.58	14487.3	GD	1.8			KS
6887.62	6889.52	14514.8	GD	1.5			KS
6878.53	6880.43	14534.0	GD	0.8			KS
6861.89	6863.78	14569.2	GD	0.3			KS
6858.50	6860.39	14576.4	GD	0.6			KS
6857.12	6859.01	14579.4	GD	2.3			MIT
6852.79	6854.68	14588.6	GD	0.7			ED
6849.89	6851.78	14594.7	GD	0.7			KS
6846.60	6848.49	14601.8	GD	2.7			MIT
6836.60	6838.49	14623.1	GD	0.5			KS
6834.82	6836.71	14626.9	GD	1.0			KS
6828.25	6830.13	14641.0	GD	2.2			MIT
6822.73	6824.61	14652.8	GD	1.2			ED
6820.91	6822.79	14656.8	GD	1.7			MIT
6816.47	6818.35	14666.3	GD	1.6			MIT
6814.55	6816.43	14670.4	GD	1.5			KS
6800.70	6802.58	14700.3	GD	0.5			KS
6787.14	6789.01	14729.7	GD	1.3			KS
6786.32	6788.19	14731.5	GD	2.1			KS
6783.35	6785.22	14737.9	GD	1.6			MIT
6777.77	6779.64	14750.0	GD	0.5H			ED
6772.03	6773.90	14762.6	GD	1.3			KS
6766.91	6768.78	14773.7	GD	1.2			KS
6761.54	6763.41	14785.4	GD	0.7H			KS
6754.77	6756.63	14800.3	GD	0.7			ED
6753.90	6755.76	14802.2	GD	2.0			KS
6752.68	6754.54	14804.8	GD	2.5			MIT
6730.76	6732.62	14853.1	GD	2.0			MIT
6727.84	6729.70	14859.5	GD	1.7			KS
6718.14	6719.99	14881.0	GD	1.6			KS
6704.17	6706.02	14912.0	GD	1.4			KS
6702.09	6703.94	14916.6	GD	1.4			KS
6694.91	6696.76	14932.6	GD	1.3			KS
6692.85	6694.70	14937.2	GD	1.4			KS
6690.88	6692.73	14941.6	GD	0.7H			KS
6687.82	6689.67	14948.4	GD	0.7			KS
6681.22	6683.06	14963.2	GD	2.0			KS
6679.54	6681.38	14967.0	GD	1.4			KS
6656.56	6658.40	15018.6	GD	1.2			KS
6651.08	6652.92	15031.0	GD	1.2			KS
6646.84	6648.68	15040.6	GD	1.0			KS

AIR WAVELENGTH (ANGSTRMS)	VACUUM WAVELENGTH (ANGSTRMS)	WAVENUMBER (KAYSERS)	LMENT	LOG ARC	INTENSITY SPK	DIS	REF
6645.18	6647.02	15044.3	GD	2.2			ED
6644.04	6645.87	15046.9	GD	0.9			MIT
6642.75	6644.58	15049.8	GD	0.7			KS
6640.09	6641.92	15055.9	GD	1.6			MIT
6634.35	6636.18	15068.9	GD	2.2			KS
6628.42	6630.25	15082.4	GD	1.0H			KS
6622.27	6624.10	15096.4	GD	1.2			KS
6610.05	6611.88	15124.3	GD	1.2			KS
6600.60	6602.42	15145.9	GD	0.9			KS
6593.40	6595.22	15162.5	GD	0.9			MIT
6591.60	6593.42	15166.6	GD	1.6			MIT
6573.80	6575.62	15207.7	GD	1.0			KS
6567.98	6569.79	15221.2	GD	1.4			KS
6564.78	6566.59	15228.6	GD	1.4			KS
6563.66	6565.47	15231.2	GD	0.8			KS
6549.24	6551.05	15264.7	GD	1.4			MIT
6547.28	6549.09	15269.3	GD	0.8			KS
6542.04	6543.85	15281.5	GD	0.7			ED
6538.14	6539.95	15290.7	GD	1.7			MIT
6530.98	6532.78	15307.4	GD	1.0			MIT
6530.01	6531.81	15309.7	GD	1.0			KS
6526.21	6528.01	15318.6	GD	1.0			KS
6517.72	6519.52	15338.6	GD	1.1			KS
6517.23	6519.03	15339.7	GD	1.2			MIT
6514.219	6516.019	15346.79	GD	1.2			MIT
6508.70	6510.50	15359.8	GD	0.3			ED
6505.41	6507.21	15367.6	GD	1.2			KS
6494.13	6495.92	15394.3	GD	1.3			KS
6483.99	6485.78	15418.3	GD	1.0			KS
6482.95	6484.74	15420.8	GD	0.7H			KS
6480.107	6481.898	15427.58	GD	1.6			MIT
6472.72	6474.51	15445.2	GD	0.9			KS
6470.270	6472.058	15451.04	GD	1.0			MIT
6466.565	6468.352	15459.89	GD	1.4			MIT
6462.581	6464.367	15469.42	GD	1.7			MIT
6456.73	6458.51	15483.4	GD	1.2			KS
6444.860	6446.641	15511.95	GD	0.3			MIT
6424.51	6426.29	15561.1	GD	1.0			MIT
6422.415	6424.190	15566.16	GD	1.7			MIT
6421.88	6423.66	15567.5	GD	0.3			KS
6408.555	6410.326	15599.83	GD	0.3			MIT
6402.31	6404.08	15615.1	GD	0.5			KS
6382.188	6383.952	15664.28	GD	1.8			MIT
6380.974	6382.738	15667.26	GD	2.0			MIT
6368.78	6370.54	15697.2	GD	1.0R			KS
6363.22	6364.98	15711.0	GD	0.5			KS
6358.26	6360.02	15723.2	GD	0.6H			KS
6351.70	6353.46	15739.5	GD	0.7			KS
6350.44	6352.20	15742.6	GD	1.0			KS
6346.64	6348.39	15752.0	GD	1.5			KS
6336.31	6338.06	15777.7	GD	1.0			KS
6331.36	6333.11	15790.0	GD	1.2			KS
6317.18	6318.93	15825.5	GD	1.0			KS
6309.10	6310.84	15845.7	GD	1.0			KS
6305.14	6306.88	15855.7	GD	2.0			KS
6299.05	6300.79	15871.0	GD	1.2			KS
6292.86	6294.60	15886.6	GD	1.0			KS
6289.74	6291.48	15894.5	GD	1.2			KS
6265.32	6267.05	15956.5	GD	0.6			KS
6260.27	6262.00	15969.3	GD	0.8			KS
6259.76	6261.49	15970.6	GD	0.7			KS
6238.44	6240.17	16025.2	GD	0.3			KS
6220.94	6222.66	16070.3	GD	0.7			KS
6203.60	6205.32	16115.2	GD	0.6			KS
6180.455	6182.165	16175.56	GD	1.5			MIT
6164.51	6166.22	16217.4	GD	1.0			MIT
6151.13	6152.83	16252.7	GD	1.2			KS
6138.353	6140.052	16286.51	GD	1.0			MIT
6136.434	6138.132	16291.60	GD	0.7			MIT
6114.077	6115.769	16351.17	GD	1.4			MIT

AIR WAVELENGTH (ANGSTRMS)	VACUUM WAVELENGTH (ANGSTRMS)	WAVENUMBER (KAYSERS)	LMENT	LOG ARC	INTENSITY SPK	DIS	REF
6109.078	6110.769	16364.55	GD	1.0			MIT
6106.183	6107.873	16372.31	GD	1.2			MIT
6082.383	6084.067	16436.37	GD	0.7			MIT
6080.657	6082.340	16441.04	GD	1.4			MIT
6064.249	6065.928	16485.52	GD	0.8			MIT
6049.507	6051.182	16525.70	GD	0.8			MIT
6021.131	6022.798	16603.58	GD	1.0			MIT
6010.93	6012.59	16631.7	GD	1.0			KS
6008.715	6010.379	16637.89	GD	1.2			MIT
6004.571	6006.234	16649.37	GD	1.5	0.9		MIT
5999.069	6000.731	16664.64	GD	0.9			MIT
5991.849	5993.509	16684.72	GD	0.5			MIT
5988.008	5989.667	16695.42	GD	0.6			MIT
5987.097	5988.755	16697.96	GD	0.9			MIT
5982.428	5984.085	16710.99	GD	0.9			MIT
5977.249	5978.905	16725.47	GD	0.9			MIT
5970.309	5971.963	16744.91	GD	0.9			MIT
5956.457	5958.107	16783.85	GD	0.9			MIT
5951.580	5953.229	16797.61	GD	0.5H			MIT
5940.88	5942.53	16827.9	GD	0.5H			KS
5937.706	5939.351	16836.86	GD	0.9			MIT
5936.818	5938.463	16839.37	GD	0.9			MIT
5930.276	5931.919	16857.95	GD	0.8			MIT
5922.027	5923.668	16881.43	GD	0.6			MIT
5921.343	5922.984	16883.38	GD	0.8			MIT
5916.754	5918.393	16896.48	GD	0.9			MIT
5913.543	5915.182	16905.65	GD	1.0			MIT
5911.417	5913.055	16911.73	GD	0.5			MIT
5904.54	5906.18	16931.4	GD	0.9			KS
5904.04	5905.68	16932.9	GD	0.9			KS
5897.59	5899.22	16951.4	GD	0.8			KS
5886.44	5888.07	16983.5	GD	0.8			KS
5882.16	5883.79	16995.9	GD	1.0			KS
5877.24	5878.87	17010.1	GD	0.8			KS
5876.72	5878.35	17011.6	GD	0.5H			KS
5870.60	5872.23	17029.3	GD	0.6H			KS
5860.71	5862.33	17058.1	GD	1.0			KS
5857.06	5858.68	17068.7	GD	0.8			KS
5856.216	5857.839	17071.14	GD	1.0			MIT
5855.220	5856.843	17074.04	GD	0.9			MIT
5851.631	5853.253	17084.52	GD	1.3			MIT
5848.467	5850.088	17093.76	GD	0.8			MIT
5845.69	5847.31	17101.9	GD	0.8			KS
5840.471	5842.090	17117.16	GD	1.0			MIT
5823.959	5825.574	17165.69	GD	0.8			MIT
5820.986	5822.600	17174.46	GD	1.0			MIT
5819.513	5821.126	17178.81	GD	0.3			MIT
5815.851	5817.463	17189.62	GD	1.0			MIT
5809.229	5810.840	17209.22	GD	0.9			MIT
5807.724	5809.334	17213.68	GD	0.6			MIT
5807.051	5808.661	17215.67	GD	0.8			MIT
5802.932	5804.541	17227.89	GD	0.8			MIT
5796.805	5798.412	17246.10	GD	0.9			MIT
5791.386	5792.992	17262.24	GD	1.0			MIT
5776.029	5777.631	17308.13	GD	1.0			MIT
5769.754	5771.354	17326.96	GD	0.9			MIT
5765.29	5766.89	17340.4	GD	0.9			KS
5754.154	5755.750	17373.93	GD	1.4R			MIT
5751.878	5753.473	17380.80	GD	0.9			MIT
5749.406	5751.001	17388.28	GD	1.0			MIT
5746.351	5747.945	17397.52	GD	0.9			MIT
5744.664	5746.257	17402.63	GD	0.8			MIT
5737.623	5739.215	17423.99	GD	0.6			MIT
5735.972	5737.563	17429.00	GD	0.9			MIT
5733.863	5735.454	17435.41	GD	1.3			MIT
5732.145	5733.735	17440.64	GD	0.8			MIT
5728.312	5729.901	17452.31	GD	0.8			MIT
5724.736	5726.324	17463.21	GD	0.9			MIT
5721.974	5723.561	17471.64	GD	1.0			MIT
5711.353	5712.938	17504.13	GD	0.9			MIT

AIR WAVELENGTH (ANGSTRMS)	VACUUM WAVELENGTH (ANGSTRMS)	WAVENUMBER (KAYSERS)	LMENT	LOG ARC	INTENSITY SPK	DIS	REF
5710.329	5711.913	17507.27	GD	0.9			MIT
5709.407	5710.991	17510.10	GD	1.0			MIT
5705.987	5707.570	17520.59	GD	0.5H			MIT
5701.351	5702.933	17534.84	GD	1.3			MIT
5697.994	5699.575	17545.17	GD	0.8			MIT
5696.215	5697.795	17550.65	GD	1.3			MIT
5692.122	5693.701	17563.27	GD	0.9			MIT
5686.671	5688.249	17580.10	GD	0.6			MIT
5684.108	5685.685	17588.03	GD	0.3			MIT
5683.339	5684.916	17590.41	GD	0.5H			MIT
5680.857	5682.433	17598.09	GD	0.8			MIT
5677.442	5679.017	17608.68	GD	0.9			MIT
5653.315	5654.884	17683.83	GD	0.9			MIT
5652.793	5654.362	17685.46	GD	0.8			MIT
5644.833	5646.400	17710.40	GD	1.1			MIT
5643.677	5645.243	17714.03	GD	0.8			MIT
5643.245	5644.811	17715.38	GD	1.3			MIT
5633.483	5635.047	17746.08	GD	1.1			MIT
5632.245	5633.808	17749.98	GD	1.0			MIT
5631.965	5633.528	17750.86	GD	0.8			MIT
5630.309	5631.872	17756.09	GD	0.8			MIT
5629.546	5631.109	17758.49	GD	1.3			MIT
5622.912	5624.473	17779.44	GD	0.9			MIT
5621.422	5622.982	17784.16	GD	1.0			MIT
5619.494	5621.054	17790.26	GD	0.8			MIT
5617.915	5619.475	17795.26	GD	1.3			MIT
5616.196	5617.755	17800.71	GD	1.0			MIT
5614.433	5615.992	17806.29	GD	1.0			MIT
5610.926	5612.484	17817.42	GD	0.8			MIT
5605.89	5607.45	17833.4	GD	1.0			KS
5604.888	5606.444	17836.62	GD	1.0			MIT
5597.207	5598.761	17861.09	GD	1.1			MIT
5594.120	5595.673	17870.95	GD	0.9			MIT
5593.474	5595.027	17873.01	GD	0.8			MIT
5591.883	5593.436	17878.10	GD	1.3R			MIT
5588.935	5590.487	17887.53	GD	0.9			MIT
5586.149	5587.700	17896.45	GD	0.9			MIT
5583.681	5585.231	17904.36	GD	1.4			MIT
5581.599	5583.149	17911.04	GD	0.8			MIT
5579.658	5581.207	17917.27	GD	0.9			MIT
5576.129	5577.677	17928.61	GD	1.1			MIT
5572.526	5574.073	17940.20	GD	1.0			MIT
5560.690	5562.234	17978.39	GD	1.3			MIT
5559.726	5561.270	17981.50	GD	0.9			MIT
5552.207	5553.749	18005.85	GD	0.3			MIT
5550.216	5551.757	18012.31	GD	0.9			MIT
5548.201	5549.742	18018.86	GD	0.9			MIT
5544.999	5546.539	18029.26	GD	1.1			MIT
5539.817	5541.356	18046.13	GD	1.1			MIT
5538.333	5539.871	18050.96	GD	1.3			MIT
5535.16	5536.70	18061.3	GD	0.9			KS
5534.548	5536.085	18063.31	GD	0.9			MIT
5534.291	5535.828	18064.14	GD	0.9			MIT
5533.33	5534.87	18067.3	GD	0.9			KS
5524.57	5526.10	18095.9	GD	1.0			KS
5521.71	5523.24	18105.3	GD	0.9			KS
5516.21	5517.74	18123.4	GD	1.0			ED
5515.63	5517.16	18125.3	GD	0.8			KS
5514.56	5516.09	18128.8	GD	0.9			KS
5513.76	5515.29	18131.4	GD	0.6			KS
5505.08	5506.61	18160.0	GD	0.9			KS
5500.41	5501.94	18175.4	GD	1.0			KS
5499.92	5501.45	18177.0	GD	1.0			KS
5498.68	5500.21	18181.1	GD	0.9			KS
5493.38	5494.91	18198.7	GD	0.9			KS
5491.65	5493.18	18204.4	GD	0.8			KS
5491.37	5492.90	18205.3	GD	0.8			KS
5490.58	5492.11	18207.9	GD	0.6H			KS
5488.18	5489.70	18215.9	GD	0.5H			KS
5486.57	5488.09	18221.3	GD	0.8			KS

AIR WAVELENGTH (ANGSTRMS)	VACUUM WAVELENGTH (ANGSTRMS)	WAVENUMBER (KAYSERS)	LMENT	LOG ARC	INTENSITY SPK	DIS	REF
5481.97	5483.49	18236.6	GD	0.8			KS
5480.18	5481.70	18242.5	GD	0.8			KS
5476.29	5477.81	18255.5	GD	0.6			KS
5475.66	5477.18	18257.6	GD	0.9			KS
5470.47	5471.99	18274.9	GD	0.9			ED
5470.29	5471.81	18275.5	GD	0.3			KS
5469.729	5471.249	18277.36	GD	1.1			MIT
5469.041	5470.561	18279.66	GD	0.9			MIT
5463.277	5464.795	18298.95	GD	0.9			MIT
5455.28	5456.80	18325.8	GD	1.0			ED
5453.464	5454.980	18331.87	GD	0.9			MIT
5452.66	5454.18	18334.6	GD	0.6			ED
5447.735	5449.249	18351.15	GD	1.1			MIT
5441.56	5443.07	18372.0	GD	0.9			ED
5441.13	5442.64	18373.4	GD	0.9			ED
5439.45	5440.96	18379.1	GD	0.5			ED
5436.305	5437.816	18389.74	GD	1.0			MIT
5424.65	5426.16	18429.2	GD	0.8			ED
5423.61	5425.12	18432.8	GD	0.9			ED
5421.195	5422.702	18440.99	GD	1.1			MIT
5419.870	5421.377	18445.50	GD	0.5D			MIT
5417.08	5418.59	18455.0	GD	1.0			ED
5415.693	5417.199	18459.73	GD	1.1			MIT
5413.21	5414.71	18468.2	GD	1.0			ED
5412.629	5414.134	18470.18	GD	0.9			MIT
5411.190	5412.694	18475.09	GD	0.9			MIT
5399.53	5401.03	18515.0	GD	0.9			ED
5394.347	5395.847	18532.77	GD	1.0			MIT
5393.665	5395.165	18535.12	GD	1.0			MIT
5389.513	5391.012	18549.40	GD	1.3			MIT
5385.38	5386.88	18563.6	GD	1.0			ED
5384.174	5385.671	18567.79	GD	1.0			MIT
5380.26	5381.76	18581.3	GD	0.9			ED
5375.405	5376.900	18598.08	GD	1.0			MIT
5372.224	5373.718	18609.09	GD	1.3			MIT
5370.69	5372.18	18614.4	GD	1.4R			ED
5369.911	5371.404	18617.10	GD	1.1			MIT
5369.628	5371.121	18618.09	GD	0.9			MIT
5368.77	5370.26	18621.1	GD	1.0			ED
5367.68	5369.17	18624.8	GD	0.9			ED
5366.35	5367.84	18629.5	GD	0.8			ED
5365.404	5366.896	18632.74	GD	1.0			MIT
5362.57	5364.06	18642.6	GD	0.8			ED
5361.641	5363.132	18645.82	GD	1.0			MIT
5359.188	5360.678	18654.35	GD	0.9			MIT
5357.757	5359.247	18659.34	GD	0.9			MIT
5355.958	5357.448	18665.60	GD	0.6			MIT
5354.423	5355.912	18670.96	GD	1.0			MIT
5353.283	5354.772	18674.93	GD	1.4			MIT
5350.406	5351.894	18684.97	GD	1.4			MIT
5349.62	5351.11	18687.7	GD	0.9			ED
5348.689	5350.177	18690.97	GD	1.3			MIT
5347.83	5349.32	18694.0	GD	1.3			ED
5345.692	5347.179	18701.45	GD	1.0			MIT
5345.14	5346.63	18703.4	GD	1.0			MIT
5343.022	5344.508	18710.80	GD	1.0			MIT
5341.823	5343.309	18715.00	GD	0.9			MIT
5337.548	5339.033	18729.99	GD	1.3			MIT
5333.25	5334.73	18745.1	GD	1.3			ED
5331.941	5333.424	18749.68	GD	0.9			MIT
5331.02	5332.50	18752.9	GD	0.9			ED
5328.27	5329.75	18762.6	GD	1.1			ED
5327.330	5328.812	18765.91	GD	1.3			MIT
5322.710	5324.191	18782.20	GD	0.9			MIT
5321.791	5323.271	18785.44	GD	1.0			MIT
5321.508	5322.988	18786.44	GD	0.9			MIT
5321.265	5322.745	18787.30	GD	0.9			MIT
5316.809	5318.288	18803.04	GD	0.9			MIT
5315.806	5317.285	18806.59	GD	0.8			MIT
5311.856	5313.334	18820.57	GD	1.0			MIT

AIR WAVELENGTH (ANGSTRMS)	VACUUM WAVELENGTH (ANGSTRMS)	WAVENUMBER (KAYSERS)	LMENT	LOG ARC	INTENSITY SPK	DIS	REF
5307.322	5308.799	18836.65	GD	1.4			MIT
5306.717	5308.193	18838.80	GD	1.0			MIT
5302.768	5304.243	18852.83	GD	1.4			MIT
5301.684	5303.159	18856.69	GD	1.4			MIT
5298.592	5300.066	18867.69	GD	1.2			MIT
5283.865	5285.335	18920.27	GD	1.0			MIT
5283.089	5284.559	18923.05	GD	1.0			MIT
5282.493	5283.963	18925.19	GD	1.0			MIT
5276.537	5278.005	18946.55	GD	1.3			MIT
5272.930	5274.397	18959.51	GD	1.0			MIT
5272.374	5273.841	18961.51	GD	0.8			MIT
5271.75	5273.22	18963.8	GD	0.6			ED
5268.794	5270.260	18974.40	GD	1.3			MIT
5263.809	5265.274	18992.36	GD	1.0			MIT
5255.821	5257.284	19021.23	GD	1.1			MIT
5254.759	5256.222	19025.07	GD	1.2			MIT
5252.164	5253.626	19034.47	GD	1.2			MIT
5251.175	5252.637	19038.06	GD	1.5			MIT
5246.87	5248.33	19053.7	GD	1.3			ED
5233.962	5235.419	19100.67	GD	1.3			MIT
5220.27	5221.72	19150.8	GD	1.3			ED
5219.40	5220.85	19154.0	GD	1.4			ED
5217.49	5218.94	19161.0	GD	1.4			ED
5210.99	5212.44	19184.9	GD	1.1			ED
5197.77	5199.22	19233.7	GD	1.4			ED
5191.07	5192.52	19258.5	GD	1.5			ED
5187.85	5189.29	19270.4	GD	1.2			ED
5187.24	5188.68	19272.7	GD	1.2			ED
5186.91	5188.35	19273.9	GD	1.3			ED
5179.91	5181.35	19300.0	GD	1.3			ED
5178.84	5180.28	19304.0	GD	1.3			ED
5176.28	5177.72	19313.5	GD	1.5			ED
5163.68	5165.12	19360.6	GD	1.2			ED
5160.88	5162.32	19371.1	GD	1.2			ED
5156.74	5158.18	19386.7	GD	1.3			ED
5155.84	5157.28	19390.1	GD	1.4			ED
5142.65	5144.08	19439.8	GD	1.2			ED
5141.49	5142.92	19444.2	GD	1.3			ED
5140.84	5142.27	19446.7	GD	1.4			ED
5136.02	5137.45	19464.9	GD	1.3			ED
5135.58	5137.01	19466.6	GD	1.2			ED
5132.21	5133.64	19479.4	GD	1.3			ED
5130.26	5131.69	19486.8	GD	1.3			ED
5125.54	5126.97	19504.7	GD	1.7			ED
5123.68	5125.11	19511.8	GD	1.3			ED
5108.91	5110.33	19568.2	GD	1.5			ED
5103.46	5104.88	19589.1	GD	1.4			ED
5102.79	5104.21	19591.7	GD	0.9			ED
5100.96	5102.38	19598.7	GD	1.3			ED
5099.27	5100.69	19605.2	GD	1.0			ED
5098.40	5099.82	19608.5	GD	1.3			ED
5096.06	5097.48	19617.5	GD	1.2			ED
5092.25	5093.67	19632.2	GD	1.5			ED
5071.02	5072.43	19714.4	GD	1.2			ED
5050.87	5052.28	19793.1	GD	1.7			ED
5031.287	5032.690	19870.09	GD	1.5			MIT
5023.131	5024.532	19902.35	GD	1.3			MIT
5020.376	5021.776	19913.27	GD	1.3			MIT
5015.057	5016.456	19934.39	GD	2.0O			MIT
5011.755	5013.153	19947.53	GD	1.0			MIT
5010.833	5012.231	19951.20	GD	1.7			MIT
4999.085	5000.480	19998.08	GD	1.0H			MIT
4998.389	4999.783	20000.87	GD	1.4	0.3		MIT
4993.81	4995.20	20019.2	GD	1.0H			ED
4986.208	4987.599	20049.73	GD	1.4			MIT
4973.85	4975.24	20099.5	GD	1.4			ED
4972.622	4974.010	20104.50	GD	1.4			MIT
4969.164	4970.551	20118.50	GD	2.0			MIT
4968.587	4969.973	20120.83	GD	1.3	1.0		MIT
4965.064	4966.449	20135.11	GD	2.0	1.0H		MIT

AIR WAVELENGTH (ANGSTRMS)	VACUUM WAVELENGTH (ANGSTRMS)	WAVENUMBER (KAYSERS)	LMENT	LOG ARC	INTENSITY SPK	INTENSITY DIS	REF
4961.497	4962.882	20149.58	GD	2.0			MIT
4958.793	4960.177	20160.57	GD	2.1	0.6		MIT
4953.166	4954.548	20183.48	GD	1.0			MIT
4952.493	4953.875	20186.22	GD	1.3			MIT
4942.569	4943.949	20226.75	GD	2.0			MIT
4939.550	4940.929	20239.11	GD	1.5			MIT
4938.618	4939.996	20242.93	GD	2.2	0.3		MIT
4936.345	4937.723	20252.25	GD	1.5	0.3		MIT
4930.708	4932.084	20275.40	GD	1.9			MIT
4923.51	4924.88	20305.0	GD	1.0	0.3		ED
4918.66	4920.03	20325.1	GD	1.0			ED
4916.602	4917.975	20333.57	GD	1.5			MIT
4915.836	4917.208	20336.74	GD	1.6			MIT
4910.124	4911.495	20360.40	GD	1.4			MIT
4894.320	4895.687	20426.14	GD	2.3	0.6		MIT
4889.203	4890.568	20447.52	GD	1.8			MIT
4883.205	4884.569	20472.64	GD	2.0			MIT
4881.940	4883.303	20477.94	GD	2.0	1.3		MIT
4881.376	4882.739	20480.31	GD	1.7			MIT
4881.08	4882.44	20481.6	GD	1.8			ED
4875.973	4877.335	20503.00	GD	2.0	1.0		MIT
4873.350	4874.711	20514.04	GD	2.3	0.3		MIT
4871.524	4872.885	20521.73	GD	2.0			MIT
4870.049	4871.409	20527.94	GD	2.3	1.0		MIT
4866.408	4867.767	20543.30	GD	2.0			MIT
4865.042	4866.401	20549.07	GD	2.6	1.0		MIT
4862.608	4863.966	20559.35	GD	2.0	0.3		MIT
4861.802	4863.160	20562.76	GD	2.0			MIT
4859.232	4860.589	20573.64	GD	1.7			MIT
4856.738	4858.095	20584.20	GD	1.9			MIT
4856.193	4857.550	20586.51	GD	1.9			MIT
4848.108	4849.462	20620.84	GD	1.8	1.0		MIT
4835.274	4836.625	20675.57	GD	2.0	1.0		MIT
4834.244	4835.595	20679.98	GD	2.1	1.4		MIT
4829.950	4831.300	20698.36	GD	1.5			MIT
4823.08	4824.43	20727.9	GD	0.7			ED
4821.708	4823.055	20733.75	GD	2.2	1.9		MIT
4819.803	4821.150	20741.94	GD	1.4			MIT
4816.845	4818.191	20754.68	GD	1.7	1.3		MIT
4813.768	4815.113	20767.94	GD	1.2			MIT
4808.019	4809.363	20792.77	GD	1.2			MIT
4807.465	4808.809	20795.17	GD	2.0	1.6		MIT
4805.824	4807.167	20802.27	GD	2.3	1.9		MIT
4803.551	4804.894	20812.11	GD	2.0	1.6		MIT
4802.583	4803.925	20816.31	GD	2.0	1.6H		MIT
4801.077	4802.419	20822.84	GD	2.3	2.3		MIT
4800.111	4801.453	20827.03	GD	1.5	1.9		MIT
4799.869	4801.211	20828.08	GD	1.4	1.8		MIT
4793.454	4794.794	20855.95	GD	0.7H			MIT
4791.602	4792.941	20864.01	GD	2.2			MIT
4786.806	4788.144	20884.92	GD	1.6	0.3		MIT
4784.641	4785.979	20894.37	GD	2.0	1.7		MIT
4783.565	4784.902	20899.07	GD	1.3			MIT
4781.935	4783.272	20906.19	GD	2.3	1.7		MIT
4781.06	4782.40	20910.0	GD	1.4			ED
4779.939	4781.275	20914.92	GD	0.3H			MIT
4773.094	4774.429	20944.92	GD	1.3			MIT
4767.251	4768.584	20970.59	GD	2.0	1.4		MIT
4763.841	4765.173	20985.60	GD	1.4	1.0		MIT
4760.751	4762.082	20999.22	GD	1.6	1.6		MIT
4758.708	4760.039	21008.23	GD	1.5			MIT
4758.272	4759.603	21010.16	GD	1.4			MIT
4755.517	4756.847	21022.33	GD	0.9	1.3		MIT
4749.150	4750.478	21050.51	GD	1.2			MIT
4745.806	4747.133	21065.34	GD	1.3	0.3		MIT
4743.655	4744.982	21074.90	GD	2.5	2.5		MIT
4739.908	4741.234	21091.56	GD	1.0H			MIT
4738.131	4739.456	21099.47	GD	1.7	1.7H		MIT
4735.763	4737.088	21110.02	GD	2.2	2.2		MIT
4734.441	4735.765	21115.91	GD	2.0	1.6		MIT

AIR WAVELENGTH (ANGSTRMS)	VACUUM WAVELENGTH (ANGSTRMS)	WAVENUMBER (KAYSERS)	LMENT	LOG ARC	INTENSITY SPK	INTENSITY DIS	REF
4732.612	4733.936	21124.07	GD	2.5	2.5		MIT
4730.135	4731.458	21135.13	GD	1.5			MIT
4728.474	4729.797	21142.56	GD	2.2	2.0		MIT
4726.741	4728.063	21150.31	GD	1.6	2.0		MIT
4723.735	4725.056	21163.77	GD	2.0	2.3		MIT
4721.466	4722.787	21173.94	GD	1.3	1.3H		MIT
4719.054	4720.374	21184.76	GD	1.7	1.3		MIT
4717.29	4718.61	21192.7	GD	1.0			ED
4716.586	4717.906	21195.85	GD	1.2	1.2		MIT
4715.51	4716.83	21200.7	GD	0.9			MIT
4712.819	4714.138	21212.79	GD	1.4	1.6		MIT
4711.983	4713.301	21216.55	GD	1.5	1.5		MIT
4709.783	4711.101	21226.46	GD	2.0	1.4		MIT
4707.891	4709.208	21234.99	GD	1.5	1.5		MIT
4705.767	4707.084	21244.58	GD	1.4			MIT
4704.289	4705.605	21251.25	GD	0.6			MIT
4703.136	4704.452	21256.46	GD	1.7	1.3		MIT
4702.323	4703.639	21260.14	GD	1.7	2.0		MIT
4697.488	4698.803	21282.02	GD	2.0	0.6		MIT
4696.613	4697.927	21285.98	GD	1.4	0.3		MIT
4695.486	4696.800	21291.09	GD	1.4	1.0		MIT
4695.14	4696.45	21292.7	GD	1.2			MIT
4694.339	4695.653	21296.29	GD	1.7	0.9		MIT
4691.17	4692.48	21310.7	GD	1.4	0.3		MIT
4688.906	4690.218	21320.97	GD	0.7			MIT
4688.136	4689.448	21324.47	GD	1.3	1.0		MIT
4686.403	4687.715	21332.36	GD	1.0	0.7H		MIT
4683.343	4684.654	21346.29	GD	2.0	1.7		MIT
4683.071	4684.382	21347.53	GD	1.3			MIT
4680.65	4681.96	21358.6	GD	0.5			MIT
4680.056	4681.366	21361.29	GD	1.7	1.4		MIT
4679.179	4680.489	21365.29	GD	1.4	1.4		MIT
4678.247	4679.556	21369.55	GD	1.0H	1.0H		MIT
4677.621	4678.930	21372.41	GD	0.7			MIT
4676.98	4678.29	21375.3	GD	0.9			M
4670.849	4672.156	21403.39	GD	0.5	0.3H		MIT
4668.23	4669.54	21415.4	GD	1.1			KN
4666.443	4667.749	21423.60	GD	1.4	1.0		MIT
4664.24	4665.55	21433.7	GD	1.4	0.3		M
4659.847	4661.152	21453.93	GD	1.0	1.0		MIT
4658.610	4659.914	21459.62	GD	0.7			MIT
4654.990	4656.293	21476.31	GD	1.2	1.2		MIT
4654.765	4656.068	21477.35	GD	0.5			KN
4653.548	4654.851	21482.97	GD	1.2	0.3		MIT
4652.323	4653.626	21488.62	GD	0.9	0.9H		MIT
4648.704	4650.006	21505.35	GD	0.9			KN
4648.585	4649.887	21505.90	GD	0.9			KN
4647.658	4648.959	21510.19	GD	0.9	0.9		MIT
4646.336	4647.637	21516.31	GD	1.1	0.8		MIT
4646.01	4647.31	21517.8	GD	1.1	0.0		MIT
4641.315	4642.615	21539.59	GD	0.8			MIT
4640.53	4641.83	21543.2	GD	0.6			MIT
4640.047	4641.346	21545.47	GD	1.0	0.7		MIT
4639.01	4640.31	21550.3	GD	1.4	0.5		MIT
4638.599	4639.898	21552.20	GD	0.9			MIT
4636.655	4637.953	21561.23	GD	1.4	1.1		MIT
4630.51	4631.81	21589.9	GD	0.7			MIT
4627.63	4628.93	21603.3	GD	0.9	0.0		MIT
4624.428	4625.723	21618.24	GD	1.2	0.9		MIT
4622.311	4623.606	21628.14	GD	0.6			KN
4622.187	4623.482	21628.72	GD	0.6			KN
4620.470	4621.764	21636.76	GD	0.5	0.5H		MIT
4619.63	4620.92	21640.7	GD	0.5H			ED
4619.130	4620.424	21643.04	GD	0.7	0.5		MIT
4614.505	4615.798	21664.73	GD	1.7	1.4		MIT
4611.045	4612.337	21680.99	GD	0.6	0.6H		MIT
4608.591	4609.882	21692.53	GD	0.8			MIT
4608.02	4609.31	21695.2	GD	0.9	0.5		MIT
4606.650	4607.940	21701.67	GD	0.5	0.5		MIT
4606.056	4607.346	21704.47	GD	0.7	0.3		MIT

AIR WAVELENGTH (ANGSTRMS)	VACUUM WAVELENGTH (ANGSTRMS)	WAVENUMBER (KAYSERS)	LMENT	LOG ARC	INTENSITY SPK	INTENSITY DIS	REF
4602.944	4604.234	21719.14	GD	1.0	0.3		MIT
4601.05	4602.34	21728.1	GD	1.6	0.8		MIT
4598.905	4600.193	21738.22	GD	1.3	1.0		MIT
4597.922	4599.210	21742.86	GD	1.4	1.6		MIT
4596.994	4598.282	21747.25	GD	1.4	1.4		MIT
4595.368	4596.656	21754.95	GD	0.6	0.0		MIT
4589.542	4590.828	21782.56	GD	0.5	0.5		MIT
4586.98	4588.27	21794.7	GD	1.2	0.3		MIT
4584.282	4585.567	21807.56	GD	0.5	0.3		MIT
4583.087	4584.371	21813.24	GD	1.6	0.3		MIT
4582.51	4583.79	21816.0	GD	1.0	0.5		MIT
4582.382	4583.666	21816.60	GD	0.8	0.3		MIT
4581.301	4582.585	21821.75	GD	1.5	0.6		MIT
4581.09	4582.37	21822.7	GD	0.7			MIT
4579.606	4580.889	21829.82	GD	1.2	0.6		MIT
4575.912	4577.194	21847.45	GD	0.6	0.0		KN
4575.84	4577.12	21847.8	GD	0.5			MIT
4573.81	4575.09	21857.5	GD	0.7			MIT
4572.215	4573.496	21865.11	GD	0.9	0.9		MIT
4564.57	4565.85	21901.7	GD	0.6			KN
4561.05	4562.33	21918.6	GD	0.9	0.6		KN
4558.094	4559.372	21932.85	GD	1.5	0.5		MIT
4554.99	4556.27	21947.8	GD	0.8	0.3		MIT
4551.466	4552.742	21964.79	GD	0.3	0.3		MIT
4550.964	4552.240	21967.21	GD	1.0	1.0		MIT
4548.010	4549.285	21981.48	GD	1.7	1.7		MIT
4545.11	4546.38	21995.5	GD	0.3			KN
4544.246	4545.520	21999.68	GD	0.8H	0.8H		MIT
4542.727	4544.001	22007.04	GD	0.5	0.5		MIT
4542.033	4543.306	22010.40	GD	1.7	1.7		MIT
4540.025	4541.298	22020.14	GD	1.9	2.3		MIT
4537.821	4539.093	22030.83	GD	2.2	2.0		MIT
4536.980	4538.252	22034.92	GD	1.3	0.3		MIT
4533.516	4534.787	22051.75	GD	0.5			MIT
4531.81	4533.08	22060.1	GD	0.5	0.0		MIT
4531.12	4532.39	22063.4	GD	0.7			M
4524.33	4525.60	22096.5	GD	0.5	0.0		KN
4524.099	4525.368	22097.65	GD	0.9	0.3		MIT
4523.838	4525.107	22098.93	GD	0.9			MIT
4522.838	4524.106	22103.81	GD	1.7	0.6		MIT
4522.43	4523.70	22105.8	GD	0.5			KN
4521.956	4523.224	22108.12	GD	1.4	0.7		MIT
4521.298	4522.566	22111.34	GD	1.4	0.3		MIT
4520.084	4521.352	22117.28	GD	1.7	0.5		MIT
4519.657	4520.925	22119.37	GD	2.2	2.0		MIT
4517.075	4518.342	22132.01	GD	0.5	0.0		KN
4516.989	4518.256	22132.43	GD	0.7	0.3		MIT
4514.517	4515.783	22144.55	GD	1.8	1.0		MIT
4513.809	4515.075	22148.03	GD	0.5			MIT
4510.386	4511.651	22164.83	GD	1.0H	1.0H		MIT
4509.093	4510.358	22171.19	GD	1.0	0.3		MIT
4508.995	4510.260	22171.67	GD	0.8			MIT
4507.675	4508.939	22178.17	GD	0.8	0.8		MIT
4507.065	4508.329	22181.17	GD	0.6H			KN
4506.941	4508.205	22181.78	GD	1.7	1.3		MIT
4506.346	4507.610	22184.71	GD	1.9	0.7		MIT
4506.221	4507.485	22185.32	GD	2.0	1.7		MIT
4504.962	4506.226	22191.52	GD	0.5			MIT
4503.803	4505.066	22197.23	GD	0.5			KN
4503.21	4504.47	22200.1	GD	0.7	0.3		MIT
4501.461	4502.724	22208.78	GD	1.0H			MIT
4500.877	4502.140	22211.66	GD	0.5	0.5		MIT
4500.678	4501.940	22212.64	GD	0.7	0.3		MIT
4498.294	4499.556	22224.42	GD	2.0			MIT
4497.55	4498.81	22228.1	GD	0.3			KN
4497.328	4498.590	22229.19	GD	1.0	0.7		MIT
4497.133	4498.395	22230.15	GD	2.2	1.9		MIT
4496.620	4497.881	22232.69	GD	0.6	0.6		MIT
4494.858	4496.119	22241.40	GD	0.7			MIT
4492.038	4493.298	22255.37	GD	0.3			MIT

AIR WAVELENGTH (ANGSTRMS)	VACUUM WAVELENGTH (ANGSTRMS)	WAVENUMBER (KAYSERS)	LMENT	LOG ARC	INTENSITY SPK	INTENSITY DIS	REF
4491.264	4492.524	22259.20	GD	0.3			MIT
4489.742	4491.002	22266.75	GD	0.5			MIT
4488.560	4489.819	22272.61	GD	0.6			MIT
4488.415	4489.674	22273.33	GD	1.4			MIT
4486.908	4488.167	22280.81	GD	2.0	1.2		MIT
4486.364	4487.623	22283.51	GD	1.4	1.9		MIT
4485.479	4486.737	22287.91	GD	0.7	0.5		MIT
4484.710	4485.968	22291.73	GD	1.4	1.0		MIT
4484.477	4485.735	22292.89	GD	0.7	0.7		MIT
4483.339	4484.597	22298.55	GD	1.9	2.1		MIT
4481.068	4482.325	22309.85	GD	1.8			MIT
4478.812	4480.069	22321.09	GD	1.9			MIT
4476.144	4477.400	22334.39	GD	2.0			AB
4474.138	4475.393	22344.40	GD	2.2	2.2		MIT
4473.283	4474.538	22348.68	GD	1.0	1.0		MIT
4471.303	4472.558	22358.57	GD	1.7			MIT
4470.430	4471.685	22362.94	GD	0.7			MIT
4467.17	4468.42	22379.3	GD	1.3			KN
4467.088	4468.342	22379.67	GD	1.7			KN
4466.553	4467.806	22382.35	GD	2.3	2.2		MIT
4465.809	4467.062	22386.08	GD	0.7			MIT
4464.748	4466.001	22391.40	GD	1.3	1.3		MIT
4463.249	4464.502	22398.92	GD	1.2			MIT
4462.799	4464.051	22401.18	GD	1.0			MIT
4461.372	4462.624	22408.34	GD	0.6			MIT
4456.69	4457.94	22431.9	GD	0.5H			MIT
4454.697	4455.947	22441.92	GD	0.5			MIT
4453.929	4455.179	22445.79	GD	1.0	1.0		MIT
4452.728	4453.978	22451.84	GD	1.5			MIT
4449.946	4451.195	22465.88	GD	0.6	1.6		MIT
4449.019	4450.268	22470.56	GD	0.6	0.6H		MIT
4447.345	4448.593	22479.01	GD	0.3			MIT
4446.48	4447.73	22483.4	GD	1.9	0.3		M
4445.501	4446.749	22488.34	GD	0.5H	0.0		MIT
4444.93	4446.18	22491.2	GD	0.7			MIT
4438.463	4439.709	22524.00	GD	0.5			KN
4438.268	4439.514	22524.99	GD	1.6	2.0		MIT
4438.143	4439.389	22525.62	GD	0.7			KN
4437.82	4439.07	22527.3	GD	0.6			MIT
4436.225	4437.470	22535.36	GD	1.5	2.0		MIT
4436.103	4437.348	22535.98	GD	0.6			KN
4433.638	4434.883	22548.51	GD	1.3	1.1		MIT
4431.769	4433.013	22558.02	GD	1.8	0.0		KN
4430.627	4431.871	22563.83	GD	2.2	1.6		MIT
4428.94	4430.18	22572.4	GD	0.5			MIT
4427.605	4428.848	22579.23	GD	1.3	1.2		MIT
4426.146	4427.389	22586.68	GD	1.7	0.3		MIT
4425.009	4426.252	22592.48	GD	0.9	0.6		MIT
4424.102	4425.344	22597.11	GD	1.4	1.0		MIT
4422.413	4423.655	22605.74	GD	2.0	1.6		MIT
4421.24	4422.48	22611.7	GD	2.0	0.9		MIT
4420.64	4421.88	22614.8	GD	1.0			KN
4419.036	4420.277	22623.02	GD	2.3	2.3		MIT
4415.97	4417.21	22638.7	GD	0.5			KN
4415.63	4416.87	22640.5	GD	0.5			KN
4415.435	4416.675	22641.47	GD	0.5H	0.0		MIT
4415.029	4416.269	22643.55	GD	0.5			KN
4414.735	4415.975	22645.06	GD	2.0	1.7		MIT
4414.162	4415.402	22648.00	GD	2.0	1.8		MIT
4413.47	4414.71	22651.6	GD	0.7	0.3		MIT
4411.159	4412.398	22663.41	GD	2.0	1.7		MIT
4409.256	4410.494	22673.20	GD	1.0	1.0		MIT
4408.261	4409.499	22678.31	GD	2.0	2.2		MIT
4407.19	4408.43	22683.8	GD	0.5			KN
4406.672	4407.910	22686.49	GD	1.8	2.3		MIT
4403.231	4404.468	22704.22	GD	0.5			MIT
4403.136	4404.373	22704.71	GD	2.0	2.0		MIT
4401.853	4403.089	22711.33	GD	2.3	2.0		MIT
4400.764	4402.000	22716.95	GD	1.2	0.9		MIT
4400.183	4401.419	22719.95	GD	1.0	0.7		MIT

AIR WAVELENGTH (ANGSTRMS)	VACUUM WAVELENGTH (ANGSTRMS)	WAVENUMBER (KAYSERS)	LMENT	LOG ARC	INTENSITY SPK	DIS	REF
4397.517	4398.752	22733.72	GD	2.0	0.7		MIT
4394.73	4395.96	22748.1	GD	1.0			MIT
4392.071	4393.305	22761.91	GD	2.0	2.0		MIT
4391.652	4392.886	22764.08	GD	0.5			KN
4391.440	4392.674	22765.18	GD	1.2	1.4		MIT
4390.955	4392.189	22767.69	GD	2.0	2.0		MIT
4389.995	4391.228	22772.67	GD	1.5	1.9		MIT
4389.885	4391.118	22773.24	GD	1.6	1.6H		MIT
4388.991	4390.224	22777.88	GD	0.6	0.6		MIT
4387.678	4388.911	22784.70	GD	2.2			MIT
4387.16	4388.39	22787.4	GD	0.5			MIT
4386.201	4387.433	22792.37	GD	0.6			MIT
4383.137	4384.368	22808.30	GD	1.5	1.6		MIT
4382.059	4383.290	22813.91	GD	1.6			MIT
4380.640	4381.871	22821.30	GD	2.0	2.1		MIT
4379.52	4380.75	22827.1	GD	0.3			KN
4378.570	4379.800	22832.09	GD	1.6			MIT
4376.076	4377.306	22845.10	GD	0.9	0.9		MIT
4374.986	4376.215	22850.79	GD	1.4			KN
4374.254	4375.483	22854.62	GD	1.3	1.3		MIT
4373.839	4375.068	22856.79	GD	2.3	1.9		MIT
4372.01	4373.24	22866.4	GD	0.5			KN
4370.195	4371.423	22875.85	GD	1.6			MIT
4369.775	4371.003	22878.04	GD	2.4	2.2		MIT
4369.168	4370.396	22881.22	GD	1.7	1.3		MIT
4368.53	4369.76	22884.6	GD	0.7			KN
4364.14	4365.37	22907.6	GD	1.0			MIT
4360.926	4362.152	22924.47	GD	2.3			MIT
4360.12	4361.35	22928.7	GD	0.5			MIT
4359.644	4360.869	22931.21	GD	1.2			MIT
4359.162	4360.387	22933.74	GD	1.3			MIT
4357.80	4359.02	22940.9	GD	0.3			KN
4354.02	4355.24	22960.8	GD	1.3			KN
4353.79	4355.01	22962.0	GD	1.0			MIT
4347.320	4348.542	22996.21	GD	2.0	2.0		MIT
4346.626	4347.848	22999.89	GD	1.7	1.3		MIT
4346.459	4347.681	23000.77	GD	2.2	1.8		MIT
4344.45	4345.67	23011.4	GD	1.4	1.4		KN
4344.305	4345.526	23012.17	GD	1.7			MIT
4342.191	4343.412	23023.38	GD	2.3	2.3		MIT
4341.9	4343.1	23025.	GD	0.3H			KN
4341.292	4342.512	23028.14	GD	2.3	2.1		MIT
4337.510	4338.729	23048.22	GD	1.7	1.7		MIT
4336.50	4337.72	23053.6	GD	0.7			KN
4335.292	4336.511	23060.01	GD	1.0			MIT
4333.248	4334.466	23070.89	GD	1.0			MIT
4331.381	4332.599	23080.83	GD	1.2	0.9		MIT
4330.616	4331.834	23084.91	GD	2.0			MIT
4330.320	4331.538	23086.49	GD	0.7			MIT
4329.584	4330.801	23090.42	GD	2.0	2.0		MIT
4328.941	4330.158	23093.84	GD	0.5			MIT
4327.104	4328.321	23103.65	GD	2.7G	2.0		MIT
4326.291	4327.508	23107.99	GD	0.5			MIT
4325.690	4326.906	23111.20	GD I	2.7G	2.4		MIT
4325.568	4326.784	23111.85	GD	0.7			MIT
4324.568	4325.784	23117.20	GD	0.9	0.6		MIT
4324.074	4325.290	23119.84	GD	2.0			MIT
4323.609	4324.825	23122.32	GD	0.7			MIT
4322.195	4323.410	23129.89	GD	1.7	1.4		MIT
4322.008	4323.223	23130.89	GD	0.8			MIT
4321.207	4322.422	23135.18	GD	1.6			MIT
4321.108	4322.323	23135.71	GD	1.7	1.3		MIT
4320.527	4321.742	23138.82	GD	1.8	1.3		MIT
4316.275	4317.489	23161.61	GD	1.6			MIT
4316.061	4317.275	23162.76	GD	2.2	1.8		MIT
4314.403	4315.616	23171.66	GD	2.0	1.3		MIT
4313.851	4315.064	23174.63	GD	2.3	1.9		MIT
4310.988	4312.201	23190.02	GD	2.0			MIT
4309.294	4310.506	23199.13	GD	1.3	1.0		MIT
4308.210	4309.422	23204.97	GD	1.0			MIT

AIR WAVELENGTH (ANGSTRMS)	VACUUM WAVELENGTH (ANGSTRMS)	WAVENUMBER (KAYSERS)	LMENT	LOG ARC	INTENSITY SPK	DIS	REF
4306.350	4307.561	23214.99	GD	2.3	1.9		MIT
4304.897	4306.108	23222.83	GD	2.0	2.0		MIT
4303.456	4304.667	23230.60	GD	0.8	0.8		MIT
4299.301	4300.510	23253.05	GD	1.3			KN
4298.424	4299.633	23257.80	GD	1.4	0.0		MIT
4297.179	4298.388	23264.54	GD	2.0	0.6		MIT
4296.90	4298.11	23266.1	GD	0.6			KN
4296.291	4297.500	23269.34	GD	1.6	1.6		MIT
4296.079	4297.288	23270.49	GD	2.3	0.6		MIT
4292.749	4293.957	23288.54	GD	0.7	0.6		MIT
4290.067	4291.274	23303.10	GD	0.8			MIT
4289.901	4291.108	23304.00	GD	1.6	2.0		MIT
4286.122	4287.328	23324.55	GD	1.8	1.3		MIT
4285.827	4287.033	23326.16	GD	1.8	1.3		MIT
4284.948	4286.154	23330.94	GD	0.6			MIT
4282.792	4283.997	23342.69	GD	1.2	1.2		MIT
4280.501	4281.705	23355.18	GD	2.3	2.0		MIT
4278.202	4279.406	23367.73	GD	0.8	0.8		MIT
4274.171	4275.374	23389.77	GD	2.0			MIT
4273.25	4274.45	23394.8	GD	0.6			KN
4270.775	4271.977	23408.37	GD	0.6H	0.0H		MIT
4270.277	4271.479	23411.10	GD	0.7			MIT
4268.739	4269.940	23419.53	GD	1.6	1.6		MIT
4267.016	4268.217	23428.99	GD	1.6	1.6		MIT
4266.601	4267.802	23431.27	GD	1.7			MIT
4262.095	4263.295	23456.04	GD	2.2	1.0		MIT
4260.112	4261.311	23466.96	GD	1.5	1.0		MIT
4255.57	4256.77	23492.0	GD	1.0			MIT
4255.44	4256.64	23492.7	GD	1.0			KN
4254.03	4255.23	23500.5	GD	0.5			KN
4253.619	4254.816	23502.78	GD	1.7	1.7		MIT
4253.370	4254.567	23504.15	GD	1.7	0.6		MIT
4251.736	4252.933	23513.19	GD	2.5	1.0		MIT
4250.275	4251.472	23521.27	GD	0.5			MIT
4246.547	4247.743	23541.92	GD	2.2	0.5		MIT
4245.345	4246.540	23548.58	GD	1.4			MIT
4243.845	4245.040	23556.91	GD	1.8	2.0		MIT
4241.282	4242.476	23571.14	GD	1.0			MIT
4240.680	4241.874	23574.49	GD	0.6			MIT
4238.785	4239.979	23585.02	GD	2.3	2.3		MIT
4235.890	4237.083	23601.14	GD	0.7			KN
4235.06	4236.25	23605.8	GD	0.8			MIT
4232.941	4234.133	23617.59	GD	0.6			KN
4232.46	4233.65	23620.3	GD	1.2			KN
4229.808	4230.999	23635.08	GD	1.2	1.3		MIT
4227.145	4228.335	23649.97	GD	1.7	1.3		MIT
4225.853	4227.043	23657.20	GD	2.2	1.7		MIT
4225.264	4226.454	23660.50	GD	0.5			KN
4225.148	4226.338	23661.15	GD	1.3	1.0		MIT
4225.028	4226.218	23661.82	GD I	1.2			AB
4224.22	4225.41	23666.3	GD	0.7			KN
4222.98	4224.17	23673.3	GD	1.0			KN
4217.195	4218.383	23705.77	GD	2.0	2.0		MIT
4215.024	4216.211	23717.98	GD	2.3	2.2		MIT
4212.019	4213.205	23734.90	GD	2.2	1.7		MIT
4208.064	4209.249	23757.20	GD	0.7			KN
4206.78	4207.97	23764.5	GD	0.6H			MIT
4204.839	4206.024	23775.43	GD	1.4			KN
4202.512	4203.696	23788.59	GD	1.2			MIT
4197.696	4198.879	23815.88	GD	1.6	1.7		MIT
4197.054	4198.237	23819.52	GD	1.0	1.0		MIT
4195.398	4196.580	23828.93	GD	0.9			MIT
4193.158	4194.340	23841.66	GD	1.4			MIT
4191.624	4192.805	23850.38	GD I	1.6	1.2		KN
4190.785	4191.966	23855.16	GD I	2.0	1.6		MIT
4190.147	4191.328	23858.79	GD	2.0O			MIT
4188.099	4189.279	23870.46	GD	1.3			MIT
4184.264	4185.443	23892.33	GD	2.2	2.2		MIT
4183.619	4184.798	23896.02	GD	1.5			MIT
4182.770	4183.949	23900.87	GD	0.7			MIT

AIR WAVELENGTH (ANGSTRMS)	VACUUM WAVELENGTH (ANGSTRMS)	WAVENUMBER (KAYSERS)	LMENT	LOG ARC	INTENSITY SPK	DIS	REF
4177.83	4179.01	23929.1	GD	0.7			ED
4175.539	4176.716	23942.26	GD	1.7			MIT
4173.561	4174.737	23953.60	GD	1.5	1.5		MIT
4171.71	4172.89	23964.2	GD	1.4	0.0		MIT
4170.110	4171.285	23973.43	GD	1.7	1.7		MIT
4167.271	4168.446	23989.76	GD I	1.0			MIT
4167.157	4168.332	23990.41	GD	1.3			KN
4163.111	4164.285	24013.73	GD	1.4			KN
4162.74	4163.91	24015.9	GD	1.7			MIT
4158.469	4159.641	24040.53	GD	0.7H			KN
4157.788	4158.960	24044.47	GD I	1.4			MIT
4154.87	4156.04	24061.4	GD	1.0			MIT
4153.510	4154.681	24069.24	GD	1.0			MIT
4151.59	4152.76	24080.4	GD	0.5			KN
4150.61	4151.78	24086.1	GD	0.7			KN
4149.475	4150.645	24092.64	GD	0.7			KN
4148.861	4150.031	24096.21	GD	1.0			MIT
4144.240	4145.409	24123.07	GD	0.5			KN
4141.02	4142.19	24141.8	GD	0.7			MIT
4140.454	4141.622	24145.13	GD	1.3	1.3		MIT
4138.020	4139.187	24159.33	GD	0.7			KN
4137.083	4138.250	24164.80	GD	1.3	1.6		MIT
4135.439	4136.605	24174.41	GD	0.6	0.6		MIT
4134.167	4135.333	24181.85	GD I	1.4	0.0		MIT
4132.281	4133.447	24192.89	GD	1.4	1.3		MIT
4131.482	4132.647	24197.56	GD	1.7	1.7		MIT
4130.378	4131.543	24204.03	GD	2.3	1.0		MIT
4127.685	4128.849	24219.82	GD	1.0	1.3		MIT
4125.761	4126.925	24231.12	GD	0.5			MIT
4123.59	4124.75	24243.9	GD	0.5			KN
4123.010	4124.173	24247.28	GD	0.9	1.0		MIT
4121.07	4122.23	24258.7	GD	0.5	0.0		MIT
4119.380	4120.542	24268.65	GD	0.9			KN
4119.199	4120.361	24269.72	GD	0.5			KN
4115.353	4116.514	24292.40	GD	1.4			MIT
4113.769	4114.930	24301.75	GD	0.6	0.6H		MIT
4112.964	4114.125	24306.51	GD	0.7H			MIT
4111.77	4112.93	24313.6	GD	1.0	0.3		MIT
4111.440	4112.600	24315.52	GD	1.2	1.2		MIT
4110.603	4111.763	24320.47	GD	1.0	1.2		MIT
4110.431	4111.591	24321.49	GD	0.7			MIT
4108.42	4109.58	24333.4	GD	0.6	0.3		KN
4105.79	4106.95	24349.0	GD	0.5	0.0H		KN
4104.986	4106.144	24353.75	GD	0.7			MIT
4103.410	4104.568	24363.10	GD	0.7H			MIT
4100.261	4101.418	24381.81	GD I	1.2	0.7		MIT
4098.912	4100.069	24389.83	GD	2.0	2.0		MIT
4098.613	4099.770	24391.61	GD	1.4H	0.8		MIT
4098.04	4099.20	24395.0	GD	1.0			KN
4094.491	4095.647	24416.17	GD	1.4	1.2		MIT
4093.725	4094.880	24420.74	GD	1.2	1.2		MIT
4092.712	4093.867	24426.78	GD I	2.0	1.6		MIT
4091.754	4092.909	24432.50	GD	0.7H			KN
4090.747	4091.902	24438.51	GD	0.7			KN
4090.422	4091.577	24440.46	GD I	2.0	1.3		MIT
4088.804	4089.958	24450.13	GD	0.8			MIT
4087.709	4088.863	24456.68	GD	1.9	2.0		MIT
4087.332	4088.486	24458.93	GD	0.7			MIT
4085.648	4086.801	24469.01	GD	2.0	1.9		MIT
4084.69	4085.84	24474.7	GD	1.0			MIT
4083.95	4085.10	24479.2	GD	0.7			KN
4083.707	4084.860	24480.64	GD	1.0	0.7		MIT
4080.781	4081.933	24498.20	GD	0.7H			MIT
4080.534	4081.686	24499.68	GD I	1.0	1.0		MIT
4078.709	4079.861	24510.64	GD I	1.3	1.0		MIT
4078.465	4079.616	24512.11	GD	1.2	1.0		MIT
4075.472	4076.623	24530.11	GD	0.7			MIT
4073.771	4074.921	24540.35	GD	2.0	1.7		MIT
4073.212	4074.362	24543.72	GD	1.9	1.9		MIT
4070.399	4071.548	24560.68	GD	1.6	1.0		MIT

AIR WAVELENGTH (ANGSTRMS)	VACUUM WAVELENGTH (ANGSTRMS)	WAVENUMBER (KAYSERS)	LMENT	LOG ARC	INTENSITY SPK	DIS	REF
4070.287	4071.436	24561.36	GD	1.9			MIT
4068.77	4069.92	24570.5	GD	0.7			KN
4068.357	4069.506	24573.01	GD	1.0			MIT
4066.11	4067.26	24586.6	GD	0.5	0.5H		KN
4065.637	4066.785	24589.45	GD	1.7L	1.7H		MIT
4063.43	4064.58	24602.8	GD	1.0	0.7		MIT
4062.594	4063.741	24607.86	GD	1.5	0.8		MIT
4061.858	4063.005	24612.32	GD	0.5			MIT
4061.298	4062.445	24615.72	GD	0.9	1.1		MIT
4061.164	4062.311	24616.53	GD	0.5			MIT
4059.881	4061.028	24624.31	GD	1.7	1.3		MIT
4059.346	4060.492	24627.55	GD	1.0W	0.5		MIT
4058.231	4059.377	24634.32	GD I	2.0	1.8		MIT
4056.04	4057.19	24647.6	GD	0.5			KN
4054.734	4055.879	24655.57	GD I	1.9	1.3		MIT
4053.648	4054.793	24662.17	GD I	2.0	1.6		MIT
4053.302	4054.447	24664.28	GD	2.0	1.9		MIT
4051.619	4052.763	24674.52	GD	0.6			MIT
4050.370	4051.514	24682.13	GD	0.9			MIT
4049.90	4051.04	24685.0	GD	2.0	1.8		M
4049.548	4050.692	24687.14	GD	0.5			KN
4049.441	4050.585	24687.79	GD	1.9	1.3		MIT
4049.196	4050.340	24689.29	GD	0.7			MIT
4048.82	4049.96	24691.6	GD	1.0O			KN
4048.590	4049.734	24692.98	GD	0.9	0.9		MIT
4047.846	4048.989	24697.52	GD	2.2L	1.7		MIT
4047.093	4048.236	24702.11	GD	0.8	0.7		MIT
4046.842	4047.985	24703.65	GD	1.0	1.0		MIT
4045.862	4047.005	24709.63	GD			1.5	KN
4045.010	4046.153	24714.84	GD I	1.3	0.7		MIT
4044.030	4045.172	24720.82	GD	0.5H			KN
4043.710	4044.852	24722.78	GD	0.7	0.7		MIT
4043.09	4044.23	24726.6	GD	0.6			ED
4042.761	4043.903	24728.58	GD	0.7	0.7		MIT
4039.669	4040.810	24747.51	GD	1.0	0.7		MIT
4039.50	4040.64	24748.6	GD	1.0			KN
4037.906	4039.047	24758.32	GD	2.0	1.5		MIT
4037.338	4038.479	24761.80	GD	2.0	1.5		MIT
4036.840	4037.981	24764.85	GD	1.0	0.9		MIT
4035.403	4036.543	24773.67	GD I	0.9	0.7		MIT
4034.38	4035.52	24779.9	GD	0.7			KN
4033.491	4034.631	24785.42	GD	1.0	0.7		MIT
4030.881	4032.020	24801.46	GD	0.9			KN
4028.158	4029.296	24818.23	GD	1.4	0.9		MIT
4027.613	4028.751	24821.59	GD I	1.0	0.9		MIT
4023.74	4024.88	24845.5	GD	0.5H			MIT
4023.350	4024.487	24847.89	GD I	1.3	1.0		MIT
4023.154	4024.291	24849.10	GD	1.3	1.0		MIT
4022.345	4023.482	24854.09	GD	1.2			MIT
4019.732	4020.868	24870.25	GD I	1.2	1.0		MIT
4017.717	4018.853	24882.72	GD	1.0	0.9		MIT
4017.252	4018.387	24885.60	GD	0.7	0.5		MIT
4015.596	4016.731	24895.87	GD I	0.6			MIT
4015.22	4016.35	24898.2	GD	0.7			MIT
4013.92	4015.05	24906.3	GD	1.0			M
4013.817	4014.952	24906.90	GD	1.4	0.5		MIT
4013.430	4014.564	24909.30	GD	0.7			MIT
4009.21	4010.34	24935.5	GD	1.7	0.3		M
4008.922	4010.055	24937.31	GD	1.3	0.5		MIT
4008.331	4009.464	24940.99	GD I	1.2	1.0		MIT
4006.969	4008.102	24949.47	GD I	0.5			MIT
4004.93	4006.06	24962.2	GD	0.9	0.5		M
4003.855	4004.987	24968.87	GD	0.9	0.9		MIT
4003.38	4004.51	24971.8	GD	0.5H			MIT
4001.97	4003.10	24980.6	GD	0.9	0.0		MIT
4001.63	4002.76	24982.7	GD	1.2			KN
4001.24	4002.37	24985.2	GD	1.9	0.5		MIT
4000.16	4001.29	24991.9	GD	1.0H			KN
3997.775	3998.905	25006.84	GD	1.3	1.5		MIT
3996.325	3997.455	25015.92	GD	2.0	2.0		MIT

AIR WAVELENGTH (ANGSTRMS)	VACUUM WAVELENGTH (ANGSTRMS)	WAVENUMBER (KAYSERS)	LMENT		LOG ARC	INTENSITY SPK	DIS	REF
3994.179	3995.308	25029.36	GD		0.9	0.8		MIT
3993.227	3994.356	25035.32	GD		1.2	1.0		MIT
3992.698	3993.827	25038.64	GD		1.2	1.2		MIT
3991.705	3992.834	25044.87	GD		0.7	0.7		MIT
3989.249	3990.377	25060.29	GD		1.0	1.3		MIT
3987.842	3988.970	25069.13	GD		1.7	1.4		MIT
3987.218	3988.346	25073.05	GD		2.0	2.0		MIT
3985.67	3986.80	25082.8	GD		0.3			KN
3984.803	3985.930	25088.25	GD		0.3			MIT
3983.028	3984.155	25099.43	GD		1.6	1.6		MIT
3980.127	3981.253	25117.72	GD		0.3			MIT
3979.749	3980.875	25120.11	GD		0.6 H			MIT
3979.336	3980.462	25122.72	GD	I	2.0	1.5		MIT
3976.864	3977.989	25138.33	GD		0.7			MIT
3975.121	3976.245	25149.35	GD		0.9	1.2		MIT
3974.814	3975.938	25151.29	GD	I	1.0	1.0		MIT
3974.243	3975.367	25154.91	GD		0.7			MIT
3974.066	3975.190	25156.03	GD		2.0	1.9		MIT
3972.710	3973.834	25164.61	GD	I	1.3	1.0		MIT
3972.181	3973.305	25167.97	GD		0.6	0.6		MIT
3971.762	3972.886	25170.62	GD		1.4	1.3		MIT
3971.077	3972.200	25174.96	GD		1.3	1.3		MIT
3970.198	3971.321	25180.54	GD		0.6 H			MIT
3969.261	3970.384	25186.48	GD		2.3			MIT
3969.005	3970.128	25188.11	GD	I	1.6			MIT
3968.35	3969.47	25192.3	GD		1.3			MIT
3966.854	3967.976	25201.76	GD		0.9			MIT
3966.279	3967.401	25205.42	GD		1.4	1.1		MIT
3965.04	3966.16	25213.3	GD		0.7 H			MIT
3963.658	3964.779	25222.08	GD		1.7	1.8		MIT
3962.105	3963.226	25231.97	GD		1.4	1.4		MIT
3960.115	3961.236	25244.65	GD		1.3	1.3		MIT
3959.524	3960.644	25248.42	GD		1.4	0.7		MIT
3959.435	3960.555	25248.98	GD		1.4			MIT
3958.682	3959.802	25253.79	GD		1.2			MIT
3957.681	3958.801	25260.17	GD		2.5 O	2.3		MIT
3953.371	3954.490	25287.71	GD	I	2.0	1.7		MIT
3952.009	3953.127	25296.43	GD		2.0	1.8		MIT
3949.208	3950.326	25314.37	GD		1.0	1.3		MIT
3947.05	3948.17	25328.2	GD		0.6			KN
3945.74	3946.86	25336.6	GD		0.5			MIT
3945.544	3946.661	25337.88	GD		2.3 O	2.2		MIT
3943.631	3944.747	25350.17	GD		1.3	1.3		MIT
3943.246	3944.362	25352.64	GD		1.6	1.6		MIT
3942.642	3943.758	25356.52	GD		1.8	1.5		MIT
3941.807	3942.923	25361.90	GD		0.7	0.7		MIT
3938.982	3940.097	25380.09	GD		1.3	1.6		MIT
3938.153	3939.268	25385.43	GD		1.0	1.3		MIT
3935.393	3936.507	25403.23	GD	I	1.0	1.0		MIT
3934.801	3935.915	25407.05	GD	I	2.0	1.7		MIT
3932.97	3934.08	25418.9	GD		1.0	1.0		MIT
3926.680	3927.792	25459.60	GD		0.7			MIT
3924.25	3925.36	25475.4	GD		1.0			MIT
3923.345	3924.456	25481.24	GD		0.8	0.6		MIT
3923.255	3924.366	25481.82	GD		1.0			MIT
3918.25	3919.36	25514.4	GD		1.0	1.0		MIT
3918.051	3919.161	25515.67	GD		1.0	1.0		MIT
3916.666	3917.775	25524.69	GD		0.3			MIT
3916.589	3917.698	25525.19	GD		2.2 W	2.0		MIT
3913.791	3914.899	25543.44	GD		1.0	1.0		MIT
3912.745	3913.853	25550.27	GD		0.7			MIT
3911.66	3912.77	25557.4	GD		0.7			KN
3910.239	3911.347	25566.64	GD		1.0 H	1.0		MIT
3909.943	3911.050	25568.58	GD		0.7			MIT
3909.247	3910.354	25573.13	GD		0.5			MIT
3908.146	3909.253	25580.33	GD		0.7	0.5 H		MIT
3907.125	3908.232	25587.02	GD		2.0 O	2.0		MIT
3905.651	3906.757	25596.68	GD		1.7	1.7		MIT
3904.289	3905.395	25605.60	GD	I	1.0	1.0		MIT
3902.717	3903.823	25615.92	GD	I	1.4			MIT

AIR WAVELENGTH (ANGSTRMS)	VACUUM WAVELENGTH (ANGSTRMS)	WAVENUMBER (KAYSERS)	LMENT	LOG ARC	INTENSITY SPK	DIS	REF
3902.404	3903.510	25617.97	GD	2.0	1.9		MIT
3898.48	3899.58	25643.8	GD	1.0	1.0		M
3897.348	3898.452	25651.21	GD	0.6			MIT
3896.479	3897.583	25656.93	GD	0.7	0.7		MIT
3895.968	3897.072	25660.29	GD	1.0			MIT
3895.784	3896.888	25661.50	GD	1.4	1.0		MIT
3895.232	3896.336	25665.14	GD	1.3	1.3		MIT
3894.708	3895.811	25668.59	GD	2.2 O	1.9		MIT
3892.75	3893.85	25681.5	GD	0.7 H			KN
3890.884	3891.987	25693.82	GD	1.2	1.2		MIT
3890.421	3891.523	25696.88	GD	0.7			MIT
3888.934	3890.036	25706.70	GD	0.8			MIT
3887.745	3888.847	25714.56	GD	0.7			MIT
3887.177	3888.279	25718.32	GD	1.0	1.0		MIT
3884.71	3885.81	25734.6	GD	1.0 H	1.0		KN
3875.311	3876.409	25797.07	GD	0.6			MIT
3874.474	3875.572	25802.64	GD	0.5 H			MIT
3867.619	3868.715	25848.37	GD	1.0			MIT
3866.981	3868.077	25852.64	GD	1.0			MIT
3863.065	3864.160	25878.84	GD	0.9	1.0		MIT
3855.581	3856.674	25929.08	GD	1.2	0.8		MIT
3852.497	3853.590	25949.83	GD	2.0	0.9		MIT
3850.981	3852.073	25960.05	GD	1.0	0.8		MIT
3850.703	3851.795	25961.92	GD	0.7	0.3		MIT
3844.584	3845.674	26003.24	GD	1.2	1.2		MIT
3843.261	3844.351	26012.19	GD	1.0			MIT
3842.213	3843.303	26019.29	GD	1.0	1.2		MIT
3840.254	3841.343	26032.56	GD	0.3			MIT
3839.642	3840.731	26036.71	GD	1.4			MIT
3836.923	3838.011	26055.16	GD	1.2	1.3		MIT
3836.514	3837.602	26057.94	GD	2.0 J	1.8		MIT
3834.997	3836.085	26068.24	GD	0.9	1.0		MIT
3833.151	3834.238	26080.80	GD	0.3 H			MIT
3829.411	3830.498	26106.27	GD	1.4 H			MIT
3827.365	3828.451	26120.23	GD	1.0	1.0		MIT
3826.056	3827.142	26129.16	GD	0.9	0.9		MIT
3825.013	3826.098	26136.29	GD	0.6	0.9		MIT
3818.760	3819.844	26179.08	GD	1.0	1.2		MIT
3816.640	3817.723	26193.62	GD	1.0	1.0		MIT
3814.755	3815.838	26206.56	GD	0.9	1.0		MIT
3813.981	3815.063	26211.88	GD	2.0 W	1.8		MIT
3811.339	3812.421	26230.05	GD	0.7 H			MIT
3805.526	3806.606	26270.12	GD	1.0	1.0		MIT
3805.107	3806.187	26273.01	GD	1.0	1.0		MIT
3804.735	3805.815	26275.58	GD	0.9			MIT
3803.087	3804.167	26286.97	GD	0.7			MIT
3801.334	3802.413	26299.09	GD	1.2	1.2		MIT
3796.393	3797.471	26333.32	GD	2.2 W	2.2		MIT
3792.396	3793.473	26361.07	GD	1.3	1.4		MIT
3791.135	3792.212	26369.84	GD	1.4	1.4		MIT
3790.621	3791.697	26373.41	GD	0.9	1.2		MIT
3787.56	3788.64	26394.7	GD	1.4	1.4		MIT
3783.811	3784.886	26420.88	GD	0.7	0.7		MIT
3783.062	3784.136	26426.11	GD	1.4	1.3		MIT
3782.284	3783.358	26431.54	GD	1.4	1.7		MIT
3780.492	3781.566	26444.07	GD	0.7			MIT
3771.725	3772.797	26505.54	GD	1.0			MIT
3770.699	3771.770	26512.75	GD	1.7	1.8		MIT
3769.453	3770.524	26521.51	GD	1.7	1.6		MIT
3768.457	3769.528	26528.52	GD	1.0			MIT
3768.405	3769.476	26528.89	GD	1.3	1.3		MIT
3767.036	3768.106	26538.53	GD	1.2	1.0		MIT
3764.634	3765.704	26555.46	GD	1.0	1.0		MIT
3764.202	3765.272	26558.51	GD	0.7	0.9		MIT
3763.328	3764.397	26564.68	GD	0.9	0.9		MIT
3763.004	3764.073	26566.96	GD	1.0	0.7		MIT
3762.212	3763.281	26572.56	GD	1.0	0.5		MIT
3760.931	3762.000	26581.61	GD	1.0	1.0		MIT
3760.715	3761.784	26583.13	GD	1.4	1.4		MIT
3759.011	3760.079	26595.18	GD	1.3	1.2		MIT

AIR WAVELENGTH (ANGSTRMS)	VACUUM WAVELENGTH (ANGSTRMS)	WAVENUMBER (KAYSERS)	LMENT	LOG ARC	INTENSITY SPK	DIS	REF
3758.237	3759.305	26600.66	GD	1.0H	0.9		MIT
3757.947	3759.015	26602.71	GD	1.2	0.7		MIT
3757.73	3758.80	26604.2	GD	1.2	1.3		MIT
3755.256	3756.323	26621.78	GD	0.0	0.3		MIT
3746.423	3747.488	26684.54	GD	0.7	1.0H		MIT
3744.805	3745.870	26696.07	GD	1.4	1.1		MIT
3743.484	3744.548	26705.49	GD	1.4	1.4		MIT
3740.048	3741.111	26730.02	GD	1.7	1.7		MIT
3739.764	3740.827	26732.05	GD	2.0	1.7		MIT
3733.092	3734.153	26779.83	GD	1.3	1.3		MIT
3732.668	3733.729	26782.87	GD	0.7			MIT
3730.857	3731.918	26795.87	GD	2.0O	2.0		MIT
3725.487	3726.547	26834.50	GD	1.7	1.7		MIT
3723.686	3724.745	26847.48	GD	0.7	1.0H		MIT
3722.035	3723.094	26859.38	GD	1.7	1.5		MIT
3719.464	3720.522	26877.95	GD	1.6	1.6		MIT
3717.487	3718.544	26892.24	GD	2.0W	1.7		MIT
3716.369	3717.426	26900.33	GD	2.2W	2.1		MIT
3715.914	3716.971	26903.63	GD	1.0			MIT
3713.583	3714.639	26920.51	GD	2.0O	1.9		MIT
3712.714	3713.770	26926.81	GD	2.3O	2.4		MIT
3700.921	3701.974	27012.61	GD	1.0			MIT
3699.748	3700.801	27021.18	GD	2.3O	2.4		MIT
3697.739	3698.791	27035.86	GD	2.3W	2.3W		MIT
3696.765	3697.817	27042.98	GD	1.0	1.2		MIT
3694.011	3695.062	27063.14	GD	1.4	1.4		MIT
3687.759	3688.809	27109.02	GD	2.3W	2.3		MIT
3686.340	3687.389	27119.46	GD	2.2O	2.3		MIT
3684.124	3685.173	27135.77	GD	2.3O	2.2		MIT
3682.742	3683.790	27145.95	GD	1.0	1.3		MIT
3674.063	3675.109	27210.08	GD	2.0O	1.5		MIT
3671.216	3672.261	27231.18	GD	2.2W	2.0		MIT
3668.31	3669.35	27252.7	GD	0.7	1.0		MIT
3664.621	3665.665	27280.18	GD	2.3W	2.3		MIT
3662.268	3663.311	27297.71	GD	2.3W	2.3		MIT
3661.668	3662.711	27302.18	GD	1.0	1.3		MIT
3658.186	3659.228	27328.17	GD	0.3			MIT
3656.163	3657.205	27343.29	GD	2.3O	2.3		MIT
3654.637	3655.678	27354.71	GD	2.3O	2.3		MIT
3652.560	3653.601	27370.26	GD	1.3	1.4		MIT
3651.119	3652.159	27381.06	GD	1.0	1.2		MIT
3650.968	3652.008	27382.19	GD	1.4	1.5		MIT
3649.441	3650.481	27393.65	GD	0.7	0.7		MIT
3648.475	3649.515	27400.90	GD	1.3			MIT
3646.196	3647.235	27418.03	GD	2.3W	2.2		MIT
3645.634	3646.673	27422.26	GD	1.3	1.3		MIT
3641.38	3642.42	27454.3	GD	1.1	1.1		MIT
3640.189	3641.226	27463.27	GD	1.0	0.9		MIT
3639.053	3640.090	27471.85	GD	1.0	1.2		MIT
3637.742	3638.779	27481.75	GD	0.6			MIT
3634.760	3635.796	27504.29	GD	1.3	1.3		MIT
3629.517	3630.552	27544.02	GD	1.6	1.8		MIT
3626.42	3627.45	27567.6	GD	0.7	0.7		MIT
3625.267	3626.301	27576.31	GD	1.2	1.2		MIT
3624.901	3625.935	27579.10	GD	1.3	1.5		MIT
3622.807	3623.840	27595.04	GD	1.3	1.3		MIT
3620.461	3621.493	27612.92	GD	1.4	1.4		MIT
3614.41	3615.44	27659.1	GD	0.5			M
3613.400	3614.431	27666.88	GD	0.9	0.8		MIT
3611.049	3612.079	27684.89	GD	0.7	0.7H		MIT
3610.766	3611.796	27687.06	GD	1.4	0.8		MIT
3608.758	3609.787	27702.46	GD	2.0	2.1		MIT
3607.066	3608.095	27715.46	GD	1.2	1.3		MIT
3605.62	3606.65	27726.6	GD	1.3	1.2		M
3605.249	3606.277	27729.43	GD	1.2	1.2		MIT
3604.884	3605.912	27732.23	GD	1.7	1.1		MIT
3600.966	3601.993	27762.41	GD	1.5	1.5		MIT
3593.432	3594.457	27820.61	GD	1.2	1.2		MIT
3592.696	3593.721	27826.31	GD	1.7	1.8		MIT
3591.907	3592.932	27832.42	GD	0.9	0.6		MIT

AIR WAVELENGTH (ANGSTRMS)	VACUUM WAVELENGTH (ANGSTRMS)	WAVENUMBER (KAYSERS)	LMENT	LOG ARC	INTENSITY SPK	DIS	REF
3591.429	3592.454	27836.13	GD	1.0	0.7		MIT
3590.472	3591.497	27843.55	GD	1.2	1.2		MIT
3587.191	3588.215	27869.01	GD	1.4	1.2		MIT
3584.964	3585.987	27886.32	GD	2.0	2.0		MIT
3581.916	3582.938	27910.05	GD	1.2	1.2		MIT
3580.629	3581.651	27920.08	GD	0.7			MIT
3579.550	3580.572	27928.50	GD	0.7	0.7		MIT
3578.58	3579.60	27936.1	GD	0.9			MIT
3578.34	3579.36	27937.9	GD	0.7	0.7		MIT
3576.759	3577.780	27950.29	GD	0.7	0.5H		MIT
3574.738	3575.759	27966.09	GD	1.4	1.4		MIT
3571.941	3572.961	27987.99	GD	1.4	1.3		MIT
3564.64	3565.66	28045.3	GD	0.3	0.6		MIT
3564.053	3565.071	28049.93	GD	1.0	1.0		MIT
3558.519	3559.535	28093.55	GD	2.0R	1.7		MIT
3558.189	3559.205	28096.16	GD	1.4	1.4		MIT
3557.062	3558.078	28105.06	GD	1.4	1.4		MIT
3553.744	3554.759	28131.30	GD	1.2			MIT
3549.365	3550.379	28166.01	GD	2.1	2.1		MIT
3545.794	3546.807	28194.37	GD	2.1	2.1		MIT
3544.981	3545.994	28200.84	GD	0.5	0.5		MIT
3542.768	3543.780	28218.45	GD	1.4	1.4		MIT
3537.15	3538.16	28263.3	GD	0.3	0.3		MIT
3528.549	3529.558	28332.16	GD	1.0	1.0		MIT
3524.198	3525.206	28367.14	GD	1.4	1.2		MIT
3513.66	3514.66	28452.2	GD	0.6			MIT
3512.511	3513.516	28461.52	GD	1.5	1.5		MIT
3512.227	3513.231	28463.82	GD	1.5	1.3		MIT
3505.520	3506.523	28518.28	GD	1.8	1.8		MIT
3494.418	3495.418	28608.88	GD	1.8	1.8		MIT
3491.967	3492.966	28628.96	GD	1.7	1.4		MIT
3491.738	3492.737	28630.84	GD	0.6	0.9		MIT
3486.196	3487.194	28676.35	GD	0.5	0.5		MIT
3482.613	3483.610	28705.85	GD	1.2	1.2		MIT
3481.818	3482.815	28712.41	GD	1.9	1.8		MIT
3481.359	3482.356	28716.20	GD	2.2	2.2		MIT
3476.347	3477.342	28757.59	GD	0.7	0.7		MIT
3473.229	3474.223	28783.41	GD	1.7	1.6		MIT
3468.994	3469.987	28818.55	GD	2.2	2.2		MIT
3468.087	3469.080	28826.08	GD	0.5	0.3		MIT
3467.280	3468.273	28832.79	GD	2.0	2.0		MIT
3466.961	3467.954	28835.45	GD	1.2	1.0		MIT
3464.139	3465.131	28858.94	GD	0.6	0.5		MIT
3463.984	3464.976	28860.23	GD	2.0	2.1		MIT
3463.00	3463.99	28868.4	GD	0.8	0.6		ED
3461.956	3462.948	28877.13	GD	0.7	0.7		MIT
3457.053	3458.043	28918.09	GD	0.7	0.7		MIT
3454.900	3455.890	28936.11	GD	1.4	1.3		MIT
3454.149	3455.139	28942.40	GD	1.2	1.2		MIT
3451.241	3452.230	28966.78	GD	1.7	1.6		MIT
3450.385	3451.374	28973.97	GD	2.0	2.0		MIT
3449.628	3450.616	28980.33	GD	1.0	0.9		MIT
3441.793	3442.779	29046.30	GD	0.5	0.3		MIT
3439.990	3440.976	29061.52	GD	1.8	1.7		MIT
3439.785	3440.771	29063.25	GD	1.7	1.7		MIT
3439.212	3440.198	29068.10	GD	1.8	1.5		MIT
3432.997	3433.981	29120.72	GD	1.7	1.6		MIT
3430.982	3431.966	29137.82	GD	0.6	0.3		MIT
3428.469	3429.452	29159.18	GD	0.7	0.5		MIT
3426.327	3427.309	29177.41	GD	0.5	0.5		MIT
3425.918	3426.900	29180.89	GD	1.0	0.9		MIT
3424.601	3425.583	29192.11	GD	1.5			MIT
3423.928	3424.910	29197.85	GD	1.5	1.3		MIT
3422.713	3423.695	29208.21	GD	0.9	0.8		MIT
3422.466	3423.447	29210.32	GD	1.9	2.0		MIT
3418.735	3419.716	29242.20	GD	1.7	1.4		MIT
3416.957	3417.937	29257.41	GD	1.7	1.5		MIT
3413.279	3414.258	29288.94	GD	0.7	0.3		MIT
3412.013	3412.992	29299.81	GD	0.3	0.3		MIT
3409.301	3410.279	29323.11	GD	0.7	0.7		MIT

AIR WAVELENGTH (ANGSTRMS)	VACUUM WAVELENGTH (ANGSTRMS)	WAVENUMBER (KAYSERS)	LMENT	LOG ARC	INTENSITY SPK	INTENSITY DIS	REF
3407.596	3408.574	29337.78	GD	2.0R	2.0		MIT
3402.074	3403.050	29385.40	GD	1.2	1.3		MIT
3401.060	3402.036	29394.16	GD	0.3	0.0		MIT
3399.993	3400.969	29403.39	GD	1.2	1.2		MIT
3395.126	3396.100	29445.54	GD	1.0	1.1		MIT
3393.629	3394.603	29458.52	GD	0.3	0.3		MIT
3392.534	3393.508	29468.03	GD	1.6	1.4		MIT
3380.519	3381.490	29572.76	GD	0.3	0.3		MIT
3379.763	3380.734	29579.38	GD	0.3	0.3		MIT
3374.693	3375.662	29623.82	GD	0.7	0.5		MIT
3369.616	3370.584	29668.45	GD	0.3	0.3		MIT
3364.249	3365.216	29715.78	GD	0.5	0.0		MIT
3362.244	3363.210	29733.50	GD	2.2	2.3		MIT
3360.716	3361.682	29747.02	GD	1.4	1.4		MIT
3358.628	3359.593	29765.51	GD	2.0	2.0		MIT
3350.482	3351.445	29837.87	GD	2.2	2.3		MIT
3350.099	3351.062	29841.29	GD	0.5	0.5H		MIT
3345.991	3346.953	29877.92	GD	1.3	1.2		MIT
3336.183	3337.142	29965.76	GD	1.5	1.5		MIT
3332.137	3333.095	30002.14	GD	1.0	1.1		MIT
3331.388	3332.346	30008.89	GD	2.0	1.9		MIT
3330.341	3331.299	30018.32	GD	1.0	1.0		MIT
3329.349	3330.307	30027.26	GD	0.3	0.3		MIT
3320.437	3321.392	30107.85	GD	0.5	0.5		MIT
3317.229	3318.184	30136.97	GD	0.3			MIT
3315.597	3316.551	30151.80	GD	0.3	0.0		MIT
3313.734	3314.688	30168.75	GD	0.7	0.7		MIT
3294.086	3295.035	30348.69	GD	0.3			MIT
3292.215	3293.163	30365.94	GD	1.0	1.0		MIT
3291.490	3292.438	30372.63	GD	0.3			MIT
3282.257	3283.203	30458.06	GD	0.7R	0.7		MIT
3281.611	3282.557	30464.06	GD	0.7	0.7		MIT
3279.923	3280.868	30479.74	GD	0.3			MIT
3279.527	3280.472	30483.42	GD	0.7	0.9		MIT
3271.683	3272.626	30556.50	GD	0.3			MIT
3268.341	3269.283	30587.75	GD	0.6	0.3		MIT
3267.64	3268.58	30594.3	GD	0.3			MIT
3266.733	3267.675	30602.80	GD	0.5			MIT
3250.234	3251.172	30758.14	GD	0.3	0.6		MIT
3226.326	3227.258	30986.06	GD	0.5	0.5		MIT
3225.44	3226.37	30994.6	GD II	0.3	0.0		MIT
3223.740	3224.671	31010.92	GD	0.7	0.3		MIT
3215.265	3216.194	31092.66	GD	0.7	0.5		MIT
3190.282	3191.204	31336.13	GD	0.7			MIT
3169.474	3170.391	31541.85	GD	0.3	0.3		MIT
3161.379	3162.294	31622.61	GD	1.5	1.5		MIT
3160.660	3161.575	31629.81	GD	0.3	0.0		MIT
3156.539	3157.453	31671.10	GD	1.4	1.4		MIT
3146.884	3147.795	31768.26	GD	0.7	0.7		MIT
3145.521	3146.432	31782.03	GD	1.0	0.9		MIT
3145.006	3145.917	31787.23	GD	1.7	1.5		MIT
3143.141	3144.051	31806.09	GD	0.7	0.5		MIT
3142.889	3143.799	31808.65	GD	0.3	0.0		MIT
3135.039	3135.947	31888.29	GD	0.7	0.3		MIT
3133.859	3134.767	31900.30	GD	1.4	1.4		MIT
3133.092	3134.000	31908.10	GD	0.5	0.0		MIT
3130.814	3131.721	31931.32	GD	0.5	0.0		MIT
3129.965	3130.872	31939.98	GD	0.3	0.0		MIT
3128.569	3129.476	31954.23	GD	0.3	0.3		MIT
3124.264	3125.170	31998.26	GD	0.6			MIT
3124.002	3124.908	32000.95	GD	1.0	0.9		MIT
3120.186	3121.091	32040.08	GD	0.3	0.0		MIT
3119.946	3120.851	32042.55	GD	1.2	1.0		MIT
3119.018	3119.922	32052.08	GD	0.3	0.0		MIT
3118.603	3119.507	32056.34	GD	0.3	0.0		MIT
3113.179	3114.082	32112.19	GD	0.5	0.5		MIT
3108.364	3109.266	32161.93	GD	0.5	0.5		MIT
3102.557	3103.457	32222.13	GD	1.4	1.4		MIT
3101.925	3102.825	32228.69	GD	0.7	0.5		MIT
3101.186	3102.086	32236.37	GD	0.5	0.3		MIT

AIR WAVELENGTH (ANGSTRMS)	VACUUM WAVELENGTH (ANGSTRMS)	WAVENUMBER (KAYSERS)	LMENT	LOG ARC	INTENSITY SPK	INTENSITY DIS	REF
3100.508	3101.408	32243.42	GD	2.0	1.9		MIT
3098.903	3099.802	32260.12	GD	0.7	0.3		MIT
3098.651	3099.550	32262.75	GD	1.2	0.9		MIT
3092.064	3092.962	32331.47	GD	0.5	0.3		MIT
3085.081	3085.977	32404.65	GD	0.3	0.0		MIT
3084.003	3084.899	32415.98	GD	0.3	0.3		MIT
3082.000	3082.895	32437.04	GD	2.0	1.8		MIT
3077.093	3077.987	32488.77	GD	0.7	0.5H		MIT
3076.971	3077.865	32490.05	GD	1.4	1.4		MIT
3072.574	3073.467	32536.55	GD	1.3	1.4		MIT
3068.649	3069.541	32578.16	GD	1.7	1.7		MIT
3059.909	3060.799	32671.21	GD	0.3			MIT
3053.577	3054.465	32738.96	GD	0.7	0.5		MIT
3046.480	3047.366	32815.22	GD	0.3			MIT
3034.059	3034.942	32949.56	GD	2.0	1.8		MIT
3032.850	3033.733	32962.69	GD	2.0	2.0		MIT
3028.979	3029.861	33004.82	GD	0.5	0.5		MIT
3027.612	3028.494	33019.72	GD	2.0	1.8		MIT
3012.196	3013.074	33188.70	GD	0.6	0.3		MIT
3010.139	3011.016	33211.38	GD	2.0	2.0		MIT
3002.869	3003.744	33291.78	GD	1.2	1.3		MIT
2999.056	2999.930	33334.11	GD	1.6	1.5		MIT
2993.038	2993.911	33401.13	GD	0.5	0.5H		MIT
2980.160	2981.030	33545.46	GD	1.4	1.4		MIT
2972.17	2973.04	33635.6	GD II				ME
2969.267	2970.134	33668.52	GD II				ME
2965.420	2966.286	33712.19	GD	0.5	0.5		MIT
2960.893	2961.758	33763.73	GD	0.3	0.3		MIT
2955.60	2956.46	33824.2	GD	0.3	1.4		MIT
2923.371	2924.227	34197.08	GD	0.3	0.3		MIT
2918.44	2919.29	34254.9	GD	0.3	0.5		MIT
2910.526	2911.378	34347.99	GD	0.9	0.6		MIT
2905.313	2906.164	34409.62	GD	1.3	1.3		MIT
2904.71	2905.56	34416.8	GD		1.5		MIT
2885.600	2886.446	34644.68	GD	0.7	1.0		MIT
2881.328	2882.173	34696.04	GD	0.3	0.5		MIT
2862.493	2863.334	34924.33	GD	0.7	0.7		MIT
2841.335	2842.170	35184.38	GD	0.6	0.6		MIT
2840.236	2841.071	35197.99	GD	1.7	1.8		MIT
2836.955	2837.789	35238.70	GD	0.3			MIT
2833.753	2834.587	35278.51	GD	1.3	1.3		MIT
2809.720	2810.548	35580.25	GD	1.8	1.9		MIT
2796.94	2797.76	35742.8	GD	1.8	1.9		MIT
2791.97	2792.79	35806.4	GD	1.4	1.2		MIT
2781.405	2782.226	35942.45	GD	1.5	1.6		MIT
2770.18	2771.00	36088.1	GD	1.2	0.3		MIT
2769.81	2770.63	36092.9	GD	1.3	1.3		MIT
2768.50	2769.32	36110.0	GD	0.5	0.5		MIT
2764.080	2764.897	36167.72	GD	1.4	1.5		MIT
2717.30	2718.11	36790.3	GD		0.7		EX
2692.832	2693.631	37124.61	GD	0.0H	0.3H		MIT
2679.42	2680.22	37310.4	GD	0.5			MIT
2661.48	2662.27	37561.9	GD	0.3	0.0		MIT
2655.56	2656.35	37645.6	GD		0.8		MIT
2638.50	2639.29	37889.0	GD	0.3			MIT
2588.26	2589.03	38624.4	GD	0.0	0.3		MIT
2586.13	2586.90	38656.2	GD	0.3			MIT
2583.65	2584.42	38693.4	GD		0.5		MIT
2573.61	2574.38	38844.3	GD		0.6		MIT
2566.01	2566.78	38959.3	GD		0.5		MIT
2564.51	2565.28	38982.1	GD		0.7		EX
2554.08	2554.85	39141.3	GD		0.5H		MIT
2553.91	2554.68	39143.9	GD		0.6		EX
2551.63	2552.40	39178.9	GD		0.3		MIT
2550.19	2550.96	39201.0	GD	0.3			MIT
2543.71	2544.47	39300.9	GD	0.3			MIT
2524.54	2525.30	39599.3	GD	0.5			MIT
2520.38	2521.14	39664.6	GD		0.3		MIT
2499.03	2499.78	40003.5	GD	0.3			MIT
2496.40	2497.15	40045.6	GD	0.3	0.0		MIT

AIR WAVELENGTH (ANGSTRMS)	VACUUM WAVELENGTH (ANGSTRMS)	WAVENUMBER (KAYSERS)	LMENT	LOG ARC	INTENSITY SPK	DIS	REF
2485.62	2486.37	40219.3	GD	0.3	0.0		MIT
2474.19	2474.94	40405.1	GD	0.3	0.0		MIT
2468.22	2468.97	40502.8	GD	0.0H	0.3		MIT
2424.662	2425.398	41230.34	GD	0.3			MIT
2363.31	2364.03	42300.6	GD	0.0	0.6		MIT
2361.97	2362.69	42324.6	GD		0.5		EX
2354.0	2354.7	42468.	GD	0.3			DS
2339.02	2339.74	42739.9	GD		0.5		EX
2329.40	2330.11	42916.3	GD		0.5		EX
2323.85	2324.56	43018.8	GD		0.5		EX
2321.007	2321.720	43071.52	GD	0.3	0.0		MIT
2307.11	2307.82	43330.9	GD		0.3		MIT
2261.195	2261.895	44210.72	GD	1.4	0.5		MIT
2236.83	2237.52	44692.2	GD		0.7		EX
2224.02	2224.71	44949.6	GD		0.5		EX
7147.0	7149.0	13988.	GE II		0.3		LG
7050.1	7052.0	14180.	GE II		0.3		LG
6967.5	6969.4	14348.	GE II		0.3		LG
6484.32	6486.11	15417.6	GE II		1.2		LG
6336.31	6338.06	15777.7	GE II		1.0		LG
6021.14	6022.81	16603.6	GE II		1.4		LG
5893.46	5895.09	16963.3	GE II		2.0		LG
5178.58	5180.02	19304.9	GE II		2.0		LG
5131.7	5133.1	19481.	GE II		2.0		LG
4824.20	4825.55	20723.0	GE II		1.0		LG
4814.80	4816.15	20763.5	GE II		2.3		LG
4741.937	4743.263	21082.53	GE II		1.7		MIT
4685.837	4687.148	21334.93	GE I	1.3			MIT
4260.802	4262.001	23463.16	GE	1.3	1.0H		MIT
4226.570	4227.760	23653.19	GE I	2.3	1.7		MIT
4179.041	4180.219	23922.19	GE		1.4J		MIT
4074.368	4075.518	24536.76	GE	0.3H			MIT
3269.494	3270.436	30576.96	GE I	2.5	2.5		MIT
3255.343	3256.282	30709.87	GE		2.0V		MIT
3124.817	3125.723	31992.60	GE I	2.3	1.9		MIT
3067.132	3068.023	32594.28	GE I	0.3			MIT
3067.007	3067.898	32595.60	GE	1.8	1.6		MIT
3039.064	3039.948	32895.29	GE I	3.0	3.0		MIT
2845.48	2846.32	35133.1	GE II		1.1		MIT
2831.77	2832.60	35303.2	GE II		0.9		MIT
2829.011	2829.843	35337.65	GE I	0.9	0.7		MIT
2793.938	2794.762	35781.23	GE I	0.9	0.7		M
2754.592	2755.406	36292.29	GE	1.5	1.3		MIT
2740.431	2741.242	36479.82	GE	1.0	1.0		M
2709.626	2710.429	36894.52	GE	1.5	1.3		MIT
2691.344	2692.143	37145.13	GE	1.4	1.2		MIT
2651.575	2652.364	37702.21	GE I	1.5	1.3		MIT
2651.178	2651.967	37707.86	GE I	1.6	1.3		MIT
2644.190	2644.978	37807.50	GE	0.6	0.5		MIT
2592.537	2593.312	38560.72	GE	1.3	1.2		MIT
2589.188	2589.963	38610.60	GE I	0.8	0.8		MIT
2556.305	2557.072	39107.23	GE	1.3	0.3		MIT
2533.229	2533.990	39463.45	GE I	0.7	0.7		MIT
2497.963	2498.716	40020.55	GE I	0.9	1.0		MIT
2436.39	2437.13	41031.9	GE I	0.3	0.0		MIT
2417.366	2418.101	41354.77	GE	1.0	1.1		MIT
2397.89	2398.62	41690.6	GE	0.3	0.0		MIT
2394.10	2394.83	41756.6	GE I	0.3	0.0		MIT
2389.470	2390.198	41837.53	GE I	0.3	0.0		MIT
2379.14	2379.87	42019.2	GE I	0.5	0.6		MIT
2359.21	2359.93	42374.1	GE	0.3			MIT
2338.67	2339.39	42746.2	GE	0.3	0.0		MIT
2327.90	2328.61	42944.0	GE I	0.5	0.7		MIT
2314.20	2314.91	43198.2	GE	0.5	0.5		MIT
2256.00	2256.70	44312.5	GE I	0.3	0.0		MIT
2247.04	2247.74	44489.2	GE	0.3			MIT
2242.47	2243.17	44579.9	GE I	0.3			MIT
2222.73	2223.42	44975.7	GE	0.3			GT
2220.37	2221.06	45023.5	GE I	0.5			MIT
2198.701	2199.388	45467.20	GE	1.0G	0.3H		MIT

AIR WAVELENGTH (ANGSTRMS)	VACUUM WAVELENGTH (ANGSTRMS)	WAVENUMBER (KAYSERS)	LMENT	LOG ARC	INTENSITY SPK	DIS	REF
2186.443	2187.127	45722.08	GE	0.3			MIT
2124.75	2125.42	47049.5	GE	0.5			MIT
2123.82	2124.49	47070.1	GE I	0.3			MIT
2105.80	2106.47	47472.8	GE	0.5			MIT
2094.23	2094.90	47735.1	GE	0.9G			MIT
2086.00	2086.66	47923.4	GE I	0.6			MIT
2068.65	2069.31	48325.3	GE	0.7R	2.3G		MIT
2067.63	2068.29	48349.1	GE	0.3			GT
2065.20	2065.86	48406.0	GE	0.6G	1.7		MIT
2058.59	2059.25	48561.4	GE I	0.3			GT
2057.23	2057.89	48593.5	GE	0.5R			MIT
2054.45	2055.11	48659.2	GE	0.6			MIT
2043.76	2044.42	48913.7	GE	0.6R	1.8H		MIT
2041.69	2042.35	48963.3	GE	0.7R	2.2R		MIT
2019.05	2019.70	49512.3	GE	0.3G			MIT
2011.31	2011.96	49702.8	GE I	0.3G			GT
9546.2	9548.8	10473.	H			0.7	PK
9229.7	9232.2	10832.	H			0.6	PK
9015.3	9017.8	11089.	H			0.5	PK
8863.4	8865.8	11279.	H			0.3	PK
6562.849	6564.662	15233.08	H			3.3	MS
6562.725	6564.538	15233.36	H			3.0	MS
4861.327	4862.685	20564.77	H			2.7	M
4340.465	4341.685	23032.53	H			2.3	M
4101.735	4102.893	24373.05	H I			2.0	M
3970.074	3971.197	25181.32	H I			1.9	M
3889.055	3890.157	25705.90	H I			1.8	M
3835.397	3836.485	26065.53	H I			1.6	RK
3797.910	3798.988	26322.80	H I			1.3	RK
3770.634	3771.705	26513.21	H I			1.2	RK
3750.152	3751.218	26658.01	H I			1.0	RK
3734.372	3735.434	26770.65	H I			0.9	RK
3721.948	3723.007	26860.01	H I			0.8	RK
3711.98	3713.04	26932.1	H I			0.7	RK
3703.86	3704.91	26991.2	H I			0.6	RK
3697.15	3698.20	27040.2	H I			0.5	RK
3691.55	3692.60	27081.2	H I			0.3	RK
9702.66	9705.32	10303.6	HE I			1.0	ME
9625.80	9628.44	10385.9	HE I			0.5	ME
9603.50	9606.13	10410.0	HE I			0.8	ME
9529.27	9531.88	10491.1	HE I			0.6	ME
9526.17	9528.78	10494.5	HE I			1.0	ME
9516.70	9519.31	10505.0	HE I			1.5	ME
9463.66	9466.26	10563.8	HE I			1.8	ME
9210.28	9212.81	10854.4	HE I			0.8	ME
9074.6	9077.1	11017.	HE I			0.3	ME
9063.40	9065.89	11030.4	HE I			0.8	ME
8996.7	8999.2	11112.	HE I			0.3	ME
8914.7	8917.1	11214.	HE I			0.3	ME
8361.7	8364.0	11956.	HE I			0.5	ME
7816.16	7818.31	12790.5	HE I			1.1	PS
7281.349	7283.355	13729.94	HE I			1.5	IMR
7065.705	7067.653	14148.97	HE I			1.0	MIT
7065.188	7067.136	14150.00	HE I			1.8	IMR
6678.149	6679.993	14970.08	HE I			2.0	IMR
6560.13	6561.94	15239.4	HE II			2.0	PS
5875.989	5877.618	17013.70	HE I			1.0	MIT
5875.618	5877.246	17014.77	HE I			3.0	IMR
5411.55	5413.05	18473.9	HE II			1.7	PS
5047.736	5049.143	19805.34	HE I			1.2	IMR
5015.675	5017.074	19931.94	HE I			2.0	IMR
4921.929	4923.303	20311.57	HE I			1.7	I
4859.34	4860.70	20573.2	HE II			0.8	PS
4713.373	4714.692	21210.29	HE I			0.8	PS
4713.143	4714.462	21211.33	HE I			1.6	IMR
4685.75	4687.06	21335.3	HE II			2.5	PS
4541.61	4542.88	22012.4	HE			0.7	PS
4471.681	4472.936	22356.68	HE I			0.8	PS
4471.477	4472.732	22357.70	HE I			2.0	IMR
4437.549	4438.795	22528.64	HE I			1.0	IMR

AIR WAVELENGTH (ANGSTRMS)	VACUUM WAVELENGTH (ANGSTRMS)	WAVENUMBER (KAYSERS)	LMENT	LOG ARC	INTENSITY SPK	INTENSITY DIS	REF
4387.928	4389.161	22783.40	HE I			1.5	IMR
4338.69	4339.91	23041.9	HE II			0.5	PS
4199.85	4201.03	23803.7	HE II			0.3	PS
4168.967	4170.142	23980.00	HE I			0.8	IOF
4143.759	4144.928	24125.87	HE I			1.2	IMR
4120.989	4122.152	24259.18	HE I			0.7	PS
4120.812	4121.975	24260.22	HE I			1.4	IMR
4100.00	4101.16	24383.4	HE II			0.3	PS
4026.363	4027.501	24829.29	HE I			0.7	PS
4026.189	4027.327	24830.37	HE I			1.8	IMR
4023.973	4025.110	24844.04	HE I			0.7	PS
4009.270	4010.403	24935.15	HE I			1.0	PS
3964.727	3965.849	25215.28	HE I			1.7	IMR
3935.914	3937.028	25399.87	HE I			0.6	PS
3926.530	3927.642	25460.57	HE I			0.8	PS
3888.646	3889.748	25708.61	HE I			3.0	IMR
3878.183	3879.282	25777.97	HE I			0.5	PS
3871.819	3872.917	25820.33	HE I			0.7	PS
3867.630	3868.726	25848.30	HE I			0.6	PS
3867.477	3868.573	25849.32	HE I			1.2	PS
3838.094	3839.183	26047.21	HE I			0.3	PS
3833.574	3834.662	26077.92	HE I			0.6	PS
3819.761	3820.845	26172.22	HE I			0.6	PS
3819.606	3820.690	26173.28	HE I			1.7	IMR
3805.765	3806.845	26268.47	HE I			0.5	PS
3784.886	3785.961	26413.37	HE I			0.3	PS
3768.81	3769.88	26526.0	HE I			0.3	PS
3732.987	3734.048	26780.58	HE I			0.5	PS
3732.861	3733.922	26781.49	HE I			1.0	PS
3705.140	3706.194	26981.86	HE I			0.5	PS
3705.003	3706.057	26982.85	HE I			1.5	IMR
3652.104	3653.145	27373.68	HE I			0.3	PS
3651.981	3653.022	27374.60	HE I			0.8	PS
3634.367	3635.403	27507.27	HE I			0.3	PS
3634.235	3635.271	27508.27	HE I			1.2	PS
3613.641	3614.672	27665.03	HE I			1.5	IMR
3599.442	3600.469	27774.16	HE I			0.3	PS
3599.304	3600.331	27775.23	HE I			0.7	PS
3587.396	3588.420	27867.42	HE I			0.3	PS
3587.252	3588.276	27868.54	HE I			1.0	PS
3562.950	3563.968	28058.62	HE I			0.6	PS
3554.394	3555.409	28126.16	HE I			0.8	PS
3536.820	3537.831	28265.91	HE I			0.5	PS
3530.487	3531.496	28316.61	HE I			0.7	PS
3517.327	3518.333	28422.55	HE I			0.3	PS
3512.511	3513.516	28461.52	HE I			0.6	PS
3502.381	3503.383	28543.84	HE I			0.3	PS
3498.641	3499.642	28574.35	HE I			0.5	PS
3487.721	3488.719	28663.81	HE I			0.3	PS
3478.97	3479.97	28735.9	HE I			0.3	PS
3447.590	3448.578	28997.46	HE I			1.2	PS
3354.550	3355.514	29801.69	HE I			1.0	PS
3296.786	3297.735	30323.84	HE I			0.8	PS
3258.275	3259.215	30682.24	HE I			0.7	PS
3231.266	3232.199	30938.69	HE I			0.5	PS
3211.496	3212.424	31129.14	HE I			0.3	PS
3203.14	3204.07	31210.4	HE II			2.0	PS
3196.760	3197.684	31272.63	HE I			0.3	PS
3187.743	3188.665	31361.09	HE I			2.3	IMR
2945.104	2945.965	33944.73	HE I			2.0	IMR
2829.073	2829.905	35336.87	HE I			1.6	PS
2763.800	2764.616	36171.38	HE I			1.3	PS
2733.32	2734.13	36574.7	HE II			2.0	PS
2723.191	2723.998	36710.75	HE I			1.0	PS
2696.119	2696.919	37079.35	HE I			0.8	PS
2677.135	2677.931	37342.27	HE I			0.7	PS
2663.271	2664.063	37536.65	HE I			0.6	PS
2652.848	2653.638	37684.12	HE I			0.5	PS
2644.802	2645.590	37798.75	HE I			0.3	PS
2511.22	2511.98	39809.3	HE II			1.7	PS

AIR WAVELENGTH (ANGSTRMS)	VACUUM WAVELENGTH (ANGSTRMS)	WAVENUMBER (KAYSERS)	LMENT	LOG ARC	INTENSITY SPK	INTENSITY DIS	REF
2385.42	2386.15	41908.6	HE II			1.5	PS
2306.22	2306.93	43347.7	HE II			1.3	PS
2252.71	2253.41	44377.2	HE II			1.0	PS
2214.69	2215.38	45139.0	HE II			0.7	PS
2186.62	2187.30	45718.4	HE II			0.3	PS
9742.28	9744.95	10261.7	HF II	1.0			ME
9453.55	9456.14	10575.1	HF II	0.3			ME
9250.27	9252.81	10807.5	HF	0.5			ME
9182.07	9184.59	10887.8	HF	0.3			ME
9004.73	9007.20	11102.2	HF	0.5			ME
8715.62	8718.01	11470.5	HF	0.3			ME
8711.24	8713.63	11476.3	HF	0.8			ME
8640.06	8642.43	11570.8	HF	1.2			ME
8603.24	8605.60	11620.3	HF	0.3			ME
8584.12	8586.48	11646.2	HF	0.7H			ME
8581.88	8584.24	11649.3	HF II	0.0	0.7		ME
8569.0	8571.4	11667.	HF	0.5D			ME
8546.48	8548.83	11697.5	HF	1.3			ME
8460.01	8462.33	11817.1	HF	1.0			ME
8382.98	8385.28	11925.7	HF	0.3			ME
8361.78	8364.08	11955.9	HF	0.6			ME
8344.25	8346.54	11981.0	HF	0.9			ME
8338.12	8340.41	11989.8	HF	0.3			ME
8305.91	8308.19	12036.3	HF II	0.3	0.6		ME
8276.95	8279.23	12078.4	HF	1.3			ME
8248.81	8251.08	12119.6	HF	0.5			ME
8236.3	8238.6	12138.	HF II	0.7	1.0		ME
8216.32	8218.58	12167.6	HF	0.5			ME
8204.58	8206.84	12185.0	HF	1.3			ME
8194.78	8197.03	12199.5	HF	0.9			ME
8173.89	8176.14	12230.7	HF	0.7			ME
8080.32	8082.54	12372.3	HF	0.6			ME
8059.10	8061.32	12404.9	HF	0.5			ME
8056.52	8058.74	12408.9	HF	0.6			ME
8010.58	8012.78	12480.1	HF	0.3			ME
7994.76	7996.96	12504.7	HF I	3.7	2.9		ME
7983.66	7985.86	12522.1	HF II	0.5	0.7		ME
7938.06	7940.24	12594.1	HF	0.3			ME
7920.75	7922.93	12621.6	HF I	3.3	2.6		ME
7861.22	7863.38	12717.2	HF II	0.5	0.9		ME
7845.35	7847.51	12742.9	HF	1.3			ME
7814.57	7816.72	12793.1	HF I	2.6	1.9		ME
7801.53	7803.68	12814.5	HF II	0.3	0.8		ME
7790.90	7793.04	12832.0	HF	0.7			ME
7757.89	7760.02	12886.6	HF II	0.7	1.2		ME
7740.17	7742.30	12916.1	HF I	3.3	1.6		ME
7720.72	7722.84	12948.6	HF II	0.0	0.3		ME
7663.09	7665.20	13046.0	HF II	0.3	1.5		ME
7645.64	7647.74	13075.7	HF	0.3			ME
7624.387	7626.486	13112.20	HF I	3.7	3.0		ME
7608.59	7610.68	13139.4	HF	0.3			ME
7592.96	7595.05	13166.5	HF	0.3			ME
7576.95	7579.04	13194.3	HF	0.3			ME
7565.35	7567.43	13214.5	HF II		0.3		ME
7564.22	7566.30	13216.5	HF	0.3			ME
7562.93	7565.01	13218.7	HF I	3.2	2.5		ME
7561.08	7563.16	13222.0	HF II	0.0	1.0		ME
7556.37	7558.45	13230.2	HF	0.3			ME
7514.88	7516.95	13303.3	HF II		0.3		ME
7484.56	7486.62	13357.2	HF	0.5			ME
7463.86	7465.92	13394.2	HF	1.0			ME
7442.33	7444.38	13432.9	HF	0.0	0.5		ME
7437.56	7439.61	13441.6	HF	0.5	1.3		ME
7423.69	7425.73	13466.7	HF	0.7	1.2		ME
7398.96	7401.00	13511.7	HF II		1.0		ME
7390.70	7392.74	13526.8	HF	0.5	1.0		ME
7365.28	7367.31	13573.5	HF	1.0			ME
7356.10	7358.13	13590.4	HF	1.0			ME
7328.64	7330.66	13641.3	HF II	0.5	1.5		ME
7321.76	7323.78	13654.2	HF	0.5	1.0		ME

AIR WAVELENGTH (ANGSTRMS)	VACUUM WAVELENGTH (ANGSTRMS)	WAVENUMBER (KAYSERS)	LMENT		LOG ARC	INTENSITY SPK	DIS	REF
7320.06	7322.08	13657.3	HF	I	3.0	2.3		ME
7292.33	7294.34	13709.3	HF	II	0.5	0.5		ME
7278.72	7280.73	13734.9	HF	II		0.8		ME
7277.67	7279.68	13736.9	HF	II	0.7	1.7		ME
7262.62	7264.62	13765.3	HF		0.5			ME
7240.87	7242.87	13806.7	HF	I	3.7	2.8		ME
7237.101	7239.095	13813.88	HF	I	3.9	3.0		ME
7189.30	7191.28	13905.7	HF	II	0.0	0.5		ME
7164.52	7166.49	13953.8	HF	II	0.0	0.7		ME
7131.82	7133.79	14017.8	HF	I	3.8	3.0		ME
7119.52	7121.48	14042.0	HF		1.2	1.7		ME
7115.65	7117.61	14049.7	HF			0.5		ME
7100.54	7102.50	14079.6	HF		0.5	1.2		ME
7094.40	7096.36	14091.7	HF		0.5	1.2		ME
7063.82	7065.77	14152.7	HF	I	3.5	2.6		ME
7062.87	7064.82	14154.7	HF		0.5	1.2		ME
7061.90	7063.85	14156.6	HF		1.0	1.5		ME
7052.62	7054.56	14175.2	HF			0.5		ME
7035.13	7037.07	14210.5	HF		0.5	1.0		ME
7030.33	7032.27	14220.2	HF	II	1.5	2.2		ME
7021.23	7023.17	14238.6	HF	II	0.5	1.5		ME
7019.25	7021.19	14242.6	HF			1.0		ME
7016.99	7018.93	14247.2	HF	II		0.8		ME
7010.68	7012.61	14260.0	HF	II	0.7	1.0		ME
6997.83	6999.76	14286.2	HF	II	0.3	1.3		ME
6981.60	6983.53	14319.4	HF		0.7	1.0		ME
6980.91	6982.84	14320.8	HF	II	2.0	2.3		ME
6979.59	6981.52	14323.5	HF		0.5	1.3		ME
6976.18	6978.10	14330.5	HF			0.3		ME
6970.44	6972.36	14342.3	HF		0.7	1.0		ME
6965.80	6967.72	14351.9	HF		0.0	0.5		ME
6935.16	6937.07	14415.3	HF	II	0.7	1.7		ME
6926.19	6928.10	14434.0	HF		0.7	1.2		ME
6911.40	6913.31	14464.9	HF		1.2	1.7		ME
6874.95	6876.85	14541.6	HF		0.7	1.0	D	ME
6858.70	6860.59	14576.0	HF		1.2	1.7		ME
6857.03	6858.92	14579.6	HF	II	0.7	1.0		ME
6855.29	6857.18	14583.2	HF	II	0.8	1.7		ME
6850.07	6851.96	14594.4	HF		1.3	1.8		ME
6833.77	6835.66	14629.2	HF		0.7	1.0		ME
6826.56	6828.44	14644.6	HF		1.0	1.4		ME
6818.94	6820.82	14661.0	HF		2.0	2.3		ME
6789.28	6791.15	14725.0	HF	I	3.0	2.0		ME
6773.07	6774.94	14760.3	HF		0.7	1.2		ME
6769.95	6771.82	14767.1	HF	II	1.0	1.3		ME
6767.91	6769.78	14771.5	HF	II	0.3	0.3		ME
6754.61	6756.47	14800.6	HF	II	1.8	2.0		ME
6719.40	6721.26	14878.2	HF	II	0.3	1.7		ME
6716.00	6717.85	14885.7	HF		0.0	0.7		ME
6713.5	6715.4	14891.	HF		1.0	1.3		ME
6709.41	6711.26	14900.3	HF		0.0	0.8		ME
6708.33	6710.18	14902.7	HF			0.5		ME
6691.67	6693.52	14939.8	HF		0.3	0.5		ME
6671.29	6673.13	14985.5	HF		0.0	0.5		ME
6669.31	6671.15	14989.9	HF		0.0	0.5		ME
6659.40	6661.24	15012.2	HF		0.0	0.7		ME
6647.06	6648.90	15040.1	HF	II	1.5	2.0		ME
6644.60	6646.43	15045.7	HF	II	2.0	2.3		ME
6635.38	6637.21	15066.6	HF		0.3	0.6		ME
6616.14	6617.97	15110.4	HF		0.3	0.6		ME
6609.20	6611.03	15126.2	HF	II	0.0	0.9		ME
6599.76	6601.58	15147.9	HF	II	0.0	0.6		ME
6587.23	6589.05	15176.7	HF		0.7	1.0		ME
6584.53	6586.35	15182.9	HF	II	0.6	1.6		ME
6581.15	6582.97	15190.7	HF		0.3	0.3		ME
6567.39	6569.20	15222.5	HF	II	0.8	1.8		ME
6565.76	6567.57	15226.3	HF	II		0.5		ME
6562.86	6564.67	15233.1	HF	II	0.3	1.0		ME
6557.91	6559.72	15244.6	HF	II	1.0	2.0		ME
6557.00	6558.81	15246.7	HF		0.3	0.3		ME

AIR WAVELENGTH (ANGSTRMS)	VACUUM WAVELENGTH (ANGSTRMS)	WAVENUMBER (KAYSERS)	LMENT		LOG ARC	INTENSITY SPK	DIS	REF
6556.50	6558.31	15247.8	HF		0.7	1.0		ME
6555.98	6557.79	15249.0	HF		0.0	0.5		ME
6552.92	6554.73	15256.2	HF		0.0	0.5		ME
6550.01	6551.82	15262.9	HF	II	0.0	1.0		ME
6548.72	6550.53	15265.9	HF		0.0	1.0		ME
6548.24	6550.05	15267.1	HF	II	0.3	0.5		ME
6546.02	6547.83	15272.2	HF		0.3	0.7		ME
6542.80	6544.61	15279.8	HF	II	0.5	1.7		ME
6541.01	6542.82	15283.9	HF		0.3	0.7		ME
6537.06	6538.87	15293.2	HF		0.3	0.7		ME
6536.56	6538.37	15294.3	HF		0.3			ME
6531.66	6533.46	15305.8	HF	II	0.3	1.5		ME
6526.20	6528.00	15318.6	HF	II		0.7		ME
6523.94	6525.74	15323.9	HF		0.0	0.5		ME
6512.61	6514.41	15350.6	HF	II	0.0	1.0		ME
6511.62	6513.42	15352.9	HF	II		0.8		ME
6494.58	6496.37	15393.2	HF		0.0	0.7		ME
6486.5	6488.3	15412.	HF		0.3			ME
6482.57	6484.36	15421.7	HF		0.0	0.3		ME
6478.7	6480.5	15431.	HF		0.3			ME
6473.89	6475.68	15442.4	HF	II	0.5	1.3		ME
6467.90	6469.69	15456.7	HF	II	0.3	0.3		ME
6462.30	6464.09	15470.1	HF	II	0.0	0.6		ME
6456.97	6458.75	15482.9	HF		0.6	0.9		ME
6455.85	6457.63	15485.6	HF	II	0.3	1.3		ME
6449.9	6451.7	15500.	HF		0.3			ME
6448.43	6450.21	15503.4	HF		0.0	0.3		ME
6444.98	6446.76	15511.7	HF		0.0	0.3		ME
6431.8	6433.6	15543.	HF		0.3			ME
6429.49	6431.27	15549.0	HF		0.0	0.3		ME
6409.52	6411.29	15597.5	HF		0.5	0.7		ME
6390.99	6392.76	15642.7	HF		0.0	0.3		ME
6390.30	6392.07	15644.4	HF		0.0	0.3		ME
6388.91	6390.68	15647.8	HF		0.0	0.3		ME
6388.20	6389.97	15649.5	HF		0.0	0.3		ME
6386.23	6388.00	15654.4	HF		1.2	1.3		ME
6380.19	6381.95	15669.2	HF		0.5	0.8		ME
6376.20	6377.96	15679.0	HF		0.0	0.5		ME
6374.60	6376.36	15682.9	HF		0.0	0.3		ME
6359.83	6361.59	15719.3	HF		0.0	0.3		ME
6357.36	6359.12	15725.4	HF		0.0	0.3		ME
6347.62	6349.38	15749.6	HF		0.0	0.3		ME
6343.77	6345.52	15759.1	HF	II	0.0	0.6		ME
6340.71	6342.46	15766.7	HF		0.0	0.3		ME
6338.10	6339.85	15773.2	HF		0.6	0.8		ME
6334.55	6336.30	15782.1	HF		0.0	0.3		ME
6328.86	6330.61	15796.3	HF		0.0	0.3		ME
6318.33	6320.08	15822.6	HF		0.5	0.7		ME
6315.94	6317.69	15828.6	HF	II	0.0	0.7		ME
6313.46	6315.21	15834.8	HF		0.3	0.5		ME
6311.85	6313.60	15838.8	HF		0.5	0.8		ME
6310.75	6312.50	15841.6	HF		0.0	0.3		ME
6308.67	6310.41	15846.8	HF		0.0	0.3		ME
6306.17	6307.91	15853.1	HF	II	0.0	0.7		ME
6299.54	6301.28	15869.8	HF		0.3	0.7		ME
6291.48	6293.22	15890.1	HF		0.3	0.6		ME
6279.84	6281.58	15919.6	HF	II	1.2	1.3		ME
6271.05	6272.78	15941.9	HF	II	0.0	0.5		ME
6258.81	6260.54	15973.1	HF		1.0	1.3		ME
6257.00	6258.73	15977.7	HF		1.0	1.5		ME
6248.95	6250.68	15998.3	HF	II	1.9	2.0		ME
6238.58	6240.31	16024.9	HF		0.8	0.9		ME
6230.84	6232.56	16044.8	HF	II	1.0	1.3		ME
6229.66	6231.38	16047.8	HF	II	0.3	0.5		ME
6222.81	6224.53	16065.5	HF	II	0.8	1.0		ME
6216.82	6218.54	16080.9	HF		0.3	0.8		ME
6211.94	6213.66	16093.6	HF		0.3			ME
6210.70	6212.42	16096.8	HF		0.8	0.9		ME
6209.42	6211.14	16100.1	HF		0.3			ME
6206.95	6208.67	16106.5	HF	II	0.0	0.5		ME

AIR WAVELENGTH (ANGSTRMS)	VACUUM WAVELENGTH (ANGSTRMS)	WAVENUMBER (KAYSERS)	LMENT	LOG ARC	INTENSITY SPK DIS	REF
6202.85	6204.57	16117.2	HF II	0.3	0.3	ME
6192.50	6194.21	16144.1	HF	0.3	0.6	ME
6185.13	6186.84	16163.3	HF	1.0	1.1	ME
6160.70	6162.40	16227.4	HF		0.6	ME
6158.72	6160.42	16232.7	HF II	0.0	0.3	ME
6156.25	6157.95	16239.2	HF II	0.5	0.5D	ME
6135.10	6136.80	16295.1	HF II	1.0	1.3	ME
6118.16	6119.85	16340.3	HF	0.3	0.0	ME
6098.67	6100.36	16392.5	HF	0.7	0.3	ME
6097.47	6099.16	16395.7	HF II		0.3	ME
6093.15	6094.84	16407.3	HF II		0.3	ME
6054.17	6055.85	16513.0	HF	0.7		ME
6048.00	6049.67	16529.8	HF II	0.5	0.7	ME
6043.19	6044.86	16543.0	HF	0.7	0.0	ME
6041.44	6043.11	16547.8	HF II	0.3	0.8	ME
6031.96	6033.63	16573.8	HF II	0.0	0.5	ME
6027.57	6029.24	16585.8	HF II	1.0	1.3	ME
6016.79	6018.46	16615.6	HF	0.6	0.0	ME
6006.39	6008.05	16644.3	HF II	0.3	0.3	ME
5992.96	5994.62	16681.6	HF	0.3	0.0	ME
5978.66	5980.32	16721.5	HF	0.8	0.0	ME
5974.72	5976.38	16732.6	HF	0.3		ME
5974.28	5975.93	16733.8	HF	1.0	0.3	ME
5969.38	5971.03	16747.5	HF II	0.6	0.7	ME
5933.69	5935.33	16848.2	HF	0.8	0.0	ME
5929.35	5930.99	16860.6	HF II	0.5	0.7	ME
5926.47	5928.11	16868.8	HF	0.3		ME
5902.94	5904.58	16936.0	HF	1.2	0.5	ME
5896.61	5898.24	16954.2	HF	0.3		ME
5890.45	5892.08	16971.9	HF	0.9		ME
5885.58	5887.21	16986.0	HF II		0.3	ME
5883.66	5885.29	16991.5	HF	0.5	0.0	ME
5858.35	5859.97	17064.9	HF	0.3		ME
5847.77	5849.39	17095.8	HF	0.5		ME
5845.87	5847.49	17101.4	HF	0.6		ME
5842.23	5843.85	17112.0	HF II	1.7	1.9	ME
5838.87	5840.49	17121.9	HF	0.3		ME
5817.47	5819.08	17184.8	HF	0.5	0.0	ME
5809.50	5811.11	17208.4	HF II	1.3	1.5	ME
5808.42	5810.03	17211.6	HF	0.3		ME
5801.71	5803.32	17231.5	HF II	0.8	1.2	ME
5799.78	5801.39	17237.2	HF	0.3		ME
5788.08	5789.69	17272.1	HF	0.3		ME
5767.18	5768.78	17334.7	HF II	1.2	1.5	ME
5765.99	5767.59	17338.3	HF	0.3		ME
5765.38	5766.98	17340.1	HF	0.3	0.3	ME
5752.53	5754.13	17378.8	HF	0.5	0.7	ME
5748.72	5750.31	17390.4	HF	0.6	0.0	ME
5742.52	5744.11	17409.1	HF	0.3	0.3	ME
5734.54	5736.13	17433.4	HF	0.5	0.0	ME
5719.18	5720.77	17480.2	HF	1.6	1.0	ME
5713.28	5714.87	17498.2	HF	0.6	0.0	ME
5673.58	5675.15	17620.7	HF II	0.6	1.0	ME
5658.83	5660.40	17666.6	HF II	0.0	0.5	ME
5654.64	5656.21	17679.7	HF	0.6	0.0	ME
5650.83	5652.40	17691.6	HF	0.5	0.0	ME
5631.36	5632.92	17752.8	HF II	0.0	0.6	ME
5628.27	5629.83	17762.5	HF	0.3		ME
5614.01	5615.57	17807.6	HF	0.5		ME
5613.27	5614.83	17810.0	HF	1.0	0.5	ME
5600.77	5602.32	17849.7	HF	0.3		ME
5590.73	5592.28	17881.8	HF II	0.5	0.7	ME
5575.86	5577.41	17929.5	HF	0.8	0.7	ME
5565.56	5567.11	17962.7	HF II	0.3	0.7	ME
5552.12	5553.66	18006.1	HF	1.6	0.7	ME
5550.60	5552.14	18011.1	HF	1.5	0.7	ME
5538.26	5539.80	18051.2	HF	0.5		ME
5538.02	5539.56	18052.0	HF	0.6	0.0	ME
5530.29	5531.83	18077.2	HF	0.3		ME
5524.35	5525.88	18096.6	HF II	1.6	1.7	ME

AIR WAVELENGTH (ANGSTRMS)	VACUUM WAVELENGTH (ANGSTRMS)	WAVENUMBER (KAYSERS)	LMENT	LOG ARC	INTENSITY SPK DIS	REF
5514.96	5516.49	18127.5	HF II	0.0	0.5	ME
5510.12	5511.65	18143.4	HF	0.5		ME
5497.30	5498.83	18185.7	HF	0.5	0.0	ME
5493.22	5494.75	18199.2	HF	0.0	0.8	ME
5463.38	5464.90	18298.6	HF II	1.0	1.0	ME
5452.92	5454.44	18333.7	HF	0.9	0.3	ME
5444.07	5445.58	18363.5	HF II	1.3	1.5	ME
5438.74	5440.25	18381.5	HF	0.7	0.3	ME
5438.59	5440.10	18382.0	HF	0.3		ME
5435.78	5437.29	18391.5	HF	0.5		ME
5430.06	5431.57	18410.9	HF	0.0	0.6	ME
5420.44	5421.95	18443.6	HF II	0.3	0.5	ME
5404.47	5405.97	18498.1	HF	0.6	0.0	ME
5394.89	5396.39	18530.9	HF	0.3		ME
5391.36	5392.86	18543.0	HF II	0.8	1.0	ME
5383.04	5384.54	18571.7	HF	0.5		ME
5373.86	5375.35	18603.4	HF	1.2	0.5	ME
5371.80	5373.29	18610.6	HF II	0.3	0.6	ME
5371.12	5372.61	18612.9	HF	0.3		ME
5361.38	5362.87	18646.7	HF II	0.0	0.3	ME
5360.75	5362.24	18648.9	HF	0.3H		ME
5358.33	5359.82	18657.3	HF	0.3		ME
5354.73	5356.22	18669.9	HF	1.1	0.3	ME
5348.40	5349.89	18692.0	HF II	1.0	1.2	ME
5346.30	5347.79	18699.3	HF II	1.0	1.6	ME
5324.26	5325.74	18776.7	HF II	1.3	1.5	ME
5311.60	5313.08	18821.5	HF II	2.0	2.2	ME
5309.68	5311.16	18828.3	HF	0.8	0.0	ME
5307.82	5309.30	18834.9	HF	0.6		ME
5304.19	5305.67	18847.8	HF	0.3		ME
5299.85	5301.32	18863.2	HF II	0.9	1.0	ME
5298.06	5299.53	18869.6	HF II	1.9	2.0	ME
5294.87	5296.34	18880.9	HF	1.1	0.3	ME
5292.78	5294.25	18888.4	HF	0.3		ME
5289.98	5291.45	18898.4	HF II	0.5	1.0	ME
5286.09	5287.56	18912.3	HF	0.6		ME
5276.39	5277.86	18947.1	HF II	0.0	0.5	ME
5275.04	5276.51	18951.9	HF	0.8	0.0	ME
5264.95	5266.42	18988.2	HF II	1.7	1.9	ME
5260.44	5261.90	19004.5	HF II	1.5	1.6	ME
5258.75	5260.21	19010.6	HF	0.3		ME
5254.49	5255.95	19026.1	HF	0.3		ME
5247.10	5248.56	19052.8	HF II	1.6	1.8	ME
5244.68	5246.14	19061.6	HF	0.3		ME
5243.99	5245.45	19064.1	HF	1.2	0.3	ME
5235.12	5236.58	19096.4	HF	0.3		ME
5194.57	5196.02	19245.5	HF II	0.5	0.8	ME
5187.75	5189.19	19270.8	HF II	1.3	1.5	ME
5186.84	5188.28	19274.2	HF	0.6		ME
5181.86	5183.30	19292.7	HF	1.4	1.0	ME
5170.18	5171.62	19336.3	HF	1.0	0.7	ME
5167.42	5168.86	19346.6	HF	0.5		ME
5166.38	5167.82	19350.5	HF	0.3		ME
5164.56	5166.00	19357.3	HF II	0.5	0.9	ME
5164.43	5165.87	19357.8	HF	0.6		ME
5157.96	5159.40	19382.1	HF	1.1	1.0	ME
5156.06	5157.50	19389.2	HF II	0.0	0.7	ME
5154.64	5156.08	19394.6	HF	0.3D		ME
5153.12	5154.56	19400.3	HF	0.5	0.0	ME
5128.53	5129.96	19493.3	HF II	1.0	1.3	ME
5118.46	5119.89	19531.7	HF	0.3		ME
5113.57	5114.99	19550.4	HF	0.6		ME
5112.13	5113.55	19555.9	HF	0.8	0.0	ME
5110.61	5112.03	19561.7	HF II		0.8	ME
5101.68	5103.10	19595.9	HF	0.3		ME
5100.67	5102.09	19599.8	HF	0.5		ME
5093.88	5095.30	19625.9	HF	0.7	0.3	ME
5080.44	5081.86	19677.9	HF II	0.3	1.0	ME
5079.65	5081.07	19680.9	HF II	1.6	1.8	ME
5075.92	5077.33	19695.4	HF II	1.0	1.3	ME

AIR WAVELENGTH (ANGSTRMS)	VACUUM WAVELENGTH (ANGSTRMS)	WAVENUMBER (KAYSERS)	LMENT	LOG ARC	INTENSITY SPK	DIS	REF
5071.23	5072.64	19713.6	HF	0.8	0.9		ME
5069.80	5071.21	19719.1	HF	0.3			ME
5058.18	5059.59	19764.4	HF II	0.9	1.0		ME
5057.03	5058.44	19768.9	HF II	1.3	1.5		ME
5051.32	5052.73	19791.3	HF	0.5	0.0		ME
5047.45	5048.86	19806.5	HF	1.2	0.7		ME
5040.82	5042.23	19832.5	HF II	2.0	2.2		ME
5034.90	5036.30	19855.8	HF	0.5			ME
5034.33	5035.73	19858.1	HF II	0.6	0.9		ME
5025.91	5027.31	19891.4	HF	0.3	0.3		ME
5023.076	5024.477	19902.57	HF	0.6			MIT
5021.75	5023.15	19907.8	HF	0.6			ME
5021.11	5022.51	19910.4	HF	0.6	0.3		ME
5018.20	5019.60	19921.9	HF	1.3	0.3		ME
5000.54	5001.93	19992.3	HF	0.3	0.3		ME
4999.68	5001.07	19995.7	HF II	1.5	1.6		M
4992.36	4993.75	20025.0	HF	0.5H	0.3		ME
4984.769	4986.160	20055.51	HF II	0.3	0.6		MIT
4975.25	4976.64	20093.9	HF	1.4	0.6		ME
4965.312	4966.698	20134.10	HF	0.3H	0.3		MIT
4962.37	4963.75	20146.0	HF	0.8	0.3		ME
4948.94	4950.32	20200.7	HF	1.0	0.3		ME
4945.37	4946.75	20215.3	HF II	1.0	1.1		M
4934.45	4935.83	20260.0	HF II	1.6	1.7		M
4928.38	4929.76	20285.0	HF	0.3	0.7		MIT
4926.97	4928.35	20290.8	HF II	0.9	1.0		M
4926.14	4927.52	20294.2	HF	0.3H			ME
4920.975	4922.349	20315.50	HF	0.3	0.7		MIT
4915.26	4916.63	20339.1	HF	0.7	0.5		M
4910.93	4912.30	20357.1	HF	0.5			ME
4910.046	4911.417	20360.72	HF	0.5	0.3		MIT
4907.34	4908.71	20371.9	HF II	0.3	0.6		ME
4906.33	4907.70	20376.1	HF	0.3H	0.3H		ME
4904.52	4905.89	20383.7	HF II	1.1	1.4		MIT
4903.04	4904.41	20389.8	HF	0.3	0.5		ME
4896.32	4897.69	20417.8	HF	0.6	0.6		ME
4889.89	4891.26	20444.6	HF	0.3H	0.0H		ME
4885.74	4887.10	20462.0	HF II	0.3	0.6		M
4883.265	4884.629	20472.39	HF II		0.6		MIT
4878.15	4879.51	20493.9	HF	0.5	0.5		ME
4877.58	4878.94	20496.2	HF	1.3	0.3		ME
4872.94	4874.30	20515.8	HF	0.8	0.3		ME
4865.43	4866.79	20547.4	HF II	0.7	1.0		M
4863.27	4864.63	20556.6	HF	1.3	0.5		ME
4861.49	4862.85	20564.1	HF	0.5	0.3		ME
4859.237	4860.594	20573.61	HF	1.5	0.7		MIT
4858.414	4859.771	20577.10	HF	1.2	0.6		MIT
4850.607	4851.962	20610.22	HF	1.0	0.5		MIT
4848.46	4849.81	20619.4	HF II	0.7	1.3		M
4844.00	4845.35	20638.3	HF II	1.0	1.2		ME
4837.23	4838.58	20667.2	HF	1.5	0.6		MIT
4834.84	4836.19	20677.4	HF II		0.5		ME
4834.187	4835.538	20680.22	HF	1.3	0.5		MIT
4818.867	4820.214	20745.97	HF	1.4	0.6		MIT
4817.21	4818.56	20753.1	HF II	1.2	1.6		M
4811.14	4812.48	20779.3	HF	0.8	0.3		ME
4809.18	4810.52	20787.8	HF II	0.9	1.0		ME
4807.14	4808.48	20796.6	HF II	0.8	0.9		ME
4800.499	4801.841	20825.35	HF	1.7	0.8		MIT
4795.963	4797.304	20845.04	HF	0.5	0.3		MIT
4790.725	4792.064	20867.83	HF II	1.3	1.3		MIT
4782.739	4784.076	20902.68	HF	1.6	0.7		MIT
4777.20	4778.54	20926.9	HF	0.5	0.5H		ME
4774.893	4776.228	20937.02	HF	0.8	0.3		MIT
4773.715	4775.050	20942.19	HF	1.4	0.6		MIT
4769.36	4770.69	20961.3	HF	0.6			ME
4766.51	4767.84	20973.9	HF	1.3	0.6		M
4765.78	4767.11	20977.1	HF II	0.5H	1.2		M
4760.589	4761.920	20999.93	HF II	0.7	1.1		MIT
4757.584	4758.914	21013.20	HF	1.2	0.6		MIT

AIR WAVELENGTH (ANGSTRMS)	VACUUM WAVELENGTH (ANGSTRMS)	WAVENUMBER (KAYSERS)	LMENT	LOG ARC	INTENSITY SPK	DIS	REF
4746.732	4748.060	21061.24	HF	0.7H			MIT
4739.812	4741.138	21091.98	HF	1.1			MIT
4738.575	4739.900	21097.49	HF	1.3	0.5		MIT
4738.13	4739.46	21099.5	HF	0.6			ME
4735.77	4737.09	21110.0	HF II	0.3	1.0		MIT
4735.668	4736.993	21110.44	HF II	1.0	1.3		MIT
4733.74	4735.06	21119.0	HF II		0.3		ME
4731.366	4732.689	21129.64	HF II	1.2	1.3		MIT
4721.711	4723.032	21172.84	HF	1.0	0.6		MIT
4719.102	4720.422	21184.55	HF II	1.5	1.6		MIT
4714.99	4716.31	21203.0	HF II	0.5	0.6		ME
4713.48	4714.80	21209.8	HF II	0.8	0.3		MIT
4708.84	4710.16	21230.7	HF	0.3	0.6		MIT
4703.610	4704.926	21254.32	HF II	0.6	1.1		MIT
4703.12	4704.44	21256.5	HF	1.3			MIT
4699.715	4701.030	21271.93	HF II	1.3	1.4		MIT
4699.013	4700.328	21275.11	HF	1.5	0.6		MIT
4695.965	4697.279	21288.92	HF	0.7			MIT
4688.39	4689.70	21323.3	HF	1.5	0.6		ME
4686.38	4687.69	21332.5	HF	0.8			ME
4683.934	4685.245	21343.60	HF	1.1			MIT
4682.667	4683.978	21349.38	HF II	0.3	0.9		MIT
4675.455	4676.764	21382.31	HF II	1.0	1.0		MIT
4670.91	4672.22	21403.1	HF	0.7	0.3		ME
4669.24	4670.55	21410.8	HF	0.8	0.8		MIT
4664.123	4665.429	21434.26	HF II	1.7	2.0		MIT
4659.21	4660.51	21456.9	HF II	0.9	0.9		M
4655.186	4656.489	21475.41	HF	1.7	0.6		MIT
4652.109	4653.412	21489.61	HF	0.8H	0.8H		MIT
4650.603	4651.905	21496.57	HF	1.0	0.5H		MIT
4648.338	4649.640	21507.04	HF	1.0	0.3H		MIT
4647.438	4648.739	21511.21	HF	0.8			MIT
4642.241	4643.541	21535.29	HF	1.2	0.6		MIT
4640.094	4641.393	21545.25	HF II	1.1	1.3		MIT
4630.610	4631.907	21589.38	HF	1.1			MIT
4628.19	4629.49	21600.7	HF	0.6			ME
4626.42	4627.72	21608.9	HF II	1.2	0.8		MIT
4622.699	4623.994	21626.33	HF II	1.3	1.8		MIT
4620.864	4622.158	21634.92	HF	1.7	0.6		MIT
4614.206	4615.498	21666.13	HF	1.0			MIT
4613.738	4615.030	21668.33	HF II	1.1	1.4J		MIT
4608.092	4609.383	21694.88	HF	1.4	0.6		MIT
4605.774	4607.064	21705.80	HF II	1.3	1.5		MIT
4602.713	4604.002	21720.23	HF	0.8	0.5		MIT
4599.44	4600.73	21735.7	HF II	1.0	1.4		MIT
4598.800	4600.088	21738.71	HF	1.3	0.5		MIT
4597.904	4599.192	21742.95	HF	0.9	0.3		MIT
4586.248	4587.533	21798.21	HF II	0.9	1.0		MIT
4573.788	4575.070	21857.59	HF II	1.2	1.2		MIT
4570.690	4571.971	21872.41	HF II	0.9	1.3		MIT
4565.936	4567.216	21895.18	HF	1.6	0.8		MIT
4563.81	4565.09	21905.4	HF		0.9H		ME
4553.776	4555.053	21953.64	HF	1.0			MIT
4553.239	4554.515	21956.23	HF	0.7H			MIT
4550.14	4551.42	21971.2	HF	0.6			ME
4546.83	4548.10	21987.2	HF	1.0	0.3		ME
4544.019	4545.293	22000.78	HF	1.3	0.3		MIT
4543.01	4544.28	22005.7	HF	0.7	0.3		MIT
4541.716	4542.989	22011.94	HF	1.1	0.3		MIT
4541.297	4542.570	22013.97	HF II	1.2	1.3		MIT
4540.932	4542.205	22015.74	HF	1.7	0.3		MIT
4539.758	4541.031	22021.43	HF II	0.5	0.8		MIT
4535.366	4536.638	22042.76	HF II	1.0	1.3		MIT
4533.152	4534.423	22053.52	HF II	1.3	1.6		MIT
4524.74	4526.01	22094.5	HF II	1.0	1.2		ME
4520.57	4521.84	22114.9	HF	1.0H	0.6		MIT
4519.031	4520.298	22122.43	HF	0.0	0.9		MIT
4518.294	4519.561	22126.04	HF	1.0	0.6		MIT
4499.650	4500.912	22217.72	HF	1.2	0.8		MIT
4490.595	4491.855	22262.52	HF II	0.7	1.2		MIT

AIR WAVELENGTH (ANGSTRMS)	VACUUM WAVELENGTH (ANGSTRMS)	WAVENUMBER (KAYSERS)	LMENT	LOG ARC	INTENSITY SPK	INTENSITY DIS	REF
4486.648	4487.907	22282.10	HF II	0.5	1.2		MIT
4486.132	4487.391	22284.67	HF II	1.4	1.5		MIT
4483.29	4484.55	22298.8	HF II		0.7		ME
4466.40	4467.65	22383.1	HF II	1.0	1.3		MIT
4461.183	4462.435	22409.29	HF	1.4	0.3		MIT
4457.344	4458.595	22428.59	HF	1.4	0.3		MIT
4453.00	4454.25	22450.5	HF	0.9H	0.3		MIT
4452.70	4453.95	22452.0	HF II		1.0		ME
4443.07	4444.32	22500.6	HF	1.2	1.3		ME
4438.04	4439.29	22526.1	HF	1.5	0.3		MIT
4434.52	4435.77	22544.0	HF	0.8H			ME
4432.93	4434.17	22552.1	HF	0.8H	0.6H		MIT
4426.178	4427.421	22586.51	HF II	0.5H	0.9		MIT
4422.74	4423.98	22604.1	HF II	1.2	1.4		MIT
4422.23	4423.47	22606.7	HF	1.1	0.8		MIT
4418.25	4419.49	22627.0	HF	0.9			MIT
4417.91	4419.15	22628.8	HF	1.4	0.0		MIT
4417.35	4418.59	22631.6	HF II	1.4	1.7		MIT
4408.81	4410.05	22675.5	HF	0.9			MIT
4397.15	4398.39	22735.6	HF		0.5H		ME
4395.01	4396.24	22746.7	HF	0.9H			MIT
4392.69	4393.92	22758.7	HF	0.3			ME
4391.26	4392.49	22766.1	HF	0.5			ME
4390.70	4391.93	22769.0	HF	0.5			MIT
4385.474	4386.706	22796.15	HF II	0.3H	0.6		MIT
4379.167	4380.397	22828.98	HF	1.0H	0.6H		MIT
4370.97	4372.20	22871.8	HF II	1.5	1.6		MIT
4367.905	4369.132	22887.84	HF II	1.4	1.3		MIT
4365.37	4366.60	22901.1	HF	0.9	0.9		MIT
4356.99	4358.21	22945.2	HF	1.0			MIT
4356.327	4357.551	22948.67	HF	1.5	0.6		MIT
4353.34	4354.56	22964.4	HF	0.9	0.8H		MIT
4352.566	4353.789	22968.50	HF	1.0	0.8		MIT
4351.15	4352.37	22976.0	HF	0.9			MIT
4350.508	4351.731	22979.36	HF II	1.3	1.6		MIT
4349.74	4350.96	22983.4	HF	0.9	0.5		MIT
4336.66	4337.88	23052.7	HF II	1.5	1.8H		MIT
4335.148	4336.367	23060.78	HF II	0.9	0.9		MIT
4334.638	4335.857	23063.49	HF II	1.0	1.3		MIT
4330.27	4331.49	23086.8	HF	1.2	0.5		ME
4327.78	4329.00	23100.0	HF	0.6	1.1		ME
4327.55	4328.77	23101.3	HF II		0.6		ME
4326.24	4327.46	23108.3	HF	0.8H			ME
4321.36	4322.58	23134.4	HF II	0.6	1.0		ME
4320.672	4321.887	23138.04	HF II	1.2	1.3		MIT
4319.51	4320.72	23144.3	HF II	0.3	1.0J		ME
4318.139	4319.353	23151.61	HF	1.1	0.8		MIT
4304.407	4305.618	23225.47	HF	0.8	0.3		MIT
4303.596	4304.807	23229.85	HF	1.0	0.8		MIT
4296.41	4297.62	23268.7	HF	1.0	0.8		M
4294.787	4295.995	23277.49	HF	1.4	0.3		MIT
4283.47	4284.68	23339.0	HF	0.7			MIT
4272.848	4274.050	23397.01	HF II	1.1	1.3		MIT
4270.56	4271.76	23409.5	HF	0.9	0.6		MIT
4269.65	4270.85	23414.5	HF II	1.0	1.3		MIT
4268.10	4269.30	23423.0	HF II		0.7		ME
4263.389	4264.589	23448.92	HF	1.3	1.0		MIT
4262.72	4263.92	23452.6	HF II	1.0	1.2		ME
4260.98	4262.18	23462.2	HF	1.3			MIT
4249.34	4250.54	23526.4	HF	0.6	1.2		MIT
4245.842	4247.037	23545.82	HF II	1.0	1.1		MIT
4245.16	4246.36	23549.6	HF	1.0	0.3		MIT
4241.93	4243.12	23567.5	HF II	0.3	1.0		ME
4232.443	4233.635	23620.36	HF II	1.3	0.8		MIT
4231.61	4232.80	23625.0	HF II		0.6		ME
4231.20	4232.39	23627.3	HF	0.7			MIT
4228.082	4229.273	23644.73	HF	0.9	0.3		MIT
4225.10	4226.29	23661.4	HF II		0.7		ME
4221.76	4222.95	23680.1	HF	0.3			MIT
4218.84	4220.03	23696.5	HF II	0.5			ME

AIR WAVELENGTH (ANGSTRMS)	VACUUM WAVELENGTH (ANGSTRMS)	WAVENUMBER (KAYSERS)	LMENT	LOG ARC	INTENSITY SPK	INTENSITY DIS	REF
4209.70	4210.89	23748.0	HF	0.7	0.3		M
4207.40	4208.59	23760.9	HF II		0.6		MIT
4206.577	4207.762	23765.60	HF II	1.0	1.0H		MIT
4188.24	4189.42	23869.6	HF	0.9			MIT
4187.66	4188.84	23873.0	HF II	0.9	1.0		MIT
4179.53	4180.71	23919.4	HF II	0.7	0.9		MIT
4177.505	4178.682	23930.99	HF II	0.7	0.9		MIT
4174.34	4175.52	23949.1	HF	1.4	0.8		MIT
4170.904	4172.080	23968.86	HF	1.0	0.0		MIT
4170.41	4171.59	23971.7	HF II	0.6	0.6		ME
4166.64	4167.81	23993.4	HF II		0.6		ME
4163.444	4164.618	24011.81	HF II	0.7	1.0		MIT
4162.688	4163.862	24016.17	HF	1.0	0.5		MIT
4162.36	4163.53	24018.1	HF II	0.7	1.0		MIT
4158.88	4160.05	24038.2	HF II	0.7	0.8		MIT
4156.74	4157.91	24050.5	HF II		0.5		MIT
4145.759	4146.928	24114.24	HF	1.0	0.0		MIT
4141.83	4143.00	24137.1	HF II		0.9		ME
4140.21	4141.38	24146.6	HF II		0.6		ME
4138.704	4139.871	24155.34	HF II		0.8		MIT
4136.361	4137.528	24169.02	HF	1.0	0.8		MIT
4127.795	4128.959	24219.18	HF II	1.0	1.0		MIT
4125.10	4126.26	24235.0	HF	0.3	0.8		ME
4123.534	4124.697	24244.20	HF II	0.0	1.0		MIT
4118.913	4120.075	24271.40	HF	0.5			MIT
4118.60	4119.76	24273.2	HF	0.5	0.0		ME
4113.531	4114.692	24303.16	HF II	0.7	1.0		MIT
4106.58	4107.74	24344.3	HF	1.0	0.3		MIT
4104.233	4105.391	24358.21	HF	0.9	0.6		MIT
4100.933	4102.090	24377.81	HF	1.0	0.0		MIT
4097.24	4098.40	24399.8	HF II	0.6	1.0		MIT
4095.493	4096.649	24410.20	HF	0.7	0.0		MIT
4093.161	4094.316	24424.10	HF II	1.4	1.3		MIT
4087.959	4089.113	24455.18	HF	0.8	0.0		MIT
4083.352	4084.505	24482.77	HF	1.0	0.3		MIT
4082.68	4083.83	24486.8	HF II	0.3	0.3		MIT
4080.442	4081.594	24500.23	HF II	1.2	1.2		MIT
4071.223	4072.373	24555.71	HF II	0.3	0.6		MIT
4067.838	4068.987	24576.14	HF	0.8	0.3		MIT
4066.22	4067.37	24585.9	HF	1.0	0.0H		MIT
4062.842	4063.989	24606.36	HF	1.0	0.5		MIT
4060.199	4061.346	24622.38	HF	1.0	0.6		MIT
4057.43	4058.58	24639.2	HF	0.5	0.3		M
4051.899	4053.044	24672.82	HF	0.9	0.6		MIT
4050.885	4052.029	24678.99	HF	1.3	0.0H		MIT
4050.667	4051.811	24680.32	HF II	0.6	0.5		MIT
4049.74	4050.88	24686.0	HF	0.8	0.3		ME
4049.45	4050.59	24687.7	HF II	1.0	1.0		MIT
4047.96	4049.10	24696.8	HF	0.9	1.4		ME
4044.39	4045.53	24718.6	HF	1.0	0.6		MIT
4033.88	4035.02	24783.0	HF II	0.7	0.9		
4032.266	4033.405	24792.95	HF	0.7	0.3		MIT
4029.173	4030.312	24811.98	HF II	1.0	1.1		MIT
4022.84	4023.98	24851.0	HF	0.3			ME
4020.247	4021.383	24867.06	HF II	0.8	0.3H		MIT
4011.944	4013.078	24918.53	HF	0.5			MIT
4009.78	4010.91	24932.0	HF	0.7			ME
4008.46	4009.59	24940.2	HF II	0.7	0.9		M
4007.35	4008.48	24947.1	HF II	0.7	0.6H		M
3998.51	3999.64	25002.2	HF II		0.6D		ME
3996.80	3997.93	25012.9	HF II	0.0	0.7		ME
3984.85	3985.98	25087.9	HF II	0.5	0.6H		ME
3984.03	3985.16	25093.1	HF II	0.6	0.9		M
3979.40	3980.53	25122.3	HF II	0.8	1.6		ME
3975.15	3976.27	25149.2	HF II	0.5	0.5		ME
3970.051	3971.174	25181.47	HF	1.0	0.5		MIT
3968.01	3969.13	25194.4	HF	0.7	0.0		M
3964.95	3966.07	25213.9	HF II	0.9	1.2		M
3951.826	3952.944	25297.60	HF II	1.2	0.6		MIT
3950.759	3951.877	25304.43	HF	1.0H			MIT

AIR WAVELENGTH (ANGSTRMS)	VACUUM WAVELENGTH (ANGSTRMS)	WAVENUMBER (KAYSERS)	LMENT	LOG ARC	INTENSITY SPK	DIS	REF
3946.012	3947.129	25334.87	HF II		0.6H		MIT
3945.36	3946.48	25339.1	HF II	0.9	0.8H		ME
3943.06	3944.18	25353.8	HF	0.5			ME
3939.044	3940.159	25379.69	HF	0.9			MIT
3935.647	3936.761	25401.59	HF II	1.2	1.2		MIT
3933.664	3934.778	25414.40	HF II	1.3	1.2		MIT
3932.40	3933.51	25422.6	HF II	0.5	1.0		ME
3931.75	3932.86	25426.8	HF	0.3	0.3		ME
3931.377	3932.490	25429.18	HF	1.0	0.5		MIT
3929.535	3930.648	25441.10	HF II	0.5	0.0H		MIT
3929.212	3930.324	25443.19	HF	0.3	0.0		MIT
3927.570	3928.682	25453.83	HF	1.0	0.0		MIT
3926.424	3927.536	25461.26	HF	1.0			MIT
3923.904	3925.015	25477.61	HF II	1.1	1.1		MIT
3918.091	3919.201	25515.41	HF II	1.3	1.1		MIT
3917.448	3918.557	25519.59	HF II	0.7	1.2		MIT
3909.18	3910.29	25573.6	HF	0.5	0.5D		ME
3906.89	3908.00	25588.6	HF	0.5	0.5D		M
3900.651	3901.756	25629.49	HF II	0.3	1.2		MIT
3899.936	3901.041	25634.18	HF	1.2	0.8		MIT
3892.472	3893.575	25683.34	HF	0.7	0.0H		MIT
3889.328	3890.430	25704.10	HF	0.7	0.0		MIT
3889.232	3890.334	25704.73	HF	0.6	0.0H		MIT
3883.767	3884.868	25740.90	HF II	0.5	1.0		MIT
3880.818	3881.918	25760.46	HF II	1.3	1.5		MIT
3877.104	3878.203	25785.14	HF II	0.6	1.5		MIT
3872.55	3873.65	25815.5	HF II	0.8	1.3		M
3867.337	3868.433	25850.26	HF II	0.5	1.1		MIT
3864.75	3865.85	25867.6	HF II	0.3	1.3		ME
3860.910	3862.005	25893.29	HF	0.8	0.8		MIT
3858.309	3859.403	25910.74	HF	0.7J	0.7J		MIT
3849.524	3850.616	25969.87	HF II	0.3H	0.6		MIT
3849.183	3850.275	25972.17	HF	1.2	1.2		MIT
3840.026	3841.115	26034.10	HF	1.0	0.8		MIT
3838.37	3839.46	26045.3	HF II	0.3	0.5		MIT
3833.669	3834.757	26077.27	HF	1.0	0.8		MIT
3831.135	3832.222	26094.52	HF	1.4	1.4		MIT
3830.018	3831.105	26102.13	HF	1.0	0.8		MIT
3829.67	3830.76	26104.5	HF	0.5	0.0		ME
3823.511	3824.596	26146.55	HF II	0.3H	0.0H		MIT
3820.728	3821.812	26165.60	HF I	3.0	2.0		ME
3819.925	3821.009	26171.10	HF	1.1	1.0		MIT
3819.376	3820.460	26174.86	HF	1.2	1.3		MIT
3817.205	3818.288	26189.75	HF II	0.7H	1.2		MIT
3816.08	3817.16	26197.5	HF	0.3			M
3811.780	3812.862	26227.02	HF	0.5	0.0		MIT
3810.611	3811.693	26235.06	HF II	1.2	1.2		MIT
3806.074	3807.154	26266.34	HF II	0.7	1.3		MIT
3804.529	3805.609	26277.00	HF	0.5	0.3		MIT
3804.165	3805.245	26279.52	HF	0.7	0.6		MIT
3800.39	3801.47	26305.6	HF II	1.3	1.1		ME
3799.52	3800.60	26311.6	HF II		0.6		ME
3798.662	3799.741	26317.59	HF	0.7	0.5		MIT
3797.923	3799.001	26322.71	HF II	1.4	1.4		MIT
3793.372	3794.449	26354.29	HF II	1.0	1.3		MIT
3788.208	3789.284	26390.21	HF	0.5	0.0		MIT
3787.372	3788.448	26396.04	HF	0.5	0.3		MIT
3785.460	3786.535	26409.37	HF I	3.5	2.4		ME
3782.78	3783.85	26428.1	HF II	0.6	0.7		ME
3782.43	3783.50	26430.5	HF	0.7	0.5		M
3780.09	3781.16	26446.9	HF II	0.6	1.0		ME
3777.658	3778.731	26463.91	HF I	3.3	2.3		ME
3773.12	3774.19	26495.7	HF	0.3			ME
3772.852	3773.924	26497.62	HF II	0.0	0.3		MIT
3771.365	3772.436	26508.07	HF II	0.3	0.8		MIT
3770.64	3771.71	26513.2	HF II		0.3		ME
3768.254	3769.325	26529.95	HF	0.6	0.0		MIT
3766.924	3767.994	26539.32	HF II	0.7	1.4		MIT
3762.51	3763.58	26570.4	HF II	0.3	1.4		ME
3758.05	3759.12	26602.0	HF II	0.7			ME

AIR WAVELENGTH (ANGSTRMS)	VACUUM WAVELENGTH (ANGSTRMS)	WAVENUMBER (KAYSERS)	LMENT	LOG ARC	INTENSITY SPK	DIS	REF
3757.209	3758.277	26607.94	HF	0.8	0.3		MIT
3747.49	3748.56	26676.9	HF II	0.8	0.9		M
3746.803	3747.868	26681.84	HF	1.3	0.9		MIT
3744.98	3746.04	26694.8	HF II	1.2	1.3		M
3741.945	3743.009	26716.48	HF	0.3	0.3H		MIT
3739.041	3740.104	26737.22	HF	0.5	0.0		MIT
3737.876	3738.939	26745.56	HF II	1.2	1.4		MIT
3733.788	3734.850	26774.84	HF	1.1	0.7		MIT
3729.104	3730.164	26808.47	HF	0.9			MIT
3726.49	3727.55	26827.3	HF	0.8	0.5		M
3722.65	3723.71	26854.9	HF		0.3		ME
3721.507	3722.565	26863.19	HF	1.0			MIT
3719.279	3720.337	26879.29	HF II	1.2	1.5		MIT
3718.842	3719.900	26882.44	HF		0.5		MIT
3717.800	3718.858	26889.98	HF	1.3	0.9		MIT
3707.363	3708.418	26965.68	HF	0.5	0.3		MIT
3705.405	3706.459	26979.93	HF II	1.2	1.4		MIT
3704.92	3705.97	26983.5	HF	0.3			ME
3702.36	3703.41	27002.1	HF II		0.3H		M
3701.148	3702.201	27010.96	HF II	1.2	1.6		MIT
3699.716	3700.769	27021.41	HF II	1.3	1.4		MIT
3699.56	3700.61	27022.6	HF II		0.5		ME
3698.395	3699.448	27031.06	HF II	1.0	1.4		MIT
3696.51	3697.56	27044.9	HF	0.9	0.3		MIT
3682.247	3683.295	27149.60	HF I	3.5	2.3		ME
3681.39	3682.44	27155.9	HF II	0.5	0.6		M
3678.086	3679.133	27180.31	HF II		0.7H		MIT
3676.06	3677.11	27195.3	HF		0.5		ME
3675.740	3676.787	27197.66	HF	1.0	0.8		MIT
3672.269	3673.315	27223.37	HF	1.4	0.5		MIT
3668.207	3669.252	27253.51	HF	1.0	0.0		MIT
3666.775	3667.819	27264.16	HF II	1.4	0.6		MIT
3665.346	3666.390	27274.79	HF II	1.3	1.4		MIT
3661.73	3662.77	27301.7	HF II	0.0	0.3		ME
3661.046	3662.089	27306.82	HF II	1.0	1.4		MIT
3659.045	3660.087	27321.75	HF II	0.5	0.9		MIT
3656.679	3657.721	27339.43	HF II		0.3		MIT
3651.772	3652.812	27376.17	HF	1.2	0.0		MIT
3650.53	3651.57	27385.5	HF	0.3	0.3		ME
3649.105	3650.145	27396.17	HF	1.3	0.7		MIT
3648.35	3649.39	27401.8	HF II	0.7	0.9		ME
3647.492	3648.531	27408.29	HF II	1.3	0.3		MIT
3644.355	3645.394	27431.88	HF II	1.4	1.7		MIT
3637.59	3638.63	27482.9	HF	0.8	0.3		MIT
3633.185	3634.221	27516.22	HF II	0.8	0.9		MIT
3632.691	3633.727	27519.96	HF	0.7	0.3		MIT
3630.874	3631.909	27533.73	HF II	1.2	0.8		MIT
3627.832	3628.866	27556.82	HF	0.7			MIT
3624.000	3625.033	27585.95	HF II	1.2	1.3		MIT
3622.46	3623.49	27597.7	HF II	0.3	0.7		ME
3620.041	3621.073	27616.12	HF	0.9			MIT
3619.966	3620.998	27616.69	HF	0.5			MIT
3617.685	3618.717	27634.11	HF	0.6	0.5		MIT
3616.892	3617.923	27640.17	HF I	3.3	2.3		ME
3615.045	3616.076	27654.29	HF	0.6	0.0		MIT
3600.055	3601.082	27769.43	HF II	0.3	0.9H		MIT
3599.870	3600.897	27770.86	HF	1.0	0.9		MIT
3599.139	3600.166	27776.50	HF II	0.6	1.0		MIT
3597.510	3598.536	27789.07	HF	0.5			MIT
3597.42	3598.45	27789.8	HF II	0.7	1.2		ME
3594.427	3595.453	27812.91	HF II		0.3		MIT
3583.28	3584.30	27899.4	HF	0.3			MIT
3580.45	3581.47	27921.5	HF II	0.3	0.5		M
3579.905	3580.927	27925.73	HF	1.2	1.0		MIT
3569.041	3570.060	28010.73	HF II	1.3	1.7		MIT
3567.364	3568.383	28023.90	HF	1.3	1.0		MIT
3565.32	3566.34	28040.0	HF	0.3			ME
3564.314	3565.332	28047.88	HF	0.3	0.0		MIT
3561.664	3562.681	28068.75	HF II	1.3	1.5		MIT
3558.483	3559.499	28093.84	HF	0.5	0.0		MIT

AIR WAVELENGTH (ANGSTRMS)	VACUUM WAVELENGTH (ANGSTRMS)	WAVENUMBER (KAYSERS)	LMENT	LOG ARC	INTENSITY SPK	DIS	REF
3554.00	3555.02	28129.3	HF	0.7			ME
3552.696	3553.711	28139.60	HF II	1.3	1.5		MIT
3548.812	3549.826	28170.40	HF	0.7	0.3		MIT
3536.62	3537.63	28267.5	HF	1.0	0.3		ME
3535.545	3536.555	28276.10	HF II	1.2	1.7		MIT
3531.227	3532.236	28310.68	HF	0.7	0.5		MIT
3530.874	3531.883	28313.51	HF	0.7	0.0		MIT
3526.857	3527.865	28345.75	HF	1.0	0.0H		MIT
3523.021	3524.028	28376.62	HF	1.3	1.0		MIT
3521.558	3522.565	28388.41	HF	0.5	0.3		MIT
3518.752	3519.758	28411.04	HF II	0.7	1.2		MIT
3518.38	3519.39	28414.1	HF II		0.5		ME
3513.276	3514.281	28455.32	HF	1.0	0.0		MIT
3511.88	3512.88	28466.6	HF II	0.7	1.0		ME
3505.227	3506.230	28520.66	HF II	1.3	1.7		MIT
3498.985	3499.986	28571.54	HF	0.8			MIT
3497.491	3498.492	28583.75	HF	1.3	1.0		MIT
3497.160	3498.161	28586.45	HF	1.0	0.6		MIT
3495.928	3496.928	28596.53	HF II	1.0	1.0		MIT
3495.748	3496.748	28598.00	HF II	1.0	1.2		MIT
3487.57	3488.57	28665.1	HF II	0.8	1.2		MIT
3485.17	3486.17	28684.8	HF II	0.9	0.5		MIT
3484.71	3485.71	28688.6	HF	0.8			MIT
3481.854	3482.851	28712.11	HF	0.7			MIT
3479.285	3480.281	28733.31	HF II	1.0	1.2		MIT
3478.990	3479.986	28735.75	HF II	1.5	1.6		MIT
3472.405	3473.399	28790.24	HF	1.4	1.0		MIT
3469.23	3470.22	28816.6	HF II	0.7			ME
3468.848	3469.841	28819.76	HF II	0.5H	0.5		MIT
3467.603	3468.596	28830.11	HF	0.9			MIT
3465.92	3466.91	28844.1	HF II	0.7	0.3		ME
3462.641	3463.633	28871.42	HF II	1.2	1.1		MIT
3452.31	3453.30	28957.8	HF	0.5			ME
3441.84	3442.83	29045.9	HF	1.0			MIT
3440.88	3441.87	29054.0	HF	0.5			ME
3438.432	3439.418	29074.69	HF	1.1	0.5		MIT
3438.235	3439.220	29076.36	HF II	1.4	1.4		MIT
3434.89	3435.87	29104.7	HF	0.7			MIT
3428.366	3429.349	29160.05	HF II	1.2	1.2		MIT
3421.45	3422.43	29219.0	HF II	0.5	0.6		ME
3419.175	3420.156	29238.44	HF	1.2	0.6		MIT
3417.345	3418.325	29254.09	HF	1.0	0.3		MIT
3413.74	3414.72	29285.0	HF II	1.1	1.0		MIT
3412.338	3413.317	29297.02	HF	1.0			MIT
3410.171	3411.149	29315.63	HF II	1.4	1.8		MIT
3407.759	3408.737	29336.38	HF II	1.3	1.4		MIT
3407.143	3408.121	29341.69	HF	1.0	0.0		MIT
3402.514	3403.490	29381.60	HF	1.3	0.5		MIT
3400.211	3401.187	29401.50	HF	1.2	0.3		MIT
3399.795	3400.771	29405.10	HF II	1.8	2.0		MIT
3397.597	3398.572	29424.12	HF	1.3	0.5		MIT
3397.257	3398.232	29427.07	HF	1.3	0.5		MIT
3395.94	3396.91	29438.5	HF	1.0			MIT
3394.983	3395.957	29446.78	HF II	1.3	1.7		MIT
3394.589	3395.563	29450.19	HF II	1.3	1.4		MIT
3392.811	3393.785	29465.63	HF	1.3	0.5		MIT
3389.833	3390.806	29491.51	HF II	1.5	1.6		MIT
3386.10	3387.07	29524.0	HF	0.7	0.8		ME
3385.23	3386.20	29531.6	HF		0.5H		MIT
3384.702	3385.674	29536.22	HF II	1.0	1.1		MIT
3384.14	3385.11	29541.1	HF II	1.2	1.2		M
3378.928	3379.898	29586.69	HF	1.4R	1.3		MIT
3376.68	3377.65	29606.4	HF II	0.8	0.3		MIT
3370.69	3371.66	29659.0	HF II	0.5	0.5H		MIT
3367.09	3368.06	29690.7	HF II	0.3	0.5		ME
3366.68	3367.65	29694.3	HF	1.2	0.0		MIT
3365.80	3366.77	29702.1	HF	0.3	0.3J		ME
3360.055	3361.021	29752.87	HF	0.9	0.0		MIT
3358.906	3359.871	29763.05	HF	1.4	0.3		MIT
3358.303	3359.268	29768.39	HF II	1.0	1.2		MIT

AIR WAVELENGTH (ANGSTRMS)	VACUUM WAVELENGTH (ANGSTRMS)	WAVENUMBER (KAYSERS)	LMENT	LOG ARC	INTENSITY SPK	DIS	REF
3356.78	3357.74	29781.9	HF	1.0			ME
3356.092	3357.056	29788.00	HF II	1.2	0.7		MIT
3352.055	3353.018	29823.87	HF II	1.5	1.7		MIT
3349.17	3350.13	29849.6	HF II	0.7	0.6		ME
3333.49	3334.45	29990.0	HF II		0.5H		ME
3332.729	3333.688	29996.81	HF	1.6	1.0		MIT
3331.887	3332.845	30004.39	HF	1.1			MIT
3328.213	3329.170	30037.51	HF II	1.3	1.2		MIT
3327.812	3328.769	30041.13	HF	0.7			MIT
3324.18	3325.14	30073.9	HF II	0.9	0.8		ME
3323.36	3324.32	30081.4	HF II	1.1	1.2H		MIT
3318.74	3319.69	30123.2	HF II	0.5			ME
3318.28	3319.23	30127.4	HF	0.8			ME
3317.988	3318.943	30130.08	HF II	1.4	1.3		MIT
3317.224	3318.179	30137.02	HF II	0.8	0.0		MIT
3316.19	3317.14	30146.4	HF	1.2			MIT
3314.03	3314.98	30166.1	HF II	0.7	0.0		ME
3312.865	3313.818	30176.67	HF	1.5	1.0		MIT
3310.856	3311.809	30194.98	HF II	1.2	0.9		MIT
3310.274	3311.227	30200.29	HF	1.3	0.7		MIT
3309.187	3310.140	30210.21	HF	1.2	0.0		MIT
3306.118	3307.070	30238.25	HF	1.2	0.5		MIT
3298.956	3299.906	30303.89	HF	0.9	0.8		MIT
3294.68	3295.63	30343.2	HF II	0.6	0.7H		ME
3291.050	3291.998	30376.69	HF	1.3	0.8H		MIT
3289.74	3290.69	30388.8	HF II	0.7	0.5		ME
3283.384	3284.330	30447.61	HF II	1.0	0.8		MIT
3279.98	3280.93	30479.2	HF II	1.4	1.4		ME
3273.655	3274.599	30538.09	HF II	1.3	1.0		MIT
3267.175	3268.117	30598.66	HF	1.0	0.0		MIT
3267.009	3267.951	30600.22	HF	0.9			MIT
3266.53	3267.47	30604.7	HF II	0.0	0.5		ME
3265.292	3266.233	30616.31	HF	1.0	0.0		MIT
3262.474	3263.415	30642.75	HF	1.0	0.0		MIT
3255.28	3256.22	30710.5	HF II	1.3	1.5		ME
3254.857	3255.796	30714.46	HF	1.0			MIT
3253.702	3254.640	30725.36	HF II	1.5	1.5		MIT
3250.07	3251.01	30759.7	HF II	0.7			ME
3249.534	3250.471	30764.77	HF	1.3	0.0		MIT
3247.67	3248.61	30782.4	HF I	2.6	1.9		ME
3243.353	3244.289	30823.40	HF	1.2			MIT
3243.00	3243.94	30826.7	HF II	0.8	0.9		ME
3239.438	3240.373	30860.65	HF	1.2			MIT
3233.80	3234.73	30914.4	HF II	0.5	0.6		ME
3230.061	3230.993	30950.23	HF	1.0			MIT
3226.996	3227.928	30979.63	HF II	0.8	0.5		MIT
3225.026	3225.957	30998.55	HF II	0.6	0.5		MIT
3220.606	3221.536	31041.09	HF II	1.4	1.5		MIT
3218.20	3219.13	31064.3	HF II	0.9	0.9		ME
3217.305	3218.234	31072.94	HF II	1.5	1.2		MIT
3213.719	3214.647	31107.61	HF	1.0			MIT
3210.976	3211.904	31134.19	HF	1.1	0.0		MIT
3206.774	3207.701	31174.98	HF II	0.8	0.7		MIT
3206.108	3207.034	31181.46	HF	1.3	0.3		MIT
3203.674	3204.600	31205.15	HF II	0.7	1.0		MIT
3202.16	3203.09	31219.9	HF II	0.7	0.7		ME
3199.994	3200.919	31241.03	HF II	1.3	1.5		MIT
3196.926	3197.850	31271.01	HF	1.2	0.0		MIT
3195.614	3196.538	31283.85	HF II	0.8	0.9		MIT
3194.47	3195.39	31295.1	HF II	0.8	0.5		ME
3194.193	3195.116	31297.76	HF II	1.6	1.6		MIT
3193.526	3194.449	31304.30	HF II	1.4	1.4		MIT
3189.625	3190.547	31342.59	HF	1.2	0.0		MIT
3181.756	3182.676	31420.10	HF II	0.6	0.3		MIT
3181.153	3182.073	31426.05	HF	1.0	0.0		MIT
3181.012	3181.932	31427.45	HF	1.0	0.3		MIT
3176.856	3177.775	31468.56	HF II	1.5	1.5		MIT
3172.942	3173.860	31507.38	HF	1.5	0.9		MIT
3168.390	3169.307	31552.64	HF	1.5	0.3		MIT
3167.446	3168.363	31562.04	HF II	0.3	0.0		MIT

AIR WAVELENGTH (ANGSTRMS)	VACUUM WAVELENGTH (ANGSTRMS)	WAVENUMBER (KAYSERS)	LMENT	LOG ARC	INTENSITY SPK	INTENSITY DIS	REF
3162.611	3163.526	31610.29	HF II	1.6	1.5		MIT
3159.93	3160.84	31637.1	HF II	0.7	0.0		ME
3159.823	3160.738	31638.18	HF	1.3	0.3		MIT
3156.626	3157.540	31670.23	HF	1.3	0.5		MIT
3152.960	3153.873	31707.05	HF	0.9			MIT
3151.629	3152.542	31720.44	HF	1.3	0.0		MIT
3148.414	3149.326	31752.83	HF	1.3	0.0		MIT
3145.319	3146.230	31784.07	HF II	1.7	1.3		MIT
3140.763	3141.673	31830.18	HF II	1.4	1.4		MIT
3139.653	3140.563	31841.43	HF II	1.4	1.3		MIT
3137.512	3138.421	31863.16	HF	1.5	1.0		MIT
3134.718	3135.626	31891.55	HF	1.9	2.1		MIT
3133.50	3134.41	31903.9	HF II	1.2	0.7		ME
3133.10	3134.01	31908.0	HF II	0.7	0.5		ME
3131.814	3132.722	31921.12	HF	1.6	1.0		MIT
3129.584	3130.491	31943.87	HF	1.2			MIT
3128.755	3129.662	31952.33	HF	1.3	0.0		MIT
3126.291	3127.197	31977.52	HF II	1.3	0.7		MIT
3123.953	3124.859	32001.45	HF	0.9			MIT
3122.547	3123.452	32015.86	HF	0.9	0.0		MIT
3119.980	3120.885	32042.20	HF	1.4	0.0		MIT
3116.950	3117.854	32073.34	HF II	1.3	1.3		MIT
3112.182	3113.085	32122.48	HF	0.5			MIT
3110.874	3111.776	32135.99	HF II	1.5	1.6		MIT
3109.117	3110.019	32154.15	HF II	1.7	2.0		MIT
3107.841	3108.743	32167.35	HF II	0.0	0.3		MIT
3106.023	3106.924	32186.17	HF	1.1	0.0		MIT
3105.668	3106.569	32189.85	HF	1.0			MIT
3103.69	3104.59	32210.4	HF	1.0			ME
3102.358	3103.258	32224.20	HF	1.2	0.5H		MIT
3102.147	3103.047	32226.39	HF	0.8	0.0		MIT
3101.397	3102.297	32234.18	HF II	1.8	2.0		MIT
3099.123	3100.022	32257.83	HF II		0.5		MIT
3098.593	3099.492	32263.35	HF	1.0			MIT
3096.758	3097.657	32282.47	HF	1.4	0.0		MIT
3092.245	3093.143	32329.58	HF II	1.3	1.3		MIT
3091.79	3092.69	32334.3	HF II	0.9	0.5H		ME
3091.366	3092.263	32338.77	HF	1.1			MIT
3083.647	3084.543	32419.72	HF	0.8			MIT
3080.845	3081.740	32449.20	HF	1.4	0.7		MIT
3080.657	3081.552	32451.18	HF II	1.5	2.0		MIT
3076.874	3077.768	32491.08	HF	0.7	0.9		MIT
3076.70	3077.59	32492.9	HF II	0.6			ME
3075.30	3076.19	32507.7	HF	1.2			ME
3074.791	3075.684	32513.09	HF	1.5	0.6		MIT
3074.097	3074.990	32520.43	HF	1.4	0.0		MIT
3072.877	3073.770	32533.34	HF I	1.9	1.3		MIT
3070.481	3071.373	32558.73	HF	0.7			MIT
3070.00	3070.89	32563.8	HF II		0.7H		MIT
3069.676	3070.568	32567.26	HF II	0.5			MIT
3069.182	3070.074	32572.51	HF	1.2			MIT
3067.414	3068.305	32591.28	HF	1.5	1.0		MIT
3064.68	3065.57	32620.4	HF II	1.0	1.5		MIT
3064.38	3065.27	32623.6	HF II		0.3		ME
3063.78	3064.67	32629.9	HF	1.4	0.0		MIT
3057.016	3057.905	32702.13	HF	1.8	1.0		MIT
3055.44	3056.33	32719.0	HF II	1.2	1.2		MIT
3054.52	3055.41	32728.9	HF II	1.2	1.2		MIT
3050.758	3051.645	32769.21	HF	1.7	1.0		MIT
3049.29	3050.18	32785.0	HF	1.2			MIT
3046.08	3046.97	32819.5	HF II	1.4	1.4		ME
3031.16	3032.04	32981.1	HF II	1.8	2.0		MIT
3025.29	3026.17	33045.1	HF II	1.5	1.5		MIT
3024.76	3025.64	33050.9	HF II	1.3	1.4		MIT
3022.114	3022.994	33079.79	HF II	0.9	0.3		MIT
3020.529	3021.409	33097.14	HF	1.2	0.3		MIT
3018.314	3019.193	33121.43	HF	1.8	1.0		MIT
3016.94	3017.82	33136.5	HF II	1.2	0.8		ME
3016.78	3017.66	33138.3	HF	1.4	1.0		MIT
3012.902	3013.780	33180.92	HF II	1.9	2.0		MIT

AIR WAVELENGTH (ANGSTRMS)	VACUUM WAVELENGTH (ANGSTRMS)	WAVENUMBER (KAYSERS)	LMENT	LOG ARC	INTENSITY SPK	INTENSITY DIS	REF
3012.20	3013.08	33188.7	HF II		0.5		MIT
3011.24	3012.12	33199.2	HF II	1.2	1.3		ME
3005.557	3006.433	33262.01	HF	1.7	0.9		MIT
3001.85	3002.73	33303.1	HF II	0.7	0.0		M
3000.096	3000.971	33322.55	HF II	1.6	1.5		MIT
2996.77	2997.64	33359.5	HF II		0.3		MIT
2992.00	2992.87	33412.7	HF	0.9	0.0		MIT
2990.811	2991.683	33426.00	HF II	0.8	0.3		MIT
2984.363	2985.234	33498.21	HF II		0.3		MIT
2984.05	2984.92	33501.7	HF	1.3	0.7		MIT
2982.721	2983.591	33516.65	HF	1.5	0.0		MIT
2980.810	2981.680	33538.14	HF	1.7	1.0		MIT
2980.21	2981.08	33544.9	HF II	0.7	0.0		MIT
2979.280	2980.149	33555.36	HF	1.3	0.0		MIT
2977.595	2978.464	33574.35	HF II	1.5	1.4		MIT
2977.119	2977.988	33579.72	HF		0.5		MIT
2975.886	2976.755	33593.63	HF II	2.8	3.5		ME
2974.09	2974.96	33613.9	HF II	0.5	0.8		ME
2973.373	2974.241	33622.02	HF	1.2	1.2		MIT
2970.57	2971.44	33653.7	HF		0.5H		MIT
2968.946	2969.813	33672.15	HF II	1.1	0.8H		MIT
2968.812	2969.679	33673.67	HF II	1.5	1.5		MIT
2967.231	2968.097	33691.62	HF II	1.3	1.4		MIT
2966.929	2967.795	33695.04	HF	1.4	0.6		MIT
2964.885	2965.751	33718.27	HF I	2.6	1.9		ME
2961.797	2962.662	33753.43	HF I	1.3	1.5		MIT
2960.823	2961.688	33764.53	HF II	1.3	1.3		MIT
2959.163	2960.027	33783.47	HF II		0.3		MIT
2958.018	2958.882	33796.55	HF	1.5	0.7		MIT
2954.201	2955.064	33840.21	HF I	2.7	2.0		ME
2953.65	2954.51	33846.5	HF II	0.7	0.0		ME
2951.90	2952.76	33866.6	HF	0.9			MIT
2950.671	2951.533	33880.69	HF I	2.8	2.0		ME
2949.62	2950.48	33892.8	HF II		0.5		ME
2947.130	2947.991	33921.40	HF II	1.2	1.2		MIT
2944.711	2945.572	33949.27	HF	1.3	0.0		MIT
2940.762	2941.622	33994.85	HF I	2.9	2.0		ME
2937.782	2938.641	34029.33	HF II	2.6	3.5		ME
2935.37	2936.23	34057.3	HF	0.7			MIT
2929.898	2930.755	34120.90	HF I	2.3	1.7		ME
2929.642	2930.499	34123.88	HF II	2.6	3.3		ME
2929.009	2929.866	34131.25	HF	1.2			MIT
2926.32	2927.18	34162.6	HF II	1.0	0.3		MIT
2926.06	2926.92	34165.6	HF		0.3		MIT
2924.615	2925.471	34182.53	HF	1.4	0.3		MIT
2919.60	2920.45	34241.2	HF II	2.5	3.3		ME
2918.576	2919.430	34253.26	HF	1.5	0.9		MIT
2917.49	2918.34	34266.0	HF II	1.0	0.9H		MIT
2916.481	2917.335	34277.86	HF I	2.9	2.0		ME
2913.219	2914.072	34316.24	HF II	0.5	0.5H		MIT
2912.775	2913.628	34321.47	HF II	0.5	0.3		MIT
2909.912	2910.764	34355.24	HF II	2.3	2.5		ME
2908.858	2909.710	34367.69	HF II	0.5	0.7		MIT
2904.760	2905.611	34416.17	HF I	2.5	1.8		ME
2904.519	2905.370	34419.02	HF II	0.9	0.9		MIT
2904.412	2905.263	34420.29	HF I	2.6	1.8		ME
2903.577	2904.428	34430.19	HF	0.0	0.3		MIT
2898.709	2899.559	34488.01	HF II	1.7	2.3		ME
2898.256	2899.105	34493.40	HF I	2.7	1.7		ME
2894.009	2894.857	34544.02	HF II	0.3	0.3		MIT
2892.561	2893.409	34561.31	HF	1.1	0.7		MIT
2891.03	2891.88	34579.6	HF	0.8			ME
2889.619	2890.466	34596.49	HF I	2.5	1.7		ME
2888.012	2888.859	34615.74	HF II	0.7	0.5		MIT
2887.541	2888.388	34621.39	HF	1.2	0.3		MIT
2887.135	2887.982	34626.26	HF	1.4	0.5		MIT
2885.473	2886.319	34646.20	HF II	1.1	1.2		MIT
2879.112	2879.957	34722.75	HF II	1.2	1.0		MIT
2877.16	2878.00	34746.3	HF	1.1			ME
2876.32	2877.16	34756.5	HF II	2.3	3.0		MIT

AIR WAVELENGTH (ANGSTRMS)	VACUUM WAVELENGTH (ANGSTRMS)	WAVENUMBER (KAYSERS)	LMENT	LOG ARC	INTENSITY SPK	DIS	REF
2873.650	2874.493	34788.74	HF	1.0			MIT
2869.825	2870.667	34835.10	HF II	1.4	1.3		MIT
2867.702	2868.544	34860.89	HF	1.0	0.3		MIT
2866.373	2867.215	34877.06	HF	1.7	1.1		MIT
2865.571	2866.412	34886.82	HF II	0.5	0.0H		MIT
2861.695	2862.535	34934.07	HF II	2.8	3.4		ME
2861.009	2861.849	34942.44	HF II	2.7	3.3		ME
2860.557	2861.397	34947.96	HF	1.3	0.3		MIT
2860.312	2861.152	34950.96	HF II	1.2	1.5		MIT
2858.661	2859.501	34971.14	HF II	0.7	0.6		MIT
2857.650	2858.489	34983.51	HF II	1.3	1.3		MIT
2856.98	2857.82	34991.7	HF II	0.3	0.0		MIT
2852.012	2852.850	35052.67	HF II	1.3	1.7		MIT
2851.206	2852.044	35062.58	HF II	1.4	1.7		MIT
2850.960	2851.798	35065.60	HF	1.4	0.7		MIT
2850.148	2850.986	35075.59	HF II	1.3	1.3		MIT
2849.215	2850.052	35087.08	HF II	2.5	3.0		ME
2846.40	2847.24	35121.8	HF		0.5		ME
2845.828	2846.664	35128.83	HF	1.4	0.7		MIT
2841.95	2842.79	35176.8	HF		0.7J		ME
2841.492	2842.327	35182.44	HF	1.0			MIT
2835.18	2836.01	35260.8	HF II	0.0	0.5H		ME
2834.134	2834.968	35273.77	HF	1.3	1.1		MIT
2833.276	2834.109	35284.45	HF	1.4	0.6		MIT
2829.322	2830.154	35333.76	HF II	1.2	1.5		MIT
2828.517	2829.349	35343.82	HF		0.3		MIT
2828.149	2828.981	35348.42	HF II	0.5	0.0		MIT
2822.678	2823.509	35416.92	HF II	2.5	3.2		ME
2820.419	2821.249	35445.29	HF II	1.0	0.7		MIT
2820.227	2821.057	35447.70	HF II	2.7	3.3		ME
2819.738	2820.568	35453.85	HF	1.3	0.0		MIT
2818.940	2819.770	35463.89	HF	1.2			MIT
2817.675	2818.505	35479.81	HF	1.3	0.0		MIT
2816.069	2816.898	35500.04	HF II	1.0	0.8		MIT
2815.815	2816.644	35503.24	HF	1.0			MIT
2814.758	2815.587	35516.57	HF II	1.2	1.5		MIT
2814.475	2815.304	35520.15	HF II	1.4	1.6		MIT
2813.865	2814.694	35527.85	HF II	1.4	1.5		MIT
2809.606	2810.434	35581.70	HF II	0.5	0.3		MIT
2808.002	2808.829	35602.02	HF II	1.4	1.5		MIT
2808.00	2808.83	35602.0	HF II	2.3	3.0		ME
2807.527	2808.354	35608.04	HF II		0.3		MIT
2799.784	2800.609	35706.52	HF II		0.3		MIT
2792.802	2793.626	35795.78	HF II	0.3	0.3		MIT
2790.45	2791.27	35826.0	HF II	0.5	0.5		MIT
2789.804	2790.627	35834.24	HF II	1.2	0.7H		MIT
2789.73	2790.55	35835.2	HF II	1.3	1.5		MIT
2789.50	2790.32	35838.1	HF	0.7	1.2		MIT
2786.301	2787.123	35879.29	HF II	1.0	1.2		MIT
2783.692	2784.513	35912.92	HF	1.2			MIT
2779.366	2780.186	35968.81	HF	1.3	0.6		MIT
2775.266	2776.085	36021.95	HF II	1.2	1.2		MIT
2774.016	2774.835	36038.18	HF II	1.4	1.7		MIT
2773.500	2774.319	36044.88	HF II	1.0	1.0H		MIT
2773.360	2774.179	36046.70	HF II	2.5	3.3		ME
2773.017	2773.836	36051.16	HF	1.3	0.5		MIT
2772.324	2773.143	36060.17	HF II	1.2	1.3		MIT
2770.457	2771.275	36084.47	HF II	1.2	1.3		MIT
2769.672	2770.490	36094.70	HF		0.3		MIT
2767.34	2768.16	36125.1	HF II	0.8	0.9		MIT
2766.963	2767.780	36130.04	HF	1.2	0.0		MIT
2764.885	2765.702	36157.19	HF II	1.0	1.0		MIT
2762.687	2763.503	36185.95	HF	1.0	0.6		MIT
2761.631	2762.447	36199.79	HF	1.4	0.5		MIT
2761.20	2762.02	36205.4	HF II		0.7H		ME
2758.309	2759.124	36243.39	HF	1.0			MIT
2756.911	2757.726	36261.76	HF II	1.2	1.6		MIT
2753.61	2754.42	36305.2	HF		1.8		ME
2752.548	2753.362	36319.24	HF		0.6		MIT
2751.812	2752.626	36328.95	HF II	1.4	1.9		MIT
2746.617	2747.429	36397.66	HF	0.5			MIT
2743.638	2744.450	36437.18	HF	1.2	0.0		MIT
2740.417	2741.228	36480.00	HF II	0.6	0.3J		MIT
2738.760	2739.570	36502.07	HF II				MIT
2738.460	2739.270	36506.07	HF II	2.5	3.0		ME
2737.832	2738.642	36514.45	HF	1.0	0.6		MIT
2735.76	2736.57	36542.1	HF II		0.7H		ME
2735.12	2735.93	36550.6	HF II	0.8	0.5		ME
2732.98	2733.79	36579.3	HF II		1.0V		ME
2732.68	2733.49	36583.3	HF II	1.0	1.2		ME
2731.142	2731.951	36603.88	HF II	0.9	1.0		MIT
2730.852	2731.660	36607.77	HF	1.2			MIT
2730.71	2731.52	36609.7	HF	0.6			MIT
2729.100	2729.908	36631.27	HF	1.0	0.9		MIT
2726.699	2727.506	36663.52	HF	1.2			MIT
2718.588	2719.393	36772.91	HF	1.2	0.0		MIT
2718.51	2719.32	36774.0	HF II	1.0	1.3		ME
2717.90	2718.71	36782.2	HF II		0.8		ME
2713.839	2714.643	36837.25	HF	1.0	0.0		MIT
2713.51	2714.31	36841.7	HF II	0.3	0.7		ME
2712.425	2713.229	36856.45	HF II	1.4	1.7		MIT
2712.137	2712.941	36860.37	HF II	1.0	1.0		MIT
2711.993	2712.797	36862.33	HF II	1.0	1.0		MIT
2710.66	2711.46	36880.5	HF		0.5		MIT
2709.997	2710.800	36889.47	HF II		0.6		MIT
2706.727	2707.530	36934.04	HF II	1.0	1.7		MIT
2705.611	2706.413	36949.27	HF	0.7	0.0		MIT
2703.164	2703.966	36982.72	HF II	0.0	0.7		MIT
2699.63	2700.43	37031.1	HF	1.0			MIT
2696.180	2696.980	37078.51	HF	1.0			MIT
2688.35	2689.15	37186.5	HF	0.5			ME
2687.22	2688.02	37202.1	HF		1.4		ME
2685.22	2686.02	37229.8	HF II	1.0	1.3		ME
2684.02	2684.82	37246.5	HF II	0.0	0.5		ME
2683.350	2684.147	37255.78	HF II	2.3	3.0		ME
2678.426	2679.222	37324.27	HF II	1.0	1.1		MIT
2677.578	2678.374	37336.09	HF II	1.0	0.0		MIT
2676.63	2677.43	37349.3	HF II	1.0	1.1		ME
2671.25	2672.04	37424.5	HF II	1.0	1.0		ME
2669.56	2670.35	37448.2	HF II	0.3	0.6		MIT
2669.002	2669.796	37456.05	HF II	1.2	0.9		MIT
2668.279	2669.072	37466.20	HF	0.5	0.3		MIT
2665.974	2666.767	37498.59	HF II	2.0	2.5		ME
2661.883	2662.675	37556.22	HF II	2.3	2.7		ME
2660.52	2661.31	37575.5	HF II		0.3		MIT
2659.69	2660.48	37587.2	HF		0.5H		ME
2657.847	2658.638	37613.25	HF II	2.0	2.3		ME
2657.50	2658.29	37618.2	HF II	1.0	1.3		ME
2655.42	2656.21	37647.6	HF II	0.3			MIT
2652.86	2653.65	37684.0	HF II	0.6	0.5		ME
2652.33	2653.12	37691.5	HF II		0.5		ME
2651.165	2651.954	37708.04	HF II	2.0	2.5		ME
2650.28	2651.07	37720.6	HF		0.6H		ME
2649.15	2649.94	37736.7	HF II	1.0	1.3		ME
2647.294	2648.082	37763.18	HF II	2.3	2.8		ME
2642.753	2643.540	37828.06	HF	1.2	0.0		MIT
2642.08	2642.87	37837.7	HF	0.7			MIT
2641.410	2642.197	37847.29	HF II	2.5	3.0		ME
2638.712	2639.498	37885.99	HF II	2.3	2.6		ME
2637.001	2637.787	37910.57	HF	1.0	0.0		MIT
2635.79	2636.58	37928.0	HF II	1.1	1.3		ME
2630.907	2631.691	37998.38	HF I	2.0	0.5		ME
2626.95	2627.73	38055.6	HF II	1.0	1.0		MIT
2625.56	2626.34	38075.8	HF II	0.5	0.7		ME
2622.737	2623.519	38116.74	HF II	2.6	3.3		ME
2621.83	2622.61	38129.9	HF II		0.3H		MIT
2620.953	2621.735	38142.68	HF II	0.3	0.5		MIT
2616.606	2617.387	38206.04	HF	1.0	0.9		MIT
2614.59	2615.37	38235.5	HF		0.5H		ME
2614.293	2615.073	38239.84	HF II	0.8	0.8		MIT

AIR WAVELENGTH (ANGSTRMS)	VACUUM WAVELENGTH (ANGSTRMS)	WAVENUMBER (KAYSERS)	LMENT	LOG ARC	INTENSITY SPK	DIS	REF
2613.604	2614.384	38249.92	HF II	2.3	3.2		ME
2612.59	2613.37	38264.8	HF	0.6	0.5		MIT
2609.962	2610.741	38303.29	HF	0.7			MIT
2609.264	2610.043	38313.54	HF II		0.3		MIT
2608.451	2609.230	38325.48	HF	1.0	0.7		MIT
2607.243	2608.022	38343.24	HF II	1.0	1.3		MIT
2607.032	2607.811	38346.34	HF II	2.4	3.3		ME
2606.375	2607.154	38356.01	HF II	2.3	3.2		ME
2602.868	2603.646	38407.68	HF	0.6	0.6		MIT
2602.668	2603.446	38410.63	HF	0.8	0.7		MIT
2599.215	2599.992	38461.66	HF II	1.0	1.0		MIT
2598.851	2599.628	38467.04	HF II		0.3		MIT
2596.68	2597.46	38499.2	HF		0.5H		ME
2595.61	2596.39	38515.1	HF II	0.0	0.5		ME
2591.326	2592.101	38578.74	HF II	1.0	1.1		MIT
2582.923	2583.696	38704.24	HF	0.0	0.3H		MIT
2582.508	2583.281	38710.46	HF II	2.3	2.8		ME
2580.064	2580.836	38747.13	HF		0.3		MIT
2579.052	2579.824	38762.33	HF		0.5H		MIT
2578.155	2578.927	38775.82	HF II	2.2	2.8		ME
2576.830	2577.602	38795.75	HF II	2.2	2.9		ME
2575.511	2576.282	38815.62	HF II		0.9		MIT
2574.889	2575.660	38825.00	HF	1.0	1.3		MIT
2574.291	2575.062	38834.02	HF	0.3	0.3H		MIT
2573.912	2574.683	38839.73	HF II	3.1	3.0		ME
2572.96	2573.73	38854.1	HF II	0.7	0.5		MIT
2571.678	2572.448	38873.47	HF II	2.5	3.3		ME
2571.13	2571.90	38881.8	HF II		0.5		ME
2570.712	2571.482	38888.08	HF II	1.0	1.1		MIT
2567.46	2568.23	38937.3	HF		1.3		ME
2566.63	2567.40	38949.9	HF		0.5H		MIT
2563.613	2564.381	38995.76	HF II	2.0	2.6		ME
2560.74	2561.51	39039.5	HF		1.4		ME
2559.191	2559.958	39063.13	HF II	1.3	1.6		MIT
2559.02	2559.79	39065.7	HF	0.9			MIT
2555.33	2556.10	39122.1	HF		0.3H		MIT
2552.378	2553.144	39167.40	HF		0.7H		MIT
2551.847	2552.613	39175.55	HF II	0.5	0.0H		MIT
2551.428	2552.194	39181.98	HF II	1.6	2.5		ME
2551.38	2552.15	39182.7	HF II	1.9	2.7		ME
2548.51	2549.27	39226.8	HF II	0.0	0.3		MIT
2548.187	2548.952	39231.81	HF II	1.7	2.5		ME
2544.80	2545.56	39284.0	HF		0.5H		MIT
2543.13	2543.89	39309.8	HF II		0.3H		MIT
2537.337	2538.099	39399.56	HF II	1.7	2.6		ME
2534.332	2535.094	39446.27	HF		0.7H		MIT
2532.97	2533.73	39467.5	HF II	0.5	0.6		ME
2531.196	2531.957	39495.14	HF II	2.3	3.0		ME
2529.78	2530.54	39517.2	HF II		0.5H		ME
2523.654	2524.413	39613.17	HF		0.6H		MIT
2521.72	2522.48	39643.5	HF II		0.5H		ME
2521.50	2522.26	39647.0	HF	1.0	1.3		ME
2521.485	2522.244	39647.24	HF II	0.7	1.0		MIT
2520.53	2521.29	39662.3	HF	0.0	0.3H		MIT
2517.859	2518.617	39704.33	HF	0.7			MIT
2516.881	2517.639	39719.76	HF II	1.5	2.0		MIT
2515.486	2516.243	39741.79	HF II	1.7	2.6		ME
2513.033	2513.790	39780.58	HF II	2.3	3.0		ME
2512.963	2513.720	39781.68	HF II	2.0	2.5		ME
2512.702	2513.459	39785.82	HF II	2.3	2.9		ME
2505.17	2505.92	39905.4	HF II		0.3H		MIT
2502.656	2503.410	39945.51	HF	0.9	0.8		MIT
2500.736	2501.490	39976.18	HF II	1.1	1.3		MIT
2498.57	2499.32	40010.8	HF II		0.3H		MIT
2496.986	2497.739	40036.21	HF II	1.5	1.6		MIT
2494.37	2495.12	40078.2	HF II	0.0	0.7H		MIT
2487.16	2487.91	40194.4	HF	0.7			ME
2483.33	2484.08	40256.4	HF II	0.8	0.7		MIT
2481.438	2482.187	40287.05	HF II	0.8	1.0		MIT
2478.563	2479.312	40333.78	HF II	2.0	2.5J		MIT

AIR WAVELENGTH (ANGSTRMS)	VACUUM WAVELENGTH (ANGSTRMS)	WAVENUMBER (KAYSERS)	LMENT	LOG ARC	INTENSITY SPK	DIS	REF
2476.04	2476.79	40374.9	HF II	0.0H	0.3J		MIT
2474.10	2474.85	40406.5	HF II	0.3	0.3H		MIT
2473.915	2474.663	40409.55	HF II	1.2	1.4		MIT
2469.179	2469.925	40487.05	HF II	2.2	2.7		ME
2467.967	2468.713	40506.93	HF II	1.2	1.0		MIT
2465.06	2465.81	40554.7	HF II	1.0	1.2		MIT
2464.193	2464.938	40568.97	HF II	2.3	2.8		ME
2463.967	2464.712	40572.69	HF II	1.2	1.2		MIT
2461.72	2462.46	40609.7	HF		1.5		MIT
2460.493	2461.237	40629.97	HF II	1.3	1.7		MIT
2459.46	2460.20	40647.0	HF II	0.0	0.5		ME
2455.16	2455.90	40718.2	HF II	0.5	0.5		MIT
2453.979	2454.722	40737.81	HF II	0.7	0.9		MIT
2453.338	2454.081	40748.45	HF II	1.1	1.4		MIT
2452.47	2453.21	40762.9	HF		1.2		ME
2452.30	2453.04	40765.7	HF II	1.0	0.9		MIT
2450.065	2450.807	40802.89	HF II		0.3		MIT
2449.440	2450.182	40813.30	HF II	1.9	2.3		ME
2447.251	2447.992	40849.80	HF II	2.3	2.7		ME
2441.02	2441.76	40954.1	HF	0.0	0.7B		MIT
2434.74	2435.48	41059.7	HF II	0.9	1.2		MIT
2433.890	2434.628	41074.03	HF II		0.3H		MIT
2433.564	2434.302	41079.53	HF II	1.8	2.5		ME
2428.993	2429.730	41156.83	HF II	1.7	2.3		ME
2426.87	2427.61	41192.8	HF		0.5		MIT
2425.975	2426.712	41208.03	HF II	2.0	2.3		ME
2424.003	2424.739	41241.55	HF II		0.6H		MIT
2420.76	2421.50	41296.8	HF II	0.0	0.3		ME
2417.690	2418.425	41349.23	HF II	2.3	2.5		ME
2417.23	2417.96	41357.1	HF		0.5H		ME
2415.959	2416.693	41378.85	HF II	0.7	0.7		MIT
2413.33	2414.06	41423.9	HF II	0.7	0.9		ME
2410.137	2410.870	41478.80	HF II	2.3	2.5		ME
2406.442	2407.174	41542.49	HF II	1.1	1.4		MIT
2405.421	2406.153	41560.12	HF II	2.4	2.6		ME
2404.562	2405.294	41574.96	HF II	0.9	1.0		MIT
2403.60	2404.33	41591.6	HF II	0.3	0.7		ME
2402.03	2402.76	41618.8	HF II		0.3H		MIT
2400.78	2401.51	41640.5	HF II	1.2	1.2		MIT
2397.21	2397.94	41702.5	HF		0.3H		MIT
2394.96	2395.69	41741.6	HF		0.3H		ME
2393.828	2394.557	41761.37	HF II	2.0	2.2		ME
2393.362	2394.091	41769.50	HF II	1.7	1.9		MIT
2393.183	2393.912	41772.63	HF II	1.3	1.6		MIT
2392.008	2392.737	41793.15	HF		0.5H		MIT
2385.78	2386.51	41902.2	HF II		0.5		ME
2385.18	2385.91	41912.8	HF		0.6H		ME
2381.00	2381.73	41986.4	HF II	1.3	1.6		ME
2380.302	2381.028	41998.66	HF II	1.5	1.8		MIT
2378.81	2379.54	42025.0	HF II		0.3H		ME
2377.993	2378.719	42039.44	HF		0.8		MIT
2372.33	2373.05	42139.8	HF II	0.5	0.6		ME
2371.93	2372.65	42146.9	HF		0.5H		ME
2371.428	2372.152	42155.81	HF II	0.8	0.9		MIT
2368.47	2369.19	42208.5	HF II		0.6H		ME
2367.618	2368.341	42223.64	HF II	0.3	0.5		MIT
2365.98	2366.70	42252.9	HF II	0.9	1.0		ME
2362.342	2363.064	42317.94	HF		0.3		MIT
2353.025	2353.745	42485.48	HF	0.7			MIT
2351.215	2351.935	42518.19	HF II	2.0	2.2		MIT
2349.32	2350.04	42552.5	HF II		0.3H		MIT
2347.444	2348.163	42586.48	HF II	1.9	2.1		MIT
2345.41	2346.13	42623.4	HF II	0.6	0.7		ME
2343.324	2344.042	42661.35	HF II	1.8	1.9		MIT
2340.39	2341.11	42714.8	HF		0.9H		ME
2338.25	2338.97	42753.9	HF II		1.0H		MIT
2337.59	2338.31	42766.0	HF		0.5H		MIT
2337.33	2338.05	42770.7	HF II	1.5	1.5		MIT
2335.250	2335.966	42808.84	HF II		0.3		MIT
2333.77	2334.49	42836.0	HF II		0.6H		MIT

AIR WAVELENGTH (ANGSTRMS)	VACUUM WAVELENGTH (ANGSTRMS)	WAVENUMBER (KAYSERS)	LMENT	LOG ARC	INTENSITY SPK	INTENSITY DIS	REF
2332.968	2333.684	42850.71	HF II	1.6	1.7		MIT
2332.44	2333.16	42860.4	HF II	0.8	1.0		ME
2332.17	2332.89	42865.4	HF II	0.3			MIT
2329.09	2329.80	42922.0	HF II	0.3	0.3H		ME
2324.89	2325.60	42999.6	HF II	1.6	1.9		ME
2324.50	2325.21	43006.8	HF II	1.3	1.6		MIT
2323.25	2323.96	43029.9	HF II	1.6	1.8		MIT
2322.47	2323.18	43044.4	HF II	1.8	1.8		ME
2321.14	2321.85	43069.0	HF II	1.7	1.8		ME
2318.49	2319.20	43118.3	HF II	0.9	1.0		ME
2316.49	2317.20	43155.5	HF	1.0	1.3		ME
2305.34	2306.05	43364.2	HF II	0.8	0.9		ME
2304.50	2305.21	43380.0	HF	0.5	0.5H		ME
2303.21	2303.92	43404.3	HF II	0.6	0.6H		ME
2302.11	2302.82	43425.0	HF II	0.7	0.8		ME
2298.78	2299.49	43487.9	HF II	0.8	0.9		ME
2298.34	2299.05	43496.3	HF II	1.4	1.7		ME
2296.13	2296.84	43538.1	HF II	0.7	0.8H		ME
2292.77	2293.48	43601.9	HF II	0.5	0.5H		ME
2292.34	2293.05	43610.1	HF	0.3	0.3H		ME
2291.64	2292.35	43623.4	HF II	1.5	1.6		MIT
2288.63	2289.34	43680.8	HF II	0.7	0.8		ME
2285.43	2286.14	43742.0	HF II	0.8	0.8H		ME
2284.60	2285.30	43757.8	HF II	1.3	1.5		ME
2281.81	2282.51	43811.3	HF	0.7	0.8H		ME
2277.166	2277.869	43900.67	HF II	1.6	2.2		ME
2274.64	2275.34	43949.4	HF II	1.2	1.3		ME
2273.15	2273.85	43978.2	HF II	1.6	1.8		ME
2271.14	2271.84	44017.1	HF II	0.3	0.3H		ME
2270.68	2271.38	44026.1	HF II	0.8	0.8		ME
2269.86	2270.56	44042.0	HF	0.9	0.9H		ME
2268.47	2269.17	44069.0	HF	0.3	0.3H		ME
2267.91	2268.61	44079.8	HF	0.5	0.5H		ME
2267.58	2268.28	44086.2	HF	0.5	0.5H		ME
2267.35	2268.05	44090.7	HF II	0.5	0.5H		ME
2266.83	2267.53	44100.8	HF II	1.8	1.9		ME
2266.52	2267.22	44106.9	HF II	1.5	1.6		ME
2263.36	2264.06	44168.4	HF II	0.3	0.3H		ME
2261.54	2262.24	44204.0	HF	0.5	0.5H		ME
2260.25	2260.95	44229.2	HF	0.3	0.3H		ME
2258.68	2259.38	44259.9	HF II	1.2	1.3		ME
2257.89	2258.59	44275.4	HF II	1.0	1.2		ME
2256.26	2256.96	44307.4	HF II	0.0	0.3		ME
2255.15	2255.85	44329.2	HF II	1.6	1.8		ME
2254.008	2254.706	44351.67	HF II	1.8	1.9		MIT
2251.85	2252.55	44394.2	HF II	0.9	1.0		ME
2249.82	2250.52	44434.2	HF	0.7	0.7H		ME
2248.7	2249.4	44456.	HF		0.3		MD
2244.23	2244.93	44544.9	HF	0.6	0.6H		ME
2243.18	2243.88	44565.7	HF II	1.2	1.3		ME
2241.1	2241.8	44607.	HF		0.3		MD
2239.86	2240.56	44631.8	HF II	0.7	0.8		ME
2236.34	2237.03	44702.0	HF	0.6	0.6H		ME
2235.41	2236.10	44720.6	HF II	0.5	1.3H		ME
2231.41	2232.10	44800.8	HF II	0.5	0.5H		ME
2228.78	2229.47	44853.6	HF II	0.8	0.9		ME
2228.45	2229.14	44860.3	HF	0.6	0.6H		ME
2228.18	2228.87	44865.7	HF II	0.8	0.9		ME
2225.35	2226.04	44922.8	HF II	0.8	0.8		ME
2224.29	2224.98	44944.2	HF II	1.0	1.0J		ME
2223.36	2224.05	44963.0	HF II	0.6	0.7		ME
2220.48	2221.17	45021.3	HF	0.8	0.9H		ME
2218.37	2219.06	45064.1	HF II	1.0	1.2		ME
2217.66	2218.35	45078.5	HF II	0.3	0.3H		ME
2212.45	2213.14	45184.7	HF II	1.4	1.5		ME
2211.86	2212.55	45196.7	HF II	0.7	0.8		ME
2210.82	2211.51	45218.0	HF II	0.3	0.3H		ME
2206.59	2207.28	45304.7	HF	0.3	0.3H		ME
2206.11	2206.80	45314.5	HF II	1.0	1.0H		ME
2203.85	2204.54	45361.0	HF II	0.5	0.5		ME
2202.99	2203.68	45378.7	HF II	0.5	0.5H		ME
2201.67	2202.36	45405.9	HF II	0.5	0.5		ME
2199.84	2200.53	45443.7	HF	0.6	0.6H		ME
2199.57	2200.26	45449.2	HF II	0.9	0.9H		ME
2199.40	2200.09	45452.7	HF II	0.8	0.9		ME
2191.79	2192.48	45610.5	HF II	1.0	1.2		ME
2191.27	2191.96	45621.4	HF	0.7	0.7H		ME
2190.22	2190.90	45643.2	HF II	1.5	1.5		ME
2188.87	2189.55	45671.4	HF	0.7	0.8H		ME
2184.95	2185.63	45753.3	HF II	0.8	0.8H		ME
2184.31	2184.99	45766.7	HF II	0.8	0.9		ME
2181.94	2182.62	45816.4	HF	0.5	0.5H		ME
2180.22	2180.90	45852.6	HF	0.3	0.3H		ME
2178.90	2179.58	45880.4	HF II	1.8	1.9		ME
2177.7	2178.4	45906.	HF		0.3		MD
2175.36	2176.04	45955.0	HF II	1.4	1.5		ME
2173.44	2174.12	45995.6	HF II	1.2	1.3		ME
2172.96	2173.64	46005.7	HF II	0.7	0.7H		ME
2171.89	2172.57	46028.4	HF II	0.8	0.8H		ME
2170.22	2170.90	46063.8	HF II	1.3	1.5		ME
2166.5	2167.2	46143.	HF		0.3		MD
2164.8	2165.5	46179.	HF		0.3		MD
2162.48	2163.16	46228.7	HF II	1.2	1.2		ME
2161.61	2162.29	46247.3	HF II	1.0	1.0H		ME
2158.12	2158.80	46322.1	HF	1.2	1.2H		ME
2156.44	2157.12	46358.1	HF II	1.4	1.4		ME
2154.66	2155.34	46396.5	HF II	0.6	0.6H		ME
2150.30	2150.98	46490.5	HF	1.0	1.0H		ME
2150.07	2150.75	46495.5	HF II	0.7	0.7		ME
2149.86	2150.54	46500.0	HF	0.8	0.8H		ME
2144.73	2145.41	46611.2	HF II	0.5	0.5H		ME
2144.17	2144.85	46623.4	HF II	0.8	0.8H		ME
2141.84	2142.51	46674.1	HF II	1.3	1.3		ME
2139.24	2139.91	46730.8	HF II	1.5	1.6		ME
2137.22	2137.89	46775.0	HF II	0.8	0.8		ME
2134.53	2135.20	46833.9	HF II	1.2	1.3		ME
2133.36	2134.03	46859.6	HF	0.6	0.6H		ME
2132.28	2132.95	46883.4	HF II	1.0	1.0		ME
2130.09	2130.76	46931.6	HF	0.9	0.9H		ME
2129.10	2129.77	46953.4	HF II	1.8	2.0		ME
2127.50	2128.17	46988.7	HF	0.7	0.7H		ME
2126.61	2127.28	47008.4	HF	0.8	0.9H		ME
2124.59	2125.26	47053.0	HF II	1.7	1.9		ME
2123.68	2124.35	47073.2	HF II	1.6	1.6		ME
2122.94	2123.61	47089.6	HF II	1.2	1.2H		ME
2119.9	2120.6	47157.	HF		0.3		MD
2119.3	2120.0	47170.	HF		0.3		MD
2117.95	2118.62	47200.5	HF II	0.7	0.7H		ME
2116.80	2117.47	47226.2	HF	0.6			ME
2115.61	2116.28	47252.7	HF	0.6			ME
2115.02	2115.69	47265.9	HF II	1.3	1.3		ME
2114.62	2115.29	47274.9	HF II	0.3			ME
2114.2	2114.9	47284.	HF		0.3		MD
2110.42	2111.09	47368.9	HF II	1.2	1.2		ME
2108.50	2109.17	47412.0	HF	1.2	1.0H		ME
2107.47	2108.14	47435.2	HF II	1.8	1.8		ME
2106.4	2107.1	47459.	HF		0.3		MD
2105.0	2105.7	47491.	HF		0.6		MD
2102.9	2103.6	47538.	HF		0.8		MD
2098.91	2099.58	47628.7	HF II	0.3			ME
2096.18	2096.85	47690.7	HF II	2.0	2.2		ME
2094.33	2095.00	47732.8	HF II	0.5	0.3		ME
2093.22	2093.89	47758.1	HF II	0.8			ME
2090.83	2091.49	47812.7	HF II	1.6	1.6		ME
2089.95	2090.61	47832.8	HF II	1.5	1.5		ME
2089.38	2090.04	47845.9	HF	0.5			ME
2088.77	2089.43	47859.8	HF II	1.7	1.7		ME
2087.66	2088.32	47885.3	HF	0.6			ME
2086.80	2087.46	47905.0	HF II	0.9	0.6		ME
2086.29	2086.95	47916.7	HF	1.1			ME

AIR WAVELENGTH (ANGSTRMS)	VACUUM WAVELENGTH (ANGSTRMS)	WAVENUMBER (KAYSERS)	LMENT	LOG ARC	INTENSITY SPK	INTENSITY DIS	REF
2084.45	2085.11	47959.0	HF	0.3			ME
2083.80	2084.46	47974.0	HF II	1.6	1.5		ME
2082.80	2083.46	47997.0	HF II	1.2	0.9		ME
2079.72	2080.38	48068.1	HF	0.8			ME
2079.11	2079.77	48082.2	HF II	0.5D			ME
2076.70	2077.36	48138.0	HF	0.5			ME
2076.04	2076.70	48153.3	HF	0.6			ME
2071.1	2071.8	48268.	HF		0.8		MD
2068.84	2069.50	48320.8	HF	1.3	0.6H		ME
2068.76	2069.42	48322.7	HF II	0.5			ME
2068.58	2069.24	48326.9	HF	0.7			ME
2067.6	2068.3	48350.	HF		0.3		MD
2066.20	2066.86	48382.6	HF	0.3			ME
2065.55	2066.21	48397.8	HF II	0.3	0.6		ME
2064.78	2065.44	48415.8	HF II	1.2	0.6		ME
2064.43	2065.09	48424.0	HF	0.3			ME
2063.23	2063.89	48452.2	HF II	0.5			ME
2061.39	2062.05	48495.5	HF	0.6	0.3		ME
2060.49	2061.15	48516.6	HF II	0.8			ME
2057.39	2058.05	48589.7	HF II	0.5			ME
2055.30	2055.96	48639.1	HF	0.6			ME
2054.86	2055.52	48649.5	HF	0.3			ME
2054.3	2055.0	48663.	HF		0.8		MD
2053.7	2054.4	48677.	HF		0.3		MD
2051.7	2052.4	48724.	HF		0.3		MD
2049.6	2050.3	48774.	HF		0.6		MD
2045.2	2045.9	48879.	HF		0.3		MD
2039.08	2039.74	49026.0	HF	0.8	0.3		ME
2037.32	2037.97	49068.3	HF II	0.3	0.3		ME
2036.24	2036.89	49094.3	HF II	0.5			ME
2035.4	2036.1	49115.	HF		0.3		MD
2031.4	2032.1	49211.	HF		0.3		MD
2029.88	2030.53	49248.1	HF II	0.7	0.6		ME
2029.3	2030.0	49262.	HF		0.3		MD
2028.18	2028.83	49289.4	HF II	1.4	1.2		ME
2025.9	2026.6	49345.	HF		0.6		MD
2023.36	2024.01	49406.8	HF	0.3			ME
2022.0	2022.7	49440.	HF		0.3		MD
2020.07	2020.72	49487.3	HF	0.5	0.6		ME
2013.38	2014.03	49651.7	HF II	0.3	0.3		ME
2012.78	2013.43	49666.5	HF II	1.0	0.6		ME
2011.3	2012.0	49703.	HF		0.3		MD
2010.5	2011.1	49723.	HF		0.3		MD
2008.0	2008.6	49785.	HF		0.6		MD
2002.0	2002.6	49934.	HF		0.8		MD
9999.00	10001.7	9998.26	HG I			1.0	SU
9980.90	9983.64	10016.4	HG I			1.0	SU
9969.34	9972.07	10028.0	HG I			1.0	SU
9962.15	9964.88	10035.2	HG			0.3	SU
9946.0	9948.7	10051.	HG II			0.6	RS
9945.01	9947.74	10052.5	HG I			0.3	SU
9838.08	9840.78	10161.8	HG I			1.0	SU
9821.75	9824.44	10178.7	HG I			0.5	SU
9792.90	9795.59	10208.7	HG I			0.3	SU
9775.30	9777.98	10227.1	HG I			0.5	SU
9639.0	9641.6	10372.	HG II			0.8	RS
9574.26	9576.89	10441.8	HG I			0.5	SU
9554.1	9556.7	10464.	HG I			0.3	SU
9526.21	9528.82	10494.5	HG I			0.8	SU
9515.70	9518.31	10506.1	HG			0.3	SU
9508.42	9511.03	10514.1	HG			0.3	SU
9495.82	9498.42	10528.1	HG I			1.0	SU
9442.75	9445.34	10587.2	HG I			1.0	SU
9438.26	9440.85	10592.3	HG I			1.0	SU
9432.06	9434.65	10599.2	HG I			1.0	SU
9425.64	9428.23	10606.4	HG I			1.0	SU
9419.38	9421.96	10613.5	HG I			0.5	SU
9342.46	9345.02	10700.9	HG I			0.3	SU
9338.38	9340.94	10705.6	HG I			0.9	SU
9327.24	9329.80	10718.3	HG I			0.3	SU

AIR WAVELENGTH (ANGSTRMS)	VACUUM WAVELENGTH (ANGSTRMS)	WAVENUMBER (KAYSERS)	LMENT	LOG ARC	INTENSITY SPK	INTENSITY DIS	REF
9298.62	9301.17	10751.3	HG I			0.3	SU
9279.12	9281.67	10773.9	HG I			0.3	SU
9263.69	9266.23	10791.9	HG I			0.3	SU
9253.90	9256.44	10803.3	HG I			0.9	SU
9243.10	9245.64	10815.9	HG I			1.0	SU
9200.72	9203.25	10865.7	HG I			0.5	SU
9157.85	9160.36	10916.6	HG I			0.3	SU
9039.20	9041.68	11059.9	HG I			0.6	SU
8991.36	8993.83	11118.7	HG I			0.8	SU
8988.86	8991.33	11121.8	HG I			0.9	SU
8973.65	8976.11	11140.7	HG I			0.9	SU
8893.73	8896.17	11240.8	HG I			0.8	SU
8887.50	8889.94	11248.7	HG I			0.6	SU
8783.71	8786.12	11381.6	HG I			1.0	SU
8778.45	8780.86	11388.4	HG I			0.8	SU
8773.11	8775.52	11395.3	HG I			1.0	SU
8763.02	8765.43	11408.5	HG I			1.0	SU
8758.06	8760.47	11414.9	HG I			1.0	SU
8751.91	8754.31	11422.9	HG I			0.8	SU
8746.09	8748.49	11430.5	HG I			0.5	SU
8716.43	8718.82	11469.4	HG			0.7	SU
8706.41	8708.80	11482.6	HG I			0.7	SU
8704.51	8706.90	11485.1	HG I			0.8	SU
8686.67	8689.06	11508.7	HG I			0.8	SU
8652.74	8655.12	11553.9	HG I			1.0	SU
8613.58	8615.95	11606.4	HG I			0.8	SU
8590.07	8592.43	11638.2	HG I			0.3	SU
8582.50	8584.86	11648.4	HG I			0.8	SU
8557.04	8559.39	11683.1	HG I			0.8	SU
8548.2	8550.5	11695.	HG II			1.0	RS
8538.87	8541.22	11707.9	HG I			0.5	SU
8511.85	8514.19	11745.1	HG			0.6	SU
8505.51	8507.85	11753.9	HG I			1.0	SU
8492.47	8494.80	11771.9	HG I			0.7	SU
8464.07	8466.40	11811.4	HG			0.6	SU
8443.58	8445.90	11840.1	HG I			0.6	SU
8423.05	8425.36	11868.9	HG I			0.7	SU
8418.16	8420.47	11875.8	HG I			0.7	SU
8406.2	8408.5	11893.	HG II			0.7	RS
8401.56	8403.87	11899.3	HG I			0.8	SU
8373.21	8375.51	11939.6	HG I			0.7	SU
8353.97	8356.27	11967.1	HG I			0.3	SU
8315.01	8317.30	12023.1	HG I			0.3	SU
8287.46	8289.74	12063.1	HG I			0.3	SU
8271.76	8274.03	12086.0	HG I			0.3	SU
8252.63	8254.90	12114.0	HG I			0.3	SU
8214.12	8216.38	12170.8	HG I			1.1	SU
8206.14	8208.40	12182.7	HG I			0.5	SU
8200.84	8203.09	12190.5	HG I			0.8	SU
8197.4	8199.7	12196.	HG II			0.3	WD
8195.61	8197.86	12198.3	HG I			0.9	SU
8177.6	8179.8	12225.	HG II			0.8	RS
8172.90	8175.15	12232.2	HG I			0.7	SU
8167.22	8169.47	12240.7	HG I			0.7	MU
8164.52	8166.76	12244.7	HG I			1.0	SU
8153.55	8155.79	12261.2	HG I			0.6	SU
8145.40	8147.64	12273.5	HG I			0.3	SU
8094.27	8096.50	12351.0	HG			0.5	SU
8090.48	8092.70	12356.8	HG I			0.3	SU
8070.59	8072.81	12387.3	HG I			0.6	SU
8022.28	8024.49	12461.9	HG I			0.8	SU
8013.31	8015.51	12475.8	HG I			0.3	SU
7978.70	7980.89	12529.9	HG I			1.8	SU
7972.76	7974.95	12539.3	HG			0.5	SU
7944.66	7946.85	12583.6	HG II			0.9	PS
7929.64	7931.82	12607.4	HG I			0.6	SU
7821.15	7823.30	12782.3	HG I			0.9	SU
7820.73	7822.88	12783.0	HG I			0.3	MU
7728.49	7730.62	12935.6	HG I			1.0	SU
7718.48	7720.60	12952.3	HG			0.3	SU

AIR WAVELENGTH (ANGSTRMS)	VACUUM WAVELENGTH (ANGSTRMS)	WAVENUMBER (KAYSERS)	LMENT		LOG ARC	INTENSITY SPK	DIS	REF
7687.21	7689.33	13005.0	HG	I			0.3	SU
7674.37	7676.48	13026.8	HG	I			0.8	SU
7666.92	7669.03	13039.5	HG	I			0.6	SU
7646.15	7648.25	13074.9	HG	I			0.5	SU
7621.2	7623.3	13118.	HG	II			0.3	SU
7602.18	7604.27	13150.5	HG	I			0.7	SU
7551.79	7553.87	13238.2	HG	I			0.8	SU
7542.81	7544.89	13254.0	HG	I			0.5	SU
7519.96	7522.03	13294.3	HG	I			0.5	SU
7454.33	7456.38	13411.3	HG	I			1.7	SU
7418.1	7420.1	13477.	HG	II			1.6	PS
7410.46	7412.50	13490.7	HG	I			1.0	SU
7395.8	7397.8	13517.	HG				1.0	LF
7373.5	7375.5	13558.	HG				1.0	WD
7371.71	7373.74	13561.6	HG	I			1.9	SU
7367.3	7369.3	13570.	HG	I			0.9	SU
7346.37	7348.39	13608.4	HG	II			3.0	PS
7344.16	7346.18	13612.5	HG	II			0.5	PS
7332.4	7334.4	13634.	HG				0.7	WD
7327.47	7329.49	13643.5	HG	I			1.0	SU
7316.9	7318.9	13663.	HG	I			0.5	SU
7313.	7315.	13670.	HG	I			0.7	SU
7310.87	7312.88	13674.5	HG				1.2	SU
7301.68	7303.69	13691.7	HG				1.4	SU
7294.76	7296.77	13704.7	HG	I			1.3	SU
7269.2	7271.2	13753.	HG	I			1.2	SU
7245.7	7247.7	13797.	HG				0.7	WD
7177.9	7179.9	13928.	HG	I			0.5	SU
7175.96	7177.94	13931.6	HG	I			1.2	SU
7126.29	7128.25	14028.7	HG	I			1.0	SU
7119.92	7121.88	14041.2	HG				0.7	SU
7110.9	7112.9	14059.	HG				1.0	LF
7091.99	7093.95	14096.5	HG	I			2.0	SU
7081.88	7083.83	14116.7	HG	I			2.1	SU
7044.54	7046.48	14191.5	HG	I			1.0	SU
7009.31	7011.24	14262.8	HG	I			0.7	SU
6969.	6971.	14350.	HG				1.0	SU
6943.69	6945.61	14397.6	HG				1.0	SU
6938.1	6940.0	14409.	HG	II			1.4	RS
6924.	6926.	14440.	HG	I			0.7	SU
6907.16	6909.07	14473.7	HG	I			2.1	SU
6897.	6899.	14500.	HG				0.7	SU
6888.74	6890.64	14512.4	HG	I			1.4	SU
6870.80	6872.70	14550.3	HG				1.2	SU
6838.9	6840.8	14618.	HG				0.7	LF
6836.6	6838.5	14623.	HG				0.7	LF
6716.17	6718.02	14885.3	HG	I			1.9	SU
6715.2	6717.1	14887.	HG	II			1.4	RS
6709.5	6711.4	14900.	HG				0.7	LF
6689.85	6691.70	14943.9	HG				0.7	LF
6675.5	6677.3	14976.	HG	II			0.7	RS
6646.7	6648.5	15041.	HG	II			1.0	NU
6639.5	6641.3	15057.	HG				1.0	LF
6609.8	6611.6	15125.	HG	II			0.7	PS
6602.4	6604.2	15142.	HG				1.0	LF
6594.3	6596.1	15160.	HG				1.0	LF
6584.6	6586.4	15183.	HG				1.0	LF
6553.6	6555.4	15255.	HG				1.0	LF
6541.2	6543.0	15283.	HG		0.5		0.7	LF
6521.13	6522.93	15330.5	HG	II			1.1	PS
6512.95	6514.75	15349.8	HG				1.0	LF
6501.47	6503.27	15376.9	HG				0.7	PS
6501.25	6503.05	15377.4	HG				0.7	LF
6501.03	6502.83	15377.9	HG				0.7	LF
6499.01	6500.81	15382.7	HG				0.7	LF
6418.95	6420.72	15574.6	HG				1.4	LF
6394.94	6396.71	15633.0	HG	II			1.4	PS
6383.34	6385.10	15661.4	HG				1.2	LF
6362.10	6363.86	15713.7	HG				1.2	LF
6348.95	6350.71	15746.3	HG				0.7	LF

AIR WAVELENGTH (ANGSTRMS)	VACUUM WAVELENGTH (ANGSTRMS)	WAVENUMBER (KAYSERS)	LMENT		LOG ARC	INTENSITY SPK	DIS	REF
6343.80	6345.55	15759.1	HG				1.2	LF
6316.47	6318.22	15827.2	HG				1.2	LF
6314.61	6316.36	15831.9	HG				1.0	LF
6300.09	6301.83	15868.4	HG				0.7	LF
6291.26	6293.00	15890.7	HG	II			1.7	PS
6242.24	6243.97	16015.5	HG				1.0	LF
6234.37	6236.09	16035.7	HG	I			1.2	LF
6220.30	6222.02	16071.9	HG				1.2	LF
6196.43	6198.14	16133.9	HG	II			1.1	LF
6186.48	6188.19	16159.8	HG				1.0	LF
6170.98	6172.69	16200.4	HG				1.2	LF
6166.50	6168.21	16212.2	HG				0.7	LF
6149.50	6151.20	16257.0	HG	II			2.3	PS
6146.93	6148.63	16263.8	HG				1.2	PS
6123.27	6124.96	16326.6	HG				1.2	PS
6122.3	6124.0	16329.	HG				0.7	LF
6100.36	6102.05	16387.9	HG				1.4	LF
6089.79	6091.48	16416.4	HG	II			1.4	LF
6072.64	6074.32	16462.7	HG				1.0	WT
6046.01	6047.68	16535.3	HG				0.7	LF
6040.12	6041.79	16551.4	HG	II			0.7	PS
6023.62	6025.29	16596.7	HG				0.7	LF
6017.15	6018.82	16614.6	HG				1.0	LF
5981.42	5983.08	16713.8	HG				0.6	LF
5961.49	5963.14	16769.7	HG				1.1	LF
5956.30	5957.95	16784.3	HG				0.6	LF
5948.67	5950.32	16805.8	HG				0.6	LF
5947.8	5949.4	16808.	HG	II			0.6	PS
5943.98	5945.63	16819.1	HG				0.6	LF
5935.55	5937.19	16843.0	HG				0.6	LF
5890.16	5891.79	16972.8	HG				1.6	WD
5888.94	5890.57	16976.3	HG	II			1.3	PS
5872.03	5873.66	17025.2	HG	I			1.0	WT
5871.73	5873.36	17026.0	HG	II			1.6	PS
5868.06	5869.69	17036.7	HG				0.6	WD
5860.10	5861.72	17059.8	HG				0.7	WD
5859.38	5861.00	17061.9	HG	I			1.5	WT
5851.89	5853.51	17083.8	HG	II			0.9	PS
5838.77	5840.39	17122.1	HG				0.7	WD
5821.46	5823.07	17173.1	HG				1.2	WD
5803.65	5805.26	17225.8	HG	I			1.8	WT
5790.654	5792.260	17264.42	HG	I			3.0	WT
5789.66	5791.27	17267.4	HG	I	2.7			ME
5769.59	5771.19	17327.4	HG	I	2.8	2.3		WD
5698.69	5700.27	17543.0	HG	II			1.1	PS
5677.17	5678.75	17609.5	HG	II			2.5	PS
5675.86	5677.44	17613.6	HG	I			1.9	WD
5644.55	5646.12	17711.3	HG	II			0.3	PS
5595.40	5596.95	17866.9	HG	II			1.3	PS
5571.0	5572.5	17945.	HG	II			0.6	PS
5549.79	5551.33	18013.7	HG	I			1.2	WD
5499.8	5501.3	18177.	HG	II			0.7	ML
5460.753	5462.271	18307.41	HG	I	3.3			CN
5457.8	5459.3	18317.	HG	II			0.9	PS
5425.25	5426.76	18427.2	HG	II			2.3	PS
5420.15	5421.66	18444.6	HG				0.9	WD
5405.79	5407.29	18493.5	HG				0.8 J	WD
5393.50	5395.00	18535.7	HG	I			0.7	WD
5389.01	5390.51	18551.1	HG	I			0.9	WD
5385.79	5387.29	18562.2	HG				0.6	WD
5384.70	5386.20	18566.0	HG	I			1.2	WD
5371.42	5372.91	18611.9	HG	II			1.0	NU
5365.06	5366.55	18633.9	HG				1.3 J	WD
5354.05	5355.54	18672.3	HG	I			1.5 J	WD
5342.4	5343.9	18713.	HG	II			1.1	PS
5316.69	5318.17	18803.5	HG	I			1.2 J	WD
5299.53	5301.00	18864.4	HG	II			1.0	NU
5293.95	5295.42	18884.2	HG	II			1.1	PS
5290.1	5291.6	18898.	HG	I			1.0	WD
5286.11	5287.58	18912.2	HG	II			0.3	PS

AIR WAVELENGTH (ANGSTRMS)	VACUUM WAVELENGTH (ANGSTRMS)	WAVENUMBER (KAYSERS)	LMENT		LOG ARC	INTENSITY SPK	DIS	REF
5277.3	5278.8	18944.	HG				0.8	PS
5271.33	5272.80	18965.3	HG	II			0.6	PS
5238.7	5240.2	19083.	HG	II			0.6	PS
5222.81	5224.26	19141.4	HG	II			1.9	PS
5218.9	5220.4	19156.	HG	I			1.0	WD
5216.39	5217.84	19165.0	HG				1.3	PS
5210.79	5212.24	19185.6	HG				1.8	PS
5204.78	5206.23	19207.8	HG				1.6	PS
5196.15	5197.60	19239.7	HG	II			1.3	PS
5165.8	5167.2	19353.	HG	I			0.7	WD
5162.15	5163.59	19366.4	HG	II			0.7	NU
5146.26	5147.69	19426.2	HG				1.5	PS
5140.10	5141.53	19449.5	HG	I			0.7	WD
5138.09	5139.52	19457.1	HG	I			1.0	WD
5128.45	5129.88	19493.6	HG	II			2.2	PS
5120.55	5121.98	19523.7	HG	I			1.3	WD
5102.42	5103.84	19593.1	HG	I			1.0	WD
5045.82	5047.23	19812.9	HG				1.2J	WD
5025.64	5027.04	19892.4	HG	I			1.3	WD
4991.5	4992.9	20028.	HG	I	1.2			WD
4980.82	4982.21	20071.4	HG	I			0.8	WD
4980.57	4981.96	20072.4	HG				1.8	PS
4970.13	4971.52	20114.6	HG	I	0.7			WD
4961.89	4963.27	20148.0	HG	II		1.0		NU
4960.32	4961.70	20154.4	HG				0.7H	WD
4951.9	4953.3	20189.	HG	II			0.3	PS
4916.036	4917.408	20335.92	HG	I			1.7	CN
4896.9	4898.3	20415.	HG	I			0.7	WD
4890.27	4891.64	20443.1	HG	I			1.2L	WD
4883.1	4884.5	20473.	HG	I			0.9	WD
4855.72	4857.08	20588.5	HG	II			2.0	PS
4832.2	4833.6	20689.	HG	I			0.7	WD
4827.1	4828.4	20711.	HG	I			1.2	WD
4825.62	4826.97	20716.9	HG				1.8	PS
4822.3	4823.6	20731.	HG	I			0.7	WD
4797.01	4798.35	20840.5	HG				2.5	PS
4782.1	4783.4	20905.	HG	I			0.7	WD
4762.22	4763.55	20992.7	HG	II			2.0	PS
4748.1	4749.4	21055.	HG	I			0.7	WD
4722.8	4724.1	21168.	HG	I			0.7	WD
4704.63	4705.95	21249.7	HG	II			2.3	PS
4701.8	4703.1	21262.	HG	I			0.7	WD
4696.25	4697.56	21287.6	HG	II			0.7	NU
4685.3	4686.6	21337.	HG	I			0.7	WD
4672.7	4674.0	21395.	HG	I			0.7	WD
4662.4	4663.7	21442.	HG	I			0.7	WD
4660.28	4661.58	21451.9	HG	II			2.3	PS
4653.4	4654.7	21484.	HG	I			0.7	WD
4640.66	4641.96	21542.6	HG	II			0.3	PS
4630.14	4631.44	21591.6	HG				1.5	PS
4609.72	4611.01	21687.2	HG				1.0	PS
4606.60	4607.89	21701.9	HG				0.7	PS
4597.72	4599.01	21743.8	HG				1.3	PS
4589.91	4591.20	21780.8	HG				1.0	PS
4563.27	4564.55	21908.0	HG				0.7	PS
4552.89	4554.17	21957.9	HG	II			1.5	PS
4537.01	4538.28	22034.8	HG				1.0	PS
4536.43	4537.70	22037.6	HG				0.7	PS
4522.68	4523.95	22104.6	HG				0.3	PS
4492.81	4494.07	22251.5	HG	II			1.5	PS
4487.48	4488.74	22278.0	HG				2.5	PS
4463.80	4465.05	22396.1	HG	II			0.3	NU
4455.49	4456.74	22437.9	HG	II			1.5	PS
4454.33	4455.58	22443.8	HG				1.5	PS
4451.637	4452.887	22457.34	HG				0.8	ST
4425.22	4426.46	22591.4	HG	II			1.5	PS
4404.86	4406.10	22695.8	HG	II			1.7	PS
4402.06	4403.30	22710.3	HG				1.7	PS
4398.62	4399.86	22728.0	HG	II			2.5	PS
4382.16	4383.39	22813.4	HG	II			1.0	PS

AIR WAVELENGTH (ANGSTRMS)	VACUUM WAVELENGTH (ANGSTRMS)	WAVENUMBER (KAYSERS)	LMENT		LOG ARC	INTENSITY SPK	DIS	REF
4376.194	4377.424	22844.49	HG				1.7	ST
4358.35	4359.57	22938.0	HG	I	3.5W	2.7		MIT
4347.59	4348.81	22994.8	HG		0.3H	0.3H		MIT
4347.496	4348.718	22995.28	HG	I	2.3	1.7		MIT
4343.634	4344.855	23015.73	HG	I	1.3	0.7		MIT
4339.235	4340.455	23039.06	HG	I	2.2	1.3		MIT
4313.1	4314.3	23179.	HG	I			0.7	CN
4284.6	4285.8	23333.	HG				1.0	PS
4282.6	4283.8	23344.	HG				1.0	PS
4281.94	4283.14	23347.3	HG				1.3	PS
4277.80	4279.00	23369.9	HG	II			1.0	PS
4273.72	4274.92	23392.2	HG				0.3H	ST
4264.65	4265.85	23442.0	HG				1.3	PS
4261.88	4263.08	23457.2	HG				1.8	PS
4259.30	4260.50	23471.4	HG	II			1.3	PS
4237.49	4238.68	23592.2	HG				1.3	PS
4233.985	4235.177	23611.76	HG				1.3	ST
4230.126	4231.317	23633.30	HG				1.3	ST
4227.87	4229.06	23645.9	HG				1.8	PS
4227.29	4228.48	23649.2	HG				2.0	PS
4216.725	4217.913	23708.41	HG				1.7H	ST
4212.53	4213.72	23732.0	HG	II			1.7	PS
4212.22	4213.41	23733.8	HG				1.5	PS
4206.10	4207.28	23768.3	HG				1.5	PS
4180.95	4182.13	23911.3	HG				2.0	PS
4178.025	4179.203	23928.01	HG				1.7H	ST
4175.64	4176.82	23941.7	HG				1.3	MIT
4156.68	4157.85	24050.9	HG				1.7	PS
4140.38	4141.55	24145.6	HG				2.3	PS
4140.03	4141.20	24147.6	HG	I			0.7	WD
4124.91	4126.07	24236.1	HG	II			0.7	PS
4124.071	4125.234	24241.05	HG				1.5	CN
4122.121	4123.284	24252.51	HG				1.3	ST
4120.6	4121.8	24261.	HG	II			1.7	PS
4115.41	4116.57	24292.1	HG	II			1.7	PS
4108.066	4109.225	24335.49	HG	I	1.3	0.7		MIT
4103.87	4105.03	24360.4	HG				1.7	PS
4081.20	4082.35	24495.7	HG	II			1.0	PS
4077.811	4078.962	24516.04	HG	I	2.2	2.2		MIT
4060.89	4062.04	24618.2	HG				1.0	PS
4046.561	4047.704	24705.36	HG	I	2.3	2.5		MIT
4044.10	4045.24	24720.4	HG	II	0.7	1.0		MIT
4040.40	4041.54	24743.0	HG	II			1.3	PS
4025.954	4027.092	24831.82	HG				1.3	ST
4006.27	4007.40	24953.8	HG				1.5	PS
4003.10	4004.23	24973.6	HG				1.3	PS
3984.071	3985.198	25092.86	HG				0.7	CN
3983.977	3985.104	25093.45	HG	II			2.6	CN
3983.920	3985.047	25093.81	HG				0.7	CN
3983.845	3984.972	25094.28	HG				1.2	CN
3968.03	3969.15	25194.3	HG				1.7	PS
3960.24	3961.36	25243.9	HG				1.5	PS
3948.29	3949.41	25320.2	HG				2.0	PS
3945.09	3946.21	25340.8	HG				2.0	PS
3942.59	3943.71	25356.9	HG				2.0	PS
3942.24	3943.36	25359.1	HG				2.0	PS
3940.40	3941.52	25370.9	HG	II			0.3	PS
3925.65	3926.76	25466.3	HG				2.0	PS
3918.92	3920.03	25510.0	HG	II			2.3	PS
3916.25	3917.36	25527.4	HG	II			1.3	PS
3914.29	3915.40	25540.2	HG	II			2.0	PS
3906.410	3907.517	25591.70	HG	I	1.4	1.2		MIT
3903.638	3904.744	25609.87	HG	I	0.5			MIT
3901.903	3903.008	25621.26	HG	I	1.2	0.7		MIT
3893.89	3894.99	25674.0	HG	I			0.3H	W
3874.977	3876.075	25799.29	HG				1.5H	ST
3869.14	3870.24	25838.2	HG	II			1.0	NU
3861.08	3862.17	25892.1	HG				1.0H	WD
3849.274	3850.366	25971.56	HG				0.3	ST
3845.149	3846.240	25999.42	HG				1.5	ST

AIR WAVELENGTH (ANGSTRMS)	VACUUM WAVELENGTH (ANGSTRMS)	WAVENUMBER (KAYSERS)	LMENT	LOG ARC	INTENSITY SPK	DIS	REF
3839.26	3840.35	26039.3	HG			1.7	PS
3832.46	3833.55	26085.5	HG			1.3	PS
3826.613	3827.699	26125.36	HG			1.5	ST
3820.13	3821.21	26169.7	HG			0.3	WD
3815.84	3816.92	26199.1	HG I			0.3	CN
3806.38	3807.46	26264.2	HG II			2.3	PS
3801.658	3802.737	26296.85	HG I	1.2	0.3		MIT
3790.23	3791.31	26376.1	HG			0.3H	CN
3789.818	3790.894	26379.00	HG			0.3H	ST
3777.82	3778.89	26462.8	HG II			1.0	PS
3776.26	3777.33	26473.7	HG II			1.5	PS
3775.697	3776.770	26477.66	HG			0.3	ST
3774.52	3775.59	26485.9	HG			1.5	PS
3773.540	3774.612	26492.79	HG			0.3	ST
3771.010	3772.081	26510.56	HG			1.0	ST
3755.25	3756.32	26621.8	HG			0.3	WD
3751.737	3752.803	26646.75	HG			0.7	ST
3728.213	3729.273	26814.88	HG			0.3	ST
3704.18	3705.23	26988.9	HG I	1.3	1.3		MIT
3702.36	3703.41	27002.1	HG I			0.3	WD
3701.442	3702.495	27008.81	HG I	1.2	0.3		MIT
3684.91	3685.96	27130.0	HG			1.3	PS
3680.008	3681.056	27166.12	HG I			1.6	ST
3673.02	3674.07	27217.8	HG			0.7	WD
3666.55	3667.59	27265.8	HG	1.0			MIT
3663.276	3664.319	27290.20	HG I	2.7	2.6		CN
3663.09	3664.13	27291.6	HG	0.7D	0.5		MIT
3662.878	3663.921	27293.16	HG I	1.7	2.6		CN
3654.833	3655.874	27353.24	HG I	2.3		2.3	CN
3650.146	3651.186	27388.36	HG I	2.3	2.7		CN
3644.320	3645.359	27432.14	HG			1.6	ST
3638.34	3639.38	27477.2	HG			2.0	PS
3632.38	3633.42	27522.3	HG II			1.0	PS
3630.65	3631.68	27535.4	HG			2.0	PS
3618.53	3619.56	27627.6	HG			1.7	PS
3613.607	3614.638	27665.29	HG			1.6	ST
3607.60	3608.63	27711.4	HG			1.3	PS
3606.92	3607.95	27716.6	HG			1.5	PS
3605.80	3606.83	27725.2	HG II			2.3	PS
3604.09	3605.12	27738.3	HG II			1.7	PS
3594.94	3595.97	27808.9	HG II			0.3	PS
3593.48	3594.51	27820.2	HG			1.0	PS
3592.97	3594.00	27824.2	HG I		0.3H		ST
3591.48	3592.50	27835.7	HG I			0.5	WD
3590.95	3591.97	27839.8	HG I			0.6	WD
3578.747	3579.769	27934.77	HG			1.6	ST
3562.882	3563.900	28059.15	HG			0.3	ST
3561.737	3562.754	28068.17	HG			0.3	ST
3561.203	3562.220	28072.38	HG	1.0	0.8H		ST
3549.42	3550.43	28165.6	HG II			2.3	PS
3545.06	3546.07	28200.2	HG		0.3		ST
3543.70	3544.71	28211.0	HG		0.5H		ST
3543.079	3544.091	28215.98	HG		1.6		ST
3532.63	3533.64	28299.4	HG			2.3	PS
3524.27	3525.28	28366.6	HG I			1.2	WD
3524.19	3525.20	28367.2	HG II			2.0	PS
3523.00	3524.01	28376.8	HG I	0.3			WD
3510.68	3511.68	28476.4	HG II			1.0	PS
3493.85	3494.85	28613.5	HG II			2.0	PS
3492.77	3493.77	28622.4	HG			1.7	PS
3488.35	3489.35	28658.6	HG		0.3H		ST
3484.87	3485.87	28687.3	HG II			0.7	PS
3478.90	3479.90	28736.5	HG II	1.0H		1.3	PS
3477.85	3478.85	28745.2	HG I	0.3			WD
3473.01	3474.00	28785.2	HG			1.5	PS
3465.75	3466.74	28845.5	HG		0.3		ST
3456.17	3457.16	28925.5	HG			1.0	PS
3451.69	3452.68	28963.0	HG II			2.3	PS
3451.04	3452.03	28968.5	HG		0.3		ST
3447.22	3448.21	29000.6	HG I	0.3			WD

AIR WAVELENGTH (ANGSTRMS)	VACUUM WAVELENGTH (ANGSTRMS)	WAVENUMBER (KAYSERS)	LMENT	LOG ARC	INTENSITY SPK	DIS	REF
3445.43	3446.42	29015.6	HG		0.3		ST
3434.730	3435.715	29106.03	HG		1.2		ST
3434.16	3435.14	29110.9	HG		0.3		ST
3402.77	3403.75	29379.4	HG			1.3	PS
3396.315	3397.290	29435.23	HG		1.0		ST
3391.04	3392.01	29481.0	HG		0.3H		ST
3390.06	3391.03	29489.5	HG		1.7J		CN
3385.25	3386.22	29531.4	HG II			2.3	PS
3358.78	3359.75	29764.2	HG			1.3	PS
3351.61	3352.57	29827.8	HG		0.3		ST
3351.30	3352.26	29830.6	HG	1.0	1.3H		CN
3341.478	3342.439	29918.27	HG I	2.0	2.0		CN
3338.42	3339.38	29945.7	HG II			0.7	PS
3317.62	3318.57	30133.4	HG			0.7	PS
3317.30	3318.25	30136.3	HG II			1.0	PS
3312.31	3313.26	30181.7	HG II			1.3	PS
3305.09	3306.04	30247.6	HG I	0.3			CN
3277.87	3278.81	30498.8	HG II			1.7	PS
3264.06	3265.00	30627.9	HG II			2.3	PS
3257.59	3258.53	30688.7	HG II			1.0	PS
3252.21	3253.15	30739.5	HG II			1.5	PS
3227.840	3228.772	30971.53	HG		0.3		ST
3211.98	3212.91	31124.4	HG			0.7	PS
3208.20	3209.13	31161.1	HG II			1.3	PS
3201.36	3202.29	31227.7	HG			0.7	PS
3191.03	3191.95	31328.8	HG II			2.0	PS
3179.32	3180.24	31444.2	HG		0.3		CN
3144.48	3145.39	31792.6	HG I	1.0			CN
3135.76	3136.67	31881.0	HG	1.0			CN
3131.833	3132.741	31920.93	HG I	2.3	2.0		MIT
3131.546	3132.454	31923.86	HG I	2.6	2.5		MIT
3125.663	3126.569	31983.94	HG I	2.3	2.2		MIT
3116.24	3117.14	32080.6	HG II			2.0	PS
3109.84	3110.74	32146.7	HG			1.3	PS
3090.98	3091.88	32342.8	HG			0.7	PS
3090.60	3091.50	32346.8	HG II			2.3	PS
3085.29	3086.19	32402.4	HG I	0.3H			CN
3056.32	3057.21	32709.6	HG II			1.0	PS
3050.46	3051.35	32772.4	HG I	0.3H			CN
3045.56	3046.45	32825.1	HG			1.0	PS
3035.10	3035.98	32938.3	HG			1.5	PS
3027.496	3028.377	33020.98	HG I	1.4	1.2		MIT
3025.617	3026.498	33041.49	HG I	1.2	1.0		MIT
3024.603	3025.484	33052.56	HG	1.2			MIT
3023.476	3024.356	33064.88	HG I	1.8	1.0H		MIT
3021.499	3022.379	33086.52	HG I	1.9	1.6		CN
3011.05	3011.93	33201.3	HG I	0.5H			CN
3006.57	3007.45	33250.8	HG II			1.7	PS
3004.47	3005.35	33274.0	HG			1.5	PS
2991.41	2992.28	33419.3	HG II		0.7		NU
2982.46	2983.33	33519.6	HG II			1.2	PS
2981.74	2982.61	33527.7	HG II		0.3		NU
2980.61	2981.48	33540.4	HG II			0.5	PS
2975.19	2976.06	33601.5	HG			1.7	PS
2974.12	2974.99	33613.6	HG II			1.0	PS
2967.592	2968.459	33687.52	HG I	1.0			MIT
2967.517	2968.384	33688.37	HG I	0.7			MIT
2967.278	2968.144	33691.08	HG	2.0W	2.0		CN
2955.13	2955.99	33829.6	HG II			2.0	PS
2947.08	2947.94	33922.0	HG II			1.4	PS
2939.03	2939.89	34014.9	HG			1.0	PS
2935.92	2936.78	34050.9	HG II			0.3	PS
2925.406	2926.262	34173.29	HG I	1.8	1.7		CN
2916.27	2917.12	34280.3	HG II			1.5	PS
2909.36	2910.21	34361.8	HG II			1.4	PS
2893.595	2894.443	34548.96	HG I	1.6	1.7		MIT
2873.24	2874.08	34793.7	HG			1.3	PS
2856.969	2857.808	34991.85	HG I	1.3	1.0		MIT
2852.42	2853.26	35047.6	HG			0.9	PS
2847.83	2848.67	35104.1	HG I	1.2	2.0		CN

AIR WAVELENGTH (ANGSTRMS)	VACUUM WAVELENGTH (ANGSTRMS)	WAVENUMBER (KAYSERS)	LMENT	LOG INTENSITY ARC	SPK	DIS	REF
2847.67	2848.51	35106.1	HG II			2.5	PS
2836.67	2837.50	35242.2	HG		0.7		CN
2820.0	2820.8	35451.	HG	1.0H	1.3H		CN
2814.93	2815.76	35514.4	HG II			2.3	PS
2810.51	2811.34	35570.2	HG	0.3			CN
2806.42	2807.25	35622.1	HG I			1.2	PS
2804.462	2805.288	35646.96	HG I	1.3	1.0		CN
2803.472	2804.298	35659.55	HG I	1.3	1.3H		MIT
2802.76	2803.59	35668.6	HG	0.7	1.2		CN
2801.22	2802.05	35688.2	HG	0.3			CN
2799.76	2800.59	35706.8	HG I	1.0	1.0H		CN
2797.43	2798.25	35736.6	HG			0.9	PS
2796.13	2796.95	35753.2	HG II			0.9	PS
2791.04	2791.86	35818.4	HG	0.8	1.0		CN
2790.46	2791.28	35825.8	HG II			0.5	PS
2788.40	2789.22	35852.3	HG II			1.1	PS
2781.93	2782.75	35935.7	HG			0.9	PS
2769.26	2770.08	36100.1	HG II			0.5	NU
2761.97	2762.79	36195.4	HG			1.6	PS
2759.712	2760.527	36224.96	HG I	1.3	1.2		MIT
2752.84	2753.65	36315.4	HG	1.6	1.0H		MIT
2752.775	2753.589	36316.24	HG I	2.0G	1.7		MIT
2749.83	2750.64	36355.1	HG II			0.5	PS
2732.60	2733.41	36584.4	HG II			1.1	PS
2732.145	2732.954	36590.45	HG II			0.3	PS
2729.57	2730.38	36625.0	HG II			1.1	PS
2724.417	2725.224	36694.23	HG I		1.3		SS
2710.45	2711.25	36883.3	HG			1.1	DJ
2705.36	2706.16	36952.7	HG II			1.5	PS
2702.48	2703.28	36992.1	HG			1.2	PS
2700.92	2701.72	37013.4	HG I			0.3	DJ
2699.50	2700.30	37032.9	HG I	1.4		0.7	DJ
2698.851	2699.652	37041.81	HG I	1.4	1.5		MIT
2697.29	2698.09	37063.2	HG I			0.7	DJ
2691.13	2691.93	37148.1	HG II			0.7	PS
2686.6	2687.4	37211.	HG	0.7H		0.3	DJ
2685.52	2686.32	37225.7	HG			0.9	DJ
2682.98	2683.78	37260.9	HG			0.5	DJ
2674.99	2675.78	37372.2	HG I	0.7	0.7		M
2672.67	2673.46	37404.6	HG I	0.7		0.3	CN
2670.94	2671.73	37428.9	HG			0.7	DJ
2660.4	2661.2	37577.	HG	0.7H		0.3	DJ
2658.51	2659.30	37603.9	HG	0.7H	1.0H		DJ
2655.121	2655.911	37651.86	HG I	1.9	1.6		MIT
2653.681	2654.471	37672.29	HG I	1.9	1.6		CN
2652.042	2652.831	37695.57	HG I	2.0	1.8		CN
2642.84	2643.63	37826.8	HG I	1.2			DI
2642.60	2643.39	37830.2	HG I	1.0J			DI
2642.48	2643.27	37832.0	HG I	0.7H		0.3	DI
2639.93	2640.72	37868.5	HG I	0.7J	1.0		M
2638.85	2639.64	37884.0	HG I	0.3	0.0		DI
2634.771	2635.556	37942.65	HG I	1.3	1.1		CN
2627.90	2628.68	38041.9	HG II			0.7	PS
2625.24	2626.02	38080.4	HG I	1.0	1.2		M
2603.84	2604.62	38393.4	HG I	0.5			DI
2603.15	2603.93	38403.5	HG I	1.3J	1.3		M
2598.45	2599.23	38473.0	HG			0.7	DJ
2593.41	2594.19	38547.7	HG I	0.7	0.5		DJ
2589.79	2590.56	38601.6	HG			0.7D	DJ
2587.79	2588.56	38631.5	HG			0.3	DJ
2584.74	2585.51	38677.0	HG		0.7H		DJ
2580.07	2580.84	38747.0	HG II			0.5	PS
2578.91	2579.68	38764.5	HG I			0.3	DJ
2578.44	2579.21	38771.5	HG I	0.7J		0.3	DJ
2576.295	2577.066	38803.81	HG I	1.3	1.2		MIT
2574.86	2575.63	38825.4	HG			1.4	PS
2572.15	2572.92	38866.3	HG			1.3	PS
2571.75	2572.52	38872.4	HG I	0.5		0.0	DJ
2563.90	2564.67	38991.4	HG I		0.5		DJ
2561.18	2561.95	39032.8	HG I	0.5H	0.5		DI

AIR WAVELENGTH (ANGSTRMS)	VACUUM WAVELENGTH (ANGSTRMS)	WAVENUMBER (KAYSERS)	LMENT	LOG INTENSITY ARC	SPK	DIS	REF
2557.75	2558.52	39085.1	HG			0.7	DJ
2556.30	2557.07	39107.3	HG I	0.5			DI
2552.87	2553.64	39159.9	HG			1.4	PS
2551.67	2552.44	39178.3	HG			0.3	DJ
2548.55	2549.31	39226.2	HG I	0.6H	0.3		DI
2547.25	2548.01	39246.2	HG			0.3	DJ
2545.33	2546.09	39275.8	HG II			0.9	PS
2544.87	2545.63	39282.9	HG I	0.5		0.7	DI
2543.37	2544.13	39306.1	HG II			0.5	PS
2539.01	2539.77	39373.6	HG I		0.5H	0.5	PS
2536.519	2537.281	39412.27	HG I	3.3G	3.0G		CN
2534.775	2535.537	39439.38	HG I	1.5	1.5		MIT
2531.69	2532.45	39487.4	HG I	0.5H		0.5	PS
2530.04	2530.80	39513.2	HG			0.3	DJ
2529.53	2530.29	39521.1	HG I	0.3		0.0	DI
2525.98	2526.74	39576.7	HG I	0.5H	0.7		PS
2524.11	2524.87	39606.0	HG I	0.7			DI
2523.37	2524.13	39617.6	HG II			0.3	PS
2521.69	2522.45	39644.0	HG			0.3	DJ
2521.23	2521.99	39651.2	HG I	0.5H		0.0	DI
2517.45	2518.21	39710.8	HG I	0.5H		0.0	DI
2514.79	2515.55	39752.8	HG II			1.3	PS
2514.26	2515.02	39761.2	HG I	0.3H			DI
2512.03	2512.79	39796.5	HG		0.7		CN
2511.64	2512.40	39802.6	HG I	0.3H		0.0	DI
2509.47	2510.23	39837.0	HG I	0.3H		0.0	DI
2509.09	2509.85	39843.1	HG		0.5		CN
2501.18	2501.93	39969.1	HG			0.3H	DI
2499.37	2500.12	39998.0	HG II		0.5		DJ
2497.77	2498.52	40023.6	HG		0.5		CN
2496.78	2497.53	40039.5	HG		0.5		CN
2494.12	2494.87	40082.2	HG			0.3	DJ
2493.295	2494.047	40095.48	HG			0.5	PS
2492.09	2492.84	40114.9	HG II			1.0H	PS
2486.32	2487.07	40208.0	HG			0.5	PS
2483.829	2484.579	40248.27	HG I	1.5	0.7H		CN
2482.721	2483.471	40266.23	HG I	1.4	1.0		CN
2482.008	2482.757	40277.80	HG I	1.5	0.7		CN
2481.18	2481.93	40291.2	HG			0.7	DJ
2480.62	2481.37	40300.3	HG			1.6	PS
2478.66	2479.41	40332.2	HG I	1.2H	1.2H		DJ
2472.31	2473.06	40435.8	HG			0.7	DJ
2468.95	2469.70	40490.8	HG II		0.7H		NU
2468.73	2469.48	40494.4	HG		1.0H		DJ
2467.52	2468.27	40514.3	HG			1.3	DJ
2464.068	2464.813	40571.02	HG I	1.2	1.2		MIT
2458.85	2459.59	40657.1	HG			1.0	DJ
2453.53	2454.27	40745.3	HG II		0.3		NU
2446.905	2447.646	40855.58	HG I	1.0	1.2		MIT
2441.03	2441.77	40953.9	HG I	0.6		0.0	DI
2436.03	2436.77	41038.0	HG			0.3	DJ
2417.16	2417.89	41358.3	HG			0.3	DJ
2416.25	2416.98	41373.9	HG			0.3	DJ
2414.13	2414.86	41410.2	HG II			1.4H	PS
2407.35	2408.08	41526.8	HG II			1.4H	PS
2404.09	2404.82	41583.1	HG			1.0D	DJ
2400.52	2401.25	41645.0	HG I	0.3	0.5		CN
2399.74	2400.47	41658.5	HG I	1.0	1.2		CN
2399.379	2400.109	41664.77	HG I	1.3J	1.0		CN
2390.16	2390.89	41825.5	HG			0.7	PS
2384.42	2385.15	41926.1	HG			0.7	DJ
2382.06	2382.79	41967.7	HG			0.9	PS
2379.990	2380.716	42004.17	HG I	1.0	1.0		CN
2378.33	2379.06	42033.5	HG	1.3	1.3		MIT
2374.02	2374.74	42109.8	HG I	1.0H			DI
2358.94	2359.66	42379.0	HG			0.7D	DI
2358.48	2359.20	42387.2	HG	0.3			DI
2354.216	2354.936	42463.99	HG		1.3		ST
2352.48	2353.20	42495.3	HG I	1.1H	1.2H		CN
2350.45	2351.17	42532.0	HG			0.3	DJ

AIR WAVELENGTH (ANGSTRMS)	VACUUM WAVELENGTH (ANGSTRMS)	WAVENUMBER (KAYSERS)	LMENT		LOG ARC	INTENSITY SPK	INTENSITY DIS	REF
2345.55	2346.27	42620.9	HG	II	1.0	1.5		NU
2345.431	2346.149	42623.03	HG	I	0.5	0.7		CN
2341.54	2342.26	42693.9	HG				1.0	DJ
2340.55	2341.27	42711.9	HG	I	0.5	0.3		DJ
2339.37	2340.09	42733.5	HG				0.9	PS
2336.05	2336.77	42794.2	HG				0.3W	DJ
2327.30	2328.01	42955.1	HG	I			0.3	DJ
2323.20	2323.91	43030.9	HG	I	1.0H	0.7H		DJ
2322.53	2323.24	43043.3	HG	II			0.3	NU
2320.67	2321.38	43077.8	HG				0.3	DJ
2314.39	2315.10	43194.6	HG			0.7H		DJ
2302.09	2302.80	43425.4	HG	I	1.2H	1.2		CN
2296.23	2296.94	43536.2	HG	II		1.0		NU
2291.92	2292.63	43618.1	HG			0.7H		DJ
2269.92	2270.62	44040.8	HG	I			1.0	DJ
2263.64	2264.34	44163.0	HG				1.0	PS
2262.233	2262.933	44190.44	HG	II			2.0	PS
2260.260	2260.960	44229.01	HG	II			2.3	PS
2258.77	2259.47	44258.2	HG	I	0.7	0.5		DJ
2252.780	2253.478	44375.85	HG	II			1.7	PS
2247.70	2248.40	44476.1	HG				0.7H	DJ
2234.06	2234.75	44747.6	HG	II		0.5		NU
2230.04	2230.73	44828.3	HG			0.7H		DJ
2229.86	2230.55	44831.9	HG	II		0.5		NU
2224.82	2225.51	44933.5	HG			1.0H		CN
2224.710	2225.402	44935.70	HG	II			1.5	PS
2224.03	2224.72	44949.4	HG	II		0.7		NU
2216.31	2217.00	45106.0	HG				0.3	DJ
2195.48	2196.17	45533.9	HG				0.3	DJ
2190.74	2191.42	45632.4	HG	II			1.2	DJ
2183.15	2183.83	45791.0	HG				0.3	DJ
2180.87	2181.55	45838.9	HG				0.3	DJ
2175.40	2176.08	45954.2	HG				0.3	DJ
2171.07	2171.75	46045.8	HG	II		0.3		NU
2166.24	2166.92	46148.5	HG				0.3	DJ
2154.72	2155.40	46395.1	HG				0.3	DJ
2148.003	2148.679	46540.22	HG	II			1.8	PS
2144.67	2145.35	46612.5	HG				0.7	DJ
2141.13	2141.80	46689.6	HG				0.7H	DJ
2138.25	2138.92	46752.5	HG				0.7	DJ
2122.38	2123.05	47102.0	HG				0.3	DJ
2116.36	2117.03	47236.0	HG				0.3	DJ
2113.36	2114.03	47303.0	HG				0.3H	DJ
2100.84	2101.51	47584.9	HG				0.3H	DJ
2091.57	2092.23	47795.8	HG				0.7H	DJ
2066.65	2067.31	48372.0	HG				0.3	DJ
2058.49	2059.15	48563.8	HG				1.0	DJ
2054.61	2055.27	48655.5	HG				1.3	DJ
2052.929	2053.587	48695.29	HG	II			2.0	PS
2043.8	2044.5	48913.	HG				1.0	DJ
2042.18	2042.84	48951.6	HG				1.0	DJ
2037.62	2038.27	49061.1	HG				0.5	DJ
2029.46	2030.11	49258.3	HG				0.9	DJ
2026.971	2027.624	49318.81	HG	II			2.0	PS
2022.19	2022.84	49435.4	HG				1.5	DJ
2020.90	2021.55	49467.0	HG				0.3	DJ
2017.13	2017.78	49559.4	HG				0.3	DJ
2008.55	2009.20	49771.1	HG				1.0	DJ
2004.57	2005.22	49869.9	HG				0.9	DJ
2001.91	2002.56	49936.1	HG				1.0	DJ
2000.73	2001.38	49965.6	HG				0.3	DJ
6694.32	6696.17	14933.9	HO		0.8			ED
6667.93	6669.77	14993.0	HO		0.3			ED
6628.99	6630.82	15081.1	HO		1.0			ED
6607.47	6609.29	15130.2	HO		0.6			ED
6604.94	6606.76	15136.0	HO		1.3			ED
6550.97	6552.78	15260.7	HO		1.3			ED
6471.77	6473.56	15447.4	HO		0.7			ED
6373.86	6375.62	15684.7	HO		0.3			ED
6372.59	6374.35	15687.9	HO		1.0			ED

AIR WAVELENGTH (ANGSTRMS)	VACUUM WAVELENGTH (ANGSTRMS)	WAVENUMBER (KAYSERS)	LMENT	LOG ARC	INTENSITY SPK	INTENSITY DIS	REF
6354.35	6356.11	15732.9	HO	0.9			ED
6347.31	6349.06	15750.3	HO	0.3			ED
6306.68	6308.42	15851.8	HO	0.5			ED
6305.36	6307.10	15855.1	HO	1.3			ED
6282.78	6284.52	15912.1	HO	0.5			ED
6274.90	6276.64	15932.1	HO	0.3			ED
6255.75	6257.48	15980.9	HO	1.0			ED
6234.17	6235.89	16036.2	HO	1.0			ED
6208.65	6210.37	16102.1	HO	0.7			ED
6191.68	6193.39	16146.2	HO	0.9			ED
6183.32	6185.03	16168.1	HO	0.5			ED
6162.54	6164.25	16222.6	HO	0.5			ED
6162.20	6163.91	16223.5	HO	0.3			ED
6158.44	6160.14	16233.4	HO	0.5			ED
6156.58	6158.28	16238.3	HO	0.5			ED
6156.38	6158.08	16238.8	HO	0.5			ED
6133.60	6135.30	16299.1	HO	1.0			ED
6114.92	6116.61	16348.9	HO	0.5			ED
6088.22	6089.91	16420.6	HO	0.5			ED
6081.79	6083.47	16438.0	HO	0.9			ED
6067.97	6069.65	16475.4	HO	0.3			ED
6062.94	6064.62	16489.1	HO	0.7			ED
6060.31	6061.99	16496.2	HO	0.6			ED
6058.23	6059.91	16501.9	HO	0.3			ED
6050.71	6052.39	16522.4	HO	0.5			ED
6024.30	6025.97	16594.8	HO	0.3			ED
6023.47	6025.14	16597.1	HO	0.5			ED
6021.43	6023.10	16602.7	HO	0.7			ED
6005.33	6006.99	16647.3	HO	0.5			ED
6002.04	6003.70	16656.4	HO	0.9			ED
5993.04	5994.70	16681.4	HO	0.9			ED
5982.90	5984.56	16709.7	HO	2.3			ED
5981.43	5983.09	16713.8	HO	0.9			ED
5973.52	5975.17	16735.9	HO	2.1			ED
5962.67	5964.32	16766.4	HO	1.1			ED
5955.98	5957.63	16785.2	HO	2.0			ED
5948.48	5950.13	16806.4	HO	1.1			ED
5948.03	5949.68	16807.6	HO	2.3			ED
5933.71	5935.35	16848.2	HO	2.3			ED
5921.76	5923.40	16882.2	HO	2.3			ED
5904.29	5905.93	16932.1	HO	1.1			ED
5900.00	5901.64	16944.5	HO	0.9			ED
5892.56	5894.19	16965.9	HO	1.7			ED
5887.58	5889.21	16980.2	HO	1.1			ED
5882.99	5884.62	16993.4	HO	2.3			ED
5870.85	5872.48	17028.6	HO	1.3			ED
5860.28	5861.90	17059.3	HO	2.3			ED
5839.47	5841.09	17120.1	HO	1.5			ED
5839.14	5840.76	17121.1	HO	1.1			ED
5821.90	5823.51	17171.8	HO	1.1			ED
5813.13	5814.74	17197.7	HO	0.9			ED
5798.72	5800.33	17240.4	HO	1.1			ED
5786.07	5787.67	17278.1	HO	0.9			ED
5784.64	5786.24	17282.4	HO	0.9			ED
5780.01	5781.61	17296.2	HO	1.1			ED
5768.41	5770.01	17331.0	HO	0.9			ED
5766.64	5768.24	17336.3	HO	0.9			ED
5757.70	5759.30	17363.2	HO	0.9			ED
5751.12	5752.72	17383.1	HO	1.5			ED
5749.58	5751.17	17387.7	HO	1.1			ED
5743.83	5745.42	17405.2	HO	1.1			ED
5739.24	5740.83	17419.1	HO	1.1			ED
5735.13	5736.72	17431.6	HO	0.9			ED
5728.95	5730.54	17450.4	HO	0.9			ED
5726.72	5728.31	17457.2	HO	0.9			ED
5722.63	5724.22	17469.6	HO	0.9			ED
5719.53	5721.12	17479.1	HO	0.9			ED
5709.14	5710.72	17510.9	HO	0.9			ED
5706.88	5708.46	17517.9	HO	1.7			ED
5696.57	5698.15	17549.6	HO	0.9			ED

AIR WAVELENGTH (ANGSTRMS)	VACUUM WAVELENGTH (ANGSTRMS)	WAVENUMBER (KAYSERS)	LMENT	LOG ARC	INTENSITY SPK	DIS	REF
5693.64	5695.22	17558.6	HO	1.1			ED
5691.47	5693.05	17565.3	HO	2.3			ED
5685.76	5687.34	17582.9	HO	0.9			ED
5682.12	5683.70	17594.2	HO	1.5			ED
5681.41	5682.99	17596.4	HO	1.3			ED
5676.77	5678.35	17610.8	HO	1.1			ED
5674.70	5676.27	17617.2	HO	2.3			ED
5659.58	5661.15	17664.2	HO	1.3			ED
5648.10	5649.67	17700.2	HO	0.9			ED
5645.97	5647.54	17706.8	HO	1.1			ED
5642.91	5644.48	17716.4	HO	0.9			ED
5640.62	5642.19	17723.6	HO	2.0			ED
5628.24	5629.80	17762.6	HO	1.1			ED
5613.64	5615.20	17808.8	HO	1.1			ED
5593.70	5595.25	17872.3	HO	0.3			ED
5579.07	5580.62	17919.2	HO	0.9			ED
5578.84	5580.39	17919.9	HO	0.9			ED
5573.96	5575.51	17935.6	HO	1.1			ED
5566.52	5568.07	17959.6	HO	2.0			ED
5560.94	5562.48	17977.6	HO	1.3			ED
5553.14	5554.68	18002.8	HO	1.5			ED
5547.29	5548.83	18021.8	HO	0.9			ED
5530.57	5532.11	18076.3	HO	0.9			ED
5530.42	5531.96	18076.8	HO	0.9			ED
5516.45	5517.98	18122.6	HO	1.1			ED
5504.51	5506.04	18161.9	HO	1.1			ED
5498.57	5500.10	18181.5	HO	1.3			ED
5485.28	5486.80	18225.6	HO	0.9			ED
5484.61	5486.13	18227.8	HO	0.9			ED
5479.99	5481.51	18243.1	HO	1.1			ED
5471.95	5473.47	18269.9	HO	0.9			ED
5468.46	5469.98	18281.6	HO	1.3			ED
4979.97	4981.36	20074.8	HO	0.3			EX
4967.21	4968.60	20126.4	HO	0.7	0.3		EX
4939.01	4940.39	20241.3	HO	0.5			EX
4906.99	4908.36	20373.4	HO	0.3	0.3		EX
4868.22	4869.58	20535.6	HO	0.3			EX
4832.31	4833.66	20688.3	HO	0.5	0.3		EX
4820.36	4821.71	20739.5	HO	0.3	0.0		EX
4777.48	4778.82	20925.7	HO	0.3	0.0		EX
4762.39	4763.72	20992.0	HO	0.3	0.0H		EX
4759.67	4761.00	21004.0	HO	0.3			EX
4757.01	4758.34	21015.7	HO	0.3			EX
4749.02	4750.35	21051.1	HO	0.3	0.0H		EX
4742.04	4743.37	21082.1	HO	1.0	0.5		EX
4717.52	4718.84	21191.6	HO	0.3	0.0H		EX
4709.84	4711.16	21226.2	HO	0.5	0.3		EX
4701.69	4703.01	21263.0	HO	0.3	0.3		EX
4701.16	4702.48	21265.4	HO	0.3	0.0H		EX
4688.20	4689.51	21324.2	HO	0.3	0.3		EX
4685.83	4687.14	21335.0	HO	0.5	0.3		EX
4675.63	4676.94	21381.5	HO	0.5	0.0		EX
4674.62	4675.93	21386.1	HO	0.6	0.5		EX
4661.33	4662.63	21447.1	HO	0.6	0.5		KN
4651.39	4652.69	21492.9	HO	0.3			EX
4649.77	4651.07	21500.4	HO	0.9	0.0H		EX
4647.25	4648.55	21512.1	HO	0.3			EX
4641.7	4643.0	21538.	HO	0.3			KN
4640.61	4641.91	21542.9	HO	0.3			KN
4638.30	4639.60	21553.6	HO	0.3	0.0H		EX
4638.19	4639.49	21554.1	HO	0.3			EX
4629.10	4630.40	21596.4	HO	0.8	0.6		KN
4628.22	4629.52	21600.5	HO	0.3			KN
4618.84	4620.13	21644.4	HO	0.3			KN
4613.37	4614.66	21670.1	HO	0.6			KN
4612.27	4613.56	21675.2	HO	0.5			EX
4609.52	4610.81	21688.2	HO	0.8			KN
4609.32	4610.61	21689.1	HO	0.6	0.3H		EX
4608.67	4609.96	21692.2	HO	0.5	0.0		EX
4608.0	4609.3	21695.	HO	0.3			KN

AIR WAVELENGTH (ANGSTRMS)	VACUUM WAVELENGTH (ANGSTRMS)	WAVENUMBER (KAYSERS)	LMENT	LOG ARC	INTENSITY SPK	DIS	REF
4589.40	4590.69	21783.2	HO	0.5	0.0		EX
4572.43	4573.71	21864.1	HO	0.3	0.0		KN
4562.52	4563.80	21911.6	HO	0.8	0.0H		KN
4548.93	4550.21	21977.0	HO	0.5	0.3		EX
4534.58	4535.85	22046.6	HO	0.6			KN
4531.62	4532.89	22061.0	HO	0.7	0.3		EX
4531.28	4532.55	22062.6	HO	0.3	0.3H		KN
4530.08	4531.35	22068.5	HO	0.5	0.3		EX
4526.14	4527.41	22087.7	HO	0.3	0.3H		KN
4512.54	4513.81	22154.2	HO	0.3	0.0		KN
4510.81	4512.08	22162.7	HO	0.3	0.0		KN
4497.7	4499.0	22227.	HO	0.3			KN
4484.57	4485.83	22292.4	HO	0.6	0.3		EX
4481.0	4482.3	22310.	HO		0.5H		EX
4477.63	4478.89	22327.0	HO	0.5	0.5		KN
4473.58	4474.84	22347.2	HO	0.5	0.3		EX
4455.59	4456.84	22437.4	HO	0.7			KN
4449.69	4450.94	22467.2	HO	0.5	0.0		EX
4447.24	4448.49	22479.6	HO	0.5	0.5H		EX
4429.80	4431.04	22568.1	HO	0.5	0.0H		EX
4422.56	4423.80	22605.0	HO	0.5			EX
4420.54	4421.78	22615.3	HO	0.3	0.3		KN
4419.60	4420.84	22620.1	HO	0.5	0.3		EX
4409.36	4410.60	22672.7	HO	0.5	0.0		EX
4403.27	4404.51	22704.0	HO	0.5	0.0H		KN
4401.24	4402.48	22714.5	HO	0.5	0.5		EX
4398.05	4399.29	22731.0	HO	0.6			EX
4390.53	4391.76	22769.9	HO	0.3			KN
4384.94	4386.17	22798.9	HO	0.5			KN
4384.76	4385.99	22799.9	HO	0.5	0.5H		EX
4379.64	4380.87	22826.5	HO		0.3		EX
4379.15	4380.38	22829.1	HO	0.5			EX
4371.43	4372.66	22869.4	HO	0.5	0.0H		EX
4364.18	4365.41	22907.4	HO	0.5			KN
4363.93	4365.16	22908.7	HO	0.3	0.5		KN
4358.74	4359.97	22936.0	HO	0.5			EX
4356.74	4357.96	22946.5	HO	0.9	0.9		EX
4350.73	4351.95	22978.2	HO	1.6	1.2		KN
4346.96	4348.18	22998.1	HO	0.5	0.0H		EX
4346.66	4347.88	22999.7	HO	0.5			EX
4337.13	4338.35	23050.2	HO	0.6	0.6		KN
4332.55	4333.77	23074.6	HO	0.5	0.0H		EX
4330.64	4331.86	23084.8	HO	0.3	0.3		KN
4328.99	4330.21	23093.6	HO	0.5			KN
4326.39	4327.61	23107.5	HO	0.3	0.0		KN
4325.14	4326.36	23114.1	HO		0.3H		EX
4324.19	4325.41	23119.2	HO		0.3		EX
4315.02	4316.23	23168.4	HO		0.3H		EX
4311.04	4312.25	23189.7	HO	0.3	0.0		KN
4308.65	4309.86	23202.6	HO	0.6	0.3		EX
4301.62	4302.83	23240.5	HO	0.5	0.0		EX
4301.09	4302.30	23243.4	HO	0.3	0.3H		KN
4295.01	4296.22	23276.3	HO	0.5			EX
4290.17	4291.38	23302.5	HO	0.3	0.3		KN
4286.54	4287.75	23322.3	HO	0.5	0.0		EX
4284.54	4285.75	23333.2	HO	0.3	0.3		KN
4273.63	4274.83	23392.7	HO	0.3	0.3		KN
4266.04	4267.24	23434.4	HO	0.7	0.3		KN
4264.07	4265.27	23445.2	HO	1.2	0.9		KN
4258.62	4259.82	23475.2	HO	0.5	0.0H		EX
4254.43	4255.63	23498.3	HO	2.0	1.3		KN
4245.39	4246.59	23548.3	HO	0.5	0.3		EX
4243.76	4244.95	23557.4	HO	0.5	0.3		KN
4234.79	4235.98	23607.3	HO	0.3			KN
4229.49	4230.68	23636.9	HO	0.8	0.5		KN
4225.13	4226.32	23661.2	HO	0.5	0.3		EX
4223.64	4224.83	23669.6	HO		0.3		EX
4223.47	4224.66	23670.6	HO	0.6			KN
4222.26	4223.45	23677.3	HO	0.5	0.0H		EX
4221.07	4222.26	23684.0	HO	0.6	0.3		EX

AIR WAVELENGTH (ANGSTRMS)	VACUUM WAVELENGTH (ANGSTRMS)	WAVENUMBER (KAYSERS)	LMENT	LOG ARC	INTENSITY SPK DIS	REF
4218.07	4219.26	23700.9	HO	0.5	0.3	EX
4213.15	4214.34	23728.5	HO	0.5	0.0	EX
4211.73	4212.92	23736.5	HO	0.7	0.5	EX
4211.24	4212.43	23739.3	HO	0.5	0.0H	EX
4194.88	4196.06	23831.9	HO	0.5	0.3H	EX
4194.34	4195.52	23834.9	HO	1.5	1.2	KN
4190.0	4191.2	23860.	HO		0.3H	EX
4186.84	4188.02	23877.6	HO	0.5	0.5	EX
4173.23	4174.41	23955.5	HO	1.7		KN
4172.23	4173.41	23961.2	HO	0.3		KN
4169.41	4170.59	23977.4	HO	0.3H		EX
4168.02	4169.19	23985.4	HO	0.5		EX
4163.03	4164.20	24014.2	HO	2.0	2.0	KN
4160.0	4161.2	24032.	HO	0.5		KN
4152.75	4153.92	24073.6	HO	1.5	1.6	KN
4152.54	4153.71	24074.9	HO	1.5	1.5	KN
4151.14	4152.31	24083.0	HO	0.6	0.6	KN
4150.8	4152.0	24085.	HO	0.3		KN
4148.97	4150.14	24095.6	HO	0.5	0.0	KN
4142.86	4144.03	24131.1	HO		0.5	EX
4142.19	4143.36	24135.0	HO	0.3	0.0H	KN
4136.24	4137.41	24169.7	HO	1.6	1.4	KN
4134.56	4135.73	24179.6	HO	0.7	0.0H	KN
4129.44	4130.60	24209.5	HO	0.5	0.0	EX
4128.32	4129.48	24216.1	HO		0.5	EX
4127.16	4128.32	24222.9	HO	2.2	1.8	KN
4125.65	4126.81	24231.8	HO	1.3	1.2	KN
4120.20	4121.36	24263.8	HO	1.7	1.4	KN
4119.35	4120.51	24268.8	HO	0.5	0.3H	EX
4116.63	4117.79	24284.9	HO	0.5	0.0H	KN
4112.63	4113.79	24308.5	HO		0.3H	EX
4111.99	4113.15	24312.3	HO	0.6	0.3H	KN
4111.37	4112.53	24315.9	HO		0.3H	EX
4108.63	4109.79	24332.1	HO	2.0	1.6	KN
4106.6	4107.8	24344.	HO	0.3		KN
4106.50	4107.66	24344.8	HO	0.6	0.0H	KN
4105.85	4107.01	24348.6	HO	0.3		KN
4105.6	4106.8	24350.	HO		0.3H	EX
4104.30	4105.46	24357.8	HO	0.3		KN
4103.84	4105.00	24360.6	HO	2.6	2.6	KN
4103.30	4104.46	24363.7	HO	0.5	0.5	EX
4102.40	4103.56	24369.1	HO		0.5	EX
4101.09	4102.25	24376.9	HO	1.6	1.6	KN
4100.2	4101.4	24382.	HO	0.6		KN
4087.65	4088.80	24457.0	HO	0.6	0.3	KN
4087.35	4088.50	24458.8	HO	0.6		KN
4083.67	4084.82	24480.9	HO	0.5		EX
4078.00	4079.15	24514.9	HO	0.5	0.5	EX
4073.13	4074.28	24544.2	HO	0.5	0.3	EX
4068.05	4069.20	24574.9	HO	0.8	0.3	KN
4067.8	4068.9	24576.	HO	0.3		KN
4067.04	4068.19	24581.0	HO	0.3		KN
4065.10	4066.25	24592.7	HO	1.0	0.7	KN
4060.30	4061.45	24621.8	HO	0.5	0.3	EX
4057.55	4058.70	24638.5	HO	0.3	0.3H	KN
4055.50	4056.65	24650.9	HO		0.3	EX
4054.49	4055.64	24657.1	HO	0.5	0.3	KN
4053.92	4055.07	24660.5	HO	2.6	2.3	KN
4048.79	4049.93	24691.8	HO	0.3		KN
4047.50	4048.64	24699.6	HO	0.5	0.5	EX
4045.95	4047.09	24709.1	HO	1.0	0.3	EX
4045.43	4046.57	24712.3	HO	2.3	1.9	KN
4040.84	4041.98	24740.3	HO	2.2	1.5	KN
4038.86	4040.00	24752.5	HO	0.6	0.3D	EX
4037.60	4038.74	24760.2	HO	0.5	0.0	EX
4036.10	4037.24	24769.4	HO	0.3		KN
4031.75	4032.89	24796.1	HO	0.6	0.0H	KN
4028.85	4029.99	24814.0	HO	0.6	0.0	KN
4027.20	4028.34	24824.1	HO	0.7	0.0	KN
4025.39	4026.53	24835.3	HO	0.5		KN

AIR WAVELENGTH (ANGSTRMS)	VACUUM WAVELENGTH (ANGSTRMS)	WAVENUMBER (KAYSERS)	LMENT	LOG ARC	INTENSITY SPK DIS	REF
4023.93	4025.07	24844.3	HO	0.6	0.0H	KN
4018.10	4019.24	24880.4	HO	0.5	0.0H	EX
4014.18	4015.31	24904.6	HO	0.3	0.6	KN
4003.36	4004.49	24972.0	HO	0.6	0.0	KN
4002.70	4003.83	24976.1	HO	0.5	0.0H	KN
4000.62	4001.75	24989.1	HO	0.6		KN
3999.56	4000.69	24995.7	HO	0.8	0.3	EX
3998.28	3999.41	25003.7	HO	1.6	0.8	EX
3997.17	3998.30	25010.6	HO	0.8	0.6H	EX
3993.71	3994.84	25032.3	HO	1.0	0.6	EX
3992.55	3993.68	25039.6	HO	0.6		KN
3987.55	3988.68	25071.0	HO	0.5		KN
3986.5	3987.6	25078.	HO	0.7		KN
3985.80	3986.93	25082.0	HO	0.8	0.6H	EX
3985.64	3986.77	25083.0	HO	0.9	0.6H	EX
3984.11	3985.24	25092.6	HO	0.8	0.3	EX
3983.65	3984.78	25095.5	HO	0.8	0.3	EX
3982.03	3983.16	25105.7	HO		0.8	EX
3978.55	3979.68	25127.7	HO	0.8	0.8	EX
3977.05	3978.17	25137.1	HO	0.7	0.6	EX
3976.86	3977.98	25138.4	HO	0.8	0.6	EX
3975.88	3977.00	25144.6	HO	0.5	0.3	EX
3973.59	3974.71	25159.0	HO	0.0	0.6	EX
3972.60	3973.72	25165.3	HO	0.3	0.3H	KN
3963.27	3964.39	25224.6	HO	0.7	0.6	EX
3959.84	3960.96	25246.4	HO		0.6	EX
3959.68	3960.80	25247.4	HO	0.8		KN
3959.51	3960.63	25248.5	HO		0.6	EX
3955.74	3956.86	25272.6	HO	1.2	0.6	EX
3955.05	3956.17	25277.0	HO	0.6		KN
3950.50	3951.62	25306.1	HO	0.9		KN
3940.55	3941.67	25370.0	HO	1.1	0.6H	EX
3940.37	3941.49	25371.1	HO	0.8	0.6H	EX
3938.84	3939.96	25381.0	HO	0.3	0.3H	KN
3938.62	3939.73	25382.4	HO		0.6	EX
3937.00	3938.11	25392.9	HO	0.3	0.3	KN
3936.51	3937.62	25396.0	HO	1.0	0.6H	EX
3936.35	3937.46	25397.1	HO		0.6H	EX
3924.54	3925.65	25473.5	HO		0.6H	EX
3923.37	3924.48	25481.1	HO	0.7	0.3H	EX
3919.45	3920.56	25506.6	HO	0.8	0.6	KN
3913.96	3915.07	25542.3	HO	0.5		KN
3912.44	3913.55	25552.3	HO	0.6	0.3	EX
3911.80	3912.91	25556.4	HO	0.5		KN
3909.56	3910.67	25571.1	HO	0.3		KN
3906.26	3907.37	25592.7	HO	0.5	0.5	EX
3905.78	3906.89	25595.8	HO	1.5	0.8	EX
3905.55	3906.66	25597.3	HO	1.2	0.9	EX
3904.45	3905.56	25604.6	HO	0.6	0.3	KN
3902.24	3903.35	25619.1	HO	0.9	0.8	EX
3899.64	3900.74	25636.1	HO		0.8	EX
3897.27	3898.37	25651.7	HO	0.6	0.6	KN
3896.75	3897.85	25655.1	HO	0.7	0.6	KN
3896.24	3897.34	25658.5	HO	0.6	0.8	KN
3895.53	3896.63	25663.2	HO	0.6		KN
3893.54	3894.64	25676.3	HO	0.7	0.3H	KN
3893.10	3894.20	25679.2	HO	0.6	0.6	EX
3892.95	3894.05	25680.2	HO	0.8	0.6	EX
3892.70	3893.80	25681.8	HO	0.3	0.3H	EX
3891.02	3892.12	25692.9	HO	2.3	1.6	EX
3890.42	3891.52	25696.9	HO	0.7		KN
3888.95	3890.05	25706.6	HO	1.6	1.3	EX
3886.4	3887.5	25723.	HO	0.8		KN
3881.61	3882.71	25755.2	HO		0.8	EX
3880.65	3881.75	25761.6	HO		0.8	EX
3879.59	3880.69	25768.6	HO		0.8	EX
3874.70	3875.80	25801.1	HO	0.8	0.8	EX
3874.11	3875.21	25805.1	HO	0.3	0.9	EX
3872.14	3873.24	25818.2	HO	0.8	0.9	EX
3861.68	3862.77	25888.1	HO	1.6	1.3	EX

AIR WAVELENGTH (ANGSTRMS)	VACUUM WAVELENGTH (ANGSTRMS)	WAVENUMBER (KAYSERS)	LMENT	LOG ARC	INTENSITY SPK	INTENSITY DIS	REF
3857.65	3858.74	25915.2	HO		0.6		EX
3856.98	3858.07	25919.7	HO	0.9	0.8 H		EX
3855.60	3856.69	25928.9	HO		0.6		EX
3854.05	3855.14	25939.4	HO	1.0	1.3		KN
3851.4	3852.5	25957.	HO	0.6			KN
3846.68	3847.77	25989.1	HO	1.0	1.0		KN
3843.86	3844.95	26008.1	HO	0.9	1.0		EX
3842.05	3843.14	26020.4	HO	0.6	0.9		EX
3837.60	3838.69	26050.6	HO	0.3	0.8 H		EX
3837.45	3838.54	26051.6	HO	1.2	0.8 H		EX
3835.35	3836.44	26065.9	HO	0.3	0.8 H		EX
3831.9	3833.0	26089.	HO		0.8 H		EX
3830.49	3831.58	26098.9	HO	0.3	0.8		EX
3818.69	3819.77	26179.6	HO		0.8		EX
3816.79	3817.87	26192.6	HO		0.6		EX
3813.24	3814.32	26217.0	HO	1.2	0.9 H		EX
3810.70	3811.78	26234.4	HO	1.3	1.6		EX
3809.49	3810.57	26242.8	HO		0.6		EX
3803.07	3804.15	26287.1	HO		0.6 H		EX
3801.30	3802.38	26299.3	HO	0.8	0.6 H		EX
3800.67	3801.75	26303.7	HO		0.6 H		EX
3799.06	3800.14	26314.8	HO		0.6 H		EX
3798.25	3799.33	26320.4	HO		0.6		EX
3796.73	3797.81	26331.0	HO	1.3	1.6		EX
3792.95	3794.03	26357.2	HO		0.6 H		EX
3791.00	3792.08	26370.8	HO		0.8		EX
3788.45	3789.53	26388.5	HO		0.6 H		EX
3788.10	3789.18	26391.0	HO		0.6		EX
3786.85	3787.93	26399.7	HO		0.6		EX
3786.15	3787.23	26404.6	HO	0.8	0.6		EX
3785.42	3786.50	26409.6	HO		0.6 H		EX
3785.06	3786.13	26412.2	HO		0.6 H		EX
3782.65	3783.72	26429.0	HO		0.6		EX
3780.97	3782.04	26440.7	HO		0.6		EX
3780.37	3781.44	26444.9	HO		0.8		EX
3777.65	3778.72	26464.0	HO		0.6 H		EX
3766.22	3767.29	26544.3	HO	0.6	0.3		EX
3762.85	3763.92	26568.1	HO		0.6 H		EX
3759.26	3760.33	26593.4	HO		0.6		EX
3757.26	3758.33	26607.6	HO	1.6	1.5		EX
3755.5	3756.6	26620.	HO		0.6 H		EX
3753.75	3754.82	26632.5	HO	1.3	1.0 H		EX
3752.07	3753.14	26644.4	HO		0.6 H		EX
3748.17	3749.24	26672.1	HO	1.8	1.6		EX
3737.65	3738.71	26747.2	HO	0.8	1.0 H		EX
3736.35	3737.41	26756.5	HO	0.8	0.3		EX
3732.59	3733.65	26783.4	HO	0.8	0.3 H		EX
3732.09	3733.15	26787.0	HO	0.8	0.3		EX
3731.40	3732.46	26792.0	HO	0.9	0.8		EX
3729.53	3730.59	26805.4	HO	0.8	0.6		EX
3725.99	3727.05	26830.9	HO	0.8	0.6		EX
3724.45	3725.51	26842.0	HO	0.8	0.6		EX
3721.83	3722.89	26860.9	HO	0.9	0.6 H		EX
3721.33	3722.39	26864.5	HO	0.9	0.6		EX
3720.74	3721.80	26868.7	HO	0.8	0.6		EX
3712.90	3713.96	26925.5	HO	0.8	0.6		EX
3711.30	3712.36	26937.1	HO	0.8	0.6		EX
3710.75	3711.81	26941.1	HO	0.9	0.9 H		EX
3706.03	3707.08	26975.4	HO		1.0		EX
3702.35	3703.40	27002.2	HO	1.0	0.8		EX
3701.80	3702.85	27006.2	HO	0.6	0.8		EX
3694.81	3695.86	27057.3	HO		0.6		EX
3694.24	3695.29	27061.5	HO	1.0	1.3		EX
3691.45	3692.50	27081.9	HO	0.8	0.6		EX
3691.26	3692.31	27083.3	HO	0.8			EX
3690.65	3691.70	27087.8	HO		0.6		EX
3686.65	3687.70	27117.2	HO	0.8	0.8 H		EX
3685.16	3686.21	27128.1	HO	0.8	1.2		EX
3682.65	3683.70	27146.6	HO	0.8	0.6		EX
3679.70	3680.75	27168.4	HO	0.8	0.6 H		EX

AIR WAVELENGTH (ANGSTRMS)	VACUUM WAVELENGTH (ANGSTRMS)	WAVENUMBER (KAYSERS)	LMENT	LOG ARC	INTENSITY SPK	INTENSITY DIS	REF
3679.19	3680.24	27172.2	HO	0.8	0.6		EX
3678.61	3679.66	27176.4	HO	0.8	0.6		EX
3677.62	3678.67	27183.8	HO	0.8	0.8		EX
3676.59	3677.64	27191.4	HO	0.8	0.6		EX
3674.77	3675.82	27204.8	HO	0.9	1.2		EX
3674.36	3675.41	27207.9	HO	0.8	0.3		EX
3672.30	3673.35	27223.1	HO	0.8	0.3		EX
3671.65	3672.70	27228.0	HO	0.8	0.6		EX
3670.29	3671.34	27238.1	HO	0.8	0.8		EX
3669.52	3670.57	27243.8	HO	0.8	0.6		EX
3669.05	3670.09	27247.2	HO	0.8	0.8		EX
3667.97	3669.01	27255.3	HO	1.0	0.8		EX
3667.05	3668.09	27262.1	HO	0.8	1.0		EX
3666.65	3667.69	27265.1	HO	0.8	0.6		EX
3662.98	3664.02	27292.4	HO	0.8	0.6		EX
3662.27	3663.31	27297.7	HO	1.3	1.0		EX
3658.55	3659.59	27325.4	HO		0.6		EX
3654.45	3655.49	27356.1	HO	0.8	0.8		EX
3649.52	3650.56	27393.1	HO	0.3	0.6 H		EX
3645.90	3646.94	27420.3	HO		0.6		EX
3644.39	3645.43	27431.6	HO		0.6		EX
3638.32	3639.36	27477.4	HO	0.9	1.0		EX
3635.35	3636.39	27499.8	HO		0.6		EX
3634.68	3635.72	27504.9	HO	0.8	0.6 H		EX
3631.75	3632.79	27527.1	HO	0.9	0.6 H		EX
3630.20	3631.23	27538.8	HO		0.6		EX
3627.18	3628.21	27561.8	HO	1.2	1.2		EX
3626.70	3627.73	27565.4	HO	1.3	1.2		EX
3625.46	3626.49	27574.8	HO	0.8	0.6 H		EX
3620.91	3621.94	27609.5	HO		0.6		EX
3619.43	3620.46	27620.8	HO	0.8	0.8		EX
3616.56	3617.59	27642.7	HO	0.8	0.8		EX
3613.33	3614.36	27667.4	HO	0.8	0.9		EX
3606.09	3607.12	27723.0	HO		0.6		EX
3602.55	3603.58	27750.2	HO		0.6 H		EX
3601.87	3602.90	27755.4	HO	0.8			EX
3600.95	3601.98	27762.5	HO	0.8	1.0		EX
3599.81	3600.84	27771.3	HO		0.6		EX
3599.49	3600.52	27773.8	HO		0.6		EX
3598.77	3599.80	27779.4	HO	1.6	1.5		EX
3596.11	3597.14	27799.9	HO		0.6 H		EX
3595.00	3596.03	27808.5	HO		0.6		EX
3593.13	3594.16	27822.9	HO	0.8	0.8		EX
3592.95	3593.98	27824.3	HO	0.8			EX
3592.22	3593.25	27830.0	HO	0.8	1.0		EX
3591.33	3592.35	27836.9	HO		0.8		EX
3580.75	3581.77	27919.1	HO		0.6		EX
3579.12	3580.14	27931.9	HO		0.6		EX
3576.85	3577.87	27949.6	HO		0.6		EX
3576.25	3577.27	27954.3	HO	0.6	0.8		EX
3574.78	3575.80	27965.8	HO	1.0	1.3		EX
3574.20	3575.22	27970.3	HO		0.6 H		EX
3573.22	3574.24	27978.0	HO	0.9	1.0		EX
3567.25	3568.27	28024.8	HO		0.6		EX
3566.20	3567.22	28033.1	HO		0.6 H		EX
3560.15	3561.17	28080.7	HO	0.8	0.9		EX
3558.15	3559.17	28096.5	HO		0.6 H		EX
3556.76	3557.78	28107.4	HO	1.2	1.6		EX
3551.60	3552.61	28148.3	HO		0.6		EX
3550.20	3551.21	28159.4	HO		0.6		EX
3548.95	3549.96	28169.3	HO	0.8			EX
3548.55	3549.56	28172.5	HO		0.6		EX
3545.97	3546.98	28193.0	HO	1.3	1.3		EX
3544.92	3545.93	28201.3	HO	0.9			EX
3544.07	3545.08	28208.1	HO	0.9			EX
3542.33	3543.34	28221.9	HO		0.6		EX
3540.76	3541.77	28234.5	HO	0.8	0.9		EX
3540.04	3541.05	28240.2	HO		0.8 H		EX
3538.97	3539.98	28248.7	HO	0.6			EX
3538.52	3539.53	28252.3	HO		0.8		EX

AIR WAVELENGTH (ANGSTRMS)	VACUUM WAVELENGTH (ANGSTRMS)	WAVENUMBER (KAYSERS)	LMENT	LOG ARC	INTENSITY SPK	DIS	REF
3536.04	3537.05	28272.1	HO	0.6	0.8		EX
3534.93	3535.94	28281.0	HO		0.6		EX
3531.74	3532.75	28306.6	HO	1.0	1.3		EX
3523.97	3524.98	28369.0	HO		0.6		EX
3520.16	3521.17	28399.7	HO	0.9	1.0	H	EX
3519.92	3520.93	28401.6	HO	1.0	1.0	H	EX
3519.65	3520.66	28403.8	HO		0.6	H	EX
3515.58	3516.59	28436.7	HO	1.6	1.6		EX
3511.76	3512.76	28467.6	HO		0.6		EX
3510.75	3511.75	28475.8	HO		0.6		EX
3509.35	3510.35	28487.2	HO	0.8	0.3		EX
3508.35	3509.35	28495.3	HO		0.6	H	EX
3506.94	3507.94	28506.7	HO	1.0	1.0		EX
3505.42	3506.42	28519.1	HO		0.6		EX
3499.08	3500.08	28570.8	HO	1.0	1.0		EX
3498.87	3499.87	28572.5	HO	0.9	1.0		EX
3494.77	3495.77	28606.0	HO	1.5	1.6		EX
3493.10	3494.10	28619.7	HO	1.0	1.0		EX
3490.95	3491.95	28637.3	HO	0.8	0.6	H	EX
3489.59	3490.59	28648.5	HO	1.0	1.0		EX
3486.34	3487.34	28675.2	HO	0.3	0.8		EX
3485.86	3486.86	28679.1	HO	0.8	0.8		EX
3484.73	3485.73	28688.4	HO	1.6	1.5		EX
3483.88	3484.88	28695.4	HO		0.6	H	EX
3478.05	3479.05	28743.5	HO	0.8	1.0		EX
3477.74	3478.74	28746.1	HO	0.6	0.8		EX
3474.25	3475.24	28774.9	HO	1.6	1.3		EX
3473.92	3474.91	28777.7	HO	1.0	1.0		EX
3472.25	3473.24	28791.5	HO	0.8	0.9		EX
3469.40	3470.39	28815.2	HO		0.6		EX
3467.07	3468.06	28834.5	HO	0.9	0.8		EX
3461.96	3462.95	28877.1	HO	1.3	1.3		EX
3461.36	3462.35	28882.1	HO		0.6		EX
3460.95	3461.94	28885.5	HO	0.8	0.6		EX
3456.6	3457.6	28922.	HO		0.6	H	EX
3456.00	3456.99	28926.9	HO	1.8	1.8		EX
3455.70	3456.69	28929.4	HO	0.8	0.8	H	EX
3453.13	3454.12	28950.9	HO	1.5	1.3		EX
3449.01	3450.00	28985.5	HO		0.6	H	EX
3445.56	3446.55	29014.5	HO	0.8	0.8		EX
3437.05	3438.04	29086.4	HO		0.6	H	EX
3436.32	3437.31	29092.6	HO		0.6	H	EX
3435.61	3436.59	29098.6	HO		0.6	H	EX
3434.75	3435.73	29105.9	HO		0.6	H	EX
3432.09	3433.07	29128.4	HO	0.8	0.9		EX
3429.19	3430.17	29153.1	HO	1.0	1.2		EX
3428.13	3429.11	29162.1	HO	1.6	1.6		EX
3426.76	3427.74	29173.7	HO	0.8	0.8	H	EX
3425.35	3426.33	29185.7	HO	1.6	1.6		EX
3421.64	3422.62	29217.4	HO	1.3	1.3		EX
3416.46	3417.44	29261.7	HO	1.5	1.6		EX
3414.92	3415.90	29274.9	HO	1.5	1.5		EX
3414.26	3415.24	29280.5	HO		1.0		EX
3412.86	3413.84	29292.5	HO		0.6		EX
3411.56	3412.54	29303.7	HO	0.8	0.8		EX
3410.65	3411.63	29311.5	HO	1.0	1.0		EX
3410.25	3411.23	29314.9	HO	1.3	1.2		EX
3409.06	3410.04	29325.2	HO	0.8	0.6	H	EX
3408.21	3409.19	29332.5	HO	0.8	0.6	H	EX
3407.83	3408.81	29335.8	HO	0.8	0.6		EX
3406.27	3407.25	29349.2	HO		0.6	H	EX
3402.17	3403.15	29384.6	HO		0.6		EX
3401.59	3402.57	29389.6	HO		0.6		EX
3400.57	3401.55	29398.4	HO		0.6		EX
3398.98	3399.96	29412.1	HO	1.6	1.8		EX
3397.34	3398.32	29426.4	HO		0.6	H	EX
3396.13	3397.10	29436.8	HO		0.6		EX
3394.63	3395.60	29449.8	HO	0.9	0.9		EX
3393.60	3394.57	29458.8	HO	0.8	0.6		EX
3392.05	3393.02	29472.2	HO	0.8	0.9		EX

AIR WAVELENGTH (ANGSTRMS)	VACUUM WAVELENGTH (ANGSTRMS)	WAVENUMBER (KAYSERS)	LMENT	LOG ARC	INTENSITY SPK	DIS	REF
3390.75	3391.72	29483.5	HO	0.8	0.9		EX
3389.55	3390.52	29494.0	HO		0.6		EX
3385.05	3386.02	29533.2	HO	1.0	1.0		EX
3374.16	3375.13	29628.5	HO	0.8	0.9	H	EX
3372.79	3373.76	29640.5	HO	1.1	1.2		EX
3370.86	3371.83	29657.5	HO	0.8	0.9		EX
3368.06	3369.03	29682.2	HO		0.6		EX
3364.26	3365.23	29715.7	HO		1.0		EX
3363.41	3364.38	29723.2	HO		0.6		EX
3360.86	3361.83	29745.7	HO		0.6	H	EX
3359.49	3360.46	29757.9	HO		0.6		EX
3357.90	3358.86	29772.0	HO	0.8	0.9		EX
3354.57	3355.53	29801.5	HO	0.8	0.9		EX
3353.57	3354.53	29810.4	HO	0.8	0.9		EX
3353.45	3354.41	29811.5	HO	0.8			M
3352.08	3353.04	29823.6	HO	0.8	0.8		M
3350.46	3351.42	29838.1	HO	0.8	0.9		EX
3348.58	3349.54	29854.8	HO		0.6	H	EX
3346.06	3347.02	29877.3	HO		0.6		EX
3344.46	3345.42	29891.6	HO	0.6	0.8		EX
3343.56	3344.52	29899.6	HO	1.3	1.3		EX
3342.71	3343.67	29907.2	HO		0.6		EX
3340.46	3341.42	29927.4	HO		0.6	H	EX
3338.76	3339.72	29942.6	HO	1.1	1.3		EX
3337.20	3338.16	29956.6	HO	1.1	1.1		EX
3335.11	3336.07	29975.4	HO		0.6	H	EX
3333.16	3334.12	29992.9	HO		0.6		EX
3332.71	3333.67	29997.0	HO		0.6		EX
3331.91	3332.87	30004.2	HO	0.6	0.8	H	EX
3331.01	3331.97	30012.3	HO		0.6	H	EX
3329.66	3330.62	30024.5	HO		0.6	H	EX
3329.02	3329.98	30030.2	HO	0.6	0.8	H	EX
3323.21	3324.17	30082.7	HO		0.6		EX
3321.11	3322.07	30101.7	HO		0.6		EX
3320.24	3321.20	30109.6	HO	0.8	0.9		EX
3319.87	3320.83	30113.0	HO II	0.8	0.8		ME
3319.87	3320.83	30113.0	HO	0.8	0.8		EX
3316.37	3317.32	30144.8	HO		0.6		EX
3315.66	3316.61	30151.2	HO	0.6	0.8		EX
3314.01	3314.96	30166.2	HO		0.6		EX
3308.86	3309.81	30213.2	HO		0.6		EX
3305.16	3306.11	30247.0	HO	0.8	0.6	H	EX
3297.06	3298.01	30321.3	HO		0.6	H	EX
3290.96	3291.91	30377.5	HO	0.9	0.8		EX
3288.46	3289.41	30400.6	HO	0.8	0.8		EX
3283.07	3284.02	30450.5	HO		0.6	H	EX
3281.98	3282.93	30460.6	HO	1.1	1.2		EX
3281.17	3282.12	30468.1	HO		0.6	H	EX
3279.26	3280.20	30485.9	HO	0.9	1.0		EX
3278.15	3279.09	30496.2	HO	1.0	1.2		EX
3277.16	3278.10	30505.4	HO		0.6	H	EX
3264.77	3265.71	30621.2	HO	0.8	0.9		EX
3259.17	3260.11	30673.8	HO		0.8		EX
3258.8	3259.7	30677.	HO		0.6	H	EX
3258.47	3259.41	30680.4	HO		0.6	H	EX
3258.06	3259.00	30684.3	HO		0.6		EX
3257.39	3258.33	30690.6	HO	0.8	0.8		EX
3256.27	3257.21	30701.1	HO		0.6		EX
3239.02	3239.95	30864.6	HO		0.6		EX
3237.38	3238.31	30880.3	HO		0.6		EX
3236.56	3237.49	30888.1	HO		0.6		EX
3235.62	3236.55	30897.1	HO		0.6	H	EX
3234.52	3235.45	30907.6	HO	0.3	0.8		EX
3233.36	3234.29	30918.7	HO	0.8	0.3	H	EX
3230.57	3231.50	30945.4	HO	0.8	0.8		EX
3224.27	3225.20	31005.8	HO		0.6		EX
3223.27	3224.20	31015.4	HO		0.6	H	EX
3221.42	3222.35	31033.2	HO		0.6	H	EX
3216.67	3217.60	31079.1	HO		0.9		ME
3215.37	3216.30	31091.6	HO		0.6	H	EX

AIR WAVELENGTH (ANGSTRMS)	VACUUM WAVELENGTH (ANGSTRMS)	WAVENUMBER (KAYSERS)	LMENT	LOG ARC	INTENSITY SPK DIS	REF
3213.30	3214.23	31111.7	HO		0.6	EX
3211.87	3212.80	31125.5	HO		0.6H	EX
3210.39	3211.32	31139.9	HO	0.8	0.8H	EX
3206.86	3207.79	31174.1	HO	0.8	0.6	EX
3204.27	3205.20	31199.3	HO		0.6	EX
3203.33	3204.26	31208.5	HO		0.8	EX
3201.77	3202.70	31223.7	HO	0.8	0.8	EX
3197.83	3198.75	31262.2	HO	0.8	0.6	EX
3196.07	3196.99	31279.4	HO		0.6H	EX
3195.56	3196.48	31284.4	HO		0.6	EX
3190.97	3191.89	31329.4	HO		0.6H	EX
3187.37	3188.29	31364.8	HO		0.6	EX
3184.13	3185.05	31396.7	HO		0.6H	EX
3183.85	3184.77	31399.4	HO	0.8	0.8	EX
3181.52	3182.44	31422.4	HO	0.9	1.6	EX
3176.96	3177.88	31467.5	HO	0.8	0.6H	EX
3174.86	3175.78	31488.3	HO	0.8	0.8	EX
3173.79	3174.71	31499.0	HO	0.8	1.0	EX
3171.71	3172.63	31519.6	HO	1.0	1.2	EX
3167.89	3168.81	31557.6	HO		0.6	EX
3166.62	3167.54	31570.3	HO	1.0	1.2	EX
3165.67	3166.59	31579.7	HO		0.6H	EX
3164.03	3164.95	31596.1	HO		0.6H	EX
3159.67	3160.58	31639.7	HO	0.8		EX
3156.96	3157.87	31666.9	HO	0.8	0.8	EX
3156.18	3157.09	31674.7	HO	0.8	0.6	EX
3154.22	3155.13	31694.4	HO		0.6	EX
3149.92	3150.83	31737.6	HO		0.6	EX
3148.85	3149.76	31748.4	HO	0.6	0.8	EX
3144.35	3145.26	31793.9	HO	0.9	0.8	MIT
3134.40	3135.31	31894.8	HO	0.8	0.6H	EX
3130.99	3131.90	31929.5	HO	0.8	0.8	EX
3130.38	3131.29	31935.7	HO		0.6H	EX
3122.64	3123.55	32014.9	HO		0.6	EX
3118.51	3119.41	32057.3	HO	0.9	1.0	EX
3113.48	3114.38	32109.1	HO		0.6H	EX
3109.93	3110.83	32145.7	HO	0.8	0.3H	EX
3108.63	3109.53	32159.2	HO		0.6	EX
3108.33	3109.23	32162.3	HO	0.8	0.6	EX
3102.68	3103.58	32220.9	HO		0.6	EX
3086.54	3087.44	32389.3	HO	0.9	0.8	EX
3084.36	3085.26	32412.2	HO	0.9	1.0	EX
3074.30	3075.19	32518.3	HO		0.6	EX
3064.19	3065.08	32625.6	HO		0.6	EX
3054.88	3055.77	32725.0	HO		0.6H	EX
3053.99	3054.88	32734.5	HO	0.8	0.6H	EX
3050.73	3051.62	32769.5	HO		0.6	EX
3049.38	3050.27	32784.0	HO	0.8	0.6	EX
3044.41	3045.30	32837.5	HO		0.6H	EX
3038,69	3039.57	32899.3	HO	0.6	0.8	EX
3023.14	3024.02	33068.6	HO		0.6	EX
3014.61	3015.49	33162.1	HO		0.6	EX
3009.48	3010.36	33218.6	HO		0.6H	EX
2990.27	2991.14	33432.0	HO		1.0	EX
2985.64	2986.51	33483.9	HO		1.0	EX
2979.63	2980.50	33551.4	HO	1.3	1.6	EX
2973.00	2973.87	33626.2	HO	1.3		EX
2964.48	2965.35	33722.9	HO		1.0H	EX
2953.11	2953.97	33852.7	HO	1.3	1.0	EX
2949.19	2950.05	33897.7	HO		1.3	EX
2945.83	2946.69	33936.4	HO	0.5	1.8H	EX
2944.50	2945.36	33951.7	HO	1.0	1.3	EX
2942.05	2942.91	33980.0	HO		1.0	EX
2939.29	2940.15	34011.9	HO		1.0	EX
2936.77	2937.63	34041.1	HO		3.0G	EX
2933.05	2933.91	34084.2	HO		1.0	EX
2928.79	2929.65	34133.8	HO		2.0	EX
2928.29	2929.15	34139.6	HO		1.0H	EX
2919.59	2920.44	34241.4	HO		1.0	EX
2915.80	2916.65	34285.9	HO		1.0	EX
2910.35	2911.20	34350.1	HO		1.0	EX
2909.42	2910.27	34361.0	HO	1.6	1.0	EX
2900.83	2901.68	34462.8	HO		1.0	EX
2897.36	2898.21	34504.1	HO		1.3H	EX
2894.99	2895.84	34532.3	HO	1.3	1.0	EX
2891.39	2892.24	34575.3	HO		1.0	EX
2880.99	2881.84	34700.1	HO	1.3	1.0	EX
2880.27	2881.11	34708.8	HO	1.3	1.0	EX
2867.82	2868.66	34859.5	HO		1.6H	EX
2861.49	2862.33	34936.6	HO		1.0	EX
2858.34	2859.18	34975.1	HO		1.0	EX
2849.10	2849.94	35088.5	HO	1.0	1.3	EX
2847.50	2848.34	35108.2	HO		1.6	EX
2845.64	2846.48	35131.1	HO		1.8H	EX
2844.18	2845.02	35149.2	HO		1.0H	EX
2843.39	2844.23	35159.0	HO		1.0H	EX
2842.04	2842.88	35175.6	HO		1.0	EX
2834.99	2835.82	35263.1	HO		1.0	EX
2831.60	2832.43	35305.3	HO		1.8	EX
2830.79	2831.62	35315.4	HO		1.0A	EX
2828.14	2828.97	35348.5	HO		1.3H	EX
2826.63	2827.46	35367.4	HO	0.5	1.3H	EX
2824.79	2825.62	35390.5	HO		1.0H	EX
2824.19	2825.02	35398.0	HO	1.3	0.5	EX
2819.74	2820.57	35453.8	HO		1.0H	EX
2819.08	2819.91	35462.1	HO		1.0H	EX
2818.69	2819.52	35467.0	HO		1.0H	EX
2818.14	2818.97	35474.0	HO		1.0H	EX
2816.38	2817.21	35496.1	HO		1.0	EX
2814.74	2815.57	35516.8	HO	1.0	1.3	EX
2813.77	2814.60	35529.0	HO		1.0H	EX
2812.87	2813.70	35540.4	HO		1.3	EX
2812.00	2812.83	35551.4	HO		1.0	EX
2811.36	2812.19	35559.5	HO		1.0	EX
2809.99	2810.82	35576.8	HO		1.0H	EX
2809.08	2809.91	35588.4	HO		1.0	EX
2806.72	2807.55	35618.3	HO		1.0	EX
2799.99	2800.82	35703.9	HO		1.0	EX
2792.53	2793.35	35799.3	HO		1.0H	EX
2786.95	2787.77	35870.9	HO		1.0	EX
2778.19	2779.01	35984.0	HO		1.0H	EX
2777.10	2777.92	35998.2	HO		1.0	EX
2774.70	2775.52	36029.3	HO		2.5	EX
2773.84	2774.66	36040.5	HO		1.0	EX
2759.30	2760.12	36230.4	HO		1.0H	EX
2750.44	2751.25	36347.1	HO		1.0H	EX
2733.95	2734.76	36566.3	HO		1.0	EX
2684.75	2685.55	37236.4	HO		1.0	EX
2681.18	2681.98	37285.9	HO		1.3	EX
2677.95	2678.75	37330.9	HO		1.3	EX
2673.95	2674.74	37386.7	HO		1.0	EX
2666.10	2666.89	37496.8	HO		1.0H	EX
2653.72	2654.51	37671.7	HO		1.0	EX
2642.26	2643.05	37835.1	HO		1.0H	EX
2636.50	2637.29	37917.8	HO		1.8	EX
2597.51	2598.29	38486.9	HO		1.3	EX
2545.57	2546.33	39272.1	HO		1.0	EX
2533.61	2534.37	39457.5	HO		1.0	EX
2533.21	2533.97	39463.7	HO		1.0	EX
2514.72	2515.48	39753.9	HO		1.0	EX
2511.12	2511.88	39810.9	HO		1.3	EX
2489.24	2489.99	40160.8	HO		1.0	EX
2480.61	2481.36	40300.5	HO		1.0	EX
2473.76	2474.51	40412.1	HO		1.0H	EX
2472.21	2472.96	40437.4	HO		1.0	EX
2469.92	2470.67	40474.9	HO		1.0	EX
2466.53	2467.28	40530.5	HO		1.0	EX
2464.63	2465.38	40561.8	HO		1.0	EX
2462.32	2463.06	40599.8	HO		1.0H	EX
2458.74	2459.48	40658.9	HO		1.0H	EX

AIR WAVELENGTH (ANGSTRMS)	VACUUM WAVELENGTH (ANGSTRMS)	WAVENUMBER (KAYSERS)	LMENT		LOG ARC	INTENSITY SPK	DIS	REF
2455.21	2455.95	40717.4	HO			1.0H		EX
2454.42	2455.16	40730.5	HO			1.0H		EX
2452.69	2453.43	40759.2	HO			1.3		EX
2450.97	2451.71	40787.8	HO			1.0		EX
2448.95	2449.69	40821.5	HO			1.0H		EX
2442.76	2443.50	40924.9	HO			1.3		EX
2439.33	2440.07	40982.4	HO			1.3		EX
2437.07	2437.81	41020.4	HO			1.0		EX
2433.00	2433.74	41089.0	HO			1.3		EX
2431.03	2431.77	41122.4	HO			1.3		EX
9731.57	9734.24	10273.0	I				0.3	EV
9335.99	9338.55	10708.3	I	I			0.3	EV
9113.88	9116.38	10969.3	I	I			0.8	EV
9058.38	9060.87	11036.5	I	I			0.8	EV
9022.43	9024.91	11080.4	I	I			0.6	EV
8969.54	8972.00	11145.8	I	II			1.0	MU
8898.44	8900.88	11234.8	I	I			0.7	EV
8858.62	8861.05	11285.3	I	I			0.6	DB
8857.46	8859.89	11286.8	I	I			0.6	EV
8853.39	8855.82	11292.0	I				0.6	EV
8704.12	8706.51	11485.7	I	II			0.9	MU
8700.80	8703.19	11490.0	I	I			0.3	EV
8664.93	8667.31	11537.6	I	I			0.8	EV
8623.40	8625.77	11593.2	I	II			0.8	MU
8486.16	8488.49	11780.7	I	I			0.6	EV
8393.42	8395.73	11910.8	I	I			0.8	EV
8251.90	8254.17	12115.1	I	II			0.6	MU
8240.13	8242.40	12132.4	I	I			0.8	EV
8222.65	8224.91	12158.2	I	I			0.7	EV
8169.51	8171.76	12237.3	I	I			0.8	EV
8123.98	8126.21	12305.8	I	II			0.8	MU
8043.74	8045.95	12428.6	I	I			1.0	EV
8003.70	8005.90	12490.8	I	I			0.7	EV
7969.53	7971.72	12544.3	I	I			0.9	MU
7875.55	7877.72	12694.0	I	I			0.3	MU
7700.20	7702.32	12983.1	I	I			0.9	EV
7657.99	7660.10	13054.7	I	II			1.4	MU
7618.56	7620.66	13122.2	I	II			1.0	MU
7595.40	7597.49	13162.2	I	II			1.4	MU
7588.70	7590.79	13173.9	I	I			0.5	EV
7556.69	7558.77	13229.7	I	I			0.7	EV
7554.25	7556.33	13233.9	I	I			0.8	EV
7490.58	7492.64	13346.4	I	I			1.8	EV
7469.04	7471.10	13384.9	I	I			2.7	EV
7416.55	7418.59	13479.7	I				1.7	EV
7414.61	7416.65	13483.2	I	I			0.7	EV
7413.64	7415.68	13484.9	I				1.5	EV
7411.25	7413.29	13489.3	I	I			1.7	EV
7402.10	7404.14	13506.0	I	I			2.5	EV
7351.36	7353.39	13599.2	I	II			1.8	KE
7318.12	7320.14	13660.9	I	I			1.2	DB
7300.30	7302.31	13694.3	I	I			1.7	DB
7277.15	7279.16	13737.9	I				1.2H	BL
7243.66	7245.66	13801.4	I	I			0.3	MU
7237.88	7239.87	13812.4	I	I			1.5	BL
7236.80	7238.79	13814.5	I	I			2.2	EV
7231.86	7233.85	13823.9	I				1.3	EV
7227.34	7229.33	13832.5	I				1.0	EV
7225.85	7227.84	13835.4	I				1.0	KE
7213.11	7215.10	13859.8	I	I			0.7	DB
7191.73	7193.71	13901.0	I	I			0.7	EV
7168.87	7170.85	13945.4	I	I			1.2	BL
7166.47	7168.45	13950.0	I	I			1.5	DB
7164.82	7166.79	13953.2	I				1.5	EV
7148.63	7150.60	13984.8	I	I			1.5	EV
7146.57	7148.54	13988.9	I				1.2	BL
7142.09	7144.06	13997.7	I	I			2.0	EV
7138.99	7140.96	14003.7	I	II			1.8H	BL
7122.09	7124.05	14036.9	I				1.8	EV
7120.11	7122.07	14040.9	I				1.3	EV

AIR WAVELENGTH (ANGSTRMS)	VACUUM WAVELENGTH (ANGSTRMS)	WAVENUMBER (KAYSERS)	LMENT		LOG ARC	INTENSITY SPK	DIS	REF
7107.66	7109.62	14065.4	I				1.2	BL
7102.65	7104.61	14075.4	I	I			1.7	DB
7101.69	7103.65	14077.3	I				1.5	BL
7100.16	7102.12	14080.3	I	I			1.2	DB
7088.00	7089.95	14104.5	I	I			2.0	DB
7087.69	7089.64	14105.1	I				1.2	BL
7085.22	7087.17	14110.0	I				2.2	BL
7080.34	7082.29	14119.7	I	I			1.5	BL
7067.24	7069.19	14145.9	I				1.2	BL
7063.60	7065.55	14153.2	I	I			1.7	EV
7057.27	7059.22	14165.9	I	I			0.7	BL
7042.24	7044.18	14196.1	I				1.8H	BL
7032.99	7034.93	14214.8	I	II			1.8	KE
7027.70	7029.64	14225.5	I				1.2	BL
7021.64	7023.58	14237.8	I	II			1.0	KE
7018.94	7020.88	14243.2	I	II			1.7	KE
7018.32	7020.26	14244.5	I				1.7	EV
6993.45	6995.38	14295.2	I	I			1.0	EV
6989.84	6991.77	14302.5	I				1.3	EV
6986.35	6988.28	14309.7	I				1.2	EV
6958.781	6960.700	14366.37	I	II			3.0	KE
6951.29	6953.21	14381.8	I				1.2	BL
6941.03	6942.94	14403.1	I	I			0.3	MU
6928.86	6930.77	14428.4	I				1.0	EV
6910.13	6912.04	14467.5	I	II			0.3	MU
6904.82	6906.72	14478.6	I	II			1.5	BL
6902.127	6904.031	14484.29	I	II			2.2	KE
6888.66	6890.56	14512.6	I				1.2	BL
6879.22	6881.12	14532.5	I				1.2	BL
6869.88	6871.78	14552.3	I				1.2	BL
6856.82	6858.71	14580.0	I				1.2	EV
6835.70	6837.59	14625.0	I	I			1.2	DB
6828.02	6829.90	14641.5	I	I			1.2	DB
6825.17	6827.05	14647.6	I				0.7	BL
6821.04	6822.92	14656.5	I				1.5	BL
6813.19	6815.07	14673.4	I				1.2	BL
6812.562	6814.442	14674.72	I	II			2.0	KE
6812.19	6814.07	14675.5	I	I			1.7	EV
6801.00	6802.88	14699.7	I	I			1.7	DB
6792.45	6794.32	14718.2	I	II			1.2	BL
6789.32	6791.19	14724.9	I	I			1.7	EV
6788.93	6790.80	14725.8	I	I			2.0	DB
6773.56	6775.43	14759.2	I	I			1.7	DB
6764.14	6766.01	14779.8	I				1.5	BL
6755.72	6757.58	14798.2	I				1.2H	BL
6741.56	6743.42	14829.3	I	I			1.7	EV
6739.50	6741.36	14833.8	I	I			1.7	EV
6738.12	6739.98	14836.8	I				1.8	EV
6736.60	6738.46	14840.2	I				1.7	EV
6732.10	6733.96	14850.1	I				2.2	EV
6727.00	6728.86	14861.4	I				1.8	EV
6723.52	6725.38	14869.1	I				1.2	BL
6720.67	6722.53	14875.4	I				1.2	BL
6718.803	6720.658	14879.50	I	I			1.7	KE
6718.803	6720.658	14879.50	I	II			1.7	KE
6713.13	6714.98	14892.1	I	I			0.7	DB
6698.53	6700.38	14924.5	I				1.7	EV
6697.33	6699.18	14927.2	I	I			2.0	EV
6687.74	6689.59	14948.6	I				1.2	BL
6685.32	6687.17	14954.0	I	II			0.7	MU
6672.33	6674.17	14983.1	I				1.5	BL
6665.970	6667.811	14997.43	I	II			1.8	KE
6662.14	6663.98	15006.1	I				2.0	EV
6661.16	6663.00	15008.3	I				2.0	EV
6650.81	6652.65	15031.6	I	I			1.5	EV
6633.94	6635.77	15069.8	I				1.5	BL
6622.49	6624.32	15095.9	I	II			1.5	BL
6619.69	6621.52	15102.3	I	I			2.3	EV
6618.41	6620.24	15105.2	I	II			0.3	MU
6585.190	6587.009	15181.40	I	II			1.8	KE

AIR WAVELENGTH (ANGSTRMS)	VACUUM WAVELENGTH (ANGSTRMS)	WAVENUMBER (KAYSERS)	LMENT	LOG INTENSITY ARC	SPK	DIS	REF
6583.81	6585.63	15184.6	I			1.8	KE
6582.87	6584.69	15186.7	I I			1.3	KE
6581.68	6583.50	15189.5	I			1.2	BL
6580.58	6582.40	15192.0	I			1.2	EV
6570.81	6572.63	15214.6	I I			1.5	BL
6569.27	6571.08	15218.2	I I			0.7	DB
6566.48	6568.29	15224.7	I I			2.6	KE
6564.80	6566.61	15228.6	I			1.5	BL
6560.87	6562.68	15237.7	I I			1.5	EV
6550.58	6552.39	15261.6	I I			0.7	DB
6549.35	6551.16	15264.5	I			1.5	BL
6546.94	6548.75	15270.1	I II			0.7	MU
6538.34	6540.15	15290.2	I			1.8	BL
6524.80	6526.60	15321.9	I II			0.3	MU
6521.45	6523.25	15329.8	I			1.2	BL
6518.86	6520.66	15335.9	I II			0.7	BL
6518.86	6520.66	15335.9	I I			0.7	BL
6517.88	6519.68	15338.2	I I			1.5	BL
6516.19	6517.99	15342.2	I II			1.0	KE
6496.01	6497.80	15389.8	I II			1.2	BL
6496.01	6497.80	15389.8	I I			1.2	BL
6488.18	6489.97	15408.4	I I			2.2	EV
6478.94	6480.73	15430.4	I I			0.7	DB
6477.48	6479.27	15433.8	I I			1.4	EV
6475.91	6477.70	15437.6	I			1.8	BL
6458.41	6460.19	15479.4	I II			1.2	BL
6448.41	6450.19	15503.4	I			1.2H	BL
6444.51	6446.29	15512.8	I I			2.0	DB
6443.26	6445.04	15515.8	I		1.5H		BL
6440.22	6442.00	15523.1	I			2.0	BL
6430.97	6432.75	15545.5	I		1.2H		BL
6428.54	6430.32	15551.3	I			1.5	BL
6417.66	6419.43	15577.7	I			1.2	BL
6415.79	6417.56	15582.2	I			1.3	EV
6411.29	6413.06	15593.2	I			1.5	EV
6409.54	6411.31	15597.4	I			1.2	BL
6405.87	6407.64	15606.4	I I			0.7	BL
6395.26	6397.03	15632.3	I			1.5	BL
6388.94	6390.71	15647.7	I I			1.5	BL
6378.80	6380.56	15672.6	I			1.5	EV
6371.76	6373.52	15689.9	I I			2.0	EV
6367.34	6369.10	15700.8	I I			1.8	EV
6363.26	6365.02	15710.9	I			1.2	BL
6359.19	6360.95	15720.9	I I			1.8	EV
6355.66	6357.42	15729.7	I I			1.2	EV
6348.34	6350.10	15747.8	I			1.7	BL
6341.09	6342.84	15765.8	I			1.5	BL
6339.97	6341.72	15768.6	I II			2.0H	BL
6339.52	6341.27	15769.7	I			2.5	EV
6338.97	6340.72	15771.1	I I			2.0	DB
6338.02	6339.77	15773.4	I			2.0	EV
6333.58	6335.33	15784.5	I			1.5	EV
6330.45	6332.20	15792.3	I			1.7	EV
6322.38	6324.13	15812.5	I I			1.5	BL
6320.82	6322.57	15816.4	I			1.7	BL
6320.60	6322.35	15816.9	I I			2.0	DB
6320.41	6322.16	15817.4	I I			1.7	DB
6313.12	6314.87	15835.7	I I			1.5	DB
6312.55	6314.30	15837.1	I I			0.7	EV
6300.50	6302.24	15867.4	I			1.5	BL
6297.08	6298.82	15876.0	I			1.5	EV
6294.08	6295.82	15883.6	I I			2.5	EV
6291.402	6293.142	15890.31	I II			1.5	KE
6290.68	6292.42	15892.1	I I			0.7	EV
6280.32	6282.06	15918.3	I			1.7	BL
6277.02	6278.76	15926.7	I			1.2	BL
6276.66	6278.40	15927.6	I			1.2	BL
6268.55	6270.28	15948.2	I			1.5	KE
6267.67	6269.40	15950.5	I			1.2	BL
6257.489	6259.220	15976.43	I			1.6	KE
6256.42	6258.15	15979.2	I II			0.3	MU
6255.54	6257.27	15981.4	I II			1.5	KE
6250.59	6252.32	15994.1	I II			1.0	KE
6246.26	6247.99	16005.2	I			1.2	EV
6245.41	6247.14	16007.3	I I			0.7	EV
6244.55	6246.28	16009.5	I			1.5	EV
6244.11	6245.84	16010.7	I			1.2	EV
6243.28	6245.01	16012.8	I			1.2	EV
6242.81	6244.54	16014.0	I			1.5	EV
6240.93	6242.66	16018.8	I			1.2	EV
6238.74	6240.47	16024.4	I I			1.2	DB
6236.40	6238.13	16030.5	I			1.7	BL
6233.59	6235.31	16037.7	I			1.2	EV
6232.85	6234.57	16039.6	I			1.7	BL
6229.39	6231.11	16048.5	I			1.7	BL
6213.91	6215.63	16088.5	I I			1.8	DB
6213.17	6214.89	16090.4	I I			1.8	EV
6204.869	6206.586	16111.92	I II			1.8	KE
6200.52	6202.24	16123.2	I I			1.7	DB
6195.51	6197.22	16136.3	I			2.0	BL
6191.97	6193.68	16145.5	I I			2.2	EV
6186.92	6188.63	16158.7	I			1.2	BL
6161.90	6163.61	16224.3	I II			0.3	MU
6148.39	6150.09	16259.9	I II			1.0	MU
6132.94	6134.64	16300.9	I			1.7	BL
6127.464	6129.160	16315.45	I II			2.1	KE
6125.53	6127.23	16320.6	I I			2.0	DB
6115.04	6116.73	16348.6	I			1.2	EV
6086.77	6088.46	16424.5	I			2.2	BL
6084.80	6086.48	16429.9	I			1.6	KE
6082.46	6084.14	16436.2	I I			3.0	EV
6077.91	6079.59	16448.5	I II			0.3	MU
6074.99	6076.67	16456.4	I II			1.9	KE
6068.948	6070.628	16472.76	I II			1.7	KE
6058.85	6060.53	16500.2	I			1.2	BL
6056.10	6057.78	16507.7	I			1.5	EV
6055.13	6056.81	16510.4	I			1.2	EV
6054.16	6055.84	16513.0	I			1.2	EV
6053.56	6055.24	16514.6	I			1.5	EV
6048.72	6050.39	16527.9	I I			1.8	EV
6046.59	6048.26	16533.7	I			1.2	BL
6044.09	6045.76	16540.5	I			1.5	BL
6042.78	6044.45	16544.1	I			1.5	EV
6041.57	6043.24	16547.4	I			1.5	BL
6039.44	6041.11	16553.2	I			1.2H	BL
6036.52	6038.19	16561.2	I			1.2	BL
6029.39	6031.06	16580.8	I I			0.7	BL
6024.13	6025.80	16595.3	I I			2.5	EV
6021.85	6023.52	16601.6	I			1.2	BL
6015.87	6017.54	16618.1	I			1.7	BL
6007.51	6009.17	16641.2	I			1.7	BL
6005.09	6006.75	16647.9	I			1.2	EV
5990.86	5992.52	16687.5	I			0.9	EV
5985.03	5986.69	16703.7	I			1.2	EV
5984.36	5986.02	16705.6	I			1.2	EV
5981.42	5983.08	16713.8	I I			1.2	EV
5980.96	5982.62	16715.1	I			1.2	EV
5971.85	5973.50	16740.6	I I			1.5	DB
5968.41	5970.06	16750.2	I			1.2	EV
5956.98	5958.63	16782.4	I			1.2	EV
5954.49	5956.14	16789.4	I			0.9	EV
5950.26	5951.91	16801.3	I II			1.7	KE
5949.92	5951.57	16802.3	I II			0.3	MU
5949.20	5950.85	16804.3	I I			0.3	MU
5928.61	5930.25	16862.7	I			0.9L	BL
5920.90	5922.54	16884.6	I II			1.2	BL
5920.61	5922.25	16885.5	I II			1.2	MU
5908.60	5910.24	16919.8	I			0.9	BL
5899.02	5900.65	16947.3	I I			1.4	DB
5894.05	5895.68	16961.6	I I			1.8	MU

AIR WAVELENGTH (ANGSTRMS)	VACUUM WAVELENGTH (ANGSTRMS)	WAVENUMBER (KAYSERS)	LMENT	LOG ARC	INTENSITY SPK	INTENSITY DIS	REF
5893.43	5895.06	16963.4	I II			0.9	BL
5882.33	5883.96	16995.4	I			0.9	EV
5875.13	5876.76	17016.2	I			1.2	BL
5849.10	5850.72	17091.9	I II			0.3	MU
5832.66	5834.28	17140.1	I			0.9	BL
5830.16	5831.78	17147.4	I			1.2	KE
5819.721	5821.335	17178.19	I II			1.2	MU
5812.25	5813.86	17200.3	I			0.9	BL
5807.35	5808.96	17214.8	I			0.9	BL
5805.57	5807.18	17220.1	I II			0.7	MU
5803.58	5805.19	17226.0	I I			1.4	DB
5787.067	5788.672	17275.12	I II			1.5	KE
5785.75	5787.35	17279.1	I			0.9H	BL
5784.46	5786.06	17282.9	I			0.9H	BL
5781.23	5782.83	17292.6	I I			1.2	DB
5780.74	5782.34	17294.0	I			0.9	EV
5775.110	5776.712	17310.89	I			1.5	KE
5774.861	5776.463	17311.63	I II			0.9	KE
5774.50	5776.10	17312.7	I			1.4	KE
5764.33	5765.93	17343.3	I I			2.0	MU
5760.726	5762.324	17354.11	I II			1.6	KE
5750.36	5751.95	17385.4	I			0.9	BL
5739.917	5741.509	17417.02	I II			0.7	MU
5739.86	5741.45	17417.2	I			1.0	KE
5739.434	5741.026	17418.49	I I			1.4	KE
5739.208	5740.800	17419.18	I II			1.3	MU
5739.02	5740.61	17419.7	I II			1.0	KE
5738.846	5740.438	17420.27	I II			0.3	MU
5738.289	5739.881	17421.97	I II			1.7	KE
5710.531	5712.115	17506.65	I II			2.2	KE
5702.075	5703.657	17532.61	I II			1.3	KE
5690.915	5692.494	17566.99	I II			1.5	KE
5679.97	5681.55	17600.8	I II			0.3	KE
5678.056	5679.632	17606.78	I II			1.9	KE
5673.66	5675.23	17620.4	I			1.2	KE
5643.38	5644.95	17715.0	I			1.2	KE
5625.704	5627.266	17770.62	I II			2.2	KE
5612.890	5614.448	17811.19	I			1.4	KE
5603.14	5604.70	17842.2	I II			0.3	KE
5600.310	5601.865	17851.20	I			1.7	KE
5598.597	5600.151	17856.66	I			1.4	KE
5598.362	5599.916	17857.41	I			1.5	KE
5593.133	5594.686	17874.10	I			1.4	KE
5590.27	5591.82	17883.3	I			0.7	KE
5586.41	5587.96	17895.6	I I			1.3	MU
5568.10	5569.65	17954.5	I II			0.3	MU
5551.651	5553.193	18007.66	I II			1.2	KE
5546.36	5547.90	18024.8	I I			0.3	MU
5525.833	5527.368	18091.79	I			1.4	KE
5525.71	5527.24	18092.2	I			1.2	BL
5522.047	5523.581	18104.20	I			1.5	KE
5504.721	5506.250	18161.18	I II			1.8	KE
5500.60	5502.13	18174.8	I I			0.7	MU
5497.661	5499.188	18184.50	I II			1.2	KE
5496.92	5498.45	18186.9	I II			3.0	KE
5496.60	5498.13	18188.0	I			1.5	BL
5494.07	5495.60	18196.4	I			1.2	KE
5493.50	5495.03	18198.3	I			1.3	KE
5491.567	5493.093	18204.68	I			2.0	KE
5479.60	5481.12	18244.4	I II			1.2	KE
5470.53	5472.05	18274.7	I I			1.4	DB
5470.24	5471.76	18275.7	I			0.3	BL
5468.16	5469.68	18282.6	I			0.3	KE
5464.61	5466.13	18294.5	I			3.0	KE
5457.06	5458.58	18319.8	I			1.2	KE
5454.49	5456.01	18328.4	I II			0.3	MU
5448.78	5450.29	18347.6	I			0.3	BL
5438.005	5439.516	18383.99	I II			1.5	KE
5435.843	5437.354	18391.30	I II			2.1	KE
5435.48	5436.99	18392.5	I I			1.5	DB

AIR WAVELENGTH (ANGSTRMS)	VACUUM WAVELENGTH (ANGSTRMS)	WAVENUMBER (KAYSERS)	LMENT	LOG ARC	INTENSITY SPK	INTENSITY DIS	REF
5427.593	5429.102	18419.25	I			0.9	KE
5427.10	5428.61	18420.9	I I			1.7	MU
5426.752	5428.260	18422.11	I II			0.3	MU
5422.799	5424.306	18435.54	I			1.0	KE
5422.11	5423.62	18437.9	I			1.0	KE
5414.87	5416.38	18462.5	I			1.2	BL
5411.72	5413.22	18473.3	I			0.9	KE
5407.362	5408.865	18488.17	I II			1.8	KE
5405.65	5407.15	18494.0	I			1.6	KE
5405.445	5406.948	18494.72	I			0.9	KE
5405.28	5406.78	18495.3	I II			0.9	KE
5405.14	5406.64	18495.8	I			1.6	KE
5380.043	5381.539	18582.04	I II			1.2	KE
5374.49	5375.98	18601.2	I			1.2H	BL
5372.500	5373.994	18608.13	I II			1.0	KE
5370.43	5371.92	18615.3	I I			0.3	DB
5369.87	5371.36	18617.2	I II			1.6	KE
5367.60	5369.09	18625.1	I II			0.7	KE
5356.176	5357.666	18664.84	I			0.9	KE
5351.95	5353.44	18679.6	I II			0.3	MU
5349.65	5351.14	18687.6	I			1.2	BL
5345.146	5346.633	18703.36	I			2.5	KE
5341.58	5343.07	18715.9	I II			0.3	BL
5338.192	5339.677	18727.73	I II			2.5	KE
5330.16	5331.64	18755.9	I I			0.3	DB
5326.44	5327.92	18769.0	I			0.3	KE
5322.819	5324.300	18781.81	I II			1.3	KE
5314.594	5316.073	18810.88	I			0.9	KE
5309.146	5310.623	18830.18	I			1.2	KE
5299.79	5301.26	18863.4	I			1.0	KE
5296.52	5297.99	18875.1	I I			2.2	BL
5288.65	5290.12	18903.2	I I			1.2	BL
5269.35	5270.82	18972.4	I			1.5	BL
5266.94	5268.41	18981.1	I			0.9H	BL
5265.17	5266.64	18987.4	I			1.3	KE
5264.713	5266.178	18989.10	I			0.9	KE
5245.696	5247.156	19057.94	I II			1.9	KE
5234.63	5236.09	19098.2	I I			1.9	KE
5229.01	5230.47	19118.8	I			1.3	KE
5216.26	5217.71	19165.5	I II			0.7	KE
5214.08	5215.53	19173.5	I I			1.3	KE
5205.54	5206.99	19205.0	I II			0.9	BL
5204.20	5205.65	19209.9	I I			1.7	EV
5198.88	5200.33	19229.6	I			1.4	BL
5189.38	5190.83	19264.8	I			0.3	BL
5185.213	5186.657	19280.24	I II			1.5	KE
5178.131	5179.573	19306.61	I			1.4	KE
5175.40	5176.84	19316.8	I I			0.9	DB
5174.74	5176.18	19319.3	I			0.3	BL
5161.188	5162.626	19369.99	I II			2.5	KE
5156.45	5157.89	19387.8	I II			1.4	BL
5155.06	5156.50	19393.0	I			1.2	BL
5149.732	5151.167	19413.08	I			0.9	BL
5147.52	5148.95	19421.4	I			0.3H	KE
5124.56	5125.99	19508.4	I II			0.7	MU
5119.285	5120.712	19528.54	I I			2.7	KE
5118.01	5119.44	19533.4	I I			0.9	DB
5114.400	5115.825	19547.19	I			1.4	KE
5090.772	5092.191	19637.91	I I			0.9	KE
5065.587	5066.999	19735.55	I			0.9	KE
5065.39	5066.80	19736.3	I II			1.0	KE
5061.884	5063.295	19749.98	I			0.7	KE
5061.42	5062.83	19751.8	I I			1.4	DB
5057.374	5058.784	19767.60	I			0.9	KE
5046.416	5047.823	19810.52	I II			1.0	KE
5036.185	5037.589	19850.76	I			1.4	KE
5030.90	5032.30	19871.6	I			0.9	DB
5028.88	5030.28	19879.6	I			0.3	KE
5008.363	5009.760	19961.04	I I			0.7	KE
4997.29	4998.68	20005.3	I			0.9	KE

AIR WAVELENGTH (ANGSTRMS)	VACUUM WAVELENGTH (ANGSTRMS)	WAVENUMBER (KAYSERS)	LMENT		LOG ARC	INTENSITY SPK	DIS	REF
4995.15	4996.54	20013.8	I	I			1.4	DB
4986.934	4988.325	20046.81	I	II			1.5	KE
4974.51	4975.90	20096.9	I	I			0.5	DB
4968.447	4969.833	20121.40	I	II			1.0	KE
4957.86	4959.24	20164.4	I				0.9	KE
4943.20	4944.58	20224.2	I	II			1.0	MU
4943.06	4944.44	20224.7	I				0.9	KE
4917.03	4918.40	20331.8	I	I			2.0	MU
4896.78	4898.15	20415.9	I	I			1.5	MU
4893.97	4895.34	20427.6	I				0.9	BL
4891.54	4892.91	20437.7	I				1.2	BL
4884.82	4886.18	20465.9	I	II			1.4	DB
4884.82	4886.18	20465.9	I	I			1.4	DB
4884.13	4885.49	20468.8	I	II			1.2	MU
4882.18	4883.54	20476.9	I				1.2	BL
4867.08	4868.44	20540.5	I	I			0.5	DB
4864.51	4865.87	20551.3	I	II			1.4	MU
4862.31	4863.67	20560.6	I	I			2.8	KE
4862.13	4863.49	20561.4	I	II			1.4	BL
4862.13	4863.49	20561.4	I	I			1.4	BL
4853.19	4854.55	20599.2	I	II			1.2	BL
4851.05	4852.41	20608.3	I				0.9	BL
4850.46	4851.82	20610.8	I				1.4	BL
4850.26	4851.62	20611.7	I	I			1.4	DB
4835.37	4836.72	20675.2	I	I			1.2	DB
4835.18	4836.53	20676.0	I				1.4	BL
4828.450	4829.799	20704.79	I	II			0.7	MU
4828.28	4829.63	20705.5	I	II			0.5	MU
4828.150	4829.499	20706.08	I	II			0.3	MU
4828.07	4829.42	20706.4	I				0.9	BL
4808.02	4809.36	20792.8	I	I			1.2	BL
4808.02	4809.36	20792.8	I	II			1.2	BL
4806.377	4807.720	20799.88	I	II			1.2	KE
4800.07	4801.41	20827.2	I				0.9	BL
4796.58	4797.92	20842.4	I	I			0.9	BL
4790.69	4792.03	20868.0	I				0.9	BL
4788.25	4789.59	20878.6	I			0.3H		BL
4787.20	4788.54	20883.2	I	II			0.3	MU
4784.76	4786.10	20893.9	I				0.7	KE
4768.160	4769.493	20966.59	I	II			1.3	KE
4763.790	4765.122	20985.82	I				1.4	KE
4763.38	4764.71	20987.6	I	I			1.9	MU
4752.67	4754.00	21034.9	I				1.2	BL
4730.98	4732.30	21131.4	I	II			1.0	MU
4730.38	4731.70	21134.0	I				1.4	KE
4726.55	4727.87	21151.2	I	II			0.7	MU
4711.40	4712.72	21219.2	I				1.2H	BL
4709.80	4711.12	21226.4	I				0.9H	BL
4707.83	4709.15	21235.3	I				0.9	BL
4702.46	4703.78	21259.5	I	II			0.9	MU
4687.43	4688.74	21327.7	I	I			1.2	BL
4678.98	4680.29	21366.2	I				0.9	BL
4677.94	4679.25	21370.9	I	II			1.0	KE
4676.941	4678.250	21375.51	I	II			1.9	KE
4676.05	4677.36	21379.6	I	II			0.7	MU
4675.528	4676.837	21381.97	I	II			1.7	KE
4668.15	4669.46	21415.8	I	II			1.0	KE
4666.52	4667.83	21423.2	I	II			2.4	KE
4663.672	4664.978	21436.33	I				1.2	KE
4657.35	4658.65	21465.4	I	II			0.9	KE
4656.67	4657.97	21468.6	I	I			1.4	DB
4642.02	4643.32	21536.3	I	II			0.5	MU
4640.88	4642.18	21541.6	I	I			1.7	BL
4638.87	4640.17	21550.9	I				0.9H	BL
4634.867	4636.165	21569.55	I				1.5	KE
4633.363	4634.661	21576.55	I	II			1.2	KE
4632.433	4633.730	21580.89	I	II			1.5	KE
4632.32	4633.62	21581.4	I				1.7	BL
4627.34	4628.64	21604.6	I	II			0.3	MU
4621.887	4623.182	21630.13	I				1.5	KE

AIR WAVELENGTH (ANGSTRMS)	VACUUM WAVELENGTH (ANGSTRMS)	WAVENUMBER (KAYSERS)	LMENT		LOG ARC	INTENSITY SPK	DIS	REF
4620.97	4622.26	21634.4	I	I			1.2	DB
4620.74	4622.03	21635.5	I				1.2	BL
4611.22	4612.51	21680.2	I	II			1.3	KE
4603.85	4605.14	21714.9	I				0.9	BL
4599.806	4601.095	21733.96	I				1.5	KE
4590.95	4592.24	21775.9	I	II			1.4	MU
4586.58	4587.87	21796.6	I	II			0.9	KE
4579.992	4581.275	21827.98	I				1.5	KE
4576.58	4577.86	21844.3	I	II			1.4	KE
4574.107	4575.389	21856.07	I	II			1.5	KE
4571.51	4572.79	21868.5	I	II			1.0	KE
4570.24	4571.52	21874.6	I				0.5	KE
4560.60	4561.88	21920.8	I	II			0.7	MU
4559.57	4560.85	21925.7	I				0.9H	BL
4558.012	4559.290	21933.24	I				1.2	KE
4544.29	4545.56	21999.5	I	II			0.9	KE
4543.86	4545.13	22001.6	I				1.0	KE
4540.63	4541.90	22017.2	I	II			0.9	KE
4530.22	4531.49	22067.8	I	I			0.9	DB
4529.28	4530.55	22072.4	I				1.4	BL
4528.097	4529.367	22078.14	I				1.6	KE
4527.15	4528.42	22082.8	I				1.2H	BL
4513.212	4514.478	22150.96	I	II			1.2	MU
4512.632	4513.898	22153.80	I				1.5	BL
4512.57	4513.84	22154.1	I				1.5	BL
4499.58	4500.84	22218.1	I				1.5	BL
4497.54	4498.80	22228.1	I				1.2	BL
4496.86	4498.12	22231.5	I	II			0.5	KE
4496.52	4497.78	22233.2	I				0.9	BL
4495.72	4496.98	22237.1	I	II			1.0	MU
4492.68	4493.94	22252.2	I	II			0.3	MU
4490.774	4492.034	22261.63	I	II			1.0	MU
4488.545	4489.804	22272.69	I	II			1.2	KE
4487.30	4488.56	22278.9	I	II			1.0	MU
4480.95	4482.21	22310.4	I				1.4	KE
4479.74	4481.00	22316.5	I				1.4H	BL
4478.64	4479.90	22321.9	I	I			1.2	MU
4476.05	4477.31	22334.9	I	II			1.8	MU
4473.44	4474.70	22347.9	I	II			1.9	KE
4464.318	4465.571	22393.55	I	II			1.5	KE
4462.69	4463.94	22401.7	I				1.0	BL
4460.19	4461.44	22414.3	I				0.9	BL
4458.47	4459.72	22422.9	I				1.5H	BL
4456.614	4457.865	22432.26	I	II			1.0	KE
4456.26	4457.51	22434.1	I				0.9	BL
4452.881	4454.131	22451.07	I	II			2.8	KE
4451.21	4452.46	22459.5	I				0.9	BL
4451.08	4452.33	22460.1	I				0.9	BL
4446.78	4448.03	22481.9	I	II			1.5	MU
4444.908	4446.156	22491.34	I	II			1.4	KE
4442.560	4443.807	22503.23	I	II			1.5	KE
4441.21	4442.46	22510.1	I				0.9	BL
4440.80	4442.05	22512.1	I				0.9	BL
4434.233	4435.478	22545.49	I				1.3	KE
4431.730	4432.974	22558.22	I				1.3	KE
4428.225	4429.468	22576.07	I	II			1.5	KE
4423.759	4425.001	22598.86	I	II			1.9	KE
4422.02	4423.26	22607.7	I	II			1.2	KE
4417.58	4418.82	22630.5	I				0.9	BL
4416.84	4418.08	22634.3	I	II			1.2	KE
4412.37	4413.61	22657.2	I	II			1.4	KE
4409.31	4410.55	22672.9	I				1.2	KE
4408.962	4410.200	22674.71	I	II			2.4	KE
4404.56	4405.80	22697.4	I				0.9	BL
4403.55	4404.79	22702.6	I				1.3	KE
4399.086	4400.322	22725.61	I				1.3	KE
4398.592	4399.828	22728.16	I				0.9	KE
4394.55	4395.78	22749.1	I				0.9	BL
4392.00	4393.23	22762.3	I				0.9	BL
4390.34	4391.57	22770.9	I				1.2H	BL

AIR WAVELENGTH (ANGSTRMS)	VACUUM WAVELENGTH (ANGSTRMS)	WAVENUMBER (KAYSERS)	LMENT	LOG INTENSITY ARC	SPK	DIS	REF
4388.51	4389.74	22780.4	I			1.2H	BL
4385.06	4386.29	22798.3	I			0.9	KE
4376.16	4377.39	22844.7	I			1.3	KE
4362.477	4363.703	22916.32	I II			1.2	KE
4346.92	4348.14	22998.3	I I			1.5	DB
4342.105	4343.326	23023.83	I II			1.5	KE
4338.52	4339.74	23042.9	I			0.5	KE
4338.40	4339.62	23043.5	I			0.9	BL
4338.12	4339.34	23045.0	I			0.5	KE
4337.49	4338.71	23048.3	I II			0.3	MU
4335.029	4336.248	23061.41	I			1.0	KE
4333.00	4334.22	23072.2	I I			1.2	DB
4322.75	4323.97	23126.9	I II			0.9	KE
4316.88	4318.09	23158.4	I			0.9H	BL
4315.66	4316.87	23164.9	I			0.9H	BL
4313.34	4314.55	23177.4	I II			0.7	MU
4310.59	4311.80	23192.2	I			0.9	KE
4296.35	4297.56	23269.0	I			0.7	KE
4291.95	4293.16	23292.9	I			1.3	KE
4287.98	4289.19	23314.4	I II			0.7	MU
4281.86	4283.06	23347.8	I II			0.7	KE
4271.53	4272.73	23404.2	I			0.7	BL
4263.772	4264.972	23446.81	I			1.2	KE
4260.14	4261.34	23466.8	I II			1.2	MU
4258.933	4260.132	23473.45	I			1.3	KE
4243.56	4244.75	23558.5	I I			0.5	DB
4240.40	4241.59	23576.0	I II			0.5	BL
4235.47	4236.66	23603.5	I II			1.4	KE
4225.54	4226.73	23658.9	I II			1.2	KE
4224.98	4226.17	23662.1	I			0.9	BL
4223.327	4224.516	23671.35	I			1.2	KE
4220.958	4222.147	23684.63	I			1.9	KE
4219.18	4220.37	23694.6	I II			1.0	KE
4217.08	4218.27	23706.4	I I			0.5	DB
4208.98	4210.17	23752.0	I I			1.0	MU
4173.795	4174.971	23952.26	I II			1.5	KE
4171.56	4172.74	23965.1	I I			0.9	DB
4170.468	4171.644	23971.37	I			1.4	KE
4145.782	4146.951	24114.10	I I			1.2	KE
4138.022	4139.189	24159.32	I			0.9	KE
4136.328	4137.495	24169.22	I			1.2	KE
4133.246	4134.412	24187.24	I			1.2	KE
4129.13	4130.29	24211.4	I I			0.3	MU
4128.690	4129.855	24213.93	I			1.5	KE
4126.101	4127.265	24229.12	I			0.9	KE
4120.177	4121.339	24263.96	I			1.2	KE
4116.791	4117.953	24283.91	I II			0.5	KE
4116.178	4117.339	24287.53	I			0.9	KE
4108.281	4109.440	24334.21	I I			1.3	KE
4102.883	4104.041	24366.23	I			1.0	KE
4093.513	4094.668	24422.00	I			1.0	KE
4089.55	4090.70	24445.7	I II			0.5	BL
4083.84	4084.99	24479.9	I			0.9	BL
4081.40	4082.55	24494.5	I II			0.5H	BL
4079.901	4081.053	24503.48	I II			0.7	MU
4075.56	4076.71	24529.6	I I			1.4	DB
4072.01	4073.16	24551.0	I II			0.3	MU
4070.748	4071.897	24558.57	I			2.2	KE
4060.175	4061.322	24622.53	I II			0.7	KE
4056.321	4057.467	24645.92	I			1.2	KE
4050.09	4051.23	24683.8	I I			1.5	DB
4049.885	4051.029	24685.09	I			1.5	KE
4043.879	4045.021	24721.75	I			1.3	KE
4041.937	4043.079	24733.62	I			1.2	KE
4036.080	4037.220	24769.52	I II			1.7	KE
4032.092	4033.231	24794.01	I			1.0	KE
4031.96	4033.10	24794.8	I			1.0	BL
4025.076	4026.214	24837.23	I			1.5	KE
4023.82	4024.96	24845.0	I II			0.3	MU
4017.21	4018.35	24885.9	I II			1.4	KE

AIR WAVELENGTH (ANGSTRMS)	VACUUM WAVELENGTH (ANGSTRMS)	WAVENUMBER (KAYSERS)	LMENT	LOG INTENSITY ARC	SPK	DIS	REF
4017.21	4018.35	24885.9	I I			1.4	KE
4016.12	4017.26	24892.6	I			0.5	KE
4015.94	4017.08	24893.7	I I			0.9	DB
4013.794	4014.929	24907.04	I			1.2	KE
4006.10	4007.23	24954.9	I			0.9	BL
3998.12	3999.25	25004.7	I			0.7	BL
3996.65	3997.78	25013.9	I I			0.7	BL
3994.979	3996.109	25024.34	I			1.5	KE
3985.68	3986.81	25082.7	I I			0.7	BL
3985.15	3986.28	25086.1	I II			0.3	MU
3983.948	3985.075	25093.63	I			1.4	KE
3973.75	3974.87	25158.0	I I			0.7	BL
3972.823	3973.947	25163.90	I II			1.2	KE
3965.756	3966.878	25208.74	I			0.8	KE
3965.526	3966.648	25210.20	I II			1.2	KE
3964.57	3965.69	25216.3	I I			0.7	DB
3957.10	3958.22	25263.9	I I			1.0	BL
3955.57	3956.69	25273.6	I			0.3	KE
3949.904	3951.022	25309.91	I			1.3	KE
3945.52	3946.64	25338.0	I I			0.3	BL
3942.46	3943.58	25357.7	I			1.2	KE
3941.82	3942.94	25361.8	I I			0.7	DB
3940.24	3941.36	25372.0	I			2.7	KE
3937.906	3939.021	25387.02	I			1.4	KE
3937.223	3938.338	25391.42	I II			1.1	KE
3931.425	3932.538	25428.87	I			0.5	KE
3931.014	3932.127	25431.53	I			2.6	KE
3924.013	3925.124	25476.90	I II			0.7	KE
3920.00	3921.11	25503.0	I			0.7	KE
3915.216	3916.325	25534.14	I			1.3	KE
3912.477	3913.585	25552.02	I			1.4	KE
3909.477	3910.584	25571.63	I I			0.8	KE
3907.186	3908.293	25586.62	I II			1.3	KE
3905.721	3906.827	25596.22	I I			1.0	KE
3901.13	3902.24	25626.3	I			0.5	KE
3898.44	3899.54	25644.0	I I			1.4	BL
3897.260	3898.364	25651.78	I			1.6	KE
3892.988	3894.091	25679.93	I I			1.2	KE
3884.05	3885.15	25739.0	I II			0.3	MU
3877.192	3878.291	25784.55	I II			0.7	KE
3875.823	3876.922	25793.66	I			1.2	KE
3850.82	3851.91	25961.1	I II			1.2	KS
3842.92	3844.01	26014.5	I II			0.8	KE
3840.92	3842.01	26028.1	I			0.3	KE
3838.24	3839.33	26046.2	I			1.0	KE
3833.40	3834.49	26079.1	I II			0.8	KE
3821.353	3822.437	26161.32	I			1.4	KE
3809.64	3810.72	26241.7	I			0.3	KE
3808.075	3809.156	26252.53	I I			1.6	KE
3803.49	3804.57	26284.2	I			0.7D	BL
3801.450	3802.529	26298.29	I			0.7	KE
3799.819	3800.898	26309.57	I			0.8	KE
3793.42	3794.50	26353.9	I			0.5	KE
3781.656	3782.730	26435.93	I I			0.3	KE
3781.412	3782.486	26437.64	I			0.8	KE
3779.373	3780.447	26451.90	I II			0.5	KE
3778.897	3779.970	26455.23	I II			0.3	MU
3773.758	3774.830	26491.26	I			1.0	KE
3771.436	3772.507	26507.57	I II			0.3	MU
3748.06	3749.13	26672.9	I I			0.7	BL
3742.126	3743.190	26715.18	I			1.6	KE
3741.710	3742.774	26718.15	I I			1.6	KE
3739.86	3740.92	26731.4	I			0.8	KE
3739.34	3740.40	26735.1	I			1.3	BL
3734.353	3735.415	26770.79	I I			1.3	KE
3730.88	3731.94	26795.7	I I			0.7	KE
3724.812	3725.871	26839.36	I			1.7	KE
3716.162	3717.219	26901.83	I II			0.7	KE
3711.441	3712.497	26936.05	I I			1.3	KE
3709.514	3710.569	26950.04	I II			0.5	KE

AIR WAVELENGTH (ANGSTRMS)	VACUUM WAVELENGTH (ANGSTRMS)	WAVENUMBER (KAYSERS)	LMENT	LOG ARC	INTENSITY SPK	INTENSITY DIS	REF
3702.027	3703.080	27004.54	I			1.0	KE
3696.282	3697.334	27046.51	I			0.5	KE
3695.646	3696.698	27051.17	I			0.7	KE
3695.510	3696.562	27052.17	I II			0.3	KE
3695.510	3696.562	27052.17	I I			0.3	KE
3692.264	3693.315	27075.95	I II			0.7	KE
3690.53	3691.58	27088.7	I			0.7	BL
3688.213	3689.263	27105.69	I			2.1	KE
3686.547	3687.596	27117.94	I			1.8	KE
3683.64	3684.69	27139.3	I I			0.3	BL
3678.611	3679.658	27176.44	I			0.8	KE
3678.55	3679.60	27176.9	I			0.5	BL
3661.773	3662.816	27301.40	I II			1.6D	KE
3657.056	3658.098	27336.61	I II			1.2	KE
3652.04	3653.08	27374.2	I II			0.3	MU
3637.837	3638.874	27481.03	I			1.3	KE
3632.48	3633.52	27521.6	I II			0.3	MU
3631.88	3632.92	27526.1	I II			0.3	MU
3627.499	3628.533	27559.35	I			0.5	KE
3626.759	3627.793	27564.97	I			0.5	KE
3615.877	3616.908	27647.92	I II			0.7	KE
3613.81	3614.84	27663.7	I			1.0H	BL
3607.511	3608.540	27712.04	I			1.3	KE
3603.65	3604.68	27741.7	I			0.5	BL
3602.50	3603.53	27750.6	I II			0.3	MU
3600.769	3601.796	27763.92	I			0.7	KE
3600.209	3601.236	27768.24	I			1.0	KE
3584.46	3585.48	27890.2	I I			0.7	DB
3583.324	3584.347	27899.09	I			1.3	KE
3577.53	3578.55	27944.3	I			0.7	BL
3575.856	3576.877	27957.35	I II			1.2	KE
3575.856	3576.877	27957.35	I I			1.2	KE
3573.666	3574.686	27974.48	I II			1.3	KE
3571.72	3572.74	27989.7	I			0.7	BL
3571.35	3572.37	27992.6	I II			0.7	BL
3566.24	3567.26	28032.7	I II			0.7	MU
3561.177	3562.194	28072.59	I			1.6	KE
3560.51	3561.53	28077.8	I			0.3	BL
3559.81	3560.83	28083.4	I			0.7	BL
3558.013	3559.029	28097.55	I			0.8	BL
3552.76	3553.77	28139.1	I II			0.7	MU
3552.19	3553.20	28143.6	I II			1.3	KE
3545.48	3546.49	28196.9	I			1.0	BL
3541.60	3542.61	28227.8	I			0.3	KE
3536.52	3537.53	28268.3	I			0.7	BL
3535.859	3536.870	28273.59	I			0.8	KE
3533.36	3534.37	28293.6	I			0.7	BL
3526.893	3527.901	28345.47	I			1.3	KE
3526.23	3527.24	28350.8	I II			0.7	KE
3524.277	3525.285	28366.50	I			0.5D	KE
3516.51	3517.52	28429.2	I II			0.7	KE
3516.023	3517.028	28433.09	I			1.0	KE
3512.701	3513.706	28459.98	I			1.3	KE
3512.150	3513.154	28464.45	I			1.0	KE
3511.347	3512.351	28470.96	I			0.5	KE
3502.75	3503.75	28540.8	I			1.0	BL
3498.028	3499.029	28579.36	I			1.4	KE
3497.377	3498.378	28584.68	I II			0.8	KE
3483.86	3484.86	28695.6	I			0.8	KE
3481.833	3482.830	28712.29	I			1.6	KE
3474.200	3475.195	28775.37	I			0.5	KE
3467.99	3468.98	28826.9	I II			0.8	BL
3461.007	3461.998	28885.05	I II			1.4	KE
3450.86	3451.85	28970.0	I II			0.3	MU
3441.33	3442.32	29050.2	I			0.7	BL
3441.07	3442.06	29052.4	I			1.0	BL
3436.94	3437.93	29087.3	I			1.0	BL
3435.075	3436.060	29103.10	I II			0.8	KE
3426.45	3427.43	29176.4	I			0.8	KE
3424.971	3425.953	29188.96	I			1.4	KE

AIR WAVELENGTH (ANGSTRMS)	VACUUM WAVELENGTH (ANGSTRMS)	WAVENUMBER (KAYSERS)	LMENT	LOG ARC	INTENSITY SPK	INTENSITY DIS	REF
3420.42	3421.40	29227.8	I			1.0	BL
3420.17	3421.15	29229.9	I			0.5	BL
3419.04	3420.02	29239.6	I			0.7	BL
3413.891	3414.870	29283.69	I			1.2	KE
3409.929	3410.907	29317.71	I II			0.7	MU
3407.77	3408.75	29336.3	I			1.0	BL
3401.49	3402.47	29390.4	I II			1.0	MU
3394.76	3395.73	29448.7	I			0.7	BL
3383.857	3384.829	29543.59	I II			0.5	KE
3381.806	3382.777	29561.51	I			1.0	KE
3380.98	3381.95	29568.7	I			0.5	KE
3377.89	3378.86	29595.8	I			1.0	BL
3374.86	3375.83	29622.4	I			0.3	KE
3374.44	3375.41	29626.0	I			0.8	KE
3374.00	3374.97	29629.9	I			0.7	BL
3367.24	3368.21	29689.4	I I			0.3	DB
3367.06	3368.03	29691.0	I			0.7	BL
3355.525	3356.489	29793.03	I II			1.0	KE
3350.48	3351.44	29837.9	I			0.7	BL
3350.240	3351.203	29840.03	I			1.2	KE
3349.793	3350.756	29844.01	I			1.0	KE
3342.561	3343.522	29908.58	I			1.3	KE
3342.26	3343.22	29911.3	I			0.7	BL
3341.277	3342.238	29920.07	I II			0.7	KE
3326.389	3327.346	30053.98	I			0.5	KE
3323.50	3324.46	30080.1	I			0.5	BL
3321.515	3322.471	30098.08	I			0.8	KE
3320.73	3321.69	30105.2	I			0.7	BL
3309.99	3310.94	30202.9	I			0.7	BL
3303.080	3304.031	30266.06	I II			1.4	KE
3302.563	3303.514	30270.80	I II			1.0	MU
3302.489	3303.440	30271.48	I II			0.7	MU
3302.432	3303.383	30272.00	I II			0.3	MU
3295.90	3296.85	30332.0	I			0.3	KE
3294.80	3295.75	30342.1	I			0.3	KE
3288.858	3289.805	30396.94	I II			0.9	KE
3288.355	3289.302	30401.58	I			1.5	KE
3279.46	3280.40	30484.0	I			1.0	BL
3276.54	3277.48	30511.2	I II			0.3	KE
3275.115	3276.059	30524.48	I			1.4	KE
3273.45	3274.39	30540.0	I			0.5	KE
3266.84	3267.78	30601.8	I			0.7	KE
3260.69	3261.63	30659.5	I			1.0	BL
3257.93	3258.87	30685.5	I			0.7	BL
3256.463	3257.402	30699.31	I			1.3	KE
3253.803	3254.741	30724.41	I			1.0	KE
3244.21	3245.15	30815.2	I			1.0	BL
3241.27	3242.21	30843.2	I			0.7	BL
3239.32	3240.25	30861.8	I			1.0	BL
3235.694	3236.628	30896.35	I			1.0	KE
3234.73	3235.66	30905.6	I			0.5	KE
3232.78	3233.71	30924.2	I			0.7	BL
3230.572	3231.505	30945.34	I			1.0	KE
3229.993	3230.925	30950.89	I			1.2	KE
3209.61	3210.54	31147.4	I I			0.7	BL
3209.61	3210.54	31147.4	I II			0.7	BL
3208.88	3209.81	31154.5	I			0.7	BL
3206.62	3207.55	31176.5	I			0.7	BL
3199.74	3200.66	31243.5	I II			0.7	BL
3196.92	3197.84	31271.1	I			0.7	BL
3194.008	3194.931	31299.58	I			1.3	KE
3193.95	3194.87	31300.1	I			2.0	BL
3189.62	3190.54	31342.6	I I			0.7	BL
3186.35	3187.27	31374.8	I			0.7	BL
3183.58	3184.50	31402.1	I			1.0	BL
3182.07	3182.99	31417.0	I			1.0	BL
3175.047	3175.966	31486.49	I II			1.0	KE
3173.09	3174.01	31505.9	I			1.0	BL
3170.12	3171.04	31535.4	I II			0.7	MU
3168.02	3168.94	31556.3	I			1.3	BL

AIR WAVELENGTH (ANGSTRMS)	VACUUM WAVELENGTH (ANGSTRMS)	WAVENUMBER (KAYSERS)	LMENT	LOG INTENSITY ARC	SPK	DIS	REF
3165.72	3166.64	31579.2	I			1.0	BL
3161.47	3162.39	31621.7	I			0.7	BL
3160.95	3161.86	31626.9	I			0.7	BL
3157.178	3158.092	31664.69	I			0.7	KE
3149.46	3150.37	31742.3	I			1.5	BL
3139.780	3140.690	31840.14	I			1.3	KE
3137.56	3138.47	31862.7	I II			0.3	MU
3136.480	3137.389	31873.64	I			0.7	KE
3133.40	3134.31	31905.0	I II			1.2D	MU
3127.37	3128.28	31966.5	I			0.7	BL
3125.92	3126.83	31981.3	I			1.0	BL
3121.57	3122.48	32025.9	I			0.7	BL
3116.60	3117.50	32076.9	I			0.8	KE
3112.96	3113.86	32114.4	I			1.3	BL
3106.63	3107.53	32179.9	I			0.7	BL
3103.736	3104.637	32209.89	I			0.5	KE
3102.72	3103.62	32220.4	I II			0.3	MU
3100.00	3100.90	32248.7	I II			0.3	MU
3097.81	3098.71	32271.5	I			0.7	BL
3091.626	3092.524	32336.05	I			1.4	KE
3088.187	3089.084	32372.06	I			1.5	KE
3082.875	3083.770	32427.84	I			1.4	KE
3081.664	3082.559	32440.58	I			2.0	KE
3078.774	3079.668	32471.03	I			2.5	KE
3077.923	3078.817	32480.01	I			1.0	KE
3077.648	3078.542	32482.91	I			1.3	KE
3073.820	3074.713	32523.36	I			1.0	KE
3073.243	3074.136	32529.47	I			0.8	KE
3072.254	3073.147	32539.94	I			0.8	KE
3068.98	3069.87	32574.6	I			1.3	BL
3068.047	3068.939	32584.55	I			0.7	KE
3063.925	3064.816	32628.39	I			1.0	KE
3059.11	3060.00	32679.7	I			0.7	BL
3055.370	3056.258	32719.75	I			2.5	KE
3050.582	3051.469	32771.10	I			0.8	KE
3049.71	3050.60	32780.5	I			0.5	KE
3047.14	3048.03	32808.1	I			1.0	BL
3038.39	3039.27	32902.6	I			1.3	BL
3038.367	3039.251	32902.84	I			1.4	KE
3037.888	3038.772	32908.03	I			0.5	KE
3036.33	3037.21	32924.9	I			0.7	BL
3033.275	3034.158	32958.07	I			0.7	KE
3032.057	3032.940	32971.31	I			0.5	KE
3030.507	3031.389	32988.17	I			1.3	KE
3021.219	3022.099	33089.58	I			1.3	KE
3019.94	3020.82	33103.6	I			1.0	BL
3018.17	3019.05	33123.0	I			1.0	BL
3015.642	3016.521	33150.78	I			1.2	KE
3013.18	3014.06	33177.9	I			1.0	BL
3007.01	3007.89	33245.9	I			1.0	BL
3000.79	3001.66	33314.9	I			0.7	BL
3000.18	3001.05	33321.6	I			1.3	BL
2996.726	2997.600	33360.02	I II			1.2	KE
2993.864	2994.737	33391.91	I			1.8	KE
2992.66	2993.53	33405.4	I			1.1	BL
2991.39	2992.26	33419.5	I			1.3	BL
2988.395	2989.267	33453.02	I			1.2	KE
2987.92	2988.79	33458.3	I			1.1	BL
2983.343	2984.213	33509.67	I			1.3	KE
2979.06	2979.93	33557.8	I			1.1	BL
2978.124	2978.993	33568.39	I			1.2	KE
2973.55	2974.42	33620.0	I			1.3	BL
2970.86	2971.73	33650.5	I			1.1	BL
2970.027	2970.894	33659.90	I			1.3	KE
2958.364	2959.228	33792.59	I			1.4	KE
2957.66	2958.52	33800.6	I II			1.0D	MU
2957.49	2958.35	33802.6	I II			1.5	BL
2954.39	2955.25	33838.0	I			1.1	BL
2949.105	2949.967	33898.69	I			1.5	KE
2942.27	2943.13	33977.4	I			1.3	BL

AIR WAVELENGTH (ANGSTRMS)	VACUUM WAVELENGTH (ANGSTRMS)	WAVENUMBER (KAYSERS)	LMENT	LOG INTENSITY ARC	SPK	DIS	REF
2941.55	2942.41	33985.7	I			1.1	BL
2937.00	2937.86	34038.4	I			1.3	BL
2934.09	2934.95	34072.1	I			1.1	BL
2931.727	2932.585	34099.61	I			1.3	KE
2930.18	2931.04	34117.6	I			1.1	BL
2928.468	2929.325	34137.56	I			1.1	KE
2925.091	2925.947	34176.97	I II			1.1	KE
2924.74	2925.60	34181.1	I			1.1	BL
2923.67	2924.53	34193.6	I			1.1	BL
2922.70	2923.56	34204.9	I			1.1	BL
2921.972	2922.827	34213.45	I II			0.3	MU
2921.89	2922.75	34214.4	I II			0.6	BL
2917.733	2918.587	34263.15	I			1.2	KE
2915.73	2916.58	34286.7	I			1.1	BL
2906.603	2907.454	34394.35	I			1.1L	KE
2903.013	2903.864	34436.88	I			1.1	KE
2901.84	2902.69	34450.8	I			1.1	BL
2900.22	2901.07	34470.0	I			1.1	BL
2897.83	2898.68	34498.5	I			1.3	BL
2889.54	2890.39	34597.4	I			1.3	BL
2888.48	2889.33	34610.1	I			1.1	B
2887.36	2888.21	34623.6	I			1.3	BL
2882.557	2883.403	34681.25	I			0.9	KE
2882.01	2882.86	34687.8	I			1.3	BL
2878.643	2879.488	34728.40	I II			2.6	KE
2875.25	2876.09	34769.4	I			1.1	BL
2872.888	2873.731	34797.97	I			1.8	KE
2868.767	2869.609	34847.95	I			1.5	KE
2865.17	2866.01	34891.7	I II			0.3	MU
2861.833	2862.673	34932.38	I			1.1	KE
2860.92	2861.76	34943.5	I			1.1	BL
2855.87	2856.71	35005.3	I			1.1	BL
2854.155	2854.994	35026.35	I			1.5	KE
2852.846	2853.684	35042.42	I			0.6	KE
2851.29	2852.13	35061.5	I			1.1	BL
2849.70	2850.54	35081.1	I			0.6	BL
2847.718	2848.555	35105.52	I			1.2	KE
2846.988	2847.825	35114.52	I			1.2	KE
2841.72	2842.56	35179.6	I			1.3	BL
2836.922	2837.756	35239.11	I			1.6	KE
2832.917	2833.750	35288.92	I			1.3	KE
2830.73	2831.56	35316.2	I			1.2	BL
2828.51	2829.34	35343.9	I			1.3	BL
2826.800	2827.632	35365.28	I			0.9	KE
2824.836	2825.667	35389.87	I			0.9	KE
2824.15	2824.98	35398.5	I			0.6	BL
2821.15	2821.98	35436.1	I			1.3	BL
2820.65	2821.48	35442.4	I			1.1	BL
2819.21	2820.04	35460.5	I			1.1H	BL
2818.48	2819.31	35469.7	I			1.2	BL
2817.56	2818.39	35481.3	I			1.1	BL
2816.82	2817.65	35490.6	I			0.6	KE
2810.012	2810.840	35576.56	I			1.1	KE
2809.18	2810.01	35587.1	I			1.1	BL
2808.584	2809.411	35594.65	I			1.6	KE
2804.76	2805.59	35643.2	I			1.3	BL
2799.676	2800.501	35707.89	I			1.1L	KE
2797.20	2798.02	35739.5	I			1.1	BL
2793.51	2794.33	35786.7	I			1.3	BL
2790.47	2791.29	35825.7	I			0.6	KE
2789.430	2790.253	35839.05	I			1.5	KE
2783.97	2784.79	35909.3	I			1.1	BL
2782.10	2782.92	35933.5	I			1.1	BL
2780.98	2781.80	35947.9	I			1.1	BL
2780.28	2781.10	35957.0	I			1.3	BL
2777.42	2778.24	35994.0	I			1.1	BL
2776.13	2776.95	36010.7	I			1.3	BL
2774.96	2775.78	36025.9	I			1.3	BL
2771.912	2772.730	36065.53	I			1.2	KE
2770.93	2771.75	36078.3	I			1.1	BL

AIR WAVELENGTH (ANGSTRMS)	VACUUM WAVELENGTH (ANGSTRMS)	WAVENUMBER (KAYSERS)	LMENT	LOG ARC	INTENSITY SPK	INTENSITY DIS	REF
2766.27	2767.09	36139.1	I			1.1	BL
2765.639	2766.456	36147.33	I			1.3	KE
2763.23	2764.05	36178.8	I			1.5	BL
2759.252	2760.067	36231.00	I			1.1	KE
2757.43	2758.24	36254.9	I			1.1H	BL
2755.58	2756.39	36279.3	I			1.1	BL
2754.72	2755.53	36290.6	I			1.3	BL
2752.42	2753.23	36320.9	I			1.1	BL
2751.94	2752.75	36327.3	I			1.1	BL
2751.42	2752.23	36334.1	I			1.3	BL
2746.50	2747.31	36399.2	I			1.1	BL
2744.50	2745.31	36425.7	I			1.1	BL
2738.44	2739.25	36506.3	I			1.3	BL
2737.49	2738.30	36519.0	I			1.3	BL
2733.08	2733.89	36577.9	I			1.1	BL
2730.131	2730.939	36617.44	I			1.7	KE
2729.48	2730.29	36626.2	I			1.1	BL
2724.18	2724.99	36697.4	I			1.5L	BL
2723.06	2723.87	36712.5	I			1.3	BL
2720.43	2721.24	36748.0	I			1.3	BL
2719.77	2720.58	36756.9	I			1.3	BL
2712.235	2713.039	36859.04	I			2.0	KE
2711.66	2712.46	36866.9	I			1.3	BL
2708.156	2708.959	36914.55	I			1.5	KE
2705.38	2706.18	36952.4	I			0.9	KE
2704.64	2705.44	36962.5	I			1.1	BL
2701.90	2702.70	37000.0	I			1.1	BL
2698.321	2699.122	37049.09	I II			1.7	KE
2694.066	2694.866	37107.60	I II			1.5	KE
2690.50	2691.30	37156.8	I			1.3	BL
2688.989	2689.787	37177.66	I			2.0	KE
2686.18	2686.98	37216.5	I			1.1	BL
2684.72	2685.52	37236.8	I			1.1	BL
2674.804	2675.599	37374.81	I			1.8	KE
2671.26	2672.05	37424.4	I			1.1	BL
2670.95	2671.74	37428.7	I			1.1	BL
2669.59	2670.38	37447.8	I			1.3	BL
2665.65	2666.44	37503.1	I			1.1	BL
2665.31	2666.10	37507.9	I			1.5	BL
2661.22	2662.01	37565.6	I			1.1	BL
2660.75	2661.54	37572.2	I			1.3	BL
2659.27	2660.06	37593.1	I			1.3	BL
2655.84	2656.63	37641.7	I			1.3	BL
2654.85	2655.64	37655.7	I			1.1	BL
2643.31	2644.10	37820.1	I			1.3	BL
2643.07	2643.86	37823.5	I			1.1	BL
2641.39	2642.18	37847.6	I			1.6	BL
2637.62	2638.41	37901.7	I			1.3H	BL
2636.749	2637.535	37914.19	I II			1.4	KE
2636.30	2637.09	37920.6	I			1.3	BL
2635.30	2636.09	37935.0	I			1.8	BL
2631.99	2632.77	37982.7	I			1.1	BL
2627.04	2627.82	38054.3	I			1.8	BL
2625.00	2625.78	38083.9	I			1.6	BL
2622.84	2623.62	38115.2	I			1.5	BL
2619.88	2620.66	38158.3	I			1.3	BL
2617.45	2618.23	38193.7	I			1.3	BL
2615.01	2615.79	38229.4	I			1.1	BL
2614.70	2615.48	38233.9	I			1.5	BL
2612.91	2613.69	38260.1	I			1.1	BL
2610.757	2611.537	38291.63	I			1.5	KE
2607.32	2608.10	38342.1	I			1.5	BL
2605.55	2606.33	38368.1	I			1.3	BL
2605.07	2605.85	38375.2	I			1.3	BL
2603.68	2604.46	38395.7	I			1.1	BL
2602.96	2603.74	38406.3	I			1.1	BL
2597.89	2598.67	38481.3	I			0.6	BL
2596.48	2597.26	38502.2	I			1.1	BL
2593.466	2594.242	38546.91	I II			2.2	KE
2591.02	2591.79	38583.3	I			1.3	BL

AIR WAVELENGTH (ANGSTRMS)	VACUUM WAVELENGTH (ANGSTRMS)	WAVENUMBER (KAYSERS)	LMENT	LOG ARC	INTENSITY SPK	INTENSITY DIS	REF
2588.683	2589.457	38618.13	I II			1.6	KE
2586.734	2587.508	38647.22	I II			1.6	KE
2584.843	2585.617	38675.50	I II			0.3	MU
2584.765	2585.538	38676.66	I II			1.3	MU
2582.808	2583.581	38705.97	I II			2.6	KE
2581.60	2582.37	38724.1	I			1.3	BL
2580.62	2581.39	38738.8	I			1.3	BL
2575.47	2576.24	38816.2	I			1.3	BL
2566.53	2567.30	38951.4	I			1.0	BL
2566.258	2567.027	38955.57	I II			2.5	KE
2564.401	2565.170	38983.78	I II			1.8	KE
2562.47	2563.24	39013.1	I			1.5	KE
2561.49	2562.26	39028.1	I			2.2	BL
2559.718	2560.486	39055.09	I II			1.6	MU
2559.277	2560.044	39061.82	I II			1.4	MU
2557.28	2558.05	39092.3	I II			1.3	BL
2553.98	2554.75	39142.8	I			1.3	BL
2551.43	2552.20	39182.0	I			1.6	BL
2548.12	2548.88	39232.8	I			1.1	BL
2540.13	2540.89	39356.2	I			1.1	BL
2535.57	2536.33	39427.0	I			1.5	BL
2534.45	2535.21	39444.4	I			1.1	BL
2534.24	2535.00	39447.7	I			1.1H	BL
2533.619	2534.380	39457.37	I II			2.0	MU
2530.977	2531.738	39498.56	I II			1.8	MU
2527.72	2528.48	39549.5	I			1.3	BL
2527.28	2528.04	39556.3	I			1.1	BL
2526.88	2527.64	39562.6	I			1.6	BL
2526.60	2527.36	39567.0	I			1.1	BL
2515.012	2515.769	39749.28	I II			1.3	MU
2513.21	2513.97	39777.8	I			1.3	BL
2511.50	2512.26	39804.9	I			1.3	BL
2508.68	2509.44	39849.6	I			1.1	BL
2502.985	2503.739	39940.26	I II			2.2	MU
2499.56	2500.31	39995.0	I			1.5	BL
2499.319	2500.072	39998.84	I II			1.8	MU
2497.01	2497.76	40035.8	I			1.3	BL
2494.71	2495.46	40072.7	I			2.0	BL
2491.61	2492.36	40122.6	I			1.6	BL
2484.64	2485.39	40235.1	I			1.3	BL
2482.05	2482.80	40277.1	I			1.3	BL
2481.09	2481.84	40292.7	I			1.1	BL
2478.04	2478.79	40342.3	I			1.1	BL
2471.01	2471.76	40457.0	I			1.1	BL
2464.95	2465.70	40556.5	I			1.1	BL
2464.679	2465.424	40560.97	I II			2.0	MU
2463.98	2464.73	40572.5	I			1.1H	BL
2461.128	2461.873	40619.49	I II			1.8	MU
2457.27	2458.01	40683.3	I			1.3	BL
2454.46	2455.20	40729.8	I			1.3	BL
2450.36	2451.10	40798.0	I			1.3	BL
2448.48	2449.22	40829.3	I			1.3	BL
2447.81	2448.55	40840.5	I			1.5	BL
2444.12	2444.86	40902.1	I			1.8	BL
2438.28	2439.02	41000.1	I II			0.7	MU
2437.61	2438.35	41011.4	I			1.3	BL
2436.20	2436.94	41035.1	I			1.3	BL
2430.92	2431.66	41124.2	I			1.3	BL
2423.40	2424.14	41251.8	I			1.3	BL
2423.21	2423.95	41255.0	I			1.3	BL
2422.32	2423.06	41270.2	I			1.3	BL
2419.41	2420.15	41319.8	I			1.3L	BL
2419.16	2419.89	41324.1	I			1.8	BL
2412.52	2413.25	41437.8	I			1.5	BL
2411.95	2412.68	41447.6	I			1.1	BL
2407.98	2408.71	41516.0	I			1.8	BL
2407.57	2408.30	41523.0	I			1.1	BL
2404.03	2404.76	41584.2	I			1.1	BL
2398.15	2398.88	41686.1	I			1.1	BL
2397.21	2397.94	41702.5	I			1.1	BL

AIR WAVELENGTH (ANGSTRMS)	VACUUM WAVELENGTH (ANGSTRMS)	WAVENUMBER (KAYSERS)	LMENT	LOG ARC	INTENSITY SPK	DIS	REF
2395.20	2395.93	41737.5	I			1.1	BL
2392.92	2393.65	41777.2	I			1.1H	BL
2392.35	2393.08	41787.2	I			1.3	BL
2384.03	2384.76	41933.0	I			1.5	BL
2381.79	2382.52	41972.4	I			1.3	BL
2380.99	2381.72	41986.5	I			1.1	BL
2377.99	2378.72	42039.5	I			1.5	BL
2372.23	2372.95	42141.6	I			1.1	BL
2360.13	2360.85	42357.6	I			1.1	BL
2354.35	2355.07	42461.6	I			1.3	BL
2348.56	2349.28	42566.2	I			1.1	BL
2346.32	2347.04	42606.9	I			1.5	BL
2344.36	2345.08	42642.5	I			1.5	BL
2342.38	2343.10	42678.5	I			2.2	BL
2335.52	2336.24	42803.9	I II			1.2D	MU
2335.45	2336.17	42805.2	I II			1.8	BL
2334.13	2334.85	42829.4	I			1.5	BL
2331.63	2332.35	42875.3	I II			0.7D	MU
2331.54	2332.26	42877.0	I II			1.6	BL
2328.40	2329.11	42934.8	I			1.5L	BL
2327.29	2328.00	42955.2	I			1.3	BL
2324.70	2325.41	43003.1	I			1.1	BL
2320.20	2320.91	43086.5	I			1.3	BL
2314.88	2315.59	43185.5	I			1.1	BL
2308.57	2309.28	43303.5	I			1.3	BL
2307.75	2308.46	43318.9	I II			0.3D	MU
2293.68	2294.39	43584.6	I			1.3	BL
2293.24	2293.95	43593.0	I			1.1	BL
2292.64	2293.35	43604.4	I			1.1	BL
2292.44	2293.15	43608.2	I			1.6	BL
2288.72	2289.43	43679.1	I II			0.3	MU
2287.89	2288.60	43694.9	I			1.1	BL
2284.90	2285.61	43752.1	I II			0.6	BL
2265.69	2266.39	44123.0	I			1.5L	BL
2263.82	2264.52	44159.5	I			1.1	BL
2255.27	2255.97	44326.9	I			1.5	LC
2250.03	2250.73	44430.1	I			2.0	LC
2240.14	2240.84	44626.2	I			1.3	LC
2238.81	2239.51	44652.7	I			1.3	LC
2229.97	2230.66	44829.7	I			2.6	LC
2229.28	2229.97	44843.6	I			1.3	BL
2225.27	2225.96	44924.4	I			1.5	LC
2220.64	2221.33	45018.0	I			1.1	LC
2204.12	2204.81	45355.4	I			1.3	LC
2183.27	2183.95	45788.5	I			1.5	LC
2172.93	2173.61	46006.4	I			1.1	LC
2169.59	2170.27	46077.2	I			2.0	LC
2122.95	2123.62	47089.4	I			1.3	LC
2110.49	2111.16	47367.4	I			1.3	LC
2107.68	2108.35	47430.5	I			1.3	LC
2106.76	2107.43	47451.2	I			1.3	LC
2105.20	2105.87	47486.4	I II			0.3	MU
2097.55	2098.22	47659.5	I			1.3	LC
2094.56	2095.23	47727.6	I			1.3	LC
2087.60	2088.26	47886.7	I			1.3	LC
2081.29	2081.95	48031.8	I			1.6	LC
2075.59	2076.25	48163.7	I			1.1	LC
2069.24	2069.90	48311.5	I			1.3	LC
2062.38	2063.04	48472.2	I			3.0	LC
2054.47	2055.13	48658.8	I			1.3	LC
2054.36	2055.02	48661.4	I			1.3	LC
2049.45	2050.11	48777.9	I			2.2	LC
2042.05	2042.71	48954.7	I			1.3	LC
2037.98	2038.63	49052.4	I II			0.3	MU
2034.86	2035.51	49127.6	I			2.0	LC
2023.80	2024.45	49396.1	I			2.0	LC
2022.82	2023.47	49420.0	I			1.1	LC
2015.86	2016.51	49590.6	I			1.1	LC
9245.380	9247.917	10813.25	IN II			0.3	PS
9242.245	9244.781	10816.91	IN II			1.0	PS

AIR WAVELENGTH (ANGSTRMS)	VACUUM WAVELENGTH (ANGSTRMS)	WAVENUMBER (KAYSERS)	LMENT	LOG ARC	INTENSITY SPK	DIS	REF
9241.985	9244.521	10817.22	IN II			0.8	PS
9213.658	9216.187	10850.47	IN II			0.9	PS
9213.278	9215.806	10850.92	IN II			0.8	PS
9212.950	9215.478	10851.31	IN II			0.6	PS
9212.688	9215.216	10851.62	IN II			0.5	PS
9212.468	9214.996	10851.88	IN II			0.3	PS
9202.140	9204.665	10864.06	IN II			0.8	PS
9201.938	9204.463	10864.29	IN II			0.8	PS
9198.016	9200.540	10868.93	IN II			0.5	PS
9197.707	9200.231	10869.29	IN II			0.6	PS
9197.332	9199.856	10869.73	IN II			0.7	PS
8832.565	8834.990	11318.63	IN II			0.6	PS
8832.376	8834.801	11318.87	IN II			0.5	PS
8814.333	8816.753	11342.04	IN II			0.6	PS
8813.925	8816.345	11342.57	IN II			0.5	PS
8813.543	8815.963	11343.06	IN II			0.3	PS
8803.988	8806.406	11355.37	IN II			0.3	PS
8803.764	8806.182	11355.66	IN II			0.5	PS
8685.910	8688.296	11509.74	IN II			0.3	PS
8461.894	8464.219	11814.44	IN II			0.3	PS
8434.776	8437.094	11852.42	IN II			0.3	PS
8434.324	8436.642	11853.06	IN II			0.5	PS
8414.130	8416.442	11881.51	IN II			0.3	PS
8413.623	8415.935	11882.22	IN II			0.3	PS
8413.083	8415.395	11882.98	IN II			0.7	PS
8412.319	8414.631	11884.06	IN II			0.5	PS
8234.640	8236.904	12140.48	IN II			0.6	PS
8233.291	8235.554	12142.47	IN II			0.3	PS
8232.600	8234.863	12143.49	IN II			0.3	PS
8227.354	8229.616	12151.24	IN II			0.8	PS
8227.274	8229.536	12151.35	IN II			0.3	PS
8226.977	8229.239	12151.79	IN II			0.5	PS
8226.918	8229.180	12151.88	IN II			0.6	PS
8226.871	8229.133	12151.95	IN II			0.3	PS
8226.640	8228.902	12152.29	IN II			0.5	PS
8226.581	8228.843	12152.38	IN II			0.6	PS
8158.976	8161.219	12253.07	IN II			0.3	PS
7841.180	7843.337	12749.68	IN II			0.6	PS
7840.913	7843.070	12750.11	IN II			0.5	PS
7840.695	7842.852	12750.46	IN II			0.3	PS
7829.396	7831.550	12768.86	IN II			0.5	PS
7815.190	7817.340	12792.07	IN II			0.6	PS
7814.850	7817.000	12792.63	IN II			0.3	PS
7806.990	7809.138	12805.51	IN II			0.5	PS
7806.819	7808.967	12805.79	IN II			0.5	PS
7789.318	7791.461	12834.56	IN II			0.7	PS
7789.045	7791.188	12835.01	IN II			0.8	PS
7788.721	7790.864	12835.55	IN II			0.8	PS
7787.516	7789.659	12837.53	IN II			0.6	PS
7787.179	7789.322	12838.09	IN II			0.3	PS
7786.834	7788.977	12838.66	IN II			0.3	PS
7786.439	7788.582	12839.31	IN II			0.3	PS
7786.157	7788.300	12839.77	IN II			0.5	PS
7785.995	7788.137	12840.04	IN II			0.3	PS
7776.958	7779.098	12854.96	IN II			1.0	PS
7776.751	7778.891	12855.30	IN II			0.7	PS
7776.571	7778.711	12855.60	IN II			0.3	PS
7740.940	7743.070	12914.77	IN II			0.5	PS
7740.733	7742.863	12915.12	IN II			0.6	PS
7740.481	7742.611	12915.54	IN II			0.7	PS
7740.194	7742.324	12916.02	IN II			0.8	PS
7683.865	7685.980	13010.70	IN II			0.5	PS
7683.455	7685.570	13011.40	IN II			0.8	PS
7683.401	7685.516	13011.49	IN II			0.7	PS
7683.027	7685.142	13012.12	IN II			0.8	PS
7682.925	7685.040	13012.30	IN II			1.0	PS
7682.876	7684.991	13012.38	IN II			0.9	PS
7682.262	7684.376	13013.42	IN II			0.5	PS
7681.678	7683.792	13014.41	IN II			0.3	PS
7603.312	7605.405	13148.54	IN II			0.5	PS

AIR WAVELENGTH (ANGSTRMS)	VACUUM WAVELENGTH (ANGSTRMS)	WAVENUMBER (KAYSERS)	LMENT	LOG ARC	INTENSITY SPK	INTENSITY DIS	REF
7602.775	7604.868	13149.47	IN II			0.5	PS
7602.282	7604.375	13150.32	IN II			0.3	PS
7455.16	7457.21	13409.8	IN		1.3		SQ
7453.845	7455.898	13412.20	IN II			1.3	PS
7452.899	7454.952	13413.90	IN II			1.1	PS
7452.082	7454.134	13415.37	IN II			0.7	PS
7425.82	7427.87	13462.8	IN		1.1		SQ
7367.24	7369.27	13569.9	IN		1.1		SQ
7355.124	7357.150	13592.22	IN II			0.7	PS
7354.931	7356.957	13592.58	IN II			1.1	PS
7354.699	7356.725	13593.01	IN II			1.3	PS
7351.585	7353.610	13598.76	IN II			1.7	PS
7351.487	7353.512	13598.94	IN II			1.7	PS
7350.906	7352.931	13600.02	IN II			1.1	PS
7350.371	7352.396	13601.01	IN II			1.7	PS
7350.004	7352.029	13601.69	IN II			1.5	PS
7349.567	7351.592	13602.50	IN II			1.5	PS
7303.754	7305.766	13687.82	IN II			1.7	PS
7303.348	7305.360	13688.58	IN II			1.6	PS
7303.012	7305.024	13689.21	IN II			1.5	PS
7297.91	7299.92	13698.8	IN II			1.1	PS
7293.00	7295.01	13708.0	IN II			1.1H	PS
7279.76	7281.77	13732.9	IN		0.7		SQ
7277.586	7279.591	13737.04	IN II			1.8	PS
7276.413	7278.418	13739.25	IN II			1.7	PS
7275.455	7277.460	13741.06	IN II			1.6	PS
7272.04	7274.04	13747.5	IN II			0.3H	PS
7256.069	7258.069	13777.77	IN II			1.5	PS
7255.182	7257.181	13779.45	IN II			1.6	PS
7254.104	7256.103	13781.50	IN II			1.7	PS
7241.459	7243.455	13805.57	IN II			0.7	PS
7229.216	7231.208	13828.95	IN II			1.5	PS
7226.08	7228.07	13834.9	IN		0.7		SQ
7222.048	7224.038	13842.67	IN II			1.3H	PS
7221.32	7223.31	13844.1	IN		1.5		SQ
7183.958	7185.938	13916.07	IN II			1.7	PS
7183.450	7185.430	13917.05	IN II			1.6	PS
7183.190	7185.170	13917.56	IN II			1.9	PS
7182.523	7184.503	13918.85	IN II			1.3	PS
7182.048	7184.028	13919.77	IN II			1.6	PS
7181.878	7183.858	13920.10	IN II			1.6	PS
7181.10	7183.08	13921.6	IN		0.7		SQ
7171.92	7173.90	13939.4	IN		1.1		SQ
7167.86	7169.84	13947.3	IN		0.7		SQ
6932.70	6934.61	14420.4	IN		0.7		SQ
6902.98	6904.88	14482.5	IN		0.7		SQ
6900.37	6902.27	14488.0	IN I	1.7			PS
6892.207	6894.109	14505.14	IN II			1.3	PS
6891.994	6893.895	14505.59	IN II			0.7	PS
6891.664	6893.565	14506.28	IN II			1.8	PS
6891.626	6893.527	14506.36	IN II			1.6	PS
6891.434	6893.335	14506.77	IN II			1.3	PS
6891.155	6893.056	14507.35	IN II			1.5	PS
6890.767	6892.668	14508.17	IN II			0.7	PS
6875.02	6876.92	14541.4	IN		1.5		SQ
6867.713	6869.608	14556.87	IN II			0.7	PS
6860.403	6862.296	14572.38	IN II			0.3	PS
6849.523	6851.413	14595.53	IN II			0.7	PS
6847.44	6849.33	14600.0	IN I	1.8			PS
6783.718	6785.590	14737.11	IN II			2.0	PS
6767.405	6769.273	14772.64	IN II			0.7	PS
6766.334	6768.202	14774.97	IN II			1.7	PS
6765.964	6767.832	14775.78	IN II			1.7	PS
6765.378	6767.245	14777.06	IN II			1.5	PS
6752.042	6753.906	14806.25	IN II			1.6	PS
6751.880	6753.744	14806.60	IN II			1.8	PS
6751.615	6753.479	14807.18	IN II			1.5	PS
6750.520	6752.383	14809.59	IN II			1.3	PS
6666.00	6667.84	14997.4	IN II			0.7	PS
6627.12	6628.95	15085.3	IN II			1.3H	PS

AIR WAVELENGTH (ANGSTRMS)	VACUUM WAVELENGTH (ANGSTRMS)	WAVENUMBER (KAYSERS)	LMENT	LOG ARC	INTENSITY SPK	INTENSITY DIS	REF
6582.82	6584.64	15186.9	IN		0.7		SQ
6578.16	6579.98	15197.6	IN		1.1		SQ
6541.455	6543.262	15282.90	IN II			1.6	PS
6541.224	6543.031	15283.44	IN II			1.7	PS
6540.959	6542.766	15284.06	IN II			1.8	PS
6517.22	6519.02	15339.7	IN		0.7		SQ
6512.18	6513.98	15351.6	IN		1.5		SQ
6469.326	6471.114	15453.29	IN II			1.1	PS
6469.248	6471.036	15453.48	IN II			1.3	PS
6468.994	6470.782	15454.08	IN II			1.9	PS
6468.889	6470.677	15454.33	IN II			1.7	PS
6468.557	6470.345	15455.13	IN II			1.1	PS
6468.477	6470.265	15455.32	IN II			1.1	PS
6438.96	6440.74	15526.2	IN		0.7		SQ
6437.540	6439.319	15529.59	IN II			1.1	PS
6436.914	6438.693	15531.10	IN II			0.7	PS
6436.412	6438.191	15532.31	IN II			0.7	PS
6362.958	6364.717	15711.62	IN II			1.6	PS
6362.896	6364.655	15711.77	IN II			1.6	PS
6362.373	6364.132	15713.06	IN II			1.1	PS
6362.133	6363.892	15713.66	IN II			1.6	PS
6361.740	6363.499	15714.62	IN II			1.3	PS
6361.492	6363.251	15715.24	IN II			1.3	PS
6354.738	6356.495	15731.94	IN II			0.7	PS
6354.318	6356.075	15732.98	IN II			1.3	PS
6338.001	6339.753	15773.48	IN II			0.7	PS
6337.364	6339.116	15775.07	IN II			1.1	PS
6336.566	6338.318	15777.06	IN II			1.3	PS
6305.737	6307.481	15854.19	IN II			1.6	PS
6305.555	6307.299	15854.65	IN II			1.5	PS
6304.846	6306.590	15856.43	IN II			1.5	PS
6304.540	6306.284	15857.20	IN II			1.3	PS
6303.965	6305.708	15858.65	IN II			1.3	PS
6303.830	6305.573	15858.99	IN II			1.1	PS
6303.173	6304.916	15860.64	IN II			0.7	PS
6302.526	6304.269	15862.27	IN II			0.7	PS
6283.428	6285.166	15910.48	IN II			0.7	PS
6283.216	6284.954	15911.02	IN II			1.1	PS
6231.905	6233.629	16042.02	IN II			0.7	PS
6231.480	6233.204	16043.11	IN II			1.3	PS
6230.425	6232.149	16045.83	IN II			1.1	PS
6229.854	6231.577	16047.30	IN II			1.1	PS
6228.852	6230.575	16049.88	IN II			1.1	PS
6228.762	6230.485	16050.11	IN II			1.6	PS
6228.533	6230.256	16050.70	IN II			0.7	PS
6228.315	6230.038	16051.27	IN II			1.5	PS
6228.038	6229.761	16051.98	IN II			1.3	PS
6227.919	6229.642	16052.29	IN II			1.3	PS
6227.815	6229.538	16052.56	IN II			1.3	PS
6224.275	6225.997	16061.69	IN II			1.8	PS
6198.06	6199.77	16129.6	IN		1.3		SQ
6163.008	6164.714	16221.35	IN II			1.3	PS
6162.754	6164.459	16222.02	IN II			1.3	PS
6162.533	6164.238	16222.60	IN II			1.3	PS
6162.340	6164.045	16223.11	IN II			1.1	PS
6161.862	6163.567	16224.37	IN II			1.3	PS
6161.149	6162.854	16226.25	IN II			1.8	PS
6149.988	6151.690	16255.70	IN II			0.7	PS
6149.675	6151.377	16256.52	IN II			1.1	PS
6149.375	6151.077	16257.32	IN II			1.3	PS
6149.096	6150.798	16258.05	IN II			1.5	PS
6148.834	6150.536	16258.75	IN II			1.5	PS
6148.416	6150.118	16259.85	IN II			1.6	PS
6148.255	6149.957	16260.28	IN II			1.3	PS
6148.108	6149.810	16260.67	IN II			0.7	PS
6143.229	6144.929	16273.58	IN II			1.9	PS
6141.248	6142.948	16278.83	IN II			0.7	PS
6141.084	6142.784	16279.27	IN II			1.3	PS
6140.659	6142.359	16280.39	IN II			1.5	PS
6140.357	6142.056	16281.19	IN II			1.3	PS

AIR WAVELENGTH (ANGSTRMS)	VACUUM WAVELENGTH (ANGSTRMS)	WAVENUMBER (KAYSERS)	LMENT	LOG ARC	INTENSITY SPK	INTENSITY DIS	REF
6140.026	6141.725	16282.07	IN II			1.1	PS
6139.682	6141.381	16282.98	IN II			0.7	PS
6137.188	6138.887	16289.60	IN II			1.7	PS
6132.742	6134.439	16301.41	IN II			1.7	PS
6132.419	6134.116	16302.27	IN II			1.3	PS
6132.133	6133.830	16303.03	IN II			1.6	PS
6131.508	6133.205	16304.69	IN II			1.3	PS
6131.275	6132.972	16305.31	IN II			1.6	PS
6130.948	6132.645	16306.18	IN II			1.5	PS
6129.754	6131.451	16309.35	IN II			1.3	PS
6129.696	6131.393	16309.51	IN II			1.8	PS
6129.474	6131.171	16310.10	IN II			0.7	PS
6129.096	6130.792	16311.10	IN II			1.6	PS
6128.992	6130.688	16311.38	IN II			1.1	PS
6128.721	6130.417	16312.10	IN II			1.6	PS
6128.360	6130.056	16313.06	IN II			1.5	PS
6128.053	6129.749	16313.88	IN II			1.3	PS
6127.760	6129.456	16314.66	IN II			0.7	PS
6127.513	6129.209	16315.32	IN II			1.6	PS
6116.267	6117.960	16345.32	IN II			1.6	PS
6115.863	6117.556	16346.40	IN II			1.3	PS
6115.630	6117.323	16347.02	IN II			1.3	PS
6115.427	6117.120	16347.56	IN II			1.3	PS
6115.312	6117.005	16347.87	IN II			0.3	PS
6108.995	6110.686	16364.78	IN II			1.8	PS
6108.655	6110.346	16365.69	IN II			1.7	PS
6108.334	6110.025	16366.55	IN II			1.6	PS
6097.02	6098.71	16396.9	IN		0.7		SQ
6096.117	6097.805	16399.34	IN II			1.3	PS
6095.960	6097.648	16399.77	IN II			1.9	PS
6095.846	6097.534	16400.07	IN II			1.7	PS
6095.782	6097.469	16400.25	IN II			0.7	PS
6095.725	6097.412	16400.40	IN II			0.7	PS
6063.477	6065.156	16487.62	IN II			1.5	PS
6062.856	6064.535	16489.31	IN II			1.3	PS
6062.311	6063.990	16490.79	IN II			1.3	PS
5918.899	5920.539	16890.35	IN II			1.5	PS
5918.784	5920.424	16890.68	IN II			1.7	PS
5918.650	5920.290	16891.06	IN II			1.8	PS
5915.966	5917.605	16898.73	IN II			2.0	PS
5915.634	5917.273	16899.68	IN II			1.7	PS
5915.449	5917.088	16900.21	IN II			1.7	PS
5914.832	5916.471	16901.97	IN II			1.7	PS
5914.679	5916.318	16902.41	IN II			1.8	PS
5914.397	5916.036	16903.21	IN II			1.5	PS
5905.04	5906.68	16930.0	IN		1.2		SQ
5903.872	5905.508	16933.34	IN II			1.2	PS
5903.755	5905.391	16933.68	IN II			2.2	PS
5903.626	5905.262	16934.05	IN II			2.7	PS
5903.472	5905.108	16934.49	IN II			1.8	PS
5903.367	5905.003	16934.79	IN II			2.0	PS
5903.243	5904.879	16935.15	IN II			2.0	PS
5903.137	5904.773	16935.45	IN II			2.0	PS
5903.045	5904.681	16935.72	IN II			1.8	PS
5902.972	5904.608	16935.93	IN II			1.5	PS
5896.02	5897.65	16955.9	IN		0.7		SQ
5890.26	5891.89	16972.5	IN		1.0		SQ
5854.58	5856.20	17075.9	IN		0.7		SQ
5853.434	5855.057	17079.25	IN II			2.5	PS
5853.107	5854.729	17080.21	IN II			2.2	PS
5852.829	5854.451	17081.02	IN II			2.0	PS
5820.08	5821.69	17177.1	IN		1.2		SQ
5727.69	5729.28	17454.2	IN I	1.7			PS
5723.66	5725.25	17466.5	IN		0.7		SQ
5722.041	5723.628	17471.44	IN II			1.8	PS
5721.747	5723.334	17472.33	IN II			1.5	PS
5721.522	5723.109	17473.02	IN II			1.2	PS
5709.75	5711.33	17509.0	IN I	1.7J			PS
5708.687	5710.271	17512.30	IN II			1.8	PS
5708.522	5710.106	17512.81	IN II			2.0	PS

AIR WAVELENGTH (ANGSTRMS)	VACUUM WAVELENGTH (ANGSTRMS)	WAVENUMBER (KAYSERS)	LMENT	LOG ARC	INTENSITY SPK	INTENSITY DIS	REF
5708.312	5709.896	17513.46	IN II			2.0	PS
5644.86	5646.43	17710.3	IN		1.8		SQ
5636.963	5638.528	17735.13	IN II			1.5	PS
5636.747	5638.312	17735.81	IN II			2.5	PS
5636.663	5638.228	17736.07	IN II			2.2	PS
5636.379	5637.943	17736.96	IN II			1.2	PS
5636.045	5637.609	17738.01	IN II			0.3	PS
5609.15	5610.71	17823.1	IN		1.5		SQ
5577.038	5578.587	17925.69	IN II			2.0	PS
5576.910	5578.459	17926.10	IN II			2.2	PS
5576.750	5578.299	17926.61	IN II			2.5	PS
5556.045	5557.588	17993.42	IN II			2.0	PS
5555.607	5557.150	17994.84	IN II			1.2	PS
5555.432	5556.975	17995.40	IN II			1.8	PS
5555.140	5556.683	17996.35	IN II			1.5	PS
5554.946	5556.489	17996.98	IN II	1.5			PS
5540.44	5541.98	18044.1	IN		0.7		SQ
5537.032	5538.570	18055.20	IN II			1.7	PS
5536.553	5538.091	18056.76	IN II			1.8	PS
5535.940	5537.478	18058.76	IN II			1.8	PS
5530.16	5531.70	18077.6	IN II			1.1	PS
5528.743	5530.279	18082.27	IN II			1.2	PS
5525.97	5527.50	18091.4	IN II			1.2	PS
5523.915	5525.449	18098.07	IN II			2.0	PS
5523.861	5525.395	18098.25	IN II			1.8	PS
5523.613	5525.147	18099.06	IN II			1.7	PS
5523.287	5524.821	18100.13	IN II			1.7	PS
5523.001	5524.535	18101.07	IN II			1.7	PS
5522.579	5524.113	18102.45	IN II			1.2	PS
5519.359	5520.892	18113.01	IN II			2.7	PS
5519.253	5520.786	18113.36	IN II			1.2	PS
5514.45	5515.98	18129.1	IN		0.7		SQ
5513.208	5514.740	18133.22	IN II			1.2	PS
5513.156	5514.688	18133.39	IN II			1.5	PS
5513.098	5514.630	18133.58	IN II			1.7	PS
5513.060	5514.592	18133.71	IN II			1.8	PS
5512.998	5514.529	18133.91	IN II			1.8	PS
5512.916	5514.447	18134.18	IN II			2.0	PS
5512.824	5514.355	18134.49	IN II			2.2	PS
5510.974	5512.505	18140.57	IN II			1.2	PS
5510.881	5512.412	18140.88	IN II			1.8	PS
5510.800	5512.331	18141.15	IN II			1.5	PS
5510.772	5512.303	18141.24	IN II			1.2	PS
5507.335	5508.865	18152.56	IN II			1.8	PS
5507.107	5508.637	18153.31	IN II			1.5	PS
5506.822	5508.352	18154.25	IN II			1.2	PS
5506.713	5508.243	18154.61	IN II			1.5	PS
5503.295	5504.824	18165.89	IN II			1.2	PS
5503.152	5504.681	18166.36	IN II			1.5	PS
5498.048	5499.575	18183.22	IN II			1.5	PS
5497.645	5499.172	18184.55	IN II			1.7	PS
5497.553	5499.080	18184.86	IN II			1.8	PS
5497.377	5498.904	18185.44	IN II			1.5	PS
5496.900	5498.427	18187.02	IN II			1.5H	PS
5496.691	5498.218	18187.71	IN II			1.5H	PS
5437.402	5438.913	18386.03	IN II			1.7	PS
5436.929	5438.440	18387.62	IN II			2.2	PS
5436.279	5437.790	18389.82	IN II			2.0	PS
5436.007	5437.518	18390.74	IN II			1.7	PS
5435.635	5437.146	18392.00	IN II			1.5	PS
5418.734	5420.240	18449.37	IN II			2.2H	PS
5418.480	5419.986	18450.23	IN II			1.7	PS
5418.216	5419.722	18451.13	IN II			1.5	PS
5411.406	5412.910	18474.35	IN II			2.5	PS
5402.944	5404.446	18503.28	IN II			1.7	PS
5402.443	5403.945	18505.00	IN II			1.5	PS
5309.828	5311.305	18827.76	IN II			2.0	PS
5309.398	5310.875	18829.29	IN II			1.8	PS
5309.035	5310.512	18830.58	IN II			1.8	PS
5262.74	5264.20	18996.2	IN I	1.1			PS

AIR WAVELENGTH (ANGSTRMS)	VACUUM WAVELENGTH (ANGSTRMS)	WAVENUMBER (KAYSERS)	LMENT	LOG ARC	INTENSITY SPK	INTENSITY DIS	REF
5253.97	5255.43	19027.9	IN I	1.5			PS
5253.375	5254.837	19030.08	IN II			1.2	PS
5248.52	5249.98	19047.7	IN		0.7		SQ
5184.660	5186.104	19282.30	IN II			1.8	PS
5184.438	5185.882	19283.12	IN II			2.5	PS
5175.558	5176.999	19316.21	IN II			2.2	PS
5175.422	5176.863	19316.72	IN II			2.5	PS
5175.292	5176.733	19317.20	IN II			2.6	PS
5130.363	5131.792	19486.37	IN II			1.2	PS
5129.939	5131.368	19487.98	IN II			1.8	PS
5129.763	5131.192	19488.65	IN II			1.5	PS
5121.629	5123.056	19519.60	IN II			1.5	PS
5121.498	5122.925	19520.10	IN II			1.5	PS
5121.339	5122.766	19520.70	IN II			1.7	PS
5121.154	5122.581	19521.41	IN II			1.2	PS
5121.098	5122.525	19521.62	IN II			2.6	PS
5120.963	5122.390	19522.14	IN II			2.2	PS
5120.855	5122.282	19522.55	IN II			2.0	PS
5120.534	5121.961	19523.77	IN II			1.5	PS
5117.461	5118.887	19535.50	IN II			1.2	PS
5117.412	5118.838	19535.68	IN II			2.5	PS
5117.367	5118.793	19535.85	IN II			1.7	PS
5116.755	5118.181	19538.19	IN II			1.8	PS
5115.911	5117.337	19541.42	IN II			2.0	PS
5115.628	5117.054	19542.50	IN II			1.5	PS
5115.454	5116.879	19543.16	IN II			1.5	PS
5115.250	5116.675	19543.94	IN II			1.2	PS
5115.023	5116.448	19544.81	IN II			1.2	PS
5109.360	5110.784	19566.47	IN II			2.5W	PS
5085.98	5087.40	19656.4	IN		0.7		SQ
5044.141	5045.548	19819.46	IN II			1.7	PS
5043.772	5045.178	19820.91	IN II			1.5	PS
5043.546	5044.952	19821.79	IN II			1.7	PS
5012.01	5013.41	19946.5	IN		0.7		SQ
4973.851	4975.239	20099.54	IN II			1.0	PS
4973.774	4975.162	20099.85	IN II			1.0	PS
4973.691	4975.079	20100.18	IN II			1.2	PS
4973.605	4974.993	20100.53	IN II			1.2	PS
4971.616	4973.003	20108.57	IN II			0.7	PS
4928.30	4929.68	20285.3	IN II			0.7	PS
4924.934	4926.309	20299.17	IN II			1.9H	PS
4915.085	4916.457	20339.85	IN II			1.5	PS
4908.423	4909.793	20367.46	IN II			0.7	PS
4908.316	4909.686	20367.90	IN II			1.0	PS
4907.149	4908.519	20372.74	IN II			1.7	PS
4906.973	4908.343	20373.47	IN II			1.0	PS
4902.450	4903.819	20392.27	IN II			0.7	PS
4878.8	4880.2	20491.	IN I	0.5	0.3		PS
4856.240	4857.597	20586.31	IN II			1.5	PS
4800.01	4801.35	20827.5	IN	0.7	1.0		SQ
4797.011	4798.352	20840.49	IN II			1.0	PS
4747.412	4748.740	21058.22	IN II			1.0	PS
4701.481	4702.797	21263.94	IN		1.5		MIT
4685.223	4686.534	21337.73	IN II			2.0	PS
4685.040	4686.351	21338.56	IN II			1.0	PS
4684.934	4686.245	21339.05	IN II			1.2	PS
4684.760	4686.071	21339.84	IN II			1.4H	PS
4684.587	4685.898	21340.63	IN II			1.4	PS
4684.449	4685.760	21341.25	IN II			1.3	PS
4684.316	4685.627	21341.86	IN II			1.5	PS
4682.00	4683.31	21352.4	IN		2.40		SQ
4681.108	4682.418	21356.49	IN II			2.3	PS
4678.17	4679.48	21369.9	IN		1.5		SQ
4673.772	4675.080	21390.01	IN II			0.7	PS
4667.325	4668.632	21419.55	IN		1.0		MIT
4657.075	4658.379	21466.70	IN II			1.5	PS
4656.997	4658.301	21467.05	IN II			1.5	PS
4656.812	4658.116	21467.91	IN II			1.5	PS
4656.707	4658.011	21468.39	IN II			0.7	PS
4656.544	4657.848	21469.14	IN II			1.3	PS

AIR WAVELENGTH (ANGSTRMS)	VACUUM WAVELENGTH (ANGSTRMS)	WAVENUMBER (KAYSERS)	LMENT	LOG ARC	INTENSITY SPK	INTENSITY DIS	REF
4656.407	4657.711	21469.77	IN II			1.0	PS
4656.302	4657.606	21470.26	IN II			1.0	PS
4655.790	4657.093	21472.62	IN II			2.0	PS
4655.657	4656.960	21473.23	IN II			1.7	PS
4655.523	4656.826	21473.85	IN II			1.7	PS
4655.409	4656.712	21474.38	IN II			1.7	PS
4644.648	4645.949	21524.13	IN II			1.8	PS
4644.536	4645.837	21524.65	IN II			2.1	PS
4638.861	4640.160	21550.98	IN	0.5	1.8		MIT
4638.245	4639.544	21553.84	IN II			2.1	PS
4638.100	4639.399	21554.52	IN II			2.3	PS
4637.974	4639.273	21555.10	IN II			1.3	PS
4637.112	4638.411	21559.11	IN II			1.0	PS
4637.025	4638.324	21559.51	IN II			1.3	PS
4636.915	4638.213	21560.02	IN II			1.0	PS
4628.310	4629.606	21600.11	IN II			0.7	PS
4628.182	4629.478	21600.71	IN II			1.0	PS
4628.072	4629.368	21601.22	IN II			1.0	PS
4627.784	4629.080	21602.56	IN II			0.7	PS
4627.385	4628.681	21604.43	IN II			2.2	PS
4625.74	4627.04	21612.1	IN		0.7		SQ
4620.828	4622.122	21635.08	IN II			0.7	PS
4620.667	4621.961	21635.84	IN II			0.7	PS
4620.529	4621.823	21636.48	IN II			1.0	PS
4620.413	4621.707	21637.03	IN II			1.2	PS
4620.243	4621.537	21637.82	IN II			2.3	PS
4620.055	4621.349	21638.70	IN II			1.9	PS
4617.159	4618.452	21652.27	IN II			2.3	PS
4616.585	4617.878	21654.97	IN II			1.0	PS
4616.176	4617.469	21656.89	IN II			1.3	PS
4616.033	4617.326	21657.56	IN II			1.2	PS
4615.925	4617.218	21658.06	IN II			1.2	PS
4615.570	4616.863	21659.73	IN II			0.7	PS
4612.133	4613.425	21675.87	IN		2.2		MIT
4593.698	4594.985	21762.86	IN		1.3H		MIT
4579.32	4580.60	21831.2	IN		1.0		SQ
4578.526	4579.809	21834.97	IN II			0.7	PS
4578.393	4579.676	21835.61	IN II			1.8	PS
4578.087	4579.370	21837.06	IN II			1.7	PS
4576.096	4577.378	21846.57	IN		0.5H		MIT
4572.904	4574.186	21861.81	IN II			1.0	PS
4571.402	4572.683	21869.00	IN II			0.3	PS
4571.330	4572.611	21869.34	IN II			0.7	PS
4571.215	4572.496	21869.89	IN II			1.5	PS
4571.169	4572.450	21870.11	IN II			1.0	PS
4570.979	4572.260	21871.02	IN II			0.7	PS
4570.932	4572.213	21871.25	IN II			1.5	PS
4570.841	4572.122	21871.68	IN II			1.2	PS
4570.782	4572.063	21871.97	IN II			0.7	PS
4564.97	4566.25	21899.8	IN		1.0		SQ
4549.353	4550.628	21974.99	IN II			1.0	PS
4549.23	4550.51	21975.6	IN II			1.0	PS
4549.051	4550.326	21976.45	IN II			1.2	PS
4548.738	4550.013	21977.96	IN II			1.2	PS
4535.50	4536.77	22042.1	IN		0.7		SQ
4530.050	4531.320	22068.62	IN		1.0		MIT
4523.57	4524.84	22100.2	IN		1.0		SQ
4517.42	4518.69	22130.3	IN		1.0		SQ
4511.323	4512.588	22160.23	IN I	3.7G	3.6G		MIT
4511.30	4512.57	22160.3	IN I	2.3			ME
4505.42	4506.68	22189.3	IN		1.0		SQ
4500.949	4502.212	22211.31	IN II			1.7	PS
4500.770	4502.033	22212.19	IN II			1.5	PS
4500.627	4501.889	22212.90	IN II			1.2	PS
4487.36	4488.62	22278.6	IN		1.0		SQ
4415.69	4416.93	22640.2	IN		1.2H		SQ
4389.48	4390.71	22775.3	IN		0.7H		SQ
4383.76	4384.99	22805.1	IN		0.7H		SQ
4375.08	4376.31	22850.3	IN II			0.3H	PS
4373.040	4374.269	22860.96	IN II			1.2	PS

AIR WAVELENGTH (ANGSTRMS)	VACUUM WAVELENGTH (ANGSTRMS)	WAVENUMBER (KAYSERS)	LMENT	LOG ARC	INTENSITY SPK	DIS	REF
4372.874	4374.103	22861.83	IN II			1.9	PS
4372.804	4374.033	22862.20	IN II			1.2	PS
4365.06	4366.29	22902.8	IN II			1.2	PS
4364.77	4366.00	22904.3	IN II			1.0	PS
4364.65	4365.88	22904.9	IN II			0.3	PS
4360.31	4361.54	22927.7	IN II			1.0	PS
4358.50	4359.73	22937.2	IN II			0.3	PS
4357.92	4359.14	22940.3	IN II			0.3	PS
4336.339	4337.558	23054.45	IN II			1.2	PS
4330.020	4331.238	23088.09	IN II			1.7	PS
4329.649	4330.866	23090.07	IN II			0.3	PS
4327.334	4328.551	23102.42	IN II			1.9	PS
4326.809	4328.026	23105.22	IN II			0.7	PS
4326.761	4327.978	23105.48	IN II			1.2	PS
4326.335	4327.552	23107.75	IN II			0.7	PS
4326.229	4327.446	23108.32	IN II			1.2	PS
4326.019	4327.235	23109.44	IN II			1.0	PS
4325.885	4327.101	23110.16	IN II			1.0	PS
4325.757	4326.973	23110.84	IN II			0.7	PS
4277.99	4279.19	23368.9	IN		0.3		SQ
4245.33	4246.53	23548.7	IN		0.7		SQ
4229.59	4230.78	23636.3	IN II			0.7	PS
4229.00	4230.19	23639.6	IN II			1.0	PS
4227.98	4229.17	23645.3	IN II			1.0	PS
4227.16	4228.35	23649.9	IN II			1.7H	PS
4223.57	4224.76	23670.0	IN		0.7		SQ
4219.83	4221.02	23691.0	IN II			1.7	PS
4219.50	4220.69	23692.8	IN II			1.5	PS
4219.40	4220.59	23693.4	IN II			0.7	PS
4218.72	4219.91	23697.2	IN II			0.7	PS
4215.90	4217.09	23713.1	IN		0.7		SQ
4213.73	4214.92	23725.3	IN II			1.2	PS
4213.58	4214.77	23726.1	IN II			1.0	PS
4213.10	4214.29	23728.8	IN II			1.7	PS
4212.97	4214.16	23729.5	IN II			1.2	PS
4205.217	4206.402	23773.29	IN II			1.2	PS
4205.151	4206.336	23773.66	IN II			1.5	PS
4205.079	4206.264	23774.07	IN II			1.7	PS
4196.99	4198.17	23819.9	IN		1.0		SQ
4180.94	4182.12	23911.3	IN		1.2		MIT
4158.08	4159.25	24042.8	IN		0.9		SQ
4149.710	4150.880	24091.28	IN II			1.5	PS
4141.460	4142.628	24139.27	IN II			1.2	PS
4140.964	4142.132	24142.16	IN II			0.7	PS
4140.420	4141.588	24145.33	IN II			0.7	PS
4132.904	4134.070	24189.24	IN		1.2		MIT
4122.791	4123.954	24248.57	IN II			1.5H	PS
4115.48	4116.64	24291.6	IN II			0.3	PS
4109.34	4110.50	24327.9	IN II			0.7H	PS
4101.773	4102.931	24372.82	IN I	3.3G	3.0G		MIT
4072.40	4073.55	24548.6	IN		2.3J		SQ
4057.866	4059.012	24636.54	IN	1.9	1.0		MIT
4057.188	4058.334	24640.65	IN II			1.0	PS
4057.070	4058.216	24641.37	IN II			2.0	PS
4056.936	4058.082	24642.18	IN II			2.7	PS
4056.785	4057.931	24643.10	IN II			1.5	PS
4056.747	4057.893	24643.33	IN II			1.7	PS
4056.591	4057.737	24644.28	IN II			0.7	PS
4033.066	4034.206	24788.03	IN	0.6			MIT
4027.79	4028.93	24820.5	IN II			1.7H	PS
4024.83	4025.97	24838.7	IN		2.0J		SQ
4023.76	4024.90	24845.4	IN II			1.2	PS
4021.99	4023.13	24856.3	IN II			1.0	PS
4021.66	4022.80	24858.3	IN II			1.7	PS
4016.76	4017.90	24888.6	IN II			0.3	PS
4016.24	4017.38	24891.9	IN II			1.7	PS
4015.80	4016.94	24894.6	IN II			0.3	PS
4013.93	4015.06	24906.2	IN II			1.9	PS
4013.49	4014.62	24908.9	IN II			1.5	PS
4012.96	4014.09	24912.2	IN II			1.0	PS

AIR WAVELENGTH (ANGSTRMS)	VACUUM WAVELENGTH (ANGSTRMS)	WAVENUMBER (KAYSERS)	LMENT	LOG ARC	INTENSITY SPK	DIS	REF
4007.608	4008.741	24945.49	IN II			1.0	PS
4007.543	4008.676	24945.89	IN II			1.2	PS
4004.834	4005.966	24962.77	IN II			1.0	PS
4004.709	4005.841	24963.55	IN II			1.2	PS
4004.528	4005.660	24964.67	IN II			1.5	PS
3995.158	3996.288	25023.22	IN		1.3V		MIT
3969.13	3970.25	25187.3	IN	1.2			SQ
3962.609	3963.730	25228.76	IN II			1.4	PS
3962.588	3963.709	25228.89	IN II			1.4	PS
3962.418	3963.539	25229.98	IN II			0.7	PS
3962.275	3963.396	25230.89	IN II			1.4	PS
3962.159	3963.280	25231.63	IN II			1.3	PS
3962.041	3963.162	25232.38	IN II			1.4	PS
3957.46	3958.58	25261.6	IN		1.2		SQ
3941.33	3942.45	25365.0	IN		0.5		S
3936.79	3937.90	25394.2	IN		1.3		SQ
3934.431	3935.545	25409.44	IN II			0.7	PS
3934.123	3935.237	25411.43	IN II			1.0	PS
3931.11	3932.22	25430.9	IN		0.5		SQ
3929.80	3930.91	25439.4	IN II			0.7	PS
3929.53	3930.64	25441.1	IN II			1.0	PS
3924.35	3925.46	25474.7	IN II			1.0	PS
3923.94	3925.05	25477.4	IN II			0.3	PS
3922.16	3923.27	25488.9	IN II			1.0	PS
3922.08	3923.19	25489.5	IN II			1.0	PS
3920.54	3921.65	25499.5	IN		1.3		SQ
3912.81	3913.92	25549.8	IN		1.2		SQ
3902.123	3903.228	25619.82	IN II			1.0	PS
3902.076	3903.181	25620.13	IN II			1.3	PS
3902.024	3903.129	25620.47	IN II			1.3	PS
3894.82	3895.92	25667.9	IN II			1.3	PS
3889.78	3890.88	25701.1	IN II			2.0	PS
3860.73	3861.82	25894.5	IN II			0.7	PS
3855.723	3856.816	25928.12	IN II			0.7	PS
3853.379	3854.472	25943.89	IN II		1.1		MIT
3842.272	3843.362	26018.89	IN II			1.5	PS
3842.217	3843.307	26019.26	IN II			1.4	PS
3842.168	3843.258	26019.59	IN II			1.4	PS
3842.125	3843.215	26019.88	IN II			1.3	PS
3835.183	3836.271	26066.98	IN		1.2		MIT
3834.722	3835.810	26070.11	IN II			1.3	PS
3834.686	3835.774	26070.36	IN II			1.4	PS
3834.647	3835.735	26070.62	IN II			1.5	PS
3834.606	3835.694	26070.90	IN II			1.6	PS
3834.563	3835.651	26071.20	IN II			1.6	PS
3808.11	3809.19	26252.3	IN II			0.3	PS
3801.5	3802.6	26298.	IN II			1.7	PS
3799.423	3800.502	26312.31	IN II			1.0	PS
3799.371	3800.450	26312.68	IN II			1.4	PS
3799.314	3800.393	26313.07	IN II			1.0	PS
3799.204	3800.283	26313.83	IN II			1.3	PS
3799.118	3800.197	26314.43	IN II			1.0	PS
3799.055	3800.134	26314.86	IN II			1.0	PS
3795.272	3796.350	26341.09	IN II			1.0	PS
3795.211	3796.289	26341.52	IN II			1.7	PS
3795.166	3796.244	26341.83	IN II			1.3	PS
3739.95	3741.01	26730.7	IN		0.5		MIT
3723.645	3724.704	26847.77	IN II			1.4	PS
3723.407	3724.466	26849.49	IN II			1.5	PS
3723.205	3724.264	26850.94	IN II			1.5	PS
3718.836	3719.894	26882.49	IN II			1.6	PS
3718.707	3719.765	26883.42	IN II			1.3	PS
3718.634	3719.692	26883.95	IN II			1.4	PS
3718.390	3719.448	26885.71	IN II			1.4	PS
3718.332	3719.390	26886.13	IN II			1.4	PS
3718.218	3719.276	26886.96	IN II			1.3	PS
3716.284	3717.341	26900.95	IN II			1.4	PS
3716.233	3717.290	26901.32	IN II			1.6	PS
3716.183	3717.240	26901.68	IN II			1.3	PS
3716.130	3717.187	26902.06	IN II			1.4	PS

AIR WAVELENGTH (ANGSTRMS)	VACUUM WAVELENGTH (ANGSTRMS)	WAVENUMBER (KAYSERS)	LMENT	LOG ARC	INTENSITY SPK	INTENSITY DIS	REF
3716.081	3717.138	26902.42	IN II			1.3	PS
3716.048	3717.105	26902.66	IN II			1.3	PS
3716.004	3717.061	26902.98	IN II			1.0	PS
3708.246	3709.301	26959.26	IN II			1.3	PS
3708.106	3709.161	26960.27	IN II			1.0	PS
3708.001	3709.056	26961.04	IN II			1.0	PS
3694.029	3695.080	27063.01	IN II			1.6	PS
3693.899	3694.950	27063.96	IN II			1.5	PS
3693.788	3694.839	27064.78	IN II			1.4	PS
3683.520	3684.569	27140.22	IN		1.2		MIT
3627.782	3628.816	27557.20	IN II			1.2	PS
3612.86	3613.89	27671.0	IN		1.2		MIT
3610.508	3611.538	27689.04	IN		1.3		MIT
3602.88	3603.91	27747.7	IN		0.5		SQ
3590.35	3591.37	27844.5	IN		1.1		M
3585.86	3586.88	27879.4	IN		1.1		SQ
3583.85	3584.87	27895.0	IN		1.1		SQ
3572.75	3573.77	27981.6	IN		0.8		MIT
3546.83	3547.84	28186.1	IN		0.5		SQ
3542.82	3543.83	28218.0	IN		0.5J		SQ
3535.71	3536.72	28274.8	IN		1.0J		SQ
3517.566	3518.572	28420.62	IN II			0.7	PS
3480.16	3481.16	28726.1	IN		0.5		SQ
3471.49	3472.48	28797.8	IN		0.8		SQ
3467.66	3468.65	28829.6	IN		1.1		SQ
3466.24	3467.23	28841.4	IN		1.3		SQ
3454.73	3455.72	28937.5	IN		0.5		SQ
3441.975	3442.961	29044.76	IN II			0.7	PS
3438.519	3439.505	29073.96	IN II			1.0	PS
3438.339	3439.325	29075.48	IN II			1.7	PS
3404.448	3405.425	29364.91	IN II			1.0	PS
3404.297	3405.274	29366.21	IN II			1.3	PS
3404.131	3405.108	29367.65	IN II			1.3	PS
3403.70	3404.68	29371.4	IN		1.3		SQ
3376.644	3377.614	29606.70	IN II			0.7	PS
3376.604	3377.574	29607.05	IN II			1.3	PS
3376.554	3377.524	29607.49	IN II			1.0	PS
3370.64	3371.61	29659.4	IN		0.5		SQ
3360.85	3361.82	29745.8	IN		0.8		SQ
3358.74	3359.71	29764.5	IN		0.8		SQ
3352.33	3353.29	29821.4	IN II			1.0	PS
3346.062	3347.024	29877.29	IN II			1.0	PS
3345.940	3346.902	29878.38	IN II			0.7	PS
3338.575	3339.535	29944.29	IN II			1.3	PS
3338.486	3339.446	29945.09	IN II			1.0	PS
3338.415	3339.375	29945.72	IN II			0.7	PS
3336.26	3337.22	29965.1	IN		0.5		SQ
3285.80	3286.75	30425.2	IN		0.5J		SQ
3280.692	3281.637	30472.59	IN		0.5		MIT
3274.02	3274.96	30534.7	IN II		1.2		SQ
3264.041	3264.982	30628.04	IN II			0.7	PS
3264.005	3264.946	30628.38	IN II			1.0	PS
3258.564	3259.504	30679.52	IN I	2.7G	2.5G		MIT
3256.090	3257.029	30702.83	IN I	3.2G	2.8G		MIT
3250.38	3251.32	30756.8	IN		1.0		SQ
3247.61	3248.55	30783.0	IN		1.2		SQ
3236.86	3237.79	30885.2	IN II			0.7	PS
3198.184	3199.108	31258.71	IN II			1.0	PS
3198.037	3198.961	31260.15	IN II			1.5	PS
3194.634	3195.557	31293.44	IN II			1.4	PS
3194.435	3195.358	31295.39	IN II			1.0	PS
3194.360	3195.283	31296.13	IN II			1.0	PS
3194.278	3195.201	31296.93	IN II			0.7	PS
3187.03	3187.95	31368.1	IN		1.1		SQ
3176.475	3177.394	31472.33	IN II			0.7	PS
3176.309	3177.228	31473.98	IN II			1.0	PS
3176.112	3177.031	31475.93	IN II			1.3	PS
3175.03	3175.95	31486.7	IN		0.5		MIT
3158.396	3159.310	31652.48	IN II			2.0	PS
3155.771	3156.685	31678.81	IN II			2.3	PS
3146.806	3147.717	31769.05	IN II			1.7	PS
3146.601	3147.512	31771.12	IN II			1.4	PS
3142.790	3143.700	31809.65	IN II			1.0	PS
3142.714	3143.624	31810.42	IN II			1.4	PS
3138.642	3139.551	31851.69	IN II			1.7	PS
3138.56	3139.47	31852.5	IN II			1.7	PS
3132.93	3133.84	31909.7	IN		0.5		SQ
3121.55	3122.46	32026.1	IN		0.5		SQ
3101.005	3101.905	32238.25	IN II			1.0	PS
3100.934	3101.834	32238.99	IN II			1.0	PS
3100.867	3101.767	32239.69	IN II			1.3	PS
3100.698	3101.598	32241.45	IN II			1.3	PS
3100.659	3101.559	32241.85	IN II			1.0	PS
3100.571	3101.471	32242.77	IN II			0.7	PS
3099.866	3100.766	32250.10	IN II			1.6	PS
3099.739	3100.639	32251.42	IN II			1.3	PS
3083.738	3084.634	32418.76	IN II			1.0	PS
3083.653	3084.549	32419.66	IN II			0.7	PS
3081.05	3081.94	32447.0	IN		0.5		CX
3078.27	3079.16	32476.4	IN		1.0		CX
3069.76	3070.65	32566.4	IN II			1.3	PS
3067.73	3068.62	32587.9	IN		0.5		SQ
3066.55	3067.44	32600.5	IN		0.8		MIT
3056.58	3057.47	32706.8	IN		1.0		MIT
3051.25	3052.14	32763.9	IN I			1.2	SY
3047.93	3048.82	32799.6	IN		0.5		SQ
3047.00	3047.89	32809.6	IN		0.8H		CX
3043.40	3044.29	32848.4	IN				ME
3039.356	3040.240	32892.13	IN I	3.0G	2.7G		PS
3038.67	3039.55	32899.6	IN		0.5		CX
3037.11	3037.99	32916.5	IN		1.1		CX
3034.122	3035.005	32948.87	IN	0.9	0.0		MIT
3028.565	3029.447	33009.33	IN II			0.3	PS
3023.39	3024.27	33065.8	IN		0.5		SQ
3022.372	3023.252	33076.96	IN II			1.3	PS
3017.78	3018.66	33127.3	IN		0.5		SQ
3008.31	3009.19	33231.6	IN		2.70		CX
2999.503	3000.378	33329.14	IN II			1.0	PS
2999.397	3000.271	33330.32	IN II			1.5	PS
2999.316	3000.190	33331.22	IN II			1.0	PS
2989.59	2990.46	33439.6	IN		0.3		CX
2986.92	2987.79	33469.5	IN		0.3		CX
2983.99	2984.86	33502.4	IN		0.3		SQ
2982.87	2983.74	33515.0	IN		1.5		MIT
2979.58	2980.45	33552.0	IN		1.0		CX
2966.201	2967.067	33703.31	IN II			1.3	PS
2966.147	2967.013	33703.93	IN II			0.7	PS
2959.72	2960.58	33777.1	IN		0.3		SQ
2957.012	2957.876	33808.04	IN I	1.7	1.4		UH
2941.050	2941.910	33991.52	IN II			1.9	PS
2939.36	2940.22	34011.1	IN		1.0		CX
2938.71	2939.57	34018.6	IN		1.0		CX
2933.27	2934.13	34081.7	IN		0.5		CX
2932.624	2933.482	34089.18	IN I	2.7	2.5		MIT
2928.04	2928.90	34142.5	IN		0.3		CX
2927.08	2927.94	34153.7	IN		0.3		MIT
2916.467	2917.321	34278.03	IN II			1.0	PS
2915.63	2916.48	34287.9	IN		0.3		MIT
2911.45	2912.30	34337.1	IN		0.5		CX
2910.52	2911.37	34348.1	IN		0.7		CX
2907.93	2908.78	34378.6	IN		0.3		CX
2907.08	2907.93	34388.7	IN		0.3		CX
2905.16	2906.01	34411.4	IN		0.3		CX
2900.56	2901.41	34466.0	IN		1.1		CX
2890.200	2891.047	34589.54	IN II			1.3	PS
2890.161	2891.008	34590.01	IN II			1.6	PS
2884.08	2884.93	34662.9	IN		0.3		SQ
2881.61	2882.46	34692.6	IN		0.3		SQ
2873.50	2874.34	34790.6	IN		0.3		SQ
2867.387	2868.229	34864.72	IN II			1.0	PS

AIR WAVELENGTH (ANGSTRMS)	VACUUM WAVELENGTH (ANGSTRMS)	WAVENUMBER (KAYSERS)	LMENT	LOG ARC	INTENSITY SPK	DIS	REF
2867.222	2868.064	34866.73	IN II			1.0	PS
2866.464	2867.306	34875.95	IN II			1.3	PS
2865.734	2866.575	34884.83	IN II			0.7	PS
2865.684	2866.525	34885.44	IN II			1.7	PS
2865.637	2866.478	34886.01	IN II			0.7	PS
2864.23	2865.07	34903.1	IN		0.5		CX
2860.52	2861.36	34948.4	IN		0.7		CX
2858.74	2859.58	34970.2	IN I			1.5	SY
2844.51	2845.35	35145.1	IN		0.3		CX
2839.980	2840.815	35201.17	IN		0.3		MIT
2839.33	2840.16	35209.2	IN		0.3		CX
2838.87	2839.70	35214.9	IN		0.8		SQ
2836.919	2837.753	35239.15	IN I	1.9	1.9		MIT
2835.95	2836.78	35251.2	IN		0.5		MIT
2833.06	2833.89	35287.1	IN		0.5		MIT
2831.540	2832.373	35306.08	IN II			1.3	PS
2823.99	2824.82	35400.5	IN		0.5		SQ
2819.52	2820.35	35456.6	IN II			1.6	PS
2819.039	2819.869	35462.64	IN II			1.5	PS
2818.932	2819.762	35463.99	IN II			1.0	PS
2811.93	2812.76	35552.3	IN		0.3		MIT
2810.90	2811.73	35565.3	IN		0.3		CX
2805.400	2806.227	35635.04	IN II			1.3	PS
2805.288	2806.115	35636.46	IN II			0.7H	PS
2803.18	2804.01	35663.3	IN		0.3		SQ
2802.01	2802.84	35678.1	IN		0.5		MIT
2798.760	2799.585	35719.58	IN II			1.5	PS
2792.743	2793.566	35796.54	IN		0.3		MIT
2786.25	2787.07	35880.0	IN		0.9		CX
2775.355	2776.174	36020.79	IN I	1.9	1.5		UH
2773.04	2773.86	36050.9	IN		0.5		CX
2753.878	2754.692	36301.70	IN I	2.5G	2.5J		PS
2752.834	2753.648	36315.46	IN II			0.7	PS
2752.743	2753.557	36316.67	IN II			0.3	PS
2749.803	2750.616	36355.49	IN II			1.5	PS
2749.702	2750.515	36356.83	IN II			1.7	PS
2748.72	2749.53	36369.8	IN		1.4J		MIT
2747.9	2748.7	36381.	IN		0.3		CX
2726.27	2727.08	36669.3	IN		0.3		CX
2720.00	2720.81	36753.8	IN I	1.0H			FL
2715.72	2716.52	36811.7	IN		0.3		MIT
2713.935	2714.739	36835.95	IN I	2.3G	2.1J		MIT
2710.265	2711.068	36885.83	IN I	2.9G	2.3P		MIT
2709.2	2710.0	36900.	IN		0.7		CX
2705.2	2706.0	36955.	IN		0.3		CX
2704.483	2705.285	36964.68	IN II			1.5	PS
2693.882	2694.682	37110.14	IN II			1.5	PS
2691.254	2692.053	37146.37	IN II			1.0	PS
2691.050	2691.849	37149.19	IN II			1.0	PS
2683.117	2683.914	37259.02	IN II			1.7	PS
2682.20	2683.00	37271.8	IN II			1.3	LG
2674.632	2675.427	37377.21	IN II			1.6	PS
2674.480	2675.275	37379.34	IN II			1.6	PS
2672.173	2672.967	37411.61	IN II			1.4	PS
2668.685	2669.478	37460.50	IN II			1.7	PS
2668.623	2669.416	37461.37	IN II			1.5	PS
2667.62	2668.41	37475.5	IN I			0.7	LG
2667.62	2668.41	37475.5	IN II			0.7	LG
2666.9	2667.7	37486.	IN I		0.8		CX
2662.685	2663.477	37544.91	IN II			1.5	PS
2662.583	2663.375	37546.35	IN II			1.5	PS
2654.76	2655.55	37657.0	IN II			1.5	PS
2654.65	2655.44	37658.5	IN II			1.3	PS
2650.351	2651.140	37719.62	IN II			1.3	PS
2637.646	2638.432	37901.30	IN II			0.7	PS
2634.7	2635.5	37944.	IN		0.3		CX
2630.94	2631.72	37997.9	IN		0.3		SQ
2623.433	2624.216	38106.62	IN II			1.3	PS
2623.281	2624.064	38108.83	IN II			1.3	PS
2614.31	2615.09	38239.6	IN		0.3		MIT

AIR WAVELENGTH (ANGSTRMS)	VACUUM WAVELENGTH (ANGSTRMS)	WAVENUMBER (KAYSERS)	LMENT	LOG ARC	INTENSITY SPK	DIS	REF
2612.01	2612.79	38273.3	IN		0.3		SQ
2604.045	2604.823	38390.32	IN II			1.5	PS
2601.756	2602.534	38424.10	IN I	1.7G	1.2J		PS
2598.802	2599.579	38467.77	IN II			1.3	PS
2598.692	2599.469	38469.40	IN II			1.0	PS
2589.5	2590.3	38606.	IN		0.3		CX
2585.8	2586.6	38661.	IN		0.3		CX
2584.6	2585.4	38679.	IN		0.3		CX
2580.8	2581.6	38736.	IN		0.3		CX
2578.9	2579.7	38765.	IN		0.5		MIT
2574.512	2575.283	38830.68	IN II			0.7	PS
2574.372	2575.143	38832.79	IN II			1.0	PS
2572.8	2573.6	38856.	IN I		0.3		CX
2570.3	2571.1	38894.	IN		0.3		CX
2565.397	2566.166	38968.64	IN II			0.7	PS
2565.13	2565.90	38972.7	IN II			1.6	PS
2564.725	2565.494	38978.85	IN II			1.0	PS
2560.228	2560.996	39047.31	IN I	2.2G	1.7P		MIT
2554.477	2555.243	39135.22	IN II			1.7	PS
2554.399	2555.165	39136.41	IN II			1.7	PS
2553.56	2554.33	39149.3	IN II			1.6	PS
2553.33	2554.10	39152.8	IN II			0.7	PS
2552.0	2552.8	39173.	IN		0.3		CX
2541.62	2542.38	39333.2	IN		0.3		SQ
2536.669	2537.431	39409.94	IN II			1.0	PS
2528.17	2528.93	39542.4	IN		0.7		SQ
2524.20	2524.96	39604.6	IN		0.3		SQ
2522.985	2523.744	39623.67	IN I	1.0			PS
2521.64	2522.40	39644.8	IN		0.3		SQ
2521.371	2522.130	39649.03	IN I	1.5			PS
2519.41	2520.17	39679.9	IN		0.3		MIT
2516.40	2517.16	39727.4	IN		0.5		MIT
2514.08	2514.84	39764.0	IN II			1.0	PS
2512.370	2513.126	39791.07	IN II			1.5	PS
2512.247	2513.003	39793.02	IN II			1.4	PS
2510.41	2511.17	39822.1	IN		0.6		SQ
2508.37	2509.13	39854.5	IN		0.9		MIT
2508.157	2508.913	39857.91	IN II			1.5	PS
2505.04	2505.79	39907.5	IN		0.3		SQ
2501.078	2501.832	39970.71	IN II			1.3	PS
2500.900	2501.654	39973.56	IN II			1.4	PS
2499.599	2500.353	39994.36	IN II			1.9	PS
2499.34	2500.09	39998.5	IN II			1.0	PS
2498.590	2499.343	40010.51	IN II			1.7	PS
2497.20	2497.95	40032.8	IN		0.5		SQ
2496.333	2497.086	40046.68	IN II			1.3	PS
2496.215	2496.968	40048.58	IN II			1.0	PS
2495.97	2496.72	40052.5	IN		0.3		MIT
2488.951	2489.702	40165.45	IN II			1.8	PS
2488.666	2489.417	40170.05	IN II			1.6	PS
2488.556	2489.307	40171.82	IN II			1.3	PS
2486.15	2486.90	40210.7	IN II			1.6	PS
2475.565	2476.313	40382.62	IN		1.1		MIT
2468.023	2468.769	40506.01	IN I	1.4H	0.3J		PS
2466.472	2467.218	40531.48	IN		0.3		MIT
2461.07	2461.81	40620.4	IN II			0.7	PS
2460.19	2460.93	40635.0	IN II			0.7J	PS
2460.079	2460.823	40636.81	IN I	1.0J			PS
2455.682	2456.425	40709.56	IN		0.5		MIT
2453.855	2454.598	40739.87	IN II			1.3H	PS
2453.228	2453.971	40750.28	IN II			1.5	PS
2449.14	2449.88	40818.3	IN		0.3		SQ
2447.90	2448.64	40839.0	IN II			1.9	PS
2446.02	2446.76	40870.4	IN II			0.7	LG
2442.63	2443.37	40927.1	IN II			1.5	PS
2442.46	2443.20	40929.9	IN II			1.3	PS
2438.75	2439.49	40992.2	IN		0.5		SQ
2433.9	2434.6	41074.	IN		0.3		CX
2432.730	2433.468	41093.62	IN II			1.4	PS
2430.986	2431.724	41123.09	IN I	1.06			PS

AIR WAVELENGTH (ANGSTRMS)	VACUUM WAVELENGTH (ANGSTRMS)	WAVENUMBER (KAYSERS)	LMENT	LOG ARC	INTENSITY SPK	DIS	REF
2429.864	2430.601	41142.08	IN I	1.3G			PS
2428.38	2429.12	41167.2	IN		0.5		SQ
2427.205	2427.942	41187.15	IN II			1.6H	PS
2425.964	2426.701	41208.22	IN II			1.4H	PS
2425.834	2426.571	41210.42	IN II			1.3	PS
2420.60	2421.34	41299.5	IN		0.3		SQ
2419.195	2419.930	41323.51	IN II			1.4	PS
2419.064	2419.799	41325.75	IN II			1.4	PS
2418.932	2419.667	41328.00	IN II			1.0	PS
2417.666	2418.401	41349.64	IN		0.3		MIT
2415.488	2416.222	41386.92	IN II			0.7	PS
2415.316	2416.050	41389.87	IN II			1.0	PS
2412.83	2413.56	41432.5	IN II			0.3	PS
2410.846	2411.579	41466.61	IN II			1.3	PS
2408.757	2409.490	41502.56	IN II			1.4	PS
2408.565	2409.298	41505.87	IN II			1.3	PS
2407.590	2408.322	41522.68	IN I	1.0	0.0		MIT
2406.472	2407.204	41541.97	IN II			1.4H	PS
2399.177	2399.907	41668.27	IN I	1.0			PS
2398.38	2399.11	41682.1	IN II			1.4	PS
2393.179	2393.908	41772.70	IN II			1.3	PS
2393.037	2393.766	41775.18	IN II			1.0	PS
2389.543	2390.271	41836.26	IN I	1.7G			PS
2384.557	2385.284	41923.73	IN II			0.7	PS
2382.673	2383.400	41956.87	IN II			1.3	PS
2382.593	2383.320	41958.28	IN II			1.6	PS
2378.996	2379.722	42021.72	IN I	1.0			PS
2378.141	2378.867	42036.82	IN I	1.2			PS
2377.180	2377.905	42053.82	IN II			0.3	PS
2376.66	2377.39	42063.0	IN II			1.0	PS
2375.96	2376.69	42075.4	IN II			1.0	PS
2373.04	2373.76	42127.2	IN II			1.0H	PS
2372.90	2373.62	42129.7	IN II			0.9H	PS
2364.72	2365.44	42275.4	IN II			1.0	PS
2358.698	2359.419	42383.31	IN I	0.7J			PS
2356.88	2357.60	42416.0	IN II			1.0	PS
2351.45	2352.17	42513.9	IN		0.5		MIT
2350.765	2351.485	42526.33	IN II			0.7	PS
2350.735	2351.455	42526.87	IN II			1.0	PS
2346.561	2347.280	42602.51	IN I	0.7			PS
2345.903	2346.621	42614.46	IN I	0.9			PS
2340.186	2340.903	42718.55	IN I	1.0			PS
2334.57	2335.29	42821.3	IN II			1.7H	PS
2332.755	2333.471	42854.62	IN I	0.5			PS
2328.00	2328.71	42942.1	IN II			1.3	PS
2327.90	2328.61	42944.0	IN II			1.0	PS
2324.924	2325.638	42998.96	IN I	0.5	0.0		PS
2324.407	2325.121	43008.52	IN I	0.7			PS
2323.400	2324.113	43027.16	IN II			1.0	PS
2321.83	2322.54	43056.2	IN		0.3		SD
2315.09	2315.80	43181.6	IN I	0.3			PS
2313.268	2313.979	43215.60	IN II			1.4	PS
2313.156	2313.867	43217.69	IN II			1.3	PS
2309.75	2310.46	43281.4	IN I	0.3			PS
2309.32	2310.03	43289.5	IN I	0.5			PS
2306.879	2307.589	43335.28	IN I	1.4	1.5		MIT
2306.118	2306.828	43349.58	IN II			1.0	PS
2306.062	2306.772	43350.63	IN II			1.0	PS
2305.994	2306.704	43351.91	IN II			1.3	PS
2302.49	2303.20	43417.9	IN I	0.3			PS
2298.697	2299.405	43489.51	IN I	0.3			PS
2298.33	2299.04	43496.5	IN I	0.3			PS
2297.09	2297.80	43519.9	IN		1.0		SQ
2292.8	2293.5	43601.	IN		0.3		CX
2283.75	2284.45	43774.1	IN I	0.3			PS
2281.642	2282.346	43814.56	IN II			1.3	PS
2278.3	2279.0	43879.	IN I	1.2			UH
2278.196	2278.900	43880.83	IN I	0.7			PS
2265.06	2265.76	44135.3	IN		0.8		MIT
2261.3	2262.0	44209.	IN		0.7		MIT

AIR WAVELENGTH (ANGSTRMS)	VACUUM WAVELENGTH (ANGSTRMS)	WAVENUMBER (KAYSERS)	LMENT	LOG ARC	INTENSITY SPK	DIS	REF
2259.991	2260.691	44234.27	IN I	1.0G	0.0		PS
2255.788	2256.487	44316.68	IN II			1.0	PS
2249.62	2250.32	44438.2	IN II			1.4	PS
2241.664	2242.360	44595.88	IN I	0.5			PS
2241.10	2241.80	44607.1	IN I	1.0G			MIT
2230.699	2231.392	44815.07	IN I	0.7G			PS
2229.8	2230.5	44833.	IN	0.3			UH
2227.50	2228.19	44879.4	IN		0.3		WB
2216.3	2217.0	45106.	IN		0.3		MIT
2211.137	2211.826	45211.51	IN I	0.5			PS
2205.28	2205.97	45331.6	IN II			1.4	PS
2202.240	2202.927	45394.14	IN I	0.3			PS
2197.407	2198.093	45493.97	IN I	0.3	0.0		PS
2195.67	2196.36	45530.0	IN II			0.3	PS
2190.835	2191.520	45630.43	IN I	0.3			PS
2187.40	2188.08	45702.1	IN I	0.3	0.0		PS
2182.400	2183.083	45806.77	IN I	0.3	0.0		PS
2179.90	2180.58	45859.3	IN I	0.3G			PS
2177.7	2178.4	45906.	IN	0.3			UH
2171.0	2171.7	46047.	IN	0.3			UH
2158.4	2159.1	46316.	IN		0.3		MIT
2157.0	2157.7	46346.	IN		0.3		MIT
2154.81	2155.49	46393.2	IN		0.3		SD
2079.26	2079.92	48078.7	IN II		1.2		LG
2062.0	2062.7	48481.	IN		1.1		WB
2025.44	2026.09	49356.1	IN		1.1		WB
8117.5	8119.7	12316.	IR	0.3V			IT
7834.32	7836.48	12760.8	IR I	0.7			ME
7216.3	7218.3	13854.	IR	0.6			IT
7199.00	7200.98	13887.0	IR	0.7			ED
7183.71	7185.69	13916.6	IR I	1.6			MIT
7138.77	7140.74	14004.2	IR I	0.8			MIT
7128.34	7130.31	14024.7	IR I	0.6			MIT
7101.09	7103.05	14078.5	IR	0.8			MIT
7037.87	7039.81	14204.9	IR	1.2			MIT
6998.84	7000.77	14284.1	IR I	1.2			MIT
6966.46	6968.38	14350.5	IR I	0.7			MIT
6959.10	6961.02	14365.7	IR I	1.0			MIT
6929.88	6931.79	14426.3	IR I	1.7			MIT
6893.34	6895.24	14502.7	IR	1.5			ME
6888.72	6890.62	14512.5	IR I	1.4			ME
6886.23	6888.13	14517.7	IR I	1.0			ME
6870.1	6872.0	14552.	IR	0.3H			IT
6837.8	6839.7	14621.	IR	0.6			IT
6830.01	6831.89	14637.2	IR I	1.7			MIT
6813.2	6815.1	14673.	IR	0.6			IT
6789.84	6791.71	14723.8	IR I	0.6			ME
6748.62	6750.48	14813.8	IR	0.6			ME
6733.12	6734.98	14847.9	IR I	0.5			ME
6722.5	6724.4	14871.	IR	0.7			IT
6721.54	6723.40	14873.4	IR	0.6			ME
6686.08	6687.93	14952.3	IR	1.3			ME
6668.317	6670.158	14992.15	IR I	0.6			MIT
6667.694	6669.535	14993.55	IR I	0.7			MIT
6659.189	6661.028	15012.70	IR I	0.7			MIT
6643.67	6645.50	15047.8	IR	0.5			MIT
6624.731	6626.561	15090.79	IR	1.2			MIT
6619.903	6621.731	15101.79	IR	0.6			MIT
6588.204	6590.024	15174.45	IR	0.7			MIT
6552.263	6554.073	15257.69	IR I	0.5			MIT
6540.29	6542.10	15285.6	IR	0.6			ME
6508.709	6510.507	15359.79	IR	0.6			MIT
6486.98	6488.77	15411.2	IR	0.9			ME
6455.06	6456.84	15487.4	IR	0.7			ME
6448.69	6450.47	15502.7	IR	0.5			ME
6418.340	6420.114	15576.05	IR	0.7			MIT
6403.885	6405.655	15611.21	IR	0.5			MIT
6398.259	6400.028	15624.93	IR	0.6			MIT
6334.438	6336.190	15782.36	IR	1.3			MIT
6327.49	6329.24	15799.7	IR	0.5			MIT

AIR WAVELENGTH (ANGSTRMS)	VACUUM WAVELENGTH (ANGSTRMS)	WAVENUMBER (KAYSERS)	LMENT	LOG ARC	INTENSITY SPK	DIS	REF
6303.743	6305.486	15859.20	IR	0.7			MIT
6288.277	6290.016	15898.21	IR	1.3			MIT
6242.730	6244.457	16014.20	IR	0.6			MIT
6217.374	6219.094	16079.51	IR	0.6			MIT
6211.314	6213.032	16095.20	IR	1.2			MIT
6125.79	6127.49	16319.9	IR	0.5			ME
6123.01	6124.70	16327.3	IR	0.5			ME
6114.53	6116.22	16350.0	IR	0.5			ME
6110.672	6112.363	16360.28	IR	1.5			MIT
6083.12	6084.80	16434.4	IR	0.5			ME
6067.830	6069.510	16475.79	IR	1.4			MIT
6060.544	6062.222	16495.60	IR	1.2			MIT
6036.645	6038.317	16560.91	IR	1.0			MIT
6026.097	6027.766	16589.90	IR	1.3			MIT
6023.83	6025.50	16596.1	IR	0.7			ME
5999.714	6001.376	16662.85	IR	0.7			MIT
5960.690	5962.341	16771.94	IR I	0.5			MIT
5899.576	5901.211	16945.68	IR I	0.3			MIT
5894.065	5895.698	16961.52	IR I	1.3			MIT
5887.364	5888.996	16980.82	IR	1.0			MIT
5882.30	5883.93	16995.4	IR I	1.0			MIT
5873.500	5875.128	17020.91	IR I	0.9			MIT
5872.487	5874.115	17023.84	IR I	0.6			MIT
5846.035	5847.656	17100.87	IR I	0.3H			MIT
5832.492	5834.109	17140.58	IR I	0.3H			MIT
5828.548	5830.164	17152.18	IR I	0.3			MIT
5816.371	5817.984	17188.09	IR I	0.3			MIT
5806.955	5808.565	17215.96	IR I	0.3			MIT
5778.28	5779.88	17301.4	IR	0.3			MIT
5768.909	5770.509	17329.49	IR	0.5			MIT
5747.525	5749.119	17393.97	IR	0.3			MIT
5745.403	5746.997	17400.39	IR	0.3			MIT
5736.226	5737.817	17428.23	IR	0.5			MIT
5731.77	5733.36	17441.8	IR	0.3			ME
5725.294	5726.882	17461.51	IR	0.5			MIT
5709.339	5710.923	17510.30	IR	0.7			MIT
5703.79	5705.37	17527.3	IR	0.3			ME
5687.36	5688.94	17578.0	IR	0.5			ME
5678.79	5680.37	17604.5	IR	0.3			ME
5649.519	5651.087	17695.71	IR I	0.3			MIT
5627.644	5629.206	17764.49	IR I	0.3			MIT
5625.553	5627.115	17771.10	IR	1.2			MIT
5620.042	5621.602	17788.52	IR	0.8			MIT
5540.612	5542.151	18043.54	IR	0.3			MIT
5495.173	5496.700	18192.74	IR I	0.7			MIT
5469.404	5470.924	18278.45	IR I	0.5			MIT
5459.625	5461.142	18311.19	IR I	0.5			MIT
5454.499	5456.015	18328.40	IR I	1.3			MIT
5449.500	5451.015	18345.21	IR I	1.5	0.3		MIT
5446.753	5448.267	18354.46	IR I	0.6			MIT
5431.162	5432.672	18407.15	IR	0.3			MIT
5429.003	5430.512	18414.47	IR	0.3			MIT
5417.168	5418.674	18454.70	IR	0.5			MIT
5395.422	5396.922	18529.08	IR	0.3			MIT
5390.982	5392.481	18544.34	IR	0.6			MIT
5385.600	5387.097	18562.87	IR	0.6			MIT
5382.482	5383.979	18573.62	IR	0.3			MIT
5364.940	5366.432	18634.35	IR	0.6			MIT
5364.323	5365.815	18636.50	IR	1.2			MIT
5356.859	5358.349	18662.47	IR	0.3			MIT
5340.742	5342.228	18718.78	IR I	0.7			MIT
5329.214	5330.696	18759.27	IR	0.5			MIT
5313.01	5314.49	18816.5	IR	0.3			ME
5289.329	5290.801	18900.73	IR	0.3			MIT
5283.67	5285.14	18921.0	IR	0.6			MIT
5273.777	5275.245	18956.47	IR	0.7			MIT
5273.492	5274.960	18957.49	IR	0.3			MIT
5243.692	5245.152	19065.23	IR	0.3			MIT
5234.80	5236.26	19097.6	IR	0.3H			MIT
5230.372	5231.828	19113.78	IR	0.3			MIT

AIR WAVELENGTH (ANGSTRMS)	VACUUM WAVELENGTH (ANGSTRMS)	WAVENUMBER (KAYSERS)	LMENT	LOG ARC	INTENSITY SPK	DIS	REF
5215.788	5217.240	19167.22	IR	0.3H			MIT
5180.584	5182.027	19297.47	IR	0.3H			MIT
5177.947	5179.389	19307.30	IR	0.9	0.3		MIT
5164.78	5166.22	19356.5	IR	0.3H			MIT
5150.033	5151.468	19411.94	IR	0.3H			MIT
5143.403	5144.836	19436.97	IR	0.5			MIT
5131.529	5132.959	19481.94	IR	0.5			MIT
5129.894	5131.323	19488.15	IR	0.3H			MIT
5128.966	5130.395	19491.68	IR	0.3H			MIT
5123.661	5125.089	19511.86	IR	1.0			MIT
5109.149	5110.573	19567.28	IR	0.3			MIT
5107.43	5108.85	19573.9	IR	0.5			MIT
5099.181	5100.602	19605.53	IR	0.5			MIT
5086.851	5088.269	19653.05	IR	0.3H			MIT
5082.413	5083.830	19670.21	IR	0.3H			MIT
5058.948	5060.358	19761.45	IR	0.7	0.3		MIT
5056.430	5057.840	19771.29	IR	0.3H			MIT
5046.270	5047.677	19811.09	IR	0.3	0.3		MIT
5046.060	5047.467	19811.92	IR	0.7	0.3		MIT
5035.068	5036.472	19855.17	IR	0.3H			MIT
5025.459	5026.861	19893.13	IR	0.3			MIT
5014.981	5016.380	19934.70	IR	1.0			MIT
5014.22	5015.62	19937.7	IR	0.5			ME
5009.167	5010.564	19957.83	IR	0.7	0.3H		MIT
5005.78	5007.18	19971.3	IR	0.3			MIT
5004.390	5005.786	19976.88	IR	0.5			MIT
5002.743	5004.139	19983.46	IR	0.9			MIT
4999.743	5001.138	19995.45	IR	0.8	0.3		MIT
4981.879	4983.269	20067.15	IR	0.5			MIT
4979.477	4980.866	20076.83	IR	0.5			MIT
4975.585	4976.973	20092.53	IR	0.3J			MIT
4970.476	4971.863	20113.19	IR	0.8	0.3		MIT
4969.09	4970.48	20118.8	IR	0.3			ME
4959.455	4960.839	20157.88	IR	0.5			MIT
4939.941	4941.320	20237.51	IR	0.3			MIT
4939.173	4940.552	20240.66	IR	0.6			MIT
4938.76	4940.14	20242.4	IR	0.3			ME
4938.088	4939.466	20245.10	IR	1.0	0.5		MIT
4933.497	4934.874	20263.94	IR	0.3			MIT
4927.936	4929.312	20286.81	IR	0.3	0.3		MIT
4924.418	4925.793	20301.30	IR	0.3			MIT
4913.825	4915.197	20345.06	IR	0.3			MIT
4913.36	4914.73	20347.0	IR	0.3			ME
4905.638	4907.008	20379.02	IR	0.5	0.3		MIT
4904.444	4905.813	20383.98	IR	0.3			MIT
4903.323	4904.692	20388.64	IR	0.5			MIT
4888.509	4889.874	20450.42	IR	0.3J			MIT
4879.25	4880.61	20489.2	IR I	0.3			ME
4858.001	4859.358	20578.85	IR	0.3			MIT
4851.382	4852.737	20606.93	IR	0.3			MIT
4849.771	4851.126	20613.77	IR	0.3			MIT
4845.377	4846.731	20632.47	IR	0.8	0.3		MIT
4840.768	4842.120	20652.11	IR	0.7	0.5		MIT
4826.748	4828.097	20712.09	IR	1.2			MIT
4819.734	4821.081	20742.24	IR	0.3H			MIT
4815.39	4816.74	20760.9	IR	0.3			ME
4815.10	4816.45	20762.2	IR	0.3			ME
4809.467	4810.811	20786.51	IR	0.5			MIT
4795.666	4797.007	20846.33	IR	0.3J	0.3J		MIT
4795.247	4796.587	20848.16	IR	0.3			MIT
4793.18	4794.52	20857.1	IR	0.3			ME
4787.748	4789.086	20880.81	IR	0.7			MIT
4778.157	4779.493	20922.72	IR	1.7	0.5		MIT
4770.414	4771.748	20956.68	IR	0.3			MIT
4769.306	4770.640	20961.55	IR	0.8			MIT
4768.393	4769.726	20965.56	IR	0.3			MIT
4766.549	4767.882	20973.67	IR	0.3J			MIT
4757.963	4759.294	21011.52	IR	1.3	0.3J		MIT
4756.459	4757.789	21018.17	IR	1.3	0.3		MIT
4747.07	4748.40	21059.7	IR	0.3			ME

AIR WAVELENGTH (ANGSTRMS)	VACUUM WAVELENGTH (ANGSTRMS)	WAVENUMBER (KAYSERS)	LMENT	LOG ARC	INTENSITY SPK	INTENSITY DIS	REF
4745.84	4747.17	21065.2	IR	0.3J			ME
4740.264	4741.590	21089.97	IR	0.3J			MIT
4731.855	4733.179	21127.45	IR	0.9	0.7		MIT
4728.861	4730.184	21140.83	IR	2.2	0.5		MIT
4728.023	4729.346	21144.57	IR	0.6			MIT
4721.88	4723.20	21172.1	IR	0.3			ME
4720.93	4722.25	21176.3	IR	0.3J			ME
4719.318	4720.638	21183.58	IR	0.3			MIT
4711.85	4713.17	21217.1	IR	0.3			ME
4708.879	4710.197	21230.54	IR	1.2			MIT
4708.184	4709.501	21233.67	IR	0.6			MIT
4704.054	4705.370	21252.31	IR	1.1			MIT
4702.605	4703.921	21258.86	IR	0.8			MIT
4689.49	4690.80	21318.3	IR	0.5			ME
4685.06	4686.37	21338.5	IR	0.6			ME
4684.10	4685.41	21342.8	IR	0.8			ME
4680.465	4681.775	21359.42	IR	0.3			MIT
4675.537	4676.846	21381.93	IR	1.2	0.3		MIT
4668.987	4670.294	21411.93	IR	1.3			MIT
4664.67	4665.98	21431.7	IR	0.8			ME
4662.960	4664.265	21439.60	IR	0.9			MIT
4659.42	4660.72	21455.9	IR	0.3			ME
4657.64	4658.94	21464.1	IR	0.6			ME
4656.178	4657.482	21470.83	IR	1.8			MIT
4652.14	4653.44	21489.5	IR	0.3			ME
4651.077	4652.379	21494.38	IR	0.6			MIT
4648.24	4649.54	21507.5	IR	0.5			ME
4644.36	4645.66	21525.5	IR	0.5			ME
4640.083	4641.382	21545.30	IR	1.6	0.3J		MIT
4637.882	4639.181	21555.53	IR	1.0			MIT
4636.05	4637.35	21564.1	IR	0.3			ME
4627.097	4628.393	21605.77	IR	1.3			AB
4622.135	4623.430	21628.97	IR	0.3			AB
4618.009	4619.302	21648.29	IR	0.3H			AB
4617.613	4618.906	21650.15	IR	0.5H			AB
4617.368	4618.661	21651.29	IR	0.7H			AB
4616.754	4618.047	21654.18	IR	0.7H			AB
4616.388	4617.681	21655.89	IR	2.3	0.7		MIT
4614.173	4615.465	21666.29	IR	1.0			AB
4612.244	4613.536	21675.35	IR	0.3			AB
4611.776	4613.068	21677.55	IR	0.3J			AB
4604.475	4605.765	21711.92	IR	1.0H			MIT
4603.836	4605.126	21714.93	IR	0.3			MIT
4596.42	4597.71	21750.0	IR	0.6			MIT
4591.13	4592.42	21775.0	IR	0.5			MIT
4590.91	4592.20	21776.1	IR	0.7			MIT
4587.473	4588.758	21792.39	IR	0.7			AB
4587.409	4588.694	21792.69	IR	1.3			MIT
4585.59	4586.87	21801.3	IR	0.9			MIT
4585.13	4586.41	21803.5	IR	0.3			MIT
4584.75	4586.03	21805.3	IR	0.5			MIT
4581.92	4583.20	21818.8	IR	1.3H			MIT
4579.345	4580.628	21831.07	IR	0.9H			AB
4577.175	4578.458	21841.42	IR	0.9			MIT
4571.390	4572.671	21869.06	IR	0.3			AB
4570.614	4571.895	21872.77	IR	0.9H			AB
4570.417	4571.698	21873.71	IR	0.9H			AB
4570.019	4571.300	21875.62	IR	1.7	0.3		MIT
4568.092	4569.372	21884.84	IR	2.0	0.5		MIT
4564.056	4565.335	21904.20	IR	0.3			AB
4562.909	4564.188	21909.70	IR	0.3			AB
4560.890	4562.168	21919.40	IR	1.1			AB
4558.743	4560.021	21929.72	IR	0.9			AB
4557.842	4559.120	21934.06	IR	0.7			AB
4554.778	4556.055	21948.81	IR	0.6			MIT
4550.775	4552.051	21968.12	IR	1.9			MIT
4548.804	4550.079	21977.64	IR	0.6			MIT
4548.485	4549.760	21979.18	IR	2.0	0.7		MIT
4545.680	4546.954	21992.74	IR	2.3	0.6		MIT
4542.871	4544.145	22006.34	IR	0.5	0.3J		MIT

AIR WAVELENGTH (ANGSTRMS)	VACUUM WAVELENGTH (ANGSTRMS)	WAVENUMBER (KAYSERS)	LMENT	LOG ARC	INTENSITY SPK	INTENSITY DIS	REF
4538.671	4539.944	22026.71	IR	1.4			MIT
4535.001	4536.273	22044.53	IR	0.3			MIT
4532.867	4534.138	22054.91	IR	1.9	0.3		MIT
4528.42	4529.69	22076.6	IR	0.3			MIT
4526.997	4528.266	22083.51	IR	0.5			MIT
4526.407	4527.676	22086.39	IR	0.3			MIT
4522.215	4523.483	22106.86	IR	0.3			AB
4521.339	4522.607	22111.14	IR	0.5			MIT
4517.372	4518.639	22130.56	IR	0.6			MIT
4517.008	4518.275	22132.34	IR	0.6H			MIT
4516.078	4517.345	22136.90	IR	0.8			MIT
4515.331	4516.597	22140.56	IR	0.9	0.0		MIT
4512.028	4513.293	22156.77	IR	0.5			MIT
4507.639	4508.903	22178.34	IR	0.6			MIT
4507.127	4508.391	22180.86	IR	0.3			AB
4505.703	4506.967	22187.87	IR	1.0			MIT
4505.072	4506.336	22190.98	IR	1.0			MIT
4503.282	4504.545	22199.80	IR	0.3			MIT
4500.928	4502.191	22211.41	IR	0.9			MIT
4496.032	4497.293	22235.60	IR	1.6	0.3		MIT
4495.347	4496.608	22238.99	IR	2.0	0.5		MIT
4492.158	4493.418	22254.77	IR	1.5	0.3		MIT
4491.358	4492.618	22258.74	IR	1.6	0.3J		MIT
4491.001	4492.261	22260.51	IR	0.3			AB
4489.864	4491.124	22266.14	IR	0.3			AB
4489.418	4490.678	22268.35	IR	0.8			MIT
4483.948	4485.206	22295.52	IR	0.3			MIT
4483.788	4485.046	22296.31	IR	0.5			MIT
4482.032	4483.290	22305.05	IR	0.7			MIT
4481.198	4482.455	22309.20	IR	0.6			AB
4478.476	4479.733	22322.76	IR	2.3	1.0		MIT
4471.262	4472.517	22358.78	IR	0.5			MIT
4470.475	4471.730	22362.71	IR	0.6			MIT
4467.342	4468.596	22378.40	IR	0.6			MIT
4467.182	4468.436	22379.20	IR	0.3			MIT
4466.733	4467.987	22381.45	IR	1.4			MIT
4465.038	4466.291	22389.94	IR	0.5			MIT
4463.484	4464.737	22397.74	IR	0.5			MIT
4452.808	4454.058	22451.44	IR	1.2	0.3		MIT
4452.612	4453.862	22452.43	IR	1.0			MIT
4451.318	4452.567	22458.95	IR	1.1			MIT
4451.178	4452.427	22459.66	IR	0.3			MIT
4450.790	4452.039	22461.62	IR	0.7			MIT
4450.178	4451.427	22464.71	IR	1.8	0.5		MIT
4447.456	4448.704	22478.45	IR	0.6			MIT
4445.941	4447.189	22486.11	IR	0.5			MIT
4445.851	4447.099	22486.57	IR	0.9			MIT
4443.806	4445.053	22496.92	IR	1.0			MIT
4442.141	4443.388	22505.35	IR	0.3J			MIT
4440.837	4442.084	22511.96	IR	1.0			MIT
4437.823	4439.069	22527.25	IR	0.3			MIT
4435.592	4436.837	22538.58	IR	0.9			MIT
4434.958	4436.203	22541.80	IR	0.5			MIT
4434.931	4436.176	22541.94	IR	0.5			AB
4434.474	4435.719	22544.26	IR	0.7			MIT
4433.908	4435.153	22547.14	IR	1.0			MIT
4431.904	4433.148	22557.33	IR	0.3			MIT
4428.928	4430.172	22572.49	IR	0.6			MIT
4426.269	4427.512	22586.05	IR	2.6W	1.0		MIT
4425.761	4427.004	22588.64	IR	1.0			MIT
4424.752	4425.994	22593.79	IR	0.5			MIT
4421.961	4423.203	22608.05	IR	1.3	0.3		MIT
4420.348	4421.589	22616.30	IR	1.0			MIT
4411.176	4412.415	22663.33	IR	1.6	0.3		MIT
4410.537	4411.776	22666.61	IR	1.0			MIT
4408.861	4410.099	22675.23	IR	0.6			MIT
4406.759	4407.997	22686.04	IR	1.4	0.3		MIT
4405.501	4406.738	22692.52	IR	0.5			MIT
4403.784	4405.021	22701.37	IR	2.5	1.0		MIT
4401.246	4402.482	22714.46	IR	1.1			MIT

AIR WAVELENGTH (ANGSTRMS)	VACUUM WAVELENGTH (ANGSTRMS)	WAVENUMBER (KAYSERS)	LMENT	LOG ARC	INTENSITY SPK	DIS	REF
4399.472	4400.708	22723.62	IR	2.6	2.0		MIT
4392.591	4393.825	22759.21	IR	2.0	0.6		MIT
4388.333	4389.566	22781.30	IR	0.9			MIT
4385.986	4387.218	22793.49	IR	0.9			MIT
4381.359	4382.590	22817.56	IR	0.5			MIT
4380.774	4382.005	22820.60	IR	1.0			MIT
4377.006	4378.236	22840.25	IR	2.0	0.6		MIT
4373.718	4374.947	22857.42	IR	0.7			MIT
4372.127	4373.356	22865.74	IR	1.6W			MIT
4362.121	4363.347	22918.19	IR	1.4	0.3		MIT
4361.846	4363.072	22919.63	IR	0.6			MIT
4359.580	4360.805	22931.54	IR	1.1			MIT
4358.279	4359.504	22938.39	IR	0.9			MIT
4355.842	4357.066	22951.22	IR	0.6			MIT
4353.382	4354.606	22964.19	IR	1.3			MIT
4352.562	4353.785	22968.52	IR	1.7	0.3		MIT
4351.299	4352.522	22975.19	IR	1.7			MIT
4346.556	4347.778	23000.25	IR	0.8			MIT
4332.332	4333.550	23075.77	IR	1.0			MIT
4331.551	4332.769	23079.93	IR	0.6			MIT
4329.902	4331.119	23088.72	IR	1.5	0.3		MIT
4323.426	4324.642	23123.30	IR	0.7			MIT
4319.894	4321.109	23142.21	IR	1.5			MIT
4318.643	4319.858	23148.91	IR	0.9			MIT
4316.304	4317.518	23161.46	IR	1.1			MIT
4311.498	4312.711	23187.27	IR	2.5	1.0		MIT
4310.591	4311.803	23192.15	IR	2.2	0.9		MIT
4309.767	4310.979	23196.59	IR	0.5			MIT
4308.182	4309.394	23205.12	IR	0.5			MIT
4307.799	4309.011	23207.18	IR	0.9			AB
4305.948	4307.159	23217.16	IR	1.4	0.3		MIT
4305.198	4306.409	23221.20	IR	1.2			MIT
4301.603	4302.813	23240.61	IR	2.3	1.0		MIT
4300.641	4301.851	23245.81	IR	1.0	0.3		MIT
4297.030	4298.239	23265.34	IR	0.5			MIT
4288.492	4289.699	23311.66	IR	0.3			MIT
4287.899	4289.105	23314.89	IR	1.1	0.3		MIT
4286.619	4287.825	23321.85	IR	2.3	0.5		MIT
4283.395	4284.600	23339.40	IR	0.5			MIT
4269.706	4270.908	23414.23	IR I	0.9			MIT
4268.942	4270.143	23418.42	IR	1.0	0.3		MIT
4268.761	4269.962	23419.41	IR I	0.5			MIT
4268.101	4269.302	23423.03	IR I	2.3	1.2		MIT
4266.362	4267.563	23432.58	IR I	1.0			MIT
4266.038	4267.239	23434.36	IR I	1.0			MIT
4265.305	4266.505	23438.39	IR I	1.8	0.3		MIT
4263.065	4264.265	23450.70	IR	0.5			AB
4261.888	4263.088	23457.18	IR	1.0			MIT
4261.258	4262.457	23460.65	IR	1.4			MIT
4260.898	4262.097	23462.63	IR	1.0			MIT
4260.031	4261.230	23467.40	IR	1.4			MIT
4259.113	4260.312	23472.46	IR	2.3	1.0		MIT
4258.093	4259.292	23478.08	IR	0.7			MIT
4257.373	4258.571	23482.05	IR	1.5			MIT
4252.514	4253.711	23508.88	IR	0.6			MIT
4248.999	4250.195	23528.33	IR	0.3			MIT
4248.806	4250.002	23529.40	IR	0.6			MIT
4245.117	4246.312	23549.85	IR	0.3			MIT
4244.664	4245.859	23552.36	IR	0.3			MIT
4243.804	4244.999	23557.13	IR	1.1	0.3H		MIT
4240.493	4241.687	23575.53	IR	1.2			MIT
4236.667	4237.860	23596.82	IR	1.0			MIT
4226.735	4227.925	23652.26	IR	1.5			MIT
4226.628	4227.818	23652.86	IR	0.7			MIT
4226.205	4227.395	23655.23	IR	0.3			AB
4225.496	4226.686	23659.20	IR	1.2	0.3		MIT
4224.649	4225.839	23663.94	IR	0.5			AB
4223.159	4224.348	23672.29	IR	1.2	0.3H		MIT
4220.800	4221.989	23685.52	IR	1.5	0.3		MIT
4218.842	4220.030	23696.51	IR	0.5			MIT
4218.289	4219.477	23699.62	IR	1.3	0.3		MIT
4218.085	4219.273	23700.76	IR	0.9			MIT
4217.760	4218.948	23702.59	IR	1.6	0.5		MIT
4216.584	4217.772	23709.20	IR	0.9			MIT
4214.524	4215.711	23720.79	IR	0.7			MIT
4213.285	4214.472	23727.77	IR	0.5	0.3		MIT
4210.587	4211.773	23742.97	IR	0.5J	0.3J		MIT
4202.516	4203.700	23788.57	IR	0.9			MIT
4202.426	4203.610	23789.08	IR	0.9			MIT
4199.662	4200.845	23804.73	IR	0.6			MIT
4199.305	4200.488	23806.76	IR	0.3			MIT
4198.503	4199.686	23811.30	IR	1.0			MIT
4197.536	4198.719	23816.79	IR	1.6	0.3		MIT
4192.837	4194.018	23843.48	IR	1.4	0.3		MIT
4190.547	4191.728	23856.51	IR	0.7			MIT
4189.876	4191.057	23860.33	IR I	0.3			AB
4187.967	4189.147	23871.21	IR	0.9			MIT
4186.280	4187.460	23880.83	IR	0.3			AB
4185.658	4186.838	23884.37	IR	1.4			MIT
4183.207	4184.386	23898.37	IR	1.6	0.6		MIT
4182.467	4183.646	23902.60	IR	1.7	0.8		MIT
4178.342	4179.520	23926.19	IR	0.6			MIT
4177.358	4178.535	23931.83	IR	0.7			MIT
4176.183	4177.360	23938.56	IR	1.0	0.3		MIT
4175.864	4177.041	23940.39	IR	0.5			AB
4175.216	4176.393	23944.11	IR	0.6			MIT
4172.559	4173.735	23959.35	IR I	2.2	1.1		MIT
4170.309	4171.485	23972.28	IR	0.6			AB
4169.167	4170.342	23978.85	IR	1.2	0.3H		MIT
4168.043	4169.218	23985.31	IR	0.5			AB
4167.025	4168.200	23991.17	IR	1.0			MIT
4166.040	4167.214	23996.84	IR	2.2	1.0		MIT
4163.559	4164.733	24011.14	IR	1.4	0.3		MIT
4162.112	4163.285	24019.49	IR	0.5			AB
4159.381	4160.554	24035.26	IR	0.3			AB
4158.081	4159.253	24042.78	IR	1.4			MIT
4155.702	4156.874	24056.54	IR	1.9	0.7		AB
4153.506	4154.677	24069.26	IR	1.5			AB
4151.951	4153.122	24078.27	IR	0.3			AB
4151.352	4152.523	24081.75	IR	1.5	0.3		AB
4143.140	4144.308	24129.48	IR	0.9			MIT
4141.291	4142.459	24140.25	IR	0.8			MIT
4140.818	4141.986	24143.01	IR	0.8			MIT
4140.111	4141.279	24147.13	IR	0.7			MIT
4139.967	4141.135	24147.97	IR	0.3			MIT
4137.923	4139.090	24159.90	IR	0.7			MIT
4135.916	4137.083	24171.62	IR	0.9			MIT
4135.146	4136.312	24176.12	IR	0.3			AB
4133.148	4134.314	24187.81	IR	1.0			MIT
4130.731	4131.896	24201.96	IR	0.3			AB
4130.100	4131.265	24205.66	IR	1.2			AB
4128.429	4129.594	24215.46	IR	1.0	0.3		AB
4127.917	4129.081	24218.46	IR	1.5	0.3		MIT
4126.477	4127.641	24226.91	IR	1.4			MIT
4126.067	4127.231	24229.32	IR	1.1			AB
4124.745	4125.909	24237.09	IR	0.9			AB
4124.540	4125.704	24238.29	IR	0.6			AB
4119.601	4120.763	24267.35	IR	1.2			AB
4118.663	4119.825	24272.88	IR	0.9			AB
4117.596	4118.758	24279.17	IR	1.7	0.5		MIT
4117.210	4118.372	24281.44	IR	0.3	0.5		AB
4116.838	4118.000	24283.64	IR	0.7			AB
4115.778	4116.939	24289.89	IR	2.0	1.5		MIT
4113.634	4114.795	24302.55	IR	1.1			AB
4112.287	4113.447	24310.51	IR	1.1	0.3		MIT
4111.768	4112.928	24313.58	IR	0.3			AB
4108.467	4109.626	24333.11	IR	0.6	0.3		MIT
4108.208	4109.367	24334.65	IR	0.6	0.5H		MIT
4107.823	4108.982	24336.93	IR	1.0	0.3		MIT
4107.129	4108.288	24341.04	IR	0.5			AB

AIR WAVELENGTH (ANGSTRMS)	VACUUM WAVELENGTH (ANGSTRMS)	WAVENUMBER (KAYSERS)	LMENT	LOG ARC	INTENSITY SPK	INTENSITY DIS	REF
4105.755	4106.914	24349.19	IR	1.2	0.3		MIT
4105.203	4106.361	24352.46	IR	1.4	0.3		MIT
4104.235	4105.393	24358.20	IR	1.3	0.3		MIT
4103.667	4104.825	24361.57	IR I	1.0			AB
4101.667	4102.825	24373.45	IR	0.5			AB
4100.148	4101.305	24382.48	IR	2.0	0.5H		MIT
4099.482	4100.639	24386.44	IR	0.6			AB
4097.250	4098.406	24399.73	IR	0.5			AB
4095.110	4096.266	24412.48	IR	0.5			AB
4094.881	4096.037	24413.84	IR	0.9			AB
4093.634	4094.789	24421.28	IR	0.3			AB
4092.610	4093.765	24427.39	IR	1.8	1.3		MIT
4090.973	4092.128	24437.17	IR	0.5			AB
4089.639	4090.793	24445.14	IR	1.2	0.3		AB
4086.957	4088.111	24461.18	IR	0.3			AB
4085.931	4087.084	24467.32	IR	1.2	1.2H		AB
4082.377	4083.530	24488.62	IR	1.4	0.5		MIT
4081.397	4082.549	24494.50	IR	1.2	0.3		MIT
4080.479	4081.631	24500.01	IR	0.7			AB
4079.897	4081.049	24503.50	IR	1.4			AB
4079.170	4080.322	24507.87	IR	0.5			MIT
4076.806	4077.957	24522.08	IR	0.3H			AB
4076.553	4077.704	24523.60	IR	1.4			AB
4076.370	4077.521	24524.71	IR	0.3	0.3		MIT
4075.634	4076.785	24529.13	IR	1.0			MIT
4072.386	4073.536	24548.70	IR	1.0	0.5		MIT
4070.677	4071.826	24559.00	IR	1.5	1.0		MIT
4069.919	4071.068	24563.58	IR	1.5	1.3		MIT
4062.673	4063.820	24607.39	IR	0.9			MIT
4059.234	4060.380	24628.23	IR	1.5	0.6		MIT
4056.473	4057.619	24645.00	IR	1.1	0.3		AB
4055.717	4056.863	24649.59	IR	0.8			AB
4055.353	4056.498	24651.80	IR	0.8			AB
4055.034	4056.179	24653.74	IR	0.6			MIT
4051.190	4052.334	24677.14	IR	0.7			AB
4047.329	4048.472	24700.68	IR	0.6			AB
4046.539	4047.682	24705.50	IR	0.9			AB
4044.892	4046.035	24715.56	IR	0.3	0.8		AB
4040.732	4041.874	24741.00	IR	0.3			AB
4040.440	4041.582	24742.79	IR	0.7	0.3		AB
4040.078	4041.219	24745.01	IR	1.6	0.7		MIT
4035.335	4036.475	24774.09	IR I	0.8			AB
4033.762	4034.902	24783.75	IR	2.0	1.4		MIT
4032.214	4033.353	24793.26	IR	1.0			AB
4031.558	4032.697	24797.30	IR I	0.5			AB
4029.143	4030.282	24812.16	IR I	0.8			AB
4020.028	4021.164	24868.42	IR	1.9	2.0		MIT
4015.789	4016.924	24894.67	IR	0.6	0.3		MIT
4014.658	4015.793	24901.68	IR	0.9	0.3		MIT
4014.177	4015.312	24904.67	IR	0.5			AB
4013.997	4015.132	24905.78	IR I	0.6			AB
4009.854	4010.988	24931.52	IR I	0.9	0.3H		AB
4008.967	4010.100	24937.03	IR I	0.7			AB
4008.052	4009.185	24942.73	IR I	1.1	0.3		AB
4005.580	4006.712	24958.12	IR I	1.3	0.5		AB
4005.020	4006.152	24961.61	IR	1.4			AB
4002.128	4003.260	24979.65	IR	0.6			AB
3998.828	3999.959	25000.26	IR	0.3			AB
3996.974	3998.104	25011.85	IR	0.3			MIT
3996.450	3997.580	25015.13	IR I	0.9	0.3		MIT
3995.291	3996.421	25022.39	IR I	0.8			MIT
3995.196	3996.326	25022.99	IR I	0.5			MIT
3992.121	3993.250	25042.26	IR I	2.2	1.8		MIT
3989.435	3990.563	25059.12	IR I	1.4	0.3		MIT
3988.978	3990.106	25061.99	IR I	0.6			MIT
3987.829	3988.957	25069.21	IR	1.1			MIT
3987.657	3988.785	25070.29	IR I	0.6			MIT
3987.376	3988.504	25072.06	IR	0.7			MIT
3986.381	3987.508	25078.32	IR	0.3	0.3		AB
3985.866	3986.993	25081.56	IR	0.3			MIT

AIR WAVELENGTH (ANGSTRMS)	VACUUM WAVELENGTH (ANGSTRMS)	WAVENUMBER (KAYSERS)	LMENT	LOG ARC	INTENSITY SPK	INTENSITY DIS	REF
3984.278	3985.405	25091.55	IR	0.3			MIT
3982.516	3983.642	25102.66	IR	0.3H			MIT
3978.107	3979.232	25130.48	IR I	1.0	0.3		MIT
3976.869	3977.994	25138.30	IR I	0.8	0.3		MIT
3976.311	3977.436	25141.83	IR I	1.0	1.8		MIT
3970.517	3971.640	25178.51	IR I	0.7			MIT
3969.172	3970.295	25187.05	IR	1.5	1.0		MIT
3968.475	3969.598	25191.47	IR	1.4			MIT
3967.507	3968.629	25197.62	IR I	0.5			AB
3966.580	3967.702	25203.50	IR	0.5			AB
3966.088	3967.210	25206.63	IR I	1.5	0.6		MIT
3962.783	3963.904	25227.65	IR I	1.3			MIT
3962.443	3963.564	25229.82	IR	0.8			MIT
3960.555	3961.676	25241.84	IR I	0.6			MIT
3956.682	3957.802	25266.55	IR I	0.8			MIT
3956.115	3957.235	25270.17	IR I	0.8			MIT
3952.616	3953.735	25292.54	IR I	1.3	0.5		MIT
3951.952	3953.070	25296.79	IR I	1.3	0.9		MIT
3948.506	3949.624	25318.87	IR	0.6			MIT
3948.317	3949.434	25320.08	IR I	1.0	0.5		MIT
3946.269	3947.386	25333.22	IR I	1.7	1.2		MIT
3944.375	3945.491	25345.39	IR I	1.3	0.6		MIT
3944.058	3945.174	25347.42	IR	0.3			AB
3943.205	3944.321	25352.91	IR	0.8			MIT
3941.761	3942.877	25362.19	IR	0.5			AB
3941.110	3942.226	25366.38	IR	0.8			MIT
3934.844	3935.958	25406.77	IR I	2.3	1.7		MIT
3933.901	3935.015	25412.87	IR	1.3	0.9		MIT
3933.664	3934.778	25414.40	IR	1.3			MIT
3932.971	3934.084	25418.87	IR I	0.8			MIT
3928.438	3929.550	25448.21	IR I	1.1	0.3		MIT
3928.229	3929.341	25449.56	IR I	0.3			MIT
3926.038	3927.150	25463.76	IR I	0.7			MIT
3924.427	3925.538	25474.21	IR I	1.2	0.3		MIT
3923.522	3924.633	25480.09	IR I	1.5			MIT
3919.062	3920.172	25509.09	IR I	0.6			MIT
3915.384	3916.493	25533.05	IR	2.2	1.7		MIT
3914.913	3916.022	25536.12	IR	1.3	0.5		MIT
3912.410	3913.518	25552.46	IR	0.3			AB
3908.834	3909.941	25575.83	IR	0.7			MIT
3907.896	3909.003	25581.97	IR I	1.0	0.3		MIT
3907.753	3908.860	25582.91	IR I	0.6			MIT
3905.451	3906.557	25597.99	IR I	0.9			MIT
3902.849	3903.955	25615.05	IR I	0.9	0.3		MIT
3902.662	3903.768	25616.28	IR I	0.9	0.9		MIT
3902.506	3903.612	25617.30	IR I	1.0	1.2		MIT
3897.128	3898.232	25652.65	IR	0.7			MIT
3895.592	3896.696	25662.77	IR	1.5	1.4		MIT
3889.577	3890.679	25702.45	IR	0.3	0.5		MIT
3887.381	3888.483	25716.97	IR	0.5			AB
3887.050	3888.152	25719.16	IR	0.9			MIT
3886.390	3887.491	25723.53	IR	0.8			AB
3885.418	3886.519	25729.97	IR I	0.9			MIT
3883.364	3884.465	25743.57	IR	0.6			MIT
3882.519	3883.619	25749.18	IR	0.3			MIT
3878.877	3879.976	25773.35	IR I	0.3			MIT
3878.574	3879.673	25775.37	IR	0.6			MIT
3873.153	3874.251	25811.44	IR I	1.4	1.0		MIT
3872.943	3874.041	25812.84	IR	0.8			MIT
3871.851	3872.949	25820.12	IR I	0.9	0.3		MIT
3871.546	3872.643	25822.16	IR I	0.3			MIT
3870.131	3871.228	25831.60	IR I	0.9	0.0		MIT
3869.873	3870.970	25833.32	IR I	0.3H			MIT
3869.468	3870.565	25836.02	IR I	1.2	0.3		MIT
3865.976	3867.072	25859.36	IR I	0.5			MIT
3865.639	3866.735	25861.61	IR I	1.4	1.3		MIT
3861.945	3863.040	25886.35	IR I	1.2	1.0		MIT
3861.318	3862.413	25890.55	IR	0.5	0.3		MIT
3856.072	3857.165	25925.77	IR I	1.4	0.6		AB
3851.818	3852.910	25954.41	IR	0.3			AB

AIR WAVELENGTH (ANGSTRMS)	VACUUM WAVELENGTH (ANGSTRMS)	WAVENUMBER (KAYSERS)	LMENT	LOG ARC	INTENSITY SPK	DIS	REF
3851.259	3852.351	25958.17	IR I	0.9			AB
3850.667	3851.759	25962.16	IR I	0.6			AB
3848.160	3849.251	25979.08	IR	1.0	0.3		MIT
3846.689	3847.780	25989.01	IR	1.1			MIT
3845.962	3847.053	25993.92	IR	0.6H	0.3H		MIT
3845.435	3846.526	25997.49	IR	0.9			MIT
3845.014	3846.105	26000.33	IR I	1.2	0.3H		MIT
3844.449	3845.539	26004.15	IR	0.8			MIT
3843.004	3844.094	26013.93	IR	0.9			AB
3841.703	3842.793	26022.74	IR I	1.0	0.3H		MIT
3840.295	3841.384	26032.28	IR	0.3			AB
3839.317	3840.406	26038.91	IR	0.6			MIT
3839.134	3840.223	26040.15	IR	0.8			AB
3837.723	3838.812	26049.73	IR	1.2	0.6		MIT
3837.527	3838.616	26051.06	IR I	0.9			AB
3836.501	3837.589	26058.02	IR	1.2			AB
3835.703	3836.791	26063.45	IR	0.3			AB
3835.363	3836.451	26065.76	IR	0.5			MIT
3833.883	3834.971	26075.82	IR	1.7			AB
3833.279	3834.367	26079.93	IR	0.3			AB
3832.179	3833.266	26087.41	IR	1.2			AB
3831.033	3832.120	26095.22	IR	1.0			MIT
3830.650	3831.737	26097.83	IR I	0.6			AB
3830.516	3831.603	26098.74	IR	0.3H			MIT
3830.341	3831.428	26099.93	IR	0.3	0.3H		MIT
3830.051	3831.138	26101.91	IR	0.3			MIT
3829.676	3830.763	26104.46	IR I	0.6H			AB
3828.465	3829.551	26112.72	IR	1.5			MIT
3826.960	3828.046	26122.99	IR	0.9	0.0H		AB
3826.626	3827.712	26125.27	IR I	0.3			AB
3824.524	3825.609	26139.63	IR	1.4	0.9		MIT
3823.377	3824.462	26147.47	IR	1.3	0.5		MIT
3822.930	3824.015	26150.53	IR	0.3H			AB
3822.577	3823.662	26152.94	IR	0.3			AB
3821.710	3822.795	26158.87	IR	0.5H			AB
3820.088	3821.172	26169.98	IR	0.6			MIT
3817.244	3818.327	26189.48	IR I	1.2	1.5		MIT
3815.516	3816.599	26201.34	IR I	1.4	0.5		MIT
3813.786	3814.868	26213.22	IR I	0.7			AB
3812.657	3813.739	26220.99	IR	0.5			MIT
3811.782	3812.864	26227.00	IR I	0.5			MIT
3811.539	3812.621	26228.68	IR	0.5			AB
3810.608	3811.690	26235.08	IR I	0.5	0.3		MIT
3810.387	3811.469	26236.61	IR I	1.0	0.6		MIT
3808.690	3809.771	26248.30	IR	0.6			AB
3807.003	3808.084	26259.93	IR I	0.5			AB
3806.573	3807.654	26262.89	IR	0.9	0.7		MIT
3806.294	3807.374	26264.82	IR	0.3			AB
3806.095	3807.175	26266.19	IR	0.6			AB
3803.447	3804.527	26284.48	IR	0.3			AB
3802.982	3804.062	26287.69	IR	0.3			AB
3802.778	3803.858	26289.10	IR I	0.5			AB
3802.678	3803.758	26289.79	IR	1.0	0.0H		AB
3800.123	3801.202	26307.47	IR	2.2	2.0		MIT
3798.666	3799.745	26317.56	IR I	0.7			AB
3798.272	3799.350	26320.29	IR	0.9			MIT
3798.059	3799.137	26321.76	IR	0.8			MIT
3797.924	3799.002	26322.70	IR	0.3			MIT
3796.790	3797.868	26330.56	IR I	0.6			AB
3796.223	3797.301	26334.50	IR	0.3			MIT
3794.056	3795.133	26349.53	IR	1.5	0.8		MIT
3793.951	3795.028	26350.26	IR	1.3			MIT
3793.792	3794.869	26351.37	IR	1.5	1.0		MIT
3792.255	3793.332	26362.05	IR	0.9			AB
3791.967	3793.044	26364.05	IR I	0.6			AB
3790.928	3792.005	26371.28	IR	0.3			MIT
3790.709	3791.785	26372.80	IR I	0.5			AB
3790.527	3791.603	26374.07	IR	1.7			MIT
3790.300	3791.376	26375.65	IR I	0.5			MIT
3789.474	3790.550	26381.40	IR I	1.0	0.3H		MIT

AIR WAVELENGTH (ANGSTRMS)	VACUUM WAVELENGTH (ANGSTRMS)	WAVENUMBER (KAYSERS)	LMENT	LOG ARC	INTENSITY SPK	DIS	REF
3788.736	3789.812	26386.53	IR	0.3H			MIT
3788.159	3789.235	26390.55	IR	0.3H			MIT
3787.276	3788.352	26396.71	IR	0.6			MIT
3786.058	3787.133	26405.20	IR	1.5			AB
3784.929	3786.004	26413.07	IR	0.5			AB
3784.717	3785.792	26414.55	IR	0.3H	0.5H		AB
3781.133	3782.207	26439.59	IR	0.7			AB
3779.644	3780.718	26450.00	IR	1.2			MIT
3779.163	3780.236	26453.37	IR	0.3H			AB
3776.934	3778.007	26468.98	IR I	1.0	0.3		MIT
3775.206	3776.278	26481.10	IR	0.9			MIT
3771.604	3772.675	26506.39	IR I	1.6	0.6		MIT
3770.730	3771.801	26512.53	IR	1.6	0.7		MIT
3768.997	3770.068	26524.72	IR	0.6			MIT
3768.676	3769.747	26526.98	IR I	1.8	1.0		MIT
3766.553	3767.623	26541.93	IR	0.5			MIT
3766.157	3767.227	26544.72	IR I	0.6	0.0H		AB
3765.412	3766.482	26549.98	IR	0.5			MIT
3764.939	3766.009	26553.31	IR	1.0	0.3		MIT
3763.198	3764.267	26565.59	IR	0.5			MIT
3762.208	3763.277	26572.58	IR	0.3			AB
3761.941	3763.010	26574.47	IR	0.3	0.6		MIT
3761.555	3762.624	26577.20	IR I	0.7			MIT
3760.839	3761.908	26582.26	IR	0.3H			AB
3757.185	3758.253	26608.11	IR	1.0	0.0H		AB
3754.510	3755.577	26627.07	IR I	1.4	0.3		MIT
3754.042	3755.109	26630.39	IR	0.5			AB
3753.324	3754.391	26635.48	IR I	1.2			MIT
3752.865	3753.932	26638.74	IR I	0.5			AB
3752.515	3753.582	26641.22	IR	0.6			AB
3750.768	3751.834	26653.63	IR	1.2	0.3		MIT
3750.404	3751.470	26656.22	IR	1.5			MIT
3748.564	3749.630	26669.30	IR I	0.5			AB
3747.204	3748.269	26678.98	IR I	2.0	1.8		MIT
3745.43	3746.49	26691.6	IR I	0.5			MIT
3744.954	3746.019	26695.01	IR I	0.7			AB
3742.94	3744.00	26709.4	IR	0.3	0.5		MIT
3742.56	3743.62	26712.1	IR I	0.3			MIT
3742.27	3743.33	26714.1	IR	1.4	0.3		MIT
3741.731	3742.795	26718.00	IR I	1.4	0.3		MIT
3740.049	3741.112	26730.02	IR I	0.3			AB
3738.531	3739.594	26740.87	IR	1.8	1.0		MIT
3736.751	3737.813	26753.61	IR	0.3			MIT
3736.653	3737.715	26754.31	IR I	0.3			AB
3736.480	3737.542	26755.55	IR	0.3			AB
3735.499	3736.561	26762.58	IR	0.9			AB
3734.770	3735.832	26767.80	IR	2.0	1.5		MIT
3734.332	3735.394	26770.94	IR	0.7			MIT
3732.579	3733.640	26783.51	IR	1.3	0.5		MIT
3731.359	3732.420	26792.27	IR	1.7	1.7		MIT
3730.44	3731.50	26798.9	IR	1.4	0.6		MIT
3730.196	3731.257	26800.62	IR	0.3			AB
3728.855	3729.915	26810.26	IR I	0.5			AB
3728.457	3729.517	26813.12	IR I	0.6			MIT
3728.032	3729.092	26816.18	IR	1.8	1.0		MIT
3727.845	3728.905	26817.52	IR I	0.9			AB
3727.390	3728.450	26820.80	IR	0.3			AB
3726.928	3727.988	26824.12	IR	1.5	0.3		MIT
3725.385	3726.445	26835.23	IR I	1.7	1.3		MIT
3722.747	3723.806	26854.25	IR I	1.5	0.7		MIT
3722.434	3723.493	26856.50	IR	0.3	0.8		AB
3722.021	3723.080	26859.49	IR I	1.0	0.3		MIT
3721.483	3722.541	26863.37	IR I	1.2	0.6		MIT
3720.821	3721.879	26868.15	IR I	0.8			MIT
3718.211	3719.269	26887.01	IR	0.5			MIT
3715.676	3716.733	26905.35	IR	0.7			AB
3714.045	3715.102	26917.17	IR	0.3H	0.3H		MIT
3713.026	3714.082	26924.55	IR	1.3			MIT
3712.476	3713.532	26928.54	IR I	1.5	1.0		MIT
3711.758	3712.814	26933.75	IR	0.3			MIT

AIR WAVELENGTH (ANGSTRMS)	VACUUM WAVELENGTH (ANGSTRMS)	WAVENUMBER (KAYSERS)	LMENT	LOG ARC	INTENSITY SPK	INTENSITY DIS	REF
3711.153	3712.209	26938.14	IR I	0.8			MIT
3708.683	3709.738	26956.08	IR	0.9	0.3		MIT
3708.090	3709.145	26960.39	IR	0.9	0.3		MIT
3707.172	3708.227	26967.07	IR	0.8			AB
3706.982	3708.037	26968.45	IR	1.5	0.9		MIT
3705.263	3706.317	26980.96	IR	1.0	0.3		MIT
3700.550	3701.603	27015.32	IR I	0.3H			AB
3700.090	3701.143	27018.68	IR	0.3H			AB
3698.102	3699.154	27033.21	IR I	1.5	0.9		MIT
3697.698	3698.750	27036.16	IR I	0.5			AB
3696.158	3697.210	27047.42	IR	1.3	0.3		MIT
3694.454	3695.505	27059.90	IR	0.3			AB
3692.694	3693.745	27072.79	IR I	1.2	0.6		MIT
3692.20	3693.25	27076.4	IR	0.3	0.7		MIT
3689.691	3690.741	27094.83	IR	0.3			AB
3689.307	3690.357	27097.65	IR I	0.8	1.0		MIT
3688.935	3689.985	27100.38	IR I	0.3			M
3688.168	3689.218	27106.02	IR I	1.0			AB
3687.080	3688.130	27114.01	IR I	1.6	0.6		MIT
3685.318	3686.367	27126.98	IR	0.3			MIT
3684.327	3685.376	27134.27	IR I	0.8	0.7		MIT
3683.523	3684.572	27140.20	IR	0.7			MIT
3679.711	3680.759	27168.31	IR	0.3			AB
3678.286	3679.333	27178.84	IR	1.0			AB
3677.089	3678.136	27187.68	IR	0.5			AB
3676.656	3677.703	27190.89	IR	1.2	0.3		MIT
3674.984	3676.030	27203.26	IR I	2.0	1.7		MIT
3673.143	3674.189	27216.89	IR	1.0			MIT
3668.184	3669.229	27253.68	IR	0.9			MIT
3665.986	3667.030	27270.02	IR	0.3			MIT
3665.878	3666.922	27270.83	IR	0.5			AB
3665.069	3666.113	27276.85	IR	0.7	0.3		AB
3664.618	3665.662	27280.20	IR I	1.8	1.2		MIT
3664.267	3665.311	27282.82	IR I	0.5			MIT
3661.708	3662.751	27301.88	IR I	1.7	1.5		MIT
3657.616	3658.658	27332.43	IR I	0.7	0.3		MIT
3656.677	3657.719	27339.45	IR I	0.7			MIT
3656.416	3657.458	27341.40	IR	0.8			MIT
3653.759	3654.800	27361.28	IR I	0.3H			MIT
3653.323	3654.364	27364.54	IR I	0.5			AB
3653.191	3654.232	27365.53	IR	1.2	1.7		MIT
3647.702	3648.741	27406.71	IR I	1.3			MIT
3647.105	3648.144	27411.20	IR I	0.5			AB
3645.303	3646.342	27424.75	IR	1.4	0.3H		MIT
3643.481	3644.519	27438.46	IR	0.3			AB
3640.869	3641.907	27458.15	IR I	1.5	0.6		MIT
3639.854	3640.891	27465.80	IR	0.3			MIT
3639.336	3640.373	27469.71	IR	0.5			MIT
3639.165	3640.202	27471.00	IR	0.5			MIT
3636.997	3638.034	27487.38	IR	0.3H			MIT
3636.199	3637.235	27493.41	IR I	1.7	1.4		MIT
3635.488	3636.524	27498.78	IR I	1.5	0.6H		MIT
3633.862	3634.898	27511.09	IR I	0.5			MIT
3632.557	3633.592	27520.97	IR	0.3			MIT
3631.714	3632.749	27527.36	IR I	1.0	0.3		MIT
3629.755	3630.790	27542.22	IR I	1.2	0.6		MIT
3629.158	3630.193	27546.75	IR I	1.0	0.3		MIT
3628.670	3629.704	27550.45	IR I	2.0	1.5		MIT
3626.286	3627.320	27568.56	IR I	1.5	1.0		MIT
3625.706	3626.740	27572.97	IR I	1.5	0.8		MIT
3623.803	3624.836	27587.45	IR I	1.3	0.6		MIT
3622.787	3623.820	27595.19	IR	0.3			MIT
3622.501	3623.534	27597.37	IR	0.5			AB
3621.887	3622.920	27602.05	IR I	0.6			AB
3620.550	3621.582	27612.24	IR I	0.5			AB
3619.163	3620.195	27622.82	IR I	1.5	0.5		MIT
3617.212	3618.244	27637.72	IR I	1.7	1.2		MIT
3614.454	3615.485	27658.81	IR	0.3	0.3H		MIT
3609.772	3610.802	27694.68	IR	1.5	1.4		MIT
3608.354	3609.383	27705.56	IR I	0.8	0.3		MIT

AIR WAVELENGTH (ANGSTRMS)	VACUUM WAVELENGTH (ANGSTRMS)	WAVENUMBER (KAYSERS)	LMENT	LOG ARC	INTENSITY SPK	INTENSITY DIS	REF
3607.124	3608.153	27715.01	IR I	0.5			MIT
3604.315	3605.343	27736.61	IR I	0.7			MIT
3602.064	3603.092	27753.94	IR I	0.5			AB
3601.403	3602.430	27759.04	IR I	1.5	0.9		MIT
3598.763	3599.790	27779.40	IR I	1.4	0.7		MIT
3596.417	3597.443	27797.52	IR I	0.5			MIT
3595.491	3596.517	27804.68	IR	0.5			AB
3594.391	3595.417	27813.19	IR	1.2	1.5		MIT
3594.137	3595.163	27815.15	IR I	1.0	0.6		MIT
3591.583	3592.608	27834.93	IR	0.3H			MIT
3591.283	3592.308	27837.26	IR I	0.9	0.3		AB
3588.234	3589.258	27860.91	IR	0.7			MIT
3584.663	3585.686	27888.67	IR I	1.0	0.0H		MIT
3584.532	3585.555	27889.68	IR I	0.5			AB
3583.388	3584.411	27898.59	IR	0.8	0.5		MIT
3580.861	3581.883	27918.27	IR I	1.2	0.5		AB
3577.622	3578.643	27943.55	IR I	0.6			AB
3577.093	3578.114	27947.68	IR	0.5			MIT
3576.737	3577.758	27950.46	IR I	1.0	0.3		AB
3576.233	3577.254	27954.40	IR	0.3H			MIT
3573.724	3574.744	27974.03	IR I	0.9	2.0		MIT
3573.612	3574.632	27974.91	IR I	0.7H			AB
3569.833	3570.852	28004.52	IR I	0.8	0.5		MIT
3567.995	3569.014	28018.94	IR	1.2	0.7		MIT
3566.830	3567.849	28028.10	IR I	0.3			MIT
3566.491	3567.509	28030.76	IR I	0.9			MIT
3566.409	3567.427	28031.40	IR I	0.8	0.7		AB
3562.822	3563.840	28059.62	IR I	0.8	0.3		MIT
3561.373	3562.390	28071.04	IR	0.7			MIT
3561.172	3562.189	28072.63	IR I	0.7	0.3		AB
3558.992	3560.009	28089.82	IR I	1.7	1.7		AB
3557.170	3558.186	28104.21	IR I	1.4	1.5		MIT
3555.603	3556.619	28116.59	IR I	1.0	0.3		MIT
3555.327	3556.343	28118.78	IR	0.3			MIT
3552.100	3553.115	28144.32	IR I	1.3	0.3		AB
3550.395	3551.409	28157.84	IR	0.3			MIT
3550.271	3551.285	28158.82	IR I	0.5			MIT
3549.291	3550.305	28166.59	IR	0.9	0.3		MIT
3548.648	3549.662	28171.70	IR	1.4	0.0H		MIT
3546.096	3547.109	28191.97	IR I	0.7			AB
3542.575	3543.587	28219.99	IR I	1.0	1.0H		MIT
3540.686	3541.698	28235.05	IR I	0.7	0.3H		MIT
3538.152	3539.163	28255.27	IR	1.3	0.0H		AB
3535.787	3536.798	28274.17	IR	1.1	1.0		MIT
3534.984	3535.994	28280.59	IR I	1.0			MIT
3532.267	3533.277	28302.34	IR	1.1	0.5		MIT
3532.06	3533.07	28304.0	IR	0.3D			MIT
3531.798	3532.808	28306.10	IR I	0.3			AB
3530.739	3531.748	28314.59	IR	1.0	0.3		MIT
3527.641	3528.649	28339.45	IR I	0.7			MIT
3526.765	3527.773	28346.49	IR	1.3	0.3		MIT
3524.372	3525.380	28365.74	IR	0.3H			MIT
3522.028	3523.035	28384.62	IR I				ME
3518.906	3519.912	28409.80	IR I	0.3H			AB
3515.949	3516.954	28433.69	IR I	1.5	1.2		MIT
3513.645	3514.650	28452.34	IR I	2.0H	2.0		MIT
3512.200	3513.204	28464.04	IR I	1.1	0.7		MIT
3511.893	3512.897	28466.53	IR I	1.3	0.8		MIT
3510.642	3511.646	28476.67	IR I	1.3	1.0		MIT
3509.241	3510.245	28488.04	IR I	0.9	0.3		AB
3508.719	3509.723	28492.28	IR I	0.3			MIT
3508.576	3509.580	28493.44	IR I	0.8	0.0		MIT
3507.506	3508.509	28502.13	IR	0.0	0.3		MIT
3507.223	3508.226	28504.43	IR I	0.3			MIT
3504.866	3505.869	28523.60	IR	0.3			MIT
3504.124	3505.126	28529.64	IR I	0.7			MIT
3502.936	3503.938	28539.32	IR I	0.9	0.6		MIT
3502.616	3503.618	28541.92	IR I	0.3			AB
3500.230	3501.231	28561.38	IR	0.7			MIT
3499.764	3500.765	28565.18	IR I	0.5			MIT

AIR WAVELENGTH (ANGSTRMS)	VACUUM WAVELENGTH (ANGSTRMS)	WAVENUMBER (KAYSERS)	LMENT	LOG ARC	INTENSITY SPK	INTENSITY DIS	REF
3499.368	3500.369	28568.42	IR I	0.3			MIT
3499.112	3500.113	28570.50	IR I	0.7			MIT
3498.951	3499.952	28571.82	IR	1.4			MIT
3498.737	3499.738	28573.57	IR	1.2			MIT
3496.439	3497.439	28592.35	IR I	0.5			MIT
3496.010	3497.010	28595.85	IR I	0.9			MIT
3494.624	3495.624	28607.20	IR	0.7	0.3		MIT
3492.060	3493.059	28628.20	IR I	0.3			MIT
3489.572	3490.571	28648.61	IR	0.3			MIT
3488.575	3489.573	28656.80	IR I	1.2	0.5		MIT
3488.052	3489.050	28661.09	IR	0.3			MIT
3486.879	3487.877	28670.74	IR I	0.5			MIT
3485.502	3486.500	28682.06	IR I	1.2	0.5		MIT
3484.481	3485.478	28690.47	IR I	1.2	0.6		MIT
3484.083	3485.080	28693.74	IR I	1.2	0.5		MIT
3482.598	3483.595	28705.98	IR I	1.2	0.6		MIT
3482.039	3483.036	28710.59	IR I	0.3			MIT
3481.085	3482.081	28718.46	IR I	0.8	0.3		MIT
3479.202	3480.198	28734.00	IR I	0.3			MIT
3477.767	3478.763	28745.85	IR I	1.2			MIT
3476.457	3477.452	28756.69	IR I	1.2	0.3		MIT
3476.015	3477.010	28760.34	IR	0.7			MIT
3475.535	3476.530	28764.31	IR I	0.3	0.5H		AB
3468.593	3469.586	28821.88	IR I	0.3	0.3		MIT
3465.225	3466.217	28849.89	IR I	1.1	0.0		MIT
3463.120	3464.112	28867.43	IR	0.6			AB
3461.582	3462.573	28880.25	IR	0.6			AB
3460.544	3461.535	28888.92	IR	0.5			AB
3455.038	3456.028	28934.95	IR	0.5	0.0H		MIT
3448.971	3449.959	28985.85	IR I	1.8	1.0		MIT
3446.636	3447.624	29005.49	IR I	0.9	0.3		MIT
3446.301	3447.289	29008.31	IR I	1.5	0.6		MIT
3439.418	3440.404	29066.36	IR	0.6			AB
3438.094	3439.079	29077.55	IR I	0.8			AB
3437.498	3438.483	29082.59	IR I	1.5	0.5		MIT
3437.015	3438.000	29086.68	IR I	1.3	1.2		MIT
3434.757	3435.742	29105.80	IR I	1.0			MIT
3433.314	3434.298	29118.03	IR I	0.9			MIT
3432.029	3433.013	29128.93	IR	0.5			AB
3431.556	3432.540	29132.95	IR	0.9			AB
3431.297	3432.281	29135.15	IR	0.6			MIT
3430.408	3431.391	29142.70	IR	0.5H			AB
3430.035	3431.018	29145.86	IR I	0.5			AB
3428.868	3429.851	29155.78	IR I	0.7			AB
3425.367	3426.349	29185.58	IR	0.5			MIT
3424.693	3425.675	29191.33	IR I	1.3	0.3		MIT
3423.699	3424.681	29199.80	IR	0.7			MIT
3421.760	3422.741	29216.35	IR	0.8			MIT
3420.487	3421.468	29227.22	IR I	1.3	0.3		MIT
3419.423	3420.404	29236.31	IR I	1.5	0.7		MIT
3418.363	3419.343	29245.38	IR	0.5			MIT
3417.336	3418.316	29254.17	IR I	0.5			AB
3415.744	3416.724	29267.80	IR I	0.8	0.0H		MIT
3415.241	3416.221	29272.11	IR	1.0	0.3H		MIT
3412.595	3413.574	29294.81	IR I	1.0	0.3H		MIT
3411.577	3412.556	29303.55	IR I	0.8	0.0		AB
3410.112	3411.090	29316.14	IR I	0.5			MIT
3410.022	3411.000	29316.91	IR I	0.9	0.3		MIT
3409.775	3410.753	29319.04	IR	0.6			AB
3402.791	3403.767	29379.21	IR I	1.0	0.3		MIT
3402.683	3403.659	29380.14	IR I	0.5			MIT
3401.770	3402.746	29388.03	IR I	1.3	0.5		MIT
3392.589	3393.563	29467.55	IR I	0.5			MIT
3392.475	3393.449	29468.55	IR	0.3	0.6H		MIT
3390.868	3391.841	29482.51	IR I	0.7			AB
3387.988	3388.961	29507.57	IR	0.7			MIT
3387.866	3388.839	29508.64	IR	0.5			MIT
3386.173	3387.145	29523.39	IR	1.2	0.3		MIT
3385.597	3386.569	29528.41	IR I	0.7			MIT
3385.094	3386.066	29532.80	IR I	0.7			AB
3381.370	3382.341	29565.32	IR I	0.5			AB
3380.994	3381.965	29568.61	IR I	1.2	0.0		MIT
3379.835	3380.806	29578.75	IR	0.6	0.3H		MIT
3379.409	3380.379	29582.48	IR	0.5			MIT
3377.949	3378.919	29595.26	IR I	0.6			MIT
3377.126	3378.096	29602.48	IR	0.6H			MIT
3375.998	3376.968	29612.37	IR I	0.5			MIT
3374.453	3375.422	29625.92	IR I	0.6			MIT
3371.441	3372.409	29652.39	IR I	1.3	0.3		MIT
3370.633	3371.601	29659.50	IR	0.9	0.3H		MIT
3368.481	3369.449	29678.45	IR I	1.4	1.3		MIT
3367.050	3368.017	29691.06	IR I	0.5			MIT
3366.904	3367.871	29692.35	IR I	0.6	0.0H		MIT
3366.476	3367.443	29696.12	IR	0.5			MIT
3366.231	3367.198	29698.28	IR I	0.6			MIT
3365.524	3366.491	29704.52	IR I	0.7			MIT
3364.231	3365.198	29715.94	IR	0.7	0.3H		MIT
3362.791	3363.757	29728.66	IR	0.3			MIT
3360.313	3361.279	29750.58	IR	0.5			MIT
3360.066	3361.032	29752.77	IR	0.3H			MIT
3359.738	3360.703	29755.68	IR	0.7	0.8		MIT
3359.474	3360.439	29758.01	IR I	0.9	0.6		MIT
3359.286	3360.251	29759.68	IR	0.3			MIT
3358.895	3359.860	29763.14	IR	0.3			MIT
3358.292	3359.257	29768.49	IR	0.3			MIT
3357.734	3358.699	29773.43	IR	0.3			MIT
3356.534	3357.499	29784.08	IR	0.3			MIT
3356.193	3357.158	29787.10	IR I	0.6			MIT
3355.793	3356.757	29790.65	IR I	1.0	0.5		MIT
3355.585	3356.549	29792.50	IR I	0.6			MIT
3355.026	3355.990	29797.47	IR	0.6	0.6H		MIT
3354.726	3355.690	29800.13	IR	0.3			MIT
3353.544	3354.508	29810.63	IR I	0.3			MIT
3352.844	3353.808	29816.86	IR		0.3H		MIT
3347.863	3348.825	29861.22	IR I	0.5			MIT
3347.547	3348.509	29864.03	IR I	0.8	0.0		MIT
3346.819	3347.781	29870.53	IR	0.3H			MIT
3346.461	3347.423	29873.73	IR I	0.3			MIT
3346.085	3347.047	29877.08	IR	0.3			MIT
3344.218	3345.179	29893.76	IR I	1.0	0.0		MIT
3343.591	3344.552	29899.37	IR I	0.3			MIT
3342.797	3343.758	29906.47	IR I	0.3			MIT
3340.351	3341.311	29928.37	IR I	1.0			MIT
3339.392	3340.352	29936.96	IR I	1.2	0.3		MIT
3338.374	3339.334	29946.09	IR I	1.2	0.3		MIT
3336.056	3337.015	29966.90	IR I	0.3			MIT
3334.159	3335.118	29983.95	IR I	1.6	0.5		MIT
3326.900	3327.857	30049.37	IR I	1.0	0.0		MIT
3326.531	3327.488	30052.70	IR I	0.3			MIT
3326.403	3327.360	30053.86	IR	0.3			MIT
3326.100	3327.057	30056.59	IR I	0.3			MIT
3325.453	3326.410	30062.44	IR	0.5	0.0H		MIT
3322.871	3323.827	30085.80	IR I	1.3	1.2		MIT
3322.603	3323.559	30088.23	IR I	1.5	1.3		MIT
3320.359	3321.314	30108.56	IR I	0.3			MIT
3319.185	3320.140	30119.21	IR	0.5			MIT
3319.092	3320.047	30120.05	IR I	0.5			MIT
3318.672	3319.627	30123.87	IR	0.3H			MIT
3318.448	3319.403	30125.90	IR I	0.5			MIT
3317.525	3318.480	30134.28	IR I	0.3			MIT
3317.319	3318.274	30136.15	IR I	0.3			MIT
3316.636	3317.590	30142.36	IR	0.9			MIT
3312.134	3313.087	30183.33	IR I	1.4	1.2		MIT
3311.023	3311.976	30193.46	IR I	0.3			MIT
3310.525	3311.478	30198.00	IR I	1.5	1.1		MIT
3309.395	3310.348	30208.31	IR I	0.7	0.0H		MIT
3308.442	3309.394	30217.01	IR I	0.3			MIT
3307.630	3308.582	30224.43	IR I	0.7			MIT
3305.845	3306.797	30240.75	IR	0.5	0.0		MIT
3305.643	3306.595	30242.59	IR I	0.3			MIT

AIR WAVELENGTH (ANGSTRMS)	VACUUM WAVELENGTH (ANGSTRMS)	WAVENUMBER (KAYSERS)	LMENT	LOG ARC	INTENSITY SPK	DIS	REF
3304.923	3305.874	30249.18	IR I	0.9			MIT
3303.628	3304.579	30261.04	IR	0.6			MIT
3303.091	3304.042	30265.96	IR I	0.5			MIT
3301.757	3302.708	30278.19	IR I	0.3			MIT
3298.410	3299.360	30308.91	IR I	0.3			MIT
3297.925	3298.875	30313.37	IR	0.3H			MIT
3297.508	3298.458	30317.20	IR	0.5	0.0H		MIT
3295.087	3296.036	30339.48	IR I	0.9			MIT
3290.896	3291.844	30378.11	IR	0.3H			MIT
3290.258	3291.206	30384.00	IR I	0.3			MIT
3287.589	3288.536	30408.67	IR I	1.1	0.0		MIT
3287.058	3288.005	30413.58	IR I	1.1	0.0		MIT
3284.568	3285.514	30436.64	IR I	0.5	0.0		MIT
3283.586	3284.532	30445.74	IR I	1.3P			MIT
3282.327	3283.273	30457.42	IR I	0.3	0.3		MIT
3281.899	3282.845	30461.39	IR I	0.5			MIT
3277.283	3278.227	30504.29	IR I	0.9	0.0		MIT
3276.167	3277.111	30514.68	IR I	0.3			MIT
3275.603	3276.547	30519.94	IR I	0.5			MIT
3275.031	3275.975	30525.26	IR	0.3H			MIT
3274.561	3275.505	30529.65	IR I	0.6			MIT
3271.984	3272.927	30553.69	IR	0.5			MIT
3271.807	3272.750	30555.34	IR I	1.0	0.0		MIT
3271.445	3272.388	30558.72	IR I	0.3			MIT
3271.235	3272.178	30560.69	IR I	1.0	0.0		MIT
3267.100	3268.042	30599.36	IR I	0.3			MIT
3266.444	3267.386	30605.51	IR I	1.7	1.0		MIT
3263.309	3264.250	30634.91	IR I	0.5	0.0H		MIT
3262.930	3263.871	30638.47	IR I	0.6			MIT
3262.718	3263.659	30640.46	IR I	0.5			MIT
3262.010	3262.951	30647.11	IR I	1.3	0.3		MIT
3260.998	3261.938	30656.62	IR I	0.3H			MIT
3257.888	3258.828	30685.88	IR	0.3			MIT
3256.784	3257.723	30696.28	IR I	0.6			MIT
3255.202	3256.141	30711.20	IR I	0.3			MIT
3255.058	3255.997	30712.56	IR I	0.7			MIT
3254.403	3255.342	30718.74	IR I	1.3	0.0		MIT
3253.37	3254.31	30728.5	IR	0.3	0.0		MIT
3249.730	3250.667	30762.91	IR I	1.0	0.5		MIT
3249.508	3250.445	30765.01	IR I	1.0	1.0		MIT
3247.667	3248.604	30782.45	IR I	0.5	0.0		AB
3246.293	3247.230	30795.48	IR I	0.3			MIT
3242.608	3243.544	30830.48	IR I	0.5			MIT
3242.323	3243.259	30833.19	IR I	0.9			MIT
3241.517	3242.452	30840.85	IR I	2.0	1.7		MIT
3240.562	3241.497	30849.94	IR I	0.5			MIT
3240.214	3241.149	30853.26	IR I	1.2	0.0		MIT
3237.866	3238.800	30875.63	IR I	0.3			MIT
3234.510	3235.444	30907.66	IR I	0.5			MIT
3232.001	3232.934	30931.66	IR	1.3	0.0		MIT
3230.763	3231.696	30943.51	IR I	1.3	0.0		MIT
3229.282	3230.214	30957.70	IR I	1.5	0.3		MIT
3226.710	3227.642	30982.37	IR I	1.3	0.0		MIT
3223.886	3224.817	31009.51	IR I	0.5	0.0H		MIT
3223.512	3224.443	31013.11	IR I	0.5			MIT
3223.009	3223.940	31017.95	IR I	0.3			MIT
3221.281	3222.211	31034.59	IR	1.0	0.0H		MIT
3220.780	3221.710	31039.42	IR I	2.0	1.5		MIT
3219.514	3220.444	31051.62	IR I	1.5	0.3		MIT
3218.460	3219.390	31061.79	IR I	1.3	0.0		MIT
3216.798	3217.727	31077.84	IR	0.5H			MIT
3213.548	3214.476	31109.27	IR I	1.0	0.3		MIT
3212.221	3213.149	31122.12	IR I	0.7			MIT
3212.119	3213.047	31123.11	IR I	1.4	1.2		MIT
3208.151	3209.078	31161.60	IR I	1.0			MIT
3207.094	3208.021	31171.87	IR I	0.7			MIT
3205.091	3206.017	31191.35	IR I	1.3	0.7		MIT
3204.453	3205.379	31197.56	IR	0.3			MIT
3200.887	3201.812	31232.31	IR I	1.0	0.7		MIT
3200.037	3200.962	31240.61	IR I	0.8			MIT
3198.924	3199.849	31251.48	IR I	1.5	1.0		MIT
3198.097	3199.021	31259.56	IR I	0.8	0.7H		MIT
3194.852	3195.776	31291.31	IR	0.3			MIT
3193.967	3194.890	31299.98	IR	0.8			MIT
3193.116	3194.039	31308.32	IR	0.3			MIT
3189.335	3190.257	31345.44	IR I	0.8			MIT
3187.133	3188.055	31367.09	IR I	0.3H			MIT
3186.113	3187.034	31377.13	IR	0.5			AB
3185.886	3186.807	31379.37	IR I	0.3H			MIT
3182.798	3183.718	31409.81	IR I	0.3			MIT
3180.354	3181.274	31433.95	IR I	1.4	0.0		MIT
3180.174	3181.094	31435.73	IR I	0.8			MIT
3179.476	3180.396	31442.63	IR I	0.3			MIT
3179.201	3180.121	31445.35	IR I	1.1	0.3H		MIT
3178.689	3179.608	31450.41	IR	1.0	1.0		MIT
3177.582	3178.501	31461.37	IR I	1.4	0.3		MIT
3173.337	3174.255	31503.45	IR I	0.3			MIT
3172.792	3173.710	31508.86	IR I	1.4	0.0		MIT
3168.880	3169.797	31547.76	IR I	1.4	0.3		MIT
3168.178	3169.095	31554.75	IR I	1.3	0.3		MIT
3167.203	3168.120	31564.47	IR I	1.1	0.0		MIT
3166.749	3167.665	31568.99	IR I	0.9	0.0		MIT
3166.341	3167.257	31573.06	IR	0.6			MIT
3165.195	3166.111	31584.49	IR I	0.5			MIT
3163.87	3164.79	31597.7	IR		0.3		MIT
3162.824	3163.739	31608.17	IR	0.3			MIT
3162.739	3163.654	31609.01	IR	0.3			MIT
3161.822	3162.737	31618.18	IR I	0.8	0.0		MIT
3161.357	3162.272	31622.83	IR I	0.8	0.0H		MIT
3159.856	3160.771	31637.85	IR I	0.3			MIT
3159.522	3160.437	31641.20	IR I	0.8	0.0		MIT
3159.149	3160.064	31644.93	IR I	1.7R	0.3H		MIT
3157.494	3158.408	31661.52	IR I	0.8			MIT
3156.148	3157.062	31675.02	IR I	0.6			MIT
3154.737	3155.650	31689.19	IR I	1.3	0.3		MIT
3154.551	3155.464	31691.06	IR I	1.3	0.0		MIT
3152.615	3153.528	31710.52	IR I	1.0	0.0H		MIT
3152.301	3153.214	31713.68	IR	0.3			MIT
3152.125	3153.038	31715.45	IR	0.3			MIT
3151.627	3152.540	31720.46	IR	0.3	0.0H		MIT
3150.999	3151.911	31726.78	IR	0.5			MIT
3150.727	3151.639	31729.52	IR	0.7			MIT
3150.607	3151.519	31730.73	IR I	1.3	0.0		MIT
3149.995	3150.907	31736.89	IR	0.3H			MIT
3147.739	3148.651	31759.64	IR I	1.0			MIT
3147.458	3148.370	31762.47	IR	0.5			MIT
3145.068	3145.979	31786.61	IR	1.3	0.3		MIT
3143.802	3144.713	31799.41	IR	0.5			MIT
3143.008	3143.918	31807.44	IR	0.3			MIT
3142.515	3143.425	31812.43	IR I	0.3H			MIT
3141.810	3142.720	31819.57	IR	0.3H			MIT
3140.740	3141.650	31830.41	IR	0.3			MIT
3140.412	3141.322	31833.73	IR I	1.7R	0.0		MIT
3139.588	3140.498	31842.09	IR I	0.7			MIT
3139.169	3140.078	31846.34	IR	0.5			MIT
3137.721	3138.630	31861.03	IR	0.9	0.5		MIT
3137.275	3138.184	31865.56	IR	0.3H			MIT
3136.896	3137.805	31869.41	IR	0.5			MIT
3136.476	3137.385	31873.68	IR	0.5			AB
3136.300	3137.209	31875.47	IR I	0.3			MIT
3136.100	3137.009	31877.50	IR I	0.3H			MIT
3135.227	3136.135	31886.38	IR	0.3H			MIT
3133.321	3134.229	31905.77	IR I	1.6	0.3H		MIT
3133.086	3133.994	31908.17	IR I	1.3	0.0		MIT
3130.578	3131.485	31933.73	IR	0.5			MIT
3130.285	3131.192	31936.72	IR	0.6			MIT
3129.236	3130.143	31947.42	IR	0.3H	0.0H		MIT
3128.390	3129.297	31956.06	IR I	1.3	0.0		MIT
3127.693	3128.600	31963.18	IR	0.5			MIT
3125.254	3126.160	31988.13	IR	0.3H			MIT

AIR WAVELENGTH (ANGSTRMS)	VACUUM WAVELENGTH (ANGSTRMS)	WAVENUMBER (KAYSERS)	LMENT	LOG ARC	INTENSITY SPK	DIS	REF
3124.927	3125.833	31991.47	IR	0.6			MIT
3124.081	3124.987	32000.14	IR I	0.6			MIT
3123.921	3124.827	32001.77	IR	0.3			MIT
3123.217	3124.122	32008.99	IR I	0.3			MIT
3122.384	3123.289	32017.53	IR I	1.4	0.0		MIT
3121.778	3122.683	32023.74	IR I	1.5	0.0		MIT
3120.758	3121.663	32034.21	IR I	1.7	0.3		MIT
3119.674	3120.579	32045.34	IR	0.7	0.0		MIT
3118.836	3119.740	32053.95	IR	0.3	0.3		MIT
3117.856	3118.760	32064.02	IR	0.7			MIT
3117.760	3118.664	32065.01	IR	0.7			MIT
3117.523	3118.427	32067.45	IR	0.8			MIT
3117.332	3118.236	32069.41	IR	0.3	0.3		MIT
3117.185	3118.089	32070.93	IR	0.5			MIT
3114.549	3115.452	32098.07	IR I	1.4	1.0		MIT
3114.047	3114.950	32103.24	IR I	1.4	1.2		MIT
3113.788	3114.691	32105.91	IR I	0.3			MIT
3113.355	3114.258	32110.38	IR	0.8			MIT
3113.131	3114.034	32112.69	IR I	0.3H			MIT
3112.356	3113.259	32120.68	IR I	0.5			MIT
3112.178	3113.081	32122.52	IR	0.5			MIT
3112.087	3112.990	32123.46	IR	0.6	0.3		MIT
3110.227	3111.129	32142.67	IR	0.6	0.0		MIT
3109.357	3110.259	32151.66	IR I	0.9	0.9		MIT
3108.327	3109.229	32162.32	IR	0.3			MIT
3107.821	3108.723	32167.55	IR	0.3H			MIT
3107.556	3108.458	32170.30	IR	0.6			MIT
3106.021	3106.922	32186.19	IR	0.6			MIT
3105.946	3106.847	32186.97	IR I	0.6			MIT
3105.671	3106.572	32189.82	IR	0.3			MIT
3104.169	3105.070	32205.40	IR I	0.3			MIT
3103.754	3104.655	32209.70	IR I	1.0	0.0		MIT
3102.355	3103.255	32224.23	IR	0.6			MIT
3102.144	3103.044	32226.42	IR	0.6			MIT
3101.167	3102.067	32236.57	IR I	0.7	0.0		MIT
3100.446	3101.346	32244.07	IR I	1.3	0.5		MIT
3100.286	3101.186	32245.73	IR I	1.5	0.3		MIT
3098.928	3099.827	32259.86	IR I	1.0R			MIT
3098.590	3099.489	32263.38	IR	0.7			MIT
3098.436	3099.335	32264.98	IR	0.3			MIT
3097.815	3098.714	32271.45	IR I	1.1	1.0		MIT
3097.349	3098.248	32276.31	IR	0.3H			MIT
3096.941	3097.840	32280.56	IR I	0.3H			MIT
3096.823	3097.722	32281.79	IR	0.7			MIT
3096.340	3097.239	32286.82	IR	0.3H			MIT
3096.128	3097.027	32289.03	IR	0.5			MIT
3095.344	3096.242	32297.21	IR	0.6			MIT
3094.617	3095.515	32304.80	IR	0.5			MIT
3094.366	3095.264	32307.42	IR I	0.7			MIT
3094.009	3094.907	32311.15	IR I	1.3	1.0		MIT
3092.401	3093.299	32327.95	IR	0.7			MIT
3091.366	3092.263	32338.77	IR	0.5			MIT
3090.372	3091.269	32349.17	IR	0.3			MIT
3090.163	3091.060	32351.36	IR I	1.0	0.0		MIT
3088.038	3088.935	32373.62	IR I	1.7	0.3		MIT
3086.440	3087.336	32390.38	IR I	1.5	0.3		MIT
3085.198	3086.094	32403.42	IR	0.6	0.0		MIT
3084.952	3085.848	32406.00	IR	0.3H			MIT
3084.736	3085.632	32408.27	IR	0.3H			MIT
3083.985	3084.881	32416.17	IR	0.7			AB
3083.224	3084.119	32424.17	IR I	1.4	0.3H		MIT
3079.993	3080.888	32458.18	IR	0.3H			MIT
3078.576	3079.470	32473.12	IR	0.7	0.7		MIT
3077.882	3078.776	32480.44	IR	0.3H			MIT
3077.648	3078.542	32482.91	IR I	0.9	0.0		MIT
3076.686	3077.580	32493.06	IR I	1.0H	0.3		MIT
3074.762	3075.655	32513.40	IR I	1.1			MIT
3073.281	3074.174	32529.06	IR I	1.0			MIT
3070.918	3071.810	32554.09	IR	0.5			AB
3070.637	3071.529	32557.07	IR	0.5			MIT

AIR WAVELENGTH (ANGSTRMS)	VACUUM WAVELENGTH (ANGSTRMS)	WAVENUMBER (KAYSERS)	LMENT	LOG ARC	INTENSITY SPK	DIS	REF
3069.706	3070.598	32566.95	IR I	1.4	0.7		MIT
3069.094	3069.986	32573.44	IR I	1.4	1.0		MIT
3068.890	3069.782	32575.60	IR I	1.6	1.3		MIT
3067.939	3068.831	32585.70	IR	0.3H			MIT
3064.790	3065.681	32619.18	IR I	0.7			MIT
3064.509	3065.400	32622.17	IR I	1.3	0.0H		MIT
3063.726	3064.617	32630.51	IR	0.6			MIT
3061.408	3062.298	32655.22	IR I	1.4	0.3		MIT
3060.838	3061.728	32661.30	IR I	1.4	0.0		MIT
3059.743	3060.633	32672.99	IR	0.5			MIT
3058.653	3059.542	32684.63	IR	0.7			MIT
3057.951	3058.840	32692.13	IR	0.5			MIT
3057.278	3058.167	32699.33	IR I	1.5	0.3		MIT
3056.635	3057.524	32706.21	IR I	0.3H			MIT
3054.460	3055.348	32729.49	IR	0.3			MIT
3053.596	3054.484	32738.75	IR I	1.1	0.0H		MIT
3052.240	3053.128	32753.30	IR	0.5	0.0		MIT
3052.161	3053.049	32754.15	IR I	0.9	0.0		MIT
3051.149	3052.036	32765.01	IR I	0.7	0.0		MIT
3051.093	3051.980	32765.61	IR I	0.7			MIT
3050.018	3050.905	32777.16	IR I	0.7			MIT
3049.435	3050.322	32783.42	IR I	1.4	1.0		MIT
3048.666	3049.553	32791.69	IR I	1.1			MIT
3047.788	3048.675	32801.14	IR I	1.2			MIT
3047.157	3048.043	32807.93	IR I	1.7	1.3		MIT
3045.654	3046.540	32824.12	IR	0.9	0.3H		MIT
3043.831	3044.717	32843.78	IR	0.3			MIT
3042.646	3043.531	32856.57	IR	1.4	1.2		MIT
3041.853	3042.738	32865.14	IR	0.8	0.0H		MIT
3040.937	3041.822	32875.04	IR I	0.7			AB
3040.467	3041.352	32880.12	IR I	1.5	0.3		MIT
3039.768	3040.653	32887.68	IR	0.3			MIT
3039.714	3040.599	32888.26	IR	0.3			MIT
3039.260	3040.144	32893.17	IR I	1.4	0.3		MIT
3037.748	3038.632	32909.55	IR	1.6	0.3		MIT
3035.534	3036.417	32933.55	IR	0.5			MIT
3034.556	3035.439	32944.16	IR I	1.3			MIT
3034.039	3034.922	32949.77	IR I	0.6			MIT
3033.622	3034.505	32954.30	IR I	1.4	0.0		MIT
3032.410	3033.293	32967.47	IR I	1.7	0.0		MIT
3031.878	3032.761	32973.26	IR	0.3H			MIT
3030.328	3031.210	32990.12	IR	0.3H			MIT
3030.238	3031.120	32991.10	IR I	0.9			MIT
3029.362	3030.244	33000.64	IR I	1.8	0.5		MIT
3028.881	3029.763	33005.88	IR I	0.6			MIT
3028.464	3029.346	33010.43	IR	0.5			AB
3026.363	3027.244	33033.34	IR		0.5		MIT
3025.822	3026.703	33039.25	IR I	1.7	0.5		MIT
3024.782	3025.663	33050.61	IR I	0.6			MIT
3024.296	3025.177	33055.92	IR I	1.5			MIT
3022.688	3023.568	33073.50	IR I	1.3	0.0		MIT
3022.410	3023.290	33076.55	IR	1.5	0.0		MIT
3020.009	3020.889	33102.84	IR	1.5	0.0		MIT
3019.229	3020.108	33111.39	IR	1.5	0.3H		MIT
3018.032	3018.911	33124.53	IR I	1.2			MIT
3017.726	3018.605	33127.88	IR	0.3H			MIT
3017.313	3018.192	33132.42	IR	1.5	0.3H		MIT
3016.426	3017.305	33142.16	IR I	1.5	0.3H		MIT
3014.727	3015.605	33160.84	IR	0.9			MIT
3014.466	3015.344	33163.71	IR I	1.0			MIT
3012.858	3013.736	33181.41	IR	0.7			MIT
3012.574	3013.452	33184.54	IR I	1.3			MIT
3011.690	3012.568	33194.28	IR I	1.5	0.0		MIT
3009.901	3010.778	33214.00	IR I	1.4	1.0		MIT
3009.572	3010.449	33217.64	IR	0.3			MIT
3008.625	3009.502	33228.09	IR I	0.6			MIT
3007.618	3008.495	33239.22	IR I	0.8	0.3H		MIT
3007.282	3008.158	33242.93	IR I	1.4			MIT
3005.212	3006.088	33265.83	IR I	1.5	1.3		MIT
3005.039	3005.915	33267.74	IR I	0.5			MIT

AIR WAVELENGTH (ANGSTRMS)	VACUUM WAVELENGTH (ANGSTRMS)	WAVENUMBER (KAYSERS)	LMENT	LOG ARC	INTENSITY SPK	INTENSITY DIS	REF
3003.632	3004.508	33283.33	IR I	1.8	1.5		MIT
3002.485	3003.360	33296.04	IR	0.8			AB
3002.253	3003.128	33298.61	IR I	1.7	1.0		MIT
3001.256	3002.131	33309.67	IR I	0.7			MIT
3000.027	3000.902	33323.32	IR	1.5			MIT
2999.541	3000.416	33328.72	IR I	1.5	0.0		MIT
2999.014	2999.888	33334.57	IR I	0.3			MIT
2997.968	2998.842	33346.20	IR	0.8			MIT
2997.408	2998.282	33352.43	IR I	1.4	0.3		MIT
2997.192	2998.066	33354.84	IR I	1.4	0.0		MIT
2996.662	2997.536	33360.74	IR	0.5			MIT
2996.078	2996.952	33367.24	IR I	1.7	0.3		MIT
2993.629	2994.502	33394.53	IR	0.3			MIT
2993.058	2993.931	33400.90	IR	0.9			MIT
2991.396	2992.268	33419.46	IR I	0.8			MIT
2990.617	2991.489	33428.17	IR I	1.5	0.5		MIT
2988.979	2989.851	33446.48	IR I	1.5			MIT
2988.195	2989.067	33455.26	IR	0.5			MIT
2987.646	2988.518	33461.41	IR	0.7	0.0		MIT
2986.616	2987.487	33472.95	IR	0.5			AB
2986.173	2987.044	33477.91	IR I	1.4	0.7H		MIT
2985.800	2986.671	33482.09	IR I	1.4	0.3		MIT
2984.656	2985.527	33494.93	IR	0.3H			MIT
2982.603	2983.473	33517.98	IR I	0.7			MIT
2982.414	2983.284	33520.10	IR I	1.2			MIT
2980.649	2981.519	33539.95	IR I	1.2H	1.0		MIT
2980.222	2981.092	33544.76	IR I	0.5			AB
2978.426	2979.295	33564.98	IR	0.5	0.7		MIT
2977.935	2978.804	33570.52	IR I	0.9			MIT
2977.681	2978.550	33573.38	IR	0.9			MIT
2976.727	2977.596	33584.14	IR I	0.8			MIT
2976.445	2977.314	33587.32	IR	0.5	0.7		MIT
2974.947	2975.815	33604.23	IR I	1.4	1.0		MIT
2974.702	2975.570	33607.00	IR	0.5			MIT
2974.540	2975.408	33608.83	IR I	0.6			MIT
2974.101	2974.969	33613.79	IR I	1.0	0.7		MIT
2971.39	2972.26	33644.5	IR		1.2H		MIT
2971.082	2971.949	33647.95	IR I	1.0			MIT
2969.473	2970.340	33666.18	IR I	1.0			MIT
2968.954	2969.821	33672.06	IR I	1.1	0.7		MIT
2968.487	2969.354	33677.36	IR I	1.1	0.3H		MIT
2968.207	2969.074	33680.54	IR I	1.1	0.7H		MIT
2967.239	2968.105	33691.52	IR I	1.0			MIT
2967.004	2967.870	33694.19	IR	0.8			MIT
2966.127	2966.993	33704.15	IR I	1.0			MIT
2965.202	2966.068	33714.67	IR I	1.4H	0.7		MIT
2964.629	2965.495	33721.19	IR	0.3H			MIT
2964.099	2964.965	33727.21	IR		0.6		MIT
2962.994	2963.859	33739.79	IR I	1.4	0.7		MIT
2962.455	2963.320	33745.93	IR I	1.0			MIT
2961.472	2962.337	33757.13	IR	1.1			MIT
2960.886	2961.751	33763.81	IR I	1.3	0.7		MIT
2959.446	2960.311	33780.24	IR I	0.7			MIT
2957.364	2958.228	33804.02	IR I	1.3			MIT
2956.561	2957.425	33813.20	IR	0.5	0.7		MIT
2954.779	2955.642	33833.59	IR		0.7H		MIT
2952.732	2953.595	33857.05	IR	1.0			MIT
2952.574	2953.437	33858.86	IR	0.3H			MIT
2951.222	2952.084	33874.37	IR I	1.0H	0.5		MIT
2950.766	2951.628	33879.60	IR I	1.1	0.3		MIT
2950.482	2951.344	33882.87	IR I	1.1			MIT
2949.762	2950.624	33891.13	IR I	1.4	1.0		MIT
2949.197	2950.059	33897.63	IR	0.6			MIT
2949.042	2949.904	33899.41	IR	0.6			MIT
2948.432	2949.294	33906.42	IR I	0.7			MIT
2947.367	2948.229	33918.67	IR I	1.1	0.3		MIT
2946.972	2947.833	33923.22	IR I	1.3H	1.0		MIT
2943.87	2944.73	33959.0	IR	0.6D	0.7D		MIT
2943.725	2944.586	33960.64	IR	0.5H			MIT
2943.628	2944.489	33961.75	IR	0.5H			AB

AIR WAVELENGTH (ANGSTRMS)	VACUUM WAVELENGTH (ANGSTRMS)	WAVENUMBER (KAYSERS)	LMENT	LOG ARC	INTENSITY SPK	INTENSITY DIS	REF
2943.260	2944.121	33966.00	IR	0.6			MIT
2943.151	2944.011	33967.26	IR I	1.5	1.3		MIT
2941.079	2941.939	33991.19	IR I	1.3	0.7		MIT
2940.542	2941.402	33997.40	IR I	1.2	1.0		MIT
2940.428	2941.288	33998.71	IR I	0.7			MIT
2939.271	2940.131	34012.10	IR I	1.3	1.2		MIT
2938.757	2939.616	34018.04	IR I	1.0			MIT
2938.469	2939.328	34021.38	IR I	1.3H	1.1		MIT
2937.956	2938.815	34027.32	IR I	0.7			MIT
2937.260	2938.119	34035.38	IR I	0.8			MIT
2936.908	2937.767	34039.46	IR	0.5H			MIT
2936.682	2937.541	34042.08	IR I	1.2H	0.7		MIT
2936.625	2937.484	34042.74	IR	1.6			AB
2936.105	2936.964	34048.77	IR I	1.0	0.0		MIT
2935.192	2936.051	34059.36	IR I	1.1			MIT
2934.638	2935.496	34065.79	IR	1.2J	0.8		MIT
2934.003	2934.861	34073.16	IR I	0.8			MIT
2933.657	2934.515	34077.18	IR I	0.3			MIT
2933.138	2933.996	34083.21	IR I	1.3	0.7		MIT
2930.628	2931.485	34112.40	IR	1.2	0.7		MIT
2930.176	2931.033	34117.66	IR I	1.1	0.7		MIT
2929.751	2930.608	34122.61	IR		0.5H		MIT
2927.015	2927.872	34154.50	IR I	1.0			MIT
2925.113	2925.969	34176.71	IR	0.3H	1.0		MIT
2924.792	2925.648	34180.46	IR I	1.4J	1.2		MIT
2923.851	2924.707	34191.46	IR I	1.0	0.7		MIT
2921.119	2921.974	34223.44	IR		0.3H		MIT
2919.786	2920.641	34239.06	IR	0.5	0.7		MIT
2919.189	2920.044	34246.06	IR	0.5H	1.0		MIT
2918.568	2919.422	34253.35	IR I	1.3	0.3		MIT
2918.035	2918.889	34259.61	IR I	1.1			MIT
2917.770	2918.624	34262.72	IR I	1.0			MIT
2917.229	2918.083	34269.07	IR I	0.3H			MIT
2916.365	2917.219	34279.22	IR I	1.4	0.3		MIT
2915.494	2916.348	34289.46	IR	0.3H			MIT
2909.943	2910.795	34354.87	IR I	0.8			MIT
2909.789	2910.641	34356.69	IR	0.3			MIT
2909.558	2910.410	34359.42	IR I	1.3	0.7		MIT
2907.235	2908.087	34386.87	IR I	1.4	1.0		MIT
2906.156	2907.007	34399.64	IR I	0.5			MIT
2905.638	2906.489	34405.77	IR	1.3	0.6		MIT
2904.799	2905.650	34415.71	IR I	1.4	1.0		MIT
2903.873	2904.724	34426.68	IR I	0.9			MIT
2903.745	2904.596	34428.20	IR I	0.9			MIT
2902.951	2903.802	34437.62	IR I	0.5			MIT
2902.317	2903.167	34445.14	IR I	0.6			MIT
2901.951	2902.801	34449.48	IR I	1.4	1.2		MIT
2901.002	2901.852	34460.75	IR I	0.7			MIT
2900.389	2901.239	34468.03	IR I	1.3	0.7		MIT
2899.629	2900.479	34477.07	IR I	1.4	0.7		MIT
2898.942	2899.792	34485.24	IR I	0.9			MIT
2898.350	2899.199	34492.28	IR I	1.0			MIT
2897.653	2898.502	34500.58	IR	0.5			MIT
2897.152	2898.001	34506.54	IR I	1.3	1.0		MIT
2896.960	2897.809	34508.83	IR I	0.7			MIT
2893.682	2894.530	34547.92	IR I	1.0			MIT
2892.261	2893.109	34564.89	IR	0.7	0.5H		MIT
2890.535	2891.382	34585.53	IR I	0.9			MIT
2889.581	2890.428	34596.95	IR I	1.0	0.7		MIT
2889.234	2890.081	34601.10	IR I	0.7			MIT
2887.125	2887.972	34626.38	IR I	1.0			MIT
2882.635	2883.481	34680.31	IR I	1.6	0.8		MIT
2881.737	2882.582	34691.12	IR	0.5			AB
2881.643	2882.488	34692.25	IR I	0.8			MIT
2881.357	2882.202	34695.69	IR	0.6			MIT
2881.158	2882.003	34698.09	IR	1.2	0.5		MIT
2880.211	2881.056	34709.50	IR	1.0	0.3H		MIT
2880.067	2880.912	34711.23	IR I	0.8			MIT
2879.407	2880.252	34719.19	IR I	1.3	0.7		MIT
2878.514	2879.359	34729.96	IR	0.6			MIT

AIR WAVELENGTH (ANGSTRMS)	VACUUM WAVELENGTH (ANGSTRMS)	WAVENUMBER (KAYSERS)	LMENT	LOG ARC	INTENSITY SPK	DIS	REF
2877.678	2878.522	34740.05	IR I	1.3	1.0		MIT
2875.983	2876.827	34760.52	IR I	1.4	1.2		MIT
2875.605	2876.449	34765.09	IR I	1.4	1.2		MIT
2873.796	2874.639	34786.97	IR I	0.3	0.7		MIT
2873.334	2874.177	34792.56	IR I	1.3	0.5		MIT
2872.111	2872.954	34807.38	IR	0.3	0.5H		MIT
2870.561	2871.404	34826.17	IR I	0.5			AB
2870.220	2871.062	34830.31	IR	0.7			AB
2869.704	2870.546	34836.57	IR I	1.2	0.3		MIT
2867.630	2868.472	34861.77	IR I	1.1	0.3		MIT
2866.689	2867.531	34873.21	IR I	1.3	0.3		MIT
2865.252	2866.093	34890.70	IR	0.9			MIT
2863.844	2864.685	34907.85	IR I	1.2			MIT
2862.485	2863.326	34924.42	IR I	1.0			MIT
2862.349	2863.190	34926.08	IR I	1.0			MIT
2860.665	2861.505	34946.64	IR I	1.1	0.3		MIT
2859.024	2859.864	34966.70	IR	0.6			MIT
2856.941	2857.780	34992.19	IR I	1.0	0.3H		MIT
2855.929	2856.768	35004.59	IR I	1.0			MIT
2855.823	2856.662	35005.89	IR I	1.0	0.3H		MIT
2853.314	2854.152	35036.67	IR I	0.9			MIT
2852.479	2853.317	35046.93	IR	0.3			MIT
2852.131	2852.969	35051.20	IR	1.3			MIT
2851.410	2852.248	35060.07	IR	0.9			MIT
2851.035	2851.873	35064.68	IR I	0.3			MIT
2850.785	2851.623	35067.75	IR		0.7H		MIT
2850.254	2851.092	35074.29	IR I	0.7			MIT
2849.725	2850.562	35080.80	IR I	1.6H	1.3H		MIT
2848.441	2849.278	35096.61	IR	0.3			MIT
2846.646	2847.483	35118.74	IR I	1.0			MIT
2844.852	2845.688	35140.88	IR	0.7			AB
2842.283	2843.119	35172.64	IR I	1.1	0.3		MIT
2841.691	2842.526	35179.97	IR I	0.8	0.3		MIT
2840.219	2841.054	35198.20	IR I	1.2	1.0		MIT
2839.245	2840.080	35210.28	IR I	1.1			MIT
2839.158	2839.993	35211.36	IR I	1.4	1.2		MIT
2837.327	2838.161	35234.08	IR I	1.3	1.0		MIT
2836.404	2837.238	35245.54	IR I	1.4	1.0		MIT
2836.097	2836.931	35249.36	IR I	0.7	0.0		MIT
2835.656	2836.490	35254.84	IR I	1.1			MIT
2835.305	2836.139	35259.20	IR I	0.3			MIT
2833.671	2834.505	35279.54	IR	0.5			AB
2833.236	2834.069	35284.95	IR	0.8	1.3		MIT
2832.774	2833.607	35290.71	IR	0.7			MIT
2831.811	2832.644	35302.71	IR I	0.8	0.7		MIT
2831.360	2832.193	35308.33	IR I	0.8	0.3		MIT
2830.864	2831.697	35314.52	IR I	1.5			MIT
2830.744	2831.577	35316.01	IR I	0.6			AB
2830.509	2831.342	35318.94	IR I	0.9	0.7		MIT
2830.168	2831.001	35323.20	IR I	1.1			MIT
2829.612	2830.445	35330.14	IR I	0.7			MIT
2827.163	2827.995	35360.74	IR I	0.8	0.6		MIT
2826.219	2827.051	35372.55	IR	0.5			AB
2825.503	2826.335	35381.52	IR I	0.9	0.0		MIT
2824.448	2825.279	35394.73	IR I	1.3	1.2		MIT
2823.179	2824.010	35410.64	IR I	1.1	0.7		MIT
2822.561	2823.392	35418.39	IR I	0.6	0.0		MIT
2821.827	2822.658	35427.60	IR	0.3			MIT
2820.632	2821.462	35442.61	IR I	0.6			MIT
2816.936	2817.765	35489.12	IR I	0.9	0.0		MIT
2814.877	2815.706	35515.07	IR I	0.7	0.0		MIT
2814.430	2815.259	35520.71	IR	0.6	0.6		MIT
2812.800	2813.628	35541.30	IR I	1.0	0.3		MIT
2812.09	2812.92	35550.3	IR		0.3H		MIT
2809.847	2810.675	35578.65	IR I	0.5			AB
2807.654	2808.481	35606.44	IR I	0.7	0.0		MIT
2806.396	2807.223	35622.40	IR I	0.8	0.0		MIT
2806.051	2806.878	35626.77	IR I	0.5			MIT
2800.820	2801.645	35693.31	IR I	1.3	0.7		MIT
2799.825	2800.650	35705.99	IR I	0.5			MIT

AIR WAVELENGTH (ANGSTRMS)	VACUUM WAVELENGTH (ANGSTRMS)	WAVENUMBER (KAYSERS)	LMENT	LOG ARC	INTENSITY SPK	DIS	REF
2799.738	2800.563	35707.10	IR I	0.8	0.7		MIT
2799.417	2800.242	35711.20	IR I	0.5			MIT
2799.153	2799.978	35714.57	IR	0.3			MIT
2798.182	2799.007	35726.96	IR I	1.2	0.7		MIT
2797.702	2798.527	35733.09	IR I	1.3	1.0		MIT
2797.35	2798.17	35737.6	IR I				ME
2796.456	2797.280	35749.01	IR I	1.0	0.7		MIT
2794.553	2795.377	35773.35	IR I	0.5			MIT
2794.086	2794.910	35779.33	IR I	0.9			MIT
2793.809	2794.633	35782.88	IR		0.6H		MIT
2790.688	2791.511	35822.89	IR I	0.5	0.3		MIT
2789.805	2790.628	35834.23	IR I	0.5	0.0		MIT
2788.962	2789.785	35845.06	IR	0.3			MIT
2785.224	2786.046	35893.17	IR I	1.4	1.0		MIT
2782.229	2783.050	35931.80	IR I	0.3			MIT
2781.289	2782.110	35943.95	IR I	1.3	1.0		MIT
2780.943	2781.764	35948.42	IR I	0.7	0.7		MIT
2780.409	2781.229	35955.32	IR I	0.5			MIT
2779.654	2780.474	35965.09	IR I	0.5			MIT
2777.877	2778.697	35988.09	IR I	0.6			AB
2777.530	2778.350	35992.59	IR I	0.5			MIT
2777.434	2778.254	35993.83	IR I	0.9	0.7		MIT
2775.554	2776.373	36018.21	IR I	1.1	0.3		MIT
2774.968	2775.787	36025.82	IR	0.3	0.7		MIT
2774.585	2775.404	36030.79	IR I	0.7	0.0		MIT
2773.901	2774.720	36039.67	IR I	0.8	0.7		MIT
2773.594	2774.413	36043.66	IR I	0.6			MIT
2772.456	2773.275	36058.46	IR I	1.2	1.0		MIT
2771.612	2772.430	36069.44	IR I	1.1	0.6		MIT
2768.434	2769.252	36110.84	IR I	0.3			MIT
2767.654	2768.471	36121.02	IR I	0.8	0.7H		MIT
2767.328	2768.145	36125.27	IR	0.3			MIT
2762.028	2762.844	36194.59	IR I	0.3	0.6		MIT
2760.586	2761.402	36213.49	IR	0.6	0.3		MIT
2759.910	2760.726	36222.36	IR I	1.0	0.7		MIT
2759.316	2760.131	36230.16	IR I	1.0	0.7		MIT
2759.002	2759.817	36234.28	IR	0.7	0.7		MIT
2758.478	2759.293	36241.17	IR I	0.5			MIT
2758.226	2759.041	36244.48	IR I	1.0	0.7H		MIT
2756.112	2756.927	36272.27	IR I	0.7	0.7		MIT
2755.739	2756.553	36277.19	IR	0.5			MIT
2752.875	2753.689	36314.92	IR	0.3	0.6H		MIT
2749.323	2750.136	36361.84	IR I	0.8			MIT
2748.294	2749.107	36375.45	IR I	0.3			MIT
2747.513	2748.325	36385.79	IR	0.7	0.0H		MIT
2743.998	2744.810	36432.40	IR I	1.1	0.3		MIT
2743.674	2744.486	36436.70	IR I	0.5			MIT
2743.378	2744.189	36440.63	IR	0.3			MIT
2743.156	2743.967	36443.58	IR		0.3H		MIT
2740.330	2741.141	36481.16	IR	0.0	0.6H		MIT
2740.177	2740.988	36483.20	IR I	0.6			MIT
2740.077	2740.888	36484.53	IR	0.3			MIT
2739.997	2740.808	36485.60	IR I	0.7			MIT
2739.318	2740.129	36494.64	IR I	0.9	0.3H		MIT
2737.309	2738.119	36521.42	IR	1.1	0.7H		MIT
2736.414	2737.224	36533.37	IR I	0.5			MIT
2735.052	2735.861	36551.56	IR I	0.3			MIT
2734.513	2735.322	36558.76	IR I	0.5			MIT
2732.670	2733.479	36583.42	IR	1.3	0.3H		MIT
2732.407	2733.216	36586.94	IR		0.7H		MIT
2731.869	2732.678	36594.14	IR I	0.5			MIT
2730.712	2731.520	36609.65	IR I	1.3	0.7		MIT
2730.407	2731.215	36613.74	IR I	0.3			MIT
2729.557	2730.365	36625.14	IR I	1.1	0.3		MIT
2728.410	2729.218	36640.53	IR I	0.6			MIT
2726.475	2727.282	36666.54	IR I	0.9	0.7		MIT
2726.255	2727.062	36669.50	IR I	0.6			AB
2724.796	2725.603	36689.13	IR	0.6	0.7H		MIT
2723.755	2724.562	36703.15	IR I	1.2	0.5		MIT
2721.001	2721.807	36740.30	IR I	0.5			AB

AIR WAVELENGTH (ANGSTRMS)	VACUUM WAVELENGTH (ANGSTRMS)	WAVENUMBER (KAYSERS)	LMENT	LOG ARC	INTENSITY SPK	DIS	REF
2720.451	2721.257	36747.72	IR I	1.1	0.7		MIT
2719.854	2720.660	36755.79	IR	0.3			MIT
2717.655	2718.460	36785.53	IR I	0.3			MIT
2716.534	2717.339	36800.71	IR	0.8			MIT
2714.550	2715.355	36827.60	IR I	0.6			MIT
2714.103	2714.907	36833.67	IR		1.0H		MIT
2712.740	2713.544	36852.17	IR I	1.6	1.0		MIT
2711.572	2712.376	36868.05	IR		0.8H		MIT
2710.083	2710.886	36888.30	IR I	1.0	0.7		MIT
2708.675	2709.478	36907.48	IR		1.0H		MIT
2707.175	2707.978	36927.92	IR I	0.3			MIT
2706.882	2707.685	36931.92	IR I	0.3			MIT
2705.537	2706.339	36950.28	IR I	1.0	0.3		MIT
2705.360	2706.162	36952.70	IR	0.3			MIT
2705.119	2705.921	36955.99	IR I	0.5	0.9H		MIT
2704.931	2705.733	36958.56	IR I	1.1	0.3		MIT
2704.623	2705.425	36962.77	IR I	0.5			MIT
2704.034	2704.836	36970.82	IR I	1.2			MIT
2701.115	2701.916	37010.77	IR	0.6	0.0		MIT
2698.607	2699.408	37045.16	IR	0.7	0.0		MIT
2695.930	2696.730	37081.95	IR I	0.9	0.3		MIT
2695.468	2696.268	37088.30	IR I	0.7	0.7		MIT
2694.533	2695.333	37101.17	IR I	0.3H			MIT
2694.233	2695.033	37105.30	IR I	2.2H	1.7		MIT
2693.490	2694.289	37115.54	IR I	1.0			MIT
2692.876	2693.675	37124.00	IR I	1.0			MIT
2692.344	2693.143	37131.33	IR I	1.2	0.3		MIT
2692.190	2692.989	37133.46	IR I	0.7			MIT
2691.904	2692.703	37137.40	IR	0.3			MIT
2691.064	2691.863	37149.00	IR I	1.2			MIT
2690.576	2691.375	37155.73	IR		1.0H		MIT
2689.679	2690.478	37168.12	IR I	0.7	0.0		MIT
2684.036	2684.833	37246.26	IR I	1.2			MIT
2683.833	2684.630	37249.08	IR		1.0H		MIT
2682.455	2683.252	37268.21	IR I	1.0	0.3		MIT
2681.096	2681.892	37287.10	IR I	1.0	0.3		MIT
2679.418	2680.214	37310.45	IR I	0.9	0.7		MIT
2679.063	2679.859	37315.40	IR I	1.2	0.5		MIT
2676.828	2677.623	37346.55	IR I	1.5	1.0		MIT
2673.607	2674.402	37391.54	IR I	1.6	1.0		MIT
2672.803	2673.597	37402.79	IR I	0.6			MIT
2671.838	2672.632	37416.30	IR I	1.7	1.0		MIT
2669.913	2670.707	37443.27	IR I	1.8	1.0		MIT
2669.462	2670.256	37449.60	IR I	1.0	0.7		MIT
2668.993	2669.787	37456.18	IR I	1.7	0.7		MIT
2667.458	2668.251	37477.73	IR I	1.0	0.3		MIT
2666.580	2667.373	37490.07	IR I	0.6			MIT
2666.412	2667.205	37492.43	IR I	1.0	0.3		MIT
2665.067	2665.860	37511.35	IR I	0.9			MIT
2664.786	2665.579	37515.31	IR I	2.3H	1.7		MIT
2664.454	2665.246	37519.98	IR		1.0H		MIT
2663.669	2664.461	37531.04	IR I	0.3	0.7H		MIT
2663.314	2664.106	37536.04	IR I	1.0	0.7		MIT
2662.626	2663.418	37545.74	IR I	1.6	1.0		MIT
2661.983	2662.775	37554.81	IR I	2.2H	1.2		MIT
2661.261	2662.053	37565.00	IR	0.3	0.6		MIT
2660.993	2661.785	37568.78	IR	0.7			MIT
2660.076	2660.867	37581.73	IR I	0.5			MIT
2659.946	2660.737	37583.57	IR I	0.3			MIT
2659.144	2659.935	37594.90	IR	0.6	0.0		MIT
2657.712	2658.503	37615.16	IR I	1.0	0.7		MIT
2657.498	2658.289	37618.19	IR	0.5	0.5		MIT
2656.812	2657.603	37627.90	IR I	1.2	0.5		MIT
2654.573	2655.363	37659.63	IR I	0.5			MIT
2653.946	2654.736	37668.53	IR I	1.0	0.7H		MIT
2653.765	2654.555	37671.10	IR I	1.2	0.7H		MIT
2653.029	2653.819	37681.55	IR	0.9	0.7		MIT
2651.768	2652.557	37699.47	IR		0.7		MIT
2646.246	2647.034	37778.13	IR I	1.0	0.7		MIT
2645.410	2646.198	37790.07	IR I	1.0	0.7		MIT

AIR WAVELENGTH (ANGSTRMS)	VACUUM WAVELENGTH (ANGSTRMS)	WAVENUMBER (KAYSERS)	LMENT	LOG ARC	INTENSITY SPK	DIS	REF
2644.186	2644.974	37807.56	IR I	1.5	0.7		MIT
2640.377	2641.164	37862.10	IR I	0.7	0.0		MIT
2639.712	2640.499	37871.64	IR I	2.0H	1.2		MIT
2639.424	2640.210	37875.77	IR I	1.2	0.7		MIT
2638.969	2639.755	37882.30	IR I	0.7			MIT
2638.312	2639.098	37891.73	IR		0.9H		MIT
2637.312	2638.098	37906.10	IR	0.5			MIT
2636.876	2637.662	37912.37	IR I	0.5			MIT
2635.271	2636.056	37935.45	IR I	1.2	0.5		MIT
2634.253	2635.038	37950.11	IR I	1.2	0.7		MIT
2634.167	2634.952	37951.35	IR I	1.0			MIT
2629.409	2630.193	38020.02	IR I	1.0	0.3		MIT
2628.202	2628.986	38037.48	IR I	0.7			MIT
2626.826	2627.609	38057.41	IR I	0.3			MIT
2626.761	2627.544	38058.35	IR	1.3	0.7		MIT
2625.674	2626.457	38074.10	IR I	1.3			MIT
2625.316	2626.099	38079.29	IR I	1.3			MIT
2623.641	2624.424	38103.60	IR	1.0	0.7		MIT
2621.527	2622.309	38134.33	IR I	0.3			MIT
2620.700	2621.482	38146.36	IR I	0.3	0.3H		MIT
2619.883	2620.665	38158.26	IR I	1.5	0.7		MIT
2618.268	2619.049	38181.79	IR I	0.3			MIT
2617.781	2618.562	38188.89	IR I	1.4	1.0		MIT
2617.417	2618.198	38194.21	IR I	0.6			MIT
2615.998	2616.779	38214.92	IR I	1.0	0.3		MIT
2615.881	2616.662	38216.63	IR I	0.9	0.3		MIT
2614.984	2615.765	38229.74	IR I	1.4	0.7		MIT
2614.198	2614.978	38241.23	IR I	1.0	0.3		MIT
2612.259	2613.039	38269.62	IR I	1.0	0.3		MIT
2612.040	2612.820	38272.83	IR I	1.0			MIT
2611.295	2612.075	38283.74	IR I	1.9	1.0		MIT
2610.103	2610.882	38301.23	IR	0.3			MIT
2609.905	2610.684	38304.13	IR I	0.6	0.0H		MIT
2608.907	2609.686	38318.78	IR I	0.3			MIT
2608.246	2609.025	38328.49	IR I	1.7	1.0		MIT
2607.519	2608.298	38339.18	IR I	1.0	0.3		MIT
2604.552	2605.330	38382.85	IR I	1.0	0.3		MIT
2602.035	2602.813	38419.98	IR I	1.2	0.3		MIT
2599.400	2600.177	38458.92	IR	1.6			MIT
2599.040	2599.817	38464.25	IR I	1.4	1.0		MIT
2598.284	2599.061	38475.44	IR I	0.9			MIT
2598.000	2598.777	38479.64	IR I	0.3			MIT
2597.862	2598.639	38481.69	IR	0.5			MIT
2597.353	2598.129	38489.23	IR I	1.1	0.6		MIT
2595.829	2596.605	38511.82	IR	0.5H	1.4H		MIT
2593.134	2593.909	38551.85	IR	0.5			MIT
2592.056	2592.831	38567.88	IR I	2.0	1.3		MIT
2591.845	2592.620	38571.02	IR I	0.5			MIT
2591.646	2592.421	38573.98	IR	0.3			MIT
2591.040	2591.815	38583.00	IR	0.3	0.6		MIT
2590.199	2590.974	38595.53	IR I	0.3			MIT
2589.381	2590.156	38607.72	IR I	0.7			MIT
2589.118	2589.893	38611.64	IR	0.3			MIT
2588.964	2589.738	38613.94	IR I	0.5			MIT
2587.426	2588.200	38636.89	IR I	0.3			MIT
2584.778	2585.551	38676.47	IR I	0.3			MIT
2583.177	2583.950	38700.44	IR I	0.8			MIT
2581.411	2582.184	38726.91	IR I	0.3			MIT
2580.907	2581.680	38734.47	IR	0.3	0.6		MIT
2579.759	2580.531	38751.71	IR I	0.5			MIT
2579.488	2580.260	38755.78	IR		1.2H		MIT
2578.911	2579.683	38764.45	IR I	1.1	0.3		MIT
2578.712	2579.484	38767.44	IR I	0.7	0.0		MIT
2577.529	2578.301	38785.23	IR I	0.5			MIT
2577.265	2578.037	38789.21	IR I	1.8	1.2		MIT
2575.743	2576.514	38812.13	IR I	1.0	0.3		MIT
2575.481	2576.252	38816.07	IR I	0.5	0.3		MIT
2573.218	2573.989	38850.21	IR	0.3			MIT
2572.701	2573.472	38858.02	IR I	1.4	0.7		MIT
2572.440	2573.211	38861.96	IR I	0.3	0.3		MIT

AIR WAVELENGTH (ANGSTRMS)	VACUUM WAVELENGTH (ANGSTRMS)	WAVENUMBER (KAYSERS)	LMENT	LOG ARC	INTENSITY SPK	INTENSITY DIS	REF
2572.367	2573.138	38863.06	IR I	0.9			MIT
2572.069	2572.839	38867.56	IR I	1.0	0.6		MIT
2570.622	2571.392	38889.44	IR I	1.3	0.6		MIT
2569.877	2570.647	38900.71	IR I	1.3	0.5		MIT
2569.660	2570.430	38904.00	IR	0.5			MIT
2568.315	2569.085	38924.37	IR	1.5			MIT
2566.912	2567.681	38945.64	IR	0.6			MIT
2566.752	2567.521	38948.07	IR I	0.3			MIT
2566.347	2567.116	38954.22	IR I	0.3			MIT
2564.827	2565.596	38977.30	IR I	0.3			MIT
2564.558	2565.327	38981.39	IR I	0.5			MIT
2564.177	2564.946	38987.18	IR I	1.6	0.9		MIT
2563.284	2564.052	39000.76	IR I	1.2	0.5		MIT
2562.907	2563.675	39006.50	IR		1.0H		MIT
2561.98	2562.75	39020.6	IR		0.7D		MIT
2561.782	2562.550	39023.63	IR	0.3	0.6H		MIT
2560.539	2561.307	39042.57	IR I	1.0	0.3H		MIT
2560.300	2561.068	39046.21	IR I	0.9	0.3H		MIT
2559.553	2560.321	39057.61	IR I	0.7			MIT
2558.739	2559.506	39070.03	IR I	0.3			MIT
2558.545	2559.312	39073.00	IR	0.3	0.8D		MIT
2557.200	2557.967	39093.55	IR	0.3			MIT
2556.774	2557.541	39100.06	IR I	0.9	0.3H		MIT
2555.884	2556.651	39113.67	IR I	1.3	0.6		MIT
2555.347	2556.114	39121.89	IR I	1.4	0.7		MIT
2554.399	2555.165	39136.41	IR I	1.2	1.2H		MIT
2551.399	2552.165	39182.42	IR I	1.3	0.6		MIT
2550.903	2551.669	39190.04	IR I	0.3			MIT
2550.756	2551.521	39192.30	IR	0.3			MIT
2550.212	2550.977	39200.66	IR I	0.3			MIT
2549.690	2550.455	39208.69	IR I	0.7			MIT
2547.694	2548.459	39239.40	IR I	1.2	0.7		MIT
2547.205	2547.970	39246.94	IR I	1.4	0.7		MIT
2546.034	2546.798	39264.98	IR I	2.0H	1.3		MIT
2545.778	2546.542	39268.93	IR I	0.3			MIT
2545.538	2546.302	39272.64	IR I	1.0	0.3		MIT
2544.561	2545.325	39287.71	IR		0.7		MIT
2543.971	2544.735	39296.83	IR I	2.3H	2.0		MIT
2542.796	2543.560	39314.98	IR	1.0	0.3H		MIT
2542.301	2543.064	39322.64	IR I	0.3			MIT
2542.016	2542.779	39327.04	IR I	1.5	1.0		MIT
2541.475	2542.238	39335.42	IR I	1.1	0.3		MIT
2540.658	2541.421	39348.06	IR I	0.6			MIT
2540.404	2541.167	39352.00	IR I	1.0	0.3		MIT
2539.610	2540.373	39364.30	IR	0.3	0.7		MIT
2538.884	2539.647	39375.56	IR I	0.9			MIT
2538.453	2539.216	39382.24	IR	0.5			MIT
2537.675	2538.437	39394.31	IR I	1.2	0.5		MIT
2537.225	2537.987	39401.30	IR I	1.5	1.0		MIT
2536.665	2537.427	39410.00	IR	0.5			MIT
2536.127	2536.889	39418.36	IR	0.3	0.0		MIT
2534.457	2535.219	39444.33	IR I	2.0	1.0		MIT
2534.006	2534.768	39451.35	IR I	0.3			MIT
2533.131	2533.892	39464.98	IR I	2.0	1.3		MIT
2532.520	2533.281	39474.50	IR I	1.3	0.5		MIT
2532.274	2533.035	39478.33	IR	0.3			MIT
2532.199	2532.960	39479.50	IR I	0.6	0.0		MIT
2530.410	2531.171	39507.41	IR		1.0		MIT
2529.781	2530.542	39517.23	IR I	0.7	0.0		MIT
2529.474	2530.234	39522.03	IR I	1.4	0.7		MIT
2528.392	2529.152	39538.94	IR		1.0H		MIT
2527.938	2528.698	39546.04	IR	0.5			MIT
2527.784	2528.544	39548.45	IR	0.5			MIT
2526.766	2527.526	39564.38	IR I	0.5	0.0		MIT
2525.941	2526.701	39577.30	IR	0.3	0.7		MIT
2525.052	2525.811	39591.24	IR I	1.3	0.5		MIT
2524.876	2525.635	39594.00	IR	0.7	0.0		MIT
2521.212	2521.971	39651.53	IR		0.7		MIT
2521.089	2521.848	39653.47	IR I	0.3			MIT
2518.542	2519.300	39693.57	IR I	0.3			MIT
2517.94	2518.70	39703.1	IR		1.0H		MIT
2515.364	2516.121	39743.71	IR I	1.3	0.5		MIT
2513.708	2514.465	39769.90	IR I	1.2			MIT
2512.651	2513.408	39786.62	IR I	0.5			MIT
2512.580	2513.337	39787.75	IR	0.7	1.2H		MIT
2512.103	2512.859	39795.30	IR		1.4H		MIT
2511.937	2512.693	39797.93	IR	1.3			MIT
2511.157	2511.913	39810.29	IR I	0.3			MIT
2511.052	2511.808	39811.96	IR	0.5			MIT
2509.779	2510.535	39832.15	IR I	0.3			MIT
2509.710	2510.466	39833.24	IR I	1.2	0.3		MIT
2508.351	2509.107	39854.82	IR I	0.9	0.3		MIT
2507.628	2508.383	39866.31	IR I	1.0	0.3		MIT
2506.601	2507.356	39882.65	IR	1.0	0.3		MIT
2505.740	2506.495	39896.35	IR I	1.2	0.3		MIT
2505.229	2505.984	39904.49	IR I	0.7			MIT
2504.369	2505.124	39918.19	IR I	1.0	0.7		MIT
2502.983	2503.737	39940.29	IR I	2.0W	0.7		MIT
2502.628	2503.382	39945.96	IR I	1.0	0.3		MIT
2500.271	2501.025	39983.61	IR I	0.6	0.0H		MIT
2498.891	2499.644	40005.69	IR I	0.3			MIT
2496.271	2497.024	40047.68	IR I	1.0	0.3		MIT
2495.292	2496.045	40063.39	IR	0.6			MIT
2493.080	2493.832	40098.93	IR I	1.3	0.7		MIT
2492.705	2493.457	40104.96	IR	0.3H			MIT
2492.321	2493.073	40111.14	IR I	0.3			MIT
2489.204	2489.955	40161.37	IR I	0.3	0.0		MIT
2488.284	2489.035	40176.21	IR		0.7		MIT
2486.747	2487.498	40201.04	IR I	0.3	0.0		MIT
2486.371	2487.121	40207.12	IR I	0.3	0.0		MIT
2485.382	2486.132	40223.12	IR I	0.8	0.0		MIT
2484.25	2485.00	40241.5	IR		1.0H		MIT
2481.183	2481.932	40291.19	IR I	1.7	1.0		MIT
2480.589	2481.338	40300.84	IR	0.3	0.0		MIT
2479.165	2479.914	40323.98	IR	0.3	0.0		MIT
2478.109	2478.858	40341.17	IR I	0.8	0.0		MIT
2475.122	2475.870	40389.85	IR I	2.0	1.0		MIT
2474.077	2474.825	40406.90	IR	0.3	0.0H		MIT
2472.942	2473.689	40425.45	IR	0.7H	0.5H		MIT
2472.623	2473.370	40430.66	IR I	0.5			MIT
2470.060	2470.807	40472.61	IR I	0.3	0.0		MIT
2469.51	2470.26	40481.6	IR		1.0		MIT
2468.944	2469.690	40490.90	IR I	0.3			MIT
2468.619	2469.365	40496.23	IR I	0.3			MIT
2468.176	2468.922	40503.50	IR I	0.3	0.0		MIT
2467.767	2468.513	40510.22	IR	0.3			AB
2467.51	2468.26	40514.4	IR		0.5H		MIT
2467.302	2468.048	40517.85	IR I	1.6	0.7		MIT
2465.089	2465.835	40554.22	IR I	1.0			MIT
2464.901	2465.646	40557.31	IR I	0.7	0.0		MIT
2464.351	2465.096	40566.37	IR	0.3	0.3		MIT
2463.034	2463.779	40588.06	IR I	0.6	0.0		MIT
2462.364	2463.109	40599.10	IR I	0.5	0.0		MIT
2462.230	2462.975	40601.31	IR I	0.3H			MIT
2461.449	2462.194	40614.19	IR I	0.3H			MIT
2461.209	2461.954	40618.15	IR	0.3	0.9H		MIT
2460.713	2461.458	40626.34	IR I	0.3H			MIT
2458.972	2459.716	40655.10	IR I	0.6			MIT
2457.227	2457.971	40683.97	IR I	1.0	0.5		MIT
2457.030	2457.774	40687.23	IR I	1.1	0.3		MIT
2456.769	2457.513	40691.55	IR	0.3			MIT
2455.868	2456.611	40706.48	IR I	1.0	0.3		MIT
2455.609	2456.352	40710.77	IR I	1.5	0.7		MIT
2454.584	2455.327	40727.77	IR I	0.5	0.0		MIT
2454.120	2454.863	40735.47	IR I	0.6	0.0		MIT
2452.807	2453.550	40757.27	IR I	1.5	0.3		MIT
2452.258	2453.001	40766.40	IR		0.3D		MIT
2449.816	2450.558	40807.03	IR I	0.3			MIT
2449.024	2449.766	40820.23	IR I	0.7	0.0		MIT
2448.229	2448.971	40833.48	IR I	1.0	0.0		MIT

AIR WAVELENGTH (ANGSTRMS)	VACUUM WAVELENGTH (ANGSTRMS)	WAVENUMBER (KAYSERS)	LMENT	LOG ARC	INTENSITY SPK	INTENSITY DIS	REF
2447.763	2448.505	40841.26	IR I	1.0	0.0		MIT
2447.488	2448.229	40845.84	IR I	0.3	0.0		MIT
2446.836	2447.577	40856.73	IR	0.3	0.0		MIT
2445.341	2446.082	40881.70	IR I	1.0	0.6		MIT
2445.092	2445.833	40885.87	IR I	0.6	0.5		MIT
2440.969	2441.709	40954.92	IR I	0.3H			MIT
2440.183	2440.923	40968.11	IR		0.7		MIT
2436.424	2437.163	41031.31	IR I	0.5			MIT
2436.29	2437.03	41033.6	IR		0.5		MIT
2435.140	2435.879	41052.95	IR I	0.9	0.0		MIT
2434.014	2434.752	41071.94	IR	0.3	0.0		MIT
2432.579	2433.317	41096.17	IR I	0.7	0.0H		MIT
2432.355	2433.093	41099.95	IR I	0.7			MIT
2431.938	2432.676	41107.00	IR I	1.7	1.7		MIT
2431.241	2431.979	41118.78	IR I	1.4	0.7		MIT
2429.734	2430.471	41144.28	IR I	0.3	0.0		MIT
2428.360	2429.097	41167.56	IR I	0.3			MIT
2427.963	2428.700	41174.29	IR	0.7			MIT
2427.613	2428.350	41180.23	IR I	1.4	0.3		MIT
2427.086	2427.823	41189.17	IR I	0.3			MIT
2426.882	2427.619	41192.63	IR I	0.3			MIT
2426.776	2427.513	41194.43	IR I	0.3			MIT
2426.532	2427.269	41198.57	IR	0.5	0.0		MIT
2426.310	2427.047	41202.34	IR	0.3	0.6		MIT
2425.77	2426.51	41211.5	IR		0.3		MIT
2425.661	2426.397	41213.36	IR I	1.2			MIT
2424.990	2425.726	41224.77	IR I	1.2	0.3		MIT
2424.886	2425.622	41226.53	IR I	1.0			MIT
2424.660	2425.396	41230.38	IR	1.0	0.9H		MIT
2422.180	2422.916	41272.59	IR	0.3			MIT
2421.214	2421.949	41289.05	IR I	0.3			MIT
2418.553	2419.288	41334.48	IR I	0.3			MIT
2418.106	2418.841	41342.12	IR I	1.5	0.7		MIT
2417.903	2418.638	41345.59	IR		0.7		MIT
2416.240	2416.974	41374.04	IR	0.3	0.0H		MIT
2415.862	2416.596	41380.52	IR I	1.2			MIT
2414.416	2415.150	41405.30	IR I	0.3	0.0H		MIT
2413.310	2414.044	41424.27	IR I	1.4			MIT
2413.097	2413.831	41427.93	IR		0.7H		MIT
2412.809	2413.543	41432.87	IR		0.7H		MIT
2412.442	2413.175	41439.17	IR		0.6		MIT
2410.731	2411.464	41468.58	IR I	0.9	0.0		MIT
2410.469	2411.202	41473.09	IR I	0.3			MIT
2410.167	2410.900	41478.29	IR I	1.2			MIT
2409.371	2410.104	41491.99	IR I	1.2	0.0		MIT
2408.36	2409.09	41509.4	IR		1.2H		MIT
2406.027	2406.759	41549.65	IR I	0.5	0.0H		MIT
2405.865	2406.597	41552.45	IR	0.3			MIT
2403.029	2403.760	41601.48	IR	0.3	0.0		MIT
2402.282	2403.013	41614.42	IR I	0.3			MIT
2401.774	2402.505	41623.22	IR I	0.7	0.0		MIT
2401.168	2401.899	41633.73	IR I	0.0	0.3		MIT
2399.575	2400.306	41661.36	IR	1.2	0.3H		MIT
2398.746	2399.476	41675.76	IR	1.0	2.2		MIT
2398.06	2398.79	41687.7	IR	0.3			MIT
2397.22	2397.95	41702.3	IR	0.5	1.2		MIT
2396.23	2396.96	41719.5	IR		1.0		MIT
2396.094	2396.824	41721.88	IR I	1.3	0.7		MIT
2395.886	2396.616	41725.50	IR I	1.2	0.9		MIT
2394.56	2395.29	41748.6	IR	1.0			MIT
2394.326	2395.055	41752.69	IR	1.5			MIT
2393.93	2394.66	41759.6	IR		1.4		MIT
2393.660	2394.389	41764.31	IR	0.7			MIT
2393.03	2393.76	41775.3	IR	0.5J	1.4		MIT
2392.22	2392.95	41789.4	IR		1.4		MIT
2391.25	2391.98	41806.4	IR		1.4		MIT
2391.178	2391.907	41807.65	IR	1.7			MIT
2391.055	2391.784	41809.80	IR	0.3			MIT
2390.801	2391.530	41814.24	IR	0.7			MIT
2390.617	2391.346	41817.46	IR	1.6	0.8		MIT

AIR WAVELENGTH (ANGSTRMS)	VACUUM WAVELENGTH (ANGSTRMS)	WAVENUMBER (KAYSERS)	LMENT	LOG ARC	INTENSITY SPK	INTENSITY DIS	REF
2390.46	2391.19	41820.2	IR		1.3		MIT
2388.916	2389.644	41847.23	IR		0.5J		MIT
2388.410	2389.138	41856.10	IR	1.0	0.3H		MIT
2387.86	2388.59	41865.7	IR		1.5		MIT
2386.892	2387.620	41882.72	IR	1.7L	1.2		MIT
2386.58	2387.31	41888.2	IR	1.4	1.3		MIT
2386.153	2386.881	41895.69	IR	1.2			MIT
2385.863	2386.590	41900.78	IR	1.3	0.5		MIT
2384.81	2385.54	41919.3	IR	0.6	1.9		MIT
2383.789	2384.516	41937.23	IR	1.2	0.5H		MIT
2383.450	2384.177	41943.20	IR	0.3			MIT
2383.168	2383.895	41948.16	IR	1.0	0.7		MIT
2381.82	2382.55	41971.9	IR	0.9	1.7		MIT
2381.622	2382.348	41975.39	IR	1.5	0.3H		MIT
2379.379	2380.105	42014.95	IR	1.4	0.9		MIT
2378.297	2379.023	42034.07	IR	0.3			MIT
2377.983	2378.709	42039.62	IR	1.5	1.2		MIT
2377.938	2378.664	42040.41	IR	0.6			MIT
2377.275	2378.000	42052.14	IR I	1.4	0.7W		MIT
2377.06	2377.79	42055.9	IR		0.9		MIT
2375.84	2376.57	42077.5	IR		1.0		MIT
2375.58	2376.31	42082.1	IR	1.1			MIT
2375.090	2375.815	42090.82	IR	1.0	1.3		MIT
2373.172	2373.897	42124.83	IR	0.7	1.0J		MIT
2372.774	2373.498	42131.90	IR I	2.0	1.6		MIT
2370.67	2371.39	42169.3	IR		0.8		MIT
2370.372	2371.096	42174.59	IR I	1.2	0.7W		MIT
2369.564	2370.288	42188.97	IR	0.7			MIT
2369.22	2369.94	42195.1	IR		1.5		MIT
2368.409	2369.133	42209.54	IR	0.9			MIT
2368.33	2369.05	42211.0	IR		1.1		MIT
2368.04	2368.76	42216.1	IR	1.4	2.1		MIT
2367.683	2368.406	42222.48	IR	0.6			MIT
2366.06	2366.78	42251.4	IR		0.9		MIT
2365.764	2366.487	42256.73	IR I	1.2	0.5		MIT
2365.29	2366.01	42265.2	IR		0.7		MIT
2364.640	2365.363	42276.81	IR	1.0	0.3		MIT
2364.486	2365.209	42279.57	IR	0.9			MIT
2363.996	2364.719	42288.33	IR	1.0	0.3		MIT
2363.042	2363.764	42305.40	IR	1.7	1.4		MIT
2362.24	2362.96	42319.8	IR		0.6		MIT
2361.67	2362.39	42330.0	IR	0.5	1.4		MIT
2360.730	2361.452	42346.83	IR	1.6	1.4		MIT
2360.684	2361.406	42347.66	IR	0.6			MIT
2359.578	2360.300	42367.50	IR	1.0			MIT
2359.372	2360.093	42371.20	IR	0.3	1.4		MIT
2359.04	2359.76	42377.2	IR		0.5H		MIT
2358.754	2359.475	42382.30	IR		1.4		MIT
2358.165	2358.886	42392.89	IR	1.4	0.3		MIT
2357.954	2358.675	42396.68	IR	0.9	1.7		MIT
2357.31	2358.03	42408.3	IR		1.5		MIT
2356.615	2357.336	42420.77	IR	1.0			MIT
2356.552	2357.273	42421.90	IR	1.6	0.7		MIT
2356.316	2357.037	42426.15	IR	1.2	0.8		MIT
2356.031	2356.752	42431.28	IR	0.6			MIT
2355.96	2356.68	42432.6	IR		1.5		MIT
2355.52	2356.24	42440.5	IR	1.0	1.7		MIT
2355.004	2355.725	42449.79	IR	1.3			MIT
2353.99	2354.71	42468.1	IR		0.7		MIT
2353.640	2354.360	42474.38	IR	1.0	0.7		MIT
2353.117	2353.837	42483.82	IR	0.6	1.7W		MIT
2352.625	2353.345	42492.71	IR	1.2	0.3		MIT
2351.99	2352.71	42504.2	IR	0.7H	1.0J		MIT
2351.406	2352.126	42514.73	IR	1.1			MIT
2350.61	2351.33	42529.1	IR		1.4		MIT
2350.054	2350.773	42539.19	IR	1.2	0.7J		MIT
2349.662	2350.381	42546.29	IR	0.7			MIT
2349.309	2350.028	42552.68	IR	0.7	1.0J		MIT
2348.301	2349.020	42570.94	IR	1.3			MIT
2347.246	2347.965	42590.08	IR	0.6	0.3H		MIT

AIR WAVELENGTH (ANGSTRMS)	VACUUM WAVELENGTH (ANGSTRMS)	WAVENUMBER (KAYSERS)	LMENT	LOG ARC	INTENSITY SPK	DIS	REF
2346.14	2346.86	42610.1	IR	0.5	0.9		MIT
2343.88	2344.60	42651.2	IR I	0.3			AB
2343.659	2344.377	42655.26	IR	0.5			AB
2343.606	2344.324	42656.22	IR I	1.3	0.9		MIT
2343.176	2343.894	42664.05	IR I	1.6	1.0		MIT
2342.973	2343.691	42667.74	IR I	0.7	0.0		MIT
2342.697	2343.415	42672.77	IR I	1.2			MIT
2342.499	2343.217	42676.38	IR I	1.2	0.7		MIT
2341.697	2342.415	42690.99	IR		1.3		MIT
2341.526	2342.243	42694.11	IR		0.7		MIT
2340.40	2341.12	42714.6	IR		0.7		MIT
2340.269	2340.986	42717.04	IR	0.6			MIT
2340.042	2340.759	42721.18	IR	1.0	1.5		MIT
2339.31	2340.03	42734.5	IR		1.5		MIT
2337.535	2338.252	42767.00	IR I	1.0	0.3		MIT
2335.285	2336.001	42808.20	IR I	0.7			MIT
2334.958	2335.674	42814.19	IR		1.4		MIT
2334.505	2335.221	42822.50	IR I	1.4	0.7		MIT
2334.330	2335.046	42825.71	IR I	1.0	0.3		MIT
2333.839	2334.555	42834.72	IR I	1.6	1.0		MIT
2333.298	2334.014	42844.65	IR I	1.4	0.9		MIT
2332.239	2332.954	42864.10	IR	0.6	1.5		MIT
2330.768	2331.483	42891.15	IR	0.5			MIT
2330.50	2331.22	42896.1	IR		1.7		MIT
2329.82	2330.53	42908.6	IR	0.7J	0.9		MIT
2329.415	2330.130	42916.06	IR	0.7	1.7		MIT
2328.98	2329.69	42924.1	IR	0.5	1.2		MIT
2328.714	2329.429	42928.98	IR I	0.7			MIT
2328.543	2329.258	42932.13	IR I	1.2	0.0H		MIT
2328.248	2328.963	42937.57	IR	1.0	0.3H		AB
2327.981	2328.695	42942.50	IR	1.4	0.7H		MIT
2327.256	2327.970	42955.87	IR	0.0H	1.5		AB
2326.946	2327.660	42961.60	IR	0.7			AB
2326.050	2326.764	42978.14	IR	0.8	1.7		MIT
2325.82	2326.53	42982.4	IR	0.6			MIT
2325.321	2326.035	42991.62	IR	0.7			MIT
2324.941	2325.655	42998.64	IR	1.0			MIT
2324.705	2325.419	43003.01	IR	0.7	1.9		MIT
2324.11	2324.82	43014.0	IR		1.5		MIT
2323.63	2324.34	43022.9	IR		1.7L		MIT
2322.73	2323.44	43039.6	IR		1.0		MIT
2322.31	2323.02	43047.4	IR		1.5		MIT
2321.73	2322.44	43058.1	IR		0.7		MIT
2321.580	2322.293	43060.89	IR	1.1			MIT
2321.451	2322.164	43063.28	IR	1.0S	0.9		MIT
2321.371	2322.084	43064.76	IR	0.3			MIT
2319.313	2320.026	43102.97	IR	1.4	0.9		MIT
2318.320	2319.032	43121.43	IR	0.5			AB
2318.136	2318.848	43124.86	IR	1.1			MIT
2317.900	2318.612	43129.25	IR	1.1			AB
2317.442	2318.154	43137.77	IR	0.3	1.4		MIT
2315.378	2316.090	43176.22	IR	1.5			MIT
2315.146	2315.858	43180.55	IR	0.3	1.6		AB
2314.90	2315.61	43185.1	IR	1.0	1.6		MIT
2314.111	2314.822	43199.86	IR	0.3	1.6		MIT
2312.51	2313.22	43229.8	IR	0.7	0.3		MIT
2311.207	2311.918	43254.13	IR	0.5			AB
2310.533	2311.244	43266.75	IR	0.3	1.2		AB
2309.656	2310.366	43283.18	IR	1.0			MIT
2309.57	2310.28	43284.8	IR		0.7H		MIT
2308.934	2309.644	43296.71	IR	1.3	0.7		AB
2306.77	2307.48	43337.3	IR		1.5		MIT
2305.466	2306.176	43361.83	IR I	1.5	0.7L		MIT
2304.215	2304.924	43385.37	IR I	2.0			MIT
2303.582	2304.291	43397.29	IR	0.9			MIT
2303.389	2304.098	43400.93	IR I	1.3	0.0		MIT
2303.076	2303.785	43406.83	IR		0.5H		AB
2302.409	2303.118	43419.40	IR I	1.2			MIT
2301.675	2302.384	43433.25	IR		1.4J		MIT
2300.908	2301.617	43447.72	IR	0.9	1.0		MIT
2300.64	2301.35	43452.8	IR		0.7		MIT
2300.499	2301.207	43455.45	IR I	1.5	1.4		MIT
2300.342	2301.050	43458.41	IR I	0.5			MIT
2299.67	2300.38	43471.1	IR		1.0		A
2299.526	2300.234	43473.83	IR	1.4	0.5		MIT
2298.252	2298.960	43497.93	IR	0.5	1.5		AB
2298.159	2298.867	43499.69	IR I	1.1			MIT
2298.048	2298.756	43501.79	IR I	1.0			MIT
2297.559	2298.267	43511.05	IR	0.7			MIT
2297.138	2297.846	43519.02	IR	0.3	1.3		MIT
2296.209	2296.916	43536.63	IR I	1.4			MIT
2296.000	2296.707	43540.59	IR I	1.2	0.5		MIT
2295.56	2296.27	43548.9	IR		0.6		MIT
2295.084	2295.791	43557.97	IR I	1.6	0.7		MIT
2293.279	2293.986	43592.25	IR I	0.9			MIT
2293.02	2293.73	43597.2	IR	0.5W			A
2291.78	2292.49	43620.8	IR		0.7		MIT
2291.655	2292.361	43623.14	IR	1.0			MIT
2290.80	2291.51	43639.4	IR	0.5	1.7		MIT
2290.450	2291.156	43646.08	IR	0.5			MIT
2289.393	2290.099	43666.23	IR	0.7	1.6		MIT
2288.920	2289.626	43675.26	IR	0.3			MIT
2288.09	2288.80	43691.1	IR		0.7W		MIT
2287.878	2288.584	43695.15	IR	1.3	0.7		MIT
2286.221	2286.926	43726.81	IR	1.1			MIT
2285.74	2286.45	43736.0	IR		1.0W		MIT
2285.465	2286.170	43741.28	IR I	0.9			MIT
2284.605	2285.310	43757.74	IR	1.3	0.7H		MIT
2282.496	2283.200	43798.17	IR	0.9			MIT
2282.171	2282.875	43804.40	IR I	0.5			AB
2281.907	2282.611	43809.47	IR I	1.4	0.5		MIT
2281.021	2281.725	43826.49	IR	0.3	1.7		MIT
2280.574	2281.278	43835.08	IR	1.5	1.3J		AB
2279.999	2280.703	43846.13	IR	1.3	0.6		MIT
2279.622	2280.326	43853.38	IR	0.5J	0.3J		MIT
2278.49	2279.19	43875.2	IR		1.4		MIT
2277.873	2278.576	43887.05	IR	1.0	1.2		MIT
2277.485	2278.188	43894.53	IR	1.0			MIT
2275.66	2276.36	43929.7	IR		1.4		MIT
2273.092	2273.794	43979.35	IR	0.7	0.6		AB
2272.51	2273.21	43990.6	IR		1.5		MIT
2271.39	2272.09	44012.3	IR		1.6		MIT
2270.019	2270.721	44038.88	IR I	1.3	0.5		MIT
2269.598	2270.300	44047.05	IR	1.1	0.5		AB
2268.898	2269.600	44060.64	IR I	1.4	1.5		AB
2268.40	2269.10	44070.3	IR	1.3			MIT
2267.71	2268.41	44083.7	IR		0.7		MIT
2266.929	2267.630	44098.90	IR I	0.7			MIT
2266.634	2267.335	44104.64	IR	0.3			AB
2266.331	2267.032	44110.54	IR I	1.4	1.0		MIT
2265.454	2266.155	44127.61	IR I	1.1	0.5		MIT
2265.163	2265.864	44133.28	IR	0.7	1.7		MIT
2264.607	2265.308	44144.12	IR I	1.5	1.2		MIT
2264.172	2264.873	44152.60	IR I	1.1	0.5		MIT
2263.970	2264.670	44156.53	IR	0.6	1.0		AB
2263.804	2264.504	44159.77	IR I	1.0			AB
2262.43	2263.13	44186.6	IR		0.8		MIT
2260.65	2261.35	44221.4	IR	0.5	1.6		MIT
2259.27	2259.97	44248.4	IR		1.5		MIT
2258.856	2259.555	44256.49	IR I	1.5	1.0		MIT
2258.508	2259.207	44263.31	IR I	1.2	1.7		MIT
2257.50	2258.20	44283.1	IR	0.3	1.7		MIT
2256.99	2257.69	44293.1	IR		0.7H		MIT
2256.849	2257.548	44295.85	IR I	0.8			MIT
2256.752	2257.451	44297.75	IR I	0.3			MIT
2256.43	2257.13	44304.1	IR		1.0		MIT
2256.227	2256.926	44308.06	IR	0.7			AB
2255.810	2256.509	44316.25	IR I	1.4	1.0		MIT
2255.625	2256.324	44319.88	IR I	1.0			AB
2255.594	2256.293	44320.49	IR I	1.0			MIT

AIR WAVELENGTH (ANGSTRMS)	VACUUM WAVELENGTH (ANGSTRMS)	WAVENUMBER (KAYSERS)	LMENT		LOG ARC	INTENSITY SPK	INTENSITY DIS	REF
2255.101	2255.800	44330.18	IR	I	1.4	1.0		MIT
2254.91	2255.61	44333.9	IR			1.0		MIT
2254.47	2255.17	44342.6	IR		0.3H	1.0		MIT
2253.487	2254.185	44361.93	IR		1.0	0.5		MIT
2253.375	2254.073	44364.13	IR		0.6			AB
2253.23	2253.93	44367.0	IR			1.4		MIT
2253.040	2253.738	44370.73	IR	I	1.0	0.3		MIT
2252.298	2252.996	44385.34	IR		0.7			MIT
2251.223	2251.921	44406.54	IR		0.3			MIT
2249.406	2250.103	44442.40	IR	I	0.6	1.0		MIT
2247.66	2248.36	44476.9	IR		0.0	1.3		MIT
2246.90	2247.60	44492.0	IR		1.0	2.0		MIT
2245.76	2246.46	44514.5	IR		1.0	2.2		MIT
2245.502	2246.199	44519.66	IR		0.6			MIT
2244.932	2245.628	44530.96	IR		1.0	0.3H		MIT
2244.23	2244.93	44544.9	IR			1.0		MIT
2242.68	2243.38	44575.7	IR		1.7	2.5		MIT
2242.351	2243.047	44582.22	IR		0.9			MIT
2242.24	2242.94	44584.4	IR			1.2		MIT
2240.96	2241.66	44609.9	IR		0.7	0.5H		A
2240.64	2241.34	44616.3	IR			1.2		MIT
2240.423	2241.118	44620.58	IR		1.3	1.5		MIT
2238.818	2239.513	44652.56	IR	I	1.5	0.7		MIT
2238.68	2239.38	44655.3	IR			1.1		MIT
2238.288	2238.983	44663.14	IR			1.9		MIT
2237.43	2238.12	44680.3	IR		0.3			MIT
2237.09	2237.78	44687.0	IR			2.0		MIT
2237.038	2237.733	44688.09	IR		0.3			MIT
2237.001	2237.696	44688.83	IR		1.0			MIT
2235.750	2236.444	44713.83	IR		1.4	0.3H		MIT
2235.289	2235.983	44723.05	IR		1.4	0.6		MIT
2234.86	2235.55	44731.6	IR			1.0		MIT
2234.38	2235.07	44741.2	IR		0.7	1.9		MIT
2234.154	2234.848	44745.77	IR	I	1.3			MIT
2233.370	2234.064	44761.48	IR	I	1.0	2.0		AB
2232.471	2233.165	44779.50	IR	I	1.4			AB
2232.248	2232.942	44783.97	IR		1.5	1.7		AB
2232.179	2232.873	44785.36	IR	I	0.7			AB
2231.657	2232.351	44795.83	IR	I	1.4	0.7		MIT
2230.99	2231.68	44809.2	IR			1.4		MIT
2230.441	2231.134	44820.25	IR		1.3	1.3L		AB
2229.780	2230.473	44833.54	IR	I	1.3	0.7L		MIT
2229.53	2230.22	44838.6	IR			0.8		MIT
2228.671	2229.364	44855.84	IR		1.0	0.3		AB
2224.80	2225.49	44933.9	IR			1.4		MIT
2223.653	2224.345	44957.06	IR	I	0.8	0.5		AB
2223.099	2223.791	44968.26	IR	I	1.0	0.6		MIT
2222.969	2223.661	44970.89	IR		0.3			AB
2221.07	2221.76	45009.3	IR		0.3	2.0		MIT
2220.373	2221.064	45023.46	IR	I	1.7	1.0		MIT
2220.182	2220.873	45027.34	IR		0.9	0.3H		AB
2219.21	2219.90	45047.1	IR		0.3	1.3		MIT
2218.739	2219.430	45056.62	IR	I	0.9	0.3		AB
2217.739	2218.430	45076.93	IR		1.0	1.3		AB
2217.05	2217.74	45090.9	IR			0.7		MIT
2216.031	2216.721	45111.67	IR	I	1.4	0.7		MIT
2215.741	2216.431	45117.58	IR	I	0.9			AB
2214.305	2214.995	45146.83	IR	I	1.0	0.3		MIT
2213.77	2214.46	45157.7	IR			1.4		MIT
2212.32	2213.01	45187.3	IR			1.7		MIT
2211.25	2211.94	45209.2	IR		0.3J	0.6		MIT
2208.96	2209.65	45256.1	IR		0.3	1.7		MIT
2208.72	2209.41	45261.0	IR			1.6		MIT
2208.088	2208.777	45273.93	IR		1.6	0.5J		AB
2205.68	2206.37	45323.4	IR		1.4	0.7		MIT
2205.122	2205.810	45334.82	IR	I	1.0			AB
2204.961	2205.649	45338.13	IR		0.7	2.5		MIT
2203.68	2204.37	45364.5	IR			1.4		MIT
2198.854	2199.541	45464.04	IR		1.7	1.2		AB
2197.50	2198.19	45492.0	IR		0.7	2.0		MIT

AIR WAVELENGTH (ANGSTRMS)	VACUUM WAVELENGTH (ANGSTRMS)	WAVENUMBER (KAYSERS)	LMENT		LOG ARC	INTENSITY SPK	INTENSITY DIS	REF
2196.44	2197.13	45514.0	IR		0.7	1.7		MIT
2195.744	2196.430	45528.43	IR	I	1.4			AB
2194.27	2194.96	45559.0	IR			1.6		MIT
2192.22	2192.91	45601.6	IR			1.4W		MIT
2191.64	2192.33	45613.7	IR		1.5	0.7		MIT
2191.351	2192.036	45619.69	IR	I	1.1	0.7		MIT
2190.38	2191.06	45639.9	IR			2.2		MIT
2187.427	2188.111	45701.52	IR	I	1.3	1.3W		AB
2184.66	2185.34	45759.4	IR			1.2		MIT
2182.999	2183.682	45794.21	IR		1.2			AB
2182.83	2183.51	45797.7	IR			1.4		MIT
2182.349	2183.032	45807.85	IR		1.2	1.2		AB
2180.59	2181.27	45844.8	IR		0.5			MIT
2178.98	2179.66	45878.7	IR		1.0	1.7		MIT
2178.171	2178.853	45895.70	IR	I	1.4	0.7		MIT
2175.245	2175.927	45957.43	IR	I	1.5	0.6		MIT
2175.014	2175.696	45962.31	IR			1.7		MIT
2172.21	2172.89	46021.6	IR		1.5	1.3		MIT
2170.046	2170.727	46067.52	IR		1.0	0.5		AB
2169.76	2170.44	46073.6	IR			0.5		MIT
2169.42	2170.10	46080.8	IR		0.3	1.7		MIT
2166.48	2167.16	46143.3	IR			0.5		MIT
2163.70	2164.38	46202.6	IR			1.0		MIT
2163.314	2163.993	46210.87	IR		1.3	1.0		MIT
2162.92	2163.60	46219.3	IR			1.5		MIT
2162.88	2163.56	46220.1	IR		1.4	0.0		MIT
2161.38	2162.06	46252.2	IR		1.2	1.4		MIT
2161.02	2161.70	46259.9	IR		1.2	0.5		MIT
2160.74	2161.42	46265.9	IR		1.0	1.7		MIT
2160.02	2160.70	46281.3	IR		1.0	1.0W		A
2159.381	2160.059	46295.02	IR		0.3			AB
2158.40	2159.08	46316.1	IR		0.7	1.0		MIT
2158.05	2158.73	46323.6	IR		1.7	1.7		MIT
2155.99	2156.67	46367.8	IR		0.6	0.7		MIT
2155.81	2156.49	46371.7	IR		1.4	1.0		MIT
2154.59	2155.27	46398.0	IR			1.5		MIT
2154.18	2154.86	46406.8	IR		1.2	0.7		MIT
2154.02	2154.70	46410.2	IR		1.0	0.3		A
2152.68	2153.36	46439.1	IR		1.7	2.3		MIT
2151.62	2152.30	46462.0	IR		0.3W	1.9		MIT
2150.54	2151.22	46485.3	IR		1.3	0.6H		MIT
2148.97	2149.65	46519.3	IR		1.0			A
2148.22	2148.90	46535.5	IR		1.4	1.7		MIT
2147.74	2148.42	46545.9	IR		0.3	1.0		A
2144.58	2145.26	46614.5	IR			1.2		A
2144.28	2144.96	46621.0	IR		1.0	0.5		MIT
2144.07	2144.75	46625.6	IR			0.7		MIT
2143.20	2143.88	46644.5	IR			0.9		A
2142.82	2143.50	46652.8	IR		0.5	0.0		MIT
2142.25	2142.92	46665.2	IR		1.3	1.0		MIT
2141.65	2142.32	46678.3	IR		0.0H	0.6		A
2141.11	2141.78	46690.0	IR		0.7W			A
2140.53	2141.20	46702.7	IR		0.5	1.0		A
2140.08	2140.75	46712.5	IR		0.7			A
2139.78	2140.45	46719.0	IR			1.2		A
2139.56	2140.23	46723.9	IR		0.6			A
2139.22	2139.89	46731.3	IR		0.6			A
2138.57	2139.24	46745.5	IR		1.2			A
2137.02	2137.69	46779.4	IR		0.9			A
2135.14	2135.81	46820.6	IR		0.7	0.0H		A
2134.68	2135.35	46830.6	IR		0.6			A
2133.72	2134.39	46851.7	IR		0.6	1.3		A
2133.37	2134.04	46859.4	IR		0.6	0.5H		A
2132.95	2133.62	46868.6	IR		0.7	0.3		MIT
2132.56	2133.23	46877.2	IR		0.5	1.9		MIT
2132.29	2132.96	46883.1	IR		0.6W			A
2131.66	2132.33	46897.0	IR			1.7		MIT
2131.33	2132.00	46904.2	IR		0.7	0.7		A
2131.04	2131.71	46910.6	IR		0.7	0.3W		A
2130.45	2131.12	46923.6	IR		0.7	1.9		MIT

AIR WAVELENGTH (ANGSTRMS)	VACUUM WAVELENGTH (ANGSTRMS)	WAVENUMBER (KAYSERS)	LMENT	LOG ARC	INTENSITY SPK	DIS	REF
2130.09	2130.76	46931.6	IR	0.6			A
2129.85	2130.52	46936.8	IR	1.0	0.7J		A
2129.46	2130.13	46945.4	IR	0.6	1.0		A
2128.62	2129.29	46964.0	IR	0.3	0.7		A
2127.94	2128.61	46979.0	IR	1.3	1.2		MIT
2127.52	2128.19	46988.2	IR	1.2			MIT
2127.26	2127.93	46994.0	IR	0.7	0.9		A
2126.81	2127.48	47003.9	IR	1.4	2.3		MIT
2126.24	2126.91	47016.5	IR	0.6			A
2125.97	2126.64	47022.5	IR	0.7			A
2125.44	2126.11	47034.2	IR	1.4	1.0		A
2125.15	2125.82	47040.6	IR	0.6	0.8		A
2123.92	2124.59	47067.9	IR	0.6			A
2123.64	2124.31	47074.1	IR	0.3	0.3		MIT
2123.21	2123.88	47083.6	IR	1.0			A
2121.88	2122.55	47113.1	IR	1.1	0.6		A
2121.68	2122.35	47117.6	IR	0.6	0.7		A
2121.25	2121.92	47127.1	IR		0.7		A
2120.87	2121.54	47135.6	IR	0.9			A
2119.54	2120.21	47165.1	IR	0.7	1.2		A
2119.34	2120.01	47169.6	IR		0.9		A
2119.12	2119.79	47174.5	IR	0.6			A
2118.76	2119.43	47182.5	IR	0.6	0.3		MIT
2118.52	2119.19	47187.8	IR		1.3		A
2118.14	2118.81	47196.3	IR	0.3	0.5		A
2117.72	2118.39	47205.7	IR	1.0	1.7		A
2116.66	2117.33	47229.3	IR	1.0			A
2115.97	2116.64	47244.7	IR		1.0		A
2115.00	2115.67	47266.4	IR		0.9		A
2114.09	2114.76	47286.7	IR	0.9	0.8		A
2113.83	2114.50	47292.5	IR		0.9		A
2112.68	2113.35	47318.3	IR	1.3	1.0		A
2112.45	2113.12	47323.4	IR	0.6	1.2		A
2111.96	2112.63	47334.4	IR	0.3	1.6		MIT
2111.34	2112.01	47348.3	IR	0.7	0.3H		A
2110.98	2111.65	47356.4	IR	0.3	0.7R		A
2110.68	2111.35	47363.1	IR	1.1	1.0		A
2109.38	2110.05	47392.3	IR	0.5H	1.7		A
2108.80	2109.47	47405.3	IR	0.5			A
2108.62	2109.29	47409.4	IR	1.2			A
2107.60	2108.27	47432.3	IR	0.5			A
2107.22	2107.89	47440.9	IR		1.2		A
2106.92	2107.59	47447.6	IR	1.3	1.0L		A
2106.69	2107.36	47452.8	IR	0.6			A
2106.35	2107.02	47460.4	IR	1.3	1.3W		A
2106.08	2106.75	47466.5	IR		1.2		A
2104.66	2105.33	47498.5	IR		1.3		A
2104.33	2105.00	47506.0	IR	0.3	1.2		A
2103.87	2104.54	47516.4	IR		1.7		A
2103.58	2104.25	47522.9	IR	1.0	0.5		A
2102.85	2103.52	47539.4	IR	0.7			A
2102.45	2103.12	47548.5	IR	1.2	0.8		A
2101.86	2102.53	47561.8	IR	1.3	0.7		A
2101.63	2102.30	47567.0	IR	0.5			A
2100.96	2101.63	47582.2	IR	0.3H	1.7		A
2100.20	2100.87	47599.4	IR	1.3	1.3		A
2099.97	2100.64	47604.6	IR	1.2	0.6		A
2099.62	2100.29	47612.5	IR		1.6		A
2099.37	2100.04	47618.2	IR	0.7W	0.5		A
2097.55	2098.22	47659.5	IR	0.7			A
2097.30	2097.97	47665.2	IR		1.4		A
2097.10	2097.77	47669.8	IR	0.6	1.7		A
2096.70	2097.37	47678.9	IR		0.6		MIT
2096.20	2096.87	47690.2	IR	0.0	1.9		A
2095.04	2095.71	47716.6	IR		1.2		A
2094.85	2095.52	47721.0	IR	0.5			A
2094.18	2094.85	47736.2	IR	0.9	1.2		A
2094.05	2094.72	47739.2	IR		0.5		MIT
2093.97	2094.64	47741.0	IR		1.0		A
2092.77	2093.44	47768.4	IR	1.2	0.7		A
2092.63	2093.30	47771.6	IR	1.3	1.3		MIT
2091.86	2092.53	47789.1	IR	1.0	1.3S		MIT
2091.59	2092.25	47795.3	IR	0.9	0.0		A
2091.44	2092.10	47798.7	IR		1.1		A
2090.80	2091.46	47813.4	IR		1.5		MIT
2090.52	2091.18	47819.8	IR	0.9	0.8		A
2090.32	2090.98	47824.4	IR		1.4		MIT
2089.93	2090.59	47833.3	IR		0.7		MIT
2088.82	2089.48	47858.7	IR	1.7	1.7		MIT
2088.51	2089.17	47865.8	IR	0.0H	1.3		A
2086.92	2087.58	47902.3	IR	0.5	1.3		A
2086.80	2087.46	47905.0	IR	1.0	0.5		A
2086.58	2087.24	47910.1	IR		1.3		A
2086.06	2086.72	47922.0	IR	0.6	0.0		A
2085.74	2086.40	47929.4	IR	1.5	1.3		A
2084.49	2085.15	47958.1	IR		1.9		A
2084.38	2085.04	47960.6	IR	0.7			A
2084.22	2084.88	47964.3	IR		1.1		A
2083.85	2084.51	47972.8	IR	1.2	0.7H		A
2083.43	2084.09	47982.5	IR		1.4		MIT
2083.22	2083.88	47987.3	IR	1.4	1.0		A
2082.79	2083.45	47997.2	IR	0.0	1.1		A
2082.22	2082.88	48010.4	IR	0.8	0.7H		A
2081.83	2082.49	48019.4	IR	0.3	1.0		A
2080.95	2081.61	48039.7	IR	0.9	1.2		MIT
2080.64	2081.30	48046.8	IR	0.3	1.8		MIT
2079.71	2080.37	48068.3	IR		1.6		MIT
2079.33	2079.99	48077.1	IR	1.2	0.7		A
2078.88	2079.54	48087.5	IR		1.4		A
2078.37	2079.03	48099.3	IR		1.4		A
2077.67	2078.33	48115.5	IR		1.1		A
2076.76	2077.42	48136.6	IR	0.3	1.6		MIT
2076.61	2077.27	48140.1	IR	0.7			A
2076.42	2077.08	48144.5	IR	0.3	0.7		A
2076.07	2076.73	48152.6	IR		1.3		A
2075.90	2076.56	48156.5	IR	0.9	0.5		A
2075.43	2076.09	48167.4	IR		1.4		MIT
2075.05	2075.71	48176.2	IR	0.8	1.2		A
2074.59	2075.25	48186.9	IR	1.0	0.6		A
2074.33	2074.99	48193.0	IR	0.9			A
2074.15	2074.81	48197.1	IR	0.6			A
2073.70	2074.36	48207.6	IR	0.6J			A
2072.96	2073.62	48224.8	IR	1.3	1.0		A
2072.29	2072.95	48240.4	IR	0.6	0.3		A
2071.94	2072.60	48248.5	IR	1.2	0.9		A
2071.81	2072.47	48251.6	IR	0.8	0.6		A
2070.51	2071.17	48281.9	IR	0.8	1.8		MIT
2070.21	2070.87	48288.9	IR	0.3W	0.7		A
2069.48	2070.14	48305.9	IR		1.0		A
2069.07	2069.73	48315.5	IR	1.0	0.9		A
2068.73	2069.39	48323.4	IR	0.5W	0.3H		A
2068.25	2068.91	48334.6	IR	1.0	0.3		A
2067.87	2068.53	48343.5	IR	0.7	0.6D		A
2067.17	2067.83	48359.9	IR		1.5		A
2066.98	2067.64	48364.3	IR	0.6			A
2066.21	2066.87	48382.3	IR		1.6		A
2066.00	2066.66	48387.2	IR	0.6			A
2065.79	2066.45	48392.2	IR		1.9		MIT
2065.59	2066.25	48396.9	IR	0.5	1.1		A
2065.18	2065.84	48406.5	IR	0.5W			A
2065.09	2065.75	48408.6	IR		1.4		A
2064.77	2065.43	48416.1	IR		1.2		A
2064.55	2065.21	48421.2	IR	0.7	0.7		A
2064.38	2065.04	48425.2	IR	0.6			A
2063.89	2064.55	48436.7	IR		1.4		A
2063.59	2064.25	48443.7	IR	0.7	1.2		A
2063.37	2064.03	48448.9	IR	1.1	0.6		A
2063.03	2063.69	48456.9	IR		1.9		A
2062.54	2063.20	48468.4	IR	0.3H	0.3		MIT
2062.28	2062.94	48474.5	IR	0.3	0.7		A

AIR WAVELENGTH (ANGSTRMS)	VACUUM WAVELENGTH (ANGSTRMS)	WAVENUMBER (KAYSERS)	LMENT	LOG ARC	INTENSITY SPK	DIS	REF
2061.99	2062.65	48481.3	IR	0.6	0.3		MIT
2061.86	2062.52	48484.4	IR	0.3	0.7J		A
2061.47	2062.13	48493.6	IR	1.0	0.6		A
2060.90	2061.56	48507.0	IR	0.7	0.3		A
2060.64	2061.30	48513.1	IR	1.3	1.2		A
2060.09	2060.75	48526.0	IR	0.6	0.3H		A
2059.70	2060.36	48535.2	IR	1.0	1.2		A
2058.93	2059.59	48553.4	IR	0.6W			A
2058.05	2058.71	48574.1	IR		1.0		A
2057.93	2058.59	48577.0	IR	1.0	0.6		A
2057.78	2058.44	48580.5	IR	0.9	0.3		A
2057.63	2058.29	48584.0	IR	0.3	1.2		A
2057.23	2057.89	48593.5	IR	0.8	1.7		A
2057.01	2057.67	48598.7	IR	0.5W	0.6		MIT
2056.73	2057.39	48605.3	IR	0.5			A
2056.52	2057.18	48610.3	IR	0.6W			A
2056.40	2057.06	48613.1	IR	0.3			A
2056.17	2056.83	48618.5	IR		1.3		A
2055.72	2056.38	48629.2	IR	1.2	1.4		A
2055.44	2056.10	48635.8	IR	1.0	0.6		A
2054.77	2055.43	48651.7	IR	0.3	1.5		A
2053.87	2054.53	48673.0	IR	0.3	1.4		A
2053.72	2054.38	48676.5	IR	0.6			A
2052.22	2052.88	48712.1	IR	1.1	1.0		A
2051.16	2051.82	48737.3	IR	1.0	1.7		MIT
2050.72	2051.38	48747.7	IR	0.7	0.0		A
2050.49	2051.15	48753.2	IR	0.7			A
2050.22	2050.88	48759.6	IR		1.4		MIT
2049.56	2050.22	48775.3	IR	0.3	1.5		MIT
2048.82	2049.48	48792.9	IR	0.5			A
2048.55	2049.21	48799.4	IR		1.0		A
2047.96	2048.62	48813.4	IR	0.5	1.0		A
2047.74	2048.40	48818.7	IR	0.5			A
2047.57	2048.23	48822.7	IR	0.7	0.0H		A
2047.22	2047.88	48831.1	IR	0.8	0.6H		A
2046.68	2047.34	48844.0	IR	0.5	0.3H		A
2046.18	2046.84	48855.9	IR		0.5		MIT
2045.88	2046.54	48863.0	IR	0.7	1.2		A
2045.48	2046.14	48872.6	IR	1.0	0.9		A
2045.31	2045.97	48876.7	IR	1.0	0.5		A
2044.83	2045.49	48888.1	IR	0.6	1.9		MIT
2044.19	2044.85	48903.4	IR	1.0	2.0		MIT
2044.03	2044.69	48907.3	IR	1.3			A
2043.21	2043.87	48926.9	IR	1.0	0.3		A
2042.96	2043.62	48932.9	IR	0.6	1.0		A
2042.13	2042.79	48952.8	IR	1.1			A
2041.69	2042.35	48963.3	IR	0.3	1.7		MIT
2041.17	2041.83	48975.8	IR	0.3	1.0		A
2040.76	2041.42	48985.6	IR	0.8	1.3		A
2040.48	2041.14	48992.3	IR	0.3	0.7W		A
2039.79	2040.45	49008.9	IR	0.9	1.6		A
2039.43	2040.09	49017.6	IR	1.4	1.1		A
2039.14	2039.80	49024.5	IR	0.7	0.7		A
2038.99	2039.65	49028.1	IR	0.0	1.4		A
2038.60	2039.25	49037.5	IR		1.3		A
2038.11	2038.76	49049.3	IR		1.5		A
2037.33	2037.98	49068.1	IR	0.5	0.9D		A
2036.84	2037.49	49079.9	IR	0.3	0.3H		A
2036.13	2036.78	49097.0	IR	0.3	1.0		A
2036.00	2036.65	49100.1	IR	0.3H	0.9		A
2035.58	2036.23	49110.3	IR	1.0	0.6		A
2035.39	2036.04	49114.8	IR	0.8	0.5W		A
2034.80	2035.45	49129.1	IR	1.0	0.3H		A
2034.60	2035.25	49133.9	IR	0.5$			A
2034.01	2034.66	49148.2	IR		1.5		A
2033.57	2034.22	49158.8	IR	1.3	0.8D		A
2032.93	2033.58	49174.3	IR	0.5W	0.6		A
2032.66	2033.31	49180.8	IR	0.6W			A
2032.12	2032.77	49193.9	IR	0.9D	1.0		A
2031.50	2032.15	49208.9	IR	0.6	1.4		A

AIR WAVELENGTH (ANGSTRMS)	VACUUM WAVELENGTH (ANGSTRMS)	WAVENUMBER (KAYSERS)	LMENT	LOG ARC	INTENSITY SPK	DIS	REF
2031.26	2031.91	49214.7	IR	0.0	0.7W		A
2030.68	2031.33	49228.7	IR	0.3H	0.7		A
2030.30	2030.95	49238.0	IR	0.5	1.0		A
2029.88	2030.53	49248.1	IR	0.9W			A
2029.74	2030.39	49251.5	IR		0.7D		A
2029.64	2030.29	49254.0	IR		0.7D		A
2029.20	2029.85	49264.6	IR	0.6	0.9		A
2028.65	2029.30	49278.0	IR	0.7	1.7		MIT
2027.92	2028.57	49295.7	IR	0.3	1.4		A
2027.78	2028.43	49299.1	IR	1.1	0.7		A
2027.26	2027.91	49311.8	IR	0.3	1.0		A
2026.10	2026.75	49340.0	IR	0.6W			A
2025.96	2026.61	49343.4	IR		1.4		A
2025.49	2026.14	49354.9	IR	0.6	1.0		A
2024.47	2025.12	49379.7	IR	0.5	1.4		MIT
2024.07	2024.72	49389.5	IR	0.9	0.0H		A
2023.97	2024.62	49391.9	IR	1.6			A
2023.48	2024.13	49403.9	IR	0.5W	0.0		A
2023.24	2023.89	49409.7	IR	0.5	0.0		A
2022.83	2023.48	49419.8	IR	0.6	1.7		MIT
2022.35	2023.00	49431.5	IR	1.2	0.6		A
2022.20	2022.85	49435.1	IR		1.1		MIT
2021.51	2022.16	49452.0	IR		1.7		A
2021.46	2022.11	49453.2	IR	0.0	0.5		A
2020.82	2021.47	49468.9	IR	0.0	1.4		A
2019.78	2020.43	49494.4	IR		1.5		A
2019.13	2019.78	49510.3	IR		1.0D		A
2019.06	2019.71	49512.0	IR		1.0D		A
2017.99	2018.64	49538.3	IR	0.3H	1.6		MIT
2017.79	2018.44	49543.2	IR	0.5W			A
2017.57	2018.22	49548.6	IR	0.7W			A
2017.41	2018.06	49552.5	IR		1.7		A
2017.32	2017.97	49554.7	IR	0.3H	0.5		MIT
2015.90	2016.55	49589.6	IR	1.0	0.7		A
2015.48	2016.13	49600.0	IR	0.3	1.0		A
2015.30	2015.95	49604.4	IR		1.4W		A
2015.16	2015.81	49607.8	IR	0.8			A
2014.75	2015.40	49617.9	IR	0.3	0.6		A
2014.38	2015.03	49627.0	IR	0.5	1.0W		A
2013.83	2014.48	49640.6	IR	0.5	0.7		A
2013.24	2013.89	49655.1	IR	1.0	1.0		A
2012.71	2013.36	49668.2	IR	0.7J	0.5		MIT
2012.40	2013.05	49675.9	IR	0.3H	0.6		A
2012.17	2012.82	49681.5	IR		1.2		A
2011.96	2012.61	49686.7	IR	0.7	0.6		A
2011.26	2011.91	49704.0	IR	0.3H	1.2		A
2011.04	2011.69	49709.5	IR	0.6	0.9		A
2010.91	2011.56	49712.7	IR	0.7			A
2010.65	2011.30	49719.1	IR	1.2	1.2		A
2009.91	2010.56	49737.4	IR	0.5W			A
2009.76	2010.41	49741.1	IR	0.5W	1.3		MIT
2009.36	2010.01	49751.0	IR		1.4		MIT
2009.01	2009.66	49759.7	IR	0.6W			A
2008.29	2008.94	49777.5	IR		1.2		A
2007.18	2007.83	49805.0	IR	0.6	1.2		A
2006.44	2007.09	49823.4	IR	1.1	0.3		A
2005.86	2006.51	49837.8	IR	0.5	1.5		MIT
2005.57	2006.22	49845.0	IR	0.6	0.8		A
2005.24	2005.89	49853.2	IR	0.7			A
2004.85	2005.50	49862.9	IR	0.8	0.8		A
2004.52	2005.17	49871.1	IR		1.4		A
2004.35	2005.00	49875.3	IR	0.7			A
2004.08	2004.73	49882.1	IR	0.6	1.2		A
2003.56	2004.21	49895.0	IR	0.3	1.4		A
2001.92	2002.57	49935.9	IR	1.0	1.4		A
2001.23	2001.88	49953.1	IR		1.0		A
2000.70	2001.35	49966.3	IR	0.6	0.3H		A
9955.2	9957.9	10042.	K I	1.0H			ME
9950.5	9953.2	10047.	K I	1.3H			ME
9597.76	9600.39	10416.2	K I	0.3			EN

AIR WAVELENGTH (ANGSTRMS)	VACUUM WAVELENGTH (ANGSTRMS)	WAVENUMBER (KAYSERS)	LMENT		LOG ARC	INTENSITY SPK	DIS	REF
9595.60	9598.23	10418.6	K	I	0.5			EN
9591.8	9594.4	10423.	K	I	1.7X			ME
8904.04	8906.48	11227.8	K	I	0.5			EN
8902.20	8904.64	11230.1	K	I			0.7	EN
8505.19	8507.53	11754.3	K	I	0.3			EN
8503.51	8505.85	11756.6	K	I	0.5			EN
8250.22	8252.49	12117.6	K	I	0.3			EN
7698.979	7701.098	12985.16	K	I	3.7G	2.3		IHZ
7664.907	7667.017	13042.88	K	I	4.0G	2.6		IHZ
6964.69	6966.61	14354.2	K	I			0.7	EN
6964.18	6966.10	14355.2	K	I			0.5	EN
6938.98	6940.89	14407.4	K	I	2.7			ME
6911.30	6913.21	14465.1	K		2.5			ME
6595.00	6596.82	15158.8	K	II			0.7	BN
6427.690	6429.467	15553.39	K	II			1.3	DM
6307.240	6308.984	15850.41	K	II			1.6	DM
6246.465	6248.193	16004.63	K	II			1.5	DM
6120.219	6121.913	16334.76	K				1.8	DM
6119.95	6121.64	16335.5	K				1.0	SG
6050.	6052.	16520.	K				1.1	ML
6012.41	6014.08	16627.7	K	II			0.3	BN
5969.64	5971.29	16746.8	K	II			0.7	BN
5832.09	5833.71	17141.8	K	I	1.7			ME
5812.52	5814.13	17199.5	K	I	1.5			ME
5801.96	5803.57	17230.8	K	I	1.7H	1.3		ME
5782.60	5784.20	17288.5	K	I	1.8			ME
5772.32	5773.92	17319.2	K	II			1.2	BN
5642.674	5644.240	17717.18	K	II				DM
5536.01	5537.55	18058.5	K	II			1.0	BN
5512.69	5514.22	18134.9	K	II			0.7	BN
5488.06	5489.58	18216.3	K	II			0.7	BN
5470.065	5471.585	18276.24	K	II			1.6	DM
5359.521	5361.012	18653.20	K	I	1.6L			DA
5342.974	5344.460	18710.96	K	I	1.5L			DA
5339.670	5341.155	18722.54	K	I	1.6L			DA
5323.228	5324.709	18780.37	K	I	1.6L			DA
5310.208	5311.685	18826.42	K	II			1.4	DM
5112.204	5113.629	19555.58	K	I	1.5L			DA
5099.180	5100.601	19605.53	K	I	1.4L			DA
5097.144	5098.565	19613.36	K	I	1.4H			DA
5084.212	5085.629	19663.25	K	I	1.3L			DA
5056.182	5057.592	19772.26	K	II			1.8	DM
5005.34	5006.74	19973.1	K	II			1.2	SG
4965.038	4966.423	20135.21	K	I	1.2			DA
4956.043	4957.426	20171.76	K	I	1.0			DA
4950.816	4952.198	20193.05	K	I	1.0			DA
4943.24	4944.62	20224.0	K	II			1.5	BN
4941.964	4943.343	20229.22	K	I	0.7			DA
4938.75	4940.13	20242.4	K	II			1.0	BN
4869.70	4871.06	20529.4	K	I	1.0			FL
4863.61	4864.97	20555.1	K	I	0.8			FL
4856.03	4857.39	20587.2	K	I	0.8			FL
4849.88	4851.23	20613.3	K	I	0.5			FL
4829.212	4830.561	20701.53	K	II			2.0	DM
4805.19	4806.53	20805.0	K	I	0.6			FL
4800.16	4801.50	20826.8	K	I	0.5			FL
4791.08	4792.42	20866.3	K	I	0.3			FL
4786.89	4788.23	20884.6	K	I	0.3			FL
4744.92	4746.25	21069.3	K	II			1.2	BN
4659.317	4660.621	21456.37	K	II			1.6	DM
4642.172	4643.472	21535.61	K	I	0.7			DA
4641.585	4642.885	21538.33	K	I	0.5			DA
4608.425	4609.716	21693.31	K	II			1.6	DM
4595.613	4596.901	21753.79	K	II			1.6	DM
4505.340	4506.604	22189.66	K	II			1.5	DM
4466.658	4467.912	22381.82	K	II			1.3	DM
4455.00	4456.25	22440.4	K	II			0.7	BN
4423.716	4424.958	22599.08	K	II			1.0	DM
4388.129	4389.362	22782.35	K	II			1.6	DM
4374.870	4376.099	22851.40	K	II			0.3	DM
4362.96	4364.19	22913.8	K	II			1.3	BN
4339.977	4341.197	23035.12	K	II			1.3	DM
4332.024	4333.242	23077.41	K				1.0	DM
4317.85	4319.06	23153.2	K	II			0.7	BN
4309.076	4310.288	23200.31	K	II			1.6	DM
4305.265	4306.476	23220.84	K	II			0.3	DM
4304.937	4306.148	23222.61	K	II			1.6	DM
4288.651	4289.858	23310.80	K	II			1.2	DM
4284.853	4286.059	23331.46	K	II			1.0	DM
4263.312	4264.512	23449.34	K	II			1.6	DM
4225.605	4226.795	23658.59	K	II			1.6	DM
4222.975	4224.164	23673.32	K	II			1.6	DM
4209.498	4210.684	23749.11	K	II			1.2	DM
4186.226	4187.406	23881.14	K	II			1.8	DM
4149.171	4150.341	24094.41	K	II			1.3	DM
4134.721	4135.887	24178.61	K	II			1.6	DM
4114.952	4116.113	24294.76	K	II			1.5	DM
4112.071	4113.231	24311.79	K	II			1.2	DM
4093.697	4094.852	24420.90	K	II			1.3	DM
4065.209	4066.357	24592.04	K	II			1.2	DM
4047.201	4048.344	24701.46	K	I	2.6	2.3		DA
4044.140	4045.283	24720.15	K	I	2.9	2.6		DA
4042.594	4043.736	24729.61	K	II			1.5	DM
4039.692	4040.833	24747.37	K	II			1.2	DM
4024.920	4026.057	24838.20	K	II			1.2	DM
4017.509	4018.645	24884.01	K	II			1.2	DM
4012.10	4013.23	24917.6	K	II			1.3	BN
4001.200	4002.331	24985.44	K	II			1.6	DM
3995.10	3996.23	25023.6	K	II		1.5		BN
3991.775	3992.904	25044.43	K	II			1.2	DM
3972.551	3973.675	25165.62	K	II			1.5	DM
3966.687	3967.809	25202.82	K	II			1.5	DM
3959.777	3960.897	25246.80	K	II			1.0	DM
3956.099	3957.219	25270.27	K	II			1.0	DM
3955.207	3956.326	25275.97	K	II			1.5	DM
3942.505	3943.621	25357.41	K	II			1.5	DM
3934.46	3935.57	25409.3	K				1.3	BN
3926.338	3927.450	25461.81	K	II			1.3	DM
3923.053	3924.164	25483.14	K	II			1.3	DM
3900.11	3901.21	25633.0	K	II			1.0	BN
3899.242	3900.347	25638.75	K	II			1.0	DM
3897.870	3898.974	25647.77	K	II			1.8	DM
3886.84	3887.94	25720.6	K	II			0.7	BN
3883.358	3884.459	25743.61	K	II			1.0	DM
3878.62	3879.72	25775.1	K	II			1.2	BN
3873.747	3874.845	25807.48	K	II			1.3	DM
3861.412	3862.507	25889.92	K	II			1.0	DM
3844.02	3845.11	26007.1	K	II			1.0	BN
3821.30	3822.38	26161.7	K	II			1.0	BN
3817.541	3818.624	26187.44	K	II			1.6	DM
3816.549	3817.632	26194.25	K	II			1.5	DM
3800.141	3801.220	26307.34	K	II			1.5	DM
3783.192	3784.266	26425.20	K	II			1.5	DM
3767.372	3768.442	26536.16	K	II			1.5	DM
3756.635	3757.703	26612.00	K	II			1.0	DM
3744.404	3745.468	26698.93	K	II			1.3	DM
3739.125	3740.188	26736.62	K	II			1.3	DM
3721.344	3722.402	26864.37	K	II			1.3	DM
3716.594	3717.651	26898.70	K	II			1.3	DM
3681.525	3682.573	27154.93	K	II			1.5	DM
3676.05	3677.10	27195.4	K	II			1.0	BN
3668.627	3669.672	27250.39	K	II			1.0	DM
3647.95	3648.99	27404.9	K	II			0.7	BN
3637.00	3638.04	27487.4	K	II			1.0	BN
3626.42	3627.45	27567.6	K				1.2	BN
3618.429	3619.461	27628.42	K	II			1.3	DM
3608.871	3609.900	27701.60	K	II			1.0	DM
3593.22	3594.25	27822.2	K	II			0.7	BN
3586.60	3587.62	27873.6	K	II			0.7	BN
3562.15	3563.17	28064.9	K	II			1.2	BN

AIR WAVELENGTH (ANGSTRMS)	VACUUM WAVELENGTH (ANGSTRMS)	WAVENUMBER (KAYSERS)	LMENT	LOG INTENSITY ARC	SPK	DIS	REF
3530.707	3531.716	28314.85	K II			1.6	DM
3529.53	3530.54	28324.3	K II			1.0	BN
3528.51	3529.52	28332.5	K II			0.3	BN
3481.11	3482.11	28718.2	K II			1.5	BN
3457.85	3458.84	28911.4	K II			0.7	BN
3447.375	3448.363	28999.27	K I	2.0G	1.9G		RG
3446.372	3447.360	29007.71	K I	2.2G	2.0G		RG
3440.05	3441.04	29061.0	K II			1.6	BN
3427.13	3428.11	29170.6	K II			0.7	BN
3416.76	3417.74	29259.1	K			0.7	BN
3404.24	3405.22	29366.7	K II			1.5	BN
3392.63	3393.60	29467.2	K II			1.0	BN
3384.86	3385.83	29534.8	K II			1.5	BN
3380.62	3381.59	29571.9	K II			1.5	BN
3373.60	3374.57	29633.4	K			1.5	BN
3356.51	3357.47	29784.3	K II			0.7	BN
3345.32	3346.28	29883.9	K			1.5	BN
3337.67	3338.63	29952.4	K II			0.3	BN
3301.60	3302.55	30279.6	K II			1.0	BN
3290.65	3291.60	30380.4	K II			1.0H	BN
3258.81	3259.75	30677.2	K II			1.0	BN
3220.60	3221.53	31041.1	K II			1.2	BN
3217.621	3218.550	31069.89	K I	1.7G	1.4		RG
3217.155	3218.084	31074.39	K I	2.0G	1.3H		RG
3190.07	3190.99	31338.2	K II			1.3	BN
3189.28	3190.20	31346.0	K			0.7	BN
3171.81	3172.73	31518.6	K II			0.7	BN
3169.80	3170.72	31538.6	K II			1.0	BN
3157.15	3158.06	31665.0	K II			0.7	BN
3142.75	3143.66	31810.1	K II			0.7	BN
3105.00	3105.90	32196.8	K II			1.5	BN
3102.790	3103.690	32219.71	K I	1.7G			RG
3102.043	3102.943	32227.47	K I	1.3G			RG
3075.00	3075.89	32510.9	K II			1.0	BN
3062.18	3063.07	32647.0	K II			1.3	BN
3047.16	3048.05	32807.9	K II			0.7	BN
3034.82	3035.70	32941.3	K I	1.5H			FL
3030.43	3031.31	32989.0	K II			0.7	BN
2993.4	2994.3	33397.	K			1.0	ML
2992.21	2993.08	33410.4	K I	1.2G			FL
2985.9	2986.8	33481.	K			0.7	ML
2966.9	2967.8	33695.	K			0.3	ML
2963.24	2964.11	33737.0	K I	0.9G			FL
2955.8	2956.7	33822.	K			0.3	ML
2950.88	2951.74	33878.3	K II		0.7		BN
2942.7	2943.6	33972.	K I	0.7G			FL
2939.5	2940.4	34009.	K			0.7	ML
2927.9	2928.8	34144.	K I	0.3G			FL
2853.95	2854.79	35028.9	K			0.7	S
2836.2	2837.0	35248.	K			1.0	ML
2833.14	2833.97	35286.1	K	0.3H		0.3	SG
2819.26	2820.09	35459.9	K II			1.0	BN
2808.99	2809.82	35589.5	K II			1.0	BN
2803.8	2804.6	35655.	K			1.5	ML
2780.1	2780.9	35959.	K			0.3	SG
2777.89	2778.71	35987.9	K II			0.3	BN
2768.1	2768.9	36115.	K			0.3	ML
2743.55	2744.36	36438.4	K II			0.7	BN
2736.0	2736.8	36539.	K			1.3	SG
2732.0	2732.8	36592.	K			0.3	ML
2662.4	2663.2	37549.	K			1.0	SG
2655.8	2656.6	37642.	K			0.3	ML
2614.1	2614.9	38243.	K			0.7	SG
2569.8	2570.6	38902.	K			0.3	ML
2559.2	2560.0	39063.	K			0.7	ML
2550.02	2550.79	39203.6	K II			1.3	BN
2538.6	2539.4	39380.	K			1.0	SG
2504.60	2505.35	39914.5	K II			0.7	BN
2474.2	2474.9	40405.	K			1.0	SG
2470.4	2471.1	40467.	K			1.0	ML

AIR WAVELENGTH (ANGSTRMS)	VACUUM WAVELENGTH (ANGSTRMS)	WAVENUMBER (KAYSERS)	LMENT	LOG INTENSITY ARC	SPK	DIS	REF
2451.25	2451.99	40783.2	K			0.7	ML
2446.0	2446.7	40871.	K			1.3	ML
2440.9	2441.6	40956.	K			0.7	ML
2440.6	2441.3	40961.	K			0.3	SG
2436.6	2437.3	41028.	K			1.3	ML
2410.4	2411.1	41474.	K			1.0	ML
2402.0	2402.7	41619.	K			1.0	ML
2393.4	2394.1	41769.	K			1.0	ML
2379.5	2380.2	42013.	K			1.3	SG
2370.25	2370.97	42176.8	K			1.5	ML
2365.8	2366.5	42256.	K			0.7	ML
2358.70	2359.42	42383.3	K II			1.7	BN
2350.2	2350.9	42536.	K			1.0	SG
2348.3	2349.0	42571.	K			1.3	ML
2342.30	2343.02	42680.0	K II			1.0	BN
2340.15	2340.87	42719.2	K			1.0	ML
2327.9	2328.6	42944.	K			0.7	SG
2324.33	2325.04	43009.9	K			1.0	ML
2319.15	2319.86	43106.0	K			1.5	ML
2315.22	2315.93	43179.2	K			1.3	ML
2306.58	2307.29	43340.9	K			1.0	ML
2304.0	2304.7	43389.	K			0.3	ML
2300.9	2301.6	43448.	K			1.0	ML
2296.79	2297.50	43525.6	K II			0.7	BN
2284.32	2285.02	43763.2	K			0.7	ML
2280.05	2280.75	43845.1	K			1.0	ML
2270.9	2271.6	44022.	K			1.5	ML
2265.04	2265.74	44135.7	K			1.5	ML
2262.1	2262.8	44193.	K			1.0	SG
2261.7	2262.4	44201.	K			0.3	SG
2260.6	2261.3	44222.	K			1.5	SG
2260.32	2261.02	44227.8	K			1.5	ML
2257.8	2258.5	44277.	K			0.3	ML
2255.29	2255.99	44326.5	K II			1.0	ML
2253.0	2253.7	44371.	K			0.7	ML
2250.92	2251.62	44412.5	K			1.5	ML
2246.32	2247.02	44503.5	K			0.7	ML
2243.4	2244.1	44561.	K II			0.7	SG
2240.89	2241.59	44611.3	K			1.6	ML
2231.67	2232.36	44795.6	K			0.7	ML
2226.0	2226.7	44910.	K			0.7	ML
2221.58	2222.27	44999.0	K			1.5	ML
2217.23	2217.92	45087.3	K			1.0	ML
2213.1	2213.8	45171.	K			1.3	ML
2210.53	2211.22	45223.9	K II			1.3	ML
2200.89	2201.58	45422.0	K			0.7	ML
2196.7	2197.4	45509.	K			1.0	SG
2190.0	2190.7	45648.	K II			1.6	ML
2186.93	2187.61	45711.9	K			1.6	ML
2182.4	2183.1	45807.	K			0.3	ML
2177.92	2178.60	45901.0	K			1.0	ML
2175.12	2175.80	45960.1	K			1.0	ML
2171.03	2171.71	46046.6	K			0.7	ML
2163.31	2163.99	46211.0	K			0.3	ML
2155.3	2156.0	46383.	K			1.3	ML
2149.42	2150.10	46509.5	K			1.0	ML
2144.5	2145.2	46616.	K			1.3	SQ
2141.0	2141.7	46692.	K			0.3	ML
2140.49	2141.16	46703.5	K			0.3	ML
2137.56	2138.23	46767.6	K			1.5	ML
2132.8	2133.5	46872.	K			1.0	ML
2130.75	2131.42	46917.0	K			1.3	ML
2128.9	2129.6	46958.	K			1.3	ML
2127.25	2127.92	46994.2	K			1.3	ML
2122.0	2122.7	47110.	K			1.3	ML
2114.2	2114.9	47284.	K			1.0	ML
2110.05	2110.72	47377.2	K			1.0	ML
2105.45	2106.12	47480.7	K			0.3	ML
2101.05	2101.72	47580.1	K			0.3	ML
2100.0	2100.7	47604.	K			1.0	ML

AIR WAVELENGTH (ANGSTRMS)	VACUUM WAVELENGTH (ANGSTRMS)	WAVENUMBER (KAYSERS)	LMENT	LOG ARC	INTENSITY SPK	INTENSITY DIS	REF
2097.65	2098.32	47657.3	K			1.0	ML
2093.3	2094.0	47756.	K			0.7	ML
2089.8	2090.5	47836.	K			1.6	ML
2085.2	2085.9	47942.	K			1.3	ML
2082.1	2082.8	48013.	K			1.0	ML
2077.9	2078.6	48110.	K			1.9	ML
2070.0	2070.7	48294.	K			0.7	ML
2067.2	2067.9	48359.	K			0.3	ML
2062.05	2062.71	48479.9	K			0.7	ML
2054.9	2055.6	48649.	K			1.3	ML
2053.1	2053.8	48691.	K			0.3	ML
2049.8	2050.5	48770.	K			0.7	ML
2045.7	2046.4	48867.	K			0.3	ML
2043.1	2043.8	48929.	K			0.3	ML
2036.9	2037.6	49078.	K			1.6	ML
2027.8	2028.5	49299.	K			0.3	ML
2019.6	2020.3	49499.	K			0.3	ML
2010.7	2011.3	49718.	K			0.7	ML
2006.8	2007.4	49814.	K			0.7	ML
9954.75	9957.48	10042.7	KR II			0.7H	ME
9917.60	9920.32	10080.3	KR I			0.5	ME
9916.37	9919.09	10081.6	KR I			0.6	ME
9897.08	9899.79	10101.2	KR I			0.3	ME
9892.97	9895.68	10105.4	KR II			0.3H	ME
9862.95	9865.65	10136.2	KR I			0.6	ME
9856.24	9858.94	10143.1	KR I			2.7	ME
9838.33	9841.03	10161.5	KR I			0.7	ME
9826.58	9829.27	10173.7	KR			1.4	ME
9823.39	9826.08	10177.0	KR			1.40	ME
9810.27	9812.96	10190.6	KR I			0.3H	ME
9803.14	9805.83	10198.0	KR			2.1	ME
9794.89	9797.58	10206.6	KR I			0.5	ME
9768.69	9771.37	10234.0	KR I			0.3	ME
9751.759	9754.433	10251.75	KR I			3.3	ME
9743.11	9745.78	10260.8	KR I			1.7	ME
9727.51	9730.18	10277.3	KR I			0.3V	ME
9717.16	9719.82	10288.2	KR			0.3V	ME
9714.85	9717.51	10290.7	KR I			1.2	ME
9711.60	9714.26	10294.1	KR			1.7V	ME
9704.22	9706.88	10302.0	KR I			1.7	ME
9687.83	9690.49	10319.4	KR I			1.0	ME
9682.26	9684.92	10325.3	KR I			0.3	ME
9663.34	9665.99	10345.6	KR II			1.7	ME
9619.61	9622.25	10392.6	KR			2.0V	ME
9615.63	9618.27	10396.9	KR I			0.5	ME
9613.80	9616.44	10398.9	KR II			1.4V	ME
9605.80	9608.43	10407.5	KR			2.1V	ME
9594.24	9596.87	10420.1	KR II			1.4	ME
9577.52	9580.15	10438.2	KR II			2.1	ME
9552.85	9555.47	10465.2	KR			0.3V	ME
9543.64	9546.26	10475.3	KR II			0.3V	ME
9540.89	9543.51	10478.3	KR I			1.5	ME
9504.70	9507.31	10518.2	KR II			1.4	ME
9500.60	9503.21	10522.8	KR			1.4V	ME
9475.06	9477.66	10551.1	KR II			1.4V	ME
9470.93	9473.53	10555.7	KR II			1.7V	ME
9450.88	9453.47	10578.1	KR I			1.3	ME
9440.02	9442.61	10590.3	KR II			1.3V	ME
9437.21	9439.80	10593.4	KR II			0.7J	ME
9414.94	9417.52	10618.5	KR II			1.4	ME
9402.82	9405.40	10632.2	KR II			1.7Y	ME
9388.08	9390.66	10648.9	KR			1.1X	ME
9362.03	9364.60	10678.5	KR I			2.0	ME
9361.95	9364.52	10678.6	KR II			1.9	ME
9352.23	9354.80	10689.7	KR I			2.0	ME
9349.08	9351.65	10693.3	KR II			1.5V	ME
9345.11	9347.67	10697.8	KR II			1.5V	ME
9330.66	9333.22	10714.4	KR II			0.3H	ME
9326.03	9328.59	10719.7	KR I			1.0	ME
9320.99	9323.55	10725.5	KR II			1.8H	ME

AIR WAVELENGTH (ANGSTRMS)	VACUUM WAVELENGTH (ANGSTRMS)	WAVENUMBER (KAYSERS)	LMENT	LOG ARC	INTENSITY SPK	INTENSITY DIS	REF
9317.84	9320.40	10729.2	KR II			0.9H	ME
9296.1	9298.7	10754.	KR			1.2V	ME
9293.82	9296.37	10756.9	KR II			2.0X	ME
9289.95	9292.50	10761.4	KR			0.7X	ME
9279.9	9282.4	10773.	KR I			0.3	ME
9273.02	9275.56	10781.0	KR I			0.9	ME
9271.99	9274.53	10782.2	KR II			1.0	ME
9270.96	9273.50	10783.4	KR			1.0	ME
9245.45	9247.99	10813.2	KR			0.7X	ME
9243.54	9246.08	10815.4	KR			1.5	ME
9238.48	9241.02	10821.3	KR II			2.1	ME
9233.18	9235.71	10827.5	KR II			1.1	ME
9207.27	9209.80	10858.0	KR			0.3V	ME
9188.69	9191.21	10880.0	KR I			0.3	ME
9181.23	9183.75	10888.8	KR			0.5X	ME
9175.42	9177.94	10895.7	KR			1.0X	ME
9133.4	9135.9	10946.	KR			0.3X	ME
9122.49	9124.99	10958.9	KR I			1.3	ME
9115.00	9117.50	10967.9	KR			0.7X	ME
9111.69	9114.19	10971.9	KR I			1.3	ME
9099.72	9102.22	10986.3	KR			0.6H	ME
9094.33	9096.83	10992.8	KR I			0.6H	ME
9044.55	9047.03	11053.3	KR II			0.3H	ME
9044.47	9046.95	11053.4	KR I			0.5	ME
9039.95	9042.43	11059.0	KR			0.6B	ME
9025.67	9028.15	11076.5	KR II			0.5B	ME
9006.15	9008.62	11100.5	KR II			0.5	ME
8999.19	9001.66	11109.1	KR I			1.5	ME
8999.11	9001.58	11109.2	KR II			0.3	ME
8978.70	8981.16	11134.4	KR II			0.6B	ME
8977.99	8980.45	11135.3	KR I			1.7	ME
8967.53	8969.99	11148.3	KR			1.0	ME
8928.692	8931.143	11196.78	KR I			3.3	IME
8925.3	8927.8	11201.	KR			0.3	ME
8870.32	8872.76	11270.5	KR			0.6	ME
8842.46	8844.89	11306.0	KR I			0.5	ME
8805.78	8808.20	11353.1	KR I			1.3	ME
8780.25	8782.66	11386.1	KR I			1.5	ME
8776.749	8779.159	11390.61	KR I			3.7	IME
8774.05	8776.46	11394.1	KR I			1.7	ME
8773.00	8775.41	11395.5	KR I			0.6	ME
8764.112	8766.519	11407.04	KR I			2.2	IME
8755.20	8757.60	11418.7	KR I			1.5	ME
8747.29	8749.69	11429.0	KR I			0.3	ME
8746.43	8748.83	11430.1	KR I			0.5	ME
8726.54	8728.94	11456.2	KR I			0.9	ME
8713.62	8716.01	11473.1	KR I			0.3	ME
8707.61	8710.00	11481.1	KR II			0.3H	ME
8697.50	8699.89	11494.4	KR I			1.6	ME
8690.19	8692.58	11504.1	KR II			1.3C	ME
8673.48	8675.86	11526.2	KR I			0.3	ME
8651.49	8653.87	11555.5	KR I			0.9	ME
8624.82	8627.19	11591.3	KR I			0.6H	ME
8605.85	8608.21	11616.8	KR I			1.6	ME
8595.91	8598.27	11630.2	KR			0.3J	ME
8593.1	8595.5	11634.	KR I			1.0	ME
8569.02	8571.37	11666.7	KR I			1.3	ME
8560.89	8563.24	11677.8	KR I			1.7	ME
8537.93	8540.28	11709.2	KR I			1.6	ME
8508.870	8511.208	11749.21	KR I			3.5	IME
8498.21	8500.54	11763.9	KR I			1.5	ME
8477.20	8479.53	11793.1	KR I			0.3	ME
8473.31	8475.64	11798.5	KR II			1.3B	ME
8469.96	8472.29	11803.2	KR I			0.3H	ME
8464.92	8467.25	11810.2	KR II			0.6Q	ME
8412.428	8414.740	11883.91	KR I			2.0	IME
8409.24	8411.55	11888.4	KR			0.5	GR
8384.90	8387.20	11922.9	KR I			1.2	ME
8375.93	8378.23	11935.7	KR			0.7	ME
8321.09	8323.38	12014.3	KR			0.3	ME

AIR WAVELENGTH (ANGSTRMS)	VACUUM WAVELENGTH (ANGSTRMS)	WAVENUMBER (KAYSERS)	LMENT	LOG ARC	INTENSITY SPK	DIS	REF
8303.20	8305.48	12040.2	KR I			1.0	ME
8301.39	8303.67	12042.9	KR I			1.3	ME
8298.108	8300.389	12047.63	KR I			3.7	IME
8287.56	8289.84	12063.0	KR I			0.6H	ME
8281.049	8283.325	12072.45	KR			3.0	IME
8272.355	8274.629	12085.13	KR I			2.0	IME
8263.240	8265.511	12098.46	KR I			3.3	IME
8228.89	8231.15	12149.0	KR I			1.0H	ME
8222.69	8224.95	12158.1	KR I			0.8	ME
8218.40	8220.66	12164.5	KR I			1.9	ME
8206.62	8208.88	12181.9	KR I			1.6	ME
8205.22	8207.48	12184.0	KR I			1.3	ME
8202.72	8204.98	12187.7	KR II			1.6H	ME
8195.070	8197.323	12199.10	KR I			1.7	IME
8192.4	8194.7	12203.	KR I			0.3	ME
8190.054	8192.306	12206.58	KR I			3.5	IME
8157.25	8159.49	12255.7	KR II			0.5C	ME
8145.15	8147.39	12273.9	KR II			1.3V	ME
8144.96	8147.20	12274.2	KR I			1.2	ME
8132.98	8135.22	12292.2	KR I			1.8	ME
8130.03	8132.27	12296.7	KR II			0.5H	ME
8112.902	8115.133	12322.66	KR I			3.7	IHU
8104.364	8106.592	12335.64	KR I			3.7	I
8104.02	8106.25	12336.2	KR I			2.7	ME
8059.504	8061.720	12404.30	KR I			3.0	IME
8040.50	8042.71	12433.6	KR I			0.9H	ME
8033.52	8035.73	12444.4	KR I			0.3H	ME
7993.22	7995.42	12507.2	KR II			1.7H	ME
7993.12	7995.32	12507.3	KR I			0.7	ME
7990.78	7992.98	12511.0	KR			0.3	ME
7982.406	7984.602	12524.11	KR I			2.0	IME
7981.82	7984.02	12525.0	KR I			1.5	ME
7981.19	7983.39	12526.0	KR I			1.3	ME
7973.62	7975.81	12537.9	KR II			1.5C	ME
7957.67	7959.86	12563.0	KR I			0.3	ME
7946.99	7949.18	12579.9	KR I			1.3	ME
7938.34	7940.52	12593.6	KR I			0.3	ME
7931.41	7933.59	12604.6	KR II			1.0	ME
7928.600	7930.781	12609.10	KR I			2.2	IME
7920.47	7922.65	12622.0	KR I			1.6	ME
7913.424	7915.601	12633.28	KR I			2.3	IME
7904.62	7906.79	12647.3	KR I			1.5	ME
7882.36	7884.53	12683.1	KR I			1.0	ME
7881.76	7883.93	12684.0	KR I			1.5	ME
7871.93	7874.10	12699.9	KR I			0.3	ME
7863.91	7866.07	12712.8	KR I			1.3	ME
7856.52	7858.68	12724.8	KR II		1.5V		ME
7854.821	7856.982	12727.53	KR I			2.9	I
7840.40	7842.56	12750.9	KR I			0.6	ME
7840.01	7842.17	12751.6	KR I			0.9H	ME
7830.21	7832.36	12767.5	KR I			0.3H	ME
7806.52	7808.67	12806.3	KR I			1.7	ME
7791.90	7794.04	12830.3	KR II			0.8B	ME
7786.66	7788.80	12838.9	KR I			0.3	ME
7781.97	7784.11	12846.7	KR			2.0H	E
7776.28	7778.42	12856.1	KR I			1.6	ME
7772.40	7774.54	12862.5	KR I			0.7H	ME
7768.43	7770.57	12869.1	KR I			0.7	ME
7756.52	7758.65	12888.8	KR II			1.5V	ME
7749.16	7751.29	12901.1	KR II			0.5	ME
7749.16	7751.29	12901.1	KR I			0.5	ME
7746.828	7748.960	12904.96	KR I			2.2	IME
7741.39	7743.52	12914.0	KR I			1.6	ME
7735.69	7737.82	12923.5	KR II			2.3H	ME
7703.41	7705.53	12977.7	KR I			0.3H	ME
7694.539	7696.657	12992.65	KR I			3.0	IME
7685.246	7687.361	13008.36	KR I			3.0	IME
7663.75	7665.86	13044.8	KR			0.3	ME
7652.16	7654.27	13064.6	KR I			0.6	ME
7641.16	7643.26	13083.4	KR II			2.2	ME
7635.13	7637.23	13093.7	KR II			0.7	ME
7629.46	7631.56	13103.5	KR II			0.7H	ME
7615.69	7617.79	13127.2	KR			0.5H	ME
7601.544	7603.637	13151.60	KR I			3.7	IME
7587.413	7589.502	13176.09	KR I			3.0	IME
7550.63	7552.71	13240.3	KR			0.5	ME
7543.10	7545.18	13253.5	KR			0.5	ME
7525.48	7527.55	13284.5	KR			1.3H	ME
7524.46	7526.53	13286.3	KR II			2.5H	ME
7517.52	7519.59	13298.6	KR			0.3H	ME
7494.15	7496.21	13340.1	KR I			1.5	ME
7493.58	7495.64	13341.1	KR I			1.3	ME
7486.862	7488.924	13353.05	KR I			2.0	ME
7467.99	7470.05	13386.8	KR			0.8J	ME
7465.01	7467.07	13392.1	KR I			0.5H	ME
7444.64	7446.69	13428.8	KR			0.3H	ME
7435.78	7437.83	13444.8	KR II			2.3H	ME
7434.74	7436.79	13446.7	KR II			1.2H	ME
7425.54	7427.59	13463.3	KR I			1.8	ME
7407.02	7409.06	13497.0	KR II			2.6H	ME
7367.02	7369.05	13570.3	KR I			0.3H	ME
7366.80	7368.83	13570.7	KR I			0.3H	ME
7362.83	7364.86	13578.0	KR			0.6	ME
7361.34	7363.37	13580.7	KR I			0.3	RS
7359.96	7361.99	13583.3	KR I			0.7	ME
7355.48	7357.51	13591.6	KR			0.6	ME
7341.16	7343.18	13618.1	KR			0.3	ME
7334.33	7336.35	13630.8	KR			0.6	ME
7327.00	7329.02	13644.4	KR			0.7	ME
7301.25	7303.26	13692.5	KR I			0.7	ME
7289.78	7291.79	13714.1	KR II			2.6H	ME
7287.262	7289.270	13718.80	KR I			1.9	ME
7272.97	7274.97	13745.7	KR II			0.6	ME
7268.28	7270.28	13754.6	KR I			0.7	RS
7241.56	7243.56	13805.4	KR II			0.3B	ME
7234.58	7236.57	13818.7	KR			0.3	ME
7227.34	7229.33	13832.5	KR			0.3	ME
7224.103	7226.094	13838.73	KR I			2.0	ME
7213.13	7215.12	13859.8	KR II			2.4	ME
7200.59	7202.57	13883.9	KR I			0.3H	ME
7180.47	7182.45	13922.8	KR			0.5	ME
7152.21	7154.18	13977.8	KR I			0.7	ME
7143.45	7145.42	13995.0	KR I			0.9	ME
7139.99	7141.96	14001.8	KR II			1.8	ME
7128.14	7130.11	14025.0	KR			0.3B	ME
7089.48	7091.43	14101.5	KR I			0.3	RS
7078.44	7080.39	14123.5	KR II			0.5	ME
7073.97	7075.92	14132.4	KR II			1.8	ME
7057.27	7059.22	14165.9	KR I			1.0	ME
7022.56	7024.50	14235.9	KR II			0.3H	ME
7008.62	7010.55	14264.2	KR			0.3	ME
7001.62	7003.55	14278.5	KR			0.3	ME
7000.79	7002.72	14280.2	KR I			0.8	ME
6993.05	6994.98	14296.0	KR I			0.3	ME
6977.95	6979.87	14326.9	KR II			0.5H	ME
6944.06	6945.98	14396.8	KR II			1.0B	ME
6911.29	6913.20	14465.1	KR I			0.3	RS
6904.68	6906.58	14478.9	KR I			2.0	ME
6904.22	6906.12	14479.9	KR I			1.2	ME
6870.85	6872.75	14550.2	KR II			1.6	ME
6869.63	6871.53	14552.8	KR I			1.3	ME
6862.82	6864.71	14567.2	KR			0.5	ME
6853.32	6855.21	14587.4	KR			0.3	ME
6846.40	6848.29	14602.2	KR I			1.3	ME
6829.09	6830.97	14639.2	KR I			0.9	ME
6813.10	6814.98	14673.6	KR I			1.7	ME
6795.40	6797.28	14711.8	KR I			0.6	ME
6776.15	6778.02	14753.6	KR I			0.5	ME
6771.22	6773.09	14764.3	KR II			1.7	ME
6764.51	6766.38	14779.0	KR I			0.7	RS

AIR WAVELENGTH (ANGSTRMS)	VACUUM WAVELENGTH (ANGSTRMS)	WAVENUMBER (KAYSERS)	LMENT	LOG ARC	INTENSITY SPK	DIS	REF
6764.43	6766.30	14779.1	KR II			1.9	ME
6763.61	6765.48	14780.9	KR II			2.0	ME
6740.10	6741.96	14832.5	KR I			1.3	ME
6723.36	6725.22	14869.4	KR I			0.6	ME
6699.228	6701.078	14922.97	KR I			1.8	ME
6652.239	6654.076	15028.38	KR I			1.6	ME
6647.94	6649.78	15038.1	KR I			0.5	RS
6634.36	6636.19	15068.9	KR II			1.2H	ME
6627.96	6629.79	15083.4	KR II			0.3B	ME
6624.22	6626.05	15091.9	KR II			0.3B	ME
6605.12	6606.94	15135.6	KR I			0.3	ME
6605.00	6606.82	15135.9	KR II			1.2	ME
6602.90	6604.72	15140.7	KR II			1.0H	ME
6576.42	6578.24	15201.6	KR I			1.3	ME
6570.07	6571.88	15216.3	KR II			2.2	ME
6565.32	6567.13	15227.3	KR II			0.8H	ME
6555.69	6557.50	15249.7	KR I			0.8	ME
6555.56	6557.37	15250.0	KR I			0.3	ME
6536.55	6538.36	15294.4	KR I			0.9	ME
6510.95	6512.75	15354.5	KR II			2.0	ME
6510.14	6511.94	15356.4	KR II			0.9B	ME
6508.37	6510.17	15360.6	KR I			0.5	ME
6504.89	6506.69	15368.8	KR I			1.0	ME
6493.7	6495.5	15395.	KR II			0.3H	ME
6488.07	6489.86	15408.7	KR I			1.2	ME
6470.89	6472.68	15449.6	KR II			1.7	ME
6456.291	6458.075	15484.49	KR I			2.3	S
6448.78	6450.56	15502.5	KR I			1.0	ME
6440.74	6442.52	15521.9	KR II			0.7B	ME
6431.92	6433.70	15543.2	KR			0.3H	ME
6421.029	6422.804	15569.52	KR I			2.0	S
6420.18	6421.95	15571.6	KR II			2.5	ME
6416.61	6418.38	15580.2	KR II			1.8C	ME
6415.65	6417.42	15582.6	KR I			1.3	ME
6412.53	6414.30	15590.2	KR II			0.6H	ME
6410.17	6411.94	15595.9	KR I			0.7	ME
6409.84	6411.61	15596.7	KR II			1.0C	ME
6404.69	6406.46	15609.2	KR			0.5J	ME
6394.28	6396.05	15634.7	KR II			0.6C	ME
6391.14	6392.91	15642.3	KR II			1.5	ME
6373.58	6375.34	15685.4	KR I			1.5	ME
6368.26	6370.02	15698.5	KR I			0.6	ME
6351.90	6353.66	15739.0	KR I			0.9	ME
6346.66	6348.41	15752.0	KR I			1.3	ME
6344.61	6346.36	15757.1	KR II			0.6H	ME
6322.42	6324.17	15812.4	KR II			0.6	ME
6303.66	6305.40	15859.4	KR II			2.0	ME
6290.96	6292.70	15891.4	KR II			0.5B	ME
6277.52	6279.26	15925.4	KR			0.3H	ME
6267.33	6269.06	15951.3	KR I			0.3	ME
6264.91	6266.64	15957.5	KR			0.3J	ME
6257.84	6259.57	15975.5	KR II			0.6N	ME
6243.53	6245.26	16012.2	KR			0.3K	ME
6241.39	6243.12	16017.6	KR I			1.0	ME
6236.354	6238.079	16030.58	KR I			1.5	JA
6230.74	6232.46	16045.0	KR II			1.0B	ME
6222.71	6224.43	16065.7	KR I			1.3	ME
6196.14	6197.85	16134.6	KR			0.5K	ME
6185.35	6187.06	16162.8	KR II			0.8B	ME
6172.08	6173.79	16197.5	KR I			0.3	ME
6171.77	6173.48	16198.3	KR II			0.8C	ME
6168.80	6170.51	16206.1	KR II			1.7	ME
6163.65	6165.36	16219.7	KR I			0.8	ME
6151.38	6153.08	16252.0	KR I			1.3	ME
6119.56	6121.25	16336.5	KR			1.0K	ME
6115.23	6116.92	16348.1	KR I			0.5	ME
6112.61	6114.30	16355.1	KR II			0.6H	ME
6108.34	6110.03	16366.5	KR I			0.5	ME
6107.61	6109.30	16368.5	KR			0.7K	ME
6094.50	6096.19	16403.7	KR II			1.5B	ME

AIR WAVELENGTH (ANGSTRMS)	VACUUM WAVELENGTH (ANGSTRMS)	WAVENUMBER (KAYSERS)	LMENT	LOG ARC	INTENSITY SPK	DIS	REF
6094.31	6096.00	16404.2	KR I			0.3	ME
6091.81	6093.50	16410.9	KR I			0.8	ME
6088.00	6089.69	16421.2	KR I			0.3	ME
6082.85	6084.53	16435.1	KR I			1.6	ME
6079.71	6081.39	16443.6	KR II			1.3C	ME
6075.24	6076.92	16455.7	KR I			1.3	ME
6056.128	6057.805	16507.63	KR I			1.8	ME
6049.35	6051.03	16526.1	KR I			0.5	RS
6046.06	6047.73	16535.1	KR			1.0J	ME
6044.85	6046.52	16538.4	KR			0.3H	ME
6040.7	6042.4	16550.	KR			1.0J	ME
6035.82	6037.49	16563.2	KR I			1.2	ME
6022.39	6024.06	16600.1	KR II			1.6	ME
6012.157	6013.822	16628.36	KR I			1.7	JA
6009.99	6011.65	16634.4	KR II			1.0H	ME
6008.10	6009.76	16639.6	KR II			0.5C	ME
6002.19	6003.85	16656.0	KR I			0.5	ME
5993.8503	5995.5105	16679.147	KR I			1.8	S
5992.22	5993.88	16683.7	KR II			2.3	ME
5977.65	5979.31	16724.4	KR I			0.6	ME
5974.82	5976.48	16732.3	KR II			0.3	ME
5969.57	5971.22	16747.0	KR			0.3Q	ME
5967.54	5969.19	16752.7	KR II			1.2T	ME
5955.14	5956.79	16787.6	KR I			0.3	ME
5949.93	5951.58	16802.3	KR II			0.5H	ME
5945.44	5947.09	16814.9	KR I			0.7	ME
5942.13	5943.78	16824.3	KR I			0.3	ME
5941.82	5943.47	16825.2	KR			0.6T	ME
5935.03	5936.67	16844.4	KR II			0.9H	ME
5919.4	5921.0	16889.	KR			0.3J	ME
5918.81	5920.45	16890.6	KR II			0.3V	ME
5911.72	5913.36	16910.9	KR			1.0Q	ME
5900.89	5902.53	16941.9	KR II			0.9Q	ME
5897.47	5899.10	16951.7	KR			0.3Q	ME
5894.56	5896.19	16960.1	KR II			0.9Q	ME
5887.68	5889.31	16979.9	KR I			0.5	ME
5882.67	5884.30	16994.4	KR			0.3Q	ME
5881.18	5882.81	16998.7	KR I			0.3	ME
5879.900	5881.530	17002.38	KR I			1.7	IJA
5870.9158	5872.5430	17028.398	KR I			3.5	S
5866.752	5868.378	17040.48	KR I			1.7	ME
5860.75	5862.37	17057.9	KR			1.0Q	ME
5854.04	5855.66	17077.5	KR			0.6Q	ME
5852.86	5854.48	17080.9	KR I			0.7	ME
5849.66	5851.28	17090.3	KR I			0.3	ME
5841.44	5843.06	17114.3	KR I			0.6	RS
5832.859	5834.476	17139.50	KR I			2.0	JA
5827.07	5828.69	17156.5	KR I			1.3	ME
5824.50	5826.11	17164.1	KR I			1.6	ME
5823.51	5825.12	17167.0	KR I			0.5	ME
5820.10	5821.71	17177.1	KR I			1.2	ME
5810.80	5812.41	17204.6	KR I			0.9	ME
5805.53	5807.14	17220.2	KR I			1.3	ME
5801.17	5802.78	17233.1	KR I			0.3	ME
5800.16	5801.77	17236.1	KR II			0.8T	ME
5788.24	5789.85	17271.6	KR I			0.8	ME
5787.29	5788.89	17274.4	KR I			0.8	ME
5783.89	5785.49	17284.6	KR I			1.0	ME
5777.72	5779.32	17303.1	KR II			0.3J	ME
5775.56	5777.16	17309.5	KR			0.3	ME
5771.41	5773.01	17322.0	KR II			2.0	ME
5762.90	5764.50	17347.6	KR I			0.6	ME
5755.60	5757.20	17369.6	KR II			0.3J	ME
5755.04	5756.64	17371.3	KR I			0.3	ME
5752.98	5754.58	17377.5	KR II			1.8	ME
5750.57	5752.17	17384.8	KR I			1.0	ME
5749.27	5750.86	17388.7	KR II			0.7Q	ME
5749.02	5750.61	17389.4	KR I			0.7	ME
5730.86	5732.45	17444.6	KR I			0.6	ME
5726.59	5728.18	17457.6	KR I			1.3	ME

AIR WAVELENGTH (ANGSTRMS)	VACUUM WAVELENGTH (ANGSTRMS)	WAVENUMBER (KAYSERS)	LMENT	LOG ARC	INTENSITY SPK	DIS	REF
5723.56	5725.15	17466.8	KR I			1.2	ME
5721.88	5723.47	17471.9	KR I			1.0	ME
5717.61	5719.20	17485.0	KR I			0.5	ME
5714.11	5715.70	17495.7	KR I			0.3	ME
5707.512	5709.095	17515.91	KR I			1.6	IJA
5702.19	5703.77	17532.3	KR I			1.0	ME
5699.84	5701.42	17539.5	KR II			1.0	ME
5696.54	5698.12	17549.6	KR I			0.5	ME
5692.11	5693.69	17563.3	KR			0.7Q	ME
5690.35	5691.93	17568.7	KR II			2.3T	ME
5681.89	5683.47	17594.9	KR II			2.6	ME
5674.52	5676.09	17617.7	KR II			1.5C	ME
5672.78	5674.35	17623.1	KR II			1.6C	ME
5672.452	5674.026	17624.17	KR I			1.7	IJA
5662.67	5664.24	17654.6	KR I			0.5	ME
5650.37	5651.94	17693.1	KR II			1.0J	ME
5649.5628	5651.1308	17695.573	KR I			2.0	S
5641.07	5642.64	17722.2	KR II			0.5B	ME
5633.02	5634.58	17747.5	KR II			2.0H	ME
5617.63	5619.19	17796.2	KR II			0.3	ME
5611.82	5613.38	17814.6	KR I			0.6	ME
5608.37	5609.93	17825.5	KR			0.5	ME
5591.41	5592.96	17879.6	KR I			0.3	ME
5580.388	5581.937	17914.93	KR I			1.9	IJA
5577.64	5579.19	17923.7	KR I			0.5H	ME
5575.56	5577.11	17930.4	KR I			1.0	ME
5573.13	5574.68	17938.3	KR I			0.3	ME
5570.2895	5571.8363	17947.405	KR I			3.3	S
5568.65	5570.20	17952.7	KR II			2.0	ME
5562.2257	5563.7703	17973.423	KR I			2.7	S
5559.26	5560.80	17983.0	KR I			0.3	RS
5552.99	5554.53	18003.3	KR II			2.0T	ME
5546.11	5547.65	18025.6	KR II			0.7J	ME
5541.65	5543.19	18040.2	KR II			0.6T	ME
5532.29	5533.83	18070.7	KR II			0.7	ME
5528.63	5530.17	18082.6	KR I			0.3	ME
5523.47	5525.00	18099.5	KR II			1.5	ME
5522.94	5524.47	18101.3	KR II			1.8	ME
5521.17	5522.70	18107.1	KR I			0.5	ME
5520.52	5522.05	18109.2	KR I			1.6	ME
5516.66	5518.19	18121.9	KR I			1.3	ME
5504.34	5505.87	18162.4	KR I			1.3	ME
5504.02	5505.55	18163.5	KR I			1.2	ME
5500.71	5502.24	18174.4	KR I			1.7	ME
5499.54	5501.07	18178.3	KR II			1.7	ME
5496.21	5497.74	18189.3	KR I			0.5	ME
5491.43	5492.96	18205.1	KR II			0.6H	ME
5491.33	5492.86	18205.5	KR I			0.3H	ME
5490.94	5492.47	18206.8	KR I			1.7	ME
5488.86	5490.39	18213.7	KR I			0.7	ME
5476.58	5478.10	18254.5	KR I			0.3	ME
5476.46	5477.98	18254.9	KR II			0.6Q	ME
5468.17	5469.69	18282.6	KR II			2.3C	ME
5462.65	5464.17	18301.1	KR I			0.3	ME
5459.47	5460.99	18311.7	KR I			0.6	ME
5458.80	5460.32	18313.9	KR I			0.8	ME
5456.39	5457.91	18322.0	KR I			0.3H	ME
5449.61	5451.12	18344.8	KR II			0.3J	ME
5447.86	5449.37	18350.7	KR I			0.5H	ME
5446.34	5447.85	18355.9	KR II			1.9	ME
5438.63	5440.14	18381.9	KR II			1.6	ME
5433.24	5434.75	18400.1	KR II			0.3C	ME
5418.43	5419.94	18450.4	KR II			1.5T	ME
5403.03	5404.53	18503.0	KR I			0.3H	ME
5379.64	5381.14	18583.4	KR I			1.2	ME
5374.19	5375.68	18602.3	KR II			0.5H	ME
5371.74	5373.23	18610.8	KR I			0.3	ME
5355.45	5356.94	18667.4	KR II			1.0H	ME
5347.37	5348.86	18695.6	KR I			0.3	ME
5346.76	5348.25	18697.7	KR II			1.8B	ME

AIR WAVELENGTH (ANGSTRMS)	VACUUM WAVELENGTH (ANGSTRMS)	WAVENUMBER (KAYSERS)	LMENT	LOG ARC	INTENSITY SPK	DIS	REF
5339.13	5340.62	18724.4	KR I			1.3	ME
5334.78	5336.26	18739.7	KR I			1.0	ME
5333.41	5334.89	18744.5	KR II			2.7H	ME
5331.08	5332.56	18752.7	KR I			0.3	ME
5329.15	5330.63	18759.5	KR II			0.6H	ME
5327.87	5329.35	18764.0	KR I			0.3	ME
5322.77	5324.25	18782.0	KR II			1.8B	ME
5322.02	5323.50	18784.6	KR I			0.3	ME
5317.41	5318.89	18800.9	KR II			1.5B	ME
5310.26	5311.74	18826.2	KR II			0.6H	ME
5308.66	5310.14	18831.9	KR II			2.3	ME
5300.74	5302.21	18860.0	KR I			0.5	ME
5299.79	5301.26	18863.4	KR I			0.3H	ME
5286.86	5288.33	18909.6	KR			0.3H	ME
5279.84	5281.31	18934.7	KR I			1.0	ME
5276.50	5277.97	18946.7	KR II			2.0H	ME
5274.61	5276.08	18953.5	KR I			0.6	ME
5256.75	5258.21	19017.9	KR II			1.5	ME
5249.06	5250.52	19045.7	KR II			0.6B	ME
5245.25	5246.71	19059.6	KR			0.6T	ME
5241.29	5242.75	19074.0	KR II			0.3B	ME
5232.06	5233.52	19107.6	KR			0.3	ME
5230.15	5231.61	19114.6	KR II			0.5H	ME
5229.52	5230.98	19116.9	KR II			1.8	ME
5228.18	5229.64	19121.8	KR I			1.3	ME
5225.05	5226.50	19133.2	KR II			0.5	ME
5224.56	5226.01	19135.0	KR II			0.8	ME
5223.57	5225.02	19138.7	KR I			0.7	ME
5222.38	5223.83	19143.0	KR I			0.5	ME
5217.93	5219.38	19159.4	KR II			1.1	ME
5217.45	5218.90	19161.1	KR II			1.5	ME
5215.81	5217.26	19167.1	KR I			0.9	ME
5208.32	5209.77	19194.7	KR II			2.7	ME
5201.56	5203.01	19219.6	KR II			0.3H	ME
5200.22	5201.67	19224.6	KR II			1.8T	ME
5186.99	5188.43	19273.6	KR II			1.8T	ME
5177.71	5179.15	19308.2	KR II			0.8T	ME
5172.36	5173.80	19328.1	KR I			0.3	ME
5168.06	5169.50	19344.2	KR I			0.6	ME
5166.80	5168.24	19348.9	KR II			1.9	ME
5152.01	5153.45	19404.5	KR			0.5B	ME
5149.61	5151.04	19413.5	KR II			0.5B	ME
5145.28	5146.71	19429.9	KR II			0.6	ME
5145.04	5146.47	19430.8	KR I			0.3H	ME
5143.05	5144.48	19438.3	KR II			2.8B	ME
5142.7	5144.1	19440.	KR I			0.6	ME
5125.73	5127.16	19504.0	KR II			2.6Q	ME
5123.16	5124.59	19513.8	KR II			1.2Q	ME
5109.81	5111.23	19564.7	KR			0.3	ME
5089.12	5090.54	19644.3	KR I			0.3H	ME
5086.52	5087.94	19654.3	KR II			2.4B	ME
5078.19	5079.61	19686.6	KR II			0.3Q	ME
5077.23	5078.65	19690.3	KR II			1.6	ME
5075.92	5077.33	19695.4	KR II			0.6Q	ME
5072.55	5073.96	19708.5	KR II			1.6	ME
5067.41	5068.82	19728.4	KR			0.5H	ME
5065.58	5066.99	19735.6	KR II			1.3Q	ME
5058.08	5059.49	19764.8	KR I			0.6	ME
5054.53	5055.94	19778.7	KR II			1.5Q	ME
5047.52	5048.93	19806.2	KR II			0.6B	ME
5046.31	5047.72	19810.9	KR II			1.9Q	ME
5040.34	5041.75	19834.4	KR I			0.8	ME
5033.85	5035.25	19860.0	KR II			2.0Q	ME
5030.96	5032.36	19871.4	KR			0.5T	ME
5029.15	5030.55	19878.5	KR I			0.7	ME
5028.36	5029.76	19881.7	KR II			1.5B	ME
5022.40	5023.80	19905.2	KR II			2.3	ME
5021.88	5023.28	19907.3	KR II			2.0	ME
5020.43	5021.83	19913.1	KR II			0.6T	ME
5013.29	5014.69	19941.4	KR II			2.0	ME

AIR WAVELENGTH (ANGSTRMS)	VACUUM WAVELENGTH (ANGSTRMS)	WAVENUMBER (KAYSERS)	LMENT	LOG ARC	INTENSITY SPK	INTENSITY DIS	REF
5002.14	5003.54	19985.9	KR I			0.3	ME
4998.54	4999.93	20000.3	KR II			0.7Q	ME
4982.83	4984.22	20063.3	KR II			1.7B	ME
4978.89	4980.28	20079.2	KR II			2.0B	ME
4969.36	4970.75	20117.7	KR I			1.2	ME
4969.08	4970.47	20118.8	KR I			1.3	ME
4960.25	4961.63	20154.6	KR II			2.0B	ME
4955.27	4956.65	20174.9	KR I			1.2	ME
4948.50	4949.88	20202.5	KR II			1.7B	ME
4945.59	4946.97	20214.4	KR II			2.5	ME
4938.38	4939.76	20243.9	KR I			0.3H	ME
4934.48	4935.86	20259.9	KR I			0.6H	ME
4930.38	4931.76	20276.7	KR I			0.6H	ME
4915.94	4917.31	20336.3	KR II			2.0B	ME
4914.62	4915.99	20341.8	KR II			0.3H	ME
4910.39	4911.76	20359.3	KR I			0.3	ME
4908.34	4909.71	20367.8	KR II			0.3B	ME
4897.2	4898.6	20414.	KR			0.5	ME
4870.14	4871.50	20527.6	KR II			1.3T	ME
4864.91	4866.27	20549.6	KR I			0.3H	ME
4862.1	4863.5	20561.	KR			0.3H	ME
4861.84	4863.20	20562.6	KR I			0.3H	ME
4857.20	4858.56	20582.2	KR II			2.2	ME
4852.61	4853.97	20601.7	KR II			0.3	ME
4846.60	4847.95	20627.3	KR II			2.8	ME
4845.14	4846.49	20633.5	KR II			0.3H	ME
4839.04	4840.39	20659.5	KR II		0.6H		ME
4836.56	4837.91	20670.1	KR II			1.3B	ME
4833.68	4835.03	20682.4	KR II			0.6H	ME
4832.07	4833.42	20689.3	KR II			2.9	ME
4825.18	4826.53	20718.8	KR II			2.5	ME
4812.637	4813.982	20772.82	KR I			1.6	IHU
4811.76	4813.10	20776.6	KR II			2.5	ME
4810.51	4811.85	20782.0	KR I			0.5	ME
4802.97	4804.31	20814.6	KR II			0.6	ME
4796.33	4797.67	20843.4	KR II			1.8B	ME
4791.15	4792.49	20866.0	KR II			0.5	ME
4788.76	4790.10	20876.4	KR II			0.7H	ME
4774.46	4775.79	20938.9	KR II			0.3	ME
4773.01	4774.34	20945.3	KR II			1.6H	ME
4765.74	4767.07	20977.2	KR II			3.0	ME
4762.43	4763.76	20991.8	KR II			2.5	ME
4752.02	4753.35	21037.8	KR II			2.0B	ME
4739.00	4740.33	21095.6	KR II			3.5	ME
4724.89	4726.21	21158.6	KR I			1.3	ME
4722.16	4723.48	21170.8	KR			0.5	ME
4706.31	4707.63	21242.1	KR II			0.5	ME
4705.44	4706.76	21246.1	KR II			0.3B	ME
4699.69	4701.01	21272.1	KR II			1.5Q	ME
4695.66	4696.97	21290.3	KR II			1.7B	ME
4694.84	4696.15	21294.0	KR I			0.6	ME
4694.44	4695.75	21295.8	KR II			2.3B	ME
4691.28	4692.59	21310.2	KR II			2.0	ME
4688.3	4689.6	21324.	KR II			0.5H	ME
4687.28	4688.59	21328.4	KR II			1.0B	ME
4686.30	4687.61	21332.8	KR II			0.9Q	ME
4683.68	4684.99	21344.8	KR II			0.7	ME
4680.41	4681.72	21359.7	KR II			2.7	ME
4673.80	4675.11	21389.9	KR II			0.5	ME
4672.09	4673.40	21397.7	KR II			0.3Q	ME
4671.61	4672.92	21399.9	KR I			1.0	ME
4658.87	4660.17	21458.4	KR II			3.3	ME
4650.17	4651.47	21498.6	KR II			1.5	ME
4640.20	4641.50	21544.8	KR			0.3J	ME
4636.14	4637.44	21563.6	KR I			1.3	ME
4635.42	4636.72	21567.0	KR II			0.9	ME
4633.88	4635.18	21574.1	KR II			2.9	ME
4619.99	4621.28	21639.0	KR II			0.7H	ME
4619.15	4620.44	21642.9	KR II			3.0	ME
4615.28	4616.57	21661.1	KR II			2.7	ME

AIR WAVELENGTH (ANGSTRMS)	VACUUM WAVELENGTH (ANGSTRMS)	WAVENUMBER (KAYSERS)	LMENT	LOG ARC	INTENSITY SPK	INTENSITY DIS	REF
4614.50	4615.79	21664.7	KR II			1.2	ME
4613.79	4615.08	21668.1	KR II			0.3H	ME
4610.65	4611.94	21682.8	KR II			1.8B	ME
4609.72	4611.01	21687.2	KR II			1.3C	ME
4604.02	4605.31	21714.1	KR II			1.8B	ME
4600.69	4601.98	21729.8	KR			0.3	ME
4598.49	4599.78	21740.2	KR II			1.7B	ME
4596.81	4598.10	21748.1	KR			0.3	ME
4592.80	4594.09	21767.1	KR II			2.2Q	ME
4582.85	4584.13	21814.4	KR II			2.5B	ME
4580.11	4581.39	21827.4	KR II			0.3H	ME
4577.20	4578.48	21841.3	KR II			2.9	ME
4573.33	4574.61	21859.8	KR II			1.5B	ME
4560.38	4561.66	21921.9	KR II			0.5B	ME
4556.61	4557.89	21940.0	KR II			2.3B	ME
4552.77	4554.05	21958.5	KR II			0.5	ME
4550.298	4551.574	21970.42	KR I			1.6	I
4538.06	4539.33	22029.7	KR I			0.5	ME
4528.62	4529.89	22075.6	KR II			0.5B	ME
4523.14	4524.41	22102.3	KR II			2.6B	ME
4502.3547	4503.6176	22204.372	KR I			2.8	S
4489.88	4491.14	22266.1	KR II			2.6B	ME
4488.22	4489.48	22274.3	KR			0.5J	ME
4481.85	4483.11	22306.0	KR II			1.7Q	ME
4479.86	4481.12	22315.9	KR II			0.7Q	ME
4475.00	4476.26	22340.1	KR II			2.9C	ME
4463.6902	4464.9429	22396.703	KR I			2.9	S
4461.61	4462.86	22407.1	KR			0.3B	ME
4459.99	4461.24	22415.3	KR II			0.9H	ME
4457.25	4458.50	22429.1	KR II			1.6Q	ME
4454.37	4455.62	22443.6	KR II			1.0Q	ME
4453.9179	4455.1681	22445.842	KR I			2.8	S
4453.21	4454.46	22449.4	KR II			1.7T	ME
4450.34	4451.59	22463.9	KR			0.6C	ME
4443.72	4444.97	22497.4	KR II			0.5	ME
4436.81	4438.06	22532.4	KR II			2.8	ME
4431.67	4432.91	22558.5	KR II			2.7	ME
4425.191	4426.434	22591.55	KR I			2.0	I
4422.70	4423.94	22604.3	KR II			2.0C	ME
4418.763	4420.004	22624.41	KR I			1.7	IHU
4417.24	4418.48	22632.2	KR II			1.6	ME
4416.884	4418.124	22634.04	KR I			1.3	IHU
4412.39	4413.63	22657.1	KR I			0.8	ME
4410.369	4411.608	22667.47	KR I			1.7	I
4408.89	4410.13	22675.1	KR II			1.6C	ME
4404.33	4405.57	22698.6	KR II			1.5H	ME
4400.87	4402.11	22716.4	KR II			2.0B	ME
4399.9670	4401.2029	22721.061	KR I			2.3	S
4399.39	4400.63	22724.0	KR			1.2C	ME
4389.72	4390.95	22774.1	KR II			1.3B	ME
4388.90	4390.13	22778.4	KR II			0.5B	ME
4386.54	4387.77	22790.6	KR II			2.5B	ME
4385.27	4386.50	22797.2	KR II			1.7Q	ME
4381.52	4382.75	22816.7	KR II			2.0H	ME
4380.11	4381.34	22824.1	KR			0.3	ME
4377.71	4378.94	22836.6	KR			1.6H	ME
4376.1220	4377.3517	22844.863	KR I			2.9	S
4371.25	4372.48	22870.3	KR II			1.3B	ME
4369.69	4370.92	22878.5	KR II			2.3	ME
4366.26	4367.49	22896.5	KR II			0.8B	ME
4364.61	4365.84	22905.1	KR II			0.6B	ME
4362.6423	4363.8684	22915.448	KR I			2.7	S
4355.478	4356.702	22953.14	KR II			3.5	ME
4354.23	4355.45	22959.7	KR			0.3	ME
4353.90	4355.12	22961.5	KR			0.3	ME
4351.3607	4352.5838	22974.859	KR I			2.0	S
4351.02	4352.24	22976.7	KR II			1.6J	ME
4349.55	4350.77	22984.4	KR			0.3	ME
4341.33	4342.55	23027.9	KR II			0.9Q	ME
4333.34	4334.56	23070.4	KR II			1.7J	ME

AIR WAVELENGTH (ANGSTRMS)	VACUUM WAVELENGTH (ANGSTRMS)	WAVENUMBER (KAYSERS)	LMENT	LOG ARC	INTENSITY SPK	DIS	REF
4331.24	4332.46	23081.6	KR II			1.9J	ME
4322.98	4324.20	23125.7	KR II			2.2Q	ME
4319.5797	4320.7945	23143.892	KR I			3.0	S
4319.12	4320.33	23146.4	KR II			0.6	ME
4318.5525	4319.7670	23149.397	KR I			2.6	S
4317.81	4319.02	23153.4	KR II			2.7Q	ME
4305.81	4307.02	23217.9	KR			0.5H	ME
4302.446	4303.656	23236.06	KR			1.0	IHU
4301.53	4302.74	23241.0	KR II			1.6	ME
4300.4877	4301.6974	23246.637	KR I			2.3	S
4300.4877	4301.6974	23246.637	KR II			2.3	S
4295.21	4296.42	23275.2	KR			0.9B	ME
4292.92	4294.13	23287.6	KR II			2.8	ME
4292.64	4293.85	23289.1	KR I			0.8	ME
4290.78	4291.99	23299.2	KR			0.6	ME
4288.02	4289.23	23314.2	KR I			0.7	ME
4287.45	4288.66	23317.3	KR			0.6J	ME
4286.4873	4287.6934	23322.563	KR I			1.6	S
4285.40	4286.61	23328.5	KR			0.6H	ME
4282.9683	4284.1734	23341.725	KR I			2.0	S
4280.61	4281.81	23354.6	KR II			0.7B	ME
4280.05	4281.25	23357.6	KR			0.3H	ME
4275.75	4276.95	23381.1	KR			0.3	ME
4273.9700	4275.1728	23390.867	KR I			3.0	S
4273.48	4274.68	23393.6	KR			0.6	ME
4268.81	4270.01	23419.1	KR II			2.0Q	ME
4268.57	4269.77	23420.5	KR II			1.8Q	ME
4263.288	4264.488	23449.47	KR I			1.3	IHU
4260.85	4262.05	23462.9	KR II			0.7B	ME
4259.44	4260.64	23470.7	KR			1.9C	ME
4254.85	4256.05	23496.0	KR II			2.0B	ME
4252.67	4253.87	23508.0	KR II			1.7C	ME
4250.58	4251.78	23519.6	KR II			2.2	ME
4236.64	4237.83	23597.0	KR II			2.0H	ME
4233.43	4234.62	23614.9	KR			0.3H	ME
4229.21	4230.40	23638.4	KR			0.9Q	ME
4228.79	4229.98	23640.8	KR II			1.3B	ME
4222.20	4223.39	23677.7	KR II			1.3B	ME
4217.88	4219.07	23701.9	KR II			0.3	ME
4210.67	4211.86	23742.5	KR			1.4Q	ME
4204.31	4205.49	23778.4	KR			0.5Q	ME
4201.42	4202.60	23794.8	KR II			1.5Q	ME
4185.12	4186.30	23887.4	KR II			1.7	ME
4184.473	4185.652	23891.14	KR I			1.3	IHU
4179.58	4180.76	23919.1	KR II			1.3Q	ME
4177.02	4178.20	23933.8	KR			0.5C	ME
4172.83	4174.01	23957.8	KR I			0.5	ME
4172.51	4173.69	23959.6	KR II			1.3B	ME
4167.28	4168.45	23989.7	KR I			0.7D	ME
4164.48	4165.65	24005.8	KR I			0.3	ME
4163.82	4164.99	24009.6	KR II			0.3	ME
4159.00	4160.17	24037.5	KR			0.6C	ME
4145.12	4146.29	24117.9	KR II			2.4	ME
4139.11	4140.28	24153.0	KR			2.0Q	ME
4137.96	4139.13	24159.7	KR			1.7J	ME
4135.86	4137.03	24171.9	KR II			0.5H	ME
4133.68	4134.85	24184.7	KR II			0.7Q	ME
4118.14	4119.30	24276.0	KR II			1.5Q	ME
4113.73	4114.89	24302.0	KR II			0.9Q	ME
4110.16	4111.32	24323.1	KR II			0.7Q	ME
4109.23	4110.39	24328.6	KR II			2.0C	ME
4108.43	4109.59	24333.3	KR I			0.5	ME
4099.71	4100.87	24385.1	KR II			0.5	ME
4098.72	4099.88	24391.0	KR II			2.4	ME
4088.33	4089.48	24453.0	KR II			2.7	ME
4086.90	4088.05	24461.5	KR I			0.3	ME
4066.09	4067.24	24586.7	KR			0.8Q	ME
4065.11	4066.26	24592.6	KR II			2.5	ME
4057.01	4058.16	24641.7	KR II			2.5C	ME
4056.57	4057.72	24644.4	KR I			0.5	ME

AIR WAVELENGTH (ANGSTRMS)	VACUUM WAVELENGTH (ANGSTRMS)	WAVENUMBER (KAYSERS)	LMENT	LOG ARC	INTENSITY SPK	DIS	REF
4050.42	4051.56	24681.8	KR II			1.7Q	ME
4044.67	4045.81	24716.9	KR II			1.9	ME
4037.83	4038.97	24758.8	KR II			1.5	ME
4029.66	4030.80	24809.0	KR I			0.3	ME
4008.48	4009.61	24940.1	KR II			1.0Q	ME
4008.08	4009.21	24942.6	KR II			1.4	ME
4005.57	4006.70	24958.2	KR II			1.5B	ME
4000.72	4001.85	24988.4	KR I			0.3	ME
3997.95	3999.08	25005.7	KR II			2.0Q	ME
3996.69	3997.82	25013.6	KR II			0.5	ME
3994.83	3995.96	25025.3	KR II			2.0	ME
3994.82	3995.95	25025.3	KR I			0.5	ME
3991.94	3993.07	25043.4	KR II			1.2	ME
3991.258	3992.387	25047.67	KR I			1.0	IHU
3991.079	3992.208	25048.80	KR I			1.3	HU
3990.66	3991.79	25051.4	KR II			1.2B	ME
3987.78	3988.91	25069.5	KR II			1.4	ME
3987.09	3988.22	25073.9	KR II			0.7T	ME
3982.170	3983.296	25104.84	KR I			0.8	IHU
3964.89	3966.01	25214.2	KR II			1.5B	ME
3962.34	3963.46	25230.5	KR II			1.0B	ME
3954.78	3955.90	25278.7	KR II			2.0Q	ME
3953.59	3954.71	25286.3	KR II			1.3	ME
3947.66	3948.78	25324.3	KR II			0.7B	ME
3945.48	3946.60	25338.3	KR II			0.7B	ME
3942.93	3944.05	25354.7	KR			1.3Q	ME
3940.92	3942.04	25367.6	KR			0.7Q	ME
3938.88	3940.00	25380.7	KR II			1.3Q	ME
3929.26	3930.37	25442.9	KR II			1.3B	ME
3921.68	3922.79	25492.1	KR II			0.8B	ME
3920.14	3921.25	25502.1	KR II			2.3B	ME
3917.64	3918.75	25518.3	KR II			1.7Q	ME
3916.90	3918.01	25523.2	KR II			0.5B	ME
3912.88	3913.99	25549.4	KR II			0.7B	ME
3912.59	3913.70	25551.3	KR II			1.8	ME
3906.25	3907.36	25592.7	KR II			2.2B	ME
3901.15	3902.26	25626.2	KR II			1.0B	ME
3894.71	3895.81	25668.6	KR II			1.8Q	ME
3887.54	3888.64	25715.9	KR II			0.7Q	ME
3880.07	3881.17	25765.4	KR II			0.3Q	ME
3875.44	3876.54	25796.2	KR II			2.2Q	ME
3858.78	3859.87	25907.6	KR II			0.7Q	ME
3857.32	3858.41	25917.4	KR II			1.3Q	ME
3846.83	3847.92	25988.1	KR II			0.7H	ME
3846.12	3847.21	25992.9	KR I			0.3	ME
3845.978	3847.069	25993.82	KR I			1.2	IHU
3844.45	3845.54	26004.1	KR II			1.7Q	ME
3842.28	3843.37	26018.8	KR II			1.3Q	ME
3840.92	3842.01	26028.1	KR II			0.7Q	ME
3839.37	3840.46	26038.6	KR II			0.6Q	ME
3837.816	3838.905	26049.10	KR I			1.5	I
3837.703	3838.792	26049.86	KR I			1.5	IHU
3836.54	3837.63	26057.8	KR II			1.5Q	ME
3831.17	3832.26	26094.3	KR II			0.3Q	ME
3826.15	3827.24	26128.5	KR II			0.3H	ME
3817.11	3818.19	26190.4	KR II			1.2Q	ME
3812.216	3813.298	26224.02	KR I			1.3	I
3806.17	3807.25	26265.7	KR II			0.9Q	ME
3804.67	3805.75	26276.0	KR II			1.5B	ME
3800.544	3801.623	26304.55	KR I			1.5	I
3796.884	3797.962	26329.91	KR I			1.3	I
3783.13	3784.20	26425.6	KR II			2.7B	ME
3778.09	3779.16	26460.9	KR II			2.7B	ME
3773.424	3774.496	26493.60	KR I			1.7	I
3771.34	3772.41	26508.2	KR II			1.5H	ME
3765.88	3766.95	26546.7	KR II			0.3J	ME
3758.93	3760.00	26595.8	KR II			0.8Q	ME
3754.24	3755.31	26629.0	KR II			1.9	ME
3747.50	3748.57	26676.9	KR			0.3J	ME
3744.80	3745.86	26696.1	KR II			2.2C	ME

AIR WAVELENGTH (ANGSTRMS)	VACUUM WAVELENGTH (ANGSTRMS)	WAVENUMBER (KAYSERS)	LMENT	LOG ARC	INTENSITY SPK	DIS	REF
3741.69	3742.75	26718.3	KR II			2.3B	ME
3740.73	3741.79	26725.1	KR II			0.8	ME
3735.78	3736.84	26760.6	KR II			1.6B	ME
3732.92	3733.98	26781.1	KR II			0.8	ME
3732.61	3733.67	26783.3	KR II			1.2H	ME
3731.67	3732.73	26790.0	KR II			0.3H	ME
3728.04	3729.10	26816.1	KR II			0.8B	ME
3721.35	3722.41	26864.3	KR II			2.2B	ME
3718.63	3719.69	26884.0	KR II			2.3B	ME
3718.02	3719.08	26888.4	KR II			2.5B	ME
3716.15	3717.21	26901.9	KR II			0.6	ME
3715.04	3716.10	26910.0	KR II			1.1H	ME
3698.045	3699.097	27033.62	KR I			0.8	IHU
3690.65	3691.70	27087.8	KR II			1.5	ME
3686.15	3687.20	27120.9	KR II			1.9Q	ME
3680.37	3681.42	27163.4	KR II			2.0Q	ME
3679.611	3680.659	27169.05	KR I			1.7	IHU
3679.561	3680.609	27169.42	KR I			1.7	IHU
3678.66	3679.71	27176.1	KR			0.8B	ME
3669.01	3670.05	27247.6	KR			2.2B	ME
3668.737	3669.782	27249.58	KR I			1.0	I
3668.59	3669.63	27250.7	KR II			0.8	ME
3666.01	3667.05	27269.9	KR II			0.7	ME
3665.326	3666.370	27274.93	KR I			1.9	I
3663.44	3664.48	27289.0	KR II			1.3	ME
3661.00	3662.04	27307.2	KR II			1.2	ME
3653.97	3655.01	27359.7	KR II			2.4B	ME
3651.02	3652.06	27381.8	KR II			1.4B	ME
3648.61	3649.65	27399.9	KR II			1.6B	ME
3637.93	3638.97	27480.3	KR II			0.6Q	ME
3637.48	3638.52	27483.7	KR II			1.3B	ME
3634.42	3635.46	27506.9	KR II			0.5Q	ME
3633.54	3634.58	27513.5	KR II			0.5H	ME
3632.490	3633.525	27521.48	KR I			0.6	IHU
3631.87	3632.91	27526.2	KR II			2.3B	ME
3628.157	3629.191	27554.35	KR I			1.0	I
3626.91	3627.94	27563.8	KR I			0.3	ME
3623.61	3624.64	27588.9	KR II			1.5B	ME
3615.475	3616.506	27651.00	KR I			1.3	I
3607.88	3608.91	27709.2	KR II			2.0Q	ME
3602.12	3603.15	27753.5	KR II			0.3H	ME
3599.90	3600.93	27770.6	KR II			1.6B	ME
3599.21	3600.24	27775.9	KR II			1.4H	ME
3596.86	3597.89	27794.1	KR II			0.3B	ME
3589.65	3590.67	27849.9	KR II			1.8Q	ME
3586.25	3587.27	27876.3	KR II			1.1B	ME
3577.60	3578.62	27943.7	KR II			0.6B	ME
3572.68	3573.70	27982.2	KR II			1.2Q	ME
3569.68	3570.70	28005.7	KR II			0.3H	ME
3555.54	3556.56	28117.1	KR II			0.9Q	ME
3553.49	3554.51	28133.3	KR II			1.3B	ME
3548.71	3549.72	28171.2	KR II			0.8	ME
3546.46	3547.47	28189.1	KR I			0.5	ME
3544.54	3545.55	28204.4	KR II			1.5Q	ME
3544.14	3545.15	28207.5	KR II			1.5Q	ME
3540.954	3541.966	28232.91	KR I			0.7	IHU
3539.542	3540.554	28244.17	KR I			1.2	IHU
3535.35	3536.36	28277.7	KR II			1.7B	ME
3527.42	3528.43	28341.2	KR			0.5B	ME
3522.675	3523.682	28379.40	KR I			1.2	IHU
3517.37	3518.38	28422.2	KR II			0.7C	ME
3511.896	3512.900	28466.51	KR I			0.6	IHU
3507.84	3508.84	28499.4	KR I			0.5	ME
3506.66	3507.66	28509.0	KR I			0.5	ME
3503.898	3504.900	28531.48	KR I			1.2	IHU
3503.25	3504.25	28536.8	KR II			1.7Q	ME
3502.554	3503.556	28542.43	KR I			1.3	IHU
3498.92	3499.92	28572.1	KR II			0.3J	ME
3498.50	3499.50	28575.5	KR II			0.6J	ME
3497.45	3498.45	28584.1	KR II			0.5H	ME

AIR WAVELENGTH (ANGSTRMS)	VACUUM WAVELENGTH (ANGSTRMS)	WAVENUMBER (KAYSERS)	LMENT	LOG ARC	INTENSITY SPK	DIS	REF
3495.990	3496.990	28596.02	KR I			1.0	I
3493.04	3494.04	28620.2	KR II			0.9J	ME
3488.65	3489.65	28656.2	KR II			1.5H	ME
3487.49	3488.49	28665.7	KR II			0.8H	ME
3479.00	3480.00	28735.7	KR II			0.5H	ME
3477.89	3478.89	28744.8	KR II			0.7	ME
3475.31	3476.31	28766.2	KR			0.5B	ME
3470.05	3471.04	28809.8	KR II			1.5J	ME
3465.41	3466.40	28848.4	KR II			0.8Q	ME
3460.90	3461.89	28885.9	KR			0.3	ME
3460.13	3461.12	28892.4	KR I			0.3	ME
3460.09	3461.08	28892.7	KR II			1.7	ME
3456.87	3457.86	28919.6	KR I			0.5	ME
3453.46	3454.45	28948.2	KR			0.5H	ME
3446.51	3447.50	29006.6	KR II			1.7J	ME
3443.29	3444.28	29033.7	KR			0.7H	ME
3438.88	3439.87	29070.9	KR II			0.5H	ME
3434.28	3435.26	29109.8	KR			0.3	ME
3434.142	3435.126	29111.01	KR I			0.9	IHU
3431.722	3432.706	29131.54	KR I			1.3	IHU
3431.45	3432.43	29133.9	KR I			0.3	ME
3431.03	3432.01	29137.4	KR			0.9B	ME
3429.91	3430.89	29146.9	KR II			0.5H	ME
3427.71	3428.69	29165.6	KR II			1.5	ME
3426.27	3427.25	29177.9	KR I			0.3	ME
3424.943	3425.925	29189.20	KR I			1.2	IHU
3423.73	3424.71	29199.5	KR II			1.3C	ME
3414.80	3415.78	29275.9	KR II			1.0	ME
3413.78	3414.76	29284.6	KR			0.3H	ME
3409.89	3410.87	29318.1	KR I			0.3	ME
3408.97	3409.95	29326.0	KR I			0.3	ME
3405.16	3406.14	29358.8	KR II			1.9Q	ME
3402.79	3403.77	29379.2	KR II			0.3	ME
3401.40	3402.38	29391.2	KR I			0.7	ME
3389.67	3390.64	29492.9	KR II			0.7	ME
3387.11	3388.08	29515.2	KR			0.8H	ME
3385.23	3386.20	29531.6	KR II			1.2Q	ME
3381.11	3382.08	29567.6	KR II			1.3J	ME
3379.03	3380.00	29585.8	KR II			1.2J	ME
3375.78	3376.75	29614.3	KR II			0.5H	ME
3361.74	3362.71	29738.0	KR I			0.3	ME
3357.58	3358.54	29774.8	KR II			0.3	ME
3347.50	3348.46	29864.4	KR I			0.3	ME
3345.73	3346.69	29880.2	KR I			0.6	ME
3335.16	3336.12	29974.9	KR II			0.6B	ME
3328.00	3328.96	30039.4	KR I			0.3	ME
3321.16	3322.12	30101.3	KR II			0.9	ME
3315.72	3316.67	30150.7	KR II			1.2H	ME
3306.17	3307.12	30237.8	KR I			0.8	ME
3302.54	3303.49	30271.0	KR I			1.0	ME
3302.28	3303.23	30273.4	KR			0.6B	ME
3301.75	3302.70	30278.2	KR II			0.7H	ME
3300.18	3301.13	30292.7	KR II			0.6J	ME
3295.29	3296.24	30337.6	KR			0.5H	ME
3294.43	3295.38	30345.5	KR			0.3H	ME
3287.69	3288.64	30407.7	KR II			0.3J	ME
3287.38	3288.33	30410.6	KR II			0.3H	ME
3282.08	3283.03	30459.7	KR II			1.2H	ME
3264.33	3265.27	30625.3	KR II			0.7J	ME
3261.58	3262.52	30651.1	KR II			0.9H	ME
3256.67	3257.61	30697.4	KR II			0.6	ME
3248.03	3248.97	30779.0	KR II			0.8J	ME
3247.00	3247.94	30788.8	KR II			1.1J	ME
3246.18	3247.12	30796.6	KR			0.3	ME
3240.20	3241.14	30853.4	KR II			0.3	ME
3232.80	3233.73	30924.0	KR I			0.3	ME
3232.15	3233.08	30930.2	KR II			0.3H	ME
3230.68	3231.61	30944.3	KR I			0.3	ME
3226.57	3227.50	30983.7	KR II			0.7Q	ME
3223.52	3224.45	31013.0	KR II			1.1H	ME

AIR WAVELENGTH (ANGSTRMS)	VACUUM WAVELENGTH (ANGSTRMS)	WAVENUMBER (KAYSERS)	LMENT	LOG ARC	INTENSITY SPK	DIS	REF
3223.00	3223.93	31018.0	KR II			0.8	ME
3220.25	3221.18	31044.5	KR			0.8H	ME
3216.25	3217.18	31083.1	KR II			0.8H	ME
3210.89	3211.82	31135.0	KR II			0.8	ME
3210.64	3211.57	31137.4	KR II			0.3H	ME
3209.17	3210.10	31151.7	KR II			0.8B	ME
3208.28	3209.21	31160.4	KR II			1.6H	ME
3205.44	3206.37	31187.9	KR II			0.3	ME
3205.26	3206.19	31189.7	KR II			0.6	ME
3202.54	3203.47	31216.2	KR II			1.2B	ME
3200.40	3201.32	31237.1	KR II			1.7H	ME
3197.65	3198.57	31263.9	KR II			0.6B	ME
3195.50	3196.42	31285.0	KR II			0.3	ME
3192.54	3193.46	31314.0	KR II			0.3	ME
3187.61	3188.53	31362.4	KR II			0.6	ME
3181.25	3182.17	31425.1	KR			0.7J	ME
3176.94	3177.86	31467.7	KR II			1.2Q	ME
3175.67	3176.59	31480.3	KR II			1.6J	ME
3170.63	3171.55	31530.4	KR II			0.3	ME
3164.94	3165.86	31587.0	KR II			0.5Q	ME
3153.34	3154.25	31703.2	KR			0.3J	ME
3150.93	3151.84	31727.5	KR II			1.9H	ME
3140.44	3141.35	31833.4	KR II			0.5Q	ME
3139.86	3140.77	31839.3	KR II			0.6	ME
3139.58	3140.49	31842.2	KR II			1.3	ME
3138.24	3139.15	31855.8	KR			0.3	ME
3135.10	3136.01	31887.7	KR II			0.9	ME
3132.84	3133.75	31910.7	KR			0.6J	ME
3126.02	3126.93	31980.3	KR			0.8H	ME
3113.92	3114.82	32104.6	KR II			0.3	ME
3111.45	3112.35	32130.0	KR			0.3H	ME
3096.52	3097.42	32284.9	KR II			1.3C	ME
3095.14	3096.04	32299.3	KR II			1.5C	ME
3080.20	3081.09	32456.0	KR			0.3H	ME
3066.72	3067.61	32598.6	KR II			0.3	ME
3063.57	3064.46	32632.2	KR II			0.5B	ME
3061.51	3062.40	32654.1	KR			0.8H	ME
3060.84	3061.73	32661.3	KR II			1.5T	ME
3056.01	3056.90	32712.9	KR II			1.5J	ME
3055.31	3056.20	32720.4	KR II			0.5	ME
3049.23	3050.12	32785.6	KR II			0.9T	ME
3038.38	3039.26	32902.7	KR			0.5Q	ME
3034.16	3035.04	32948.5	KR II			0.3H	ME
3032.77	3033.65	32963.6	KR			0.7J	ME
3031.59	3032.47	32976.4	KR			0.7T	ME
3030.01	3030.89	32993.6	KR II			0.6	ME
3022.49	3023.37	33075.7	KR II			0.7H	ME
3017.65	3018.53	33128.7	KR II			1.3T	ME
3008.42	3009.30	33230.4	KR II			0.9H	ME
3002.48	3003.36	33296.1	KR II			0.3B	ME
2999.84	3000.71	33325.4	KR II			1.6H	ME
2996.60	2997.47	33361.4	KR II			1.3	ME
2988.69	2989.56	33449.7	KR II			0.5Q	ME
2985.33	2986.20	33487.4	KR			0.6Q	ME
2983.94	2984.81	33503.0	KR II			0.3Q	ME
2979.81	2980.68	33549.4	KR II			1.3	ME
2978.87	2979.74	33560.0	KR II			1.4	ME
2976.28	2977.15	33589.2	KR II			0.5H	ME
2975.92	2976.79	33593.2	KR II			0.5H	ME
2974.04	2974.91	33614.5	KR II			1.4H	ME
2972.34	2973.21	33633.7	KR II			0.3H	ME
2971.80	2972.67	33639.8	KR II			0.6H	ME
2968.11	2968.98	33681.6	KR			0.3	ME
2967.25	2968.12	33691.4	KR II			1.9J	ME
2966.13	2967.00	33704.1	KR			0.5T	ME
2965.11	2965.98	33715.7	KR II			0.3	ME
2963.11	2963.98	33738.5	KR			0.3	ME
2961.05	2961.91	33761.9	KR II			0.6	ME
2960.78	2961.64	33765.0	KR II			0.7	ME
2960.14	2961.00	33772.3	KR II			1.6J	ME

AIR WAVELENGTH (ANGSTRMS)	VACUUM WAVELENGTH (ANGSTRMS)	WAVENUMBER (KAYSERS)	LMENT	LOG ARC	INTENSITY SPK	DIS	REF
2958.35	2959.21	33792.7	KR II			1.3J	ME
2956.30	2957.16	33816.2	KR II			0.5H	ME
2954.28	2955.14	33839.3	KR II			1.1H	ME
2950.21	2951.07	33886.0	KR			1.5H	ME
2949.54	2950.40	33893.7	KR II			1.2H	ME
2939.70	2940.56	34007.1	KR II			0.3	ME
2930.40	2931.26	34115.0	KR II			0.3Q	ME
2921.92	2922.78	34214.1	KR			0.6Q	ME
2919.07	2919.92	34247.5	KR			0.3Q	ME
2913.23	2914.08	34316.1	KR II			0.6	ME
2908.62	2909.47	34370.5	KR II			0.7	ME
2894.63	2895.48	34536.6	KR II			0.3Q	ME
2884.21	2885.06	34661.4	KR II			0.3	ME
2875.71	2876.55	34763.8	KR			0.3H	ME
2873.72	2874.56	34787.9	KR II			0.6Q	ME
2862.17	2863.01	34928.3	KR II			0.3Q	ME
2847.36	2848.20	35109.9	KR II			1.4H	ME
2844.46	2845.30	35145.7	KR II			1.3	ME
2839.20	2840.03	35210.8	KR II			0.3	ME
2838.79	2839.62	35215.9	KR II			1.3	ME
2835.35	2836.18	35258.6	KR II			0.9B	ME
2833.00	2833.83	35287.9	KR II			2.0	ME
2832.39	2833.22	35295.5	KR II			0.3	ME
2830.43	2831.26	35319.9	KR II			0.5B	ME
2823.03	2823.86	35412.5	KR II			0.3H	ME
2822.63	2823.46	35417.5	KR II			0.7	ME
2816.87	2817.70	35490.0	KR II			1.5	ME
2816.46	2817.29	35495.1	KR II			1.8	ME
2803.60	2804.43	35657.9	KR II			0.6H	ME
2803.20	2804.03	35663.0	KR II			1.3H	ME
2801.23	2802.06	35688.1	KR II			0.3Q	ME
2800.98	2801.81	35691.3	KR II			0.3Q	ME
2796.26	2797.08	35751.5	KR II			0.3	ME
2795.81	2796.63	35757.3	KR			1.9H	ME
2789.83	2790.65	35833.9	KR			0.5Q	ME
2779.51	2780.33	35967.0	KR II			0.6	ME
2779.11	2779.93	35972.1	KR II			1.3	ME
2778.99	2779.81	35973.7	KR II			0.3	ME
2774.59	2775.41	36030.7	KR II			0.5	ME
2772.60	2773.42	36056.6	KR II			1.0H	ME
2759.02	2759.84	36234.0	KR II			0.6Q	ME
2751.59	2752.40	36331.9	KR II			0.7Q	ME
2747.41	2748.22	36387.2	KR			0.3Q	ME
2746.31	2747.12	36401.7	KR II			1.2	ME
2742.56	2743.37	36451.5	KR II			1.6	ME
2733.26	2734.07	36575.5	KR			1.7	ME
2732.33	2733.14	36588.0	KR II			0.6H	ME
2729.46	2730.27	36626.4	KR II			1.5H	ME
2719.90	2720.71	36755.2	KR			0.7Q	ME
2716.16	2716.96	36805.8	KR II			1.0H	ME
2714.49	2715.29	36828.4	KR II			0.5H	ME
2712.40	2713.20	36856.8	KR II			1.9H	ME
2711.11	2711.91	36874.3	KR II			0.3	ME
2710.27	2711.07	36885.8	KR II			0.5	ME
2701.34	2702.14	37007.7	KR II			1.2H	ME
2700.60	2701.40	37017.8	KR			0.5H	ME
2695.70	2696.50	37085.1	KR II			1.5H	ME
2691.20	2692.00	37147.1	KR II			0.3	ME
2688.37	2689.17	37186.2	KR			0.6H	ME
2683.55	2684.35	37253.0	KR II			1.2	ME
2677.20	2678.00	37341.4	KR II			0.8	ME
2675.31	2676.11	37367.7	KR II			0.6H	ME
2672.79	2673.58	37403.0	KR II			0.5	ME
2666.61	2667.40	37489.6	KR II			0.8H	ME
2664.37	2665.16	37521.2	KR II			0.6	ME
2664.00	2664.79	37526.4	KR II			0.9	ME
2662.57	2663.36	37546.5	KR II			0.3B	ME
2661.47	2662.26	37562.0	KR II			0.7	ME
2660.97	2661.76	37569.1	KR II			0.9B	ME
2659.60	2660.39	37588.5	KR II			0.3Q	ME

AIR WAVELENGTH (ANGSTRMS)	VACUUM WAVELENGTH (ANGSTRMS)	WAVENUMBER (KAYSERS)	LMENT	LOG ARC	INTENSITY SPK	DIS	REF
2656.38	2657.17	37634.0	KR			1.2H	ME
2653.95	2654.74	37668.5	KR II			0.8	ME
2649.67	2650.46	37729.3	KR II			0.6B	ME
2649.27	2650.06	37735.0	KR II			1.3	ME
2648.15	2648.94	37751.0	KR			1.3J	ME
2643.06	2643.85	37823.7	KR II			1.3H	ME
2642.08	2642.87	37837.7	KR II			0.6B	ME
2640.74	2641.53	37856.9	KR II			0.3B	ME
2638.32	2639.11	37891.6	KR II			0.3H	ME
2636.51	2637.30	37917.6	KR II			0.5B	ME
2634.41	2635.20	37947.9	KR II			0.8H	ME
2627.75	2628.53	38044.0	KR II			0.8	ME
2627.22	2628.00	38051.7	KR			0.5J	ME
2624.78	2625.56	38087.1	KR			0.8Q	ME
2622.82	2623.60	38115.5	KR II			0.3	ME
2620.65	2621.43	38147.1	KR			0.8H	ME
2620.44	2621.22	38150.1	KR II			1.6H	ME
2616.71	2617.49	38204.5	KR II			1.0H	ME
2610.98	2611.76	38288.4	KR			1.0H	ME
2602.11	2602.89	38418.9	KR II			0.8	ME
2597.73	2598.51	38483.6	KR II			0.8	ME
2596.73	2597.51	38498.5	KR			0.7J	ME
2595.36	2596.14	38518.8	KR II			0.6J	ME
2594.40	2595.18	38533.0	KR II			0.6	ME
2592.48	2593.26	38561.6	KR II			1.8	ME
2590.74	2591.51	38587.5	KR II			0.3H	ME
2589.08	2589.85	38612.2	KR			1.5	ME
2584.15	2584.92	38685.9	KR II			0.5H	ME
2581.74	2582.51	38722.0	KR			0.7Q	ME
2580.12	2580.89	38746.3	KR			0.3	ME
2578.98	2579.75	38763.4	KR II			0.3	ME
2574.25	2575.02	38834.6	KR			1.4	ME
2572.03	2572.80	38868.1	KR			1.0H	ME
2561.94	2562.71	39021.2	KR II			0.5B	ME
2559.10	2559.87	39064.5	KR II			0.9H	ME
2556.36	2557.13	39106.4	KR			0.8	ME
2555.91	2556.68	39113.3	KR			0.8	ME
2550.13	2550.90	39201.9	KR			0.3	ME
2538.34	2539.10	39384.0	KR II			0.7Q	ME
2527.16	2527.92	39558.2	KR			0.5B	ME
2511.74	2512.50	39801.0	KR II			0.5	ME
2510.56	2511.32	39819.8	KR II			0.7	ME
2506.56	2507.32	39883.3	KR			0.7Q	ME
2503.87	2504.62	39926.1	KR II			0.8Q	ME
2497.42	2498.17	40029.2	KR			0.3H	ME
2489.39	2490.14	40158.4	KR II			0.9H	ME
2487.62	2488.37	40186.9	KR II			0.6	ME
2487.50	2488.25	40188.9	KR II			0.5	ME
2478.85	2479.60	40329.1	KR II			0.5	ME
2474.69	2475.44	40396.9	KR II			0.3H	ME
2470.45	2471.20	40466.2	KR			1.0H	ME
2464.77	2465.52	40559.5	KR II			2.0H	ME
2463.27	2464.02	40584.2	KR II			0.3	ME
2462.33	2463.07	40599.7	KR II			0.3	ME
2456.07	2456.81	40703.1	KR			0.8	ME
2455.31	2456.05	40715.7	KR II			0.3	ME
2455.04	2455.78	40720.2	KR II			0.3	ME
2446.44	2447.18	40863.3	KR II			0.9	ME
2432.74	2433.48	41093.5	KR			0.9	ME
2428.35	2429.09	41167.7	KR			1.3	ME
2426.36	2427.10	41201.5	KR II			1.0	ME
2418.41	2419.14	41336.9	KR			0.6	ME
2414.89	2415.62	41397.2	KR II			1.0	ME
2413.81	2414.54	41415.7	KR			1.0H	ME
2409.06	2409.79	41497.4	KR II			0.7	ME
2408.52	2409.25	41506.6	KR			0.7H	ME
2392.78	2393.51	41779.7	KR II			1.0	ME
2390.50	2391.23	41819.5	KR II			0.6	ME
2375.52	2376.25	42083.2	KR II			1.3	ME
2373.68	2374.40	42115.8	KR II			0.6	ME

AIR WAVELENGTH (ANGSTRMS)	VACUUM WAVELENGTH (ANGSTRMS)	WAVENUMBER (KAYSERS)	LMENT	LOG ARC	INTENSITY SPK	DIS	REF
2368.94	2369.66	42200.1	KR II			0.5	ME
2365.52	2366.24	42261.1	KR II			0.5	ME
2362.74	2363.46	42310.8	KR II			0.8	ME
2353.68	2354.40	42473.7	KR			1.7	ME
2352.86	2353.58	42488.5	KR II			0.3	ME
2344.38	2345.10	42642.1	KR II			1.0	ME
2316.32	2317.03	43158.7	KR			1.0	ME
2315.52	2316.23	43173.6	KR II			0.9	ME
2314.24	2314.95	43197.5	KR			0.8	ME
2312.00	2312.71	43239.3	KR II			0.8	ME
2302.67	2303.38	43414.5	KR II			0.5	ME
2301.73	2302.44	43432.2	KR II			0.8	ME
2300.38	2301.09	43457.7	KR II			0.8	ME
2287.79	2288.50	43696.8	KR II			1.5	ME
2283.07	2283.77	43787.2	KR			1.5	ME
2273.24	2273.94	43976.5	KR			0.9	ME
2250.32	2251.02	44424.4	KR			0.9	ME
2245.39	2246.09	44521.9	KR			1.0	ME
2240.89	2241.59	44611.3	KR			0.3	ME
2237.15	2237.84	44685.9	KR			0.6	ME
2236.43	2237.12	44700.2	KR			0.3	ME
2234.34	2235.03	44742.0	KR			0.3	ME
2227.92	2228.61	44871.0	KR II			1.5	ME
2225.18	2225.87	44926.2	KR			0.3	ME
2221.86	2222.55	44993.3	KR			0.6	ME
2213.96	2214.65	45153.9	KR II			0.7	ME
2212.29	2212.98	45188.0	KR			0.8	ME
2211.71	2212.40	45199.8	KR II			0.7	ME
2199.17	2199.86	45457.5	KR			0.3	ME
2197.25	2197.94	45497.2	KR			0.3	ME
2191.91	2192.60	45608.0	KR			0.3	ME
2185.52	2186.20	45741.4	KR			0.6	ME
2177.79	2178.47	45903.7	KR			0.3H	ME
2170.83	2171.51	46050.9	KR			0.3	ME
2164.38	2165.06	46188.1	KR			0.6H	ME
2159.02	2159.70	46302.8	KR			0.3H	ME
2145.08	2145.76	46603.6	KR II			1.0	ME
2130.55	2131.22	46921.4	KR			0.3	ME
2130.43	2131.10	46924.1	KR			0.3	ME
2125.56	2126.23	47031.6	KR			0.3	ME
2123.48	2124.15	47077.6	KR			0.5	ME
2118.83	2119.50	47180.9	KR II			1.1	ME
2109.81	2110.48	47382.6	KR II			0.7	ME
2096.24	2096.91	47689.3	KR II			1.2	ME
2093.37	2094.04	47754.7	KR II			0.5	ME
2088.16	2088.82	47873.8	KR II			1.3	ME
2086.73	2087.39	47906.6	KR II			0.7	ME
2082.60	2083.26	48001.6	KR			0.3	ME
9988.47	9991.21	10008.8	LA	1.0			ME
9981.24	9983.98	10016.1	LA I	0.8			ME
9980.38	9983.12	10016.9	LA	1.0			ME
9965.70	9968.43	10031.7	LA	0.5			ME
9932.72	9935.44	10065.0	LA I	0.3			ME
9920.82	9923.54	10077.1	LA I	2.2			ME
9911.08	9913.80	10086.9	LA	0.5			ME
9893.82	9896.53	10104.6	LA II	0.6			ME
9881.24	9883.95	10117.4	LA I	2.0			ME
9862.60	9865.30	10136.5	LA I	0.5			ME
9852.58	9855.28	10146.8	LA I	0.8			ME
9848.70	9851.40	10150.8	LA	0.6			ME
9842.0	9844.7	10158.	LA	0.3			ME
9833.30	9836.00	10166.7	LA	0.5H			ME
9804.20	9806.89	10196.9	LA	0.3			ME
9775.09	9777.77	10227.3	LA	0.9			ME
9772.24	9774.92	10230.3	LA I	1.3			ME
9768.82	9771.50	10233.8	LA	0.5H			ME
9737.09	9739.76	10267.2	LA I	2.0			ME
9713.52	9716.18	10292.1	LA	0.5			ME
9709.45	9712.11	10296.4	LA I	1.0			ME
9706.48	9709.14	10299.6	LA I	1.3			ME

AIR WAVELENGTH (ANGSTRMS)	VACUUM WAVELENGTH (ANGSTRMS)	WAVENUMBER (KAYSERS)	LMENT	LOG ARC	INTENSITY SPK	DIS	REF
9699.64	9702.30	10306.8	LA I	1.3			ME
9692.6	9695.3	10314.	LA	0.3			ME
9672.94	9675.59	10335.3	LA II	0.5			ME
9672.04	9674.69	10336.2	LA I	0.9			ME
9657.00	9659.65	10352.3	LA II	1.3			ME
9646.47	9649.12	10363.6	LA I	0.5			ME
9640.81	9643.45	10369.7	LA	1.5			ME
9633.72	9636.36	10377.4	LA I	1.6			ME
9631.84	9634.48	10379.4	LA	0.3			ME
9570.38	9573.01	10446.0	LA I	0.7			ME
9563.60	9566.22	10453.4	LA II	0.6			ME
9560.72	9563.34	10456.6	LA I	0.9			MIT
9542.09	9544.71	10477.0	LA I	1.6			MIT
9541.23	9543.85	10477.9	LA I	1.3			ME
9485.14	9487.74	10539.9	LA I	1.1			MIT
9476.98	9479.58	10549.0	LA I	0.5			ME
9474.45	9477.05	10551.8	LA I	0.7			ME
9467.25	9469.85	10559.8	LA	0.3			ME
9461.79	9464.39	10565.9	LA I	1.7			MIT
9457.62	9460.21	10570.6	LA	0.3			ME
9438.29	9440.88	10592.2	LA I	2.0			MIT
9415.64	9418.22	10617.7	LA	0.5			ME
9412.64	9415.22	10621.1	LA I	1.9			MIT
9390.50	9393.08	10646.1	LA	0.6			MIT
9377.71	9380.28	10660.7	LA I	0.5			ME
9376.10	9378.67	10662.5	LA	0.5			ME
9372.58	9375.15	10666.5	LA I	1.3			MIT
9346.69	9349.25	10696.0	LA II	1.2			ME
9328.87	9331.43	10716.5	LA	0.3			ME
9293.3	9295.9	10757.	LA	0.3H			ME
9260.42	9262.96	10795.7	LA II	0.5			ME
9254.72	9257.26	10802.3	LA I	0.9			MIT
9250.06	9252.60	10807.8	LA	1.0			MIT
9226.60	9229.13	10835.3	LA I	1.0W			MIT
9219.63	9222.16	10843.4	LA I	0.8			MIT
9172.88	9175.40	10898.7	LA	0.5			ME
9172.39	9174.91	10899.3	LA I	0.8			ME
9157.11	9159.62	10917.5	LA	0.8			MIT
9151.63	9154.14	10924.0	LA I	0.6			MIT
9146.75	9149.26	10929.8	LA II	0.3			ME
9143.77	9146.28	10933.4	LA I	0.7			MIT
9142.18	9144.69	10935.3	LA I	0.6H			MIT
9119.17	9121.67	10962.9	LA I	0.90			MIT
9101.10	9103.60	10984.7	LA II	0.3			ME
9096.71	9099.21	10990.0	LA II	0.5	0.5		ME
9079.08	9081.57	11011.3	LA I	1.7			MIT
9058.63	9061.12	11036.2	LA I	0.3			ME
9056.48	9058.97	11038.8	LA I	0.6			MIT
9046.97	9049.45	11050.4	LA	0.3			ME
9025.05	9027.53	11077.2	LA I	0.3			ME
9016.80	9019.28	11087.4	LA II	0.3			ME
9008.26	9010.73	11097.9	LA	0.8			MIT
8977.39	8979.85	11136.0	LA I	0.3			ME
8970.07	8972.53	11145.1	LA I	0.5			ME
8965.41	8967.87	11150.9	LA	0.3			ME
8963.65	8966.11	11153.1	LA I	0.8			MIT
8957.73	8960.19	11160.5	LA I	1.4			MIT
8948.89	8951.35	11171.5	LA	0.3			ME
8884.24	8886.68	11252.8	LA I	0.3			ME
8879.56	8882.00	11258.7	LA	0.5			ME
8875.05	8877.49	11264.4	LA	0.3			ME
8871.02	8873.46	11269.6	LA	0.3H			MIT
8867.35	8869.78	11274.2	LA	0.3H			ME
8839.63	8842.06	11309.6	LA I	1.0			MIT
8825.82	8828.24	11327.3	LA I	1.4W			MIT
8818.93	8821.35	11336.1	LA	1.0			MIT
8810.57	8812.99	11346.9	LA	0.5			ME
8797.6	8800.0	11364.	LA	0.3H			ME
8781.98	8784.39	11383.8	LA II	0.3			ME
8767.92	8770.33	11402.1	LA	0.6			ME

AIR WAVELENGTH (ANGSTRMS)	VACUUM WAVELENGTH (ANGSTRMS)	WAVENUMBER (KAYSERS)	LMENT	LOG ARC	INTENSITY SPK	DIS	REF
8748.38	8750.78	11427.6	LA I	1.4			MIT
8720.41	8722.81	11464.2	LA I	1.1W			MIT
8703.11	8705.50	11487.0	LA I	0.7			MIT
8674.43	8676.81	11525.0	LA I	1.7W			MIT
8672.11	8674.49	11528.1	LA I	1.4W			MIT
8650.82	8653.20	11556.4	LA II	0.0H	0.3		ME
8638.47	8640.84	11572.9	LA I	1.0			MIT
8624.22	8626.59	11592.1	LA I	0.7			ME
8621.55	8623.92	11595.7	LA	0.3			ME
8590.94	8593.30	11637.0	LA I	0.7			MIT
8545.44	8547.79	11698.9	LA I	1.5			MIT
8543.46	8545.81	11701.6	LA I	1.3			ME
8529.67	8532.01	11720.6	LA	0.5			MIT
8514.65	8516.99	11741.2	LA II	0.5	0.5		ME
8513.57	8515.91	11742.7	LA I	0.9			MIT
8507.37	8509.71	11751.3	LA I	0.8			MIT
8484.01	8486.34	11783.6	LA II	0.3			ME
8476.48	8478.81	11794.1	LA I	1.4			MIT
8467.62	8469.95	11806.4	LA I	0.7			ME
8440.06	8442.38	11845.0	LA	0.5			ME
8379.80	8382.10	11930.2	LA I	0.8			ME
8346.53	8348.82	11977.7	LA I	1.8			MIT
8334.41	8336.70	11995.2	LA	0.7			MIT
8324.69	8326.98	12009.2	LA I	1.9			MIT
8316.04	8318.33	12021.7	LA I	1.2			MIT
8302.88	8305.16	12040.7	LA I	0.8			MIT
8247.44	8249.71	12121.6	LA I	1.6			MIT
8237.90	8240.16	12135.7	LA I	0.5K			ME
8211.60	8213.86	12174.6	LA	0.6W			MIT
8203.32	8205.58	12186.8	LA I	0.5			MIT
8159.05	8161.29	12253.0	LA II	0.7W	0.9		ME
8086.05	8088.27	12363.6	LA I	0.8N			MIT
8084.50	8086.72	12365.9	LA I	0.3			MIT
8059.5	8061.7	12404.	LA II	0.5H	0.5H		ME
8051.39	8053.60	12416.8	LA I	0.6	0.0		MIT
8001.89	8004.09	12493.6	LA I	0.6	0.0		MIT
7964.83	7967.02	12551.7	LA I	0.5	0.0		MIT
7948.30	7950.49	12577.8	LA I	0.7			RL
7947.92	7950.11	12578.4	LA I			1.0	KS
7927.83	7930.01	12610.3	LA II	0.0	0.3		ME
7891.69	7893.86	12668.1	LA II	0.5	0.3		ME
7879.93	7882.10	12687.0	LA II	0.3	0.6		ME
7841.80	7843.96	12748.7	LA	1.0	0.3		MIT
7740.54	7742.67	12915.4	LA II		0.3H		ME
7664.34	7666.45	13043.8	LA I	0.9	0.5		MIT
7612.94	7615.04	13131.9	LA II	0.5	0.5		ME
7539.23	7541.31	13260.3	LA I	1.0	0.5		MIT
7533.59	7535.66	13270.2	LA I	0.3			MIT
7498.84	7500.90	13331.7	LA I	0.7			MIT
7483.52	7485.58	13359.0	LA II	1.3	1.2		MIT
7463.04	7465.10	13395.7	LA I	1.0			MIT
7382.72	7384.75	13541.4	LA I	0.7			MIT
7379.71	7381.74	13546.9	LA I	1.0			KS
7345.31	7347.33	13610.4	LA I	2.1			MIT
7344.42	7346.44	13612.0	LA I	0.3			RL
7334.17	7336.19	13631.1	LA I	2.0			MIT
7320.91	7322.93	13655.7	LA I	0.3			MIT
7308.46	7310.47	13679.0	LA	0.3H			ME
7282.34	7284.35	13728.1	LA II	1.8	2.1		MIT
7270.30	7272.30	13750.8	LA I	0.7			RL
7270.11	7272.11	13751.2	LA I	1.2			MIT
7263.68	7265.68	13763.3	LA	0.5			ME
7262.83	7264.83	13764.9	LA	0.5			ME
7219.90	7221.89	13846.8	LA I	1.3			MIT
7217.16	7219.15	13852.1	LA	0.3			ME
7161.25	7163.22	13960.2	LA I	2.0			MIT
7158.05	7160.02	13966.4	LA I	2.1			MIT
7149.78	7151.75	13982.6	LA I	0.7			MIT
7116.8	7118.8	14047.	LA II		0.3		ME
7104.7	7106.7	14071.	LA II	1.7	0.3		ME

AIR WAVELENGTH (ANGSTRMS)	VACUUM WAVELENGTH (ANGSTRMS)	WAVENUMBER (KAYSERS)	LMENT	LOG ARC	INTENSITY SPK	DIS	REF
7098.17	7100.13	14084.3	LA	0.5			MIT
7092.40	7094.36	14095.7	LA I	0.3			RL
7079.74	7081.69	14120.9	LA II		0.3		ME
7076.39	7078.34	14127.6	LA	0.5			MIT
7068.36	7070.31	14143.7	LA I	2.0			MIT
7066.21	7068.16	14148.0	LA II	2.6	2.2		MIT
7045.96	7047.90	14188.6	LA I	2.1			MIT
7032.03	7033.97	14216.7	LA I	1.6			MIT
7023.67	7025.61	14233.7	LA I	2.0H			MIT
7013.20	7015.13	14254.9	LA I	0.6			MIT
6978.06	6979.98	14326.7	LA I	0.7			MIT
6976.85	6978.77	14329.2	LA	0.9			MIT
6968.73	6970.65	14345.9	LA II	0.5	1.1		MIT
6958.08	6960.00	14367.8	LA II	1.4	1.7		MIT
6954.50	6956.42	14375.2	LA II	1.2	1.0		MIT
6952.49	6954.41	14379.4	LA II	0.8	0.7		MIT
6934.99	6936.90	14415.7	LA I	1.7			MIT
6925.21	6927.12	14436.0	LA I	2.0			MIT
6918.26	6920.17	14450.5	LA I	1.3			MIT
6917.22	6919.13	14452.7	LA I	1.4			MIT
6859.02	6860.91	14575.3	LA II	0.6	0.5		MIT
6837.88	6839.77	14620.4	LA II	1.2	0.9		MIT
6834.02	6835.91	14628.6	LA II	1.4	1.0		MIT
6830.83	6832.71	14635.5	LA II		0.5		ME
6823.75	6825.63	14650.7	LA I	1.7			MIT
6821.51	6823.39	14655.5	LA I	0.7H			MIT
6813.64	6815.52	14672.4	LA II	0.7	1.4		MIT
6808.84	6810.72	14682.7	LA II	1.1	1.2		MIT
6801.38	6803.26	14698.8	LA II		0.5		ME
6796.73	6798.61	14708.9	LA I	0.7			MIT
6774.23	6776.10	14757.7	LA II	2.0	2.1		MIT
6753.02	6754.88	14804.1	LA I	2.0			MIT
6748.125	6749.988	14814.84	LA I	1.4			MIT
6741.20	6743.06	14830.1	LA	0.3			ME
6732.78	6734.64	14848.6	LA II	0.0	1.3		MIT
6718.67	6720.52	14879.8	LA II	0.5	1.5		MIT
6714.09	6715.94	14889.9	LA II	0.6	1.6		MIT
6709.496	6711.348	14900.14	LA I	2.2			KS
6699.84	6701.69	14921.6	LA	0.9			MIT
6699.23	6701.08	14923.0	LA	0.9			MIT
6692.87	6694.72	14937.2	LA I	1.7			MIT
6676.14	6677.98	14974.6	LA II		0.3		ME
6671.382	6673.224	14985.26	LA II	1.0	1.4		MIT
6661.399	6663.238	15007.72	LA I	1.8			MIT
6650.801	6652.638	15031.63	LA I	2.0			MIT
6645.160	6646.995	15044.39	LA I	0.9			MIT
6644.407	6646.242	15046.10	LA I	1.6			MIT
6642.787	6644.621	15049.77	LA II	1.0	1.4		MIT
6636.531	6638.364	15063.95	LA	0.0	0.5		MIT
6631.206	6633.037	15076.05	LA I	0.5			MIT
6628.413	6630.244	15082.40	LA I	0.7			MIT
6616.577	6618.404	15109.38	LA I	2.1			MIT
6612.44	6614.27	15118.8	LA	0.8			MIT
6608.257	6610.082	15128.41	LA I	1.7			MIT
6607.749	6609.574	15129.57	LA I	0.5			MIT
6600.168	6601.991	15146.95	LA I	1.6			MIT
6593.467	6595.288	15162.34	LA I	1.4			MIT
6582.191	6584.009	15188.31	LA I	0.9			MIT
6578.513	6580.330	15196.81	LA I	2.0			MIT
6570.960	6572.775	15214.27	LA II	0.7			MIT
6565.434	6567.248	15227.08	LA I	1.2			MIT
6555.95	6557.76	15249.1	LA I	0.7			KS
6555.118	6556.929	15251.04	LA I	0.6			MIT
6549.185	6550.994	15264.86	LA I	0.5			MIT
6543.151	6544.959	15278.94	LA I	2.1			MIT
6529.731	6531.535	15310.34	LA II	0.6	0.3H		MIT
6526.986	6528.789	15316.78	LA II	2.1	2.0		MIT
6523.86	6525.66	15324.1	LA I	0.5			ME
6520.770	6522.572	15331.38	LA I	1.2			MIT
6519.343	6521.144	15334.73	LA I	0.8			MIT

AIR WAVELENGTH (ANGSTRMS)	VACUUM WAVELENGTH (ANGSTRMS)	WAVENUMBER (KAYSERS)	LMENT	LOG ARC	INTENSITY SPK	DIS	REF
6506.211	6508.009	15365.68	LA I	1.2			MIT
6498.188	6499.984	15384.65	LA II	1.4	2.1		MIT
6492.848	6494.642	15397.31	LA I	1.0			MIT
6485.549	6487.341	15414.64	LA I	1.7			MIT
6468.434	6470.222	15455.42	LA I	1.4			MIT
6455.988	6457.772	15485.22	LA I	2.0			MIT
6454.531	6456.315	15488.71	LA I	2.1			MIT
6450.331	6452.114	15498.80	LA I	1.3			MIT
6449.94	6451.72	15499.7	LA	0.3			ME
6448.25	6450.03	15503.8	LA I	0.7			ME
6448.109	6449.891	15504.14	LA I	1.0			MIT
6446.604	6448.386	15507.76	LA II	1.2	2.0		MIT
6443.05	6444.83	15516.3	LA II	0.9	1.4H		ME
6426.614	6428.390	15555.99	LA I	0.6			MIT
6417.220	6418.994	15578.77	LA I	0.8			MIT
6410.995	6412.767	15593.89	LA I	2.0			MIT
6399.053	6400.822	15622.99	LA II	1.2	2.3		MIT
6394.234	6396.002	15634.77	LA I	2.2			MIT
6390.484	6392.251	15643.94	LA II	1.8	2.0		MIT
6375.56	6377.32	15680.6	LA I	0.5			MIT
6375.056	6376.818	15681.80	LA	0.6			MIT
6374.107	6375.869	15684.14	LA II	0.8	1.2		MIT
6360.238	6361.996	15718.34	LA I	1.2			MIT
6358.137	6359.895	15723.53	LA II	1.0	1.2		MIT
6356.434	6358.191	15727.74	LA I	0.7			MIT
6339.26	6341.01	15770.3	LA I	0.5			MIT
6337.88	6339.63	15773.8	LA II	0.5	0.3		MIT
6333.820	6335.571	15783.90	LA I	0.5			MIT
6333.211	6334.962	15785.41	LA I	0.3			MIT
6330.448	6332.198	15792.30	LA I	0.5			MIT
6325.923	6327.672	15803.60	LA I	1.5			MIT
6320.392	6322.140	15817.43	LA II	1.8	2.0		MIT
6318.260	6320.007	15822.77	LA I	1.0			MIT
6315.817	6317.564	15828.89	LA II	0.6	1.4		MIT
6310.926	6312.671	15841.15	LA II	1.0	2.0		MIT
6310.145	6311.890	15843.11	LA	0.8			MIT
6308.87	6310.61	15846.3	LA	0.5			ME
6308.243	6309.987	15847.89	LA	0.5			MIT
6307.266	6309.010	15850.35	LA II	0.5	1.0H		MIT
6305.45	6307.19	15854.9	LA II	1.1	0.7		MIT
6296.096	6297.837	15878.47	LA II	1.7	2.2		MIT
6293.565	6295.306	15884.85	LA I	1.5			MIT
6288.56	6290.30	15897.5	LA I	0.6			ME
6287.750	6289.489	15899.54	LA I	0.7			MIT
6273.763	6275.498	15934.99	LA II	0.7	1.7		MIT
6266.030	6267.763	15954.66	LA I	1.9			MIT
6262.296	6264.028	15964.17	LA II	2.1	2.2		MIT
6249.929	6251.658	15995.76	LA I	2.5			MIT
6238.587	6240.313	16024.84	LA I	1.4			MIT
6236.713	6238.438	16029.65	LA I	1.2			MIT
6234.855	6236.580	16034.43	LA I	1.4			MIT
6233.501	6235.225	16037.91	LA I	1.2			MIT
6225.279	6227.001	16059.09	LA I	0.5			MIT
6219.472	6221.193	16074.09	LA	0.6			MIT
6218.211	6219.931	16077.35	LA I	0.9			MIT
6214.35	6216.07	16087.3	LA I	0.5			ME
6206.76	6208.48	16107.0	LA	0.5			ME
6203.510	6205.226	16115.45	LA II	0.9	1.4		MIT
6188.091	6189.803	16155.60	LA II	0.9	1.7		MIT
6174.194	6175.903	16191.96	LA II	0.7	0.5		MIT
6172.730	6174.438	16195.81	LA II	1.2	0.7		MIT
6167.649	6169.356	16209.15	LA I	0.5			MIT
6165.699	6167.405	16214.27	LA I	2.0			MIT
6165.006	6166.712	16216.10	LA	0.7			MIT
6146.531	6148.232	16264.84	LA I	1.3	0.9		MIT
6145.309	6147.010	16268.07	LA I	1.0			MIT
6143.99	6145.69	16271.6	LA	0.5			KS
6142.981	6144.681	16274.24	LA I	1.7			MIT
6136.531	6138.229	16291.34	LA I	0.3			MIT
6134.39	6136.09	16297.0	LA I	1.8G			ME

AIR WAVELENGTH (ANGSTRMS)	VACUUM WAVELENGTH (ANGSTRMS)	WAVENUMBER (KAYSERS)	LMENT	LOG ARC	INTENSITY SPK	INTENSITY DIS	REF
6129.554	6131.251	16309.89	LA II	1.6	1.4		MIT
6127.057	6128.753	16316.53	LA I	1.4			MIT
6126.086	6127.782	16319.12	LA II	1.3	1.4		MIT
6123.758	6125.453	16325.32	LA	0.7			MIT
6120.34	6122.03	16334.4	LA I	0.3	0.0		ME
6120.34	6122.03	16334.4	LA II	0.3	0.0		ME
6111.723	6113.415	16357.47	LA I	1.7			MIT
6108.492	6110.183	16366.12	LA I	1.8			MIT
6107.280	6108.971	16369.37	LA I	1.4			MIT
6100.379	6102.068	16387.89	LA II	1.5	1.2		MIT
6092.278	6093.965	16409.68	LA I	0.6			MIT
6087.961	6089.646	16421.31	LA	0.5			MIT
6085.485	6087.170	16428.00	LA II	0.5	0.7		MIT
6084.888	6086.573	16429.61	LA I	1.2			MIT
6074.045	6075.727	16458.94	LA I	0.5			MIT
6074.045	6075.727	16458.94	LA II	0.5			MIT
6072.061	6073.742	16464.31	LA I	0.3			MIT
6068.663	6070.343	16473.53	LA I	1.0			MIT
6067.140	6068.820	16477.67	LA II	0.0	0.5		MIT
6061.459	6063.137	16493.11	LA	0.8	0.0		MIT
6046.079	6047.753	16535.07	LA	0.5	0.0		MIT
6044.819	6046.493	16538.51	LA I	0.5			MIT
6041.551	6043.224	16547.46	LA	0.8			MIT
6038.606	6040.278	16555.53	LA I	1.7			MIT
6034.546	6036.217	16566.67	LA I	0.3			MIT
6031.540	6033.210	16574.92	LA	0.5			MIT
6025.093	6026.762	16592.66	LA	0.5			MIT
6017.162	6018.828	16614.53	LA I	0.9			MIT
6007.381	6009.045	16641.58	LA I	1.8			MIT
5992.360	5994.020	16683.29	LA I	0.5			MIT
5991.98	5993.64	16684.4	LA II		0.6H		ME
5982.357	5984.014	16711.19	LA I	1.0			MIT
5975.754	5977.409	16729.66	LA I	0.9			MIT
5973.527	5975.182	16735.89	LA II	1.2	2.1		MIT
5971.09	5972.74	16742.7	LA II	0.0	0.9		ME
5962.604	5964.256	16766.55	LA I	0.7			MIT
5961.43	5963.08	16769.9	LA II		0.5		ME
5960.586	5962.237	16772.23	LA I	1.0			MIT
5957.90	5959.55	16779.8	LA II		0.6		ME
5948.236	5949.884	16807.05	LA II	0.3	1.3		KS
5940.838	5942.484	16827.98	LA I	0.6			KS
5936.218	5937.863	16841.08	LA II	1.2	1.3		MIT
5935.288	5936.932	16843.72	LA I	1.3			MIT
5930.667	5932.310	16856.84	LA I	2.0			MIT
5930.628	5932.271	16856.95	LA I	2.2			MIT
5928.503	5930.146	16862.99	LA I	1.0			MIT
5927.71	5929.35	16865.2	LA II		1.5		ME
5918.26	5919.90	16892.2	LA II		0.6		ME
5917.629	5919.269	16893.98	LA I	1.4			MIT
5904.308	5905.944	16932.09	LA I	0.8			MIT
5901.953	5903.589	16938.85	LA II	0.3	1.6		MIT
5900.75	5902.39	16942.3	LA I	0.5			ME
5894.847	5896.481	16959.27	LA I	1.4			MIT
5892.66	5894.29	16965.6	LA II	0.0	0.6		ME
5880.647	5882.277	17000.22	LA II	1.5	1.7		MIT
5878.008	5879.637	17007.85	LA I	0.8			MIT
5877.634	5879.263	17008.94	LA I	0.9			MIT
5874.736	5876.364	17017.32	LA I	1.3			MIT
5874.003	5875.631	17019.45	LA II	0.3	0.8		MIT
5869.971	5871.598	17031.14	LA I	0.5			MIT
5863.709	5865.334	17049.33	LA II	1.6	1.9		MIT
5855.591	5857.214	17072.96	LA I	1.5			MIT
5852.283	5853.905	17082.61	LA I	1.0			MIT
5848.949	5850.570	17092.35	LA II	0.3	0.5		KS
5848.385	5850.006	17094.00	LA I	1.3			MIT
5845.040	5846.660	17103.78	LA I	1.2			MIT
5839.791	5841.410	17119.16	LA I	0.5			MIT
5829.726	5831.342	17148.71	LA I	1.4			MIT
5828.494	5830.110	17152.33	LA II	0.5	0.6		KS
5827.556	5829.172	17155.10	LA I	1.2			MIT

AIR WAVELENGTH (ANGSTRMS)	VACUUM WAVELENGTH (ANGSTRMS)	WAVENUMBER (KAYSERS)	LMENT	LOG ARC	INTENSITY SPK	INTENSITY DIS	REF
5823.827	5825.442	17166.08	LA I	1.2			MIT
5821.998	5823.612	17171.47	LA I	1.6			MIT
5808.63	5810.24	17211.0	LA II		1.0		ME
5808.332	5809.942	17211.87	LA II	1.4	1.8		MIT
5806.56	5808.17	17217.1	LA II		0.9		ME
5805.783	5807.393	17219.43	LA II	1.8	2.1		MIT
5797.587	5799.195	17243.77	LA II	1.9	2.2		MIT
5791.345	5792.951	17262.36	LA I	2.3			MIT
5789.248	5790.853	17268.61	LA I	2.1			MIT
5781.02	5782.62	17293.2	LA II		0.5		ME
5779.91	5781.51	17296.5	LA II		0.6		ME
5770.006	5771.606	17326.20	LA I	1.4			MIT
5769.349	5770.949	17328.17	LA I	1.8			MIT
5769.07	5770.67	17329.0	LA II	1.5	1.8		MIT
5761.845	5763.443	17350.74	LA I	1.8			MIT
5744.412	5746.005	17403.40	LA I	1.9			MIT
5742.94	5744.53	17407.9	LA I	0.7			MIT
5740.662	5742.254	17414.76	LA I	2.0	0.0		MIT
5734.959	5736.550	17432.08	LA I	1.2			MIT
5727.293	5728.882	17455.41	LA II	1.1			MIT
5720.019	5721.606	17477.61	LA I	1.4			MIT
5714.02	5715.61	17496.0	LA I	1.2			MIT
5712.402	5713.987	17500.92	LA II	1.3	1.6		MIT
5703.327	5704.909	17528.76	LA I	1.3	1.3		MIT
5703.327	5704.909	17528.76	LA II	1.3	1.3		MIT
5702.53	5704.11	17531.2	LA I	0.7			MIT
5699.386	5700.967	17540.88	LA I	0.7			MIT
5696.193	5697.773	17550.72	LA I	1.7			MIT
5671.550	5673.124	17626.97	LA II	1.0	2.0		MIT
5657.731	5659.301	17670.03	LA I	1.7			MIT
5656.54	5658.11	17673.7	LA I	0.5			ME
5654.87	5656.44	17679.0	LA I	0.7			MIT
5648.254	5649.822	17699.67	LA I	2.2			MIT
5639.301	5640.866	17727.77	LA I	1.2			MIT
5632.025	5633.588	17750.68	LA I	1.3			MIT
5631.214	5632.777	17753.23	LA I	1.7			MIT
5610.53	5612.09	17818.7	LA II	0.3	1.6		ME
5588.342	5589.894	17889.43	LA I	1.6			MIT
5570.386	5571.933	17947.09	LA I	0.7			MIT
5568.474	5570.020	17953.26	LA I	1.6			MIT
5566.943	5568.489	17958.19	LA II	0.9	1.3		MIT
5565.727	5567.273	17962.12	LA I	1.0			MIT
5565.452	5566.998	17963.00	LA I	1.3			MIT
5547.56	5549.10	18020.9	LA	0.3H	0.6H		ME
5544.907	5546.447	18029.56	LA I	0.7			MIT
5541.258	5542.797	18041.43	LA I	1.7O			MIT
5535.671	5537.209	18059.64	LA II	1.7	2.0		MIT
5532.17	5533.71	18071.1	LA I	0.3	1.0		ME
5532.059	5533.596	18071.43	LA	0.5			MIT
5529.882	5531.418	18078.55	LA I	0.7			MIT
5517.342	5518.875	18119.64	LA I	1.5			MIT
5515.280	5516.812	18126.41	LA I	1.2			MIT
5510.34	5511.87	18142.7	LA I	2.3			ME
5507.33	5508.86	18152.6	LA I	0.7			KN
5506.001	5507.531	18156.96	LA I	1.7			MIT
5503.809	5505.338	18164.19	LA I	2.0			MIT
5502.669	5504.198	18167.95	LA I	1.0			MIT
5502.245	5503.774	18169.35	LA I	1.0			MIT
5501.340	5502.868	18172.34	LA	2.3	1.7		MIT
5493.452	5494.978	18198.43	LA II	1.2	1.3		MIT
5491.067	5492.593	18206.34	LA I	1.0			MIT
5486.837	5488.362	18220.37	LA II	0.3	0.6		MIT
5482.270	5483.793	18235.55	LA II	1.4	1.7		MIT
5480.739	5482.262	18240.65	LA II	0.7	1.6		MIT
5475.161	5476.682	18259.23	LA	0.9			MIT
5466.927	5468.446	18286.73	LA I	0.5			MIT
5464.381	5465.900	18295.25	LA II	1.4	1.5		MIT
5458.687	5460.204	18314.33	LA II	0.7	1.7		MIT
5455.146	5456.662	18326.22	LA I	2.3	0.0		MIT
5447.59	5449.10	18351.6	LA II	0.3	1.0		ME

AIR WAVELENGTH (ANGSTRMS)	VACUUM WAVELENGTH (ANGSTRMS)	WAVENUMBER (KAYSERS)	LMENT	LOG ARC	INTENSITY SPK	INTENSITY DIS	REF
5437.523	5439.034	18385.62	LA I	0.6H			MIT
5429.864	5431.373	18411.55	LA I	1.0			MIT
5423.82	5425.33	18432.1	LA II	0.5H	0.6		ME
5422.132	5423.639	18437.80	LA I	0.5			MIT
5415.694	5417.200	18459.72	LA I	0.6			MIT
5381.925	5383.421	18575.55	LA II	1.2	2.0		MIT
5381.77	5383.27	18576.1	LA II	0.7	1.3		ME
5380.997	5382.493	18578.75	LA II	1.7	2.0		MIT
5380.022	5381.518	18582.12	LA I	0.5			MIT
5377.095	5378.590	18592.23	LA II	1.5	2.3		MIT
5365.897	5367.389	18631.03	LA I	0.7			MIT
5357.873	5359.363	18658.93	LA I	1.6			MIT
5340.67	5342.16	18719.0	LA II	1.9	2.0		MIT
5333.42	5334.90	18744.5	LA II	0.5	0.7		ME
5323.573	5325.054	18779.15	LA I	0.5			MIT
5320.165	5321.645	18791.18	LA I	0.5			MIT
5307.54	5309.02	18835.9	LA	0.5			MIT
5304.03	5305.51	18848.3	LA I	1.4			MIT
5303.556	5305.032	18850.03	LA II	2.0	2.1		MIT
5302.62	5304.10	18853.4	LA II	1.7	2.2		ME
5301.984	5303.459	18855.62	LA II	2.5R	2.3		MIT
5290.836	5292.308	18895.35	LA II	1.8	2.0		MIT
5279.15	5280.62	18937.2	LA II	0.7	1.2		MIT
5276.43	5277.90	18946.9	LA I	1.3			MIT
5271.199	5272.666	18965.74	LA I	2.0	1.3		MIT
5259.389	5260.853	19008.32	LA II	1.6	1.7		MIT
5257.857	5259.320	19013.86	LA I	0.7			MIT
5253.458	5254.920	19029.78	LA I	2.0	0.7		MIT
5240.84	5242.30	19075.6	LA	0.7			MIT
5239.56	5241.02	19080.3	LA I	0.7			MIT
5234.274	5235.731	19099.53	LA I	2.3R	0.90		MIT
5226.216	5227.671	19128.98	LA II	0.8	1.3L		MIT
5217.83	5219.28	19159.7	LA	0.3H	1.0H		ME
5211.870	5213.321	19181.63	LA I	2.5R	0.7		MIT
5204.155	5205.604	19210.07	LA II	1.7	2.5		MIT
5191.50	5192.95	19256.9	LA	0.6	0.5H		ME
5190.34	5191.79	19261.2	LA	0.6H			ME
5188.229	5189.674	19269.03	LA II	1.7	2.7		MIT
5183.923	5185.367	19285.04	LA I	1.4			MIT
5183.422	5184.866	19286.90	LA II	2.5	2.6		MIT
5179.133	5180.575	19302.87	LA I	0.7			MIT
5177.311	5178.753	19309.67	LA I	2.2	1.5		MIT
5173.854	5175.295	19322.57	LA II	1.3	1.4H		MIT
5172.924	5174.365	19326.04	LA II	0.7	1.3L		MIT
5168.971	5170.411	19340.82	LA I	0.5			MIT
5167.791	5169.230	19345.24	LA I	1.3			MIT
5167.28	5168.72	19347.1	LA II	0.5	1.0		ME
5163.615	5165.053	19360.88	LA II	1.4	1.6		MIT
5162.68	5164.12	19364.4	LA II	0.3	0.5		ME
5161.566	5163.004	19368.57	LA	0.5B			MIT
5158.693	5160.130	19379.36	LA I	1.7			MIT
5157.431	5158.868	19384.10	LA II	1.6	2.0		MIT
5156.741	5158.177	19386.69	LA II	1.6	1.6		MIT
5145.422	5146.855	19429.34	LA I	2.0	1.0		MIT
5139.146	5140.578	19453.07	LA	0.6			MIT
5135.445	5136.876	19467.08	LA	0.5H			MIT
5122.990	5124.417	19514.41	LA II	2.2	2.3		MIT
5120.875	5122.302	19522.47	LA I	0.8			MIT
5114.573	5115.998	19546.53	LA II	2.2	2.3		MIT
5109.133	5110.557	19567.34	LA I	0.6			MIT
5107.54	5108.96	19573.4	LA II	0.3H	0.8H		ME
5106.233	5107.656	19578.45	LA I	2.0	1.0		MIT
5103.135	5104.557	19590.34	LA	0.5			MIT
5096.614	5098.034	19615.40	LA	0.5H			MIT
5090.58	5092.00	19638.6	LA II	0.5	0.3H		MIT
5086.71	5088.13	19653.6	LA II	0.6	0.3		ME
5086.24	5087.66	19655.4	LA I	0.5			MIT
5080.21	5081.63	19678.7	LA II	0.90	1.0H		ME
5079.393	5080.809	19681.91	LA	0.8			MIT
5078.923	5080.339	19683.73	LA	0.7			MIT
5072.14	5073.55	19710.1	LA I	0.5			MIT
5067.898	5069.311	19726.55	LA I	1.5			MIT
5066.99	5068.40	19730.1	LA II	0.3	1.3H		ME
5063.76	5065.17	19742.7	LA II	0.3	0.5		ME
5062.925	5064.337	19745.92	LA II	1.2	1.3		MIT
5060.85	5062.26	19754.0	LA II	0.3	0.5		ME
5056.461	5057.871	19771.17	LA I	1.9			MIT
5050.569	5051.977	19794.23	LA I	1.9	0.6		MIT
5048.040	5049.448	19804.15	LA II	0.8	1.3		MIT
5046.884	5048.291	19808.68	LA I	1.9			MIT
5019.516	5020.916	19916.69	LA	1.2			MIT
5014.45	5015.85	19936.8	LA II	0.3	1.5B		ME
5002.133	5003.528	19985.90	LA II	1.0	1.2		MIT
5001.790	5003.185	19987.27	LA I	1.0			MIT
4999.468	5000.863	19996.55	LA II	2.6	2.5		MIT
4996.823	4998.217	20007.14	LA II	1.4	1.5		MIT
4994.64	4996.03	20015.9	LA I	0.3			M
4993.876	4995.269	20018.94	LA I	1.9			MIT
4991.284	4992.676	20029.34	LA II	2.0	1.9		MIT
4986.834	4988.225	20047.21	LA II	2.2	2.0		MIT
4984.919	4986.310	20054.91	LA I	0.60			MIT
4983.571	4984.961	20060.34	LA I	0.6			MIT
4977.953	4979.342	20082.98	LA I	1.0			MIT
4974.20	4975.59	20098.1	LA II		0.6H		ME
4970.392	4971.779	20113.52	LA II	2.1	2.0		MIT
4968.59	4969.98	20120.8	LA I	0.6D			ME
4964.838	4966.223	20136.02	LA I	0.7			MIT
4957.791	4959.175	20164.65	LA	0.5			MIT
4956.060	4957.443	20171.69	LA	0.6			MIT
4952.069	4953.451	20187.95	LA II	1.7	1.6		MIT
4949.769	4951.150	20197.33	LA I	2.3			MIT
4946.466	4947.847	20210.81	LA II	2.0	1.7		MIT
4945.848	4947.228	20213.34	LA I	0.8			MIT
4935.618	4936.996	20255.23	LA II	0.9			MIT
4934.825	4936.202	20258.49	LA II	2.2	2.0		MIT
4925.405	4926.780	20297.23	LA	0.8			MIT
4921.783	4923.157	20312.17	LA II	2.7	2.6		MIT
4920.971	4922.345	20315.52	LA II	2.7	2.6		MIT
4916.618	4917.991	20333.51	LA	0.5			MIT
4911.337	4912.708	20355.37	LA II	0.6	1.3		MIT
4905.126	4906.496	20381.15	LA I	0.9	0.3		MIT
4904.43	4905.80	20384.0	LA II		0.3H		ME
4901.862	4903.231	20394.72	LA I	1.7			MIT
4899.924	4901.292	20402.78	LA II	2.6	2.3		MIT
4894.232	4895.599	20426.51	LA	0.6			MIT
4891.43	4892.80	20438.2	LA II		1.0		ME
4887.617	4888.982	20454.16	LA I	0.3			MIT
4886.822	4888.187	20457.48	LA	0.3			MIT
4880.20	4881.56	20485.2	LA II		1.0H		ME
4878.848	4880.211	20490.92	LA I	1.3	1.5		MIT
4870.557	4871.917	20525.80	LA I	0.9			MIT
4868.907	4870.267	20532.75	LA	0.6			MIT
4867.375	4868.735	20539.22	LA I	0.3H			MIT
4860.908	4862.266	20566.54	LA II	2.0	2.0		MIT
4859.18	4860.54	20573.9	LA II		0.7H		ME
4854.949	4856.305	20591.79	LA I	1.2			MIT
4850.817	4852.172	20609.33	LA I	1.3	0.5		MIT
4850.584	4851.939	20610.32	LA II	1.4	1.3		MIT
4843.304	4844.657	20641.29	LA	0.7	1.0		MIT
4840.001	4841.353	20655.38	LA II	1.5	1.0		MIT
4839.516	4840.868	20657.45	LA I	1.6	1.0		MIT
4830.518	4831.868	20695.93	LA II	0.9	0.6		MIT
4826.886	4828.235	20711.50	LA II	1.2	1.5		MIT
4824.066	4825.414	20723.61	LA II	2.2	2.2		MIT
4817.55	4818.90	20751.6	LA	0.6			ME
4817.167	4818.513	20753.29	LA I	1.0			MIT
4809.015	4810.359	20788.47	LA II	2.2	2.2		MIT
4804.041	4805.384	20809.99	LA II	2.2	2.2		MIT
4800.258	4801.600	20826.39	LA I	1.2			MIT
4800.005	4801.347	20827.49	LA I	1.2			MIT

AIR WAVELENGTH (ANGSTRMS)	VACUUM WAVELENGTH (ANGSTRMS)	WAVENUMBER (KAYSERS)	LMENT	LOG ARC	INTENSITY SPK	DIS	REF
4796.688	4798.029	20841.89	LA II	1.1	1.0		MIT
4794.561	4795.901	20851.14	LA	1.0			MIT
4792.465	4793.805	20860.26	LA I	0.5			MIT
4791.397	4792.736	20864.91	LA I	0.8			MIT
4780.559	4781.896	20912.21	LA	0.6			MIT
4779.895	4781.231	20915.11	LA I	0.8			MIT
4775.152	4776.487	20935.89	LA I	1.3			MIT
4770.433	4771.767	20956.60	LA U	1.4	0.7		MIT
4766.894	4768.227	20972.16	LA I	2.0	1.0		MIT
4759.718	4761.049	21003.77	LA I	0.8			MIT
4757.162	4758.492	21015.06	LA	0.7			MIT
4756.981	4758.311	21015.86	LA I	0.6			MIT
4753.138	4754.467	21032.85	LA	0.7H			MIT
4752.416	4753.745	21036.05	LA I	0.5			MIT
4750.414	4751.743	21044.91	LA I	1.6	1.6		MIT
4748.729	4750.057	21052.38	LA II	2.0	2.3		MIT
4743.085	4744.412	21077.43	LA II	2.5R	2.5		MIT
4740.277	4741.603	21089.92	LA II	2.2	2.5		MIT
4739.792	4741.118	21092.07	LA II	0.6	0.9		MIT
4733.806	4735.130	21118.74	LA I	0.9			MIT
4730.771	4732.094	21132.29	LA II	0.3	0.3H		MIT
4728.417	4729.740	21142.81	LA II	2.6R	2.5		MIT
4724.426	4725.748	21160.67	LA II	1.2	1.3		MIT
4723.716	4725.037	21163.85	LA I	0.8			MIT
4722.14	4723.46	21170.9	LA II		0.3H		ME
4719.947	4721.267	21180.75	LA II	2.3R	2.5		MIT
4717.586	4718.906	21191.35	LA II	1.2	1.4		MIT
4716.443	4717.763	21196.49	LA II	2.0	2.3		MIT
4714.152	4715.471	21206.79	LA I	0.7	0.3		MIT
4712.934	4714.253	21212.27	LA II	2.0R	2.2		MIT
4708.186	4709.503	21233.66	LA I	1.4	0.3		MIT
4703.278	4704.594	21255.82	LA II	2.3R	2.5R		MIT
4702.644	4703.960	21258.69	LA I	1.0	0.5		MIT
4700.263	4701.578	21269.45	LA	0.9	0.5		MIT
4699.627	4700.942	21272.33	LA II	2.3R	2.3R		MIT
4695.305	4696.619	21291.91	LA I	0.6			MIT
4692.502	4693.815	21304.63	LA II	2.3	2.5		MIT
4691.177	4692.490	21310.65	LA II	1.0	1.4		MIT
4688.656	4689.968	21322.11	LA II	0.8	1.3		MIT
4682.125	4683.435	21351.85	LA II	0.5	0.5		MIT
4671.833	4673.141	21398.89	LA II	2.0	2.2		MIT
4668.914	4670.221	21412.26	LA II	2.3R	2.5R		MIT
4663.765	4665.071	21435.90	LA II	2.0	2.3		MIT
4662.511	4663.816	21441.67	LA II	2.2	2.3		MIT
4660.701	4662.006	21450.00	LA I	0.7	0.3		MIT
4655.497	4656.800	21473.97	LA II	2.2	2.5		MIT
4653.900	4655.203	21481.34	LA I	0.6	0.3H		MIT
4652.084	4653.387	21489.73	LA I	0.7	0.6		MIT
4650.331	4651.633	21497.83	LA I	0.9	0.6		MIT
4648.649	4649.951	21505.60	LA I	1.2	0.9		MIT
4647.509	4648.810	21510.88	LA II	1.0	1.7		MIT
4646.343	4647.644	21516.28	LA I	0.9	0.5		MIT
4645.281	4646.582	21521.20	LA II	1.4	1.6		MIT
4643.124	4644.424	21531.20	LA I	0.9	0.7H		MIT
4641.40	4642.70	21539.2	LA II		0.3H		ME
4636.425	4637.723	21562.30	LA II	1.0	1.3		MIT
4634.971	4636.269	21569.07	LA II	0.5	0.7		MIT
4627.347	4628.643	21604.60	LA I	0.5			MIT
4619.882	4621.176	21639.51	LA II	2.2	2.3		MIT
4615.065	4616.358	21662.10	LA I	1.0	0.3		MIT
4613.393	4614.685	21669.95	LA II	2.0	2.0		MIT
4605.782	4607.072	21705.76	LA II	2.0	2.0		MIT
4605.093	4606.383	21709.01	LA I	0.7	0.3		MIT
4604.239	4605.529	21713.03	LA I	0.9	0.3		MIT
4602.046	4603.335	21723.38	LA I	1.3			MIT
4601.65	4602.94	21725.2	LA II		0.5		ME
4600.59	4601.88	21730.2	LA II		0.7H		ME
4596.198	4597.486	21751.02	LA I	0.6	0.3		MIT
4595.053	4596.340	21756.44	LA II	0.5			MIT
4589.902	4591.188	21780.85	LA I	0.6			MIT
4587.130	4588.415	21794.02	LA II	0.5H			MIT
4581.210	4582.494	21822.18	LA I	1.2	0.7		MIT
4580.057	4581.340	21827.67	LA II	1.9	2.0		MIT
4574.875	4576.157	21852.40	LA II	2.5	2.5		MIT
4570.971	4572.252	21871.06	LA II	0.9	1.0		MIT
4570.027	4571.308	21875.58	LA I	1.9	1.4		MIT
4567.909	4569.189	21885.72	LA I	2.0	1.4		MIT
4564.863	4566.142	21900.32	LA I	1.0	0.5		MIT
4562.5	4563.8	21912.	LA II		0.7H		ME
4559.295	4560.573	21927.07	LA II	2.0	2.2		MIT
4558.464	4559.742	21931.07	LA II	2.0	2.3		MIT
4552.486	4553.762	21959.86	LA I	0.7			MIT
4550.777	4552.053	21968.11	LA I	0.6	0.5		MIT
4550.176	4551.452	21971.01	LA I	0.5	0.3		MIT
4549.502	4550.777	21974.27	LA I	1.2	0.8		MIT
4541.788	4543.061	22011.59	LA I	1.5	0.7		MIT
4540.707	4541.980	22016.83	LA II	0.6	1.4		MIT
4538.87	4540.14	22025.7	LA II		0.9B		ME
4537.578	4538.850	22032.01	LA I	0.9			MIT
4530.562	4531.832	22066.13	LA II	1.2	1.4		MIT
4528.915	4530.185	22074.15	LA I	0.3			MIT
4526.109	4527.378	22087.84	LA II	2.0	2.2		MIT
4525.303	4526.572	22091.77	LA II	2.0	2.0		MIT
4522.372	4523.640	22106.09	LA II	2.3	2.6		MIT
4516.38	4517.65	22135.4	LA		0.7B		ME
4508.478	4509.743	22174.22	LA II	0.5	0.3		MIT
4507.405	4508.669	22179.49	LA I	0.6	0.5		MIT
4505.844	4507.108	22187.18	LA II	0.3	0.8B		MIT
4502.16	4503.42	22205.3	LA II		1.0B		ME
4501.577	4502.840	22208.21	LA I	0.7			MIT
4500.219	4501.481	22214.91	LA I	1.2	0.6		MIT
4499.052	4500.314	22220.67	LA I	1.0			MIT
4498.751	4500.013	22222.16	LA II	0.5	0.6H		MIT
4497.005	4498.267	22230.79	LA II	0.3			MIT
4494.706	4495.967	22242.16	LA I	1.0			MIT
4493.809	4495.070	22246.60	LA I	0.7			MIT
4493.111	4494.371	22250.05	LA I	1.0			MIT
4491.756	4493.016	22256.76	LA I	0.9	0.5		MIT
4486.056	4487.315	22285.04	LA I	1.2	0.6		MIT
4484.502	4485.760	22292.76	LA II	1.0H	1.0H		MIT
4481.198	4482.455	22309.20	LA II	0.6	0.3H		MIT
4479.822	4481.079	22316.05	LA I	1.0			MIT
4474.543	4475.799	22342.38	LA I	0.6			MIT
4474.029	4475.284	22344.95	LA II	0.5	1.3		MIT
4468.964	4470.218	22370.27	LA I	1.0			MIT
4459.120	4460.372	22419.66	LA II	0.9	0.5		MIT
4455.795	4457.046	22436.39	LA II	1.6	1.4		MIT
4455.211	4456.461	22439.33	LA I	0.5			MIT
4453.853	4455.103	22446.17	LA I	0.5	0.5		MIT
4452.153	4453.403	22454.74	LA I	1.2	0.7		MIT
4445.12	4446.37	22490.3	LA I	0.5			M
4443.949	4445.197	22496.19	LA I	0.5	0.5		MIT
4443.949	4445.197	22496.19	LA II	0.5	0.5		MIT
4442.675	4443.922	22502.64	LA I	0.8			MIT
4435.847	4437.092	22537.28	LA II	0.7	0.5		MIT
4432.949	4434.194	22552.01	LA II	1.4	0.7		MIT
4429.904	4431.148	22567.52	LA II	2.3	2.5		MIT
4427.567	4428.810	22579.43	LA I	1.5	1.7		MIT
4423.898	4425.140	22598.15	LA I	0.7	0.7		MIT
4419.161	4420.402	22622.38	LA II	0.6	0.8		MIT
4417.14	4418.38	22632.7	LA I	0.3H	0.3H		ME
4417.14	4418.38	22632.7	LA II	0.3H	0.3H		ME
4413.436	4414.675	22651.72	LA I	0.5			MIT
4412.207	4413.446	22658.03	LA II	0.3	0.3H		MIT
4411.204	4412.443	22663.18	LA II	0.6	1.0H		MIT
4403.02	4404.26	22705.3	LA I	0.6	0.3		ME
4403.02	4404.26	22705.3	LA II	0.6	0.3		ME
4402.649	4403.886	22707.22	LA I	0.7			MIT
4397.04	4398.28	22736.2	LA I	0.5			KN
4396.79	4398.03	22737.5	LA I	0.6	0.3H		KN

AIR WAVELENGTH (ANGSTRMS)	VACUUM WAVELENGTH (ANGSTRMS)	WAVENUMBER (KAYSERS)	LMENT	LOG ARC	INTENSITY SPK	DIS	REF
4396.318	4397.553	22739.92	LA	0.6			MIT
4393.505	4394.739	22754.48	LA	0.6	0.3		MIT
4389.868	4391.101	22773.33	LA I	0.7			MIT
4385.204	4386.436	22797.55	LA II	0.9	1.0		MIT
4383.45	4384.68	22806.7	LA II	1.0	1.7		M
4380.555	4381.786	22821.75	LA I	0.6			MIT
4378.097	4379.327	22834.56	LA II	1.6	1.5		MIT
4364.666	4365.893	22904.82	LA II	1.7	1.7		MIT
4363.062	4364.288	22913.24	LA II	0.9	0.5		MIT
4360.847	4362.073	22924.88	LA I	0.5			MIT
4360.496	4361.722	22926.73	LA I	0.7			MIT
4357.917	4359.142	22940.29	LA I	0.7	0.3	H	MIT
4354.804	4356.028	22956.69	LA I	1.6	0.5		MIT
4354.399	4355.623	22958.83	LA II	1.9	2.0		MIT
4340.727	4341.947	23031.14	LA I	1.7	0.5		MIT
4339.934	4341.154	23035.35	LA	1.3			MIT
4337.777	4338.997	23046.80	LA II	0.8	0.5	H	MIT
4334.964	4336.183	23061.76	LA II	1.6	1.7		MIT
4333.734	4334.952	23068.30	LA II	2.9	2.7		MIT
4326.185	4327.402	23108.56	LA I	0.7			MIT
4322.503	4323.719	23128.24	LA II	2.2	2.2		MIT
4315.902	4317.116	23163.61	LA II	1.7	0.5	H	
4311.743	4312.956	23185.96	LA I	1.2	0.7	H	MIT
4306.002	4307.213	23216.87	LA I	1.5			MIT
4304.117	4305.328	23227.04	LA II	0.5	0.3		MIT
4300.618	4301.828	23245.93	LA I	0.7			MIT
4300.435	4301.645	23246.92	LA II	1.4	1.3		MIT
4296.048	4297.257	23270.66	LA II	2.3	2.3		MIT
4291.019	4292.226	23297.93	LA I	0.6			MIT
4289.01	4290.22	23308.9	LA I	0.6			RL
4286.973	4288.179	23319.92	LA II	2.6	2.5		MIT
4280.261	4281.465	23356.49	LA I	1.7	1.5		MIT
4275.642	4276.845	23381.72	LA II	1.6	2.7		MIT
4271.157	4272.359	23406.27	LA I	1.5			MIT
4269.494	4270.696	23415.39	LA II	2.2	2.2		MIT
4267.741	4268.942	23425.01	LA I	0.9			MIT
4263.580	4264.780	23447.87	LA II	2.2	2.2		MIT
4262.334	4263.534	23454.72	LA I	1.2	0.3	H	MIT
4259.51	4260.71	23470.3	LA II		0.3	H	ME
4256.915	4258.113	23484.58	LA I	1.7			MIT
4256.490	4257.688	23486.92	LA	0.3	0.3		MIT
4252.925	4254.122	23506.61	LA II	0.7	0.3		MIT
4249.991	4251.187	23522.84	LA II	2.0	1.7		MIT
4248.32	4249.52	23532.1	LA II		0.3		ME
4241.20	4242.39	23571.6	LA II		1.2	B	ME
4238.593	4239.786	23586.09	LA I	0.8			MIT
4238.379	4239.572	23587.28	LA II	2.7	2.5		MIT
4230.953	4232.144	23628.68	LA II	2.2	1.7		MIT
4217.554	4218.742	23703.75	LA II	2.3	2.0		MIT
4216.549	4217.737	23709.40	LA I	1.0			MIT
4210.22	4211.41	23745.0	LA		1.0	H	ME
4207.617	4208.802	23759.73	LA II	0.7	0.5	H	MIT
4204.038	4205.222	23779.96	LA II	2.3	1.4		MIT
4201.50	4202.68	23794.3	LA	0.3	1.1	H	ME
4196.547	4197.729	23822.40	LA II	2.3	2.2		MIT
4194.354	4195.536	23834.86	LA II	1.2	1.2		MIT
4193.371	4194.553	23840.45	LA II	1.2	1.0		MIT
4192.730	4193.911	23844.09	LA I	1.0			MIT
4192.358	4193.539	23846.21	LA II	1.7	1.7		MIT
4187.316	4188.496	23874.92	LA I	1.7	1.6		MIT
4180.97	4182.15	23911.2	LA II		1.0	H	ME
4177.486	4178.663	23931.10	LA I	1.2			MIT
4172.316	4173.492	23960.75	LA I	0.9			MIT
4171.132	4172.308	23967.55	LA I	0.9			MIT
4163.305	4164.479	24012.61	LA I	0.9	0.6		MIT
4161.963	4163.136	24020.35	LA		1.0		MIT
4160.264	4161.437	24030.16	LA I	1.6	1.7		MIT
4157.513	4158.685	24046.06	LA I	1.0	0.5		MIT
4154.59	4155.76	24063.0	LA II	0.3	0.3		ME
4152.775	4153.946	24073.50	LA II	1.6	1.7		MIT

AIR WAVELENGTH (ANGSTRMS)	VACUUM WAVELENGTH (ANGSTRMS)	WAVENUMBER (KAYSERS)	LMENT	LOG ARC	INTENSITY SPK	DIS	REF
4151.955	4153.126	24078.25	LA II	2.3	2.5		MIT
4150.236	4151.406	24088.22	LA I	0.3			MIT
4148.155	4149.325	24100.31	LA II		0.6		MIT
4144.354	4145.523	24122.41	LA I	0.5			MIT
4143.936	4145.105	24124.84	LA I	1.4	1.2	H	MIT
4143.742	4144.911	24125.97	LA II		1.0	H	MIT
4141.740	4142.908	24137.63	LA II	2.3	2.3		MIT
4137.925	4139.092	24159.89	LA II	0.6	0.3		MIT
4137.025	4138.192	24165.14	LA I	1.2			MIT
4133.334	4134.500	24186.72	LA	1.2	1.4		MIT
4132.501	4133.667	24191.60	LA II		1.0	H	ME
4131.74	4132.91	24196.1	LA II	0.5	0.5		ME
4123.228	4124.391	24246.00	LA II	2.7	2.7		MIT
4117.687	4118.849	24278.63	LA	1.3	0.9		MIT
4113.28	4114.44	24304.6	LA II	1.0	1.3		ME
4109.805	4110.965	24325.19	LA I	1.2	1.0		MIT
4109.485	4110.645	24327.08	LA I	1.2	0.7		MIT
4104.877	4106.035	24354.39	LA I	1.6	0.3		MIT
4101.01	4102.17	24377.4	LA II	0.3	0.3	H	ME
4099.542	4100.699	24386.09	LA II	2.0	2.0		MIT
4098.73	4099.89	24390.9	LA II		0.6		ME
4090.407	4091.562	24440.55	LA I	0.5			MIT
4089.612	4090.766	24445.30	LA I	1.6	0.7		MIT
4086.714	4087.868	24462.63	LA II	2.7	2.7		MIT
4079.178	4080.330	24507.82	LA I	1.4	0.5		MIT
4077.340	4078.491	24518.87	LA II	2.8	2.6		MIT
4076.711	4077.862	24522.65	LA II	1.2	0.7		MIT
4067.392	4068.541	24578.84	LA II	2.2	1.9		MIT
4065.580	4066.728	24589.79	LA I	1.5	0.3		MIT
4064.783	4065.931	24594.61	LA I	1.6	0.5		MIT
4060.321	4061.468	24621.64	LA I	1.9	0.7		MIT
4058.085	4059.231	24635.21	LA II	0.6			MIT
4050.079	4051.223	24683.90	LA II	1.8	1.8		MIT
4042.911	4044.053	24727.67	LA II	2.6	2.5		MIT
4040.973	4042.115	24739.53	LA I	0.9			MIT
4037.214	4038.355	24762.56	LA I	1.7	0.5	H	MIT
4036.596	4037.737	24766.35	LA II	0.7	1.2		MIT
4031.692	4032.831	24796.48	LA II	2.6	2.5		MIT
4025.882	4027.020	24832.26	LA II	1.7	1.7		MIT
4023.588	4024.725	24846.42	LA II	1.7	1.2		MIT
4015.393	4016.528	24897.12	LA I	2.0	0.3	H	MIT
4007.662	4008.795	24945.15	LA	0.5	0.3	H	MIT
4001.38	4002.51	24984.3	LA I	0.5			ME
3995.750	3996.880	25019.52	LA II	2.8	2.5		MIT
3994.474	3995.604	25027.51	LA II		0.7		MIT
3988.518	3989.646	25064.88	LA II	3.0	2.9		MIT
3981.35	3982.48	25110.0	LA II		1.0	L	MIT
3979.08	3980.21	25124.3	LA II		0.3		ME
3963.04	3964.16	25226.0	LA		0.7	L	ME
3962.03	3963.15	25232.4	LA	0.3	0.5		ME
3958.54	3959.66	25254.7	LA II	0.3	0.3	H	MIT
3957.25	3958.37	25262.9	LA		0.3	H	ME
3956.07	3957.19	25270.5	LA II		0.6		MIT
3955.19	3956.31	25276.1	LA	0.5	0.3	H	MIT
3953.680	3954.799	25285.74	LA I	1.5			MIT
3953.36	3954.48	25287.8	LA	0.3	0.3		ME
3951.43	3952.55	25300.1	LA II	0.6	0.3	H	ME
3949.106	3950.224	25315.02	LA II	3.0	2.9		MIT
3944.15	3945.27	25346.8	LA II		0.3		ME
3939.85	3940.97	25374.5	LA II	0.3	0.5	H	ME
3936.221	3937.335	25397.89	LA II	2.0	1.7		MIT
3932.523	3933.636	25421.77	LA II		0.3	H	MIT
3930.47	3931.58	25435.1	LA		0.5		ME
3929.216	3930.328	25443.17	LA II	2.6	2.5		MIT
3927.558	3928.670	25453.91	LA I	1.6			MIT
3925.09	3926.20	25469.9	LA II	0.6	0.5	H	ME
3924.69	3925.80	25472.5	LA		0.5		ME
3921.535	3922.645	25493.00	LA II	2.6	2.3		MIT
3916.045	3917.154	25528.74	LA II	2.6	2.6		MIT
3910.806	3911.914	25562.94	LA II	1.0	0.7	H	MIT

AIR WAVELENGTH (ANGSTRMS)	VACUUM WAVELENGTH (ANGSTRMS)	WAVENUMBER (KAYSERS)	LMENT	LOG ARC	INTENSITY SPK	DIS	REF
3902.576	3903.682	25616.84	LA I	1.3			MIT
3898.604	3899.709	25642.94	LA I	1.5	0.7		MIT
3897.429	3898.533	25650.67	LA II	0.5	0.6		MIT
3895.655	3896.759	25662.35	LA	1.0			MIT
3892.47	3893.57	25683.4	LA II		0.5H		ME
3892.05	3893.15	25686.1	LA II		0.5H		ME
3886.368	3887.469	25723.68	LA II	2.6	2.3		MIT
3885.09	3886.19	25732.1	LA II	0.6	0.6		ME
3871.631	3872.728	25821.59	LA II	2.3	1.2		MIT
3868.35	3869.45	25843.5	LA		0.5H		ME
3864.488	3865.584	25869.31	LA II	2.0	2.2		MIT
3863.11	3864.21	25878.5	LA II	0.3	0.3		ME
3860.31	3861.40	25897.3	LA II		0.3		ME
3854.910	3856.003	25933.59	LA II		1.6		MIT
3849.013	3850.105	25973.32	LA II	2.3	2.2		MIT
3845.997	3847.088	25993.69	LA II	1.6	1.7		MIT
3840.709	3841.798	26029.48	LA II	1.7	1.8		MIT
3835.074	3836.162	26067.72	LA II	1.1	1.2		MIT
3817.224	3818.307	26189.61	LA II		0.7		MIT
3816.168	3817.251	26196.86	LA II		1.0H		MIT
3814.02	3815.10	26211.6	LA II	0.3	0.3		MIT
3808.781	3809.862	26247.67	LA II		1.0		MIT
3804.76	3805.84	26275.4	LA II		0.3H		MIT
3798.19	3799.27	26320.9	LA II		0.3		ME
3794.773	3795.851	26344.56	LA II	2.6	2.3		MIT
3790.822	3791.898	26372.01	LA II	2.6	2.5		MIT
3784.797	3785.872	26414.00	LA II	0.7	0.3		MIT
3780.674	3781.748	26442.80	LA II	1.7	1.7		MIT
3780.515	3781.589	26443.91	LA II		1.3		MIT
3773.117	3774.189	26495.76	LA II	0.3	1.3L		MIT
3768.98	3770.05	26524.8	LA II		0.3		ME
3767.05	3768.12	26538.4	LA II	0.3H	1.0H		ME
3766.56	3767.63	26541.9	LA II		0.5H		MIT
3759.080	3760.148	26594.70	LA II	2.6	2.2		MIT
3753.04	3754.11	26637.5	LA II	0.3	0.0		ME
3747.99	3749.06	26673.4	LA	0.3	0.6L		MIT
3744.85	3745.91	26695.7	LA	0.0	0.3H		MIT
3736.415	3737.477	26756.01	LA II		0.8		MIT
3735.851	3736.913	26760.05	LA II	1.3	1.0		MIT
3731.42	3732.48	26791.8	LA II	0.3	0.9		MIT
3728.98	3730.04	26809.4	LA II		0.3H		MIT
3725.052	3726.111	26837.63	LA II	1.4	1.2		MIT
3724.772	3725.831	26839.65	LA	0.3	0.3		MIT
3720.75	3721.81	26868.7	LA II		0.5H		ME
3717.98	3719.04	26888.7	LA II	0.5	0.5		MIT
3715.524	3716.581	26906.45	LA II	2.0	1.7		MIT
3714.858	3715.915	26911.27	LA II	1.8	1.6		MIT
3714.293	3715.350	26915.37	LA I	1.0	0.3		MIT
3713.544	3714.600	26920.80	LA II	2.3	1.8		MIT
3710.59	3711.65	26942.2	LA II	0.5	0.3		MIT
3705.818	3706.872	26976.92	LA II	1.7	1.9H		MIT
3704.531	3705.585	26986.29	LA I	1.4	0.6		MIT
3701.807	3702.860	27006.15	LA II	0.9	1.0		MIT
3699.556	3700.609	27022.58	LA I	1.0	0.3		MIT
3696.11	3697.16	27047.8	LA II		0.3		ME
3695.2	3696.3	27054.	LA	0.3H	0.3		ME
3694.27	3695.32	27061.2	LA II	0.6	1.0		ME
3692.31	3693.36	27075.6	LA II		0.3		ME
3678.24	3679.29	27179.2	LA II	0.3H	0.3H		MIT
3672.016	3673.062	27225.24	LA I	1.4	0.5		MIT
3670.23	3671.28	27238.5	LA II	0.3	0.5		ME
3669.27	3670.31	27245.6	LA		0.5H		ME
3665.22	3666.26	27275.7	LA II	0.6	0.6		ME
3662.073	3663.116	27299.16	LA II	1.8	1.6		MIT
3658.41	3659.45	27326.5	LA II	0.5	0.5		MIT
3650.174	3651.214	27388.15	LA II	2.0	1.8		MIT
3649.509	3650.549	27393.14	LA I	1.6	0.9		MIT
3645.414	3646.453	27423.91	LA II	2.0	1.8		MIT
3641.656	3642.694	27452.21	LA II		0.6		MIT
3641.522	3642.560	27453.22	LA I	1.4	0.6		MIT

AIR WAVELENGTH (ANGSTRMS)	VACUUM WAVELENGTH (ANGSTRMS)	WAVENUMBER (KAYSERS)	LMENT	LOG ARC	INTENSITY SPK	DIS	REF
3641.10	3642.14	27456.4	LA II	0.3	0.5		ME
3639.25	3640.29	27470.4	LA II	0.3	0.3		ME
3637.148	3638.185	27486.24	LA II	1.7	1.6		MIT
3636.660	3637.697	27489.92	LA I	1.3	0.3		MIT
3629.99	3631.02	27540.4	LA II	0.3	0.3		ME
3628.822	3629.857	27549.30	LA II	1.9	1.6H		MIT
3621.77	3622.80	27602.9	LA II	0.7	0.9		MIT
3613.077	3614.107	27669.35	LA I	1.1	0.5		MIT
3612.334	3613.364	27675.04	LA II	0.9	1.2H		MIT
3611.103	3612.133	27684.47	LA II		0.5		MIT
3610.245	3611.275	27691.05	LA	0.8	0.7		MIT
3609.23	3610.26	27698.8	LA II	0.3	0.5		MIT
3608.16	3609.19	27707.1	LA II	0.3	0.8		MIT
3606.432	3607.461	27720.33	LA II	0.3	0.5H		MIT
3601.054	3602.081	27761.73	LA II	0.7	1.2		MIT
3598.927	3599.954	27778.13	LA II		0.5B		MIT
3596.65	3597.68	27795.7	LA II	0.3	0.5		ME
3593.29	3594.32	27821.7	LA	0.3			M
3585.523	3586.546	27881.98	LA II		0.8H		MIT
3581.68	3582.70	27911.9	LA II		1.3B		ME
3580.099	3581.121	27924.22	LA II	0.3	0.5H		MIT
3578.89	3579.91	27933.6	LA II	0.3H	0.5H		ME
3576.581	3577.602	27951.68	LA II		0.5D		MIT
3574.426	3575.446	27968.53	LA I	1.5	0.9		MIT
3570.095	3571.114	28002.46	LA II	1.0	1.0		MIT
3557.251	3558.267	28103.57	LA II	1.0	0.9		MIT
3550.815	3551.829	28154.50	LA II	1.0	0.6		MIT
3536.37	3537.38	28269.5	LA II	0.5	0.5		ME
3533.67	3534.68	28291.1	LA II		0.5H		ME
3530.661	3531.670	28315.22	LA II	1.2	0.8		MIT
3526.77	3527.78	28346.4	LA II		0.3H		ME
3520.711	3521.718	28395.24	LA II	0.6	0.6		MIT
3517.11	3518.12	28424.3	LA	1.4H			KN
3514.87	3515.88	28442.4	LA II	0.5	0.3		ME
3514.063	3515.068	28448.95	LA I	0.9	0.5		MIT
3512.917	3513.922	28458.23	LA II	1.7	1.2		MIT
3509.989	3510.993	28481.97	LA II	1.0	1.0		MIT
3507.90	3508.90	28498.9	LA		0.6B		ME
3484.38	3485.38	28691.3	LA II	0.7	0.7		MIT
3480.607	3481.603	28722.40	LA I	0.8			MIT
3474.84	3475.83	28770.1	LA II	0.6	0.9		ME
3462.32	3463.31	28874.1	LA	0.3	0.3H		ME
3461.184	3462.175	28883.57	LA I	1.0	0.3		MIT
3460.31	3461.30	28890.9	LA II	0.3	0.5		ME
3453.168	3454.157	28950.62	LA II	1.7	1.6		MIT
3452.184	3453.173	28958.87	LA II	1.7	1.6		MIT
3451.12	3452.11	28967.8	LA II	0.3	0.6H		ME
3450.641	3451.630	28971.82	LA I	1.0	0.3		MIT
3432.81	3433.79	29122.3	LA II	0.5	0.5		ME
3427.57	3428.55	29166.8	LA II	0.3	0.5		ME
3423.9	3424.9	29198.	LA II		0.3		ME
3420.54	3421.52	29226.8	LA II		0.3		ME
3411.76	3412.74	29302.0	LA II	0.7	1.4B		ME
3407.00	3407.98	29342.9	LA II	0.3	0.6		ME
3404.52	3405.50	29364.3	LA I	1.0	0.3		MIT
3398.29	3399.27	29418.1	LA II	0.3	0.5		ME
3397.76	3398.74	29422.7	LA II	0.7	0.9		MIT
3392.94	3393.91	29464.5	LA II	0.6	0.5		ME
3390.40	3391.37	29486.6	LA II	0.3	0.3		ME
3388.612	3389.585	29502.14	LA I	1.0	0.3		MIT
3381.438	3382.409	29564.73	LA I	0.7	0.3		MIT
3380.910	3381.881	29569.34	LA II	2.3	2.0H		MIT
3376.329	3377.299	29609.46	LA II	2.0	1.7		MIT
3374.89	3375.86	29622.1	LA	0.3	0.5		ME
3368.36	3369.33	29679.5	LA I	0.7			MIT
3364.900	3365.867	29710.03	LA I	0.6	0.3H		MIT
3362.040	3363.006	29735.30	LA I	1.0	0.5		MIT
3357.49	3358.45	29775.6	LA I	0.8			MIT
3351.89	3352.85	29825.3	LA II	0.3	0.3		ME
3349.834	3350.797	29843.65	LA I	0.6			MIT

AIR WAVELENGTH (ANGSTRMS)	VACUUM WAVELENGTH (ANGSTRMS)	WAVENUMBER (KAYSERS)	LMENT	LOG ARC	INTENSITY SPK	DIS	REF
3344.560	3345.522	29890.71	LA II	2.5	2.3J		MIT
3342.224	3343.185	29911.60	LA I	1.9	0.7		MIT
3337.488	3338.448	29954.04	LA II	2.9	2.5J		MIT
3329.07	3330.03	30029.8	LA II	0.6	0.6		ME
3326.21	3327.17	30055.6	LA II	0.5	0.5		ME
3325.33	3326.29	30063.6	LA II	0.3	0.3		ME
3310.62	3311.57	30197.1	LA II	0.3	0.5		ME
3306.98	3307.93	30230.4	LA II	1.0	0.9		MIT
3303.11	3304.06	30265.8	LA II	2.6	2.2		MIT
3298.72	3299.67	30306.1	LA	0.3H	0.5H		ME
3297.15	3298.10	30320.5	LA II	0.3	0.3		MIT
3294.44	3295.39	30345.4	LA II	0.3	0.7		ME
3283.95	3284.90	30442.4	LA II	0.5	0.6		ME
3277.83	3278.77	30499.2	LA II	0.3H	0.5H		ME
3267.31	3268.25	30597.4	LA II	0.3	0.5		ME
3265.67	3266.61	30612.8	LA II	2.5	2.3		ME
3263.99	3264.93	30628.5	LA II	0.5	0.6		MIT
3256.601	3257.540	30698.01	LA	0.5			MIT
3253.42	3254.36	30728.0	LA II	0.5	0.6		MIT
3249.351	3250.288	30766.50	LA II	2.5	1.9		MIT
3247.04	3247.98	30788.4	LA I	0.9	0.3		MIT
3245.120	3246.056	30806.61	LA II	2.6	2.5		MIT
3235.65	3236.58	30896.8	LA I	0.9			MIT
3226.008	3226.939	30989.12	LA II	0.3	0.3		MIT
3217.12	3218.05	31074.7	LA II	0.3	1.0H		ME
3215.813	3216.742	31087.36	LA I	1.0	0.5		MIT
3212.56	3213.49	31118.8	LA II	0.3	0.3		ME
3209.13	3210.06	31152.1	LA II	0.5	0.5		ME
3208.13	3209.06	31161.8	LA	0.3H	1.0B		ME
3205.76	3206.69	31184.8	LA II	0.6	0.7		MIT
3204.55	3205.48	31196.6	LA II	0.3	0.3		ME
3194.70	3195.62	31292.8	LA II		0.3		ME
3193.02	3193.94	31309.3	LA II	1.2	1.8		MIT
3191.39	3192.31	31325.2	LA II	0.5	0.9B		ME
3179.78	3180.70	31439.6	LA I	0.9	0.3		MIT
3175.987	3176.906	31477.17	LA I	1.2	0.5		MIT
3174.88	3175.80	31488.1	LA II	0.5	1.0B		ME
3171.668	3172.586	31520.03	LA	0.3H	3.0J		MIT
3166.26	3167.18	31573.9	LA II	0.5	0.5		ME
3165.19	3166.11	31584.5	LA II				ME
3160.56	3161.47	31630.8	LA II	0.5	0.5H		ME
3157.58	3158.49	31660.7	LA II	0.3	0.3		ME
3148.51	3149.42	31751.9	LA I	0.8			ME
3145.7	3146.6	31780.	LA II	0.3H	0.3H		ME
3142.762	3143.672	31809.93	LA II	2.2	1.7		MIT
3132.14	3133.05	31917.8	LA II	0.3	0.5H		ME
3130.25	3131.16	31937.1	LA II	0.5	0.6		ME
3125.72	3126.63	31983.4	LA II		0.5		ME
3112.63	3113.53	32117.9	LA II	0.5	1.0B		ME
3109.436	3110.338	32150.85	LA I	1.4	0.3		MIT
3108.462	3109.364	32160.92	LA II	1.0	0.9		MIT
3104.589	3105.490	32201.04	LA II	2.3	1.7		MIT
3101.99	3102.89	32228.0	LA		0.6B		ME
3096.023	3096.922	32290.13	LA I	1.2	0.3		MIT
3094.76	3095.66	32303.3	LA	0.3	0.5		ME
3088.53	3089.43	32368.5	LA II	0.3	0.5H		ME
3081.42	3082.31	32443.1	LA II	0.6	0.8		ME
3075.51	3076.40	32505.5	LA II	0.3H	0.3		ME
3069.45	3070.34	32569.7	LA II	0.3	0.5		ME
3068.96	3069.85	32574.9	LA II	0.3H	0.5		ME
3059.91	3060.80	32671.2	LA II	0.3	0.8		ME
3054.02	3054.91	32734.2	LA II	0.3	0.8		ME
3049.39	3050.28	32783.9	LA II	0.5	0.5		ME
3025.88	3026.76	33038.6	LA II		0.5		ME
3022.26	3023.14	33078.2	LA II	0.3	0.5		ME
3018.95	3019.83	33114.5	LA	0.6	0.5		ME
3010.78	3011.66	33204.3	LA I	0.7			ME
3007.32	3008.20	33242.5	LA II	0.6	0.6		ME
2985.75	2986.62	33482.6	LA II	0.3	0.3		MIT
2985.432	2986.303	33486.22	LA II	0.3	0.5		MIT

AIR WAVELENGTH (ANGSTRMS)	VACUUM WAVELENGTH (ANGSTRMS)	WAVENUMBER (KAYSERS)	LMENT	LOG ARC	INTENSITY SPK	DIS	REF
2984.336	2985.207	33498.52	LA II	0.3	0.3		MIT
2983.45	2984.32	33508.5	LA II		0.3		MIT
2976.83	2977.70	33583.0	LA II		0.5		MIT
2966.57	2967.44	33699.1	LA II	0.3	0.6		MIT
2966.05	2966.92	33705.0	LA II	0.0H	0.3		MIT
2962.910	2963.775	33740.75	LA II	0.3	1.2		MIT
2959.850	2960.715	33775.63	LA II	0.6	0.7		MIT
2951.45	2952.31	33871.7	LA II	0.0	0.5		MIT
2950.492	2951.354	33882.75	LA II	0.5	1.7		MIT
2943.551	2944.412	33962.64	LA II	0.3	0.8B		ME
2939.618	2940.478	34008.08	LA II		0.5H		ME
2929.88	2930.74	34121.1	LA II	0.3	0.8		ME
2925.148	2926.004	34176.30	LA II	0.0H	0.7H		MIT
2923.90	2924.76	34190.9	LA II	0.5	1.3		MIT
2913.61	2914.46	34311.6	LA II		0.3		MIT
2909.631	2910.483	34358.56	LA	0.3			MIT
2905.52	2906.37	34407.2	LA II	0.3	0.6B		MIT
2899.80	2900.65	34475.0	LA II	0.3	0.6B		MIT
2897.76	2898.61	34499.3	LA II	0.3	0.7B		MIT
2893.071	2893.919	34555.22	LA II	0.8	1.8		MIT
2885.141	2885.987	34650.19	LA II	0.7	1.7		MIT
2880.644	2881.489	34704.23	LA II	0.6	1.6		MIT
2874.28	2875.12	34781.1	LA II	0.7	0.5		ME
2873.18	2874.02	34794.4	LA II	0.0	0.3		MIT
2862.967	2863.808	34918.55	LA II	0.3	1.2B		MIT
2862.373	2863.214	34925.79	LA II	0.3	0.8		MIT
2859.757	2860.597	34957.74	LA II	0.0	0.7		MIT
2855.902	2856.741	35004.92	LA II	0.5	1.7B		MIT
2853.72	2854.56	35031.7	LA II		0.6H		ME
2848.34	2849.18	35097.9	LA II	0.3	0.8		MIT
2846.69	2847.53	35118.2	LA II		0.7		MIT
2843.659	2844.495	35155.63	LA II	0.6	0.6		MIT
2840.50	2841.34	35194.7	LA II	0.5	1.4B		MIT
2838.438	2839.273	35220.29	LA II	0.3	0.7		MIT
2832.55	2833.38	35293.5	LA II	0.6	0.7		MIT
2825.51	2826.34	35381.4	LA II	0.7	0.7		ME
2821.040	2821.870	35437.49	LA II	0.0	0.7		MIT
2819.73	2820.56	35454.0	LA II		0.3		MIT
2818.39	2819.22	35470.8	LA II		0.5		MIT
2817.500	2818.330	35482.01	LA	0.3			MIT
2815.35	2816.18	35509.1	LA II	0.3	0.8		MIT
2813.72	2814.55	35529.7	LA II	0.3H	0.7		MIT
2813.03	2813.86	35538.4	LA II	0.0H	0.5		MIT
2809.34	2810.17	35585.1	LA II	0.3	0.3		MIT
2808.39	2809.22	35597.1	LA II	1.0D	2.2		MIT
2805.59	2806.42	35632.6	LA II	0.0D	0.7		MIT
2804.53	2805.36	35646.1	LA II		0.6		MIT
2798.546	2799.371	35722.31	LA II	0.3	1.6B		MIT
2796.366	2797.190	35750.16	LA II	0.3	0.7		MIT
2794.02	2794.84	35780.2	LA	0.6			MIT
2791.513	2792.336	35812.31	LA II	0.3	1.4		MIT
2780.234	2781.054	35957.58	LA II	1.3	1.0H		MIT
2779.778	2780.598	35963.48	LA II	0.0	1.0		MIT
2778.757	2779.577	35976.70	LA II	0.0	1.0		MIT
2767.40	2768.22	36124.3	LA II	0.6	0.9		ME
2766.458	2767.275	36136.63	LA I	0.6			MIT
2761.542	2762.358	36200.96	LA I	0.8			MIT
2761.111	2761.927	36206.61	LA II		0.7		MIT
2760.504	2761.320	36214.57	LA II		0.5		MIT
2759.545	2760.360	36227.15	LA	0.6			MIT
2759.156	2759.971	36232.26	LA II		0.5		MIT
2758.652	2759.467	36238.88	LA II		0.5		MIT
2756.57	2757.38	36266.2	LA	0.3			MIT
2752.858	2753.672	36315.15	LA II	0.0	1.0		MIT
2749.51	2750.32	36359.4	LA	0.5			MIT
2748.317	2749.130	36375.15	LA II	0.0	0.9		MIT
2739.269	2740.079	36495.29	LA	0.5			MIT
2737.50	2738.31	36518.9	LA	0.3			MIT
2736.90	2737.71	36526.9	LA II	0.5	0.3		ME
2736.41	2737.22	36533.4	LA II		0.5		ME

AIR WAVELENGTH (ANGSTRMS)	VACUUM WAVELENGTH (ANGSTRMS)	WAVENUMBER (KAYSERS)	LMENT	LOG ARC	INTENSITY SPK	INTENSITY DIS	REF
2732.415	2733.224	36586.83	LA II	0.0	1.0		MIT
2730.138	2730.946	36617.34	LA	0.3H			MIT
2729.877	2730.685	36620.85	LA I	0.6			MIT
2727.54	2728.35	36652.2	LA II	0.3	0.3		MIT
2725.586	2726.393	36678.50	LA I	1.0			MIT
2722.321	2723.127	36722.48	LA I	0.7			MIT
2717.360	2718.165	36789.52	LA	0.3			MIT
2715.416	2716.221	36815.86	LA II	0.0	1.0B		MIT
2714.538	2715.343	36827.77	LA I	0.8			MIT
2710.685	2711.489	36880.11	LA	0.5			MIT
2709.92	2710.72	36890.5	LA II	0.0H	0.5		ME
2707.07	2707.87	36929.4	LA I	0.5			ME
2706.49	2707.29	36937.3	LA II	0.7	0.3		MIT
2702.159	2702.961	36996.47	LA II	0.3	0.9		MIT
2695.458	2696.258	37088.44	LA II	0.5	1.5		MIT
2694.207	2695.007	37105.66	LA II	0.0H	0.7		MIT
2687.72	2688.52	37195.2	LA II	0.3	0.3		MIT
2684.142	2684.939	37244.79	LA	0.7			MIT
2681.501	2682.298	37281.47	LA II	0.3	1.0		MIT
2679.86	2680.66	37304.3	LA II		0.6H		MIT
2677.76	2678.56	37333.5	LA I	0.3H			MIT
2675.655	2676.450	37362.92	LA II	0.3	0.7		MIT
2673.743	2674.538	37389.64	LA II		0.5		MIT
2672.906	2673.700	37401.35	LA II	0.5	1.5		MIT
2671.892	2672.686	37415.54	LA	0.3			MIT
2666.54	2667.33	37490.6	LA II	0.0	0.5		ME
2666.178	2666.971	37495.72	LA II	0.0	0.8		MIT
2665.62	2666.41	37503.6	LA II		0.6		ME
2664.75	2665.54	37515.8	LA II	0.0	0.5		ME
2661.665	2662.457	37559.29	LA II		0.5		MIT
2661.36	2662.15	37563.6	LA II	0.0H	0.6		MIT
2653.495	2654.285	37674.93	LA I	0.3			MIT
2651.71	2652.50	37700.3	LA	0.0	0.9		MIT
2647.36	2648.15	37762.2	LA II		0.6		ME
2647.144	2647.932	37765.32	LA	0.3			MIT
2638.988	2639.774	37882.02	LA II		0.7		MIT
2631.93	2632.71	37983.6	LA II	0.5	0.9		MIT
2631.52	2632.30	37989.5	LA II		0.6		ME
2620.03	2620.81	38156.1	LA II	0.0	0.8		MIT
2616.314	2617.095	38210.31	LA II	0.0	0.8B		MIT
2613.097	2613.877	38257.34	LA II		0.6		MIT
2610.335	2611.115	38297.82	LA II	1.0	2.2		MIT
2601.777	2602.555	38423.79	LA II		0.7		MIT
2600.869	2601.646	38437.20	LA II		0.3		MIT
2600.347	2601.124	38444.92	LA II		0.6		MIT
2596.32	2597.10	38504.5	LA II		0.5		ME
2596.091	2596.867	38507.94	LA II		1.3		MIT
2592.87	2593.65	38555.8	LA II		0.5H		ME
2586.37	2587.14	38652.7	LA II		1.0		MIT
2582.960	2583.733	38703.69	LA II		0.9		MIT
2582.56	2583.33	38709.7	LA II		0.8		MIT
2580.823	2581.596	38735.73	LA II		0.9B		MIT
2577.921	2578.693	38779.34	LA II		0.3		MIT
2573.481	2574.252	38846.24	LA		0.3H		MIT
2566.078	2566.847	38958.30	LA II		1.0H		MIT
2561.848	2562.616	39022.62	LA II		1.3		MIT
2560.374	2561.142	39045.09	LA II		1.7		MIT
2558.991	2559.758	39066.19	LA II		0.5		MIT
2553.398	2554.164	39151.75	LA II		0.5H		MIT
2552.600	2553.366	39163.99	LA II		0.8		MIT
2546.40	2547.16	39259.3	LA II		1.3B		ME
2542.40	2543.16	39321.1	LA II		0.8		ME
2541.59	2542.35	39333.6	LA II		0.6		MIT
2538.40	2539.16	39383.1	LA II		0.3		MIT
2536.76	2537.52	39408.5	LA II		0.5D		MIT
2534.98	2535.74	39436.2	LA II		0.8		ME
2533.14	2533.90	39464.8	LA II		1.2		ME
2531.60	2532.36	39488.8	LA II		0.9		ME
2527.84	2528.60	39547.6	LA II		0.5		ME
2523.07	2523.83	39622.3	LA II		0.7B		ME

AIR WAVELENGTH (ANGSTRMS)	VACUUM WAVELENGTH (ANGSTRMS)	WAVENUMBER (KAYSERS)	LMENT	LOG ARC	INTENSITY SPK	INTENSITY DIS	REF
2519.215	2519.973	39682.96	LA II		1.7		MIT
2515.79	2516.55	39737.0	LA II		0.6		ME
2514.56	2515.32	39756.4	LA II		0.5		MIT
2501.18	2501.93	39969.1	LA II		1.2H		ME
2487.588	2488.339	40187.46	LA II		1.6		MIT
2483.00	2483.75	40261.7	LA II		0.3H		MIT
2479.84	2480.59	40313.0	LA II		1.0		MIT
2476.75	2477.50	40363.3	LA		1.5H		MIT
2474.51	2475.26	40399.8	LA II		0.5		MIT
2472.43	2473.18	40433.8	LA II		1.0		MIT
2471.900	2472.647	40442.49	LA II		1.3		MIT
2471.06	2471.81	40456.2	LA II		0.7		ME
2470.55	2471.30	40464.6	LA II		0.5		ME
2458.15	2458.89	40668.7	LA II		0.3J		MIT
2455.88	2456.62	40706.3	LA II		1.0		MIT
2452.74	2453.48	40758.4	LA II		0.9		MIT
2445.57	2446.31	40877.9	LA		1.0H		MIT
2443.16	2443.90	40918.2	LA II		0.3H		MIT
2442.80	2443.54	40924.2	LA		0.5		ME
2438.39	2439.13	40998.2	LA II		1.0		MIT
2437.999	2438.738	41004.81	LA II		1.3		MIT
2437.13	2437.87	41019.4	LA II		1.0W		MIT
2436.42	2437.16	41031.4	LA II		1.2O		MIT
2431.40	2432.14	41116.1	LA		0.8		ME
2424.524	2425.260	41232.69	LA		0.3H		MIT
2421.61	2422.35	41282.3	LA II		0.7H		ME
2420.01	2420.75	41309.6	LA		0.7H		MIT
2417.61	2418.34	41350.6	LA II		0.5H		ME
2410.10	2410.83	41479.4	LA		0.7B		ME
2407.81	2408.54	41518.9	LA		0.7B		MIT
2404.643	2405.375	41573.56	LA II		0.8		MIT
2403.28	2404.01	41597.1	LA II		0.8		MIT
2401.47	2402.20	41628.5	LA		0.3H		MIT
2399.66	2400.39	41659.9	LA II		1.3B		MIT
2398.70	2399.43	41676.6	LA		0.5H		ME
2397.237	2397.967	41701.99	LA II		0.8B		MIT
2394.98	2395.71	41741.3	LA II		0.6		ME
2389.84	2390.57	41831.1	LA		0.5B		ME
2384.28	2385.01	41928.6	LA		0.5H		ME
2379.42	2380.15	42014.2	LA		1.8		MIT
2370.49	2371.21	42172.5	LA II		0.3		MIT
2369.18	2369.90	42195.8	LA	0.3	0.3		ME
2365.50	2366.22	42261.5	LA II	0.3	0.5		MIT
2358.02	2358.74	42395.5	LA	0.3	0.5H		ME
2355.81	2356.53	42435.3	LA	0.3	0.7H		ME
2353.40	2354.12	42478.7	LA		0.3		ME
2348.86	2349.58	42560.8	LA II	0.3H	0.3		ME
2341.85	2342.57	42688.2	LA II		0.6		ME
2328.753	2329.468	42928.26	LA	0.3	0.3		MIT
2325.75	2326.46	42983.7	LA II	0.3	1.3B		ME
2322.78	2323.49	43038.6	LA	0.3	0.5H		ME
2319.441	2320.154	43100.60	LA II	0.0	1.3		MIT
2317.82	2318.53	43130.7	LA II	0.0	1.3B		MIT
2297.78	2298.49	43506.9	LA	0.3H	2.2		MIT
2292.32	2293.03	43610.5	LA II		0.5		ME
2280.94	2281.64	43828.0	LA II	0.3	0.6H		ME
2265.58	2266.28	44125.2	LA II		0.5		MIT
2256.76	2257.46	44297.6	LA II	0.3H	1.5		MIT
2230.74	2231.43	44814.2	LA II	0.3H	0.8		ME
2216.11	2216.80	45110.1	LA	0.3	1.4		MIT
2195.83	2196.52	45526.6	LA II	0.3H	0.6		MIT
2187.87	2188.55	45692.3	LA II	0.5	1.6		MIT
2163.66	2164.34	46203.5	LA	0.5	1.3B		ME
2161.36	2162.04	46252.6	LA II	0.3H	0.6		ME
2142.81	2143.49	46653.0	LA	0.3H	1.3B		ME
8126.52	8128.75	12302.0	LI I	3.0			ME
6707.85	6709.70	14903.8	LI	2.5W			MIT
6707.844	6709.696	14903.81	LI I	3.5G	2.3		HZ
6240.1	6241.8	16021.	LI I	2.5			SD
6103.642	6105.332	16379.13	LI I	3.3G	2.5		HZ

AIR WAVELENGTH (ANGSTRMS)	VACUUM WAVELENGTH (ANGSTRMS)	WAVENUMBER (KAYSERS)	LMENT	LOG ARC	INTENSITY SPK	INTENSITY DIS	REF
5484.7	5486.2	18227.	LI II			0.9	WR
4971.990	4973.377	20107.06	LI I	2.7			MIT
4881.3	4882.7	20481.	LI II			0.5	WR
4671.8	4673.1	21399.	LI II			0.6	WR
4636.1	4637.4	21564.	LI I	0.5			FL
4602.863	4604.152	21719.52	LI I	2.9			DA
4325.7	4326.9	23111.	LI II			0.5	WR
4273.28	4274.48	23394.6	LI I	2.3R	2.0H		FL
4132.29	4133.46	24192.8	LI I	2.6J			FL
3985.79	3986.92	25082.0	LI I	2.0			FL
3921.65	3922.76	25492.2	LI I	0.3			FL
3915.0	3916.1	25536.	LI I	2.3J			FL
3838.15	3839.24	26046.8	LI I	0.7			FL
3794.722	3795.799	26344.91	LI I	1.8			MIT
3718.7	3719.8	26883.	LI I	1.5			FL
3684.1	3685.1	27136.	LI II			0.3	WR
3670.4	3671.4	27237.	LI I	0.7			FL
3232.61	3233.54	30925.8	LI I	3.0G	2.7		FL
3195.8	3196.7	31282.	LI II			0.5	WR
3155.4	3156.3	31682.	LI II			0.3	WR
3143.7	3144.6	31800.	LI		0.5		AN
3047.1	3048.0	32808.	LI		0.8		AN
2988.5	2989.4	33452.	LI		1.2		AN
2959.5	2960.4	33780.	LI		0.3		AN
2899.66	2900.51	34476.7	LI		1.8		AN
2869.90	2870.74	34834.2	LI		0.5		AN
2866.45	2867.29	34876.1	LI		0.3		AN
2741.31	2742.12	36468.1	LI I	2.3			FL
2741.188	2741.999	36469.74	LI II		1.5		DA
2728.4	2729.2	36641.	LI II			0.3	WR
2704.60	2705.40	36963.1	LI		0.3		AN
2674.4	2675.2	37380.	LI II			0.3	WR
2635.20	2635.99	37936.5	LI		0.5		AN
2562.537	2563.305	39012.13	LI I	2.2	1.2		MIT
2475.29	2476.04	40387.1	LI I	2.0W	0.7		FL
2425.68	2426.42	41213.0	LI I	1.5W			FL
2394.48	2395.21	41750.0	LI I	0.7G			FL
9914.92	9917.64	10083.1	LU	2.0			ME
9892.00	9894.71	10106.4	LU	0.7			ME
9841.32	9844.02	10158.4	LU	0.3			ME
9696.03	9698.69	10310.7	LU	1.5			ME
9661.69	9664.34	10347.3	LU	1.0			ME
9504.50	9507.11	10518.4	LU	0.3			ME
9273.79	9276.33	10780.1	LU	0.3			ME
9203.92	9206.45	10861.9	LU	0.9			ME
9116.26	9118.76	10966.4	LU	1.3			ME
8950.18	8952.64	11169.9	LU	1.0			ME
8949.63	8952.09	11170.6	LU	1.2			ME
8948.93	8951.39	11171.4	LU	1.3			ME
8795.6	8798.0	11366.	LU	0.3			ME
8788.83	8791.24	11375.0	LU	0.6			ME
8744.03	8746.43	11433.2	LU	0.5			ME
8712.78	8715.17	11474.2	LU	0.5			ME
8690.23	8692.62	11504.0	LU	0.5H			ME
8610.98	8613.35	11609.9	LU	2.1			ME
8508.08	8510.42	11750.3	LU	2.0			ME
8478.50	8480.83	11791.3	LU	1.7			ME
8459.19	8461.51	11818.2	LU	2.2			ME
8408.21	8410.52	11889.9	LU	0.9			ME
8382.08	8384.38	11926.9	LU	1.5			ME
8329.45	8331.74	12002.3	LU	0.7			ME
8248.68	8250.95	12119.8	LU	0.3			ME
8178.16	8180.41	12224.3	LU	1.6			ME
8082.36	8084.58	12369.2	LU	0.3			ME
7965.31	7967.50	12551.0	LU	1.0			ME
7939.63	7941.81	12591.6	LU	0.3			ME
7927.58	7929.76	12610.7	LU	0.3			ME
7917.52	7919.70	12626.7	LU	1.2			ME
7911.62	7913.80	12636.2	LU	1.5			ME
7816.14	7818.29	12790.5	LU	0.8			ME

AIR WAVELENGTH (ANGSTRMS)	VACUUM WAVELENGTH (ANGSTRMS)	WAVENUMBER (KAYSERS)	LMENT	LOG ARC	INTENSITY SPK	INTENSITY DIS	REF
7815.91	7818.06	12790.9	LU	1.4			ME
7815.84	7817.99	12791.0	LU	1.2			ME
7758.30	7760.44	12885.9	LU	1.3			ME
7686.81	7688.93	13005.7	LU	0.3H			ME
7659.01	7661.12	13052.9	LU	0.6			ME
7649.79	7651.90	13068.7	LU	0.5			ME
7649.54	7651.65	13069.1	LU	0.9			ME
7649.28	7651.39	13069.5	LU	0.5			ME
7640.08	7642.18	13085.3	LU	1.0			ME
7543.06	7545.14	13253.6	LU	0.3			ME
7456.96	7459.01	13406.6	LU	0.7			ME
7441.52	7443.57	13434.4	LU	1.3			ME
7409.70	7411.74	13492.1	LU	0.8			ME
7237.98	7239.97	13812.2	LU	1.6			ME
7196.40	7198.38	13892.0	LU	0.7			ME
7165.94	7167.92	13951.1	LU	1.0			ME
7143.10	7145.07	13995.7	LU	0.7			ME
7142.79	7144.76	13996.3	LU	0.8			ME
7125.84	7127.80	14029.6	LU	2.1			ME
7096.34	7098.30	14087.9	LU	1.5			ME
7031.24	7033.18	14218.3	LU	1.7			ME
6943.96	6945.88	14397.0	LU	0.7			ME
6917.31	6919.22	14452.5	LU	1.7			ME
6826.59	6828.47	14644.6	LU	0.6			ME
6793.77	6795.65	14715.3	LU	1.6	0.3		ME
6735.76	6737.62	14842.0	LU	0.7			ME
6677.14	6678.98	14972.3	LU	1.6	0.0		ME
6668.74	6670.58	14991.2	LU	0.6			ME
6612.04	6613.87	15119.7	LU	1.0			ME
6611.95	6613.78	15120.0	LU	1.2			ME
6611.80	6613.63	15120.3	LU	1.3			ME
6611.71	6613.54	15120.5	LU	2.0	2.2		ME
6611.58	6613.41	15120.8	LU	1.4			ME
6611.28	6613.11	15121.5	LU	1.5			ME
6523.18	6524.98	15325.7	LU	1.9	0.7		ME
6477.67	6479.46	15433.4	LU	1.5	0.3		ME
6463.12	6464.91	15468.1	LU	2.6	2.9		ME
6444.89	6446.67	15511.9	LU	0.7	1.2		ME
6441.14	6442.92	15520.9	LU	1.6			ME
6366.00	6367.76	15704.1	LU	1.2			ME
6365.79	6367.55	15704.6	LU	1.0			ME
6354.85	6356.61	15731.7	LU	1.3	0.0		ME
6345.35	6347.10	15755.2	LU	1.8	0.6		ME
6248.80	6250.53	15998.7	LU	1.0			ME
6242.34	6244.07	16015.2	LU	1.6	2.3		ME
6235.36	6237.08	16033.1	LU	1.4	2.0		ME
6228.14	6229.86	16051.7	LU	1.0	1.6		ME
6221.87	6223.59	16067.9	LU	2.7	3.0		ME
6199.66	6201.38	16125.5	LU	1.6	2.1		ME
6195.12	6196.83	16137.3	LU	1.2			ME
6159.94	6161.64	16229.4	LU	1.7	2.3		ME
6140.71	6142.41	16280.3	LU	1.0			ME
6084.14	6085.82	16431.6	LU	0.9W			ME
6055.03	6056.71	16510.6	LU	2.2D	1.0		ME
6041.66	6043.33	16547.2	LU	1.3	0.0		ME
6004.52	6006.18	16649.5	LU	2.6	1.6		ME
5997.13	5998.79	16670.0	LU	1.7	0.7		ME
5984.15	5985.81	16706.2	LU	1.3	1.6		KN
5984.00	5985.66	16706.6	LU	1.3			ME
5983.80	5985.46	16707.2	LU	1.3	1.6		KN
5983.59	5985.25	16707.7	LU	1.5	1.8		KN
5961.39	5963.04	16770.0	LU	0.5			ME
5887.23	5888.86	16981.2	LU	0.0	0.8H		ME
5866.30	5867.93	17041.8	LU	0.5			ME
5860.79	5862.41	17057.8	LU	1.3	0.3		ME
5800.59	5802.20	17234.9	LU	1.5	0.3		ME
5775.40	5777.00	17310.0	LU	1.7	0.7		ME
5736.55	5738.14	17427.2	LU	2.2	1.2		ME
5713.49	5715.08	17497.6	LU	0.5	1.2		ME
5664.89	5666.46	17647.7	LU	0.3	1.3		ME

AIR WAVELENGTH (ANGSTRMS)	VACUUM WAVELENGTH (ANGSTRMS)	WAVENUMBER (KAYSERS)	LMENT	LOG ARC	INTENSITY SPK	INTENSITY DIS	REF
5602.56	5604.12	17844.0	LU	0.0	0.7		ME
5476.69	5478.21	18254.1	LU	2.7	3.0		ME
5465.50	5467.02	18291.5	LU	0.5			ME
5453.57	5455.09	18331.5	LU	0.9	0.0		ME
5437.88	5439.39	18384.4	LU	1.5	0.5		ME
5421.90	5423.41	18438.6	LU	1.7	0.7		ME
5402.57	5404.07	18504.6	LU	2.2	1.0		ME
5349.12	5350.61	18689.5	LU	1.4	0.3		ME
5304.40	5305.88	18847.0	LU	1.3	0.3		ME
5278.48	5279.95	18939.6	LU		0.3H		ME
5206.47	5207.92	19201.5	LU	1.0			ME
5196.61	5198.06	19238.0	LU	1.0			ME
5135.09	5136.52	19468.4	LU	2.3	1.3		ME
5134.05	5135.48	19472.4	LU	1.3	0.3		ME
5057.60	5059.01	19766.7	LU	1.2	0.0		ME
5001.14	5002.54	19989.9	LU	2.0			ME
4994.13	4995.52	20017.9	LU	2.4	2.6		ME
4942.34	4943.72	20227.7	LU	1.6	0.5		ME
4921.70	4923.07	20312.5	LU	0.5	0.9		ME
4904.88	4906.25	20382.2	LU	1.8	0.7		ME
4888.14	4889.51	20452.0	LU	0.3H			ME
4882.40	4883.76	20476.0	LU	0.5H			ME
4865.36	4866.72	20547.7	LU	0.6	1.3		ME
4858.75	4860.11	20575.7	LU	0.3	0.9		ME
4844.44	4845.79	20636.5	LU	0.5			ME
4843.03	4844.38	20642.5	LU	0.3	0.7		ME
4839.62	4840.97	20657.0	LU	1.7	2.0		ME
4815.05	4816.40	20762.4	LU	1.3	0.3		ME
4785.42	4786.76	20891.0	LU	2.0	2.3		ME
4748.38	4749.71	21053.9	LU		1.0B		ME
4733.50	4734.82	21120.1	LU		0.5H		ME
4726.20	4727.52	21152.7	LU	0.6			ME
4716.70	4718.02	21195.3	LU	0.7			ME
4689.77	4691.08	21317.0	LU	0.6			ME
4675.29	4676.60	21383.1	LU	0.6	0.0		ME
4659.03	4660.33	21457.7	LU	1.0	0.0		ME
4658.02	4659.32	21462.3	LU	2.0	1.2		ME
4648.85	4650.15	21504.7	LU	1.4H			ME
4648.21	4649.51	21507.6	LU	1.4H	0.3		ME
4645.47	4646.77	21520.3	LU	1.4H	0.3		ME
4643.29	4644.59	21530.4	LU	0.9H			ME
4610.3	4611.6	21684.	LU	1.0			KN
4605.39	4606.68	21707.6	LU	1.0H			ME
4602.60	4603.89	21720.8	LU	0.5H			ME
4586.93	4588.22	21795.0	LU	0.3	0.8		ME
4525.48	4526.75	22090.9	LU	0.3H			ME
4521.08	4522.35	22112.4	LU	0.3H			ME
4518.57	4519.84	22124.7	LU	2.5	1.6		ME
4505.22	4506.48	22190.2	LU	0.0	0.7		ME
4498.85	4500.11	22221.7	LU	1.0	0.0		ME
4490.53	4491.79	22262.8	LU	0.3			ME
4480.15	4481.41	22314.4	LU	0.5			ME
4471.55	4472.80	22357.3	LU	0.8			ME
4450.81	4452.06	22461.5	LU	1.6	0.3		ME
4438.79	4440.04	22522.3	LU	0.6			ME
4430.48	4431.72	22564.6	LU	1.5	0.3		ME
4420.96	4422.20	22613.2	LU	1.2			ME
4416.45	4417.69	22636.3	LU	0.8			ME
4404.86	4406.10	22695.8	LU	0.7			ME
4397.31	4398.55	22734.8	LU	0.8			ME
4396.16	4397.39	22740.7	LU	0.3			ME
4374.90	4376.13	22851.2	LU	0.7			ME
4341.98	4343.20	23024.5	LU	0.5	1.5B		ME
4332.72	4333.94	23073.7	LU	1.0	0.0		ME
4309.57	4310.78	23197.6	LU	1.4	0.3		ME
4296.114	4297.323	23270.30	LU	0.8	0.0		KN
4295.950	4297.159	23271.19	LU	1.5	0.5		KN
4281.03	4282.23	23352.3	LU	1.6	0.6		ME
4277.50	4278.70	23371.6	LU	1.5	0.5		ME
4266.40	4267.60	23432.4	LU		0.3H		ME

AIR WAVELENGTH (ANGSTRMS)	VACUUM WAVELENGTH (ANGSTRMS)	WAVENUMBER (KAYSERS)	LMENT	LOG ARC	INTENSITY SPK	INTENSITY DIS	REF
4262.02	4263.22	23456.4	LU		0.9		ME
4223.99	4225.18	23667.6	LU	0.3			ME
4223.09	4224.28	23672.7	LU	0.0	0.6		ME
4184.25	4185.43	23892.4	LU	2.0	2.3		ME
4167.50	4168.67	23988.4	LU	0.3			ME
4154.08	4155.25	24065.9	LU	1.6	0.5		ME
4131.79	4132.96	24195.8	LU	1.0			ME
4124.73	4125.89	24237.2	LU	2.3	1.0		ME
4122.49	4123.65	24250.3	LU	1.2	0.3		ME
4112.67	4113.83	24308.2	LU	0.7			ME
4054.45	4055.60	24657.3	LU	1.4	0.5		ME
4030.86	4032.00	24801.6	LU		0.7B		ME
3991.38	3992.51	25046.9	LU	0.5			ME
3981.01	3982.14	25112.1	LU		0.3H		ME
3968.464	3969.587	25191.54	LU	1.7			KN
3937.61	3938.72	25388.9	LU		0.7B		ME
3926.62	3927.73	25460.0	LU	0.3			ME
3925.30	3926.41	25468.6	LU	1.3	0.0		ME
3918.86	3919.97	25510.4	LU	0.6			ME
3911.77	3912.88	25556.6	LU	0.5H			ME
3899.54	3900.64	25636.8	LU		0.5H		ME
3876.69	3877.79	25787.9	LU	1.2			ME
3876.63	3877.73	25788.3	LU	1.7	2.0		KN
3876.55	3877.65	25788.8	LU	0.5			KN
3874.61	3875.71	25801.7	LU	0.6			ME
3870.88	3871.98	25826.6	LU	0.7			ME
3853.29	3854.38	25944.5	LU	1.0			ME
3843.61	3844.70	26009.8	LU	1.2H			ME
3841.18	3842.27	26026.3	LU	2.0	0.9		ME
3829.07	3830.16	26108.6	LU	1.0H			ME
3814.02	3815.10	26211.6	LU	0.3			ME
3802.62	3803.70	26290.2	LU	0.5			ME
3800.67	3801.75	26303.7	LU	0.7			ME
3793.24	3794.32	26355.2	LU	0.6			ME
3786.18	3787.26	26404.4	LU	0.6			ME
3777.15	3778.22	26467.5	LU		0.3H		ME
3756.79	3757.86	26610.9	LU	0.8			ME
3756.70	3757.77	26611.5	LU	0.9			ME
3742.08	3743.14	26715.5	LU	0.3			ME
3724.32	3725.38	26842.9	LU	0.3H			ME
3722.58	3723.64	26855.4	LU	0.5			ME
3710.95	3712.01	26939.6	LU	0.5			ME
3706.16	3707.21	26974.4	LU	0.3			ME
3704.79	3705.84	26984.4	LU		0.5H		ME
3684.32	3685.37	27134.3	LU	1.2	0.0		ME
3647.77	3648.81	27406.2	LU	2.0	0.7		ME
3642.35	3643.39	27447.0	LU	0.3			ME
3636.25	3637.29	27493.0	LU	1.4	0.5		ME
3628.90	3629.93	27548.7	LU	0.3			ME
3623.99	3625.02	27586.0	LU	1.3	1.6		ME
3620.31	3621.34	27614.1	LU	0.6			ME
3596.34	3597.37	27798.1	LU	0.8			ME
3580.26	3581.28	27923.0	LU	0.5			ME
3567.84	3568.86	28020.2	LU	2.0	0.8		ME
3554.43	3555.45	28125.9	LU	1.7	2.2		ME
3546.39	3547.40	28189.6	LU	0.8			ME
3525.94	3526.95	28353.1	LU	0.3			ME
3521.18	3522.19	28391.4	LU	0.5			ME
3508.42	3509.42	28494.7	LU	1.5	0.5		ME
3507.47	3508.47	28502.4	LU	1.5			KN
3507.39	3508.39	28503.1	LU	2.0	2.2		KN
3491.92	3492.92	28629.4	LU		0.5H		ME
3472.48	3473.47	28789.6	LU	1.7	2.2		ME
3454.77	3455.76	28937.2	LU		0.5H		ME
3397.07	3398.04	29428.7	LU	1.7	1.3R		ME
3396.82	3397.79	29430.9	LU	1.5	0.0		ME
3391.55	3392.52	29476.6	LU	1.0	0.3		ME
3385.50	3386.47	29529.3	LU	1.5	0.6		ME
3376.50	3377.47	29608.0	LU	2.0	1.0		ME
3359.56	3360.53	29757.2	LU	2.2	1.2		ME

AIR WAVELENGTH (ANGSTRMS)	VACUUM WAVELENGTH (ANGSTRMS)	WAVENUMBER (KAYSERS)	LMENT	LOG ARC	INTENSITY SPK	DIS	REF
3332.61	3333.57	29997.9	LU		0.9H		ME
3319.63	3320.59	30115.2	LU		0.5H		ME
3312.11	3313.06	30183.6	LU	2.0	1.0		ME
3305.68	3306.63	30242.3	LU	0.3	0.0		ME
3281.74	3282.69	30462.9	LU	1.8	0.7		ME
3280.50	3281.45	30474.4	LU	1.0			ME
3278.97	3279.91	30488.6	LU	1.7	0.7		ME
3265.62	3266.56	30613.2	LU	0.3			ME
3265.00	3265.94	30619.0	LU		1.0B		ME
3254.31	3255.25	30719.6	LU	1.7	2.2		ME
3249.47	3250.41	30765.4	LU		0.6		ME
3242.93	3243.87	30827.4	LU		0.6H		ME
3222.57	3223.50	31022.2	LU		0.9H		ME
3219.09	3220.02	31055.7	LU		0.5H		ME
3198.12	3199.04	31259.3	LU	1.6	1.9		ME
3191.80	3192.72	31321.2	LU	0.5	1.8B		ME
3183.73	3184.65	31400.6	LU	0.5	0.8		ME
3171.36	3172.28	31523.1	LU	1.6	0.7		ME
3167.33	3168.25	31563.2	LU	0.0	1.2B		ME
3161.62	3162.54	31620.2	LU		1.0B		ME
3118.43	3119.33	32058.1	LU	1.6	0.7		ME
3104.98	3105.88	32197.0	LU		1.4B		ME
3089.02	3089.92	32363.3	LU		0.9H		ME
3081.47	3082.36	32442.6	LU	1.9	0.9		ME
3080.11	3081.00	32456.9	LU	1.2	0.3		ME
3077.60	3078.49	32483.4	LU	2.0	2.3		ME
3067.83	3068.72	32586.9	LU		0.5H		ME
3063.51	3064.40	32632.8	LU	1.3H			ME
3062.65	3063.54	32642.0	LU		0.3H		ME
3057.90	3058.79	32692.7	LU	0.5	2.2H		ME
3056.72	3057.61	32705.3	LU	1.7	2.0		ME
3047.36	3048.25	32805.7	LU	0.5H			ME
3040.04	3040.92	32884.7	LU		1.3B		ME
3027.29	3028.17	33023.2	LU		0.9		ME
3020.54	3021.42	33097.0	LU		2.0		ME
3016.37	3017.25	33142.8	LU		1.2B		ME
3003.65	3004.53	33283.1	LU		0.5H		ME
2995.84	2996.71	33369.9	LU		1.2B		ME
2989.27	2990.14	33443.2	LU	1.7	0.6		ME
2969.82	2970.69	33662.2	LU	1.5	2.0		ME
2963.32	2964.19	33736.1	LU	1.7	2.2		ME
2955.78	2956.64	33822.1	LU	0.3	1.8B		ME
2951.69	2952.55	33869.0	LU	1.3	1.9		ME
2949.73	2950.59	33891.5	LU	1.3H	0.0		ME
2931.53	2932.39	34101.9	LU	0.8H			ME
2912.70	2913.55	34322.4	LU	1.2H			ME
2911.39	2912.24	34337.8	LU	2.0	2.5		ME
2903.05	2903.90	34436.4	LU	1.3	0.0		ME
2900.30	2901.15	34469.1	LU	1.7	2.2		ME
2894.84	2895.69	34534.1	LU	1.8	2.3		ME
2886.04	2886.89	34639.4	LU	0.5			ME
2885.14	2885.99	34650.2	LU	1.6H	0.5		ME
2860.31	2861.15	34951.0	LU		0.5		ME
2847.51	2848.35	35108.1	LU	1.6	2.1		ME
2845.13	2845.97	35137.5	LU	1.5H	0.3		ME
2835.25	2836.08	35259.9	LU		1.0A		ME
2834.35	2835.18	35271.1	LU	0.7	1.6H		ME
2829.42	2830.25	35332.5	LU		0.5H		ME
2826.82	2827.65	35365.0	LU		0.8B		ME
2821.23	2822.06	35435.1	LU	0.3	1.7B		ME
2819.50	2820.33	35456.8	LU		0.8B		ME
2796.63	2797.45	35746.8	LU	1.4	2.0		ME
2784.80	2785.62	35898.6	LU		0.9H		ME
2781.47	2782.29	35941.6	LU	0.6H			ME
2772.58	2773.40	36056.8	LU	0.7	2.2H		ME
2765.74	2766.56	36146.0	LU	1.3	0.5		ME
2754.17	2754.98	36297.9	LU	1.6	2.1		ME
2738.17	2738.98	36509.9	LU	0.3	1.4B		ME
2728.95	2729.76	36633.3	LU	1.6	0.3		ME
2724.81	2725.62	36688.9	LU		0.6		ME
2719.09	2719.90	36766.1	LU	1.0H			ME
2715.91	2716.71	36809.2	LU	0.6H			ME
2703.13	2703.93	36983.2	LU	0.8H			ME
2701.71	2702.51	37002.6	LU	1.6	2.2		ME
2699.74	2700.54	37029.6	LU		0.5		ME
2692.34	2693.14	37131.4	LU	0.7H			ME
2685.54	2686.34	37225.4	LU	1.0H	0.0		ME
2685.08	2685.88	37231.8	LU	1.7H	0.5		ME
2677.77	2678.57	37333.4	LU	0.0	1.0H		ME
2677.25	2678.05	37340.7	LU	1.0H			ME
2670.78	2671.57	37431.1	LU	0.7H			ME
2657.80	2658.59	37613.9	LU	1.7	2.2		ME
2619.26	2620.04	38167.3	LU	1.5	2.0		ME
2615.42	2616.20	38223.4	LU	2.0	2.4		ME
2613.40	2614.18	38252.9	LU	1.5	2.0		ME
2612.86	2613.64	38260.8	LU	0.5H			ME
2607.32	2608.10	38342.1	LU		0.6H		ME
2603.39	2604.17	38400.0	LU	0.7			ME
2603.28	2604.06	38401.6	LU	0.7			ME
2592.34	2593.12	38563.6	LU		0.7H		ME
2590.03	2590.80	38598.0	LU		0.5		ME
2582.13	2582.90	38716.1	LU	0.5	1.3		ME
2578.79	2579.56	38766.3	LU	1.6	2.1		ME
2571.23	2572.00	38880.2	LU	1.5	2.0		ME
2563.52	2564.29	38997.2	LU	0.0	1.9H		ME
2561.80	2562.57	39023.4	LU		0.8H		ME
2558.42	2559.19	39074.9	LU		0.8H		ME
2550.67	2551.44	39193.6	LU		0.5		ME
2549.52	2550.29	39211.3	LU	1.3H			ME
2548.63	2549.39	39225.0	LU		1.0H		ME
2546.87	2547.63	39252.1	LU		1.0H		ME
2546.32	2547.08	39260.6	LU		0.7H		ME
2538.12	2538.88	39387.4	LU		0.5		ME
2536.95	2537.71	39405.6	LU	1.0	1.3		ME
2530.55	2531.31	39505.2	LU		0.9H		ME
2528.59	2529.35	39535.8	LU		0.7H		ME
2526.83	2527.59	39563.4	LU	0.0	1.0H		ME
2523.27	2524.03	39619.2	LU		0.3		ME
2518.04	2518.80	39701.5	LU		0.6		ME
2509.04	2509.80	39843.9	LU		0.5		ME
2500.86	2501.61	39974.2	LU	0.7H	0.0		ME
2498.99	2499.74	40004.1	LU		0.5		ME
2494.64	2495.39	40073.9	LU		0.8		ME
2487.96	2488.71	40181.5	LU		0.6		ME
2481.72	2482.47	40282.5	LU	1.3	2.0		ME
2475.34	2476.09	40386.3	LU		0.5		ME
2469.27	2470.02	40485.6	LU	1.0	1.6		ME
2464.82	2465.57	40558.6	LU	0.7			ME
2463.95	2464.70	40573.0	LU		0.5		ME
2459.64	2460.38	40644.1	LU	0.5	0.9		ME
2457.56	2458.30	40678.5	LU		0.7H		ME
2455.60	2456.34	40710.9	LU		0.6		ME
2430.26	2431.00	41135.4	LU	0.5	1.2		ME
2427.21	2427.95	41187.1	LU		0.5		ME
2421.42	2422.16	41285.5	LU		0.6		ME
2419.21	2419.94	41323.2	LU	0.9	1.6		ME
2415.92	2416.65	41379.5	LU		0.7		ME
2403.69	2404.42	41590.0	LU		0.6H		ME
2399.14	2399.87	41668.9	LU	1.0	1.7		ME
2392.19	2392.92	41790.0	LU	1.5	2.0		ME
2381.69	2382.42	41974.2	LU		1.5H		ME
2380.57	2381.30	41993.9	LU		0.3		ME
2369.08	2369.80	42197.6	LU		0.7		ME
2368.22	2368.94	42212.9	LU		0.3		ME
2359.38	2360.10	42371.1	LU		0.3		ME
2357.28	2358.00	42408.8	LU		0.6		ME
2350.43	2351.15	42532.4	LU		0.5		ME
2348.60	2349.32	42565.5	LU		0.3		ME
2343.07	2343.79	42666.0	LU	0.3	1.2		ME
2341.19	2341.91	42700.2	LU		0.5		ME

AIR WAVELENGTH (ANGSTRMS)	VACUUM WAVELENGTH (ANGSTRMS)	WAVENUMBER (KAYSERS)	LMENT	LOG ARC	INTENSITY SPK	DIS	REF
2337.05	2337.77	42775.9	LU		0.5		ME
2313.98	2314.69	43202.3	LU	0.5	1.2B		ME
2313.03	2313.74	43220.0	LU		0.7		ME
2310.66	2311.37	43264.4	LU		0.3H		ME
2297.41	2298.12	43513.9	LU	1.2	2.0		ME
2293.78	2294.49	43582.7	LU		0.3		ME
2276.94	2277.64	43905.0	LU	0.9	1.3		ME
2271.97	2272.67	44001.1	LU		0.3		ME
2270.61	2271.31	44027.4	LU		0.6H		ME
2264.63	2265.33	44143.7	LU		0.3		ME
2263.98	2264.68	44156.3	LU		0.5		ME
2260.79	2261.49	44218.6	LU		0.3		ME
2260.40	2261.10	44226.3	LU		0.6		ME
2249.53	2250.23	44440.0	LU		0.3H		ME
2236.17	2236.86	44705.4	LU		2.2H		ME
2224.02	2224.71	44949.6	LU		0.8H		ME
2217.32	2218.01	45085.5	LU		0.5H		ME
2207.64	2208.33	45283.1	LU		0.3H		ME
2206.82	2207.51	45299.9	LU		0.7H		ME
2204.60	2205.29	45345.5	LU		0.3H		ME
2200.31	2201.00	45434.0	LU		0.8		ME
2197.90	2198.59	45483.8	LU		0.3		ME
2196.98	2197.67	45502.8	LU		0.6		ME
2196.54	2197.23	45511.9	LU		0.3		ME
2195.54	2196.23	45532.7	LU	1.5	2.0		ME
2194.80	2195.49	45548.0	LU		0.8		ME
2191.37	2192.06	45619.3	LU	0.8	1.8		ME
2190.77	2191.45	45631.8	LU		0.6		ME
2190.14	2190.82	45644.9	LU		0.8		ME
2185.42	2186.10	45743.5	LU		0.3		ME
2184.80	2185.48	45756.5	LU		0.8		ME
2180.79	2181.47	45840.6	LU		0.7		ME
2178.02	2178.70	45898.9	LU		0.9		ME
2176.93	2177.61	45921.9	LU		0.6		ME
2164.33	2165.01	46189.2	LU		1.2		ME
2159.91	2160.59	46283.7	LU		0.3		ME
2158.04	2158.72	46323.8	LU		0.3		ME
2156.73	2157.41	46351.9	LU		0.6		ME
2155.92	2156.60	46369.3	LU		0.5		ME
2154.36	2155.04	46402.9	LU		0.8		ME
2147.59	2148.27	46549.2	LU		0.3H		ME
2143.84	2144.52	46630.6	LU		0.5		ME
2139.19	2139.86	46731.9	LU		0.8		ME
2135.18	2135.85	46819.7	LU		0.7		ME
2132.31	2132.98	46882.7	LU		0.7		ME
2131.66	2132.33	46897.0	LU		0.8		ME
2128.39	2129.06	46969.0	LU		0.7		ME
2127.43	2128.10	46990.2	LU		0.6		ME
2124.64	2125.31	47051.9	LU		0.3		ME
2122.76	2123.43	47093.6	LU		0.8H		ME
2117.14	2117.81	47218.6	LU		0.6		ME
2110.37	2111.04	47370.0	LU		0.6		ME
2110.22	2110.89	47373.4	LU		0.8		ME
2109.14	2109.81	47397.7	LU		0.7		ME
2107.84	2108.51	47426.9	LU		0.6		ME
2104.40	2105.07	47504.4	LU		1.6		ME
2100.40	2101.07	47594.9	LU		0.3		ME
2099.54	2100.21	47614.4	LU		1.0H		ME
2092.14	2092.81	47782.8	LU		0.8		ME
2087.50	2088.16	47889.0	LU		0.7		ME
2086.45	2087.11	47913.0	LU		0.6		ME
2086.06	2086.72	47922.0	LU		0.3		ME
2085.69	2086.35	47930.5	LU		0.7		ME
2071.95	2072.61	48248.3	LU		0.3		ME
2070.66	2071.32	48278.4	LU		0.5H		ME
2065.42	2066.08	48400.8	LU		1.5H		ME
9257.9	9260.4	10799.	MG I	0.3			PS
8928.97	8931.42	11196.4	MG I	0.3			PS
8806.79	8809.21	11351.8	MG I	2.0			QB
7779.9	7782.0	12850.	MG	0.7			PS

AIR WAVELENGTH (ANGSTRMS)	VACUUM WAVELENGTH (ANGSTRMS)	WAVENUMBER (KAYSERS)	LMENT	LOG ARC	INTENSITY SPK	DIS	REF
7657.60	7659.71	13055.3	MG I	0.3			PS
6545.44	6547.25	15273.6	MG II	1.0	0.3		LR
6347.06	6348.81	15751.0	MG II	1.0	0.3		LR
6332.21	6333.96	15787.9	MG	0.3H	0.8		MIT
6319.08	6320.83	15820.7	MG	0.3H			LR
6318.55	6320.30	15822.0	MG I	0.3H			LR
6021.761	6023.429	16601.84	MG	0.7			MIT
5711.115	5712.699	17504.86	MG I	0.6	0.0		MIT
5528.461	5529.997	18083.19	MG I	1.8	1.5		MIT
5401.05	5402.55	18509.8	MG II	0.3	0.7		FL
5264.14	5265.61	18991.2	MG II	0.3	0.7		FL
5183.618	5185.062	19286.17	MG I	2.7J	2.5		MIT
5172.699	5174.140	19326.88	MG I	2.3J	2.0J		MIT
5167.343	5168.782	19346.92	MG I	2.0J	1.7		MIT
4851.10	4852.46	20608.1	MG II	0.7			FL
4739.59	4740.92	21093.0	MG II	0.7			FL
4730.16	4731.48	21135.0	MG I	0.3			FL
4703.02	4704.34	21257.0	MG I	0.9R	0.5		MIT
4571.15	4572.43	21870.2	MG I	1.3	0.3		MIT
4534.26	4535.53	22048.1	MG II	0.6			FL
4481.327	4482.584	22308.56	MG II	2.0			FL
4481.130	4482.387	22309.54	MG II		1.7		ME
4436.48	4437.73	22534.1	MG II	0.7			FL
4433.991	4435.236	22546.72	MG II	0.9			FL
4427.995	4429.238	22577.25	MG II	0.8			FL
4393.24	4394.47	22755.9	MG		0.3		MIT
4390.585	4391.818	22769.61	MG II	1.0			FL
4384.643	4385.875	22800.47	MG II	0.9			FL
4351.91	4353.13	22972.0	MG I	1.2	0.3		FL
4331.93	4333.15	23077.9	MG II	0.5			FL
4242.47	4243.66	23564.5	MG II	0.6			FL
4193.44	4194.62	23840.1	MG II	0.3			FL
4167.65	4168.82	23987.6	MG	0.7			MIT
4167.389	4168.564	23989.08	MG I	0.8	0.6		MIT
4109.54	4110.70	24326.8	MG II	0.3			FL
4058.96	4060.11	24629.9	MG	0.3			MIT
4057.632	4058.778	24637.96	MG I	1.0W			MIT
4013.80	4014.93	24907.0	MG II	0.3			FL
3986.733	3987.861	25076.10	MG I	1.2W	0.5		MIT
3938.423	3939.538	25383.69	MG I	1.0W	0.5		MIT
3904.035	3905.141	25607.27	MG I	0.3	1.1		MIT
3898.120	3899.224	25646.13	MG I	0.7	0.9		PS
3895.662	3896.766	25662.31	MG I		1.2		PS
3893.376	3894.479	25677.37	MG I		0.9		PS
3892.118	3893.221	25685.67	MG	0.5	1.1		MIT
3890.241	3891.343	25698.07	MG I	0.5	0.9		PS
3878.573	3879.672	25775.37	MG I	1.0			MIT
3859.211	3860.305	25904.69	MG I	0.3			MIT
3854.538	3855.631	25936.09	MG I	0.3	0.7		MIT
3854.117	3855.210	25938.93	MG I	0.3	0.9		MIT
3850.40	3851.49	25964.0	MG II	0.9	0.7		FL
3848.92	3850.01	25973.9	MG		0.3		MIT
3848.773	3849.865	25974.94	MG I	0.3	1.1		MIT
3848.24	3849.33	25978.5	MG II	1.0	1.0		FL
3848.090	3849.181	25979.55	MG I	0.3	0.5		MIT
3844.973	3846.064	26000.61	MG I	0.3	1.0		MIT
3838.258	3839.347	26046.10	MG I	2.5	2.3		MIT
3832.306	3833.393	26086.55	MG I	2.4	2.3		MIT
3829.350	3830.436	26106.69	MG I	2.0W	2.2		MIT
3615.64	3616.67	27649.7	MG II	0.3			FL
3613.80	3614.83	27663.8	MG II	0.6			FL
3553.51	3554.53	28133.1	MG II	0.9			FL
3549.61	3550.62	28164.1	MG II	0.6			FL
3538.86	3539.87	28249.6	MG II	0.9			FL
3535.04	3536.05	28280.1	MG II	0.9			FL
3336.680	3337.640	29961.29	MG I	2.1	1.8		MIT
3332.153	3333.111	30002.00	MG I	2.0	1.4		MIT
3329.930	3330.888	30022.03	MG I	1.9	0.9		MIT
3229.371	3230.303	30956.85	MG I	1.4			MIT
3175.84	3176.76	31478.6	MG II	0.7			FL

AIR WAVELENGTH (ANGSTRMS)	VACUUM WAVELENGTH (ANGSTRMS)	WAVENUMBER (KAYSERS)	LMENT	LOG ARC	INTENSITY SPK	INTENSITY DIS	REF
3172.79	3173.71	31508.9	MG II	0.6			FL
3168.98	3169.90	31546.8	MG II	0.9			FL
3165.94	3166.86	31577.1	MG II	0.7			FL
3139.5	3140.4	31843.	MG	0.5	0.3		ED
3135.0	3135.9	31889.	MG	0.6	0.3H		ED
3107.02	3107.92	32175.9	MG		0.7		ED
3104.809	3105.710	32198.76	MG II	0.7	1.2		RG
3104.722	3105.623	32199.66	MG II		1.3		RG
3096.899	3097.798	32281.00	MG I	2.2	1.4		MIT
3092.991	3093.889	32321.78	MG I	2.1	1.3		MIT
3091.077	3091.974	32341.79	MG I	1.9	1.0		MIT
3073.987	3074.880	32521.59	MG	0.9	1.0		MIT
3051.987	3052.875	32756.01	MG I	0.3			MIT
3043.75	3044.64	32844.6	MG I	0.3			FL
2971.70	2972.57	33641.0	MG II	1.0			FL
2969.02	2969.89	33671.3	MG II	0.8			FL
2967.87	2968.74	33684.4	MG II	1.0			FL
2965.19	2966.06	33714.8	MG II	0.8			FL
2942.11	2942.97	33979.3	MG I	1.3	0.3H		MIT
2938.538	2939.397	34020.58	MG I	1.4			MIT
2936.895	2937.754	34039.61	MG I	1.1			MIT
2936.537	2937.396	34043.76	MG II	1.3			MIT
2936.509	2937.368	34044.08	MG II	0.3	1.0		RG
2928.75	2929.61	34134.3	MG II	1.4	2.0		MIT
2915.52	2916.37	34289.2	MG	1.3	1.1		M
2852.129	2852.967	35051.23	MG I	2.5G	2.0G		MIT
2851.65	2852.49	35057.1	MG I	1.4			FL
2848.42	2849.26	35096.9	MG I	1.3			MIT
2848.37	2849.21	35097.5	MG		0.5		MIT
2847.00	2847.84	35114.4	MG	0.9	0.9		M
2846.75	2847.59	35117.5	MG I	1.3	0.6		MIT
2817.22	2818.05	35485.5	MG	0.5	0.9		MIT
2815.54	2816.37	35506.7	MG	0.5	0.9		MIT
2811.27	2812.10	35560.6	MG		0.9		MIT
2809.78	2810.61	35579.5	MG	0.9	0.9		MIT
2802.695	2803.521	35669.43	MG II	2.2	2.5		MIT
2798.06	2798.88	35728.5	MG II	1.5	1.9		MIT
2795.53	2796.35	35760.9	MG II	2.2	2.5		MIT
2790.787	2791.610	35821.62	MG II	1.6	1.9		MIT
2782.974	2783.795	35922.18	MG	1.2	1.2		MIT
2781.417	2782.238	35942.29	MG	1.3	0.9		MIT
2779.834	2780.654	35962.76	MG	1.6	1.7		MIT
2778.288	2779.108	35982.77	MG	1.4	1.3		MIT
2776.690	2777.510	36003.48	MG	1.5	1.3		MIT
2768.46	2769.28	36110.5	MG I	0.9			MIT
2765.34	2766.16	36151.2	MG I	0.9	0.5		FL
2736.72	2737.53	36529.3	MG	0.9	0.3		M
2736.533	2737.343	36531.78	MG I	1.5	0.5		MIT
2733.57	2734.38	36571.4	MG I	1.4			MIT
2732.08	2732.89	36591.3	MG I	1.1	0.5		MIT
2698.14	2698.94	37051.6	MG I	1.1			MIT
2695.214	2696.014	37091.80	MG I	1.0			MIT
2693.71	2694.51	37112.5	MG I	1.0			MIT
2672.45	2673.24	37407.7	MG I	1.3			MIT
2669.54	2670.33	37448.5	MG I	1.1			MIT
2668.11	2668.90	37468.6	MG I	1.0H			MIT
2660.821	2661.613	37571.21	MG II	1.6	0.8		FL
2660.755	2661.547	37572.14	MG II	1.6	0.8		FL
2649.02	2649.81	37738.6	MG I	1.1			FL
2646.16	2646.95	37779.4	MG I	0.7			MIT
2644.78	2645.57	37799.1	MG I	0.6	0.6		MIT
2632.88	2633.66	37969.9	MG I	1.0	0.6		MIT
2630.020	2630.804	38011.19	MG I	0.9			MIT
2628.63	2629.41	38031.3	MG I	0.9	0.6		MIT
2617.48	2618.26	38193.3	MG I	0.7	0.7		FL
2614.67	2615.45	38234.3	MG I	0.7	0.3H		MIT
2606.654	2607.433	38351.90	MG I	0.9	0.3H		MIT
2603.88	2604.66	38392.8	MG I	0.9	0.7		MIT
2602.45	2603.23	38413.9	MG I	0.8			MIT
2595.940	2596.716	38510.18	MG I	0.9	0.7H		MIT

AIR WAVELENGTH (ANGSTRMS)	VACUUM WAVELENGTH (ANGSTRMS)	WAVENUMBER (KAYSERS)	LMENT	LOG ARC	INTENSITY SPK	INTENSITY DIS	REF
2593.182	2593.957	38551.13	MG I	0.7	0.3		MIT
2588.28	2589.05	38624.1	MG I	0.9	0.9		MIT
2585.57	2586.34	38664.6	MG I	0.7			MIT
2584.30	2585.07	38683.6	MG I	0.9	0.9		MIT
2574.91	2575.68	38824.7	MG I	0.9			MIT
2572.325	2573.096	38863.69	MG I	0.9			MIT
2570.885	2571.655	38885.46	MG I	0.6			MIT
2564.83	2565.60	38977.3	MG I	0.8			MIT
2562.225	2562.993	39016.88	MG I	0.5			MIT
2560.86	2561.63	39037.7	MG I	0.6			MIT
2557.264	2558.031	39092.57	MG I	0.8			MIT
2554.62	2555.39	39133.0	MG I	0.6			MIT
2551.096	2551.862	39187.08	MG I	0.8			MIT
2548.49	2549.25	39227.1	MG I	0.3			MIT
2449.573	2450.315	40811.08	MG II	1.3			FL
2329.58	2330.29	42913.0	MG II	1.1			FL
2253.87	2254.57	44354.4	MG II	0.9			FL
2202.68	2203.37	45385.1	MG II	0.7			FL
2166.28	2166.96	46147.6	MG II	0.3			FL
2025.82	2026.47	49346.8	MG I	0.9H			PS
9686.3	9689.0	10321.	MN	1.2			ME
9684.9	9687.6	10322.	MN I	1.2			ME
9676.50	9679.15	10331.5	MN I	1.6			ME
9633.02	9635.66	10378.1	MN I	0.6			ME
9608.56	9611.20	10404.5	MN I	2.0			ME
9606.71	9609.34	10406.5	MN I	0.7			ME
9598.7	9601.3	10415.	MN I	0.5			ME
9584.0	9586.6	10431.	MN	1.0H			ME
9550.80	9553.42	10467.5	MN I	1.3H			ME
9542.14	9544.76	10477.0	MN	1.0H			ME
9535.72	9538.34	10484.0	MN I	0.7H			ME
9502.12	9504.73	10521.1	MN I	0.9H			ME
9476.57	9479.17	10549.4	MN I	0.6H			ME
9444.90	9447.49	10584.8	MN I	1.6			ME
9429.58	9432.17	10602.0	MN I	1.5H			ME
9412.78	9415.36	10620.9	MN	1.0H			ME
9408.38	9410.96	10625.9	MN	0.6H			ME
9336.47	9339.03	10707.7	MN I	1.6H			ME
9331.90	9334.46	10713.0	MN I	1.3H			ME
9325.16	9327.72	10720.7	MN I	0.7			ME
9323.76	9326.32	10722.3	MN	0.6H			ME
9243.29	9245.83	10815.7	MN I	2.2			ME
9240.81	9243.35	10818.6	MN	0.6			ME
9234.40	9236.93	10826.1	MN I	1.0			ME
9172.09	9174.61	10899.7	MN I	2.0			ME
9155.85	9158.36	10919.0	MN I	0.7			ME
9125.0	9127.5	10956.	MN	0.3			ME
9114.02	9116.52	10969.1	MN I	1.6			ME
9084.29	9086.78	11005.0	MN I	1.5			ME
9066.50	9068.99	11026.6	MN	0.3			ME
9021.80	9024.28	11081.2	MN	0.6H			ME
8932.96	8935.41	11191.4	MN	0.3			ME
8929.72	8932.17	11195.5	MN I	1.7H			ME
8926.06	8928.51	11200.1	MN I	1.2H			ME
8895.36	8897.80	11238.7	MN	0.6			ME
8859.08	8861.51	11284.8	MN	0.3H			ME
8842.48	8844.91	11305.9	MN I	0.5H			ME
8827.83	8830.25	11324.7	MN	0.5			ME
8820.26	8822.68	11334.4	MN I	0.6H			ME
8798.66	8801.08	11362.2	MN I	0.5H			ME
8796.83	8799.25	11364.6	MN I	0.8H			ME
8767.96	8770.37	11402.0	MN I	0.7H			ME
8740.93	8743.33	11437.3	MN I	2.70			ME
8737.32	8739.72	11442.0	MN I	2.2W			ME
8734.60	8737.00	11445.6	MN I	1.5			ME
8729.80	8732.20	11451.9	MN I	0.3H			ME
8717.29	8719.68	11468.3	MN I	0.3H			ME
8710.21	8712.60	11477.6	MN	1.0			ME
8703.76	8706.15	11486.1	MN I	2.3W			ME
8701.05	8703.44	11489.7	MN I	2.2W			ME

AIR WAVELENGTH (ANGSTRMS)	VACUUM WAVELENGTH (ANGSTRMS)	WAVENUMBER (KAYSERS)	LMENT	LOG INTENSITY ARC	SPK	DIS	REF
8699.13	8701.52	11492.2	MN I	2.0W			ME
8680.24	8682.62	11517.3	MN I	0.3H			ME
8673.97	8676.35	11525.6	MN I	2.0W			ME
8672.06	8674.44	11528.1	MN I	1.9W			ME
8670.92	8673.30	11529.6	MN I	1.8K			ME
8666.31	8668.69	11535.8	MN I	0.3H			ME
8664.66	8667.04	11538.0	MN I	0.3H			ME
8659.38	8661.76	11545.0	MN I	1.0H			ME
8654.63	8657.01	11551.3	MN I	1.6H			ME
8603.03	8605.39	11620.6	MN	0.5			ME
8600.34	8602.70	11624.2	MN I	0.7V			ME
8558.63	8560.98	11680.9	MN I	0.9H			ME
8521.57	8523.91	11731.7	MN I	1.0H			ME
8481.70	8484.03	11786.8	MN	0.5			ME
8431.20	8433.52	11857.4	MN I	1.2H			ME
8409.88	8412.19	11887.5	MN I	1.0H			ME
8395.87	8398.18	11907.3	MN I	1.0H			ME
8380.77	8383.07	11928.8	MN	1.4			ME
8373.93	8376.23	11938.5	MN	0.6H			ME
8353.79	8356.09	11967.3	MN I	0.3V			ME
8344.80	8347.09	11980.2	MN	0.3			ME
8304.42	8306.70	12038.5	MN	0.7H			ME
8284.48	8286.76	12067.4	MN I	0.6H			ME
8251.64	8253.91	12115.5	MN	0.3			ME
8229.99	8232.25	12147.3	MN	0.3			ME
8212.43	8214.69	12173.3	MN	1.7H			ME
8210.16	8212.42	12176.7	MN I	0.3			ME
8204.50	8206.76	12185.1	MN	0.3			ME
8148.78	8151.02	12268.4	MN	0.3			ME
8119.36	8121.59	12312.9	MN	0.3			ME
8110.51	8112.74	12326.3	MN	0.5			ME
8043.37	8045.58	12429.2	MN I	0.6			ME
8032.54	8034.75	12445.9	MN	0.3			ME
7948.10	7950.29	12578.2	MN	0.3			ME
7942.93	7945.11	12586.3	MN	1.2			MIT
7928.45	7930.63	12609.3	MN I	0.3			ME
7854.26	7856.42	12728.4	MN	0.3			MIT
7834.34	7836.50	12760.8	MN I	0.8H			ME
7821.25	7823.40	12782.2	MN I	1.1H			ME
7816.61	7818.76	12789.7	MN I	1.2H			ME
7806.00	7808.15	12807.1	MN I	0.6L			ME
7790.82	7792.96	12832.1	MN I	1.0H			ME
7764.72	7766.86	12875.2	MN I	2.3H			ME
7755.15	7757.28	12891.1	MN I	0.7H			ME
7752.67	7754.80	12895.2	MN I	0.7B			ME
7737.20	7739.33	12921.0	MN	0.3			MIT
7734.43	7736.56	12925.6	MN I	1.3H			ME
7733.24	7735.37	12927.6	MN I	1.8B			ME
7712.42	7714.54	12962.5	MN I	2.0H			ME
7709.98	7712.10	12966.6	MN I	1.2V			ME
7706.52	7708.64	12972.5	MN I	0.7B			ME
7680.20	7682.31	13016.9	MN I	2.3			MIT
7670.42	7672.53	13033.5	MN	0.7H			ME
7667.89	7670.00	13037.8	MN	1.3H			ME
7655.99	7658.10	13058.1	MN I	0.5B			ME
7651.62	7653.73	13065.5	MN I	0.6B			ME
7646.02	7648.12	13075.1	MN I	0.7B			ME
7603.43	7605.52	13148.3	MN	0.3			ME
7515.74	7517.81	13301.7	MN	0.3			ME
7443.50	7445.55	13430.8	MN	0.3			ME
7415.81	7417.85	13481.0	MN	0.3			ME
7376.85	7378.88	13552.2	MN	0.3			ME
7342.80	7344.82	13615.0	MN	0.3			ME
7326.51	7328.53	13645.3	MN I	2.7			ME
7302.89	7304.90	13689.4	MN	2.5H			ME
7301.38	7303.39	13692.3	MN	0.3			ME
7283.815	7285.822	13725.29	MN	2.6H			MIT
7247.82	7249.82	13793.4	MN	1.6			ME
7239.62	7241.62	13809.1	MN	0.5			MIT
7184.25	7186.23	13915.5	MN	1.6			SL

AIR WAVELENGTH (ANGSTRMS)	VACUUM WAVELENGTH (ANGSTRMS)	WAVENUMBER (KAYSERS)	LMENT	LOG INTENSITY ARC	SPK	DIS	REF
7174.46	7176.44	13934.5	MN	0.5			SL
7158.07	7160.04	13966.4	MN	0.8H			SL
7151.28	7153.25	13979.7	MN	1.5H			SL
7103.01	7104.97	14074.7	MN	0.7H			SL
7069.84	7071.79	14140.7	MN	1.8			SL
7055.65	7057.60	14169.1	MN	0.7			SL
7025.81	7027.75	14229.3	MN	0.3			SL
7012.25	7014.18	14256.8	MN	1.0H			SL
7002.52	7004.45	14276.6	MN	0.3			SL
6989.812	6991.740	14302.59	MN	1.9			SL
6968.00	6969.92	14347.4	MN	0.7H			SL
6942.518	6944.433	14400.02	MN	2.0H			SL
6931.13	6933.04	14423.7	MN	1.3H			SL
6887.755	6889.655	14514.51	MN	1.5			MIT
6863.08	6864.97	14566.7	MN	0.3H			ME
6833.92	6835.81	14628.8	MN	1.6			SL
6827.12	6829.00	14643.4	MN	0.3H			MIT
6691.581	6693.429	14940.03	MN	1.4			FU
6610.03	6611.86	15124.3	MN	0.3			SL
6605.53	6607.35	15134.7	MN	1.5H			SL
6598.24	6600.06	15151.4	MN	0.3H			MIT
6586.343	6588.162	15178.74	MN	1.3H			SL
6570.834	6572.649	15214.57	MN	0.3			MIT
6569.23	6571.04	15218.3	MN	0.3			SL
6552.010	6553.820	15258.28	MN	0.5			SL
6534.08	6535.89	15300.2	MN	1.0H			SL
6533.494	6535.299	15301.52	MN	1.3			MIT
6526.54	6528.34	15317.8	MN	1.0			MIT
6519.371	6521.172	15334.67	MN	1.3			MIT
6518.753	6520.554	15336.12	MN	1.3			MIT
6491.712	6493.506	15400.00	MN	2.0			MIT
6490.624	6492.417	15402.58	MN	0.5			SL
6475.32	6477.11	15439.0	MN	0.6H			SL
6443.492	6445.273	15515.25	MN	1.0			MIT
6440.974	6442.754	15521.31	MN	1.8			MIT
6413.950	6415.723	15586.71	MN	1.4			SL
6391.215	6392.982	15642.15	MN	0.5			SL
6388.973	6390.739	15647.64	MN	0.3			MIT
6384.669	6386.434	15658.19	MN	1.4			MIT
6382.169	6383.933	15664.32	MN	1.3			MIT
6378.956	6380.719	15672.21	MN	1.3			MIT
6370.814	6372.575	15692.24	MN	0.6H			SL
6356.082	6357.839	15728.61	MN	0.9			SL
6349.766	6351.522	15744.26	MN	1.2			SL
6344.110	6345.864	15758.30	MN	1.3			SL
6315.063	6316.809	15830.78	MN	0.7			MIT
6265.627	6267.360	15955.68	MN	0.6			SL
6243.145	6244.872	16013.14	MN	0.5H			SL
6237.42	6239.15	16027.8	MN	0.3			SL
6176.585	6178.294	16185.70	MN	0.3			SL
6135.37	6137.07	16294.4	MN	0.7			ME
6131.917	6133.614	16303.60	MN II			1.0	CZ
6131.005	6132.702	16306.03	MN II			1.0	CZ
6130.794	6132.491	16306.59	MN II			1.5	CZ
6129.022	6130.718	16311.30	MN II			1.3	CZ
6128.725	6130.421	16312.09	MN II			1.6	CZ
6126.210	6127.906	16318.79	MN II			1.3	CZ
6125.855	6127.551	16319.73	MN II			1.7	CZ
6122.799	6124.494	16327.88	MN II			1.2	CZ
6122.438	6124.133	16328.84	MN II			1.9	CZ
6081.455	6083.139	16438.88	MN	0.3			SL
6057.110	6058.787	16504.95	MN	0.7			SL
6041.715	6043.388	16547.01	MN	0.5			SL
6021.798	6023.466	16601.74	MN	1.9H	0.7		IKS
6016.639	6018.305	16615.97	MN	1.9H	0.7		IKS
6014.356	6016.022	16622.28	MN	0.9			FU
6013.498	6015.163	16624.65	MN	2.0H	0.7		MIT
5848.955	5850.576	17092.33	MN	0.9			MIT
5816.844	5818.457	17186.69	MN	1.0			MIT
5780.189	5781.792	17295.68	MN	1.0			MIT

AIR WAVELENGTH (ANGSTRMS)	VACUUM WAVELENGTH (ANGSTRMS)	WAVENUMBER (KAYSERS)	LMENT	LOG ARC	INTENSITY SPK	DIS	REF
5738.286	5739.878	17421.97	MN	1.0			MIT
5700.583	5702.165	17537.20	MN	0.5			MIT
5626.239	5627.801	17768.93	MN	0.5			MIT
5602.046	5603.601	17845.67	MN	0.7			FU
5573.680	5575.228	17936.49	MN	1.0			MIT
5573.013	5574.561	17938.63	MN	1.0			MIT
5567.762	5569.308	17955.55	MN	1.1			MIT
5552.45	5553.99	18005.1	MN	0.6			MIT
5551.985	5553.527	18006.57	MN	1.0			MIT
5537.756	5539.294	18052.84	MN	1.6			MIT
5516.771	5518.303	18121.51	MN	1.7			MIT
5505.869	5507.399	18157.39	MN	1.6			MIT
5504.21	5505.74	18162.9	MN	0.8			MIT
5497.38	5498.91	18185.4	MN	0.6			MIT
5481.396	5482.919	18238.46	MN	1.7			MIT
5470.638	5472.158	18274.33	MN	1.7			MIT
5457.471	5458.988	18318.42	MN	1.4			MIT
5433.418	5434.928	18399.51	MN	0.8			MIT
5432.548	5434.058	18402.45	MN	1.6			MIT
5420.362	5421.869	18443.82	MN	1.8			MIT
5413.687	5415.192	18466.57	MN	1.5			MIT
5407.424	5408.927	18487.95	MN	1.8			MIT
5406.018	5407.521	18492.76	MN	0.7			MIT
5399.489	5400.990	18515.12	MN	1.6			MIT
5394.674	5396.174	18531.65	MN	1.7			MIT
5388.521	5390.019	18552.81	MN	1.0			MIT
5377.628	5379.123	18590.39	MN	1.6			MIT
5377.209	5378.704	18591.84	MN	0.9			MIT
5375.30	5376.79	18598.4	MN	0.3			MIT
5364.484	5365.976	18635.94	MN	0.6			MIT
5349.881	5351.369	18686.81	MN	1.3			MIT
5348.069	5349.556	18693.14	MN	1.0			MIT
5344.470	5345.957	18705.73	MN	1.1			MIT
5341.065	5342.551	18717.65	MN	2.3	2.0		MIT
5317.095	5318.574	18802.03	MN	0.8			MIT
5308.924	5310.401	18830.97	MN	0.7H			MIT
5302.320	5303.795	18854.42	MN II			1.8	CZ
5299.278	5300.752	18865.25	MN II			1.7	CZ
5298.854	5300.328	18866.76	MN	0.6			MIT
5296.968	5298.442	18873.47	MN II			1.6	CZ
5295.292	5296.765	18879.45	MN II			1.5	CZ
5294.216	5295.689	18883.28	MN II			1.3	CZ
5289.51	5290.98	18900.1	MN	0.3H			MIT
5260.771	5262.235	19003.33	MN	1.0			MIT
5255.325	5256.788	19023.02	MN	1.7			MIT
5197.216	5198.663	19235.71	MN	1.0			MIT
5196.591	5198.038	19238.03	MN	1.5			MIT
5150.890	5152.325	19408.71	MN	1.6			MIT
5149.13	5150.56	19415.4	MN	0.9			MIT
5117.937	5119.363	19533.68	MN	1.5			MIT
5074.794	5076.209	19699.74	MN	1.2			MIT
5042.589	5043.995	19825.55	MN	0.7			MIT
5030.639	5032.042	19872.65	MN	0.9			MIT
5029.812	5031.215	19875.92	MN	1.1			MIT
5022.038	5023.439	19906.68	MN	0.6			MIT
5017.615	5019.014	19924.23	MN	1.0			MIT
5010.364	5011.762	19953.06	MN	1.0			MIT
5004.907	5006.303	19974.82	MN	1.3			MIT
4985.769	4987.160	20051.49	MN	1.1			MIT
4965.881	4967.267	20131.80	MN	1.7	0.3		MIT
4952.415	4953.797	20186.53	MN	0.5			MIT
4952.02	4953.40	20188.1	MN	0.7			MIT
4942.418	4943.797	20227.37	MN	0.5			MIT
4934.15	4935.53	20261.3	MN	1.4	0.7		MIT
4917.495	4918.868	20329.88	MN	1.1			MIT
4900.88	4902.25	20398.8	MN	1.0D	0.7H		MIT
4900.718	4902.086	20399.48	MN	0.9			MIT
4881.572	4882.935	20479.49	MN	1.0			MIT
4862.054	4863.412	20561.70	MN	1.6	0.7		MIT
4854.810	4856.166	20592.38	MN	1.5	0.7		MIT

AIR WAVELENGTH (ANGSTRMS)	VACUUM WAVELENGTH (ANGSTRMS)	WAVENUMBER (KAYSERS)	LMENT	LOG ARC	INTENSITY SPK	DIS	REF
4854.604	4855.960	20593.25	MN	1.2	0.7		MIT
4844.315	4845.668	20636.99	MN	1.9	0.7		MIT
4843.19	4844.54	20641.8	MN	1.2			MIT
4840.148	4841.500	20654.75	MN	1.7			MIT
4838.244	4839.596	20662.88	MN	1.7			MIT
4826.896	4828.245	20711.46	MN	1.0	0.7		MIT
4825.593	4826.941	20717.05	MN	1.3	0.7		MIT
4823.516	4824.864	20725.97	MN	2.6	1.9		MIT
4818.331	4819.678	20748.28	MN	0.5			MIT
4815.114	4816.460	20762.14	MN	1.2H			MIT
4808.723	4810.067	20789.73	MN	1.3			MIT
4807.180	4808.524	20796.40	MN	1.1			MIT
4797.700	4799.041	20837.50	MN	1.4			MIT
4796.709	4798.050	20841.80	MN	1.2			MIT
4793.004	4794.344	20857.91	MN	0.9			MIT
4783.420	4784.757	20899.70	MN	2.6	1.8		MIT
4779.150	4780.486	20918.37	MN	1.3O			MIT
4774.096	4775.431	20940.52	MN	1.7			MIT
4771.672	4773.006	20951.16	MN	0.9			MIT
4766.430	4767.763	20974.20	MN	1.9	1.5		MIT
4765.859	4767.192	20976.71	MN	1.8	1.4		MIT
4762.376	4763.708	20992.05	MN	2.0	1.6		MIT
4761.526	4762.857	20995.80	MN	1.8	1.2		MIT
4755.72	4757.05	21021.4	MN	1.0	0.7		MIT
4754.042	4755.372	21028.85	MN	2.6	1.8		MIT
4743.847	4745.174	21074.04	MN	0.3			MIT
4739.108	4740.434	21095.12	MN	2.2	1.2		MIT
4738.29	4739.62	21098.8	MN	1.1	0.7		MIT
4730.361	4731.684	21134.12	MN	1.2	0.7		MIT
4727.476	4728.798	21147.02	MN	2.2	1.3		MIT
4712.02	4713.34	21216.4	MN	0.7			MIT
4709.715	4711.033	21226.77	MN	2.2	1.2		MIT
4701.159	4702.474	21265.40	MN	2.0	0.7		MIT
4696.909	4698.223	21284.64	MN	1.1			MIT
4671.688	4672.996	21399.55	MN	2.0	0.7		MIT
4666.49	4667.80	21423.4	MN	1.1			MIT
4661.87	4663.18	21444.6	MN	1.2			MIT
4660.73	4662.03	21449.9	MN	0.7			MIT
4659.34	4660.64	21456.3	MN	0.9			MIT
4658.778	4660.082	21458.85	MN	1.4			MIT
4658.38	4659.68	21460.7	MN	1.1			MIT
4657.460	4658.764	21464.92	MN	1.0			MIT
4656.237	4657.541	21470.56	MN	1.0			MIT
4649.01	4650.31	21503.9	MN	0.3H			MIT
4647.585	4648.886	21510.53	MN II			0.6	CZ
4645.035	4646.336	21522.34	MN	1.6			MIT
4642.812	4644.112	21532.64	MN	1.7			MIT
4642.29	4643.59	21535.1	MN	1.1			MIT
4627.744	4629.040	21602.75	MN	1.7			MIT
4626.544	4627.840	21608.35	MN	1.9	1.2		MIT
4625.38	4626.68	21613.8	MN	1.3			MIT
4624.33	4625.63	21618.7	MN	0.9			MIT
4623.33	4624.62	21623.4	MN	0.9			MIT
4622.74	4624.03	21626.1	MN	0.9			MIT
4621.04	4622.33	21634.1	MN	0.7H			MIT
4620.21	4621.50	21638.0	MN	0.7H			MIT
4619.62	4620.91	21640.7	MN	0.5			MIT
4619.30	4620.59	21642.2	MN	1.3			MIT
4617.82	4619.11	21649.2	MN	0.5			MIT
4615.65	4616.94	21659.4	MN	0.9			MIT
4611.43	4612.72	21679.2	MN	0.0			MIT
4611.02	4612.31	21681.1	MN	0.3			MIT
4609.931	4611.222	21686.22	MN	1.3			MIT
4607.625	4608.916	21697.08	MN	1.7			MIT
4605.363	4606.653	21707.73	MN	2.2	1.2		MIT
4596.380	4597.668	21750.16	MN	1.0			MIT
4594.108	4595.395	21760.91	MN	1.1			MIT
4591.70	4592.99	21772.3	MN	0.9			MIT
4591.513	4592.799	21773.21	MN	1.0			MIT
4586.108	4587.393	21798.87	MN	1.5L			MIT

AIR WAVELENGTH (ANGSTRMS)	VACUUM WAVELENGTH (ANGSTRMS)	WAVENUMBER (KAYSERS)	LMENT	LOG ARC	INTENSITY SPK	INTENSITY DIS	REF
4584.921	4586.206	21804.52	MN	1.3			MIT
4583.050	4584.334	21813.42	MN	0.9			MIT
4582.805	4584.089	21814.58	MN	1.3			MIT
4581.833	4583.117	21819.21	MN	2.1			MIT
4579.699	4580.982	21829.38	MN	1.7			MIT
4578.041	4579.324	21837.28	MN	1.2			MIT
4575.410	4576.692	21849.84	MN	1.7			MIT
4574.522	4575.804	21854.08	MN	0.9			MIT
4573.635	4574.917	21858.32	MN	1.0			MIT
4572.74	4574.02	21862.6	MN	1.3			MIT
4572.39	4573.67	21864.3	MN	0.7			MIT
4571.55	4572.83	21868.3	MN	0.9			MIT
4571.233	4572.514	21869.81	MN	1.3			MIT
4565.73	4567.01	21896.2	MN	1.0			MIT
4563.62	4564.90	21906.3	MN	0.3			MIT
4562.74	4564.02	21910.5	MN	0.9H			MIT
4561.13	4562.41	21918.2	MN	0.3			MIT
4560.483	4561.761	21921.36	MN	1.3H			MIT
4553.31	4554.59	21955.9	MN	1.1H			MIT
4549.718	4550.993	21973.22	MN	1.1			MIT
4548.582	4549.857	21978.71	MN	1.9	0.7		MIT
4547.33	4548.60	21984.8	MN	0.9			MIT
4545.310	4546.584	21994.53	MN	1.1			MIT
4544.414	4545.688	21998.87	MN	1.8	0.7		MIT
4542.436	4543.710	22008.45	MN	1.9	0.7		MIT
4538.457	4539.729	22027.74	MN	1.6			MIT
4534.463	4535.734	22047.15	MN	1.5			MIT
4532.31	4533.58	22057.6	MN	0.5H			MIT
4530.034	4531.304	22068.70	MN II			1.0	CZ
4529.790	4531.060	22069.89	MN	1.7			MIT
4523.389	4524.657	22101.12	MN	1.7			MIT
4520.037	4521.305	22117.51	MN	1.1			MIT
4511.238	4512.503	22160.65	MN	0.3			MIT
4510.210	4511.475	22165.70	MN II			0.8	CZ
4506.726	4507.990	22182.83	MN	0.9			MIT
4503.869	4505.132	22196.91	MN	1.8	0.7		MIT
4502.223	4503.486	22205.02	MN I	0.9			ME
4502.220	4503.483	22205.04	MN	2.1	1.6		MIT
4498.897	4500.159	22221.44	MN	2.2	1.6		MIT
4497.714	4498.976	22227.28	MN	0.3			MIT
4496.989	4498.251	22230.86	MN II			0.6	CZ
4496.631	4497.892	22232.64	MN	1.6	0.7		MIT
4491.651	4492.911	22257.28	MN	1.7	0.7		MIT
4490.081	4491.341	22265.07	MN	2.0	1.4		MIT
4479.399	4480.656	22318.16	MN	1.8			MIT
4472.792	4474.047	22351.13	MN	2.0	1.4		MIT
4470.138	4471.392	22364.40	MN	1.9	1.6		MIT
4464.677	4465.930	22391.75	MN	1.8	1.7		MIT
4462.022	4463.274	22405.08	MN	1.6	1.8		MIT
4461.085	4462.337	22409.78	MN	1.5	1.4		MIT
4460.377	4461.629	22413.34	MN	1.3	0.7		MIT
4458.262	4459.513	22423.97	MN	1.4	1.3		MIT
4457.549	4458.800	22427.56	MN	1.3	1.2		MIT
4457.045	4458.296	22430.09	MN	1.3	1.0		MIT
4455.821	4457.072	22436.26	MN	1.4	1.2		MIT
4455.318	4456.569	22438.79	MN	1.4	1.2		MIT
4455.012	4456.262	22440.33	MN	1.4	1.0		MIT
4453.005	4454.255	22450.44	MN	1.7	1.3		MIT
4452.524	4453.774	22452.87	MN	0.9			MIT
4451.586	4452.836	22457.60	MN	2.1	2.0		MIT
4449.372	4450.621	22468.77	MN	0.5			MIT
4447.146	4448.394	22480.02	MN	1.8			MIT
4444.91	4446.16	22491.3	MN	1.0			MIT
4439.830	4441.076	22517.06	MN	1.0			MIT
4436.352	4437.598	22534.72	MN	1.9	1.7		MIT
4436.025	4437.270	22536.38	MN	1.1			MIT
4433.683	4434.928	22548.28	MN	1.1			MIT
4431.922	4433.166	22557.24	MN	0.9			MIT
4419.776	4421.017	22619.23	MN	2.0	1.3		MIT
4419.25	4420.49	22621.9	MN	0.9			MIT
4414.879	4416.119	22644.32	MN	2.2	1.8		MIT
4411.878	4413.117	22659.72	MN	2.0	1.3		MIT
4410.487	4411.726	22666.87	MN	1.7			MIT
4408.085	4409.323	22679.22	MN	1.8	0.7		MIT
4393.37	4394.60	22755.2	MN	0.9	0.7H		MIT
4389.752	4390.985	22773.93	MN	1.7			MIT
4388.082	4389.315	22782.60	MN	1.8			MIT
4383.067	4384.298	22808.67	MN	1.0			MIT
4382.626	4383.857	22810.96	MN	1.9H			MIT
4381.700	4382.931	22815.78	MN	1.9	1.3		MIT
4381.26	4382.49	22818.1	MN	0.7			MIT
4374.947	4376.176	22851.00	MN	2.2	1.3		MIT
4370.875	4372.103	22872.29	MN	1.5			MIT
4368.878	4370.106	22882.74	MN	1.7			MIT
4365.24	4366.47	22901.8	MN	1.4	0.7		MIT
4363.25	4364.48	22912.3	MN	1.3	0.7		MIT
4359.805	4361.030	22930.36	MN	1.4			FU
4359.627	4360.852	22931.30	MN	1.2			MIT
4356.626	4357.851	22947.09	MN	1.3	0.7		MIT
4356.13	4357.35	22949.7	MN	1.0			MIT
4347.520	4348.742	22995.16	MN	1.0			MIT
4343.97	4345.19	23013.9	MN	2.0	1.5		MIT
4338.824	4340.044	23041.24	MN	1.2			MIT
4338.139	4339.359	23044.88	MN	1.1			MIT
4337.423	4338.642	23048.68	MN	1.9			MIT
4329.420	4330.637	23091.29	MN	1.2			MIT
4326.756	4327.973	23105.51	MN	1.9	1.5		MIT
4323.401	4324.617	23123.44	MN	1.7			MIT
4321.161	4322.376	23135.42	MN	1.8	0.7		MIT
4312.550	4313.763	23181.62	MN	2.0	1.3		MIT
4305.660	4306.871	23218.71	MN	1.7			MIT
4300.197	4301.407	23248.21	MN	1.8	0.7		MIT
4290.112	4291.319	23302.86	MN	0.9			MIT
4287.75	4288.96	23315.7	MN	0.7H			MIT
4284.084	4285.289	23335.65	MN	1.9	1.3		MIT
4281.099	4282.304	23351.92	MN	2.0	1.7		MIT
4278.676	4279.880	23365.14	MN	1.2	0.7		MIT
4265.924	4267.125	23434.98	MN	2.0	1.7		MIT
4261.296	4262.495	23460.44	MN	1.3	0.7		MIT
4259.338	4260.537	23471.22	MN	0.9	1.0		MIT
4257.659	4258.857	23480.48	MN	2.0	1.6		MIT
4239.725	4240.919	23579.80	MN	2.0	1.7		MIT
4235.290	4236.483	23604.49	MN	1.9	2.0		MIT
4235.140	4236.333	23605.32	MN	1.9			MIT
4220.610	4221.799	23686.59	MN	1.8	1.3		MIT
4211.748	4212.934	23736.43	MN	1.5	1.3		MIT
4201.757	4202.941	23792.86	MN	1.6	1.3		MIT
4189.988	4191.169	23859.69	MN	1.9	1.6		MIT
4176.602	4177.779	23936.16	MN	2.0	1.6		MIT
4157.019	4158.191	24048.92	MN	1.6	0.3D		MIT
4155.525	4156.697	24057.56	MN	1.6	0.7		MIT
4148.797	4149.967	24096.58	MN	1.7	1.5		MIT
4147.532	4148.702	24103.93	MN	1.6	1.3		MIT
4141.063	4142.231	24141.58	MN	1.7	1.5		MIT
4140.189	4141.357	24146.68	MN	1.2	0.7		MIT
4137.257	4138.424	24163.79	MN	1.6	0.7		MIT
4135.036	4136.202	24176.77	MN	1.7	1.5		MIT
4131.430	4132.595	24197.87	MN	1.0			MIT
4131.117	4132.282	24199.70	MN	1.7	1.6		MIT
4128.14	4129.30	24217.1	MN	1.2	1.2S		MIT
4123.543	4124.706	24244.15	MN	1.2	0.7		MIT
4123.279	4124.442	24245.70	MN	1.1	0.7		MIT
4116.649	4117.810	24284.75	MN	1.2H	0.7		MIT
4114.384	4115.545	24298.12	MN	1.3	1.3		MIT
4113.876	4115.037	24301.12	MN	1.3	0.7		MIT
4113.236	4114.397	24304.90	MN	1.6	1.3		MIT
4110.903	4112.063	24318.69	MN	1.9R	1.6		MIT
4107.871	4109.030	24336.64	MN	1.3	0.7		MIT
4105.365	4106.524	24351.50	MN	1.7	1.3		MIT
4103.549	4104.707	24362.27	MN	1.1	1.0		MIT

AIR WAVELENGTH (ANGSTRMS)	VACUUM WAVELENGTH (ANGSTRMS)	WAVENUMBER (KAYSERS)	LMENT	LOG ARC	INTENSITY SPK	DIS	REF
4103.463	4104.621	24362.78	MN	1.1			FU
4102.965	4104.123	24365.74	MN	2.0	1.3		MIT
4095.248	4096.404	24411.66	MN	1.2	1.2		MIT
4095.053	4096.209	24412.82	MN	1.1	0.7		MIT
4092.388	4093.543	24428.72	MN	1.5	1.3		MIT
4090.606	4091.761	24439.36	MN	1.1	0.7S		MIT
4089.938	4091.092	24443.35	MN	1.9	1.3		MIT
4083.628	4084.781	24481.12	MN	1.9	1.8		MIT
4083.215	4084.368	24483.59	MN	0.0	0.3		MIT
4082.944	4084.097	24485.22	MN	1.9	1.8		MIT
4079.422	4080.574	24506.36	MN	1.7	1.6		MIT
4079.241	4080.393	24507.45	MN	1.7	1.6		MIT
4075.254	4076.405	24531.42	MN	1.4	0.7		MIT
4070.279	4071.428	24561.41	MN	1.9	1.5		MIT
4068.003	4069.152	24575.15	MN	1.7	1.3		MIT
4066.223	4067.371	24585.90	MN	1.1	0.7		MIT
4065.082	4066.230	24592.80	MN	1.3	1.3		MIT
4063.528	4064.676	24602.21	MN	2.0	1.8		MIT
4061.742	4062.889	24613.03	MN	1.9	1.5		MIT
4059.392	4060.538	24627.27	MN	1.3	1.2		MIT
4058.930	4060.076	24630.08	MN	1.9	1.8		MIT
4057.950	4059.096	24636.03	MN	1.9	1.3		MIT
4055.543	4056.688	24650.65	MN	1.9	1.9		MIT
4055.214	4056.359	24652.65	MN	1.0	0.7		MIT
4052.472	4053.617	24669.33	MN	1.3	1.3		MIT
4051.731	4052.875	24673.84	MN	1.2	1.3		MIT
4049.431	4050.575	24687.85	MN	0.7	0.7		MIT
4048.755	4049.899	24691.98	MN	1.8	1.8		MIT
4045.206	4046.349	24713.64	MN	1.2	1.2		MIT
4045.133	4046.276	24714.08	MN	1.2			MIT
4041.361	4042.503	24737.15	MN	2.0	1.7		MIT
4038.732	4039.873	24753.25	MN	1.2	1.2		MIT
4037.554	4038.695	24760.48	MN	0.5	0.5		MIT
4036.560	4037.701	24766.57	MN	0.3	0.3		MIT
4036.249	4037.389	24768.48	MN	0.3	0.3		MIT
4035.728	4036.868	24771.68	MN	1.7	1.8		MIT
4034.490	4035.630	24779.28	MN I	2.4R	1.3		MIT
4033.630	4034.770	24784.56	MN	0.7	0.7		MIT
4033.073	4034.213	24787.99	MN I	2.6R	1.3		MIT
4031.790	4032.929	24795.87	MN	0.9	1.0		MIT
4030.755	4031.894	24802.24	MN I	2.7R	1.3		MIT
4028.593	4029.731	24815.55	MN	0.3	0.3		MIT
4026.435	4027.573	24828.85	MN	1.7	1.6		MIT
4025.910	4027.048	24832.09	MN	0.5	0.5		MIT
4020.092	4021.228	24868.02	MN	1.0	0.7		MIT
4018.102	4019.238	24880.34	MN	1.9	1.8		MIT
4016.665	4017.800	24889.24	MN	0.9	0.7		MIT
4011.905	4013.039	24918.77	MN	1.1	1.0		MIT
4011.531	4012.665	24921.09	MN	1.2	1.2		MIT
4008.020	4009.153	24942.92	MN	1.2	0.7		MIT
4007.037	4008.170	24949.04	MN	1.0	0.7		MIT
4003.255	4004.387	24972.61	MN	1.3	0.7H		MIT
4002.162	4003.294	24979.43	MN	1.2	0.7		MIT
4001.913	4003.044	24980.99	MN	1.2	1.0		MIT
4001.183	4002.314	24985.54	MN	1.1	0.7		MIT
3997.214	3998.344	25010.35	MN	1.1	1.4		MIT
3992.491	3993.620	25039.94	MN	1.6	1.9		MIT
3991.596	3992.725	25045.55	MN	1.3	1.4		MIT
3988.671	3989.799	25063.92	MN	1.1	1.1		MIT
3987.464	3988.592	25071.51	MN	1.2	1.2		MIT
3987.098	3988.226	25073.81	MN	1.5	1.8		MIT
3986.826	3987.954	25075.52	MN	1.6	1.9		MIT
3985.241	3986.368	25085.49	MN	1.9	2.0		MIT
3984.177	3985.304	25092.19	MN	1.3	1.3		MIT
3982.907	3984.034	25100.19	MN	1.3	1.5		MIT
3982.583	3983.709	25102.23	MN	1.3	1.5		MIT
3982.169	3983.295	25104.84	MN	1.1	1.4		MIT
3980.882	3982.008	25112.96	MN	0.9	0.9		MIT
3980.139	3981.265	25117.65	MN	1.0	1.0		MIT
3977.080	3978.205	25136.97	MN	1.7	2.0		MIT

AIR WAVELENGTH (ANGSTRMS)	VACUUM WAVELENGTH (ANGSTRMS)	WAVENUMBER (KAYSERS)	LMENT	LOG ARC	INTENSITY SPK	DIS	REF
3975.888	3977.013	25144.50	MN	1.6	1.7		MIT
3952.842	3953.961	25291.10	MN	1.8	1.9		MIT
3951.962	3953.080	25296.73	MN	1.6	1.7		MIT
3942.855	3943.971	25355.16	MN	1.9	1.9		MIT
3937.762	3938.877	25387.95	MN	1.2	1.2		MIT
3936.761	3937.875	25394.40	MN	1.4	1.7		MIT
3929.652	3930.765	25440.34	MN	1.1	1.4		MIT
3929.241	3930.353	25443.00	MN	1.5	1.5		MIT
3926.467	3927.579	25460.98	MN	1.6	1.7L		MIT
3924.075	3925.186	25476.50	MN	1.3	1.3H		MIT
3923.327	3924.438	25481.36	MN	1.0	1.3		MIT
3922.908	3924.019	25484.08	MN	0.7	0.7W		MIT
3922.668	3923.779	25485.64	MN	1.3			MIT
3921.762	3922.873	25491.52	MN	1.3H	1.3J		MIT
3918.315	3919.425	25513.95	MN	1.6	1.7		MIT
3916.609	3917.718	25525.06	MN	1.1			MIT
3912.748	3913.856	25550.25	MN	0.7	0.7		MIT
3911.419	3912.527	25558.93	MN	1.2	1.2		MIT
3911.123	3912.231	25560.86	MN	1.3			MIT
3908.162	3909.269	25580.23	MN	0.5	0.5		MIT
3904.965	3906.071	25601.17	MN	1.0	1.0D		MIT
3904.305	3905.411	25605.50	MN	0.7	0.7		MIT
3899.333	3900.438	25638.15	MN	1.1	1.4		MIT
3898.362	3899.466	25644.53	MN	1.5			MIT
3896.333	3897.437	25657.89	MN	1.1	1.1		MIT
3894.708	3895.811	25668.59	MN	1.6	1.6		MIT
3892.615	3893.718	25682.39	MN	1.3	1.3		MIT
3889.450	3890.552	25703.29	MN	1.4	1.7		MIT
3873.196	3874.294	25811.16	MN	1.1	1.1		MIT
3872.125	3873.223	25818.29	MN	1.0	1.0		MIT
3872.056	3873.154	25818.75	MN	1.0			MIT
3865.660	3866.756	25861.47	MN	1.1	1.1		MIT
3856.535	3857.629	25922.66	MN	1.2	1.5		MIT
3853.471	3854.564	25943.27	MN	1.4	1.3		MIT
3852.397	3853.489	25950.50	MN	0.9			MIT
3843.983	3845.073	26007.31	MN	1.9	2.0		MIT
3841.082	3842.172	26026.95	MN	1.7	1.7		MIT
3839.777	3840.866	26035.79	MN	2.0	2.1		MIT
3838.247	3839.336	26046.17	MN	1.0			MIT
3834.364	3835.452	26072.55	MN	1.9R	1.9		MIT
3833.862	3834.950	26075.96	MN	1.9	1.9		MIT
3829.945	3831.032	26102.63	MN	0.9			MIT
3829.683	3830.770	26104.42	MN	1.8	1.8		MIT
3823.893	3824.978	26143.94	MN	1.7H	1.7		MIT
3823.513	3824.598	26146.54	MN	1.9H	1.9		MIT
3816.753	3817.836	26192.85	MN	1.8	1.7		MIT
3810.693	3811.775	26234.50	MN	1.2	1.2		MIT
3809.592	3810.673	26242.08	MN	2.2	2.2		MIT
3809.480	3810.561	26242.85	MN	2.2			MIT
3809.096	3810.177	26245.50	MN	0.9	0.9		MIT
3806.719	3807.800	26261.89	MN	1.7H	1.3		MIT
3804.757	3805.837	26275.43	MN	0.7	0.7		MIT
3801.907	3802.986	26295.12	MN	1.3	1.3		MIT
3801.633	3802.712	26297.02	MN II			0.8	CZ
3800.552	3801.631	26304.50	MN	1.8	1.8		MIT
3800.240	3801.319	26306.66	MN II			0.6	CZ
3799.259	3800.338	26313.45	MN	1.7	1.7		MIT
3790.215	3791.291	26376.24	MN	2.0	2.1		MIT
3785.421	3786.496	26409.64	MN	0.6	0.6		MIT
3776.527	3777.600	26471.84	MN	1.4	1.4		MIT
3774.645	3775.717	26485.03	MN	1.3			MIT
3768.157	3769.228	26530.64	MN	1.0	1.0		MIT
3767.696	3768.766	26533.88	MN	1.1	1.1		MIT
3763.800	3764.869	26561.35	MN	1.6	1.4		MIT
3763.377	3764.446	26564.33	MN	1.3S	1.3		MIT
3756.639	3757.707	26611.98	MN	1.3	1.3		MIT
3750.763	3751.829	26653.67	MN	1.8	1.5		MIT
3746.616	3747.681	26683.17	MN	1.4	1.4		MIT
3741.030	3742.094	26723.01	MN	1.2	1.2		MIT
3736.899	3737.961	26752.55	MN	1.4	1.4		MIT

AIR WAVELENGTH (ANGSTRMS)	VACUUM WAVELENGTH (ANGSTRMS)	WAVENUMBER (KAYSERS)	LMENT	LOG ARC	INTENSITY SPK	DIS	REF
3731.932	3732.993	26788.16	MN	1.9	2.0		MIT
3728.889	3729.949	26810.01	MN	1.9	2.0		MIT
3726.931	3727.991	26824.10	MN	0.7	0.7		MIT
3718.930	3719.988	26881.81	MN	1.9	2.0		MIT
3711.340	3712.396	26936.78	MN	1.0			MIT
3706.658	3707.713	26970.81	MN	0.7	0.7		MIT
3706.078	3707.133	26975.03	MN	1.9			MIT
3701.730	3702.783	27006.71	MN	1.8	1.5		MIT
3700.296	3701.349	27017.18	MN	0.7			MIT
3700.130	3701.183	27018.39	MN	1.0			MIT
3696.568	3697.620	27044.42	MN	2.0	1.7		MIT
3694.115	3695.166	27062.38	MN	0.7	0.7		MIT
3693.667	3694.718	27065.66	MN	1.7	1.8		MIT
3692.812	3693.863	27071.93	MN	1.7	1.7		MIT
3685.556	3686.605	27125.23	MN	1.1	1.1		MIT
3685.212	3686.261	27127.76	MN	1.2			MIT
3684.868	3685.917	27130.29	MN	1.2	1.2		MIT
3684.521	3685.570	27132.85	MN	0.7	0.7		MIT
3683.474	3684.523	27140.56	MN	1.1			MIT
3682.087	3683.135	27150.78	MN	1.4	1.6		MIT
3680.148	3681.196	27165.09	MN	0.7	0.7		MIT
3676.959	3678.006	27188.65	MN	1.8	2.0		MIT
3675.673	3676.720	27198.16	MN	1.3			MIT
3670.517	3671.562	27236.36	MN	1.4	1.2		MIT
3669.838	3670.883	27241.40	MN	1.5	1.5		MIT
3669.399	3670.444	27244.66	MN	0.5	0.5		MIT
3660.404	3661.447	27311.61	MN	1.9	1.9		MIT
3658.518	3659.560	27325.69	MN	0.7	0.7		MIT
3657.902	3658.944	27330.29	MN	1.1	1.0		MIT
3641.391	3642.429	27454.21	MN	1.7	1.7H		MIT
3640.104	3641.141	27463.92	MN	0.3	0.3		MIT
3635.682	3636.718	27497.32	MN	1.0	0.7		MIT
3629.741	3630.776	27542.32	MN	2.0	1.5		MIT
3623.792	3624.825	27587.54	MN	1.9	1.6		MIT
3619.284	3620.316	27621.90	MN	1.9	1.7		MIT
3610.299	3611.329	27690.64	MN	1.8	1.6		MIT
3608.494	3609.523	27704.49	MN	1.8	1.6		MIT
3607.537	3608.566	27711.84	MN	1.9	1.6		MIT
3605.691	3606.720	27726.03	MN	1.0			MIT
3604.689	3605.717	27733.73	MN	1.1			MIT
3601.268	3602.295	27760.08	MN	1.2	1.2H		MIT
3595.119	3596.145	27807.56	MN	1.7	1.4		MIT
3591.809	3592.834	27833.18	MN	1.2			MIT
3586.543	3587.567	27874.05	MN	1.7H	1.6		MIT
3583.676	3584.699	27896.35	MN	0.7			MIT
3582.431	3583.454	27906.04	MN	1.0			FU
3580.102	3581.124	27924.19	MN	0.3			MIT
3579.661	3580.683	27927.63	MN	0.5			MIT
3577.880	3578.901	27941.54	MN	1.7	1.4		MIT
3570.100	3571.119	28002.42	MN	1.3G	1.2		MIT
3570.041	3571.060	28002.89	MN	1.3G			MIT
3569.804	3570.823	28004.75	MN	1.1	0.7		MIT
3569.493	3570.512	28007.19	MN	1.4H	0.9		MIT
3552.721	3553.736	28139.40	MN	1.1			MIT
3548.202	3549.216	28175.24	MN	1.6H	1.6		MIT
3548.029	3549.043	28176.61	MN	1.6H	1.2		MIT
3547.802	3548.816	28178.42	MN	1.6H	1.2		MIT
3537.895	3538.906	28257.32	MN	1.1			MIT
3535.307	3536.317	28278.00	MN	0.7			MIT
3532.121	3533.131	28303.51	MN	1.7H	1.5		MIT
3531.998	3533.008	28304.50	MN	1.7H	0.9		MIT
3531.848	3532.858	28305.70	MN	1.6G	1.5R		MIT
3524.540	3525.548	28364.39	MN	1.2			MIT
3511.833	3512.837	28467.02	MN	0.6	0.6		MIT
3502.280	3503.282	28544.66	MN	0.3H			MIT
3500.852	3501.854	28556.30	MN	0.7			MIT
3497.538	3498.539	28583.36	MN II	1.2	2.2		MIT
3496.807	3497.808	28589.34	MN II	1.0	1.5		MIT
3495.839	3496.839	28597.25	MN II	1.4	2.2		MIT
3492.960	3493.960	28620.82	MN	1.0			MIT

AIR WAVELENGTH (ANGSTRMS)	VACUUM WAVELENGTH (ANGSTRMS)	WAVENUMBER (KAYSERS)	LMENT	LOG ARC	INTENSITY SPK	DIS	REF
3488.680	3489.678	28655.94	MN	1.7	2.3		MIT
3482.909	3483.906	28703.42	MN II	1.7	2.4		MIT
3475.762	3476.757	28762.44	MN	0.9			MIT
3474.133	3475.128	28775.92	MN II	1.1	2.6		MIT
3474.044	3475.039	28776.66	MN II	1.3			MIT
3470.010	3471.004	28810.11	MN	0.3	0.3		MIT
3466.336	3467.329	28840.65	MN II			1.3	CZ
3465.195	3466.187	28850.14	MN	0.3			MIT
3465.037	3466.029	28851.46	MN II			1.3	CZ
3464.043	3465.035	28859.74	MN II			1.2	CZ
3463.330	3464.322	28865.68	MN II			1.1	CZ
3462.878	3463.870	28869.45	MN II			1.0	CZ
3460.328	3461.319	28890.72	MN II	1.8	2.7		MIT
3460.018	3461.009	28893.31	MN II	0.7	0.7		MIT
3451.479	3452.468	28964.79	MN	0.3			MIT
3450.605	3451.594	28972.12	MN	0.6			MIT
3441.988	3442.974	29044.65	MN II	1.9	1.9		MIT
3440.055	3441.041	29060.97	MN	0.3			MIT
3438.974	3439.960	29070.11	MN II	1.3	1.3		MIT
3433.565	3434.549	29115.90	MN	1.2			MIT
3420.795	3421.776	29224.59	MN	0.9			MIT
3366.212	3367.179	29698.45	MN	0.7W	0.7W		MIT
3355.487	3356.451	29793.37	MN	0.5			MIT
3351.662	3352.625	29827.37	MN	0.9			MIT
3351.427	3352.390	29829.46	MN	0.6			FU
3350.401	3351.364	29838.60	MN	0.6			MIT
3345.352	3346.314	29883.63	MN	1.2			MIT
3343.731	3344.692	29898.12	MN	1.5			MIT
3330.668	3331.626	30015.37	MN	1.9			MIT
3320.693	3321.648	30105.53	MN	1.8	1.5H		MIT
3317.305	3318.260	30136.28	MN	2.0	1.5H		MIT
3316.48	3317.43	30143.8	MN	1.3H			FU
3316.324	3317.278	30145.19	MN	1.1			MIT
3314.904	3315.858	30158.11	MN	1.5			FU
3314.424	3315.378	30162.47	MN	1.5H			FU
3313.524	3314.478	30170.67	MN	1.1			MIT
3313.223	3314.177	30173.41	MN	1.7			FU
3311.905	3312.858	30185.42	MN	1.9			MIT
3308.785	3309.737	30213.88	MN	1.3			MIT
3307.005	3307.957	30230.14	MN	1.2			MIT
3304.894	3305.845	30249.45	MN	0.9	0.9H		MIT
3303.278	3304.229	30264.25	MN	1.6			MIT
3300.948	3301.898	30285.61	MN	0.5			MIT
3298.224	3299.174	30310.62	MN	1.7	1.4H		MIT
3296.882	3297.831	30322.96	MN	1.8	1.5		MIT
3296.027	3296.976	30330.82	MN	1.3			MIT
3290.988	3291.936	30377.26	MN	1.0			MIT
3280.756	3281.701	30472.00	MN	1.8	1.5		MIT
3278.553	3279.498	30492.47	MN	1.8	1.5		MIT
3278.066	3279.011	30497.00	MN	1.0			MIT
3273.015	3273.958	30544.07	MN	1.3	1.3		MIT
3270.351	3271.294	30568.95	MN	1.5	1.5		MIT
3268.722	3269.664	30584.18	MN	1.5	1.5		MIT
3267.794	3268.736	30592.86	MN	1.6	1.6		MIT
3264.711	3265.652	30621.75	MN	1.9	1.7		MIT
3260.231	3261.171	30663.83	MN	1.9	1.7		MIT
3258.413	3259.353	30680.94	MN	1.9	1.6		MIT
3256.137	3257.076	30702.38	MN	1.9	1.7		MIT
3254.039	3254.978	30722.18	MN	1.7	1.4		MIT
3252.948	3253.886	30732.48	MN	1.9	1.7		MIT
3251.135	3252.073	30749.62	MN	1.7	1.4		MIT
3248.516	3249.453	30774.41	MN	2.0	2.0		MIT
3247.542	3248.479	30783.64	MN	2.1			MIT
3243.780	3244.716	30819.34	MN	2.0	1.9		MIT
3240.886	3241.821	30846.86	MN	0.5			MIT
3240.616	3241.551	30849.43	MN	1.8	1.5		MIT
3240.399	3241.334	30851.49	MN	1.8	1.5		MIT
3237.414	3238.348	30879.94	MN	1.5H			MIT
3236.778	3237.712	30886.01	MN	1.9	1.9		MIT
3235.003	3235.937	30902.95	MN	1.5H			MIT

AIR WAVELENGTH (ANGSTRMS)	VACUUM WAVELENGTH (ANGSTRMS)	WAVENUMBER (KAYSERS)	LMENT	LOG ARC	INTENSITY SPK	INTENSITY DIS	REF
3233.968	3234.901	30912.84	MN	1.9J			MIT
3230.719	3231.652	30943.93	MN	1.9	1.9		MIT
3229.711	3230.643	30953.59	MN	0.3			MIT
3228.090	3229.022	30969.13	MN	2.0	2.0		MIT
3226.034	3226.965	30988.87	MN	1.6	1.3		MIT
3224.761	3225.692	31001.10	MN	1.9	1.6		MIT
3223.224	3224.155	31015.88	MN	0.6H			MIT
3216.946	3217.875	31076.41	MN	1.9	1.9		MIT
3212.884	3213.812	31115.70	MN	2.0	2.0		MIT
3206.908	3207.835	31173.68	MN	1.8			MIT
3203.742	3204.668	31204.48	MN	1.1			MIT
3202.555	3203.480	31216.05	MN	1.1			MIT
3201.121	3202.046	31230.03	MN	0.9			MIT
3178.495	3179.414	31452.33	MN	2.2	1.7		MIT
3177.049	3177.968	31466.65	MN	0.9			MIT
3175.714	3176.633	31479.87	MN	0.9			MIT
3174.651	3175.569	31490.42	MN	1.2H			MIT
3161.039	3161.954	31626.01	MN	2.2	1.7		MIT
3160.145	3161.060	31634.96	MN	0.7			MIT
3159.948	3160.863	31636.93	MN	1.1			MIT
3158.740	3159.654	31649.03	MN	0.9			MIT
3155.779	3156.693	31678.73	MN	0.7			MIT
3148.857	3149.769	31748.36	MN	0.7			MIT
3148.179	3149.091	31755.20	MN	2.2	1.6		MIT
3146.325	3147.236	31773.91	MN	0.6			MIT
3142.670	3143.580	31810.86	MN	1.7			MIT
3141.818	3142.728	31819.49	MN	0.5			MIT
3138.219	3139.128	31855.98	MN	0.5			MIT
3136.958	3137.867	31868.78	MN	1.0	1.0		MIT
3132.792	3133.700	31911.16	MN	0.9			MIT
3132.285	3133.193	31916.33	MN	1.2			MIT
3128.323	3129.230	31956.75	MN	0.3H			MIT
3120.336	3121.241	32038.54	MN	1.7			MIT
3118.128	3119.032	32061.23	MN	0.6H			MIT
3115.465	3116.369	32088.63	MN	1.7	1.4		MIT
3114.121	3115.024	32102.48	MN	0.5			MIT
3113.130	3114.033	32112.70	MN	1.2			MIT
3110.677	3111.579	32138.02	MN	1.5	1.5		MIT
3108.770	3109.672	32157.73	MN	0.7			MIT
3108.635	3109.537	32159.13	MN	0.9			MIT
3107.774	3108.676	32168.04	MN	1.1			MIT
3106.744	3107.645	32178.70	MN	0.7			MIT
3103.073	3103.973	32216.77	MN	0.3			MIT
3101.557	3102.457	32232.52	MN	1.7	1.7		MIT
3100.302	3101.202	32245.56	MN	1.8	1.8		MIT
3097.063	3097.962	32279.29	MN	1.9W	1.6		MIT
3082.703	3083.598	32429.65	MN	1.1	1.1		MIT
3082.052	3082.947	32436.50	MN	1.7			MIT
3081.330	3082.225	32444.09	MN	1.9	1.4		MIT
3079.627	3080.522	32462.04	MN	2.1	1.6		MIT
3073.126	3074.019	32530.70	MN	1.9	1.3		MIT
3070.266	3071.158	32561.01	MN	2.0	1.4		MIT
3066.019	3066.910	32606.11	MN	1.9			MIT
3062.119	3063.009	32647.63	MN	1.9	1.3		MIT
3056.990	3057.879	32702.41	MN	0.3			MIT
3056.456	3057.345	32708.12	MN	0.5			MIT
3054.362	3055.250	32730.54	MN	1.9	1.6		MIT
3048.864	3049.751	32789.56	MN	1.4			MIT
3047.035	3047.921	32809.25	MN	1.8	1.8		MIT
3046.596	3047.482	32813.97	MN	0.7			MIT
3046.266	3047.152	32817.53	MN II	0.0	0.8H		CZ
3045.808	3046.694	32822.46	MN	1.4	1.1		MIT
3045.593	3046.479	32824.78	MN	1.6	1.3		MIT
3044.567	3045.453	32835.84	MN	2.0J	1.6		MIT
3043.770	3044.656	32844.44	MN	1.2			MIT
3043.356	3044.241	32848.91	MN	1.6	1.6		MIT
3043.143	3044.028	32851.21	MN	1.1			MIT
3042.733	3043.618	32855.63	MN	1.4R	1.4		MIT
3041.224	3042.109	32871.93	MN	1.4	1.4		MIT
3040.603	3041.488	32878.65	MN	1.7	1.4		MIT

AIR WAVELENGTH (ANGSTRMS)	VACUUM WAVELENGTH (ANGSTRMS)	WAVENUMBER (KAYSERS)	LMENT	LOG ARC	INTENSITY SPK	INTENSITY DIS	REF
3039.551	3040.435	32890.02	MN II	0.0	0.8H		CZ
3038.602	3039.486	32900.30	MN	0.5			MIT
3038.503	3039.387	32901.37	MN	0.6	0.6		MIT
3035.365	3036.248	32935.38	MN	0.7	0.9		MIT
3033.563	3034.446	32954.94	MN	0.0	0.3		MIT
3031.063	3031.945	32982.12	MN	0.9	1.4		MIT
3029.694	3030.576	32997.03	MN	1.1			MIT
3029.041	3029.923	33004.14	MN II	0.3	0.8H		CZ
3022.749	3023.629	33072.84	MN	1.7	1.4		MIT
3016.454	3017.333	33141.85	MN	1.4	1.4		MIT
3014.668	3015.546	33161.49	MN	1.2	1.2		MIT
3012.853	3013.731	33181.46	MN	0.9			MIT
3011.376	3012.253	33197.74	MN	1.4	1.4		MIT
3011.162	3012.039	33200.10	MN	1.4	1.4		MIT
3008.265	3009.142	33232.07	MN	1.2	1.2		MIT
3007.655	3008.532	33238.81	MN	1.6	1.6		MIT
3002.489	3003.364	33296.00	MN	1.1			MIT
2992.110	2992.983	33411.49	MN	0.7			MIT
2986.001	2986.872	33479.84	MN	1.3			MIT
2978.563	2979.432	33563.44	MN	1.1			MIT
2978.121	2978.990	33568.42	MN	1.0H			MIT
2974.105	2974.973	33613.75	MN	1.2			MIT
2970.967	2971.834	33649.25	MN	0.5			MIT
2963.610	2964.476	33732.78	MN	1.3			MIT
2963.259	2964.124	33736.77	MN	1.1			MIT
2961.676	2962.541	33754.81	MN	0.0	0.3		MIT
2956.971	2957.835	33808.51	MN	1.1	0.0		MIT
2956.102	2956.966	33818.45	MN	1.3	0.0		MIT
2953.008	2953.871	33853.88	MN	1.0			MIT
2951.160	2952.022	33875.08	MN	0.6	0.6		MIT
2949.205	2950.067	33897.54	MN	2.0	1.5		ME
2943.908	2944.769	33958.52	MN	0.3	0.3		MIT
2943.132	2943.992	33967.48	MN	0.0	0.5		MIT
2942.745	2943.605	33971.94	MN	1.0	0.0		MIT
2941.041	2941.901	33991.63	MN	1.4	0.0		MIT
2940.483	2941.343	33998.08	MN	1.6V			MIT
2940.392	2941.252	33999.13	MN	1.6V			FU
2939.904	2940.764	34004.77	MN	1.1			MIT
2939.304	2940.164	34011.71	MN	1.7			MIT
2937.924	2938.783	34027.69	MN	1.4V			MIT
2936.156	2937.015	34048.18	MN	1.0			MIT
2935.656	2936.515	34053.98	MN	1.3			MIT
2934.022	2934.880	34072.94	MN	1.4			MIT
2933.442	2934.300	34079.68	MN	0.6			MIT
2933.063	2933.921	34084.08	MN	1.9	1.2		MIT
2930.247	2931.104	34116.83	MN	1.4			MIT
2928.680	2929.537	34135.09	MN	1.4			MIT
2925.57	2926.43	34171.4	MN	2.2V	0.0V		MIT
2924.632	2925.488	34182.33	MN	0.7			MIT
2924.437	2925.293	34184.61	MN	1.1			MIT
2919.124	2919.979	34246.83	MN	1.1			MIT
2914.603	2915.456	34299.95	MN	2.2V			MIT
2913.804	2914.657	34309.35	MN	0.7			MIT
2913.110	2913.963	34317.52	MN	0.3H	0.3		MIT
2910.242	2911.094	34351.34	MN	0.7			MIT
2908.879	2909.731	34367.44	MN	1.0			MIT
2907.993	2908.845	34377.91	MN	1.3			MIT
2907.216	2908.068	34387.10	MN	1.7			MIT
2902.401	2903.251	34444.14	MN	1.1			MIT
2902.205	2903.055	34446.47	MN	1.7			MIT
2900.547	2901.397	34466.16	MN	1.7			MIT
2900.163	2901.013	34470.72	MN	1.1	0.5		MIT
2898.693	2899.543	34488.20	MN	1.1	0.5		MIT
2897.990	2898.839	34496.56	MN	1.2			MIT
2897.797	2898.646	34498.86	MN	1.2			MIT
2897.427	2898.276	34503.27	MN	0.9			MIT
2897.065	2897.914	34507.58	MN	0.7	0.3		MIT
2895.190	2896.039	34529.93	MN	1.1			MIT
2892.658	2893.506	34560.15	MN	1.3			MIT
2892.485	2893.333	34562.22	MN	0.7			MIT

AIR WAVELENGTH (ANGSTRMS)	VACUUM WAVELENGTH (ANGSTRMS)	WAVENUMBER (KAYSERS)	LMENT		LOG ARC	INTENSITY SPK	DIS	REF
2892.392	2893.240	34563.33	MN		0.9	0.6		MIT
2891.321	2892.169	34576.13	MN		0.5	0.5		MIT
2889.58	2890.43	34597.0	MN		1.1D	1.0		MIT
2886.678	2887.525	34631.74	MN		1.2	0.8		MIT
2885.125	2885.971	34650.38	MN		0.0	0.3		MIT
2882.902	2883.748	34677.10	MN		1.4			MIT
2879.488	2880.333	34718.21	MN		1.1	0.7		MIT
2872.914	2873.757	34797.65	MN		0.0	0.5		MIT
2872.581	2873.424	34801.69	MN		1.5			MIT
2870.083	2870.925	34831.97	MN		0.6	0.6		MIT
2863.260	2864.101	34914.97	MN		0.5			MIT
2858.659	2859.499	34971.17	MN		1.7			MIT
2854.805	2855.644	35018.37	MN		0.7			MIT
2842.086	2842.922	35175.08	MN		1.1V			MIT
2840.001	2840.836	35200.90	MN		1.3			MIT
2836.313	2837.147	35246.67	MN		1.2			MIT
2830.794	2831.627	35315.39	MN		1.7			MIT
2828.762	2829.594	35340.75	MN		1.7J			MIT
2826.718	2827.550	35366.31	MN		1.1V			MIT
2822.550	2823.381	35418.53	MN		1.1	0.0		MIT
2821.452	2822.283	35432.31	MN		1.6			MIT
2819.728	2820.558	35453.98	MN		1.0			MIT
2819.377	2820.207	35458.39	MN		0.7			MIT
2818.919	2819.749	35464.15	MN		1.3			MIT
2818.771	2819.601	35466.01	MN		1.4			MIT
2817.972	2818.802	35476.07	MN		1.7	0.0		MIT
2817.669	2818.499	35479.88	MN		1.2			MIT
2817.168	2817.997	35486.19	MN		1.0			MIT
2816.324	2817.153	35496.83	MN		0.0	0.3		MIT
2815.596	2816.425	35506.00	MN		1.1H			MIT
2815.018	2815.847	35513.29	MN		1.4	1.9		MIT
2814.460	2815.289	35520.33	MN		0.3H			MIT
2813.992	2814.821	35526.24	MN		1.1H			MIT
2813.473	2814.302	35532.79	MN		1.5	0.0		MIT
2812.845	2813.673	35540.73	MN		1.3			MIT
2812.61	2813.44	35543.7	MN		0.0	0.5		MIT
2812.270	2813.098	35547.99	MN		0.0	0.3		MIT
2811.343	2812.171	35559.71	MN		0.5H			MIT
2809.106	2809.933	35588.03	MN		1.4			MIT
2808.384	2809.211	35597.18	MN		0.9			MIT
2808.016	2808.843	35601.85	MN		1.3			MIT
2806.793	2807.620	35617.36	MN		1.1			MIT
2806.136	2806.963	35625.69	MN		1.4			MIT
2805.362	2806.189	35635.52	MN		0.0	0.7		MIT
2805.207	2806.034	35637.49	MN	II	0.3	0.8		CZ
2804.924	2805.750	35641.09	MN		1.0			MIT
2804.362	2805.188	35648.23	MN		0.9			MIT
2804.095	2804.921	35651.62	MN		1.2			MIT
2803.620	2804.446	35657.66	MN		1.1			MIT
2802.800	2803.626	35668.10	MN		1.1			MIT
2802.65	2803.48	35670.0	MN		0.3D			MIT
2802.414	2803.240	35673.01	MN		1.1			MIT
2802.168	2802.994	35676.14	MN		0.7			MIT
2801.064	2801.890	35690.20	MN		2.8G	1.8		MIT
2799.842	2800.667	35705.78	MN		1.3			MIT
2798.271	2799.096	35725.82	MN		2.9G	1.9		MIT
2797.094	2797.919	35740.85	MN		0.7			MIT
2796.942	2797.767	35742.80	MN		0.9			MIT
2794.817	2795.641	35769.97	MN		3.0G	0.7		MIT
2791.584	2792.407	35811.40	MN		0.5			MIT
2791.080	2791.903	35817.86	MN		1.2			MIT
2790.915	2791.738	35819.98	MN		0.9			MIT
2790.356	2791.179	35827.15	MN		1.3			MIT
2789.731	2790.554	35835.18	MN		0.3			MIT
2789.351	2790.174	35840.06	MN		1.1H			MIT
2789.197	2790.020	35842.04	MN		1.2			MIT
2788.681	2789.504	35848.67	MN		0.7			MIT
2787.819	2788.641	35859.76	MN		1.1			MIT
2783.082	2783.903	35920.79	MN		1.1			MIT
2782.734	2783.555	35925.28	MN		1.7			MIT

AIR WAVELENGTH (ANGSTRMS)	VACUUM WAVELENGTH (ANGSTRMS)	WAVENUMBER (KAYSERS)	LMENT		LOG ARC	INTENSITY SPK	DIS	REF
2779.998	2780.818	35960.64	MN		1.4			MIT
2778.560	2779.380	35979.25	MN		1.8			MIT
2777.464	2778.284	35993.44	MN		1.0			MIT
2776.231	2777.050	36009.43	MN		1.9			MIT
2775.654	2776.473	36016.91	MN	II	0.6	1.0H		CZ
2773.664	2774.483	36042.75	MN		1.2			MIT
2773.025	2773.844	36051.06	MN		1.0			MIT
2771.435	2772.253	36071.74	MN		1.4			MIT
2769.629	2770.447	36095.26	MN		0.5	1.0		MIT
2768.447	2769.265	36110.67	MN		0.5	1.3		MIT
2765.443	2766.260	36149.89	MN		0.5			MIT
2762.090	2762.906	36193.77	MN	II	0.3	1.0H		CZ
2761.357	2762.173	36203.38	MN		0.7			MIT
2760.930	2761.746	36208.98	MN		1.9			MIT
2758.952	2759.767	36234.94	MN		1.0			MIT
2755.649	2756.463	36278.37	MN		0.5			MIT
2752.317	2753.131	36322.29	MN		1.3D			MIT
2745.550	2746.362	36411.81	MN		1.2			MIT
2740.563	2741.374	36478.06	MN		0.5H			MIT
2738.862	2739.672	36500.71	MN		1.4			MIT
2737.006	2737.816	36525.46	MN		0.9	1.0		MIT
2734.729	2735.538	36555.87	MN		1.0			MIT
2729.417	2730.225	36627.02	MN		1.1			MIT
2728.608	2729.416	36637.87	MN			1.7		MIT
2728.082	2728.890	36644.94	MN			0.7		MIT
2726.139	2726.946	36671.06	MN		2.5V			MIT
2725.922	2726.729	36673.98	MN			1.7		MIT
2724.449	2725.256	36693.80	MN		0.3	1.9		MIT
2724.167	2724.974	36697.60	MN			0.5		MIT
2722.081	2722.887	36725.72	MN		0.0	1.7H		MIT
2719.893	2720.699	36755.26	MN			1.0		MIT
2719.723	2720.529	36757.56	MN		0.3	1.2		MIT
2719.300	2720.106	36763.28	MN			1.3H		MIT
2719.003	2719.809	36767.29	MN			1.3		MIT
2716.787	2717.592	36797.28	MN		0.0	1.6		MIT
2714.734	2715.539	36825.11	MN			0.7		MIT
2713.843	2714.647	36837.20	MN			1.2		MIT
2713.331	2714.135	36844.15	MN		2.5V			MIT
2711.584	2712.388	36867.88	MN		0.3	2.1H		MIT
2710.625	2711.429	36880.93	MN			1.4H		MIT
2710.333	2711.136	36884.90	MN		1.1	1.6H		MIT
2709.956	2710.759	36890.03	MN			1.4		MIT
2709.610	2710.413	36894.74	MN			1.1		MIT
2708.807	2709.610	36905.68	MN			1.1		MIT
2708.451	2709.254	36910.53	MN		1.2	1.7H		MIT
2707.530	2708.333	36923.08	MN		1.0	1.7H		MIT
2706.634	2707.437	36935.31	MN			1.1		MIT
2705.735	2706.537	36947.58	MN		1.4	1.4H		MIT
2705.561	2706.363	36949.95	MN	II	0.6	1.3		CZ
2703.990	2704.792	36971.42	MN		2.0J	1.4		MIT
2703.453	2704.255	36978.76	MN			1.1		MIT
2702.999	2703.801	36984.97	MN			1.0H		MIT
2701.700	2702.501	37002.76	MN		2.2	1.6H		MIT
2701.169	2701.970	37010.03	MN		0.9	1.0H		MIT
2701.005	2701.806	37012.28	MN		0.7	1.1H		MIT
2698.968	2699.769	37040.21	MN			1.6		MIT
2695.958	2696.758	37081.56	MN			1.3		MIT
2695.958	2696.758	37081.56	MN			1.3		MIT
2695.360	2696.160	37089.79	MN		2.0G	1.7		MIT
2695.048	2695.848	37094.08	MN			1.2		MIT
2694.093	2694.893	37107.23	MN		0.9	1.7		MIT
2693.574	2694.373	37114.38	MN			1.2		MIT
2693.194	2693.993	37119.62	MN		0.9	1.3O		MIT
2692.657	2693.456	37127.02	MN		2.2			MIT
2692.447	2693.246	37129.91	MN			1.2		MIT
2691.974	2692.773	37136.44	MN			1.3		MIT
2690.977	2691.776	37150.20	MN			1.4W		MIT
2690.405	2691.204	37158.09	MN			1.0		MIT
2689.798	2690.597	37166.48	MN			1.6		MIT
2688.246	2689.044	37187.94	MN		0.5	2.0H		MIT

AIR WAVELENGTH (ANGSTRMS)	VACUUM WAVELENGTH (ANGSTRMS)	WAVENUMBER (KAYSERS)	LMENT	LOG ARC	INTENSITY SPK	DIS	REF
2687.412	2688.210	37199.48	MN	1.4	1.9R		MIT
2686.779	2687.577	37208.24	MN	1.0			MIT
2685.940	2686.738	37219.86	MN	1.1H	2.0W		MIT
2684.546	2685.343	37239.19	MN	0.3	1.9		MIT
2683.824	2684.621	37249.20	MN		1.4		MIT
2683.746	2684.543	37250.29	MN	1.1			MIT
2683.016	2683.813	37260.42	MN	1.2	1.2		MIT
2682.371	2683.168	37269.38	MN		1.1		MIT
2681.725	2682.522	37278.36	MN	1.3	1.2		MIT
2681.235	2682.031	37285.17	MN		1.7		MIT
2679.95	2680.75	37303.0	MN		0.7		MIT
2679.161	2679.957	37314.03	MN		1.6		MIT
2677.846	2678.642	37332.35	MN		1.6		MIT
2677.246	2678.042	37340.72	MN		1.2		MIT
2676.747	2677.542	37347.68	MN		1.0		MIT
2676.331	2677.126	37353.49	MN	1.2			MIT
2675.506	2676.301	37365.00	MN		1.4		MIT
2674.979	2675.774	37372.36	MN	0.0	1.2		MIT
2674.744	2675.539	37375.65	MN		1.3		MIT
2674.434	2675.229	37379.98	MN	0.3	1.4		MIT
2673.368	2674.163	37394.88	MN		1.7		MIT
2672.586	2673.380	37405.83	MN	1.2	2.1H		MIT
2671.805	2672.599	37416.76	MN		1.7		MIT
2670.810	2671.604	37430.70	MN	0.7	0.0		MIT
2670.205	2670.999	37439.18	MN	1.2H			MIT
2669.313	2670.107	37451.69	MN		1.3		MIT
2667.826	2668.619	37472.56	MN		1.2		MIT
2667.005	2667.798	37484.10	MN	0.7	1.7H		MIT
2666.768	2667.561	37487.43	MN	0.9	1.7H		MIT
2666.546	2667.339	37490.55	MN	0.9			MIT
2665.185	2665.978	37509.69	MN		1.2		MIT
2665.068	2665.861	37511.34	MN	1.0			MIT
2664.031	2664.823	37525.94	MN		1.4		MIT
2662.536	2663.328	37547.01	MN		1.7		MIT
2660.619	2661.411	37574.06	MN		1.0		MIT
2660.34	2661.13	37578.0	MN		0.7		MIT
2659.21	2660.00	37594.0	MN		1.0D		MIT
2659.084	2659.875	37595.75	MN		1.4		MIT
2657.890	2658.681	37612.64	MN	1.3J			MIT
2656.170	2656.960	37636.99	MN		1.1		MIT
2655.914	2656.704	37640.62	MN	0.7	1.3		MIT
2655.792	2656.582	37642.35	MN	0.7	1.0		MIT
2653.561	2654.351	37674.00	MN		1.2		MIT
2652.485	2653.275	37689.28	MN	0.5	2.0		MIT
2651.861	2652.650	37698.15	MN		1.1		MIT
2650.994	2651.783	37710.47	MN		2.2		MIT
2648.938	2649.727	37739.74	MN	0.0	1.7		MIT
2648.806	2649.595	37741.62	MN	1.3W			MIT
2648.044	2648.833	37752.48	MN		1.4		MIT
2647.615	2648.403	37758.60	MN		1.2		MIT
2643.750	2644.538	37813.79	MN		1.2		MIT
2643.051	2643.838	37823.79	MN		1.1		MIT
2642.766	2643.553	37827.87	MN		0.9		MIT
2642.231	2643.018	37835.53	MN		1.2		MIT
2639.835	2640.622	37869.87	MN	1.1	1.9H		MIT
2638.62	2639.41	37887.3	MN		1.0D		MIT
2638.548	2639.334	37888.34	MN		0.9		MIT
2638.171	2638.957	37893.76	MN	1.4	1.9H		MIT
2637.861	2638.647	37898.21	MN		1.3		MIT
2637.157	2637.943	37908.33	MN		1.2		MIT
2636.866	2637.652	37912.51	MN		0.7		MIT
2635.600	2636.386	37930.72	MN		1.8		MIT
2633.801	2634.586	37956.63	MN		1.1		MIT
2633.325	2634.110	37963.49	MN		1.3		MIT
2633.249	2634.034	37964.58	MN	0.9H			MIT
2632.967	2633.752	37968.65	MN		0.3		MIT
2632.352	2633.137	37977.52	MN	1.1	1.9H		MIT
2631.981	2632.766	37982.87	MN		1.1		MIT
2630.566	2631.350	38003.30	MN	2.0	1.2		MIT
2630.260	2631.044	38007.72	MN	1.2	0.3		MIT

AIR WAVELENGTH (ANGSTRMS)	VACUUM WAVELENGTH (ANGSTRMS)	WAVENUMBER (KAYSERS)	LMENT	LOG ARC	INTENSITY SPK	DIS	REF
2629.548	2630.332	38018.01	MN		1.5		MIT
2627.050	2627.834	38054.16	MN		0.9		MIT
2626.640	2627.423	38060.10	MN	2.1	0.7		MIT
2626.453	2627.236	38062.81	MN		1.3		MIT
2625.585	2626.368	38075.39	MN		2.0H		MIT
2625.120	2625.903	38082.14	MN	0.5	0.3		MIT
2624.804	2625.587	38086.72	MN	1.4	1.1		MIT
2624.646	2625.429	38089.01	MN	0.5			MIT
2624.045	2624.828	38097.74	MN	1.4L	1.8		MIT
2623.33	2624.11	38108.1	MN	1.1D	0.0		MIT
2622.898	2623.681	38114.40	MN	2.3	1.2		MIT
2619.980	2620.762	38156.84	MN	1.7	1.1		MIT
2619.509	2620.291	38163.70	MN	2.1	1.0		MIT
2619.296	2620.078	38166.81	MN		1.2		MIT
2618.914	2619.696	38172.37	MN	1.6	1.1		MIT
2618.143	2618.924	38183.62	MN	1.7	2.0H		MIT
2617.444	2618.225	38193.81	MN		1.1		MIT
2616.918	2617.699	38201.49	MN		1.1		MIT
2616.508	2617.289	38207.47	MN		1.8		MIT
2614.037	2614.817	38243.59	MN		0.7		MIT
2613.589	2614.369	38250.14	MN	1.1	0.3		MIT
2612.855	2613.635	38260.89	MN	1.2	0.3		MIT
2612.627	2613.407	38264.23	MN		1.1		MIT
2610.849	2611.629	38290.28	MN		1.0		MIT
2610.572	2611.352	38294.35	MN		0.9		MIT
2610.200	2610.980	38299.80	MN	1.2	2.0H		MIT
2609.552	2610.331	38309.31	MN		1.2		MIT
2608.806	2609.585	38320.27	MN		1.3		MIT
2608.432	2609.211	38325.76	MN		1.3		MIT
2608.094	2608.873	38330.73	MN		0.3		MIT
2607.854	2608.633	38334.25	MN		0.3		MIT
2607.284	2608.063	38342.63	MN		0.5		MIT
2606.583	2607.362	38352.95	MN		0.9		MIT
2606.252	2607.031	38357.82	MN		0.7		MIT
2605.688	2606.466	38366.12	MN II	2.0G	2.7G		MIT
2604.356	2605.134	38385.74	MN		1.0		MIT
2603.719	2604.497	38395.13	MN	0.7	1.7		MIT
2603.253	2604.031	38402.00	MN		0.9		MIT
2603.114	2603.892	38404.05	MN		0.9		MIT
2602.720	2603.498	38409.87	MN		1.9		MIT
2602.135	2602.913	38418.50	MN	1.1	0.0		MIT
2601.973	2602.751	38420.89	MN		1.0		MIT
2601.830	2602.608	38423.00	MN		1.0		MIT
2600.588	2601.365	38441.35	MN		1.0		MIT
2600.267	2601.044	38446.10	MN	0.0	1.0		MIT
2599.871	2600.648	38451.95	MN	1.0	0.0		MIT
2599.036	2599.813	38464.31	MN II	0.3	1.1		CZ
2598.902	2599.679	38466.29	MN	0.7	2.0$		MIT
2598.174	2598.951	38477.07	MN	1.1			MIT
2597.510	2598.287	38486.90	MN		1.0		MIT
2596.827	2597.603	38497.02	MN		1.7R		MIT
2595.761	2596.537	38512.83	MN	2.3	1.4		MIT
2595.649	2596.425	38514.50	MN		1.4		MIT
2594.720	2595.496	38528.28	MN	0.0	1.1		MIT
2593.729	2594.505	38543.00	MN II	2.3R	3.0P		MIT
2593.281	2594.057	38549.66	MN		0.3		MIT
2592.944	2593.719	38554.67	MN	2.2	0.5		MIT
2592.298	2593.073	38564.28	MN	1.1	0.3		MIT
2592.029	2592.804	38568.28	MN		0.9		MIT
2591.822	2592.597	38571.36	MN		0.7		MIT
2591.424	2592.199	38577.28	MN	0.3	1.1		MIT
2591.229	2592.004	38580.19	MN		1.1		MIT
2590.691	2591.466	38588.20	MN		0.3		MIT
2590.143	2590.918	38596.36	MN		0.7		MIT
2590.017	2590.792	38598.24	MN		0.5		MIT
2589.708	2590.483	38602.85	MN	1.0	1.7H		MIT
2588.957	2589.731	38614.04	MN		1.9		MIT
2587.593	2588.367	38634.40	MN		0.7		MIT
2587.493	2588.267	38635.89	MN		0.7		MIT
2587.277	2588.051	38639.11	MN		1.1		MIT

AIR WAVELENGTH (ANGSTRMS)	VACUUM WAVELENGTH (ANGSTRMS)	WAVENUMBER (KAYSERS)	LMENT	LOG ARC	INTENSITY SPK	DIS	REF
2586.567	2587.341	38649.72	MN		1.0		MIT
2585.889	2586.663	38659.85	MN II	0.6	0.8H		CZ
2585.483	2586.257	38665.92	MN		1.0J		MIT
2584.533	2585.306	38680.13	MN	1.1	1.1		MIT
2584.308	2585.081	38683.50	MN	2.2W	1.2		MIT
2584.110	2584.883	38686.47	MN	1.2	0.7W		MIT
2583.277	2584.050	38698.94	MN	1.2			MIT
2583.092	2583.865	38701.71	MN	0.7	0.0		MIT
2582.965	2583.738	38703.61	MN		1.7		MIT
2581.649	2582.422	38723.34	MN		1.2		MIT
2580.183	2580.955	38745.34	MN	1.0			MIT
2579.670	2580.442	38753.05	MN	2.1	0.0		MIT
2579.388	2580.160	38757.28	MN		0.7		MIT
2578.910	2579.682	38764.47	MN		1.4J		MIT
2578.354	2579.126	38772.83	MN		1.0H		MIT
2577.950	2578.722	38778.90	MN		0.7		MIT
2577.468	2578.240	38786.15	MN	1.1			MIT
2576.104	2576.875	38806.69	MN II	2.5G	3.3G		MIT
2575.509	2576.280	38815.65	MN	2.2	0.0		MIT
2572.758	2573.529	38857.15	MN	2.3	1.7		MIT
2572.428	2573.199	38862.14	MN		1.0		MIT
2571.892	2572.662	38870.24	MN		1.1		MIT
2570.940	2571.710	38884.63	MN		1.9		MIT
2570.088	2570.858	38897.52	MN		1.0		MIT
2569.323	2570.093	38909.10	MN		1.0		MIT
2568.718	2569.488	38918.26	MN		1.2		MIT
2568.519	2569.289	38921.28	MN		1.1		MIT
2568.308	2569.078	38924.48	MN		1.0		MIT
2566.785	2567.554	38947.57	MN	0.3			MIT
2565.953	2566.722	38960.20	MN	1.2H	1.0		MIT
2565.223	2565.992	38971.29	MN	0.3	1.9		MIT
2564.120	2564.889	38988.05	MN		0.9		MIT
2563.648	2564.417	38995.23	MN	1.4	1.7J		MIT
2563.246	2564.014	39001.34	MN		0.7		MIT
2562.247	2563.015	39016.55	MN		1.0		MIT
2561.550	2562.318	39027.16	MN		0.9		MIT
2560.760	2561.528	39039.20	MN		1.1		MIT
2559.664	2560.432	39055.92	MN		1.6D		MIT
2559.412	2560.180	39059.76	MN	0.3	1.7		MIT
2558.857	2559.624	39068.23	MN		1.1		MIT
2558.588	2559.355	39072.34	MN		1.9		MIT
2558.289	2559.056	39076.91	MN		1.1		MIT
2557.537	2558.304	39088.40	MN	0.0	1.7		MIT
2556.895	2557.662	39098.21	MN	0.5	1.6		MIT
2556.573	2557.340	39103.13	MN	1.0	1.9H		MIT
2554.513	2555.279	39134.66	MN		1.3		MIT
2553.981	2554.747	39142.82	MN		1.0		MIT
2553.256	2554.022	39153.93	MN		1.7		MIT
2552.987	2553.753	39158.06	MN		0.5		MIT
2551.88	2552.65	39175.0	MN	0.0	2.0D		MIT
2551.564	2552.330	39179.89	MN		0.7		MIT
2549.819	2550.584	39206.70	MN		0.9		MIT
2549.610	2550.375	39209.92	MN		0.9		MIT
2549.327	2550.092	39214.27	MN		0.9		MIT
2548.744	2549.509	39223.24	MN	0.9	2.2		MIT
2548.335	2549.100	39229.53	MN		1.1		MIT
2547.901	2548.666	39236.21	MN		0.5H		MIT
2547.457	2548.222	39243.05	MN		1.1J		MIT
2546.607	2547.371	39256.15	MN	1.0H			MIT
2545.163	2545.927	39278.42	MN		1.4		MIT
2544.295	2545.059	39291.82	MN		0.5		MIT
2543.973	2544.737	39296.79	MN		0.5H		MIT
2543.452	2544.216	39304.84	MN	0.6	2.0		MIT
2542.925	2543.689	39312.99	MN	0.0	2.0		MIT
2542.648	2543.412	39317.27	MN		1.2		MIT
2542.495	2543.259	39319.64	MN	1.2			MIT
2541.109	2541.872	39341.08	MN		1.9		MIT
2540.765	2541.528	39346.41	MN		1.0		MIT
2540.188	2540.951	39355.34	MN		0.7		MIT
2539.796	2540.559	39361.42	MN	1.0			MIT

AIR WAVELENGTH (ANGSTRMS)	VACUUM WAVELENGTH (ANGSTRMS)	WAVENUMBER (KAYSERS)	LMENT	LOG ARC	INTENSITY SPK	DIS	REF
2539.648	2540.411	39363.71	MN	1.0			MIT
2539.400	2540.163	39367.56	MN		1.3V		MIT
2538.915	2539.678	39375.08	MN		0.9		MIT
2538.047	2538.809	39388.54	MN		1.4		MIT
2537.353	2538.115	39399.31	MN		0.3		MIT
2536.830	2537.592	39407.44	MN		0.7		MIT
2536.08	2536.84	39419.1	MN		1.1		MIT
2535.645	2536.407	39425.85	MN		1.9		MIT
2535.031	2535.793	39435.40	MN		0.5		MIT
2534.210	2534.972	39448.17	MN II		1.4		MIT
2534.15	2534.91	39449.1	MN		1.4D		MIT
2533.317	2534.078	39462.08	MN		1.0		MIT
2533.061	2533.822	39466.07	MN	1.3	0.0		MIT
2532.763	2533.524	39470.71	MN		1.3		MIT
2529.132	2529.892	39527.37	MN	1.9	0.7D		MIT
2528.702	2529.462	39534.09	MN	1.1	0.3		MIT
2527.436	2528.196	39553.90	MN	2.2	1.1		MIT
2525.669	2526.429	39581.57	MN	1.0	1.0		MIT
2524.473	2525.232	39600.32	MN	1.1	0.0		MIT
2520.584	2521.342	39661.41	MN		1.4D		MIT
2518.147	2518.905	39699.79	MN		1.6		MIT
2516.73	2517.49	39722.1	MN		1.1		MIT
2514.645	2515.402	39755.08	MN		1.1J		MIT
2510.68	2511.44	39817.9	MN		1.0		MIT
2502.474	2503.228	39948.42	MN		0.5		MIT
2499.426	2500.179	39997.13	MN	1.2			MIT
2497.780	2498.533	40023.48	MN		0.5H		MIT
2494.400	2495.152	40077.71	MN	1.2			MIT
2492.720	2493.472	40104.72	MN		0.9		MIT
2490.918	2491.669	40133.73	MN		1.0H		MIT
2488.460	2489.211	40173.37	MN		1.1		MIT
2486.975	2487.726	40197.36	MN		0.7		MIT
2486.163	2486.913	40210.49	MN		1.3		MIT
2485.056	2485.806	40228.40	MN		0.7		MIT
2482.662	2483.412	40267.19	MN		0.3		MIT
2482.147	2482.896	40275.54	MN		0.9W		MIT
2480.954	2481.703	40294.91	MN		1.0		MIT
2475.460	2476.208	40384.33	MN		0.5		MIT
2474.973	2475.721	40392.28	MN		1.2		MIT
2473.657	2474.404	40413.76	MN		1.0J		MIT
2469.408	2470.155	40483.30	MN	1.0			MIT
2469.294	2470.040	40485.17	MN		0.7		MIT
2468.777	2469.523	40493.64	MN	0.3	1.4		MIT
2467.982	2468.728	40506.69	MN		1.2		MIT
2466.419	2467.165	40532.35	MN		1.6		MIT
2466.211	2466.957	40535.77	MN		1.7		MIT
2464.007	2464.752	40572.03	MN		1.0H		MIT
2463.002	2463.747	40588.58	MN	1.2	0.0		MIT
2462.778	2463.523	40592.27	MN	1.0	0.0		MIT
2461.011	2461.756	40621.42	MN	1.4	0.7		MIT
2460.887	2461.632	40623.46	MN	1.4	0.5		MIT
2459.698	2460.442	40643.10	MN	0.7			MIT
2458.687	2459.431	40659.81	MN		1.9H		MIT
2453.659	2454.402	40743.12	MN		1.2J		MIT
2453.175	2453.918	40751.16	MN		1.6J		MIT
2452.526	2453.269	40761.94	MN	0.5H	2.0J		MIT
2449.747	2450.489	40808.18	MN		1.0		MIT
2446.452	2447.193	40863.14	MN		1.3J		MIT
2441.16	2441.90	40951.7	MN		1.2J		MIT
2440.454	2441.194	40963.56	MN		0.9		MIT
2438.222	2438.961	41001.06	MN	0.0	1.2J		MIT
2437.914	2438.653	41006.24	MN		1.4J		MIT
2437.424	2438.163	41014.48	MN		1.6J		MIT
2435.138	2435.877	41052.98	MN	0.0	0.9		MIT
2434.205	2434.943	41068.72	MN	1.1			MIT
2434.070	2434.808	41070.99	MN	1.0			MIT
2431.520	2432.258	41114.06	MN	1.1	0.0		MIT
2430.396	2431.134	41133.08	MN	1.1	0.0		MIT
2429.234	2429.971	41152.75	MN	1.2			MIT
2428.422	2429.159	41166.51	MN	1.1			MIT

AIR WAVELENGTH (ANGSTRMS)	VACUUM WAVELENGTH (ANGSTRMS)	WAVENUMBER (KAYSERS)	LMENT		LOG ARC	INTENSITY SPK	DIS	REF
2427.978	2428.715	41174.04	MN			1.7J		MIT
2427.753	2428.490	41177.85	MN			1.7J		MIT
2427.410	2428.147	41183.67	MN			1.7J		MIT
2424.260	2424.996	41237.18	MN		1.0			MIT
2424.154	2424.890	41238.98	MN			0.5		MIT
2420.405	2421.140	41302.85	MN		1.1			MIT
2420.115	2420.850	41307.80	MN		1.1			MIT
2419.812	2420.547	41312.97	MN			0.7		MIT
2417.929	2418.664	41345.14	MN			0.9		MIT
2416.351	2417.085	41372.14	MN			1.2		MIT
2412.740	2413.474	41434.06	MN			1.2		MIT
2410.57	2411.30	41471.4	MN	II		0.8		CZ
2408.846	2409.579	41501.03	MN			1.0		MIT
2401.728	2402.459	41624.02	MN			0.7		MIT
2398.88	2399.61	41673.4	MN			0.8W		MIT
2397.97	2398.70	41689.2	MN			0.5W		MIT
2391.40	2392.13	41803.8	MN			0.8W		MIT
2389.59	2390.32	41835.4	MN			0.8W		MIT
2389.08	2389.81	41844.4	MN			0.8J		MIT
2387.02	2387.75	41880.5	MN		0.3H	1.5		MIT
2385.28	2386.01	41911.0	MN		0.3	0.6		MIT
2384.05	2384.78	41932.6	MN		1.6	0.5		MIT
2376.73	2377.46	42061.8	MN			1.3		MIT
2374.30	2375.02	42104.8	MN			1.5		MIT
2373.37	2374.09	42121.3	MN	II	0.6	1.7		MIT
2372.117	2372.841	42143.57	MN		1.1			MIT
2367.10	2367.82	42232.9	MN		0.8			MIT
2366.91	2367.63	42236.3	MN			1.6		MIT
2361.76	2362.48	42328.4	MN		0.3H	1.5		MIT
2360.09	2360.81	42358.3	MN			0.8		MIT
2359.46	2360.18	42369.6	MN		0.5	0.9		MIT
2358.45	2359.17	42387.8	MN		0.9	1.4		MIT
2357.64	2358.36	42402.3	MN		0.3	1.0		MIT
2356.81	2357.53	42417.3	MN		0.8	1.0		MIT
2352.933	2353.653	42487.15	MN		1.1	0.8		MIT
2349.22	2349.94	42554.3	MN		0.5	1.2		MIT
2348.83	2349.55	42561.4	MN		0.5	1.2		MIT
2345.540	2346.258	42621.05	MN		0.5			MIT
2341.25	2341.97	42699.1	MN		0.3H			MIT
2340.59	2341.31	42711.2	MN			1.5		MIT
2337.474	2338.191	42768.11	MN		0.3			MIT
2334.94	2335.66	42814.5	MN			1.4		MIT
2333.26	2333.98	42845.4	MN		0.3	1.4		MIT
2332.14	2332.86	42865.9	MN			1.4		MIT
2329.94	2330.65	42906.4	MN		0.3H			MIT
2328.85	2329.56	42926.5	MN			1.5		MIT
2328.67	2329.38	42929.8	MN			1.4		MIT
2325.92	2326.63	42980.5	MN			1.1		MIT
2325.81	2326.52	42982.6	MN		1.0			MIT
2320.98	2321.69	43072.0	MN		0.3H			MIT
2320.45	2321.16	43081.9	MN			1.8		MIT
2318.92	2319.63	43110.3	MN			1.5		MIT
2317.14	2317.85	43143.4	MN		0.7			MIT
2315.68	2316.39	43170.6	MN			1.4		MIT
2313.94	2314.65	43203.0	MN		0.5	1.3		MIT
2312.69	2313.40	43226.4	MN			1.6		MIT
2312.31	2313.02	43233.5	MN		1.1			MIT
2310.96	2311.67	43258.8	MN		1.4			MIT
2308.18	2308.89	43310.9	MN			1.4		MIT
2304.997	2305.706	43370.66	MN		1.6	1.8		MIT
2301.868	2302.577	43429.61	MN			0.6J		MIT
2298.95	2299.66	43484.7	MN		1.3S	1.6		MIT
2289.97	2290.68	43655.2	MN			0.5		MIT
2288.42	2289.13	43684.8	MN		0.8	0.5		A
2281.37	2282.07	43819.8	MN			1.2		A
2278.05	2278.75	43883.6	MN			1.3		A
2269.869	2270.571	44041.79	MN		0.3H	1.8		MIT
2266.58	2267.28	44105.7	MN			1.0		A
2256.93	2257.63	44294.3	MN		0.5	1.1		A
2243.67	2244.37	44556.0	MN			0.7		A

AIR WAVELENGTH (ANGSTRMS)	VACUUM WAVELENGTH (ANGSTRMS)	WAVENUMBER (KAYSERS)	LMENT	LOG ARC	INTENSITY SPK	DIS	REF
2231.662	2232.356	44795.73	MN		1.8		MIT
2229.97	2230.66	44829.7	MN		1.8		MIT
2228.41	2229.10	44861.1	MN	0.5	0.9		A
2227.44	2228.13	44880.6	MN		1.0		MIT
2220.54	2221.23	45020.1	MN		1.6W		A
2215.21	2215.90	45128.4	MN		0.8		MIT
2213.82	2214.51	45156.7	MN	1.8R	1.0		A
2210.88	2211.57	45216.8	MN	0.9	0.9		A
2209.64	2210.33	45242.1	MN	0.8	0.3H		A
2194.85	2195.54	45547.0	MN		1.1		A
2191.39	2192.08	45618.9	MN	1.4	1.0		A
2184.88	2185.56	45754.8	MN	0.9	1.5		A
2175.56	2176.24	45950.8	MN		1.6		A
2148.07	2148.75	46538.8	MN		1.0		A
2144.80	2145.48	46609.7	MN		1.0		A
2131.31	2131.98	46904.7	MN	0.3	0.3		A
2129.86	2130.53	46936.6	MN	0.5	1.0		A
2117.25	2117.92	47216.1	MN	0.5	0.8		A
2109.58	2110.25	47387.8	MN	1.6G	1.1		A
2106.04	2106.71	47467.4	MN	1.6R			MIT
2098.96	2099.63	47627.5	MN	1.0			A
2097.54	2098.21	47659.8	MN	1.3	1.5		MIT
2093.39	2094.06	47754.2	MN	1.5R	1.3		MIT
2092.48	2093.15	47775.0	MN	1.2			MIT
2092.13	2092.80	47783.0	MN	1.5R	1.3		MIT
2072.89	2073.55	48226.4	MN	1.0	0.8		ME
2012.12	2012.77	49682.8	MN		1.5		A
2003.82	2004.47	49888.5	MN	1.7R	1.4		A
2001.95	2002.60	49935.1	MN	0.3	1.3		A
9460.60	9463.20	10567.3	MO	1.0			MIT
9424.71	9427.30	10607.5	MO	0.9			MIT
9347.96	9350.52	10694.6	MO	1.3			MIT
9252.88	9255.42	10804.5	MO	0.6			MIT
9067.46	9069.95	11025.4	MO	0.3			KS
8483.39	8485.72	11784.5	MO	0.3			MIT
8389.32	8391.63	11916.6	MO	2.2			MIT
8351.15	8353.45	11971.1	MO	0.7			KS
8328.44	8330.73	12003.7	MO	2.0			MIT
8245.06	8247.33	12125.1	MO	1.5			MIT
8192.60	8194.85	12202.8	MO	0.3			MIT
8153.45	8155.69	12261.4	MO	0.5			KS
8104.67	8106.90	12335.2	MO	0.5			MIT
8058.22	8060.44	12406.3	MO	0.3			MIT
8027.32	8029.53	12454.0	MO	0.8			MIT
7992.55	7994.75	12508.2	MO	0.3			MIT
7986.60	7988.80	12517.5	MO	1.2			MIT
7968.85	7971.04	12545.4	MO	0.9			MIT
7949.03	7951.22	12576.7	MO	0.3			MIT
7923.15	7925.33	12617.8	MO	1.2			MIT
7917.62	7919.80	12626.6	MO	0.6			MIT
7887.74	7889.91	12674.4	MO	0.9			MIT
7869.40	7871.57	12703.9	MO	0.3			MIT
7854.45	7856.61	12728.1	MO	1.2			MIT
7829.65	7831.80	12768.4	MO	1.3			MIT
7811.31	7813.46	12798.4	MO	0.5			MIT
7795.76	7797.91	12824.0	MO	0.5			MIT
7752.34	7754.47	12895.8	MO	0.5			MIT
7732.49	7734.62	12928.9	MO				MIT
7730.61	7732.74	12932.0	MO	0.5			MIT
7723.63	7725.76	12943.7	MO	1.2			MIT
7720.77	7722.89	12948.5	MO	1.7			MIT
7709.54	7711.66	12967.4	MO	1.2			MIT
7707.78	7709.90	12970.3	MO	0.5			MIT
7679.49	7681.60	13018.1	MO	1.2			MIT
7656.76	7658.87	13056.8	MO	1.4H			MIT
7653.26	7655.37	13062.7	MO	0.5			KS
7649.52	7651.63	13069.1	MO	0.3			KS
7631.46	7633.56	13100.1	MO	0.3			KS
7621.72	7623.82	13116.8	MO	0.3			KS
7617.39	7619.49	13124.2	MO	0.3			KS

AIR WAVELENGTH (ANGSTRMS)	VACUUM WAVELENGTH (ANGSTRMS)	WAVENUMBER (KAYSERS)	LMENT	LOG ARC	INTENSITY SPK	INTENSITY DIS	REF
7601.84	7603.93	13151.1	MO	0.9			MIT
7595.16	7597.25	13162.7	MO	1.2			KS
7591.66	7593.75	13168.7	MO	0.6			MIT
7579.58	7581.67	13189.7	MO	0.5			KS
7572.64	7574.72	13201.8	MO	0.9			MIT
7571.53	7573.61	13203.7	MO	0.5			MIT
7556.21	7558.29	13230.5	MO	0.3			KS
7504.47	7506.54	13321.7	MO	1.2			MIT
7501.62	7503.69	13326.8	MO	0.5			MIT
7485.74	7487.80	13355.1	MO	2.0			MIT
7475.43	7477.49	13373.5	MO	0.7			MIT
7456.67	7458.72	13407.1	MO	0.7H			MIT
7452.85	7454.90	13414.0	MO	1.2H			MIT
7447.34	7449.39	13423.9	MO	1.0H			MIT
7434.10	7436.15	13447.8	MO	1.0			MIT
7427.55	7429.60	13459.7	MO	0.5			MIT
7413.32	7415.36	13485.5	MO	0.6			MIT
7404.50	7406.54	13501.6	MO	0.7			MIT
7391.36	7393.40	13525.6	MO	1.7			MIT
7383.14	7385.17	13540.6	MO	0.8			MIT
7365.25	7367.28	13573.5	MO	0.8			MIT
7364.41	7366.44	13575.1	MO	0.7			MIT
7361.65	7363.68	13580.2	MO	1.0			MIT
7360.38	7362.41	13582.5	MO	0.9			MIT
7348.49	7350.51	13604.5	MO	1.2			MIT
7341.79	7343.81	13616.9	MO	0.5			MIT
7333.71	7335.73	13631.9	MO	0.9			MIT
7331.53	7333.55	13636.0	MO	1.1H			MIT
7329.02	7331.04	13640.6	MO	1.2H			MIT
7325.34	7327.36	13647.5	MO	0.7H			MIT
7322.53	7324.55	13652.7	MO	0.8			KS
7315.33	7317.35	13666.2	MO	0.6W			MIT
7300.19	7302.20	13694.5	MO	1.3			MIT
7295.01	7297.02	13704.2	MO	0.6			MIT
7281.53	7283.54	13729.6	MO	0.7			MIT
7267.62	7269.62	13755.9	MO	1.6			MIT
7245.85	7247.85	13797.2	MO	1.8			MIT
7244.58	7246.58	13799.6	MO	0.6			MIT
7242.50	7244.50	13803.6	MO	1.9			MIT
7240.46	7242.46	13807.5	MO	0.9			MIT
7228.28	7230.27	13830.7	MO	0.5			MS
7216.94	7218.93	13852.5	MO	0.9			MIT
7182.78	7184.76	13918.3	MO	0.7			MIT
7171.78	7173.76	13939.7	MO	0.3			MIT
7159.75	7161.72	13963.1	MO	0.5			MIT
7154.10	7156.07	13974.2	MO	0.5			MIT
7149.42	7151.39	13983.3	MO	0.5			MIT
7146.25	7148.22	13989.5	MO	0.6			MIT
7134.08	7136.05	14013.4	MO	1.6			MIT
7124.32	7126.28	14032.6	MO	0.7			KS
7122.65	7124.61	14035.8	MO	0.9			KS
7109.87	7111.83	14061.1	MO	1.9			MIT
7105.59	7107.55	14069.6	MO	0.6			KS
7102.65	7104.61	14075.4	MO	0.9			KS
7098.18	7100.14	14084.2	MO	0.7			KS
7088.78	7090.73	14102.9	MO	0.7H			KS
7081.22	7083.17	14118.0	MO	1.1			KS
7063.34	7065.29	14153.7	MO	1.1			KS
7060.21	7062.16	14160.0	MO	1.4			MIT
7045.29	7047.23	14190.0	MO	0.8			KS
7037.98	7039.92	14204.7	MO	1.4			MIT
7032.28	7034.22	14216.2	MO	0.5H			MIT
7025.32	7027.26	14230.3	MO	1.1			KS
7019.61	7021.55	14241.9	MO	0.5H			KS
7018.43	7020.37	14244.3	MO	0.9			MIT
7016.44	7018.37	14248.3	MO	1.0			MIT
7013.97	7015.90	14253.3	MO	0.7			MIT
7011.80	7013.73	14257.7	MO	0.5			MIT
7001.60	7003.53	14278.5	MO	1.3			MIT
6999.88	7001.81	14282.0	MO	0.7			MIT

AIR WAVELENGTH (ANGSTRMS)	VACUUM WAVELENGTH (ANGSTRMS)	WAVENUMBER (KAYSERS)	LMENT	LOG ARC	INTENSITY SPK	INTENSITY DIS	REF
6999.13	7001.06	14283.6	MO	0.9			MIT
6991.69	6993.62	14298.7	MO	1.1			MIT
6988.94	6990.87	14304.4	MO	1.5			MIT
6984.67	6986.60	14313.1	MO	0.7			MIT
6980.37	6982.30	14321.9	MO	1.0			MIT
6978.71	6980.63	14325.3	MO	1.4			MIT
6961.48	6963.40	14360.8	MO	1.0			MIT
6960.64	6962.56	14362.5	MO	1.1			MIT
6959.44	6961.36	14365.0	MO	0.5			MIT
6957.03	6958.95	14370.0	MO	0.7			MIT
6953.78	6955.70	14376.7	MO	1.1			MIT
6947.39	6949.31	14389.9	MO	1.2			MIT
6946.75	6948.67	14391.2	MO	0.8			MIT
6934.10	6936.01	14417.5	MO	1.2			MIT
6931.40	6933.31	14423.1	MO	1.2			MIT
6925.85	6927.76	14434.7	MO	0.7			MIT
6914.01	6915.92	14459.4	MO	1.6			MIT
6908.20	6910.11	14471.6	MO	1.3			MIT
6904.70	6906.60	14478.9	MO	0.9			MIT
6900.55	6902.45	14487.6	MO	0.6H			KS
6899.097	6901.000	14490.65	MO	1.1			MIT
6898.01	6899.91	14492.9	MO	1.2			MIT
6892.36	6894.26	14504.8	MO	1.1			MIT
6889.10	6891.00	14511.7	MO	0.7			MIT
6886.28	6888.18	14517.6	MO	1.4			MIT
6865.44	6867.33	14561.7	MO	0.5			MIT
6863.66	6865.55	14565.5	MO	0.5			MIT
6848.92	6850.81	14596.8	MO	1.2			MIT
6838.88	6840.77	14618.2	MO	1.5	0.3		MIT
6834.23	6836.12	14628.2	MO	0.7			MIT
6828.98	6830.86	14639.4	MO	1.7			MIT
6763.50	6765.37	14781.2	MO	0.9			MIT
6753.97	6755.83	14802.0	MO	1.4	0.3		MIT
6746.268	6748.130	14818.92	MO	1.7	0.5		MIT
6746.075	6747.937	14819.34	MO	1.3			MIT
6733.978	6735.837	14845.96	MO	2.0	0.6		MIT
6728.04	6729.90	14859.1	MO	0.8			MIT
6691.08	6692.93	14941.2	MO	0.7			MIT
6690.47	6692.32	14942.5	MO	1.3	0.3		MIT
6687.866	6689.713	14948.32	MO	0.8			MIT
6679.66	6681.50	14966.7	MO	0.5			MIT
6678.892	6680.736	14968.41	MO	0.7H			MIT
6659.680	6661.519	15011.59	MO	1.3	0.3		MIT
6650.375	6652.211	15032.59	MO	1.9	0.8		MIT
6649.703	6651.539	15034.11	MO	0.8			MIT
6649.35	6651.19	15034.9	MO	0.7			MIT
6637.165	6638.998	15062.51	MO	0.9			MIT
6624.573	6626.403	15091.15	MO	1.2	0.3		MIT
6619.130	6620.958	15103.55	MO	2.5	1.2		MIT
6611.203	6613.029	15121.66	MO	1.3			MIT
6590.897	6592.717	15168.25	MO	1.1			MIT
6590.395	6592.215	15169.41	MO	0.5			MIT
6571.088	6572.903	15213.98	MO	0.7	0.3		MIT
6569.776	6571.591	15217.02	MO	1.0	0.3		MIT
6552.596	6554.406	15256.91	MO	0.7	0.3		MIT
6537.621	6539.427	15291.86	MO	0.8	0.3		MIT
6536.307	6538.113	15294.93	MO	0.8			MIT
6527.634	6529.437	15315.26	MO	0.6			MIT
6519.836	6521.637	15333.57	MO	1.4	0.8		MIT
6512.680	6514.479	15350.42	MO	0.5	0.3		MIT
6493.129	6494.923	15396.64	MO	1.2	0.6		MIT
6492.490	6494.284	15398.16	MO	1.0	0.3		MIT
6485.832	6487.624	15413.96	MO	1.0H	0.7W		MIT
6479.201	6480.991	15429.74	MO	0.5	0.3		MIT
6473.991	6475.780	15442.16	MO	1.3	0.6		MIT
6471.201	6472.989	15448.81	MO	1.7	0.6		MIT
6466.970	6468.757	15458.92	MO	0.6	0.3		MIT
6463.516	6465.302	15467.18	MO	0.9	0.3		MIT
6453.293	6455.076	15491.68	MO	1.0	0.3		MIT
6446.342	6448.124	15508.39	MO	1.3	0.6		MIT

AIR WAVELENGTH (ANGSTRMS)	VACUUM WAVELENGTH (ANGSTRMS)	WAVENUMBER (KAYSERS)	LMENT	LOG ARC	INTENSITY SPK	DIS	REF
6429.04	6430.82	15550.1	MO	0.6	0.3		MIT
6424.368	6426.144	15561.43	MO	2.0	1.3		MIT
6419.763	6421.537	15572.59	MO	0.8	0.3		MIT
6418.992	6420.766	15574.46	MO	0.7	0.3		MIT
6412.389	6414.161	15590.50	MO	1.2	0.6		MIT
6409.109	6410.881	15598.48	MO	1.4	0.6		MIT
6406.966	6408.737	15603.70	MO	0.5	0.3		MIT
6406.462	6408.233	15604.93	MO	0.9	0.3		MIT
6401.295	6403.064	15617.52	MO	0.5			MIT
6401.070	6402.839	15618.07	MO	1.3	0.8		MIT
6391.118	6392.885	15642.39	MO	1.1	0.6		MIT
6389.111	6390.877	15647.30	MO	1.2	0.6		MIT
6383.731	6385.496	15660.49	MO	0.5	0.3		MIT
6372.357	6374.119	15688.44	MO	0.7	0.3		MIT
6357.215	6358.973	15725.81	MO	1.6	1.0		MIT
6339.904	6341.657	15768.75	MO	1.2	0.3		MIT
6335.107	6336.859	15780.69	MO	0.6	0.3		MIT
6323.542	6325.291	15809.55	MO	1.1	0.3		MIT
6316.857	6318.604	15826.28	MO	0.6	0.3		MIT
6306.399	6308.143	15852.53	MO	0.6			MIT
6301.753	6303.496	15864.21	MO	1.2			MIT
6290.743	6292.483	15891.98	MO	1.2	0.6		MIT
6286.609	6288.348	15902.43	MO	0.8	0.3		MIT
6281.935	6283.672	15914.26	MO	0.9	0.3		MIT
6278.43	6280.17	15923.1	MO	0.3			MIT
6265.878	6267.611	15955.04	MO	1.4	0.3		MIT
6264.269	6266.002	15959.14	MO	1.2	0.3		MIT
6228.430	6230.153	16050.97	MO	0.7J			MIT
6217.894	6219.614	16078.17	MO	1.3	0.3		MIT
6214.678	6216.397	16086.49	MO	0.8	0.3		MIT
6213.248	6214.967	16090.19	MO	1.1			MIT
6209.478	6211.196	16099.96	MO	0.9	0.3		MIT
6208.273	6209.991	16103.08	MO	1.1	0.3		MIT
6197.662	6199.377	16130.65	MO	1.2	0.3		MIT
6193.201	6194.915	16142.27	MO	0.5	0.3		MIT
6188.987	6190.699	16153.26	MO	1.1	0.3		MIT
6175.530	6177.239	16188.46	MO	0.5	0.3		MIT
6170.751	6172.459	16201.00	MO	0.9	0.3		MIT
6158.442	6160.146	16233.38	MO	1.0	0.5		MIT
6148.991	6150.693	16258.33	MO	0.5H			MIT
6139.351	6141.050	16283.86	MO	0.5			MIT
6138.608	6140.307	16285.83	MO	1.0	0.3		MIT
6134.589	6136.287	16296.50	MO	0.6	0.3		MIT
6130.864	6132.561	16306.40	MO	0.6			MIT
6130.628	6132.325	16307.03	MO	1.1	0.3		MIT
6128.168	6129.864	16313.57	MO	1.1	0.3		MIT
6123.69	6125.38	16325.5	MO	1.1S			KS
6123.528	6125.223	16325.94	MO	1.1L			MIT
6119.027	6120.721	16337.94	MO	0.9	0.3		MIT
6110.676	6112.367	16360.27	MO	0.6H			MIT
6110.241	6111.932	16361.44	MO	1.0	0.3		MIT
6101.868	6103.557	16383.89	MO	1.6	0.6		MIT
6096.194	6097.882	16399.14	MO	0.6	0.5		MIT
6086.627	6088.312	16424.91	MO	0.7	0.5		MIT
6081.275	6082.959	16439.37	MO	0.9	0.3		MIT
6079.579	6081.262	16443.96	MO	1.4	0.6		MIT
6075.568	6077.250	16454.81	MO	0.5	0.3		MIT
6063.827	6065.506	16486.67	MO	0.8	1.0		MIT
6058.014	6059.691	16502.49	MO	1.1	0.3		MIT
6054.806	6056.482	16511.23	MO	1.3	0.3		MIT
6050.895	6052.570	16521.91	MO	1.0	0.3		MIT
6047.829	6049.504	16530.28	MO	1.2	0.3		WN
6039.771	6041.443	16552.34	MO	1.1			MIT
6030.662	6032.332	16577.34	MO	2.5	2.1		MIT
6027.268	6028.937	16586.67	MO	1.3	0.3		MIT
6025.486	6027.155	16591.58	MO	1.4	0.3		MIT
6013.034	6014.699	16625.94	MO	0.9	0.3		MIT
6011.343	6013.008	16630.61	MO	1.1	0.3		MIT
5991.348	5993.007	16686.11	MO	1.0			MIT
5990.006	5991.665	16689.85	MO	1.0			MIT

AIR WAVELENGTH (ANGSTRMS)	VACUUM WAVELENGTH (ANGSTRMS)	WAVENUMBER (KAYSERS)	LMENT	LOG ARC	INTENSITY SPK	DIS	REF
5989.474	5991.133	16691.33	MO	1.1			MIT
5988.173	5989.832	16694.96	MO	1.5			MIT
5982.927	5984.584	16709.60	MO	1.3			MIT
5974.257	5975.912	16733.85	MO	1.3			MIT
5968.481	5970.134	16750.04	MO	0.6			MIT
5965.574	5967.227	16758.20	MO	1.2			MIT
5949.175	5950.823	16804.40	MO	0.6			MIT
5937.911	5939.556	16836.27	MO	1.3			MIT
5928.882	5930.525	16861.91	MO	2.0H			MIT
5926.359	5928.001	16869.09	MO	1.8H	1.3		MIT
5923.792	5925.433	16876.40	MO	0.9H			MIT
5914.293	5915.932	16903.51	MO	0.6			MIT
5912.122	5913.760	16909.72	MO	0.8			MIT
5901.472	5903.107	16940.23	MO	1.5			MIT
5899.678	5901.313	16945.38	MO	1.1			MIT
5898.825	5900.460	16947.83	MO	0.9			MIT
5898.785	5900.420	16947.95	MO	0.9			MIT
5897.865	5899.499	16950.59	MO	0.7			MIT
5893.738	5895.371	16962.46	MO	0.7			MIT
5893.376	5895.009	16963.50	MO	1.8H	1.2		MIT
5892.294	5893.927	16966.62	MO	1.3			MIT
5891.562	5893.195	16968.73	MO	1.4H			MIT
5889.978	5891.610	16973.29	MO	1.7H			MIT
5888.326	5889.958	16978.05	MO	2.2	2.0		MIT
5884.332	5885.963	16989.57	MO	1.1			MIT
5882.724	5884.354	16994.22	MO	1.2			MIT
5881.526	5883.156	16997.68	MO	1.3			MIT
5876.587	5878.216	17011.97	MO	1.4			MIT
5874.225	5875.853	17018.80	MO	0.7			MIT
5869.780	5871.407	17031.69	MO	0.7			MIT
5869.329	5870.956	17033.00	MO	1.7			MIT
5868.763	5870.390	17034.64	MO	1.3			MIT
5867.459	5869.085	17038.43	MO	0.6			MIT
5861.379	5863.004	17056.10	MO	1.3			MIT
5858.269	5859.893	17065.16	MO	2.3	2.3		MIT
5851.516	5853.138	17084.85	MO	1.6H	1.3H		MIT
5849.728	5851.350	17090.07	MO	1.8H	1.0H		MIT
5848.860	5850.481	17092.61	MO	1.7H			MIT
5839.988	5841.607	17118.58	MO	1.2			MIT
5835.588	5837.206	17131.48	MO	1.3			MIT
5826.894	5828.509	17157.04	MO	0.5			MIT
5825.205	5826.820	17162.02	MO	1.3			MIT
5825.022	5826.637	17162.56	MO	1.2			MIT
5820.692	5822.306	17175.33	MO	0.9			MIT
5815.735	5817.347	17189.97	MO	1.1			MIT
5815.518	5817.130	17190.61	MO	1.3			MIT
5813.861	5815.473	17195.51	MO	0.8			MIT
5812.509	5814.121	17199.51	MO	0.6			MIT
5809.030	5810.641	17209.81	MO	0.5			MIT
5806.689	5808.299	17216.74	MO	1.3			MIT
5806.188	5807.798	17218.23	MO	1.2			MIT
5803.980	5805.589	17224.78	MO	1.0			MIT
5802.668	5804.277	17228.68	MO	1.5			MIT
5800.459	5802.067	17235.24	MO	1.4			MIT
5795.773	5797.380	17249.17	MO	1.3			MIT
5791.847	5793.453	17260.86	MO	2.0	1.8		MIT
5785.692	5787.296	17279.23	MO	1.0			MIT
5784.005	5785.609	17284.26	MO	0.7			MIT
5783.33	5784.93	17286.3	MO	1.3H			KS
5780.638	5782.241	17294.33	MO	1.0H			MIT
5780.111	5781.714	17295.91	MO	1.1			MIT
5779.362	5780.965	17298.15	MO	1.3			MIT
5778.190	5779.792	17301.66	MO	1.1			MIT
5774.546	5776.147	17312.58	MO	1.3			MIT
5771.053	5772.653	17323.06	MO	1.2			MIT
5769.748	5771.348	17326.97	MO	1.1			MIT
5766.59	5768.19	17336.5	MO	0.7			KS
5765.283	5766.882	17340.39	MO	0.9			MIT
5757.486	5759.083	17363.88	MO	1.0			MIT
5751.402	5752.997	17382.24	MO	2.1	2.0		MIT

AIR WAVELENGTH (ANGSTRMS)	VACUUM WAVELENGTH (ANGSTRMS)	WAVENUMBER (KAYSERS)	LMENT	LOG ARC	INTENSITY SPK	DIS	REF
5747.670	5749.264	17393.53	MO	1.2			MIT
5741.708	5743.301	17411.59	MO	1.0			MIT
5739.656	5741.248	17417.82	MO	1.0			MIT
5734.057	5735.648	17434.82	MO	1.3			MIT
5729.589	5731.178	17448.42	MO	1.0			MIT
5729.450	5731.039	17448.84	MO	1.0			MIT
5728.769	5730.358	17450.92	MO	1.2			MIT
5727.052	5728.641	17456.15	MO	0.5H			MIT
5723.112	5724.700	17468.17	MO	1.0			MIT
5722.735	5724.323	17469.32	MO	1.9	1.8		MIT
5711.798	5713.383	17502.77	MO	1.3			MIT
5707.992	5709.576	17514.44	MO	0.9			MIT
5705.720	5707.303	17521.41	MO	1.6	1.6		MIT
5702.106	5703.688	17532.52	MO	1.2			MIT
5699.285	5700.866	17541.19	MO	1.3			MIT
5698.274	5699.855	17544.31	MO	1.0			MIT
5696.027	5697.607	17551.23	MO	1.1			MIT
5694.390	5695.970	17556.27	MO	1.0			MIT
5689.519	5691.098	17571.30	MO	1.1			MIT
5689.136	5690.715	17572.49	MO	1.9L	1.6		MIT
5687.635	5689.213	17577.12	MO	0.7			MIT
5682.894	5684.471	17591.79	MO	1.3			MIT
5677.893	5679.469	17607.28	MO	1.4			MIT
5674.466	5676.041	17617.92	MO	1.5			MIT
5673.633	5675.207	17620.50	MO	1.3			MIT
5672.069	5673.643	17625.36	MO	1.0			MIT
5667.296	5668.869	17640.20	MO	1.2O			MIT
5664.382	5665.954	17649.28	MO	0.8			MIT
5664.337	5665.909	17649.42	MO	0.8			MIT
5655.419	5656.989	17677.25	MO	0.7	1.3H		MIT
5652.166	5653.735	17687.42	MO	1.0	0.5		MIT
5651.866	5653.435	17688.36	MO	1.0	0.8		MIT
5651.281	5652.849	17690.19	MO	1.0	0.7		MIT
5650.130	5651.698	17693.80	MO	2.0	1.7		MIT
5634.858	5636.422	17741.75	MO	1.5	1.3		MIT
5632.471	5634.034	17749.27	MO	2.0	1.7		MIT
5627.224	5628.786	17765.82	MO	0.6	0.3		MIT
5619.381	5620.941	17790.62	MO	1.2	1.1		MIT
5618.771	5620.331	17792.55	MO	1.0	0.7		MIT
5618.447	5620.007	17793.57	MO	1.3	1.0		MIT
5613.070	5614.628	17810.62	MO	1.3	1.1		MIT
5610.933	5612.491	17817.40	MO	1.5	1.3		MIT
5609.541	5611.098	17821.82	MO	0.7	0.3		MIT
5609.233	5610.790	17822.80	MO	1.1	0.9		MIT
5608.624	5610.181	17824.74	MO	1.3	1.1		MIT
5602.756	5604.311	17843.41	MO	1.3	0.7		MIT
5601.046	5602.601	17848.85	MO	1.0	1.0		MIT
5596.322	5597.876	17863.92	MO	1.0			MIT
5591.577	5593.129	17879.08	MO	1.3	1.1		MIT
5575.186	5576.734	17931.64	MO	1.3	1.2		MIT
5570.450	5571.997	17946.89	MO	2.3	2.0		MIT
5569.476	5571.023	17950.03	MO	1.2	1.0		MIT
5568.620	5570.166	17952.78	MO	1.5	1.2		MIT
5566.663	5568.209	17959.10	MO	0.5	0.5		MIT
5564.046	5565.591	17967.54	MO	1.3	1.0		MIT
5563.367	5564.912	17969.74	MO	1.0	0.3		MIT
5562.49	5564.03	17972.6	MO	0.5	0.5		MIT
5556.723	5558.266	17991.22	MO	1.3	1.1		MIT
5556.282	5557.825	17992.65	MO	1.5	1.2		MIT
5552.188	5553.730	18005.92	MO	1.1	0.5		MIT
5544.489	5546.029	18030.92	MO	1.3	1.1		MIT
5543.121	5544.661	18035.37	MO	1.3	1.2		MIT
5541.647	5543.186	18040.17	MO	1.1	0.7		MIT
5539.410	5540.949	18047.45	MO	1.4	1.1		MIT
5534.542	5536.079	18063.32	MO	0.7	0.6		MIT
5533.046	5534.583	18068.21	MO	2.3	2.0		MIT
5531.735	5533.271	18072.49	MO	0.7H	0.3H		MIT
5529.286	5530.822	18080.50	MO	0.5H	0.3H		MIT
5526.968	5528.503	18088.08	MO	1.4	1.2		MIT
5526.525	5528.060	18089.53	MO	1.4	1.3		MIT

AIR WAVELENGTH (ANGSTRMS)	VACUUM WAVELENGTH (ANGSTRMS)	WAVENUMBER (KAYSERS)	LMENT	LOG ARC	INTENSITY SPK	DIS	REF
5521.170	5522.704	18107.07	MO	0.5	0.3		MIT
5520.645	5522.179	18108.79	MO	1.3	1.1		MIT
5520.038	5521.571	18110.79	MO	1.3	1.1		MIT
5517.429	5518.962	18119.35	MO	0.9	0.0		MIT
5513.808	5515.340	18131.25	MO	0.5	0.0		MIT
5511.490	5513.021	18138.87	MO	1.2	0.6		MIT
5508.243	5509.773	18149.57	MO	1.3			MIT
5506.491	5508.021	18155.34	MO	2.3R	2.0		MIT
5503.537	5505.066	18165.09	MO	0.7H	0.7H		MIT
5501.874	5503.403	18170.58	MO	1.3	1.0		MIT
5501.544	5503.072	18171.67	MO	1.3	1.2		MIT
5499.491	5501.019	18178.45	MO	0.7H	0.5H		MIT
5498.494	5500.022	18181.75	MO	1.3	1.0		MIT
5496.942	5498.469	18186.88	MO	0.7H	0.7H		MIT
5493.804	5495.330	18197.27	MO	1.3H	1.1H		MIT
5492.166	5493.692	18202.70	MO	1.2	0.9		MIT
5490.279	5491.804	18208.95	MO	1.3	1.0		MIT
5489.398	5490.923	18211.87	MO	0.5	0.0		MIT
5488.671	5490.196	18214.29	MO	0.5	0.0		MIT
5475.897	5477.419	18256.77	MO	1.3	1.1		MIT
5473.370	5474.891	18265.20	MO	1.7	1.4		MIT
5465.570	5467.089	18291.27	MO	1.3	0.7		MIT
5460.529	5462.046	18308.16	MO	1.3	1.2H		MIT
5456.459	5457.975	18321.81	MO	1.3	1.1		MIT
5453.026	5454.541	18333.35	MO	1.2	0.8		MIT
5450.513	5452.028	18341.80	MO	1.5	1.3		MIT
5448.548	5450.062	18348.41	MO	0.5	0.0		MIT
5439.711	5441.223	18378.22	MO	1.1	0.8		MIT
5437.750	5439.261	18384.85	MO	1.5	1.2		MIT
5435.684	5437.195	18391.84	MO	1.5	1.1		MIT
5431.018	5432.528	18407.64	MO	1.2	0.7		MIT
5427.547	5429.056	18419.41	MO	0.9	0.5		MIT
5426.887	5428.395	18421.65	MO	1.3	0.7		MIT
5425.996	5427.504	18424.68	MO	0.5	0.3		MIT
5417.383	5418.889	18453.97	MO	1.3	1.0		MIT
5414.672	5416.177	18463.21	MO	0.8	0.3		MIT
5406.386	5407.889	18491.50	MO	1.2	0.9		MIT
5405.79	5407.29	18493.5	MO	0.7	0.5		MIT
5400.471	5401.972	18511.76	MO	1.3	1.2		MIT
5398.731	5400.232	18517.72	MO	0.7	0.0		MIT
5397.378	5398.879	18522.36	MO	1.3	1.0		MIT
5394.517	5396.017	18532.19	MO	1.3	1.2		MIT
5388.688	5390.186	18552.23	MO	1.1H	0.7H		MIT
5372.404	5373.898	18608.47	MO	1.3	1.0		MIT
5367.111	5368.604	18626.82	MO	1.0H			MIT
5367.008	5368.501	18627.18	MO	1.1	0.7		MIT
5364.284	5365.776	18636.63	MO	1.8H	1.4H		MIT
5360.898	5362.389	18648.41	MO	1.0	0.7		MIT
5360.557	5362.048	18649.59	MO	2.0H	1.8H		MIT
5356.477	5357.967	18663.80	MO	1.4	1.1		MIT
5355.513	5357.002	18667.16	MO	1.1H	0.5H		MIT
5354.879	5356.368	18669.37	MO	1.4	1.2		MIT
5352.347	5353.836	18678.20	MO	1.0	0.6		MIT
5349.786	5351.274	18687.14	MO	0.8	0.6		MIT
5348.811	5350.299	18690.54	MO	0.6H	0.0H		MIT
5339.083	5340.568	18724.60	MO	0.6	0.0		MIT
5337.201	5338.686	18731.20	MO	1.0	0.5		MIT
5334.787	5336.271	18739.68	MO	0.6	0.0		MIT
5327.063	5328.545	18766.85	MO	1.3	1.1		MIT
5325.863	5327.345	18771.08	MO	1.0	0.3		MIT
5324.467	5325.948	18776.00	MO	1.2	1.0		MIT
5319.886	5321.366	18792.17	MO	1.1	0.9		MIT
5317.948	5319.427	18799.02	MO	1.1	0.5		MIT
5317.130	5318.609	18801.91	MO	0.7	0.0		MIT
5315.820	5317.299	18806.54	MO	1.0	0.5		MIT
5315.044	5316.523	18809.29	MO	1.3	1.1		MIT
5313.893	5315.371	18813.36	MO	1.4	1.2		MIT
5312.756	5314.234	18817.39	MO	1.0	0.3		MIT
5306.260	5307.736	18840.42	MO	1.2	0.9		MIT
5298.060	5299.534	18869.58	MO	1.2	0.5		MIT

AIR WAVELENGTH (ANGSTRMS)	VACUUM WAVELENGTH (ANGSTRMS)	WAVENUMBER (KAYSERS)	LMENT	LOG ARC	INTENSITY SPK	INTENSITY DIS	REF
5295.466	5296.939	18878.83	MO	1.3	1.1		MIT
5293.458	5294.931	18885.99	MO	1.3	1.0		MIT
5292.084	5293.557	18890.89	MO	1.3	1.2		MIT
5284.103	5285.573	18919.42	MO	1.1	0.6		MIT
5283.843	5285.313	18920.35	MO	1.1	0.8		MIT
5280.864	5282.334	18931.03	MO	1.7	1.4		MIT
5279.841	5281.310	18934.70	MO	1.1	0.3		MIT
5279.650	5281.119	18935.38	MO	1.3	1.1		MIT
5279.358	5280.827	18936.43	MO	1.0	0.0		MIT
5277.358	5278.827	18943.60	MO	1.0	0.3		MIT
5276.275	5277.743	18947.49	MO	1.3	1.0		MIT
5275.220	5276.688	18951.28	MO	0.6	0.0		MIT
5271.797	5273.264	18963.59	MO	1.3	1.0		MIT
5268.946	5270.412	18973.85	MO	1.1	0.7		MIT
5261.569	5263.033	19000.45	MO	0.5H	0.3H		MIT
5261.140	5262.604	19002.00	MO	1.3	1.2		MIT
5260.168	5261.632	19005.51	MO	1.0H	0.5H		MIT
5259.040	5260.504	19009.59	MO	1.8	1.3		MIT
5251.142	5252.604	19038.18	MO	0.7	0.0		MIT
5250.339	5251.800	19041.09	MO	1.0	0.3		MIT
5245.510	5246.970	19058.62	MO	1.4	0.3D		MIT
5242.812	5244.271	19068.43	MO	1.7H	1.3H		MIT
5240.880	5242.339	19075.46	MO	1.9H	1.6H		MIT
5240.310	5241.769	19077.53	MO	0.8	0.0		MIT
5238.197	5239.655	19085.23	MO	1.9H	1.5H		MIT
5237.967	5239.425	19086.06	MO	1.2	1.0		MIT
5234.265	5235.722	19099.56	MO	1.4	1.2		MIT
5232.362	5233.819	19106.51	MO	1.3	1.0		MIT
5231.065	5232.521	19111.25	MO	1.3	1.2		MIT
5230.170	5231.626	19114.52	MO	0.8	0.3		MIT
5219.405	5220.858	19153.94	MO	1.4	1.3		MIT
5211.864	5213.315	19181.65	MO	1.3	1.1		MIT
5210.439	5211.890	19186.90	MO	1.2	0.6		MIT
5203.943	5205.392	19210.85	MO	1.1	0.7		MIT
5200.744	5202.192	19222.67	MO	1.3	1.0		MIT
5200.168	5201.616	19224.79	MO	1.6	1.3		MIT
5198.563	5200.011	19230.73	MO	0.7	0.5		MIT
5196.872	5198.319	19236.99	MO	0.7	0.3		MIT
5181.591	5183.034	19293.72	MO		1.3		WN
5180.216	5181.659	19298.84	MO	1.1	0.6		MIT
5179.975	5181.418	19299.74	MO	1.0	0.5		MIT
5179.405	5180.848	19301.86	MO	0.7	0.0		MIT
5177.09	5178.53	19310.5	MO	0.7	0.0		MIT
5174.178	5175.619	19321.36	MO	1.8H	1.4H		MIT
5172.941	5174.382	19325.98	MO	1.8H	1.4H		MIT
5171.246	5172.686	19332.31	MO	1.1	0.6		MIT
5171.076	5172.516	19332.95	MO	1.5H	0.8H		MIT
5170.694	5172.134	19334.38	MO	0.7	0.0		MIT
5167.757	5169.196	19345.37	MO	1.4	1.3		MIT
5163.192	5164.630	19362.47	MO	1.4	1.3		MIT
5156.205	5157.641	19388.71	MO	1.0	0.0		MIT
5155.263	5156.699	19392.25	MO	1.0	0.0		MIT
5149.652	5151.087	19413.38	MO	0.6	0.0		MIT
5148.438	5149.872	19417.96	MO	0.9	0.6		MIT
5147.388	5148.822	19421.92	MO	1.4	1.3		MIT
5146.703	5148.137	19424.50	MO	0.7	0.3		MIT
5145.385	5146.818	19429.48	MO	1.4	1.3		MIT
5142.247	5143.680	19441.33	MO	1.0	0.3		MIT
5141.262	5142.694	19445.06	MO	1.1	0.7		MIT
5134.967	5136.398	19468.90	MO	1.0	0.3		MIT
5127.111	5128.540	19498.73	MO	0.6	0.3		MIT
5126.725	5128.153	19500.20	MO	1.0	0.7		MIT
5123.828	5125.256	19511.22	MO	1.1	0.7		MIT
5121.774	5123.201	19519.05	MO	1.1	0.6		MIT
5116.967	5118.393	19537.38	MO	1.1	0.7		MIT
5115.644	5117.070	19542.44	MO	0.9	0.3		MIT
5114.974	5116.399	19545.00	MO	1.3	1.1		MIT
5109.714	5111.138	19565.11	MO	1.4	1.3		MIT
5109.258	5110.682	19566.86	MO	1.0	0.0		MIT
5104.701	5106.124	19584.33	MO	0.5	0.0		MIT
5100.335	5101.756	19601.09	MO	0.8	0.3		MIT
5098.034	5099.455	19609.94	MO	1.3	1.0		MIT
5097.522	5098.943	19611.91	MO	1.6	1.3		MIT
5096.647	5098.067	19615.28	MO	1.5	1.3		MIT
5095.888	5097.308	19618.20	MO	1.3	1.0		MIT
5092.749	5094.168	19630.29	MO	1.1W	0.5		MIT
5092.26	5093.68	19632.2	MO	0.3			MIT
5092.164	5093.583	19632.54	MO	0.8	0.8		MIT
5091.342	5092.761	19635.71	MO	1.1	0.5		MIT
5090.966	5092.385	19637.16	MO	1.3	0.8		MIT
5090.615	5092.034	19638.52	MO	1.0	0.5		MIT
5084.23	5085.65	19663.2	MO	1.0	0.5		MIT
5081.260	5082.676	19674.67	MO	1.3	0.7		MIT
5080.018	5081.434	19679.48	MO	1.3	1.1		MIT
5079.867	5081.283	19680.07	MO	1.0	0.5		MIT
5077.042	5078.457	19691.02	MO	0.6	0.0		MIT
5064.637	5066.049	19739.25	MO	1.3	0.7		MIT
5062.521	5063.932	19747.50	MO	1.3	0.8		MIT
5060.349	5061.760	19755.98	MO	1.0	0.5		MIT
5059.877	5061.288	19757.82	MO	1.6	1.3		MIT
5058.073	5059.483	19764.86	MO	1.1	0.8		MIT
5057.289	5058.699	19767.93	MO	1.1	0.3		MIT
5055.005	5056.414	19776.86	MO	1.3	1.0		MIT
5054.045	5055.454	19780.62	MO	0.9	0.3		MIT
5053.59	5055.00	19782.4	MO	0.8	0.3		MIT
5047.706	5049.113	19805.46	MO	1.4	1.3		MIT
5046.525	5047.932	19810.09	MO	1.3	0.8		MIT
5044.382	5045.789	19818.51	MO	0.5H	0.0H		MIT
5038.907	5040.312	19840.04	MO	1.0	0.6		MIT
5037.178	5038.583	19846.85	MO	0.7	0.3		MIT
5034.175	5035.579	19858.69	MO	0.6	0.0		MIT
5032.380	5033.783	19865.77	MO	0.5	0.0		MIT
5030.775	5032.178	19872.11	MO	1.3	1.0		MIT
5029.002	5030.404	19879.12	MO	1.3	1.1		MIT
5026.413	5027.815	19889.36	MO	0.7	0.0		MIT
5025.397	5026.799	19893.38	MO	0.7	0.0		MIT
5023.601	5025.002	19900.49	MO	0.7	0.0		MIT
5021.229	5022.629	19909.89	MO	0.6	0.0		MIT
5019.850	5021.250	19915.36	MO	1.1	1.0		MIT
5017.524	5018.923	19924.59	MO	0.6H	0.0H		MIT
5016.779	5018.178	19927.55	MO	1.3	0.9		MIT
5014.595	5015.994	19936.23	MO	1.5	1.3		MIT
5012.331	5013.729	19945.23	MO	1.0	0.5		MIT
5010.814	5012.212	19951.27	MO	1.1	0.6		MIT
5009.032	5010.429	19958.37	MO	0.6	0.0		MIT
5007.616	5009.013	19964.01	MO	1.0	0.3		MIT
4999.912	5001.307	19994.77	MO	1.7	1.4		MIT
4995.318	4996.712	20013.16	MO	1.3	0.7		MIT
4992.236	4993.629	20025.52	MO	0.9	0.5		MIT
4987.989	4989.381	20042.57	MO	1.2	0.3		MIT
4985.562	4986.953	20052.32	MO	1.3	0.8		MIT
4983.461	4984.851	20060.78	MO	0.9	0.3		MIT
4981.827	4983.217	20067.36	MO	1.0	0.5		MIT
4980.000	4981.389	20074.72	MO	0.5	0.3		MIT
4979.115	4980.504	20078.29	MO	2.0	1.5		MIT
4978.404	4979.793	20081.16	MO	0.9	0.0		MIT
4977.694	4979.083	20084.02	MO	1.7	0.5		MIT
4975.983	4977.371	20090.93	MO	1.3	0.5		MIT
4973.360	4974.748	20101.52	MO	1.4	0.7		MIT
4970.768	4972.155	20112.00	MO	0.6	0.0		MIT
4970.367	4971.754	20113.63	MO	0.5	0.3		MIT
4969.833	4971.220	20115.79	MO	0.5	0.0		MIT
4969.639	4971.026	20116.57	MO	1.0	0.6		MIT
4964.414	4965.799	20137.75	MO	1.4	0.9		MIT
4964.192	4965.577	20138.65	MO	1.2	0.8		MIT
4962.910	4964.295	20143.85	MO	0.7H	0.3H		MIT
4959.619	4961.003	20157.21	MO	0.6	0.0		MIT
4957.539	4958.922	20165.67	MO	1.8	1.4		MIT
4956.578	4957.961	20169.58	MO	1.2	0.7		MIT
4953.790	4955.173	20180.93	MO	0.8	0.3		MIT

AIR WAVELENGTH (ANGSTRMS)	VACUUM WAVELENGTH (ANGSTRMS)	WAVENUMBER (KAYSERS)	LMENT	LOG ARC	INTENSITY SPK	INTENSITY DIS	REF
4951.959	4953.341	20188.39	MO	1.1	0.6		MIT
4950.618	4952.000	20193.86	MO	1.9	1.5		MIT
4949.019	4950.400	20200.39	MO	1.2	0.5		MIT
4945.275	4946.655	20215.68	MO	0.8	0.0		MIT
4943.874	4945.254	20221.41	MO	0.8	0.5		MIT
4942.965	4944.345	20225.13	MO	0.6	0.0		MIT
4942.367	4943.746	20227.57	MO	0.7	0.5		MIT
4941.656	4943.035	20230.49	MO	1.6	1.4		MIT
4939.665	4941.044	20238.64	MO	1.2	0.7		MIT
4937.433	4938.811	20247.79	MO	0.5	0.3		MIT
4933.732	4935.109	20262.98	MO	1.1	0.6		MIT
4933.461	4934.838	20264.09	MO	0.6	0.3		MIT
4933.333	4934.710	20264.61	MO	1.2	0.5		MIT
4933.102	4934.479	20265.56	MO	1.5	1.2		MIT
4931.149	4932.525	20273.59	MO	1.3	0.6		MIT
4927.047	4928.422	20290.47	MO	0.5	0.0		MIT
4926.430	4927.805	20293.01	MO	1.4	1.1		MIT
4926.193	4927.568	20293.99	MO	1.4	1.1		MIT
4924.783	4926.158	20299.80	MO	1.3	0.9		MIT
4918.155	4919.528	20327.15	MO	0.6			MIT
4916.190	4917.563	20335.28	MO	0.6	0.3		MIT
4909.186	4910.557	20364.29	MO	1.5	1.2		MIT
4908.224	4909.594	20368.28	MO	0.7H			MIT
4907.426	4908.796	20371.59	MO	1.5	1.3		MIT
4906.911	4908.281	20373.73	MO	0.6			MIT
4904.387	4905.756	20384.22	MO	1.0	0.6		MIT
4903.811	4905.180	20386.61	MO	1.9	1.5		MIT
4900.456	4901.824	20400.57	MO		0.7		MIT
4899.580	4900.948	20404.22	MO	1.4	1.4		MIT
4889.216	4890.581	20447.47	MO	1.4	0.7		MIT
4887.158	4888.523	20456.08	MO	1.3	1.0		MIT
4886.473	4887.838	20458.95	MO	1.4	1.2		MIT
4885.640	4887.004	20462.43	MO	0.7	0.5		MIT
4885.324	4886.688	20463.76	MO	0.7	0.6		MIT
4884.949	4886.313	20465.33	MO	0.7	0.7		MIT
4884.332	4885.696	20467.91	MO	1.0	0.6		MIT
4881.867	4883.230	20478.25	MO	0.7	0.5		MIT
4880.791	4882.154	20482.76	MO	1.0	1.0		MIT
4878.367	4879.729	20492.94	MO	1.4	1.0		MIT
4875.114	4876.476	20506.61	MO	0.6	0.6		MIT
4869.198	4870.558	20531.53	MO	1.4	1.3		MIT
4868.892	4870.252	20532.82	MO	0.5	0.3		MIT
4868.713	4870.073	20533.57	MO	0.5	0.3		MIT
4868.004	4869.364	20536.56	MO	1.7	1.6		MIT
4865.823	4867.182	20545.77	MO	0.9	0.9		MIT
4864.743	4866.102	20550.33	MO	0.5	0.5		MIT
4860.755	4862.113	20567.19	MO	1.0	0.9		MIT
4860.556	4861.914	20568.03	MO	0.7	0.6		MIT
4860.050	4861.408	20570.17	MO	1.4	1.3		MIT
4858.222	4859.579	20577.91	MO	1.3	1.3		MIT
4854.469	4855.825	20593.82	MO	0.5	0.5		MIT
4851.700	4853.055	20605.58	MO	1.2	1.2		MIT
4849.831	4851.186	20613.52	MO	1.1	1.0		MIT
4848.175	4849.529	20620.56	MO	0.6	0.5		MIT
4847.234	4848.588	20624.56	MO	0.5	0.5		MIT
4845.172	4846.526	20633.34	MO	1.3	1.3		MIT
4844.949	4846.303	20634.29	MO	0.6	0.6		MIT
4839.587	4840.939	20657.15	MO	1.3	1.3		MIT
4838.113	4839.465	20663.44	MO	1.2	1.2		MIT
4835.763	4837.114	20673.48	MO	1.1	1.2		MIT
4833.965	4835.316	20681.17	MO	1.4	1.4		MIT
4832.921	4834.271	20685.64	MO	1.2	1.0		MIT
4832.808	4834.158	20686.12	MO	1.2	1.0		MIT
4830.513	4831.863	20695.95	MO	2.1	2.0		MIT
4829.936	4831.286	20698.42	MO	1.3	1.0		MIT
4828.467	4829.816	20704.72	MO	1.4	1.4		MIT
4822.931	4824.279	20728.49	MO	0.8	0.7		MIT
4822.424	4823.772	20730.67	MO	1.2	1.1		MIT
4819.249	4820.596	20744.32	MO	1.9	1.8		MIT
4817.699	4819.045	20751.00	MO	1.4	1.4		MIT
4816.964	4818.310	20754.16	MO	0.5	0.5		MIT
4814.468	4815.814	20764.92	MO	1.0	0.9		MIT
4814.099	4815.444	20766.51	MO	0.3	0.7H		MIT
4813.161	4814.506	20770.56	MO	0.7	0.7		MIT
4812.484	4813.829	20773.48	MO	0.5	0.5		MIT
4811.062	4812.407	20779.62	MO	1.7	1.7		MIT
4808.459	4809.803	20790.87	MO	1.0	0.9		MIT
4808.087	4809.431	20792.48	MO	1.4	1.4		MIT
4807.712	4809.056	20794.10	MO	0.6	0.6		MIT
4806.362	4807.705	20799.94	MO	0.5	1.3H		MIT
4805.579	4806.922	20803.33	MO	1.5	1.5		MIT
4804.912	4806.255	20806.22	MO	1.0	1.0		MIT
4801.012	4802.354	20823.12	MO	0.5	0.6		MIT
4796.522	4797.863	20842.61	MO	1.6	1.6		MIT
4795.371	4796.711	20847.62	MO	0.6	0.5		MIT
4794.604	4795.944	20850.95	MO	1.1	1.0		MIT
4793.821	4795.161	20854.36	MO	1.2	1.2		MIT
4793.411	4794.751	20856.14	MO	1.5	1.5		MIT
4792.742	4794.082	20859.05	MO	1.6	1.5		MIT
4791.826	4793.166	20863.04	MO	0.7	0.7		MIT
4790.977	4792.316	20866.74	MO	1.0G			PU
4789.343	4790.682	20873.85	MO	0.9	0.9		MIT
4788.181	4789.520	20878.92	MO	1.2	1.2		MIT
4787.626	4788.964	20881.34	MO	1.0	0.9		MIT
4786.457	4787.795	20886.44	MO	1.4	1.4		MIT
4785.122	4786.460	20892.27	MO	1.5	1.5		MIT
4784.414	4785.752	20895.36	MO	0.7	0.7		MIT
4782.941	4784.278	20901.79	MO	1.6	1.6		MIT
4777.881	4779.217	20923.93	MO	0.7	0.7		MIT
4776.343	4777.678	20930.67	MO	1.6	1.6		MIT
4775.665	4777.000	20933.64	MO	1.4	1.4		MIT
4774.222	4775.557	20939.97	MO	1.3	1.3		MIT
4773.437	4774.772	20943.41	MO	1.3	1.3		MIT
4773.286	4774.621	20944.07	MO	1.0	1.0		MIT
4770.872	4772.206	20954.67	MO	0.9	0.9		MIT
4764.419	4765.751	20983.05	MO	1.7	1.7		MIT
4761.278	4762.609	20996.89	MO	0.5	0.3		MIT
4761.087	4762.418	20997.74	MO	0.6	0.5		MIT
4760.186	4761.517	21001.71	MO	2.1	2.1		MIT
4759.660	4760.991	21004.03	MO	0.7	0.7		MIT
4758.502	4759.833	21009.14	MO	1.6	1.6		MIT
4757.272	4758.602	21014.57	MO	0.5	0.5		MIT
4757.150	4758.480	21015.11	MO	0.6	0.6		MIT
4755.886	4757.216	21020.70	MO	0.9	1.0		MIT
4755.327	4756.657	21023.17	MO	0.5	0.6		MIT
4754.934	4756.264	21024.91	MO	0.3	0.6		MIT
4753.345	4754.674	21031.94	MO	0.8	0.7		MIT
4751.111	4752.440	21041.82	MO	0.8	0.6		MIT
4750.393	4751.722	21045.00	MO	1.5	1.5		MIT
4748.865	4750.193	21051.78	MO	0.6	0.7		MIT
4743.614	4744.941	21075.08	MO	0.5	0.5		MIT
4743.085	4744.412	21077.43	MO	1.0	1.0		MIT
4742.589	4743.915	21079.63	MO		1.0		MIT
4740.359	4741.685	21089.55	MO	0.7	0.7		MIT
4736.637	4737.962	21106.12	MO	1.0	1.0		MIT
4735.297	4736.622	21112.09	MO	0.8	0.9		MIT
4734.120	4735.444	21117.34	MO	0.8	0.9		MIT
4733.395	4734.719	21120.58	MO	0.5	0.7		MIT
4731.443	4732.767	21129.29	MO	2.0	2.0		MIT
4729.141	4730.464	21139.58	MO	1.5	1.5		MIT
4728.235	4729.558	21143.63	MO	0.3	0.5		MIT
4725.339	4726.661	21156.58	MO	0.9	0.9		MIT
4723.449	4724.770	21165.05	MO	0.6	0.6		MIT
4723.315	4724.636	21165.65	MO	0.6	0.6		MIT
4723.063	4724.384	21166.78	MO	1.0	1.3		MIT
4718.879	4720.199	21185.55	MO	1.4H	1.4H		MIT
4717.924	4719.244	21189.83	MO	1.7	1.7		MIT
4716.674	4717.994	21195.45	MO	0.7	0.9		MIT
4714.510	4715.829	21205.18	MO	1.2	1.2		MIT
4709.970	4711.288	21225.62	MO	0.6	0.7		MIT

AIR WAVELENGTH (ANGSTRMS)	VACUUM WAVELENGTH (ANGSTRMS)	WAVENUMBER (KAYSERS)	LMENT	LOG ARC	INTENSITY SPK	DIS	REF
4708.225	4709.542	21233.49	MO	1.5	1.5		MIT
4707.255	4708.572	21237.86	MO	2.1	2.1		MIT
4706.200	4707.517	21242.62	MO	0.6	0.6		MIT
4706.057	4707.374	21243.27	MO	1.4	1.4		MIT
4700.490	4701.805	21268.43	MO	1.4	1.4		MIT
4698.784	4700.099	21276.15	MO	0.3	0.6		MIT
4696.508	4697.822	21286.46	MO	0.9	1.0		MIT
4695.862	4697.176	21289.39	MO	0.6	0.8		MIT
4693.931	4695.245	21298.15	MO	1.3	1.4		MIT
4693.342	4694.655	21300.82	MO	0.3	0.5		MIT
4692.694	4694.007	21303.76	MO	0.3	0.6		MIT
4692.008	4693.321	21306.87	MO	0.7	0.8		MIT
4690.859	4692.172	21312.09	MO	1.3	1.4		MIT
4688.220	4689.532	21324.09	MO	1.4	1.5		MIT
4686.095	4687.407	21333.76	MO	0.9	0.8		MIT
4685.813	4687.124	21335.04	MO	1.1	1.1		MIT
4684.336	4685.647	21341.77	MO	0.7	0.6		MIT
4683.827	4685.138	21344.09	MO	0.8	0.9		MIT
4683.719	4685.030	21344.58	MO	0.7	0.7		MIT
4682.244	4683.554	21351.30	MO	0.6	0.7		MIT
4681.933	4683.243	21352.72	MO	0.5	0.5		MIT
4681.626	4682.936	21354.12	MO	0.6	0.7		MIT
4681.046	4682.356	21356.77	MO	0.8	0.9		MIT
4678.204	4679.513	21369.74	MO	0.5	0.7		MIT
4675.705	4677.014	21381.16	MO	0.3	0.6		MIT
4673.032	4674.340	21393.39	MO	0.5	0.5		MIT
4671.898	4673.206	21398.59	MO	1.5	1.5		MIT
4670.909	4672.216	21403.12	MO	0.3			MIT
4670.241	4671.548	21406.18	MO	0.7	0.7		MIT
4670.000	4671.307	21407.28	MO	0.3			MIT
4669.637	4670.944	21408.95	MO	0.3	0.5		MIT
4668.805	4670.112	21412.76	MO	0.7	0.7		MIT
4667.420	4668.727	21419.12	MO		1.3L		MIT
4665.380	4666.686	21428.48	MO	0.7	0.7		MIT
4663.108	4664.413	21438.92	MO	0.5	0.5		MIT
4662.762	4664.067	21440.51	MO	1.6	1.6		MIT
4661.934	4663.239	21444.32	MO	1.4	1.4		MIT
4658.544	4659.848	21459.93	MO	0.6	0.7		MIT
4657.478	4658.782	21464.84	MO	1.0	1.0		MIT
4656.368	4657.672	21469.96	MO	0.7	0.9		MIT
4656.215	4657.519	21470.66	MO	0.5	0.5		MIT
4656.047	4657.351	21471.44	MO	0.7	0.7		MIT
4653.935	4655.238	21481.18	MO		0.9		MIT
4653.688	4654.991	21482.32	MO	0.5	0.6		MIT
4652.285	4653.588	21488.80	MO	0.7H			MIT
4651.049	4652.351	21494.51	MO	1.2	1.2		MIT
4649.117	4650.419	21503.44	MO	1.2	1.2		MIT
4647.814	4649.115	21509.47	MO	1.4	1.4		MIT
4646.484	4647.785	21515.62	MO	0.5	0.6		MIT
4645.379	4646.680	21520.74	MO	0.6	0.6		MIT
4642.700	4644.000	21533.16	MO	0.9	0.8		MIT
4641.570	4642.870	21538.40	MO	0.7	0.7		MIT
4640.916	4642.216	21541.44	MO	0.6	0.7		MIT
4639.358	4640.657	21548.67	MO	0.5	0.5		MIT
4638.360	4639.659	21553.31	MO	0.3	0.3		MIT
4637.93	4639.23	21555.3	MO		1.0		MIT
4637.745	4639.044	21556.17	MO	0.7	0.5		MIT
4635.011	4636.309	21568.88	MO	0.7	0.7		MIT
4633.681	4634.979	21575.07	MO	0.6	0.6		MIT
4633.104	4634.401	21577.76	MO	1.4H	1.4H		MIT
4632.564	4633.861	21580.27	MO	0.7	0.7		MIT
4631.549	4632.846	21585.00	MO	0.7	0.7		MIT
4630.918	4632.215	21587.95	MO	0.5	0.5		MIT
4630.016	4631.313	21592.15	MO	1.2	1.1		MIT
4628.717	4630.013	21598.21	MO	0.3	0.3		MIT
4628.449	4629.745	21599.46	MO	0.5	0.6		MIT
4627.475	4628.771	21604.01	MO	1.9	1.9		MIT
4626.466	4627.762	21608.72	MO	2.0	1.9		MIT
4624.240	4625.535	21619.12	MO	1.4	1.4		MIT
4623.676	4624.971	21621.76	MO		0.3		MIT

AIR WAVELENGTH (ANGSTRMS)	VACUUM WAVELENGTH (ANGSTRMS)	WAVENUMBER (KAYSERS)	LMENT	LOG ARC	INTENSITY SPK	DIS	REF
4623.464	4624.759	21622.75	MO	1.2	1.2		MIT
4622.798	4624.093	21625.86	MO		1.4		MIT
4622.567	4623.862	21626.94	MO	0.3	0.3		MIT
4621.375	4622.669	21632.52	MO	1.5	1.4		MIT
4619.718	4621.012	21640.28	MO	0.3	0.6		MIT
4617.949	4619.242	21648.57	MO	1.1	1.1		MIT
4617.638	4618.931	21650.03	MO	0.7	1.0		MIT
4616.618	4617.911	21654.81	MO	1.2	1.2		MIT
4615.476	4616.769	21660.17	MO	0.9			MIT
4614.748	4616.041	21663.59	MO	1.0	1.1		MIT
4613.208	4614.500	21670.82	MO	0.7			MIT
4613.071	4614.363	21671.46	MO		1.1		MIT
4612.609	4613.901	21673.63	MO		1.0		MIT
4612.242	4613.534	21675.36	MO	0.7			MIT
4611.955	4613.247	21676.71	MO	0.7			MIT
4611.154	4612.446	21680.47	MO	1.3	1.2		MIT
4610.843	4612.135	21681.93	MO	1.0	0.9		MIT
4610.481	4611.773	21683.64	MO		0.9		MIT
4609.879	4611.170	21686.47	MO	1.6	1.6		MIT
4608.709	4610.000	21691.97	MO	1.0	1.0		MIT
4608.116	4609.407	21694.76	MO	0.7	0.7		MIT
4607.075	4608.366	21699.67	MO	0.9	1.0		MIT
4606.511	4607.801	21702.32	MO	0.6	0.5		MIT
4604.234	4605.524	21713.06	MO		1.3		MIT
4603.560	4604.850	21716.24	MO	0.8	0.8H		MIT
4600.881	4602.170	21728.88	MO	0.5	0.5		MIT
4599.157	4600.446	21737.02	MO	1.4	1.3		MIT
4598.246	4599.534	21741.33	MO	1.2	1.2		MIT
4597.882	4599.170	21743.05	MO	1.2	1.3		MIT
4595.159	4596.446	21755.94	MO	1.6	1.6		MIT
4593.643	4594.930	21763.12	MO	0.9	0.9		MIT
4592.209	4593.496	21769.91	MO	1.3	1.3		MIT
4590.380	4591.666	21778.59	MO	1.2J	1.0J		MIT
4589.338	4590.624	21783.53	MO	0.7J	0.5J		MIT
4588.147	4589.433	21789.19	MO	1.4	1.5		MIT
4587.400	4588.685	21792.73	MO	0.7	0.7		MIT
4586.789	4588.074	21795.64	MO	1.2	1.2		MIT
4586.574	4587.859	21796.66	MO	1.2	1.2		MIT
4586.061	4587.346	21799.10	MO	1.3	1.3		MIT
4584.450	4585.735	21806.76	MO	0.5	0.5		MIT
4582.498	4583.782	21816.05	MO	1.0	1.0		MIT
4582.346	4583.630	21816.77	MO	1.0	1.0		MIT
4580.965	4582.249	21823.35	MO		1.3H		MIT
4579.715	4580.998	21829.30	MO	0.7	0.6		MIT
4578.783	4580.066	21833.75	MO		1.1H		MIT
4578.483	4579.766	21835.18	MO	0.8	0.8		MIT
4577.778	4579.061	21838.54	MO	1.0	1.2		MIT
4576.501	4577.784	21844.63	MO	1.6	1.6		MIT
4575.863	4577.145	21847.68	MO	1.2	1.2		MIT
4575.169	4576.451	21850.99	MO	0.7	0.9		MIT
4574.608	4575.890	21853.67	MO	1.0	0.9		MIT
4574.485	4575.767	21854.26	MO	1.0	1.0		MIT
4571.759	4573.040	21867.29	MO	0.8J	0.8J		MIT
4571.062	4572.343	21870.62	MO	0.8J	0.8J		MIT
4570.595	4571.876	21872.86	MO	0.7	0.7		MIT
4570.128	4571.409	21875.09	MO	1.4	1.4		MIT
4569.020	4570.301	21880.40	MO	0.7	0.9		MIT
4567.683	4568.963	21886.80	MO	1.4	1.4		MIT
4567.397	4568.677	21888.17	MO	0.9	0.8		MIT
4566.879	4568.159	21890.66	MO	0.3H	0.5H		MIT
4565.703	4566.983	21896.29	MO	0.3	1.4H		MIT
4564.668	4565.947	21901.26	MO		1.0H		MIT
4560.132	4561.410	21923.04	MO	1.4	1.4		MIT
4559.752	4561.030	21924.87	MO	0.7	0.7		MIT
4559.034	4560.312	21928.32	MO	0.5	0.5		MIT
4558.744	4560.022	21929.72	MO	1.1	1.2		MIT
4558.107	4559.385	21932.78	MO	1.5	1.5		MIT
4556.028	4557.305	21942.79	MO	0.3	0.6L		MIT
4554.028	4555.305	21952.43	MO	0.0	0.6		MIT
4553.798	4555.075	21953.54	MO	1.3	1.3		MIT

AIR WAVELENGTH (ANGSTRMS)	VACUUM WAVELENGTH (ANGSTRMS)	WAVENUMBER (KAYSERS)	LMENT	LOG ARC	INTENSITY SPK	DIS	REF
4553.503	4554.779	21954.96	MO		1.4		MIT
4553.320	4554.596	21955.84	MO	1.1	0.6		MIT
4553.220	4554.496	21956.32	MO	1.1	0.8		MIT
4552.803	4554.079	21958.33	MO	0.7	0.7		MIT
4549.420	4550.695	21974.66	MO	0.9	0.9		MIT
4549.159	4550.434	21975.92	MO	0.8	0.7		MIT
4547.532	4548.807	21983.79	MO	0.6S	0.6S		MIT
4547.205	4548.480	21985.37	MO	0.6L	0.6L		MIT
4545.99	4547.26	21991.2	MO		0.6H		MIT
4545.044	4546.318	21995.82	MO	0.6	0.8		MIT
4543.400	4544.674	22003.78	MO	0.3	1.2		MIT
4542.886	4544.160	22006.27	MO	0.5	0.6		MIT
4541.556	4542.829	22012.71	MO	1.3	1.3		MIT
4541.022	4542.295	22015.30	MO	0.6	0.6		MIT
4540.755	4542.028	22016.60	MO	1.4D	1.4D		MIT
4539.638	4540.911	22022.01	MO	0.8	0.8		MIT
4539.206	4540.479	22024.11	MO	0.6	0.5		MIT
4538.877	4540.150	22025.71	MO	0.7	0.6		MIT
4538.412	4539.684	22027.96	MO	0.7	0.7		MIT
4537.325	4538.597	22033.24	MO	0.0	0.5		MIT
4536.800	4538.072	22035.79	MO	1.6	1.9		MIT
4535.543	4536.815	22041.90	MO	0.8	0.7		MIT
4535.385	4536.657	22042.66	MO	1.3	1.3		MIT
4534.881	4536.153	22045.11	MO	1.3H	1.3H		MIT
4534.429	4535.700	22047.31	MO	1.3H	1.5H		MIT
4529.398	4530.668	22071.80	MO	1.4	1.3		MIT
4528.618	4529.888	22075.60	MO	1.2L	1.0L		MIT
4527.869	4529.139	22079.25	MO	0.6H	0.6H		MIT
4526.369	4527.638	22086.57	MO	1.4	1.4		MIT
4525.861	4527.130	22089.05	MO	0.3	0.5		MIT
4525.324	4526.593	22091.67	MO	0.8	0.9		MIT
4524.726	4525.995	22094.59	MO	0.6	0.7		MIT
4524.338	4525.607	22096.49	MO	1.5	1.5		MIT
4522.897	4524.165	22103.52	MO	0.5	0.5		MIT
4522.190	4523.458	22106.98	MO	1.4	1.5		MIT
4521.554	4522.822	22110.09	MO	0.7	0.7		MIT
4521.149	4522.417	22112.07	MO	0.9	1.2		MIT
4520.285	4521.553	22116.30	MO	0.5	0.5		MIT
4519.586	4520.853	22119.72	MO		1.6		MIT
4518.668	4519.935	22124.21	MO	0.6H	0.6H		MIT
4518.440	4519.707	22125.33	MO	0.7	0.7		MIT
4517.409	4518.676	22130.38	MO	1.1	0.9		MIT
4517.132	4518.399	22131.73	MO	1.5	1.5		MIT
4515.185	4516.451	22141.28	MO	1.2	1.2		MIT
4515.036	4516.302	22142.01	MO	0.7	0.9		MIT
4514.407	4515.673	22145.09	MO	0.9	1.0		MIT
4512.148	4513.414	22156.18	MO	1.4	1.4		MIT
4508.675	4509.940	22173.25	MO		0.7H		MIT
4508.209	4509.473	22175.54	MO	0.7	0.6		MIT
4506.673	4507.937	22183.10	MO	1.4	1.4		MIT
4506.051	4507.315	22186.16	MO	1.3	1.3		MIT
4505.958	4507.222	22186.62	MO	1.3	1.3		MIT
4504.897	4506.161	22191.84	MO	0.3	0.6		MIT
4503.558	4504.821	22198.44	MO	1.4W			MIT
4501.290	4502.553	22209.62	MO	1.4	1.4		MIT
4500.523	4501.785	22213.41	MO	0.7H			MIT
4499.444	4500.706	22218.74	MO	1.3	1.3		MIT
4496.11	4497.37	22235.2	MO	0.3	0.5		MIT
4494.090	4495.351	22245.21	MO	0.7	0.7		MIT
4493.528	4494.789	22247.99	MO		1.0H		MIT
4492.066	4493.326	22255.23	MO	0.5	0.5		MIT
4491.656	4492.916	22257.26	MO	0.7	0.8		MIT
4491.282	4492.542	22259.11	MO	1.6	1.5		MIT
4490.192	4491.452	22264.52	MO	1.3	1.3		MIT
4489.28	4490.54	22269.0	MO		0.7H		MIT
4488.999	4490.258	22270.43	MO	1.2	1.1		MIT
4487.047	4488.306	22280.12	MO	1.4	1.4		MIT
4485.790	4487.049	22286.36	MO	0.6	0.6		MIT
4484.969	4486.227	22290.44	MO	1.5	1.5		MIT
4481.276	4482.533	22308.81	MO	1.0	0.9		MIT

AIR WAVELENGTH (ANGSTRMS)	VACUUM WAVELENGTH (ANGSTRMS)	WAVENUMBER (KAYSERS)	LMENT	LOG ARC	INTENSITY SPK	DIS	REF
4479.041	4480.298	22319.95	MO	0.5	0.7		MIT
4477.16	4478.42	22329.3	MO	0.5	0.6		MIT
4475.618	4476.874	22337.01	MO	1.4	1.4		MIT
4475.25	4476.51	22338.9	MO	0.0	0.5		MIT
4474.653	4475.909	22341.83	MO	1.9			MIT
4474.564	4475.820	22342.28	MO	2.1	2.1		MIT
4473.182	4474.437	22349.18	MO	1.5	1.5		MIT
4472.571	4473.826	22352.23	MO	0.7	0.7		MIT
4472.045	4473.300	22354.86	MO	1.3	1.2		MIT
4471.658	4472.913	22356.80	MO	1.3	1.3		MIT
4471.051	4472.306	22359.83	MO	0.3	0.6		M
4469.38	4470.63	22368.2	MO	0.3	0.5		MIT
4468.276	4469.530	22373.72	MO	1.4	1.4		MIT
4468.087	4469.341	22374.66	MO	0.7	0.7		MIT
4467.40	4468.65	22378.1	MO	0.3	0.6		MIT
4464.769	4466.022	22391.29	MO	1.3	1.3		MIT
4464.48	4465.73	22392.7	MO	0.6	0.7		MIT
4463.85	4465.10	22395.9	MO		1.4		MIT
4463.45	4464.70	22397.9	MO	0.3	0.6		MIT
4461.84	4463.09	22406.0	MO	0.7H	0.7H		MIT
4461.55	4462.80	22407.4	MO	0.6	0.7		MIT
4460.625	4461.877	22412.09	MO	1.3	1.3		MIT
4458.646	4459.897	22422.04	MO	0.9	1.0		MIT
4457.356	4458.607	22428.53	MO	1.7	1.7		MIT
4455.305	4456.556	22438.85	MO	1.0	1.1		MIT
4453.87	4455.12	22446.1	MO	0.6	0.7		EX
4452.74	4453.99	22451.8	MO		1.2		MIT
4452.558	4453.808	22452.70	MO	1.1	0.9		MIT
4451.968	4453.218	22455.67	MO		1.5		MIT
4449.738	4450.987	22466.93	MO	1.6	1.6		MIT
4447.233	4448.481	22479.58	MO	1.0	0.8		MIT
4446.74	4447.99	22482.1	MO	0.6	0.3		EX
4446.427	4447.675	22483.66	MO	1.4	1.4		MIT
4445.38	4446.63	22488.9	MO		1.2		EX
4444.73	4445.98	22492.2	MO	0.6	0.6		EX
4444.438	4445.686	22493.72	MO	0.5	0.5		MIT
4444.005	4445.253	22495.91	MO	1.0H	1.0H		M
4443.069	4444.316	22500.65	MO	1.4	1.3		MIT
4442.203	4443.450	22505.03	MO	1.6	1.5		MIT
4440.71	4441.96	22512.6	MO	0.3	0.7		EX
4440.183	4441.430	22515.27	MO	0.6	0.7		MIT
4439.489	4440.735	22518.79	MO	0.5	0.5		MIT
4438.959	4440.205	22521.48	MO	1.2	1.2		MIT
4437.434	4438.680	22529.22	MO	0.6	0.6		MIT
4437.151	4438.397	22530.66	MO	0.9	0.8		MIT
4436.888	4438.134	22531.99	MO	1.3	1.2		MIT
4436.653	4437.899	22533.19	MO	0.6	0.5		MIT
4434.953	4436.198	22541.82	MO	1.9	1.9		MIT
4434.23	4435.47	22545.5	MO		0.8H		EX
4433.501	4434.746	22549.21	MO	0.6	2.1		MIT
4432.880	4434.125	22552.37	MO	0.7	0.5		MIT
4430.875	4432.119	22562.57	MO	0.7	0.6		MIT
4430.48	4431.72	22564.6	MO		0.9		EX
4429.111	4430.355	22571.56	MO	0.6	0.6		MIT
4428.211	4429.454	22576.14	MO	0.7	0.6		MIT
4426.671	4427.914	22584.00	MO	1.5	1.5		MIT
4424.197	4425.439	22596.63	MO	0.7	0.5		MIT
4423.618	4424.860	22599.58	MO	1.6	1.6		MIT
4423.057	4424.299	22602.45	MO	0.9	0.8		MIT
4422.850	4424.092	22603.51	MO	0.8	0.8		MIT
4422.063	4423.305	22607.53	MO	0.9	0.8		MIT
4420.742	4421.983	22614.29	MO	0.8	0.7		MIT
4419.721	4420.962	22619.51	MO	0.5	0.3		MIT
4417.228	4418.468	22632.28	MO	0.7	0.7		MIT
4415.670	4416.910	22640.26	MO	0.3	0.5		MIT
4414.35	4415.59	22647.0	MO	0.7	0.6		EX
4413.672	4414.912	22650.51	MO	0.8	0.8		MIT
4412.770	4414.009	22655.14	MO	1.5	1.3		MIT
4412.225	4413.464	22657.94	MO		1.4		MIT
4411.702	4412.941	22660.62	MO	1.4	1.4		MIT

AIR WAVELENGTH (ANGSTRMS)	VACUUM WAVELENGTH (ANGSTRMS)	WAVENUMBER (KAYSERS)	LMENT	LOG ARC	INTENSITY SPK	DIS	REF
4411.567	4412.806	22661.32	MO	1.0	0.9		MIT
4409.950	4411.189	22669.63	MO	1.2	1.2		MIT
4409.443	4410.681	22672.23	MO	1.0	1.1		MIT
4407.444	4408.682	22682.52	MO	1.5			MIT
4406.866	4408.104	22685.49	MO	0.9	1.0		MIT
4405.027	4406.264	22694.96	MO	0.7	0.6		MIT
4404.546	4405.783	22697.44	MO	1.2	0.9		MIT
4404.18	4405.42	22699.3	MO	0.0	1.2H		EX
4403.34	4404.58	22703.7	MO		1.3		EX
4402.900	4404.137	22705.93	MO	1.3	1.3		MIT
4402.491	4403.728	22708.03	MO	1.2	1.2		MIT
4400.658	4401.894	22717.49	MO	0.5	0.5		MIT
4399.22	4400.46	22724.9	MO	0.6	0.6		EX
4398.491	4399.727	22728.69	MO	0.8	1.1		MIT
4397.292	4398.527	22734.88	MO	1.5	1.5		MIT
4396.850	4398.085	22737.17	MO	0.7	0.8		MIT
4396.661	4397.896	22738.15	MO	1.4	1.4		MIT
4396.369	4397.604	22739.66	MO	0.6	0.6		MIT
4394.471	4395.705	22749.48	MO	0.9	0.9		MIT
4394.322	4395.556	22750.25	MO	1.3	1.2		MIT
4393.709	4394.943	22753.42	MO	0.9H	0.7H		MIT
4392.123	4393.357	22761.64	MO	1.2	1.2		MIT
4391.535	4392.769	22764.69	MO	1.2	1.2		MIT
4390.93	4392.16	22767.8	MO	0.0	0.9H		EX
4389.570	4390.803	22774.88	MO	0.8	0.8		MIT
4388.275	4389.508	22781.60	MO	0.8	1.3		MIT
4387.299	4388.532	22786.67	MO	0.5	0.3		MIT
4385.890	4387.122	22793.99	MO	1.1	0.9		MIT
4385.56	4386.79	22795.7	MO		1.3		EX
4384.192	4385.424	22802.81	MO	0.9	0.9		MIT
4382.413	4383.644	22812.07	MO	1.0	1.3		MIT
4381.640	4382.871	22816.09	MO	2.2	2.2		MIT
4380.592	4381.823	22821.55	MO	1.2	1.2		MIT
4380.293	4381.524	22823.11	MO	1.5	1.4		MIT
4377.765	4378.995	22836.29	MO	0.7	2.3		MIT
4376.684	4377.914	22841.93	MO	0.9	0.9		MIT
4375.014	4376.243	22850.65	MO	0.9	0.9		MIT
4374.888	4376.117	22851.31	MO	0.9	0.7		MIT
4373.321	4374.550	22859.50	MO	1.0	1.1		MIT
4372.128	4373.357	22865.73	MO	1.0	1.0		MIT
4370.138	4371.366	22876.14	MO	0.7	0.5		MIT
4369.045	4370.273	22881.87	MO	1.6	1.4		MIT
4368.772	4370.000	22883.30	MO	0.7	0.6		MIT
4366.536	4367.763	22895.01	MO	1.4	1.4		MIT
4365.896	4367.123	22898.37	MO	0.5	0.6		MIT
4364.698	4365.925	22904.66	MO	0.6	0.6		MIT
4364.474	4365.701	22905.83	MO	0.9	0.8		MIT
4363.644	4364.870	22910.19	MO	0.7	2.3		MIT
4363.017	4364.243	22913.48	MO	0.6	0.5		MIT
4362.817	4364.043	22914.53	MO	0.5	0.3		MIT
4362.710	4363.936	22915.09	MO	0.7	0.7		MIT
4362.023	4363.249	22918.70	MO	0.9	0.7		MIT
4361.99	4363.22	22918.9	MO		1.3		MIT
4361.404	4362.630	22921.95	MO	0.8	0.8		MIT
4361.099	4362.325	22923.56	MO	0.5	0.3		MIT
4359.621	4360.846	22931.33	MO	1.2	1.2		MIT
4358.551	4359.776	22936.96	MO	1.3W	1.0W		MIT
4358.324	4359.549	22938.15	MO		1.6		MIT
4357.867	4359.092	22940.56	MO	0.8	0.3		MIT
4357.335	4358.560	22943.36	MO	0.7	1.2		MIT
4356.07	4357.29	22950.0	MO		1.5		MIT
4355.21	4356.43	22954.6	MO		1.3		MIT
4354.692	4355.916	22957.28	MO	0.7	0.7		MIT
4353.310	4354.534	22964.57	MO	1.4	1.4		MIT
4352.883	4354.107	22966.82	MO	0.7	0.5		MIT
4351.549	4352.772	22973.86	MO	1.2	1.2		MIT
4351.02	4352.24	22976.7	MO	0.7H	0.7H		MIT
4350.339	4351.562	22980.25	MO	1.7	1.6		MIT
4349.224	4350.447	22986.15	MO	0.8	0.6		MIT
4346.895	4348.117	22998.46	MO	0.3	0.5		MIT

AIR WAVELENGTH (ANGSTRMS)	VACUUM WAVELENGTH (ANGSTRMS)	WAVENUMBER (KAYSERS)	LMENT	LOG ARC	INTENSITY SPK	DIS	REF
4346.204	4347.426	23002.12	MO	1.2	1.0		MIT
4345.681	4346.903	23004.89	MO	0.3	0.5		MIT
4344.660	4345.881	23010.29	MO	1.3	1.3		MIT
4341.964	4343.185	23024.58	MO	0.7	0.6		MIT
4341.421	4342.642	23027.46	MO	1.4	1.4		MIT
4340.746	4341.966	23031.04	MO	1.3	1.3		MIT
4339.823	4341.043	23035.94	MO	1.2	1.2		MIT
4339.234	4340.454	23039.06	MO	1.2	1.1		MIT
4338.714	4339.934	23041.83	MO	1.1	1.1		MIT
4338.564	4339.784	23042.62	MO	0.8	0.7		MIT
4334.811	4336.030	23062.57	MO	1.4	1.4		MIT
4334.470	4335.689	23064.39	MO	0.8	0.7		MIT
4333.211	4334.429	23071.09	MO	1.0	1.1		MIT
4332.744	4333.962	23073.57	MO	0.8	0.7		MIT
4332.506	4333.724	23074.84	MO	1.2	1.2		MIT
4331.396	4332.614	23080.75	MO	1.1D	1.2D		MIT
4330.088	4331.306	23087.73	MO	1.0	0.9		MIT
4329.629	4330.846	23090.18	MO	1.2	1.0		MIT
4329.335	4330.552	23091.74	MO	1.0	1.0		MIT
4328.695	4329.912	23095.16	MO	0.5	0.7		MIT
4328.02	4329.24	23098.8	MO		1.5		MIT
4326.743	4327.960	23105.58	MO	1.7	1.7		MIT
4326.137	4327.353	23108.81	MO	1.7	1.6		MIT
4325.823	4327.039	23110.49	MO	1.4	1.3		PU
4325.256	4326.472	23113.52	MO	1.2	1.3		MIT
4324.542	4325.758	23117.33	MO	0.7	0.6		MIT
4322.471	4323.687	23128.41	MO	0.9	1.0		MIT
4321.968	4323.183	23131.10	MO	1.2	1.2		MIT
4321.781	4322.996	23132.10	MO	1.1	1.1		MIT
4319.512	4320.727	23144.25	MO	0.7H	0.6H		MIT
4318.550	4319.764	23149.41	MO	0.7	0.8		MIT
4318.251	4319.465	23151.01	MO	0.7	0.6		MIT
4317.929	4319.143	23152.74	MO	1.5	1.4		MIT
4317.265	4318.479	23156.30	MO	0.7D	0.7D		MIT
4316.366	4317.580	23161.12	MO	0.6	0.6		MIT
4315.799	4317.013	23164.17	MO	0.7	0.8		MIT
4315.386	4316.600	23166.38	MO	0.8	0.6		MIT
4315.25	4316.46	23167.1	MO		1.0		EX
4313.547	4314.760	23176.26	MO	0.8	0.8		MIT
4312.967	4314.180	23179.38	MO	1.0	1.0		MIT
4312.798	4314.011	23180.28	MO	1.2	0.8		MIT
4312.452	4313.665	23182.14	MO		1.2		MIT
4311.653	4312.866	23186.44	MO	0.3	1.6		MIT
4311.023	4312.236	23189.83	MO	0.0	1.5		MIT
4310.390	4311.602	23193.23	MO	1.2	1.1		MIT
4308.651	4309.863	23202.59	MO	0.8	0.8		MIT
4306.646	4307.857	23213.40	MO	0.6	0.7		MIT
4304.916	4306.127	23222.73	MO	1.2	1.2		MIT
4304.551	4305.762	23224.69	MO	0.6	0.7		MIT
4304.020	4305.231	23227.56	MO	1.1	1.0		MIT
4303.891	4305.102	23228.25	MO		1.0		MIT
4301.932	4303.142	23238.83	MO	1.0	1.0		MIT
4301.264	4302.474	23242.44	MO	1.3	1.3		MIT
4298.897	4300.106	23255.24	MO	1.3	1.0		MIT
4297.638	4298.847	23262.05	MO	0.8	0.8		MIT
4296.624	4297.833	23267.54	MO	1.2	1.2		MIT
4296.160	4297.369	23270.05	MO	1.2	1.2		MIT
4295.368	4296.576	23274.34	MO	0.7			MIT
4294.599	4295.807	23278.51	MO	0.8	0.8		MIT
4293.880	4295.088	23282.41	MO	1.3	1.3		MIT
4293.753	4294.961	23283.10	MO	1.0	0.6		MIT
4293.210	4294.418	23286.04	MO	2.1	2.0		MIT
4292.132	4293.340	23291.89	MO	2.0	1.9		MIT
4291.671	4292.878	23294.39	MO	0.5	0.3		MIT
4291.202	4292.409	23296.94	MO	1.2	1.2		MIT
4290.184	4291.391	23302.47	MO	1.5H	1.4H		MIT
4289.415	4290.622	23306.65	MO	1.3	1.2		MIT
4288.637	4289.844	23310.87	MO	1.9	2.0		MIT
4288.247	4289.454	23312.99	MO	0.6	0.6		MIT
4287.075	4288.281	23319.37	MO	1.0	1.0		MIT

AIR WAVELENGTH (ANGSTRMS)	VACUUM WAVELENGTH (ANGSTRMS)	WAVENUMBER (KAYSERS)	LMENT	LOG ARC	INTENSITY SPK	DIS	REF
4285.598	4286.804	23327.40	MO	0.6	0.6		MIT
4284.597	4285.803	23332.85	MO	2.1H	1.9H		MIT
4283.284	4284.489	23340.00	MO	0.3	0.6		MIT
4282.44	4283.64	23344.6	MO	0.6	0.5		MIT
4281.826	4283.031	23347.95	MO	1.2	1.0		MIT
4281.173	4282.378	23351.51	MO	0.7	1.5		MIT
4280.875	4282.080	23353.14	MO	0.6	0.5		MIT
4280.578	4281.783	23354.76	MO	0.3	1.2		MIT
4280.001	4281.205	23357.91	MO	0.7	0.7		MIT
4279.390	4280.594	23361.24	MO	0.5			MIT
4279.023	4280.227	23363.25	MO	0.9	2.0		MIT
4278.584	4279.788	23365.64	MO	0.7	0.3		MIT
4277.410	4278.614	23372.06	MO	1.2	1.2		MIT
4277.242	4278.446	23372.97	MO	1.5	1.5		MIT
4276.912	4278.116	23374.78	MO	1.5	1.5		MIT
4275.676	4276.879	23381.53	MO	1.1	1.1		MIT
4274.41	4275.61	23388.5	MO		1.5		EX
4274.047	4275.250	23390.45	MO	0.8	0.8		MIT
4273.066	4274.269	23395.82	MO	1.3	1.0		MIT
4272.057	4273.259	23401.34	MO	1.2	1.2		MIT
4269.282	4270.484	23416.55	MO	1.5	1.5		MIT
4268.075	4269.276	23423.17	MO	1.3	1.3		MIT
4266.608	4267.809	23431.23	MO	0.3	1.2H		MIT
4266.181	4267.382	23433.57	MO	1.4	1.3		MIT
4265.117	4266.317	23439.42	MO	1.2	1.2		MIT
4264.631	4265.831	23442.09	MO	1.2	1.2		MIT
4263.103	4264.303	23450.49	MO	0.6	0.3		MIT
4261.443	4262.642	23459.63	MO	1.3	1.3		MIT
4260.977	4262.176	23462.19	MO	1.2	1.2		MIT
4260.655	4261.854	23463.97	MO	1.3	1.3		MIT
4260.359	4261.558	23465.59	MO	1.3	1.3		MIT
4258.660	4259.859	23474.96	MO	1.3	1.2		MIT
4256.99	4258.19	23484.2	MO	0.5	0.3		MIT
4256.80	4258.00	23485.2	MO	0.5	0.5		MIT
4256.60	4257.80	23486.3	MO		1.2		MIT
4254.955	4256.153	23495.40	MO	1.4	1.4		MIT
4254.429	4255.627	23498.30	MO	1.0	0.7		MIT
4253.576	4254.773	23503.01	MO	0.9	0.7		MIT
4252.494	4253.691	23508.99	MO	0.8	0.7		MIT
4251.873	4253.070	23512.43	MO	1.8	1.8		MIT
4251.389	4252.586	23515.10	MO	0.7	0.5		MIT
4250.689	4251.886	23518.98	MO	0.7	2.1		MIT
4249.494	4250.690	23525.59	MO	0.7	0.7		MIT
4246.616	4247.812	23541.53	MO		1.4		MIT
4246.022	4247.217	23544.83	MO	1.5	1.5		MIT
4244.800	4245.995	23551.60	MO	0.6	1.9		MIT
4243.14	4244.33	23560.8	MO		1.4		MIT
4242.803	4243.998	23562.69	MO	1.2	0.6		MIT
4240.831	4242.025	23573.65	MO	1.5	1.4		MIT
4240.279	4241.473	23576.72	MO	1.4	1.3		MIT
4240.076	4241.270	23577.84	MO	1.4	1.3		MIT
4239.194	4240.388	23582.75	MO	1.2	1.0		MIT
4239.073	4240.267	23583.42	MO	1.2	1.1		MIT
4238.028	4239.221	23589.24	MO		0.7		MIT
4237.162	4238.355	23594.06	MO	1.3	1.0		MIT
4235.034	4236.227	23605.91	MO	1.0	0.9		MIT
4233.976	4235.168	23611.81	MO	0.7	0.7		MIT
4233.491	4234.683	23614.52	MO	0.9	0.9		MIT
4233.368	4234.560	23615.20	MO	0.6	0.6		MIT
4232.586	4233.778	23619.57	MO	2.1	2.0		MIT
4231.625	4232.817	23624.93	MO	0.7	0.6		MIT
4229.525	4230.716	23636.66	MO	0.6	0.6		MIT
4228.787	4229.978	23640.78	MO	0.6	0.6		MIT
4227.085	4228.275	23650.30	MO	0.5	1.4		MIT
4226.549	4227.739	23653.30	MO	1.0	0.7		MIT
4226.291	4227.481	23654.75	MO	1.3	1.3		MIT
4225.248	4226.438	23660.59	MO	0.7	0.5		MIT
4224.929	4226.119	23662.37	MO	0.7	0.7L		MIT
4224.766	4225.956	23663.29	MO	0.7	0.5		MIT
4223.918	4225.108	23668.04	MO	0.7	0.6		MIT
4222.961	4224.150	23673.40	MO	1.2	1.0		MIT
4222.411	4223.600	23676.48	MO	1.3	1.2		MIT
4219.989	4221.178	23690.07	MO	0.7	0.7		MIT
4219.395	4220.583	23693.41	MO	1.2	1.1		MIT
4219.015	4220.203	23695.54	MO	0.9	1.0		MIT
4216.841	4218.029	23707.76	MO	1.0	0.9		MIT
4214.065	4215.252	23723.37	MO	1.3	1.2		MIT
4211.024	4212.210	23740.50	MO	1.2	1.1		MIT
4210.209	4211.395	23745.10	MO	0.7	0.7		MIT
4209.649	4210.835	23748.26	MO	0.6	1.9		MIT
4208.775	4209.961	23753.19	MO	0.7	0.7		MIT
4207.561	4208.746	23760.04	MO	1.0	0.7		MIT
4207.401	4208.586	23760.95	MO	1.0	0.7		MIT
4207.249	4208.434	23761.81	MO	1.0	1.0		MIT
4205.809	4206.994	23769.94	MO	0.3	0.0		MIT
4204.809	4205.994	23775.59	MO	1.4	1.3		MIT
4204.609	4205.794	23776.73	MO	1.2	1.0		MIT
4203.12	4204.30	23785.1	MO	0.6	0.3		MIT
4202.219	4203.403	23790.25	MO	0.7	0.7		MIT
4201.318	4202.502	23795.35	MO	0.7	0.7		MIT
4201.178	4202.362	23796.14	MO	0.6	0.5		MIT
4200.570	4201.753	23799.59	MO	1.2	1.2		MIT
4200.199	4201.382	23801.69	MO	0.6	0.3		MIT
4199.830	4201.013	23803.78	MO	0.7	0.7		MIT
4199.653	4200.836	23804.78	MO	0.7	0.7		MIT
4198.91	4200.09	23809.0	MO	0.0	1.4		MIT
4198.61	4199.79	23810.7	MO	0.6	0.6		MIT
4197.393	4198.576	23817.60	MO	0.5	0.5		MIT
4197.05	4198.23	23819.6	MO	0.5	0.5		MIT
4194.561	4195.743	23833.68	MO	1.5	1.5		MIT
4194.009	4195.191	23836.82	MO	0.7	0.5		MIT
4193.88	4195.06	23837.6	MO	0.7	0.5		MIT
4192.29	4193.47	23846.6	MO		1.6		MIT
4192.177	4193.358	23847.24	MO	0.6			MIT
4191.03	4192.21	23853.8	MO		1.5		MIT
4190.420	4191.601	23857.23	MO	0.8	0.5		MIT
4190.005	4191.186	23859.60	MO	1.3	1.2		MIT
4188.321	4189.501	23869.19	MO	2.0	1.9		MIT
4187.609	4188.789	23873.25	MO	0.5	0.5		MIT
4186.281	4187.461	23880.82	MO	1.2	1.1		MIT
4185.825	4187.005	23883.42	MO	1.6	1.6		MIT
4184.391	4185.570	23891.61	MO	0.6	0.6		MIT
4184.170	4185.349	23892.87	MO	0.6	0.6		MIT
4183.06	4184.24	23899.2	MO	0.6H	0.3H		MIT
4181.049	4182.227	23910.70	MO	1.4	1.2		MIT
4180.503	4181.681	23913.83	MO	0.9	0.6		MIT
4179.936	4181.114	23917.07	MO	0.9	0.7		MIT
4178.535	4179.713	23925.09	MO	0.7	0.3		MIT
4178.272	4179.450	23926.59	MO	1.4	1.3		MIT
4177.917	4179.095	23928.63	MO		1.2H		MI
4177.257	4178.434	23932.41	MO	1.3	1.3		MIT
4176.900	4178.077	23934.45	MO	1.0	1.0		MIT
4175.124	4176.301	23944.64	MO	1.1	0.9		MIT
4174.080	4175.257	23950.62	MO	1.1	1.0		MIT
4172.751	4173.927	23958.25	MO	0.5	0.3		MIT
4171.90	4173.08	23963.1	MO		1.4		MIT
4171.454	4172.630	23965.70	MO	1.0	0.5		MIT
4171.074	4172.250	23967.88	MO	1.2	1.0		MIT
4170.348	4171.524	23972.06	MO	0.9	0.3		MIT
4169.820	4170.995	23975.09	MO	1.3	1.3		MIT
4168.494	4169.669	23982.72	MO	1.2	1.2		MIT
4166.282	4167.456	23995.45	MO	1.3	1.2		MIT
4165.759	4166.933	23998.46	MO	0.8	0.6		MIT
4164.079	4165.253	24008.15	MO	1.2	1.0		MIT
4162.682	4163.856	24016.20	MO	1.4	1.5		MIT
4161.298	4162.471	24024.19	MO	0.0	1.4		MIT
4160.87	4162.04	24026.7	MO	0.6	0.3		MIT
4160.251	4161.424	24030.24	MO	0.6	0.7		MIT
4158.081	4159.253	24042.78	MO	0.5	0.5		MIT
4157.403	4158.575	24046.70	MO	1.4	1.4		MIT

AIR WAVELENGTH (ANGSTRMS)	VACUUM WAVELENGTH (ANGSTRMS)	WAVENUMBER (KAYSERS)	LMENT	LOG ARC	INTENSITY SPK	DIS	REF
4156.79	4157.96	24050.2	MO	1.2	1.0		MIT
4155.578	4156.750	24057.26	MO	1.4	1.3		MIT
4155.283	4156.455	24058.97	MO	1.5	1.4		MIT
4154.29	4155.46	24064.7	MO		1.0H		MIT
4153.173	4154.344	24071.19	MO	1.0	0.3		MIT
4151.879	4153.050	24078.69	MO	1.3	1.2		MIT
4151.373	4152.544	24081.63	MO	0.6	0.3		MIT
4150.85	4152.02	24084.7	MO		1.3H		MIT
4149.695	4150.865	24091.36	MO	0.8	0.7		MIT
4149.582	4150.752	24092.02	MO	0.7			MIT
4149.41	4150.58	24093.0	MO	0.6	0.5		MIT
4148.945	4150.115	24095.72	MO	1.6	1.4		MIT
4148.680	4149.850	24097.26	MO	0.8	0.7		MIT
4146.846	4148.015	24107.91	MO		1.5		MIT
4146.479	4147.648	24110.05	MO	0.6	0.3		MIT
4145.596	4146.765	24115.18	MO	0.7	0.5		MIT
4143.549	4144.718	24127.10	MO	2.0	2.0		MIT
4142.162	4143.330	24135.18	MO	0.9	0.6		MIT
4142.047	4143.215	24135.84	MO	0.9	0.6		MIT
4141.489	4142.657	24139.10	MO	0.0	1.5		MIT
4139.528	4140.695	24150.53	MO	1.0	0.5		MIT
4138.65	4139.82	24155.7	MO		0.7		MIT
4138.543	4139.710	24156.28	MO	1.3	1.1		MIT
4138.185	4139.352	24158.37	MO	1.3	1.2		MIT
4136.95	4138.12	24165.6	MO	1.2	0.9		MIT
4135.383	4136.549	24174.74	MO	1.0	1.0		MIT
4135.212	4136.378	24175.74	MO	0.7	0.6		MIT
4133.002	4134.168	24188.67	MO	1.2			MIT
4132.750	4133.916	24190.14	MO	1.1	1.0		MIT
4132.230	4133.396	24193.18	MO	1.3	1.3		MIT
4131.920	4133.085	24195.00	MO	1.3	1.0		MIT
4130.846	4132.011	24201.29	MO		1.2		MIT
4130.257	4131.422	24204.74	MO		0.9		MIT
4130.105	4131.270	24205.63	MO	1.0	0.9		MIT
4129.695	4130.860	24208.03	MO	0.9	0.6		MIT
4128.833	4129.998	24213.09	MO	1.6	1.6		MIT
4128.284	4129.449	24216.31	MO	1.7	1.4		MIT
4127.645	4128.809	24220.06	MO	0.8	0.5		MIT
4127.322	4128.486	24221.95	MO		1.3		MIT
4126.529	4127.693	24226.61	MO	1.4	1.3		MIT
4125.619	4126.783	24231.95	MO		1.7		MIT
4124.545	4125.709	24238.26	MO	1.5	1.4		MIT
4123.650	4124.813	24243.52	MO	0.7	1.4		MIT
4122.395	4123.558	24250.90	MO	1.2	1.7		MIT
4120.697	4121.860	24260.90	MO	0.9			MIT
4120.096	4121.258	24264.43	MO	1.4	1.7		MIT
4119.633	4120.795	24267.16	MO	0.7	1.7		MIT
4118.965	4120.127	24271.10	MO	1.0	0.9		MIT
4118.549	4119.711	24273.55	MO		1.0		MIT
4115.782	4116.943	24289.87	MO		0.7H		MIT
4114.928	4116.089	24294.91	MO	1.0	1.1		MIT
4113.605	4114.766	24302.72	MO	0.8	0.8		MIT
4112.120	4113.280	24311.50	MO	0.6	0.5		MIT
4110.704	4111.864	24319.87	MO	0.8	0.7		MIT
4110.287	4111.447	24322.34	MO	0.5	0.7		MIT
4108.695	4109.854	24331.76	MO		0.9H		MIT
4108.124	4109.283	24335.14	MO	0.7	0.8		MIT
4107.468	4108.627	24339.03	MO	1.5	1.6		MIT
4105.526	4106.685	24350.54	MO	1.0	1.0		MIT
4105.083	4106.241	24353.17	MO	1.2	1.3		MIT
4103.753	4104.911	24361.06	MO	0.7	0.7		MIT
4102.153	4103.311	24370.56	MO	1.5	1.4		MIT
4100.322	4101.479	24381.45	MO		1.3		MIT
4098.743	4099.900	24390.84	MO	1.3	1.3		MIT
4098.183	4099.340	24394.17	MO	1.3	1.2		MIT
4097.708	4098.865	24397.00	MO	0.5	0.3		MIT
4096.810	4097.966	24402.35	MO	1.3	1.2		MIT
4095.318	4096.474	24411.24	MO	0.3	0.5		MIT
4094.908	4096.064	24413.68	MO		1.5		MIT
4094.467	4095.623	24416.31	MO	0.7	0.6		MIT

AIR WAVELENGTH (ANGSTRMS)	VACUUM WAVELENGTH (ANGSTRMS)	WAVENUMBER (KAYSERS)	LMENT	LOG ARC	INTENSITY SPK	DIS	REF
4094.347	4095.503	24417.03	MO	0.5	0.3		MIT
4093.161	4094.316	24424.10	MO	0.7	0.7		MIT
4090.953	4092.108	24437.28	MO		1.3		MIT
4090.877	4092.032	24437.74	MO	1.0	1.3		MIT
4089.731	4090.885	24444.59	MO	0.8	0.8		MIT
4088.660	4089.814	24450.99	MO	0.7H	0.7H		MIT
4087.743	4088.897	24456.47	MO	0.5	0.6		MIT
4087.463	4088.617	24458.15	MO	0.5	0.5		MIT
4086.025	4087.178	24466.76	MO	1.2	1.2		MIT
4085.520	4086.673	24469.78	MO	1.0H	0.8H		MIT
4084.383	4085.536	24476.59	MO	1.6	1.6		MIT
4084.01	4085.16	24478.8	MO		0.7		MIT
4082.116	4083.268	24490.19	MO	0.5	0.3		MIT
4081.764	4082.916	24492.30	MO	0.7	0.7		MIT
4081.438	4082.590	24494.25	MO	1.7	1.7		MIT
4081.078	4082.230	24496.41	MO		1.5		MIT
4079.342	4080.494	24506.84	MO	0.6	0.6		MIT
4078.381	4079.532	24512.61	MO	0.7	0.5		MIT
4078.074	4079.225	24514.46	MO	0.6	0.6		MIT
4077.682	4078.833	24516.81	MO	0.9	1.0		MIT
4076.506	4077.657	24523.89	MO	0.7	0.7		MIT
4076.194	4077.345	24525.76	MO	1.4	1.4		MIT
4075.937	4077.088	24527.31	MO	1.0	0.7		MIT
4075.541	4076.692	24529.69	MO	1.3	1.3		MIT
4075.246	4076.397	24531.47	MO	1.4	1.4		MIT
4074.438	4075.588	24536.33	MO		1.3		MIT
4070.002	4071.151	24563.08	MO	1.4	1.4		MIT
4069.882	4071.031	24563.80	MO	1.4	1.4		MIT
4067.720	4068.869	24576.86	MO	1.0	1.0		MIT
4066.975	4068.123	24581.36	MO	1.0H	1.0H		MIT
4066.367	4067.515	24585.03	MO	1.2	1.3		MIT
4064.667	4065.815	24595.32	MO	1.4	1.4D		MIT
4064.171	4065.319	24598.32	MO	0.7	0.9		MIT
4063.904	4065.052	24599.93	MO	0.7	0.7		MIT
4062.592	4063.739	24607.88	MO				ME
4062.076	4063.223	24611.00	MO	1.9	1.9		MIT
4059.608	4060.755	24625.97	MO	1.0	1.0		MIT
4058.61	4059.76	24632.0	MO		1.0		MIT
4057.584	4058.730	24638.25	MO	1.0	0.6		MIT
4057.438	4058.584	24639.14	MO	0.6	0.6		MIT
4056.318	4057.464	24645.94	MO	1.2	1.2		MIT
4056.012	4057.158	24647.80	MO	1.4	1.5		MIT
4055.545	4056.690	24650.64	MO	0.3	1.0		MIT
4052.293	4053.438	24670.42	MO	1.2H	0.7H		MIT
4051.936	4053.081	24672.59	MO	0.6	0.7		MIT
4051.184	4052.328	24677.17	MO	0.7	1.0		MIT
4050.092	4051.236	24683.82	MO	0.8	0.9		MIT
4047.563	4048.706	24699.25	MO	0.6	0.6		MIT
4047.398	4048.541	24700.25	MO	0.6	0.6		MIT
4047.204	4048.347	24701.44	MO	0.6	0.5		MIT
4046.886	4048.029	24703.38	MO	0.5	1.3		MIT
4045.543	4046.686	24711.58	MO	0.5	0.5		MIT
4045.125	4046.268	24714.13	MO	0.3	0.5		MIT
4043.738	4044.880	24722.61	MO	0.9	0.9		M
4043.270	4044.412	24725.47	MO	0.5	0.3		MIT
4042.873	4044.015	24727.90	MO	1.2	1.2		MIT
4041.122	4042.264	24738.61	MO	0.8	0.7		MIT
4040.637	4041.779	24741.58	MO	0.7	0.5		MIT
4039.00	4040.14	24751.6	MO	0.5	0.5		MIT
4038.631	4039.772	24753.87	MO		1.0H		MIT
4038.084	4039.225	24757.23	MO	1.3	1.2		MIT
4037.778	4038.919	24759.10	MO	1.0	1.2		MIT
4037.303	4038.444	24762.01	MO	0.9	0.9		MIT
4036.667	4037.808	24765.92	MO	0.7	0.7		MIT
4035.661	4036.801	24772.09	MO	0.5	1.4		MIT
4033.999	4035.139	24782.29	MO	0.5	0.5		MIT
4033.631	4034.771	24784.56	MO	0.8	0.8		MIT
4032.502	4033.641	24791.49	MO	0.9	0.9		MIT
4030.915	4032.054	24801.25	MO	0.5	0.7		MIT
4029.941	4031.080	24807.25	MO	0.0	1.5		MIT

AIR WAVELENGTH (ANGSTRMS)	VACUUM WAVELENGTH (ANGSTRMS)	WAVENUMBER (KAYSERS)	LMENT	LOG ARC	INTENSITY SPK	DIS	REF
4029.51	4030.65	24809.9	MO	0.5	0.5		MIT
4028.82	4029.96	24814.1	MO	0.5	0.5		MIT
4028.646	4029.784	24815.22	MO	0.8	0.8		MIT
4026.924	4028.062	24825.83	MO	0.7	0.7		MIT
4025.99	4027.13	24831.6	MO	1.5H	1.5H		MIT
4025.487	4026.625	24834.70	MO	0.8	0.7		MIT
4024.091	4025.228	24843.31	MO	1.5	1.4		MIT
4023.56	4024.70	24846.6	MO		1.4		MIT
4021.015	4022.151	24862.32	MO	1.2	1.4		MIT
4020.668	4021.804	24864.46	MO	0.7			MIT
4020.455	4021.591	24865.78	MO	1.0	1.0		MIT
4019.789	4020.925	24869.90	MO	1.0			MIT
4019.046	4020.182	24874.50	MO	0.3	0.6		MIT
4017.382	4018.518	24884.80	MO	0.8	0.8		MIT
4016.702	4017.837	24889.01	MO	0.7	0.7		MIT
4016.062	4017.197	24892.98	MO		1.0		PU
4013.21	4014.34	24910.7	MO	0.0	1.6		EX
4012.80	4013.93	24913.2	MO	0.5	0.3		EX
4012.51	4013.64	24915.0	MO	0.5	0.5		EX
4012.270	4013.404	24916.50	MO	0.5	0.5		MIT
4011.966	4013.100	24918.39	MO	1.4	1.4		MIT
4010.30	4011.43	24928.7	MO		1.0		EX
4010.136	4011.270	24929.76	MO	0.7	0.7		MIT
4009.366	4010.499	24934.55	MO	1.3	1.4		MIT
4008.667	4009.800	24938.90	MO		1.3		MIT
4008.054	4009.187	24942.71	MO	0.6	0.7		MIT
4007.45	4008.58	24946.5	MO	0.6	0.7		EX
4006.95	4008.08	24949.6	MO		0.7		EX
4006.70	4007.83	24951.1	MO	0.7	0.7		EX
4006.049	4007.182	24955.20	MO	1.3	1.3		MIT
4005.75	4006.88	24957.1	MO	0.5	0.5		EX
4005.12	4006.25	24961.0	MO		1.3		EX
4003.452	4004.584	24971.38	MO	0.5	0.5		MIT
4002.97	4004.10	24974.4	MO		1.3		MIT
4000.497	4001.628	24989.83	MO	0.9	0.9		MIT
4000.386	4001.517	24990.52	MO	0.8	0.7		MIT
3999.864	4000.995	24993.78	MO		0.7		MIT
3998.631	3999.762	25001.49	MO		1.4		MIT
3998.286	3999.417	25003.65	MO	0.9	0.7		MIT
3996.251	3997.381	25016.38	MO	0.5	0.3		MIT
3995.482	3996.612	25021.19	MO	0.5	0.3		MIT
3994.628	3995.758	25026.54	MO	0.5	0.5		MIT
3993.934	3995.063	25030.89	MO	0.7	0.7		MIT
3993.051	3994.180	25036.43	MO	0.7	0.7		MIT
3991.854	3992.983	25043.93	MO	0.3	0.5		MIT
3991.391	3992.520	25046.84	MO	0.8	0.9		MIT
3990.838	3991.967	25050.31	MO		1.5		MIT
3989.922	3991.050	25056.06	MO	0.7	0.8		MIT
3989.506	3990.634	25058.67	MO		1.4H		MIT
3987.373	3988.501	25072.08	MO	0.6	0.7		MIT
3986.978	3988.106	25074.56	MO	0.6	0.0		MIT
3986.201	3987.328	25079.45	MO	0.7	1.3		MIT
3985.723	3986.850	25082.46	MO	0.7	0.5		MIT
3982.605	3983.731	25102.09	MO	0.7	0.7		MIT
3982.052	3983.178	25105.58	MO	0.9	0.9		MIT
3981.640	3982.766	25108.18	MO	0.3	0.7		MIT
3980.709	3981.835	25114.05	MO	0.7	0.7		MIT
3980.204	3981.330	25117.24	MO	1.2	1.2		MIT
3979.221	3980.347	25123.44	MO	0.8	1.1		MIT
3977.902	3979.027	25131.77	MO	1.0	1.0		MIT
3975.956	3977.081	25144.07	MO	1.0	0.9		MIT
3974.859	3975.983	25151.01	MO		1.3		MIT
3973.928	3975.052	25156.90	MO	0.7	0.5		MIT
3973.771	3974.895	25157.90	MO	1.4	1.4		MIT
3972.948	3974.072	25163.11	MO	0.6	1.3		MIT
3971.374	3972.497	25173.08	MO	0.3	0.5		MIT
3970.959	3972.082	25175.71	MO	0.7	0.7		MIT
3969.015	3970.138	25188.04	MO	0.5	0.5		MIT
3968.748	3969.871	25189.74	MO	0.9	1.7		MIT
3966.149	3967.271	25206.24	MO	0.7D	0.8		MIT

AIR WAVELENGTH (ANGSTRMS)	VACUUM WAVELENGTH (ANGSTRMS)	WAVENUMBER (KAYSERS)	LMENT	LOG ARC	INTENSITY SPK	DIS	REF
3965.757	3966.879	25208.73	MO	1.0	1.0		MIT
3963.986	3965.108	25220.00	MO	1.0	1.0		MIT
3963.526	3964.647	25222.92	MO	0.8	0.8		MIT
3961.988	3963.109	25232.72	MO	0.5			MIT
3961.503	3962.624	25235.80	MO	0.7	2.7		MIT
3960.137	3961.258	25244.51	MO		0.7		MIT
3959.953	3961.074	25245.68	MO	0.5	0.3		MIT
3959.673	3960.793	25247.47	MO	0.3	0.5		MIT
3958.860	3959.980	25252.65	MO	0.5	0.5		MIT
3958.604	3959.724	25254.28	MO	0.9	0.9		MIT
3957.971	3959.091	25258.32	MO	0.5	0.5		MIT
3957.651	3958.771	25260.36	MO	0.3	0.5		MIT
3955.493	3956.612	25274.15	MO	1.0	1.2		MIT
3953.928	3955.047	25284.15	MO	1.2	1.3		MIT
3952.992	3954.111	25290.14	MO	0.0	0.7		MIT
3952.273	3953.392	25294.74	MO	0.6	0.3		MIT
3951.533	3952.651	25299.47	MO	0.6	0.6		MIT
3951.348	3952.466	25300.66	MO	0.6	0.6		MIT
3950.986	3952.104	25302.98	MO	1.2	1.2		MIT
3950.257	3951.375	25307.65	MO	0.6	0.6		MIT
3948.649	3949.767	25317.95	MO	0.8	1.0		MIT
3947.173	3948.290	25327.42	MO	1.0	1.0		MIT
3946.886	3948.003	25329.26	MO	0.6	0.5		MIT
3945.250	3946.367	25339.76	MO	1.0	0.8		MIT
3943.507	3944.623	25350.96	MO	0.7	0.6		MIT
3943.086	3944.202	25353.67	MO	1.0	1.0		MIT
3943.044	3944.160	25353.94	MO	1.0	0.8		MIT
3942.713	3943.829	25356.07	MO		1.3		MIT
3942.183	3943.299	25359.48	MO	0.6	0.3		MIT
3941.757	3942.873	25362.22	MO	0.8	0.3		MIT
3941.478	3942.594	25364.01	MO	0.7	2.2		MIT
3939.495	3940.610	25376.78	MO	0.6	0.6		MIT
3939.138	3940.253	25379.08	MO	0.6	0.6		MIT
3938.722	3939.837	25381.76	MO	0.9	1.0		MIT
3938.426	3939.541	25383.67	MO	0.3	0.5		MIT
3936.728	3937.842	25394.62	MO	0.6	0.6		MIT
3936.143	3937.257	25398.39	MO	0.5	0.5		MIT
3935.697	3936.811	25401.27	MO	0.8	0.8		MIT
3935.183	3936.297	25404.59	MO	0.7	0.7		MIT
3935.024	3936.138	25405.61	MO	0.7	0.7		MIT
3934.956	3936.070	25406.05	MO	0.7	0.8		MIT
3934.260	3935.374	25410.55	MO	0.7	0.7		MIT
3931.403	3932.516	25429.01	MO	0.7	0.7		MIT
3930.203	3931.316	25436.78	MO	0.7	0.7		MIT
3929.584	3930.697	25440.78	MO		1.4		MIT
3928.788	3929.900	25445.94	MO	0.7	0.7		MIT
3928.703	3929.815	25446.49	MO	0.7	0.7		MIT
3928.281	3929.393	25449.22	MO	0.9	0.9		MIT
3927.615	3928.727	25453.54	MO		1.0		MIT
3926.955	3928.067	25457.81	MO		1.3		MIT
3926.441	3927.553	25461.15	MO	0.6	0.5		MIT
3926.327	3927.439	25461.89	MO	0.5	0.5		MIT
3925.832	3926.944	25465.10	MO	0.0	1.5		MIT
3925.649	3926.761	25466.28	MO		1.3		MIT
3923.748	3924.859	25478.62	MO	1.0	1.3		MIT
3922.661	3923.772	25485.68	MO	0.5	0.5		MIT
3922.320	3923.431	25487.90	MO	1.0	1.0		MIT
3921.767	3922.878	25491.49	MO	0.5	0.0		MIT
3921.541	3922.651	25492.96	MO		1.3		MIT
3920.923	3922.033	25496.98	MO	0.7	0.7		MIT
3920.079	3921.189	25502.47	MO	0.5	0.5		MIT
3919.453	3920.563	25506.54	MO		1.3		MIT
3917.779	3918.889	25517.44	MO	1.2	0.7		MIT
3917.545	3918.654	25518.96	MO	1.2	1.0		MIT
3916.925	3918.034	25523.00	MO	0.9	0.7		MIT
3916.433	3917.542	25526.21	MO	0.7	0.7		MIT
3915.661	3916.770	25531.24	MO	0.5	0.0		MIT
3915.438	3916.547	25532.70	MO	0.8	1.7		MIT
3915.014	3916.123	25535.46	MO	0.5	0.0		MIT
3913.678	3914.786	25544.18	MO	0.6	0.5		MIT

AIR WAVELENGTH (ANGSTRMS)	VACUUM WAVELENGTH (ANGSTRMS)	WAVENUMBER (KAYSERS)	LMENT	LOG ARC	INTENSITY SPK	DIS	REF
3913.531	3914.639	25545.14	MO	0.6	0.6		MIT
3913.362	3914.470	25546.24	MO	0.9	0.9		MIT
3911.945	3913.053	25555.49	MO	0.7	0.7		MIT
3911.092	3912.200	25561.07	MO	1.3	1.3		MIT
3909.548	3910.655	25571.16	MO	0.8	0.8		MIT
3908.612	3909.719	25577.28	MO		1.5		MIT
3908.249	3909.356	25579.66	MO	0.7	0.5		MIT
3906.976	3908.083	25588.00	MO	0.7	0.7		MIT
3906.916	3908.023	25588.39	MO	0.7	0.7		MIT
3906.480	3907.587	25591.24	MO	0.7	1.0		MIT
3905.019	3906.125	25600.82	MO		0.7		MIT
3902.963	3904.069	25614.30	MO I	3.0G	2.7G		MIT
3901.769	3902.874	25622.14	MO	1.2	1.3		MIT
3900.665	3901.770	25629.39	MO	0.3			MIT
3900.227	3901.332	25632.27	MO	0.0	0.5		MIT
3897.507	3898.611	25650.16	MO	0.7	0.9		MIT
3896.852	3897.956	25654.47	MO	0.7	0.7		MIT
3896.379	3897.483	25657.58	MO	0.9	1.0		MIT
3894.015	3895.118	25673.16	MO	0.0	0.5		MIT
3893.325	3894.428	25677.71	MO	0.7	0.7		MIT
3892.292	3893.395	25684.53	MO	0.7	1.2		MIT
3891.877	3892.980	25687.26	MO	0.3	1.3D		MIT
3890.706	3891.808	25695.00	MO	0.7	0.9		MIT
3890.455	3891.557	25696.65	MO	0.3	0.5		MIT
3889.953	3891.055	25699.97	MO	0.3	0.6		MIT
3889.285	3890.387	25704.38	MO	0.3	0.5		MIT
3888.875	3889.977	25707.09	MO	0.9	0.9		MIT
3888.178	3889.280	25711.70	MO	1.0	0.9		MIT
3887.960	3889.062	25713.14	MO	0.5	0.6		MIT
3887.672	3888.774	25715.05	MO	0.5	0.5		MIT
3886.822	3887.923	25720.67	MO	1.5	1.5		MIT
3885.508	3886.609	25729.37	MO	0.7	0.7		MIT
3883.698	3884.799	25741.36	MO		1.3		MIT
3883.462	3884.563	25742.92	MO	0.6	0.3		MIT
3882.950	3884.050	25746.32	MO	0.7	1.2		MIT
3882.306	3883.406	25750.59	MO		0.6		MIT
3881.598	3882.698	25755.29	MO		0.6		MIT
3879.984	3881.084	25766.00	MO	0.6	0.6		MIT
3879.522	3880.622	25769.07	MO	1.0	0.9		MIT
3879.020	3880.119	25772.40	MO	0.7	0.7		MIT
3874.285	3875.383	25803.90	MO	0.0	0.8		MIT
3874.154	3875.252	25804.77	MO	0.7	0.8		MIT
3873.259	3874.357	25810.74	MO		1.0		MIT
3871.882	3872.980	25819.91	MO		1.4		MIT
3871.446	3872.543	25822.82	MO		1.7		MIT
3870.593	3871.690	25828.51	MO	0.7	0.7		MIT
3870.436	3871.533	25829.56	MO	0.7	0.8		MIT
3869.080	3870.177	25838.61	MO	1.4	1.5		MIT
3868.808	3869.905	25840.43	MO	0.5	0.8		MIT
3867.673	3868.769	25848.01	MO	0.5	0.6		MIT
3866.787	3867.883	25853.94	MO		1.0		MIT
3866.691	3867.787	25854.58	MO	0.7	0.5		MIT
3865.153	3866.249	25864.86	MO		1.3		MIT
3864.110	3865.206	25871.84	MO I	3.0G	2.7G		MIT
3861.288	3862.383	25890.75	MO		0.6		MIT
3858.84	3859.93	25907.2	MO		1.2		MIT
3858.321	3859.415	25910.66	MO	0.7	0.7		MIT
3857.203	3858.297	25918.17	MO		1.8		MIT
3856.373	3857.467	25923.75	MO	0.7	0.6		MIT
3855.972	3857.065	25926.45	MO	0.7	0.7		MIT
3854.905	3855.998	25933.62	MO	0.6	0.7		MIT
3853.487	3854.580	25943.17	MO	0.3			MIT
3851.990	3853.082	25953.25	MO	1.0	1.2		MIT
3851.393	3852.485	25957.27	MO	0.8	0.7		MIT
3850.819	3851.911	25961.14	MO		1.4		MIT
3849.785	3850.877	25968.11	MO	0.7	0.7		MIT
3848.301	3849.392	25978.13	MO	1.4	1.3		MIT
3847.248	3848.339	25985.24	MO	1.4	1.4		MIT
3846.180	3847.271	25992.45	MO	0.7	0.7		MIT
3845.954	3847.045	25993.98	MO	1.3	1.3		MIT

AIR WAVELENGTH (ANGSTRMS)	VACUUM WAVELENGTH (ANGSTRMS)	WAVENUMBER (KAYSERS)	LMENT	LOG ARC	INTENSITY SPK	DIS	REF
3843.895	3844.985	26007.90	MO	0.7	0.8		MIT
3842.958	3844.048	26014.24	MO	0.5	0.5		MIT
3842.58	3843.67	26016.8	MO		1.4		MIT
3840.58	3841.67	26030.4	MO	0.8W	0.8W		MIT
3839.475	3840.564	26037.84	MO	0.7	0.7		MIT
3837.882	3838.971	26048.65	MO	0.5	0.5		MIT
3837.29	3838.38	26052.7	MO		2.0W		MIT
3837.141	3838.230	26053.68	MO		1.3		MIT
3836.905	3837.993	26055.28	MO	0.6	0.0		MIT
3835.312	3836.400	26066.10	MO	0.9	1.4		MIT
3834.974	3836.062	26068.40	MO	0.9	0.8		MIT
3834.639	3835.727	26070.68	MO	0.9	0.8		MIT
3833.748	3834.836	26076.74	MO	1.9	1.4		MIT
3833.47	3834.56	26078.6	MO		1.4		MIT
3832.37	3833.46	26086.1	MO		1.0H		MIT
3832.110	3833.197	26087.88	MO	1.0	0.9		MIT
3831.765	3832.852	26090.23	MO	0.6	0.5		MIT
3831.071	3832.158	26094.96	MO	0.6	0.6		MIT
3830.815	3831.902	26096.70	MO	0.7	0.7		MIT
3830.055	3831.142	26101.88	MO	0.7	0.8		MIT
3829.914	3831.001	26102.84	MO	0.6	0.5		MIT
3829.791	3830.878	26103.68	MO	0.7	0.7		MIT
3828.873	3829.959	26109.94	MO	1.6	1.5		MIT
3827.156	3828.242	26121.65	MO	1.4	1.4		MIT
3826.695	3827.781	26124.80	MO	1.5	1.4		MIT
3825.461	3826.546	26133.23	MO	0.5	0.6		MIT
3825.323	3826.408	26134.17	MO	0.6	0.6		MIT
3824.779	3825.864	26137.89	MO	0.6	0.6		MIT
3824.168	3825.253	26142.06	MO	0.7	0.7		MIT
3823.15	3824.23	26149.0	MO		1.3H		MIT
3822.984	3824.069	26150.16	MO	1.3	1.2		MIT
3821.950	3823.035	26157.23	MO	0.7	0.7		MIT
3821.642	3822.726	26159.34	MO	0.7	0.7		MIT
3820.921	3822.005	26164.28	MO	0.8	0.6		MIT
3819.873	3820.957	26171.45	MO	0.8	0.8		MIT
3819.778	3820.862	26172.10	MO	0.8	0.8		MIT
3819.22	3820.30	26175.9	MO		0.7		MIT
3818.663	3819.747	26179.75	MO	0.9	1.2		MIT
3817.974	3819.058	26184.47	MO	0.7	1.0		MIT
3817.175	3818.258	26189.95	MO	0.5	0.7		MIT
3816.61	3817.69	26193.8	MO		1.3		MIT
3815.36	3816.44	26202.4	MO		0.9L		MIT
3815.055	3816.138	26204.50	MO	0.7	0.6		MIT
3814.49	3815.57	26208.4	MO	0.6	0.6		MIT
3814.099	3815.182	26211.07	MO	0.6	0.5		MIT
3813.902	3814.984	26212.43	MO	0.3	1.3		MIT
3812.472	3813.554	26222.26	MO	0.7	0.6		MIT
3812.276	3813.358	26223.61	MO		1.4		MIT
3811.84	3812.92	26226.6	MO		1.4D		MIT
3811.390	3812.472	26229.70	MO	0.7	0.7		MIT
3810.816	3811.898	26233.65	MO	0.6	0.6		MIT
3810.129	3811.210	26238.38	MO	0.6	0.7		MIT
3808.623	3809.704	26248.76	MO	0.7	0.7		MIT
3807.854	3808.935	26254.06	MO	0.6	0.6		MIT
3807.654	3808.735	26255.44	MO	0.5	0.3		MIT
3807.008	3808.089	26259.89	MO	0.0	1.3W		MIT
3805.993	3807.073	26266.90	MO	1.0	1.0		MIT
3805.930	3807.010	26267.33	MO	0.7	0.7		MIT
3805.42	3806.50	26270.9	MO		0.7		MIT
3804.520	3805.600	26277.06	MO	1.3	1.3		MIT
3803.799	3804.879	26282.05	MO	0.5	0.7		MIT
3803.41	3804.49	26284.7	MO		1.3W		MIT
3802.170	3803.249	26293.31	MO	0.7	0.7		MIT
3801.837	3802.916	26295.61	MO	1.3	1.4		MIT
3800.942	3802.021	26301.80	MO	0.0	0.6		MIT
3800.090	3801.169	26307.70	MO	0.3	0.5		MIT
3798.252	3799.330	26320.43	MO I	3.0G	3.0G		MIT
3797.300	3798.378	26327.03	MO	0.7	0.7		MIT
3797.031	3798.109	26328.89	MO	0.7	0.7		MIT
3796.570	3797.648	26332.09	MO		1.3		MIT

AIR WAVELENGTH (ANGSTRMS)	VACUUM WAVELENGTH (ANGSTRMS)	WAVENUMBER (KAYSERS)	LMENT	LOG ARC	INTENSITY SPK DIS	REF
3796.043	3797.121	26335.74	MO	1.0	0.7	MIT
3795.59	3796.67	26338.9	MO		1.3D	MIT
3794.432	3795.509	26346.93	MO	0.7	1.0	MIT
3793.621	3794.698	26352.56	MO	0.7	0.7	MIT
3792.321	3793.398	26361.59	MO	0.6	0.7	MIT
3792.094	3793.171	26363.17	MO	0.0	1.2	MIT
3791.41	3792.49	26367.9	MO		1.0H	MIT
3788.256	3789.332	26389.88	MO	1.2	1.2	MIT
3786.361	3787.436	26403.08	MO	0.3	2.1	MIT
3785.514	3786.589	26408.99	MO	0.7	0.7	MIT
3785.032	3786.107	26412.35	MO	0.9	1.0	MIT
3783.532	3784.607	26422.83	MO	0.9	0.0	MIT
3783.181	3784.255	26425.28	MO		1.6	MIT
3782.673	3783.747	26428.83	MO	0.5	0.0	MIT
3782.189	3783.263	26432.21	MO	0.7		MIT
3782.075	3783.149	26433.00	MO		2.0	MIT
3781.593	3782.667	26436.37	MO	1.4	1.3	MIT
3781.211	3782.285	26439.04	MO	0.6	0.6H	MIT
3780.617	3781.691	26443.20	MO	0.5	0.6	MIT
3779.769	3780.843	26449.13	MO	1.4	1.5	MIT
3777.964	3779.037	26461.77	MO	0.7	0.6H	MIT
3777.724	3778.797	26463.45	MO	0.7	0.9	MIT
3776.550	3777.623	26471.67	MO	0.7	0.9	MIT
3776.104	3777.177	26474.80	MO	0.7	0.7	MIT
3775.647	3776.720	26478.00	MO	1.2	1.3	MIT
3773.325	3774.397	26494.30	MO	0.3	0.5	MIT
3772.822	3773.894	26497.83	MO	1.3	1.3	MIT
3771.952	3773.024	26503.94	MO	1.5	1.5	MIT
3771.424	3772.495	26507.65	MO	0.5	0.5	MIT
3770.523	3771.594	26513.99	MO	0.7	0.7	MIT
3770.447	3771.518	26514.52	MO	0.7	0.7	MIT
3770.353	3771.424	26515.18	MO	0.5	0.5	MIT
3769.996	3771.067	26517.69	MO	0.7	1.0	MIT
3768.742	3769.813	26526.52	MO	0.6	0.7	MIT
3768.624	3769.695	26527.35	MO	0.6	0.6	MIT
3767.732	3768.802	26533.63	MO	0.6	1.4	MIT
3766.398	3767.468	26543.02	MO	0.5	0.7	MIT
3765.737	3766.807	26547.68	MO	0.7	1.0H	MIT
3765.225	3766.295	26551.29	MO	0.8	0.8	MIT
3765.035	3766.105	26552.63	MO	0.7	0.6	MIT
3764.435	3765.505	26556.87	MO	0.7	1.0	MIT
3764.013	3765.083	26559.84	MO	0.5	0.6	MIT
3763.351	3764.420	26564.51	MO	1.0	1.0	MIT
3762.086	3763.155	26573.45	MO	1.0	1.5	MIT
3761.755	3762.824	26575.78	MO	0.8	0.8	MIT
3760.885	3761.954	26581.93	MO	0.8	1.0	MIT
3759.603	3760.671	26591.00	MO	1.0	1.0	MIT
3758.523	3759.591	26598.64	MO	1.4	1.4	MIT
3757.850	3758.918	26603.40	MO	0.3	0.5	MIT
3755.842	3756.909	26617.62	MO	0.7	0.5	MIT
3755.54	3756.61	26619.8	MO	0.0	1.7	MIT
3755.102	3756.169	26622.87	MO	1.5D	1.3	MIT
3754.682	3755.749	26625.85	MO		1.4	MIT
3753.53	3754.60	26634.0	MO		1.4	MIT
3752.378	3753.445	26642.20	MO	0.7	0.0	MIT
3752.261	3753.327	26643.03	MO		1.0	MIT
3751.940	3753.006	26645.30	MO	0.6	0.7	MIT
3751.203	3752.269	26650.54	MO	0.8	1.6D	MIT
3748.490	3749.555	26669.83	MO	1.2	1.0	MIT
3748.13	3749.20	26672.4	MO	0.0	1.7	MIT
3747.192	3748.257	26679.07	MO	1.2	1.0	MIT
3746.41	3747.47	26684.6	MO		1.6W	MIT
3745.48	3746.54	26691.3	MO	1.5A	1.3D	MIT
3744.945	3746.010	26695.07	MO	0.7	0.9	MIT
3744.366	3745.430	26699.20	MO	1.3	1.9	MIT
3743.810	3744.874	26703.17	MO	0.7	0.7	MIT
3742.285	3743.349	26714.05	MO	1.3		MIT
3741.81	3742.87	26717.4	MO		1.3	MIT
3741.121	3742.185	26722.36	MO	1.0H		MIT
3740.764	3741.827	26724.91	MO	1.0	0.6	MIT

AIR WAVELENGTH (ANGSTRMS)	VACUUM WAVELENGTH (ANGSTRMS)	WAVENUMBER (KAYSERS)	LMENT	LOG ARC	INTENSITY SPK DIS	REF
3740.571	3741.634	26726.29	MO	0.5	0.5	MIT
3738.841	3739.904	26738.65	MO	0.0	1.0	MIT
3737.906	3738.969	26745.34	MO	1.3	1.3	MIT
3736.40	3737.46	26756.1	MO		1.3L	MIT
3736.174	3737.236	26757.74	MO	0.6	0.0	MIT
3735.909	3736.971	26759.64	MO	0.7	0.5	MIT
3735.622	3736.684	26761.70	MO	0.7	0.7	MIT
3734.370	3735.432	26770.67	MO	1.2	0.7	MIT
3733.84	3734.90	26774.5	MO		1.4	MIT
3733.749	3734.811	26775.12	MO	0.6	0.0	MIT
3733.404	3734.466	26777.59	MO	1.0	1.0	MIT
3733.030	3734.091	26780.28	MO	1.0	1.0	MIT
3732.805	3733.866	26781.89	MO	1.0	1.0	MIT
3732.708	3733.769	26782.59	MO	1.0	1.0	MIT
3730.559	3731.620	26798.01	MO	0.7	1.5	MIT
3730.01	3731.07	26802.0	MO		1.3	MIT
3728.501	3729.561	26812.80	MO	1.0	1.0	MIT
3728.301	3729.361	26814.24	MO	1.0	0.7	MIT
3727.69	3728.75	26818.6	MO	1.4W	1.4W	MIT
3726.590	3727.650	26826.55	MO	0.5	0.5	MIT
3726.319	3727.379	26828.50	MO	0.6	0.7	MIT
3726.220	3727.280	26829.22	MO	0.6	0.7	MIT
3725.556	3726.616	26834.00	MO	1.3	1.3	MIT
3724.321	3725.380	26842.90	MO	0.3	0.5	MIT
3723.814	3724.873	26846.55	MO	1.3	1.5H	MIT
3723.514	3724.573	26848.72	MO	0.7	0.9	MIT
3723.007	3724.066	26852.37	MO		1.3	MIT
3722.306	3723.365	26857.43	MO	0.5	0.6	MIT
3720.254	3721.312	26872.24	MO	1.0	1.6	MIT
3719.74	3720.80	26875.9	MO		1.5	MIT
3719.692	3720.750	26876.30	MO	0.5		MIT
3719.553	3720.611	26877.31	MO	0.7	0.5	MIT
3719.047	3720.105	26880.96	MO		1.3	MIT
3718.479	3719.537	26885.07	MO	0.7	0.7	MIT
3716.97	3718.03	26896.0	MO		0.5H	MIT
3716.870	3717.927	26896.71	MO	0.6	1.4	MIT
3716.070	3717.127	26902.50	MO	0.7	0.7	MIT
3715.647	3716.704	26905.56	MO	0.7	0.7	MIT
3714.553	3715.610	26913.48	MO	0.7	0.6	MIT
3713.96	3715.02	26917.8	MO		1.6	MIT
3713.470	3714.526	26921.33	MO	0.9	0.9	MIT
3712.947	3714.003	26925.12	MO	0.9	0.9	MIT
3712.052	3713.108	26931.62	MO	0.6	0.6	MIT
3711.509	3712.565	26935.56	MO	0.7	0.7	MIT
3710.142	3711.198	26945.48	MO	1.3	1.2	MIT
3708.555	3709.610	26957.01	MO	0.7	0.7	MIT
3708.067	3709.122	26960.56	MO	0.3	0.5	MIT
3707.171	3708.226	26967.07	MO	0.7	0.7	MIT
3705.411	3706.465	26979.88	MO	0.6	0.5	MIT
3704.14	3705.19	26989.1	MO		1.3	MIT
3702.553	3703.607	27000.71	MO	1.0	2.2	MIT
3702.178	3703.231	27003.44	MO	0.6	0.0	MIT
3702.032	3703.085	27004.51	MO	1.0	0.7	MIT
3701.517	3702.570	27008.26	MO	0.7	0.3	MIT
3701.43	3702.48	27008.9	MO		0.5O	MIT
3700.012	3701.065	27019.25	MO	0.7	0.5	MIT
3699.85	3700.90	27020.4	MO		1.4D	MIT
3699.113	3700.166	27025.82	MO	0.6	0.6	MIT
3698.530	3699.583	27030.08	MO	0.7	0.7	MIT
3698.166	3699.218	27032.74	MO	1.2H	0.7	MIT
3697.034	3698.086	27041.01	MO	0.0	1.4	MIT
3696.045	3697.097	27048.25	MO	0.7	0.9	MIT
3694.945	3695.997	27056.30	MO	1.6	1.5	MIT
3694.421	3695.472	27060.14	MO		1.3	MIT
3693.375	3694.426	27067.80	MO	0.7	0.5	MIT
3693.232	3694.283	27068.85	MO	0.5		MIT
3692.645	3693.696	27073.15	MO	0.5	2.2	MIT
3692.081	3693.132	27077.29	MO	0.7	0.5	MIT
3690.593	3691.643	27088.21	MO	1.0	0.7	MIT
3690.359	3691.409	27089.92	MO		1.4	MIT

AIR WAVELENGTH (ANGSTRMS)	VACUUM WAVELENGTH (ANGSTRMS)	WAVENUMBER (KAYSERS)	LMENT	LOG ARC	INTENSITY SPK	DIS	REF
3690.118	3691.168	27091.69	MO	0.6	0.0		MIT
3688.973	3690.023	27100.10	MO	1.3	0.7		MIT
3688.307	3689.357	27105.00	MO	0.6	2.2		MIT
3687.962	3689.012	27107.53	MO	0.7	0.5		MIT
3686.964	3688.014	27114.87	MO	0.6	0.5		MIT
3686.575	3687.624	27117.73	MO	0.7	0.6		MIT
3686.114	3687.163	27121.12	MO	0.7	0.7		MIT
3685.773	3686.822	27123.63	MO	0.5	0.3		MIT
3684.327	3685.376	27134.27	MO	0.7	0.0		MIT
3684.22	3685.27	27135.1	MO	0.0	1.4D		MIT
3681.950	3682.998	27151.79	MO	0.6	0.0		MIT
3681.725	3682.773	27153.45	MO	0.9	1.4		MIT
3681.553	3682.601	27154.72	MO	0.6	0.5		MIT
3680.684	3681.732	27161.13	MO	1.3	1.3		MIT
3680.602	3681.650	27161.74	MO	1.3	1.3		MIT
3680.209	3681.257	27164.64	MO	0.5	0.7		MIT
3679.227	3680.275	27171.89	MO	0.6	0.6		MIT
3677.85	3678.90	27182.1	MO		0.7		MIT
3677.700	3678.747	27183.17	MO	0.8	0.8		MIT
3676.235	3677.282	27194.00	MO	0.9	1.0		MIT
3675.981	3677.028	27195.88	MO	1.0	0.6		MIT
3675.355	3676.402	27200.51	MO	1.4R	1.4		MIT
3673.871	3674.917	27211.50	MO	0.5	0.6		MIT
3673.216	3674.262	27216.35	MO	0.6	0.6		MIT
3672.820	3673.866	27219.28	MO	1.3	1.3		MIT
3671.65	3672.70	27228.0	MO	0.7	0.5		MIT
3670.668	3671.713	27235.24	MO		1.4		MIT
3669.344	3670.389	27245.07	MO	0.9	0.9		MIT
3668.488	3669.533	27251.43	MO	0.7	0.7		MIT
3668.002	3669.047	27255.04	MO	0.6	0.5		MIT
3666.935	3667.979	27262.97	MO	0.8	0.8		MIT
3666.715	3667.759	27264.60	MO	0.9	1.0		MIT
3665.747	3666.791	27271.80	MO	1.2	1.2		MIT
3664.813	3665.857	27278.75	MO	1.3	1.6		MIT
3664.305	3665.349	27282.53	MO	0.7	0.7		MIT
3663.965	3665.009	27285.07	MO	0.5	0.5		MIT
3663.884	3664.928	27285.67	MO	0.5	0.5		MIT
3663.659	3664.703	27287.34	MO	0.6	0.6		MIT
3663.30	3664.34	27290.0	MO	0.9	0.9		MIT
3662.990	3664.033	27292.33	MO	0.9	1.0		MIT
3662.152	3663.195	27298.57	MO	0.8	0.9		MIT
3661.782	3662.825	27301.33	MO	1.0	1.0		MIT
3661.076	3662.119	27306.60	MO	0.7	0.7		MIT
3660.917	3661.960	27307.78	MO	0.8	0.8		MIT
3659.359	3660.401	27319.41	MO	1.5	1.5		MIT
3658.92	3659.96	27322.7	MO		1.3		MIT
3658.333	3659.375	27327.07	MO	0.6	1.4		MIT
3658.147	3659.189	27328.46	MO	0.7	0.7		MIT
3657.354	3658.396	27334.38	MO	1.5	1.5		MIT
3655.788	3656.829	27346.09	MO		1.4		MIT
3655.076	3656.117	27351.42	MO	0.9	1.3		MIT
3654.583	3655.624	27355.11	MO	1.4	1.3		MIT
3653.897	3654.938	27360.25	MO	0.5	0.3		MIT
3653.55	3654.59	27362.8	MO	0.6D	0.6		MIT
3652.474	3653.515	27370.90	MO		1.4		MIT
3652.332	3653.373	27371.97	MO		1.4		MIT
3651.352	3652.392	27379.31	MO	1.0	0.3		MIT
3651.107	3652.147	27381.15	MO	0.3	1.7		MIT
3650.583	3651.623	27385.08	MO	0.5	0.3		MIT
3650.046	3651.086	27389.11	MO	0.5	1.4		MIT
3649.47	3650.51	27393.4	MO	0.7	0.7		MIT
3648.607	3649.647	27399.91	MO	1.0	1.0		MIT
3646.869	3647.908	27412.97	MO	0.5	0.7		MIT
3645.589	3646.628	27422.59	MO	0.5	0.5		MIT
3643.47	3644.51	27438.5	MO	0.0H	1.3		MIT
3642.723	3643.761	27444.17	MO	0.6	0.3		MIT
3642.413	3643.451	27446.51	MO	0.5	0.3		MIT
3642.204	3643.242	27448.08	MO	0.7	1.0		MIT
3641.825	3642.863	27450.94	MO	0.5	0.5		MIT
3641.390	3642.428	27454.22	MO	0.5	0.5		MIT

AIR WAVELENGTH (ANGSTRMS)	VACUUM WAVELENGTH (ANGSTRMS)	WAVENUMBER (KAYSERS)	LMENT	LOG ARC	INTENSITY SPK	DIS	REF
3640.989	3642.027	27457.24	MO	0.6	0.6		MIT
3640.911	3641.949	27457.83	MO	0.6	0.6		MIT
3640.623	3641.661	27460.00	MO	0.9	0.7		MIT
3639.64	3640.68	27467.4	MO		1.0		MIT
3639.546	3640.583	27468.13	MO	0.6	0.5		MIT
3639.007	3640.044	27472.19	MO	0.5	0.6		MIT
3638.56	3639.60	27475.6	MO	0.5	0.5		MIT
3638.425	3639.462	27476.59	MO	0.5	0.5		MIT
3638.205	3639.242	27478.25	MO	1.3	1.0		MIT
3637.791	3638.828	27481.38	MO		1.3		MIT
3637.518	3638.555	27483.44	MO	0.9	0.7		MIT
3636.827	3637.864	27488.66	MO	0.5	0.3		MIT
3636.64	3637.68	27490.1	MO		1.0		MIT
3635.612	3636.648	27497.85	MO	0.6	0.5		MIT
3635.429	3636.465	27499.23	MO	1.4	0.7		MIT
3635.144	3636.180	27501.39	MO	2.0H	1.0		MIT
3633.29	3634.33	27515.4	MO		1.3		MIT
3629.309	3630.344	27545.60	MO	0.8	0.8		MIT
3628.655	3629.689	27550.57	MO	0.7	0.8		MIT
3628.350	3629.384	27552.88	MO	1.0	0.7		MIT
3627.35	3628.38	27560.5	MO		1.5		EX
3626.184	3627.218	27569.34	MO	1.3	1.3		MIT
3625.56	3626.59	27574.1	MO		1.4		MIT
3624.625	3625.658	27581.20	MO	0.7	0.7		MIT
3624.464	3625.497	27582.42	MO	1.4	1.4		MIT
3623.70	3624.73	27588.2	MO		1.7		MIT
3623.231	3624.264	27591.81	MO	1.2	1.2		MIT
3622.850	3623.883	27594.71	MO		1.3		MIT
3621.618	3622.651	27604.10	MO	0.6	0.5		MIT
3618.35	3619.38	27629.0	MO		1.3		MIT
3617.554	3618.586	27635.11	MO		1.3		MIT
3616.84	3617.87	27640.6	MO	1.2	1.2		MIT
3615.740	3616.771	27648.97	MO	0.7	0.8		MIT
3615.15	3616.18	27653.5	MO	0.6	0.7		MIT
3614.686	3615.717	27657.03	MO	0.0	1.3		MIT
3614.253	3615.284	27660.35	MO	1.7D	1.5		MIT
3613.642	3614.673	27665.02	MO	0.5	0.6		MIT
3613.374	3614.405	27667.07	MO	0.7	0.7		MIT
3612.454	3613.484	27674.12	MO	0.8	0.8		MIT
3612.20	3613.23	27676.1	MO		1.3		MIT
3611.998	3613.028	27677.61	MO	0.9	0.7		MIT
3610.619	3611.649	27688.19	MO	0.5	0.7		MIT
3609.493	3610.523	27696.82	MO	0.7	1.0		MIT
3608.369	3609.398	27705.45	MO	1.2	1.2		MIT
3607.412	3608.441	27712.80	MO	0.6	0.6		MIT
3606.908	3607.937	27716.67	MO		1.5		MIT
3606.674	3607.703	27718.47	MO	0.9	1.3		MIT
3606.295	3607.324	27721.38	MO	0.0	1.0		MIT
3605.018	3606.046	27731.20	MO	0.3	0.7		MIT
3604.564	3605.592	27734.70	MO	0.7	0.7		MIT
3604.074	3605.102	27738.47	MO	0.8	1.2		MIT
3603.724	3604.752	27741.16	MO	0.6	0.7		MIT
3603.592	3604.620	27742.18	MO	0.5	0.6		MIT
3602.941	3603.969	27747.19	MO	1.3	1.4		MIT
3602.470	3603.498	27750.81	MO	0.3	0.5		MIT
3601.84	3602.87	27755.7	MO	0.0	0.5		MIT
3600.737	3601.764	27764.17	MO	0.3	0.6		MIT
3600.284	3601.311	27767.67	MO	0.7	0.7		MIT
3600.204	3601.231	27768.28	MO	0.5	0.5		MIT
3599.829	3600.856	27771.17	MO	0.5	0.7		MIT
3598.881	3599.908	27778.49	MO	1.0	1.2		MIT
3597.372	3598.398	27790.14	MO	0.3	0.5		PU
3596.351	3597.377	27798.03	MO		1.5		MIT
3595.706	3596.732	27803.02	MO	0.5	0.6		MIT
3595.551	3596.577	27804.22	MO	0.9	0.7		MIT
3595.40	3596.43	27805.4	MO		1.0		MIT
3594.550	3595.576	27811.96	MO	0.5	0.6		MIT
3592.65	3593.68	27826.7	MO		1.3		MIT
3592.24	3593.27	27829.8	MO		1.2		MIT
3592.015	3593.04C	27831.58	MO	0.7	0.5		MIT

AIR WAVELENGTH (ANGSTRMS)	VACUUM WAVELENGTH (ANGSTRMS)	WAVENUMBER (KAYSERS)	LMENT	LOG ARC	INTENSITY SPK	INTENSITY DIS	REF
3591.69	3592.71	27834.1	MO		1.4		MIT
3591.404	3592.429	27836.32	MO	0.6	0.3		MIT
3590.738	3591.763	27841.48	MO	1.0	1.0		MIT
3590.306	3591.331	27844.83	MO	0.5	0.3		MIT
3590.20	3591.22	27845.6	MO		1.2D		MIT
3589.52	3590.54	27850.9	MO		1.5D		MIT
3588.949	3589.973	27855.36	MO	1.0	1.0		MIT
3588.106	3589.130	27861.91	MO		1.2		MIT
3586.861	3587.885	27871.58	MO	0.8	0.8		MIT
3586.78	3587.80	27872.2	MO		0.7		MIT
3585.90	3586.92	27879.0	MO		1.2		M
3585.700	3586.723	27880.60	MO		1.3		MIT
3585.571	3586.594	27881.60	MO	0.6			MIT
3584.256	3585.279	27891.83	MO	0.7	1.0		MIT
3583.871	3584.894	27894.83	MO	0.3	0.5		MIT
3583.151	3584.174	27900.43	MO	0.5	0.6		MIT
3582.676	3583.699	27904.13	MO	1.3	0.0		MIT
3582.42	3583.44	27906.1	MO		1.2		MIT
3581.891	3582.913	27910.25	MO	1.0	1.2		MIT
3581.805	3582.827	27910.92	MO	1.0	0.7		MIT
3581.00	3582.02	27917.2	MO	0.3	0.5		MIT
3580.543	3581.565	27920.75	MO	0.7	1.0		MIT
3580.29	3581.31	27922.7	MO		1.0		MIT
3579.073	3580.095	27932.22	MO		1.4		MIT
3576.851	3577.872	27949.57	MO	0.0	0.5		MIT
3576.174	3577.195	27954.86	MO	0.5	0.7		MIT
3575.67	3576.69	27958.8	MO	0.5D	0.6D		MIT
3575.607	3576.628	27959.30	MO	0.5	0.6		MIT
3574.457	3575.477	27968.29	MO	0.7	0.7		MIT
3573.882	3574.902	27972.79	MO	1.3	1.4		MIT
3572.592	3573.612	27982.89	MO		1.3		MIT
3571.262	3572.282	27993.31	MO	0.7	0.9		MIT
3570.647	3571.667	27998.14	MO	1.2	1.3		MIT
3570.462	3571.481	27999.58	MO	0.3	0.6		MIT
3569.505	3570.524	28007.09	MO	0.3	0.7		MIT
3568.879	3569.898	28012.00	MO	0.6	0.3		MIT
3568.181	3569.200	28017.48	MO	0.5	0.6		MIT
3568.067	3569.086	28018.38	MO	0.5	0.5		MIT
3567.735	3568.754	28020.99	MO	0.3	0.5		MIT
3567.465	3568.484	28023.11	MO	0.0	0.5		MIT
3567.068	3568.087	28026.23	MO	0.7	0.7		MIT
3566.747	3567.766	28028.75	MO	0.5	0.5		MIT
3566.309	3567.327	28032.19	MO	0.5	0.5		MIT
3566.054	3567.072	28034.20	MO	1.0	1.0		MIT
3564.516	3565.534	28046.29	MO	0.0	0.9		MIT
3564.291	3565.309	28048.06	MO	0.7	0.7		MIT
3563.755	3564.773	28052.28	MO	1.0	1.0		MIT
3563.138	3564.156	28057.14	MO	1.2	1.2		MIT
3562.106	3563.123	28065.26	MO	0.7	0.7		MIT
3561.361	3562.378	28071.14	MO	0.0	1.3		MIT
3559.882	3560.899	28082.80	MO	0.7	0.7		MIT
3559.720	3560.737	28084.08	MO		1.4		MIT
3559.245	3560.262	28087.82	MO	0.5	0.5		MIT
3558.795	3559.811	28091.37	MO	0.5	0.7		MIT
3558.100	3559.116	28096.86	MO	1.2	1.2		MIT
3556.940	3557.956	28106.02	MO	0.3D	1.4		MIT
3556.327	3557.343	28110.87	MO		1.4		MIT
3555.433	3556.449	28117.94	MO	0.5	0.5		MIT
3554.195	3555.210	28127.73	MO	0.7	1.0		MIT
3553.835	3554.850	28130.58	MO	1.0	0.0		MIT
3552.720	3553.735	28139.41	MO	0.5	0.7		MIT
3552.397	3553.412	28141.97	MO	0.5	0.5		MIT
3550.962	3551.976	28153.34	MO	0.6	0.8		MIT
3549.139	3550.153	28167.80	MO	0.0	0.5		MIT
3548.730	3549.744	28171.05	MO	0.3	0.5		MIT
3548.439	3549.453	28173.36	MO	0.5H	0.3H		MIT
3547.404	3548.418	28181.58	MO	0.5	0.6		MIT
3545.997	3547.010	28192.76	MO		1.4		MIT
3544.617	3545.630	28203.73	MO	1.6H	0.3		MIT
3543.115	3544.127	28215.69	MO	0.6	0.5		MIT
3542.775	3543.787	28218.40	MO	0.5	0.3		MIT
3542.166	3543.178	28223.25	MO	0.7	0.7		MIT
3540.454	3541.466	28236.90	MO	0.3	0.5		MIT
3540.212	3541.224	28238.83	MO	0.5	0.5		MIT
3539.465	3540.476	28244.79	MO	0.5	0.5		MIT
3538.923	3539.934	28249.11	MO	0.5	0.6		MIT
3537.275	3538.286	28262.27	MO	1.2	1.2		MIT
3537.09	3538.10	28263.7	MO		1.7		MIT
3536.126	3537.137	28271.46	MO	0.3	0.5		MIT
3535.841	3536.852	28273.73	MO	0.0	0.5		MIT
3535.271	3536.281	28278.29	MO	0.5	0.5		MIT
3534.688	3535.698	28282.96	MO	1.0	1.4		MIT
3533.07	3534.08	28295.9	MO		1.4D		MIT
3532.55	3533.56	28300.1	MO		1.0		MIT
3531.303	3532.312	28310.07	MO	0.5	0.5		MIT
3530.809	3531.818	28314.03	MO	0.0	0.5		MIT
3527.87	3528.88	28337.6	MO	0.0	1.5D		MIT
3526.537	3527.545	28348.33	MO	0.3	0.5		MIT
3526.35	3527.36	28349.8	MO	0.3	0.5		MIT
3525.939	3526.947	28353.13	MO	0.7	0.5		MIT
3524.981	3525.989	28360.84	MO	0.7	0.7		MIT
3524.646	3525.654	28363.53	MO	0.7	1.7H		MIT
3524.228	3525.236	28366.90	MO	0.9	0.5		MIT
3523.89	3524.90	28369.6	MO		1.0		MIT
3523.316	3524.323	28374.24	MO	0.6	0.5		MIT
3522.371	3523.378	28381.85	MO	0.7	0.5		MIT
3522.063	3523.070	28384.33	MO		1.0		MIT
3521.413	3522.420	28389.57	MO	0.9	0.5		MIT
3521.175	3522.182	28391.49	MO	0.5	1.0W		MIT
3520.198	3521.205	28399.37	MO	0.0	1.3		MIT
3518.570	3519.576	28412.51	MO	0.0	0.5		MIT
3518.221	3519.227	28415.33	MO	0.8	1.2		MIT
3517.557	3518.563	28420.70	MO	0.3	1.3		MIT
3515.689	3516.694	28435.79	MO		0.8		MIT
3515.053	3516.058	28440.94	MO		0.8		MIT
3514.782	3515.787	28443.13	MO	0.6	0.5		MIT
3513.708	3514.713	28451.83	MO	0.5	0.0		MIT
3511.793	3512.797	28467.34	MO	0.0	0.5		MIT
3510.780	3511.784	28475.55	MO	0.6	0.8		MI
3509.986	3510.990	28482.00	MO	0.0	0.5		MIT
3509.23	3510.23	28488.1	MO		0.5		MIT
3508.476	3509.480	28494.25	MO	0.3	0.7		MIT
3508.115	3509.118	28497.19	MO	1.3	1.3		MIT
3507.287	3508.290	28503.91	MO	0.5	0.5		MIT
3507.024	3508.027	28506.05	MO	0.5	0.3		MIT
3506.704	3507.707	28508.65	MO		1.3		MIT
3505.315	3506.318	28519.95	MO	1.0	1.3		MIT
3504.413	3505.415	28527.29	MO	1.3	1.3		MIT
3503.502	3504.504	28534.71	MO	0.5	0.5		MIT
3502.673	3503.675	28541.46	MO		1.3		M
3501.963	3502.965	28547.25	MO		1.5		MIT
3499.985	3500.986	28563.38	MO		1.4		MIT
3499.199	3500.200	28569.79	MO	0.5			MIT
3499.071	3500.072	28570.84	MO		1.3		MIT
3498.923	3499.924	28572.05	MO	0.5	0.0		MIT
3496.704	3497.704	28590.18	MO	0.5	0.5		MIT
3493.337	3494.337	28617.74	MO	0.8	1.0		MIT
3492.825	3493.824	28621.93	MO	0.5	0.6		MIT
3491.871	3492.870	28629.75	MO	0.5	0.5		MIT
3491.768	3492.767	28630.59	MO	0.5	0.5		MIT
3491.136	3492.135	28635.78	MO	0.5	0.7		MIT
3489.430	3490.429	28649.78	MO		0.5		MIT
3488.14	3489.14	28660.4	MO		1.3		MIT
3485.928	3486.926	28678.56	MO	0.9	0.6		MIT
3485.73	3486.73	28680.2	MO		1.3		MIT
3485.483	3486.481	28682.22	MO		1.4		MIT
3484.56	3485.56	28689.8	MO II				ME
3483.843	3484.840	28695.72	MO	1.3W	1.0		MIT
3483.673	3484.670	28697.12	MO	0.7	0.7		MIT
3482.76	3483.76	28704.6	MO		1.0		MIT

AIR WAVELENGTH (ANGSTRMS)	VACUUM WAVELENGTH (ANGSTRMS)	WAVENUMBER (KAYSERS)	LMENT	LOG ARC	INTENSITY SPK	DIS	REF
3482.395	3483.392	28707.65	MO	0.7	0.5		MIT
3482.340	3483.337	28708.10	MO	0.7	0.5		MIT
3481.787	3482.784	28712.67	MO	0.7	0.7		MIT
3480.089	3481.085	28726.67	MO	0.7	0.7		MIT
3479.426	3480.422	28732.15	MO	1.3	1.3		MIT
3476.830	3477.825	28753.60	MO	0.3	0.5		MIT
3475.032	3476.027	28768.48	MO	0.7	1.0		MIT
3474.643	3475.638	28771.70	MO	0.3	0.5		MIT
3474.203	3475.198	28775.34	MO	0.3	0.5		MIT
3473.223	3474.217	28783.46	MO	0.6	0.6		MIT
3470.924	3471.918	28802.52	MO	0.7	0.9		MIT
3469.634	3470.628	28813.23	MO	0.5	0.6		MIT
3469.219	3470.212	28816.68	MO	1.3	1.0		MIT
3467.853	3468.846	28828.03	MO	0.7	1.0		MIT
3466.969	3467.962	28835.38	MO	0.7	0.9		MIT
3466.826	3467.819	28836.57	MO	0.8	1.0		MIT
3465.86	3466.85	28844.6	MO	0.7	0.7		MIT
3465.661	3466.654	28846.26	MO	0.8	0.8		MIT
3463.573	3464.565	28863.65	MO	1.0W	1.3W		MIT
3463.032	3464.024	28868.16	MO		1.3		MIT
3462.06	3463.05	28876.3	MO		1.4		MIT
3460.784	3461.775	28886.91	MO	1.4	1.4		MIT
3460.226	3461.217	28891.57	MO	0.7	0.7		MIT
3459.923	3460.914	28894.10	MO	0.9	0.9		MIT
3458.863	3459.854	28902.96	MO		1.4		MIT
3458.152	3459.143	28908.90	MO	0.7	0.9		MIT
3456.524	3457.514	28922.51	MO	0.5	0.6		MIT
3456.387	3457.377	28923.66	MO	1.2W	1.0		MIT
3456.152	3457.142	28925.63	MO	0.7	0.8		MIT
3455.86	3456.85	28928.1	MO		1.0		MIT
3454.223	3455.213	28941.78	MO	0.7	0.7		MIT
3453.73	3454.72	28945.9	MO		1.2		MIT
3453.488	3454.477	28947.94	MO	0.9	0.3		ME
3453.37	3454.36	28948.9	MO	0.6	0.0		ME
3452.795	3453.784	28953.75	MO		1.3		MIT
3452.603	3453.592	28955.36	MO	1.0	0.9		MIT
3451.749	3452.738	28962.52	MO	1.0	1.3		MIT
3450.590	3451.579	28972.25	MO		1.3		MIT
3449.851	3450.839	28978.46	MO	0.7	0.7		MIT
3449.074	3450.062	28984.98	MO	1.0	1.0		MIT
3448.542	3449.530	28989.46	MO	0.3	1.3		MIT
3447.123	3448.111	29001.39	MO	1.4R	1.3		MIT
3446.085	3447.073	29010.12	MO	0.0	1.6H		MIT
3445.808	3446.795	29012.46	MO	0.5			MIT
3445.475	3446.462	29015.26	MO	0.3	1.4		MIT
3445.260	3446.247	29017.07	MO	0.9	0.3		MIT
3445.036	3446.023	29018.96	MO	0.7	0.7		MIT
3443.260	3444.247	29033.92	MO	1.3	1.2		MIT
3442.665	3443.652	29038.94	MO	0.7	0.7		MIT
3442.470	3443.457	29040.59	MO		0.8		MIT
3441.872	3442.858	29045.63	MO	0.7	0.9		MIT
3441.445	3442.431	29049.24	MO	1.0	1.0		MIT
3439.832	3440.818	29062.86	MO	0.5	0.5		MIT
3439.553	3440.539	29065.22	MO	0.5	0.5		MIT
3439.230	3440.216	29067.94	MO	0.3	0.3		MIT
3438.871	3439.857	29070.98	MO	1.3	1.3		MIT
3437.216	3438.201	29084.98	MO	1.4	1.4		MIT
3435.451	3436.436	29099.92	MO	0.8			MIT
3435.40	3436.38	29100.4	MO		1.8H		MIT
3434.790	3435.775	29105.52	MO	1.7	1.1		MIT
3434.496	3435.481	29108.01	MO	0.7	0.6		MIT
3434.045	3435.029	29111.83	MO	0.6	0.6		MIT
3433.377	3434.361	29117.50	MO	0.5	0.0		MIT
3433.279	3434.263	29118.33	MO		0.7		MIT
3432.870	3433.854	29121.80	MO	0.7	0.7		MIT
3432.232	3433.216	29127.21	MO		1.4		MIT
3430.222	3431.205	29144.28	MO	0.5	0.6		MIT
3428.890	3429.873	29155.60	MO	0.0	1.4		MIT
3427.902	3428.885	29164.00	MO	0.6	0.6		MIT
3426.790	3427.773	29173.46	MO	1.0	0.3		MIT

AIR WAVELENGTH (ANGSTRMS)	VACUUM WAVELENGTH (ANGSTRMS)	WAVENUMBER (KAYSERS)	LMENT	LOG ARC	INTENSITY SPK	DIS	REF
3426.002	3426.984	29180.17	MO	1.0	1.0		MIT
3425.479	3426.461	29184.63	MO	0.9	0.9		MIT
3424.758	3425.740	29190.77	MO	0.7	0.5		MIT
3424.600	3425.582	29192.12	MO	0.9	0.7		MIT
3422.777	3423.759	29207.67	MO		1.5H		MIT
3422.309	3423.290	29211.66	MO	1.0	1.0		MIT
3421.250	3422.231	29220.70	MO	0.8	0.8		MIT
3420.037	3421.018	29231.07	MO	1.2	1.2		MIT
3418.958	3419.939	29240.29	MO	0.9	1.0H		MIT
3418.519	3419.499	29244.05	MO	1.0	1.1		MIT
3418.341	3419.321	29245.57	MO	0.6	0.0		MIT
3417.512	3418.492	29252.66	MO	0.7	0.6		MIT
3417.00	3417.98	29257.1	MO		0.8		MIT
3416.143	3417.123	29264.39	MO	0.6	1.4		MIT
3416.025	3417.005	29265.40	MO	0.3	1.0		MIT
3415.633	3416.613	29268.75	MO	1.0	0.7		MIT
3415.273	3416.253	29271.84	MO	0.7	0.7		MIT
3414.422	3415.401	29279.14	MO		1.3		MIT
3413.657	3414.636	29285.70	MO	1.1	0.8		MIT
3413.370	3414.349	29288.16	MO	0.9	0.9		MIT
3412.955	3413.934	29291.72	MO		1.2		MIT
3412.295	3413.274	29297.39	MO	0.0	1.3		MIT
3412.016	3412.995	29299.78	MO	0.7	0.7		MIT
3411.094	3412.073	29307.70	MO	0.6	0.5		MIT
3410.62	3411.60	29311.8	MO		1.4		MIT
3408.634	3409.612	29328.85	MO	0.0	1.4		MIT
3407.639	3408.617	29337.42	MO	0.6	0.7		MIT
3405.937	3406.914	29352.07	MO	1.4	1.4		MIT
3405.68	3406.66	29354.3	MO	0.0D	0.5D		MIT
3405.204	3406.181	29358.39	MO	0.9	0.7		MIT
3404.864	3405.841	29361.32	MO	0.8	0.6		MIT
3404.342	3405.319	29365.83	MO	1.3	1.4		MIT
3403.353	3404.330	29374.36	MO	1.3	0.5		MIT
3402.812	3403.788	29379.03	MO	0.7D	2.0		MIT
3402.315	3403.291	29383.32	MO	0.7	0.0		MIT
3400.100	3401.076	29402.46	MO	0.0	1.0		MIT
3397.688	3398.663	29423.33	MO	1.3	1.3		MIT
3396.825	3397.800	29430.81	MO	0.6	0.5		MIT
3395.615	3396.590	29441.30	MO	0.6			MIT
3395.360	3396.335	29443.51	MO		1.4		MIT
3394.807	3395.781	29448.30	MO		1.0		MIT
3393.653	3394.627	29458.32	MO	1.3	1.0		MIT
3393.106	3394.080	29463.07	MO		1.0		MIT
3392.173	3393.147	29471.17	MO	1.2	0.3		MIT
3391.851	3392.825	29473.97	MO	0.0	1.5		MIT
3391.536	3392.510	29476.70	MO	0.6			MIT
3390.105	3391.078	29489.15	MO	0.7	0.5		MIT
3389.799	3390.772	29491.81	MO	1.2	1.3		MIT
3388.952	3389.925	29499.18	MO	0.5	0.5		MIT
3388.730	3389.703	29501.11	MO	0.5	0.5		MIT
3388.451	3389.424	29503.54	MO	1.0	0.9		MIT
3387.750	3388.723	29509.65	MO	1.0	0.9		MIT
3386.307	3387.279	29522.22	MO	0.0	0.9		MIT
3385.876	3386.848	29525.98	MO	0.8	1.0		MIT
3385.246	3386.218	29531.47	MO	0.7	0.3		MIT
3384.616	3385.588	29536.97	MO	1.5R	1.4		MIT
3383.981	3384.953	29542.51	MO	0.0	1.5		MIT
3383.554	3384.526	29546.24	MO	0.7	0.0		MIT
3382.484	3383.455	29555.58	MO	1.2	1.2		MIT
3382.292	3383.263	29557.26	MO	1.0	0.8		MIT
3381.761	3382.732	29561.90	MO	0.6	0.6		MIT
3381.119	3382.090	29567.52	MO	0.5	0.3		MIT
3380.215	3381.186	29575.42	MO	0.0	1.4		MIT
3379.966	3380.937	29577.60	MO	1.4	0.0		MIT
3379.762	3380.733	29579.39	MO	0.0	1.4		MIT
3378.461	3379.431	29590.78	MO	1.0	1.0		MIT
3378.198	3379.168	29593.08	MO	1.0	1.0		MIT
3376.773	3377.743	29605.57	MO	0.5			MIT
3376.689	3377.659	29606.31	MO		1.4		MIT
3375.648	3376.617	29615.44	MO	1.3	1.0		MIT

AIR WAVELENGTH (ANGSTRMS)	VACUUM WAVELENGTH (ANGSTRMS)	WAVENUMBER (KAYSERS)	LMENT	LOG ARC	INTENSITY SPK	DIS	REF
3375.222	3376.191	29619.17	MO	1.3	1.3		MIT
3374.768	3375.737	29623.16	MO	0.3	1.3		MIT
3373.811	3374.780	29631.56	MO	0.8	0.8		MIT
3373.126	3374.095	29637.58	MO	0.7	1.4		MIT
3372.920	3373.889	29639.39	MO	0.7	0.0		MIT
3372.065	3373.034	29646.90	MO	0.7	0.3		MIT
3371.692	3372.660	29650.18	MO		1.4		MIT
3370.520	3371.488	29660.49	MO	0.3	1.4		MIT
3369.939	3370.907	29665.61	MO	0.7	1.3		MIT
3369.253	3370.221	29671.65	MO	1.3	1.2		MIT
3367.969	3368.937	29682.96	MO	0.3	2.0		MIT
3367.630	3368.597	29685.95	MO	0.7			MIT
3367.400	3368.367	29687.97	MO	0.7	0.0		MIT
3367.162	3368.129	29690.07	MO	0.7	0.0		MIT
3366.91	3367.88	29692.3	MO		1.2		MIT
3366.115	3367.082	29699.31	MO		1.3		MIT
3365.476	3366.443	29704.95	MO	0.7	0.6		MIT
3365.399	3366.366	29705.62	MO	0.7	0.3		MIT
3365.303	3366.270	29706.47	MO		0.7		MIT
3363.783	3364.749	29719.90	MO	1.6	1.5		MIT
3363.001	3363.967	29726.80	MO	0.0	1.5		MIT
3362.725	3363.691	29729.25	MO	0.9	0.0		MIT
3362.366	3363.332	29732.42	MO	1.4	1.4		MIT
3361.373	3362.339	29741.20	MO	1.5	1.5		MIT
3360.28	3361.25	29750.9	MO		1.4		MIT
3359.538	3360.503	29757.45	MO	0.5	0.3		MIT
3359.200	3360.165	29760.44	MO	0.6	0.7		MIT
3358.124	3359.089	29769.98	MO	1.8 O	1.5		MIT
3357.778	3358.743	29773.04	MO	1.4	0.3		MIT
3357.078	3358.043	29779.25	MO		1.5		MIT
3356.453	3357.418	29784.80	MO	0.5	1.0		MIT
3354.989	3355.953	29797.79	MO	0.7			MIT
3354.760	3355.724	29799.83	MO	0.9			MIT
3354.488	3355.452	29802.24	MO	1.0			MIT
3354.124	3355.088	29805.48	MO	1.0			MIT
3351.512	3352.475	29828.71	MO	0.7	0.9		MIT
3351.118	3352.081	29832.21	MO	0.5	0.7		MIT
3350.299	3351.262	29839.50	MO	0.7	0.8		MIT
3349.193	3350.156	29849.36	MO	0.8	0.7		MIT
3348.940	3349.903	29851.61	MO	0.0	1.5		MIT
3348.075	3349.037	29859.33	MO	0.8	0.0		MIT
3347.269	3348.231	29866.51	MO	0.0	1.5		MIT
3347.018	3347.980	29868.75	MO	1.6	0.6		MIT
3346.403	3347.365	29874.24	MO	0.3	1.7		MIT
3346.142	3347.104	29876.57	MO		1.3		MIT
3345.572	3346.534	29881.66	MO	0.7	0.3		MIT
3344.746	3345.708	29889.04	MO	1.7	1.6		MIT
3343.722	3344.683	29898.20	MO	1.0	0.7		MIT
3342.59	3343.55	29908.3	MO		1.3		MIT
3341.846	3342.807	29914.98	MO	0.9	0.7		MIT
3340.508	3341.469	29926.96	MO	0.0	1.4		MIT
3340.167	3341.127	29930.02	MO	1.5	1.4		MIT
3339.35	3340.31	29937.3	MO	0.5D	0.5H		MIT
3338.431	3339.391	29945.58	MO	0.7	0.8		MIT
3338.112	3339.072	29948.44	MO	0.0	1.3		MIT
3336.571	3337.531	29962.27	MO	0.3			MIT
3336.507	3337.467	29962.85	MO	1.4D	1.4D		MIT
3336.077	3337.036	29966.71	MO	0.7	1.3		MIT
3335.503	3336.462	29971.87	MO	0.5	1.3		MIT
3335.076	3336.035	29975.70	MO	0.6	1.3		MIT
3334.966	3335.925	29976.69	MO	0.9	0.5		MIT
3333.334	3334.293	29991.37	MO	0.7	0.5		MIT
3332.52	3333.48	29998.7	MO		1.9		MIT
3331.59	3332.55	30007.1	MO		0.8D		MIT
3331.405	3332.363	30008.73	MO	1.2	1.2		MIT
3330.90	3331.86	30013.3	MO		1.3		MIT
3330.663	3331.621	30015.42	MO		1.0		MIT
3330.294	3331.252	30018.74	MO		1.4		MIT
3329.215	3330.173	30028.47	MO	0.3	2.0		MIT
3329.035	3329.993	30030.10	MO	1.2			MIT

AIR WAVELENGTH (ANGSTRMS)	VACUUM WAVELENGTH (ANGSTRMS)	WAVENUMBER (KAYSERS)	LMENT	LOG ARC	INTENSITY SPK	DIS	REF
3328.563	3329.520	30034.35	MO	0.0	1.3		MIT
3327.661	3328.618	30042.50	MO	0.7	0.6		MIT
3327.301	3328.258	30045.75	MO	1.6	1.3		MIT
3326.109	3327.066	30056.51	MO	0.7	0.6		MIT
3325.674	3326.631	30060.45	MO	1.7	1.4		MIT
3325.226	3326.183	30064.50	MO		1.6		MIT
3325.131	3326.088	30065.35	MO	0.6			MIT
3323.949	3324.905	30076.04	MO	1.6	1.4		MIT
3322.170	3323.126	30092.15	MO	0.6	1.5		MIT
3321.196	3322.152	30100.97	MO	0.0	1.2		MIT
3320.902	3321.858	30103.64	MO	0.5	1.9		MIT
3320.285	3321.240	30109.23	MO	0.9	0.3		MIT
3319.593	3320.548	30115.51	MO	1.0	0.7		MIT
3318.366	3319.321	30126.64	MO	0.6	0.5H		MIT
3317.04	3317.99	30138.7	MO		1.3		MIT
3316.69	3317.64	30141.9	MO		0.9		MIT
3316.50	3317.45	30143.6	MO	0.6D	0.0		MIT
3314.463	3315.417	30162.12	MO		1.3		MIT
3313.893	3314.847	30167.31	MO	0.5			MIT
3313.624	3314.578	30169.76	MO	0.7	1.7		MIT
3312.94	3313.89	30176.0	MO	0.0	1.7		MIT
3312.330	3313.283	30181.54	MO	1.2	0.9		MIT
3310.771	3311.724	30195.75	MO	1.3	1.3		MIT
3310.405	3311.358	30199.09	MO	0.5	0.5		MIT
3310.057	3311.010	30202.27	MO	0.6	0.9		MIT
3309.42	3310.37	30208.1	MO	0.5	1.4		MIT
3307.428	3308.380	30226.27	MO		1.7		MIT
3307.122	3308.074	30229.07	MO	1.5	1.2		MIT
3306.648	3307.600	30233.40	MO	0.9	0.7		MIT
3305.905	3306.857	30240.20	MO	1.2	1.0		MIT
3305.565	3306.517	30243.31	MO	1.6	1.5		MIT
3305.25	3306.20	30246.2	MO		1.2		MIT
3304.52	3305.47	30252.9	MO		1.3		MIT
3304.224	3305.175	30255.58	MO	1.4	1.0		MIT
3303.957	3304.908	30258.03	MO	0.3	1.2		MIT
3303.54	3304.49	30261.9	MO		1.3		MIT
3303.342	3304.293	30263.66	MO	1.4	0.7		MIT
3303.113	3304.064	30265.76	MO	0.6	0.3		MIT
3302.716	3303.667	30269.40	MO		1.4		MIT
3301.707	3302.658	30278.65	MO	0.6	0.6		MIT
3300.689	3301.639	30287.98	MO	0.8	0.8		MIT
3300.597	3301.547	30288.83	MO		1.2		MIT
3299.784	3300.734	30296.29	MO	0.5			MIT
3299.059	3300.009	30302.95	MO	0.7	0.0		MIT
3297.684	3298.634	30315.58	MO	0.0	1.8		MIT
3297.543	3298.493	30316.88	MO	0.7			MIT
3296.582	3297.531	30325.72	MO		1.4		MIT
3296.404	3297.353	30327.35	MO	1.2	0.7		MIT
3296.29	3297.24	30328.4	MO		0.7		MIT
3295.432	3296.381	30336.30	MO	0.7	0.7		MIT
3295.181	3296.130	30338.61	MO	0.5	0.0		MIT
3294.851	3295.800	30341.65	MO	1.2	0.7		MIT
3293.844	3294.793	30350.92	MO	0.6	0.0		MIT
3293.656	3294.605	30352.66	MO	0.8	0.0		MIT
3292.983	3293.931	30358.86	MO	0.0	1.0		MIT
3292.312	3293.260	30365.05	MO	1.0	2.5		MIT
3291.595	3292.543	30371.66	MO	0.9	0.6		MIT
3290.823	3291.771	30378.78	MO	1.6	2.0		MIT
3289.844	3290.792	30387.82	MO	1.0	1.0		MIT
3289.015	3289.962	30395.48	MO	1.6	1.5		MIT
3287.383	3288.330	30410.57	MO	1.3	0.6		MIT
3287.202	3288.149	30412.25	MO	0.0	1.4		MIT
3285.996	3286.943	30423.41	MO	0.6	0.0		MIT
3285.607	3286.554	30427.01	MO	0.0	0.6		MIT
3285.359	3286.305	30429.31	MO	1.3	1.0		MIT
3285.024	3285.970	30432.41	MO	1.2	0.7		MIT
3284.622	3285.568	30436.14	MO		1.6		MIT
3284.560	3285.506	30436.71	MO	0.5			MIT
3283.365	3284.311	30447.79	MO	0.7	0.0		MIT
3282.909	3283.855	30452.02	MO	0.0	1.5		MIT

AIR WAVELENGTH (ANGSTRMS)	VACUUM WAVELENGTH (ANGSTRMS)	WAVENUMBER (KAYSERS)	LMENT	LOG ARC	INTENSITY SPK	DIS	REF
3282.481	3283.427	30455.99	MO	0.3	0.9		MIT
3281.617	3282.563	30464.00	MO		1.3		MIT
3281.345	3282.290	30466.53	MO	0.7	0.0		MIT
3281.068	3282.013	30469.10	MO	1.4	0.5		MIT
3280.877	3281.822	30470.87	MO		1.4		MIT
3280.318	3281.263	30476.07	MO	0.7	0.7		MIT
3279.442	3280.387	30484.21	MO	1.2	0.5		MIT
3278.881	3279.826	30489.42	MO	0.0	1.7		MIT
3278.655	3279.600	30491.52	MO	0.6	0.0		MIT
3277.168	3278.112	30505.36	MO	0.6	0.3		MIT
3276.336	3277.280	30513.11	MO	0.3	1.6		MIT
3276.076	3277.020	30515.53	MO	0.9	0.0		MIT
3274.628	3275.572	30529.02	MO		1.5		MIT
3274.204	3275.148	30532.98	MO		0.7H		MIT
3273.582	3274.525	30538.78	MO		1.4		MIT
3272.898	3273.841	30545.16	MO	0.6	0.0		MIT
3272.382	3273.325	30549.97	MO		1.4		MIT
3271.666	3272.609	30556.66	MO		1.8		MIT
3270.904	3271.847	30563.78	MO	1.7	1.4		MIT
3268.979	3269.921	30581.78	MO		1.4		MIT
3268.193	3269.135	30589.13	MO	0.8	0.3		MIT
3267.639	3268.581	30594.32	MO	0.6	1.5		MIT
3266.887	3267.829	30601.36	MO		1.6		MIT
3266.289	3267.231	30606.96	MO	0.8	0.5		MIT
3266.164	3267.106	30608.13	MO	0.8	0.7		MIT
3265.139	3266.080	30617.74	MO	1.4	1.4		MIT
3264.404	3265.345	30624.63	MO	1.4	1.3		MIT
3263.832	3264.773	30630.00	MO	1.2	1.2		MIT
3262.628	3263.569	30641.30	MO	1.2	0.6		MIT
3262.353	3263.294	30643.89	MO	0.9	0.0		MIT
3262.188	3263.129	30645.44	MO	0.0	1.3		MIT
3261.84	3262.78	30648.7	MO	0.3			MIT
3260.484	3261.424	30661.45	MO	1.0	0.7		MIT
3259.539	3260.479	30670.34	MO	0.7	0.5		MIT
3259.161	3260.101	30673.90	MO	0.7	0.6		MIT
3258.687	3259.627	30678.36	MO	0.0	1.4		MIT
3256.728	3257.667	30696.81	MO	0.6	0.3		MIT
3256.210	3257.149	30701.70	MO	1.6	1.4		MIT
3255.246	3256.185	30710.79	MO		1.6		MIT
3254.680	3255.619	30716.13	MO	0.9	1.7		MIT
3253.78	3254.72	30724.6	MO	0.0D	1.7D		MIT
3253.510	3254.448	30727.17	MO	0.5			MIT
3252.33	3253.27	30738.3	MO		1.5		MIT
3251.944	3252.882	30741.97	MO	0.5	0.0		MIT
3251.649	3252.587	30744.76	MO	0.8	0.5		MIT
3251.361	3252.299	30747.48	MO	0.6	0.3		MIT
3250.747	3251.685	30753.29	MO	0.3	2.0		MIT
3249.924	3250.861	30761.08	MO	0.8	0.5		MIT
3247.621	3248.558	30782.89	MO	1.5	1.3		MIT
3245.923	3246.859	30798.99	MO	1.0	0.8		MIT
3244.472	3245.408	30812.77	MO	0.7	0.7		MIT
3243.202	3244.138	30824.83	MO		1.4		MIT
3240.713	3241.648	30848.50	MO	0.5	1.8		MIT
3240.493	3241.428	30850.60	MO	1.4	0.7		MIT
3238.399	3239.334	30870.55	MO		1.6		MIT
3237.985	3238.919	30874.49	MO	1.2	0.5		MIT
3237.84	3238.77	30875.9	MO		1.3		MIT
3237.075	3238.009	30883.17	MO	1.6	1.4		MIT
3236.420	3237.354	30889.42	MO	0.6	0.3		MIT
3235.878	3236.812	30894.60	MO	0.5	0.3		MIT
3235.385	3236.319	30899.30	MO	1.3	1.4		MIT
3235.020	3235.954	30902.79	MO		1.4		MIT
3234.180	3235.113	30910.82	MO	0.9	0.7		MIT
3233.140	3234.073	30920.76	MO	1.7	1.5		MIT
3231.123	3232.056	30940.06	MO	0.7	0.0		MIT
3230.786	3231.719	30943.29	MO	0.6	0.3		MIT
3229.794	3230.726	30952.79	MO	1.4	0.6		MIT
3229.710	3230.642	30953.60	MO		1.7		MIT
3229.36	3230.29	30956.9	MO		0.7		MIT
3228.215	3229.147	30967.93	MO	1.6	1.4		MIT

AIR WAVELENGTH (ANGSTRMS)	VACUUM WAVELENGTH (ANGSTRMS)	WAVENUMBER (KAYSERS)	LMENT	LOG ARC	INTENSITY SPK	DIS	REF
3226.795	3227.727	30981.56	MO	0.3	0.5H		MIT
3226.55	3227.48	30983.9	MO		1.3		MIT
3224.408	3225.339	31004.49	MO	0.6			MIT
3223.494	3224.425	31013.28	MO	0.8	0.8		MIT
3222.90	3223.83	31019.0	MO		1.6		MIT
3221.735	3222.665	31030.22	MO	1.3	1.3		MIT
3220.855	3221.785	31038.69	MO	0.9	0.9		MIT
3219.402	3220.332	31052.70	MO	0.7H	1.4		MIT
3218.508	3219.438	31061.33	MO	0.7	0.0		MIT
3216.779	3217.708	31078.02	MO	0.7	0.7		MIT
3216.068	3216.997	31084.89	MO		1.7		MIT
3215.281	3216.210	31092.50	MO	0.6	0.3		MIT
3215.073	3216.002	31094.51	MO	1.4	1.3		MIT
3214.442	3215.370	31100.61	MO	1.3	1.7		MIT
3213.317	3214.245	31111.50	MO	1.0	0.0		MIT
3212.593	3213.521	31118.51	MO	1.2	1.0		MIT
3211.510	3212.438	31129.01	MO	0.7	0.3		MIT
3210.966	3211.894	31134.28	MO	1.2	1.3		MIT
3210.652	3211.580	31137.33	MO	0.6	0.0		MIT
3210.555	3211.483	31138.27	MO	0.6	0.0		MIT
3210.456	3211.383	31139.23	MO	0.6	0.0		MIT
3209.712	3210.639	31146.45	MO		1.5		MIT
3208.834	3209.761	31154.97	MO	2.2R	1.8		MIT
3207.170	3208.097	31171.13	MO		1.0		MIT
3206.804	3207.731	31174.69	MO		1.3		MIT
3205.883	3206.809	31183.64	MO	1.5	1.3		MIT
3205.541	3206.467	31186.97	MO	0.7	0.3		MIT
3205.217	3206.143	31190.12	MO	1.3	0.6		MIT
3204.935	3205.861	31192.87	MO		1.4		MIT
3202.133	3203.058	31220.16	MO	0.6	0.0		MIT
3201.500	3202.425	31226.33	MO	0.0	1.5		MIT
3200.889	3201.814	31232.29	MO	0.7			MIT
3200.213	3201.138	31238.89	MO	0.7	1.5		MIT
3199.241	3200.166	31248.38	MO		1.2		MIT
3198.852	3199.777	31252.18	MO	0.8	0.0		MIT
3198.411	3199.335	31256.49	MO	0.0	1.4		MIT
3197.182	3198.106	31268.51	MO	0.6	0.5		MIT
3195.958	3196.882	31280.48	MO	1.1	1.7		MIT
3195.235	3196.159	31287.56	MO	0.0	1.3		MIT
3194.873	3195.797	31291.10	MO	0.5	0.5		MIT
3193.973	3194.896	31299.92	MO	3.0R	1.7R		MIT
3192.797	3193.720	31311.45	MO	0.8	0.3		MIT
3192.104	3193.027	31318.25	MO	0.5	1.7		MIT
3191.519	3192.442	31323.99	MO	0.6	0.5		MIT
3191.024	3191.947	31328.84	MO	0.6			MIT
3190.232	3191.154	31336.62	MO	0.6	0.6		MIT
3189.281	3190.203	31345.97	MO	0.5	1.4		MIT
3188.403	3189.325	31354.60	MO	0.7	0.5		MIT
3188.094	3189.016	31357.64	MO	0.7	0.5		MIT
3187.683	3188.605	31361.68	MO	0.7			MIT
3187.592	3188.514	31362.57	MO	0.0	1.7		MIT
3186.386	3187.307	31374.45	MO		1.4		MIT
3185.711	3186.632	31381.09	MO	0.9	0.5		MIT
3185.104	3186.025	31387.07	MO	1.3	0.9		MIT
3184.572	3185.493	31392.32	MO	0.8	0.6		MIT
3184.402	3185.323	31393.99	MO	0.5	0.0		MIT
3183.322	3184.243	31404.64	MO	0.8D			MIT
3183.033	3183.954	31407.49	MO	1.0			MIT
3181.985	3182.905	31417.84	MO	0.6	0.5		MIT
3181.173	3182.093	31425.86	MO		0.9		MIT
3179.774	3180.694	31439.68	MO	0.8	0.7		MIT
3179.322	3180.242	31444.15	MO	0.8	0.7		MIT
3178.869	3179.789	31448.63	MO		0.6		MIT
3177.902	3178.821	31458.20	MO	0.7			MIT
3176.814	3177.733	31468.98	MO	0.7	0.7		MIT
3176.33	3177.25	31473.8	MO		1.7		MIT
3175.587	3176.506	31481.13	MO	0.7	0.5		MIT
3175.53	3176.45	31481.7	MO	0.7	0.0		MIT
3175.049	3175.968	31486.47	MO	0.3	1.8		MIT
3174.672	3175.590	31490.21	MO	0.6	0.0		MIT

AIR WAVELENGTH (ANGSTRMS)	VACUUM WAVELENGTH (ANGSTRMS)	WAVENUMBER (KAYSERS)	LMENT	LOG ARC	INTENSITY SPK	DIS	REF
3173.777	3174.695	31499.09	MO	0.0	1.3		MIT
3172.742	3173.660	31509.36	MO	0.5	1.7		MIT
3172.370	3173.288	31513.06	MO	0.7	1.3		MIT
3172.026	3172.944	31516.47	MO	0.3	1.4		MIT
3171.375	3172.293	31522.94	MO	0.7	0.6		MIT
3170.347	3171.264	31533.16	MO	3.0G	1.4R		MIT
3169.958	3170.875	31537.03	MO		1.3		MIT
3169.165	3170.082	31544.93	MO	0.6	0.0		MIT
3168.726	3169.643	31549.29	MO		1.3		MIT
3167.725	3168.642	31559.26	MO	0.0	1.3		MIT
3167.03	3167.95	31566.2	MO		1.3		MIT
3164.526	3165.442	31591.17	MO	1.0	1.0		MIT
3163.905	3164.821	31597.37	MO	1.0	1.3		MIT
3163.288	3164.204	31603.53	MO	0.6	0.5		MIT
3162.606	3163.521	31610.34	MO	0.3	1.3		MIT
3162.00	3162.92	31616.4	MO	0.5	0.0		MIT
3160.498	3161.413	31631.43	MO	0.5	0.3		MIT
3160.12	3161.03	31635.2	MO		1.2		MIT
3159.343	3160.258	31642.99	MO	1.3	0.7		MIT
3159.23	3160.14	31644.1	MO		1.4		MIT
3158.994	3159.908	31646.49	MO	0.3	1.0		MIT
3158.165	3159.079	31654.79	MO	2.5G	1.5R		MIT
3157.328	3158.242	31663.18	MO		1.3		MIT
3156.846	3157.760	31668.02	MO	0.0	1.5		MIT
3156.513	3157.427	31671.36	MO	0.9			MIT
3155.963	3156.877	31676.88	MO		0.7		MIT
3155.644	3156.558	31680.08	MO		1.7		MIT
3155.191	3156.105	31684.63	MO	0.5			MIT
3152.819	3153.732	31708.47	MO	0.8	1.9		MIT
3152.341	3153.254	31713.27	MO		0.6		MIT
3151.630	3152.543	31720.43	MO	0.3	1.6		MIT
3148.299	3149.211	31753.99	MO		1.0		MIT
3148.040	3148.952	31756.60	MO	0.5			MIT
3147.954	3148.866	31757.47	MO		1.4		MIT
3147.351	3148.263	31763.55	MO	1.3	1.0		MIT
3146.683	3147.594	31770.29	MO	1.4	0.7H		MIT
3145.905	3146.816	31778.15	MO	0.7	0.0		MIT
3145.751	3146.662	31779.71	MO		1.1		MIT
3145.286	3146.197	31784.41	MO	0.8	0.5		MIT
3144.806	3145.717	31789.26	MO	0.5			MIT
3144.613	3145.524	31791.21	MO	0.5	1.7		MIT
3144.337	3145.248	31794.00	MO	0.8	0.0		MIT
3143.903	3144.814	31798.39	MO	0.5	0.0		MIT
3143.054	3143.964	31806.98	MO	0.0	1.4		MIT
3142.750	3143.660	31810.05	MO	1.3			MIT
3141.730	3142.640	31820.38	MO		1.4		MIT
3141.415	3142.325	31823.57	MO		1.5		MIT
3140.648	3141.558	31831.34	MO	0.5	0.0		MIT
3139.871	3140.781	31839.22	MO	0.8	0.0		MIT
3138.715	3139.624	31850.94	MO	0.0	1.5		MIT
3138.43	3139.34	31853.8	MO II				ME
3138.355	3139.264	31854.60	MO	0.7			MIT
3137.850	3138.759	31859.72	MO	0.5	0.0		MIT
3136.749	3137.658	31870.91	MO	0.0	0.7		MIT
3136.466	3137.375	31873.78	MO II		2.0		KS
3136.465	3137.374	31873.79	MO	0.0	1.6		MIT
3135.894	3136.803	31879.59	MO	0.7	0.6		MIT
3135.602	3136.511	31882.56	MO	0.6	0.5		MIT
3132.594	3133.502	31913.18	MO	3.0G	2.5G		MIT
3131.195	3132.102	31927.44	MO		0.7		MIT
3130.060	3130.967	31939.01	MO		1.3		MIT
3128.51	3129.42	31954.8	MO	0.0D	1.3D		MIT
3127.808	3128.715	31962.01	MO		1.5		MIT
3127.022	3127.928	31970.04	MO	0.0D	0.3		MIT
3126.777	3127.683	31972.55	MO		1.3		MIT
3126.026	3126.932	31980.23	MO		1.0		MIT
3123.365	3124.271	32007.47	MO	0.7	1.4		MIT
3122.758	3123.663	32013.69	MO	0.7	0.0		MIT
3121.999	3122.904	32021.48	MO	0.7	2.2		MIT
3121.228	3122.133	32029.39	MO		0.7H		MIT

AIR WAVELENGTH (ANGSTRMS)	VACUUM WAVELENGTH (ANGSTRMS)	WAVENUMBER (KAYSERS)	LMENT	LOG ARC	INTENSITY SPK	DIS	REF
3119.807	3120.712	32043.97	MO	0.5	0.0		MIT
3118.812	3119.716	32054.20	MO		1.0		MIT
3117.466	3118.370	32068.03	MO II		1.3		KS
3116.092	3116.996	32082.18	MO	0.0	1.8		MIT
3114.964	3115.867	32093.79	MO II		1.4		KS
3113.480	3114.383	32109.09	MO	0.3	0.8		MIT
3112.124	3113.027	32123.08	MO	1.6	1.0		MIT
3111.640	3112.543	32128.07	MO		1.4		MIT
3111.224	3112.126	32132.37	MO	0.0	1.2		MIT
3110.844	3111.746	32136.29	MO		1.5		MIT
3110.644	3111.546	32138.36	MO	0.5			MIT
3107.549	3108.451	32170.37	MO		1.4		MIT
3106.859	3107.760	32177.51	MO	0.0	0.7		MIT
3106.460	3107.361	32181.65	MO	0.3			MIT
3106.341	3107.242	32182.88	MO	1.0	1.4D		MIT
3104.381	3105.282	32203.20	MO	1.2			MIT
3104.067	3104.968	32206.46	MO	0.3	0.5		MIT
3103.916	3104.817	32208.02	MO	0.5			MIT
3101.915	3102.815	32228.80	MO	0.7			MIT
3101.345	3102.245	32234.72	MO	1.9	1.0		MIT
3101.218	3102.118	32236.04	MO	0.6			MIT
3101.11	3102.01	32237.2	MO	0.0	1.2		MIT
3100.875	3101.775	32239.61	MO	1.6	0.3		MIT
3100.507	3101.407	32243.43	MO	0.3	0.8		MIT
3099.932	3100.832	32249.41	MO	1.4	0.5		MIT
3099.569	3100.469	32253.19	MO	1.0			MIT
3099.24	3100.14	32256.6	MO		1.3		MIT
3099.159	3100.058	32257.46	MO	1.0			MIT
3098.465	3099.364	32264.68	MO	1.3	1.4		MIT
3097.689	3098.588	32272.76	MO	1.0	1.5		MIT
3097.385	3098.284	32275.93	MO	1.0			MIT
3097.201	3098.100	32277.85	MO	1.3			MIT
3095.701	3096.600	32293.49	MO	1.4			MIT
3094.664	3095.562	32304.31	MO	2.2	1.4		MIT
3093.680	3094.578	32314.58	MO	1.0			MIT
3092.92	3093.82	32322.5	MO		1.0D		MIT
3092.70	3093.60	32324.8	MO	1.3	0.3		MIT
3092.074	3092.972	32331.37	MO	1.5	2.0		MIT
3091.840	3092.738	32333.81	MO	1.0			MIT
3090.647	3091.544	32346.29	MO	0.6			MIT
3089.706	3090.603	32356.15	MO	1.6	0.7		MIT
3089.122	3090.019	32362.26	MO	1.5	0.5		MIT
3088.318	3089.215	32370.69	MO	1.3			MIT
3087.621	3088.518	32377.99	MO	1.5	2.3		MIT
3086.879	3087.775	32385.78	MO	0.7			MIT
3086.363	3087.259	32391.19	MO	1.3			MIT
3085.615	3086.511	32399.04	MO	2.1	1.4		MIT
3084.239	3085.135	32413.50	MO	1.4	0.0		MIT
3082.220	3083.115	32434.73	MO	0.6	1.6		MIT
3081.950	3082.845	32437.57	MO	1.4			MIT
3081.65	3082.55	32440.7	MO		1.3		MIT
3081.155	3082.050	32445.94	MO	1.4			MIT
3080.635	3081.530	32451.41	MO	0.9			MIT
3080.408	3081.303	32453.81	MO	1.8	0.8		MIT
3079.879	3080.774	32459.38	MO	1.6	0.5		MIT
3078.267	3079.161	32476.38	MO	1.0			MIT
3077.661	3078.555	32482.77	MO	1.3	2.1		MIT
3074.929	3075.822	32511.63	MO	1.3			MIT
3074.63	3075.52	32514.8	MO	0.0	0.9		MIT
3074.374	3075.267	32517.50	MO	1.8	1.2		MIT
3073.382	3074.275	32528.00	MO	1.1	1.0		MIT
3073.24	3074.13	32529.5	MO		1.3		MIT
3071.439	3072.331	32548.57	MO	1.4	0.5		MIT
3070.895	3071.787	32554.34	MO	1.6	0.5		MIT
3070.619	3071.511	32557.26	MO		1.4		MIT
3069.970	3070.862	32564.15	MO	1.3	0.0		MIT
3069.790	3070.682	32566.05	MO	1.2			MIT
3069.517	3070.409	32568.95	MO	1.3			MIT
3069.050	3069.942	32573.91	MO	0.7			MIT
3068.642	3069.534	32578.24	MO	1.2			MIT

AIR WAVELENGTH (ANGSTRMS)	VACUUM WAVELENGTH (ANGSTRMS)	WAVENUMBER (KAYSERS)	LMENT	LOG ARC	INTENSITY SPK	DIS	REF
3067.997	3068.889	32585.09	MO	1.5	0.0		MIT
3067.642	3068.534	32588.86	MO	1.0	1.7		MIT
3066.399	3067.290	32602.07	MO	1.2			MIT
3065.042	3065.933	32616.50	MO	1.5	1.0		MIT
3064.555	3065.446	32621.68	MO	1.2			MIT
3064.279	3065.170	32624.62	MO	1.9	1.0		MIT
3062.438	3063.328	32644.23	MO	1.2			MIT
3062.102	3062.992	32647.81	MO		0.5		MIT
3061.586	3062.476	32653.32	MO	1.7	0.5		MIT
3060.928	3061.818	32660.34	MO	0.8	0.8		MIT
3060.777	3061.667	32661.95	MO	1.4	1.2		MIT
3058.597	3059.486	32685.23	MO	1.3			MIT
3057.88	3058.77	32692.9	MO		1.5		MIT
3057.559	3058.448	32696.32	MO	1.4	0.0		MIT
3056.731	3057.620	32705.18	MO	1.3			MIT
3055.323	3056.211	32720.25	MO	1.7	0.7		MIT
3054.763	3055.651	32726.25	MO		0.5		MIT
3053.630	3054.518	32738.39	MO	1.3 O			MIT
3053.464	3054.352	32740.17	MO	0.6	0.8		MIT
3053.289	3054.177	32742.04	MO	1.3			MIT
3053.02	3053.91	32744.9	MO	0.9			MIT
3052.559	3053.447	32749.87	MO	1.3			MIT
3052.320	3053.208	32752.44	MO	1.0	1.7		MIT
3050.214	3051.101	32775.05	MO	1.0	0.0		MIT
3049.29	3050.18	32785.0	MO	0.5			MIT
3048.83	3049.72	32789.9	MO		0.7		MIT
3048.05	3048.94	32798.3	MO	0.0	1.4		MIT
3047.821	3048.708	32800.79	MO	1.2	0.0		MIT
3047.312	3048.198	32806.26	MO	1.7	0.6		MIT
3047.117	3048.003	32808.36	MO	1.2			MIT
3046.805	3047.691	32811.72	MO	1.7	0.3		MIT
3046.404	3047.290	32816.04	MO		0.5		MIT
3045.719	3046.605	32823.42	MO	0.0	1.0		MIT
3045.231	3046.117	32828.68	MO	0.9			MIT
3044.733	3045.619	32834.05	MO	1.0			MIT
3043.899	3044.785	32843.05	MO	1.3			MIT
3043.692	3044.578	32845.28	MO	1.3			MIT
3043.438	3044.323	32848.02	MO II		1.5		KS
3042.957	3043.842	32853.21	MO	1.0			MIT
3041.702	3042.587	32866.77	MO	1.6	0.7		MIT
3041.01	3041.89	32874.2	MO	0.0	1.4		MIT
3039.824	3040.709	32887.07	MO	1.3	0.3		MIT
3039.058	3039.942	32895.36	MO		1.4		MIT
3038.790	3039.674	32898.26	MO		0.5		MIT
3036.314	3037.198	32925.09	MO	1.5	0.3		MIT
3035.333	3036.216	32935.73	MO	1.5	0.3		MIT
3034.925	3035.808	32940.16	MO	0.7	1.4		MIT
3033.33	3034.21	32957.5	MO	0.0	1.5		MIT
3033.234	3034.117	32958.52	MO	0.9	1.0		MIT
3032.545	3033.428	32966.01	MO	0.7			MIT
3031.213	3032.095	32980.49	MO	1.3			MIT
3030.325	3031.207	32990.16	MO	0.3	1.3		MIT
3027.771	3028.653	33017.98	MO	1.3	1.5		MIT
3025.004	3025.885	33048.18	MO	1.7	0.6		MIT
3023.690	3024.571	33062.54	MO	1.3			MIT
3023.300	3024.180	33066.81	MO	0.7	2.0		MIT
3022.994	3023.874	33070.16	MO	1.4	0.0		MIT
3022.740	3023.620	33072.94	MO	0.0	0.8		MIT
3021.617	3022.497	33085.23	MO		1.6		MIT
3020.693	3021.573	33095.35	MO	0.7	1.0		MIT
3018.55	3019.43	33118.8	MO	0.0	2.0		MIT
3016.777	3017.656	33138.31	MO	1.4	0.3		MIT
3015.418	3016.296	33153.24	MO	0.7	1.0		MIT
3014.780	3015.658	33160.25	MO	1.2			MIT
3014.691	3015.569	33161.23	MO	1.3	0.5		MIT
3014.159	3015.037	33167.09	MO	0.7	1.7		MIT
3013.763	3014.641	33171.44	MO	2.0 H	0.7		MIT
3013.388	3014.266	33175.57	MO	1.4	0.7		MIT
3008.15	3009.03	33233.3	MO II				ME
3007.82	3008.70	33237.0	MO	0.0	0.7		MIT

AIR WAVELENGTH (ANGSTRMS)	VACUUM WAVELENGTH (ANGSTRMS)	WAVENUMBER (KAYSERS)	LMENT	LOG ARC	INTENSITY SPK	DIS	REF
3007.713	3008.590	33238.17	MO	1.2			MIT
3004.460	3005.336	33274.15	MO	0.7	1.6		MIT
3004.14	3005.02	33277.7	MO	0.7			MIT
3003.172	3004.047	33288.42	MO	1.0			MIT
3002.74	3003.62	33293.2	MO	1.0			MIT
3002.213	3003.088	33299.06	MO	1.6	0.5		MIT
3001.435	3002.310	33307.69	MO	1.2			MIT
3000.942	3001.817	33313.16	MO	0.3	0.5		MIT
3000.855	3001.730	33314.12	MO	1.4	0.0		MIT
3000.45	3001.32	33318.6	MO	1.3			MIT
3000.232	3001.107	33321.04	MO	1.4	1.5		MIT
2998.149	2999.023	33344.19	MO	1.0			MIT
2997.666	2998.540	33349.56	MO	0.3	0.5		MIT
2997.413	2998.287	33352.38	MO	1.3			MIT
2997.346	2998.220	33353.12	MO	0.5	1.4		MIT
2996.293	2997.167	33364.84	MO	0.0	1.3		MIT
2995.533	2996.407	33373.31	MO		1.2		MIT
2993.515	2994.388	33395.81	MO	1.2	1.4		MIT
2992.838	2993.711	33403.36	MO	1.3	1.5		MIT
2992.645	2993.518	33405.51	MO	0.0	1.0		MIT
2992.268	2993.141	33409.72	MO	0.0	1.2		MIT
2989.89	2990.76	33436.3	MO		1.2		MIT
2989.801	2990.673	33437.29	MO	1.4	0.5		MIT
2988.679	2989.551	33449.84	MO	1.4	0.3		MIT
2988.228	2989.100	33454.89	MO	1.4	0.0		MIT
2987.920	2988.792	33458.34	MO	1.4	1.5		MIT
2987.35	2988.22	33464.7	MO		1.4		MIT
2986.912	2987.783	33469.63	MO	0.5	1.4		MIT
2986.162	2987.033	33478.04	MO	0.6	1.7		MIT
2985.841	2986.712	33481.63	MO	1.0	0.6		MIT
2985.455	2986.326	33485.96	MO	0.8			MIT
2985.308	2986.179	33487.61	MO		0.9		MIT
2985.162	2986.033	33489.25	MO	1.2	0.9		MIT
2984.88	2985.75	33492.4	MO		1.4		MIT
2983.988	2984.859	33502.42	MO		1.4		MIT
2983.809	2984.680	33504.43	MO	1.3 R			MIT
2983.626	2984.497	33506.49	MO	0.0	1.4		MIT
2983.045	2983.915	33513.01	MO	1.0	0.0		MIT
2982.78	2983.65	33516.0	MO		1.0		MIT
2982.45	2983.32	33519.7	MO		1.0		MIT
2982.134	2983.004	33523.25	MO	1.3			MIT
2979.873	2980.743	33548.69	MO	0.5	1.3		MIT
2979.704	2980.574	33550.59	MO	0.6			MIT
2979.28	2980.15	33555.4	MO	0.5			MIT
2978.609	2979.478	33562.92	MO	0.3	1.3		MIT
2978.285	2979.154	33566.57	MO	1.5	0.3		MIT
2977.765	2978.634	33572.44	MO	0.0	1.4		MIT
2977.270	2978.139	33578.02	MO	1.2	0.0		MIT
2976.923	2977.792	33581.93	MO		1.3		MIT
2975.614	2976.483	33596.70	MO		1.3		MIT
2975.404	2976.273	33599.07	MO	0.5	1.5		MIT
2974.002	2974.870	33614.91	MO	0.0	0.7		MIT
2972.961	2973.829	33626.68	MO	1.3	0.0		MIT
2972.614	2973.482	33630.61	MO	1.3	1.7		MIT
2971.906	2972.774	33638.62	MO	0.3	1.6		MIT
2971.145	2972.012	33647.23	MO		0.5		MIT
2969.736	2970.603	33663.20	MO		0.7		MIT
2969.39	2970.26	33667.1	MO		1.2		MIT
2968.774	2969.641	33674.11	MO		1.5		MIT
2967.00	2967.87	33694.2	MO		1.2		MIT
2966.697	2967.563	33697.68	MO	0.3	0.7		MIT
2965.270	2966.136	33713.90	MO	0.9	1.6		MIT
2964.958	2965.824	33717.44	MO	0.7	1.0		MIT
2964.344	2965.210	33724.43	MO	0.0	0.8		MIT
2963.794	2964.660	33730.69	MO	1.0	1.6		MIT
2962.887	2963.752	33741.01	MO	0.9	0.7		MIT
2962.212	2963.077	33748.70	MO		1.3		MIT
2961.320	2962.185	33758.86	MO	0.6	1.5		MIT
2961.036	2961.901	33762.10	MO		1.0		MIT
2960.241	2961.106	33771.17	MO	0.5	1.2		MIT

AIR WAVELENGTH (ANGSTRMS)	VACUUM WAVELENGTH (ANGSTRMS)	WAVENUMBER (KAYSERS)	LMENT	LOG ARC	INTENSITY SPK	DIS	REF
2959.804	2960.669	33776.15	MO	0.9	0.0		MIT
2959.476	2960.341	33779.90	MO	0.7	0.3		MIT
2958.73	2959.59	33788.4	MO		1.0		MIT
2957.749	2958.613	33799.62	MO	0.5	0.0		MIT
2956.902	2957.766	33809.30	MO	0.0	1.5		MIT
2956.055	2956.919	33818.99	MO	1.0	1.4		MIT
2955.838	2956.702	33821.47	MO	0.3	1.5		MIT
2955.156	2956.019	33829.28	MO	0.0	1.4		MIT
2954.514	2955.377	33836.63	MO		0.5		MIT
2954.170	2955.033	33840.57	MO	0.5			MIT
2953.947	2954.810	33843.12	MO	0.0	1.3		MIT
2953.560	2954.423	33847.56	MO	0.5	0.0		MIT
2951.814	2952.677	33867.58	MO		0.6H		MIT
2947.285	2948.147	33919.62	MO	0.3	1.4		MIT
2946.690	2947.551	33926.46	MO	0.5	1.4		MIT
2946.422	2947.283	33929.55	MO	1.0	0.8		MIT
2946.009	2946.870	33934.31	MO	1.3	1.6		MIT
2945.665	2946.526	33938.27	MO	1.3	0.3		MIT
2945.428	2946.289	33941.00	MO	0.7	0.0		MIT
2944.821	2945.682	33948.00	MO	0.3	1.7H		MIT
2944.213	2945.074	33955.01	MO	1.4	0.3		MIT
2943.989	2944.850	33957.59	MO		0.5		MIT
2943.380	2944.241	33964.62	MO	0.0	1.4		MIT
2942.850	2943.710	33970.73	MO	1.0	0.0		MIT
2942.222	2943.082	33977.98	MO		0.5		MIT
2941.222	2942.082	33989.54	MO	0.3	1.6		MIT
2940.978	2941.838	33992.35	MO	1.0	0.0		MIT
2940.101	2940.961	34002.49	MO	0.3	1.6		MIT
2940.022	2940.882	34003.41	MO	0.5			MIT
2938.886	2939.745	34016.55	MO		0.5		MIT
2938.768	2939.627	34017.92	MO	0.3	0.7		MIT
2938.59	2939.45	34020.0	MO	0.0	1.0		MIT
2938.300	2939.159	34023.33	MO	0.0	1.5		MIT
2937.665	2938.524	34030.69	MO	1.3	0.0		MIT
2936.781	2937.640	34040.93	MO	0.3	1.4		MIT
2935.695	2936.554	34053.52	MO	0.0	1.2		MIT
2935.195	2936.054	34059.33	MO	0.3	1.0		MIT
2934.840	2935.698	34063.44	MO	0.7	0.0		MIT
2934.299	2935.157	34069.72	MO	1.5	1.7H		MIT
2933.204	2934.062	34082.44	MO		0.9		MIT
2932.189	2933.047	34094.24	MO		1.2		MIT
2931.081	2931.939	34107.13	MO	1.2	0.0		MIT
2930.771	2931.628	34110.73	MO	0.7	1.4		MIT
2930.503	2931.360	34113.85	MO	1.4	1.7H		MIT
2930.400	2931.257	34115.05	MO	0.7			MIT
2930.064	2930.921	34118.96	MO	0.0	1.3		MIT
2929.497	2930.354	34125.57	MO		0.7		MIT
2928.493	2929.350	34137.27	MO		1.0		MIT
2927.539	2928.396	34148.39	MO	0.5	1.6		MIT
2926.733	2927.589	34157.79	MO		1.3		MIT
2926.22	2927.08	34163.8	MO		1.6		MIT
2925.405	2926.261	34173.30	MO	0.7	1.5		MIT
2924.318	2925.174	34186.00	MO	1.0	1.4		MIT
2923.391	2924.247	34196.84	MO	1.5	1.5		MIT
2923.220	2924.076	34198.84	MO	0.5	0.6		MIT
2922.738	2923.593	34204.48	MO	0.0	1.3		MIT
2921.905	2922.760	34214.23	MO	0.3	0.6		MIT
2920.263	2921.118	34233.47	MO	0.0	1.0		MIT
2919.383	2920.238	34243.79	MO	0.9	0.0		MIT
2919.202	2920.057	34245.91	MO	0.9	0.0		MIT
2918.828	2919.682	34250.30	MO	1.2	1.4L		MIT
2917.151	2918.005	34269.99	MO		1.3		MIT
2916.100	2916.954	34282.34	MO	1.3			MIT
2915.380	2916.234	34290.81	MO	1.0	0.0		MIT
2915.256	2916.110	34292.26	MO	0.8			MIT
2914.428	2915.281	34302.01	MO	0.0	0.7		MIT
2914.314	2915.167	34303.35	MO		1.0		MIT
2913.808	2914.661	34309.31	MO	0.0	1.7W		MIT
2913.519	2914.372	34312.71	MO	1.3	0.0		MIT
2911.915	2912.768	34331.61	MO	1.5	1.7H		MIT

AIR WAVELENGTH (ANGSTRMS)	VACUUM WAVELENGTH (ANGSTRMS)	WAVENUMBER (KAYSERS)	LMENT	LOG ARC	INTENSITY SPK	DIS	REF
2911.765	2912.618	34333.38	MO	0.7			MIT
2910.933	2911.786	34343.19	MO		1.0		MIT
2909.116	2909.968	34364.64	MO II	1.4	1.6H		MIT
2908.164	2909.016	34375.89	MO	0.7	0.9		MIT
2907.784	2908.636	34380.38	MO	0.8			MIT
2907.116	2907.968	34388.28	MO	1.0	1.5		MIT
2906.056	2906.907	34400.82	MO	1.2	0.0		MIT
2905.833	2906.684	34403.46	MO		1.2		MIT
2905.266	2906.117	34410.17	MO	1.5	0.6		MIT
2904.333	2905.184	34421.23	MO		1.0		MIT
2903.069	2903.920	34436.21	MO	1.3	2.0H		MIT
2902.622	2903.472	34441.52	MO	0.7			MIT
2902.243	2903.093	34446.02	MO	0.7			MIT
2901.794	2902.644	34451.35	MO		1.2		MIT
2900.795	2901.645	34463.21	MO	0.3	1.6		MIT
2898.652	2899.501	34488.69	MO	1.3	0.8		MIT
2898.481	2899.330	34490.72	MO		1.2		MIT
2898.386	2899.235	34491.85	MO	0.3			MIT
2897.628	2898.477	34500.87	MO	1.3	1.4		MIT
2897.421	2898.270	34503.34	MO		1.3		MIT
2896.971	2897.820	34508.70	MO	0.5H	0.0		MIT
2896.444	2897.293	34514.98	MO	0.0	1.3		MIT
2894.855	2895.704	34533.92	MO		0.9		MIT
2894.451	2895.299	34538.74	MO	1.7	1.9H		MIT
2893.778	2894.626	34546.77	MO		0.5		MIT
2893.226	2894.074	34553.37	MO	1.0	0.0		MIT
2892.812	2893.660	34558.31	MO	1.4	1.5		MIT
2892.565	2893.413	34561.26	MO	0.7			MIT
2892.028	2892.876	34567.68	MO		1.2		MIT
2891.275	2892.123	34576.68	MO	1.2	1.3		MIT
2890.994	2891.842	34580.04	MO II	1.5	1.7H		MIT
2889.837	2890.684	34593.88	MO	1.1			MIT
2888.690	2889.537	34607.62	MO		1.0		MIT
2888.530	2889.377	34609.54	MO	1.0			MIT
2888.153	2889.000	34614.05	MO	0.0	1.6		MIT
2887.619	2888.466	34620.46	MO	1.3D			MIT
2886.972	2887.819	34628.21	MO	0.0	1.4		MIT
2886.606	2887.453	34632.60	MO	1.5	0.0		MIT
2885.736	2886.582	34643.04	MO	1.0	1.4		MIT
2884.788	2885.634	34654.43	MO		0.8		MIT
2884.590	2885.436	34656.81	MO	0.7			MIT
2883.957	2884.803	34664.41	MO		1.0		MIT
2883.297	2884.143	34672.35	MO		0.8		MIT
2882.543	2883.389	34681.42	MO	0.8			MIT
2882.378	2883.223	34683.40	MO	0.0	1.4		MIT
2882.036	2882.881	34687.52	MO		0.5		MIT
2881.937	2882.782	34688.71	MO		0.5		MIT
2881.370	2882.215	34695.54	MO		1.0		MIT
2879.689	2880.534	34715.79	MO	0.0	1.4		MIT
2879.047	2879.892	34723.53	MO	1.2	2.0H		MIT
2878.382	2879.226	34731.55	MO	1.3			MIT
2876.537	2877.381	34753.83	MO	1.2	0.3		MIT
2874.847	2875.691	34774.25	MO	0.3	1.8		MIT
2873.637	2874.480	34788.90	MO	1.0	0.0		MIT
2872.884	2873.727	34798.02	MO	0.3	1.7		MIT
2871.892	2872.735	34810.03	MO	1.0			MIT
2871.508	2872.351	34814.69	MO II	2.0	2.0H		MIT
2871.183	2872.026	34818.63	MO	1.0			MIT
2870.903	2871.746	34822.02	MO	1.2			MIT
2870.179	2871.021	34830.81	MO	1.0	0.0		MIT
2869.969	2870.811	34833.36	MO		0.7		MIT
2869.565	2870.407	34838.26	MO	1.2			MIT
2869.217	2870.059	34842.49	MO		1.2		MIT
2868.316	2869.158	34853.43	MO	0.3	1.3		MIT
2868.110	2868.952	34855.93	MO	0.5	1.3		MIT
2867.050	2867.892	34868.82	MO	1.0			MIT
2866.693	2867.535	34873.16	MO	1.5	1.5		MIT
2866.261	2867.103	34878.42	MO	0.3	0.6		MIT
2866.084	2866.925	34880.57	MO		0.5		MIT
2865.620	2866.461	34886.22	MO	0.7	1.3		MIT

AIR WAVELENGTH (ANGSTRMS)	VACUUM WAVELENGTH (ANGSTRMS)	WAVENUMBER (KAYSERS)	LMENT	LOG ARC	INTENSITY SPK	DIS	REF
2865.505	2866.346	34887.62	MO	0.0	0.9		MIT
2865.118	2865.959	34892.33	MO	0.6			MIT
2864.656	2865.497	34897.96	MO	1.6	0.5		MIT
2864.306	2865.147	34902.22	MO	1.6	0.3		MIT
2863.811	2864.652	34908.25	MO	1.5	2.0H		MIT
2863.121	2863.962	34916.67	MO		1.3		MIT
2862.839	2863.680	34920.11	MO	1.0			MIT
2861.858	2862.698	34932.08	MO		1.0		MIT
2861.561	2862.401	34935.70	MO		0.5H		MIT
2859.115	2859.955	34965.59	MO	0.5			MIT
2858.995	2859.835	34967.06	MO		1.2		MIT
2858.054	2858.893	34978.57	MO		0.6		MIT
2856.880	2857.719	34992.94	MO	0.0	0.8		MIT
2855.995	2856.834	35003.78	MO	0.3	1.4		MIT
2855.706	2856.545	35007.33	MO		1.3		MIT
2854.869	2855.708	35017.59	MO	0.7			MIT
2854.111	2854.950	35026.89	MO		1.0		MIT
2853.584	2854.422	35033.36	MO	0.0	1.4		MIT
2853.229	2854.067	35037.72	MO	1.4	2.0H		MIT
2852.131	2852.969	35051.20	MO	1.0H	1.0		MIT
2851.179	2852.017	35062.91	MO	1.2			MIT
2850.898	2851.736	35066.36	MO	1.0			MIT
2850.787	2851.625	35067.73	MO	1.3			MIT
2850.674	2851.512	35069.12	MO	0.0	1.6		MIT
2849.381	2850.218	35085.03	MO	1.7	0.7		MIT
2848.381	2849.218	35097.35	MO	0.7			MIT
2848.232	2849.069	35099.18	MO II	2.1	2.3H		MIT
2847.683	2848.520	35105.95	MO		0.8		MIT
2846.620	2847.457	35119.06	MO	0.0	1.0		MIT
2845.647	2846.483	35131.07	MO		1.0		MIT
2844.814	2845.650	35141.35	MO		1.3		MIT
2844.390	2845.226	35146.59	MO	1.5	0.7		MIT
2843.728	2844.564	35154.77	MO	0.0	0.9		MIT
2842.908	2843.744	35164.91	MO	0.6			MIT
2842.457	2843.293	35170.49	MO	0.5	1.5		MIT
2842.369	2843.205	35171.58	MO	1.0			MIT
2842.151	2842.987	35174.28	MO	0.3	1.6		MIT
2841.773	2842.608	35178.96	MO		1.2		MIT
2839.585	2840.420	35206.06	MO	1.4	0.0		MIT
2839.164	2839.999	35211.28	MO		1.4		MIT
2837.901	2838.736	35226.95	MO	1.2			MIT
2837.320	2838.154	35234.17	MO	0.7H	0.0		MIT
2836.965	2837.799	35238.57	MO	0.9			MIT
2836.705	2837.539	35241.80	MO	0.0	1.4H		MIT
2836.295	2837.129	35246.90	MO	0.3	1.0		MIT
2836.028	2836.862	35250.22	MO	0.9			MIT
2835.914	2836.748	35251.63	MO	1.2			MIT
2835.331	2836.165	35258.88	MO	1.3	1.6		MIT
2834.780	2835.614	35265.73	MO	1.2	1.0		MIT
2834.394	2835.228	35270.54	MO	1.3	1.6		MIT
2833.792	2834.626	35278.03	MO	0.0	0.9		MIT
2832.657	2833.490	35292.16	MO		0.7		MIT
2832.073	2832.906	35299.44	MO	0.0	1.3		MIT
2831.956	2832.789	35300.90	MO	0.3	0.9		MIT
2831.441	2832.274	35307.32	MO	0.0H	1.5		MIT
2829.942	2830.775	35326.02	MO	1.4	0.7		MIT
2829.793	2830.626	35327.88	MO	1.2			MIT
2829.031	2829.863	35337.40	MO	0.0	0.8		MIT
2828.789	2829.621	35340.42	MO	1.0	0.0		MIT
2827.743	2828.575	35353.49	MO	0.9	1.6		MIT
2827.172	2828.004	35360.63	MO		0.7		MIT
2826.746	2827.578	35365.96	MO	1.2	0.0		MIT
2826.545	2827.377	35368.47	MO	1.6	0.7		MIT
2825.994	2826.826	35375.37	MO	1.2	0.6		MIT
2825.671	2826.503	35379.41	MO	1.4	0.0		MIT
2824.172	2825.003	35398.19	MO		0.5		MIT
2822.863	2823.694	35414.60	MO	1.3	0.8		MIT
2822.429	2823.260	35420.05	MO	1.2	0.6		MIT
2822.029	2822.860	35425.07	MO	1.2	1.3		MIT
2821.834	2822.665	35427.52	MO	0.0	1.4		MIT

AIR WAVELENGTH (ANGSTRMS)	VACUUM WAVELENGTH (ANGSTRMS)	WAVENUMBER (KAYSERS)	LMENT	LOG ARC	INTENSITY SPK	DIS	REF
2820.633	2821.463	35442.60	MO	0.9			MIT
2820.003	2820.833	35450.52	MO		0.9		MIT
2819.581	2820.411	35455.83	MO		0.9		MIT
2818.300	2819.130	35471.94	MO	1.4	0.0		MIT
2817.500	2818.330	35482.01	MO	1.2	1.4		MIT
2817.440	2818.270	35482.77	MO	0.9	1.4		MIT
2816.941	2817.770	35489.05	MO	0.7			MIT
2816.154	2816.983	35498.97	MO II	2.3	2.5H		MIT
2815.908	2816.737	35502.07	MO	1.3			MIT
2815.542	2816.371	35506.69	MO	1.0			MIT
2814.993	2815.822	35513.61	MO		0.5		MIT
2814.672	2815.501	35517.66	MO	0.0	1.3		MIT
2812.585	2813.413	35544.01	MO	0.3	1.5		MIT
2812.147	2812.975	35549.55	MO	0.6			MIT
2811.503	2812.331	35557.69	MO	1.3			MIT
2811.374	2812.202	35559.32	MO		0.5		MIT
2811.153	2811.981	35562.12	MO		1.0		MIT
2810.433	2811.261	35571.23	MO	1.0	1.0		MIT
2810.233	2811.061	35573.76	MO		0.9		MIT
2809.955	2810.783	35577.28	MO	1.3			MIT
2809.396	2810.224	35584.36	MO	0.7			MIT
2808.955	2809.782	35589.94	MO		0.8		MIT
2808.374	2809.201	35597.31	MO	1.4	0.0		MIT
2807.755	2808.582	35605.15	MO	1.8	1.9H		MIT
2807.355	2808.182	35610.23	MO	1.3			MIT
2806.181	2807.008	35625.12	MO		1.2		MIT
2803.131	2803.957	35663.88	MO	0.5			MIT
2802.354	2803.180	35673.77	MO	1.2	1.4		MIT
2801.546	2802.372	35684.06	MO	1.3	0.0		MIT
2801.467	2802.293	35685.07	MO	1.3	0.5		MIT
2800.731	2801.556	35694.44	MO	0.3	1.2		MIT
2800.344	2801.169	35699.38	MO		1.0H		MIT
2798.899	2799.724	35717.81	MO		1.0H		MIT
2798.011	2798.836	35729.14	MO	1.2	1.5		MIT
2797.931	2798.756	35730.16	MO	1.2	0.3		MIT
2796.89	2797.71	35743.5	MO		1.2J		MIT
2796.777	2797.601	35744.90	MO	2.0	1.0		MIT
2794.573	2795.397	35773.10	MO	0.5	0.5		MIT
2793.972	2794.796	35780.79	MO	0.5	1.2J		MIT
2792.964	2793.788	35793.70	MO	1.5	0.3		MIT
2791.540	2792.363	35811.96	MO	0.0	1.5		MIT
2790.409	2791.232	35826.48	MO	0.5	1.4		MIT
2790.308	2791.131	35827.77	MO	1.5	0.0		MIT
2790.010	2790.833	35831.60	MO	1.2			MIT
2789.133	2789.956	35842.86	MO	1.0			MIT
2787.830	2788.652	35859.62	MO	1.6	0.6		MIT
2787.315	2788.137	35866.24	MO	0.0	0.7		MIT
2786.230	2787.052	35880.21	MO		0.6		MIT
2785.619	2786.441	35888.08	MO	0.0	1.2		MIT
2784.992	2785.814	35896.16	MO	2.0	2.3		MIT
2784.008	2784.829	35908.84	MO	1.3	0.0		MIT
2783.129	2783.950	35920.18	MO	0.5			MIT
2782.004	2782.825	35934.71	MO		0.5H		MIT
2780.994	2781.815	35947.76	MO	1.2	0.0H		MIT
2780.036	2780.856	35960.15	MO	1.8	2.0H		MIT
2779.476	2780.296	35967.39	MO	1.4	0.0		MIT
2779.239	2780.059	35970.46	MO		1.4		MIT
2777.856	2778.676	35988.36	MO	1.2	1.0		MIT
2777.743	2778.563	35989.83	MO	1.5	0.3		MIT
2776.672	2777.492	36003.71	MO	0.0	1.3		MIT
2775.551	2776.370	36018.25	MO	1.0			MIT
2775.400	2776.219	36020.21	MO	1.9	2.0H		MIT
2774.392	2775.211	36033.30	MO	1.5	1.7H		MIT
2773.784	2774.603	36041.19	MO	1.2	1.4		MIT
2771.775	2772.593	36067.31	MO	0.7			MIT
2771.689	2772.507	36068.43	MO		1.3		MIT
2771.356	2772.174	36072.77	MO	1.3			MIT
2770.908	2771.726	36078.60	MO		0.5		MIT
2770.582	2771.400	36082.85	MO		1.4		MIT
2770.158	2770.976	36088.37	MO		0.9		MIT

AIR WAVELENGTH (ANGSTRMS)	VACUUM WAVELENGTH (ANGSTRMS)	WAVENUMBER (KAYSERS)	LMENT	LOG ARC	INTENSITY SPK	INTENSITY DIS	REF
2769.862	2770.680	36092.22	MO	0.7			MIT
2769.762	2770.580	36093.53	MO	1.0	2.0		MIT
2768.860	2769.678	36105.28	MO	1.0			MIT
2768.094	2768.911	36115.27	MO	1.0			MIT
2767.224	2768.041	36126.63	MO	1.3			MIT
2766.722	2767.539	36133.18	MO	1.3	0.0		MIT
2766.258	2767.075	36139.24	MO	1.5	0.5		MIT
2766.157	2766.974	36140.56	MO		1.0		MIT
2765.096	2765.913	36154.43	MO	1.2	0.0		MIT
2763.934	2764.750	36169.63	MO	1.4	0.0		MIT
2763.620	2764.436	36173.74	MO	1.4	1.7H		MIT
2763.298	2764.114	36177.95	MO	0.0	1.3		MIT
2763.027	2763.843	36181.50	MO	1.4	0.0		MIT
2762.697	2763.513	36185.82	MO	1.4	0.3		MIT
2762.442	2763.258	36189.16	MO	0.0	1.0		MIT
2761.787	2762.603	36197.75	MO		1.0		MIT
2761.533	2762.349	36201.08	MO	1.6	1.3		MIT
2760.526	2761.342	36214.28	MO	0.0	1.5		MIT
2760.006	2760.822	36221.10	MO	1.0			MIT
2759.582	2760.397	36226.67	MO	1.3	0.0		MIT
2759.183	2759.998	36231.91	MO		0.7		MIT
2758.634	2759.449	36239.12	MO	1.0	1.0		MIT
2758.506	2759.321	36240.80	MO	0.3	1.3		MIT
2757.095	2757.910	36259.34	MO	0.7			MIT
2756.259	2757.074	36270.34	MO	1.2			MIT
2756.072	2756.887	36272.80	MO	1.0	1.7H		MIT
2755.367	2756.181	36282.08	MO	1.2	1.0		MIT
2754.294	2755.108	36296.22	MO	1.3	1.3		MIT
2753.922	2754.736	36301.12	MO		0.7		MIT
2753.817	2754.631	36302.50	MO	0.0	1.0		MIT
2752.583	2753.397	36318.78	MO		0.6H		MIT
2752.150	2752.964	36324.49	MO	0.5	0.5H		MIT
2751.472	2752.285	36333.44	MO	1.7	0.7		MIT
2750.029	2750.842	36352.50	MO	0.3	1.7		MIT
2748.494	2749.307	36372.81	MO	1.3			MIT
2748.414	2749.227	36373.87	MO		1.0		MIT
2747.145	2747.957	36390.67	MO		1.0		MIT
2746.892	2747.704	36394.02	MO	0.9	0.3		MIT
2746.301	2747.113	36401.85	MO	1.5	1.4		MIT
2745.384	2746.196	36414.01	MO	1.3			MIT
2745.272	2746.084	36415.49	MO	0.7	0.7H		MIT
2745.088	2745.900	36417.93	MO	1.3	0.0		MIT
2744.193	2745.005	36429.81	MO	0.3	1.4		MIT
2743.714	2744.526	36436.17	MO	1.5	0.0		MIT
2743.068	2743.879	36444.75	MO	1.2			MIT
2742.890	2743.701	36447.12	MO	0.0	1.3		MIT
2741.618	2742.429	36464.02	MO	0.7	1.4		MIT
2741.319	2742.130	36468.00	MO	0.3	1.3		MIT
2740.965	2741.776	36472.71	MO	0.9			MIT
2740.068	2740.879	36484.65	MO	0.0H	0.6Q		MIT
2738.605	2739.415	36504.14	MO	0.0	1.6		MIT
2737.880	2738.690	36513.81	MO	1.1	1.3		MIT
2736.960	2737.770	36526.08	MO	1.4	1.5		MIT
2736.648	2737.458	36530.24	MO		0.6		MIT
2736.417	2737.227	36533.33	MO	0.7	1.0		MIT
2735.883	2736.693	36540.46	MO	1.0	0.3		MIT
2735.653	2736.463	36543.53	MO	0.5	0.7		MIT
2735.307	2736.117	36548.15	MO		1.0		MIT
2735.155	2735.965	36550.18	MO	0.9			MIT
2734.374	2735.183	36560.62	MO		0.7		MIT
2733.837	2734.646	36567.80	MO	1.0			MIT
2733.388	2734.197	36573.81	MO	1.5	0.9		MIT
2733.231	2734.040	36575.91	MO	1.0			MIT
2733.117	2733.926	36577.44	MO		0.7		MIT
2732.880	2733.689	36580.61	MO	1.3	1.5		MIT
2732.650	2733.459	36583.69	MO		1.0		MIT
2730.929	2731.737	36606.74	MO	0.0	1.0		MIT
2730.197	2731.005	36616.55	MO	0.3	1.8		MIT
2729.683	2730.491	36623.45	MO	1.2	1.8		MIT
2729.134	2729.942	36630.81	MO	1.3	0.0		MIT

AIR WAVELENGTH (ANGSTRMS)	VACUUM WAVELENGTH (ANGSTRMS)	WAVENUMBER (KAYSERS)	LMENT	LOG ARC	INTENSITY SPK	INTENSITY DIS	REF
2728.704	2729.512	36636.59	MO	1.3	1.4		MIT
2728.337	2729.145	36641.51	MO	1.2			MIT
2728.178	2728.986	36643.65	MO		1.0		MIT
2727.801	2728.609	36648.71	MO		0.5		MIT
2726.973	2727.781	36659.84	MO	1.0	2.0		MIT
2726.648	2727.455	36664.21	MO	1.1			MIT
2725.950	2726.757	36673.60	MO	1.2			MIT
2725.147	2725.954	36684.40	MO	1.3	0.6		MIT
2724.413	2725.220	36694.29	MO	0.6			MIT
2724.018	2724.825	36699.61	MO	0.0	1.4H		MIT
2723.355	2724.162	36708.54	MO	0.3	1.0		MIT
2722.559	2723.365	36719.27	MO	0.3H	1.0		MIT
2720.171	2720.977	36751.51	MO	1.4	0.0		MIT
2719.685	2720.491	36758.07	MO		0.7		MIT
2719.177	2719.983	36764.94	MO		0.7		MIT
2718.644	2719.450	36772.15	MO	1.2			MIT
2718.145	2718.950	36778.90	MO		0.3		MIT
2717.785	2718.590	36783.77	MO	1.0			MIT
2717.352	2718.157	36789.63	MO	1.3	2.0		MIT
2717.157	2717.962	36792.27	MO	1.3			MIT
2715.592	2716.397	36813.47	MO	0.6	0.8V		MIT
2715.170	2715.975	36819.19	MO	1.3	0.0		MIT
2713.506	2714.310	36841.77	MO	1.3	1.6		MIT
2713.093	2713.897	36847.38	MO	0.7			MIT
2712.941	2713.745	36849.44	MO		0.7		MIT
2712.689	2713.493	36852.87	MO	0.5			MIT
2712.348	2713.152	36857.50	MO	0.0	1.6		MIT
2711.917	2712.721	36863.36	MO	1.0			MIT
2711.486	2712.290	36869.22	MO	0.0	1.4		MIT
2710.927	2711.731	36876.82	MO	0.0	1.4		MIT
2710.742	2711.546	36879.34	MO	1.3	0.0		MIT
2710.194	2710.997	36886.79	MO	1.0	1.3		MIT
2709.757	2710.560	36892.74	MO		0.8		MIT
2709.527	2710.330	36895.87	MO	1.2			MIT
2709.246	2710.049	36899.70	MO	1.3	0.0		MIT
2707.006	2707.809	36930.23	MO		0.8		MIT
2706.119	2706.921	36942.33	MO	1.3	1.3V		MIT
2705.242	2706.044	36954.31	MO	1.3	0.0		MIT
2704.934	2705.736	36958.52	MO	0.0	1.7		MIT
2703.844	2704.646	36973.42	MO		0.5		MIT
2703.610	2704.412	36976.62	MO	0.0	1.2		MIT
2702.979	2703.781	36985.25	MO	0.7H			MIT
2702.772	2703.574	36988.08	MO	0.9			MIT
2702.612	2703.414	36990.27	MO	0.9			MIT
2701.873	2702.674	37000.39	MO	0.3	1.5		MIT
2701.417	2702.218	37006.63	MO	1.3	2.0		MIT
2701.030	2701.831	37011.93	MO	1.2	0.0		MIT
2700.746	2701.547	37015.83	MO	1.2			MIT
2700.430	2701.231	37020.16	MO		0.5V		MIT
2700.213	2701.014	37023.13	MO	1.3	0.0		MIT
2699.988	2700.789	37026.22	MO		1.2		MIT
2699.407	2700.208	37034.19	MO	0.5	1.5		MIT
2697.806	2698.606	37056.16	MO	1.4	0.6		MIT
2697.506	2698.306	37060.28	MO		0.6		MIT
2697.262	2698.062	37063.64	MO	0.6	1.0		MIT
2696.833	2697.633	37069.53	MO	0.0	1.6		MIT
2696.072	2696.872	37079.99	MO	1.4	0.0		MIT
2695.217	2696.017	37091.76	MO	0.7	1.6		MIT
2694.705	2695.505	37098.80	MO		0.5		MIT
2693.532	2694.331	37114.96	MO	0.9			MIT
2693.177	2693.976	37119.85	MO	0.0	1.4		MIT
2693.041	2693.840	37121.73	MO	1.2			MIT
2692.943	2693.742	37123.08	MO		1.2		MIT
2692.613	2693.412	37127.62	MO	0.3	1.6		MIT
2692.182	2692.981	37133.57	MO	0.5			MIT
2691.812	2692.611	37138.67	MO	0.7	0.3		MIT
2691.702	2692.501	37140.19	MO		0.7		MIT
2690.972	2691.771	37150.27	MO		0.7		MIT
2690.717	2691.516	37153.79	MO	1.0			MIT
2688.643	2689.441	37182.44	MO	1.3			MIT

AIR WAVELENGTH (ANGSTRMS)	VACUUM WAVELENGTH (ANGSTRMS)	WAVENUMBER (KAYSERS)	LMENT	LOG INTENSITY ARC	SPK	DIS	REF
2687.993	2688.791	37191.44	MO	1.5	2.0		MIT
2685.790	2686.588	37221.94	MO		1.4		MIT
2685.589	2686.387	37224.73	MO	1.0			MIT
2684.143	2684.940	37244.78	MO	1.6	2.2		MIT
2683.231	2684.028	37257.44	MO	1.3	2.2		MIT
2682.624	2683.421	37265.87	MO	1.3			MIT
2682.430	2683.227	37268.56	MO		1.0		MIT
2681.523	2682.320	37281.17	MO	1.2			MIT
2681.359	2682.156	37283.45	MO	1.0	2.0		MIT
2679.854	2680.650	37304.38	MO	1.5H	1.3		MIT
2678.666	2679.462	37320.93	MO	1.3			MIT
2676.484	2677.279	37351.35	MO	0.5	1.4		MIT
2673.844	2674.639	37388.23	MO	0.7			MIT
2673.273	2674.068	37396.21	MO	0.0	2.0		MIT
2672.843	2673.637	37402.23	MO	1.2	2.0		MIT
2671.834	2672.628	37416.35	MO	0.0	2.0		MIT
2671.471	2672.265	37421.44	MO	1.0			MIT
2670.935	2671.729	37428.95	MO		1.4		MIT
2670.322	2671.116	37437.54	MO	1.4	0.0		MIT
2669.98	2670.77	37442.3	MO	0.0	1.2		MIT
2666.751	2667.544	37487.67	MO	1.0	1.0		MIT
2665.487	2666.280	37505.44	MO	0.7			MIT
2665.096	2665.889	37510.94	MO	1.4	0.3		MIT
2663.326	2664.118	37535.87	MO	1.0	0.0H		MIT
2662.833	2663.625	37542.82	MO	1.0			MIT
2660.579	2661.371	37574.62	MO	1.4	2.1		MIT
2658.902	2659.693	37598.32	MO	1.0			MIT
2658.110	2658.901	37609.52	MO	1.6	0.7		MIT
2656.983	2657.774	37625.48	MO		1.4		MIT
2656.490	2657.281	37632.46	MO	1.0			MIT
2655.934	2656.724	37640.34	MO	1.2	0.0		MIT
2655.812	2656.602	37642.06	MO		1.3		MIT
2655.027	2655.817	37653.19	MO	1.7H	0.9		MIT
2653.790	2654.580	37670.74	MO	0.0	1.3		MIT
2653.349	2654.139	37677.00	MO	1.4	2.2		MIT
2652.35	2653.14	37691.2	MO	0.0	1.0		MIT
2650.682	2651.471	37714.91	MO	1.0	0.0		MIT
2649.999	2650.788	37724.63	MO	0.0	1.3		MIT
2649.460	2650.249	37732.31	MO	1.5H	1.0		MIT
2649.250	2650.039	37735.30	MO I	1.0	1.4		ME
2648.216	2649.005	37750.03	MO		0.3		MIT
2647.253	2648.041	37763.76	MO	1.3			MIT
2646.488	2647.276	37774.68	MO	1.4	2.0		MIT
2645.793	2646.581	37784.60	MO	1.2			MIT
2644.353	2645.141	37805.17	MO	1.5	1.8		MIT
2643.808	2644.596	37812.97	MO	1.3			MIT
2642.408	2643.195	37833.00	MO		1.5		MIT
2641.15	2641.94	37851.0	MO		1.3		MIT
2640.987	2641.774	37853.35	MO	1.6H	1.6J		MIT
2640.282	2641.069	37863.46	MO	1.3	0.0		MIT
2638.758	2639.544	37885.33	MO	1.3	2.1		MIT
2638.304	2639.090	37891.85	MO	1.4	0.7		MIT
2636.670	2637.456	37915.33	MO	1.4	2.2		MIT
2635.566	2636.352	37931.21	MO	1.4	1.2		MIT
2635.189	2635.974	37936.63	MO	1.1	0.6		MIT
2634.878	2635.663	37941.11	MO	0.8			MIT
2634.755	2635.540	37942.88	MO	1.0			MIT
2633.507	2634.292	37960.86	MO	0.3	1.4		MIT
2631.503	2632.288	37989.77	MO	1.4	1.0		MIT
2630.726	2631.510	38000.99	MO	0.7	1.6		MIT
2629.850	2630.634	38013.65	MO	1.7	0.8		MIT
2628.740	2629.524	38029.70	MO	1.6	0.3		MIT
2627.550	2628.334	38046.92	MO	1.3	1.2		MIT
2626.081	2626.864	38068.20	MO	0.5	1.5		MIT
2624.64	2625.42	38089.1	MO		1.0		MIT
2623.42	2624.20	38106.8	MO		1.5		MIT
2621.071	2621.853	38140.96	MO	1.6H	0.0		MIT
2620.815	2621.597	38144.69	MO	1.0			MIT
2620.060	2620.842	38155.68	MO		1.2		MIT
2619.341	2620.123	38166.15	MO	1.3	1.6		MIT

AIR WAVELENGTH (ANGSTRMS)	VACUUM WAVELENGTH (ANGSTRMS)	WAVENUMBER (KAYSERS)	LMENT	LOG INTENSITY ARC	SPK	DIS	REF
2617.878	2618.659	38187.48	MO	1.2			MIT
2616.782	2617.563	38203.47	MO	1.5	0.7		MIT
2615.391	2616.172	38223.79	MO	1.4	0.5		MIT
2613.719	2614.499	38248.24	MO	1.2			MIT
2613.079	2613.859	38257.61	MO	1.6H	1.3		MIT
2612.289	2613.069	38269.18	MO	1.0	0.8H		MIT
2611.627	2612.407	38278.88	MO	1.2			MIT
2611.199	2611.979	38285.15	MO	1.4	1.2		MIT
2610.259	2611.039	38298.94	MO	1.2			MIT
2609.22	2610.00	38314.2	MO		1.6J		MIT
2608.864	2609.643	38319.42	MO	1.2	1.0		MIT
2607.374	2608.153	38341.31	MO	1.7H	0.7		MIT
2606.58	2607.36	38353.0	MO		1.3		MIT
2605.933	2606.712	38362.51	MO	0.7	1.0		MIT
2605.075	2605.853	38375.15	MO	1.2	0.7		MIT
2603.316	2604.094	38401.07	MO	1.2	0.0		MIT
2602.798	2603.576	38408.71	MO	1.4	2.0		MIT
2601.96	2602.74	38421.1	MO	0.0	1.3		MIT
2601.825	2602.603	38423.08	MO	1.2			MIT
2601.691	2602.468	38425.06	MO	1.3	0.0		MIT
2601.345	2602.122	38430.17	MO	1.3			MIT
2600.366	2601.143	38444.63	MO	1.3	0.0		MIT
2600.12	2600.90	38448.3	MO		1.0		MIT
2599.643	2600.420	38455.33	MO	1.3	0.0		MIT
2599.182	2599.959	38462.15	MO	0.0	1.2		MIT
2597.376	2598.152	38488.89	MO	0.3	1.5		MIT
2597.225	2598.001	38491.13	MO	1.3			MIT
2597.13	2597.91	38492.5	MO		1.3H		MIT
2596.768	2597.544	38497.90	MO	1.3			MIT
2595.403	2596.179	38518.15	MO	1.4	1.3		MIT
2594.005	2594.781	38538.90	MO	1.2			MIT
2593.705	2594.481	38543.36	MO	1.3	1.6		MIT
2593.378	2594.154	38548.22	MO	0.5	1.3		MIT
2592.792	2593.567	38556.93	MO	0.5	1.2		MIT
2591.975	2592.750	38569.08	MO	1.6	0.0		MIT
2591.770	2592.545	38572.13	MO	0.3	1.4		MIT
2588.783	2589.557	38616.64	MO	1.0	1.5		MIT
2587.31	2588.08	38638.6	MO	0.0H	1.5J		MIT
2587.04	2587.81	38642.6	MO		1.0		MIT
2585.948	2586.722	38658.97	MO	1.2	1.5		MIT
2584.18	2584.95	38685.4	MO		0.7		MIT
2582.292	2583.065	38713.70	MO	1.0			MIT
2582.157	2582.930	38715.72	MO	1.4	0.5		MIT
2579.442	2580.214	38756.47	MO	1.0	1.3		MIT
2578.909	2579.681	38764.48	MO	0.3	1.8		MIT
2578.766	2579.538	38766.63	MO	1.4	0.0		MIT
2578.356	2579.128	38772.79	MO	0.0	1.4		MIT
2576.563	2577.335	38799.77	MO	0.3	1.4		MIT
2574.424	2575.195	38832.01	MO	1.2	1.3		MIT
2573.313	2574.084	38848.77	MO	1.4			MIT
2572.341	2573.112	38863.45	MO	1.8H	1.0		MIT
2572.25	2573.02	38864.8	MO		1.2		MIT
2571.45	2572.22	38876.9	MO	0.3	1.3		MIT
2571.24	2572.01	38880.1	MO		1.0		MIT
2570.80	2571.57	38886.7	MO		1.4J		MIT
2567.99	2568.76	38929.3	MO	1.0H	0.3H		MIT
2567.050	2567.819	38943.55	MO	1.6H	0.5		MIT
2566.259	2567.028	38955.55	MO	1.0	1.3		MIT
2564.769	2565.538	38978.18	MO	1.0	0.0H		MIT
2564.338	2565.107	38984.73	MO	0.5	1.6		MIT
2562.081	2562.849	39019.07	MO	0.0	1.5		MIT
2561.623	2562.391	39026.05	MO	1.4			MIT
2559.694	2560.462	39055.46	MO		1.3		MIT
2559.15	2559.92	39063.8	MO	0.0	1.0H		MIT
2558.936	2559.703	39067.03	MO	1.2			MIT
2558.88	2559.65	39067.9	MO	0.7H	1.5H		MIT
2557.39	2558.16	39090.6	MO	0.0	1.0		MIT
2556.753	2557.520	39100.38	MO	0.3	1.3		MIT
2555.42	2556.19	39120.8	MO	0.5	1.7		MIT
2553.698	2554.464	39147.15	MO	1.2			MIT

AIR WAVELENGTH (ANGSTRMS)	VACUUM WAVELENGTH (ANGSTRMS)	WAVENUMBER (KAYSERS)	LMENT	LOG ARC	INTENSITY SPK	DIS	REF
2552.873	2553.639	39159.80	MO	1.3	0.3		MIT
2550.851	2551.616	39190.84	MO	1.5H	0.5H		MIT
2550.75	2551.52	39192.4	MO		1.2		MIT
2548.225	2548.990	39231.23	MO	1.5H	0.7		MIT
2547.571	2548.336	39241.30	MO	1.2	1.3		MIT
2547.351	2548.116	39244.69	MO	0.0	1.3		MIT
2545.371	2546.135	39275.21	MO	1.0			MIT
2543.611	2544.375	39302.39	MO	0.5	1.2		MIT
2543.349	2544.113	39306.43	MO	1.4	0.3		MIT
2542.792	2543.556	39315.04	MO	1.0	1.0		MIT
2542.671	2543.435	39316.92	MO	1.3	1.4		MIT
2540.451	2541.214	39351.27	MO	1.6H	0.0		MIT
2540.14	2540.90	39356.1	MO		1.2		MIT
2539.44	2540.20	39366.9	MO	0.0	1.3		MIT
2538.455	2539.218	39382.21	MO	1.5	2.1		MIT
2536.849	2537.611	39407.14	MO	1.4	0.6		MIT
2533.556	2534.317	39458.36	MO		1.3		MIT
2532.69	2533.45	39471.9	MO		0.6*		MIT
2532.31	2533.07	39477.8	MO	0.9	1.5		MIT
2530.338	2531.099	39508.54	MO	0.5	1.4		MIT
2529.839	2530.600	39516.33	MO	1.4	1.3		MIT
2528.866	2529.626	39531.53	MO	0.0	1.4		MIT
2527.14	2527.90	39558.5	MO	0.0	1.4		MIT
2524.811	2525.570	39595.02	MO	1.6H			MIT
2524.71	2525.47	39596.6	MO		1.4		MIT
2523.18	2523.94	39620.6	MO	0.0	1.4		MIT
2522.59	2523.35	39629.9	MO	0.0	1.3		MIT
2521.69	2522.45	39644.0	MO	0.0	1.0		MIT
2521.12	2521.88	39653.0	MO	0.0	1.3		MIT
2518.44	2519.20	39695.2	MO		1.0		MIT
2517.832	2518.590	39704.76	MO	1.3			MIT
2517.464	2518.222	39710.56	MO	1.4			MIT
2516.109	2516.866	39731.95	MO	1.4	1.4		MIT
2515.66	2516.42	39739.0	MO	1.0	1.0		MIT
2515.082	2515.839	39748.17	MO	0.9	1.5		MIT
2514.628	2515.385	39755.35	MO	1.4			MIT
2514.181	2514.938	39762.41	MO		1.0H		MIT
2513.332	2514.089	39775.84	MO	1.4			MIT
2512.048	2512.804	39796.17	MO		1.0		MIT
2509.557	2510.313	39835.67	MO	1.3			MIT
2508.667	2509.423	39849.80	MO	1.5H			MIT
2506.19	2506.95	39889.2	MO		1.6H		MIT
2505.105	2505.860	39906.46	MO	1.4			MIT
2503.59	2504.34	39930.6	MO		0.7		MIT
2502.839	2503.593	39942.59	MO	1.0	1.2		MIT
2502.222	2502.976	39952.44	MO		1.3		MIT
2501.609	2502.363	39962.23	MO	0.3	0.5		MIT
2500.44	2501.19	39980.9	MO		1.4		MIT
2499.259	2500.012	39999.80	MO	0.0	0.7		MIT
2498.281	2499.034	40015.46	MO	1.0	1.4		MIT
2498.092	2498.845	40018.49	MO		0.9		MIT
2497.863	2498.616	40022.15	MO	1.3	1.2		MIT
2497.580	2498.333	40026.69	MO	1.5H			MIT
2497.376	2498.129	40029.96	MO		1.2		MIT
2497.02	2497.77	40035.7	MO		0.9		MIT
2496.06	2496.81	40051.1	MO		0.7H		MIT
2494.689	2495.441	40073.07	MO	0.0	0.8		MIT
2491.59	2492.34	40122.9	MO		0.6H		MIT
2490.02	2490.77	40148.2	MO		1.0H		MIT
2487.668	2488.419	40186.16	MO	0.5	1.0H		MIT
2485.313	2486.063	40224.24	MO	1.5H			MIT
2484.753	2485.503	40233.30	MO	0.3	1.3		MIT
2482.572	2483.322	40268.65	MO	0.7	1.4		MIT
2481.814	2482.563	40280.94	MO	1.7H			MIT
2481.191	2481.940	40291.06	MO	1.0	1.5		MIT
2479.21	2479.96	40323.2	MO		0.9		MIT
2478.219	2478.968	40339.37	MO		0.6		MIT
2477.990	2478.738	40343.10	MO		0.7		MIT
2477.567	2478.315	40349.99	MO	0.3	1.7		MIT
2474.709	2475.457	40396.59	MO	1.3	0.3H		MIT

AIR WAVELENGTH (ANGSTRMS)	VACUUM WAVELENGTH (ANGSTRMS)	WAVENUMBER (KAYSERS)	LMENT	LOG ARC	INTENSITY SPK	DIS	REF
2474.24	2474.99	40404.2	MO		1.0H		MIT
2471.967	2472.714	40441.39	MO	1.7H	0.3		MIT
2470.043	2470.790	40472.89	MO	0.0	1.4		MIT
2469.104	2469.850	40488.28	MO	1.3			MIT
2467.345	2468.091	40517.14	MO	0.7	1.3		MIT
2466.968	2467.714	40523.33	MO	1.0	1.3		MIT
2466.678	2467.424	40528.10	MO	0.7	1.4		MIT
2465.577	2466.323	40546.20	MO	1.2			MIT
2462.482	2463.227	40597.15	MO	0.0	0.8		MIT
2461.810	2462.555	40608.23	MO	0.3	1.4		MIT
2459.761	2460.505	40642.06	MO	0.0	1.5		MIT
2458.714	2459.458	40659.36	MO	1.0	0.7		MIT
2457.766	2458.510	40675.05	MO	1.0	1.8		MIT
2457.404	2458.148	40681.04	MO	1.2			MIT
2455.453	2456.196	40713.36	MO		0.9H		MIT
2455.00	2455.74	40720.9	MO	1.2			MIT
2454.266	2455.009	40733.05	MO	0.7			MIT
2454.088	2454.831	40736.00	MO	0.9H			MIT
2453.80	2454.54	40740.8	MO		0.5		MIT
2453.333	2454.076	40748.54	MO	1.0	0.8		MIT
2453.133	2453.876	40751.86	MO	0.3	1.3		MIT
2452.22	2452.96	40767.0	MO		1.0		MIT
2444.735	2445.476	40891.84	MO	0.0	0.9		MIT
2444.203	2444.944	40900.74	MO	1.0H			MIT
2443.442	2444.183	40913.47	MO	0.6	1.3		MIT
2443.188	2443.928	40917.73	MO	0.3	1.3		MIT
2441.644	2442.384	40943.60	MO	1.0G			PU
2440.277	2441.017	40966.54	MO	0.6	1.3		MIT
2440.060	2440.800	40970.18	MO	0.0	0.9		MIT
2438.57	2439.31	40995.2	MO	0.0	1.0		MIT
2438.343	2439.082	40999.02	MO	0.0	1.2O		MIT
2437.736	2438.475	41009.23	MO	1.0	1.9O		MIT
2435.963	2436.702	41039.08	MO	1.0	1.7H		MIT
2434.189	2434.927	41068.99	MO		0.5H		MIT
2433.96	2434.70	41072.9	MO		1.4D		MIT
2430.429	2431.167	41132.52	MO	1.3	0.6		MIT
2430.274	2431.012	41135.14	MO		1.2		MIT
2429.41	2430.15	41149.8	MO	0.6	1.4		MIT
2429.08	2429.82	41155.4	MO		0.7H		MIT
2428.78	2429.52	41160.4	MO		1.3H		MIT
2428.18	2428.92	41170.6	MO		1.2		MIT
2427.295	2428.032	41185.62	MO	0.3	1.3		MIT
2426.974	2427.711	41191.07	MO		0.3		MIT
2426.37	2427.11	41201.3	MO	0.5	0.7		MIT
2425.73	2426.47	41212.2	MO		1.2P		MIT
2425.10	2425.84	41222.9	MO		0.7		MIT
2424.258	2424.994	41237.21	MO		1.2		MIT
2423.995	2424.731	41241.69	MO	0.8	1.6		MIT
2423.716	2424.452	41246.43	MO	0.3	1.0		MIT
2422.179	2422.915	41272.60	MO	0.3	1.8		MIT
2421.650	2422.386	41281.62	MO		1.4		MIT
2420.717	2421.452	41297.53	MO	1.2	0.5		MIT
2420.529	2421.264	41300.74	MO	1.2	0.0		MIT
2420.176	2420.911	41306.76	MO	0.0	1.6		MIT
2419.008	2419.743	41326.70	MO	0.5	1.6		MIT
2417.958	2418.693	41344.65	MO	0.7	1.5		MIT
2417.373	2418.108	41354.65	MO		1.0		MIT
2416.572	2417.306	41368.36	MO	1.2	0.0		MIT
2416.18	2416.91	41375.1	MO		0.3		MIT
2415.326	2416.060	41389.70	MO	1.3	0.8		MIT
2414.247	2414.981	41408.20	MO	0.3	1.2		MIT
2413.010	2413.744	41429.42	MO	1.2	1.5H		MIT
2412.76	2413.49	41433.7	MO	0.3	1.3H		MIT
2411.815	2412.548	41449.95	MO	0.3	1.5		MIT
2411.266	2411.999	41459.38	MO	0.0	1.5		MIT
2410.916	2411.649	41465.40	MO		1.0		MIT
2410.09	2410.82	41479.6	MO		1.8H		MIT
2408.893	2409.626	41500.22	MO		1.2		MIT
2408.67	2409.40	41504.1	MO		1.0		MIT
2408.390	2409.123	41508.89	MO	1.0			MIT

AIR WAVELENGTH (ANGSTRMS)	VACUUM WAVELENGTH (ANGSTRMS)	WAVENUMBER (KAYSERS)	LMENT	LOG ARC	INTENSITY SPK	DIS	REF
2408.320	2409.053	41510.10	MO	0.0	1.2		MIT
2407.135	2407.867	41530.53	MO	0.5	1.6		MIT
2405.858	2406.590	41552.57	MO	1.4	1.2		MIT
2405.19	2405.92	41564.1	MO	0.5	0.6H		MIT
2404.657	2405.389	41573.32	MO	0.7	1.5		MIT
2403.610	2404.341	41591.43	MO	1.4	1.4		MIT
2403.418	2404.149	41594.75	MO	0.8	1.4		MIT
2401.927	2402.658	41620.57	MO	1.0	1.6		MIT
2399.76	2400.49	41658.1	MO		0.6		MIT
2399.32	2400.05	41665.8	MO		1.0		MIT
2398.95	2399.68	41672.2	MO		0.5		MIT
2398.09	2398.82	41687.2	MO	0.7			MIT
2397.82	2398.55	41691.9	MO		0.9		MIT
2397.58	2398.31	41696.0	MO	0.8			MIT
2396.24	2396.97	41719.3	MO		0.9		MIT
2395.98	2396.71	41723.9	MO		0.5		MIT
2395.25	2395.98	41736.6	MO	1.1			MIT
2395.10	2395.83	41739.2	MO		0.8		MIT
2394.74	2395.47	41745.5	MO	1.3	1.0		MIT
2393.97	2394.70	41758.9	MO		0.6		MIT
2393.20	2393.93	41772.3	MO	0.7	1.0		MIT
2392.75	2393.48	41780.2	MO		0.8		MIT
2392.63	2393.36	41782.3	MO		0.8		MIT
2392.33	2393.06	41787.5	MO	0.3	1.3		MIT
2391.71	2392.44	41798.4	MO	0.5	1.3		MIT
2390.78	2391.51	41814.6	MO		1.4		MIT
2390.43	2391.16	41820.7	MO		0.9		MIT
2390.10	2390.83	41826.5	MO	1.0	1.3		MIT
2389.20	2389.93	41842.3	MO	1.3	1.3		MIT
2388.96	2389.69	41846.5	MO		1.0		MIT
2388.70	2389.43	41851.0	MO II		1.5		MIT
2388.05	2388.78	41862.4	MO		0.5		MIT
2387.81	2388.54	41866.6	MO		0.8		MIT
2387.18	2387.91	41877.7	MO	1.1			MIT
2386.96	2387.69	41881.5	MO		1.5		MIT
2386.24	2386.97	41894.2	MO		0.9		MIT
2386.05	2386.78	41897.5	MO		1.3		MIT
2384.65	2385.38	41922.1	MO		1.1		MIT
2383.87	2384.60	41935.8	MO		1.3		MIT
2383.52	2384.25	41942.0	MO	1.1			MIT
2383.36	2384.09	41944.8	MO		1.0		MIT
2383.05	2383.78	41950.2	MO		0.6		MIT
2382.39	2383.12	41961.9	MO		1.1		MIT
2381.47	2382.20	41978.1	MO		1.1		MIT
2380.41	2381.14	41996.8	MO	1.1			MIT
2380.00	2380.73	42004.0	MO		1.1		MIT
2378.65	2379.38	42027.8	MO	0.8	0.5		MIT
2377.85	2378.58	42042.0	MO		0.7		MIT
2377.12	2377.85	42054.9	MO		1.5		MIT
2375.44	2376.17	42084.6	MO		0.3		MIT
2374.87	2375.59	42094.7	MO		0.9		MIT
2373.93	2374.65	42111.4	MO		0.7		MIT
2372.96	2373.68	42128.6	MO		1.5		MIT
2372.27	2372.99	42140.9	MO	1.4	0.8		MIT
2372.07	2372.79	42144.4	MO	0.7			MIT
2371.54	2372.26	42153.8	MO		0.9		MIT
2371.26	2371.98	42158.8	MO		1.2		MIT
2370.39	2371.11	42174.3	MO	0.5	0.9		MIT
2370.23	2370.95	42177.1	MO	0.3H	1.0		MIT
2369.97	2370.69	42181.7	MO		1.0		MIT
2368.95	2369.67	42199.9	MO		1.2		MIT
2366.96	2367.68	42235.4	MO	0.3H	1.3		MIT
2366.38	2367.10	42245.7	MO	0.8	1.3		MIT
2366.09	2366.81	42250.9	MO	0.8	1.0		MIT
2365.85	2366.57	42255.2	MO		0.9		MIT
2364.37	2365.09	42281.6	MO	0.9			MIT
2363.69	2364.41	42293.8	MO		1.0		MIT
2362.98	2363.70	42306.5	MO		0.9		MIT
2362.42	2363.14	42316.5	MO		0.8		MIT
2361.25	2361.97	42337.5	MO		0.5		MIT

AIR WAVELENGTH (ANGSTRMS)	VACUUM WAVELENGTH (ANGSTRMS)	WAVENUMBER (KAYSERS)	LMENT	LOG ARC	INTENSITY SPK	DIS	REF
2359.73	2360.45	42364.8	MO		1.6		MIT
2359.34	2360.06	42371.8	MO		0.7		MIT
2357.73	2358.45	42400.7	MO	0.9			MIT
2357.57	2358.29	42403.6	MO		1.1		MIT
2357.33	2358.05	42407.9	MO		0.6		MIT
2357.16	2357.88	42411.0	MO		0.8		MIT
2356.02	2356.74	42431.5	MO		0.6		MIT
2355.42	2356.14	42442.3	MO	0.7	1.4		MIT
2355.28	2356.00	42444.8	MO		1.0		MIT
2355.22	2355.94	42445.9	MO	1.0			MIT
2355.01	2355.73	42449.7	MO		1.0		MIT
2354.67	2355.39	42455.8	MO		0.8		MIT
2354.46	2355.18	42459.6	MO	1.0			MIT
2354.16	2354.88	42465.0	MO		1.1		MIT
2353.72	2354.44	42472.9	MO		1.3		MIT
2352.61	2353.33	42493.0	MO	1.2	0.3H		MIT
2351.28	2352.00	42517.0	MO	1.0			MIT
2350.53	2351.25	42530.6	MO		0.7		MIT
2350.11	2350.83	42538.2	MO		1.1		MIT
2349.89	2350.61	42542.2	MO		1.4		MIT
2349.78	2350.50	42544.1	MO	1.0W			MIT
2348.98	2349.70	42558.6	MO		0.8		MIT
2348.84	2349.56	42561.2	MO	0.5	1.0		MIT
2348.58	2349.30	42565.9	MO		1.0		MIT
2347.79	2348.51	42580.2	MO		1.3		MIT
2347.46	2348.18	42586.2	MO		0.9		MIT
2346.65	2347.37	42600.9	MO		0.9		MIT
2344.73	2345.45	42635.8	MO		1.1		A
2344.653	2345.371	42637.17	MO	0.5	1.1		MIT
2343.72	2344.44	42654.1	MO		0.8		MIT
2343.13	2343.85	42664.9	MO		1.1		MIT
2342.63	2343.35	42674.0	MO		0.8		MIT
2342.27	2342.99	42680.5	MO		0.8		MIT
2341.59	2342.31	42692.9	MO	1.0	1.8		MIT
2341.42	2342.14	42696.0	MO	1.0			MIT
2340.47	2341.19	42713.4	MO I	1.5	1.7		ME
2339.36	2340.08	42733.6	MO		0.9		MIT
2339.18	2339.90	42736.9	MO		0.3		MIT
2338.95	2339.67	42741.1	MO		1.0		MIT
2338.47	2339.19	42749.9	MO		0.8		MIT
2337.72	2338.44	42763.6	MO	1.1	0.5		MIT
2337.42	2338.14	42769.1	MO	0.8	1.1		MIT
2337.04	2337.76	42776.0	MO		0.9		MIT
2336.65	2337.37	42783.2	MO		1.0		MIT
2336.42	2337.14	42787.4	MO		1.0		MIT
2336.19	2336.91	42791.6	MO		1.0		MIT
2335.80	2336.52	42798.8	MO		1.0		MIT
2335.38	2336.10	42806.5	MO		0.7		MIT
2334.92	2335.64	42814.9	MO		1.3		MIT
2334.80	2335.52	42817.1	MO		1.0		MIT
2334.38	2335.10	42824.8	MO		0.5		MIT
2334.05	2334.77	42830.9	MO		0.5		MIT
2333.63	2334.35	42838.6	MO		0.6		MIT
2332.68	2333.40	42856.0	MO	0.8	1.3		MIT
2332.26	2332.98	42863.7	MO	0.5			MIT
2332.12	2332.84	42866.3	MO	0.9	1.9		MIT
2331.16	2331.88	42883.9	MO		0.8		MIT
2330.93	2331.65	42888.2	MO	1.0	1.5		MIT
2330.46	2331.18	42896.8	MO	1.3	0.3H		MIT
2330.04	2330.75	42904.5	MO		1.2		MIT
2329.69	2330.40	42911.0	MO	0.5	1.4		MIT
2328.65	2329.36	42930.2	MO		0.5		MIT
2328.11	2328.82	42940.1	MO		1.0		MIT
2327.79	2328.50	42946.0	MO		1.0		MIT
2326.75	2327.46	42965.2	MO	0.5	1.3		MIT
2326.05	2326.76	42978.1	MO	0.5			MIT
2325.81	2326.52	42982.6	MO	1.2	0.7		MIT
2325.54	2326.25	42987.6	MO		1.4		MIT
2324.85	2325.56	43000.3	MO	0.5	1.4		MIT
2323.90	2324.61	43017.9	MO		1.5		MIT

AIR WAVELENGTH (ANGSTRMS)	VACUUM WAVELENGTH (ANGSTRMS)	WAVENUMBER (KAYSERS)	LMENT	LOG ARC	INTENSITY SPK	DIS	REF
2322.95	2323.66	43035.5	MO		0.8		MIT
2322.48	2323.19	43044.2	MO		1.3		MIT
2321.05	2321.76	43070.7	MO	0.7	0.6		MIT
2320.72	2321.43	43076.8	MO	1.1			MIT
2318.53	2319.24	43117.5	MO	1.1			MIT
2317.89	2318.60	43129.4	MO		1.2		MIT
2316.45	2317.16	43156.2	MO		1.1		MIT
2316.17	2316.88	43161.5	MO		0.9		MIT
2315.62	2316.33	43171.7	MO	0.5	1.2		MIT
2314.42	2315.13	43194.1	MO		1.3		MIT
2313.88	2314.59	43204.2	MO		1.0		MIT
2312.22	2312.93	43235.2	MO		0.8		MIT
2309.85	2310.56	43279.5	MO		1.5		MIT
2309.45	2310.16	43287.0	MO		1.3		MIT
2307.98	2308.69	43314.6	MO	0.7	1.3L		MIT
2306.97	2307.68	43333.6	MO	0.9	1.4		MIT
2306.506	2307.216	43342.28	MO		1.3		MIT
2306.06	2306.77	43350.7	MO	0.6			A
2305.87	2306.58	43354.2	MO		1.0		MIT
2305.658	2306.368	43358.22	MO	0.7	1.3		MIT
2305.25	2305.96	43365.9	MO		0.9		MIT
2304.251	2304.960	43384.70	MO	0.8	1.7		MIT
2302.842	2303.551	43411.24	MO		1.3		MIT
2302.130	2302.839	43424.66	MO		0.9		MIT
2301.16	2301.87	43443.0	MO		0.5		MIT
2299.874	2300.582	43467.26	MO		1.0		MIT
2299.27	2299.98	43478.7	MO		0.7		MIT
2298.86	2299.57	43486.4	MO		0.3		MIT
2298.379	2299.087	43495.53	MO	0.5	1.0		MIT
2298.22	2298.93	43498.5	MO		1.1		MIT
2297.64	2298.35	43509.5	MO		0.8		MIT
2296.549	2297.257	43530.18	MO		1.4		MIT
2295.308	2296.015	43553.72	MO		0.9		MIT
2294.960	2295.667	43560.32	MO		1.7		MIT
2292.983	2293.690	43597.87	MO		0.5		MIT
2290.87	2291.58	43638.1	MO	0.5	1.3		MIT
2290.32	2291.03	43648.6	MO		1.3		MIT
2290.29	2291.00	43649.1	MO	0.9			A
2290.06	2290.77	43653.5	MO		1.3		MIT
2289.89	2290.60	43656.8	MO	0.7	1.2		MIT
2289.19	2289.90	43670.1	MO II	0.6	1.3		KS
2286.42	2287.13	43723.0	MO	0.3H	1.0		MIT
2285.14	2285.85	43747.5	MO		1.3		MIT
2284.43	2285.13	43761.1	MO	0.8	1.5		MIT
2283.28	2283.98	43783.1	MO		1.0		MIT
2282.89	2283.59	43790.6	MO		1.3		MIT
2282.21	2282.91	43803.7	MO		0.7		MIT
2281.44	2282.14	43818.4	MO		0.5		MIT
2280.82	2281.52	43830.4	MO		1.5L		MIT
2279.58	2280.28	43854.2	MO	0.6	1.0		MIT
2279.04	2279.74	43864.6	MO		0.8		MIT
2276.94	2277.64	43905.0	MO		0.8		MIT
2276.21	2276.91	43919.1	MO		1.3		MIT
2275.64	2276.34	43930.1	MO		1.2		MIT
2275.47	2276.17	43933.4	MO		1.2		MIT
2275.00	2275.70	43942.5	MO	0.5	1.5		MIT
2274.35	2275.05	43955.0	MO		0.7		MIT
2273.26	2273.96	43976.1	MO		2.0		MIT
2272.672	2273.374	43987.48	MO		0.7		MIT
2272.37	2273.07	43993.3	MO		0.9		MIT
2271.783	2272.485	44004.69	MO		0.7H		MIT
2270.68	2271.38	44026.1	MO		1.0		MIT
2270.332	2271.034	44032.81	MO		0.7		MIT
2269.690	2270.392	44045.26	MO	1.1	1.5		MIT
2269.42	2270.12	44050.5	MO	0.5	1.0		MIT
2268.47	2269.17	44069.0	MO		1.3		MIT
2267.12	2267.82	44095.2	MO		0.9		MIT
2266.22	2266.92	44112.7	MO		1.3		MIT
2265.52	2266.22	44126.3	MO		0.9		MIT
2265.17	2265.87	44133.1	MO		0.3		MIT
2264.95	2265.65	44137.4	MO		0.6		MIT
2264.742	2265.443	44141.48	MO		1.3		MIT
2264.35	2265.05	44149.1	MO	0.6	1.0		MIT
2264.20	2264.90	44152.0	MO		0.7		MIT
2263.75	2264.45	44160.8	MO		0.6		MIT
2261.976	2262.676	44195.46	MO	0.7H	1.2		MIT
2260.751	2261.451	44219.40	MO		0.7		MIT
2260.52	2261.22	44223.9	MO		0.7		MIT
2258.22	2258.92	44269.0	MO		0.7		MIT
2258.06	2258.76	44272.1	MO		0.6		MIT
2257.705	2258.404	44279.06	MO		0.9		MIT
2257.19	2257.89	44289.2	MO		1.1		MIT
2256.983	2257.682	44293.22	MO		1.1		MIT
2254.70	2255.40	44338.1	MO		0.9		MIT
2253.18	2253.88	44368.0	MO		1.4		MIT
2252.88	2253.58	44373.9	MO		0.7		A
2252.406	2253.104	44383.22	MO		1.3		MIT
2252.03	2252.73	44390.6	MO		0.7		MIT
2251.33	2252.03	44404.4	MO		1.5		MIT
2250.25	2250.95	44425.7	MO		1.3		MIT
2249.93	2250.63	44432.0	MO	1.3	1.3		MIT
2249.33	2250.03	44443.9	MO	0.6	1.3		MIT
2249.07	2249.77	44449.0	MO		1.3		MIT
2247.11	2247.81	44487.8	MO		1.2		MIT
2246.95	2247.65	44491.0	MO		1.3		MIT
2246.05	2246.75	44508.8	MO		1.2		MIT
2243.623	2244.319	44556.94	MO		1.0		MIT
2242.21	2242.91	44585.0	MO		1.4		MIT
2241.63	2242.33	44596.5	MO		0.6		MIT
2241.17	2241.87	44605.7	MO		0.6		MIT
2240.58	2241.28	44617.5	MO		0.6		MIT
2239.39	2240.09	44641.2	MO	0.5	1.3		MIT
2238.88	2239.58	44651.3	MO		0.3H		MIT
2238.41	2239.10	44660.7	MO		1.2		MIT
2236.04	2236.73	44708.0	MO II	0.5	1.0W		MIT
2235.66	2236.35	44715.6	MO		1.2		MIT
2234.500	2235.194	44738.84	MO		0.9		MIT
2232.75	2233.44	44773.9	MO	1.0	1.0		A
2232.434	2233.128	44780.24	MO	0.6	1.1		MIT
2232.12	2232.81	44786.5	MO		1.0		MIT
2231.483	2232.177	44799.33	MO		0.8		MIT
2231.08	2231.77	44807.4	MO		1.2		MIT
2230.56	2231.25	44817.9	MO		1.0		MIT
2230.08	2230.77	44827.5	MO	0.6	1.4		MIT
2227.39	2228.08	44881.6	MO	1.1			A
2226.93	2227.62	44890.9	MO		0.8		MIT
2226.78	2227.47	44893.9	MO		0.8		MIT
2226.07	2226.76	44908.2	MO		1.0		MIT
2224.66	2225.35	44936.7	MO		0.9		MIT
2223.92	2224.61	44951.7	MO	1.1	0.5		A
2223.193	2223.885	44966.36	MO		1.3		MIT
2222.77	2223.46	44974.9	MO		0.8		MIT
2221.67	2222.36	44997.2	MO	0.5			MIT
2220.97	2221.66	45011.4	MO	0.7	1.4		MIT
2220.25	2220.94	45026.0	MO	1.1			A
2219.82	2220.51	45034.7	MO		0.6		MIT
2217.876	2218.567	45074.15	MO		0.8		MIT
2217.17	2217.86	45088.5	MO		0.7		MIT
2216.61	2217.30	45099.9	MO	0.6	1.1		MIT
2215.52	2216.21	45122.1	MO		0.5		MIT
2214.27	2214.96	45147.5	MO	0.3	1.4		MIT
2214.01	2214.70	45152.9	MO		0.6		MIT
2213.44	2214.13	45164.5	MO		0.5		MIT
2211.00	2211.69	45214.3	MO		1.2		MIT
2210.88	2211.57	45216.8	MO	0.6			MIT
2209.85	2210.54	45237.8	MO	0.3	0.8		MIT
2209.56	2210.25	45243.8	MO	0.3	1.2		MIT
2209.07	2209.76	45253.8	MO		0.6		MIT
2208.18	2208.87	45272.0	MO		1.0		MIT
2207.35	2208.04	45289.1	MO		1.0		MIT

AIR WAVELENGTH (ANGSTRMS)	VACUUM WAVELENGTH (ANGSTRMS)	WAVENUMBER (KAYSERS)	LMENT	LOG ARC	INTENSITY SPK	INTENSITY DIS	REF
2206.84	2207.53	45299.5	MO		1.3		MIT
2206.450	2207.138	45307.54	MO	0.5	1.3		MIT
2206.06	2206.75	45315.5	MO		0.8		MIT
2204.87	2205.56	45340.0	MO		1.0		MIT
2200.96	2201.65	45420.5	MO	0.5	1.2		MIT
2200.60	2201.29	45428.0	MO		0.5		MIT
2200.25	2200.94	45435.2	MO	0.3H	0.8		MIT
2198.14	2198.83	45478.8	MO	0.3	1.3		MIT
2197.47	2198.16	45492.7	MO	0.7	1.5		MIT
2197.26	2197.95	45497.0	MO	0.5	1.0		MIT
2196.97	2197.66	45503.0	MO		0.7		MIT
2195.25	2195.94	45538.7	MO		0.8		MIT
2194.95	2195.64	45544.9	MO		0.9		MIT
2191.45	2192.14	45617.6	MO		0.7		MIT
2190.43	2191.11	45638.9	MO		1.0		MIT
2190.23	2190.91	45643.0	MO	0.5	0.8		A
2190.006	2190.691	45647.70	MO	0.5	1.0		MIT
2189.74	2190.42	45653.2	MO		0.6		MIT
2189.42	2190.10	45659.9	MO	0.6	1.5		MIT
2188.95	2189.63	45669.7	MO	0.6	1.2		MIT
2188.64	2189.32	45676.2	MO		0.8		MIT
2188.03	2188.71	45688.9	MO		1.3		MIT
2187.62	2188.30	45697.5	MO		0.9		MIT
2184.89	2185.57	45754.6	MO		1.1		MIT
2184.35	2185.03	45765.9	MO	0.3	1.3		MIT
2183.60	2184.28	45781.6	MO		0.8		MIT
2182.535	2183.218	45803.94	MO		1.2		MIT
2181.98	2182.66	45815.6	MO		0.8		MIT
2181.37	2182.05	45828.4	MO	0.3J	1.3		MIT
2180.25	2180.93	45851.9	MO		1.2		MIT
2180.03	2180.71	45856.6	MO		0.9		MIT
2179.36	2180.04	45870.7	MO		1.3		MIT
2177.86	2178.54	45902.2	MO		0.6		MIT
2177.51	2178.19	45909.6	MO		0.8		MIT
2175.40	2176.08	45954.2	MO		0.7		MIT
2174.08	2174.76	45982.0	MO		0.9		MIT
2172.917	2173.598	46006.66	MO	0.7	1.3		MIT
2172.46	2173.14	46016.3	MO		0.9		MIT
2171.09	2171.77	46045.4	MO		0.5		MIT
2170.57	2171.25	46056.4	MO		1.3		MIT
2169.51	2170.19	46078.9	MO		1.1		MIT
2168.75	2169.43	46095.0	MO		1.1		MIT
2168.293	2168.973	46104.76	MO		1.1		MIT
2167.65	2168.33	46118.4	MO		1.1L		MIT
2167.122	2167.802	46129.67	MO		1.0		MIT
2165.80	2166.48	46157.8	MO		1.3		MIT
2165.17	2165.85	46171.3	MO		1.4		MIT
2163.644	2164.323	46203.82	MO		1.0		MIT
2161.05	2161.73	46259.3	MO		1.4L		MIT
2157.02	2157.70	46345.7	MO		1.0		MIT
2154.02	2154.70	46410.2	MO		0.8		MIT
2153.36	2154.04	46424.5	MO		1.0		MIT
2152.50	2153.18	46443.0	MO		1.1		MIT
2149.77	2150.45	46502.0	MO		1.0		MIT
2147.80	2148.48	46544.6	MO		1.0		A
2147.66	2148.34	46547.6	MO		0.7		MIT
2147.51	2148.19	46550.9	MO		0.8		A
2145.45	2146.13	46595.6	MO		1.1		MIT
2144.07	2144.75	46625.6	MO		1.3		MIT
2143.25	2143.93	46643.4	MO	0.3H	1.2		MIT
2142.44	2143.12	46661.0	MO		1.0		MIT
2142.22	2142.89	46665.8	MO		1.1		MIT
2141.37	2142.04	46684.4	MO		0.8		MIT
2139.44	2140.11	46726.5	MO	0.3	1.1		MIT
2136.06	2136.73	46800.4	MO		0.9		MIT
2134.99	2135.66	46823.9	MO		0.8		MIT
2133.37	2134.04	46859.4	MO	0.3H	1.2		MIT
2133.06	2133.73	46866.2	MO		1.2		MIT
2131.87	2132.54	46892.4	MO		1.1		MIT
2128.63	2129.30	46963.7	MO		1.2		MIT

AIR WAVELENGTH (ANGSTRMS)	VACUUM WAVELENGTH (ANGSTRMS)	WAVENUMBER (KAYSERS)	LMENT	LOG ARC	INTENSITY SPK	INTENSITY DIS	REF
2127.26	2127.93	46994.0	MO	0.5	1.2		MIT
2125.91	2126.58	47023.8	MO	0.3H	1.5		MIT
2124.82	2125.49	47047.9	MO	0.3H	1.2		MIT
2124.10	2124.77	47063.9	MO	0.8	1.3		MIT
2123.30	2123.97	47081.6	MO		1.4		MIT
2121.88	2122.55	47113.1	MO		1.1		MIT
2119.69	2120.36	47161.8	MO	0.3	1.3		MIT
2117.83	2118.50	47203.2	MO		1.2		MIT
2116.78	2117.45	47226.6	MO	0.6	1.3		MIT
2115.43	2116.10	47256.7	MO	0.3H	1.0		MIT
2115.11	2115.78	47263.9	MO	0.6	1.3		MIT
2114.30	2114.97	47282.0	MO	0.5	1.1		MIT
2113.64	2114.31	47296.8	MO		0.6		MIT
2113.36	2114.03	47303.0	MO		0.7		MIT
2111.18	2111.85	47351.9	MO	0.5	1.4		MIT
2110.57	2111.24	47365.6	MO		0.8		MIT
2109.94	2110.61	47379.7	MO	0.6	1.3		MIT
2108.73	2109.40	47406.9	MO	0.3H	1.1		MIT
2108.02	2108.69	47422.9	MO	0.5	1.5		MIT
2105.02	2105.69	47490.4	MO	0.7	1.4		MIT
2104.29	2104.96	47506.9	MO	0.6	1.5		MIT
2101.69	2102.36	47565.7	MO	0.6	1.3		MIT
2101.19	2101.86	47577.0	MO		1.0		MIT
2100.84	2101.51	47584.9	MO	0.8	1.4		MIT
2095.28	2095.95	47711.2	MO	0.6	1.4		MIT
2093.11	2093.78	47760.6	MO	0.8	1.4		MIT
2092.50	2093.17	47774.5	MO	0.8	1.3		MIT
2091.21	2091.87	47804.0	MO	0.5	1.2		MIT
2089.52	2090.18	47842.7	MO	0.5	1.3		MIT
2088.28	2088.94	47871.1	MO	0.5	1.0		MIT
2086.78	2087.44	47905.5	MO	0.3H	1.0		A
2074.29	2074.95	48193.9	MO	0.6	1.2		A
2071.08	2071.74	48268.6	MO	0.8	1.3		A
2063.96	2064.62	48435.1	MO	0.7			A
2049.36	2050.02	48780.1	MO	0.6J			A
2045.59	2046.25	48870.0	MO	0.8			A
2045.47	2046.13	48872.8	MO	0.3	1.1		A
2043.36	2044.02	48923.3	MO	0.6	0.3P		A
2041.45	2042.11	48969.1	MO	0.3	1.5		A
2038.44	2039.09	49041.4	MO	1.3			A
2038.01	2038.66	49051.7	MO	0.3	1.3		A
2037.68	2038.33	49059.6	MO	0.3	1.3		A
2036.21	2036.86	49095.1	MO	0.5			A
2035.57	2036.22	49110.5	MO	0.5			A
2032.05	2032.70	49195.6	MO	0.3H	1.1		A
2028.14	2028.79	49290.4	MO	0.3H	1.0		A
2026.75	2027.40	49324.2	MO		1.1		A
2020.30	2020.95	49481.6	MO	1.3	1.7R		A
2015.11	2015.76	49609.1	MO	1.1	1.6		A
2000.21	2000.86	49978.6	MO	0.3	1.1		A
2000.04	2000.69	49982.8	MO	1.0			A
9862.5	9865.2	10137.	N I			1.3	IG
9821.8	9824.5	10179.	N I			0.5	IG
9460.0	9462.6	10568.	N I			0.9	IG
9392.5	9395.1	10644.	N I			2.1	IG
9386.5	9389.1	10651.	N I			1.8	IG
9060.6	9063.1	11034.	N I			2.1	IG
9045.1	9047.6	11053.	N	2.2			IG
9028.9	9031.4	11072.	N I			1.2	IG
8729.07	8731.47	11452.8	N I			0.3	KS
8718.99	8721.38	11466.1	N I			0.5	KS
8711.78	8714.17	11475.6	N I			0.5	KS
8703.42	8705.81	11486.6	N I			0.5	KS
8686.38	8688.77	11509.1	N I			0.7	KS
8683.61	8686.00	11512.8	N I			0.9	KS
8680.35	8682.73	11517.1	N I			1.1	KS
8656.32	8658.70	11549.1	N I			0.5	KS
8629.61	8631.98	11584.8	N I			1.0	KS
8594.34	8596.70	11632.4	N I			0.5	KS
8568.04	8570.39	11668.1	N I			0.5	KS

AIR WAVELENGTH (ANGSTRMS)	VACUUM WAVELENGTH (ANGSTRMS)	WAVENUMBER (KAYSERS)	LMENT	LOG ARC	INTENSITY SPK	INTENSITY DIS	REF
8242.47	8244.74	12128.9	N I			1.2	KS
8223.28	8225.54	12157.3	N I			1.2	KS
8216.46	8218.72	12167.3	N I			1.5	KS
8210.94	8213.20	12175.5	N I			1.0	KS
8200.59	8202.84	12190.9	N I			0.3	KS
8188.16	8190.41	12209.4	N I			1.2	KS
8185.05	8187.30	12214.0	N I			1.2	KS
7468.79	7470.85	13385.4	N I			2.3	KS
7442.56	7444.61	13432.5	N I			2.0	KS
7423.88	7425.92	13466.3	N I			1.7	KS
7259.3	7261.3	13772.	N II			1.2	FM
7217.0	7219.0	13852.	N II			1.2	FM
7139.8	7141.8	14002.	N II			1.5	FM
7015.3	7017.2	14251.	N II			0.7	FM
6979.10	6981.02	14324.5	N I			0.7	DU
6976.8	6978.7	14329.	N II			1.2	FM
6967.6	6969.5	14348.	N II			0.7	FM
6951.50	6953.42	14381.4	N I			0.7	DU
6945.22	6947.14	14394.4	N I			1.6	DU
6942.9	6944.8	14399.	N II			1.5	FM
6926.90	6928.81	14432.5	N I			0.7	DU
6888.7	6890.6	14512.	N II			1.2	FM
6874.30	6876.20	14542.9	N			0.7	DU
6870.8	6872.7	14550.	N II			0.7	FM
6857.6	6859.5	14578.	N II			0.7	FM
6836.2	6838.1	14624.	N II			0.7	FL
6812.26	6814.14	14675.4	N II			1.2	FL
6758.60	6760.47	14791.9	N I			1.7	DU
6752.40	6754.26	14805.5	N I			1.7	DU
6741.29	6743.15	14829.9	N I			1.5	DU
6733.48	6735.34	14847.1	N I			2.0	DU
6723.12	6724.98	14869.9	N I			2.7	DU
6713.12	6714.97	14892.1	N			0.7	DU
6708.81	6710.66	14901.7	N			1.7	DU
6706.20	6708.05	14907.5	N I			1.7	DU
6656.61	6658.45	15018.5	N I			0.7	DU
6653.41	6655.25	15025.7	N I			1.8	DU
6646.52	6648.36	15041.3	N I			1.2	MT
6644.96	6646.79	15044.8	N I			2.7	DU
6637.01	6638.84	15062.9	N I			1.7	DU
6630.5	6632.3	15078.	N II			1.2	FL
6622.53	6624.36	15095.8	N I			1.5	DU
6610.58	6612.41	15123.1	N II			2.0	FL
6533.0	6534.8	15303.	N II			0.7	FL
6504.9	6506.7	15369.	N II			1.2	FL
6499.52	6501.32	15381.5	N I			1.4	DU
6491.28	6493.07	15401.0	N I			1.4	DU
6484.88	6486.67	15416.2	N I			2.7	DU
6483.75	6485.54	15418.9	N I			1.5	DU
6482.74	6484.53	15421.3	N I			2.7	MT
6482.07	6483.86	15422.9	N II			2.5	FL
6481.73	6483.52	15423.7	N I			1.2	MT
6471.03	6472.82	15449.2	N I			0.5	DU
6468.32	6470.11	15455.7	N I			1.5	DU
6457.93	6459.71	15480.6	N I			1.4	DU
6452.75	6454.53	15493.0	N I			0.7	DU
6441.70	6443.48	15519.6	N I			1.8	MT
6440.95	6442.73	15521.4	N			1.4	DU
6437.01	6438.79	15530.9	N I			1.5	DU
6422.93	6424.71	15564.9	N			1.0	DU
6420.47	6422.24	15570.9	N I			1.5	DU
6417.05	6418.82	15579.2	N			1.0	DU
6379.63	6381.39	15670.6	N II			1.8	FL
6357.0	6358.8	15726.	N II			1.5	FL
6347.1	6348.9	15751.	N II			0.7	FL
6340.67	6342.42	15766.8	N II			1.7	FL
6328.6	6330.3	15797.	N II			0.7	FL
6285.78	6287.52	15904.5	N			0.5	DU
6284.30	6286.04	15908.3	N II			1.5	FL
6279.42	6281.16	15920.6	N			0.3	MT

AIR WAVELENGTH (ANGSTRMS)	VACUUM WAVELENGTH (ANGSTRMS)	WAVENUMBER (KAYSERS)	LMENT	LOG ARC	INTENSITY SPK	INTENSITY DIS	REF
6275.43	6277.17	15930.8	N			0.5	MT
6272.83	6274.56	15937.4	N I			0.9	DU
6242.52	6244.25	16014.7	N II			1.8	FL
6173.40	6175.11	16194.1	N II			1.5	FL
6170.26	6171.97	16202.3	N I			0.7	FL
6167.82	6169.53	16208.7	N II			1.7	FL
6075.83	6077.51	16454.1	N			1.5	DU
6017.70	6019.37	16613.0	N			1.0	DU
6015.40	6017.07	16619.4	N			0.7	DU
6008.48	6010.14	16638.5	N I			2.9	DU
5999.47	6001.13	16663.5	N I			2.0	DU
5952.39	5954.04	16795.3	N II			1.5	FL
5941.67	5943.32	16825.6	N II			2.3	FL
5940.25	5941.90	16829.6	N II			1.2	FL
5931.79	5933.43	16853.6	N II			2.2	FL
5927.82	5929.46	16864.9	N II			1.7	FL
5856.23	5857.85	17071.1	N I			0.7	DU
5854.16	5855.78	17077.1	N I			1.2	DU
5841.01	5842.63	17115.6	N I			1.2	DU
5834.71	5836.33	17134.1	N I			0.7	DU
5829.53	5831.15	17149.3	N I			1.8	DU
5816.48	5818.09	17187.8	N I			1.2	DU
5793.51	5795.12	17255.9	N			0.7	DU
5767.43	5769.03	17333.9	N II			1.5	FL
5752.64	5754.24	17378.5	N I			1.5	DU
5747.36	5748.95	17394.5	N I			1.2	DU
5747.29	5748.88	17394.7	N II			1.7	FL
5740.65	5742.24	17414.8	N			1.2	DU
5735.63	5737.22	17430.0	N			0.7	DU
5730.67	5732.26	17445.1	N II			1.2	FL
5710.76	5712.34	17505.9	N II			2.0	FL
5686.21	5687.79	17581.5	N II			2.0	FL
5679.56	5681.14	17602.1	N II			2.7	FL
5676.02	5677.60	17613.1	N II			2.0	FL
5667.04	5668.61	17641.0	N			0.7	DU
5666.64	5668.21	17642.2	N II			2.5	FL
5625.43	5626.99	17771.5	N I			1.0	DU
5623.20	5624.76	17778.5	N I			1.6	DU
5618.18	5619.74	17794.4	N I			0.5	MT
5616.54	5618.10	17799.6	N I			1.8	DU
5611.36	5612.92	17816.1	N I			0.7	MT
5567.63	5569.18	17956.0	N			0.7	MT
5564.37	5565.92	17966.5	N I			2.3	DU
5563.84	5565.39	17968.2	N			1.5H	MT
5560.37	5561.91	17979.4	N I			2.3	DU
5557.44	5558.98	17988.9	N			1.0	DU
5551.95	5553.49	18006.7	N II			1.5	FL
5545.11	5546.65	18028.9	N I			1.3	MT
5543.49	5545.03	18034.2	N II			1.5	FL
5540.36	5541.90	18044.4	N			0.7	DU
5540.16	5541.70	18045.0	N II			0.7	FL
5535.39	5536.93	18060.6	N II			1.8	FL
5530.27	5531.81	18077.3	N II			1.7	FL
5526.26	5527.80	18090.4	N II			1.2	FL
5495.70	5497.23	18191.0	N II			1.8	FL
5480.10	5481.62	18242.8	N II			1.5	FL
5478.13	5479.65	18249.3	N II			1.2	FL
5462.62	5464.14	18301.1	N II			1.5	FL
5454.26	5455.78	18329.2	N II			1.2	FL
5378.45	5379.95	18587.6	N I			1.5	MT
5372.66	5374.15	18607.6	N I			1.3H	MT
5371.10	5372.59	18613.0	N I			0.7	DU
5367.27	5368.76	18626.3	N I			0.5	DU
5356.77	5358.26	18662.8	N I			1.7	DU
5351.21	5352.70	18682.2	N II			1.5	MIT
5340.15	5341.64	18720.9	N II			0.7	FL
5338.77	5340.25	18725.7	N II			1.2	FL
5334.42	5335.90	18741.0	N I			0.7	DU
5328.70	5330.18	18761.1	N I			1.8	DU
5320.75	5322.23	18789.1	N II			1.7	FL

AIR WAVELENGTH (ANGSTRMS)	VACUUM WAVELENGTH (ANGSTRMS)	WAVENUMBER (KAYSERS)	LMENT		LOG ARC	INTENSITY SPK	INTENSITY DIS	REF
5314.47	5315.95	18811.3	N	II			0.7	FL
5310.52	5312.00	18825.3	N	I			0.5	DU
5309.48	5310.96	18829.0	N	I			0.5	MT
5281.18	5282.65	18929.9	N	I			1.3	DU
5201.71	5203.16	19219.1	N	I			1.0	DU
5190.42	5191.87	19260.9	N	II			1.2	FL
5189.51	5190.96	19264.3	N				0.7	DU
5184.97	5186.41	19281.1	N	II			1.2	FL
5183.21	5184.65	19287.7	N	II			1.2	FL
5181.80	5183.24	19292.9	N	I			1.2	RY
5181.0	5182.4	19296.	N	I			0.7	RY
5180.34	5181.78	19298.4	N	II			0.7	FM
5179.50	5180.94	19301.5	N	II			1.8	FL
5175.89	5177.33	19315.0	N	II			1.5	FL
5174.46	5175.90	19320.3	N	II			0.7	FL
5173.37	5174.81	19324.4	N	II			1.2	FL
5172.32	5173.76	19328.3	N	II			0.7	FL
5171.46	5172.90	19331.5	N	II			0.7	FL
5170.08	5171.52	19336.7	N	II			0.7	FL
5169.45	5170.89	19339.0	N	II			0.7	DU
5168.24	5169.68	19343.6	N	II			0.7	FL
5162.78	5164.22	19364.0	N				0.7	DU
5104.45	5105.87	19585.3	N	II			1.2	FL
5073.60	5075.01	19704.4	N	II			1.5	FL
5045.098	5046.505	19815.70	N	II			2.3	FL
5025.665	5027.067	19892.32	N	II			2.0	FL
5023.11	5024.51	19902.4	N	II			1.2	FL
5016.387	5017.786	19929.11	N	II			1.8	FL
5012.026	5013.424	19946.45	N	II			1.2	FL
5011.24	5012.64	19949.6	N	II			0.7	FM
5010.620	5012.018	19952.04	N	II			2.0	FL
5007.316	5008.713	19965.21	N	II			2.2	FL
5005.140	5006.536	19973.89	N	II			2.7	FL
5002.692	5004.087	19983.66	N	II			1.2	FL
5001.469	5002.864	19988.55	N	II			2.3	FL
5001.128	5002.523	19989.91	N	II			2.2	FL
4994.358	4995.751	20017.01	N	II			1.5	FL
4991.22	4992.61	20029.6	N	II			0.7	FL
4987.377	4988.768	20045.03	N	II			1.2	FL
4935.03	4936.41	20257.6	N	I			2.4	MT
4914.90	4916.27	20340.6	N	I			1.3	DU
4886.30	4887.66	20459.7	N				0.7	DU
4881.79	4883.15	20478.6	N				0.6	DU
4860.35	4861.71	20568.9	N	II			0.7	FL
4847.38	4848.73	20623.9	N				0.7	DU
4837.93	4839.28	20664.2	N				0.3	DU
4831.16	4832.51	20693.2	N				0.3H	DU
4810.286	4811.630	20782.98	N	II			0.7	FL
4803.272	4804.615	20813.32	N	II			1.5	FL
4793.656	4794.996	20855.07	N	II			0.7	FL
4788.126	4789.465	20879.16	N	II			1.4	FL
4781.168	4782.505	20909.55	N	II			0.7	FL
4779.710	4781.046	20915.92	N	II			1.2	FL
4774.222	4775.557	20939.97	N	II			0.7	FL
4753.13	4754.46	21032.9	N				0.7	DU
4750.26	4751.59	21045.6	N				0.7	DU
4744.04	4745.37	21073.2	N				1.0	DU
4742.90	4744.23	21078.2	N				0.6	DU
4718.43	4719.75	21187.6	N	II			0.7	FL
4709.45	4710.77	21228.0	N	II			0.3	FL
4694.55	4695.86	21295.3	N	II			1.0H	FL
4685.74	4687.05	21335.4	N				1.0	DU
4677.94	4679.25	21370.9	N	II			1.0W	FL
4674.98	4676.29	21384.5	N	II			0.7	FL
4669.77	4671.08	21408.3	N				1.0	DU
4667.28	4668.59	21419.8	N	II			0.7	FL
4660.05	4661.35	21453.0	N				0.7	DU
4654.57	4655.87	21478.2	N	II			0.7	FL
4643.106	4644.406	21531.28	N	II			2.0	FL
4630.551	4631.848	21589.66	N	II			2.5	FL

AIR WAVELENGTH (ANGSTRMS)	VACUUM WAVELENGTH (ANGSTRMS)	WAVENUMBER (KAYSERS)	LMENT		LOG ARC	INTENSITY SPK	INTENSITY DIS	REF
4621.405	4622.699	21632.38	N	II			1.7	FL
4613.887	4615.179	21667.63	N	II			1.5	FL
4609.60	4610.89	21687.8	N	II			1.5	FL
4607.167	4608.458	21699.23	N	II			1.7	FL
4601.490	4602.779	21726.00	N	II			2.0	FL
4564.78	4566.06	21900.7	N	II			0.3	FL
4552.50	4553.78	21959.8	N	II			1.2	FL
4530.37	4531.64	22067.1	N				1.4	FL
4507.58	4508.84	22178.6	N	II			1.0	FL
4502.27	4503.53	22204.8	N				0.7	DU
4494.67	4495.93	22242.3	N				1.4	DU
4492.40	4493.66	22253.6	N				1.6	DU
4477.74	4479.00	22326.4	N	II			0.7	FL
4459.96	4461.21	22415.4	N	II			0.3	FL
4447.035	4448.283	22480.58	N	II			2.5	FL
4441.99	4443.24	22506.1	N	II			1.0H	FL
4433.48	4434.72	22549.3	N				0.7H	FL
4432.71	4433.95	22553.2	N	II			1.5H	FL
4427.97	4429.21	22577.4	N				0.7	FL
4427.21	4428.45	22581.2	N	II			0.7	FL
4358.27	4359.49	22938.4	N				2.4	DU
4336.48	4337.70	23053.7	N				1.3	DU
4317.70	4318.91	23154.0	N				1.3	DU
4313.11	4314.32	23178.6	N				1.2	MT
4305.46	4306.67	23219.8	N				1.5	MT
4284.92	4286.13	23331.1	N				0.7	DU
4281.39	4282.59	23350.3	N				0.7	DU
4254.75	4255.95	23496.5	N	I	1.2			RY
4253.28	4254.48	23504.6	N	I			1.2	DU
4241.80	4242.99	23568.3	N	II			2.0H	FL
4236.98	4238.17	23595.1	N				1.5H	FL
4230.35	4231.54	23632.1	N	I			1.2	DU
4229.59	4230.78	23636.3	N				0.7	DU
4227.83	4229.02	23646.1	N	II			1.0H	FL
4224.74	4225.93	23663.4	N				1.2	DU
4223.04	4224.23	23673.0	N	I			1.4	DU
4220.79	4221.98	23685.6	N				0.7	DU
4215.92	4217.11	23712.9	N	I			0.7	DU
4214.73	4215.92	23719.6	N	I			1.4	DU
4205.65	4206.83	23770.8	N				0.7H	DU
4193.49	4194.67	23839.8	N				1.0	DU
4179.68	4180.86	23918.5	N	II			0.3H	FL
4176.17	4177.35	23938.6	N				1.0H	FL
4171.63	4172.81	23964.7	N				0.7H	FL
4151.46	4152.63	24081.1	N	I			3.0	DU
4145.759	4146.928	24114.24	N	II			1.0	FL
4143.65	4144.82	24126.5	N	I			1.5	RY
4137.63	4138.80	24161.6	N	I			1.7	DU
4133.654	4134.820	24184.85	N	II			0.7	FL
4124.10	4125.26	24240.9	N	II			0.3	FL
4114.00	4115.16	24300.4	N	I			1.5	DU
4109.98	4111.14	24324.1	N	I			3.0	DU
4102.18	4103.34	24370.4	N				0.7	DU
4099.94	4101.10	24383.7	N	I			2.2	DU
4082.28	4083.43	24489.2	N	II			0.7H	FL
4073.04	4074.19	24544.8	N	II			0.7H	FL
4043.54	4044.68	24723.8	N				1.0H	FL
4041.325	4042.467	24737.37	N	II			1.3H	FL
4035.090	4036.230	24775.59	N	II			1.2H	FL
4033.64	4034.78	24784.5	N				0.3	DU
4026.087	4027.225	24831.00	N				1.0H	FL
4010.99	4012.12	24924.4	N				0.7	DU
3999.98	4001.11	24993.1	N				1.2	U
3994.995	3996.125	25024.24	N	II			2.5	FL
3969.95	3971.07	25182.1	N				0.3	DU
3957.20	3958.32	25263.2	N				1.0	DU
3955.851	3956.970	25271.86	N	II			1.5	FL
3952.21	3953.33	25295.1	N				1.2	DU
3919.003	3920.113	25509.47	N	II			1.5	FL
3869.10	3870.20	25838.5	N				1.2	DU

AIR WAVELENGTH (ANGSTRMS)	VACUUM WAVELENGTH (ANGSTRMS)	WAVENUMBER (KAYSERS)	LMENT	LOG ARC	INTENSITY SPK	DIS	REF
3856.07	3857.16	25925.8	N II			1.0	FL
3855.08	3856.17	25932.4	N II			0.7	FL
3847.38	3848.47	25984.3	N II			1.0	FL
3842.20	3843.29	26019.4	N II			1.0	FL
3838.39	3839.48	26045.2	N II			1.4	FL
3834.84	3835.93	26069.3	N			0.7	DU
3834.24	3835.33	26073.4	N I			1.2	DU
3830.39	3831.48	26099.6	N I			2.2	DU
3829.80	3830.89	26103.6	N II			1.0	FL
3822.07	3823.15	26156.4	N I			1.5	DU
3818.27	3819.35	26182.4	N I			0.7	DU
3687.88	3688.93	27108.1	N			0.7	DU
3681.10	3682.15	27158.1	N			1.0	DU
3650.19	3651.23	27388.0	N			1.8	DU
3615.88	3616.91	27647.9	N II			0.3	FL
3609.09	3610.12	27699.9	N II			0.7	FL
3593.60	3594.63	27819.3	N II			1.0	FL
3545.62	3546.63	28195.8	N			0.7	DU
3532.65	3533.66	28299.3	N			1.2	DU
3437.162	3438.147	29085.43	N II			1.5	FL
3408.136	3409.114	29333.14	N II			1.0	FL
3331.32	3332.28	30009.5	N II			1.0	FL
3330.30	3331.26	30018.7	N II			0.7	FL
3328.79	3329.75	30032.3	N II			1.2	FL
3324.58	3325.54	30070.3	N II			0.7	FL
3318.14	3319.09	30128.7	N II			0.7	FL
3023.80	3024.68	33061.3	N II			0.7	FL
3006.86	3007.74	33247.6	N II			1.7	FL
2897.49	2898.34	34502.5	N			1.2H	FL
2892.86	2893.71	34557.7	N			1.4H	FL
2885.25	2886.10	34648.9	N			1.7H	FL
2884.25	2885.10	34660.9	N			0.9H	FL
2879.73	2880.57	34715.3	N			1.4H	FL
2877.66	2878.50	34740.3	N			0.9H	FL
2823.67	2824.50	35404.5	N II			1.4	FL
2799.20	2800.03	35714.0	N II			1.4	FL
2735.0	2735.8	36552.	N II			0.5H	FL
2709.82	2710.62	36891.9	N II			1.7	FL
2590.91	2591.68	38584.9	N II			1.4	FL
2524.55	2525.31	39599.1	N II			0.5	FL
2522.51	2523.27	39631.1	N II			0.5	FL
2522.27	2523.03	39634.9	N II				FL
2520.85	2521.61	39657.2	N II			1.2	FL
2520.27	2521.03	39666.4	N II			0.9	FL
2496.88	2497.63	40037.9	N II			1.4	FL
2494.02	2494.77	40083.8	N II			0.5	FL
2490.37	2491.12	40142.6	N II			0.9	FL
2488.82	2489.57	40167.6	N II			0.5	FL
2461.30	2462.04	40616.6	N II			1.2	FL
2390.90	2391.63	41812.5	N II			0.9	FL
2388.24	2388.97	41859.1	N II			0.5	FL
2321.62	2322.33	43060.1	N II			0.5	FM
2319.90	2320.61	43092.1	N II			0.5	FM
2317.01	2317.72	43145.8	N II			1.2	FL
2316.66	2317.37	43152.3	N II			0.5	FL
2316.46	2317.17	43156.1	N II			0.9	FL
2311.60	2312.31	43246.8	N			0.5	FL
2310.61	2311.32	43265.3	N			0.5	FL
2308.90	2309.61	43297.4	N			1.5	FL
2308.55	2309.26	43303.9	N			1.5	FL
2293.40	2294.11	43590.0	N II			0.9	FL
2292.72	2293.43	43602.9	N II			0.5	FL
2291.67	2292.38	43622.9	N II			0.9	FL
2290.31	2291.02	43648.7	N II			0.5	FL
2288.47	2289.18	43683.8	N II			1.2	FL
2286.73	2287.44	43717.1	N II			1.4	FL
2285.3	2286.0	43744.	N II			1.2	FL
2283.70	2284.40	43775.1	N II			0.9	FL
2238.91	2239.61	44650.7	N II			0.9	FL
2235.18	2235.87	44725.2	N II			0.5	FL

AIR WAVELENGTH (ANGSTRMS)	VACUUM WAVELENGTH (ANGSTRMS)	WAVENUMBER (KAYSERS)	LMENT	LOG ARC	INTENSITY SPK	DIS	REF
2206.10	2206.79	45314.7	N II			1.2	FL
2203.72	2204.41	45363.7	N II			0.5	FL
2197.58	2198.27	45490.4	N II			0.9	FL
2097.58	2098.25	47658.9	N II			0.9	FL
2096.79	2097.46	47676.8	N II			1.2	FL
2096.16	2096.83	47691.1	N II			0.9	FL
2095.47	2096.14	47706.8	N II			1.7	FL
2094.12	2094.79	47737.6	N II			0.9	FL
2091.20	2091.86	47804.2	N II			1.2	FL
2063.99	2064.65	48434.4	N II			1.2	FM
2063.50	2064.16	48445.9	N II			1.2	FM
2035.02	2035.67	49123.8	N			0.7	FM
8194.811	8197.064	12199.49	NA I	3.0G			IHZ
8183.270	8185.520	12216.69	NA I	2.7G			IHZ
6160.760	6162.465	16227.27	NA I	2.7	2.0		HZ
6154.229	6155.932	16244.49	NA I	2.7	2.0		HZ
5895.923	5897.557	16956.17	NA I	3.7G	2.7G		HZ
5889.953	5891.585	16973.36	NA I	4.0G	3.0G		HZ
5688.61	5690.19	17574.1	NA	1.0			ME
5688.224	5689.802	17575.30	NA I	2.5			HZ
5682.657	5684.234	17592.52	NA I	1.9			HZ
5675.70	5677.27	17614.1	NA I	2.2J			FL
5670.18	5671.75	17631.2	NA I	2.0V			FL
5532.0	5533.5	18072.	NA I	1.2			FL
5153.645	5155.081	19398.34	NA I	2.8			ME
5149.090	5150.524	19415.50	NA I	2.6			ME
4982.845	4984.235	20063.26	NA I	2.3J	2.0		HZ
4978.585	4979.974	20080.43	NA I	1.2	1.0		HZ
4975.9	4977.3	20091.	NA I	0.5			FL
4972.8	4974.2	20104.	NA I	0.5			FL
4751.891	4753.220	21038.37	NA I	1.3			DA
4748.016	4749.344	21055.54	NA I	1.2			DA
4668.597	4669.904	21413.72	NA I	2.3	2.0		DA
4664.858	4666.164	21430.88	NA I	1.9			DA
4545.218	4546.492	21994.98	NA I	1.2			DA
4541.671	4542.944	22012.16	NA I	1.0			DA
4497.724	4498.986	22227.23	NA I	1.8			DA
4494.266	4495.527	22244.33	NA I	1.8			DA
4423.31	4424.55	22601.2	NA I	0.5			DA
4419.94	4421.18	22618.4	NA I	0.5			DA
4393.45	4394.68	22754.8	NA I	1.3			DA
4390.14	4391.37	22771.9	NA I	1.2			DA
4344.5	4345.7	23011.	NA I	0.5			FO
4341.3	4342.5	23028.	NA I	0.5			FO
4324.3	4325.5	23119.	NA I	1.0			FO
4321.1	4322.3	23136.	NA I	0.5			FO
4290.6	4291.8	23300.	NA I	0.5			FO
4287.5	4288.7	23317.	NA I	0.5			FO
4276.3	4277.5	23378.	NA I	0.5			FO
4273.2	4274.4	23395.	NA I	0.5			FO
4252.4	4253.6	23509.	NA I	0.5			FO
4249.4	4250.6	23526.	NA I	0.5			FO
4241.6	4242.8	23569.	NA I	0.5			FO
4238.6	4239.8	23586.	NA I	0.5			FO
4222.8	4224.0	23674.	NA I	0.5			FO
4220.3	4221.5	23688.	NA I	1.0			FL
4215.4	4216.6	23716.	NA I	0.5			FO
4212.4	4213.6	23733.	NA I	0.5			FO
4198.3	4199.5	23812.	NA I	1.0			FL
4195.5	4196.7	23828.	NA I	0.5			FO
4192.6	4193.8	23845.	NA I	0.5			FO
4180.2	4181.4	23916.	NA I	0.5			FO
4177.4	4178.6	23932.	NA I	0.5			FO
4167.8	4169.0	23987.	NA I	0.5			FO
4165.0	4166.2	24003.	NA I	0.5			FO
4156.0	4157.2	24055.	NA I	0.5			FO
4148.0	4149.2	24101.	NA I	0.5			FO
4141.0	4142.2	24142.	NA I	0.5			FO
4138.6	4139.8	24156.	NA	0.5			FO
4123.069	4124.232	24246.94	NA II	1.0		1.2	FR

AIR WAVELENGTH (ANGSTRMS)	VACUUM WAVELENGTH (ANGSTRMS)	WAVENUMBER (KAYSERS)	LMENT	LOG ARC	INTENSITY SPK	DIS	REF
4114.95	4116.11	24294.8	NA II		1.2		SO
3870.7	3871.8	25828.	NA			0.5	NM
3816.8	3817.9	26192.	NA			0.5	NM
3711.074	3712.130	26938.71	NA II	0.9		1.8	FR
3631.266	3632.301	27530.76	NA II	1.1		2.0	FR
3533.010	3534.020	28296.39	NA II	1.7		2.3	MIT
3462.494	3463.486	28872.65	NA II	0.3		1.2	FR
3427.121	3428.104	29170.65	NA	1.0			MIT
3400.110	3401.086	29402.38	NA II			0.7	FR
3327.685	3328.642	30042.28	NA II	0.8		1.3	FR
3318.032	3318.987	30129.68	NA II	0.8		1.3	FR
3302.988	3303.939	30266.90	NA I	2.5G	2.2G		MIT
3302.323	3303.274	30273.00	NA I	2.8G	2.5G		MIT
3301.346	3302.297	30281.96	NA II			0.7	FR
3285.748	3286.695	30425.71	NA II	1.6		2.0	MIT
3274.220	3275.164	30532.82	NA II	1.2		1.6	FR
3260.218	3261.158	30663.95	NA II			1.2	FR
3257.965	3258.905	30685.16	NA II	1.5		1.8	FR
3250.949	3251.887	30751.38	NA II			1.2	FR
3234.926	3235.860	30903.69	NA II	0.8		1.3	FR
3225.976	3226.907	30989.42	NA II	0.3		1.3	FR
3216.284	3217.213	31082.80	NA II			0.7	FR
3212.186	3213.114	31122.46	NA II	1.5		1.8	FR
3189.783	3190.705	31341.03	NA II	1.5		1.8	FR
3179.055	3179.975	31446.79	NA II		0.8	1.6	FR
3175.088	3176.007	31486.08	NA II			1.2	FR
3167.487	3168.404	31561.64	NA II			0.7	FR
3163.731	3164.647	31599.10	NA II	1.5		1.8	FR
3149.267	3150.179	31744.23	NA II	1.1		1.6	FR
3145.697	3146.608	31780.25	NA II			1.2	FR
3137.852	3138.761	31859.70	NA II			1.2	FR
3135.483	3136.392	31883.77	NA II	1.1		1.6	FR
3129.368	3130.275	31946.07	NA II	1.5		1.8	FR
3124.414	3125.320	31996.73	NA II	0.3		1.2	FR
3104.396	3105.297	32203.04	NA II	0.3		1.3	FR
3092.729	3093.627	32324.52	NA II	1.7		2.3	FR
3087.047	3087.943	32384.01	NA II			0.7	FR
3080.250	3081.145	32455.47	NA I	0.3		1.2	FR
3078.315	3079.209	32475.87	NA II	1.1		1.8	FR
3074.334	3075.227	32517.92	NA II	0.9		1.8	FR
3066.536	3067.427	32600.61	NA II	0.3		1.3	FR
3064.372	3065.263	32623.63	NA II	0.3		1.3	FR
3056.157	3057.046	32711.32	NA II	1.5		1.8	FR
3053.664	3054.552	32738.02	NA II	0.9		1.8	FR
3045.593	3046.479	32824.78	NA II			1.6	FR
3037.071	3037.955	32916.88	NA II	0.9		1.6	FR
3029.068	3029.950	33003.85	NA II			1.8	FR
3015.400	3016.278	33153.44	NA II			1.8	FR
3009.138	3010.015	33222.43	NA II	0.5		1.3	FR
3007.442	3008.318	33241.16	NA II			1.3	FR
2984.430	2985.301	33497.46	NA II	1.3		1.9	MIT
2980.614	2981.484	33540.35	NA II		1.1		MIT
2979.662	2980.532	33551.06	NA II	1.5		1.6	FR
2979.050	2979.919	33557.95	NA			0.7	FR
2977.132	2978.001	33579.57	NA II	0.5		1.1	FR
2974.991	2975.859	33603.74	NA II	1.2		1.8	FR
2974.236	2975.104	33612.27	NA II			0.7	FR
2970.725	2971.592	33651.99	NA II	0.0		0.3	FR
2965.750	2966.616	33708.44	NA II			0.7	FR
2960.112	2960.977	33772.64	NA II	0.0		0.3	FR
2952.395	2953.258	33860.91	NA II			1.1	FR
2951.231	2952.094	33874.27	NA II	1.6		2.0	FR
2947.441	2948.303	33917.82	NA II	0.8		1.6	FR
2945.695	2946.556	33937.92	NA II	0.3		1.3	FR
2937.725	2938.584	34029.99	NA II	0.8		1.6	FR
2934.065	2934.923	34072.44	NA II	0.0		0.7	FR
2930.883	2931.740	34109.43	NA II	0.0		0.3	FR
2923.661	2924.517	34193.68	NA II	0.3		1.1	MIT
2920.940	2921.795	34225.54	NA II	0.8		1.3	FR
2919.845	2920.700	34238.37	NA II			0.7	FR

AIR WAVELENGTH (ANGSTRMS)	VACUUM WAVELENGTH (ANGSTRMS)	WAVENUMBER (KAYSERS)	LMENT	LOG ARC	INTENSITY SPK	DIS	REF
2919.048	2919.903	34247.72	NA II	0.8		1.6	FR
2917.516	2918.370	34265.70	NA II	0.9		1.6	FR
2904.914	2905.765	34414.35	NA II	1.3		1.9	FR
2901.381	2902.231	34456.25	NA II	0.5		1.3	MIT
2893.946	2894.794	34544.77	NA II	0.9		1.8	FR
2886.250	2887.096	34636.88	NA II	0.8		1.3	FP
2881.140	2881.985	34698.31	NA II	0.9		1.8	FR
2871.270	2872.113	34817.58	NA II	0.8		1.6	FR
2870.89	2871.73	34822.2	NA	0.8	0.8		FO
2861.011	2861.851	34942.42	NA II	0.0		0.3	FR
2859.481	2860.321	34961.11	NA II	0.3		1.6	FR
2853.031	2853.869	35040.15	NA I	1.9G	1.2		FL
2852.828	2853.666	35042.64	NA I	2.0G	1.3		FL
2841.721	2842.556	35179.60	NA II	1.3		1.9	FR
2839.555	2840.390	35206.43	NA II	0.3		1.3	FR
2829.854	2830.687	35327.12	NA II	0.3		0.7	FR
2818.271	2819.101	35472.31	NA II	0.3		0.7	FR
2809.514	2810.342	35582.86	NA II	0.9		1.6	FR
2808.685	2809.512	35593.37	NA II			0.3	FR
2680.443	2681.239	37296.19	NA I	1.6G			FL
2680.335	2681.131	37297.69	NA I	1.8G	1.0		FL
2678.086	2678.882	37329.01	NA II	0.8		1.6	FR
2671.829	2672.623	37416.42	NA II	1.1		1.8	FR
2660.996	2661.788	37568.74	NA II			1.9	FR
2611.815	2612.595	38276.12	NA II			0.9	FP
2594.965	2595.741	38524.65	NA II			0.3	FR
2593.927	2594.703	38540.06	NA I	1.2G			FL
2593.828	2594.604	38541.53	NA I	1.3R			FL
2586.312	2587.086	38653.53	NA II			0.7	FR
2563.329	2564.097	39000.08	NA			0.3	FR
2553.586	2554.352	39148.87	NA			0.3	FR
2543.875	2544.639	39298.31	NA I	1.1G			FL
2543.817	2544.581	39299.20	NA I	0.8G			FL
2531.548	2532.309	39489.65	NA II	0.8		1.8	FR
2515.460	2516.217	39742.20	NA II	0.5		0.7	FR
2512.210	2512.966	39793.61	NA I	0.6G			FL
2512.128	2512.884	39794.91	NA I	0.3G			FL
2510.331	2511.087	39823.39	NA II			0.3	FR
2506.295	2507.050	39887.52	NA II			0.7	FR
2497.059	2497.812	40035.04	NA			0.7	FR
2493.152	2493.904	40097.77	NA II			1.6	FR
2490.733	2491.484	40136.71	NA I	0.5G			FL
2475.533	2476.281	40383.14	NA I	0.3G			FL
2474.771	2475.519	40395.57	NA			0.3	FR
2464.397	2465.142	40565.61	NA I	0.3G			FL
2459.329	2460.073	40649.20	NA			0.7	FR
2455.915	2456.658	40705.70	NA I	0.3G			FL
2449.393	2450.135	40814.08	NA I	0.3G			FL
2444.195	2444.936	40900.87	NA I	0.3R			FL
2440.046	2440.786	40970.41	NA I	0.3R			FL
2394.051	2394.780	41757.48	NA			0.7	FR
2387.026	2387.754	41880.37	NA			0.7	FR
2317.98	2318.69	43127.8	NA	0.5			MIT
8838.96	8841.39	11310.4	ND	0.7			KS
8643.48	8645.85	11566.2	ND	0.3			MIT
8594.85	8597.21	11631.7	ND	0.3			MIT
8456.80	8459.12	11821.6	ND	0.3			KS
8400.82	8403.13	11900.3	ND	0.3			MIT
8394.74	8397.05	11908.9	ND	0.3			MIT
8386.68	8388.98	11920.4	ND	0.3			KS
8375.21	8377.51	11936.7	ND	0.3			MIT
8362.39	8364.69	11955.0	ND	0.3			MIT
8346.36	8348.65	11978.0	ND	0.9			MIT
8336.63	8338.92	11992.0	ND	0.3			KS
8332.00	8334.29	11998.6	ND	0.5			MIT
8324.49	8326.78	12009.4	ND	0.7			MIT
8307.75	8310.03	12033.7	ND	0.7			MIT
8302.73	8305.01	12040.9	ND	0.5			MIT
8283.52	8285.80	12068.8	ND	0.3			KS
8272.77	8275.04	12084.5	ND	0.3			MIT

AIR WAVELENGTH (ANGSTRMS)	VACUUM WAVELENGTH (ANGSTRMS)	WAVENUMBER (KAYSERS)	LMENT	LOG INTENSITY ARC	SPK	DIS	REF
8266.71	8268.98	12093.4	ND	0.5			MIT
8262.79	8265.06	12099.1	ND	0.3			MIT
8249.66	8251.93	12118.4	ND	0.3			MIT
8248.74	8251.01	12119.7	ND	0.3			MIT
8231.51	8233.77	12145.1	ND	0.8			MIT
8227.26	8229.52	12151.4	ND	0.3			MIT
8216.97	8219.23	12166.6	ND	0.3			MIT
8213.14	8215.40	12172.3	ND	0.3			MIT
8205.38	8207.64	12183.8	ND	0.3			MIT
8185.65	8187.90	12213.1	ND	0.3H			MIT
8182.39	8184.64	12218.0	ND	0.9			MIT
8179.78	8182.03	12221.9	ND	0.8			MIT
8172.54	8174.79	12232.7	ND	0.6			MIT
8169.81	8172.06	12236.8	ND	0.3			MIT
8164.92	8167.16	12244.2	ND	0.5			MIT
8143.29	8145.53	12276.7	ND	0.8			KS
8141.74	8143.98	12279.0	ND	0.8			MIT
8122.05	8124.28	12308.8	ND	0.8			MIT
8120.89	8123.12	12310.5	ND	0.5			MIT
8118.70	8120.93	12313.9	ND	0.3			MIT
8108.70	8110.93	12329.0	ND	0.3			MIT
8099.11	8101.34	12343.6	ND	0.6			MIT
8071.50	8073.72	12385.9	ND	0.5			MIT
8063.94	8066.16	12397.5	ND	0.5			MIT
8051.29	8053.50	12417.0	ND	0.5			MIT
8043.22	8045.43	12429.4	ND	0.8			MIT
8020.06	8022.27	12465.3	ND	0.3			MIT
8007.68	8009.88	12484.6	ND	0.8			MIT
8005.41	8007.61	12488.1	ND	1.0			MIT
8000.74	8002.94	12495.4	ND	0.8			MIT
7982.090	7984.285	12524.60	ND	0.6			MIT
7970.64	7972.83	12542.6	ND	0.3			KS
7965.70	7967.89	12550.4	ND	0.8			MIT
7958.92	7961.11	12561.1	ND	0.8			MIT
7957.45	7959.64	12563.4	ND	0.3			MIT
7955.38	7957.57	12566.7	ND	0.3			KS
7949.66	7951.85	12575.7	ND	0.5			MIT
7927.41	7929.59	12611.0	ND	0.5			MIT
7925.01	7927.19	12614.8	ND	0.5			MIT
7923.05	7925.23	12617.9	ND	0.3			MIT
7916.98	7919.16	12627.6	ND	0.6			MIT
7906.03	7908.20	12645.1	ND	0.3			KS
7900.39	7902.56	12654.1	ND	0.3			MIT
7896.51	7898.68	12660.3	ND	0.3			MIT
7886.59	7888.76	12676.3	ND	0.3			MIT
7872.01	7874.18	12699.7	ND	0.3			MIT
7862.84	7865.00	12714.6	ND	0.6			KS
7853.18	7855.34	12730.2	ND	0.3			MIT
7845.51	7847.67	12742.6	ND	0.3			KS
7844.25	7846.41	12744.7	ND	0.3			MIT
7837.22	7839.38	12756.1	ND	0.5			MIT
7825.20	7827.35	12775.7	ND	0.3			MIT
7818.82	7820.97	12786.1	ND	0.3			MIT
7814.35	7816.50	12793.4	ND	0.3H			MIT
7808.45	7810.60	12803.1	ND	0.5			MIT
7797.30	7799.45	12821.4	ND	0.3			MIT
7796.42	7798.57	12822.9	ND	0.3			MIT
7792.24	7794.38	12829.7	ND	0.3			MIT
7787.31	7789.45	12837.9	ND	0.3			MIT
7773.03	7775.17	12861.5	ND	0.5			MIT
7767.91	7770.05	12869.9	ND	0.3			MIT
7750.95	7753.08	12898.1	ND	0.5			MIT
7749.10	7751.23	12901.2	ND	0.5			MIT
7743.88	7746.01	12909.9	ND	0.3			MIT
7740.84	7742.97	12914.9	ND	0.3			MIT
7724.91	7727.04	12941.6	ND	0.3			MIT
7718.15	7720.27	12952.9	ND	0.3			MIT
7705.70	7707.82	12973.8	ND	0.3			KS
7698.94	7701.06	12985.2	ND	0.3H			MIT
7696.53	7698.65	12989.3	ND	0.5			MIT

AIR WAVELENGTH (ANGSTRMS)	VACUUM WAVELENGTH (ANGSTRMS)	WAVENUMBER (KAYSERS)	LMENT	LOG INTENSITY ARC	SPK	DIS	REF
7684.86	7686.98	13009.0	ND	0.3H			MIT
7663.52	7665.63	13045.2	ND	0.3			MIT
7645.92	7648.02	13075.3	ND	0.3			MIT
7639.76	7641.86	13085.8	ND	0.5			MIT
7632.82	7634.92	13097.7	ND	0.3			KS
7614.70	7616.80	13128.9	ND	0.3			MIT
7605.91	7608.00	13144.1	ND	0.3			MIT
7603.75	7605.84	13147.8	ND	0.3			KS
7595.93	7598.02	13161.3	ND	0.3			KS
7590.49	7592.58	13170.7	ND	0.3			KS
7587.66	7589.75	13175.7	ND	0.3			MIT
7579.18	7581.27	13190.4	ND	0.3			KS
7577.49	7579.58	13193.3	ND	0.3			MIT
7566.00	7568.08	13213.4	ND	0.3			KS
7563.21	7565.29	13218.3	ND	0.3			KS
7546.97	7549.05	13246.7	ND	0.5			MIT
7540.98	7543.06	13257.2	ND	0.3			MIT
7538.26	7540.34	13262.0	ND	0.6			MIT
7535.74	7537.81	13266.4	ND	0.3			MIT
7528.98	7531.05	13278.4	ND	0.6			MIT
7526.46	7528.53	13282.8	ND	0.5			MIT
7516.03	7518.10	13301.2	ND	0.3			MIT
7513.77	7515.84	13305.2	ND	0.7			MIT
7511.11	7513.18	13309.9	ND	0.6			MIT
7451.11	7453.16	13417.1	ND	0.7			KS
7448.76	7450.81	13421.4	ND	0.5			KS
7437.56	7439.61	13441.6	ND	0.7			KS
7427.43	7429.48	13459.9	ND	0.3			MIT
7418.16	7420.20	13476.7	ND	0.3			MIT
7409.73	7411.77	13492.1	ND	0.7			KS
7406.63	7408.67	13497.7	ND	0.3			MIT
7401.29	7403.33	13507.4	ND	0.7			MIT
7381.82	7383.85	13543.1	ND	0.3			MIT
7363.28	7365.31	13577.2	ND	0.7			KS
7323.12	7325.14	13651.6	ND	0.3			MIT
7321.38	7323.40	13654.9	ND	0.7			MIT
7316.85	7318.87	13663.3	ND	0.3			MIT
7272.26	7274.26	13747.1	ND	0.7			KS
7263.43	7265.43	13763.8	ND	0.3			ED
7236.56	7238.55	13814.9	ND	0.3			MIT
7192.01	7193.99	13900.5	ND	0.3			KS
7189.45	7191.43	13905.4	ND	0.5			MIT
7151.03	7153.00	13980.2	ND	0.3			MIT
7129.38	7131.35	14022.6	ND	0.3			MIT
7066.89	7068.84	14146.6	ND	0.3			KS
7037.37	7039.31	14205.9	ND	0.5			MIT
7024.59	7026.53	14231.8	ND	0.3			MIT
7022.99	7024.93	14235.0	ND	0.3			ED
7020.92	7022.86	14239.2	ND	0.6			KS
6995.27	6997.20	14291.4	ND	0.3			KN
6985.25	6987.18	14311.9	ND	0.6			MIT
6964.67	6966.59	14354.2	ND	0.7			MIT
6941.39	6943.30	14402.4	ND	0.3			KS
6940.20	6942.11	14404.8	ND	0.3			MIT
6923.86	6925.77	14438.8	ND	0.6			KS
6906.07	6907.98	14476.0	ND	1.0			MIT
6900.43	6902.33	14487.8	ND	1.0			MIT
6896.68	6898.58	14495.7	ND	0.3			MIT
6874.62	6876.52	14542.2	ND	0.3			MIT
6857.00	6858.89	14579.6	ND	0.9			MIT
6853.72	6855.61	14586.6	ND	0.7			KS
6849.30	6851.19	14596.0	ND	0.5			KS
6846.715	6848.604	14601.52	ND	1.0			MIT
6842.66	6844.55	14610.2	ND	1.0			MIT
6825.34	6827.22	14647.2	ND	0.3			MIT
6816.02	6817.90	14667.3	ND	1.0			MIT
6812.30	6814.18	14675.3	ND	0.5			KS
6804.00	6805.88	14693.2	ND	1.2			MIT
6803.06	6804.94	14695.2	ND	0.6			KS
6801.34	6803.22	14698.9	ND	0.6			KS

AIR WAVELENGTH (ANGSTRMS)	VACUUM WAVELENGTH (ANGSTRMS)	WAVENUMBER (KAYSERS)	LMENT	LOG INTENSITY ARC	SPK	DIS	REF
6792.28	6794.15	14718.5	ND	0.3			KS
6790.41	6792.28	14722.6	ND	1.0D			MIT
6763.78	6765.65	14780.6	ND	0.5			KS
6763.01	6764.88	14782.2	ND	0.5			KS
6749.60	6751.46	14811.6	ND	1.0			KN
6745.18	6747.04	14821.3	ND	0.7			KS
6742.537	6744.398	14827.12	ND	1.0			MIT
6740.107	6741.968	14832.46	ND	1.0			MIT
6737.787	6739.647	14837.57	ND	1.0			MIT
6735.04	6736.90	14843.6	ND	0.7			KS
6732.06	6733.92	14850.2	ND	1.0			KN
6728.93	6730.79	14857.1	ND	0.9			KS
6727.740	6729.597	14859.73	ND	0.3			MIT
6680.143	6681.987	14965.61	ND	0.3			MIT
6679.771	6681.615	14966.44	ND	0.3			MIT
6678.523	6680.367	14969.24	ND	0.3			MIT
6670.391	6672.233	14987.49	ND	0.3			MIT
6664.61	6666.45	15000.5	ND	0.9			KS
6655.672	6657.510	15020.63	ND	1.0			MIT
6650.57	6652.41	15032.2	ND	1.0			KS
6637.962	6639.795	15060.71	ND	1.4			MIT
6637.191	6639.024	15062.45	ND	1.2			MIT
6630.138	6631.969	15078.48	ND	1.5			MIT
6624.26	6626.09	15091.9	ND	0.6			KS
6619.354	6621.182	15103.04	ND	1.0			MIT
6618.529	6620.357	15104.93	ND	0.9			MIT
6615.88	6617.71	15111.0	ND	0.7			KS
6611.988	6613.814	15119.87	ND	0.9			MIT
6609.685	6611.510	15125.14	ND	0.7			MIT
6609.305	6611.130	15126.01	ND	0.6			MIT
6604.44	6606.26	15137.2	ND	0.9D			KS
6603.694	6605.518	15138.86	ND	0.5			MIT
6601.756	6603.579	15143.30	ND	1.0			MIT
6595.013	6596.835	15158.79	ND	0.7			MIT
6592.66	6594.48	15164.2	ND	0.3			KN
6591.429	6593.250	15167.03	ND	0.7			MIT
6590.945	6592.765	15168.14	ND	0.5			MIT
6589.639	6591.459	15171.15	ND	1.0			MIT
6589.067	6590.887	15172.46	ND	0.7			MIT
6588.033	6589.853	15174.85	ND	0.8			MIT
6585.706	6587.525	15180.21	ND	0.9			MIT
6584.59	6586.41	15182.8	ND	0.7			KS
6582.617	6584.435	15187.33	ND	0.5			MIT
6580.943	6582.761	15191.19	ND	1.0			MIT
6579.640	6581.457	15194.20	ND	0.5			MIT
6577.732	6579.549	15198.61	ND	0.6			MIT
6575.935	6577.751	15202.76	ND	0.6			MIT
6572.651	6574.467	15210.36	ND	0.9			MIT
6571.864	6573.679	15212.18	ND	0.7			MIT
6569.56	6571.37	15217.5	ND	0.7			KS
6568.469	6570.283	15220.04	ND	0.7			MIT
6563.927	6565.740	15230.57	ND	0.6H			MIT
6560.292	6562.104	15239.01	ND	0.7			MIT
6558.966	6560.778	15242.09	ND	0.7			MIT
6553.070	6554.880	15255.81	ND	0.9			MIT
6550.191	6552.001	15262.51	ND	0.9J			MIT
6549.542	6551.351	15264.03	ND	1.0			MIT
6541.073	6542.880	15283.79	ND	0.5			MIT
6539.937	6541.744	15286.44	ND	1.0O			MIT
6523.151	6524.953	15325.78	ND	1.0			MIT
6520.279	6522.080	15332.53	ND	0.7			MIT
6519.858	6521.659	15333.52	ND	1.0			MIT
6514.961	6516.761	15345.05	ND	1.2			MIT
6504.463	6506.260	15369.81	ND	1.0			MIT
6501.864	6503.661	15375.96	ND	0.7			MIT
6501.072	6502.868	15377.83	ND	0.5			MIT
6500.159	6501.955	15379.99	ND	0.9			MIT
6495.591	6497.386	15390.81	ND	0.9			MIT
6492.348	6494.142	15398.49	ND	1.1			MIT
6490.945	6492.739	15401.82	ND	0.7			MIT

AIR WAVELENGTH (ANGSTRMS)	VACUUM WAVELENGTH (ANGSTRMS)	WAVENUMBER (KAYSERS)	LMENT	LOG INTENSITY ARC	SPK	DIS	REF
6489.199	6490.992	15405.96	ND	0.7			MIT
6487.531	6489.324	15409.93	ND	0.7			MIT
6485.687	6487.479	15414.31	ND	1.3			MIT
6484.365	6486.157	15417.45	ND	0.5			MIT
6483.445	6485.237	15419.64	ND	0.7H			MIT
6482.282	6484.073	15422.40	ND	1.0			MIT
6480.214	6482.005	15427.33	ND	1.0			MIT
6479.849	6481.640	15428.19	ND	0.8			MIT
6478.78	6480.57	15430.7	ND	0.6			KS
6476.201	6477.991	15436.89	ND	0.5			MIT
6474.231	6476.020	15441.58	ND	0.5			MIT
6471.71	6473.50	15447.6	ND	0.3			KS
6465.236	6467.023	15463.07	ND	0.9			MIT
6463.582	6465.368	15467.02	ND	1.2			MIT
6461.170	6462.956	15472.80	ND	0.5			MIT
6457.953	6459.738	15480.51	ND	0.6			MIT
6457.126	6458.910	15482.49	ND	0.7			MIT
6455.058	6456.842	15487.45	ND	0.5			MIT
6454.795	6456.579	15488.08	ND	0.9			MIT
6451.234	6453.017	15496.63	ND	0.7D			MIT
6445.786	6447.567	15509.72	ND	0.9			MIT
6439.86	6441.64	15524.0	ND	0.5			KN
6439.171	6440.951	15525.66	ND	1.3			MIT
6433.236	6435.014	15539.98	ND	1.1			MIT
6432.653	6434.431	15541.39	ND	1.1			MIT
6431.711	6433.489	15543.67	ND	0.3			MIT
6431.177	6432.955	15544.96	ND	0.3			MIT
6429.840	6431.617	15548.19	ND	0.3			MIT
6428.645	6430.422	15551.08	ND	1.4			MIT
6425.790	6427.566	15557.99	ND	1.2			MIT
6424.841	6426.617	15560.29	ND	0.6			MIT
6414.029	6415.802	15586.52	ND	0.5			MIT
6410.598	6412.370	15594.86	ND	0.3			MIT
6407.606	6409.377	15602.14	ND	0.7			MIT
6403.196	6404.966	15612.89	ND	0.5			MIT
6402.758	6404.528	15613.95	ND	0.5			MIT
6396.876	6398.644	15628.31	ND	0.5			MIT
6394.80	6396.57	15633.4	ND	0.6			MIT
6389.997	6391.763	15645.14	ND	1.2			MIT
6385.196	6386.961	15656.90	ND	2.0			MIT
6384.633	6386.398	15658.28	ND	0.8			MIT
6382.069	6383.833	15664.57	ND	1.3			MIT
6377.256	6379.019	15676.39	ND	0.6			MIT
6375.970	6377.733	15679.55	ND	0.7			MIT
6372.776	6374.538	15687.41	ND	0.5			MIT
6366.753	6368.513	15702.25	ND	0.6			MIT
6365.554	6367.314	15705.21	ND	1.2			MIT
6363.935	6365.694	15709.20	ND	0.3			MIT
6362.093	6363.852	15713.75	ND	1.0			MIT
6361.430	6363.189	15715.39	ND	1.0			MIT
6360.869	6362.628	15716.78	ND	0.7			MIT
6356.545	6358.302	15727.47	ND	0.3			MIT
6355.947	6357.704	15728.95	ND	0.5			MIT
6354.754	6356.511	15731.90	ND	0.6			MIT
6351.958	6353.714	15738.83	ND	0.6			MIT
6348.743	6350.498	15746.80	ND	0.3			MIT
6347.824	6349.579	15749.07	ND	0.3			MIT
6346.539	6348.294	15752.26	ND	0.7			MIT
6346.320	6348.075	15752.81	ND	0.5			MIT
6343.057	6344.811	15760.91	ND	0.3			MIT
6341.507	6343.260	15764.76	ND	1.2			MIT
6340.001	6341.754	15768.51	ND	0.7			MIT
6338.838	6340.591	15771.40	ND	0.5			MIT
6333.896	6335.647	15783.71	ND	0.3			MIT
6331.889	6333.640	15788.71	ND	0.5			MIT
6330.167	6331.917	15793.01	ND	1.2			MIT
6328.465	6330.215	15797.25	ND	0.5			MIT
6326.952	6328.702	15801.03	ND	0.5			MIT
6324.781	6326.530	15806.45	ND	0.3			MIT
6322.867	6324.615	15811.24	ND	0.5			MIT

AIR WAVELENGTH (ANGSTRMS)	VACUUM WAVELENGTH (ANGSTRMS)	WAVENUMBER (KAYSERS)	LMENT	LOG ARC	INTENSITY SPK	INTENSITY DIS	REF
6321.218	6322.966	15815.36	ND	0.3			MIT
6319.686	6321.434	15819.20	ND	0.7			MIT
6317.703	6319.450	15824.16	ND	0.5			MIT
6313.235	6314.981	15835.36	ND	0.6			MIT
6312.718	6314.464	15836.66	ND	0.3			MIT
6310.487	6312.232	15842.26	ND	1.7			MIT
6308.257	6310.002	15847.86	ND	1.0			MIT
6307.043	6308.787	15850.91	ND	1.0			MIT
6305.784	6307.528	15854.07	ND	0.5			MIT
6305.231	6306.975	15855.46	ND	0.5			MIT
6301.972	6303.715	15863.66	ND	0.6			MIT
6300.970	6302.713	15866.18	ND	0.7			MIT
6298.418	6300.160	15872.61	ND	1.3			MIT
6297.072	6298.813	15876.01	ND	1.2			MIT
6294.425	6296.166	15882.68	ND	0.3			MIT
6292.843	6294.583	15886.68	ND	1.3			MIT
6292.004	6293.744	15888.79	ND	0.6			MIT
6288.024	6289.763	15898.85	ND	0.5			MIT
6287.574	6289.313	15899.99	ND	0.5			MIT
6287.207	6288.946	15900.92	ND	0.7			MIT
6285.794	6287.532	15904.49	ND	1.2			MIT
6284.726	6286.464	15907.19	ND	0.5			MIT
6282.515	6284.253	15912.79	ND	0.5			MIT
6281.997	6283.734	15914.10	ND	1.0			MIT
6278.253	6279.989	15923.59	ND	0.3			MIT
6277.286	6279.022	15926.05	ND	0.7			MIT
6272.771	6274.506	15937.51	ND	0.3			MIT
6271.737	6273.472	15940.14	ND	0.6			MIT
6270.273	6272.007	15943.86	ND	0.7			MIT
6269.417	6271.151	15946.04	ND	0.6			MIT
6263.227	6264.959	15961.80	ND	0.3			MIT
6258.733	6260.464	15973.26	ND	0.9			MIT
6257.517	6259.248	15976.36	ND	1.2			MIT
6257.308	6259.039	15976.89	ND	0.6			MIT
6256.715	6258.446	15978.41	ND	0.7			MIT
6250.431	6252.160	15994.47	ND	1.0			MIT
6248.276	6250.004	15999.99	ND	1.2			MIT
6246.828	6248.556	16003.70	ND	0.7			MIT
6244.084	6245.811	16010.73	ND	1.5			MIT
6238.499	6240.225	16025.06	ND	1.3			MIT
6237.338	6239.063	16028.05	ND	0.7			MIT
6230.322	6232.046	16046.10	ND	0.3			MIT
6230.022	6231.746	16046.87	ND	0.5			MIT
6227.955	6229.678	16052.19	ND	0.3			MIT
6227.206	6228.929	16054.12	ND	0.3			MIT
6226.502	6228.225	16055.94	ND	1.3			MIT
6225.250	6226.972	16059.17	ND	0.3			MIT
6223.394	6225.116	16063.96	ND	1.4			MIT
6222.148	6223.869	16067.18	ND	0.9			MIT
6221.150	6222.871	16069.75	ND	0.3			MIT
6219.794	6221.515	16073.26	ND	0.3			MIT
6218.185	6219.905	16077.42	ND	0.3			MIT
6216.688	6218.408	16081.29	ND	0.9			MIT
6212.794	6214.513	16091.37	ND	0.3			MIT
6210.681	6212.399	16096.84	ND	1.0			MIT
6208.239	6209.957	16103.17	ND	1.0			MIT
6208.009	6209.727	16103.77	ND	0.7			MIT
6206.110	6207.827	16108.70	ND	0.7			MIT
6201.744	6203.460	16120.04	ND	1.2			MIT
6198.610	6200.325	16128.19	ND	0.6			MIT
6197.706	6199.421	16130.54	ND	0.3			MIT
6196.204	6197.918	16134.45	ND	0.6			MIT
6191.684	6193.397	16146.23	ND	0.5			MIT
6189.460	6191.173	16152.03	ND	0.5			MIT
6186.979	6188.691	16158.51	ND	0.3			MIT
6183.907	6185.618	16166.53	ND	1.3			MIT
6182.975	6184.686	16168.97	ND	0.8			MIT
6181.626	6183.337	16172.50	ND	0.5			MIT
6180.764	6182.474	16174.75	ND	0.5			MIT
6178.591	6180.301	16180.44	ND	1.2			MIT

AIR WAVELENGTH (ANGSTRMS)	VACUUM WAVELENGTH (ANGSTRMS)	WAVENUMBER (KAYSERS)	LMENT	LOG ARC	INTENSITY SPK	INTENSITY DIS	REF
6178.198	6179.908	16181.47	ND	0.5			MIT
6176.085	6177.794	16187.01	ND	0.6			MIT
6174.301	6176.010	16191.69	ND	0.3			MIT
6170.486	6172.194	16201.69	ND	1.5			MIT
6169.168	6170.875	16205.16	ND	0.3			MIT
6168.096	6169.803	16207.97	ND	0.7			MIT
6166.673	6168.380	16211.71	ND	1.3			MIT
6165.656	6167.362	16214.39	ND	0.9			MIT
6165.332	6167.038	16215.24	ND	0.3			MIT
6165.132	6166.838	16215.77	ND	0.3			MIT
6164.422	6166.128	16217.63	ND	0.6			MIT
6162.377	6164.082	16223.02	ND	0.9			MIT
6161.508	6163.213	16225.30	ND	1.0			MIT
6159.281	6160.986	16231.17	ND	0.5			MIT
6157.834	6159.538	16234.98	ND	1.4			MIT
6156.912	6158.616	16237.42	ND	0.3			MIT
6156.578	6158.282	16238.30	ND	0.9			MIT
6156.163	6157.867	16239.39	ND	0.5			MIT
6155.062	6156.765	16242.30	ND	1.2			MIT
6152.788	6154.491	16248.30	ND	0.3			MIT
6150.281	6151.983	16254.92	ND	0.5			MIT
6149.277	6150.979	16257.57	ND	1.3			MIT
6148.574	6150.276	16259.43	ND	0.5			MIT
6144.512	6146.213	16270.18	ND	0.3			MIT
6142.439	6144.139	16275.67	ND	0.3			MIT
6141.891	6143.591	16277.13	ND	0.3			MIT
6137.383	6139.082	16289.08	ND	0.5			MIT
6133.964	6135.662	16298.16	ND	0.7			MIT
6133.468	6135.166	16299.48	ND	0.7			MIT
6131.790	6133.487	16303.94	ND	0.7			MIT
6130.604	6132.301	16307.09	ND	0.6			MIT
6129.203	6130.899	16310.82	ND	0.5			MIT
6122.964	6124.659	16327.44	ND	0.5			MIT
6122.224	6123.919	16329.41	ND	1.3			MIT
6114.454	6116.146	16350.16	ND	0.7			MIT
6113.468	6115.160	16352.80	ND	0.9			MIT
6111.99	6113.68	16356.8	ND	0.3			MIT
6109.688	6111.379	16362.92	ND	0.9			MIT
6109.058	6110.749	16364.61	ND	0.5			MIT
6108.410	6110.101	16366.34	ND	1.3			MIT
6106.135	6107.825	16372.44	ND	0.3			MIT
6105.518	6107.208	16374.09	ND	0.7			MIT
6104.106	6105.796	16377.88	ND	0.5			MIT
6101.747	6103.436	16384.21	ND	1.0			MIT
6099.014	6100.702	16391.55	ND	1.0			MIT
6098.203	6099.891	16393.74	ND	0.7			MIT
6097.787	6099.475	16394.85	ND	0.5			MIT
6092.058	6093.744	16410.27	ND	0.6			MIT
6091.602	6093.288	16411.50	ND	0.5			MIT
6088.01	6089.70	16421.2	ND	0.3			MIT
6087.919	6089.604	16421.43	ND	0.3			MIT
6087.509	6089.194	16422.53	ND	0.5			MIT
6086.964	6088.649	16424.00	ND	0.6			MIT
6084.622	6086.306	16430.33	ND	0.9			MIT
6082.896	6084.580	16434.99	ND	0.3			MIT
6082.032	6083.716	16437.32	ND	0.3			MIT
6081.569	6083.253	16438.57	ND	0.6			MIT
6080.607	6082.290	16441.18	ND	0.5			MIT
6077.222	6078.905	16450.33	ND	1.0			MIT
6074.432	6076.114	16457.89	ND	0.5			MIT
6073.967	6075.649	16459.15	ND	1.3			MIT
6073.114	6074.795	16461.46	ND	0.3			MIT
6071.701	6073.382	16465.29	ND	1.3			MIT
6069.698	6071.378	16470.72	ND	0.6			MIT
6066.031	6067.711	16480.68	ND	1.3			MIT
6065.743	6067.422	16481.46	ND	0.3			MIT
6065.142	6066.821	16483.10	ND	0.3			MIT
6061.051	6062.729	16494.22	ND	0.6			MIT
6058.307	6059.984	16501.69	ND	0.3			MIT
6057.642	6059.319	16503.50	ND	0.5			MIT

AIR WAVELENGTH (ANGSTRMS)	VACUUM WAVELENGTH (ANGSTRMS)	WAVENUMBER (KAYSERS)	LMENT	LOG ARC	INTENSITY SPK	DIS	REF
6056.993	6058.670	16505.27	ND	0.3			MIT
6054.482	6056.158	16512.12	ND	0.3			MIT
6051.875	6053.551	16519.23	ND	0.6			MIT
6050.473	6052.148	16523.06	ND	0.3			MIT
6049.861	6051.536	16524.73	ND	0.6			MIT
6043.783	6045.457	16541.35	ND	0.5			MIT
6040.958	6042.631	16549.08	ND	0.5			MIT
6035.608	6037.279	16563.75	ND	0.5			MIT
6034.239	6035.910	16567.51	ND	1.3			MIT
6033.292	6034.963	16570.11	ND	1.0			MIT
6031.266	6032.936	16575.68	ND	1.4			MIT
6025.536	6027.205	16591.44	ND	0.9			MIT
6020.236	6021.903	16606.05	ND	0.3			MIT
6018.933	6020.600	16609.64	ND	0.3			MIT
6017.708	6019.375	16613.02	ND	0.3			MIT
6016.295	6017.961	16616.92	ND	0.3			MIT
6015.329	6016.995	16619.59	ND	0.3			MIT
6011.533	6013.198	16630.09	ND	0.3			MIT
6010.884	6012.549	16631.88	ND	0.6			MIT
6009.304	6010.968	16636.25	ND	0.7			MIT
6007.669	6009.333	16640.78	ND	1.4			MIT
6006.399	6008.063	16644.30	ND	0.3			MIT
6004.332	6005.995	16650.03	ND	0.3			MIT
6003.978	6005.641	16651.01	ND	0.3			MIT
6002.528	6004.190	16655.03	ND	0.3			MIT
6002.316	6003.978	16655.62	ND	0.3			MIT
6001.448	6003.110	16658.03	ND	0.3			MIT
6000.043	6001.705	16661.93	ND	0.3			MIT
5996.465	5998.126	16671.87	ND	1.2			MIT
5994.756	5996.416	16676.63	ND	1.2H			MIT
5994.035	5995.695	16678.63	ND	0.5			MIT
5990.689	5992.348	16687.95	ND	0.5			MIT
5990.244	5991.903	16689.19	ND	0.7			MIT
5989.341	5991.000	16691.70	ND	1.2			MIT
5988.142	5989.801	16695.05	ND	0.3			MIT
5979.951	5981.607	16717.91	ND	0.5			MIT
5977.415	5979.071	16725.01	ND	0.5			MIT
5976.354	5978.009	16727.98	ND	0.5			MIT
5975.503	5977.158	16730.36	ND	0.3			MIT
5974.563	5976.218	16732.99	ND	0.6			MIT
5974.248	5975.903	16733.87	ND	0.7			MIT
5972.499	5974.153	16738.77	ND	0.3			MIT
5968.789	5970.442	16749.18	ND	0.3			MIT
5968.282	5969.935	16750.60	ND	0.5			MIT
5966.710	5968.363	16755.01	ND	0.3			MIT
5966.379	5968.032	16755.94	ND	0.3			MIT
5963.258	5964.910	16764.71	ND	0.5			MIT
5962.39	5964.04	16767.1	ND	0.3			KS
5961.156	5962.807	16770.62	ND	0.6			MIT
5957.589	5959.239	16780.67	ND	0.5			MIT
5955.869	5957.519	16785.51	ND	1.2			MIT
5953.080	5954.729	16793.37	ND	0.3			MIT
5950.470	5952.119	16800.74	ND	0.9			MIT
5949.636	5951.284	16803.10	ND	1.0			MIT
5947.406	5949.054	16809.40	ND	0.9			MIT
5943.216	5944.863	16821.25	ND	1.2	0.0		MIT
5939.744	5941.390	16831.08	ND	0.5			MIT
5935.475	5937.120	16843.19	ND	0.5			MIT
5934.747	5936.391	16845.25	ND	1.0			MIT
5932.140	5933.784	16852.65	ND	0.5			MIT
5931.421	5933.064	16854.70	ND	0.5			MIT
5928.919	5930.562	16861.81	ND	0.6			MIT
5927.958	5929.601	16864.54	ND	0.3			MIT
5922.790	5924.431	16879.26	ND	0.7			MIT
5922.364	5924.005	16880.47	ND	0.3			MIT
5921.820	5923.461	16882.02	ND	0.3			MIT
5921.217	5922.858	16883.74	ND	1.0			MIT
5919.811	5921.451	16887.75	ND	0.5			MIT
5919.310	5920.950	16889.18	ND	0.7			MIT
5917.465	5919.105	16894.45	ND	0.3			MIT

AIR WAVELENGTH (ANGSTRMS)	VACUUM WAVELENGTH (ANGSTRMS)	WAVENUMBER (KAYSERS)	LMENT	LOG ARC	INTENSITY SPK	DIS	REF
5914.401	5916.040	16903.20	ND	0.7			MIT
5910.268	5911.906	16915.02	ND	0.3L			MIT
5909.874	5911.512	16916.15	ND	1.0	0.0		MIT
5906.646	5908.283	16925.39	ND	0.8			MIT
5906.276	5907.913	16926.45	ND	0.5			MIT
5905.808	5907.445	16927.79	ND	0.7			MIT
5902.119	5903.755	16938.37	ND	0.5			MIT
5900.430	5902.065	16943.22	ND	0.7			MIT
5899.491	5901.126	16945.92	ND	0.5			MIT
5899.171	5900.806	16946.84	ND	0.3			MIT
5898.867	5900.502	16947.71	ND	0.5			MIT
5895.578	5897.212	16957.17	ND	0.3			MIT
5892.670	5894.303	16965.53	ND	0.3			MIT
5891.528	5893.161	16968.82	ND	1.3			MIT
5890.503	5892.135	16971.77	ND	0.7			MIT
5887.907	5889.539	16979.26	ND	1.4			MIT
5886.235	5887.866	16984.08	ND	0.5			MIT
5883.292	5884.923	16992.58	ND	1.2			MIT
5882.784	5884.414	16994.04	ND	1.0			MIT
5881.33	5882.96	16998.2	ND	0.5			KS
5880.25	5881.88	17001.4	ND	0.5			KS
5877.827	5879.456	17008.38	ND	0.5			MIT
5876.344	5877.973	17012.67	ND	0.3			MIT
5875.813	5877.442	17014.21	ND	0.3			MIT
5874.392	5876.020	17018.32	ND	0.3			MIT
5873.342	5874.970	17021.36	ND	0.3			MIT
5871.042	5872.669	17028.03	ND	0.7			MIT
5869.606	5871.233	17032.20	ND	0.3H			MIT
5868.895	5870.522	17034.26	ND	1.0			MIT
5867.611	5869.237	17037.99	ND	0.5			MIT
5867.083	5868.709	17039.52	ND	0.9			MIT
5865.057	5866.683	17045.41	ND	1.2			MIT
5861.146	5862.771	17056.78	ND	0.3			MIT
5860.250	5861.874	17059.39	ND	0.5			MIT
5859.676	5861.300	17061.06	ND	0.3			MIT
5858.907	5860.531	17063.30	ND	1.2			MIT
5857.522	5859.146	17067.33	ND	1.0			MIT
5856.687	5858.310	17069.77	ND	0.3			MIT
5848.374	5849.995	17094.03	ND	0.3			MIT
5847.604	5849.225	17096.28	ND	0.3			MIT
5846.360	5847.981	17099.92	ND	0.3			MIT
5845.945	5847.566	17101.13	ND	0.3			MIT
5844.656	5846.276	17104.91	ND	0.6			MIT
5843.232	5844.852	17109.07	ND	0.3			MIT
5842.391	5844.011	17111.54	ND	1.4			MIT
5839.127	5840.746	17121.10	ND	0.3			MIT
5837.139	5838.757	17126.93	ND	0.3			MIT
5835.28	5836.90	17132.4	ND	0.3			MIT
5834.549	5836.166	17134.53	ND	0.3			MIT
5830.727	5832.343	17145.77	ND	0.3H			MIT
5829.598	5831.214	17149.09	ND	0.3			MIT
5826.743	5828.358	17157.49	ND	1.2			MIT
5825.872	5827.487	17160.05	ND	1.4			MIT
5824.00	5825.61	17165.6	ND	0.7			KN
5823.75	5825.36	17166.3	ND	0.6H			MIT
5823.371	5824.985	17167.43	ND	1.0			MIT
5820.955	5822.569	17174.55	ND	0.3			MIT
5820.365	5821.979	17176.29	ND	0.3			MIT
5815.442	5817.054	17190.83	ND	0.3H			MIT
5813.889	5815.501	17195.42	ND	1.5			MIT
5811.572	5813.183	17202.28	ND	1.2	0.0		MIT
5809.247	5810.858	17209.16	ND	0.9			MIT
5806.433	5808.043	17217.50	ND	0.3			MIT
5804.020	5805.629	17224.66	ND	2.0			MIT
5800.087	5801.695	17236.34	ND	0.5			MIT
5795.172	5796.779	17250.96	ND	1.0			MIT
5793.976	5795.583	17254.52	ND	0.3H			MIT
5792.411	5794.017	17259.18	ND	0.3			MIT
5788.222	5789.827	17271.67	ND	1.5			MIT
5784.958	5786.562	17281.42	ND	1.3			MIT

AIR WAVELENGTH (ANGSTRMS)	VACUUM WAVELENGTH (ANGSTRMS)	WAVENUMBER (KAYSERS)	LMENT	LOG ARC	INTENSITY SPK	DIS	REF
5783.685	5785.289	17285.22	ND	0.5H			MIT
5776.118	5777.720	17307.87	ND	1.2			MIT
5773.89	5775.49	17314.5	ND	0.6			KS
5772.160	5773.761	17319.73	ND	0.7			MIT
5771.480	5773.081	17321.77	ND	0.3			MIT
5770.500	5772.100	17324.72	ND	1.3			MIT
5769.866	5771.466	17326.62	ND	1.0			MIT
5767.330	5768.929	17334.24	ND	0.7			MIT
5764.225	5765.824	17343.58	ND	0.5			MIT
5762.388	5763.986	17349.10	ND	0.3			MIT
5762.077	5763.675	17350.04	ND	1.0			MIT
5761.695	5763.293	17351.19	ND	1.0			MIT
5759.999	5761.597	17356.30	ND	0.7			MIT
5755.202	5756.798	17370.77	ND	0.3			MIT
5753.530	5755.126	17375.81	ND	1.2			MIT
5749.656	5751.251	17387.52	ND	1.0			MIT
5749.186	5750.781	17388.94	ND	1.3			MIT
5749.056	5750.651	17389.34	ND	1.0			MIT
5748.154	5749.748	17392.06	ND	0.6			MIT
5746.88	5748.47	17395.9	ND	0.3			MIT
5744.766	5746.359	17402.32	ND	1.4	0.0		MIT
5744.138	5745.731	17404.23	ND	0.9			MIT
5743.196	5744.789	17407.08	ND	1.3J	0.0		MIT
5742.763	5744.356	17408.39	ND	0.7			MIT
5742.092	5743.685	17410.43	ND	1.6			MIT
5741.276	5742.869	17412.90	ND	1.0			MIT
5740.862	5742.454	17414.16	ND	1.5			MIT
5739.955	5741.547	17416.91	ND	1.2	0.0		MIT
5738.912	5740.504	17420.07	ND	0.5			MIT
5731.810	5733.400	17441.66	ND	0.3			MIT
5731.050	5732.640	17443.97	ND	0.3			MIT
5729.294	5730.883	17449.32	ND	1.5			MIT
5727.872	5729.461	17453.65	ND	0.6			MIT
5726.825	5728.414	17456.84	ND	1.4			MIT
5723.860	5725.448	17465.88	ND	0.3			MIT
5719.086	5720.673	17480.46	ND	1.2			MIT
5718.120	5719.706	17483.42	ND	1.5	0.0		MIT
5708.280	5709.864	17513.55	ND	1.8			MIT
5707.372	5708.955	17516.34	ND	0.6			MIT
5706.206	5707.789	17519.92	ND	1.6			MIT
5702.244	5703.826	17532.09	ND	1.4			MIT
5701.566	5703.148	17534.18	ND	0.7			MIT
5698.927	5700.508	17542.30	ND	1.0	0.0		MIT
5695.228	5696.808	17553.69	ND	0.7			MIT
5690.251	5691.830	17569.04	ND	0.9			MIT
5689.506	5691.085	17571.34	ND	1.0			MIT
5688.525	5690.103	17574.37	ND	2.2			MIT
5686.639	5688.217	17580.20	ND	1.0			MIT
5679.64	5681.22	17601.9	ND	0.3			MIT
5677.805	5679.381	17607.55	ND	0.9			MIT
5676.334	5677.909	17612.12	ND	1.0			MIT
5675.971	5677.546	17613.24	ND	1.5			MIT
5669.770	5671.343	17632.51	ND	1.6			MIT
5669.320	5670.893	17633.91	ND	0.5			MIT
5668.868	5670.441	17635.31	ND	1.2			MIT
5665.747	5667.319	17645.03	ND	0.7			MIT
5665.261	5666.833	17646.54	ND	0.5			MIT
5662.455	5664.026	17655.28	ND	0.6			MIT
5659.781	5661.352	17663.63	ND	1.2	0.0		MIT
5653.566	5655.135	17683.04	ND	1.0			MIT
5651.514	5653.083	17689.46	ND	0.3			MIT
5651.269	5652.837	17690.23	ND	0.7			MIT
5647.98	5649.55	17700.5	ND	0.6			KS
5644.127	5645.694	17712.62	ND	1.0			MIT
5643.270	5644.836	17715.31	ND	0.6			MIT
5639.544	5641.109	17727.01	ND	1.4			MIT
5639.046	5640.611	17728.57	ND	0.7			MIT
5637.827	5639.392	17732.41	ND	0.3H			MIT
5635.761	5637.325	17738.91	ND	1.4			MIT
5631.64	5633.20	17751.9	ND	0.3			MIT

AIR WAVELENGTH (ANGSTRMS)	VACUUM WAVELENGTH (ANGSTRMS)	WAVENUMBER (KAYSERS)	LMENT	LOG ARC	INTENSITY SPK	DIS	REF
5627.246	5628.808	17765.75	ND	0.7			MIT
5626.661	5628.223	17767.60	ND	0.3			MIT
5625.722	5627.284	17770.56	ND	1.0	0.0		MIT
5625.368	5626.930	17771.68	ND	0.5			MIT
5624.332	5625.893	17774.96	ND	0.5			MIT
5622.303	5623.864	17781.37	ND	0.3			MIT
5620.538	5622.098	17786.95	ND	2.3	0.7		MIT
5619.000	5620.560	17791.82	ND	1.0			MIT
5617.705	5619.264	17795.92	ND	1.2	0.0		MIT
5615.353	5616.912	17803.38	ND	0.7			MIT
5614.303	5615.862	17806.71	ND	1.0			MIT
5611.185	5612.743	17816.60	ND	0.7			MIT
5605.646	5607.202	17834.21	ND	0.5H			MIT
5603.651	5605.207	17840.55	ND	0.7			MIT
5603.198	5604.754	17842.00	ND	0.5			MIT
5602.678	5604.233	17843.65	ND	0.7			MIT
5601.918	5603.473	17846.07	ND	1.3			MIT
5601.433	5602.988	17847.62	ND	1.2			MIT
5598.489	5600.043	17857.00	ND	0.6			MIT
5597.852	5599.406	17859.04	ND	0.5			MIT
5595.811	5597.365	17865.55	ND	0.9			MIT
5594.425	5595.978	17869.98	ND	2.2	0.7		MIT
5592.671	5594.224	17875.58	ND	0.5			MIT
5589.919	5591.471	17884.38	ND	0.3			MIT
5588.441	5589.993	17889.11	ND	0.3			MIT
5587.962	5589.514	17890.65	ND	1.0			MIT
5587.612	5589.163	17891.76	ND	0.7			MIT
5581.601	5583.151	17911.03	ND	0.5			MIT
5578.657	5580.206	17920.49	ND	1.0			MIT
5577.703	5579.252	17923.55	ND	1.0			MIT
5576.700	5578.249	17926.77	ND	0.7			MIT
5575.503	5577.051	17930.62	ND	0.7			MIT
5574.665	5576.213	17933.32	ND	0.5			MIT
5569.956	5571.503	17948.48	ND	0.6	0.3		MIT
5565.708	5567.254	17962.18	ND	0.5			MIT
5561.168	5562.712	17976.84	ND	1.0			MIT
5559.006	5560.550	17983.83	ND	0.7			MIT
5557.62	5559.16	17988.3	ND	0.6			KN
5555.765	5557.308	17994.32	ND	0.7			MIT
5553.33	5554.87	18002.2	ND	0.3			MIT
5552.863	5554.405	18003.73	ND	0.9			MIT
5550.092	5551.633	18012.72	ND	1.0	0.0		MIT
5548.706	5550.247	18017.22	ND	0.7			MIT
5548.474	5550.015	18017.97	ND	1.0	0.0		MIT
5547.678	5549.219	18020.55	ND	0.3			MIT
5546.979	5548.520	18022.83	ND	0.3			MIT
5545.910	5547.450	18026.30	ND	0.6			MIT
5544.848	5546.388	18029.75	ND	0.3			MIT
5543.241	5544.781	18034.98	ND	1.3			MIT
5541.791	5543.330	18039.70	ND	0.7			MIT
5539.26	5540.80	18047.9	ND	1.0			MIT
5538.806	5540.344	18049.42	ND	0.7			MIT
5537.768	5539.306	18052.80	ND	0.3			MIT
5537.279	5538.817	18054.40	ND	0.5			MIT
5536.444	5537.982	18057.12	ND	0.3			MIT
5535.476	5537.013	18060.28	ND	0.5			MIT
5535.271	5536.808	18060.95	ND	0.3			MIT
5533.840	5535.377	18065.62	ND	1.0			MIT
5533.396	5534.933	18067.07	ND	1.2			MIT
5532.971	5534.508	18068.45	ND	1.0H			MIT
5532.087	5533.624	18071.34	ND	1.2			MIT
5530.485	5532.021	18076.58	ND	0.5			MIT
5529.069	5530.605	18081.21	ND	1.0			MIT
5528.333	5529.869	18083.61	ND	1.3			MIT
5526.143	5527.678	18090.78	ND	0.7			MIT
5525.721	5527.256	18092.16	ND	1.3			MIT
5523.938	5525.472	18098.00	ND	0.7			MIT
5522.165	5523.699	18103.81	ND	1.0			MIT
5520.602	5522.136	18108.94	ND	1.2			MIT
5520.034	5521.567	18110.80	ND	0.3			MIT

AIR WAVELENGTH (ANGSTRMS)	VACUUM WAVELENGTH (ANGSTRMS)	WAVENUMBER (KAYSERS)	LMENT	LOG ARC	INTENSITY SPK	INTENSITY DIS	REF
5519.354	5520.887	18113.03	ND	0.7			MIT
5518.685	5520.218	18115.23	ND	1.0			MIT
5518.392	5519.925	18116.19	ND	0.7			MIT
5516.291	5517.823	18123.09	ND	1.0			MIT
5516.008	5517.540	18124.02	ND	1.0			MIT
5515.674	5517.206	18125.12	ND	0.3			MIT
5514.752	5516.284	18128.15	ND	0.3			MIT
5511.947	5513.478	18137.37	ND	0.7			MIT
5509.132	5510.662	18146.64	ND	1.0			MIT
5508.398	5509.928	18149.06	ND	1.2			MIT
5507.660	5509.190	18151.49	ND	0.9			MIT
5507.194	5508.724	18153.02	ND	0.3			MIT
5504.978	5506.507	18160.33	ND	1.0			MIT
5503.704	5505.233	18164.53	ND	0.5			MIT
5501.470	5502.998	18171.91	ND	1.3			MIT
5498.859	5500.387	18180.54	ND	0.7			MIT
5496.418	5497.945	18188.61	ND	1.0			MIT
5494.01	5495.54	18196.6	ND	1.2	0.0		KS
5493.339	5494.865	18198.81	ND	0.6			MIT
5492.301	5493.827	18202.25	ND	0.9			MIT
5487.026	5488.551	18219.75	ND	0.3	0.0		MIT
5485.699	5487.223	18224.15	ND	1.3	0.5		MIT
5485.101	5486.625	18226.14	ND	0.9			MIT
5484.001	5485.525	18229.80	ND	0.5			MIT
5478.61	5480.13	18247.7	ND	1.2			KN
5474.734	5476.255	18260.65	ND	0.5	0.3		MIT
5473.083	5474.604	18266.16	ND	0.8			MIT
5470.410	5471.930	18275.09	ND	0.9			MIT
5469.399	5470.919	18278.47	ND	0.3			MIT
5468.021	5469.540	18283.07	ND	0.3			MIT
5467.497	5469.016	18284.82	ND	0.3			MIT
5458.060	5459.577	18316.44	ND	0.6			MIT
5455.815	5457.331	18323.98	ND	1.5			MIT
5453.876	5455.392	18330.49	ND	0.5			MIT
5451.115	5452.630	18339.77	ND	1.4			MIT
5449.236	5450.750	18346.10	ND	1.0			MIT
5448.606	5450.120	18348.22	ND	0.3			MIT
5447.556	5449.070	18351.75	ND	0.9			MIT
5447.283	5448.797	18352.68	ND	1.1			MIT
5445.175	5446.688	18359.78	ND	0.3			MIT
5444.923	5446.436	18360.63	ND	0.6			MIT
5442.274	5443.787	18369.57	ND	1.4			MIT
5441.490	5443.002	18372.21	ND	0.3			MIT
5441.260	5442.772	18372.99	ND	0.3	0.0		MIT
5437.576	5439.087	18385.44	ND	0.3			MIT
5436.881	5438.392	18387.79	ND	0.3			MIT
5434.483	5435.994	18395.90	ND	0.6			MIT
5432.360	5433.870	18403.09	ND	0.5	0.0		MIT
5431.526	5433.036	18405.92	ND	1.4	0.3		MIT
5430.792	5432.302	18408.40	ND	0.5			MIT
5429.298	5430.807	18413.47	ND	0.7			MIT
5424.068	5425.576	18431.22	ND	0.5			MIT
5423.549	5425.057	18432.99	ND	0.5			MIT
5422.489	5423.996	18436.59	ND	0.3			MIT
5421.559	5423.066	18439.75	ND	1.2	0.0		MIT
5420.657	5422.164	18442.82	ND	0.8			MIT
5416.879	5418.385	18455.68	ND	0.3			MIT
5416.381	5417.887	18457.38	ND	1.2			MIT
5415.307	5416.812	18461.04	ND	0.7			MIT
5414.739	5416.244	18462.98	ND	0.5			MIT
5412.940	5414.445	18469.11	ND	0.3			MIT
5411.929	5413.434	18472.56	ND	0.7			MIT
5406.174	5407.677	18492.23	ND	0.3			MIT
5402.896	5404.398	18503.45	ND	0.7			MIT
5400.195	5401.696	18512.70	ND	0.7			MIT
5398.114	5399.615	18519.84	ND	0.3	0.0		MIT
5385.896	5387.394	18561.85	ND	1.2	0.0		MIT
5383.848	5385.345	18568.91	ND	0.3			MIT
5381.741	5383.237	18576.18	ND	0.3			MIT
5380.611	5382.107	18580.08	ND	0.5			MIT

AIR WAVELENGTH (ANGSTRMS)	VACUUM WAVELENGTH (ANGSTRMS)	WAVENUMBER (KAYSERS)	LMENT	LOG ARC	INTENSITY SPK	INTENSITY DIS	REF
5378.227	5379.723	18588.32	ND	0.3			MIT
5377.785	5379.280	18589.85	ND	0.9			MIT
5373.675	5375.169	18604.06	ND	0.3			MIT
5371.935	5373.429	18610.09	ND	1.3	0.0		MIT
5371.388	5372.882	18611.99	ND	0.3			MIT
5365.121	5366.613	18633.73	ND	0.3			MIT
5363.659	5365.151	18638.81	ND	0.3			MIT
5362.857	5364.348	18641.59	ND	0.3			MIT
5361.891	5363.382	18644.95	ND	0.3			MIT
5361.474	5362.965	18646.40	ND	1.4	0.0		MIT
5361.174	5362.665	18647.45	ND	0.5			MIT
5356.976	5358.466	18662.06	ND	1.2			MIT
5349.577	5351.065	18687.87	ND	0.6			MIT
5349.261	5350.749	18688.97	ND	0.3			MIT
5345.709	5347.196	18701.39	ND	0.8			MIT
5343.646	5345.132	18708.61	ND	0.5			MIT
5338.008	5339.493	18728.37	ND	0.3			MIT
5336.851	5338.335	18732.43	ND	0.3			MIT
5336.552	5338.036	18733.48	ND	0.9			MIT
5334.326	5335.810	18741.30	ND	0.3			MIT
5332.428	5333.911	18747.97	ND	0.3			MIT
5324.592	5326.073	18775.56	ND	0.3			MIT
5320.790	5322.270	18788.98	ND	0.3			MIT
5319.818	5321.298	18792.41	ND	1.8	0.3		MIT
5319.110	5320.590	18794.91	ND	0.3			MIT
5316.604	5318.083	18803.77	ND	0.7			MIT
5311.461	5312.939	18821.98	ND	1.2	0.0		MIT
5308.276	5309.753	18833.27	ND	0.3			MIT
5306.472	5307.948	18839.67	ND	0.7			MIT
5303.205	5304.680	18851.28	ND	0.7			MIT
5302.607	5304.082	18853.40	ND	0.3H			MIT
5302.279	5303.754	18854.57	ND	1.1	0.0		MIT
5300.577	5302.052	18860.62	ND	0.3			MIT
5298.876	5300.350	18866.68	ND	0.5			MIT
5293.168	5294.641	18887.02	ND	1.8	0.6		MIT
5291.665	5293.137	18892.39	ND	1.0			MIT
5288.161	5289.632	18904.91	ND	0.3			MIT
5286.683	5288.154	18910.19	ND	0.5			MIT
5284.350	5285.820	18918.54	ND	0.3			MIT
5276.879	5278.347	18945.32	ND	1.0	0.0		MIT
5273.431	5274.899	18957.71	ND	1.4	0.3		MIT
5270.694	5272.161	18967.55	ND	0.7			MIT
5269.778	5271.245	18970.85	ND	0.8	0.0		MIT
5264.224	5265.689	18990.87	ND	0.3			MIT
5255.510	5256.973	19022.35	ND	1.7	0.5		MIT
5250.816	5252.278	19039.36	ND	1.0	0.0		MIT
5249.585	5251.046	19043.82	ND	1.8	0.3		MIT
5249.294	5250.755	19044.88	ND	0.3			MIT
5239.792	5241.251	19079.42	ND	1.2			MIT
5234.195	5235.652	19099.82	ND	1.7			MIT
5228.427	5229.883	19120.89	ND	0.8			MIT
5225.045	5226.500	19133.26	ND	0.7			MIT
5221.574	5223.028	19145.98	ND	0.6			MIT
5215.654	5217.106	19167.71	ND	0.7			MIT
5213.232	5214.684	19176.62	ND	0.7			MIT
5212.365	5213.816	19179.81	ND	1.5			MIT
5209.904	5211.355	19188.87	ND	0.3			MIT
5204.381	5205.830	19209.23	ND	0.3			MIT
5200.119	5201.567	19224.98	ND	1.0			MIT
5199.727	5201.175	19226.43	ND	0.6			MIT
5198.067	5199.514	19232.56	ND	0.3			MIT
5195.601	5197.048	19241.69	ND	0.3			MIT
5192.621	5194.067	19252.74	ND	1.6			MIT
5191.448	5192.894	19257.09	ND	1.6			MIT
5187.053	5188.498	19273.40	ND	0.3			MIT
5182.603	5184.046	19289.95	ND	0.9			MIT
5181.165	5182.608	19295.30	ND	0.7			MIT
5179.781	5181.224	19300.46	ND	1.4			MIT
5178.750	5180.192	19304.30	ND	1.0			MIT
5176.791	5178.233	19311.61	ND	0.6			MIT

AIR WAVELENGTH (ANGSTRMS)	VACUUM WAVELENGTH (ANGSTRMS)	WAVENUMBER (KAYSERS)	LMENT	LOG ARC	INTENSITY SPK	INTENSITY DIS	REF
5167.923	5169.362	19344.75	ND	0.7			MIT
5166.087	5167.526	19351.62	ND	0.6			MIT
5165.140	5166.579	19355.17	ND	1.0			MIT
5161.706	5163.144	19368.04	ND	0.5			MIT
5153.449	5154.885	19399.08	ND	0.6			MIT
5151.783	5153.218	19405.35	ND	0.7			MIT
5149.563	5150.998	19413.72	ND	1.0			MIT
5143.332	5144.765	19437.23	ND	0.9	0.0		MIT
5132.330	5133.760	19478.90	ND	1.0	0.0		MIT
5130.596	5132.026	19485.48	ND	1.6	0.3		MIT
5123.785	5125.213	19511.39	ND	1.5	0.0		MIT
5107.585	5109.008	19573.27	ND	1.3	0.0		MIT
5106.637	5108.060	19576.90	ND	0.9			MIT
5105.354	5106.777	19581.82	ND	1.0			MIT
5105.211	5106.634	19582.37	ND	0.7	0.0		MIT
5103.111	5104.533	19590.43	ND	1.2			MIT
5102.393	5103.815	19593.19	ND	1.2	0.0		MIT
5096.524	5097.944	19615.75	ND	0.9			MIT
5092.797	5094.216	19630.10	ND	1.3			MIT
5089.837	5091.256	19641.52	ND	1.0			MIT
5079.085	5080.501	19683.10	ND	0.7			MIT
5077.162	5078.577	19690.55	ND	0.5			MIT
5076.586	5078.001	19692.79	ND	1.4			MIT
5071.866	5073.280	19711.11	ND	0.7			MIT
5063.726	5065.138	19742.80	ND	1.1			MIT
5056.890	5058.300	19769.49	ND	1.2			MIT
5051.064	5052.472	19792.29	ND	0.9			MIT
5040.195	5041.600	19834.97	ND	1.0			MIT
5033.517	5034.921	19861.29	ND	0.8			MIT
5029.450	5030.853	19877.35	ND	1.2			MIT
5028.500	5029.902	19881.10	ND	0.5D			MIT
5027.852	5029.254	19883.66	ND	0.6			MIT
5027.150	5028.552	19886.44	ND	0.6			MIT
5014.545	5015.944	19936.43	ND	0.7			MIT
5011.666	5013.064	19947.88	ND	0.6			MIT
5000.436	5001.831	19992.68	ND	1.0			MIT
4998.547	4999.941	20000.23	ND	1.0			MIT
4996.147	4997.541	20009.84	ND	0.3			MIT
4989.937	4991.329	20034.74	ND	1.5			MIT
4988.649	4990.041	20039.92	ND	0.3			MIT
4987.166	4988.557	20045.88	ND	1.0			MIT
4982.899	4984.289	20063.04	ND	0.3			MIT
4981.283	4982.673	20069.55	ND	1.0			MIT
4980.893	4982.283	20071.12	ND	0.3H			MIT
4975.494	4976.882	20092.90	ND	0.8			MIT
4970.922	4972.309	20111.38	ND	0.7			MIT
4969.749	4971.136	20116.13	ND	0.7			MIT
4963.334	4964.719	20142.13	ND	1.2			MIT
4961.396	4962.781	20149.99	ND	1.2			MIT
4959.130	4960.514	20159.20	ND	1.5			MIT
4958.139	4959.523	20163.23	ND	0.7			MIT
4954.783	4956.166	20176.89	ND	1.7			MIT
4952.509	4953.891	20186.15	ND	0.9			MIT
4950.719	4952.101	20193.45	ND	0.7			MIT
4950.285	4951.667	20195.22	ND	0.5			MIT
4947.023	4948.404	20208.54	ND	0.3H			MIT
4944.835	4946.215	20217.48	ND	1.7			MIT
4943.901	4945.281	20221.30	ND	1.0			MIT
4942.961	4944.341	20225.14	ND	0.7			MIT
4940.320	4941.699	20235.96	ND	0.3H			MIT
4930.729	4932.105	20275.32	ND	0.7			MIT
4924.531	4925.906	20300.83	ND	1.9			MIT
4924.255	4925.630	20301.97	ND	0.3			MIT
4922.468	4923.842	20309.34	ND	0.6			MIT
4921.157	4922.531	20314.75	ND	0.5			MIT
4920.692	4922.066	20316.67	ND	1.8			MIT
4917.385	4918.758	20330.34	ND	0.6			MIT
4914.385	4915.757	20342.75	ND	1.2			MIT
4913.422	4914.794	20346.73	ND	1.8			MIT
4910.060	4911.431	20360.67	ND	0.9			MIT

AIR WAVELENGTH (ANGSTRMS)	VACUUM WAVELENGTH (ANGSTRMS)	WAVENUMBER (KAYSERS)	LMENT	LOG ARC	INTENSITY SPK	INTENSITY DIS	REF
4907.793	4909.163	20370.07	ND	0.5			MIT
4907.276	4908.646	20372.22	ND	0.5			MIT
4902.041	4903.410	20393.97	ND	1.2			MIT
4901.850	4903.219	20394.77	ND	1.5			MIT
4901.545	4902.914	20396.03	ND	1.5			MIT
4896.934	4898.301	20415.24	ND	1.8			MIT
4893.221	4894.587	20430.73	ND	1.0D			MIT
4891.063	4892.429	20439.75	ND	1.6			MIT
4890.701	4892.067	20441.26	ND	1.5			MIT
4889.106	4890.471	20447.93	ND	0.9			MIT
4889.04	4890.41	20448.2	ND		0.5		KN
4887.333	4888.698	20455.34	ND	0.3			MIT
4885.008	4886.372	20465.08	ND	0.9			MIT
4883.815	4885.179	20470.08	ND	1.8			MIT
4882.885	4884.249	20473.98	ND	0.8			MIT
4881.709	4883.072	20478.91	ND	0.7			MIT
4880.452	4881.815	20484.19	ND	0.3			MIT
4879.792	4881.155	20486.96	ND	0.9			MIT
4876.112	4877.474	20502.42	ND	0.5			MIT
4875.837	4877.199	20503.57	ND	0.8			MIT
4875.718	4877.080	20504.07	ND	0.8			MIT
4874.362	4875.723	20509.78	ND	0.3			MIT
4869.279	4870.639	20531.19	ND	0.6			MIT
4867.839	4869.199	20537.26	ND	0.8			MIT
4866.735	4868.094	20541.92	ND	1.5			MIT
4864.201	4865.560	20552.62	ND	0.3			MIT
4863.550	4864.909	20555.37	ND	0.3H			MIT
4861.771	4863.129	20562.89	ND	0.3			MIT
4859.592	4860.949	20572.11	ND	1.0			MIT
4859.030	4860.387	20574.49	ND	1.8	1.8		MIT
4855.318	4856.674	20590.22	ND	1.0			MIT
4854.719	4856.075	20592.76	ND	0.9			MIT
4853.336	4854.692	20598.63	ND	1.0			MIT
4849.064	4850.419	20616.78	ND	1.2			MIT
4836.625	4837.976	20669.80	ND	1.4			MIT
4835.982	4837.333	20672.55	ND	1.2			MIT
4835.665	4837.016	20673.90	ND	1.0			MIT
4835.321	4836.672	20675.37	ND	0.3			MIT
4833.513	4834.864	20683.11	ND	0.3			MIT
4832.276	4833.626	20688.40	ND	1.3			MIT
4831.080	4832.430	20693.52	ND	0.7			MIT
4828.574	4829.923	20704.26	ND	1.0			MIT
4827.740	4829.089	20707.84	ND	0.3	0.0		MIT
4827.562	4828.911	20708.60	ND	0.7			MIT
4825.482	4826.830	20717.53	ND	2.0	0.9		MIT
4824.185	4825.533	20723.10	ND	1.0			MIT
4820.336	4821.683	20739.65	ND	1.6			MIT
4819.643	4820.990	20742.63	ND	1.0			MIT
4818.972	4820.319	20745.52	ND	1.1			MIT
4817.180	4818.526	20753.23	ND	1.1			MIT
4812.670	4814.015	20772.68	ND	0.5			MIT
4811.343	4812.688	20778.41	ND	1.8	1.8		MIT
4810.509	4811.853	20782.01	ND	0.3	0.3J		MIT
4806.624	4807.967	20798.81	ND	1.1			MIT
4805.103	4806.446	20805.39	ND	0.8			MIT
4803.745	4805.088	20811.27	ND	0.5			MIT
4799.423	4800.765	20830.02	ND	1.2			MIT
4798.031	4799.372	20836.06	ND	0.6			MIT
4797.157	4798.498	20839.85	ND	1.5			MIT
4796.861	4798.202	20841.14	ND	0.8			MIT
4792.623	4793.963	20859.57	ND	0.3			MIT
4789.428	4790.767	20873.49	ND	1.6			MIT
4788.339	4789.678	20878.23	ND	0.5			MIT
4787.433	4788.771	20882.18	ND	0.9			MIT
4786.457	4787.795	20886.44	ND	0.3			MIT
4786.112	4787.450	20887.95	ND	1.0			MIT
4783.828	4785.165	20897.92	ND	1.3			MIT
4781.462	4782.799	20908.26	ND	0.5			MIT
4780.537	4781.874	20912.30	ND	0.3			MIT
4779.458	4780.794	20917.03	ND	1.5			MIT

AIR WAVELENGTH (ANGSTRMS)	VACUUM WAVELENGTH (ANGSTRMS)	WAVENUMBER (KAYSERS)	LMENT	LOG ARC	INTENSITY SPK	DIS	REF
4778.401	4779.737	20921.65	ND	0.7			MIT
4777.713	4779.049	20924.67	ND	1.2D			MIT
4774.138	4775.473	20940.33	ND	1.0			MIT
4772.884	4774.218	20945.84	ND	1.0			MIT
4772.263	4773.597	20948.56	ND	0.9			MIT
4771.728	4773.062	20950.91	ND	0.6			MIT
4771.063	4772.397	20953.83	ND	0.3			MIT
4770.192	4771.526	20957.66	ND	1.3			MIT
4769.315	4770.649	20961.51	ND	0.7			MIT
4768.946	4770.279	20963.13	ND	0.3			MIT
4763.865	4765.197	20985.49	ND	1.3			MIT
4763.624	4764.956	20986.55	ND	1.0			MIT
4761.761	4763.093	20994.76	ND	0.6			MIT
4760.457	4761.788	21000.51	ND	0.3			MIT
4759.341	4760.672	21005.44	ND	0.7			MIT
4759.100	4760.431	21006.50	ND	1.3S			MIT
4758.634	4759.965	21008.56	ND	0.3			MIT
4758.505	4759.836	21009.13	ND	0.6			MIT
4757.516	4758.846	21013.50	ND	0.7			MIT
4755.852	4757.182	21020.85	ND	1.2			MIT
4752.535	4753.864	21035.52	ND	0.3			MIT
4750.020	4751.348	21046.66	ND	0.6			MIT
4749.750	4751.078	21047.85	ND	1.6			MIT
4749.556	4750.884	21048.71	ND	0.8			MIT
4749.033	4750.361	21051.03	ND	0.5			MIT
4745.380	4746.707	21067.24	ND	0.5			MIT
4744.903	4746.230	21069.35	ND	0.5			MIT
4742.392	4743.718	21080.51	ND	0.6			MIT
4739.888	4741.214	21091.65	ND	0.5			MIT
4736.203	4737.528	21108.06	ND	1.0			MIT
4734.908	4736.232	21113.83	ND	0.9			MIT
4734.791	4736.115	21114.35	ND	0.6			MIT
4733.917	4735.241	21118.25	ND	0.8			MIT
4731.783	4733.107	21127.77	ND	1.6			MIT
4727.916	4729.239	21145.05	ND	0.3			MIT
4726.556	4727.878	21151.14	ND	1.2			MIT
4724.358	4725.680	21160.98	ND	1.5	0.3		MIT
4720.910	4722.231	21176.43	ND	0.6			MIT
4719.032	4720.352	21184.86	ND	1.7			MIT
4717.081	4718.401	21193.62	ND	1.6			MIT
4715.589	4716.908	21200.33	ND	1.5	0.0		MIT
4714.228	4715.547	21206.45	ND	0.6			MIT
4713.059	4714.378	21211.71	ND	1.5	0.3		MIT
4709.714	4711.032	21226.77	ND	1.5	0.0		MIT
4709.556	4710.874	21227.49	ND	0.3			MIT
4708.370	4709.687	21232.83	ND	0.5			MIT
4708.040	4709.357	21234.32	ND	0.3			MIT
4707.580	4708.897	21236.40	ND	1.2			MIT
4706.960	4708.277	21239.19	ND	1.6			MIT
4706.542	4707.859	21241.08	ND	1.7	0.7		MIT
4703.864	4705.180	21253.17	ND	1.2			MIT
4703.576	4704.892	21254.47	ND	1.3			MIT
4703.074	4704.390	21256.74	ND	0.5			MIT
4698.297	4699.612	21278.35	ND	1.2	0.3		MIT
4697.289	4698.603	21282.92	ND	0.8			MIT
4696.445	4697.759	21286.74	ND	1.5			MIT
4693.630	4694.944	21299.51	ND	1.3	0.0		MIT
4692.968	4694.281	21302.51	ND	0.9			MIT
4690.344	4691.657	21314.43	ND	1.5			MIT
4689.588	4690.900	21317.87	ND	1.4			MIT
4688.557	4689.869	21322.56	ND	0.7	0.0		MIT
4684.890	4686.201	21339.25	ND	0.5			MIT
4684.358	4685.669	21341.67	ND	0.5			MIT
4684.039	4685.350	21343.12	ND	1.6			MIT
4683.441	4684.752	21345.85	ND	1.7			MIT
4681.765	4683.075	21353.49	ND	1.4			MIT
4681.255	4682.565	21355.82	ND	0.7			MIT
4680.734	4682.044	21358.19	ND	1.7	0.0		MIT
4678.908	4680.218	21366.53	ND	0.3H	0.0		MIT
4676.262	4677.571	21378.62	ND	0.3	0.3H		MIT

AIR WAVELENGTH (ANGSTRMS)	VACUUM WAVELENGTH (ANGSTRMS)	WAVENUMBER (KAYSERS)	LMENT	LOG ARC	INTENSITY SPK	DIS	REF
4675.523	4676.832	21382.00	ND	0.7			MIT
4674.595	4675.903	21386.24	ND	1.7	1.0		MIT
4674.184	4675.492	21388.12	ND	0.7			MIT
4673.975	4675.283	21389.08	ND	0.7			MIT
4672.687	4673.995	21394.97	ND	0.7	0.3		MIT
4672.46	4673.77	21396.0	ND	0.3H			KN
4671.094	4672.402	21402.27	ND	1.3			MIT
4670.560	4671.867	21404.72	ND	1.3			MIT
4669.133	4670.440	21411.26	ND	0.7			MIT
4664.446	4665.752	21432.77	ND	1.2			MIT
4664.012	4665.318	21434.77	ND	0.3			MIT
4662.968	4664.273	21439.57	ND	0.7			MIT
4661.281	4662.586	21447.33	ND	0.3H			MIT
4660.472	4661.777	21451.05	ND	0.5			MIT
4660.277	4661.582	21451.95	ND	1.0			MIT
4659.373	4660.677	21456.11	ND	1.0			MIT
4658.187	4659.491	21461.57	ND	0.7			MIT
4657.219	4658.523	21466.03	ND	0.6			MIT
4654.728	4656.031	21477.52	ND	1.5D			MIT
4654.291	4655.594	21479.54	ND	0.8			MIT
4653.301	4654.604	21484.11	ND	0.7			MIT
4652.678	4653.981	21486.98	ND	0.7			MIT
4652.390	4653.693	21488.31	ND	1.2			MIT
4651.563	4652.865	21492.13	ND	0.9			MIT
4651.194	4652.496	21493.84	ND	1.1			MIT
4651.017	4652.319	21494.66	ND	1.2			MIT
4650.234	4651.536	21498.27	ND	1.4			MIT
4649.669	4650.971	21500.89	ND	1.6			MIT
4647.946	4649.247	21508.86	ND	0.3			MIT
4647.759	4649.060	21509.72	ND	0.9			MIT
4646.692	4647.993	21514.66	ND	1.8	0.6		MIT
4646.399	4647.700	21516.02	ND	1.7			MIT
4645.765	4647.066	21518.96	ND	1.3	0.3		MIT
4644.979	4646.280	21522.60	ND	0.7			MIT
4644.845	4646.146	21523.22	ND	0.6	0.3W		KN
4641.102	4642.402	21540.57	ND	1.9			MIT
4640.172	4641.471	21544.89	ND	0.9			MIT
4639.593	4640.892	21547.58	ND	0.7			MIT
4639.377	4640.676	21548.58	ND	1.0			MIT
4639.141	4640.440	21549.68	ND	1.3			MIT
4638.712	4640.011	21551.67	ND	1.2	0.5		MIT
4637.206	4638.505	21558.67	ND	1.0			MIT
4636.569	4637.867	21561.63	ND	0.7	0.3		MIT
4636.288	4637.586	21562.94	ND	1.2			MIT
4634.229	4635.527	21572.52	ND	1.7			MIT
4632.688	4633.985	21579.70	ND	1.1			MIT
4631.614	4632.911	21584.70	ND	0.7			MIT
4631.295	4632.592	21586.19	ND	1.0			MIT
4630.301	4631.598	21590.82	ND	1.3			MIT
4629.903	4631.200	21592.68	ND	1.0			MIT
4629.252	4630.548	21595.71	ND	0.5			MIT
4628.503	4629.799	21599.21	ND	0.5D			MIT
4628.414	4629.710	21599.62	ND	0.5D			MIT
4627.986	4629.282	21601.62	ND	1.1			MIT
4626.497	4627.793	21608.57	ND	1.0			MIT
4624.198	4625.493	21619.32	ND	1.2			MIT
4622.214	4623.509	21628.60	ND	0.5			MIT
4621.937	4623.232	21629.89	ND	1.6			MIT
4618.111	4619.405	21647.81	ND	0.5			MIT
4618.032	4619.326	21648.18	ND	0.5			MIT
4616.68	4617.97	21654.5	ND	0.7			EX
4615.688	4616.981	21659.18	ND	1.5	0.7		MIT
4613.845	4615.137	21667.83	ND	0.6	1.4H		MIT
4612.473	4613.765	21674.27	ND	1.1			MIT
4610.509	4611.801	21683.50	ND	0.7			MIT
4609.874	4611.165	21686.49	ND	1.4			MIT
4609.809	4611.100	21686.80	ND	1.1			MIT
4609.584	4610.875	21687.86	ND	0.5			MIT
4609.230	4610.521	21689.52	ND	0.7D			MIT
4607.381	4608.672	21698.23	ND	1.4			MIT

AIR WAVELENGTH (ANGSTRMS)	VACUUM WAVELENGTH (ANGSTRMS)	WAVENUMBER (KAYSERS)	LMENT	LOG ARC	INTENSITY SPK	DIS	REF
4605.829	4607.119	21705.54	ND	0.6H			MIT
4603.823	4605.113	21715.00	ND	1.4			MIT
4602.242	4603.531	21722.46	ND	1.0			MIT
4600.522	4601.811	21730.58	ND	1.0			MIT
4599.342	4600.631	21736.15	ND	0.5			MIT
4598.432	4599.720	21740.45	ND	0.6			MIT
4597.013	4598.301	21747.16	ND	1.4	0.3		MIT
4594.675	4595.962	21758.23	ND	0.6L			MIT
4594.447	4595.734	21759.31	ND	1.0			MIT
4593.936	4595.223	21761.73	ND	0.3H			MIT
4590.505	4591.791	21777.99	ND	0.5			MIT
4586.967	4588.252	21794.79	ND	1.0			MIT
4586.614	4587.899	21796.47	ND	1.7			MIT
4584.043	4585.328	21808.69	ND	1.1	0.3		MIT
4581.202	4582.486	21822.22	ND	0.5			MIT
4579.311	4580.594	21831.23	ND	1.4	0.5		MIT
4579.158	4580.441	21831.96	ND	0.3H	0.3		MIT
4578.882	4580.165	21833.27	ND	1.5	0.3		MIT
4571.045	4572.326	21870.71	ND	0.7			MIT
4569.849	4571.130	21876.43	ND	0.9			MIT
4567.606	4568.886	21887.17	ND	1.3	0.0		MIT
4567.344	4568.624	21888.43	ND	0.8			MIT
4564.959	4566.238	21899.86	ND	0.6			MIT
4563.219	4564.498	21908.21	ND	1.7	0.6		MIT
4562.344	4563.623	21912.42	ND	0.7			MIT
4561.847	4563.126	21914.80	ND	1.3			MIT
4561.181	4562.459	21918.00	ND	1.4			MIT
4560.416	4561.694	21921.68	ND	1.7	0.5		MIT
4559.673	4560.951	21925.25	ND	1.6			MIT
4559.342	4560.620	21926.84	ND	0.6			MIT
4559.183	4560.461	21927.61	ND	0.9			MIT
4557.768	4559.046	21934.42	ND	0.6			MIT
4557.380	4558.657	21936.28	ND	1.0			MIT
4556.735	4558.012	21939.39	ND	1.2			MIT
4556.136	4557.413	21942.27	ND	1.3	0.0		MIT
4555.140	4556.417	21947.07	ND	1.2			MIT
4554.967	4556.244	21947.90	ND	0.7	0.0		MIT
4550.394	4551.670	21969.96	ND	0.5H			MIT
4549.017	4550.292	21976.61	ND	0.7			MIT
4548.239	4549.514	21980.37	ND	1.3			MIT
4545.328	4546.602	21994.45	ND	1.0			MIT
4544.257	4545.531	21999.63	ND	0.8			MIT
4542.603	4543.877	22007.64	ND	1.7	0.7		MIT
4542.050	4543.323	22010.32	ND	1.7	0.7		MIT
4541.269	4542.542	22014.10	ND	1.7	0.6		MIT
4539.423	4540.696	22023.06	ND	0.5			MIT
4539.260	4540.533	22023.85	ND	1.0	0.3		MIT
4537.569	4538.841	22032.05	ND	0.3			MIT
4535.943	4537.215	22039.95	ND	0.7			MIT
4533.799	4535.070	22050.37	ND	1.5			MIT
4532.308	4533.579	22057.63	ND	0.7			MIT
4530.328	4531.598	22067.27	ND	1.0			MIT
4529.935	4531.205	22069.18	ND	1.6			MIT
4529.764	4531.034	22070.02	ND	0.5			MIT
4529.154	4530.424	22072.99	ND	0.3			MIT
4527.243	4528.513	22082.31	ND	1.2	0.3		MIT
4526.380	4527.649	22086.52	ND	0.9			MIT
4523.569	4524.838	22100.24	ND	1.1	0.3		MIT
4522.825	4524.093	22103.88	ND	1.0	0.3		MIT
4521.245	4522.513	22111.60	ND	0.8			MIT
4520.252	4521.520	22116.46	ND	0.7			MIT
4517.779	4519.046	22128.56	ND	0.3H			MIT
4516.645	4517.912	22134.12	ND	0.3H			MIT
4516.381	4517.648	22135.41	ND	1.5			KN
4516.349	4517.616	22135.57	ND	1.6	0.5		MIT
4515.382	4516.648	22140.31	ND	0.7			MIT
4513.333	4514.599	22150.36	ND	1.4	0.3		MIT
4512.289	4513.555	22155.49	ND	0.5			MIT
4511.823	4513.088	22157.77	ND	1.7	1.0		MIT
4511.290	4512.555	22160.39	ND	1.4			MIT

AIR WAVELENGTH (ANGSTRMS)	VACUUM WAVELENGTH (ANGSTRMS)	WAVENUMBER (KAYSERS)	LMENT	LOG ARC	INTENSITY SPK	DIS	REF
4509.824	4511.089	22167.60	ND	0.8			MIT
4509.426	4510.691	22169.55	ND	0.7			MIT
4507.981	4509.245	22176.66	ND	0.9			MIT
4506.582	4507.846	22183.54	ND	1.7	0.5		MIT
4505.044	4506.308	22191.12	ND	1.3	0.3		MIT
4502.440	4503.703	22203.95	ND	0.6			MIT
4501.934	4503.197	22206.45	ND	0.6			MIT
4501.808	4503.071	22207.07	ND	1.7	0.9		MIT
4500.923	4502.186	22211.44	ND	0.7			MIT
4499.282	4500.544	22219.54	ND	1.2	0.3		MIT
4498.741	4500.003	22222.21	ND	0.3			MIT
4497.915	4499.177	22226.29	ND	1.2	0.9		MIT
4497.380	4498.642	22228.93	ND	0.6J			MIT
4497.259	4498.521	22229.53	ND	1.0	0.7		MIT
4494.758	4496.019	22241.90	ND	0.7			MIT
4493.420	4494.681	22248.52	ND	1.1	0.9		MIT
4492.474	4493.734	22253.21	ND	0.9	0.5		MIT
4492.056	4493.316	22255.28	ND	0.7			MIT
4491.644	4492.904	22257.32	ND	0.8	0.3		MIT
4490.773	4492.033	22261.64	ND	0.3			MIT
4487.618	4488.877	22277.29	ND	0.9	0.5		MIT
4485.952	4487.211	22285.56	ND	1.1	0.5		MIT
4484.948	4486.206	22290.55	ND	0.7	0.3		MIT
4481.893	4483.151	22305.74	ND	1.3			MIT
4480.970	4482.227	22310.34	ND	1.4			MIT
4478.525	4479.782	22322.52	ND	0.6			MIT
4477.885	4479.141	22325.71	ND	1.2			MIT
4477.452	4478.708	22327.87	ND	1.2	0.7		MIT
4475.834	4477.090	22335.94	ND	0.6			MIT
4475.566	4476.822	22337.27	ND	1.1	0.6		MIT
4474.718	4475.974	22341.51	ND	0.7	0.0		MIT
4471.412	4472.667	22358.03	ND	1.2	0.7		MIT
4470.965	4472.220	22360.26	ND	0.9D	0.6D		MIT
4470.689	4471.944	22361.64	ND	1.0			MIT
4469.261	4470.515	22368.79	ND	1.1	1.0		MIT
4469.084	4470.338	22369.67	ND	0.9	0.3		MIT
4467.853	4469.107	22375.84	ND	1.0	0.7		MIT
4465.834	4467.087	22385.95	ND	0.9	0.6		MIT
4465.601	4466.854	22387.12	ND	1.3	1.0		MIT
4465.075	4466.328	22389.76	ND	1.3	0.9		MIT
4462.985	4464.238	22400.24	ND	1.8	1.3		MIT
4462.407	4463.659	22403.14	ND	1.5	1.2		MIT
4460.138	4461.390	22414.54	ND	0.5	0.0		MIT
4458.934	4460.185	22420.59	ND	0.5			MIT
4458.519	4459.770	22422.68	ND	1.4	1.3		MIT
4457.179	4458.430	22429.42	ND	0.7	0.0		MIT
4456.901	4458.152	22430.82	ND	0.7	0.0		MIT
4456.394	4457.645	22433.37	ND	1.3	1.2		MIT
4456.134	4457.385	22434.68	ND	1.2	0.7		MIT
4455.623	4456.874	22437.25	ND	1.0	0.5		MIT
4451.978	4453.228	22455.62	ND	1.7	1.3		MIT
4451.566	4452.816	22457.70	ND	2.0	1.7		MIT
4451.474	4452.724	22458.17	ND	1.0	1.2		KN
4447.994	4449.243	22475.74	ND	1.0	0.3		MIT
4446.387	4447.635	22483.86	ND	2.0	1.7		MIT
4444.981	4446.229	22490.97	ND	1.3			MIT
4444.277	4445.525	22494.53	ND	1.3	0.9		MIT
4444.115	4445.363	22495.35	ND	0.7			MIT
4443.400	4444.647	22498.97	ND	1.0	0.5		MIT
4442.485	4443.732	22503.61	ND	0.8	0.8		MIT
4441.301	4442.548	22509.61	ND	0.5D			MIT
4440.882	4442.129	22511.73	ND	0.7			MIT
4439.952	4441.198	22516.45	ND	0.7			MIT
4438.999	4440.245	22521.28	ND	1.2	0.7		MIT
4438.057	4439.303	22526.06	ND	0.6			MIT
4435.095	4436.340	22541.10	ND	1.2	0.6		MIT
4432.293	4433.537	22555.35	ND	1.1	0.3		MIT
4432.218	4433.462	22555.73	ND	0.3			MIT
4431.872	4433.116	22557.50	ND	0.7			MIT
4431.725	4432.969	22558.24	ND	0.9			MIT

AIR WAVELENGTH (ANGSTRMS)	VACUUM WAVELENGTH (ANGSTRMS)	WAVENUMBER (KAYSERS)	LMENT	LOG ARC	INTENSITY SPK	INTENSITY DIS	REF
4428.992	4430.236	22572.16	ND	0.7			MIT
4426.822	4428.065	22583.23	ND	1.1	0.3		MIT
4426.103	4427.346	22586.90	ND	0.7			MIT
4424.343	4425.585	22595.88	ND	1.7	1.7		MIT
4422.439	4423.681	22605.61	ND	1.0			MIT
4416.884	4418.124	22634.04	ND	1.3	0.9		MIT
4416.563	4417.803	22635.68	ND	1.2	0.6		MIT
4415.982	4417.222	22638.66	ND	1.1	0.6		MIT
4414.432	4415.672	22646.61	ND	1.1	0.5		MIT
4413.784	4415.024	22649.94	ND	0.7	0.0		MIT
4413.326	4414.565	22652.29	ND	1.0	0.3		MIT
4412.265	4413.504	22657.73	ND	1.6	1.2		MIT
4411.052	4412.291	22663.96	ND	1.7	1.3		MIT
4410.247	4411.486	22668.10	ND	1.0	0.6		MIT
4409.520	4410.758	22671.84	ND	0.9	0.3		MIT
4408.819	4410.057	22675.44	ND	1.2	0.7		MIT
4407.072	4408.310	22684.43	ND	1.2	0.9		MIT
4406.530	4407.768	22687.22	ND	0.7			MIT
4405.906	4407.143	22690.43	ND	0.9	0.3		MIT
4403.767	4405.004	22701.46	ND	0.7			MIT
4402.470	4403.707	22708.14	ND	1.0	0.0		MIT
4400.828	4402.064	22716.62	ND	1.7	1.3		MIT
4399.579	4400.815	22723.06	ND	1.0	0.5		MIT
4398.030	4399.265	22731.07	ND	1.2	0.7		MIT
4397.670	4398.905	22732.93	ND	0.9	0.3		MIT
4397.312	4398.547	22734.78	ND	1.2	0.3		MIT
4396.370	4397.605	22739.65	ND	0.3			KN
4395.500	4396.735	22744.15	ND	1.1	0.6		MIT
4394.195	4395.429	22750.91	ND	1.2	0.7		MIT
4393.348	4394.582	22755.29	ND	1.2	0.3		MIT
4392.117	4393.351	22761.67	ND	1.0	0.6		MIT
4391.110	4392.344	22766.89	ND	1.0	0.6		MIT
4390.662	4391.895	22769.21	ND	1.3	1.2		MIT
4390.051	4391.284	22772.38	ND	1.5	0.5		MIT
4385.663	4386.895	22795.17	ND	1.6	1.3		MIT
4384.514	4385.746	22801.14	ND	0.7	0.3		MIT
4382.737	4383.968	22810.38	ND	1.2	1.0		MIT
4381.942	4383.173	22814.52	ND	0.6			MIT
4381.873	4383.104	22814.88	ND	0.6			MIT
4381.290	4382.521	22817.92	ND	1.0	0.5		MIT
4379.876	4381.107	22825.28	ND	0.6			MIT
4379.111	4380.341	22829.27	ND	0.3	0.0		KN
4377.396	4378.626	22838.22	ND	1.3	0.5		MIT
4377.099	4378.329	22839.76	ND	1.1	0.5		MIT
4376.440	4377.670	22843.20	ND	1.2	0.6		MIT
4375.039	4376.268	22850.52	ND	1.5	0.7		MIT
4374.923	4376.152	22851.12	ND	1.3	0.7		MIT
4373.657	4374.886	22857.74	ND	0.3H			MIT
4372.729	4373.958	22862.59	ND	1.1	0.8		MIT
4372.276	4373.505	22864.96	ND	1.1			MIT
4372.138	4373.367	22865.68	ND	0.3H			MIT
4371.069	4372.297	22871.27	ND	1.0	0.5		MIT
4370.756	4371.984	22872.91	ND	0.5			MIT
4370.303	4371.531	22875.28	ND	1.0	0.5		MIT
4370.109	4371.337	22876.30	ND	1.0	0.3		MIT
4369.437	4370.665	22879.81	ND	0.7			MIT
4368.632	4369.860	22884.03	ND	1.7	1.2		MIT
4367.160	4368.387	22891.74	ND	1.0	0.3		MIT
4366.391	4367.618	22895.77	ND	1.1D	0.5		MIT
4366.315	4367.542	22896.17	ND	1.1D	0.5D		MIT
4365.225	4366.452	22901.89	ND	1.0	0.3		MIT
4364.940	4366.167	22903.39	ND	0.9	0.3		MIT
4364.143	4365.369	22907.57	ND	1.2	0.9		MIT
4361.865	4363.091	22919.53	ND	0.7			MIT
4361.404	4362.630	22921.95	ND	1.0	0.5		MIT
4360.818	4362.044	22925.03	ND	1.4	0.9		MIT
4359.244	4360.469	22933.31	ND	1.2	0.5		MIT
4358.699	4359.924	22936.18	ND	1.2	0.9		MIT
4358.169	4359.394	22938.97	ND	1.7	1.3		MIT
4357.572	4358.797	22942.11	ND	0.5			MIT
4357.222	4358.447	22943.95	ND	0.9			MIT
4356.922	4358.147	22945.53	ND	1.0	0.3		MIT
4356.018	4357.242	22950.30	ND	1.5	1.2		MIT
4355.493	4356.717	22953.06	ND	0.7			MIT
4355.349	4356.573	22953.82	ND	1.1	0.6		MIT
4353.615	4354.839	22962.96	ND	0.7	0.0		MIT
4353.246	4354.470	22964.91	ND	0.9			MIT
4353.033	4354.257	22966.03	ND	0.7	0.0		MIT
4352.486	4353.709	22968.92	ND	1.0	0.3		MIT
4351.295	4352.518	22975.21	ND	1.5	1.0		MIT
4351.185	4352.408	22975.79	ND	1.2	0.7		MIT
4350.203	4351.426	22980.97	ND	1.0	0.7		MIT
4349.566	4350.789	22984.34	ND	1.2			MIT
4349.101	4350.324	22986.80	ND	1.0	0.3		MIT
4348.526	4349.748	22989.84	ND	1.4	1.2		MIT
4347.221	4348.443	22996.74	ND	0.7	0.0		MIT
4346.882	4348.104	22998.53	ND	0.8	0.3		MIT
4345.489	4346.711	23005.90	ND	0.5			MIT
4344.927	4346.148	23008.88	ND	0.8			MIT
4343.974	4345.195	23013.93	ND	0.3			MIT
4343.770	4344.991	23015.01	ND	0.9	0.5		MIT
4343.497	4344.718	23016.45	ND	1.0			MIT
4342.071	4343.292	23024.01	ND	1.3	0.8		MIT
4341.757	4342.978	23025.68	ND	0.8	0.3		MIT
4338.697	4339.917	23041.92	ND	1.6	1.2		MIT
4337.232	4338.451	23049.70	ND	1.0			MIT
4335.200	4336.419	23060.50	ND	0.6			MIT
4334.538	4335.757	23064.02	ND	0.7	0.0		MIT
4333.394	4334.612	23070.11	ND	0.7			MIT
4333.217	4334.435	23071.06	ND	0.5			MIT
4332.575	4333.793	23074.47	ND	0.8	0.3		MIT
4332.470	4333.688	23075.03	ND	0.6			MIT
4331.422	4332.640	23080.62	ND	1.1			MIT
4331.181	4332.399	23081.90	ND	0.3			KN
4330.877	4332.095	23083.52	ND	0.5			MIT
4328.161	4329.378	23098.01	ND	1.0	0.5		MIT
4327.932	4329.149	23099.23	ND	1.5	1.2		MIT
4325.766	4326.982	23110.79	ND	2.0	1.5		MIT
4323.902	4325.118	23120.76	ND	1.1	0.8		MIT
4323.279	4324.495	23124.09	ND	1.3	1.2		MIT
4322.538	4323.754	23128.05	ND	0.7	0.0		MIT
4321.829	4323.044	23131.85	ND	0.6	0.6		MIT
4319.948	4321.163	23141.92	ND	1.3			MIT
4319.336	4320.551	23145.20	ND	0.9	0.3		MIT
4317.249	4318.463	23156.39	ND	0.6D			MIT
4316.518	4317.732	23160.31	ND	0.7			MIT
4316.091	4317.305	23162.60	ND	1.0	0.5		MIT
4314.511	4315.724	23171.08	ND	1.3	1.1		MIT
4314.369	4315.582	23171.84	ND	1.2	1.0		MIT
4313.355	4314.568	23177.29	ND	1.3	1.0		MIT
4312.075	4313.288	23184.17	ND	0.7	0.0		MIT
4311.254	4312.467	23188.59	ND	1.2	0.0		MIT
4310.506	4311.718	23192.61	ND	1.3	0.8		MIT
4309.560	4310.772	23197.70	ND	0.7	0.3		MIT
4307.778	4308.990	23207.30	ND	0.7	0.8		MIT
4306.743	4307.954	23212.87	ND	1.2	1.0		MIT
4305.812	4307.023	23217.89	ND	0.7	0.3		MIT
4305.467	4306.678	23219.75	ND	0.6	0.3		MIT
4304.444	4305.655	23225.27	ND	1.3	1.1		MIT
4303.573	4304.784	23229.97	ND	2.0	1.6		MIT
4301.252	4302.462	23242.51	ND	1.2	0.9		MIT
4300.902	4302.112	23244.40	ND	1.0	0.6		MIT
4299.696	4300.906	23250.92	ND	1.2	0.7		MIT
4299.015	4300.224	23254.60	ND	1.0	0.6		MIT
4298.465	4299.674	23257.58	ND	0.7	0.3		MIT
4297.351	4298.560	23263.60	ND	1.1	0.6		MIT
4296.363	4297.572	23268.96	ND	1.1	0.5		MIT
4295.226	4296.434	23275.11	ND	1.0	0.3		MIT
4294.735	4295.943	23277.77	ND	1.2	0.3		MIT
4293.216	4294.424	23286.01	ND	0.5D			MIT

AIR WAVELENGTH (ANGSTRMS)	VACUUM WAVELENGTH (ANGSTRMS)	WAVENUMBER (KAYSERS)	LMENT	LOG ARC	INTENSITY SPK	DIS	REF
4292.895	4294.103	23287.75	ND	0.8	0.0		MIT
4291.469	4292.676	23295.49	ND	0.7			MIT
4290.957	4292.164	23298.27	ND	1.2	0.9		MIT
4289.363	4290.570	23306.93	ND	1.2	0.5H		MIT
4287.400	4288.606	23317.60	ND	1.0	0.3		MIT
4284.518	4285.724	23333.28	ND	1.4	1.3		MIT
4283.042	4284.247	23341.32	ND	1.2	0.3		MIT
4282.570	4283.775	23343.90	ND	1.0	0.9		MIT
4282.443	4283.648	23344.59	ND	1.2			MIT
4281.342	4282.547	23350.59	ND	0.7	0.0		MIT
4280.774	4281.979	23353.69	ND	1.3	1.3		MIT
4280.380	4281.584	23355.84	ND	1.4	1.2		MIT
4280.168	4281.372	23357.00	ND	0.8	0.3		MIT
4277.279	4278.483	23372.77	ND	1.3	0.7		MIT
4276.295	4277.498	23378.15	ND	0.7	0.5H		MIT
4275.762	4276.965	23381.06	ND	1.2	0.5		MIT
4275.083	4276.286	23384.78	ND	1.3	1.0		MIT
4273.739	4274.942	23392.13	ND	1.0	0.3		MIT
4272.789	4273.991	23397.33	ND	1.2	0.7		MIT
4270.565	4271.767	23409.52	ND	1.2	0.6		MIT
4268.283	4269.484	23422.03	ND	0.5D			MIT
4267.493	4268.694	23426.37	ND		2.3J		MIT
4266.716	4267.917	23430.63	ND	1.3			MIT
4263.995	4265.195	23445.59	ND	0.3			MIT
4263.908	4265.108	23446.06	ND	0.5	0.0		MIT
4263.440	4264.640	23448.64	ND	1.2	0.6		MIT
4262.239	4263.439	23455.25	ND	1.2	0.5		MIT
4261.837	4263.037	23457.46	ND	1.3	1.0		MIT
4259.608	4260.807	23469.73	ND	1.1	0.3		MIT
4258.101	4259.300	23478.04	ND	0.7			MIT
4257.783	4258.982	23479.79	ND	1.3	1.0		MIT
4257.276	4258.474	23482.59	ND	1.1	0.6		MIT
4256.820	4258.018	23485.10	ND	1.2	0.5		MIT
4256.479	4257.677	23486.99	ND	1.6	1.3		MIT
4256.239	4257.437	23488.31	ND	1.0			MIT
4255.858	4257.056	23490.41	ND	1.1	0.3		MIT
4253.868	4255.065	23501.40	ND	1.1	0.5		MIT
4252.437	4253.634	23509.31	ND	1.5	1.2		MIT
4250.219	4251.416	23521.58	ND	0.6			KN
4248.147	4249.343	23533.05	ND	1.0	0.6		MIT
4247.367	4248.563	23537.37	ND	1.7	1.3		MIT
4246.879	4248.075	23540.08	ND	1.0	0.6		MIT
4244.971	4246.166	23550.66	ND	1.0	0.3		MIT
4244.562	4245.757	23552.93	ND	1.1	0.3		MIT
4241.208	4242.402	23571.55	ND	1.1	0.6		MIT
4239.828	4241.022	23579.22	ND	1.3	1.0		MIT
4237.273	4238.466	23593.44	ND	1.0	0.3		MIT
4235.229	4236.422	23604.83	ND	1.2	0.8		MIT
4234.196	4235.388	23610.59	ND	1.3	1.0		MIT
4233.147	4234.339	23616.44	ND	1.1	0.5		MIT
4232.378	4233.570	23620.73	ND	1.6	1.2		MIT
4231.151	4232.343	23627.58	ND	1.0	0.3		MIT
4229.515	4230.706	23636.72	ND	1.0	0.3		MIT
4228.837	4230.028	23640.51	ND	1.0	0.5		MIT
4228.572	4229.763	23641.99	ND	1.0	0.3		MIT
4228.200	4229.391	23644.07	ND	1.2	0.8		MIT
4228.025	4229.216	23645.05	ND	1.1	0.7		MIT
4227.719	4228.910	23646.76	ND	1.3	0.9		MIT
4226.992	4228.182	23650.82	ND	0.8	0.7		MIT
4225.556	4226.746	23658.86	ND	0.5	0.3		MIT
4224.847	4226.037	23662.83	ND	1.2	0.7		MIT
4223.208	4224.397	23672.01	ND	1.2	0.5		MIT
4222.058	4223.247	23678.46	ND	0.7	0.3		MIT
4221.715	4222.904	23680.39	ND	0.9	0.3		MIT
4221.138	4222.327	23683.62	ND	1.2	0.9		MIT
4220.258	4221.447	23688.56	ND	1.0D	0.7D		MIT
4219.568	4220.756	23692.44	ND	0.9	0.6		MIT
4218.548	4219.736	23698.16	ND	1.0	0.6		MIT
4217.282	4218.470	23705.28	ND	1.3	0.5		MIT
4214.603	4215.790	23720.35	ND	1.0	1.0		MIT

AIR WAVELENGTH (ANGSTRMS)	VACUUM WAVELENGTH (ANGSTRMS)	WAVENUMBER (KAYSERS)	LMENT	LOG ARC	INTENSITY SPK	DIS	REF
4214.227	4215.414	23722.46	ND	1.0	0.9		MIT
4213.210	4214.397	23728.19	ND	1.0			MIT
4213.057	4214.244	23729.05	ND	0.3			MIT
4212.746	4213.933	23730.80	ND	0.3	0.0		MIT
4211.286	4212.472	23739.03	ND	1.5	1.2		MIT
4210.978	4212.164	23740.76	ND	1.4	0.7H		MIT
4209.804	4210.990	23747.39	ND	0.7	0.6		MIT
4209.116	4210.302	23751.27	ND	0.7	0.7		MIT
4208.380	4209.566	23755.42	ND	0.9	0.7		MIT
4205.595	4206.780	23771.15	ND	1.3	1.2		MIT
4205.255	4206.440	23773.07	ND	0.7	0.6		MIT
4204.351	4205.535	23778.19	ND	1.0D	0.5D		MIT
4203.434	4204.618	23783.37	ND	1.0D	0.9		MIT
4200.031	4201.214	23802.64	ND	0.8	0.6		MIT
4199.099	4200.282	23807.93	ND	1.2	1.2		MIT
4198.168	4199.351	23813.20	ND		0.6		MIT
4195.029	4196.211	23831.02	ND	1.2	1.0		MIT
4190.461	4191.642	23857.00	ND	0.7	0.5		MIT
4186.311	4187.491	23880.65	ND	1.4	0.5		MIT
4186.033	4187.213	23882.24	ND	0.9	0.6		MIT
4185.766	4186.946	23883.76	ND	1.2	0.9S		MIT
4184.980	4186.159	23888.25	ND	1.2	1.0		MIT
4183.134	4184.313	23898.79	ND	0.8	0.3		KN
4182.513	4183.692	23902.33	ND	0.9	0.9		MIT
4182.292	4183.471	23903.60	ND	0.9	0.0H		MIT
4179.585	4180.763	23919.08	ND	1.0	1.0		MIT
4178.639	4179.817	23924.49	ND	1.0	0.8		MIT
4178.532	4179.710	23925.11	ND	0.7			MIT
4178.441	4179.619	23925.63	ND		0.7		MIT
4177.321	4178.498	23932.04	ND	1.2	1.4		MIT
4175.606	4176.783	23941.87	ND	1.6	1.0		MIT
4174.701	4175.878	23947.06	ND	0.5	0.3		MIT
4174.460	4175.637	23948.44	ND	0.7	1.0		MIT
4173.379	4174.555	23954.65	ND	1.1	0.3		MIT
4170.751	4171.927	23969.74	ND	0.8	0.7		MIT
4170.453	4171.629	23971.45	ND	0.9	0.9		MIT
4168.756	4169.931	23981.21	ND	0.7	0.7		MIT
4167.999	4169.174	23985.57	ND	0.9	1.0		MIT
4167.523	4168.698	23988.31	ND	0.7	0.5		MIT
4166.560	4167.735	23993.85	ND	0.9	0.6		MIT
4165.039	4166.213	24002.61	ND	1.2	1.2		MIT
4164.854	4166.028	24003.68	ND	0.6	0.5		MIT
4164.412	4165.586	24006.23	ND	1.0	0.7		MIT
4161.458	4162.631	24023.27	ND	0.3	0.6		MIT
4160.859	4162.032	24026.73	ND	1.2	0.3		MIT
4160.565	4161.738	24028.42	ND	1.0	1.0		MIT
4159.558	4160.731	24034.24	ND	1.0	1.2		MIT
4157.575	4158.747	24045.70	ND	0.5	0.7		MIT
4156.265	4157.437	24053.28	ND	0.7	0.9		MIT
4156.083	4157.255	24054.33	ND	1.0	1.3		MIT
4153.730	4154.901	24067.96	ND	0.5	0.6		MIT
4151.679	4152.850	24079.85	ND	0.7	1.0		MIT
4150.788	4151.958	24085.02	ND	0.5	0.5		MIT
4147.138	4148.307	24106.22	ND	1.2	0.5		MIT
4146.600	4147.769	24109.34	ND	1.0	0.6		MIT
4146.132	4147.301	24112.07	ND	1.0	1.0		MIT
4144.923	4146.092	24119.10	ND	0.7D	0.0		MIT
4144.553	4145.722	24121.25	ND	1.0	1.0		MIT
4140.587	4141.755	24144.36	ND	0.3			KN
4137.989	4139.156	24159.51	ND	1.0	0.6		MIT
4136.752	4137.919	24166.74	ND	0.5	0.7		MIT
4136.232	4137.399	24169.78	ND	0.5	0.7		MIT
4135.932	4137.099	24171.53	ND	0.6	0.5		MIT
4135.793	4136.959	24172.34	ND	0.7	0.5		MIT
4135.325	4136.491	24175.08	ND	1.0	1.2		MIT
4134.713	4135.879	24178.66	ND	0.8			MIT
4133.361	4134.527	24186.56	ND	1.2	1.0		MIT
4132.554	4133.720	24191.29	ND	0.6	0.7		MIT
4131.306	4132.471	24198.59	ND	0.9	0.6		MIT
4130.722	4131.887	24202.02	ND	1.2	0.9		MIT

AIR WAVELENGTH (ANGSTRMS)	VACUUM WAVELENGTH (ANGSTRMS)	WAVENUMBER (KAYSERS)	LMENT	LOG ARC	INTENSITY SPK	DIS	REF
4129.873	4131.038	24206.99	ND	0.3			KN
4128.705	4129.870	24213.84	ND	1.0	1.0		MIT
4128.132	4129.296	24217.20	ND	0.9	0.6		MIT
4126.564	4127.728	24226.40	ND	1.0	0.3		MIT
4125.500	4126.664	24232.65	ND	0.9	0.0H		MIT
4125.048	4126.212	24235.30	ND	0.6D	0.6		KN
4124.115	4125.278	24240.79	ND	0.3			MIT
4123.881	4125.044	24242.16	ND	1.6	1.3		MIT
4123.005	4124.168	24247.31	ND	1.2	0.9		MIT
4121.936	4123.099	24253.60	ND	0.6	0.9		MIT
4120.654	4121.817	24261.15	ND	0.8	0.6		MIT
4117.541	4118.703	24279.49	ND	0.9	0.6		MIT
4116.764	4117.926	24284.07	ND	1.0	1.0		MIT
4114.458	4115.619	24297.68	ND	0.3H			KN
4113.826	4114.987	24301.42	ND	1.0D	0.7D		MIT
4113.112	4114.273	24305.63	ND	1.3H	0.3		MIT
4112.745	4113.905	24307.80	ND	1.2			MIT
4111.791	4112.951	24313.44	ND	1.2	0.5H		MIT
4110.472	4111.632	24321.24	ND	1.0	1.0		MIT
4109.455	4110.615	24327.26	ND	1.5	1.5		MIT
4109.073	4110.232	24329.52	ND	1.2	1.2		MIT
4107.955	4109.114	24336.15	ND	0.6	0.9		MIT
4107.744	4108.903	24337.40	ND	0.3H			KN
4106.582	4107.741	24344.28	ND	1.4	1.2		MIT
4105.595	4106.754	24350.13	ND	0.9	0.5		MIT
4104.536	4105.694	24356.42	ND	0.9	0.7		MIT
4104.227	4105.385	24358.25	ND	1.0	0.6		MIT
4102.563	4103.721	24368.13	ND	1.0	0.7		MIT
4102.563	4103.721	24368.13	ND	1.0	0.7		MIT
4101.682	4102.840	24373.36	ND	0.9			MIT
4101.456	4102.614	24374.71	ND	0.9	0.8		MIT
4100.240	4101.397	24381.94	ND	1.0	0.9		MIT
4098.936	4100.093	24389.69	ND	0.9	0.9		MIT
4098.513	4099.670	24392.21	ND	1.0	0.7		MIT
4098.175	4099.332	24394.22	ND	1.2	0.7		MIT
4096.705	4097.861	24402.97	ND	1.0	0.5		MIT
4096.131	4097.287	24406.39	ND	1.0	0.9		MIT
4095.778	4096.934	24408.50	ND	0.7	0.5		MIT
4095.420	4096.576	24410.63	ND	0.7D	0.7		MIT
4094.615	4095.771	24415.43	ND	0.9	0.3		MIT
4093.10	4094.26	24424.5	ND	0.3	0.3H		EX
4092.652	4093.807	24427.14	ND	0.6			MIT
4090.993	4092.148	24437.04	ND	0.7D	0.7		MIT
4089.678	4090.832	24444.90	ND	1.0	0.6		MIT
4088.790	4089.944	24450.21	ND	1.0	0.3		MIT
4088.554	4089.708	24451.62	ND	0.7	0.0		MIT
4087.471	4088.625	24458.10	ND	1.0	0.3		MIT
4085.815	4086.968	24468.01	ND	1.0	0.8		MIT
4085.15	4086.30	24472.0	ND	0.3	0.6		EX
4084.268	4085.421	24477.28	ND	0.7			MIT
4083.918	4085.071	24479.38	ND	0.6D	0.3		MIT
4082.584	4083.737	24487.38	ND	1.0	1.0		MIT
4082.429	4083.582	24488.31	ND	1.2	1.2		MIT
4080.227	4081.379	24501.52	ND	1.3	1.0		MIT
4077.620	4078.771	24517.19	ND	0.9	0.5		MIT
4077.150	4078.301	24520.01	ND	1.0	0.6		MIT
4075.559	4076.710	24529.59	ND	0.7D	0.3		MIT
4075.272	4076.423	24531.31	ND	1.1	0.6		MIT
4075.116	4076.267	24532.25	ND	1.2	1.0		MIT
4074.416	4075.566	24536.47	ND	0.9	0.5		MIT
4070.897	4072.047	24557.68	ND	0.6D	0.0		MIT
4069.267	4070.416	24567.51	ND	1.2	1.1		MIT
4068.899	4070.048	24569.74	ND	1.1	0.9		MIT
4067.727	4068.876	24576.81	ND	0.3H	0.0		MIT
4066.854	4068.002	24582.09	ND	0.3			MIT
4065.766	4066.914	24588.67	ND	0.7	0.3		MIT
4063.285	4064.433	24603.68	ND	0.7	0.5		MIT
4062.899	4064.046	24606.02	ND	0.9			MIT
4061.715	4062.862	24613.19	ND	0.7	0.5		MIT
4061.085	4062.232	24617.01	ND	1.6	1.5		MIT
4060.562	4061.709	24620.18	ND	1.0	0.7		MIT
4059.961	4061.108	24623.82	ND	1.3	1.1		MIT
4056.842	4057.988	24642.75	ND	0.9	0.9		MIT
4055.671	4056.817	24649.87	ND	1.0	0.7		MIT
4054.724	4055.869	24655.63	ND	1.1	0.6		MIT
4053.837	4054.982	24661.02	ND	1.1	0.5		MIT
4053.525	4054.670	24662.92	ND	0.7			MIT
4051.145	4052.289	24677.41	ND	1.2	1.2		MIT
4049.859	4051.003	24685.25	ND	1.0D	0.3D		MIT
4048.814	4049.958	24691.62	ND	1.2	1.0		MIT
4047.158	4048.301	24701.72	ND	1.1	1.0		MIT
4046.702	4047.845	24704.50	ND	0.9	0.3		MIT
4044.347	4045.490	24718.89	ND	0.5			KN
4043.596	4044.738	24723.48	ND	1.2	0.7		MIT
4043.046	4044.188	24726.84	ND	1.2	0.0		MIT
4042.507	4043.649	24730.14	ND	0.3			MIT
4041.065	4042.207	24738.96	ND	1.2	0.7		MIT
4040.796	4041.938	24740.61	ND	1.6	1.6		MIT
4039.944	4041.085	24745.83	ND	0.9			MIT
4039.544	4040.685	24748.28	ND	0.7	0.0		MIT
4038.124	4039.265	24756.98	ND	1.2	1.0		MIT
4037.643	4038.784	24759.93	ND	0.7	0.3		KN
4037.380	4038.521	24761.54	ND		0.7		MIT
4036.555	4037.696	24766.60	ND	1.0	0.0		MIT
4036.005	4037.145	24769.98	ND	1.0	0.5		MIT
4035.399	4036.539	24773.70	ND	0.9D	0.3		MIT
4035.168	4036.308	24775.11	ND	0.3			KN
4034.147	4035.287	24781.39	ND	1.0D	0.3		MIT
4034.012	4035.152	24782.22	ND	0.6	0.3		MIT
4033.900	4035.040	24782.90	ND	1.0	0.6		MIT
4033.504	4034.644	24785.34	ND	1.0	0.7		MIT
4032.508	4033.647	24791.46	ND	0.3			MIT
4031.807	4032.946	24795.77	ND	1.2	1.2		MIT
4031.545	4032.684	24797.38	ND	1.0	0.5		MIT
4030.470	4031.609	24803.99	ND	1.3	1.2		MIT
4029.914	4031.053	24807.42	ND	1.0	0.5		MIT
4028.754	4029.892	24814.56	ND	1.0	0.7		MIT
4026.657	4027.795	24827.48	ND	1.0	0.5		MIT
4026.084	4027.222	24831.01	ND	1.0	0.3		MIT
4024.785	4025.922	24839.03	ND	1.3	1.0		MIT
4023.824	4024.961	24844.96	ND	1.1	0.5		MIT
4023.002	4024.139	24850.04	ND	1.2	1.2		MIT
4022.032	4023.169	24856.03	ND	0.6	0.3		MIT
4021.795	4022.932	24857.49	ND	1.1	1.0		MIT
4021.330	4022.467	24860.37	ND	1.1	1.1		MIT
4020.872	4022.008	24863.20	ND	1.2	1.2		MIT
4020.062	4021.198	24868.21	ND	0.8D	0.5D		MIT
4019.809	4020.945	24869.77	ND	1.0	0.9		MIT
4018.826	4019.962	24875.86	ND	1.2	1.0		MIT
4017.311	4018.446	24885.24	ND	0.8	0.5		MIT
4017.080	4018.215	24886.67	ND	0.6	0.3		MIT
4015.557	4016.692	24896.11	ND	1.0	0.9		MIT
4013.935	4015.070	24906.17	ND	0.7	0.5		MIT
4013.359	4014.493	24909.74	ND	0.5	0.5		MIT
4013.224	4014.358	24910.58	ND	1.0	0.7		MIT
4012.704	4013.838	24913.81	ND	1.2	1.0		MIT
4012.250	4013.384	24916.63	ND	1.9	1.6		MIT
4011.069	4012.203	24923.96	ND	1.2	1.0		MIT
4010.756	4011.890	24925.91	ND	0.6			MIT
4010.454	4011.588	24927.79	ND	1.3	0.8		MIT
4009.367	4010.500	24934.54	ND	0.9	0.7		MIT
4008.754	4009.887	24938.36	ND	1.1	1.0		MIT
4008.416	4009.549	24940.46	ND	0.7	0.5		MIT
4007.435	4008.568	24946.56	ND	1.3	1.3		MIT
4006.761	4007.894	24950.76	ND	0.7D	0.3D		MIT
4006.386	4007.519	24953.10	ND	0.7	0.5		MIT
4004.830	4005.962	24962.79	ND	1.0	0.5		MIT
4004.264	4005.396	24966.32	ND	1.2	1.0		MIT
4004.010	4005.142	24967.90	ND	1.3	1.2		MIT
4003.171	4004.303	24973.14	ND	1.0	0.5		MIT

AIR WAVELENGTH (ANGSTRMS)	VACUUM WAVELENGTH (ANGSTRMS)	WAVENUMBER (KAYSERS)	LMENT	LOG ARC	INTENSITY SPK	DIS	REF
4000.562	4001.693	24989.42	ND	0.9D	0.5D		MIT
4000.493	4001.624	24989.85	ND	1.0D	0.7D		MIT
3999.383	4000.514	24996.79	ND	1.3	0.8		MIT
3998.689	3999.820	25001.13	ND	1.6	1.2		MIT
3998.155	3999.285	25004.47	ND	1.3	1.1		MIT
3997.929	3999.059	25005.88	ND	1.3	1.0		MIT
3997.133	3998.263	25010.86	ND	0.3	0.0		MIT
3995.244	3996.374	25022.69	ND	1.3			MIT
3994.684	3995.814	25026.19	ND	1.9	1.6		MIT
3994.456	3995.586	25027.62	ND	0.3			MIT
3994.099	3995.228	25029.86	ND	1.3	0.3		MIT
3992.981	3994.110	25036.87	ND	0.8D			KN
3992.574	3993.703	25039.42	ND	1.5	1.3		MIT
3992.16	3993.29	25042.0	ND	1.3	0.9		MIT
3991.743	3992.872	25044.63	ND	1.8	1.6		MIT
3991.412	3992.541	25046.71	ND		0.6		MIT
3990.103	3991.231	25054.92	ND	1.6	1.3		MIT
3990.020	3991.148	25055.45	ND	1.3	1.2		MIT
3988.812	3989.940	25063.03	ND	1.3	0.8		MIT
3987.810	3988.938	25069.33	ND	1.4	0.8		MIT
3987.426	3988.554	25071.75	ND	1.3	1.3		MIT
3986.233	3987.360	25079.25	ND	1.5	1.3		MIT
3985.034	3986.161	25086.79	ND	0.6	0.3		MIT
3984.208	3985.335	25092.00	ND	0.6D	0.6		MIT
3983.58	3984.71	25095.9	ND	1.0	0.7		MIT
3983.398	3984.525	25097.10	ND	1.3	0.9		MIT
3983.22	3984.35	25098.2	ND	0.8	0.3		MIT
3982.974	3984.101	25099.77	ND	1.2	0.8		MIT
3982.269	3983.395	25104.21	ND	0.9	0.6		MIT
3981.216	3982.342	25110.85	ND	1.5	1.2		MIT
3980.981	3982.107	25112.33	ND	1.3	0.9		MIT
3980.293	3981.419	25116.67	ND	1.3	0.9		MIT
3979.479	3980.605	25121.81	ND	1.6	1.5		MIT
3979.033	3980.158	25124.63	ND	1.3	0.9		MIT
3978.448	3979.573	25128.32	ND		0.3		MIT
3978.284	3979.409	25129.36	ND	1.0	0.8		MIT
3977.994	3979.119	25131.19	ND	1.2	0.9		MIT
3977.325	3978.450	25135.42	ND	1.1D	0.8D		MIT
3976.836	3977.961	25138.51	ND	1.6	1.5		MIT
3976.112	3977.237	25143.08	ND	0.8D	0.6D		MIT
3975.837	3976.962	25144.82	ND	1.3	1.0		MIT
3975.203	3976.327	25148.83	ND	1.6	1.5		MIT
3974.663	3975.787	25152.25	ND	1.3	0.3		MIT
3974.484	3975.608	25153.38	ND	1.3	1.3		MIT
3973.912	3975.036	25157.00	ND	0.9			MIT
3973.650	3974.774	25158.66	ND	1.5	1.3		MIT
3973.269	3974.393	25161.07	ND	1.6	1.4		MIT
3972.393	3973.517	25166.62	ND	1.3	0.9		MIT
3971.669	3972.793	25171.21	ND	1.0	0.6		MIT
3969.666	3970.789	25183.91	ND	1.3	0.6		MIT
3968.876	3969.999	25188.92	ND	1.3	0.6		MIT
3967.705	3968.828	25196.36	ND	1.3	0.8		MIT
3967.313	3968.435	25198.85	ND	1.0	0.3		MIT
3967.068	3968.190	25200.40	ND	1.5	1.0		MIT
3964.899	3966.021	25214.19	ND	1.3	0.8		MIT
3964.304	3965.426	25217.97	ND	1.2	0.8		MIT
3963.912	3965.034	25220.47	ND	1.5	1.3		MIT
3963.114	3964.235	25225.55	ND	1.5	1.4		MIT
3962.216	3963.337	25231.26	ND	1.5	1.3		MIT
3959.088	3960.208	25251.20	ND	1.0	0.6		MIT
3958.001	3959.121	25258.13	ND	1.5	1.3		MIT
3957.465	3958.585	25261.55	ND	1.8	1.6S		MIT
3956.818	3957.938	25265.68	ND	1.0	0.6		MIT
3956.362	3957.482	25268.59	ND	1.3	0.3		MIT
3955.966	3957.085	25271.12	ND	1.3	1.0		MIT
3955.562	3956.681	25273.71	ND	1.0	0.6		MIT
3955.093	3956.212	25276.70	ND	1.3	0.8		MIT
3954.69	3955.81	25279.3	ND	1.3	0.6		MIT
3954.406	3955.525	25281.09	ND	1.3	1.1		MIT
3954.209	3955.328	25282.35	ND	1.1	0.6		MIT

AIR WAVELENGTH (ANGSTRMS)	VACUUM WAVELENGTH (ANGSTRMS)	WAVENUMBER (KAYSERS)	LMENT	LOG ARC	INTENSITY SPK	DIS	REF
3953.525	3954.644	25286.73	ND	1.8	1.8		MIT
3953.395	3954.514	25287.56	ND	1.3	0.9		MIT
3952.870	3953.989	25290.92	ND	1.4	1.3		MIT
3952.195	3953.313	25295.24	ND	1.5	1.3		MIT
3951.89	3953.01	25297.2	ND	0.9			MIT
3951.154	3952.272	25301.90	ND	1.6	1.5		MIT
3950.748	3951.866	25304.50	ND	1.0	0.8		MIT
3950.417	3951.535	25306.62	ND	1.3	1.0		MIT
3949.457	3950.575	25312.77	ND	0.3			MIT
3948.778	3949.896	25317.12	ND		0.9		MIT
3948.325	3949.442	25320.03	ND	1.3	1.0		MIT
3947.608	3948.725	25324.63	ND	0.7			MIT
3946.811	3947.928	25329.74	ND	1.3	0.9		MIT
3944.421	3945.537	25345.09	ND	1.3			MIT
3942.631	3943.747	25356.60	ND	1.4	1.4		MIT
3942.144	3943.260	25359.73	ND	1.2	0.6		MIT
3941.512	3942.628	25363.79	ND	1.8	1.5		MIT
3939.835	3940.950	25374.59	ND	1.3	0.6		MIT
3939.548	3940.663	25376.44	ND	1.3	0.5		MIT
3938.874	3939.989	25380.78	ND	1.5	1.4		MIT
3937.575	3938.690	25389.15	ND	1.3	0.8		MIT
3936.991	3938.106	25392.92	ND	1.2D	0.8D		MIT
3936.721	3937.835	25394.66	ND	1.3	0.8		MIT
3936.136	3937.250	25398.44	ND	1.5	1.3		MIT
3935.917	3937.031	25399.85	ND	0.7	0.0		KN
3934.823	3935.937	25406.91	ND	1.8	1.5		MIT
3934.107	3935.221	25411.53	ND	0.7			KN
3934.093	3935.207	25411.62	ND	1.3			MIT
3931.227	3932.340	25430.15	ND	0.8	0.3H		MIT
3930.506	3931.619	25434.82	ND	1.3	1.5		MIT
3929.953	3931.066	25438.40	ND	1.0	0.8		MIT
3929.260	3930.373	25442.88	ND	1.5	1.2		MIT
3927.106	3928.218	25456.84	ND	1.9	1.7S		MIT
3926.608	3927.720	25460.06	ND	1.2	1.1		MIT
3924.984	3926.095	25470.60	ND	1.3	1.3		MIT
3924.485	3925.596	25473.84	ND	1.3	0.9		MIT
3923.110	3924.221	25482.76	ND	0.5	0.0		MIT
3920.965	3922.075	25496.71	ND	1.6	1.2		MIT
3919.923	3921.033	25503.48	ND	1.3D	0.8D		MIT
3918.90	3920.01	25510.1	ND	1.3	0.6		MIT
3918.264	3919.374	25514.28	ND	1.0	0.8		MIT
3918.046	3919.156	25515.70	ND	1.6	0.8		MIT
3917.648	3918.757	25518.29	ND	1.3	1.2		MIT
3915.948	3917.057	25529.37	ND	1.4	1.3		MIT
3915.131	3916.240	25534.70	ND	1.3	1.1		MIT
3914.940	3916.049	25535.94	ND	0.3			MIT
3913.690	3914.798	25544.10	ND	0.9	0.3H		MIT
3913.552	3914.660	25545.00	ND	0.6			MIT
3912.228	3913.336	25553.65	ND	1.3	1.3		MIT
3911.909	3913.017	25555.73	ND	1.1	0.8		MIT
3911.169	3912.277	25560.56	ND	1.4	1.4		MIT
3909.276	3910.383	25572.94	ND	1.2	1.0		MIT
3907.843	3908.950	25582.32	ND	1.3	1.1		MIT
3906.096	3907.202	25593.76	ND	1.2	1.0		MIT
3905.886	3906.992	25595.14	ND	1.6	1.5		MIT
3905.555	3906.661	25597.30	ND	1.0	1.0		MIT
3903.510	3904.616	25610.71	ND	1.0	0.3		MIT
3902.453	3903.559	25617.65	ND	1.0	0.9		MIT
3901.850	3902.955	25621.61	ND	1.5	1.5		MIT
3900.226	3901.331	25632.28	ND	1.5	1.5		MIT
3897.638	3898.742	25649.30	ND	1.3	1.0		MIT
3896.143	3897.247	25659.14	ND	1.3	1.0		MIT
3895.905	3897.009	25660.71	ND	0.9	0.6		MIT
3895.376	3896.480	25664.19	ND	1.0	1.0		MIT
3895.027	3896.131	25666.49	ND		0.7		MIT
3894.627	3895.730	25669.13	ND	1.3	1.3		MIT
3893.703	3894.806	25675.22	ND	1.1	0.6		MIT
3892.065	3893.168	25686.02	ND	1.3	1.3		MIT
3891.512	3892.615	25689.67	ND	1.3	1.2		MIT
3890.940	3892.043	25693.45	ND	1.3	1.3		MIT

AIR WAVELENGTH (ANGSTRMS)	VACUUM WAVELENGTH (ANGSTRMS)	WAVENUMBER (KAYSERS)	LMENT	LOG ARC	INTENSITY SPK	DIS	REF
3890.580	3891.682	25695.83	ND	1.5	1.4		MIT
3890.221	3891.323	25698.20	ND	1.1	0.9		MIT
3889.929	3891.031	25700.13	ND	1.5	1.3		MIT
3889.664	3890.766	25701.88	ND	1.5	1.4		MIT
3889.218	3890.320	25704.83	ND	1.0	1.2		MIT
3887.866	3888.968	25713.76	ND	1.4	1.3		MIT
3886.049	3887.150	25725.79	ND	1.0	1.0		MIT
3884.741	3885.842	25734.45	ND	0.9	0.6		MIT
3884.076	3885.177	25738.85	ND	0.8	0.9		MIT
3883.757	3884.858	25740.97	ND	0.8	1.3		MIT
3880.779	3881.879	25760.72	ND		1.3		MIT
3880.376	3881.476	25763.40	ND	1.4	0.6		MIT
3879.543	3880.643	25768.93	ND	1.5	0.7		MIT
3878.582	3879.681	25775.31	ND	1.5	0.3		KN
3876.728	3877.827	25787.64	ND	0.5			MIT
3875.866	3876.965	25793.37	ND	1.5	0.6		MIT
3875.738	3876.837	25794.23	ND	1.2	0.8		MIT
3869.045	3870.142	25838.85	ND	1.3	0.6		MIT
3866.796	3867.892	25853.87	ND	1.2	0.3		MIT
3866.524	3867.620	25855.69	ND	1.0	0.3		MIT
3865.985	3867.081	25859.30	ND	1.5	1.5		MIT
3864.896	3865.992	25866.58	ND	0.6			KN
3863.409	3864.504	25876.54	ND	1.3	1.3		MIT
3863.327	3864.422	25877.09	ND	1.0	0.6		MIT
3862.487	3863.582	25882.72	ND	1.3	1.1		MIT
3860.942	3862.037	25893.07	ND	0.6	0.6		MIT
3859.670	3860.764	25901.61	ND	1.4	1.4		MIT
3859.422	3860.516	25903.27	ND	1.0	1.1		MIT
3858.55	3859.64	25909.1	ND	0.5	0.5		MIT
3857.85	3858.94	25913.8	ND	1.3	1.3		MIT
3857.14	3858.23	25918.6	ND	1.5	1.3H		MIT
3856.95	3858.04	25919.9	ND	1.3	1.2		MIT
3853.48	3854.57	25943.2	ND	1.6	1.3		MIT
3852.90	3853.99	25947.1	ND	1.0	0.9		MIT
3852.38	3853.47	25950.6	ND	1.8	1.7		MIT
3851.748	3852.840	25954.88	ND	0.9	1.2		MIT
3851.657	3852.749	25955.49	ND	1.3			MIT
3850.227	3851.319	25965.13	ND	1.0	0.3		MIT
3848.524	3849.615	25976.62	ND	1.0	1.3		MIT
3848.309	3849.400	25978.07	ND	1.6			KN
3848.233	3849.324	25978.58	ND	1.0D	1.0D		MIT
3847.85	3848.94	25981.2	ND	1.8	1.7		MIT
3847.25	3848.34	25985.2	ND	1.3	1.0		MIT
3846.969	3848.060	25987.12	ND	1.3	1.3		MIT
3846.71	3847.80	25988.9	ND	1.5D	1.5D		MIT
3845.994	3847.085	25993.71	ND	1.5	1.4		MIT
3843.51	3844.60	26010.5	ND		1.3		MIT
3842.988	3844.078	26014.04	ND	1.5	1.6		MIT
3842.695	3843.785	26016.02	ND	1.5	1.1		KN
3842.432	3843.522	26017.80	ND	1.0	1.0		MIT
3841.95	3843.04	26021.1	ND	1.5	1.5		MIT
3839.81	3840.90	26035.6	ND	1.5	1.3		MIT
3839.497	3840.586	26037.69	ND	1.0	1.1		MIT
3838.981	3840.070	26041.19	ND	1.3	1.5		MIT
3838.724	3839.813	26042.94	ND	1.7	1.4		MIT
3838.33	3839.42	26045.6	ND	1.6	1.4		MIT
3837.909	3838.998	26048.47	ND	1.0	0.3		KN
3836.541	3837.629	26057.75	ND	1.9	2.0		MIT
3836.108	3837.196	26060.70	ND	1.6	1.2		MIT
3834.404	3835.492	26072.28	ND	0.3			KN
3833.604	3834.692	26077.72	ND	1.3	1.1		MIT
3833.035	3834.122	26081.59	ND	1.8	1.5		MIT
3831.030	3832.117	26095.24	ND	1.8D	1.5		MIT
3830.915	3832.002	26096.02	ND	0.5			KN
3830.476	3831.563	26099.01	ND	1.0	1.0		MIT
3830.000	3831.087	26102.25	ND	1.3	0.9		KN
3829.63	3830.72	26104.8	ND	1.6	1.5		MIT
3829.41	3830.50	26106.3	ND	1.7	1.6		MIT
3829.151	3830.237	26108.04	ND	1.0	1.2		MIT
3828.842	3829.928	26110.15	ND	1.3	1.3		MIT
3828.17	3829.26	26114.7	ND	1.7	1.6		MIT
3827.986	3829.072	26115.99	ND	0.8	0.9		MIT
3826.91	3828.00	26123.3	ND	1.6W	1.4W		MIT
3826.416	3827.502	26126.70	ND	1.0	1.3		MIT
3825.27	3826.36	26134.5	ND	1.3D	1.0D		MIT
3824.789	3825.874	26137.82	ND	1.6	1.5		MIT
3824.08	3825.17	26142.7	ND	0.8D	0.6D		MIT
3823.80	3824.89	26144.6	ND	1.3D	1.0D		MIT
3823.255	3824.340	26148.30	ND	0.6	0.8		MIT
3822.474	3823.559	26153.65	ND	0.6	1.0		KN
3822.29	3823.37	26154.9	ND	1.3D	1.0D		MIT
3821.77	3822.85	26158.5	ND	1.7W	1.6W		MIT
3821.132	3822.216	26162.83	ND	0.3			KN
3820.866	3821.950	26164.65	ND	0.6	0.3		KN
3820.431	3821.515	26167.63	ND		1.6		MIT
3819.965	3821.049	26170.82	ND	1.2	0.6		MIT
3819.703	3820.787	26172.62	ND	0.3	0.3		MIT
3818.837	3819.921	26178.55	ND	0.6	0.8		MIT
3817.841	3818.924	26185.38	ND	1.3	1.0		MIT
3817.675	3818.758	26186.52	ND	1.6	1.6		MIT
3817.373	3818.456	26188.59	ND	1.5	1.4		MIT
3815.339	3816.422	26202.55	ND	0.8	0.6		MIT
3814.890	3815.973	26205.64	ND	1.0D	0.8D		MIT
3814.725	3815.808	26206.77	ND	0.8	0.9		MIT
3812.852	3813.934	26219.65	ND	0.3	0.0		KN
3812.533	3813.615	26221.84	ND	1.0	1.2		MIT
3812.053	3813.135	26225.14	ND	1.3D	1.3D		MIT
3811.774	3812.856	26227.06	ND	0.9	1.0		MIT
3811.34	3812.42	26230.1	ND	1.3D	1.0D		MIT
3811.073	3812.155	26231.88	ND	1.0	1.0		MIT
3810.480	3811.562	26235.97	ND	1.3D	1.3		MIT
3809.90	3810.98	26240.0	ND	1.3	1.3		MIT
3809.046	3810.127	26245.84	ND	1.2	1.3D		MIT
3808.772	3809.853	26247.73	ND	1.3	1.1		MIT
3808.251	3809.332	26251.32	ND	0.6	0.3		MIT
3807.935	3809.016	26253.50	ND	1.5	1.3		MIT
3807.803	3808.884	26254.41	ND	0.6	0.3		MIT
3807.227	3808.308	26258.38	ND	1.2	1.3		MIT
3806.540	3807.621	26263.12	ND	1.3	1.0D		KN
3805.547	3806.627	26269.97	ND	0.9	0.9		MIT
3805.359	3806.439	26271.27	ND	1.7	1.5		MIT
3804.768	3805.848	26275.35	ND	1.0	1.0		MIT
3804.102	3805.182	26279.95	ND	1.2D	1.0D		MIT
3803.474	3804.554	26284.29	ND	1.3	1.3		MIT
3801.373	3802.452	26298.82	ND	1.8	1.6		MIT
3801.118	3802.197	26300.58	ND	1.2	1.2		MIT
3800.619	3801.698	26304.04	ND	1.3	0.8		MIT
3800.026	3801.105	26308.14	ND	1.5D	1.3		MIT
3799.236	3800.315	26313.61	ND	1.0D	1.0D		MIT
3797.89	3798.97	26322.9	ND	1.3D	1.0D		MIT
3796.892	3797.970	26329.85	ND	1.3D	1.0D		MIT
3796.038	3797.116	26335.78	ND	1.2	0.0		MIT
3795.801	3796.879	26337.42	ND	1.3	1.2		MIT
3795.454	3796.532	26339.83	ND	0.8	1.0		MIT
3794.689	3795.766	26345.14	ND	1.0	0.3		MIT
3793.640	3794.717	26352.43	ND	1.3D	0.8D		KN
3792.55	3793.63	26360.0	ND	1.3D	1.0D		MIT
3792.317	3793.394	26361.62	ND	0.8	0.0		MIT
3791.503	3792.580	26367.28	ND	1.3	1.0		MIT
3791.243	3792.320	26369.09	ND	1.0	0.6		KN
3790.840	3791.916	26371.89	ND	1.3D	1.0D		MIT
3790.644	3791.720	26373.25	ND	1.0			MIT
3788.967	3790.043	26384.93	ND	1.2D	1.0D		MIT
3788.48	3789.56	26388.3	ND	1.3D	0.8D		MIT
3787.644	3788.720	26394.14	ND	0.8	0.6		MIT
3787.519	3788.595	26395.01	ND	1.0	0.8		MIT
3787.158	3788.234	26397.53	ND	1.8D	1.7		MIT
3786.895	3787.970	26399.36	ND	1.0	0.8		MIT
3785.992	3787.067	26405.66	ND	1.3	0.6		KN
3785.094	3786.169	26411.92	ND	1.0D	1.0		MIT

AIR WAVELENGTH (ANGSTRMS)	VACUUM WAVELENGTH (ANGSTRMS)	WAVENUMBER (KAYSERS)	LMENT	LOG ARC	INTENSITY SPK	DIS	REF
3784.850	3785.925	26413.62	ND	1.3	1.3		MIT
3784.250	3785.325	26417.81	ND	1.4	1.4		MIT
3783.777	3784.852	26421.11	ND	1.3D	1.3D		MIT
3783.070	3784.144	26426.05	ND	1.4D	1.0		MIT
3782.310	3783.384	26431.36	ND	0.9D	0.6D		MIT
3781.771	3782.845	26435.13	ND	0.3			MIT
3781.315	3782.389	26438.32	ND	1.3	1.3		MIT
3780.922	3781.996	26441.06	ND		1.3		MIT
3780.391	3781.465	26444.78	ND	1.3D	1.2		MIT
3779.767	3780.841	26449.15	ND	1.3D	1.0D		MIT
3779.466	3780.540	26451.25	ND	1.2	1.2		MIT
3778.826	3779.899	26455.73	ND	1.0D	0.9D		KN
3778.140	3779.213	26460.53	ND	1.6	1.6		MIT
3777.41	3778.48	26465.6	ND	1.0	0.3		MIT
3776.338	3777.411	26473.16	ND	1.2	1.1		MIT
3775.501	3776.573	26479.03	ND	1.0	1.3		MIT
3775.24	3776.31	26480.9	ND	1.2D	0.9D		MIT
3774.928	3776.000	26483.05	ND	1.0	0.8		MIT
3774.322	3775.394	26487.30	ND	1.3	1.5		MIT
3774.08	3775.15	26489.0	ND	1.3D	1.0D		MIT
3773.335	3774.407	26494.23	ND	0.6			MIT
3773.175	3774.247	26495.35	ND	1.0D	1.0		MIT
3772.932	3774.004	26497.06	ND	1.3D	1.0		MIT
3772.402	3773.474	26500.78	ND	1.0	0.8		KN
3772.138	3773.210	26502.64	ND	1.3D	1.0D		KN
3772.056	3773.128	26503.21	ND	0.7	0.0		MIT
3771.799	3772.871	26505.02	ND	1.2D	0.8D		MIT
3771.325	3772.396	26508.35	ND	1.0D	0.8D		MIT
3771.103	3772.174	26509.91	ND	1.0	0.9		MIT
3769.816	3770.887	26518.96	ND	1.3	0.6		MIT
3769.644	3770.715	26520.17	ND	2.0	1.3		MIT
3769.343	3770.414	26522.29	ND	1.1	0.8		MIT
3767.759	3768.829	26533.44	ND	1.2	1.3		MIT
3766.590	3767.660	26541.67	ND	1.0	0.3		MIT
3765.924	3766.994	26546.37	ND	0.8	0.6		MIT
3765.344	3766.414	26550.46	ND	1.0	0.6		KN
3764.382	3765.452	26557.24	ND	1.0D	0.6		MIT
3763.475	3764.544	26563.64	ND	1.3	1.3		MIT
3762.594	3763.663	26569.86	ND	1.2	1.3		MIT
3762.361	3763.430	26571.50	ND	1.0	0.6		MIT
3761.575	3762.644	26577.06	ND	1.3	1.0		MIT
3760.942	3762.011	26581.53	ND	1.0D	0.6D		MIT
3760.360	3761.429	26585.64	ND	1.0D	0.8D		MIT
3759.795	3760.863	26589.64	ND	1.0	1.1		MIT
3758.944	3760.012	26595.66	ND	1.3	1.5		MIT
3757.944	3759.012	26602.74	ND	0.8	0.3		KN
3757.820	3758.888	26603.61	ND	1.2	1.1		MIT
3756.580	3757.648	26612.39	ND	1.2	1.1		MIT
3756.094	3757.161	26615.84	ND	0.6H	0.3H		KN
3755.602	3756.669	26619.32	ND	1.2	1.1		MIT
3754.830	3755.897	26624.80	ND	1.3	1.3		MIT
3754.353	3755.420	26628.18	ND	0.9D	0.6D		KN
3753.916	3754.983	26631.28	ND	0.6	0.9		MIT
3752.679	3753.746	26640.06	ND	1.0	1.1		MIT
3752.496	3753.563	26641.36	ND	1.0	1.2		MIT
3752.287	3753.353	26642.84	ND	1.0D	0.6D		KN
3751.62	3752.69	26647.6	ND	1.0W	0.9W		MIT
3750.735	3751.801	26653.86	ND	1.0	0.8		MIT
3750.307	3751.373	26656.91	ND	0.9	1.0		MIT
3749.853	3750.919	26660.13	ND	0.8	0.9		KN
3746.128	3747.193	26686.64	ND	0.3			KN
3744.804	3745.869	26696.08	ND	0.9	0.6		MIT
3742.156	3743.220	26714.97	ND	0.8	0.6		MIT
3741.427	3742.491	26720.17	ND	1.3	1.2		MIT
3740.964	3742.028	26723.48	ND	0.8	0.9		MIT
3738.057	3739.120	26744.26	ND	1.4			MIT
3736.44	3737.50	26755.8	ND		0.6		MIT
3735.599	3736.661	26761.86	ND	1.0	1.7		KN
3735.538	3736.600	26762.30	ND	1.0			MIT
3735.224	3736.286	26764.55	ND	0.3			MIT

AIR WAVELENGTH (ANGSTRMS)	VACUUM WAVELENGTH (ANGSTRMS)	WAVENUMBER (KAYSERS)	LMENT	LOG ARC	INTENSITY SPK	DIS	REF
3734.862	3735.924	26767.14	ND	0.8	0.6		MIT
3733.054	3734.115	26780.10	ND	0.3	0.3H		MIT
3732.761	3733.822	26782.21	ND	0.8	0.9		MIT
3731.218	3732.279	26793.28	ND	0.8	0.0		MIT
3730.578	3731.639	26797.88	ND	0.9	1.0		MIT
3729.501	3730.562	26805.62	ND	0.9	0.6		MIT
3729.137	3730.197	26808.23	ND	1.0D	0.6D		MIT
3728.130	3729.190	26815.47	ND	0.8	0.9		MIT
3726.345	3727.405	26828.32	ND	1.0	0.3		MIT
3724.877	3725.936	26838.89	ND	1.2	0.3		MIT
3723.894	3724.953	26845.98	ND	0.9			MIT
3723.506	3724.565	26848.77	ND	1.5	1.3		MIT
3721.746	3722.805	26861.47	ND	0.6	0.0		MIT
3721.330	3722.388	26864.47	ND	1.2	1.2		MIT
3720.54	3721.60	26870.2	ND	0.9	0.9		MIT
3719.595	3720.653	26877.00	ND	1.0	0.9		KN
3717.482	3718.539	26892.28	ND	0.3			KN
3716.577	3717.634	26898.83	ND	0.9	0.8		MIT
3715.680	3716.737	26905.32	ND	1.3	1.3		MIT
3715.393	3716.450	26907.40	ND	1.1	0.9		MIT
3715.047	3716.104	26909.91	ND	1.0	1.0		MIT
3714.808	3715.865	26911.64	ND	0.9D	0.8D		MIT
3714.727	3715.784	26912.22	ND	0.9D	0.8D		MIT
3714.194	3715.251	26916.08	ND	1.3	1.2		MIT
3713.695	3714.751	26919.70	ND	1.4	1.3		MIT
3712.369	3713.425	26929.32	ND	1.0	0.6		MIT
3712.120	3713.176	26931.12	ND	1.4	1.0		MIT
3711.545	3712.601	26935.29	ND	0.3	0.0		KN
3707.735	3708.790	26962.97	ND	0.9	0.6		MIT
3707.174	3708.229	26967.05	ND	0.3			MIT
3703.145	3704.199	26996.39	ND	1.5	0.8		MIT
3701.981	3703.034	27004.88	ND	0.3	0.0		MIT
3701.751	3702.804	27006.56	ND	1.0	0.6H		MIT
3701.487	3702.540	27008.48	ND	0.9	0.8		MIT
3700.925	3701.978	27012.58	ND	1.3	1.2		MIT
3700.65	3701.70	27014.6	ND	0.9	0.3		MIT
3700.444	3701.497	27016.10	ND	0.3	0.0		MIT
3699.740	3700.793	27021.24	ND	1.3	1.0		MIT
3699.404	3700.457	27023.69	ND	1.0	0.3		KN
3697.536	3698.588	27037.34	ND	1.3	1.0		MIT
3697.165	3698.217	27040.06	ND	1.3	1.0		MIT
3694.792	3695.844	27057.42	ND	1.5	1.3		MIT
3693.133	3694.184	27069.58	ND	1.3	0.9		MIT
3692.768	3693.819	27072.25	ND	1.1	0.9		MIT
3690.543	3691.593	27088.57	ND	0.9	0.6		MIT
3690.086	3691.136	27091.93	ND	0.9			MIT
3689.692	3690.742	27094.82	ND	1.4	1.2		MIT
3689.166	3690.216	27098.68	ND	0.8	0.6		MIT
3688.331	3689.381	27104.82	ND	1.0	0.9		MIT
3688.050	3689.100	27106.88	ND	1.2	0.8		MIT
3687.294	3688.344	27112.44	ND	1.3	1.0		MIT
3685.804	3686.853	27123.40	ND	1.5	1.3		KN
3684.909	3685.958	27129.99	ND	1.0	1.0		MIT
3684.64	3685.69	27132.0	ND	0.9	0.0		MIT
3684.293	3685.342	27134.52	ND	1.0	0.8		MIT
3684.131	3685.180	27135.72	ND	0.5			KN
3681.873	3682.921	27152.36	ND	0.8	0.0		MIT
3679.404	3680.452	27170.58	ND	0.6H	0.3H		KN
3678.189	3679.236	27179.55	ND	1.0	1.1		MIT
3674.645	3675.691	27205.77	ND	1.2	0.8		MIT
3674.055	3675.101	27210.14	ND	1.5	1.0		MIT
3673.542	3674.588	27213.94	ND	1.3	1.3		KN
3672.363	3673.409	27222.67	ND	1.5	1.1		KN
3671.660	3672.706	27227.88	ND	1.0	0.8		MIT
3671.449	3672.495	27229.45	ND	1.0	0.8		MIT
3670.915	3671.960	27233.41	ND	0.6			MIT
3670.51	3671.56	27236.4	ND	1.0	0.6		MIT
3670.022	3671.067	27240.03	ND	0.6	0.3		MIT
3668.789	3669.834	27249.19	ND	1.4	0.6		MIT
3666.565	3667.609	27265.72	ND	1.0	0.6		MIT

AIR WAVELENGTH (ANGSTRMS)	VACUUM WAVELENGTH (ANGSTRMS)	WAVENUMBER (KAYSERS)	LMENT	LOG ARC	INTENSITY SPK	INTENSITY DIS	REF
3665.751	3666.795	27271.77	ND	1.0	0.6		MIT
3665.180	3666.224	27276.02	ND	1.4	1.1		MIT
3664.649	3665.693	27279.97	ND	1.3	1.0		KN
3663.033	3664.076	27292.01	ND	0.8	0.3		MIT
3662.263	3663.306	27297.75	ND	1.5	1.5		KN
3661.691	3662.734	27302.01	ND	1.0	0.6		MIT
3661.345	3662.388	27304.59	ND	1.6	1.4		MIT
3660.974	3662.017	27307.36	ND	1.0	0.3		MIT
3659.948	3660.991	27315.01	ND	1.2	0.6		MIT
3658.104	3659.146	27328.78	ND	0.3			MIT
3657.991	3659.033	27329.62	ND	0.6H	0.3H		MIT
3655.963	3657.005	27344.78	ND	1.0	0.6		MIT
3655.031	3656.072	27351.76	ND	0.7	0.0		KN
3654.156	3655.197	27358.31	ND	1.0	1.0		MIT
3653.150	3654.191	27365.84	ND	1.0	0.6		KN
3651.592	3652.632	27377.51	ND	1.0	0.8H		MIT
3650.694	3651.734	27384.25	ND	1.2	0.8		MIT
3650.415	3651.455	27386.34	ND	1.1	0.8		MIT
3648.188	3649.228	27403.06	ND	1.1	0.9		MIT
3645.776	3646.815	27421.19	ND	0.9	0.0		MIT
3645.626	3646.665	27422.32	ND	1.2	0.8		MIT
3645.167	3646.206	27425.77	ND	0.8	0.3		MIT
3644.66	3645.70	27429.6	ND	1.0	0.6		MIT
3644.426	3645.465	27431.35	ND		1.2		MIT
3643.630	3644.668	27437.34	ND	0.7	0.0		MIT
3642.46	3643.50	27446.1	ND	1.3	0.8		MIT
3642.205	3643.243	27448.07	ND	0.3			MIT
3641.502	3642.540	27453.37	ND	1.3	1.0		MIT
3641.114	3642.152	27456.30	ND	0.8	0.6		MIT
3640.235	3641.272	27462.93	ND	1.2	1.0		MIT
3639.175	3640.212	27470.93	ND	0.6D	0.0		MIT
3637.783	3638.820	27481.44	ND	1.0	0.9		MIT
3637.229	3638.266	27485.62	ND	1.3	1.1		MIT
3636.990	3638.027	27487.43	ND	1.3	1.1		MIT
3635.106	3636.142	27501.68	ND	0.8	0.8		MIT
3634.871	3635.907	27503.45	ND	1.5	1.2		MIT
3634.282	3635.318	27507.91	ND	1.4	1.5		MIT
3633.469	3634.505	27514.06	ND	1.1	0.6		MIT
3631.020	3632.055	27532.62	ND	1.0	0.9		MIT
3629.932	3630.967	27540.87	ND	1.0	0.8		MIT
3628.445	3629.479	27552.16	ND	0.8	0.6		KN
3627.41	3628.44	27560.0	ND	1.3	1.2H		MIT
3626.859	3627.893	27564.21	ND	1.0	0.0H		MIT
3624.654	3625.687	27580.98	ND	1.3	0.8		KN
3624.319	3625.352	27583.53	ND	0.8	0.3		KN
3620.74	3621.77	27610.8	ND	1.0	0.6		MIT
3618.969	3620.001	27624.30	ND	1.3	1.2		MIT
3616.325	3617.356	27644.50	ND	1.2	0.0H		MIT
3615.817	3616.848	27648.38	ND	1.3	1.0		MIT
3613.244	3614.275	27668.07	ND	1.0	0.6		MIT
3612.44	3613.47	27674.2	ND	0.8	1.3		MIT
3609.788	3610.818	27694.56	ND	1.2	1.0		MIT
3609.495	3610.525	27696.81	ND	1.4	1.5		MIT
3607.722	3608.751	27710.42	ND	1.0	0.8		MIT
3606.401	3607.430	27720.57	ND	1.1	0.3		MIT
3606.285	3607.314	27721.46	ND	1.1	0.6		MIT
3604.04	3605.07	27738.7	ND		0.8		MIT
3601.321	3602.348	27759.67	ND	1.0	0.3		MIT
3600.915	3601.942	27762.80	ND	1.3	1.0		MIT
3600.118	3601.145	27768.95	ND	1.2	1.0		MIT
3599.070	3600.097	27777.03	ND	1.0			MIT
3598.026	3599.053	27785.09	ND	1.0	0.8		MIT
3592.595	3593.620	27827.09	ND	1.3	1.5		MIT
3592.073	3593.098	27831.14	ND	0.9	0.9		MIT
3587.504	3588.528	27866.58	ND	0.9	1.3		MIT
3576.156	3577.177	27955.00	ND	1.2	0.8		MIT
3574.977	3575.998	27964.22	ND	1.5	0.9		MIT
3574.337	3575.357	27969.23	ND	1.3	0.9		MIT
3573.183	3574.203	27978.26	ND	1.0	0.6		MIT
3572.751	3573.771	27981.65	ND	1.0	1.0		MIT

AIR WAVELENGTH (ANGSTRMS)	VACUUM WAVELENGTH (ANGSTRMS)	WAVENUMBER (KAYSERS)	LMENT	LOG ARC	INTENSITY SPK	INTENSITY DIS	REF
3572.405	3573.425	27984.36	ND	1.4	0.8		MIT
3570.98	3572.00	27995.5	ND	0.8	0.3		MIT
3570.408	3571.427	28000.01	ND	0.3			MIT
3569.498	3570.517	28007.15	ND	1.1			MIT
3569.229	3570.248	28009.26	ND	1.3	0.6		MIT
3568.879	3569.898	28012.00	ND	0.8	1.1		MIT
3566.551	3567.569	28030.29	ND	0.5			MIT
3566.336	3567.354	28031.98	ND	0.9			KN
3564.094	3565.112	28049.61	ND	1.5	0.6		MIT
3563.403	3564.421	28055.05	ND	1.7	1.0		MIT
3561.594	3562.611	28069.30	ND	0.9	0.3		MIT
3560.729	3561.746	28076.12	ND	1.5	1.3		MIT
3555.723	3556.739	28115.64	ND	2.0	1.2		MIT
3553.988	3555.003	28129.37	ND	1.0	0.6		MIT
3553.57	3554.59	28132.7	ND	1.5	1.5		MIT
3551.748	3552.763	28147.11	ND	1.2	0.8		MIT
3549.370	3550.384	28165.97	ND		1.3		MIT
3543.468	3544.481	28212.88	ND	0.9			MIT
3543.352	3544.364	28213.80	ND	1.0	0.3		MIT
3541.626	3542.638	28227.55	ND	1.0D	0.6D		MIT
3540.956	3541.968	28232.89	ND		0.6		MIT
3539.372	3540.383	28245.53	ND	0.6	0.6		MIT
3539.183	3540.194	28247.04	ND	1.0	0.6		MIT
3538.858	3539.869	28249.63	ND	1.3	0.8		MIT
3536.676	3537.687	28267.06	ND	0.8	0.6		MIT
3534.54	3535.55	28284.1	ND	1.2D	0.6		MIT
3534.176	3535.186	28287.05	ND	0.8	0.6H		KN
3533.588	3534.598	28291.76	ND	0.8	0.8		KN
3531.712	3532.721	28306.79	ND	0.8	0.6		MIT
3531.39	3532.40	28309.4	ND	0.6	0.0		MIT
3529.19	3530.20	28327.0	ND	0.9	0.3		MIT
3529.055	3530.064	28328.10	ND	1.0D			MIT
3527.529	3528.537	28340.35	ND	1.3	1.0		MIT
3523.625	3524.632	28371.75	ND	1.2	0.9		MIT
3522.991	3523.998	28376.86	ND	0.3	0.3		MIT
3522.044	3523.051	28384.49	ND	1.1	0.9		MIT
3519.77	3520.78	28402.8	ND	1.0D	0.3		MIT
3516.899	3517.905	28426.01	ND	1.0			KN
3512.909	3513.914	28458.30	ND	0.9	1.0		MIT
3512.040	3513.044	28465.34	ND	1.3			MIT
3510.694	3511.698	28476.25	ND	1.0	0.3		MIT
3509.235	3510.239	28488.09	ND	1.0D			MIT
3506.722	3507.725	28508.51	ND	0.3			MIT
3506.028	3507.031	28514.15	ND	0.8D			MIT
3505.297	3506.300	28520.09	ND	1.2	0.8		MIT
3504.215	3505.217	28528.90	ND	0.8	1.3		MIT
3504.077	3505.079	28530.02	ND	1.0D	0.7		MIT
3502.705	3503.707	28541.20	ND	0.9D	0.6		MIT
3502.01	3503.01	28546.9	ND	0.9	0.6		MIT
3501.535	3502.537	28550.74	ND	1.0D	0.6		MIT
3499.97	3500.97	28563.5	ND	0.9	0.6		MIT
3497.25	3498.25	28585.7	ND	1.0	0.8		MIT
3496.05	3497.05	28595.5	ND	0.6	0.6		MIT
3495.48	3496.48	28600.2	ND	0.6	0.3		MIT
3494.26	3495.26	28610.2	ND	1.0	0.3		MIT
3493.600	3494.600	28615.58	ND	0.8D	0.8		MIT
3491.364	3492.363	28633.91	ND	0.6D	0.7H		MIT
3489.58	3490.58	28648.6	ND	1.0			MIT
3489.293	3490.292	28650.90	ND	1.0			MIT
3488.331	3489.329	28658.80	ND	0.8D	0.6		MIT
3486.90	3487.90	28670.6	ND	0.9D			MIT
3486.83	3487.83	28671.1	ND	1.2	0.6		MIT
3486.32	3487.32	28675.3	ND	0.8D	0.9		MIT
3484.89	3485.89	28687.1	ND	0.5D	0.0		MIT
3484.367	3485.364	28691.41	ND	0.8	0.3		MIT
3484.09	3485.09	28693.7	ND	1.1	0.8		MIT
3482.49	3483.49	28706.9	ND	0.9	0.6		MIT
3481.438	3482.435	28715.54	ND	0.6	0.3		MIT
3480.560	3481.556	28722.79	ND	0.8	0.6		MIT
3479.854	3480.850	28728.61	ND		0.6		MIT

AIR WAVELENGTH (ANGSTRMS)	VACUUM WAVELENGTH (ANGSTRMS)	WAVENUMBER (KAYSERS)	LMENT	LOG ARC	INTENSITY SPK	DIS	REF
3478.93	3479.93	28736.2	ND	0.7			MIT
3477.256	3478.252	28750.08	ND	1.0J			MIT
3476.364	3477.359	28757.45	ND	0.6			MIT
3475.102	3476.097	28767.90	ND	1.0	0.3		MIT
3473.428	3474.423	28781.76	ND	1.0	0.3		MIT
3472.258	3473.252	28791.46	ND	1.1			MIT
3470.866	3471.860	28803.01	ND	1.0	0.3		MIT
3470.59	3471.58	28805.3	ND	0.9D	0.3D		MIT
3468.420	3469.413	28823.32	ND	1.0	0.3		MIT
3468.16	3469.15	28825.5	ND	0.8D	0.3D		MIT
3465.44	3466.43	28848.1	ND	0.8J			MIT
3464.934	3465.926	28852.31	ND	1.0	0.3H		MIT
3460.581	3461.572	28888.61	ND	1.4	0.8		MIT
3460.13	3461.12	28892.4	ND	0.8	0.3H		MIT
3458.949	3459.940	28902.24	ND	0.8	0.3		MIT
3458.779	3459.770	28903.66	ND	1.1	0.3		MIT
3458.000	3458.991	28910.17	ND	1.1	0.6		MIT
3456.004	3456.994	28926.86	ND	0.6			MIT
3455.767	3456.757	28928.85	ND	0.3			MIT
3454.389	3455.379	28940.39	ND	0.8	0.3		MIT
3450.73	3451.72	28971.1	ND	0.9	0.3		MIT
3450.18	3451.17	28975.7	ND	1.0			MIT
3449.300	3450.288	28983.08	ND	1.0	0.6		MIT
3448.27	3449.26	28991.7	ND	0.9D	0.3H		MIT
3447.623	3448.611	28997.18	ND	0.8			MIT
3446.88	3447.87	29003.4	ND	1.1			MIT
3446.18	3447.17	29009.3	ND	0.9	0.6		MIT
3445.895	3446.882	29011.72	ND	0.3			MIT
3443.603	3444.590	29031.03	ND	0.9	0.3		MIT
3443.308	3444.295	29033.52	ND	0.9	0.3		MIT
3442.82	3443.81	29037.6	ND	0.9			MIT
3442.203	3443.190	29042.84	ND	1.2	0.8		MIT
3441.41	3442.40	29049.5	ND	1.0	0.3		MIT
3440.061	3441.047	29060.92	ND	0.6			MIT
3439.210	3440.196	29068.11	ND	1.5	1.1		MIT
3438.97	3439.96	29070.1	ND	0.8			MIT
3438.474	3439.460	29074.33	ND	0.8	0.0		MIT
3435.434	3436.419	29100.06	ND	1.3	0.6		MIT
3432.985	3433.969	29120.82	ND	1.3	1.0		MIT
3431.57	3432.55	29132.8	ND		0.8		MIT
3428.925	3429.908	29155.30	ND	1.3	0.9		MIT
3425.209	3426.191	29186.93	ND	0.9	0.6		MIT
3424.075	3425.057	29196.60	ND	0.3	0.3		MIT
3423.63	3424.61	29200.4	ND	0.8			MIT
3420.09	3421.07	29230.6	ND	1.2	0.3		MIT
3417.528	3418.508	29252.53	ND	0.9	0.8		MIT
3416.19	3417.17	29264.0	ND	0.9			MIT
3415.553	3416.533	29269.44	ND	1.2	0.8		MIT
3414.298	3415.277	29280.20	ND	1.0	0.3		MIT
3411.028	3412.007	29308.27	ND	0.9	0.3		MIT
3410.234	3411.212	29315.09	ND	0.9	0.6		MIT
3409.675	3410.653	29319.90	ND	0.6			MIT
3409.25	3410.23	29323.6	ND	1.3	0.8		MIT
3408.024	3409.002	29334.10	ND	0.6			MIT
3406.994	3407.971	29342.97	ND	0.9	0.6		MIT
3406.392	3407.369	29348.15	ND	0.9	0.3		MIT
3405.681	3406.658	29354.28	ND	0.6			MIT
3404.763	3405.740	29362.20	ND	0.6	0.8		MIT
3403.459	3404.436	29373.45	ND	1.0	0.6		MIT
3403.15	3404.13	29376.1	ND	1.3	1.2		MIT
3400.351	3401.327	29400.29	ND	1.0	0.6		MIT
3399.41	3400.39	29408.4	ND	1.0	0.6		MIT
3399.193	3400.169	29410.31	ND	0.9	0.3		MIT
3398.881	3399.856	29413.01	ND	1.0	0.3		MIT
3398.327	3399.302	29417.80	ND	1.1	0.6		MIT
3397.453	3398.428	29425.37	ND	0.3	0.3		MIT
3396.178	3397.153	29436.42	ND	1.0	0.9		MIT
3395.534	3396.509	29442.00	ND	0.9	0.3		MIT
3395.345	3396.320	29443.64	ND	1.0	0.3		MIT
3394.987	3395.961	29446.74	ND	0.9	0.3		MIT

AIR WAVELENGTH (ANGSTRMS)	VACUUM WAVELENGTH (ANGSTRMS)	WAVENUMBER (KAYSERS)	LMENT	LOG ARC	INTENSITY SPK	DIS	REF
3393.641	3394.615	29458.42	ND	1.0	0.6		MIT
3391.948	3392.922	29473.12	ND	0.8			MIT
3389.325	3390.298	29495.93	ND	1.3	1.3		MIT
3388.025	3388.998	29507.25	ND	1.0	0.6		MIT
3387.960	3388.933	29507.82	ND	0.5			MIT
3387.075	3388.047	29515.53	ND	0.6	0.3H		MIT
3386.518	3387.490	29520.38	ND	1.2	0.6		MIT
3386.285	3387.257	29522.41	ND	1.0	0.6		MIT
3384.666	3385.638	29536.53	ND	1.3	1.3		MIT
3384.127	3385.099	29541.24	ND	0.6	0.6		MIT
3383.449	3384.420	29547.16	ND	0.3			MIT
3382.093	3383.064	29559.00	ND	0.6			MIT
3381.052	3382.023	29568.10	ND	0.6D	0.6		MIT
3379.803	3380.774	29579.03	ND	1.1	0.6		MIT
3379.19	3380.16	29584.4	ND	1.0	0.6		MIT
3377.86	3378.83	29596.0	ND	0.6D	0.3D		MIT
3377.146	3378.116	29602.30	ND	0.8	0.6H		MIT
3376.49	3377.46	29608.1	ND	1.3	1.0		MIT
3375.634	3376.603	29615.56	ND	1.1	0.8H		MIT
3375.244	3376.213	29618.98	ND	1.0	0.6		MIT
3374.936	3375.905	29621.68	ND	1.0	0.3		MIT
3373.300	3374.269	29636.05	ND	0.9			MIT
3373.095	3374.064	29637.85	ND	0.6			MIT
3369.58	3370.55	29668.8	ND	0.6D	0.3D		MIT
3369.242	3370.210	29671.74	ND	0.8			MIT
3369.050	3370.018	29673.43	ND	1.0	0.9		MIT
3368.876	3369.844	29674.97	ND	0.8	0.3		MIT
3365.707	3366.674	29702.91	ND	1.3	1.2		MIT
3364.950	3365.917	29709.59	ND	1.0	0.8		MIT
3364.37	3365.34	29714.7	ND	0.9	0.3		MIT
3362.262	3363.228	29733.34	ND	1.1	0.9		MIT
3361.773	3362.739	29737.66	ND	0.9	0.3		MIT
3359.767	3360.732	29755.42	ND	0.9	0.0		MIT
3358.72	3359.69	29764.7	ND	1.3	1.0		MIT
3358.158	3359.123	29769.68	ND	0.8	0.3		MIT
3356.554	3357.519	29783.90	ND	1.0	0.3		MIT
3355.920	3356.884	29789.53	ND	1.0	0.6		MIT
3355.68	3356.64	29791.7	ND	1.0	0.6		MIT
3354.621	3355.585	29801.06	ND	1.0	0.3		MIT
3353.592	3354.556	29810.21	ND	0.7	0.5		MIT
3353.521	3354.485	29810.84	ND	1.0	0.8		MIT
3353.39	3354.35	29812.0	ND	0.9			MIT
3352.221	3353.185	29822.40	ND	0.9	0.3		MIT
3351.80	3352.76	29826.1	ND	0.9D	0.0H		MIT
3348.586	3349.549	29854.77	ND	0.3			MIT
3348.161	3349.123	29858.56	ND	1.0	0.3		MIT
3347.89	3348.85	29861.0	ND	0.9	0.8		MIT
3347.569	3348.531	29863.84	ND	0.9	0.3		MIT
3347.18	3348.14	29867.3	ND	1.0D	0.6D		MIT
3346.589	3347.551	29872.58	ND	0.8			KN
3345.711	3346.673	29880.42	ND	1.3	0.6		MIT
3345.522	3346.484	29882.11	ND	1.0	0.3		MIT
3345.088	3346.050	29885.99	ND	1.1	0.3		MIT
3343.38	3344.34	29901.2	ND	0.8	0.3H		MIT
3339.740	3340.700	29933.84	ND	1.2	0.6		MIT
3339.440	3340.400	29936.53	ND	1.0	0.3		MIT
3339.063	3340.023	29939.91	ND	1.0	0.6		MIT
3335.771	3336.730	29969.46	ND	1.3			MIT
3334.98	3335.94	29976.6	ND	0.8D			MIT
3334.471	3335.430	29981.14	ND	1.2	0.8		MIT
3333.058	3334.017	29993.85	ND	0.8D	0.3D		MIT
3332.174	3333.132	30001.81	ND	1.5	0.6		MIT
3331.567	3332.525	30007.27	ND	0.9	0.3		MIT
3330.804	3331.762	30014.15	ND	0.6			MIT
3330.09	3331.05	30020.6	ND	0.9D	0.0H		MIT
3328.991	3329.949	30030.49	ND	0.8			MIT
3328.270	3329.227	30037.00	ND	1.2	0.8		MIT
3328.037	3328.994	30039.10	ND	0.9	0.3		MIT
3327.688	3328.645	30042.25	ND	1.1	0.3		MIT
3327.560	3328.517	30043.41	ND	1.2	0.3		MIT

AIR WAVELENGTH (ANGSTRMS)	VACUUM WAVELENGTH (ANGSTRMS)	WAVENUMBER (KAYSERS)	LMENT	LOG ARC	INTENSITY SPK	DIS	REF
3327.147	3328.104	30047.14	ND	0.9	0.3H		MIT
3326.26	3327.22	30055.1	ND	0.8	0.3		MIT
3325.889	3326.846	30058.50	ND	1.0	0.6		MIT
3325.732	3326.689	30059.92	ND	0.6H			MIT
3325.146	3326.103	30065.22	ND	1.0			MIT
3322.936	3323.892	30085.21	ND	1.4	0.3		MIT
3321.395	3322.351	30099.17	ND	0.9			MIT
3319.04	3320.00	30120.5	ND	1.0	0.3		MIT
3318.413	3319.368	30126.22	ND	0.9	0.3		MIT
3316.917	3317.872	30139.80	ND	0.6			MIT
3316.20	3317.15	30146.3	ND	1.1	0.8H		MIT
3316.011	3316.965	30148.04	ND	1.0	0.6		MIT
3315.446	3316.400	30153.18	ND	1.1			MIT
3315.01	3315.96	30157.1	ND	1.3			MIT
3313.146	3314.100	30174.11	ND	1.0	0.3		MIT
3312.743	3313.696	30177.78	ND	1.0	0.3		MIT
3310.900	3311.853	30194.58	ND	1.4	0.3		MIT
3310.360	3311.313	30199.50	ND	0.9	0.3		MIT
3307.97	3308.92	30221.3	ND	1.3	0.3		MIT
3307.802	3308.754	30222.86	ND	1.2	0.3		MIT
3305.33	3306.28	30245.5	ND	0.9	0.3		MIT
3304.649	3305.600	30251.69	ND	0.9			MIT
3300.908	3301.858	30285.97	ND	0.9D	0.3D		MIT
3300.148	3301.098	30292.95	ND	1.0	0.8		MIT
3298.609	3299.559	30307.08	ND	0.9	0.8		MIT
3294.661	3295.610	30343.40	ND	0.9	0.3		MIT
3293.820	3294.769	30351.15	ND	1.1	0.6		MIT
3290.643	3291.591	30380.45	ND	1.0	0.8		MIT
3287.41	3288.36	30410.3	ND	0.8	0.3H		MIT
3286.607	3287.554	30417.75	ND	0.8	0.3		MIT
3285.093	3286.039	30431.77	ND	1.0	0.6		MIT
3282.777	3283.723	30453.24	ND	0.9	0.3		MIT
3281.487	3282.432	30465.21	ND	0.6	0.8		MIT
3280.202	3281.147	30477.15	ND	0.3H	0.6H		MIT
3276.132	3277.076	30515.01	ND	1.1	0.3		MIT
3275.842	3276.786	30517.71	ND	1.0	0.6		MIT
3275.218	3276.162	30523.52	ND	1.2	0.9		MIT
3273.173	3274.116	30542.59	ND	0.9	0.3		MIT
3267.251	3268.193	30597.95	ND	1.0	0.6		MIT
3265.373	3266.314	30615.55	ND	1.0	0.3		MIT
3265.124	3266.065	30617.88	ND	1.0	0.6		MIT
3262.36	3263.30	30643.8	ND	1.0D	0.3H		MIT
3261.33	3262.27	30653.5	ND		0.6H		MIT
3260.655	3261.595	30659.84	ND	1.0	0.6		MIT
3259.231	3260.171	30673.24	ND	1.0	0.6		MIT
3256.904	3257.843	30695.15	ND	1.0	0.3		MIT
3255.625	3256.564	30707.21	ND	0.9	0.6		MIT
3254.066	3255.005	30721.92	ND	0.9	0.3		MIT
3251.804	3252.742	30743.29	ND	0.6D			MIT
3250.974	3251.912	30751.14	ND	0.6	0.3H		MIT
3245.703	3246.639	30801.08	ND	0.9	0.6		MIT
3242.455	3243.391	30831.93	ND	0.6	0.3		MIT
3238.474	3239.409	30869.83	ND	1.0			MIT
3238.259	3239.194	30871.88	ND	0.8	0.3		MIT
3237.911	3238.845	30875.20	ND	1.0	0.6		MIT
3237.68	3238.61	30877.4	ND	0.3	0.6H		MIT
3236.499	3237.433	30888.67	ND	0.8D			MIT
3229.872	3230.804	30952.04	ND		1.0		MIT
3228.048	3228.980	30969.53	ND	0.9	0.3		MIT
3223.776	3224.707	31010.57	ND	0.5	0.3		MIT
3222.624	3223.555	31021.66	ND	0.9	0.6		MIT
3222.173	3223.103	31026.00	ND	0.8			MIT
3220.529	3221.459	31041.83	ND	0.3H	0.6H		MIT
3217.533	3218.462	31070.74	ND	0.6			MIT
3217.116	3218.045	31074.77	ND	0.3D			MIT
3213.286	3214.214	31111.80	ND	0.6			KN
3210.973	3211.901	31134.21	ND	0.8	0.3		MIT
3205.917	3206.843	31183.31	ND	0.9			MIT
3203.462	3204.388	31207.21	ND	1.0	0.6		MIT
3200.623	3201.548	31234.89	ND	0.9	0.6		MIT

AIR WAVELENGTH (ANGSTRMS)	VACUUM WAVELENGTH (ANGSTRMS)	WAVENUMBER (KAYSERS)	LMENT	LOG ARC	INTENSITY SPK	DIS	REF
3197.678	3198.602	31263.66	ND	0.8			MIT
3193.034	3193.957	31309.12	ND	0.6			MIT
3191.855	3192.778	31320.69	ND	0.6			MIT
3191.093	3192.016	31328.17	ND	0.6			MIT
3190.707	3191.629	31331.96	ND	0.8			MIT
3188.734	3189.656	31351.34	ND	1.0			MIT
3181.516	3182.436	31422.47	ND	0.8	0.3		MIT
3178.925	3179.845	31448.08	ND	0.8			MIT
3175.987	3176.906	31477.17	ND	0.9	0.3		MIT
3170.015	3170.932	31536.47	ND	1.0			MIT
3162.632	3163.547	31610.08	ND	0.8D	0.3		MIT
3159.216	3160.131	31644.26	ND	0.8	0.6		MIT
3156.259	3157.173	31673.91	ND	0.9			KN
3155.763	3156.677	31678.89	ND	0.6			MIT
3149.511	3150.423	31741.77	ND	0.9	0.6		KN
3149.296	3150.208	31743.94	ND	0.9	0.8		MIT
3148.525	3149.437	31751.71	ND	0.8	0.3		MIT
3146.016	3146.927	31777.03	ND	0.8			MIT
3144.839	3145.750	31788.92	ND	1.3	0.6		MIT
3144.564	3145.475	31791.70	ND	1.2	0.6		MIT
3143.310	3144.221	31804.39	ND	0.7	0.6		MIT
3142.444	3143.354	31813.15	ND	0.9	0.9		MIT
3141.476	3142.386	31822.95	ND	1.0	1.0		MIT
3137.253	3138.162	31865.79	ND	1.0	0.6		MIT
3136.679	3137.588	31871.62	ND	0.8			KN
3134.897	3135.805	31889.73	ND	1.6	1.5		MIT
3133.603	3134.511	31902.90	ND	1.2	1.0		MIT
3127.175	3128.081	31968.48	ND	0.6	0.3H		MIT
3124.567	3125.473	31995.16	ND	1.1	1.0		MIT
3124.074	3124.980	32000.21	ND	0.6	0.3D		MIT
3123.069	3123.974	32010.50	ND	0.9	0.8		MIT
3119.755	3120.660	32044.51	ND	0.6	0.3		MIT
3116.141	3117.045	32081.67	ND	1.0	0.6		MIT
3115.172	3116.075	32091.65	ND	0.6	0.3		MIT
3114.860	3115.763	32094.86	ND	0.6			MIT
3113.228	3114.131	32111.69	ND	0.6	0.0		MIT
3108.008	3108.910	32165.62	ND	0.8	0.3		MIT
3106.173	3107.074	32184.62	ND	0.8	0.6		MIT
3105.419	3106.320	32192.43	ND	0.9	0.3		MIT
3099.515	3100.415	32253.75	ND	0.8	0.6		MIT
3098.476	3099.375	32264.57	ND	0.9	0.3		MIT
3096.829	3097.728	32281.73	ND	0.3	0.3		MIT
3093.883	3094.781	32312.46	ND	0.8			MIT
3092.915	3093.813	32322.58	ND	0.9	0.8		MIT
3089.677	3090.574	32356.45	ND	1.0			KN
3080.930	3081.825	32448.31	ND	1.0	0.6		MIT
3080.099	3080.994	32457.06	ND	0.6	0.3		MIT
3079.365	3080.259	32464.80	ND	0.9	0.3		MIT
3075.380	3076.273	32506.86	ND	0.9	0.8		MIT
3073.085	3073.978	32531.14	ND	0.6	0.3		MIT
3071.482	3072.374	32548.12	ND	0.3			MIT
3071.426	3072.318	32548.71	ND	0.6D	0.3		MIT
3069.716	3070.608	32566.84	ND	0.9	0.6		MIT
3056.698	3057.587	32705.53	ND	0.9	0.6		MIT
3052.147	3053.035	32754.30	ND	1.0	0.8		MIT
3051.108	3051.995	32765.45	ND	0.8	0.3		MIT
3038.962	3039.846	32896.40	ND	0.6	0.3		MIT
3018.333	3019.212	33121.22	ND	0.9	0.6		MIT
3014.735	3015.613	33160.75	ND	0.6	0.3H		MIT
3014.165	3015.043	33167.02	ND	0.9	0.3		MIT
3007.975	3008.852	33235.27	ND	0.8	0.3		MIT
2994.734	2995.607	33382.21	ND	0.5	0.3		MIT
2993.179	2994.052	33399.55	ND	1.6	0.3		MIT
2986.950	2987.821	33469.20	ND	0.5	0.3		MIT
2985.33	2986.20	33487.4	ND	0.3			MIT
2977.96	2978.83	33570.2	ND		0.7		MIT
2970.88	2971.75	33650.2	ND		0.7		MIT
2970.03	2970.90	33659.9	ND		0.70		MIT
2965.70	2966.57	33709.0	ND			0.7	MIT
2963.560	2964.426	33733.35	ND	1.5			MIT

AIR WAVELENGTH (ANGSTRMS)	VACUUM WAVELENGTH (ANGSTRMS)	WAVENUMBER (KAYSERS)	LMENT	LOG ARC	INTENSITY SPK	INTENSITY DIS	REF
2963.37	2964.24	33735.5	ND		0.7		MIT
2962.865	2963.730	33741.26	ND	1.3	0.3		MIT
2961.45	2962.32	33757.4	ND		0.7		MIT
2960.29	2961.15	33770.6	ND		0.7		MIT
2946.72	2947.58	33926.1	ND		1.0		MIT
2945.913	2946.774	33935.41	ND		0.7		MIT
2943.500	2944.361	33963.23	ND	0.5	0.7		MIT
2939.535	2940.395	34009.04	ND		0.7		MIT
2932.15	2933.01	34094.7	ND		0.7H		MIT
2928.67	2929.53	34135.2	ND		0.7		MIT
2928.11	2928.97	34141.7	ND		0.7H		MIT
2923.439	2924.295	34196.28	ND		0.7H		MIT
2921.484	2922.339	34219.16	ND		0.7		MIT
2921.26	2922.12	34221.8	ND	0.7D	0.7		MIT
2917.624	2918.478	34264.43	ND	0.3H	0.7H		MIT
2916.735	2917.589	34274.88	ND	0.3			MIT
2914.499	2915.352	34301.17	ND		1.2		MIT
2914.25	2915.10	34304.1	ND	0.7D			MIT
2908.98	2909.83	34366.2	ND		0.7H		MIT
2908.598	2909.450	34370.76	ND		0.7		MIT
2908.140	2908.992	34376.17	ND		0.7		MIT
2906.44	2907.29	34396.3	ND		1.0		MIT
2904.302	2905.153	34421.60	ND		1.0		MIT
2903.571	2904.422	34430.26	ND		1.0		MIT
2902.66	2903.51	34441.1	ND		0.7		MIT
2899.85	2900.70	34474.4	ND		1.0		MIT
2899.60	2900.45	34477.4	ND		1.0		MIT
2896.98	2897.83	34508.6	ND		0.7H		MIT
2894.12	2894.97	34542.7	ND		0.7		MIT
2892.68	2893.53	34559.9	ND	0.7			MIT
2890.84	2891.69	34581.9	ND	0.7	0.3H		MIT
2889.19	2890.04	34601.6	ND		0.7		MIT
2887.56	2888.41	34621.2	ND		0.7		MIT
2885.82	2886.67	34642.0	ND		0.7		MIT
2879.36	2880.20	34719.7	ND	0.7H	0.3J		MIT
2876.046	2876.890	34759.76	ND		1.0		MIT
2871.44	2872.28	34815.5	ND	0.7			MIT
2868.85	2869.69	34846.9	ND		0.7		MIT
2868.053	2868.895	34856.63	ND		0.7		MIT
2865.63	2866.47	34886.1	ND	0.7			MIT
2864.34	2865.18	34901.8	ND		0.7H		MIT
2863.949	2864.790	34906.57	ND	0.7	0.3		MIT
2855.24	2856.08	35013.0	ND	0.7			MIT
2848.415	2849.252	35096.93	ND		0.7		MIT
2846.58	2847.42	35119.5	ND	0.7			MIT
2839.69	2840.52	35204.8	ND		0.7		MIT
2838.702	2839.537	35217.01	ND		0.7H		MIT
2837.51	2838.34	35231.8	ND	0.7			MIT
2837.10	2837.93	35236.9	ND	0.7			MIT
2835.65	2836.48	35254.9	ND	0.7			MIT
2832.20	2833.03	35297.9	ND	0.7			MIT
2828.545	2829.377	35343.47	ND		0.7		MIT
2823.133	2823.964	35411.22	ND		0.7		MIT
2819.64	2820.47	35455.1	ND	0.7H			MIT
2814.65	2815.48	35517.9	ND	0.7H			MIT
2810.06	2810.89	35576.0	ND		0.7		MIT
2808.652	2809.479	35593.78	ND	0.7			MIT
2801.887	2802.713	35679.72	ND		0.3		MIT
2797.22	2798.04	35739.2	ND	0.7			MIT
2785.791	2786.613	35885.86	ND	0.7			MIT
2782.730	2783.551	35925.33	ND	0.7	0.3		MIT
2780.76	2781.58	35950.8	ND	0.7			MIT
2772.32	2773.14	36060.2	ND	0.7			MIT
2765.96	2766.78	36143.1	ND	0.7D			MIT
2765.49	2766.31	36149.3	ND	0.7			MIT
2764.982	2765.799	36155.92	ND	1.0			MIT
2751.50	2752.31	36333.1	ND	0.7	0.3		MIT
2747.73	2748.54	36382.9	ND	0.7			MIT
2733.36	2734.17	36574.2	ND	0.7	0.3		MIT
2710.04	2710.84	36888.9	ND	0.7	0.3		MIT

AIR WAVELENGTH (ANGSTRMS)	VACUUM WAVELENGTH (ANGSTRMS)	WAVENUMBER (KAYSERS)	LMENT	LOG ARC	INTENSITY SPK	INTENSITY DIS	REF
2704.54	2705.34	36963.9	ND	1.0D	0.3		MIT
2702.455	2703.257	36992.42	ND	1.0	0.3		MIT
2683.73	2684.53	37250.5	ND	0.7			MIT
2670.47	2671.26	37435.5	ND	0.7	0.3		MIT
2662.42	2663.21	37548.6	ND	1.0	0.3		MIT
2655.41	2656.20	37647.8	ND	0.7	0.3		MIT
2648.89	2649.68	37740.4	ND	1.0	0.3		MIT
2637.17	2637.96	37908.1	ND	0.7			MIT
2634.59	2635.38	37945.3	ND	0.7	0.3		MIT
2622.657	2623.439	38117.90	ND	1.0	0.3		MIT
2565.11	2565.88	38973.0	ND	0.7			MIT
2466.83	2467.58	40525.6	ND		0.7		MIT
2435.57	2436.31	41045.7	ND		0.7		MIT
2426.38	2427.12	41201.1	ND		0.7		MIT
2409.25	2409.98	41494.1	ND	0.7			MIT
2400.71	2401.44	41641.7	ND		1.0		MIT
2398.38	2399.11	41682.1	ND		1.1		MIT
2396.12	2396.85	41721.4	ND		1.2		MIT
2394.20	2394.93	41754.9	ND		1.3		MIT
2389.92	2390.65	41829.7	ND		1.3		MIT
2384.85	2385.58	41918.6	ND		1.0		MIT
2382.82	2383.55	41954.3	ND		1.0		MIT
2378.06	2378.79	42038.3	ND		0.3		MIT
2373.17	2373.89	42124.9	ND		0.8		MIT
2369.98	2370.70	42181.6	ND		1.0		MIT
2368.14	2368.86	42214.3	ND		1.0		MIT
2363.87	2364.59	42290.6	ND		1.3		MIT
2362.21	2362.93	42320.3	ND		0.6		MIT
2358.57	2359.29	42385.6	ND		1.0		MIT
2358.07	2358.79	42394.6	ND		0.5		MIT
2357.56	2358.28	42403.8	ND		0.3		MIT
2353.77	2354.49	42472.0	ND		1.3		MIT
2352.60	2353.32	42493.2	ND		0.6		MIT
2352.17	2352.89	42500.9	ND		0.6		MIT
2349.30	2350.02	42552.8	ND		1.2		MIT
2348.46	2349.18	42568.1	ND		1.3		MIT
2344.49	2345.21	42640.1	ND		0.9		MIT
2340.47	2341.19	42713.4	ND		0.8		MIT
2337.71	2338.43	42763.8	ND		0.9		MIT
2336.37	2337.09	42788.3	ND		0.6		MIT
2333.45	2334.17	42841.9	ND		0.5		MIT
2332.97	2333.69	42850.7	ND		1.3		MIT
2323.89	2324.60	43018.1	ND		1.0		MIT
2323.39	2324.10	43027.3	ND		0.9		MIT
2316.69	2317.40	43151.8	ND		0.5		MIT
2316.41	2317.12	43157.0	ND		0.6		MIT
2314.82	2315.53	43186.6	ND		0.6		MIT
2310.82	2311.53	43261.4	ND		0.7		MIT
2307.10	2307.81	43331.1	ND		0.7		MIT
2301.16	2301.87	43443.0	ND		0.7		MIT
2299.81	2300.52	43468.5	ND		1.0		MIT
2295.46	2296.17	43550.8	ND		0.7		MIT
2292.69	2293.40	43603.5	ND		0.9		MIT
2285.86	2286.57	43733.7	ND		0.5		MIT
2285.08	2285.79	43748.6	ND		0.5		MIT
2282.73	2283.43	43793.7	ND		0.9		MIT
2264.08	2264.78	44154.4	ND		1.2		MIT
2261.51	2262.21	44204.6	ND		0.3		MIT
2260.54	2261.24	44223.5	ND		1.0		MIT
2258.19	2258.89	44269.5	ND		0.3		MIT
2257.26	2257.96	44287.8	ND		1.2		MIT
2253.50	2254.20	44361.7	ND		1.4		MIT
2252.59	2253.29	44379.6	ND		1.0		A
2250.87	2251.57	44413.5	ND		1.2		MIT
2249.88	2250.58	44433.0	ND		0.8		MIT
2248.72	2249.42	44456.0	ND		0.6		MIT
2246.47	2247.17	44500.5	ND		1.1		MIT
2241.75	2242.45	44594.2	ND		1.1		MIT
2240.14	2240.84	44626.2	ND		1.0		MIT
2238.47	2239.17	44659.5	ND		1.0		MIT

AIR WAVELENGTH (ANGSTRMS)	VACUUM WAVELENGTH (ANGSTRMS)	WAVENUMBER (KAYSERS)	LMENT	LOG ARC	INTENSITY SPK	INTENSITY DIS	REF
2237.37	2238.06	44681.5	ND		0.8		MIT
2233.75	2234.44	44753.9	ND		0.7		MIT
2228.41	2229.10	44861.1	ND		1.1		MIT
2225.48	2226.17	44920.1	ND		1.0		MIT
2208.60	2209.29	45263.4	ND		1.3		MIT
2203.56	2204.25	45367.0	ND		1.5		MIT
2203.13	2203.82	45375.8	ND		1.0		MIT
2193.89	2194.58	45566.9	ND		1.4		MIT
2193.18	2193.87	45581.6	ND		1.3		MIT
2187.45	2188.13	45701.0	ND		1.2		MIT
2186.36	2187.04	45723.8	ND		0.5		MIT
2184.62	2185.30	45760.2	ND		1.3		MIT
2178.33	2179.01	45892.4	ND		1.0		MIT
2176.47	2177.15	45931.6	ND		1.2		MIT
2154.59	2155.27	46398.0	ND		0.5		MIT
2117.62	2118.29	47207.9	ND		1.4		A
2110.79	2111.46	47360.6	ND		0.3H		MIT
2098.61	2099.28	47635.5	ND		1.2		A
2091.98	2092.65	47786.4	ND		1.1		A
2089.60	2090.26	47840.8	ND		1.0		A
2088.96	2089.62	47855.5	ND		1.1		A
2085.15	2085.81	47942.9	ND		1.3		A
2079.64	2080.30	48069.9	ND		1.2		A
2078.53	2079.19	48095.6	ND		1.0		A
9974.2	9976.9	10023.	NE I			0.3	ME
9963.55	9966.28	10033.8	NE I			0.8	ME
9947.94	9950.67	10049.6	NE I			1.2	ME
9944.9	9947.6	10053.	NE I			0.3	ME
9944.1	9946.8	10053.	NE I			0.8H	ME
9938.35	9941.07	10059.3	NE I			1.2	ME
9936.83	9939.55	10060.8	NE I			1.0	ME
9918.52	9921.24	10079.4	NE I			0.6	ME
9915.13	9917.85	10082.8	NE I			1.3	ME
9902.31	9905.03	10095.9	NE I			1.5	ME
9900.58	9903.29	10097.7	NE I			1.6	ME
9899.06	9901.77	10099.2	NE I			0.3	ME
9897.30	9900.01	10101.0	NE I			0.5	ME
9875.90	9878.61	10122.9	NE I			0.3	ME
9837.47	9840.17	10162.4	NE I			1.3	ME
9823.42	9826.11	10177.0	NE I			0.7	ME
9788.1	9790.8	10214.	NE I			0.3	ME
9760.57	9763.25	10242.5	NE			0.3	ME
9702.40	9705.06	10303.9	NE I			0.5	ME
9665.424	9668.075	10343.32	NE I			3.0	IME
9592.19	9594.82	10422.3	NE			0.7	ME
9584.79	9587.42	10430.3	NE			0.5	ME
9573.99	9576.62	10442.1	NE I			0.3	ME
9547.40	9550.02	10471.2	NE I			2.5	ME
9534.167	9536.782	10485.72	NE I			2.7	IME
9508.4	9511.0	10514.	NE I			0.7	ME
9506.59	9509.20	10516.1	NE			0.5	ME
9497.9	9500.5	10526.	NE			0.3	ME
9486.680	9489.282	10538.20	NE I			2.7	IME
9467.8	9470.4	10559.	NE I			0.3H	ME
9459.21	9461.81	10568.8	NE I			2.5	ME
9452.08	9454.67	10576.8	NE			1.0	ME
9445.26	9447.85	10584.4	NE I			0.5	ME
9443.8	9446.4	10586.	NE I			0.3	ME
9432.94	9435.53	10598.2	NE I			1.6	ME
9425.38	9427.97	10606.7	NE I			2.7	ME
9412.32	9414.90	10621.5	NE I			0.6	ME
9410.75	9413.33	10623.2	NE I			0.8	ME
9405.75	9408.33	10628.9	NE I			0.9	ME
9393.8	9396.4	10642.	NE			0.3	ME
9377.2	9379.8	10661.	NE I			0.7	ME
9373.28	9375.85	10665.7	NE I			2.3	ME
9353.3	9355.9	10688.	NE I			0.5	ME
9340.5	9343.1	10703.	NE I			0.3	ME
9326.52	9329.08	10719.2	NE I			2.8	ME
9313.98	9316.54	10733.6	NE I			2.5	ME

AIR WAVELENGTH (ANGSTRMS)	VACUUM WAVELENGTH (ANGSTRMS)	WAVENUMBER (KAYSERS)	LMENT	LOG ARC	INTENSITY SPK	INTENSITY DIS	REF
9310.58	9313.13	10737.5	NE I			2.2	ME
9300.85	9303.40	10748.8	NE I			2.8	ME
9275.53	9278.08	10778.1	NE I			2.0	ME
9226.67	9229.20	10835.2	NE I			2.3	ME
9221.88	9224.41	10840.8	NE I			1.2	MS
9221.59	9224.12	10841.1	NE I			2.3	ME
9220.05	9222.58	10842.9	NE I			2.6	ME
9212.9	9215.4	10851.	NE I			0.3	ME
9201.76	9204.29	10864.5	NE I			2.8	ME
9191.8	9194.3	10876.	NE			0.5	ME
9148.68	9151.19	10927.5	NE I			2.8	ME
9121.14	9123.64	10960.5	NE I			1.3	ME
9103.53	9106.03	10981.7	NE I			0.5	ME
9073.04	9075.53	11018.6	NE I			0.9	ME
9069.7	9072.2	11023.	NE			0.3	ME
9052.54	9055.02	11043.6	NE I			0.8	ME
9049.06	9051.54	11047.8	NE I			0.5	ME
9039.0	9041.5	11060.	NE I			0.5	ME
9036.98	9039.46	11062.6	NE I			0.8	ME
8988.58	8991.05	11122.2	NE I			2.3	ME
8968.6	8971.1	11147.	NE I			0.3	ME
8962.34	8964.80	11154.7	NE I			0.5	ME
8948.12	8950.58	11172.5	NE I			0.8	ME
8941.47	8943.92	11180.8	NE I			0.8	ME
8929.24	8931.69	11196.1	NE I			1.0	ME
8927.4	8929.9	11198.	NE I			0.3	ME
8919.50	8921.95	11208.3	NE I			2.5	ME
8915.44	8917.89	11213.4	NE I			0.5	ME
8913.0	8915.4	11216.	NE			0.5	ME
8895.6	8898.0	11238.	NE I			0.3	ME
8892.22	8894.66	11242.7	NE I			1.0	ME
8865.759	8868.193	11276.25	NE I			2.7	IME
8865.33	8867.76	11276.8	NE I			2.0	ME
8853.866	8856.297	11291.40	NE I			2.8	IME
8842.1	8844.5	11306.	NE			0.3	ME
8830.92	8833.34	11320.7	NE I			1.7	ME
8820.36	8822.78	11334.3	NE			0.8	ME
8792.51	8794.92	11370.2	NE I			1.5	ME
8783.755	8786.167	11381.53	NE I			3.0	IME
8782.01	8784.42	11383.8	NE I			1.7	ME
8780.622	8783.033	11385.59	NE I			3.0	IME
8778.75	8781.16	11388.0	NE I			2.2	ME
8771.70	8774.11	11397.2	NE I			2.6	ME
8767.55	8769.96	11402.6	NE I			1.2	ME
8714.52	8716.91	11471.9	NE I			0.7	GR
8704.15	8706.54	11485.6	NE I			2.3	ME
8681.920	8684.305	11515.03	NE I			2.7	IME
8679.491	8681.875	11518.25	NE I			2.7	IME
8654.51	8656.89	11551.5	NE I			2.6	ME
8654.383	8656.760	11551.67	NE I			3.3	IME
8647.05	8649.43	11561.5	NE I			2.5	ME
8635.31	8637.68	11577.2	NE I			1.7	ME
8634.648	8637.020	11578.07	NE I			2.8	IME
8591.258	8593.618	11636.54	NE I			2.6	IME
8582.91	8585.27	11647.9	NE I			1.8	ME
8571.36	8573.71	11663.6	NE I			2.0	ME
8544.70	8547.05	11699.9	NE I			1.8	ME
8495.360	8497.694	11767.90	NE I			2.7	IME
8484.45	8486.78	11783.0	NE I			1.9	ME
8470.72	8473.05	11802.1	NE I			0.7	GR
8463.37	8465.70	11812.4	NE I			2.2	ME
8418.427	8420.740	11875.44	NE I			2.6	IME
8417.18	8419.49	11877.2	NE I			2.0	ME
8377.607	8379.909	11933.30	NE I			2.9	IME
8376.41	8378.71	11935.0	NE I			2.3	ME
8365.75	8368.05	11950.2	NE I			2.2	ME
8301.54	8303.82	12042.7	NE I			2.2	ME
8300.326	8302.607	12044.41	NE I			2.8	IME
8267.11	8269.38	12092.8	NE I			1.9	ME
8266.076	8268.348	12094.31	NE I			2.3	IME

AIR WAVELENGTH (ANGSTRMS)	VACUUM WAVELENGTH (ANGSTRMS)	WAVENUMBER (KAYSERS)	LMENT	LOG ARC	INTENSITY SPK	INTENSITY DIS	REF
8259.380	8261.650	12104.12	NE I			2.2	IME
8248.70	8250.97	12119.8	NE I			1.5	ME
8136.406	8138.643	12287.06	NE I			2.5	IME
8128.93	8131.17	12298.4	NE I			1.8	ME
8118.549	8120.781	12314.09	NE I			2.0	IME
8093.08	8095.31	12352.8	NE I			0.3	ME
8082.458	8084.681	12369.07	NE I			2.3	IME
8041.79	8044.00	12431.6	NE I			0.3	ME
8024.11	8026.32	12459.0	NE I			0.3	ME
7944.16	7946.35	12584.4	NE I			1.3	ME
7943.180	7945.365	12585.95	NE I			2.3	IME
7937.010	7939.193	12595.74	NE I			1.8	IMS
7927.13	7929.31	12611.4	NE I			1.6	ME
7839.08	7841.24	12753.1	NE I			1.5	ME
7833.06	7835.22	12762.9	NE I			0.8	ME
7724.63	7726.76	12942.0	NE I			1.0	ME
7670.85	7672.96	13032.8	NE I			0.7	GR
7621.33	7623.43	13117.5	NE I			0.7	GR
7572.06	7574.14	13202.8	NE I			0.7	GR
7544.046	7546.123	13251.84	NE I			2.0	IME
7535.775	7537.850	13266.38	NE I			2.5	IME
7488.872	7490.934	13349.47	NE I			2.7	IME
7472.425	7474.483	13378.85	NE I			1.7	PS
7438.899	7440.948	13439.15	NE I			2.5	IME
7325.57	7327.59	13647.1	NE I			1.2	GR
7307.93	7309.94	13680.0	NE			1.2	GR
7304.82	7306.83	13685.8	NE I			1.5	MS
7245.167	7247.164	13798.50	NE I			3.0	IME
7173.939	7175.916	13935.50	NE I			3.0	IME
7138.70	7140.67	14004.3	NE			1.5	PS
7112.2	7114.2	14056.	NE I			1.0	PS
7059.109	7061.055	14162.19	NE I			2.3	IME
7051.292	7053.236	14177.89	NE I			1.8	IME
7032.4127	7034.3520	14215.951	NE I			3.0	S
7024.051	7025.988	14232.87	NE I			2.7	IME
6929.468	6931.380	14427.14	NE I			3.0	IME
6759.586	6761.452	14789.72	NE I			1.2	PS
6738.058	6739.918	14836.98	NE I			1.8	PS
6717.0428	6718.8972	14883.395	NE I			1.8	S
6678.2764	6680.1203	14969.790	NE I			2.7	S
6666.893	6668.734	14995.35	NE I			2.0	PS
6652.093	6653.930	15028.71	NE I			2.2	PS
6640.80	6642.63	15054.3	NE I			0.7	GR
6640.012	6641.846	15056.06	NE I			1.0	PS
6602.907	6604.731	15140.66	NE I			2.0	PS
6598.9529	6600.7755	15149.735	NE I			3.0	S
6532.8824	6534.6872	15302.951	NE I			2.0	S
6506.5279	6508.3257	15364.935	NE I			3.0	S
6444.721	6446.502	15512.29	NE I			2.2	PS
6421.708	6423.483	15567.88	NE I			2.0	PS
6409.753	6411.525	15596.91	NE I			2.2	PS
6402.246	6404.016	15615.20	NE I			3.3	I
6401.076	6402.845	15618.06	NE I			2.0	PS
6382.9914	6384.7560	15662.306	NE I			3.0	S
6365.013	6366.773	15706.55	NE I			2.0	PS
6351.873	6353.629	15739.04	NE I			2.0	PS
6334.4279	6336.1794	15782.381	NE I			3.0	S
6330.901	6332.652	15791.17	NE I			2.2	PS
6328.173	6329.923	15797.98	NE I			2.5	PS
6313.692	6315.438	15834.21	NE I			2.2	PS
6304.7892	6306.5328	15856.573	NE I			2.0	S
6293.766	6295.507	15884.34	NE I			2.0	PS
6276.039	6277.775	15929.21	NE I			1.7	PS
6273.018	6274.753	15936.88	NE I			1.8	PS
6266.4950	6268.2283	15953.471	NE I			3.0	S
6258.796	6260.527	15973.10	NE I			2.0	PS
6249.593	6251.322	15996.62	NE I			0.7	PS
6246.734	6248.462	16003.94	NE I			2.0	PS
6225.742	6227.464	16057.90	NE I			1.7	PS
6217.2813	6219.0014	16079.752	NE I			3.0	S

AIR WAVELENGTH (ANGSTRMS)	VACUUM WAVELENGTH (ANGSTRMS)	WAVENUMBER (KAYSERS)	LMENT	LOG ARC	INTENSITY SPK	INTENSITY DIS	REF
6213.878	6215.597	16088.56	NE I			2.2	PS
6205.787	6207.504	16109.54	NE I			2.0	PS
6202.981	6204.697	16116.82	NE I			1.2	PS
6193.078	6194.792	16142.59	NE I			1.7	PS
6189.076	6190.789	16153.03	NE I			1.8	PS
6183.169	6184.880	16168.46	NE I			0.7	PS
6182.146	6183.857	16171.14	NE I			2.2	IME
6175.291	6177.000	16189.09	NE I			1.7	PS
6174.888	6176.597	16190.15	NE I			1.8	PS
6172.821	6174.529	16195.57	NE I			1.2	PS
6163.5939	6165.2996	16219.812	NE I			3.0	S
6156.145	6157.849	16239.44	NE I			1.7	PS
6154.09	6155.79	16244.9	NE			1.8W	WA
6150.303	6152.005	16254.86	NE I			2.0	PS
6143.0623	6144.7625	16274.022	NE I			3.0	S
6142.508	6144.208	16275.49	NE I			2.0	PS
6128.451	6130.147	16312.82	NE I			2.0	IME
6118.027	6119.720	16340.62	NE I			1.2	PS
6096.1630	6097.8506	16399.221	NE I			2.5	S
6074.3377	6076.0194	16458.144	NE I			3.0	S
6064.552	6066.231	16484.70	NE I			1.7	PS
6046.158	6047.832	16534.85	NE I			1.7	PS
6042.013	6043.686	16546.19	NE I			1.2	PS
6029.9971	6031.6669	16579.165	NE I			3.0	S
6000.951	6002.613	16659.41	NE I			2.0	PS
5991.675	5993.335	16685.20	NE I			1.9	PS
5987.907	5989.566	16695.70	NE I			2.2	IME
5982.401	5984.058	16711.07	NE I			0.9	PS
5975.5340	5977.1892	16730.272	NE I			2.8	S
5974.628	5976.283	16732.81	NE I			2.7	IME
5966.171	5967.824	16756.53	NE I			1.5	PS
5965.474	5967.127	16758.49	NE I			2.7	IME
5961.626	5963.278	16769.30	NE I			1.8	PS
5944.8342	5946.4812	16816.668	NE I			2.7	S
5939.319	5940.965	16832.28	NE I			1.7	PS
5934.458	5936.102	16846.07	NE I			1.9	PS
5933.958	5935.602	16847.49	NE			0.9	PS
5922.709	5924.350	16879.49	NE I			1.4	PS
5919.037	5920.677	16889.96	NE I			0.9	PS
5918.914	5920.554	16890.31	NE I			2.4	PS
5913.633	5915.272	16905.39	NE I			2.4	IME
5906.429	5908.066	16926.01	NE I			1.7	IME
5902.792	5904.428	16936.44	NE I			0.7	PS
5902.463	5904.099	16937.39	NE I			1.7	IME
5902.097	5903.733	16938.44	NE I			0.5	PS
5898.406	5900.041	16949.04	NE I			1.3	PS
5881.8950	5883.5252	16996.613	NE I			3.0	S
5872.828	5874.456	17022.85	NE			1.5	IME
5872.149	5873.777	17024.82	NE I			1.9	PS
5870.971	5872.598	17028.24	NE I			0.5	PS
5868.417	5870.044	17035.65	NE I			1.9	PS
5852.4878	5854.1101	17082.016	NE I			3.3	S
5828.910	5830.526	17151.11	NE I			1.9	PS
5820.155	5821.769	17176.91	NE I			2.7	IME
5816.645	5818.258	17187.28	NE I			1.7	PS
5812.400	5814.012	17199.83	NE			1.2	PS
5811.417	5813.028	17202.74	NE I			2.5	PS
5804.449	5806.058	17223.39	NE I			2.7	IME
5804.098	5805.707	17224.43	NE I			1.9	PS
5770.307	5771.907	17325.30	NE I			1.7	PS
5764.418	5766.017	17343.00	NE I			2.8	IME
5764.063	5765.662	17344.06	NE I			0.5	PS
5760.585	5762.183	17354.53	NE I			1.8	PS
5748.650	5750.244	17390.56	NE I			1.8	PS
5748.299	5749.893	17391.63	NE I			2.7	IME
5719.532	5721.119	17479.10	NE I			1.9	PS
5719.225	5720.812	17480.04	NE I			2.7	IME
5718.899	5720.486	17481.03	NE I			2.2	PS
5715.339	5716.925	17491.92	NE I			1.5	PS
5689.817	5691.396	17570.38	NE I			2.2	IME

AIR WAVELENGTH (ANGSTRMS)	VACUUM WAVELENGTH (ANGSTRMS)	WAVENUMBER (KAYSERS)	LMENT	LOG ARC	INTENSITY SPK	DIS	REF
5684.647	5686.224	17586.36	NE I			1.4	PS
5662.547	5664.118	17655.00	NE I			1.9	IME
5656.659	5658.229	17673.37	NE I			2.7	IME
5656.030	5657.600	17675.34	NE I			1.9	PS
5652.570	5654.139	17686.16	NE I			1.9	PS
5591.15	5592.70	17880.4	NE I			0.9	GR
5589.378	5590.930	17886.11	NE I			1.7	PS
5585.905	5587.456	17897.23	NE I			0.7	PS
5576.049	5577.597	17928.87	NE I			1.5	PS
5563.047	5564.592	17970.77	NE I			1.9	PS
5562.769	5564.314	17971.67	NE I			2.7	IME
5562.441	5563.986	17972.73	NE I			2.2	PS
5559.087	5560.631	17983.57	NE I			1.5	PS
5547.120	5548.661	18022.37	NE I			0.9	PS
5538.641	5540.179	18049.96	NE I			1.7	PS
5533.678	5535.215	18066.15	NE I			1.9	PS
5520.63	5522.16	18108.8	NE I			0.5	GR
5511.485	5513.016	18138.89	NE I			1.2	PS
5511.176	5512.707	18139.91	NE I			0.5	PS
5507.339	5508.869	18152.55	NE I			1.4	PS
5494.407	5495.934	18195.27	NE I			1.7	PS
5448.508	5450.022	18348.55	NE I			2.2	IME
5433.649	5435.159	18398.73	NE I			2.4	IME
5420.155	5421.662	18444.53	NE I			1.7	PS
5418.555	5420.061	18449.98	NE I			2.2	PS
5412.655	5414.160	18470.09	NE I			2.4	PS
5410.12	5411.62	18478.7	NE I			0.7	PS
5400.562	5402.063	18511.45	NE I			3.3	I
5383.257	5384.754	18570.95	NE I			1.4	PS
5374.975	5376.470	18599.57	NE I			1.7	IME
5372.314	5373.808	18608.78	NE I			1.9	PS
5366.222	5367.714	18629.90	NE I			1.4	PS
5362.248	5363.739	18643.71	NE I			1.4	PS
5360.442	5361.933	18649.99	NE I			1.5	PS
5360.012	5361.503	18651.49	NE I			2.2	IME
5358.020	5359.510	18658.42	NE I			1.0	PS
5355.422	5356.911	18667.47	NE I			2.2	IME
5355.176	5356.665	18668.33	NE I			2.2	PS
5349.210	5350.698	18689.15	NE I			2.2	PS
5343.284	5344.770	18709.88	NE I			2.8	IME
5341.093	5342.579	18717.55	NE I			3.0	I
5335.710	5337.194	18736.44	NE I			1.0	PS
5333.323	5334.807	18744.82	NE I			1.7	PS
5330.777	5332.260	18753.77	NE I			2.8	IME
5326.396	5327.878	18769.20	NE I			1.9	IME
5316.806	5318.285	18803.05	NE			1.4	PS
5314.781	5316.260	18810.22	NE I			1.5	PS
5304.756	5306.232	18845.76	NE I			1.8	IME
5298.190	5299.664	18869.12	NE I			2.2	IME
5280.070	5281.539	18933.87	NE I			1.7	PS
5274.043	5275.511	18955.51	NE I			1.6	PS
5234.028	5235.485	19100.43	NE I			1.7	IME
5222.351	5223.805	19143.13	NE I			1.7	IME
5214.337	5215.789	19172.55	NE I			1.5	PS
5210.573	5212.024	19186.41	NE I			1.7	IME
5208.863	5210.313	19192.70	NE I			1.8	IME
5206.565	5208.015	19201.18	NE I			0.5	PS
5203.895	5205.344	19211.03	NE I			2.2	IME
5193.224	5194.670	19250.50	NE I			2.2	IME
5193.130	5194.576	19250.85	NE			2.2	IME
5191.327	5192.773	19257.53	NE I			1.5	PS
5188.612	5190.057	19267.61	NE I			2.2	IME
5163.474	5164.912	19361.41	NE I			1.0	PS
5158.894	5160.331	19378.60	NE I			1.7	PS
5156.664	5158.100	19386.98	NE I			1.7	IME
5154.422	5155.858	19395.41	NE I			1.7	IME
5151.963	5153.398	19404.67	NE I			1.9	IME
5150.077	5151.512	19411.78	NE I			1.5	PS
5145.122	5146.555	19430.47	NE I			1.5	PS
5145.011	5146.444	19430.89	NE I			2.7	PS

AIR WAVELENGTH (ANGSTRMS)	VACUUM WAVELENGTH (ANGSTRMS)	WAVENUMBER (KAYSERS)	LMENT	LOG ARC	INTENSITY SPK	DIS	REF
5144.938	5146.371	19431.17	NE I			2.7	IME
5143.265	5144.698	19437.49	NE I			0.7	PS
5122.337	5123.764	19516.90	NE I			2.2	PS
5122.257	5123.684	19517.21	NE I			2.2	IME
5120.506	5121.933	19523.88	NE I			1.4	PS
5117.011	5118.437	19537.22	NE I			1.5	PS
5116.503	5117.929	19539.15	NE I			2.2	IME
5113.675	5115.100	19549.96	NE I			1.9	IME
5104.705	5106.128	19584.31	NE I			1.5	IME
5099.042	5100.463	19606.06	NE I			1.4	PS
5090.321	5091.740	19639.65	NE I			0.9	PS
5083.968	5085.385	19664.19	NE I			1.4	PS
5081.360	5082.776	19674.29	NE I			1.2	PS
5080.383	5081.799	19678.07	NE I			2.2	IME
5078.762	5080.178	19684.35	NE I			1.2	PS
5076.581	5077.996	19692.81	NE I			1.5	PS
5074.200	5075.615	19702.05	NE I			1.5	IME
5074.062	5075.476	19702.58	NE I			0.5	PS
5052.930	5054.339	19784.98	NE I			1.4	PS
5046.608	5048.015	19809.77	NE I			0.5	PS
5045.816	5047.223	19812.87	NE I			1.2	PS
5042.853	5044.259	19824.52	NE I			1.2	PS
5037.751	5039.156	19844.59	NE I			2.7	IME
5037.577	5038.982	19845.28	NE I			0.5	PS
5035.989	5037.393	19851.54	NE I			1.5	PS
5031.348	5032.751	19869.85	NE I			2.4	IME
5022.870	5024.271	19903.39	NE I			1.4	IME
5015.187	5016.586	19933.88	NE I			0.7	PS
5011.003	5012.401	19950.52	NE I			1.4	IME
5005.333	5006.729	19973.12	NE I			1.7	PS
5005.160	5006.556	19973.81	NE I			2.7	IME
5000.395	5001.790	19992.84	NE I			0.5	PS
4998.502	4999.896	20000.41	NE I			1.0	PS
4997.482	4998.876	20004.50	NE I			1.2	PS
4994.930	4996.323	20014.72	NE I			2.2L	IME
4979.625	4981.014	20076.23	NE I			0.7	PS
4975.961	4977.349	20091.01	NE I			1.0	PS
4974.760	4976.148	20095.86	NE I			1.7	PS
4973.538	4974.926	20100.80	NE I			2.0	PS
4957.122	4958.505	20167.37	NE I			2.2	IME
4957.033	4958.416	20167.73	NE I			3.0	IME
4955.382	4956.765	20174.45	NE I			2.2	PS
4944.987	4946.367	20216.86	NE I			2.0	IME
4939.041	4940.420	20241.20	NE I			2.0	IME
4930.944	4932.320	20274.43	NE I			1.7	PS
4928.235	4929.611	20285.58	NE I			1.8	IME
4899.013	4900.381	20406.58	NE I			1.7	PS
4897.924	4899.292	20411.11	NE I			1.8	PS
4892.228	4893.594	20434.88	NE I			1.0	PS
4892.090	4893.456	20435.46	NE I			2.7	IME
4888.365	4889.730	20451.03	NE I			0.7	PS
4885.084	4886.448	20464.76	NE I			2.0	PS
4884.915	4886.279	20465.47	NE I			3.0	IME
4883.403	4884.767	20471.81	NE I			1.2	PS
4868.268	4869.628	20535.45	NE I			1.8	PS
4867.010	4868.369	20540.76	NE I			1.8	PS
4866.476	4867.835	20543.01	NE I			1.9	IME
4865.505	4866.864	20547.11	NE I			2.0	IME
4864.351	4865.710	20551.99	NE I			1.5	PS
4863.079	4864.437	20557.36	NE I			2.0	IME
4859.604	4860.961	20572.06	NE I			1.2	PS
4852.655	4854.011	20601.52	NE I			2.0	IME
4851.501	4852.856	20606.42	NE I			1.8	PS
4849.530	4850.885	20614.80	NE I			1.5	PS
4845.767	4847.121	20630.80	NE I			0.7	PS
4845.145	4846.499	20633.45	NE I			1.2	PS
4842.941	4844.294	20642.84	NE I			1.7	PS
4842.566	4843.919	20644.44	NE I			1.0	PS
4837.312	4838.664	20666.86	NE I			2.7	IME
4829.288	4830.637	20701.20	NE I			0.7	PS

AIR WAVELENGTH (ANGSTRMS)	VACUUM WAVELENGTH (ANGSTRMS)	WAVENUMBER (KAYSERS)	LMENT	LOG INTENSITY ARC	SPK	DIS	REF
4827.587	4828.936	20708.50	NE I			2.5	IME
4827.338	4828.687	20709.56	NE I			3.0	IME
4825.529	4826.877	20717.33	NE I			1.7	PS
4823.370	4824.718	20726.60	NE I			1.7	PS
4823.174	4824.522	20727.44	NE I			2.0	PS
4821.924	4823.271	20732.82	NE I			2.5	IME
4819.937	4821.284	20741.36	NE I			1.8	PS
4818.789	4820.136	20746.30	NE I			2.2	PS
4817.636	4818.982	20751.27	NE I			2.5	IME
4814.338	4815.683	20765.48	NE I			1.7	PS
4810.634	4811.979	20781.47	NE I			2.0	PS
4810.063	4811.407	20783.94	NE I			2.2	IME
4809.500	4810.844	20786.37	NE I			1.0	PS
4802.363	4803.705	20817.26	NE I			1.0	PS
4800.111	4801.453	20827.03	NE			1.2	IME
4795.62	4796.96	20846.5	NE II			1.2	BN
4790.728	4792.067	20867.82	NE I			1.5	PS
4790.218	4791.557	20870.04	NE			1.7	IME
4789.600	4790.939	20872.74	NE I			2.0	PS
4788.926	4790.265	20875.67	NE			2.5	IME
4781.95	4783.29	20906.1	NE II			0.7	BN
4781.239	4782.576	20909.24	NE I			0.3	PS
4780.884	4782.221	20910.79	NE I			1.5	PS
4780.338	4781.674	20913.18	NE I			1.7	IME
4758.728	4760.059	21008.14	NE I			2.2	IME
4754.440	4755.770	21027.09	NE I			2.0	PS
4752.731	4754.060	21034.65	NE I			3.0	IME
4751.802	4753.131	21038.76	NE I			1.5	PS
4750.686	4752.015	21043.71	NE I			1.5	PS
4749.572	4750.900	21048.64	NE I			2.5	IME
4732.53	4733.85	21124.4	NE II			0.7	BN
4732.3	4733.6	21125.	NE			1.2	BL
4725.145	4726.467	21157.45	NE I			1.8	IME
4724.162	4725.484	21161.85	NE I			0.7	PS
4723.810	4725.131	21163.43	NE I			1.3	PS
4722.714	4724.035	21168.34	NE I			1.2	PS
4722.150	4723.471	21170.87	NE I			0.7	PS
4721.536	4722.857	21173.62	NE I			1.8	PS
4719.37	4720.69	21183.3	NE II			1.0	BN
4717.608	4718.928	21191.25	NE I			1.8	PS
4715.344	4716.663	21201.43	NE I			3.2	IME
4715.246	4716.565	21201.87	NE I			1.5	PS
4715.132	4716.451	21202.38	NE I			1.5	PS
4714.336	4715.655	21205.96	NE I			1.8	PS
4713.13	4714.45	21211.4	NE			2.0	WA
4712.800	4714.119	21212.87	NE I			1.0	PS
4712.135	4713.453	21215.87	NE I			1.2	PS
4712.060	4713.378	21216.21	NE I			3.0	PS
4710.478	4711.796	21223.33	NE I			1.5	PS
4710.058	4711.376	21225.22	NE I			3.0	PS
4710.058	4711.376	21225.22	NE II			3.0	PS
4708.854	4710.172	21230.65	NE I			3.1	IME
4704.395	4705.711	21250.77	NE I			3.2	IME
4702.526	4703.842	21259.22	NE I			2.2	PS
4700.469	4701.784	21268.52	NE I			0.7	PS
4696.943	4698.257	21284.49	NE I			0.7	PS
4695.363	4696.677	21291.65	NE I			1.3	PS
4691.580	4692.893	21308.82	NE I			1.2	PS
4688.191	4689.503	21324.22	NE I			0.3	PS
4687.671	4688.983	21326.59	NE I			2.0	IME
4683.764	4685.075	21344.38	NE I			1.5	PS
4683.238	4684.549	21346.77	NE I			0.7	PS
4682.910	4684.221	21348.27	NE I			1.0	PS
4682.146	4683.456	21351.75	NE I			1.3	PS
4681.930	4683.240	21352.74	NE I			1.3	PS
4681.200	4682.510	21356.07	NE I			1.7	PS
4680.363	4681.673	21359.89	NE I			2.0	PS
4679.135	4680.445	21365.49	NE I			2.2	IME
4678.604	4679.914	21367.92	NE I			1.7	PS
4678.218	4679.527	21369.68	NE I			2.5	IME

AIR WAVELENGTH (ANGSTRMS)	VACUUM WAVELENGTH (ANGSTRMS)	WAVENUMBER (KAYSERS)	LMENT	LOG INTENSITY ARC	SPK	DIS	REF
4670.884	4672.191	21403.23	NE I			1.8	IME
4669.02	4670.33	21411.8	NE			1.7	BL
4667.356	4668.663	21419.41	NE I			2.0	IME
4666.654	4667.960	21422.63	NE I			1.7	PS
4663.518	4664.824	21437.04	NE I			1.3	PS
4663.092	4664.397	21439.00	NE I			1.6	PS
4661.104	4662.409	21448.14	NE I			2.2	IME
4656.392	4657.696	21469.84	NE I			2.5	IME
4653.699	4655.002	21482.27	NE I			1.7	PS
4652.101	4653.404	21489.65	NE I			1.5	PS
4649.904	4651.206	21499.80	NE I			1.8	IME
4645.416	4646.717	21520.57	NE I			2.5	IME
4644.833	4646.134	21523.27	NE I			1.6	L
4643.182	4644.482	21530.93	NE I			0.7	PS
4640.443	4641.742	21543.63	NE I			1.8	PS
4639.591	4640.890	21547.59	NE I			1.5	PS
4636.974	4638.273	21559.75	NE I			1.7	PS
4636.634	4637.932	21561.33	NE I			1.8	IME
4636.125	4637.423	21563.70	NE I			1.8	IME
4634.73	4636.03	21570.2	NE II			1.2	BN
4628.460	4629.756	21599.41	NE I			1.5	PS
4628.309	4629.605	21600.11	NE I			2.2	IME
4627.799	4629.095	21602.49	NE II			1.3	PS
4627.799	4629.095	21602.49	NE I			1.3	PS
4617.837	4619.130	21649.10	NE I			1.8	IME
4616.911	4618.204	21653.44	NE I			0.7	PS
4615.98	4617.27	21657.8	NE II			1.7	BN
4614.391	4615.684	21665.26	NE I			2.0	IME
4612.89	4614.18	21672.3	NE II			0.7	BN
4609.910	4611.201	21686.32	NE I			2.2	IME
4609.365	4610.656	21688.89	NE I			1.5	PS
4604.938	4606.228	21709.74	NE I			0.7	PS
4604.095	4605.385	21713.71	NE I			1.2	PS
4600.16	4601.45	21732.3	NE II			0.7	BL
4595.249	4596.536	21755.51	NE I			1.7	PS
4593.243	4594.530	21765.01	NE I			1.7	PS
4588.13	4589.42	21789.3	NE II			1.5	BN
4585.876	4587.161	21799.98	NE I			1.0	PS
4582.980	4584.264	21813.75	NE I			0.7	PS
4582.556	4583.840	21815.77	NE I			1.2	PS
4582.450	4583.734	21816.27	NE I			2.2	IME
4582.105	4583.389	21817.92	NE I			1.2	PS
4582.035	4583.319	21818.25	NE I			2.2	IME
4580.35	4581.63	21826.3	NE II			1.5	BN
4575.858	4577.140	21847.70	NE I			1.3	PS
4575.060	4576.342	21851.51	NE I			2.5	IME
4574.49	4575.77	21854.2	NE II			0.7	BN
4573.759	4575.041	21857.73	NE I			1.5	IME
4573.557	4574.839	21858.69	NE I			1.7	PS
4573.066	4574.348	21861.04	NE I			0.7	PS
4569.01	4570.29	21880.4	NE II			1.8	BN
4567.845	4569.125	21886.03	NE I			1.0	PS
4567.139	4568.419	21889.41	NE I			1.2	PS
4566.830	4568.110	21890.89	NE I			1.6	PS
4565.888	4567.168	21895.41	NE I			1.8	IME
4565.49	4566.77	21897.3	NE II			0.7	BN
4562.05	4563.33	21913.8	NE II			0.7	BN
4556.698	4557.975	21939.57	NE I			0.3	PS
4555.392	4556.669	21945.86	NE I			1.5	PS
4554.824	4556.101	21948.59	NE I			1.6	PS
4554.561	4555.838	21949.86	NE I			0.7	PS
4554.415	4555.692	21950.56	NE I			1.0	PS
4553.16	4554.44	21956.6	NE II			1.7	BN
4552.598	4553.874	21959.32	NE I			1.5	IME
4547.728	4549.003	21982.84	NE I			1.2	PS
4547.218	4548.493	21985.30	NE I			1.0	PS
4544.502	4545.776	21998.44	NE I			1.7	PS
4544.11	4545.38	22000.3	NE II			0.7	BN
4540.376	4541.649	22018.43	NE I			1.7	IME
4539.168	4540.441	22024.29	NE I			1.7	PS

AIR WAVELENGTH (ANGSTRMS)	VACUUM WAVELENGTH (ANGSTRMS)	WAVENUMBER (KAYSERS)	LMENT	LOG INTENSITY ARC	SPK	DIS	REF
4538.309	4539.581	22028.46	NE I			2.5	PS
4537.751	4539.023	22031.17	NE I			3.0	IME
4537.683	4538.955	22031.50	NE I			2.5	PS
4536.312	4537.584	22038.16	NE I			2.2	PS
4535.47	4536.74	22042.2	NE II			1.5	BN
4534.66	4535.93	22046.2	NE II			1.2	BN
4529.476	4530.746	22071.42	NE I			1.5	PS
4527.725	4528.995	22079.96	NE I			1.2	PS
4526.685	4527.954	22085.03	NE I			1.2	PS
4526.177	4527.446	22087.51	NE I			1.7	PS
4525.764	4527.033	22089.52	NE I			1.8	IME
4522.66	4523.93	22104.7	NE II			1.7	BN
4517.736	4519.003	22128.77	NE II			2.0	IME
4517.736	4519.003	22128.77	NE I			2.0	IME
4517.29	4518.56	22131.0	NE			1.2	BL
4516.936	4518.203	22132.69	NE I			1.7	PS
4515.411	4516.677	22140.17	NE I			1.5	PS
4515.022	4516.288	22142.08	NE I			0.3	PS
4514.891	4516.157	22142.72	NE I			1.8	PS
4514.80	4516.07	22143.2	NE II			1.2	BN
4511.509	4512.774	22159.32	NE I			1.3	PS
4511.37	4512.64	22160.0	NE II			1.7	BL
4511.29	4512.56	22160.4	NE II			1.2	BN
4510.170	4511.435	22165.90	NE I			1.2	PS
4508.21	4509.47	22175.5	NE II			1.5	BN
4502.52	4503.78	22203.6	NE II			0.3	BN
4500.182	4501.444	22215.09	NE I			1.7	IME
4499.843	4501.105	22216.76	NE I			0.7	PS
4499.000	4500.262	22220.93	NE I			1.3	PS
4499.000	4500.262	22220.93	NE II			1.3	PS
4493.699	4494.960	22247.14	NE I			1.7	PS
4493.108	4494.368	22250.07	NE I			0.7	PS
4492.689	4493.949	22252.14	NE I			1.2	PS
4492.412	4493.672	22253.51	NE I			1.5	PS
4492.132	4493.392	22254.90	NE I			0.7	PS
4491.838	4493.098	22256.36	NE I			1.7	PS
4491.771	4493.031	22256.69	NE I			1.9	PS
4488.093	4489.352	22274.93	NE I			2.5	IME
4483.190	4484.448	22299.29	NE I			2.2	IME
4480.823	4482.080	22311.07	NE I			1.2	PS
4475.656	4476.912	22336.83	NE I			2.0	IME
4475.131	4476.387	22339.45	NE I			0.7	PS
4475.131	4476.387	22339.45	NE II			0.7	PS
4471.52	4472.77	22357.5	NE II			1.5	BN
4470.971	4472.226	22360.23	NE I			0.7	PS
4468.91	4470.16	22370.5	NE II			1.8	BN
4466.807	4468.061	22381.07	NE I			1.8	IME
4466.503	4467.756	22382.60	NE I			0.3	PS
4466.045	4467.298	22384.89	NE I			0.7	PS
4465.651	4466.904	22386.87	NE I			1.7	PS
4462.856	4464.109	22400.89	NE I			0.3	PS
4460.175	4461.427	22414.35	NE I			2.0	IME
4456.95	4458.20	22430.6	NE II			1.8	BN
4455.564	4456.815	22437.55	NE I			1.2	PS
4454.285	4455.535	22443.99	NE I			0.7	PS
4453.324	4454.574	22448.84	NE I			0.3	PS
4453.253	4454.503	22449.19	NE I			0.7	PS
4452.983	4454.233	22450.55	NE I			1.2	PS
4452.56	4453.81	22452.7	NE			0.3	BL
4446.46	4447.71	22483.5	NE II			1.5	BN
4444.978	4446.226	22490.99	NE I			1.5	PS
4442.67	4443.92	22502.7	NE II			1.5	BN
4440.812	4442.059	22512.08	NE I			0.3	PS
4440.363	4441.610	22514.36	NE I			1.2	PS
4439.95	4441.20	22516.4	NE II			1.2	BL
4439.30	4440.55	22519.7	NE II			1.5	BL
4435.094	4436.339	22541.11	NE I			0.7	PS
4433.721	4434.966	22548.09	NE I			1.8	IME
4433.398	4434.643	22549.73	NE I			1.0	PS
4432.526	4433.771	22554.17	NE I			1.3	PS

AIR WAVELENGTH (ANGSTRMS)	VACUUM WAVELENGTH (ANGSTRMS)	WAVENUMBER (KAYSERS)	LMENT	LOG INTENSITY ARC	SPK	DIS	REF
4432.26	4433.50	22555.5	NE II			0.7	BL
4431.67	4432.91	22558.5	NE II			0.7	BL
4430.90	4432.14	22562.4	NE II			1.7	BN
4429.60	4430.84	22569.1	NE II			1.2	BN
4428.54	4429.78	22574.5	NE II			2.0	BN
4427.981	4429.224	22577.32	NE I			1.2	PS
4427.755	4428.998	22578.47	NE I			1.5	PS
4425.400	4426.643	22590.48	NE I			2.2	IME
4424.800	4426.042	22593.55	NE I			2.5	IME
4422.519	4423.761	22605.20	NE I			2.5	IME
4421.559	4422.801	22610.11	NE I			1.7	PS
4421.38	4422.62	22611.0	NE II			1.5	BL
4416.817	4418.057	22634.38	NE I			1.7	PS
4416.817	4418.057	22634.38	NE II			1.7	PS
4415.141	4416.381	22642.97	NE I			0.7	PS
4413.561	4414.801	22651.08	NE I			1.2	PS
4413.20	4414.44	22652.9	NE II			1.7	BN
4412.54	4413.78	22656.3	NE II			1.2	BL
4412.285	4413.524	22657.63	NE I			1.3	PS
4409.620	4410.858	22671.32	NE I			1.3	PS
4409.30	4410.54	22673.0	NE II			2.2	BN
4402.374	4403.611	22708.64	NE I			0.3	PS
4398.136	4399.371	22730.52	NE I			0.7	PS
4397.94	4399.18	22731.5	NE II			2.0	BL
4395.556	4396.791	22743.86	NE I			1.7	IME
4394.773	4396.008	22747.91	NE I			1.2	PS
4394.370	4395.604	22750.00	NE I			1.2	PS
4391.94	4393.17	22762.6	NE II			2.2	BL
4385.00	4386.23	22798.6	NE II			1.2	BN
4384.08	4385.31	22803.4	NE II			0.7	BL
4381.220	4382.451	22818.28	NE I			1.5	IME
4379.50	4380.73	22827.2	NE II			2.0	BN
4377.95	4379.18	22835.3	NE II			1.2	BL
4377.754	4378.984	22836.35	NE I			0.3	PS
4374.997	4376.226	22850.74	NE I			0.3	PS
4372.157	4373.386	22865.58	NE I			1.5	PS
4371.796	4373.025	22867.47	NE I			0.3	PS
4369.77	4371.00	22878.1	NE II			1.8	BN
4365.72	4366.95	22899.3	NE II			1.2	BL
4363.524	4364.750	22910.82	NE I			1.8	IME
4363.228	4364.454	22912.37	NE I			0.3	PS
4362.690	4363.916	22915.20	NE I			1.5	PS
4358.816	4360.041	22935.56	NE I			0.3	PS
4357.918	4359.143	22940.29	NE I			0.7	PS
4357.298	4358.523	22943.55	NE I			0.3	PS
4346.036	4347.258	23003.01	NE I			1.2	PS
4345.762	4346.984	23004.46	NE I			0.3	PS
4345.479	4346.701	23005.96	NE I			0.3	PS
4341.42	4342.64	23027.5	NE II			1.2	BL
4340.420	4341.640	23032.77	NE I			0.3	PS
4340.256	4341.476	23033.64	NE I			0.3	PS
4339.78	4341.00	23036.2	NE II			1.2	BN
4338.200	4339.420	23044.56	NE I			0.3	PS
4336.221	4337.440	23055.07	NE I			1.7	PS
4334.125	4335.344	23066.22	NE I			1.8	IME
4327.265	4328.482	23102.79	NE I			1.0	PS
4322.66	4323.88	23127.4	NE II			0.7	BL
4322.26	4323.48	23129.5	NE			1.2	BL
4321.492	4322.707	23133.65	NE I			0.3	PS
4318.834	4320.049	23147.89	NE I			0.7	PS
4316.008	4317.222	23163.04	NE I			1.2	PS
4314.695	4315.908	23170.09	NE I			1.5	PS
4310.130	4311.342	23194.63	NE I			0.3	S
4306.244	4307.455	23215.56	NE I			1.8	PS
4303.955	4305.166	23227.91	NE I			0.7	PS
4303.248	4304.458	23231.73	NE I			1.5	PS
4291.976	4293.184	23292.74	NE I			0.3	PS
4290.40	4291.61	23301.3	NE II			2.0	BL
4289.799	4291.006	23304.56	NE I			0.3	PS
4288.541	4289.748	23311.40	NE I			0.7	PS

AIR WAVELENGTH (ANGSTRMS)	VACUUM WAVELENGTH (ANGSTRMS)	WAVENUMBER (KAYSERS)	LMENT	LOG INTENSITY ARC	SPK	DIS	REF
4283.242	4284.447	23340.23	NE I			1.0	PS
4279.279	4280.483	23361.85	NE I			1.2	PS
4278.850	4280.054	23364.19	NE I			0.7	PS
4275.560	4276.763	23382.17	NE I			1.8	PS
4274.656	4275.859	23387.11	NE I			1.7	PS
4270.227	4271.429	23411.37	NE I			1.7	PS
4269.724	4270.926	23414.13	NE I			1.8	PS
4268.009	4269.210	23423.54	NE I			1.8	PS
4267.724	4268.925	23425.10	NE I			0.7	PS
4267.286	4268.487	23427.50	NE I			0.3	PS
4262.479	4263.679	23453.93	NE I			0.3	PS
4257.82	4259.02	23479.6	NE II			1.5	BL
4256.498	4257.696	23486.88	NE I			0.3	PS
4252.775	4253.972	23507.44	NE I			0.3	PS
4252.418	4253.615	23509.41	NE I			0.3	PS
4250.68	4251.88	23519.0	NE II			1.7	BL
4249.538	4250.734	23525.35	NE I			0.3	PS
4242.20	4243.39	23566.0	NE II			0.7	BN
4239.95	4241.14	23578.5	NE II			1.2	BN
4233.86	4235.05	23612.5	NE			1.5	BN
4231.60	4232.79	23625.1	NE II			1.7	BN
4224.57	4225.76	23664.4	NE II			0.7	BL
4220.92	4222.11	23684.9	NE II			1.2	BN
4219.76	4220.95	23691.4	NE II			2.0	BN
4217.15	4218.34	23706.0	NE II			1.5	BL
4206.43	4207.62	23766.4	NE II			1.2	BL
4203.270	4204.454	23784.30	NE I			0.3	PS
4198.099	4199.282	23813.60	NE I			1.8	PS
4196.415	4197.597	23823.15	NE I			1.2	PS
4175.488	4176.665	23942.55	NE I			1.6	PS
4175.223	4176.400	23944.07	NE I			1.8	PS
4174.369	4175.546	23948.97	NE I			1.8	PS
4173.966	4175.143	23951.28	NE I			0.3	PS
4166.091	4167.265	23996.55	NE I			1.5	PS
4164.802	4165.976	24003.98	NE I			1.7	PS
4150.67	4151.84	24085.7	NE II			1.5	BN
4133.65	4134.82	24184.9	NE II			1.5	BL
4131.054	4132.219	24200.07	NE I			1.8	PS
4130.512	4131.677	24203.25	NE I			1.3	PS
4128.072	4129.236	24217.55	NE I			1.5	PS
4126.941	4128.105	24224.19	NE I			0.3	PS
4112.865	4114.025	24307.09	NE I			1.0	PS
4112.694	4113.854	24308.10	NE I			1.3	PS
4112.100	4113.260	24311.61	NE I			1.2	PS
4100.30	4101.46	24381.6	NE II			0.7H	BN
4098.77	4099.93	24390.7	NE II			1.7	BN
4086.69	4087.84	24462.8	NE II			0.7	BL
4080.48	4081.63	24500.0	NE II			1.2H	BN
4080.148	4081.300	24502.00	NE I			1.7	PS
4079.359	4080.511	24506.74	NE I			0.3	PS
4069.389	4070.538	24566.78	NE I			0.7	PS
4069.243	4070.392	24567.66	NE I			1.5	PS
4069.049	4070.198	24568.83	NE			0.3	PS
4068.835	4069.984	24570.12	NE I			1.5	PS
4064.829	4065.977	24594.33	NE I			1.2	PS
4064.036	4065.184	24599.13	NE I			1.7	PS
4062.90	4064.05	24606.0	NE II			1.5	BN
4045.662	4046.805	24710.85	NE I			0.3	PS
4042.642	4043.784	24729.31	NE I			1.7	PS
4042.327	4043.469	24731.24	NE I			1.0	PS
4037.696	4038.837	24759.60	NE I			0.7	PS
4037.615	4038.756	24760.10	NE I			1.2	PS
4037.262	4038.403	24762.27	NE I			0.7	PS
4020.015	4021.151	24868.50	NE I			0.3	PS
4013.995	4015.130	24905.80	NE I			0.3	PS
3984.253	3985.380	25091.71	NE I			0.8	PS
3942.19	3943.31	25359.4	NE			0.8	BN
3829.77	3830.86	26103.8	NE II			1.6	BN
3818.44	3819.52	26181.3	NE II			1.4	BN
3806.30	3807.38	26264.8	NE II			0.6	BN

AIR WAVELENGTH (ANGSTRMS)	VACUUM WAVELENGTH (ANGSTRMS)	WAVENUMBER (KAYSERS)	LMENT	LOG INTENSITY ARC	SPK	DIS	REF
3800.02	3801.10	26308.2	NE II			1.3	BN
3777.16	3778.23	26467.4	NE II			1.9	BN
3769.449	3770.520	26521.54	NE I			0.8	PS
3766.29	3767.36	26543.8	NE II			1.9	BN
3754.216	3755.283	26629.15	NE I			1.7	IHU
3753.83	3754.90	26631.9	NE II			1.3	BN
3751.26	3752.33	26650.1	NE II			1.3	BN
3744.66	3745.72	26697.1	NE II			1.1	BL
3734.94	3736.00	26766.6	NE II			1.6	BL
3727.08	3728.14	26823.0	NE II			2.1	BN
3721.86	3722.92	26860.6	NE II			0.6	BN
3713.084	3714.140	26924.13	NE II			2.4	PS
3709.64	3710.70	26949.1	NE II			1.6	BN
3701.81	3702.86	27006.1	NE II			1.1	BN
3701.225	3702.278	27010.40	NE			1.6	IHU
3697.09	3698.14	27040.6	NE II			0.6	BN
3694.197	3695.248	27061.78	NE II			2.4	PS
3685.736	3686.785	27123.90	NE I			1.9	IHU
3682.243	3683.291	27149.63	NE I			1.9	IHU
3679.80	3680.85	27167.6	NE II			0.6	BN
3664.112	3665.156	27283.97	NE II			2.4	PS
3659.93	3660.97	27315.1	NE II			0.8	BN
3644.86	3645.90	27428.1	NE II			1.1	BN
3643.89	3644.93	27435.4	NE II			1.3	BN
3633.665	3634.701	27512.58	NE I			1.9	IHU
3632.75	3633.79	27519.5	NE II			0.6	BN
3628.06	3629.09	27555.1	NE II			1.1	BN
3612.35	3613.38	27674.9	NE II			0.8	BL
3609.179	3610.208	27699.23	NE I			1.7	IHU
3600.169	3601.196	27768.55	NE I			1.9	IHU
3594.18	3595.21	27814.8	NE II			1.1	BN
3593.640	3594.665	27819.00	NE I			2.4	IHU
3593.526	3594.551	27819.88	NE I			2.7	IHU
3590.47	3591.49	27843.6	NE II			0.6	BL
3574.64	3575.66	27966.9	NE II			1.3	BN
3571.26	3572.28	27993.3	NE II			1.1	BN
3568.53	3569.55	28014.7	NE II			1.4	BN
3565.84	3566.86	28035.9	NE II			1.1	BN
3562.942	3563.960	28058.68	NE I			1.2	PS
3561.23	3562.25	28072.2	NE II			1.1	BN
3557.84	3558.86	28098.9	NE II			1.1	BN
3542.90	3543.91	28217.4	NE II			1.6	BN
3542.28	3543.29	28222.3	NE II			0.6	BN
3537.99	3539.00	28256.6	NE II			0.8	BN
3520.472	3521.479	28397.16	NE I			3.0	IHU
3515.191	3516.196	28439.82	NE I			2.2	IHU
3510.721	3511.725	28476.03	NE I			1.7	IHU
3503.61	3504.61	28533.8	NE II			1.3	BN
3501.217	3502.219	28553.33	NE I			2.2	IHU
3498.064	3499.065	28579.06	NE I			1.9	IHU
3481.96	3482.96	28711.2	NE II			1.4	BN
3480.75	3481.75	28721.2	NE II			0.6	BL
3477.69	3478.69	28746.5	NE II			0.8	BN
3472.571	3473.565	28788.86	NE I			2.7	IHU
3466.579	3467.572	28838.62	NE I			2.2	IHU
3464.339	3465.331	28857.27	NE I			1.9	IHU
3460.525	3461.516	28889.07	NE I			1.9	IHU
3459.38	3460.37	28898.6	NE II			0.6	BL
3456.68	3457.67	28921.2	NE II			1.1L	BN
3456.52	3457.51	28922.6	NE			0.8	BL
3454.195	3455.185	28942.01	NE I			1.9	IHU
3453.10	3454.09	28951.2	NE II			0.8	BN
3450.765	3451.754	28970.78	NE I			1.7	IHU
3447.703	3448.691	28996.51	NE I			2.2	IHU
3443.70	3444.69	29030.2	NE II			0.6	BN
3438.97	3439.96	29070.1	NE II			0.6	BL
3428.76	3429.74	29156.7	NE II			1.3	BN
3423.913	3424.895	29197.98	NE I			1.7	IHU
3418.007	3418.987	29248.43	NE I			1.7	IHU
3417.904	3418.884	29249.31	NE I			2.7	IHU

AIR WAVELENGTH (ANGSTRMS)	VACUUM WAVELENGTH (ANGSTRMS)	WAVENUMBER (KAYSERS)	LMENT	LOG ARC	INTENSITY SPK	DIS	REF
3417.71	3418.69	29251.0	NE II			1.3	BN
3416.87	3417.85	29258.2	NE II			1.1	BN
3414.82	3415.80	29275.7	NE II			0.6	BN
3413.13	3414.11	29290.2	NE II			0.8	BL
3406.88	3407.86	29343.9	NE II			1.3	BL
3404.77	3405.75	29362.1	NE II			1.1	BL
3392.78	3393.75	29465.9	NE II			1.3	BN
3390.56	3391.53	29485.2	NE II			0.6	BL
3388.46	3389.43	29503.5	NE II			1.4	BN
3386.24	3387.21	29522.8	NE II			0.6	BN
3378.28	3379.25	29592.4	NE II			1.3	BN
3377.23	3378.20	29601.6	NE II			0.6	BN
3375.650	3376.619	29615.42	NE I			1.7	IHU
3374.10	3375.07	29629.0	NE II			0.6	BN
3371.87	3372.84	29648.6	NE II			1.1	BN
3369.908	3370.876	29665.88	NE I			2.8	IHU
3369.809	3370.777	29666.75	NE I			2.7	IHU
3367.20	3368.17	29689.7	NE II			1.4	BN
3362.89	3363.86	29727.8	NE II			0.6	BN
3360.63	3361.60	29747.8	NE II			1.3	BL
3357.90	3358.86	29772.0	NE II			0.8	BN
3356.35	3357.31	29785.7	NE II			0.6	BN
3355.05	3356.01	29797.2	NE II			1.6	BN
3353.63	3354.59	29809.9	NE II			0.6	BN
3351.744	3352.707	29826.64	NE I			1.4	PS
3345.49	3346.45	29882.4	NE II			0.8	BL
3344.43	3345.39	29891.9	NE II			1.3	BN
3336.12	3337.08	29966.3	NE II			0.6	BN
3334.87	3335.83	29977.6	NE II			2.4	BN
3330.78	3331.74	30014.4	NE II			0.6	BN
3329.20	3330.16	30028.6	NE II			1.1	BN
3327.16	3328.12	30047.0	NE II			1.3	BN
3323.75	3324.71	30077.9	NE II			1.6	BN
3320.29	3321.25	30109.2	NE II			0.6	BN
3319.75	3320.71	30114.1	NE II			0.8	BN
3311.30	3312.25	30190.9	NE II			0.8	BN
3309.78	3310.73	30204.8	NE II			0.8	BL
3297.74	3298.69	30315.1	NE II			1.6	BL
3275.20	3276.14	30523.7	NE II			0.6	BN
3270.79	3271.73	30564.8	NE II			0.6	BL
3269.86	3270.80	30573.5	NE II			0.8	BN
3263.43	3264.37	30633.8	NE II			0.8	BN
3255.39	3256.33	30709.4	NE II			0.6	BN
3248.15	3249.09	30777.9	NE II			0.8L	BN
3244.15	3245.09	30815.8	NE II			1.3	BL
3243.34	3244.28	30823.5	NE II			0.6	BN
3232.38	3233.31	30928.0	NE II			0.8	BL
3230.16	3231.09	30949.3	NE II			1.3	BN
3229.50	3230.43	30955.6	NE II			0.8	BN
3224.82	3225.75	31000.5	NE II			1.1	BN
3218.21	3219.14	31064.2	NE II			1.9	BN
3214.38	3215.31	31101.2	NE II			1.3	BN
3213.70	3214.63	31107.8	NE II			0.8	BN
3209.38	3210.31	31149.7	NE II			0.8	BN
3208.99	3209.92	31153.4	NE II			0.6	BN
3207.906	3208.833	31163.98	NE			1.0	PS
3198.62	3199.54	31254.4	NE II			1.3	BN
3194.61	3195.53	31293.7	NE II			1.1	BN
3190.86	3191.78	31330.5	NE II			0.6	BL
3188.74	3189.66	31351.3	NE II			0.8	BN
3187.60	3188.52	31362.5	NE II			0.6	BN
3176.16	3177.08	31475.4	NE II			0.8	BN
3173.58	3174.50	31501.0	NE II			0.8	BL
3167.568	3168.485	31560.83	NE I			1.7	PS
3165.70	3166.62	31579.4	NE II			1.1	BN
3164.46	3165.38	31591.8	NE II			0.8	BN
3153.404	3154.317	31702.58	NE I			1.7	PS
3151.16	3152.07	31725.2	NE II			0.6	BN
3148.603	3149.515	31750.92	NE I			1.9	PS
3147.701	3148.613	31760.02	NE I			1.4	PS

AIR WAVELENGTH (ANGSTRMS)	VACUUM WAVELENGTH (ANGSTRMS)	WAVENUMBER (KAYSERS)	LMENT	LOG ARC	INTENSITY SPK	DIS	REF
3143.74	3144.65	31800.0	NE II			0.6	BN
3141.35	3142.26	31824.2	NE II			0.8	BN
3132.22	3133.13	31917.0	NE II			0.6	BN
3126.190	3127.096	31978.55	NE I			2.2	PS
3118.02	3118.92	32062.3	NE II			1.1	BN
3101.407	3102.307	32234.08	NE			0.8	PS
3097.15	3098.05	32278.4	NE II			0.8	BL
3094.08	3094.98	32310.4	NE II			1.1	BN
3092.91	3093.81	32322.6	NE II			0.6	BL
3088.23	3089.13	32371.6	NE II			0.8	BN
3079.175	3080.069	32466.80	NE I			1.9	PS
3078.875	3079.769	32469.96	NE I			1.9	PS
3076.971	3077.865	32490.05	NE I			2.2	PS
3071.08	3071.97	32552.4	NE II			0.6	BN
3067.214	3068.105	32593.40	NE			0.7	PS
3065.668	3066.559	32609.84	NE			0.7	PS
3063.695	3064.586	32630.84	NE I			2.2	PS
3063.36	3064.25	32634.4	NE			0.6	BN
3062.58	3063.47	32642.7	NE			0.6	BL
3059.16	3060.05	32679.2	NE II			0.8	BN
3057.388	3058.277	32698.15	NE I			2.4	PS
3054.69	3055.58	32727.0	NE II			1.3	BN
3047.57	3048.46	32803.5	NE II			1.4	BN
3045.949	3046.835	32820.94	NE I			0.8	PS
3045.58	3046.47	32824.9	NE II			1.1	BN
3044.16	3045.05	32840.2	NE II			0.6	BN
3039.65	3040.53	32889.0	NE II			0.8	BN
3037.73	3038.61	32909.7	NE II			1.1	BN
3035.98	3036.86	32928.7	NE II			0.8	BN
3034.48	3035.36	32945.0	NE II			1.3	BN
3030.85	3031.73	32984.4	NE II			0.6	BN
3030.313	3031.195	32990.29	NE I			1.7	PS
3028.84	3029.72	33006.3	NE II			1.1	BN
3027.04	3027.92	33026.0	NE II			1.1	BN
3026.913	3027.794	33027.34	NE			1.2	PS
3017.348	3018.227	33132.04	NE I			1.7	PS
3017.348	3018.227	33132.04	NE II			1.7	PS
3012.955	3013.833	33180.34	NE I			1.7	PS
3012.129	3013.007	33189.44	NE I			1.7	PS
3001.65	3002.53	33305.3	NE II			1.4	BN
2994.250	2995.123	33387.61	NE I			0.5	PS
2992.438	2993.311	33407.83	NE I			2.2	PS
2982.663	2983.533	33517.31	NE I			2.4	PS
2980.922	2981.792	33536.88	NE I			1.7	PS
2980.642	2981.512	33540.03	NE I			1.6	PS
2979.806	2980.676	33549.44	NE I			1.7	PS
2975.518	2976.387	33597.79	NE I			1.5	PS
2974.714	2975.582	33606.87	NE I			2.4	PS
2967.20	2968.07	33692.0	NE II			0.9	BL
2963.29	2964.16	33736.4	NE II			0.6	BL
2957.293	2958.157	33804.83	NE I			0.9	PS
2955.73	2956.59	33822.7	NE II			1.6	BN
2952.527	2953.390	33859.40	NE I			0.7	PS
2951.10	2951.96	33875.8	NE II			0.6	BN
2949.316	2950.178	33896.26	NE I			1.2	PS
2949.043	2949.905	33899.40	NE I			1.0	PS
2947.297	2948.159	33919.48	NE I			2.2	PS
2935.30	2936.16	34058.1	NE II			0.3	BN
2933.70	2934.56	34076.7	NE II			0.6	BN
2932.721	2933.579	34088.06	NE I			1.9	PS
2929.312	2930.169	34127.72	NE I			1.2	PS
2925.63	2926.49	34170.7	NE II			1.0	BN
2913.168	2914.021	34316.84	NE I			2.2	PS
2911.461	2912.314	34336.96	NE I			1.4	PS
2910.44	2911.29	34349.0	NE			0.6	BL
2910.10	2910.95	34353.0	NE II			1.3	BN
2906.85	2907.70	34391.4	NE II			1.0	BL
2897.03	2897.88	34508.0	NE II			0.7	BL
2888.43	2889.28	34610.7	NE II			0.3	BN
2880.290	2881.135	34708.54	NE I			0.5	PS

AIR WAVELENGTH (ANGSTRMS)	VACUUM WAVELENGTH (ANGSTRMS)	WAVENUMBER (KAYSERS)	LMENT	LOG ARC	INTENSITY SPK	DIS	REF
2876.43	2877.27	34755.1	NE II			1.3L	BN
2873.00	2873.84	34796.6	NE II			1.0	BN
2872.663	2873.506	34800.69	NE I			1.5	PS
2869.95	2870.79	34833.6	NE II			0.7	BN
2862.070	2862.910	34929.49	NE I			0.9	PS
2858.01	2858.85	34979.1	NE II			0.7	BN
2842.57	2843.41	35169.1	NE I			1.2	PS
2835.233	2836.067	35260.10	NE I			1.2	PS
2832.921	2833.754	35288.87	NE I			0.9	PS
2827.584	2828.416	35355.48	NE I			0.5	PS
2825.609	2826.441	35380.19	NE I			0.9	PS
2825.259	2826.090	35384.57	NE I			1.0	PS
2814.685	2815.514	35517.50	NE I			1.3	PS
2809.50	2810.33	35583.0	NE II			1.3	BN
2795.963	2796.787	35755.31	NE I			0.9	PS
2795.101	2795.925	35766.34	NE I			1.5	PS
2794.592	2795.416	35772.85	NE I			0.7	PS
2794.26	2795.08	35777.1	NE II			1.0	BN
2792.660	2793.483	35797.60	NE I			0.5	PS
2792.318	2793.141	35801.98	NE I			1.3	PS
2792.05	2792.87	35805.4	NE II			1.4	BN
2781.68	2782.50	35938.9	NE I			0.5	PS
2781.42	2782.24	35942.2	NE			0.5	PS
2780.06	2780.88	35959.8	NE II			0.7	BN
2775.049	2775.868	36024.77	NE I			0.7	PS
2770.06	2770.88	36089.6	NE II			0.5	BN
2767.28	2768.10	36125.9	NE I			0.5	PS
2766.353	2767.170	36138.00	NE I			0.5	PS
2762.97	2763.79	36182.2	NE II			1.0	BL
2762.324	2763.140	36190.71	NE I			0.5	PS
2758.64	2759.46	36239.0	NE I			0.5	PS
2756.68	2757.49	36264.8	NE II			1.0	BL
2755.82	2756.63	36276.1	NE I			1.2	PS
2743.53	2744.34	36438.6	NE I			1.2	PS
2736.177	2736.987	36536.53	NE I			0.7	PS
2735.69	2736.50	36543.0	NE I			0.9	PS
2735.168	2735.978	36550.01	NE I			0.5	PS
2731.528	2732.337	36598.71	NE I			0.5	PS
2731.358	2732.167	36600.99	NE I			0.5	PS
2724.765	2725.572	36689.55	NE I			0.5	PS
2702.554	2703.356	36991.06	NE I			0.5	PS
2700.555	2701.356	37018.44	NE I			0.9	PS
2686.75	2687.55	37208.6	NE			1.2	PS
2677.36	2678.16	37339.1	NE			1.2	PS
2675.64	2676.44	37363.1	NE I			2.2	PS
2675.24	2676.04	37368.7	NE I			2.2	PS
2669.36	2670.15	37451.0	NE			0.7	PS
2669.13	2669.92	37454.3	NE I			0.5	PS
2663.28	2664.07	37536.5	NE			0.9	PS
2657.52	2658.31	37617.9	NE I			1.2	PS
2651.01	2651.80	37710.2	NE			1.7	PS
2648.56	2649.35	37745.1	NE I			1.4	PS
2648.21	2649.00	37750.1	NE I			1.2	PS
2647.76	2648.55	37756.5	NE I			0.9	PS
2647.42	2648.21	37761.4	NE I			2.2	PS
2646.19	2646.98	37778.9	NE I			1.2	PS
2645.70	2646.49	37785.9	NE I			1.5	PS
2645.51	2646.30	37788.6	NE			1.7	PS
2644.16	2644.95	37807.9	NE I			0.7	PS
2642.47	2643.26	37832.1	NE I			0.9	PS
2639.97	2640.76	37867.9	NE I			1.2	PS
2635.98	2636.77	37925.2	NE I			1.4	PS
2622.90	2623.68	38114.4	NE I			1.2	PS
2621.10	2621.88	38140.5	NE			0.9	PS
2619.02	2619.80	38170.8	NE			0.7	PS
2616.62	2617.40	38205.8	NE I			1.4	PS
2614.26	2615.04	38240.3	NE I			0.7	PS
2613.59	2614.37	38250.1	NE I			1.5	PS
2595.21	2595.99	38521.0	NE			1.7	PS
2591.15	2591.92	38581.4	NE I			0.5	PS

AIR WAVELENGTH (ANGSTRMS)	VACUUM WAVELENGTH (ANGSTRMS)	WAVENUMBER (KAYSERS)	LMENT	LOG ARC	INTENSITY SPK	DIS	REF
2590.67	2591.44	38588.5	NE I			1.0	PS
2574.55	2575.32	38830.1	NE I			0.9	PS
2561.79	2562.56	39023.5	NE I			0.9	PS
9966.19	9968.92	10031.2	NI	0.5			ME
9898.90	9901.61	10099.4	NI I	1.3			ME
9842.04	9844.74	10157.7	NI	0.3			ME
9689.35	9692.01	10317.8	NI I	0.5			ME
9520.06	9522.67	10501.2	NI I	1.7			ME
9447.29	9449.88	10582.1	NI I	0.7			ME
9396.57	9399.15	10639.3	NI I	0.3H			ME
9385.62	9388.20	10651.7	NI I	0.3H			ME
9258.47	9261.01	10798.0	NI I	0.7			ME
9196.18	9198.70	10871.1	NI I	0.3			ME
9106.40	9108.90	10978.3	NI I	0.7H			ME
9085.25	9087.74	11003.8	NI I	0.5			ME
9078.67	9081.16	11011.8	NI I	0.3			ME
9058.55	9061.04	11036.3	NI I	0.3			ME
9005.14	9007.61	11101.7	NI I	0.7			ME
8982.35	8984.82	11129.9	NI I	0.3			ME
8968.20	8970.66	11147.4	NI I	0.9			ME
8965.99	8968.45	11150.2	NI I	0.9			MIT
8954.65	8957.11	11164.3	NI I	0.3			ME
8877.07	8879.51	11261.9	NI I	0.3			ME
8862.58	8865.01	11280.3	NI I	1.0			MIT
8809.47	8811.89	11348.3	NI I	0.9			ME
8770.68	8773.09	11398.5	NI I	0.3			ME
8716.58	8718.97	11469.2	NI I	0.5			ME
8702.49	8704.88	11487.8	NI I	0.6			ME
8680.2	8682.6	11517.	NI	0.3			ME
8637.04	8639.41	11574.9	NI I	0.7			ME
8606.45	8608.81	11616.0	NI I	0.3			ME
8598.84	8601.20	11626.3	NI I	0.3			ME
8580.04	8582.40	11651.8	NI I	0.3			ME
8501.81	8504.15	11759.0	NI I	0.3			ME
8417.24	8419.55	11877.1	NI I	0.9			MIT
8214.30	8216.56	12170.6	NI	0.3H			SL
8034.56	8036.77	12442.8	NI I	0.3H			MIT
8012.96	8015.16	12476.3	NI I	0.3			MIT
7960.72	7962.91	12558.2	NI	0.3H			MIT
7953.11	7955.30	12570.2	NI I	0.3H			MIT
7917.48	7919.66	12626.8	NI I	1.5			MIT
7890.22	7892.39	12670.4	NI I	1.0			MIT
7863.79	7865.95	12713.0	NI I	1.3			MIT
7861.10	7863.26	12717.4	NI I	1.0			MIT
7855.12	7857.28	12727.1	NI I	0.9H			MIT
7826.81	7828.96	12773.1	NI I	1.0			MIT
7797.62	7799.77	12820.9	NI I	1.9			MIT
7788.95	7791.09	12835.2	NI I	1.8			MIT
7748.93	7751.06	12901.5	NI I	0.9			MIT
7727.66	7729.79	12937.0	NI I	1.9			MIT
7715.63	7717.75	12957.1	NI I	1.6			MIT
7714.32	7716.44	12959.3	NI I	1.8			MIT
7657.30	7659.41	13055.8	NI I	0.3H			MIT
7654.28	7656.39	13061.0	NI	1.0H			SL
7619.21	7621.31	13121.1	NI I	1.3			MIT
7617.00	7619.10	13124.9	NI I	1.8			MIT
7609.97	7612.06	13137.0	NI	0.3H			MIT
7574.08	7576.17	13199.3	NI I	1.5			MIT
7559.62	7561.70	13224.5	NI I	0.3			MIT
7555.60	7557.68	13231.6	NI I	1.6			MIT
7525.14	7527.21	13285.1	NI I	1.5			MIT
7522.78	7524.85	13289.3	NI I	1.6			MIT
7488.73	7490.79	13349.7	NI I	0.3H			MIT
7433.48	7435.53	13448.9	NI I	0.3			MIT
7422.30	7424.34	13469.2	NI I	2.8			MIT
7419.35	7421.39	13474.6	NI I	0.3			MIT
7414.51	7416.55	13483.3	NI I	2.3			MIT
7409.39	7411.43	13492.7	NI I	2.6			MIT
7401.13	7403.17	13507.7	NI I	1.2			MIT
7393.63	7395.67	13521.4	NI I	2.8			MIT

AIR WAVELENGTH (ANGSTRMS)	VACUUM WAVELENGTH (ANGSTRMS)	WAVENUMBER (KAYSERS)	LMENT	LOG INTENSITY ARC	SPK	DIS	REF
7386.21	7388.24	13535.0	NI I	2.0			MIT
7385.24	7387.27	13536.8	NI I	2.2			MIT
7381.94	7383.97	13542.8	NI I	1.6			MIT
7351.40	7353.43	13599.1	NI I	0.5H			MIT
7333.49	7335.51	13632.3	NI I	0.3			MIT
7327.67	7329.69	13643.1	NI I	1.4			MIT
7309.64	7311.65	13676.8	NI I	0.5			MIT
7297.75	7299.76	13699.1	NI I	0.6			MIT
7291.48	7293.49	13710.9	NI I	2.0			MIT
7290.87	7292.88	13712.0	NI I	1.3			MIT
7286.56	7288.57	13720.1	NI I	1.0			MIT
7266.22	7268.22	13758.5	NI I	1.2			MIT
7261.93	7263.93	13766.7	NI I	2.5			MIT
7225.13	7227.12	13836.8	NI I	0.3			MIT
7220.79	7222.78	13845.1	NI I	0.5H			MIT
7197.03	7199.01	13890.8	NI I	2.3			MIT
7182.00	7183.98	13919.9	NI I	2.3			MIT
7173.73	7175.71	13935.9	NI I	0.3			MIT
7170.14	7172.12	13942.9	NI I	0.7			MIT
7167.01	7168.99	13949.0	NI I	1.5			MIT
7157.39	7159.36	13967.7	NI	0.3H			SL
7150.93	7152.90	13980.3	NI	0.7H			MIT
7146.74	7148.71	13988.5	NI	0.3			MIT
7141.62	7143.59	13998.6	NI I	0.3			MIT
7126.71	7128.67	14027.8	NI I	0.5			MIT
7122.24	7124.20	14036.7	NI I	3.00			MIT
7112.72	7114.68	14055.4	NI	0.3			SL
7110.91	7112.87	14059.0	NI I	2.0			MIT
7101.95	7103.91	14076.8	NI	0.5			MIT
7095.40	7097.36	14089.7	NI I	1.2			MIT
7082.22	7084.17	14116.0	NI I	0.3			ME
7068.33	7070.28	14143.7	NI I	0.3			MIT
7067.50	7069.45	14145.4	NI I	0.3			MIT
7063.57	7065.52	14153.2	NI I	1.0			MIT
7062.97	7064.92	14154.4	NI I	1.5			MIT
7049.68	7051.62	14181.1	NI I	0.3			MIT
7037.37	7039.31	14205.9	NI I	0.6			MIT
7034.42	7036.36	14211.9	NI I	1.5			MIT
7030.06	7032.00	14220.7	NI I	2.0			MIT
7028.79	7030.73	14223.3	NI I	0.7			ME
7024.86	7026.80	14231.2	NI	1.7			MIT
7023.76	7025.70	14233.5	NI	1.0			MIT
7022.37	7024.31	14236.3	NI	0.6			MIT
7004.46	7006.39	14272.7	NI I	1.2			MIT
7001.57	7003.50	14278.6	NI I	1.5			MIT
6973.57	6975.49	14335.9	NI I	0.3			MIT
6955.06	6956.98	14374.1	NI I	1.9			MIT
6928.25	6930.16	14429.7	NI I	0.3			MIT
6914.57	6916.48	14458.2	NI I	2.5			MIT
6902.84	6904.74	14482.8	NI I	0.3			MIT
6901.88	6903.78	14484.8	NI I	0.3			MIT
6876.71	6878.61	14537.8	NI I	1.4			MIT
6870.13	6872.03	14551.7	NI I	0.5			MIT
6861.24	6863.13	14570.6	NI I	0.7			MIT
6850.48	6852.37	14593.5	NI I	0.3			MIT
6842.07	6843.96	14611.4	NI I	1.8			MIT
6813.598	6815.478	14672.48	NI I	1.5			MIT
6791.19	6793.06	14720.9	NI I	0.3			ME
6782.46	6784.33	14739.8	NI I	0.7W			MIT
6772.36	6774.23	14761.8	NI I	2.3			MIT
6767.778	6769.646	14771.82	NI I	2.5			IKS
6759.41	6761.28	14790.1	NI I	0.3			ME
6758.73	6760.60	14791.6	NI	0.3			SL
6690.80	6692.65	14941.8	NI I	0.3			MIT
6661.39	6663.23	15007.7	NI I	0.3H			MIT
6647.80	6649.64	15038.4	NI	0.7L			ME
6643.641	6645.476	15047.83	NI I	2.5W			IKS
6635.15	6636.98	15067.1	NI I	0.7H			MIT
6629.52	6631.35	15079.9	NI	0.3			MIT
6621.24	6623.07	15098.7	NI I	0.3H			ME

AIR WAVELENGTH (ANGSTRMS)	VACUUM WAVELENGTH (ANGSTRMS)	WAVENUMBER (KAYSERS)	LMENT	LOG INTENSITY ARC	SPK	DIS	REF
6598.594	6600.417	15150.56	NI I	0.7H			MIT
6592.517	6594.338	15164.53	NI I	0.7H			MIT
6586.328	6588.147	15178.77	NI I	1.6			MIT
6580.22	6582.04	15192.9	NI	0.3H			MIT
6576.38	6578.20	15201.7	NI	0.3			MIT
6532.891	6534.696	15302.93	NI I	0.5			MIT
6482.811	6484.602	15421.15	NI I	1.5			MIT
6455.07	6456.85	15487.4	NI	1.0			MIT
6451.580	6453.363	15495.80	NI I	0.3H			MIT
6424.905	6426.681	15560.13	NI I	0.3H			MIT
6421.507	6423.282	15568.36	NI I	0.5H			MIT
6414.603	6416.376	15585.12	NI I	0.7			MIT
6384.697	6386.462	15658.12	NI I	0.7H			MIT
6378.263	6380.026	15673.92	NI I	1.3H			MIT
6370.383	6372.144	15693.31	NI I	0.5			MIT
6366.483	6368.243	15702.92	NI I	1.2			MIT
6360.798	6362.557	15716.95	NI I	0.7H			MIT
6339.148	6340.901	15770.63	NI I	1.7			MIT
6327.603	6329.353	15799.40	NI I	1.4			MIT
6322.165	6323.913	15812.99	NI I	0.5			MIT
6314.675	6316.421	15831.75	NI I	2.5			MIT
6259.615	6261.346	15971.01	NI I	0.3			MIT
6258.591	6260.322	15973.62	NI I	0.8			MIT
6256.365	6258.096	15979.30	NI I	2.8W	1.0		IKS
6230.115	6231.839	16046.63	NI I	0.7			MIT
6223.994	6225.716	16062.41	NI I	1.5			MIT
6204.640	6206.357	16112.51	NI I	1.0			MIT
6191.186	6192.899	16147.53	NI I	2.7	0.0		MIT
6186.740	6188.452	16159.13	NI I	1.5			MIT
6180.093	6181.803	16176.51	NI I	0.3			MIT
6177.258	6178.967	16183.93	NI I	0.7			MIT
6176.814	6178.523	16185.10	NI I	2.6W			MIT
6175.424	6177.133	16188.74	NI I	2.5			MIT
6170.568	6172.276	16201.48	NI I	0.7			MIT
6163.42	6165.13	16220.3	NI I	2.0			MIT
6130.174	6131.871	16308.24	NI I	1.2			MIT
6128.990	6130.686	16311.39	NI I	1.0			MIT
6119.780	6121.474	16335.94	NI I	0.3			MIT
6116.181	6117.874	16345.55	NI I	2.2			MIT
6111.06	6112.75	16359.2	NI I	1.4			MIT
6108.123	6109.814	16367.11	NI I	2.3			MIT
6086.290	6087.975	16425.82	NI I	2.0			MIT
6053.680	6055.356	16514.30	NI I	0.7			MIT
6025.73	6027.40	16590.9	NI I	0.3H			ME
6012.251	6013.916	16628.10	NI	0.5			MIT
6007.258	6008.922	16641.92	NI I	1.3			MIT
5997.610	5999.271	16668.69	NI I	0.7			MIT
5996.74	5998.40	16671.1	NI I	1.0			MIT
5923.930	5925.571	16876.01	NI I	0.3			MIT
5906.486	5908.123	16925.85	NI I	0.3H			MIT
5892.878	5894.511	16964.94	NI I	0.8			IKS
5857.755	5859.379	17066.66	NI I	1.7			MIT
5847.010	5848.631	17098.02	NI I	0.7			MIT
5831.624	5833.241	17143.13	NI I	0.9			MIT
5805.233	5806.843	17221.06	NI I	1.7			MIT
5796.078	5797.685	17248.26	NI I	0.5			MIT
5760.847	5762.445	17353.75	NI I	1.7			MIT
5754.675	5756.271	17372.36	NI I	2.2W			MIT
5748.343	5749.937	17391.49	NI I	1.6			MIT
5715.086	5716.672	17492.70	NI I	1.7			MIT
5711.905	5713.490	17502.44	NI I	1.7			MIT
5709.559	5711.143	17509.63	NI I	2.0W		0.0	MIT
5694.998	5696.578	17554.40	NI I	1.6			MIT
5682.204	5683.781	17593.92	NI I	1.7			MIT
5669.945	5671.518	17631.96	NI I	1.0			MIT
5664.017	5665.589	17650.42	NI I	1.2			MIT
5649.697	5651.265	17695.15	NI I	1.2			MIT
5643.099	5644.665	17715.84	NI I	0.7			MIT
5642.660	5644.226	17717.22	NI I	0.3			MIT
5641.880	5643.446	17719.67	NI I	1.4			MIT

AIR WAVELENGTH (ANGSTRMS)	VACUUM WAVELENGTH (ANGSTRMS)	WAVENUMBER (KAYSERS)	LMENT	LOG ARC	INTENSITY SPK	DIS	REF
5637.121	5638.686	17734.63	NI I	1.2			MIT
5628.347	5629.909	17762.27	NI I	0.7			MIT
5625.326	5626.888	17771.81	NI I	1.5			MIT
5614.790	5616.349	17805.16	NI I	1.7			MIT
5600.038	5601.593	17852.06	NI I	1.0			MIT
5593.735	5595.288	17872.18	NI I	1.6	0.3H		MIT
5592.283	5593.836	17876.82	NI I	2.2G	0.0		MIT
5589.384	5590.936	17886.09	NI I	1.3			MIT
5587.865	5589.416	17890.96	NI I	1.7			MIT
5578.734	5580.283	17920.24	NI I	1.7			MIT
5553.693	5555.235	18001.04	NI I	0.9			MIT
5510.001	5511.532	18143.78	NI I	1.3			MIT
5504.120	5505.649	18163.16	NI I	0.3			MIT
5494.890	5496.417	18193.67	NI I	0.7			MIT
5480.893	5482.416	18240.13	NI I	0.3			MIT
5476.906	5478.428	18253.41	NI I	2.6W	0.9		MIT
5468.101	5469.620	18282.80	NI I	0.3			MIT
5462.487	5464.005	18301.59	NI I	1.3			MIT
5453.255	5454.771	18332.58	NI I	0.5			MIT
5435.871	5437.382	18391.20	NI I	1.7			MIT
5424.654	5426.162	18429.23	NI I	1.5			MIT
5411.227	5412.731	18474.96	NI I	1.6	0.3		MIT
5392.371	5393.870	18539.56	NI I	0.7			MIT
5388.350	5389.848	18553.40	NI I	0.5			MIT
5371.35	5372.84	18612.1	NI I	1.5			MIT
5353.415	5354.904	18674.47	NI I	1.6			MIT
5281.692	5283.162	18928.06	NI I	0.5			MIT
5268.348	5269.814	18976.00	NI I	1.0			MIT
5265.748	5267.214	18985.37	NI I	0.7			MIT
5235.35	5236.81	19095.6	NI I	1.5	0.3		MIT
5220.307	5221.760	19150.63	NI I	1.2			MIT
5216.494	5217.946	19164.63	NI I	1.0			MIT
5197.165	5198.612	19235.90	NI I	1.0			MIT
5192.524	5193.970	19253.09	NI I	1.0			MIT
5186.592	5188.036	19275.11	NI I	0.8			MIT
5184.585	5186.029	19282.58	NI I	1.7			MIT
5179.136	5180.578	19302.86	NI I	0.6			MIT
5176.565	5178.007	19312.45	NI I	1.8	0.3		MIT
5168.660	5170.100	19341.99	NI I	1.8			MIT
5157.993	5159.430	19381.99	NI I	0.9			MIT
5155.764	5157.200	19390.37	NI I	1.9	0.0		MIT
5155.140	5156.576	19392.71	NI I	1.7			MIT
5146.478	5147.912	19425.35	NI I	2.2	0.0		MIT
5142.771	5144.204	19439.35	NI I	2.0			MIT
5139.255	5140.687	19452.65	NI I	1.7			MIT
5137.075	5138.506	19460.91	NI I	2.2	0.0		MIT
5129.383	5130.812	19490.09	NI I	1.9			MIT
5125.211	5126.639	19505.96	NI I	1.7			MIT
5121.570	5122.997	19519.82	NI I	1.3			MIT
5115.397	5116.822	19543.38	NI I	1.9	0.7J		MIT
5102.971	5104.393	19590.97	NI I	1.6			MIT
5099.946	5101.367	19602.59	NI I	2.2W			MIT
5099.322	5100.743	19604.99	NI I	1.9			MIT
5096.874	5098.295	19614.40	NI I	1.7			MIT
5094.416	5095.836	19623.87	NI I	1.4			MIT
5088.956	5090.374	19644.92	NI I	1.3			MIT
5088.534	5089.952	19646.55	NI I	1.3			MIT
5085.479	5086.897	19658.35	NI I	1.0			MIT
5084.081	5085.498	19663.76	NI I	2.5W	0.3		MIT
5082.354	5083.771	19670.44	NI I	2.0W			MIT
5081.111	5082.527	19675.25	NI I	2.2W	0.3		MIT
5080.523	5081.939	19677.53	NI I	2.3W	0.5		MIT
5079.961	5081.377	19679.71	NI I	1.5			MIT
5076.321	5077.736	19693.82	NI I	1.0			MIT
5058.03	5059.44	19765.0	NI I	1.3			MIT
5051.527	5052.935	19790.48	NI I	1.7			MIT
5048.851	5050.259	19800.97	NI I	1.9	0.3H		MIT
5048.082	5049.490	19803.98	NI I	1.3			MIT
5042.195	5043.601	19827.10	NI I	1.9			MIT
5041.077	5042.483	19831.50	NI I	1.5			MIT

AIR WAVELENGTH (ANGSTRMS)	VACUUM WAVELENGTH (ANGSTRMS)	WAVENUMBER (KAYSERS)	LMENT	LOG ARC	INTENSITY SPK	DIS	REF
5039.259	5040.664	19838.66	NI I	1.3			MIT
5038.599	5040.004	19841.25	NI I	1.7			MIT
5035.961	5037.365	19851.65	NI I	1.8W			MIT
5035.374	5036.778	19853.96	NI I	2.5W	0.7		MIT
5032.748	5034.151	19864.32	NI I	0.8			MIT
5018.294	5019.694	19921.53	NI I	1.8W			MIT
5017.591	5018.990	19924.33	NI I	2.0W	0.0		MIT
5014.236	5015.635	19937.66	NI I	1.4			MIT
5012.464	5013.862	19944.71	NI I	1.8	0.3H		MIT
5010.961	5012.359	19950.69	NI I	1.7			MIT
5010.045	5011.442	19954.33	NI I	1.4			MIT
5003.751	5005.147	19979.43	NI I	1.3			MIT
5000.335	5001.730	19993.08	NI I	2.2W			MIT
4998.233	4999.627	20001.49	NI I	2.2			MIT
4996.850	4998.244	20007.03	NI I	1.9			MIT
4984.126	4985.517	20058.10	NI I	2.7O	0.0		MIT
4980.161	4981.551	20074.07	NI I	2.7O	0.0		MIT
4976.345	4977.733	20089.46	NI I	1.6			MIT
4976.155	4977.543	20090.23	NI I	1.0			MIT
4971.354	4972.741	20109.63	NI I	2.0			MIT
4953.204	4954.586	20183.32	NI I	2.2			MIT
4946.037	4947.417	20212.57	NI I	0.7			MIT
4945.458	4946.838	20214.93	NI I	2.0			MIT
4941.920	4943.299	20229.40	NI I	0.3			MIT
4937.337	4938.715	20248.18	NI I	2.6W			MIT
4935.830	4937.208	20254.36	NI I	2.2	0.0		MIT
4934.000	4935.377	20261.88	NI I	0.5			MIT
4925.578	4926.953	20296.52	NI I	2.0			MIT
4918.712	4920.085	20324.85	NI I	1.6			MIT
4918.363	4919.736	20326.29	NI I	2.3O	0.0		MIT
4913.970	4915.342	20344.47	NI I	2.3			MIT
4912.030	4913.401	20352.50	NI I	2.0			MIT
4904.413	4905.782	20384.11	NI I	2.6O	0.0		MIT
4886.992	4888.357	20456.77	NI I	1.5			MIT
4886.725	4888.090	20457.89	NI I	0.3			MIT
4874.809	4876.171	20507.90	NI I	1.4			MIT
4873.437	4874.798	20513.67	NI I	2.3	0.3H		MIT
4870.845	4872.205	20524.59	NI I	2.0			MIT
4866.267	4867.626	20543.89	NI I	2.5W	0.0		MIT
4864.282	4865.641	20552.28	NI I	1.3			MIT
4863.931	4865.290	20553.76	NI I	1.5			MIT
4857.382	4858.739	20581.47	NI I	2.0			MIT
4855.414	4856.770	20589.81	NI I	2.6W	0.0		MIT
4852.560	4853.916	20601.92	NI I	2.2			MIT
4843.165	4844.518	20641.89	NI I	1.3			MIT
4838.651	4840.003	20661.14	NI I	2.2	0.6		MIT
4832.704	4834.054	20686.57	NI I	1.8			MIT
4831.183	4832.533	20693.08	NI I	2.3	0.3		MIT
4829.028	4830.377	20702.32	NI I	2.5W	0.3H		MIT
4821.143	4822.490	20736.17	NI I	1.4	0.3		MIT
4817.847	4819.193	20750.36	NI I	1.2			MIT
4814.617	4815.963	20764.28	NI I	0.7			MIT
4811.999	4813.344	20775.58	NI I	1.0			MIT
4808.864	4810.208	20789.12	NI I	1.4			MIT
4806.996	4808.340	20797.20	NI I	2.2O	0.0		MIT
4786.541	4787.879	20886.07	NI I	2.5O	0.3		MIT
4786.293	4787.631	20887.16	NI I	1.4			MIT
4773.412	4774.747	20943.52	NI I	1.2			MIT
4763.950	4765.282	20985.12	NI I	2.2	0.0		MIT
4762.627	4763.959	20990.95	NI I	2.2	0.0H		MIT
4756.519	4757.849	21017.90	NI I	2.4	0.5		MIT
4756.259	4757.589	21019.05	NI	0.8			HA
4754.768	4756.098	21025.64	NI I	2.0			MIT
4752.426	4753.755	21036.00	NI I	2.2			MIT
4752.124	4753.453	21037.34	NI I	1.5			MIT
4740.165	4741.491	21090.41	NI I	1.2			MIT
4732.465	4733.789	21124.73	NI I	2.0			MIT
4731.809	4733.133	21127.66	NI I	2.0	0.3		MIT
4729.291	4730.614	21138.91	NI I	1.0			MIT
4727.851	4729.174	21145.34	NI I	0.7			MIT

AIR WAVELENGTH (ANGSTRMS)	VACUUM WAVELENGTH (ANGSTRMS)	WAVENUMBER (KAYSERS)	LMENT	LOG ARC	INTENSITY SPK	INTENSITY DIS	REF
4715.778	4717.097	21199.48	NI I	2.3	0.3		MIT
4714.421	4715.740	21205.58	NI I	3.0	0.9		MIT
4712.069	4713.387	21216.16	NI I	1.5			MIT
4705.50	4706.82	21245.8	NI I	0.3			MIT
4703.808	4705.124	21253.42	NI I	2.3			MIT
4701.536	4702.852	21263.70	NI I	2.2			MIT
4701.336	4702.652	21264.60	NI I	2.0			MIT
4698.408	4699.723	21277.85	NI I	1.5			MIT
4686.218	4687.530	21333.20	NI I	2.3	0.0		MIT
4675.639	4676.948	21381.47	NI I	0.9			MIT
4667.766	4669.073	21417.53	NI I	2.0			MIT
4666.994	4668.300	21421.07	NI I	1.7			MIT
4655.661	4656.964	21473.22	NI I	1.6			MIT
4648.659	4649.961	21505.56	NI I	2.6W	0.5		MIT
4606.231	4607.521	21703.64	NI I	2.0			MIT
4604.994	4606.284	21709.47	NI I	2.5	1.0H		MIT
4600.372	4601.661	21731.28	NI I	2.3			MIT
4595.951	4597.239	21752.19	NI I	1.2			MIT
4594.908	4596.195	21757.13	NI	1.2	1.0H		MIT
4592.529	4593.816	21768.40	NI I	2.3	0.3		MIT
4580.619	4581.903	21825.00	NI I	0.7			MIT
4559.945	4561.223	21923.94	NI I	1.0			MIT
4553.175	4554.451	21956.54	NI I	1.2R			MIT
4551.236	4552.512	21965.90	NI I	0.7			MIT
4547.234	4548.509	21985.23	NI I	1.5			MIT
4546.930	4548.205	21986.70	NI I	1.7	0.3		MIT
4519.986	4521.254	22117.76	NI I	1.4			MIT
4512.995	4514.261	22152.02	NI I	0.3			MIT
4490.541	4491.801	22262.79	NI I	0.5			MIT
4480.570	4481.827	22312.33	NI I	0.5			MIT
4470.483	4471.738	22362.67	NI I	2.2	1.3		ME
4467.969	4469.223	22375.25	NI	0.3			MIT
4466.394	4467.647	22383.15	NI I	0.5			MIT
4463.427	4464.680	22398.02	NI I	1.0			MIT
4462.460	4463.712	22402.88	NI I	2.2	1.3		MIT
4459.037	4460.289	22420.07	NI I	2.6	1.3		MIT
4450.301	4451.550	22464.08	NI I	0.7			MIT
4450.15	4451.40	22464.9	NI I	0.3			M
4442.441	4443.688	22503.83	NI I	0.7			MIT
4441.446	4442.693	22508.87	NI I	0.7			MIT
4437.570	4438.816	22528.53	NI I	1.0			MIT
4436.981	4438.227	22531.52	NI I	1.4	0.7		MIT
4423.000	4424.242	22602.74	NI I	0.5			MIT
4410.516	4411.755	22666.72	NI I	1.4	0.6		MIT
4401.547	4402.783	22712.91	NI I	3.0O	1.5		MIT
4400.870	4402.106	22716.40	NI I	1.2	0.5		MIT
4399.607	4400.843	22722.92	NI I	1.0	0.7		MIT
4398.625	4399.861	22727.99	NI I	0.3	0.5		MIT
4390.322	4391.555	22770.98	NI I	0.3			MIT
4389.870	4391.103	22773.32	NI I	0.7	0.7		MIT
4386.461	4387.693	22791.02	NI I	0.5	0.6		MIT
4384.543	4385.775	22800.99	NI I	1.4	0.0		MIT
4370.041	4371.269	22876.65	NI I	0.7			MIT
4368.312	4369.540	22885.71	NI I	0.7			MIT
4359.585	4360.810	22931.52	NI I	2.0	1.0		MIT
4355.911	4357.135	22950.86	NI I	1.2			MIT
4331.645	4332.863	23079.43	NI I	2.3	1.1		MIT
4330.720	4331.938	23084.36	NI I	0.7			MIT
4325.607	4326.823	23111.64	NI I	1.8			MIT
4325.361	4326.577	23112.96	NI I	1.0			MIT
4307.286	4308.498	23209.95	NI I	0.6			MIT
4298.767	4299.976	23255.94	NI I	0.5			MIT
4298.515	4299.724	23257.31	NI I	0.7			MIT
4296.984	4298.193	23265.59	NI I	0.3			MIT
4295.888	4297.097	23271.53	NI I	2.0	0.3H		MIT
4288.005	4289.211	23314.31	NI I	2.2			MIT
4284.683	4285.889	23332.38	NI I	1.4	1.0		MIT
4252.107	4253.304	23511.13	NI I	0.3			MIT
4236.372	4237.565	23598.46	NI I	0.7			MIT
4231.040	4232.231	23628.20	NI I	1.2			MIT

AIR WAVELENGTH (ANGSTRMS)	VACUUM WAVELENGTH (ANGSTRMS)	WAVENUMBER (KAYSERS)	LMENT	LOG ARC	INTENSITY SPK	INTENSITY DIS	REF
4221.696	4222.885	23680.49	NI I	0.7			MIT
4202.154	4203.338	23790.62	NI I	0.7			MIT
4201.723	4202.907	23793.06	NI I	1.5			MIT
4200.464	4201.647	23800.19	NI I	1.6			MIT
4195.531	4196.713	23828.17	NI I	1.5			MIT
4184.475	4185.654	23891.13	NI I	0.8			MIT
4166.970	4168.145	23991.49	NI I	0.3			MIT
4150.366	4151.536	24087.47	NI I	0.5			MIT
4142.320	4143.488	24134.25	NI	1.2			MIT
4142.184	4143.352	24135.05	NI I	0.3			MIT
4123.778	4124.941	24242.77	NI	0.3			MIT
4115.982	4117.143	24288.69	NI I	0.8			MIT
4086.151	4087.304	24466.00	NI	0.6			MIT
4074.897	4076.048	24533.57	NI I	1.0			MIT
4072.913	4074.063	24545.52	NI I	0.3			MIT
4069.243	4070.392	24567.66	NI	0.3			MIT
4067.051	4068.199	24580.90	NI II		1.5		MIT
4064.374	4065.522	24597.09	NI I	1.0			MIT
4057.347	4058.493	24639.69	NI I	0.3			MIT
4046.761	4047.904	24704.14	NI I	0.3			MIT
4044.75	4045.89	24716.4	NI II			0.3	FL
4025.114	4026.252	24837.00	NI I	0.3			MIT
4022.052	4023.189	24855.91	NI I	0.5			MIT
4019.046	4020.182	24874.50	NI I	0.7			MIT
4017.462	4018.598	24884.30	NI I	1.2V			MIT
4009.984	4011.118	24930.71	NI I	0.5			MIT
4006.136	4007.269	24954.65	NI	1.0			MIT
3993.952	3995.081	25030.78	NI I	1.5H			MIT
3984.140	3985.267	25092.42	NI I	1.5J			MIT
3974.650	3975.774	25152.33	NI I	1.6H			MIT
3973.562	3974.686	25159.22	NI I	2.9	1.0		MIT
3972.171	3973.295	25168.03	NI I	2.0	0.8		MIT
3970.503	3971.626	25178.60	NI I	1.7W			MIT
3962.12	3963.24	25231.9	NI I	1.0H			MIT
3944.126	3945.242	25346.99	NI I	0.7J			MIT
3912.979	3914.087	25548.74	NI I	0.7			MIT
3912.310	3913.418	25553.11	NI	0.3			MIT
3908.931	3910.038	25575.20	NI I	0.3			MIT
3889.671	3890.773	25701.83	NI I	1.5	1.0H		MIT
3863.072	3864.167	25878.80	NI I	0.6	0.7		MIT
3858.301	3859.395	25910.80	NI I	2.9R	1.8H		MIT
3849.58	3850.67	25969.5	NI II				MIT
3844.276	3845.366	26005.32	NI I	0.5H			MIT
3832.873	3833.960	26082.69	NI I	1.4			MIT
3831.690	3832.777	26090.74	NI	2.5	1.0		MIT
3807.144	3808.225	26258.96	NI I	2.9O	1.6H		MIT
3793.608	3794.685	26352.65	NI I	1.7			MIT
3792.337	3793.414	26361.48	NI I	1.4	0.7		MIT
3783.530	3784.605	26422.84	NI I	2.7	1.6H		MIT
3778.063	3779.136	26461.07	NI I	1.4	0.7		MIT
3775.572	3776.645	26478.53	NI I	2.7H	1.6H		MIT
3772.530	3773.602	26499.88	NI I	1.2	0.7		MIT
3769.455	3770.526	26521.50	NI II	0.3	1.7H		MIT
3762.618	3763.687	26569.69	NI	0.3			MIT
3749.045	3750.111	26665.88	NI I	1.7	0.7		MIT
3744.562	3745.626	26697.80	NI I	0.8			MIT
3739.782	3740.845	26731.93	NI I	0.3			MIT
3739.229	3740.292	26735.88	NI I	2.0	1.0		MIT
3736.813	3737.875	26753.17	NI	2.5	1.2		MIT
3730.751	3731.812	26796.64	NI I	1.4			MIT
3729.23	3730.29	26807.6	NI I	0.3			HA
3728.933	3729.993	26809.70	NI I	0.3			MIT
3724.827	3725.886	26839.25	NI I	0.6			MIT
3722.484	3723.543	26856.14	NI I	2.3	1.3		MIT
3715.499	3716.556	26906.63	NI I	0.3			MIT
3696.913	3697.965	27041.90	NI I	0.5			MIT
3693.932	3694.983	27063.72	NI I	1.7			MIT
3689.305	3690.355	27097.66	NI I	0.3			MIT
3688.415	3689.465	27104.20	NI I	2.2	1.2		MIT
3674.15	3675.20	27209.4	NI I	2.3	1.7R		MIT

AIR WAVELENGTH (ANGSTRMS)	VACUUM WAVELENGTH (ANGSTRMS)	WAVENUMBER (KAYSERS)	LMENT	LOG ARC	INTENSITY SPK	DIS	REF
3670.427	3671.472	27237.03	NI I	2.2	1.3		MIT
3669.241	3670.286	27245.83	NI I	2.2	1.0		MIT
3668.216	3669.261	27253.45	NI I	0.5	0.6		MIT
3665.924	3666.968	27270.49	NI	0.5			MIT
3664.095	3665.139	27284.10	NI I	2.5	1.5		MIT
3661.951	3662.994	27300.07	NI I	1.7	0.8		MIT
3643.941	3644.979	27435.00	NI I	0.3			MIT
3642.387	3643.425	27446.70	NI I	0.3			MIT
3641.641	3642.679	27452.32	NI I	0.8			MIT
3634.941	3635.977	27502.92	NI I	1.7	1.0		MIT
3629.906	3630.941	27541.07	NI I	0.6	0.5H		MIT
3624.733	3625.766	27580.37	NI I	2.2	1.2		MIT
3619.392	3620.424	27621.07	NI I	3.3G	2.2H		MIT
3612.741	3613.771	27671.92	NI I	2.6	1.7H		MIT
3610.462	3611.492	27689.39	NI I	3.0R			MIT
3609.314	3610.343	27698.20	NI I	2.3	1.2		MIT
3606.852	3607.881	27717.10	NI I	2.0R			MIT
3602.281	3603.309	27752.27	NI I	2.2	1.2		MIT
3597.705	3598.731	27787.57	NI I	3.0R	1.7H		MIT
3587.931	3588.955	27863.26	NI I	2.3	1.1		MIT
3577.240	3578.261	27946.53	NI I	0.6			MIT
3576.762	3577.783	27950.27	NI II	0.3	1.6H		MIT
3575.952	3576.973	27956.60	NI I	0.5			MIT
3571.869	3572.889	27988.56	NI I	3.0G	1.6H		MIT
3566.372	3567.390	28031.70	NI I	3.3G	2.0J		MIT
3561.751	3562.768	28068.06	NI I	1.8	1.1		MIT
3559.930	3560.947	28082.42	NI I	0.7			MIT
3553.483	3554.498	28133.37	NI I	1.7	1.0		MIT
3551.534	3552.549	28148.81	NI I	1.7	1.1		MIT
3548.185	3549.199	28175.37	NI I	2.6	1.4		MIT
3530.595	3531.604	28315.74	NI I	1.5			MIT
3528.891	3529.900	28329.42	NI I	1.2			MIT
3527.982	3528.991	28336.72	NI I	2.3	1.2		MIT
3526.540	3527.548	28348.30	NI I	0.7			MIT
3524.541	3525.549	28364.38	NI I	3.0G	2.0J		MIT
3523.444	3524.451	28373.21	NI I	2.0			MIT
3523.074	3524.081	28376.19	NI I	0.7			MIT
3519.766	3520.772	28402.86	NI I	2.7H	1.5		MIT
3518.634	3519.640	28412.00	NI I	2.0	0.9		MIT
3516.234	3517.240	28431.39	NI I	0.7J			MIT
3515.054	3516.059	28440.93	NI I	3.0G	1.7H		MIT
3513.933	3514.938	28450.00	NI I	2.3	1.6H		MIT
3513.484	3514.489	28453.64	NI	0.5			MIT
3510.338	3511.342	28479.14	NI I	3.0G	1.7H		MIT
3507.694	3508.697	28500.61	NI I	2.0	1.1		MIT
3502.595	3503.597	28542.09	NI I	2.0			MIT
3500.852	3501.854	28556.30	NI I	2.7J	1.9		MIT
3496.350	3497.350	28593.07	NI I	1.2			MIT
3492.956	3493.956	28620.86	NI I	3.0G	2.0H		MIT
3488.293	3489.291	28659.11	NI I	0.3			MIT
3485.888	3486.886	28678.89	NI I	2.2	1.5		MIT
3485.110	3486.108	28685.29	NI I	0.8			MIT
3483.774	3484.771	28696.29	NI I	2.7G	1.5		MIT
3480.183	3481.179	28725.90	NI I	0.7			MIT
3479.264	3480.260	28733.49	NI I	0.7			MIT
3478.292	3479.288	28741.51	NI I	0.5			MIT
3477.864	3478.860	28745.05	NI I	0.7	0.5		MIT
3476.63	3477.63	28755.2	NI I	0.6			MIT
3472.545	3473.539	28789.08	NI I	2.9G	1.6		MIT
3469.486	3470.480	28814.46	NI I	2.5	1.3		MIT
3467.732	3468.725	28829.04	NI I	0.9			MIT
3467.502	3468.495	28830.95	NI I	2.5	1.2		MIT
3462.804	3463.796	28870.06	NI I	0.8	1.0		MIT
3461.652	3462.643	28879.67	NI I	2.9G	1.7H		MIT
3458.474	3459.465	28906.21	NI I	2.9G	1.7H		MIT
3454.161	3455.151	28942.30	NI II		0.3		MIT
3452.890	3453.879	28952.95	NI I	2.8G	1.7		MIT
3446.263	3447.251	29008.62	NI I	3.0G	1.7H		MIT
3444.251	3445.238	29025.57	NI I	1.0			MIT
3442.559	3443.546	29039.84	NI I	0.6			MIT

AIR WAVELENGTH (ANGSTRMS)	VACUUM WAVELENGTH (ANGSTRMS)	WAVENUMBER (KAYSERS)	LMENT	LOG ARC	INTENSITY SPK	DIS	REF
3442.044	3443.030	29044.18	NI I	1.2			MIT
3437.280	3438.265	29084.43	NI I	2.8G	1.6		MIT
3435.489	3436.474	29099.60	NI I	0.5			MIT
3433.558	3434.542	29115.96	NI I	2.9G	1.7J		MIT
3423.711	3424.693	29199.70	NI I	2.8G	1.4		MIT
3422.878	3423.860	29206.80	NI I	1.0	0.7		MIT
3422.332	3423.313	29211.47	NI I	1.0	0.7		MIT
3421.342	3422.323	29219.92	NI I	1.5	0.9		MIT
3421.204	3422.185	29221.10	NI I	0.6			MIT
3420.741	3421.722	29225.05	NI I	1.5	0.5		MIT
3414.765	3415.744	29276.20	NI I	3.0G	1.7J		MIT
3413.939	3414.918	29283.28	NI I	2.5	1.0		MIT
3413.478	3414.457	29287.23	NI I	2.7	1.2		MIT
3409.578	3410.556	29320.73	NI I	2.5			MIT
3407.304	3408.282	29340.30	NI II		1.4		MIT
3403.432	3404.409	29373.68	NI I	1.6			MIT
3401.766	3402.742	29388.06	NI II		1.2		MIT
3401.166	3402.142	29393.25	NI I	1.6	0.0		MIT
3397.248	3398.223	29427.15	NI	0.3			MIT
3396.184	3397.159	29436.36	NI I	0.9			MIT
3392.992	3393.966	29464.05	NI I	2.8G			MIT
3391.050	3392.023	29480.93	NI I	2.6	1.6		MIT
3387.466	3388.439	29512.12	NI I	0.3			MIT
3380.885	3381.856	29569.56	NI I	2.3	1.1		MIT
3380.574	3381.545	29572.28	NI I	2.8G	2.0		MIT
3376.331	3377.301	29609.45	NI	1.4			MIT
3375.561	3376.530	29616.20	NI I	0.6			MIT
3374.642	3375.611	29624.26	NI I	1.2	0.8		MIT
3374.221	3375.190	29627.96	NI I	2.6	0.8		MIT
3373.973	3374.942	29630.14	NI II		0.6		MIT
3371.993	3372.962	29647.54	NI I	2.6	1.0		MIT
3369.573	3370.541	29668.83	NI I	2.7G	2.0		MIT
3367.892	3368.860	29683.64	NI I	1.9			MIT
3366.807	3367.774	29693.20	NI I	1.8	0.0		MIT
3366.168	3367.135	29698.84	NI I	2.6O	1.1		MIT
3365.766	3366.733	29702.39	NI I	2.6W	1.1		MIT
3364.591	3365.558	29712.76	NI I	1.2			MIT
3363.613	3364.579	29721.40	NI I	1.3			MIT
3362.806	3363.772	29728.53	NI I	2.0			MIT
3361.556	3362.522	29739.58	NI I	2.7O	1.3		MIT
3361.241	3362.207	29742.37	NI I	1.0			MIT
3359.106	3360.071	29761.27	NI I	1.8			MIT
3350.42	3351.38	29838.4	NI II		0.3		MIT
3339.050	3340.010	29940.03	NI I	0.7			MIT
3338.758	3339.718	29942.65	NI I	0.5			MIT
3337.014	3337.974	29958.29	NI I	0.8			MIT
3332.180	3333.138	30001.75	NI	0.9			MIT
3328.714	3329.672	30032.99	NI I	1.2			MIT
3327.392	3328.349	30044.93	NI I	0.7			MIT
3326.670	3327.627	30051.45	NI I	0.7			MIT
3322.310	3323.266	30090.88	NI I	2.6	1.0		MIT
3321.242	3322.198	30100.56	NI I	0.6			MIT
3320.779	3321.734	30104.75	NI I	1.0			MIT
3320.257	3321.212	30109.49	NI I	2.6W	1.2		MIT
3315.663	3316.617	30151.20	NI I	2.6G	1.3		MIT
3312.992	3313.946	30175.51	NI I	1.2			MIT
3312.320	3313.273	30181.63	NI I	1.8	0.3		MIT
3310.202	3311.155	30200.94	NI I	1.7			MIT
3309.428	3310.381	30208.01	NI	0.6			MIT
3307.013	3307.965	30230.07	NI I	0.5			MIT
3304.950	3305.901	30248.94	NI I	1.4			MIT
3287.221	3288.168	30412.07	NI I	0.5			MIT
3286.946	3287.893	30414.62	NI I	2.0	0.0		MIT
3284.340	3285.286	30438.75	NI	0.3			MIT
3282.827	3283.773	30452.78	NI I	1.4	0.0		MIT
3282.696	3283.642	30453.99	NI I	2.0			MIT
3281.880	3282.826	30461.56	NI I	1.3			MIT
3274.90	3275.84	30526.5	NI II		0.7		ME
3271.118	3272.061	30561.78	NI I	2.1	0.0		MIT
3268.971	3269.913	30581.85	NI I	0.3			MIT

AIR WAVELENGTH (ANGSTRMS)	VACUUM WAVELENGTH (ANGSTRMS)	WAVENUMBER (KAYSERS)	LMENT	LOG ARC	INTENSITY SPK	DIS	REF
3268.064	3269.006	30590.34	NI	0.7			MIT
3250.743	3251.681	30753.33	NI I	2.1	0.0		MIT
3249.440	3250.377	30765.66	NI I	1.5			MIT
3248.457	3249.394	30774.97	NI I	2.2	0.3		MIT
3245.370	3246.306	30804.24	NI I	0.3			MIT
3243.058	3243.994	30826.20	NI I	2.6G	1.2		MIT
3235.753	3236.687	30895.79	NI I	1.2			MIT
3234.649	3235.583	30906.33	NI I	2.5	1.2		MIT
3233.174	3234.107	30920.43	NI I	0.6			MIT
3232.963	3233.896	30922.45	NI I	2.5G	1.5		MIT
3226.984	3227.916	30979.74	NI I	2.0			MIT
3225.020	3225.951	30998.61	NI I	2.5	0.8		MIT
3223.534	3224.465	31012.90	NI I	0.7			MIT
3221.652	3222.582	31031.01	NI I	2.5	0.6		MIT
3221.273	3222.203	31034.67	NI I	1.5			MIT
3219.811	3220.741	31048.76	NI I	0.6			MIT
3217.830	3218.759	31067.87	NI I	1.0	0.7		MIT
3216.821	3217.750	31077.61	NI I	0.8			MIT
3214.059	3214.987	31104.32	NI I	1.3			MIT
3213.423	3214.351	31110.48	NI I	1.1			MIT
3209.912	3210.839	31144.50	NI I	1.3			MIT
3206.952	3207.879	31173.25	NI I	0.9			MIT
3202.142	3203.067	31220.07	NI I	1.4			MIT
3200.423	3201.348	31236.84	NI	1.5			MIT
3199.342	3200.267	31247.40	NI	0.7			MIT
3197.113	3198.037	31269.18	NI I	2.5			MIT
3195.573	3196.497	31284.25	NI I	2.1			MIT
3191.875	3192.798	31320.49	NI I	0.6			MIT
3184.367	3185.288	31394.34	NI I	2.2	0.5		MIT
3183.251	3184.172	31405.34	NI I	1.4			MIT
3183.038	3183.959	31407.44	NI I	0.7			MIT
3181.740	3182.660	31420.26	NI I	1.7	0.0H		MIT
3176.292	3177.211	31474.15	NI I	0.6			MIT
3170.715	3171.632	31529.50	NI I	0.6			MIT
3165.508	3166.424	31581.37	NI I	1.2			MIT
3164.166	3165.082	31594.76	NI I	0.6			MIT
3159.521	3160.436	31641.21	NI I	1.0			MIT
3154.585	3155.498	31690.72	NI I	0.6			MIT
3151.259	3152.172	31724.16	NI	0.8			MIT
3145.719	3146.630	31780.03	NI I	2.3	0.5		MIT
3145.121	3146.032	31786.07	NI I	0.8			MIT
3134.108	3135.016	31897.76	NI I	3.0G	2.2		MIT
3129.314	3130.221	31946.63	NI I	2.1			MIT
3116.714	3117.618	32075.77	NI I	0.3			MIT
3114.124	3115.027	32102.45	NI I	2.5	1.7		MIT
3107.714	3108.616	32168.66	NI I	1.4			MIT
3105.469	3106.370	32191.92	NI I	2.3	1.5		MIT
3101.879	3102.779	32229.17	NI I	2.6G	2.2		MIT
3101.554	3102.454	32232.55	NI I	3.0G	2.2		MIT
3099.115	3100.014	32257.91	NI I	2.3	1.7		MIT
3097.118	3098.017	32278.71	NI I	2.3	1.7		MIT
3087.077	3087.973	32383.70	NI II		2.2		MIT
3080.755	3081.650	32450.15	NI I	2.3	1.8		MIT
3066.442	3067.333	32601.61	NI I	0.7			MIT
3064.623	3065.514	32620.96	NI I	2.3R	1.7		MIT
3063.937	3064.828	32628.26	NI II		0.3		MIT
3057.638	3058.527	32695.48	NI I	2.6G	2.1		MIT
3054.316	3055.204	32731.04	NI I	2.6G	2.0		MIT
3050.819	3051.706	32768.55	NI I	3.0G			MIT
3045.006	3045.892	32831.11	NI I	2.3	1.0		MIT
3037.935	3038.819	32907.52	NI I	2.9G	2.0		MIT
3031.870	3032.753	32973.35	NI I	2.3			MIT
3029.297	3030.179	33001.35	NI I	0.5			MIT
3019.143	3020.022	33112.34	NI I	2.3G	1.5		MIT
3012.004	3012.882	33190.82	NI I	2.9G	2.10		MIT
3003.629	3004.505	33283.36	NI I	2.7G	1.9		MIT
3002.491	3003.366	33295.97	NI I	3.0G	2.0		MIT
2994.460	2995.333	33385.27	NI I	2.1G	1.0		MIT
2992.595	2993.468	33406.07	NI I	1.9G	1.0		MIT
2991.095	2991.967	33422.82	NI I	1.2			MIT

AIR WAVELENGTH (ANGSTRMS)	VACUUM WAVELENGTH (ANGSTRMS)	WAVENUMBER (KAYSERS)	LMENT	LOG ARC	INTENSITY SPK	DIS	REF
2988.053	2988.925	33456.85	NI II		0.6		MIT
2984.131	2985.002	33500.82	NI I	1.7G	1.0		MIT
2983.422	2984.292	33508.78	NI I	1.0			MIT
2981.651	2982.521	33528.68	NI I	1.9R	1.3		MIT
2947.453	2948.315	33917.68	NI II		1.0		MIT
2943.914	2944.775	33958.46	NI I	1.7R	1.3		MIT
2932.623	2933.481	34089.19	NI	1.0			MIT
2914.007	2914.860	34306.96	NI I	1.3			MIT
2913.586	2914.439	34311.92	NI II		1.3		MIT
2907.459	2908.311	34384.22	NI I	1.6			MIT
2881.247	2882.092	34697.02	NI II		0.9		MIT
2865.502	2866.343	34887.66	NI I	0.7			MIT
2864.15	2864.99	34904.1	NI II		2.5J		MIT
2863.700	2864.541	34909.61	NI II		2.4		MIT
2842.420	2843.256	35170.95	NI II		2.2		MIT
2838.958	2839.793	35213.84	NI I	1.4			MIT
2834.553	2835.387	35268.56	NI I	1.6	1.2		MIT
2825.236	2826.067	35384.86	NI II		2.1		MIT
2821.293	2822.123	35434.31	NI I	2.1	2.1		MIT
2814.361	2815.190	35521.58	NI I	1.2			MIT
2808.359	2809.186	35597.50	NI II		1.6		MIT
2805.674	2806.501	35631.56	NI II		2.3H		MIT
2805.083	2805.910	35639.07	NI I	1.7	1.2		MIT
2803.148	2803.974	35663.67	NI I	0.7			MIT
2802.274	2803.100	35674.79	NI	1.7	1.2		MIT
2798.653	2799.478	35720.95	NI I	2.1			MIT
2775.327	2776.146	36021.16	NI II		2.4J		MIT
2768.785	2769.603	36106.26	NI II		2.4J		MIT
2760.671	2761.487	36212.38	NI II		1.6		MIT
2759.019	2759.834	36234.06	NI II		2.7J		MIT
2746.746	2747.558	36395.95	NI I	2.1	1.7		MIT
2742.994	2743.805	36445.73	NI II		2.7		MIT
2708.791	2709.594	36905.90	NI II		2.7		MIT
2696.491	2697.291	37074.23	NI I	0.3			ME
2690.640	2691.439	37154.85	NI II	0.0	2.4H		MIT
2684.412	2685.209	37241.04	NI II		2.8J		MIT
2679.241	2680.037	37312.92	NI II		2.7J		MIT
2670.328	2671.122	37437.45	NI II		1.9		MIT
2665.249	2666.042	37508.79	NI II		2.1		MIT
2655.907	2656.697	37640.72	NI II		2.7V		MIT
2655.47	2656.26	37646.9	NI II		2.6J		MIT
2648.719	2649.508	37742.86	NI II		1.9		MIT
2647.056	2647.844	37766.57	NI II		2.7J		MIT
2632.891	2633.676	37969.74	NI II		3.3J		MIT
2631.520	2632.305	37989.52	NI II		2.0J		MIT
2630.278	2631.062	38007.46	NI II		2.2W		MIT
2626.561	2627.344	38061.25	NI II		2.7H		MIT
2615.193	2615.974	38226.68	NI II		3.0H		MIT
2611.654	2612.434	38278.48	NI II		2.1		MIT
2610.092	2610.871	38301.39	NI II		3.0H		MIT
2606.387	2607.166	38355.83	NI II		2.8H		MIT
2605.347	2606.125	38371.14	NI II		2.4J		MIT
2601.126	2601.903	38433.40	NI II		3.3H		ME
2588.314	2589.088	38623.63	NI II		1.9		MIT
2587.246	2588.020	38639.58	NI II		1.7		MIT
2583.995	2584.768	38688.19	NI II		2.3		MIT
2578.470	2579.242	38771.08	NI I	1.3			MIT
2566.081	2566.850	38958.25	NI II		2.8H		MIT
2565.372	2566.141	38969.02	NI II		2.2J		MIT
2562.125	2562.893	39018.40	NI	1.2			MIT
2561.424	2562.192	39029.08	NI I	1.6			MIT
2560.300	2561.068	39046.21	NI II		2.7H		MIT
2559.664	2560.432	39055.92	NI	1.3			MIT
2557.870	2558.637	39083.31	NI II		1.9		MIT
2555.113	2555.879	39125.48	NI II		3.0H		MIT
2553.377	2554.143	39152.07	NI I	1.3			MIT
2551.033	2551.799	39188.05	NI II		1.9		MIT
2549.558	2550.323	39210.72	NI II		2.2		MIT
2547.413	2548.178	39243.73	NI I	1.3			MIT
2547.188	2547.953	39247.20	NI II		2.0		MIT

AIR WAVELENGTH (ANGSTRMS)	VACUUM WAVELENGTH (ANGSTRMS)	WAVENUMBER (KAYSERS)	LMENT	LOG ARC	INTENSITY SPK	DIS	REF
2545.900	2546.664	39267.05	NI II	1.3	3.0H		MIT
2540.020	2540.783	39357.95	NI I	1.6	0.0		MIT
2539.100	2539.863	39372.21	NI II		2.4W		MIT
2535.967	2536.729	39420.85	NI	1.4			MIT
2532.076	2532.837	39481.42	NI I	1.5			MIT
2528.976	2529.736	39529.81	NI	1.3	0.3		MIT
2528.050	2528.810	39544.29	NI I	1.3			MIT
2525.388	2526.148	39585.97	NI II		2.5V		MIT
2524.216	2524.975	39604.35	NI I	1.7G			MIT
2520.351	2521.109	39665.08	NI II		0.3		MIT
2514.761	2515.518	39753.24	NI II		0.3H		MIT
2510.873	2511.629	39814.79	NI II	1.7H	2.4H		MIT
2505.837	2506.592	39894.81	NI II		1.7		MIT
2501.132	2501.886	39969.85	NI	1.3			MIT
2497.817	2498.570	40022.89	NI II		0.7		MIT
2491.187	2491.939	40129.40	NI	1.3			MIT
2490.700	2491.451	40137.25	NI	1.6			MIT
2489.507	2490.258	40156.48	NI I	1.6	0.8		MIT
2484.327	2485.077	40240.20	NI II		1.7J		MIT
2484.034	2484.784	40244.95	NI I	1.3			MIT
2479.481	2480.230	40318.84	NI	0.9			MIT
2476.874	2477.622	40361.28	NI I	1.60	0.3		MIT
2473.151	2473.898	40422.03	NI II	1.9	2.7		MIT
2472.229	2472.976	40437.11	NI I	0.7	0.0		MIT
2472.065	2472.812	40439.79	NI I	1.2			MIT
2466.958	2467.704	40523.50	NI I	1.60			MIT
2465.264	2466.010	40551.34	NI I	1.3	0.7		MIT
2459.315	2460.059	40649.43	NI II		0.3		MIT
2455.526	2456.269	40712.15	NI II		1.6		MIT
2453.987	2454.730	40737.68	NI I	1.60	0.9		MIT
2450.986	2451.728	40787.55	NI	2.0H	1.5		MIT
2450.466	2451.208	40796.21	NI I	1.3			MIT
2441.822	2442.562	40940.62	NI I	1.3	0.8		MIT
2441.667	2442.407	40943.21	NI I	1.3	0.6		MIT
2437.888	2438.627	41006.68	NI II	1.6W	2.3		MIT
2436.672	2437.411	41027.14	NI	1.5L	1.3		MIT
2434.415	2435.153	41065.17	NI I	1.6W			MIT
2433.567	2434.305	41079.48	NI II		1.9		MIT
2431.570	2432.308	41113.22	NI II		1.3		MIT
2429.096	2429.833	41155.09	NI I	1.2	0.0		MIT
2424.028	2424.764	41241.12	NI I	1.3			MIT
2423.656	2424.392	41247.46	NI I	1.3	0.7		MIT
2423.326	2424.062	41253.07	NI I	1.4	0.6		MIT
2421.227	2421.962	41288.83	NI I	1.3S	0.6		MIT
2419.313	2420.048	41321.49	NI I	1.3R	0.6		MIT
2416.138	2416.872	41375.79	NI II	1.6	2.4H		MIT
2413.049	2413.783	41428.75	NI II		1.7		MIT
2412.643	2413.377	41435.72	NI I	1.3	0.9		MIT
2412.273	2413.006	41442.08	NI II		0.9		MIT
2410.752	2411.485	41468.22	NI II		0.8		MIT
2406.882	2407.614	41534.89	NI II		1.3		MIT
2406.395	2407.127	41543.30	NI II		0.9		MIT
2405.169	2405.901	41564.47	NI II		1.9		MIT
2401.841	2402.572	41622.06	NI I	1.6R	1.0		MIT
2398.63	2399.36	41677.8	NI II		0.3		MIT
2396.63	2397.36	41712.5	NI I	1.1	0.3		MIT
2396.37	2397.10	41717.1	NI I	1.1	0.5		MIT
2395.61	2396.34	41730.3	NI	1.0	0.6		A
2394.841	2395.570	41743.71	NI II	0.3	0.8H		MIT
2394.516	2395.245	41749.38	NI II	1.3	1.3		MIT
2393.11	2393.84	41773.9	NI I	1.0			MIT
2392.96	2393.69	41776.5	NI I	1.0	0.5		MIT
2392.58	2393.31	41783.2	NI II	0.6	1.0		MIT
2392.10	2392.83	41791.5	NI II	1.0	1.0		MIT
2388.919	2389.647	41847.18	NI	1.0	1.3		MIT
2387.762	2388.490	41867.46	NI II	0.8	1.5		MIT
2387.54	2388.27	41871.4	NI I	1.2	0.3		MIT
2386.58	2387.31	41888.2	NI I	1.3	0.7		MIT
2385.01	2385.74	41915.8	NI	1.0	0.3		MIT
2384.84	2385.57	41918.7	NI	0.9	0.5H		MIT

AIR WAVELENGTH (ANGSTRMS)	VACUUM WAVELENGTH (ANGSTRMS)	WAVENUMBER (KAYSERS)	LMENT	LOG ARC	INTENSITY SPK	DIS	REF
2384.38	2385.11	41926.8	NI I	1.2	0.6		MIT
2380.82	2381.55	41989.5	NI	0.3H			MIT
2380.79	2381.52	41990.0	NI I	1.0	0.3		MIT
2379.72	2380.45	42008.9	NI I	1.2	0.3		MIT
2376.03	2376.76	42074.2	NI I	1.2	0.7		MIT
2375.421	2376.146	42084.96	NI II	1.0	1.5		MIT
2369.22	2369.94	42195.1	NI II	0.0	1.3		MIT
2367.389	2368.112	42227.73	NI II	0.9	1.2		MIT
2366.54	2367.26	42242.9	NI II		1.0		MIT
2365.66	2366.38	42258.6	NI I	1.0	0.3		MIT
2362.06	2362.78	42323.0	NI I	1.2	1.0		MIT
2360.63	2361.35	42348.6	NI I	1.0	0.5		MIT
2358.85	2359.57	42380.6	NI I	1.2	0.5		MIT
2356.87	2357.59	42416.2	NI I	1.0	0.5		MIT
2356.63	2357.35	42420.5	NI	1.0			MIT
2356.41	2357.13	42424.5	NI II	0.6	1.3		MIT
2355.06	2355.78	42448.8	NI I	1.0	0.3		MIT
2353.708	2354.428	42473.16	NI	0.9			MIT
2353.38	2354.10	42479.1	NI	0.7	0.8		MIT
2350.85	2351.57	42524.8	NI II		0.7		MIT
2350.467	2351.186	42531.72	NI	1.0			MIT
2350.05	2350.77	42539.3	NI		0.5H		MIT
2348.74	2349.46	42563.0	NI I	1.0	0.0		MIT
2347.52	2348.24	42585.1	NI I	1.1	0.3		MIT
2346.63	2347.35	42601.3	NI I	1.0	0.5		MIT
2346.090	2346.809	42611.06	NI	0.8			MIT
2345.54	2346.26	42621.0	NI I	1.3R			MIT
2345.444	2346.162	42622.80	NI I	1.3R	0.8		MIT
2345.26	2345.98	42626.1	NI II		1.0		MIT
2343.95	2344.67	42650.0	NI II		0.6		MIT
2343.49	2344.21	42658.3	NI II	0.3	1.2		MIT
2341.18	2341.90	42700.4	NI II	0.5	1.3		MIT
2338.49	2339.21	42749.5	NI I	0.6			MIT
2337.82	2338.54	42761.8	NI I	0.6D	0.5		MIT
2337.49	2338.21	42767.8	NI I	0.9D	0.7		MIT
2337.09	2337.81	42775.1	NI I	1.0	0.6		MIT
2336.68	2337.40	42782.6	NI II		1.5		MIT
2336.586	2337.302	42784.37	NI II		2.2		MIT
2336.22	2336.94	42791.1	NI		0.3		MIT
2334.580	2335.296	42821.12	NI II	0.9	1.3		MIT
2331.70	2332.42	42874.0	NI I	1.0H	0.3		MIT
2329.964	2330.679	42905.95	NI I	1.1G	1.0		MIT
2326.448	2327.162	42970.79	NI II	0.6	1.2		MIT
2325.79	2326.50	42983.0	NI I	1.5G	1.0		MIT
2324.64	2325.35	43004.2	NI I	1.2	0.3		MIT
2322.68	2323.39	43040.5	NI	1.1	0.5		MIT
2321.96	2322.67	43053.8	NI I	0.7D	0.6		MIT
2321.38	2322.09	43064.6	NI I	1.3G	1.0		MIT
2320.03	2320.74	43089.6	NI I	1.5G	0.7L		MIT
2319.756	2320.469	43094.74	NI II		1.0		MIT
2318.77	2319.48	43113.1	NI I	1.0	0.3		MIT
2318.48	2319.19	43118.5	NI II	0.0	1.3		MIT
2317.162	2317.874	43142.98	NI I	1.5G	1.1		MIT
2316.037	2316.749	43163.94	NI II	1.2	1.4W		MIT
2313.982	2314.693	43202.27	NI I	1.0G	1.0		MIT
2313.655	2314.366	43208.37	NI	1.0G	0.8		MIT
2312.916	2313.627	43222.18	NI II		1.3		MIT
2312.34	2313.05	43232.9	NI I	1.50	0.8		MIT
2312.24	2312.95	43234.8	NI II		1.0		MIT
2310.965	2311.676	43258.66	NI I	1.7G	1.0S		MIT
2310.024	2310.735	43276.28	NI	0.9			MIT
2309.486	2310.196	43286.36	NI	1.1			MIT
2308.519	2309.229	43304.49	NI II		1.3		MIT
2308.171	2308.881	43311.02	NI	1.0			MIT
2307.785	2308.495	43318.27	NI II	0.3	1.2		MIT
2307.354	2308.064	43326.36	NI I	1.0	0.3		MIT
2306.40	2307.11	43344.3	NI	1.0	0.6		MIT
2305.244	2305.953	43366.01	NI II	0.3	1.3		MIT
2303.857	2304.566	43392.12	NI II		1.2		MIT
2302.997	2303.706	43408.32	NI II	1.0S	1.5		MIT

AIR WAVELENGTH (ANGSTRMS)	VACUUM WAVELENGTH (ANGSTRMS)	WAVENUMBER (KAYSERS)	LMENT	LOG ARC	INTENSITY SPK	DIS	REF
2302.997	2303.706	43408.32	NI I	1.0S	1.5		MIT
2302.472	2303.181	43418.21	NI II	0.3	1.3		MIT
2301.57	2302.28	43435.2	NI	1.0			MIT
2301.019	2301.728	43445.63	NI II		0.5		MIT
2300.778	2301.486	43450.18	NI I	1.3R	0.6		MIT
2300.088	2300.796	43463.21	NI II	0.5	1.3		MIT
2299.646	2300.354	43471.56	NI II	0.5	1.2		MIT
2298.499	2299.207	43493.26	NI II		1.0		MIT
2298.275	2298.983	43497.50	NI II	1.1	1.2		MIT
2297.491	2298.199	43512.34	NI II	1.0	1.2		MIT
2297.138	2297.846	43519.02	NI II	1.2	1.3		MIT
2296.553	2297.261	43530.11	NI II	1.0	1.3		MIT
2293.116	2293.823	43595.35	NI I	1.0	0.5		MIT
2289.976	2290.682	43655.12	NI I	1.3R	1.3		MIT
2288.39	2289.10	43685.4	NI I	1.1	0.3		MIT
2287.65	2288.36	43699.5	NI II	0.3	1.3		MIT
2287.323	2288.029	43705.75	NI I	0.7			MIT
2287.084	2287.789	43710.31	NI II	2.0	2.7		MIT
2286.80	2287.51	43715.7	NI	0.3H			MIT
2279.53	2280.23	43855.1	NI	1.3			MIT
2278.773	2279.477	43869.72	NI II	1.1	1.4		MIT
2277.76	2278.46	43889.2	NI	1.0			MIT
2277.28	2277.98	43898.5	NI II	0.7	1.4		MIT
2276.445	2277.148	43914.58	NI II		1.0		MIT
2275.689	2276.392	43929.17	NI II	0.3	1.2		MIT
2274.73	2275.43	43947.7	NI II		1.2		MIT
2274.66	2275.36	43949.0	NI I	1.1			MIT
2273.81	2274.51	43965.5	NI	0.90	0.6H		MIT
2271.950	2272.652	44001.45	NI I	1.0	0.5		MIT
2270.213	2270.915	44035.12	NI II	2.0	2.6		MIT
2267.556	2268.257	44086.71	NI I	1.0			MIT
2266.352	2267.053	44110.13	NI I	1.3	0.3		MIT
2265.354	2266.055	44129.56	NI II		0.7		MIT
2264.457	2265.158	44147.04	NI II	2.2	2.6		MIT
2262.900	2263.600	44177.41	NI II		0.6		MIT
2261.424	2262.124	44206.24	NI I	1.1	0.5		MIT
2259.571	2260.271	44242.49	NI I	2.5W			MIT
2258.146	2258.845	44270.41	NI I	1.2	0.3		MIT
2257.89	2258.59	44275.4	NI		1.1		MIT
2256.146	2256.845	44309.65	NI II		1.0		MIT
2255.878	2256.577	44314.91	NI I	1.0			MIT
2254.80	2255.50	44336.1	NI I	1.0	0.5		MIT
2253.86	2254.56	44354.6	NI II	2.0	2.5		MIT
2253.66	2254.36	44358.5	NI II		1.0		MIT
2253.55	2254.25	44360.7	NI I	1.0			MIT
2252.886	2253.584	44373.76	NI	0.0	0.7H		MIT
2251.48	2252.18	44401.5	NI I	1.0	0.5		MIT
2247.23	2247.93	44485.4	NI II	0.3	1.1		MIT
2245.08	2245.78	44528.0	NI	0.0	0.6		MIT
2244.90	2245.60	44531.6	NI	0.0	0.9		MIT
2244.53	2245.23	44538.9	NI I	1.6			MIT
2244.48	2245.18	44539.9	NI I	1.2	0.5		MIT
2243.21	2243.91	44565.1	NI I	0.5			MIT
2242.88	2243.58	44571.7	NI I	0.5H			MIT
2242.76	2243.46	44574.1	NI		0.9H		MIT
2242.15	2242.85	44586.2	NI II		0.6		MIT
2241.66	2242.36	44596.0	NI		0.9H		MIT
2236.08	2236.77	44707.2	NI II		0.6		MIT
2230.954	2231.647	44809.95	NI I	1.0	0.0		MIT
2229.858	2230.551	44831.97	NI II		0.8H		MIT
2226.332	2227.024	44902.96	NI II	1.1	1.2		MIT
2225.82	2226.51	44913.3	NI		0.9		MIT
2225.342	2226.034	44922.94	NI I	0.7			MIT
2224.867	2225.559	44932.53	NI II	1.2	1.4		MIT
2224.512	2225.204	44939.70	NI II		0.9		MIT
2224.358	2225.050	44942.81	NI II	0.8	1.0		MIT
2222.944	2223.636	44971.40	NI II	1.2	1.4W		MIT
2222.33	2223.02	44983.8	NI		0.6W		MIT
2221.940	2222.631	44991.71	NI I	1.0			MIT
2221.85	2222.54	44993.5	NI		0.7		MIT

AIR WAVELENGTH (ANGSTRMS)	VACUUM WAVELENGTH (ANGSTRMS)	WAVENUMBER (KAYSERS)	LMENT	LOG ARC	INTENSITY SPK	DIS	REF
2221.11	2221.80	45008.5	NI		1.2H		MIT
2220.400	2221.091	45022.92	NI II	1.0	1.4		MIT
2219.53	2220.22	45040.6	NI		0.9H		MIT
2217.768	2218.459	45076.34	NI I	1.0	0.0		MIT
2216.473	2217.163	45102.68	NI II	1.3	1.6J		MIT
2213.849	2214.539	45156.13	NI	0.7			MIT
2213.193	2213.883	45169.51	NI II	0.3	1.1		MIT
2212.15	2212.84	45190.8	NI I	1.1	0.3H		MIT
2211.290	2211.979	45208.38	NI I	1.1			MIT
2211.10	2211.79	45212.3	NI II		0.9		MIT
2211.02	2211.71	45213.9	NI I	0.9	1.1		MIT
2210.39	2211.08	45226.8	NI II	1.1	1.3W		MIT
2208.99	2209.68	45255.5	NI	0.7			MIT
2208.67	2209.36	45262.0	NI I	0.5			MIT
2207.74	2208.43	45281.1	NI I	0.6			MIT
2207.44	2208.13	45287.2	NI		0.5H		MIT
2206.70	2207.39	45302.4	NI II	1.3	1.5H		MIT
2203.52	2204.21	45367.8	NI		0.9H		MIT
2201.56	2202.25	45408.2	NI I	0.8H			MIT
2201.41	2202.10	45411.3	NI II	1.0	1.5		MIT
2200.70	2201.39	45425.9	NI I	1.0	0.3		MIT
2197.35	2198.04	45495.1	NI I	1.3			MIT
2192.359	2193.044	45598.71	NI II		0.9H		MIT
2192.12	2192.81	45603.7	NI		0.8H		MIT
2191.56	2192.25	45615.3	NI I	0.3			MIT
2191.19	2191.88	45623.0	NI I	1.2			MIT
2190.969	2191.654	45627.64	NI II		1.0		MIT
2190.57	2191.25	45636.0	NI		0.9H		MIT
2190.23	2190.91	45643.0	NI I	1.3	0.5		MIT
2188.88	2189.56	45671.2	NI		0.6		MIT
2188.049	2188.733	45688.52	NI II	0.6	1.4		MIT
2187.600	2188.284	45697.90	NI I	0.6			MIT
2187.21	2187.89	45706.0	NI	0.3			MIT
2186.94	2187.62	45711.7	NI I	0.9			MIT
2185.498	2186.182	45741.85	NI II	0.6	1.5		MIT
2184.606	2185.290	45760.52	NI II	1.2	1.3H		MIT
2183.904	2184.588	45775.23	NI I	1.0			MIT
2183.378	2184.061	45786.26	NI	1.0	0.3H		MIT
2182.37	2183.05	45807.4	NI I	1.2	0.5		MIT
2180.470	2181.153	45847.31	NI II	0.6	1.6		MIT
2179.992	2180.675	45857.37	NI II		1.1		MIT
2179.464	2180.147	45868.48	NI II	0.6	0.9		MIT
2179.350	2180.033	45870.87	NI II	0.3	1.0		MIT
2177.360	2178.042	45912.79	NI II	0.5	1.3		MIT
2177.088	2177.770	45918.53	NI II	0.5	1.3		MIT
2175.156	2175.838	45959.31	NI II	1.2	1.4		MIT
2174.669	2175.351	45969.60	NI II	1.1	1.5		MIT
2174.472	2175.154	45973.77	NI I	1.0			MIT
2173.547	2174.228	45993.33	NI I	1.2			MIT
2170.06	2170.74	46067.2	NI		0.6		MIT
2169.61	2170.29	46076.8	NI		0.3H		MIT
2169.100	2169.780	46087.61	NI II	1.3	1.3		MIT
2166.156	2166.836	46150.24	NI I	1.0			MIT
2165.559	2166.239	46162.96	NI II	1.3	1.6G		MIT
2161.224	2161.903	46255.55	NI II	1.0	1.5		MIT
2161.03	2161.71	46259.7	NI I	1.2	0.0		MIT
2158.739	2159.417	46308.79	NI II	1.0	1.3		MIT
2158.312	2158.990	46317.95	NI I	1.2R	0.6		MIT
2157.83	2158.51	46328.3	NI I	1.1	0.3		MIT
2152.22	2152.90	46449.0	NI I	1.0	0.0		MIT
2151.922	2152.599	46455.47	NI I	1.0			MIT
2147.803	2148.479	46544.55	NI I	1.2	0.9		MIT
2145.18	2145.86	46601.5	NI I	0.9	1.0		MIT
2140.09	2140.76	46712.3	NI	0.3			MIT
2139.06	2139.73	46734.8	NI II		0.5		MIT
2138.58	2139.25	46745.3	NI II	1.0	1.2		MIT
2138.05	2138.72	46756.9	NI II		0.7		MIT
2135.335	2136.009	46816.29	NI I	1.0			MIT
2134.928	2135.602	46825.21	NI I	1.1	0.3		MIT
2134.28	2134.95	46839.4	NI II	0.3	1.0		MIT

AIR WAVELENGTH (ANGSTRMS)	VACUUM WAVELENGTH (ANGSTRMS)	WAVENUMBER (KAYSERS)	LMENT	LOG ARC	INTENSITY SPK	INTENSITY DIS	REF
2131.266	2131.939	46905.66	NI II	0.9	1.3H		MIT
2131.04	2131.71	46910.6	NI II		1.0		MIT
2129.95	2130.62	46934.6	NI I	1.2	0.5		MIT
2129.14	2129.81	46952.5	NI II		0.6		MIT
2128.57	2129.24	46965.1	NI II	0.7	1.5		MIT
2128.41	2129.08	46968.6	NI I	1.2			MIT
2127.91	2128.58	46979.6	NI	1.2			MIT
2127.77	2128.44	46982.7	NI II		1.3		MIT
2126.80	2127.47	47004.1	NI II	0.3	1.4W		MIT
2125.90	2126.57	47024.0	NI II	0.6	1.4		MIT
2125.62	2126.29	47030.2	NI I	1.2	0.3		MIT
2125.11	2125.78	47041.5	NI II	0.6	1.4		MIT
2124.81	2125.48	47048.2	NI I	1.4	0.5		MIT
2124.11	2124.78	47063.7	NI	1.5			MIT
2122.25	2122.92	47104.9	NI I	1.0			MIT
2121.39	2122.06	47124.0	NI I	1.3	0.9		MIT
2120.70	2121.37	47139.3	NI	0.5	0.5H		MIT
2118.56	2119.23	47186.9	NI I	1.0			MIT
2114.41	2115.08	47279.5	NI I	1.3	0.3H		MIT
2113.51	2114.18	47299.7	NI II	0.5	1.7		MIT
2113.29	2113.96	47304.6	NI II	0.3	0.5		MIT
2112.41	2113.08	47324.3	NI II		0.7		MIT
2111.72	2112.39	47339.8	NI II	1.4	1.0		MIT
2109.776	2110.445	47383.38	NI I	1.0			MIT
2108.99	2109.66	47401.0	NI II		1.3		MIT
2107.96	2108.63	47424.2	NI II	0.3	1.7		MIT
2107.194	2107.862	47441.44	NI I	1.1			MIT
2105.836	2106.504	47472.03	NI I	1.1	0.5		MIT
2103.38	2104.05	47527.5	NI II		1.4		MIT
2102.81	2103.48	47540.3	NI II		1.0		MIT
2100.20	2100.87	47599.4	NI II		0.6		MIT
2097.096	2097.762	47669.85	NI II	0.6	1.5		MIT
2095.73	2096.40	47700.9	NI I	1.4	0.3		MIT
2095.123	2095.789	47714.73	NI I	1.2	0.3		MIT
2093.56	2094.23	47750.4	NI II	0.8	1.3		MIT
2091.71	2092.37	47792.6	NI I	1.0			MIT
2090.389	2091.054	47822.78	NI I	1.2			MIT
2090.10	2090.76	47829.4	NI II	0.5	1.4		MIT
2089.07	2089.73	47853.0	NI I	1.2R	1.2H		MIT
2088.98	2089.64	47855.0	NI I	1.5D			MIT
2087.736	2088.400	47883.54	NI	0.3	0.8H		MIT
2085.550	2086.214	47933.73	NI I	0.8			MIT
2085.348	2086.012	47938.37	NI I	1.2	0.3		MIT
2084.86	2085.52	47949.6	NI II		1.4		MIT
2084.12	2084.78	47966.6	NI	0.9			MIT
2083.75	2084.41	47975.1	NI II		1.0		MIT
2083.643	2084.306	47977.59	NI II	0.3	1.0		MIT
2082.856	2083.519	47995.72	NI I	1.4	0.3		MIT
2080.849	2081.512	48042.00	NI II	0.5	1.5		MIT
2078.76	2079.42	48090.3	NI II	0.0	1.1		MIT
2077.18	2077.84	48126.9	NI	0.3			MIT
2076.19	2076.85	48149.8	NI II		0.6		MIT
2076.043	2076.705	48153.21	NI I	1.0			MIT
2074.58	2075.24	48187.2	NI	0.3			MIT
2074.12	2074.78	48197.8	NI II		1.2		MIT
2072.238	2072.899	48241.61	NI I	1.0			MIT
2069.499	2070.160	48305.45	NI I	1.3	0.3		MIT
2069.023	2069.684	48316.56	NI	1.2	0.6		MIT
2068.606	2069.267	48326.30	NI I	1.0			MIT
2068.35	2069.01	48332.3	NI	0.3			MIT
2066.410	2067.070	48377.65	NI II	0.3	1.4		MIT
2064.366	2065.026	48425.55	NI I	1.2	0.3		MIT
2063.39	2064.05	48448.5	NI I	1.3	0.3		MIT
2062.341	2063.000	48473.09	NI I	1.2	0.3		MIT
2060.73	2061.39	48511.0	NI I	0.7			MIT
2060.189	2060.848	48523.71	NI I	1.0	0.3		MIT
2059.895	2060.554	48530.64	NI I	1.2R	0.8		MIT
2057.83	2058.49	48579.3	NI II	0.8	1.0S		MIT
2057.37	2058.03	48590.2	NI II	0.3	1.3		MIT
2055.50	2056.16	48634.4	NI I	1.0	0.3		MIT

AIR WAVELENGTH (ANGSTRMS)	VACUUM WAVELENGTH (ANGSTRMS)	WAVENUMBER (KAYSERS)	LMENT	LOG ARC	INTENSITY SPK	INTENSITY DIS	REF
2054.32	2054.98	48662.3	NI II		1.4		MIT
2053.90	2054.56	48672.3	NI I	0.8	0.0		MIT
2053.290	2053.948	48686.73	NI II	0.5	1.4		MIT
2052.428	2053.086	48707.18	NI I	0.8			MIT
2052.06	2052.72	48715.9	NI I	1.1D	0.8		MIT
2050.827	2051.484	48745.20	NI I	1.2	0.3		MIT
2048.30	2048.96	48805.3	NI	0.3			MIT
2047.333	2047.990	48828.37	NI I	1.1	0.8		MIT
2044.39	2045.05	48898.7	NI	0.3			MIT
2041.139	2041.794	48976.53	NI I	1.0			MIT
2038.75	2039.40	49033.9	NI I	0.3H	0.5		MIT
2035.35	2036.00	49115.8	NI II		1.0		MIT
2035.08	2035.73	49122.3	NI I	1.0H			MIT
2034.85	2035.50	49127.9	NI I	0.8H	0.3		MIT
2034.397	2035.051	49138.81	NI I	1.2	0.3		MIT
2033.538	2034.192	49159.57	NI I	0.9			MIT
2033.45	2034.10	49161.7	NI II		0.7		MIT
2032.30	2032.95	49189.5	NI II	0.3	1.4		MIT
2029.83	2030.48	49249.4	NI	0.7			MIT
2029.251	2029.904	49263.41	NI I	1.0			MIT
2029.19	2029.84	49264.9	NI II		1.4		MIT
2026.63	2027.28	49327.1	NI I	1.0	0.7H		MIT
2026.36	2027.01	49333.7	NI	0.9			MIT
2025.81	2026.46	49347.1	NI	0.8			MIT
2025.38	2026.03	49357.5	NI I	1.0R	0.8		MIT
2021.28	2021.93	49457.6	NI I	0.7			MIT
2020.97	2021.62	49465.2	NI II		1.5		MIT
2019.04	2019.69	49512.5	NI II	0.3	1.5		MIT
2016.35	2017.00	49578.6	NI I	0.7			MIT
2014.228	2014.879	49630.78	NI I	1.2R	1.0		MIT
2007.676	2008.325	49792.73	NI I	1.0	0.5		MIT
2006.996	2007.645	49809.60	NI I	1.2	0.5		MIT
2004.29	2004.94	49876.8	NI II		1.5		MIT
2001.817	2002.465	49938.44	NI I	1.2	0.5		MIT
2000.46	2001.11	49972.3	NI	1.0	0.3		MIT
9265.99	9268.53	10789.2	O I	1.2			EN
9262.73	9265.27	10793.0	O I	1.2			EN
9260.88	9263.42	10795.2	O I	1.1			EN
8819.60	8822.02	11335.3	O			1.8	FH
8586.00	8588.36	11643.7	O			1.2	FH
8446.38	8448.70	11836.1	O I			3.3	PS
8428.342	8430.658	11861.47	O I			1.2	FH
8426.326	8428.641	11864.31	O I			1.7	FH
8235.408	8237.672	12139.35	O I			1.7	FH
8233.085	8235.348	12142.78	O I			3.0	FH
8230.016	8232.278	12147.31	O I			3.0	FH
8227.680	8229.942	12150.75	O I			3.0	FH
8221.829	8224.089	12159.40	O I			3.3	FH
7995.086	7997.285	12504.24	O I			1.7	FH
7987.036	7989.233	12516.85	O I			1.2	FH
7952.182	7954.369	12571.71	O I			1.7	FH
7950.824	7953.011	12573.85	O I			2.0	FH
7947.566	7949.752	12579.01	O I			3.0	FH
7947.204	7949.390	12579.58	O I			1.5	FH
7943.178	7945.363	12585.96	O I			1.5	FH
7775.433	7777.573	12857.48	O I			2.0	ME
7774.138	7776.277	12859.62	O I			2.5	ME
7771.928	7774.067	12863.28	O I			3.0	ME
7480.652	7482.712	13364.14	O I			1.5	FH
7479.148	7481.208	13366.83	O I			1.5	FH
7477.264	7479.323	13370.19	O I			1.2	FH
7476.473	7478.532	13371.61	O I			1.8	FH
7473.226	7475.284	13377.42	O I			1.2	FH
7471.374	7473.432	13380.73	O I			1.2	FH
7254.05	7256.05	13781.6	O I			1.2	PS
7157.360	7159.333	13967.78	O I			1.8	FH
7002.22	7004.15	14277.2	O I			1.7	PS
6910.75	6912.66	14466.2	O II			1.5	MH
6908.11	6910.02	14471.7	O II			1.2	MH
6906.54	6908.45	14475.0	O II			1.7	MH

AIR WAVELENGTH (ANGSTRMS)	VACUUM WAVELENGTH (ANGSTRMS)	WAVENUMBER (KAYSERS)	LMENT	LOG ARC	INTENSITY SPK	DIS	REF
6895.29	6897.19	14498.7	0 II			1.8	MH
6727.866	6729.723	14859.45	0 I			1.8	FH
6721.35	6723.21	14873.9	0 II			1.8	FL
6721.21	6723.07	14874.2	0 II			2.5	MH
6654.77	6656.61	15022.7	0			1.0	FL
6654.121	6655.958	15024.13	0 I			1.7	FH
6640.90	6642.73	15054.0	0 II			1.8	FL
6627.62	6629.45	15084.2	0 II			1.5	MH
6456.07	6457.85	15485.0	0 I			2.7	PS
6454.55	6456.33	15488.7	0 I			2.2	PS
6453.69	6455.47	15490.7	0 I			2.0	PS
6374.292	6376.054	15683.68	0			1.8	FH
6366.282	6368.042	15703.41	0			1.7	FH
6324.682	6326.431	15806.70	0 I			1.5	FH
6264.346	6266.079	15958.94	0			1.2	FH
6261.314	6263.046	15966.67	0			1.8	FH
6256.616	6258.347	15978.66	0			1.5	FH
6158.20	6159.90	16234.0	0 I			3.0	PS
6156.78	6158.48	16237.8	0 I			2.5	PS
6155.99	6157.69	16239.8	0 I			2.2	PS
6106.398	6108.088	16371.73	0			1.5	FH
6046.34	6048.01	16534.4	0 I			2.2	PS
5995.198	5996.859	16675.40	0			1.6	FH
5992.45	5994.11	16683.0	0			1.5	PS
5991.852	5993.512	16684.71	0			1.2	FH
5958.53	5960.18	16778.0	0 I			2.0H	PS
5950.60	5952.25	16800.4	0			1.8	PS
5750.424	5752.019	17385.20	0			1.8	FH
5731.103	5732.693	17443.81	0			1.5	FH
5554.94	5556.48	17997.0	0 I			2.0H	PS
5512.70	5514.23	18134.9	0 I			1.8H	PS
5436.83	5438.34	18388.0	0 I			2.3	PS
5435.76	5437.27	18391.6	0 I			2.0	PS
5435.16	5436.67	18393.6	0 I			1.8	PS
5410.76	5412.26	18476.6	0			1.5	PS
5408.59	5410.09	18484.0	0			1.7	PS
5404.87	5406.37	18496.7	0			1.5	PS
5330.66	5332.14	18754.2	0 I			2.7	PS
5329.59	5331.07	18757.9	0 I			2.2	PS
5328.98	5330.46	18760.1	0 I			2.0	PS
5299.00	5300.47	18866.2	0 I			1.8	PS
5275.08	5276.55	18951.8	0 I			1.7	PS
5206.61	5208.06	19201.0	0 II			1.8	MH
5190.45	5191.90	19260.8	0 II			1.5	MH
5175.86	5177.30	19315.1	0 II			1.2	MH
5159.88	5161.32	19374.9	0 II			1.6	MH
5146.06	5147.49	19426.9	0 I			1.8	PS
5130.53	5131.96	19485.7	0 I			1.5	PS
5047.70	5049.11	19805.5	0 I			1.2H	PS
5037.16	5038.56	19846.9	0 I			1.2H	PS
5020.13	5021.53	19914.2	0 I			1.8	PS
5019.34	5020.74	19917.4	0 I			1.7	PS
5018.78	5020.18	19919.6	0 I			1.5	PS
4968.76	4970.15	20120.1	0 I			2.0	PS
4967.86	4969.25	20123.8	0 I			1.9	PS
4967.40	4968.79	20125.6	0 I			1.7	PS
4955.78	4957.16	20172.8	0 II			1.5	FL
4942.97	4944.35	20225.1	0 II			2.0	MH
4941.02	4942.40	20233.1	0 II			1.7	MH
4924.50	4925.87	20301.0	0 II			1.8	MH
4906.80	4908.17	20374.2	0 II			1.7	MH
4890.85	4892.22	20440.6	0 II			1.5	MH
4871.48	4872.84	20521.9	0 II			1.6	MH
4864.95	4866.31	20549.5	0 II			1.5	FL
4860.93	4862.29	20566.4	0 II			1.3	MH
4856.76	4858.12	20584.1	0 II			1.3	MH
4856.49	4857.85	20585.2	0 II			1.2	FL
4803.00	4804.34	20814.5	0 I			1.7	PS
4802.20	4803.54	20818.0	0 I			1.5	PS
4801.80	4803.14	20819.7	0 I			1.2	PS

AIR WAVELENGTH (ANGSTRMS)	VACUUM WAVELENGTH (ANGSTRMS)	WAVENUMBER (KAYSERS)	LMENT	LOG ARC	INTENSITY SPK	DIS	REF
4773.76	4775.09	20942.0	0 I			1.8	PS
4772.89	4774.22	20945.8	0 I			1.7	PS
4772.54	4773.87	20947.4	0 I			1.5	PS
4752.70	4754.03	21034.8	0 II			1.2	FL
4751.34	4752.67	21040.8	0 II			1.7	FL
4741.71	4743.04	21083.5	0 II			1.3	FL
4710.00	4711.32	21225.5	0 II			1.8	MH
4705.32	4706.64	21246.6	0 II			2.5	MH
4703.14	4704.46	21256.4	0 II			1.5	MH
4701.16	4702.48	21265.4	0 II			1.3	MH
4699.21	4700.52	21274.2	0 II			2.0	FL
4698.99	4700.30	21275.2	0 II			1.5	MH
4696.32	4697.63	21287.3	0 II			1.5	MH
4691.37	4692.68	21309.8	0 II			1.2	MH
4676.246	4677.555	21378.69	0 II			2.1	FL
4673.71	4675.02	21390.3	0 I			1.5	MH
4673.71	4675.02	21390.3	0 II			1.5	MH
4672.75	4674.06	21394.7	0 I			1.5	PS
4661.650	4662.955	21445.63	0 II			2.1	FL
4655.36	4656.66	21474.6	0 I			1.7	PS
4654.56	4655.86	21478.3	0 I			1.5	PS
4654.23	4655.53	21479.8	0 I			1.2	PS
4650.853	4652.155	21495.41	0 II			1.8	FL
4649.148	4650.450	21503.30	0 II			2.5	FL
4641.827	4643.127	21537.21	0 II			2.2	FL
4638.865	4640.164	21550.96	0 II			1.8	FL
4610.14	4611.43	21685.2	0 II			1.2H	FL
4609.39	4610.68	21688.8	0 II			1.8H	MH
4602.08	4603.37	21723.2	0 II			1.0H	MH
4596.13	4597.42	21751.3	0 II			2.2	MH
4590.94	4592.23	21775.9	0 II			2.5	MH
4589.89	4591.18	21780.9	0 I			1.5	PS
4588.98	4590.27	21785.2	0 I			1.2D	PS
4577.66	4578.94	21839.1	0 I			1.5	PS
4576.79	4578.07	21843.2	0 I			1.2D	PS
4506.50	4507.76	22183.9	0			1.2H	FL
4491.23	4492.49	22259.4	0 II			1.5H	MH
4489.47	4490.73	22268.1	0 II			1.0H	MH
4488.17	4489.43	22274.6	0 II			1.2H	MH
4477.92	4479.18	22325.5	0 II			1.0H	MH
4469.41	4470.66	22368.0	0 II			1.6	MH
4467.83	4469.08	22375.9	0 II			1.7	MH
4466.28	4467.53	22383.7	0 II			1.5H	MH
4465.45	4466.70	22387.9	0 II			1.7	MH
4452.41	4453.66	22453.4	0 II			1.8	MH
4448.20	4449.45	22474.7	0 II			1.8	MH
4443.04	4444.29	22500.8	0 II			1.7	MH
4416.974	4418.214	22633.58	0 II			2.2	FL
4414.888	4416.128	22644.27	0 II			2.5	FL
4395.95	4397.18	22741.8	0 II			1.9	FL
4379.55	4380.78	22827.0	0			1.2H	FL
4378.41	4379.64	22832.9	0 II			1.0H	MH
4371.59	4372.82	22868.6	0 II			1.0H	MH
4369.28	4370.51	22880.6	0 II			1.7	FL
4368.30	4369.53	22885.8	0 I			3.0	PS
4366.906	4368.133	22893.07	0 II			2.0	FL
4351.275	4352.498	22975.31	0 II			2.1	FL
4349.435	4350.658	22985.03	0 II			2.5	FL
4347.429	4348.651	22995.64	0 II			1.8	FL
4345.570	4346.792	23005.47	0 II			2.1	FL
4341.94	4343.16	23024.7	0 II			1.5H	MH
4340.29	4341.51	23033.5	0 II			1.0H	MH
4336.85	4338.07	23051.7	0 II			1.8	MH
4331.84	4333.06	23078.4	0 II			1.2	MH
4328.62	4329.84	23095.6	0 II			1.2	FL
4327.47	4328.69	23101.7	0 II			1.3	MH
4325.76	4326.98	23110.8	0 II			1.3	MH
4319.647	4320.862	23143.53	0 II			2.2	FL
4317.160	4318.374	23156.86	0 II			2.2	FL
4303.78	4304.99	23228.9	0 II			1.8H	MH

AIR WAVELENGTH (ANGSTRMS)	VACUUM WAVELENGTH (ANGSTRMS)	WAVENUMBER (KAYSERS)	LMENT		LOG ARC	INTENSITY SPK	DIS	REF
4294.74	4295.95	23277.7	O	II			1.5H	MH
4291.19	4292.40	23297.0	O	II			1.2H	MH
4285.62	4286.83	23327.3	O	II			1.3H	MH
4282.82	4284.03	23342.5	O	II			1.0H	MH
4276.64	4277.84	23376.3	O	II			1.2H	MH
4275.47	4276.67	23382.7	O	II			1.7H	MH
4253.98	4255.18	23500.8	O	II			2.0H	FL
4253.74	4254.94	23502.1	O	II			1.7H	MH
4233.32	4234.51	23615.5	O				2.0	PS
4222.78	4223.97	23674.4	O				1.7	PS
4217.09	4218.28	23706.4	O				1.5	PS
4192.53	4193.71	23845.2	O	II			1.2	MH
4189.793	4190.974	23860.80	O	II			2.7	FL
4185.453	4186.633	23885.54	O	II			2.2	FL
4169.23	4170.41	23978.5	O	II			1.7	FL
4156.54	4157.71	24051.7	O	II			1.5	FL
4153.310	4154.481	24070.40	O	II			2.3	FL
4146.06	4147.23	24112.5	O	II			1.6	MH
4143.76	4144.93	24125.9	O	II			1.2	MH
4132.82	4133.99	24189.7	O	II			2.0	FL
4129.34	4130.50	24210.1	O	II			1.2	FL
4121.46	4122.62	24256.4	O	II			1.7	MH
4120.55	4121.71	24261.8	O	II			1.3	FL
4120.27	4121.43	24263.4	O	II			1.7	MH
4119.222	4120.384	24269.58	O	II			2.5	FL
4113.82	4114.98	24301.4	O	II			1.2	MH
4112.02	4113.18	24312.1	O	II			1.7	MH
4110.79	4111.95	24319.4	O	II			1.6	MH
4105.001	4106.159	24353.66	O	II			2.1	FL
4104.73	4105.89	24355.3	O	II			1.7	MH
4103.01	4104.17	24365.5	O	II			1.7	FL
4098.24	4099.40	24393.8	O	II			0.7H	MH
4097.24	4098.40	24399.8	O	II			1.8	MH
4096.53	4097.69	24404.0	O	II			1.5	MH
4095.63	4096.79	24409.4	O	II			1.0H	FL
4092.94	4094.10	24425.4	O	II			1.9	FL
4089.27	4090.42	24447.3	O	II			1.8H	MH
4087.14	4088.29	24460.1	O	II			1.6H	MH
4085.12	4086.27	24472.2	O	II			1.8	FL
4084.66	4085.81	24474.9	O	II			1.5	MH
4083.91	4085.06	24479.4	O	II			1.6H	MH
4078.85	4080.00	24509.8	O	II			1.8	MH
4075.869	4077.020	24527.72	O	II			2.9	FL
4072.156	4073.306	24550.08	O	II			2.5	FL
4071.24	4072.39	24555.6	O	II			0.7	MH
4069.903	4071.052	24563.67	O	II			2.1	FL
4069.635	4070.784	24565.29	O	II			1.8	FL
4062.94	4064.09	24605.8	O	II			1.5H	MH
4061.00	4062.15	24617.5	O	II			1.2H	MH
4060.60	4061.75	24619.9	O	II			1.5H	MH
4048.22	4049.36	24695.2	O	II			1.0H	FL
3982.725	3983.851	25101.34	O	II			1.3	FL
3973.266	3974.390	25161.09	O	II			2.1	FL
3954.687	3955.806	25279.30	O	I	1.0			EN
3954.596	3955.715	25279.88	O	I			1.6	FH
3953.056	3954.175	25289.73	O	I			0.7	FH
3951.987	3953.105	25296.57	O	I			1.0	FH
3947.61	3948.73	25324.6	O	I			1.3	PS
3947.51	3948.63	25325.3	O	I			1.7	PS
3947.33	3948.45	25326.4	O	I			2.5	PS
3945.04	3946.16	25341.1	O	II			1.3	MH
3919.279	3920.389	25507.67	O	II			1.5	FL
3912.088	3913.196	25554.56	O	II			0.7	FL
3911.951	3913.059	25555.45	O	II			2.2	FL
3907.44	3908.55	25585.0	O	II			1.3	MH
3893.52	3894.62	25676.4	O	II			0.7	MH
3883.12	3884.22	25745.2	O	II			0.8	MH
3882.43	3883.53	25749.8	O	II			1.0	MH
3882.194	3883.294	25751.33	O	II			1.5	FL
3875.80	3876.90	25793.8	O	II			1.1	MH
3874.07	3875.17	25805.3	O	II			0.7	MH
3864.66	3865.76	25868.2	O	II			0.7	MH
3864.42	3865.52	25869.8	O	II			1.3	MH
3863.49	3864.59	25876.0	O	II			0.7	MH
3857.16	3858.25	25918.5	O	II			1.0	MH
3856.16	3857.25	25925.2	O	II			1.3	FL
3851.47	3852.56	25956.7	O	II			0.3	MH
3851.04	3852.13	25959.6	O	II			1.0	FL
3850.81	3851.90	25961.2	O	II			0.7	FL
3847.89	3848.98	25980.9	O	II			1.0	FL
3843.58	3844.67	26010.0	O	II			1.0	FL
3842.82	3843.91	26015.2	O	II			1.0	FL
3833.10	3834.19	26081.1	O	II			1.0	FL
3830.45	3831.54	26099.2	O	II			1.3	FL
3825.249	3826.334	26134.67	O	I			1.3	FH
3825.090	3826.175	26135.76	O	I			1.3	FH
3824.425	3825.510	26140.30	O	I			1.0	FH
3823.469	3824.554	26146.84	O	I			2.1	FH
3821.64	3822.72	26159.4	O	II			1.0L	MH
3803.13	3804.21	26286.7	O	II			1.3L	MH
3794.48	3795.56	26346.6	O	II			1.0L	FL
3777.60	3778.67	26464.3	O	II			1.3L	FL
3762.51	3763.58	26570.4	O	II			1.1L	MH
3749.47	3750.54	26662.9	O	II			2.1	MH
3739.92	3740.98	26730.9	O	II			1.5L	FL
3739.79	3740.85	26731.9	O	II			1.0L	MH
3735.94	3737.00	26759.4	O	II			1.0	FL
3729.34	3730.40	26806.8	O	II			0.7	FL
3727.30	3728.36	26821.4	O	II			1.7	MH
3712.74	3713.80	26926.6	O	II			1.4	MH
3692.44	3693.49	27074.7	O	I			1.7	PS
3638.70	3639.74	27474.5	O				1.0	FL
3496.32	3497.32	28593.3	O	II			0.7	FL
3470.77	3471.76	28803.8	O	II			2.0	MH
3470.37	3471.36	28807.1	O	II			1.4	MH
3447.92	3448.91	28994.7	O	II			1.3	MH
3440.40	3441.39	29058.1	O				1.3	MH
3427.42	3428.40	29168.1	O				1.0H	MH
3420.63	3421.61	29226.0	O				0.8H	MH
3419.91	3420.89	29232.1	O				0.5H	MH
3409.77	3410.75	29319.1	O	II			1.4L	MH
3407.35	3408.33	29339.9	O	II			1.6L	MH
3390.26	3391.23	29487.8	O	II			2.0	MH
3382.68	3383.65	29553.9	O	I			1.1	MH
3377.20	3378.17	29601.8	O	II			1.6	FL
3371.85	3372.82	29648.8	O	II			0.7H	FL
3306.54	3307.49	30234.4	O	II			1.3L	MH
3305.09	3306.04	30247.6	O	II			1.3L	MH
3301.56	3302.51	30280.0	O	II			1.0L	FL
3295.09	3296.04	30339.4	O	II			1.0L	MH
3290.08	3291.03	30385.6	O	II			1.3L	MH
3287.57	3288.52	30408.8	O	II			1.8L	MH
3277.66	3278.60	30500.8	O	II			1.4L	MH
3273.54	3274.48	30539.2	O	II			1.5L	MH
3270.96	3271.90	30563.2	O	II			1.4L	MH
3218.10	3219.03	31065.3	O	II			0.8	MH
3216.76	3217.69	31078.2	O	II			0.5	MH
3216.08	3217.01	31084.8	O	II			0.3	MH
3139.75	3140.66	31840.4	O	II			1.0L	MH
3138.40	3139.31	31854.1	O	II			1.5L	MH
3134.79	3135.70	31890.8	O	II			2.0L	MH
3134.32	3135.23	31895.6	O	II			1.0L	FL
3129.41	3130.32	31945.6	O	II			1.4L	MH
3124.02	3124.93	32000.8	O	II			0.7L	FL
3122.62	3123.53	32015.1	O	II			1.3L	FL
3081.46	3082.35	32442.7	O				0.7H	FL
3032.47	3033.35	32966.8	O	II			0.5H	MH
3032.09	3032.97	32971.0	O	II			0.8H	MH
3013.41	3014.29	33175.3	O	II			1.1H	MH
3008.83	3009.71	33225.8	O	II			1.0H	MH

AIR WAVELENGTH (ANGSTRMS)	VACUUM WAVELENGTH (ANGSTRMS)	WAVENUMBER (KAYSERS)	LMENT	LOG INTENSITY ARC	SPK	DIS	REF
3007.74	3008.62	33237.9	O II			1.0H	MH
3007.28	3008.16	33243.0	O			0.3	FI
3007.08	3007.96	33245.2	O II			1.0	FL
3006.90	3007.78	33247.1	O II			1.1H	MH
3006.01	3006.89	33257.0	O II			0.7H	MH
3005.62	3006.50	33261.3	O II			0.7	FL
2997.74	2998.61	33348.7	O II			0.8H	MH
2992.59	2993.46	33406.1	O			1.0H	MH
2943.00	2943.86	33969.0	O			0.7H	MH
2942.88	2943.74	33970.4	O			0.7H	FL
2921.54	2922.40	34218.5	O			1.5	MH
2916.40	2917.25	34278.8	O			1.2	MH
2911.85	2912.70	34332.4	O II			0.8H	MH
2911.20	2912.05	34340.0	O II			0.7H	MH
2906.62	2907.47	34394.1	O II			1.2H	MH
2905.00	2905.85	34413.3	O II			0.7H	FL
2904.29	2905.14	34421.7	O II			0.7H	MH
2903.30	2904.15	34433.5	O			1.2H	MH
2894.98	2895.83	34532.4	O			1.0H	MH
2893.70	2894.55	34547.7	O			1.0H	MH
2892.47	2893.32	34562.4	O II			0.7H	MH
2887.91	2888.76	34617.0	O II			1.0H	MH
2885.36	2886.21	34647.6	O			1.0H	MH
2883.82	2884.67	34666.1	O I			1.2	FH
2881.70	2882.55	34691.6	O			1.0	MH
2879.04	2879.88	34723.6	O			0.8	MH
2863.57	2864.41	34911.2	O			1.0H	MH
2862.52	2863.36	34924.0	O			1.0H	MH
2862.26	2863.10	34927.2	O			1.0H	MH
2861.38	2862.22	34937.9	O			1.0H	MH
2859.89	2860.73	34956.1	O			0.7D	MH
2857.89	2858.73	34980.6	O			1.0H	MH
2856.78	2857.62	34994.2	O			1.0H	MH
2848.91	2849.75	35090.8	O			1.5	MH
2845.84	2846.68	35128.7	O			1.0H	MH
2836.35	2837.18	35246.2	O II			0.7	MH
2816.52	2817.35	35494.4	O			1.4	MH
2808.84	2809.67	35591.4	O II			0.7	MH
2805.96	2806.79	35627.9	O			1.0	MH
2783.15	2783.97	35919.9	O II			0.5	MH
2782.46	2783.28	35928.8	O			0.7	MH
2761.30	2762.12	36204.1	O			1.0H	MH
2747.50	2748.31	36386.0	O II			1.4	MH
2733.40	2734.21	36573.6	O II			2.2	MH
2718.90	2719.71	36768.7	O II			0.8	MH
2715.45	2716.25	36815.4	O II			1.2	MH
2696.11	2696.91	37079.5	O			0.8	MH
2681.49	2682.29	37281.6	O			0.8	MH
2641.53	2642.32	37845.6	O			1.0H	MH
2632.78	2633.56	37971.3	O			1.0	MH
2623.05	2623.83	38112.2	O			0.7H	MH
2622.41	2623.19	38121.5	O			0.8	MH
2620.04	2620.82	38156.0	O			0.9	MH
2597.572	2598.349	38485.98	O	1.0	0.3		MIT
2582.99	2583.76	38703.2	O			0.7	MH
2575.300	2576.071	38818.80	O II			2.0	FL
2571.476	2572.246	38876.52	O II			1.5	FL
2538.94	2539.70	39374.7	O			0.9	MH
2530.30	2531.06	39509.1	O II			1.6	FL
2529.92	2530.68	39515.1	O			0.9	MH
2526.91	2527.67	39562.1	O II			1.4	MH
2523.20	2523.96	39620.3	O II			1.2	MH
2517.97	2518.73	39702.6	O II			1.4	FL
2517.41	2518.17	39711.4	O			1.7	MH
2507.77	2508.53	39864.1	O			1.3	MH
2506.56	2507.32	39883.3	O			0.9	MH
2504.70	2505.45	39912.9	O			1.2	MH
2501.80	2502.55	39959.2	O			1.5	MH
2499.29	2500.04	39999.3	O			1.3	MH
2497.10	2497.85	40034.4	O			0.9	MH

AIR WAVELENGTH (ANGSTRMS)	VACUUM WAVELENGTH (ANGSTRMS)	WAVENUMBER (KAYSERS)	LMENT	LOG INTENSITY ARC	SPK	DIS	REF
2493.75	2494.50	40088.2	O			1.7	MH
2493.33	2494.08	40094.9	O			1.5	MH
2459.45	2460.19	40647.2	O			0.9	MH
2445.55	2446.29	40878.2	O II			2.5	FL
2444.26	2445.00	40899.8	O II			1.5	FL
2441.67	2442.41	40943.2	O II			0.7	MH
2436.06	2436.80	41037.5	O II			1.3H	MH
2433.53	2434.27	41080.1	O II			2.4	MH
2425.55	2426.29	41215.2	O II			1.4	MH
2418.46	2419.19	41336.1	O II			1.6	MH
2415.13	2415.86	41393.1	O II			1.2	FL
2411.60	2412.33	41453.6	O II			1.3	MH
2407.49	2408.22	41524.4	O II			1.4	FL
2406.41	2407.14	41543.0	O II			1.3	FL
2375.73	2376.46	42079.5	O II			1.2	FL
2365.15	2365.87	42267.7	O II			1.0	FL
2339.31	2340.03	42734.5	O II			0.8	MH
2327.97	2328.68	42942.7	O II			0.7	MH
2325.95	2326.66	42980.0	O II			0.7	MH
2322.15	2322.86	43050.3	O II			0.8H	MH
2319.68	2320.39	43096.1	O II			1.2	MH
2316.79	2317.50	43149.9	O II			0.8H	MH
2316.12	2316.83	43162.4	O II			1.0H	MH
2313.05	2313.76	43219.7	O II			0.8H	MH
2302.83	2303.54	43411.5	O II			1.4	MH
2300.36	2301.07	43458.1	O II			1.8	MH
2293.33	2294.04	43591.3	O II			1.7	MH
2290.88	2291.59	43637.9	O II			1.4	MH
2285.68	2286.39	43737.2	O			0.8	MH
2284.89	2285.60	43752.3	O II			1.0	MH
2283.42	2284.12	43780.5	O II			0.8	MH
2259.66	2260.36	44240.7	O II			0.7	MH
2252.90	2253.60	44373.5	O II			1.0	FL
2252.74	2253.44	44376.6	O II			1.0	MH
2250.00	2250.70	44430.7	O II			0.3	FL
2218.70	2219.39	45057.4	O II			0.7	MH
2195.70	2196.39	45529.3	O II			0.7	FL
2195.43	2196.12	45534.9	O II			0.7	MH
2191.44	2192.13	45617.8	O II			0.7	MH
2190.66	2191.34	45634.1	O II			0.7	FL
2190.42	2191.10	45639.1	O II			1.0	MH
2189.20	2189.88	45664.5	O II			0.7	MH
2182.64	2183.32	45801.7	O II			1.2	M
2165.32	2166.00	46168.1	O			1.0	MH
2162.88	2163.56	46220.1	O			1.4	MH
2131.99	2132.66	46889.7	O II			1.4	MH
2131.76	2132.43	46894.8	O II			1.4	MH
2107.13	2107.80	47442.9	O			1.4	MH
2106.07	2106.74	47466.7	O			1.4	MH
2101.29	2101.96	47574.7	O II			1.2	MH
2100.69	2101.36	47588.3	O II			0.3	MH
2092.86	2093.53	47766.3	O II			0.5	MH
2075.14	2075.80	48174.2	O II			1.2	MH
2072.23	2072.89	48241.8	O II			1.5	FL
2052.53	2053.19	48704.8	O			1.2	MH
2045.41	2046.07	48874.3	O			1.4	MH
2020.44	2021.09	49478.2	O II			0.7	FL
2020.27	2020.92	49482.4	O			0.7	FL
2016.60	2017.25	49572.4	O II			0.7	FL
8041.29	8043.50	12432.4	OS	0.3			MIT
7981.20	7983.40	12526.0	OS	0.3			MIT
7974.72	7976.91	12536.2	OS	0.3			MIT
7957.31	7959.50	12563.6	OS	0.3			MIT
7901.57	7903.74	12652.2	OS	0.3			MIT
7852.17	7854.33	12731.8	OS	0.5			MIT
7789.96	7792.10	12833.5	OS	0.5			MIT
7701.46	7703.58	12981.0	OS	0.3			ME
7618.97	7621.07	13121.5	OS	0.3			MIT
7602.95	7605.04	13149.2	OS	1.3			MIT
7407.95	7409.99	13495.3	OS	1.0			MIT

AIR WAVELENGTH (ANGSTRMS)	VACUUM WAVELENGTH (ANGSTRMS)	WAVENUMBER (KAYSERS)	LMENT	LOG ARC	INTENSITY SPK	DIS	REF
7364.05	7366.08	13575.7	OS	1.0			MIT
7253.49	7255.49	13782.7	OS	1.2			MIT
7251.16	7253.16	13787.1	OS	1.0			MIT
7224.43	7226.42	13838.1	OS	0.5			MIT
7209.96	7211.95	13865.9	OS	0.9			MIT
7206.33	7208.32	13872.9	OS	1.2			MIT
7184.10	7186.08	13915.8	OS	1.0			MIT
7157.86	7159.83	13966.8	OS	0.5H			MIT
7148.91	7150.88	13984.3	OS	1.2			MIT
7145.54	7147.51	13990.9	OS	1.5			MIT
7104.86	7106.82	14071.0	OS	0.5			MIT
7060.67	7062.62	14159.1	OS	1.2			MIT
7054.86	7056.81	14170.7	OS	0.9			MIT
7034.31	7036.25	14212.1	OS	0.6			MIT
7029.20	7031.14	14222.4	OS	0.5			MIT
7007.04	7008.97	14267.4	OS	0.5			MIT
6984.95	6986.88	14312.6	OS	1.0			MIT
6956.02	6957.94	14372.1	OS	0.5			MIT
6901.58	6903.48	14485.4	OS	0.5			MIT
6878.74	6880.64	14533.5	OS	0.5			MIT
6806.61	6808.49	14687.6	OS	0.9			ME
6791.60	6793.47	14720.0	OS	1.0			MIT
6729.556	6731.414	14855.72	OS	1.5			MIT
6661.81	6663.65	15006.8	OS	0.3			MIT
6614.56	6616.39	15114.0	OS	0.5			MIT
6576.828	6578.645	15200.70	OS	0.5			MIT
6538.295	6540.101	15290.28	OS	1.0			MIT
6533.139	6534.944	15302.35	OS	0.5H			MIT
6528.871	6530.675	15312.35	OS	0.3			MIT
6520.855	6522.657	15331.18	OS	0.3			MIT
6448.126	6449.908	15504.10	OS	0.5			MIT
6403.150	6404.920	15613.00	OS	1.2			MIT
6398.857	6400.626	15623.47	OS	0.5			MIT
6286.831	6288.570	15901.87	OS	0.7			MIT
6274.937	6276.673	15932.01	OS	0.5			MIT
6269.414	6271.148	15946.04	OS	0.9			MIT
6241.704	6243.431	16016.83	OS	0.6			MIT
6227.701	6229.424	16052.85	OS	1.5			MIT
6158.03	6159.73	16234.5	OS	0.6H			ME
6154.019	6155.722	16245.05	OS	0.3H			MIT
6144.526	6146.227	16270.15	OS	0.9			MIT
6015.793	6017.459	16618.31	OS	0.9			MIT
5995.998	5997.659	16673.17	OS	1.7			MIT
5906.84	5908.48	16924.8	OS	0.5			MIT
5903.982	5905.618	16933.03	OS	0.5			MIT
5882.916	5884.546	16993.66	OS	0.8			MIT
5860.641	5862.265	17058.25	OS	0.9			MIT
5857.760	5859.384	17066.64	OS	1.9			MIT
5842.49	5844.11	17111.2	OS	0.7			MIT
5830.986	5832.603	17145.00	OS	0.5			MIT
5800.603	5802.211	17234.81	OS	1.7			MIT
5780.815	5782.418	17293.80	OS	1.7			MIT
5765.047	5766.646	17341.10	OS	1.2			MIT
5751.492	5753.087	17381.97	OS	0.7			MIT
5739.717	5741.309	17417.63	OS	0.5			MIT
5737.888	5739.480	17423.18	OS	0.5			MIT
5731.076	5732.666	17443.89	OS	0.3			MIT
5721.931	5723.518	17471.77	OS	1.9			MIT
5709.370	5710.954	17510.21	OS	0.6			MIT
5680.884	5682.460	17598.01	OS	1.3			MIT
5645.253	5646.820	17709.08	OS	0.8			MIT
5642.558	5644.124	17717.54	OS	0.6			MIT
5620.085	5621.645	17788.39	OS	1.5			MIT
5584.444	5585.995	17901.92	OS	1.7			MIT
5580.659	5582.209	17914.06	OS	0.9			MIT
5560.622	5562.166	17978.61	OS	0.8			MIT
5557.134	5558.677	17989.89	OS	0.7			MIT
5552.883	5554.425	18003.66	OS	1.1			MIT
5548.76	5550.30	18017.0	OS	0.5			MIT
5546.819	5548.360	18023.34	OS	0.9			MIT

AIR WAVELENGTH (ANGSTRMS)	VACUUM WAVELENGTH (ANGSTRMS)	WAVENUMBER (KAYSERS)	LMENT	LOG ARC	INTENSITY SPK	DIS	REF
5523.531	5525.065	18099.33	OS	2.0			MIT
5516.011	5517.543	18124.01	OS	0.8			MIT
5509.330	5510.861	18145.99	OS	1.2			MIT
5491.28	5492.81	18205.6	OS	0.9			MIT
5481.852	5483.375	18236.94	OS	1.2			MIT
5477.268	5478.790	18252.21	OS	1.5			MIT
5475.13	5476.65	18259.3	OS	1.2			MIT
5474.58	5476.10	18261.2	OS	0.9			MIT
5469.998	5471.518	18276.46	OS	1.5			MIT
5457.298	5458.815	18319.00	OS	1.4			MIT
5453.395	5454.911	18332.11	OS	1.3			MIT
5449.37	5450.88	18345.6	OS	1.3			MIT
5447.76	5449.27	18351.1	OS	0.5			MIT
5443.310	5444.823	18366.07	OS	1.7			MIT
5441.821	5443.333	18371.10	OS	1.1			MIT
5419.73	5421.24	18446.0	OS	0.9			MIT
5417.507	5419.013	18453.54	OS	1.6			MIT
5416.693	5418.199	18456.32	OS	1.7			MIT
5416.345	5417.851	18457.50	OS	1.9			MIT
5412.14	5413.64	18471.8	OS	1.2			MIT
5403.432	5404.934	18501.61	OS	0.9			MIT
5391.48	5392.98	18542.6	OS	0.5			MIT
5386.926	5388.424	18558.30	OS	0.6			MIT
5385.72	5387.22	18562.5	OS	0.9H			MIT
5379.308	5380.804	18584.58	OS	0.8			MIT
5376.792	5378.287	18593.28	OS	1.7			MIT
5362.893	5364.384	18641.47	OS	0.9			MIT
5352.25	5353.74	18678.5	OS	1.3			MIT
5346.03	5347.52	18700.3	OS	0.9			MIT
5345.102	5346.589	18703.51	OS	0.7			MIT
5336.23	5337.71	18734.6	OS	1.3			MIT
5302.58	5304.06	18853.5	OS	1.2			MIT
5298.781	5300.255	18867.02	OS	1.3			MIT
5295.65	5297.12	18878.2	OS	1.3			MIT
5287.70	5289.17	18906.6	OS	1.0			MIT
5283.889	5285.359	18920.19	OS	0.9			MIT
5282.282	5283.752	18925.95	OS	0.7			MIT
5271.84	5273.31	18963.4	OS	0.9W			ME
5265.150	5266.615	18987.53	OS	1.5			MIT
5259.797	5261.261	19006.85	OS	1.2			MIT
5255.823	5257.286	19021.22	OS	1.3			MIT
5250.465	5251.926	19040.63	OS	0.9			MIT
5241.20	5242.66	19074.3	OS	0.7			MIT
5233.23	5234.69	19103.3	OS	0.7			MIT
5232.89	5234.35	19104.6	OS	0.7			MIT
5226.57	5228.03	19127.7	OS	0.5			MIT
5203.232	5204.681	19213.47	OS	1.1			MIT
5202.634	5204.083	19215.68	OS	1.5H			MIT
5193.525	5194.971	19249.39	OS	1.5			MIT
5182.282	5183.725	19291.15	OS	0.7			MIT
5171.72	5173.16	19330.5	OS	0.9			ME
5168.975	5170.415	19340.81	OS	0.9			MIT
5157.508	5158.945	19383.81	OS	0.8			MIT
5152.013	5153.448	19404.48	OS	0.8			MIT
5149.736	5151.171	19413.06	OS	1.9			MIT
5146.86	5148.29	19423.9	OS	0.7			ME
5145.543	5146.976	19428.88	OS	1.2			MIT
5143.285	5144.718	19437.41	OS	0.8			MIT
5135.898	5137.329	19465.37	OS	0.7			MIT
5134.898	5136.329	19469.16	OS	0.8			MIT
5124.349	5125.777	19509.24	OS	1.1			MIT
5123.381	5124.809	19512.92	OS	0.9			MIT
5122.229	5123.656	19517.31	OS	1.4			MIT
5108.264	5109.688	19570.67	OS	0.5			MIT
5104.740	5106.163	19584.18	OS	0.9			MIT
5103.495	5104.917	19588.96	OS	1.5			MIT
5093.092	5094.512	19628.97	OS	0.7			MIT
5079.088	5080.504	19683.09	OS	1.2			MIT
5077.59	5079.01	19688.9	OS	0.5			ME
5076.906	5078.321	19691.55	OS	0.7			MIT

AIR WAVELENGTH (ANGSTRMS)	VACUUM WAVELENGTH (ANGSTRMS)	WAVENUMBER (KAYSERS)	LMENT	LOG ARC	INTENSITY SPK	DIS	REF
5076.223	5077.638	19694.20	OS	0.9			MIT
5074.769	5076.184	19699.84	OS	1.1			MIT
5072.884	5074.298	19707.16	OS	1.1			MIT
5056.01	5057.42	19772.9	OS	0.5H			MIT
5050.417	5051.825	19794.83	OS	0.7			MIT
5047.154	5048.561	19807.62	OS	0.5			MIT
5039.122	5040.527	19839.20	OS	1.7			MIT
5034.172	5035.576	19858.70	OS	1.1			MIT
5031.831	5033.234	19867.94	OS	1.3			MIT
5029.247	5030.650	19878.15	OS	0.9			MIT
5020.892	5022.292	19911.23	OS	0.8			MIT
5019.641	5021.041	19916.19	OS	0.9			MIT
5012.525	5013.923	19944.46	OS	1.1			MIT
4990.103	4991.495	20034.08	OS	0.8			MIT
4982.895	4984.285	20063.06	OS	0.8			MIT
4979.318	4980.707	20077.47	OS	1.3			MIT
4959.034	4960.418	20159.59	OS	0.3			MIT
4950.590	4951.972	20193.98	OS	0.5			MIT
4942.940	4944.320	20225.23	OS	1.3			MIT
4940.208	4941.587	20236.41	OS	0.7			MIT
4935.813	4937.191	20254.43	OS	1.1			MIT
4926.385	4927.760	20293.20	OS	0.5			MIT
4921.804	4923.178	20312.08	OS	0.8			MIT
4912.605	4913.977	20350.12	OS	1.9			MIT
4910.463	4911.834	20358.99	OS	0.9			MIT
4902.180	4903.549	20393.39	OS	0.3			MIT
4899.215	4900.583	20405.74	OS	1.8			MIT
4895.268	4896.635	20422.19	OS	0.9			MIT
4882.285	4883.649	20476.49	OS	1.1			MIT
4878.518	4879.881	20492.30	OS	1.3			MIT
4877.34	4878.70	20497.2	OS	0.5			ME
4867.177	4868.537	20540.05	OS	0.9			MIT
4865.598	4866.957	20546.72	OS	1.9	0.0		MIT
4864.009	4865.368	20553.43	OS	0.8			MIT
4859.471	4860.828	20572.62	OS	0.5			MIT
4858.470	4859.827	20576.86	OS	0.3			MIT
4854.940	4856.296	20591.82	OS	0.5			MIT
4851.879	4853.234	20604.81	OS	0.9			MIT
4849.217	4850.572	20616.13	OS	1.3			MIT
4843.868	4845.221	20638.89	OS	1.4			MIT
4838.122	4839.474	20663.40	OS	0.3			MIT
4828.441	4829.790	20704.83	OS	0.5			MIT
4826.656	4828.005	20712.49	OS	1.3			MIT
4823.43	4824.78	20726.3	OS	0.9H			MIT
4815.957	4817.303	20758.50	OS	1.8			MIT
4815.495	4816.841	20760.50	OS	1.3			MIT
4813.798	4815.143	20767.81	OS	1.3			MIT
4812.620	4813.965	20772.90	OS	0.3			MIT
4807.091	4808.435	20796.79	OS	0.8			MIT
4793.994	4795.334	20853.60	OS	2.5	0.8		MIT
4763.099	4764.431	20988.87	OS	1.3			MIT
4760.770	4762.101	20999.13	OS	0.7			MIT
4755.155	4756.485	21023.93	OS	1.0	0.0		MIT
4752.158	4753.487	21037.19	OS	1.5			MIT
4743.890	4745.217	21073.85	OS	1.8			MIT
4742.626	4743.952	21079.47	OS	0.3			MIT
4738.351	4739.676	21098.49	OS	1.3			MIT
4738.040	4739.365	21099.87	OS	1.1			MIT
4734.410	4735.734	21116.05	OS	0.5			MIT
4732.800	4734.124	21123.23	OS	1.2			MIT
4721.284	4722.605	21174.75	OS	1.1			MIT
4718.989	4720.309	21185.05	OS	0.8			MIT
4710.018	4711.336	21225.40	OS	0.3			MIT
4705.936	4707.253	21243.81	OS	1.2	0.0		MIT
4692.062	4693.375	21306.63	OS	1.9	0.5		MIT
4686.491	4687.803	21331.96	OS	0.3			MIT
4682.312	4683.623	21351.00	OS	1.3			MIT
4681.428	4682.738	21355.03	OS	0.3			MIT
4674.014	4675.322	21388.90	OS	0.5			MIT
4670.686	4671.993	21404.14	OS	0.8			MIT

AIR WAVELENGTH (ANGSTRMS)	VACUUM WAVELENGTH (ANGSTRMS)	WAVENUMBER (KAYSERS)	LMENT	LOG ARC	INTENSITY SPK	DIS	REF
4667.542	4668.849	21418.56	OS	0.7			MIT
4663.822	4665.128	21435.64	OS	2.0	0.7		MIT
4658.567	4659.871	21459.82	OS	0.8H			MIT
4657.251	4658.555	21465.88	OS	0.8			MIT
4655.202	4656.505	21475.33	OS	0.5			MIT
4650.58	4651.88	21496.7	OS	0.3			ME
4641.831	4643.131	21537.19	OS	1.5	0.5H		MIT
4638.625	4639.924	21552.08	OS	1.4			MIT
4634.768	4636.066	21570.01	OS	1.5			MIT
4633.173	4634.471	21577.44	OS	0.8			MIT
4631.828	4633.125	21583.70	OS	2.0	0.7		MIT
4629.356	4630.652	21595.23	OS	0.3			MIT
4629.028	4630.324	21596.76	OS	0.3			MIT
4628.612	4629.908	21598.70	OS	1.0			MIT
4616.783	4618.076	21654.04	OS	2.2	0.8		MIT
4608.280	4609.571	21693.99	OS	1.1			MIT
4605.036	4606.326	21709.27	OS	1.2			MIT
4597.863	4599.151	21743.14	OS	0.9			MIT
4597.160	4598.448	21746.47	OS	2.0	0.6		MIT
4595.040	4596.327	21756.50	OS	1.9	0.6		MIT
4590.946	4592.232	21775.90	OS	0.9	0.0H		MIT
4588.591	4589.877	21787.08	OS	0.3			MIT
4584.937	4586.222	21804.44	OS	0.3			MIT
4579.041	4580.324	21832.52	OS	1.2			MIT
4572.953	4574.235	21861.58	OS	0.7			MIT
4570.501	4571.782	21873.31	OS	0.3			MIT
4567.48	4568.76	21887.8	OS	0.7S			MIT
4566.484	4567.764	21892.55	OS	1.0			MIT
4562.604	4563.883	21911.17	OS	0.7			MIT
4551.298	4552.574	21965.60	OS	2.2	0.9		MIT
4550.410	4551.686	21969.88	OS	2.2	1.0		MIT
4548.662	4549.937	21978.32	OS	2.0	0.7		MIT
4545.800	4547.074	21992.16	OS	0.7	0.0		MIT
4539.917	4541.190	22020.66	OS	2.0	0.3		MIT
4537.615	4538.887	22031.83	OS	1.7			MIT
4529.674	4530.944	22070.46	OS	1.9	0.3		MIT
4524.869	4526.138	22093.89	OS	1.9	0.3		MIT
4520.318	4521.586	22116.14	OS	1.3	0.0		MIT
4519.864	4521.132	22118.36	OS	1.0			MIT
4518.889	4520.156	22123.13	OS	1.2			MIT
4514.902	4516.168	22142.67	OS	0.7			MIT
4510.096	4511.361	22166.26	OS	0.3			MIT
4504.040	4505.303	22196.06	OS	0.6			MIT
4500.729	4501.992	22212.39	OS	0.7			MIT
4488.601	4489.860	22272.41	OS	1.8			MIT
4484.763	4486.021	22291.47	OS	2.0	0.0		MIT
4479.808	4481.065	22316.12	OS	1.9L	0.0		MIT
4474.834	4476.090	22340.93	OS	0.3			MIT
4473.08	4474.34	22349.7	OS	0.5			MIT
4465.940	4467.193	22385.42	OS	1.0			MIT
4462.287	4463.539	22403.75	OS	0.9	0.0		MIT
4459.527	4460.779	22417.61	OS	1.2	0.3		MIT
4458.331	4459.582	22423.62	OS	0.5L			MIT
4451.82	4453.07	22456.4	OS	0.5			MIT
4447.354	4448.602	22478.97	OS	2.3	0.5		MIT
4445.69	4446.94	22487.4	OS	1.0	0.3		MIT
4439.639	4440.885	22518.03	OS	1.5	0.3		MIT
4437.088	4438.334	22530.98	OS	1.1			MIT
4436.317	4437.563	22534.89	OS	1.9	0.5		MIT
4432.412	4433.656	22554.75	OS	1.5	0.0		MIT
4420.468	4421.709	22615.69	OS I	2.6G	2.0		MIT
4404.213	4405.450	22699.16	OS	1.3	0.0		MIT
4402.736	4403.973	22706.77	OS	1.7	0.5		MIT
4400.579	4401.815	22717.90	OS	1.3	0.0		MIT
4397.263	4398.498	22735.03	OS	1.3	0.5		MIT
4394.858	4396.093	22747.47	OS	2.2	0.8		MIT
4391.080	4392.314	22767.04	OS	1.2	0.5		MIT
4376.904	4378.134	22840.78	OS	0.8			MIT
4370.661	4371.889	22873.41	OS	1.7	0.5		MIT
4365.674	4366.901	22899.53	OS	1.8	0.6		MIT

AIR WAVELENGTH (ANGSTRMS)	VACUUM WAVELENGTH (ANGSTRMS)	WAVENUMBER (KAYSERS)	LMENT	LOG ARC	INTENSITY SPK	DIS	REF
4358.141	4359.366	22939.12	OS	1.0	0.0		MIT
4357.980	4359.205	22939.96	OS	1.1			MIT
4354.464	4355.688	22958.49	OS	1.0	0.0		MIT
4351.528	4352.751	22973.98	OS	1.0	0.3		MIT
4342.521	4343.742	23021.63	OS	0.6			MIT
4338.753	4339.973	23041.62	OS	1.0	0.0		MIT
4328.677	4329.894	23095.25	OS	1.8	0.6		MIT
4326.254	4327.471	23108.19	OS	1.5	1.1		MIT
4319.339	4320.554	23145.18	OS	1.0	0.0		MIT
4311.399	4312.612	23187.81	OS	2.2	1.0		MIT
4308.877	4310.089	23201.38	OS	1.3	0.3		MIT
4296.218	4297.427	23269.74	OS	1.1	0.0		MIT
4293.949	4295.157	23282.04	OS I	1.8	0.8		MIT
4285.895	4287.101	23325.79	OS	1.5	0.5		MIT
4277.147	4278.351	23373.49	OS	1.1	0.0		MIT
4274.905	4276.108	23385.75	OS	1.0	0.0		MIT
4270.788	4271.990	23408.29	OS	1.1	0.0		MIT
4269.610	4270.812	23414.75	OS	1.5	0.5		MIT
4269.364	4270.566	23416.10	OS	1.1	0.0		MIT
4264.746	4265.946	23441.46	OS	1.3	0.5		MIT
4260.854	4262.053	23462.87	OS I	2.3	2.3		MIT
4252.538	4253.735	23508.75	OS	1.0			MIT
4251.170	4252.367	23516.32	OS	0.8			MIT
4241.521	4242.715	23569.81	OS	1.0			MIT
4233.460	4234.652	23614.69	OS	1.1	0.5		MIT
4226.527	4227.717	23653.43	OS	1.1	0.8		MIT
4215.155	4216.342	23717.24	OS	0.9	0.0		MIT
4213.859	4215.046	23724.53	OS	1.5	0.5		MIT
4211.855	4213.041	23735.82	OS	2.2	1.7		MIT
4205.222	4206.407	23773.26	OS	1.0	0.0		MIT
4204.560	4205.745	23777.00	OS	1.1			MIT
4202.062	4203.246	23791.14	OS	2.0	0.6		MIT
4201.449	4202.633	23794.61	OS	1.5	0.5		MIT
4195.144	4196.326	23830.37	OS	2.0S	0.0		MIT
4192.637	4193.818	23844.62	OS	1.0	0.0		MIT
4192.190	4193.371	23847.16	OS	0.6			MIT
4189.906	4191.087	23860.16	OS	1.8	0.5		MIT
4184.131	4185.310	23893.09	OS	1.4	0.3		MIT
4182.450	4183.629	23902.70	OS	0.8			MIT
4175.627	4176.804	23941.75	OS	2.0	0.6		MIT
4173.234	4174.410	23955.48	OS	2.0	0.8		MIT
4172.57	4173.75	23959.3	OS	1.8	0.5		MIT
4160.268	4161.441	24030.14	OS	0.8			MIT
4159.965	4161.138	24031.89	OS	1.3	0.0		MIT
4158.784	4159.957	24038.71	OS	1.7	0.0		MIT
4153.374	4154.545	24070.02	OS	1.0			MIT
4150.699	4151.869	24085.54	OS	0.9	0.0H		MIT
4147.331	4148.501	24105.09	OS	1.0			MIT
4137.840	4139.007	24160.38	OS	2.0	0.5		MIT
4135.784	4136.950	24172.40	OS	2.3	1.7		MIT
4135.038	4136.204	24176.75	OS	0.8	0.0		MIT
4131.049	4132.214	24200.10	OS	1.1	0.0		MIT
4128.961	4130.126	24212.34	OS	1.8	0.5		MIT
4124.599	4125.763	24237.94	OS	1.5	1.0		MIT
4116.585	4117.746	24285.13	OS	0.8	0.0		MIT
4116.27	4117.43	24287.0	OS	1.3	0.0H		MIT
4112.018	4113.178	24312.10	OS	2.2	1.0		MIT
4111.028	4112.188	24317.95	OS	0.8	0.0		MIT
4107.974	4109.133	24336.03	OS	1.4			MIT
4105.42	4106.58	24351.2	OS	0.8			MIT
4103.616	4104.774	24361.88	OS	1.3	0.5		MIT
4100.300	4101.457	24381.58	OS	1.8	0.5		MIT
4098.10	4099.26	24394.7	OS	1.0	0.5		MIT
4091.817	4092.972	24432.12	OS	2.0	1.1		MIT
4090.769	4091.924	24438.38	OS	0.5	0.5		MIT
4088.442	4089.596	24452.29	OS	2.0H	0.5		MIT
4074.682	4075.832	24534.86	OS	1.9	0.8		MIT
4073.62	4074.77	24541.3	OS	0.8	0.5		MIT
4071.56	4072.71	24553.7	OS	1.5	0.5		MIT
4071.013	4072.163	24556.98	OS	1.1	0.6		MIT

AIR WAVELENGTH (ANGSTRMS)	VACUUM WAVELENGTH (ANGSTRMS)	WAVENUMBER (KAYSERS)	LMENT	LOG ARC	INTENSITY SPK	DIS	REF
4070.855	4072.004	24557.93	OS	1.3	0.6		MIT
4066.693	4067.841	24583.06	OS	2.0	2.0		MIT
4066.313	4067.461	24585.36	OS	1.0	0.6		MIT
4055.496	4056.641	24650.93	OS	1.5	0.5		MIT
4051.429	4052.573	24675.68	OS	1.1	0.5		MIT
4048.054	4049.198	24696.25	OS	1.3	0.3		MIT
4041.917	4043.059	24733.75	OS	2.0L	0.8		MIT
4038.640	4039.781	24753.82	OS	1.0	0.5		MIT
4037.841	4038.982	24758.72	OS	1.9L	0.6		MIT
4018.257	4019.393	24879.38	OS	1.8	0.6		MIT
4015.037	4016.172	24899.33	OS	1.3	0.7		MIT
4005.155	4006.287	24960.77	OS	1.5	1.3		MIT
4004.024	4005.156	24967.82	OS	1.7	0.8		MIT
4003.48	4004.61	24971.2	OS	1.7	0.8		MIT
3998.933	4000.064	24999.60	OS	1.9	1.1		MIT
3996.805	3997.935	25012.91	OS	1.7	1.0		MIT
3994.929	3996.059	25024.66	OS	1.5	0.7		MIT
3991.487	3992.616	25046.24	OS	1.6	1.0		MIT
3988.179	3989.307	25067.01	OS	1.7	1.1		MIT
3977.231	3978.356	25136.01	OS I	2.5	1.6		MIT
3975.441	3976.566	25147.33	OS	1.7	1.1		MIT
3969.671	3970.794	25183.88	OS	2.0	2.0		MIT
3964.965	3966.087	25213.77	OS	1.8	1.1		MIT
3963.628	3964.749	25222.27	OS I	2.7	1.7		MIT
3961.016	3962.137	25238.91	OS	2.1	1.3		MIT
3960.506	3961.627	25242.16	OS	1.7	1.2		MIT
3955.367	3956.486	25274.95	OS	1.1	1.0		MIT
3952.768	3953.887	25291.57	OS	1.6	1.2		MIT
3949.784	3950.902	25310.68	OS	1.7	1.0		MIT
3939.566	3940.681	25376.32	OS	1.7	1.1		MIT
3938.593	3939.708	25382.59	OS	2.1	1.3		MIT
3931.525	3932.638	25428.22	OS	1.6	1.1		MIT
3929.997	3931.110	25438.11	OS	1.9	1.1		MIT
3928.541	3929.653	25447.54	OS	1.7	1.0		MIT
3928.409	3929.521	25448.39	OS	1.6	1.0		MIT
3926.770	3927.882	25459.01	OS	1.5	1.0		MIT
3925.103	3926.214	25469.83	OS	1.5	1.0		MIT
3922.033	3923.144	25489.76	OS	1.5	1.0		MIT
3920.870	3921.980	25497.32	OS	1.3	1.1		MIT
3918.971	3920.081	25509.68	OS	1.5	1.0		MIT
3911.812	3912.920	25556.36	OS	1.5	0.7		MIT
3901.708	3902.813	25622.54	OS	2.2	1.3		MIT
3901.005	3902.110	25627.16	OS	1.0	0.7		MIT
3900.394	3901.499	25631.17	OS	1.7	1.1		MIT
3895.177	3896.281	25665.50	OS	1.3	1.0		MIT
3886.750	3887.851	25721.15	OS	1.5	0.9		MIT
3885.749	3886.850	25727.77	OS	1.3	0.9		MIT
3881.858	3882.958	25753.56	OS	2.1	1.3		MIT
3880.768	3881.868	25760.79	OS	1.0	1.1		MIT
3878.572	3879.671	25775.38	OS	1.7	1.1		MIT
3877.307	3878.406	25783.79	OS	1.3	1.0		MIT
3876.768	3877.867	25787.37	OS I	2.5	1.7		MIT
3873.724	3874.822	25807.64	OS	1.3	1.3		MIT
3868.686	3869.783	25841.24	OS	1.5	1.0		MIT
3866.480	3867.576	25855.99	OS	1.0	0.7		MIT
3865.469	3866.565	25862.75	OS I	2.1	2.3		MIT
3865.045	3866.141	25865.59	OS	1.0	0.7		MIT
3857.089	3858.183	25918.94	OS I	2.2	1.2		MIT
3854.705	3855.798	25934.97	OS	1.5	1.1		MIT
3853.589	3854.682	25942.48	OS	0.9	0.9		MIT
3853.439	3854.532	25943.49	OS	2.0	1.2		MIT
3849.944	3851.036	25967.04	OS I	2.1	1.3		MIT
3846.411	3847.502	25990.89	OS	1.2	1.1		MIT
3843.665	3844.755	26009.46	OS	1.9	1.1		MIT
3841.292	3842.382	26025.52	OS	1.5	1.0		MIT
3840.304	3841.393	26032.22	OS	2.2	1.3		MIT
3838.59	3839.68	26043.8	OS I	0.5			AB
3836.056	3837.144	26061.05	OS I	2.2	1.3		MIT
3832.176	3833.263	26087.43	OS	1.5L	1.0		MIT
3827.136	3828.222	26121.79	OS	1.7	1.3		MIT

AIR WAVELENGTH (ANGSTRMS)	VACUUM WAVELENGTH (ANGSTRMS)	WAVENUMBER (KAYSERS)	LMENT	LOG ARC	INTENSITY SPK	INTENSITY DIS	REF
3826.634	3827.720	26125.21	OS	1.5	1.1		MIT
3821.642	3822.726	26159.34	OS	0.9	0.7		MIT
3818.641	3819.725	26179.90	OS	0.9	1.2		MIT
3818.082	3819.166	26183.73	OS	1.1	1.2		MIT
3814.259	3815.342	26209.97	OS	1.5	1.0		MIT
3812.289	3813.371	26223.52	OS	1.3	0.7		MIT
3802.612	3803.692	26290.25	OS	0.7H	0.7		MIT
3801.604	3802.683	26297.22	OS	1.3	0.7		MIT
3800.437	3801.516	26305.29	OS	1.7	1.2		MIT
3795.667	3796.745	26338.35	OS	1.6	1.1		MIT
3794.662	3795.739	26345.33	OS	1.6	1.0		MIT
3793.909	3794.986	26350.56	OS I	2.1	2.5		MIT
3790.731	3791.807	26372.65	OS	2.0	1.3		MIT
3790.14	3791.22	26376.8	OS I	1.9	1.5		M
3789.106	3790.182	26383.96	OS	1.5	1.1		MIT
3783.658	3784.733	26421.95	OS	1.3	1.0		MIT
3782.195	3783.269	26432.17	OS I	2.6G	2.3		MIT
3780.214	3781.288	26446.02	OS	1.3	1.0		MIT
3776.992	3778.065	26468.58	OS	2.2	1.3		MIT
3776.254	3777.327	26473.75	OS	1.7	1.2		MIT
3774.619	3775.691	26485.22	OS	1.8	1.1		MIT
3774.400	3775.472	26486.75	OS	1.8	1.2		MIT
3773.780	3774.852	26491.10	OS	1.1	0.7		MIT
3771.944	3773.016	26504.00	OS	1.6S	1.1		MIT
3771.640	3772.711	26506.14	OS	1.3	1.0		MIT
3768.138	3769.209	26530.77	OS	1.9	1.2		MIT
3766.301	3767.371	26543.71	OS	2.0	1.3		MIT
3760.274	3761.343	26586.25	OS	1.3	1.0		MIT
3757.116	3758.184	26608.60	OS	1.6	1.2		MIT
3752.775	3753.842	26639.38	OS	0.6D			MIT
3752.524	3753.591	26641.16	OS I	2.6G	2.0		MIT
3751.943	3753.009	26645.28	OS	1.2	1.0		MIT
3751.315	3752.381	26649.74	OS	1.3	1.0		MIT
3750.804	3751.870	26653.37	OS	1.2	0.7		MIT
3750.590	3751.656	26654.90	OS	1.2	0.7		MIT
3749.073	3750.139	26665.68	OS	0.6	0.3		MIT
3746.466	3747.531	26684.24	OS	2.0	1.3		MIT
3741.529	3742.593	26719.45	OS	1.3	0.7		MIT
3741.085	3742.149	26722.62	OS	1.6	1.2		MIT
3735.536	3736.598	26762.31	OS	1.3	1.0		MIT
3732.850	3733.911	26781.57	OS	2.3G	0.7H		MIT
3731.802	3732.863	26789.09	OS	1.2	1.0		MIT
3730.731	3731.792	26796.78	OS	1.6	1.1		MIT
3729.220	3730.280	26807.64	OS	1.5	1.1		MIT
3728.411	3729.471	26813.45	OS	1.3	0.7		MIT
3725.284	3726.343	26835.96	OS	1.2	1.0		MIT
3721.959	3723.018	26859.93	OS	1.5	1.2		MIT
3720.132	3721.190	26873.12	OS I	1.9	1.6		MIT
3719.522	3720.580	26877.53	OS I	1.6	1.1		MIT
3718.339	3719.397	26886.08	OS	1.5	1.0		MIT
3717.926	3718.984	26889.07	OS	1.0	1.1		MIT
3716.336	3717.393	26900.57	OS	1.1	1.0		MIT
3713.726	3714.782	26919.48	OS	2.0	1.3		MIT
3712.838	3713.894	26925.92	OS	1.7	1.1		MIT
3711.851	3712.907	26933.07	OS	1.2	1.0		MIT
3709.145	3710.200	26952.72	OS	0.6	0.0		MIT
3706.556	3707.611	26971.55	OS	1.7	1.2		MIT
3703.247	3704.301	26995.65	OS	2.0	1.5		MIT
3702.763	3703.817	26999.18	OS	0.6	0.3		MIT
3701.596	3702.649	27007.69	OS	1.1	1.2		MIT
3700.299	3701.352	27017.16	OS	0.7H	0.7H		MIT
3698.832	3699.885	27027.87	OS	1.3	1.1		MIT
3690.724	3691.775	27087.25	OS	1.2	1.3		MIT
3689.058	3690.108	27099.48	OS	2.3	1.5		MIT
3684.546	3685.595	27132.66	OS	1.2	0.9		MIT
3681.568	3682.616	27154.61	OS	1.5	0.9		MIT
3678.016	3679.063	27180.83	OS	1.0	1.0		MIT
3675.451	3676.498	27199.80	OS	1.6	1.2		MIT
3670.891	3671.936	27233.59	OS I	2.3	1.3		MIT
3666.309	3667.353	27267.62	OS	1.6	1.2		MIT
3661.252	3662.295	27305.28	OS	1.0	1.0		•MIT
3656.902	3657.944	27337.76	OS I	2.2	1.5		MIT
3654.492	3655.533	27355.79	OS	2.0	1.2		MIT
3653.725	3654.766	27361.53	OS	1.5	1.0		MIT
3653.201	3654.242	27365.46	OS	1.0O	0.7O		MIT
3650.381	3651.421	27386.60	OS	1.2	0.6		MIT
3648.806	3649.846	27398.42	OS	2.0	1.0		MIT
3648.304	3649.344	27402.19	OS	1.0	0.7		MIT
3642.505	3643.543	27445.81	OS	1.3	1.0		MIT
3641.227	3642.265	27455.45	OS	1.5	1.0		MIT
3640.329	3641.366	27462.22	OS I	2.3	1.6		MIT
3629.953	3630.988	27540.72	OS	1.6	1.1		MIT
3620.241	3621.273	27614.60	OS	1.2	1.1		MIT
3619.431	3620.463	27620.78	OS	1.8	1.4		MIT
3616.574	3617.605	27642.59	OS I	2.2	1.3		MIT
3613.329	3614.360	27667.42	OS	1.5	1.1		MIT
3609.147	3610.176	27699.48	OS	1.3	1.1		MIT
3604.475	3605.503	27735.38	OS	1.2	2.0		MIT
3602.484	3603.512	27750.71	OS	1.2	1.0		MIT
3601.833	3602.861	27755.72	OS	1.8	1.3		MIT
3598.111	3599.138	27784.43	OS I	2.5	1.5		MIT
3597.519	3598.545	27789.01	OS	1.2	1.0		MIT
3595.794	3596.820	27802.34	OS	0.0	0.7		MIT
3592.317	3593.342	27829.25	OS	1.3	1.1		MIT
3591.602	3592.627	27834.79	OS	0.3	0.7		MIT
3590.107	3591.132	27846.38	OS	1.5	1.2		MIT
3587.873	3588.897	27863.71	OS	0.3	1.0		MIT
3587.315	3588.339	27868.05	OS	1.8	1.2		MIT
3586.506	3587.530	27874.33	OS	1.5	1.2		MIT
3584.404	3585.427	27890.68	OS	1.2	1.3		MIT
3583.398	3584.421	27898.51	OS	1.2	1.0		MIT
3583.091	3584.114	27900.90	OS	1.3	1.0		MIT
3581.794	3582.816	27911.00	OS	0.7	0.9		MIT
3580.530	3581.552	27920.86	OS	0.3	0.6		MIT
3579.867	3580.889	27926.03	OS	0.3	0.7		MIT
3577.487	3578.508	27944.60	OS	0.5			MIT
3574.079	3575.099	27971.25	OS	1.5	0.7		MIT
3569.775	3570.794	28004.97	OS	2.0	1.5		MIT
3568.566	3569.585	28014.46	OS	0.3	0.7		MIT
3564.092	3565.110	28049.63	OS	0.7	1.0		MIT
3562.337	3563.354	28063.45	OS	1.7	1.3		MIT
3560.855	3561.872	28075.12	OS I	2.2G	2.0		MIT
3560.468	3561.485	28078.18	OS	0.0	0.7H		MIT
3559.786	3560.803	28083.55	OS I	2.2	1.7		MIT
3558.805	3559.821	28091.30	OS	0.6H	0.3		MIT
3555.973	3556.989	28113.67	OS	1.5	1.1		MIT
3555.707	3556.723	28115.77	OS	1.0	0.9		MIT
3554.544	3555.559	28124.97	OS	0.9	0.7		MIT
3554.048	3555.063	28128.90	OS	0.9	1.0		MIT
3550.939	3551.953	28153.52	OS	0.8	1.0		MIT
3550.714	3551.728	28155.31	OS	1.2	1.0		MIT
3549.684	3550.698	28163.48	OS	1.3	1.1		MIT
3547.876	3548.890	28177.83	OS	0.0	0.9		MIT
3546.133	3547.146	28191.68	OS	0.9	0.7		MIT
3544.582	3545.595	28204.01	OS	1.1	1.0		MIT
3544.582	3545.595	28204.01	OS	1.1	1.0		MIT
3543.724	3544.737	28210.84	OS	1.2	1.1		MIT
3542.711	3543.723	28218.91	OS	2.2	1.0		MIT
3541.906	3542.918	28225.32	OS	1.3	0.7		MIT
3539.860	3540.872	28241.63	OS	0.0	0.5		MIT
3537.988	3538.999	28256.58	OS	0.0	0.6		MIT
3537.498	3538.509	28260.49	OS	0.3	0.5		MIT
3533.408	3534.418	28293.20	OS	1.6	1.3		MIT
3532.800	3533.810	28298.07	OS I	2.0	1.3		MIT
3531.129	3532.138	28311.46	OS	1.2	1.1		MIT
3530.062	3531.071	28320.02	OS	2.0	1.3		MIT
3528.602	3529.611	28331.74	OS I	2.6G	1.7		MIT
3526.035	3527.043	28352.36	OS	1.9	1.3		MIT
3525.290	3526.298	28358.35	OS	0.6	0.7		MIT
3523.636	3524.643	28371.66	OS	2.2	1.5		MIT

AIR WAVELENGTH (ANGSTRMS)	VACUUM WAVELENGTH (ANGSTRMS)	WAVENUMBER (KAYSERS)	LMENT	LOG ARC	INTENSITY SPK	DIS	REF
3523.167	3524.174	28375.44	OS	1.0	0.7		MIT
3521.965	3522.972	28385.12	OS	1.2	1.0		MIT
3519.998	3521.004	28400.99	OS	1.5	1.3		MIT
3519.176	3520.182	28407.62	OS	1.2	0.9		MIT
3518.942	3519.948	28409.51	OS	1.3	0.7		MIT
3518.725	3519.731	28411.26	OS	2.3	1.5		MIT
3517.292	3518.298	28422.84	OS	1.2	1.0		MIT
3517.164	3518.170	28423.87	OS	0.9	0.9		MIT
3516.630	3517.636	28428.19	OS	1.3	1.3		MIT
3513.822	3514.827	28450.90	OS	1.0	0.5		MIT
3512.988	3513.993	28457.66	OS	1.5	1.5		MIT
3511.225	3512.229	28471.95	OS	1.5	1.1		MIT
3508.851	3509.855	28491.21	OS	0.9	1.0		MIT
3507.049	3508.052	28505.85	OS	1.0	0.7		MIT
3505.014	3506.017	28522.40	OS		0.7H		MIT
3504.659	3505.662	28525.29	OS I	2.5	1.3		MIT
3503.451	3504.453	28535.12	OS	1.2	0.9		MIT
3501.687	3502.689	28549.50	OS	1.5	0.7		MIT
3501.162	3502.164	28553.78	OS	2.0	1.2		MIT
3499.535	3500.536	28567.05	OS	1.0	0.5		MIT
3499.266	3500.267	28569.25	OS	0.5	0.3		MIT
3498.536	3499.537	28575.21	OS	1.9	1.2		MIT
3498.074	3499.075	28578.98	OS	1.2	1.0		MIT
3495.848	3496.848	28597.18	OS		1.0H		MIT
3495.615	3496.615	28599.09	OS	1.3	1.1		MIT
3491.495	3492.494	28632.83	OS	1.5	1.2		MIT
3490.331	3491.330	28642.38	OS	1.6	1.5		MIT
3488.87	3489.87	28654.4	OS	1.0D	0.9		MIT
3488.765	3489.763	28655.24	OS	1.5	1.0		MIT
3487.463	3488.461	28665.94	OS	1.7	1.2		MIT
3487.248	3488.246	28667.70	OS	1.6	1.1		MIT
3482.231	3483.228	28709.00	OS	1.3	1.0		MIT
3482.112	3483.109	28709.99	OS	1.6	1.2		MIT
3478.529	3479.525	28739.56	OS	2.0	1.2		MIT
3477.642	3478.638	28746.89	OS	1.3	0.5		MIT
3476.852	3477.847	28753.42	OS	1.3	0.0		MIT
3474.125	3475.120	28775.99	OS	0.6	1.0		MIT
3469.368	3470.361	28815.44	OS	1.1	0.7		MIT
3465.435	3466.427	28848.14	OS	1.8	1.1		MIT
3464.881	3465.873	28852.76	OS	1.2	0.7		MIT
3462.191	3463.183	28875.17	OS	1.3	1.0		MIT
3459.018	3460.009	28901.66	OS	2.0	1.0		MIT
3458.385	3459.376	28906.95	OS	2.3	1.1		MIT
3456.139	3457.129	28925.74	OS	1.3	1.0		MIT
3455.031	3456.021	28935.01	OS	1.7	1.2		MIT
3453.054	3454.043	28951.58	OS	1.3S	1.0		MIT
3450.390	3451.379	28973.93	OS	1.0	1.0		MIT
3449.196	3450.184	28983.96	OS	2.0	1.3		MIT
3445.551	3446.538	29014.62	OS	1.9	1.2		MIT
3444.460	3445.447	29023.81	OS	1.7	1.1		MIT
3439.831	3440.817	29062.87	OS	1.2	0.9		MIT
3439.486	3440.472	29065.78	OS	1.6	1.1		MIT
3438.611	3439.597	29073.18	OS	0.6	1.0H		MIT
3435.256	3436.241	29101.57	OS	1.3	1.0		MIT
3434.889	3435.874	29104.68	OS	1.3	0.5		MIT
3430.066	3431.049	29145.60	OS	0.6	0.7		MIT
3429.987	3430.970	29146.27	OS		0.9		MIT
3427.671	3428.654	29165.97	OS	1.9	1.2		MIT
3427.438	3428.421	29167.95	OS	1.5	1.1		MIT
3422.275	3423.256	29211.95	OS	1.3	1.0		MIT
3421.687	3422.668	29216.97	OS	1.5	1.0		MIT
3415.224	3416.204	29272.26	OS	0.7	0.9		MIT
3414.245	3415.224	29280.65	OS	1.0	0.9		MIT
3412.740	3413.719	29293.56	OS	1.2			MIT
3408.756	3409.734	29327.80	OS	1.6	1.0		MIT
3406.667	3407.644	29345.78	OS	1.5	1.0		MIT
3406.279	3407.256	29349.13	OS	1.5	1.1		MIT
3402.511	3403.487	29381.63	OS I	2.3	1.2		MIT
3401.859	3402.835	29387.26	OS I	2.3	1.3		MIT
3401.174	3402.150	29393.18	OS	1.6	1.2		MIT

AIR WAVELENGTH (ANGSTRMS)	VACUUM WAVELENGTH (ANGSTRMS)	WAVENUMBER (KAYSERS)	LMENT	LOG ARC	INTENSITY SPK	DIS	REF
3400.117	3401.093	29402.31	OS	1.3	1.0		MIT
3398.575	3399.550	29415.66	OS	1.3	1.0		MIT
3397.758	3398.733	29422.73	OS	1.1	1.0		MIT
3395.719	3396.694	29440.39	OS	1.6	1.1		MIT
3394.593	3395.567	29450.16	OS	1.3	1.1		MIT
3391.286	3392.259	29478.88	OS	0.3	0.7		MIT
3389.524	3390.497	29494.20	OS	1.0	0.7		MIT
3388.639	3389.612	29501.90	OS	0.5	0.5		MIT
3387.836	3388.809	29508.90	OS I	2.0	1.2		MIT
3386.627	3387.599	29519.43	OS	1.0	0.6		MIT
3386.138	3387.110	29523.69	OS	1.5	0.9		MIT
3385.940	3386.912	29525.42	OS	1.6	1.0		MIT
3384.596	3385.568	29537.14	OS	1.3	0.5		MIT
3384.003	3384.975	29542.32	OS	1.9	0.7		MIT
3381.669	3382.640	29562.71	OS	1.2	0.0		MIT
3378.679	3379.649	29588.87	OS	1.7	1.0		MIT
3377.593	3378.563	29598.38	OS	0.8	0.7		MIT
3376.957	3377.927	29603.96	OS	1.2	1.0		MIT
3375.131	3376.100	29619.97	OS	0.5	0.7		MIT
3374.221	3375.190	29627.96	OS	0.0	0.5		MIT
3373.204	3374.173	29636.89	OS	0.8	0.9		MIT
3373.038	3374.007	29638.35	OS	1.2	0.7		MIT
3372.524	3373.493	29642.87	OS	0.5	0.3		MIT
3372.083	3373.052	29646.75	OS	1.6	0.9		MIT
3371.516	3372.484	29651.73	OS	0.5	0.5		MIT
3370.588	3371.556	29659.89	OS I	2.5G	1.5		MIT
3370.202	3371.170	29663.29	OS	1.7	1.0		MIT
3365.872	3366.839	29701.45	OS	1.2	1.0		MIT
3364.362	3365.329	29714.78	OS	1.0	0.9		MIT
3364.122	3365.089	29716.90	OS	2.0	1.1		MIT
3362.926	3363.892	29727.47	OS	0.7	0.7		MIT
3362.559	3363.525	29730.71	OS	0.3	0.7		MIT
3361.149	3362.115	29743.18	OS	1.9	1.3		MIT
3359.743	3360.708	29755.63	OS	0.9	1.0		MIT
3357.969	3358.934	29771.35	OS	2.0	1.2		MIT
3357.528	3358.493	29775.26	OS	0.6	0.9		MIT
3353.912	3354.876	29807.36	OS	1.6	1.2		MIT
3351.735	3352.698	29826.72	OS	1.7	1.1		MIT
3348.662	3349.625	29854.09	OS	1.5	1.2		MIT
3341.914	3342.875	29914.37	OS	1.0	1.0S		MIT
3340.689	3341.650	29925.34	OS	0.3	0.5		MIT
3337.138	3338.098	29957.18	OS	0.3	0.5		MIT
3336.150	3337.109	29966.05	OS I	2.3G	1.7		MIT
3334.151	3335.110	29984.02	OS	1.2	1.0		MIT
3333.849	3334.808	29986.74	OS	1.1	1.0		MIT
3329.217	3330.175	30028.46	OS	0.5	0.5		MIT
3329.126	3330.084	30029.28	OS	1.3	0.7		MIT
3327.425	3328.382	30044.63	OS	1.9	1.2		MIT
3326.517	3327.474	30052.83	OS	1.1	0.9		MIT
3326.386	3327.343	30054.01	OS	1.2	1.0		MIT
3324.749	3325.706	30068.81	OS	1.2	0.9		MIT
3324.33	3325.29	30072.6	OS	1.7D	1.2D		MIT
3323.172	3324.128	30083.08	OS	0.5H	0.5H		MIT
3322.051	3323.007	30093.23	OS	1.0	1.0		MIT
3320.425	3321.380	30107.96	OS	1.0	0.9		MIT
3319.902	3320.857	30112.71	OS	0.6	0.7		MIT
3318.590	3319.545	30124.61	OS	0.8	0.7		MIT
3318.167	3319.122	30128.45	OS	0.6	0.6		MIT
3317.857	3318.812	30131.27	OS	0.9	0.9		MIT
3317.276	3318.231	30136.54	OS	0.8	0.7		MIT
3316.688	3317.642	30141.89	OS	1.5	1.2		MIT
3315.688	3316.642	30150.98	OS	1.6	1.2		MIT
3315.421	3316.375	30153.40	OS	1.7	1.2		MIT
3314.77	3315.72	30159.3	OS	1.2H	1.1		MIT
3313.459	3314.413	30171.26	OS	1.0	0.7		MIT
3312.027	3312.980	30184.30	OS	0.9	0.7		MIT
3310.912	3311.865	30194.47	OS	2.3	1.5		MIT
3309.667	3310.620	30205.83	OS	1.0	0.9		MIT
3306.233	3307.185	30237.20	OS	1.9	1.1		MIT
3305.377	3306.329	30245.03	OS	1.3	1.0		MIT

AIR WAVELENGTH (ANGSTRMS)	VACUUM WAVELENGTH (ANGSTRMS)	WAVENUMBER (KAYSERS)	LMENT	LOG ARC	INTENSITY SPK	DIS	REF
3301.559	3302.510	30280.00	OS I	2.7G	1.7		MIT
3293.162	3294.110	30357.21	OS	0.6	0.5		MIT
3291.127	3292.075	30375.98	OS	1.0	0.7		MIT
3290.263	3291.211	30383.96	OS	2.3	1.3		MIT
3288.837	3289.784	30397.13	OS	1.5	1.2		MIT
3286.674	3287.621	30417.13	OS	1.5	0.7		MIT
3284.537	3285.483	30436.92	OS	0.8	0.0		MIT
3280.915	3281.860	30470.52	OS	0.7			MIT
3277.966	3278.911	30497.93	OS	1.9	0.9		MIT
3276.411	3277.355	30512.41	OS	0.5	0.0		MIT
3275.201	3276.145	30523.68	OS	2.3	1.2		MIT
3273.384	3274.327	30540.62	OS	1.2	0.7		MIT
3272.479	3273.422	30549.07	OS	1.3	0.0		MIT
3272.104	3273.047	30552.57	OS	1.6	1.0		MIT
3271.984	3272.927	30553.69	OS	1.2	0.9		MIT
3270.851	3271.794	30564.27	OS	1.2H	0.6		MIT
3269.887	3270.830	30573.28	OS	0.9	0.3		MIT
3269.209	3270.151	30579.62	OS	2.3	1.3		MIT
3267.945	3268.887	30591.45	OS I	2.6G	1.5		MIT
3267.200	3268.142	30598.43	OS	1.3	1.2		MIT
3266.747	3267.689	30602.67	OS	1.0	0.7		MIT
3264.686	3265.627	30621.99	OS	2.0H	1.0		MIT
3262.751	3263.692	30640.15	OS	2.0	1.3		MIT
3262.290	3263.231	30644.48	OS I	2.7G	1.7		MIT
3260.568	3261.508	30660.66	OS	1.2	0.7		MIT
3260.299	3261.239	30663.19	OS	1.8	1.0		MIT
3256.924	3257.863	30694.97	OS	1.9	1.1		MIT
3255.270	3256.209	30710.56	OS	1.0H	0.5		MIT
3254.914	3255.853	30713.92	OS	1.8	1.1		MIT
3252.01	3252.95	30741.4	OS	1.7	1.2		MIT
3247.996	3248.933	30779.34	OS	0.0	0.7		MIT
3245.678	3246.614	30801.32	OS	0.7	0.5		MIT
3241.982	3242.917	30836.43	OS	1.2	1.1		MIT
3241.796	3242.731	30838.20	OS	1.2	1.0		MIT
3241.426	3242.361	30841.72	OS	1.2	1.0		MIT
3241.044	3241.979	30845.35	OS	1.9	1.3		MIT
3238.628	3239.563	30868.36	OS	2.0	1.3		MIT
3234.731	3235.665	30905.55	OS	2.0	1.0		MIT
3234.196	3235.129	30910.66	OS	2.2	1.1		MIT
3232.540	3233.473	30926.50	OS	2.2	1.0		MIT
3232.055	3232.988	30931.14	OS I	2.7G	1.3		MIT
3231.417	3232.350	30937.25	OS	2.2	1.1		MIT
3230.399	3231.332	30947.00	OS	1.6	0.7		MIT
3229.206	3230.138	30958.43	OS	2.1	0.7		MIT
3227.280	3228.212	30976.90	OS	2.1	1.1		MIT
3223.858	3224.789	31009.78	OS	2.0	0.9		MIT
3221.383	3222.313	31033.61	OS	1.2	0.7		MIT
3220.189	3221.119	31045.11	OS	1.5	1.0		MIT
3219.135	3220.065	31055.28	OS	1.4	0.9		MIT
3218.023	3218.952	31066.01	OS	1.3	0.6		MIT
3217.057	3217.986	31075.34	OS	1.3	1.0		MIT
3213.463	3214.391	31110.09	OS	1.3	0.3		MIT
3213.312	3214.240	31111.55	OS	1.7	1.6		MIT
3212.720	3213.648	31117.28	OS	1.6	1.0		MIT
3205.777	3206.703	31184.68	OS	1.0	0.7		MIT
3204.515	3205.441	31196.96	OS	0.9	0.7		MIT
3203.327	3204.253	31208.52	OS	0.7	0.7		MIT
3202.834	3203.760	31213.33	OS	1.6	1.0		MIT
3200.782	3201.707	31233.34	OS	1.0	0.6		MIT
3197.200	3198.124	31268.33	OS	1.6S	1.0		MIT
3195.966	3196.890	31280.40	OS	1.5	1.0		MIT
3195.382	3196.306	31286.12	OS	2.0	1.1		MIT
3194.687	3195.610	31292.93	OS	1.7	1.0		MIT
3194.234	3195.157	31297.36	OS	2.1	1.2		MIT
3193.867	3194.790	31300.96	OS	1.6L	1.0		MIT
3191.191	3192.114	31327.21	OS	1.3	0.9		MIT
3189.459	3190.381	31344.22	OS	2.1	1.2		MIT
3187.335	3188.257	31365.10	OS	1.9	1.1		MIT
3186.980	3187.902	31368.60	OS	2.0	1.2		MIT
3186.533	3187.454	31373.00	OS	1.5	0.9		MIT

AIR WAVELENGTH (ANGSTRMS)	VACUUM WAVELENGTH (ANGSTRMS)	WAVENUMBER (KAYSERS)	LMENT	LOG ARC	INTENSITY SPK	DIS	REF
3186.402	3187.323	31374.29	OS	1.2	1.5		MIT
3185.327	3186.248	31384.87	OS	2.2	1.1		MIT
3184.337	3185.258	31394.63	OS	1.0	0.7		MIT
3182.803	3183.723	31409.76	OS	1.6	0.9		MIT
3182.567	3183.487	31412.09	OS	2.0	1.2		MIT
3182.229	3183.149	31415.43	OS	1.3	0.9		MIT
3181.879	3182.799	31418.88	OS	2.0	1.1		MIT
3179.259	3180.179	31444.77	OS	1.3	0.6		MIT
3178.240	3179.159	31454.86	OS	1.9	1.0		MIT
3178.065	3178.984	31456.59	OS	2.2	1.3		MIT
3177.389	3178.308	31463.28	OS	1.0	0.7		MIT
3175.648	3176.567	31480.53	OS	1.3	0.9		MIT
3173.926	3174.844	31497.61	OS	1.9	1.5		MIT
3173.196	3174.114	31504.85	OS	2.0	1.2		MIT
3172.800	3173.718	31508.79	OS	0.9	0.7		MIT
3171.616	3172.534	31520.55	OS	1.0	0.7		MIT
3168.281	3169.198	31553.73	OS	2.0	1.2		MIT
3166.512	3167.428	31571.35	OS	2.3	1.3		MIT
3165.659	3166.575	31579.86	OS	1.3	1.3		MIT
3164.612	3165.528	31590.31	OS	1.8	1.1		MIT
3161.728	3162.643	31619.12	OS	2.0	1.0		MIT
3161.445	3162.360	31621.95	OS	1.9	1.1		MIT
3160.427	3161.342	31632.14	OS	1.3	0.9		MIT
3160.286	3161.201	31633.55	OS	1.5	1.0		MIT
3159.356	3160.271	31642.86	OS	1.3	0.7		MIT
3157.238	3158.152	31664.09	OS	2.0	0.3		MIT
3156.775	3157.689	31668.73	OS	2.0	0.5		MIT
3156.248	3157.162	31674.02	OS I	2.7G	1.2J		MIT
3155.328	3156.242	31683.25	OS	1.3	0.7		MIT
3153.611	3154.524	31700.50	OS	2.1	1.3		MIT
3152.673	3153.586	31709.93	OS	2.2	1.3		MIT
3152.070	3152.983	31716.00	OS	1.9	1.2		MIT
3149.807	3150.719	31738.79	OS	1.2	0.9		MIT
3145.956	3146.867	31777.64	OS	1.8	1.0		MIT
3144.338	3145.249	31793.99	OS	1.3	0.9		MIT
3143.043	3143.953	31807.09	OS	1.5	1.0		MIT
3140.938	3141.848	31828.40	OS	1.7	1.1		MIT
3140.314	3141.224	31834.73	OS	1.8	1.1		MIT
3137.519	3138.428	31863.08	OS	1.4	1.0		MIT
3131.481	3132.389	31924.52	OS	1.3	1.0		MIT
3131.115	3132.022	31928.25	OS	2.1	1.5		MIT
3130.002	3130.909	31939.60	OS	1.5	0.9		MIT
3129.229	3130.136	31947.49	OS	1.8	1.2		MIT
3128.439	3129.346	31955.56	OS	1.3	0.7		MIT
3124.939	3125.845	31991.35	OS	2.2H	1.0J		MIT
3124.004	3124.910	32000.93	OS	1.2	0.9H		MIT
3120.650	3121.555	32035.32	OS	1.3	0.7		MIT
3119.895	3120.800	32043.07	OS	1.5	1.2		MIT
3118.328	3119.232	32059.17	OS	2.2	1.3		MIT
3118.122	3119.026	32061.29	OS	1.9	1.2		MIT
3117.891	3118.795	32063.66	OS	1.5	1.0		MIT
3116.475	3117.379	32078.23	OS	1.7	1.2		MIT
3114.814	3115.717	32095.34	OS	1.7	1.1		MIT
3111.09	3111.99	32133.7	OS	2.0	1.3		MIT
3110.618	3111.520	32138.63	OS	1.5	1.0		MIT
3109.679	3110.581	32148.33	OS	1.3	1.3		MIT
3109.381	3110.283	32151.42	OS	2.1	1.3		MIT
3108.981	3109.883	32155.55	OS	2.1	1.2		MIT
3107.976	3108.878	32165.95	OS	1.4	0.9		MIT
3107.864	3108.766	32167.11	OS	1.5	0.9		MIT
3107.378	3108.279	32172.14	OS	1.6	1.0		MIT
3105.992	3106.893	32186.50	OS	2.2	1.3		MIT
3104.98	3105.88	32197.0	OS	2.3D	1.2D		MIT
3103.415	3104.315	32213.22	OS	1.3	0.9		MIT
3102.716	3103.616	32220.48	OS	1.3	0.9		MIT
3101.528	3102.428	32232.82	OS	2.1	1.3		MIT
3099.261	3100.160	32256.40	OS	1.6	0.9		MIT
3094.074	3094.972	32310.47	OS	1.5	0.9		MIT
3093.587	3094.485	32315.55	OS	2.1	1.2		MIT
3091.248	3092.145	32340.00	OS	1.6	1.2		MIT

AIR WAVELENGTH (ANGSTRMS)	VACUUM WAVELENGTH (ANGSTRMS)	WAVENUMBER (KAYSERS)	LMENT		LOG ARC	INTENSITY SPK	INTENSITY DIS	REF
3090.492	3091.389	32347.92	OS		1.9	1.2		MIT
3090.296	3091.193	32349.97	OS		2.0	1.1		MIT
3090.085	3090.982	32352.18	OS		2.0	1.2		MIT
3088.266	3089.163	32371.23	OS		1.8	1.1		MIT
3087.748	3088.645	32376.66	OS		1.7	1.0		MIT
3086.274	3087.170	32392.12	OS		1.7	1.0		MIT
3084.596	3085.492	32409.74	OS		1.8	1.0		MIT
3079.560	3080.455	32462.74	OS		1.6	1.0		MIT
3078.383	3079.277	32475.15	OS		2.1	1.2		MIT
3078.113	3079.007	32478.00	OS		2.1	1.2		MIT
3077.720	3078.614	32482.15	OS		2.0	1.5		MIT
3077.437	3078.331	32485.14	OS		1.9	0.9		MIT
3077.056	3077.950	32489.16	OS		2.0	1.1		MIT
3076.751	3077.645	32492.38	OS		1.5	0.7		MIT
3074.96	3075.85	32511.3	OS		2.1	1.3		MIT
3074.08	3074.97	32520.6	OS		2.1	1.3		MIT
3069.936	3070.828	32564.51	OS		2.1	1.2		MIT
3069.122	3070.014	32573.14	OS		1.5	1.0		MIT
3066.833	3067.724	32597.45	OS		1.5	1.0		MIT
3066.598	3067.489	32599.95	OS		1.3	0.9		MIT
3066.116	3067.007	32605.08	OS		1.7	1.0		MIT
3062.695	3063.585	32641.49	OS		1.3	0.9		MIT
3062.468	3063.358	32643.91	OS		1.3	0.9		MIT
3062.192	3063.082	32646.85	OS		2.0	1.5		MIT
3060.305	3061.195	32666.99	OS		2.0	1.5		MIT
3058.66	3059.55	32684.6	OS	I	2.7G	2.7		MIT
3056.901	3057.790	32703.36	OS		1.3	0.9		MIT
3055.211	3056.099	32721.45	OS		1.9	1.2		MIT
3054.968	3055.856	32724.05	OS		1.7	1.0		MIT
3052.418	3053.306	32751.39	OS		1.6	1.1		MIT
3051.166	3052.053	32764.83	OS		1.9	1.2		MIT
3050.386	3051.273	32773.20	OS		2.0	1.7		MIT
3049.456	3050.343	32783.20	OS		1.9	1.2		MIT
3049.042	3049.929	32787.65	OS		1.6	0.9		MIT
3045.777	3046.663	32822.80	OS		1.5	1.0		MIT
3045.323	3046.209	32827.69	OS		1.5	1.1		MIT
3044.910	3045.796	32832.14	OS		2.0	1.1		MIT
3044.408	3045.294	32837.56	OS		1.7	1.0		MIT
3044.067	3044.953	32841.23	OS		1.5	0.9		MIT
3043.638	3044.524	32845.86	OS		1.8	1.1		MIT
3043.505	3044.390	32847.30	OS		2.0	1.2		MIT
3042.739	3043.624	32855.57	OS		1.3	1.7		MIT
3040.900	3041.785	32875.44	OS		2.3	2.0		MIT
3032.806	3033.689	32963.17	OS		1.7	1.2		MIT
3031.295	3032.177	32979.60	OS		1.6	1.0		MIT
3031.006	3031.888	32982.75	OS		1.5	1.0		MIT
3030.695	3031.577	32986.13	OS	I	2.7	1.6		MIT
3028.917	3029.799	33005.49	OS		1.6	1.1		MIT
3020.496	3021.376	33097.50	OS			0.5		MIT
3019.375	3020.254	33109.79	OS		2.0	1.3		MIT
3018.039	3018.918	33124.45	OS	I	2.5G	1.7		MIT
3017.247	3018.126	33133.14	OS		2.0	1.4		MIT
3015.653	3016.532	33150.66	OS		2.0	1.2		MIT
3013.074	3013.952	33179.03	OS		2.2	1.3		MIT
3009.896	3010.773	33214.06	OS		1.3	0.9		MIT
3007.896	3008.773	33236.14	OS		1.6	0.9		MIT
3003.482	3004.357	33284.99	OS		1.8	1.1		MIT
2997.647	2998.521	33349.77	OS		1.6	0.9		MIT
2993.571	2994.444	33395.18	OS		1.2	0.9		MIT
2992.112	2992.985	33411.46	OS		1.3	0.9		MIT
2989.526	2990.398	33440.37	OS		0.7	0.6		MIT
2989.127	2989.999	33444.83	OS		1.1	1.6		MIT
2988.262	2989.134	33454.51	OS		0.9	0.0		MIT
2985.614	2986.485	33484.18	OS		1.0	0.6		MIT
2984.290	2985.161	33499.03	OS		0.7	0.7		MIT
2982.902	2983.772	33514.62	OS		1.6	1.0		MIT
2982.557	2983.427	33518.50	OS		1.0	0.8		MIT
2982.125	2982.995	33523.35	OS		0.9	0.6		MIT
2979.432	2980.302	33553.65	OS		1.1	0.8		MIT
2978.528	2979.397	33563.83	OS		1.0	0.9		MIT

AIR WAVELENGTH (ANGSTRMS)	VACUUM WAVELENGTH (ANGSTRMS)	WAVENUMBER (KAYSERS)	LMENT		LOG ARC	INTENSITY SPK	INTENSITY DIS	REF
2978.208	2979.077	33567.44	OS		1.3	0.7		MIT
2977.637	2978.506	33573.88	OS		1.5	1.1		MIT
2975.343	2976.211	33599.76	OS		1.2	0.9		MIT
2972.251	2973.119	33634.71	OS		1.1	0.9		MIT
2970.972	2971.839	33649.19	OS		1.6	1.0		MIT
2970.689	2971.556	33652.40	OS		0.9	0.7		MIT
2968.446	2969.313	33677.83	OS		0.7	0.7		MIT
2964.615	2965.481	33721.34	OS		0.9	0.7		MIT
2964.062	2964.928	33727.63	OS		1.4	1.1		MIT
2962.327	2963.192	33747.39	OS		1.3	0.9		MIT
2962.148	2963.013	33749.43	OS		1.5	1.2		MIT
2961.012	2961.877	33762.37	OS		1.6	1.0		MIT
2958.341	2959.205	33792.86	OS		0.9	0.6		MIT
2957.084	2957.948	33807.22	OS		1.0	0.8		MIT
2955.000	2955.863	33831.06	OS		0.9	0.7		MIT
2952.344	2953.207	33861.50	OS		1.2	0.9		MIT
2949.896	2950.758	33889.60	OS		0.7	0.6		MIT
2949.810	2950.672	33890.58	OS		0.9	0.7		MIT
2949.532	2950.394	33893.78	OS		1.5	1.0		MIT
2948.229	2949.091	33908.76	OS		1.1	0.7		MIT
2942.848	2943.708	33970.76	OS		1.5	0.9		MIT
2942.204	2943.064	33978.19	OS		0.9	0.7		MIT
2934.642	2935.500	34065.74	OS		1.5	0.7		MIT
2931.280	2932.138	34104.81	OS		1.6	1.0		MIT
2930.566	2931.423	34113.12	OS		1.2	0.6		MIT
2930.191	2931.048	34117.49	OS		1.0	0.6		MIT
2929.507	2930.364	34125.45	OS		1.5	0.5		MIT
2925.568	2926.424	34171.40	OS		1.6	1.0		MIT
2925.320	2926.176	34174.29	OS		0.9	0.6		MIT
2919.794	2920.649	34238.97	OS		2.0	1.2		MIT
2917.827	2918.681	34262.05	OS		1.0	0.5		MIT
2917.258	2918.112	34268.73	OS		1.6	1.3		MIT
2914.710	2915.563	34298.69	OS		1.0	0.8		MIT
2913.844	2914.697	34308.88	OS		1.5	0.9		MIT
2912.334	2913.187	34326.67	OS		1.7	1.7		MIT
2911.341	2912.194	34338.38	OS		0.7	0.7		MIT
2909.669	2910.521	34358.11	OS		1.1	0.7		MIT
2909.061	2909.913	34365.29	OS	I	2.7G	2.6		MIT
2908.026	2908.878	34377.52	OS		0.7	0.5		MIT
2905.970	2906.821	34401.84	OS		1.1	0.9		MIT
2905.730	2906.581	34404.68	OS		1.6	0.9		MIT
2903.119	2903.970	34435.62	OS		0.9	0.7		MIT
2901.322	2902.172	34456.95	OS		1.0	0.6		MIT
2896.063	2896.912	34519.52	OS		1.6	0.9		MIT
2895.064	2895.913	34531.43	OS		1.4	0.9		MIT
2890.849	2891.697	34581.77	OS		1.0	0.6		MIT
2886.542	2887.389	34633.37	OS		0.7	0.7		MIT
2884.408	2885.254	34658.99	OS		1.3	0.7		MIT
2878.400	2879.244	34731.33	OS		1.6	1.1		MIT
2877.351	2878.195	34744.00	OS		1.5	0.3		MIT
2874.955	2875.799	34772.95	OS		1.7	1.2		MIT
2874.635	2875.479	34776.82	OS		1.0	0.6		MIT
2872.405	2873.248	34803.82	OS		1.7	0.9		MIT
2865.680	2866.521	34885.49	OS		1.2	0.7		MIT
2860.956	2861.796	34943.09	OS		2.0	1.4		MIT
2860.063	2860.903	34954.00	OS		1.4	1.0		MIT
2857.537	2858.376	34984.90	OS		0.7	0.7		MIT
2855.337	2856.176	35011.85	OS		1.4	0.9		MIT
2850.762	2851.600	35068.04	OS		1.9	1.4		MIT
2849.302	2850.139	35086.00	OS		0.9	0.6		MIT
2849.046	2849.883	35089.16	OS		1.2	0.7		MIT
2848.247	2849.084	35099.00	OS		1.5	1.2		MIT
2846.554	2847.391	35119.87	OS		1.1	0.7		MIT
2846.391	2847.228	35121.88	OS		1.6	1.0		MIT
2844.680	2845.516	35143.01	OS		1.2	0.7		MIT
2844.396	2845.232	35146.52	OS		1.7	1.4		MIT
2838.626	2839.461	35217.96	OS	I	2.0G	2.0		MIT
2838.173	2839.008	35223.58	OS		1.5	1.1		MIT
2837.422	2838.256	35232.90	OS		1.4	1.0		MIT
2829.269	2830.101	35334.42	OS		1.6	0.8		MIT

AIR WAVELENGTH (ANGSTRMS)	VACUUM WAVELENGTH (ANGSTRMS)	WAVENUMBER (KAYSERS)	LMENT	LOG ARC	INTENSITY SPK	DIS	REF
2824.166	2824.997	35398.27	OS	1.3	0.6		MIT
2821.251	2822.081	35434.84	OS	1.3	0.8		MIT
2820.555	2821.385	35443.58	OS	1.0	0.7		MIT
2820.178	2821.008	35448.32	OS	1.1	0.7		MIT
2815.780	2816.609	35503.68	OS	1.6	0.6		MIT
2815.272	2816.101	35510.09	OS	0.9	0.5		MIT
2814.839	2815.668	35515.55	OS	1.0	0.5		MIT
2814.200	2815.029	35523.62	OS	1.7	1.4		MIT
2813.839	2814.668	35528.17	OS	1.0	0.6		MIT
2808.935	2809.762	35590.20	OS	1.6	0.9		MIT
2806.906	2807.733	35615.92	OS	2.0W	0.0		MIT
2804.067	2804.893	35651.98	OS	1.9	1.3		MIT
2796.727	2797.551	35745.54	OS	2.0	1.2		MIT
2786.798	2787.620	35872.90	OS	1.6	0.9		MIT
2786.306	2787.128	35879.23	OS	1.4	0.7		MIT
2782.552	2783.373	35927.63	OS	1.6	1.2		MIT
2776.910	2777.730	36000.62	OS	1.4	0.9		MIT
2774.900	2775.719	36026.70	OS	0.9S	0.6W		MIT
2774.375	2775.194	36033.52	OS	0.9	0.7		MIT
2774.151	2774.970	36036.43	OS	0.7	0.3		MIT
2774.016	2774.835	36038.18	OS	1.1	0.8		MIT
2773.069	2773.888	36050.49	OS	1.1	0.7		MIT
2770.098	2770.916	36089.15	OS	0.7	0.5		MIT
2769.875	2770.693	36092.05	OS	1.3	0.9		MIT
2767.122	2767.939	36127.96	OS	1.0	0.5		MIT
2765.447	2766.264	36149.84	OS	1.0	0.6		MIT
2764.058	2764.875	36168.01	OS	0.3	0.0		MIT
2763.935	2764.751	36169.62	OS	1.3	0.9		MIT
2763.273	2764.089	36178.28	OS	1.4	1.0		MIT
2761.418	2762.234	36202.58	OS	1.7	1.0		MIT
2761.082	2761.898	36206.99	OS	1.3	0.7		MIT
2758.821	2759.636	36236.66	OS	1.2	0.8		MIT
2757.808	2758.623	36249.97	OS	1.4	0.9		MIT
2753.722	2754.536	36303.75	OS	0.9	0.6		MIT
2751.147	2751.960	36337.73	OS	0.9	0.7		MIT
2749.180	2749.993	36363.73	OS	1.2	0.8		MIT
2748.863	2749.676	36367.92	OS	1.1	0.7		MIT
2740.753	2741.564	36475.53	OS	0.9	0.7		MIT
2740.607	2741.418	36477.48	OS	1.0	0.9		MIT
2740.321	2741.132	36481.28	OS	1.2	0.9		MIT
2732.805	2733.614	36581.61	OS	1.9	1.2		MIT
2728.272	2729.080	36642.39	OS	1.1	0.7		MIT
2721.862	2722.668	36728.68	OS	1.9	1.0		MIT
2720.044	2720.850	36753.22	OS	1.9	1.2		MIT
2715.636	2716.441	36812.88	OS	0.9	0.3		MIT
2715.364	2716.169	36816.56	OS	1.1	0.7		MIT
2714.642	2715.447	36826.36	OS	1.7R	1.0		MIT
2709.864	2710.667	36891.28	OS	1.3	0.7		MIT
2708.179	2708.982	36914.24	OS	1.0	1.2		MIT
2707.421	2708.224	36924.57	OS	1.1	0.6		MIT
2706.702	2707.505	36934.38	OS	1.7	0.9		MIT
2704.448	2705.250	36965.16	OS	1.0	0.0		MIT
2702.830	2703.632	36987.29	OS	0.9	0.3		MIT
2702.401	2703.203	36993.16	OS	0.3			MIT
2700.746	2701.547	37015.83	OS	1.0	0.0		MIT
2699.589	2700.390	37031.69	OS	1.7	0.9		MIT
2697.241	2698.041	37063.92	OS	0.9	0.0		MIT
2694.748	2695.548	37098.21	OS	0.7	0.0		MIT
2694.523	2695.323	37101.31	OS	1.0	0.5		MIT
2692.701	2693.500	37126.41	OS	0.9	0.3		MIT
2689.816	2690.615	37166.23	OS	1.7	1.0		MIT
2689.350	2690.148	37172.67	OS	0.7	0.3		MIT
2688.077	2688.875	37190.27	OS	0.9	0.3		MIT
2682.187	2682.984	37271.94	OS	1.0	0.5		MIT
2679.735	2680.531	37306.04	OS	0.9	0.0		MIT
2674.885	2675.680	37373.68	OS	1.4	0.7		MIT
2674.568	2675.363	37378.11	OS	1.2	0.7		MIT
2670.527	2671.321	37434.66	OS	0.3	0.0		MIT
2669.529	2670.323	37448.66	OS	0.9	0.5		MIT
2666.211	2667.004	37495.26	OS	0.7	0.0		MIT

AIR WAVELENGTH (ANGSTRMS)	VACUUM WAVELENGTH (ANGSTRMS)	WAVENUMBER (KAYSERS)	LMENT	LOG ARC	INTENSITY SPK	DIS	REF
2665.986	2666.779	37498.42	OS	0.7	0.3L		MIT
2661.179	2661.971	37566.15	OS	1.3	0.7		MIT
2660.918	2661.710	37569.84	OS	0.7	0.0		MIT
2659.833	2660.624	37585.16	OS	1.5	0.9		MIT
2658.600	2659.391	37602.59	OS	1.7	1.0		MIT
2656.681	2657.472	37629.75	OS	1.3	0.9		MIT
2655.780	2656.570	37642.52	OS	0.9	0.0		MIT
2655.19	2655.98	37650.9	OS	1.2	0.3		MIT
2653.776	2654.566	37670.94	OS	1.1	0.0		MIT
2652.977	2653.767	37682.29	OS	1.0	0.3		MIT
2649.335	2650.124	37734.09	OS	1.4	0.8		MIT
2647.730	2648.518	37756.96	OS	1.4	0.8		MIT
2646.892	2647.680	37768.91	OS	1.2	0.7		MIT
2644.114	2644.902	37808.59	OS	1.9	1.0		MIT
2643.629	2644.416	37815.53	OS	0.8	0.0		MIT
2641.174	2641.961	37850.67	OS	1.0			MIT
2640.518	2641.305	37860.08	OS	0.3	0.7H		MIT
2637.133	2637.919	37908.67	OS I	2.2	1.5		MIT
2634.439	2635.224	37947.43	OS	0.6	0.0		MIT
2634.291	2635.076	37949.57	OS	0.7	0.0		MIT
2632.892	2633.677	37969.73	OS	0.7	0.3		MIT
2628.479	2629.263	38033.47	OS	1.3	0.7		MIT
2621.818	2622.600	38130.10	OS	1.4	0.6		MIT
2621.371	2622.153	38136.60	OS	0.7	0.0		MIT
2620.619	2621.401	38147.54	OS	1.0	0.5		MIT
2619.944	2620.726	38157.37	OS	1.5	0.9		MIT
2617.796	2618.577	38188.68	OS	0.7	1.5		MIT
2615.962	2616.743	38215.45	OS	0.7	0.0		MIT
2613.058	2613.838	38257.92	OS	1.7	1.0		MIT
2612.630	2613.410	38264.18	OS	1.3	0.7		MIT
2611.330	2612.110	38283.23	OS	0.9	0.0		MIT
2610.782	2611.562	38291.27	OS	1.3	0.8		MIT
2609.560	2610.339	38309.19	OS	1.3	0.6		MIT
2609.203	2609.982	38314.44	OS	0.9	0.6		MIT
2604.604	2605.382	38382.08	OS	1.3	0.5		MIT
2603.220	2603.998	38402.49	OS	0.7	0.0		MIT
2600.749	2601.526	38438.97	OS	0.7	0.0		MIT
2600.452	2601.229	38443.36	OS	0.9	0.3		MIT
2599.907	2600.684	38451.42	OS	1.1	0.5		MIT
2599.129	2599.906	38462.93	OS	0.7	0.0		MIT
2597.291	2598.067	38490.15	OS	0.6	0.0		MIT
2597.201	2597.977	38491.48	OS	0.7	0.3		MIT
2596.692	2597.468	38499.03	OS	1.0	0.5		MIT
2596.005	2596.781	38509.21	OS	1.1	0.5		MIT
2594.143	2594.919	38536.85	OS	1.0	0.5		MIT
2590.755	2591.530	38587.25	OS	1.9	0.9L		MIT
2589.507	2590.282	38605.84	OS	0.7	0.0		MIT
2589.391	2590.166	38607.57	OS	0.9	0.3		MIT
2587.486	2588.260	38635.99	OS	0.9	0.3		MIT
2581.958	2582.731	38718.71	OS	1.9S	0.7		MIT
2581.052	2581.825	38732.30	OS	1.4	0.6		MIT
2580.026	2580.798	38747.70	OS	1.2	2.0W		MIT
2578.321	2579.093	38773.32	OS	1.0	1.0		MIT
2578.164	2578.936	38775.68	OS	0.9	0.0		MIT
2571.782	2572.552	38871.90	OS	1.4	0.7		MIT
2568.834	2569.604	38916.51	OS	1.4L	1.0L		MIT
2566.490	2567.259	38952.05	OS	1.4	0.6		MIT
2565.167	2565.936	38972.13	OS	1.1	0.0		MIT
2564.369	2565.138	38984.26	OS	0.9	0.0		MIT
2563.164	2563.932	39002.59	OS	0.9	1.4		MIT
2562.665	2563.433	39010.18	OS	1.1	0.0		MIT
2558.093	2558.860	39079.90	OS	1.1	0.3		MIT
2557.768	2558.535	39084.87	OS	1.0	0.0		MIT
2556.077	2556.844	39110.72	OS	1.0	0.3		MIT
2555.803	2556.570	39114.91	OS	1.0	0.3		MIT
2555.273	2556.040	39123.03	OS	1.0	0.5		MIT
2554.465	2555.231	39135.40	OS	1.3	0.6		MIT
2548.832	2549.597	39221.88	OS	0.5	1.2		MIT
2547.700	2548.465	39239.31	OS	0.7	0.3		MIT
2543.804	2544.568	39299.40	OS	1.0	0.0		MIT

AIR WAVELENGTH (ANGSTRMS)	VACUUM WAVELENGTH (ANGSTRMS)	WAVENUMBER (KAYSERS)	LMENT	LOG ARC	INTENSITY SPK	DIS	REF
2542.512	2543.276	39319.37	OS	1.7	0.9		MIT
2540.741	2541.504	39346.78	OS	1.6	0.5		MIT
2538.102	2538.864	39387.69	OS	0.7	0.0		MIT
2537.997	2538.759	39389.32	OS		1.2		MIT
2534.166	2534.928	39448.86	OS	0.9	0.3		MIT
2532.436	2533.197	39475.81	OS	1.1	0.5		MIT
2531.985	2532.746	39482.84	OS	0.9	1.3		MIT
2527.756	2528.516	39548.89	OS	0.5	1.0H		MIT
2527.087	2527.847	39559.36	OS	1.5R	0.7R		MIT
2526.013	2526.773	39576.18	OS	1.0	0.0		MIT
2519.788	2520.546	39673.94	OS	1.1	0.7		MIT
2519.028	2519.786	39685.91	OS	0.0	1.5H		MIT
2518.441	2519.199	39695.16	OS	1.3	0.7		MIT
2517.920	2518.678	39703.37	OS	1.1	0.7		MIT
2515.044	2515.801	39748.77	OS	1.6S	0.7		MIT
2513.246	2514.003	39777.21	OS	1.7	0.9		MIT
2512.873	2513.630	39783.11	OS	1.6	0.5		MIT
2509.936	2510.692	39829.66	OS	1.4	0.0D		MIT
2509.708	2510.464	39833.28	OS	0.9	1.3		MIT
2508.610	2509.366	39850.71	OS	0.9S	0.0		MIT
2504.507	2505.262	39915.99	OS	1.0	0.0		MIT
2504.387	2505.142	39917.90	OS	0.9	0.3		MIT
2503.670	2504.424	39929.33	OS	1.2H	1.0H		MIT
2502.290	2503.044	39951.35	OS	1.0S	0.0		MIT
2500.911	2501.665	39973.38	OS	1.1	0.6		MIT
2499.924	2500.678	39989.16	OS	0.7H	0.3		MIT
2498.414	2499.167	40013.33	OS	1.3B	0.7		MIT
2493.616	2494.368	40090.31	OS	0.9S	0.0		MIT
2492.367	2493.119	40110.40	OS	1.7R	0.9R		MIT
2491.689	2492.441	40121.32	OS	1.0	0.5		MIT
2491.021	2491.773	40132.07	OS	0.9W	0.9		MIT
2489.280	2490.031	40160.14	OS	1.0	1.0H		MIT
2488.548	2489.299	40171.95	OS I	1.7W	1.2		MIT
2486.244	2486.994	40209.18	OS	0.9H	1.2H		MIT
2482.427	2483.177	40271.00	OS	1.1W	0.3		MIT
2481.791	2482.540	40281.32	OS	1.0S	0.0		MIT
2476.836	2477.584	40361.90	OS	1.4	0.9		MIT
2472.281	2473.028	40436.26	OS	1.2S	0.6		MIT
2470.067	2470.814	40472.50	OS		0.3		MIT
2468.898	2469.644	40491.66	OS	1.0	1.5		MIT
2465.162	2465.908	40553.02	OS	1.4	1.6		MIT
2464.501	2465.246	40563.90	OS	0.6	0.0		MIT
2461.417	2462.162	40614.72	OS	1.7	1.0		MIT
2457.71	2458.45	40676.0	OS	0.3	1.0		MIT
2456.462	2457.206	40696.64	OS	1.1S	0.5		MIT
2453.895	2454.638	40739.21	OS	1.1	0.5		MIT
2451.726	2452.468	40775.24	OS	1.1S	0.9		MIT
2450.738	2451.480	40791.68	OS	1.4	0.9		MIT
2446.023	2446.764	40870.31	OS	1.1	0.9		MIT
2445.880	2446.621	40872.70	OS	0.9	1.2		MIT
2440.815	2441.555	40957.51	OS	0.5	1.2		MIT
2431.607	2432.345	41112.59	OS	1.6	0.6		MIT
2431.193	2431.931	41119.59	OS	1.6	0.6		MIT
2427.900	2428.637	41175.36	OS	0.9	1.6		MIT
2426.812	2427.549	41193.82	OS	1.2	0.5		MIT
2424.970	2425.706	41225.11	OS	1.7	0.9		MIT
2424.717	2425.453	41229.41	OS	0.5	0.0		MIT
2424.565	2425.301	41231.99	OS	1.3	0.5		MIT
2424.024	2424.760	41241.19	OS	0.9	1.0		MIT
2423.071	2423.807	41257.41	OS	1.4	1.9		MIT
2418.529	2419.264	41334.89	OS	1.2	0.5		MIT
2417.986	2418.721	41344.17	OS	1.3S	0.5		MIT
2414.517	2415.251	41403.56	OS	1.4	0.5D		MIT
2408.667	2409.400	41504.12	OS	1.6	0.9		MIT
2405.960	2406.692	41550.81	OS	1.0	0.3		MIT
2405.078	2405.810	41566.05	OS	0.9	1.3		MIT
2403.853	2404.585	41587.23	OS	1.4	0.5		MIT
2402.233	2402.964	41615.27	OS	1.0	0.0		MIT
2401.127	2401.858	41634.44	OS	1.4	0.9		MIT
2399.65	2400.38	41660.1	OS	1.3			A

AIR WAVELENGTH (ANGSTRMS)	VACUUM WAVELENGTH (ANGSTRMS)	WAVENUMBER (KAYSERS)	LMENT	LOG ARC	INTENSITY SPK	DIS	REF
2398.18	2398.91	41685.6	OS	1.4	0.5		MIT
2397.61	2398.34	41695.5	OS	1.3	0.5		A
2396.77	2397.50	41710.1	OS	2.0	1.1		MIT
2396.43	2397.16	41716.0	OS		0.6		MIT
2395.39	2396.12	41734.1	OS	1.0	1.0		MIT
2394.58	2395.31	41748.3	OS	0.5	0.7		MIT
2394.29	2395.02	41753.3	OS	1.7	1.0		MIT
2393.86	2394.59	41760.8	OS	1.5	0.7		MIT
2393.59	2394.32	41765.5	OS	1.0	0.3H		A
2393.34	2394.07	41769.9	OS	0.7			A
2392.82	2393.55	41779.0	OS	0.7			A
2392.56	2393.29	41783.5	OS	1.0			A
2392.46	2393.19	41785.2	OS	0.3	0.6		MIT
2392.22	2392.95	41789.4	OS	1.1			A
2391.78	2392.51	41797.1	OS	0.7L	1.3		MIT
2391.12	2391.85	41808.7	OS	1.1			A
2390.95	2391.68	41811.6	OS	1.5	0.5		A
2390.11	2390.84	41826.3	OS	0.3			MIT
2389.08	2389.81	41844.4	OS	1.2	0.5		MIT
2388.87	2389.60	41848.0	OS	0.8			A
2387.99	2388.72	41863.5	OS	1.1	0.7		A
2387.29	2388.02	41875.7	OS	1.6	1.2		MIT
2386.75	2387.48	41885.2	OS	1.0			A
2386.04	2386.77	41897.7	OS	0.3	1.3		MIT
2385.49	2386.22	41907.3	OS	1.2	0.3		A
2384.62	2385.35	41922.6	OS	1.5	0.7		MIT
2383.21	2383.94	41947.4	OS	1.2			A
2382.46	2383.19	41960.6	OS	1.5	0.7		A
2380.82	2381.55	41989.5	OS	1.5	1.3		A
2379.84	2380.57	42006.8	OS	1.4	0.9		MIT
2379.64	2380.37	42010.4	OS	1.2	0.7		MIT
2379.39	2380.12	42014.8	OS	1.6	1.0		MIT
2378.74	2379.47	42026.2	OS	1.4	0.7		A
2378.54	2379.27	42029.8	OS	1.4	0.7W		A
2378.14	2378.87	42036.8	OS	1.5	0.7		A
2377.61	2378.34	42046.2	OS	1.7	1.2		MIT
2377.03	2377.76	42056.5	OS	1.7	1.5		MIT
2376.51	2377.24	42065.7	OS	1.0	0.6W		A
2376.29	2377.02	42069.6	OS	1.4	0.7		A
2376.15	2376.88	42072.0	OS	0.3	1.0		MIT
2376.02	2376.75	42074.4	OS	0.7			A
2375.45	2376.18	42084.4	OS	1.0			A
2375.06	2375.78	42091.4	OS	1.0	1.4		MIT
2374.51	2375.23	42101.1	OS	1.4			A
2374.33	2375.05	42104.3	OS	1.4	0.5		A
2373.62	2374.34	42116.9	OS	0.6	0.7		MIT
2373.12	2373.84	42125.8	OS		1.0		MIT
2372.91	2373.63	42129.5	OS	1.4	0.5		A
2372.65	2373.37	42134.1	OS	1.3	0.7W		MIT
2371.85	2372.57	42148.3	OS		0.6		MIT
2371.18	2371.90	42160.2	OS	1.7	1.2		MIT
2370.70	2371.42	42168.8	OS	1.4	1.0		MIT
2370.53	2371.25	42171.8	OS	1.2			A
2369.24	2369.96	42194.7	OS	1.7	1.1		MIT
2367.93	2368.65	42218.1	OS	0.3	0.5		MIT
2367.66	2368.38	42222.9	OS	0.7			A
2367.35	2368.07	42228.4	OS	1.7	1.9		MIT
2366.57	2367.29	42242.3	OS	0.9			A
2366.28	2367.00	42247.5	OS	1.1	0.3		A
2366.12	2366.84	42250.4	OS	1.1	0.3		A
2365.86	2366.58	42255.0	OS	1.2	0.3W		A
2365.07	2365.79	42269.1	OS	1.2	0.0		A
2364.85	2365.57	42273.1	OS	1.2	1.5		A
2364.71	2365.43	42275.6	OS	1.0			A
2364.51	2365.23	42279.1	OS	0.7			A
2363.90	2364.62	42290.0	OS	1.4	1.0W		MIT
2363.53	2364.25	42296.7	OS	1.0	0.3		A
2363.33	2364.05	42300.2	OS	1.4	1.2W		MIT
2362.77	2363.49	42310.3	OS	1.7	1.1		MIT
2362.41	2363.13	42316.7	OS	1.4	0.9		MIT

AIR WAVELENGTH (ANGSTRMS)	VACUUM WAVELENGTH (ANGSTRMS)	WAVENUMBER (KAYSERS)	LMENT	LOG ARC	INTENSITY SPK	INTENSITY DIS	REF
2362.21	2362.93	42320.3	OS	1.0	0.3		A
2361.66	2362.38	42330.2	OS	1.2			MIT
2360.44	2361.16	42352.0	OS	1.0			A
2359.23	2359.95	42373.7	OS	0.9			A
2358.68	2359.40	42383.6	OS	0.8	1.5		MIT
2358.50	2359.22	42386.9	OS	0.6W			A
2357.64	2358.36	42402.3	OS	1.2	0.6		A
2357.25	2357.97	42409.3	OS	1.4	0.7		MIT
2356.92	2357.64	42415.3	OS	1.2	1.0W		MIT
2356.05	2356.77	42430.9	OS	1.0	0.3		A
2355.64	2356.36	42438.3	OS	0.7			A
2355.284	2356.005	42444.74	OS	1.2	1.4		MIT
2352.99	2353.71	42486.1	OS	1.5L	0.9		MIT
2352.17	2352.89	42500.9	OS		0.6W		MIT
2351.97	2352.69	42504.5	OS	0.0	0.7		MIT
2351.72	2352.44	42509.1	OS	1.4	0.6		A
2351.55	2352.27	42512.1	OS	1.4	0.80		A
2350.23	2350.95	42536.0	OS	1.5	1.7		MIT
2350.04	2350.76	42539.5	OS	1.1			A
2349.81	2350.53	42543.6	OS	1.6	0.8L		MIT
2349.48	2350.20	42549.6	OS		0.6W		MIT
2348.61	2349.33	42565.3	OS	0.5			A
2347.38	2348.10	42587.6	OS	1.5	0.7		MIT
2346.49	2347.21	42603.8	OS	1.0	0.0		A
2346.21	2346.93	42608.9	OS	0.9	0.6W		A
2345.75	2346.47	42617.2	OS	1.4	0.7W		MIT
2345.58	2346.30	42620.3	OS	0.8			A
2345.34	2346.06	42624.7	OS	0.7			A
2345.16	2345.88	42628.0	OS	0.6			A
2344.90	2345.62	42632.7	OS	0.6			MIT
2344.74	2345.46	42635.6	OS	1.2	0.5		A
2344.31	2345.03	42643.4	OS	1.3			A
2343.89	2344.61	42651.0	OS	1.3			MIT
2343.74	2344.46	42653.8	OS	1.2			A
2342.78	2343.50	42671.3	OS	1.2	0.8		MIT
2342.654	2343.372	42673.55	OS	0.3H			MIT
2342.48	2343.20	42676.7	OS	1.3	0.3		A
2341.92	2342.64	42686.9	OS	1.5L	0.7		A
2340.69	2341.41	42709.4	OS	1.5L	1.3		MIT
2340.44	2341.16	42713.9	OS	1.0			A
2339.81	2340.53	42725.4	OS	0.7	1.3		MIT
2339.22	2339.94	42736.2	OS	0.8			A
2338.63	2339.35	42747.0	OS	1.5L	0.8		A
2338.01	2338.73	42758.3	OS	1.3			A
2337.78	2338.50	42762.5	OS	1.0			A
2337.64	2338.36	42765.1	OS	1.1	0.6		A
2336.80	2337.52	42780.5	OS	1.7	1.9		MIT
2336.50	2337.22	42785.9	OS	1.2	0.5L		A
2336.19	2336.91	42791.6	OS	0.7	0.9		MIT
2334.25	2334.97	42827.2	OS		0.3W		MIT
2333.50	2334.22	42840.9	OS	1.0	0.3		A
2332.64	2333.36	42856.7	OS	1.0			A
2332.14	2332.86	42865.9	OS	1.5	0.6W		A
2331.45	2332.17	42878.6	OS	0.3	0.6		MIT
2330.91	2331.63	42888.5	OS	0.3H	0.6		MIT
2330.22	2330.93	42901.2	OS		0.9		MIT
2329.71	2330.42	42910.6	OS	1.0	0.3		A
2329.56	2330.27	42913.4	OS	0.5	1.0		A
2329.29	2330.00	42918.4	OS	1.1	0.8W		MIT
2328.17	2328.88	42939.0	OS	0.5			A
2327.90	2328.61	42944.0	OS	1.4	0.5		A
2325.65	2326.36	42985.5	OS	0.8	1.2		MIT
2325.02	2325.73	42997.2	OS	1.2	0.6		A
2324.89	2325.60	42999.6	OS	0.6			A
2324.24	2324.95	43011.6	OS	1.5	1.0		MIT
2323.98	2324.69	43016.4	OS	1.5	1.3		MIT
2321.66	2322.37	43059.4	OS	1.2W	0.5W		A
2321.19	2321.90	43068.1	OS	0.5	1.0W		MIT
2320.95	2321.66	43072.6	OS	0.7			A
2320.18	2320.89	43086.9	OS	1.4	1.1		MIT

AIR WAVELENGTH (ANGSTRMS)	VACUUM WAVELENGTH (ANGSTRMS)	WAVENUMBER (KAYSERS)	LMENT	LOG ARC	INTENSITY SPK	INTENSITY DIS	REF
2317.63	2318.34	43134.3	OS	0.5H			A
2316.38	2317.09	43157.5	OS	1.0	0.0		A
2316.12	2316.83	43162.4	OS	0.7			A
2315.58	2316.29	43172.5	OS	1.0			A
2315.45	2316.16	43174.9	OS	0.9	1.3		MIT
2315.16	2315.87	43180.3	OS	1.0	1.4		MIT
2314.84	2315.55	43186.2	OS	1.3	0.3		A
2314.41	2315.12	43194.3	OS	0.9	0.0		A
2313.75	2314.46	43206.6	OS	1.2	1.5		MIT
2312.11	2312.82	43237.2	OS	0.7	0.50		A
2310.16	2310.87	43273.7	OS	1.0	0.8		MIT
2309.81	2310.52	43280.3	OS		0.7		MIT
2309.40	2310.11	43288.0	OS	1.6	0.5		A
2308.44	2309.15	43306.0	OS	1.1			A
2308.31	2309.02	43308.4	OS	1.3	0.8		MIT
2307.61	2308.32	43321.5	OS	0.7	0.7L		MIT
2306.87	2307.58	43335.5	OS	1.1	0.3		A
2306.48	2307.19	43342.8	OS	0.7W			A
2306.30	2307.01	43346.1	OS	1.0	0.3		A
2306.046	2306.756	43350.93	OS	0.8	1.5		MIT
2305.84	2306.55	43354.8	OS	0.5	0.8		MIT
2304.83	2305.54	43373.8	OS		1.0		MIT
2304.69	2305.40	43376.4	OS	1.0			A
2304.38	2305.09	43382.3	OS	1.6			A
2303.69	2304.40	43395.3	OS	1.1	0.5		A
2303.51	2304.22	43398.6	OS	1.0	0.6		A
2302.62	2303.33	43415.4	OS	1.3	0.3		A
2301.88	2302.59	43429.4	OS	1.4	0.6		A
2301.14	2301.85	43443.3	OS	1.2	0.5		A
2300.86	2301.57	43448.6	OS		1.3		MIT
2299.82	2300.53	43468.3	OS	0.3	1.0		MIT
2299.12	2299.83	43481.5	OS	1.2			A
2298.47	2299.18	43493.8	OS	1.1	0.3		A
2298.15	2298.86	43499.9	OS	0.7	0.7		MIT
2295.56	2296.27	43548.9	OS		0.9		MIT
2293.54	2294.25	43587.3	OS	1.3	1.4		MIT
2293.26	2293.97	43592.6	OS	1.0			A
2292.21	2292.92	43612.6	OS	1.2	0.7W		MIT
2291.93	2292.64	43617.9	OS	0.6			A
2290.13	2290.84	43652.2	OS	1.0			A
2286.517	2287.222	43721.15	OS	1.5	1.0		MIT
2284.92	2285.63	43751.7	OS	1.3			MIT
2284.90	2285.61	43752.1	OS	1.3	0.7		A
2284.76	2285.46	43754.8	OS	1.1			A
2283.67	2284.37	43775.6	OS	1.7	1.2		MIT
2283.17	2283.87	43785.2	OS		1.5		MIT
2283.04	2283.74	43787.7	OS	1.0			A
2282.83	2283.53	43791.8	OS	1.4	0.7		A
2282.26	2282.96	43802.7	OS	2.0	2.1		MIT
2280.93	2281.63	43828.2	OS	1.0			A
2279.85	2280.55	43849.0	OS	1.3	0.8		A
2279.11	2279.81	43863.2	OS	2.0	1.4		A
2278.44	2279.14	43876.1	OS	1.4			A
2276.43	2277.13	43914.9	OS	1.5	0.7		A
2274.28	2274.98	43956.4	OS		1.4		MIT
2272.538	2273.240	43990.07	OS	0.6W	1.7		MIT
2270.61	2271.31	44027.4	OS	1.4	1.0		A
2270.17	2270.87	44036.0	OS	1.8	1.2		A
2269.68	2270.38	44045.5	OS	1.2			A
2269.54	2270.24	44048.2	OS	0.5	1.2		MIT
2268.28	2268.98	44072.6	OS	1.6	1.0L		A
2266.57	2267.27	44105.9	OS	1.3	0.6		A
2265.39	2266.09	44128.9	OS	0.7			MIT
2263.91	2264.61	44157.7	OS		1.2		MIT
2263.367	2264.067	44168.30	OS	0.8	1.2		MIT
2262.32	2263.02	44188.7	OS	1.4	0.8		A
2261.73	2262.43	44200.3	OS	1.4L	1.7L		MIT
2260.35	2261.05	44227.2	OS	0.7			A
2260.11	2260.81	44231.9	OS	1.4	1.7		MIT
2259.63	2260.33	44241.3	OS	1.0	0.3		A

AIR WAVELENGTH (ANGSTRMS)	VACUUM WAVELENGTH (ANGSTRMS)	WAVENUMBER (KAYSERS)	LMENT	LOG ARC	INTENSITY SPK	DIS	REF
2258.975	2259.674	44254.16	OS	0.5	1.4		MIT
2258.332	2259.031	44266.76	OS	1.1	1.5		MIT
2255.847	2256.546	44315.52	OS	2.1	0.3		MIT
2255.41	2256.11	44324.1	OS	1.1			A
2254.69	2255.39	44338.3	OS	1.1	0.5L		A
2254.45	2255.15	44343.0	OS		1.0		MIT
2253.00	2253.70	44371.5	OS	1.0			MIT
2252.68	2253.38	44377.8	OS	1.0			A
2252.22	2252.92	44386.9	OS		0.6		MIT
2252.15	2252.85	44388.3	OS	1.0			A
2250.86	2251.56	44413.7	OS	0.6			MIT
2249.51	2250.21	44440.4	OS	0.7			A
2248.49	2249.19	44460.5	OS	0.8	0.3		A
2247.68	2248.38	44476.5	OS	1.6	0.8		A
2245.25	2245.95	44524.7	OS	1.3			A
2243.94	2244.64	44550.6	OS	0.9	1.3		MIT
2243.54	2244.24	44558.6	OS	1.0			A
2243.32	2244.02	44563.0	OS	0.3	1.0		A
2242.10	2242.80	44587.2	OS	1.5	0.7		A
2241.62	2242.32	44596.7	OS	1.4	0.3W		A
2240.14	2240.84	44626.2	OS	0.9			A
2239.72	2240.42	44634.6	OS	1.3	0.3		A
2239.12	2239.82	44646.5	OS	0.7			A
2235.89	2236.58	44711.0	OS	1.0	0.5		A
2235.65	2236.34	44715.8	OS	1.2	0.7		A
2234.61	2235.30	44736.6	OS	1.5	1.0		A
2232.82	2233.51	44772.5	OS	1.0	0.0		A
2231.16	2231.85	44805.8	OS	1.1	1.7		MIT
2230.25	2230.94	44824.1	OS	1.3			A
2229.11	2229.80	44847.0	OS	0.7	0.5		A
2228.53	2229.22	44858.7	OS	1.4	0.7		A
2227.98	2228.67	44869.7	OS	1.5	1.5		A
2226.83	2227.52	44892.9	OS	1.4	1.2		MIT
2226.23	2226.92	44905.0	OS	1.3	0.7		A
2225.44	2226.13	44921.0	OS	1.4	0.7		A
2225.27	2225.96	44924.4	OS	0.9	1.3		MIT
2224.57	2225.26	44938.5	OS	1.1	0.5		A
2223.85	2224.54	44953.1	OS	1.5	1.0		A
2223.08	2223.77	44968.6	OS	1.0			A
2222.53	2223.22	44979.8	OS	1.0	0.0		A
2222.37	2223.06	44983.0	OS	0.9			A
2221.36	2222.05	45003.5	OS	1.2	0.3		A
2220.99	2221.68	45011.0	OS	1.2	0.3		A
2219.82	2220.51	45034.7	OS	0.8	0.6		A
2218.46	2219.15	45062.3	OS	1.0			A
2218.30	2218.99	45065.5	OS	0.7	1.3		MIT
2217.23	2217.92	45087.3	OS	1.3	0.5		A
2216.52	2217.21	45101.7	OS	0.7	0.7		A
2215.37	2216.06	45125.1	OS	1.3	0.7		A
2214.76	2215.45	45137.6	OS	0.7	1.5		MIT
2214.16	2214.85	45149.8	OS	1.4	1.0		A
2213.56	2214.25	45162.0	OS	1.0	0.7J		A
2213.34	2214.03	45166.5	OS		1.2		A
2212.22	2212.91	45189.4	OS	1.0			A
2212.11	2212.80	45191.6	OS		0.8		A
2211.96	2212.65	45194.7	OS	1.6	0.7		A
2211.64	2212.33	45201.2	OS	0.7			A
2211.51	2212.20	45203.9	OS	0.7	0.6W		A
2210.53	2211.22	45223.9	OS	1.2	0.0		A
2209.50	2210.19	45245.0	OS	0.7	1.0		A
2208.46	2209.15	45266.3	OS	1.1	0.3		A
2208.27	2208.96	45270.2	OS	1.0	1.2		MIT
2207.48	2208.17	45286.4	OS	1.2	0.6		A
2206.27	2206.96	45311.2	OS	1.4	1.7		MIT
2205.74	2206.43	45322.1	OS	1.3	1.7		MIT
2205.17	2205.86	45333.8	OS	1.2			A
2204.85	2205.54	45340.4	OS	0.3	1.6		MIT
2203.91	2204.60	45359.7	OS	1.4	0.7		A
2203.75	2204.44	45363.0	OS	0.7			A
2202.49	2203.18	45389.0	OS	1.4	0.7		A

AIR WAVELENGTH (ANGSTRMS)	VACUUM WAVELENGTH (ANGSTRMS)	WAVENUMBER (KAYSERS)	LMENT	LOG ARC	INTENSITY SPK	DIS	REF
2201.93	2202.62	45400.5	OS	1.4	0.5		A
2200.80	2201.49	45423.8	OS	1.0	1.1		A
2199.43	2200.12	45452.1	OS	1.2	0.7		A
2198.16	2198.85	45478.4	OS	1.0	0.7		A
2196.49	2197.18	45513.0	OS	0.9			A
2194.39	2195.08	45556.5	OS	1.6	2.0		MIT
2194.11	2194.80	45562.3	OS	1.0			A
2193.32	2194.01	45578.7	OS	1.0	0.3		A
2189.54	2190.22	45657.4	OS	1.0			A
2188.97	2189.65	45669.3	OS	1.7			A
2188.06	2188.74	45688.3	OS	1.1	0.7		A
2184.68	2185.36	45759.0	OS	1.4	1.2		A
2184.14	2184.82	45770.3	OS	1.4	0.8		A
2183.94	2184.62	45774.5	OS	1.0	0.7		A
2183.69	2184.37	45779.7	OS	1.3	0.8		A
2181.50	2182.18	45825.7	OS	1.4	0.7		A
2178.68	2179.36	45885.0	OS	0.7			A
2178.17	2178.85	45895.7	OS	0.9			A
2177.10	2177.78	45918.3	OS	1.0	0.6		A
2176.67	2177.35	45927.4	OS	1.2	0.5		A
2174.24	2174.92	45978.7	OS	1.0			A
2173.83	2174.51	45987.3	OS	1.0	0.7J		A
2173.49	2174.17	45994.5	OS	1.0	1.1		A
2172.84	2173.52	46008.3	OS	1.2	1.3		A
2172.31	2172.99	46019.5	OS	0.7	1.3		A
2171.65	2172.33	46033.5	OS	1.3	0.7		A
2169.81	2170.49	46072.5	OS	0.7	0.0		A
2169.53	2170.21	46078.5	OS	1.0			A
2167.75	2168.43	46116.3	OS	1.7	1.0		A
2167.09	2167.77	46130.4	OS	1.0	0.3		A
2166.90	2167.58	46134.4	OS	1.3	1.0		A
2166.47	2167.15	46143.5	OS	1.0	0.3		A
2166.31	2166.99	46147.0	OS	0.9	0.3		A
2166.05	2166.73	46152.5	OS	1.3	0.3J		A
2165.19	2165.87	46170.8	OS	1.6	0.7		A
2164.85	2165.53	46178.1	OS	1.4	1.7		MIT
2164.49	2165.17	46185.8	OS	1.4	0.5		A
2162.28	2162.96	46233.0	OS	1.0	0.0		A
2161.00	2161.68	46260.3	OS	1.7	0.0		A
2159.98	2160.66	46282.2	OS	1.8	0.7		A
2159.52	2160.20	46292.0	OS	1.8			A
2158.53	2159.21	46313.3	OS	1.7	1.4		A
2158.04	2158.72	46323.8	OS	0.5	1.4		A
2157.84	2158.52	46328.1	OS	1.5	0.6		A
2157.08	2157.76	46344.4	OS	1.4	0.5		A
2156.88	2157.56	46348.7	OS	0.6	1.3		A
2156.31	2156.99	46361.0	OS	1.7	0.5		A
2155.79	2156.47	46372.1	OS	1.6	0.3		A
2154.83	2155.51	46392.8	OS	1.0			A
2154.59	2155.27	46398.0	OS	1.4	1.0		A
2153.63	2154.31	46418.6	OS	1.0	0.0		A
2150.42	2151.10	46487.9	OS	0.7	1.2		A
2150.13	2150.81	46494.2	OS	1.0	0.3		A
2149.97	2150.65	46497.6	OS	1.2	1.0		A
2149.81	2150.49	46501.1	OS	1.0	0.3		A
2149.20	2149.88	46514.3	OS		1.0		A
2148.62	2149.30	46526.9	OS	0.5	1.2		A
2148.27	2148.95	46534.4	OS	0.8			A
2147.36	2148.04	46554.1	OS	0.7	0.9D		A
2147.27	2147.95	46556.1	OS	0.3	0.9D		A
2145.26	2145.94	46599.7	OS	1.0			A
2143.23	2143.91	46643.9	OS	1.1	0.5		A
2142.73	2143.41	46654.7	OS	0.6	0.5		A
2142.38	2143.06	46662.4	OS	0.6			A
2142.13	2142.80	46667.8	OS		0.7S		A
2141.26	2141.93	46686.8	OS	0.6			A
2141.10	2141.77	46690.2	OS	0.9			A
2139.97	2140.64	46714.9	OS	0.5			A
2139.17	2139.84	46732.4	OS		1.0		A
2138.75	2139.42	46741.5	OS	0.5	0.0		A

AIR WAVELENGTH (ANGSTRMS)	VACUUM WAVELENGTH (ANGSTRMS)	WAVENUMBER (KAYSERS)	LMENT	LOG ARC	INTENSITY SPK	INTENSITY DIS	REF
2138.61	2139.28	46744.6	OS	0.9			A
2138.40	2139.07	46749.2	OS	0.5	0.6		A
2137.94	2138.61	46759.2	OS	0.9			A
2137.11	2137.78	46777.4	OS	2.0	1.5		A
2136.81	2137.48	46784.0	OS	0.8			A
2136.69	2137.36	46786.6	OS	1.3	0.0H		A
2134.66	2135.33	46831.1	OS	1.0	0.3		A
2134.38	2135.05	46837.2	OS		0.9		A
2133.58	2134.25	46854.8	OS	0.5	0.7		A
2133.31	2133.98	46860.7	OS	0.5	0.3		A
2132.67	2133.34	46874.8	OS	1.0	0.5		A
2132.49	2133.16	46878.7	OS	0.3	1.0		A
2132.26	2132.93	46883.8	OS	1.0	0.3		A
2131.48	2132.15	46901.0	OS	0.6	1.4		A
2130.58	2131.25	46920.8	OS	1.0	0.3		A
2129.48	2130.15	46945.0	OS	0.5	0.5		A
2128.92	2129.59	46957.3	OS	0.7			A
2128.78	2129.45	46960.4	OS	0.9	1.0L		A
2128.28	2128.95	46971.5	OS		0.7		A
2127.97	2128.64	46978.3	OS	0.7	1.7		A
2125.34	2126.01	47036.4	OS	1.4	0.6		A
2124.82	2125.49	47047.9	OS	0.7			A
2124.39	2125.06	47057.5	OS	1.4	0.7		A
2123.84	2124.51	47069.6	OS	1.6	1.4W		A
2123.34	2124.01	47080.7	OS	1.0			A
2123.15	2123.82	47084.9	OS	0.7	0.9		A
2122.83	2123.50	47092.0	OS	1.1	0.5		A
2122.02	2122.69	47110.0	OS	0.7			A
2120.01	2120.68	47154.7	OS	1.4	0.6		A
2119.79	2120.46	47159.6	OS	2.1	1.5		A
2119.31	2119.98	47170.2	OS	1.2	0.6		A
2118.82	2119.49	47181.1	OS	1.2	1.0		A
2117.96	2118.63	47200.3	OS	1.9	1.3		A
2117.66	2118.33	47207.0	OS	1.6	1.0		A
2117.49	2118.16	47210.8	OS	0.5			A
2117.03	2117.70	47221.0	OS	1.0	0.9		A
2116.34	2117.01	47236.4	OS		1.5		A
2114.97	2115.64	47267.0	OS	0.7			A
2114.00	2114.67	47288.7	OS	0.5	0.3		A
2113.59	2114.26	47297.9	OS	1.3	0.6		A
2113.45	2114.12	47301.0	OS	0.5			A
2112.58	2113.25	47320.5	OS	0.7	0.6		A
2111.40	2112.07	47346.9	OS	1.0	0.3H		A
2110.95	2111.62	47357.0	OS		0.7		A
2110.74	2111.41	47361.7	OS	1.2	0.3		A
2109.25	2109.92	47395.2	OS	0.5	1.0		A
2108.44	2109.11	47413.4	OS	0.7			A
2107.52	2108.19	47434.1	OS	1.2	0.5		A
2107.30	2107.97	47439.0	OS	0.9	0.5		A
2107.01	2107.68	47445.6	OS	0.3			A
2104.41	2105.08	47504.2	OS	0.8			A
2102.91	2103.58	47538.1	OS	0.7	1.4		A
2102.63	2103.30	47544.4	OS	0.7			A
2101.38	2102.05	47572.7	OS	1.4	0.5W		A
2101.18	2101.85	47577.2	OS	0.6			A
2100.63	2101.30	47589.7	OS	1.2	0.8		A
2099.92	2100.59	47605.7	OS	1.0	0.5		A
2099.51	2100.18	47615.0	OS	1.0	0.3		A
2099.25	2099.92	47620.9	OS	0.6			A
2097.60	2098.27	47658.4	OS	1.4	0.9W		A
2096.57	2097.24	47681.8	OS	1.0	0.0		A
2096.28	2096.95	47688.4	OS	1.1	1.2		A
2092.12	2092.79	47783.2	OS	0.5			A
2091.90	2092.57	47788.2	OS	0.0	1.0		A
2091.34	2092.00	47801.0	OS	0.6	0.3H		A
2090.71	2091.37	47815.4	OS	1.1			A
2089.21	2089.87	47849.8	OS	1.3	0.6		A
2089.03	2089.69	47853.9	OS	1.3	0.6		A
2088.31	2088.97	47870.4	OS	0.6			A
2087.84	2088.50	47881.2	OS	1.3	0.6		A

AIR WAVELENGTH (ANGSTRMS)	VACUUM WAVELENGTH (ANGSTRMS)	WAVENUMBER (KAYSERS)	LMENT	LOG ARC	INTENSITY SPK	INTENSITY DIS	REF
2086.62	2087.28	47909.1	OS	0.7			A
2086.23	2086.89	47918.1	OS	0.3	0.7		A
2084.51	2085.17	47957.6	OS	0.9	0.0		MIT
2083.58	2084.24	47979.0	OS	1.2	0.3		A
2083.20	2083.86	47987.8	OS	1.2	0.3		A
2082.54	2083.20	48003.0	OS	1.3	0.7		A
2081.72	2082.38	48021.9	OS	0.5			A
2081.46	2082.12	48027.9	OS	1.3	1.0		A
2080.28	2080.94	48055.1	OS	0.6	0.3		A
2079.97	2080.63	48062.3	OS	1.6	1.0		A
2079.21	2079.87	48079.9	OS	1.0	0.0		A
2078.99	2079.65	48085.0	OS	0.7	1.3		A
2078.49	2079.15	48096.5	OS	1.0	0.0		A
2078.09	2078.75	48105.8	OS	1.4	1.0		A
2077.33	2077.99	48123.4	OS	1.2	0.5		A
2076.95	2077.61	48132.2	OS	1.4	1.1		A
2076.10	2076.76	48151.9	OS	0.6			A
2075.85	2076.51	48157.7	OS	0.7			A
2075.65	2076.31	48162.3	OS	0.8			A
2074.95	2075.61	48178.6	OS	1.2	0.7		A
2074.10	2074.76	48198.3	OS	0.5			A
2073.92	2074.58	48202.5	OS	1.0			A
2073.84	2074.50	48204.4	OS		1.0		A
2071.28	2071.94	48263.9	OS	0.6	1.0		A
2070.67	2071.33	48278.1	OS	1.2	1.5		A
2070.42	2071.08	48284.0	OS	1.1	0.3		A
2069.75	2070.41	48299.6	OS	0.7	0.3		A
2069.61	2070.27	48302.9	OS	1.4	0.6		A
2068.38	2069.04	48331.6	OS	1.0	1.0		A
2067.71	2068.37	48347.2	OS	1.0	0.6		A
2067.21	2067.87	48358.9	OS	1.0	1.6		A
2067.11	2067.77	48361.3	OS	0.5			A
2066.35	2067.01	48379.1	OS	0.3			A
2065.80	2066.46	48391.9	OS	0.7	0.3W		A
2064.24	2064.90	48428.5	OS	0.6			A
2063.55	2064.21	48444.7	OS	1.4	0.3W		A
2063.20	2063.86	48452.9	OS	0.7			A
2062.16	2062.82	48477.3	OS	1.1	0.5		A
2061.69	2062.35	48488.4	OS	1.3	0.8		A
2060.40	2061.06	48518.7	OS	1.0			A
2060.08	2060.74	48526.3	OS	1.0			A
2059.79	2060.45	48533.1	OS	0.6			A
2059.65	2060.31	48536.4	OS	1.0	0.6		A
2059.50	2060.16	48540.0	OS	0.5			A
2058.78	2059.44	48556.9	OS	1.0	0.5		A
2058.69	2059.35	48559.0	OS	1.0	0.6		A
2057.48	2058.14	48587.6	OS	0.8			A
2055.53	2056.19	48633.7	OS	0.7	1.3		A
2055.40	2056.06	48636.8	OS	0.5	0.7		A
2054.85	2055.51	48649.8	OS	1.0	0.3		A
2054.60	2055.26	48655.7	OS	0.5			A
2053.84	2054.50	48673.7	OS	0.8	0.7		A
2052.78	2053.44	48698.8	OS	0.5	0.7		A
2052.40	2053.06	48707.8	OS	1.0	0.3		A
2050.61	2051.27	48750.4	OS	0.5H	0.3H		A
2049.95	2050.61	48766.0	OS	0.8	0.3		A
2049.42	2050.08	48778.7	OS	1.2	0.7		A
2048.80	2049.46	48793.4	OS	0.8	1.4		A
2048.62	2049.28	48797.7	OS	0.7			A
2048.28	2048.94	48805.8	OS	1.4	0.7		A
2048.02	2048.68	48812.0	OS	0.6			A
2047.77	2048.43	48818.0	OS	0.7			A
2046.81	2047.47	48840.9	OS	0.7	0.7		A
2046.28	2046.94	48853.5	OS	0.5	1.0		A
2045.36	2046.02	48875.5	OS	1.7	1.3		A
2044.89	2045.55	48886.7	OS	1.1	0.5		A
2044.76	2045.42	48889.8	OS	0.7			A
2044.36	2045.02	48899.4	OS	0.5	0.7		A
2043.68	2044.34	48915.6	OS	0.9			A
2043.50	2044.16	48920.0	OS	1.0			A

AIR WAVELENGTH (ANGSTRMS)	VACUUM WAVELENGTH (ANGSTRMS)	WAVENUMBER (KAYSERS)	LMENT	LOG ARC	INTENSITY SPK	DIS	REF
2043.18	2043.84	48927.6	OS	0.6			A
2043.00	2043.66	48931.9	OS	0.5	1.2		A
2042.75	2043.41	48937.9	OS	0.7			A
2042.44	2043.10	48945.3	OS	0.8	0.3		A
2040.68	2041.34	48987.5	OS	1.0	0.6		A
2038.72	2039.37	49034.6	OS	1.2	0.7		A
2038.16	2038.81	49048.1	OS	0.6			A
2037.62	2038.27	49061.1	OS	0.3	0.7W		A
2036.49	2037.14	49088.3	OS	1.0	0.5		A
2035.93	2036.58	49101.8	OS	1.2	0.6		A
2034.44	2035.09	49137.8	OS	1.5	0.7		A
2034.29	2034.94	49141.4	OS	0.7	0.5W		A
2033.53	2034.18	49159.8	OS	1.0	1.4		A
2032.73	2033.38	49179.1	OS	1.1	0.5		A
2032.51	2033.16	49184.4	OS	0.7	0.0		A
2032.39	2033.04	49187.3	OS	0.7	0.5		A
2030.19	2030.84	49240.6	OS	0.8			A
2029.18	2029.83	49265.1	OS	0.7	0.0		A
2028.23	2028.88	49288.2	OS	1.3	0.7		A
2027.45	2028.10	49307.2	OS	1.0	0.3		A
2026.54	2027.19	49329.3	OS	0.7	0.0		A
2026.16	2026.81	49338.5	OS	0.7			A
2025.69	2026.34	49350.0	OS	0.8			A
2025.12	2025.77	49363.9	OS	0.8	0.3		A
2024.49	2025.14	49379.2	OS	1.0	0.3		A
2023.72	2024.37	49398.0	OS	1.0	0.3		A
2023.09	2023.74	49413.4	OS	0.7			A
2022.76	2023.41	49421.5	OS	1.2	0.8		A
2022.16	2022.81	49436.1	OS	0.7			A
2021.61	2022.26	49449.6	OS	0.0	0.3		A
2021.28	2021.93	49457.6	OS	0.9			A
2020.26	2020.91	49482.6	OS	1.4	1.0		A
2020.04	2020.69	49488.0	OS	1.0	0.3		A
2019.87	2020.52	49492.2	OS	0.5	0.3		A
2019.40	2020.05	49503.7	OS	0.7	0.3		A
2019.08	2019.73	49511.5	OS	0.3	0.7		A
2018.14	2018.79	49534.6	OS	1.3	0.7		A
2018.00	2018.65	49538.0	OS	0.7	0.5W		A
2017.34	2017.99	49554.2	OS	0.3			A
2017.13	2017.78	49559.4	OS	0.3	0.8		A
2016.62	2017.27	49571.9	OS	0.5	1.2		A
2015.98	2016.63	49587.7	OS	0.8	0.3		A
2015.78	2016.43	49592.6	OS	0.9	0.7		A
2015.36	2016.01	49602.9	OS	0.7	0.3		A
2013.42	2014.07	49650.7	OS	0.7	0.3		A
2013.09	2013.74	49658.8	OS	0.7	0.5W		A
2012.46	2013.11	49674.4	OS	0.6			A
2010.15	2010.80	49731.5	OS	1.0	0.6		A
2009.65	2010.30	49743.8	OS	0.6	1.3		A
2009.35	2010.00	49751.2	OS	0.5			A
2007.97	2008.62	49785.4	OS	0.7	0.6		A
2006.16	2006.81	49830.4	OS	0.7			A
2005.87	2006.52	49837.5	OS	0.6	0.9		A
2004.78	2005.43	49864.6	OS	1.0			A
2003.92	2004.57	49886.0	OS	0.7	0.5		A
2003.73	2004.38	49890.8	OS	0.7	1.0		A
2003.51	2004.16	49896.2	OS	0.7	0.5		A
2001.89	2002.54	49936.6	OS	1.2	0.3W		A
2001.45	2002.10	49947.6	OS	1.0	0.5		A
2001.26	2001.91	49952.3	OS	0.5			A
2001.09	2001.74	49956.6	OS	0.5	0.5		A
9976.65	9979.39	10020.7	P I	1.0			KS
9903.74	9906.46	10094.4	P I	1.3			KS
9834.61	9837.31	10165.4	P	0.3			KS
9796.79	9799.48	10204.6	P I	2.0			KS
9790.08	9792.76	10211.6	P I	0.7			KS
9750.73	9753.40	10252.8	P I	1.8			KS
9734.74	9737.41	10269.7	P I	1.7			KS
9706.80	9709.46	10299.2	P I	0.3			KS
9676.27	9678.92	10331.7	P	0.5			KS

AIR WAVELENGTH (ANGSTRMS)	VACUUM WAVELENGTH (ANGSTRMS)	WAVENUMBER (KAYSERS)	LMENT	LOG ARC	INTENSITY SPK	DIS	REF
9639.06	9641.70	10371.6	P	0.3			KS
9608.97	9611.61	10404.1	P I	0.7			KS
9593.54	9596.17	10420.8	P I	1.8			KS
9563.45	9566.07	10453.6	P I	1.4			KS
9545.27	9547.89	10473.5	P I	1.3			KS
9525.78	9528.39	10494.9	P I	2.0			KS
9493.48	9496.08	10530.7	P I	0.8			KS
9452.87	9455.46	10575.9	P	0.3			KS
9441.76	9444.35	10588.3	P	0.7			KS
9435.07	9437.66	10595.8	P I	0.5			KS
9323.55	9326.11	10722.6	P I	0.5			KS
9304.88	9307.43	10744.1	P I	0.7			KS
9278.82	9281.37	10774.3	P	0.3			KS
9193.86	9196.38	10873.8	P I	0.3			KS
6752.1	6754.0	14806.	P	1.2			SA
6460.1	6461.9	15475.	P II			1.5	DJ
6435.5	6437.3	15534.	P II			0.7	DJ
6165.2	6166.9	16216.	P II			1.2	DJ
6087.4	6089.1	16423.	P II			1.2	DJ
6055.2	6056.9	16510.	P II			0.7	DJ
6043.05	6044.72	16543.4	P II			2.2	SA
6033.9	6035.6	16568.	P II			1.5	DJ
6024.14	6025.81	16595.3	P II			1.7	SA
5727.68	5729.27	17454.2	P II			1.2L	GU
5676.89	5678.47	17610.4	P			1.3	GU
5588.25	5589.80	17889.7	P II			1.8L	GU
5583.33	5584.88	17905.5	P II			1.8L	GU
5552.63	5554.17	18004.5	P			1.2	GU
5544.49	5546.03	18030.9	P			1.7	GU
5541.18	5542.72	18041.7	P II			1.7	GU
5533.20	5534.74	18067.7	P			1.2	GU
5517.08	5518.61	18120.5	P			1.5	GU
5514.77	5516.30	18128.1	P			1.5	GU
5507.13	5508.66	18153.2	P II			1.8L	GU
5502.83	5504.36	18167.4	P			1.5H	GU
5499.71	5501.24	18177.7	P II			2.2	GU
5483.55	5485.07	18231.3	P II			1.8L	GU
5477.74	5479.26	18250.6	P			1.5	GU
5460.85	5462.37	18307.1	P			2.0	GU
5458.34	5459.86	18315.5	P			1.5	GU
5450.65	5452.16	18341.3	P II			2.0L	GU
5447.09	5448.60	18353.3	P			1.2	GU
5437.28	5438.79	18386.4	P			1.8	GU
5433.18	5434.69	18400.3	P			1.2	GU
5425.92	5427.43	18424.9	P II			2.2W	GU
5409.65	5411.15	18480.4	P II			2.2W	GU
5386.87	5388.37	18558.5	P			2.2W	GU
5378.11	5379.61	18588.7	P II			1.5L	GU
5375.15	5376.64	18599.0	P			1.2	GU
5371.4	5372.9	18612.	P			1.5L	GU
5368.71	5370.20	18621.3	P			1.5	GU
5364.47	5365.96	18636.0	P			1.2	GU
5345.81	5347.30	18701.0	P II			1.7	GU
5344.72	5346.21	18704.9	P			2.2W	GU
5316.07	5317.55	18805.7	P II			2.2W	GU
5303.21	5304.69	18851.3	P			1.7	GU
5296.09	5297.56	18876.6	P II			2.5W	GU
5293.63	5295.10	18885.4	P II			1.5	GU
5287.97	5289.44	18905.6	P			1.2	GU
5254.93	5256.39	19024.4	P			1.2	GU
5253.48	5254.94	19029.7	P II			2.5W	GU
5242.63	5244.09	19069.1	P			1.5	GU
5235.52	5236.98	19095.0	P			1.8	GU
5191.40	5192.85	19257.3	P II			2.0	GU
5176.38	5177.82	19313.1	P			1.8	GU
5162.27	5163.71	19365.9	P			1.5	GU
5161.97	5163.41	19367.1	P			1.5	GU
5156.72	5158.16	19386.8	P			1.7	GU
5154.84	5156.28	19393.8	P			1.0	GU
5152.20	5153.64	19403.8	P II			1.7	GU

AIR WAVELENGTH (ANGSTRMS)	VACUUM WAVELENGTH (ANGSTRMS)	WAVENUMBER (KAYSERS)	LMENT		LOG ARC	INTENSITY SPK	INTENSITY DIS	REF
5149.41	5150.84	19414.3	P				1.2	GU
5141.49	5142.92	19444.2	P				1.7	GU
5120.12	5121.55	19525.4	P	II			0.7	GU
5111.95	5113.37	19556.6	P				1.5	GU
5109.5	5110.9	19566.	P				1.2	GU
5108.6	5110.0	19569.	P				1.2	GU
5104.13	5105.55	19586.5	P				1.5	GU
5100.91	5102.33	19598.9	P				1.0	GU
5079.27	5080.69	19682.4	P				1.2	GU
5067.52	5068.93	19728.0	P				1.2	GU
5055.69	5057.10	19774.2	P				1.2	GU
5051.59	5053.00	19790.2	P				1.5	GU
5040.74	5042.15	19832.8	P				1.8L	GU
4969.64	4971.03	20116.6	P	II			2.2L	GU
4954.32	4955.70	20178.8	P	II			1.8L	GU
4943.41	4944.79	20223.3	P	II			2.2L	GU
4935.55	4936.93	20255.5	P			1.2		GU
4927.16	4928.54	20290.0	P	II			1.7L	GU
4888.52	4889.89	20450.4	P				1.5D	GU
4883.65	4885.01	20470.8	P				1.5	GU
4876.98	4878.34	20498.8	P				1.7L	GU
4872.33	4873.69	20518.3	P				1.7	GU
4864.38	4865.74	20551.9	P	II			1.7L	GU
4862.83	4864.19	20558.4	P				1.2	GU
4854.69	4856.05	20592.9	P				1.8	GU
4844.25	4845.60	20637.3	P				1.5	GU
4819.34	4820.69	20743.9	P				1.2	GU
4814.2	4815.5	20766.	P	II			1.5J	GU
4805.96	4807.30	20801.7	P	II			0.7H	GU
4792.06	4793.40	20862.0	P	II			1.8L	GU
4778.80	4780.14	20919.9	P				1.5	GU
4764.84	4766.17	20981.2	P				1.2	GU
4763.91	4765.24	20985.3	P				1.2	GU
4755.79	4757.12	21021.1	P				1.5H	GU
4739.49	4740.82	21093.4	P	II			1.5L	GU
4734.27	4735.59	21116.7	P				1.2H	GU
4727.46	4728.78	21147.1	P				2.0	GU
4724.25	4725.57	21161.5	P				1.9	GU
4720.26	4721.58	21179.4	P	II			1.5	GU
4717.00	4718.32	21194.0	P	II			1.2H	GU
4700.79	4702.11	21267.1	P	II			1.7L	GU
4698.56	4699.87	21277.2	P				1.5	GU
4678.94	4680.25	21366.4	P	II			2.0L	GU
4675.78	4677.09	21380.8	P				1.8	GU
4661.78	4663.09	21445.0	P				1.2	GU
4658.11	4659.41	21461.9	P	II			2.0L	GU
4649.05	4650.35	21503.7	P				1.7	GU
4641.72	4643.02	21537.7	P				1.7	GU
4637.16	4638.46	21558.9	P				1.7	GU
4628.70	4630.00	21598.3	P	II			1.7	GU
4626.60	4627.90	21608.1	P	II			1.8	GU
4622.70	4623.99	21626.3	P	II			1.7H	GU
4613.8	4615.1	21668.	P				1.5	GU
4612.83	4614.12	21672.6	P	II			1.5	GU
4601.96	4603.25	21723.8	P	II			2.5W	GU
4595.98	4597.27	21752.1	P	II			1.5	GU
4595.53	4596.82	21754.2	P				1.0	GU
4589.78	4591.07	21781.4	P	II			2.5W	GU
4587.90	4589.19	21790.4	P	II			2.5W	GU
4581.76	4583.04	21819.6	P	II			1.5H	GU
4574.90	4576.18	21852.3	P				1.5	GU
4565.21	4566.49	21898.7	P	II			2.0L	GU
4561.93	4563.21	21914.4	P	II			1.2	GU
4558.03	4559.31	21933.1	P	II			2.0L	GU
4554.80	4556.08	21948.7	P	II			2.0	GU
4548.40	4549.68	21979.6	P				1.7	GU
4546.03	4547.30	21991.1	P				1.8	GU
4541.12	4542.39	22014.8	P				1.8	GU
4540.20	4541.47	22019.3	P				1.8	GU
4533.81	4535.08	22050.3	P	II			1.2	GU

AIR WAVELENGTH (ANGSTRMS)	VACUUM WAVELENGTH (ANGSTRMS)	WAVENUMBER (KAYSERS)	LMENT		LOG ARC	INTENSITY SPK	INTENSITY DIS	REF
4530.78	4532.05	22065.1	P	II			2.2L	GU
4523.73	4525.00	22099.4	P				1.7	GU
4522.92	4524.19	22103.4	P	II			1.7	GU
4520.49	4521.76	22115.3	P				1.2	GU
4512.71	4513.98	22153.4	P				1.2	GU
4507.55	4508.81	22178.8	P				1.5	GU
4504.06	4505.32	22196.0	P				1.5	GU
4501.2	4502.5	22210.	P				1.2	GU
4499.17	4500.43	22220.1	P	II			2.2L	GU
4485.29	4486.55	22288.9	P				1.7	GU
4483.66	4484.92	22296.9	P	II			1.5	GU
4479.74	4481.00	22316.5	P				1.8	GU
4475.26	4476.52	22338.8	P	II			2.2L	GU
4467.97	4469.22	22375.2	P	II			1.7	GU
4466.10	4467.35	22384.6	P	II			1.5	GU
4463.70	4464.95	22396.6	P	II			1.8	GU
4462.94	4464.19	22400.5	P				1.8	GU
4452.44	4453.69	22453.3	P	II			2.2L	GU
4443.87	4445.12	22496.6	P				1.7	GU
4428.15	4429.39	22576.5	P				1.8	GU
4425.94	4427.18	22587.7	P	II			1.2	GU
4423.9	4425.1	22598.	P	II			1.5	GU
4423.55	4424.79	22599.9	P				1.5	GU
4420.64	4421.88	22614.8	P				1.8	GU
4417.30	4418.54	22631.9	P	II			1.5	GU
4414.60	4415.84	22645.7	P	II			1.8	GU
4414.28	4415.52	22647.4	P				2.0	GU
4411.94	4413.18	22659.4	P				1.2H	GU
4400.99	4402.23	22715.8	P				1.7L	GU
4385.33	4386.56	22796.9	P				2.0L	GU
4352.22	4353.44	22970.3	P				1.5	GU
4316.84	4318.05	23158.6	P				1.5	GU
4294.11	4295.32	23281.2	P	II			1.2	GU
4288.52	4289.73	23311.5	P	II			1.7	GU
4249.57	4250.77	23525.2	P				2.0	GU
4246.88	4248.08	23540.1	P		1.8		2.2W	SA
4244.55	4245.75	23553.0	P	II			1.5	GU
4224.43	4225.62	23665.2	P	II			1.2	GU
4222.15	4223.34	23677.9	P		2.5		2.2W	GU
4216.56	4217.75	23709.3	P	II			1.2	GU
4202.24	4203.42	23790.1	P				1.5H	GU
4199.6	4200.8	23805.	P	II			0.7H	GU
4193.42	4194.60	23840.2	P	II			0.7H	GU
4189.08	4190.26	23864.9	P				1.5H	GU
4188.07	4189.25	23870.6	P				1.5H	GU
4178.36	4179.54	23926.1	P				2.5W	GU
4172.79	4173.97	23958.0	P	II			0.7	GU
4166.73	4167.90	23992.9	P	II			1.2H	GU
4161.64	4162.81	24022.2	P				1.2	GU
4160.56	4161.73	24028.4	P	II			1.5	GU
4157.33	4158.50	24047.1	P				1.5H	GU
4152.4	4153.6	24076.	P				1.2	GU
4151.19	4152.36	24082.7	P				1.2	GU
4143.84	4145.01	24125.4	P				1.7	GU
4137.3	4138.5	24163.	P				1.2H	GU
4130.77	4131.94	24201.7	P	II			1.5	GU
4127.49	4128.65	24221.0	P	II			1.8	GU
4120.78	4121.94	24260.4	P	II			1.2H	GU
4118.96	4120.12	24271.1	P	II			1.2	GU
4117.09	4118.25	24282.1	P	II			1.7	GU
4109.19	4110.35	24328.8	P	II			1.8	GU
4094.43	4095.59	24416.5	P				1.5	GU
4091.53	4092.68	24433.8	P	II			1.5L	GU
4089.25	4090.40	24447.5	P				1.5	GU
4080.04	4081.19	24502.6	P				2.2	GU
4075.74	4076.89	24528.5	P				1.2	GU
4072.13	4073.28	24550.2	P	II			1.5	GU
4064.64	4065.79	24595.5	P	II			1.5	GU
4062.08	4063.23	24611.0	P	II			1.2H	GU
4059.27	4060.42	24628.0	P				2.0	GU

AIR WAVELENGTH (ANGSTRMS)	VACUUM WAVELENGTH (ANGSTRMS)	WAVENUMBER (KAYSERS)	LMENT	LOG ARC	INTENSITY SPK	INTENSITY DIS	REF
4057.39	4058.54	24639.4	P			1.7	GU
4044.49	4045.63	24718.0	P II			2.2W	GU
4036.22	4037.36	24768.7	P II			1.2H	GU
4033.68	4034.82	24784.2	P II			1.2	GU
4019.45	4020.59	24872.0	P II			1.7	GU
4013.80	4014.93	24907.0	P			1.5	GU
3997.16	3998.29	25010.7	P			1.8	GU
3985.86	3986.99	25081.6	P			1.5	GU
3973.10	3974.22	25162.1	P			1.2	GU
3967.54	3968.66	25197.4	P			1.2	GU
3957.62	3958.74	25260.6	P			2.0	GU
3951.50	3952.62	25299.7	P			1.8	GU
3933.37	3934.48	25416.3	P			1.7	GU
3927.29	3928.40	25455.6	P			1.5	GU
3922.71	3923.82	25485.4	P			1.7	GU
3914.26	3915.37	25540.4	P			2.0L	GU
3904.78	3905.89	25602.4	P			2.0	GU
3895.02	3896.12	25666.5	P			2.0	GU
3885.17	3886.27	25731.6	P			2.2L	GU
3837.14	3838.23	26053.7	P			1.5	GU
3827.44	3828.53	26119.7	P II			2.2L	GU
3802.07	3803.15	26294.0	P			2.0	GU
3795.09	3796.17	26342.4	P II			1.5	GU
3793.60	3794.68	26352.7	P II			1.5	GU
3788.06	3789.14	26391.2	P II			1.2H	GU
3786.69	3787.77	26400.8	P II			1.2	GU
3775.02	3776.09	26482.4	P II			1.5	GU
3768.70	3769.77	26526.8	P II			1.7	GU
3761.81	3762.88	26575.4	P II			1.5	GU
3744.21	3745.27	26700.3	P			1.8	GU
3733.26	3734.32	26778.6	P			1.7	GU
3728.66	3729.72	26811.7	P II			1.7D	GU
3723.62	3724.68	26847.9	P II			1.5	GU
3717.62	3718.68	26891.3	P II			1.8	GU
3715.85	3716.91	26904.1	P II			1.7	GU
3710.45	3711.51	26943.2	P II			1.5H	GU
3706.05	3707.10	26975.2	P II			2.2W	GU
3676.26	3677.31	27193.8	P II			2.0O	GU
3668.59	3669.63	27250.7	P			1.7	GU
3664.19	3665.23	27283.4	P II			2.0W	GU
3659.26	3660.30	27320.1	P			1.7	GU
3653.38	3654.42	27364.1	P			2.0W	GU
3631.40	3632.44	27529.7	P II			1.7	GU
3624.68	3625.71	27580.8	P II			0.7	GU
3623.10	3624.13	27592.8	P			1.2	GU
3620.65	3621.68	27611.5	P			1.2	GU
3617.09	3618.12	27638.6	P			2.0W	GU
3587.35	3588.37	27867.8	P			1.5	GU
3583.60	3584.62	27896.9	P			1.7	GU
3580.35	3581.37	27922.3	P			1.5	GU
3577.60	3578.62	27943.7	P			1.7D	GU
3570.33	3571.35	28000.6	P II			1.5D	GU
3566.42	3567.44	28031.3	P II			1.8	GU
3562.47	3563.49	28062.4	P II			1.5	GU
3559.92	3560.94	28082.5	P II			1.5	GU
3556.48	3557.50	28109.7	P II			2.0	GU
3551.15	3552.16	28151.9	P II			1.5H	GU
3536.29	3537.30	28270.1	P II			1.5	GU
3533.66	3534.67	28291.2	P II			1.5	GU
3533.06	3534.07	28296.0	P II			1.2	GU
3530.24	3531.25	28318.6	P II			1.8	GU
3527.10	3528.11	28343.8	P II			1.5	GU
3519.22	3520.23	28407.3	P II			1.2	GU
3518.60	3519.61	28412.3	P II			1.7H	GU
3516.15	3517.16	28432.1	P			1.8	GU
3507.36	3508.36	28503.3	P II			2.0W	GU
3504.50	3505.50	28526.6	P			1.2	GU
3502.99	3503.99	28538.9	P II			1.8	GU
3490.44	3491.44	28641.5	P II			1.8	GU
3488.77	3489.77	28655.2	P			1.8	GU

AIR WAVELENGTH (ANGSTRMS)	VACUUM WAVELENGTH (ANGSTRMS)	WAVENUMBER (KAYSERS)	LMENT	LOG ARC	INTENSITY SPK	INTENSITY DIS	REF
3485.00	3486.00	28686.2	P			1.7	GU
3478.73	3479.73	28737.9	P II			1.5	GU
3477.44	3478.44	28748.6	P			1.5	GU
3474.14	3475.13	28775.9	P			1.8	GU
3472.87	3473.86	28786.4	P II			1.8	GU
3470.82	3471.81	28803.4	P II			1.7	GU
3430.09	3431.07	29145.4	P			1.5	GU
3426.19	3427.17	29178.6	P II			1.7	GU
3424.87	3425.85	29189.8	P II			2.0	GU
3419.24	3420.22	29237.9	P II			2.0	GU
3413.51	3414.49	29287.0	P			1.8	GU
3406.93	3407.91	29343.5	P			1.7	GU
3404.33	3405.31	29365.9	P II			1.7	GU
3395.35	3396.32	29443.6	P			1.7	GU
3378.76	3379.73	29588.2	P			1.7	GU
3377.52	3378.49	29599.0	P II			1.7H	GU
3372.70	3373.67	29641.3	P II			1.7	GU
3371.10	3372.07	29655.4	P			1.8	GU
3364.43	3365.40	29714.2	P			2.0	GU
3360.49	3361.46	29749.0	P			1.5	GU
3308.85	3309.80	30213.3	P II			2.0W	GU
3283.20	3284.15	30449.3	P			1.2	GU
3280.20	3281.15	30477.2	P			1.5H	GU
3277.80	3278.74	30499.5	P			1.5	GU
3270.43	3271.37	30568.2	P			1.2	GU
3233.61	3234.54	30916.3	P			1.3	GU
3219.30	3220.23	31053.7	P			2.0W	GU
3204.04	3204.97	31201.6	P			1.5	GU
3186.24	3187.16	31375.9	P			1.7	GU
3184.82	3185.74	31389.9	P			1.7	GU
3175.14	3176.06	31485.6	P II			1.8	GU
3171.84	3172.76	31518.3	P			1.7	GU
3163.87	3164.79	31597.7	P			1.7	GU
3162.34	3163.26	31613.0	P			1.7	GU
3157.73	3158.64	31659.1	P II			1.2	GU
3144.38	3145.29	31793.6	P			1.5	GU
3130.38	3131.29	31935.7	P II			1.5	GU
3124.30	3125.21	31997.9	P II			1.2	GU
3080.6	3081.5	32452.	P II			1.2H	DJ
3075.14	3076.03	32509.4	P			1.2H	GU
3071.83	3072.72	32544.4	P			1.5	GU
3038.2	3039.1	32905.	P			1.2	GU
3027.2	3028.1	33024.	P			1.2H	GU
2980.6	2981.5	33540.	P II			0.7H	GU
2977.6	2978.5	33574.	P			0.7H	GU
2948.14	2949.00	33909.8	P			0.7H	GU
2945.52	2946.38	33939.9	P			0.7	GU
2912.8	2913.7	34321.	P II			0.7	DJ
2912.04	2912.89	34330.1	P II			0.7	GU
2910.39	2911.24	34349.6	P			0.7H	GU
2895.32	2896.17	34528.4	P			1.4W	GU
2883.90	2884.75	34665.1	P			1.4	GU
2882.75	2883.60	34678.9	P			1.3	GU
2877.53	2878.37	34741.8	P			1.0	GU
2875.9	2876.7	34761.	P			0.7H	GU
2866.14	2866.98	34879.9	P			1.3	GU
2862.06	2862.90	34929.6	P			1.0	GU
2856.96	2857.80	34992.0	P			0.7	GU
2826.5	2827.3	35369.	P			1.3B	GU
2825.16	2825.99	35385.8	P			0.7	GU
2822.99	2823.82	35413.0	P			1.3B	GU
2780.79	2781.61	35950.4	P			1.0	GU
2758.52	2759.34	36240.6	P			1.3	GU
2739.90	2740.71	36486.9	P			1.0	GU
2739.32	2740.13	36494.6	P			1.4	GU
2729.19	2730.00	36630.1	P			1.0	GU
2728.80	2729.61	36635.3	P			1.3	GU
2724.86	2725.67	36688.3	P			1.3	GU
2715.80	2716.60	36810.6	P			0.7H	GU
2711.28	2712.08	36872.0	P			0.7	GU

AIR WAVELENGTH (ANGSTRMS)	VACUUM WAVELENGTH (ANGSTRMS)	WAVENUMBER (KAYSERS)	LMENT	LOG INTENSITY ARC	INTENSITY SPK	INTENSITY DIS	REF
2710.40	2711.20	36884.0	P			1.0	GU
2688.02	2688.82	37191.1	P I	0.6		0.0	KS
2684.98	2685.78	37233.2	P			0.7	GU
2680.36	2681.16	37297.3	P			1.0J	GU
2677.12	2677.92	37342.5	P I	1.0		0.3	KS
2676.28	2677.08	37354.2	P			1.0	GU
2663.99	2664.78	37526.5	P			1.0	GU
2644.20	2644.99	37807.4	P			1.4	GU
2638.18	2638.97	37893.6	P II			0.7	GU
2628.54	2629.32	38032.6	P II			0.7	GU
2626.15	2626.93	38067.2	P II			1.3	GU
2624.76	2625.54	38087.4	P II			1.0	GU
2606.02	2606.80	38361.2	P II			1.0	GU
2605.45	2606.23	38369.6	P			1.3	GU
2603.72	2604.50	38395.1	P II			0.7	GU
2554.93	2555.70	39128.3	P I	1.8		1.3	KS
2553.28	2554.05	39153.6	P I	1.9		1.3	KS
2535.65	2536.41	39425.8	P I	2.0		1.5	KS
2534.01	2534.77	39451.3	P I	1.7		1.3	KS
2500.922	2501.676	39973.21	P II			1.7	RI
2497.328	2498.081	40030.73	P II			2.0	RI
2496.003	2496.756	40051.98	P II			1.7	RI
2485.775	2486.525	40216.76	P II			0.3	RI
2484.152	2484.902	40243.04	P II			1.3	RI
2481.983	2482.732	40278.20	P II			1.0	RI
2478.22	2478.97	40339.4	P			1.0	GU
2477.77	2478.52	40346.7	P			1.0	GU
2454.484	2455.227	40729.43	P II			0.3	RI
2440.89	2441.63	40956.2	P			0.7	GU
2424.34	2425.08	41235.8	P			0.6	GU
2323.449	2324.162	43026.25	P II			0.3	RI
2314.725	2315.437	43188.40	P II			0.8	RI
2298.325	2299.033	43496.55	P II			0.8	RI
2287.492	2288.198	43702.52	P II			1.3	RI
2285.114	2285.819	43748.00	P II			1.3	RI
2281.003	2281.707	43826.83	P II			1.3	RI
2243.189	2243.885	44565.56	P I			0.3	RI
2242.492	2243.188	44579.41	P I			0.3	RI
2235.734	2236.428	44714.15	P I	1.0		0.0	RI
2234.99	2235.68	44729.0	P I	0.8			KS
2223.35	2224.04	44963.2	P I	0.8			KS
2154.081	2154.758	46408.92	P I	1.2		1.4	RI
2152.950	2153.627	46433.29	P I	1.1		1.4	RI
2149.108	2149.784	46516.29	P I	1.2		1.4	RI
2136.199	2136.873	46797.36	P I	1.2		1.4	RI
2135.466	2136.140	46813.42	P I	1.1		1.3	RI
2033.489	2034.143	49160.75	P I	1.2		1.3	RI
2032.447	2033.101	49185.95	P I	1.1		1.2	RI
2024.546	2025.198	49377.88	P I	1.1		1.1	RI
2023.471	2024.123	49404.11	P I	1.2		1.1	RI
9807.3	9810.0	10194.	PB	0.7V			ME
9674.56	9677.21	10333.6	PB	2.0X			ME
9604.06	9606.69	10409.4	PB	1.7X			ME
9470.32	9472.92	10556.4	PB	1.0X			ME
9440.1	9442.7	10590.	PB II			0.3	EA
9438.38	9440.97	10592.1	PB	1.3X			ME
9293.98	9296.53	10756.7	PB	0.7V			ME
9136.65	9139.16	10941.9	PB	0.9V			ME
9063.7	9066.2	11030.	PB II			2.0	EA
9050.7	9053.2	11046.	PB II			2.0	EA
8858.4	8860.8	11286.	PB	0.7V			ME
8723.1	8725.5	11461.	PB II		0.9		EA
8719.7	8722.1	11465.	PB II			1.2	EA
8710.1	8712.5	11478.	PB II			1.5	EA
8550.6	8552.9	11692.	PB II			0.7	EA
8545.0	8547.3	11699.	PB II			1.5	EA
8478.88	8481.21	11790.8	PB	1.0&			ME
8409.30	8411.61	11888.3	PB	0.7V			ME
8395.6	8397.9	11908.	PB II			2.0	EA
8335.0	8337.3	11994.	PB II			1.4	EA

AIR WAVELENGTH (ANGSTRMS)	VACUUM WAVELENGTH (ANGSTRMS)	WAVENUMBER (KAYSERS)	LMENT	LOG INTENSITY ARC	INTENSITY SPK	INTENSITY DIS	REF
8272.84	8275.11	12084.4	PB	2.3K			ME
7818.21	7820.36	12787.1	PB	0.7B			ME
7809.45	7811.60	12801.5	PB	1.0B			ME
7777.8	7779.9	12854.	PB II			0.7	EA
7739.8	7741.9	12917.	PB II			0.7	EA
7732.3	7734.4	12929.	PB II			1.1	EA
7679.4	7681.5	13018.	PB II			0.3	EA
7632.2	7634.3	13099.	PB II			2.0	EA
7558.7	7560.8	13226.	PB II			2.0	EA
7553.7	7555.8	13235.	PB II			0.3	EA
7330.12	7332.14	13638.6	PB	1.0			WT
7229.11	7231.10	13829.2	PB	1.7			WT
7193.6	7195.6	13897.	PB II			2.0	EA
7187.1	7189.1	13910.	PB II			1.0	EA
7165.1	7167.1	13953.	PB II			1.0	EA
7158.7	7160.7	13965.	PB II			1.0	EA
7114.7	7116.7	14051.	PB II			0.3	EA
7099.78	7101.74	14081.1	PB	1.0			WT
7050.7	7052.6	14179.	PB II			1.6	EA
7013.2	7015.1	14255.	PB II			1.7	EA
6872.0	6873.9	14548.	PB II			0.3	EA
6791.7	6793.6	14720.	PB II			1.0	EA
6790.8	6792.7	14722.	PB II			1.7	EA
6785.9	6787.8	14732.	PB II			1.2	EA
6660.0	6661.8	15011.	PB II			2.7	EA
6569.4	6571.2	15218.	PB II			1.4	EA
6558.7	6560.5	15243.	PB II			1.2	EA
6533.1	6534.9	15302.	PB II			1.3	EA
6527.8	6529.6	15315.	PB II			1.4	EA
6518.2	6520.0	15337.	PB II			1.5	EA
6422.9	6424.7	15565.	PB II			0.3	EA
6345.0	6346.8	15756.	PB II			1.4	EA
6339.8	6341.6	15769.	PB II			1.2	EA
6311.5	6313.2	15840.	PB II			1.6	EA
6235.44	6237.16	16032.9	PB	1.3			WT
6229.7	6231.4	16048.	PB II			1.7	EA
6203.7	6205.4	16115.	PB II			0.7D	EA
6198.8	6200.5	16128.	PB II			0.7D	EA
6181.9	6183.6	16172.	PB II			1.5	EA
6169.40	6171.11	16204.6	PB	1.0H			WT
6160.0	6161.7	16229.	PB II			1.7	EA
6110.787	6112.479	16359.98	PB	1.2B			MIT
6094.0	6095.7	16405.	PB II			0.3D	EA
6089.4	6091.1	16417.	PB II			0.7	EA
6081.5	6083.2	16439.	PB II			2.3	EA
6075.8	6077.5	16454.	PB II			2.3	EA
6059.71	6061.39	16497.9	PB	1.3B			WT
6041.4	6043.1	16548.	PB II			1.5D	EA
6011.98	6013.65	16628.9	PB	1.4B			WT
6009.7	6011.4	16635.	PB II			1.6	EA
6007.5	6009.2	16641.	PB II			1.0	EA
6001.884	6003.546	16656.82	PB	1.6B	0.5H		HZ
5895.70	5897.33	16956.8	PB	1.3B	0.3		WT
5876.7	5878.3	17012.	PB II			1.6	EA
5857.67	5859.29	17066.9	PB I		1.3		RO
5767.9	5769.5	17332.	PB II			1.6	EA
5713.8	5715.4	17497.	PB II			1.0	EA
5692.26	5693.84	17562.8	PB	1.3			WT
5660.1	5661.7	17663.	PB II			0.6	EA
5608.8	5610.4	17824.	PB II			1.6	EA
5562.694	5564.239	17971.91	PB I	0.3			MIT
5545.1	5546.6	18029.	PB		0.9		KL
5544.6	5546.1	18031.	PB II			1.6	EA
5523.5	5525.0	18099.	PB I		1.4		RO
5496.851	5498.378	18187.18	PB I	0.8			MIT
5472.4	5473.9	18268.	PB II			1.4	EA
5372.5	5374.0	18608.	PB			1.0H	KL
5372.1	5373.6	18609.	PB II			1.6	EA
5367.3	5368.8	18626.	PB II			1.6	EA
5308.2	5309.7	18833.	PB II			0.8	EA

AIR WAVELENGTH (ANGSTRMS)	VACUUM WAVELENGTH (ANGSTRMS)	WAVENUMBER (KAYSERS)	LMENT	LOG ARC	INTENSITY SPK	INTENSITY DIS	REF
5306.8	5308.3	18838.	PB II			1.3	EA
5201.444	5202.892	19220.08	PB	1.0	0.3H		HZ
5191.4	5192.8	19257.	PB II			0.3	EA
5189.2	5190.6	19265.	PB		1.3		RO
5163.8	5165.2	19360.	PB II			1.4	EA
5155.8	5157.2	19390.	PB II			1.4	EA
5143.14	5144.57	19438.0	PB		0.6H		ED
5111.9	5113.3	19557.	PB II			1.6	EA
5109.6	5111.0	19566.	PB II			0.8	EA
5081.2	5082.6	19675.	PB II			0.8	EA
5074.6	5076.0	19700.	PB II			1.6	EA
5070.7	5072.1	19716.	PB II			1.6	EA
5063.1	5064.5	19745.	PB I		1.0		RO
5049.3	5050.7	19799.	PB II			1.5	EA
5042.5	5043.9	19826.	PB II			2.3	EA
5032.2	5033.6	19866.	PB II			1.3	EA
5005.433	5006.829	19972.72	PB	1.3	0.6		MIT
4912.7	4914.1	20350.	PB II			0.5	EA
4895.6	4897.0	20421.	PB II			0.3	EA
4836.3	4837.7	20671.	PB II			0.8	EA
4833.7	4835.1	20682.	PB II			0.7	EA
4804.5	4805.8	20808.	PB II			0.9	EA
4802.23	4803.57	20817.8	PB		1.0		SX
4798.52	4799.86	20833.9	PB		1.3		SX
4798.4	4799.7	20834.	PB I		0.7		KL
4788.1	4789.4	20879.	PB II			0.8	EA
4760.98	4762.31	20998.2	PB I		0.8		RO
4684.9	4686.2	21339.	PB II			0.7	EA
4677.8	4679.1	21372.	PB II			0.8	EA
4665.5	4666.8	21428.	PB II			0.7	EA
4605.43	4606.72	21707.4	PB		0.3		SX
4582.34	4583.62	21816.8	PB II			1.0	GS
4581.3	4582.6	21822.	PB II			0.6	EA
4579.15	4580.43	21832.0	PB II			1.0	GS
4571.72	4573.00	21867.5	PB I		0.8		RO
4571.35	4572.63	21869.2	PB		1.5		SX
4557.2	4558.5	21937.	PB II			0.3	EA
4544.8	4546.1	21997.	PB II			0.6	EA
4534.69	4535.96	22046.0	PB		0.7		SX
4476.3	4477.6	22334.	PB II			0.5	EA
4453.8	4455.1	22446.	PB II			0.5	EA
4428.7	4429.9	22574.	PB II			0.3	EA
4400.89	4402.13	22716.3	PB		1.0		SX
4386.58	4387.81	22790.4	PB II			1.3	SX
4386.0	4387.2	22793.	PB II			0.3	EA
4352.7	4353.9	22968.	PB II			1.0	EA
4351.5	4352.7	22974.	PB II			0.5	EA
4340.432	4341.652	23032.71	PB	1.0			KL
4296.71	4297.92	23267.1	PB II			0.8	GS
4293.84	4295.05	23282.6	PB II			0.8	GS
4272.63	4273.83	23398.2	PB I		1.5		SX
4272.55	4273.75	23398.6	PB II		0.3		KL
4244.99	4246.19	23550.6	PB II				GS
4242.47	4243.66	23564.5	PB II		1.0		KL
4242.20	4243.39	23566.0	PB II		0.3		SX
4232.43	4233.62	23620.4	PB II			0.3	SX
4195.5	4196.7	23828.	PB II			0.5	EA
4182.161	4183.340	23904.35	PB		0.7H		MIT
4168.045	4169.220	23985.30	PB	1.3	1.0		MIT
4152.93	4154.10	24072.6	PB II		0.7		SX
4141.42	4142.59	24139.5	PB		1.0		SX
4128.21	4129.37	24216.7	PB		0.7		SX
4113.27	4114.43	24304.7	PB II			0.6	GS
4110.77	4111.93	24319.5	PB II			0.7	GS
4094.68	4095.84	24415.0	PB		0.7		SX
4077.61	4078.76	24517.2	PB		0.3		SX
4062.144	4063.291	24610.59	PB	1.3	1.3		MIT
4057.820	4058.966	24636.82	PB I	3.3G	2.5G		MIT
4041.3	4042.4	24737.	PB		0.7H		KL
4031.33	4032.47	24798.7	PB I		0.7		SX

AIR WAVELENGTH (ANGSTRMS)	VACUUM WAVELENGTH (ANGSTRMS)	WAVENUMBER (KAYSERS)	LMENT	LOG ARC	INTENSITY SPK	INTENSITY DIS	REF
4019.639	4020.775	24870.83	PB	0.8	0.8		MIT
3987.5	3988.6	25071.	PB II		0.7		GS
3985.2	3986.3	25086.	PB II			1.0	EA
3971.3	3972.4	25173.	PB II			1.5	EA
3951.94	3953.06	25296.9	PB I		1.7H		SX
3943.80	3944.92	25349.1	PB		0.7		SX
3927.79	3928.90	25452.4	PB		0.3		KL
3927.41	3928.52	25454.9	PB II		0.7H		SX
3925.23	3926.34	25469.0	PB		0.3		SX
3909.17	3910.28	25573.6	PB II		1.6		SX
3896.9	3898.0	25654.	PB II			0.3	EA
3894.6	3895.7	25669.	PB II			0.7	EA
3880.5	3881.6	25763.	PB II			0.3	EA
3874.65	3875.75	25801.5	PB		0.3		SX
3872.64	3873.74	25814.9	PB		0.3		SX
3854.053	3855.146	25939.35	PB I		2.0		MIT
3841.91	3843.00	26021.3	PB		0.7H		KL
3841.62	3842.71	26023.3	PB		1.8		SX
3832.904	3833.991	26082.48	PB		0.7		MIT
3832.83	3833.92	26083.0	PB		1.7		SX
3829.2	3830.3	26108.	PB II			0.3	EA
3827.89	3828.98	26116.6	PB		0.3		KL
3827.2	3828.3	26121.	PB II			1.3	EA
3786.243	3787.318	26403.91	PB II		1.0H		KL
3786.00	3787.08	26405.6	PB II		1.6		SX
3784.0	3785.1	26420.	PB II			1.0	EA
3749.15	3750.22	26665.1	PB		0.3		SX
3746.9	3748.0	26681.	PB II			1.5	EA
3739.947	3741.010	26730.75	PB	2.2	1.8H		MIT
3734.8	3735.9	26768.	PB II			1.0	EA
3728.83	3729.89	26810.4	PB I		1.3		SX
3723.87	3724.93	26846.1	PB I		0.3		SX
3718.34	3719.40	26886.1	PB II		0.3		SX
3714.05	3715.11	26917.1	PB II		1.0		SX
3706.13	3707.18	26974.6	PB I		1.0		SX
3699.2	3700.3	27025.	PB II			1.0	EA
3689.309	3690.359	27097.63	PB II		1.6		MIT
3689.309	3690.359	27097.63	PB I		1.6		MIT
3689.201	3690.251	27098.43	PB		0.7		MIT
3689.0	3690.1	27100.	PB II			0.3	EA
3683.471	3684.520	27140.58	PB I	2.5	1.7		MIT
3678.541	3679.588	27176.95	PB		0.3		MIT
3676.377	3677.424	27192.95	PB		0.3		MIT
3674.95	3676.00	27203.5	PB		0.3		SX
3671.503	3672.549	27229.05	PB	1.7	0.8		MIT
3671.39	3672.44	27229.9	PB		1.8		SX
3665.485	3666.529	27273.75	PB II		0.3		MIT
3665.05	3666.09	27277.0	PB I		0.3		SX
3649.0	3650.0	27397.	PB II			1.3	EA
3639.580	3640.617	27467.87	PB I	2.5	1.7H		MIT
3620.85	3621.88	27609.9	PB		0.7		SX
3601.8	3602.8	27756.	PB II			1.3	EA
3593.12	3594.15	27823.0	PB		1.5		SX
3592.921	3593.946	27824.57	PB		0.5		MIT
3589.92	3590.94	27847.8	PB I		1.6		SX
3586.44	3587.46	27874.9	PB		1.3		SX
3572.734	3573.754	27981.78	PB	2.3	1.3		MIT
3567.1	3568.1	28026.	PB		0.7		SX
3562.89	3563.91	28059.1	PB		1.3		SX
3533.91	3534.92	28289.2	PB		0.3		SX
3530.35	3531.36	28317.7	PB		0.7		SX
3505.15	3506.15	28521.3	PB		0.7		SX
3501.9	3502.9	28548.	PB II			1.0	EA
3483.39	3484.39	28699.4	PB		1.5		SX
3476.25	3477.25	28758.4	PB		0.3		SX
3463.6	3464.6	28863.	PB II			1.7	EA
3455.49	3456.48	28931.2	PB I		1.8		RO
3455.0	3456.0	28935.	PB II			1.0	EA
3453.0	3454.0	28952.	PB II			0.3	EA
3451.90	3452.89	28961.3	PB I		1.0		SX

AIR WAVELENGTH (ANGSTRMS)	VACUUM WAVELENGTH (ANGSTRMS)	WAVENUMBER (KAYSERS)	LMENT	LOG ARC	INTENSITY SPK	DIS	REF
3451.7	3452.7	28963.	PB II			1.9	EA
3451.619	3452.608	28963.61	PB II		1.0		MIT
3450.0	3451.0	28977.	PB II			1.0	EA
3437.36	3438.35	29083.8	PB I		1.0		RO
3431.35	3432.33	29134.7	PB		0.3		SX
3429.6	3430.6	29150.	PB II			0.7	EA
3389.4	3390.4	29495.	PB II			1.0	EA
3361.58	3362.55	29739.4	PB I		1.2		SX
3309.2	3310.2	30210.	PB II			1.2	EA
3297.64	3298.59	30316.0	PB		1.5		SX
3279.85	3280.80	30480.4	PB II		1.0		GS
3279.33	3280.27	30485.2	PB		1.0		SX
3276.44	3277.38	30512.1	PB		0.7H		KL
3276.19	3277.13	30514.5	PB		1.8		SX
3262.353	3263.294	30643.89	PB	1.3H	0.7H		MIT
3261.21	3262.15	30654.6	PB		0.3		SX
3261.0	3261.9	30657.	PB II			1.7	EA
3242.86	3243.80	30828.1	PB I		1.6		SX
3240.192	3241.127	30853.47	PB	1.5H			MIT
3232.353	3233.286	30928.29	PB	1.5			KL
3231.23	3232.16	30939.0	PB		1.0		SX
3227.08	3228.01	30978.8	PB		0.7		SX
3220.538	3221.468	31041.75	PB	1.7H	0.7		MIT
3217.9	3218.8	31067.	PB II			1.2	EA
3190.89	3191.81	31330.2	PB		0.3		SX
3190.06	3190.98	31338.3	PB		0.3		MIT
3176.54	3177.46	31471.7	PB I		2.0		KL
3173.5	3174.4	31502.	PB II			0.3	EA
3145.60	3146.51	31781.2	PB I		1.0		SX
3137.83	3138.74	31859.9	PB I		2.0		KL
3129.63	3130.54	31943.4	PB I		0.5		SX
3125.6	3126.5	31985.	PB II			1.7	EA
3118.920	3119.824	32053.09	PB	0.7H			MIT
3118.19	3119.09	32060.6	PB		0.3		MIT
3117.7	3118.6	32066.	PB II			2.0	EA
3111.38	3112.28	32130.8	PB		0.3		SX
3102.874	3103.774	32218.84	PB I		1.0		MIT
3089.093	3089.990	32362.56	PB I		1.5		MIT
3087.039	3087.935	32384.10	PB		1.3		MIT
3068.5	3069.4	32580.	PB II			1.0	EA
3062.44	3063.33	32644.2	PB		1.0		SX
3061.32	3062.21	32656.1	PB		0.3		SX
3043.90	3044.79	32843.0	PB I		2.0		MIT
3031.65	3032.53	32975.7	PB		1.3		SX
3025.51	3026.39	33042.7	PB		0.3		SX
3017.46	3018.34	33130.8	PB		1.0		MIT
3016.4	3017.3	33142.	PB II		1.4		EA
3010.19	3011.07	33210.8	PB		0.3		SX
3002.74	3003.62	33293.2	PB II		1.0		GS
2986.9	2987.8	33470.	PB II			1.3	EA
2980.160	2981.030	33545.46	PB	0.3			MIT
2973.00	2973.87	33626.2	PB	0.3			KL
2966.514	2967.380	33699.76	PB	0.3			MIT
2948.719	2949.581	33903.12	PB II		2.1		MIT
2929.848	2930.705	34121.48	PB		0.3H		MIT
2926.646	2927.502	34158.81	PB	0.3H			KL
2914.5	2915.4	34301.	PB II			1.0	EA
2887.19	2888.04	34625.6	PB II			1.0	GS
2873.316	2874.159	34792.78	PB	2.0G	1.8		MIT
2873.0	2873.8	34797.	PB II			1.3	EA
2868.148	2868.990	34855.47	PB		0.7		MIT
2864.514	2865.355	34899.69	PB		1.8		MIT
2864.257	2865.098	34902.82	PB		1.8		MIT
2860.64	2861.48	34947.0	PB		0.3		SX
2845.2	2846.0	35137.	PB II			0.3	EA
2840.66	2841.50	35192.7	PB II		1.3		MIT
2833.069	2833.902	35287.03	PB I	2.7G	1.9G		MIT
2823.189	2824.020	35410.52	PB	2.2G	1.6		MIT
2802.003	2802.829	35678.24	PB	2.4P	2.0H		MIT
2772.71	2773.53	36055.1	PB II		0.3		MIT
2745.47	2746.28	36412.9	PB		0.7H		SX
2719.80	2720.61	36756.5	PB II			1.3	SX
2717.5	2718.3	36788.	PB II			1.3	EA
2717.37	2718.18	36789.4	PB		0.3H		SX
2697.527	2698.327	37060.00	PB	1.2J	0.7H		KL
2684.76	2685.56	37236.2	PB II		0.7		MIT
2663.166	2663.958	37538.13	PB	2.5J	1.6		MIT
2661.143	2661.935	37566.66	PB II		2.0		MIT
2660.19	2660.98	37580.1	PB II			0.7	BX
2658.722	2659.513	37600.87	PB II	1.3	2.5		MIT
2657.563	2658.354	37617.27	PB II		2.3		MIT
2657.106	2657.897	37623.73	PB		0.5		MIT
2653.60	2654.39	37673.4	PB		0.5		MIT
2650.4	2651.2	37719.	PB	2.0	1.9		KL
2650.26	2651.05	37720.9	PB		0.7H		SX
2638.36	2639.15	37891.0	PB		0.3		MIT
2637.72	2638.51	37900.2	PB		0.3		MIT
2634.3	2635.1	37949.	PB II			1.3	EA
2630.795	2631.579	37999.99	PB II		0.3H		MIT
2628.263	2629.047	38036.60	PB	1.7	1.0		HZ
2614.178	2614.958	38241.52	PB	2.3R	1.9		HZ
2613.653	2614.433	38249.21	PB	1.7G	0.7		MIT
2608.4	2609.2	38326.	PB II			0.7	EA
2577.263	2578.035	38789.24	PB	2.0J	1.6		MIT
2576.55	2577.32	38800.0	PB II			2.0	GS
2568.43	2569.20	38922.6	PB		1.0		MIT
2562.28	2563.05	39016.0	PB		2.0		MIT
2550.29	2551.06	39199.5	PB		0.3		SX
2534.78	2535.54	39439.3	PB		0.3		SX
2527.03	2527.79	39560.2	PB		0.3		SX
2526.62	2527.38	39566.7	PB II			2.0	GS
2508.90	2509.66	39846.1	PB		0.7		SX
2500.50	2501.25	39980.0	PB		0.3		SX
2498.90	2499.65	40005.5	PB II		1.2		MIT
2495.58	2496.33	40058.8	PB		0.3		SX
2478.594	2479.343	40333.27	PB		0.3		KL
2476.379	2477.127	40369.35	PB	2.2J	1.4		MIT
2446.188	2446.929	40867.55	PB	2.2W	1.2		MIT
2445.1	2445.8	40886.	PB II			0.3	EA
2443.838	2444.579	40906.85	PB	2.0W	1.2		MIT
2428.644	2429.381	41162.75	PB	1.0H	0.9		KL
2411.735	2412.468	41451.32	PB	1.9	1.2		HZ
2401.946	2402.677	41620.24	PB	1.7	1.6		MIT
2399.583	2400.314	41661.22	PB	1.5	1.1		MIT
2393.794	2394.523	41761.97	PB	3.4	3.0		MIT
2388.774	2389.502	41849.72	PB	1.6	1.3		MIT
2359.58	2360.30	42367.5	PB		0.5		MIT
2356.99	2357.71	42414.0	PB II		0.6		MIT
2332.426	2333.141	42860.67	PB	1.8	1.5		MIT
2296.85	2297.56	43524.5	PB		0.5		MIT
2280.85	2281.55	43829.8	PB II		0.3H		MIT
2253.95	2254.65	44352.8	PB	1.6	0.7		MIT
2246.888	2247.585	44492.20	PB	1.5G	2.0G		MIT
2242.610	2243.306	44577.07	PB	1.2	1.2		MIT
2237.426	2238.121	44680.34	PB	1.7W	1.5W		MIT
2218.08	2218.77	45070.0	PB	1.7	1.6		MIT
2203.505	2204.193	45368.09	PB II	1.7W	3.7G		MIT
2187.85	2188.53	45692.7	PB	1.7	0.5		MIT
2181.66	2182.34	45822.3	PB II	0.8			MIT
2175.579	2176.261	45950.37	PB	1.6W			MIT
2169.99	2170.67	46068.7	PB I	2.7G	2.7G		HZ
2159.52	2160.20	46292.0	PB	1.2			MIT
2115.02	2115.69	47265.9	PB	1.5G	0.9W		MIT
2112.08	2112.75	47331.7	PB		1.2		MIT
2111.74	2112.41	47339.3	PB	1.3W	0.9W		MIT
2088.43	2089.09	47867.6	PB	1.5G	1.60		MIT
2087.58	2088.24	47887.1	PB	1.1			MIT
2074.0	2074.7	48201.	PB		1.3		LG
2061.74	2062.40	48487.2	PB	0.9R	1.6		MIT
2059.63	2060.29	48536.9	PB	2.7U			MIT

AIR WAVELENGTH (ANGSTRMS)	VACUUM WAVELENGTH (ANGSTRMS)	WAVENUMBER (KAYSERS)	LMENT	LOG ARC	INTENSITY SPK	DIS	REF
2053.32	2053.98	48686.0	PB I	1.1			MIT
2049.15	2049.81	48785.1	PB I	0.9S			MIT
2022.07	2022.72	49438.3	PB	0.7	1.1		MIT
2014.43	2015.08	49625.8	PB I	0.5			MIT
8761.35	8763.76	11410.6	PD I	0.3			MIT
8599.10	8601.46	11625.9	PD I	0.3			MIT
8582.08	8584.44	11649.0	PD I	0.3			MIT
8532.74	8535.08	11716.3	PD I	0.3			MIT
8353.58	8355.88	11967.6	PD I	0.3			MIT
8346.3	8348.6	11978.	PD		0.3		IT
8300.83	8303.11	12043.7	PD I	0.8			MIT
8295.5	8297.8	12051.	PD		0.5		IT
8132.82	8135.06	12292.5	PD I	0.9			MIT
8003.97	8006.17	12490.4	PD		0.3		MIT
8001.0	8003.2	12495.	PD		0.3		IT
7961.08	7963.27	12557.7	PD I	0.7			MIT
7915.80	7917.98	12629.5	PD I	1.0			MIT
7904.4	7906.6	12648.	PD		0.3		IT
7786.67	7788.81	12838.9	PD I	1.0			MIT
7764.03	7766.17	12876.4	PD I	1.4			MIT
7486.90	7488.96	13353.0	PD I	0.9			MIT
7408.2	7410.2	13495.	PD		0.3		IT
7391.92	7393.96	13524.6	PD I	1.0			MIT
7388.3	7390.3	13531.	PD		0.5		IT
7368.12	7370.15	13568.2	PD I	1.3			MIT
7310.06	7312.07	13676.0	PD I	0.7			ME
7278.43	7280.44	13735.4	PD I	0.3			MIT
7242.90	7244.90	13802.8	PD I	0.3			ME
7149.11	7151.08	13983.9	PD I	0.8			ME
7115.84	7117.80	14049.3	PD I	0.5			ME
7060.29	7062.24	14159.8	PD I	0.7			ME
7052.04	7053.98	14176.4	PD I	0.3			ME
7037.58	7039.52	14205.5	PD I	0.5			ME
7033.7	7035.6	14213.	PD		0.3		IT
7016.44	7018.37	14248.3	PD I	1.0			ME
7013.3	7015.2	14255.	PD		0.3		IT
6981.2	6983.1	14320.	PD		0.3		IT
6947.43	6949.35	14389.8	PD		0.3		MIT
6917.54	6919.45	14452.0	PD I	0.3			MIT
6916.55	6918.46	14454.1	PD I	1.0			MIT
6914.98	6916.89	14457.4	PD I	0.3H			ME
6878.35	6880.25	14534.4	PD I	0.3			ME
6833.42	6835.31	14629.9	PD I	1.0			ME
6784.52	6786.39	14735.4	PD I	1.1	0.3		ME
6782.3	6784.2	14740.	PD		0.7		IT
6774.54	6776.41	14757.1	PD I	1.2			ME
6686.79	6688.64	14950.7	PD I	0.5			ME
6685.71	6687.56	14953.2	PD I	0.3			ME
6681.56	6683.40	14962.4	PD I	0.5			ME
6662.86	6664.70	15004.4	PD I	0.6			ME
6625.28	6627.11	15089.5	PD I	0.6			ME
6623.26	6625.09	15094.1	PD I	0.6			ME
6591.44	6593.26	15167.0	PD I	0.5			ME
6508.415	6510.213	15360.48	PD I	0.8			MIT
6444.89	6446.67	15511.9	PD I	0.3			ME
6243.97	6245.70	16011.0	PD I	0.3			ME
6195.61	6197.32	16136.0	PD I	0.3			ME
6188.024	6189.736	16155.78	PD I	0.8			MIT
6176.168	6177.877	16186.79	PD I	0.7			MIT
6170.955	6172.663	16200.46	PD I	0.7			MIT
6130.557	6132.254	16307.22	PD I	1.0			MIT
5868.14	5869.77	17036.4	PD I	0.3			ME
5774.25	5775.85	17313.5	PD I	0.5			ME
5759.912	5761.510	17356.56	PD I	0.6			MIT
5739.676	5741.268	17417.75	PD I	0.9			MIT
5737.636	5739.228	17423.95	PD I	0.6			MIT
5736.609	5738.200	17427.07	PD I	1.1			MIT
5695.090	5696.670	17554.11	PD I	1.7			MIT
5690.139	5691.718	17569.39	PD I	1.0			MIT
5687.478	5689.056	17577.61	PD I	0.5			MIT
5680.80	5682.38	17598.3	PD I	0.3			ME
5674.243	5675.818	17618.61	PD I	0.5			MIT
5670.068	5671.641	17631.58	PD I	2.0			MIT
5668.399	5669.972	17636.77	PD I	0.5			MIT
5655.425	5656.995	17677.23	PD I	1.1			MIT
5642.690	5644.256	17717.13	PD I	1.0	0.3H		MIT
5621.322	5622.882	17784.47	PD I	0.3			MIT
5619.442	5621.002	17790.42	PD I	1.7	0.3		MIT
5608.022	5609.579	17826.65	PD I	1.0			MIT
5602.996	5604.552	17842.64	PD I	0.6			MIT
5602.310	5603.865	17844.83	PD I	0.3			MIT
5601.623	5603.178	17847.01	PD I	0.9H			MIT
5547.020	5548.561	18022.69	PD I	1.7	0.3		MIT
5542.799	5544.338	18036.42	PD I	2.0	0.3		MIT
5529.455	5530.991	18079.94	PD I	1.0			MIT
5459.164	5460.681	18312.73	PD I	0.3			MIT
5435.161	5436.672	18393.61	PD I	0.8			MIT
5427.692	5429.201	18418.92	PD I	0.5			MIT
5406.611	5408.114	18490.73	PD I	0.6			MIT
5395.244	5396.744	18529.69	PD I	1.7	0.3		MIT
5394.842	5396.342	18531.07	PD	0.7H			MIT
5377.645	5379.140	18590.33	PD I	0.3			MIT
5363.283	5364.775	18640.11	PD I	0.3			MIT
5362.659	5364.150	18642.28	PD I	1.2			MIT
5346.785	5348.272	18697.63	PD I	0.3H			MIT
5345.105	5346.592	18703.50	PD I	1.0			MIT
5312.573	5314.051	18818.03	PD I	1.2			MIT
5295.629	5297.102	18878.25	PD I	2.3	1.0		MIT
5294.143	5295.616	18883.54	PD I	0.8			MIT
5256.175	5257.638	19019.95	PD I	1.1			MIT
5238.416	5239.874	19084.43	PD I	0.3H			MIT
5234.860	5236.317	19097.39	PD I	1.7	0.3		MIT
5208.914	5210.364	19192.52	PD I	1.0			MIT
5163.842	5165.280	19360.03	PD I	2.5	0.9		MIT
5161.358	5162.796	19369.35	PD I	0.6			MIT
5127.711	5129.140	19496.45	PD I	0.9			MIT
5117.015	5118.441	19537.20	PD I	1.7	0.6		MIT
5114.38	5115.81	19547.3	PD I	0.8			MIT
5110.813	5112.237	19560.91	PD I	2.0	0.3		MIT
5063.408	5064.820	19744.04	PD I	1.2			MIT
4971.958	4973.345	20107.19	PD	1.0			MIT
4930.002	4931.378	20278.31	PD	0.5			MIT
4924.162	4925.537	20302.36	PD	0.3			MIT
4919.859	4921.232	20320.11	PD I	1.1			MIT
4875.430	4876.792	20505.28	PD	1.4	0.3		MIT
4836.428	4837.779	20670.64	PD	0.9			MIT
4819.151	4820.498	20744.75	PD I	0.3			MIT
4817.509	4818.855	20751.82	PD	1.6	0.9		MIT
4817.007	4818.353	20753.98	PD	0.6	0.3H		MIT
4806.37	4807.71	20799.9	PD I	0.3			ME
4790.832	4792.171	20867.37	PD I	0.5			MIT
4788.175	4789.514	20878.95	PD I	2.3H	0.6H		MIT
4776.567	4777.902	20929.69	PD I	0.7	1.0		MIT
4761.873	4763.205	20994.27	PD I	0.5	0.3H		MIT
4724.001	4725.323	21162.58	PD I	1.2	0.3H		MIT
4708.043	4709.360	21234.31	PD I	0.3			MIT
4677.461	4678.770	21373.14	PD I	0.9	0.3		MIT
4632.632	4633.929	21579.96	PD I	0.5			MIT
4631.379	4632.676	21585.80	PD I	0.3H			MIT
4589.978	4591.264	21780.49	PD I	0.8	0.5H		MIT
4552.892	4554.168	21957.91	PD I	0.3			MIT
4541.137	4542.410	22014.75	PD I	1.2	0.5		MIT
4516.177	4517.444	22136.41	PD I	1.0	0.0		MIT
4497.669	4498.931	22227.50	PD I	0.5			MIT
4489.479	4490.739	22268.05	PD I	1.1	0.3		MIT
4473.590	4474.845	22347.14	PD I	1.8	0.8		MIT
4443.039	4444.286	22500.80	PD I	0.7	0.3H		MIT
4421.040	4422.281	22612.76	PD I	0.8			MIT
4406.546	4407.784	22687.14	PD I	1.5	0.3		MIT
4388.620	4389.853	22779.81	PD I	0.9	0.3		MIT

AIR WAVELENGTH (ANGSTRMS)	VACUUM WAVELENGTH (ANGSTRMS)	WAVENUMBER (KAYSERS)	LMENT	LOG ARC	INTENSITY SPK	DIS	REF
4386.420	4387.652	22791.23	PD I	1.0			MIT
4379.561	4380.792	22826.93	PD I	0.8	0.3H		MIT
4360.282	4361.507	22927.85	PD I	1.0H			MIT
4358.599	4359.824	22936.71	PD I	1.4			MIT
4351.008	4352.231	22976.72	PD I	0.7H			MIT
4344.672	4345.893	23010.23	PD I	0.7			MIT
4328.086	4329.303	23098.41	PD I	0.3H			MIT
4314.985	4316.199	23168.54	PD I	0.5			MIT
4281.902	4283.107	23347.54	PD I	0.5H			MIT
4268.330	4269.531	23421.77	PD I	1.0H			MIT
4241.7	4242.9	23569.	PD I	0.3H			EX
4212.950	4214.137	23729.65	PD I	2.70	2.50		MIT
4169.842	4171.017	23974.97	PD I	2.3	1.7		MIT
4166.279	4167.453	23995.47	PD I	1.0			MIT
4162.885	4164.059	24015.03	PD I	0.6H			MIT
4156.955	4158.127	24049.29	PD I	0.5	0.5H		MIT
4152.12	4153.29	24077.3	PD	0.3			SH
4151.357	4152.528	24081.72	PD I	0.3			MIT
4140.827	4141.995	24142.96	PD I	2.0R			MIT
4128.368	4129.533	24215.82	PD I	0.7H			MIT
4123.617	4124.780	24243.72	PD I	1.0	0.3		MIT
4099.27	4100.43	24387.7	PD I		0.6H		SH
4098.904	4100.061	24389.88	PD I	0.3			MIT
4090.071	4091.226	24442.55	PD I	0.6			MIT
4087.343	4088.497	24458.87	PD I	2.7	2.0		MIT
4084.35	4085.50	24476.8	PD I	0.7H			SH
4081.666	4082.818	24492.89	PD I	0.3			MIT
4020.679	4021.815	24864.39	PD	0.7H			MIT
4020.225	4021.361	24867.20	PD I	1.2J			MIT
4007.470	4008.603	24946.35	PD I	0.7H			MIT
3992.330	3993.459	25040.95	PD I	0.7H			MIT
3985.490	3986.617	25083.92	PD I	0.3			MIT
3978.912	3980.037	25125.39	PD I	0.5H			MIT
3958.642	3959.762	25254.04	PD I	2.7W	2.3		MIT
3958.315	3959.435	25256.13	PD		0.3H		DN
3926.922	3928.034	25458.03	PD I	0.3			MIT
3913.095	3914.203	25547.98	PD I	0.3			MIT
3894.201	3895.304	25671.94	PD I	2.30	2.30		MIT
3873.564	3874.662	25808.70	PD I	0.8			MIT
3832.291	3833.378	26086.65	PD I	2.2	2.2		MIT
3827.135	3828.221	26121.79	PD I	0.6	1.4J		MIT
3822.022	3823.107	26156.74	PD I	0.5H			SH
3807.87	3808.95	26253.9	PD I	0.7H			SH
3802.525	3803.605	26290.85	PD		0.5		MIT
3799.190	3800.269	26313.93	PD I	2.3W	2.2		MIT
3778.313	3779.386	26459.32	PD I	0.3			MIT
3738.86	3739.92	26738.5	PD II		1.2		MIT
3718.909	3719.967	26881.96	PD I	2.5	2.3		MIT
3690.341	3691.391	27090.06	PD I	2.5H	3.0W		MIT
3654.425	3655.466	27356.29	PD I		0.6		MIT
3645.968	3647.007	27419.75	PD I	1.2			MIT
3634.695	3635.731	27504.78	PD I	3.3G	3.0G		MIT
3609.548	3610.578	27696.40	PD I	3.0G	2.8G		MIT
3596.661	3597.687	27795.64	PD	0.7	0.0		MIT
3589.108	3590.132	27854.13	PD I	0.7			MIT
3584.09	3585.11	27893.1	PD	0.5			MIT
3571.155	3572.175	27994.15	PD I	1.6H	1.6H		MIT
3566.626	3567.644	28029.70	PD	1.8			MIT
3553.082	3554.097	28136.54	PD I	2.0G	1.2J		MIT
3528.723	3529.732	28330.76	PD I	1.0			MIT
3521.16	3522.17	28391.6	PD I	0.3			MIT
3516.943	3517.949	28425.66	PD I	3.0G	2.7G		MIT
3507.95	3508.95	28498.5	PD II		1.5H		EX
3489.772	3490.771	28646.97	PD I	2.20	1.5		MIT
3481.152	3482.149	28717.90	PD I	2.7R	0.3H		MIT
3468.54	3469.53	28822.3	PD II		1.7H		MIT
3460.774	3461.765	28887.00	PD I	2.5R	2.8H		MIT
3451.346	3452.335	28965.90	PD II		2.6H		MIT
3442.403	3443.390	29041.15	PD I	0.7	0.3H		MIT
3441.396	3442.382	29049.65	PD I	2.9H	0.3H		MIT

AIR WAVELENGTH (ANGSTRMS)	VACUUM WAVELENGTH (ANGSTRMS)	WAVENUMBER (KAYSERS)	LMENT	LOG ARC	INTENSITY SPK	DIS	REF
3433.449	3434.433	29116.89	PD I	3.0H	2.7H		MIT
3421.24	3422.22	29220.8	PD I	3.3G	3.06		MIT
3419.652	3420.633	29234.36	PD I	1.0	0.3		MIT
3406.042	3407.019	29351.17	PD I	0.5	0.3		MIT
3404.580	3405.557	29363.77	PD I	3.3G	3.06		MIT
3396.780	3397.755	29431.20	PD I	1.0			MIT
3389.039	3390.012	29498.42	PD I	1.0			MIT
3382.566	3383.537	29554.87	PD II		0.3H		MIT
3380.674	3381.645	29571.41	PD I	2.2W	0.3H		MIT
3377.35	3378.32	29600.5	PD			0.3	BX
3373.001	3373.970	29638.68	PD I	2.9R	2.7J		MIT
3371.670	3372.638	29650.38	PD		0.3H		MIT
3355.93	3356.89	29789.4	PD			0.7	BX
3352.19	3353.15	29822.7	PD			0.5	BX
3346.77	3347.73	29871.0	PD II		0.6H		EX
3327.23	3328.19	30046.4	PD II		1.7H		EX
3320.987	3321.943	30102.87	PD I	1.2			MIT
3312.985	3313.939	30175.57	PD I	1.2			MIT
3311.023	3311.976	30193.46	PD	0.6	0.3H		MIT
3310.123	3311.076	30201.67	PD I	0.8			MIT
3307.07	3308.02	30229.5	PD	0.6L			MIT
3302.128	3303.079	30274.78	PD I	3.0J	2.3H		MIT
3287.249	3288.196	30411.81	PD I	2.5W	1.4		MIT
3280.678	3281.623	30472.72	PD		0.3H		MIT
3278.05	3278.99	30497.1	PD II		0.3H		BX
3272.56	3273.50	30548.3	PD II		1.8H		EX
3267.35	3268.29	30597.0	PD II		2.3H		MIT
3261.73	3262.67	30649.7	PD		0.3H		MIT
3258.780	3259.720	30677.48	PD I	2.5	2.3H		MIT
3251.640	3252.578	30744.84	PD I	2.3	2.7		MIT
3243.131	3244.067	30825.51	PD II		1.8H		DN
3242.703	3243.639	30829.57	PD I	3.3J	2.86		MIT
3241.79	3242.73	30838.3	PD		0.3		BX
3232.32	3233.25	30928.6	PD I	0.3			MIT
3220.467	3221.397	31042.43	PD II		0.3		MIT
3218.974	3219.904	31056.83	PD I	2.5	0.9		MIT
3210.448	3211.375	31139.30	PD II		1.8H		MIT
3209.412	3210.339	31149.36	PD II		0.3H		MIT
3206.885	3207.812	31173.90	PD II		0.3		MIT
3189.34	3190.26	31345.4	PD		0.3H		BX
3184.59	3185.51	31392.1	PD II		0.3		BX
3179.410	3180.330	31443.28	PD I	1.2	0.0H		MIT
3178.77	3179.69	31449.6	PD II		1.0		EX
3173.865	3174.783	31498.21	PD II		0.3		MIT
3170.26	3171.18	31534.0	PD II		1.7J		SH
3162.86	3163.78	31607.8	PD II		0.6		EX
3161.947	3162.862	31616.93	PD II		2.0H		MIT
3155.59	3156.50	31680.6	PD II		1.6		EX
3153.14	3154.05	31705.2	PD II		0.3H		MIT
3142.812	3143.722	31809.42	PD I	2.5	2.0		MIT
3138.32	3139.23	31854.9	PD		0.3H		MIT
3136.67	3137.58	31871.7	PD		0.3H		BX
3135.78	3136.69	31880.7	PD II		0.7H		SH
3132.513	3133.421	31914.00	PD II		1.2H		MIT
3114.040	3114.943	32103.31	PD I	2.6W	2.7W		MIT
3111.52	3112.42	32129.3	PD II		0.5H		MIT
3109.156	3110.058	32153.74	PD II	0.7	1.5H		MIT
3086.54	3087.44	32389.3	PD II		0.3H		MIT
3086.272	3087.168	32392.15	PD II		0.3H		MIT
3079.97	3080.86	32458.4	PD		0.3H		BX
3077.01	3077.90	32489.6	PD II		0.3H		BX
3075.166	3076.059	32509.12	PD I	0.3	0.3H		MIT
3069.184	3070.076	32572.48	PD II		0.3H		MIT
3068.885	3069.777	32575.66	PD II		0.3H		MIT
3066.103	3066.994	32605.21	PD I	2.2	0.3		MIT
3065.306	3066.197	32613.69	PD I	1.0	2.0		MIT
3059.431	3060.320	32676.32	PD II		2.2W		MIT
3055.29	3056.18	32720.6	PD II		0.5H		MIT
3052.15	3053.04	32754.3	PD II		2.2H		MIT
3050.084	3050.971	32776.45	PD II		2.0H		MIT

AIR WAVELENGTH (ANGSTRMS)	VACUUM WAVELENGTH (ANGSTRMS)	WAVENUMBER (KAYSERS)	LMENT	LOG ARC	INTENSITY SPK	INTENSITY DIS	REF
3046.50	3047.39	32815.0	PD II	0.3H	0.3H		MIT
3041.664	3042.549	32867.18	PD II		1.5H		ML
3032.20	3033.08	32969.8	PD II	0.3	2.0H		MIT
3028.762	3029.644	33007.18	PD	0.5H			MIT
3027.910	3028.792	33016.47	PD I	2.2	2.3H		MIT
3024.29	3025.17	33056.0	PD		0.6H		MIT
3021.749	3022.629	33083.78	PD I	0.3			MIT
3019.616	3020.496	33107.15	PD II		1.5H		MIT
3018.95	3019.83	33114.5	PD II		0.3H		BX
3018.498	3019.377	33119.41	PD II		1.7H		MIT
3015.06	3015.94	33157.2	PD II		0.3		MIT
3014.743	3015.621	33160.66	PD II		0.3H		MIT
3009.777	3010.654	33215.37	PD I	1.7R	1.0		MIT
3002.652	3003.527	33294.19	PD I	2.0R	1.8		MIT
2999.550	3000.425	33328.62	PD II		2.0H		MIT
2997.491	2998.365	33351.51	PD II		0.3J		MIT
2980.655	2981.525	33539.88	PD II		2.3G		MIT
2956.495	2957.359	33813.96	PD II		0.8H		MIT
2955.28	2956.14	33827.9	PD II		0.7		SH
2954.394	2955.257	33838.00	PD II		1.8H		MIT
2953.774	2954.637	33845.10	PD II		0.3H		MIT
2951.66	2952.52	33869.3	PD II		0.3H		BX
2949.069	2949.931	33899.10	PD II		0.6H		MIT
2946.44	2947.30	33929.3	PD II		0.3		MIT
2939.43	2940.29	34010.3	PD II		0.3H		MIT
2936.77	2937.63	34041.1	PD I	0.3	0.0		SH
2934.84	2935.70	34063.4	PD II		1.0H		MIT
2931.462	2932.320	34102.69	PD I	1.0	0.3		MIT
2927.304	2928.161	34151.13	PD II		1.0		MIT
2925.417	2926.273	34173.16	PD II		0.3J		MIT
2922.492	2923.347	34207.36	PD I	2.3	1.4		MIT
2920.36	2921.21	34232.3	PD II		0.6J		BX
2911.03	2911.88	34342.0	PD II		0.3J		BX
2910.708	2911.560	34345.84	PD		0.3J		MIT
2900.25	2901.10	34469.7	PD		0.3J		BX
2893.089	2893.937	34555.00	PD II		2.0		MIT
2889.97	2890.82	34592.3	PD II		0.3H		BX
2880.32	2881.16	34708.2	PD II			0.5	BX
2877.87	2878.71	34737.7	PD II		1.2J		MIT
2875.744	2876.588	34763.41	PD	0.3			MIT
2871.37	2872.21	34816.4	PD II		1.0J		MIT
2870.16	2871.00	34831.0	PD II		0.3H		MIT
2859.316	2860.156	34963.13	PD II		1.2		MIT
2857.726	2858.565	34982.58	PD II		2.0		MIT
2854.581	2855.420	35021.12	PD II	0.6	2.7H		MIT
2853.658	2854.496	35032.45	PD II		0.7J		MIT
2850.73	2851.57	35068.4	PD II		0.7J		BX
2846.762	2847.599	35117.31	PD II		1.2		MIT
2841.029	2841.864	35188.17	PD II		2.0		MIT
2839.892	2840.727	35202.26	PD II		2.0		MIT
2837.655	2838.489	35230.00	PD II		1.0J		MIT
2837.12	2837.95	35236.6	PD II			1.6	BX
2829.213	2830.045	35335.12	PD II		0.3H		MIT
2826.336	2827.168	35371.09	PD II		0.3J		MIT
2822.943	2823.774	35413.60	PD II		0.7J		MIT
2821.853	2822.684	35427.28	PD II		1.0J		MIT
2814.004	2814.833	35526.09	PD II		1.0		MIT
2811.593	2812.421	35556.55	PD II		0.5H		MIT
2808.331	2809.158	35597.85	PD II		0.3H		MIT
2807.55	2808.38	35607.7	PD II		1.2H		MIT
2807.327	2808.154	35610.58	PD II		1.2H		MIT
2805.31	2806.14	35636.2	PD			0.3H	BX
2802.473	2803.299	35672.26	PD II		1.0		MIT
2801.547	2802.373	35684.05	PD II		0.5		MIT
2800.74	2801.57	35694.3	PD II			1.0J	BX
2800.638	2801.463	35695.63	PD II		1.0J		MIT
2796.606	2797.430	35747.09	PD II		0.7		MIT
2791.62	2792.44	35810.9	PD II			0.5	BX
2791.11	2791.93	35817.5	PD			0.7	BX
2787.921	2788.743	35858.44	PD II		1.2J		MIT

AIR WAVELENGTH (ANGSTRMS)	VACUUM WAVELENGTH (ANGSTRMS)	WAVENUMBER (KAYSERS)	LMENT	LOG ARC	INTENSITY SPK	INTENSITY DIS	REF
2781.48	2782.30	35941.5	PD II			0.3J	BX
2779.701	2780.521	35964.48	PD II		0.7		MIT
2776.843	2777.663	36001.49	PD II		1.4J		MIT
2771.884	2772.702	36065.90	PD II		0.7		MIT
2765.833	2766.650	36144.80	PD		0.3H		MIT
2763.092	2763.908	36180.65	PD II	2.5R	1.5R		MIT
2762.932	2763.748	36182.75	PD		1.2		MIT
2754.943	2755.757	36287.67	PD		0.7		MIT
2751.55	2752.36	36332.4	PD II			0.3H	BX
2751.06	2751.87	36338.9	PD II			0.3J	BX
2750.51	2751.32	36346.1	PD			0.3J	BX
2742.603	2743.414	36450.93	PD II		2.0		MIT
2739.548	2740.359	36491.58	PD		0.3		MIT
2732.36	2733.17	36587.6	PD II			0.5	BX
2732.07	2732.88	36591.5	PD II			0.7	BX
2731.810	2732.619	36594.93	PD II		0.7		MIT
2731.61	2732.42	36597.6	PD II			1.7	BX
2731.36	2732.17	36601.0	PD II			0.7	BX
2727.891	2728.699	36647.50	PD II		2.3		MIT
2726.78	2727.59	36662.4	PD II			1.0J	BX
2714.902	2715.707	36822.83	PD II		2.3		MIT
2714.319	2715.123	36830.74	PD		2.2		MIT
2709.82	2710.62	36891.9	PD II			0.7	BX
2709.206	2710.009	36900.24	PD II		1.5J		MIT
2703.397	2704.199	36979.53	PD II		0.7H		MIT
2698.547	2699.348	37045.99	PD II		2.3		MIT
2696.41	2697.21	37075.4	PD II		1.0J		MIT
2693.882	2694.682	37110.14	PD II		1.2J		MIT
2688.551	2689.349	37183.72	PD II		2.3		MIT
2687.658	2688.456	37196.07	PD II		2.2		MIT
2686.284	2687.082	37215.10	PD I	0.5	0.0		MIT
2684.761	2685.558	37236.21	PD II		2.0		MIT
2680.67	2681.47	37293.0	PD II			0.5	BX
2679.578	2680.374	37308.23	PD II		2.0		MIT
2679.12	2679.92	37314.6	PD II		1.2J		MIT
2677.851	2678.647	37332.29	PD II		1.5		MIT
2677.025	2677.820	37343.80	PD II		1.0J		MIT
2651.09	2651.88	37709.1	PD II		1.0H		BX
2650.39	2651.18	37719.1	PD II		0.8		MIT
2649.470	2650.259	37732.16	PD II		2.3		MIT
2644.057	2644.845	37809.40	PD II		0.3		MIT
2642.168	2642.955	37836.44	PD II		2.0		MIT
2641.650	2642.437	37843.85	PD		0.3		MIT
2640.184	2640.971	37864.87	PD II		1.8		MIT
2637.069	2637.855	37909.59	PD II		2.0		MIT
2635.942	2636.728	37925.80	PD II	1.7	2.5		MIT
2632.467	2633.252	37975.86	PD II		1.0J		MIT
2630.330	2631.114	38006.71	PD II		1.2J		MIT
2629.014	2629.798	38025.73	PD II		1.0J		MIT
2628.254	2629.038	38036.73	PD II		2.3		MIT
2625.08	2625.86	38082.7	PD II			1.5	BX
2624.63	2625.41	38089.2	PD II			0.5	BX
2620.557	2621.339	38148.44	PD II		1.0J		MIT
2619.053	2619.835	38170.35	PD II		1.0H		MIT
2618.01	2618.79	38185.5	PD II			0.6	BX
2615.12	2615.90	38227.7	PD			1.4	BX
2613.434	2614.214	38252.41	PD II		2.0		MIT
2613.03	2613.81	38258.3	PD II		1.0H		MIT
2609.862	2610.641	38304.76	PD II		2.3		MIT
2606.343	2607.122	38356.48	PD II		0.7H		MIT
2606.20	2606.98	38358.6	PD II			0.6	BX
2605.678	2606.456	38366.27	PD		0.3		MIT
2605.057	2605.835	38375.41	PD	0.5			MIT
2602.760	2603.538	38409.28	PD II	1.3	2.3		MIT
2595.969	2596.745	38509.75	PD II	0.5	2.2		MIT
2594.273	2595.049	38534.92	PD II		1.4J		MIT
2593.269	2594.044	38549.84	PD I	0.5	2.0		MIT
2592.09	2592.87	38567.4	PD II			0.7H	BX
2587.289	2588.063	38638.94	PD II		1.2H		MIT
2586.56	2587.33	38649.8	PD II		0.3		MIT

AIR WAVELENGTH (ANGSTRMS)	VACUUM WAVELENGTH (ANGSTRMS)	WAVENUMBER (KAYSERS)	LMENT	LOG ARC	INTENSITY SPK	INTENSITY DIS	REF
2584.132	2584.905	38686.14	PD II		1.9		MIT
2583.849	2584.622	38690.37	PD II		2.3		MIT
2583.029	2583.802	38702.65	PD II		0.8		MIT
2577.097	2577.869	38791.74	PD II	0.5	2.2		MIT
2576.399	2577.171	38802.24	PD II		2.0		MIT
2575.490	2576.261	38815.94	PD II		2.0		MIT
2572.643	2573.414	38858.89	PD II		0.3H		MIT
2569.554	2570.324	38905.60	PD II	1.3	2.2		MIT
2568.03	2568.80	38928.7	PD II		1.2J		MIT
2567.20	2567.97	38941.3	PD II			0.3H	BX
2565.507	2566.276	38966.97	PD II	0.3	2.3		MIT
2564.507	2565.276	38982.17	PD II		1.0J		MIT
2561.67	2562.44	39025.3	PD			0.5H	BX
2561.44	2562.21	39028.8	PD		0.3H		MIT
2561.024	2561.792	39035.18	PD II		2.3		MIT
2553.74	2554.51	39146.5	PD			1.3	BX
2552.28	2553.05	39168.9	PD II		0.3H		MIT
2551.848	2552.614	39175.53	PD II		2.0H		MIT
2550.989	2551.755	39188.72	PD II		1.5		MIT
2550.656	2551.421	39193.84	PD II		2.2		MIT
2544.833	2545.597	39283.52	PD II		2.3		MIT
2543.35	2544.11	39306.4	PD II			0.3	BX
2541.998	2542.761	39327.32	PD II		1.2J		MIT
2539.358	2540.121	39368.21	PD II		1.7J		MIT
2537.974	2538.736	39389.67	PD II		2.0		MIT
2537.169	2537.931	39402.17	PD II		2.0		MIT
2535.94	2536.70	39421.3	PD			0.5H	BX
2534.599	2535.361	39442.12	PD II		2.0		MIT
2533.966	2534.728	39451.97	PD		0.7H		MIT
2532.21	2532.97	39479.3	PD			0.7H	BX
2526.76	2527.52	39564.5	PD			0.5H	BX
2521.917	2522.676	39640.45	PD II		0.3H		MIT
2521.496	2522.255	39647.07	PD II		0.3H		MIT
2521.30	2522.06	39650.1	PD II			0.7H	BX
2518.291	2519.049	39697.52	PD		0.5H		DN
2518.117	2518.875	39700.27	PD II		1.2J		MIT
2515.33	2516.09	39744.2	PD II			1.0J	BX
2514.478	2515.235	39757.72	PD II		2.3		MIT
2511.22	2511.98	39809.3	PD II			0.5H	BX
2509.114	2509.870	39842.71	PD II		0.3H		MIT
2508.92	2509.68	39845.8	PD II			1.7	BX
2508.06	2508.82	39859.5	PD II		1.2J		MIT
2505.739	2506.494	39896.37	PD II	0.5	1.5		MIT
2500.036	2500.790	39987.37	PD		0.6H		DN
2499.11	2499.86	40002.2	PD II			0.8	BX
2498.784	2499.537	40007.40	PD II	0.6	2.2H		MIT
2498.22	2498.97	40016.4	PD II		0.5H		MIT
2496.686	2497.439	40041.02	PD II		2.0		MIT
2489.611	2490.362	40154.80	PD II		1.9		MIT
2488.921	2489.672	40165.93	PD II	1.0	1.5		MIT
2488.412	2489.163	40174.15	PD II		1.0		MIT
2487.804	2488.555	40183.97	PD II		0.3H		MIT
2486.528	2487.278	40204.58	PD II	0.7	1.5		MIT
2484.01	2484.76	40245.3	PD			0.3	BX
2481.865	2482.614	40280.12	PD		0.8		MIT
2479.115	2479.864	40324.80	PD II		0.7H		DN
2478.802	2479.551	40329.89	PD II		1.2H		MIT
2478.565	2479.314	40333.74	PD		1.4		MIT
2477.565	2478.313	40350.02	PD		0.7H		MIT
2477.46	2478.21	40351.7	PD II			0.8	BX
2477.005	2477.753	40359.14	PD II		1.4		MIT
2476.57	2477.32	40366.2	PD			1.3	BX
2476.418	2477.166	40368.71	PD I	2.5R	1.7		MIT
2475.905	2476.653	40377.07	PD		0.3H		MIT
2473.532	2474.279	40415.81	PD		0.3H		MIT
2472.512	2473.259	40432.48	PD II		2.2		MIT
2471.152	2471.899	40454.73	PD II		2.2		MIT
2470.011	2470.758	40473.42	PD II		2.2		MIT
2469.77	2470.52	40477.4	PD II			0.8	BX
2469.254	2470.000	40485.82	PD II		2.2		MIT

AIR WAVELENGTH (ANGSTRMS)	VACUUM WAVELENGTH (ANGSTRMS)	WAVENUMBER (KAYSERS)	LMENT	LOG ARC	INTENSITY SPK	INTENSITY DIS	REF
2469.013	2469.759	40489.77	PD		0.3		MIT
2467.92	2468.67	40507.7	PD II			0.3	BX
2465.458	2466.204	40548.15	PD		0.3H		MIT
2463.970	2464.715	40572.64	PD		1.2H		MIT
2461.89	2462.63	40606.9	PD		0.3H		MIT
2461.51	2462.25	40613.2	PD		1.3		SX
2459.702	2460.446	40643.03	PD		0.3H		MIT
2457.764	2458.508	40675.08	PD II	0.3	0.9		MIT
2457.260	2458.004	40683.42	PD II	0.3	1.0		MIT
2457.02	2457.76	40687.4	PD			0.5	BX
2456.268	2457.011	40699.85	PD		0.3		MIT
2454.763	2455.506	40724.80	PD II		1.2H		MIT
2453.474	2454.217	40746.20	PD II		0.5		MIT
2452.758	2453.501	40758.09	PD II		0.3H		MIT
2452.433	2453.176	40763.49	PD		1.0		MIT
2451.043	2451.785	40786.61	PD		1.0H		MIT
2450.958	2451.700	40788.02	PD		1.4J		MIT
2450.039	2450.781	40803.32	PD		1.2H		MIT
2448.164	2448.906	40834.57	PD II		2.0		MIT
2447.909	2448.651	40838.82	PD I	2.3R	2.0		MIT
2446.714	2447.455	40858.77	PD II		1.7		MIT
2446.182	2446.923	40867.65	PD II		2.0		MIT
2444.375	2445.116	40897.86	PD		0.7H		MIT
2444.206	2444.947	40900.69	PD II		1.2H		MIT
2443.614	2444.355	40910.60	PD		0.5H		DN
2443.45	2444.19	40913.3	PD II			0.7	BX
2442.01	2442.75	40937.5	PD			0.3	BX
2440.014	2440.754	40970.95	PD		0.3H		MIT
2437.930	2438.669	41005.97	PD II		1.2J		MIT
2437.797	2438.536	41008.21	PD II		1.2H		MIT
2437.76	2438.50	41008.8	PD		0.5H		MIT
2436.499	2437.238	41030.05	PD		1.4J		MIT
2436.348	2437.087	41032.60	PD II		0.3H		MIT
2435.322	2436.061	41049.88	PD II		1.7		MIT
2433.104	2433.842	41087.30	PD II		1.7		MIT
2431.777	2432.515	41109.72	PD II		1.2		MIT
2431.46	2432.20	41115.1	PD II			1.4	BX
2430.929	2431.667	41124.06	PD II		1.5H		MIT
2430.538	2431.276	41130.67	PD II		1.4		MIT
2430.252	2430.990	41135.51	PD II		1.0		MIT
2430.078	2430.815	41138.46	PD II		1.0		MIT
2426.868	2427.605	41192.87	PD II		1.7		MIT
2426.735	2427.472	41195.12	PD		0.5		MIT
2426.08	2426.82	41206.2	PD			1.5	BX
2425.798	2426.535	41211.04	PD II		1.2		MIT
2425.02	2425.76	41224.3	PD II			0.5	BX
2424.480	2425.216	41233.44	PD II		2.0		MIT
2423.390	2424.126	41251.98	PD II		0.3H		MIT
2422.642	2423.378	41264.72	PD II		0.3H		MIT
2422.394	2423.130	41268.94	PD II		1.3H		MIT
2422.048	2422.784	41274.84	PD II		1.2H		MIT
2419.410	2420.145	41319.84	PD II		0.3H		MIT
2418.727	2419.462	41331.50	PD II		1.7		MIT
2418.56	2419.29	41334.4	PD			0.5	BX
2417.09	2417.82	41359.5	PD			0.5	BX
2416.684	2417.418	41366.44	PD II		1.4J		MIT
2416.349	2417.083	41372.18	PD		0.3		MIT
2415.614	2416.348	41384.76	PD II		1.5		MIT
2415.306	2416.040	41390.04	PD		0.7H		MIT
2414.95	2415.68	41396.1	PD II			0.3H	BX
2414.732	2415.466	41399.88	PD II		2.2		MIT
2413.91	2414.64	41414.0	PD II			0.5	BX
2413.390	2414.124	41422.90	PD II		1.6		MIT
2411.756	2412.489	41450.96	PD II			1.4H	MIT
2410.514	2411.247	41472.32	PD		0.3		MIT
2410.186	2410.919	41477.96	PD II		1.4J		MIT
2408.736	2409.469	41502.93	PD II		2.0		MIT
2406.743	2407.475	41537.29	PD II	0.0	2.2		MIT
2406.34	2407.07	41544.2	PD II			1.8	BX
2406.071	2406.803	41548.89	PD II		0.6		MIT

AIR WAVELENGTH (ANGSTRMS)	VACUUM WAVELENGTH (ANGSTRMS)	WAVENUMBER (KAYSERS)	LMENT	LOG ARC	INTENSITY SPK	INTENSITY DIS	REF
2401.479	2402.210	41628.33	PD II		0.7H		MIT
2401.383	2402.114	41630.00	PD		1.4H		MIT
2401.130	2401.861	41634.38	PD		1.0		MIT
2400.97	2401.70	41637.2	PD II			1.1	BX
2400.331	2401.062	41648.24	PD		0.3		MIT
2399.37	2400.10	41664.9	PD		0.5H		MIT
2398.830	2399.560	41674.30	PD		1.2H		MIT
2396.40	2397.13	41716.6	PD II			0.6	BX
2396.11	2396.84	41721.6	PD			0.7	BX
2391.35	2392.08	41804.6	PD			0.9	BX
2390.84	2391.57	41813.6	PD II		1.1H		MIT
2390.13	2390.86	41826.0	PD	0.3	0.6H		MIT
2389.86	2390.59	41830.7	PD II		0.3		MIT
2388.289	2389.017	41858.22	PD II	0.3	1.6		MIT
2387.58	2388.31	41870.6	PD		0.7H		MIT
2386.24	2386.97	41894.2	PD II		0.7J		MIT
2385.17	2385.90	41913.0	PD		1.2H		MIT
2385.01	2385.74	41915.8	PD II		1.7J		MIT
2383.40	2384.13	41944.1	PD II		1.7J		MIT
2383.24	2383.97	41946.9	PD II		0.3		MIT
2382.756	2383.483	41955.41	PD II	0.3	0.3H		MIT
2382.57	2383.30	41958.7	PD		1.6H		MIT
2382.25	2382.98	41964.3	PD		0.3H		MIT
2381.78	2382.51	41972.6	PD II			0.3	BX
2381.024	2381.750	41985.93	PD	1.4	1.0		MIT
2379.638	2380.364	42010.38	PD		1.2H		MIT
2379.50	2380.23	42012.8	PD II		0.3H		MIT
2378.721	2379.447	42026.58	PD		1.0H		MIT
2377.916	2378.642	42040.80	PD II		1.5		MIT
2376.71	2377.44	42062.1	PD II	0.3	0.3		MIT
2374.97	2375.69	42093.0	PD II			0.5	BX
2374.162	2374.887	42107.27	PD II		0.5		MIT
2373.92	2374.64	42111.6	PD II		0.3H		MIT
2372.98	2373.70	42128.2	PD		1.0		MIT
2372.154	2372.878	42142.91	PD II	0.7	1.7		MIT
2371.51	2372.23	42154.4	PD		0.3H		MIT
2369.07	2369.79	42197.8	PD II			0.6	BX
2368.54	2369.26	42207.2	PD II			0.6	BX
2367.960	2368.683	42217.55	PD	1.0	1.8		MIT
2367.94	2368.66	42217.9	PD II		0.3		MIT
2367.59	2368.31	42224.1	PD II			0.3H	BX
2367.047	2367.770	42233.83	PD II	1.0	1.1		MIT
2366.292	2367.015	42247.30	PD II		1.2		MIT
2365.614	2366.337	42259.41	PD		0.6H		MIT
2365.48	2366.20	42261.8	PD II		1.2H		MIT
2364.81	2365.53	42273.8	PD II		1.0H		MIT
2364.61	2365.33	42277.4	PD			0.8	BX
2364.06	2364.78	42287.2	PD		0.3H		MIT
2363.93	2364.65	42289.5	PD	0.7	0.8		MIT
2363.540	2364.262	42296.49	PD II		1.4W		MIT
2362.95	2363.67	42307.0	PD			0.9	BX
2362.77	2363.49	42310.3	PD	0.8			MIT
2362.322	2363.044	42318.29	PD II	0.7	1.5		MIT
2361.484	2362.206	42333.31	PD	1.0	1.0		MIT
2361.31	2362.03	42336.4	PD II		1.4H		MIT
2361.123	2361.845	42339.78	PD II		0.7		MIT
2360.95	2361.67	42342.9	PD II		0.3H		MIT
2360.797	2361.519	42345.63	PD		0.6		MIT
2360.53	2361.25	42350.4	PD	1.1			MIT
2357.630	2358.351	42402.51	PD II	0.5	1.6		MIT
2357.432	2358.153	42406.07	PD	0.3	1.2		MIT
2355.728	2356.449	42436.74	PD	0.3	1.4H		MIT
2355.510	2356.231	42440.67	PD II	0.8	1.4V		MIT
2355.33	2356.05	42443.9	PD II		0.7H		MIT
2354.797	2355.517	42453.52	PD II		1.4		MIT
2354.220	2354.940	42463.92	PD		1.3		MIT
2352.99	2353.71	42486.1	PD	0.6	0.8		MIT
2351.857	2352.577	42506.58	PD II		1.5		MIT
2351.338	2352.058	42515.96	PD II	1.0	1.8		MIT
2350.70	2351.42	42527.5	PD	0.8	1.4H		MIT

AIR WAVELENGTH (ANGSTRMS)	VACUUM WAVELENGTH (ANGSTRMS)	WAVENUMBER (KAYSERS)	LMENT	LOG ARC	INTENSITY SPK	INTENSITY DIS	REF
2349.38	2350.10	42551.4	PD			0.3	BX
2349.02	2349.74	42557.9	PD		1.2		MIT
2347.76	2348.48	42580.7	PD II			1.4	BX
2347.527	2348.246	42584.98	PD	1.3	1.3		MIT
2347.306	2348.025	42588.99	PD II	0.5	1.4W		MIT
2347.17	2347.89	42591.5	PD II			1.0	BX
2346.463	2347.182	42604.29	PD II	0.3	1.4		MIT
2345.52	2346.24	42621.4	PD II			0.9	BX
2344.937	2345.655	42632.01	PD II		1.3		MIT
2344.74	2345.46	42635.6	PD II	0.3			MIT
2342.46	2343.18	42677.1	PD	0.7	0.5		MIT
2341.93	2342.65	42686.7	PD		0.3		MIT
2341.13	2341.85	42701.3	PD		0.5		MIT
2340.26	2340.98	42717.2	PD II			0.5	BX
2340.039	2340.756	42721.24	PD II	1.0	0.7		MIT
2339.64	2340.36	42728.5	PD	1.0	0.3		MIT
2339.26	2339.98	42735.5	PD		0.5		MIT
2338.10	2338.82	42756.7	PD II			0.3	BX
2337.77	2338.49	42762.7	PD II		1.4		MIT
2336.600	2337.316	42784.11	PD II	0.5	1.5		MIT
2336.411	2337.127	42787.57	PD II	0.3	1.5		MIT
2335.57	2336.29	42803.0	PD			0.7	BX
2335.43	2336.15	42805.5	PD			0.5	BX
2334.066	2334.782	42830.55	PD II		0.6		MIT
2333.26	2333.98	42845.4	PD	0.5			MIT
2332.81	2333.53	42853.6	PD			0.3	BX
2332.544	2333.259	42858.50	PD II		1.4W		MIT
2331.413	2332.128	42879.29	PD II		1.6		MIT
2330.04	2330.75	42904.5	PD II		0.9H		MIT
2329.79	2330.50	42909.2	PD	0.6			MIT
2328.56	2329.27	42931.8	PD II		1.0H		BX
2327.51	2328.22	42951.2	PD I	0.7	0.3H		MIT
2327.35	2328.06	42954.1	PD			0.9	BX
2326.11	2326.82	42977.0	PD II			0.8	BX
2322.79	2323.50	43038.5	PD II			0.5	BX
2322.576	2323.289	43042.42	PD		1.4		MIT
2321.896	2322.609	43055.03	PD II		1.4		MIT
2320.56	2321.27	43079.8	PD II		0.7H		BX
2319.49	2320.20	43099.7	PD	0.3	1.4W		MIT
2319.25	2319.96	43104.1	PD	1.2			MIT
2318.06	2318.77	43126.3	PD II	0.5	1.0H		MIT
2317.96	2318.67	43128.1	PD II			0.7	BX
2317.37	2318.08	43139.1	PD	0.3			MIT
2316.47	2317.18	43155.9	PD	1.1			MIT
2315.873	2316.585	43166.99	PD II	0.3	1.4		MIT
2315.65	2316.36	43171.1	PD			0.7	BX
2314.46	2315.17	43193.3	PD II	0.5	0.3H		MIT
2313.80	2314.51	43205.7	PD			0.6	BX
2312.29	2313.00	43233.9	PD II			0.8	BX
2310.57	2311.28	43266.1	PD II			1.0	BX
2309.52	2310.23	43285.7	PD		0.8		MIT
2308.604	2309.314	43302.90	PD II		1.5		MIT
2307.50	2308.21	43323.6	PD II		1.7		MIT
2306.99	2307.70	43333.2	PD			0.3	BX
2306.75	2307.46	43337.7	PD	0.5	0.3		MIT
2305.71	2306.42	43357.2	PD		0.7		MIT
2304.86	2305.57	43373.2	PD II	0.5	0.5		MIT
2304.35	2305.06	43382.8	PD			0.8	BX
2303.47	2304.18	43399.4	PD	0.6			MIT
2302.319	2303.028	43421.10	PD II		1.2		MIT
2302.14	2302.85	43424.5	PD			0.5	BX
2302.012	2302.721	43426.89	PD II		1.8		MIT
2299.598	2300.306	43472.47	PD II		1.2		MIT
2299.432	2300.140	43475.61	PD II		1.4		MIT
2296.513	2297.221	43530.87	PD II	1.3	1.8		MIT
2295.77	2296.48	43545.0	PD		1.0		MIT
2295.51	2296.22	43549.9	PD II			1.1	BX
2294.01	2294.72	43578.4	PD II		1.0H		MIT
2293.70	2294.41	43584.2	PD II		1.0		MIT
2293.59	2294.30	43586.3	PD II		1.4H		MIT

AIR WAVELENGTH (ANGSTRMS)	VACUUM WAVELENGTH (ANGSTRMS)	WAVENUMBER (KAYSERS)	LMENT	LOG ARC	INTENSITY SPK	INTENSITY DIS	REF
2293.357	2294.064	43590.77	PD II		1.4		MIT
2292.98	2293.69	43597.9	PD	0.3			MIT
2291.44	2292.15	43627.2	PD		1.3		MIT
2288.59	2289.30	43681.5	PD		0.6		MIT
2287.55	2288.26	43701.4	PD			1.0	BX
2286.78	2287.49	43716.1	PD II			0.9	BX
2283.09	2283.79	43786.8	PD			0.5	BX
2282.977	2283.682	43788.94	PD II		1.2D		MIT
2282.45	2283.15	43799.0	PD II		1.2		MIT
2282.100	2282.804	43805.77	PD II		1.5		MIT
2281.37	2282.07	43819.8	PD II			0.6D	BX
2280.827	2281.531	43830.21	PD II	1.2	1.5		MIT
2280.43	2281.13	43837.8	PD			1.4	BX
2276.020	2276.723	43922.78	PD		1.5H		MIT
2274.811	2275.514	43946.12	PD II		1.0W		MIT
2274.454	2275.157	43953.02	PD II		1.3		MIT
2273.31	2274.01	43975.1	PD II		1.5		MIT
2270.209	2270.911	44035.19	PD		1.6J		MIT
2270.11	2270.81	44037.1	PD II			1.6	BX
2266.95	2267.65	44098.5	PD II	0.5	1.3		MIT
2266.44	2267.14	44108.4	PD II		0.5H		MIT
2265.75	2266.45	44121.9	PD II		0.5H		MIT
2264.68	2265.38	44142.7	PD II			0.5	BX
2264.282	2264.983	44150.45	PD II		1.4		MIT
2263.63	2264.33	44163.2	PD II			0.3	BX
2262.484	2263.184	44185.53	PD II		1.5		MIT
2262.30	2263.00	44189.1	PD			1.2	BX
2262.141	2262.841	44192.23	PD II		1.4		MIT
2261.29	2261.99	44208.9	PD			1.3	BX
2260.53	2261.23	44223.7	PD		1.4		MIT
2260.45	2261.15	44225.3	PD II			1.3	BX
2260.13	2260.83	44231.5	PD II		1.4		MIT
2259.32	2260.02	44247.4	PD II			0.7	BX
2259.03	2259.73	44253.1	PD II			0.5	BX
2258.55	2259.25	44262.5	PD		0.7		MIT
2257.50	2258.20	44283.1	PD	0.6			MIT
2254.448	2255.146	44343.02	PD II		1.0		MIT
2254.26	2254.96	44346.7	PD I	1.4	0.9		MIT
2253.74	2254.44	44357.0	PD			0.3	BX
2253.655	2254.353	44358.62	PD	0.3	1.4		MIT
2252.618	2253.316	44379.04	PD II		1.0H		MIT
2252.03	2252.73	44390.6	PD II		1.5		MIT
2251.50	2252.20	44401.1	PD II	0.8	1.5		MIT
2249.51	2250.21	44440.4	PD II		0.9		MIT
2248.03	2248.73	44469.6	PD		0.9		MIT
2247.40	2248.10	44482.1	PD II			0.5	BX
2247.01	2247.71	44489.8	PD II		0.9		MIT
2245.12	2245.82	44527.2	PD II		0.7H		MIT
2243.90	2244.60	44551.4	PD	0.5	1.3		MIT
2243.541	2244.237	44558.57	PD		1.3		MIT
2243.17	2243.87	44565.9	PD			0.3	BX
2242.54	2243.24	44578.5	PD			0.8	BX
2241.51	2242.21	44598.9	PD II			0.6	BX
2240.76	2241.46	44613.9	PD I	0.6			ME
2240.58	2241.28	44617.5	PD II	0.5			MIT
2237.721	2238.416	44674.45	PD II		1.3		MIT
2236.84	2237.53	44692.0	PD II	0.5	0.9		MIT
2236.619	2237.314	44696.46	PD		0.8		MIT
2236.38	2237.07	44701.2	PD	1.0			MIT
2235.236	2235.930	44724.11	PD		0.9		MIT
2235.05	2235.74	44727.8	PD			0.5	BX
2234.25	2234.94	44743.9	PD			1.2	BX
2232.63	2233.32	44776.3	PD			0.6	BX
2231.59	2232.28	44797.2	PD II	1.0	1.8		MIT
2231.317	2232.010	44802.66	PD II		1.0		MIT
2230.75	2231.44	44814.0	PD			0.3	BX
2230.43	2231.12	44820.5	PD II			0.6	BX
2229.76	2230.45	44833.9	PD II			0.6	BX
2229.220	2229.913	44844.80	PD II		1.5		MIT
2228.40	2229.09	44861.3	PD II			0.7	BX

AIR WAVELENGTH (ANGSTRMS)	VACUUM WAVELENGTH (ANGSTRMS)	WAVENUMBER (KAYSERS)	LMENT	LOG ARC	INTENSITY SPK	INTENSITY DIS	REF
2225.29	2225.98	44924.0	PD	1.2	0.7		MIT
2224.14	2224.83	44947.2	PD			0.3	BX
2223.726	2224.418	44955.58	PD II		1.5		MIT
2223.57	2224.26	44958.7	PD			0.3	BX
2223.079	2223.771	44968.67	PD II		1.1		MIT
2222.29	2222.98	44984.6	PD		1.5		MIT
2220.20	2220.89	45027.0	PD			0.7	BX
2218.158	2218.849	45068.42	PD II		1.6		MIT
2217.537	2218.228	45081.04	PD II	0.3	1.4		MIT
2217.324	2218.015	45085.37	PD II		1.4		MIT
2216.48	2217.17	45102.5	PD I	0.5W			SH
2216.48	2217.17	45102.5	PD II	0.5W			SH
2214.038	2214.728	45152.27	PD II		1.6		MIT
2212.81	2213.50	45177.3	PD		1.0		MIT
2212.44	2213.13	45184.9	PD	0.3H			MIT
2212.14	2212.83	45191.0	PD II	0.7	1.7		MIT
2210.57	2211.26	45223.1	PD			0.3	BX
2207.47	2208.16	45286.6	PD II	0.6	1.5		MIT
2206.32	2207.01	45310.2	PD			0.6	BX
2205.41	2206.10	45328.9	PD II			1.4	BX
2204.80	2205.49	45341.4	PD			0.3	BX
2204.55	2205.24	45346.6	PD			0.6	BX
2203.478	2204.166	45368.64	PD II		1.4		MIT
2202.355	2203.042	45391.77	PD II		1.6		MIT
2200.09	2200.78	45438.5	PD			0.3	BX
2198.240	2198.926	45476.74	PD II		1.6		MIT
2197.41	2198.10	45493.9	PD		1.3		MIT
2197.14	2197.83	45499.5	PD II		0.8		MIT
2196.29	2196.98	45517.1	PD II		0.5		MIT
2195.51	2196.20	45533.3	PD I	1.0			MIT
2194.56	2195.25	45553.0	PD II		0.5H		SH
2193.91	2194.60	45566.5	PD II		1.0		MIT
2193.47	2194.16	45575.6	PD			0.6	BX
2193.26	2193.95	45580.0	PD II		1.2		MIT
2191.92	2192.61	45607.9	PD II			0.5	BX
2190.55	2191.23	45636.4	PD II			0.6	BX
2190.45	2191.13	45638.5	PD II			0.8	BX
2184.34	2185.02	45766.1	PD			0.3	BX
2182.344	2183.027	45807.95	PD II		1.5		MIT
2180.86	2181.54	45839.1	PD II		0.7		MIT
2180.67	2181.35	45843.1	PD II		0.8		MIT
2179.40	2180.08	45869.8	PD II		1.0		MIT
2178.27	2178.95	45893.6	PD II	1.2	0.5		MIT
2178.27	2178.95	45893.6	PD I	1.2	0.5		MIT
2176.90	2177.58	45922.5	PD II		1.4		MIT
2174.66	2175.34	45969.8	PD	0.6			MIT
2172.911	2173.592	46006.79	PD I	1.1	0.6		MIT
2172.383	2173.064	46017.97	PD II		1.3		MIT
2166.36	2167.04	46145.9	PD			0.5	BX
2165.458	2166.138	46165.12	PD II		1.2		ML
2165.29	2165.97	46168.7	PD II	0.5	1.2		MIT
2165.18	2165.86	46171.0	PD II		1.4		MIT
2162.271	2162.950	46233.15	PD II	0.5	1.6		MIT
2155.91	2156.59	46369.5	PD			0.3	BX
2152.75	2153.43	46437.6	PD II		1.5		MIT
2151.02	2151.70	46475.0	PD	1.0			MIT
2149.13	2149.81	46515.8	PD		1.5		MIT
2148.28	2148.96	46534.2	PD II		1.4		MIT
2147.96	2148.64	46541.1	PD		1.3		MIT
2146.70	2147.38	46568.5	PD II		0.9		MIT
2144.25	2144.93	46621.7	PD	0.3	1.4		MIT
2142.57	2143.25	46658.2	PD	0.5			A
2142.13	2142.80	46667.8	PD I	1.2			MIT
2141.30	2141.97	46685.9	PD II		0.7H		MIT
2140.69	2141.36	46699.2	PD	0.3H	0.5		MIT
2140.26	2140.93	46708.6	PD II		1.5		MIT
2137.24	2137.91	46774.6	PD II		1.6		MIT
2135.646	2136.320	46809.47	PD II		1.4		MIT
2135.078	2135.752	46821.93	PD II		1.4		MIT
2134.86	2135.53	46826.7	PD II		1.1		MIT

AIR WAVELENGTH (ANGSTRMS)	VACUUM WAVELENGTH (ANGSTRMS)	WAVENUMBER (KAYSERS)	LMENT	LOG ARC	INTENSITY SPK	DIS	REF
2133.30	2133.97	46860.9	PD II	0.5	0.3		MIT
2130.677	2131.350	46918.63	PD I	0.8			MIT
2126.652	2127.324	47007.42	PD II		1.3		MIT
2124.818	2125.489	47047.99	PD	0.3H	1.0		MIT
2124.33	2125.00	47058.8	PD II		0.6H		BX
2123.76	2124.43	47071.4	PD	0.5			MIT
2123.669	2124.340	47073.44	PD		1.4		MIT
2122.636	2123.307	47096.34	PD	0.5			MIT
2120.88	2121.55	47135.3	PD		1.0		MIT
2119.590	2120.260	47164.02	PD		1.0		MIT
2118.069	2118.739	47197.88	PD I	1.0			MIT
2117.805	2118.475	47203.77	PD II		1.2		MIT
2117.079	2117.749	47219.95	PD	0.3			MIT
2116.888	2117.558	47224.21	PD II		1.2		MIT
2115.55	2116.22	47254.1	PD	0.3			MIT
2110.80	2111.47	47360.4	PD	0.3	0.5		A
2108.10	2108.77	47421.0	PD I	1.0			MIT
2107.68	2108.35	47430.5	PD II		1.0		MIT
2105.87	2106.54	47471.3	PD I	1.3	0.6		MIT
2103.66	2104.33	47521.1	PD II		1.3		MIT
2102.43	2103.10	47548.9	PD II		1.3		MIT
2098.62	2099.29	47635.2	PD II		1.3		MIT
2095.51	2096.18	47705.9	PD II		1.1		MIT
2092.62	2093.29	47771.8	PD I	1.1			MIT
2092.38	2093.05	47777.3	PD II		1.4		MIT
2092.01	2092.68	47785.7	PD II		1.5		MIT
2091.29	2091.95	47802.2	PD	0.3	0.5		MIT
2090.135	2090.800	47828.59	PD II		1.3		MIT
2089.99	2090.65	47831.9	PD I	0.8			MIT
2089.59	2090.25	47841.1	PD II	0.3	1.4		MIT
2088.49	2089.15	47866.3	PD I	0.8			MIT
2087.94	2088.60	47878.9	PD I	1.2	0.7		MIT
2086.49	2087.15	47912.1	PD II		1.4		MIT
2082.27	2082.93	48009.2	PD I	1.3	0.3		MIT
2081.42	2082.08	48028.8	PD II		1.4		MIT
2081.12	2081.78	48035.7	PD I	1.3			MIT
2079.32	2079.98	48077.3	PD	0.3			MIT
2077.95	2078.61	48109.0	PD I	1.0			MIT
2076.47	2077.13	48143.3	PD	0.6			MIT
2072.03	2072.69	48246.5	PD II		1.4		MIT
2068.80	2069.46	48321.8	PD I	1.1			MIT
2068.67	2069.33	48324.8	PD		0.9		MIT
2062.56	2063.22	48467.9	PD II		1.0		MIT
2061.91	2062.57	48483.2	PD		1.1		MIT
2057.76	2058.42	48581.0	PD I	0.5			SH
2057.42	2058.08	48589.0	PD	0.7			MIT
2055.49	2056.15	48634.6	PD		1.1		MIT
2054.43	2055.09	48659.7	PD		1.5		MIT
2045.62	2046.28	48869.3	PD II		2.0		MIT
2045.58	2046.24	48870.2	PD		1.4		MIT
2043.80	2044.46	48912.8	PD II	0.3	1.2		MIT
2042.98	2043.64	48932.4	PD	0.7			MIT
2040.18	2040.84	48999.5	PD	0.9D			MIT
2039.83	2040.49	49008.0	PD I	0.6			MIT
2038.93	2039.59	49029.6	PD		1.3		MIT
2032.02	2032.67	49196.3	PD		1.5		MIT
2026.69	2027.34	49325.6	PD		1.4		MIT
2019.79	2020.44	49494.1	PD I	1.1	0.6		MIT
2019.54	2020.19	49500.3	PD I	1.2	1.0		MIT
2017.34	2017.99	49554.2	PD	0.3H			MIT
2014.20	2014.85	49631.5	PD		1.5		MIT
2013.96	2014.61	49637.4	PD I	0.8			MIT
2013.81	2014.46	49641.1	PD		1.5		MIT
2008.21	2008.86	49779.5	PD II		1.4		MIT
2006.97	2007.62	49810.2	PD	1.3	0.9		MIT
2003.80	2004.45	49889.0	PD		1.6		MIT
2001.49	2002.14	49946.6	PD		1.6		MIT
2000.73	2001.38	49965.6	PD I	0.3	1.0		MIT
2558.1	2558.9	39080.	PO			1.3	KA
2450.0	2450.7	40804.	PO			1.0	KA

AIR WAVELENGTH (ANGSTRMS)	VACUUM WAVELENGTH (ANGSTRMS)	WAVENUMBER (KAYSERS)	LMENT	LOG ARC	INTENSITY SPK	DIS	REF
8543.9	8546.2	11701.	PR	0.5			IT
8510.6	8512.9	11747.	PR	0.5			IT
8024.2	8026.4	12459.	PR	0.5			IT
7988.72	7990.92	12514.2	PR	0.5			IT
7886.0	7888.2	12677.	PR	0.7			IT
7871.65	7873.82	12700.3	PR	0.7			MIT
7786.11	7788.25	12839.8	PR	0.6			MIT
7726.02	7728.15	12939.7	PR	0.3			MIT
7721.81	7723.94	12946.8	PR	0.6			MIT
7714.72	7716.84	12958.7	PR	0.5			MIT
7706.0	7708.1	12973.	PR	0.5			IT
7704.96	7707.08	12975.1	PR	0.5			MIT
7645.67	7647.77	13075.7	PR	0.3			MIT
7543.5	7545.6	13253.	PR	0.7			IT
7541.02	7543.10	13257.2	PR	0.3			MIT
7503.9	7506.0	13323.	PR	0.5			IT
7499.45	7501.52	13330.6	PR	0.5			MIT
7495.62	7497.68	13337.4	PR	0.6			MIT
7451.71	7453.76	13416.0	PR	0.7			MIT
7289.14	7291.15	13715.3	PR	0.3			MIT
7287.71	7289.72	13717.9	PR	0.3			MIT
7285.71	7287.72	13721.7	PR	0.3H			MIT
7259.22	7261.22	13771.8	PR	0.3			MIT
7231.49	7233.48	13824.6	PR	0.6			MIT
7227.71	7229.70	13831.8	PR	0.6			MIT
7208.81	7210.80	13868.1	PR	0.3			MIT
7159.87	7161.84	13962.9	PR	0.6			MIT
7114.55	7116.51	14051.8	PR	0.7			MIT
7099.54	7101.50	14081.5	PR	0.3			MIT
7080.02	7081.97	14120.4	PR	0.5			MIT
7051.06	7053.00	14178.4	PR	0.5			MIT
7042.54	7044.48	14195.5	PR	0.5			MIT
7024.59	7026.53	14231.8	PR	0.3			MIT
7021.54	7023.48	14238.0	PR	0.8			MIT
6980.18	6982.11	14322.3	PR	0.3			MIT
6892.76	6894.66	14504.0	PR	0.5			ED
6887.9	6889.8	14514.	PR	0.3			IT
6884.72	6886.62	14520.9	PR	0.3			ED
6881.23	6883.13	14528.3	PR	0.6			ED
6870.55	6872.45	14550.9	PR	0.3			ED
6852.90	6854.79	14588.3	PR	0.5			ED
6850.55	6852.44	14593.3	PR	0.6			ED
6845.54	6847.43	14604.0	PR	0.3			ED
6830.57	6832.45	14636.0	PR	0.3			ED
6827.70	6829.58	14642.2	PR	0.7			ED
6798.68	6800.56	14704.7	PR	1.3W			MIT
6788.27	6790.14	14727.2	PR	0.3H			ED
6749.28	6751.14	14812.3	PR	0.8			ED
6747.2	6749.1	14817.	PR	1.30			KN
6736.89	6738.75	14839.6	PR	0.7			ED
6691.33	6693.18	14940.6	PR	0.30			ED
6673.778	6675.621	14979.88	PR	1.6			MIT
6673.410	6675.253	14980.71	PR	1.5			MIT
6656.829	6658.667	15018.02	PR	1.5			MIT
6647.119	6648.955	15039.96	PR	0.7			MIT
6632.04	6633.87	15074.2	PR	0.3			ED
6616.650	6618.477	15109.21	PR	1.5			MIT
6609.850	6611.676	15124.76	PR	0.6			MIT
6595.470	6597.292	15157.73	PR	0.5			MIT
6578.001	6579.818	15197.99	PR	0.5			MIT
6571.044	6572.859	15214.08	PR	0.3H			MIT
6566.747	6568.561	15224.04	PR	1.30			MIT
6564.632	6566.445	15228.94	PR	1.0	0.0		MIT
6560.747	6562.559	15237.96	PR	0.6			MIT
6553.301	6555.111	15255.27	PR	0.6W			MIT
6540.483	6542.290	15285.17	PR	1.2W			MIT
6534.64	6536.45	15298.8	PR	0.5W			ED
6518.807	6520.608	15335.99	PR	0.6			MIT
6517.145	6518.946	15339.90	PR	0.5			MIT
6504.095	6505.892	15370.68	PR	0.6			MIT

AIR WAVELENGTH (ANGSTRMS)	VACUUM WAVELENGTH (ANGSTRMS)	WAVENUMBER (KAYSERS)	LMENT	LOG ARC	INTENSITY SPK	DIS	REF
6500.730	6502.526	15378.64	PR	1.0			MIT
6498.939	6500.735	15382.88	PR	0.5			MIT
6497.18	6498.98	15387.0	PR	0.6W			MIT
6494.899	6496.694	15392.44	PR	0.6			MIT
6493.494	6495.288	15395.78	PR	0.7			MIT
6491.759	6493.553	15399.89	PR	1.3			MIT
6486.557	6488.349	15412.24	PR	1.5			MIT
6478.026	6479.816	15432.54	PR	1.2W			MIT
6475.287	6477.076	15439.06	PR	0.5			MIT
6470.256	6472.044	15451.07	PR	0.5			MIT
6467.757	6469.544	15457.04	PR	0.8			MIT
6462.682	6464.468	15469.18	PR	0.7			MIT
6461.467	6463.253	15472.09	PR	0.3			MIT
6460.30	6462.09	15474.9	PR	0.5W			ED
6456.200	6457.984	15484.71	PR	0.70			MIT
6454.865	6456.649	15487.91	PR	0.5			MIT
6453.441	6455.224	15491.33	PR	1.0			MIT
6443.937	6445.718	15514.18	PR	0.3			MIT
6431.863	6433.641	15543.30	PR	0.9			MIT
6429.645	6431.422	15548.66	PR	0.9	0.0		MIT
6421.52	6423.29	15568.3	PR	0.3H			KN
6419.633	6421.407	15572.91	PR	0.3			MIT
6415.531	6417.304	15582.87	PR	0.50			MIT
6413.699	6415.472	15587.32	PR	1.0	0.0		MIT
6410.660	6412.432	15594.71	PR	0.8W			MIT
6404.395	6406.165	15609.96	PR	0.6W			MIT
6397.996	6399.765	15625.57	PR	1.1			MIT
6393.191	6394.958	15637.32	PR	1.4			MIT
6392.103	6393.870	15639.98	PR	1.00			MIT
6389.589	6391.355	15646.13	PR	0.90			MIT
6378.623	6380.386	15673.03	PR	0.90			MIT
6377.617	6379.380	15675.50	PR	0.7			MIT
6359.043	6360.801	15721.29	PR	1.6W			MIT
6357.237	6358.995	15725.76	PR	1.0W			MIT
6350.984	6352.740	15741.24	PR	1.3W			MIT
6347.721	6349.476	15749.33	PR	0.5			MIT
6347.127	6348.882	15750.81	PR	0.7			MIT
6346.748	6348.503	15751.74	PR	0.5W			MIT
6343.938	6345.692	15758.72	PR	1.0W			MIT
6322.367	6324.115	15812.49	PR	1.5W			MIT
6305.262	6307.006	15855.38	PR	1.0			MIT
6304.026	6305.769	15858.49	PR	0.9			MIT
6302.362	6304.105	15862.68	PR	1.0			MIT
6289.026	6290.765	15896.32	PR	1.2W			MIT
6287.026	6288.765	15901.37	PR	0.5			MIT
6281.306	6283.043	15915.85	PR	1.4	0.0		MIT
6278.675	6280.412	15922.52	PR	1.0			MIT
6262.539	6264.271	15963.55	PR	1.0			MIT
6262.338	6264.070	15964.06	PR	0.3			MIT
6251.022	6252.751	15992.96	PR	0.5W			MIT
6244.344	6246.071	16010.06	PR	1.0	0.0		MIT
6241.110	6242.836	16018.36	PR	0.9W			MIT
6236.828	6238.553	16029.36	PR	0.7			MIT
6218.082	6219.802	16077.68	PR	0.7W			MIT
6212.727	6214.446	16091.54	PR	1.0W			MIT
6210.587	6212.305	16097.08	PR	0.5			MIT
6191.553	6193.266	16146.57	PR	0.7W			MIT
6187.977	6189.689	16155.90	PR	0.6W			MIT
6187.498	6189.210	16157.15	PR	0.5			MIT
6182.343	6184.054	16170.62	PR	1.1	0.0		MIT
6174.332	6176.041	16191.60	PR	0.6W			MIT
6165.945	6167.651	16213.63	PR	1.7	0.3		MIT
6165.256	6166.962	16215.44	PR	0.3			MIT
6162.188	6163.893	16223.51	PR	0.7			MIT
6161.194	6162.899	16226.13	PR	1.7	0.3		MIT
6159.093	6160.797	16231.67	PR	0.5			MIT
6157.804	6159.508	16235.06	PR	0.5			MIT
6148.250	6149.952	16260.29	PR	1.0			MIT
6142.385	6144.085	16275.82	PR	0.3			MIT
6141.508	6143.208	16278.14	PR	0.8			MIT

AIR WAVELENGTH (ANGSTRMS)	VACUUM WAVELENGTH (ANGSTRMS)	WAVENUMBER (KAYSERS)	LMENT	LOG ARC	INTENSITY SPK	DIS	REF
6122.244	6123.939	16329.36	PR	0.7			MIT
6118.022	6119.715	16340.63	PR	0.80			MIT
6114.396	6116.088	16350.32	PR	1.0	0.0		MIT
6109.057	6110.748	16364.61	PR	0.6			MIT
6108.546	6110.237	16365.98	PR	0.3			MIT
6106.759	6108.449	16370.77	PR	0.6			MIT
6105.548	6107.238	16374.01	PR	0.8W			MIT
6096.346	6098.034	16398.73	PR	0.5			MIT
6093.037	6094.724	16407.64	PR	0.5			MIT
6090.372	6092.058	16414.81	PR	0.6			MIT
6090.149	6091.835	16415.42	PR	0.5D	0.3		MIT
6087.505	6089.190	16422.54	PR	1.2			MIT
6086.148	6087.833	16426.21	PR	0.7			MIT
6085.853	6087.538	16427.00	PR	0.9			MIT
6075.192	6076.874	16455.83	PR	0.7W			MIT
6067.236	6068.916	16477.41	PR	0.5			MIT
6055.133	6056.810	16510.34	PR	2.0W			MIT
6050.878	6052.553	16521.95	PR	0.5			MIT
6050.029	6051.704	16524.27	PR	0.6			MIT
6049.257	6050.932	16526.38	PR	0.7			MIT
6046.684	6048.358	16533.41	PR	1.1			MIT
6042.874	6044.547	16543.84	PR	0.7			MIT
6025.723	6027.392	16590.92	PR	1.4	0.3		MIT
6019.903	6021.570	16606.96	PR	0.9W			MIT
6017.803	6019.470	16612.76	PR	1.6	0.3		MIT
6016.492	6018.158	16616.38	PR	1.1			MIT
6006.347	6008.011	16644.45	PR	1.1	0.0		MIT
6002.450	6004.112	16655.25	PR	0.8			MIT
5996.054	5997.715	16673.02	PR	0.9W			MIT
5987.148	5988.806	16697.82	PR	0.6			MIT
5986.140	5987.798	16700.63	PR	1.4W			MIT
5981.209	5982.866	16714.40	PR	1.1	0.0		MIT
5978.882	5980.538	16720.90	PR	0.6			MIT
5976.957	5978.613	16726.29	PR	0.5			MIT
5968.301	5969.954	16750.55	PR	0.6			MIT
5967.837	5969.490	16751.85	PR	1.2			MIT
5963.022	5964.674	16765.38	PR	1.0			MIT
5962.208	5963.860	16767.67	PR	0.8			MIT
5959.308	5960.959	16775.82	PR	0.5W			MIT
5956.617	5958.267	16783.40	PR	1.1	0.0		MIT
5951.783	5953.432	16797.03	PR	0.6			MIT
5951.282	5952.931	16798.45	PR	1.0	0.0		MIT
5949.79	5951.44	16802.7	PR	1.0W			MIT
5947.20	5948.85	16810.0	PR	0.7			MIT
5941.648	5943.294	16825.69	PR	0.7			MIT
5940.733	5942.379	16828.28	PR	1.0	0.0		MIT
5939.913	5941.559	16830.60	PR	1.3	0.3		MIT
5935.436	5937.081	16843.29	PR	0.5			MIT
5924.172	5925.813	16875.32	PR	0.5			MIT
5920.762	5922.403	16885.04	PR	1.0			MIT
5915.984	5917.623	16898.68	PR	0.6			MIT
5915.306	5916.945	16900.61	PR	0.7			MIT
5908.67	5910.31	16919.6	PR	0.9			MIT
5904.448	5906.084	16931.69	PR	1.0			MIT
5903.127	5904.763	16935.48	PR	0.9			MIT
5894.291	5895.924	16960.87	PR	1.2W			MIT
5892.231	5893.864	16966.80	PR	1.0	0.0		MIT
5884.701	5886.332	16988.51	PR	0.6			MIT
5879.253	5880.882	17004.25	PR	1.0	0.0		MIT
5878.111	5879.740	17007.55	PR	1.0W			MIT
5874.732	5876.360	17017.34	PR	0.6			MIT
5868.832	5870.459	17034.44	PR	1.0	0.0		MIT
5859.691	5861.315	17061.02	PR	1.2	0.0		MIT
5856.92	5858.54	17069.1	PR	0.9			MIT
5856.08	5857.70	17071.5	PR	1.0			MIT
5852.63	5854.25	17081.6	PR	0.9			MIT
5850.648	5852.270	17087.39	PR	1.2H			MIT
5847.125	5848.746	17097.68	PR	1.0J			MIT
5845.003	5846.623	17103.89	PR	0.7			MIT
5844.64	5846.26	17104.9	PR	0.6D	0.0		MIT

AIR WAVELENGTH (ANGSTRMS)	VACUUM WAVELENGTH (ANGSTRMS)	WAVENUMBER (KAYSERS)	LMENT	LOG ARC	INTENSITY SPK	DIS	REF
5835.134	5836.752	17132.82	PR	1.2			MIT
5823.723	5825.338	17166.39	PR	1.8W	0.0		MIT
5822.63	5824.24	17169.6	PR	0.9			MIT
5821.351	5822.965	17173.38	PR	0.8W			MIT
5820.615	5822.229	17175.55	PR	0.7			MIT
5818.572	5820.185	17181.58	PR	1.0	0.0		MIT
5815.367	5816.979	17191.05	PR	1.0			MIT
5815.176	5816.788	17191.62	PR	1.2	0.0		MIT
5813.593	5815.205	17196.30	PR	0.5			MIT
5810.622	5812.233	17205.09	PR	1.0			MIT
5800.918	5802.526	17233.87	PR	0.6D			MIT
5792.955	5794.561	17257.56	PR	0.5			MIT
5791.382	5792.988	17262.25	PR	0.9			MIT
5786.183	5787.788	17277.76	PR	0.9			MIT
5785.303	5786.907	17280.39	PR	0.6			MIT
5779.293	5780.896	17298.36	PR	1.7			MIT
5777.285	5778.887	17304.37	PR	0.6			MIT
5775.920	5777.522	17308.46	PR	0.6H			MIT
5769.778	5771.378	17326.88	PR	0.5			MIT
5769.148	5770.748	17328.78	PR	0.5			MIT
5760.205	5761.803	17355.68	PR	0.7			MIT
5756.173	5757.770	17367.84	PR	1.4	0.0		MIT
5753.022	5754.618	17377.35	PR	0.7			MIT
5747.88	5749.47	17392.9	PR	1.0V			MIT
5747.142	5748.736	17395.13	PR	0.6			MIT
5744.928	5746.521	17401.83	PR	0.7H			MIT
5731.87	5733.46	17441.5	PR	1.0			MIT
5730.51	5732.10	17445.6	PR	1.2W			MIT
5728.36	5729.95	17452.2	PR	0.7			MIT
5722.61	5724.20	17469.7	PR	0.5H			MIT
5722.231	5723.818	17470.85	PR	0.5H			MIT
5719.81	5721.40	17478.2	PR	0.8			MIT
5719.63	5721.22	17478.8	PR	0.7	0.0		MIT
5719.090	5720.677	17480.45	PR	1.0	0.0		MIT
5716.08	5717.67	17489.6	PR	0.6			MIT
5715.37	5716.96	17491.8	PR	0.6			MIT
5713.830	5715.415	17496.54	PR	0.8	0.0		MIT
5711.65	5713.23	17503.2	PR	0.9	0.0		MIT
5707.614	5709.198	17515.60	PR	2.0W			MIT
5704.386	5705.969	17525.51	PR	0.9			MIT
5702.22	5703.80	17532.2	PR	0.5			MIT
5697.98	5699.56	17545.2	PR	0.7			MIT
5695.93	5697.51	17551.5	PR	0.9			MIT
5695.480	5697.060	17552.91	PR	0.9			MIT
5692.86	5694.44	17561.0	PR	0.7			MIT
5691.04	5692.62	17566.6	PR	1.3	0.0		MIT
5690.95	5692.53	17566.9	PR	1.0	0.0		MIT
5688.460	5690.038	17574.57	PR	0.9	0.0		MIT
5687.19	5688.77	17578.5	PR	0.6H			MIT
5686.50	5688.08	17580.6	PR	0.7H			MIT
5685.61	5687.19	17583.4	PR	0.5			MIT
5681.896	5683.473	17594.88	PR	0.8	0.0		MIT
5677.04	5678.62	17609.9	PR	0.5			MIT
5674.13	5675.70	17619.0	PR	0.7W	0.0		MIT
5670.017	5671.590	17631.74	PR	0.6	0.0		MIT
5669.553	5671.126	17633.18	PR	0.7	0.0		MIT
5668.449	5670.022	17636.62	PR	1.4W			MIT
5661.564	5663.135	17658.06	PR	1.0W			MIT
5659.844	5661.415	17663.43	PR	0.7	0.0		MIT
5654.247	5655.816	17680.91	PR	0.6	0.3		MIT
5643.195	5644.761	17715.54	PR	0.5H			MIT
5640.342	5641.908	17724.50	PR	0.5			MIT
5638.794	5640.359	17729.37	PR	1.5O	0.0		MIT
5636.455	5638.019	17736.73	PR	0.5	0.0		MIT
5633.024	5634.588	17747.53	PR	0.6			MIT
5624.434	5625.995	17774.63	PR	1.1	0.3		MIT
5623.046	5624.607	17779.02	PR	1.2	0.3		MIT
5621.854	5623.415	17782.79	PR	1.2O	0.0		MIT
5620.251	5621.811	17787.86	PR	0.7	0.0		MIT
5617.478	5619.037	17796.64	PR	0.5			MIT

AIR WAVELENGTH (ANGSTRMS)	VACUUM WAVELENGTH (ANGSTRMS)	WAVENUMBER (KAYSERS)	LMENT	LOG ARC	INTENSITY SPK	DIS	REF
5610.219	5611.776	17819.67	PR	0.8	0.0		MIT
5605.639	5607.195	17834.23	PR	1.0	0.3		MIT
5597.292	5598.846	17860.82	PR	0.5	0.0		MIT
5594.918	5596.471	17868.40	PR	0.8			MIT
5578.813	5580.362	17919.98	PR	0.5			MIT
5571.841	5573.388	17942.41	PR	1.0			MIT
5570.247	5571.794	17947.54	PR	0.3W			MIT
5566.915	5568.461	17958.28	PR	0.5	0.0		MIT
5565.535	5567.081	17962.74	PR	1.0W			MIT
5562.055	5563.600	17973.98	PR	1.2W			MIT
5561.459	5563.003	17975.90	PR	0.5	0.0		MIT
5553.399	5554.941	18001.99	PR	0.7D			MIT
5548.312	5549.853	18018.50	PR	0.5	0.0		MIT
5545.008	5546.548	18029.23	PR	0.7	0.0		MIT
5540.723	5542.262	18043.18	PR	0.5W			MIT
5538.767	5540.305	18049.55	PR	0.5	0.0		MIT
5538.373	5539.911	18050.83	PR	1.0			MIT
5535.176	5536.713	18061.26	PR	1.0	0.3		MIT
5531.153	5532.689	18074.39	PR	1.3O			MIT
5530.193	5531.729	18077.53	PR	0.6	0.0		MIT
5527.876	5529.411	18085.11	PR	0.7W			MIT
5526.569	5528.104	18089.38	PR	0.7W			MIT
5525.900	5527.435	18091.57	PR	0.6D	0.0		MIT
5524.141	5525.675	18097.33	PR	1.0O			MIT
5522.804	5524.338	18101.72	PR	1.2W	0.0		MIT
5520.300	5521.833	18109.93	PR	0.6W			MIT
5515.110	5516.642	18126.97	PR	0.7	0.0		MIT
5513.583	5515.115	18131.99	PR	1.0	0.3		MIT
5511.656	5513.187	18138.33	PR	0.9W			MIT
5509.146	5510.676	18146.59	PR	1.7	0.3		MIT
5508.781	5510.311	18147.79	PR	0.8	0.3		MIT
5497.238	5498.765	18185.90	PR	1.0O			MIT
5492.368	5493.894	18202.03	PR	0.6	0.0		MIT
5491.671	5493.197	18204.34	PR	0.5H			MIT
5490.563	5492.088	18208.01	PR	0.6			MIT
5488.934	5490.459	18213.41	PR	0.5			MIT
5487.577	5489.102	18217.92	PR	0.9			MIT
5487.413	5488.938	18218.46	PR	0.9			MIT
5485.528	5487.052	18224.72	PR	0.6			MIT
5481.759	5483.282	18237.25	PR	0.5			MIT
5479.730	5481.253	18244.00	PR	1.0			MIT
5475.669	5477.191	18257.53	PR	1.3W	0.0		MIT
5471.951	5473.472	18269.94	PR	0.5			MIT
5471.750	5473.270	18270.61	PR	0.5W			MIT
5469.920	5471.440	18276.72	PR	0.9			MIT
5469.708	5471.228	18277.43	PR	0.5			MIT
5460.250	5461.767	18309.09	PR	0.7			MIT
5458.972	5460.489	18313.38	PR	0.5W			MIT
5457.064	5458.581	18319.78	PR	0.5			MIT
5453.263	5454.779	18332.55	PR	0.6			MIT
5445.428	5446.941	18358.93	PR	0.8W			MIT
5437.365	5438.876	18386.15	PR	0.9O			MIT
5432.913	5434.423	18401.22	PR	0.9			MIT
5432.068	5433.578	18404.08	PR	0.5	0.0		MIT
5431.135	5432.645	18407.24	PR	0.5			MIT
5427.266	5428.775	18420.36	PR	0.8			MIT
5422.423	5423.930	18436.81	PR	0.5	0.0		MIT
5419.042	5420.548	18448.32	PR	0.5O			MIT
5418.781	5420.287	18449.21	PR	0.5			MIT
5413.398	5414.903	18467.55	PR	0.9			MIT
5413.188	5414.693	18468.27	PR	1.2W			MIT
5411.555	5413.059	18473.84	PR	1.4	0.0		MIT
5404.808	5406.311	18496.90	PR	0.5			MIT
5402.585	5404.087	18504.51	PR	1.2W			MIT
5400.946	5402.448	18510.13	PR	0.6			MIT
5395.885	5397.385	18527.49	PR	1.2			MIT
5391.248	5392.747	18543.43	PR	0.5H			MIT
5390.080	5391.579	18547.44	PR	0.5			MIT
5388.802	5390.300	18551.84	PR	0.6H			MIT
5388.041	5389.539	18554.46	PR	0.6H			MIT

AIR WAVELENGTH (ANGSTRMS)	VACUUM WAVELENGTH (ANGSTRMS)	WAVENUMBER (KAYSERS)	LMENT	LOG ARC	INTENSITY SPK	DIS	REF
5381.703	5383.199	18576.31	PR	0.5			MIT
5381.262	5382.758	18577.84	PR	1.8	0.3		MIT
5377.441	5378.936	18591.04	PR	0.5			MIT
5374.242	5375.736	18602.10	PR	0.6			MIT
5372.371	5373.865	18608.58	PR	0.5			MIT
5368.830	5370.323	18620.85	PR	0.7D			MIT
5358.989	5360.479	18655.05	PR	0.6			MIT
5354.307	5355.796	18671.36	PR	0.5			MIT
5352.403	5353.892	18678.00	PR	1.9	0.3		MIT
5343.862	5345.348	18707.85	PR	0.9			MIT
5342.545	5344.031	18712.47	PR	0.5			MIT
5341.920	5343.406	18714.66	PR	0.7			MIT
5336.490	5337.974	18733.70	PR	0.7R			MIT
5331.483	5332.966	18751.29	PR	0.8	0.0		MIT
5322.778	5324.259	18781.96	PR	1.5	0.5		MIT
5321.086	5322.566	18787.93	PR	0.9	0.0		MIT
5314.71	5316.19	18810.5	PR	0.6W			MIT
5313.39	5314.87	18815.1	PR	0.5			MIT
5312.327	5313.805	18818.91	PR	0.7	0.0		MIT
5311.119	5312.597	18823.19	PR	1.0	0.0		MIT
5308.96	5310.44	18830.8	PR	0.5			MIT
5298.106	5299.580	18869.42	PR	1.4			MIT
5295.91	5297.38	18877.2	PR	0.5			MIT
5292.630	5294.103	18888.94	PR	1.7	0.3		MIT
5292.10	5293.57	18890.8	PR	1.8W			MIT
5291.94	5293.41	18891.4	PR	0.5			MIT
5289.937	5291.409	18898.56	PR	0.8W			MIT
5289.346	5290.818	18900.67	PR	1.0			MIT
5288.06	5289.53	18905.3	PR	0.5W			MIT
5286.67	5288.14	18910.2	PR	1.0W			MIT
5285.629	5287.100	18913.96	PR	1.6	0.0		MIT
5277.32	5278.79	18943.7	PR	0.7O	0.0		MIT
5272.72	5274.19	18960.3	PR	0.5			MIT
5267.66	5269.13	18978.5	PR	0.5			MIT
5263.878	5265.343	18992.11	PR	1.4	0.3		MIT
5259.743	5261.207	19007.05	PR	2.1	0.5		MIT
5251.738	5253.200	19036.02	PR	1.0W	0.0		MIT
5249.835	5251.296	19042.92	PR	0.8			MIT
5243.685	5245.145	19065.25	PR	0.5			MIT
5230.272	5231.728	19114.14	PR	1.0W			MIT
5228.005	5229.460	19122.43	PR	1.2			MIT
5223.279	5224.733	19139.73	PR	0.5			MIT
5220.113	5221.566	19151.34	PR	1.9	0.5		MIT
5219.053	5220.506	19155.23	PR	1.7	0.3		MIT
5216.761	5218.213	19163.65	PR	0.9			MIT
5214.368	5215.820	19172.44	PR	0.5			MIT
5212.495	5213.946	19179.33	PR	0.5			MIT
5211.917	5213.368	19181.46	PR	0.5D			MIT
5207.905	5209.355	19196.23	PR	1.2	0.0		MIT
5206.562	5208.012	19201.19	PR	1.5	0.3		MIT
5195.481	5196.928	19242.14	PR	1.3			MIT
5195.307	5196.754	19242.78	PR	1.3	0.0		MIT
5195.110	5196.557	19243.51	PR	1.0	0.0		MIT
5194.413	5195.860	19246.09	PR	1.3			MIT
5191.335	5192.781	19257.50	PR	1.3	0.0		MIT
5188.239	5189.684	19269.00	PR	0.5			MIT
5183.848	5185.292	19285.32	PR	0.7			MIT
5177.369	5178.811	19309.45	PR	0.7			MIT
5175.833	5177.275	19315.18	PR	1.0W	0.0		MIT
5175.186	5176.627	19317.60	PR	1.5W	0.0		MIT
5173.898	5175.339	19322.41	PR	2.0	0.6		MIT
5168.314	5169.754	19343.28	PR	0.5			MIT
5161.743	5163.181	19367.91	PR	1.6	0.0		MIT
5159.210	5160.647	19377.42	PR	0.5			MIT
5156.488	5157.924	19387.64	PR	1.3W	0.0		MIT
5152.21	5153.65	19403.7	PR	1.2W			MIT
5151.361	5152.796	19406.94	PR	0.5			MIT
5149.872	5151.307	19412.55	PR	0.7			MIT
5147.455	5148.889	19421.67	PR	0.6			MIT
5139.799	5141.231	19450.59	PR	1.2	0.0		MIT

AIR WAVELENGTH (ANGSTRMS)	VACUUM WAVELENGTH (ANGSTRMS)	WAVENUMBER (KAYSERS)	LMENT	LOG ARC	INTENSITY SPK	DIS	REF
5135.125	5136.556	19468.30	PR	1.7	0.3		MIT
5133.423	5134.853	19474.75	PR	1.8	0.0		MIT
5129.520	5130.949	19489.57	PR	2.0			MIT
5118.021	5119.447	19533.36	PR	0.6W			MIT
5117.272	5118.698	19536.22	PR	0.6			MIT
5110.768	5112.192	19561.08	PR	1.9	0.5		MIT
5110.382	5111.806	19562.56	PR	1.9	0.3		MIT
5087.111	5088.529	19652.04	PR	1.4			MIT
5076.701	5078.116	19692.34	PR	0.5			MIT
5075.681	5077.096	19696.30	PR	1.1W			MIT
5070.009	5071.422	19718.33	PR	0.5			MIT
5063.389	5064.801	19744.11	PR	0.7			MIT
5053.400	5054.809	19783.14	PR	1.5	0.0		MIT
5045.526	5046.933	19814.01	PR	1.6	0.0		MIT
5043.829	5045.235	19820.68	PR	1.2	0.0		MIT
5037.458	5038.863	19845.75	PR	1.0W			MIT
5034.415	5035.819	19857.74	PR	1.5			MIT
5033.380	5034.784	19861.83	PR	1.0			MIT
5029.000	5030.402	19879.12	PR	0.5			MIT
5026.968	5028.370	19887.16	PR	1.9	0.0		MIT
5019.755	5021.155	19915.74	PR	1.7	0.0		MIT
5018.583	5019.983	19920.39	PR	1.7	0.0		MIT
5015.543	5016.942	19932.46	PR	0.6			MIT
5009.406	5010.803	19956.88	PR	0.5H			MIT
5007.630	5009.027	19963.96	PR	0.6			MIT
5004.584	5005.980	19976.11	PR	0.7	0.0		MIT
5002.454	5003.849	19984.61	PR	0.9	0.0		MIT
5000.593	5001.988	19992.05	PR	0.5			MIT
4995.66	4997.05	20011.8	PR	0.3$			MIT
4992.311	4993.704	20025.22	PR	0.3			MIT
4991.774	4993.167	20027.37	PR	0.5			MIT
4989.268	4990.660	20037.43	PR	1.7			MIT
4987.706	4989.098	20043.71	PR	0.3			MIT
4981.996	4983.386	20066.68	PR	0.3			MIT
4980.522	4981.912	20072.62	PR	0.5R			MIT
4979.125	4980.514	20078.25	PR	0.5			MIT
4978.361	4979.750	20081.33	PR	0.5			MIT
4977.578	4978.967	20084.49	PR	0.3H			MIT
4976.398	4977.787	20089.25	PR	1.3			MIT
4975.752	4977.140	20091.86	PR	1.4			MIT
4974.920	4976.308	20095.22	PR	1.2			MIT
4972.478	4973.865	20105.09	PR	0.3			MIT
4971.544	4972.931	20108.86	PR	0.3			MIT
4970.918	4972.305	20111.40	PR	0.7			MIT
4967.886	4969.272	20123.67	PR	1.0			MIT
4965.441	4966.827	20133.58	PR	0.3			MIT
4964.567	4965.952	20137.12	PR	0.3			MIT
4964.185	4965.570	20138.67	PR	0.3			MIT
4961.040	4962.424	20151.44	PR	0.3			MIT
4960.260	4961.644	20154.61	PR	1.0			MIT
4956.645	4958.028	20169.31	PR	1.6W			MIT
4956.060	4957.443	20171.69	PR	1.0			MIT
4951.357	4952.739	20190.85	PR	2.2			MIT
4950.645	4952.027	20193.75	PR	0.3			MIT
4948.202	4949.583	20203.72	PR	0.3			MIT
4947.197	4948.578	20207.83	PR	0.6			MIT
4943.735	4945.115	20221.98	PR	0.5			MIT
4942.310	4943.689	20227.81	PR	0.6O			MIT
4940.296	4941.675	20236.05	PR	1.7			MIT
4939.735	4941.114	20238.35	PR	2.0			MIT
4938.897	4940.276	20241.79	PR	0.9			MIT
4938.427	4939.805	20243.71	PR	0.7			MIT
4936.004	4937.382	20253.65	PR	1.7			MIT
4935.380	4936.758	20256.21	PR	0.3			MIT
4935.136	4936.514	20257.21	PR	0.3			MIT
4934.071	4935.448	20261.58	PR	0.5			MIT
4932.809	4934.186	20266.77	PR	0.3			MIT
4932.168	4933.545	20269.40	PR	0.6			MIT
4929.371	4930.747	20280.90	PR	0.5			MIT
4928.398	4929.774	20284.91	PR	0.6			MIT

AIR WAVELENGTH (ANGSTRMS)	VACUUM WAVELENGTH (ANGSTRMS)	WAVENUMBER (KAYSERS)	LMENT	LOG ARC	INTENSITY SPK	INTENSITY DIS	REF
4927.055	4928.430	20290.44	PR	0.5			MIT
4925.630	4927.005	20296.31	PR	1.0W			MIT
4925.321	4926.696	20297.58	PR	0.7			MIT
4924.815	4926.190	20299.66	PR	0.7			MIT
4924.588	4925.963	20300.60	PR	1.9			MIT
4922.455	4923.829	20309.40	PR	0.3W			MIT
4921.765	4923.139	20312.24	PR	0.6			MIT
4919.590	4920.963	20321.22	PR	0.6			MIT
4917.001	4918.374	20331.92	PR	0.3			MIT
4915.418	4916.790	20338.47	PR	0.6			MIT
4914.709	4916.081	20341.41	PR	0.6			MIT
4914.418	4915.790	20342.61	PR	0.5			MIT
4914.029	4915.401	20344.22	PR	1.8			MIT
4913.127	4914.499	20347.96	PR	0.5			MIT
4912.629	4914.001	20350.02	PR	1.0			MIT
4911.468	4912.839	20354.83	PR	0.3			MIT
4906.983	4908.353	20373.43	PR	1.7			MIT
4904.412	4905.781	20384.11	PR	0.5			MIT
4903.367	4904.736	20388.46	PR	0.7			MIT
4901.483	4902.852	20396.29	PR	0.7			MIT
4899.921	4901.289	20402.79	PR	0.6			MIT
4899.462	4900.830	20404.71	PR	0.3			MIT
4897.886	4899.254	20411.27	PR	0.5			MIT
4897.285	4898.652	20413.78	PR	0.5			MIT
4896.135	4897.502	20418.57	PR	1.4			MIT
4894.913	4896.280	20423.67	PR	0.5			MIT
4892.941	4894.307	20431.90	PR	0.3H			MIT
4892.585	4893.951	20433.39	PR	0.5			MIT
4892.053	4893.419	20435.61	PR	1.0W			MIT
4890.264	4891.630	20443.08	PR	1.2			MIT
4889.663	4891.028	20445.60	PR	0.7			MIT
4886.976	4888.341	20456.84	PR	0.5			MIT
4886.045	4887.410	20460.74	PR	1.3	0.0		MIT
4884.455	4885.819	20467.40	PR	1.1			MIT
4884.315	4885.679	20467.98	PR	0.8			MIT
4882.245	4883.609	20476.66	PR	1.3			MIT
4880.957	4882.320	20482.06	PR	0.6			MIT
4879.522	4880.885	20488.09	PR	0.7			MIT
4879.121	4880.484	20489.77	PR	1.5W			MIT
4877.823	4879.185	20495.23	PR	1.7W	0.7W		MIT
4876.257	4877.619	20501.81	PR	1.3			MIT
4875.016	4876.378	20507.03	PR	1.0			MIT
4874.812	4876.174	20507.88	PR	0.6			MIT
4873.688	4875.049	20512.61	PR	1.2R			MIT
4872.958	4874.319	20515.69	PR	1.0W			MIT
4870.832	4872.192	20524.64	PR	0.8			MIT
4870.395	4871.755	20526.48	PR	0.7			MIT
4870.108	4871.468	20527.69	PR	0.7			MIT
4869.333	4870.693	20530.96	PR	1.0			MIT
4868.234	4869.594	20535.59	PR	0.6			MIT
4867.391	4868.751	20539.15	PR	1.2			MIT
4866.119	4867.478	20544.52	PR	0.5			MIT
4865.237	4866.596	20548.24	PR	1.6	0.3		MIT
4863.868	4865.227	20554.03	PR	1.2			MIT
4863.126	4864.484	20557.16	PR	0.6			MIT
4862.728	4864.086	20558.85	PR	0.6			MIT
4862.310	4863.668	20560.61	PR	0.5			MIT
4861.965	4863.323	20562.07	PR	0.7			MIT
4859.953	4861.311	20570.58	PR	1.0			MIT
4859.038	4860.395	20574.46	PR	1.6W			MIT
4858.585	4859.942	20576.38	PR	1.0			MIT
4857.365	4858.722	20581.54	PR	1.2			MIT
4856.080	4857.437	20586.99	PR	1.0			MIT
4855.322	4856.678	20590.20	PR	0.8			MIT
4853.684	4855.040	20597.15	PR	1.3			MIT
4851.481	4852.836	20606.51	PR	1.0			MIT
4850.666	4852.021	20609.97	PR	1.0			MIT
4850.277	4851.632	20611.62	PR	1.0			MIT
4848.547	4849.902	20618.98	PR	2.1W			MIT
4846.401	4847.755	20628.10	PR	0.8			MIT

AIR WAVELENGTH (ANGSTRMS)	VACUUM WAVELENGTH (ANGSTRMS)	WAVENUMBER (KAYSERS)	LMENT	LOG ARC	INTENSITY SPK	INTENSITY DIS	REF
4845.973	4847.327	20629.93	PR	1.3			MIT
4845.610	4846.964	20631.47	PR	0.7			MIT
4845.184	4846.538	20633.29	PR	0.6			MIT
4844.218	4845.571	20637.40	PR	0.9			MIT
4842.588	4843.941	20644.35	PR	0.9			MIT
4841.876	4843.229	20647.38	PR	0.5			MIT
4841.553	4842.906	20648.76	PR	0.7			MIT
4840.738	4842.090	20652.24	PR	1.3L			MIT
4840.476	4841.828	20653.35	PR	0.7			MIT
4839.540	4840.892	20657.35	PR	1.4	0.0		MIT
4839.336	4840.688	20658.22	PR	0.9			MIT
4838.616	4839.968	20661.29	PR	1.2			MIT
4837.041	4838.393	20668.02	PR	1.9W			MIT
4836.002	4837.353	20672.46	PR	0.6			MIT
4835.437	4836.788	20674.88	PR	1.0D			MIT
4834.713	4836.064	20677.97	PR	0.9			MIT
4833.14	4834.49	20684.7	PR	1.2W			MIT
4832.734	4834.084	20686.44	PR	0.5			MIT
4832.070	4833.420	20689.28	PR	2.0W			MIT
4831.435	4832.785	20692.00	PR	0.8			MIT
4830.194	4831.544	20697.32	PR	1.0			MIT
4829.468	4830.818	20700.43	PR	0.8W			MIT
4828.065	4829.414	20706.45	PR	1.2			MIT
4827.250	4828.599	20709.94	PR	1.3D			MIT
4826.649	4827.998	20712.52	PR	1.6J			MIT
4826.310	4827.659	20713.98	PR	1.0			MIT
4825.822	4827.171	20716.07	PR	1.0			MIT
4825.474	4826.822	20717.56	PR	0.7			MIT
4824.655	4826.003	20721.08	PR	0.7			MIT
4822.980	4824.328	20728.28	PR	2.1	1.0W		MIT
4821.949	4823.297	20732.71	PR	1.2			MIT
4821.153	4822.500	20736.13	PR	0.6			MIT
4820.577	4821.924	20738.61	PR	0.6			MIT
4820.064	4821.411	20740.82	PR	0.5			MIT
4819.735	4821.082	20742.23	PR	0.5			MIT
4818.987	4820.334	20745.45	PR	0.8			MIT
4818.325	4819.672	20748.30	PR	0.5			MIT
4817.545	4818.891	20751.66	PR	0.7			MIT
4816.133	4817.479	20757.75	PR	0.8			MIT
4815.672	4817.018	20759.73	PR	0.8			MIT
4814.344	4815.689	20765.46	PR	1.5	0.5		MIT
4811.735	4813.080	20776.72	PR	0.8			MIT
4811.064	4812.409	20779.61	PR	0.7			MIT
4808.191	4809.535	20792.03	PR	1.3			MIT
4807.419	4808.763	20795.37	PR	0.5			MIT
4807.239	4808.583	20796.15	PR	1.0			MIT
4805.833	4807.176	20802.23	PR	0.6			MIT
4804.039	4805.382	20810.00	PR	1.2W			MIT
4802.675	4804.017	20815.91	PR	0.6			MIT
4802.034	4803.376	20818.69	PR	0.7			MIT
4801.501	4802.843	20821.00	PR	1.0			MIT
4801.150	4802.492	20822.52	PR	1.6	0.5		MIT
4799.418	4800.760	20830.04	PR	0.6			MIT
4798.753	4800.094	20832.92	PR	1.4			MIT
4798.315	4799.656	20834.82	PR	0.6H			MIT
4797.957	4799.298	20836.38	PR	0.5			MIT
4796.843	4798.184	20841.22	PR	0.8			MIT
4796.424	4797.765	20843.04	PR	0.6			MIT
4795.828	4797.169	20845.63	PR	0.7			MIT
4795.249	4796.589	20848.15	PR	1.3			MIT
4794.903	4796.243	20849.65	PR	0.8			MIT
4794.321	4795.661	20852.18	PR	0.6			MIT
4792.921	4794.261	20858.27	PR	1.0H			MIT
4791.952	4793.292	20862.49	PR	0.9			MIT
4791.597	4792.936	20864.04	PR	0.5			MIT
4791.266	4792.605	20865.48	PR	0.8			MIT
4790.842	4792.181	20867.32	PR	0.7			MIT
4789.367	4790.706	20873.75	PR	0.8H			MIT
4788.279	4789.618	20878.49	PR	1.0			MIT
4787.573	4788.911	20881.57	PR	0.7			MIT

AIR WAVELENGTH (ANGSTRMS)	VACUUM WAVELENGTH (ANGSTRMS)	WAVENUMBER (KAYSERS)	LMENT	LOG ARC	INTENSITY SPK	DIS	REF
4787.229	4788.567	20883.07	PR	0.5			MIT
4786.593	4787.931	20885.85	PR	0.8			MIT
4785.617	4786.955	20890.11	PR	1.0W			MIT
4784.635	4785.973	20894.40	PR	0.5			MIT
4783.354	4784.691	20899.99	PR	2.1	1.0W		MIT
4782.838	4784.175	20902.25	PR	0.5			MIT
4782.100	4783.437	20905.47	PR	0.5			MIT
4779.196	4780.532	20918.17	PR	1.2			MIT
4778.303	4779.639	20922.08	PR	1.5			MIT
4778.165	4779.501	20922.69	PR	1.5			MIT
4775.996	4777.331	20932.19	PR	1.0			MIT
4775.419	4776.754	20934.72	PR	1.0			MIT
4775.170	4776.505	20935.81	PR	1.0			MIT
4774.537	4775.872	20938.58	PR	0.9			MIT
4773.524	4774.859	20943.03	PR	1.0			MIT
4772.531	4773.865	20947.39	PR	0.6			MIT
4772.296	4773.630	20948.42	PR	0.7			MIT
4771.866	4773.200	20950.30	PR	1.0			MIT
4770.455	4771.789	20956.50	PR	0.6			MIT
4769.896	4771.230	20958.96	PR	0.6			MIT
4769.593	4770.927	20960.29	PR	0.7			MIT
4767.804	4769.137	20968.15	PR	1.3			MIT
4766.906	4768.239	20972.10	PR	1.5			MIT
4765.928	4767.261	20976.41	PR	1.0			MIT
4765.222	4766.554	20979.51	PR	1.9W	0.7W		MIT
4764.446	4765.778	20982.93	PR	1.1			MIT
4763.571	4764.903	20986.79	PR	1.2			MIT
4762.727	4764.059	20990.50	PR	1.8	1.0		MIT
4762.402	4763.734	20991.94	PR	0.5			MIT
4761.969	4763.301	20993.85	PR	0.8			MIT
4761.125	4762.456	20997.57	PR	1.4			MIT
4759.911	4761.242	21002.92	PR	0.8			MIT
4759.237	4760.568	21005.90	PR	0.5			MIT
4758.044	4759.375	21011.16	PR	1.0	0.0		MIT
4757.937	4759.268	21011.64	PR	2.0W	0.7W		MIT
4757.119	4758.449	21015.25	PR	1.0D			MIT
4756.13	4757.46	21019.6	PR	1.7D	0.3H		MIT
4755.978	4757.308	21020.29	PR	1.7	0.3		MIT
4755.355	4756.685	21023.04	PR	0.7			MIT
4754.635	4755.965	21026.23	PR	1.2			MIT
4754.456	4755.786	21027.02	PR	0.9			MIT
4753.904	4755.233	21029.46	PR	0.9			MIT
4752.848	4754.177	21034.13	PR	0.9			MIT
4751.671	4753.000	21039.34	PR	0.5			MIT
4751.318	4752.647	21040.91	PR	0.7D			MIT
4750.556	4751.885	21044.28	PR	0.6			MIT
4749.733	4751.061	21047.93	PR	1.0			MIT
4748.603	4749.931	21052.94	PR	0.6			MIT
4748.181	4749.509	21054.81	PR	0.5			MIT
4747.587	4748.915	21057.44	PR	0.5			MIT
4746.93	4748.26	21060.4	PR	2.0	1.4W		MIT
4744.925	4746.252	21069.26	PR	2.0	1.0W		MIT
4744.164	4745.491	21072.64	PR	1.9			MIT
4743.565	4744.892	21075.30	PR	0.8			MIT
4743.076	4744.403	21077.47	PR	0.7			MIT
4742.325	4743.651	21080.81	PR	0.8			MIT
4741.503	4742.829	21084.46	PR	1.5			MIT
4741.005	4742.331	21086.68	PR	0.8			MIT
4739.886	4741.212	21091.65	PR	0.6			MIT
4739.325	4740.651	21094.15	PR	0.5			MIT
4738.622	4739.947	21097.28	PR	0.9D			MIT
4736.688	4738.013	21105.89	PR	2.1	0.0		MIT
4734.177	4735.501	21117.09	PR	2.0W	0.9W		MIT
4733.751	4735.075	21118.99	PR	0.9			MIT
4732.191	4733.515	21125.95	PR	0.5W			MIT
4730.686	4732.009	21132.67	PR	1.8	0.0		MIT
4729.130	4730.453	21139.62	PR	0.6			MIT
4728.630	4729.953	21141.86	PR	1.6W	0.8W		MIT
4726.518	4727.840	21151.31	PR	1.0			MIT
4726.355	4727.677	21152.04	PR	1.6W			MIT

AIR WAVELENGTH (ANGSTRMS)	VACUUM WAVELENGTH (ANGSTRMS)	WAVENUMBER (KAYSERS)	LMENT	LOG ARC	INTENSITY SPK	DIS	REF
4724.765	4726.087	21159.16	PR	0.6			MIT
4724.438	4725.760	21160.62	PR	0.3			MIT
4723.899	4725.221	21163.03	PR	0.6			MIT
4722.670	4723.991	21168.54	PR	0.7			MIT
4721.910	4723.231	21171.95	PR	0.5			MIT
4721.675	4722.996	21173.00	PR	0.5			MIT
4721.471	4722.792	21173.92	PR	0.3			MIT
4720.235	4721.556	21179.46	PR	0.6			MIT
4719.646	4720.966	21182.10	PR	0.6			MIT
4719.304	4720.624	21183.64	PR	0.9D			MIT
4718.680	4720.000	21186.44	PR	0.9D			MIT
4717.076	4718.396	21193.64	PR	0.5			MIT
4715.588	4716.907	21200.33	PR	0.8			MIT
4715.230	4716.549	21201.94	PR	0.6			MIT
4714.852	4716.171	21203.64	PR	0.6			MIT
4714.146	4715.465	21206.82	PR	1.5			MIT
4713.100	4714.419	21211.52	PR	1.5			MIT
4712.285	4713.603	21215.19	PR	0.7			MIT
4711.840	4713.158	21217.20	PR	1.2			MIT
4711.343	4712.661	21219.43	PR	0.3			MIT
4709.519	4710.837	21227.65	PR	1.6	0.0		MIT
4708.155	4709.472	21233.80	PR	1.7	1.0W		MIT
4707.939	4709.256	21234.77	PR	1.7	1.0W		MIT
4707.541	4708.858	21236.57	PR	1.9	1.0W		MIT
4706.548	4707.865	21241.05	PR	0.3			MIT
4706.228	4707.545	21242.50	PR	0.8			MIT
4705.573	4706.890	21245.45	PR	0.7H			MIT
4703.998	4705.314	21252.57	PR	0.6			MIT
4703.258	4704.574	21255.91	PR	0.5			MIT
4702.886	4704.202	21257.59	PR	0.5			MIT
4702.450	4703.766	21259.56	PR	0.6			MIT
4700.645	4701.960	21267.73	PR	1.0			MIT
4700.118	4701.433	21270.11	PR	0.6			MIT
4699.582	4700.897	21272.53	PR	0.9			MIT
4698.858	4700.173	21275.81	PR	0.6			MIT
4698.383	4699.698	21277.96	PR	0.6			MIT
4698.006	4699.321	21279.67	PR	0.7			MIT
4697.295	4698.609	21282.89	PR	0.6	0.0		MIT
4696.440	4697.754	21286.77	PR	0.6			MIT
4695.767	4697.081	21289.82	PR	1.8	0.3		MIT
4694.667	4695.981	21294.81	PR	0.6			MIT
4693.751	4695.065	21298.96	PR	1.0W			MIT
4692.658	4693.971	21303.92	PR	1.0			MIT
4692.019	4693.332	21306.82	PR	0.6			MIT
4691.789	4693.102	21307.87	PR	1.0			MIT
4690.576	4691.889	21313.38	PR	1.0			MIT
4689.557	4690.869	21318.01	PR	1.2			MIT
4689.341	4690.653	21318.99	PR	0.6			MIT
4689.039	4690.351	21320.36	PR	0.5			MIT
4688.559	4689.871	21322.55	PR	0.6			MIT
4687.809	4689.121	21325.96	PR	1.7	0.0		MIT
4687.233	4688.545	21328.58	PR	0.5			MIT
4686.285	4687.597	21332.89	PR	0.6	0.0		MIT
4685.447	4686.758	21336.71	PR	0.7			MIT
4684.936	4686.247	21339.04	PR	2.1	1.0W		MIT
4684.541	4685.852	21340.84	PR	0.8			MIT
4683.836	4685.147	21344.05	PR	0.5			MIT
4683.430	4684.741	21345.90	PR	0.5			MIT
4683.225	4684.536	21346.83	PR	0.3			MIT
4682.692	4684.003	21349.26	PR	0.3			MIT
4682.445	4683.756	21350.39	PR	0.3			MIT
4682.213	4683.523	21351.45	PR	0.5			MIT
4682.056	4683.366	21352.16	PR	0.5			MIT
4680.472	4681.782	21359.39	PR	0.6			MIT
4680.318	4681.628	21360.09	PR	0.3			MIT
4679.831	4681.141	21362.31	PR	0.5			MIT
4679.498	4680.808	21363.83	PR	0.3			MIT
4679.112	4680.422	21365.60	PR	0.9	0.5		MIT
4679.039	4680.349	21365.93	PR	0.7			MIT
4678.168	4679.477	21369.91	PR	1.1	0.3		MIT

AIR WAVELENGTH (ANGSTRMS)	VACUUM WAVELENGTH (ANGSTRMS)	WAVENUMBER (KAYSERS)	LMENT	LOG ARC	INTENSITY SPK	INTENSITY DIS	REF
4677.780	4679.089	21371.68	PR	0.3			MIT
4676.994	4678.303	21375.27	PR	0.5			MIT
4676.725	4678.034	21376.50	PR	0.7			MIT
4676.181	4677.490	21378.99	PR	1.0	0.9H		MIT
4675.472	4676.781	21382.23	PR	0.5			MIT
4675.038	4676.347	21384.22	PR	0.7			MIT
4674.800	4676.109	21385.30	PR	1.4			MIT
4674.238	4675.546	21387.87	PR	0.5			MIT
4672.595	4673.903	21395.40	PR	0.5			MIT
4672.081	4673.389	21397.75	PR	2.0	1.4W		MIT
4670.543	4671.850	21404.79	PR	1.0			MIT
4668.454	4669.761	21414.37	PR	0.8			MIT
4668.230	4669.537	21415.40	PR	1.0	0.0		MIT
4666.50	4667.81	21423.3	PR	0.3			MIT
4666.171	4667.477	21424.85	PR	0.3			MIT
4665.945	4667.251	21425.89	PR	0.5			MIT
4664.647	4665.953	21431.85	PR	2.0	1.2		MIT
4663.723	4665.029	21436.10	PR	0.5			MIT
4663.153	4664.458	21438.72	PR	0.3			MIT
4662.77	4664.08	21440.5	PR	0.3			MIT
4660.915	4662.220	21449.01	PR	1.4			MIT
4660.156	4661.461	21452.50	PR	0.5			MIT
4659.845	4661.150	21453.94	PR	0.7			MIT
4659.710	4661.015	21454.56	PR	0.5			MIT
4659.50	4660.80	21455.5	PR	0.3			MIT
4658.733	4660.037	21459.06	PR	1.2			MIT
4658.361	4659.665	21460.77	PR	0.5			MIT
4658.09	4659.39	21462.0	PR	1.3	0.0		MIT
4657.622	4658.926	21464.17	PR	0.3			MIT
4657.235	4658.539	21465.96	PR	0.5			MIT
4656.764	4658.068	21468.13	PR	0.7			MIT
4655.993	4657.297	21471.68	PR	0.7			MIT
4654.035	4655.338	21480.72	PR	0.9	0.0		MIT
4653.81	4655.11	21481.8	PR	0.5			MIT
4653.64	4654.94	21482.5	PR	0.6	0.0		MIT
4652.678	4653.981	21486.98	PR	0.7			MIT
4651.517	4652.819	21492.34	PR	2.1	1.6W		MIT
4650.88	4652.18	21495.3	PR	0.6	0.9H		MIT
4650.44	4651.74	21497.3	PR	0.9W			MIT
4650.08	4651.38	21499.0	PR	0.5			MIT
4649.840	4651.142	21500.10	PR	0.8			MIT
4649.42	4650.72	21502.0	PR	0.6			MIT
4649.088	4650.390	21503.57	PR	0.5	1.0H		MIT
4648.546	4649.848	21506.08	PR	0.5			MIT
4647.943	4649.244	21508.87	PR	0.9W			MIT
4646.991	4648.292	21513.28	PR	1.4			MIT
4646.059	4647.360	21517.59	PR	1.7	0.9		MIT
4645.485	4646.786	21520.25	PR	0.6W			MIT
4645.111	4646.412	21521.99	PR	0.5			MIT
4643.505	4644.805	21529.43	PR	1.8W	0.7W		MIT
4642.029	4643.329	21536.27	PR	0.6H			MIT
4640.206	4641.505	21544.73	PR	1.3			MIT
4639.877	4641.176	21546.26	PR	0.7			MIT
4639.555	4640.854	21547.76	PR	1.8	0.3		MIT
4638.970	4640.269	21550.48	PR	0.3			MIT
4636.727	4638.025	21560.90	PR	0.5			MIT
4636.26	4637.56	21563.1	PR	0.6			MIT
4635.692	4636.990	21565.71	PR	1.6			MIT
4635.050	4636.348	21568.70	PR	0.7			MIT
4634.717	4636.015	21570.25	PR	0.5			MIT
4634.012	4635.310	21573.53	PR	0.3			MIT
4633.47	4634.77	21576.1	PR	0.5			MIT
4632.913	4634.210	21578.65	PR	0.6			MIT
4632.280	4633.577	21581.60	PR	1.6	0.0		MIT
4631.673	4632.970	21584.43	PR	0.6			MIT
4630.712	4632.009	21588.91	PR	0.5			MIT
4629.675	4630.972	21593.74	PR	0.8W			MIT
4629.455	4630.752	21594.77	PR	0.3			MIT
4628.751	4630.047	21598.05	PR	2.3	1.7W		MIT
4627.05	4628.35	21606.0	PR	1.2	0.5		MIT

AIR WAVELENGTH (ANGSTRMS)	VACUUM WAVELENGTH (ANGSTRMS)	WAVENUMBER (KAYSERS)	LMENT	LOG ARC	INTENSITY SPK	INTENSITY DIS	REF
4625.616	4626.912	21612.69	PR	0.7H			MIT
4625.298	4626.593	21614.18	PR	0.6	0.0		MIT
4624.048	4625.343	21620.02	PR	0.9			MIT
4623.633	4624.928	21621.96	PR	0.7			MIT
4623.190	4624.485	21624.03	PR	0.5			MIT
4622.726	4624.021	21626.20	PR	0.8			MIT
4622.26	4623.55	21628.4	PR	0.8	0.0		MIT
4621.341	4622.635	21632.68	PR	0.8			MIT
4619.846	4621.140	21639.68	PR	0.5			MIT
4619.660	4620.954	21640.55	PR	0.8			MIT
4618.257	4619.551	21647.13	PR	0.5			MIT
4618.011	4619.304	21648.28	PR	1.0	0.3		MIT
4617.727	4619.020	21649.61	PR	1.0			MIT
4617.573	4618.866	21650.33	PR	0.3			MIT
4617.326	4618.619	21651.49	PR	0.3			MIT
4617.121	4618.414	21652.45	PR	0.7			MIT
4615.362	4616.655	21660.71	PR	1.0	0.0		MIT
4614.930	4616.223	21662.73	PR	0.5			MIT
4613.81	4615.10	21668.0	PR	0.6			MIT
4612.754	4614.046	21672.95	PR	0.3			MIT
4612.553	4613.845	21673.90	PR	0.6			MIT
4612.072	4613.364	21676.16	PR	1.8	1.2		MIT
4611.673	4612.965	21678.03	PR	0.3			MIT
4610.62	4611.91	21683.0	PR	1.0			MIT
4610.186	4611.477	21685.02	PR	0.5			MIT
4609.61	4610.90	21687.7	PR	0.7D			MIT
4609.13	4610.42	21690.0	PR	0.8D			MIT
4608.005	4609.296	21695.29	PR	0.7			MIT
4606.920	4608.211	21700.40	PR	0.5			MIT
4606.446	4607.736	21702.63	PR	1.5W	0.5W		MIT
4605.214	4606.504	21708.44	PR	0.7			MIT
4604.89	4606.18	21710.0	PR	1.0	0.3		MIT
4604.205	4605.495	21713.19	PR	0.5			MIT
4603.810	4605.100	21715.06	PR	1.0	0.3		MIT
4603.478	4604.768	21716.62	PR	0.6			MIT
4602.562	4603.851	21720.94	PR	1.0			MIT
4601.075	4602.364	21727.96	PR	0.5			MIT
4600.391	4601.680	21731.20	PR	1.3	0.5		MIT
4599.48	4600.77	21735.5	PR	0.5			MIT
4598.951	4600.239	21738.00	PR	1.0			MIT
4598.743	4600.031	21738.98	PR	0.6			MIT
4596.948	4598.236	21747.47	PR	1.3	0.6		MIT
4595.87	4597.16	21752.6	PR	0.8	0.5H		MIT
4594.963	4596.250	21756.86	PR	0.5			MIT
4594.572	4595.859	21758.72	PR	0.5			MIT
4593.926	4595.213	21761.78	PR	1.5	0.3		MIT
4593.571	4594.858	21763.46	PR	0.9D	0.3D		MIT
4592.153	4593.440	21770.18	PR	0.9	0.3		MIT
4591.923	4593.210	21771.27	PR	0.6			MIT
4591.555	4592.841	21773.01	PR	0.5			MIT
4590.932	4592.218	21775.97	PR	0.9	1.0H		MIT
4590.705	4591.991	21777.04	PR	0.6			MIT
4590.145	4591.431	21779.70	PR	0.5			MIT
4589.76	4591.05	21781.5	PR	0.7D			MIT
4589.16	4590.45	21784.4	PR	0.7			MIT
4588.00	4589.29	21789.9	PR	0.7D			MIT
4587.135	4588.420	21793.99	PR	0.9			MIT
4586.982	4588.267	21794.72	PR	0.8			MIT
4586.532	4587.817	21796.86	PR	0.7W			MIT
4586.16	4587.45	21798.6	PR	0.3			MIT
4585.94	4587.23	21799.7	PR	0.6			MIT
4584.427	4585.712	21806.87	PR	0.3			MIT
4582.571	4583.855	21815.70	PR	0.7W			MIT
4581.584	4582.868	21820.40	PR	1.0W			MIT
4580.821	4582.105	21824.03	PR	0.6W			MIT
4580.460	4581.744	21825.75	PR	0.5			MIT
4579.845	4581.128	21828.68	PR	0.5			MIT
4578.71	4579.99	21834.1	PR	0.5			MIT
4578.139	4579.422	21836.82	PR	1.6W	0.5W		MIT
4577.53	4578.81	21839.7	PR	0.5			MIT

AIR WAVELENGTH (ANGSTRMS)	VACUUM WAVELENGTH (ANGSTRMS)	WAVENUMBER (KAYSERS)	LMENT	LOG ARC	INTENSITY SPK	DIS	REF
4577.006	4578.289	21842.22	PR	0.6			MIT
4576.64	4577.92	21844.0	PR	0.5			MIT
4576.320	4577.602	21845.50	PR	1.7	1.2		MIT
4574.906	4576.188	21852.25	PR	0.6			MIT
4574.636	4575.918	21853.54	PR	0.6			MIT
4574.203	4575.485	21855.61	PR	0.7W			MIT
4573.281	4574.563	21860.01	PR	0.6	0.0		MIT
4572.30	4573.58	21864.7	PR	0.5			MIT
4572.129	4573.410	21865.52	PR	1.0			MIT
4571.613	4572.894	21867.99	PR	1.3	0.7		MIT
4570.565	4571.846	21873.00	PR	1.6	1.2		MIT
4570.350	4571.631	21874.03	PR	0.5			MIT
4570.116	4571.397	21875.15	PR	1.3W	0.3H		MIT
4569.624	4570.905	21877.51	PR	1.0			MIT
4569.372	4570.653	21878.71	PR	0.5			MIT
4568.545	4569.825	21882.67	PR	1.5	0.5H		MIT
4567.481	4568.761	21887.77	PR	0.5			MIT
4566.845	4568.125	21890.82	PR	0.7			MIT
4566.422	4567.702	21892.85	PR	0.5			MIT
4565.929	4567.209	21895.21	PR	1.0			MIT
4564.80	4566.08	21900.6	PR	1.3W			MIT
4564.335	4565.614	21902.86	PR	0.6			MIT
4563.941	4565.220	21904.75	PR	0.5			MIT
4563.780	4565.059	21905.52	PR	0.7			MIT
4563.126	4564.405	21908.66	PR	2.0	1.6		MIT
4561.461	4562.740	21916.66	PR	0.8			MIT
4561.100	4562.378	21918.39	PR	0.5			MIT
4560.958	4562.236	21919.07	PR	0.7			MIT
4559.70	4560.98	21925.1	PR	0.5H			MIT
4559.49	4560.77	21926.1	PR	0.5			MIT
4559.135	4560.413	21927.84	PR	0.5			MIT
4558.485	4559.763	21930.97	PR	0.6	0.0		MIT
4557.855	4559.133	21934.00	PR	0.9	0.0		MIT
4557.43	4558.71	21936.0	PR	0.5			MIT
4556.94	4558.22	21938.4	PR	0.3			MIT
4556.656	4557.933	21939.77	PR	0.3			MIT
4556.26	4557.54	21941.7	PR	0.6			MIT
4555.982	4557.259	21943.01	PR	0.5			MIT
4554.790	4556.067	21948.76	PR	0.6	0.0		MIT
4554.498	4555.775	21950.16	PR	0.3			MIT
4553.498	4554.774	21954.98	PR	0.6			MIT
4553.256	4554.532	21956.15	PR	0.7H			MIT
4552.829	4554.105	21958.21	PR	0.7			MIT
4552.262	4553.538	21960.95	PR	1.8			MIT
4551.538	4552.814	21964.44	PR	0.6			MIT
4551.177	4552.453	21966.18	PR	0.5			MIT
4550.882	4552.158	21967.60	PR	1.0	0.3		MIT
4550.769	4552.045	21968.15	PR	1.0			MIT
4550.670	4551.946	21968.63	PR	1.0D			MIT
4550.06	4551.34	21971.6	PR	1.2	0.5		MIT
4549.838	4551.113	21972.65	PR	1.2	0.6		MIT
4548.539	4549.814	21978.92	PR	1.8	1.5		MIT
4546.631	4547.906	21988.14	PR	0.7W			MIT
4546.329	4547.604	21989.60	PR	0.5			MIT
4545.853	4547.127	21991.91	PR	0.5			MIT
4545.335	4546.609	21994.41	PR	0.8	0.3		MIT
4545.103	4546.377	21995.53	PR	0.6	0.0		MIT
4544.588	4545.862	21998.03	PR	0.5			MIT
4543.965	4545.239	22001.04	PR	1.0	0.5		MIT
4543.534	4544.808	22003.13	PR	1.4W	0.5		MIT
4542.537	4543.811	22007.96	PR	1.3	0.9		MIT
4541.993	4543.266	22010.60	PR	0.5H			MIT
4541.267	4542.540	22014.11	PR	1.3	0.0		MIT
4541.048	4542.321	22015.18	PR	0.6			MIT
4540.151	4541.424	22019.53	PR	0.7	0.0		MIT
4539.288	4540.561	22023.71	PR	1.5D	0.5W		MIT
4538.018	4539.290	22029.87	PR	1.2W			MIT
4537.386	4538.658	22032.94	PR	0.3			MIT
4537.153	4538.425	22034.07	PR	0.3			MIT
4535.921	4537.193	22040.06	PR	2.1	2.0		MIT

AIR WAVELENGTH (ANGSTRMS)	VACUUM WAVELENGTH (ANGSTRMS)	WAVENUMBER (KAYSERS)	LMENT	LOG ARC	INTENSITY SPK	DIS	REF
4535.242	4536.514	22043.36	PR	0.3			MIT
4534.885	4536.157	22045.09	PR	0.6			MIT
4534.154	4535.425	22048.65	PR	2.2	1.9		MIT
4533.933	4535.204	22049.72	PR	0.3			MIT
4533.791	4535.062	22050.41	PR	0.7	0.3		MIT
4533.303	4534.574	22052.79	PR	0.7			MIT
4532.788	4534.059	22055.29	PR	0.6			MIT
4532.336	4533.607	22057.49	PR	1.2			MIT
4531.846	4533.117	22059.88	PR	0.3			MIT
4531.511	4532.782	22061.51	PR	0.5			MIT
4531.089	4532.360	22063.56	PR	1.0W	0.3W		MIT
4529.930	4531.200	22069.21	PR	1.4W	0.6W		MIT
4528.910	4530.180	22074.18	PR	0.5			MIT
4528.068	4529.338	22078.28	PR	0.3			MIT
4526.465	4527.734	22086.10	PR	0.5W			MIT
4524.695	4525.964	22094.74	PR	0.7	0.0		MIT
4524.365	4525.634	22096.35	PR	1.0W	0.3W		MIT
4523.991	4525.260	22098.18	PR	0.6			MIT
4523.209	4524.477	22102.00	PR	0.5			MIT
4522.143	4523.411	22107.21	PR	0.5			MIT
4520.778	4522.046	22113.89	PR	1.6	1.2		MIT
4520.386	4521.654	22115.80	PR	0.5	0.0		MIT
4519.629	4520.896	22119.51	PR	1.0			MIT
4519.115	4520.382	22122.02	PR	0.5			MIT
4518.687	4519.954	22124.12	PR	0.3			MIT
4517.595	4518.862	22129.47	PR	1.6W	1.2W		MIT
4516.955	4518.222	22132.60	PR	0.7			MIT
4516.461	4517.728	22135.02	PR	0.9	0.5		MIT
4514.215	4515.481	22146.03	PR	1.0W			MIT
4513.723	4514.989	22148.45	PR	1.0	0.3		MIT
4513.231	4514.497	22150.86	PR	0.8	0.3		MIT
4512.272	4513.538	22155.57	PR	0.9			MIT
4511.815	4513.080	22157.81	PR	0.5			MIT
4511.455	4512.720	22159.58	PR	0.7			MIT
4511.349	4512.614	22160.10	PR	0.6			MIT
4511.091	4512.356	22161.37	PR	0.5			MIT
4510.160	4511.425	22165.95	PR	2.3	2.1		MIT
4509.381	4510.646	22169.77	PR	0.6	0.3		MIT
4508.926	4510.191	22172.01	PR	0.6	0.0		MIT
4508.370	4509.635	22174.75	PR	0.9	0.3		MIT
4507.765	4509.029	22177.72	PR	0.6	0.7		MIT
4506.948	4508.212	22181.74	PR	1.0	0.3		MIT
4505.876	4507.140	22187.02	PR	0.3			MIT
4505.096	4506.360	22190.86	PR	0.3			MIT
4504.591	4505.855	22193.35	PR	1.3	0.3		MIT
4503.742	4505.005	22197.53	PR	1.0			MIT
4503.273	4504.536	22199.84	PR	1.2W			MIT
4502.856	4504.119	22201.90	PR	1.0			MIT
4502.283	4503.546	22204.73	PR	0.3			MIT
4501.828	4503.091	22206.97	PR	1.3	0.0		MIT
4501.367	4502.630	22209.24	PR	0.3			MIT
4501.213	4502.476	22210.00	PR	0.6			MIT
4500.518	4501.780	22213.43	PR	1.3	0.3W		MIT
4500.102	4501.364	22215.49	PR	0.5			MIT
4499.800	4501.062	22216.98	PR	0.7W			MIT
4499.545	4500.807	22218.24	PR	0.7W			MIT
4499.377	4500.639	22219.07	PR	1.0W			MIT
4498.992	4500.254	22220.97	PR	0.7W			MIT
4498.633	4499.895	22222.74	PR	0.6			MIT
4498.295	4499.557	22224.41	PR	0.7W			MIT
4497.272	4498.534	22229.47	PR	0.3			MIT
4496.429	4497.690	22233.63	PR	2.3	2.1		MIT
4494.342	4495.603	22243.96	PR	0.6			MIT
4494.193	4495.454	22244.70	PR	1.5W			MIT
4493.709	4494.970	22247.09	PR	1.3	0.6		MIT
4493.119	4494.379	22250.01	PR	1.4W			MIT
4492.927	4494.187	22250.96	PR	1.1			MIT
4492.427	4493.687	22253.44	PR	1.4	0.3W		MIT
4491.478	4492.738	22258.14	PR	0.5			MIT
4491.137	4492.397	22259.83	PR	0.5			MIT

AIR WAVELENGTH (ANGSTRMS)	VACUUM WAVELENGTH (ANGSTRMS)	WAVENUMBER (KAYSERS)	LMENT	LOG ARC	INTENSITY SPK	INTENSITY DIS	REF
4490.389	4491.649	22263.54	PR	0.9W			MIT
4488.601	4489.860	22272.41	PR	1.0			MIT
4488.173	4489.432	22274.53	PR	1.6	1.0		MIT
4487.821	4489.080	22276.28	PR	1.6	0.5W		MIT
4487.033	4488.292	22280.19	PR	0.7			MIT
4486.588	4487.847	22282.40	PR	0.5			MIT
4485.977	4487.236	22285.44	PR	0.7			MIT
4485.544	4486.802	22287.59	PR	1.6	0.7W		MIT
4485.311	4486.569	22288.75	PR	0.6			MIT
4484.883	4486.141	22290.87	PR	0.6			MIT
4484.714	4485.972	22291.71	PR	0.3			KN
4483.491	4484.749	22297.79	PR	1.5	0.3		MIT
4483.083	4484.341	22299.82	PR	1.4	0.3		MIT
4482.694	4483.952	22301.76	PR	0.9W			MIT
4481.410	4482.667	22308.15	PR	1.5	0.6		MIT
4480.688	4481.945	22311.74	PR	0.5			MIT
4480.275	4481.532	22313.80	PR	0.7			MIT
4480.133	4481.390	22314.50	PR	0.7			MIT
4479.618	4480.875	22317.07	PR	1.5			MIT
4479.098	4480.355	22319.66	PR	0.5			MIT
4478.638	4479.895	22321.95	PR	0.5			KN
4477.991	4479.248	22325.18	PR	0.3			MIT
4477.259	4478.515	22328.83	PR	2.1	1.5W		MIT
4475.697	4476.953	22336.62	PR	0.8			MIT
4475.375	4476.631	22338.23	PR	0.9			MIT
4474.840	4476.096	22340.90	PR	1.0	0.0		MIT
4474.463	4475.719	22342.78	PR	1.3			MIT
4473.835	4475.090	22345.92	PR	1.5	0.8		MIT
4473.498	4474.753	22347.60	PR	1.0			MIT
4473.323	4474.578	22348.48	PR	1.0			MIT
4472.926	4474.181	22350.46	PR	1.5	1.0		MIT
4471.745	4473.000	22356.36	PR	0.9			MIT
4471.428	4472.683	22357.95	PR	1.0D			MIT
4470.575	4471.830	22362.21	PR	0.6			MIT
4470.430	4471.685	22362.94	PR	0.7H			MIT
4469.655	4470.909	22366.81	PR	1.0			MIT
4469.027	4470.281	22369.96	PR	0.9	0.3		MIT
4468.712	4469.966	22371.53	PR	2.1	2.0		MIT
4468.196	4469.450	22374.12	PR	1.0			MIT
4466.785	4468.039	22381.19	PR	0.3			KN
4466.431	4467.684	22382.96	PR	0.6	0.0		MIT
4465.981	4467.234	22385.22	PR	2.0	1.5		MIT
4465.365	4466.618	22388.30	PR	0.7			MIT
4465.075	4466.328	22389.76	PR	1.5	0.5W		MIT
4463.925	4465.178	22395.52	PR	0.3H			MIT
4463.762	4465.015	22396.34	PR	1.0			KN
4463.317	4464.570	22398.57	PR	0.8			MIT
4463.08	4464.33	22399.8	PR	1.4	0.6		MIT
4462.627	4463.879	22402.04	PR	1.0			MIT
4461.292	4462.544	22408.74	PR	1.2	0.3		MIT
4459.434	4460.686	22418.08	PR	0.5			KN
4458.336	4459.587	22423.60	PR	2.0	1.0W		MIT
4455.687	4456.938	22436.93	PR	0.9			MIT
4455.462	4456.713	22438.06	PR	0.9			MIT
4455.035	4456.285	22440.21	PR	1.0D			MIT
4454.695	4455.945	22441.93	PR	1.5	1.0		MIT
4454.382	4455.632	22443.50	PR	1.8	1.2		MIT
4453.149	4454.399	22449.72	PR	0.5			MIT
4452.529	4453.779	22452.84	PR	0.3			KN
4451.949	4453.199	22455.77	PR	1.9	1.3		MIT
4450.806	4452.055	22461.54	PR	0.7			MIT
4450.214	4451.463	22464.52	PR	1.6W	1.3W		MIT
4449.867	4451.116	22466.27	PR	2.1	1.9		MIT
4448.702	4449.951	22472.16	PR	0.7			MIT
4447.910	4449.159	22476.16	PR	0.6			KN
4446.985	4448.233	22480.83	PR	1.7	1.5J		MIT
4445.866	4447.114	22486.49	PR	1.4	0.7		MIT
4445.354	4446.602	22489.08	PR	0.3			MIT
4445.155	4446.403	22490.09	PR	0.7	0.0		MIT
4444.006	4445.254	22495.91	PR	1.5W	0.5W		MIT

AIR WAVELENGTH (ANGSTRMS)	VACUUM WAVELENGTH (ANGSTRMS)	WAVENUMBER (KAYSERS)	LMENT	LOG ARC	INTENSITY SPK	INTENSITY DIS	REF
4442.913	4444.160	22501.44	PR	0.3			KN
4441.414	4442.661	22509.03	PR	0.5			MIT
4441.086	4442.333	22510.70	PR	1.2	0.0		MIT
4440.616	4441.863	22513.08	PR	1.2			MIT
4439.988	4441.234	22516.26	PR	0.7			MIT
4439.144	4440.390	22520.54	PR	0.7			MIT
4438.178	4439.424	22525.45	PR	1.7	1.3		MIT
4437.688	4438.934	22527.93	PR	0.7			MIT
4437.451	4438.697	22529.14	PR	0.3			MIT
4436.454	4437.700	22534.20	PR	0.3			MIT
4435.730	4436.975	22537.88	PR	1.2W			MIT
4435.252	4436.497	22540.30	PR	0.7			MIT
4434.852	4436.097	22542.34	PR	1.4	0.5W		MIT
4432.336	4433.580	22555.13	PR	1.9	1.0W		MIT
4431.890	4433.134	22557.40	PR	1.6	0.7W		MIT
4431.559	4432.803	22559.09	PR	1.0			MIT
4430.126	4431.370	22566.39	PR	0.7			MIT
4429.238	4430.482	22570.91	PR	2.3	2.1		MIT
4428.458	4429.701	22574.89	PR	0.5			MIT
4427.883	4429.126	22577.82	PR	1.0W			MIT
4426.961	4428.204	22582.52	PR	0.3			MIT
4426.508	4427.751	22584.83	PR	0.5			MIT
4426.325	4427.568	22585.76	PR	0.5			MIT
4425.865	4427.108	22588.11	PR	0.3			MIT
4425.498	4426.741	22589.98	PR	0.5			MIT
4425.220	4426.463	22591.40	PR	0.3			MIT
4424.595	4425.837	22594.59	PR	2.0	1.5		MIT
4423.933	4425.175	22597.98	PR	0.7	0.0		MIT
4423.685	4424.927	22599.24	PR	0.3			MIT
4423.32	4424.56	22601.1	PR	1.0W			MIT
4422.092	4423.334	22607.38	PR	0.5			MIT
4421.231	4422.473	22611.78	PR	2.0	1.5W		MIT
4420.304	4421.545	22616.53	PR	0.6			MIT
4420.023	4421.264	22617.97	PR	0.5			MIT
4419.667	4420.908	22619.79	PR	2.0	1.7		MIT
4419.058	4420.299	22622.90	PR	1.9	1.5		MIT
4418.456	4419.697	22625.99	PR	0.3			MIT
4418.23	4419.47	22627.1	PR	0.6			MIT
4417.873	4419.114	22628.97	PR	1.0D			MIT
4417.185	4418.425	22632.50	PR	0.6			MIT
4416.586	4417.826	22635.57	PR	1.0	0.0		MIT
4415.977	4417.217	22638.69	PR	0.5			MIT
4415.587	4416.827	22640.69	PR	0.5			MIT
4414.403	4415.643	22646.76	PR	1.3	0.6		MIT
4413.765	4415.005	22650.03	PR	2.0	1.6		MIT
4413.249	4414.488	22652.68	PR	0.7			MIT
4412.766	4414.005	22655.16	PR	0.5			MIT
4412.155	4413.394	22658.30	PR	1.7	1.2		MIT
4411.802	4413.041	22660.11	PR	0.6	0.0		MIT
4411.334	4412.573	22662.51	PR	1.4			MIT
4410.855	4412.094	22664.98	PR	0.3			MIT
4410.628	4411.867	22666.14	PR	0.5			MIT
4410.244	4411.483	22668.12	PR	0.6			MIT
4409.683	4410.921	22671.00	PR	1.4W	0.3W		MIT
4408.844	4410.082	22675.31	PR	2.1	2.0		KN
4408.163	4409.401	22678.82	PR	1.0	0.3		MIT
4406.665	4407.903	22686.53	PR	1.4	0.9		MIT
4405.849	4407.086	22690.73	PR	2.0	2.0		MIT
4405.145	4406.382	22694.35	PR	1.4	1.3		MIT
4404.707	4405.944	22696.61	PR	1.4W	0.6		KN
4403.605	4404.842	22702.29	PR	2.0	1.6		MIT
4403.282	4404.519	22703.96	PR	1.4	1.0		MIT
4402.580	4403.817	22707.58	PR	0.6			MIT
4402.158	4403.395	22709.75	PR	0.3			MIT
4400.838	4402.074	22716.56	PR	0.5			MIT
4400.254	4401.490	22719.58	PR	1.4	1.0		MIT
4400.028	4401.264	22720.75	PR	1.5	1.3		MIT
4399.326	4400.562	22724.37	PR	1.6	1.3		MIT
4398.265	4399.500	22729.85	PR	1.4	1.0		MIT
4397.197	4398.432	22735.37	PR	0.3			MIT

AIR WAVELENGTH (ANGSTRMS)	VACUUM WAVELENGTH (ANGSTRMS)	WAVENUMBER (KAYSERS)	LMENT	LOG ARC	INTENSITY SPK	DIS	REF
4397.121	4398.356	22735.77	PR	0.3			KN
4396.872	4398.107	22737.05	PR	1.1	0.3		MIT
4396.117	4397.352	22740.96	PR	1.9	1.7		MIT
4395.788	4397.023	22742.66	PR	1.5	0.3W		MIT
4395.005	4396.240	22746.71	PR	1.4	1.2		MIT
4394.746	4395.981	22748.05	PR	1.0	0.3		MIT
4394.043	4395.277	22751.69	PR	0.9			MIT
4393.492	4394.726	22754.55	PR	0.9	0.3		MIT
4393.045	4394.279	22756.86	PR	0.5			MIT
4392.713	4393.947	22758.58	PR	0.3			MIT
4392.442	4393.676	22759.99	PR	0.3			MIT
4392.079	4393.313	22761.87	PR	0.7			MIT
4391.988	4393.222	22762.34	PR	1.5W	0.7W		MIT
4391.514	4392.748	22764.79	PR	1.3	0.6		MIT
4391.111	4392.345	22766.88	PR	1.0			MIT
4390.145	4391.378	22771.89	PR	0.9	0.0		MIT
4389.511	4390.744	22775.18	PR	0.9			MIT
4388.729	4389.962	22779.24	PR	1.0	0.0		MIT
4387.577	4388.810	22785.22	PR	0.3			MIT
4387.444	4388.677	22785.91	PR	0.7	0.0		MIT
4386.646	4387.878	22790.06	PR	0.5			MIT
4385.481	4386.713	22796.11	PR	0.9	0.0		MIT
4385.238	4386.470	22797.37	PR	1.2	0.7		MIT
4384.797	4386.029	22799.67	PR	1.6	1.2		MIT
4384.136	4385.368	22803.10	PR	1.5W	1.4W		MIT
4382.817	4384.048	22809.97	PR	1.4W	0.9W		MIT
4382.418	4383.649	22812.04	PR	1.5	1.3		MIT
4381.622	4382.853	22816.19	PR	1.2			MIT
4381.082	4382.313	22819.00	PR	1.3W			MIT
4380.316	4381.547	22822.99	PR	1.7	1.3		MIT
4379.335	4380.565	22828.10	PR	2.0W	0.3		MIT
4378.836	4380.066	22830.70	PR	0.5			MIT
4378.635	4379.865	22831.75	PR	0.5			MIT
4378.263	4379.493	22833.69	PR	1.2	0.5		MIT
4377.846	4379.076	22835.87	PR	0.6			MIT
4377.614	4378.844	22837.08	PR	0.7			MIT
4376.609	4377.839	22842.32	PR	0.8			MIT
4375.726	4376.956	22846.93	PR	1.0	0.3		MIT
4374.410	4375.639	22853.80	PR	1.7	1.2		MIT
4373.815	4375.044	22856.91	PR	1.6	1.0		MIT
4372.945	4374.174	22861.46	PR	0.5			MIT
4372.720	4373.949	22862.64	PR	1.2H			MIT
4372.385	4373.614	22864.39	PR	1.0			MIT
4371.614	4372.842	22868.42	PR	2.1	1.6W		MIT
4371.218	4372.446	22870.49	PR	0.9	0.3		MIT
4370.800	4372.028	22872.68	PR	1.3	0.8		MIT
4369.866	4371.094	22877.57	PR	0.7			MIT
4369.050	4370.278	22881.84	PR	0.7W			MIT
4368.327	4369.555	22885.63	PR	2.1	2.0		MIT
4367.233	4368.460	22891.36	PR	1.4W	0.7W		MIT
4366.693	4367.920	22894.19	PR	0.7			MIT
4366.088	4367.315	22897.36	PR	1.5W			MIT
4365.760	4366.987	22899.08	PR	0.7			MIT
4365.328	4366.555	22901.35	PR	0.8			MIT
4364.983	4366.210	22903.16	PR	0.7			MIT
4364.773	4366.000	22904.26	PR	1.3W			MIT
4364.510	4365.737	22905.64	PR	0.5			MIT
4364.163	4365.390	22907.46	PR	0.9D			MIT
4363.988	4365.214	22908.38	PR	1.0			MIT
4363.219	4364.445	22912.42	PR	1.4	0.9		MIT
4362.981	4364.207	22913.67	PR	1.7	1.3		MIT
4362.146	4363.372	22918.05	PR	1.0W			MIT
4361.818	4363.044	22919.78	PR	1.2	0.6		MIT
4361.527	4362.753	22921.31	PR	1.0	0.3		MIT
4361.261	4362.487	22922.71	PR	1.3	0.6		MIT
4360.956	4362.182	22924.31	PR	0.3			MIT
4360.568	4361.794	22926.35	PR	0.6			MIT
4359.795	4361.020	22930.41	PR	2.0	1.6		MIT
4359.107	4360.332	22934.03	PR	1.8	1.4		KN
4357.503	4358.728	22942.47	PR	1.4	0.7		MIT

AIR WAVELENGTH (ANGSTRMS)	VACUUM WAVELENGTH (ANGSTRMS)	WAVENUMBER (KAYSERS)	LMENT	LOG ARC	INTENSITY SPK	DIS	REF
4356.672	4357.897	22946.85	PR	0.7			MIT
4356.222	4357.446	22949.22	PR	0.5			MIT
4355.753	4356.977	22951.69	PR	0.7	0.0		MIT
4355.190	4356.414	22954.66	PR	1.4	0.7		MIT
4354.912	4356.136	22956.12	PR	1.9	1.5		MIT
4353.798	4355.022	22962.00	PR	1.0	0.3		MIT
4353.559	4354.783	22963.26	PR	0.7			MIT
4353.175	4354.399	22965.28	PR	1.0	0.0		MIT
4352.513	4353.736	22968.78	PR	0.7	0.0		MIT
4352.047	4353.270	22971.24	PR	0.6	0.3		MIT
4351.849	4353.072	22972.28	PR	1.9	1.8		MIT
4350.910	4352.133	22977.24	PR	0.6			MIT
4350.399	4351.622	22979.94	PR	1.8	1.4		MIT
4349.672	4350.895	22983.78	PR	1.2			MIT
4348.168	4349.390	22991.73	PR	0.9	0.3H		MIT
4347.490	4348.712	22995.31	PR	2.0	1.6		MIT
4346.892	4348.114	22998.48	PR	1.5	0.3H		MIT
4346.245	4347.467	23001.90	PR	1.4			MIT
4345.503	4346.725	23005.83	PR	0.6	0.7H		MIT
4344.895	4346.116	23009.05	PR	0.8	0.3		MIT
4344.334	4345.555	23012.02	PR	2.2W	1.9W		MIT
4343.891	4345.112	23014.37	PR	1.0	0.5		MIT
4343.389	4344.610	23017.03	PR	0.5			MIT
4342.810	4344.031	23020.09	PR	1.6	0.7W		MIT
4341.375	4342.596	23027.70	PR	1.0	0.0		MIT
4340.888	4342.108	23030.29	PR	0.7			MIT
4340.255	4341.475	23033.65	PR	0.6			MIT
4339.68	4340.90	23036.7	PR	1.3	0.3		MIT
4338.694	4339.914	23041.93	PR	2.0	1.7		MIT
4338.285	4339.505	23044.10	PR	0.9			MIT
4337.515	4338.734	23048.20	PR	0.6			MIT
4335.969	4337.188	23056.41	PR	0.9	0.0		MIT
4335.748	4336.967	23057.59	PR	1.9	1.3		MIT
4334.616	4335.835	23063.61	PR	1.6	0.8W		MIT
4333.913	4335.132	23067.35	PR	2.2	2.0		KN
4333.148	4334.366	23071.42	PR	1.6	1.0		MIT
4332.487	4333.705	23074.94	PR	1.5	0.3		MIT
4331.472	4332.690	23080.35	PR	0.7	0.0		MIT
4331.294	4332.512	23081.30	PR	1.0	0.5		MIT
4330.440	4331.658	23085.85	PR	1.7	0.5W		MIT
4329.415	4330.632	23091.32	PR	1.7	1.3		MIT
4328.985	4330.202	23093.61	PR	1.7	1.0		MIT
4328.421	4329.638	23096.62	PR	1.7	1.2		MIT
4328.202	4329.419	23097.79	PR	1.5	0.6W		MIT
4327.698	4328.915	23100.48	PR	0.5			MIT
4327.511	4328.728	23101.48	PR	0.9			MIT
4325.358	4326.574	23112.97	PR	1.3D	0.3H		MIT
4324.802	4326.018	23115.95	PR	0.5			MIT
4324.318	4325.534	23118.53	PR	0.5			MIT
4323.922	4325.138	23120.65	PR	1.5	0.5W		MIT
4323.551	4324.767	23122.63	PR	2.0	1.5		MIT
4323.071	4324.287	23125.20	PR	0.9	0.0		MIT
4322.765	4323.981	23126.84	PR	0.8	0.0		MIT
4322.410	4323.626	23128.74	PR	1.0W			MIT
4322.036	4323.251	23130.74	PR	0.7			MIT
4321.329	4322.544	23134.52	PR	1.2	0.3		MIT
4320.754	4321.969	23137.60	PR	1.3	0.3		MIT
4320.367	4321.582	23139.67	PR	0.7			MIT
4320.168	4321.383	23140.74	PR	1.3	0.3H		MIT
4319.723	4320.938	23143.12	PR	0.9	0.3H		MIT
4319.003	4320.218	23146.98	PR	1.4	0.9		MIT
4318.440	4319.654	23150.00	PR	0.5	0.0		MIT
4317.836	4319.050	23153.24	PR	1.6	0.7W		MIT
4317.045	4318.259	23157.48	PR	1.5	1.0J		MIT
4316.057	4317.271	23162.78	PR	1.7	0.9		MIT
4315.523	4316.737	23165.65	PR	1.5	0.9		MIT
4314.691	4315.904	23170.11	PR	1.0	0.0		MIT
4314.356	4315.569	23171.91	PR	1.4	0.5		MIT
4313.260	4314.473	23177.80	PR	1.2W			MIT
4313.148	4314.361	23178.40	PR	0.3			MIT

AIR WAVELENGTH (ANGSTRMS)	VACUUM WAVELENGTH (ANGSTRMS)	WAVENUMBER (KAYSERS)	LMENT	LOG ARC	INTENSITY SPK	DIS	REF
4312.355	4313.568	23182.67	PR	0.9	0.0		MIT
4311.919	4313.132	23185.01	PR	1.0	0.3		MIT
4311.54	4312.75	23187.1	PR	1.2	0.0		MIT
4311.102	4312.315	23189.40	PR	1.7	1.0W		MIT
4310.203	4311.415	23194.24	PR	0.6			MIT
4309.873	4311.085	23196.02	PR	0.7			MIT
4309.00	4310.21	23200.7	PR	0.9	0.3		MIT
4308.888	4310.100	23201.32	PR	1.4	0.7		MIT
4307.673	4308.885	23207.86	PR	1.5	1.0		MIT
4307.24	4308.45	23210.2	PR	1.5	0.3W		MIT
4306.808	4308.019	23212.52	PR	0.6			MIT
4306.535	4307.746	23214.00	PR	0.7	0.0		MIT
4306.083	4307.294	23216.43	PR	1.4	1.2		MIT
4305.763	4306.974	23218.16	PR	2.2	2.0		MIT
4305.595	4306.806	23219.06	PR	0.5	0.0		KN
4304.651	4305.862	23224.15	PR	0.6	0.0		MIT
4303.594	4304.805	23229.86	PR	2.0	1.8		MIT
4303.139	4304.349	23232.31	PR	1.3	0.7		MIT
4302.10	4303.31	23237.9	PR	1.8	0.7W		MIT
4301.583	4302.793	23240.72	PR	1.3	1.5		MIT
4298.920	4300.129	23255.11	PR	1.5	1.5		MIT
4297.764	4298.973	23261.37	PR	1.7	1.6		MIT
4296.55	4297.76	23267.9	PR	1.2	0.3W		MIT
4296.18	4297.39	23269.9	PR	1.0	0.0		MIT
4295.11	4296.32	23275.7	PR	1.1	0.7		MIT
4294.700	4295.908	23277.97	PR	1.4	1.0		MIT
4293.582	4294.790	23284.03	PR	1.3	0.9		MIT
4293.135	4294.343	23286.45	PR	1.4	0.9		MIT
4291.63	4292.84	23294.6	PR	1.3	0.3		MIT
4290.99	4292.20	23298.1	PR	0.7	0.0		MIT
4290.512	4291.719	23300.69	PR	0.6			KN
4289.89	4291.10	23304.1	PR	1.2	0.6		MIT
4289.42	4290.63	23306.6	PR	1.1	0.5		MIT
4288.46	4289.67	23311.8	PR	1.0	0.3W		MIT
4286.98	4288.19	23319.9	PR	0.5	0.0		MIT
4286.38	4287.59	23323.1	PR	0.7D			MIT
4285.85	4287.06	23326.0	PR	0.7			MIT
4285.540	4286.746	23327.72	PR	0.7	0.3		MIT
4284.48	4285.69	23333.5	PR	1.0			MIT
4283.78	4284.99	23337.3	PR	1.4W	0.0		MIT
4283.01	4284.22	23341.5	PR	0.7	0.0		MIT
4282.440	4283.645	23344.60	PR	1.9	1.6		MIT
4281.82	4283.02	23348.0	PR	1.3	0.3		MIT
4281.05	4282.25	23352.2	PR	0.9	0.0		MIT
4280.105	4281.309	23357.34	PR	1.8	1.5		KN
4278.99	4280.19	23363.4	PR	1.2	0.5		MIT
4278.62	4279.82	23365.4	PR	1.0	0.3		MIT
4278.04	4279.24	23368.6	PR	1.5	1.2		MIT
4277.23	4278.43	23373.0	PR	0.5			MIT
4276.19	4277.39	23378.7	PR	1.1	0.5		MIT
4275.82	4277.02	23380.7	PR	1.3	0.6		MIT
4275.32	4276.52	23383.5	PR	0.7	0.0		MIT
4275.17	4276.37	23384.3	PR	1.1	0.3		MIT
4274.959	4276.162	23385.46	PR	0.5			MIT
4274.27	4275.47	23389.2	PR	0.9D	0.0D		MIT
4273.21	4274.41	23395.0	PR	0.5	0.0		MIT
4272.93	4274.13	23396.6	PR	0.7	0.0		MIT
4272.271	4273.473	23400.17	PR	1.7	1.5		MIT
4271.764	4272.966	23402.95	PR	1.3	1.2		MIT
4269.67	4270.87	23414.4	PR	1.3	0.0		MIT
4269.100	4270.301	23417.55	PR	1.1	0.6		MIT
4267.78	4268.98	23424.8	PR	1.1			MIT
4266.73	4267.93	23430.6	PR	0.5			MIT
4266.45	4267.65	23432.1	PR	0.6			MIT
4265.66	4266.86	23436.4	PR	0.5			MIT
4265.490	4266.691	23437.37	PR	0.5			KN
4263.805	4265.005	23446.63	PR	1.7	0.7W		MIT
4263.15	4264.35	23450.2	PR	0.6	0.0		MIT
4262.80	4264.00	23452.2	PR	1.1	0.3		MIT
4262.314	4263.514	23454.83	PR	1.0	0.6		KN

AIR WAVELENGTH (ANGSTRMS)	VACUUM WAVELENGTH (ANGSTRMS)	WAVENUMBER (KAYSERS)	LMENT	LOG ARC	INTENSITY SPK	DIS	REF
4261.796	4262.996	23457.68	PR	1.2	0.3		MIT
4261.04	4262.24	23461.9	PR	0.6	0.0		MIT
4260.06	4261.26	23467.2	PR	0.5	0.0		MIT
4259.64	4260.84	23469.6	PR	0.7D			MIT
4257.94	4259.14	23478.9	PR	0.5			MIT
4257.69	4258.89	23480.3	PR	0.7	0.0		MIT
4256.33	4257.53	23487.8	PR	0.6	0.0		MIT
4255.66	4256.86	23491.5	PR	0.6	0.0		MIT
4255.00	4256.20	23495.1	PR	0.6	0.0		MIT
4254.420	4255.618	23498.35	PR	1.5	1.3		MIT
4253.032	4254.229	23506.02	PR	1.1	0.3		KN
4252.637	4253.834	23508.20	PR	0.5			MIT
4252.07	4253.27	23511.3	PR	0.6	0.0		MIT
4251.490	4252.687	23514.55	PR	1.6W	1.2W		MIT
4250.401	4251.598	23520.57	PR	1.3	1.1		MIT
4249.484	4250.680	23525.65	PR	1.3	0.3		MIT
4249.083	4250.279	23527.87	PR	1.1	0.3		MIT
4247.662	4248.858	23535.74	PR	1.8	1.5		MIT
4246.15	4247.35	23544.1	PR	1.0	0.3		MIT
4245.46	4246.66	23547.9	PR	1.0	0.0		MIT
4245.14	4246.34	23549.7	PR	1.0W	0.0W		MIT
4244.490	4245.685	23553.32	PR	0.6	0.0		MIT
4243.528	4244.723	23558.67	PR	1.3	0.9		MIT
4241.30	4242.49	23571.0	PR	1.0	0.3		MIT
4241.019	4242.213	23572.60	PR	1.7	1.1		MIT
4240.033	4241.227	23578.08	PR	1.1	0.5		MIT
4239.47	4240.66	23581.2	PR	0.7	0.0		MIT
4237.010	4238.203	23594.91	PR	0.7	0.0		KN
4236.635	4237.828	23596.99	PR	1.0	0.5		MIT
4236.210	4237.403	23599.36	PR	1.3W	0.5W		MIT
4233.134	4234.326	23616.51	PR	1.2	0.7		KN
4232.65	4233.84	23619.2	PR	0.9	0.0H		MIT
4232.330	4233.522	23621.00	PR	0.7	0.0		MIT
4232.04	4233.23	23622.6	PR	0.6	0.0		MIT
4230.667	4231.858	23630.28	PR	0.9	0.3		MIT
4229.80	4230.99	23635.1	PR	0.9	0.0		MIT
4229.10	4230.29	23639.0	PR	1.2W	0.3		MIT
4228.50	4229.69	23642.4	PR	1.2W	0.5W		MIT
4225.327	4226.517	23660.14	PR	1.7	1.6		MIT
4223.51	4224.70	23670.3	PR	1.1	0.5		MIT
4222.98	4224.17	23673.3	PR	2.1	1.6		MIT
4221.45	4222.64	23681.9	PR	0.6	0.0		MIT
4220.14	4221.33	23689.2	PR	1.0	0.3		MIT
4219.652	4220.840	23691.96	PR	1.5W	0.9W		MIT
4217.81	4219.00	23702.3	PR	1.1	0.9		MIT
4217.19	4218.38	23705.8	PR	0.7	0.0		MIT
4216.773	4217.961	23708.14	PR	0.5			MIT
4215.98	4217.17	23712.6	PR	0.6	0.5		MIT
4215.14	4216.33	23717.3	PR	0.9	0.5		MIT
4214.478	4215.665	23721.05	PR	0.5	0.3		KN
4213.96	4215.15	23724.0	PR	1.1	0.5		MIT
4213.573	4214.760	23726.14	PR	1.3	1.0		MIT
4213.26	4214.45	23727.9	PR	0.6	0.3		MIT
4212.70	4213.89	23731.1	PR	0.7	0.3		MIT
4211.858	4213.044	23735.80	PR	1.7D	1.4D		KN
4211.24	4212.43	23739.3	PR	0.7	0.3		MIT
4208.305	4209.491	23755.84	PR	1.3	1.1		MIT
4207.810	4208.995	23758.64	PR	1.1	0.7		KN
4206.739	4207.924	23764.69	PR	1.7	1.7		MIT
4205.72	4206.90	23770.4	PR	0.7	0.0		MIT
4204.58	4205.76	23776.9	PR	0.7	0.0		MIT
4203.49	4204.67	23783.1	PR	0.6	0.3		MIT
4202.70	4203.88	23787.5	PR	0.9	0.5		MIT
4202.41	4203.59	23789.2	PR	0.5	0.3		MIT
4201.529	4202.713	23794.16	PR	1.5W	1.1W		MIT
4201.184	4202.368	23796.11	PR	1.2	1.0		MIT
4200.52	4201.70	23799.9	PR	0.6	0.0		MIT
4199.58	4200.76	23805.2	PR	0.5	0.3		MIT
4196.79	4197.97	23821.0	PR	1.0	0.6		MIT
4195.96	4197.14	23825.7	PR	1.1	0.7		MIT

AIR WAVELENGTH (ANGSTRMS)	VACUUM WAVELENGTH (ANGSTRMS)	WAVENUMBER (KAYSERS)	LMENT	LOG ARC	INTENSITY SPK	DIS	REF
4195.52	4196.70	23828.2	PR	0.9H	0.5H		MIT
4194.62	4195.80	23833.4	PR	1.2	0.7		MIT
4194.24	4195.42	23835.5	PR	1.0	0.5		MIT
4193.88	4195.06	23837.6	PR	0.7	0.3		MIT
4192.50	4193.68	23845.4	PR	1.3	0.7		MIT
4191.615	4192.796	23850.43	PR	1.6	1.4		MIT
4190.64	4191.82	23856.0	PR	1.0	0.5		MIT
4189.518	4190.699	23862.37	PR	2.0	1.7		KN
4187.79	4188.97	23872.2	PR	0.3	0.3		MIT
4186.395	4187.575	23880.17	PR	1.1	0.5		MIT
4185.82	4187.00	23883.4	PR	1.2	0.7		MIT
4185.154	4186.333	23887.25	PR	0.7	0.3		MIT
4184.242	4185.421	23892.46	PR	0.6	0.0		MIT
4182.91	4184.09	23900.1	PR	0.7	0.3		MIT
4182.33	4183.51	23903.4	PR	1.0	0.7		MIT
4180.68	4181.86	23912.8	PR	0.9	0.0		MIT
4180.40	4181.58	23914.4	PR	0.7	0.3		MIT
4179.422	4180.600	23920.01	PR	2.3	1.6		MIT
4178.641	4179.819	23924.48	PR	1.3	1.0		MIT
4177.89	4179.07	23928.8	PR	1.1	0.5		MIT
4177.50	4178.68	23931.0	PR	0.6	0.3		MIT
4176.33	4177.51	23937.7	PR	0.7	0.3		MIT
4175.644	4176.821	23941.65	PR	1.3	1.0		MIT
4175.299	4176.476	23943.63	PR	1.3	1.0		KN
4174.63	4175.81	23947.5	PR	0.7	0.0		MIT
4173.70	4174.88	23952.8	PR	0.7	0.3		MIT
4172.273	4173.449	23961.00	PR	1.9	1.6		MIT
4171.824	4173.000	23963.58	PR	1.9	1.6		KN
4171.04	4172.22	23968.1	PR	1.1	0.6		MIT
4169.678	4170.853	23975.91	PR	0.5			KN
4169.459	4170.634	23977.17	PR	1.2	1.0		MIT
4168.08	4169.25	23985.1	PR	1.3	0.9		MIT
4167.69	4168.86	23987.4	PR	1.0	0.5		MIT
4166.84	4168.01	23992.2	PR	0.6	0.3		MIT
4166.36	4167.53	23995.0	PR	0.7	0.3		MIT
4165.50	4166.67	24000.0	PR	0.3	0.0		MIT
4165.187	4166.361	24001.76	PR	0.7	0.3		MIT
4164.192	4165.366	24007.49	PR	2.3	1.7		KN
4163.01	4164.18	24014.3	PR	0.7	0.0		MIT
4162.51	4163.68	24017.2	PR	0.7	0.3		MIT
4161.65	4162.82	24022.2	PR	0.9	0.3		MIT
4161.30	4162.47	24024.2	PR	0.7	0.3		MIT
4160.86	4162.03	24026.7	PR	1.1	0.5		MIT
4160.47	4161.64	24029.0	PR	1.3W	0.7W		MIT
4159.83	4161.00	24032.7	PR	1.2	0.5		MIT
4159.47	4160.64	24034.7	PR	0.9	0.5		MIT
4158.07	4159.24	24042.8	PR	1.0	0.7		MIT
4157.77	4158.94	24044.6	PR	0.9	0.3		MIT
4156.90	4158.07	24049.6	PR	1.0	0.6		MIT
4156.518	4157.690	24051.82	PR	1.1	1.0		MIT
4154.05	4155.22	24066.1	PR	1.0	0.3		MIT
4153.663	4154.834	24068.35	PR	0.8			KN
4152.34	4153.51	24076.0	PR	0.7	0.6		MIT
4151.01	4152.18	24083.7	PR	0.8D	0.5D		MIT
4150.04	4151.21	24089.4	PR	0.7	0.3		MIT
4148.457	4149.627	24098.55	PR	1.2	1.1		MIT
4147.55	4148.72	24103.8	PR	0.6	0.3		MIT
4147.14	4148.31	24106.2	PR	1.1	0.6		MIT
4146.537	4147.706	24109.71	PR	1.3	1.1		MIT
4144.75	4145.92	24120.1	PR	0.8	0.5		MIT
4144.23	4145.40	24123.1	PR	0.6	0.5		MIT
4143.136	4144.304	24129.50	PR	2.3	1.7		MIT
4142.434	4143.602	24133.59	PR	0.5			MIT
4142.00	4143.17	24136.1	PR	0.9	0.5		MIT
4141.257	4142.425	24140.45	PR	2.2	1.7		MIT
4140.573	4141.741	24144.44	PR	0.5			MIT
4140.32	4141.49	24145.9	PR	0.7	0.0		MIT
4138.19	4139.36	24158.3	PR	1.3	0.9		MIT
4137.49	4138.66	24162.4	PR	0.3	0.0		MIT
4137.16	4138.33	24164.4	PR	0.5	0.0		MIT

AIR WAVELENGTH (ANGSTRMS)	VACUUM WAVELENGTH (ANGSTRMS)	WAVENUMBER (KAYSERS)	LMENT	LOG ARC	INTENSITY SPK	DIS	REF
4133.618	4134.784	24185.06	PR	1.0	0.7		MIT
4132.230	4133.396	24193.18	PR	1.2	0.9		MIT
4131.82	4132.99	24195.6	PR	1.1	0.6		MIT
4130.770	4131.935	24201.74	PR	1.4	1.3		MIT
4129.148	4130.313	24211.24	PR	1.3	1.2		MIT
4126.15	4127.31	24228.8	PR	1.0	0.7		MIT
4125.06	4126.22	24235.2	PR	1.1	0.7		MIT
4124.35	4125.51	24239.4	PR	0.7	0.3		MIT
4124.06	4125.22	24241.1	PR	0.7	0.3		MIT
4122.98	4124.14	24247.5	PR	0.9	0.3		MIT
4122.31	4123.47	24251.4	PR	0.9	0.3		MIT
4122.09	4123.25	24252.7	PR	1.1	0.5		MIT
4120.96	4122.12	24259.4	PR	0.5	0.0		MIT
4120.00	4121.16	24265.0	PR	0.9D	0.3A		MIT
4119.80	4120.96	24266.2	PR	0.7	0.0		MIT
4119.37	4120.53	24268.7	PR	0.7	0.3		MIT
4118.481	4119.643	24273.95	PR	2.4D	1.7D		MIT
4117.21	4118.37	24281.4	PR	0.5	0.0		MIT
4115.83	4116.99	24289.6	PR	1.1	0.7		MIT
4114.84	4116.00	24295.4	PR	0.9	0.3		MIT
4114.31	4115.47	24298.6	PR	0.5	0.0		MIT
4113.887	4115.048	24301.05	PR	1.5W	1.8W		MIT
4113.381	4114.542	24304.04	PR	0.5			MIT
4112.74	4113.90	24307.8	PR	1.0	0.3		MIT
4111.98	4113.14	24312.3	PR	0.5	0.0		MIT
4111.871	4113.031	24312.97	PR	1.5W	0.9W		MIT
4110.92	4112.08	24318.6	PR	1.0	0.3		MIT
4110.47	4111.63	24321.3	PR	0.7	0.0		MIT
4110.11	4111.27	24323.4	PR	0.5	0.0		MIT
4109.41	4110.57	24327.5	PR	1.2	0.3		MIT
4109.09	4110.25	24329.4	PR	0.7	0.0		MIT
4108.34	4109.50	24333.9	PR	1.2	0.5		MIT
4107.754	4108.913	24337.34	PR	1.2	0.7		MIT
4107.11	4108.27	24341.1	PR	1.0	0.0		MIT
4105.733	4106.892	24349.32	PR	0.6	0.3		KN
4104.86	4106.02	24354.5	PR	0.6	0.0		MIT
4104.48	4105.64	24356.7	PR	0.7	0.3		MIT
4103.20	4104.36	24364.4	PR	0.3	0.0		MIT
4100.746	4101.903	24378.93	PR	2.3	1.7		MIT
4100.22	4101.38	24382.1	PR	1.2	1.0		MIT
4099.88	4101.04	24384.1	PR	0.6	0.0		MIT
4098.65	4099.81	24391.4	PR	0.9	0.5		MIT
4098.410	4099.567	24392.82	PR	1.3	1.1		MIT
4096.822	4097.978	24402.28	PR	1.5	1.2		MIT
4096.342	4097.498	24405.14	PR	1.3	0.9		MIT
4095.97	4097.13	24407.4	PR	0.5	0.0		MIT
4094.970	4096.126	24413.31	PR	1.7	1.1		MIT
4093.79	4094.95	24420.4	PR	0.9	0.0		MIT
4092.85	4094.01	24426.0	PR	0.7	0.3		MIT
4092.63	4093.79	24427.3	PR	1.2	0.7		MIT
4091.112	4092.267	24436.33	PR	0.5			MIT
4090.74	4091.89	24438.6	PR	1.2	0.9		MIT
4089.87	4091.02	24443.8	PR	0.7	0.3		MIT
4089.47	4090.62	24446.1	PR	0.5	0.0		MIT
4088.87	4090.02	24449.7	PR	0.6	0.3		MIT
4088.35	4089.50	24452.8	PR	0.5	0.0		MIT
4087.206	4088.360	24459.69	PR	1.7	0.5		MIT
4086.24	4087.39	24465.5	PR	1.4	0.3		MIT
4085.34	4086.49	24470.9	PR	1.2	0.9		MIT
4085.13	4086.28	24472.1	PR	0.6	0.3		MIT
4084.74	4085.89	24474.4	PR	1.0	0.7		MIT
4083.336	4084.489	24482.87	PR	1.3	1.3		MIT
4082.78	4083.93	24486.2	PR	1.0	0.3		MIT
4081.901	4083.053	24491.48	PR	1.9	1.5		KN
4081.54	4082.69	24493.6	PR	0.7	0.0		MIT
4081.018	4082.170	24496.77	PR	1.7	1.4		MIT
4079.786	4080.938	24504.17	PR	1.7	1.5		MIT
4078.16	4079.31	24513.9	PR	0.7	0.0		MIT
4077.98	4079.13	24515.0	PR	1.0	0.3		MIT
4077.69	4078.84	24516.8	PR	0.6	0.3		MIT

AIR WAVELENGTH (ANGSTRMS)	VACUUM WAVELENGTH (ANGSTRMS)	WAVENUMBER (KAYSERS)	LMENT	LOG ARC	INTENSITY SPK	DIS	REF
4076.21	4077.36	24525.7	PR	1.0	0.0		MIT
4075.33	4076.48	24531.0	PR	0.7	0.0		MIT
4074.84	4075.99	24533.9	PR	0.7	0.0		MIT
4074.65	4075.80	24535.1	PR	0.5	0.0		MIT
4074.34	4075.49	24536.9	PR	0.5			MIT
4073.94	4075.09	24539.3	PR	0.7	0.0		MIT
4072.52	4073.67	24547.9	PR	1.0	0.3		MIT
4072.22	4073.37	24549.7	PR	0.3			MIT
4071.99	4073.14	24551.1	PR	0.7	0.0		MIT
4071.38	4072.53	24554.8	PR	0.6	0.0		MIT
4070.259	4071.408	24561.53	PR	1.3	0.9		MIT
4068.804	4069.953	24570.31	PR	1.6	1.2		MIT
4068.09	4069.24	24574.6	PR	0.7	0.0		MIT
4066.01	4067.16	24587.2	PR	1.3	0.5		MIT
4065.404	4066.552	24590.86	PR	0.3			MIT
4062.817	4063.964	24606.51	PR	2.2	1.7		MIT
4062.225	4063.372	24610.10	PR	0.7	0.5		MIT
4061.336	4062.483	24615.49	PR	0.6	0.3		MIT
4059.37	4060.52	24627.4	PR	0.6	0.5		MIT
4058.778	4059.924	24631.00	PR	1.4	1.2		MIT
4058.19	4059.34	24634.6	PR	0.8	0.9		MIT
4056.543	4057.689	24644.57	PR	2.0	1.8		MIT
4056.13	4057.28	24647.1	PR	0.7	0.3		MIT
4055.76	4056.91	24649.3	PR	0.8J	0.5		MIT
4054.845	4055.990	24654.89	PR	1.7	1.6		MIT
4053.48	4054.62	24663.2	PR	0.9	0.6		MIT
4052.575	4053.720	24668.70	PR	1.0	0.5		MIT
4051.152	4052.296	24677.37	PR	1.7	1.5		MIT
4049.83	4050.97	24685.4	PR	0.9	0.3		MIT
4049.61	4050.75	24686.8	PR	0.9	0.3		MIT
4048.14	4049.28	24695.7	PR	0.9	0.6		MIT
4047.098	4048.241	24702.08	PR	1.3	1.1		MIT
4046.635	4047.778	24704.91	PR	0.9	0.5		MIT
4044.818	4045.961	24716.01	PR	1.7	1.5		MIT
4041.87	4043.01	24734.0	PR	0.6	0.0		MIT
4041.37	4042.51	24737.1	PR	0.7			MIT
4040.67	4041.81	24741.4	PR	0.3			MIT
4039.357	4040.498	24749.42	PR	1.7	1.3		MIT
4038.90	4040.04	24752.2	PR	0.6	0.0		MIT
4038.467	4039.608	24754.88	PR	1.2	1.0		MIT
4038.154	4039.295	24756.80	PR	1.1	0.5		MIT
4037.22	4038.36	24762.5	PR	0.9	0.3		MIT
4036.05	4037.19	24769.7	PR	1.2	0.0		MIT
4035.79	4036.93	24771.3	PR	0.5	0.0		MIT
4035.42	4036.56	24773.6	PR	1.0	0.0		MIT
4035.07	4036.21	24775.7	PR	0.3	0.0		MIT
4034.30	4035.44	24780.4	PR	1.3	0.7		MIT
4033.857	4034.997	24783.17	PR	1.7	1.5		MIT
4033.24	4034.38	24787.0	PR	0.5	0.3		MIT
4032.974	4034.114	24788.59	PR	1.2	1.0		MIT
4032.492	4033.631	24791.56	PR	1.3	1.1		MIT
4031.755	4032.894	24796.09	PR	1.7	1.5		MIT
4031.09	4032.23	24800.2	PR	1.1	0.9		MIT
4029.73	4030.87	24808.6	PR	1.2	1.1		MIT
4029.041	4030.180	24812.79	PR	1.1	0.7		MIT
4028.69	4029.83	24814.9	PR	0.7	0.0		MIT
4027.64	4028.78	24821.4	PR	1.2	0.7		MIT
4026.839	4027.977	24826.36	PR	1.3	0.7		KN
4025.546	4026.684	24834.33	PR	1.6	1.4		MIT
4025.19	4026.33	24836.5	PR	1.2	0.9		MIT
4024.41	4025.55	24841.3	PR	1.2	0.3		MIT
4024.07	4025.21	24843.4	PR	0.7	0.0		MIT
4023.737	4024.874	24845.50	PR	0.5			MIT
4022.738	4023.875	24851.67	PR	1.1	0.8		MIT
4022.199	4023.336	24855.00	PR	0.9	0.6		MIT
4020.991	4022.127	24862.46	PR	1.6	1.5		MIT
4020.28	4021.42	24866.9	PR	1.0	0.3		MIT
4019.84	4020.98	24869.6	PR	1.2	0.9		MIT
4019.44	4020.58	24872.1	PR	1.0	0.5		MIT
4016.748	4017.883	24888.73	PR	1.4	1.3		MIT

AIR WAVELENGTH (ANGSTRMS)	VACUUM WAVELENGTH (ANGSTRMS)	WAVENUMBER (KAYSERS)	LMENT	LOG ARC	INTENSITY SPK	DIS	REF
4015.389	4016.524	24897.15	PR	1.7	1.5		MIT
4014.66	4015.79	24901.7	PR	1.0	0.3		MIT
4014.35	4015.48	24903.6	PR	1.0	0.5		MIT
4013.43	4014.56	24909.3	PR	1.2	0.9		MIT
4013.23	4014.36	24910.5	PR	1.0	0.6		MIT
4012.91	4014.04	24912.5	PR	0.9	0.0		MIT
4011.532	4012.666	24921.09	PR	0.7	0.0		MIT
4010.643	4011.777	24926.61	PR	0.9R	0.7R		KN
4009.97	4011.10	24930.8	PR	1.0	0.5		MIT
4009.24	4010.37	24935.3	PR	0.7	0.3		MIT
4008.714	4009.847	24938.61	PR	2.2	1.7		MIT
4007.78	4008.91	24944.4	PR	0.9	0.5		MIT
4006.704	4007.837	24951.12	PR	1.0	0.7		KN
4004.714	4005.846	24963.51	PR	1.3	1.4		MIT
4003.26	4004.39	24972.6	PR	0.5	0.0		MIT
4002.89	4004.02	24974.9	PR	0.5	0.0		MIT
4002.003	4003.134	24980.43	PR	0.5	0.0		KN
4001.47	4002.60	24983.7	PR	1.2	0.7		MIT
4000.91	4002.04	24987.2	PR	1.3	0.9		MIT
4000.478	4001.609	24989.95	PR	0.9	0.6		KN
4000.190	4001.321	24991.75	PR	1.7	1.4		MIT
3999.391	4000.522	24996.74	PR	0.3			MIT
3999.188	4000.319	24998.01	PR	1.7D	1.6D		MIT
3998.690	3999.821	25001.12	PR	0.6	0.3		MIT
3998.441	3999.572	25002.68	PR	1.0	0.5		MIT
3997.963	3999.093	25005.67	PR	1.3	1.0		MIT
3997.054	3998.184	25011.35	PR	2.0	1.6		MIT
3996.686	3997.816	25013.66	PR	1.3	0.6		MIT
3995.846	3996.976	25018.92	PR	0.6	0.3		MIT
3994.834	3995.964	25025.25	PR	2.5	1.4		MIT
3994.342	3995.471	25028.33	PR	0.6			MIT
3994.01	3995.14	25030.4	PR	1.3	0.5		MIT
3992.92	3994.05	25037.2	PR	1.2	0.5		MIT
3992.181	3993.310	25041.88	PR	1.4	1.0		MIT
3991.890	3993.019	25043.71	PR	1.4	1.0		MIT
3989.718	3990.846	25057.34	PR	2.3	2.2		MIT
3989.137	3990.265	25060.99	PR	1.0	0.0		MIT
3988.019	3989.147	25068.02	PR	1.4	0.8		MIT
3987.371	3988.499	25072.09	PR	1.4	1.2		MIT
3987.171	3988.299	25073.35	PR	1.2	0.5		MIT
3986.171	3987.298	25079.64	PR	1.6	1.2		MIT
3985.659	3986.786	25082.86	PR	1.4	1.0		MIT
3984.267	3985.394	25091.62	PR	1.3	0.9		MIT
3983.580	3984.707	25095.95	PR	1.4	0.9		MIT
3982.503	3983.629	25102.74	PR	1.2W	0.8W		MIT
3982.063	3983.189	25105.51	PR	2.1	2.0		MIT
3981.204	3982.330	25110.93	PR	0.3	0.6		MIT
3980.230	3981.356	25117.07	PR	0.7	1.0		MIT
3979.687	3980.813	25120.50	PR	0.7D	0.3D		MIT
3979.317	3980.443	25122.83	PR	1.0	0.5		MIT
3978.646	3979.771	25127.07	PR	1.3	0.6		MIT
3977.417	3978.542	25134.83	PR	1.1	0.8		MIT
3976.826	3977.951	25138.57	PR	1.1	0.7		MIT
3976.56	3977.68	25140.2	PR	1.2	0.9		MIT
3976.297	3977.422	25141.92	PR	1.5	1.2		MIT
3976.287	3977.412	25141.98	PR	1.7	1.0		MIT
3975.686	3976.811	25145.78	PR	1.3	0.6		MIT
3974.976	3976.100	25150.27	PR	0.7	0.3		MIT
3974.858	3975.982	25151.02	PR	1.2	1.0		MIT
3974.37	3975.49	25154.1	PR	1.5	0.9		MIT
3973.900	3975.024	25157.08	PR	1.0	0.5		MIT
3973.124	3974.248	25161.99	PR	0.6	0.5		MIT
3972.164	3973.288	25168.07	PR	2.1	1.9		MIT
3971.693	3972.817	25171.06	PR	1.8	1.6		MIT
3971.164	3972.287	25174.41	PR	2.0	1.8		MIT
3970.070	3971.193	25181.35	PR	1.2	0.6		MIT
3969.755	3970.878	25183.35	PR	1.1	0.5		MIT
3969.510	3970.633	25184.90	PR	0.9	0.5		MIT
3968.158	3969.281	25193.48	PR	1.4	1.0		MIT
3967.661	3968.784	25196.64	PR	1.0	0.5		MIT

AIR WAVELENGTH (ANGSTRMS)	VACUUM WAVELENGTH (ANGSTRMS)	WAVENUMBER (KAYSERS)	LMENT	LOG ARC	INTENSITY SPK	DIS	REF
3967.131	3968.253	25200.00	PR	1.6D	1.4D		MIT
3966.573	3967.695	25203.55	PR	2.0D	1.8D		MIT
3965.832	3966.954	25208.26	PR	1.0	0.6		MIT
3965.619	3966.741	25209.61	PR	1.0	0.7		MIT
3965.263	3966.385	25211.87	PR	2.0	1.7		MIT
3964.825	3965.947	25214.66	PR	2.1D	1.9D		MIT
3964.261	3965.383	25218.25	PR	1.8	1.7		MIT
3963.713	3964.834	25221.73	PR	1.5	1.2		MIT
3962.445	3963.566	25229.80	PR	1.8	1.7		MIT
3962.186	3963.307	25231.45	PR	1.0	0.5		MIT
3961.284	3962.405	25237.20	PR	1.0	0.7		MIT
3960.599	3961.720	25241.56	PR	1.7	1.4H		MIT
3959.780	3960.900	25246.78	PR	1.4	0.7		MIT
3959.517	3960.637	25248.46	PR	1.4	1.0		MIT
3959.412	3960.532	25249.13	PR	1.4D	1.0D		MIT
3958.495	3959.615	25254.98	PR	1.4	1.0		MIT
3958.185	3959.305	25256.96	PR	0.8	0.5		MIT
3957.682	3958.802	25260.17	PR	1.3	0.9		MIT
3956.762	3957.882	25266.04	PR	1.3	1.1		MIT
3955.421	3956.540	25274.61	PR	0.8	0.3		MIT
3954.995	3956.114	25277.33	PR	1.0	0.3		MIT
3953.516	3954.635	25286.78	PR	2.2	2.0		MIT
3953.201	3954.320	25288.80	PR	1.2D	0.7D		MIT
3952.357	3953.476	25294.20	PR	0.9	0.5		MIT
3952.020	3953.138	25296.36	PR	0.7	0.6		MIT
3951.843	3952.961	25297.49	PR	1.1	0.7		MIT
3950.659	3951.777	25305.07	PR	1.0	0.6		MIT
3949.438	3950.556	25312.89	PR	2.2	2.0		MIT
3948.621	3949.739	25318.13	PR	1.0	0.3		MIT
3947.633	3948.750	25324.47	PR	2.1D	1.8D		MIT
3946.953	3948.070	25328.83	PR	0.9	0.7		MIT
3946.224	3947.341	25333.51	PR	0.9	0.0		MIT
3945.662	3946.779	25337.12	PR	1.5	1.0		MIT
3945.424	3946.541	25338.65	PR	1.1	0.6		MIT
3944.899	3946.016	25342.02	PR	1.5	1.1		MIT
3944.617	3945.734	25343.83	PR	0.9	0.5		MIT
3944.137	3945.253	25346.91	PR	1.0	0.7		MIT
3943.753	3944.869	25349.38	PR	1.0	0.9		MIT
3943.371	3944.487	25351.84	PR	0.7	0.5		MIT
3942.918	3944.034	25354.75	PR	1.2	0.8		MIT
3942.339	3943.455	25358.47	PR	0.7	0.5		MIT
3942.260	3943.376	25358.98	PR	1.0	0.5		MIT
3941.507	3942.623	25363.83	PR	1.0	0.5		MIT
3940.377	3941.492	25371.10	PR	1.0D	0.5D		MIT
3940.152	3941.267	25372.55	PR	1.9	1.2		MIT
3938.915	3940.030	25380.52	PR	0.6	0.0		MIT
3938.312	3939.427	25384.40	PR	1.6	1.5		MIT
3937.694	3938.809	25388.39	PR	1.0	0.3		MIT
3937.030	3938.145	25392.67	PR	1.7	0.9		MIT
3936.689	3937.803	25394.87	PR	0.7	0.3		MIT
3935.823	3936.937	25400.46	PR	2.1	1.7		MIT
3935.134	3936.248	25404.90	PR	1.0D	0.5D		MIT
3934.883	3935.997	25406.52	PR	0.9	0.3		MIT
3933.300	3934.414	25416.75	PR	1.0	0.3		MIT
3932.978	3934.091	25418.83	PR	1.4	0.9		MIT
3932.130	3933.243	25424.31	PR	1.4	0.7		MIT
3930.965	3932.078	25431.85	PR	1.0	0.5		MIT
3930.619	3931.732	25434.08	PR	1.0	0.5		MIT
3929.877	3930.990	25438.89	PR	0.9	0.8		MIT
3929.256	3930.369	25442.91	PR	1.6	1.5		MIT
3928.911	3930.023	25445.14	PR	1.5	0.9		MIT
3928.615	3929.727	25447.06	PR	1.5	1.2		MIT
3928.295	3929.407	25449.13	PR	0.6	0.3		MIT
3927.712	3928.824	25452.91	PR	0.9	0.9		MIT
3927.454	3928.566	25454.58	PR	1.9	1.5		MIT
3926.591	3927.703	25460.18	PR	1.0	0.5		MIT
3926.397	3927.509	25461.43	PR	0.7D	0.7		MIT
3925.989	3927.101	25464.08	PR	0.5	0.5		MIT
3925.456	3926.568	25467.54	PR	2.1	2.0		MIT
3924.995	3926.106	25470.53	PR	0.3	0.3		MIT

AIR WAVELENGTH (ANGSTRMS)	VACUUM WAVELENGTH (ANGSTRMS)	WAVENUMBER (KAYSERS)	LMENT	LOG ARC	INTENSITY SPK	DIS	REF
3924.507	3925.618	25473.69	PR	0.7			MIT
3924.144	3925.255	25476.05	PR	2.0	1.2		MIT
3923.821	3924.932	25478.15	PR	1.0	0.5		MIT
3923.562	3924.673	25479.83	PR	0.8	0.9		MIT
3922.961	3924.072	25483.73	PR	0.6	0.5H		MIT
3922.245	3923.356	25488.39	PR	0.6	0.5		MIT
3921.406	3922.516	25493.84	PR	1.0	0.3		MIT
3920.524	3921.634	25499.57	PR	1.5	1.0		MIT
3919.620	3920.730	25505.45	PR	1.5	1.2		MIT
3918.856	3919.966	25510.43	PR	2.0	1.5		MIT
3917.915	3919.025	25516.55	PR	1.0H	0.7H		MIT
3917.229	3918.338	25521.02	PR	1.4	1.0		MIT
3916.802	3917.911	25523.80	PR	0.6	0.6		MIT
3916.458	3917.567	25526.05	PR	0.8	0.3		MIT
3915.467	3916.576	25532.51	PR	1.6	1.3		MIT
3914.872	3915.981	25536.39	PR	0.8	0.5		MIT
3914.764	3915.873	25537.09	PR	1.0	0.9		MIT
3913.561	3914.669	25544.94	PR	1.9	1.5		MIT
3912.898	3914.006	25549.27	PR	2.2	1.9		MIT
3912.614	3913.722	25551.12	PR	1.0	0.7		MIT
3912.268	3913.376	25553.38	PR	1.0	0.5		MIT
3911.989	3913.097	25555.21	PR	1.3D	0.8		MIT
3911.798	3912.906	25556.45	PR	0.9	0.3		MIT
3911.307	3912.415	25559.66	PR	0.7	0.5		MIT
3910.572	3911.680	25564.47	PR	1.0			MIT
3909.620	3910.727	25570.69	PR	1.3	1.0		MIT
3909.312	3910.419	25572.71	PR	0.6	0.3		MIT
3908.431	3909.538	25578.47	PR	2.0	1.8		MIT
3908.033	3909.140	25581.07	PR	2.0	1.7		MIT
3907.296	3908.403	25585.90	PR	0.9	0.5		MIT
3906.093	3907.199	25593.78	PR	0.8	0.9		MIT
3904.800	3905.906	25602.25	PR	0.8	0.3		MIT
3903.908	3905.014	25608.10	PR	1.0	0.9		MIT
3902.470	3903.576	25617.54	PR	1.8	1.6		MIT
3901.679	3902.784	25622.73	PR	1.0	0.5		MIT
3900.108	3901.213	25633.05	PR	0.5	0.0		MIT
3899.555	3900.660	25636.69	PR	1.0	0.9		MIT
3898.840	3899.945	25641.39	PR	1.2	1.0		MIT
3897.717	3898.821	25648.78	PR	0.9	0.5		MIT
3897.284	3898.388	25651.63	PR	1.3	0.9		MIT
3897.039	3898.143	25653.24	PR	1.0	0.5		MIT
3896.843	3897.947	25654.53	PR	1.1	0.6		MIT
3895.078	3896.182	25666.16	PR	0.7	0.5		MIT
3894.943	3896.047	25667.04	PR	1.2D	0.5D		MIT
3894.141	3895.244	25672.33	PR	0.9	0.3		MIT
3892.523	3893.626	25683.00	PR	1.2	0.6		MIT
3891.704	3892.807	25688.41	PR	1.0	0.6		MIT
3890.170	3891.272	25698.54	PR	0.9	0.3		MIT
3889.969	3891.071	25699.86	PR	0.7	0.5		MIT
3889.419	3890.521	25703.50	PR	1.0	0.8		MIT
3889.330	3890.432	25704.09	PR	2.2	1.8		MIT
3888.291	3889.393	25710.95	PR	0.7	0.6		MIT
3885.190	3886.291	25731.48	PR	2.0W	1.6W		MIT
3884.741	3885.842	25734.45	PR	0.6	0.0		MIT
3884.585	3885.686	25735.48	PR	0.3	0.0		MIT
3884.039	3885.140	25739.10	PR	0.7	0.6H		MIT
3882.315	3883.415	25750.53	PR	0.3			MIT
3881.876	3882.976	25753.44	PR	0.6			MIT
3880.466	3881.566	25762.80	PR	1.9	1.8		MIT
3879.214	3880.313	25771.11	PR	2.0	1.9		MIT
3878.307	3879.406	25777.14	PR		0.6		MIT
3877.929	3879.028	25779.65	PR		0.6		MIT
3877.225	3878.324	25784.33	PR	2.1W	1.9W		MIT
3876.183	3877.282	25791.26	PR	1.9	1.5		MIT
3875.314	3876.412	25797.05	PR		0.7		MIT
3874.76	3875.86	25800.7	PR	0.3	0.7		MIT
3874.45	3875.55	25802.8	PR	1.3	0.6		MIT
3870.725	3871.822	25827.63	PR	0.6	0.5		MIT
3868.784	3869.881	25840.59	PR	0.3	0.7		MIT
3867.55	3868.65	25848.8	PR	1.3	0.3		M

AIR WAVELENGTH (ANGSTRMS)	VACUUM WAVELENGTH (ANGSTRMS)	WAVENUMBER (KAYSERS)	LMENT	LOG ARC	INTENSITY SPK	DIS	REF
3866.984	3868.080	25852.62	PR	0.5	0.5		MIT
3866.818	3867.914	25853.73	PR	0.3	0.6		MIT
3865.987	3867.083	25859.28	PR		0.3		MIT
3865.458	3866.554	25862.82	PR	2.3R	2.1R		MIT
3865.147	3866.243	25864.90	PR		0.3		MIT
3864.534	3865.630	25869.01	PR	0.3	0.6		MIT
3864.059	3865.155	25872.19	PR	0.5	0.7		MIT
3861.313	3862.408	25890.59	PR	1.2	0.7		MIT
3859.416	3860.510	25903.31	PR	0.5	0.5		MIT
3859.142	3860.236	25905.15	PR	0.6	0.6		MIT
3858.256	3859.350	25911.10	PR	1.3	0.9		MIT
3855.884	3856.977	25927.04	PR	0.3	0.6		MIT
3854.963	3856.056	25933.23	PR II	1.9W	1.5		ZA
3854.574	3855.667	25935.85	PR	0.3	0.6		MIT
3853.492	3854.585	25943.13	PR	1.0	0.9		MIT
3852.805	3853.898	25947.76	PR	2.0	1.7		MIT
3851.617	3852.709	25955.76	PR	2.3W	2.2W		MIT
3850.825	3851.917	25961.10	PR	1.7	1.2		MIT
3848.332	3849.423	25977.92	PR	0.5	0.3		MIT
3846.605	3847.696	25989.58	PR	1.8D	1.5		MIT
3844.558	3845.648	26003.42	PR	1.7	1.0		MIT
3842.363	3843.453	26018.27	PR	1.9	1.6		MIT
3841.006	3842.096	26027.46	PR	1.5	1.0		MIT
3838.341	3839.430	26045.53	PR	0.7	0.5		MIT
3834.918	3836.006	26068.78	PR	1.5	1.2		MIT
3834.574	3835.662	26071.12	PR		0.3		MIT
3834.199	3835.287	26073.67	PR		0.6		MIT
3833.604	3834.692	26077.72	PR	1.0	0.9		MIT
3833.042	3834.129	26081.54	PR	1.3	1.3		MIT
3832.648	3833.735	26084.22	PR	1.0	0.9		MIT
3831.394	3832.481	26092.76	PR	0.8	0.3		MIT
3830.719	3831.806	26097.36	PR	2.0	1.8		MIT
3830.365	3831.452	26099.77	PR	1.0	0.8		MIT
3829.647	3830.734	26104.66	PR	1.0	1.0		MIT
3829.435	3830.522	26106.11	PR	1.0	1.0		MIT
3826.708	3827.794	26124.71	PR		0.6		MIT
3826.292	3827.378	26127.55	PR	1.9D	1.7D		MIT
3825.68	3826.77	26131.7	PR		0.7		MIT
3824.071	3825.156	26142.72	PR	0.3	0.3		MIT
3823.184	3824.269	26148.79	PR	2.1	1.4		MIT
3822.316	3823.401	26154.73	PR	0.3	0.3		MIT
3821.817	3822.902	26158.14	PR	1.7	1.7		MIT
3820.807	3821.891	26165.06	PR	0.6	0.8		MIT
3820.411	3821.495	26167.77	PR	0.3	0.3		MIT
3819.068	3820.152	26176.97	PR	0.5	0.5		MIT
3818.707	3819.791	26179.45	PR	0.3	0.3		MIT
3818.281	3819.365	26182.36	PR	2.1	2.0		MIT
3817.873	3818.957	26185.16	PR	1.3	1.0		MIT
3816.166	3817.249	26196.88	PR	1.6	1.6		MIT
3812.032	3813.114	26225.28	PR	1.2D	1.0		MIT
3811.845	3812.927	26226.57	PR	1.4	1.0		MIT
3811.348	3812.430	26229.99	PR	1.0D	1.3		MIT
3809.962	3811.043	26239.53	PR	1.3R	1.0R		MIT
3809.162	3810.243	26245.04	PR	1.9	1.6		MIT
3808.143	3809.224	26252.07	PR	0.7	0.5		MIT
3807.612	3808.693	26255.73	PR	0.9D	0.3		MIT
3806.857	3807.938	26260.93	PR	1.0	0.6		MIT
3806.415	3807.496	26263.98	PR	1.4D	1.2		MIT
3806.125	3807.205	26265.99	PR	0.9	0.6		MIT
3805.508	3806.588	26270.24	PR	1.0	0.9		MIT
3804.852	3805.932	26274.77	PR	1.4W	1.7W		MIT
3803.110	3804.190	26286.81	PR	1.7D	1.3		MIT
3802.763	3803.843	26289.21	PR	1.5D	1.2		MIT
3802.299	3803.378	26292.41	PR	1.0	0.3H		MIT
3801.348	3802.427	26298.99	PR	1.5D	1.0		MIT
3800.303	3801.382	26306.22	PR	2.0	1.7		MIT
3799.679	3800.758	26310.54	PR	1.7	1.4		MIT
3799.356	3800.435	26312.78	PR	0.8	0.5		MIT
3797.231	3798.309	26327.50	PR	1.2	0.9		MIT
3796.925	3798.003	26329.63	PR	1.6D	1.3D		MIT
3796.292	3797.370	26334.02	PR	0.6	0.5		MIT
3796.292	3797.370	26334.02	PR	0.6	0.5		MIT
3795.765	3796.843	26337.67	PR	1.5D	1.3D		MIT
3795.340	3796.418	26340.62	PR	1.4D	1.4D		MIT
3794.947	3796.025	26343.35	PR	0.3	0.5		MIT
3794.771	3795.849	26344.57	PR	0.6	0.3		MIT
3794.379	3795.456	26347.29	PR	1.7D	1.4D		MIT
3793.789	3794.866	26351.39	PR	1.6D	1.5D		MIT
3792.933	3794.010	26357.34	PR	1.1	0.6		MIT
3792.524	3793.601	26360.18	PR	2.0	0.9		MIT
3792.435	3793.512	26360.80	PR	0.6	0.5		MIT
3792.163	3793.240	26362.69	PR	1.0	0.7		MIT
3790.823	3791.899	26372.01	PR	1.2	0.7		MIT
3790.595	3791.671	26373.59	PR	1.1	0.8		MIT
3789.991	3791.067	26377.80	PR	1.4	1.3		MIT
3789.716	3790.792	26379.71	PR	1.0			KN
3789.039	3790.115	26384.42	PR	1.2	1.0		MIT
3788.930	3790.006	26385.18	PR	1.7D	1.4D		MIT
3788.493	3789.569	26388.23	PR	0.3	0.3		MIT
3788.058	3789.134	26391.26	PR	0.8	0.5		MIT
3786.882	3787.957	26399.45	PR	1.5D	1.2		MIT
3785.969	3787.044	26405.82	PR	0.6	0.3		MIT
3785.498	3786.573	26409.10	PR	1.7	1.3		MIT
3785.071	3786.146	26412.08	PR	0.5	0.3		MIT
3784.310	3785.385	26417.39	PR	1.0D	0.5D		MIT
3783.985	3785.060	26419.66	PR	0.5			MIT
3783.859	3784.934	26420.54	PR	2.0D	1.5D		MIT
3783.509	3784.584	26422.99	PR	1.5D	1.3D		MIT
3782.346	3783.420	26431.11	PR	1.4	1.0		MIT
3781.926	3783.000	26434.05	PR	1.3	0.9		MIT
3781.645	3782.719	26436.01	PR	2.0D	1.7D		MIT
3781.421	3782.495	26437.58	PR	1.0	0.6		MIT
3780.664	3781.738	26442.87	PR	1.3	0.7		MIT
3780.268	3781.342	26445.64	PR	1.1	0.7		MIT
3779.735	3780.809	26449.37	PR	1.2	0.9		MIT
3779.468	3780.542	26451.24	PR	1.0	0.3		MIT
3779.315	3780.388	26452.31	PR	0.5	0.3		MIT
3778.747	3779.820	26456.28	PR	1.6	1.3		MIT
3777.842	3778.915	26462.62	PR	1.0	0.9		MIT
3777.633	3778.706	26464.08	PR	1.2	0.8		MIT
3777.102	3778.175	26467.81	PR	1.3	1.0		MIT
3776.089	3777.162	26474.91	PR	0.9	0.5		MIT
3775.211	3776.283	26481.06	PR	1.0	0.6		MIT
3774.588	3775.660	26485.43	PR	1.0	0.3		MIT
3774.064	3775.136	26489.11	PR	2.0	1.7		MIT
3773.476	3774.548	26493.24	PR	0.9	0.5		MIT
3773.198	3774.270	26495.19	PR	1.0	0.7		MIT
3772.854	3773.926	26497.61	PR	1.9D	1.3		MIT
3772.552	3773.624	26499.73	PR	1.0	1.1		MIT
3772.077	3773.149	26503.06	PR	0.0	0.5		MIT
3771.766	3772.838	26505.25	PR	1.6D	1.3D		MIT
3771.373	3772.444	26508.01	PR	1.4D	1.0D		MIT
3770.484	3771.555	26514.26	PR	1.5	1.0		MIT
3770.211	3771.282	26516.18	PR	1.4D	1.2D		MIT
3769.695	3770.766	26519.81	PR	1.3	0.8		MIT
3768.934	3770.005	26525.17	PR	1.7	1.5		MIT
3768.388	3769.459	26529.01	PR	0.3	0.3		KN
3766.478	3767.548	26542.46	PR	0.9	0.6		MIT
3765.993	3767.063	26545.88	PR	0.9	0.5		MIT
3764.987	3766.057	26552.97	PR	1.0	0.6		MIT
3764.811	3765.881	26554.21	PR	2.0	1.7		MIT
3764.086	3765.156	26559.33	PR	1.1D	0.7D		MIT
3763.573	3764.642	26562.95	PR	0.6	0.3		MIT
3763.033	3764.102	26566.76	PR	1.0	0.7		MIT
3762.563	3763.632	26570.08	PR	0.9	0.3		MIT
3762.366	3763.435	26571.47	PR	0.7	0.5		MIT
3761.867	3762.936	26574.99	PR	2.2	2.0		MIT
3761.606	3762.675	26576.84	PR	1.0	0.8		MIT
3761.381	3762.450	26578.43	PR	0.9	0.3		MIT
3760.956	3762.025	26581.43	PR	0.7	0.5		MIT

AIR WAVELENGTH (ANGSTRMS)	VACUUM WAVELENGTH (ANGSTRMS)	WAVENUMBER (KAYSERS)	LMENT	LOG ARC	INTENSITY SPK	DIS	REF
3760.297	3761.366	26586.09	PR	0.3	0.0		MIT
3760.080	3761.148	26587.62	PR	1.3	0.9		MIT
3759.609	3760.677	26590.95	PR	1.5	1.0		MIT
3756.811	3757.879	26610.76	PR	0.7	0.5		MIT
3756.657	3757.725	26611.85	PR	1.1D	0.8D		MIT
3756.143	3757.210	26615.49	PR	0.3			MIT
3755.486	3756.553	26620.15	PR	0.5	0.3		MIT
3754.976	3756.043	26623.76	PR	1.3	0.8		MIT
3754.411	3755.478	26627.77	PR	0.7	0.3		MIT
3753.375	3754.442	26635.12	PR	1.0	0.3		MIT
3752.692	3753.759	26639.97	PR	1.0	0.3		MIT
3752.291	3753.357	26642.81	PR	1.6D	1.5D		MIT
3751.601	3752.667	26647.71	PR	0.3	0.0		MIT
3751.001	3752.067	26651.98	PR	1.6	1.5		MIT
3750.486	3751.552	26655.64	PR	1.3	1.0		MIT
3750.125	3751.191	26658.20	PR	1.5	1.2		MIT
3749.796	3750.862	26660.54	PR	0.7	0.3		MIT
3749.005	3750.071	26666.16	PR	0.3			MIT
3748.822	3749.888	26667.47	PR	1.0	0.5		MIT
3748.504	3749.569	26669.73	PR	0.7	0.5		MIT
3748.058	3749.123	26672.90	PR	1.0	0.6		MIT
3747.473	3748.538	26677.07	PR	1.0	0.5		MIT
3747.265	3748.330	26678.55	PR	0.7	0.3		MIT
3744.817	3745.882	26695.99	PR	0.5			MIT
3743.989	3745.053	26701.89	PR	0.8	0.6		MIT
3742.260	3743.324	26714.23	PR	0.6	0.3		MIT
3741.823	3742.887	26717.35	PR	0.7	0.3		MIT
3741.007	3742.071	26723.17	PR	1.6	1.2		MIT
3739.589	3740.652	26733.31	PR	1.0	0.5		MIT
3739.193	3740.256	26736.14	PR	1.9	1.5		MIT
3737.670	3738.733	26747.03	PR	1.0	0.8		MIT
3736.497	3737.559	26755.43	PR	1.3	1.3		MIT
3735.760	3736.822	26760.71	PR	0.3	0.3		MIT
3734.407	3735.469	26770.40	PR	1.6	1.5		MIT
3734.026	3735.088	26773.13	PR	0.6	0.3H		MIT
3733.027	3734.088	26780.30	PR	1.6D	1.3		MIT
3732.886	3733.947	26781.31	PR	0.7	0.3		MIT
3732.55	3733.61	26783.7	PR	0.5			MIT
3732.479	3733.540	26784.23	PR	0.7			KN
3731.496	3732.557	26791.28	PR	0.5	0.3		MIT
3730.936	3731.997	26795.31	PR	1.2	0.3		MIT
3730.578	3731.639	26797.88	PR	1.5O	1.0		MIT
3729.404	3730.465	26806.31	PR	1.5	0.6		MIT
3729.108	3730.168	26808.44	PR	1.6	0.8		MIT
3726.308	3727.368	26828.58	PR	1.3	0.6		MIT
3725.034	3726.093	26837.76	PR	1.2	0.9		MIT
3724.755	3725.814	26839.77	PR	0.3	0.3		MIT
3724.380	3725.439	26842.47	PR	0.3			MIT
3724.175	3725.234	26843.95	PR	0.5			KN
3723.583	3724.642	26848.22	PR	0.8	0.3		MIT
3722.066	3723.125	26859.16	PR	0.6	0.3		MIT
3721.671	3722.730	26862.01	PR	0.3			MIT
3721.515	3722.573	26863.14	PR	1.0	0.5		MIT
3721.27	3722.33	26864.9	PR	0.9	0.3		MIT
3720.83	3721.89	26868.1	PR	1.0	0.9		MIT
3720.222	3721.280	26872.47	PR	1.2	0.8		MIT
3719.435	3720.493	26878.16	PR	1.2	1.0		MIT
3718.877	3719.935	26882.19	PR	0.5	0.5		MIT
3718.016	3719.074	26888.42	PR	0.5H	0.3		MIT
3717.841	3718.899	26889.68	PR	0.9	0.5		MIT
3716.980	3718.037	26895.91	PR	1.0H	0.3H		MIT
3716.198	3717.255	26901.57	PR	0.6	0.5		MIT
3715.615	3716.672	26905.79	PR	1.2	0.3		MIT
3714.057	3715.114	26917.08	PR	1.7	1.3		MIT
3713.278	3714.334	26922.72	PR	1.1	0.5		MIT
3712.345	3713.401	26929.49	PR	1.1	0.5		MIT
3711.218	3712.274	26937.67	PR	0.5	0.3		MIT
3711.099	3712.155	26938.53	PR	0.9	0.5		MIT
3710.012	3711.068	26946.42	PR	0.8	0.3		MIT
3709.665	3710.720	26948.95	PR	0.5			MIT

AIR WAVELENGTH (ANGSTRMS)	VACUUM WAVELENGTH (ANGSTRMS)	WAVENUMBER (KAYSERS)	LMENT	LOG ARC	INTENSITY SPK	DIS	REF
3708.687	3709.742	26956.05	PR	0.3			MIT
3707.24	3708.29	26966.6	PR	0.9	0.8		MIT
3706.766	3707.821	26970.02	PR	1.4	1.1		MIT
3705.338	3706.392	26980.41	PR		0.5D		MIT
3704.972	3706.026	26983.08	PR	1.2	0.9		MIT
3704.319	3705.373	26987.84	PR		0.3		MIT
3703.342	3704.396	26994.96	PR	1.1	0.0		MIT
3701.812	3702.865	27006.11	PR	1.5	1.0		MIT
3701.476	3702.529	27008.56	PR	1.2	0.6		MIT
3700.988	3702.041	27012.12	PR	1.3	0.9		MIT
3700.661	3701.714	27014.51	PR	0.8	0.0		MIT
3700.248	3701.301	27017.53	PR	0.8			MIT
3699.952	3701.005	27019.69	PR	1.1	0.0		MIT
3699.506	3700.559	27022.95	PR	1.6	1.0		MIT
3698.727	3699.780	27028.64	PR	0.3			MIT
3698.069	3699.121	27033.45	PR	1.4	1.1		MIT
3696.658	3697.710	27043.76	PR	1.0	0.3		MIT
3695.462	3696.514	27052.52	PR	0.6			MIT
3694.695	3695.747	27058.13	PR	1.6	0.6H		MIT
3694.321	3695.372	27060.87	PR	0.5	0.0		MIT
3693.493	3694.544	27066.94	PR	0.3	0.0		MIT
3693.363	3694.414	27067.89	PR	1.5	0.9		MIT
3692.293	3693.344	27075.74	PR	0.6	0.0		MIT
3691.484	3692.535	27081.67	PR	1.0	0.5		MIT
3690.450	3691.500	27089.26	PR	1.2	0.3		MIT
3689.985	3691.035	27092.67	PR	1.0	0.5		MIT
3689.714	3690.764	27094.66	PR	1.3	1.0		MIT
3689.399	3690.449	27096.97	PR	1.6	1.0		MIT
3687.662	3688.712	27109.74	PR	0.5			MIT
3687.200	3688.250	27113.13	PR	1.7D	1.2D		MIT
3687.039	3688.089	27114.32	PR	1.8D	1.3D		MIT
3686.478	3687.527	27118.44	PR	1.4	0.8		MIT
3685.688	3686.737	27124.25	PR	0.8	0.3		MIT
3685.265	3686.314	27127.37	PR	1.8	0.9		MIT
3684.924	3685.973	27129.88	PR	0.9	0.0		MIT
3684.759	3685.808	27131.09	PR	0.6	0.0		MIT
3683.846	3684.895	27137.82	PR	0.6	0.3		MIT
3683.196	3684.245	27142.61	PR	0.5	0.0		MIT
3681.856	3682.904	27152.48	PR	1.0	0.7		MIT
3681.648	3682.696	27154.02	PR	0.3			MIT
3680.787	3681.835	27160.37	PR	0.5	0.0		MIT
3679.985	3681.033	27166.29	PR	0.7	0.3		MIT
3678.302	3679.349	27178.72	PR	0.8			MIT
3677.720	3678.767	27183.02	PR	1.0	0.3		MIT
3676.826	3677.873	27189.63	PR		0.3		MIT
3675.915	3676.962	27196.37	PR	0.8	0.3		MIT
3675.518	3676.565	27199.30	PR	0.6			MIT
3674.883	3675.929	27204.00	PR	0.7	0.3		MIT
3674.140	3675.186	27209.50	PR	1.0	0.3		MIT
3673.493	3674.539	27214.30	PR	0.8	0.3		MIT
3672.616	3673.662	27220.80	PR	0.8	0.0		MIT
3671.915	3672.961	27225.99	PR	1.3	0.7		MIT
3671.563	3672.609	27228.60	PR	0.9	0.3		MIT
3671.370	3672.416	27230.03	PR	0.9	0.3		MIT
3670.794	3671.839	27234.31	PR	0.9	0.3		MIT
3670.263	3671.308	27238.25	PR	0.9	0.5		MIT
3669.817	3670.862	27241.56	PR	1.0	0.3		MIT
3669.523	3670.568	27243.74	PR	1.0	0.5		MIT
3668.830	3669.875	27248.89	PR	2.0	1.6		MIT
3668.477	3669.522	27251.51	PR	0.5			MIT
3667.674	3668.719	27257.47	PR	1.0	0.6		MIT
3667.143	3668.187	27261.42	PR	1.0	0.3		MIT
3664.640	3665.684	27280.04	PR	1.0	0.5		MIT
3662.457	3663.500	27296.30	PR	0.5	0.0		MIT
3661.624	3662.667	27302.51	PR	1.4	0.7		MIT
3660.375	3661.418	27311.82	PR	1.6	1.3		MIT
3660.075	3661.118	27314.06	PR	1.0	0.5		MIT
3659.038	3660.080	27321.80	PR	1.0	0.9		MIT
3658.347	3659.389	27326.97	PR	0.8	0.0		MIT
3658.213	3659.255	27327.97	PR	1.2	0.3		MIT

AIR WAVELENGTH (ANGSTRMS)	VACUUM WAVELENGTH (ANGSTRMS)	WAVENUMBER (KAYSERS)	LMENT	LOG INTENSITY ARC	SPK	DIS	REF
3657.625	3658.667	27332.36	PR	1.0	0.0		MIT
3657.421	3658.463	27333.88	PR	1.4	0.6		MIT
3656.271	3657.313	27342.48	PR	1.0	0.0		MIT
3655.118	3656.159	27351.11	PR	1.0	0.3		MIT
3654.609	3655.650	27354.92	PR	0.8			MIT
3654.341	3655.382	27356.92	PR	0.9	0.3		MIT
3653.653	3654.694	27362.07	PR	0.6	0.0		MIT
3652.380	3653.421	27371.61	PR	1.3	0.5		MIT
3651.038	3652.078	27381.67	PR	1.0	0.3		MIT
3650.176	3651.216	27388.14	PR	1.5	0.8		MIT
3649.405	3650.445	27393.92	PR	1.0	0.5		MIT
3648.299	3649.339	27402.23	PR	1.5	1.0		MIT
3647.542	3648.581	27407.91	PR	0.6	0.3		MIT
3646.878	3647.917	27412.90	PR	0.6	0.3H		MIT
3646.299	3647.338	27417.26	PR	1.7	1.2		MIT
3645.660	3646.699	27422.06	PR	1.5	1.3		MIT
3645.539	3646.578	27422.97	PR	1.3	0.9		MIT
3645.114	3646.153	27426.17	PR	0.7	0.3		MIT
3644.897	3645.936	27427.80	PR	0.7	0.3		MIT
3644.543	3645.582	27430.47	PR	0.9	0.5		MIT
3643.317	3644.355	27439.70	PR	1.4	0.9		MIT
3642.923	3643.961	27442.66	PR	0.9	0.3		MIT
3641.873	3642.911	27450.57	PR	0.9	0.3		MIT
3641.616	3642.654	27452.51	PR	1.7	0.7		MIT
3641.114	3642.152	27456.30	PR	0.9	0.0		MIT
3640.889	3641.927	27457.99	PR	1.0	0.3		MIT
3639.770	3640.807	27466.44	PR	1.0	0.3		MIT
3638.865	3639.902	27473.27	PR	0.5			MIT
3638.576	3639.613	27475.45	PR	0.9	0.0		MIT
3638.300	3639.337	27477.53	PR	1.0	0.3		MIT
3637.647	3638.684	27482.47	PR	1.1	0.7		MIT
3637.163	3638.200	27486.12	PR	0.6	0.0		MIT
3636.499	3637.536	27491.14	PR	0.9	0.3		MIT
3635.284	3636.320	27500.33	PR	1.4	0.8		MIT
3634.467	3635.503	27506.51	PR	1.3	0.6		MIT
3633.348	3634.384	27514.98	PR	0.8	0.3		MIT
3630.967	3632.002	27533.02	PR	1.7	1.3		MIT
3629.715	3630.750	27542.52	PR	1.3	0.5		MIT
3628.818	3629.853	27549.33	PR	0.9	0.0		MIT
3628.177	3629.211	27554.20	PR	0.8	0.0		MIT
3627.799	3628.833	27557.07	PR	1.1	0.3		MIT
3626.485	3627.519	27567.05	PR	0.9	0.3		MIT
3625.405	3626.439	27575.26	PR	0.7	0.3		MIT
3624.306	3625.339	27583.62	PR	0.5			MIT
3622.737	3623.770	27595.57	PR	0.8	0.3		MIT
3622.384	3623.417	27598.26	PR	1.4	0.7		MIT
3621.181	3622.214	27607.43	PR	1.0	0.0		MIT
3621.088	3622.121	27608.14	PR	1.3H	0.3H		KN
3620.426	3621.458	27613.19	PR	0.7	0.3		MIT
3619.590	3620.622	27619.56	PR	0.9	0.3		MIT
3619.098	3620.130	27623.32	PR	1.0	0.5		MIT
3618.082	3619.114	27631.07	PR	1.2	0.5		MIT
3616.676	3617.707	27641.81	PR	1.3	0.9		MIT
3616.055	3617.086	27646.56	PR	0.7	0.3H		MIT
3615.159	3616.190	27653.41	PR	1.6	0.6		MIT
3613.70	3614.73	27664.6	PR	0.8A	0.3J		MIT
3612.715	3613.745	27672.12	PR	1.0	0.3		MIT
3611.942	3612.972	27678.04	PR	1.2	0.8		MIT
3611.159	3612.189	27684.04	PR	0.7	0.3		MIT
3610.684	3611.714	27687.69	PR	0.5	0.3		MIT
3608.468	3609.497	27704.69	PR	1.0	0.5		MIT
3607.820	3608.849	27709.67	PR	0.7	0.5		MIT
3607.225	3608.254	27714.24	PR	1.0	0.3		MIT
3605.962	3606.991	27723.94	PR	1.0	0.5		MIT
3605.054	3606.082	27730.93	PR	1.3	1.0		MIT
3603.732	3604.760	27741.10	PR	0.3	0.0		MIT
3601.916	3602.944	27755.08	PR	0.5			MIT
3601.023	3602.050	27761.97	PR	1.0	0.3		MIT
3600.750	3601.777	27764.07	PR	1.4	0.7		MIT
3599.51	3600.54	27773.6	PR	0.6D	0.3A		MIT

AIR WAVELENGTH (ANGSTRMS)	VACUUM WAVELENGTH (ANGSTRMS)	WAVENUMBER (KAYSERS)	LMENT	LOG INTENSITY ARC	SPK	DIS	REF
3598.914	3599.941	27778.24	PR	0.9	0.3		MIT
3596.193	3597.219	27799.25	PR	1.4	0.9		MIT
3593.685	3594.710	27818.65	PR	0.6	0.3		MIT
3593.036	3594.061	27823.68	PR	0.7	0.3H		MIT
3591.209	3592.234	27837.83	PR	0.6	0.0		KN
3589.47	3590.49	27851.3	PR	0.6	0.3		MIT
3588.305	3589.329	27860.36	PR	0.8	0.5		MIT
3587.865	3588.889	27863.78	PR	0.9	0.8		MIT
3587.342	3588.366	27867.84	PR	1.0	0.7		MIT
3587.14	3588.16	27869.4	PR	0.9	0.9		MIT
3586.531	3587.555	27874.14	PR	1.0	0.6		MIT
3586.25	3587.27	27876.3	PR	0.3	0.0		MIT
3584.257	3585.280	27891.82	PR	1.2	0.9		KN
3583.45	3584.47	27898.1	PR	0.6	0.5		MIT
3583.22	3584.24	27899.9	PR	0.7	0.5		MIT
3582.972	3583.995	27901.83	PR	0.6	0.5		KN
3582.704	3583.727	27903.91	PR	0.6	0.3		MIT
3582.25	3583.27	27907.4	PR		0.3H		MIT
3581.840	3582.862	27910.65	PR	1.0	0.8		MIT
3579.95	3580.97	27925.4	PR	0.8	0.5		MIT
3578.75	3579.77	27934.7	PR	0.9	0.7		MIT
3578.571	3579.593	27936.14	PR	0.3	0.0		MIT
3578.424	3579.446	27937.29	PR	1.0	0.5		MIT
3578.106	3579.127	27939.77	PR	1.0	0.7		MIT
3577.846	3578.867	27941.80	PR	0.8			MIT
3577.465	3578.486	27944.78	PR	1.5	0.9		MIT
3576.83	3577.85	27949.7	PR	0.3			MIT
3576.32	3577.34	27953.7	PR	1.0	0.5		MIT
3575.544	3576.565	27959.79	PR	1.0	0.5		MIT
3574.958	3575.979	27964.37	PR	1.0	0.6		MIT
3572.24	3573.26	27985.6	PR	0.7	0.3		MIT
3571.787	3572.807	27989.20	PR	1.0	0.5		MIT
3570.56	3571.58	27998.8	PR	0.9	0.6		MIT
3569.56	3570.58	28006.7	PR	1.0	0.7		MIT
3568.29	3569.31	28016.6	PR	0.7	0.6		MIT
3566.742	3567.761	28028.79	PR	0.8	0.3		MIT
3566.089	3567.107	28033.92	PR	1.0	0.8		MIT
3564.20	3565.22	28048.8	PR	0.5H			MIT
3563.771	3564.789	28052.15	PR	1.5	0.9		MIT
3563.490	3564.508	28054.36	PR	0.5			MIT
3562.980	3563.998	28058.38	PR	0.9	0.5		MIT
3562.548	3563.565	28061.78	PR	1.0	0.6		MIT
3562.22	3563.24	28064.4	PR	1.0	0.6		MIT
3561.68	3562.70	28068.6	PR	0.8	0.5		MIT
3561.27	3562.29	28071.9	PR	1.4	0.6		MIT
3559.081	3560.098	28089.12	PR	0.9	0.6		MIT
3558.813	3559.829	28091.23	PR	1.0	0.3		KN
3558.007	3559.023	28097.60	PR	1.0	0.5		MIT
3557.70	3558.72	28100.0	PR	0.3	0.3		MIT
3557.14	3558.16	28104.4	PR	1.2	0.7		MIT
3555.20	3556.22	28119.8	PR	1.0E	0.6E		MIT
3554.385	3555.400	28126.23	PR	0.7	0.5		MIT
3553.849	3554.864	28130.47	PR	0.6	0.3		MIT
3553.64	3554.66	28132.1	PR	0.5D	0.3D		MIT
3553.340	3554.355	28134.50	PR	0.6	0.3		MIT
3552.677	3553.692	28139.75	PR	1.2	0.8		MIT
3552.07	3553.08	28144.6	PR	1.0	0.3		MIT
3551.96	3552.97	28145.4	PR	0.9A			MIT
3551.405	3552.420	28149.83	PR	0.9	0.3		KN
3551.05	3552.06	28152.6	PR	0.7D	0.3D		MIT
3550.22	3551.23	28159.2	PR	1.5E	0.9E		MIT
3549.72	3550.73	28163.2	PR	0.7	0.0		MIT
3549.526	3550.540	28164.73	PR	1.0	0.5		MIT
3548.09	3549.10	28176.1	PR	0.6D	0.7D		MIT
3546.28	3547.29	28190.5	PR	0.8	0.3		MIT
3546.02	3547.03	28192.6	PR	0.9	0.0		MIT
3544.96	3545.97	28201.0	PR	1.5	1.0		MIT
3544.00	3545.01	28208.6	PR	1.6R	1.0R		MIT
3542.98	3543.99	28216.8	PR	0.5	0.0		M
3542.63	3543.64	28219.6	PR	0.9	0.3		MIT

AIR WAVELENGTH (ANGSTRMS)	VACUUM WAVELENGTH (ANGSTRMS)	WAVENUMBER (KAYSERS)	LMENT	LOG ARC	INTENSITY SPK	DIS	REF
3542.391	3543.403	28221.46	PR	0.9	0.5		MIT
3539.922	3540.934	28241.14	PR	1.4	0.7		MIT
3539.63	3540.64	28243.5	PR	0.7	0.3		MIT
3538.85	3539.86	28249.7	PR	1.0	0.3		MIT
3538.31	3539.32	28254.0	PR	0.5	0.3		MIT
3537.47	3538.48	28260.7	PR	0.9	0.5		MIT
3537.31	3538.32	28262.0	PR	0.8	0.3		MIT
3535.51	3536.52	28276.4	PR	0.5	0.5		MIT
3534.90	3535.91	28281.3	PR	0.7	0.3		MIT
3534.52	3535.53	28284.3	PR	1.4	0.6		MIT
3533.749	3534.759	28290.47	PR	1.3	0.7		MIT
3532.815	3533.825	28297.95	PR	0.6	0.3		MIT
3531.623	3532.632	28307.50	PR	0.6	0.3		MIT
3530.837	3531.846	28313.80	PR	1.0	0.3		MIT
3529.74	3530.75	28322.6	PR	0.7	0.3		MIT
3528.71	3529.72	28330.9	PR	0.5	0.3		M
3525.49	3526.50	28356.7	PR	0.6	0.3		MIT
3524.47	3525.48	28364.9	PR	0.5	0.3		MIT
3522.980	3523.987	28376.95	PR	0.3			MIT
3522.70	3523.71	28379.2	PR	0.5	0.3		M
3522.11	3523.12	28384.0	PR	0.6	0.3		EX
3519.128	3520.134	28408.01	PR	1.0	0.3		MIT
3518.508	3519.514	28413.01	PR	0.6			MIT
3513.27	3514.27	28455.4	PR	0.7			MIT
3512.22	3513.22	28463.9	PR	0.5D			MIT
3511.436	3512.440	28470.24	PR	1.2	0.5		MIT
3510.405	3511.409	28478.60	PR	0.7	0.3		MIT
3508.212	3509.215	28496.40	PR	1.1	0.5		MIT
3507.34	3508.34	28503.5	PR	0.8	0.3		MIT
3504.783	3505.786	28524.28	PR	0.7	0.3		MIT
3504.496	3505.498	28526.61	PR	0.6	0.0		MIT
3504.270	3505.272	28528.45	PR	1.1	0.5		MIT
3503.430	3504.432	28535.29	PR	1.0	0.3		MIT
3503.060	3504.062	28538.31	PR	1.2	0.6		MIT
3500.328	3501.329	28560.58	PR	0.8			MIT
3499.568	3500.569	28566.78	PR	1.0	0.6		MIT
3499.088	3500.089	28570.70	PR	1.6	0.5		MIT
3496.18	3497.18	28594.5	PR	0.9	0.3		MIT
3494.30	3495.30	28609.9	PR	0.8	0.3		MIT
3493.158	3494.158	28619.20	PR	1.0	0.3		MIT
3492.73	3493.73	28622.7	PR	0.6	0.0		MIT
3491.942	3492.941	28629.17	PR	1.0	0.3		MIT
3491.53	3492.53	28632.6	PR	1.0D	0.3		MIT
3489.013	3490.012	28653.20	PR	1.2	0.6		MIT
3488.235	3489.233	28659.59	PR	0.7			MIT
3487.574	3488.572	28665.02	PR	1.4	0.7		MIT
3487.39	3488.39	28666.5	PR	0.5	0.0		MIT
3487.021	3488.019	28669.57	PR	0.8	0.0		MIT
3486.483	3487.481	28673.99	PR	1.1	0.3		MIT
3484.80	3485.80	28687.8	PR	1.0	0.3		MIT
3484.32	3485.32	28691.8	PR	0.8	0.0		MIT
3483.790	3484.787	28696.16	PR	0.7			MIT
3483.518	3484.515	28698.40	PR	1.2	0.5		MIT
3482.248	3483.245	28708.86	PR	1.2			MIT
3481.028	3482.024	28718.93	PR	1.0	0.0		MIT
3479.46	3480.46	28731.9	PR	0.9D	0.0H		MIT
3479.283	3480.279	28733.33	PR	1.0			MIT
3474.450	3475.445	28773.30	PR	1.0	0.0		MIT
3473.854	3474.849	28778.23	PR	1.5	0.6		MIT
3472.357	3473.351	28790.64	PR	0.9	0.0		MIT
3469.318	3470.311	28815.86	PR	0.9	0.0		MIT
3469.028	3470.021	28818.27	PR	0.9	0.0		MIT
3468.659	3469.652	28821.33	PR	1.0	0.0		MIT
3468.065	3469.058	28826.27	PR	0.9	0.0		MIT
3467.872	3468.865	28827.87	PR	0.9	0.3		MIT
3467.040	3468.033	28834.79	PR	0.9	0.3		MIT
3466.749	3467.742	28837.21	PR	1.0	0.3		MIT
3466.237	3467.230	28841.47	PR	0.3	0.0H		MIT
3465.757	3466.750	28845.46	PR	1.0	0.6		MIT
3463.111	3464.103	28867.50	PR	0.9	0.0		MIT

AIR WAVELENGTH (ANGSTRMS)	VACUUM WAVELENGTH (ANGSTRMS)	WAVENUMBER (KAYSERS)	LMENT	LOG ARC	INTENSITY SPK	DIS	REF
3462.36	3463.35	28873.8	PR	0.5			MIT
3462.10	3463.09	28875.9	PR	0.8			MIT
3461.873	3462.865	28877.83	PR	0.8	0.3		MIT
3461.244	3462.235	28883.07	PR	0.6			MIT
3461.056	3462.047	28884.64	PR	1.0	0.3		MIT
3460.663	3461.654	28887.92	PR	0.5			MIT
3458.796	3459.787	28903.51	PR	0.5			MIT
3458.24	3459.23	28908.2	PR	0.8	0.3		MIT
3455.972	3456.962	28927.13	PR	1.5	0.5		MIT
3454.47	3455.46	28939.7	PR	0.8	0.0H		MIT
3453.784	3454.773	28945.46	PR	0.8			MIT
3452.93	3453.92	28952.6	PR	0.8	0.0		MIT
3451.48	3452.47	28964.8	PR	1.4	0.5		MIT
3449.826	3450.814	28978.67	PR	1.5	0.6		MIT
3448.205	3449.193	28992.29	PR	1.2	0.6		MIT
3447.844	3448.832	28995.32	PR	0.9			MIT
3444.307	3445.294	29025.10	PR	0.7			MIT
3442.759	3443.746	29038.15	PR	0.9	0.3		MIT
3442.41	3443.40	29041.1	PR	0.5			MIT
3440.131	3441.117	29060.33	PR	1.0	0.3		MIT
3438.31	3439.30	29075.7	PR	1.4W	0.0H		MIT
3437.603	3438.588	29081.70	PR	0.8			MIT
3436.629	3437.614	29089.94	PR	1.2			MIT
3436.46	3437.45	29091.4	PR		0.5H		MIT
3434.757	3435.742	29105.80	PR	1.3	0.3		MIT
3434.486	3435.471	29108.09	PR	0.5			MIT
3433.552	3434.536	29116.01	PR	1.5	0.5		MIT
3432.051	3433.035	29128.75	PR	0.7			MIT
3431.62	3432.60	29132.4	PR	1.2	0.3		MIT
3430.510	3431.494	29141.83	PR	1.4	0.5		MIT
3430.278	3431.261	29143.80	PR	1.4	0.6		MIT
3427.02	3428.00	29171.5	PR		0.3		MIT
3422.26	3423.24	29212.1	PR	0.7	0.5		MIT
3421.112	3422.093	29221.88	PR	1.0	0.0		MIT
3419.241	3420.222	29237.87	PR	1.0	0.3		MIT
3418.467	3419.447	29244.49	PR	1.5	0.5		KN
3417.98	3418.96	29248.7	PR	0.6			MIT
3415.708	3416.688	29268.11	PR	1.4	0.6		MIT
3415.079	3416.059	29273.50	PR	0.6	0.5		MIT
3414.767	3415.746	29276.18	PR	0.8			MIT
3413.14	3414.12	29290.1	PR		0.5		MIT
3409.984	3410.962	29317.24	PR	0.8	0.0		MIT
3403.568	3404.545	29372.50	PR	0.9	0.3		MIT
3400.877	3401.853	29395.75	PR	1.0	0.0		MIT
3397.56	3398.54	29424.4	PR	0.5	0.5		MIT
3396.58	3397.55	29432.9	PR	0.6	0.6		MIT
3395.43	3396.40	29442.9	PR	0.8			MIT
3394.613	3395.587	29449.99	PR	1.4	0.7		MIT
3394.144	3395.118	29454.05	PR	0.7	0.0		MIT
3390.911	3391.884	29482.14	PR	1.0	0.3		MIT
3388.036	3389.009	29507.15	PR	1.3	0.5		MIT
3386.578	3387.550	29519.86	PR	0.3			MIT
3386.025	3386.997	29524.68	PR	1.0			MIT
3383.73	3384.70	29544.7	PR	1.1	0.5		MIT
3383.376	3384.347	29547.79	PR	1.2	0.5		MIT
3382.660	3383.631	29554.05	PR	1.0	0.3		MIT
3381.77	3382.74	29561.8	PR	0.5	0.0		MIT
3381.376	3382.347	29565.27	PR	0.5			MIT
3380.11	3381.08	29576.3	PR	0.5	0.0		MIT
3379.759	3380.730	29579.41	PR	1.0	0.3		MIT
3378.93	3379.90	29586.7	PR	0.6	0.0H		MIT
3377.449	3378.419	29599.65	PR	1.0	0.0		MIT
3377.04	3378.01	29603.2	PR	0.9	0.0		MIT
3376.660	3377.630	29606.56	PR	1.0	0.3		MIT
3375.630	3376.599	29615.59	PR	0.9	0.3		MIT
3372.797	3373.766	29640.47	PR	1.0	0.6		MIT
3372.510	3373.479	29642.99	PR	1.3	0.5		MIT
3370.30	3371.27	29662.4	PR	1.0D	0.7D		MIT
3369.125	3370.093	29672.77	PR	0.5			MIT
3368.957	3369.925	29674.25	PR	0.7	0.0		MIT

AIR WAVELENGTH (ANGSTRMS)	VACUUM WAVELENGTH (ANGSTRMS)	WAVENUMBER (KAYSERS)	LMENT	LOG ARC	INTENSITY SPK	DIS	REF
3367.53	3368.50	29686.8	PR		1.0D		MIT
3367.16	3368.13	29690.1	PR	0.3	0.0		MIT
3365.931	3366.898	29700.93	PR	1.0	0.3		MIT
3364.92	3365.89	29709.9	PR	1.2D	0.6D		MIT
3363.945	3364.911	29718.46	PR	0.8			MIT
3363.249	3364.215	29724.61	PR	0.7	0.3		MIT
3360.900	3361.866	29745.39	PR	0.9	0.3		MIT
3360.162	3361.128	29751.92	PR	0.6	0.0		MIT
3359.49	3360.46	29757.9	PR		0.6H		MIT
3359.269	3360.234	29759.83	PR	0.9	0.3		MIT
3358.501	3359.466	29766.64	PR	1.0	0.3		MIT
3357.692	3358.657	29773.81	PR	0.9			MIT
3355.845	3356.809	29790.19	PR	1.0D	0.3		MIT
3355.668	3356.632	29791.76	PR	1.4	0.6		MIT
3354.90	3355.86	29798.6	PR	0.3	0.3		MIT
3353.49	3354.45	29811.1	PR	0.8	0.0		MIT
3351.252	3352.215	29831.02	PR	0.7	0.0		MIT
3350.275	3351.238	29839.72	PR	1.0	0.5		MIT
3344.758	3345.720	29888.94	PR	0.3			MIT
3344.564	3345.526	29890.67	PR	0.9	0.0		MIT
3343.934	3344.895	29896.30	PR	1.0	0.0		MIT
3341.473	3342.434	29918.32	PR	1.0	0.5		MIT
3340.955	3341.916	29922.96	PR	1.0	0.3		MIT
3333.945	3334.904	29985.87	PR	0.8	0.0		MIT
3333.450	3334.409	29990.32	PR	0.9	0.0		MIT
3331.643	3332.601	30006.59	PR	1.0	0.0		MIT
3330.677	3331.635	30015.29	PR	0.9	0.0		MIT
3327.50	3328.46	30043.9	PR	0.6D			MIT
3325.686	3326.643	30060.34	PR	1.0	0.0		MIT
3324.551	3325.507	30070.60	PR	1.0	0.0		MIT
3324.136	3325.092	30074.35	PR	1.0	0.0		MIT
3322.266	3323.222	30091.28	PR	0.8	0.0		MIT
3319.975	3320.930	30112.04	PR	0.9	0.3		MIT
3319.643	3320.598	30115.06	PR	0.9			MIT
3317.59	3318.54	30133.7	PR	0.6			MIT
3314.978	3315.932	30157.43	PR	1.0	0.0		MIT
3314.376	3315.330	30162.91	PR	1.1	0.3		MIT
3312.918	3313.871	30176.19	PR	1.0	0.3		MIT
3312.321	3313.274	30181.62	PR	1.0	0.3		MIT
3309.128	3310.081	30210.75	PR	1.0	0.0		MIT
3308.342	3309.294	30217.92	PR	1.0	0.0		MIT
3306.785	3307.737	30232.15	PR	0.7			MIT
3306.275	3307.227	30236.81	PR	0.8	0.3		MIT
3302.66	3303.61	30269.9	PR	1.2	0.3		MIT
3301.898	3302.849	30276.89	PR	1.0			MIT
3300.355	3301.305	30291.05	PR	0.7	0.0		MIT
3296.390	3297.339	30327.48	PR	1.0	0.3		MIT
3295.528	3296.477	30335.42	PR	1.0	0.3		MIT
3294.942	3295.891	30340.81	PR	0.5			MIT
3288.650	3289.597	30398.86	PR	1.1	0.0		MIT
3285.849	3286.796	30424.77	PR	1.1	0.3		MIT
3283.121	3284.067	30450.05	PR	1.2	0.3		MIT
3282.297	3283.243	30457.69	PR	0.8	0.0		MIT
3279.028	3279.973	30488.06	PR	0.8			MIT
3278.839	3279.784	30489.81	PR	1.0	0.3		MIT
3276.667	3277.611	30510.02	PR	1.0	0.3		MIT
3274.576	3275.520	30529.51	PR	0.9			MIT
3272.897	3273.840	30545.17	PR	1.0	0.0		MIT
3268.233	3269.175	30588.76	PR	1.3	0.0		MIT
3264.620	3265.561	30622.61	PR	0.8	0.0H		MIT
3264.31	3265.25	30625.5	PR	1.0			MIT
3262.39	3263.33	30643.5	PR	0.3			MIT
3258.94	3259.88	30676.0	PR	1.0	0.0		MIT
3258.306	3259.246	30681.95	PR	0.6			MIT
3257.589	3258.528	30688.70	PR	1.0			MIT
3253.195	3254.133	30730.15	PR	0.7	0.0		MIT
3249.747	3250.684	30762.75	PR	1.0	0.3		MIT
3248.75	3249.69	30772.2	PR	0.6			MIT
3245.462	3246.398	30803.37	PR	1.6	0.6		MIT
3244.31	3245.25	30814.3	PR	0.5			MIT

AIR WAVELENGTH (ANGSTRMS)	VACUUM WAVELENGTH (ANGSTRMS)	WAVENUMBER (KAYSERS)	LMENT	LOG ARC	INTENSITY SPK	DIS	REF
3240.027	3240.962	30855.04	PR	1.0	0.0		MIT
3238.867	3239.802	30866.09	PR	1.1	0.3		MIT
3235.432	3236.366	30898.86	PR	1.0	0.0		MIT
3234.223	3235.157	30910.41	PR	1.1	0.3		MIT
3225.20	3226.13	30996.9	PR	0.5			MIT
3224.999	3225.930	30998.81	PR	1.0	0.0		MIT
3224.419	3225.350	31004.39	PR	1.0	0.0		MIT
3219.551	3220.481	31051.26	PR	1.3	0.7		KN
3219.177	3220.107	31054.87	PR	0.8	0.0		MIT
3218.245	3219.174	31063.86	PR	0.9			MIT
3214.400	3215.328	31101.02	PR	0.9	0.0		MIT
3213.578	3214.506	31108.98	PR	1.0	0.0		MIT
3207.894	3208.821	31164.10	PR	1.8	0.3		MIT
3205.018	3205.944	31192.06	PR	0.9	0.3		MIT
3204.808	3205.734	31194.10	PR	1.0	0.3		MIT
3199.049	3199.974	31250.26	PR	1.3	0.5		MIT
3196.036	3196.960	31279.72	PR	1.2	0.3		MIT
3191.414	3192.337	31325.02	PR	1.4	0.5		MIT
3182.447	3183.367	31413.28	PR	1.2	0.5		MIT
3179.148	3180.068	31445.87	PR	1.0	0.0		MIT
3172.314	3173.232	31513.61	PR	1.4	0.7		MIT
3169.367	3170.284	31542.91	PR	1.1	0.0		MIT
3168.242	3169.159	31554.11	PR	1.4	0.7		MIT
3164.985	3165.901	31586.58	PR	0.9	0.0		MIT
3164.823	3165.739	31588.20	PR	1.0	0.5		MIT
3163.735	3164.651	31599.06	PR	0.6			MIT
3158.646	3159.560	31649.97	PR	1.1	0.3		MIT
3153.833	3154.746	31698.27	PR	1.1	0.3		MIT
3151.539	3152.452	31721.34	PR	1.0	0.5		MIT
3149.562	3150.474	31741.25	PR	1.2	0.0		MIT
3146.466	3147.377	31772.49	PR	1.2	0.5		MIT
3141.212	3142.122	31825.63	PR	1.0	0.3		MIT
3140.692	3141.602	31830.90	PR	1.0	0.3		MIT
3137.053	3137.962	31867.82	PR	1.0	0.3		MIT
3136.789	3137.698	31870.50	PR	1.0	0.3		MIT
3135.351	3136.260	31885.12	PR	1.0	0.3		MIT
3129.204	3130.111	31947.75	PR	1.1	0.3		MIT
3123.004	3123.909	32011.17	PR	1.0	0.3		MIT
3121.571	3122.476	32025.87	PR	1.7	0.6		MIT
3119.028	3119.932	32051.98	PR	1.2	0.3		MIT
3114.137	3115.040	32102.31	PR	0.5R			MIT
3111.337	3112.239	32131.20	PR	1.2	0.3		MIT
3110.580	3111.482	32139.02	PR	1.0	0.0		MIT
3109.757	3110.659	32147.53	PR	1.0	0.3		MIT
3105.400	3106.301	32192.63	PR	1.3	0.0		MIT
3101.27	3102.17	32235.5	PR	1.6	0.3		MIT
3098.503	3099.402	32264.29	PR	1.3	0.5		MIT
3097.78	3098.68	32271.8	PR	1.2	0.3		MIT
3096.94	3097.84	32280.6	PR	0.3			MIT
3091.34	3092.24	32339.0	PR	1.0	0.3		MIT
3085.029	3085.925	32405.20	PR	1.0	0.0		MIT
3080.27	3081.16	32455.3	PR		0.7D		MIT
3066.74	3067.63	32598.4	PR	0.0	0.5		MIT
3058.97	3059.86	32681.2	PR		0.7		MIT
3050.35	3051.24	32773.6	PR	0.3	0.6		MIT
3045.82	3046.71	32822.3	PR		0.5		MIT
3034.20	3035.08	32948.0	PR		0.5		MIT
3033.31	3034.19	32957.7	PR		0.6H		MIT
3029.393	3030.275	33000.31	PR		1.0		MIT
3025.23	3026.11	33045.7	PR		0.9		MIT
3021.68	3022.56	33084.5	PR		1.0		MIT
3015.17	3016.05	33156.0	PR		1.0		MIT
3014.503	3015.381	33163.30	PR	1.0	0.9		MIT
3010.68	3011.56	33205.4	PR	0.5	1.3		MIT
3007.98	3008.86	33235.2	PR	0.7	1.2		MIT
3003.180	3004.055	33288.33	PR	0.7	1.3		MIT
3000.454	3001.329	33318.58	PR	0.7	1.2		MIT
2997.07	2997.94	33356.2	PR		1.0		MIT
2985.77	2986.64	33482.4	PR		1.4		MIT
2980.52	2981.39	33541.4	PR		1.4		MIT

AIR WAVELENGTH (ANGSTRMS)	VACUUM WAVELENGTH (ANGSTRMS)	WAVENUMBER (KAYSERS)	LMENT	LOG ARC	INTENSITY SPK	DIS	REF
2976.975	2977.844	33581.34	PR		1.3		MIT
2968.80	2969.67	33673.8	PR	0.5	0.9		MIT
2964.76	2965.63	33719.7	PR	1.0	1.0		MIT
2953.537	2954.400	33847.82	PR	1.0	1.0		MIT
2942.302	2943.162	33977.06	PR		1.0		MIT
2930.08	2930.94	34118.8	PR	0.5	0.7		MIT
2914.513	2915.366	34301.01	PR		1.0		MIT
2911.70	2912.55	34334.1	PR		1.2		MIT
2841.97	2842.81	35176.5	PR		0.9		MIT
2679.50	2680.30	37309.3	PR		1.0		MIT
2667.55	2668.34	37476.4	PR		1.0		MIT
2654.75	2655.54	37657.1	PR		0.8		MIT
2644.65	2645.44	37800.9	PR		1.3		MIT
2639.92	2640.71	37868.6	PR		0.3J		MIT
2615.75	2616.53	38218.5	PR		0.5		MIT
2596.85	2597.63	38496.7	PR		0.8		MIT
2596.10	2596.88	38507.8	PR		0.3		MIT
2595.30	2596.08	38519.7	PR		0.7		MIT
2590.39	2591.16	38592.7	PR		0.7		MIT
2587.76	2588.53	38631.9	PR		1.2L		MIT
2580.13	2580.90	38746.1	PR		0.3J		MIT
2575.02	2575.79	38823.0	PR		0.5		MIT
2571.83	2572.60	38871.2	PR		0.6		MIT
2569.23	2570.00	38910.5	PR		1.1H		MIT
2558.55	2559.32	39072.9	PR		0.8		MIT
2555.73	2556.50	39116.0	PR		0.6		MIT
2551.52	2552.29	39180.6	PR		1.2S		MIT
2545.96	2546.72	39266.1	PR		0.5		MIT
2538.38	2539.14	39383.4	PR		1.0		MIT
2536.71	2537.47	39409.3	PR		0.9		MIT
2534.92	2535.68	39437.1	PR		0.3J		MIT
2531.35	2532.11	39492.7	PR		0.7		MIT
2530.46	2531.22	39506.6	PR		0.3		MIT
2511.13	2511.89	39810.7	PR		0.8		MIT
2488.75	2489.50	40168.7	PR		1.5		MIT
2485.17	2485.92	40226.5	PR		0.6		MIT
2468.98	2469.73	40490.3	PR		0.7		MIT
2468.22	2468.97	40502.8	PR		1.2		MIT
2462.90	2463.65	40590.3	PR		0.8		MIT
2460.73	2461.47	40626.1	PR		1.1		MIT
2454.85	2455.59	40723.4	PR		1.1		MIT
2448.19	2448.93	40834.1	PR		0.8		MIT
2446.78	2447.52	40857.7	PR		0.9		MIT
2445.52	2446.26	40878.7	PR		1.6		MIT
2438.64	2439.38	40994.0	PR		1.4		MIT
2437.08	2437.82	41020.3	PR		0.6		MIT
2434.41	2435.15	41065.3	PR		0.8		MIT
2431.74	2432.48	41110.3	PR		0.7H		MIT
2431.26	2432.00	41118.5	PR		0.7		MIT
2418.97	2419.70	41327.4	PR		1.3		MIT
2409.84	2410.57	41483.9	PR		0.9		MIT
2408.20	2408.93	41512.2	PR		0.9		MIT
2405.60	2406.33	41557.0	PR		1.5		MIT
2400.87	2401.60	41638.9	PR		0.9H		MIT
2400.78	2401.51	41640.5	PR		0.8		MIT
2399.71	2400.44	41659.0	PR		1.1		MIT
2394.06	2394.79	41757.3	PR		0.8		MIT
2378.98	2379.71	42022.0	PR		1.0S		MIT
2378.09	2378.82	42037.7	PR		1.0		MIT
2377.67	2378.40	42045.1	PR		0.8		MIT
2376.12	2376.85	42072.6	PR		0.9		MIT
2373.25	2373.97	42123.5	PR		0.9		MIT
2372.13	2372.85	42143.3	PR		0.8		MIT
2371.42	2372.14	42156.0	PR		1.2		MIT
2368.80	2369.52	42202.6	PR		1.2		MIT
2365.55	2366.27	42260.5	PR		1.0		MIT
2353.77	2354.49	42472.0	PR		1.2		MIT
2352.54	2353.26	42494.2	PR		0.8		MIT
2350.07	2350.79	42538.9	PR		1.3		MIT
2346.72	2347.44	42599.6	PR		1.0		MIT

AIR WAVELENGTH (ANGSTRMS)	VACUUM WAVELENGTH (ANGSTRMS)	WAVENUMBER (KAYSERS)	LMENT	LOG ARC	INTENSITY SPK	DIS	REF
2339.76	2340.48	42726.3	PR		1.2		MIT
2337.52	2338.24	42767.3	PR		1.2		MIT
2336.23	2336.95	42790.9	PR		1.0		MIT
2328.58	2329.29	42931.5	PR		1.1		MIT
2325.84	2326.55	42982.0	PR		0.8		MIT
2320.43	2321.14	43082.2	PR		0.8		MIT
2319.42	2320.13	43101.0	PR		1.1		MIT
2318.83	2319.54	43112.0	PR		1.0		MIT
2318.17	2318.88	43124.2	PR		0.7		MIT
2314.19	2314.90	43198.4	PR		1.0		MIT
2311.47	2312.18	43249.2	PR		0.7		MIT
2307.79	2308.50	43318.2	PR		1.0		MIT
2305.10	2305.81	43368.7	PR		1.0		MIT
2303.82	2304.53	43392.8	PR		1.0		MIT
2301.62	2302.33	43434.3	PR		0.3H		MIT
2297.77	2298.48	43507.0	PR		2.0		MIT
2294.686	2295.393	43565.52	PR		1.0		MIT
2290.26	2290.97	43649.7	PR		0.9		MIT
2288.15	2288.86	43690.0	PR		0.8		MIT
2286.61	2287.32	43719.4	PR		0.8		MIT
2284.64	2285.34	43757.1	PR		1.0		MIT
2273.86	2274.56	43964.5	PR		1.0		MIT
2273.21	2273.91	43977.1	PR		1.0		MIT
2242.17	2242.87	44585.8	PR		1.0		A
2230.37	2231.06	44821.7	PR		1.6		MIT
2223.25	2223.94	44965.2	PR		1.4		MIT
2218.11	2218.80	45069.4	PR		1.3		A
2215.27	2215.96	45127.2	PR		1.3		MIT
2205.50	2206.19	45327.0	PR		1.2		MIT
2194.24	2194.93	45559.6	PR		1.0		MIT
2189.88	2190.56	45650.3	PR		0.7		A
2187.26	2187.94	45705.0	PR		0.8		A
2169.48	2170.16	46079.5	PR		0.3H		MIT
2142.89	2143.57	46651.2	PR		1.5		MIT
2136.66	2137.33	46787.3	PR		1.5		A
2130.65	2131.32	46919.2	PR		0.9		MIT
2122.35	2123.02	47102.7	PR		1.5		A
2111.35	2112.02	47348.1	PR		1.0		A
2090.75	2091.41	47814.5	PR		1.3		A
2071.88	2072.54	48250.0	PR		0.3		MIT
2064.07	2064.73	48432.5	PR		1.1		A
2052.87	2053.53	48696.7	PR		1.1		A
8301.87	8304.15	12042.2	PT I	0.3			ME
8259.00	8261.27	12104.7	PT I	0.3			MIT
8227.55	8229.81	12150.9	PT I	0.5			MIT
8224.74	8227.00	12155.1	PT I	1.0			MIT
8204.45	8206.71	12185.2	PT I	0.6			MIT
8012.8	8015.0	12477.	PT		0.7		IT
7977.32	7979.51	12532.1	PT	0.5			ME
7911.26	7913.44	12636.7	PT	0.3			ME
7790.22	7792.36	12833.1	PT	0.3			ME
7786.78	7788.92	12838.7	PT I	0.6			ME
7780.51	7782.65	12849.1	PT	0.3			MIT
7749.74	7751.87	12900.1	PT I	0.3			ME
7723.4	7725.5	12944.	PT		0.6		IT
7618.20	7620.30	13122.8	PT	0.3			ME
7515.5	7517.6	13302.	PT		0.3		IT
7486.05	7488.11	13354.5	PT	0.7			MIT
7384.6	7386.6	13538.	PT		0.3		IT
7217.58	7219.57	13851.2	PT	0.8			ME
7131.68	7133.65	14018.1	PT I	0.7			MIT
7113.73	7115.69	14053.4	PT I	1.9			MIT
7094.78	7096.74	14091.0	PT I	1.0			MIT
7078.08	7080.03	14124.2	PT	0.5			MIT
7065.60	7067.55	14149.2	PT	1.0			MIT
7056.30	7058.25	14167.8	PT	0.3			MIT
7030.09	7032.03	14220.7	PT	0.5			MIT
6989.83	6991.76	14302.6	PT	0.5			MIT
6975.70	6977.62	14331.5	PT	0.7			MIT
6957.51	6959.43	14369.0	PT	0.5			MIT

AIR WAVELENGTH (ANGSTRMS)	VACUUM WAVELENGTH (ANGSTRMS)	WAVENUMBER (KAYSERS)	LMENT	LOG ARC	INTENSITY SPK	INTENSITY DIS	REF
6908.80	6910.71	14470.3	PT	0.5			ME
6896.74	6898.64	14495.6	PT	0.5			ME
6842.60	6844.49	14610.3	PT	1.0			ME
6838.08	6839.97	14619.9	PT	0.6			ME
6820.23	6822.11	14658.2	PT	0.5			ME
6760.020	6761.886	14788.77	PT	2.0			MIT
6710.416	6712.269	14898.09	PT	1.7			MIT
6648.312	6650.148	15037.26	PT	1.0			MIT
6597.93	6599.75	15152.1	PT	0.6			ME
6592.65	6594.47	15164.2	PT	0.6			ME
6523.453	6525.255	15325.07	PT	1.9			MIT
6490.45	6492.24	15403.0	PT	0.8			MIT
6435.94	6437.72	15533.4	PT	0.5			MIT
6398.858	6400.627	15623.47	PT	0.8			MIT
6326.577	6328.326	15801.97	PT	1.7			MIT
6318.369	6320.116	15822.49	PT	1.5			MIT
6291.85	6293.59	15889.2	PT	0.7			ME
6283.478	6285.216	15910.35	PT	1.0			MIT
6282.278	6284.016	15913.39	PT	0.7			MIT
6263.670	6265.403	15960.67	PT	0.6			MIT
6237.655	6239.381	16027.23	PT	0.7			MIT
6215.999	6217.719	16083.07	PT	1.3			MIT
6172.531	6174.239	16196.33	PT	1.2			MIT
6141.66	6143.36	16277.7	PT	0.5			ME
6127.08	6128.78	16316.5	PT	0.3			ME
6113.345	6115.037	16353.13	PT	0.7			MIT
6111.661	6113.353	16357.64	PT	0.9			MIT
6076.866	6078.548	16451.30	PT I	1.0			MIT
6033.674	6035.345	16569.06	PT	0.3			MIT
6026.035	6027.704	16590.06	PT	1.3			MIT
6024.248	6025.916	16594.99	PT	1.0			MIT
6005.45	6007.11	16646.9	PT	0.3			ME
5988.094	5989.753	16695.18	PT	0.3			MIT
5979.106	5980.762	16720.28	PT I	0.5			MIT
5860.818	5862.443	17057.74	PT	1.5			MIT
5844.837	5846.457	17104.38	PT	1.6	0.3		MIT
5840.124	5841.743	17118.18	PT	1.9			MIT
5763.572	5765.170	17345.54	PT	1.5			MIT
5762.706	5764.304	17348.15	PT	0.8			MIT
5750.195	5751.790	17385.89	PT	0.5H			MIT
5728.143	5729.732	17452.82	PT	0.5H			MIT
5700.471	5702.053	17537.54	PT	0.3H			MIT
5698.974	5700.555	17542.15	PT	0.8			MIT
5684.716	5686.293	17586.15	PT	0.8			MIT
5560.179	5561.723	17980.04	PT	0.3			MIT
5560.015	5561.559	17980.57	PT	0.8			MIT
5525.846	5527.381	18091.75	PT	1.3			MIT
5521.68	5523.21	18105.4	PT	0.6			MIT
5514.096	5515.628	18130.30	PT	1.2			MIT
5478.495	5480.017	18248.12	PT	1.7	0.3		MIT
5475.766	5477.288	18257.21	PT	1.8	0.3		MIT
5469.490	5471.010	18278.16	PT I	0.8			MIT
5390.787	5392.286	18545.01	PT	1.7	0.3		MIT
5387.88	5389.38	18555.0	PT	1.2			MIT
5368.987	5370.480	18620.31	PT	1.7	0.0		MIT
5328.60	5330.08	18761.4	PT	0.3			ME
5325.90	5327.38	18770.9	PT	0.5			ME
5324.583	5326.064	18775.59	PT	0.6			MIT
5319.337	5320.817	18794.11	PT	1.0			MIT
5301.02	5302.49	18859.1	PT	2.2	1.0		ME
5286.12	5287.59	18912.2	PT	1.0			MIT
5268.24	5269.71	18976.4	PT	0.3H			ME
5260.84	5262.30	19003.1	PT	1.3			MIT
5257.479	5258.942	19015.23	PT	1.0			MIT
5238.47	5239.93	19084.2	PT	0.3H			ME
5227.66	5229.12	19123.7	PT	1.9	0.3		ME
5208.59	5210.04	19193.7	PT	0.3			ME
5193.90	5195.35	19248.0	PT	0.5			MIT
5164.970	5166.409	19355.80	PT	0.3			MIT
5155.38	5156.82	19391.8	PT	1.2			ME

AIR WAVELENGTH (ANGSTRMS)	VACUUM WAVELENGTH (ANGSTRMS)	WAVENUMBER (KAYSERS)	LMENT	LOG ARC	INTENSITY SPK	INTENSITY DIS	REF
5130.915	5132.345	19484.27	PT	0.5			MIT
5118.431	5119.857	19531.79	PT	0.6			MIT
5108.426	5109.850	19570.05	PT	0.3			MIT
5095.794	5097.214	19618.56	PT	1.0			MIT
5082.35	5083.77	19670.4	PT	0.7			MIT
5077.81	5079.23	19688.0	PT	0.7			MIT
5059.481	5060.892	19759.36	PT	1.8	0.5		MIT
5055.37	5056.78	19775.4	PT	0.6			MIT
5053.864	5055.273	19781.32	PT	0.6			MIT
5044.044	5045.450	19819.84	PT	1.8	0.0		MIT
5038.54	5039.95	19841.5	PT	0.7			ME
5033.522	5034.926	19861.27	PT	1.5	0.0		MIT
5002.630	5004.025	19983.91	PT	1.2	0.0		MIT
4997.972	4999.366	20002.53	PT	0.8			MIT
4986.834	4988.225	20047.21	PT	0.6			MIT
4980.380	4981.770	20073.19	PT	0.6			MIT
4940.138	4941.517	20236.70	PT	0.5			MIT
4879.533	4880.896	20488.04	PT	1.3			MIT
4862.393	4863.751	20560.26	PT	0.5			MIT
4853.928	4855.284	20596.12	PT I	1.2			MIT
4831.956	4833.306	20689.77	PT I	0.3	0.0H		MIT
4831.217	4832.567	20692.94	PT I	0.5			MIT
4824.219	4825.567	20722.95	PT	0.3			MIT
4804.241	4805.584	20809.13	PT	0.3			MIT
4772.320	4773.654	20948.31	PT I	0.7			MIT
4768.091	4769.424	20966.89	PT I	0.5			MIT
4747.880	4749.208	21056.14	PT I	0.5			MIT
4739.758	4741.084	21092.22	PT I	0.7			MIT
4737.561	4738.886	21102.00	PT I	0.6	0.3		MIT
4736.089	4737.414	21108.56	PT	0.3			MIT
4684.107	4685.418	21342.81	PT I	0.7	0.0		MIT
4657.956	4659.260	21462.64	PT I	1.0	0.6		MIT
4650.07	4651.37	21499.0	PT	0.5			ME
4640.822	4642.122	21541.87	PT I	1.2			MIT
4639.73	4641.03	21546.9	PT	0.3			MIT
4631.12	4632.42	21587.0	PT	0.3			ME
4617.074	4618.367	21652.67	PT	0.3			MIT
4580.668	4581.952	21824.76	PT I	0.5	0.0		MIT
4580.546	4581.830	21825.34	PT I	0.5	0.0		MIT
4577.422	4578.705	21840.24	PT I	0.7	0.3		MIT
4560.072	4561.350	21923.33	PT	0.0	0.3		MIT
4554.593	4555.870	21949.71	PT	1.0	0.7		MIT
4552.423	4553.699	21960.17	PT I	1.8	1.0		MIT
4551.950	4553.226	21962.45	PT I	0.3	0.0H		MIT
4547.972	4549.247	21981.66	PT I	0.5	0.0H		MIT
4523.006	4524.274	22102.99	PT I	1.0	0.0		MIT
4520.901	4522.169	22113.28	PT I	1.6	0.3H		MIT
4514.14	4515.41	22146.4	PT II	0.3	0.7H		M
4511.257	4512.522	22160.55	PT I	0.3	0.0H		MIT
4498.761	4500.023	22222.11	PT I	2.0	0.7H		MIT
4493.187	4494.448	22249.68	PT	0.3	0.0		MIT
4484.695	4485.953	22291.81	PT I	0.7	0.0H		MIT
4481.645	4482.902	22306.98	PT I	0.3	0.0H		MIT
4473.461	4474.716	22347.78	PT I	0.3	0.3		MIT
4471.65	4472.90	22356.8	PT		0.3H		MIT
4445.547	4446.795	22488.11	PT I	1.3	0.3		MIT
4442.552	4443.799	22503.27	PT I	2.9	1.4		MIT
4437.283	4438.529	22529.99	PT I	1.4	0.3H		MIT
4430.20	4431.44	22566.0	PT I	0.6	0.0H		MIT
4414.254	4415.494	22647.52	PT I	0.3	0.0		MIT
4411.401	4412.640	22662.17	PT I	0.3	0.0H		MIT
4391.826	4393.060	22763.18	PT I	1.7	0.5		MIT
4371.921	4373.150	22866.81	PT		0.3H		MIT
4364.463	4365.690	22905.89	PT I	0.5	0.3		MIT
4358.336	4359.561	22938.09	PT I	0.3			MIT
4343.663	4344.884	23015.57	PT I	0.5			MIT
4334.658	4335.877	23063.39	PT I	0.3			MIT
4327.065	4328.282	23103.86	PT I	1.9	0.6		MIT
4309.172	4310.384	23199.79	PT	0.5			MIT
4302.43	4303.64	23236.1	PT		0.3		MIT

AIR WAVELENGTH (ANGSTRMS)	VACUUM WAVELENGTH (ANGSTRMS)	WAVENUMBER (KAYSERS)	LMENT	LOG ARC	INTENSITY SPK	DIS	REF
4290.887	4292.094	23298.65	PT	0.3	0.0		MIT
4288.428	4289.635	23312.01	PT II		0.3		MIT
4288.058	4289.264	23314.02	PT I	1.9	0.0H		MIT
4273.909	4275.112	23391.20	PT	0.3			MIT
4269.259	4270.461	23416.68	PT I	0.3	0.0H		MIT
4263.511	4264.711	23448.25	PT I	1.0	0.0H		MIT
4259.938	4261.137	23467.91	PT	0.7			MIT
4223.69	4224.88	23669.3	PT II		0.5		SH
4201.216	4202.400	23795.93	PT I	0.3	0.3		MIT
4192.429	4193.610	23845.80	PT I	2.0	0.3		MIT
4164.557	4165.731	24005.39	PT I	2.0	1.9		MIT
4148.30	4149.47	24099.5	PT II		0.7		SH
4118.686	4119.848	24272.74	PT I	2.6	1.0		MIT
4092.261	4093.416	24429.47	PT I	0.6	0.0		MIT
4081.474	4082.626	24494.04	PT	0.5	0.0H		MIT
4065.941	4067.089	24587.61	PT	0.7	0.3		MIT
4061.66	4062.81	24613.5	PT II		1.0		SH
4054.775	4055.920	24655.32	PT I	0.7	0.3		MIT
4046.45	4047.59	24706.0	PT II		1.3		SH
4034.17	4035.31	24781.2	PT II		0.7		SH
4023.81	4024.95	24845.1	PT II		0.5		SH
4014.31	4015.44	24903.8	PT II		0.3		SH
3996.574	3997.704	25014.36	PT I	1.7			MIT
3970.06	3971.18	25181.4	PT II	0.6	1.2		M
3966.54	3967.66	25203.8	PT	0.5	0.3		EX
3966.361	3967.483	25204.90	PT I	1.9	1.6		MIT
3953.644	3954.763	25285.97	PT I	0.3	0.0		MIT
3948.395	3949.512	25319.58	PT I	1.8	0.7		MIT
3925.340	3926.451	25468.29	PT I	1.8	0.5		MIT
3922.960	3924.071	25483.74	PT I	2.0	1.3R		MIT
3910.906	3912.014	25562.28	PT I	0.7	0.3		MIT
3906.291	3907.398	25592.48	PT I	0.3	0.0H		MIT
3904.393	3905.499	25604.92	PT I	0.7	0.0		MIT
3903.722	3904.828	25609.32	PT I	0.3	0.0		MIT
3900.734	3901.839	25628.94	PT I	1.6	0.5		MIT
3898.747	3899.852	25642.00	PT I	1.3	0.3H		MIT
3875.72	3876.82	25794.4	PT		0.3H		MIT
3868.412	3869.509	25843.07	PT		0.3		MIT
3819.88	3820.96	26171.4	PT	0.5			M
3818.691	3819.775	26179.55	PT	1.6	1.0		MIT
3806.91	3807.99	26260.6	PT II		0.7		SH
3801.05	3802.13	26301.1	PT		0.6		EX
3800.503	3801.582	26304.84	PT	0.6			MIT
3800.21	3801.29	26306.9	PT	0.5			EX
3799.87	3800.95	26309.2	PT	0.7			EX
3768.39	3769.46	26529.0	PT II		1.0		SH
3766.385	3767.455	26543.12	PT II		1.3		MIT
3720.74	3721.80	26868.7	PT	0.7			M
3706.527	3707.582	26971.76	PT	1.2	0.6		MIT
3700.14	3701.19	27018.3	PT II		1.0		SH
3699.914	3700.967	27019.97	PT I	1.9	0.7		MIT
3687.426	3688.476	27111.47	PT I	1.5	0.5		MIT
3682.983	3684.032	27144.18	PT I	0.9	0.3		MIT
3674.045	3675.091	27210.21	PT I	1.9	0.6		MIT
3671.994	3673.040	27225.41	PT I	1.9	1.0		MIT
3663.095	3664.138	27291.55	PT I	1.7	0.3		MIT
3659.419	3660.461	27318.96	PT I	0.7	0.3		MIT
3653.992	3655.033	27359.53	PT I	0.3	0.0H		MIT
3652.252	3653.293	27372.57	PT	0.3	0.0H		MIT
3643.172	3644.210	27440.79	PT I	1.8	0.9		MIT
3638.793	3639.830	27473.81	PT I	2.4	1.0		MIT
3628.870	3629.905	27548.93	PT	0.8	0.3H		MIT
3628.113	3629.147	27554.68	PT I	2.50	1.3		MIT
3621.656	3622.689	27603.81	PT I	0.6	0.0		MIT
3610.909	3611.939	27685.96	PT	0.6	0.0		MIT
3607.89	3608.92	27709.1	PT II		0.7		SH
3594.01	3595.04	27816.1	PT	0.5			EX
3587.405	3588.429	27867.35	PT I	0.6H			MIT
3577.20	3578.22	27946.9	PT II		1.0		SH
3571.999	3573.019	27987.54	PT II		1.2		MIT

AIR WAVELENGTH (ANGSTRMS)	VACUUM WAVELENGTH (ANGSTRMS)	WAVENUMBER (KAYSERS)	LMENT	LOG ARC	INTENSITY SPK	DIS	REF
3551.37	3552.38	28150.1	PT II		1.2		SH
3548.494	3549.508	28172.92	PT II		0.9		SH
3546.443	3547.456	28189.21	PT	0.5			MIT
3535.848	3536.859	28273.68	PT II		1.0		MIT
3528.544	3529.553	28332.20	PT	0.7	0.3H		MIT
3514.715	3515.720	28443.68	PT	0.3	0.3		MIT
3500.318	3501.319	28560.66	PT	0.3H			MIT
3498.17	3499.17	28578.2	PT I	0.0	0.3		MIT
3490.995	3491.994	28636.93	PT	0.3	0.3		SF
3485.267	3486.265	28684.00	PT I	2.2	2.36		MIT
3483.428	3484.425	28699.14	PT	1.8	1.0		MIT
3477.67	3478.67	28746.6	PT II		0.3		SH
3476.755	3477.750	28754.22	PT I	1.0	0.3		ML
3463.941	3464.933	28860.59	PT	0.3	0.0H		MIT
3454.134	3455.124	28942.52	PT I	0.3	0.3		MIT
3453.86	3454.85	28944.8	PT	0.0	0.9		SH
3447.78	3448.77	28995.9	PT II	0.0	1.2		SH
3447.67	3448.66	28996.8	PT II		0.7		SH
3431.848	3432.832	29130.47	PT I	0.7	0.5		MIT
3427.930	3428.913	29163.76	PT I	1.7	0.8		MIT
3426.722	3427.705	29174.04	PT I	0.9	0.0		MIT
3421.72	3422.70	29216.7	PT	0.3H			M
3417.067	3418.047	29256.47	PT	0.5			SF
3408.134	3409.112	29333.15	PT I	2.40	1.8		MIT
3383.82	3384.79	29543.9	PT	0.0	0.9		MIT
3372.996	3373.965	29638.72	PT	1.0			MIT
3372.791	3373.760	29640.52	PT	0.3			MIT
3366.998	3367.965	29691.52	PT I	1.4	0.7		MIT
3343.900	3344.861	29896.60	PT I	2.0	1.9		MIT
3340.08	3341.04	29930.8	PT	0.0	1.2		SH
3338.180	3339.140	29947.83	PT	0.3			MIT
3335.835	3336.794	29968.88	PT I	0.3			MIT
3327.171	3328.128	30046.92	PT I	0.5			ML
3325.729	3326.686	30059.95	PT	0.3			SF
3323.795	3324.751	30077.44	PT I	2.2	1.0		MIT
3315.046	3316.000	30156.81	PT I	2.3	1.0		MIT
3312.488	3313.441	30180.10	PT	0.3	0.3H		MIT
3302.12	3303.07	30274.9	PT	0.3H			MIT
3301.861	3302.812	30277.23	PT I	2.5	2.40		MIT
3290.220	3291.168	30384.35	PT I	2.2	1.0		MIT
3283.312	3284.258	30448.28	PT I	0.9	0.5		MIT
3283.209	3284.155	30449.23	PT	0.9			MIT
3281.969	3282.915	30460.74	PT I	1.0	0.5		MIT
3273.058	3274.001	30543.67	PT II		0.7		MIT
3268.424	3269.366	30586.97	PT	1.2	0.6		MIT
3261.692	3262.632	30650.10	PT	0.9	0.3		MIT
3261.072	3262.012	30655.92	PT I	0.5			MIT
3259.733	3260.673	30668.52	PT I	0.5	0.8		MIT
3256.433	3257.372	30699.59	PT	0.3			SF
3255.916	3256.855	30704.47	PT I	0.5	1.5		MIT
3251.982	3252.920	30741.61	PT I	2.0	0.0		MIT
3250.355	3251.293	30757.00	PT	1.6	0.9		MIT
3248.502	3249.439	30774.54	PT I	0.3			MIT
3243.71	3244.65	30820.0	PT II	0.6	1.3		SH
3240.203	3241.138	30853.36	PT I	1.6	0.8		MIT
3233.422	3234.355	30918.06	PT I	1.6	1.0		MIT
3230.287	3231.220	30948.07	PT I	2.0	0.8		MIT
3227.165	3228.097	30978.01	PT	0.6	0.5		MIT
3220.778	3221.708	31039.44	PT	0.3			MIT
3212.374	3213.302	31120.64	PT I	0.8			MIT
3208.851	3209.778	31154.80	PT	0.5			MIT
3204.040	3204.966	31201.58	PT I	2.4	2.0		MIT
3200.714	3201.639	31234.00	PT I	2.0	1.6		MIT
3198.929	3199.854	31251.43	PT	0.5			MIT
3197.956	3198.880	31260.94	PT II	0.6	1.3H		SH
3192.505	3193.428	31314.31	PT I	0.8			MIT
3188.078	3189.000	31357.79	PT II	0.0	0.7		SH
3179.002	3179.922	31447.32	PT	0.0	0.9		SH
3174.824	3175.742	31488.70	PT I	0.3			MIT
3174.526	3175.444	31491.66	PT		0.3		SH

AIR WAVELENGTH (ANGSTRMS)	VACUUM WAVELENGTH (ANGSTRMS)	WAVENUMBER (KAYSERS)	LMENT	LOG ARC	INTENSITY SPK	DIS	REF
3159.080	3159.995	31645.62	PT II	0.5	1.2		SH
3156.565	3157.479	31670.84	PT I	2.2	1.7		MIT
3145.030	3145.941	31786.99	PT II	0.0	0.7H		SH
3144.097	3145.008	31796.42	PT II	0.8	1.0H		SH
3141.658	3142.568	31821.11	PT	2.2	0.7H		MIT
3139.386	3140.296	31844.14	PT I	2.5	1.9		MIT
3126.925	3127.831	31971.03	PT	0.0	0.9		SH
3119.802	3120.707	32044.02	PT	0.8	0.5		MIT
3118.012	3118.916	32062.42	PT II	0.0	1.2H		SH
3114.039	3114.942	32103.32	PT	0.5			MIT
3100.961	3101.861	32238.71	PT I	1.2	0.6		MIT
3100.038	3100.938	32248.31	PT I	0.3	1.4		MIT
3084.860	3085.756	32406.97	PT	0.3			MIT
3084.116	3085.012	32414.79	PT	1.0	0.3		MIT
3079.572	3080.467	32462.61	PT	0.5	0.3		MIT
3076.719	3077.613	32492.72	PT	0.0	0.9		MIT
3075.933	3076.827	32501.02	PT II		0.7		MIT
3074.146	3075.039	32519.91	PT II		0.7		SH
3071.937	3072.830	32543.29	PT I	1.8	1.2		MIT
3070.260	3071.152	32561.07	PT	0.5			MIT
3064.712	3065.603	32620.01	PT I	3.3G	2.5G		MIT
3059.642	3060.532	32674.06	PT I	1.4	0.7		MIT
3056.071	3056.960	32712.24	PT II	0.0	0.7		SH
3055.318	3056.206	32720.30	PT I	0.6			MIT
3042.637	3043.522	32856.67	PT I	2.3G	2.4G		MIT
3041.217	3042.102	32872.01	PT I	0.5			MIT
3041.079	3041.964	32873.50	PT II	0.7	1.3H		SH
3036.450	3037.334	32923.61	PT I	2.3	1.0		MIT
3031.216	3032.098	32980.46	PT II	1.0	1.6H		MIT
3024.299	3025.180	33055.89	PT I	0.7			MIT
3022.847	3023.727	33071.77	PT	0.7	0.3		MIT
3017.883	3018.762	33126.16	PT	1.8	1.0		MIT
3017.246	3018.125	33133.15	PT	0.3	0.9		MIT
3012.523	3013.401	33185.10	PT	0.0	0.5		MIT
3012.380	3013.258	33186.67	PT I	0.3			MIT
3011.547	3012.425	33195.85	PT	0.5			MIT
3005.783	3006.659	33259.51	PT	0.5			MIT
3002.269	3003.144	33298.43	PT I	2.3	1.5		MIT
3001.169	3002.044	33310.64	PT II	0.5	1.7W		MIT
2997.967	2998.841	33346.21	PT I	3.0G	2.3R		MIT
2983.747	2984.618	33505.13	PT I	0.8			MIT
2979.806	2980.676	33549.44	PT	0.0	0.9		SH
2979.183	2980.052	33556.46	PT II	0.0	0.3		MIT
2960.751	2961.616	33765.35	PT	1.4	0.6		MIT
2959.095	2959.959	33784.25	PT I	1.3	0.5		MIT
2958.491	2959.355	33791.14	PT II	0.0H	0.5H		SH
2951.218	2952.080	33874.42	PT	0.3H			MIT
2944.754	2945.615	33948.77	PT I	1.2	0.3		MIT
2942.755	2943.615	33971.83	PT I	1.3	0.5		MIT
2941.090	2941.950	33991.06	PT	0.3			MIT
2938.813	2939.672	34017.40	PT	1.2	0.3		MIT
2930.786	2931.643	34110.56	PT	1.2	0.5H		MIT
2929.794	2930.651	34122.11	PT I	2.9G	2.3W		MIT
2928.648	2929.505	34135.46	PT II	0.0	0.8H		SH
2928.107	2928.964	34141.77	PT I	0.9	0.8		MIT
2921.383	2922.238	34220.35	PT I	2.0	0.8		MIT
2919.344	2920.199	34244.25	PT I	2.2	1.6		MIT
2914.127	2914.980	34305.55	PT	0.0	1.0		MIT
2913.542	2914.395	34312.44	PT I	2.5	1.4		MIT
2913.249	2914.102	34315.89	PT I	1.4			MIT
2912.256	2913.109	34327.59	PT I	2.5	1.4		MIT
2910.448	2911.300	34348.91	PT I	0.5			MIT
2907.899	2908.751	34379.02	PT I	1.2	0.6		MIT
2905.902	2906.753	34402.65	PT I	2.0	1.2		MIT
2899.645	2900.495	34476.88	PT II	0.3H	1.4H		MIT
2897.873	2898.722	34497.96	PT I	2.6	1.2		MIT
2893.865	2894.713	34545.73	PT I	2.7	1.4		MIT
2893.220	2894.068	34553.44	PT I	1.4	0.7		MIT
2890.374	2891.221	34587.46	PT II	0.7R	1.4		MIT
2888.196	2889.043	34613.54	PT I	1.7			MIT

AIR WAVELENGTH (ANGSTRMS)	VACUUM WAVELENGTH (ANGSTRMS)	WAVENUMBER (KAYSERS)	LMENT	LOG ARC	INTENSITY SPK	DIS	REF
2877.520	2878.364	34741.95	PT II	1.6	2.3H		SH
2875.849	2876.693	34762.14	PT II	1.3	1.9H		SH
2870.468	2871.311	34827.30	PT	1.0	0.3		MIT
2866.891	2867.733	34870.75	PT II	0.8	1.2		MIT
2866.076	2866.917	34880.67	PT	0.0	0.5		SH
2865.051	2865.892	34893.15	PT II	1.3	1.9H		SH
2860.678	2861.518	34946.48	PT II	1.5	2.2H		SH
2853.384	2854.222	35035.81	PT	0.3			MIT
2853.107	2853.945	35039.21	PT I	1.2	0.3		MIT
2842.030	2842.866	35175.77	PT	0.0	1.0		MIT
2837.230	2838.064	35235.28	PT I	0.3			MIT
2834.713	2835.547	35266.57	PT I	1.9	0.7		MIT
2831.558	2832.391	35305.86	PT II		0.5		MIT
2830.295	2831.128	35321.61	PT I	3.0G	2.8R		MIT
2824.41	2825.24	35395.2	PT	0.3	0.3		MIT
2822.492	2823.323	35419.26	PT II	0.3H	1.2		MIT
2822.270	2823.101	35422.04	PT II	1.0	1.8H		SH
2818.873	2819.703	35464.73	PT II	0.0	1.0		MIT
2818.248	2819.078	35472.59	PT	1.8	0.6		MIT
2814.001	2814.830	35526.13	PT II	0.6	1.2		SH
2812.975	2813.803	35539.08	PT	0.3			MIT
2808.835	2809.662	35591.46	PT	0.0	1.2		MIT
2808.507	2809.334	35595.62	PT I	1.3	0.3		MIT
2803.239	2804.065	35662.51	PT I	2.6	0.7		MIT
2799.981	2800.806	35704.00	PT II	1.3	1.9H		SH
2797.807	2798.632	35731.75	PT II	0.3	1.3		SH
2794.208	2795.032	35777.77	PT II	1.0	2.0J		MIT
2793.649	2794.473	35784.93	PT I	1.0			MIT
2793.266	2794.090	35789.83	PT I	2.0	0.7		MIT
2788.625	2789.447	35849.39	PT II	0.8	1.3		MIT
2774.779	2775.598	36028.27	PT II	1.0	2.0J		MIT
2774.205	2775.024	36035.73	PT	0.6			MIT
2773.996	2774.815	36038.44	PT I	1.7	0.3		MIT
2773.597	2774.416	36043.62	PT I	0.8			MIT
2773.241	2774.060	36048.25	PT I	1.7	0.7		MIT
2772.829	2773.648	36053.61	PT I	1.2	0.3		MIT
2771.666	2772.484	36068.73	PT I	2.7	1.2		MIT
2769.837	2770.655	36092.55	PT I	1.7	0.3		MIT
2768.957	2769.775	36104.02	PT	0.0	0.6		MIT
2766.660	2767.477	36133.99	PT	1.0	0.3		MIT
2763.224	2764.040	36178.92	PT II	0.8	1.2J		MIT
2757.694	2758.509	36251.47	PT I	1.2			MIT
2754.920	2755.734	36287.97	PT I	2.3	0.7H		MIT
2753.856	2754.670	36301.99	PT I	2.0	0.6		MIT
2753.760	2754.574	36303.25	PT I	1.2			MIT
2747.609	2748.422	36384.52	PT I	2.2	0.3		MIT
2744.830	2745.642	36421.36	PT	0.7R	0.0H		MIT
2743.491	2744.303	36439.13	PT II	0.8	1.6H		MIT
2740.39	2741.20	36480.4	PT II		0.6H		SH
2738.484	2739.294	36505.75	PT I	2.0	0.7		MIT
2735.755	2736.565	36542.17	PT II	0.0	1.0		MIT
2734.498	2735.307	36558.96	PT I	2.0	1.0		MIT
2733.961	2734.770	36566.14	PT I	3.0H	2.3H		MIT
2733.691	2734.500	36569.75	PT	1.2	0.7H		MIT
2729.915	2730.723	36620.33	PT I	2.7	1.7		MIT
2726.410	2727.217	36667.41	PT II	0.6	1.3		MIT
2721.839	2722.645	36728.99	PT	0.0	1.0		MIT
2719.038	2719.844	36766.82	PT I	3.0W	2.00		MIT
2717.624	2718.429	36785.95	PT II	0.6	1.6		MIT
2715.770	2716.575	36811.06	PT	0.7			MIT
2713.127	2713.931	36846.92	PT	2.3	1.0		MIT
2710.922	2711.726	36876.89	PT II	0.0	1.2		MIT
2705.894	2706.696	36945.41	PT I	3.0J	2.3J		MIT
2702.399	2703.201	36993.19	PT I	3.0	2.5		MIT
2699.389	2700.190	37034.43	PT II		0.8		MIT
2698.428	2699.229	37047.62	PT I	2.7	1.7		MIT
2694.239	2695.039	37105.22	PT	0.3			MIT
2692.236	2693.035	37132.82	PT II	0.6	1.6		MIT
2689.398	2690.196	37172.01	PT II	0.0	1.0		MIT
2681.791	2682.588	37277.44	PT	0.0	1.0		MIT

AIR WAVELENGTH (ANGSTRMS)	VACUUM WAVELENGTH (ANGSTRMS)	WAVENUMBER (KAYSERS)	LMENT	LOG ARC	INTENSITY SPK	DIS	REF
2680.059	2680.855	37301.53	PT II		0.5		SH
2679.129	2679.925	37314.48	PT II	0.8	1.7		MIT
2677.149	2677.945	37342.07	PT I	2.9W	2.3W		MIT
2674.574	2675.369	37378.02	PT	2.3	1.0		MIT
2664.640	2665.433	37517.36	PT I	1.5			MIT
2659.454	2660.245	37590.52	PT I	3.3G	2.7G		MIT
2658.701	2659.492	37601.17	PT I	1.6	0.7		MIT
2658.174	2658.965	37608.62	PT I	2.0	1.0		MIT
2651.503	2652.292	37703.23	PT II		0.6		SH
2650.857	2651.646	37712.42	PT I	2.8	2.0		MIT
2646.886	2647.674	37769.00	PT I	3.0H	2.0		MIT
2645.373	2646.161	37790.60	PT I	1.6	0.7		MIT
2639.350	2640.136	37876.83	PT I	2.7	1.7		MIT
2634.897	2635.682	37940.84	PT II	0.6	1.2		MIT
2628.029	2628.813	38039.99	PT I	3.0W	2.0		MIT
2627.392	2628.176	38049.21	PT I	1.6	0.7		MIT
2625.330	2626.113	38079.09	PT II	1.5	1.8		MIT
2621.503	2622.285	38134.68	PT	0.0	1.4		SH
2621.032	2621.814	38141.53	PT II	0.0	1.4		SH
2619.567	2620.349	38162.86	PT I	2.5	0.7		MIT
2616.755	2617.536	38203.87	PT II	1.0	1.8		MIT
2614.610	2615.391	38235.21	PT I	1.0			MIT
2608.065	2608.844	38331.15	PT II		0.7		MIT
2603.681	2604.459	38395.69	PT II		0.7		MIT
2603.142	2603.920	38403.64	PT I	2.5	1.3		MIT
2599.916	2600.693	38451.29	PT I	0.7			MIT
2596.00	2596.78	38509.3	PT	2.3	1.3		MIT
2591.013	2591.788	38583.40	PT II		1.0		SH
2587.790	2588.564	38631.45	PT	1.0			MIT
2583.156	2583.929	38700.75	PT II		1.0		MIT
2581.084	2581.857	38731.82	PT II		0.7		SH
2578.398	2579.170	38772.16	PT II	0.3	1.3		MIT
2574.490	2575.261	38831.02	PT I	1.2	0.7H		MIT
2572.618	2573.389	38859.27	PT II	1.2	1.7		MIT
2568.587	2569.357	38920.25	PT II	0.0	1.6		MIT
2552.250	2553.016	39169.36	PT I	2.2	1.3		MIT
2549.464	2550.229	39212.16	PT I	1.9	1.0		MIT
2541.352	2542.115	39337.32	PT I	1.2	0.5		MIT
2539.205	2539.968	39370.58	PT I	2.6	1.3		MIT
2536.487	2537.249	39412.76	PT I	2.0	1.0		MIT
2535.967	2536.729	39420.85	PT	1.4	0.7		MIT
2529.407	2530.167	39523.08	PT I	1.9			MIT
2524.852	2525.611	39594.37	PT II		0.5H		SH
2524.304	2525.063	39602.97	PT I	2.5	1.0		MIT
2517.185	2517.943	39714.96	PT I	1.2			MIT
2515.576	2516.333	39740.37	PT I	2.7	1.3		MIT
2515.032	2515.789	39748.96	PT I	2.2	1.3		MIT
2514.072	2514.829	39764.14	PT I	2.2	1.0H		MIT
2513.881	2514.638	39767.16	PT II	0.8	1.7		MIT
2508.496	2509.252	39852.52	PT I	2.5	1.3		MIT
2505.926	2506.681	39893.39	PT	2.2	1.0		MIT
2504.042	2504.797	39923.40	PT I	1.8	0.7		MIT
2498.500	2499.253	40011.95	PT I	2.6	1.7		MIT
2497.922	2498.675	40021.21	PT II		0.5		MIT
2495.817	2496.570	40054.96	PT I	1.6	1.0		MIT
2492.531	2493.283	40107.76	PT II		0.5		MIT
2491.413	2492.165	40125.76	PT II		0.3		SH
2490.125	2490.876	40146.51	PT I	2.5	1.3		MIT
2488.88	2489.63	40166.6	PT		1.4		MIT
2488.736	2489.487	40168.92	PT	1.0	1.4		MIT
2487.168	2487.919	40194.24	PT I	2.8R	1.3		MIT
2486.989	2487.740	40197.13	PT II	0.8	1.0		MIT
2483.367	2484.117	40255.76	PT	1.6	0.3		MIT
2482.042	2482.791	40277.25	PT II	0.3	1.4		MIT
2481.187	2481.936	40291.12	PT	0.3	0.0		MIT
2471.007	2471.754	40457.10	PT I	2.0	1.3		MIT
2467.593	2468.339	40513.07	PT II	0.7	2.0		MIT
2467.443	2468.189	40515.54	PT	2.9G	2.0		MIT
2455.140	2455.883	40718.55	PT		0.5		MIT
2452.86	2453.60	40756.4	PT		0.9		MIT
2450.970	2451.712	40787.82	PT I	2.6	1.0		MIT
2450.440	2451.182	40796.64	PT II	1.4	1.7		MIT
2443.099	2443.839	40919.22	PT II		0.5		MIT
2442.626	2443.366	40927.14	PT II	1.3	1.6		MIT
2440.057	2440.797	40970.23	PT I	2.9W	2.0J		MIT
2436.689	2437.428	41026.85	PT I	2.5	1.3		MIT
2434.458	2435.196	41064.45	PT II	1.3	1.6		MIT
2433.495	2434.233	41080.70	PT II	0.0	1.0H		SH
2429.357	2430.094	41150.67	PT II	0.6	1.0		MIT
2429.096	2429.833	41155.09	PT I	2.0			MIT
2428.203	2428.940	41170.22	PT I	2.0	1.3		MIT
2428.035	2428.772	41173.07	PT I	2.0	1.0		MIT
2424.869	2425.605	41226.82	PT II	1.7	2.0		MIT
2420.817	2421.552	41295.82	PT II	0.6	1.0		MIT
2418.058	2418.793	41342.94	PT I	2.5	1.7		MIT
2417.736	2418.471	41348.44	PT II		1.0		MIT
2413.045	2413.779	41428.82	PT I	1.8	1.0		MIT
2410.332	2411.065	41475.45	PT II		1.7		MIT
2405.728	2406.460	41554.82	PT II		2.0J		MIT
2403.089	2403.820	41600.45	PT I	2.6	1.7		MIT
2401.874	2402.605	41621.49	PT I	2.5	1.5		MIT
2401.002	2401.733	41636.60	PT I	1.4	1.0		MIT
2400.577	2401.308	41643.98	PT II		1.3H		SH
2396.686	2397.416	41711.58	PT II	0.3	1.5		MIT
2396.167	2396.897	41720.61	PT I	1.4	1.3		MIT
2392.03	2392.76	41792.8	PT		0.9		MIT
2391.76	2392.49	41797.5	PT	0.8			MIT
2390.800	2391.529	41814.26	PT II		1.0		MIT
2390.14	2390.87	41825.8	PT	0.5			MIT
2390.064	2390.792	41827.14	PT II	1.2			MIT
2389.533	2390.261	41836.43	PT	1.4	1.3		MIT
2387.358	2388.086	41874.54	PT	0.6	1.0H		MIT
2386.808	2387.536	41884.19	PT I	1.4	0.7		MIT
2386.506	2387.234	41889.49	PT II	0.3	1.4		MIT
2384.46	2385.19	41925.4	PT II		1.2H		MIT
2383.641	2384.368	41939.84	PT I	1.5	1.3		MIT
2381.36	2382.09	41980.0	PT		1.1		MIT
2378.063	2378.789	42038.20	PT		1.0		MIT
2377.958	2378.684	42040.06	PT	1.0	0.6		MIT
2377.275	2378.000	42052.14	PT II	1.2	1.4		MIT
2373.32	2374.04	42122.2	PT		1.0		MIT
2371.617	2372.341	42152.45	PT II		1.3		MIT
2368.276	2368.999	42211.91	PT I	1.5	1.4		MIT
2366.470	2367.193	42244.12	PT II		1.2H		MIT
2365.36	2366.08	42264.0	PT		0.9		MIT
2365.238	2365.961	42266.13	PT II		0.9		MIT
2364.58	2365.30	42277.9	PT		0.9H		MIT
2363.86	2364.58	42290.8	PT		1.2		MIT
2358.776	2359.497	42381.91	PT II		0.7		MIT
2357.577	2358.298	42403.46	PT	1.3	1.0		MIT
2357.104	2357.825	42411.97	PT I	1.5	1.3		MIT
2356.33	2357.05	42425.9	PT	1.3	1.0		MIT
2351.408	2352.128	42514.70	PT		1.1		MIT
2348.548	2349.267	42566.47	PT II	0.5	1.3		MIT
2347.160	2347.879	42591.64	PT	1.3	1.1		MIT
2346.722	2347.441	42599.59	PT	0.9	0.5		MIT
2343.390	2344.108	42660.15	PT I	1.4			MIT
2343.236	2343.954	42662.96	PT II		1.0		MIT
2342.777	2343.495	42671.31	PT II		1.0		MIT
2341.525	2342.242	42694.13	PT II		0.6H		MIT
2340.993	2341.710	42703.83	PT II		0.9H		SH
2340.178	2340.895	42718.70	PT I	1.4	1.3		MIT
2339.52	2340.24	42730.7	PT		1.2		MIT
2339.179	2339.896	42736.94	PT II		1.2J		MIT
2338.05	2338.77	42757.6	PT		1.0		MIT
2335.190	2335.906	42809.94	PT II	0.9	1.4		SH
2331.791	2332.506	42872.34	PT II		0.6		MIT
2330.968	2331.683	42887.47	PT I	1.4	1.0		MIT
2326.331	2327.045	42972.95	PT II		1.2		SH
2326.101	2326.815	42977.20	PT	1.5	0.9		MIT

AIR WAVELENGTH (ANGSTRMS)	VACUUM WAVELENGTH (ANGSTRMS)	WAVENUMBER (KAYSERS)	LMENT	LOG ARC	INTENSITY SPK	INTENSITY DIS	REF
2323.30	2324.01	43029.0	PT	0.3			MIT
2322.21	2322.92	43049.2	PT		0.6		MIT
2319.888	2320.601	43092.29	PT II	0.6	1.4		MIT
2318.294	2319.006	43121.92	PT	1.4	1.0		MIT
2317.908	2318.620	43129.10	PT		0.7		SH
2315.495	2316.207	43174.04	PT	1.4	0.9		MIT
2313.037	2313.748	43219.92	PT II	0.5	1.3		MIT
2310.958	2311.669	43258.79	PT II	1.2	1.4		MIT
2308.040	2308.750	43313.48	PT I	1.5	1.3		MIT
2305.637	2306.347	43358.62	PT	1.0	0.8		MIT
2304.26	2304.97	43384.5	PT		1.0		MIT
2303.200	2303.909	43404.49	PT I	1.4	0.9		MIT
2302.42	2303.13	43419.2	PT		1.3W		MIT
2300.39	2301.10	43457.5	PT		0.3		MIT
2299.515	2300.223	43474.04	PT II		0.5J		SH
2298.781	2299.489	43487.92	PT	1.2	0.6		MIT
2298.361	2299.069	43495.87	PT	1.1	0.5		MIT
2298.135	2298.843	43500.15	PT		1.1		MIT
2297.31	2298.02	43515.8	PT		0.9W		MIT
2295.864	2296.571	43543.17	PT II		1.2		SH
2295.738	2296.445	43545.56	PT	1.3	0.3		MIT
2294.565	2295.272	43567.82	PT II		0.3		MIT
2294.140	2294.847	43575.89	PT II		0.5J		SH
2293.87	2294.58	43581.0	PT	0.3			MIT
2293.467	2294.174	43588.67	PT II		0.8		MIT
2292.395	2293.102	43609.06	PT I	2.6	2.0		MIT
2291.72	2292.43	43621.9	PT		1.3W		MIT
2289.27	2289.98	43668.6	PT I	1.4	1.1		MIT
2288.192	2288.898	43689.15	PT II	1.2	1.5		MIT
2287.499	2288.205	43702.39	PT II		1.4W		SH
2286.572	2287.277	43720.10	PT II		1.0W		SH
2286.190	2286.895	43727.41	PT II		0.5W		SH
2285.65	2286.36	43737.7	PT		0.9W		MIT
2281.27	2281.97	43821.7	PT	0.6	1.3W		MIT
2280.48	2281.18	43836.9	PT	1.3	1.0		MIT
2278.763	2279.467	43869.91	PT II		1.2		MIT
2276.86	2277.56	43906.6	PT	1.4	1.0		MIT
2276.41	2277.11	43915.2	PT	1.4	1.1		MIT
2274.84	2275.54	43945.6	PT	1.4	1.0		MIT
2274.38	2275.08	43954.5	PT I	1.5	1.3		MIT
2274.07	2274.77	43960.4	PT	1.0			MIT
2271.718	2272.420	44005.95	PT II		1.3W		SH
2268.84	2269.54	44061.8	PT	1.4	1.3		MIT
2267.240	2267.941	44092.85	PT	0.7	1.3		MIT
2266.417	2267.118	44108.86	PT		1.3		MIT
2263.987	2264.687	44156.20	PT II		1.3W		SH
2263.323	2264.023	44169.16	PT II		1.4		SH
2262.900	2263.600	44177.41	PT II		0.7H		SH
2262.662	2263.362	44182.06	PT II	0.6	1.4		SH
2260.497	2261.197	44224.37	PT II		1.4		MIT
2259.45	2260.15	44244.9	PT		1.0W		MIT
2258.717	2259.416	44259.22	PT II		1.3		SH
2257.13	2257.83	44290.3	PT	1.1	0.8W		MIT
2256.104	2256.803	44310.47	PT II		1.3		SH
2253.130	2253.828	44368.96	PT II		1.4		MIT
2251.926	2252.624	44392.67	PT II		1.2W		MIT
2251.521	2252.219	44400.66	PT II	0.5	1.5		MIT
2250.631	2251.329	44418.22	PT II	0.9	1.3		MIT
2249.90	2250.60	44432.6	PT	1.4	0.9		MIT
2249.30	2250.00	44444.5	PT	1.4	1.1		MIT
2247.593	2248.290	44478.25	PT II		1.0W		SH
2246.505	2247.202	44499.79	PT II	0.3	1.0		SH
2245.518	2246.215	44519.35	PT II	1.4	1.5		MIT
2244.97	2245.67	44530.2	PT I	1.4	1.0		MIT
2241.21	2241.91	44604.9	PT	1.1			MIT
2240.31	2241.01	44622.8	PT I	1.0	0.9		MIT
2237.56	2238.25	44677.7	PT	0.3			MIT
2235.45	2236.14	44719.8	PT I	0.8			MIT
2235.301	2235.995	44722.81	PT II	0.7	1.4		MIT
2234.915	2235.609	44730.54	PT I	1.4	1.2		MIT

AIR WAVELENGTH (ANGSTRMS)	VACUUM WAVELENGTH (ANGSTRMS)	WAVENUMBER (KAYSERS)	LMENT	LOG ARC	INTENSITY SPK	INTENSITY DIS	REF
2233.110	2233.804	44766.69	PT II		0.9W		SH
2229.31	2230.00	44843.0	PT		1.4		MIT
2229.218	2229.911	44844.84	PT II	0.5	1.4		MIT
2228.498	2229.191	44859.33	PT II		1.2		MIT
2227.51	2228.20	44879.2	PT	0.3	0.6		MIT
2224.52	2225.21	44939.5	PT	1.0			MIT
2224.177	2224.869	44946.47	PT II		1.0		SH
2223.481	2224.173	44960.54	PT II		1.0		SH
2222.61	2223.30	44978.1	PT I	1.4	1.2		MIT
2222.216	2222.908	44986.13	PT	0.5	1.2		MIT
2217.340	2218.031	45085.04	PT I	1.4	1.3		MIT
2215.383	2216.073	45124.87	PT II		1.0		SH
2214.28	2214.97	45147.3	PT		1.3		MIT
2213.93	2214.62	45154.5	PT	0.8			MIT
2211.510	2212.199	45203.88	PT II		0.5W		SH
2209.848	2210.537	45237.88	PT II		1.2		SH
2208.77	2209.46	45260.0	PT	0.7	0.3		MIT
2207.570	2208.258	45284.55	PT II		0.3		SH
2206.730	2207.418	45301.79	PT II		1.3		MIT
2203.882	2204.570	45360.33	PT II		1.3		MIT
2202.582	2203.269	45387.10	PT II		1.1W		MIT
2202.22	2202.91	45394.6	PT I	1.4	0.9		MIT
2202.011	2202.698	45398.86	PT II	0.3	1.3		MIT
2201.289	2201.976	45413.75	PT II		0.8		SH
2201.01	2201.70	45419.5	PT	1.3	1.0		MIT
2199.70	2200.39	45446.6	PT	1.2	1.3		MIT
2197.893	2198.579	45483.92	PT II		1.4		MIT
2196.916	2197.602	45504.14	PT I	1.2	0.5		MIT
2195.835	2196.521	45526.54	PT		1.2		MIT
2192.842	2193.527	45588.67	PT II		1.1		MIT
2192.50	2193.19	45595.8	PT	0.8	0.6		MIT
2190.77	2191.45	45631.8	PT	0.9			MIT
2190.315	2191.000	45641.26	PT II	0.8	1.5		SH
2190.17	2190.85	45644.3	PT I	1.3			MIT
2188.345	2189.029	45682.35	PT	0.5	0.3		MIT
2184.140	2184.824	45770.29	PT II		0.5		SH
2182.77	2183.45	45799.0	PT I	1.3	1.1		MIT
2180.49	2181.17	45846.9	PT I	1.4	1.1		MIT
2180.311	2180.994	45850.66	PT	2.2	1.2		MIT
2176.872	2177.554	45923.08	PT II	0.5	1.4		SH
2176.46	2177.14	45931.8	PT I	1.0	0.3		MIT
2174.67	2175.35	45969.6	PT I	1.5	1.5		MIT
2172.388	2173.069	46017.86	PT	0.3	1.4		SH
2170.72	2171.40	46053.2	PT	1.0	0.6		MIT
2169.563	2170.244	46077.78	PT II		1.1		MIT
2169.26	2169.94	46084.2	PT	0.9	0.5		MIT
2166.63	2167.31	46140.1	PT I	1.4	1.2		MIT
2165.949	2166.629	46154.65	PT II	0.3	1.4		SH
2165.17	2165.85	46171.3	PT I	2.3G	1.4		ME
2164.288	2164.967	46190.07	PT	1.3	0.6		MIT
2157.278	2157.956	46340.15	PT II		1.1		MIT
2153.55	2154.23	46420.4	PT I	1.4	0.9		MIT
2153.38	2154.06	46424.0	PT	0.7			MIT
2152.67	2153.35	46439.3	PT	0.3			MIT
2152.08	2152.76	46452.1	PT I	1.3	1.0		MIT
2150.648	2151.325	46482.99	PT	1.2	0.9		MIT
2150.231	2150.908	46492.00	PT II	0.5	1.4		MIT
2149.703	2150.379	46503.42	PT II		1.3		MIT
2147.39	2148.07	46553.5	PT	0.8			MIT
2145.026	2145.702	46604.80	PT	1.1	0.6		MIT
2144.231	2144.906	46622.08	PT I	1.5	2.0		MIT
2144.231	2144.906	46622.08	PT II	1.5	2.0		MIT
2142.499	2143.174	46659.77	PT II	0.3	1.1		SH
2141.166	2141.841	46688.81	PT II		1.2		MIT
2135.15	2135.82	46820.4	PT	1.1	1.0		MIT
2134.62	2135.29	46832.0	PT	1.1	0.5		MIT
2132.46	2133.13	46879.4	PT I	0.7			MIT
2131.07	2131.74	46910.0	PT	0.5			MIT
2130.689	2131.362	46918.36	PT II	0.6	1.5		SH
2128.61	2129.28	46964.2	PT I	1.5	1.4		MIT

AIR WAVELENGTH (ANGSTRMS)	VACUUM WAVELENGTH (ANGSTRMS)	WAVENUMBER (KAYSERS)	LMENT	LOG ARC	INTENSITY SPK	DIS	REF
2127.417	2128.089	46990.52	PT II	0.3	1.4		MIT
2122.56	2123.23	47098.0	PT	1.2	1.0		MIT
2119.88	2120.55	47157.6	PT	1.2	1.0		MIT
2115.578	2116.248	47253.45	PT II	0.7	1.0		MIT
2114.22	2114.89	47283.8	PT	0.8	0.5		MIT
2110.355	2111.024	47370.38	PT II		1.3		SH
2109.66	2110.33	47386.0	PT I	1.2	1.0		MIT
2109.49	2110.16	47389.8	PT	1.2	1.0		MIT
2107.651	2108.319	47431.15	PT II		0.9		SH
2105.069	2105.737	47489.32	PT II	0.3	1.4		MIT
2103.761	2104.428	47518.84	PT II	0.3	1.4		MIT
2103.48	2104.15	47525.2	PT	1.0			MIT
2103.33	2104.00	47528.6	PT I	1.4	1.4		MIT
2101.67	2102.34	47566.1	PT	1.3			MIT
2101.585	2102.252	47568.04	PT II		1.5		MIT
2098.74	2099.41	47632.5	PT	1.1	0.5		MIT
2097.440	2098.106	47662.03	PT II	1.0	1.5		MIT
2097.27	2097.94	47665.9	PT	1.0			MIT
2094.70	2095.37	47724.4	PT		0.6		MIT
2094.08	2094.75	47738.5	PT	0.3			MIT
2089.048	2089.712	47853.47	PT II		1.2		SH
2088.93	2089.59	47856.2	PT	1.0	0.0		MIT
2088.712	2089.376	47861.17	PT II	0.3	1.4		SH
2088.29	2088.95	47870.8	PT	1.1	0.6		MIT
2086.866	2087.530	47903.50	PT	0.5	1.3		MIT
2085.418	2086.082	47936.76	PT II		1.2		SH
2084.59	2085.25	47955.8	PT I	1.4	1.5		MIT
2083.34	2084.00	47984.6	PT	1.3	1.0		MIT
2082.51	2083.17	48003.7	PT I	1.2	0.8		MIT
2079.762	2080.425	48067.11	PT II	0.0	1.2		MIT
2079.496	2080.159	48073.26	PT II		1.2		MIT
2077.435	2078.097	48120.94	PT II	1.0	1.3		SH
2076.21	2076.87	48149.3	PT	0.7	0.6		MIT
2075.386	2076.048	48168.45	PT II	0.7	1.4		MIT
2073.58	2074.24	48210.4	PT	1.0	1.0		MIT
2071.553	2072.214	48257.56	PT II		1.3		MIT
2070.93	2071.59	48272.1	PT I	1.3	1.3		MIT
2070.36	2071.02	48285.4	PT	0.3			LV
2068.629	2069.290	48325.77	PT II	0.7	1.2		MIT
2068.160	2068.820	48336.72	PT II	0.3	1.4		MIT
2067.50	2068.16	48352.1	PT I	1.3	1.4		MIT
2066.92	2067.58	48365.7	PT	1.2	0.8		MIT
2062.78	2063.44	48462.8	PT I	1.3	1.3		MIT
2061.715	2062.374	48487.81	PT II		0.7		SH
2061.630	2062.289	48489.80	PT II		1.4		MIT
2060.75	2061.41	48510.5	PT II	1.2	1.1		MIT
2059.894	2060.553	48530.66	PT II		1.4		SH
2059.68	2060.34	48535.7	PT I	1.2	0.7		MIT
2058.973	2059.632	48552.37	PT II		1.3		SH
2058.38	2059.04	48566.4	PT	1.0	0.9		MIT
2057.003	2057.661	48598.86	PT II	0.9	1.5		SH
2056.057	2056.715	48621.22	PT II		1.1		SH
2049.83	2050.49	48768.9	PT I	1.0	1.0		MIT
2049.373	2050.030	48779.77	PT I	1.3	1.4R		MIT
2049.168	2049.825	48784.65	PT II		1.4		MIT
2041.570	2042.225	48966.19	PT II	0.3	1.5		MIT
2041.45	2042.11	48969.1	PT	1.0			MIT
2040.33	2040.99	48995.9	PT	1.2	1.2		MIT
2039.70	2040.36	49011.1	PT	1.1	1.3		MIT
2039.57	2040.23	49014.2	PT	0.7			A
2036.463	2037.118	49088.97	PT II	1.2	1.6		MIT
2035.78	2036.43	49105.4	PT	1.2	1.2		MIT
2032.41	2033.06	49186.9	PT	1.3	1.4		MIT
2031.446	2032.100	49210.19	PT II		1.3		MIT
2030.63	2031.28	49230.0	PT	1.2	1.4		MIT
2028.58	2029.23	49279.7	PT I	1.0	0.9		MIT
2028.31	2028.96	49286.3	PT	1.1	0.7		MIT
2027.24	2027.89	49312.3	PT	0.5	0.5		MIT
2025.86	2026.51	49345.9	PT	1.0	0.5		MIT
2025.159	2025.812	49362.93	PT II	0.3	1.0		SH
2024.63	2025.28	49375.8	PT	0.9	0.0		MIT
2023.53	2024.18	49402.7	PT	0.9	0.3		A
2020.92	2021.57	49466.5	PT	0.3H			MIT
2018.329	2018.980	49529.95	PT II	0.7W	1.4		MIT
2016.891	2017.542	49565.26	PT II		1.3		SH
2016.71	2017.36	49569.7	PT	0.9	0.9		MIT
2014.927	2015.578	49613.57	PT II	0.7	1.5		MIT
2013.98	2014.63	49636.9	PT	0.5	1.3		A
2012.58	2013.23	49671.4	PT	1.0	0.8		A
2011.04	2011.69	49709.5	PT	0.8			A
2007.55	2008.20	49795.9	PT		0.8		A
2007.36	2008.01	49800.6	PT	1.0			A
2004.335	2004.984	49875.71	PT II		1.4		MIT
2004.13	2004.78	49880.8	PT I	1.1	0.7		MIT
2003.146	2003.795	49905.31	PT II	0.7	1.2		SH
9932.21	9934.93	10065.5	RA I			0.8	RS
8693.94	8696.33	11499.1	RA I			0.7	RS
8335.07	8337.36	11994.2	RA			0.8	RS
8269.03	8271.30	12090.0	RA			0.8	RS
8248.70	8250.97	12119.8	RA			0.8	RS
8177.31	8179.56	12225.6	RA I			1.0	RS
8019.70	8021.91	12465.9	RA II			2.3	RS
8019.25	8021.46	12466.6	RA			0.5	RS
8005.13	8007.33	12488.6	RA			0.8	RS
7896.43	7898.60	12660.5	RA			0.8	RS
7877.08	7879.25	12691.6	RA			0.5	RS
7838.12	7840.28	12754.7	RA I			2.0	RS
7565.49	7567.57	13214.3	RA			0.5	RS
7499.87	7501.94	13329.9	RA			1.2	RS
7310.27	7312.28	13675.6	RA I			2.7	RS
7225.16	7227.15	13836.7	RA I			3.0	RS
7141.21	7143.18	13999.4	RA I			3.3	RS
7118.50	7120.46	14044.0	RA I			3.0	RS
7078.02	7079.97	14124.3	RA II			1.8	RS
6980.22	6982.15	14322.2	RA I			3.0	RS
6903.1	6905.0	14482.	RA I			1.5	RS
6758.2	6760.1	14793.	RA I			1.7	RS
6719.32	6721.17	14878.3	RA II			2.7	RS
6653.33	6655.17	15025.9	RA			1.2	RS
6645.95	6647.79	15042.6	RA			1.2	RS
6599.47	6601.29	15148.6	RA I			1.5	RS
6593.34	6595.16	15162.6	RA II			2.7	RS
6585.41	6587.23	15180.9	RA			1.7	RS
6545.93	6547.74	15272.4	RA			1.5	RS
6532.08	6533.88	15304.8	RA I			1.5	RS
6528.92	6530.72	15312.2	RA I			1.2	RS
6487.32	6489.11	15410.4	RA I			3.0	RS
6474.47	6476.26	15441.0	RA			1.2	RS
6446.20	6447.98	15508.7	RA I			3.0	RS
6438.9	6440.7	15526.	RA I			1.5	RS
6336.90	6338.65	15776.2	RA I			2.7	RS
6247.16	6248.89	16002.8	RA II			1.7	RS
6200.30	6202.02	16123.8	RA I			3.0	RS
6167.03	6168.74	16210.8	RA I			1.8	RS
6151.19	6152.89	16252.5	RA I			1.5	RS
5957.67	5959.32	16780.4	RA			1.5	RS
5957.17	5958.82	16781.9	RA			0.9	RS
5907.2	5908.8	16924.	RA I			0.9	RS
5823.49	5825.10	17167.1	RA I			1.2	RS
5813.63	5815.24	17196.2	RA II			2.7	RS
5811.58	5813.19	17202.3	RA I			1.5	RS
5795.78	5797.39	17249.1	RA I			1.5	RS
5778.28	5779.88	17301.4	RA I			1.5	RS
5755.45	5757.05	17370.0	RA			1.4	RS
5728.83	5730.42	17450.7	RA II			1.4	RS
5690.16	5691.74	17569.3	RA I			1.2	RS
5661.73	5663.30	17657.6	RA II			1.7	RS
5660.81	5662.38	17660.4	RA I			3.0	RS
5623.43	5624.99	17777.8	RA II			1.2	RS
5620.47	5622.03	17787.2	RA I			1.2	RS

AIR WAVELENGTH (ANGSTRMS)	VACUUM WAVELENGTH (ANGSTRMS)	WAVENUMBER (KAYSERS)	LMENT	LOG INTENSITY ARC	SPK	DIS	REF
5616.66	5618.22	17799.2	RA I			2.4	RS
5601.5	5603.1	17847.	RA I			1.5	RS
5591.15	5592.70	17880.4	RA I			0.9	RS
5555.85	5557.39	17994.1	RA I			2.7	RS
5553.57	5555.11	18001.4	RA I			2.4	RS
5544.25	5545.79	18031.7	RA			1.2	RS
5505.50	5507.03	18158.6	RA I			1.4	RS
5501.98	5503.51	18170.2	RA I			2.4	RS
5497.83	5499.36	18183.9	RA			0.9	RS
5488.32	5489.84	18215.4	RA I			1.4	RS
5482.13	5483.65	18236.0	RA I			2.0	RS
5406.81	5408.31	18490.1	RA I			2.7	RS
5400.23	5401.73	18512.6	RA I			2.7	RS
5399.80	5401.30	18514.1	RA I			2.4	RS
5320.29	5321.77	18790.7	RA I			2.4	RS
5283.28	5284.75	18922.4	RA I			2.4	RS
5277.91	5279.38	18941.6	RA I			0.9	RS
5263.96	5265.43	18991.8	RA			1.4	RS
5205.93	5207.38	19203.5	RA I			2.4	RS
5097.56	5098.98	19611.8	RA			2.4	RS
5081.03	5082.45	19675.6	RA I			1.7	RS
5066.57	5067.98	19731.7	RA II			1.2	RS
5041.56	5042.97	19829.6	RA I			1.5	RS
4997.26	4998.65	20005.4	RA II			0.8	RS
4982.03	4983.42	20066.5	RA			1.0	RS
4971.77	4973.16	20107.9	RA			1.2	RS
4927.53	4928.91	20288.5	RA II			2.0	RS
4903.24	4904.61	20389.0	RA I			1.0	RS
4882.28	4883.64	20476.5	RA I			0.8	RS
4862.27	4863.63	20560.8	RA I			0.6	RS
4859.41	4860.77	20572.9	RA II			2.0	RS
4856.57	4857.93	20584.9	RA I			2.0	RS
4837.27	4838.62	20667.0	RA I			0.8	RS
4825.91	4827.26	20715.7	RA I			2.9	RS
4803.11	4804.45	20814.0	RA			0.8	RS
4740.07	4741.40	21090.8	RA			0.6	RS
4701.97	4703.29	21261.7	RA I			0.6	RS
4699.28	4700.59	21273.9	RA I			1.6	RS
4682.28	4683.59	21351.1	RA II			2.9	RS
4663.52	4664.83	21437.0	RA II			0.8	RS
4641.29	4642.59	21539.7	RA I			1.6	RS
4533.11	4534.38	22053.7	RA II			2.5	RS
4436.27	4437.52	22535.1	RA II			2.3	RS
4426.35	4427.59	22585.6	RA			0.6	RS
4366.30	4367.53	22896.2	RA			0.6	RS
4340.64	4341.86	23031.6	RA II			3.0	RS
4305.00	4306.21	23222.3	RA I			1.0	RS
4265.12	4266.32	23439.4	RA I			0.8	RS
4244.72	4245.92	23552.1	RA II			1.6	RS
4195.56	4196.74	23828.0	RA II			0.6	RS
4194.09	4195.27	23836.4	RA II			1.9	RS
4177.98	4179.16	23928.3	RA I			1.0	RS
3894.55	3895.65	25669.6	RA II			1.4	RS
3851.90	3852.99	25953.9	RA II			1.4	RS
3814.42	3815.50	26208.9	RA II			3.3	RS
3771.57	3772.64	26506.6	RA			1.0	RS
3686.16	3687.21	27120.8	RA II			1.0	RS
3649.55	3650.59	27392.8	RA II			3.0	RS
3550.2	3551.2	28159.	RA II			0.7	RS
3514.8	3515.8	28443.	RA II			0.7	RS
3456.2	3457.2	28925.	RA II			0.7	RS
3423.1	3424.1	29205.	RA II			0.7	RS
3101.80	3102.70	32230.0	RA I			1.9	RS
3033.44	3034.32	32956.3	RA II			2.2	RS
2836.46	2837.29	35244.9	RA II			1.4	RS
2813.76	2814.59	35529.2	RA II			2.6	RS
2795.21	2796.03	35764.9	RA II			2.1	RS
2708.96	2709.76	36903.6	RA II			2.3	RS
2643.73	2644.52	37814.1	RA II			2.1	RS
2595.15	2595.93	38521.9	RA II			0.6	RS

AIR WAVELENGTH (ANGSTRMS)	VACUUM WAVELENGTH (ANGSTRMS)	WAVENUMBER (KAYSERS)	LMENT	LOG INTENSITY ARC	SPK	DIS	REF
2586.61	2587.38	38649.1	RA II			1.7	RS
2480.11	2480.86	40308.6	RA II			1.1	RS
2475.50	2476.25	40383.7	RA II			2.1	RS
2460.55	2461.29	40629.0	RA II			1.7	RS
2377.10	2377.83	42055.2	RA II			0.9	RS
2369.73	2370.45	42186.0	RA II			1.7	RS
2223.3	2224.0	44964.	RA II			0.9	RS
2197.8	2198.5	45486.	RA II			0.6	RS
2177.3	2178.0	45914.	RA II			1.1	RS
2169.9	2170.6	46071.	RA II			2.1	RS
2131.0	2131.7	46911.	RA II			1.3	RS
2107.6	2108.3	47432.	RA II			1.1	RS
2070.6	2071.3	48280.	RA II			0.6	RS
2012.75	2013.40	49667.2	RA II			0.9	RS
2006.4	2007.0	49824.	RA II			0.6	RS
9540.8	9543.4	10478.	RB I	0.7X			ME
9523.4	9526.0	10498.	RB I	1.0X			ME
8868.846	8871.281	11272.33	RB I	1.5			IRZ
8868.508	8870.943	11272.76	RB I	1.8			IRZ
8861.5	8863.9	11282.	RB I	1.3X			ME
8271.707	8273.981	12086.08	RB I	2.0			IRZ
8271.410	8273.684	12086.52	RB I	2.3			IRZ
7947.60	7949.79	12578.9	RB I	3.7G			IHZ
7925.537	7927.717	12613.97	RB I	1.8			IRZ
7925.260	7927.440	12614.41	RB I	2.0			IRZ
7800.227	7802.373	12816.61	RB I	4.0G			IHZ
7759.434	7761.569	12883.99	RB I	2.6			IRZ
7757.651	7759.786	12886.95	RB I	3.0			IRZ
7618.933	7621.030	13121.58	RB I	3.0			IRZ
7408.170	7410.211	13494.89	RB I	2.7			IRZ
7316.505	7318.521	13663.96	RB II			1.7	RR
7279.997	7282.003	13732.48	RB I	2.6	1.7		IRZ
7042.450	7044.392	14195.69	RB II		2.2		RR
6805.646	6807.524	14689.63	RB II		1.7		RR
6775.062	6776.932	14755.94	RB II		2.3		LP
6560.837	6562.649	15237.75	RB		2.2		RR
6555.625	6557.436	15249.86	RB II		2.0		LP
6498.314	6500.110	15384.36	RB		1.7		RR
6458.347	6460.132	15479.56	RB II		2.6		LP
6310.043	6311.788	15843.37	RB		1.7		RR
6299.225	6300.967	15870.58	RB I	2.5	1.7		IRZ
6298.327	6300.069	15872.84	RB I	3.0	2.2		IRZ
6206.309	6208.026	16108.18	RB I	2.9	2.0		IRZ
6199.093	6200.808	16126.93	RB II		2.0		LP
6159.622	6161.327	16230.27	RB I	2.6			IRZ
6070.751	6072.432	16467.87	RB I	2.8	1.7		IRZ
5724.947	5726.535	17462.57	RB I	1.7			IRZ
5724.453	5726.041	17464.07	RB I	2.8			IRZ
5699.159	5700.740	17541.58	RB II		2.0		RR
5653.744	5655.313	17682.49	RB I	2.3			IRZ
5648.097	5649.665	17700.17	RB I	2.6			IRZ
5635.994	5637.558	17738.18	RB II		2.0		RR
5578.783	5580.332	17920.08	RB I	2.2			IRZ
5522.789	5524.323	18101.76	RB II		2.0		RR
5512.566	5514.097	18135.33	RB		1.5		RR
5431.828	5433.338	18404.89	RB I	1.2			IRZ
5431.529	5433.039	18405.91	RB I	2.0			IRZ
5390.562	5392.061	18545.78	RB I	1.4H			IRZ
5362.598	5364.089	18642.49	RB I	1.7			IRZ
5322.374	5323.855	18783.38	RB I	1.0H			IRZ
5270.508	5271.975	18968.22	RB II		1.2		RR
5260.224	5261.688	19005.31	RB I	0.7			IRZ
5260.031	5261.495	19006.00	RB I	1.3			IRZ
5233.961	5235.418	19100.67	RB I	1.2			IRZ
5195.273	5196.720	19242.91	RB I	1.3H			IRZ
5169.651	5171.091	19338.28	RB I	0.7			IRZ
5164.592	5166.031	19357.22	RB II		0.7		RR
5152.094	5153.529	19404.18	RB II		2.0		RR
4885.627	4886.991	20462.49	RB II		1.0		RR
4855.361	4856.717	20590.04	RB II		1.3		RR

AIR WAVELENGTH (ANGSTRMS)	VACUUM WAVELENGTH (ANGSTRMS)	WAVENUMBER (KAYSERS)	LMENT	LOG ARC	INTENSITY SPK	INTENSITY DIS	REF
4842.2	4843.6	20646.	RB			0.6	DR
4782.871	4784.208	20902.10	RB II		1.4		RR
4776.410	4777.745	20930.37	RB II			2.0	RR
4775.998	4777.333	20932.18	RB II		1.4		RR
4757.853	4759.184	21012.01	RB II		1.0		RR
4755.329	4756.659	21023.16	RB II			1.0	RR
4736.6	4737.9	21106.	RB			1.1	DR
4731.3	4732.6	21130.	RB			0.9	DR
4730.479	4731.802	21133.60	RB II		1.0		RR
4727.2	4728.5	21148.	RB			0.7	DR
4659.320	4660.624	21456.35	RB II			0.9	RR
4657.9	4659.2	21463.	RB			0.9	DR
4648.562	4649.864	21506.01	RB II		1.3		RR
4631.918	4633.215	21583.28	RB II		0.7		RR
4622.447	4623.742	21627.51	RB II		1.7		RR
4609.5	4610.8	21688.	RB			1.2	DR
4590.3	4591.6	21779.	RB			0.9	DA
4572.162	4573.443	21865.36	RB II		0.5		RR
4571.790	4573.071	21867.14	RB II		1.5		RR
4566.8	4568.1	21891.	RB			0.6	DR
4540.771	4542.044	22016.52	RB II		1.0		RR
4533.824	4535.095	22050.25	RB II		0.8		RR
4530.358	4531.628	22067.12	RB II		1.2		RR
4519.071	4520.338	22122.24	RB		1.0		RR
4493.949	4495.210	22245.90	RB II		0.9		RR
4482.1	4483.4	22305.	RB			0.6	DR
4469.516	4470.770	22367.51	RB II		1.0		RR
4440.130	4441.377	22515.54	RB		1.0		RR
4430.7	4431.9	22563.	RB			1.3	DR
4426.1	4427.3	22587.	RB			1.5	DR
4420.8	4422.0	22614.	RB			0.3	DR
4401.4	4402.6	22714.	RB			1.5	DR
4380.7	4381.9	22821.	RB			1.3	DR
4377.150	4378.380	22839.50	RB II		1.0		RR
4371.8	4373.0	22867.	RB			1.3	DR
4352.5	4353.7	22969.	RB			0.9	DR
4348.3	4349.5	22991.	RB			1.9	DR
4346.996	4348.218	22997.93	RB II		0.9		RR
4346.582	4347.804	23000.12	RB II		1.3		RR
4331.5	4332.7	23080.	RB			1.0	DR
4316.949	4318.163	23158.00	RB		0.3		RR
4306.299	4307.510	23215.27	RB II		1.0H		RR
4298.9	4300.1	23255.	RB			0.8	DR
4294.567	4295.775	23278.69	RB II		0.3		RR
4294.362	4295.570	23279.80	RB II		0.5		RR
4293.994	4295.202	23281.79	RB II		1.6		RR
4293.484	4294.692	23284.56	RB II		0.3		RR
4282.8	4284.0	23343.	RB			0.9	DR
4277.3	4278.5	23373.	RB			0.9	DR
4273.703	4274.906	23392.33	RB II		0.3		RR
4273.524	4274.727	23393.31	RB II		0.3		RR
4273.176	4274.379	23395.21	RB II		1.4		RR
4272.640	4273.842	23398.15	RB II		0.3		RR
4270.303	4271.505	23410.95	RB II		0.9		RR
4266.622	4267.823	23431.15	RB II		1.0		RR
4249.085	4250.281	23527.85	RB II		1.2		RR
4247.7	4248.9	23535.	RB			0.9	DR
4244.981	4246.176	23550.60	RB II		0.3		RR
4244.800	4245.995	23551.60	RB II		0.3		RR
4244.436	4245.631	23553.62	RB II		1.4		RR
4243.888	4245.083	23556.67	RB II		0.3		RR
4242.6	4243.8	23564.	RB			2.2	DR
4237.2	4238.4	23594.	RB			0.9	DR
4226.1	4227.3	23656.	RB			0.9	DR
4215.556	4216.743	23714.98	RB I	3.06	2.5		RR
4201.851	4203.035	23792.33	RB I	3.36	2.7		RR
4193.612	4194.794	23839.07	RB II		0.3		RR
4193.467	4194.649	23839.90	RB II		0.3		RR
4193.097	4194.279	23842.00	RB II		1.6		RR
4192.566	4193.747	23845.02	RB II		0.3		RR
4136.439	4137.606	24168.57	RB		0.5		RR
4136.125	4137.292	24170.40	RB II		1.2		RR
4104.823	4105.981	24354.71	RB		0.5		RR
4104.665	4105.823	24355.65	RB		0.3		RR
4104.313	4105.471	24357.74	RB II		1.3		RR
4083.927	4085.080	24479.32	RB II		1.1		RR
4071.9	4073.0	24552.	RB			1.1	DR
4047.4	4048.5	24700.	RB			0.6	DR
4041.4	4042.5	24737.	RB			0.6	DR
4029.562	4030.701	24809.58	RB II		1.2		RR
4026.9	4028.0	24826.	RB			1.4	DR
4011.9	4013.0	24919.	RB			1.2	DR
3978.207	3979.332	25129.84	RB II			1.6	RR
3940.568	3941.683	25369.87	RB II			2.3	RR
3926.489	3927.601	25460.84	RB II			1.2	RR
3922.259	3923.370	25488.29	RB II			1.0	RR
3861.2	3862.3	25891.	RB			0.3	DR
3860.796	3861.891	25894.05	RB II			0.7	RR
3851.2	3852.3	25959.	RB			1.3	DR
3837.910	3838.999	26048.46	RB II			0.3	RR
3827.2	3828.3	26121.	RB			1.6	DR
3826.708	3827.794	26124.71	RB II			1.2	RR
3809.6	3810.7	26242.	RB			0.3	DR
3801.925	3803.004	26295.00	RB II			1.3	RR
3797.276	3798.354	26327.19	RB II			0.3	RR
3797.170	3798.248	26327.93	RB II			0.3	RR
3796.823	3797.901	26330.33	RB II			1.6	RR
3796.393	3797.471	26333.32	RB II			0.3	RR
3781.1	3782.2	26440.	RB			0.5	DR
3746.381	3747.446	26684.84	RB II			1.0	RR
3728.666	3729.726	26811.62	RB			1.3	RR
3715.640	3716.697	26905.61	RB II			0.3	RR
3699.622	3700.675	27022.10	RB II			1.2	RR
3675.718	3676.765	27197.82	RB			0.7	RR
3666.774	3667.818	27264.16	RB II			0.3	RR
3663.859	3664.903	27285.85	RB II			1.2	RR
3662.784	3663.827	27293.86	RB II			1.2	RR
3647.616	3648.655	27407.36	RB II			0.3	RR
3646.321	3647.360	27417.09	RB II			1.0	RR
3640.225	3641.262	27463.00	RB II			0.3	RR
3639.860	3640.897	27465.76	RB II			1.2	RR
3636.831	3637.868	27488.63	RB			0.5	RR
3604.6	3605.6	27734.	RB			0.3	DR
3600.678	3601.705	27764.63	RB			1.3	RR
3595.91	3596.94	27801.4	RB II			0.3	OK
3591.59	3592.61	27834.9	RB I	1.9	1.3		FL
3588.8	3589.8	27856.	RB			0.8	DR
3587.08	3588.10	27869.9	RB I	2.3	1.6		FL
3577.959	3578.980	27940.92	RB II			1.2	RR
3577.41	3578.43	27945.2	RB			1.3	OK
3576.8	3577.8	27950.	RB			0.8	DR
3557.800	3558.816	28099.23	RB I			1.2	RR
3545.37	3546.38	28197.7	RB			0.3	OK
3541.216	3542.228	28230.82	RB II			2.0	RR
3531.602	3532.611	28307.67	RB II			2.0	RR
3521.439	3522.446	28389.36	RB II			2.3	RR
3516.53	3517.54	28429.0	RB II			0.7	OK
3513.88	3514.88	28450.4	RB II			1.0	OK
3511.19	3512.19	28472.2	RB			1.8	OK
3492.765	3493.764	28622.42	RB			2.5	RR
3483.84	3484.84	28695.7	RB			0.7	OK
3480.71	3481.71	28721.6	RB II			1.0	OK
3474.46	3475.45	28773.2	RB			1.3	OK
3461.574	3462.565	28880.32	RB II			2.3	RR
3439.340	3440.326	29067.01	RB			2.3	RR
3434.72	3435.70	29106.1	RB			1.6	OK
3434.263	3435.247	29109.98	RB II			1.8	RR
3421.16	3422.14	29221.5	RB			1.3	OK
3415.648	3416.628	29268.63	RB II			1.5	RR
3400.18	3401.16	29401.8	RB			0.7	OK

AIR WAVELENGTH (ANGSTRMS)	VACUUM WAVELENGTH (ANGSTRMS)	WAVENUMBER (KAYSERS)	LMENT		LOG INTENSITY ARC	SPK	DIS	REF
3393.111	3394.085	29463.02	RB	II			1.6	RR
3372.61	3373.58	29642.1	RB				1.5	OK
3361.49	3362.46	29740.2	RB				1.0	OK
3353.98	3354.94	29806.8	RB				1.5	OK
3350.89	3351.85	29834.2	RB	I	2.2			FL
3348.72	3349.68	29853.6	RB	I	2.0			FL
3347.00	3347.96	29868.9	RB				1.8	OK
3344.73	3345.69	29889.2	RB				1.2	OK
3340.605	3341.566	29926.09	RB	II			1.8	RR
3333.61	3334.57	29988.9	RB				1.6	OK
3331.62	3332.58	30006.8	RB				1.6	OK
3330.27	3331.23	30019.0	RB				1.6	OK
3329.91	3330.87	30022.2	RB	II			0.7	OK
3327.17	3328.13	30046.9	RB				1.2	OK
3323.34	3324.30	30081.6	RB				1.0	OK
3322.3	3323.3	30091.	RB				1.2	DR
3321.545	3322.501	30097.81	RB	II			1.8	RR
3306.93	3307.88	30230.8	RB				1.2	OK
3304.42	3305.37	30253.8	RB				1.3	OK
3300.73	3301.68	30287.6	RB	II			1.0	OK
3286.47	3287.42	30419.0	RB				1.8	OK
3281.49	3282.44	30465.2	RB	II			1.3	OK
3271.03	3271.97	30562.6	RB	II			1.6	OK
3268.33	3269.27	30587.9	RB				0.3	OK
3264.85	3265.79	30620.4	RB				1.3	OK
3262.15	3263.09	30645.8	RB				1.5	OK
3253.82	3254.76	30724.2	RB				1.0	OK
3234.64	3235.57	30906.4	RB				0.7	OK
3230.55	3231.48	30945.6	RB				1.2	OK
3229.112	3230.044	30959.33	RB	I	1.0H			MIT
3228.001	3228.933	30969.98	RB	I	1.3			MIT
3223.4	3224.3	31014.	RB				1.3	DR
3222.64	3223.57	31021.5	RB				1.6	OK
3221.70	3222.63	31030.6	RB				1.0	OK
3200.04	3200.96	31240.6	RB				1.0	OK
3198.77	3199.69	31253.0	RB				1.8	OK
3169.43	3170.35	31542.3	RB				1.2	OK
3161.11	3162.03	31625.3	RB	II			0.3	OK
3158.105	3159.019	31655.39	RB	I	0.9			MIT
3157.56	3158.47	31660.9	RB	I	1.3			BV
3151.66	3152.57	31720.1	RB				0.3	OK
3148.98	3149.89	31747.1	RB	II			1.3	OK
3134.90	3135.81	31889.7	RB				1.0	OK
3122.03	3122.94	32021.2	RB				0.3	OK
3119.00	3119.90	32052.3	RB				1.3	OK
3114.88	3115.78	32094.7	RB				1.3	OK
3112.83	3113.73	32115.8	RB	I	1.2			BV
3111.43	3112.33	32130.2	RB				1.5	OK
3106.50	3107.40	32181.2	RB				0.3	OK
3099.73	3100.63	32251.5	RB				1.0	OK
3098.55	3099.45	32263.8	RB	II			0.7	OK
3086.91	3087.81	32385.4	RB				1.3	OK
3082.27	3083.17	32434.2	RB	I	1.0			BV
3078.50	3079.39	32473.9	RB				0.3	OK
3070.76	3071.65	32555.8	RB				1.2	OK
3069.82	3070.71	32565.7	RB				0.7	OK
3065.25	3066.14	32614.3	RB				0.3	OK
3060.50	3061.39	32664.9	RB	I	1.1			BV
3051.43	3052.32	32762.0	RB	II			0.3	OK
3044.21	3045.10	32839.7	RB	I	0.8			BV
3039.68	3040.56	32888.6	RB				1.2	OK
3032.08	3032.96	32971.1	RB	I	0.6			BV
3024.89	3025.77	33049.4	RB				0.3	OK
3023.66	3024.54	33062.9	RB				1.3	OK
3022.60	3023.48	33074.5	RB	I			1.0	BV
3005.88	3006.76	33258.4	RB				0.3	OK
2999.56	3000.43	33328.5	RB	I			0.3	OK
2996.391	2997.265	33363.75	RB	I	0.7			MIT
2970.80	2971.67	33651.1	RB				1.6	OK
2968.16	2969.03	33681.1	RB				1.3	OK

AIR WAVELENGTH (ANGSTRMS)	VACUUM WAVELENGTH (ANGSTRMS)	WAVENUMBER (KAYSERS)	LMENT		LOG INTENSITY ARC	SPK	DIS	REF
2967.51	2968.38	33688.5	RB				1.6	OK
2956.12	2956.98	33818.2	RB				1.8	OK
2951.05	2951.91	33876.3	RB				0.7	OK
2949.68	2950.54	33892.1	RB				1.0	OK
2904.18	2905.03	34423.0	RB				0.3	OK
2903.72	2904.57	34428.5	RB				1.0	OK
2876.73	2877.57	34751.5	RB	II			0.7	OK
2873.93	2874.77	34785.4	RB				0.3	OK
2869.82	2870.66	34835.2	RB				1.0	OK
2845.52	2846.36	35132.6	RB				1.0	OK
2827.00	2827.83	35362.8	RB				0.7	OK
2816.81	2817.64	35490.7	RB				0.7	OK
2812.18	2813.01	35549.1	RB				0.7	OK
2807.63	2808.46	35606.7	RB				1.8	OK
2800.32	2801.15	35699.7	RB				1.0	OK
2798.91	2799.74	35717.7	RB				1.6	OK
2783.84	2784.66	35911.0	RB				1.0	OK
2773.9	2774.7	36040.	RB				0.3	DR
2763.14	2763.96	36180.0	RB				0.3	OK
2761.13	2761.95	36206.4	RB				0.7	OK
2745.57	2746.38	36411.5	RB				0.3	OK
2742.42	2743.23	36453.4	RB				0.3	OK
2733.61	2734.42	36570.8	RB				0.7	OK
2732.25	2733.06	36589.0	RB				0.3	OK
2716.54	2717.34	36800.6	RB				0.3	OK
2713.92	2714.72	36836.1	RB				1.0	OK
2711.82	2712.62	36864.7	RB				1.0	OK
2706.08	2706.88	36942.9	RB				0.3	OK
2693.91	2694.71	37109.7	RB				0.3	OK
2688.08	2688.88	37190.2	RB				0.3	OK
2684.17	2684.97	37244.4	RB				0.7	OK
2656.70	2657.49	37629.5	RB				1.0	OK
2636.87	2637.66	37912.5	RB				1.0	OK
2635.61	2636.40	37930.6	RB				0.3	OK
2631.8	2632.6	37985.	RB				1.6	OK
2586.91	2587.68	38644.6	RB				0.7	OK
2561.92	2562.69	39021.5	RB				1.0	OK
2524.0	2524.8	39608.	RB				0.3	DR
2495.9	2496.7	40054.	RB				1.3	DR
2484.4	2485.1	40239.	RB				0.7	DR
2472.0	2472.7	40441.	RB				1.3	DR
2418.6	2419.3	41334.	RB				0.7	DR
2385.6	2386.3	41905.	RB				0.3	DR
2381.30	2382.03	41981.1	RB				2.0	FA
2380.44	2381.17	41996.2	RB				2.1	FA
2377.85	2378.58	42042.0	RB				0.7	FA
2373.22	2373.94	42124.0	RB				1.8	FA
2370.96	2371.68	42164.1	RB				0.3	FA
2367.52	2368.24	42225.4	RB				1.6	FA
2366.63	2367.35	42241.3	RB				0.3	FA
2365.97	2366.69	42253.0	RB				1.9	FA
2365.08	2365.80	42269.0	RB				1.8	FA
2364.28	2365.00	42283.2	RB				2.2	FA
2362.34	2363.06	42318.0	RB				1.0	FA
2358.05	2358.77	42395.0	RB				0.7	FA
2356.98	2357.70	42414.2	RB				1.6	FA
2353.10	2353.82	42484.1	RB				1.6	FA
2352.56	2353.28	42493.9	RB				0.3	FA
2351.50	2352.22	42513.0	RB				0.7	FA
2349.80	2350.52	42543.8	RB				1.9	FA
2345.35	2346.07	42624.5	RB				2.0	FA
2341.89	2342.61	42687.5	RB				1.6	FA
2337.05	2337.77	42775.9	RB				2.1	FA
2335.42	2336.14	42805.7	RB				0.7	FA
2333.38	2334.10	42843.1	RB				1.9	FA
2333.03	2333.75	42849.6	RB				0.3	FA
2331.29	2332.01	42881.5	RB				0.7	FA
2330.04	2330.75	42904.5	RB				0.7	FA
2323.74	2324.45	43020.9	RB				1.6	FA
2314.56	2315.27	43191.5	RB				1.0	FA

AIR WAVELENGTH (ANGSTRMS)	VACUUM WAVELENGTH (ANGSTRMS)	WAVENUMBER (KAYSERS)	LMENT	LOG INTENSITY ARC	SPK	DIS	REF
2313.82	2314.53	43205.3	RB			0.7	FA
2312.45	2313.16	43230.9	RB			2.0	FA
2308.42	2309.13	43306.4	RB			0.7	FA
2307.27	2307.98	43327.9	RB			1.3	FA
2304.8	2305.5	43374.	RB			0.7	DR
2304.27	2304.98	43384.3	RB			1.3	FA
2304.14	2304.85	43386.8	RB			2.1	FA
2301.0	2301.7	43446.	RB			0.3	DR
2300.14	2300.85	43462.2	RB			1.6	FA
2298.86	2299.57	43486.4	RB			1.3	FA
2294.99	2295.70	43559.7	RB			0.7	FA
2292.2	2292.9	43613.	RB			0.3	DR
2291.76	2292.47	43621.1	RB			1.9	FA
2286.97	2287.68	43712.5	RB			0.7D	FA
2268.04	2268.74	44077.3	RB			1.3	FA
2264.00	2264.70	44156.0	RB			1.0	FA
2263.62	2264.32	44163.4	RB			0.3	FA
2254.24	2254.94	44347.1	RB			1.8	FA
2251.48	2252.18	44401.5	RB			0.3	FA
2250.73	2251.43	44416.3	RB			0.7	FA
2249.01	2249.71	44450.2	RB			0.3	FA
2234.37	2235.06	44741.5	RB			0.3	FA
2226.15	2226.84	44906.6	RB			0.7D	FA
2217.12	2217.81	45089.5	RB			2.0	FA
2198.29	2198.98	45475.7	RB			0.7	FA
2198.01	2198.70	45481.5	RB			0.7	FA
2182.2	2182.9	45811.	RB			0.7	FA
2180.6	2181.3	45845.	RB			0.3	FA
2172.0	2172.7	46026.	RB			0.3	FA
2164.6	2165.3	46183.	RB			0.7	FA
9976.45	9979.19	10020.9	RE	0.6			ME
9964.90	9967.63	10032.5	RE	0.3			ME
9955.45	9958.18	10042.0	RE I	1.8	O		ME
9953.02	9955.75	10044.4	RE	0.5	W		ME
9949.90	9952.63	10047.6	RE I	2.3	O		ME
9943.70	9946.43	10053.9	RE	1.3			ME
9908.97	9911.69	10089.1	RE	0.7			ME
9903.30	9906.02	10094.9	RE	0.8			ME
9884.10	9886.81	10114.5	RE	0.3	W		ME
9872.38	9875.09	10126.5	RE	1.2	I		ME
9842.63	9845.33	10157.1	RE	1.3	O		ME
9831.35	9834.05	10168.7	RE	1.0			ME
9813.75	9816.44	10187.0	RE	0.3			ME
9762.65	9765.33	10240.3	RE I	1.3			ME
9752.07	9754.74	10251.4	RE	0.6			ME
9749.67	9752.34	10253.9	RE I	0.8			ME
9748.60	9751.27	10255.1	RE	0.5			ME
9722.76	9725.43	10282.3	RE	0.6			ME
9710.52	9713.18	10295.3	RE	1.7			ME
9704.94	9707.60	10301.2	RE	0.5			ME
9635.30	9637.94	10375.7	RE	0.7			ME
9591.32	9593.95	10423.2	RE	0.5	W		ME
9581.12	9583.75	10434.3	RE I	0.9			ME
9571.75	9574.38	10444.6	RE	0.8	W		ME
9523.40	9526.01	10497.6	RE	0.8	W		ME
9504.34	9506.95	10518.6	RE	0.5			ME
9500.49	9503.10	10522.9	RE	0.9			ME
9470.14	9472.74	10556.6	RE I	1.5			ME
9427.53	9430.12	10604.3	RE	0.5			ME
9423.44	9426.03	10608.9	RE	1.2			ME
9391.12	9393.70	10645.4	RE I	0.3			ME
9383.74	9386.31	10653.8	RE	1.6			ME
9380.24	9382.81	10657.8	RE	1.0			ME
9370.87	9373.44	10668.4	RE	0.3			ME
9363.13	9365.70	10677.3	RE	1.3			ME
9340.8	9343.4	10703.	RE	0.3	V		ME
9337.73	9340.29	10706.3	RE	0.3	O		ME
9332.47	9335.03	10712.3	RE	0.3	W		ME
9325.90	9328.46	10719.9	RE	0.5			ME
9311.59	9314.15	10736.4	RE	0.7			ME

AIR WAVELENGTH (ANGSTRMS)	VACUUM WAVELENGTH (ANGSTRMS)	WAVENUMBER (KAYSERS)	LMENT	LOG INTENSITY ARC	SPK	DIS	REF
9268.46	9271.00	10786.3	RE	1.2	W		ME
9262.28	9264.82	10793.5	RE	0.6			ME
9255.71	9258.25	10801.2	RE	0.3	W		ME
9250.02	9252.56	10807.8	RE	1.0			ME
9236.50	9239.03	10823.6	RE	0.6			ME
9213.58	9216.11	10850.6	RE	0.3			ME
9209.66	9212.19	10855.2	RE	0.3	W		ME
9205.12	9207.65	10860.5	RE	0.3	W		ME
9165.38	9167.90	10907.6	RE	0.3			ME
9144.86	9147.37	10932.1	RE I	0.8	W		ME
9132.08	9134.59	10947.4	RE	0.3	W		ME
9083.03	9085.52	11006.5	RE	0.3			ME
9081.45	9083.94	11008.4	RE	0.6			ME
9070.74	9073.23	11021.4	RE	0.7			ME
9066.6	9069.1	11026.	RE	0.3	O		ME
9063.31	9065.80	11030.5	RE	0.3			ME
9059.85	9062.34	11034.7	RE	0.6			ME
9058.53	9061.02	11036.3	RE	0.5			ME
9044.38	9046.86	11053.6	RE	0.8			ME
9032.46	9034.94	11068.1	RE	0.5			ME
9009.15	9011.62	11096.8	RE	0.3	W		ME
8988.40	8990.87	11122.4	RE	0.3			ME
8971.14	8973.60	11143.8	RE I	0.3			ME
8969.21	8971.67	11146.2	RE	0.6			ME
8966.63	8969.09	11149.4	RE I	0.8	W		ME
8962.07	8964.53	11155.1	RE	0.3			ME
8944.56	8947.02	11176.9	RE	0.3			ME
8927.28	8929.73	11198.6	RE	0.7			ME
8901.36	8903.80	11231.2	RE	0.5	O		ME
8899.2	8901.6	11234.	RE	0.3			ME
8886.58	8889.02	11249.8	RE I	1.2			ME
8882.95	8885.39	11254.4	RE	1.2	N		ME
8851.32	8853.75	11294.7	RE I	0.5	X		ME
8845.23	8847.66	11302.4	RE	0.5	O		ME
8844.57	8847.00	11303.3	RE	0.3	S		ME
8797.70	8800.12	11363.5	RE	1.5	X		ME
8792.0	8794.4	11371.	RE	0.3	H		ME
8786.77	8789.18	11377.6	RE	1.6	X		ME
8772.49	8774.90	11396.1	RE	0.3			ME
8750.45	8752.85	11424.8	RE	0.5			ME
8721.64	8724.04	11462.6	RE	0.8	O		ME
8715.7	8718.1	11470.	RE	0.3	O		ME
8697.26	8699.65	11494.7	RE	1.3	W		ME
8683.75	8686.14	11512.6	RE	0.9			ME
8680.33	8682.71	11517.1	RE	0.5			ME
8675.65	8678.03	11523.3	RE I	1.7	X		ME
8648.97	8651.35	11558.9	RE I	1.5	X		ME
8643.61	8645.98	11566.1	RE I	1.2			ME
8633.10	8635.47	11580.1	RE	0.5	X		ME
8630.66	8633.03	11583.4	RE	0.9			ME
8628.44	8630.81	11586.4	RE	0.7			ME
8603.88	8606.24	11619.5	RE	0.8			ME
8593.62	8595.98	11633.3	RE	0.5			ME
8588.52	8590.88	11640.2	RE	0.5			ME
8570.71	8573.06	11664.4	RE	1.2			ME
8557.50	8559.85	11682.4	RE	0.3			ME
8554.61	8556.96	11686.4	RE	0.6			ME
8537.38	8539.73	11710.0	RE	0.3			ME
8534.99	8537.33	11713.3	RE	0.5	O		ME
8527.73	8530.07	11723.2	RE I	2.5	X		ME
8507.72	8510.06	11750.8	RE	0.8	S		ME
8503.35	8505.69	11756.8	RE	0.5			ME
8433.90	8436.22	11853.7	RE I	0.6	O		ME
8430.87	8433.19	11857.9	RE	0.3	D		ME
8417.14	8419.45	11877.3	RE I	2.5			ME
8403.58	8405.89	11896.4	RE	0.5			ME
8399.79	8402.10	11901.8	RE	0.9	W		ME
8357.59	8359.89	11961.9	RE	1.4			ME
8354.00	8356.30	11967.0	RE	0.3			ME
8313.02	8315.30	12026.0	RE	0.7	O		ME

AIR WAVELENGTH (ANGSTRMS)	VACUUM WAVELENGTH (ANGSTRMS)	WAVENUMBER (KAYSERS)	LMENT	LOG ARC	INTENSITY SPK DIS	REF
8301.01	8303.29	12043.4	RE	1.3W		ME
8293.73	8296.01	12054.0	RE	1.3		ME
8285.13	8287.41	12066.5	RE	0.5Y		ME
8262.87	8265.14	12099.0	RE	0.8W		ME
8256.25	8258.52	12108.7	RE	0.5W		ME
8254.0	8256.3	12112.	RE I	0.6X		ME
8246.93	8249.20	12122.4	RE	0.3		ME
8224.30	8226.56	12155.7	RE	0.3		ME
8190.75	8193.00	12205.5	RE	0.3H		ME
8166.63	8168.88	12241.6	RE	0.9W		ME
8088.25	8090.47	12360.2	RE	1.2		ME
8064.18	8066.40	12397.1	RE	0.3		ME
8060.03	8062.25	12403.5	RE	1.5		ME
8055.98	8058.20	12409.7	RE	0.6		ME
8052.11	8054.32	12415.7	RE	1.0N		ME
8027.53	8029.74	12453.7	RE	0.5		ME
8004.63	8006.83	12489.3	RE	0.5		ME
7985.28	7987.48	12519.6	RE	0.6		ME
7980.75	7982.95	12526.7	RE I	2.5K		ME
7979.04	7981.23	12529.4	RE I	1.3		ME
7971.26	7973.45	12541.6	RE	1.3		ME
7970.87	7973.06	12542.2	RE	1.2		ME
7938.57	7940.75	12593.3	RE	0.9		ME
7912.94	7915.12	12634.1	RE I	2.6K		ME
7898.47	7900.64	12657.2	RE I	1.6		ME
7888.93	7891.10	12672.5	RE	0.8		ME
7882.09	7884.26	12683.5	RE	1.4K		ME
7880.72	7882.89	12685.7	RE	0.9		ME
7872.09	7874.26	12699.6	RE	0.3		ME
7869.60	7871.77	12703.6	RE I	2.0K		ME
7861.85	7864.01	12716.2	RE	0.7		ME
7844.14	7846.30	12744.9	RE I	1.0		ME
7828.13	7830.28	12770.9	RE	0.9		ME
7825.85	7828.00	12774.7	RE	1.0		ME
7801.05	7803.20	12815.3	RE	0.6		ME
7799.58	7801.73	12817.7	RE	0.5		ME
7794.67	7796.81	12825.7	RE	0.6		ME
7789.88	7792.02	12833.6	RE	0.9		ME
7743.12	7745.25	12911.1	RE	1.2W		MIT
7733.59	7735.72	12927.1	RE	1.4		MIT
7712.58	7714.70	12962.3	RE	0.3		ME
7705.92	7708.04	12973.5	RE I	1.4		ME
7693.63	7695.75	12994.2	RE	1.3		ME
7684.83	7686.95	13009.1	RE	0.5		ME
7683.45	7685.56	13011.4	RE	0.3		ME
7656.81	7658.92	13056.7	RE	0.7		ME
7652.49	7654.60	13064.1	RE	0.3		ME
7640.93	7643.03	13083.8	RE I	2.6K		ME
7620.25	7622.35	13119.3	RE I	2.3K		ME
7618.32	7620.42	13122.6	RE	0.5		ME
7611.90	7614.00	13133.7	RE I	2.0		ME
7596.30	7598.39	13160.7	RE	0.3		ME
7587.66	7589.75	13175.7	RE	0.3		ME
7583.26	7585.35	13183.3	RE	0.8W		ME
7578.72	7580.81	13191.2	RE I	2.3W		ME
7573.47	7575.56	13200.3	RE	0.3		ME
7567.82	7569.90	13210.2	RE	0.8		ME
7548.71	7550.79	13243.7	RE	1.5		ME
7545.37	7547.45	13249.5	RE	0.7K		ME
7536.42	7538.50	13265.2	RE	0.3		ME
7526.50	7528.57	13282.7	RE	0.5		ME
7524.48	7526.55	13286.3	RE	1.3W		ME
7494.04	7496.10	13340.3	RE	0.6		ME
7489.13	7491.19	13349.0	RE	0.6		ME
7488.12	7490.18	13350.8	RE	1.0		ME
7481.15	7483.21	13363.2	RE	1.1		ME
7478.63	7480.69	13367.7	RE	0.9		ME
7447.58	7449.63	13423.5	RE	0.9		ME
7442.75	7444.80	13432.2	RE	0.3		ME
7440.71	7442.76	13435.9	RE	0.8		ME

AIR WAVELENGTH (ANGSTRMS)	VACUUM WAVELENGTH (ANGSTRMS)	WAVENUMBER (KAYSERS)	LMENT	LOG ARC	INTENSITY SPK DIS	REF
7437.78	7439.83	13441.2	RE	0.6		ME
7431.93	7433.98	13451.7	RE	0.6		ME
7416.44	7418.48	13479.8	RE	0.8		ME
7413.41	7415.45	13485.3	RE	0.6		ME
7409.47	7411.51	13492.5	RE I	1.5		ME
7404.27	7406.31	13502.0	RE	0.8		ME
7396.04	7398.08	13517.0	RE	0.6		ME
7390.71	7392.75	13526.8	RE	0.9		ME
7386.35	7388.38	13534.8	RE	1.5		ME
7384.07	7386.10	13538.9	RE I	0.3		ME
7382.68	7384.71	13541.5	RE	0.6		ME
7362.12	7364.15	13579.3	RE	0.6		ME
7360.95	7362.98	13581.5	RE	0.6		ME
7352.03	7354.06	13597.9	RE	1.7		ME
7346.88	7348.90	13607.5	RE	0.3		ME
7329.37	7331.39	13640.0	RE	1.0		ME
7324.20	7326.22	13649.6	RE	1.1		ME
7318.02	7320.04	13661.1	RE	0.6		ME
7316.1	7318.1	13665.	RE	0.9		ME
7307.55	7309.56	13680.7	RE I	0.9		ME
7292.67	7294.68	13708.6	RE	2.5		ME
7273.84	7275.84	13744.1	RE I	2.2		ME
7265.45	7267.45	13760.0	RE	0.8		ME
7263.87	7265.87	13763.0	RE	0.9		ME
7246.67	7248.67	13795.6	RE I	2.5		ME
7237.28	7239.27	13813.5	RE	0.8		ME
7198.41	7200.39	13888.1	RE	0.8		ME
7172.21	7174.19	13938.9	RE	1.0		ME
7139.64	7141.61	14002.4	RE	0.3		ME
7129.25	7131.22	14022.9	RE	1.0		MIT
7125.30	7127.26	14030.6	RE	0.9		ME
7108.8	7110.8	14063.	RE	0.3		ME
7066.46	7068.41	14147.5	RE I	1.0		ME
7059.98	7061.93	14160.4	RE	0.6		ME
7058.23	7060.18	14163.9	RE	0.9		ME
7028.50	7030.44	14223.9	RE	0.9		ME
7027.14	7029.08	14226.6	RE	0.3		ME
7024.13	7026.07	14232.7	RE I	2.1		MIT
7012.52	7014.45	14256.3	RE	0.6		ME
7006.65	7008.58	14268.2	RE I	2.0		MIT
6985.19	6987.12	14312.1	RE	1.3		MIT
6971.53	6973.45	14340.1	RE I	2.2		MIT
6967.68	6969.60	14348.0	RE	0.8		MIT
6874.38	6876.28	14542.7	RE	0.3W		ME
6844.44	6846.33	14606.4	RE	0.8		ME
6844.06	6845.95	14607.2	RE	0.8		ME
6829.96	6831.84	14637.3	RE I	2.3W		ME
6813.42	6815.30	14672.9	RE	2.3W		ME
6811.37	6813.25	14677.3	RE	0.5		ME
6805.36	6807.24	14690.2	RE	0.5H		ME
6801.65	6803.53	14698.3	RE	0.7		ME
6799.57	6801.45	14702.7	RE	0.5W		ME
6761.19	6763.06	14786.2	RE	1.0W		ME
6752.03	6753.89	14806.3	RE	1.7		SJ
6751.22	6753.08	14808.1	RE	1.7		ME
6711.29	6713.14	14896.2	RE I	1.0		ME
6683.30	6685.15	14958.5	RE	1.20		ME
6665.29	6667.13	14999.0	RE	1.4W		ME
6652.399	6654.236	15028.02	RE	1.90		MIT
6645.02	6646.85	15044.7	RE	1.4		SJ
6637.23	6639.06	15062.4	RE	1.3W		ME
6628.17	6630.00	15083.0	RE	1.4		SJ
6623.981	6625.810	15092.49	RE I	1.5W		MIT
6616.09	6617.92	15110.5	RE	1.4		SJ
6612.548	6614.374	15118.59	RE	0.3		MIT
6611.585	6613.411	15120.79	RE	0.5		MIT
6608.52	6610.35	15127.8	RE	1.4		SJ
6605.19	6607.01	15135.4	RE I	2.00		ME
6592.54	6594.36	15164.5	RE I	1.8W		MIT
6577.151	6578.968	15199.95	RE I	1.70		MIT

AIR WAVELENGTH (ANGSTRMS)	VACUUM WAVELENGTH (ANGSTRMS)	WAVENUMBER (KAYSERS)	LMENT	LOG INTENSITY ARC	SPK	DIS	REF
6557.65	6559.46	15245.2	RE	0.6H			ME
6557.151	6558.962	15246.31	RE I	1.0W			MIT
6554.678	6556.489	15252.07	RE	0.5			MIT
6545.740	6547.548	15272.89	RE	0.8H			MIT
6544.933	6546.741	15274.78	RE	1.2			MIT
6529.225	6531.029	15311.52	RE	0.6			MIT
6515.256	6517.056	15344.35	RE	1.3			MIT
6511.478	6513.277	15353.25	RE I	1.8			MIT
6510.74	6512.54	15355.0	RE	0.3			ME
6502.237	6504.034	15375.07	RE	0.8			MIT
6501.188	6502.984	15377.56	RE	0.7			MIT
6449.28	6451.06	15501.3	RE	0.6H			ME
6411.467	6413.239	15592.74	RE	1.3			MIT
6406.16	6407.93	15605.7	RE	1.3O			ME
6382.944	6384.709	15662.42	RE	1.2O			MIT
6374.812	6376.574	15682.40	RE	0.3W			MIT
6373.424	6375.186	15685.82	RE	0.3H			MIT
6362.942	6364.701	15711.66	RE	0.3			MIT
6362.693	6364.452	15712.27	RE	0.3			MIT
6350.75	6352.51	15741.8	RE I	2.0G			ME
6321.89	6323.64	15813.7	RE I	2.0W			ME
6307.720	6309.464	15849.21	RE I	2.0W			MIT
6303.454	6305.197	15859.93	RE	1.0			MIT
6290.105	6291.845	15893.59	RE I	0.3H			MIT
6286.417	6288.156	15902.91	RE	1.3W			MIT
6278.761	6280.498	15922.31	RE I	1.3			MIT
6271.409	6273.144	15940.97	RE I	1.2			MIT
6265.07	6266.80	15957.1	RE	0.3H			ME
6260.24	6261.97	15969.4	RE	1.3			ME
6260.017	6261.749	15969.98	RE	1.0			MIT
6253.043	6254.773	15987.79	RE	0.3			MIT
6252.06	6253.79	15990.3	RE I	0.3			ME
6243.220	6244.947	16012.95	RE I	1.7W			MIT
6229.44	6231.16	16048.4	RE I	1.6W			ME
6218.76	6220.48	16075.9	RE I	0.3H			ME
6217.987	6219.707	16077.93	RE I	1.6W			MIT
6203.246	6204.962	16116.13	RE I	1.4			MIT
6200.743	6202.459	16122.64	RE	0.5			MIT
6195.435	6197.149	16136.45	RE	1.2W			MIT
6187.914	6189.626	16156.06	RE	0.3H			MIT
6169.58	6171.29	16204.1	RE	0.3			ME
6146.817	6148.518	16264.08	RE I	1.7			MIT
6145.80	6147.50	16266.8	RE	1.7W			ME
6144.65	6146.35	16269.8	RE	1.4			SJ
6132.97	6134.67	16300.8	RE	0.3H			ME
6125.455	6127.150	16320.80	RE	1.0			MIT
6114.198	6115.890	16350.85	RE I	1.5			MIT
6051.176	6052.852	16521.14	RE	0.7			MIT
6039.638	6041.310	16552.70	RE	0.5W			MIT
6036.325	6037.997	16561.78	RE	0.5H			MIT
5995.756	5997.417	16673.85	RE	1.3W			MIT
5990.00	5991.66	16689.9	RE	0.5H			MIT
5983.15	5984.81	16709.0	RE	0.5H			MIT
5982.662	5984.319	16710.34	RE	0.5H			MIT
5981.683	5983.340	16713.07	RE	0.3H			MIT
5969.796	5971.450	16746.35	RE	1.3W			MIT
5967.640	5969.293	16752.40	RE	0.7			MIT
5954.319	5955.969	16789.88	RE	0.5			MIT
5950.229	5951.877	16801.42	RE	0.7W			MIT
5943.242	5944.889	16821.17	RE	2.0			MIT
5927.38	5929.02	16866.2	RE	0.3			ME
5919.855	5921.495	16887.63	RE	0.8H			MIT
5911.131	5912.769	16912.55	RE	0.3H			MIT
5909.992	5911.630	16915.81	RE	0.5H			MIT
5904.007	5905.643	16932.96	RE	0.3H			MIT
5870.734	5872.361	17028.93	RE	0.6			MIT
5868.043	5869.669	17036.73	RE	1.0			MIT
5866.590	5868.216	17040.95	RE	0.6			MIT
5852.751	5854.373	17081.25	RE	0.3			MIT
5852.024	5853.646	17083.37	RE	1.6W			MIT

AIR WAVELENGTH (ANGSTRMS)	VACUUM WAVELENGTH (ANGSTRMS)	WAVENUMBER (KAYSERS)	LMENT	LOG INTENSITY ARC	SPK	DIS	REF
5835.48	5837.10	17131.8	RE	0.3			ME
5834.33	5835.95	17135.2	RE	2.3			MIT
5818.26	5819.87	17182.5	RE	0.3			MIT
5818.075	5819.688	17183.05	RE	0.3			MIT
5815.866	5817.478	17189.58	RE	1.7W			MIT
5808.884	5810.495	17210.24	RE	0.8			MIT
5806.985	5808.595	17215.87	RE	1.3			MIT
5791.605	5793.211	17261.58	RE	1.4			MIT
5786.23	5787.83	17277.6	RE	0.7			MIT
5783.03	5784.63	17287.2	RE	0.5			MIT
5778.33	5779.93	17301.2	RE	0.3			MIT
5776.84	5778.44	17305.7	RE	2.5W			MIT
5752.951	5754.547	17377.56	RE	2.3W			MIT
5740.30	5741.89	17415.9	RE	1.7W			MIT
5739.434	5741.026	17418.49	RE	1.0			MIT
5733.058	5734.648	17437.86	RE	0.3H			MIT
5727.907	5729.496	17453.54	RE	0.5			MIT
5725.633	5727.221	17460.47	RE	1.2			MIT
5716.965	5718.551	17486.95	RE	1.3			MIT
5714.037	5715.622	17495.91	RE	0.8			MIT
5711.429	5713.014	17503.90	RE	1.4			MIT
5689.739	5691.318	17570.62	RE	0.3H			MIT
5684.323	5685.900	17587.36	RE	0.3			MIT
5680.950	5682.526	17597.81	RE	0.3H			MIT
5671.046	5672.620	17628.54	RE	0.3			MIT
5667.902	5669.475	17638.32	RE	2.0			MIT
5664.724	5666.296	17648.21	RE	0.3H			MIT
5662.878	5664.450	17653.97	RE	0.6W			MIT
5653.035	5654.604	17684.71	RE	1.0			MIT
5644.472	5646.039	17711.53	RE	0.7			MIT
5635.443	5637.007	17739.91	RE	0.3			MIT
5625.446	5627.008	17771.44	RE	1.2W			MIT
5621.219	5622.779	17784.80	RE	0.3			MIT
5619.760	5621.320	17789.42	RE	0.5			MIT
5612.264	5613.822	17813.18	RE	1.3W			MIT
5610.519	5612.077	17818.72	RE	0.8			MIT
5608.812	5610.369	17824.14	RE	0.3			MIT
5607.217	5608.774	17829.21	RE	1.0			MIT
5601.936	5603.491	17846.02	RE	0.6			MIT
5594.857	5596.410	17868.60	RE	0.3H			MIT
5584.755	5586.306	17900.92	RE	1.6W			MIT
5582.912	5584.462	17906.83	RE	0.3			MIT
5580.794	5582.344	17913.62	RE	0.3			MIT
5577.715	5579.264	17923.51	RE	0.7O			MIT
5573.485	5575.033	17937.11	RE	1.5W			MIT
5564.143	5565.688	17967.23	RE	1.2			MIT
5563.212	5564.757	17970.24	RE	2.2W			MIT
5557.214	5558.757	17989.63	RE	1.2			MIT
5532.663	5534.200	18069.46	RE	2.0			MIT
5523.395	5524.929	18099.78	RE	1.0W			MIT
5521.104	5522.638	18107.29	RE	1.3			MIT
5520.06	5521.59	18110.7	RE	1.2H			MIT
5519.520	5521.053	18112.49	RE	0.5H			MIT
5505.61	5507.14	18158.2	RE	0.3H			MIT
5501.919	5503.448	18170.43	RE	1.0			MIT
5495.57	5497.10	18191.4	RE	0.3H			ME
5481.001	5482.524	18239.77	RE	0.5			MIT
5460.63	5462.15	18307.8	RE	1.5O			MIT
5456.31	5457.83	18322.3	RE	0.3H			MIT
5447.915	5449.429	18350.55	RE	1.3			MIT
5437.414	5438.925	18385.99	RE	0.3			MIT
5437.062	5438.573	18387.18	RE	1.3			MIT
5431.885	5433.395	18404.70	RE	1.5			MIT
5423.805	5425.313	18432.12	RE	0.8			MIT
5423.603	5425.111	18432.80	RE	0.7			MIT
5423.274	5424.782	18433.92	RE	0.3			MIT
5409.512	5411.016	18480.82	RE	0.3			MIT
5396.490	5397.990	18525.41	RE	1.2			MIT
5379.712	5381.208	18583.19	RE	0.7			MIT
5378.14	5379.64	18588.6	RE	0.5			MIT

AIR WAVELENGTH (ANGSTRMS)	VACUUM WAVELENGTH (ANGSTRMS)	WAVENUMBER (KAYSERS)	LMENT	LOG ARC	INTENSITY SPK	DIS	REF
5377.045	5378.540	18592.41	RE	2.5O			MIT
5374.709	5376.204	18600.49	RE	0.7			MIT
5369.81	5371.30	18617.5	RE	1.6			MIT
5369.458	5370.951	18618.68	RE	1.5			MIT
5366.659	5368.151	18628.39	RE	0.3			MIT
5353.31	5354.80	18674.8	RE	0.3H			MIT
5350.392	5351.880	18685.02	RE	0.3W			MIT
5349.247	5350.735	18689.02	RE	0.5			MIT
5345.22	5346.71	18703.1	RE	0.3			MIT
5339.413	5340.898	18723.44	RE	1.0			MIT
5338.629	5340.114	18726.19	RE	1.0			MIT
5333.854	5335.338	18742.96	RE	1.5			MIT
5332.761	5334.244	18746.80	RE	1.0			MIT
5332.482	5333.965	18747.78	RE	0.3			MIT
5331.896	5333.379	18749.84	RE	1.9			MIT
5327.459	5328.941	18765.46	RE	2.0			MIT
5321.52	5323.00	18786.4	RE	0.3			MIT
5321.263	5322.743	18787.30	RE	1.6			MIT
5317.278	5318.757	18801.38	RE	1.6W			MIT
5311.55	5313.03	18821.7	RE	0.6H			MIT
5306.73	5308.21	18838.7	RE	0.3			MIT
5305.562	5307.038	18842.90	RE	1.4			MIT
5300.77	5302.24	18859.9	RE	0.5H			MIT
5292.283	5293.756	18890.18	RE	1.0			MIT
5284.583	5286.054	18917.71	RE	0.3			MIT
5278.242	5279.711	18940.43	RE	2.0			MIT
5275.53	5277.00	18950.2	RE	2.7W			MIT
5270.984	5272.451	18966.51	RE	2.3O			MIT
5265.140	5266.605	18987.56	RE	0.5H			MIT
5264.415	5265.880	18990.18	RE	0.9			MIT
5260.61	5262.07	19003.9	RE	0.3			MIT
5252.351	5253.813	19033.79	RE	0.5			MIT
5248.855	5250.316	19046.47	RE	1.4			MIT
5244.342	5245.802	19062.86	RE	1.3			MIT
5236.668	5238.126	19090.80	RE	1.3W			MIT
5234.328	5235.785	19099.33	RE	1.4O			MIT
5224.697	5226.152	19134.54	RE	1.4W			MIT
5222.13	5223.58	19143.9	RE	0.8			MIT
5221.132	5222.586	19147.60	RE	0.5H			MIT
5218.23	5219.68	19158.2	RE	0.3H			MIT
5217.45	5218.90	19161.1	RE	0.3			MIT
5214.727	5216.179	19171.12	RE	1.0			MIT
5200.807	5202.255	19222.43	RE	0.3H			MIT
5199.893	5201.341	19225.81	RE	1.2W			MIT
5186.442	5187.886	19275.67	RE	0.3			MIT
5186.147	5187.591	19276.77	RE	0.3			MIT
5185.89	5187.33	19277.7	RE	0.3			MIT
5181.761	5183.204	19293.08	RE	1.4			MIT
5178.910	5180.352	19303.71	RE	2.0O			MIT
5172.72	5174.16	19326.8	RE	0.3			ME
5172.377	5173.818	19328.09	RE	0.3			MIT
5165.171	5166.610	19355.05	RE	0.5			MIT
5161.655	5163.093	19368.24	RE	1.4			MIT
5156.257	5157.693	19388.51	RE	1.4W			MIT
5146.880	5148.314	19423.83	RE	1.4W			MIT
5145.604	5147.038	19428.65	RE	1.2			MIT
5141.285	5142.717	19444.97	RE	0.9			MIT
5140.04	5141.47	19449.7	RE	1.0			MIT
5126.69	5128.12	19500.3	RE	1.7W			MIT
5124.600	5126.028	19508.28	RE	1.3			MIT
5120.324	5121.751	19524.57	RE	1.6			MIT
5112.272	5113.697	19555.32	RE	1.4			MIT
5108.780	5110.204	19568.69	RE	1.3			MIT
5105.157	5106.580	19582.58	RE	1.3			MIT
5104.626	5106.049	19584.62	RE	1.7W			MIT
5100.700	5102.122	19599.69	RE	0.3			MIT
5096.504	5097.924	19615.83	RE	2.0			MIT
5095.77	5097.19	19618.6	RE	1.2W			MIT
5094.030	5095.450	19625.35	RE	0.3H			MIT
5093.185	5094.605	19628.61	RE	0.3H			MIT

AIR WAVELENGTH (ANGSTRMS)	VACUUM WAVELENGTH (ANGSTRMS)	WAVENUMBER (KAYSERS)	LMENT	LOG ARC	INTENSITY SPK	DIS	REF
5090.569	5091.988	19638.70	RE	1.0			MIT
5078.420	5079.836	19685.68	RE	1.2			MIT
5065.312	5066.724	19736.62	RE	0.5			MIT
5063.76	5065.17	19742.7	RE	0.7			MIT
5060.430	5061.841	19755.66	RE	0.7			MIT
5058.556	5059.966	19762.98	RE	1.6			MIT
5056.079	5057.489	19772.66	RE	0.7H			MIT
5052.204	5053.613	19787.82	RE	0.3			MIT
5042.620	5044.026	19825.43	RE	1.0			MIT
5027.947	5029.349	19883.29	RE	1.4			MIT
5026.453	5027.855	19889.20	RE	1.4			MIT
5024.022	5025.423	19898.82	RE	0.3			MIT
5010.62	5012.02	19952.0	RE	0.3			MIT
5003.618	5005.014	19979.97	RE	0.3			MIT
4985.992	4987.383	20050.59	RE I	1.6O			MIT
4981.541	4982.931	20068.51	RE I	1.2			MIT
4980.686	4982.076	20071.96	RE I	0.6			MIT
4972.854	4974.242	20103.57	RE	0.5			MIT
4969.450	4970.837	20117.34	RE	0.6			MIT
4967.826	4969.212	20123.91	RE	0.6			MIT
4963.080	4964.465	20143.16	RE	0.3			MIT
4961.389	4962.774	20150.02	RE	0.3			MIT
4956.769	4958.152	20168.80	RE	1.5			MIT
4949.836	4951.217	20197.05	RE	1.2			MIT
4946.736	4948.117	20209.71	RE	2.0			MIT
4944.88	4946.26	20217.3	RE	0.6			ME
4943.742	4945.122	20221.95	RE	0.3W			MIT
4935.85	4937.23	20254.3	RE	1.2			M
4933.740	4935.117	20262.94	RE	1.2W			MIT
4926.907	4928.282	20291.04	RE	0.3			MIT
4923.929	4925.304	20303.32	RE	2.2			MIT
4922.366	4923.740	20309.76	RE	0.3			MIT
4917.84	4919.21	20328.5	RE	0.6			ME
4915.025	4916.397	20340.10	RE	1.5			MIT
4908.572	4909.942	20366.84	RE	1.3			MIT
4906.205	4907.575	20376.66	RE	1.2			MIT
4903.744	4905.113	20386.89	RE	1.0W			MIT
4896.173	4897.540	20418.41	RE	0.3			MIT
4889.17	4890.54	20447.7	RE I	3.3W			MIT
4876.938	4878.300	20498.94	RE	0.3			MIT
4874.844	4876.206	20507.75	RE	0.7			MIT
4869.703	4871.063	20529.40	RE	0.7X			MIT
4857.15	4858.51	20582.5	RE	0.8			ME
4852.598	4853.954	20601.76	RE	0.5J			MIT
4848.435	4849.790	20619.45	RE	1.4W			MIT
4845.67	4847.02	20631.2	RE	0.3			MIT
4839.629	4840.981	20656.97	RE	0.7			MIT
4834.830	4836.181	20677.47	RE	1.0W			MIT
4829.89	4831.24	20698.6	RE	0.6W			ME
4820.60	4821.95	20738.5	RE	1.3			MIT
4811.208	4812.553	20778.99	RE	0.6			MIT
4810.104	4811.448	20783.76	RE	0.6			MIT
4803.34	4804.68	20813.0	RE	0.3			MIT
4799.497	4800.839	20829.69	RE	0.5			MIT
4799.114	4800.455	20831.36	RE	1.0			MIT
4796.908	4798.249	20840.94	RE	0.3			MIT
4793.570	4794.910	20855.45	RE	0.3			MIT
4791.419	4792.758	20864.81	RE	2.3W			MIT
4789.212	4790.551	20874.43	RE	1.0			MIT
4771.805	4773.139	20950.57	RE	0.7			MIT
4763.664	4764.996	20986.38	RE	1.3O			MIT
4758.855	4760.186	21007.58	RE	1.5			MIT
4752.100	4753.429	21037.45	RE	0.3			MIT
4751.354	4752.683	21040.75	RE	0.3			MIT
4751.004	4752.333	21042.30	RE	0.3H			MIT
4749.029	4750.357	21051.05	RE	1.4W			MIT
4748.379	4749.707	21053.93	RE	1.7W			MIT
4748.072	4749.400	21055.29	RE	0.6H			MIT
4743.52	4744.85	21075.5	RE	0.6			ME
4733.873	4735.197	21118.45	RE	1.0			MIT

AIR WAVELENGTH (ANGSTRMS)	VACUUM WAVELENGTH (ANGSTRMS)	WAVENUMBER (KAYSERS)	LMENT	LOG INTENSITY ARC	SPK	DIS	REF
4727.619	4728.942	21146.38	RE	1.6			MIT
4725.937	4727.259	21153.91	RE	1.3			MIT
4725.02	4726.34	21158.0	RE	0.6W			ME
4712.783	4714.102	21212.95	RE	1.3W			MIT
4711.240	4712.558	21219.90	RE	0.3			MIT
4705.034	4706.351	21247.89	RE	1.6W			MIT
4703.767	4705.083	21253.61	RE	0.3			MIT
4700.433	4701.748	21268.68	RE	1.3			MIT
4699.696	4701.011	21272.02	RE	0.8			MIT
4695.011	4696.325	21293.25	RE	1.6			MIT
4693.403	4694.716	21300.54	RE	0.7W			MIT
4689.545	4690.857	21318.06	RE	0.5			MIT
4687.858	4689.170	21325.74	RE	1.2			MIT
4682.332	4683.643	21350.90	RE	1.5O			MIT
4681.888	4683.198	21352.93	RE	0.3H			MIT
4679.475	4680.785	21363.94	RE	1.3			MIT
4675.390	4676.699	21382.60	RE	0.3H			MIT
4674.305	4675.613	21387.57	RE	1.2			MIT
4666.729	4668.035	21422.29	RE	0.5			MIT
4665.230	4666.536	21429.17	RE	0.3			MIT
4663.010	4664.315	21439.37	RE	0.6H			MIT
4662.496	4663.801	21441.74	RE	1.4			MIT
4661.478	4662.783	21446.42	RE	1.0			MIT
4660.544	4661.849	21450.72	RE	1.2			MIT
4656.49	4657.79	21469.4	RE	1.0			MIT
4654.553	4655.856	21478.33	RE	1.0H			MIT
4652.323	4653.626	21488.62	RE	1.5			MIT
4651.823	4653.125	21490.93	RE	0.5			MIT
4650.833	4652.135	21495.51	RE	0.3H			MIT
4649.70	4651.00	21500.7	RE	0.3			MIT
4649.183	4650.485	21503.14	RE	0.3			MIT
4648.58	4649.88	21505.9	RE	0.6			M
4647.47	4648.77	21511.1	RE	1.0H			M
4644.95	4646.25	21522.7	RE	1.2			M
4635.802	4637.100	21565.20	RE	0.3			MIT
4634.404	4635.702	21571.71	RE	0.6H			MIT
4630.839	4632.136	21588.31	RE	1.7			MIT
4630.24	4631.54	21591.1	RE	0.8			M
4625.98	4627.28	21611.0	RE	1.6			M
4623.331	4624.626	21623.37	RE	0.3H			MIT
4621.384	4622.678	21632.48	RE	1.5			MIT
4616.595	4617.888	21654.92	RE	1.3W			MIT
4614.696	4615.989	21663.83	RE	1.7			MIT
4613.960	4615.252	21667.29	RE	0.5			MIT
4608.810	4610.101	21691.50	RE	0.5H			MIT
4605.734	4607.024	21705.99	RE	1.7			MIT
4598.14	4599.43	21741.8	RE	0.3H			M
4597.33	4598.62	21745.7	RE	0.3H			MIT
4595.249	4596.536	21755.51	RE	0.3H			MIT
4592.55	4593.84	21768.3	RE	0.3H			MIT
4591.677	4592.964	21772.44	RE	1.4			MIT
4590.558	4591.844	21777.74	RE	0.3H			MIT
4587.135	4588.420	21793.99	RE	1.2			MIT
4586.054	4587.339	21799.13	RE	0.3W			MIT
4580.668	4581.952	21824.76	RE	1.5			MIT
4578.244	4579.527	21836.32	RE	0.3			MIT
4577.554	4578.837	21839.61	RE	0.3			MIT
4572.189	4573.470	21865.23	RE	0.3			MIT
4565.50	4566.78	21897.3	RE	1.0			M
4565.31	4566.59	21898.2	RE	1.4			M
4564.622	4565.901	21901.48	RE	0.8			MIT
4563.63	4564.91	21906.2	RE	0.6			ME
4559.700	4560.978	21925.12	RE	1.5W			MIT
4559.263	4560.541	21927.22	RE	1.4			MIT
4557.31	4558.59	21936.6	RE	0.3H			M
4556.017	4557.294	21942.84	RE	0.3			MIT
4551.483	4552.759	21964.70	RE	0.3H			MIT
4545.532	4546.806	21993.46	RE	1.0			MIT
4545.164	4546.438	21995.24	RE	1.5			MIT
4541.797	4543.070	22011.55	RE	1.4			MIT

AIR WAVELENGTH (ANGSTRMS)	VACUUM WAVELENGTH (ANGSTRMS)	WAVENUMBER (KAYSERS)	LMENT	LOG INTENSITY ARC	SPK	DIS	REF
4537.587	4538.859	22031.97	RE	0.5			MIT
4536.014	4537.286	22039.61	RE	1.3			MIT
4530.890	4532.160	22064.53	RE	1.3			MIT
4529.928	4531.198	22069.22	RE	1.6W			MIT
4528.943	4530.213	22074.02	RE	0.3			MIT
4526.962	4528.231	22083.68	RE	0.7H			MIT
4525.986	4527.255	22088.44	RE	1.5			MIT
4523.885	4525.154	22098.70	RE	1.6			MIT
4522.72	4523.99	22104.4	RE	2.0			MIT
4519.744	4521.012	22118.94	RE	1.6D			MIT
4516.63	4517.90	22134.2	RE	1.9			M
4515.148	4516.414	22141.46	RE	0.9			MIT
4514.275	4515.541	22145.74	RE	1.2H			MIT
4513.30	4514.57	22150.5	RE	2.5			MIT
4507.99	4509.25	22176.6	RE	1.5W			ME
4507.03	4508.29	22181.3	RE	1.6W			ME
4504.42	4505.68	22194.2	RE	0.3H			M
4498.389	4499.651	22223.95	RE	0.9			MIT
4496.446	4497.707	22233.55	RE	1.3D			MIT
4494.102	4495.363	22245.15	RE	0.3H			MIT
4486.995	4488.254	22280.38	RE	0.9			MIT
4486.59	4487.85	22282.4	RE	0.3			M
4481.447	4482.704	22307.96	RE	1.2			MIT
4478.387	4479.644	22323.20	RE	1.6W			MIT
4477.99	4479.25	22325.2	RE	1.6W			M
4477.646	4478.902	22326.90	RE	0.5			MIT
4475.080	4476.336	22339.70	RE	1.5			MIT
4473.241	4474.496	22348.89	RE	0.5H			MIT
4467.937	4469.191	22375.42	RE	1.7			MIT
4467.546	4468.800	22377.37	RE	1.4			MIT
4465.831	4467.084	22385.97	RE	0.8			MIT
4464.620	4465.873	22392.04	RE	0.5			MIT
4463.527	4464.780	22397.52	RE	1.3			MIT
4457.51	4458.76	22427.7	RE	0.3H			MIT
4456.710	4457.961	22431.78	RE	1.0W			MIT
4456.380	4457.631	22433.44	RE	0.8			MIT
4454.668	4455.918	22442.06	RE	2.0			MIT
4453.875	4455.125	22446.06	RE	1.3			MIT
4448.265	4449.514	22474.37	RE	0.6			MIT
4448.124	4449.373	22475.08	RE	0.6			MIT
4440.44	4441.69	22514.0	RE	1.6			M
4436.627	4437.873	22533.32	RE	0.8			MIT
4432.840	4434.085	22552.57	RE	0.3			MIT
4427.953	4429.196	22577.46	RE	1.4W			MIT
4427.672	4428.915	22578.89	RE	0.5			MIT
4425.773	4427.016	22588.58	RE	0.5			MIT
4423.817	4425.059	22598.57	RE	0.3			MIT
4423.004	4424.246	22602.72	RE	0.3			MIT
4418.812	4420.053	22624.16	RE	1.0			MIT
4417.29	4418.53	22632.0	RE	1.3			M
4415.825	4417.065	22639.47	RE	1.6			MIT
4413.506	4414.746	22651.36	RE	0.3			MIT
4406.404	4407.642	22687.87	RE	1.8			MIT
4403.678	4404.915	22701.91	RE	1.1			MIT
4402.605	4403.842	22707.45	RE	1.5			MIT
4399.84	4401.08	22721.7	RE	1.1W			M
4396.79	4398.03	22737.5	RE	1.4			M
4396.686	4397.921	22738.02	RE	0.8			MIT
4396.087	4397.322	22741.11	RE	1.3W			MIT
4394.384	4395.618	22749.93	RE	2.0R			MIT
4392.486	4393.720	22759.76	RE	2.0			MIT
4392.17	4393.40	22761.4	RE	0.7			M
4391.33	4392.56	22765.7	RE	1.8W			MIT
4387.412	4388.645	22786.08	RE	1.4W			MIT
4385.454	4386.686	22796.25	RE	1.3			MIT
4373.20	4374.43	22860.1	RE	0.7			M
4369.772	4371.000	22878.06	RE	0.6			MIT
4369.662	4370.890	22878.64	RE	0.3			MIT
4367.582	4368.809	22889.53	RE	1.9			MIT
4364.842	4366.069	22903.90	RE	1.2			MIT

AIR WAVELENGTH (ANGSTRMS)	VACUUM WAVELENGTH (ANGSTRMS)	WAVENUMBER (KAYSERS)	LMENT	LOG INTENSITY ARC	SPK	DIS	REF
4364.129	4365.355	22907.64	RE	1.2			MIT
4363.359	4364.585	22911.68	RE	0.6W			MIT
4361.276	4362.502	22922.63	RE	0.5W			MIT
4360.375	4361.601	22927.36	RE	1.2			MIT
4359.311	4360.536	22932.96	RE	1.1			MIT
4358.688	4359.913	22936.24	RE I	1.9			MIT
4357.97	4359.19	22940.0	RE	1.2			MIT
4357.091	4358.316	22944.64	RE	1.0			MIT
4352.417	4353.640	22969.28	RE	0.3W			MIT
4351.712	4352.935	22973.00	RE	0.3			MIT
4345.513	4346.735	23005.77	RE	0.3			MIT
4344.655	4345.876	23010.32	RE	0.3			MIT
4342.159	4343.380	23023.54	RE	0.5			MIT
4339.681	4340.901	23036.69	RE I	1.3			MIT
4335.859	4337.078	23057.00	RE	1.0			MIT
4332.261	4333.479	23076.15	RE I	1.7			MIT
4331.376	4332.594	23080.86	RE	0.7			MIT
4329.079	4330.296	23093.11	RE	0.3			MIT
4328.72	4329.94	23095.0	RE	0.3			MIT
4324.343	4325.559	23118.40	RE	1.0			MIT
4319.530	4320.745	23144.16	RE I	1.3			MIT
4318.577	4319.792	23149.26	RE	1.3			MIT
4317.947	4319.161	23152.64	RE	0.3H			MIT
4317.667	4318.881	23154.14	RE	0.3H			MIT
4317.117	4318.331	23157.09	RE	0.8W			MIT
4316.176	4317.390	23162.14	RE	0.7			MIT
4315.71	4316.92	23164.6	RE I	1.6W			MIT
4314.582	4315.795	23170.70	RE	1.3			MIT
4310.44	4311.65	23193.0	RE	0.7			MIT
4309.50	4310.71	23198.0	RE	1.0			MIT
4308.78	4309.99	23201.9	RE	0.6			MIT
4305.322	4306.533	23220.53	RE	1.5			MIT
4304.784	4305.995	23223.44	RE	0.7			MIT
4304.411	4305.622	23225.45	RE I	0.7W			MIT
4301.053	4302.263	23243.58	RE	0.5			MIT
4299.923	4301.133	23249.69	RE	1.2			MIT
4298.217	4299.426	23258.92	RE	0.5H			MIT
4291.80	4293.01	23293.7	RE	0.5			MIT
4291.646	4292.853	23294.53	RE I	1.4			MIT
4291.183	4292.390	23297.04	RE	2.0W			MIT
4286.472	4287.678	23322.65	RE	1.1			MIT
4286.090	4287.296	23324.73	RE	0.7			MIT
4282.740	4283.945	23342.97	RE I	1.2			MIT
4280.600	4281.805	23354.64	RE	0.8			MIT
4279.200	4280.404	23362.28	RE	0.5			MIT
4274.34	4275.54	23388.8	RE	1.3W			MIT
4269.82	4271.02	23413.6	RE	0.3H			MIT
4268.75	4269.95	23419.5	RE	0.3H			MIT
4263.769	4264.969	23446.83	RE	0.3H			MIT
4263.355	4264.555	23449.11	RE	0.3H			MIT
4261.26	4262.46	23460.6	RE	0.7			MIT
4260.915	4262.114	23462.53	RE I	1.0			MIT
4259.883	4261.082	23468.22	RE I	0.5			MIT
4257.593	4258.791	23480.84	RE I	2.1W			MIT
4255.752	4256.950	23491.00	RE I	1.3			MIT
4255.339	4256.537	23493.28	RE	0.3H			MIT
4255.110	4256.308	23494.54	RE	0.3H			MIT
4247.694	4248.890	23535.56	RE	0.5			MIT
4246.813	4248.009	23540.44	RE	1.4			MIT
4244.143	4245.338	23555.25	RE I	1.5			MIT
4241.822	4243.016	23568.14	RE I	0.3			MIT
4241.387	4242.581	23570.56	RE I	1.5			MIT
4241.156	4242.350	23571.84	RE I	1.3			MIT
4239.963	4241.157	23578.47	RE	0.5			MIT
4238.584	4239.777	23586.14	RE I	1.3			MIT
4238.032	4239.225	23589.22	RE	1.2			MIT
4236.239	4237.432	23599.20	RE	1.4			MIT
4234.991	4236.184	23606.15	RE I	1.2W			MIT
4234.422	4235.614	23609.33	RE I	1.2W			MIT
4233.277	4234.469	23615.71	RE I	1.3			MIT
4232.958	4234.150	23617.49	RE I	1.3			MIT
4227.46	4228.65	23648.2	RE I	2.30			MIT
4224.176	4225.366	23666.59	RE I	1.3			MIT
4221.084	4222.273	23683.93	RE I	2.0			MIT
4215.506	4216.693	23715.26	RE	1.3			MIT
4204.53	4205.71	23777.2	RE	1.4W			MIT
4203.323	4204.507	23784.00	RE	1.0H			MIT
4195.755	4196.937	23826.90	RE	1.3			MIT
4194.670	4195.852	23833.06	RE I	1.6			MIT
4191.891	4193.072	23848.86	RE	0.5H			MIT
4187.921	4189.101	23871.47	RE I	0.8			MIT
4185.722	4186.902	23884.01	RE	0.3H			MIT
4182.979	4184.158	23899.67	RE I	2.2R			MIT
4179.75	4180.93	23918.1	RE	0.3H			MIT
4178.597	4179.775	23924.73	RE	0.5			MIT
4176.90	4178.08	23934.4	RE	0.7			MIT
4176.56	4177.74	23936.4	RE I	1.1W			MIT
4175.17	4176.35	23944.4	RE	0.8			MIT
4173.987	4175.164	23951.16	RE	1.3			MIT
4173.682	4174.858	23952.91	RE	0.3			MIT
4170.394	4171.570	23971.79	RE	1.6			MIT
4168.586	4169.761	23982.19	RE I	1.2			MIT
4166.412	4167.587	23994.70	RE	1.2W			MIT
4165.978	4167.152	23997.20	RE	1.0			MIT
4165.332	4166.506	24000.92	RE	0.3			MIT
4161.763	4162.936	24021.51	RE	0.3			MIT
4160.763	4161.936	24027.28	RE	1.2W			MIT
4159.918	4161.091	24032.16	RE	1.3			MIT
4157.26	4158.43	24047.5	RE	0.3			MIT
4152.631	4153.802	24074.33	RE I	1.5			MIT
4152.285	4153.456	24076.34	RE	1.1			MIT
4151.047	4152.217	24083.52	RE	0.3H			MIT
4150.793	4151.963	24084.99	RE	0.3			MIT
4150.531	4151.701	24086.51	RE	1.0			MIT
4149.972	4151.142	24089.75	RE I	1.6W			MIT
4148.288	4149.458	24099.53	RE	1.0			MIT
4146.239	4147.408	24111.44	RE	1.3			MIT
4144.359	4145.528	24122.38	RE I	2.1W			MIT
4142.771	4143.939	24131.63	RE	0.9			MIT
4138.530	4139.697	24156.36	RE	0.3			MIT
4137.606	4138.773	24161.75	RE I	1.2			MIT
4136.446	4137.613	24168.53	RE I	2.2W			MIT
4133.417	4134.583	24186.24	RE I	2.3			MIT
4132.282	4133.448	24192.88	RE I	1.3			MIT
4130.462	4131.627	24203.54	RE	1.0			MIT
4128.094	4129.258	24217.42	RE	0.8			MIT
4124.790	4125.954	24236.82	RE	0.6			MIT
4123.505	4124.668	24244.37	RE	0.3			MIT
4122.763	4123.926	24248.74	RE	1.0			MIT
4121.63	4122.79	24255.4	RE I	1.7			MIT
4114.680	4115.841	24296.37	RE	0.6			MIT
4113.395	4114.556	24303.96	RE I	1.3			MIT
4112.259	4113.419	24310.68	RE	0.5H			MIT
4111.597	4112.757	24314.59	RE	0.5			MIT
4110.899	4112.059	24318.72	RE I	1.6			MIT
4108.624	4109.783	24332.18	RE	1.0			MIT
4107.960	4109.119	24336.12	RE	0.5			MIT
4106.842	4108.001	24342.74	RE	1.0			MIT
4106.444	4107.603	24345.10	RE	1.2			MIT
4104.422	4105.580	24357.09	RE	1.5			MIT
4103.779	4104.937	24360.91	RE	0.3H			MIT
4102.159	4103.317	24370.53	RE	0.6H			MIT
4089.916	4091.070	24443.48	RE I	1.4			MIT
4089.43	4090.58	24446.4	RE	0.3H			MIT
4089.006	4090.160	24448.92	RE	0.3H			MIT
4088.034	4089.188	24454.73	RE I	0.5			MIT
4083.594	4084.747	24481.32	RE I	1.4			MIT
4083.372	4084.525	24482.65	RE I	1.4			MIT
4081.428	4082.580	24494.31	RE I	1.5			MIT
4079.363	4080.515	24506.71	RE	1.3			MIT

AIR WAVELENGTH (ANGSTRMS)	VACUUM WAVELENGTH (ANGSTRMS)	WAVENUMBER (KAYSERS)	LMENT	LOG INTENSITY ARC	SPK	DIS	REF
4078.124	4079.275	24514.16	RE	1.0			MIT
4073.062	4074.212	24544.62	RE I	0.7H			MIT
4069.13	4070.28	24568.3	RE	0.6W			MIT
4064.325	4065.473	24597.39	RE I	0.9			MIT
4061.862	4063.009	24612.30	RE I	1.3			MIT
4050.926	4052.070	24678.74	RE	1.0			MIT
4048.992	4050.136	24690.53	RE I	1.5			MIT
4045.762	4046.905	24710.24	RE	1.4			MIT
4040.178	4041.319	24744.39	RE	1.2			MIT
4037.511	4038.652	24760.74	RE	1.5			MIT
4033.307	4034.447	24786.55	RE I	1.6			MIT
4032.146	4033.285	24793.68	RE	1.0			MIT
4031.64	4032.78	24796.8	RE	0.3			MIT
4029.639	4030.778	24809.11	RE I	1.9			MIT
4028.563	4029.701	24815.73	RE	1.4			MIT
4027.128	4028.266	24824.58	RE	0.5			MIT
4026.723	4027.861	24827.07	RE	0.3			MIT
4025.614	4026.752	24833.91	RE I	1.2			MIT
4023.353	4024.490	24847.87	RE I	1.6W			MIT
4022.965	4024.102	24850.26	RE I	1.4D			MIT
4019.130	4020.266	24873.98	RE I	1.2			MIT
4018.409	4019.545	24878.44	RE	1.4			MIT
4014.85	4015.98	24900.5	RE	0.3H			MIT
4012.260	4013.394	24916.57	RE	1.4			MIT
4011.514	4012.648	24921.20	RE	1.5			MIT
4004.932	4006.064	24962.16	RE	1.5			MIT
4001.19	4002.32	24985.5	RE I	0.7			MIT
3996.92	3998.05	25012.2	RE	0.5H			MIT
3995.68	3996.81	25019.9	RE	1.0H			MIT
3994.14	3995.27	25029.6	RE	0.7H			MIT
3992.725	3993.854	25038.47	RE I	1.3			MIT
3991.580	3992.709	25045.65	RE	1.3			MIT
3991.039	3992.168	25049.05	RE I	1.4			MIT
3990.663	3991.792	25051.41	RE I	1.2			MIT
3987.929	3989.057	25068.58	RE	0.5			MIT
3987.083	3988.211	25073.90	RE	0.8			MIT
3984.225	3985.352	25091.89	RE	1.3			MIT
3983.93	3985.06	25093.7	RE	0.7			M
3983.422	3984.549	25096.95	RE	0.7W			MIT
3978.63	3979.76	25127.2	RE	0.9H			M
3975.654	3976.779	25145.98	RE I	1.3			MIT
3975.384	3976.509	25147.69	RE	0.3			MIT
3967.424	3968.546	25198.14	RE	1.4			MIT
3964.811	3965.933	25214.75	RE	1.3			MIT
3963.700	3964.821	25221.82	RE I	1.0			MIT
3963.270	3964.391	25224.55	RE I	1.4			MIT
3962.476	3963.597	25229.61	RE I	2.0			MIT
3961.033	3962.154	25238.80	RE I	1.5			MIT
3960.584	3961.705	25241.66	RE	1.2			MIT
3958.363	3959.483	25255.82	RE	1.0			MIT
3957.366	3958.486	25262.19	RE	1.1			MIT
3954.965	3956.084	25277.52	RE	0.3			MIT
3954.433	3955.552	25280.92	RE	1.3			MIT
3953.246	3954.365	25288.51	RE	0.5			MIT
3950.610	3951.728	25305.39	RE	1.5D			MIT
3945.910	3947.027	25335.53	RE I	1.6			MIT
3944.727	3945.844	25343.12	RE	1.5			MIT
3944.349	3945.465	25345.55	RE	1.2			MIT
3944.016	3945.132	25347.69	RE	1.2			MIT
3942.561	3943.677	25357.05	RE	1.0			MIT
3941.539	3942.655	25363.62	RE	1.3D			MIT
3936.903	3938.017	25393.49	RE I	1.6			MIT
3931.199	3932.312	25430.33	RE I	1.3			MIT
3930.522	3931.635	25434.71	RE	0.7W			MIT
3929.843	3930.956	25439.11	RE I	2.0			MIT
3929.205	3930.317	25443.24	RE	1.0			MIT
3928.707	3929.819	25446.46	RE I	1.1			MIT
3927.60	3928.71	25453.6	RE I	1.3W			M
3926.840	3927.952	25458.56	RE	1.0H			MIT
3926.518	3927.630	25460.65	RE	1.0			MIT
3924.674	3925.785	25472.61	RE	0.7W			MIT
3924.089	3925.200	25476.41	RE	1.0			MIT
3923.580	3924.691	25479.71	RE I	1.3			MIT
3922.272	3923.383	25488.21	RE	1.0			MIT
3921.785	3922.896	25491.37	RE	0.7			MIT
3920.879	3921.989	25497.26	RE	1.60			MIT
3919.11	3920.22	25508.8	RE	0.9			MIT
3917.275	3918.384	25520.72	RE I	2.0W			MIT
3915.233	3916.342	25534.03	RE	0.3			MIT
3913.918	3915.027	25542.61	RE I	1.5			MIT
3911.775	3912.883	25556.60	RE	0.5			MIT
3908.202	3909.309	25579.97	RE	1.4			MIT
3905.120	3906.226	25600.16	RE	1.3			MIT
3902.821	3903.927	25615.24	RE	0.5			MIT
3902.583	3903.689	25616.80	RE	0.8			MIT
3901.096	3902.201	25626.56	RE I	1.0			MIT
3900.901	3902.006	25627.84	RE I	1.3			MIT
3896.111	3897.215	25659.35	RE	1.4			MIT
3895.409	3896.513	25663.97	RE	1.0			MIT
3893.917	3895.020	25673.81	RE	0.3			MIT
3891.398	3892.501	25690.43	RE	1.2			MIT
3889.957	3891.059	25699.94	RE	1.4			MIT
3887.952	3889.054	25713.20	RE	1.3			MIT
3887.482	3888.584	25716.30	RE	1.3			MIT
3881.890	3882.990	25753.35	RE I	1.4			MIT
3878.852	3879.951	25773.52	RE	1.0			MIT
3876.893	3877.992	25786.54	RE I	1.8			MIT
3875.248	3876.346	25797.49	RE I	1.6			MIT
3869.922	3871.019	25832.99	RE I	1.6			MIT
3864.764	3865.860	25867.47	RE	0.5			MIT
3863.750	3864.845	25874.26	RE I	1.2			MIT
3863.153	3864.248	25878.25	RE	1.0			MIT
3862.116	3863.211	25885.20	RE	1.0			MIT
3860.857	3861.952	25893.64	RE I	0.5			MIT
3855.935	3857.028	25926.70	RE	1.3			MIT
3852.706	3853.799	25948.42	RE	0.9			MIT
3851.983	3853.075	25953.29	RE	1.3			MIT
3848.96	3850.05	25973.7	RE	0.5			M
3846.86	3847.95	25987.9	RE	0.5			M
3843.453	3844.543	26010.89	RE I	1.7			MIT
3841.493	3842.583	26024.16	RE	1.2			MIT
3836.324	3837.412	26059.23	RE I	1.6			MIT
3834.228	3835.316	26073.47	RE I	1.7			MIT
3833.704	3834.792	26077.04	RE I	1.5			MIT
3832.406	3833.493	26085.87	RE	1.3			MIT
3829.806	3830.893	26103.58	RE I	1.4			MIT
3828.328	3829.414	26113.66	RE I	1.5			MIT
3827.621	3828.707	26118.48	RE	1.2			MIT
3827.041	3828.127	26122.44	RE I	1.5			MIT
3823.754	3824.839	26144.89	RE	1.4W			MIT
3817.561	3818.644	26187.30	RE I	1.6			MIT
3815.66	3816.74	26200.4	RE I	1.4			M
3812.826	3813.908	26219.82	RE	0.5			MIT
3812.260	3813.342	26223.72	RE	1.2			MIT
3810.103	3811.184	26238.56	RE I	1.3			MIT
3808.219	3809.300	26251.54	RE	1.2			MIT
3807.742	3808.823	26254.83	RE I	1.3			MIT
3805.417	3806.497	26270.87	RE	0.3H			MIT
3800.948	3802.027	26301.76	RE	0.5			MIT
3797.591	3798.669	26325.01	RE I	1.6			MIT
3796.600	3797.678	26331.88	RE I	1.8			MIT
3795.800	3796.878	26337.43	RE	1.60			MIT
3788.11	3789.19	26390.9	RE	0.5			MIT
3787.526	3788.602	26394.96	RE I	1.9			MIT
3787.213	3788.289	26397.15	RE	1.0			MIT
3784.177	3785.252	26418.32	PE	1.3			MIT
3780.361	3781.435	26444.99	PE	1.3			MIT
3777.663	3778.736	26463.87	RE I	1.6			MIT
3775.464	3776.536	26479.29	RE	1.0			MIT
3770.736	3771.807	26512.49	RE I	0.9			MIT

AIR WAVELENGTH (ANGSTRMS)	VACUUM WAVELENGTH (ANGSTRMS)	WAVENUMBER (KAYSERS)	LMENT		LOG INTENSITY ARC	SPK	DIS	REF
3769.307	3770.378	26522.54	RE		0.8			MIT
3768.253	3769.324	26529.96	RE		0.8			MIT
3766.476	3767.546	26542.48	RE	I	1.3			MIT
3763.495	3764.564	26563.50	RE	I	1.2K			MIT
3761.025	3762.094	26580.94	RE		0.3H			MIT
3757.628	3758.696	26604.97	RE	I	1.4W			MIT
3755.625	3756.692	26619.16	RE		1.2			MIT
3752.78	3753.85	26639.3	RE		0.5D			MIT
3750.680	3751.746	26654.26	RE		1.0			MIT
3749.020	3750.086	26666.06	RE		0.7			MIT
3746.934	3747.999	26680.90	RE		1.0			MIT
3745.44	3746.50	26691.6	RE	I	1.6O			MIT
3742.269	3743.333	26714.16	RE		1.2			MIT
3740.429	3741.492	26727.30	RE	I	1.3			MIT
3740.096	3741.159	26729.68	RE	I	1.7O			MIT
3739.620	3740.683	26733.08	RE		0.5			MIT
3736.834	3737.896	26753.01	RE		1.2			MIT
3735.329	3736.391	26763.79	RE	I	1.6			MIT
3735.00	3736.06	26766.1	RE	I	1.5O			M
3732.275	3733.336	26785.69	RE	I	1.5			MIT
3731.871	3732.932	26788.59	RE	I	1.4			MIT
3727.492	3728.552	26820.06	RE		0.9			MIT
3726.746	3727.806	26825.43	RE		1.2D			MIT
3726.500	3727.560	26827.20	RE		0.9			MIT
3725.767	3726.827	26832.48	RE	I	1.6H			MIT
3717.29	3718.35	26893.7	RE	I	2.2O			M
3715.02	3716.08	26910.1	RE		0.7			M
3713.18	3714.24	26923.4	RE		0.7			MIT
3712.94	3714.00	26925.2	RE		0.8			M
3711.526	3712.582	26935.43	RE	I	1.2			MIT
3709.937	3710.993	26946.97	RE	I	1.6			MIT
3708.764	3709.819	26955.49	RE		1.2			MIT
3705.69	3706.74	26977.9	RE	I	0.7			MIT
3704.994	3706.048	26982.92	RE	I	1.3			MIT
3704.845	3705.899	26984.00	RE	I	1.3			MIT
3704.462	3705.516	26986.79	RE		1.4			MIT
3703.241	3704.295	26995.69	RE	I	1.6			MIT
3702.100	3703.153	27004.01	RE		1.0			MIT
3701.18	3702.23	27010.7	RE		0.7D			MIT
3700.366	3701.419	27016.67	RE		0.3H			MIT
3697.704	3698.756	27036.11	RE		2.2W			MIT
3692.126	3693.177	27076.96	RE		1.2			MIT
3691.500	3692.551	27081.55	RE	I	2.0W			MIT
3691.381	3692.432	27082.42	RE		1.1			MIT
3690.384	3691.434	27089.74	RE		0.9			MIT
3689.522	3690.572	27096.07	RE	I	2.0O			MIT
3688.894	3689.944	27100.68	RE		1.0			MIT
3688.648	3689.698	27102.49	RE		1.0			MIT
3686.458	3687.507	27118.59	RE		0.9			MIT
3681.311	3682.359	27156.50	RE		1.2			MIT
3680.216	3681.264	27164.58	RE		1.2			MIT
3678.39	3679.44	27178.1	RE		0.8			MIT
3676.572	3677.619	27191.51	RE		0.9			MIT
3676.002	3677.049	27195.72	RE		1.3			MIT
3672.400	3673.446	27222.40	RE	I	1.3			MIT
3670.522	3671.567	27236.32	RE	I	1.3			MIT
3670.361	3671.406	27237.52	RE		1.2			MIT
3670.025	3671.070	27240.01	RE	I	1.2			MIT
3669.774	3670.819	27241.88	RE	I	1.4			MIT
3669.433	3670.478	27244.41	RE		1.2			MIT
3663.029	3664.072	27292.04	RE		0.7			MIT
3662.121	3663.164	27298.80	RE		1.0			MIT
3660.528	3661.571	27310.68	RE		1.4			MIT
3658.74	3659.78	27324.0	RE		0.6			MIT
3654.930	3655.971	27352.51	RE		1.2			MIT
3654.358	3655.399	27356.79	RE	I	0.9			MIT
3653.615	3654.656	27362.36	RE	I	1.2			MIT
3653.161	3654.202	27365.76	RE		0.6H			MIT
3651.967	3653.008	27374.70	RE	I	1.6			MIT
3651.661	3652.701	27377.00	RE		1.4O			MIT

AIR WAVELENGTH (ANGSTRMS)	VACUUM WAVELENGTH (ANGSTRMS)	WAVENUMBER (KAYSERS)	LMENT		LOG INTENSITY ARC	SPK	DIS	REF
3649.507	3650.547	27393.16	RE		0.5H			MIT
3646.628	3647.667	27414.78	RE		1.0			MIT
3645.590	3646.629	27422.59	RE		0.8			MIT
3642.987	3644.025	27442.18	RE	I	2.0			MIT
3639.148	3640.185	27471.13	RE		1.0H			MIT
3637.843	3638.880	27480.98	RE	I	1.7			MIT
3637.235	3638.272	27485.58	RE		0.8			MIT
3637.062	3638.099	27486.89	RE	I	1.4			MIT
3636.761	3637.798	27489.16	RE		0.5			MIT
3629.200	3630.235	27546.43	RE		0.6			MIT
3625.906	3626.940	27571.45	RE	I	1.0			MIT
3621.465	3622.498	27605.26	RE	I	1.5			MIT
3617.88	3618.91	27632.6	RE		1.1W			MIT
3617.247	3618.279	27637.45	RE	I	1.2			MIT
3617.080	3618.111	27638.73	RE	I	1.7			MIT
3610.493	3611.523	27689.15	RE		1.6			MIT
3607.229	3608.258	27714.21	RE		0.5			MIT
3606.558	3607.587	27719.36	RE		0.5W			MIT
3604.396	3605.424	27735.99	RE		1.2W			MIT
3598.769	3599.796	27779.35	RE		1.3			MIT
3596.390	3597.416	27797.73	RE	I	1.4			MIT
3596.222	3597.248	27799.03	RE	I	1.3H			MIT
3595.166	3596.192	27807.19	RE		1.3H			MIT
3593.397	3594.422	27820.88	RE		1.2O			MIT
3590.879	3591.904	27840.39	RE		1.2			MIT
3585.321	3586.344	27883.55	RE		1.5			MIT
3585.025	3586.048	27885.85	RE		1.2			MIT
3583.022	3584.045	27901.44	RE	I	2.0W			MIT
3580.968	3581.990	27917.44	RE	I	1.6W			MIT
3580.133	3581.155	27923.95	RE		1.9			MIT
3579.127	3580.149	27931.80	RE	I	1.7			MIT
3577.350	3578.371	27945.68	RE		0.3			MIT
3572.805	3573.825	27981.22	RE	I	0.3H			MIT
3571.706	3572.726	27989.83	RE		0.7			MIT
3570.256	3571.275	28001.20	RE		1.5			MIT
3568.231	3569.250	28017.09	RE		1.6			MIT
3565.202	3566.220	28040.89	RE		0.3			MIT
3564.730	3565.748	28044.61	RE		1.3			MIT
3562.456	3563.473	28062.51	RE		0.3H			MIT
3559.409	3560.426	28086.53	RE	I	0.3			MIT
3558.946	3559.963	28090.18	RE	I	1.7W			MIT
3553.652	3554.667	28132.03	RE	I	1.4W			MIT
3551.595	3552.610	28148.32	RE		1.4			MIT
3551.306	3552.321	28150.61	RE		1.6W			MIT
3549.889	3550.903	28161.85	RE	I	1.6			MIT
3549.323	3550.337	28166.34	RE		0.5			MIT
3548.111	3549.125	28175.96	RE		0.5			MIT
3544.369	3545.382	28205.71	RE		0.3			MIT
3543.682	3544.695	28211.18	RE		0.5			MIT
3539.947	3540.959	28240.94	RE		1.2			MIT
3539.329	3540.340	28245.87	RE		1.4			MIT
3537.466	3538.477	28260.75	RE	I	1.9W			MIT
3536.158	3537.169	28271.20	RE		0.9			MIT
3534.826	3535.836	28281.85	RE	I	1.4			MIT
3529.816	3530.825	28321.99	RE		1.0			MIT
3529.212	3530.221	28326.84	RE		1.4			MIT
3526.735	3527.743	28346.73	RE		1.3			MIT
3522.324	3523.331	28382.23	RE		0.5W			MIT
3522.174	3523.181	28383.44	RE		0.9			MIT
3521.746	3522.753	28386.89	RE		0.7			MIT
3520.732	3521.739	28395.06	RE	I	1.5			MIT
3517.326	3518.332	28422.56	RE		1.7			MIT
3516.648	3517.654	28428.04	RE		1.8			MIT
3516.154	3517.159	28432.03	RE		0.3			MIT
3515.754	3516.759	28435.27	RE		0.5			MIT
3515.003	3516.008	28441.34	RE		0.3			MIT
3512.287	3513.291	28463.34	RE	I	1.7			MIT
3510.894	3511.898	28474.63	RE	I	1.2			MIT
3506.392	3507.395	28511.19	RE	I	1.2			MIT
3503.718	3504.720	28532.95	RE		1.2O			MIT

AIR WAVELENGTH (ANGSTRMS)	VACUUM WAVELENGTH (ANGSTRMS)	WAVENUMBER (KAYSERS)	LMENT	LOG ARC	INTENSITY SPK	DIS	REF
3503.059	3504.061	28538.31	RE I	1.9			MIT
3502.731	3503.733	28540.99	RE	1.3			MIT
3495.889	3496.889	28596.84	RE	1.2			MIT
3494.727	3495.727	28606.35	RE I	1.3			MIT
3491.439	3492.438	28633.29	RE	0.5			MIT
3490.862	3491.861	28638.02	RE	1.4			MIT
3488.846	3489.844	28654.57	RE	0.7			MIT
3487.521	3488.519	28665.46	RE	0.6			MIT
3484.733	3485.730	28688.39	RE	0.5			MIT
3484.354	3485.351	28691.51	RE	0.5			MIT
3482.237	3483.234	28708.95	RE	1.6			MIT
3480.853	3481.849	28720.37	RE I	1.7			MIT
3480.381	3481.377	28724.26	RE I	1.7			MIT
3478.556	3479.552	28739.33	RE I	0.3			MIT
3477.143	3478.138	28751.01	RE	1.2			MIT
3476.441	3477.436	28756.82	RE	1.5			MIT
3476.023	3477.018	28760.27	RE	0.5H			MIT
3475.185	3476.180	28767.21	RE	0.9			MIT
3474.208	3475.203	28775.30	RE I	1.4			MIT
3473.475	3474.470	28781.37	RE I	1.0			MIT
3472.874	3473.868	28786.35	RE I	1.2			MIT
3472.724	3473.718	28787.60	RE I	1.4			MIT
3472.653	3473.647	28788.19	RE	1.1	0.5		MIT
3471.994	3472.988	28793.65	RE	1.0H			MIT
3468.648	3469.641	28821.42	RE	1.0			MIT
3467.955	3468.948	28827.18	RE I	2.0W			MIT
3465.983	3466.976	28843.58	RE I	1.3			MIT
3464.722	3465.714	28854.08	RE I	2.0			MIT
3460.47	3461.46	28889.5	RE I	3.0O			MIT
3458.884	3459.875	28902.78	RE I	1.4			MIT
3453.502	3454.491	28947.82	RE I	1.6			MIT
3453.285	3454.274	28949.64	RE	1.3			MIT
3451.808	3452.797	28962.03	RE I	2.0			MIT
3450.084	3451.073	28976.50	RE	1.5			MIT
3449.371	3450.359	28982.49	RE	2.0R			MIT
3446.576	3447.564	29005.99	RE	0.3W			MIT
3442.965	3443.952	29036.41	RE I	1.3			MIT
3441.256	3442.242	29050.83	RE	1.6			MIT
3437.720	3438.705	29080.71	RE I	2.0			MIT
3433.728	3434.712	29114.52	RE	0.5			MIT
3431.815	3432.799	29130.75	RE I	0.6			MIT
3428.830	3429.813	29156.11	RE	0.5			MIT
3428.511	3429.494	29158.82	RE I	1.0			MIT
3427.622	3428.605	29166.38	RE I	1.7			MIT
3426.191	3427.173	29178.56	RE I	1.5			MIT
3424.604	3425.586	29192.09	RE I	2.5O			MIT
3421.579	3422.560	29217.89	RE	1.2			MIT
3420.758	3421.739	29224.91	RE I	1.6			MIT
3419.400	3420.381	29236.51	RE I	1.9			MIT
3419.244	3420.225	29237.85	RE I	1.2			MIT
3417.799	3418.779	29250.21	RE I	1.6R			MIT
3413.737	3414.716	29285.01	RE	1.4			MIT
3409.829	3410.807	29318.57	RE I	1.5			MIT
3408.676	3409.654	29328.49	RE	2.0			MIT
3405.890	3406.867	29352.48	RE I	2.2			MIT
3404.724	3405.701	29362.53	RE I	2.0			MIT
3401.195	3402.171	29393.00	RE	1.0W			MIT
3399.299	3400.275	29409.39	RE I	2.3W			MIT
3398.790	3399.765	29413.79	RE	1.2			MIT
3397.683	3398.658	29423.38	RE	0.8			MIT
3397.210	3398.185	29427.47	RE	1.2			MIT
3394.127	3395.101	29454.20	RE	1.3			MIT
3392.384	3393.358	29469.34	RE	0.5H			MIT
3391.237	3392.210	29479.30	RE	0.3H			MIT
3390.242	3391.215	29487.96	RE	1.6			MIT
3389.764	3390.737	29492.11	RE	1.2			MIT
3389.425	3390.398	29495.06	RE	1.4			MIT
3385.774	3386.746	29526.87	RE	1.3			MIT
3379.719	3380.690	29579.76	RE I	1.9			MIT
3379.046	3380.016	29585.66	RE	1.7			MIT

AIR WAVELENGTH (ANGSTRMS)	VACUUM WAVELENGTH (ANGSTRMS)	WAVENUMBER (KAYSERS)	LMENT	LOG ARC	INTENSITY SPK	DIS	REF
3377.741	3378.711	29597.09	RE	1.4			MIT
3370.894	3371.862	29657.20	RE	0.8			MIT
3368.619	3369.587	29677.23	RE	0.3H			MIT
3367.687	3368.654	29685.44	RE	0.7			MIT
3367.489	3368.456	29687.19	RE I	1.2			MIT
3366.878	3367.845	29692.58	RE I	0.5			MIT
3366.184	3367.151	29698.70	RE I	1.2W			MIT
3365.840	3366.807	29701.73	RE	1.3			MIT
3365.740	3366.707	29702.61	RE	1.3			MIT
3363.032	3363.998	29726.53	RE	1.3			MIT
3362.748	3363.714	29729.04	RE I	1.6			MIT
3361.833	3362.799	29737.13	RE	0.6			MIT
3361.142	3362.108	29743.25	RE I	1.4			MIT
3359.814	3360.779	29755.00	RE I	1.4O			MIT
3359.205	3360.170	29760.40	RE I	1.6			MIT
3358.568	3359.533	29766.04	RE I	1.0			MIT
3358.032	3358.997	29770.79	RE	1.4			MIT
3356.825	3357.790	29781.50	RE	0.6			MIT
3356.461	3357.426	29784.73	RE I	1.3			MIT
3356.328	3357.293	29785.91	RE	1.3			MIT
3355.912	3356.876	29789.60	RE	1.2W			MIT
3355.663	3356.627	29791.81	RE	1.3			MIT
3355.626	3356.590	29792.14	RE	1.0			MIT
3355.290	3356.254	29795.12	RE	1.8H			MIT
3353.211	3354.175	29813.59	RE I	1.6			MIT
3349.910	3350.873	29842.97	RE I	1.3			MIT
3347.586	3348.548	29863.69	RE	0.6H			MIT
3346.608	3347.570	29872.41	RE	0.7			MIT
3346.199	3347.161	29876.06	RE I	2.0			MIT
3344.347	3345.308	29892.61	RE I	2.2			MIT
3342.263	3343.224	29911.25	RE I	2.3			MIT
3340.308	3341.268	29928.75	RE	0.3			MIT
3339.692	3340.652	29934.27	RE	1.5			MIT
3338.174	3339.134	29947.89	RE I	2.2			MIT
3337.254	3338.214	29956.14	RE	0.7D			MIT
3335.372	3336.331	29973.04	RE I	2.0			MIT
3331.518	3332.476	30007.72	RE I	1.5			MIT
3328.175	3329.132	30037.86	RE	0.5			MIT
3327.725	3328.682	30041.92	RE	1.2H			MIT
3327.204	3328.161	30046.62	RE	1.0			MIT
3326.876	3327.833	30049.58	RE	0.5H			MIT
3324.933	3325.890	30067.14	RE I	1.4			MIT
3322.478	3323.434	30089.36	RE I	2.2			MIT
3322.205	3323.161	30091.83	RE I	1.4O			MIT
3321.462	3322.418	30098.56	RE	1.0			MIT
3320.651	3321.606	30105.91	RE	0.3			MIT
3318.671	3319.626	30123.88	RE	1.5			MIT
3317.017	3317.972	30138.90	RE	0.7			MIT
3313.950	3314.904	30166.79	RE I	1.6			MIT
3312.294	3313.247	30181.87	RE I	1.4			MIT
3309.321	3310.274	30208.98	RE	0.8			MIT
3308.876	3309.828	30213.05	RE	1.2			MIT
3308.475	3309.427	30216.71	RE	0.6			MIT
3308.250	3309.202	30218.76	RE	1.3			MIT
3307.010	3307.962	30230.09	RE	1.2H			MIT
3306.481	3307.433	30234.93	RE	0.3			MIT
3303.753	3304.704	30259.90	RE I	1.6			MIT
3303.212	3304.163	30264.85	RE	1.5			MIT
3302.227	3303.178	30273.88	RE	1.5			MIT
3301.600	3302.551	30279.63	RE I	1.7			MIT
3300.970	3301.920	30285.41	RE I	1.4			MIT
3296.991	3297.940	30321.96	RE I	1.8			MIT
3296.698	3297.647	30324.65	RE I	1.9			MIT
3294.831	3295.780	30341.83	RE I	1.5			MIT
3290.098	3291.046	30385.48	RE I	0.7H			MIT
3287.134	3288.081	30412.88	RE	1.3H			MIT
3285.654	3286.601	30426.58	RE	1.4			MIT
3277.721	3278.666	30500.21	RE	1.2			MIT
3271.089	3272.032	30562.05	RE	0.5			MIT
3270.048	3270.991	30571.78	RE	0.7H			MIT

AIR WAVELENGTH (ANGSTRMS)	VACUUM WAVELENGTH (ANGSTRMS)	WAVENUMBER (KAYSERS)	LMENT	LOG ARC	INTENSITY SPK	DIS	REF
3269.037	3269.979	30581.23	RE	1.5			MIT
3268.897	3269.839	30582.54	RE I	1.5			MIT
3268.479	3269.421	30586.45	RE I	1.6			MIT
3268.079	3269.021	30590.20	RE	1.6			MIT
3266.850	3267.792	30601.71	RE I	1.5			MIT
3262.765	3263.706	30640.02	RE I	1.4			MIT
3261.554	3262.494	30651.39	RE	1.7			MIT
3259.549	3260.489	30670.25	RE	2.0			MIT
3258.850	3259.790	30676.82	RE I	2.0			MIT
3258.067	3259.007	30684.20	RE I	0.6W			MIT
3256.289	3257.228	30700.95	RE	0.9			MIT
3255.801	3256.740	30705.55	RE	0.6			MIT
3253.948	3254.887	30723.04	RE	1.2			MIT
3253.185	3254.123	30730.24	RE	0.9W			MIT
3252.257	3253.195	30739.01	RE I	1.6			MIT
3248.547	3249.484	30774.11	RE	1.4			MIT
3246.311	3247.248	30795.31	RE	1.3			MIT
3241.464	3242.399	30841.36	RE I	1.4			MIT
3239.169	3240.104	30863.21	RE	1.0H			MIT
3238.364	3239.299	30870.88	RE	0.3			MIT
3237.511	3238.445	30879.01	RE I	1.5			MIT
3236.998	3237.932	30883.91	RE	0.3O			MIT
3235.942	3236.876	30893.99	RE	1.7			MIT
3228.732	3229.664	30962.97	RE	1.4			MIT
3228.330	3229.262	30966.83	RE	0.7			MIT
3227.456	3228.388	30975.21	RE	1.6W			MIT
3219.917	3220.847	31047.74	RE	0.3			MIT
3214.822	3215.751	31096.94	RE	0.5			MIT
3214.107	3215.035	31103.86	RE	1.4			MIT
3213.891	3214.819	31105.95	RE	1.0			MIT
3213.489	3214.417	31109.84	RE	1.3W			MIT
3212.945	3213.873	31115.10	RE	1.7			MIT
3212.379	3213.307	31120.59	RE	0.3			MIT
3211.756	3212.684	31126.62	RE I	1.6			MIT
3208.224	3209.151	31160.89	RE	1.2			MIT
3206.461	3207.387	31178.02	RE	0.3H			MIT
3205.402	3206.328	31188.32	RE	1.2H			MIT
3204.685	3205.611	31195.30	RE	0.3			MIT
3204.202	3205.128	31200.00	RE	2.5			MIT
3202.221	3203.146	31219.30	RE	0.3			MIT
3200.724	3201.649	31233.91	RE I	1.4			MIT
3200.039	3200.964	31240.59	RE I	1.7W			MIT
3199.518	3200.443	31245.68	RE I	0.5H			MIT
3198.582	3199.506	31254.82	RE	1.5			MIT
3194.477	3195.400	31294.98	RE I	1.7W			MIT
3193.202	3194.125	31307.48	RE	1.4			MIT
3192.368	3193.291	31315.66	RE I	1.6			MIT
3190.789	3191.712	31331.15	RE I	1.5			MIT
3190.173	3191.095	31337.20	RE	1.4			MIT
3186.291	3187.212	31375.38	RE	1.0			MIT
3185.563	3186.484	31382.55	RE I	2.3			MIT
3184.754	3185.675	31390.52	RE I	2.2			MIT
3182.868	3183.789	31409.12	RE I	2.0			MIT
3182.667	3183.587	31411.10	RE	1.2			MIT
3178.612	3179.531	31451.18	RE I	1.5			MIT
3178.492	3179.411	31452.36	RE	1.2			MIT
3177.715	3178.634	31460.05	RE	1.9			MIT
3174.780	3175.698	31489.14	RE	1.3W			MIT
3174.619	3175.537	31490.73	RE I	1.5			MIT
3173.091	3174.009	31505.90	RE	1.4			MIT
3168.384	3169.301	31552.70	RE I	2.2W			MIT
3167.594	3168.511	31560.57	RE	0.5			MIT
3167.157	3168.074	31564.92	RE I	1.3			MIT
3166.904	3167.820	31567.45	RE	0.9			MIT
3166.473	3167.389	31571.74	RE I	0.9			MIT
3165.803	3166.719	31578.42	RE	0.3			MIT
3164.866	3165.782	31587.77	RE	0.5			MIT
3164.520	3165.436	31591.23	RE I	1.5			MIT
3159.311	3160.226	31643.31	RE I	1.2			MIT
3158.315	3159.229	31653.29	RE I	2.3			MIT

AIR WAVELENGTH (ANGSTRMS)	VACUUM WAVELENGTH (ANGSTRMS)	WAVENUMBER (KAYSERS)	LMENT	LOG ARC	INTENSITY SPK	DIS	REF
3157.711	3158.625	31659.34	RE	0.3			MIT
3153.787	3154.700	31698.73	RE I	1.9			MIT
3151.633	3152.546	31720.40	RE I	2.2W			MIT
3151.168	3152.081	31725.08	RE	1.4			MIT
3150.520	3151.432	31731.60	RE I	0.6			MIT
3150.050	3150.962	31736.34	RE	0.3			MIT
3148.204	3149.116	31754.95	RE	0.5			MIT
3146.544	3147.455	31771.70	RE	1.0			MIT
3142.616	3143.526	31811.41	RE	1.4W			MIT
3141.384	3142.294	31823.88	RE I	1.5			MIT
3139.948	3140.858	31838.44	RE	1.2			MIT
3139.797	3140.707	31839.97	RE	1.4W			MIT
3135.068	3135.976	31887.99	RE I	1.2			MIT
3134.018	3134.926	31898.68	RE I	1.5			MIT
3132.025	3132.933	31918.98	RE I	0.3			MIT
3128.952	3129.859	31950.32	RE I	2.0O			MIT
3125.518	3126.424	31985.42	RE I	1.5			MIT
3123.168	3124.073	32009.49	RE	1.4			MIT
3121.741	3122.646	32024.12	RE	0.7			MIT
3121.370	3122.275	32027.93	RE I	2.0			MIT
3119.203	3120.107	32050.18	RE	0.7			MIT
3118.193	3119.097	32060.56	RE	2.3			MIT
3114.631	3115.534	32097.22	RE	0.5			MIT
3113.214	3114.117	32111.83	RE	1.0			MIT
3111.569	3112.472	32128.81	RE I	1.4			MIT
3110.862	3111.764	32136.11	RE I	2.0			MIT
3109.759	3110.661	32147.51	RE I	1.3			MIT
3108.811	3109.713	32157.31	RE I	2.1			MIT
3107.871	3108.773	32167.04	RE	0.3			MIT
3104.651	3105.552	32200.40	RE I	1.5			MIT
3103.256	3104.156	32214.87	RE	0.5			MIT
3101.703	3102.603	32231.00	RE	0.5			MIT
3101.003	3101.903	32238.27	RE	0.3H			MIT
3100.666	3101.566	32241.78	RE I	2.0			MIT
3096.411	3097.310	32286.08	RE I	1.3			MIT
3095.81	3096.71	32292.4	RE I	1.5W			MIT
3095.061	3095.959	32300.17	RE	1.6			MIT
3093.652	3094.550	32314.88	RE I	1.8			MIT
3089.937	3090.834	32353.73	RE	1.3			MIT
3088.767	3089.664	32365.98	RE	1.8			MIT
3087.161	3088.057	32382.82	RE I	1.4			MIT
3085.542	3086.438	32399.81	RE	0.3			MIT
3085.352	3086.248	32401.80	RE	0.3			MIT
3084.776	3085.672	32407.85	RE	0.5			MIT
3084.214	3085.110	32413.76	RE I	1.6			MIT
3082.771	3083.666	32428.93	RE	0.3			MIT
3082.432	3083.327	32432.50	RE I	2.0W			MIT
3080.923	3081.818	32448.38	RE	0.7W			MIT
3078.857	3079.751	32470.15	RE I	1.4W			MIT
3076.306	3077.200	32497.08	RE	0.7			MIT
3076.147	3077.041	32498.76	RE	1.4			MIT
3075.03	3075.92	32510.6	RE	0.7H			M
3072.965	3073.858	32532.41	RE	1.6			MIT
3072.453	3073.346	32537.83	RE	0.7			MIT
3071.753	3072.646	32545.24	RE	1.0B			MIT
3071.165	3072.057	32551.48	RE	1.7			MIT
3069.945	3070.837	32564.41	RE	2.1			MIT
3068.765	3069.657	32576.93	RE	1.2			MIT
3067.390	3068.281	32591.53	RE I	1.8			MIT
3065.578	3066.469	32610.80	RE I	0.7			MIT
3065.282	3066.173	32613.95	RE	0.7			MIT
3064.600	3065.491	32621.20	RE	1.3			MIT
3061.614	3062.504	32653.02	RE	1.3			MIT
3061.223	3062.113	32657.19	RE	0.5			MIT
3060.324	3061.214	32666.78	RE	1.4			MIT
3058.786	3059.675	32683.21	RE I	1.7			MIT
3057.853	3058.742	32693.18	RE	0.9			MIT
3057.659	3058.548	32695.25	RE I	1.4			MIT
3056.459	3057.348	32708.09	RE	0.3			MIT
3054.904	3055.792	32724.74	RE I	1.6			MIT

AIR WAVELENGTH (ANGSTRMS)	VACUUM WAVELENGTH (ANGSTRMS)	WAVENUMBER (KAYSERS)	LMENT	LOG INTENSITY ARC	SPK	DIS	REF
3053.631	3054.519	32738.38	RE	1.5			MIT
3052.84	3053.73	32746.9	RE	0.3W			MIT
3052.232	3053.120	32753.38	RE	0.3			MIT
3049.803	3050.690	32779.47	RE I	0.6W			MIT
3048.591	3049.478	32792.50	RE	0.9			MIT
3047.280	3048.166	32806.61	RE	1.5W			MIT
3046.004	3046.890	32820.35	RE	1.0			MIT
3044.085	3044.971	32841.04	RE I	1.4			MIT
3042.297	3043.182	32860.34	RE	1.0			MIT
3041.995	3042.880	32863.60	RE	1.2			MIT
3041.004	3041.889	32874.31	RE I	1.3			MIT
3040.04	3040.92	32884.7	RE	1.3W			MIT
3037.960	3038.844	32907.25	RE	1.5			MIT
3036.550	3037.434	32922.53	RE	1.3			MIT
3034.536	3035.419	32944.38	RE I	1.5			MIT
3032.789	3033.672	32963.35	RE	1.3			MIT
3031.279	3032.161	32979.77	RE	1.0			MIT
3030.450	3031.332	32988.80	RE I	2.0			MIT
3026.26	3027.14	33034.5	RE	0.3			MIT
3025.060	3025.941	33047.57	RE I	1.0			MIT
3023.584	3024.465	33063.70	RE	0.3			MIT
3022.994	3023.874	33070.16	RE	1.4			MIT
3021.890	3022.770	33082.24	RE	1.5			MIT
3016.976	3017.855	33136.12	RE	1.3			MIT
3016.491	3017.370	33141.45	RE	2.0			MIT
3016.02	3016.90	33146.6	RE I	1.9O			MIT
3013.151	3014.029	33178.18	RE I	1.5			MIT
3011.945	3012.823	33191.47	RE	1.3W			MIT
3007.200	3008.076	33243.84	RE	0.7			MIT
3007.047	3007.923	33245.53	RE	1.2			MIT
3006.429	3007.305	33252.36	RE I	1.6			MIT
3005.966	3006.842	33257.48	RE I	1.0O			MIT
3004.339	3005.215	33275.49	RE	1.4			MIT
3004.138	3005.014	33277.72	RE I	1.6			MIT
3001.132	3002.007	33311.05	RE	1.6			MIT
3000.576	3001.451	33317.22	RE	0.9			MIT
2999.592	3000.467	33328.15	RE I	2.1			MIT
2998.553	2999.427	33339.70	RE	0.9			MIT
2997.703	2998.577	33349.15	RE I	1.2			MIT
2995.378	2996.251	33375.04	RE	1.3			MIT
2993.183	2994.056	33399.51	RE	1.2			MIT
2992.817	2993.690	33403.59	RE	0.5O			MIT
2992.372	2993.245	33408.56	RE I	2.0			MIT
2991.839	2992.712	33414.51	RE	0.3			MIT
2988.479	2989.351	33452.08	RE I	1.6			MIT
2986.051	2986.922	33479.28	RE	0.6			MIT
2984.762	2985.633	33493.74	RE	0.5			MIT
2982.183	2983.053	33522.70	RE	1.5O			MIT
2981.012	2981.882	33535.87	RE	1.2H			MIT
2980.831	2981.701	33537.90	RE	1.6			MIT
2978.159	2979.028	33567.99	RE	1.6			MIT
2977.306	2978.175	33577.61	RE	1.3			MIT
2976.292	2977.161	33589.05	RE I	1.7W			MIT
2975.249	2976.117	33600.82	RE	1.3			MIT
2975.026	2975.894	33603.34	RE I	1.2			MIT
2968.973	2969.840	33671.85	RE	1.3			MIT
2968.051	2968.918	33682.31	RE I	1.6			MIT
2967.250	2968.116	33691.40	RE	1.2			MIT
2965.758	2966.624	33708.35	RE I	2.2W			MIT
2965.125	2965.991	33715.54	RE I	2.0			MIT
2962.870	2963.735	33741.20	RE	1.3			MIT
2962.269	2963.134	33748.05	RE	1.4			MIT
2961.757	2962.622	33753.88	RE I	1.5W			MIT
2958.906	2959.770	33786.40	RE	0.9			MIT
2957.883	2958.747	33798.09	RE	0.3			MIT
2954.616	2955.479	33835.46	RE	1.0			MIT
2954.344	2955.207	33838.57	RE	1.4			MIT
2950.828	2951.690	33878.89	RE I	1.4			MIT
2949.880	2950.742	33889.78	RE I	1.3			MIT
2949.263	2950.125	33896.87	RE	1.2			MIT
2949.098	2949.960	33898.77	RE	1.3H			MIT
2946.576	2947.437	33927.78	RE	1.0			MIT
2945.895	2946.756	33935.62	RE	0.3			MIT
2945.706	2946.567	33937.80	RE	0.5			MIT
2944.321	2945.182	33953.76	RE	1.0			MIT
2943.380	2944.241	33964.62	RE I	1.2			MIT
2943.144	2944.004	33967.34	RE I	1.8			MIT
2941.557	2942.417	33985.66	RE	1.2			MIT
2937.814	2938.673	34028.96	RE	0.9			MIT
2936.498	2937.357	34044.21	RE	1.4			MIT
2934.009	2934.867	34073.09	RE I	0.5			MIT
2933.436	2934.294	34079.75	RE	0.3			MIT
2932.310	2933.168	34092.83	RE	1.0			MIT
2930.616	2931.473	34112.54	RE	1.7			MIT
2929.536	2930.393	34125.11	RE	1.5			MIT
2928.580	2929.437	34136.25	RE	0.8			MIT
2927.716	2928.573	34146.33	RE	1.0			MIT
2927.404	2928.261	34149.97	RE	2.1W			MIT
2926.940	2927.796	34155.38	RE	1.0W			MIT
2925.200	2926.056	34175.69	RE	1.4			MIT
2924.600	2925.456	34182.71	RE I	1.6O			MIT
2919.411	2920.266	34243.46	RE	1.4			MIT
2918.878	2919.732	34249.71	RE	1.0			MIT
2913.802	2914.655	34309.37	RE	0.3			MIT
2913.157	2914.010	34316.97	RE	1.3			MIT
2912.584	2913.437	34323.72	RE	0.6			MIT
2911.231	2912.084	34339.67	RE I	0.9			MIT
2910.084	2910.936	34353.21	RE I	1.2			MIT
2909.818	2910.670	34356.35	RE	1.6			MIT
2908.343	2909.195	34373.77	RE	1.3			MIT
2907.105	2907.957	34388.41	RE	1.2			MIT
2906.018	2906.869	34401.27	RE I	1.5			MIT
2905.582	2906.433	34406.43	RE I	1.7			MIT
2905.396	2906.247	34408.64	RE	1.3			MIT
2902.481	2903.331	34443.19	RE I	2.1O			MIT
2898.787	2899.637	34487.08	RE	1.2			MIT
2897.592	2898.441	34501.30	RE	1.0D			MIT
2896.448	2897.297	34514.93	RE	0.9			MIT
2896.016	2896.865	34520.08	RE I	2.1W			MIT
2895.657	2896.506	34524.36	RE I	1.4			MIT
2894.326	2895.174	34540.23	RE	1.3			MIT
2892.644	2893.492	34560.32	RE	1.4			MIT
2891.882	2892.730	34569.42	RE	1.6			MIT
2891.483	2892.331	34574.19	RE	1.4			MIT
2889.463	2890.310	34598.36	RE I	1.5W			MIT
2888.013	2888.860	34615.73	RE	0.5			MIT
2887.666	2888.513	34619.89	RE I	2.1			MIT
2887.312	2888.159	34624.14	RE	1.4			MIT
2886.938	2887.785	34628.62	RE	1.3			MIT
2885.931	2886.777	34640.70	RE	1.2			MIT
2885.167	2886.013	34649.88	RE	0.3			MIT
2884.634	2885.480	34656.28	RE	1.4			MIT
2884.055	2884.901	34663.24	RE	1.3			MIT
2883.450	2884.296	34670.51	RE	1.8			MIT
2882.235	2883.080	34685.12	RE	0.8			MIT
2879.276	2880.121	34720.77	RE	1.4			MIT
2876.872	2877.716	34749.78	RE	1.2			MIT
2876.028	2876.872	34759.98	RE	0.5			MIT
2875.286	2876.130	34768.95	RE I	1.9			MIT
2872.673	2873.516	34800.57	RE	1.0			MIT
2872.303	2873.146	34805.05	RE	1.2			MIT
2871.814	2872.657	34810.98	RE I	1.7			MIT
2868.151	2868.993	34855.44	RE	0.7			MIT
2867.198	2868.040	34867.02	RE	1.6			MIT
2864.834	2865.675	34895.79	RE	0.9			MIT
2864.569	2865.410	34899.02	RE	1.0			MIT
2857.436	2858.275	34986.13	RE	1.2			MIT
2855.531	2856.370	35009.47	RE	1.2			MIT
2852.857	2853.695	35042.29	RE	1.5W			MIT
2852.399	2853.237	35047.91	RE	1.0			MIT

AIR WAVELENGTH (ANGSTRMS)	VACUUM WAVELENGTH (ANGSTRMS)	WAVENUMBER (KAYSERS)	LMENT	LOG ARC	INTENSITY SPK	DIS	REF
2850.977	2851.815	35065.39	RE I	1.6			MIT
2850.279	2851.117	35073.98	RE	0.3			MIT
2847.748	2848.585	35105.15	RE	0.5			MIT
2846.976	2847.813	35114.67	RE	1.5W			MIT
2846.478	2847.315	35120.81	RE	0.5			MIT
2844.165	2845.001	35149.37	RE	1.4			MIT
2843.001	2843.837	35163.76	RE I	1.5			MIT
2840.348	2841.183	35196.60	RE I	1.6			MIT
2839.186	2840.021	35211.01	RE	0.8			MIT
2838.120	2838.955	35224.23	RE	0.9			MIT
2837.549	2838.383	35231.32	RE	1.6			MIT
2837.015	2837.849	35237.95	RE	0.3H			MIT
2834.615	2835.449	35267.79	RE I	1.5			MIT
2834.060	2834.894	35274.69	RE	2.0R			MIT
2830.834	2831.667	35314.89	RE	0.3H			MIT
2830.091	2830.924	35324.16	RE	0.5H			MIT
2829.877	2830.710	35326.83	RE	0.5H			MIT
2827.834	2828.666	35352.35	RE	0.8			MIT
2827.528	2828.360	35356.18	RE	1.5			MIT
2826.764	2827.596	35365.73	RE	0.5H			MIT
2825.461	2826.292	35382.04	RE	1.3			MIT
2824.247	2825.078	35397.25	RE	1.3			MIT
2823.193	2824.024	35410.46	RE	1.4			MIT
2822.409	2823.240	35420.30	RE I	1.0			MIT
2822.124	2822.955	35423.88	RE	1.3			MIT
2819.955	2820.785	35451.12	RE I	2.2O			MIT
2816.963	2817.792	35488.77	RE I	1.5			MIT
2816.329	2817.158	35496.76	RE	1.6			MIT
2815.642	2816.471	35505.42	RE	1.0			MIT
2814.677	2815.506	35517.60	RE	1.7			MIT
2813.962	2814.791	35526.62	RE	1.5			MIT
2813.115	2813.943	35537.32	RE	1.3			MIT
2812.360	2813.188	35546.86	RE	1.4			MIT
2812.070	2812.898	35550.52	RE	1.3			MIT
2807.870	2808.697	35603.70	RE	1.7W			MIT
2805.976	2806.803	35627.73	RE	0.8			MIT
2803.236	2804.062	35662.55	RE	0.9H			MIT
2802.840	2803.666	35667.59	RE	0.7W			MIT
2802.251	2803.077	35675.08	RE	1.0H			MIT
2800.695	2801.520	35694.90	RE	1.2H			MIT
2798.111	2798.936	35727.87	RE	1.3			MIT
2797.465	2798.290	35736.12	RE	0.5			MIT
2796.620	2797.444	35746.91	RE	0.9			MIT
2796.081	2796.905	35753.80	RE	1.0			MIT
2794.163	2794.987	35778.34	RE	0.3			MIT
2793.663	2794.487	35784.75	RE	0.6H			MIT
2791.631	2792.454	35810.79	RE	1.0			MIT
2791.293	2792.116	35815.13	RE	1.8			MIT
2790.945	2791.768	35819.60	RE	1.6			MIT
2789.273	2790.096	35841.06	RE	1.3			MIT
2787.391	2788.213	35865.26	RE	0.3			MIT
2786.560	2787.382	35875.96	RE I	1.4			MIT
2786.139	2786.961	35881.38	RE	1.3			MIT
2785.718	2786.540	35886.80	RE	1.2			MIT
2785.428	2786.250	35890.54	RE	1.3			MIT
2785.208	2786.030	35893.37	RE I	1.5			MIT
2783.573	2784.394	35914.45	RE I	2.2W			MIT
2783.145	2783.966	35919.98	RE	1.0			MIT
2781.448	2782.269	35941.89	RE	1.6			MIT
2780.826	2781.647	35949.93	RE	0.5W			MIT
2779.299	2780.119	35969.68	RE	0.7			MIT
2778.502	2779.322	35980.00	RE	1.3			MIT
2778.091	2778.911	35985.32	RE	1.3			MIT
2777.723	2778.543	35990.09	RE	1.3W			MIT
2777.226	2778.046	35996.53	RE	1.0			MIT
2776.962	2777.782	35999.95	RE	1.2			MIT
2775.646	2776.465	36017.02	RE	1.2			MIT
2774.387	2775.206	36033.36	RE	0.8			MIT
2773.115	2773.934	36049.89	RE	1.5			MIT
2772.427	2773.246	36058.83	RE	0.5			MIT

AIR WAVELENGTH (ANGSTRMS)	VACUUM WAVELENGTH (ANGSTRMS)	WAVENUMBER (KAYSERS)	LMENT	LOG ARC	INTENSITY SPK	DIS	REF
2771.619	2772.437	36069.35	RE	0.5			MIT
2770.418	2771.236	36084.98	RE I	1.8			MIT
2769.332	2770.150	36099.13	RE	1.5W			MIT
2768.856	2769.674	36105.34	RE I	1.4			MIT
2768.282	2769.100	36112.82	RE	0.5			MIT
2767.751	2768.568	36119.75	RE	0.8			MIT
2766.556	2767.373	36135.35	RE	1.0			MIT
2766.403	2767.220	36137.35	RE	1.7W			MIT
2766.058	2766.875	36141.86	RE	1.0			MIT
2763.803	2764.619	36171.34	RE	1.7			MIT
2763.306	2764.122	36177.85	RE I	1.6W			MIT
2761.932	2762.748	36195.85	RE	1.4			MIT
2758.711	2759.526	36238.10	RE	1.2			MIT
2758.000	2758.815	36247.45	RE	1.8W			MIT
2757.485	2758.300	36254.22	RE	1.3			MIT
2755.223	2756.037	36283.98	RE I	1.4			MIT
2753.688	2754.502	36304.20	RE	0.3			MIT
2753.046	2753.860	36312.67	RE I	1.6			MIT
2752.857	2753.671	36315.16	RE	1.4			MIT
2747.438	2748.250	36386.79	RE I	1.5			MIT
2745.868	2746.680	36407.59	RE	1.4			MIT
2744.212	2745.024	36429.56	RE I	0.8W			MIT
2743.871	2744.683	36434.08	RE	1.4			MIT
2742.850	2743.661	36447.65	RE	1.3			MIT
2742.739	2743.550	36449.12	RE	1.0			MIT
2741.967	2742.778	36459.38	RE	1.0			MIT
2739.950	2740.761	36486.22	RE	1.4			MIT
2738.749	2739.559	36502.22	RE	0.8			MIT
2738.328	2739.138	36507.83	RE	1.3			MIT
2734.885	2735.694	36553.79	RE	1.2			MIT
2734.318	2735.127	36561.37	RE	1.3			MIT
2733.042	2733.851	36578.44	RE	1.6W			MIT
2732.211	2733.020	36589.56	RE	1.6			MIT
2731.559	2732.368	36598.30	RE	1.0W			MIT
2730.829	2731.637	36608.08	RE	1.2			MIT
2729.639	2730.447	36624.04	RE	1.3			MIT
2728.627	2729.435	36637.62	RE	1.3			MIT
2727.546	2728.354	36652.14	RE	1.5			MIT
2723.858	2724.665	36701.76	RE	1.3			MIT
2723.361	2724.168	36708.46	RE	1.2			MIT
2722.709	2723.515	36717.25	RE I	1.7			MIT
2722.205	2723.011	36724.05	RE	1.5			MIT
2720.658	2721.464	36744.93	RE	1.2			MIT
2719.537	2720.343	36760.07	RE	1.3			MIT
2716.761	2717.566	36797.63	RE	1.4			MIT
2715.771	2716.576	36811.05	RE	1.5			MIT
2715.470	2716.275	36815.13	RE I	2.0			MIT
2713.670	2714.474	36839.55	RE	1.2			MIT
2713.165	2713.969	36846.40	RE	1.4			MIT
2713.033	2713.837	36848.19	RE	1.4			MIT
2712.483	2713.287	36855.67	RE	1.5			MIT
2710.234	2711.037	36886.25	RE	1.2W			MIT
2707.860	2708.663	36918.58	RE	0.9			MIT
2707.411	2708.214	36924.71	RE	1.4			MIT
2706.489	2707.292	36937.29	RE	1.0			MIT
2706.056	2706.858	36943.19	RE	1.4			MIT
2704.373	2705.175	36966.18	RE	1.4			MIT
2703.250	2704.052	36981.54	RE	0.7			MIT
2702.665	2703.467	36989.54	RE	1.4			MIT
2699.583	2700.384	37031.77	RE	1.4			MIT
2698.801	2699.602	37042.50	RE	1.3			MIT
2697.266	2698.066	37063.58	RE I	1.7			MIT
2695.562	2696.362	37087.01	RE	1.5			MIT
2694.399	2695.199	37103.02	RE				MIT
2694.394	2695.194	37103.08	RE I	1.3	1.0		ME
2693.724	2694.523	37112.31	RE	1.0			MIT
2690.793	2691.592	37152.74	RE	1.5H			MIT
2690.260	2691.059	37160.10	RE	1.7H			MIT
2689.329	2690.127	37172.96	RE	1.4			MIT
2688.527	2689.325	37184.05	RE	2.0			MIT

AIR WAVELENGTH (ANGSTRMS)	VACUUM WAVELENGTH (ANGSTRMS)	WAVENUMBER (KAYSERS)	LMENT	LOG INTENSITY ARC	SPK	DIS	REF
2685.317	2686.114	37228.50	RE	1.2			MIT
2683.565	2684.362	37252.80	RE	1.7			MIT
2681.012	2681.808	37288.27	RE	1.2			MIT
2679.916	2680.712	37303.52	RE	1.5			MIT
2679.077	2679.873	37315.20	RE	1.4			MIT
2677.764	2678.560	37333.50	RE	1.4			MIT
2677.037	2677.832	37343.63	RE	1.4			MIT
2674.337	2675.132	37381.33	RE I	2.0			MIT
2672.774	2673.568	37403.19	RE	1.4			MIT
2671.841	2672.635	37416.25	RE	1.8			MIT
2670.797	2671.591	37430.88	RE	1.7			MIT
2670.236	2671.030	37438.74	RE I	1.5			MIT
2667.796	2668.589	37472.98	RE	1.2			MIT
2667.133	2667.926	37482.30	RE	1.5			MIT
2664.810	2665.603	37514.97	RE	1.4			MIT
2664.220	2665.012	37523.28	RE	1.4			MIT
2663.632	2664.424	37531.56	RE	2.2			MIT
2661.034	2661.826	37568.20	RE	1.0			MIT
2660.536	2661.328	37575.23	RE	0.8			MIT
2659.792	2660.583	37585.74	RE	1.2			MIT
2659.023	2659.814	37596.61	RE	1.4			MIT
2658.688	2659.479	37601.35	RE	1.2			MIT
2657.445	2658.236	37618.94	RE	1.3			MIT
2655.816	2656.606	37642.01	RE	1.4 O			MIT
2655.182	2655.972	37651.00	RE	1.4			MIT
2654.117	2654.907	37666.10	RE	1.7			MIT
2653.738	2654.528	37671.48	RE	1.2			MIT
2652.912	2653.702	37683.21	RE	1.5			MIT
2651.902	2652.691	37697.56	RE I	2.0			MIT
2649.583	2650.372	37730.55	RE	1.3			MIT
2649.049	2649.838	37738.16	RE	2.0			MIT
2648.505	2649.294	37745.91	RE	1.0			MIT
2647.124	2647.912	37765.60	RE I	2.0			MIT
2646.380	2647.168	37776.22	RE	1.3			MIT
2645.802	2646.590	37784.47	RE	1.3			MIT
2644.768	2645.556	37799.24	RE	1.0			MIT
2642.760	2643.547	37827.96	RE	2.1			MIT
2641.170	2641.957	37850.73	RE	1.3			MIT
2641.023	2641.810	37852.84	RE	0.7			MIT
2637.012	2637.798	37910.41	RE	1.3			MIT
2636.637	2637.423	37915.80	RE I	2.1			MIT
2635.824	2636.610	37927.50	RE	1.3 W			MIT
2633.622	2634.407	37959.21	RE	1.6			MIT
2633.008	2633.793	37968.06	RE	1.3			MIT
2631.584	2632.369	37988.60	RE	1.9			MIT
2630.757	2631.541	38000.54	RE	0.3			MIT
2630.153	2630.937	38009.27	RE	1.2			MIT
2625.814	2626.597	38072.07	RE	1.2			MIT
2623.306	2624.089	38108.47	RE	1.7 W			MIT
2622.758	2623.541	38116.43	RE	1.5			MIT
2621.976	2622.758	38127.80	RE	1.5			MIT
2621.147	2621.929	38139.86	RE	1.3			MIT
2620.349	2621.131	38151.47	RE	1.7			MIT
2620.023	2620.805	38156.22	RE	1.8			MIT
2617.445	2618.226	38193.80	RE	1.4			MIT
2617.118	2617.899	38198.57	RE	1.4			MIT
2616.711	2617.492	38204.51	RE	0.7			MIT
2615.681	2616.462	38219.55	RE	1.0			MIT
2614.561	2615.342	38235.92	RE	1.8			MIT
2613.756	2614.536	38247.70	RE	1.4			MIT
2613.233	2614.013	38255.35	RE	1.2			MIT
2612.925	2613.705	38259.86	RE	1.0			MIT
2612.724	2613.504	38262.81	RE	1.3			MIT
2611.604	2612.384	38279.21	RE	1.5			MIT
2609.399	2610.178	38311.56	RE	1.3			MIT
2608.501	2609.280	38324.75	RE	1.4			MIT
2607.316	2608.095	38342.16	RE	1.3			MIT
2603.870	2604.648	38392.90	RE	2.0			MIT
2603.463	2604.241	38398.90	RE	1.4			MIT
2602.927	2603.705	38406.81	RE	1.4			MIT
2602.550	2603.328	38412.37	RE	1.0			MIT
2601.880	2602.658	38422.27	RE	1.5			MIT
2600.871	2601.648	38437.17	RE	1.4			MIT
2599.859	2600.636	38452.13	RE	1.9			MIT
2598.593	2599.370	38470.86	RE	0.8			MIT
2597.964	2598.741	38480.18	RE	1.3			MIT
2596.941	2597.717	38495.33	RE	1.4			MIT
2596.781	2597.557	38497.71	RE	1.4			MIT
2596.402	2597.178	38503.33	RE	1.4			MIT
2595.237	2596.013	38520.61	RE I	1.8			MIT
2594.853	2595.629	38526.31	RE I	1.6			MIT
2592.868	2593.643	38555.80	RE	1.7 W			MIT
2591.588	2592.363	38574.84	RE	1.5			MIT
2591.430	2592.205	38577.19	RE	1.2			MIT
2591.138	2591.913	38581.54	RE	1.2			MIT
2587.137	2587.911	38641.20	RE	1.6 W			MIT
2586.998	2587.772	38643.28	RE	1.3			MIT
2586.788	2587.562	38646.42	RE	2.0			MIT
2584.768	2585.541	38676.62	RE	1.4			MIT
2582.772	2583.545	38706.51	RE	1.2			MIT
2581.425	2582.198	38726.70	RE	1.4			MIT
2580.314	2581.086	38743.37	RE	1.3			MIT
2579.021	2579.793	38762.80	RE	1.5 G			MIT
2576.319	2577.090	38803.45	RE	1.2			MIT
2574.215	2574.986	38835.16	RE	1.2			MIT
2573.774	2574.545	38841.82	RE	1.8			MIT
2571.82	2572.59	38871.3	RE	0.7 W			MIT
2571.258	2572.028	38879.82	RE	1.4			MIT
2568.644	2569.414	38919.38	RE	1.5 W			MIT
2565.843	2566.612	38961.87	RE	1.2			MIT
2564.188	2564.957	38987.01	RE	1.7			MIT
2563.009	2563.777	39004.95	RE	1.5			MIT
2561.467	2562.235	39028.43	RE	1.4			MIT
2559.885	2560.653	39052.54	RE	1.2			MIT
2559.712	2560.480	39055.18	RE	1.2			MIT
2559.079	2559.846	39064.84	RE	1.5			MIT
2558.049	2558.816	39080.57	RE	1.8			MIT
2556.511	2557.278	39104.08	RE	2.0			MIT
2555.661	2556.428	39117.09	RE	1.2			MIT
2554.931	2555.697	39128.26	RE	1.2			MIT
2554.631	2555.397	39132.86	RE	0.7			MIT
2554.162	2554.928	39140.04	RE	1.5 G			MIT
2553.556	2554.322	39149.33	RE	0.7			MIT
2552.701	2553.467	39162.44	RE	1.3			MIT
2552.026	2552.792	39172.80	RE	1.9			MIT
2550.097	2550.862	39202.43	RE	0.7			MIT
2548.889	2549.654	39221.01	RE	1.4			MIT
2548.138	2548.903	39232.57	RE	1.4			MIT
2545.494	2546.258	39273.31	RE	1.8			MIT
2544.886	2545.650	39282.70	RE	1.3			MIT
2544.739	2545.503	39284.97	RE	1.4			MIT
2544.214	2544.978	39293.07	RE	1.4			MIT
2543.831	2544.595	39298.99	RE	1.3			MIT
2543.667	2544.431	39301.52	RE	1.3 R			MIT
2541.090	2541.853	39341.37	RE	1.3 H			MIT
2540.519	2541.282	39350.22	RE I	1.4			MIT
2539.346	2540.109	39368.39	RE	1.2			MIT
2534.801	2535.563	39438.98	RE	2.0 W			MIT
2534.101	2534.863	39449.87	RE	1.0			MIT
2533.312	2534.073	39462.16	RE	1.4			MIT
2532.962	2533.723	39467.61	RE	1.2			MIT
2529.503	2530.263	39521.58	RE	1.5			MIT
2526.819	2527.579	39563.55	RE	1.3			MIT
2525.561	2526.321	39583.26	RE	1.3			MIT
2521.586	2522.345	39645.65	RE	2.0 G			MIT
2520.009	2520.767	39670.46	RE I	1.7			MIT
2516.116	2516.873	39731.84	RE	2.1			MIT
2515.471	2516.228	39742.02	RE	1.0			MIT
2514.504	2515.261	39757.31	RE	1.3			MIT
2512.557	2513.314	39788.11	RE	1.4			MIT

AIR WAVELENGTH (ANGSTRMS)	VACUUM WAVELENGTH (ANGSTRMS)	WAVENUMBER (KAYSERS)	LMENT	LOG ARC	INTENSITY SPK	INTENSITY DIS	REF
2508.988	2509.744	39844.71	RE I	2.1			MIT
2507.412	2508.167	39869.75	RE	1.5			MIT
2505.936	2506.691	39893.23	RE	1.3			MIT
2505.391	2506.146	39901.91	RE	1.3			MIT
2504.598	2505.353	39914.54	RE	1.9			MIT
2502.377	2503.131	39949.96	RE	1.8			MIT
2501.722	2502.476	39960.42	RE	1.9			MIT
2500.575	2501.329	39978.75	RE	1.3			MIT
2498.220	2498.973	40016.44	RE	1.5			MIT
2496.70	2497.45	40040.8	RE	1.3			MIT
2492.847	2493.599	40102.68	RE	1.3			MIT
2492.552	2493.304	40107.43	RE	1.2			MIT
2491.236	2491.988	40128.61	RE	0.3			MIT
2487.23	2487.98	40193.2	RE	1.7			MIT
2486.776	2487.527	40200.58	RE	1.7W			MIT
2485.809	2486.559	40216.21	RE	1.7			MIT
2483.918	2484.668	40246.83	RE	2.2			MIT
2478.988	2479.737	40326.86	RE	0.7			MIT
2476.278	2477.026	40370.99	RE	1.2			MIT
2475.172	2475.920	40389.03	RE	1.4			MIT
2474.727	2475.475	40396.29	RE	1.4			MIT
2473.720	2474.468	40412.73	RE	0.7			MIT
2470.428	2471.175	40466.58	RE	0.7			MIT
2467.702	2468.448	40511.28	RE	0.5			MIT
2463.314	2464.059	40583.44	RE	1.4			MIT
2461.860	2462.605	40607.41	RE	1.9			MIT
2461.179	2461.924	40618.64	RE	2.1			MIT
2460.240	2460.984	40634.15	RE	1.4			MIT
2455.851	2456.594	40706.76	RE	0.9			MIT
2453.100	2453.843	40752.41	RE	1.2			MIT
2450.892	2451.634	40789.12	RE	1.5			MIT
2449.713	2450.455	40808.75	RE	1.6			MIT
2446.988	2447.729	40854.19	RE	2.00			MIT
2442.52	2443.26	40928.9	RE	1.4			MIT
2441.48	2442.22	40946.4	RE	1.6			MIT
2438.473	2439.212	40996.84	RE	1.4			MIT
2436.061	2436.800	41037.43	RE	1.4			MIT
2433.291	2434.029	41084.14	RE	1.5			MIT
2432.693	2433.431	41094.24	RE	1.2			MIT
2432.173	2432.911	41103.02	RE	2.00			MIT
2431.534	2432.272	41113.83	RE	1.7			MIT
2429.653	2430.390	41145.65	RE	1.4			MIT
2429.50	2430.24	41148.2	RE	1.5			MIT
2421.762	2422.498	41279.71	RE	1.7			MIT
2421.39	2422.13	41286.0	RE	1.4			MIT
2416.257	2416.991	41373.75	RE	0.3			MIT
2410.37	2411.10	41474.8	RE	1.3			MIT
2405.602	2406.334	41556.99	RE	1.9			MIT
2405.054	2405.786	41566.46	RE	2.0			MIT
2399.67	2400.40	41659.7	RE	1.0	0.7		A
2398.73	2399.46	41676.0	RE	1.1			A
2397.36	2398.09	41699.9	RE	1.5	0.9		A
2396.81	2397.54	41709.4	RE	1.5	0.8		A
2396.04	2396.77	41722.8	RE	1.0	0.5		A
2393.91	2394.64	41759.9	RE	1.3	1.3W		A
2393.67	2394.40	41764.1	RE	1.3			A
2393.12	2393.85	41773.7	RE	1.2			A
2392.71	2393.44	41780.9	RE	1.0			A
2392.38	2393.11	41786.6	RE	1.3	0.8		A
2391.30	2392.03	41805.5	RE	1.3	0.8		A
2390.44	2391.17	41820.6	RE	1.3	0.8		A
2389.28	2390.01	41840.9	RE	0.7			A
2389.13	2389.86	41843.5	RE	1.5	0.8		A
2387.75	2388.48	41867.7	RE	1.2			A
2387.48	2388.21	41872.4	RE	1.3	0.8		A
2385.50	2386.23	41907.2	RE	1.1			A
2383.97	2384.70	41934.0	RE	1.0	0.5		A
2383.48	2384.21	41942.7	RE	1.4	0.7		A
2381.14	2381.87	41983.9	RE	1.6	0.8		A
2380.90	2381.63	41988.1	RE	1.1	0.7		A

AIR WAVELENGTH (ANGSTRMS)	VACUUM WAVELENGTH (ANGSTRMS)	WAVENUMBER (KAYSERS)	LMENT	LOG ARC	INTENSITY SPK	INTENSITY DIS	REF
2380.24	2380.97	41999.8	RE	1.2	0.7		A
2379.97	2380.70	42004.5	RE	0.8	0.8		A
2379.79	2380.52	42007.7	RE	1.2	0.9		A
2379.11	2379.84	42019.7	RE	0.7	1.1		A
2378.42	2379.15	42031.9	RE	0.6			A
2377.38	2378.11	42050.3	RE	1.3	0.8		A
2377.12	2377.85	42054.9	RE	1.0	0.3H		A
2375.96	2376.69	42075.4	RE	0.8			A
2375.83	2376.56	42077.7	RE	1.3	0.8		A
2375.07	2375.80	42091.2	RE	1.4	0.8		A
2373.52	2374.24	42118.7	RE	0.8	1.2		A
2373.29	2374.01	42122.7	RE	0.8	0.5		A
2371.52	2372.24	42154.2	RE	1.5	0.8		A
2370.78	2371.50	42167.3	RE	1.1	1.5		A
2369.29	2370.01	42193.9	RE	1.5	0.8		A
2368.90	2369.62	42200.8	RE	1.1	0.8		A
2368.54	2369.26	42207.2	RE	0.8	1.6		A
2368.14	2368.86	42214.3	RE	0.8	0.8		A
2367.90	2368.62	42218.6	RE	1.0			A
2367.69	2368.41	42222.4	RE	1.5	0.9		A
2367.02	2367.74	42234.3	RE	0.8	0.6		A
2365.69	2366.41	42258.0	RE	1.2	0.6		A
2365.32	2366.04	42264.7	RE	1.3	0.8		A
2363.82	2364.54	42291.5	RE	1.2	0.3H		A
2362.90	2363.62	42307.9	RE	1.3	0.8		A
2362.58	2363.30	42313.7	RE	1.3	0.8		A
2361.48	2362.20	42333.4	RE	0.6	0.8		A
2360.93	2361.65	42343.2	RE	0.8	0.7		A
2359.47	2360.19	42369.4	RE	0.9			A
2358.59	2359.31	42385.2	RE	1.0			A
2358.08	2358.80	42394.4	RE	0.9			A
2357.61	2358.33	42402.9	RE	0.7			A
2357.06	2357.78	42412.8	RE	1.1	0.5		A
2356.51	2357.23	42422.7	RE	1.5	0.8		A
2355.89	2356.61	42433.8	RE	1.1			A
2355.45	2356.17	42441.7	RE	0.8	1.3		A
2355.06	2355.78	42448.8	RE	1.0			A
2354.75	2355.47	42454.4	RE	1.2			A
2353.21	2353.93	42482.1	RE	0.8			A
2352.10	2352.82	42502.2	RE	1.5	0.7		A
2350.63	2351.35	42528.8	RE	1.3	0.8		A
2350.48	2351.20	42531.5	RE	1.3	0.8		A
2349.61	2350.33	42547.2	RE	1.1	1.0		A
2349.40	2350.12	42551.0	RE	1.1	0.8		A
2348.80	2349.52	42561.9	RE	1.0			A
2347.64	2348.36	42582.9	RE	0.5	1.0		A
2346.79	2347.51	42598.4	RE	0.9			A
2346.62	2347.34	42601.4	RE	1.2	0.6		A
2345.64	2346.36	42619.2	RE	1.2	0.8		A
2345.30	2346.02	42625.4	RE	1.2	0.8		A
2344.83	2345.55	42634.0	RE	1.4	1.3		A
2344.21	2344.93	42645.2	RE	1.0			A
2344.03	2344.75	42648.5	RE	0.7			A
2343.81	2344.53	42652.5	RE	0.7	0.7		A
2343.32	2344.04	42661.4	RE		0.8		A
2342.75	2343.47	42671.8	RE	1.2	0.5		A
2342.06	2342.78	42684.4	RE	0.7	0.7		A
2340.31	2341.03	42716.3	RE	0.5H	1.0		A
2338.47	2339.19	42749.9	RE	1.5	0.8		A
2337.12	2337.84	42774.6	RE	1.0			A
2336.74	2337.46	42781.5	RE	1.1	0.3		A
2336.48	2337.20	42786.3	RE	0.5	1.0		A
2336.13	2336.85	42792.7	RE	1.4	1.0		A
2335.75	2336.47	42799.7	RE	1.6	1.0		A
2335.03	2335.75	42812.9	RE	1.0			A
2334.49	2335.21	42822.8	RE	1.0			A
2334.36	2335.08	42825.2	RE	1.3	0.9		A
2333.33	2334.05	42844.1	RE	1.2	0.6		A
2333.04	2333.76	42849.4	RE	1.2	0.5		A
2331.74	2332.46	42873.3	RE	1.4			A

AIR WAVELENGTH (ANGSTRMS)	VACUUM WAVELENGTH (ANGSTRMS)	WAVENUMBER (KAYSERS)	LMENT	LOG ARC	INTENSITY SPK	INTENSITY DIS	REF
2331.40	2332.12	42879.5	RE	1.3	0.3H		A
2331.17	2331.89	42883.8	RE	1.5	1.0		A
2330.56	2331.28	42895.0	RE	0.7			A
2330.46	2331.18	42896.8	RE	0.8			A
2330.20	2330.91	42901.6	RE	1.0	1.1		A
2328.69	2329.40	42929.4	RE	1.4	0.8		A
2327.28	2327.99	42955.4	RE	0.8	1.6		A
2327.13	2327.84	42958.2	RE	0.7			A
2325.91	2326.62	42980.7	RE	0.7			A
2325.08	2325.79	42996.1	RE	1.3	0.6		A
2323.44	2324.15	43026.4	RE	0.5	1.5		A
2322.99	2323.70	43034.7	RE	0.8	1.2		A
2322.51	2323.22	43043.6	RE	1.4	0.7		A
2321.73	2322.44	43058.1	RE	0.9	1.1		A
2321.43	2322.14	43063.7	RE	1.2	0.7		A
2321.28	2321.99	43066.5	RE	1.0			A
2319.22	2319.93	43104.7	RE	1.5			A
2317.89	2318.60	43129.4	RE	0.3H	1.2		A
2317.27	2317.98	43141.0	RE	0.8			A
2316.54	2317.25	43154.6	RE	1.5	0.7		A
2315.97	2316.68	43165.2	RE	1.5	1.0		A
2315.28	2315.99	43178.0	RE	1.0			A
2313.36	2314.07	43213.9	RE	1.3	0.9		A
2312.98	2313.69	43221.0	RE	1.2	0.5		A
2312.82	2313.53	43224.0	RE	1.2	0.7		A
2312.03	2312.74	43238.7	RE	1.3			A
2310.16	2310.87	43273.7	RE	1.2			A
2308.62	2309.33	43302.6	RE	1.0	1.0		A
2307.44	2308.15	43324.7	RE	1.0			A
2306.63	2307.34	43340.0	RE	1.2			A
2306.56	2307.27	43341.3	RE	1.3	0.9		A
2305.76	2306.47	43356.3	RE	1.2	0.3H		A
2305.57	2306.28	43359.9	RE	1.2			A
2304.45	2305.16	43381.0	RE	0.8			A
2303.71	2304.42	43394.9	RE	1.0	0.3H		A
2301.823	2302.532	43430.46	RE		1.6		MIT
2301.34	2302.05	43439.6	RE	1.0			A
2299.78	2300.49	43469.0	RE	1.3	0.9J		A
2298.61	2299.32	43491.2	RE		0.8		A
2298.36	2299.07	43495.9	RE	1.0			A
2298.12	2298.83	43500.4	RE	0.8	1.5		A
2296.24	2296.95	43536.0	RE	1.2	0.3H		A
2295.06	2295.77	43558.4	RE		1.0		A
2294.11	2294.82	43576.5	RE	0.9			A
2293.64	2294.35	43585.4	RE	1.0	1.0		A
2291.33	2292.04	43629.3	RE	1.3	0.7		A
2290.81	2291.52	43639.2	RE	1.0	0.8		A
2286.62	2287.33	43719.2	RE	0.6	1.4		A
2285.31	2286.02	43744.2	RE	1.4	0.8		A
2282.21	2282.91	43803.7	RE		0.6		A
2281.66	2282.36	43814.2	RE	1.5	0.8		A
2281.31	2282.01	43820.9	RE	1.0			A
2279.60	2280.30	43853.8	RE	0.5	1.1		A
2278.76	2279.46	43870.0	RE	1.3	0.8		A
2278.57	2279.27	43873.6	RE	1.2			A
2276.65	2277.35	43910.6	RE	1.6	0.5		A
2275.25	2275.95	43937.6	RE	2.5R	2.5R		A
2274.65	2275.35	43949.2	RE	1.0	0.8		A
2272.66	2273.36	43987.7	RE	0.5	1.3		A
2271.64	2272.34	44007.5	RE	1.0	0.6		A
2271.33	2272.03	44013.5	RE	1.4	0.8		A
2269.79	2270.49	44043.3	RE	0.5H	1.0		A
2266.09	2266.79	44115.2	RE	1.4	0.3H		A
2263.97	2264.67	44156.5	RE	1.1	0.3H		A
2263.76	2264.46	44160.6	RE		0.9		A
2260.51	2261.21	44224.1	RE	1.0	1.4		A
2259.01	2259.71	44253.5	RE	1.0			A
2256.22	2256.92	44308.2	RE	1.6	0.8		A
2255.75	2256.45	44317.4	RE	1.5	0.8		A
2254.82	2255.52	44335.7	RE	1.5	0.8		A

AIR WAVELENGTH (ANGSTRMS)	VACUUM WAVELENGTH (ANGSTRMS)	WAVENUMBER (KAYSERS)	LMENT	LOG ARC	INTENSITY SPK	INTENSITY DIS	REF
2254.22	2254.92	44347.5	RE	1.3	1.0		A
2254.05	2254.75	44350.9	RE	1.2			A
2252.56	2253.26	44380.2	RE	1.0	0.3H		A
2251.60	2252.30	44399.1	RE	0.3H	1.4		A
2251.29	2251.99	44405.2	RE	1.2			A
2250.33	2251.03	44424.2	RE	0.5	0.8		A
2249.46	2250.16	44441.3	RE	1.3			A
2248.58	2249.28	44458.7	RE	1.1			A
2247.51	2248.21	44479.9	RE	0.6	0.8		A
2246.47	2247.17	44500.5	RE	1.4	0.7		A
2245.43	2246.13	44521.1	RE	1.1	0.7		A
2245.14	2245.84	44526.8	RE	1.2	0.3H		A
2244.12	2244.82	44547.1	RE	1.2	0.6		A
2243.29	2243.99	44563.6	RE	1.2	0.8		A
2242.41	2243.11	44581.0	RE	1.4	1.0		A
2240.09	2240.79	44627.2	RE	1.3	0.7		A
2238.96	2239.66	44649.7	RE	1.0			A
2238.64	2239.34	44656.1	RE	1.6	0.6H		A
2235.79	2236.48	44713.0	RE	0.3H	1.5		A
2235.46	2236.15	44719.6	RE	1.2	0.8		A
2233.90	2234.59	44750.9	RE	1.3	0.8		A
2233.02	2233.71	44768.5	RE	0.3	0.7		A
2232.31	2233.00	44782.7	RE	0.8	1.4		A
2230.83	2231.52	44812.4	RE I	1.0			A
2230.61	2231.30	44816.9	RE I	0.7			A
2228.05	2228.74	44868.3	RE I	1.5	1.0		A
2226.92	2227.61	44891.1	RE	1.5			A
2226.45	2227.14	44900.6	RE	1.8	1.1		A
2226.06	2226.75	44908.5	RE		1.1		A
2224.44	2225.13	44941.1	RE	0.8	0.7		A
2224.21	2224.90	44945.8	RE	1.0	0.5		A
2223.18	2223.87	44966.6	RE	1.1	0.3H		A
2222.76	2223.45	44975.1	RE I	1.2	0.8		ME
2222.465	2223.157	44981.09	RE I	1.0	0.5		ME
2220.60	2221.29	45018.9	RE	1.2	0.3H		A
2219.75	2220.44	45036.1	RE		1.2		ME
2219.29	2219.98	45045.4	RE	1.2	0.6		A
2218.97	2219.66	45051.9	RE	1.0	0.8		ME
2218.71	2219.40	45057.2	RE	1.2	0.8		A
2216.161	2216.851	45109.02	RE II	0.0	1.7		ME
2215.535	2216.225	45121.77	RE	0.3	0.3		ME
2214.261	2214.951	45147.73	RE II	2.7G	3.0W		ME
2213.68	2214.37	45159.6	RE		1.3		ME
2212.809	2213.499	45177.35	RE I	1.0	0.6		ME
2212.088	2212.777	45192.07	RE I	1.0	0.0		ME
2211.21	2211.90	45210.0	RE	1.1	0.8		A
2209.81	2210.50	45238.7	RE	1.2	0.8		A
2207.680	2208.368	45282.30	RE I	1.3	0.7		ME
2206.870	2207.558	45298.92	RE II	0.3	2.0		ME
2206.78	2207.47	45300.8	RE	0.6	0.3H		A
2206.154	2206.842	45313.62	RE II	0.3	1.6		ME
2205.43	2206.12	45328.5	RE	0.6	0.9		A
2204.55	2205.24	45346.6	RE I	1.3	0.8		ME
2203.010	2203.697	45378.28	RE I	1.6	1.3		ME
2202.06	2202.75	45397.9	RE I	1.0			ME
2201.68	2202.37	45405.7	RE I	0.8			ME
2198.914	2199.601	45462.80	RE I	2.3G	1.6		ME
2198.617	2199.304	45468.94	RE I	0.7	0.0		ME
2197.44	2198.13	45493.3	RE I	1.0			ME
2195.28	2195.97	45538.0	RE	0.3H	1.3		A
2193.85	2194.54	45567.7	RE I	0.7			ME
2193.507	2194.193	45574.85	RE I	1.0	0.9		ME
2192.44	2193.13	45597.0	RE I	1.0			ME
2191.471	2192.156	45617.19	RE I	1.6	1.0		ME
2190.518	2191.203	45637.03	RE I	1.0	1.0		ME
2189.87	2190.55	45650.5	RE I	0.8	0.7		ME
2188.58	2189.26	45677.4	RE	0.8			A
2188.51	2189.19	45678.9	RE	0.8			A
2188.23	2188.91	45684.7	RE		1.0		A
2188.02	2188.70	45689.1	RE	1.1			A

AIR WAVELENGTH (ANGSTRMS)	VACUUM WAVELENGTH (ANGSTRMS)	WAVENUMBER (KAYSERS)	LMENT	LOG ARC	INTENSITY SPK	DIS	REF
2187.76	2188.44	45694.6	RE	1.2	0.8		A
2185.19	2185.87	45748.3	RE	1.0	0.5		A
2184.540	2185.224	45761.91	RE I	1.0	1.0		ME
2183.72	2184.40	45779.1	RE I	2.0G	1.0		ME
2181.81	2182.49	45819.2	RE I	0.6			ME
2181.02	2181.70	45835.7	RE I	1.3	1.0		ME
2179.09	2179.77	45876.4	RE II		0.7		ME
2178.79	2179.47	45882.7	RE		1.1		A
2178.59	2179.27	45886.9	RE I	1.0	0.7		ME
2177.851	2178.533	45902.44	RE II	0.3	2.0		ME
2177.579	2178.261	45908.18	RE II	0.0	1.3		ME
2176.206	2176.888	45937.14	RE I	2.3G	1.7		ME
2175.806	2176.488	45945.58	RE I	1.3	1.0		ME
2175.346	2176.028	45955.30	RE I	1.3	0.8		ME
2170.806	2171.487	46051.40	RE II	0.3	1.7		ME
2170.340	2171.021	46061.28	RE II	0.7	1.0		ME
2169.62	2170.30	46076.6	RE II		0.7		ME
2168.78	2169.46	46094.4	RE	1.0			A
2168.42	2169.10	46102.1	RE I	1.2	1.0		ME
2167.938	2168.618	46112.31	RE I	2.3G	1.7		ME
2165.409	2166.089	46166.16	RE II	0.3	1.5		ME
2163.10	2163.78	46215.4	RE I	1.0			ME
2162.81	2163.49	46221.6	RE I	1.2	1.0		ME
2160.42	2161.10	46272.8	RE I	1.0	0.9		ME
2159.14	2159.82	46300.2	RE I	0.8	0.6		ME
2158.73	2159.41	46309.0	RE I	1.2	0.9		ME
2156.673	2157.351	46353.15	RE I	2.0G	1.7		ME
2155.30	2155.98	46382.7	RE	0.5	0.7		ME
2154.20	2154.88	46406.4	RE I	1.0	0.3		ME
2153.32	2154.00	46425.3	RE	1.2	0.3		A
2151.74	2152.42	46459.4	RE I	0.8	0.3		ME
2150.82	2151.50	46479.3	RE II		1.0		ME
2149.59	2150.27	46505.9	RE II	0.0	0.6		ME
2147.97	2148.65	46540.9	RE I	0.8	0.6		ME
2147.483	2148.159	46551.49	RE I	0.9	0.6		ME
2145.90	2146.58	46585.8	RE II	0.0	1.5		ME
2145.71	2146.39	46590.0	RE	0.8	0.9		A
2145.44	2146.12	46595.8	RE		1.1		A
2145.41	2146.09	46596.5	RE II		1.0		ME
2144.085	2144.760	46625.26	RE II	0.3	1.5		ME
2143.48	2144.16	46638.4	RE	0.8	0.3		A
2143.00	2143.68	46648.9	RE	1.0	0.6		A
2142.736	2143.411	46654.61	RE II	1.0	1.3		ME
2141.78	2142.45	46675.4	RE	1.0	0.6		A
2140.95	2141.62	46693.5	RE	1.1			A
2140.66	2141.33	46699.9	RE II		0.8		ME
2139.48	2140.15	46725.6	RE		0.5		A
2139.06	2139.73	46734.8	RE		1.3		A
2137.77	2138.44	46763.0	RE		0.9		A
2137.56	2138.23	46767.6	RE	0.8			A
2136.73	2137.40	46785.7	RE		0.9		A
2135.75	2136.42	46807.2	RE		0.8		A
2135.07	2135.74	46822.1	RE I	0.7	0.3		ME
2133.83	2134.50	46849.3	RE	0.3H	1.1		A
2133.42	2134.09	46858.3	RE	1.0			A
2133.15	2133.82	46864.2	RE	0.5	0.8		A
2130.513	2131.186	46922.24	RE II	0.6	1.0		ME
2130.16	2130.83	46930.0	RE I	0.8			ME
2129.58	2130.25	46942.8	RE II	0.3	1.0		ME
2128.773	2129.445	46960.59	RE II	0.6	0.8		ME
2128.67	2129.34	46962.9	RE	0.8			A
2126.85	2127.52	47003.0	RE	1.5	0.8J		A
2126.73	2127.40	47005.7	RE II	0.5	0.6		ME
2126.47	2127.14	47011.4	RE	1.0	0.3H		A
2125.21	2125.88	47039.3	RE	0.3H	1.0		A
2122.081	2122.752	47108.66	RE II	0.3	1.3		ME
2121.62	2122.29	47118.9	RE	0.5H			A
2121.16	2121.83	47129.1	RE	0.9	0.5		A
2120.60	2121.27	47141.6	RE	0.7			A
2116.16	2116.83	47240.5	RE	0.8			A

AIR WAVELENGTH (ANGSTRMS)	VACUUM WAVELENGTH (ANGSTRMS)	WAVENUMBER (KAYSERS)	LMENT	LOG ARC	INTENSITY SPK	DIS	REF
2115.77	2116.44	47249.2	RE	0.6			A
2115.51	2116.18	47255.0	RE II	0.0	1.3		ME
2114.25	2114.92	47283.1	RE II	1.0	2.0		ME
2112.245	2112.914	47328.00	RE	0.6	2.2		ME
2111.854	2112.523	47336.77	RE II	1.0	1.8		ME
2110.91	2111.58	47357.9	RE	1.3			A
2110.26	2110.93	47372.5	RE	0.8J			A
2110.24	2110.91	47373.0	RE I	1.3?	0.7H		ME
2109.89	2110.56	47380.8	RE	0.6			A
2109.22	2109.89	47395.9	RE I	2.0G	1.5		ME
2108.48	2109.15	47412.5	RE	0.9	0.3H		A
2107.452	2108.120	47435.63	RE I	1.8R	1.0		ME
2106.96	2107.63	47446.7	RE	0.8J	0.3H		A
2106.51	2107.18	47456.8	RE II	0.5	1.8		ME
2106.13	2106.80	47465.4	RE I	0.6			ME
2104.13	2104.80	47510.5	RE	1.3	0.8		A
2102.69	2103.36	47543.0	RE	0.6	0.3H		A
2101.72	2102.39	47565.0	RE I	1.0	0.7		ME
2100.10	2100.77	47601.7	RE	0.5	0.5		ME
2098.62	2099.29	47635.2	RE I	0.8	0.3		ME
2098.44	2099.11	47639.3	RE	0.6			A
2097.64	2098.31	47657.5	RE	1.1			A
2097.16	2097.83	47668.4	RE	1.6	1.0		A
2096.87	2097.54	47675.0	RE	0.5H	0.5H		ME
2095.62	2096.29	47703.4	RE	1.3	0.3H		A
2095.33	2096.00	47710.0	RE	0.8			A
2094.53	2095.20	47728.2	RE	0.8			A
2092.41	2093.08	47776.6	RE II	0.7	2.0		ME
2092.26	2092.93	47780.0	RE	1.3			A
2091.725	2092.390	47792.24	RE I	0.8			ME
2091.535	2092.200	47796.58	RE II		1.5		ME
2087.855	2088.519	47880.81	RE I	0.9	0.5		ME
2087.66	2088.32	47885.3	RE	0.3	1.3		A
2087.16	2087.82	47896.8	RE I	0.6	0.0		ME
2085.80	2086.46	47928.0	RE	0.8	1.4		A
2085.62	2086.28	47932.1	RE	1.5	0.8		A
2084.53	2085.19	47957.2	RE I	1.2	0.7		ME
2083.925	2084.588	47971.10	RE I	1.6R	0.9		ME
2083.342	2084.005	47984.52	RE II	0.6	1.3		ME
2082.26	2082.92	48009.5	RE	0.8	1.3W		A
2078.68	2079.34	48092.1	RE	0.9			A
2077.30	2077.96	48124.1	RE	1.4	0.9		A
2075.69	2076.35	48161.4	RE	0.8	1.1		A
2075.20	2075.86	48172.8	RE I	0.7	0.3		ME
2074.82	2075.48	48181.6	RE I	0.7	0.3		ME
2074.72	2075.38	48183.9	RE	1.0	0.6		A
2073.898	2074.560	48203.00	RE I	1.0	0.8		ME
2073.273	2073.934	48217.53	RE I	1.3	1.0		ME
2072.62	2073.28	48232.7	RE	0.8	1.0		A
2069.44	2070.10	48306.8	RE	0.7	1.0		A
2067.89	2068.55	48343.0	RE I	1.3R	0.9		ME
2067.60	2068.26	48349.8	RE	0.7	0.7		A
2066.90	2067.56	48366.2	RE	0.8	0.7		A
2064.762	2065.422	48416.26	RE I	1.5	0.7		ME
2063.79	2064.45	48439.1	RE	0.6			A
2057.14	2057.80	48595.6	RE	0.5	0.3H		A
2056.33	2056.99	48614.8	RE	0.9	0.3H		A
2056.06	2056.72	48621.1	RE	0.3	0.8		A
2055.27	2055.93	48639.8	RE	0.6	1.0		A
2049.11	2049.77	48786.0	RE	1.5R	0.8		A
2046.96	2047.62	48837.3	RE	0.5	0.3		A
2045.78	2046.44	48865.4	RE	0.8			A
2042.67	2043.33	48939.8	RE	0.5	1.0		A
2041.00	2041.66	48979.9	RE I	0.8	0.0		ME
2040.81	2041.47	48984.4	RE II	0.5	1.0		ME
2039.91	2040.57	49006.0	RE	0.5	1.0		A
2039.23	2039.89	49022.4	RE	1.3			A
2039.01	2039.67	49027.7	RE	1.3			A
2037.96	2038.61	49052.9	RE II	0.6	1.2		ME
2037.17	2037.82	49071.9	RE	0.9	0.8		A

AIR WAVELENGTH (ANGSTRMS)	VACUUM WAVELENGTH (ANGSTRMS)	WAVENUMBER (KAYSERS)	LMENT	LOG ARC	INTENSITY SPK	INTENSITY DIS	REF
2036.20	2036.85	49095.3	RE II	0.5	1.2		ME
2035.96	2036.61	49101.1	RE I	1.0			ME
2035.52	2036.17	49111.7	RE	0.6	0.8		A
2033.78	2034.43	49153.7	RE I	1.3R	1.2		ME
2032.944	2033.598	49173.93	RE I	1.0	0.0		ME
2032.64	2033.29	49181.3	RE	0.3H	0.9		A
2031.97	2032.62	49197.5	RE	0.7	1.0		A
2031.655	2032.309	49205.12	RE I	0.6			ME
2030.92	2031.57	49222.9	RE	0.9	0.6		A
2027.93	2028.58	49295.5	RE I	0.8	0.5		ME
2023.82	2024.47	49395.6	RE	0.6			A
2023.66	2024.31	49399.5	RE	1.0	1.7		A
2023.14	2023.79	49412.2	RE	0.8J	0.8		A
2021.11	2021.76	49461.8	RE	0.6			A
2019.67	2020.32	49497.1	RE	0.8			A
2019.30	2019.95	49506.1	RE	0.5H	1.1		A
2018.55	2019.20	49524.5	RE	0.5	1.1		A
2017.88	2018.53	49541.0	RE	1.6	0.8		A
2016.59	2017.24	49572.7	RE	1.0	0.8J		A
2016.19	2016.84	49582.5	RE	1.0			A
2015.51	2016.16	49599.2	RE	0.5			A
2014.69	2015.34	49619.4	RE	0.8			A
2014.36	2015.01	49627.5	RE	0.3H	0.8		A
2013.48	2014.13	49649.2	RE	0.6	0.3		A
2012.76	2013.41	49667.0	RE	1.0	0.3H		A
2011.07	2011.72	49708.7	RE	0.5	0.7		A
2009.92	2010.57	49737.1	RE	0.5	1.2		A
2008.06	2008.71	49783.2	RE I	1.3R	1.2		ME
2007.88	2008.53	49787.7	RE II	0.3H	0.9		ME
2004.69	2005.34	49866.9	RE	0.6	1.3		A
2004.39	2005.04	49874.4	RE I	0.3	0.0		ME
2003.56	2004.21	49895.0	RE	1.3	0.8		A
2001.47	2002.12	49947.1	RE	0.3	0.3		ME
2000.47	2001.12	49972.1	RE II		0.8		ME
8425.51	8427.83	11865.5	RH I	0.5			ME
8193.63	8195.88	12201.2	RH I	1.0			ME
8136.20	8138.44	12287.4	RH I	1.7			ME
8063.50	8065.72	12398.2	RH I	1.0			ME
8045.40	8047.61	12426.1	RH I	2.1			ME
8043.92	8046.13	12428.3	RH I	0.9			ME
8036.11	8038.32	12440.4	RH I	2.0			ME
8029.91	8032.12	12450.0	RH I	2.2			ME
8016.59	8018.79	12470.7	RH I	0.8			ME
7981.99	7984.19	12524.8	RH I	0.6			ME
7846.50	7848.66	12741.0	RH I	1.7			ME
7830.05	7832.20	12767.8	RH I	1.9			ME
7824.91	7827.06	12776.2	RH	2.3			ME
7791.61	7793.75	12830.8	RH I	2.0			ME
7772.90	7775.04	12861.7	RH I	2.0			ME
7707.93	7710.05	12970.1	RH	0.3			ME
7690.05	7692.17	13000.2	RH I	1.3			ME
7579.54	7581.63	13189.8	RH	0.5			ME
7577.22	7579.31	13193.8	RH I	1.0			ME
7557.67	7559.75	13227.9	RH I	1.0			ME
7542.02	7544.10	13255.4	RH I	0.7			ME
7524.29	7526.36	13286.6	RH I	0.3			ME
7513.98	7516.05	13304.9	RH	0.7			MIT
7496.94	7499.00	13335.1	RH	0.3			MIT
7495.24	7497.30	13338.1	RH I	2.0			MIT
7492.64	7494.70	13342.8	RH I	0.5			ME
7475.74	7477.80	13372.9	RH I	2.0			ME
7446.77	7448.82	13424.9	RH I	0.7			MIT
7442.39	7444.44	13432.8	RH I	2.0			MIT
7430.80	7432.85	13453.8	RH	1.0			MIT
7423.68	7425.72	13466.7	RH	0.3H			ME
7386.64	7388.67	13534.2	RH I	0.6			MIT
7375.57	7377.60	13554.5	RH	1.2			MIT
7299.82	7301.83	13695.2	RH I	0.3			ME
7293.20	7295.21	13707.6	RH I	0.3			ME
7290.53	7292.54	13712.7	RH I	0.3H			ME

AIR WAVELENGTH (ANGSTRMS)	VACUUM WAVELENGTH (ANGSTRMS)	WAVENUMBER (KAYSERS)	LMENT	LOG ARC	INTENSITY SPK	INTENSITY DIS	REF
7282.82	7284.83	13727.2	RH I	0.3			MIT
7273.61	7275.61	13744.5	RH I	0.3			ME
7273.03	7275.03	13745.6	RH I	0.8			MIT
7271.94	7273.94	13747.7	RH I	1.9			MIT
7270.82	7272.82	13749.8	RH I	2.3			MIT
7268.18	7270.18	13754.8	RH I	2.1			MIT
7262.62	7264.62	13765.3	RH I	0.3			ME
7240.35	7242.35	13807.7	RH I	0.5			MIT
7227.72	7229.71	13831.8	RH	0.3			ME
7219.06	7221.05	13848.4	RH I	1.3			ME
7187.04	7189.02	13910.1	RH	0.7			ME
7166.15	7168.13	13950.7	RH I	0.3H			MIT
7142.55	7144.52	13996.7	RH I	0.7			MIT
7104.45	7106.41	14071.8	RH I	1.9			MIT
7101.64	7103.60	14077.4	RH I	1.8			MIT
7038.76	7040.70	14203.1	RH I	0.7			MIT
7001.58	7003.51	14278.6	RH I	1.3			MIT
6994.64	6996.57	14292.7	RH	0.3H			ME
6979.15	6981.07	14324.4	RH I	1.4			MIT
6972.91	6974.83	14337.3	RH I	0.8			ME
6965.67	6967.59	14352.2	RH	2.3			MIT
6925.49	6927.40	14435.4	RH	0.3			ME
6919.02	6920.93	14448.9	RH	0.6			ME
6911.85	6913.76	14463.9	RH I	0.3H			MIT
6879.94	6881.84	14531.0	RH I	1.5			ME
6857.68	6859.57	14578.2	RH I	0.9			ME
6843.51	6845.40	14608.3	RH I	0.7			ME
6837.80	6839.69	14620.6	RH	0.3			ME
6829.33	6831.21	14638.7	RH I	0.5			ME
6827.33	6829.21	14643.0	RH I	1.2			ME
6796.65	6798.53	14709.1	RH I	0.8			ME
6752.35	6754.21	14805.6	RH I	2.2			MIT
6721.33	6723.19	14873.9	RH I	0.3			ME
6686.65	6688.50	14951.0	RH I	0.3H			ME
6645.01	6646.84	15044.7	RH I	0.3			ME
6632.13	6633.96	15073.9	RH I	0.5			ME
6630.16	6631.99	15078.4	RH I	1.6			MIT
6627.80	6629.63	15083.8	RH I	1.5			ME
6519.703	6521.504	15333.89	RH I	1.6			MIT
6517.562	6519.363	15338.92	RH I	0.9			MIT
6510.409	6512.208	15355.78	RH I	1.4			MIT
6485.660	6487.452	15414.37	RH I	0.8			MIT
6452.730	6454.513	15493.04	RH	0.3			MIT
6445.15	6446.93	15511.3	RH	0.3			ME
6431.258	6433.036	15544.76	RH	0.3			MIT
6418.60	6420.37	15575.4	RH	0.3			ME
6414.724	6416.497	15584.83	RH I	1.7			MIT
6412.995	6414.768	15589.03	RH I	0.9			MIT
6410.216	6411.988	15595.79	RH I	0.6			MIT
6402.33	6404.10	15615.0	RH I	0.5			MIT
6386.56	6388.33	15653.6	RH I	0.6H			ME
6377.00	6378.76	15677.0	RH	0.3			ME
6363.77	6365.53	15709.6	RH I	0.3			ME
6332.977	6334.728	15786.00	RH I	0.9			MIT
6319.530	6321.278	15819.59	RH I	1.7			MIT
6307.18	6308.92	15850.6	RH I	0.5			ME
6294.17	6295.91	15883.3	RH I	0.5			ME
6293.383	6295.124	15885.31	RH I	1.0			MIT
6290.39	6292.13	15892.9	RH I	0.5			ME
6287.397	6289.136	15900.44	RH	0.6			MIT
6280.67	6282.41	15917.5	RH I	0.5			ME
6278.08	6279.82	15924.0	RH	0.5			ME
6277.464	6279.200	15925.59	RH I	1.2			MIT
6276.665	6278.401	15927.62	RH I	0.9			MIT
6258.964	6260.695	15972.67	RH	0.6			MIT
6253.718	6255.448	15986.07	RH I	1.5			MIT
6219.587	6221.308	16073.79	RH	0.6			MIT
6199.992	6201.707	16124.59	RH I	1.4			MIT
6186.892	6188.604	16158.73	RH I	1.5			MIT
6180.365	6182.075	16175.80	RH I	0.3			MIT

AIR WAVELENGTH (ANGSTRMS)	VACUUM WAVELENGTH (ANGSTRMS)	WAVENUMBER (KAYSERS)	LMENT		LOG ARC	INTENSITY SPK	INTENSITY DIS	REF
6173.930	6175.638	16192.66	RH		1.0			MIT
6158.87	6160.57	16232.2	RH	I	0.3			ME
6147.70	6149.40	16261.7	RH	I	0.3			ME
6144.340	6146.041	16270.64	RH		0.5			MIT
6128.062	6129.758	16313.86	RH	I	1.2			MIT
6118.874	6120.568	16338.35	RH	I	0.5			MIT
6116.146	6117.839	16345.64	RH	I	1.0			MIT
6102.721	6104.410	16381.60	RH		2.0			MIT
6087.337	6089.022	16423.00	RH	I	0.5			MIT
6072.96	6074.64	16461.9	RH	I	0.3			ME
6027.91	6029.58	16584.9	RH		0.3			ME
5991.186	5992.845	16686.56	RH	I	1.0			MIT
5983.599	5985.256	16707.72	RH	I	2.3	0.3		MIT
5970.69	5972.34	16743.8	RH	I	0.3			ME
5952.43	5954.08	16795.2	RH	I	0.5			MIT
5941.46	5943.11	16826.2	RH	I	0.6			MIT
5923.74	5925.38	16876.6	RH	I	0.5			MIT
5918.544	5920.184	16891.37	RH	I	1.3			MIT
5908.18	5909.82	16921.0	RH		0.3			MIT
5907.313	5908.950	16923.48	RH	I	0.7			MIT
5898.94	5900.57	16947.5	RH		0.5			ME
5871.77	5873.40	17025.9	RH	I	0.5			MIT
5831.582	5833.199	17143.25	RH	I	1.9	0.0		MIT
5821.839	5823.453	17171.94	RH	I	0.6			MIT
5806.909	5808.519	17216.09	RH	I	2.0	0.0		MIT
5803.339	5804.948	17226.68	RH		0.9			MIT
5797.52	5799.13	17244.0	RH		0.5			MIT
5795.789	5797.396	17249.12	RH		0.9			MIT
5792.767	5794.373	17258.12	RH	I	1.6			MIT
5755.695	5757.291	17369.28	RH	I	0.3			MIT
5730.431	5732.021	17445.85	RH		0.5			MIT
5727.295	5728.884	17455.41	RH	I	0.8			MIT
5726.676	5728.265	17457.29	RH	I	0.5			MIT
5713.632	5715.217	17497.15	RH		0.3			MIT
5702.470	5704.052	17531.40	RH	I	0.9			MIT
5695.661	5697.241	17552.35	RH	I	0.3			MIT
5686.377	5687.955	17581.01	RH	I	2.0	0.0		MIT
5660.761	5662.332	17660.57	RH		0.6			MIT
5659.621	5661.192	17664.12	RH	I	1.0			MIT
5651.308	5652.876	17690.11	RH	I	0.3			MIT
5634.652	5636.216	17742.40	RH		0.5			MIT
5632.769	5634.332	17748.33	RH	I	0.6			MIT
5626.071	5627.633	17769.46	RH	I	0.5			MIT
5608.346	5609.903	17825.62	RH	I	1.0	0.0		MIT
5607.712	5609.269	17827.64	RH	I	0.9			MIT
5599.421	5600.976	17854.03	RH	I	2.5W	0.5		MIT
5594.879	5596.432	17868.53	RH	I	0.7			MIT
5568.31	5569.86	17953.8	RH	I	0.3			ME
5557.18	5558.72	17989.7	RH	I	0.3			MIT
5556.802	5558.345	17990.97	RH		0.5			MIT
5555.08	5556.62	17996.5	RH	I	0.3			MIT
5544.585	5546.125	18030.61	RH	I	1.7	0.0		MIT
5535.04	5536.58	18061.7	RH		1.9	0.0		MIT
5533.84	5535.38	18065.6	RH	I	0.5			MIT
5504.652	5506.181	18161.41	RH	I	1.0			MIT
5503.593	5505.122	18164.90	RH	I	0.7			MIT
5496.97	5498.50	18186.8	RH	I	0.5			ME
5491.867	5493.393	18203.69	RH	I	0.7			MIT
5484.229	5485.753	18229.04	RH	I	1.2	0.0		MIT
5481.419	5482.942	18238.38	RH	I	1.0	0.0		MIT
5476.123	5477.645	18256.02	RH	I	1.0			MIT
5470.852	5472.372	18273.61	RH	I	1.4	0.0		MIT
5468.735	5470.255	18280.69	RH	I	0.8			MIT
5468.110	5469.629	18282.77	RH	I	1.2	0.0		MIT
5445.228	5446.741	18359.60	RH		1.4	0.0		MIT
5444.320	5445.833	18362.66	RH		1.2			MIT
5441.365	5442.877	18372.64	RH		1.2			MIT
5439.523	5441.035	18378.86	RH	I	1.3	0.0		MIT
5432.048	5433.558	18404.15	RH	I	0.7			MIT
5425.446	5426.954	18426.54	RH		1.4	0.0		MIT

AIR WAVELENGTH (ANGSTRMS)	VACUUM WAVELENGTH (ANGSTRMS)	WAVENUMBER (KAYSERS)	LMENT		LOG ARC	INTENSITY SPK	INTENSITY DIS	REF
5424.719	5426.227	18429.01	RH	I	1.4	0.0		MIT
5424.068	5425.576	18431.22	RH		2.0	0.3		MIT
5423.29	5424.80	18433.9	RH	I	1.2			MIT
5408.780	5410.284	18483.32	RH	I	0.5			MIT
5404.726	5406.229	18497.18	RH		1.7	0.0		MIT
5390.440	5391.939	18546.21	RH	I	2.1	0.5		MIT
5381.48	5382.98	18577.1	RH	I	2.0			ME
5379.095	5380.591	18585.32	RH	I	2.0	0.5		MIT
5369.298	5370.791	18619.23	RH	I	0.7			MIT
5364.073	5365.565	18637.37	RH	I	0.6			MIT
5356.467	5357.957	18663.83	RH	I	1.5	0.0		MIT
5354.399	5355.888	18671.04	RH	I	2.5	0.7		MIT
5349.308	5350.796	18688.81	RH		1.3			MIT
5339.65	5341.14	18722.6	RH	I	0.3			ME
5336.640	5338.124	18733.17	RH	I	0.7			MIT
5331.08	5332.56	18752.7	RH		1.2			MIT
5329.74	5331.22	18757.4	RH	I	1.5	0.3		ME
5329.43	5330.91	18758.5	RH	I	0.5			ME
5314.786	5316.265	18810.20	RH	I	1.6	0.0		MIT
5292.144	5293.617	18890.68	RH	I	1.9	0.0		MIT
5280.124	5281.593	18933.68	RH	I	1.0			MIT
5269.266	5270.732	18972.70	RH	I	1.7	0.0		MIT
5267.93	5269.40	18977.5	RH	I	0.7			ME
5261.95	5263.41	18999.1	RH		0.5			ME
5259.27	5260.73	19008.8	RH		0.9	0.0		MIT
5251.41	5252.87	19037.2	RH	I	1.4			ME
5248.74	5250.20	19046.9	RH	I	0.5			ME
5237.805	5239.263	19086.65	RH		0.9			MIT
5237.156	5238.614	19089.02	RH	I	2.0	0.3		MIT
5230.625	5232.081	19112.85	RH	I	1.4	0.0		MIT
5225.58	5227.03	19131.3	RH	I	1.0			ME
5222.663	5224.117	19141.99	RH	I	1.5	0.0		MIT
5214.787	5216.239	19170.90	RH	I	1.0			MIT
5213.36	5214.81	19176.1	RH	I	0.7			ME
5212.729	5214.180	19178.47	RH	I	1.2	0.0		MIT
5211.516	5212.967	19182.93	RH	I	0.9	0.0		MIT
5206.951	5208.401	19199.75	RH	I	0.8	0.0		MIT
5203.329	5204.778	19213.12	RH	I	0.9			MIT
5197.55	5199.00	19234.5	RH	I	0.5			ME
5193.137	5194.583	19250.82	RH		2.3	0.5		MIT
5185.023	5186.467	19280.95	RH	I	0.9			MIT
5184.194	5185.638	19284.03	RH	I	2.0	0.0		MIT
5178.178	5179.620	19306.44	RH	I	0.7			MIT
5177.267	5178.709	19309.83	RH	I	1.4			MIT
5175.969	5177.411	19314.67	RH		2.3	0.0		MIT
5174.71	5176.15	19319.4	RH		0.5			ME
5165.41	5166.85	19354.2	RH	I	1.2			ME
5164.10	5165.54	19359.1	RH	I	0.5			ME
5160.35	5161.79	19373.1	RH	I	0.6			ME
5158.69	5160.13	19379.4	RH		1.9	0.0		ME
5157.09	5158.53	19385.4	RH		1.4			ME
5155.54	5156.98	19391.2	RH	I	2.2	0.0		ME
5153.34	5154.78	19399.5	RH	I	0.3			ME
5144.97	5146.40	19431.1	RH	I	0.5			ME
5130.76	5132.19	19484.9	RH	I	1.3			ME
5120.69	5122.12	19523.2	RH	I	1.3			ME
5109.97	5111.39	19564.1	RH	I	1.0			ME
5090.63	5092.05	19638.5	RH	I	2.2	0.0		ME
5088.78	5090.20	19645.6	RH	I	0.3			ME
5085.52	5086.94	19658.2	RH		1.0			ME
5064.32	5065.73	19740.5	RH		0.9			ME
5057.42	5058.83	19767.4	RH	I	0.3			ME
5046.43	5047.84	19810.5	RH	I	0.5			ME
5028.358	5029.760	19881.66	RH	I	0.7			MIT
5026.30	5027.70	19889.8	RH		0.3			ME
5025.58	5026.98	19892.6	RH	I	0.6			ME
5012.39	5013.79	19945.0	RH	I	0.5			ME
4997.793	4999.187	20003.25	RH		0.6			MIT
4995.875	4997.269	20010.93	RH	I	0.3			MIT
4985.352	4986.743	20053.17	RH		0.3H			MIT

AIR WAVELENGTH (ANGSTRMS)	VACUUM WAVELENGTH (ANGSTRMS)	WAVENUMBER (KAYSERS)	LMENT	LOG ARC	INTENSITY SPK	DIS	REF
4984.985	4986.376	20054.65	RH I	0.7			MIT
4979.175	4980.564	20078.05	RH I	1.3			MIT
4977.748	4979.137	20083.80	RH I	1.4			MIT
4966.386	4967.772	20129.75	RH	0.9			MIT
4963.707	4965.092	20140.61	RH I	2.0			MIT
4961.137	4962.521	20151.05	RH	0.3			MIT
4960.173	4961.557	20154.96	RH	0.3			MIT
4944.858	4946.238	20217.39	RH	0.7			MIT
4922.513	4923.887	20309.16	RH I	1.0			MIT
4919.693	4921.066	20320.80	RH I	1.0			MIT
4918.835	4920.208	20324.34	RH I	0.7			MIT
4913.520	4914.892	20346.33	RH I	0.8			MIT
4908.620	4909.991	20366.64	RH	0.7			MIT
4897.88	4899.25	20411.3	RH I	0.7			ME
4887.92	4889.29	20452.9	RH	0.3			ME
4870.19	4871.55	20527.4	RH	0.3			ME
4866.75	4868.11	20541.9	RH	0.3			ME
4865.783	4867.142	20545.94	RH I	1.0			MIT
4861.378	4862.736	20564.55	RH	0.6			MIT
4851.634	4852.989	20605.86	RH	1.9	1.5		MIT
4843.99	4845.34	20638.4	RH	2.0	1.8		ME
4842.43	4843.78	20645.0	RH I	1.7			MIT
4833.47	4834.82	20683.3	RH	0.3	0.3		ME
4817.07	4818.42	20753.7	RH	0.3			ME
4813.40	4814.75	20769.5	RH	0.3			ME
4810.487	4811.831	20782.11	RH	1.2	1.4		MIT
4798.673	4800.014	20833.27	RH	0.5			MIT
4791.478	4792.817	20864.55	RH I	0.3H			ME
4791.021	4792.360	20866.54	RH	0.5	0.5		MIT
4777.171	4778.507	20927.04	RH	0.7	0.3		MIT
4771.563	4772.897	20951.64	RH	0.3			MIT
4770.792	4772.126	20955.02	RH	0.3	1.2		MIT
4768.121	4769.454	20966.76	RH	0.3			MIT
4755.577	4756.907	21022.06	RH I	0.6	1.2		MIT
4745.110	4746.437	21068.44	RH I	1.3	1.0		MIT
4739.223	4740.549	21094.60	RH I	1.2	0.5		MIT
4731.18	4732.50	21130.5	RH	0.3			M
4724.342	4725.664	21161.05	RH I	0.3	0.3H		MIT
4721.000	4722.321	21176.03	RH I	0.9	0.7		MIT
4719.405	4720.725	21183.19	RH I	0.3H			MIT
4706.94	4708.26	21239.3	RH	0.3H			M
4704.081	4705.397	21252.19	RH I	1.0	0.6		MIT
4700.711	4702.026	21267.43	RH I	0.3			MIT
4696.322	4697.636	21287.30	RH I	1.5			MIT
4689.459	4690.771	21318.46	RH	0.3			MIT
4682.959	4684.270	21348.04	RH I	0.5	0.3		MIT
4677.407	4678.716	21373.38	RH I	0.8	0.3		MIT
4675.026	4676.335	21384.27	RH I	2.0	1.7		MIT
4666.126	4667.432	21425.06	RH I	0.3H			MIT
4664.963	4666.269	21430.40	RH	0.3H			ME
4643.184	4644.484	21530.92	RH I	1.2	1.0		MIT
4639.37	4640.67	21548.6	RH I	0.6	0.3		M
4633.856	4635.154	21574.26	RH	0.3			MIT
4625.937	4627.233	21611.19	RH I	0.3			MIT
4619.907	4621.201	21639.40	RH	0.9	0.7		MIT
4608.119	4609.410	21694.75	RH I	1.2	0.7		MIT
4601.636	4602.925	21725.31	RH I	0.5	0.3		MIT
4595.571	4596.859	21753.99	RH	0.3	0.0H		MIT
4572.641	4573.923	21863.07	RH I	0.6	0.3		MIT
4571.306	4572.587	21869.46	RH	0.9	0.7		MIT
4570.32	4571.60	21874.2	RH	0.3	0.0		M
4569.005	4570.286	21880.47	RH I	2.0	1.4		MIT
4568.373	4569.653	21883.50	RH I	0.3			MIT
4565.189	4566.469	21898.76	RH I	1.2	0.6		MIT
4560.892	4562.170	21919.39	RH I	1.3	0.8		MIT
4558.728	4560.006	21929.80	RH I	1.0	0.5		MIT
4557.161	4558.438	21937.34	RH I	0.5	0.0		MIT
4551.638	4552.914	21963.96	RH I	1.4	1.0		MIT
4548.727	4550.002	21978.01	RH I	1.4	0.7		MIT
4544.271	4545.545	21999.56	RH I	1.4	0.6		MIT

AIR WAVELENGTH (ANGSTRMS)	VACUUM WAVELENGTH (ANGSTRMS)	WAVENUMBER (KAYSERS)	LMENT	LOG ARC	INTENSITY SPK	DIS	REF
4530.595	4531.865	22065.97	RH	0.6	0.3		MIT
4528.725	4529.995	22075.08	RH I	2.7R	1.8		MIT
4525.208	4526.477	22092.24	RH I	0.7	0.3		MIT
4506.680	4507.944	22183.06	RH I	0.5	0.0		MIT
4503.784	4505.047	22197.32	RH I	1.5	1.0		MIT
4492.469	4493.729	22253.23	RH I	1.5	0.9		MIT
4483.831	4485.089	22296.10	RH	0.7	0.3		MIT
4433.320	4434.565	22550.13	RH I	1.2	0.7		MIT
4424.047	4425.289	22597.39	RH I	0.7	0.3		MIT
4379.917	4381.148	22825.07	RH I	1.8	1.4		MIT
4376.184	4377.414	22844.54	RH I	0.5	0.0		MIT
4374.80	4376.03	22851.8	RH I	3.0O	2.7		MIT
4373.042	4374.271	22860.95	RH I	1.8	1.0		MIT
4349.175	4350.398	22986.41	RH I	0.6	0.3		MIT
4345.466	4346.688	23006.02	RH I	1.0	0.6		MIT
4345.086	4346.307	23008.04	RH I	0.5	0.0		MIT
4342.444	4343.665	23022.03	RH I	1.2	0.6		MIT
4336.020	4337.239	23056.14	RH I	0.3	0.0		MIT
4308.831	4310.043	23201.62	RH I	0.6	0.3		MIT
4296.770	4297.979	23266.75	RH I	1.6	1.3		MIT
4288.706	4289.913	23310.50	RH I	2.6	2.0		MIT
4288.005	4289.211	23314.31	RH II		1.3		RR
4278.598	4279.802	23365.57	RH I	1.4	1.0		MIT
4276.822	4278.026	23375.27	RH I	1.0	0.5		MIT
4273.426	4274.629	23393.84	RH I	1.4	1.0		MIT
4258.466	4259.665	23476.03	RH I	0.3	0.0		MIT
4244.443	4245.638	23553.59	RH I	1.2	0.7		MIT
4230.202	4231.393	23632.88	RH I	1.0	0.5		MIT
4218.001	4219.189	23701.24	RH I	0.3	0.0		MIT
4211.141	4212.327	23739.85	RH I	2.3	1.2		MIT
4206.617	4207.802	23765.38	RH I	0.9	0.7		MIT
4196.504	4197.686	23822.65	RH I	2.0	1.7		MIT
4177.635	4178.812	23930.24	RH I	1.4	0.7		MIT
4154.369	4155.540	24064.26	RH	1.8	0.0		MIT
4135.272	4136.438	24175.39	RH I	2.5	2.2		MIT
4128.870	4130.035	24212.87	RH I	2.5	2.2		MIT
4124.901	4126.065	24236.17	RH I	0.7	0.5		MIT
4121.682	4122.845	24255.10	RH I	2.2	1.7		MIT
4119.680	4120.842	24266.88	RH I	2.0	1.4		MIT
4116.331	4117.492	24286.63	RH I	1.5	1.0		MIT
4107.49	4108.65	24338.9	RH I	1.4	0.9		MIT
4097.521	4098.677	24398.11	RH I	1.4	1.0		MIT
4088.50	4089.65	24451.9	RH I	0.7	0.6		M
4087.79	4088.94	24456.2	RH	0.5	0.6		MIT
4087.52	4088.67	24457.8	RH I	0.5			MIT
4084.285	4085.438	24477.18	RH I	0.3	0.3		MIT
4082.78	4083.93	24486.2	RH I	2.0	1.7		MIT
4081.813	4082.965	24492.00	RH I	0.3	0.3		MIT
4080.539	4081.691	24499.65	RH I	0.3	0.3		MIT
4077.57	4078.72	24517.5	RH I	0.7	0.6		MIT
4056.342	4057.488	24645.79	RH I	0.5	0.3		MIT
4053.443	4054.588	24663.42	RH I	0.6	0.5		MIT
4049.039	4050.183	24690.24	RH I	0.3	0.3		MIT
4048.409	4049.553	24694.09	RH	0.6	0.5		MIT
4025.949	4027.087	24831.85	RH I	0.5	0.0		MIT
4023.144	4024.281	24849.16	RH I	1.0	0.7		MIT
3996.154	3997.284	25016.99	RH I	1.4	1.0		MIT
3995.607	3996.737	25020.41	RH I	1.2	1.0		MIT
3984.40	3985.53	25090.8	RH I	1.4	1.0		MIT
3976.110	3977.235	25143.10	RH I	0.3	0.3		MIT
3975.314	3976.439	25148.13	RH I	1.3	1.0		MIT
3968.164	3969.287	25193.44	RH I	0.3			MIT
3964.542	3965.664	25216.46	RH		0.5		MIT
3958.865	3959.985	25252.62	RH I	2.3	2.0		MIT
3958.240	3959.360	25256.61	RH	0.7	0.6		MIT
3953.067	3954.186	25289.66	RH I	0.5			MIT
3944.019	3945.135	25347.67	RH	0.7	0.6		MIT
3942.716	3943.832	25356.05	RH I	1.8	1.4		MIT
3935.838	3936.952	25400.36	RH I	1.6	1.1		MIT
3934.982	3936.096	25405.88	RH I	0.5	0.3		MIT

AIR WAVELENGTH (ANGSTRMS)	VACUUM WAVELENGTH (ANGSTRMS)	WAVENUMBER (KAYSERS)	LMENT	LOG ARC	INTENSITY SPK	INTENSITY DIS	REF
3934.229	3935.343	25410.75	RH I	2.0	0.3		MIT
3922.194	3923.305	25488.72	RH I	1.2	0.9		MIT
3913.513	3914.621	25545.25	RH I	0.6	0.3		MIT
3912.826	3913.934	25549.74	RH I	0.3	0.0		MIT
3904.224	3905.330	25606.03	RH I	0.5	0.3		MIT
3888.335	3889.437	25710.66	RH I	0.7	0.3		MIT
3877.345	3878.444	25783.54	RH I	1.3	0.7		MIT
3872.392	3873.490	25816.51	RH I	1.7	0.5		MIT
3870.011	3871.108	25832.40	RH I	1.2	0.5		MIT
3856.515	3857.609	25922.80	RH I	1.7	1.3		MIT
3856.025	3857.118	25926.09	RH	0.8	0.6		MIT
3854.68	3855.77	25935.1	RH	0.6	0.6		MIT
3849.005	3850.097	25973.37	RH I	0.3	0.3		MIT
3834.754	3835.842	26069.90	RH I	0.6	0.7		MIT
3834.22	3835.31	26073.5	RH	0.5			MIT
3833.889	3834.977	26075.78	RH I	1.7	1.4		MIT
3828.479	3829.565	26112.62	RH I	2.0	1.8		MIT
3822.262	3823.347	26155.10	RH I	2.0	2.0		MIT
3818.187	3819.271	26183.01	RH I	1.7	1.4		MIT
3816.470	3817.553	26194.79	RH I	1.2	1.2		MIT
3815.012	3816.095	26204.80	RH	1.3	1.3		MIT
3812.451	3813.533	26222.40	RH I	0.7	0.6		MIT
3809.498	3810.579	26242.73	RH I	0.5	0.5		MIT
3806.758	3807.839	26261.62	RH I	1.7	1.7		MIT
3805.917	3806.997	26267.42	RH I	1.4	1.7		MIT
3799.311	3800.390	26313.09	RH I	2.0	1.4		MIT
3793.217	3794.294	26355.36	RH I	2.3	1.8		MIT
3792.182	3793.259	26362.56	RH I	1.0	0.6		MIT
3788.57	3789.65	26387.7	RH I	0.3			MIT
3788.474	3789.550	26388.36	RH	1.7	1.4		MIT
3778.126	3779.199	26460.63	RH I	1.4	1.0		MIT
3775.716	3776.789	26477.52	RH I	0.9	0.5		MIT
3771.632	3772.703	26506.19	RH I	0.7	1.1		MIT
3770.240	3771.311	26515.98	RH	0.5	0.3		MIT
3769.974	3771.045	26517.85	RH I	1.4	1.5		MIT
3765.078	3766.148	26552.33	RH I	2.0	1.8		MIT
3760.405	3761.474	26585.33	RH I	0.8	0.3		MIT
3755.584	3756.651	26619.45	RH I	0.6	0.3		MIT
3754.273	3755.340	26628.75	RH I	1.0	1.0		MIT
3754.123	3755.190	26629.81	RH I	0.9	0.9		MIT
3748.217	3749.282	26671.77	RH I	2.3	2.0		MIT
3744.167	3745.231	26700.62	RH I	1.5	1.2		MIT
3737.273	3738.336	26749.87	RH	1.7	1.0		MIT
3737.12	3738.18	26751.0	RH I	1.7	0.0		MIT
3735.281	3736.343	26764.14	RH I	1.8	0.3		MIT
3724.942	3726.001	26838.42	RH I	0.7	0.3		MIT
3714.828	3715.885	26911.49	RH I	0.5	0.3		MIT
3713.429	3714.485	26921.63	RH I	0.6	0.5		MIT
3713.022	3714.078	26924.58	RH I	2.0	2.0R		MIT
3700.909	3701.962	27012.70	RH I	2.2D	2.2D		MIT
3700.909	3701.962	27012.70	RH I	2.2G			ME
3699.309	3700.362	27024.39	RH I	0.6	0.3		MIT
3698.595	3699.648	27029.60	RH I	1.3	1.2		MIT
3698.260	3699.312	27032.05	RH I	1.2	1.0		MIT
3695.525	3696.577	27052.06	RH	1.2	1.0		MIT
3694.953	3696.005	27056.24	RH I	0.5	0.3		MIT
3692.357	3693.408	27075.26	RH I	2.7G			ME
3692.357	3693.408	27075.26	RH I	2.7A	2.2$		MIT
3691.333	3692.384	27082.78	RH I	0.6	0.3		MIT
3690.704	3691.755	27087.39	RH I	2.1	1.7		MIT
3681.039	3682.087	27158.51	RH I	1.3	1.4		MIT
3674.765	3675.811	27204.88	RH I	1.0	0.6		MIT
3673.558	3674.604	27213.82	RH I	0.3			MIT
3666.911	3667.955	27263.15	RH I	1.0	0.6		MIT
3666.215	3667.259	27268.32	RH I	1.8	1.5		MIT
3661.863	3662.906	27300.73	RH I	1.0	0.7		MIT
3661.601	3662.644	27302.68	RH I	0.3			MIT
3657.987	3659.029	27329.66	RH I	2.7G	2.3G		ME
3657.987	3659.029	27329.66	RH I	2.70	2.30		MIT
3654.873	3655.914	27352.94	RH I	1.6	1.0		MIT

AIR WAVELENGTH (ANGSTRMS)	VACUUM WAVELENGTH (ANGSTRMS)	WAVENUMBER (KAYSERS)	LMENT	LOG ARC	INTENSITY SPK	INTENSITY DIS	REF
3651.356	3652.396	27379.28	RH I	0.3	0.0		MIT
3639.512	3640.549	27468.38	RH I	2.1	1.8		MIT
3627.803	3628.837	27557.04	RH I	0.7	0.5		MIT
3627.183	3628.217	27561.75	RH I	0.6	0.3		MIT
3626.590	3627.624	27566.25	RH I	2.2	1.8		MIT
3620.456	3621.488	27612.96	RH I	1.3	1.0		MIT
3614.775	3615.806	27656.35	RH I	1.2	1.0		MIT
3614.511	3615.542	27658.37	RH	0.6	0.3		MIT
3613.939	3614.970	27662.75	RH I	0.3			MIT
3612.470	3613.500	27674.00	RH I	2.3	1.7		MIT
3608.086	3609.115	27707.62	RH I	1.0	0.5		MIT
3605.863	3606.892	27724.70	RH I	1.4	1.5		MIT
3600.754	3601.781	27764.04	RH I	0.9	0.3		MIT
3597.892	3598.919	27786.12	RH	0.8	0.5		MIT
3597.147	3598.173	27791.88	RH I	2.3	2.0		MIT
3596.194	3597.220	27799.24	RH I	2.3	1.7		MIT
3596.016	3597.042	27800.62	RH	1.4	0.7		MIT
3593.889	3594.915	27817.07	RH	0.5			MIT
3593.530	3594.555	27819.85	RH I	1.0	0.3		MIT
3590.519	3591.544	27843.18	RH I	0.6	0.5		MIT
3583.528	3584.551	27897.50	RH I	1.0	0.7		MIT
3583.098	3584.121	27900.85	RH I	2.3	2.1		MIT
3580.262	3581.284	27922.95	RH I	1.0	0.3		MIT
3570.182	3571.201	28001.78	RH I	2.6R	2.2		MIT
3564.132	3565.150	28049.31	RH I	1.4	0.0		MIT
3549.992	3551.006	28161.03	RH I	1.0	0.3		MIT
3549.543	3550.557	28164.59	RH I	2.2	1.7		MIT
3543.948	3544.961	28209.06	RH I	2.2	1.6		MIT
3541.912	3542.924	28225.27	RH I	1.7	1.0		MIT
3538.256	3539.267	28254.44	RH I	1.7	0.6		MIT
3538.142	3539.153	28255.35	RH I	2.0	1.0		MIT
3528.024	3529.033	28336.38	RH I	3.0W	2.2		MIT
3525.658	3526.666	28355.39	RH I	1.7	0.3		MIT
3519.541	3520.547	28404.67	RH I	1.6	0.3		MIT
3513.102	3514.107	28456.73	RH I	1.7	0.5		MIT
3511.785	3512.789	28467.41	RH I	1.7	0.5		MIT
3511.538	3512.542	28469.41	RH I	1.4	0.3		MIT
3508.602	3509.606	28493.23	RH I	0.5			MIT
3508.51	3509.51	28494.0	RH	0.3			MIT
3507.316	3508.319	28503.68	RH I	2.7	2.1		MIT
3505.409	3506.412	28519.18	RH I	1.5	0.5		MIT
3502.524	3503.526	28542.67	RH I	3.0	2.2		MIT
3498.730	3499.731	28573.62	RH I	2.7	1.8		MIT
3494.442	3495.442	28608.69	RH I	1.7	0.5		MIT
3491.201	3492.200	28635.24	RH I	1.2	0.0		MIT
3491.072	3492.071	28636.30	RH I	1.6	0.3		MIT
3487.461	3488.459	28665.95	RH I	1.0	0.0		MIT
3487.214	3488.212	28667.98	RH I	1.0	0.0		MIT
3484.036	3485.033	28694.13	RH I	1.6	0.6		MIT
3481.152	3482.149	28717.90	RH	1.0			MIT
3478.906	3479.902	28736.44	RH I	2.7	2.0		MIT
3478.492	3479.488	28739.86	RH I	0.8	0.0		MIT
3477.779	3478.775	28745.75	RH	0.3	1.0		MIT
3474.780	3475.775	28770.56	RH I	2.8	2.1		MIT
3472.251	3473.245	28791.52	RH I	2.0	0.9		MIT
3471.302	3472.296	28799.39	RH	0.3			MIT
3470.657	3471.651	28804.74	RH I	2.7	2.1		MIT
3469.624	3470.618	28813.32	RH I	2.0	1.0		MIT
3462.040	3463.032	28876.43	RH I	3.0	2.2		MIT
3459.218	3460.209	28899.99	RH	0.3	0.3		MIT
3457.926	3458.917	28910.79	RH I	2.1	1.0		MIT
3457.071	3458.061	28917.94	RH I	2.0	0.6		MIT
3455.422	3456.412	28931.74	RH I	1.7	0.3		MIT
3455.219	3456.209	28933.44	RH I	2.5	1.1		MIT
3451.149	3452.138	28967.56	RH	1.7	0.3		MIT
3450.288	3451.277	28974.79	RH I	2.0	1.0		MIT
3448.575	3449.563	28989.18	RH I	1.4	0.3		MIT
3447.736	3448.724	28996.23	RH I	1.7	0.7		MIT
3442.631	3443.618	29039.23	RH I	0.5	0.6		MIT
3440.534	3441.520	29056.93	RH	2.0	0.3		MIT

AIR WAVELENGTH (ANGSTRMS)	VACUUM WAVELENGTH (ANGSTRMS)	WAVENUMBER (KAYSERS)	LMENT	LOG ARC	INTENSITY SPK	DIS	REF
3434.893	3435.878	29104.65	RH	3.0R	2.3R		MIT
3432.091	3433.075	29128.41	RH I	0.5			MIT
3428.405	3429.388	29159.72	RH	0.7			MIT
3424.382	3425.364	29193.98	RH I	1.5	0.7		MIT
3422.291	3423.272	29211.81	RH I	1.1			MIT
3412.274	3413.253	29297.57	RH	2.5	1.8		MIT
3407.742	3408.720	29336.53	RH I	0.8			MIT
3407.238	3408.216	29340.87	RH I	0.7			MIT
3406.546	3407.523	29346.83	RH I	1.7	0.9		MIT
3400.980	3401.956	29394.85	RH I	1.0	0.3		MIT
3399.696	3400.672	29405.96	RH I	2.7	1.8		MIT
3396.85	3397.82	29430.6	RH I	3.0W	2.7		MIT
3394.891	3395.865	29447.57	RH I	1.2	0.7		MIT
3391.73	3392.70	29475.0	RH I	0.5D	0.0		MIT
3389.219	3390.192	29496.85	RH I	0.8			MIT
3385.781	3386.753	29526.81	RH I	1.5	1.0		MIT
3381.445	3382.416	29564.67	RH	0.6			MIT
3380.636	3381.607	29571.74	RH I	0.6			MIT
3377.708	3378.678	29597.37	RH I	1.4	0.7		MIT
3377.138	3378.108	29602.37	RH	1.6	1.0		MIT
3372.526	3373.495	29642.85	RH I	1.0			MIT
3372.254	3373.223	29645.24	RH I	2.5	2.3		MIT
3369.679	3370.647	29667.90	RH I	1.4	0.7		MIT
3368.775	3369.743	29675.86	RH I	1.2			MIT
3368.375	3369.343	29679.38	RH I	2.5	1.7		MIT
3362.184	3363.150	29734.03	RH I	2.0	1.3		MIT
3360.803	3361.769	29746.25	RH I	0.3	0.0		MIT
3359.895	3360.860	29754.28	RH I	2.0	1.7		MIT
3357.849	3358.814	29772.41	RH I	0.5			MIT
3354.707	3355.671	29800.30	RH I	1.0	0.3		MIT
3353.687	3354.651	29809.36	RH I	1.0			MIT
3352.373	3353.337	29821.04	RH I	0.7			MIT
3344.198	3345.159	29893.94	RH I	2.0	1.3		MIT
3343.43	3344.39	29900.8	RH I	0.3			MIT
3342.904	3343.865	29905.51	RH I	0.3	0.0		MIT
3338.545	3339.505	29944.56	RH I	2.3	1.7		MIT
3331.245	3332.203	30010.18	RH	1.6	1.0		MIT
3331.093	3332.051	30011.54	RH I	1.7	1.0		MIT
3323.092	3324.048	30083.80	RH I	3.0	2.3		MIT
3314.541	3315.495	30161.41	RH I	0.6			MIT
3309.54	3310.49	30207.0	RH	0.3			MIT
3307.948	3308.900	30221.52	RH I	0.7			MIT
3305.172	3306.124	30246.90	RH I	1.6	1.0		MIT
3304.140	3305.091	30256.35	RH I	0.6			MIT
3300.465	3301.415	30290.04	RH I	2.0	1.3		MIT
3300.356	3301.306	30291.04	RH I	0.7			MIT
3298.940	3299.890	30304.04	RH I	0.5			MIT
3297.286	3298.236	30319.24	RH I	0.7			MIT
3296.717	3297.666	30324.48	RH I	1.6	1.0		MIT
3294.278	3295.227	30346.93	RH I	1.8	1.4		MIT
3289.636	3290.584	30389.75	RH	1.7R	0.7		MIT
3289.138	3290.085	30394.35	RH I	2.2	1.7		MIT
3288.039	3288.986	30404.51	RH I	0.6			MIT
3286.396	3287.343	30419.71	RH I	0.7			MIT
3283.573	3284.519	30445.86	RH I	2.2			MIT
3281.701	3282.647	30463.23	RH I	0.7	0.0		MIT
3280.55	3281.50	30473.9	RH I	1.5G	1.0		MIT
3275.986	3276.930	30516.37	RH I	0.6			MIT
3274.778	3275.722	30527.62	RH	0.3			MIT
3271.612	3272.555	30557.16	RH I	2.3	1.8		MIT
3270.576	3271.519	30566.84	RH I	0.5			MIT
3268.474	3269.416	30586.50	RH I	1.4	0.7		MIT
3267.48	3268.42	30595.8	RH	0.3	1.2		MIT
3263.821	3264.762	30630.10	RH	0.3			MIT
3263.144	3264.085	30636.46	RH I	2.3	1.6		MIT
3254.97	3255.91	30713.4	RH I	1.0			MIT
3253.343	3254.281	30728.75	RH	0.3			MIT
3250.034	3250.972	30760.04	RH I	0.3			MIT
3237.662	3238.596	30877.57	RH I	1.8	1.3		MIT
3235.785	3236.719	30895.49	RH	0.7			MIT

AIR WAVELENGTH (ANGSTRMS)	VACUUM WAVELENGTH (ANGSTRMS)	WAVENUMBER (KAYSERS)	LMENT	LOG ARC	INTENSITY SPK	DIS	REF
3233.32	3234.25	30919.0	RH	0.3	0.7		MIT
3232.504	3233.437	30926.84	RH I	0.8			MIT
3220.78	3221.71	31039.4	RH	0.6	0.3		MIT
3218.277	3219.206	31063.56	RH I	1.8			MIT
3217.882	3218.811	31067.37	RH I	1.8	1.3		MIT
3214.872	3215.801	31096.46	RH I	1.5			MIT
3214.325	3215.253	31101.75	RH I	1.8	1.3		MIT
3211.387	3212.315	31130.20	RH I	1.0			MIT
3207.29	3208.22	31170.0	RH	0.5	0.8		MIT
3206.086	3207.012	31181.67	RH	1.0			MIT
3197.133	3198.057	31268.99	RH I	1.7	1.0		MIT
3194.55	3195.47	31294.3	RH I	1.7	1.0H		MIT
3193.845	3194.768	31301.18	RH I	0.7			MIT
3191.187	3192.110	31327.25	RH I	2.5	1.7		MIT
3190.342	3191.264	31335.54	RH I	1.6	1.0		MIT
3189.048	3189.970	31348.26	RH I	2.0	1.3		MIT
3188.292	3189.214	31355.69	RH I	0.9			MIT
3187.88	3188.80	31359.7	RH	0.3	0.6		MIT
3185.593	3186.514	31382.25	RH I	2.0	1.3		MIT
3181.222	3182.142	31425.37	RH I	1.0			MIT
3179.729	3180.649	31440.13	RH I	1.7			MIT
3178.41	3179.33	31453.2	RH I	1.2			MIT
3177.092	3178.011	31466.22	RH	1.2			MIT
3172.27	3173.19	31514.1	RH I	1.4			MIT
3171.53	3172.45	31521.4	RH	0.5			MIT
3166.95	3167.87	31567.0	PH	0.3	0.5		MIT
3162.260	3163.175	31613.80	RH I	0.3	0.3		MIT
3159.25	3160.16	31643.9	RH	0.7	1.0		MIT
3157.57	3158.48	31660.8	RH I	0.8			M
3155.775	3156.689	31678.76	RH	2.2	0.3		MIT
3155.35	3156.26	31683.0	RH	0.8			M
3152.603	3153.516	31710.64	RH	1.9	0.5		MIT
3151.363	3152.276	31723.11	RH	1.9	0.3		MIT
3150.273	3151.185	31734.09	RH I	0.9			MIT
3147.627	3148.539	31760.77	RH I	0.7			MIT
3145.61	3146.52	31781.1	RH	0.3			MIT
3137.707	3138.616	31861.18	RH I	2.0			MIT
3137.323	3138.232	31865.07	RH I	1.2			MIT
3135.47	3136.38	31883.9	RH I	0.7			MIT
3130.790	3131.697	31931.56	RH I	1.8	0.3		MIT
3124.402	3125.308	31996.85	RH I	0.7			MIT
3123.697	3124.603	32004.07	RH I	2.2	0.3		MIT
3121.755	3122.660	32023.98	RH I	2.2			MIT
3117.96	3118.86	32062.9	RH	0.6	1.0		MIT
3114.909	3115.812	32094.36	RH I	2.0	0.3		MIT
3108.29	3109.19	32162.7	RH	0.5			MIT
3104.995	3105.896	32196.83	RH I	0.6			MIT
3102.527	3103.427	32222.44	RH I	1.0			MIT
3093.48	3094.38	32316.7	RH I	0.3	0.5		MIT
3090.39	3091.29	32349.0	RH	0.6			MIT
3088.320	3089.217	32370.67	RH I	0.5			MIT
3087.415	3088.311	32380.15	RH I	1.3S			MIT
3085.67	3086.57	32398.5	RH	0.3H			MIT
3083.965	3084.861	32416.37	RH I	2.2	0.3		MIT
3076.642	3077.536	32493.53	RH I	0.5			MIT
3071.027	3071.919	32552.94	RH I	1.3			MIT
3068.887	3069.779	32575.64	RH	0.3	0.0		MIT
3067.305	3068.196	32592.44	RH	1.9	0.0		MIT
3062.19	3063.08	32646.9	RH		0.3		MIT
3059.356	3060.245	32677.12	RH I	0.6			MIT
3057.889	3058.778	32692.79	RH	1.0			MIT
3055.65	3056.54	32716.7	RH	0.3H	0.3		MIT
3053.872	3054.760	32735.79	RH I	0.9			MIT
3051.679	3052.567	32759.32	RH I	0.5			MIT
3050.818	3051.705	32768.56	RH	0.3			MIT
3049.217	3050.104	32785.77	RH I	0.7			MIT
3048.89	3049.78	32789.3	RH I	0.3	0.5		MIT
3047.97	3048.86	32799.2	RH	0.3			MIT
3047.153	3048.039	32807.98	RH	1.2	0.5		MIT
3046.758	3047.644	32812.23	RH I	0.6			MIT

AIR WAVELENGTH (ANGSTRMS)	VACUUM WAVELENGTH (ANGSTRMS)	WAVENUMBER (KAYSERS)	LMENT	LOG ARC	INTENSITY SPK	DIS	REF
3046.19	3047.08	32818.4	RH I	0.3			MIT
3045.773	3046.659	32822.84	RH I	1.5			MIT
3034.99	3035.87	32939.5	RH	0.3	0.6		MIT
3028.87	3029.75	33006.0	RH	0.3H			MIT
3028.426	3029.308	33010.84	RH I	1.9			MIT
3026.94	3027.82	33027.0	RH I	0.3			MIT
3025.397	3026.278	33043.89	RH I	0.8			MIT
3023.911	3024.792	33060.13	RH I	2.0	0.3		MIT
3019.544	3020.423	33107.94	RH I	0.7			MIT
3014.23	3015.11	33166.3	RH I	0.3			MIT
3009.57	3010.45	33217.7	RH	0.3	0.3		MIT
3008.98	3009.86	33224.2	RH	0.3	0.5		MIT
3005.823	3006.699	33259.06	RH	0.8			MIT
3004.457	3005.333	33274.19	RH I	0.6	0.0		MIT
2991.743	2992.616	33415.58	RH I	0.7			MIT
2988.36	2989.23	33453.4	RH	0.3	0.3		MIT
2987.449	2988.321	33463.61	RH I	1.6	1.3		MIT
2986.989	2987.860	33468.77	RH I	1.9	1.3		MIT
2986.202	2987.073	33477.59	RH I	2.2	1.8		MIT
2983.085	2983.955	33512.56	RH I	1.5			MIT
2982.398	2983.268	33520.28	RH	0.3			MIT
2981.116	2981.986	33534.70	RH I	0.7			MIT
2977.679	2978.548	33573.40	RH I	2.1	1.5		MIT
2974.030	2974.898	33614.60	RH I	1.0			MIT
2968.663	2969.530	33675.37	RH I	2.1	1.5		MIT
2965.161	2966.027	33715.13	RH	0.9			MIT
2963.536	2964.402	33733.62	RH	1.2	0.7		MIT
2961.69	2962.56	33754.6	RH	0.3			MIT
2959.646	2960.511	33777.96	RH I	0.7			MIT
2955.412	2956.276	33826.35	RH I	1.3			MIT
2955.284	2956.148	33827.81	RH	0.8			MIT
2949.380	2950.242	33895.52	RH I	0.3			MIT
2939.47	2940.33	34009.8	RH I	0.3			MIT
2931.941	2932.799	34097.12	RH I	1.9	1.3		MIT
2929.107	2929.964	34130.11	RH	2.0	1.0		MIT
2924.09	2924.95	34188.7	RH	0.5	0.5		MIT
2924.024	2924.880	34189.44	RH I	2.0			MIT
2923.105	2923.961	34200.19	RH I	1.6			MIT
2915.419	2916.273	34290.35	RH I	1.9	1.6		MIT
2913.987	2914.840	34307.20	RH I	1.3			MIT
2913.593	2914.446	34311.84	RH I	0.8			MIT
2912.616	2913.469	34323.35	RH I	1.7	1.3		MIT
2910.171	2911.023	34352.18	RH	1.7	1.1		MIT
2907.209	2908.061	34387.18	RH I	2.0	1.5		MIT
2903.31	2904.16	34433.4	RH I	0.3			MIT
2899.955	2900.805	34473.19	RH I	1.8	1.5		MIT
2899.69	2900.54	34476.3	RH I	0.7R			MIT
2897.633	2898.482	34500.81	RH	0.3	1.0		MIT
2896.08	2896.93	34519.3	RH	0.3	1.0		MIT
2892.216	2893.064	34565.43	RH I	1.5			MIT
2889.839	2890.686	34593.86	RH I	1.8	1.5		MIT
2889.106	2889.953	34602.64	RH I	1.9	0.0H		MIT
2885.974	2886.820	34640.19	RH I	0.3	0.0		MIT
2884.556	2885.402	34657.22	RH I	0.7			MIT
2882.366	2883.211	34683.55	RH	1.9	1.0		MIT
2881.254	2882.099	34696.93	RH I	1.3			MIT
2880.648	2881.493	34704.23	RH I	1.0	0.3J		MIT
2878.655	2879.500	34728.26	RH I	1.7	1.0		MIT
2873.99	2874.83	34784.6	RH I	0.8			MIT
2873.623	2874.466	34789.07	RH	1.8	1.0		MIT
2871.354	2872.197	34816.56	RH	2.0	1.0H		MIT
2869.91	2870.75	34834.1	RH	0.9	0.0H		MIT
2868.276	2869.118	34853.92	RH I	1.0			MIT
2864.404	2865.245	34901.03	RH	1.8	1.0		MIT
2862.935	2863.776	34918.94	RH I	2.2	1.8		MIT
2860.762	2861.602	34945.46	RH I	1.5	1.0H		MIT
2860.675	2861.515	34946.52	RH I	1.5			MIT
2859.785	2860.625	34957.40	RH	0.8			MIT
2859.615	2860.455	34959.48	RH	0.8			MIT
2856.164	2857.003	35001.71	RH I	1.8	1.5H		MIT

AIR WAVELENGTH (ANGSTRMS)	VACUUM WAVELENGTH (ANGSTRMS)	WAVENUMBER (KAYSERS)	LMENT	LOG ARC	INTENSITY SPK	DIS	REF
2849.343	2850.180	35085.50	RH	0.5			MIT
2845.753	2846.589	35129.76	RH	0.9	0.6		MIT
2842.126	2842.962	35174.59	RH	1.2			MIT
2836.690	2837.524	35241.99	RH I	1.8			MIT
2834.117	2834.951	35273.98	RH I	1.8	1.5		MIT
2832.769	2833.602	35290.77	RH I	0.7			MIT
2829.306	2830.138	35333.96	RH I	1.5			MIT
2827.310	2828.142	35358.90	RH I	1.7			MIT
2826.675	2827.507	35366.85	RH I	2.0	1.7D		MIT
2823.370	2824.201	35408.25	RH I	1.4			MIT
2822.856	2823.687	35414.69	RH I	0.9			MIT
2820.836	2821.666	35440.05	RH I	0.5			MIT
2819.626	2820.456	35455.26	RH I	1.2			MIT
2810.892	2811.720	35565.42	RH I	1.2	0.0H		MIT
2807.142	2807.969	35612.93	RH I	0.9			MIT
2805.785	2806.612	35630.15	RH I	1.3	0.0		MIT
2803.903	2804.729	35654.07	RH I	0.6	0.3		MIT
2796.632	2797.456	35746.76	RH I	2.0	0.0		MIT
2795.701	2796.525	35758.66	RH I	1.2			MIT
2792.78	2793.60	35796.1	RH	0.3	0.5		MIT
2791.158	2791.981	35816.86	RH I	2.0	0.0H		MIT
2790.755	2791.578	35822.03	RH	0.5	0.5		MIT
2783.029	2783.850	35921.47	RH I	2.2	1.0		MIT
2781.80	2782.62	35937.3	RH	0.0	2.2		MIT
2779.537	2780.357	35966.60	RH I	2.0	0.8		MIT
2778.15	2778.97	35984.6	RH	0.3	2.0		MIT
2778.055	2778.875	35985.79	RH I	2.0	0.5		MIT
2775.769	2776.588	36015.42	RH	0.7	2.1		MIT
2774.17	2774.99	36036.2	RH	0.0	1.6		MIT
2773.31	2774.13	36047.4	RH I	0.5	0.3H		MIT
2773.11	2773.93	36050.0	RH		1.2		MIT
2771.510	2772.328	36070.76	RH I	2.0	0.9		MIT
2768.229	2769.047	36113.51	RH	1.7	0.6		MIT
2767.728	2768.545	36120.05	RH I	2.0	0.6		MIT
2766.54	2767.36	36135.6	RH	0.7	2.2		MIT
2764.83	2765.65	36157.9	RH	1.2R	2.1		MIT
2762.828	2763.644	36184.11	RH I	1.0			MIT
2761.26	2762.08	36204.6	RH	0.0	1.7		MIT
2756.83	2757.64	36262.8	RH		1.0		MIT
2754.09	2754.90	36298.9	RH	0.0	1.4		MIT
2752.839	2753.653	36315.40	RH I	1.7	0.3		MIT
2747.63	2748.44	36384.2	RH	0.0	2.0		MIT
2740.552	2741.363	36478.21	RH I	1.6	0.3		MIT
2740.208	2741.019	36482.79	RH I	1.6	0.3		MIT
2739.923	2740.734	36486.58	RH	1.0	2.5		MIT
2738.268	2739.078	36508.63	RH I	1.0	0.3		MIT
2737.605	2738.415	36517.47	RH	1.5	0.3		MIT
2737.400	2738.210	36520.21	RH	1.5	2.6J		MIT
2736.761	2737.571	36528.73	RH I	2.0	0.5		MIT
2730.71	2731.52	36609.7	RH	0.3	1.0		MIT
2728.945	2729.753	36633.35	RH I	2.3	2.3		MIT
2720.523	2721.329	36746.75	RH I	1.7	0.3		MIT
2720.137	2720.943	36751.97	RH	2.0	0.8		MIT
2718.544	2719.349	36773.50	RH I	2.2	1.3		MIT
2717.98	2718.79	36781.1	RH		1.7		MIT
2717.512	2718.317	36787.46	RH I	2.0	0.7		MIT
2716.819	2717.624	36796.85	RH I	1.7	0.5		MIT
2715.308	2716.113	36817.32	RH	1.7	2.7J		MIT
2715.045	2715.850	36820.89	RH I	1.7	0.3		MIT
2714.410	2715.214	36829.50	RH I	2.2	0.7		MIT
2713.32	2714.12	36844.3	RH		1.4		MIT
2709.523	2710.326	36895.93	RH I	1.7	0.3		MIT
2707.226	2708.029	36927.23	RH I	2.0	0.6		MIT
2705.629	2706.431	36949.02	RH	2.0	2.5J		MIT
2704.96	2705.76	36958.2	RH	0.5	1.6		MIT
2703.732	2704.534	36974.95	RH I	2.2	1.4		MIT
2702.233	2703.035	36995.46	RH I	0.7	0.3		MIT
2700.59	2701.39	37018.0	RH	0.3	1.9		MIT
2699.75	2700.55	37029.5	RH		1.4		MIT
2695.78	2696.58	37084.0	RH	0.0	1.2		MIT

AIR WAVELENGTH (ANGSTRMS)	VACUUM WAVELENGTH (ANGSTRMS)	WAVENUMBER (KAYSERS)	LMENT	LOG ARC	INTENSITY SPK	DIS	REF
2694.308	2695.108	37104.27	RH I	0.7	0.6		MIT
2692.36	2693.16	37131.1	RH I	0.3			MIT
2692.308	2693.107	37131.83	RH I	0.8	0.5		MIT
2691.12	2691.92	37148.2	RH	0.0	1.8		MIT
2689.62	2690.42	37168.9	RH	0.0	2.0		MIT
2688.095	2688.893	37190.02	RH I	0.6	0.8		MIT
2687.317	2688.115	37200.79	RH I	0.3			MIT
2684.214	2685.011	37243.79	RH	0.3	2.2		MIT
2683.563	2684.360	37252.83	RH	0.0	2.3		MIT
2682.540	2683.337	37267.03	RH I	0.3	0.3		MIT
2681.784	2682.581	37277.54	RH I	0.3	0.5		MIT
2681.60	2682.40	37280.1	RH		2.0		MIT
2680.631	2681.427	37293.57	RH I	1.8	1.0		MIT
2680.28	2681.08	37298.5	RH	0.5	0.3		MIT
2676.483	2677.278	37351.37	RH I	0.3	0.3		MIT
2676.25	2677.05	37354.6	RH	0.0	2.0		MIT
2676.110	2676.905	37356.57	RH I	1.0	0.7		MIT
2674.441	2675.236	37379.88	RH	0.0	2.3W		MIT
2674.200	2674.995	37383.25	RH I	0.5	0.3		MIT
2671.055	2671.849	37427.26	RH I	0.5	0.6		MIT
2669.20	2669.99	37453.3	RH	0.0	1.2		MIT
2666.41	2667.20	37492.5	RH	0.3	0.3		MIT
2664.49	2665.28	37519.5	RH	0.0	1.5		MIT
2663.72	2664.51	37530.3	RH	0.0	0.9		MIT
2659.472	2660.263	37590.27	RH I	0.5	0.5		MIT
2659.11	2659.90	37595.4	RH		2.0		MIT
2659.011	2659.802	37596.78	RH I	0.3	0.3		MIT
2658.36	2659.15	37606.0	RH	0.3	1.4		MIT
2657.32	2658.11	37620.7	RH		1.8		MIT
2655.902	2656.692	37640.79	RH I	0.3	0.3		MIT
2654.771	2655.561	37656.83	RH		1.4		MIT
2652.660	2653.450	37686.79	RH I	2.0	1.4		MIT
2643.00	2643.79	37824.5	RH	1.0	0.7		MIT
2642.76	2643.55	37828.0	RH	0.0	1.3		MIT
2641.65	2642.44	37843.9	RH	0.0	1.0		MIT
2639.25	2640.04	37878.3	RH	0.3	2.0		MIT
2638.74	2639.53	37885.6	RH	0.3	2.0		MIT
2635.322	2636.107	37934.72	RH	0.5	1.6		MIT
2634.990	2635.775	37939.50	RH	1.0	0.6		SY
2633.419	2634.204	37962.13	RH	0.3	0.3		MIT
2633.288	2634.073	37964.02	RH I	0.5	0.5		MIT
2630.418	2631.202	38005.44	RH I	1.0	0.6		MIT
2630.334	2631.118	38006.65	RH	0.3	1.7		MIT
2629.912	2630.696	38012.75	RH I	0.5	0.5		MIT
2628.125	2628.909	38038.60	RH	0.0	2.2		MIT
2627.79	2628.57	38043.5	RH		0.9		MIT
2626.61	2627.39	38060.5	RH	0.0	1.4		MIT
2625.884	2626.667	38071.06	RH I	1.8	1.4		MIT
2625.405	2626.188	38078.00	RH	0.3	2.3V		MIT
2625.214	2625.997	38080.77	RH	0.3			MIT
2622.575	2623.357	38119.09	RH I	1.7	1.4		MIT
2621.000	2621.782	38142.00	RH	0.3	0.3		MIT
2616.073	2616.854	38213.83	RH	0.3	0.5		MIT
2611.76	2612.54	38276.9	RH		1.0		MIT
2609.17	2609.95	38314.9	RH	0.5	2.3		MIT
2606.440	2607.219	38355.05	RH	0.9	0.7		MIT
2603.321	2604.099	38401.00	RH		2.0		MIT
2598.072	2598.849	38478.58	RH	0.6	0.5		MIT
2597.683	2598.460	38484.34	RH I	0.3	0.3		MIT
2597.07	2597.85	38493.4	RH	0.5	2.2		MIT
2596.924	2597.700	38495.59	RH	0.3	0.3		MIT
2593.618	2594.394	38544.65	RH		1.2		MIT
2592.160	2592.935	38566.33	RH		1.9		MIT
2587.287	2588.061	38638.96	RH	0.3	2.0		MIT
2582.83	2583.60	38705.6	RH	0.3	1.0		MIT
2581.692	2582.465	38722.70	RH	0.0	2.2		MIT
2576.229	2577.000	38804.81	RH	0.3	0.3		MIT
2575.75	2576.52	38812.0	RH I	0.3	0.6		MIT
2574.657	2575.428	38828.50	RH	0.7	0.7		MIT
2570.111	2570.881	38897.17	RH I	0.6	0.3		MIT
2569.07	2569.84	38912.9	RH	0.5	2.1		MIT
2568.83	2569.60	38916.6	RH	0.3	2.0		MIT
2566.92	2567.69	38945.5	RH		1.0		MIT
2566.043	2566.812	38958.83	RH I	0.7	0.6		MIT
2565.787	2566.556	38962.72	RH I	0.7	0.6		MIT
2565.11	2565.88	38973.0	RH	0.0	1.5		MIT
2561.92	2562.69	39021.5	RH		1.7		MIT
2560.21	2560.98	39047.6	RH I	0.3R			MIT
2559.905	2560.673	39052.24	RH	0.7	2.0		MIT
2558.62	2559.39	39071.9	RH	0.7			MIT
2557.92	2558.69	39082.5	RH	0.0	1.7		MIT
2557.20	2557.97	39093.5	RH	0.0	2.0		MIT
2555.360	2556.127	39121.69	RH I	2.0	1.8		MIT
2551.192	2551.958	39185.60	RH I	0.9	0.3		MIT
2550.751	2551.516	39192.38	RH	0.0	1.3		MIT
2549.66	2550.43	39209.1	RH	0.0	1.3		MIT
2545.695	2546.459	39270.21	RH	1.7	1.0		MIT
2545.354	2546.118	39275.47	RH	0.9	2.2		MIT
2544.223	2544.987	39292.93	RH	0.9	0.6		MIT
2543.941	2544.705	39297.29	RH	1.2	2.0R		MIT
2542.156	2542.919	39324.88	RH	0.0	1.7		MIT
2541.153	2541.916	39340.40	RH	0.3	1.2		MIT
2541.029	2541.792	39342.32	RH I	0.7	0.5		MIT
2539.721	2540.484	39362.58	RH	0.7	0.3		MIT
2537.729	2538.491	39393.48	RH	0.6	1.7		MIT
2537.039	2537.801	39404.19	RH	1.2	2.0		MIT
2536.706	2537.468	39409.36	RH	1.2	0.7		MIT
2534.57	2535.33	39442.6	RH	0.3W	2.0W		MIT
2534.072	2534.834	39450.32	RH	0.6	0.6		MIT
2533.591	2534.352	39457.81	RH	0.3	0.3		MIT
2532.661	2533.422	39472.30	RH	0.6	0.7		MIT
2531.742	2532.503	39486.63	RH	0.3			MIT
2527.292	2528.052	39556.15	RH	0.0	1.4		MIT
2523.482	2524.241	39615.87	RH		1.4		MIT
2520.533	2521.291	39662.21	RH	1.0	3.0J		MIT
2517.523	2518.281	39709.63	RH		2.2		MIT
2515.746	2516.503	39737.68	RH I	1.8	1.0		MIT
2515.325	2516.082	39744.33	RH	0.3	1.4		MIT
2513.356	2514.113	39775.46	RH	0.9	0.8		MIT
2511.027	2511.783	39812.35	RH	1.2	1.2		MIT
2510.655	2511.411	39818.25	RH	0.7	2.3J		MIT
2509.697	2510.453	39833.45	RH I	1.6	1.3		MIT
2505.673	2506.428	39897.42	RH I	1.0	0.9		MIT
2505.104	2505.859	39906.48	RH	0.3	2.3		MIT
2504.294	2505.049	39919.38	RH	0.7	1.0		MIT
2503.843	2504.597	39926.58	RH	0.3	2.2		MIT
2502.456	2503.210	39948.70	RH I	0.6	0.8		MIT
2501.272	2502.026	39967.61	RH	1.7R	1.7		MIT
2499.017	2499.770	40003.67	RH	1.2W	0.3H		MIT
2498.21	2498.96	40016.6	RH		0.5H		EX
2495.951	2496.704	40052.81	RH	0.3	1.7		MIT
2492.299	2493.051	40111.50	RH I	2.0	1.0		MIT
2491.855	2492.607	40118.64	RH	0.3	2.6H		MIT
2490.770	2491.521	40136.12	RH	2.0	3.0J		MIT
2487.470	2488.221	40189.36	RH	2.0	0.9		MIT
2485.817	2486.567	40216.08	RH	0.3	1.7		MIT
2483.333	2484.083	40256.31	RH I	2.0R	0.7		MIT
2482.734	2483.484	40266.02	RH	0.3	2.0		MIT
2482.044	2482.793	40277.21	RH	1.4	0.5		MIT
2480.44	2481.19	40303.3	RH	0.3W	1.4W		MIT
2479.756	2480.505	40314.37	RH	0.8			MIT
2477.535	2478.283	40350.51	RH	0.8	1.0J		MIT
2477.190	2477.938	40356.13	RH	0.3	0.9		MIT
2472.51	2473.26	40432.5	RH	1.6W	0.6W		MIT
2471.768	2472.515	40444.65	RH	0.5	2.0		MIT
2471.472	2472.219	40449.49	RH I	1.8	0.9		MIT
2470.391	2471.138	40467.19	RH I	1.8	0.7		MIT
2467.228	2467.974	40519.06	RH		1.7		MIT
2466.149	2466.895	40536.79	RH	0.0	2.0		MIT
2463.609	2464.354	40578.58	RH	1.5	0.3		MIT

AIR WAVELENGTH (ANGSTRMS)	VACUUM WAVELENGTH (ANGSTRMS)	WAVENUMBER (KAYSERS)	LMENT	LOG ARC	INTENSITY SPK	DIS	REF
2463.44	2464.19	40581.4	RH	0.3	2.2		MIT
2461.036	2461.781	40621.00	RH	1.9	2.3J		MIT
2459.17	2459.91	40651.8	RH	0.0	1.0J		MIT
2458.904	2459.648	40656.22	RH	1.7	2.5		MIT
2456.180	2456.923	40701.31	RH	0.7	2.2		MIT
2455.705	2456.448	40709.18	RH	1.2	2.30		MIT
2450.56	2451.30	40794.6	RH	1.0D	0.7D		MIT
2449.038	2449.780	40820.00	RH	1.5	0.7		MIT
2448.835	2449.577	40823.38	RH I	1.5	0.7		MIT
2448.280	2449.022	40832.63	RH	0.3	2.0		MIT
2447.85	2448.59	40839.8	RH	0.6	2.3W		MIT
2446.66	2447.40	40859.7	RH		1.4J		MIT
2445.643	2446.384	40876.66	RH I	1.4	0.3		MIT
2445.119	2445.860	40885.42	RH	0.3H	0.9		MIT
2444.742	2445.483	40891.72	RH	0.6	1.4		MIT
2444.270	2445.011	40899.62	RH I	2.0W	0.5		MIT
2444.058	2444.799	40903.16	RH	0.3	2.0		MIT
2443.711	2444.452	40908.97	RH	0.6	2.0		MIT
2442.727	2443.467	40925.45	RH	0.0	1.0		MIT
2438.786	2439.525	40991.58	RH	0.3	2.0		MIT
2431.847	2432.585	41108.54	RH	0.5	1.0		MIT
2430.813	2431.551	41126.02	RH	0.3	1.3		MIT
2429.516	2430.253	41147.97	RH I	0.6	0.6		MIT
2429.105	2429.842	41154.94	RH		1.0		MIT
2427.339	2428.076	41184.87	RH	0.3	1.7		MIT
2426.465	2427.202	41199.71	RH	0.5	0.3		MIT
2423.076	2423.812	41257.33	RH		1.2		MIT
2422.14	2422.88	41273.3	RH	0.0	0.3		MIT
2420.979	2421.714	41293.06	RH	1.5	2.0		MIT
2420.177	2420.912	41306.74	RH	0.6	0.6		MIT
2419.950	2420.685	41310.62	RH	0.0	0.3		MIT
2415.841	2416.575	41380.88	RH	2.0	2.3		MIT
2414.39	2415.12	41405.7	RH	0.3	0.6		MIT
2411.94	2412.67	41447.8	RH	0.3	1.7		EX
2410.688	2411.421	41469.32	RH	0.3	1.7		MIT
2408.722	2409.455	41503.17	RH	0.3	1.5		MIT
2407.884	2408.616	41517.61	RH I	1.8	0.7		MIT
2405.221	2405.953	41563.58	RH	0.0	1.7		MIT
2403.471	2404.202	41593.84	RH	0.0	1.0		MIT
2400.708	2401.439	41641.70	RH		1.5		MIT
2398.95	2399.68	41672.2	RH	1.0			MIT
2397.97	2398.70	41689.2	RH	0.3			MIT
2396.93	2397.66	41707.3	RH		0.7		MIT
2396.55	2397.28	41714.0	RH	0.3	2.3		MIT
2393.89	2394.62	41760.3	RH		1.1		MIT
2392.42	2393.15	41786.0	RH		1.6		MIT
2392.23	2392.96	41789.3	RH		1.2		MIT
2390.62	2391.35	41817.4	RH	1.0	1.7W		MIT
2389.84	2390.57	41831.1	RH		1.0		MIT
2389.27	2390.00	41841.0	RH		1.1		MIT
2388.02	2388.75	41862.9	RH	0.6			A
2387.82	2388.55	41866.4	RH		1.0		A
2386.40	2387.13	41891.4	RH	0.8	0.3		MIT
2386.14	2386.87	41895.9	RH	1.9	0.9		MIT
2385.60	2386.33	41905.4	RH		1.0		MIT
2385.44	2386.17	41908.2	RH	0.5	1.4		MIT
2384.65	2385.38	41922.1	RH	1.4			MIT
2383.59	2384.32	41940.7	RH		1.4		MIT
2383.40	2384.13	41944.1	RH	1.7	1.0		MIT
2382.89	2383.62	41953.0	RH	1.7	0.7		MIT
2382.64	2383.37	41957.5	RH		1.5		MIT
2378.93	2379.66	42022.9	RH I	0.9			MIT
2377.81	2378.54	42042.7	RH		1.7		MIT
2375.57	2376.30	42082.3	RH	0.7	0.3H		A
2374.84	2375.56	42095.2	RH		1.0		MIT
2371.66	2372.38	42151.7	RH	0.5			A
2371.48	2372.20	42154.9	RH	0.5	1.4J		MIT
2371.07	2371.79	42162.2	RH	0.5	1.6		MIT
2370.58	2371.30	42170.9	RH	1.0			MIT
2369.58	2370.30	42188.7	RH I	1.0	1.3J		MIT

AIR WAVELENGTH (ANGSTRMS)	VACUUM WAVELENGTH (ANGSTRMS)	WAVENUMBER (KAYSERS)	LMENT	LOG ARC	INTENSITY SPK	DIS	REF
2368.87	2369.59	42201.3	RH	0.8			MIT
2368.34	2369.06	42210.8	RH	1.7	0.3		MIT
2367.09	2367.81	42233.1	RH	0.7	1.5		MIT
2366.88	2367.60	42236.8	RH	0.5	1.7		MIT
2365.99	2366.71	42252.7	RH	0.5			A
2364.66	2365.38	42276.5	RH	1.0	1.4		MIT
2364.14	2364.86	42285.8	RH		1.4		MIT
2363.05	2363.77	42305.3	RH	0.9	0.6J		MIT
2361.92	2362.64	42325.5	RH	1.5	0.8		MIT
2361.45	2362.17	42333.9	RH	0.5W	0.8		MIT
2361.17	2361.89	42338.9	RH	1.0	0.5		MIT
2359.57	2360.29	42367.6	RH		1.7		MIT
2359.18	2359.90	42374.6	RH	1.2	2.0		MIT
2358.79	2359.51	42381.7	RH		1.0W		MIT
2358.47	2359.19	42387.4	RH I	1.0	0.5H		MIT
2357.43	2358.15	42406.1	RH	0.3	1.7J		MIT
2356.30	2357.02	42426.4	RH		1.5		MIT
2352.47	2353.19	42495.5	RH I	1.3	1.3W		MIT
2351.64	2352.36	42510.5	RH		0.7		MIT
2350.35	2351.07	42533.8	RH		1.7		MIT
2349.68	2350.40	42546.0	RH	0.5	2.1		MIT
2347.90	2348.62	42578.2	RH		1.2		MIT
2347.85	2348.57	42579.1	RH	0.9			A
2347.69	2348.41	42582.0	RH		1.0J		MIT
2346.77	2347.49	42598.7	RH	0.3	1.5		MIT
2346.44	2347.16	42604.7	RH	0.3	2.0		MIT
2345.41	2346.13	42623.4	RH	1.0	0.5H		A
2344.89	2345.61	42632.9	RH		1.3		MIT
2343.02	2343.74	42666.9	RH		1.0W		MIT
2342.47	2343.19	42676.9	RH	0.5			MIT
2342.37	2343.09	42678.7	RH	0.5	1.4		A
2338.68	2339.40	42746.1	RH	0.5	0.7		A
2337.51	2338.23	42767.5	RH	0.6			A
2336.84	2337.56	42779.7	RH	0.0	2.1		MIT
2335.28	2336.00	42808.3	RH		1.2		MIT
2334.77	2335.49	42817.6	RH	1.4L	2.7		MIT
2333.69	2334.41	42837.5	RH		0.5		MIT
2333.31	2334.03	42844.4	RH I	0.9	2.1		MIT
2328.64	2329.35	42930.4	RH I	1.3L			MIT
2327.68	2328.39	42948.0	RH	0.3	1.2		MIT
2326.47	2327.18	42970.4	RH I	1.4	0.5H		MIT
2326.29	2327.00	42973.7	RH		1.4W		MIT
2324.86	2325.57	43000.1	RH		1.4		MIT
2324.51	2325.22	43006.6	RH	0.0	0.9		MIT
2323.03	2323.74	43034.0	RH		1.2J		MIT
2322.58	2323.29	43042.4	RH	1.7	1.0		MIT
2321.86	2322.57	43055.7	RH		1.9		MIT
2321.73	2322.44	43058.1	RH I	1.3	0.3		MIT
2321.08	2321.79	43070.2	RH	0.7			A
2319.87	2320.58	43092.6	RH	1.0	0.5		MIT
2319.10	2319.81	43106.9	RH I	1.4	1.0		MIT
2318.36	2319.07	43120.7	RH	0.7	0.7		MIT
2317.68	2318.39	43133.3	RH	0.7H			A
2316.59	2317.30	43153.6	RH	1.0			A
2315.66	2316.37	43171.0	RH	0.8			A
2313.88	2314.59	43204.2	RH	0.7	1.3		MIT
2312.65	2313.36	43227.1	RH	0.3	2.0		MIT
2312.32	2313.03	43233.3	RH	1.0			A
2311.66	2312.37	43245.7	RH		1.0J		MIT
2311.47	2312.18	43249.2	RH		1.2		MIT
2311.07	2311.78	43256.7	RH	0.9			MIT
2310.53	2311.24	43266.8	RH		1.4		MIT
2309.823	2310.533	43280.05	RH I	1.4	1.0		MIT
2306.95	2307.66	43333.9	RH		1.2		MIT
2306.79	2307.50	43337.0	RH	0.3	0.6		MIT
2305.95	2306.66	43352.7	RH		2.0J		MIT
2305.06	2305.77	43369.5	RH	0.3	1.0		A
2303.468	2304.177	43399.44	RH		1.3		MIT
2301.944	2302.653	43428.17	RH		1.4		MIT
2300.914	2301.623	43447.61	RH	0.3	1.5		MIT

AIR WAVELENGTH (ANGSTRMS)	VACUUM WAVELENGTH (ANGSTRMS)	WAVENUMBER (KAYSERS)	LMENT	LOG ARC	INTENSITY SPK	DIS	REF
2300.29	2301.00	43459.4	RH		1.0		MIT
2298.258	2298.966	43497.82	RH		2.2		MIT
2295.119	2295.826	43557.30	RH	0.7	1.7		MIT
2294.656	2295.363	43566.09	RH		1.5		MIT
2294.49	2295.20	43569.2	RH	1.0			MIT
2294.30	2295.01	43572.9	RH	0.5	0.7		A
2294.119	2294.826	43576.29	RH	0.0	1.7		MIT
2293.29	2294.00	43592.0	RH	1.0			MIT
2291.77	2292.48	43621.0	RH	0.7			A
2290.36	2291.07	43647.8	RH	0.9			A
2290.03	2290.74	43654.1	RH	1.4	2.7		MIT
2289.58	2290.29	43662.7	RH		1.2		MIT
2288.57	2289.28	43681.9	RH	1.4	1.0		MIT
2288.25	2288.96	43688.0	RH		1.6		MIT
2285.011	2285.716	43749.97	RH		1.5		MIT
2284.08	2284.78	43767.8	RH		2.3		MIT
2282.88	2283.58	43790.8	RH		0.8		MIT
2277.85	2278.55	43887.5	RH		1.2		MIT
2277.21	2277.91	43899.8	RH		1.4		MIT
2276.96	2277.66	43904.6	RH	1.3	2.2		MIT
2276.21	2276.91	43919.1	RH		1.7		MIT
2270.345	2271.047	44032.56	RH		1.3		MIT
2264.140	2264.841	44153.22	RH	1.7	1.4		MIT
2263.43	2264.13	44167.1	RH	0.7	2.3		MIT
2261.75	2262.45	44199.9	RH	0.3	2.0		MIT
2258.250	2258.949	44268.37	RH		1.4		MIT
2257.238	2257.937	44288.21	RH		1.4		MIT
2255.694	2256.393	44318.53	RH		1.2		MIT
2255.418	2256.117	44323.95	RH		1.3		MIT
2255.260	2255.959	44327.06	RH		1.0		MIT
2254.07	2254.77	44350.5	RH	0.5	1.4		MIT
2250.82	2251.52	44414.5	RH		0.7W		MIT
2248.74	2249.44	44455.6	RH	1.3	0.5		MIT
2246.38	2247.08	44502.3	RH		1.7		MIT
2246.07	2246.77	44508.4	RH		1.2		A
2242.05	2242.75	44588.2	RH	1.4	1.4		MIT
2240.00	2240.70	44629.0	RH		1.7		MIT
2238.19	2238.88	44665.1	RH	0.9			A
2237.71	2238.40	44674.7	RH		2.0		MIT
2237.10	2237.79	44686.9	RH		1.0		MIT
2236.75	2237.44	44693.8	RH	0.3	0.9		MIT
2236.38	2237.07	44701.2	RH		1.2		MIT
2235.30	2235.99	44722.8	RH	0.3	1.4		MIT
2232.98	2233.67	44769.3	RH		1.4		MIT
2231.69	2232.38	44795.2	RH	0.5	0.3W		A
2230.65	2231.34	44816.0	RH		1.4		A
2226.532	2227.224	44898.93	RH	1.7			MIT
2224.97	2225.66	44930.5	RH		1.5		MIT
2221.969	2222.660	44991.13	RH		1.4		MIT
2221.42	2222.11	45002.2	RH		1.4		MIT
2220.739	2221.430	45016.04	RH	0.3	1.3		MIT
2220.193	2220.884	45027.11	RH	1.3	0.0		MIT
2212.80	2213.49	45177.5	RH	1.2	1.0		MIT
2210.89	2211.58	45216.6	RH	1.2	0.3		A
2206.64	2207.33	45303.6	RH	1.0	1.0		A
2206.35	2207.04	45309.6	RH		1.9		MIT
2206.06	2206.75	45315.5	RH	1.0	0.7		MIT
2205.02	2205.71	45336.9	RH		2.0		MIT
2203.55	2204.24	45367.2	RH	1.2	1.7W		A
2202.71	2203.40	45384.5	RH		1.1		A
2201.89	2202.58	45401.4	RH		1.0		A
2201.75	2202.44	45404.2	RH	0.3	0.6		A
2201.02	2201.71	45419.3	RH		1.5		MIT
2199.96	2200.65	45441.2	RH		1.7		MIT
2195.62	2196.31	45531.0	RH	0.0	1.0		A
2195.22	2195.91	45539.3	RH	0.0W	0.7		A
2190.92	2191.60	45628.7	RH		1.5		MIT
2187.580	2188.264	45698.32	RH		1.7L		MIT
2186.99	2187.67	45710.6	RH		1.4		A
2182.01	2182.69	45815.0	RH		1.4		A

AIR WAVELENGTH (ANGSTRMS)	VACUUM WAVELENGTH (ANGSTRMS)	WAVENUMBER (KAYSERS)	LMENT	LOG ARC	INTENSITY SPK	DIS	REF
2178.95	2179.63	45879.3	RH	1.1	1.5		A
2177.13	2177.81	45917.6	RH	0.5	1.4L		A
2171.77	2172.45	46031.0	RH	0.5			A
2168.75	2169.43	46095.0	RH	0.6	0.5W		A
2167.322	2168.002	46125.42	RH		1.9		MIT
2163.178	2163.857	46213.77	RH		1.7		MIT
2158.194	2158.872	46320.48	RH		1.3		MIT
2143.97	2144.65	46627.8	RH		1.4		A
2142.28	2142.95	46664.5	RH	0.5	0.5		A
2140.03	2140.70	46713.6	RH		0.7		A
2139.76	2140.43	46719.5	RH		0.7		A
2139.43	2140.10	46726.7	RH		2.0		A
2138.66	2139.33	46743.5	RH		1.4		A
2137.62	2138.29	46766.2	RH		0.7		A
2137.35	2138.02	46772.2	RH		0.7		A
2136.31	2136.98	46794.9	RH	0.3	1.3S		A
2134.42	2135.09	46836.4	RH	0.7			A
2132.33	2133.00	46882.3	RH	0.5	0.6W		A
2130.26	2130.93	46927.8	RH	0.6	0.7		A
2129.28	2129.95	46949.4	RH	0.5	1.5		A
2126.06	2126.73	47020.5	RH	0.3	0.7		A
2125.37	2126.04	47035.8	RH		1.5		A
2124.13	2124.80	47063.2	RH	1.6	1.0		A
2123.99	2124.66	47066.3	RH	0.7	1.4		A
2123.13	2123.80	47085.4	RH	0.5	1.0		A
2121.67	2122.34	47117.8	RH	0.7W			A
2120.06	2120.73	47153.6	RH	0.5	1.4		A
2119.76	2120.43	47160.2	RH	0.7W			A
2119.17	2119.84	47173.4	RH	0.6	1.2		A
2118.50	2119.17	47188.3	RH	0.3W	2.3		A
2117.66	2118.33	47207.0	RH	0.5W	1.0		A
2116.87	2117.54	47224.6	RH	0.5W	1.7		A
2116.62	2117.29	47230.2	RH	0.6W	0.5		A
2116.47	2117.14	47233.5	RH		1.5		A
2114.87	2115.54	47269.3	RH		1.7		A
2113.70	2114.37	47295.4	RH		1.8		A
2112.70	2113.37	47317.8	RH	0.5	1.9		A
2111.69	2112.36	47340.4	RH		2.2		A
2109.08	2109.75	47399.0	RH		1.5		A
2107.93	2108.60	47424.9	RH		1.5		A
2105.93	2106.60	47469.9	RH		1.2		A
2104.89	2105.56	47493.4	RH		2.0		A
2103.10	2103.77	47533.8	RH	1.4	1.4		A
2102.45	2103.12	47548.5	RH		1.7		A
2101.96	2102.63	47559.5	RH		1.6		A
2101.35	2102.02	47573.4	RH		1.5		A
2100.76	2101.43	47586.7	RH	1.7			A
2100.52	2101.19	47592.1	RH	0.3	2.0		A
2100.36	2101.03	47595.8	RH		1.5		A
2098.99	2099.66	47626.8	RH	1.2	2.5		A
2097.83	2098.50	47653.2	RH	0.5H	2.3		A
2097.54	2098.21	47659.8	RH	0.5H	0.5H		A
2096.22	2096.89	47689.8	RH	0.3	1.4		A
2096.05	2096.72	47693.6	RH		1.9		A
2092.64	2093.31	47771.3	RH	1.0	1.4		A
2091.07	2091.73	47807.2	RH		2.0		A
2089.48	2090.14	47843.6	RH		1.9		A
2087.48	2088.14	47889.4	RH	1.5	1.4		A
2085.58	2086.24	47933.0	RH		2.0		A
2085.18	2085.84	47942.2	RH		1.5		A
2082.93	2083.59	47994.0	RH		0.9D		A
2082.34	2083.00	48007.6	RH	0.3	1.0		A
2081.52	2082.18	48026.5	RH		1.5		A
2080.24	2080.90	48056.1	RH	0.0	1.5		A
2079.66	2080.32	48069.5	RH	0.7			A
2079.37	2080.03	48076.2	RH	0.5H	1.4		A
2077.97	2078.63	48108.6	RH		1.2		A
2077.18	2077.84	48126.9	RH	0.3	0.8		A
2076.83	2077.49	48135.0	RH		2.0		A
2076.22	2076.88	48149.1	RH		1.0		A

AIR WAVELENGTH (ANGSTRMS)	VACUUM WAVELENGTH (ANGSTRMS)	WAVENUMBER (KAYSERS)	LMENT	LOG ARC	INTENSITY SPK	DIS	REF
2073.75	2074.41	48206.4	RH	0.0	1.5		A
2072.85	2073.51	48227.4	RH		1.3		A
2072.36	2073.02	48238.8	RH	0.3	0.3		A
2071.94	2072.60	48248.5	RH	0.5	1.4		A
2071.76	2072.42	48252.7	RH		1.3		A
2068.32	2068.98	48333.0	RH	0.9			A
2068.14	2068.80	48337.2	RH		1.2		A
2066.78	2067.44	48369.0	RH		1.1		A
2065.70	2066.36	48394.3	RH		1.4		A
2065.23	2065.89	48405.3	RH	0.7			A
2064.17	2064.83	48430.1	RH	0.5	2.0		A
2062.04	2062.70	48480.2	RH		1.3		A
2059.48	2060.14	48540.4	RH		1.3		A
2059.16	2059.82	48548.0	RH		1.7		A
2058.15	2058.81	48571.8	RH	0.5	1.2		A
2057.85	2058.51	48578.9	RH		0.8		A
2057.71	2058.37	48582.2	RH		1.1		A
2057.18	2057.84	48594.7	RH		1.2		A
2057.04	2057.70	48598.0	RH		1.0		A
2055.31	2055.97	48638.9	RH	0.3	1.6		A
2054.99	2055.65	48646.5	RH	0.5	0.5		A
2054.37	2055.03	48661.1	RH	0.5	1.4		A
2053.61	2054.27	48679.1	RH		1.5		A
2053.20	2053.86	48688.9	RH	0.3	1.5		A
2052.84	2053.50	48697.4	RH		1.0		A
2052.57	2053.23	48703.8	RH		1.3		A
2051.25	2051.91	48735.1	RH	0.3	2.0		A
2049.97	2050.63	48765.6	RH	0.3	0.5		A
2048.95	2049.61	48789.8	RH		0.7		A
2048.65	2049.31	48797.0	RH		2.3		A
2048.31	2048.97	48805.1	RH	0.7			A
2048.04	2048.70	48811.5	RH		1.0		A
2046.19	2046.85	48855.6	RH		1.3		A
2045.86	2046.52	48863.5	RH	0.3	0.3		A
2045.68	2046.34	48867.8	RH		1.3		A
2045.33	2045.99	48876.2	RH		1.4		A
2044.07	2044.73	48906.3	RH	0.3H	1.0		A
2042.46	2043.12	48944.9	RH		1.2		A
2042.32	2042.98	48948.2	RH	0.3	1.5		A
2041.72	2042.38	48962.6	RH		1.4		A
2041.04	2041.70	48978.9	RH		1.6		A
2040.19	2040.85	48999.3	RH	0.3	2.0		A
2039.67	2040.33	49011.8	RH		1.4		A
2039.35	2040.01	49019.5	RH		1.5		A
2037.85	2038.50	49055.6	RH	0.6			A
2037.60	2038.25	49061.6	RH		1.7		A
2037.05	2037.70	49074.8	RH		1.5		A
2036.70	2037.35	49083.3	RH		1.9		A
2036.27	2036.92	49093.6	RH		1.4		A
2034.63	2035.28	49133.2	RH		1.3		A
2034.09	2034.74	49146.2	RH	0.0	1.0		A
2033.59	2034.24	49158.3	RH	0.6	1.3		A
2033.34	2033.99	49164.4	RH		0.9		A
2032.51	2033.16	49184.4	RH	0.3	0.5		A
2031.52	2032.17	49208.4	RH	0.0H	0.7		A
2030.88	2031.53	49223.9	RH		1.3		A
2030.18	2030.83	49240.9	RH	0.0	0.6		A
2029.90	2030.55	49247.7	RH	0.5H	1.5		A
2029.03	2029.68	49268.8	RH	0.0	1.0		A
2028.47	2029.12	49282.4	RH	0.0	1.4		A
2028.38	2029.03	49284.6	RH	0.0H	0.6		A
2027.80	2028.45	49298.6	RH	0.3	1.4		A
2026.17	2026.82	49338.3	RH	0.0H	0.5		A
2025.15	2025.80	49363.1	RH	0.3	1.6		A
2024.93	2025.58	49368.5	RH	0.5	1.4		A
2023.74	2024.39	49397.5	RH		1.8		A
2022.37	2023.02	49431.0	RH	0.5	0.6		A
2021.45	2022.10	49453.5	RH		0.9		A
2021.18	2021.83	49460.1	RH		1.0		A
2018.45	2019.10	49527.0	RH		1.3		A

AIR WAVELENGTH (ANGSTRMS)	VACUUM WAVELENGTH (ANGSTRMS)	WAVENUMBER (KAYSERS)	LMENT	LOG ARC	INTENSITY SPK	DIS	REF
2018.18	2018.83	49533.6	RH	0.7	0.6		A
2017.77	2018.42	49543.7	RH		1.0		A
2017.45	2018.10	49551.5	RH		1.9		A
2015.20	2015.85	49606.9	RH	0.3	1.4		A
2014.45	2015.10	49625.3	RH		1.1		A
2013.74	2014.39	49642.8	RH		2.0		A
2013.40	2014.05	49651.2	RH		0.7		A
2012.72	2013.37	49668.0	RH		1.4		A
2010.47	2011.12	49723.5	RH		1.4		A
2005.85	2006.50	49838.0	RH		0.9		A
2005.61	2006.26	49844.0	RH		1.0		A
2005.37	2006.02	49850.0	RH		1.0		A
2005.12	2005.77	49856.2	RH		1.4		A
2003.53	2004.18	49895.7	RH		1.0		A
2003.39	2004.04	49899.2	RH		0.8		A
2001.80	2002.45	49938.9	RH	0.3	0.7		A
2001.43	2002.08	49948.1	RH		1.4		A
2000.76	2001.41	49964.8	RH	0.0	1.0		A
2000.56	2001.21	49969.8	RH		1.0		A
9948.57	9951.30	10048.9	RN I			0.8	RS
9855.31	9858.01	10144.0	RN I			0.6	RS
9809.67	9812.36	10191.2	RN I			0.5	RS
9747.89	9750.56	10255.8	RN I			0.7	RS
9327.02	9329.58	10718.6	RN I			1.7	RS
9300.72	9303.27	10748.9	RN I			0.5	RS
9295.90	9298.45	10754.5	RN I			0.5	RS
9206.66	9209.19	10858.7	RN I			0.3	RS
8999.56	9002.03	11108.6	RN I			0.3	RS
8807.75	8810.17	11350.5	RN I			1.0	RS
8746.08	8748.48	11430.6	RN I			0.5	RS
8719.20	8721.59	11465.8	RN I			0.6	RS
8675.83	8678.21	11523.1	RN I			1.2	RS
8639.76	8642.13	11571.2	RN I			1.0	RS
8627.96	8630.33	11587.0	RN I			0.7	RS
8600.07	8602.43	11624.6	RN I			2.0	RS
8520.95	8523.29	11732.6	RN I			1.3	RS
8494.89	8497.22	11768.6	RN I			1.0	RS
8487.48	8489.81	11778.8	RN I			1.0	RS
8450.60	8452.92	11830.2	RN I			0.7	RS
8381.05	8383.35	11928.4	RN I			1.0	RS
8349.74	8352.03	11973.1	RN I			0.8	RS
8314.51	8316.80	12023.9	RN I			0.9	RS
8270.96	8273.23	12087.2	RN I			2.0	RS
8179.58	8181.83	12222.2	RN I			0.7	RS
8173.84	8176.09	12230.8	RN I			0.8	RS
8099.51	8101.74	12343.0	RN I			2.0	RS
8049.00	8051.21	12420.5	RN I			1.3	RS
8015.30	8017.50	12472.7	RN I			0.7	RS
7972.26	7974.45	12540.1	RN I			0.3	RS
7871.3	7873.5	12701.	RN I			0.7	RS
7809.82	7811.97	12800.9	RN I			2.0	RS
7746.64	7748.77	12905.3	RN I			1.3	RS
7738.43	7740.56	12919.0	RN I			1.0	RS
7657.48	7659.59	13055.5	RN I			1.0	RS
7601.28	7603.37	13152.1	RN I			0.9	RS
7597.55	7599.64	13158.5	RN I			0.8	RS
7584.18	7586.27	13181.7	RN I			0.3	RS
7523.93	7526.00	13287.3	RN I			0.8	RS
7516.92	7518.99	13299.7	RN I			0.9	RS
7514.13	7516.20	13304.6	RN I			0.9	RS
7485.33	7487.39	13355.8	RN I			0.8	RS
7483.13	7485.19	13359.7	RN I			1.2	RS
7470.89	7472.95	13381.6	RN I			1.2	RS
7468.92	7470.98	13385.1	RN I			1.0	RS
7450.00	7452.05	13419.1	RN I			2.8	RS
7419.04	7421.08	13475.1	RN I			1.3	RS
7412.84	7414.88	13486.4	RN I			0.8	RS
7332.43	7334.45	13634.3	RN I			1.0	RS
7320.98	7323.00	13655.6	RN I			1.1	RS
7294.51	7296.52	13705.2	RN I			0.6	RS

AIR WAVELENGTH (ANGSTRMS)	VACUUM WAVELENGTH (ANGSTRMS)	WAVENUMBER (KAYSERS)	LMENT	LOG ARC	INTENSITY SPK	INTENSITY DIS	REF
7291.00	7293.01	13711.8	RN I			1.6	RS
7268.11	7270.11	13754.9	RN I			2.3	RS
7247.23	7249.23	13794.6	RN I			1.0	RS
7231.46	7233.45	13824.7	RN I			0.6	RS
7151.82	7153.79	13978.6	RN I			0.8	RS
7055.42	7057.37	14169.6	RN I			2.6	RS
6998.90	7000.83	14284.0	RN I			1.3	RS
6944.07	6945.99	14396.8	RN I			1.0	RS
6891.16	6893.06	14507.3	RN I			1.3	RS
6837.57	6839.46	14621.0	RN I			1.2	RS
6836.95	6838.84	14622.4	RN I			1.2	RS
6808.38	6810.26	14683.7	RN I			1.0	RS
6806.79	6808.67	14687.2	RN I			1.1	RS
6791.90	6793.77	14719.4	RN I			0.8	RS
6786.36	6788.23	14731.4	RN I			0.6	RS
6762.71	6764.58	14782.9	RN I			0.9	RS
6751.81	6753.67	14806.8	RN I			1.6	RS
6730.16	6732.02	14854.4	RN I			1.0	RS
6704.28	6706.13	14911.7	RN I			1.2	RS
6699.70	6701.55	14921.9	RN I			0.9	RS
6669.60	6671.44	14989.3	RN I			1.1	RS
6627.23	6629.06	15085.1	RN I			1.5	RS
6613.03	6614.86	15117.5	RN I			0.8	RS
6606.43	6608.25	15132.6	RN I			1.3	RS
6595.93	6597.75	15156.7	RN I			1.0	RS
6594.56	6596.38	15159.8	RN I			1.0	RS
6557.49	6559.30	15245.5	RN I			1.3	RS
6552.27	6554.08	15257.7	RN I			0.8	RS
6502.53	6504.33	15374.4	RN I			1.0	RS
6500.15	6501.95	15380.0	RN I			0.6	RS
6472.15	6473.94	15446.6	RN I			1.0	RS
6448.32	6450.10	15503.6	RN I			1.0	RS
6443.24	6445.02	15515.8	RN I			0.6	RS
6441.91	6443.69	15519.1	RN I			0.8	RS
6439.83	6441.61	15524.1	RN I			0.3	RS
6426.73	6428.51	15555.7	RN I			0.9	RS
6421.48	6423.25	15568.4	RN I			1.0	RS
6380.45	6382.21	15668.5	RN I			1.1	RS
6312.11	6313.86	15838.2	RN I			0.3	RS
6309.	6311.	15850.	RN			1.0	WA
6296.33	6298.07	15877.9	RN I			0.8	RS
6261.46	6263.19	15966.3	RN I			1.1	RS
6231.15	6232.87	16044.0	RN I			0.6	RS
6224.	6226.	16060.	RN			0.6	WA
6200.75	6202.47	16122.6	RN I			1.1	RS
6194.91	6196.62	16137.8	RN I			0.6	RS
6172.23	6173.94	16197.1	RN I			0.6	RS
6101.72	6103.41	16384.3	RN I			0.6	RS
6099.83	6101.52	16389.4	RN I			0.6	RS
6066.58	6068.26	16479.2	RN I			0.3	RS
6064.75	6066.43	16484.2	RN I			0.3	RS
6061.92	6063.60	16491.9	RN I			1.3	RS
6058.50	6060.18	16501.2	RN I			0.6	RS
6041.90	6043.57	16546.5	RN I			1.0	RS
5977.4	5979.1	16725.	RN			1.3	WA
5969.80	5971.45	16746.3	RN I			1.3	RS
5951.57	5953.22	16797.6	RN I			1.4	RS
5944.7	5946.3	16817.	RN			1.0	WA
5932.60	5934.24	16851.4	RN I			1.4	RS
5919.84	5921.48	16887.7	RN I			1.2	RS
5907.50	5909.14	16922.9	RN I			1.2	RS
5900.93	5902.57	16941.8	RN I			1.3	RS
5894.4	5896.0	16961.	RN			1.5	NY
5888.6	5890.2	16977.	RN			1.9	NY
5877.56	5879.19	17009.1	RN I			1.0	RS
5824.20	5825.81	17165.0	RN I			1.0	RS
5762.01	5763.61	17350.2	RN I			1.0	RS
5722.58	5724.17	17469.8	RN I			1.5	RS
5715.9	5717.5	17490.	RN			1.9	WA
5711.88	5713.46	17502.5	RN I			1.3	RS
5643.05	5644.62	17716.0	RN I			1.0	RS
5606.91	5608.47	17830.2	RN I			1.2	RS
5595.45	5597.00	17866.7	RN I			1.3	RS
5582.4	5584.0	17908.	RN			2.3	WA
5577.28	5578.83	17924.9	RN I			1.3	RS
5542.064	5543.603	18038.81	RN I	0.3			MIT
5526.98	5528.52	18088.0	RN I			1.2	RS
5520.53	5522.06	18109.2	RN I			0.7	RS
5499.63	5501.16	18178.0	RN I			0.7	RS
5402.92	5404.42	18503.4	RN I			1.0	RS
5394.2	5395.7	18533.	RN			1.3	WA
5390.40	5391.90	18546.3	RN I			0.7	RS
5386.1	5387.6	18561.	RN			1.0	WA
5371.3	5372.8	18612.	RN			1.0	WA
5362.8	5364.3	18642.	RN			1.0	WA
5119.3	5120.7	19528.	RN			1.3	WA
5084.48	5085.90	19662.2	RN			2.5	WA
5044.8	5046.2	19817.	RN			1.5	WA
4978.84	4980.23	20079.4	RN			2.5	RC
4957.8	4959.2	20165.	RN			1.0	WA
4950.0	4951.4	20196.	RN			1.5	WA
4915.5	4916.9	20338.	RN			1.5	WA
4891.1	4892.5	20440.	RN			1.5	WA
4856.2	4857.6	20586.	RN			1.5	WA
4829.2	4830.5	20702.	RN			1.5	WA
4817.15	4818.50	20753.4	RN			2.0	RC
4796.58	4797.92	20842.4	RN			0.7	RC
4793.0	4794.3	20858.	RN			1.0	WA
4768.59	4769.92	20964.7	RN			2.0	WA
4749.27	4750.60	21050.0	RN I			1.4	RS
4721.76	4723.08	21172.6	RN I			2.2	RS
4701.70	4703.02	21262.9	RN			1.7	RC
4680.83	4682.14	21357.7	RN			2.7	WA
4644.18	4645.48	21526.3	RN			2.5	WA
4631.71	4633.01	21584.2	RN			0.7	RC
4625.48	4626.78	21613.3	RN			2.7	WA
4609.38	4610.67	21688.8	RN I			2.4	RS
4604.40	4605.69	21712.3	RN			2.3	WA
4577.72	4579.00	21838.8	RN I			2.4	RS
4568.06	4569.34	21885.0	RN			0.7	RC
4546.8	4548.1	21987.	RN			1.5	WA
4527.70	4528.97	22080.1	RN			1.0	WA
4510.2	4511.5	22166.	RN			1.0	NY
4508.48	4509.74	22174.2	RN I			2.4	RS
4507.83	4509.09	22177.4	RN			1.9	WA
4503.72	4504.98	22197.6	RN			1.3	RC
4459.25	4460.50	22419.0	RN I			2.4	RS
4439.71	4440.96	22517.7	RN			1.0	RC
4435.05	4436.30	22541.3	RN I			2.3	RS
4383.30	4384.53	22807.4	RN			1.5	WA
4371.53	4372.76	22868.9	RN			1.5	RC
4349.60	4350.82	22984.2	RN I			3.7	RS
4335.78	4337.00	23057.4	RN I			1.5	RS
4307.76	4308.97	23207.4	RN I			2.6	RS
4227.3	4228.5	23649.	RN I			1.2	RS
4226.06	4227.25	23656.0	RN I			1.7	RS
4203.23	4204.41	23784.5	RN			2.3	RC
4192.9	4194.1	23843.	RN			1.0	WA
4187.81	4188.99	23872.1	RN			1.5	WA
4169.92	4171.10	23974.5	RN			1.3	WA
4166.43	4167.60	23994.6	RN			2.7	RC
4114.56	4115.72	24297.1	RN			1.9	RC
4094.3	4095.5	24417.	RN			1.0	WA
4088.0	4089.2	24455.	RN			1.0	WA
4055.3	4056.4	24652.	RN			0.7	WA
4051.0	4052.1	24678.	RN			1.0	WA
4045.3	4046.4	24713.	RN			1.5	RC
4039.4	4040.5	24749.	RN			0.7	WA
4017.75	4018.89	24882.5	RN			2.2	RC
3997.18	3998.31	25010.6	RN I			0.5	RS

AIR WAVELENGTH (ANGSTRMS)	VACUUM WAVELENGTH (ANGSTRMS)	WAVENUMBER (KAYSERS)	LMENT	LOG ARC	INTENSITY SPK	DIS	REF
3995.7	3996.8	25020.	RN I			0.5	RS
3994.76	3995.89	25025.7	RN I			1.0	RS
3988.98	3990.11	25062.0	RN I			0.9	RS
3981.68	3982.81	25107.9	RN			2.2	RC
3978.97	3980.10	25125.0	RN I			1.1	RS
3972.69	3973.81	25164.7	RN I			0.7	RS
3971.67	3972.79	25171.2	RN			1.9	WA
3964.9	3966.0	25214.	RN			0.7	WA
3957.15	3958.27	25263.6	RN			1.4	RC
3952.36	3953.48	25294.2	RN I			1.4	RS
3941.72	3942.84	25362.5	RN I			1.4	RS
3931.82	3932.93	25426.3	RN			2.4	WA
3917.20	3918.31	25521.2	RN I			1.4	RS
3905.6	3906.7	25597.	RN			0.5	RC
3867.53	3868.63	25849.0	RN			0.9	WA
3814.70	3815.78	26206.9	RN I			1.1	RS
3790.07	3791.15	26377.2	RN I			1.1	RS
3782.66	3783.73	26428.9	RN I			1.1	RS
3760.8	3761.9	26582.	RN			1.0	WO
3757.1	3758.2	26609.	RN			0.7	WO
3753.65	3754.72	26633.2	RN I			1.7	RS
3749.48	3750.55	26662.8	RN I			0.9	RS
3739.89	3740.95	26731.1	RN I			1.4	RS
3735.4	3736.5	26763.	RN			0.7	WO
3717.2	3718.3	26894.	RN			1.0	WO
3701.2	3702.3	27011.	RN			0.3	WO
3690.31	3691.36	27090.3	RN I			0.9	RS
3688.3	3689.3	27105.	RN			1.6	WO
3685.30	3686.35	27127.1	RN I			0.9	RS
3679.0	3680.0	27174.	RN			1.5	PE
3673.41	3674.46	27214.9	RN I			1.2	RS
3669.7	3670.7	27242.	RN			0.7	PE
3664.81	3665.85	27278.8	RN			1.4	RC
3657.2	3658.2	27335.	RN			1.0	PE
3642.5	3643.5	27446.	RN			0.7	PE
3634.8	3635.8	27504.	RN			2.4	WO
3629.7	3630.7	27543.	RN			1.0	PE
3626.5	3627.5	27567.	RN			1.4	PE
3625.37	3626.40	27575.5	RN I			0.5	RS
3623.53	3624.56	27589.5	RN I			0.5	RS
3621.0	3622.0	27609.	RN			2.4	WO
3618.70	3619.73	27626.4	RN I			0.9	RS
3615.0	3616.0	27655.	RN			1.5	PE
3612.61	3613.64	27672.9	RN			1.3	RC
3605.6	3606.6	27727.	RN			1.0	PE
3601.6	3602.6	27757.	RN			0.7	WO
3597.0	3598.0	27793.	RN			0.7	PE
3591.5	3592.5	27836.	RN			0.3	PE
3589.5	3590.5	27851.	RN			0.7	PE
3584.6	3585.6	27889.	RN			0.3	PE
3582.6	3583.6	27905.	RN			1.3	WO
3575.3	3576.3	27962.	RN			0.7	PE
3573.3	3574.3	27977.	RN			0.3	PE
3559.2	3560.2	28088.	RN			0.7	WO
3555.24	3556.26	28119.5	RN I			0.5	RS
3552.36	3553.37	28142.3	RN I			0.5	RS
3551.1	3552.1	28152.	RN			0.7	PE
3549.0	3550.0	28169.	RN			0.7	PE
3545.84	3546.85	28194.0	RN I			0.7	RS
3541.2	3542.2	28231.	RN			1.0	WO
3534.2	3535.2	28287.	RN			0.7	PE
3532.3	3533.3	28302.	RN			1.0	PE
3531.71	3532.72	28306.8	RN I			0.7	RS
3529.7	3530.7	28323.	RN			0.7	PE
3528.38	3529.39	28333.5	RN I			0.7	RS
3514.60	3515.61	28444.6	RN I			1.1	RS
3509.8	3510.8	28483.	RN			0.7	WO
3508.22	3509.22	28496.3	RN I			1.0	RS
3507.8	3508.8	28500.	RN			0.7	PE
3502.6	3503.6	28542.	RN			0.3	PE

AIR WAVELENGTH (ANGSTRMS)	VACUUM WAVELENGTH (ANGSTRMS)	WAVENUMBER (KAYSERS)	LMENT	LOG ARC	INTENSITY SPK	DIS	REF
3500.6	3501.6	28558.	RN			0.3	PE
3486.6	3487.6	28673.	RN			1.0	WO
3479.5	3480.5	28731.	RN			1.5	PE
3470.0	3471.0	28810.	RN			0.3	WO
3463.1	3464.1	28868.	RN			0.3	PE
3458.6	3459.6	28905.	RN			0.7	WO
3455.1	3456.1	28934.	RN			0.3	PE
3448.55	3449.54	28989.4	RN I			0.5	RS
3446.6	3447.6	29006.	RN			0.3	PE
3440.7	3441.7	29055.	RN			1.3	WO
3435.4	3436.4	29100.	RN			1.3	PE
3429.7	3430.7	29149.	RN			1.8	WO
3426.2	3427.2	29178.	RN			0.7	PE
3422.4	3423.4	29211.	RN			1.3	PE
3393.1	3394.1	29463.	RN			1.0	WO
3392.2	3393.2	29471.	RN			0.3	PE
3391.27	3392.24	29479.0	RN I			0.9	RS
3388.6	3389.6	29502.	RN			1.0	PE
3387.52	3388.49	29511.6	RN I			0.7	RS
3384.4	3385.4	29539.	RN			0.7	PE
3377.2	3378.2	29602.	RN			1.5	WO
3372.7	3373.7	29641.	RN			1.3	PE
3370.7	3371.7	29659.	RN			0.3	PE
3365.3	3366.3	29706.	RN			0.3	PE
3362.4	3363.4	29732.	RN			0.7	WO
3356.5	3357.5	29784.	RN			1.6	PE
3350.9	3351.9	29834.	RN			0.7	WO
3345.5	3346.5	29882.	RN			0.3	PE
3344.1	3345.1	29895.	RN			0.3	PE
3342.4	3343.4	29910.	RN			0.3	WO
3335.9	3336.9	29968.	RN			0.7	PE
3330.0	3331.0	30021.	RN			1.6	PE
3327.1	3328.1	30048.	RN			0.3	WO
3318.44	3319.39	30126.0	RN I			0.7	RS
3316.14	3317.09	30146.9	RN I			0.5	RS
3312.8	3313.8	30177.	RN			2.0	WO
3306.7	3307.7	30233.	RN			2.0	PE
3254.5	3255.4	30718.	RN			1.5	WO
3250.3	3251.2	30757.	RN			0.3	PE
3247.8	3248.7	30781.	RN			1.0	PE
3241.5	3242.4	30841.	RN			1.6	PE
3229.5	3230.4	30956.	RN			0.7	PE
3221.6	3222.5	31031.	RN			1.3	WO
3216.3	3217.2	31083.	RN			1.2	PE
3210.4	3211.3	31140.	RN			0.7	PE
3206.8	3207.7	31175.	RN			1.3	WO
3201.8	3202.7	31223.	RN			0.7	PE
3196.9	3197.8	31271.	RN			1.5	WO
3192.0	3192.9	31319.	RN			1.5	PE
3190.2	3191.1	31337.	RN			0.3	PE
3187.9	3188.8	31359.	RN			1.0	PE
3185.5	3186.4	31383.	RN			1.0	PE
3175.6	3176.5	31481.	RN			1.6	WO
3169.7	3170.6	31540.	RN			1.5	PE
3165.8	3166.7	31578.	RN			0.7	PE
3159.8	3160.7	31638.	RN			1.3	PE
3157.8	3158.7	31658.	RN			1.0	PE
3153.4	3154.3	31703.	RN			0.7	PE
3137.4	3138.3	31864.	RN			0.3	WO
3120.4	3121.3	32038.	RN			1.3	PE
3116.3	3117.2	32080.	RN			0.7	PE
3114.6	3115.5	32097.	RN			1.7	PE
3111.8	3112.7	32126.	RN			0.3	WO
3106.7	3107.6	32179.	RN			1.5	PE
3105.7	3106.6	32189.	RN			1.8	WO
3104.9	3105.8	32198.	RN			1.0	PE
3100.19	3101.09	32246.7	RN			1.7	RC
3097.2	3098.1	32278.	RN			1.3	WO
3093.0	3093.9	32322.	RN			1.3	PE
3091.4	3092.3	32338.	RN			1.0	PE

AIR WAVELENGTH (ANGSTRMS)	VACUUM WAVELENGTH (ANGSTRMS)	WAVENUMBER (KAYSERS)	LMENT	LOG ARC	INTENSITY SPK	DIS	REF
3082.3	3083.2	32434.	RN			1.0	WO
3077.7	3078.6	32482.	RN			1.5	PE
3070.6	3071.5	32557.	RN			0.7	PE
3068.9	3069.8	32575.	RN			2.0	WO
3064.6	3065.5	32621.	RN			1.6	PE
3059.7	3060.6	32673.	RN			1.3	PE
3056.6	3057.5	32707.	RN			0.7	PE
3054.3	3055.2	32731.	RN			2.4	PE
3050.1	3051.0	32776.	RN			1.2	PE
3045.2	3046.1	32829.	RN			1.8	PE
3043.0	3043.9	32853.	RN			0.7	WO
3037.7	3038.6	32910.	RN			1.6	PE
3036.8	3037.7	32920.	RN			1.8	WO
3033.2	3034.1	32959.	RN			1.0	PE
3032.5	3033.4	32966.	RN			1.6	PE
3027.7	3028.6	33019.	RN			0.7	WO
3019.2	3020.1	33112.	RN			1.3	WO
3015.8	3016.7	33149.	RN			1.0	PE
3014.5	3015.4	33163.	RN			1.3	PE
3010.8	3011.7	33204.	RN			2.0	WO
3006.8	3007.7	33248.	RN			2.5	PE
3005.72	3006.60	33260.2	RN			0.5	RC
2994.5	2995.4	33385.	RN			1.3	PE
2987.4	2988.3	33464.	RN			1.3	PE
2985.0	2985.9	33491.	RN			0.5	PE
2983.3	2984.2	33510.	RN			0.5	PE
2976.3	2977.2	33589.	RN			0.5	PE
2971.5	2972.4	33643.	RN			0.5	WO
2936.0	2936.9	34050.	RN			0.5	PE
2917.8	2918.7	34262.	RN			0.5	WO
2904.6	2905.5	34418.	RN			0.5	WO
2892.7	2893.5	34560.	RN			2.2	WO
2889.3	2890.1	34600.	RN			1.2	PE
2887.2	2888.0	34625.	RN			2.1	PE
2883.8	2884.6	34666.	RN			1.4	PE
2882.4	2883.2	34683.	RN			0.5	WO
2877.0	2877.8	34748.	RN			0.8	PE
2868.7	2869.5	34849.	RN			1.8	WO
2864.4	2865.2	34901.	RN			1.1	PE
2843.0	2843.8	35164.	RN			0.5	PE
2842.1	2842.9	35175.	RN			2.2	WO
2838.5	2839.3	35219.	RN			1.8	PE
2836.3	2837.1	35247.	RN			1.4	PE
2830.6	2831.4	35318.	RN			1.8	WO
2826.5	2827.3	35369.	RN			1.8	PE
2822.8	2823.6	35415.	RN			0.5	WO
2819.8	2820.6	35453.	RN			1.1	PE
2817.9	2818.7	35477.	RN			1.1	PE
2813.4	2814.2	35534.	RN			0.5	PE
2812.1	2812.9	35550.	RN			1.4	WO
2808.4	2809.2	35597.	RN			1.6	PE
2802.1	2802.9	35677.	RN			0.5	WO
2789.2	2790.0	35842.	RN			1.3	PE
2785.2	2786.0	35893.	RN			0.5	WO
2782.6	2783.4	35927.	RN			0.5	PE
2756.3	2757.1	36270.	RN			1.4	WO
2751.3	2752.1	36336.	RN			0.5	WO
2747.7	2748.5	36383.	RN			0.8	PE
2740.8	2741.6	36475.	RN			0.5	PE
2736.8	2737.6	36528.	RN			0.8	PE
2734.8	2735.6	36555.	RN			0.8	PE
2731.9	2732.7	36594.	RN			0.5	PE
2730.4	2731.2	36614.	RN			0.3	PE
2727.7	2728.5	36650.	RN			1.1	PE
2721.0	2721.8	36740.	RN			0.5	WO
2717.0	2717.8	36794.	RN			0.5	PE
2714.4	2715.2	36830.	RN			0.5	PE
2712.1	2712.9	36861.	RN			0.8	PE
2710.7	2711.5	36880.	RN			0.5	WO
2707.9	2708.7	36918.	RN			1.1	PE

AIR WAVELENGTH (ANGSTRMS)	VACUUM WAVELENGTH (ANGSTRMS)	WAVENUMBER (KAYSERS)	LMENT	LOG ARC	INTENSITY SPK	DIS	REF
2686.3	2687.1	37215.	RN			0.5	PE
2673.7	2674.5	37390.	RN			0.8	PE
2665.3	2666.1	37508.	RN			0.8	PE
2655.7	2656.5	37644.	RN			0.8	WO
2645.5	2646.3	37789.	RN			0.5	WO
2626.0	2626.8	38069.	RN			1.1	WO
2623.5	2624.3	38106.	RN			0.8	PE
2618.8	2619.6	38174.	RN			1.3	WO
2616.0	2616.8	38215.	RN			1.1	PE
2579.9	2580.7	38750.	RN			0.8	WO
2572.5	2573.3	38861.	RN			1.1	PE
2540.9	2541.7	39344.	RN			0.5	WO
2530.3	2531.1	39509.	RN			0.8	PE
2529.3	2530.1	39525.	RN			0.5	PE
2506.2	2507.0	39889.	RN			0.8	PE
2502.8	2503.6	39943.	RN			0.8	PE
2502.0	2502.8	39956.	RN			0.8	PE
2480.0	2480.7	40310.	RN			0.5	PE
2471.9	2472.6	40442.	RN			0.5	PE
2463.4	2464.1	40582.	RN			1.3	PE
2457.6	2458.3	40678.	RN			1.1	WO
2444.1	2444.8	40902.	RN			1.1	PE
2442.1	2442.8	40936.	RN			1.3	PE
2413.4	2414.1	41423.	RN			0.5	WO
2410.9	2411.6	41466.	RN			1.3	PE
2400.9	2401.6	41638.	RN			0.5	PE
2399.6	2400.3	41661.	RN			0.5	PE
2375.5	2376.2	42084.	RN			0.8	PE
2337.5	2338.2	42768.	RN			0.8	PE
2280.5	2281.2	43836.	RN			0.5	PE
9273.15	9275.69	10780.9	RU	0.6			MIT
8724.98	8727.38	11458.2	RU	0.5			ME
8473.64	8475.97	11798.1	RU I	0.3			ME
8352.94	8355.24	11968.5	RU I	1.1			MIT
8348.98	8351.27	11974.2	RU I	1.2			MIT
8283.09	8285.37	12069.5	RU	0.8			MIT
8264.96	8267.23	12095.9	RU	1.4			MIT
8220.47	8222.73	12161.4	RU	0.8			MIT
8216.36	8218.62	12167.5	RU	0.8			MIT
8181.97	8184.22	12218.6	RU	0.3			ME
8168.78	8171.03	12238.4	RU	0.5			ME
8157.69	8159.93	12255.0	RU	0.5			MIT
8135.78	8138.02	12288.0	RU	0.6			MIT
8112.47	8114.70	12323.3	RU I	1.3			MIT
8049.48	8051.69	12419.7	RU	0.6			MIT
8036.65	8038.86	12439.6	RU	0.8			MIT
8017.63	8019.84	12469.1	RU	0.7			MIT
7999.72	8001.92	12497.0	RU	0.5			MIT
7967.84	7970.03	12547.0	RU I	1.2			MIT
7948.15	7950.34	12578.1	RU I	1.2			MIT
7924.43	7926.61	12615.7	RU I	1.4			MIT
7922.95	7925.13	12618.1	RU	0.6			MIT
7906.14	7908.31	12644.9	RU I	0.5			MIT
7900.17	7902.34	12654.5	RU	0.5			ME
7890.37	7892.54	12670.2	RU I	1.3			MIT
7881.49	7883.66	12684.5	RU	2.0			MIT
7871.94	7874.11	12699.8	RU	0.5			ME
7853.81	7855.97	12729.2	RU	0.5			MIT
7847.80	7849.96	12738.9	RU I	2.0			MIT
7841.90	7844.06	12748.5	RU	1.0			MIT
7834.81	7836.97	12760.0	RU	0.8			MIT
7833.39	7835.55	12762.3	RU I	1.0			MIT
7830.52	7832.67	12767.0	RU	0.3			ME
7829.81	7831.96	12768.2	RU I	0.8			MIT
7806.82	7808.97	12805.8	RU	0.8			MIT
7797.89	7800.04	12820.4	RU	0.7			MIT
7791.86	7794.00	12830.4	RU I	2.0			MIT
7759.67	7761.81	12883.6	RU	0.3			ME
7729.91	7732.04	12933.2	RU	0.5			MIT
7722.87	7725.00	12945.0	RU	1.4			MIT

AIR WAVELENGTH (ANGSTRMS)	VACUUM WAVELENGTH (ANGSTRMS)	WAVENUMBER (KAYSERS)	LMENT		LOG ARC	INTENSITY SPK	DIS	REF
7720.44	7722.56	12949.1	RU		0.5			MIT
7621.50	7623.60	13117.2	RU	I	1.0			MIT
7612.94	7615.04	13131.9	RU	I	0.5			MIT
7570.02	7572.10	13206.4	RU		1.0			MIT
7559.61	7561.69	13224.6	RU	I	2.0			MIT
7532.07	7534.14	13272.9	RU		0.8			MIT
7529.58	7531.65	13277.3	RU		0.6			MIT
7499.75	7501.82	13330.1	RU		2.3			MIT
7485.79	7487.85	13355.0	RU		2.2			MIT
7475.40	7477.46	13373.5	RU		1.7			MIT
7468.91	7470.97	13385.2	RU		2.2			MIT
7463.70	7465.76	13394.5	RU		0.6			MIT
7453.59	7455.64	13412.7	RU		0.7			MIT
7450.65	7452.70	13417.9	RU		0.6			MIT
7445.45	7447.50	13427.3	RU		0.6			MIT
7438.33	7440.38	13440.2	RU		0.6			MIT
7430.78	7432.83	13453.8	RU		1.0			MIT
7410.25	7412.29	13491.1	RU		0.7			MIT
7407.49	7409.53	13496.1	RU	I	0.6	0.6		MIT
7393.93	7395.97	13520.9	RU		2.2			MIT
7381.08	7383.11	13544.4	RU		1.0			MIT
7377.77	7379.80	13550.5	RU		1.3			MIT
7351.90	7353.93	13598.2	RU		0.6			MIT
7348.71	7350.73	13604.1	RU		0.8			MIT
7323.56	7325.58	13650.8	RU		1.4			MIT
7315.67	7317.69	13665.5	RU		0.6			MIT
7312.63	7314.64	13671.2	RU		0.7			MIT
7297.07	7299.08	13700.4	RU		0.6			MIT
7266.96	7268.96	13757.1	RU		1.2			MIT
7238.92	7240.91	13810.4	RU		2.3			MIT
7194.48	7196.46	13895.7	RU		0.6			MIT
7183.02	7185.00	13917.9	RU		0.8			MIT
7182.22	7184.20	13919.4	RU		0.7			MIT
7167.77	7169.75	13947.5	RU		0.6			MIT
7154.19	7156.16	13974.0	RU		1.2			MIT
7141.72	7143.69	13998.4	RU		1.0			MIT
7135.19	7137.16	14011.2	RU	I	0.8			MIT
7126.83	7128.79	14027.6	RU		1.2			MIT
7110.59	7112.55	14059.7	RU		0.7			MIT
7087.35	7089.30	14105.8	RU		1.6			MIT
7086.06	7088.01	14108.3	RU		1.6			MIT
7061.20	7063.15	14158.0	RU	I	1.4			MIT
7060.37	7062.32	14159.7	RU		1.3			MIT
7027.98	7029.92	14224.9	RU	I	2.4			MIT
7001.68	7003.61	14278.3	RU		1.2			MIT
6982.01	6983.94	14318.6	RU		2.3			MIT
6934.86	6936.77	14415.9	RU		0.6			MIT
6923.23	6925.14	14440.1	RU		2.5			MIT
6918.51	6920.42	14450.0	RU		0.6			MIT
6911.48	6913.39	14464.7	RU		2.0			MIT
6905.92	6907.83	14476.3	RU		1.4			MIT
6872.95	6874.85	14545.8	RU		0.7			MIT
6831.52	6833.41	14634.0	RU		0.8			MIT
6824.09	6825.97	14649.9	RU		2.3			MIT
6813.51	6815.39	14672.7	RU		1.0			ME
6805.54	6807.42	14689.9	RU	I	0.7			MIT
6787.23	6789.10	14729.5	RU		1.3			MIT
6775.02	6776.89	14756.0	RU		1.6			MIT
6766.95	6768.82	14773.6	RU		1.5			MIT
6756.54	6758.40	14796.4	RU	I	1.0			MIT
6741.40	6743.26	14829.6	RU		1.2			MIT
6731.80	6733.66	14850.8	RU		0.6			MIT
6730.45	6732.31	14853.7	RU		1.4			MIT
6718.30	6720.15	14880.6	RU	I	1.2			MIT
6707.524	6709.376	14904.52	RU		0.7			MIT
6690.001	6691.848	14943.56	RU		2.5			MIT
6672.23	6674.07	14983.4	RU		0.7			MIT
6664.14	6665.98	15001.5	RU		0.7			MIT
6663.139	6664.979	15003.80	RU		2.0			MIT
6621.62	6623.45	15097.9	RU		0.6			MIT

AIR WAVELENGTH (ANGSTRMS)	VACUUM WAVELENGTH (ANGSTRMS)	WAVENUMBER (KAYSERS)	LMENT		LOG ARC	INTENSITY SPK	DIS	REF
6618.195	6620.023	15105.69	RU		1.3			MIT
6613.124	6614.950	15117.27	RU		0.6			MIT
6596.584	6598.406	15155.18	RU		0.6			MIT
6593.745	6595.566	15161.70	RU		1.2			MIT
6560.450	6562.262	15238.65	RU		1.0			MIT
6544.25	6546.06	15276.4	RU		0.7			MIT
6540.236	6542.043	15285.74	RU		1.2			MIT
6528.744	6530.548	15312.65	RU		1.4			MIT
6496.439	6498.234	15388.80	RU		1.4			MIT
6444.84	6446.62	15512.0	RU		1.4			MIT
6417.568	6419.342	15577.92	RU		1.2			MIT
6390.228	6391.995	15644.57	RU		1.0			MIT
6376.450	6378.213	15678.37	RU	I	1.0			MIT
6363.407	6365.166	15710.51	RU	I	0.6			MIT
6357.90	6359.66	15724.1	RU		0.6			ME
6351.891	6353.647	15738.99	RU	I	0.3			MIT
6336.121	6337.873	15778.16	RU		0.7			MIT
6330.619	6332.370	15791.88	RU		1.3			MIT
6316.730	6318.477	15826.60	RU		0.6			MIT
6296.216	6297.957	15878.16	RU		1.2			MIT
6295.217	6296.958	15880.68	RU		0.7			MIT
6235.67	6237.40	16032.3	RU		0.6			MIT
6225.203	6226.925	16059.29	RU		1.3			MIT
6192.561	6194.274	16143.94	RU	I	1.0			MIT
6176.784	6178.493	16185.18	RU		1.0			MIT
6170.609	6172.317	16201.37	RU		1.2			MIT
6152.06	6153.76	16250.2	RU	I	0.6			ME
6125.23	6126.93	16321.4	RU	I	0.7			ME
6116.772	6118.465	16343.97	RU	I	1.4			MIT
6090.58	6092.27	16414.2	RU	I	0.7			ME
6083.551	6085.235	16433.22	RU		0.6			MIT
6080.15	6081.83	16442.4	RU	I	0.6			ME
6046.377	6048.051	16534.25	RU		0.6			MIT
6041.732	6043.405	16546.96	RU		0.6			MIT
6034.434	6036.105	16566.98	RU		0.7			MIT
6013.158	6014.823	16625.59	RU		0.7			MIT
5993.653	5995.313	16679.70	RU	I	1.2			MIT
5988.672	5990.331	16693.57	RU	I	1.1			MIT
5974.167	5975.822	16734.10	RU		0.8			MIT
5973.679	5975.334	16735.47	RU		0.6			MIT
5973.379	5975.034	16736.31	RU	I	1.1			MIT
5953.836	5955.485	16791.24	RU		0.6			MIT
5951.146	5952.795	16798.83	RU	I	0.7			MIT
5936.653	5938.298	16839.84	RU	I	0.6			MIT
5932.379	5934.023	16851.98	RU	I	1.2			MIT
5930.311	5931.954	16857.85	RU		0.6			MIT
5926.868	5928.510	16867.64	RU		1.0			MIT
5921.446	5923.087	16883.09	RU		1.4			MIT
5919.342	5920.982	16889.09	RU	I	1.3			MIT
5904.425	5906.061	16931.76	RU		0.6			MIT
5898.971	5900.606	16947.41	RU		0.8			MIT
5891.303	5892.936	16969.47	RU		0.6			MIT
5833.214	5834.831	17138.46	RU	I	0.6			MIT
5832.476	5834.093	17140.62	RU		0.6			MIT
5828.065	5829.681	17153.60	RU	I	0.8			MIT
5817.756	5819.369	17183.99	RU		0.6			MIT
5814.980	5816.592	17192.20	RU	I	1.4			MIT
5811.299	5812.910	17203.09	RU		0.6			MIT
5804.388	5805.997	17223.57	RU		1.0			MIT
5803.853	5805.462	17225.16	RU		0.6			MIT
5773.057	5774.658	17317.04	RU		0.8			MIT
5767.920	5769.520	17332.47	RU	I	0.8			MIT
5767.163	5768.762	17334.74	RU		0.7			MIT
5766.597	5768.196	17336.44	RU		0.8			MIT
5762.341	5763.939	17349.25	RU	I	0.3			MIT
5759.491	5761.088	17357.83	RU		0.6			MIT
5756.831	5758.428	17365.85	RU	I	0.8			MIT
5752.023	5753.618	17380.37	RU	I	1.0			MIT
5747.473	5749.067	17394.13	RU		1.1			MIT
5745.990	5747.584	17398.61	RU	I	1.0			MIT

AIR WAVELENGTH (ANGSTRMS)	VACUUM WAVELENGTH (ANGSTRMS)	WAVENUMBER (KAYSERS)	LMENT	LOG ARC	INTENSITY SPK	INTENSITY DIS	REF
5743.263	5744.856	17406.88	RU	0.6			MIT
5733.269	5734.859	17437.22	RU	0.8			MIT
5729.965	5731.554	17447.27	RU	0.5			MIT
5725.732	5727.320	17460.17	RU I	1.2			MIT
5724.815	5726.403	17462.97	RU I	1.1			MIT
5722.793	5724.381	17469.14	RU	0.8			MIT
5720.029	5721.616	17477.58	RU	0.8			MIT
5708.128	5709.712	17514.02	RU	0.8			MIT
5702.361	5703.943	17531.73	RU I	1.2			MIT
5699.577	5701.158	17540.29	RU I	1.0			MIT
5699.047	5700.628	17541.93	RU I	2.1			MIT
5694.458	5696.038	17556.06	RU	0.6			MIT
5693.028	5694.608	17560.47	RU	0.8			MIT
5681.086	5682.662	17597.39	RU	0.6			MIT
5679.627	5681.203	17601.91	RU	1.0			MIT
5668.001	5669.574	17638.01	RU	0.6			MIT
5665.201	5666.773	17646.73	RU	1.0			MIT
5658.305	5659.875	17668.23	RU	0.7			MIT
5655.727	5657.297	17676.29	RU	0.8			MIT
5653.302	5654.871	17683.87	RU	0.6			MIT
5649.555	5651.123	17695.60	RU I	0.8			MIT
5648.705	5650.273	17698.26	RU	0.7			MIT
5641.655	5643.221	17720.38	RU I	0.8			MIT
5636.709	5638.274	17735.93	RU	0.8			MIT
5636.235	5637.799	17737.42	RU I	2.0			MIT
5629.787	5631.350	17757.73	RU	0.6			MIT
5621.109	5622.669	17785.15	RU	0.6			MIT
5620.055	5621.615	17788.48	RU	0.8			MIT
5606.732	5608.289	17830.75	RU	0.8			MIT
5605.608	5607.164	17834.33	RU	0.6			MIT
5603.550	5605.106	17840.88	RU I	0.7			MIT
5603.142	5604.698	17842.18	RU I	1.1			MIT
5578.398	5579.947	17921.32	RU	1.4			MIT
5575.615	5577.163	17930.26	RU	0.7			MIT
5569.033	5570.579	17951.45	RU I	1.1			MIT
5559.749	5561.293	17981.43	RU I	1.8			MIT
5556.769	5558.312	17991.07	RU	0.8			MIT
5556.520	5558.063	17991.88	RU I	1.0			MIT
5540.662	5542.201	18043.37	RU	1.1			MIT
5537.821	5539.359	18052.63	RU	0.7			MIT
5530.993	5532.529	18074.92	RU	0.7			MIT
5512.371	5513.902	18135.98	RU	0.8			MIT
5510.714	5512.245	18141.43	RU	2.0			MIT
5501.017	5502.545	18173.41	RU I	0.8			MIT
5496.687	5498.214	18187.72	RU I	1.2			MIT
5484.643	5486.167	18227.66	RU I	1.0			MIT
5484.323	5485.847	18228.73	RU I	1.8			MIT
5480.295	5481.818	18242.12	RU	1.4			MIT
5479.404	5480.927	18245.09	RU	1.6			MIT
5475.164	5476.685	18259.22	RU	0.7			MIT
5472.841	5474.362	18266.97	RU I	0.6			MIT
5456.134	5457.650	18322.90	RU I	1.6			MIT
5454.816	5456.332	18327.33	RU	2.0			MIT
5452.712	5454.227	18334.40	RU	0.8			MIT
5439.408	5440.920	18379.25	RU	0.7			MIT
5439.214	5440.726	18379.90	RU	1.0			MIT
5427.590	5429.099	18419.26	RU	1.4			MIT
5418.855	5420.361	18448.96	RU I	1.3			MIT
5401.394	5402.896	18508.59	RU I	1.3			MIT
5401.038	5402.540	18509.81	RU	2.1			MIT
5388.609	5390.107	18552.51	RU I	0.7			MIT
5385.879	5387.377	18561.91	RU I	1.4			MIT
5377.839	5379.334	18589.66	RU I	1.4			MIT
5362.087	5363.578	18644.27	RU	0.6			MIT
5361.774	5363.265	18645.36	RU	2.0			MIT
5335.930	5337.414	18735.66	RU	2.0			MIT
5334.704	5336.188	18739.97	RU	1.8			MIT
5332.931	5334.414	18746.20	RU I	1.6			MIT
5315.326	5316.805	18808.29	RU	1.0			MIT
5309.267	5310.744	18829.75	RU I	2.1			MIT
5307.280	5308.757	18836.80	RU	0.7			MIT
5306.452	5307.928	18839.74	RU	0.8			MIT
5305.854	5307.330	18841.86	RU	0.8			MIT
5304.860	5306.336	18845.40	RU	1.8			MIT
5294.627	5296.100	18881.82	RU	0.6			MIT
5291.165	5292.637	18894.17	RU I	1.4			MIT
5285.982	5287.453	18912.70	RU	0.8			MIT
5284.080	5285.550	18919.51	RU I	2.0			MIT
5280.822	5282.292	18931.18	RU I	1.1			MIT
5273.337	5274.805	18958.05	RU	0.6			MIT
5271.726	5273.193	18963.84	RU	0.7			MIT
5266.833	5268.299	18981.46	RU	1.2			MIT
5266.472	5267.938	18982.76	RU	1.1			MIT
5266.222	5267.688	18983.66	RU	0.6			MIT
5263.954	5265.419	18991.84	RU	0.7			MIT
5262.587	5264.052	18996.77	RU	0.7			MIT
5257.075	5258.538	19016.69	RU I	1.4			MIT
5251.667	5253.129	19036.27	RU I	1.4			MIT
5242.952	5244.411	19067.92	RU	0.8			MIT
5242.381	5243.840	19069.99	RU I	1.0			MIT
5230.177	5231.633	19114.49	RU	0.6			MIT
5224.137	5225.591	19136.59	RU I	1.2			MIT
5223.553	5225.007	19138.73	RU	1.3			MIT
5218.114	5219.567	19158.68	RU	1.0			MIT
5214.077	5215.529	19173.51	RU I	0.6			MIT
5213.428	5214.880	19175.90	RU	1.0			MIT
5209.501	5210.952	19190.35	RU	0.8			MIT
5202.122	5203.571	19217.57	RU	1.1			MIT
5199.867	5201.315	19225.91	RU	1.3			MIT
5195.019	5196.466	19243.85	RU I	2.0			MIT
5184.034	5185.478	19284.63	RU	0.6			MIT
5182.225	5183.668	19291.36	RU	0.7			MIT
5171.028	5172.468	19333.13	RU I	2.2			MIT
5159.998	5161.435	19374.46	RU I	1.1			MIT
5155.136	5156.572	19392.73	RU I	2.1			MIT
5153.205	5154.641	19400.00	RU I	0.8			MIT
5151.067	5152.502	19408.05	RU I	1.6			MIT
5147.237	5148.671	19422.49	RU I	1.8			MIT
5142.763	5144.196	19439.39	RU I	1.4			MIT
5137.528	5138.959	19459.19	RU	0.6			MIT
5136.550	5137.981	19462.90	RU I	2.1			MIT
5134.013	5135.443	19472.51	RU I	0.8			MIT
5133.886	5135.316	19473.00	RU	1.2			MIT
5127.257	5128.686	19498.17	RU I	1.3			MIT
5123.728	5125.156	19511.60	RU I	0.8			MIT
5120.512	5121.939	19523.86	RU	0.6			MIT
5107.067	5108.490	19575.25	RU I	1.6			MIT
5106.553	5107.976	19577.23	RU	1.0			MIT
5101.717	5103.139	19595.78	RU I	0.8			MIT
5101.387	5102.809	19597.05	RU I	1.3			MIT
5099.143	5100.564	19605.67	RU	0.7			MIT
5098.947	5100.368	19606.43	RU	0.7			MIT
5093.826	5095.246	19626.14	RU	1.8			MIT
5093.334	5094.754	19628.03	RU	0.6			MIT
5088.420	5089.838	19646.99	RU	0.8			MIT
5079.445	5080.861	19681.70	RU I	0.6			MIT
5076.318	5077.733	19693.83	RU I	1.6			MIT
5076.072	5077.487	19694.78	RU I	1.2			MIT
5072.971	5074.385	19706.82	RU I	1.4			MIT
5062.642	5064.053	19747.03	RU I	1.0			MIT
5057.684	5059.094	19766.39	RU	0.6			MIT
5057.490	5058.900	19767.14	RU	0.6			MIT
5057.331	5058.741	19767.76	RU I	2.0			MIT
5057.291	5058.701	19767.92	RU	1.8			MIT
5052.949	5054.358	19784.91	RU I	0.7			MIT
5047.312	5048.719	19807.00	RU I	1.0			MIT
5045.421	5046.828	19814.43	RU	0.6			MIT
5040.744	5042.150	19832.81	RU I	1.0			MIT
5040.353	5041.759	19834.35	RU	0.8			MIT
5039.629	5041.034	19837.20	RU	0.8			MIT

AIR WAVELENGTH (ANGSTRMS)	VACUUM WAVELENGTH (ANGSTRMS)	WAVENUMBER (KAYSERS)	LMENT	LOG ARC	INTENSITY SPK	INTENSITY DIS	REF
5034.638	5036.042	19856.86	RU	0.8			MIT
5028.16	5029.56	19882.4	RU I	1.1			MIT
5026.645	5028.047	19888.44	RU	0.7			MIT
5026.175	5027.577	19890.30	RU I	1.2			MIT
5025.605	5027.007	19892.55	RU	0.6			MIT
5020.305	5021.705	19913.55	RU I	0.8			MIT
5018.984	5020.384	19918.80	RU	0.7			MIT
5018.128	5019.528	19922.19	RU	0.6			MIT
5014.954	5016.353	19934.80	RU I	1.3			MIT
5014.017	5015.416	19938.53	RU	0.8			MIT
5011.227	5012.625	19949.63	RU	1.4			MIT
5010.595	5011.993	19952.14	RU	1.0			MIT
5005.246	5006.642	19973.47	RU	0.8			MIT
5003.525	5004.921	19980.34	RU	0.7			MIT
5002.958	5004.354	19982.60	RU	0.6			MIT
4999.552	5000.947	19996.21	RU	0.8			MIT
4992.736	4994.129	20023.51	RU I	1.4			MIT
4987.262	4988.653	20045.49	RU I	1.2			MIT
4983.449	4984.839	20060.83	RU I	0.8			MIT
4980.354	4981.744	20073.29	RU I	1.8			MIT
4976.197	4977.585	20090.06	RU I	1.6			MIT
4975.372	4976.760	20093.39	RU	0.7			MIT
4974.123	4975.511	20098.44	RU	1.3			MIT
4968.904	4970.291	20119.55	RU I	1.6			MIT
4959.863	4961.247	20156.22	RU	1.2			MIT
4957.842	4959.226	20164.44	RU	1.0			MIT
4955.255	4956.638	20174.97	RU	1.4			MIT
4953.738	4955.120	20181.14	RU	0.8			MIT
4950.804	4952.186	20193.10	RU	0.8			MIT
4938.434	4939.812	20243.68	RU I	1.8			MIT
4936.231	4937.609	20252.72	RU	0.8			MIT
4935.632	4937.010	20255.18	RU	1.0			MIT
4934.607	4935.984	20259.38	RU	0.6			MIT
4921.074	4922.448	20315.10	RU I	1.6			MIT
4918.998	4920.371	20323.67	RU	0.8			MIT
4917.354	4918.727	20330.46	RU I	0.6			MIT
4914.200	4915.572	20343.51	RU	0.7			MIT
4913.265	4914.637	20347.38	RU I	0.7			MIT
4911.593	4912.964	20354.31	RU I	1.0			MIT
4910.236	4911.607	20359.94	RU	0.8			MIT
4907.888	4909.258	20369.68	RU I	1.3			MIT
4905.022	4906.392	20381.58	RU I	1.1			MIT
4903.053	4904.422	20389.76	RU I	1.8			MIT
4901.862	4903.231	20394.72	RU I	0.7			MIT
4901.066	4902.434	20398.03	RU I	0.8			MIT
4899.252	4900.620	20405.58	RU I	1.1			MIT
4895.597	4896.964	20420.82	RU I	1.1			MIT
4895.320	4896.687	20421.97	RU I	1.0			MIT
4892.841	4894.207	20432.32	RU I	0.7			MIT
4889.831	4891.197	20444.90	RU	0.6			MIT
4888.607	4889.972	20450.01	RU	0.8			MIT
4885.006	4886.370	20465.09	RU	0.8			MIT
4877.882	4879.244	20494.98	RU	0.8			MIT
4877.408	4878.770	20496.97	RU	0.8			MIT
4875.025	4876.387	20506.99	RU	1.1			MIT
4874.326	4875.687	20509.93	RU I	1.0			MIT
4869.785	4871.145	20529.05	RU	0.8			MIT
4869.153	4870.513	20531.72	RU I	2.1			MIT
4868.729	4870.089	20533.51	RU	0.8			MIT
4865.088	4866.447	20548.87	RU	1.1			MIT
4863.111	4864.469	20557.23	RU I	1.0			MIT
4861.867	4863.225	20562.49	RU	1.2			MIT
4860.443	4861.801	20568.51	RU	0.8			MIT
4860.162	4861.520	20569.70	RU	0.6			MIT
4858.000	4859.357	20578.85	RU	0.6			MIT
4854.564	4855.920	20593.42	RU I	1.0			MIT
4853.504	4854.860	20597.92	RU	0.8			MIT
4851.934	4853.289	20604.58	RU	0.7			MIT
4848.16	4849.51	20620.6	RU I	0.8			MIT
4847.858	4849.212	20621.91	RU	0.7			MIT

AIR WAVELENGTH (ANGSTRMS)	VACUUM WAVELENGTH (ANGSTRMS)	WAVENUMBER (KAYSERS)	LMENT	LOG ARC	INTENSITY SPK	INTENSITY DIS	REF
4845.91	4847.26	20630.2	RU	0.6			MIT
4844.557	4845.911	20635.96	RU I	1.3			MIT
4843.746	4845.099	20639.41	RU	0.7			MIT
4841.749	4843.102	20647.93	RU	0.6			MIT
4839.767	4841.119	20656.38	RU I	1.1			MIT
4839.014	4840.366	20659.59	RU I	1.2			MIT
4838.162	4839.514	20663.23	RU I	0.7			MIT
4832.997	4834.347	20685.31	RU	1.4	0.0		MIT
4828.683	4830.032	20703.79	RU	0.8			MIT
4826.551	4827.900	20712.94	RU	0.6			MIT
4826.194	4827.543	20714.47	RU	0.6			MIT
4825.736	4827.085	20716.44	RU	0.7			MIT
4824.357	4825.705	20722.36	RU I	0.7			MIT
4822.568	4823.916	20730.05	RU I	1.0			MIT
4817.341	4818.687	20752.54	RU I	1.0			MIT
4815.523	4816.869	20760.37	RU I	1.3			MIT
4814.719	4816.065	20763.84	RU I	0.6			MIT
4813.226	4814.571	20770.28	RU I	0.8			MIT
4812.848	4814.193	20771.91	RU I	0.6			MIT
4812.203	4813.548	20774.70	RU	0.7			MIT
4809.699	4811.043	20785.51	RU	0.6			MIT
4806.194	4807.537	20800.67	RU	1.0			MIT
4804.884	4806.227	20806.34	RU I	1.3			MIT
4802.904	4804.246	20814.92	RU	0.8			MIT
4801.176	4802.518	20822.41	RU	1.0			MIT
4798.443	4799.784	20834.27	RU I	1.4			MIT
4795.568	4796.909	20846.76	RU I	1.3			MIT
4794.384	4795.724	20851.91	RU I	1.4			MIT
4790.04	4791.38	20870.8	RU I	0.3			MIT
4787.343	4788.681	20882.58	RU	0.7			MIT
4784.269	4785.607	20895.99	RU I	1.4			MIT
4783.287	4784.624	20900.28	RU I	1.0			MIT
4782.644	4783.981	20903.09	RU	1.0			MIT
4781.760	4783.097	20906.96	RU	1.1			MIT
4781.110	4782.447	20909.80	RU	0.8			MIT
4773.997	4775.332	20940.95	RU	1.2			MIT
4773.151	4774.486	20944.67	RU I	1.2			MIT
4769.303	4770.637	20961.56	RU I	1.3			MIT
4768.460	4769.793	20965.27	RU	0.8			MIT
4767.146	4768.479	20971.05	RU	1.0			MIT
4766.653	4767.986	20973.22	RU	0.6			MIT
4764.397	4765.729	20983.15	RU	1.1			MIT
4762.626	4763.958	20990.95	RU	0.6			MIT
4761.521	4762.852	20995.82	RU	0.7			MIT
4757.841	4759.172	21012.06	RU I	2.1			MIT
4756.233	4757.563	21019.16	RU	1.6			MIT
4753.825	4755.154	21029.81	RU	0.8			MIT
4752.141	4753.470	21037.26	RU	0.8			MIT
4751.008	4752.337	21042.28	RU	1.0			MIT
4749.809	4751.137	21047.59	RU I	0.3			MIT
4744.644	4745.971	21070.50	RU	0.6			MIT
4743.994	4745.321	21073.39	RU	0.6			MIT
4743.024	4744.351	21077.70	RU	1.1			MIT
4741.269	4742.595	21085.50	RU	0.6			MIT
4740.331	4741.657	21089.67	RU	0.8			MIT
4738.403	4739.728	21098.26	RU	1.0			MIT
4733.521	4734.845	21120.01	RU I	1.6			MIT
4731.334	4732.657	21129.78	RU I	1.8			MIT
4727.590	4728.912	21146.51	RU	0.8			MIT
4724.795	4726.117	21159.02	RU I	1.0			MIT
4723.222	4724.543	21166.07	RU	0.8			MIT
4720.925	4722.246	21176.36	RU	1.2			MIT
4718.910	4720.230	21185.41	RU	0.8			MIT
4718.065	4719.385	21189.20	RU	0.8			MIT
4716.045	4717.364	21198.28	RU I	1.0			MIT
4715.612	4716.931	21200.22	RU	0.6			MIT
4714.209	4715.528	21206.53	RU	0.7			MIT
4711.989	4713.307	21216.52	RU	1.1			MIT
4709.484	4710.802	21227.81	RU I	2.2	1.9		MIT
4705.159	4706.476	21247.32	RU I	0.7			MIT

AIR WAVELENGTH (ANGSTRMS)	VACUUM WAVELENGTH (ANGSTRMS)	WAVENUMBER (KAYSERS)	LMENT	LOG ARC	INTENSITY SPK	DIS	REF
4700.292	4701.607	21269.32	RU	0.8			MIT
4695.242	4696.556	21292.20	RU I	1.0			MIT
4690.106	4691.419	21315.51	RU I	1.4			MIT
4689.889	4691.202	21316.50	RU	0.7			MIT
4685.783	4687.094	21335.18	RU	1.2			MIT
4684.018	4685.329	21343.22	RU I	2.0			MIT
4683.103	4684.414	21347.39	RU	0.8			MIT
4681.786	4683.096	21353.39	RU I	2.0			MIT
4681.393	4682.703	21355.19	RU	1.0			MIT
4674.654	4675.962	21385.97	RU I	1.3			MIT
4674.519	4675.827	21386.59	RU	0.8			MIT
4671.370	4672.678	21401.01	RU	0.7			MIT
4669.977	4671.284	21407.39	RU I	1.6			MIT
4669.138	4670.445	21411.24	RU	1.2			MIT
4664.999	4666.305	21430.23	RU	1.0			MIT
4662.493	4663.798	21441.75	RU II	0.8			MIT
4662.241	4663.546	21442.91	RU I	0.7			MIT
4660.642	4661.947	21450.27	RU	0.8			MIT
4656.420	4657.724	21469.72	RU I	1.1			MIT
4654.795	4656.098	21477.21	RU	0.8			MIT
4654.315	4655.618	21479.43	RU I	2.1			MIT
4653.744	4655.047	21482.06	RU	1.0			MIT
4652.501	4653.804	21487.80	RU	0.7			MIT
4650.216	4651.518	21498.36	RU	0.7			MIT
4648.154	4649.455	21507.90	RU	0.8			MIT
4647.606	4648.907	21510.43	RU	2.1			MIT
4646.802	4648.103	21514.15	RU	0.8			MIT
4645.089	4646.390	21522.09	RU	2.0	1.5		MIT
4642.603	4643.903	21533.61	RU	0.8			MIT
4642.385	4643.685	21534.62	RU	0.8			MIT
4640.985	4642.285	21541.12	RU	1.0			MIT
4639.318	4640.617	21548.86	RU	1.1			MIT
4638.428	4639.727	21552.99	RU I	1.0			MIT
4635.690	4636.988	21565.72	RU I	2.1			MIT
4633.483	4634.781	21575.99	RU	0.7			MIT
4633.171	4634.469	21577.45	RU I	0.7			MIT
4631.738	4633.035	21584.12	RU I	0.5			MIT
4630.706	4632.003	21588.93	RU	0.6			MIT
4630.457	4631.754	21590.09	RU	0.8			MIT
4628.333	4629.629	21600.00	RU I	1.0			MIT
4626.039	4627.335	21610.71	RU	1.1			MIT
4624.438	4625.733	21618.20	RU	0.8			MIT
4624.324	4625.619	21618.73	RU	0.8			MIT
4620.034	4621.328	21638.80	RU	0.6			MIT
4618.795	4620.089	21644.61	RU	1.0			MIT
4617.665	4618.958	21649.90	RU I	1.1			MIT
4616.107	4617.400	21657.21	RU	0.8			MIT
4615.430	4616.723	21660.39	RU	0.7			MIT
4612.325	4613.617	21674.97	RU I	1.2			MIT
4610.502	4611.794	21683.54	RU	0.8			MIT
4608.675	4609.966	21692.13	RU	0.8			MIT
4606.833	4608.124	21700.81	RU	0.6			MIT
4605.665	4606.955	21706.31	RU I	1.2			MIT
4602.808	4604.097	21719.78	RU I	1.2			MIT
4601.763	4603.052	21724.72	RU I	1.3			MIT
4600.555	4601.844	21730.42	RU	0.8			MIT
4599.085	4600.373	21737.37	RU I	2.0			MIT
4596.710	4597.998	21748.60	RU	1.3			MIT
4593.213	4594.500	21765.15	RU I	0.8			MIT
4593.025	4594.312	21766.04	RU	0.8			MIT
4592.520	4593.807	21768.44	RU I	2.0			MIT
4592.015	4593.302	21770.83	RU	0.8			MIT
4591.560	4592.847	21772.99	RU I	1.3			MIT
4591.103	4592.389	21775.16	RU I	1.8			MIT
4589.585	4590.871	21782.36	RU	0.8			MIT
4587.099	4588.384	21794.16	RU I	1.0			MIT
4584.445	4585.730	21806.78	RU I	2.2G	1.9		MIT
4580.072	4581.355	21827.60	RU I	1.4			MIT
4577.416	4578.699	21840.27	RU	0.7			MIT
4576.319	4577.601	21845.50	RU	0.8			MIT

AIR WAVELENGTH (ANGSTRMS)	VACUUM WAVELENGTH (ANGSTRMS)	WAVENUMBER (KAYSERS)	LMENT	LOG ARC	INTENSITY SPK	DIS	REF
4575.751	4577.033	21848.21	RU	0.8			MIT
4573.956	4575.238	21856.79	RU	0.7			MIT
4570.248	4571.529	21874.52	RU	1.0			MIT
4567.906	4569.186	21885.74	RU I	0.8			MIT
4566.707	4567.987	21891.48	RU	0.8			MIT
4564.691	4565.970	21901.15	RU I	1.3			MIT
4562.597	4563.876	21911.20	RU I	1.2			MIT
4559.982	4561.260	21923.77	RU I	1.3			MIT
4559.070	4560.348	21928.15	RU	0.8			MIT
4557.814	4559.092	21934.19	RU I	0.7			MIT
4554.509	4555.786	21950.11	RU I	3.0G	2.3		MIT
4552.110	4553.386	21961.68	RU I	0.8			MIT
4549.957	4551.233	21972.07	RU	1.0			MIT
4549.427	4550.702	21974.63	RU I	1.0			MIT
4547.853	4549.128	21982.24	RU	1.3			MIT
4547.33	4548.60	21984.8	RU	1.4			MIT
4546.930	4548.205	21986.70	RU I	1.2			MIT
4545.720	4546.994	21992.55	RU	0.8			MIT
4543.687	4544.961	22002.39	RU I	1.2			MIT
4542.425	4543.699	22008.50	RU	1.1			MIT
4538.664	4539.937	22026.74	RU	1.0			MIT
4537.443	4538.715	22032.67	RU	0.8			MIT
4535.587	4536.859	22041.68	RU	1.1			MIT
4534.490	4535.761	22047.01	RU	0.8			MIT
4532.446	4533.717	22056.96	RU	0.8			MIT
4531.801	4533.072	22060.10	RU	0.7			MIT
4530.854	4532.124	22064.71	RU I	1.8			MIT
4529.054	4530.324	22073.48	RU	1.0			MIT
4526.432	4527.701	22086.26	RU	0.8			MIT
4525.935	4527.204	22088.69	RU	0.8			MIT
4520.950	4522.218	22113.04	RU I	1.4			MIT
4517.818	4519.085	22128.37	RU I	1.8			MIT
4516.893	4518.160	22132.91	RU I	2.0			MIT
4516.269	4517.536	22135.96	RU I	1.1			MIT
4514.826	4516.092	22143.04	RU	0.7			MIT
4511.197	4512.462	22160.85	RU I	1.4			MIT
4510.097	4511.362	22166.25	RU I	1.4			MIT
4509.698	4510.963	22168.22	RU	0.8			MIT
4508.561	4509.826	22173.81	RU	1.2			MIT
4508.039	4509.303	22176.37	RU	1.1			MIT
4502.645	4503.908	22202.94	RU	0.6			MIT
4498.145	4499.407	22225.15	RU I	2.1	1.6		MIT
4496.793	4498.054	22231.83	RU	0.8			MIT
4492.892	4494.152	22251.14	RU	0.8			MIT
4491.682	4492.942	22257.13	RU	1.3			MIT
4490.235	4491.495	22264.30	RU I	1.4			MIT
4488.392	4489.651	22273.45	RU I	1.4			MIT
4486.812	4488.071	22281.29	RU	0.8			MIT
4483.904	4485.162	22295.74	RU	1.0			MIT
4482.034	4483.292	22305.04	RU I	1.1			MIT
4480.448	4481.705	22312.94	RU	1.8			MIT
4479.410	4480.667	22318.11	RU	1.2			MIT
4478.013	4479.270	22325.07	RU	0.8			MIT
4475.330	4476.586	22338.45	RU I	1.0			MIT
4473.928	4475.183	22345.45	RU I	2.0			MIT
4471.029	4472.284	22359.94	RU	0.8			MIT
4467.260	4468.514	22378.81	RU	1.3			MIT
4466.343	4467.596	22383.40	RU I	1.2			MIT
4465.482	4466.735	22387.72	RU	1.2			MIT
4460.035	4461.287	22415.06	RU I	2.2	1.9		MIT
4449.336	4450.585	22468.96	RU I	2.1	2.0		MIT
4444.507	4445.755	22493.37	RU I	1.6			MIT
4439.761	4441.007	22517.41	RU I	2.1	1.7		MIT
4439.407	4440.653	22519.21	RU	1.0			MIT
4438.341	4439.587	22524.62	RU	0.8			MIT
4430.311	4431.555	22565.44	RU I	1.3			MIT
4428.461	4429.704	22574.87	RU I	2.1			MIT
4426.013	4427.256	22587.35	RU I	1.0			MIT
4424.781	4426.023	22593.64	RU	1.4			MIT
4422.976	4424.218	22602.86	RU	1.1			MIT

AIR WAVELENGTH (ANGSTRMS)	VACUUM WAVELENGTH (ANGSTRMS)	WAVENUMBER (KAYSERS)	LMENT	LOG ARC	INTENSITY SPK	DIS	REF
4421.456	4422.698	22610.64	RU I	1.8			MIT
4420.841	4422.082	22613.78	RU I	1.6			MIT
4414.436	4415.676	22646.59	RU	1.0			MIT
4413.291	4414.530	22652.47	RU	1.1			MIT
4410.026	4411.265	22669.24	RU I	2.2	1.9		MIT
4399.587	4400.823	22723.02	RU I	1.3			MIT
4397.797	4399.032	22732.27	RU I	2.2			MIT
4394.959	4396.194	22746.95	RU	1.2			MIT
4391.027	4392.261	22767.32	RU I	1.3			MIT
4390.435	4391.668	22770.39	RU I	2.2G	1.9		MIT
4388.990	4390.223	22777.89	RU	1.1			MIT
4386.272	4387.504	22792.00	RU	1.3			MIT
4385.648	4386.880	22795.24	RU I	2.1	1.7		MIT
4385.391	4386.623	22796.58	RU I	2.1	1.6		MIT
4383.360	4384.592	22807.14	RU I	1.1			MIT
4381.266	4382.497	22818.04	RU	1.2			MIT
4376.599	4377.829	22842.37	RU	0.8			MIT
4372.208	4373.437	22865.31	RU I	2.1	2.0		MIT
4371.201	4372.429	22870.58	RU	1.2	0.7		MIT
4370.415	4371.643	22874.69	RU I	1.2			MIT
4364.104	4365.330	22907.77	RU	0.8			MIT
4362.702	4363.928	22915.14	RU I	0.8			MIT
4361.211	4362.437	22922.97	RU	1.6	1.7		MIT
4354.802	4356.026	22956.70	RU I	1.1			MIT
4354.131	4355.355	22960.24	RU I	1.4	1.3		MIT
4349.698	4350.921	22983.64	RU I	1.3			MIT
4346.476	4347.698	23000.68	RU I	1.2			MIT
4342.068	4343.289	23024.03	RU I	1.8	1.6		MIT
4341.042	4342.262	23029.47	RU	0.8			MIT
4340.343	4341.563	23033.18	RU	1.0			MIT
4338.678	4339.898	23042.02	RU	0.8			MIT
4337.266	4338.485	23049.52	RU	1.2			MIT
4336.424	4337.643	23053.99	RU I	1.1			MIT
4332.504	4333.722	23074.85	RU	0.8			MIT
4331.165	4332.383	23081.99	RU I	1.2			MIT
4328.558	4329.775	23095.89	RU I	0.8			MIT
4327.429	4328.646	23101.91	RU I	0.8			MIT
4326.825	4328.042	23105.14	RU I	1.3			MIT
4325.052	4326.268	23114.61	RU I	1.4	1.0		MIT
4322.961	4324.177	23125.79	RU	0.8			MIT
4321.301	4322.516	23134.67	RU I	0.8			MIT
4320.816	4322.031	23137.27	RU I	0.7			MIT
4320.580	4321.795	23138.53	RU	0.7			MIT
4319.869	4321.084	23142.34	RU I	1.3	1.6		MIT
4319.125	4320.340	23146.33	RU I	0.8			MIT
4318.433	4319.647	23150.04	RU I	1.2			MIT
4316.638	4317.852	23159.66	RU I	1.1			MIT
4314.302	4315.515	23172.20	RU	1.3			MIT
4312.477	4313.690	23182.01	RU	0.8			MIT
4309.208	4310.420	23199.59	RU	1.1			MIT
4307.595	4308.807	23208.28	RU I	1.3	1.7		MIT
4301.147	4302.357	23243.07	RU	0.7			MIT
4297.711	4298.920	23261.66	RU I	1.8	1.7		MIT
4296.689	4297.898	23267.19	RU	0.8			MIT
4295.928	4297.137	23271.31	RU I	1.3	1.3		MIT
4294.791	4295.999	23277.47	RU I	1.3			MIT
4293.284	4294.492	23285.64	RU I	1.2			MIT
4290.529	4291.736	23300.59	RU I	0.8			MIT
4287.046	4288.252	23319.52	RU	1.2			MIT
4284.331	4285.536	23334.30	RU I	1.4	1.3		MIT
4281.930	4283.135	23347.39	RU	0.8			MIT
4278.689	4279.893	23365.07	RU	0.8			MIT
4277.255	4278.459	23372.90	RU I	0.8			MIT
4265.607	4266.808	23436.73	RU I	1.1			MIT
4263.396	4264.596	23448.88	RU I	1.0			MIT
4260.004	4261.203	23467.55	RU I	1.1			MIT
4258.987	4260.186	23473.15	RU I	1.2			MIT
4255.708	4256.906	23491.24	RU	0.8			MIT
4248.143	4249.339	23533.07	RU I	1.1			MIT
4246.734	4247.930	23540.88	RU	1.3			MIT

AIR WAVELENGTH (ANGSTRMS)	VACUUM WAVELENGTH (ANGSTRMS)	WAVENUMBER (KAYSERS)	LMENT	LOG ARC	INTENSITY SPK	DIS	REF
4246.334	4247.530	23543.10	RU	1.2			MIT
4244.832	4246.027	23551.43	RU I	1.4			MIT
4243.061	4244.256	23561.26	RU I	2.0	1.6		MIT
4241.053	4242.247	23572.41	RU I	2.0	1.3		MIT
4240.032	4241.226	23578.09	RU	0.8			MIT
4239.652	4240.846	23580.20	RU	0.8			MIT
4236.674	4237.867	23596.78	RU	1.3			MIT
4232.322	4233.514	23621.04	RU I	1.6			MIT
4230.312	4231.503	23632.26	RU I	1.8			MIT
4229.309	4230.500	23637.87	RU I	1.6			MIT
4226.656	4227.846	23652.70	RU I	1.2			MIT
4225.092	4226.282	23661.46	RU	1.4			MIT
4220.675	4221.864	23686.22	RU I	1.8			MIT
4217.268	4218.456	23705.36	RU I	2.0	1.3		MIT
4214.557	4215.744	23720.60	RU	0.8			MIT
4214.442	4215.629	23721.25	RU	2.0	1.6		MIT
4212.063	4213.250	23734.65	RU I	2.1	1.9		MIT
4207.636	4208.821	23759.62	RU	1.3			MIT
4206.016	4207.201	23768.77	RU I	2.0	1.6		MIT
4199.902	4201.085	23803.37	RU I	2.2	2.5		MIT
4198.875	4200.058	23809.20	RU I	1.8	2.0		MIT
4197.579	4198.762	23816.55	RU I	2.0	2.0		MIT
4196.871	4198.054	23820.56	RU	1.8	1.7		MIT
4189.462	4190.643	23862.69	RU I	1.2	1.0		MIT
4185.459	4186.639	23885.51	RU	1.1	0.3		MIT
4182.831	4184.010	23900.52	RU	1.1	0.9		MIT
4182.644	4183.823	23901.59	RU I	1.2	1.1		MIT
4182.457	4183.636	23902.66	RU	1.3	1.5		MIT
4175.433	4176.610	23942.86	RU	0.8	1.0		MIT
4170.592	4171.768	23970.65	RU	1.0			MIT
4170.051	4171.226	23973.76	RU I	1.3	1.4		MIT
4168.762	4169.937	23981.18	RU	0.8			MIT
4167.512	4168.687	23988.37	RU I	2.0	2.2		MIT
4166.876	4168.051	23992.03	RU I	1.3	1.4		MIT
4164.392	4165.566	24006.34	RU	1.0			MIT
4161.656	4162.829	24022.12	RU I	1.4	1.7		MIT
4159.172	4160.345	24036.47	RU	1.2	0.8		MIT
4158.261	4159.433	24041.74	RU	0.8			MIT
4157.736	4158.908	24044.77	RU	1.2	0.7		MIT
4156.261	4157.433	24053.30	RU	1.0	0.7J		MIT
4150.299	4151.469	24087.86	RU	1.3	1.0		MIT
4148.375	4149.545	24099.03	RU I	1.8	1.2		MIT
4146.771	4147.940	24108.35	RU I	2.0	1.8		MIT
4145.738	4146.907	24114.36	RU I	2.1	2.2		MIT
4144.164	4145.333	24123.52	RU I	2.2	2.3		MIT
4137.234	4138.401	24163.92	RU I	1.4	1.2		MIT
4134.85	4136.02	24177.9	RU	1.1	0.7		MIT
4131.225	4132.390	24199.07	RU	1.0			MIT
4127.868	4129.032	24218.75	RU I	1.4	1.5		MIT
4127.440	4128.604	24221.26	RU I	1.3	1.5		MIT
4126.412	4127.576	24227.29	RU I	1.1	0.0		MIT
4123.813	4124.976	24242.56	RU	1.3	1.0		MIT
4123.064	4124.227	24246.97	RU I	1.4	1.5		MIT
4122.787	4123.950	24248.60	RU	0.8			MIT
4120.987	4122.150	24259.19	RU I	1.4	1.5		MIT
4118.498	4119.660	24273.85	RU I	1.6			MIT
4116.858	4118.020	24283.52	RU	0.8			MIT
4114.134	4115.295	24299.60	RU I	1.3	0.7		MIT
4113.383	4114.544	24304.03	RU	1.6	1.7		MIT
4112.741	4113.901	24307.83	RU I	2.1	2.3		MIT
4109.646	4110.806	24326.13	RU I	1.3	0.8		MIT
4108.060	4109.219	24335.52	RU	1.0	0.8		MIT
4107.837	4108.996	24336.84	RU	1.4	1.3		MIT
4105.912	4107.071	24348.25	RU I	1.1	0.3		MIT
4102.285	4103.443	24369.78	RU I	1.2	1.0		MIT
4101.745	4102.903	24372.99	RU I	1.3	1.8		MIT
4101.232	4102.389	24376.04	RU	0.8	0.3H		MIT
4100.373	4101.530	24381.14	RU	1.1	1.0		MIT
4097.791	4098.948	24396.51	RU I	1.4	2.1		MIT
4097.026	4098.182	24401.06	RU I	1.2	1.2		MIT

AIR WAVELENGTH (ANGSTRMS)	VACUUM WAVELENGTH (ANGSTRMS)	WAVENUMBER (KAYSERS)	LMENT	LOG ARC	INTENSITY SPK	INTENSITY DIS	REF
4093.776	4094.931	24420.43	RU	0.7	0.0		MIT
4091.062	4092.217	24436.63	RU	1.3	1.3		MIT
4088.360	4089.514	24452.78	RU	1.0	0.0		MIT
4085.429	4086.582	24470.33	RU I	1.6	1.7		MIT
4082.794	4083.947	24486.12	RU	1.2	1.0		MIT
4080.600	4081.752	24499.28	RU	2.1	2.5		MIT
4079.277	4080.429	24507.23	RU I	1.1	0.7		MIT
4076.733	4077.884	24522.52	RU I	1.8	1.4		MIT
4075.984	4077.135	24527.03	RU	0.7	1.4		MIT
4073.116	4074.266	24544.30	RU	0.8	1.4		MIT
4071.398	4072.548	24554.65	RU I	1.1	1.3		MIT
4068.366	4069.515	24572.95	RU	1.6	1.8		MIT
4067.613	4068.762	24577.50	RU	1.4	1.5		MIT
4066.243	4067.391	24585.78	RU	0.8			MIT
4064.456	4065.604	24596.59	RU I	1.3	1.8		MIT
4064.105	4065.253	24598.72	RU	1.2	1.4		MIT
4062.988	4064.135	24605.48	RU I	1.1	1.5		MIT
4062.853	4064.000	24606.30	RU	1.0	1.0		MIT
4059.431	4060.578	24627.04	RU	0.8			MIT
4058.882	4060.028	24630.37	RU	1.0			MIT
4054.732	4055.877	24655.58	RU I	0.7			MIT
4054.051	4055.196	24659.72	RU I	1.6	2.0		MIT
4052.195	4053.340	24671.01	RU I	1.4	1.7		MIT
4051.400	4052.544	24675.85	RU I	2.1	2.3		MIT
4049.413	4050.557	24687.96	RU I	1.2	1.1		MIT
4041.998	4043.140	24733.25	RU	1.1	0.0		MIT
4040.481	4041.623	24742.54	RU	1.1	0.5		MIT
4039.210	4040.351	24750.32	RU	1.4	1.7		MIT
4037.737	4038.878	24759.35	RU	1.1	0.7		MIT
4036.492	4037.633	24766.99	RU	1.0	0.5		MIT
4032.521	4033.660	24791.38	RU	1.0	0.7		MIT
4032.205	4033.344	24793.32	RU I	1.3	1.3		MIT
4030.997	4032.136	24800.75	RU I	1.2	1.1		MIT
4030.143	4031.282	24806.01	RU	0.8			MIT
4028.435	4029.573	24816.52	RU	1.0	0.3		MIT
4026.502	4027.640	24828.44	RU	0.8	0.0		MIT
4024.695	4025.832	24839.58	RU I	1.1	0.7		MIT
4024.305	4025.442	24841.99	RU	0.8	0.6		MIT
4023.833	4024.970	24844.90	RU I	1.4	1.8		MIT
4022.692	4023.829	24851.95	RU	0.8	0.5		MIT
4022.161	4023.298	24855.23	RU I	1.6	2.0		MIT
4020.995	4022.131	24862.44	RU I	1.2	1.1		MIT
4019.553	4020.689	24871.36	RU I	1.1	0.9		MIT
4016.753	4017.888	24888.70	RU I	0.8			MIT
4014.153	4015.288	24904.82	RU	1.1	0.7		MIT
4013.738	4014.873	24907.39	RU I	1.0	0.7		MIT
4013.505	4014.639	24908.84	RU I	1.2	1.1		MIT
4011.729	4012.863	24919.86	RU	0.8			MIT
4008.269	4009.402	24941.37	RU I	1.3	1.3		MIT
4007.535	4008.668	24945.94	RU	1.3	1.0		MIT
4006.598	4007.731	24951.78	RU I	1.4	1.2		MIT
4005.640	4006.772	24957.74	RU I	1.4	1.5		MIT
3999.393	4000.524	24996.73	RU	0.5			MIT
3996.509	3997.639	25014.76	RU I	1.0	0.6		MIT
3995.977	3997.107	25018.09	RU I	1.5	1.5		MIT
3994.560	3995.690	25026.97	RU I	0.7			MIT
3993.531	3994.660	25033.42	RU	1.0	0.7		MIT
3987.795	3988.923	25069.43	RU	0.5	1.7		MIT
3984.858	3985.985	25087.90	RU I	1.8	1.8		MIT
3984.682	3985.809	25089.01	RU	0.6	0.3		MIT
3982.230	3983.356	25104.46	RU	0.6			MIT
3981.902	3983.028	25106.52	RU	0.7	0.5		MIT
3979.424	3980.550	25122.16	RU I	1.8	1.8		MIT
3978.445	3979.570	25128.34	RU I	1.8	1.8		MIT
3974.501	3975.625	25153.28	RU I	1.3	0.9		MIT
3970.385	3971.508	25179.35	RU	0.5			MIT
3969.794	3970.917	25183.10	RU I	0.9	0.6		MIT
3968.461	3969.584	25191.56	RU	1.1	2.3		MIT
3964.896	3966.018	25214.21	RU	1.7	1.6		MIT
3957.449	3958.569	25261.66	RU I	0.9	0.7		MIT

AIR WAVELENGTH (ANGSTRMS)	VACUUM WAVELENGTH (ANGSTRMS)	WAVENUMBER (KAYSERS)	LMENT	LOG ARC	INTENSITY SPK	INTENSITY DIS	REF
3957.221	3958.341	25263.11	RU	0.9	0.3		MIT
3954.890	3956.009	25278.00	RU	0.7			MIT
3953.835	3954.954	25284.75	RU	0.7			MIT
3952.678	3953.797	25292.15	RU	1.3	1.5		MIT
3952.290	3953.409	25294.63	RU	0.6	0.0		MIT
3951.207	3952.325	25301.56	RU I	1.0	0.8		MIT
3950.412	3951.530	25306.65	RU	1.0	1.0		MIT
3950.214	3951.332	25307.92	RU I	1.1	1.2		MIT
3950.004	3951.122	25309.27	RU	1.0	0.9		MIT
3949.417	3950.535	25313.03	RU I	1.0	0.7		MIT
3946.314	3947.431	25332.93	RU I	0.9	0.5		MIT
3945.572	3946.689	25337.70	RU	1.7	2.0		MIT
3944.785	3945.902	25342.75	RU	0.6			MIT
3944.190	3945.306	25346.57	RU I	1.0	0.6		MIT
3942.063	3943.179	25360.25	RU I	1.1	1.0		MIT
3941.654	3942.770	25362.88	RU	1.1	0.9		MIT
3940.563	3941.678	25369.90	RU	0.6			MIT
3939.100	3940.215	25379.32	RU	0.7			MIT
3937.903	3939.018	25387.04	RU I	1.3	1.2		MIT
3936.782	3937.896	25394.27	RU	0.6			MIT
3936.158	3937.272	25398.29	RU	0.6			MIT
3935.346	3936.460	25403.53	RU I	0.7			MIT
3933.680	3934.794	25414.29	RU	0.7	2.3		MIT
3932.287	3933.400	25423.30	RU	0.5			MIT
3931.759	3932.872	25426.71	RU	1.7	1.8		MIT
3929.851	3930.964	25439.05	RU	0.5			MIT
3926.466	3927.578	25460.99	RU	0.6	0.0		MIT
3925.925	3927.037	25464.49	RU I	1.8	2.0		MIT
3924.633	3925.744	25472.88	RU	0.8	0.7		MIT
3923.467	3924.578	25480.45	RU	1.8	2.0		MIT
3922.335	3923.446	25487.80	RU	0.6			MIT
3921.990	3923.101	25490.04	RU	0.5			MIT
3920.915	3922.025	25497.03	RU I	1.3	1.3		MIT
3917.710	3918.819	25517.89	RU	0.5			MIT
3914.853	3915.962	25536.51	RU I	1.3	1.2		MIT
3913.013	3914.121	25548.52	RU	0.5			MIT
3912.112	3913.220	25554.40	RU	1.0	0.9		MIT
3911.144	3912.252	25560.73	RU	0.8	0.3		MIT
3909.075	3910.182	25574.26	RU	1.5	1.7		MIT
3908.765	3909.872	25576.28	RU	1.1	1.1		MIT
3905.991	3907.097	25594.45	RU	0.8	0.5H		MIT
3905.158	3906.264	25599.91	RU	0.5			MIT
3902.816	3903.922	25615.27	RU	0.7			MIT
3902.072	3903.177	25620.15	RU	0.6			MIT
3901.242	3902.347	25625.60	RU I	1.7	1.1		MIT
3898.361	3899.465	25644.54	RU I	1.1	0.8		MIT
3897.236	3898.340	25651.94	RU	1.1	0.8		MIT
3895.770	3896.874	25661.60	RU	0.5			MIT
3894.245	3895.348	25671.65	RU I	0.9	0.3		MIT
3892.773	3893.876	25681.35	RU	1.0	0.6		MIT
3892.209	3893.312	25685.07	RU I	1.7	1.6		MIT
3891.410	3892.513	25690.35	RU	1.3	0.5		MIT
3890.197	3891.299	25698.36	RU I	1.5	0.9		MIT
3887.772	3888.874	25714.39	RU I	1.2	0.9		MIT
3884.676	3885.777	25734.88	RU	1.3	0.8		MIT
3884.016	3885.117	25739.25	RU	1.3	0.8		MIT
3882.006	3883.106	25752.58	RU I	1.1			MIT
3880.806	3881.906	25760.54	RU I	0.7			MIT
3880.728	3881.828	25761.06	RU	0.7			MIT
3878.753	3879.852	25774.18	RU	0.5			MIT
3877.530	3878.629	25782.31	RU	0.5	0.7H		MIT
3876.648	3877.747	25788.17	RU I	1.1	0.0		MIT
3876.082	3877.181	25791.94	RU	1.3	0.5		MIT
3873.525	3874.623	25808.96	RU I	1.5	1.7		MIT
3872.718	3873.816	25814.34	RU	0.6	0.8		MIT
3872.372	3873.470	25816.65	RU	0.6	0.9		MIT
3871.688	3872.786	25821.21	RU	0.9	0.5		MIT
3871.215	3872.312	25824.36	RU I	1.0	0.0H		MIT
3867.839	3868.936	25846.90	RU I	1.8	1.5		MIT
3867.177	3868.273	25851.33	RU	0.6			MIT

AIR WAVELENGTH (ANGSTRMS)	VACUUM WAVELENGTH (ANGSTRMS)	WAVENUMBER (KAYSERS)	LMENT	LOG ARC	INTENSITY SPK	DIS	REF
3865.743	3866.839	25860.92	RU	0.6	0.6		MIT
3865.403	3866.499	25863.19	RU	1.0	0.6		MIT
3864.857	3865.953	25866.84	RU I	0.7			MIT
3862.646	3863.741	25881.65	RU	0.3	1.8		MIT
3861.426	3862.521	25889.83	RU I	0.7			MIT
3860.723	3861.818	25894.54	RU I	1.3	0.6		MIT
3859.712	3860.806	25901.32	RU	0.8	1.2		MIT
3858.686	3859.780	25908.21	RU	0.6	0.6		MIT
3857.551	3858.645	25915.83	RU	1.7	1.4		MIT
3856.981	3858.075	25919.66	RU	0.7	0.7H		MIT
3856.459	3857.553	25923.17	RU	1.7	0.9		MIT
3855.578	3856.671	25929.10	RU	0.6	1.2		MIT
3854.727	3855.820	25934.82	RU I	1.0	0.7J		MIT
3852.837	3853.930	25947.54	RU I	0.7	0.5H		MIT
3852.564	3853.657	25949.38	RU	0.6	0.7H		MIT
3852.137	3853.229	25952.26	RU	1.1	1.0		MIT
3850.432	3851.524	25963.75	RU I	1.7	1.0		MIT
3848.945	3850.037	25973.78	RU	0.9	1.1		MIT
3847.495	3848.586	25983.57	RU	0.6			MIT
3846.805	3847.896	25988.23	RU	0.7			MIT
3846.676	3847.767	25989.10	RU	1.1	1.0		MIT
3843.159	3844.249	26012.88	RU	1.0	0.7		MIT
3842.661	3843.751	26016.25	RU	0.8	0.5H		MIT
3840.985	3842.075	26027.61	RU	0.9	0.8		MIT
3840.76	3841.85	26029.1	RU I	0.6	0.8		MIT
3839.695	3840.784	26036.35	RU	1.7	1.5		MIT
3838.728	3839.817	26042.91	RU	1.0	1.0		MIT
3838.067	3839.156	26047.39	RU I	1.1	0.7J		MIT
3836.969	3838.057	26054.85	RU	0.6			MIT
3836.703	3837.791	26056.65	RU	0.9	1.6		MIT
3835.990	3837.078	26061.50	RU	0.8	0.0H		MIT
3835.048	3836.136	26067.90	RU	1.7	0.8		MIT
3831.795	3832.882	26090.03	RU	1.8	1.7		MIT
3830.877	3831.964	26096.28	RU I	1.0	0.6		MIT
3829.479	3830.566	26105.81	RU I	0.6	0.7		MIT
3829.332	3830.418	26106.81	RU	0.9	0.5		MIT
3828.714	3829.800	26111.02	RU I	1.5	0.9		MIT
3827.554	3828.640	26118.94	RU	0.6	0.0J		MIT
3826.105	3827.191	26128.83	RU I	0.9	0.5		MIT
3824.932	3826.017	26136.84	RU I	1.5	1.4		MIT
3822.091	3823.176	26156.27	RU	1.7	1.4		MIT
3821.646	3822.730	26159.31	RU	0.6			MIT
3819.767	3820.851	26172.18	RU I	1.1	0.8		MIT
3819.033	3820.117	26177.21	RU I	1.7	1.5		MIT
3818.352	3819.436	26181.88	RU	0.7	0.3		MIT
3817.948	3819.032	26184.65	RU	0.8	0.5		MIT
3817.269	3818.352	26189.31	RU	1.7	1.8		MIT
3814.857	3815.940	26205.86	RU I	1.3	1.5		MIT
3813.177	3814.259	26217.41	RU	0.6	0.5		MIT
3812.718	3813.800	26220.57	RU I	1.1			MIT
3808.684	3809.765	26248.34	RU	1.7	1.5		MIT
3805.693	3806.773	26268.97	RU	0.6			MIT
3805.434	3806.514	26270.75	RU I	1.0	0.7		MIT
3803.197	3804.277	26286.21	RU	1.0	0.9		MIT
3800.261	3801.340	26306.51	RU I	1.1	1.6		MIT
3799.347	3800.426	26312.84	RU I	1.8R	2.0		MIT
3798.901	3799.980	26315.93	RU I	1.8	2.0		MIT
3798.052	3799.130	26321.81	RU I	1.5	1.6		MIT
3795.175	3796.253	26341.77	RU I	1.1	0.8		MIT
3794.924	3796.002	26343.51	RU	1.3	1.5		MIT
3794.269	3795.346	26348.06	RU	0.5	0.9W		MIT
3790.513	3791.589	26374.16	RU I	1.8	2.2		MIT
3786.055	3787.130	26405.22	RU I	1.8	2.0		MIT
3784.737	3785.812	26414.41	RU I	0.6			MIT
3782.740	3783.814	26428.36	RU I	0.9	0.8		MIT
3781.683	3782.757	26435.74	RU	0.7	0.5		MIT
3781.181	3782.255	26439.25	RU	1.7	1.6		MIT
3779.964	3781.038	26447.77	RU I	1.0	0.6		MIT
3779.431	3780.505	26451.50	RU	0.6			MIT
3778.701	3779.774	26456.61	RU	1.1	1.0		MIT

AIR WAVELENGTH (ANGSTRMS)	VACUUM WAVELENGTH (ANGSTRMS)	WAVENUMBER (KAYSERS)	LMENT	LOG ARC	INTENSITY SPK	DIS	REF
3777.759	3778.832	26463.20	RU	0.5	0.6		MIT
3777.586	3778.659	26464.42	RU I	1.8	1.7		MIT
3773.793	3774.865	26491.01	RU	0.5			MIT
3773.170	3774.242	26495.39	RU I	1.1	0.6		MIT
3771.075	3772.146	26510.11	RU	0.5			MIT
3767.350	3768.420	26536.32	RU	1.7	1.7		MIT
3766.480	3767.550	26542.45	RU	0.5	0.3		MIT
3765.822	3766.892	26547.08	RU	0.7			MIT
3764.032	3765.102	26559.71	RU	0.8	0.7		MIT
3763.612	3764.681	26562.67	RU	0.6			MIT
3761.508	3762.577	26577.53	RU	1.1	1.7		MIT
3760.031	3761.099	26587.97	RU I	1.3	1.7		MIT
3759.836	3760.904	26589.35	RU	1.1	1.4		MIT
3755.931	3756.998	26616.99	RU I	1.5	1.8		MIT
3755.727	3756.794	26618.44	RU	0.8	0.9		MIT
3755.090	3756.157	26622.95	RU I	1.3	1.4		MIT
3753.545	3754.612	26633.91	RU	1.5	1.8		MIT
3752.778	3753.845	26639.35	RU	0.9	0.5		MIT
3751.853	3752.919	26645.92	RU	0.7	0.5		MIT
3746.218	3747.283	26686.00	RU I	0.6	0.6		MIT
3745.592	3746.657	26690.46	RU I	1.9			ME
3744.395	3745.459	26698.99	RU	0.9	1.5		MIT
3744.219	3745.283	26700.25	RU I	0.9	1.4		MIT
3743.945	3745.009	26702.20	RU I	0.7	0.6		MIT
3743.322	3744.386	26706.65	RU	0.8	0.7		MIT
3742.784	3743.848	26710.49	RU	1.7	1.7		MIT
3742.280	3743.344	26714.08	RU I	1.8	2.0		MIT
3739.465	3740.528	26734.19	RU	1.3	1.3		MIT
3738.912	3739.975	26738.15	RU	1.0	1.0		MIT
3738.629	3739.692	26740.17	RU	0.9	0.9		MIT
3737.741	3738.804	26746.52	RU	0.8	0.7		MIT
3737.401	3738.464	26748.96	RU I	1.1	1.1		MIT
3736.798	3737.860	26753.27	RU	0.5			MIT
3735.021	3736.083	26766.00	RU	0.6			MIT
3733.049	3734.110	26780.14	RU	1.0	0.5		MIT
3732.765	3733.826	26782.18	RU	0.6			MIT
3732.029	3733.090	26787.46	RU	0.8	0.9		MIT
3730.904	3731.965	26795.54	RU	0.5	0.7		MIT
3730.592	3731.653	26797.78	RU I	0.3	0.7		MIT
3730.433	3731.494	26798.92	RU I	1.1	1.8		MIT
3728.030	3729.090	26816.19	RU I	2.0	2.2		MIT
3726.926	3727.986	26824.14	RU I	2.0	2.2		MIT
3726.096	3727.156	26830.11	RU	1.1	1.8		MIT
3725.486	3726.546	26834.50	RU I	0.9	0.7		MIT
3724.969	3726.028	26838.23	RU I	1.1	1.1		MIT
3722.312	3723.371	26857.39	RU	0.7	0.3		MIT
3721.932	3722.991	26860.13	RU	0.5			MIT
3719.329	3720.387	26878.93	RU	1.3	1.4		MIT
3718.611	3719.669	26884.11	RU	0.3	1.0		MIT
3718.400	3719.458	26885.64	RU	0.9	0.3		MIT
3717.682	3718.740	26890.83	RU	0.8	0.7		MIT
3717.004	3718.061	26895.74	RU I	1.5	1.4		MIT
3716.179	3717.236	26901.71	RU	1.1	0.9		MIT
3715.562	3716.619	26906.18	RU	1.1	0.9		MIT
3714.624	3715.681	26912.97	RU I	0.9	0.0		MIT
3712.295	3713.351	26929.85	RU	1.5	1.0		MIT
3710.316	3711.372	26944.22	RU	0.6			MIT
3709.097	3710.152	26953.07	RU I	0.7			MIT
3708.618	3709.673	26956.55	RU	0.6			MIT
3705.357	3706.411	26980.28	RU I	0.9	0.3		MIT
3703.206	3704.260	26995.95	RU I	1.1	0.5		MIT
3702.933	3703.987	26997.94	PU	0.7			MIT
3702.240	3703.294	27002.99	RU	0.8	0.5		MIT
3701.848	3702.901	27005.85	RU	0.7			MIT
3701.320	3702.373	27009.70	RU	1.0	0.5		MIT
3700.991	3702.044	27012.10	RU	1.7	1.3		MIT
3700.352	3701.405	27016.77	RU	0.9	0.5		MIT
3697.861	3698.913	27034.97	RU	0.6	0.5		MIT
3697.764	3698.816	27035.68	RU I	0.9	0.8		MIT
3696.587	3697.639	27044.28	RU I	1.7	1.2		MIT

AIR WAVELENGTH (ANGSTRMS)	VACUUM WAVELENGTH (ANGSTRMS)	WAVENUMBER (KAYSERS)	LMENT	LOG ARC	INTENSITY SPK	DIS	REF
3693.591	3694.642	27066.22	RU	1.3	0.7		MIT
3692.370	3693.421	27075.17	RU	0.8			MIT
3690.029	3691.079	27092.35	RU	0.7	2.0		MIT
3688.810	3689.860	27101.30	RU	0.6			MIT
3686.601	3687.650	27117.54	RU	0.7	0.0		MIT
3685.951	3687.000	27122.32	RU I	1.0	0.7		MIT
3685.068	3686.117	27128.82	RU	0.9	0.5		MIT
3683.592	3684.641	27139.69	RU	0.5			MIT
3680.063	3681.111	27165.71	RU	0.6			MIT
3678.315	3679.362	27178.62	RU	0.8	1.2		MIT
3678.255	3679.302	27179.07	RU	0.7			MIT
3678.059	3679.106	27180.51	RU	0.6	0.3		MIT
3677.982	3679.029	27181.08	RU	0.7	0.0		MIT
3676.955	3678.002	27188.68	RU	0.9	0.5		MIT
3676.670	3677.717	27190.78	RU	0.9	0.6		MIT
3676.408	3677.455	27192.72	RU	0.7	0.3H		MIT
3675.262	3676.309	27201.20	RU I	0.8	0.5		MIT
3672.383	3673.429	27222.52	RU I	0.8	0.6		MIT
3672.066	3673.112	27224.87	RU	0.6	0.0		MIT
3671.219	3672.264	27231.15	RU I	1.0	0.6		MIT
3669.494	3670.539	27243.96	RU	1.7	1.8		MIT
3668.729	3669.774	27249.64	RU	0.6	0.3		MIT
3663.373	3664.416	27289.48	RU	0.7	1.8		MIT
3661.577	3662.620	27302.86	RU I	0.3	0.7		MIT
3661.353	3662.396	27304.53	RU I	1.8	2.0		MIT
3660.814	3661.857	27308.55	RU	0.3	0.9		MIT
3657.551	3658.593	27332.91	RU		1.7		MIT
3657.172	3658.214	27335.75	RU I	0.3	0.6		MIT
3654.405	3655.446	27356.44	RU	0.5	1.6		MIT
3652.316	3653.357	27372.09	RU	0.3	0.7		MIT
3650.322	3651.362	27387.04	RU I	0.5	1.1		MIT
3646.114	3647.153	27418.65	RU	0.3	0.9		MIT
3640.640	3641.678	27459.87	RU I	0.5	1.1		MIT
3638.016	3639.053	27479.68	RU	0.0	0.6		MIT
3637.466	3638.503	27483.83	RU	0.3	1.0		MIT
3635.516	3636.552	27498.57	RU I	0.3	0.9		MIT
3634.929	3635.965	27503.01	RU	1.7	2.0		MIT
3633.921	3634.957	27510.64	RU	0.3	0.7		MIT
3631.711	3632.746	27527.38	RU I	0.3	0.9		MIT
3627.289	3628.323	27560.94	RU	0.3	0.6		MIT
3626.744	3627.778	27565.08	RU	0.5	1.6		MIT
3625.196	3626.230	27576.85	RU I	0.6	1.5		MIT
3623.635	3624.668	27588.73	RU	0.3	1.0		MIT
3620.285	3621.317	27614.26	RU I	0.3	0.7		MIT
3619.202	3620.234	27622.52	RU I	0.3	1.0		MIT
3616.951	3617.982	27639.71	RU I	0.3	0.6		MIT
3609.107	3610.136	27699.78	RU I	0.5	0.6		MIT
3608.727	3609.756	27702.70	RU I	0.3	0.9		MIT
3606.153	3607.182	27722.47	RU	0.3	0.0		MIT
3605.641	3606.670	27726.41	RU	0.3	1.0		MIT
3601.485	3602.512	27758.41	RU	0.3	0.7		MIT
3599.764	3600.791	27771.68	RU	1.1	2.0		MIT
3599.400	3600.427	27774.48	RU I	0.3	0.5		MIT
3596.185	3597.211	27799.31	RU I	2.5	2.0		ME
3593.022	3594.047	27823.78	RU I	1.8	2.2		MIT
3590.886	3591.911	27840.34	RU	0.3	0.6		MIT
3589.215	3590.239	27853.30	RU I	1.8	2.0		MIT
3587.203	3588.227	27868.92	RU I	2.5	2.0		ME
3585.820	3586.843	27879.67	RU		1.0H		MIT
3584.198	3585.221	27892.28	RU I	0.3	0.8		MIT
3579.768	3580.790	27926.80	RU I	0.5	0.9		MIT
3574.577	3575.598	27967.35	RU	0.6	1.2		MIT
3572.015	3573.035	27987.41	RU	0.5	0.7		MIT
3570.594	3571.614	27998.55	RU	1.1	1.8		MIT
3567.155	3568.174	28025.54	RU I	0.5	1.0		MIT
3564.798	3565.816	28044.07	RU	0.3	0.7		MIT
3564.562	3565.580	28045.93	RU I	0.6	0.7		MIT
3564.353	3565.371	28047.57	RU I	0.5	0.8		MIT
3562.616	3563.633	28061.25	RU	0.3	0.5		MIT
3561.894	3562.911	28066.94	RU	0.3	0.7		MIT

AIR WAVELENGTH (ANGSTRMS)	VACUUM WAVELENGTH (ANGSTRMS)	WAVENUMBER (KAYSERS)	LMENT	LOG ARC	INTENSITY SPK	DIS	REF
3557.058	3558.074	28105.09	RU	0.5	0.7		MIT
3556.626	3557.642	28108.51	RU	0.5	0.7		MIT
3553.848	3554.863	28130.48	RU I	0.6	1.0		MIT
3553.668	3554.683	28131.90	RU	0.7	0.0		MIT
3550.630	3551.644	28155.97	RU	0.6	0.7		MIT
3550.269	3551.283	28158.83	RU I	0.6	0.9		MIT
3549.735	3550.749	28163.07	RU	0.5	0.7		MIT
3546.982	3547.995	28184.93	RU I	0.3	0.7		MIT
3541.631	3542.643	28227.51	RU	1.8	1.0		MIT
3541.147	3542.159	28231.37	RU	0.6			MIT
3541.046	3542.058	28232.18	RU	1.0			MIT
3540.219	3541.231	28238.77	PU	1.1	0.0		MIT
3539.369	3540.380	28245.55	RU I	1.8	1.2		MIT
3539.263	3540.274	28246.40	RU	1.5	0.7		MIT
3537.951	3538.962	28256.87	RU I	1.8	1.4		MIT
3536.567	3537.578	28267.93	RU	1.7			MIT
3535.831	3536.842	28273.81	RU	1.8	1.1		MIT
3535.372	3536.382	28277.49	RU	1.5	0.9		MIT
3535.307	3536.317	28278.00	RU	0.7			MIT
3535.052	3536.062	28280.04	RU	1.0	0.0		MIT
3534.577	3535.587	28283.84	RU	0.6			MIT
3533.913	3534.923	28289.16	RU	0.8			MIT
3532.814	3533.824	28297.96	RU	1.8	1.1		MIT
3532.055	3533.065	28304.04	RU	0.7			MIT
3531.794	3532.804	28306.13	RU	0.9	0.0		MIT
3531.390	3532.399	28309.37	RU	1.8	1.0		MIT
3529.611	3530.620	28323.64	RU	0.5			MIT
3529.276	3530.285	28326.33	RU	1.5	0.5		MIT
3528.683	3529.692	28331.09	RU I	1.8	1.1		MIT
3528.014	3529.023	28336.46	RU	1.1			MIT
3526.575	3527.583	28348.02	RU	1.1	0.5		MIT
3525.817	3526.825	28354.11	RU I	0.7			MIT
3525.644	3526.652	28355.51	RU	0.7	0.0		MIT
3524.902	3525.910	28361.48	RU I	1.1	0.3		MIT
3524.454	3525.462	28365.08	RU		0.8		MIT
3524.152	3525.160	28367.51	RU	0.8			MIT
3522.276	3523.283	28382.62	RU	1.0	0.9R		MIT
3521.968	3522.975	28385.10	RU I	0.9	0.7R		MIT
3520.130	3521.137	28399.92	RU	1.8	1.6		MIT
3519.635	3520.641	28403.92	RU	1.8	1.5		MIT
3518.983	3519.989	28409.18	RU	1.5	0.5		MIT
3517.407	3518.413	28421.91	RU	1.0			MIT
3516.186	3517.192	28431.78	RU	0.5			MIT
3515.894	3516.899	28434.14	RU	1.0	0.9		MIT
3515.678	3516.683	28435.88	RU I	1.0	0.9R		MIT
3514.766	3515.771	28443.26	RU I	1.1	0.6		MIT
3514.488	3515.493	28445.51	RU I	1.8	1.6		MIT
3514.141	3515.146	28448.32	RU	0.6			MIT
3512.882	3513.887	28458.52	RU	0.9			MIT
3511.560	3512.564	28469.23	RU I	0.8			MIT
3511.126	3512.130	28472.75	RU I	0.9			MIT
3510.311	3511.315	28479.36	RU	0.8			MIT
3509.717	3510.721	28484.18	RU I	1.7	0.3		MIT
3509.201	3510.205	28488.37	RU	1.0	2.0		MIT
3502.418	3503.420	28543.54	RU I	1.3	0.6		MIT
3501.354	3502.356	28552.21	RU	1.5	0.5		MIT
3498.942	3499.943	28571.89	RU I	2.7G	2.3		MIT
3497.937	3498.938	28580.10	RU I	1.5	0.7		MIT
3496.127	3497.127	28594.90	PU I	1.1			MIT
3495.973	3496.973	28596.16	RU	1.8	1.0		MIT
3494.251	3495.251	28610.25	RU I	1.7	0.9		MIT
3493.220	3494.220	28618.69	RU	1.3	0.0		MIT
3492.100	3493.099	28627.87	RU	0.8			MIT
3490.716	3491.715	28639.22	RU	1.1	0.3		MIT
3489.749	3490.748	28647.16	RU	1.1			MIT
3487.456	3488.454	28665.99	RU	0.7			MIT
3486.789	3487.787	28671.48	RU I	1.5			MIT
3486.206	3487.204	28676.27	RU I	1.3			MIT
3483.292	3484.289	28700.26	RU I	1.8	1.0		MIT
3483.159	3484.156	28701.35	RU	1.7	0.9		MIT

AIR WAVELENGTH (ANGSTRMS)	VACUUM WAVELENGTH (ANGSTRMS)	WAVENUMBER (KAYSERS)	LMENT	LOG ARC	INTENSITY SPK	DIS	REF
3482.341	3483.338	28708.10	RU	1.5	0.5		MIT
3481.297	3482.294	28716.71	RU I	1.8	1.5		MIT
3480.131	3481.127	28726.33	RU	1.0	0.7H		MIT
3474.843	3475.838	28770.04	RU	1.3	0.6		MIT
3473.746	3474.741	28779.13	RU I	1.8	1.5		MIT
3472.693	3473.687	28787.85	RU I	0.9			SV
3472.231	3473.225	28791.68	RU	1.8	1.0		MIT
3470.651	3471.645	28804.79	RU	0.5			MIT
3467.046	3468.039	28834.74	RU	1.7	0.6		MIT
3465.286	3466.278	28849.39	RU	0.7	0.3H		MIT
3463.144	3464.136	28867.23	RU I	1.8	0.6		MIT
3462.040	3463.032	28876.43	RU	0.7			MIT
3461.924	3462.916	28877.40	RU	1.5			MIT
3459.569	3460.560	28897.06	RU I	1.5			MIT
3456.620	3457.610	28921.71	RU I	1.8	0.9		MIT
3455.732	3456.722	28929.14	RU	1.1			MIT
3455.386	3456.376	28932.04	RU	1.3			MIT
3452.903	3453.892	28952.84	RU I	1.8	0.8		MIT
3448.953	3449.941	28986.00	RU	1.8	1.3		MIT
3446.488	3447.476	29006.73	RU	1.7			MIT
3446.070	3447.058	29010.25	RU	1.7	0.8		MIT
3445.939	3446.926	29011.35	RU	1.0			MIT
3445.301	3446.288	29016.73	RU I	1.1			MIT
3443.155	3444.142	29034.81	RU I	1.5			MIT
3440.205	3441.191	29059.71	RU I	2.0	1.5		MIT
3439.676	3440.662	29064.18	RU	1.5			MIT
3438.368	3439.354	29075.23	RU I	1.8	1.5		MIT
3436.737	3437.722	29089.03	RU I	2.5G	2.2		MIT
3436.332	3437.317	29092.46	RU I	1.1	0.9		MIT
3435.186	3436.171	29102.16	RU I	1.8	1.3		MIT
3433.260	3434.244	29118.49	RU I	1.8	1.4		MIT
3432.741	3433.725	29122.89	RU	1.8	1.6		MIT
3432.472	3433.456	29125.17	RU	0.6			MIT
3432.373	3433.357	29126.01	RU	0.7			MIT
3432.209	3433.193	29127.41	RU I	1.7	1.1		MIT
3430.772	3431.756	29139.60	RU I	1.8	1.7		MIT
3429.542	3430.525	29150.05	RU	1.8	1.4		MIT
3428.634	3429.617	29157.77	RU I	1.5	1.1		MIT
3428.309	3429.292	29160.54	RU	2.0	2.0		MIT
3426.456	3427.438	29176.31	RU	0.8			MIT
3425.964	3426.946	29180.50	RU I	1.5	0.6		MIT
3420.078	3421.059	29230.72	RU	1.8	0.9		MIT
3419.252	3420.233	29237.78	RU I	1.5	0.3		MIT
3417.353	3418.333	29254.02	RU I	0.0	1.8		SV
3416.182	3417.162	29264.05	RU	1.7	0.6		MIT
3414.642	3415.621	29277.25	RU I	1.7	0.7		MIT
3414.282	3415.261	29280.34	RU	1.1			MIT
3413.980	3414.959	29282.93	RU	0.5			MIT
3413.714	3414.693	29285.21	RU	0.5			MIT
3412.800	3413.779	29293.05	RU I	1.7	0.7		MIT
3412.078	3413.057	29299.25	RU	1.5			MIT
3411.636	3412.615	29303.04	RU	1.9	1.3		MIT
3410.735	3411.713	29310.78	RU	0.6	1.5		MIT
3409.568	3410.546	29320.82	RU I	1.3	0.3		MIT
3409.277	3410.255	29323.32	RU	2.0	1.6		MIT
3406.917	3407.894	29343.63	RU	0.6			MIT
3406.591	3407.568	29346.44	RU	1.7	0.5		MIT
3405.880	3406.857	29352.57	RU	1.7	0.3		MIT
3405.277	3406.254	29357.76	RU	0.5			MIT
3403.775	3404.752	29370.72	RU I	0.9			MIT
3402.522	3403.498	29381.53	RU	0.7			MIT
3401.739	3402.715	29388.30	RU I	2.0	1.7		MIT
3401.505	3402.481	29390.32	RU	1.5	0.3		MIT
3400.754	3401.730	29396.81	RU I	1.5	0.0		MIT
3400.604	3401.580	29398.10	RU	1.0			MIT
3399.974	3400.950	29403.55	RU I	0.6			MIT
3399.375	3400.351	29408.73	RU	1.8	0.5		MIT
3396.825	3397.800	29430.81	RU I	1.3			MIT
3392.537	3393.511	29468.01	RU	2.0	1.6		MIT
3391.890	3392.864	29473.63	RU	1.7	0.8		MIT

AIR WAVELENGTH (ANGSTRMS)	VACUUM WAVELENGTH (ANGSTRMS)	WAVENUMBER (KAYSERS)	LMENT	LOG ARC	INTENSITY SPK	DIS	REF
3390.910	3391.883	29482.15	RU I	1.1			MIT
3389.500	3390.473	29494.41	RU I	1.8	1.3		MIT
3388.709	3389.682	29501.29	RU I	1.9	1.3		MIT
3387.232	3388.204	29514.16	RU I	1.5	0.3		MIT
3386.251	3387.223	29522.71	RU I	1.5	0.3		MIT
3385.707	3386.679	29527.45	RU I	1.7	0.6		MIT
3385.474	3386.446	29529.48	RU I	0.9	0.3		MIT
3385.144	3386.116	29532.36	RU	1.8	1.5		MIT
3380.915	3381.886	29569.30	RU	1.0	0.3		MIT
3380.175	3381.146	29575.77	RU I	1.8	1.2		MIT
3379.605	3380.575	29580.76	RU I	1.8	1.3		MIT
3379.264	3380.234	29583.75	RU I	1.5	0.3		MIT
3379.029	3379.999	29585.80	RU	1.3			MIT
3378.020	3378.990	29594.64	RU	1.8	1.1		MIT
3376.053	3377.023	29611.88	RU I	0.7			MIT
3375.245	3376.214	29618.97	RU	1.1			MIT
3374.903	3375.872	29621.97	RU I	1.5	0.0		MIT
3374.646	3375.615	29624.23	RU I	1.9	1.3		MIT
3373.978	3374.947	29630.09	RU I	1.8	0.6		MIT
3371.860	3372.829	29648.71	RU I	1.8	1.3		MIT
3370.145	3371.113	29663.79	RU	0.7			MIT
3370.054	3371.022	29664.59	RU	0.5	1.0		MIT
3369.669	3370.637	29667.98	RU	1.5	0.3		MIT
3369.282	3370.250	29671.39	RU	1.1	1.8		MIT
3368.451	3369.419	29678.71	RU I	2.0	1.8		MIT
3364.825	3365.792	29710.69	RU	0.5			MIT
3364.099	3365.066	29717.10	RU	1.5	0.7		MIT
3362.335	3363.301	29732.69	RU	1.7	0.7H		MIT
3362.003	3362.969	29735.63	RU I	1.8	0.9		MIT
3361.148	3362.114	29743.19	RU	1.5			MIT
3360.792	3361.758	29746.34	RU	0.6			MIT
3359.095	3360.060	29761.37	RU I	1.8	1.3		MIT
3356.461	3357.426	29784.73	RU I	1.0			MIT
3356.199	3357.164	29787.05	RU	1.5	0.5		MIT
3353.871	3354.835	29807.73	RU	0.6			MIT
3353.648	3354.612	29809.71	RU	1.7	0.6		MIT
3353.308	3354.272	29812.73	RU I	1.5	1.2		MIT
3352.976	3353.940	29815.68	RU I	0.8			MIT
3351.932	3352.895	29824.97	RU I	1.7	0.6		MIT
3350.549	3351.512	29837.28	RU	1.1			MIT
3350.251	3351.214	29839.93	RU	1.0	0.3		MIT
3350.107	3351.070	29841.22	RU	1.1			MIT
3348.705	3349.668	29853.71	RU I	1.7	0.3		MIT
3348.012	3348.974	29859.89	RU	1.7	0.5		MIT
3347.613	3348.575	29863.45	RU I	1.8	0.8		MIT
3345.317	3346.279	29883.94	RU	1.8	0.7		MIT
3344.797	3345.759	29888.59	RU	0.7			MIT
3344.532	3345.494	29890.96	RU I	1.8	0.8		MIT
3343.357	3344.318	29901.46	RU	0.6			MIT
3342.879	3343.840	29905.74	RU	0.5			MIT
3342.707	3343.668	29907.27	RU	0.5			MIT
3341.664	3342.625	29916.61	RU	1.8	1.7		MIT
3341.230	3342.191	29920.50	RU	1.0			MIT
3341.087	3342.048	29921.77	RU	1.7	0.7		MIT
3339.781	3340.741	29933.48	RU	1.0	1.8		MIT
3339.552	3340.512	29935.53	RU	2.0	1.8		MIT
3338.707	3339.667	29943.10	RU	1.5	0.5		MIT
3337.823	3338.783	29951.03	RU I	1.8	0.9		MIT
3336.639	3337.599	29961.66	RU	1.7	0.6		MIT
3335.686	3336.645	29970.22	RU I	1.8	1.1		MIT
3332.643	3333.602	29997.59	RU	1.8	0.7		MIT
3332.052	3333.010	30002.91	RU	1.8	1.0		MIT
3328.456	3329.413	30035.32	RU I	1.1			MIT
3327.708	3328.665	30042.07	RU	1.7	0.8		MIT
3325.394	3326.351	30062.98	RU	0.5	1.6		MIT
3325.233	3326.190	30064.43	RU	0.7			MIT
3324.995	3325.952	30066.58	RU I	1.8	1.1		MIT
3323.940	3324.896	30076.13	RU	0.7			MIT
3322.226	3323.182	30091.64	RU	0.9			MIT
3321.254	3322.210	30100.45	RU	1.1			MIT

AIR WAVELENGTH (ANGSTRMS)	VACUUM WAVELENGTH (ANGSTRMS)	WAVENUMBER (KAYSERS)	LMENT	LOG ARC	INTENSITY SPK	INTENSITY DIS	REF
3319.808	3320.763	30113.56	RU I	0.8			MIT
3319.524	3320.479	30116.14	RU I	1.0			MIT
3318.908	3319.863	30121.73	RU	1.1	0.5		MIT
3318.822	3319.777	30122.50	RU	1.7	0.9		MIT
3317.965	3318.920	30130.28	RU	1.5			MIT
3317.888	3318.843	30130.99	RU I	1.7	1.1		MIT
3317.597	3318.552	30133.63	RU	0.5	1.2		MIT
3316.902	3317.857	30139.94	RU	0.7	1.7		MIT
3316.386	3317.340	30144.63	RU I	1.9			MIT
3315.442	3316.396	30153.21	RU	1.5	0.7		MIT
3315.228	3316.182	30155.16	RU I	1.8	1.4		MIT
3315.047	3316.001	30156.81	RU I	1.7	1.1		MIT
3314.774	3315.728	30159.29	RU	1.3	0.0		MIT
3314.058	3315.012	30165.80	RU	0.9			MIT
3312.224	3313.177	30182.51	RU	0.6			MIT
3310.957	3311.910	30194.06	RU I	1.5	0.7		MIT
3310.086	3311.039	30202.00	RU	0.6			MIT
3309.825	3310.778	30204.38	RU	0.6			MIT
3308.626	3309.578	30215.33	RU I	1.0			MIT
3307.984	3308.936	30221.19	RU I	1.7			MIT
3307.544	3308.496	30225.21	RU	0.9			MIT
3306.172	3307.124	30237.75	RU I	1.8	1.1		MIT
3304.825	3305.776	30250.08	RU I	1.7	0.5		MIT
3304.639	3305.590	30251.78	RU I	0.8			MIT
3304.507	3305.458	30252.99	RU I	1.1	0.3		MIT
3303.995	3304.946	30257.68	RU	1.8	0.9		MIT
3301.911	3302.862	30276.77	RU	1.5	0.9		MIT
3301.587	3302.538	30279.75	RU I	1.8	1.6		MIT
3299.791	3300.741	30296.23	RU I	0.5			MIT
3299.334	3300.284	30300.42	RU I	1.7	0.6		MIT
3298.411	3299.361	30308.90	RU I	1.7	1.4G		MIT
3297.955	3298.905	30313.09	RU I	1.7	0.8		MIT
3297.258	3298.208	30319.50	RU	1.7	0.6		MIT
3296.649	3297.598	30325.10	RU I	1.7	0.7		MIT
3296.111	3297.060	30330.05	RU I	1.7	1.0		MIT
3294.110	3295.059	30348.47	RU I	1.8	2.3		MIT
3292.265	3293.213	30365.48	RU	1.1			MIT
3291.659	3292.607	30371.07	RU	1.1			MIT
3291.117	3292.065	30376.07	RU I	1.3			MIT
3289.245	3290.192	30393.36	RU	0.7			MIT
3285.910	3286.857	30424.21	RU	0.6			MIT
3284.932	3285.878	30433.26	RU I	1.5	0.7		MIT
3277.567	3278.512	30501.65	RU	1.5	1.1		MIT
3274.706	3275.650	30528.29	RU I	1.8	1.4		MIT
3273.621	3274.565	30538.41	RU	0.3	0.7		MIT
3273.078	3274.021	30543.48	RU I	1.8	1.3		MIT
3270.242	3271.185	30569.97	RU	0.5			MIT
3268.943	3269.885	30582.11	RU	0.6			MIT
3268.794	3269.736	30583.51	RU	0.6	1.8		MIT
3268.208	3269.150	30588.99	RU I	1.8	1.1		MIT
3266.445	3267.387	30605.50	RU I	1.7	1.0		MIT
3264.663	3265.604	30622.20	RU I	1.5	0.8		MIT
3264.550	3265.491	30623.26	RU I	1.5	0.7		MIT
3263.853	3264.794	30629.80	RU	1.5	0.5		MIT
3261.129	3262.069	30655.39	RU	1.5	0.5		MIT
3260.353	3261.293	30662.68	RU I	2.0	1.7		MIT
3260.167	3261.107	30664.43	RU	1.1			MIT
3259.667	3260.607	30669.14	RU I	1.8	1.0		MIT
3258.967	3259.907	30675.72	RU	1.0	1.8		MIT
3258.040	3258.980	30684.45	RU I	1.7	0.9		MIT
3256.331	3257.270	30700.55	RU	1.7	0.5		MIT
3254.708	3255.647	30715.86	RU	1.7	1.0		MIT
3254.542	3255.481	30717.43	RU I	1.7	1.0		MIT
3252.997	3253.935	30732.02	RU	1.1			MIT
3252.905	3253.843	30732.89	RU I	1.3			MIT
3252.539	3253.477	30736.35	RU I	1.1			MIT
3251.890	3252.828	30742.48	RU	1.5	0.5		MIT
3251.329	3252.267	30747.78	RU	1.5	0.5		MIT
3250.009	3250.947	30760.27	RU	1.5	0.5		MIT
3249.927	3250.864	30761.05	RU	1.5	0.5		MIT

AIR WAVELENGTH (ANGSTRMS)	VACUUM WAVELENGTH (ANGSTRMS)	WAVENUMBER (KAYSERS)	LMENT	LOG ARC	INTENSITY SPK	INTENSITY DIS	REF
3248.843	3249.780	30771.31	RU	0.7			MIT
3246.210	3247.147	30796.27	RU	0.0	1.4		MIT
3245.618	3246.554	30801.89	RU	0.5	0.6		MIT
3244.578	3245.514	30811.76	RU	0.7			MIT
3244.456	3245.392	30812.92	RU	0.8			MIT
3244.341	3245.277	30814.01	RU	0.7			MIT
3243.498	3244.434	30822.02	RU	1.8	1.1		MIT
3242.848	3243.784	30828.20	RU I	1.3			MIT
3242.165	3243.101	30834.69	RU	1.9			MIT
3241.235	3242.170	30843.54	RU I	1.8	1.1		MIT
3239.605	3240.540	30859.06	RU I	1.7	0.7		MIT
3238.775	3239.710	30866.96	RU I	1.7	0.0		MIT
3238.527	3239.462	30869.33	RU	2.0	1.7		MIT
3235.974	3236.908	30893.68	RU	1.1			MIT
3235.101	3236.035	30902.02	RU	0.8			MIT
3234.798	3235.732	30904.91	RU I	1.0	0.0H		MIT
3234.430	3235.364	30908.43	RU	0.5	1.7		MIT
3232.751	3233.684	30924.48	RU I	1.7	0.6		MIT
3228.530	3229.462	30964.91	RU I	1.7	2.2		MIT
3228.157	3229.089	30968.49	RU	1.1	0.6		MIT
3227.885	3228.817	30971.10	RU I	1.3	1.0		MIT
3226.899	3227.831	30980.56	RU	0.6	0.3		MIT
3226.374	3227.306	30985.60	RU I	1.7	1.1		MIT
3224.651	3225.582	31002.16	RU	0.6			MIT
3223.274	3224.205	31015.40	RU I	1.8	1.5		MIT
3221.190	3222.120	31035.47	RU	0.6	0.5		MIT
3220.069	3220.999	31046.27	RU	0.6	0.5		MIT
3216.523	3217.452	31080.50	RU I	1.1	0.8		MIT
3212.969	3213.897	31114.87	RU I	1.0	0.8		MIT
3205.312	3206.238	31189.20	RU	0.6	0.0		MIT
3204.060	3204.986	31201.39	RU I	0.9	0.7		MIT
3201.499	3202.424	31226.34	RU I	0.6	0.7		MIT
3201.258	3202.183	31228.70	RU	0.3	2.0		MIT
3198.327	3199.251	31257.31	RU I	0.5			MIT
3196.591	3197.515	31274.29	RU	1.7	0.3		MIT
3195.151	3196.075	31288.38	RU		2.0		MIT
3194.738	3195.662	31292.43	RU	0.6	0.0		MIT
3192.069	3192.992	31318.59	RU I	1.0	0.7		MIT
3191.776	3192.699	31321.46	RU	0.5	0.0		MIT
3189.976	3190.898	31339.14	RU I	1.7	1.7		MIT
3189.718	3190.640	31341.67	RU	0.7	0.7		MIT
3189.296	3190.218	31345.82	RU	0.5			MIT
3188.907	3189.829	31349.64	RU	0.6	0.5		MIT
3188.609	3189.531	31352.57	RU I	0.5	0.7		MIT
3188.338	3189.260	31355.24	RU I	1.8	1.7		MIT
3186.044	3186.965	31377.81	RU I	1.9	1.4		MIT
3185.442	3186.363	31383.74	RU I	1.1			MIT
3184.897	3185.818	31389.11	RU I	0.6			MIT
3181.187	3182.107	31425.72	RU I	0.6			MIT
3180.444	3181.364	31433.06	RU	0.5			MIT
3179.264	3180.184	31444.73	RU	1.7	1.7P		MIT
3178.733	3179.652	31449.98	RU I	1.5			MIT
3177.049	3177.968	31466.65	RU	1.8	2.3		MIT
3176.292	3177.211	31474.15	RU	1.7	0.5		MIT
3175.298	3176.217	31484.00	RU	1.5	1.6		MIT
3175.147	3176.066	31485.50	RU	1.3	2.0		MIT
3174.131	3175.049	31495.57	RU	1.7	0.5		MIT
3173.399	3174.317	31502.84	RU	1.5			MIT
3173.110	3174.028	31505.71	RU I	1.5			MIT
3172.672	3173.590	31510.06	RU	1.3	1.2		MIT
3171.239	3172.157	31524.29	RU	1.3			MIT
3170.093	3171.010	31535.69	RU	1.5			MIT
3168.525	3169.442	31551.30	RU	2.0	1.4R		MIT
3168.239	3169.156	31554.14	RU I	1.6			MIT
3167.467	3168.384	31561.83	RU	0.7	2.0		MIT
3167.378	3168.295	31562.72	RU I	0.5			MIT
3166.539	3167.455	31571.08	RU	0.5	1.5		MIT
3166.104	3167.020	31575.42	RU	0.9	1.6		MIT
3164.948	3165.864	31586.95	RU	0.5	1.8		MIT
3163.828	3164.744	31598.14	RU	1.0			MIT

AIR WAVELENGTH (ANGSTRMS)	VACUUM WAVELENGTH (ANGSTRMS)	WAVENUMBER (KAYSERS)	LMENT	LOG ARC	INTENSITY SPK	INTENSITY DIS	REF
3163.184	3164.100	31604.57	RU	1.1	2.0		MIT
3159.923	3160.838	31637.18	RU I	1.8	1.4		MIT
3158.889	3159.803	31647.54	RU I	1.8	1.1		MIT
3157.627	3158.541	31660.19	RU	1.5			MIT
3157.419	3158.333	31662.27	RU	1.5	0.0		MIT
3156.817	3157.731	31668.31	RU I	1.7			MIT
3154.434	3155.347	31692.23	RU	1.5			MIT
3153.823	3154.736	31698.37	RU	1.8	1.1		MIT
3150.693	3151.605	31729.86	RU	1.8	1.8		MIT
3150.164	3151.076	31735.19	RU	0.5			MIT
3148.485	3149.397	31752.11	RU	1.5			MIT
3147.451	3148.363	31762.54	RU	0.8	1.9		MIT
3147.208	3148.120	31765.00	RU I	1.7	0.5		MIT
3146.067	3146.978	31776.51	RU I	1.5	1.3		MIT
3144.720	3145.631	31790.12	RU	0.6			MIT
3144.260	3145.171	31794.78	RU	1.8	0.9		MIT
3143.655	3144.566	31800.90	RU	1.3	1.9		MIT
3143.239	3144.150	31805.10	RU	0.7	0.6		MIT
3140.973	3141.883	31828.05	RU I	1.8	0.8		MIT
3140.484	3141.394	31833.00	RU	1.7			MIT
3140.084	3140.994	31837.06	RU	0.9			MIT
3139.272	3140.182	31845.29	RU	1.3			MIT
3138.760	3139.669	31850.49	RU	1.0			MIT
3136.909	3137.818	31869.28	RU	0.8			MIT
3136.555	3137.464	31872.88	RU	1.8	0.8		MIT
3136.342	3137.251	31875.04	RU	1.1			MIT
3135.924	3136.833	31879.29	RU	1.1			MIT
3135.802	3136.711	31880.53	RU	1.0	1.9		MIT
3134.802	3135.710	31890.70	RU I	1.0	2.0		MIT
3133.697	3134.605	31901.95	RU	1.1			MIT
3132.878	3133.786	31910.28	RU	1.8	0.7		MIT
3129.837	3130.744	31941.29	RU I	1.8	0.6		MIT
3129.604	3130.511	31943.67	RU	1.7	0.0		MIT
3128.431	3129.338	31955.64	RU	1.1			MIT
3127.913	3128.820	31960.93	RU	1.0	2.0		MIT
3126.608	3127.514	31974.27	RU	1.1	1.7		MIT
3125.963	3126.869	31980.87	RU I	1.8	1.1		MIT
3124.607	3125.513	31994.75	RU	1.7	0.3		MIT
3124.366	3125.272	31997.22	RU I	1.5	0.3		MIT
3124.167	3125.073	31999.25	RU I	1.8	0.9		MIT
3122.004	3122.909	32021.42	RU	0.8			MIT
3120.543	3121.448	32036.42	RU	1.0			MIT
3119.491	3120.396	32047.22	RU	0.6			MIT
3118.685	3119.589	32055.50	RU	1.7	0.5		MIT
3118.071	3118.975	32061.81	RU I	1.7	1.7		MIT
3116.839	3117.743	32074.49	RU	1.5			MIT
3116.658	3117.562	32076.35	RU	1.5			MIT
3115.452	3116.356	32088.76	RU	0.5			MIT
3113.397	3114.300	32109.94	RU I	1.7			MIT
3112.677	3113.580	32117.37	RU	1.7	0.5		MIT
3112.304	3113.207	32121.22	RU	1.5			MIT
3111.912	3112.815	32125.27	RU I	1.7	0.7		MIT
3110.549	3111.451	32139.34	RU I	1.8	0.8		MIT
3109.403	3110.305	32151.19	RU	1.3			MIT
3108.426	3109.328	32161.29	RU I	1.5			MIT
3107.715	3108.617	32168.65	RU I	1.8	0.7		MIT
3107.579	3108.481	32170.06	RU	1.0	1.7		MIT
3106.840	3107.741	32177.71	RU I	1.7	0.5		MIT
3105.411	3106.312	32192.52	RU I	1.7	0.0		MIT
3105.283	3106.184	32193.84	RU	1.7	1.6		MIT
3104.621	3105.522	32200.71	RU	0.7			MIT
3104.464	3105.365	32202.34	RU	1.5			MIT
3103.413	3104.313	32213.24	RU	0.5	1.7		MIT
3102.396	3103.296	32223.80	RU	1.3			MIT
3100.839	3101.739	32239.98	RU I	1.8	1.7		MIT
3099.283	3100.182	32256.17	RU I	1.8	1.8		MIT
3098.836	3099.735	32260.82	RU I	0.5			MIT
3097.879	3098.778	32270.78	RU	0.8	1.2		MIT
3097.605	3098.504	32273.64	RU	0.3	1.3		MIT
3097.231	3098.130	32277.54	RU	1.5			MIT
3096.568	3097.467	32284.45	RU	1.8	1.8		MIT
3094.558	3095.456	32305.42	RU	0.8	1.7		MIT
3094.393	3095.291	32307.14	RU	1.7	0.5		MIT
3093.901	3094.799	32312.27	RU	1.5	2.0		MIT
3093.447	3094.345	32317.02	RU	0.6			MIT
3091.873	3092.771	32333.47	RU	1.7	0.7		MIT
3090.897	3091.794	32343.68	RU	1.5	0.0		MIT
3090.229	3091.126	32350.67	RU	1.7	0.8		MIT
3089.801	3090.698	32355.15	RU I	1.8	0.7		MIT
3089.145	3090.042	32362.02	RU I	1.8	1.1		MIT
3088.231	3089.128	32371.60	RU	0.6			MIT
3088.075	3088.972	32373.23	RU I	1.5	0.3J		MIT
3086.930	3087.826	32385.24	RU	1.5	0.0		MIT
3086.784	3087.680	32386.77	RU I	0.9			MIT
3086.520	3087.416	32389.54	RU	1.5	1.6		MIT
3086.067	3086.963	32394.30	RU	1.8	0.8		MIT
3085.469	3086.365	32400.57	RU	0.6			MIT
3084.526	3085.422	32410.48	RU	1.5	0.3		MIT
3083.148	3084.043	32424.97	RU I	1.7	0.5		MIT
3081.384	3082.279	32443.53	RU	0.6	1.7		MIT
3080.900	3081.795	32448.62	RU I	1.7	0.8		MIT
3080.193	3081.088	32456.07	RU	1.5			MIT
3077.552	3078.446	32483.92	RU	1.5	0.0		MIT
3077.063	3077.957	32489.08	RU	1.5	1.6		MIT
3076.777	3077.671	32492.10	RU I	1.7	0.5		MIT
3075.309	3076.202	32507.61	RU	1.0	1.6		MIT
3073.515	3074.408	32526.59	RU	1.0	1.9		MIT
3073.336	3074.229	32528.48	RU I	1.7	0.7		MIT
3072.335	3073.228	32539.08	RU	0.7	1.6		MIT
3071.499	3072.391	32547.94	PU	1.3			MIT
3069.184	3070.076	32572.48	RU I	1.5	0.0		MIT
3068.261	3069.153	32582.28	RU	1.8	0.9		MIT
3066.396	3067.287	32602.10	RU	1.0	1.4		MIT
3064.838	3065.729	32618.67	RU	1.8	1.8		MIT
3062.048	3062.938	32648.39	RU	0.9	0.8H		MIT
3060.492	3061.382	32664.99	RU	0.9	1.7		MIT
3060.230	3061.120	32667.78	RU	1.3	1.7		MIT
3059.169	3060.058	32679.11	RU	1.7			MIT
3058.786	3059.675	32683.21	RU I	1.5	0.5		MIT
3058.655	3059.544	32684.61	RU I	1.5	0.5		MIT
3057.342	3058.231	32698.64	RU I	1.5	0.3		MIT
3056.865	3057.754	32703.75	RU	1.1	2.2		MIT
3056.058	3056.947	32712.38	RU	1.5	0.0		MIT
3054.937	3055.825	32724.38	RU I	1.8	1.1		MIT
3052.342	3053.230	32752.20	RU I	0.9			MIT
3051.597	3052.485	32760.20	RU I	1.5			MIT
3050.193	3051.080	32775.28	RU	0.7			MIT
3049.246	3050.133	32785.46	RU II	0.3	1.7		ME
3049.075	3049.962	32787.29	RU I	0.8			MIT
3048.785	3049.672	32790.41	RU I	1.8	1.0		MIT
3048.495	3049.382	32793.53	RU	1.7	0.7		MIT
3047.709	3048.596	32801.99	RU	0.7	1.1		MIT
3046.242	3047.128	32817.79	RU	1.3			MIT
3046.127	3047.013	32819.02	RU	0.6	0.8		MIT
3045.710	3046.596	32823.52	RU I	1.8	1.1		MIT
3042.831	3043.716	32854.57	RU I	1.8	0.7		MIT
3042.475	3043.360	32858.42	RU	1.8	1.1		MIT
3041.919	3042.804	32864.42	RU I	1.5	0.0		MIT
3040.310	3041.195	32881.81	RU I	1.8	1.0		MIT
3039.961	3040.846	32885.59	RU I	1.5	0.3		MIT
3039.684	3040.569	32888.59	RU I	1.1			MIT
3038.784	3039.668	32898.33	RU	0.5	0.8		MIT
3038.176	3039.060	32904.91	RU I	1.9	0.7		MIT
3037.964	3038.848	32907.21	RU I	1.7	0.7		MIT
3037.733	3038.617	32909.71	RU	1.5	0.5		MIT
3036.466	3037.350	32923.44	RU	1.7	2.2		MIT
3035.771	3036.655	32930.98	RU	0.6	1.3		MIT
3035.473	3036.356	32934.21	RU I	1.8	0.6		MIT
3034.060	3034.943	32949.55	RU I	1.8	0.7		MIT
3033.894	3034.777	32951.35	RU		0.8		MIT

AIR WAVELENGTH (ANGSTRMS)	VACUUM WAVELENGTH (ANGSTRMS)	WAVENUMBER (KAYSERS)	LMENT	LOG ARC	INTENSITY SPK	INTENSITY DIS	REF
3033.451	3034.334	32956.16	RU I	1.8	1.0		MIT
3032.672	3033.555	32964.63	RU	0.5			MIT
3031.914	3032.797	32972.87	RU	1.5			MIT
3030.781	3031.663	32985.19	RU I	1.5	0.3		MIT
3027.790	3028.672	33017.78	RU	1.1	1.7		MIT
3027.084	3027.965	33025.48	RU	1.3			MIT
3025.099	3025.980	33047.15	RU	0.9			MIT
3022.947	3023.827	33070.67	RU		1.1		MIT
3020.882	3021.762	33093.28	RU I	1.8	1.6		MIT
3019.764	3020.644	33105.53	RU	0.9	1.1		MIT
3019.371	3020.250	33109.84	RU I	1.3			MIT
3017.810	3018.689	33126.96	RU	1.0	1.8		MIT
3017.236	3018.115	33133.26	RU	2.0	1.7		MIT
3016.697	3017.576	33139.18	RU I	0.6			MIT
3015.406	3016.284	33153.37	RU	0.9	1.7		MIT
3013.359	3014.237	33175.89	RU I	1.8	0.7		MIT
3012.916	3013.794	33180.77	RU I	1.8	0.6		MIT
3010.515	3011.392	33207.23	RU	1.0	1.7		MIT
3009.694	3010.571	33216.29	RU I	1.3			MIT
3008.796	3009.673	33226.20	RU	1.7	0.7		MIT
3008.259	3009.136	33232.13	RU	1.7	0.5		MIT
3006.590	3007.466	33250.58	RU	1.8	1.2		MIT
3005.973	3006.849	33257.40	RU	1.1			MIT
3005.126	3006.002	33266.78	RU		1.8		MIT
3004.600	3005.476	33272.60	RU	1.7	0.3		MIT
3004.217	3005.093	33276.84	RU	1.5			MIT
3003.485	3004.360	33284.95	RU I	1.5			MIT
3002.067	3002.942	33300.67	RU I	0.8			MIT
3001.642	3002.517	33305.39	RU I	1.8	0.7		MIT
3000.227	3001.102	33321.10	RU I	1.5	0.0		MIT
2999.808	3000.683	33325.75	RU	0.9	1.7		MIT
2998.890	2999.764	33335.95	RU	1.7	2.0		MIT
2998.348	2999.222	33341.98	RU	1.9	0.9		MIT
2997.615	2998.489	33350.13	RU	1.5	0.7		MIT
2997.426	2998.300	33352.23	RU I	1.5	0.7		MIT
2996.895	2997.769	33358.14	RU	1.8	0.7		MIT
2995.991	2996.865	33368.21	RU	0.6	1.7		MIT
2995.916	2996.790	33369.04	RU	0.9			MIT
2994.964	2995.837	33379.65	RU I	1.9	1.6		MIT
2993.272	2994.145	33398.52	RU I	1.8	1.0		MIT
2992.957	2993.830	33402.03	RU	1.1	0.0		MIT
2992.577	2993.450	33406.27	RU	0.6	1.3		MIT
2992.089	2992.962	33411.72	RU	0.3	1.6		MIT
2991.621	2992.494	33416.95	RU	1.7	2.0		MIT
2990.291	2991.163	33431.81	RU I	1.5	0.0		MIT
2989.655	2990.527	33438.92	RU I	1.5	0.3		MIT
2989.330	2990.202	33442.56	RU	1.1	0.0		MIT
2988.948	2989.820	33446.83	RU I	2.4	2.0		MIT
2988.094	2988.966	33456.39	RU	0.9			MIT
2987.705	2988.577	33460.75	RU	1.5	0.6		MIT
2986.335	2987.206	33476.10	RU	1.3	1.4 J		MIT
2985.823	2986.694	33481.83	RU	1.3			MIT
2981.935	2982.805	33525.49	RU	1.8	0.5		MIT
2980.958	2981.828	33536.48	RU	1.0	1.7		MIT
2979.959	2980.829	33547.72	RU	1.8	1.9		MIT
2979.72	2980.59	33550.4	RU	1.5	1.6		MIT
2978.641	2979.510	33562.56	RU	1.7	2.2		MIT
2978.380	2979.249	33565.50	RU	0.8			MIT
2977.482	2978.351	33575.63	RU	1.5	1.8		MIT
2977.227	2978.096	33578.50	RU	1.7	1.9		MIT
2976.925	2977.794	33581.91	RU	1.8	0.7		MIT
2976.586	2977.455	33585.73	RU	1.8	2.3		MIT
2974.859	2975.727	33605.23	RU		0.7		MIT
2974.335	2975.203	33611.15	RU I	1.5	0.3		MIT
2973.991	2974.859	33615.04	RU	1.7	0.7		MIT
2973.826	2974.694	33616.90	RU		1.4		MIT
2972.465	2973.333	33632.29	RU	1.1	1.5		MIT
2971.771	2972.639	33640.15	RU	1.0	1.2		MIT
2970.656	2971.523	33652.77	RU		1.3		MIT
2968.954	2969.821	33672.06	RU I	1.8	0.8		MIT

AIR WAVELENGTH (ANGSTRMS)	VACUUM WAVELENGTH (ANGSTRMS)	WAVENUMBER (KAYSERS)	LMENT	LOG ARC	INTENSITY SPK	INTENSITY DIS	REF
2968.482	2969.349	33677.42	RU	1.2			MIT
2967.994	2968.861	33682.96	RU		1.6		MIT
2967.341	2968.208	33690.37	RU	1.5	0.0		MIT
2966.559	2967.425	33699.25	RU	1.1			MIT
2966.396	2967.262	33701.10	RU	0.5	1.4		MIT
2965.711	2966.577	33708.88	RU	1.3			MIT
2965.546	2966.412	33710.76	RU	1.8	2.3		MIT
2965.162	2966.028	33715.12	RU I	1.9	1.3		MIT
2964.309	2965.175	33724.83	RU	0.9			MIT
2963.715	2964.581	33731.58	RU I	1.8	0.7		MIT
2963.402	2964.268	33735.15	RU	1.8	2.2		MIT
2962.091	2962.956	33750.08	RU	0.3	1.3 J		MIT
2961.689	2962.554	33754.66	RU	1.8	0.5		MIT
2961.529	2962.394	33756.48	RU	0.7	1.7		MIT
2960.975	2961.840	33762.80	RU	1.5	0.0 H		MIT
2960.228	2961.093	33771.32	RU	0.7			MIT
2959.736	2960.601	33776.93	RU I	1.1	0.3 H		MIT
2958.000	2958.864	33796.75	RU I	1.8	0.7		MIT
2956.152	2957.016	33817.88	RU	0.7			MIT
2955.842	2956.706	33821.43	RU	1.2			MIT
2955.604	2956.468	33824.15	RU	1.1			MIT
2955.359	2956.223	33826.95	RU	1.7	0.0		MIT
2954.486	2955.349	33836.95	RU I	2.0	1.3		MIT
2954.098	2954.961	33841.39	RU	0.8	1.3		MIT
2953.015	2953.878	33853.80	RU	0.9			MIT
2952.692	2953.555	33857.50	RU		1.0		MIT
2952.498	2953.361	33859.73	RU I	1.8	0.3		MIT
2952.248	2953.111	33862.60	RU	0.6	1.4		MIT
2950.536	2951.398	33882.25	RU	1.5	0.3		MIT
2949.965	2950.827	33888.80	RU	0.6	1.3		MIT
2949.500	2950.362	33894.15	RU I	1.9	1.1		MIT
2946.991	2947.852	33923.00	RU I	1.8	1.1		MIT
2945.668	2946.529	33938.24	RU	1.8	2.5		MIT
2945.101	2945.962	33944.77	RU	0.8	1.7		MIT
2944.184	2945.045	33955.34	RU	1.1			MIT
2943.921	2944.782	33958.37	RU	1.7	0.7		MIT
2943.481	2944.342	33963.45	RU I	1.5			MIT
2942.250	2943.110	33977.66	RU	1.5	2.0		MIT
2940.358	2941.218	33999.52	RU I	1.7	0.5		MIT
2939.944	2940.804	34004.31	RU I	1.5	0.5		MIT
2939.692	2940.552	34007.23	RU	1.1	0.7		MIT
2939.131	2939.991	34013.71	RU I	1.1			MIT
2937.551	2938.410	34032.01	RU	1.3			MIT
2937.342	2938.201	34034.43	RU	1.3			MIT
2936.258	2937.117	34047.00	RU	1.0			MIT
2936.016	2936.875	34049.80	RU	1.3			MIT
2935.518	2936.377	34055.58	RU	1.0	1.9		MIT
2934.185	2935.043	34071.05	RU	1.5			MIT
2933.245	2934.103	34081.97	RU	1.3	2.2		MIT
2929.930	2930.787	34120.52	RU	1.1			MIT
2929.442	2930.299	34126.21	RU	1.3			MIT
2928.492	2929.349	34137.28	RU	1.5	1.0		MIT
2927.536	2928.393	34148.43	RU	1.7	2.3		MIT
2927.119	2927.976	34153.29	RU	1.5	0.3		MIT
2925.076	2925.932	34177.14	RU	1.0			MIT
2924.650	2925.506	34182.12	RU	1.0			MIT
2923.88	2924.74	34191.1	RU		1.5		MIT
2923.114	2923.970	34200.08	RU		1.0		MIT
2922.331	2923.186	34209.25	RU		1.6		MIT
2921.156	2922.011	34223.00	RU	0.9			MIT
2920.956	2921.811	34225.35	RU	1.5	1.5		MIT
2920.257	2921.112	34233.54	RU	1.5			MIT
2919.608	2920.463	34241.15	RU	1.9	1.1		MIT
2918.498	2919.352	34254.17	RU		1.8		MIT
2917.774	2918.628	34262.67	RU	1.8	0.3		MIT
2917.137	2917.991	34270.15	RU	1.3	1.2		MIT
2916.354	2917.208	34279.35	RU		1.5		MIT
2916.255	2917.109	34280.52	RU I	2.0	1.4		MIT
2915.625	2916.479	34287.92	RU	1.3			MIT
2914.299	2915.152	34303.52	RU	1.7			MIT

AIR WAVELENGTH (ANGSTRMS)	VACUUM WAVELENGTH (ANGSTRMS)	WAVENUMBER (KAYSERS)	LMENT	LOG ARC	INTENSITY SPK	DIS	REF
2913.968	2914.821	34307.42	RU		1.7		MIT
2913.170	2914.023	34316.82	RU	1.7	0.5		MIT
2912.749	2913.602	34321.78	RU	0.3	1.3		MIT
2912.435	2913.288	34325.48	RU I	1.5	0.5		MIT
2909.740	2910.592	34357.27	RU		2.2		MIT
2909.224	2910.076	34363.36	RU	1.1	0.3		MIT
2908.883	2909.735	34367.39	RU I	2.2	0.9		ME
2906.315	2907.166	34397.76	RU	1.5	0.7		MIT
2905.828	2906.679	34403.52	RU	1.3	0.7		MIT
2905.650	2906.501	34405.63	RU I	1.7	1.1		MIT
2903.079	2903.930	34436.10	RU	0.8	0.3		MIT
2902.860	2903.711	34438.69	RU	0.6			MIT
2902.098	2902.948	34447.74	RU	0.8			MIT
2902.009	2902.859	34448.79	RU		1.4		MIT
2901.937	2902.787	34449.65	RU	1.1	0.7		MIT
2901.780	2902.630	34451.51	RU	0.6			MIT
2901.409	2902.259	34455.92	RU		0.9		MIT
2900.421	2901.271	34467.65	RU		1.7		MIT
2899.727	2900.577	34475.90	RU	0.7	0.0		MIT
2899.488	2900.338	34478.74	RU		0.9		MIT
2898.538	2899.387	34490.04	RU	1.3	0.6		MIT
2898.219	2899.068	34493.84	RU		1.8		MIT
2897.715	2898.564	34499.84	RU	0.8	1.8		MIT
2896.530	2897.379	34513.95	RU	1.5	0.6		MIT
2893.730	2894.578	34547.35	RU	0.3			MIT
2892.560	2893.408	34561.32	RU	1.3	0.8		MIT
2891.973	2892.821	34568.33	RU		0.6		MIT
2891.649	2892.497	34572.21	RU	0.9	0.7		MIT
2891.139	2891.987	34578.31	RU	0.7			MIT
2890.886	2891.734	34581.33	RU	0.3			MIT
2889.385	2890.232	34599.30	RU		1.2		MIT
2888.628	2889.475	34608.36	RU	0.6	1.0		MIT
2887.995	2888.842	34615.95	RU I	1.5	0.6		MIT
2887.264	2888.111	34624.71	RU	0.3			MIT
2887.113	2887.960	34626.52	RU	0.5			MIT
2887.053	2887.900	34627.24	RU		0.6H		MIT
2886.536	2887.383	34633.44	RU	1.8	1.7		MIT
2884.848	2885.694	34653.71	RU	0.9			MIT
2884.507	2885.353	34657.80	RU	1.3	0.7		MIT
2883.595	2884.441	34668.77	RU	1.5	0.7		MIT
2882.584	2883.430	34680.92	RU	0.6	0.0H		MIT
2882.116	2882.961	34686.56	RU	1.5	2.3		MIT
2881.276	2882.121	34696.67	RU	1.5	0.5		MIT
2880.230	2881.075	34709.27	RU	0.3			MIT
2880.080	2880.925	34711.08	RU		1.4		MIT
2879.755	2880.600	34714.99	RU	1.7	1.1		MIT
2879.360	2880.205	34719.75	RU	0.5			MIT
2879.062	2879.907	34723.35	RU		1.8		MIT
2878.040	2878.884	34735.68	RU		1.2		MIT
2877.840	2878.684	34738.09	RU I	0.5	0.0		MIT
2877.089	2877.933	34747.16	RU	0.7	0.0		MIT
2875.317	2876.161	34768.57	RU	0.6			MIT
2874.984	2875.828	34772.60	RU	1.9	1.7		MIT
2874.052	2874.895	34783.87	RU	0.7			MIT
2873.731	2874.574	34787.76	RU		1.1		MIT
2873.308	2874.151	34792.88	RU	0.6	1.8		MIT
2871.638	2872.481	34813.11	RU	1.7	0.7		MIT
2871.47	2872.31	34815.1	RU		1.5		MIT
2871.186	2872.029	34818.59	RU	0.6			MIT
2870.548	2871.391	34826.33	RU		1.7		MIT
2870.214	2871.056	34830.38	RU	0.6			MIT
2868.537	2869.379	34850.75	RU	0.7			MIT
2868.309	2869.151	34853.52	RU	0.9	0.7		MIT
2868.188	2869.030	34854.99	RU I	0.9	0.6		MIT
2867.463	2868.305	34863.80	RU	0.5			MIT
2867.094	2867.936	34868.29	RU		1.3		MIT
2866.644	2867.486	34873.76	RU I	1.8	1.4		MIT
2866.253	2867.095	34878.52	RU		1.5		MIT
2866.082	2866.923	34880.60	RU		1.5		MIT
2865.530	2866.371	34887.31	RU		1.3		MIT

AIR WAVELENGTH (ANGSTRMS)	VACUUM WAVELENGTH (ANGSTRMS)	WAVENUMBER (KAYSERS)	LMENT	LOG ARC	INTENSITY SPK	DIS	REF
2863.324	2864.165	34914.19	RU	1.5	1.9		MIT
2863.000	2863.841	34918.14	RU I	0.8			MIT
2862.881	2863.722	34919.60	RU	0.8	1.8		MIT
2861.719	2862.559	34933.77	RU I	1.1			MIT
2861.407	2862.247	34937.58	RU I	1.8	1.5		MIT
2861.099	2861.939	34941.34	RU		1.3		MIT
2860.374	2861.214	34950.20	RU	0.5			MIT
2860.016	2860.856	34954.57	RU I	1.8	1.1		MIT
2857.776	2858.615	34981.97	RU	0.6	1.8		MIT
2857.650	2858.489	34983.51	RU	0.3	0.0J		MIT
2857.220	2858.059	34988.78	RU		1.4		MIT
2856.552	2857.391	34996.96	RU		1.7		MIT
2856.047	2856.886	35003.15	RU	0.8	0.0		MIT
2855.340	2856.179	35011.81	RU		1.3		MIT
2854.870	2855.709	35017.58	RU	0.3	1.2		MIT
2854.722	2855.561	35019.39	RU	0.3	1.8		MIT
2854.074	2854.913	35027.34	RU I	1.8	1.5		MIT
2853.326	2854.164	35036.52	RU	0.3	0.9		MIT
2850.681	2851.519	35069.03	RU		1.1		MIT
2849.565	2850.402	35082.77	RU		1.3		MIT
2849.289	2850.126	35086.16	RU	0.5	2.0		MIT
2848.579	2849.416	35094.91	RU	1.7	0.5		MIT
2847.598	2848.435	35107.00	RU		0.7		MIT
2846.749	2847.586	35117.47	RU	0.5			MIT
2846.316	2847.153	35122.81	RU	1.1	0.0		MIT
2845.524	2846.360	35132.58	RU	0.9			MIT
2845.210	2846.046	35136.46	RU		0.6		MIT
2844.711	2845.547	35142.62	RU		2.2		MIT
2843.171	2844.007	35161.66	RU I	1.5	0.5		MIT
2842.755	2843.591	35166.81	RU	1.3			MIT
2842.533	2843.369	35169.55	RU	1.5			MIT
2841.677	2842.512	35180.15	RU	1.7	2.3		MIT
2841.125	2841.960	35186.98	RU		2.1		MIT
2840.539	2841.374	35194.24	RU	1.8	0.9		MIT
2839.374	2840.209	35208.68	RU		0.8		MIT
2838.846	2839.681	35215.23	RU		1.0		MIT
2838.621	2839.456	35218.02	RU	1.5			MIT
2837.274	2838.108	35234.74	RU	1.3			MIT
2836.573	2837.407	35243.44	RU	1.5	0.0H		MIT
2836.147	2836.981	35248.74	RU	1.3			MIT
2834.001	2834.835	35275.43	RU	1.5	0.7		MIT
2833.784	2834.618	35278.13	RU		1.9		MIT
2832.625	2833.458	35292.56	RU	1.3			MIT
2831.843	2832.676	35302.31	RU	1.0	1.7		MIT
2830.703	2831.536	35316.52	RU	1.3			MIT
2829.160	2829.992	35335.78	RU I	1.7	0.9		MIT
2829.081	2829.913	35336.77	RU		0.7		MIT
2827.869	2828.701	35351.92	RU	1.5	1.1		MIT
2827.522	2828.354	35356.25	RU	0.7	0.9		MIT
2827.044	2827.876	35362.23	RU		0.9		MIT
2826.677	2827.509	35366.82	RU		2.0		MIT
2826.220	2827.052	35372.54	RU		1.9		MIT
2825.463	2826.294	35382.02	RU		1.9		MIT
2825.056	2825.887	35387.11	RU		1.8		MIT
2824.772	2825.603	35390.67	RU	1.0			MIT
2823.177	2824.008	35410.67	RU	1.3	1.9		MIT
2822.552	2823.383	35418.51	RU	1.5	2.2		MIT
2822.032	2822.863	35425.03	RU	1.7	0.7		MIT
2821.424	2822.255	35432.67	RU	0.7	2.0		MIT
2821.181	2822.011	35435.72	RU	1.5			MIT
2818.952	2819.782	35463.74	RU I	1.7	0.5		MIT
2818.815	2819.645	35465.46	RU I	1.5			MIT
2818.361	2819.191	35471.17	RU I	1.7	1.1		MIT
2817.566	2818.396	35481.18	RU		1.3		MIT
2817.093	2817.922	35487.14	RU I	1.7	0.6		MIT
2814.869	2815.698	35515.17	RU	1.5			MIT
2813.711	2814.540	35529.79	RU	1.7	2.1		MIT
2813.30	2814.13	35535.0	RU		1.9		EX
2812.823	2813.651	35541.01	RU	1.1	0.0		MIT
2811.980	2812.808	35551.66	RU	0.6	1.1		MIT

AIR WAVELENGTH (ANGSTRMS)	VACUUM WAVELENGTH (ANGSTRMS)	WAVENUMBER (KAYSERS)	LMENT	LOG ARC	INTENSITY SPK	DIS	REF
2811.455	2812.283	35558.30	RU	0.7	1.1		MIT
2811.249	2812.077	35560.90	RU	0.9			MIT
2810.553	2811.381	35569.71	RU I	1.7	2.3		MIT
2810.033	2810.861	35576.29	RU I	1.7	1.1		MIT
2808.228	2809.055	35599.16	RU	1.7			MIT
2807.537	2808.364	35607.92	RU		1.3		MIT
2807.194	2808.021	35612.27	RU		1.3		MIT
2806.742	2807.569	35618.00	RU	1.7	2.0		MIT
2804.882	2805.708	35641.62	RU		1.8		MIT
2803.498	2804.324	35659.22	RU	1.7	0.6		MIT
2802.806	2803.632	35668.02	RU	1.7	2.2		MIT
2802.162	2802.988	35676.22	RU	1.5	1.6		MIT
2801.865	2802.691	35680.00	RU	1.5			MIT
2800.695	2801.520	35694.90	RU	0.9			MIT
2799.911	2800.736	35704.90	RU		1.7		MIT
2799.573	2800.398	35709.21	RU		1.1		MIT
2798.765	2799.590	35719.52	RU		1.5		MIT
2797.709	2798.534	35733.00	RU	1.1	0.6H		MIT
2796.699	2797.523	35745.90	RU	1.3			MIT
2796.554	2797.378	35747.75	RU	1.3	0.5H		MIT
2795.353	2796.177	35763.11	RU	1.5	0.9H		MIT
2792.645	2793.468	35797.79	RU	1.7	0.0		MIT
2792.332	2793.155	35801.80	RU	1.0	2.0		MIT
2791.052	2791.875	35818.22	RU	1.0	0.5		MIT
2790.216	2791.039	35828.95	RU		1.5		MIT
2788.710	2789.533	35848.30	RU		1.7		MIT
2787.826	2788.648	35859.67	RU	1.8	2.2		MIT
2787.246	2788.068	35867.13	RU		1.5		MIT
2786.366	2787.188	35878.46	RU		0.9		MIT
2785.650	2786.472	35887.68	RU I	1.8	2.3		MIT
2785.161	2785.983	35893.98	RU		1.3		MIT
2784.882	2785.704	35897.57	RU	0.8			MIT
2784.526	2785.347	35902.16	RU	1.8	2.0		MIT
2782.209	2783.030	35932.06	RU I	1.7	0.0		MIT
2780.771	2781.592	35950.64	RU	1.5	0.9		MIT
2779.392	2780.212	35968.48	RU		1.5		MIT
2778.986	2779.806	35973.73	RU	1.7	1.7		MIT
2778.379	2779.199	35981.59	RU		2.2		MIT
2777.497	2778.317	35993.02	RU	0.7	1.7		MIT
2777.392	2778.212	35994.38	RU		1.7		MIT
2776.400	2777.220	36007.24	RU	0.8			MIT
2775.907	2776.726	36013.63	RU I	1.7			MIT
2775.631	2776.450	36017.21	RU	1.7	2.2		MIT
2775.185	2776.004	36023.00	RU	1.7			MIT
2774.483	2775.302	36032.11	RU	1.8	0.3		MIT
2774.181	2775.000	36036.04	RU		0.8		MIT
2772.963	2773.782	36051.86	RU	1.5	0.9		MIT
2772.612	2773.431	36056.43	RU	1.7			MIT
2772.453	2773.272	36058.50	RU		2.2		MIT
2771.990	2772.808	36064.52	RU		0.7		MIT
2771.452	2772.270	36071.52	RU		0.6		MIT
2771.067	2771.885	36076.53	RU		2.0		MIT
2770.701	2771.519	36081.29	RU	1.8			MIT
2770.299	2771.117	36086.53	RU	1.8	0.5		MIT
2768.926	2769.744	36104.42	RU	1.8	2.3		MIT
2766.549	2767.366	36135.44	RU		2.0		MIT
2766.228	2767.045	36139.64	RU	1.5			MIT
2765.874	2766.691	36144.26	RU		1.7		MIT
2765.441	2766.258	36149.92	RU	1.7	2.2		MIT
2765.133	2765.950	36153.95	RU		1.9		MIT
2764.725	2765.542	36159.28	RU	1.7	0.0		MIT
2763.903	2764.719	36170.04	RU	1.5	0.0		MIT
2763.419	2764.235	36176.37	RU	1.7	1.2		MIT
2763.142	2763.958	36180.00	RU	1.5	0.7		MIT
2762.306	2763.122	36190.94	RU	1.7	0.5		MIT
2762.05	2762.87	36194.3	RU		0.3		MIT
2760.725	2761.541	36211.67	RU		1.3		MIT
2760.167	2760.983	36218.99	RU	1.3	1.1		MIT
2759.138	2759.953	36232.50	RU		1.3		MIT
2758.265	2759.080	36243.96	RU	0.6	0.9		MIT

AIR WAVELENGTH (ANGSTRMS)	VACUUM WAVELENGTH (ANGSTRMS)	WAVENUMBER (KAYSERS)	LMENT	LOG ARC	INTENSITY SPK	DIS	REF
2758.009	2758.824	36247.33	RU	1.3			MIT
2757.808	2758.623	36249.97	RU	1.7			MIT
2757.075	2757.890	36259.61	RU	1.5			MIT
2756.391	2757.206	36268.60	RU		1.3		MIT
2755.233	2756.047	36283.85	RU		1.5		MIT
2754.612	2755.426	36292.03	RU I	1.7	0.0		MIT
2753.444	2754.258	36307.42	RU	1.7	1.7		MIT
2752.766	2753.580	36316.36	RU	1.7	2.2		MIT
2752.451	2753.265	36320.52	RU	1.7			MIT
2752.266	2753.080	36322.96	RU I	1.5			MIT
2752.105	2752.919	36325.08	RU		1.8		MIT
2750.350	2751.163	36348.26	RU	1.7			MIT
2749.677	2750.490	36357.16	RU	1.7	1.0		MIT
2749.098	2749.911	36364.81	RU		0.9H		MIT
2747.972	2748.785	36379.71	RU	1.7	2.0		MIT
2747.692	2748.505	36383.42	RU	0.9	1.4		MIT
2746.697	2747.509	36396.60	RU		0.6H		MIT
2746.074	2746.886	36404.86	RU	1.7	0.9		MIT
2745.834	2746.646	36408.04	RU	1.7	1.9		MIT
2745.249	2746.061	36415.80	RU	1.1	2.2		MIT
2745.076	2745.888	36418.09	RU	1.5			MIT
2744.450	2745.262	36426.40	RU	1.5	1.7		MIT
2743.936	2744.748	36433.22	RU	1.7	2.0		MIT
2743.534	2744.346	36438.56	RU	1.7	1.7		MIT
2742.349	2743.160	36454.31	RU	0.6	1.7		MIT
2740.223	2741.034	36482.59	RU	1.5			MIT
2739.218	2740.028	36495.97	RU I	1.8	0.7		MIT
2738.892	2739.702	36500.31	RU	1.8	0.3		MIT
2737.792	2738.602	36514.98	RU		1.8		MIT
2737.595	2738.405	36517.61	RU		1.7		MIT
2737.475	2738.285	36519.21	RU	1.1			MIT
2736.812	2737.622	36528.05	RU	1.5	1.8		MIT
2736.484	2737.294	36532.43	RU	1.1	1.9		MIT
2736.319	2737.129	36534.63	RU	1.1			MIT
2735.718	2736.528	36542.66	RU	1.8	1.8		MIT
2734.349	2735.158	36560.96	RU	1.9	2.3		MIT
2733.592	2734.401	36571.08	RU	1.9	0.6		MIT
2733.087	2733.896	36577.84	RU	1.0			MIT
2731.901	2732.710	36593.71	RU	1.7			MIT
2730.932	2731.740	36606.70	RU I	1.9	0.7		MIT
2730.672	2731.480	36610.18	RU		0.9		MIT
2730.328	2731.136	36614.80	RU I	1.8	0.3		MIT
2729.455	2730.263	36626.51	RU	1.8			MIT
2728.834	2729.642	36634.84	RU	1.8			MIT
2726.973	2727.781	36659.84	RU	1.8	1.0		MIT
2725.466	2726.273	36680.11	RU	1.9	2.3		MIT
2724.875	2725.682	36688.06	RU		2.0		MIT
2724.063	2724.870	36699.00	RU	1.8	0.6		MIT
2723.103	2723.910	36711.94	RU		1.7		MIT
2722.824	2723.631	36715.70	RU	1.0			MIT
2722.651	2723.457	36718.03	RU I	1.8	0.0		MIT
2722.391	2723.197	36721.54	RU	1.3			MIT
2721.562	2722.368	36732.72	RU I	1.8	0.7		MIT
2719.718	2720.524	36757.63	RU	1.1	1.7		MIT
2719.515	2720.321	36760.37	RU I	2.0	1.5		MIT
2718.828	2719.634	36769.66	RU	1.5	0.7H		MIT
2717.861	2718.666	36782.74	RU		1.7		MIT
2717.399	2718.204	36789.00	RU	1.7	2.0		MIT
2717.007	2717.812	36794.30	RU	1.5			MIT
2716.584	2717.389	36800.03	RU		1.9		MIT
2716.124	2716.929	36806.26	RU		2.0		MIT
2715.776	2716.581	36810.98	RU	1.7			MIT
2715.503	2716.308	36814.68	RU	1.1	0.3H		MIT
2715.241	2716.046	36818.23	RU	1.3	0.7H		MIT
2713.737	2714.541	36838.64	RU	1.8	0.3H		MIT
2713.583	2714.387	36840.73	RU		2.0		MIT
2713.194	2713.998	36846.01	RU	1.8	0.3		MIT
2713.070	2713.874	36847.69	RU		1.9		MIT
2712.879	2713.683	36850.29	RU	1.5			MIT
2712.410	2713.214	36856.66	RU	1.9	2.5		MIT

AIR WAVELENGTH (ANGSTRMS)	VACUUM WAVELENGTH (ANGSTRMS)	WAVENUMBER (KAYSERS)	LMENT	LOG ARC	INTENSITY SPK	DIS	REF
2712.089	2712.893	36861.02	RU	1.5			MIT
2710.738	2711.542	36879.39	RU	1.3			MIT
2710.232	2711.035	36886.27	RU	1.7	2.0		MIT
2709.204	2710.007	36900.27	RU	1.8	0.9		MIT
2708.843	2709.646	36905.19	RU	1.3			MIT
2708.646	2709.449	36907.87	RU	1.3			MIT
2707.971	2708.774	36917.07	RU	1.7	0.5		MIT
2707.473	2708.276	36923.86	RU	1.5			MIT
2707.294	2708.097	36926.30	RU		1.8		MIT
2705.329	2706.131	36953.12	RU	1.1			MIT
2704.814	2705.616	36960.16	RU		1.4		MIT
2704.566	2705.368	36963.55	RU		2.0		MIT
2704.191	2704.993	36968.67	RU	1.0	1.5		MIT
2703.801	2704.603	36974.00	RU	1.8	0.5		MIT
2703.316	2704.118	36980.64	RU	1.1			MIT
2702.832	2703.634	36987.26	RU I	1.9	0.9		MIT
2702.629	2703.431	36990.04	RU		1.0		MIT
2702.416	2703.218	36992.95	RU	1.1			MIT
2701.338	2702.139	37007.71	RU	1.8	0.9		MIT
2700.990	2701.791	37012.48	RU		1.7		MIT
2700.672	2701.473	37016.84	RU	1.5	0.0H		MIT
2700.481	2701.282	37019.46	RU I	1.7			MIT
2700.154	2700.955	37023.94	RU		2.0		MIT
2699.883	2700.684	37027.66	RU I	1.5	0.5		MIT
2699.253	2700.054	37036.30	RU	0.6	0.6		MIT
2698.752	2699.553	37043.17	RU		0.5H		MIT
2698.054	2698.855	37052.76	RU	1.1	0.8		MIT
2697.512	2698.312	37060.20	RU	1.5			MIT
2697.067	2697.867	37066.31	RU	0.6	1.5		MIT
2696.550	2697.350	37073.42	RU	1.5			MIT
2696.489	2697.289	37074.26	RU		0.7		MIT
2693.663	2694.462	37113.15	RU	0.8			MIT
2693.450	2694.249	37116.09	RU	0.9	1.2		MIT
2693.287	2694.086	37118.33	RU	1.9	0.3H		MIT
2692.844	2693.643	37124.44	RU	1.5			MIT
2692.065	2692.864	37135.18	RU	0.9	2.3		MIT
2690.820	2691.619	37152.36	RU	0.9			MIT
2690.399	2691.198	37158.18	RU	1.3			MIT
2690.193	2690.992	37161.02	RU		1.3		MIT
2689.897	2690.696	37165.11	RU	1.7			MIT
2688.900	2689.698	37178.89	RU	1.5			MIT
2688.587	2689.385	37183.22	RU	1.5	0.0		MIT
2688.155	2688.953	37189.19	RU	0.9			MIT
2688.109	2688.907	37189.83	RU	0.9	2.0		MIT
2687.497	2688.295	37198.30	RU	1.1	2.0		MIT
2687.138	2687.936	37203.27	RU	1.1			MIT
2687.069	2687.867	37204.22	RU		1.7		MIT
2686.887	2687.685	37206.74	RU		1.5		MIT
2686.294	2687.092	37214.96	RU I	1.9	1.1		MIT
2685.889	2686.687	37220.57	RU		0.8		MIT
2685.157	2685.954	37230.71	RU I	1.0	1.5		MIT
2684.597	2685.394	37238.48	RU	1.0	1.5		MIT
2684.090	2684.887	37245.51	RU	1.3	0.3H		MIT
2683.682	2684.479	37251.17	RU	1.3			MIT
2682.740	2683.537	37264.25	RU		0.7		MIT
2680.543	2681.339	37294.79	RU	0.7	1.7		MIT
2679.762	2680.558	37305.66	RU	1.1	0.0H		MIT
2679.434	2680.230	37310.23	RU		1.3		MIT
2678.758	2679.554	37319.65	RU	2.0	2.5		MIT
2678.178	2678.974	37327.73	RU I	1.0			MIT
2678.045	2678.841	37329.58	RU		0.8		MIT
2677.318	2678.114	37339.72	RU	1.1			MIT
2676.972	2677.767	37344.54	RU	1.1			MIT
2676.762	2677.557	37347.47	RU		1.3		MIT
2676.353	2677.148	37353.18	RU	1.7	0.5		MIT
2676.190	2676.985	37355.45	RU	0.9	2.0		MIT
2675.523	2676.318	37364.77	RU		1.7		MIT
2675.185	2675.980	37369.49	RU		1.5		MIT
2674.195	2674.990	37383.32	RU		1.7		MIT
2673.603	2674.398	37391.60	RU	1.7	0.5		MIT

AIR WAVELENGTH (ANGSTRMS)	VACUUM WAVELENGTH (ANGSTRMS)	WAVENUMBER (KAYSERS)	LMENT	LOG ARC	INTENSITY SPK	DIS	REF
2673.477	2674.272	37393.36	RU	1.7	0.5		MIT
2673.013	2673.808	37399.85	RU	0.9	1.7		MIT
2672.362	2673.156	37408.96	RU	0.9	1.6		MIT
2672.214	2673.008	37411.03	RU		1.4		MIT
2670.717	2671.511	37432.00	RU	0.9			MIT
2670.493	2671.287	37435.14	RU	0.9	1.5		MIT
2669.417	2670.211	37450.23	RU		2.0		MIT
2668.609	2669.402	37461.57	RU		1.0		MIT
2668.345	2669.138	37465.27	RU	1.3			MIT
2667.968	2668.761	37470.57	RU	1.7			MIT
2667.790	2668.583	37473.07	RU		1.9		MIT
2667.395	2668.188	37478.62	RU	1.0	2.2		MIT
2665.722	2666.515	37502.14	RU	1.1			MIT
2665.361	2666.154	37507.21	RU		0.5		MIT
2664.757	2665.550	37515.72	RU I	1.8	0.7		MIT
2663.734	2664.526	37530.12	RU		0.9		MIT
2662.856	2663.648	37542.50	RU		1.6		MIT
2662.160	2662.952	37552.31	RU		1.5		MIT
2661.864	2662.656	37556.49	RU	1.7			MIT
2661.817	2662.609	37557.15	RU		1.0		MIT
2661.613	2662.405	37560.03	RU	1.9	2.2		MIT
2661.174	2661.966	37566.22	RU	1.3	2.0		MIT
2660.603	2661.395	37574.29	RU	1.1			MIT
2659.615	2660.406	37588.24	RU I	1.9	1.1		MIT
2658.770	2659.561	37600.19	RU	1.0			MIT
2658.395	2659.186	37605.49	RU	1.3	0.3		MIT
2658.241	2659.032	37607.67	RU		1.5		MIT
2657.192	2657.983	37622.52	RU		1.7		MIT
2657.169	2657.960	37622.84	RU	1.5			MIT
2656.692	2657.483	37629.60	RU I	1.5			MIT
2656.562	2657.353	37631.44	RU	1.5			MIT
2656.248	2657.038	37635.89	RU	1.3	2.2		MIT
2655.219	2656.009	37650.47	RU	1.3			MIT
2655.112	2655.902	37651.99	RU	1.0			MIT
2654.810	2655.600	37656.27	RU	1.1	0.7H		MIT
2654.465	2655.255	37661.17	RU	0.9			MIT
2654.327	2655.117	37663.12	RU	1.1			MIT
2653.947	2654.737	37668.52	RU		1.9		MIT
2653.697	2654.487	37672.06	RU	1.3	0.0J		MIT
2653.150	2653.940	37679.83	RU	1.1	0.7		MIT
2653.076	2653.866	37680.88	RU		1.0		MIT
2652.134	2652.924	37694.26	RU	0.9	0.3		MIT
2651.841	2652.630	37698.43	RU	2.0	1.0		MIT
2651.507	2652.296	37703.18	RU	1.3			MIT
2651.292	2652.081	37706.23	RU I	1.8	0.7		MIT
2650.591	2651.380	37716.21	RU	1.1	0.7J		MIT
2650.405	2651.194	37718.85	RU	1.7			MIT
2650.253	2651.042	37721.02	RU		1.0		MIT
2649.992	2650.781	37724.73	RU	1.3			MIT
2649.513	2650.302	37731.55	RU	1.5	0.5		MIT
2648.780	2649.569	37741.99	RU	1.5	2.2		MIT
2648.451	2649.240	37746.68	RU I	1.3	0.0		MIT
2647.935	2648.724	37754.04	RU	1.1			MIT
2647.315	2648.103	37762.88	RU	1.7	0.7		MIT
2646.630	2647.418	37772.65	RU	1.0			MIT
2646.016	2646.804	37781.41	RU	1.3	2.2		MIT
2644.614	2645.402	37801.44	RU	0.9	2.0		MIT
2643.523	2644.310	37817.04	RU	0.6	0.5		MIT
2643.393	2644.180	37818.90	RU		0.7		MIT
2643.132	2643.919	37822.64	RU	0.7	1.5		MIT
2642.960	2643.747	37825.10	RU I	2.2			MIT
2642.794	2643.581	37827.47	RU		2.2		MIT
2641.978	2642.765	37839.16	RU	1.3			MIT
2641.616	2642.403	37844.34	RU		2.0		MIT
2641.463	2642.250	37846.53	RU I	1.1			MIT
2640.327	2641.114	37862.81	RU	1.8	0.7		MIT
2639.870	2640.657	37869.37	RU	1.7			MIT
2639.712	2640.499	37871.64	RU	1.0	1.5		MIT
2639.119	2639.905	37880.15	RU	1.8	0.7		MIT
2638.510	2639.296	37888.89	RU	1.8	0.6		MIT

AIR WAVELENGTH (ANGSTRMS)	VACUUM WAVELENGTH (ANGSTRMS)	WAVENUMBER (KAYSERS)	LMENT	LOG ARC	INTENSITY SPK	DIS	REF
2638.348	2639.134	37891.21	RU	0.9	1.4		MIT
2636.821	2637.607	37913.16	RU		1.7		MIT
2636.670	2637.456	37915.33	RU	1.8			MIT
2636.542	2637.328	37917.17	RU	0.9	2.0		MIT
2635.861	2636.647	37926.96	RU	1.5	2.0		MIT
2635.354	2636.140	37934.26	RU I	1.1	1.0		MIT
2633.797	2634.582	37956.68	RU		1.9		MIT
2633.457	2634.242	37961.58	RU	1.7	0.0H		MIT
2632.711	2633.496	37972.34	RU		1.9		MIT
2632.504	2633.289	37975.33	RU	1.7	0.0		MIT
2632.127	2632.912	37980.76	RU	1.7	0.3		MIT
2631.568	2632.353	37988.83	RU	1.8	0.5		MIT
2631.09	2631.87	37995.7	RU		1.7		MIT
2630.235	2631.019	38008.08	RU	1.7	0.0H		MIT
2630.181	2630.965	38008.86	RU	0.9			MIT
2630.02	2630.80	38011.2	RU		1.7		MIT
2629.928	2630.712	38012.52	RU	1.3			MIT
2629.364	2630.148	38020.67	RU		1.0		MIT
2628.729	2629.513	38029.86	RU		2.0		MIT
2628.534	2629.318	38032.68	RU	1.1			MIT
2627.795	2628.579	38043.37	RU		1.0		MIT
2627.651	2628.435	38045.46	RU	1.8	0.0		MIT
2627.305	2628.089	38050.47	RU		1.0		MIT
2626.475	2627.258	38062.49	RU	1.7			MIT
2626.352	2627.135	38064.27	RU	1.3	1.7		MIT
2626.210	2626.993	38066.33	RU	1.7			MIT
2625.566	2626.349	38075.67	RU		0.8		MIT
2625.380	2626.163	38078.37	RU		1.6		MIT
2624.699	2625.482	38088.25	RU		0.6H		MIT
2624.155	2624.938	38096.14	RU		1.3		MIT
2623.830	2624.613	38100.86	RU	1.7			MIT
2621.738	2622.520	38131.26	RU		0.6		MIT
2621.264	2622.046	38138.15	RU		1.1		MIT
2621.081	2621.863	38140.82	RU	1.3	0.6		MIT
2620.876	2621.658	38143.80	RU	1.3			MIT
2620.610	2621.392	38147.67	RU	1.7	0.7		MIT
2620.068	2620.850	38155.56	RU	1.0	0.8		MIT
2619.671	2620.453	38161.34	RU	1.7	0.7		MIT
2619.354	2620.136	38165.96	RU		1.8		MIT
2619.020	2619.802	38170.83	RU	1.5	0.0H		MIT
2618.740	2619.522	38174.91	RU	1.0			MIT
2618.597	2619.379	38177.00	RU		0.5		MIT
2617.789	2618.570	38188.78	RU	1.7	0.6		MIT
2617.441	2618.222	38193.85	RU		0.7		MIT
2617.066	2617.847	38199.33	RU		1.7		MIT
2616.329	2617.110	38210.09	RU		1.3		MIT
2615.093	2615.874	38228.15	RU	1.8	2.0		MIT
2614.865	2615.646	38231.48	RU		1.5		MIT
2614.586	2615.367	38235.56	RU	1.7	0.6		MIT
2614.068	2614.848	38243.13	RU	1.8	0.5		MIT
2613.206	2613.986	38255.75	RU	0.7	1.0J		MIT
2612.917	2613.697	38259.98	RU	0.7			MIT
2612.515	2613.295	38265.87	RU		1.9		MIT
2612.068	2612.848	38272.41	RU	2.0	1.5		MIT
2611.510	2612.290	38280.59	RU	0.5	1.9		MIT
2611.048	2611.828	38287.37	RU	1.7	0.7		MIT
2610.078	2610.857	38301.59	RU		1.8		MIT
2609.485	2610.264	38310.30	RU	1.5	0.8		MIT
2609.057	2609.836	38316.58	RU I	1.9	1.1		MIT
2607.918	2608.697	38333.31	RU	0.9	1.5		MIT
2607.347	2608.126	38341.71	RU	1.1			MIT
2606.944	2607.723	38347.63	RU	1.0	1.2		MIT
2606.20	2606.98	38358.6	RU		1.3J		MIT
2605.857	2606.635	38363.63	RU	1.7	0.5		MIT
2605.349	2606.127	38371.11	RU	1.7	0.6		MIT
2604.318	2605.096	38386.30	RU	1.1			MIT
2604.120	2604.898	38389.22	RU		1.9		MIT
2603.802	2604.580	38393.91	RU	1.1			MIT
2602.885	2603.663	38407.43	RU		1.3		MIT
2602.249	2603.027	38416.82	RU		2.0		MIT

AIR WAVELENGTH (ANGSTRMS)	VACUUM WAVELENGTH (ANGSTRMS)	WAVENUMBER (KAYSERS)	LMENT	LOG ARC	INTENSITY SPK	DIS	REF
2601.460	2602.237	38428.47	RU	1.5	0.7		MIT
2599.658	2600.435	38455.10	RU	1.0	1.4		MIT
2598.766	2599.543	38468.30	RU	0.6	1.7		MIT
2598.581	2599.358	38471.04	RU	1.3			MIT
2597.793	2598.570	38482.71	RU		1.3		MIT
2597.518	2598.295	38486.78	RU	1.0	1.3		MIT
2597.326	2598.102	38489.63	RU	1.5			MIT
2597.147	2597.923	38492.28	RU		1.7H		MIT
2595.932	2596.708	38510.30	RU	0.9			MIT
2595.796	2596.572	38512.31	RU		2.0		MIT
2595.639	2596.415	38514.64	RU	1.3			MIT
2595.530	2596.306	38516.26	RU		0.5J		MIT
2595.425	2596.201	38517.82	RU	1.0			MIT
2594.852	2595.628	38526.32	RU	1.8	0.6		MIT
2594.392	2595.168	38533.15	RU		1.5J		MIT
2593.700	2594.476	38543.43	RU	1.3			MIT
2593.636	2594.412	38544.38	RU	1.0			MIT
2592.024	2592.799	38568.35	RU	1.8	0.8		MIT
2591.640	2592.415	38574.07	RU	1.5	0.0		MIT
2591.237	2592.012	38580.07	RU		1.7		MIT
2591.118	2591.893	38581.84	RU	1.7			MIT
2590.971	2591.746	38584.03	RU	1.1	2.0		MIT
2589.785	2590.560	38601.70	RU	0.9	0.7		MIT
2589.566	2590.341	38604.96	RU	1.8	0.5		MIT
2589.413	2590.188	38607.24	RU		1.8		MIT
2589.038	2589.812	38612.83	RU	1.3			MIT
2588.862	2589.636	38615.46	RU		0.5		MIT
2588.655	2589.429	38618.55	RU		0.5		MIT
2588.195	2588.969	38625.41	RU	1.0			MIT
2587.86	2588.63	38630.4	RU		0.9		EX
2587.458	2588.232	38636.41	RU	0.8			MIT
2586.331	2587.105	38653.25	RU	0.8			MIT
2586.084	2586.858	38656.94	RU	1.1			MIT
2585.735	2586.509	38662.15	RU	1.7			MIT
2585.647	2586.421	38663.47	RU	1.0			MIT
2585.341	2586.115	38668.05	RU	1.3			MIT
2584.138	2584.911	38686.05	RU	1.7	0.5		MIT
2583.037	2583.810	38702.54	RU	1.5	1.6		MIT
2582.818	2583.591	38705.82	RU	0.5	1.3		MIT
2582.627	2583.400	38708.68	RU	0.6	1.0		MIT
2581.910	2582.683	38719.43	RU	1.5	0.3		MIT
2581.430	2582.203	38726.63	RU		0.9		MIT
2581.137	2581.910	38731.02	RU	1.8	0.3		MIT
2580.799	2581.572	38736.10	RU	1.7	0.5		MIT
2580.444	2581.216	38741.42	RU		0.9		MIT
2580.205	2580.977	38745.01	RU		1.0		MIT
2579.533	2580.305	38755.10	RU	1.5	0.6		MIT
2579.217	2579.989	38759.85	RU	1.5	0.0		MIT
2579.017	2579.789	38762.86	RU	1.1	1.9		MIT
2578.949	2579.721	38763.88	RU	1.3			MIT
2578.567	2579.339	38769.62	RU	1.5	1.0		MIT
2577.300	2578.072	38788.68	RU		0.7		MIT
2576.954	2577.726	38793.89	RU	1.1	0.9		MIT
2576.082	2576.853	38807.02	RU		1.7		MIT
2575.242	2576.013	38819.68	RU	1.5	0.0		MIT
2574.525	2575.296	38830.49	RU		0.7		MIT
2574.063	2574.834	38837.46	RU		1.3		MIT
2573.545	2574.316	38845.27	RU	0.7	1.7		MIT
2572.660	2573.431	38858.63	RU		0.9		MIT
2572.409	2573.180	38862.43	RU I	1.5	0.0		MIT
2572.279	2573.050	38864.39	RU	1.5	0.0		MIT
2571.085	2571.855	38882.44	RU	0.8	2.0		MIT
2570.972	2571.742	38884.15	RU	1.7			MIT
2570.086	2570.856	38897.55	RU	1.0			MIT
2569.736	2570.506	38902.85	RU I	1.3			MIT
2569.656	2570.426	38904.06	RU	0.5	0.3		MIT
2568.978	2569.648	38915.84	RU		1.5		MIT
2568.767	2569.537	38917.52	RU I	1.8	0.9		MIT
2567.895	2568.665	38930.74	RU	1.5			MIT
2566.587	2567.356	38950.58	RU	1.5	1.4		MIT

AIR WAVELENGTH (ANGSTRMS)	VACUUM WAVELENGTH (ANGSTRMS)	WAVENUMBER (KAYSERS)	LMENT	LOG ARC	INTENSITY SPK	INTENSITY DIS	REF
2566.234	2567.003	38955.93	RU		1.7		MIT
2565.808	2566.577	38962.40	RU	1.1	0.7		MIT
2565.702	2566.471	38964.01	RU	0.6	1.7		MIT
2565.182	2565.951	38971.91	RU	1.7			MIT
2564.809	2565.578	38977.58	RU		0.6J		MIT
2564.577	2565.346	38981.10	RU	1.7	0.3		MIT
2564.417	2565.186	38983.53	RU	0.9			MIT
2563.893	2564.662	38991.50	RU		1.5		MIT
2563.151	2563.919	39002.79	RU	1.7	0.0		MIT
2562.840	2563.608	39007.52	RU	0.5	0.0		MIT
2562.173	2562.941	39017.67	RU	0.7			MIT
2561.803	2562.571	39023.31	RU	1.3			MIT
2560.830	2561.598	39038.13	RU	1.5	0.7		MIT
2560.262	2561.030	39046.79	RU	1.8	0.7		MIT
2559.806	2560.574	39053.75	RU	0.5	0.9		MIT
2559.405	2560.173	39059.87	RU	1.5	0.6		MIT
2558.535	2559.302	39073.15	RU	1.7	0.0		MIT
2558.016	2558.783	39081.08	RU	0.6	1.0		MIT
2557.697	2558.464	39085.95	RU I	1.5	0.7J		MIT
2557.129	2557.896	39094.63	RU	0.7	1.7		MIT
2556.697	2557.464	39101.24	RU		1.5		MIT
2556.312	2557.079	39107.12	RU	1.7			MIT
2556.049	2556.816	39111.15	RU		1.3		MIT
2555.996	2556.763	39111.96	RU	1.5			MIT
2555.864	2556.631	39113.98	RU	1.3	1.3		MIT
2555.654	2556.421	39117.19	RU	1.1	0.9		MIT
2554.687	2555.453	39132.00	RU	0.9			MIT
2553.964	2554.730	39143.08	RU	1.1			MIT
2553.310	2554.076	39153.10	RU	0.8	0.7		MIT
2552.424	2553.190	39166.69	RU	1.1			MIT
2552.305	2553.071	39168.52	RU	1.3			MIT
2551.984	2552.750	39173.44	RU	1.0	2.2		MIT
2551.726	2552.492	39177.40	RU	1.5			MIT
2551.554	2552.320	39180.04	RU		1.1		MIT
2549.792	2550.557	39207.12	RU	0.5	2.0		MIT
2549.577	2550.342	39210.42	RU	1.7	0.5		MIT
2549.479	2550.244	39211.93	RU	1.3	0.5		MIT
2549.182	2549.947	39216.50	RU	0.7	2.2		MIT
2548.689	2549.454	39224.08	RU	0.3	0.6		MIT
2548.486	2549.251	39227.21	RU	0.5	0.6		MIT
2547.941	2548.706	39235.60	RU		1.0		MIT
2547.672	2548.437	39239.74	RU	0.7	1.9		MIT
2547.508	2548.273	39242.27	RU	1.1			MIT
2546.671	2547.436	39255.17	RU	0.9	0.5		MIT
2545.769	2546.533	39269.07	RU	1.1	0.5J		MIT
2544.222	2544.986	39292.95	RU	1.8	0.8		MIT
2543.678	2544.442	39301.35	RU	1.3			MIT
2543.47	2544.23	39304.6	RU		0.9		MIT
2543.250	2544.014	39307.96	RU	1.7	2.2		MIT
2542.230	2542.993	39323.73	RU		0.7		MIT
2542.035	2542.798	39326.75	RU		1.1		MIT
2541.284	2542.047	39338.37	RU	1.7			MIT
2540.300	2541.063	39353.61	RU	1.0	2.0		MIT
2539.719	2540.482	39362.61	RU	1.1	2.0		MIT
2536.216	2536.978	39416.98	RU	1.1	0.8		MIT
2535.592	2536.354	39426.67	RU		2.0		MIT
2535.222	2535.984	39432.43	RU		1.5		MIT
2534.944	2535.706	39436.75	RU	0.5	1.6		MIT
2534.001	2534.763	39451.43	RU	0.6	1.9		MIT
2533.236	2533.997	39463.34	RU I	1.7			MIT
2533.042	2533.803	39466.36	RU		0.7		MIT
2532.015	2532.776	39482.37	RU	1.0			MIT
2530.464	2531.225	39506.57	RU	0.7	0.6		MIT
2529.699	2530.460	39518.51	RU	1.3	0.9		MIT
2528.878	2529.638	39531.34	RU	1.8	0.3		MIT
2528.715	2529.475	39533.89	RU	1.5			MIT
2528.05	2528.81	39544.3	RU II		2.0		ME
2527.897	2528.657	39546.68	RU	0.3	1.6		MIT
2527.352	2528.112	39555.21	RU	0.7	0.6		MIT
2527.053	2527.813	39559.89	RU	0.3	0.7		MIT

AIR WAVELENGTH (ANGSTRMS)	VACUUM WAVELENGTH (ANGSTRMS)	WAVENUMBER (KAYSERS)	LMENT	LOG ARC	INTENSITY SPK	INTENSITY DIS	REF
2526.828	2527.588	39563.41	RU	1.7	1.3		MIT
2525.932	2526.692	39577.44	RU I	0.9	0.3J		MIT
2525.632	2526.392	39582.15	RU	1.5			MIT
2525.174	2525.933	39589.33	RU	1.5			MIT
2524.858	2525.617	39594.28	RU	1.0	1.9		MIT
2524.385	2525.144	39601.70	RU		0.9		MIT
2523.098	2523.857	39621.90	RU	1.1	0.5		MIT
2522.582	2523.341	39630.00	RU		0.9		MIT
2522.320	2523.079	39634.12	RU	1.1	0.3		MIT
2521.613	2522.372	39645.23	RU	1.8	0.0		MIT
2520.789	2521.547	39658.19	RU	0.8	1.3		MIT
2519.949	2520.707	39671.41	RU	1.1	0.0		MIT
2519.208	2519.966	39683.07	RU	1.3	1.9		MIT
2518.400	2519.158	39695.81	RU	0.5	1.7		MIT
2517.616	2518.374	39708.17	RU	1.7			MIT
2517.317	2518.075	39712.88	RU	1.8	1.9		MIT
2516.702	2517.459	39722.59	RU	0.8	0.9		MIT
2516.006	2516.763	39733.57	RU	1.3	0.7J		MIT
2515.759	2516.516	39737.47	RU	0.8	0.9		MIT
2515.285	2516.042	39744.96	RU	1.8	0.3		MIT
2514.153	2514.910	39762.86	RU		0.6		MIT
2513.322	2514.079	39776.00	RU	1.7	1.9		MIT
2512.806	2513.563	39784.17	RU	1.9	0.3		MIT
2512.481	2513.238	39789.31	RU	0.6	0.9J		MIT
2511.987	2512.743	39797.14	RU	1.3	1.6		MIT
2511.557	2512.313	39803.95	RU	1.5	0.7J		MIT
2511.21	2511.97	39809.5	RU		0.8W		MIT
2510.971	2511.727	39813.24	RU	1.3			MIT
2510.139	2510.895	39826.44	RU	1.5			MIT
2509.069	2509.825	39843.42	RU	1.7	1.3		MIT
2508.667	2509.423	39849.80	RU		1.7		MIT
2508.426	2509.182	39853.63	RU	1.1			MIT
2508.267	2509.023	39856.16	RU I	1.7	0.3		MIT
2507.553	2508.308	39867.51	RU	0.0	0.3		MIT
2507.014	2507.769	39876.08	RU	1.8	1.9		MIT
2506.414	2507.169	39885.62	RU		1.1		MIT
2506.26	2507.02	39888.1	RU		0.5J		MIT
2505.880	2506.635	39894.12	RU		1.0		MIT
2505.672	2506.427	39897.43	RU	0.5	0.8		MIT
2504.913	2505.668	39909.52	RU		0.7		MIT
2502.377	2503.131	39949.96	RU	1.3	0.6H		MIT
2501.888	2502.642	39957.77	RU	1.5	1.5		MIT
2501.482	2502.236	39964.26	RU	1.8	0.0		MIT
2500.838	2501.592	39974.55	RU	1.5			MIT
2500.136	2500.890	39985.77	RU		1.3		MIT
2499.780	2500.534	39991.46	RU	1.7	0.6		MIT
2499.563	2500.317	39994.94	RU	0.6	0.5		MIT
2499.372	2500.125	39997.99	RU		0.9		MIT
2498.572	2499.325	40010.80	RU	1.8	1.6		MIT
2498.419	2499.172	40013.25	RU	1.8	1.6		MIT
2497.678	2498.431	40025.12	RU	1.7	0.0H		MIT
2496.846	2497.599	40038.46	RU		1.0		MIT
2495.694	2496.447	40056.94	RU	1.9	1.5		MIT
2494.481	2495.233	40076.41	RU	1.7	1.8		MIT
2494.089	2494.841	40082.71	RU	0.9	0.8		MIT
2494.022	2494.774	40083.79	RU	1.9			MIT
2493.693	2494.445	40089.08	RU	1.3	1.9		MIT
2491.776	2492.528	40119.92	RU	1.5			MIT
2491.558	2492.310	40123.42	RU		1.2		MIT
2491.338	2492.090	40126.97	RU		0.6		MIT
2491.099	2491.851	40130.82	RU	1.0	1.6		MIT
2490.531	2491.282	40139.97	RU		0.6		MIT
2489.912	2490.663	40149.95	RU	1.8			MIT
2489.336	2490.087	40159.24	RU		1.6		MIT
2488.577	2489.328	40171.48	RU		1.2		MIT
2488.096	2488.847	40179.25	RU	0.5			MIT
2487.200	2487.951	40193.72	RU		0.6		MIT
2484.666	2485.416	40234.71	RU		0.9		MIT
2483.972	2484.722	40245.95	RU	1.5	1.3		MIT
2482.776	2483.526	40265.34	RU		1.2		MIT

AIR WAVELENGTH (ANGSTRMS)	VACUUM WAVELENGTH (ANGSTRMS)	WAVENUMBER (KAYSERS)	LMENT	LOG ARC	INTENSITY SPK	INTENSITY DIS	REF
2482.541	2483.291	40269.15	RU	1.5			MIT
2482.176	2482.925	40275.07	RU	1.3			MIT
2481.859	2482.608	40280.21	RU		0.7		MIT
2481.111	2481.860	40292.36	RU I	1.1	1.9		MIT
2480.808	2481.557	40297.28	RU		1.1		MIT
2479.359	2480.108	40320.83	RU	1.3			MIT
2478.928	2479.677	40327.84	RU	1.9	1.8		MIT
2478.399	2479.148	40336.44	RU	0.9	0.7J		MIT
2477.932	2478.680	40344.05	RU	1.0	0.3H		MIT
2477.257	2478.005	40355.04	RU II		1.7		ME
2476.876	2477.624	40361.25	RU	1.8	0.3		MIT
2476.308	2477.056	40370.50	RU	1.7			MIT
2476.149	2476.897	40373.10	RU		0.5		MIT
2475.406	2476.154	40385.21	RU	2.0	0.5		MIT
2474.748	2475.496	40395.95	RU		0.8		MIT
2474.407	2475.155	40401.52	RU I	1.1			MIT
2474.036	2474.784	40407.57	RU	1.7	0.0V		MIT
2473.577	2474.324	40415.07	RU		0.7		MIT
2472.830	2473.577	40427.28	RU		0.9		MIT
2472.435	2473.182	40433.74	RU		0.7		MIT
2471.487	2472.234	40449.25	RU	1.5			MIT
2471.018	2471.765	40456.92	RU		0.6H		MIT
2470.715	2471.462	40461.88	RU	1.1			MIT
2470.517	2471.264	40465.13	RU	1.3	1.3		MIT
2469.215	2469.961	40486.46	RU	0.8	0.7H		MIT
2467.578	2468.324	40513.32	RU	1.5			MIT
2467.314	2468.060	40517.65	RU	0.3	0.7		MIT
2464.995	2465.740	40555.77	RU		0.6		MIT
2464.769	2465.514	40559.49	RU		0.8		MIT
2464.699	2465.444	40560.64	RU	1.7	0.6		MIT
2464.365	2465.110	40566.13	RU I	1.1			MIT
2463.972	2464.717	40572.60	RU	0.7			MIT
2462.937	2463.682	40589.65	RU	1.8			MIT
2461.520	2462.265	40613.02	RU		0.7H		MIT
2460.717	2461.462	40626.27	RU	1.3	0.7		MIT
2460.422	2461.166	40631.14	RU		0.7		MIT
2460.166	2460.910	40635.37	RU		0.7		MIT
2459.429	2460.173	40647.54	RU		0.6		MIT
2459.168	2459.912	40651.86	RU	0.7	0.5J		MIT
2458.621	2459.365	40660.90	RU	1.8	0.3		MIT
2457.995	2458.739	40671.26	RU	0.9	0.3H		MIT
2457.181	2457.925	40684.73	RU	1.0	1.1		MIT
2456.949	2457.693	40688.57	RU	1.0			MIT
2456.568	2457.312	40694.88	RU	1.8	1.7		MIT
2456.438	2457.182	40697.03	RU	1.8	1.7		MIT
2455.889	2456.632	40706.13	RU	0.7	0.0		MIT
2455.530	2456.273	40712.08	RU	1.9	2.0R		MIT
2454.923	2455.666	40722.15	RU	1.8	0.7		MIT
2454.235	2454.978	40733.56	RU	0.3	0.7H		MIT
2453.939	2454.682	40738.48	RU	1.0	1.4		MIT
2451.318	2452.060	40782.03	RU	0.5	1.1		MIT
2451.049	2451.791	40786.51	RU	0.5	0.9		MIT
2450.584	2451.326	40794.25	RU	1.0			MIT
2449.858	2450.600	40806.33	RU	1.3			MIT
2449.786	2450.528	40807.53	RU	0.6	1.3H		MIT
2448.862	2449.604	40822.93	RU	1.1	0.8		MIT
2448.498	2449.240	40829.00	RU	0.7			MIT
2447.451	2448.192	40846.46	RU	1.5			MIT
2446.593	2447.334	40860.79	RU	0.5	0.5		MIT
2445.579	2446.320	40877.73	RU	1.0			MIT
2444.030	2444.771	40903.63	RU	1.1			MIT
2443.225	2443.965	40917.11	RU	0.5	0.0		MIT
2441.743	2442.483	40941.94	RU	0.8			MIT
2441.440	2442.180	40947.02	RU	0.6			MIT
2440.969	2441.709	40954.92	RU	1.1			MIT
2439.709	2440.449	40976.07	RU	0.5	0.7		MIT
2436.923	2437.662	41022.91	RU	1.4			MIT
2435.509	2436.248	41046.73	RU	0.8	0.3		MIT
2434.882	2435.621	41057.30	RU	1.7			MIT
2433.867	2434.605	41074.42	RU		0.9		MIT

AIR WAVELENGTH (ANGSTRMS)	VACUUM WAVELENGTH (ANGSTRMS)	WAVENUMBER (KAYSERS)	LMENT	LOG ARC	INTENSITY SPK	INTENSITY DIS	REF
2433.772	2434.510	41076.02	RU	0.3	0.3		MIT
2432.927	2433.665	41090.29	RU	1.8			MIT
2432.250	2432.988	41101.72	RU	1.0	0.0		MIT
2430.929	2431.667	41124.06	RU		0.5J		MIT
2430.323	2431.061	41134.31	RU	0.8	1.0		MIT
2429.596	2430.333	41146.62	RU	1.8			MIT
2429.109	2429.846	41154.87	RU	0.5	0.5		MIT
2428.718	2429.455	41161.49	RU	1.1	0.5		MIT
2427.741	2428.478	41178.06	RU	0.3	1.1		MIT
2427.082	2427.819	41189.23	RU	0.0	0.8		MIT
2426.707	2427.444	41195.60	RU	1.1	0.7		MIT
2424.529	2425.265	41232.60	RU	0.3	0.8		MIT
2423.540	2424.276	41249.43	RU	0.3	0.5J		MIT
2423.092	2423.828	41257.06	RU	0.9			MIT
2422.925	2423.661	41259.90	RU	1.8	0.9		MIT
2422.571	2423.307	41265.93	RU	1.7	0.0H		MIT
2422.177	2422.913	41272.64	RU	1.1	0.9J		MIT
2420.820	2421.555	41295.77	RU	1.8	0.3		MIT
2420.251	2420.986	41305.48	RU		0.6		MIT
2420.119	2420.854	41307.73	RU	0.3	0.9		MIT
2418.908	2419.643	41328.41	RU	0.6	0.9		MIT
2416.946	2417.680	41361.96	RU	0.8	1.1		MIT
2416.490	2417.224	41369.76	RU	1.0	0.3H		MIT
2415.718	2416.452	41382.98	RU	0.6	1.0		MIT
2414.825	2415.559	41398.29	RU	1.4	1.1		MIT
2413.921	2414.655	41413.79	RU		0.3		MIT
2413.384	2414.118	41423.00	RU		0.6		MIT
2411.496	2412.229	41455.43	RU	0.3	1.0		MIT
2410.892	2411.625	41465.81	RU	1.5			MIT
2410.156	2410.889	41478.48	RU		1.1		MIT
2409.676	2410.409	41486.74	RU		0.7H		MIT
2408.668	2409.401	41504.10	RU	1.1			MIT
2408.447	2409.180	41507.91	RU	0.6	1.0		MIT
2408.299	2409.032	41510.46	RU	1.1			MIT
2407.922	2408.654	41516.96	RU	1.8	1.7		MIT
2406.007	2406.739	41550.00	RU	0.0	0.8		MIT
2405.291	2406.023	41562.37	RU		0.5		MIT
2405.182	2405.914	41564.25	RU	1.1	0.3		MIT
2403.536	2404.267	41592.71	RU	0.8			MIT
2403.176	2403.907	41598.94	RU	0.9			MIT
2402.717	2403.448	41606.89	RU	2.0	2.2F		MIT
2401.902	2402.633	41621.00	RU	0.3	0.7		MIT
2400.930	2401.661	41637.85	RU	0.6	0.8		MIT
2399.750	2400.481	41658.33	RU	1.1			A
2398.578	2399.308	41678.68	RU		0.6H		MIT
2397.697	2398.427	41693.99	RU	0.9	0.9		MIT
2396.944	2397.674	41707.09	RU	0.5	0.9		MIT
2396.710	2397.440	41711.16	RU	1.8	1.9		MIT
2394.778	2395.507	41744.81	RU	0.6	0.7		MIT
2393.971	2394.700	41758.88	RU	1.7			MIT
2393.836	2394.565	41761.23	RU		1.1		MIT
2393.253	2393.982	41771.41	RU	1.9	0.0		MIT
2392.424	2393.153	41785.88	RU	1.9	0.8		MIT
2391.760	2392.489	41797.48	RU		0.9		MIT
2391.440	2392.169	41803.07	RU		0.6		MIT
2391.176	2391.905	41807.69	RU	0.8	0.5		MIT
2390.430	2391.158	41820.73	RU	0.7	0.6		MIT
2390.120	2390.848	41826.16	RU	1.0	0.5		MIT
2389.258	2389.986	41841.25	RU	0.6			A
2388.189	2388.917	41859.97	RU	1.3	0.9		MIT
2387.900	2388.628	41865.04	RU	1.8	0.5		MIT
2386.678	2387.406	41886.47	RU	0.9	0.0		MIT
2386.216	2386.944	41894.58	RU	0.3	0.0		MIT
2383.963	2384.690	41934.17	RU	1.1			MIT
2383.442	2384.169	41943.34	RU		1.1		MIT
2382.741	2383.468	41955.68	RU	0.0	0.6		MIT
2381.993	2382.720	41968.85	RU	1.7	2.2		MIT
2379.845	2380.571	42006.73	RU	1.0	1.2		MIT
2379.57	2380.30	42011.6	RU		1.0		EX
2375.631	2376.356	42081.23	RU	1.7	1.9		MIT

AIR WAVELENGTH (ANGSTRMS)	VACUUM WAVELENGTH (ANGSTRMS)	WAVENUMBER (KAYSERS)	LMENT	LOG ARC	INTENSITY SPK	INTENSITY DIS	REF
2375.271	2375.996	42087.61	RU	1.9	0.7		MIT
2374.777	2375.502	42096.37	RU		0.6		MIT
2374.652	2375.377	42098.58	RU		0.5		MIT
2373.952	2374.677	42111.00	RU		0.6		MIT
2371.915	2372.639	42147.16	RU	0.3	1.0		MIT
2370.173	2370.897	42178.13	RU	1.8			MIT
2370.018	2370.742	42180.89	RU	0.6	0.7		MIT
2367.775	2368.498	42220.84	RU	0.6			MIT
2367.222	2367.945	42230.71	RU	0.8	1.0		MIT
2364.227	2364.950	42284.20	RU		0.7		MIT
2360.724	2361.446	42346.94	RU	0.9	0.3		MIT
2360.555	2361.277	42349.97	RU	1.5			MIT
2360.097	2360.819	42358.19	RU	1.3			MIT
2359.105	2359.826	42376.00	RU		1.1		MIT
2358.791	2359.512	42381.64	RU	1.3	1.2		MIT
2357.914	2358.635	42397.40	RU	1.8	2.0		MIT
2355.627	2356.348	42438.56	RU		0.7		MIT
2354.100	2354.820	42466.08	RU	0.6	0.0		MIT
2353.595	2354.315	42475.20	RU	0.8			MIT
2352.951	2353.671	42486.82	RU	0.8	0.9		MIT
2351.334	2352.054	42516.04	RU	1.8	0.6		MIT
2350.577	2351.297	42529.73	RU	0.5	0.8		MIT
2349.336	2350.055	42552.19	RU	1.8	0.6		MIT
2348.327	2349.046	42570.47	RU	1.7			A
2346.409	2347.128	42605.27	RU	0.3	0.9		MIT
2346.349	2347.068	42606.36	RU	0.7	0.9		A
2344.816	2345.534	42634.21	RU		0.5V		MIT
2344.534	2345.252	42639.34	RU	0.7	0.0		A
2344.080	2344.798	42647.60	RU	0.9			A
2342.846	2343.564	42670.06	RU	1.8	1.6		MIT
2342.72	2343.44	42672.4	RU	1.8	1.0		A
2342.460	2343.178	42677.09	RU	0.3	1.0H		MIT
2342.01	2342.73	42685.3	RU	1.0			A
2341.061	2341.778	42702.59	RU	0.5	0.9		MIT
2340.691	2341.408	42709.34	RU	1.8	0.6		MIT
2340.501	2341.218	42712.81	RU		0.7		MIT
2339.925	2340.642	42723.32	RU	0.0	0.7		MIT
2339.49	2340.21	42731.3	RU	1.3			A
2338.671	2339.388	42746.23	RU	0.3	0.5		MIT
2337.816	2338.533	42761.86	RU	0.3	0.8		MIT
2336.855	2337.571	42779.44	RU		1.0H		MIT
2336.377	2337.093	42788.19	RU		0.5		MIT
2335.984	2336.700	42795.39	RU	0.3	1.2		MIT
2335.840	2336.556	42798.03	RU	0.0	0.9		MIT
2335.74	2336.46	42799.9	RU	1.1			A
2333.907	2334.623	42833.47	RU	1.1	1.3		MIT
2333.565	2334.281	42839.75	RU	0.7	1.3		MIT
2332.277	2332.992	42863.40	RU		1.1H		MIT
2331.780	2332.495	42872.54	RU	1.1	1.2H		MIT
2331.43	2332.15	42879.0	RU	1.3			A
2331.055	2331.770	42885.87	RU		1.1		MIT
2330.153	2330.868	42902.47	RU	0.7	0.0H		MIT
2329.016	2329.731	42923.42	RU	0.8	1.1		MIT
2328.327	2329.042	42936.12	RU		0.3		MIT
2328.154	2328.869	42939.31	RU		0.3		MIT
2327.532	2328.246	42950.78	RU	1.0			A
2326.41	2327.12	42971.5	RU	0.8			A
2326.243	2326.957	42974.58	RU	1.3			MIT
2325.956	2326.670	42979.88	RU	1.5			A
2325.516	2326.230	42988.01	RU		0.3H		MIT
2322.851	2323.564	43037.33	RU	0.0	0.8		MIT
2322.422	2323.135	43045.28	RU	0.7			A
2322.012	2322.725	43052.88	RU	1.8	0.3		A
2321.585	2322.298	43060.79	RU	0.3	0.9		MIT
2320.701	2321.414	43077.20	RU	1.7			MIT
2320.230	2320.943	43085.94	RU	1.3			A
2319.657	2320.370	43096.58	RU	0.5	0.0		MIT
2319.207	2319.920	43104.94	RU	0.7	0.8H		MIT
2318.757	2319.469	43113.31	RU	1.0			A
2318.535	2319.247	43117.44	RU	0.9	0.6		MIT

AIR WAVELENGTH (ANGSTRMS)	VACUUM WAVELENGTH (ANGSTRMS)	WAVENUMBER (KAYSERS)	LMENT	LOG ARC	INTENSITY SPK	INTENSITY DIS	REF
2317.802	2318.514	43131.07	RU		0.6		MIT
2315.898	2316.610	43166.53	RU		0.5		MIT
2314.777	2315.489	43187.43	RU		0.5		MIT
2314.022	2314.733	43201.52	RU		0.5		MIT
2313.374	2314.085	43213.62	RU	0.8	1.0		MIT
2312.559	2313.270	43228.85	RU	0.5			A
2312.00	2312.71	43239.3	RU		0.5H		MIT
2311.195	2311.906	43254.36	RU		0.3H		MIT
2309.522	2310.232	43285.69	RU		0.5		MIT
2309.155	2309.865	43292.57	RU	0.7	0.9		MIT
2308.626	2309.336	43302.49	RU		0.5		MIT
2306.919	2307.629	43334.52	RU	1.3			MIT
2305.626	2306.336	43358.83	RU	0.0	1.3		MIT
2305.518	2306.228	43360.86	RU	1.0	1.3		A
2304.827	2305.536	43373.85	RU	1.1	0.8		MIT
2304.722	2305.431	43375.83	RU	0.6	0.5		MIT
2302.806	2303.515	43411.92	RU		0.9		MIT
2302.541	2303.250	43416.91	RU	1.9	0.5		MIT
2302.233	2302.942	43422.72	RU	0.9			A
2300.367	2301.075	43457.94	RU	1.7			A
2299.29	2300.00	43478.3	RU	1.7			A
2298.659	2299.367	43490.23	RU	0.6	0.9		MIT
2297.974	2298.682	43503.19	RU	0.3	0.5		MIT
2296.650	2297.358	43528.27	RU		0.7		MIT
2296.170	2296.877	43537.37	RU		0.7		MIT
2294.06	2294.77	43577.4	RU	1.5	0.0		A
2293.05	2293.76	43596.6	RU	1.0			A
2292.337	2293.044	43610.16	RU	1.3			MIT
2291.177	2291.883	43632.24	RU I	1.8	0.0		MIT
2287.68	2288.39	43698.9	RU	1.8	0.0		A
2287.525	2288.231	43701.89	RU	0.5			MIT
2285.380	2286.085	43742.90	RU	1.9	0.0		MIT
2281.710	2282.414	43813.25	RU	1.1	1.2		MIT
2281.609	2282.313	43815.19	RU	0.6	1.2		MIT
2279.569	2280.273	43854.40	RU	2.0	1.0		MIT
2278.19	2278.89	43880.9	RU	1.9			A
2276.417	2277.120	43915.12	RU	0.5	0.5		MIT
2276.132	2276.835	43920.62	RU	0.6	0.0		MIT
2274.989	2275.692	43942.68	RU	0.8	0.7		MIT
2272.089	2272.791	43998.76	RU	2.0	0.5		MIT
2268.132	2268.833	44075.52	RU	1.0	0.0		MIT
2263.499	2264.199	44165.72	RU	1.0	0.0		MIT
2260.080	2260.780	44232.53	RU	0.6			MIT
2255.521	2256.220	44321.93	RU	1.9	0.5		MIT
2253.64	2254.34	44358.9	RU	1.7			A
2249.43	2250.13	44441.9	RU	0.9			A
2243.220	2243.916	44564.95	RU	1.3	0.0		MIT
2241.07	2241.77	44607.7	RU	1.8			A
2239.85	2240.55	44632.0	RU	0.9			A
2238.33	2239.02	44662.3	RU	1.7			A
2235.84	2236.53	44712.0	RU	1.1			A
2232.08	2232.77	44787.3	RU	1.1			A
2230.804	2231.497	44812.96	RU	0.9	0.3		MIT
2230.31	2231.00	44822.9	RU	1.5			A
2227.64	2228.33	44876.6	RU	1.5			A
2218.527	2219.218	45060.92	RU	0.6			MIT
2209.074	2209.763	45253.73	RU	1.0			MIT
2207.317	2208.005	45289.74	RU	0.3	0.0		MIT
2200.28	2200.97	45434.6	RU	0.9			A
2195.73	2196.42	45528.7	RU	1.1			A
2194.42	2195.11	45555.9	RU	1.1			A
2192.858	2193.543	45588.34	RU	0.5			MIT
2188.754	2189.439	45673.81	RU	0.5	0.0		MIT
2140.88	2141.55	46695.0	RU	1.1			A
2124.73	2125.40	47049.9	RU	0.6			A
2124.39	2125.06	47057.5	RU	0.6			A
2123.67	2124.34	47073.4	RU	0.6			A
2119.90	2120.57	47157.1	RU	0.3			A
2119.80	2120.47	47159.4	RU	0.3			A
2117.75	2118.42	47205.0	RU	0.6			A

AIR WAVELENGTH (ANGSTRMS)	VACUUM WAVELENGTH (ANGSTRMS)	WAVENUMBER (KAYSERS)	LMENT	LOG ARC	INTENSITY SPK	INTENSITY DIS	REF
2117.42	2118.09	47212.4	RU	0.9			A
2116.95	2117.62	47222.8	RU	0.5			A
2115.21	2115.88	47261.7	RU	0.5			A
2111.02	2111.69	47355.5	RU	0.3			A
2107.91	2108.58	47425.3	RU	0.6			A
2107.33	2108.00	47438.4	RU	0.3			A
2105.53	2106.20	47478.9	RU	0.3			A
2101.92	2102.59	47560.5	RU	0.7			A
2100.68	2101.35	47588.5	RU	0.9			A
2098.82	2099.49	47630.7	RU	0.9			A
2098.46	2099.13	47638.9	RU	0.9			A
2097.24	2097.91	47666.6	RU	0.7			A
2090.60	2091.26	47818.0	RU	0.6			A
2088.86	2089.52	47857.8	RU	0.3			A
2088.38	2089.04	47868.8	RU	0.7			A
2084.76	2085.42	47951.9	RU	0.5			A
2083.77	2084.43	47974.7	RU	0.7			A
2076.43	2077.09	48144.2	RU	0.6			A
2059.84	2060.50	48531.9	RU	0.3			A
2049.39	2050.05	48779.4	RU	0.3			A
9963.02	9965.75	10034.4	S			1.3	FH
9958.90	9961.63	10038.5	S			2.2	FH
9949.84	9952.57	10047.7	S			2.2	FH
9932.26	9934.98	10065.4	S			2.2	FH
9911.85	9914.57	10086.2	S			0.7	FH
9908.80	9911.52	10089.3	S			0.7	FH
9741.93	9744.60	10262.1	S			1.8	FH
9739.74	9742.41	10264.4	S			2.2	FH
9736.95	9739.62	10267.3	S			1.3	FH
9697.33	9699.99	10309.3	S			2.2	FH
9693.68	9696.34	10313.2	S			2.3	FH
9686.11	9688.77	10321.2	S			0.7	FH
9680.80	9683.46	10326.9	S			2.3	FH
9672.34	9674.99	10335.9	S			2.3	FH
9649.94	9652.59	10359.9	S			2.4	FH
9633.78	9636.42	10377.3	S			1.8	FH
9499.16	9501.77	10524.4	S			0.7	FH
9477.86	9480.46	10548.0	S			1.3	FH
9455.43	9458.02	10573.0	S			1.6	FH
9445.03	9447.62	10584.7	S			0.7	FH
9437.60	9440.19	10593.0	S			0.7	FH
9437.11	9439.70	10593.6	S			2.2	FH
9421.93	9424.51	10610.6	S			2.2	FH
9413.46	9416.04	10620.2	S			2.2	FH
9382.94	9385.51	10654.7	S			1.3	FH
9237.49	9240.02	10822.5	S I			2.3	FH
9228.11	9230.64	10833.5	S I			2.3	FH
9212.91	9215.44	10851.4	S I			2.3	FH
9039.27	9041.75	11059.8	S I			1.3	FH
9038.72	9041.20	11060.5	S I			1.3	FH
9036.73	9039.21	11062.9	S I			0.7	FH
9036.32	9038.80	11063.4	S I			1.6	FH
9035.92	9038.40	11063.9	S I			2.0	FH
8884.23	8886.67	11252.8	S I			2.2	FH
8882.47	8884.91	11255.0	S I			1.8	FH
8880.70	8883.14	11257.3	S I			1.3	FH
8874.53	8876.97	11265.1	S I			2.2	FH
8847.28	8849.71	11299.8	S I			1.2	FH
8694.71	8697.10	11498.1	S I			2.3	FH
8694.01	8696.40	11499.0	S I			1.3	FH
8680.45	8682.83	11517.0	S I			2.3	FH
8679.61	8681.99	11518.1	S I			0.7	FH
8671.35	8673.73	11529.1	S I			0.3	FH
8670.65	8673.03	11530.0	S I			0.7	FH
8668.42	8670.80	11533.0	S I			0.7	FH
8655.17	8657.55	11550.6	S I			0.3	FH
8633.18	8635.55	11580.0	S I			1.5	FH
8626.60	8628.97	11588.9	S I			0.3	FH
8617.15	8619.52	11601.6	S I			0.7	FH
8585.60	8587.96	11644.2	S			2.3	MS

AIR WAVELENGTH (ANGSTRMS)	VACUUM WAVELENGTH (ANGSTRMS)	WAVENUMBER (KAYSERS)	LMENT	LOG ARC	INTENSITY SPK	INTENSITY DIS	REF
8452.18	8454.50	11828.0	S I			1.8	FH
8451.55	8453.87	11828.9	S I			0.5	MS
8449.57	8451.89	11831.7	S I			1.5	FH
8314.73	8317.02	12023.5	S			2.3	MS
8258.27	8260.54	12105.7	S II			0.6	BT
8225.15	8227.41	12154.5	S II			0.6	BT
8179.31	8181.56	12222.6	S II			0.7	BT
8150.76	8153.00	12265.4	S I			0.3	MS
8148.40	8150.64	12269.0	S I			0.3	MS
8147.82	8150.06	12269.8	S I			1.5	MS
8136.26	8138.50	12287.3	S I			1.5	MS
8133.02	8135.26	12292.2	S II			0.6	MS
8131.29	8133.53	12294.8	S I			1.7	MS
8125.44	8127.67	12303.6	S I			1.3	FH
8123.01	8125.24	12307.3	S I			1.5	FH
7967.43	7969.62	12547.7	S			2.3	MS
7954.31	7956.50	12568.3	S			0.6	MS
7931.63	7933.81	12604.3	S I			2.3	FH
7928.84	7931.02	12608.7	S I			1.7	MS
7923.95	7926.13	12616.5	S I			2.5	MS
7902.09	7904.26	12651.4	S I			0.6	MS
7696.73	7698.85	12989.0	S I			2.3	MS
7686.13	7688.25	13006.9	S I			2.2	MS
7679.60	7681.71	13017.9	S I			1.8	MS
7629.82	7631.92	13102.9	S			2.3	MS
7578.96	7581.05	13190.8	S			2.3	MS
7476.2	7478.3	13372.	S			1.7	BZ
7468.25	7470.31	13386.3	S			1.0	MS
7450.33	7452.38	13418.5	S I			1.2	MS
7449.09	7451.14	13420.8	S I			0.7	MS
7424.11	7426.15	13465.9	S I			0.7	MS
7281.04	7283.05	13730.5	S			1.8	MS
7249.06	7251.06	13791.1	S I			1.4	MS
7244.77	7246.77	13799.3	S I			1.9	MS
7243.06	7245.06	13802.5	S I			1.4	MS
7167.76	7169.74	13947.5	S I			1.2	FH
7166.74	7168.72	13949.5	S I			0.5	MS
7165.12	7167.10	13952.7	S I			1.2	FH
7164.56	7166.53	13953.7	S I			1.0	MS
7161.43	7163.40	13959.8	S I			1.0	FH
7139.79	7141.76	14002.2	S II			1.0	BT
7040.92	7042.86	14198.8	S			1.8	MS
6994.57	6996.50	14292.9	S I			1.5	FH
6992.80	6994.73	14296.5	S I			1.2	MS
6981.40	6983.33	14319.8	S			1.7	BL
6925.00	6926.91	14436.4	S I			0.3	MS
6907.5	6909.4	14473.	S			1.5	BZ
6848.08	6849.97	14598.6	S			0.3	MS
6776.75	6778.62	14752.3	S			1.5	BL
6760.71	6762.58	14787.3	S			1.5	BL
6757.10	6758.97	14795.2	S I			2.2	MS
6748.79	6750.65	14813.4	S I			1.9	MS
6743.58	6745.44	14824.8	S I			1.7	MS
6734.61	6736.47	14844.6	S			1.2	BL
6729.47	6731.33	14855.9	S			1.5	BL
6717.74	6719.59	14881.8	S			1.7	BL
6715.83	6717.68	14886.1	S			1.5	BL
6712.75	6714.60	14892.9	S			1.5	BL
6682.92	6684.77	14959.4	S			1.2	MS
6681.92	6683.76	14961.6	S			1.5	BL
6681.24	6683.08	14963.2	S			1.5	BL
6641.50	6643.33	15052.7	S			1.2	BL
6565.04	6566.85	15228.0	S			1.2	BL
6538.57	6540.38	15289.6	S			1.8	FH
6537.97	6539.78	15291.0	S I			0.5	FH
6536.67	6538.48	15294.1	S I			1.2	FH
6536.41	6538.22	15294.7	S I			1.7	FH
6535.63	6537.44	15296.5	S I			0.9	FH
6455.36	6457.14	15486.7	S			1.8	BL
6425.64	6427.42	15558.3	S			1.2H	BL

AIR WAVELENGTH (ANGSTRMS)	VACUUM WAVELENGTH (ANGSTRMS)	WAVENUMBER (KAYSERS)	LMENT	LOG ARC	INTENSITY SPK	INTENSITY DIS	REF
6418.90	6420.67	15574.7	S			0.8	BL
6415.50	6417.27	15582.9	S I			1.0	MS
6413.71	6415.48	15587.3	S			2.7	BL
6408.13	6409.90	15600.9	S I			0.7	MS
6403.58	6405.35	15611.9	S I			0.3	MS
6398.05	6399.82	15625.4	S			2.5	BL
6397.30	6399.07	15627.3	S			2.5	BL
6396.54	6398.31	15629.1	S I			1.2	FH
6395.07	6396.84	15632.7	S I			1.2	FH
6386.48	6388.25	15653.7	S			0.7	BT
6384.89	6386.66	15657.7	S			2.5	BL
6369.34	6371.10	15695.9	S			1.7	BL
6312.68	6314.43	15836.7	S			3.0	BL
6305.51	6307.25	15854.8	S			3.0	BL
6287.06	6288.80	15901.3	S			3.0	BL
6286.35	6288.09	15903.1	S			2.5	BL
6274.34	6276.08	15933.5	S			1.5	BL
6248.67	6250.40	15999.0	S			1.2H	BL
6175.85	6177.56	16187.6	S I			1.6	FH
6174.97	6176.68	16189.9	S			1.5	FH
6173.64	6175.35	16193.4	S I			1.6	FH
6172.81	6174.52	16195.6	S I			0.7	FH
6138.98	6140.68	16284.8	S			1.7	BL
6102.26	6103.95	16382.8	S			1.7	BL
6092.12	6093.81	16410.1	S			1.2	BL
6052.63	6054.31	16517.2	S I			2.1	MS
6046.04	6047.71	16535.2	S I			1.6	MS
6041.93	6043.60	16546.4	S I			1.2	MS
5996.16	5997.82	16672.7	S			1.4	BL
5951.50	5953.15	16797.8	S			1.2	BL
5940.69	5942.34	16828.4	S			0.9	BL
5908.25	5909.89	16920.8	S			0.9	BL
5890.98	5892.61	16970.4	S			0.9	BL
5889.75	5891.38	16973.9	S I			0.7	MS
5869.08	5870.71	17033.7	S			0.8	BL
5819.22	5820.83	17179.7	S II			1.0	IG
5797.43	5799.04	17244.2	S			0.9H	BL
5730.44	5732.03	17445.8	S			1.2	BL
5706.11	5707.69	17520.2	S I			1.7	MS
5700.24	5701.82	17538.2	S I			1.4	MS
5696.63	5698.21	17549.4	S I			0.9	MS
5664.73	5666.30	17648.2	S II			1.2	IG
5659.93	5661.50	17663.2	S			2.8	BL
5648.29	5649.86	17699.6	S			1.4	BL
5646.98	5648.55	17703.7	S II			1.5	IG
5645.70	5647.27	17707.7	S II			1.0	IG
5640.37	5641.94	17724.4	S			2.7	BL
5639.98	5641.55	17725.6	S			2.7	BL
5616.63	5618.19	17799.3	S II			0.7	IG
5614.39	5615.95	17806.4	S I			0.5	MS
5606.10	5607.66	17832.8	S			2.8	BL
5578.86	5580.41	17919.8	S			1.8	BL
5564.93	5566.48	17964.7	S			2.2	BL
5558.91	5560.45	17984.1	S II			0.7	IG
5556.01	5557.55	17993.5	S II			0.7	IG
5526.20	5527.74	18090.6	S			1.2	BL
5520.42	5521.95	18109.5	S			1.2	BL
5518.74	5520.27	18115.1	S			1.2	BL
5509.67	5511.20	18144.9	S II			1.4	IG
5507.01	5508.54	18153.6	S I			1.4	MS
5501.54	5503.07	18171.7	S I			1.2	MS
5498.18	5499.71	18182.8	S I			0.9	MS
5478.28	5479.80	18248.8	S			0.9	BL
5475.06	5476.58	18259.6	S			1.2	BL
5473.63	5475.15	18264.3	S			2.9	BL
5453.88	5455.40	18330.5	S			2.9	BL
5432.83	5434.34	18401.5	S			2.8	BL
5428.69	5430.20	18415.5	S			2.4	BL
5381.02	5382.52	18578.7	S I			0.9	MS
5345.66	5347.15	18701.6	S			1.4	BL

AIR WAVELENGTH (ANGSTRMS)	VACUUM WAVELENGTH (ANGSTRMS)	WAVENUMBER (KAYSERS)	LMENT	LOG ARC	INTENSITY SPK	INTENSITY DIS	REF
5320.70	5322.18	18789.3	S II			1.5	IG
5278.91	5280.38	18938.0	S I			1.2	FH
5278.61	5280.08	18939.1	S I			0.7	FH
5212.61	5214.06	19178.9	S II			1.3	IG
5201.35	5202.80	19220.4	S II			1.0	IG
5201.00	5202.45	19221.7	S II			1.0	IG
5141.89	5143.32	19442.7	S			0.9	MS
5126.13	5127.56	19502.5	S II			0.9	BT
5047.74	5049.15	19805.3	S			1.2	MS
5032.39	5033.79	19865.7	S			0.9	BL
5014.01	5015.41	19938.6	S			0.5	BL
5009.54	5010.94	19956.4	S			0.5	BL
4993.51	4994.90	20020.4	S			2.2	MS
4925.32	4926.69	20297.6	S II			2.0	IG
4924.08	4925.45	20302.7	S II			1.8	IG
4920.48	4921.85	20317.6	S			1.2	MS
4917.15	4918.52	20331.3	S II			1.5	IG
4902.44	4903.81	20392.3	S			1.2	BL
4885.63	4886.99	20462.5	S II			1.5	HN
4835.85	4837.20	20673.1	S II			0.9	HN
4826.77	4828.12	20712.0	S II			0.5	HN
4824.07	4825.42	20723.6	S II			1.6	HN
4819.60	4820.95	20742.8	S II			1.4	HN
4815.515	4816.861	20760.41	S II			2.9	HN
4802.81	4804.15	20815.3	S			1.3	HN
4792.02	4793.36	20862.2	S II			1.6	HN
4779.11	4780.45	20918.6	S II			1.4	HN
4763.38	4764.71	20987.6	S II			1.3	HN
4755.12	4756.45	21024.1	S II			1.5	HN
4731.15	4732.47	21130.6	S			1.2	MS
4729.45	4730.77	21138.2	S II			0.9	HN
4716.226	4717.545	21197.46	S II			2.8	HN
4700.21	4701.53	21269.7	S II			0.7	HN
4696.25	4697.56	21287.6	S I			1.2	MS
4695.45	4696.76	21291.3	S I			1.5	MS
4694.13	4695.44	21297.2	S I			1.7	MS
4681.32	4682.63	21355.5	S II			0.7	HN
4669.14	4670.45	21411.2	S			1.5	MS
4668.58	4669.89	21413.8	S II			1.7	HN
4656.74	4658.04	21468.2	S II			1.9	HN
4648.17	4649.47	21507.8	S II			1.5	HN
4624.11	4625.41	21619.7	S			1.3	HN
4591.05	4592.34	21775.4	S			1.5	HN
4561.88	4563.16	21914.6	S			1.4	HN
4552.378	4553.654	21960.39	S II			2.3	HN
4549.547	4550.822	21974.05	S			1.9	HN
4529.18	4530.45	22072.9	S			0.9	MS
4524.946	4526.215	22093.52	S II			2.2	HN
4524.68	4525.95	22094.8	S II			1.3	HN
4497.88	4499.14	22226.5	S II			0.7	HN
4486.66	4487.92	22282.0	S II			1.5	HN
4483.424	4484.682	22298.12	S II			2.0	HN
4482.48	4483.74	22302.8	S			0.9	HN
4469.96	4471.21	22365.3	S			0.9H	MS
4464.425	4465.678	22393.02	S			2.0	HN
4463.582	4464.835	22397.25	S II			2.3	HN
4456.43	4457.68	22433.2	S II			1.5	HN
4437.01	4438.26	22531.4	S			1.4	MS
4432.41	4433.65	22554.8	S II			1.7	HN
4431.02	4432.26	22561.8	S II			1.3	HN
4411.34	4412.58	22662.5	S I			0.5	HN
4402.86	4404.10	22706.1	S II			0.5	HN
4391.84	4393.07	22763.1	S II			1.5	HN
4369.96	4371.19	22877.1	S			0.5H	HN
4360.49	4361.72	22926.8	S			0.9	HN
4349.41	4350.63	22985.2	S			0.6	BL
4333.84	4335.06	23067.7	S II			0.5	HN
4330.95	4332.17	23083.1	S			0.5	HN
4318.68	4319.89	23148.7	S II			1.6	HN
4294.432	4295.640	23279.42	S II			1.9	HN

AIR WAVELENGTH (ANGSTRMS)	VACUUM WAVELENGTH (ANGSTRMS)	WAVENUMBER (KAYSERS)	ELMENT		LOG ARC	INTENSITY SPK	DIS	REF
4291.45	4292.66	23295.6	S	II			0.7	HN
4283.70	4284.91	23337.7	S				0.5	HN
4282.63	4283.84	23343.6	S	II			1.5	HN
4278.54	4279.74	23365.9	S	II			1.5	HN
4269.76	4270.96	23413.9	S	II			1.5	HN
4267.802	4269.003	23424.67	S	II			1.8	HN
4259.18	4260.38	23472.1	S	II			1.2	HN
4257.42	4258.62	23481.8	S	II			1.5	HN
4249.92	4251.12	23523.2	S	II			0.5	HN
4230.98	4232.17	23628.5	S	II			1.5	HN
4217.23	4218.42	23705.6	S	II			1.5	HN
4193.51	4194.69	23839.6	S	II			1.2	HN
4189.71	4190.89	23861.3	S	II			2.4	HN
4185.95	4187.13	23882.7	S				1.2	HN
4175.21	4176.39	23944.1	S				0.7	HN
4174.300	4175.477	23949.36	S				2.2	HN
4174.042	4175.219	23950.84	S	II			1.7	HN
4168.409	4169.584	23983.21	S	II			1.7	HN
4165.11	4166.28	24002.2	S	II			1.0	HN
4164.96	4166.13	24003.1	S				0.9	HN
4162.698	4163.872	24016.11	S	II			2.8	HN
4162.39	4163.56	24017.9	S	II			1.3	HN
4157.70	4158.87	24045.0	S	I			1.0	MS
4153.098	4154.269	24071.62	S	II			2.8	HN
4148.91	4150.08	24095.9	S				0.7	HN
4146.94	4148.11	24107.4	S	II			1.5	HN
4145.100	4146.269	24118.07	S	II			2.4	HN
4142.291	4143.459	24134.42	S	II			2.2	HN
4130.95	4132.12	24200.7	S				1.2	BL
4127.54	4128.70	24220.7	S				0.5	HN
4112.28	4113.44	24310.6	S				0.5	HN
4111.56	4112.72	24314.8	S				1.5	HN
4099.44	4100.60	24386.7	S				0.9	HN
4064.45	4065.60	24596.6	S				1.4	HN
4050.11	4051.25	24683.7	S	II			1.0	HN
4034.01	4035.15	24782.2	S				0.9	MS
4032.812	4033.952	24789.59	S	II			2.1	HN
4028.791	4029.929	24814.33	S	II			2.3	HN
4009.39	4010.52	24934.4	S	II			0.5	HN
4007.78	4008.91	24944.4	S	II			0.7	HN
4003.89	4005.02	24968.6	S	II			0.9	HN
3998.79	3999.92	25000.5	S	II			1.8	HN
3997.97	3999.10	25005.6	S				0.9	HN
3993.526	3994.655	25033.45	S	II			1.7	HN
3990.94	3992.07	25049.7	S	II			1.6	HN
3979.86	3980.99	25119.4	S	II			1.5	HN
3970.69	3971.81	25177.4	S	II			0.7	HN
3963.13	3964.25	25225.4	S	II			1.0	HN
3961.55	3962.67	25235.5	S	I			1.0	HN
3946.98	3948.10	25328.7	S	II			0.7	HN
3933.294	3934.408	25416.79	S	II			1.9	HN
3932.30	3933.41	25423.2	S	II			1.0	HN
3931.938	3933.051	25425.55	S	II			1.2	HN
3923.483	3924.594	25480.34	S	II			2.3	HN
3920.68	3921.79	25498.6	S				0.9	BL
3901.99	3903.10	25620.7	S				1.3	HN
3892.321	3893.424	25684.33	S	II			1.5	HN
3879.43	3880.53	25769.7	S				0.5	BL
3867.56	3868.66	25848.8	S				2.2	MS
3860.64	3861.73	25895.1	S	II			1.2	HN
3860.15	3861.24	25898.4	S	II			0.9	HN
3853.09	3854.18	25945.8	S	II			0.9	HN
3850.93	3852.02	25960.4	S	II			0.9	HN
3845.21	3846.30	25999.0	S	II			0.3	HN
3843.41	3844.50	26011.2	S				1.2	MS
3841.40	3842.49	26024.8	S				0.7	BL
3839.19	3840.28	26039.8	S				0.9	MS
3831.41	3832.50	26092.6	S				1.0	HN
3830.94	3832.03	26095.9	S				0.9	MS
3829.33	3830.42	26106.8	S				0.7	BL

AIR WAVELENGTH (ANGSTRMS)	VACUUM WAVELENGTH (ANGSTRMS)	WAVENUMBER (KAYSERS)	ELMENT		LOG ARC	INTENSITY SPK	DIS	REF
3818.04	3819.12	26184.0	S				0.9	BL
3808.03	3809.11	26252.9	S				0.8	BL
3799.21	3800.29	26313.8	S				0.9	MS
3792.46	3793.54	26360.6	S	II			1.5	BT
3783.49	3784.56	26423.1	S				1.2	BL
3783.16	3784.23	26425.4	S	II			0.9	HN
3782.6	3783.7	26429.	S	II			0.9	HN
3782.26	3783.33	26431.7	S				1.5	BL
3780.64	3781.71	26443.0	S				0.5	HN
3776.80	3777.87	26469.9	S	II			0.5	HN
3767.92	3768.99	26532.3	S				0.6	BL
3767.73	3768.80	26533.6	S				0.9	BL
3756.78	3757.85	26611.0	S				0.9	BL
3741.15	3742.21	26722.1	S				0.5	HN
3730.64	3731.70	26797.4	S				0.7	HN
3700.09	3701.14	27018.7	S				0.5	HN
3699.37	3700.42	27023.9	S				0.9	HN
3696.25	3697.30	27046.7	S				0.7	HN
3687.13	3688.18	27113.6	S				1.2	MS
3678.13	3679.18	27180.0	S				1.0	HN
3675.85	3676.90	27196.9	S				0.6	BL
3672.14	3673.19	27224.3	S	II			1.3	HN
3669.049	3670.094	27247.26	S	II			1.8	HN
3661.95	3662.99	27300.1	S				0.9	BL
3654.51	3655.55	27355.7	S	II			0.9	HN
3641.49	3642.53	27453.5	S				0.5	BL
3638.15	3639.19	27478.7	S				0.5	HN
3634.25	3635.29	27508.1	S				1.5H	MS
3626.53	3627.56	27566.7	S				1.0	HN
3622.68	3623.71	27596.0	S				0.5	HN
3616.916	3617.947	27639.98	S	II			1.8	HN
3613.03	3614.06	27669.7	S	II			1.1	HN
3600.08	3601.11	27769.2	S				0.5	HN
3599.88	3600.91	27770.8	S				0.7	BL
3595.991	3597.017	27800.81	S	II			1.7	HN
3594.462	3595.488	27812.64	S	II			1.5	HN
3584.18	3585.20	27892.4	S				1.0	HN
3567.171	3568.190	28025.42	S	II			1.6	HN
3549.72	3550.73	28163.2	S				0.9	HN
3531.90	3532.91	28305.3	S				0.7	HN
3503.78	3504.78	28532.4	S				0.9H	BL
3497.340	3498.341	28584.98	S				2.0	HN
3464.66	3465.65	28854.6	S				0.9	BL
3448.09	3449.08	28993.3	S				0.9	BL
3413.74	3414.72	29285.0	S				0.9	BL
3395.87	3396.84	29439.1	S				1.2	BL
3387.35	3388.32	29513.1	S				0.9	HN
3385.81	3386.78	29526.6	S				1.4	HN
3373.19	3374.16	29637.0	S				0.7	HN
3372.48	3373.45	29643.3	S				0.5	HN
3371.90	3372.87	29648.4	S				1.3	HN
3368.09	3369.06	29681.9	S				1.4	HN
3356.41	3357.37	29785.2	S				1.2	HN
3354.98	3355.94	29797.9	S				0.9	BL
3347.43	3348.39	29865.1	S				0.9	HN
3342.74	3343.70	29907.0	S				0.9	BL
3341.88	3342.84	29914.7	S				0.9	HN
3314.50	3315.45	30161.8	S	II			0.9	HN
3288.15	3289.10	30403.5	S				0.9	BL
3284.68	3285.63	30435.6	S				1.2	BL
3261.60	3262.54	30651.0	S				0.9	BL
3257.83	3258.77	30686.4	S	II			1.0	BL
3253.46	3254.40	30727.6	S				1.2	BL
3147.54	3148.45	31761.6	S				1.5	BL
3130.71	3131.62	31932.4	S				1.2	BL
3090.65	3091.55	32346.3	S				0.9	BL
3081.98	3082.88	32437.2	S				1.2	BL
3076.20	3077.09	32498.2	S				1.4	BL
3066.89	3067.78	32596.9	S				1.5	BL
3060.86	3061.75	32661.1	S				0.9	BL

AIR WAVELENGTH (ANGSTRMS)	VACUUM WAVELENGTH (ANGSTRMS)	WAVENUMBER (KAYSERS)	LMENT		LOG ARC	INTENSITY SPK	INTENSITY DIS	REF
3059.07	3059.96	32680.2	S				0.9	BL
3015.59	3016.47	33151.4	S	II			0.7	BL
2998.28	2999.15	33342.7	S				1.4	BL
2957.83	2958.69	33798.7	S				0.9	BL
2847.41	2848.25	35109.3	S				0.9	BL
2817.56	2818.39	35481.3	S				1.2	BL
2802.23	2803.06	35675.4	S				0.9	BL
2750.61	2751.42	36344.8	S				0.9	BL
2749.00	2749.81	36366.1	S				0.9	BL
2737.27	2738.08	36521.9	S				0.9	BL
2730.36	2731.17	36614.4	S				0.9	BL
2688.32	2689.12	37186.9	S				1.2	BL
2675.24	2676.04	37368.7	S				0.9	BL
2670.07	2670.86	37441.1	S				0.9	BL
2662.82	2663.61	37543.0	S				1.2	BL
2639.14	2639.93	37879.8	S				0.9	BL
2636.32	2637.11	37920.4	S				0.9	BL
2629.18	2629.96	38023.3	S	II			0.7	BL
2625.14	2625.92	38081.9	S				0.9	BL
2595.70	2596.48	38513.7	S				0.9	BL
2587.10	2587.87	38641.8	S				0.9	BL
2585.25	2586.02	38669.4	S				1.0	BL
2584.29	2585.06	38683.8	S				1.2	BL
2581.02	2581.79	38732.8	S				0.9	BL
2578.86	2579.63	38765.2	S				0.9	BL
2564.98	2565.75	38975.0	S				0.9	BL
2554.76	2555.53	39130.9	S				0.9	BL
2518.44	2519.20	39695.2	S				0.9	BL
2508.15	2508.91	39858.0	S	II			1.6	IG
2475.45	2476.20	40384.5	S				0.9	BL
2357.75	2358.47	42400.4	S				0.9	BL
2336.25	2336.97	42790.5	S				1.4	BL
2332.45	2333.17	42860.2	S				1.2	BL
2236.76	2237.45	44693.6	S				1.2	BL
2198.80	2199.49	45465.2	S				0.9	LC
2193.12	2193.81	45582.9	S				1.2	LC
2191.39	2192.08	45618.9	S				1.2	LC
2189.29	2189.97	45662.6	S				1.2	LC
2188.02	2188.70	45689.1	S				1.2	LC
2178.77	2179.45	45883.1	S				0.9	LC
2170.48	2171.16	46058.3	S				0.9	LC
2166.27	2166.95	46147.8	S				0.9	LC
2126.81	2127.48	47003.9	S				1.2	LC
2125.40	2126.07	47035.1	S				1.2	LC
2124.09	2124.76	47064.1	S				1.4	LC
2123.32	2123.99	47081.2	S				1.2	LC
2098.65	2099.32	47634.6	S				1.7	LC
2098.14	2098.81	47646.1	S				2.5	LC
2089.90	2090.56	47834.0	S				2.5	LC
2085.57	2086.23	47933.3	S				0.9	LC
2072.83	2073.49	48227.8	S				0.9	LC
2067.93	2068.59	48342.1	S				0.9	LC
2066.48	2067.14	48376.0	S				0.9	LC
2002.86	2003.51	49912.4	S				1.0	LC
2001.30	2001.95	49951.3	S				0.9	LC
9949.14	9951.87	10048.4	SB		2.6H			ME
9866.78	9869.49	10132.2	SB		1.5			ME
9775.13	9777.81	10227.2	SB		1.3			ME
9756.72	9759.40	10246.5	SB		1.4			ME
9662.2	9664.9	10347.	SB		1.5D			ME
9602.94	9605.57	10410.6	SB		1.3			ME
9578.68	9581.31	10437.0	SB		2.6			ME
9549.13	9551.75	10469.3	SB		0.9			ME
9546.12	9548.74	10472.6	SB		1.0H			ME
9518.68	9521.29	10502.8	SB		2.6			ME
9132.21	9134.72	10947.2	SB		1.5			ME
9019.24	9021.72	11084.4	SB		1.0			ME
8903.1	8905.5	11229.	SB		0.7V			WT
8860.3	8862.7	11283.	SB		0.7V			WT
8789.2	8791.6	11374.	SB		0.7V			WT
8735.7	8738.1	11444.	SB		1.6V			ME
8719.85	8722.24	11464.9	SB		1.0			ME
8700.1	8702.5	11491.	SB		1.7V			ME
8682.7	8685.1	11514.	SB		2.0H			ME
8667.1	8669.5	11535.	SB		1.3H			ME
8619.52	8621.89	11598.4	SB		2.2H			ME
8617.7	8620.1	11601.	SB		1.6H			ME
8572.61	8574.97	11661.9	SB		2.3			ME
8447.	8449.	11840.	SB	II		1.0		DV
8411.70	8414.01	11884.9	SB		1.8			ME
8314.25	8316.54	12024.2	SB		1.2H			ME
8289.0	8291.3	12061.	SB		1.5H			ME
8270.12	8272.39	12088.4	SB		1.0H			ME
8257.85	8260.12	12106.4	SB		1.3H			ME
8241.70	8243.97	12130.1	SB		0.8			ME
8062.1	8064.3	12400.	SB		1.0V			WT
7988.	7990.	12520.	SB	II		1.1		DV
7969.6	7971.8	12544.	SB		1.7H			ME
7933.38	7935.56	12601.5	SB		0.7			WT
7924.65	7926.83	12615.4	SB		2.5			ME
7906.67	7908.85	12644.1	SB		1.0			ME
7867.15	7869.31	12707.6	SB		1.3			ME
7844.41	7846.57	12744.4	SB		2.0			ME
7837.30	7839.46	12756.0	SB		1.0H			ME
7834.6	7836.8	12760.	SB		0.3V			WT
7816.9	7819.1	12789.	SB		1.0V			WT
7772.	7774.	12860.	SB	II		1.2		DV
7764.9	7767.0	12875.	SB		1.0V			WT
7755.80	7757.93	12890.0	SB		1.0			ME
7648.28	7650.39	13071.2	SB		1.0H			ME
7592.82	7594.91	13166.7	SB		0.7H			WT
7428.	7430.	13460.	SB	II		1.5		DV
7405.50	7407.54	13499.8	SB		0.7H			WT
7343.4	7345.4	13614.	SB	II	0.0	0.5		LG
7275.	7277.	13740.	SB	II		1.5		DV
7222.50	7224.49	13841.8	SB		0.7H			WT
7194.7	7196.7	13895.	SB	II		1.5		DV
7163.50	7165.47	13955.8	SB	II			0.6	LG
7145.48	7147.45	13991.0	SB				0.3	LG
7075.11	7077.06	14130.2	SB				0.6	LG
7051.26	7053.20	14177.9	SB				0.9	LG
6996.4	6998.3	14289.	SB	II		1.4		DV
6995.91	6997.84	14290.1	SB		0.7H			WT
6947.3	6949.2	14390.	SB	II		1.3		DV
6915.58	6917.49	14456.1	SB	II			0.9	LG
6874.3	6876.2	14543.	SB	II		1.4		DV
6842.98	6844.87	14609.5	SB				0.9	LG
6806.67	6808.55	14687.4	SB	II			1.1	LG
6806.35	6808.23	14688.1	SB		0.7J			WT
6806.16	6808.04	14688.5	SB	II		1.5		MIT
6778.75	6780.62	14747.9	SB	II			0.6	LG
6778.38	6780.25	14748.7	SB		0.7	1.6		WT
6738.99	6740.85	14834.9	SB		0.3			WT
6713.60	6715.45	14891.0	SB	II			0.8	LG
6688.01	6689.86	14948.0	SB				1.5	LG
6648.134	6649.970	15037.66	SB		0.5			WT
6647.44	6649.28	15039.2	SB				1.8	LG
6634.63	6636.46	15068.3	SB		0.3J			WT
6629.48	6631.31	15080.0	SB	II			0.3	LG
6621.28	6623.11	15098.7	SB	II			0.6	LG
6611.48	6613.31	15121.0	SB		0.8			WT
6610.36	6612.19	15123.6	SB		0.6			MIT
6564.052	6565.865	15230.29	SB		0.8			WT
6558.150	6559.962	15243.99	SB		1.1			WT
6546.86	6548.67	15270.3	SB	II	0.6		0.3	LG
6529.88	6531.68	15310.0	SB	II	0.6		0.6	LG
6517.10	6518.90	15340.0	SB		0.6			WT
6511.13	6512.93	15354.1	SB				1.0	LG
6503.26	6505.06	15372.7	SB	II			1.1	LG
6502.88	6504.68	15373.6	SB	II		1.3		KZ

AIR WAVELENGTH (ANGSTRMS)	VACUUM WAVELENGTH (ANGSTRMS)	WAVENUMBER (KAYSERS)	LMENT	LOG ARC	INTENSITY SPK	INTENSITY DIS	REF
6494.96	6496.75	15392.3	SB	0.6			KZ
6490.25	6492.04	15403.5	SB	0.8			WT
6474.43	6476.22	15441.1	SB II	0.6H		1.0	LG
6465.39	6467.18	15462.7	SB II		1.2		KZ
6454.95	6456.73	15487.7	SB	0.8H			WT
6415.93	6417.70	15581.9	SB	0.8B			WT
6406.08	6407.85	15605.9	SB			0.8	LG
6405.406	6407.177	15607.50	SB	0.6			WT
6396.21	6397.98	15629.9	SB	0.6H		0.3H	LG
6392.175	6393.942	15639.80	SB	0.9	0.3H		WT
6388.324	6390.090	15649.23	SB	0.3H			WT
6376.11	6377.87	15679.2	SB II	0.6		0.8	LG
6359.5	6361.3	15720.	SB		1.0		DV
6345.955	6347.710	15753.71	SB	0.6			WT
6332.220	6333.971	15787.88	SB	0.8			WT
6319.76	6321.51	15819.0	SB II			1.0	LG
6319.19	6320.94	15820.4	SB II	0.6	1.1		KZ
6302.764	6304.507	15861.67	SB II			1.3	LG
6302.35	6304.09	15862.7	SB	0.6H			KZ
6287.55	6289.29	15900.1	SB		1.0H		KZ
6284.44	6286.18	15907.9	SB			0.6	LG
6221.45	6223.17	16069.0	SB	0.9	0.3H		WT
6214.00	6215.72	16088.2	SB	0.8			WT
6208.39	6210.11	16102.8	SB II		0.9J		KZ
6204.51	6206.23	16112.8	SB	0.8			WT
6203.20	6204.92	16116.2	SB II	0.3H		1.4	LG
6196.33	6198.04	16134.1	SB	0.8H			KZ
6186.00	6187.71	16161.1	SB II			0.8	LG
6178.64	6180.35	16180.3	SB	0.6H			WT
6156.83	6158.53	16237.6	SB		1.2		KZ
6154.945	6156.648	16242.60	SB	0.9H		1.6	LG
6137.30	6139.00	16289.3	SB II	0.6H		0.3	LG
6129.98	6131.68	16308.7	SB II	1.0H	2.2H		WT
6112.70	6114.39	16354.9	SB	0.6H			WT
6097.60	6099.29	16395.4	SB	1.2			WT
6081.14	6082.82	16439.7	SB	0.6	0.8		WT
6079.797	6081.480	16443.36	SB II			1.8	LG
6079.55	6081.23	16444.0	SB	1.3	2.0H		WT
6073.932	6075.614	16459.24	SB II	1.3W		0.9	LG
6064.13	6065.81	16485.9	SB II	1.2		0.3	LG
6053.411	6055.087	16515.04	SB			1.2	LG
6051.99	6053.67	16518.9	SB		0.8		KZ
6006.105	6007.768	16645.12	SB	0.6H		1.3	LG
6005.21	6006.87	16647.6	SB II			2.3	LG
6005.00	6006.66	16648.2	SB	0.9H	1.1H		WT
6004.6	6006.3	16649.	SB II		2.3H		DV
5981.42	5983.08	16713.8	SB II			0.3	LG
5980.98	5982.64	16715.0	SB II		1.2		KZ
5948.37	5950.02	16806.7	SB	0.5			WT
5927.6	5929.2	16866.	SB II		1.2		DV
5917.77	5919.41	16893.6	SB II			0.8	LG
5912.33	5913.97	16909.1	SB	0.7			WT
5910.64	5912.28	16913.9	SB		0.3H		MIT
5901.20	5902.84	16941.0	SB II			0.7	LG
5895.09	5896.72	16958.6	SB II			2.2J	LG
5845.65	5847.27	17102.0	SB II			1.2	LG
5844.66	5846.28	17104.9	SB	0.3H			WT
5825.50	5827.12	17161.1	SB II			1.1	LG
5813.97	5815.58	17195.2	SB	0.3H			WT
5806.30	5807.91	17217.9	SB	0.3H			WT
5789.52	5791.13	17267.8	SB II			0.3	LG
5782.150	5783.753	17289.81	SB	0.8			SP
5774.557	5776.158	17312.54	SB	0.8			WT
5730.46	5732.05	17445.8	SB	1.1			WT
5707.480	5709.063	17516.01	SB	0.8			KZ
5705.50	5707.08	17522.1	SB II			1.3	LG
5700.219	5701.801	17538.32	SB	0.8			KZ
5691.26	5692.84	17565.9	SB	0.6			WT
5660.78	5662.35	17660.5	SB II	0.5	0.8		WT
5639.74	5641.31	17726.4	SB II		2.0J		WT

AIR WAVELENGTH (ANGSTRMS)	VACUUM WAVELENGTH (ANGSTRMS)	WAVENUMBER (KAYSERS)	LMENT	LOG ARC	INTENSITY SPK	INTENSITY DIS	REF
5635.18	5636.74	17740.7	SB II			1.6	LG
5631.97	5633.53	17750.9	SB II	1.2		1.6	WT
5602.30	5603.86	17844.9	SB	1.0			WT
5599.93	5601.48	17852.4	SB II	0.6		0.8	WT
5568.09	5569.64	17954.5	SB II	0.8	2.3J		WT
5567.0	5568.5	17958.	SB II		1.6		DV
5556.21	5557.75	17992.9	SB	0.8			WT
5555.993	5557.536	17993.59	SB II			0.3	LG
5531.89	5533.43	18072.0	SB II	0.6		0.7	WT
5490.33	5491.86	18208.8	SB	0.3			WT
5475.77	5477.29	18257.2	SB II			0.7	LG
5471.95	5473.47	18269.9	SB II			1.2	LG
5464.08	5465.60	18296.3	SB II			2.0	LG
5381.20	5382.70	18578.1	SB II			1.5	LG
5354.24	5355.73	18671.6	SB II			2.3	LG
5320.1	5321.6	18791.	SB II		1.0		DV
5238.94	5240.40	19082.5	SB II			1.7J	LG
5210.69	5212.14	19186.0	SB		0.6H		SP
5208.80	5210.25	19192.9	SB			0.9	LG
5179.36	5180.80	19302.0	SB		0.3H		SP
5176.55	5177.99	19312.5	SB II			1.7	LG
5172.46	5173.90	19327.8	SB II			1.2	LG
5166.32	5167.76	19350.7	SB II			1.5J	LG
5164.72	5166.16	19356.7	SB II			1.2	LG
5158.6	5160.0	19380.	SB II		1.1		DV
5152.30	5153.74	19403.4	SB II			1.1	LG
5141.2	5142.6	19445.	SB		1.6		SP
5113.86	5115.29	19549.2	SB		1.8		SP
5109.71	5111.13	19565.1	SB II			0.6	LG
5066.99	5068.40	19730.1	SB II			0.6	LG
5045.23	5046.64	19815.2	SB			1.2T	SP
5044.56	5045.97	19817.8	SB			2.0	LG
5033.03	5034.43	19863.2	SB II			1.2	LG
5025.3	5026.7	19894.	SB II		1.3		DV
5021.68	5023.08	19908.1	SB II			0.9	LG
5010.42	5011.82	19952.8	SB II			1.6	LG
5000.6	5002.0	19992.	SB II		1.3		DV
4991.66	4993.05	20027.8	SB II			1.0	LG
4975.71	4977.10	20092.0	SB II			0.5	LG
4948.52	4949.90	20202.4	SB II			1.7	LG
4947.40	4948.78	20207.0	SB II			1.5	LG
4914.2	4915.6	20343.	SB II		0.6		DV
4895.0	4896.4	20423.	SB II		0.8		DV
4880.01	4881.37	20486.0	SB II			0.3	LG
4877.24	4878.60	20497.7	SB II			1.8	LG
4850.50	4851.86	20610.7	SB II			0.3	LG
4843.74	4845.09	20639.4	SB II			1.3	LG
4838.27	4839.62	20662.8	SB II			1.0	LG
4832.82	4834.17	20686.1	SB II			0.7	LG
4829.0	4830.3	20702.	SB II		0.9		DV
4821.29	4822.64	20735.5	SB II			0.3	LG
4802.01	4803.35	20818.8	SB II			1.6	LG
4784.76	4786.10	20893.9	SB		0.3H		SP
4784.03	4785.37	20897.0	SB			1.8	LG
4779.4	4780.7	20917.	SB II		0.9		DV
4766.91	4768.24	20972.1	SB II			1.4	LG
4766.19	4767.52	20975.2	SB		0.3W		SP
4765.36	4766.69	20978.9	SB II			1.6	LG
4757.82	4759.15	21012.1	SB II		1.0		MIT
4735.386	4736.711	21111.70	SB II		1.4		MIT
4734.0	4735.3	21118.	SB II		0.5		DV
4711.96	4713.28	21216.6	SB		0.5H		SP
4711.26	4712.58	21219.8	SB II			2.0	LG
4704.48	4705.80	21250.4	SB II		1.0		MIT
4700.22	4701.54	21269.6	SB II		0.3		DV
4694.95	4696.26	21293.5	SB II			0.8	LG
4675.745	4677.054	21380.98	SL			1.2	LG
4657.907	4659.211	21462.86	SB II		1.5		SP
4656.35	4657.65	21470.0	SB II			1.2	LG
4653.32	4654.62	21484.0	SB II			1.2	LG

AIR WAVELENGTH (ANGSTRMS)	VACUUM WAVELENGTH (ANGSTRMS)	WAVENUMBER (KAYSERS)	LMENT	LOG ARC	INTENSITY SPK	INTENSITY DIS	REF
4648.49	4649.79	21506.3	SB II			0.6	LG
4647.919	4649.220	21508.98	SB II			1.3	LG
4647.704	4649.005	21509.98	SB		1.3		SP
4647.316	4648.617	21511.77	SB II			1.9	LG
4643.19	4644.49	21530.9	SB II		1.2		DV
4637.31	4638.61	21558.2	SB II			1.0	LG
4623.070	4624.365	21624.59	SB		1.3		SP
4612.92	4614.21	21672.2	SB II			1.7	LG
4604.77	4606.06	21710.5	SB II			1.5	LG
4599.68	4600.97	21734.6	SB		1.0H		SP
4599.09	4600.38	21737.3	SB II			1.6	LG
4597.08	4598.37	21746.9	SB II		0.8		DV
4596.90	4598.19	21747.7	SB II			1.8	LG
4596.096	4597.384	21751.50	SB		0.3		SP
4594.93	4596.22	21757.0	SB II			1.3	LG
4594.21	4595.50	21760.4	SB II			1.2	LG
4591.880	4593.167	21771.47	SB		1.3		MIT
4586.84	4588.13	21795.4	SB II			1.4	LG
4577.95	4579.23	21837.7	SB II			1.4	LG
4544.73	4546.00	21997.3	SB II		0.7		DV
4530.810	4532.080	22064.92	SB II		1.0		MIT
4526.04	4527.31	22088.2	SB			1.0	LG
4514.50	4515.77	22144.6	SB			1.0	LG
4506.913	4508.177	22181.92	SB II		1.1		MIT
4506.709	4507.973	22182.92	SB		0.5		SP
4498.793	4500.055	22221.95	SB		0.3		SP
4457.015	4458.266	22430.25	SB II		0.3		SP
4446.48	4447.73	22483.4	SB II			1.4	LG
4445.295	4446.543	22489.38	SB		0.3		SP
4431.87	4433.11	22557.5	SB II			0.7	LG
4428.576	4429.819	22574.28	SB		0.7H		MIT
4427.25	4428.49	22581.0	SB II			1.2	LG
4425.482	4426.725	22590.07	SB		0.3H		SP
4415.40	4416.64	22641.6	SB II		0.6		DV
4411.50	4412.74	22661.7	SB		1.0H		MIT
4377.89	4379.12	22835.6	SB		0.7J		SP
4376.85	4378.08	22841.1	SB II			0.5	LG
4369.995	4371.223	22876.89	SB		0.3		SP
4367.38	4368.61	22890.6	SB II		0.5		DV
4352.25	4353.47	22970.2	SB		0.8H		SP
4344.835	4346.056	23009.36	SB II		0.7H		MIT
4337.30	4338.52	23049.3	SB II		0.3		DV
4335.348	4336.567	23059.72	SB II		0.6J		MIT
4332.64	4333.86	23074.1	SB II			0.5	LG
4320.27	4321.48	23140.2	SB II		0.5		DV
4316.896	4318.110	23158.28	SB		0.3		SP
4314.32	4315.53	23172.1	SB II			0.5	LG
4295.008	4296.216	23276.29	SB		0.3H		SP
4289.30	4290.51	23307.3	SB II			0.8	LG
4287.057	4288.263	23319.46	SB		0.3H		SP
4283.88	4285.09	23336.8	SB II			0.7	LG
4271.54	4272.74	23404.2	SB II			1.0	LG
4264.929	4266.129	23440.45	SB		1.0H		SP
4260.55	4261.75	23464.5	SB II			1.0	LG
4260.011	4261.210	23467.51	SB		0.3		SP
4245.34	4246.54	23548.6	SB		0.3H		SP
4242.86	4244.05	23562.4	SB II		0.3H		DV
4224.30	4225.49	23665.9	SB		0.9J		MIT
4223.34	4224.53	23671.3	SB II		0.6		DV
4219.070	4220.258	23695.23	SB II		1.5		MIT
4209.67	4210.86	23748.1	SB II			1.0	LG
4200.486	4201.669	23800.06	SB II		0.3J		MIT
4195.169	4196.351	23830.23	SB II		1.7		MIT
4186.34	4187.52	23880.5	SB II		0.6		DV
4171.045	4172.221	23968.05	SB		0.5J		SP
4170.35	4171.53	23972.1	SB II		0.9		DV
4146.60	4147.77	24109.3	SB II		0.8		DV
4141.21	4142.38	24140.7	SB II		0.8		DV
4140.607	4141.775	24144.24	SB II		1.2		SP
4134.172	4135.338	24181.82	SB II		0.9		MIT

AIR WAVELENGTH (ANGSTRMS)	VACUUM WAVELENGTH (ANGSTRMS)	WAVENUMBER (KAYSERS)	LMENT	LOG ARC	INTENSITY SPK	INTENSITY DIS	REF
4133.95	4135.12	24183.1	SB		0.7		SP
4120.89	4122.05	24259.8	SB II		0.7		DV
4111.202	4112.362	24316.93	SB II		1.3		MIT
4105.863	4107.022	24348.54	SB II		1.1		MIT
4104.770	4105.928	24355.03	SB		0.5		SP
4097.87	4099.03	24396.0	SB II		0.6		DV
4090.863	4092.018	24437.82	SB		0.3		SP
4085.907	4087.060	24467.46	SB II		0.8		MIT
4078.385	4079.536	24512.59	SB		0.6		SP
4065.31	4066.46	24591.4	SB II		0.5H		MIT
4040.54	4041.68	24742.2	SB			0.3H	LG
4038.387	4039.528	24755.37	SB		0.3H		SP
4033.543	4034.683	24785.10	SB	1.8	1.8		MIT
4024.137	4025.274	24843.03	SB		0.6		SP
4004.392	4005.524	24965.52	SB		0.5		SP
3994.90	3996.03	25024.8	SB		1.0		DV
3986.031	3987.158	25080.52	SB II		0.7		SP
3982.03	3983.16	25105.7	SB II		0.5		MIT
3980.982	3982.108	25112.33	SB II	0.3H	0.3		MIT
3964.747	3965.869	25215.16	SB II	0.3H	1.2		SP
3963.96	3965.08	25220.2	SB II	0.3H	0.5		MIT
3960.55	3961.67	25241.9	SB II		1.3H		MIT
3944.58	3945.70	25344.1	SB II		0.9		DV
3939.57	3940.69	25376.3	SB II		0.5		DV
3931.785	3932.898	25426.54	SB II	0.3	0.7		MIT
3929.23	3930.34	25443.1	SB II			0.3H	LG
3914.27	3915.38	25540.3	SB	0.5	0.3H		MIT
3907.736	3908.843	25583.02	SB			0.9	LG
3893.75	3894.85	25674.9	SB II			0.8	LG
3857.32	3858.41	25917.4	SB II		0.7		MIT
3850.23	3851.32	25965.1	SB II		1.3		MIT
3841.19	3842.28	26026.2	SB		0.7H		MIT
3839.76	3840.85	26035.9	SB		0.6		KZ
3825.023	3826.108	26136.22	SB II		0.3H		SP
3797.05	3798.13	26328.8	SB II	0.0	0.6		MIT
3772.775	3773.847	26498.16	SB II		0.5H		MIT
3752.02	3753.09	26644.7	SB II		0.3H		MIT
3739.95	3741.01	26730.7	SB		0.6H		SP
3739.261	3740.324	26735.65	SB	0.3	0.7		MIT
3722.793	3723.852	26853.92	SB II	1.6Y	1.7		MIT
3719.70	3720.76	26876.2	SB II		0.9		MIT
3710.519	3711.575	26942.74	SB II	0.3	0.7		MIT
3686.86	3687.91	27115.6	SB	0.6J			MIT
3683.481	3684.530	27140.51	SB	0.5	0.3		MIT
3655.25	3656.29	27350.1	SB II	0.0	0.6		MIT
3651.704	3652.744	27376.68	SB	0.3H	0.5		SP
3640.84	3641.88	27458.4	SB II		0.5		DV
3637.829	3638.866	27481.09	SB	0.3H	1.8		MIT
3629.915	3630.950	27541.00	SB II	0.3	1.2H		MIT
3627.44	3628.47	27559.8	SB II	0.0	0.5H		MIT
3603.53	3604.56	27742.6	SB	0.3	1.4Q		MIT
3599.62	3600.65	27772.8	SB		0.6		DV
3597.51	3598.54	27789.1	SB	0.3	2.3Q		MIT
3596.96	3597.99	27793.3	SB			2.0	LG
3568.00	3569.02	28018.9	SB II	0.0	0.8		MIT
3566.64	3567.66	28029.6	SB		1.4J		SP
3559.242	3560.259	28087.85	SB	0.3H	1.7J		MIT
3533.88	3534.89	28289.4	SB	0.0	1.0V		MIT
3521.771	3522.778	28386.69	SB II	0.0		1.0	MIT
3520.474	3521.481	28397.15	SB II			2.1	LG
3520.071	3521.077	28400.40	SB	0.6	0.3H		MIT
3504.469	3505.471	28526.83	SB	1.3	1.3		MIT
3498.456	3499.457	28575.86	SB II		2.5J		MIT
3473.915	3474.910	28777.73	SB	0.5	2.5J		MIT
3467.245	3468.238	28833.08	SB	0.0	0.3H		MIT
3459.568	3460.559	28897.07	SB	0.3	0.5H		MIT
3459.26	3460.25	28899.6	SB			1.2	LG
3437.49	3438.48	29082.7	SB	0.3H	0.9		MIT
3425.95	3426.93	29180.6	SB		0.9H		MIT
3425.47	3426.45	29184.7	SB			0.9	LG

AIR WAVELENGTH (ANGSTRMS)	VACUUM WAVELENGTH (ANGSTRMS)	WAVENUMBER (KAYSERS)	LMENT	LOG ARC	INTENSITY SPK	DIS	REF
3403.91	3404.89	29369.6	SB II	0.3	1.0H		MIT
3399.93	3400.91	29403.9	SB II	0.0	0.3		MIT
3395.91	3396.88	29438.7	SB		0.6J		MIT
3383.137	3384.108	29549.88	SB	1.6	1.7		MIT
3374.65	3375.62	29624.2	SB	0.3H	0.5		MIT
3366.93	3367.90	29692.1	SB	0.5	0.9H		MIT
3354.86	3355.82	29798.9	SB		0.3H		MIT
3337.15	3338.11	29957.1	SB	0.3	1.0H		SP
3333.190	3334.149	29992.66	SB II	0.3	1.0H		SP
3332.64	3333.60	29997.6	SB II	1.0	1.3		DV
3317.54	3318.49	30134.1	SB II	0.3H	0.3H		MIT
3304.63	3305.58	30251.9	SB II		0.3		DV
3304.11	3305.06	30256.6	SB		1.6J		SP
3287.15	3288.10	30412.7	SB	0.0H	0.3H		MIT
3275.6	3276.5	30520.	SB		0.3		SP
3273.97	3274.91	30535.2	SB		0.6		SP
3267.502	3268.444	30595.60	SB I	2.2	2.2V		MIT
3247.547	3248.484	30783.59	SB	0.3H	1.0		MIT
3241.280	3242.215	30843.11	SB			2.5V	LG
3232.499	3233.432	30926.89	SB I	2.2	2.4J		MIT
3210.113	3211.040	31142.55	SB			0.8	LG
3196.98	3197.90	31270.5	SB		0.7H		MIT
3171.45	3172.37	31522.2	SB			0.8	LG
3168.417	3169.334	31552.37	SB			1.1	LG
3141.08	3141.99	31827.0	SB II			1.2	LG
3040.669	3041.554	32877.93	SB II			2.6J	LG
3029.807	3030.689	32995.80	SB	2.0	2.3J		MIT
3024.70	3025.58	33051.5	SB	0.3H	0.6H		MIT
3022.189	3023.069	33078.96	SB II		1.8		SP
3011.070	3011.947	33201.11	SB		1.8		SP
3001.70	3002.58	33304.7	SB II			1.2	LG
2980.962	2981.832	33536.43	SB			2.1A	LG
2966.22	2967.09	33703.1	SB		0.3		SP
2913.271	2914.124	34315.63	SB		0.6		SP
2911.86	2912.71	34332.3	SB			0.7	LG
2891.51	2892.36	34573.9	SB		1.3J		MIT
2891.214	2892.062	34577.41	SB			1.1	LG
2889.929	2890.776	34592.78	SB		0.7		SP
2884.07	2884.92	34663.1	SB II			0.5	LG
2877.915	2878.759	34737.19	SB I	2.40	2.2		MIT
2863.018	2863.859	34917.92	SB		0.6J		SP
2858.044	2858.883	34978.69	SB	1.0	0.7		MIT
2851.112	2851.950	35063.73	SB	1.7	1.7		MIT
2847.13	2847.97	35112.8	SB			0.5	LG
2837.311	2838.145	35234.28	SB		0.5		SP
2826.793	2827.625	35365.37	SB		0.7H		SP
2797.69	2798.51	35733.2	SB II		0.7C		MIT
2797.641	2798.466	35733.87	SB		0.6J		SP
2793.52	2794.34	35786.6	SB II			1.0	LG
2790.387	2791.210	35826.76	SB		0.8		SP
2788.87	2789.69	35846.2	SB			0.6	LG
2786.002	2786.824	35883.14	SB		0.6		MIT
2775.76	2776.58	36015.5	SB		0.5H		SP
2769.939	2770.757	36091.22	SB	2.0	1.9		MIT
2741.01	2741.82	36472.1	SB		0.3		MIT
2727.21	2728.02	36656.7	SB	1.5	1.5		MIT
2718.893	2719.699	36768.78	SB	1.7	1.7		MIT
2716.72	2717.53	36798.2	SB II			0.9	LG
2692.253	2693.052	37132.59	SB	1.6	1.6		MIT
2682.762	2683.559	37263.95	SB	1.7	1.5		MIT
2674.47	2675.26	37379.5	SB II		0.3H		MIT
2670.643	2671.437	37433.04	SB	1.7	1.5		MIT
2669.641	2670.435	37447.09	SB		0.5		SP
2663.194	2663.986	37537.73	SB	0.3	0.0		SP
2656.73	2657.52	37629.1	SB		1.0J		MIT
2652.606	2653.396	37687.56	SB	1.7	1.9		MIT
2617.320	2618.101	38195.62	SB		0.8W		SP
2614.666	2615.447	38234.39	SB	0.6	0.5		MIT
2612.301	2613.081	38269.00	SB	1.7	1.8		MIT
2598.062	2598.839	38478.73	SB I	2.3	2.0		MIT

AIR WAVELENGTH (ANGSTRMS)	VACUUM WAVELENGTH (ANGSTRMS)	WAVENUMBER (KAYSERS)	LMENT	LOG ARC	INTENSITY SPK	DIS	REF
2590.238	2591.013	38594.95	SB		0.5H		SP
2574.133	2574.904	38836.40	SB	1.5	1.6		MIT
2571.56	2572.33	38875.3	SB		0.9J		SP
2567.740	2568.509	38933.09	SB		0.7		MIT
2565.54	2566.31	38966.5	SB		1.40		M
2554.617	2555.383	39133.07	SB	1.5	1.0		MIT
2543.84	2544.60	39298.9	SB		1.4		M
2528.535	2529.295	39536.71	SB I	2.5G	2.3		LG
2520.180	2520.938	39667.77	SB	1.3	0.7		SP
2519.30	2520.06	39681.6	SB		0.3H		MIT
2514.568	2515.325	39756.29	SB	1.5	1.0		MIT
2510.529	2511.285	39820.25	SB	1.7	1.3		MIT
2507.783	2508.538	39863.85	SB		1.0H		SP
2481.736	2482.485	40282.21	SB	1.2	0.7		MIT
2480.442	2481.191	40303.22	SB	1.4	1.0		MIT
2478.311	2479.060	40337.88	SB	1.9	2.0		MIT
2474.584	2475.332	40398.63	SB	1.4	1.0		MIT
2445.515	2446.256	40878.80	SB	1.9	1.5		MIT
2429.497	2430.234	41148.29	SB		0.5		MIT
2426.352	2427.089	41201.63	SB	1.9	1.4		MIT
2422.14	2422.88	41273.3	SB	1.7	1.3		MIT
2395.20	2395.93	41737.5	SB	1.7	1.2		MIT
2383.63	2384.36	41940.0	SB	1.9	1.3		WT
2373.623	2374.348	42116.83	SB	1.9	1.4		MIT
2360.495	2361.217	42351.05	SB	1.3	0.7		MIT
2352.201	2352.921	42500.37	SB	1.1	0.6		MIT
2349.853	2350.572	42542.83	SB	0.9	0.0H		MIT
2329.08	2329.79	42922.2	SB	1.2	0.5H		MIT
2315.883	2316.595	43166.81	SB	1.3	1.0		MIT
2311.469	2312.180	43249.23	SB I	2.2G	1.7		MIT
2306.50	2307.21	43342.4	SB	1.5	1.5		MIT
2296.00	2296.71	43540.6	SB		0.5H		MIT
2293.46	2294.17	43588.8	SB	1.4	1.0		MIT
2288.986	2289.692	43674.00	SB	1.3	1.0		MIT
2288.139	2288.845	43690.16	SB	0.7			MIT
2270.08	2270.78	44037.7	SB	1.4	1.2		MIT
2262.538	2263.238	44184.48	SB	1.6	1.4		MIT
2246.99	2247.69	44490.2	SB		0.6		SP
2225.15	2225.84	44926.8	SB			1.0	LG
2224.94	2225.63	44931.1	SB	1.5	1.4		MIT
2222.08	2222.77	44988.9	SB	1.4	1.2		SP
2220.80	2221.49	45014.8	SB	1.5	1.3		SP
2208.53	2209.22	45264.9	SB	1.4	1.3		SP
2207.80	2208.49	45279.8	SB	0.6H	0.5H		MIT
2203.66	2204.35	45364.9	SB	0.6	0.3		MIT
2201.40	2202.09	45411.5	SB	1.6	1.4		M
2190.96	2191.64	45627.8	SB II		0.7		SP
2179.257	2179.940	45872.83	SB	1.5	1.6		MIT
2175.890	2176.572	45943.81	SB I	2.5	1.6		MIT
2170.23	2170.91	46063.6	SB	0.3	0.9		SP
2159.26	2159.94	46297.6	SB	1.2	1.0		WT
2158.97	2159.65	46303.8	SB	1.3	1.0		WT
2145.04	2145.72	46604.5	SB	1.3	1.3		WT
2141.76	2142.43	46675.9	SB	1.5	1.5		WT
2139.76	2140.43	46719.5	SB	1.5	1.5		WT
2137.11	2137.78	46777.4	SB	1.1	0.9		WT
2127.39	2128.06	46991.1	SB I	1.4	1.2		ME
2118.52	2119.19	47187.8	SB	1.2	1.4		SP
2117.28	2117.95	47215.5	SB	0.9	0.8		WT
2098.47	2099.14	47638.6	SB	1.5	1.5		WT
2086.0	2086.7	47923.	SB			0.9	LG
2079.63	2080.29	48070.2	SB	1.0			WT
2073.39	2074.05	48214.8	SB	0.7		0.8	LG
2068.38	2069.04	48331.6	SB I	2.5G	0.5		WT
2067.5	2068.2	48352.	SB			1.4	LG
2063.44	2064.10	48447.3	SB	1.5	1.1		WT
2061.92	2062.58	48483.0	SB	0.9			WT
2054.04	2054.70	48669.0	SB	1.5	0.7		WT
2053.1	2053.8	48691.	SB	0.0		0.9	LG
2049.57	2050.23	48775.1	SB I	2.2G			WT

AIR WAVELENGTH (ANGSTRMS)	VACUUM WAVELENGTH (ANGSTRMS)	WAVENUMBER (KAYSERS)	LMENT	LOG ARC	INTENSITY SPK	DIS	REF
2046.48	2047.14	48848.7	SB	0.5	0.3		WT
2044.56	2045.22	48894.6	SB	0.7	0.7		WT
2039.60	2040.26	49013.5	SB II	1.2	1.2		ME
2036.51	2037.16	49087.8	SB	0.9	0.6		WT
2036.21	2036.86	49095.1	SB			0.3	LG
2029.49	2030.14	49257.6	SB I	1.8R			ME
2027.3	2028.0	49311.	SB			0.6	LG
2024.06	2024.71	49389.7	SB			0.7	LG
2023.86	2024.51	49394.6	SB	0.9	0.0		WT
2014.69	2015.34	49619.4	SB	0.6		0.7	WT
2012.6	2013.3	49671.	SB		0.3		LG
2004.77	2005.42	49864.9	SB	0.9			WT
8408.01	8410.32	11890.2	SC	0.3H			ME
8241.13	8243.40	12130.9	SC I	0.7			ME
8196.98	8199.23	12196.3	SC I	0.5			ME
8017.83	8020.04	12468.8	SC	0.5			ME
8006.40	8008.60	12486.6	SC	0.3			ME
8003.27	8005.47	12491.5	SC I	0.3			ME
7848.55	7850.71	12737.7	SC	0.3			ME
7821.64	7823.79	12781.5	SC I	0.9			ME
7800.44	7802.59	12816.3	SC I	1.6			ME
7794.68	7796.82	12825.7	SC I	0.8			ME
7785.17	7787.31	12841.4	SC I	0.9			ME
7771.06	7773.20	12864.7	SC I	0.7			ME
7752.72	7754.85	12895.2	SC I	0.5			ME
7750.37	7752.50	12899.1	SC I	0.7			ME
7741.17	7743.30	12914.4	SC I	1.7			MIT
7738.18	7740.31	12919.4	SC I	0.3			ME
7729.72	7731.85	12933.5	SC I	1.0			MIT
7727.09	7729.22	12937.9	SC I	0.3			ME
7697.73	7699.85	12987.3	SC I	1.3			MIT
7665.72	7667.83	13041.5	SC I	0.7			ME
7617.45	7619.55	13124.1	SC I	0.8			ME
7584.48	7586.57	13181.2	SC	0.5			ME
7574.44	7576.53	13198.7	SC I	0.8			MIT
7553.96	7556.04	13234.4	SC I	0.8			ME
7524.13	7526.20	13286.9	SC I	0.8			MIT
7521.73	7523.80	13291.2	SC	0.6H			MIT
7323.78	7325.80	13650.4	SC I	0.3			ME
7300.62	7302.63	13693.7	SC I	0.5			MIT
7275.57	7277.57	13740.8	SC I	0.6H			MIT
7257.57	7259.57	13774.9	SC I	0.8			MIT
7169.13	7171.11	13944.8	SC I	0.9			ME
7138.14	7140.11	14005.4	SC I	0.8			MIT
7130.61	7132.58	14020.2	SC	0.5H			ME
7120.43	7122.39	14040.2	SC	0.3H			ME
7049.38	7051.32	14181.7	SC	0.3H			ME
6890.93	6892.83	14507.8	SC	0.5H			ME
6889.82	6891.72	14510.2	SC	0.8H			ME
6835.03	6836.92	14626.5	SC	1.4			ME
6829.54	6831.42	14638.2	SC I	1.2			MIT
6819.52	6821.40	14659.7	SC I	1.3			MIT
6817.08	6818.96	14665.0	SC	1.0			MIT
6739.40	6741.26	14834.0	SC	0.5			MIT
6737.87	6739.73	14837.4	SC	0.7			MIT
6717.64	6719.49	14882.1	SC	0.5			ME
6714.60	6716.45	14888.8	SC I	0.7			ME
6604.60	6606.42	15136.8	SC II	1.1			ME
6558.05	6559.86	15244.2	SC I	0.7			ME
6486.35	6488.14	15412.7	SC	0.3			ME
6448.10	6449.88	15504.2	SC I	0.6H	0.5		ME
6413.353	6415.126	15588.16	SC I	1.0	1.4		MIT
6403.151	6404.921	15612.99	SC	0.3H	0.3		MIT
6396.373	6398.141	15629.54	SC	0.3H			MIT
6378.824	6380.587	15672.54	SC I	0.9	1.2		MIT
6362.286	6364.045	15713.28	SC I	0.3	0.7		MIT
6361.801	6363.560	15714.47	SC	1.0G	0.9		MIT
6344.831	6346.585	15756.50	SC I	0.7	0.8		MIT
6342.082	6343.836	15763.33	SC II	0.3	0.7		MIT
6332.227	6333.978	15787.87	SC	1.3R	0.5		MIT

AIR WAVELENGTH (ANGSTRMS)	VACUUM WAVELENGTH (ANGSTRMS)	WAVENUMBER (KAYSERS)	LMENT	LOG ARC	INTENSITY SPK	DIS	REF
6322.737	6324.485	15811.56	SC I	0.3	0.6		MIT
6320.854	6322.602	15816.27	SC II	0.5	1.2		MIT
6309.902	6311.647	15843.72	SC II	0.7	1.4		MIT
6306.047	6307.791	15853.41	SC I	0.8			MIT
6305.671	6307.415	15854.36	SC I	1.9	1.8		MIT
6300.697	6302.439	15866.87	SC II	0.3	1.0		MIT
6297.767	6299.509	15874.25	SC I	0.3	0.5		MIT
6293.045	6294.785	15886.17	SC I	0.3H	1.0		MIT
6284.757	6286.495	15907.11	SC I	0.6	0.8		MIT
6284.178	6285.916	15908.58	SC I	0.3H	0.7		MIT
6280.169	6281.906	15918.74	SC I	0.7H	0.5		MIT
6279.757	6281.494	15919.78	SC II	1.0	1.4		MIT
6276.310	6278.046	15928.52	SC I	1.0	0.9		MIT
6273.147	6274.882	15936.55	SC I	0.5			MIT
6262.245	6263.977	15964.30	SC I	0.7	0.9		MIT
6258.962	6260.693	15972.67	SC I	1.3	1.3		MIT
6249.961	6251.690	15995.68	SC I	1.0	1.0		MIT
6245.629	6247.357	16006.77	SC II	0.8	1.5		MIT
6244.555	6246.282	16009.52	SC I	0.3			MIT
6239.778	6241.504	16021.78	SC I	1.5	0.5H		MIT
6239.410	6241.136	16022.72	SC I	0.9	0.7		MIT
6210.676	6212.394	16096.85	SC I	1.3	1.4		MIT
6198.438	6200.153	16128.63	SC I	1.0	0.8		MIT
6193.672	6195.386	16141.05	SC I	0.8	0.7		MIT
6148.693	6150.395	16259.12	SC	0.3H	1.3		MIT
6146.25	6147.95	16265.6	SC I	0.8	0.5		ME
6073.137	6074.818	16461.40	SC	0.7	1.0		MIT
6026.18	6027.85	16589.7	SC I	0.5	1.0		ME
6021.702	6023.370	16602.00	SC I	0.3	0.5		MIT
6002.342	6004.004	16655.55	SC	0.5	0.7		MIT
5988.42	5990.08	16694.3	SC I	1.0			ME
5969.19	5970.84	16748.1	SC I	0.6			ME
5968.25	5969.90	16750.7	SC	0.8			ME
5961.49	5963.14	16769.7	SC I	0.8			ME
5952.20	5953.85	16795.9	SC	0.7			ME
5940.58	5942.23	16828.7	SC	0.3			ME
5931.23	5932.87	16855.2	SC	0.6H			ME
5919.11	5920.75	16889.7	SC I	0.8H			ME
5894.63	5896.26	16959.9	SC I	0.7			ME
5887.758	5889.390	16979.69	SC	0.6J			MIT
5735.20	5736.79	17431.4	SC	0.5			ME
5724.073	5725.661	17465.23	SC I	1.2			MIT
5717.246	5718.832	17486.09	SC I	1.3			MIT
5711.754	5713.339	17502.90	SC I	2.0			MIT
5708.600	5710.184	17512.57	SC I	1.2			MIT
5700.230	5701.812	17538.29	SC I	2.6G			MIT
5686.826	5688.404	17579.62	SC I	2.3			MIT
5684.190	5685.767	17587.78	SC II	1.0			MIT
5671.805	5673.379	17626.18	SC I	2.5O			MIT
5669.030	5670.603	17634.81	SC II	1.1	1.2		MIT
5668.74	5670.31	17635.7	SC I	0.3			ME
5667.164	5668.737	17640.61	SC II	0.8			MIT
5658.334	5659.904	17668.14	SC II	0.8			MIT
5657.870	5659.440	17669.59	SC II	1.5			MIT
5649.56	5651.13	17695.6	SC I	0.6H			MIT
5647.60	5649.17	17701.7	SC I	0.5			ME
5646.344	5647.911	17705.66	SC I	0.5			MIT
5640.971	5642.537	17722.52	SC	1.2			MIT
5593.372	5594.925	17873.34	SC I	0.7			MIT
5591.322	5592.874	17879.89	SC I	1.2			MIT
5579.761	5581.310	17916.94	SC I	0.5			MIT
5571.238	5572.785	17944.35	SC I	0.5			MIT
5564.861	5566.406	17964.91	SC I	0.7			MIT
5561.10	5562.64	17977.1	SC I	0.5			ME
5553.595	5555.137	18001.35	SC I	0.7			MIT
5552.25	5553.79	18005.7	SC II		0.6		ME
5550.40	5551.94	18011.7	SC I	0.3H			ME
5549.68	5551.22	18014.1	SC I	0.5			ME
5541.030	5542.569	18042.18	SC I	0.5			MIT
5526.809	5528.344	18088.60	SC II	2.0	2.5J		MIT

AIR WAVELENGTH (ANGSTRMS)	VACUUM WAVELENGTH (ANGSTRMS)	WAVENUMBER (KAYSERS)	LMENT	LOG ARC	INTENSITY SPK	DIS	REF
5520.496	5522.029	18109.28	SC I	1.9			MIT
5514.215	5515.747	18129.91	SC I	1.8			MIT
5484.618	5486.142	18227.75	SC I	1.8			MIT
5481.989	5483.512	18236.49	SC I	1.8			MIT
5472.19	5473.71	18269.1	SC I	1.0			MIT
5468.40	5469.92	18281.8	SC I	0.3			MIT
5451.343	5452.858	18339.01	SC I	0.7			MIT
5447.38	5448.89	18352.4	SC I	0.5			MIT
5446.206	5447.720	18356.30	SC I	1.2			MIT
5442.60	5444.11	18368.5	SC I	0.6			MIT
5438.22	5439.73	18383.3	SC I	0.6			ME
5433.283	5434.793	18399.96	SC I	0.8			MIT
5429.407	5430.916	18413.10	SC I	0.5			MIT
5425.57	5427.08	18426.1	SC I	0.5			ME
5416.12	5417.63	18458.3	SC I	0.5			ME
5392.075	5393.574	18540.58	SC I	1.3			MIT
5375.346	5376.841	18598.28	SC I	1.1			MIT
5357.195	5358.685	18661.29	SC II		0.7		MIT
5356.100	5357.590	18665.11	SC I	1.6			MIT
5355.752	5357.242	18666.32	SC I	0.8			MIT
5350.296	5351.784	18685.36	SC I	0.5			MIT
5349.702	5351.190	18687.43	SC I	0.8			MIT
5349.294	5350.782	18688.86	SC I	1.5			MIT
5342.961	5344.447	18711.01	SC I	0.7			MIT
5341.040	5342.526	18717.74	SC I	0.7			MIT
5339.408	5340.893	18723.46	SC I	0.6			MIT
5334.228	5335.712	18741.64	SC II		0.6		MIT
5331.761	5333.244	18750.31	SC I	0.6			MIT
5318.337	5319.817	18797.64	SC II	0.3	1.1		MIT
5301.936	5303.411	18855.79	SC I	0.3	0.3		MIT
5285.752	5287.223	18913.52	SC I	1.0	1.2		MIT
5284.971	5286.442	18916.32	SC I	0.3	0.3		MIT
5258.333	5259.797	19012.14	SC I	1.1	1.2		MIT
5239.823	5241.282	19079.30	SC II	1.5	2.1		MIT
5219.67	5221.12	19153.0	SC I	1.0	1.1		ME
5210.522	5211.973	19186.59	SC I	1.3			MIT
5112.852	5114.277	19553.11	SC	0.6	1.0		MIT
5109.064	5110.488	19567.60	SC	0.6	1.0		MIT
5101.121	5102.543	19598.07	SC I	1.1	1.1		MIT
5099.228	5100.649	19605.35	SC I	2.0R	1.9		MIT
5096.716	5098.137	19615.01	SC I	1.1	1.2		MIT
5089.888	5091.307	19641.32	SC I	1.1	1.2		MIT
5087.148	5088.566	19651.90	SC I	1.6	1.3		MIT
5086.951	5088.369	19652.66	SC I	1.8	1.4		MIT
5085.547	5086.965	19658.09	SC I	1.9	1.8		MIT
5083.713	5085.130	19665.18	SC I	2.0	1.9		MIT
5081.554	5082.970	19673.54	SC I	2.0	2.0		MIT
5075.814	5077.229	19695.78	SC I	1.1	1.1		MIT
5070.249	5071.662	19717.40	SC I	1.3	1.2		MIT
5068.859	5070.272	19722.81	SC I	0.3	0.6		MIT
5064.321	5065.733	19740.48	SC I	1.1	1.2		MIT
5031.019	5032.422	19871.15	SC II	1.7	2.3H		MIT
5021.510	5022.910	19908.78	SC I	0.3	1.0		MIT
5020.142	5021.542	19914.20	SC	0.3	0.5		MIT
5018.379	5019.779	19921.20	SC I	0.3	0.7		MIT
5014.107	5015.506	19938.17	SC	0.0	0.3		MIT
5005.075	5006.471	19974.15	SC		1.2J		MIT
4995.006	4996.399	20014.41	SC I	0.3	0.0		MIT
4991.922	4993.315	20026.78	SC I	0.9	0.9		MIT
4983.451	4984.841	20060.82	SC I	0.8	0.6		MIT
4980.369	4981.759	20073.23	SC I	0.8	0.9		MIT
4973.667	4975.055	20100.28	SC I	0.8	0.7		MIT
4954.052	4955.435	20179.86	SC I	1.0	0.9		MIT
4951.696	4953.078	20189.47	SC I	0.3H	0.3		MIT
4941.338	4942.717	20231.79	SC I	0.5	0.7		MIT
4935.729	4937.107	20254.78	SC I	0.7	0.0		MIT
4934.242	4935.619	20260.88	SC I		0.9		MIT
4922.838	4924.212	20307.82	SC I	0.6	0.7		MIT
4909.770	4911.141	20361.87	SC I	0.6	1.0		MIT
4906.683	4908.053	20374.68	SC I	0.5	0.6		MIT

AIR WAVELENGTH (ANGSTRMS)	VACUUM WAVELENGTH (ANGSTRMS)	WAVENUMBER (KAYSERS)	LMENT	LOG ARC	INTENSITY SPK	DIS	REF
4890.335	4891.701	20442.79	SC	0.5	0.0H		MIT
4878.154	4879.516	20493.83	SC I	0.5	0.0		MIT
4858.081	4859.438	20578.51	SC	0.7	0.0H		MIT
4852.681	4854.037	20601.41	SC I	0.7	0.8		MIT
4847.688	4849.042	20622.63	SC I	1.1	0.3		MIT
4840.885	4842.238	20651.61	SC I		0.5H		MIT
4840.467	4841.819	20653.39	SC I	0.0	0.3		MIT
4839.433	4840.785	20657.81	SC I	1.0	0.8		MIT
4833.675	4835.026	20682.41	SC I	0.8	0.7		MIT
4827.283	4828.632	20709.80	SC I	0.7			MIT
4792.843	4794.183	20858.61	SC	0.5	0.3		MIT
4791.500	4792.839	20864.46	SC I	1.1	0.8		MIT
4779.347	4780.683	20917.51	SC I	1.9	1.6		MIT
4771.442	4772.776	20952.17	SC	0.3	0.5		MIT
4763.116	4764.448	20988.79	SC I	0.7	0.6H		MIT
4758.926	4760.257	21007.27	SC I	0.5	0.7		MIT
4753.152	4754.481	21032.79	SC I	1.9	1.6		MIT
4748.981	4750.309	21051.26	SC I	0.5	0.5		MIT
4746.111	4747.438	21063.99	SC I	0.7	0.5		MIT
4743.814	4745.141	21074.19	SC I	2.0	1.8H		MIT
4741.018	4742.344	21086.62	SC I	2.0	1.8H		MIT
4737.642	4738.967	21101.64	SC I	2.0	1.8H		MIT
4735.079	4736.403	21113.07	SC	1.0	0.9		MIT
4734.094	4735.418	21117.46	SC I	2.0	1.8H		MIT
4732.295	4733.619	21125.49	SC I	1.0	0.8		MIT
4729.226	4730.549	21139.20	SC I	2.0	1.7H		MIT
4728.769	4730.092	21141.24	SC I	1.7	1.4		MIT
4720.830	4722.151	21176.79	SC I	0.6	0.7		MIT
4719.347	4720.667	21183.45	SC I	1.1	0.7		MIT
4717.031	4718.351	21193.85	SC I	0.9	0.6		MIT
4716.248	4717.567	21197.36	SC I	0.8	0.5		MIT
4714.331	4715.650	21205.98	SC I	0.8	0.7		MIT
4711.732	4713.050	21217.68	SC I	0.8	0.5		MIT
4709.336	4710.654	21228.48	SC I	1.2	1.2P		MIT
4706.967	4708.284	21239.16	SC I	1.0	0.7		MIT
4698.276	4699.591	21278.45	SC II	0.7	1.2		MIT
4680.481	4681.791	21359.35	SC	0.7	0.6H		MIT
4670.404	4671.711	21405.43	SC II	2.0	2.5J		MIT
4665.834	4667.140	21426.40	SC	0.8	0.8		MIT
4610.419	4611.710	21683.93	SC I	0.7	0.5		MIT
4609.951	4611.242	21686.13	SC I	1.0	0.9		MIT
4609.527	4610.818	21688.12	SC	1.0	0.7		MIT
4604.713	4606.003	21710.80	SC	0.6			MIT
4598.436	4599.724	21740.43	SC I	0.3H			MIT
4598.135	4599.423	21741.86	SC	0.3			MIT
4592.939	4594.226	21766.45	SC I	0.3			MIT
4573.993	4575.275	21856.61	SC I	0.3			MIT
4557.237	4558.514	21936.97	SC I	0.5			MIT
4544.676	4545.950	21997.60	SC I	0.5			MIT
4542.554	4543.828	22007.88	SC I	0.3			MIT
4431.369	4432.613	22560.05	SC II	1.7	0.5		MIT
4420.665	4421.906	22614.68	SC II	1.3	0.3		MIT
4415.559	4416.799	22640.83	SC II	2.0	1.4		MIT
4400.355	4401.591	22719.06	SC II	2.2	1.5		MIT
4389.597	4390.830	22774.74	SC I	1.0			MIT
4384.813	4386.045	22799.58	SC II	1.4	1.0		MIT
4381.853	4383.084	22814.99	SC I	0.3			MIT
4381.232	4382.463	22818.22	SC I	1.0			MIT
4375.170	4376.399	22849.83	SC I	0.5			MIT
4374.455	4375.684	22853.57	SC II	2.0	1.4		MIT
4364.916	4366.143	22903.51	SC	1.1			MIT
4359.654	4360.879	22931.16	SC	1.1			MIT
4359.077	4360.302	22934.19	SC	1.1			MIT
4358.645	4359.870	22936.46	SC I	1.0			MIT
4354.609	4355.833	22957.72	SC II	1.8	1.0		MIT
4352.140	4353.363	22970.75	SC	1.0			MIT
4348.528	4349.750	22989.82	SC I	1.0			MIT
4325.010	4326.226	23114.83	SC II	1.7	1.6		MIT
4320.745	4321.960	23137.65	SC II	1.7	1.6		MIT
4314.084	4315.297	23173.37	SC II	1.7	2.2		MIT

AIR WAVELENGTH (ANGSTRMS)	VACUUM WAVELENGTH (ANGSTRMS)	WAVENUMBER (KAYSERS)	LMENT	LOG ARC	INTENSITY SPK	DIS	REF
4305.715	4306.926	23218.42	SC II	1.3	1.3		MIT
4294.767	4295.975	23277.60	SC II	1.3	1.3		MIT
4286.547	4287.753	23322.24	SC	0.6	0.5		MIT
4283.562	4284.767	23338.49	SC	0.6			MIT
4279.927	4281.131	23358.31	SC II	1.0	0.5		MIT
4259.588	4260.787	23469.84	SC	0.9			MIT
4246.829	4248.025	23540.35	SC II	1.9	2.7		MIT
4246.115	4247.310	23544.31	SC I	0.6			MIT
4239.565	4240.759	23580.69	SC I	0.7			MIT
4238.039	4239.232	23589.18	SC I	1.1			MIT
4237.798	4238.991	23590.52	SC	1.1	0.3		MIT
4235.290	4236.483	23604.49	SC I	0.3H			MIT
4233.610	4234.802	23613.85	SC I	1.0	0.3		MIT
4231.921	4233.113	23623.28	SC	1.0			MIT
4231.643	4232.835	23624.83	SC I	0.8	0.3		MIT
4225.58	4226.77	23658.7	SC I	0.6			MIT
4221.877	4223.066	23679.48	SC I	1.0			MIT
4219.729	4220.918	23691.53	SC I	0.9			MIT
4218.258	4219.446	23699.79	SC I	0.6			MIT
4216.101	4217.289	23711.92	SC I	0.5			MIT
4212.489	4213.676	23732.25	SC I	0.6			MIT
4212.34	4213.53	23733.1	SC I	0.5			M
4205.194	4206.379	23773.42	SC I	1.0	0.0		MIT
4204.538	4205.723	23777.13	SC I	0.6	0.3H		MIT
4199.488	4200.671	23805.72	SC	0.5			MIT
4171.568	4172.744	23965.05	SC I	0.5			MIT
4165.184	4166.358	24001.78	SC I	1.2	0.7		MIT
4161.881	4163.054	24020.82	SC I	1.2	0.6		MIT
4160.525	4161.698	24028.65	SC I	0.0	0.3		MIT
4154.723	4155.894	24062.21	SC	0.5	0.0		MIT
4152.355	4153.526	24075.93	SC I	1.0	0.6		MIT
4147.403	4148.573	24104.68	SC I	0.5	0.0		MIT
4140.304	4141.472	24146.01	SC I	1.0	0.3		MIT
4133.006	4134.172	24188.64	SC I	0.9	0.5		MIT
4126.564	4127.728	24226.40	SC	0.3			MIT
4100.333	4101.490	24381.38	SC I	0.7	0.0		MIT
4098.339	4099.496	24393.24	SC I	0.8			MIT
4094.837	4095.993	24414.11	SC I	0.9	0.8		MIT
4093.133	4094.288	24424.27	SC I	0.8	0.9		MIT
4087.163	4088.317	24459.94	SC I	1.0	0.0		MIT
4086.673	4087.827	24462.88	SC I	1.0	0.9J		MIT
4086.030	4087.183	24466.73	SC I	0.5			MIT
4082.396	4083.549	24488.50	SC I	1.4	1.0		MIT
4078.584	4079.736	24511.39	SC I	1.0	1.0		MIT
4074.976	4076.127	24533.09	SC I	1.0	0.5		MIT
4067.668	4068.817	24577.17	SC I	0.7			MIT
4067.000	4068.148	24581.21	SC I	0.5	0.3		MIT
4056.583	4057.729	24644.33	SC I	0.7	0.3		MIT
4054.555	4055.700	24656.66	SC I	1.0	1.0		MIT
4051.849	4052.994	24673.12	SC I	0.5			MIT
4049.951	4051.095	24684.68	SC I	0.9			MIT
4047.792	4048.935	24697.85	SC I	1.4	1.0		IT
4046.486	4047.629	24705.82	SC I	1.0			MIT
4043.804	4044.946	24722.21	SC I	1.1	0.6		MIT
4036.882	4038.023	24764.60	SC I	0.9			MIT
4034.23	4035.37	24780.9	SC I	0.9	0.3H		MIT
4031.397	4032.536	24798.29	SC I	1.0	0.3		MIT
4030.657	4031.796	24802.84	SC I	1.0	0.3		MIT
4023.688	4024.825	24845.80	SC I	2.0	1.4		MIT
4023.223	4024.360	24848.67	SC I	1.8			MIT
4020.399	4021.535	24866.12	SC I	1.7	1.3		MIT
4014.489	4015.624	24902.73	SC II	1.3	0.9		MIT
3996.607	3997.737	25014.15	SC I	1.6	1.0		MIT
3989.06	3990.19	25061.5	SC II	0.5	0.3		M
3976.29	3977.41	25142.0	SC	0.6			M
3952.278	3953.397	25294.71	SC	0.5			MIT
3933.381	3934.495	25416.23	SC I	1.8	1.8		MIT
3923.503	3924.614	25480.21	SC II	0.5	0.7		MIT
3918.213	3919.323	25514.61	SC	0.3H	0.5H		MIT
3911.810	3912.918	25556.37	SC I	2.2	1.5		MIT

AIR WAVELENGTH (ANGSTRMS)	VACUUM WAVELENGTH (ANGSTRMS)	WAVENUMBER (KAYSERS)	LMENT	LOG ARC	INTENSITY SPK	DIS	REF
3907.476	3908.583	25584.72	SC I	2.1	1.4		MIT
3894.957	3896.061	25666.95	SC	0.3H	0.0		MIT
3855.620	3856.713	25928.81	SC	1.2	1.2		MIT
3853.490	3854.583	25943.15	SC	0.9	0.9		MIT
3852.392	3853.484	25950.54	SC	1.2	1.2		MIT
3846.654	3847.745	25989.25	SC	0.9	0.9		MIT
3845.454	3846.545	25997.36	SC	1.0	1.0		MIT
3844.232	3845.322	26005.62	SC	1.0	1.0		MIT
3843.000	3844.090	26013.96	SC II	1.4	1.3		MIT
3836.519	3837.607	26057.90	SC	1.4	1.4		MIT
3833.059	3834.146	26081.42	SC II	1.0	0.9		MIT
3828.179	3829.265	26114.67	SC	1.5	1.5		MIT
3823.761	3824.846	26144.84	SC	1.2	1.0		MIT
3800.029	3801.108	26308.12	SC	0.7	1.1		MIT
3787.154	3788.230	26397.56	SC	1.5	1.0		MIT
3717.096	3718.153	26895.07	SC	0.6	0.6		MIT
3678.543	3679.590	27176.94	SC	0.5	1.2		MIT
3678.342	3679.389	27178.42	SC II	0.6	1.2		MIT
3676.604	3677.651	27191.27	SC	0.3	0.0		MIT
3675.265	3676.312	27201.18	SC II	0.7	0.3		MIT
3666.537	3667.581	27265.93	SC II	1.2	0.9		MIT
3664.254	3665.298	27282.91	SC II	0.6	0.3		MIT
3651.798	3652.838	27375.97	SC II	1.7	1.7		MIT
3645.311	3646.350	27424.69	SC II	1.7	1.7		MIT
3644.686	3645.725	27429.39	SC	0.3			MIT
3642.785	3643.823	27443.70	SC II	1.8	1.7		MIT
3636.168	3637.204	27493.64	SC	0.5			MIT
3635.310	3636.346	27500.13	SC	0.5	0.3		MIT
3630.740	3631.775	27534.75	SC II	1.7	1.8		MIT
3624.621	3625.654	27581.23	SC	0.3	0.3H		MIT
3613.836	3614.867	27663.54	SC II	1.6	1.8		MIT
3606.624	3607.653	27718.85	SC	0.6			MIT
3590.475	3591.500	27843.52	SC II	1.3	1.1		MIT
3589.635	3590.659	27850.04	SC II	0.7	1.1		MIT
3580.927	3581.949	27917.76	SC II	1.1	1.6		MIT
3576.340	3577.361	27953.57	SC II	1.3	1.7		MIT
3572.523	3573.543	27983.43	SC II	1.5	1.7		MIT
3567.701	3568.720	28021.25	SC II	1.2	1.6		MIT
3558.538	3559.554	28093.40	SC II	1.2	1.6		MIT
3535.729	3536.740	28274.63	SC II	1.2	1.5		MIT
3498.912	3499.913	28572.14	SC I	0.9	0.3		MIT
3497.030	3498.031	28587.51	SC	1.1W			MIT
3475.037	3476.032	28768.44	SC I	0.3			MIT
3470.551	3471.545	28805.62	SC I	0.9			MIT
3469.617	3470.611	28813.37	SC I	0.6	0.3H		MIT
3462.183	3463.175	28875.24	SC I	0.6	0.3H		MIT
3460.700	3461.691	28887.61	SC I	0.8	0.3H		MIT
3457.448	3458.438	28914.78	SC I	0.9	0.3		MIT
3455.898	3456.888	28927.75	SC I	0.5	0.3H		MIT
3448.503	3449.491	28989.78	SC I	0.9	0.5		MIT
3444.578	3445.565	29022.81	SC I	0.7			MIT
3443.989	3444.976	29027.78	SC I	0.5			MIT
3439.40	3440.39	29066.5	SC I	0.5	0.0		MIT
3435.555	3436.540	29099.04	SC I	1.1	0.6		MIT
3431.358	3432.342	29134.63	SC I	1.0	0.3		MIT
3429.483	3430.466	29150.56	SC I	1.1	0.3		MIT
3429.206	3430.189	29152.91	SC I	1.1	0.8		MIT
3419.358	3420.339	29236.87	SC I	0.5			MIT
3418.528	3419.508	29243.97	SC I	0.7	0.3H		MIT
3416.674	3417.654	29259.84	SC I	0.7	0.3H		MIT
3379.397	3380.367	29582.58	SC II	0.6	1.0B		MIT
3379.18	3380.15	29584.5	SC II	0.5	0.5H		MIT
3378.209	3379.179	29592.99	SC II	0.6	0.8B		MIT
3372.151	3373.120	29646.15	SC II	0.8	2.2		MIT
3368.946	3369.914	29674.35	SC II	1.7	1.3		MIT
3363.501	3364.467	29722.39	SC II	0.3	0.3		MIT
3361.935	3362.901	29736.23	SC II	1.4	1.0		MIT
3361.270	3362.236	29742.11	SC II	1.4	1.0		MIT
3359.679	3360.644	29756.20	SC II	1.7	1.4		MIT
3357.303	3358.268	29777.26	SC	0.6			MIT

AIR WAVELENGTH (ANGSTRMS)	VACUUM WAVELENGTH (ANGSTRMS)	WAVENUMBER (KAYSERS)	LMENT	LOG ARC	INTENSITY SPK	INTENSITY DIS	REF
3353.734	3354.698	29808.94	SC II	1.7	1.8		MIT
3352.048	3353.011	29823.94	SC II	1.1	0.5		MIT
3349.22	3350.18	29849.1	SC I	0.6			MIT
3343.27	3344.23	29902.2	SC II	0.8	1.1W		MIT
3331.07	3332.03	30011.7	SC II	0.7	0.6B		MIT
3320.709	3321.664	30105.39	SC II	0.7			MIT
3320.422	3321.377	30107.99	SC II	0.7	1.1		MIT
3317.693	3318.648	30132.76	SC II	0.3	0.3B		MIT
3317.038	3317.993	30138.71	SC II	0.7	0.9		MIT
3313.539	3314.493	30170.53	SC II	0.5	1.0		MIT
3312.736	3313.689	30177.84	SC II	0.7	0.9		MIT
3311.708	3312.661	30187.21	SC II	0.5	0.8		MIT
3273.619	3274.562	30538.43	SC I	1.5	1.1		MIT
3269.904	3270.847	30573.12	SC I	1.5	1.1		MIT
3255.678	3256.617	30706.71	SC I	1.2	0.9		MIT
3251.32	3252.26	30747.9	SC II	1.2	0.9		MIT
3199.37	3200.29	31247.1	SC II	0.3	0.9		MIT
3191.005	3191.928	31329.03	SC II	0.5	0.3H		MIT
3190.403	3191.325	31334.94	SC II	0.3	0.3		MIT
3158.542	3159.456	31651.01	SC	0.3			MIT
3140.731	3141.641	31830.50	SC	0.3			MIT
3139.729	3140.639	31840.66	SC II	0.8	1.1		MIT
3133.096	3134.004	31908.06	SC II	0.3	1.0		MIT
3128.286	3129.193	31957.12	SC II	0.5	1.0		MIT
3127.689	3128.596	31963.22	SC	0.5	0.0H		MIT
3124.935	3125.841	31991.39	SC	0.3	0.0H		MIT
3123.956	3124.862	32001.42	SC	0.3	0.7		MIT
3122.954	3123.859	32011.68	SC II	0.3	0.9		MIT
3122.542	3123.447	32015.91	SC II	0.3H			MIT
3119.675	3120.580	32045.33	SC I	0.5	0.3H		MIT
3117.890	3118.794	32063.67	SC I	0.3			MIT
3117.191	3118.095	32070.86	SC	0.3H			MIT
3114.778	3115.681	32095.71	SC	0.3	0.3H		MIT
3113.370	3114.273	32110.22	SC	0.5	0.0		MIT
3110.238	3111.140	32142.56	SC I	0.5	0.7		MIT
3109.341	3110.243	32151.83	SC	0.5	0.6		MIT
3108.511	3109.413	32160.41	SC II	0.6	0.7B		MIT
3107.529	3108.431	32170.58	SC II	0.9	1.1		MIT
3107.387	3108.288	32172.05	SC II	0.6	0.0		MIT
3106.546	3107.447	32180.75	SC	0.6	0.0		MIT
3106.022	3106.923	32186.18	SC	0.3H	0.0H		MIT
3105.667	3106.568	32189.86	SC I	0.3			MIT
3102.149	3103.049	32226.37	SC I	0.5	0.6		MIT
3099.422	3100.321	32254.72	SC	0.3H	0.0H		MIT
3098.597	3099.496	32263.31	SC	0.3H	0.0		MIT
3096.80	3097.70	32282.0	SC	0.8D	0.5		MIT
3095.364	3096.262	32297.00	SC I	0.3			MIT
3094.625	3095.523	32304.72	SC	0.6	0.0		MIT
3092.519	3093.417	32326.71	SC II	0.5	0.6		MIT
3092.42	3093.32	32327.7	SC	0.5			MIT
3087.345	3088.241	32380.89	SC	0.6	0.0		MIT
3083.283	3084.178	32423.54	SC	0.5	0.9		MIT
3082.56	3083.46	32431.1	SC II	0.3	0.3H		MIT
3081.548	3082.443	32441.80	SC I	0.0	0.3H		MIT
3079.953	3080.848	32458.60	SC	0.3			MIT
3078.463	3079.357	32474.31	SC	0.3			MIT
3077.034	3077.928	32489.39	SC	0.3H			MIT
3075.38	3076.27	32506.9	SC II	0.6L			ME
3065.106	3065.997	32615.82	SC II	1.1D	1.4		MIT
3063.730	3064.621	32630.47	SC	0.3H			MIT
3060.531	3061.421	32664.57	SC II	0.3H	0.0H		MIT
3056.306	3057.195	32709.73	SC	0.6			MIT
3052.929	3053.817	32745.91	SC II	1.0	1.2B		MIT
3045.714	3046.600	32823.48	SC II	1.2	1.4		MIT
3039.92	3040.80	32886.0	SC II	0.6	1.3B		MIT
3030.769	3031.651	32985.32	SC I	1.3	0.9		MIT
3019.350	3020.229	33110.07	SC I	1.3	1.0		MIT
3015.364	3016.242	33153.83	SC I	1.3	1.0		MIT
2988.952	2989.824	33446.79	SC II	1.3S	0.3D		MIT
2988.952	2989.824	33446.79	SC I	1.3S	0.3D		MIT

AIR WAVELENGTH (ANGSTRMS)	VACUUM WAVELENGTH (ANGSTRMS)	WAVENUMBER (KAYSERS)	LMENT	LOG ARC	INTENSITY SPK	INTENSITY DIS	REF
2980.752	2981.622	33538.79	SC I	1.3	1.0		MIT
2979.683	2980.553	33550.83	SC II	0.3H	1.0		MIT
2974.006	2974.874	33614.87	SC I	1.2	0.9		MIT
2913.03	2913.88	34318.5	SC II	0.3H	0.7		MIT
2870.917	2871.760	34821.86	SC II	0.3	0.7		MIT
2866.71	2867.55	34873.0	SC II	0.5			ME
2866.05	2866.89	34881.0	SC II	0.7	0.7J		MIT
2863.68	2864.52	34909.9	SC II	0.6			MIT
2859.32	2860.16	34963.1	SC II	0.8D	0.7J		ME
2837.293	2838.127	35234.50	SC	0.6H	0.8		MIT
2832.34	2833.17	35296.1	SC	0.5			ME
2827.815	2828.647	35352.59	SC II	0.7			MIT
2826.664	2827.496	35366.98	SC II	1.0	1.4		MIT
2822.131	2822.962	35423.79	SC II	1.7	1.3		MIT
2819.521	2820.351	35456.58	SC II	0.9	1.3		MIT
2801.312	2802.138	35687.04	SC II	0.8	0.7H		MIT
2789.169	2789.992	35842.40	SC II	0.6	1.0J		MIT
2782.358	2783.179	35930.14	SC II	0.6	0.7		MIT
2734.089	2734.898	36564.43	SC	0.8D	1.1		MIT
2711.343	2712.147	36871.16	SC I	1.0			MIT
2706.76	2707.56	36933.6	SC I	0.8			MIT
2699.104	2699.905	37038.34	SC	0.5	1.4		MIT
2684.233	2685.030	37243.53	SC II	0.3H			MIT
2611.209	2611.989	38285.00	SC II	0.7	1.0J		MIT
2563.183	2563.951	39002.30	SC II	1.0	1.3		MIT
2560.227	2560.995	39047.33	SC II	1.0	1.5		MIT
2555.799	2556.566	39114.97	SC II	1.0	1.3		MIT
2552.359	2553.125	39167.69	SC II	1.4	1.6		MIT
2545.204	2545.968	39277.79	SC II	1.2	1.3		MIT
2273.18	2273.88	43977.6	SC		1.0		EX
9969.73	9972.46	10027.6	SE I			0.6	RD
9825.51	9828.20	10174.8	SE I			0.8	RD
9816.63	9819.32	10184.0	SE II			0.3	MZ
9733.72	9736.39	10270.7	SE II			0.5	BT
9549.40	9552.02	10469.0	SE I			0.8	RD
9544.96	9547.58	10473.9	SE I			0.5	RD
9541.64	9544.26	10477.5	SE I			0.7	RD
9461.92	9464.52	10565.8	SE I			0.3	MS
9432.43	9435.02	10598.8	SE I			1.0	RD
9395.73	9398.31	10640.2	SE I			0.3	MS
9383.82	9386.39	10653.7	SE I			0.5	MS
9373.74	9376.31	10665.2	SE			0.3	MS
9303.15	9305.70	10746.1	SE I			0.3	RD
9297.11	9299.66	10753.1	SE I			0.7	MS
9271.02	9273.56	10783.3	SE I			0.8	RD
9265.88	9268.42	10789.3	SE I			0.6	MS
9262.72	9265.26	10793.0	SE I			0.3	MS
9252.40	9254.94	10805.0	SE I			0.3	RD
9249.06	9251.60	10808.9	SE I			0.3	MS
9218.30	9220.83	10845.0	SE I			0.3	RD
9214.30	9216.83	10849.7	SE I			0.3	MS
9211.03	9213.56	10853.6	SE I			0.3H	MS
9210.10	9212.63	10854.7	SE I			0.3H	MS
9206.19	9208.72	10859.3	SE I			0.3	MS
9203.58	9206.11	10862.4	SE I			0.5	RD
9199.81	9202.33	10866.8	SE I			0.3	MS
9185.09	9187.61	10884.2	SE I			0.3	MS
9181.80	9184.32	10888.1	SE I			0.8	RD
9140.79	9143.30	10937.0	SE I			0.9	RD
9130.71	9133.22	10949.1	SE			0.3	BT
9123.17	9125.67	10958.1	SE			0.3	MS
9095.15	9097.65	10991.8	SE I			0.3	MS
9088.70	9091.19	10999.7	SE I			1.1	RD
9088.14	9090.63	11000.3	SE I			0.8	RD
9083.05	9085.54	11006.5	SE I			0.9	RD
9038.56	9041.04	11060.7	SE I			1.3	RD
9005.99	9008.46	11100.7	SE I			0.5	RD
9005.58	9008.05	11101.2	SE I			0.8	RD
9003.44	9005.91	11103.8	SE I			0.9	RD
9001.93	9004.40	11105.7	SE I			1.3	RD

AIR WAVELENGTH (ANGSTRMS)	VACUUM WAVELENGTH (ANGSTRMS)	WAVENUMBER (KAYSERS)	LMENT	LOG ARC	INTENSITY SPK	DIS	REF
8969.63	8972.09	11145.7	SE I			1.0	RD
8969.23	8971.69	11146.2	SE I			0.9	RD
8938.14	8940.59	11184.9	SE I			1.0	RD
8924.19	8926.64	11202.4	SE I			0.5	RD
8918.80	8921.25	11209.2	SE I			1.5	RD
8915.88	8918.33	11212.9	SE I			0.6	RD
8914.58	8917.03	11214.5	SE I			0.6	RD
8872.26	8874.70	11268.0	SE I			0.5	RD
8799.79	8802.21	11360.8	SE I			0.5	RD
8742.29	8744.69	11435.5	SE I			1.2	RD
8679.13	8681.51	11518.7	SE II			0.3	MZ
8634.65	8637.02	11578.1	SE			0.3	BT
8568.39	8570.74	11667.6	SE II			0.3	RD
8450.47	8452.79	11830.4	SE I			1.2	RD
8440.55	8442.87	11844.3	SE I			1.2	RD
8375.31	8377.61	11936.6	SE I			0.9	RD
8361.99	8364.29	11955.6	SE II			0.3H	BT
8350.60	8352.90	11971.9	SE			0.3	RD
8349.14	8351.43	11974.0	SE			0.3	BT
8340.03	8342.32	11987.1	SE I			0.6	RD
8309.52	8311.80	12031.1	SE II			0.5	MZ
8272.45	8274.72	12085.0	SE			0.3	RD
8256.74	8259.01	12108.0	SE I			0.3	RD
8194.61	8196.86	12199.8	SE I			1.1	RD
8191.43	8193.68	12204.5	SE I			0.9	RD
8190.13	8192.38	12206.5	SE I			0.6	RD
8185.00	8187.25	12214.1	SE I			1.0	RD
8182.93	8185.18	12217.2	SE I			1.2	RD
8163.08	8165.32	12246.9	SE I			1.3	RD
8157.73	8159.97	12254.9	SE I			1.3	RD
8155.69	8157.93	12258.0	SE			0.5	RD
8152.02	8154.26	12263.5	SE I			1.2	RD
8149.28	8151.52	12267.7	SE I			1.3	RD
8118.56	8120.79	12314.1	SE II			0.3	BT
8098.51	8100.74	12344.6	SE			0.3	RD
8094.69	8096.92	12350.4	SE I			1.2	RD
8093.19	8095.42	12352.7	SE I			1.2	RD
8087.77	8089.99	12360.9	SE			0.3	RD
8081.14	8083.36	12371.1	SE I			1.1	RD
8065.31	8067.53	12395.4	SE I			1.1	RD
8060.91	8063.13	12402.1	SE I			1.1	RD
8058.37	8060.59	12406.1	SE I			0.6	RD
8051.81	8054.02	12416.2	SE I			1.0	RD
8036.35	8038.56	12440.0	SE I			1.3	RD
8000.96	8003.16	12495.1	SE I			1.5	RD
7963.89	7966.08	12553.2	SE I			0.5	RD
7929.48	7931.66	12607.7	SE			0.5	RD
7838.81	7840.97	12753.5	SE II			1.0	MZ
7767.87	7770.01	12870.0	SE I			0.3	RD
7750.41	7752.54	12899.0	SE I			0.7	RD
7680.61	7682.72	13016.2	SE I			0.8	RD
7677.83	7679.94	13020.9	SE I			0.5	RD
7672.17	7674.28	13030.5	SE I			0.6	RD
7670.33	7672.44	13033.7	SE I			0.9	RD
7653.59	7655.70	13062.2	SE I			0.9	RD
7606.81	7608.90	13142.5	SE I			1.1	RD
7592.19	7594.28	13167.8	SE I			1.2	RD
7583.37	7585.46	13183.1	SE I			1.4	RD
7575.08	7577.17	13197.6	SE I			1.3	RD
7549.97	7552.05	13241.4	SE I			1.0	RD
7508.97	7511.04	13313.7	SE I			1.0	RD
7506.44	7508.51	13318.2	SE I			0.7	RD
7493.47	7495.53	13341.3	SE I			1.5	RD
7429.27	7431.32	13456.6	SE I			2.3	RD
7426.81	7428.86	13461.0	SE I			2.0	RD
7426.81	7428.86	13461.0	SE II			2.0	RD
7414.02	7416.06	13484.2	SE I			1.2	RD
7391.99	7394.03	13524.4	SE II			2.1	MZ
7383.88	7385.91	13539.3	SE I			2.0	RD
7380.34	7382.37	13545.8	SE I			1.8	RD

AIR WAVELENGTH (ANGSTRMS)	VACUUM WAVELENGTH (ANGSTRMS)	WAVENUMBER (KAYSERS)	LMENT	LOG ARC	INTENSITY SPK	DIS	REF
7377.77	7379.80	13550.5	SE I			1.3	RD
7372.54	7374.57	13560.1	SE I			1.6	RD
7370.82	7372.85	13563.3	SE I			1.9	RD
7328.97	7330.99	13640.7	SE I			2.5	RD
7299.51	7301.52	13695.8	SE I			2.3	RD
7265.31	7267.31	13760.2	SE I			1.5	RD
7210.04	7212.03	13865.7	SE I			1.5	RD
7188.31	7190.29	13907.6	SE I			1.2	RD
7187.86	7189.84	13908.5	SE I			1.2	RD
7178.84	7180.82	13926.0	SE I			1.5	RD
7138.91	7140.88	14003.9	SE II			1.3	MZ
7084.03	7085.98	14112.4	SE			1.2	MS
7079.81	7081.76	14120.8	SE II			1.2H	BT
7074.63	7076.58	14131.1	SE II			1.2H	BT
7063.79	7065.74	14152.8	SE I			1.8	MS
7062.06	7064.01	14156.3	SE I			3.0	RD
7013.85	7015.78	14253.6	SE I			2.6	RD
7012.75	7014.68	14255.8	SE I			2.3	RD
7010.82	7012.75	14259.7	SE I			2.7	RD
7002.84	7004.77	14276.0	SE I			1.2	RD
6991.77	6993.70	14298.6	SE I			2.3	RD
6990.65	6992.58	14300.9	SE I			2.5	RD
6931.32	6933.23	14423.3	SE I			1.2	RD
6916.86	6918.77	14453.4	SE II			1.2	BT
6895.40	6897.30	14498.4	SE			1.5	MS
6888.61	6890.51	14512.7	SE			1.5	MS
6881.61	6883.51	14527.5	SE			1.5	BT
6879.52	6881.42	14531.9	SE I			1.2	RD
6865.09	6866.98	14562.4	SE I			1.2	RD
6831.27	6833.16	14634.5	SE I			2.6	RD
6830.89	6832.78	14635.3	SE I			2.0	RD
6830.56	6832.44	14636.1	SE I			1.2	RD
6815.28	6817.16	14668.9	SE I			1.8	RD
6805.24	6807.12	14690.5	SE I			2.3	RD
6781.22	6783.09	14742.5	SE II			1.3	RD
6781.22	6783.09	14742.5	SE I			1.3	RD
6777.56	6779.43	14750.5	SE			1.2	BT
6768.50	6770.37	14770.2	SE I			1.9	RD
6752.07	6753.93	14806.2	SE			1.5	MS
6746.43	6748.29	14818.6	SE I			2.3	RD
6721.37	6723.23	14873.8	SE II			1.8	RD
6721.37	6723.23	14873.8	SE I			1.8	RD
6701.04	6702.89	14918.9	SE I			1.5	RD
6699.56	6701.41	14922.2	SE I			2.1	RD
6682.36	6684.21	14960.6	SE I			1.3	RD
6679.43	6681.27	14967.2	SE I			1.8	RD
6667.13	6668.97	14994.8	SE			1.2	BL
6660.75	6662.59	15009.2	SE			1.7	BL
6657.83	6659.67	15015.8	SE II			1.0	RD
6657.83	6659.67	15015.8	SE I			1.0	RD
6647.43	6649.27	15039.2	SE			1.2	BT
6590.88	6592.70	15168.3	SE			0.9	BL
6545.04	6546.85	15274.5	SE			0.6	BL
6541.23	6543.04	15283.4	SE II			1.2	BT
6534.95	6536.76	15298.1	SE II			2.5	MZ
6534.44	6536.25	15299.3	SE			2.1	MS
6523.85	6525.65	15324.1	SE II			1.0	BT
6516.88	6518.68	15340.5	SE			0.5	BL
6497.85	6499.65	15385.4	SE II			0.3	BT
6490.48	6492.27	15402.9	SE II			2.7	BL
6488.34	6490.13	15408.0	SE II			2.0	BL
6483.06	6484.85	15420.6	SE II			2.3	BL
6482.70	6484.49	15421.4	SE II			0.6	KH
6474.47	6476.26	15441.0	SE II			1.2	BT
6470.15	6471.94	15451.3	SE II			1.0	RD
6470.15	6471.94	15451.3	SE I			1.0	RD
6461.39	6463.18	15472.3	SE I			1.2	MS
6448.42	6450.20	15503.4	SE			1.2	BT
6444.25	6446.03	15513.4	SE I			2.0	BL
6441.43	6443.21	15520.2	SE I			1.3	RD

AIR WAVELENGTH (ANGSTRMS)	VACUUM WAVELENGTH (ANGSTRMS)	WAVENUMBER (KAYSERS)	LMENT	LOG ARC	INTENSITY SPK	INTENSITY DIS	REF
6428.67	6430.45	15551.0	SE			1.2	BL
6422.90	6424.68	15565.0	SE II			2.1	BT
6419.27	6421.04	15573.8	SE			1.2	BT
6417.69	6419.46	15577.6	SE II			1.0	BT
6416.95	6418.72	15579.4	SE			0.8	BL
6410.99	6412.76	15593.9	SE I			1.2	RD
6404.53	6406.30	15609.6	SE I			1.2	MS
6401.17	6402.94	15617.8	SE I			1.2	MS
6399.99	6401.76	15620.7	SE I			1.2	MS
6397.69	6399.46	15626.3	SE I			1.2	RD
6391.96	6393.73	15640.3	SE			1.2	BT
6376.74	6378.50	15677.7	SE			1.2	BT
6370.62	6372.38	15692.7	SE II			1.5	BL
6368.13	6369.89	15698.9	SE II			1.5	BT
6359.19	6360.95	15720.9	SE			0.7	BL
6358.37	6360.13	15722.9	SE I			1.2	RD
6348.86	6350.62	15746.5	SE			1.5	BT
6338.14	6339.89	15773.1	SE II			1.2	BL
6332.05	6333.80	15788.3	SE			1.2	BL
6326.87	6328.62	15801.2	SE II			1.0	BL
6326.87	6328.62	15801.2	SE I			1.0	BL
6325.57	6327.32	15804.5	SE I			2.7	RD
6325.11	6326.86	15805.6	SE I			1.5	RD
6303.42	6305.16	15860.0	SE I			3.0	MS
6300.99	6302.73	15866.1	SE I			1.6	RD
6296.85	6298.59	15876.6	SE			0.5	BL
6286.99	6288.73	15901.5	SE I			1.5	MS
6284.47	6286.21	15907.8	SE I			2.5	RD
6283.96	6285.70	15909.1	SE I			2.3	RD
6283.16	6284.90	15911.2	SE I			2.3	RD
6281.72	6283.46	15914.8	SE			1.5	BT
6269.15	6270.88	15946.7	SE I			1.9	RD
6266.23	6267.96	15954.2	SE I			2.3	RD
6261.09	6262.82	15967.2	SE II			0.7H	BT
6259.83	6261.56	15970.5	SE I			1.6	BT
6259.83	6261.56	15970.5	SE II			1.6	BT
6244.61	6246.34	16009.4	SE			1.5	BT
6242.21	6243.94	16015.5	SE I			1.9	RD
6240.37	6242.10	16020.3	SE II			1.0	BT
6239.70	6241.43	16022.0	SE II			0.3	KH
6233.73	6235.45	16037.3	SE II			1.5	BL
6228.69	6230.41	16050.3	SE II			1.2	BL
6220.86	6222.58	16070.5	SE II			0.7	RD
6220.86	6222.58	16070.5	SE I			0.7	RD
6207.20	6208.92	16105.9	SE I			1.5	RD
6206.29	6208.01	16108.2	SE II			1.0	BL
6186.60	6188.31	16159.5	SE I			1.2	RD
6183.75	6185.46	16166.9	SE II			1.2	BL
6183.75	6185.46	16166.9	SE I			1.2	BL
6177.71	6179.42	16182.7	SE I			1.9	RD
6164.54	6166.25	16217.3	SE II			1.5	BT
6148.84	6150.54	16258.7	SE			1.5H	BL
6144.28	6145.98	16270.8	SE			1.2H	BL
6138.46	6140.16	16286.2	SE I			1.8	RD
6135.04	6136.74	16295.3	SE II			1.8	BL
6125.51	6127.21	16320.7	SE			0.6	BL
6123.49	6125.18	16326.0	SE II			1.8	BL
6121.54	6123.23	16331.2	SE I			1.6	RD
6121.54	6123.23	16331.2	SE II			1.6	RD
6116.06	6117.75	16345.9	SE			0.6	BL
6101.96	6103.65	16383.6	SE II			2.3	BL
6100.81	6102.50	16386.7	SE			1.7	BL
6096.12	6097.81	16399.3	SE II			1.7	BL
6085.03	6086.71	16429.2	SE			1.2	BL
6065.83	6067.51	16481.2	SE II			1.9	BL
6060.65	6062.33	16495.3	SE			0.5	BL
6055.96	6057.64	16508.1	SE II			3.0	BL
6030.12	6031.79	16578.8	SE II			1.5	BL
6029.45	6031.12	16580.7	SE			1.2	BL
6029.10	6030.77	16581.6	SE I			1.2	MS

AIR WAVELENGTH (ANGSTRMS)	VACUUM WAVELENGTH (ANGSTRMS)	WAVENUMBER (KAYSERS)	LMENT	LOG ARC	INTENSITY SPK	INTENSITY DIS	REF
6020.29	6021.96	16605.9	SE			1.5	BL
6014.17	6015.84	16622.8	SE			1.2	BT
6000.31	6001.97	16661.2	SE II			1.2	BT
5990.92	5992.58	16687.3	SE II			1.5	BL
5987.61	5989.27	16696.5	SE			0.9	BT
5984.86	5986.52	16704.2	SE			0.9	BL
5976.22	5977.88	16728.4	SE II			0.9	BT
5962.67	5964.32	16766.4	SE II			0.7	BL
5962.01	5963.66	16768.2	SE I			2.0	RD
5961.32	5962.97	16770.2	SE I			1.5	RD
5959.63	5961.28	16774.9	SE			0.7	BL
5936.32	5937.96	16840.8	SE I			0.9	RD
5925.15	5926.79	16872.5	SE I			1.5	RD
5924.98	5926.62	16873.0	SE I			0.5	RD
5924.76	5926.40	16873.6	SE I			1.6	RD
5909.38	5911.02	16917.6	SE I			1.4	RD
5909.20	5910.84	16918.1	SE I			1.5	RD
5907.08	5908.72	16924.1	SE			1.2	RD
5891.29	5892.92	16969.5	SE			0.9	BT
5889.74	5891.37	16974.0	SE II			1.2	BT
5886.34	5887.97	16983.8	SE II			1.3	BL
5878.70	5880.33	17005.9	SE I			1.2	RD
5866.27	5867.90	17041.9	SE II			1.9	BL
5843.14	5844.76	17109.3	SE I			0.9	RD
5842.68	5844.30	17110.7	SE II			1.8	BL
5831.47	5833.09	17143.6	SE II			0.5	BL
5827.80	5829.42	17154.4	SE I			0.9	RD
5824.75	5826.36	17163.4	SE			1.2	BL
5799.72	5801.33	17237.4	SE II			0.9	BL
5798.84	5800.45	17240.1	SE			0.9	BL
5790.03	5791.64	17266.3	SE II			1.2	BL
5784.76	5786.36	17282.0	SE II			1.2	BL
5769.06	5770.66	17329.0	SE II			1.2	BL
5753.32	5754.92	17376.4	SE I			1.4	RD
5752.03	5753.63	17380.4	SE I			0.9	RD
5747.62	5749.21	17393.7	SE II			1.7	BL
5738.29	5739.88	17422.0	SE			1.2	RD
5733.05	5734.64	17437.9	SE II			1.3	BL
5730.84	5732.43	17444.6	SE II			1.2	BL
5726.73	5728.32	17457.1	SE II			1.4	BT
5725.70	5727.29	17460.3	SE II			1.0	BT
5723.44	5725.03	17467.2	SE			0.9H	BT
5718.43	5720.02	17482.5	SE I			0.9	RD
5718.12	5719.71	17483.4	SE			0.9	BT
5717.99	5719.58	17483.8	SE I			1.2	RD
5704.98	5706.56	17523.7	SE I			0.9	RD
5703.73	5705.31	17527.5	SE I			1.2	RD
5702.44	5704.02	17531.5	SE			0.9	RD
5701.37	5702.95	17534.8	SE I			0.9	RD
5700.24	5701.82	17538.2	SE I			0.9	RD
5697.88	5699.46	17545.5	SE II			1.7	BL
5655.49	5657.06	17677.0	SE II			0.9	BL
5652.36	5653.93	17686.8	SE II			0.9	RD
5652.36	5653.93	17686.8	SE I			0.9	RD
5623.13	5624.69	17778.7	SE II			2.5	BL
5617.87	5619.43	17795.4	SE I			1.2	RD
5611.59	5613.15	17815.3	SE			1.2	BT
5591.16	5592.71	17880.4	SE II			2.7	BL
5586.43	5587.98	17895.6	SE I			0.9	RD
5586.34	5587.89	17895.8	SE I			1.2	RD
5586.34	5587.89	17895.8	SE II			1.2	RD
5566.93	5568.48	17958.2	SE II			2.7	BL
5560.54	5562.08	17978.9	SE II			0.9	BL
5530.12	5531.66	18077.8	SE I			0.9	RD
5522.42	5523.95	18103.0	SE II			2.9	BL
5498.78	5500.31	18180.8	SE I			0.9	RD
5484.08	5485.60	18229.5	SE II			1.3	BL
5473.99	5475.51	18263.1	SE			1.0	BT
5458.52	5460.04	18314.9	SE			1.5	BT
5455.58	5457.10	18324.8	SE II			1.2	MZ

AIR WAVELENGTH (ANGSTRMS)	VACUUM WAVELENGTH (ANGSTRMS)	WAVENUMBER (KAYSERS)	LMENT	LOG ARC	INTENSITY SPK	INTENSITY DIS	REF
5444.95	5446.46	18360.5	SE II			1.7	BL
5417.12	5418.63	18454.9	SE			1.2	BT
5401.01	5402.51	18509.9	SE II			1.9	MZ
5399.88	5401.38	18513.8	SE			0.9	BL
5382.74	5384.24	18572.7	SE			1.5	BT
5381.71	5383.21	18576.3	SE			1.2	BL
5380.21	5381.71	18581.5	SE II			1.3	MZ
5374.14	5375.63	18602.5	SE I			2.2	RD
5369.91	5371.40	18617.1	SE I			2.2	RD
5365.47	5366.96	18632.5	SE I			2.1	RD
5315.50	5316.98	18807.7	SE II			1.2	BT
5305.35	5306.83	18843.7	SE II			2.7	BL
5301.03	5302.50	18859.0	SE II			1.3	BL
5271.22	5272.69	18965.7	SE II			2.2	BL
5253.63	5255.09	19029.2	SE II			2.0	BL
5253.07	5254.53	19031.2	SE II			1.7	BL
5245.17	5246.63	19059.9	SE II			1.5	BL
5243.75	5245.21	19065.0	SE II			0.9	MZ
5237.62	5239.08	19087.3	SE II			0.7	BL
5235.19	5236.65	19096.2	SE			0.7	BT
5231.74	5233.20	19108.8	SE			0.9	BT
5227.51	5228.97	19124.2	SE II			2.8	BL
5223.87	5225.32	19137.6	SE II			0.9	BT
5187.68	5189.12	19271.1	SE II			1.3	BL
5183.01	5184.45	19288.4	SE II			1.2	BL
5175.98	5177.42	19314.6	SE II			2.8	BL
5171.54	5172.98	19331.2	SE II			1.3	BL
5142.14	5143.57	19441.7	SE II			2.7	BL
5134.31	5135.74	19471.4	SE II			1.5	BL
5117.77	5119.20	19534.3	SE			1.4	BT
5112.67	5114.09	19553.8	SE			0.9	BL
5109.59	5111.01	19565.6	SE II			1.2	BL
5109.11	5110.53	19567.4	SE II			1.2	BL
5096.57	5097.99	19615.6	SE II			2.5	BT
5093.26	5094.68	19628.3	SE II			1.7	BL
5068.65	5070.06	19723.6	SE II			2.4	BL
5031.26	5032.66	19870.2	SE II			1.6	MZ
5019.32	5020.72	19917.5	SE II			1.4	MZ
5005.69	5007.09	19971.7	SE			0.9	BL
4992.75	4994.14	20023.5	SE II			2.5	BL
4992.03	4993.42	20026.3	SE II			1.1	BL
4975.66	4977.05	20092.2	SE II			2.5	BL
4944.55	4945.93	20218.6	SE			1.3	BL
4920.98	4922.35	20315.5	SE			1.2	BT
4917.29	4918.66	20330.7	SE II			0.9	BT
4897.56	4898.93	20412.6	SE II			0.9	BL
4886.99	4888.35	20456.8	SE I			1.8	RD
4844.96	4846.31	20634.2	SE II			2.9	BL
4840.63	4841.98	20652.7	SE II			2.9	BL
4830.78	4832.13	20694.8	SE II			1.1	KH
4819.80	4821.15	20741.9	SE			1.4	BT
4809.61	4810.95	20785.9	SE II			0.6	BL
4797.66	4799.00	20837.7	SE II			1.1	BT
4792.73	4794.07	20859.1	SE I			1.3	RD
4777.85	4779.19	20924.1	SE II			1.3	BL
4765.52	4766.85	20978.2	SE II			1.6	BL
4763.65	4764.98	20986.4	SE II			2.9	BL
4761.85	4763.18	20994.4	SE II			1.3	BL
4742.25	4743.58	21081.1	SE I			2.7	RD
4740.97	4742.30	21086.8	SE II			2.8	BL
4739.03	4740.36	21095.5	SE I			2.9	RD
4733.57	4734.89	21119.8	SE II			0.9	BL
4730.78	4732.10	21132.2	SE I			3.0	RD
4718.58	4719.90	21186.9	SE II			0.6	BT
4685.45	4686.76	21336.7	SE II			1.1	BL
4675.55	4676.86	21381.9	SE			1.0	BT
4669.38	4670.69	21410.1	SE II			1.0	BT
4667.80	4669.11	21417.4	SE I			1.8	RD
4666.73	4668.04	21422.3	SE II			1.0	BT
4664.98	4666.29	21430.3	SE II			2.2	RD

AIR WAVELENGTH (ANGSTRMS)	VACUUM WAVELENGTH (ANGSTRMS)	WAVENUMBER (KAYSERS)	LMENT	LOG ARC	INTENSITY SPK	INTENSITY DIS	REF
4664.98	4666.29	21430.3	SE I			2.2	RD
4664.20	4665.51	21433.9	SE I			2.2	RD
4659.89	4661.19	21453.7	SE			1.3	BT
4656.22	4657.52	21470.6	SE			1.3	BT
4648.70	4650.00	21505.4	SE			0.3	BL
4648.44	4649.74	21506.6	SE II			2.9	BL
4636.65	4637.95	21561.3	SE			2.2	BT
4633.06	4634.36	21578.0	SE			1.0	BT
4630.56	4631.86	21589.6	SE II			1.1	BL
4629.82	4631.12	21593.1	SE II			0.6	BT
4628.12	4629.42	21601.0	SE II			1.4	BL
4625.37	4626.67	21613.8	SE II			0.9	BL
4623.77	4625.07	21621.3	SE II			2.2	RD
4623.77	4625.07	21621.3	SE I			2.2	RD
4621.58	4622.87	21631.6	SE II			0.9	BL
4620.32	4621.61	21637.5	SE II			0.9	BL
4618.77	4620.06	21644.7	SE II			2.0	BL
4604.34	4605.63	21712.6	SE II			2.5	BL
4599.96	4601.25	21733.2	SE II			1.8	BL
4597.93	4599.22	21742.8	SE II			1.4	BL
4596.59	4597.88	21749.2	SE II			0.9	BL
4592.34	4593.63	21769.3	SE II			0.9	BT
4567.19	4568.47	21889.2	SE II			1.0	BL
4563.95	4565.23	21904.7	SE II			2.3	BL
4561.75	4563.03	21915.3	SE II			1.3	BL
4559.27	4560.55	21927.2	SE			1.6	BT
4552.64	4553.92	21959.1	SE II			0.3	BL
4541.40	4542.67	22013.5	SE II			1.4	BL
4534.68	4535.95	22046.1	SE I			1.0	RD
4534.13	4535.40	22048.8	SE II			0.9	BT
4532.74	4534.01	22055.5	SE I			1.3	RD
4516.25	4517.52	22136.1	SE II			1.8	BL
4507.59	4508.85	22178.6	SE II			1.3	BT
4502.85	4504.11	22201.9	SE			1.4	BT
4500.57	4501.83	22213.2	SE II			1.0	KH
4495.34	4496.60	22239.0	SE II			1.3	BT
4476.70	4477.96	22331.6	SE II			1.0	BL
4475.02	4476.28	22340.0	SE II			1.1	BT
4467.60	4468.85	22377.1	SE II			2.5	BL
4459.63	4460.88	22417.1	SE II			1.0	BT
4454.95	4456.20	22440.6	SE II			1.7	BT
4449.15	4450.40	22469.9	SE II			2.5	BL
4447.72	4448.97	22477.1	SE II			0.6	BT
4446.02	4447.27	22485.7	SE II			2.3	BL
4434.92	4436.17	22542.0	SE II			1.6	BL
4433.74	4434.98	22548.0	SE II			1.4	BL
4432.33	4433.57	22555.2	SE II			1.8	BL
4426.12	4427.36	22586.8	SE II			1.3	BL
4421.68	4422.92	22609.5	SE			1.3	BT
4415.85	4417.09	22639.3	SE			1.0	BL
4413.16	4414.40	22653.1	SE			1.3	BL
4406.58	4407.82	22687.0	SE II			1.8	BT
4401.02	4402.26	22715.6	SE II			2.0	BL
4399.94	4401.18	22721.2	SE II			1.1	BL
4399.17	4400.41	22725.2	SE II			1.2	BL
4382.87	4384.10	22809.7	SE II			2.9	BL
4374.24	4375.47	22854.7	SE II			1.6	KH
4373.31	4374.54	22859.6	SE			1.6	BL
4357.33	4358.55	22943.4	SE			1.3	BL
4356.72	4357.94	22946.6	SE II			1.1	BT
4355.15	4356.37	22954.9	SE II			1.6	BL
4354.64	4355.86	22957.6	SE II			0.3	BL
4347.79	4349.01	22993.7	SE II			0.9	BL
4345.41	4346.63	23006.3	SE II			1.4	BL
4339.59	4340.81	23037.2	SE I			2.3	RD
4337.60	4338.82	23047.7	SE II			1.6	BL
4335.52	4336.74	23058.8	SE II			1.4	BT
4330.28	4331.50	23086.7	SE I			2.3	RD
4330.28	4331.50	23086.7	SE II			2.3	RD
4328.70	4329.92	23095.1	SE I			2.3	RD

AIR WAVELENGTH (ANGSTRMS)	VACUUM WAVELENGTH (ANGSTRMS)	WAVENUMBER (KAYSERS)	LMENT	LOG INTENSITY ARC	SPK	DIS	REF
4322.19	4323.41	23129.9	SE			1.8	BL
4320.39	4321.60	23139.6	SE II			2.0	BL
4318.91	4320.12	23147.5	SE II			0.9	BL
4318.61	4319.82	23149.1	SE			1.0	BL
4316.21	4317.42	23162.0	SE II			1.6	BT
4309.09	4310.30	23200.2	SE II			1.4	BL
4307.95	4309.16	23206.4	SE II			1.1	BL
4304.13	4305.34	23227.0	SE II			1.0	BL
4297.31	4298.52	23263.8	SE II			1.6	BL
4292.05	4293.26	23292.3	SE			1.0	BT
4290.57	4291.78	23300.4	SE II			1.3	BT
4289.65	4290.86	23305.4	SE			1.0	BL
4282.10	4283.30	23346.5	SE II			2.0	BT
4280.36	4281.56	23355.9	SE II			2.2	BL
4279.98	4281.18	23358.0	SE			1.0	BL
4248.00	4249.20	23533.9	SE II			2.0	BL
4243.91	4245.10	23556.5	SE II			0.9	BT
4230.05	4231.24	23633.7	SE II			1.6	BL
4226.37	4227.56	23654.3	SE II			1.3	BL
4221.58	4222.77	23681.1	SE II			1.3	BT
4215.02	4216.21	23718.0	SE			2.2	BT
4212.58	4213.77	23731.7	SE II			2.3	BL
4211.83	4213.02	23736.0	SE II			2.3	BT
4206.53	4207.72	23765.9	SE			1.0	BL
4198.05	4199.23	23813.9	SE II			1.6	BL
4196.24	4197.42	23824.1	SE II			1.3	BL
4195.51	4196.69	23828.3	SE II			2.0	BT
4195.37	4196.55	23829.1	SE			1.0	BL
4194.55	4195.73	23833.7	SE II			1.7	BL
4193.32	4194.50	23840.7	SE			1.3	BT
4184.26	4185.44	23892.4	SE II			1.4	BT
4184.08	4185.26	23893.4	SE			1.0	RD
4182.20	4183.38	23904.1	SE II			0.6	BL
4180.94	4182.12	23911.3	SE II			2.9	BL
4175.32	4176.50	23943.5	SE II			2.9	BL
4169.65	4170.83	23976.1	SE II			1.1	BL
4167.20	4168.37	23990.2	SE II			1.1	BT
4165.66	4166.83	23999.0	SE II			1.6	BL
4159.75	4160.92	24033.1	SE II			1.8	BL
4159.75	4160.92	24033.1	SE II			1.8	GU
4153.93	4155.10	24066.8	SE II			1.6	BL
4152.34	4153.51	24076.0	SE			1.9	BT
4141.17	4142.34	24141.0	SE I			1.0	RD
4140.40	4141.57	24145.4	SE I			1.3	RD
4139.66	4140.83	24149.8	SE I			1.6	RD
4138.99	4140.16	24153.7	SE II			1.0	BL
4137.31	4138.48	24163.5	SE II			1.7	BL
4136.28	4137.45	24169.5	SE II			2.0	BL
4132.76	4133.93	24190.1	SE			2.3	BT
4129.15	4130.31	24211.2	SE II			2.3	KH
4126.57	4127.73	24226.4	SE II			2.2	BL
4115.28	4116.44	24292.8	SE II			1.0	BT
4114.35	4115.51	24298.3	SE II			1.6	BL
4111.88	4113.04	24312.9	SE I			1.0	RD
4108.83	4109.99	24331.0	SE II			2.9	BL
4097.91	4099.07	24395.8	SE II			1.8	MZ
4091.95	4093.11	24431.3	SE II			1.8	BL
4072.11	4073.26	24550.4	SE			1.3	BL
4070.16	4071.31	24562.1	SE II			2.7	BL
4062.06	4063.21	24611.1	SE II			1.8	BL
4058.20	4059.35	24634.5	SE II			1.3	BL
4047.77	4048.91	24698.0	SE II			0.9	BL
4041.82	4042.96	24734.3	SE II			0.9	BT
4038.31	4039.45	24755.8	SE II			1.6	BL
4032.89	4034.03	24789.1	SE			1.0	BL
4032.22	4033.36	24793.2	SE II			0.9	BT
4030.07	4031.21	24806.4	SE II			2.2	BT
4023.23	4024.37	24848.6	SE I			1.3	RD
4021.26	4022.40	24860.8	SE I			1.3	RD
4019.72	4020.86	24870.3	SE I			1.3	RD

AIR WAVELENGTH (ANGSTRMS)	VACUUM WAVELENGTH (ANGSTRMS)	WAVENUMBER (KAYSERS)	LMENT	LOG INTENSITY ARC	SPK	DIS	REF
4019.50	4020.64	24871.7	SE II			1.0	BL
4018.52	4019.66	24877.7	SE II			1.8	BL
4014.77	4015.90	24901.0	SE I			1.8	RD
4012.96	4014.09	24912.2	SE I			2.2	RD
4011.88	4013.01	24918.9	SE I			2.3	RD
4007.90	4009.03	24943.7	SE II			2.2	BL
4003.08	4004.21	24973.7	SE II			1.8	BL
4002.07	4003.20	24980.0	SE			1.8	BT
3952.60	3953.72	25292.6	SE II			0.7	BT
3951.84	3952.96	25297.5	SE			1.4	BT
3948.80	3949.92	25317.0	SE II			1.4	BT
3944.07	3945.19	25347.4	SE			1.3	BT
3941.45	3942.57	25364.2	SE			1.3	B
3939.60	3940.72	25376.1	SE II			0.9	BT
3935.75	3936.86	25400.9	SE			0.8	BL
3935.36	3936.47	25403.4	SE			1.8	BT
3931.72	3932.83	25427.0	SE			0.9	BL
3931.56	3932.67	25428.0	SE			0.7	BL
3924.02	3925.13	25476.9	SE II			0.3	BL
3923.38	3924.49	25481.0	SE			0.9	BL
3917.95	3919.06	25516.3	SE II			0.7	BT
3917.06	3918.17	25522.1	SE II			1.3	BL
3916.49	3917.60	25525.8	SE II			0.9	BL
3913.81	3914.92	25543.3	SE II			0.9	BL
3911.09	3912.20	25561.1	SE II			0.3	BT
3904.85	3905.96	25601.9	SE II			1.4	BT
3901.53	3902.64	25623.7	SE			0.7	BL
3895.66	3896.76	25662.3	SE			0.3	BT
3877.28	3878.38	25784.0	SE II			1.7	BL
3867.60	3868.70	25848.5	SE			1.7	BT
3855.24	3856.33	25931.4	SE			0.9	BT
3853.28	3854.37	25944.6	SE			0.6	BL
3849.60	3850.69	25969.4	SE			0.6	BL
3845.98	3847.07	25993.8	SE II			1.1	KH
3841.93	3843.02	26021.2	SE			1.3H	BT
3836.60	3837.69	26057.4	SE			0.9	BT
3836.25	3837.34	26059.7	SE II			1.3	BL
3818.75	3819.83	26179.1	SE II			0.7	BL
3817.10	3818.18	26190.5	SE			0.9	BT
3813.85	3814.93	26212.8	SE			0.3	BT
3812.16	3813.24	26224.4	SE			0.6	BL
3811.62	3812.70	26228.1	SE II			0.6	BL
3809.88	3810.96	26240.1	SE II			0.3	BT
3793.63	3794.71	26352.5	SE II			1.4	BL
3789.68	3790.76	26380.0	SE			0.7	BT
3786.61	3787.69	26401.4	SE			1.1	BT
3779.15	3780.22	26453.5	SE			0.7	BT
3770.54	3771.61	26513.9	SE II			0.9	BT
3763.24	3764.31	26565.3	SE II			1.3	BL
3754.31	3755.38	26628.5	SE II			1.3	BL
3734.50	3735.56	26769.7	SE I			0.9	RD
3733.34	3734.40	26778.1	SE I			1.3	RD
3728.23	3729.29	26814.7	SE			1.3	BT
3686.21	3687.26	27120.4	SE II			1.5	BL
3658.46	3659.50	27326.1	SE II			0.7	BL
3654.89	3655.93	27352.8	SE			0.5	BL
3653.05	3654.09	27366.6	SE II			1.4	BT
3639.40	3640.44	27469.2	SE II			1.1	BL
3631.38	3632.42	27529.9	SE			1.4	BT
3618.73	3619.76	27626.1	SE II			1.5	BL
3610.50	3611.53	27689.1	SE II			1.5	BL
3591.03	3592.05	27839.2	SE I			0.9	RD
3588.16	3589.18	27861.5	SE II			0.9	BL
3578.93	3579.95	27933.3	SE			1.1	BT
3515.63	3516.64	28436.3	SE			1.5	BL
3501.52	3502.52	28550.9	SE I			1.7	RD
3489.03	3490.03	28653.1	SE			1.3	BL
3454.05	3455.04	28943.2	SE			0.9	BL
3444.27	3445.26	29025.4	SE II			1.5	BL
3431.65	3432.63	29132.1	SE II			0.7	KH

AIR WAVELENGTH (ANGSTRMS)	VACUUM WAVELENGTH (ANGSTRMS)	WAVENUMBER (KAYSERS)	LMENT		LOG INTENSITY ARC	SPK	DIS	REF
3384.98	3385.95	29533.8	SE	II			1.4	BL
3353.67	3354.63	29809.5	SE	II			1.3	BL
3346.56	3347.52	29872.8	SE				1.5	BL
3335.62	3336.58	29970.8	SE				0.9H	BL
3325.71	3326.67	30060.1	SE				1.3	BL
3278.96	3279.90	30488.7	SE	II			1.3	KH
3268.05	3268.99	30590.5	SE				1.3	BL
3260.64	3261.58	30660.0	SE				0.9	BL
3251.64	3252.58	30744.8	SE				1.3	BL
3242.79	3243.73	30828.7	SE				1.0	RO
3242.19	3243.13	30834.4	SE	II			1.4	BL
3238.43	3239.36	30870.2	SE	II			1.2	BL
3232.27	3233.20	30929.1	SE				0.9	BL
3228.17	3229.10	30968.4	SE	II			0.9	BL
3204.58	3205.51	31196.3	SE	II			1.6	BL
3200.93	3201.86	31231.9	SE	II			0.9	BL
3194.09	3195.01	31298.8	SE				1.3	BL
3181.85	3182.77	31419.2	SE				0.9	BL
3178.20	3179.12	31455.2	SE				1.1	RO
3150.86	3151.77	31728.2	SE				0.9	BL
3148.69	3149.60	31750.0	SE				1.3	BL
3141.13	3142.04	31826.5	SE	II			2.0	BL
3134.42	3135.33	31894.6	SE	II			1.8	BL
3132.58	3133.49	31913.3	SE				1.3	BL
3130.12	3131.03	31938.4	SE				0.9	BL
3116.63	3117.53	32076.6	SE				1.3	BL
3108.54	3109.44	32160.1	SE	II			1.2	BL
3105.50	3106.40	32191.6	SE	II			1.3	BL
3085.74	3086.64	32397.7	SE				0.9	BL
3084.20	3085.10	32413.9	SE				1.3	BL
3054.27	3055.16	32731.5	SE				1.3	BL
3046.24	3047.13	32817.8	SE	II			1.5	BL
3044.71	3045.60	32834.3	SE				0.9	BL
3041.31	3042.19	32871.0	SE	II			1.8	BL
3039.50	3040.38	32890.6	SE				1.3	BL
3038.66	3039.54	32899.7	SE	II			1.8	BL
3003.86	3004.74	33280.8	SE				0.9	BL
2972.60	2973.47	33630.8	SE				1.0	BL
2971.46	2972.33	33643.7	SE				1.0	BL
2963.91	2964.78	33729.4	SE	II			1.4	BL
2952.28	2953.14	33862.2	SE	II			1.1	MZ
2947.13	2947.99	33921.4	SE	II			0.6	MZ
2941.48	2942.34	33986.5	SE				1.2	BL
2934.18	2935.04	34071.1	SE	II			0.3	MZ
2919.22	2920.07	34245.7	SE				1.0	BL
2916.13	2916.98	34282.0	SE				1.0	BL
2914.89	2915.74	34296.6	SE				1.2	BL
2911.07	2911.92	34341.6	SE				0.7	BL
2906.71	2907.56	34393.1	SE				1.0	BL
2905.88	2906.73	34402.9	SE				0.7	BL
2905.07	2905.92	34412.5	SE	II			0.8	BL
2895.88	2896.73	34521.7	SE	II			1.4	BL
2892.76	2893.61	34558.9	SE				1.0	BL
2880.31	2881.15	34708.3	SE				1.4	BL
2873.41	2874.25	34791.6	SE				0.7	BL
2872.08	2872.92	34807.8	SE	II			0.5	BL
2865.89	2866.73	34882.9	SE				1.0	BL
2855.07	2855.91	35015.1	SE				0.7	BL
2842.48	2843.32	35170.2	SE				0.7	BL
2837.21	2838.04	35235.5	SE				1.5	BL
2821.52	2822.35	35431.5	SE	II			1.3	BL
2820.10	2820.93	35449.3	SE				1.2	BL
2817.01	2817.84	35488.2	SE				1.4	BL
2810.74	2811.57	35567.3	SE				0.7	BL
2798.50	2799.32	35722.9	SE				1.0	BL
2762.34	2763.16	36190.5	SE				1.0	BL
2752.15	2752.96	36324.5	SE				0.7	BL
2749.94	2750.75	36353.7	SE				1.2	BL
2738.16	2738.97	36510.1	SE	II			1.0	BL
2733.34	2734.15	36574.5	SE				0.7	BL

AIR WAVELENGTH (ANGSTRMS)	VACUUM WAVELENGTH (ANGSTRMS)	WAVENUMBER (KAYSERS)	LMENT		LOG INTENSITY ARC	SPK	DIS	REF
2729.15	2729.96	36630.6	SE				0.7	BL
2719.48	2720.29	36760.8	SE				1.5	BL
2718.29	2719.10	36776.9	SE				0.7	BL
2706.96	2707.76	36930.9	SE				0.7	BL
2705.39	2706.19	36952.3	SE				0.7	BL
2702.66	2703.46	36989.6	SE				1.0	BL
2692.02	2692.82	37135.8	SE				1.0	BL
2688.29	2689.09	37187.3	SE				1.0	BL
2681.36	2682.16	37283.4	SE				0.7	BL
2662.05	2662.84	37553.9	SE				1.4	BL
2651.48	2652.27	37703.6	SE				1.0	BL
2649.42	2650.21	37732.9	SE				1.7	BL
2643.43	2644.22	37818.4	SE				1.0	BL
2639.24	2640.03	37878.4	SE				0.7	BL
2633.25	2634.04	37964.6	SE				0.7	BL
2630.92	2631.70	37998.2	SE				1.5	BL
2626.85	2627.63	38057.1	SE				0.7	BL
2609.41	2610.19	38311.4	SE				1.0	BL
2606.40	2607.18	38355.6	SE				0.7	BL
2602.62	2603.40	38411.3	SE				1.2	BL
2600.52	2601.30	38442.4	SE				0.7	BL
2591.41	2592.19	38577.5	SE				2.2	BL
2586.72	2587.49	38647.4	SE				0.7	BL
2586.44	2587.21	38651.6	SE				1.2	BL
2585.25	2586.02	38669.4	SE				1.2	BL
2561.69	2562.46	39025.0	SE				1.4	BL
2560.88	2561.65	39037.4	SE				1.2	BL
2554.58	2555.35	39133.6	SE				1.2	BL
2549.19	2549.96	39216.4	SE				1.7	BL
2547.98	2548.74	39235.0	SE	I			1.8	RD
2539.54	2540.30	39365.4	SE				1.0	BL
2530.17	2530.93	39511.2	SE				0.7	BL
2527.98	2528.74	39545.4	SE				1.4	BL
2518.74	2519.50	39690.5	SE				1.2	BL
2516.57	2517.33	39724.7	SE				1.2	BL
2514.87	2515.63	39751.5	SE				0.7	BL
2505.35	2506.10	39902.6	SE				0.7	BL
2496.05	2496.80	40051.2	SE				2.0	BL
2492.56	2493.31	40107.3	SE				0.8	RO
2490.84	2491.59	40135.0	SE				1.5	BL
2483.74	2484.49	40249.7	SE				0.3H	BL
2482.35	2483.10	40272.2	SE				0.7	BL
2472.82	2473.57	40427.4	SE				1.5	BL
2469.75	2470.50	40477.7	SE				0.7	BL
2468.83	2469.58	40492.8	SE				1.0	BL
2463.97	2464.72	40572.6	SE				1.0	BL
2451.29	2452.03	40782.5	SE				0.7	BL
2450.52	2451.26	40795.3	SE				0.7	BL
2447.53	2448.27	40845.1	SE				0.7	BL
2443.76	2444.50	40908.1	SE				1.0H	BL
2442.07	2442.81	40936.5	SE	I			1.2	RD
2441.50	2442.24	40946.0	SE				1.0	BL
2440.04	2440.78	40970.5	SE				0.7	BL
2438.86	2439.60	40990.3	SE				0.7	BL
2434.10	2434.84	41070.5	SE				1.0	BL
2425.99	2426.73	41207.8	SE				1.2	BL
2424.39	2425.13	41235.0	SE				0.7	BL
2423.23	2423.97	41254.7	SE				0.7	BL
2413.517	2414.251	41420.72	SE	I			2.1	RD
2411.60	2412.33	41453.6	SE				1.0	BL
2409.36	2410.09	41492.2	SE				1.0	BL
2406.59	2407.32	41539.9	SE				1.2	BL
2402.28	2403.01	41614.5	SE				1.4	BL
2393.51	2394.24	41766.9	SE				0.7	BL
2393.27	2394.00	41771.1	SE				0.7	BL
2389.38	2390.11	41839.1	SE				1.0	BL
2380.65	2381.38	41992.5	SE				0.7	BL
2372.71	2373.43	42133.0	SE				1.4	BL
2363.35	2364.07	42299.9	SE				0.7	BL
2354.35	2355.07	42461.6	SE				1.4	BL

AIR WAVELENGTH (ANGSTRMS)	VACUUM WAVELENGTH (ANGSTRMS)	WAVENUMBER (KAYSERS)	LMENT	LOG INTENSITY ARC	SPK	DIS	REF
2349.29	2350.01	42553.0	SE			0.7	BL
2340.19	2340.91	42718.5	SE			1.0	BL
2332.81	2333.53	42853.6	SE I			1.5	RD
2325.32	2326.03	42991.6	SE			0.7	BL
2318.05	2318.76	43126.5	SE			0.7	BL
2310.21	2310.92	43272.8	SE			1.0	BL
2292.07	2292.78	43615.2	SE			1.0	BL
2279.27	2279.97	43860.1	SE			1.0	BL
2272.41	2273.11	43992.5	SE			1.0	BL
2265.72	2266.42	44122.4	SE			0.7	BL
2262.98	2263.68	44175.9	SE			0.7	BL
2254.89	2255.59	44334.3	SE			1.0	BL
2241.12	2241.82	44606.7	SE			0.7	BL
2227.31	2228.00	44883.2	SE			0.7	BL
2207.87	2208.56	45278.4	SE			1.0	BL
2164.160	2164.839	46192.80	SE I			2.0	RD
2147.19	2147.87	46557.8	SE			1.1	RD
2113.95	2114.62	47289.8	SE II			0.6	MZ
2081.08	2081.74	48036.7	SE			1.2	RD
2074.793	2075.455	48182.21	SE I			2.0	RD
2062.788	2063.447	48462.59	SE I			2.9	RD
2054.36	2055.02	48661.4	SE I			0.9	RD
2050.43	2051.09	48754.6	SE I			1.1	RD
2039.851	2040.506	49007.45	SE I			3.0	RD
2026.51	2027.16	49330.0	SE			0.6	RO
2025.13	2025.78	49363.6	SE			0.6	RO
2021.80	2022.45	49444.9	SE I			0.9	RD
9891.90	9894.61	10106.5	SI I	0.7W			KS
9886.92	9889.63	10111.6	SI I	0.3W			KS
9839.58	9842.28	10160.2	SI I	0.3W			KS
9789.24	9791.92	10212.5	SI I	0.3H			KS
9770.10	9772.78	10232.5	SI I	0.6W			KS
9768.27	9770.95	10234.4	SI I	0.7W			KS
9758.08	9760.76	10245.1	SI I	0.3W			KS
9738.60	9741.27	10265.6	SI	0.8W			KS
9689.41	9692.07	10317.7	SI I	0.9W			KS
9585.72	9588.35	10429.3	SI I	0.6			KS
9570.08	9572.71	10446.4	SI I	0.6			KS
9505.28	9507.89	10517.6	SI	0.7			KS
9421.82	9424.40	10610.7	SI I	0.6			KS
9413.59	9416.17	10620.0	SI I	2.0			KS
9393.40	9395.98	10642.8	SI I	0.3H			KS
9318.24	9320.80	10728.7	SI I	0.6			KS
9254.59	9257.13	10802.5	SI I	0.6H			KS
9238.60	9241.14	10821.2	SI I	0.3W			KS
9208.55	9211.08	10856.5	SI I	0.7W			KS
9103.37	9105.87	10981.9	SI I	0.5W			KS
9022.40	9024.88	11080.5	SI	0.3H			KS
9009.04	9011.51	11096.9	SI I	0.7B			KS
8949.33	8951.79	11170.9	SI	1.2W			KS
8925.55	8928.00	11200.7	SI	0.9W			KS
8899.50	8901.94	11233.5	SI	0.5W			KS
8898.97	8901.41	11234.2	SI	0.5W			KS
8892.97	8895.41	11241.7	SI	1.4W			KS
8883.84	8886.28	11253.3	SI	0.6W			KS
8791.28	8793.69	11371.8	SI	0.7W			KS
8790.88	8793.29	11372.3	SI	0.6W			KS
8766.68	8769.09	11403.7	SI I	0.5W			KS
8752.17	8754.57	11422.6	SI I	2.3			KS
8742.60	8745.00	11435.1	SI I	2.0			KS
8729.02	8731.42	11452.9	SI I	0.7W			KS
8728.38	8730.78	11453.7	SI	1.0W			KS
8648.89	8651.27	11559.0	SI	2.0B			KS
8597.00	8599.36	11628.8	SI I	0.3B			KS
8579.15	8581.51	11653.0	SI	0.3W			KS
8556.64	8558.99	11683.6	SI I	2.0H			KS
8536.80	8539.15	11710.8	SI	0.5W			KS
8503.17	8505.51	11757.1	SI	0.7			KS
8502.38	8504.72	11758.2	SI I	1.5W			KS
8501.50	8503.84	11759.4	SI I	1.3W			KS
8444.48	8446.80	11838.8	SI I	0.5W			KS
8444.00	8446.32	11839.5	SI I	1.2W			KS
8338.43	8340.72	11989.4	SI I	0.7W			KS
8317.45	8319.74	12019.6	SI I	0.3W			KS
8306.80	8309.08	12035.0	SI I	0.6W			KS
8230.67	8232.93	12146.3	SI I	1.2			KS
8211.48	8213.74	12174.7	SI I	0.3			KS
8093.32	8095.55	12352.5	SI	1.4H			KS
8035.39	8037.60	12441.5	SI	0.8W			KS
7970.91	7973.10	12542.2	SI	0.5W			KS
7970.26	7972.45	12543.2	SI	1.0H			KS
7943.94	7946.13	12584.7	SI	2.7W			KS
7932.20	7934.38	12603.4	SI I	2.5W			KS
7918.38	7920.56	12625.4	SI	2.3			KS
7913.47	7915.65	12633.2	SI	1.0H			KS
7912.55	7914.73	12634.7	SI	0.5W			KS
7850.5	7852.7	12734.	SI I	0.3V			KS
7801.30	7803.45	12814.8	SI	0.5H			KS
7800.0	7802.1	12817.	SI	0.6V			KS
7743.2	7745.3	12911.	SI	0.6H			KS
7742.7	7744.8	12912.	SI	0.7H			KS
7680.35	7682.46	13016.7	SI I	2.0W			KS
7424.63	7426.67	13465.0	SI I	1.3			KS
7423.54	7425.58	13466.9	SI I	2.7			KS
7416.00	7418.04	13480.7	SI I	2.3			KS
7415.37	7417.41	13481.8	SI I	1.2			KS
7409.11	7411.15	13493.2	SI I	2.0			KS
7405.85	7407.89	13499.1	SI I	2.5			KS
7373.02	7375.05	13559.2	SI I	1.0W			KS
7290.21	7292.22	13713.2	SI I	1.0			KS
7289.25	7291.26	13715.1	SI I	2.3			KS
7275.28	7277.28	13741.4	SI I	1.7			KS
7250.69	7252.69	13788.0	SI I	1.6			KS
7235.86	7237.85	13816.2	SI I	1.0			KS
7235.32	7237.31	13817.3	SI I	1.0			KS
7226.20	7228.19	13834.7	SI I	1.0			KS
7193.89	7195.87	13896.8	SI I	0.7			KS
7193.56	7195.54	13897.5	SI I	0.9			KS
7184.89	7186.87	13914.3	SI I	1.0			KS
7167.47	7169.45	13948.1	SI I	0.5H			KS
7165.62	7167.60	13951.7	SI I	2.0H			KS
7164.75	7166.72	13953.4	SI I	0.3H			KS
7084.33	7086.28	14111.8	SI I	0.3H			KS
7034.96	7036.90	14210.8	SI I	1.7H			KS
7026.61	7028.55	14227.7	SI	0.3H			KS
7017.98	7019.92	14245.2	SI I	0.6H			KS
7017.68	7019.62	14245.8	SI I	1.0H			KS
7005.84	7007.77	14269.9	SI I	1.7H			KS
7003.58	7005.51	14274.5	SI I	1.7H			KS
6976.53	6978.45	14329.8	SI	1.4H			KS
6560.68	6562.49	15238.1	SI I	0.3H			KS
6527.49	6529.29	15315.6	SI I	0.5H			KS
6415.24	6417.01	15583.6	SI I	0.6H			KS
6371.09	6372.85	15691.6	SI	0.3	1.5		KS
6363.10	6364.86	15711.3	SI	0.5	0.0		SY
6347.01	6348.76	15751.1	SI	0.3	1.7		SY
6339.39	6341.14	15770.0	SI	0.3	0.3		SY
6331.92	6333.67	15788.6	SI	0.3H			KS
6329.71	6331.46	15794.1	SI	0.3	0.0		SY
6255.60	6257.33	15981.3	SI I	0.5H			KS
6254.96	6256.69	15982.9	SI I	0.3H			KS
6254.55	6256.28	15983.9	SI I	1.2B			KS
6254.25	6255.98	15984.7	SI I	1.4H			KS
6245.11	6246.84	16008.1	SI I	0.3H			KS
6244.74	6246.47	16009.1	SI I	1.1H			KS
6244.56	6246.29	16009.5	SI I	1.0H			KS
6243.86	6245.59	16011.3	SI I	1.0H			KS
6238.42	6240.15	16025.3	SI I	0.6H			KS
6237.62	6239.35	16027.3	SI I	0.7H			KS
6237.34	6239.07	16028.0	SI I	0.7H			KS

AIR WAVELENGTH (ANGSTRMS)	VACUUM WAVELENGTH (ANGSTRMS)	WAVENUMBER (KAYSERS)	LMENT	LOG ARC	INTENSITY SPK	DIS	REF
6156.00	6157.70	16239.8	SI I	1.0H			KS
6155.73	6157.43	16240.5	SI I	0.3H			KS
6155.32	6157.02	16241.6	SI I	1.7H			KS
6155.22	6156.92	16241.9	SI I	1.3H			KS
6145.22	6146.92	16268.3	SI I	1.2H			KS
6145.08	6146.78	16268.7	SI I	1.0H			KS
6142.70	6144.40	16275.0	SI I	0.8H			KS
6142.53	6144.23	16275.4	SI I	0.7H			KS
6132.14	6133.84	16303.0	SI I	0.6H			KS
6131.86	6133.56	16303.7	SI I	0.7H			KS
6131.54	6133.24	16304.6	SI I	0.6H			KS
6125.44	6127.14	16320.8	SI I	0.3H			KS
6125.03	6126.73	16321.9	SI I	0.6H			KS
6124.85	6126.55	16322.4	SI I	0.3H			KS
6106.70	6108.39	16370.9	SI	0.0H	0.5		KS
5957.80	5959.45	16780.1	SI		0.7		SY
5948.584	5950.232	16806.07	SI I	1.7	0.7		KS
5914.7	5916.3	16902.	SI		0.3		SY
5867.33	5868.96	17038.8	SI		0.7		SY
5845.06	5846.68	17103.7	SI		0.3		SY
5806.30	5807.91	17217.9	SI		0.3		SY
5800.25	5801.86	17235.9	SI		0.3		SY
5797.912	5799.520	17242.81	SI I	1.4			KS
5793.128	5794.734	17257.05	SI I	1.3			KS
5780.452	5782.055	17294.89	SI I	1.2			KS
5772.258	5773.859	17319.44	SI I	1.5			KS
5754.258	5755.854	17373.62	SI I	1.6			KS
5739.20	5740.79	17419.2	SI		0.8		SY
5716.08	5717.67	17489.6	SI		0.3		SY
5708.437	5710.021	17513.07	SI I	1.6	0.3H		KS
5706.23	5707.81	17519.8	SI		0.3		SY
5701.138	5702.720	17535.49	SI I	1.2			KS
5690.470	5692.049	17568.37	SI I	1.4			KS
5688.83	5690.41	17573.4	SI		0.3		SY
5684.523	5686.100	17586.75	SI I	1.5			KS
5669.63	5671.20	17632.9	SI		0.7		SY
5665.601	5667.173	17645.48	SI I	1.3			KS
5645.665	5647.232	17707.79	SI I	1.3			KS
5639.13	5640.70	17728.3	SI		0.3		SY
5622.22	5623.78	17781.6	SI I	0.5	0.0		KS
5472.90	5474.42	18266.8	SI		0.3		SY
5469.18	5470.70	18279.2	SI		0.3		SY
5456.61	5458.13	18321.3	SI		0.3		SY
5451.91	5453.43	18337.1	SI		0.3		SY
5438.58	5440.09	18382.0	SI		0.5		SY
5417.36	5418.87	18454.1	SI		0.3		SY
5219.05	5220.50	19155.2	SI		0.3		SY
5202.85	5204.30	19214.9	SI		0.3		SY
5196.62	5198.07	19237.9	SI		0.3		SY
5193.21	5194.66	19250.6	SI		0.3		SY
5185.64	5187.08	19278.6	SI		0.3		SY
5182.13	5183.57	19291.7	SI		0.5J		SY
5113.70	5115.13	19549.9	SI		0.3		SY
5092.00	5093.42	19633.2	SI		0.3		SY
5056.10	5057.51	19772.6	SI		0.9		SY
5041.035	5042.441	19831.67	SI		0.6J		IJA
4958.19	4959.57	20163.0	SI	0.8			SY
4949.72	4951.10	20197.5	SI	0.3			SY
4943.16	4944.54	20224.3	SI	0.7	0.3		SY
4921.86	4923.23	20311.9	SI	0.9	0.3		SY
4907.50	4908.87	20371.3	SI	0.5	0.3		SY
4900.48	4901.85	20400.5	SI	0.3			SY
4883.51	4884.87	20471.4	SI	0.6	0.3		SY
4828.84	4830.19	20703.1	SI		1.1		SY
4819.57	4820.92	20742.9	SI		0.9		SY
4813.28	4814.63	20770.1	SI		0.8		SY
4716.71	4718.03	21195.3	SI		0.9		SY
4683.10	4684.41	21347.4	SI		0.6		SY
4654.08	4655.38	21480.5	SI		0.9		SY
4631.22	4632.52	21586.5	SI		0.6		SY
4619.60	4620.89	21640.8	SI		0.3		SY
4574.66	4575.94	21853.4	SI		1.3		SY
4567.66	4568.94	21886.9	SI		1.5		SY
4552.50	4553.78	21959.8	SI		1.6		SY
4338.57	4339.79	23042.6	SI		0.6		SY
4321.81	4323.03	23131.9	SI		0.3		SY
4314.32	4315.53	23172.1	SI		0.5		SY
4304.15	4305.36	23226.9	SI		0.3		SY
4277.95	4279.15	23369.1	SI	0.0	0.5		SY
4236.45	4237.64	23598.0	SI		0.5J		SY
4212.66	4213.85	23731.3	SI		0.3		SY
4198.25	4199.43	23812.7	SI		0.5		SY
4190.92	4192.10	23854.4	SI		0.6		SY
4141.04	4142.21	24141.7	SI		0.3		SY
4130.96	4132.13	24200.6	SI		1.4W		SY
4128.11	4129.27	24217.3	SI	1.3W			SY
4116.15	4117.31	24287.7	SI		0.3		SY
4102.946	4104.104	24365.85	SI I	1.1	1.0		MIT
4088.88	4090.03	24449.7	SI		0.5		SY
4058.49	4059.64	24632.7	SI		0.5		SY
3991.62	3992.75	25045.4	SI		0.3		SY
3924.44	3925.55	25474.1	SI		0.5		SY
3905.528	3906.634	25597.48	SI I	1.3	1.2C		MIT
3905.37	3906.48	25598.5	SI		0.6		SY
3891.55	3892.65	25689.4	SI		0.5		SY
3870.64	3871.74	25828.2	SI		0.5		SY
3862.51	3863.61	25882.6	SI		0.8		SY
3856.09	3857.18	25925.6	SI		0.9		SY
3853.01	3854.10	25946.4	SI		0.7		SY
3837.65	3838.74	26050.2	SI		0.5		SY
3806.60	3807.68	26262.7	SI		0.8		SY
3796.18	3797.26	26334.8	SI		0.8		SY
3791.13	3792.21	26369.9	SI		0.7		SY
3779.47	3780.54	26451.2	SI		0.5		SY
3774.74	3775.81	26484.4	SI		0.6		SY
3762.42	3763.49	26571.1	SI		0.5		SY
3713.23	3714.29	26923.1	SI		0.3		SY
3702.01	3703.06	27004.7	SI		0.8		SY
3681.38	3682.43	27156.0	SI		0.5W		SY
3670.49	3671.54	27236.6	SI		0.6		SY
3651.66	3652.70	27377.0	SI		0.5		SY
3590.77	3591.79	27841.2	SI		0.7		SY
3576.15	3577.17	27955.1	SI		0.3H		SY
3570.05	3571.07	28002.8	SI		0.3		SY
3563.61	3564.63	28053.4	SI		0.3		SY
3548.24	3549.25	28174.9	SI		0.3		SY
3499.61	3500.61	28566.4	SI		0.3		SY
3486.79	3487.79	28671.5	SI		0.8		SY
3481.55	3482.55	28714.6	SI		0.7		SY
3470.68	3471.67	28804.6	SI		0.3		SY
3464.14	3465.13	28858.9	SI		0.5		SY
3430.51	3431.49	29141.8	SI		0.3H		SY
3423.66	3424.64	29200.1	SI		0.3		SY
3420.32	3421.30	29228.6	SI		0.3		SY
3333.48	3334.44	29990.1	SI		0.3		SY
3283.69	3284.64	30444.8	SI		0.3		SY
3281.64	3282.59	30463.8	SI		0.5		SY
3276.10	3277.04	30515.3	SI		0.3		SY
3266.95	3267.89	30600.8	SI		0.3		SY
3258.45	3259.39	30680.6	SI		0.3H		SY
3241.80	3242.74	30838.2	SI		0.8		SY
3234.05	3234.98	30912.1	SI		0.8		SY
3230.43	3231.36	30946.7	SI		0.5		SY
3210.27	3211.20	31141.0	SI		0.8H		SY
3203.87	3204.80	31203.2	SI		0.5		SY
3199.42	3200.34	31246.6	SI		0.6		SY
3196.12	3197.04	31278.9	SI		0.3		SY
3192.69	3193.61	31312.5	SI		0.3		SY
3188.85	3189.77	31350.2	SI		0.3		SY
3185.28	3186.20	31385.3	SI		0.5		SY

AIR WAVELENGTH (ANGSTRMS)	VACUUM WAVELENGTH (ANGSTRMS)	WAVENUMBER (KAYSERS)	LMENT	LOG ARC	INTENSITY SPK	INTENSITY DIS	REF
3165.78	3166.70	31578.6	SI		0.8		SY
3149.67	3150.58	31740.2	SI		0.8		SY
3147.01	3147.92	31767.0	SI		0.3H		SY
3130.48	3131.39	31934.7	SI		0.7		SY
3121.99	3122.90	32021.6	SI		0.5		SY
3106.14	3107.04	32185.0	SI		0.3		SY
3103.80	3104.70	32209.2	SI		0.3		SY
3096.92	3097.82	32280.8	SI		0.6		SY
3093.28	3094.18	32318.8	SI		0.8		SY
3086.44	3087.34	32390.4	SI		0.8		SY
3053.02	3053.91	32744.9	SI		0.7		SY
3043.70	3044.59	32845.2	SI		0.7		SY
3002.37	3003.25	33297.3	SI		0.5		SY
2987.99	2988.86	33457.5	SI			0.7	SY
2987.648	2988.520	33461.38	SI I	2.0	2.0		MIT
2976.42	2977.29	33587.6	SI			0.3	SY
2970.347	2971.214	33656.27	SI I	1.3	1.3		MIT
2924.01	2924.87	34189.6	SI	0.5		0.7	MIT
2905.59	2906.44	34406.3	SI			0.7	SY
2904.01	2904.86	34425.1	SI			0.7	SY
2899.52	2900.37	34478.4	SI			0.7	SY
2881.595	2882.440	34692.83	SI I	2.7	2.6		MIT
2873.10	2873.94	34795.4	SI			0.3	SY
2869.73	2870.57	34836.3	SI			0.3	SY
2866.28	2867.12	34878.2	SI			0.8	SY
2858.14	2858.98	34977.5	SI			0.3	SY
2842.35	2843.19	35171.8	SI I	0.5			KS
2831.40	2832.23	35307.8	SI			0.7	SY
2813.61	2814.44	35531.1	SI			0.7	SY
2716.25	2717.05	36804.6	SI			0.7	SY
2701.53	2702.33	37005.1	SI			0.7	SY
2697.20	2698.00	37064.5	SI			0.3	SY
2695.2	2696.0	37092.	SI			0.7	SY
2687.75	2688.55	37194.8	SI			0.7	SY
2682.42	2683.22	37268.7	SI			0.7	SY
2675.25	2676.05	37368.6	SI			0.7	SY
2650.73	2651.52	37714.2	SI			0.7	SY
2645.59	2646.38	37787.5	SI			0.3	SY
2640.89	2641.68	37854.7	SI			0.7	SY
2637.92	2638.71	37897.4	SI			0.3	SY
2631.310	2632.095	37992.56	SI I	1.8	1.7		FL
2593.71	2594.49	38543.3	SI			0.3	SY
2580.35	2581.12	38742.8	SI			0.3	SY
2577.13	2577.90	38791.2	SI I	1.0			KS
2570.78	2571.55	38887.0	SI			0.3	SY
2568.64	2569.41	38919.5	SI I	1.2	1.0		MIT
2564.82	2565.59	38977.4	SI I	0.5			KS
2563.67	2564.44	38994.9	SI I	0.6			KS
2559.20	2559.97	39063.0	SI			1.2	SY
2541.91	2542.67	39328.7	SI		0.7		SY
2532.378	2533.139	39476.71	SI I	1.5	1.6		MIT
2528.516	2529.276	39537.00	SI I	2.6	2.7		FL
2524.118	2524.877	39605.89	SI I	2.6	2.6		FL
2519.207	2519.965	39683.09	SI I	2.5	2.5		MIT
2517.45	2518.21	39710.8	SI		1.0		MIT
2516.123	2516.880	39731.73	SI I	2.7	2.7		FL
2514.331	2515.088	39760.04	SI I	2.5	2.3		FL
2506.899	2507.654	39877.91	SI I	2.5	2.3		MIT
2503.64	2504.39	39929.8	SI			0.3	SY
2500.90	2501.65	39973.6	SI			0.3	SY
2495.79	2496.54	40055.4	SI			0.3	SY
2486.28	2487.03	40208.6	SI			0.7	SY
2483.29	2484.04	40257.0	SI			0.3	SY
2480.04	2480.79	40309.8	SI			0.3	SY
2452.136	2452.879	40768.43	SI I	1.3	1.3		FL
2449.70	2450.44	40809.0	SI			0.7	SY
2443.378	2444.119	40914.55	SI I	1.2	1.2		FL
2438.782	2439.521	40991.65	SI I	1.5	1.3		FL
2435.159	2435.898	41052.63	SI I	2.2	1.9		FL
2432.23	2432.97	41102.1	SI			0.7	SY

AIR WAVELENGTH (ANGSTRMS)	VACUUM WAVELENGTH (ANGSTRMS)	WAVENUMBER (KAYSERS)	LMENT	LOG ARC	INTENSITY SPK	INTENSITY DIS	REF
2419.80	2420.54	41313.2	SI			0.3	SY
2416.54	2417.27	41368.9	SI			0.3	SY
2363.82	2364.54	42291.5	SI			0.7	SY
2358.02	2358.74	42395.5	SI			0.7	SY
2357.20	2357.92	42410.2	SI			0.7	SY
2356.35	2357.07	42425.5	SI			1.0	SY
2353.19	2353.91	42482.5	SI			0.3	SY
2349.54	2350.26	42548.5	SI			0.3	SY
2346.80	2347.52	42598.2	SI			0.7	SY
2338.97	2339.69	42740.8	SI			0.7	SY
2334.48	2335.20	42823.0	SI			0.3	SY
2325.35	2326.06	42991.1	SI			0.7	SY
2308.21	2308.92	43310.3	SI			0.3	SY
2303.02	2303.73	43407.9	SI I	1.0	0.8		FL
2299.86	2300.57	43467.5	SI			0.7	SY
2291.03	2291.74	43635.0	SI I	0.9			KS
2289.61	2290.32	43662.1	SI I	1.0			KS
2287.06	2287.77	43710.8	SI			1.0	SY
2278.30	2279.00	43878.8	SI I	0.8			KS
2259.57	2260.27	44242.5	SI I	0.5			FL
2228.86	2229.55	44852.0	SI			0.7	SY
2218.917	2219.608	45053.00	SI I	0.5			FL
2218.080	2218.771	45070.00	SI I	1.0	1.2		FL
2216.685	2217.375	45098.36	SI I	1.6	1.6		FL
2211.750	2212.439	45198.98	SI I	1.1	1.1		FL
2210.912	2211.601	45216.11	SI I	1.5	1.5		FL
2207.980	2208.669	45276.15	SI I	1.3	1.2		FL
2192.22	2192.91	45601.6	SI			1.0	SY
2189.62	2190.30	45655.7	SI			0.7	SY
2179.37	2180.05	45870.5	SI			0.7	SY
2177.30	2177.98	45914.1	SI I	0.90			KS
2167.74	2168.42	46116.5	SI I	0.8			KS
2163.78	2164.46	46200.9	SI I	1.0W			KS
2150.43	2151.11	46487.7	SI I	0.7			KS
2147.91	2148.59	46542.2	SI I	0.5			KS
2135.90	2136.57	46803.9	SI			0.7	SY
2124.150	2124.821	47062.78	SI I	2.36	1.7		FL
2123.77	2124.44	47071.2	SI			0.3	SY
2122.99	2123.66	47088.5	SI I	1.0			KS
2121.22	2121.89	47127.8	SI I	0.8			KS
2114.59	2115.26	47275.5	SI I	0.6W			KS
2103.28	2103.95	47529.7	SI	0.7W			KS
2094.20	2094.87	47735.8	SI I	0.3			KS
2084.47	2085.13	47958.6	SI I	1.3			KS
2082.01	2082.67	48015.2	SI I	1.2			KS
2065.48	2066.14	48399.4	SI	0.3			FL
2061.18	2061.84	48500.4	SI I	0.9			KS
2058.13	2058.79	48572.2	SI I	1.2			KS
2054.85	2055.51	48649.8	SI	0.5			FL
2010.97	2011.62	49711.2	SI I	0.9			KS
2008.43	2009.08	49774.0	SI I	0.5			KS
8913.66	8916.11	11215.7	SM II	0.8			KS
8859.76	8862.19	11283.9	SM	0.3			KS
8788.83	8791.24	11375.0	SM II	0.3			KS
8758.28	8760.69	11414.6	SM II	0.3			KS
8717.89	8720.28	11467.5	SM II	1.7D			KN
8708.43	8710.82	11480.0	SM II	1.8D			KN
8706.32	8708.71	11482.8	SM II	1.5			KN
8677.93	8680.31	11520.3	SM	1.5D			KN
8632.83	8635.20	11580.5	SM II	1.6D			KS
8622.27	8624.64	11594.7	SM II	1.2D			KN
8617.03	8619.40	11601.7	SM II	1.7D			KN
8568.94	8571.29	11666.8	SM	0.9D			KN
8543.22	8545.57	11702.0	SM II	2.2D			KN
8539.22	8541.57	11707.5	SM II	1.0D			KN
8510.90	8513.24	11746.4	SM II	2.3D			KN
8506.77	8509.11	11752.1	SM	1.0D			KN
8485.99	8488.32	11780.9	SM II	2.6D			KN
8473.54	8475.87	11798.2	SM II	2.0D			KN
8472.9	8475.2	11799.	SM	0.6			KN

AIR WAVELENGTH (ANGSTRMS)	VACUUM WAVELENGTH (ANGSTRMS)	WAVENUMBER (KAYSERS)	LMENT	LOG ARC	INTENSITY SPK	DIS	REF
8465.08	8467.41	11810.0	SM	0.9			KN
8457.77	8460.09	11820.2	SM II	1.5 D			KN
8454.76	8457.08	11824.4	SM	1.3			KN
8449.8	8452.1	11831.	SM	1.0 D			KN
8437.65	8439.97	11848.4	SM II	1.2 D			KN
8432.64	8434.96	11855.4	SM II	2.3 D			KN
8427.1	8429.4	11863.	SM	0.9 D			KN
8425.56	8427.88	11865.4	SM II	0.9 D			KN
8393.86	8396.17	11910.2	SM	1.2 D			KN
8390.30	8392.61	11915.2	SM II	1.2 D			KN
8387.77	8390.08	11918.8	SM II	2.0 D			KN
8383.71	8386.01	11924.6	SM	2.2 D			KN
8372.26	8374.56	11940.9	SM II	1.4 D			KN
8369.08	8371.38	11945.5	SM	0.7 D			KN
8355.23	8357.53	11965.3	SM	1.0 D			KN
8348.68	8350.97	11974.7	SM II	2.2 D			KN
8344.63	8346.92	11980.5	SM	0.7			KN
8325.38	8327.67	12008.2	SM	1.0			KN
8320.09	8322.38	12015.8	SM II	0.8			KN
8315.45	8317.74	12022.5	SM	2.2			KN
8312.71	8314.99	12026.5	SM II	1.2 D			KN
8305.79	8308.07	12036.5	SM II	2.7 D			KN
8301.34	8303.62	12042.9	SM	2.0			KN
8300.88	8303.16	12043.6	SM II	1.9 D			KN
8298.46	8300.74	12047.1	SM	0.9			KN
8289.26	8291.54	12060.5	SM II	2.1 D			KN
8267.45	8269.72	12092.3	SM	1.6 D			KN
8256.5	8258.8	12108.	SM	0.6			KN
8248.2	8250.5	12120.	SM	0.7			KN
8246.51	8248.78	12123.0	SM II	1.0 D			KN
8240.98	8243.25	12131.1	SM II	2.2			KN
8230.33	8232.59	12146.8	SM	2.0			KN
8218.76	8221.02	12163.9	SM II	2.5			KN
8214.74	8217.00	12169.9	SM	0.7			KN
8208.25	8210.51	12179.5	SM	1.8			KN
8206.30	8208.56	12182.4	SM	1.9 D			KN
8203.98	8206.24	12185.9	SM	1.9			KN
8195.50	8197.75	12198.5	SM II	2.3 D			KN
8186.31	8188.56	12212.2	SM	1.6 D			KN
8184.43	8186.68	12215.0	SM	1.7 D			KN
8180.34	8182.59	12221.1	SM	0.8 D			KN
8172.96	8175.21	12232.1	SM	0.8 D			KN
8165.38	8167.62	12243.5	SM	1.4 D			KN
8161.90	8164.14	12248.7	SM II	2.3 D			KN
8153.85	8156.09	12260.8	SM	0.8			KN
8146.60	8148.84	12271.7	SM	1.6			KN
8135.27	8137.51	12288.8	SM	0.7			KN
8133.67	8135.91	12291.2	SM	1.0 D			KN
8132.01	8134.25	12293.7	SM	1.0			KN
8129.11	8131.35	12298.1	SM	1.2			KN
8125.12	8127.35	12304.1	SM II	2.1 D			KN
8122.29	8124.52	12308.4	SM	0.8			KN
8121.35	8123.58	12309.8	SM	1.2			KN
8117.16	8119.39	12316.2	SM II	2.2 D			KN
8116.20	8118.43	12317.7	SM	1.1			KN
8112.31	8114.54	12323.6	SM	0.6			KN
8102.38	8104.61	12338.7	SM	1.3			KN
8091.15	8093.37	12355.8	SM	1.9			KN
8090.80	8093.02	12356.3	SM	1.8			KN
8087.08	8089.30	12362.0	SM II	1.2 D			KN
8084.52	8086.74	12365.9	SM	1.2 D			KN
8076.26	8078.48	12378.6	SM	1.2			KN
8071.74	8073.96	12385.5	SM II	1.6 D			KN
8070.37	8072.59	12387.6	SM II	1.7 D			KN
8068.46	8070.68	12390.5	SM II	2.9			KN
8065.16	8067.38	12395.6	SM	2.2			KN
8055.50	8057.72	12410.5	SM	0.9 D			KN
8048.70	8050.91	12420.9	SM II	2.6			KN
8040.27	8042.48	12434.0	SM	1.1 D			KN
8038.38	8040.59	12436.9	SM	0.6			KN

AIR WAVELENGTH (ANGSTRMS)	VACUUM WAVELENGTH (ANGSTRMS)	WAVENUMBER (KAYSERS)	LMENT	LOG ARC	INTENSITY SPK	DIS	REF
8033.69	8035.90	12444.2	SM	0.7			KN
8032.03	8034.24	12446.7	SM II	2.3 D			KN
8029.56	8031.77	12450.6	SM II	0.8 D			KN
8026.32	8028.53	12455.6	SM II	2.7 D			KN
8025.12	8027.33	12457.4	SM II	2.6			KN
8021.54	8023.75	12463.0	SM	1.6			KN
8019.56	8021.77	12466.1	SM II	1.5			KN
8016.17	8018.37	12471.4	SM	1.9			KN
8014.92	8017.12	12473.3	SM II	2.3 D			KN
8005.00	8007.20	12488.8	SM II	1.2 D			KN
8001.61	8003.81	12494.1	SM II	2.3 D			KN
7984.07	7986.27	12521.5	SM	0.8			KN
7979.18	7981.37	12529.2	SM	0.5			KN
7968.32	7970.51	12546.2	SM	1.7			KN
7955.52	7957.71	12566.4	SM	0.8			KN
7952.22	7954.41	12571.7	SM	1.1			KN
7948.12	7950.31	12578.1	SM II	2.0 D			KN
7947.00	7949.19	12579.9	SM II	1.2 D			KN
7946.15	7948.34	12581.2	SM	1.0			KN
7940.31	7942.49	12590.5	SM	2.2			KN
7937.09	7939.27	12595.6	SM II	2.0 D			KN
7934.01	7936.19	12600.5	SM	1.0 D			KN
7931.92	7934.10	12603.8	SM I	2.3			KN
7931.15	7933.33	12605.0	SM	1.7			KN
7928.14	7930.32	12609.8	SM II	2.9			KN
7926.94	7929.12	12611.7	SM II	1.3 D			KN
7926.27	7928.45	12612.8	SM	1.5			KN
7921.34	7923.52	12620.7	SM	1.2			KN
7919.44	7921.62	12623.7	SM I	1.9			KN
7914.96	7917.14	12630.8	SM II	2.3 D			KN
7913.49	7915.67	12633.2	SM	1.8			KN
7907.48	7909.66	12642.8	SM I	0.7			KN
7906.37	7908.54	12644.6	SM II	1.5 D			KN
7895.96	7898.13	12661.2	SM I	2.3			KN
7889.34	7891.51	12671.8	SM	1.2			KN
7880.07	7882.24	12686.7	SM II	2.0 D			KN
7875.10	7877.27	12694.8	SM II	1.3 D			KN
7873.72	7875.89	12697.0	SM	1.4 D			KN
7868.28	7870.44	12705.8	SM II	1.3 D			KN
7863.65	7865.81	12713.2	SM II	2.0 D			KN
7859.53	7861.69	12719.9	SM I	2.0			KN
7856.80	7858.96	12724.3	SM	0.9 D			KN
7856.10	7858.26	12725.5	SM	0.9			KN
7854.94	7857.10	12727.3	SM	1.1 D			KN
7848.77	7850.93	12737.3	SM	1.0 D			KN
7846.10	7848.26	12741.7	SM	0.6 D			KN
7844.82	7846.98	12743.8	SM II	1.8 D			KN
7837.27	7839.43	12756.0	SM II	2.6			KN
7835.99	7838.15	12758.1	SM	0.5			KN
7835.51	7837.67	12758.9	SM	0.9			KN
7835.08	7837.24	12759.6	SM II	2.6 D			KN
7831.40	7833.55	12765.6	SM	1.8 D			KN
7823.34	7825.49	12778.7	SM	0.9			KN
7820.15	7822.30	12784.0	SM II	2.2 D			KN
7815.00	7817.15	12792.4	SM	0.8 D			KN
7813.61	7815.76	12794.7	SM II	1.3 D			KN
7812.75	7814.90	12796.1	SM	1.8 D			KN
7808.96	7811.11	12802.3	SM II	1.3 D			KN
7808.21	7810.36	12803.5	SM	0.7			KN
7806.11	7808.26	12806.9	SM	1.5 D			KN
7805.56	7807.71	12807.9	SM	0.9 D			KN
7801.54	7803.69	12814.5	SM I	2.2			KN
7798.47	7800.62	12819.5	SM	1.0			KN
7794.50	7796.64	12826.0	SM I	2.0			KN
7792.76	7794.90	12828.9	SM	0.5			KN
7792.33	7794.47	12829.6	SM	1.0			KN
7789.76	7791.90	12833.8	SM	1.2 D			KN
7787.02	7789.16	12838.3	SM I	0.9			KN
7783.77	7785.91	12843.7	SM	0.7			KN
7782.51	7784.65	12845.8	SM	0.6			KN

AIR WAVELENGTH (ANGSTRMS)	VACUUM WAVELENGTH (ANGSTRMS)	WAVENUMBER (KAYSERS)	LMENT	LOG ARC	INTENSITY SPK	DIS	REF
7778.15	7780.29	12853.0	SM	0.6			KN
7777.01	7779.15	12854.9	SM	1.5			KN
7772.83	7774.97	12861.8	SM II	1.5D			KN
7772.4	7774.5	12862.	SM	0.3			KN
7771.6	7773.7	12864.	SM	0.3			KN
7768.04	7770.18	12869.7	SM	1.3D			KN
7766.99	7769.13	12871.5	SM	0.6			KN
7758.20	7760.33	12886.0	SM	0.7			KN
7757.74	7759.87	12886.8	SM II	1.3D			KN
7757.23	7759.36	12887.7	SM	0.7			KN
7755.32	7757.45	12890.8	SM I	1.9			KN
7755.20	7757.33	12891.0	SM II	2.1D			KN
7749.30	7751.43	12900.8	SM II	2.3			KN
7746.94	7749.07	12904.8	SM	0.7			KN
7743.67	7745.80	12910.2	SM	0.7			KN
7736.26	7738.39	12922.6	SM	2.2D			KN
7735.5	7737.6	12924.	SM	0.3			KN
7732.54	7734.67	12928.8	SM	0.6D			KN
7728.56	7730.69	12935.5	SM II	2.3D			KN
7725.66	7727.79	12940.3	SM	0.5			KN
7721.38	7723.51	12947.5	SM	0.7			KN
7715.71	7717.83	12957.0	SM	0.7			KN
7714.99	7717.11	12958.2	SM	0.8			KN
7712.04	7714.16	12963.2	SM II	1.8D			KN
7710.72	7712.84	12965.4	SM	1.0			KN
7709.14	7711.26	12968.1	SM	0.9			KN
7705.37	7707.49	12974.4	SM	0.8			KN
7695.78	7697.90	12990.6	SM	2.0			KN
7692.24	7694.36	12996.5	SM	1.3D			KN
7691.34	7693.46	12998.1	SM	1.2			KN
7690.75	7692.87	12999.1	SM II	1.3D			KN
7686.70	7688.82	13005.9	SM I	0.9			KN
7686.07	7688.19	13007.0	SM	0.6			KN
7680.51	7682.62	13016.4	SM	0.7D			KN
7678.79	7680.90	13019.3	SM II	1.8D			KN
7676.19	7678.30	13023.7	SM	0.8D			KN
7675.35	7677.46	13025.1	SM II	0.8D			KN
7672.49	7674.60	13030.0	SM	1.7			KN
7670.40	7672.51	13033.5	SM	1.3D			KN
7668.21	7670.32	13037.3	SM	1.3			KN
7667.20	7669.31	13039.0	SM II	1.7D			KN
7662.94	7665.05	13046.2	SM	1.3			KN
7661.30	7663.41	13049.0	SM	1.2			KN
7655.78	7657.89	13058.4	SM II	1.7D			KN
7648.02	7650.13	13071.7	SM II	2.0D			KN
7645.82	7647.92	13075.4	SM	1.9			KN
7645.09	7647.19	13076.7	SM II	2.3D			KN
7643.91	7646.01	13078.7	SM	1.6			KN
7643.01	7645.11	13080.2	SM II	1.5D			KN
7639.64	7641.74	13086.0	SM	0.9			KN
7637.94	7640.04	13088.9	SM II	1.8D			KN
7631.77	7633.87	13099.5	SM II	1.3D			KN
7629.24	7631.34	13103.9	SM	0.5			KN
7623.16	7625.26	13114.3	SM	1.2D			KN
7619.33	7621.43	13120.9	SM	0.7			KN
7613.94	7616.04	13130.2	SM II	1.5D			KN
7610.78	7612.88	13135.6	SM	0.9D			KN
7607.74	7609.83	13140.9	SM I	1.9			KN
7607.48	7609.57	13141.3	SM	1.6D			KN
7598.01	7600.10	13157.7	SM	1.6			KN
7592.55	7594.64	13167.2	SM	0.6			KN
7590.43	7592.52	13170.9	SM	1.4			KN
7590.04	7592.13	13171.5	SM	1.3			KN
7588.31	7590.40	13174.5	SM	1.6D			KN
7585.77	7587.86	13178.9	SM II	1.8D			KN
7582.86	7584.95	13184.0	SM	0.8			KN
7580.91	7583.00	13187.4	SM II	0.8			KN
7580.82	7582.91	13187.6	SM I	0.7			MIT
7580.19	7582.28	13188.7	SM	0.9D			KN
7578.09	7580.18	13192.3	SM II	1.8D			KN

AIR WAVELENGTH (ANGSTRMS)	VACUUM WAVELENGTH (ANGSTRMS)	WAVENUMBER (KAYSERS)	LMENT	LOG ARC	INTENSITY SPK	DIS	REF
7573.8	7575.9	13200.	SM	0.5			KN
7572.29	7574.37	13202.4	SM	1.9D			KN
7570.95	7573.03	13204.7	SM II	2.2D			KN
7563.83	7565.91	13217.2	SM	1.2			MIT
7562.94	7565.02	13218.7	SM	1.7D			KN
7560.15	7562.23	13223.6	SM II	1.3D			KN
7554.08	7556.16	13234.2	SM I	0.9			KN
7546.57	7548.65	13247.4	SM	1.3			KN
7544.74	7546.82	13250.6	SM	1.7			KN
7543.62	7545.70	13252.6	SM	1.2			KN
7541.42	7543.50	13256.4	SM II	2.0D			KN
7540.04	7542.12	13258.9	SM II	1.3D			KN
7534.2	7536.3	13269.	SM	0.8D			KN
7533.3	7535.4	13271.	SM	0.8D			KN
7531.35	7533.42	13274.2	SM	1.0			KN
7530.22	7532.29	13276.2	SM	0.9D			KN
7520.90	7522.97	13292.6	SM	1.3			KN
7519.91	7521.98	13294.4	SM	1.2			KN
7518.32	7520.39	13297.2	SM II	1.2D			KN
7517.00	7519.07	13299.5	SM	1.5D			KN
7513.33	7515.40	13306.0	SM I	0.3			KN
7507.85	7509.92	13315.7	SM	1.1			KN
7505.80	7507.87	13319.4	SM	1.0D			KN
7503.38	7505.45	13323.7	SM	0.8D			KN
7502.39	7504.46	13325.4	SM II	1.7D			KN
7499.50	7501.57	13330.6	SM	0.5			KN
7498.71	7500.77	13332.0	SM I	0.6			KN
7497.26	7499.32	13334.5	SM I	0.5			KN
7494.07	7496.13	13340.2	SM II	0.9			KN
7491.84	7493.90	13344.2	SM	0.3			KN
7481.99	7484.05	13361.7	SM II	1.4			KN
7479.96	7482.02	13365.4	SM II	0.6			KN
7479.40	7481.46	13366.4	SM	1.2			KN
7470.76	7472.82	13381.8	SM	2.0			KN
7460.52	7462.57	13400.2	SM	1.0			KN
7454.70	7456.75	13410.7	SM	0.5			KN
7454.38	7456.43	13411.2	SM	0.5			KN
7453.03	7455.08	13413.7	SM II	1.5			KN
7452.80	7454.85	13414.1	SM II	1.2			KN
7451.96	7454.01	13415.6	SM	0.6			KN
7449.28	7451.33	13420.4	SM I	0.9			KN
7445.41	7447.46	13427.4	SM I	2.0			KN
7444.56	7446.61	13428.9	SM	2.2			KN
7442.00	7444.05	13433.6	SM	0.5			KN
7440.80	7442.85	13435.7	SM	0.6			KN
7440.44	7442.49	13436.4	SM	0.5			KN
7434.14	7436.19	13447.7	SM	0.9			KN
7431.18	7433.23	13453.1	SM	0.8			KN
7424.43	7426.47	13465.3	SM II	0.5			KN
7423.81	7425.85	13466.5	SM	0.5			KN
7417.02	7419.06	13478.8	SM	1.0			KN
7412.24	7414.28	13487.5	SM	1.2			KN
7410.73	7412.77	13490.2	SM	1.2			KN
7405.00	7407.04	13500.7	SM	1.7			KN
7403.30	7405.34	13503.8	SM I	0.7			KN
7402.38	7404.42	13505.4	SM II	0.6			KN
7401.23	7403.27	13507.6	SM	0.6			KN
7398.94	7400.98	13511.7	SM	0.7			KN
7397.92	7399.96	13513.6	SM	0.3			KN
7393.98	7396.02	13520.8	SM	1.3			KN
7393.37	7395.41	13521.9	SM II	0.9			KN
7389.03	7391.07	13529.8	SM I	0.7			KN
7387.12	7389.15	13533.3	SM II	0.3			KN
7385.42	7387.45	13536.5	SM	0.6			KN
7383.55	7385.58	13539.9	SM	1.2			KN
7378.69	7380.72	13548.8	SM	0.8			KN
7378.04	7380.07	13550.0	SM I	0.6			KN
7376.69	7378.72	13552.5	SM II	1.8			KN
7375.14	7377.17	13555.3	SM	0.6			KN
7373.80	7375.83	13557.8	SM II	0.5			KN

AIR WAVELENGTH (ANGSTRMS)	VACUUM WAVELENGTH (ANGSTRMS)	WAVENUMBER (KAYSERS)	LMENT		LOG ARC	INTENSITY SPK	DIS	REF
7371.51	7373.54	13562.0	SM	I	1.7			KN
7371.12	7373.15	13562.7	SM		0.3			KN
7367.9	7369.9	13569.	SM	I	0.3			KN
7366.06	7368.09	13572.0	SM		0.5			KN
7364.10	7366.13	13575.7	SM		1.7			KN
7363.01	7365.04	13577.7	SM	II	0.5			KN
7362.70	7364.73	13578.2	SM		0.5			KN
7358.37	7360.40	13586.2	SM		0.5			KN
7355.99	7358.02	13590.6	SM		0.5			KN
7354.09	7356.12	13594.1	SM	II	0.5			KN
7347.30	7349.32	13606.7	SM	I	2.0			KN
7344.76	7346.78	13611.4	SM	II	0.7			KN
7338.80	7340.82	13622.4	SM		0.3			KN
7338.04	7340.06	13623.9	SM		1.7			KN
7335.92	7337.94	13627.8	SM	II	0.5			KN
7335.10	7337.12	13629.3	SM		1.0			KN
7332.65	7334.67	13633.9	SM	I	2.0			KN
7331.33	7333.35	13636.3	SM		0.3			KN
7327.08	7329.10	13644.2	SM		1.5			KN
7325.44	7327.46	13647.3	SM	II	1.0			KN
7323.88	7325.90	13650.2	SM		0.5			KN
7318.70	7320.72	13659.9	SM		0.5			KN
7317.77	7319.79	13661.6	SM	II	0.6			KN
7316.31	7318.33	13664.3	SM		1.0			KN
7314.00	7316.02	13668.6	SM	I	0.9			KN
7312.60	7314.61	13671.3	SM		0.7			KN
7310.10	7312.11	13675.9	SM		0.6			KN
7309.35	7311.36	13677.3	SM		0.5			KN
7309.02	7311.03	13678.0	SM		1.6			KN
7306.58	7308.59	13682.5	SM		0.5			KN
7304.74	7306.75	13686.0	SM		1.0			KN
7304.16	7306.17	13687.1	SM		0.3			KN
7300.72	7302.73	13693.5	SM		1.6			KN
7298.4	7300.4	13698.	SM		0.5			KN
7298.04	7300.05	13698.5	SM		0.8			KN
7296.12	7298.13	13702.1	SM		0.3			KN
7295.48	7297.49	13703.3	SM		0.5			KN
7291.52	7293.53	13710.8	SM		0.5			KN
7290.23	7292.24	13713.2	SM		1.9			KN
7288.92	7290.93	13715.7	SM	II	1.5			KN
7286.30	7288.31	13720.6	SM		0.3			KN
7285.40	7287.41	13722.3	SM		0.3			KN
7284.61	7286.62	13723.8	SM	II	0.7			KN
7283.33	7285.34	13726.2	SM		1.5			KN
7282.94	7284.95	13726.9	SM		0.3			KN
7282.21	7284.22	13728.3	SM	I	1.3			KN
7281.47	7283.48	13729.7	SM	II	1.4			KN
7281.15	7283.16	13730.3	SM		0.3			KN
7279.25	7281.26	13733.9	SM	I	2.0			KN
7277.10	7279.11	13737.9	SM		1.0			KN
7273.88	7275.88	13744.0	SM		1.5			KN
7273.33	7275.33	13745.1	SM	II	0.5			KN
7271.32	7273.32	13748.9	SM		0.9			KN
7270.54	7272.54	13750.3	SM	II	0.7			KN
7268.45	7270.45	13754.3	SM	II	0.5			KN
7264.17	7266.17	13762.4	SM		0.9			MIT
7263.59	7265.59	13763.5	SM		0.3			KN
7262.69	7264.69	13765.2	SM	I	0.7			KN
7261.45	7263.45	13767.6	SM	II	1.0			KN
7257.11	7259.11	13775.8	SM	II	1.3			KN
7256.18	7258.18	13777.6	SM		0.5			KN
7254.95	7256.95	13779.9	SM	II	1.2			KN
7254.17	7256.17	13781.4	SM		0.6			KN
7249.02	7251.02	13791.2	SM	II	0.7D			KN
7245.60	7247.60	13797.7	SM		0.7			KN
7240.90	7242.90	13806.6	SM	II	2.3			KN
7238.17	7240.16	13811.8	SM		0.5			KN
7237.02	7239.01	13814.0	SM	II	1.2			KN
7235.73	7237.72	13816.5	SM	II	0.9			KN
7234.72	7236.71	13818.4	SM		0.7			KN

AIR WAVELENGTH (ANGSTRMS)	VACUUM WAVELENGTH (ANGSTRMS)	WAVENUMBER (KAYSERS)	LMENT		LOG ARC	INTENSITY SPK	DIS	REF
7234.40	7236.39	13819.0	SM		1.2D			KN
7228.91	7230.90	13829.5	SM		1.2			KN
7227.15	7229.14	13832.9	SM	II	0.9			KN
7225.62	7227.61	13835.8	SM		1.4			KN
7224.56	7226.55	13837.9	SM		0.7			KN
7221.14	7223.13	13844.4	SM	I	0.6			KN
7220.07	7222.06	13846.5	SM	I	1.4			KN
7219.34	7221.33	13847.9	SM		0.5			KN
7218.28	7220.27	13849.9	SM		0.3			KN
7218.09	7220.08	13850.3	SM	II	1.7			KN
7216.00	7217.99	13854.3	SM		0.8			KN
7215.08	7217.07	13856.0	SM		0.5			KN
7213.82	7215.81	13858.5	SM	I	1.8			KN
7210.95	7212.94	13864.0	SM		1.6			KN
7204.09	7206.08	13877.2	SM		1.3			KN
7193.9	7195.9	13897.	SM		0.3			KN
7191.63	7193.61	13901.2	SM		0.5			KN
7189.57	7191.55	13905.2	SM		0.5			KN
7186.35	7188.33	13911.4	SM		0.6			KN
7181.60	7183.58	13920.6	SM		0.3			KN
7180.85	7182.83	13922.1	SM		0.3			KN
7179.72	7181.70	13924.3	SM		1.0			MIT
7174.0	7176.0	13935.	SM		0.3			KN
7172.67	7174.65	13938.0	SM	I	1.3			KN
7172.24	7174.22	13938.8	SM		0.3			KN
7170.58	7172.56	13942.0	SM		0.5			KN
7157.33	7159.30	13967.8	SM		0.5			KN
7156.19	7158.16	13970.1	SM		0.6			KN
7154.83	7156.80	13972.7	SM	I	1.4			KN
7152.80	7154.77	13976.7	SM		0.5			KN
7151.03	7153.00	13980.2	SM	I	0.3			KN
7149.60	7151.57	13982.9	SM	II	2.2			KN
7148.8	7150.8	13984.	SM	I	0.3			KN
7147.70	7149.67	13986.7	SM		0.5			KN
7143.98	7145.95	13993.9	SM	II	1.5			KN
7142.09	7144.06	13997.7	SM	I	1.0			KN
7141.13	7143.10	13999.5	SM	I	1.4			KN
7139.39	7141.36	14002.9	SM		1.0			KN
7137.99	7139.96	14005.7	SM		1.0			KN
7137.58	7139.55	14006.5	SM		1.2			KN
7136.42	7138.39	14008.8	SM		1.5			KN
7136.01	7137.98	14009.6	SM		1.7			KN
7131.80	7133.77	14017.8	SM	I	1.9			KN
7129.79	7131.76	14021.8	SM		0.6			KN
7128.74	7130.71	14023.9	SM		0.5			KN
7127.37	7129.33	14026.6	SM		0.6			KN
7125.11	7127.07	14031.0	SM	II	1.5			KN
7123.29	7125.25	14034.6	SM	II	0.7			KN
7123.19	7125.15	14034.8	SM		0.6			KN
7122.40	7124.36	14036.3	SM	II	1.2			KN
7119.81	7121.77	14041.4	SM		1.8			KN
7118.25	7120.21	14044.5	SM	I	1.4			KN
7117.51	7119.47	14046.0	SM		1.4			KN
7115.96	7117.92	14049.0	SM		1.7			KN
7115.30	7117.26	14050.3	SM		0.5			KN
7114.50	7116.46	14051.9	SM	I	0.7			KN
7112.72	7114.68	14055.4	SM	II	0.5			KN
7112.34	7114.30	14056.2	SM		0.7H			KN
7111.44	7113.40	14058.0	SM		0.7			KN
7110.56	7112.52	14059.7	SM	I	0.7			KN
7109.06	7111.02	14062.7	SM	I	0.5			KN
7108.36	7110.32	14064.1	SM		0.7			KN
7106.72	7108.68	14067.3	SM		1.2			KN
7106.24	7108.20	14068.3	SM	I	1.9			MIT
7104.57	7106.53	14071.6	SM	I	2.2			MIT
7103.61	7105.57	14073.5	SM		0.7			KN
7102.00	7103.96	14076.7	SM	II	1.0			KN
7101.46	7103.42	14077.7	SM	I	1.0			KN
7100.77	7102.73	14079.1	SM		0.5			KN
7096.37	7098.33	14087.8	SM	I	1.8			MIT

AIR WAVELENGTH (ANGSTRMS)	VACUUM WAVELENGTH (ANGSTRMS)	WAVENUMBER (KAYSERS)	LMENT	LOG ARC	INTENSITY SPK	DIS	REF
7095.51	7097.47	14089.5	SM I	2.0			MIT
7092.98	7094.94	14094.6	SM	0.7			KN
7091.90	7093.86	14096.7	SM II	0.5			KN
7091.22	7093.18	14098.1	SM I	1.8			MIT
7089.94	7091.89	14100.6	SM	0.3			KN
7088.30	7090.25	14103.9	SM I	1.8			MIT
7085.54	7087.49	14109.4	SM II	1.5			MIT
7082.98	7084.93	14114.5	SM	0.3			KN
7082.51	7084.46	14115.4	SM	1.0			KN
7082.37	7084.32	14115.7	SM II	2.6H			MIT
7079.76	7081.71	14120.9	SM	0.6			KN
7079.48	7081.43	14121.4	SM I	1.4			KN
7074.67	7076.62	14131.0	SM	1.7			KN
7074.41	7076.36	14131.6	SM	1.2			KN
7071.10	7073.05	14138.2	SM I	1.0			KN
7069.32	7071.27	14141.7	SM	0.3			KN
7067.38	7069.33	14145.6	SM	1.2			KN
7065.84	7067.79	14148.7	SM II	0.8D			KN
7065.02	7066.97	14150.3	SM	1.0			KN
7064.79	7066.74	14150.8	SM II	0.6			KN
7062.25	7064.20	14155.9	SM	0.3			KN
7060.02	7061.97	14160.4	SM II	0.5			KN
7059.37	7061.32	14161.7	SM	0.3			KN
7057.28	7059.23	14165.9	SM	0.3			KN
7057.0	7058.9	14166.	SM	0.3			KN
7056.55	7058.50	14167.3	SM	1.4			KN
7054.97	7056.92	14170.5	SM II	1.2			KN
7053.92	7055.87	14172.6	SM	0.3			KN
7053.54	7055.48	14173.4	SM I	0.7			KN
7051.52	7053.46	14177.4	SM II	2.6A			MIT
7049.15	7051.09	14182.2	SM	1.0			KN
7047.82	7049.76	14184.9	SM	0.3			KN
7046.98	7048.92	14186.6	SM	0.3			KN
7046.44	7048.38	14187.7	SM	0.5			KN
7042.23	7044.17	14196.1	SM II	2.6H			MIT
7041.77	7043.71	14197.1	SM	0.3			KN
7039.22	7041.16	14202.2	SM II	2.7H			MIT
7037.9	7039.8	14205.	SM	0.3			KN
7036.73	7038.67	14207.2	SM	1.4			KN
7034.67	7036.61	14211.4	SM II	0.5			KN
7033.47	7035.41	14213.8	SM	0.6			KN
7032.31	7034.25	14216.2	SM	0.3			KN
7026.64	7028.58	14227.6	SM I	2.0			MIT
7024.44	7026.38	14232.1	SM	0.3			KN
7023.75	7025.69	14233.5	SM	0.7			KN
7022.68	7024.62	14235.7	SM	0.7			KN
7020.41	7022.35	14240.3	SM II	2.7			MIT
7018.93	7020.87	14243.3	SM	0.3			KN
7018.04	7019.98	14245.1	SM	0.3			KN
7016.72	7018.66	14247.7	SM	0.6			KN
7015.36	7017.29	14250.5	SM	0.5			KN
7012.99	7014.92	14255.3	SM II	0.3			KN
7011.86	7013.79	14257.6	SM	0.6			KN
7009.91	7011.84	14261.6	SM	0.5			MIT
7009.67	7011.60	14262.1	SM	0.7D			KN
7009.15	7011.08	14263.1	SM	0.3			KN
7008.41	7010.34	14264.6	SM	0.9			KN
7007.39	7009.32	14266.7	SM	0.7			KN
7006.04	7007.97	14269.5	SM I	0.6			KN
7003.98	7005.91	14273.7	SM	0.3			KN
7002.03	7003.96	14277.6	SM I	0.3			KN
7001.55	7003.48	14278.6	SM	0.3			KN
7000.76	7002.69	14280.2	SM	0.3			KN
6996.95	6998.88	14288.0	SM	0.5			KN
6993.42	6995.35	14295.2	SM II	1.2			MIT
6992.84	6994.77	14296.4	SM	0.3			KN
6990.07	6992.00	14302.1	SM	1.2			MIT
6988.40	6990.33	14305.5	SM	1.6			MIT
6987.36	6989.29	14307.6	SM	1.0			KN
6985.67	6987.60	14311.1	SM	0.7			KN
6985.25	6987.18	14311.9	SM	0.3			KN
6984.16	6986.09	14314.2	SM	1.0			KN
6978.46	6980.38	14325.9	SM	0.3			KN
6978.22	6980.14	14326.3	SM	0.3			KN
6977.55	6979.47	14327.7	SM I	1.0			KN
6976.42	6978.34	14330.1	SM II	0.3			KN
6975.99	6977.91	14330.9	SM	1.0			KN
6975.62	6977.54	14331.7	SM	0.8			KN
6972.16	6974.08	14338.8	SM	0.3			KN
6969.69	6971.61	14343.9	SM II	1.2D			MIT
6968.99	6970.91	14345.3	SM	0.3			KN
6968.65	6970.57	14346.0	SM II	1.4			KN
6966.49	6968.41	14350.5	SM	0.9			KN
6965.95	6967.87	14351.6	SM	0.9			KN
6964.84	6966.76	14353.9	SM	0.3			KN
6963.61	6965.53	14356.4	SM	0.3			KN
6960.98	6962.90	14361.8	SM	0.3			KN
6958.97	6960.89	14366.0	SM	1.0D			KN
6957.73	6959.65	14368.5	SM	0.3			KN
6955.63	6957.55	14372.9	SM II	1.2D			KN
6955.27	6957.19	14373.6	SM II	2.6D			MIT
6954.32	6956.24	14375.6	SM	0.5			KN
6953.61	6955.53	14377.1	SM	0.5			KN
6952.39	6954.31	14379.6	SM	0.5			KN
6950.51	6952.43	14383.5	SM II	2.0			KN
6949.23	6951.15	14386.1	SM	1.2D			KN
6948.18	6950.10	14388.3	SM	1.3D			KN
6946.60	6948.52	14391.6	SM	0.3			KN
6941.56	6943.47	14402.0	SM II	1.7			KN
6939.51	6941.42	14406.3	SM	0.5			KN
6938.74	6940.65	14407.9	SM	1.0D			KN
6937.53	6939.44	14410.4	SM	0.5			KN
6935.38	6937.29	14414.8	SM	0.5			KN
6930.41	6932.32	14425.2	SM II	1.7D			KN
6929.60	6931.51	14426.9	SM	1.6D			KN
6928.84	6930.75	14428.4	SM	0.3			KN
6928.18	6930.09	14429.8	SM	0.5			KN
6927.70	6929.61	14430.8	SM II	0.7			KN
6927.07	6928.98	14432.1	SM II	1.6D			MIT
6926.14	6928.05	14434.1	SM	0.5			KN
6925.70	6927.61	14435.0	SM	0.6			KN
6925.20	6927.11	14436.0	SM	0.5			KN
6919.03	6920.94	14448.9	SM I	1.5D			KN
6918.78	6920.69	14449.4	SM	1.3			KN
6913.54	6915.45	14460.4	SM II	0.5			KN
6912.78	6914.69	14462.0	SM I	0.5H			KN
6911.44	6913.35	14464.8	SM	0.5			KN
6909.81	6911.72	14468.2	SM II	1.7D			KN
6907.21	6909.12	14473.6	SM	0.3			KN
6906.22	6908.13	14475.7	SM	1.3			KN
6904.51	6906.41	14479.3	SM	1.6			KN
6900.28	6902.18	14488.2	SM II	1.4D			KN
6892.02	6893.92	14505.5	SM	1.4			KN
6891.48	6893.38	14506.7	SM	1.6D			KN
6888.81	6890.71	14512.3	SM I	1.0			KN
6887.42	6889.32	14515.2	SM	1.3			KN
6885.16	6887.06	14520.0	SM II	1.5			KN
6884.92	6886.82	14520.5	SM II	0.8			KN
6883.46	6885.36	14523.6	SM	0.3			KN
6883.11	6885.01	14524.3	SM	0.3H			KN
6882.79	6884.69	14525.0	SM	0.3			KN
6880.86	6882.76	14529.1	SM I	0.3			KN
6879.50	6881.40	14531.9	SM I	1.0			KN
6877.27	6879.17	14536.6	SM	1.0			KN
6877.10	6879.00	14537.0	SM	1.4D			KN
6875.27	6877.17	14540.9	SM II	1.3D			KN
6873.69	6875.59	14544.2	SM	0.3			KN
6873.26	6875.16	14545.1	SM II	1.3			KN
6872.42	6874.32	14546.9	SM II	2.0D			MIT
6870.26	6872.16	14551.5	SM	0.8			KN

AIR WAVELENGTH (ANGSTRMS)	VACUUM WAVELENGTH (ANGSTRMS)	WAVENUMBER (KAYSERS)	LMENT	LOG ARC	INTENSITY SPK	DIS	REF
6868.13	6870.03	14556.0	SM	0.3			KN
6867.11	6869.00	14558.2	SM	1.2			KN
6866.53	6868.42	14559.4	SM II	0.9D			KN
6862.82	6864.71	14567.2	SM II	2.0D			KN
6861.06	6862.95	14571.0	SM II	2.7D			MIT
6860.93	6862.82	14571.3	SM I	2.9D			KN
6858.12	6860.01	14577.2	SM	0.8			KN
6857.72	6859.61	14578.1	SM	0.8			KN
6856.03	6857.92	14581.7	SM II	2.5D			KN
6854.50	6856.39	14584.9	SM II	1.8D			KN
6853.92	6855.81	14586.2	SM I	0.7			KN
6853.54	6855.43	14587.0	SM	1.0D			KN
6852.94	6854.83	14588.2	SM	0.3			KN
6848.88	6850.77	14596.9	SM II	1.6			KN
6848.31	6850.20	14598.1	SM	1.0D			KN
6848.16	6850.05	14598.4	SM II	1.2D			KN
6846.54	6848.43	14601.9	SM II	2.0D			KN
6845.74	6847.63	14603.6	SM	0.6D			KN
6844.71	6846.60	14605.8	SM II	2.2D			KN
6841.78	6843.67	14612.1	SM I	1.4			MIT
6839.64	6841.53	14616.6	SM	1.6D			KN
6839.23	6841.12	14617.5	SM I	0.3			KN
6839.08	6840.97	14617.8	SM	1.2D			KN
6838.33	6840.22	14619.4	SM	1.0			KN
6837.20	6839.09	14621.8	SM	1.3D			KN
6835.97	6837.86	14624.5	SM	0.3			KN
6834.40	6836.29	14627.8	SM	0.3			KN
6832.13	6834.02	14632.7	SM	0.3			KN
6831.37	6833.26	14634.3	SM	0.3			KN
6830.54	6832.42	14636.1	SM I	1.2			KN
6829.81	6831.69	14637.7	SM II	2.0			KN
6829.20	6831.08	14639.0	SM	0.6			KN
6828.50	6830.38	14640.5	SM	1.5D			KN
6827.81	6829.69	14641.9	SM I	1.2			KN
6827.03	6828.91	14643.6	SM	0.6			KN
6826.65	6828.53	14644.4	SM	0.5			KN
6823.62	6825.50	14650.9	SM II	1.2D			KN
6821.90	6823.78	14654.6	SM	1.0			KN
6820.91	6822.79	14656.8	SM I	1.2			KN
6819.72	6821.60	14659.3	SM	0.5			KN
6816.16	6818.04	14667.0	SM	0.6			MIT
6815.54	6817.42	14668.3	SM	1.0D			KN
6814.86	6816.74	14669.8	SM	1.3			KN
6813.54	6815.42	14672.6	SM	1.2			KN
6813.4	6815.3	14673.	SM	1.2			KN
6809.23	6811.11	14681.9	SM	0.3			KN
6808.31	6810.19	14683.9	SM I	1.3			KN
6807.50	6809.38	14685.6	SM	0.7			KN
6805.58	6807.46	14689.8	SM	0.3			KN
6804.89	6806.77	14691.3	SM	1.0D			KN
6803.1	6805.0	14695.	SM	1.5H			KN
6802.96	6804.84	14695.4	SM I	1.4			KN
6801.72	6803.60	14698.1	SM	1.2D			KN
6801.34	6803.22	14698.9	SM	0.3			KN
6796.82	6798.70	14708.7	SM I	1.0			KN
6796.13	6798.01	14710.2	SM	0.6			KN
6795.08	6796.96	14712.5	SM	0.3			KN
6794.20	6796.08	14714.4	SM II	2.3			KN
6792.90	6794.77	14717.2	SM	0.3			KN
6792.55	6794.42	14717.9	SM II	1.5D			KN
6790.88	6792.75	14721.6	SM I	0.6			KN
6790.03	6791.90	14723.4	SM II	2.5D			MIT
6784.35	6786.22	14735.7	SM	0.7D			KS
6782.95	6784.82	14738.8	SM	1.5D			KN
6781.17	6783.04	14742.7	SM II	1.7D			KN
6780.03	6781.90	14745.1	SM	1.6			KN
6779.58	6781.45	14746.1	SM	0.3			KN
6779.16	6781.03	14747.0	SM I	1.2			KN
6778.61	6780.48	14748.2	SM II	2.3D			KN
6778.19	6780.06	14749.1	SM II	1.5			KN

AIR WAVELENGTH (ANGSTRMS)	VACUUM WAVELENGTH (ANGSTRMS)	WAVENUMBER (KAYSERS)	LMENT	LOG ARC	INTENSITY SPK	DIS	REF
6775.30	6777.17	14755.4	SM I	1.2			KN
6772.02	6773.89	14762.6	SM	0.7			KN
6771.36	6773.23	14764.0	SM	0.9			KN
6770.37	6772.24	14766.2	SM	1.0			KS
6769.44	6771.31	14768.2	SM	0.3			KN
6766.52	6768.39	14774.6	SM II	1.7D			KN
6762.58	6764.45	14783.2	SM	1.0			KS
6762.06	6763.93	14784.3	SM	1.4			KS
6761.70	6763.57	14785.1	SM	0.3			KN
6759.25	6761.12	14790.5	SM I	1.2			KN
6758.34	6760.21	14792.4	SM	0.3			KN
6756.94	6758.81	14795.5	SM II	1.7			KN
6754.85	6756.71	14800.1	SM	1.5			KN
6754.68	6756.54	14800.5	SM II	1.7			KN
6751.10	6752.96	14808.3	SM	0.3			KN
6750.86	6752.72	14808.8	SM	0.3			KN
6748.39	6750.25	14814.3	SM	0.6			KN
6746.66	6748.52	14818.1	SM	0.3			KN
6746.41	6748.27	14818.6	SM II	1.3			KN
6743.64	6745.50	14824.7	SM	0.3			KN
6741.47	6743.33	14829.5	SM II	1.6			KN
6741.2	6743.1	14830.	SM	1.4			KN
6735.82	6737.68	14841.9	SM	0.6H			KN
6735.34	6737.20	14843.0	SM	2.0D			KN
6734.81	6736.67	14844.1	SM II	2.6D			KN
6734.06	6735.92	14845.8	SM II	2.6D			KN
6731.84	6733.70	14850.7	SM II	2.7D			KN
6728.24	6730.10	14858.6	SM	0.9			KN
6727.27	6729.13	14860.8	SM	0.3			KN
6726.81	6728.67	14861.8	SM	1.2			KS
6725.90	6727.76	14863.8	SM I	1.5			KN
6724.73	6726.59	14866.4	SM I	1.5			KN
6723.26	6725.12	14869.6	SM	1.5			KN
6723.07	6724.93	14870.1	SM	1.5			KN
6721.62	6723.48	14873.3	SM	1.0			KN
6720.95	6722.81	14874.7	SM	0.7W			KN
6719.59	6721.45	14877.7	SM	0.7			KN
6718.22	6720.07	14880.8	SM	0.6			KS
6717.15	6719.00	14883.2	SM	0.3			KN
6713.86	6715.71	14890.4	SM	0.3			KN
6712.8	6714.7	14893.	SM	0.9			KN
6712.62	6714.47	14893.2	SM II	1.7			KN
6712.13	6713.98	14894.3	SM	1.4D			KS
6711.51	6713.36	14895.7	SM	0.7			KN
6710.70	6712.55	14897.5	SM	1.3D			KS
6709.33	6711.18	14900.5	SM	0.3			KN
6707.45	6709.30	14904.7	SM II	1.7D			KN
6707.1	6709.0	14905.	SM	0.3			KN
6706.85	6708.70	14906.0	SM	0.7			KN
6705.50	6707.35	14909.0	SM	0.3			KN
6704.25	6706.10	14911.8	SM	1.0			KN
6703.61	6705.46	14913.2	SM I	1.5			KN
6702.19	6704.04	14916.4	SM	0.7			KS
6699.25	6701.10	14922.9	SM	0.9			KN
6698.92	6700.77	14923.7	SM	0.3			KS
6698.11	6699.96	14925.5	SM	0.6			KN
6697.28	6699.13	14927.3	SM	0.8			KN
6696.86	6698.71	14928.2	SM	0.5			KN
6695.31	6697.16	14931.7	SM	0.5			KN
6694.69	6696.54	14933.1	SM II	1.5D			KN
6693.55	6695.40	14935.6	SM II	2.0D			KN
6692.57	6694.42	14937.8	SM	0.5			KN
6689.29	6691.14	14945.1	SM	0.5			KN
6688.81	6690.66	14946.2	SM	0.6			KN
6687.79	6689.64	14948.5	SM II	1.9D			KN
6686.36	6688.21	14951.7	SM	0.3			KN
6685.02	6686.87	14954.7	SM	0.3			KN
6684.44	6686.29	14956.0	SM II	1.4D			KN
6682.57	6684.42	14960.2	SM	0.7D			KN
6681.53	6683.37	14962.5	SM II	1.8D			KN

AIR WAVELENGTH (ANGSTRMS)	VACUUM WAVELENGTH (ANGSTRMS)	WAVENUMBER (KAYSERS)	LMENT	LOG ARC	INTENSITY SPK	INTENSITY DIS	REF
6679.24	6681.08	14967.6	SM II	1.9			MIT
6679.1	6680.9	14968.	SM	1.9			KN
6678.00	6679.84	14970.4	SM	0.5			KN
6677.46	6679.30	14971.6	SM	0.5			KN
6677.06	6678.90	14972.5	SM	1.0			KN
6675.60	6677.44	14975.8	SM	0.5			KN
6674.68	6676.52	14977.9	SM	0.3H			KS
6672.52	6674.36	14982.7	SM	0.3			KN
6672.20	6674.04	14983.4	SM	0.5			KN
6671.48	6673.32	14985.0	SM I	1.7			MIT
6670.73	6672.57	14986.7	SM	0.5			KN
6669.31	6671.15	14989.9	SM	1.0D			KN
6668.91	6670.75	14990.8	SM	0.5J			KN
6668.60	6670.44	14991.5	SM	0.3			KN
6667.83	6669.67	14993.2	SM	0.3			KN
6667.22	6669.06	14994.6	SM II	1.7D			KN
6661.63	6663.47	15007.2	SM	0.7			KN
6661.37	6663.21	15007.8	SM	0.7			KN
6656.19	6658.03	15019.5	SM II	2.0			KN
6654.30	6656.14	15023.7	SM	0.6			KN
6651.61	6653.45	15029.8	SM II	1.9D			KN
6649.02	6650.86	15035.7	SM II	1.7			KN
6648.12	6649.96	15037.7	SM I	0.5			KN
6646.22	6648.06	15042.0	SM	1.0D			KN
6643.88	6645.71	15047.3	SM	0.6			KN
6641.57	6643.40	15052.5	SM	0.5			KN
6640.74	6642.57	15054.4	SM	0.6W			KN
6640.49	6642.32	15055.0	SM	0.5			KN
6639.81	6641.64	15056.5	SM	0.6			KN
6638.49	6640.32	15059.5	SM	1.2D			KN
6637.17	6639.00	15062.5	SM II	1.8D			KN
6635.36	6637.19	15066.6	SM	0.3			KN
6635.22	6637.05	15066.9	SM	1.0			KN
6633.1	6634.9	15072.	SM	1.0D			KN
6632.28	6634.11	15073.6	SM II	2.0			KN
6630.61	6632.44	15077.4	SM II	1.7D			KN
6629.71	6631.54	15079.4	SM	0.3			KN
6628.88	6630.71	15081.3	SM II	1.7D			KN
6627.83	6629.66	15083.7	SM	0.6			KN
6627.41	6629.24	15084.7	SM	0.7			KN
6626.96	6628.79	15085.7	SM	0.3			KN
6625.28	6627.11	15089.5	SM	1.5D			KN
6622.67	6624.50	15095.5	SM	0.6			KN
6619.01	6620.84	15103.8	SM II	0.7V			KN
6617.61	6619.44	15107.0	SM II	1.7D			KN
6616.80	6618.63	15108.9	SM	0.8			KN
6614.82	6616.65	15113.4	SM II	1.7D			KN
6612.60	6614.43	15118.5	SM	0.6			KN
6608.86	6610.69	15127.0	SM	0.5			KN
6608.4	6610.2	15128.	SM	0.5			KN
6607.17	6608.99	15130.9	SM	1.2			KN
6604.56	6606.38	15136.9	SM II	2.3D			KN
6603.53	6605.35	15139.2	SM	0.6			KN
6601.83	6603.65	15143.1	SM II	2.2D			KN
6600.39	6602.21	15146.4	SM	0.5			KN
6598.95	6600.77	15149.7	SM	0.3			KN
6597.03	6598.85	15154.2	SM	1.6			KN
6596.00	6597.82	15156.5	SM	1.0			KN
6593.03	6594.85	15163.3	SM	0.6			KN
6591.50	6593.32	15166.9	SM	1.6			KN
6590.20	6592.02	15169.9	SM	0.5			KN
6589.73	6591.55	15170.9	SM II	2.6D			MIT
6588.92	6590.74	15172.8	SM	2.0			MIT
6588.41	6590.23	15174.0	SM	0.9			KN
6587.54	6589.36	15176.0	SM	1.5D			KN
6586.42	6588.24	15178.6	SM	0.3			KN
6585.21	6587.03	15181.3	SM II	2.2D			KS
6584.14	6585.96	15183.8	SM	1.2			KN
6581.25	6583.07	15190.5	SM	0.6			KN
6580.53	6582.35	15192.2	SM	1.2			KN
6579.249	6581.066	15195.11	SM	0.6			MIT
6577.36	6579.18	15199.5	SM	1.4			KN
6575.95	6577.77	15202.7	SM	1.4			KN
6574.38	6576.20	15206.4	SM II	2.0			KN
6573.03	6574.85	15209.5	SM	1.0			KN
6572.35	6574.17	15211.1	SM	0.3			KN
6570.67	6572.49	15214.9	SM II	2.3D			KN
6569.31	6571.12	15218.1	SM II	2.7D			KN
6567.50	6569.31	15222.3	SM	0.5			KN
6564.69	6566.50	15228.8	SM	0.3D			MIT
6563.52	6565.33	15231.5	SM	1.3			KN
6562.94	6564.75	15232.9	SM	1.7D			KN
6562.64	6564.45	15233.6	SM	0.3			KN
6558.39	6560.20	15243.4	SM	0.5			KN
6558.17	6559.98	15243.9	SM	1.0			KN
6556.46	6558.27	15247.9	SM	0.5			KN
6555.24	6557.05	15250.8	SM	0.5D			KN
6551.80	6553.61	15258.8	SM I	1.4			KN
6551.04	6552.85	15260.5	SM	0.7			KN
6549.77	6551.58	15263.5	SM II	2.0D			KN
6544.95	6546.76	15274.7	SM	1.0			KN
6544.66	6546.47	15275.4	SM	1.6			KN
6544.57	6546.38	15275.6	SM II	1.6			KN
6543.29	6545.10	15278.6	SM	1.6			KN
6542.76	6544.57	15279.8	SM II	2.3			KN
6541.13	6542.94	15283.7	SM	0.3			KN
6537.61	6539.42	15291.9	SM	0.5			KN
6537.515	6539.321	15292.11	SM	0.7W			MIT
6536.83	6538.64	15293.7	SM	1.0H			KN
6535.35	6537.16	15297.2	SM	0.7			KN
6533.96	6535.77	15300.4	SM	1.2			KN
6532.25	6534.05	15304.4	SM	1.2			KN
6530.74	6532.54	15308.0	SM	1.0			KN
6529.70	6531.50	15310.4	SM	1.7			KN
6528.98	6530.78	15312.1	SM	0.7			KN
6528.02	6529.82	15314.3	SM	0.5			KN
6526.64	6528.44	15317.6	SM	1.7			KN
6525.52	6527.32	15320.2	SM	0.6			KN
6520.40	6522.20	15332.2	SM	0.5			KN
6516.82	6518.62	15340.7	SM	0.6H			KN
6516.17	6517.97	15342.2	SM	1.0			KN
6514.65	6516.45	15345.8	SM	0.7			KN
6512.69	6514.49	15350.4	SM II	1.2D			KN
6512.30	6514.10	15351.3	SM	0.7			KN
6510.03	6511.83	15356.7	SM	0.7			KN
6509.44	6511.24	15358.1	SM	1.0			KN
6507.700	6509.498	15362.17	SM	1.2			KS
6507.13	6508.93	15363.5	SM	1.6			KN
6505.56	6507.36	15367.2	SM	0.9D			KN
6502.00	6503.80	15375.6	SM II	1.6			KN
6501.00	6502.80	15378.0	SM	0.6			KN
6498.67	6500.47	15383.5	SM II	2.0D			KN
6497.91	6499.71	15385.3	SM	1.2D			AB
6496.16	6497.95	15389.5	SM	0.3			MIT
6496.05	6497.84	15389.7	SM II	1.0			KN
6493.92	6495.71	15394.8	SM	0.5			KN
6493.10	6494.89	15396.7	SM	0.5			KN
6492.05	6493.84	15399.2	SM I	0.6			KN
6491.28	6493.07	15401.0	SM I	0.5			KN
6490.748	6492.542	15402.29	SM II	1.8D			MIT
6490.21	6492.00	15403.6	SM	0.8D			KN
6489.44	6491.23	15405.4	SM	0.3			MIT
6488.60	6490.39	15407.4	SM	0.6			MIT
6487.62	6489.41	15409.7	SM II	1.8			KN
6484.52	6486.31	15417.1	SM II	2.0			MIT
6483.96	6485.75	15418.4	SM	0.3			MIT
6483.37	6485.16	15419.8	SM	0.6			MIT
6479.249	6481.039	15429.62	SM	0.6			MIT
6478.36	6480.15	15431.7	SM	0.8			KN
6477.02	6478.81	15434.9	SM II	1.0D			KN

AIR WAVELENGTH (ANGSTRMS)	VACUUM WAVELENGTH (ANGSTRMS)	WAVENUMBER (KAYSERS)	LMENT	LOG ARC	INTENSITY SPK	DIS	REF
6476.43	6478.22	15436.3	SM	0.7			MIT
6474.25	6476.04	15441.5	SM II	1.4			KN
6473.34	6475.13	15443.7	SM	0.9			MIT
6472.34	6474.13	15446.1	SM II	2.2D			KN
6471.580	6473.368	15447.91	SM	1.2			MIT
6470.46	6472.25	15450.6	SM II	1.8D			KN
6469.764	6471.552	15452.24	SM	0.6			MIT
6469.42	6471.21	15453.1	SM	0.5			KN
6469.00	6470.79	15454.1	SM	0.5			KN
6468.325	6470.112	15455.68	SM	0.6			MIT
6466.30	6468.09	15460.5	SM	0.7			KN
6464.40	6466.19	15465.1	SM	0.3			MIT
6461.15	6462.94	15472.8	SM	0.7			KN
6459.78	6461.57	15476.1	SM	0.5D			KN
6459.36	6461.15	15477.1	SM	1.2			MIT
6458.44	6460.22	15479.3	SM	0.3			MIT
6457.546	6459.331	15481.48	SM	1.2			MIT
6456.48	6458.26	15484.0	SM	0.3			MIT
6456.24	6458.02	15484.6	SM	1.5D			MIT
6455.59	6457.37	15486.2	SM II	1.6			MIT
6453.37	6455.15	15491.5	SM	0.3			MIT
6452.050	6453.833	15494.67	SM	1.4			MIT
6450.442	6452.225	15498.53	SM	0.7			MIT
6448.331	6450.113	15503.60	SM	0.8			MIT
6448.051	6449.833	15504.28	SM	0.9			MIT
6447.532	6449.314	15505.53	SM	1.5			MIT
6440.54	6442.32	15522.4	SM	0.9			AB
6439.72	6441.50	15524.3	SM II	1.0D			KS
6439.318	6441.098	15525.30	SM	0.5			MIT
6437.630	6439.409	15529.37	SM	1.0			MIT
6435.318	6437.097	15534.95	SM	1.4			MIT
6433.953	6435.731	15538.25	SM	1.2			MIT
6433.309	6435.087	15539.81	SM	0.5			MIT
6431.96	6433.74	15543.1	SM II	1.7			KN
6431.03	6432.81	15545.3	SM II	1.7D			MIT
6428.956	6430.733	15550.33	SM	1.2			MIT
6428.315	6430.092	15551.88	SM	1.7D			MIT
6426.62	6428.40	15556.0	SM II	2.0D			MIT
6425.909	6427.685	15557.70	SM	1.0			MIT
6424.256	6426.032	15561.70	SM	0.5			MIT
6418.928	6420.702	15574.62	SM	0.7			MIT
6417.513	6419.287	15578.05	SM II	2.0D			MIT
6417.17	6418.94	15578.9	SM	0.9			KN
6411.403	6413.175	15592.90	SM	1.0			MIT
6410.344	6412.116	15595.48	SM II	0.6			MIT
6408.042	6409.813	15601.08	SM II	1.0D			MIT
6406.24	6408.01	15605.5	SM II	1.5D			KN
6404.618	6406.388	15609.42	SM II	0.3H			MIT
6404.117	6405.887	15610.64	SM	0.7			MIT
6403.98	6405.75	15611.0	SM	0.3			KN
6402.23	6404.00	15615.2	SM	0.3			KN
6401.45	6403.22	15617.1	SM	1.0D			KN
6399.86	6401.63	15621.0	SM	0.3			KN
6399.415	6401.184	15622.11	SM	0.5			MIT
6398.295	6400.064	15624.84	SM	1.1			MIT
6395.427	6397.195	15631.85	SM	0.3			MIT
6394.967	6396.735	15632.98	SM	0.3			MIT
6392.445	6394.212	15639.14	SM	0.3			MIT
6390.838	6392.605	15643.08	SM II	2.0			MIT
6389.870	6391.636	15645.45	SM II	2.0			MIT
6388.068	6389.834	15649.86	SM	0.5			MIT
6386.768	6388.534	15653.04	SM	1.0			MIT
6384.825	6386.590	15657.81	SM	0.3			MIT
6384.303	6386.068	15659.09	SM	0.3			MIT
6380.709	6382.473	15667.91	SM	0.5			MIT
6380.045	6381.809	15669.54	SM	0.5			MIT
6378.075	6379.838	15674.38	SM	0.3			MIT
6374.623	6376.385	15682.87	SM	0.3H			MIT
6373.474	6375.236	15685.69	SM	0.3			MIT
6372.688	6374.450	15687.63	SM	0.3			MIT

AIR WAVELENGTH (ANGSTRMS)	VACUUM WAVELENGTH (ANGSTRMS)	WAVENUMBER (KAYSERS)	LMENT	LOG ARC	INTENSITY SPK	DIS	REF
6371.90	6373.66	15689.6	SM II	1.2			MIT
6371.029	6372.790	15691.71	SM	0.7			MIT
6368.41	6370.17	15698.2	SM	0.6			KN
6368.279	6370.040	15698.49	SM II	1.7D			MIT
6367.431	6369.191	15700.58	SM	1.0			MIT
6366.267	6368.027	15703.45	SM	0.3			MIT
6363.158	6364.917	15711.12	SM	0.5			MIT
6357.269	6359.027	15725.68	SM	0.3	0.3		MIT
6357.190	6358.948	15725.87	SM	1.7	0.0		MIT
6355.357	6357.114	15730.41	SM	0.9			MIT
6353.494	6355.251	15735.02	SM	1.7	0.7		MIT
6353.437	6355.194	15735.16	SM	0.3	1.5		MIT
6353.013	6354.770	15736.21	SM II	1.3	1.3		MIT
6349.240	6350.996	15745.56	SM	0.6			MIT
6348.949	6350.704	15746.29	SM	0.6			MIT
6346.505	6348.260	15752.35	SM	0.5			MIT
6343.358	6345.112	15760.16	SM	0.3			MIT
6341.572	6343.325	15764.60	SM	0.3			MIT
6340.055	6341.808	15768.37	SM II	1.4			MIT
6339.743	6341.496	15769.15	SM	0.5			MIT
6338.898	6340.651	15771.25	SM	0.7H			MIT
6337.969	6339.721	15773.56	SM	0.3			MIT
6335.39	6337.14	15780.0	SM	0.5			KN
6331.731	6333.482	15789.10	SM	0.5			MIT
6331.339	6333.090	15790.08	SM	0.6			MIT
6331.17	6332.92	15790.5	SM	0.6			KS
6330.157	6331.907	15793.03	SM	1.0D			MIT
6329.376	6331.126	15794.98	SM II	0.9			MIT
6329.222	6330.972	15795.36	SM	0.9			MIT
6328.845	6330.595	15796.30	SM II	1.0			MIT
6327.991	6329.741	15798.44	SM	0.9			MIT
6327.47	6329.22	15799.7	SM II	2.0			KN
6326.284	6328.033	15802.70	SM	0.5			MIT
6326.08	6327.83	15803.2	SM	0.3			KN
6325.582	6327.331	15804.45	SM	1.7D			MIT
6322.519	6324.267	15812.11	SM	0.6			MIT
6321.74	6323.49	15814.1	SM	1.8			MIT
6321.43	6323.18	15814.8	SM	0.5			KN
6319.49	6321.24	15819.7	SM I	0.3			KN
6317.819	6319.566	15823.87	SM I	0.3			MIT
6317.40	6319.15	15824.9	SM	0.3			KN
6316.94	6318.69	15826.1	SM	0.5			KN
6315.779	6317.526	15828.98	SM	1.2			MIT
6314.582	6316.328	15831.98	SM	0.5			MIT
6313.20	6314.95	15835.4	SM	0.3			KN
6311.56	6313.31	15839.6	SM	0.3			KN
6310.297	6312.042	15842.73	SM	0.5			MIT
6309.89	6311.63	15843.8	SM II	0.5			KN
6308.28	6310.02	15847.8	SM II	1.0			KN
6307.06	6308.80	15850.9	SM II	1.8			KN
6306.318	6308.062	15852.73	SM	0.5			MIT
6305.19	6306.93	15855.6	SM	1.6D			KN
6304.74	6306.48	15856.7	SM	0.3			KN
6303.90	6305.64	15858.8	SM	0.3			KN
6303.15	6304.89	15860.7	SM	1.6			KN
6302.394	6304.137	15862.60	SM II	1.4			MIT
6301.873	6303.616	15863.91	SM	0.7			MIT
6301.12	6302.86	15865.8	SM II	1.7			KN
6300.19	6301.93	15868.2	SM II	1.7			KN
6299.62	6301.36	15869.6	SM	0.3H			KN
6297.642	6299.384	15874.57	SM	0.6			MIT
6295.975	6297.716	15878.77	SM	1.0			MIT
6293.908	6295.649	15883.99	SM	0.9D			MIT
6292.960	6294.700	15886.38	SM	1.0			MIT
6291.82	6293.56	15889.3	SM II	2.0D			KN
6289.928	6291.668	15894.04	SM II	1.8			MIT
6287.773	6289.512	15899.48	SM	0.3			MIT
6286.96	6288.70	15901.5	SM	0.3			KN
6285.149	6286.887	15906.12	SM	1.0			MIT
6284.12	6285.86	15908.7	SM	1.0D			KS

AIR WAVELENGTH (ANGSTRMS)	VACUUM WAVELENGTH (ANGSTRMS)	WAVENUMBER (KAYSERS)	LMENT	LOG ARC	INTENSITY SPK	DIS	REF
6282.29	6284.03	15913.4	SM	0.3			KN
6281.94	6283.68	15914.2	SM	0.9D			AB
6279.493	6281.230	15920.45	SM	1.0			MIT
6279.06	6280.80	15921.6	SM	0.3			KN
6278.23	6279.97	15923.7	SM	1.0D			KS
6276.77	6278.51	15927.4	SM	1.0D			KN
6276.48	6278.22	15928.1	SM	0.3			KN
6275.076	6276.812	15931.66	SM	0.6			MIT
6273.707	6275.442	15935.13	SM	0.7			MIT
6273.392	6275.127	15935.93	SM	1.2			MIT
6272.406	6274.141	15938.44	SM	0.3			MIT
6271.374	6273.109	15941.06	SM	0.6			MIT
6271.046	6272.781	15941.89	SM	0.9			MIT
6267.28	6269.01	15951.5	SM II	2.2			KN
6266.34	6268.07	15953.9	SM	0.3H			KN
6265.655	6267.388	15955.61	SM	0.3			MIT
6262.519	6264.251	15963.60	SM	0.5			MIT
6260.12	6261.85	15969.7	SM	0.5			KN
6259.682	6261.413	15970.83	SM	0.9			MIT
6257.93	6259.66	15975.3	SM	0.3			KN
6256.66	6258.39	15978.6	SM II	1.9			KN
6256.536	6258.267	15978.86	SM	1.4			MIT
6256.108	6257.839	15979.96	SM	0.9			MIT
6255.85	6257.58	15980.6	SM	1.0			KN
6252.24	6253.97	15989.8	SM	0.3			AB
6251.73	6253.46	15991.2	SM	0.7			KN
6248.107	6249.835	16000.42	SM	1.0			MIT
6246.76	6248.49	16003.9	SM II	1.9			KN
6245.71	6247.44	16006.6	SM	0.3			KN
6245.30	6247.03	16007.6	SM	0.3			KN
6244.21	6245.94	16010.4	SM II	1.6D			KN
6242.406	6244.133	16015.03	SM	0.7			MIT
6242.01	6243.74	16016.1	SM	0.5			KN
6241.567	6243.294	16017.19	SM	0.3			MIT
6240.486	6242.212	16019.96	SM	0.6			MIT
6238.90	6240.63	16024.0	SM	1.2			KS
6238.299	6240.025	16025.58	SM	1.0			MIT
6237.661	6239.387	16027.22	SM II	1.7			MIT
6236.24	6237.97	16030.9	SM	0.3D			MIT
6236.11	6237.84	16031.2	SM	1.2			KN
6235.36	6237.08	16033.1	SM	1.2			KN
6234.95	6236.67	16034.2	SM	0.3			KN
6234.20	6235.92	16036.1	SM II	1.0			KS
6232.68	6234.40	16040.0	SM	0.3			KN
6232.17	6233.89	16041.3	SM	0.3H			KN
6231.70	6233.42	16042.6	SM II	0.6			KN
6229.82	6231.54	16047.4	SM	0.3			KN
6229.07	6230.79	16049.3	SM	0.3			KN
6228.31	6230.03	16051.3	SM	0.3			KN
6226.713	6228.436	16055.40	SM	0.7			MIT
6225.556	6227.278	16058.38	SM	1.2	1.3		MIT
6225.48	6227.20	16058.6	SM	0.5			KN
6225.236	6226.958	16059.20	SM	0.6			MIT
6224.70	6226.42	16060.6	SM	0.6			KS
6223.641	6225.363	16063.32	SM	0.6			MIT
6222.74	6224.46	16065.7	SM	0.3			KN
6222.568	6224.290	16066.09	SM	0.5			MIT
6221.32	6223.04	16069.3	SM	0.5			KN
6220.89	6222.61	16070.4	SM	1.0D			KN
6219.17	6220.89	16074.9	SM	0.6			KN
6218.236	6219.956	16077.28	SM	0.6			MIT
6217.147	6218.867	16080.10	SM	1.1			MIT
6216.909	6218.629	16080.71	SM	1.0D	1.3		MIT
6216.44	6218.16	16081.9	SM	0.7			KS
6213.85	6215.57	16088.6	SM	0.6			KN
6212.65	6214.37	16091.7	SM	0.3			KN
6210.75	6212.47	16096.7	SM	0.5			KN
6210.454	6212.172	16097.43	SM	0.5			MIT
6208.989	6210.707	16101.23	SM	0.6			MIT
6207.116	6208.833	16106.08	SM II	0.9			MIT
6206.830	6208.547	16106.83	SM	1.4			MIT
6206.239	6207.956	16108.36	SM	0.3			MIT
6205.837	6207.554	16109.41	SM	0.7			MIT
6203.878	6205.594	16114.49	SM	1.3			MIT
6201.138	6202.854	16121.61	SM	0.7			MIT
6200.075	6201.790	16124.38	SM II	0.3			MIT
6198.385	6200.100	16128.77	SM	1.3			MIT
6198.14	6199.85	16129.4	SM	0.6			KN
6197.854	6199.569	16130.15	SM	0.3			MIT
6197.395	6199.110	16131.35	SM	0.3			MIT
6196.583	6198.298	16133.46	SM	0.5			MIT
6196.11	6197.82	16134.7	SM	0.5			KN
6195.21	6196.92	16137.0	SM	0.3H			KN
6194.407	6196.121	16139.13	SM	1.0			MIT
6193.14	6194.85	16142.4	SM	1.0			KN
6192.64	6194.35	16143.7	SM	1.7			KN
6191.23	6192.94	16147.4	SM	1.2W			KN
6190.62	6192.33	16149.0	SM	0.3			KN
6189.70	6191.41	16151.4	SM	0.7			KN
6188.00	6189.71	16155.8	SM II	0.0H	1.0		KN
6187.94	6189.65	16156.0	SM	1.5D			KN
6187.417	6189.129	16157.36	SM	0.5			MIT
6182.89	6184.60	16169.2	SM II	1.8D			KN
6181.88	6183.59	16171.8	SM	1.2D			KN
6181.05	6182.76	16174.0	SM II	1.7D			KN
6179.84	6181.55	16177.2	SM II	1.5D			KN
6179.419	6181.129	16178.27	SM	1.0			MIT
6178.67	6180.38	16180.2	SM	0.3			KN
6177.455	6179.164	16183.42	SM	0.7			MIT
6176.40	6178.11	16186.2	SM	0.9D			KN
6174.96	6176.67	16190.0	SM II	1.5D			KN
6174.458	6176.167	16191.27	SM	1.0			MIT
6173.947	6175.655	16192.61	SM	1.1			MIT
6173.13	6174.84	16194.8	SM	0.5			KN
6172.563	6174.271	16196.24	SM	1.2			MIT
6169.947	6171.654	16203.11	SM	0.5			MIT
6169.147	6170.854	16205.21	SM	0.7			MIT
6168.335	6170.042	16207.34	SM II	1.5			KN
6166.76	6168.47	16211.5	SM	0.9D			KS
6165.857	6167.563	16213.86	SM	0.8			MIT
6165.61	6167.32	16214.5	SM	0.3H			KN
6164.95	6166.66	16216.2	SM	0.5			KN
6164.51	6166.22	16217.4	SM	1.3D			KN
6163.944	6165.650	16218.89	SM	1.2			MIT
6163.16	6164.87	16220.9	SM	0.3			KN
6162.25	6163.96	16223.3	SM	0.3			KN
6160.42	6162.12	16228.2	SM II	1.6D			KN
6159.56	6161.26	16230.4	SM	1.6L	1.5		KN
6157.55	6159.25	16235.7	SM II	1.5			KN
6156.90	6158.60	16237.4	SM	1.5			KN
6154.81	6156.51	16243.0	SM	0.3			KN
6150.94	6152.64	16253.2	SM	0.5			KN
6149.94	6151.64	16255.8	SM	0.5			KN
6149.10	6150.80	16258.0	SM II	1.8			KN
6148.27	6149.97	16260.2	SM	0.6			KN
6146.520	6148.221	16264.87	SM	1.0			MIT
6145.88	6147.58	16266.6	SM	0.3			KN
6145.2	6146.9	16268.	SM	0.3			KN
6144.74	6146.44	16269.6	SM	0.5			KN
6143.961	6145.661	16271.64	SM	0.8			MIT
6143.578	6145.278	16272.66	SM	0.8			MIT
6143.09	6144.79	16273.9	SM	1.5			KN
6140.581	6142.281	16280.60	SM	1.0			MIT
6139.33	6141.03	16283.9	SM	0.8H			KN
6138.451	6140.150	16286.25	SM	0.6			MIT
6138.05	6139.75	16287.3	SM	0.6			KN
6135.85	6137.55	16293.2	SM	1.3			MIT
6134.69	6136.39	16296.2	SM	0.3			KN
6133.225	6134.923	16300.12	SM	0.6			MIT
6133.07	6134.77	16300.5	SM	0.3			KN

AIR WAVELENGTH (ANGSTRMS)	VACUUM WAVELENGTH (ANGSTRMS)	WAVENUMBER (KAYSERS)	LMENT	LOG ARC	INTENSITY SPK	INTENSITY DIS	REF
6132.113	6133.810	16303.08	SM	0.5			MIT
6131.142	6132.839	16305.66	SM	1.0			MIT
6130.608	6132.305	16307.08	SM	1.0			MIT
6130.43	6132.13	16307.6	SM	0.3			KN
6129.87	6131.57	16309.1	SM	1.2			AB
6128.71	6130.41	16312.1	SM II	1.3D			AB
6127.73	6129.43	16314.7	SM	0.3			KN
6126.989	6128.685	16316.71	SM	0.8			MIT
6126.305	6128.001	16318.54	SM	1.0			MIT
6125.84	6127.54	16319.8	SM	1.2			AB
6124.95	6126.65	16322.2	SM	0.3			KN
6124.88	6126.58	16322.3	SM II	1.6			KN
6123.60	6125.29	16325.7	SM II	1.7D			KN
6122.744	6124.439	16328.03	SM	1.0			MIT
6121.81	6123.50	16330.5	SM II	1.5D			KN
6120.26	6121.95	16334.7	SM II	1.0D			KS
6117.79	6119.48	16341.2	SM	0.9			KN
6117.6	6119.3	16342.	SM	0.5			KN
6115.743	6117.436	16346.72	SM	0.3			MIT
6114.73	6116.42	16349.4	SM	1.0			MIT
6114.58	6116.27	16349.8	SM II	1.4H			KN
6112.938	6114.630	16354.22	SM	1.0			MIT
6111.828	6113.520	16357.19	SM	0.6			MIT
6111.66	6113.35	16357.6	SM	0.5			KN
6110.66	6112.35	16360.3	SM II	1.6D			KN
6110.19	6111.88	16361.6	SM II	1.1			KN
6109.26	6110.95	16364.1	SM	0.6			KN
6109.00	6110.69	16364.8	SM	0.3			KN
6107.13	6108.82	16369.8	SM	0.3			KN
6105.360	6107.050	16374.52	SM	0.5			MIT
6104.82	6106.51	16376.0	SM II	1.5D			KN
6104.392	6106.082	16377.11	SM	0.8			MIT
6104.2	6105.9	16378.	SM	0.5			KN
6103.95	6105.64	16378.3	SM	0.3			KN
6103.723	6105.413	16378.91	SM	0.6			MIT
6103.374	6105.064	16379.85	SM	1.5			MIT
6101.96	6103.65	16383.6	SM	1.5			KN
6101.422	6103.111	16385.09	SM	0.6			MIT
6099.918	6101.607	16389.13	SM I	1.2			MIT
6099.12	6100.81	16391.3	SM	0.6			KN
6098.30	6099.99	16393.5	SM	0.7			KN
6096.790	6098.478	16397.53	SM	1.2			MIT
6096.55	6098.24	16398.2	SM	0.3			KN
6096.24	6097.93	16399.0	SM	0.5			KN
6095.24	6096.93	16401.7	SM	0.6			KN
6094.082	6095.769	16404.82	SM	0.7L			MIT
6093.699	6095.386	16405.85	SM	0.6			MIT
6092.909	6094.596	16407.98	SM	0.6			MIT
6091.950	6093.636	16410.56	SM	0.6			MIT
6091.393	6093.079	16412.06	SM	1.3			MIT
6089.659	6091.345	16416.74	SM	0.7			MIT
6089.295	6090.981	16417.72	SM	0.5			MIT
6088.619	6090.305	16419.54	SM	1.0			MIT
6088.076	6089.761	16421.00	SM	0.7			MIT
6087.704	6089.389	16422.01	SM	0.5			MIT
6084.604	6086.288	16430.37	SM	0.3			MIT
6084.128	6085.812	16431.66	SM	1.7			MIT
6082.664	6084.348	16435.61	SM	1.2			MIT
6081.99	6083.67	16437.4	SM	0.9			KS
6080.487	6082.170	16441.50	SM	0.5			MIT
6080.07	6081.75	16442.6	SM	0.5			KN
6077.856	6079.539	16448.62	SM	0.6			MIT
6075.721	6077.403	16454.40	SM	0.9			MIT
6074.963	6076.645	16456.45	SM	0.7			MIT
6074.46	6076.14	16457.8	SM	0.5			AB
6071.64	6073.32	16465.5	SM	0.3			KN
6070.074	6071.755	16469.70	SM	1.6			MIT
6068.274	6069.954	16474.59	SM	0.7			MIT
6067.782	6069.462	16475.93	SM	1.6			MIT
6067.401	6069.081	16476.96	SM	0.7			MIT

AIR WAVELENGTH (ANGSTRMS)	VACUUM WAVELENGTH (ANGSTRMS)	WAVENUMBER (KAYSERS)	LMENT	LOG ARC	INTENSITY SPK	INTENSITY DIS	REF
6066.651	6068.331	16479.00	SM	0.3			MIT
6063.656	6065.335	16487.14	SM	0.5			MIT
6061.819	6063.497	16492.13	SM	0.7			MIT
6061.20	6062.88	16493.8	SM	1.0D			KN
6060.73	6062.41	16495.1	SM II	1.3D			KN
6060.60	6062.28	16495.4	SM	0.8			KN
6059.871	6061.549	16497.43	SM	0.7			MIT
6058.45	6060.13	16501.3	SM	0.3			KN
6057.680	6059.357	16503.40	SM	0.9			MIT
6057.251	6058.928	16504.57	SM	0.7			MIT
6056.59	6058.27	16506.4	SM	0.7			KN
6053.879	6055.555	16513.76	SM	0.7			MIT
6053.11	6054.79	16515.9	SM	0.5			KN
6052.07	6053.75	16518.7	SM	0.3H			KN
6050.30	6051.98	16523.5	SM	0.3H			KN
6049.81	6051.49	16524.9	SM	1.0J			KN
6047.461	6049.136	16531.29	SM	0.6			MIT
6045.372	6047.046	16537.00	SM	1.6	1.3		MIT
6044.992	6046.666	16538.04	SM	1.8			MIT
6044.676	6046.350	16538.90	SM	0.3			MIT
6043.329	6045.002	16542.59	SM	0.5			MIT
6042.814	6044.487	16544.00	SM	1.2D			MIT
6042.545	6044.218	16544.74	SM	1.0D			MIT
6041.378	6043.051	16547.93	SM	1.6			MIT
6036.45	6038.12	16561.4	SM	0.3			KN
6035.544	6037.215	16563.93	SM	0.6			MIT
6033.834	6035.505	16568.62	SM	1.3			MIT
6033.23	6034.90	16570.3	SM II	1.4			KN
6032.443	6034.114	16572.44	SM	0.5			MIT
6030.338	6032.008	16578.23	SM	0.5			MIT
6029.828	6031.498	16579.63	SM	0.6			MIT
6028.635	6030.304	16582.91	SM	0.6			MIT
6027.524	6029.193	16585.97	SM	0.7			MIT
6027.159	6028.828	16586.97	SM	1.2			MIT
6026.83	6028.50	16587.9	SM	0.3			KN
6026.260	6027.929	16589.45	SM	0.3			MIT
6025.18	6026.85	16592.4	SM	0.6D			KN
6024.45	6026.12	16594.4	SM	0.5D			KN
6024.07	6025.74	16595.5	SM	1.0			KN
6023.68	6025.35	16596.6	SM II	1.2			KN
6023.21	6024.88	16597.9	SM	0.3			KN
6021.75	6023.42	16601.9	SM II	1.2			KN
6021.267	6022.935	16603.20	SM II	1.2			MIT
6017.39	6019.06	16613.9	SM II	1.7D			KN
6016.795	6018.461	16615.54	SM	0.6			MIT
6014.601	6016.267	16621.60	SM	0.6			MIT
6013.863	6015.529	16623.64	SM	0.3			MIT
6013.446	6015.111	16624.80	SM II	1.3			MIT
6012.842	6014.507	16626.47	SM	0.5			MIT
6011.242	6012.907	16630.89	SM II	1.8	1.3		MIT
6010.794	6012.459	16632.13	SM	0.5			MIT
6010.259	6011.924	16633.61	SM	0.5			MIT
6009.895	6011.559	16634.62	SM	0.7			MIT
6008.915	6010.579	16637.33	SM	0.3			MIT
6007.94	6009.60	16640.0	SM	1.0D			KN
6007.095	6008.759	16642.37	SM	0.3			MIT
6005.156	6006.819	16647.75	SM	0.5			MIT
6004.197	6005.860	16650.41	SM	0.3			MIT
6003.567	6005.230	16652.15	SM	0.9D			MIT
6002.853	6004.516	16654.13	SM	0.5			MIT
6001.947	6003.609	16656.65	SM	0.9			MIT
6001.678	6003.340	16657.39	SM	1.7			MIT
6001.111	6002.773	16658.97	SM	0.5			MIT
5998.844	6000.505	16665.26	SM	0.3			MIT
5998.493	6000.154	16666.24	SM	0.5	0.0		MIT
5998.085	5999.746	16667.37	SM	0.3			MIT
5997.319	5998.980	16669.50	SM II	1.7			MIT
5996.944	5998.605	16670.54	SM	0.5			MIT
5996.042	5997.703	16673.05	SM	0.6			MIT
5995.308	5996.969	16675.09	SM	0.3			MIT

AIR WAVELENGTH (ANGSTRMS)	VACUUM WAVELENGTH (ANGSTRMS)	WAVENUMBER (KAYSERS)	LMENT		LOG ARC	INTENSITY SPK	DIS	REF
5995.099	5996.759	16675.67	SM		0.9			MIT
5994.651	5996.311	16676.92	SM	II	1.8			MIT
5993.852	5995.512	16679.14	SM		1.3			MIT
5991.741	5993.401	16685.02	SM	II	1.0W			MIT
5989.692	5991.351	16690.73	SM		0.9			MIT
5989.374	5991.033	16691.61	SM		1.0D			MIT
5987.39	5989.05	16697.1	SM	II	0.5			KN
5985.488	5987.146	16702.45	SM		0.3			MIT
5984.298	5985.956	16705.77	SM		1.1			MIT
5983.874	5985.531	16706.95	SM		0.6			MIT
5983.641	5985.298	16707.60	SM		0.6D			MIT
5982.345	5984.002	16711.22	SM		0.8			MIT
5982.014	5983.671	16712.15	SM		0.7			MIT
5980.549	5982.206	16716.24	SM		0.5			MIT
5980.369	5982.026	16716.75	SM		0.3			MIT
5979.758	5981.414	16718.45	SM		0.3H			MIT
5979.391	5981.047	16719.48	SM		1.2			MIT
5978.043	5979.699	16723.25	SM		0.7			MIT
5976.481	5978.137	16727.62	SM		0.5			MIT
5972.566	5974.220	16738.59	SM		1.0			MIT
5972.103	5973.757	16739.88	SM		0.3D			MIT
5970.803	5972.457	16743.53	SM		1.2			MIT
5969.485	5971.139	16747.23	SM		0.7			MIT
5968.818	5970.471	16749.10	SM	II	1.6			MIT
5967.132	5968.785	16753.83	SM		0.7			MIT
5965.708	5967.361	16757.83	SM	II	2.1			MIT
5963.926	5965.578	16762.83	SM		0.5			MIT
5963.211	5964.863	16764.84	SM	II	1.6			MIT
5963.05	5964.70	16765.3	SM		0.5			KN
5960.103	5961.754	16773.59	SM		1.3			MIT
5959.49	5961.14	16775.3	SM		0.5			KN
5958.223	5959.874	16778.88	SM		0.3			MIT
5957.521	5959.171	16780.86	SM	II	1.6			MIT
5956.786	5958.436	16782.93	SM		0.6			MIT
5955.832	5957.482	16785.61	SM		1.4			MIT
5955.687	5957.337	16786.02	SM		0.5			MIT
5952.37	5954.02	16795.4	SM	II	0.7			KS
5950.171	5951.819	16801.58	SM		0.5			MIT
5949.138	5950.786	16804.50	SM		0.7			MIT
5947.455	5949.103	16809.26	SM		0.5			MIT
5946.369	5948.016	16812.33	SM		1.2			MIT
5944.735	5946.382	16816.95	SM		0.6D			MIT
5942.999	5944.646	16821.86	SM		0.7			MIT
5942.292	5943.938	16823.86	SM		1.0			MIT
5941.45	5943.10	16826.2	SM		0.7D			M
5940.638	5942.284	16828.55	SM		0.3			MIT
5938.845	5940.490	16833.63	SM	II	1.7D			MIT
5936.116	5937.761	16841.37	SM	II	1.5D			MIT
5935.528	5937.173	16843.03	SM		0.6			MIT
5935.109	5936.753	16844.22	SM		0.3			MIT
5932.887	5934.531	16850.53	SM	II	1.7			MIT
5932.419	5934.063	16851.86	SM	II	1.5D			MIT
5932.142	5933.786	16852.65	SM		1.0			MIT
5931.534	5933.177	16854.37	SM		0.5			MIT
5930.886	5932.529	16856.22	SM		0.3			MIT
5929.199	5930.842	16861.01	SM		0.5L			MIT
5928.361	5930.004	16863.40	SM		0.3			MIT
5927.873	5929.515	16864.78	SM		0.6			MIT
5926.450	5928.092	16868.83	SM	II	1.3			MIT
5925.937	5927.579	16870.29	SM		0.6			MIT
5924.641	5926.283	16873.98	SM		1.0			MIT
5924.184	5925.825	16875.29	SM		0.6			MIT
5923.307	5924.948	16877.78	SM		1.2			MIT
5922.394	5924.035	16880.39	SM		0.6			MIT
5921.248	5922.889	16883.65	SM		0.5			MIT
5920.992	5922.633	16884.38	SM		1.2			MIT
5919.327	5920.967	16889.13	SM		1.8			MIT
5917.951	5919.591	16893.06	SM		0.5			MIT
5916.368	5918.007	16897.58	SM		1.2			MIT
5916.030	5917.669	16898.54	SM		0.3			MIT

AIR WAVELENGTH (ANGSTRMS)	VACUUM WAVELENGTH (ANGSTRMS)	WAVENUMBER (KAYSERS)	LMENT		LOG ARC	INTENSITY SPK	DIS	REF
5915.559	5917.198	16899.89	SM		0.7			MIT
5914.919	5916.558	16901.72	SM		0.5			MIT
5913.885	5915.524	16904.67	SM		1.3			MIT
5913.567	5915.206	16905.58	SM		1.0			MIT
5912.621	5914.259	16908.29	SM		1.2			MIT
5910.757	5912.395	16913.62	SM		1.0D			MIT
5909.039	5910.676	16918.54	SM		1.0			MIT
5906.608	5908.245	16925.50	SM		0.5			MIT
5906.069	5907.706	16927.04	SM		1.0			MIT
5903.565	5905.201	16934.23	SM		1.3D			MIT
5903.094	5904.730	16935.58	SM		0.6D			MIT
5902.605	5904.241	16936.98	SM		1.2			MIT
5902.421	5904.057	16937.51	SM		1.4D	0.9		MIT
5901.421	5903.056	16940.38	SM		1.1			MIT
5899.757	5901.392	16945.16	SM		0.3			MIT
5898.962	5900.597	16947.44	SM		1.4			MIT
5897.379	5899.013	16951.99	SM		2.0			MIT
5896.872	5898.506	16953.45	SM		0.5			MIT
5896.278	5897.912	16955.15	SM		0.7			MIT
5895.154	5896.788	16958.39	SM		0.8			MIT
5894.718	5896.352	16959.64	SM		1.2D			MIT
5892.401	5894.034	16966.31	SM		0.7D			MIT
5891.906	5893.539	16967.73	SM		0.3			MIT
5891.416	5893.049	16969.15	SM		1.2			MIT
5890.626	5892.259	16971.42	SM		0.6			MIT
5889.695	5891.327	16974.10	SM		1.3			MIT
5888.248	5889.880	16978.27	SM		0.3			MIT
5887.268	5888.900	16981.10	SM		0.5			MIT
5886.471	5888.102	16983.40	SM		0.7			MIT
5886.068	5887.699	16984.56	SM		0.5			MIT
5883.663	5885.294	16991.51	SM		0.5			MIT
5882.493	5884.123	16994.89	SM		0.7			MIT
5879.585	5881.215	17003.29	SM	I	0.3			MIT
5879.264	5880.893	17004.22	SM		0.3			MIT
5878.378	5880.007	17006.78	SM		1.2			MIT
5878.070	5879.699	17007.67	SM	II	1.3			MIT
5876.894	5878.523	17011.08	SM	II	1.0			MIT
5875.930	5877.559	17013.87	SM		0.9			MIT
5875.090	5876.718	17016.30	SM		0.7			MIT
5874.194	5875.822	17018.90	SM		1.6			MIT
5871.058	5872.685	17027.99	SM		1.4			MIT
5870.315	5871.942	17030.14	SM		0.5			MIT
5868.617	5870.244	17035.07	SM		1.6			MIT
5867.787	5869.413	17037.48	SM		1.7			MIT
5867.389	5869.015	17038.63	SM		0.5			MIT
5864.491	5866.117	17047.05	SM		0.6H			MIT
5863.491	5865.116	17049.96	SM	II	1.1			MIT
5860.791	5862.416	17057.81	SM		0.7			MIT
5860.417	5862.041	17058.90	SM	II	1.6D			MIT
5860.231	5861.855	17059.45	SM		0.3			MIT
5859.258	5860.882	17062.28	SM		1.0D	0.0		MIT
5858.151	5859.775	17065.50	SM		0.3			MIT
5857.089	5858.713	17068.60	SM		0.6			MIT
5853.068	5854.690	17080.32	SM		0.6			MIT
5849.710	5851.332	17090.13	SM		0.3			MIT
5848.67	5850.29	17093.2	SM	II	1.5			KN
5846.575	5848.196	17099.29	SM		0.5			MIT
5843.760	5845.380	17107.53	SM		0.9			MIT
5842.569	5844.189	17111.01	SM	II	1.6			MIT
5841.987	5843.606	17112.72	SM		1.0			MIT
5839.885	5841.504	17118.88	SM		0.7D			MIT
5839.544	5841.163	17119.88	SM		0.5			MIT
5838.427	5840.046	17123.15	SM	II	0.7			MIT
5836.595	5838.213	17128.53	SM		0.3			MIT
5836.365	5837.983	17129.20	SM	II	2.0D			MIT
5833.393	5835.010	17137.93	SM		0.5			MIT
5832.365	5833.982	17140.95	SM		0.7D	0.0		MIT
5831.769	5833.386	17142.70	SM	II	1.6R			MIT
5831.013	5832.630	17144.93	SM	II	1.9			MIT
5830.501	5832.117	17146.43	SM		1.3			MIT

AIR WAVELENGTH (ANGSTRMS)	VACUUM WAVELENGTH (ANGSTRMS)	WAVENUMBER (KAYSERS)	LMENT	LOG ARC	INTENSITY SPK	DIS	REF
5829.718	5831.334	17148.73	SM	1.5			MIT
5827.590	5829.206	17155.00	SM	0.5			MIT
5826.297	5827.912	17158.80	SM	0.6D			MIT
5825.651	5827.266	17160.71	SM	1.0			MIT
5825.478	5827.093	17161.22	SM	0.3			MIT
5823.977	5825.592	17165.64	SM	0.5			MIT
5823.197	5824.811	17167.94	SM	0.5H			MIT
5822.614	5824.228	17169.66	SM	0.3			MIT
5820.673	5822.287	17175.38	SM	1.6			MIT
5819.355	5820.968	17179.27	SM	0.5			MIT
5818.302	5819.915	17182.38	SM	1.0			MIT
5816.337	5817.950	17188.19	SM	0.8			MIT
5814.872	5816.484	17192.52	SM	1.8			MIT
5813.065	5814.677	17197.86	SM	0.9D			MIT
5812.672	5814.284	17199.02	SM II	1.4			MIT
5810.325	5811.936	17205.97	SM	1.0W			MIT
5806.74	5808.35	17216.6	SM	1.1			MIT
5804.700	5806.309	17222.64	SM	0.5			MIT
5803.114	5804.723	17227.35	SM	0.7			MIT
5802.82	5804.43	17228.2	SM	1.9			MIT
5801.67	5803.28	17231.6	SM	1.4D			MIT
5801.230	5802.839	17232.95	SM	1.1			MIT
5800.501	5802.109	17235.11	SM	1.9			MIT
5797.873	5799.481	17242.92	SM	0.6D			MIT
5795.294	5796.901	17250.60	SM	0.5			MIT
5793.281	5794.887	17256.59	SM	1.2	0.3		MIT
5792.533	5794.139	17258.82	SM	0.5			MIT
5791.776	5793.382	17261.07	SM	0.5			MIT
5790.916	5792.522	17263.64	SM	0.5			MIT
5788.390	5789.995	17271.17	SM	1.5			MIT
5787.539	5789.144	17273.71	SM	0.9			MIT
5787.151	5788.756	17274.87	SM	1.2			MIT
5786.986	5788.591	17275.36	SM II	2.3			MIT
5784.716	5786.320	17282.14	SM	0.5			MIT
5783.527	5785.131	17285.69	SM	0.9			MIT
5781.893	5783.496	17290.58	SM	2.0			MIT
5779.248	5780.851	17298.49	SM	1.7			MIT
5778.331	5779.933	17301.24	SM	1.7			MIT
5773.772	5775.373	17314.90	SM	2.0			MIT
5771.740	5773.341	17320.99	SM	0.5			MIT
5768.784	5770.384	17329.87	SM	0.3			MIT
5768.100	5769.700	17331.93	SM	0.5			MIT
5767.63	5769.23	17333.3	SM	0.3			KN
5767.105	5768.704	17334.92	SM	0.3			MIT
5765.894	5767.493	17338.56	SM	0.5			MIT
5765.564	5767.163	17339.55	SM	0.3			MIT
5764.995	5766.594	17341.26	SM	0.7			MIT
5763.881	5765.480	17344.61	SM	1.3			MIT
5763.445	5765.043	17345.92	SM	0.6			MIT
5763.265	5764.863	17346.47	SM	0.5			MIT
5761.719	5763.317	17351.12	SM	0.6			MIT
5760.757	5762.355	17354.02	SM	0.9D			MIT
5759.503	5761.100	17357.79	SM II	1.8			MIT
5758.881	5760.478	17359.67	SM	0.3			MIT
5758.527	5760.124	17360.74	SM	0.5			MIT
5757.965	5759.562	17362.43	SM	1.3			MIT
5757.361	5758.958	17364.25	SM	0.5			MIT
5756.765	5758.362	17366.05	SM	0.5			MIT
5756.400	5757.997	17367.15	SM II	1.2			MIT
5755.930	5757.526	17368.57	SM	0.6			MIT
5754.40	5756.00	17373.2	SM	0.8W			MIT
5753.115	5754.711	17377.07	SM	0.3			MIT
5752.594	5754.190	17378.64	SM	0.5			MIT
5751.365	5752.960	17382.36	SM	1.6			MIT
5750.730	5752.325	17384.27	SM	0.5			MIT
5748.432	5750.026	17391.22	SM	0.6			MIT
5748.072	5749.666	17392.31	SM	1.3			MIT
5747.51	5749.10	17394.0	SM	0.5			MIT
5746.487	5748.081	17397.11	SM	0.6			MIT
5745.482	5747.076	17400.15	SM	1.0			MIT

AIR WAVELENGTH (ANGSTRMS)	VACUUM WAVELENGTH (ANGSTRMS)	WAVENUMBER (KAYSERS)	LMENT	LOG ARC	INTENSITY SPK	DIS	REF
5745.031	5746.625	17401.52	SM	0.7			MIT
5744.226	5745.819	17403.96	SM	0.6			MIT
5743.335	5744.928	17406.66	SM II	1.3			MIT
5743.043	5744.636	17407.54	SM	0.5			MIT
5741.186	5742.778	17413.17	SM	1.8			MIT
5740.878	5742.470	17414.11	SM	1.9			MIT
5739.981	5741.573	17416.83	SM	0.5			MIT
5738.002	5739.594	17422.84	SM II	1.2D			MIT
5736.851	5738.442	17426.33	SM	1.6			MIT
5735.004	5736.595	17431.94	SM	1.0			MIT
5733.578	5735.168	17436.28	SM	0.7			MIT
5732.947	5734.537	17438.20	SM	2.0			MIT
5730.60	5732.19	17445.3	SM	0.5	0.0		MIT
5730.125	5731.715	17446.79	SM	0.9			MIT
5729.919	5731.508	17447.41	SM II	1.6			MIT
5729.298	5730.887	17449.30	SM	1.0			MIT
5728.515	5730.104	17451.69	SM	0.5			MIT
5725.59	5727.18	17460.6	SM	1.6D			MIT
5724.488	5726.076	17463.97	SM	0.7			MIT
5721.375	5722.962	17473.47	SM	1.3			MIT
5721.18	5722.77	17474.1	SM	1.4	0.3		MIT
5720.823	5722.410	17475.16	SM	0.6			MIT
5720.188	5721.775	17477.09	SM	1.5			MIT
5719.123	5720.710	17480.35	SM	1.8	1.3		MIT
5717.916	5719.502	17484.04	SM	1.5			MIT
5714.23	5715.82	17495.3	SM	0.8			MIT
5711.459	5713.044	17503.80	SM	1.0			MIT
5710.928	5712.512	17505.43	SM	1.0			MIT
5709.737	5711.321	17509.08	SM	0.9			MIT
5706.758	5708.341	17518.22	SM	0.6			MIT
5706.203	5707.786	17519.93	SM	1.4			MIT
5705.41	5706.99	17522.4	SM	0.5D			MIT
5704.12	5705.70	17526.3	SM	0.5			MIT
5703.455	5705.037	17528.37	SM II	1.0			MIT
5702.250	5703.832	17532.07	SM	0.3			MIT
5701.117	5702.699	17535.56	SM	0.5			MIT
5700.262	5701.844	17538.19	SM	0.3			MIT
5699.615	5701.196	17540.18	SM	0.7			MIT
5699.48	5701.06	17540.6	SM	0.3			MIT
5698.919	5700.500	17542.32	SM	0.5			MIT
5696.733	5698.314	17549.05	SM	1.6			MIT
5696.237	5697.817	17550.58	SM	1.6	0.5		MIT
5694.497	5696.077	17555.94	SM	1.4	0.3		MIT
5693.664	5695.244	17558.51	SM	0.6			MIT
5692.610	5694.189	17561.76	SM II	1.2D			MIT
5692.522	5694.101	17562.03	SM	1.2D	0.3D		MIT
5692.038	5693.617	17563.53	SM	0.7			MIT
5691.623	5693.202	17564.81	SM	0.7			MIT
5690.356	5691.935	17568.72	SM	0.3			MIT
5689.257	5690.836	17572.11	SM	1.1			MIT
5688.205	5689.783	17575.36	SM	0.3			MIT
5687.659	5689.237	17577.05	SM	0.3			MIT
5686.982	5688.560	17579.14	SM	0.7			MIT
5686.791	5688.369	17579.73	SM	0.7			MIT
5686.726	5688.304	17579.93	SM	1.7	1.3		MIT
5682.422	5683.999	17593.25	SM	0.5			MIT
5680.854	5682.430	17598.10	SM	0.6			MIT
5679.700	5681.276	17601.68	SM	1.2			MIT
5679.128	5680.704	17603.45	SM	0.6			MIT
5676.061	5677.636	17612.96	SM	0.7			MIT
5674.099	5675.674	17619.05	SM	0.5			MIT
5672.881	5674.455	17622.84	SM	0.7			MIT
5672.371	5673.945	17624.42	SM	0.5			MIT
5670.748	5672.322	17629.47	SM	0.6			MIT
5668.911	5670.484	17635.18	SM	0.9H			MIT
5667.519	5669.092	17639.51	SM	1.2W			MIT
5666.038	5667.610	17644.12	SM	1.0D			MIT
5664.665	5666.237	17648.40	SM	0.5			MIT
5663.881	5665.453	17650.84	SM	0.9			MIT
5662.945	5664.517	17653.76	SM	1.2L			MIT

AIR WAVELENGTH (ANGSTRMS)	VACUUM WAVELENGTH (ANGSTRMS)	WAVENUMBER (KAYSERS)	LMENT	LOG ARC	INTENSITY SPK	DIS	REF
5661.488	5663.059	17658.30	SM	1.4			MIT
5661.170	5662.741	17659.29	SM	0.6			MIT
5659.863	5661.434	17663.37	SM	1.8			MIT
5658.15	5659.72	17668.7	SM	1.2	0.3		MIT
5657.881	5659.451	17669.56	SM	0.3			MIT
5657.209	5658.779	17671.66	SM	0.7			MIT
5656.361	5657.931	17674.31	SM	1.2			MIT
5655.503	5657.073	17676.99	SM	0.3			MIT
5652.843	5654.412	17685.30	SM	1.0			MIT
5651.945	5653.514	17688.11	SM	0.5			MIT
5651.677	5653.246	17688.95	SM	0.3			MIT
5650.782	5652.350	17691.76	SM	0.5			MIT
5650.396	5651.964	17692.96	SM	0.7			MIT
5649.008	5650.576	17697.31	SM	0.7			MIT
5647.467	5649.034	17702.14	SM	0.3			MIT
5647.297	5648.864	17702.67	SM	1.0D			MIT
5644.125	5645.692	17712.62	SM	1.6			MIT
5643.372	5644.938	17714.99	SM II	1.3			MIT
5642.916	5644.482	17716.42	SM	0.7			MIT
5642.678	5644.244	17717.16	SM	0.3			MIT
5642.248	5643.814	17718.51	SM	0.5	0.0		MIT
5642.008	5643.574	17719.27	SM	0.6			MIT
5641.620	5643.186	17720.49	SM	0.5			MIT
5640.768	5642.334	17723.16	SM	0.3			MIT
5640.272	5641.838	17724.72	SM	0.8			MIT
5638.670	5640.235	17729.76	SM	0.5			MIT
5637.273	5638.838	17734.15	SM	1.9			MIT
5636.713	5638.278	17735.91	SM	0.9			MIT
5634.244	5635.808	17743.69	SM	0.5			MIT
5633.538	5635.102	17745.91	SM	0.7			MIT
5632.352	5633.915	17749.65	SM	0.5			MIT
5632.062	5633.625	17750.56	SM	0.5			MIT
5631.780	5633.343	17751.45	SM	0.5			MIT
5630.841	5632.404	17754.41	SM	0.5			MIT
5629.120	5630.683	17759.84	SM	0.7D	0.0		MIT
5628.162	5629.724	17762.86	SM	0.5			MIT
5627.725	5629.287	17764.24	SM	0.6			MIT
5626.009	5627.571	17769.66	SM	1.8			MIT
5624.966	5626.527	17772.95	SM	1.2			MIT
5624.688	5626.249	17773.83	SM	0.3			MIT
5623.552	5625.113	17777.42	SM	0.5			MIT
5622.960	5624.521	17779.29	SM II	0.7			MIT
5622.715	5624.276	17780.07	SM	0.3			MIT
5622.012	5623.573	17782.29	SM	0.6			MIT
5621.798	5623.359	17782.97	SM	1.8			MIT
5620.885	5622.445	17785.86	SM	0.5			MIT
5617.747	5619.306	17795.79	SM	0.9D	0.0		MIT
5615.749	5617.308	17802.12	SM	0.6			MIT
5615.041	5616.600	17804.37	SM	0.6			MIT
5614.410	5615.969	17806.37	SM	0.6			MIT
5612.990	5614.548	17810.87	SM	0.5			MIT
5611.857	5613.415	17814.47	SM	0.5			MIT
5611.289	5612.847	17816.27	SM	0.3			MIT
5610.032	5611.589	17820.26	SM	1.5	1.3		MIT
5609.698	5611.255	17821.32	SM	0.7			MIT
5608.139	5609.696	17826.28	SM	0.6	0.0		MIT
5606.555	5608.111	17831.31	SM	0.3			MIT
5605.462	5607.018	17834.79	SM	0.5			MIT
5604.861	5606.417	17836.70	SM	1.5	0.3		MIT
5603.645	5605.201	17840.57	SM	0.6			MIT
5603.188	5604.744	17842.03	SM II	2.0D			MIT
5600.854	5602.409	17849.47	SM II	2.3			MIT
5599.563	5601.118	17853.58	SM	0.3			MIT
5598.677	5600.231	17856.41	SM	0.5			MIT
5598.145	5599.699	17858.10	SM	0.3			MIT
5597.556	5599.110	17859.98	SM	0.5			MIT
5592.380	5593.933	17876.51	SM	1.2			MIT
5591.919	5593.472	17877.99	SM	0.5			MIT
5591.670	5593.223	17878.78	SM	0.9			MIT
5591.166	5592.718	17880.39	SM	0.8			MIT

AIR WAVELENGTH (ANGSTRMS)	VACUUM WAVELENGTH (ANGSTRMS)	WAVENUMBER (KAYSERS)	LMENT	LOG ARC	INTENSITY SPK	DIS	REF
5589.919	5591.471	17884.38	SM	0.6			MIT
5588.191	5589.743	17889.91	SM	1.4			MIT
5587.461	5589.012	17892.25	SM	0.5			MIT
5584.107	5585.657	17903.00	SM	0.7			MIT
5583.699	5585.249	17904.30	SM	0.5			MIT
5583.298	5584.848	17905.59	SM	1.0			MIT
5581.878	5583.428	17910.14	SM	0.5			MIT
5580.436	5581.986	17914.77	SM	0.9H	0.0		MIT
5579.627	5581.176	17917.37	SM	1.0	0.0		MIT
5578.935	5580.484	17919.59	SM	1.0			MIT
5575.589	5577.137	17930.35	SM	0.7			MIT
5574.908	5576.456	17932.54	SM	1.7			MIT
5574.664	5576.212	17933.32	SM	1.2	0.3		MIT
5574.347	5575.895	17934.34	SM	0.7	0.0		MIT
5573.939	5575.487	17935.65	SM	0.6			MIT
5573.430	5574.978	17937.29	SM	1.9			MIT
5570.603	5572.150	17946.40	SM	0.7			MIT
5569.278	5570.825	17950.66	SM	0.6			MIT
5565.618	5567.164	17962.47	SM	0.3			MIT
5562.845	5564.390	17971.42	SM	0.5			MIT
5561.378	5562.922	17976.16	SM	1.3			MIT
5561.161	5562.705	17976.86	SM	0.3			MIT
5558.630	5560.174	17985.05	SM	0.3			MIT
5556.382	5557.925	17992.33	SM	0.7			MIT
5554.737	5556.280	17997.65	SM	0.7			MIT
5553.015	5554.557	18003.24	SM	0.5			MIT
5551.454	5552.996	18008.30	SM	0.5			MIT
5550.399	5551.940	18011.72	SM	2.1			MIT
5548.954	5550.495	18016.41	SM	1.9			MIT
5543.603	5545.143	18033.80	SM	0.3			MIT
5542.733	5544.272	18036.63	SM	0.3			MIT
5542.650	5544.189	18036.90	SM	1.0			MIT
5541.346	5542.885	18041.15	SM II	1.0			MIT
5538.395	5539.933	18050.76	SM	0.3			MIT
5537.071	5538.609	18055.07	SM II	1.7			MIT
5536.198	5537.736	18057.92	SM	0.3			MIT
5535.504	5537.041	18060.19	SM	0.5			MIT
5532.813	5534.350	18068.97	SM	0.3			MIT
5532.605	5534.142	18069.65	SM	0.5			MIT
5529.954	5531.490	18078.31	SM	0.7			MIT
5526.135	5527.670	18090.80	SM	0.3			MIT
5525.572	5527.107	18092.65	SM	0.7			MIT
5523.310	5524.844	18100.06	SM	0.3			MIT
5521.637	5523.171	18105.54	SM	0.5			MIT
5519.622	5521.155	18112.15	SM	0.8			MIT
5518.868	5520.401	18114.63	SM	0.6			MIT
5516.145	5517.677	18123.57	SM	2.3			MIT
5513.929	5515.461	18130.85	SM	0.3			MIT
5513.209	5514.741	18133.22	SM	0.3			MIT
5512.963	5514.494	18134.03	SM	0.3			MIT
5512.101	5513.632	18136.86	SM	1.9			MIT
5511.375	5512.906	18139.25	SM	0.5			MIT
5511.096	5512.627	18140.17	SM	0.9			MIT
5510.789	5512.320	18141.18	SM	0.5			MIT
5510.114	5511.645	18143.41	SM	0.3			MIT
5509.662	5511.193	18144.89	SM II	1.6			MIT
5505.184	5506.713	18159.65	SM	0.6			MIT
5504.051	5505.580	18163.39	SM	0.5			MIT
5502.628	5504.157	18168.09	SM	0.6			MIT
5501.743	5503.271	18171.01	SM	0.3			MIT
5500.892	5502.420	18173.82	SM	0.5			MIT
5498.213	5499.741	18182.68	SM	1.9			MIT
5495.591	5497.118	18191.35	SM	0.3			MIT
5494.309	5495.835	18195.60	SM	0.5			MIT
5493.719	5495.245	18197.55	SM	1.9			MIT
5492.498	5494.024	18201.59	SM	0.8	0.3		MIT
5491.313	5492.839	18205.52	SM	0.3			MIT
5490.643	5492.169	18207.74	SM	1.2			MIT
5490.101	5491.626	18209.54	SM	0.5			MIT
5489.397	5490.922	18211.88	SM	0.5			MIT

AIR WAVELENGTH (ANGSTRMS)	VACUUM WAVELENGTH (ANGSTRMS)	WAVENUMBER (KAYSERS)	LMENT		LOG ARC	INTENSITY SPK	INTENSITY DIS	REF
5488.939	5490.464	18213.40	SM		1.2			MIT
5488.249	5489.774	18215.69	SM		0.5			MIT
5488.038	5489.563	18216.39	SM		0.5			MIT
5485.416	5486.940	18225.09	SM		1.6			MIT
5484.569	5486.093	18227.91	SM		0.6			MIT
5484.287	5485.811	18228.85	SM		0.3			MIT
5483.098	5484.622	18232.80	SM		1.2D	0.5		MIT
5482.180	5483.703	18235.85	SM		0.5			MIT
5480.661	5482.184	18240.91	SM		0.6			MIT
5478.596	5480.118	18247.78	SM		0.3			MIT
5478.29	5479.81	18248.8	SM	II	0.3			KN
5475.183	5476.704	18259.16	SM		1.0D	0.3		MIT
5473.927	5475.448	18263.35	SM		0.6			MIT
5473.688	5475.209	18264.14	SM		0.5			MIT
5473.292	5474.813	18265.46	SM	II	0.9			MIT
5472.686	5474.207	18267.49	SM		0.6			MIT
5467.208	5468.727	18285.79	SM	II	0.9			MIT
5466.728	5468.247	18287.40	SM		1.9			MIT
5463.795	5465.313	18297.21	SM		0.3			MIT
5463.167	5464.685	18299.32	SM		0.3			MIT
5462.329	5463.847	18302.12	SM		0.5			MIT
5461.537	5463.055	18304.78	SM		0.9			MIT
5459.588	5461.105	18311.31	SM		0.3			MIT
5456.394	5457.910	18322.03	SM		0.7	0.0		MIT
5455.273	5456.789	18325.79	SM		0.7D			MIT
5453.018	5454.533	18333.37	SM		2.0			MIT
5452.259	5453.774	18335.93	SM		0.3			MIT
5450.052	5451.567	18343.35	SM		0.3			MIT
5449.587	5451.102	18344.92	SM		0.7			MIT
5448.283	5449.797	18349.31	SM		0.3			MIT
5447.648	5449.162	18351.45	SM		0.3			MIT
5443.651	5445.164	18364.92	SM		1.0D	0.0		MIT
5442.948	5444.461	18367.29	SM		0.5			MIT
5441.920	5443.433	18370.76	SM		0.5			MIT
5441.378	5442.890	18372.59	SM		0.5	0.0		MIT
5440.884	5442.396	18374.26	SM		0.3			MIT
5440.653	5442.165	18375.04	SM		0.5			MIT
5439.856	5441.368	18377.73	SM		0.3			MIT
5438.977	5440.489	18380.70	SM		0.3			MIT
5436.666	5438.177	18388.51	SM		0.3			MIT
5436.322	5437.833	18389.68	SM		1.0			MIT
5433.823	5435.333	18398.14	SM		0.9			MIT
5433.548	5435.058	18399.07	SM		0.6			MIT
5431.796	5433.306	18405.00	SM		0.5			MIT
5430.633	5432.142	18408.94	SM		0.6			MIT
5427.587	5429.096	18419.27	SM		0.5	0.0		MIT
5426.388	5427.896	18423.34	SM		0.5			MIT
5425.638	5427.146	18425.89	SM		1.0			MIT
5423.961	5425.469	18431.59	SM		0.3	0.0		MIT
5422.971	5424.478	18434.95	SM	II	0.6J			MIT
5422.198	5423.705	18437.58	SM		0.7J	0.3		MIT
5421.569	5423.076	18439.72	SM		1.4			MIT
5420.711	5422.218	18442.64	SM		1.4	0.5		MIT
5419.421	5420.928	18447.03	SM		0.5			MIT
5419.069	5420.575	18448.23	SM		0.6			MIT
5416.353	5417.859	18457.48	SM		0.6			MIT
5416.052	5417.558	18458.50	SM		2.0			MIT
5414.416	5415.921	18464.08	SM		0.8D	0.0		MIT
5412.801	5414.306	18469.59	SM		0.5			MIT
5411.408	5412.912	18474.34	SM		1.6			MIT
5411.155	5412.659	18475.21	SM		0.6			MIT
5410.106	5411.610	18478.79	SM		0.7W	0.0		MIT
5407.855	5409.358	18486.48	SM		1.1			MIT
5406.811	5408.314	18490.05	SM		0.5			MIT
5405.239	5406.742	18495.43	SM		1.9			MIT
5404.970	5406.473	18496.35	SM		0.6			MIT
5403.690	5405.192	18500.73	SM		0.9			MIT
5402.315	5403.817	18505.44	SM		0.5			MIT
5402.011	5403.513	18506.48	SM		0.3			MIT
5401.428	5402.930	18508.48	SM		0.3			MIT

AIR WAVELENGTH (ANGSTRMS)	VACUUM WAVELENGTH (ANGSTRMS)	WAVENUMBER (KAYSERS)	LMENT		LOG ARC	INTENSITY SPK	INTENSITY DIS	REF
5400.858	5402.360	18510.43	SM		0.5			MIT
5399.677	5401.178	18514.48	SM		0.3			MIT
5398.959	5400.460	18516.94	SM		0.3			MIT
5398.140	5399.641	18519.75	SM	II	1.2D			MIT
5397.914	5399.415	18520.53	SM		0.5			MIT
5397.186	5398.687	18523.02	SM		0.5W	0.0		MIT
5394.483	5395.983	18532.30	SM		0.9			MIT
5393.553	5395.053	18535.50	SM		0.3			MIT
5392.666	5394.165	18538.55	SM		0.7			MIT
5391.607	5393.106	18542.19	SM		0.5			MIT
5389.823	5391.322	18548.33	SM		0.7			MIT
5387.982	5389.480	18554.67	SM		1.7	0.5		MIT
5386.685	5388.183	18559.13	SM		0.3			MIT
5385.964	5387.462	18561.62	SM		0.3			MIT
5384.883	5386.380	18565.34	SM		0.5			MIT
5383.883	5385.380	18568.79	SM		0.5			MIT
5382.608	5384.105	18573.19	SM		0.3			MIT
5382.037	5383.534	18575.16	SM		0.5			MIT
5381.043	5382.539	18578.59	SM		0.5			MIT
5380.403	5381.899	18580.80	SM		1.0	0.5		MIT
5379.406	5380.902	18584.25	SM		0.7W	0.0		MIT
5378.073	5379.568	18588.85	SM		0.7			MIT
5374.307	5375.801	18601.88	SM		0.6			MIT
5374.026	5375.520	18602.85	SM		0.6			MIT
5373.693	5375.187	18604.00	SM		1.0D	0.3		MIT
5370.269	5371.762	18615.86	SM		0.9D	0.3		MIT
5370.048	5371.541	18616.63	SM		0.3			MIT
5369.163	5370.656	18619.70	SM		1.6			MIT
5368.360	5369.853	18622.48	SM		1.9			MIT
5366.771	5368.263	18628.00	SM		1.0D	0.3		MIT
5366.547	5368.039	18628.77	SM		0.3			MIT
5364.359	5365.851	18636.37	SM		1.7			MIT
5360.755	5362.246	18648.90	SM		0.3			MIT
5355.883	5357.373	18665.87	SM		1.0			MIT
5352.283	5353.772	18678.42	SM		0.3			MIT
5351.558	5353.046	18680.95	SM		1.2D	0.5		MIT
5350.618	5352.106	18684.23	SM		1.2			MIT
5349.137	5350.625	18689.41	SM		1.4			MIT
5348.752	5350.240	18690.75	SM		1.3			MIT
5348.087	5349.574	18693.08	SM		1.3			MIT
5342.766	5344.252	18711.69	SM		0.5			MIT
5341.289	5342.775	18716.87	SM		1.9			MIT
5340.312	5341.797	18720.29	SM		0.5			MIT
5333.338	5334.822	18744.77	SM		0.6			MIT
5332.093	5333.576	18749.15	SM	II	1.1			MIT
5326.883	5328.365	18767.48	SM		0.3			MIT
5326.474	5327.956	18768.93	SM		0.5	0.0		MIT
5324.893	5326.374	18774.50	SM		1.4D	0.5		MIT
5323.534	5325.015	18779.29	SM		1.0D			MIT
5321.871	5323.351	18785.16	SM		1.7D	0.6		MIT
5321.250	5322.730	18787.35	SM		0.5			MIT
5320.597	5322.077	18789.66	SM		2.0			MIT
5318.967	5320.447	18795.41	SM		0.5			MIT
5318.045	5319.524	18798.67	SM		0.7			MIT
5317.687	5319.166	18799.94	SM		0.9D	0.3		MIT
5313.751	5315.229	18813.86	SM		0.8			MIT
5312.210	5313.688	18819.32	SM	II	2.0			MIT
5310.191	5311.668	18826.48	SM		0.3			MIT
5309.493	5310.970	18828.95	SM		0.9			MIT
5308.810	5310.287	18831.37	SM		0.5			MIT
5304.502	5305.978	18846.67	SM		0.5			MIT
5303.232	5304.708	18851.18	SM		0.3			MIT
5302.910	5304.385	18852.32	SM		1.7			MIT
5302.658	5304.133	18853.22	SM		0.5			MIT
5299.192	5300.666	18865.55	SM		0.3			MIT
5298.350	5299.824	18868.55	SM		1.0	0.3		MIT
5296.940	5298.414	18873.57	SM		0.8	0.3		MIT
5294.654	5296.127	18881.72	SM		0.7			MIT
5291.171	5292.643	18894.15	SM		0.5			MIT
5289.942	5291.414	18898.54	SM		1.2			MIT

AIR WAVELENGTH (ANGSTRMS)	VACUUM WAVELENGTH (ANGSTRMS)	WAVENUMBER (KAYSERS)	LMENT	LOG ARC	INTENSITY SPK	DIS	REF
5288.114	5289.585	18905.07	SM	0.5			MIT
5287.439	5288.910	18907.49	SM	1.0D	0.3		MIT
5285.188	5286.659	18915.54	SM	0.6			MIT
5284.347	5285.817	18918.55	SM	0.3			MIT
5282.914	5284.384	18923.68	SM	2.0			MIT
5281.541	5283.011	18928.60	SM	1.0D	0.3		MIT
5281.04	5282.51	18930.4	SM II	0.8			MIT
5280.502	5281.971	18932.32	SM	0.5			MIT
5274.126	5275.594	18955.21	SM II	1.0			MIT
5272.815	5274.282	18959.93	SM II	1.7D			MIT
5271.400	5272.867	18965.01	SM	2.2			MIT
5269.948	5271.415	18970.24	SM	0.3			MIT
5265.656	5267.121	18985.70	SM	1.0			MIT
5264.399	5265.864	18990.24	SM	0.7			MIT
5263.264	5264.729	18994.33	SM	0.6			MIT
5260.528	5261.992	19004.21	SM	0.5			MIT
5259.099	5260.563	19009.37	SM	0.6			MIT
5257.916	5259.379	19013.65	SM II	0.5			MIT
5257.102	5258.565	19016.59	SM	0.5			MIT
5253.796	5255.258	19028.56	SM	0.6			MIT
5253.449	5254.911	19029.82	SM	0.5			MIT
5252.766	5254.228	19032.29	SM II	1.6D			MIT
5251.886	5253.348	19035.48	SM	2.2			MIT
5245.595	5247.055	19058.31	SM	0.6			MIT
5238.108	5239.566	19085.55	SM	0.7			MIT
5237.598	5239.056	19087.41	SM	1.0	0.5		MIT
5234.75	5236.21	19097.8	SM II	0.3			KN
5234.175	5235.632	19099.89	SM	1.7D			KN
5233.296	5234.753	19103.10	SM	0.6	0.3		MIT
5229.863	5231.319	19115.64	SM	0.6			MIT
5229.572	5231.028	19116.70	SM	0.5			MIT
5228.781	5230.237	19119.59	SM	1.8			MIT
5224.665	5226.120	19134.66	SM	0.7			MIT
5221.319	5222.773	19146.92	SM	0.9			MIT
5221.107	5222.561	19147.70	SM	0.9			MIT
5218.398	5219.851	19157.64	SM	1.4			MIT
5218.084	5219.537	19158.79	SM	0.9D	0.3		MIT
5216.428	5217.880	19164.87	SM	0.7			MIT
5215.609	5217.061	19167.88	SM	1.2			MIT
5212.189	5213.640	19180.46	SM	0.8			MIT
5211.726	5213.177	19182.16	SM	0.7			MIT
5210.754	5212.205	19185.74	SM	0.9			MIT
5209.928	5211.379	19188.78	SM	1.3			MIT
5207.638	5209.088	19197.22	SM	0.5			MIT
5207.151	5208.601	19199.01	SM II	1.4D			MIT
5205.395	5206.844	19205.49	SM	1.2	0.3		MIT
5203.668	5205.117	19211.86	SM	0.5			MIT
5202.733	5204.182	19215.32	SM II	1.7			MIT
5202.356	5203.805	19216.71	SM	0.5			MIT
5201.452	5202.900	19220.05	SM	1.0			MIT
5200.591	5202.039	19223.23	SM	2.3	0.0		MIT
5196.412	5197.859	19238.69	SM	0.5			MIT
5194.723	5196.170	19244.95	SM	1.0			MIT
5193.518	5194.964	19249.41	SM	0.5	0.3		MIT
5193.054	5194.500	19251.13	SM	0.7	0.0		MIT
5190.445	5191.890	19260.81	SM II	0.7			MIT
5187.864	5189.309	19270.39	SM II	0.5			MIT
5187.210	5188.655	19272.82	SM	0.3			MIT
5187.079	5188.524	19273.31	SM	0.5			MIT
5179.208	5180.650	19302.59	SM	0.5			MIT
5178.058	5179.500	19306.88	SM	2.0D			MIT
5175.418	5176.859	19316.73	SM	1.8	0.0		MIT
5173.635	5175.076	19323.39	SM	0.5			MIT
5172.742	5174.183	19326.72	SM	1.9	0.0		MIT
5170.736	5172.176	19334.22	SM	0.7	0.3		MIT
5169.602	5171.042	19338.46	SM	1.7D			MIT
5168.345	5169.785	19343.17	SM	0.8			MIT
5167.271	5168.710	19347.19	SM	0.5			MIT
5166.046	5167.485	19351.77	SM II	2.1D			MIT
5164.612	5166.051	19357.15	SM	0.5			MIT

AIR WAVELENGTH (ANGSTRMS)	VACUUM WAVELENGTH (ANGSTRMS)	WAVENUMBER (KAYSERS)	LMENT	LOG ARC	INTENSITY SPK	DIS	REF
5162.860	5164.298	19363.72	SM	1.2	0.5		MIT
5162.060	5163.498	19366.72	SM II	0.6			MIT
5161.161	5162.599	19370.09	SM	0.5			MIT
5160.572	5162.009	19372.30	SM	1.0D			MIT
5159.516	5160.953	19376.27	SM II	1.2			MIT
5157.884	5159.321	19382.40	SM	0.5			MIT
5157.557	5158.994	19383.62	SM	0.5			MIT
5157.223	5158.660	19384.88	SM	1.0			MIT
5157.043	5158.480	19385.56	SM II	1.0			MIT
5155.853	5157.289	19390.03	SM	1.0			MIT
5155.016	5156.452	19393.18	SM II	2.1	0.0		MIT
5154.271	5155.707	19395.98	SM	2.1D	0.0		MIT
5152.578	5154.013	19402.36	SM	0.5			MIT
5151.950	5153.385	19404.72	SM	0.6			MIT
5150.439	5151.874	19410.41	SM	0.5			MIT
5145.797	5147.231	19427.92	SM	1.0			MIT
5144.947	5146.380	19431.13	SM	0.6			MIT
5143.285	5144.718	19437.41	SM	1.1			MIT
5137.385	5138.816	19459.73	SM	0.6			MIT
5135.835	5137.266	19465.61	SM	0.6			MIT
5132.193	5133.623	19479.42	SM	1.0			MIT
5128.739	5130.168	19492.54	SM	0.9	0.3		MIT
5127.113	5128.542	19498.72	SM II	0.7			MIT
5125.558	5126.986	19504.64	SM II	0.5			MIT
5124.830	5126.258	19507.41	SM	1.7			MIT
5122.136	5123.563	19517.67	SM	1.9			MIT
5121.120	5122.547	19521.54	SM	1.2	0.5		MIT
5117.156	5118.582	19536.66	SM	1.9	0.0		MIT
5116.677	5118.103	19538.49	SM II	2.0	0.0		MIT
5115.750	5117.176	19542.03	SM	1.0	0.0		MIT
5115.357	5116.782	19543.53	SM	0.7			MIT
5113.888	5115.313	19549.15	SM	1.2D	0.6		MIT
5112.295	5113.720	19555.24	SM II	2.1	0.0		MIT
5111.931	5113.356	19556.63	SM	0.6D			MIT
5111.148	5112.572	19559.62	SM II	1.1			MIT
5110.527	5111.951	19562.00	SM	0.6	0.5		MIT
5108.279	5109.703	19570.61	SM	0.9	0.3		MIT
5106.988	5108.411	19575.56	SM	0.5			MIT
5106.625	5108.048	19576.95	SM	1.0D	0.5		MIT
5104.475	5105.898	19585.20	SM II	2.1	0.0		MIT
5104.060	5105.482	19586.79	SM II	1.2D			MIT
5103.082	5104.504	19590.54	SM II	2.2			MIT
5102.371	5103.793	19593.27	SM	0.9	1.6		MIT
5100.305	5101.726	19601.21	SM	1.8			MIT
5100.198	5101.619	19601.62	SM II	1.7			MIT
5091.812	5093.231	19633.90	SM	0.5			MIT
5090.017	5091.436	19640.82	SM	1.2D			MIT
5089.705	5091.124	19642.03	SM II	1.2D			MIT
5088.968	5090.386	19644.87	SM II	1.5			MIT
5088.315	5089.733	19647.40	SM	1.4			MIT
5087.645	5089.063	19649.98	SM	1.6			MIT
5087.073	5088.491	19652.19	SM II	1.5			MIT
5084.788	5086.205	19661.02	SM	0.3	0.0		MIT
5083.341	5084.758	19666.62	SM	1.0	1.5		MIT
5080.846	5082.262	19676.28	SM	0.9W	0.3		MIT
5079.856	5081.272	19680.11	SM	1.0			MIT
5079.016	5080.432	19683.37	SM	0.6	0.3		MIT
5078.065	5079.481	19687.05	SM II	1.4D			MIT
5077.608	5079.023	19688.82	SM	0.5			MIT
5076.682	5078.097	19692.42	SM	1.0			MIT
5074.935	5076.350	19699.19	SM II	0.5			MIT
5072.456	5073.870	19708.82	SM	0.6			MIT
5071.187	5072.601	19713.75	SM	2.0			MIT
5069.445	5070.858	19720.53	SM II	2.2	0.0		MIT
5068.185	5069.598	19725.43	SM	0.7	0.3		MIT
5067.157	5068.570	19729.43	SM	0.9	0.3		MIT
5066.854	5068.267	19730.61	SM II	1.0			MIT
5066.384	5067.796	19732.44	SM II	1.2			MIT
5065.926	5067.338	19734.23	SM	0.5	0.5		MIT
5064.239	5065.651	19740.80	SM II	1.5			MIT

AIR WAVELENGTH (ANGSTRMS)	VACUUM WAVELENGTH (ANGSTRMS)	WAVENUMBER (KAYSERS)	LMENT	LOG ARC	INTENSITY SPK	DIS	REF
5061.397	5062.808	19751.88	SM	0.5			MIT
5060.922	5062.333	19753.74	SM	1.5			MIT
5059.854	5061.265	19757.91	SM	1.2			MIT
5058.855	5060.265	19761.81	SM II	1.0			MIT
5057.747	5059.157	19766.14	SM II	2.0	0.0		MIT
5054.072	5055.481	19780.51	SM	0.5			MIT
5052.755	5054.164	19785.67	SM II	2.2	0.3		MIT
5050.075	5051.483	19796.17	SM	0.9	0.3		MIT
5049.504	5050.912	19798.41	SM	0.9			MIT
5046.421	5047.828	19810.50	SM	0.8			MIT
5044.279	5045.686	19818.91	SM	2.2			MIT
5041.932	5043.338	19828.14	SM	0.7			MIT
5039.948	5041.353	19835.94	SM	1.4			MIT
5039.455	5040.860	19837.88	SM	0.3			MIT
5039.115	5040.520	19839.22	SM	0.3			MIT
5036.213	5037.617	19850.65	SM	1.7			MIT
5034.758	5036.162	19856.39	SM	0.6			MIT
5031.181	5032.584	19870.51	SM II	1.5			MIT
5030.958	5032.361	19871.39	SM	0.9			MIT
5029.605	5031.008	19876.73	SM	0.6			MIT
5028.443	5029.845	19881.33	SM II	2.3			MIT
5026.197	5027.599	19890.21	SM	1.6			MIT
5023.506	5024.907	19900.87	SM II	1.8			MIT
5022.461	5023.862	19905.01	SM	0.3			MIT
5019.158	5020.558	19918.10	SM	0.3			MIT
5017.539	5018.938	19924.53	SM	0.5			MIT
5016.609	5018.008	19928.23	SM II	1.9			MIT
5013.680	5015.078	19939.87	SM	1.0			MIT
5010.875	5012.273	19951.03	SM	0.3			MIT
5010.362	5011.760	19953.07	SM	0.9			MIT
5010.162	5011.559	19953.87	SM	0.9			MIT
5003.382	5004.778	19980.91	SM	0.5			MIT
5001.231	5002.626	19989.50	SM II	1.6			MIT
5000.050	5001.445	19994.22	SM II	1.4			MIT
4994.99	4996.38	20014.5	SM	0.3			MIT
4992.031	4993.424	20026.34	SM II	1.9			MIT
4989.440	4990.832	20036.74	SM II	1.8			MIT
4986.820	4988.211	20047.27	SM	0.5			MIT
4985.886	4987.277	20051.02	SM II	1.3D			MIT
4983.373	4984.763	20061.13	SM	1.2			MIT
4981.714	4983.104	20067.81	SM II	1.7			MIT
4980.259	4981.649	20073.68	SM	0.6			MIT
4975.986	4977.374	20090.91	SM	1.9			MIT
4974.733	4976.121	20095.97	SM	0.3			MIT
4973.734	4975.122	20100.01	SM II	1.6			MIT
4972.166	4973.553	20106.35	SM II	1.9			MIT
4970.09	4971.48	20114.7	SM	0.8			MIT
4967.739	4969.125	20124.27	SM	0.5			MIT
4967.019	4968.405	20127.18	SM	0.3			MIT
4965.787	4967.173	20132.18	SM	1.2			MIT
4964.565	4965.950	20137.13	SM II	1.9			MIT
4964.224	4965.609	20138.52	SM	0.3			MIT
4961.939	4963.324	20147.79	SM II	2.0			MIT
4961.260	4962.644	20150.55	SM	0.3			MIT
4960.059	4961.443	20155.43	SM II	1.0			MIT
4958.436	4959.820	20162.02	SM	0.3			MIT
4957.69	4959.07	20165.1	SM II	0.9			MIT
4956.731	4958.114	20168.96	SM	0.3			MIT
4956.123	4957.506	20171.43	SM II	1.4			MIT
4955.961	4957.344	20172.09	SM	0.7			MIT
4953.048	4954.430	20183.96	SM II	1.7			MIT
4952.373	4953.755	20186.71	SM II	2.1			MIT
4951.298	4952.680	20191.09	SM	0.3			MIT
4948.630	4950.011	20201.98	SM II	2.1			MIT
4948.067	4949.448	20204.27	SM	0.7			MIT
4947.45	4948.83	20206.8	SM	0.3			MIT
4946.306	4947.687	20211.47	SM	1.8			MIT
4945.20	4946.58	20216.0	SM	1.0			MIT
4943.44	4944.82	20223.2	SM	0.3			MIT
4941.390	4942.769	20231.57	SM II	0.9D			MIT

AIR WAVELENGTH (ANGSTRMS)	VACUUM WAVELENGTH (ANGSTRMS)	WAVENUMBER (KAYSERS)	LMENT	LOG ARC	INTENSITY SPK	DIS	REF
4940.617	4941.996	20234.74	SM II	1.5			MIT
4938.104	4939.482	20245.04	SM II	2.1			MIT
4936.590	4937.968	20251.25	SM	0.6			MIT
4936.021	4937.399	20253.58	SM II	1.9			MIT
4935.458	4936.836	20255.89	SM	1.3			MIT
4933.303	4934.680	20264.74	SM II	1.4			MIT
4930.928	4932.304	20274.50	SM	1.0			MIT
4929.561	4930.937	20280.12	SM II	1.6			MIT
4924.056	4925.431	20302.79	SM	1.5			MIT
4923.813	4925.188	20303.79	SM II	1.5			MIT
4922.472	4923.846	20309.33	SM II	1.5			MIT
4920.37	4921.74	20318.0	SM II	2.1			MIT
4919.882	4921.255	20320.02	SM	0.3			MIT
4918.984	4920.357	20323.73	SM	2.1			MIT
4917.410	4918.783	20330.23	SM II	1.2D			MIT
4914.309	4915.681	20343.06	SM II	1.4			MIT
4913.259	4914.631	20347.41	SM II	2.2			MIT
4910.406	4911.777	20359.23	SM	2.2			MIT
4907.789	4909.159	20370.09	SM	0.3			MIT
4905.790	4907.160	20378.39	SM	0.3			MIT
4904.978	4906.348	20381.76	SM	1.8			MIT
4901.903	4903.272	20394.55	SM II	1.6D			MIT
4901.01	4902.38	20398.3	SM II	1.0D			MIT
4900.737	4902.105	20399.40	SM II	2.0			MIT
4897.827	4899.195	20411.52	SM	0.5			MIT
4894.300	4895.667	20426.23	SM II	1.8			MIT
4893.765	4895.132	20428.46	SM	0.5			MIT
4893.337	4894.703	20430.25	SM II	2.2			MIT
4891.940	4893.306	20436.08	SM	1.6			MIT
4891.466	4892.832	20438.06	SM	0.3			MIT
4890.761	4892.127	20441.01	SM	0.3			MIT
4890.329	4891.695	20442.81	SM	0.5			MIT
4886.970	4888.335	20456.86	SM	0.5			MIT
4886.613	4887.978	20458.36	SM	0.3			MIT
4886.050	4887.415	20460.72	SM	0.3			MIT
4883.983	4885.347	20469.37	SM	1.9			MIT
4883.779	4885.143	20470.23	SM	1.8			MIT
4879.75	4881.11	20487.1	SM	0.5			MIT
4879.35	4880.71	20488.8	SM II	0.7			MIT
4878.99	4880.35	20490.3	SM	0.3			MIT
4877.578	4878.940	20496.25	SM II	1.0			MIT
4875.616	4876.978	20504.50	SM	1.0D			MIT
4874.378	4875.739	20509.71	SM	0.6			MIT
4873.854	4875.215	20511.92	SM	0.9			MIT
4873.199	4874.560	20514.67	SM	1.5			MIT
4871.325	4872.686	20522.56	SM	0.3			MIT
4869.982	4871.342	20528.22	SM II	2.1			MIT
4866.581	4867.940	20542.57	SM	0.3			MIT
4865.397	4866.756	20547.57	SM II	1.0			MIT
4864.229	4865.588	20552.50	SM II	1.2D			MIT
4862.08	4863.44	20561.6	SM	0.5			MIT
4861.050	4862.408	20565.94	SM	0.6			MIT
4860.697	4862.055	20567.44	SM	0.3			MIT
4859.852	4861.210	20571.01	SM	0.7			MIT
4859.568	4860.925	20572.21	SM II	1.9			MIT
4857.165	4858.522	20582.39	SM	0.3			MIT
4854.375	4855.731	20594.22	SM II	2.1			MIT
4851.525	4852.880	20606.32	SM	0.6			MIT
4848.329	4849.684	20619.90	SM	2.0			MIT
4847.766	4849.120	20622.30	SM II	2.2			MIT
4847.069	4848.423	20625.26	SM II	0.5			MIT
4845.369	4846.723	20632.50	SM	0.5			MIT
4844.212	4845.565	20637.43	SM II	2.2			MIT
4841.704	4843.057	20648.12	SM	2.0			MIT
4838.645	4839.997	20661.17	SM	0.6			MIT
4837.649	4839.001	20665.42	SM II	2.0			MIT
4836.666	4838.017	20669.62	SM II	1.7			MIT
4835.722	4837.073	20673.66	SM	0.3			MIT
4834.626	4835.977	20678.34	SM II	2.0			MIT
4833.329	4834.680	20683.89	SM II	1.9			MIT

AIR WAVELENGTH (ANGSTRMS)	VACUUM WAVELENGTH (ANGSTRMS)	WAVENUMBER (KAYSERS)	LMENT	LOG ARC	INTENSITY SPK	INTENSITY DIS	REF
4832.331	4833.681	20688.17	SM	0.3			MIT
4830.681	4832.031	20695.23	SM II	0.9			MIT
4829.579	4830.929	20699.95	SM II	2.3			MIT
4828.736	4830.085	20703.57	SM	0.3			MIT
4827.813	4829.162	20707.53	SM	0.3			MIT
4826.575	4827.924	20712.84	SM II	1.2			MIT
4825.382	4826.730	20717.96	SM	1.0			MIT
4824.664	4826.012	20721.04	SM	1.0			MIT
4822.865	4824.213	20728.77	SM	0.7			MIT
4820.766	4822.113	20737.80	SM	0.9			MIT
4816.481	4817.827	20756.25	SM II	0.9D			MIT
4816.027	4817.373	20758.20	SM II	1.3			MIT
4815.806	4817.152	20759.15	SM II	2.1	1.9		MIT
4814.33	4815.68	20765.5	SM	0.5			MIT
4809.001	4810.345	20788.53	SM	0.3			MIT
4808.487	4809.831	20790.75	SM	0.9			MIT
4807.370	4808.714	20795.58	SM	0.3			MIT
4804.914	4806.257	20806.21	SM II	1.2			MIT
4803.407	4804.750	20812.74	SM	0.5			MIT
4803.000	4804.342	20814.50	SM	1.0			MIT
4798.875	4800.216	20832.39	SM II	1.7R			MIT
4795.905	4797.246	20845.29	SM	0.3			MIT
4794.307	4795.647	20852.24	SM	1.0			MIT
4791.584	4792.923	20864.09	SM II	2.2			MIT
4789.960	4791.299	20871.17	SM	1.7			MIT
4786.151	4787.489	20887.78	SM	0.6			MIT
4785.869	4787.207	20889.01	SM	2.0			MIT
4784.427	4785.765	20895.30	SM	0.8			MIT
4783.100	4784.437	20901.10	SM	2.2			MIT
4781.838	4783.175	20906.62	SM II	1.8			MIT
4781.566	4782.903	20907.80	SM	0.9			MIT
4779.223	4780.559	20918.05	SM II	1.0			MIT
4778.636	4779.972	20920.62	SM	0.7			MIT
4777.840	4779.176	20924.11	SM II	2.0			MIT
4775.96	4777.30	20932.4	SM	0.3			MIT
4775.528	4776.863	20934.24	SM II	1.5			MIT
4774.137	4775.472	20940.34	SM II	2.0			MIT
4772.474	4773.808	20947.64	SM	0.3			MIT
4771.667	4773.001	20951.18	SM II	0.9			MIT
4770.696	4772.030	20955.44	SM	0.7			MIT
4770.290	4771.624	20957.23	SM	0.5			MIT
4770.191	4771.525	20957.66	SM	1.6			MIT
4768.389	4769.722	20965.58	SM	0.3			MIT
4767.050	4768.383	20971.47	SM	0.3			MIT
4766.140	4767.473	20975.47	SM	1.6			MIT
4763.437	4764.769	20987.38	SM	0.3			MIT
4761.587	4762.919	20995.53	SM II	0.8			MIT
4760.724	4762.055	20999.34	SM	1.0			MIT
4760.262	4761.593	21001.37	SM	2.2			MIT
4760.028	4761.359	21002.41	SM	1.3			MIT
4758.819	4760.150	21007.74	SM II	0.9			MIT
4755.373	4756.703	21022.97	SM II	2.0			MIT
4753.71	4755.04	21030.3	SM	0.3			MIT
4752.94	4754.27	21033.7	SM	0.7			MIT
4752.286	4753.615	21036.62	SM II	0.8			MIT
4750.725	4752.054	21043.53	SM	1.8			MIT
4749.64	4750.97	21048.3	SM	0.8			MIT
4748.795	4750.123	21052.09	SM	1.0			MIT
4745.675	4747.002	21065.93	SM II	2.4			MIT
4742.481	4743.807	21080.11	SM	0.5			MIT
4742.227	4743.553	21081.24	SM	0.3			MIT
4741.726	4743.052	21083.47	SM II	1.9			MIT
4739.561	4740.887	21093.10	SM	1.0			MIT
4736.958	4738.283	21104.69	SM	0.6W			MIT
4736.490	4737.815	21106.78	SM	1.0			MIT
4736.116	4737.441	21108.44	SM	0.6			MIT
4733.884	4735.208	21118.40	SM	0.5			MIT
4732.434	4733.758	21124.87	SM	0.5			MIT
4731.355	4732.678	21129.68	SM	0.7			MIT
4730.956	4732.279	21131.47	SM	0.3			MIT

AIR WAVELENGTH (ANGSTRMS)	VACUUM WAVELENGTH (ANGSTRMS)	WAVENUMBER (KAYSERS)	LMENT	LOG ARC	INTENSITY SPK	INTENSITY DIS	REF
4729.723	4731.046	21136.98	SM	0.3			MIT
4728.438	4729.761	21142.72	SM	2.2			MIT
4727.111	4728.433	21148.65	SM	0.5			MIT
4726.495	4727.817	21151.41	SM	1.0			MIT
4726.026	4727.348	21153.51	SM II	2.0			MIT
4723.686	4725.007	21163.99	SM II	1.3			MIT
4722.947	4724.268	21167.30	SM	0.5H			MIT
4722.632	4723.953	21168.71	SM	0.5H			MIT
4721.396	4722.717	21174.25	SM II	1.3			MIT
4720.129	4721.450	21179.94	SM II	1.3			MIT
4719.842	4721.162	21181.22	SM II	2.1			MIT
4718.667	4719.987	21186.50	SM	1.4			MIT
4718.346	4719.666	21187.94	SM II	2.0			MIT
4717.736	4719.056	21190.68	SM II	2.0			MIT
4717.432	4718.752	21192.04	SM	0.7			MIT
4717.088	4718.408	21193.59	SM	1.9			MIT
4716.693	4718.013	21195.36	SM	0.3			MIT
4716.108	4717.427	21197.99	SM	2.2			MIT
4715.272	4716.591	21201.75	SM II	2.0			MIT
4714.630	4715.949	21204.64	SM II	1.7			MIT
4713.074	4714.393	21211.64	SM II	2.0			MIT
4710.651	4711.969	21222.55	SM II	1.7			MIT
4706.556	4707.873	21241.01	SM II	0.7			MIT
4705.000	4706.317	21248.04	SM II	0.5			MIT
4704.408	4705.724	21250.71	SM II	2.3			MIT
4699.345	4700.660	21273.61	SM II	1.7			MIT
4696.416	4697.730	21286.88	SM	0.6			MIT
4693.629	4694.943	21299.51	SM II	1.7			MIT
4689.578	4690.890	21317.91	SM II	1.0			MIT
4688.735	4690.047	21321.75	SM	1.7			MIT
4688.132	4689.444	21324.49	SM	0.6			MIT
4687.182	4688.494	21328.81	SM II	2.0			MIT
4684.206	4685.517	21342.36	SM	0.3			MIT
4684.035	4685.346	21343.14	SM	0.6			MIT
4682.682	4683.993	21349.31	SM II	1.3			MIT
4681.560	4682.870	21354.42	SM	1.7			MIT
4680.045	4681.355	21361.34	SM	1.0			MIT
4678.123	4679.432	21370.11	SM	0.5			MIT
4676.908	4678.217	21375.67	SM II	2.0	1.7		MIT
4674.592	4675.900	21386.25	SM II	1.9	1.6		MIT
4671.845	4673.153	21398.83	SM II	0.3			MIT
4670.833	4672.140	21403.47	SM	1.5			MIT
4670.768	4672.075	21403.76	SM	1.5			MIT
4669.647	4670.954	21408.90	SM II	1.7	1.6		MIT
4669.390	4670.697	21410.08	SM II	1.6	1.5		MIT
4665.111	4666.417	21429.72	SM II	1.3			MIT
4664.513	4665.819	21432.47	SM	0.3			MIT
4663.546	4664.852	21436.91	SM	1.8			MIT
4660.452	4661.757	21451.14	SM	0.3H			MIT
4655.361	4656.664	21474.60	SM II	0.3H			MIT
4655.124	4656.427	21475.69	SM II	1.2			MIT
4654.288	4655.591	21479.55	SM	0.5			MIT
4652.663	4653.966	21487.05	SM	0.6			MIT
4649.489	4650.791	21501.72	SM	1.4			MIT
4648.630	4649.932	21505.69	SM II	0.3			MIT
4648.164	4649.465	21507.85	SM II	1.1			MIT
4648.077	4649.378	21508.25	SM	0.8			MIT
4647.544	4648.845	21510.72	SM II	0.9			MIT
4646.93	4648.23	21513.6	SM	0.5			MIT
4646.679	4647.980	21514.72	SM II	1.7			MIT
4645.399	4646.700	21520.65	SM	1.6			MIT
4642.232	4643.532	21535.33	SM II	2.0	1.6		MIT
4641.520	4642.820	21538.64	SM	0.3			MIT
4640.512	4641.811	21543.31	SM	0.5			MIT
4639.14	4640.44	21549.7	SM	0.5			MIT
4638.65	4639.95	21552.0	SM	0.5W			MIT
4636.765	4638.063	21560.72	SM	0.3			MIT
4636.274	4637.572	21563.01	SM	0.9			MIT
4633.166	4634.464	21577.47	SM II	0.3			MIT
4632.771	4634.068	21579.31	SM	0.3			MIT

AIR WAVELENGTH (ANGSTRMS)	VACUUM WAVELENGTH (ANGSTRMS)	WAVENUMBER (KAYSERS)	LMENT	LOG ARC	INTENSITY SPK	INTENSITY DIS	REF
4630.967	4632.264	21587.72	SM	0.5W			MIT
4630.569	4631.866	21589.57	SM	0.3			MIT
4630.211	4631.508	21591.24	SM	1.6			MIT
4629.431	4630.728	21594.88	SM	1.0			MIT
4628.794	4630.090	21597.85	SM	0.3H			MIT
4624.959	4626.254	21615.76	SM II	0.6W			MIT
4624.137	4625.432	21619.60	SM	0.3H			MIT
4621.190	4622.484	21633.39	SM	0.5			MIT
4620.762	4622.056	21635.39	SM II	0.5H			MIT
4618.235	4619.529	21647.23	SM	0.3			MIT
4616.613	4617.906	21654.84	SM	0.3H			MIT
4616.484	4617.777	21655.44	SM II	0.5			MIT
4615.689	4616.982	21659.17	SM II	1.7	1.6		MIT
4615.446	4616.739	21660.31	SM II	1.4			MIT
4613.503	4614.795	21669.43	SM II	0.9			MIT
4611.669	4612.961	21678.05	SM	0.6R			MIT
4611.257	4612.549	21679.99	SM	1.7			MIT
4606.520	4607.810	21702.28	SM II	1.6			MIT
4604.812	4606.102	21710.33	SM	0.3			MIT
4604.181	4605.471	21713.31	SM II	1.8			MIT
4603.987	4605.277	21714.22	SM	0.3H			MIT
4603.116	4604.406	21718.33	SM	0.7			MIT
4602.025	4603.314	21723.48	SM	0.3			MIT
4598.365	4599.653	21740.77	SM II	0.7			MIT
4596.747	4598.035	21748.42	SM	1.3			MIT
4595.300	4596.587	21755.27	SM II	2.0	1.8		MIT
4594.577	4595.864	21758.69	SM	0.3			MIT
4593.531	4594.818	21763.65	SM II	1.7	1.7		MIT
4593.407	4594.694	21764.24	SM	0.3			MIT
4591.826	4593.113	21771.73	SM II	2.0			MIT
4589.419	4590.705	21783.15	SM	0.3			MIT
4584.834	4586.119	21804.93	SM II	1.8	1.7		MIT
4581.737	4583.021	21819.67	SM	1.6			MIT
4581.584	4582.868	21820.40	SM	0.9			MIT
4580.185	4581.468	21827.06	SM	0.5			MIT
4579.046	4580.329	21832.49	SM	0.6			MIT
4578.705	4579.988	21834.12	SM	0.6			MIT
4578.004	4579.287	21837.46	SM	0.6			MIT
4577.687	4578.970	21838.97	SM II	2.0	1.7		MIT
4574.736	4576.018	21853.06	SM	0.3			MIT
4569.585	4570.866	21877.69	SM	1.2			MIT
4566.776	4568.056	21891.15	SM	1.2			MIT
4566.205	4567.485	21893.89	SM II	2.0	1.7		MIT
4564.070	4565.349	21904.13	SM II	1.2			MIT
4561.179	4562.457	21918.01	SM II	0.5			MIT
4560.423	4561.701	21921.65	SM II	1.7			MIT
4556.626	4557.903	21939.91	SM	1.0			MIT
4556.503	4557.780	21940.50	SM	0.7			MIT
4554.443	4555.720	21950.43	SM II	1.8			MIT
4552.661	4553.937	21959.02	SM II	1.9	1.6		MIT
4550.036	4551.312	21971.69	SM	1.3			MIT
4548.022	4549.297	21981.42	SM	1.0			MIT
4545.801	4547.075	21992.16	SM II	0.6			MIT
4544.826	4546.100	21996.88	SM II	1.0			MIT
4543.945	4545.219	22001.14	SM II	2.0	1.7		MIT
4542.049	4543.322	22010.32	SM	1.7			MIT
4540.188	4541.461	22019.35	SM II	1.6			MIT
4538.548	4539.820	22027.30	SM	1.0			MIT
4537.954	4539.226	22030.19	SM II	1.7	1.4		MIT
4537.556	4538.828	22032.12	SM	0.6			MIT
4536.513	4537.785	22037.18	SM	1.6			MIT
4536.171	4537.443	22038.84	SM II	0.6			MIT
4534.856	4536.128	22045.24	SM	1.0			MIT
4533.799	4535.070	22050.37	SM	1.6			MIT
4533.217	4534.488	22053.21	SM	0.3			MIT
4532.445	4533.716	22056.96	SM	1.8			MIT
4529.949	4531.219	22069.11	SM	0.6			MIT
4527.420	4528.690	22081.44	SM	0.7			MIT
4523.912	4525.181	22098.56	SM II	2.0	1.7		MIT
4523.182	4524.450	22102.13	SM	0.9			MIT

AIR WAVELENGTH (ANGSTRMS)	VACUUM WAVELENGTH (ANGSTRMS)	WAVENUMBER (KAYSERS)	LMENT	LOG ARC	INTENSITY SPK	INTENSITY DIS	REF
4523.036	4524.304	22102.84	SM II	1.6	1.3		MIT
4522.538	4523.806	22105.28	SM	1.8			MIT
4519.635	4520.902	22119.48	SM II	2.2	1.9		MIT
4519.527	4520.794	22120.01	SM II	0.3			MIT
4517.250	4518.517	22131.16	SM	0.6			MIT
4515.098	4516.364	22141.70	SM	2.0			MIT
4514.706	4515.972	22143.63	SM	0.3			MIT
4512.305	4513.571	22155.41	SM	0.3			MIT
4511.834	4513.099	22157.72	SM II	2.0	2.0		MIT
4511.307	4512.572	22160.31	SM	1.6			MIT
4507.385	4508.649	22179.59	SM	0.3			MIT
4505.047	4506.311	22191.10	SM II	0.7	0.7		MIT
4503.381	4504.644	22199.31	SM	0.6	0.3		MIT
4501.368	4502.631	22209.24	SM II	0.7	1.0		MIT
4499.484	4500.746	22218.54	SM II	2.0	2.0		MIT
4499.105	4500.367	22220.41	SM	1.0	1.0		MIT
4495.123	4496.384	22240.09	SM II		1.0		MIT
4492.301	4493.561	22254.06	SM II	0.5			MIT
4486.296	4487.555	22283.85	SM	1.0	1.3		MIT
4485.570	4486.829	22287.46	SM	0.6	0.0		KN
4484.824	4486.082	22291.17	SM II	0.3	0.3		MIT
4484.516	4485.774	22292.70	SM	0.3			MIT
4480.795	4482.052	22311.21	SM	0.8	0.3		MIT
4480.307	4481.564	22313.64	SM	1.0	0.8		MIT
4478.658	4479.915	22321.85	SM	2.0	2.0		MIT
4477.773	4479.029	22326.26	SM	0.5	0.0		MIT
4477.493	4478.749	22327.66	SM	0.3	0.3		MIT
4475.165	4476.421	22339.28	SM II	1.0	1.0		MIT
4473.012	4474.267	22350.03	SM II	2.2	2.2		MIT
4472.416	4473.671	22353.01	SM II	2.0	2.0		MIT
4472.152	4473.407	22354.33	SM II	0.3	0.3		MIT
4470.886	4472.141	22360.66	SM	1.8	1.8		MIT
4470.480	4471.735	22362.69	SM	0.3	0.0		MIT
4469.650	4470.904	22366.84	SM	1.2	1.2		MIT
4467.968	4469.222	22375.26	SM	0.5			MIT
4467.341	4468.595	22378.40	SM II	2.3	2.3		MIT
4463.889	4465.142	22395.71	SM	0.5	0.5		MIT
4462.583	4463.835	22402.26	SM	0.7	0.3		MIT
4459.288	4460.540	22418.81	SM	0.9	0.9		MIT
4458.514	4459.765	22422.70	SM II	2.2	2.3		MIT
4457.805	4459.056	22426.27	SM II	0.5	0.5		MIT
4456.708	4457.959	22431.79	SM	0.3D			KN
4456.108	4457.359	22434.81	SM II	1.5	1.4		MIT
4455.232	4456.483	22439.22	SM	0.5			MIT
4454.629	4455.879	22442.26	SM II	2.0	2.0		MIT
4452.951	4454.201	22450.72	SM	0.9	0.8		MIT
4452.713	4453.963	22451.92	SM II	2.3	2.3		MIT
4451.76	4453.01	22456.7	SM	0.5H			MIT
4450.983	4452.232	22460.64	SM		0.5		MIT
4449.977	4451.226	22465.72	SM	1.2W	0.6		MIT
4446.915	4448.163	22481.19	SM II	1.0	0.6		MIT
4445.846	4447.094	22486.59	SM	0.6	0.3		MIT
4445.149	4446.397	22490.12	SM	1.6	1.3		MIT
4444.264	4445.512	22494.60	SM	2.0	2.0		MIT
4443.266	4444.513	22499.65	SM	0.5	0.0		MIT
4442.473	4443.720	22503.67	SM	1.1	1.1		MIT
4442.271	4443.518	22504.69	SM	1.6	1.0		MIT
4441.800	4443.047	22507.08	SM	1.5	1.3		MIT
4441.10	4442.35	22510.6	SM		0.3		MIT
4439.14	4440.39	22520.6	SM	0.7			MIT
4434.321	4435.566	22545.04	SM II	2.3	2.3		MIT
4433.884	4435.129	22547.26	SM II	2.3	2.3		MIT
4433.347	4434.592	22549.99	SM	0.6			MIT
4433.071	4434.316	22551.39	SM	1.0	0.3		MIT
4429.644	4430.888	22568.84	SM	1.4	1.2		MIT
4429.037	4430.281	22571.93	SM	0.3	0.3		MIT
4427.791	4429.034	22578.28	SM II	0.9	0.9		MIT
4427.581	4428.824	22579.36	SM II	1.4	1.4		MIT
4425.984	4427.227	22587.50	SM II	1.0	0.8		MIT
4424.99	4426.23	22592.6	SM	0.5	0.3		MIT

AIR WAVELENGTH (ANGSTRMS)	VACUUM WAVELENGTH (ANGSTRMS)	WAVENUMBER (KAYSERS)	LMENT	LOG ARC	INTENSITY SPK	DIS	REF
4424.342	4425.584	22595.89	SM II	2.5	2.5		MIT
4423.379	4424.621	22600.80	SM	1.0	0.3		MIT
4421.136	4422.378	22612.27	SM II	2.2	2.2		MIT
4420.529	4421.770	22615.38	SM II	2.3	2.3		MIT
4419.338	4420.579	22621.47	SM	1.6	1.3		MIT
4418.10	4419.34	22627.8	SM	0.6	0.3		MIT
4417.583	4418.824	22630.46	SM II	1.9	1.9		MIT
4416.00	4417.24	22638.6	SM	0.7	0.5		MIT
4411.827	4413.066	22659.98	SM II	1.2	1.0		MIT
4411.584	4412.823	22661.23	SM	1.4	0.7		MIT
4411.135	4412.374	22663.54	SM	1.0	0.7		KN
4409.337	4410.575	22672.78	SM II	2.0	2.0		MIT
4407.521	4408.759	22682.12	SM II	1.0	0.8		KN
4405.645	4406.882	22691.78	SM II	0.9	0.9		MIT
4403.364	4404.601	22703.53	SM II	1.7	1.7		MIT
4403.118	4404.355	22704.80	SM	1.4	1.0		MIT
4403.049	4404.286	22705.16	SM II	1.2	1.0		MIT
4401.166	4402.402	22714.87	SM	1.7	1.5		MIT
4400.19	4401.43	22719.9	SM	0.3	0.3		MIT
4399.865	4401.101	22721.59	SM II	1.3	1.2		MIT
4397.346	4398.581	22734.60	SM	1.3	0.7		MIT
4396.759	4397.994	22737.64	SM	0.7	0.5		MIT
4393.360	4394.594	22755.23	SM	1.3	0.8		MIT
4393.177	4394.411	22756.18	SM	0.0	0.3		MIT
4392.605	4393.839	22759.14	SM II	1.0H	0.8		MIT
4392.07	4393.30	22761.9	SM		0.3		MIT
4390.865	4392.099	22768.16	SM II	2.2	2.2		MIT
4388.987	4390.220	22777.90	SM	1.0	0.9		MIT
4386.220	4387.452	22792.27	SM	1.4	0.7		MIT
4384.294	4385.526	22802.28	SM II	1.7	1.7		MIT
4381.249	4382.480	22818.13	SM	0.3			MIT
4380.422	4381.653	22822.44	SM	1.4	0.8		MIT
4378.230	4379.460	22833.86	SM II	2.0	2.0		MIT
4376.972	4378.202	22840.43	SM	0.5	0.3		MIT
4376.046	4377.276	22845.26	SM II	0.6	0.0		MIT
4374.975	4376.204	22850.85	SM II	2.3	2.3		MIT
4373.728	4374.957	22857.37	SM II	0.3	0.3		MIT
4373.459	4374.688	22858.77	SM II	1.7	1.7		MIT
4372.419	4373.648	22864.21	SM II	0.5	0.0		MIT
4371.500	4372.728	22869.02	SM II	0.5	0.0		MIT
4370.474	4371.702	22874.39	SM II	0.9	0.8		MIT
4369.919	4371.147	22877.29	SM II	1.6	1.4		MIT
4368.032	4369.260	22887.17	SM II	1.8	1.8		MIT
4365.953	4367.180	22898.07	SM	0.8	0.0		MIT
4365.377	4366.604	22901.09	SM	0.7	0.6		MIT
4364.059	4365.285	22908.01	SM	0.6W	0.6W		MIT
4363.446	4364.672	22911.23	SM II	1.8	1.7		MIT
4362.911	4364.137	22914.04	SM	1.8	1.4		MIT
4362.409	4363.635	22916.67	SM II	0.3	0.3		MIT
4362.033	4363.259	22918.65	SM II	2.2	2.2		MIT
4361.067	4362.293	22923.73	SM II	1.5	1.3		MIT
4360.712	4361.938	22925.59	SM II	2.0	1.8		MIT
4357.891	4359.116	22940.43	SM	0.5	0.3H		MIT
4355.831	4357.055	22951.28	SM	0.5			MIT
4352.096	4353.319	22970.98	SM	2.1	2.0		MIT
4350.811	4352.034	22977.76	SM	0.5	0.0		MIT
4350.460	4351.683	22979.62	SM II	2.2	2.2		MIT
4347.801	4349.023	22993.67	SM II	2.2	1.8		MIT
4347.242	4348.464	22996.63	SM II	0.3	0.3		MIT
4346.479	4347.701	23000.66	SM II	1.5	1.2		MIT
4345.853	4347.075	23003.98	SM II	2.0	2.0		MIT
4342.384	4343.605	23022.35	SM	0.3	0.5H		MIT
4339.929	4341.149	23035.37	SM	0.9	0.5		MIT
4339.353	4340.573	23038.43	SM	0.3	0.3		MIT
4338.960	4340.180	23040.52	SM	1.0	0.3		MIT
4336.742	4337.961	23052.30	SM	1.0	0.8		MIT
4336.130	4337.349	23055.56	SM	2.0	1.5		MIT
4334.149	4335.368	23066.09	SM II	2.3	2.3		MIT
4331.443	4332.661	23080.50	SM	1.4	0.6		MIT
4330.013	4331.231	23088.13	SM	1.9	1.2		MIT

AIR WAVELENGTH (ANGSTRMS)	VACUUM WAVELENGTH (ANGSTRMS)	WAVENUMBER (KAYSERS)	LMENT	LOG ARC	INTENSITY SPK	DIS	REF
4329.016	4330.233	23093.44	SM II	2.5	2.5		MIT
4327.512	4328.729	23101.47	SM II	1.0	0.8		MIT
4326.127	4327.343	23108.87	SM	0.3			MIT
4325.148	4326.364	23114.10	SM	0.5	0.3H		MIT
4324.455	4325.671	23117.80	SM	1.3	0.5		MIT
4323.277	4324.493	23124.10	SM II	2.1	2.0		MIT
4321.868	4323.083	23131.64	SM	0.5	0.3		MIT
4319.526	4320.741	23144.18	SM	1.7	1.2		MIT
4318.935	4320.150	23147.35	SM II	2.5	2.5		MIT
4316.701	4317.915	23159.33	SM	0.5	0.5		MIT
4315.745	4316.959	23164.46	SM	0.5	0.5		MIT
4315.355	4316.569	23166.55	SM II	0.7	0.7		MIT
4313.871	4315.084	23174.52	SM	1.6	1.0		MIT
4313.715	4314.928	23175.36	SM II	1.2	1.0		MIT
4313.316	4314.529	23177.50	SM II	0.9	0.6		MIT
4312.846	4314.059	23180.03	SM	1.3	0.9		MIT
4309.004	4310.216	23200.69	SM II	2.3	2.2		MIT
4306.147	4307.358	23216.09	SM	0.3H	0.3H		MIT
4304.944	4306.155	23222.57	SM II	2.0	2.0		MIT
4301.277	4302.487	23242.37	SM	0.8	0.0		MIT
4299.338	4300.547	23252.85	SM II	1.2	1.0		MIT
4299.140	4300.349	23253.92	SM	0.3			MIT
4298.42	4299.63	23257.8	SM	0.5	0.5		MIT
4296.750	4297.959	23266.86	SM	2.0	1.7		MIT
4295.716	4296.924	23272.46	SM II	1.2	0.7		MIT
4295.446	4296.654	23273.92	SM	0.3	0.5		MIT
4293.735	4294.943	23283.20	SM	0.5	0.0		MIT
4292.182	4293.390	23291.62	SM II	2.0	1.8		MIT
4291.610	4292.817	23294.73	SM	0.9	0.9		MIT
4289.365	4290.572	23306.92	SM	1.2	0.3		MIT
4286.641	4287.847	23321.73	SM II	2.0	1.8		MIT
4285.479	4286.685	23328.05	SM II	2.3	2.3		MIT
4284.497	4285.703	23333.40	SM	1.1	0.9		MIT
4283.722	4284.927	23337.62	SM	0.7			KN
4283.496	4284.701	23338.85	SM	1.5	0.8		MIT
4282.827	4284.032	23342.50	SM	1.3	0.9		MIT
4282.200	4283.405	23345.91	SM	1.2	0.9		MIT
4281.002	4282.207	23352.45	SM	1.7	1.5		MIT
4280.779	4281.984	23353.66	SM II	2.3	2.3		MIT
4280.321	4281.525	23356.16	SM II	0.9	0.6		MIT
4279.946	4281.150	23358.21	SM II	1.7	1.6		MIT
4279.748	4280.952	23359.29	SM II	1.2	1.2		MIT
4279.666	4280.870	23359.74	SM II	2.0	2.0		MIT
4279.501	4280.705	23360.64	SM	0.6	0.3		MIT
4276.203	4277.406	23378.65	SM	0.6	0.6		MIT
4274.16	4275.36	23389.8	SM	0.7	0.0		MIT
4274.035	4275.238	23390.51	SM	0.5	0.0		MIT
4272.008	4273.210	23401.61	SM II	0.7	0.5		MIT
4271.858	4273.060	23402.43	SM	0.7	0.3		MIT
4270.834	4272.036	23408.04	SM II	1.2	1.2		MIT
4270.701	4271.903	23408.77	SM II	0.7	0.5		MIT
4269.770	4270.972	23413.88	SM II	0.8	0.5		MIT
4266.311	4267.512	23432.86	SM II	1.0	0.0		MIT
4265.237	4266.437	23438.76	SM	0.6			MIT
4265.074	4266.274	23439.66	SM II	1.8	1.4		MIT
4262.678	4263.878	23452.83	SM II	2.3	2.2		MIT
4259.393	4260.592	23470.92	SM II	0.5	0.3		MIT
4258.561	4259.760	23475.50	SM II	1.7	1.6		MIT
4256.396	4257.594	23487.44	SM II	2.2	2.2		MIT
4253.715	4254.912	23502.25	SM II	0.7	0.3		MIT
4251.788	4252.985	23512.90	SM II	2.3	2.3		MIT
4251.303	4252.500	23515.58	SM	0.3	0.0		MIT
4249.536	4250.732	23525.36	SM II	1.7	1.6		MIT
4247.395	4248.591	23537.22	SM II	1.2	1.2		MIT
4245.173	4246.368	23549.54	SM II	0.9	0.8		MIT
4244.696	4245.891	23552.18	SM II	2.0	1.9		MIT
4244.244	4245.439	23554.69	SM	0.5	0.0		MIT
4240.453	4241.647	23575.75	SM	0.7	0.0		MIT
4238.734	4239.928	23585.31	SM	0.7			KN
4237.657	4238.850	23591.30	SM II	1.8	1.7		MIT

AIR WAVELENGTH (ANGSTRMS)	VACUUM WAVELENGTH (ANGSTRMS)	WAVENUMBER (KAYSERS)	LMENT	LOG ARC	INTENSITY SPK	DIS	REF
4236.740	4237.933	23596.41	SM II	1.8	1.7		MIT
4236.576	4237.769	23597.32	SM	0.8	0.0		MIT
4235.872	4237.065	23601.24	SM	1.0	0.6		MIT
4234.574	4235.766	23608.48	SM II	1.8	1.6		MIT
4229.702	4230.893	23635.67	SM II	1.6	1.5		MIT
4226.175	4227.365	23655.40	SM	0.7	0.3		MIT
4225.318	4226.508	23660.19	SM II	1.6	1.5		MIT
4224.237	4225.427	23666.25	SM II	1.2	0.7		MIT
4223.711	4224.901	23669.20	SM II	1.7	1.3		MIT
4223.056	4224.245	23672.87	SM	0.5	0.3		MIT
4221.859	4223.048	23679.58	SM II	0.6	0.3		MIT
4220.653	4221.842	23686.34	SM II	2.0	2.0		MIT
4220.541	4221.730	23686.97	SM	0.5	0.0		MIT
4220.151	4221.340	23689.16	SM II	1.1	1.0		MIT
4219.302	4220.490	23693.93	SM	0.7	0.0		MIT
4218.634	4219.822	23697.68	SM	0.8	0.0		MIT
4213.924	4215.111	23724.17	SM	1.0	0.5		MIT
4213.031	4214.218	23729.20	SM II	0.3			MIT
4212.931	4214.118	23729.76	SM		0.3		MIT
4210.340	4211.526	23744.36	SM II	1.7	1.3		MIT
4206.619	4207.804	23765.36	SM II	1.0	1.0		MIT
4206.124	4207.309	23768.16	SM II	1.3	1.3		MIT
4205.773	4206.958	23770.15	SM	0.6	0.3		MIT
4205.361	4206.546	23772.47	SM	0.9	0.6		MIT
4204.813	4205.998	23775.57	SM II	0.9	1.0		MIT
4203.047	4204.231	23785.56	SM II	1.2	1.3		MIT
4202.912	4204.096	23786.33	SM II	1.2	0.9		MIT
4199.454	4200.637	23805.91	SM II	0.9	1.1		MIT
4197.869	4199.052	23814.90	SM II	0.3	0.3		MIT
4192.152	4193.333	23847.38	SM II	0.7	1.0		MIT
4191.923	4193.104	23848.68	SM	0.9	1.2		MIT
4188.121	4189.301	23870.33	SM II	1.0	1.4		MIT
4183.757	4184.936	23895.23	SM II	1.0	1.2		MIT
4183.335	4184.514	23897.64	SM	0.3	0.3		MIT
4181.099	4182.277	23910.42	SM II	0.9	1.4		MIT
4180.385	4181.563	23914.50	SM	1.0	0.3		MIT
4178.006	4179.184	23928.12	SM II	0.5	1.0		MIT
4174.432	4175.609	23948.60	SM II	0.7	1.0		MIT
4171.554	4172.730	23965.13	SM II	0.8	1.1		MIT
4169.480	4170.655	23977.05	SM II	1.2	1.4		MIT
4167.229	4168.404	23990.00	SM		0.5		MIT
4166.333	4167.508	23995.16	SM	1.2	0.9		MIT
4165.532	4166.706	23999.77	SM	0.3	0.5		MIT
4163.722	4164.896	24010.20	SM II	0.3	0.7		MIT
4163.147	4164.321	24013.52	SM	0.7	1.0		MIT
4159.513	4160.686	24034.50	SM II	0.3	0.6		MIT
4159.393	4160.566	24035.19	SM II	0.6	0.6		MIT
4158.425	4159.597	24040.79	SM	0.9	0.3		MIT
4156.240	4157.412	24053.43	SM II	1.0W	0.7		MIT
4155.221	4156.393	24059.32	SM II	0.9	1.0		MIT
4153.326	4154.497	24070.30	SM II	0.7	1.0		MIT
4152.211	4153.382	24076.77	SM II	0.9	1.2		MIT
4152.067	4153.238	24077.60	SM	0.5	0.5		MIT
4151.209	4152.380	24082.58	SM	0.7D			MIT
4149.834	4151.004	24090.56	SM II	1.3	1.2		MIT
4149.203	4150.373	24094.22	SM	0.6	0.0H		MIT
4147.976	4149.146	24101.35	SM	0.9H			MIT
4147.713	4148.883	24102.87	SM II	0.5	1.0		MIT
4146.743	4147.912	24108.51	SM II	0.3			MIT
4145.243	4146.412	24117.24	SM	0.6	0.0		MIT
4144.320	4145.489	24122.61	SM		0.5		MIT
4142.958	4144.126	24130.54	SM	1.2			MIT
4142.813	4143.981	24131.38	SM	1.0	0.3		KN
4139.646	4140.814	24149.84	SM	0.7D	0.3		MIT
4138.969	4140.136	24153.79	SM II	0.9	0.0		KN
4138.731	4139.898	24155.18	SM	0.9			MIT
4135.894	4137.061	24171.75	SM	0.8	0.5		MIT
4135.496	4136.662	24174.08	SM	1.2	0.3		MIT
4135.135	4136.301	24176.19	SM II	0.7	1.0		MIT
4133.171	4134.337	24187.68	SM II	0.6	0.6		MIT
4129.989	4131.154	24206.31	SM	1.0			MIT
4129.225	4130.390	24210.79	SM II	0.7	1.1		MIT
4128.116	4129.280	24217.29	SM	0.7	0.3		MIT
4126.113	4127.277	24229.05	SM	1.0			MIT
4125.868	4127.032	24230.49	SM	1.2	1.0		MIT
4125.239	4126.403	24234.18	SM	0.7	0.0		MIT
4123.951	4125.114	24241.75	SM II	1.0	1.3		MIT
4123.006	4124.169	24247.31	SM	0.7	0.3		MIT
4122.499	4123.662	24250.29	SM II	0.7	1.0		MIT
4122.017	4123.180	24253.13	SM	0.3	0.6		MIT
4121.543	4122.706	24255.92	SM II	1.0H	0.6		MIT
4121.351	4122.514	24257.04	SM II	1.2	1.2		MIT
4119.569	4120.731	24267.54	SM II	0.6	0.7		MIT
4118.546	4119.708	24273.56	SM II	1.7	1.8		MIT
4116.456	4117.617	24285.89	SM II	0.9	1.0		MIT
4113.896	4115.057	24301.00	SM II	0.7	1.0		MIT
4112.252	4113.412	24310.72	SM	1.0	0.3		MIT
4110.185	4111.345	24322.94	SM II	0.7	0.7		MIT
4109.407	4110.567	24327.55	SM II	1.0	1.0		MIT
4108.323	4109.482	24333.97	SM II	0.7	0.6		MIT
4107.804	4108.963	24337.04	SM II	1.0D	0.3		MIT
4107.394	4108.553	24339.47	SM II	0.7	0.7		MIT
4107.264	4108.423	24340.24	SM II	1.0	0.7		MIT
4106.596	4107.755	24344.20	SM II	2.0	0.7		MIT
4106.276	4107.435	24346.10	SM	0.7			MIT
4104.121	4105.279	24358.88	SM II	0.7	1.0		MIT
4103.65	4104.81	24361.7	SM	0.5	0.3		MIT
4101.315	4102.472	24375.54	SM	1.2D	0.3H		MIT
4099.954	4101.111	24383.64	SM	0.6			MIT
4098.953	4100.110	24389.59	SM	0.9			MIT
4094.042	4095.198	24418.85	SM II	1.0	1.0		MIT
4093.042	4094.197	24424.81	SM II	0.5	0.5		MIT
4092.257	4093.412	24429.50	SM II	1.0	1.4		MIT
4088.746	4089.900	24450.48	SM	0.7	0.0H		MIT
4084.403	4085.556	24476.47	SM II	1.9			KN
4083.225	4084.378	24483.53	SM II	0.7	0.7		MIT
4082.589	4083.742	24487.35	SM II	1.0	1.4		MIT
4081.963	4083.115	24491.10	SM II	0.8	0.7		MIT
4080.547	4081.699	24499.60	SM II	1.2	1.0		MIT
4079.829	4080.981	24503.91	SM	1.3	0.0		MIT
4076.854	4078.005	24521.79	SM II	1.0	0.7		MIT
4076.632	4077.783	24523.13	SM II	1.4	1.2		M
4075.835	4076.986	24527.93	SM II	1.6	1.6		MIT
4070.980	4072.130	24557.18	SM II	0.7	0.5		MIT
4068.320	4069.469	24573.23	SM II	1.6	1.6		MIT
4067.384	4068.533	24578.89	SM II	0.3	0.6		MIT
4066.728	4067.876	24582.85	SM II	1.3	1.3		MIT
4066.182	4067.330	24586.15	SM II	1.0	0.7		MIT
4065.008	4066.156	24593.25	SM II	0.8	0.5		MIT
4064.569	4065.717	24595.91	SM II	1.2	1.2		MIT
4064.315	4065.463	24597.45	SM	1.2	0.8		MIT
4063.528	4064.676	24602.21	SM II	1.0	0.6		MIT
4062.319	4063.466	24609.53	SM	0.7	0.0		MIT
4062.052	4063.199	24611.15	SM	0.8			MIT
4061.049	4062.196	24617.23	SM II	0.9	0.9		MIT
4058.867	4060.013	24630.46	SM II	1.5	1.3		MIT
4057.653	4058.799	24637.83	SM	1.0	0.5		MIT
4055.282	4056.427	24652.24	SM	0.7	0.5		MIT
4054.512	4055.657	24656.92	SM		0.3		MIT
4052.290	4053.435	24670.44	SM II	1.0	0.3		MIT
4051.817	4052.962	24673.32	SM	0.3	0.0		MIT
4049.832	4050.976	24685.41	SM	1.6	1.3		MIT
4049.573	4050.717	24686.99	SM II	0.5	0.0		MIT
4048.616	4049.760	24692.82	SM II	1.0	1.0		MIT
4047.363	4048.506	24700.47	SM II	0.9	0.8		MIT
4046.154	4047.297	24707.85	SM II	1.0	1.0		MIT
4045.047	4046.190	24714.61	SM II	1.0	0.8		MIT
4044.948	4046.091	24715.22	SM	1.0	1.0		MIT
4044.113	4045.255	24720.32	SM	0.6			MIT
4043.365	4044.507	24724.89	SM II	0.7	0.5		MIT

AIR WAVELENGTH (ANGSTRMS)	VACUUM WAVELENGTH (ANGSTRMS)	WAVENUMBER (KAYSERS)	LMENT	LOG ARC	INTENSITY SPK	DIS	REF
4042.896	4044.038	24727.76	SM II	1.0	1.0		MIT
4042.711	4043.853	24728.89	SM II	1.0	1.0		MIT
4041.667	4042.809	24735.28	SM II	1.4	1.0		MIT
4039.50	4040.64	24748.6	SM	1.0	0.3		MIT
4039.105	4040.246	24750.97	SM	0.3	0.3		MIT
4038.083	4039.224	24757.23	SM II	0.8	0.6		MIT
4037.092	4038.233	24763.31	SM II	1.0	0.6		MIT
4035.101	4036.241	24775.53	SM II	1.7	0.5		MIT
4032.977	4034.117	24788.57	SM	1.3	0.9		KN
4030.425	4031.564	24804.27	SM II	1.0	0.5		MIT
4023.223	4024.360	24848.67	SM II	1.5	1.4		MIT
4022.714	4023.851	24851.81	SM II	0.8	0.7		MIT
4021.408	4022.545	24859.89	SM II	0.6	0.0		MIT
4020.771	4021.907	24863.82	SM II	0.5	0.3		MIT
4019.976	4021.112	24868.74	SM II	1.5	1.2		MIT
4019.827	4020.963	24869.66	SM II	0.9	0.9		MIT
4018.539	4019.675	24877.63	SM II	0.6	0.7		MIT
4016.118	4017.253	24892.63	SM		0.3		MIT
4016.004	4017.139	24893.34	SM	1.2			KN
4015.765	4016.900	24894.82	SM II	0.9	0.6		MIT
4011.719	4012.853	24919.93	SM II	0.9	0.6		MIT
4008.330	4009.463	24941.00	SM	0.9	1.0H		MIT
4008.091	4009.224	24942.48	SM II	1.0	0.5		MIT
4007.486	4008.619	24946.25	SM II	1.7	1.4		MIT
4006.98	4008.11	24949.4	SM	0.5			MIT
4006.834	4007.967	24950.31	SM II	0.9	0.7		MIT
4006.585	4007.718	24951.86	SM II	0.9	0.9		MIT
4004.245	4005.377	24966.44	SM II	0.9	0.5		MIT
4003.693	4004.825	24969.88	SM	0.8	0.5		MIT
4003.454	4004.586	24971.37	SM II	1.5	1.3		MIT
3998.353	3999.484	25003.23	SM	1.0	0.3		MIT
3996.909	3998.039	25012.26	SM	0.3	0.3		MIT
3995.590	3996.720	25020.52	SM	0.9	0.5		MIT
3993.302	3994.431	25034.85	SM II	1.4	1.4		MIT
3991.017	3992.146	25049.19	SM	1.0	0.3		MIT
3990.003	3991.131	25055.55	SM II	1.6	1.4		MIT
3987.424	3988.552	25071.76	SM II	1.2	0.9		MIT
3987.070	3988.198	25073.98	SM II	0.3	0.3		MIT
3986.905	3988.033	25075.02	SM	0.9	0.9		MIT
3986.234	3987.361	25079.24	SM	0.5	0.5		MIT
3985.987	3987.114	25080.80	SM II	0.7	0.8		MIT
3984.447	3985.574	25090.49	SM II	0.6	0.6		MIT
3983.137	3984.264	25098.74	SM II	2.0	1.8		MIT
3981.097	3982.223	25111.60	SM	0.6	0.6		MIT
3980.883	3982.009	25112.95	SM	0.9	0.7		MIT
3979.195	3980.321	25123.60	SM II	1.7	1.7		MIT
3978.240	3979.365	25129.64	SM	0.8			MIT
3976.862	3977.987	25138.34	SM	0.7	0.3		MIT
3976.426	3977.551	25141.10	SM II	1.4	1.0		MIT
3976.266	3977.391	25142.11	SM II	1.6	1.2		MIT
3975.220	3976.344	25148.73	SM II	1.0	0.7		MIT
3974.660	3975.784	25152.27	SM	1.2	0.6		MIT
3973.107	3974.231	25162.10	SM II	0.7	0.3		MIT
3971.394	3972.517	25172.95	SM II	1.7	1.5		MIT
3970.527	3971.650	25178.45	SM II	1.2	1.0		MIT
3967.779	3968.902	25195.89	SM	0.8	0.3		MIT
3967.675	3968.798	25196.55	SM II	0.9	0.8		MIT
3966.332	3967.454	25205.08	SM II	0.7	0.6		MIT
3966.051	3967.173	25206.87	SM II	1.8	1.7		MIT
3964.884	3966.006	25214.28	SM	0.5			MIT
3964.356	3965.478	25217.64	SM	0.9	0.0		MIT
3964.074	3965.196	25219.44	SM II	1.0	0.3		MIT
3962.996	3964.117	25226.30	SM	1.7	1.6		MIT
3962.244	3963.365	25231.08	SM II	1.0	0.7		MIT
3962.136	3963.257	25231.77	SM	0.7	0.5		MIT
3961.808	3962.929	25233.86	SM	1.4	1.0		MIT
3960.511	3961.632	25242.12	SM	0.7	0.5		MIT
3960.121	3961.242	25244.61	SM	0.7	0.5		MIT
3959.530	3960.650	25248.38	SM	1.7	1.6		MIT
3958.709	3959.829	25253.61	SM II	1.0	0.7		MIT
3957.475	3958.595	25261.49	SM	1.2	0.5		MIT
3954.972	3956.091	25277.48	SM	0.7	0.5		MIT
3954.195	3955.314	25282.44	SM II	1.2	0.7		MIT
3953.378	3954.497	25287.67	SM	0.3	0.3		MIT
3951.885	3953.003	25297.22	SM	1.0	0.5		MIT
3949.845	3950.963	25310.29	SM	0.3	0.3		MIT
3948.108	3949.225	25321.42	SM II	1.7	1.7		MIT
3947.832	3948.949	25323.19	SM II	1.2	0.9		MIT
3946.506	3947.623	25331.70	SM II	1.6	1.5		MIT
3944.744	3945.861	25343.01	SM	0.7	0.7		MIT
3943.621	3944.737	25350.23	SM	0.9	0.5		MIT
3943.235	3944.351	25352.71	SM II	1.6	1.4		MIT
3942.659	3943.775	25356.42	SM	1.0	0.3		MIT
3941.868	3942.984	25361.50	SM II	1.7	1.6		MIT
3939.643	3940.758	25375.83	SM II	0.5	0.3		MIT
3938.424	3939.539	25383.68	SM II	0.9	0.3		MIT
3937.056	3938.171	25392.50	SM	1.2	0.8		MIT
3935.760	3936.874	25400.86	SM II	1.5	1.0		MIT
3935.461	3936.575	25402.79	SM	0.3H			MIT
3935.188	3936.302	25404.55	SM II	0.6	0.5		MIT
3934.818	3935.932	25406.94	SM	0.6	0.9		MIT
3933.592	3934.706	25414.86	SM II	2.3	2.3H		MIT
3932.966	3934.079	25418.91	SM II	0.8	0.5		MIT
3931.156	3932.269	25430.61	SM II	0.6	0.3		MIT
3929.554	3930.667	25440.98	SM	0.3			MIT
3928.914	3930.026	25445.12	SM	1.4	0.5		MIT
3928.272	3929.384	25449.28	SM II	1.8	1.8		MIT
3926.324	3927.436	25461.91	SM	1.3	0.3		MIT
3925.200	3926.311	25469.20	SM	1.2	0.5		MIT
3924.673	3925.784	25472.62	SM	0.3			MIT
3923.681	3924.792	25479.06	SM	1.0	0.3		MIT
3922.695	3923.806	25485.46	SM II	0.9	0.6		MIT
3922.385	3923.496	25487.48	SM II	1.8	1.8		MIT
3922.039	3923.150	25489.72	SM II	1.6	0.8		MIT
3920.095	3921.205	25502.36	SM	0.5	0.3		MIT
3918.612	3919.722	25512.01	SM II	1.4	0.6		MIT
3917.967	3919.077	25516.22	SM	0.6			MIT
3917.438	3918.547	25519.66	SM II	1.3	1.2		MIT
3916.359	3917.468	25526.69	SM	0.8H	0.7H		MIT
3914.389	3915.498	25539.54	SM	0.8			MIT
3914.073	3915.182	25541.60	SM	0.5	0.5		MIT
3913.638	3914.746	25544.44	SM	1.0	0.5		MIT
3913.365	3914.473	25546.22	SM	0.7	0.9		MIT
3912.970	3914.078	25548.80	SM II	1.3	0.7		MIT
3910.919	3912.027	25562.20	SM II	0.0	0.3		MIT
3910.301	3911.409	25566.24	SM	0.6	0.0		MIT
3910.086	3911.194	25567.64	SM II	0.7	0.7		MIT
3909.944	3911.051	25568.57	SM	0.5			MIT
3909.273	3910.380	25572.96	SM	0.7	0.7		MIT
3908.258	3909.365	25579.60	SM	0.8			MIT
3907.124	3908.231	25587.03	SM	1.0	1.2		MIT
3906.805	3907.912	25589.11	SM II	0.8	0.7		MIT
3904.185	3905.291	25606.29	SM	0.5	0.0		MIT
3903.412	3904.518	25611.36	SM	1.8	1.8		MIT
3902.324	3903.429	25618.50	SM II	1.0			MIT
3901.056	3902.161	25626.82	SM	0.5	0.3		MIT
3900.886	3901.991	25627.94	SM II	1.0	1.0		MIT
3898.508	3899.612	25643.57	SM		0.6		MIT
3897.252	3898.356	25651.84	SM II	0.9	0.7		MIT
3896.971	3898.075	25653.69	SM II	1.7	1.7		MIT
3895.423	3896.527	25663.88	SM	1.0	0.3		MIT
3895.093	3896.197	25666.06	SM	0.5	0.5		MIT
3894.281	3895.384	25671.41	SM	0.9	0.3H		MIT
3894.052	3895.155	25672.92	SM II	1.2	1.2		MIT
3893.384	3894.487	25677.32	SM	0.5	0.0		MIT
3892.678	3893.781	25681.98	SM		0.5H		MIT
3891.938	3893.041	25686.86	SM	0.6			MIT
3891.179	3892.282	25691.87	SM II	1.7	0.9		MIT
3890.074	3891.176	25699.17	SM II	1.0	1.0		MIT
3889.216	3890.318	25704.84	SM	0.7	0.8		MIT

AIR WAVELENGTH (ANGSTRMS)	VACUUM WAVELENGTH (ANGSTRMS)	WAVENUMBER (KAYSERS)	LMENT	LOG ARC	INTENSITY SPK	DIS	REF
3889.151	3890.253	25705.27	SM II	0.6	0.6		MIT
3887.104	3888.206	25718.80	SM	0.9	0.9		MIT
3886.038	3887.139	25725.86	SM	0.7			MIT
3885.902	3887.003	25726.76	SM II	0.7	0.7		MIT
3885.284	3886.385	25730.85	SM II	1.7	1.7		MIT
3884.363	3885.464	25736.95	SM	0.7	0.3		MIT
3883.987	3885.088	25739.45	SM	0.3	0.3		MIT
3883.804	3884.905	25740.66	SM II	0.3	0.5		MIT
3882.857	3883.957	25746.94	SM II	0.7	0.7		MIT
3882.503	3883.603	25749.28	SM	1.4	1.0		MIT
3881.787	3882.887	25754.03	SM	1.3			KN
3881.387	3882.487	25756.69	SM II	1.0	1.0		MIT
3880.989	3882.089	25759.33	SM	1.0	0.9		MIT
3880.755	3881.855	25760.88	SM II	1.6	1.5		MIT
3877.473	3878.572	25782.69	SM	1.7	0.9		MIT
3877.180	3878.279	25784.63	SM	1.5	0.7		MIT
3875.541	3876.640	25795.54	SM II	0.7	1.0		MIT
3875.174	3876.272	25797.98	SM	1.7	1.0		MIT
3874.366	3875.464	25803.36	SM	0.7H	0.6		MIT
3873.718	3874.816	25807.68	SM	1.1	0.5		MIT
3873.471	3874.569	25809.32	SM	1.0	0.5		MIT
3873.200	3874.298	25811.13	SM II	1.3	0.5		MIT
3872.223	3873.321	25817.64	SM	1.0	0.5		MIT
3871.782	3872.880	25820.58	SM II	0.8	1.4		MIT
3870.885	3871.982	25826.56	SM II	0.7	0.6		MIT
3867.623	3868.719	25848.35	SM	1.0	0.9		MIT
3865.688	3866.784	25861.28	SM II	0.7	0.7		MIT
3865.240	3866.336	25864.28	SM II	1.0			MIT
3864.051	3865.147	25872.24	SM	1.0	0.3		MIT
3863.407	3864.502	25876.55	SM	1.2D	0.9		MIT
3862.477	3863.572	25882.78	SM	1.0	0.5		MIT
3862.236	3863.331	25884.40	SM II	0.5	1.0		MIT
3862.051	3863.146	25885.64	SM II	0.6	1.0		MIT
3861.788	3862.883	25887.40	SM	1.0	0.8		MIT
3861.592	3862.687	25888.72	SM II	1.2	0.9		MIT
3861.183	3862.278	25891.46	SM II	1.2	0.8		MIT
3861.059	3862.154	25892.29	SM II	1.2D	0.6		MIT
3860.619	3861.714	25895.24	SM	1.4	1.0		MIT
3860.280	3861.375	25897.51	SM II	0.6	0.7		MIT
3860.144	3861.238	25898.43	SM	0.7	0.0		MIT
3858.741	3859.835	25907.84	SM	1.3			MIT
3858.514	3859.608	25909.37	SM	1.0D	1.0H		MIT
3857.904	3858.998	25913.46	SM II	0.9	1.0		MIT
3857.332	3858.426	25917.31	SM	0.7	0.0		MIT
3855.894	3856.987	25926.97	SM II	0.7	1.2		MIT
3854.941	3856.034	25933.38	SM II	0.7	0.7		MIT
3854.560	3855.653	25935.94	SM	0.7	0.7		MIT
3854.198	3855.291	25938.38	SM	2.3J	1.4		MIT
3853.290	3854.383	25944.49	SM		0.7		MIT
3851.882	3852.974	25953.98	SM II	1.0	0.9		MIT
3849.752	3850.844	25968.33	SM	0.9	0.5		MIT
3849.012	3850.104	25973.33	SM	1.0	0.6		MIT
3848.806	3849.898	25974.72	SM	2.2D	1.0		MIT
3847.521	3848.612	25983.39	SM II	0.8	1.0		MIT
3846.986	3848.077	25987.00	SM	1.0	0.5		MIT
3846.761	3847.852	25988.52	SM	0.7			KN
3846.282	3847.373	25991.76	SM	0.9	0.5		MIT
3845.997	3847.088	25993.69	SM	1.0	0.6		MIT
3844.504	3845.594	26003.78	SM II	0.3	0.3		MIT
3843.788	3844.878	26008.63	SM II	1.0D	1.0		MIT
3843.512	3844.602	26010.49	SM II	0.5	1.0		MIT
3842.348	3843.438	26018.37	SM II	0.7	0.7		MIT
3840.596	3841.685	26030.24	SM	0.5	0.5		MIT
3840.441	3841.530	26031.29	SM II	1.2	1.0		MIT
3839.190	3840.279	26039.77	SM	1.4	0.5		MIT
3838.931	3840.020	26041.53	SM II	1.0	1.2		MIT
3836.515	3837.603	26057.93	SM	1.7	1.4		MIT
3836.121	3837.209	26060.61	SM	1.3	0.6		MIT
3835.727	3836.815	26063.28	SM II	1.2	1.2		MIT
3834.945	3836.033	26068.60	SM	0.5			KN

AIR WAVELENGTH (ANGSTRMS)	VACUUM WAVELENGTH (ANGSTRMS)	WAVENUMBER (KAYSERS)	LMENT	LOG ARC	INTENSITY SPK	DIS	REF
3834.601	3835.689	26070.94	SM	1.4	1.2		MIT
3834.473	3835.561	26071.81	SM	0.7	0.3		MIT
3833.828	3834.916	26076.19	SM II	1.2	1.0		MIT
3833.047	3834.134	26081.51	SM	1.2	0.5		MIT
3832.802	3833.889	26083.17	SM	0.8	0.0		MIT
3831.507	3832.594	26091.99	SM II	0.8	1.2		MIT
3830.300	3831.387	26100.21	SM II	1.7	1.0		MIT
3829.406	3830.493	26106.30	SM	1.2			MIT
3829.156	3830.242	26108.01	SM	0.7	0.8		MIT
3828.409	3829.495	26113.10	SM	0.9	0.3		MIT
3828.055	3829.141	26115.52	SM		0.9		MIT
3827.356	3828.442	26120.29	SM	1.0	0.9		MIT
3826.565	3827.651	26125.69	SM	0.3	0.5		MIT
3826.210	3827.296	26128.11	SM II	1.0	1.2		MIT
3824.996	3826.081	26136.40	SM II	0.3	0.9		MIT
3824.803	3825.888	26137.72	SM	1.0	0.0		MIT
3824.533	3825.618	26139.57	SM II	0.7	0.9		MIT
3824.169	3825.254	26142.05	SM II	0.7	0.9		MIT
3823.795	3824.880	26144.61	SM	1.2	1.0		MIT
3823.510	3824.595	26146.56	SM	1.2	1.2		MIT
3822.965	3824.050	26150.29	SM	1.4			MIT
3822.317	3823.402	26154.72	SM	1.2	0.6		MIT
3821.827	3822.912	26158.07	SM II	1.4	0.9		MIT
3821.726	3822.811	26158.76	SM	1.4	0.7		MIT
3820.817	3821.901	26164.99	SM	0.9	0.8		MIT
3819.678	3820.762	26172.79	SM	1.0	1.3		MIT
3818.353	3819.437	26181.87	SM	0.3	0.5		MIT
3816.847	3817.930	26192.20	SM	0.3			MIT
3814.631	3815.714	26207.42	SM II		0.7		MIT
3813.822	3814.904	26212.98	SM	0.7			MIT
3813.632	3814.714	26214.28	SM	1.0	0.6		MIT
3812.834	3813.916	26219.77	SM	0.9			MIT
3812.068	3813.150	26225.04	SM II	1.3	1.3		MIT
3810.976	3812.058	26232.55	SM	0.9	0.0H		MIT
3810.433	3811.515	26236.29	SM	0.7	1.0		MIT
3809.953	3811.034	26239.59	SM	0.6			MIT
3809.882	3810.963	26240.08	SM II	1.0	0.9		MIT
3809.743	3810.824	26241.04	SM II	1.0	1.0		MIT
3808.463	3809.544	26249.86	SM II	0.8	0.8		MIT
3807.923	3809.004	26253.58	SM II	1.2	0.3		MIT
3806.770	3807.851	26261.53	SM II	1.0	0.9		MIT
3806.460	3807.541	26263.67	SM	1.0	0.5		MIT
3806.031	3807.111	26266.63	SM	0.3	0.3		MIT
3805.626	3806.706	26269.43	SM II	0.7	0.8		MIT
3805.187	3806.267	26272.46	SM II	0.8	0.0		MIT
3803.940	3805.020	26281.07	SM	1.2	0.3		MIT
3803.090	3804.170	26286.95	SM	1.0	0.5		MIT
3802.417	3803.496	26291.60	SM	1.0			MIT
3800.890	3801.969	26302.16	SM II	1.3	1.4		MIT
3800.783	3801.862	26302.90	SM	0.7	0.3		MIT
3800.368	3801.447	26305.77	SM II	1.3			MIT
3800.042	3801.121	26308.03	SM	0.9	0.7O		MIT
3799.546	3800.625	26311.46	SM II	1.5	1.0		MIT
3797.727	3798.805	26324.07	SM	1.4	0.7		MIT
3797.278	3798.356	26327.18	SM II	1.2			MIT
3796.15	3797.23	26335.0	SM	0.9D	0.3D		MIT
3794.787	3795.865	26344.46	SM		0.3		MIT
3794.347	3795.424	26347.51	SM	1.5W	0.3H		MIT
3793.969	3795.046	26350.14	SM II	1.0	1.5		MIT
3793.332	3794.409	26354.56	SM	1.3	0.0		MIT
3793.160	3794.237	26355.76	SM	0.5			KN
3792.60	3793.68	26359.6	SM	1.4W	0.5H		MIT
3792.023	3793.100	26363.66	SM II	1.4	0.8		MIT
3791.275	3792.352	26368.86	SM	1.0	0.8		MIT
3790.779	3791.855	26372.31	SM	0.7	0.3		MIT
3789.913	3790.989	26378.34	SM	0.5	0.0		MIT
3788.130	3789.206	26390.75	SM II	1.4	1.0		MIT
3787.356	3788.432	26396.15	SM	1.0			MIT
3787.201	3788.277	26397.23	SM II	2.0	1.5		MIT
3785.350	3786.425	26410.14	SM	1.5	0.9		MIT

AIR WAVELENGTH (ANGSTRMS)	VACUUM WAVELENGTH (ANGSTRMS)	WAVENUMBER (KAYSERS)	LMENT	LOG ARC	INTENSITY SPK	DIS	REF
3783.806	3784.881	26420.91	SM	1.7D	1.0D		MIT
3783.46	3784.53	26423.3	SM	0.9D	0.7D		MIT
3783.354	3784.429	26424.07	SM II	1.0	1.2		MIT
3783.062	3784.136	26426.11	SM II	1.4	0.9		MIT
3782.679	3783.753	26428.78	SM	1.2			MIT
3782.423	3783.497	26430.57	SM II	1.2			MIT
3782.149	3783.223	26432.49	SM	1.0	0.5		KN
3781.64	3782.71	26436.1	SM	0.7D	0.5D		MIT
3781.330	3782.404	26438.21	SM	0.3	0.3		KN
3780.922	3781.996	26441.06	SM II	1.2	1.2		MIT
3780.762	3781.836	26442.18	SM	1.3	1.3		MIT
3780.504	3781.578	26443.99	SM	1.0	0.5		MIT
3780.145	3781.219	26446.50	SM II	1.2			MIT
3779.563	3780.637	26450.57	SM II	1.2	0.6		MIT
3778.135	3779.208	26460.57	SM	1.6	2.0		MIT
3777.840	3778.913	26462.64	SM	1.0	0.6		MIT
3777.088	3778.161	26467.90	SM	1.5	1.0		MIT
3776.06	3777.13	26475.1	SM	1.2			MIT
3775.849	3776.922	26476.59	SM	0.3			MIT
3775.459	3776.531	26479.32	SM	1.0	0.5		KN
3774.678	3775.750	26484.80	SM II	1.2O			MIT
3774.305	3775.377	26487.42	SM II	1.0			MIT
3774.127	3775.199	26488.67	SM	1.3	0.7H		KN
3773.420	3774.492	26493.63	SM II	1.2	0.6		MIT
3773.342	3774.414	26494.18	SM	1.2	0.0		MIT
3772.645	3773.717	26499.07	SM	1.3	1.0		MIT
3772.138	3773.210	26502.64	SM	1.4D	0.7D		MIT
3771.350	3772.421	26508.17	SM	1.7D	0.3		MIT
3770.936	3772.007	26511.08	SM	0.5			KN
3770.729	3771.800	26512.54	SM II	1.4	1.7		MIT
3770.239	3771.310	26515.98	SM	0.9	0.0		MIT
3768.810	3769.881	26526.04	SM	0.5			MIT
3768.307	3769.378	26529.58	SM	1.0			MIT
3768.254	3769.325	26529.95	SM	1.0			MIT
3767.931	3769.002	26532.23	SM	0.5	0.3		MIT
3767.758	3768.828	26533.44	SM	1.2	1.0		MIT
3767.362	3768.432	26536.23	SM II	1.3	1.0		MIT
3766.923	3767.993	26539.33	SM	0.7			KN
3765.752	3766.822	26547.58	SM	0.7			KN
3765.615	3766.685	26548.54	SM	0.7			KN
3765.427	3766.497	26549.87	SM	0.9	0.9		MIT
3764.377	3765.447	26557.27	SM II	1.6	1.4		MIT
3763.161	3764.230	26565.86	SM	1.2			KN
3761.144	3762.213	26580.10	SM	0.8	0.3H		KN
3760.697	3761.766	26583.26	SM II	1.4	1.6		MIT
3760.175	3761.244	26586.95	SM	0.7			MIT
3760.049	3761.117	26587.84	SM II	1.4	1.0		MIT
3758.974	3760.042	26595.45	SM	1.4	1.5		MIT
3758.454	3759.522	26599.13	SM II	1.0	0.9		MIT
3757.74	3758.81	26604.2	SM	0.5	0.5		MIT
3757.530	3758.598	26605.67	SM	1.2	1.2		MIT
3756.539	3757.607	26612.69	SM	1.5	0.6		MIT
3756.409	3757.477	26613.61	SM II	1.4	1.0		MIT
3756.068	3757.135	26616.02	SM	0.9	0.0		MIT
3755.476	3756.543	26620.22	SM	1.0	0.5		MIT
3755.278	3756.345	26621.62	SM II	1.3	1.3		MIT
3754.863	3755.930	26624.56	SM II	1.2	1.0		MIT
3754.675	3755.742	26625.90	SM II	1.0	0.3		KN
3754.294	3755.361	26628.60	SM	1.0			MIT
3754.22	3755.29	26629.1	SM	1.0	0.3		MIT
3753.614	3754.681	26633.42	SM	1.2			MIT
3753.196	3754.263	26636.39	SM	1.0			MIT
3753.096	3754.163	26637.10	SM	1.2			MIT
3753.024	3754.091	26637.61	SM	0.0	0.3		MIT
3752.65	3753.72	26640.3	SM		0.3H		MIT
3752.256	3753.322	26643.06	SM	1.4	0.0H		MIT
3751.744	3752.810	26646.70	SM	1.4	0.3		MIT
3751.562	3752.628	26647.99	SM II	1.4	0.8		MIT
3751.121	3752.187	26651.12	SM	1.2	0.6		MIT
3750.670	3751.736	26654.33	SM	1.0	0.9		MIT

AIR WAVELENGTH (ANGSTRMS)	VACUUM WAVELENGTH (ANGSTRMS)	WAVENUMBER (KAYSERS)	LMENT	LOG ARC	INTENSITY SPK	DIS	REF
3749.795	3750.861	26660.55	SM		0.5H		MIT
3748.629	3749.695	26668.84	SM II	1.0	0.7		MIT
3748.515	3749.580	26669.65	SM	0.7			MIT
3748.153	3749.218	26672.23	SM	0.7	0.6		MIT
3747.753	3748.818	26675.07	SM	0.5			MIT
3747.623	3748.688	26676.00	SM II	1.4	1.4		MIT
3747.353	3748.418	26677.92	SM	0.9	0.0		MIT
3746.05	3747.11	26687.2	SM		0.3H		MIT
3745.616	3746.681	26690.29	SM II	1.6	1.5		MIT
3745.461	3746.526	26691.40	SM	1.6	1.0		MIT
3743.863	3744.927	26702.79	SM II	1.7	1.4		MIT
3741.283	3742.347	26721.20	SM	1.4	1.4		MIT
3740.754	3741.817	26724.98	SM	0.6	0.0		MIT
3739.998	3741.061	26730.38	SM	0.7			MIT
3739.360	3740.423	26734.94	SM	1.0	0.3		MIT
3739.199	3740.262	26736.09	SM II	1.0	1.0		MIT
3739.120	3740.183	26736.66	SM	1.0	1.0		MIT
3738.261	3739.324	26742.80	SM	0.7			MIT
3737.873	3738.936	26745.58	SM	0.3			KN
3737.472	3738.535	26748.45	SM II	0.3	0.3		MIT
3737.134	3738.197	26750.87	SM	1.0	0.5		MIT
3735.970	3737.032	26759.20	SM II	1.7	0.9		MIT
3731.46	3732.52	26791.5	SM		0.9D		MIT
3731.260	3732.321	26792.98	SM II	1.7	1.0		MIT
3730.731	3731.792	26796.78	SM	1.2			MIT
3729.755	3730.816	26803.79	SM	1.3	0.7		MIT
3728.926	3729.986	26809.75	SM	1.0W	0.6		MIT
3728.470	3729.530	26813.03	SM II	2.0	2.0		MIT
3728.162	3729.222	26815.24	SM	1.2	0.3		MIT
3727.371	3728.431	26820.93	SM II	0.6	0.3		MIT
3726.800	3727.860	26825.04	SM II	2.0R	0.5		MIT
3726.236	3727.296	26829.10	SM	0.5	0.0		MIT
3724.897	3725.956	26838.75	SM II	1.6	1.2		MIT
3724.017	3725.076	26845.09	SM	0.5	0.3		MIT
3722.563	3723.622	26855.57	SM	1.6			MIT
3722.024	3723.083	26859.46	SM	1.0	0.6		MIT
3721.842	3722.901	26860.78	SM	2.0	1.7		MIT
3721.020	3722.078	26866.71	SM	0.6H	0.0		MIT
3720.644	3721.702	26869.43	SM	0.3			MIT
3720.572	3721.630	26869.95	SM	0.6	0.6		MIT
3719.451	3720.509	26878.04	SM	1.7	1.0		KN
3719.300	3720.358	26879.13	SM	0.5			KN
3718.880	3719.938	26882.17	SM II	2.0	0.7		MIT
3718.698	3719.756	26883.49	SM II	0.9	0.0		MIT
3716.780	3717.837	26897.36	SM	0.7			KN
3712.759	3713.815	26926.49	SM II	2.0	2.0		MIT
3712.106	3713.162	26931.22	SM	1.3	0.9		MIT
3711.545	3712.601	26935.29	SM II	1.0	1.0		MIT
3710.869	3711.925	26940.20	SM II	2.0R	0.8		MIT
3710.300	3711.356	26944.33	SM	1.4	1.2		MIT
3709.526	3710.581	26949.96	SM	1.0	0.8		MIT
3709.024	3710.079	26953.60	SM	0.9			MIT
3708.787	3709.842	26955.32	SM	0.3			MIT
3708.660	3709.715	26956.25	SM II	1.7	1.4		MIT
3708.410	3709.465	26958.06	SM II	1.4	1.0		MIT
3707.850	3708.905	26962.14	SM	1.3			KN
3707.628	3708.683	26963.75	SM	0.5			MIT
3707.165	3708.220	26967.12	SM II	1.0	0.3H		MIT
3706.976	3708.031	26968.49	SM	1.0	0.9		MIT
3706.752	3707.807	26970.12	SM II	1.0	0.9		MIT
3705.995	3707.049	26975.63	SM	0.6			KN
3705.569	3706.623	26978.73	SM	0.9	0.0		MIT
3705.063	3706.117	26982.42	SM	1.2	0.3		MIT
3701.697	3702.750	27006.95	SM	1.3	0.0		MIT
3701.558	3702.611	27007.97	SM II	0.7	0.6		MIT
3701.178	3702.231	27010.74	SM	0.5	0.5		MIT
3700.926	3701.979	27012.58	SM	1.7	0.6		MIT
3700.593	3701.646	27015.01	SM II	1.7	1.5		MIT
3700.431	3701.484	27016.19	SM	0.3			MIT
3700.197	3701.250	27017.90	SM	1.0	1.0		KN

AIR WAVELENGTH (ANGSTRMS)	VACUUM WAVELENGTH (ANGSTRMS)	WAVENUMBER (KAYSERS)	LMENT	LOG ARC	INTENSITY SPK	DIS	REF
3696.349	3697.401	27046.02	SM	0.0	0.5		MIT
3695.82	3696.87	27049.9	SM	0.3			KN
3694.817	3695.869	27057.24	SM	1.2	0.5		MIT
3694.314	3695.365	27060.92	SM	1.2	0.9		MIT
3693.996	3695.047	27063.25	SM II	2.0	2.2		MIT
3692.899	3693.950	27071.29	SM	1.0	0.3		MIT
3692.763	3693.814	27072.29	SM	1.3	0.8		MIT
3692.221	3693.272	27076.26	SM II	2.0	1.6		MIT
3691.574	3692.625	27081.01	SM	1.4W	0.5		MIT
3690.931	3691.982	27085.73	SM II	1.2	0.3		MIT
3690.653	3691.704	27087.77	SM	0.5D			KN
3690.082	3691.132	27091.96	SM	1.0			MIT
3689.602	3690.652	27095.48	SM	0.6			MIT
3688.877	3689.927	27100.81	SM	0.8	0.8		MIT
3688.421	3689.471	27104.16	SM II	1.3	0.9		MIT
3688.126	3689.176	27106.32	SM	0.6H			MIT
3687.876	3688.926	27108.16	SM	1.7			MIT
3687.100	3688.150	27113.87	SM	0.7	0.8		MIT
3684.776	3685.825	27130.97	SM	0.3			KN
3684.539	3685.588	27132.71	SM	0.8	0.8		MIT
3684.120	3685.169	27135.80	SM II	1.3	0.7		MIT
3682.541	3683.589	27147.43	SM	0.9D			MIT
3682.212	3683.260	27149.86	SM II	1.0	0.3		MIT
3682.075	3683.123	27150.87	SM	0.7	0.5		MIT
3681.719	3682.767	27153.50	SM	1.5	0.8		MIT
3680.964	3682.012	27159.06	SM	1.4	0.7		MIT
3679.997	3681.045	27166.20	SM II	1.3	0.6		KN
3679.25	3680.30	27171.7	SM	1.4	0.0		MIT
3678.167	3679.214	27179.72	SM	1.2D			MIT
3678.081	3679.128	27180.35	SM	1.2	0.3		MIT
3677.775	3678.822	27182.61	SM	1.4	1.0		MIT
3677.248	3678.295	27186.51	SM	1.2			KN
3676.837	3677.884	27189.55	SM	1.0	0.7		MIT
3676.315	3677.362	27193.41	SM	0.9	0.5		MIT
3675.439	3676.486	27199.89	SM		0.5		MIT
3675.115	3676.161	27202.29	SM	0.3			KN
3674.335	3675.381	27208.06	SM	0.3			MIT
3674.053	3675.099	27210.15	SM II	1.7	1.0		MIT
3672.213	3673.259	27223.78	SM	0.8	0.7		MIT
3670.817	3671.862	27234.14	SM II	2.0	1.7		MIT
3670.657	3671.702	27235.32	SM	1.0	0.7		MIT
3670.503	3671.548	27236.47	SM II	1.0	0.5		MIT
3669.886	3670.931	27241.04	SM II	0.9	0.3		MIT
3668.887	3669.932	27248.46	SM	0.3			MIT
3668.170	3669.215	27253.79	SM	0.3D			KN
3667.907	3668.952	27255.74	SM II	1.4	1.0		MIT
3667.541	3668.586	27258.46	SM	1.0			KN
3666.951	3667.995	27262.85	SM	1.0	0.3H		MIT
3666.565	3667.609	27265.72	SM	0.9	0.3		MIT
3666.265	3667.309	27267.95	SM	0.8	0.5		KN
3665.377	3666.421	27274.55	SM	0.8			KN
3664.011	3665.055	27284.72	SM	1.2			KN
3663.654	3664.698	27287.38	SM	0.5	0.6		MIT
3662.897	3663.940	27293.02	SM II	1.4	1.0		MIT
3662.693	3663.736	27294.54	SM	1.4	1.0		KN
3662.254	3663.297	27297.81	SM II	1.7	1.7		MIT
3661.687	3662.730	27302.04	SM	0.8	0.6		KN
3661.350	3662.393	27304.55	SM II	2.0	1.7		MIT
3659.862	3660.905	27315.65	SM	0.0	0.3		MIT
3659.620	3660.662	27317.46	SM II	1.0	0.8		MIT
3657.808	3658.850	27330.99	SM	0.3	0.5H		MIT
3657.316	3658.358	27334.67	SM	1.2	0.0		MIT
3656.245	3657.287	27342.68	SM	0.8	0.7		MIT
3655.760	3656.801	27346.30	SM	0.7	0.5		MIT
3655.548	3656.589	27347.89	SM	0.8	0.7		MIT
3655.185	3656.226	27350.60	SM	0.8	0.5		MIT
3654.847	3655.888	27353.13	SM II	1.0	0.0		MIT
3653.479	3654.520	27363.37	SM	1.2	0.3		MIT
3653.113	3654.154	27366.12	SM	1.0	0.3		MIT
3652.210	3653.251	27372.88	SM II	0.7	0.6		MIT

AIR WAVELENGTH (ANGSTRMS)	VACUUM WAVELENGTH (ANGSTRMS)	WAVENUMBER (KAYSERS)	LMENT	LOG ARC	INTENSITY SPK	DIS	REF
3651.823	3652.863	27375.78	SM	0.3H	0.3H		MIT
3651.516	3652.556	27378.08	SM	0.8	0.7		KN
3651.423	3652.463	27378.78	SM	0.5D			KN
3650.981	3652.021	27382.10	SM II	1.4	1.0		MIT
3650.168	3651.208	27388.20	SM II	1.4	0.7		MIT
3649.506	3650.546	27393.16	SM II	2.0	1.5		MIT
3649.006	3650.046	27396.92	SM	0.3	0.0		MIT
3647.270	3648.309	27409.96	SM	0.9	0.5		MIT
3646.013	3647.052	27419.41	SM	0.3			MIT
3645.898	3646.937	27420.27	SM II	0.6	0.3		MIT
3645.789	3646.828	27421.09	SM	0.7D			KN
3645.387	3646.426	27424.11	SM II	0.9	0.8		MIT
3645.290	3646.329	27424.84	SM II	1.0	0.8		MIT
3643.991	3645.029	27434.62	SM	0.9			MIT
3643.715	3644.753	27436.70	SM	0.7	0.5		MIT
3642.739	3643.777	27444.05	SM II	1.4	0.7		MIT
3641.648	3642.686	27452.27	SM	0.8	0.3		MIT
3639.409	3640.446	27469.16	SM	0.8	0.7		MIT
3639.257	3640.294	27470.31	SM II	1.0	0.3		MIT
3638.748	3639.785	27474.15	SM	1.6	0.9		MIT
3636.235	3637.271	27493.14	SM	1.0	0.3		MIT
3636.109	3637.145	27494.09	SM	0.7	0.0		MIT
3636.044	3637.080	27494.58	SM	0.7	0.3		MIT
3634.908	3635.944	27503.17	SM II	1.3	0.3		MIT
3634.271	3635.307	27507.99	SM II	2.0	1.4		MIT
3631.139	3632.174	27531.72	SM	1.6	1.2		MIT
3630.853	3631.888	27533.89	SM	0.5	0.5		MIT
3630.669	3631.704	27535.28	SM	0.7	0.7		MIT
3629.477	3630.512	27544.33	SM	0.7			MIT
3629.217	3630.252	27546.30	SM II	0.8	0.0		MIT
3627.965	3628.999	27555.81	SM II	0.9	0.5		MIT
3627.444	3628.478	27559.76	SM	1.0	0.6		MIT
3626.994	3628.028	27563.18	SM II	1.7	1.6		MIT
3626.496	3627.530	27566.97	SM II	0.3	0.0		MIT
3624.419	3625.452	27582.76	SM	0.5D			MIT
3623.315	3624.348	27591.17	SM II	1.2	0.8		MIT
3622.646	3623.679	27596.26	SM	0.3			MIT
3622.506	3623.539	27597.33	SM II	1.0	0.6		MIT
3621.216	3622.249	27607.16	SM II	1.8	1.0		MIT
3620.938	3621.970	27609.28	SM	0.6D	0.5		MIT
3620.582	3621.614	27612.00	SM II	0.6	0.3		MIT
3620.095	3621.127	27615.71	SM II	0.5	0.5		MIT
3619.288	3620.320	27621.87	SM	0.3	0.3		MIT
3618.523	3619.555	27627.71	SM	0.6			MIT
3615.244	3616.275	27652.76	SM II	1.6	0.3		MIT
3614.743	3615.774	27656.60	SM	0.6	0.0		MIT
3613.902	3614.933	27663.03	SM	0.3			MIT
3613.595	3614.626	27665.38	SM	0.5	0.3		MIT
3613.521	3614.552	27665.95	SM	0.3	0.3		MIT
3612.431	3613.461	27674.30	SM	0.3	0.3		MIT
3611.056	3612.086	27684.83	SM	0.6	0.3		MIT
3609.484	3610.514	27696.89	SM II	1.8	2.0		MIT
3607.635	3608.664	27711.09	SM	0.5			MIT
3606.329	3607.358	27721.12	SM	0.3	0.0		MIT
3604.987	3606.015	27731.44	SM	0.5	0.0		MIT
3604.712	3605.740	27733.56	SM	0.3			MIT
3604.276	3605.304	27736.91	SM II	0.8	1.3		MIT
3603.504	3604.532	27742.85	SM	0.0	0.6		MIT
3601.984	3603.012	27754.56	SM	0.3	0.0		MIT
3601.695	3602.723	27756.79	SM II	0.7	0.7		MIT
3601.252	3602.279	27760.20	SM	0.6	0.0		MIT
3597.928	3598.955	27785.85	SM	0.3	0.6		MIT
3597.064	3598.090	27792.52	SM		0.5		MIT
3596.645	3597.671	27795.76	SM	0.6	0.3		MIT
3594.99	3596.02	27808.6	SM	0.8D			KN
3593.995	3595.021	27816.25	SM	0.3			MIT
3593.730	3594.755	27818.30	SM II	0.6	0.3		MIT
3593.402	3594.427	27820.84	SM II	0.3	0.5		MIT
3592.905	3593.930	27824.69	SM	0.6			MIT
3592.595	3593.620	27827.09	SM II	1.6	1.7		MIT

AIR WAVELENGTH (ANGSTRMS)	VACUUM WAVELENGTH (ANGSTRMS)	WAVENUMBER (KAYSERS)	LMENT	LOG ARC	INTENSITY SPK	DIS	REF
3591.741	3592.766	27833.71	SM II		0.3		MIT
3591.680	3592.705	27834.18	SM	0.3			MIT
3589.742	3590.766	27849.21	SM	0.3J			MIT
3589.296	3590.320	27852.67	SM	0.6			MIT
3587.473	3588.497	27866.82	SM	1.2	0.5		MIT
3586.54	3587.56	27874.1	SM	0.5	0.5		MIT
3586.359	3587.383	27875.48	SM	1.6R			MIT
3585.832	3586.855	27879.57	SM	0.7	0.6H		MIT
3584.255	3585.278	27891.84	SM II	1.4D	0.5		MIT
3583.920	3584.943	27894.45	SM	0.3	0.0		MIT
3583.372	3584.395	27898.71	SM II	1.2	0.6		MIT
3582.683	3583.706	27904.08	SM II	1.0			MIT
3580.906	3581.928	27917.93	SM	1.6	0.8		MIT
3580.586	3581.608	27920.42	SM II	0.3			MIT
3579.948	3580.970	27925.40	SM	0.3	0.0		MIT
3579.668	3580.690	27927.58	SM	0.6	0.5		MIT
3579.499	3580.521	27928.90	SM	0.6	0.3		MIT
3577.774	3578.795	27942.36	SM II	1.3	0.5		MIT
3575.448	3576.469	27960.54	SM	0.3	0.0		MIT
3573.888	3574.908	27972.75	SM	1.0H			MIT
3573.396	3574.416	27976.60	SM	0.7	0.3		MIT
3572.825	3573.845	27981.07	SM		0.3		MIT
3571.769	3572.789	27989.34	SM	0.7H			MIT
3571.075	3572.095	27994.78	SM	0.8	0.3		MIT
3569.494	3570.513	28007.18	SM	0.5			MIT
3568.891	3569.910	28011.91	SM II	0.6	0.0		MIT
3568.258	3569.277	28016.88	SM II	1.6	1.7		MIT
3566.830	3567.849	28028.10	SM	1.3	0.5		MIT
3566.471	3567.489	28030.92	SM	1.0	0.3		MIT
3564.208	3565.226	28048.71	SM	0.3	0.0		MIT
3563.590	3564.608	28053.58	SM	0.5	0.0		MIT
3562.203	3563.220	28064.50	SM	0.3	0.0		MIT
3561.581	3562.598	28069.40	SM	1.2	0.6		MIT
3560.266	3561.283	28079.77	SM	1.4	0.5		MIT
3559.105	3560.122	28088.93	SM	1.6D	1.3		MIT
3558.866	3559.882	28090.81	SM	0.7			MIT
3557.453	3558.469	28101.97	SM	1.2	0.0		MIT
3557.362	3558.378	28102.69	SM II	1.5	0.3		MIT
3556.728	3557.744	28107.70	SM II	1.5	0.5		MIT
3555.353	3556.369	28118.57	SM	0.3	0.3		MIT
3554.175	3555.190	28127.89	SM II	1.4	0.5		MIT
3552.284	3553.299	28142.86	SM II	1.2	0.5		MIT
3552.195	3553.210	28143.57	SM	0.3			MIT
3550.222	3551.236	28159.21	SM	1.4	0.3		MIT
3546.821	3547.834	28186.21	SM	0.5			MIT
3545.416	3546.429	28197.38	SM	0.5			KN
3544.983	3545.996	28200.82	SM	0.7	0.3		MIT
3542.458	3543.470	28220.92	SM II	1.0D	0.3		MIT
3541.373	3542.385	28229.57	SM	1.0	0.0		MIT
3539.892	3540.904	28241.38	SM	1.0			MIT
3539.255	3540.266	28246.46	SM II	0.9	0.0		MIT
3538.864	3539.875	28249.58	SM II	1.0	0.0		MIT
3536.900	3537.911	28265.27	SM	0.0	0.3		MIT
3536.765	3537.776	28266.35	SM	0.7	0.5		MIT
3536.225	3537.236	28270.66	SM		0.3		MIT
3535.633	3536.644	28275.40	SM II	1.4	0.7		MIT
3534.913	3535.923	28281.16	SM II	1.0			MIT
3534.527	3535.537	28284.25	SM	1.0			MIT
3532.541	3533.551	28300.15	SM	1.3	0.6		MIT
3532.008	3533.018	28304.42	SM	0.7			MIT
3530.598	3531.607	28315.72	SM II	0.9	0.6		MIT
3529.998	3531.007	28320.53	SM	1.2			MIT
3527.878	3528.887	28337.55	SM		0.3		MIT
3527.059	3528.067	28344.13	SM	0.9			MIT
3526.893	3527.901	28345.47	SM II	1.0	0.5		MIT
3526.775	3527.783	28346.41	SM	1.2			MIT
3525.496	3526.504	28356.70	SM II	1.2	0.3		MIT
3525.065	3526.073	28360.16	SM	0.3			MIT
3524.538	3525.546	28364.40	SM	1.0			MIT
3523.413	3524.420	28373.46	SM	1.1			MIT

AIR WAVELENGTH (ANGSTRMS)	VACUUM WAVELENGTH (ANGSTRMS)	WAVENUMBER (KAYSERS)	LMENT	LOG ARC	INTENSITY SPK	DIS	REF
3523.114	3524.121	28375.87	SM	0.9	0.9		MIT
3523.050	3524.057	28376.38	SM	0.9	0.9		MIT
3522.803	3523.810	28378.37	SM	0.5			MIT
3521.521	3522.528	28388.70	SM	0.8			MIT
3518.682	3519.688	28411.61	SM	0.7			MIT
3518.304	3519.310	28414.66	SM	0.0	0.3		MIT
3518.066	3519.072	28416.58	SM II	0.9	0.0		MIT
3517.035	3518.041	28424.91	SM		0.5H		MIT
3513.048	3514.053	28457.17	SM	1.2D	0.6		MIT
3512.909	3513.914	28458.30	SM II	0.8D	0.5		MIT
3512.650	3513.655	28460.40	SM	1.2	0.5		MIT
3511.590	3512.594	28468.99	SM II	0.0	0.3		MIT
3511.213	3512.217	28472.04	SM II	0.7	0.7		MIT
3510.596	3511.600	28477.05	SM		0.3		MIT
3510.151	3511.155	28480.66	SM	0.3	0.0		MIT
3509.801	3510.805	28483.50	SM	0.5	0.0		MIT
3509.447	3510.451	28486.37	SM	0.6	0.6		MIT
3509.117	3510.121	28489.05	SM	0.9	1.0		MIT
3508.706	3509.710	28492.39	SM	0.5	0.3		MIT
3507.698	3508.701	28500.57	SM	0.6			MIT
3507.108	3508.111	28505.37	SM	1.2	0.3		MIT
3506.847	3507.850	28507.49	SM II	0.6	0.3		MIT
3504.212	3505.214	28528.93	SM	0.3	0.7		MIT
3503.291	3504.293	28536.43	SM	1.3			MIT
3501.228	3502.230	28553.24	SM II	0.3	0.0		MIT
3500.498	3501.499	28559.19	SM II	1.5	1.2		MIT
3499.825	3500.826	28564.69	SM	0.9	0.7		MIT
3495.922	3496.922	28596.57	SM	0.9	0.3		MIT
3494.826	3495.826	28605.54	SM	0.7			MIT
3493.597	3494.597	28615.60	SM	0.7	0.6		MIT
3493.408	3494.408	28617.15	SM II	0.3	0.0		MIT
3492.895	3493.895	28621.36	SM	0.5			MIT
3492.775	3493.774	28622.34	SM	0.5	0.5		MIT
3492.622	3493.621	28623.59	SM II	0.9	0.0		MIT
3491.045	3492.044	28636.52	SM	0.3			MIT
3487.609	3488.607	28664.74	SM	0.5	0.0		MIT
3487.403	3488.401	28666.43	SM	1.0	0.6		MIT
3485.796	3486.794	28679.64	SM II	0.3H	0.0		MIT
3484.723	3485.720	28688.47	SM	0.5	0.0		MIT
3483.200	3484.197	28701.02	SM	0.3	0.3		MIT
3482.824	3483.821	28704.12	SM	0.6			MIT
3482.694	3483.691	28705.19	SM II	0.3			MIT
3480.560	3481.556	28722.79	SM	0.7	0.7		MIT
3480.252	3481.248	28725.33	SM	0.7	0.6		MIT
3479.512	3480.508	28731.44	SM II	0.9	0.5		MIT
3479.135	3480.131	28734.55	SM II	0.5	0.3		MIT
3478.951	3479.947	28736.07	SM	0.6D			MIT
3478.445	3479.441	28740.25	SM	0.7			M
3477.428	3478.424	28748.66	SM	0.9			MIT
3475.129	3476.124	28767.67	SM II	0.9	0.0		MIT
3474.371	3475.366	28773.95	SM	0.5			MIT
3473.953	3474.948	28777.41	SM II	0.9	0.7		MIT
3472.034	3473.028	28793.32	SM	1.0J			MIT
3471.270	3472.264	28799.65	SM	0.5			MIT
3471.122	3472.116	28800.88	SM	0.8	0.3		MIT
3467.874	3468.867	28827.86	SM II	0.8	0.5		MIT
3466.784	3467.777	28836.92	SM	0.6			MIT
3466.740	3467.733	28837.28	SM	0.5			MIT
3465.457	3466.449	28847.96	SM	0.7			MIT
3464.428	3465.420	28856.53	SM II	0.5	0.0		MIT
3464.082	3465.074	28859.41	SM	1.5	1.0		MIT
3463.611	3464.603	28863.34	SM	0.3			MIT
3463.184	3464.176	28866.90	SM	0.3	0.0		MIT
3462.688	3463.680	28871.03	SM II	0.7	0.3		MIT
3461.405	3462.396	28881.73	SM	0.7	0.6		MIT
3461.138	3462.129	28883.96	SM	0.9	0.5		MIT
3460.63	3461.62	28888.2	SM	0.5W	0.0		MIT
3459.396	3460.387	28898.50	SM	1.0	0.3		MIT
3459.189	3460.180	28900.23	SM	0.8	0.5		MIT
3457.921	3458.912	28910.83	SM II	0.7	0.0		MIT

AIR WAVELENGTH (ANGSTRMS)	VACUUM WAVELENGTH (ANGSTRMS)	WAVENUMBER (KAYSERS)	LMENT	LOG ARC	INTENSITY SPK	DIS	REF
3456.114	3457.104	28925.94	SM	0.6			MIT
3454.959	3455.949	28935.61	SM II	1.4	1.1		MIT
3454.777	3455.767	28937.14	SM	0.6			MIT
3453.547	3454.536	28947.44	SM	1.2	0.6		MIT
3453.226	3454.215	28950.14	SM	0.6	0.6W		MIT
3452.768	3453.757	28953.98	SM II	0.6	0.0		MIT
3451.523	3452.512	28964.42	SM	0.5	0.3H		MIT
3448.856	3449.844	28986.82	SM	0.8			MIT
3447.783	3448.771	28995.84	SM	0.3	0.0		MIT
3445.621	3446.608	29014.03	SM	2.2G	1.0G		MIT
3444.613	3445.600	29022.52	SM	0.9	1.0W		MIT
3441.233	3442.219	29051.03	SM	0.5			MIT
3440.504	3441.490	29057.18	SM II	1.6	0.5		MIT
3439.602	3440.588	29064.80	SM II	0.3			MIT
3438.978	3439.964	29070.07	SM	0.9			MIT
3438.054	3439.039	29077.89	SM II	1.0	0.5		MIT
3437.614	3438.599	29081.61	SM	0.5	0.3H		MIT
3437.115	3438.100	29085.83	SM II	1.2	0.3		MIT
3436.70	3437.69	29089.3	SM	0.5	0.3		MIT
3436.136	3437.121	29094.12	SM	0.5			MIT
3435.256	3436.241	29101.57	SM II	0.7	0.5		MIT
3433.685	3434.669	29114.88	SM	0.7	0.7		MIT
3432.630	3433.614	29123.83	SM	0.7C	0.5C		MIT
3431.897	3432.881	29130.05	SM	1.1			MIT
3429.749	3430.732	29148.30	SM II	0.9	0.3		MIT
3427.961	3428.944	29163.50	SM	0.9	0.3		MIT
3426.53	3427.51	29175.7	SM	0.5W			MIT
3426.187	3427.169	29178.60	SM	0.9	0.3		MIT
3425.254	3426.236	29186.55	SM II	0.5	0.0		MIT
3424.765	3425.747	29190.71	SM	0.7	0.3		MIT
3423.504	3424.486	29201.47	SM	0.3H			MIT
3422.760	3423.742	29207.81	SM	1.0	0.5		MIT
3422.476	3423.457	29210.24	SM	0.3			MIT
3422.189	3423.170	29212.69	SM II	0.5			MIT
3422.063	3423.044	29213.76	SM	0.7			MIT
3421.297	3422.278	29220.30	SM	0.3			MIT
3420.514	3421.495	29226.99	SM	0.5	0.3		MIT
3419.769	3420.750	29233.36	SM	1.0	0.8		MIT
3418.510	3419.490	29244.12	SM II	1.7	1.0		MIT
3418.134	3419.114	29247.34	SM	1.3	0.8		MIT
3416.015	3416.995	29265.48	SM	0.8	0.0		MIT
3414.953	3415.933	29274.58	SM II	1.0	0.3		MIT
3414.52	3415.50	29278.3	SM		0.6		MIT
3413.895	3414.874	29283.66	SM II	0.8	0.3		MIT
3412.571	3413.550	29295.02	SM	0.6	0.0		MIT
3411.275	3412.254	29306.15	SM	0.7	0.3		MIT
3411.22	3412.20	29306.6	SM	0.5	0.0		MIT
3410.381	3411.359	29313.83	SM	1.0			MIT
3410.280	3411.258	29314.70	SM	0.6	0.3H		MIT
3410.031	3411.009	29316.84	SM	0.7			MIT
3408.668	3409.646	29328.56	SM	1.3	1.0		MIT
3408.599	3409.577	29329.15	SM II	0.6	0.3		MIT
3408.575	3409.553	29329.36	SM II	0.6	0.3		MIT
3408.048	3409.026	29333.89	SM	0.3			MIT
3406.545	3407.522	29346.84	SM	0.6			MIT
3404.767	3405.744	29362.16	SM	0.3	0.3		MIT
3403.085	3404.061	29376.67	SM	1.2	0.9		MIT
3402.463	3403.439	29382.04	SM II	1.7	1.0		MIT
3399.840	3400.816	29404.71	SM	1.0	0.5		MIT
3399.688	3400.664	29406.02	SM	0.7	0.0		MIT
3398.44	3399.42	29416.8	SM	0.0	0.5H		MIT
3397.763	3398.738	29422.69	SM	1.0	0.5		MIT
3396.183	3397.158	29436.37	SM II	1.5	1.2		MIT
3394.818	3395.792	29448.21	SM	0.3			MIT
3391.841	3392.815	29474.05	SM	0.3	0.5		MIT
3391.122	3392.095	29480.30	SM II	0.9	0.3		MIT
3389.325	3390.298	29495.93	SM II	1.6	1.3		MIT
3388.180	3389.153	29505.90	SM	0.5	0.0		MIT
3387.668	3388.641	29510.36	SM	0.8	0.5		MIT
3387.065	3388.037	29515.61	SM II	0.9	0.5		MIT

AIR WAVELENGTH (ANGSTRMS)	VACUUM WAVELENGTH (ANGSTRMS)	WAVENUMBER (KAYSERS)	LMENT	LOG ARC	INTENSITY SPK	DIS	REF
3386.603	3387.575	29519.64	SM	0.3H			MIT
3385.399	3386.371	29530.14	SM II	0.8	0.6		MIT
3384.867	3385.839	29534.78	SM	0.5	0.3		MIT
3384.663	3385.635	29536.56	SM II	1.0	0.9		MIT
3382.407	3383.378	29556.26	SM II	2.0	1.6		MIT
3380.703	3381.674	29571.16	SM	0.3	0.0		MIT
3378.552	3379.522	29589.98	SM	0.9	0.3		MIT
3378.312	3379.282	29592.08	SM II	0.7	0.3		MIT
3377.820	3378.790	29596.39	SM II	0.9	0.3		MIT
3376.835	3377.805	29605.03	SM	0.7	0.3		MIT
3376.494	3377.464	29608.02	SM II	1.3	0.9		MIT
3373.788	3374.757	29631.76	SM	1.2	0.6		MIT
3371.212	3372.180	29654.40	SM II	1.0	0.9		MIT
3371.019	3371.987	29656.10	SM II	0.5	0.3		MIT
3370.586	3371.554	29659.91	SM II	1.0	0.5		MIT
3369.452	3370.420	29669.89	SM II	1.0	0.7		MIT
3369.041	3370.009	29673.51	SM	0.9	0.5		MIT
3368.572	3369.540	29677.64	SM II	1.1	0.8		MIT
3367.275	3368.242	29689.07	SM II	1.0	0.7		MIT
3365.958	3366.925	29700.69	SM	0.6	0.3		MIT
3365.865	3366.832	29701.51	SM	1.4	1.0		MIT
3364.795	3365.762	29710.96	SM	0.5	0.3		MIT
3364.747	3365.714	29711.38	SM	0.3			MIT
3363.908	3364.874	29718.79	SM	1.0	0.5		MIT
3361.435	3362.401	29740.65	SM	1.0	0.3		MIT
3360.640	3361.606	29747.69	SM	0.3	0.0		MIT
3358.273	3359.238	29768.66	SM II	0.8	0.3H		MIT
3354.719	3355.683	29800.19	SM	1.0	0.7		MIT
3354.281	3355.245	29804.08	SM	0.7	0.3		MIT
3354.182	3355.146	29804.96	SM II	1.0	0.7		MIT
3353.766	3354.730	29808.66	SM	0.3			MIT
3352.724	3353.688	29817.92	SM	0.6			MIT
3351.686	3352.649	29827.16	SM	0.5			MIT
3351.283	3352.246	29830.74	SM	0.7	0.3		MIT
3350.883	3351.846	29834.30	SM	1.0	0.6		MIT
3350.652	3351.615	29836.36	SM	0.9			MIT
3349.835	3350.798	29843.64	SM	0.5	0.0		MIT
3348.871	3349.834	29852.23	SM	0.5	0.0		MIT
3348.682	3349.645	29853.91	SM	1.0	1.0		MIT
3347.296	3348.258	29866.27	SM II	0.9	0.8		MIT
3346.907	3347.869	29869.75	SM	0.7	0.5		MIT
3346.353	3347.315	29874.69	SM II	0.7	0.5		MIT
3345.02	3345.98	29886.6	SM	0.0	0.5H		MIT
3344.349	3345.310	29892.59	SM II	1.6	1.0		MIT
3344.172	3345.133	29894.17	SM	0.6			MIT
3343.644	3344.605	29898.89	SM	1.4	0.7		MIT
3343.492	3344.453	29900.25	SM	1.2	0.9		MIT
3342.664	3343.625	29907.66	SM	0.3	0.5		MIT
3341.832	3342.793	29915.10	SM	0.7	0.0		MIT
3341.435	3342.396	29918.66	SM II	0.7	0.3		MIT
3340.585	3341.546	29926.27	SM II	1.6	1.4		MIT
3338.865	3339.825	29941.69	SM	0.8			MIT
3338.515	3339.475	29944.83	SM	0.3			MIT
3336.120	3337.079	29966.32	SM II	1.4	0.7		MIT
3335.039	3335.998	29976.04	SM	1.0	0.5		MIT
3334.679	3335.638	29979.27	SM	0.3	0.0		MIT
3334.114	3335.073	29984.35	SM	0.9	0.6		MIT
3333.632	3334.591	29988.69	SM II	1.2	0.7		MIT
3332.703	3333.662	29997.05	SM II	1.0	0.3		MIT
3331.078	3332.036	30011.68	SM	0.3			MIT
3329.623	3330.581	30024.79	SM II	0.9	0.3		MIT
3328.939	3329.897	30030.96	SM II	1.0	0.3		MIT
3327.901	3328.858	30040.33	SM	1.7	1.3		MIT
3326.977	3327.934	30048.67	SM	0.5	0.3H		MIT
3325.475	3326.432	30062.24	SM	1.2	0.6		MIT
3325.258	3326.215	30064.21	SM	1.2	0.7		MIT
3324.870	3325.827	30067.71	SM	0.5			MIT
3323.769	3324.725	30077.67	SM	1.2	0.9		MIT
3321.184	3322.140	30101.08	SM II	1.7	1.2		MIT
3320.587	3321.542	30106.50	SM	0.6	0.3		MIT

AIR WAVELENGTH (ANGSTRMS)	VACUUM WAVELENGTH (ANGSTRMS)	WAVENUMBER (KAYSERS)	LMENT	LOG ARC	INTENSITY SPK	DIS	REF
3320.163	3321.118	30110.34	SM II	1.6	0.9		MIT
3319.563	3320.518	30115.78	SM	0.9	0.3		MIT
3319.122	3320.077	30119.78	SM	0.9	0.3		MIT
3317.866	3318.821	30131.18	SM	0.9	0.3		MIT
3316.582	3317.536	30142.85	SM	1.4	1.0		MIT
3316.195	3317.149	30146.37	SM II	0.7	0.3		MIT
3314.93	3315.88	30157.9	SM	0.5D	0.0H		MIT
3312.415	3313.368	30180.77	SM II	1.1	1.0		MIT
3310.655	3311.608	30196.81	SM II	1.4	1.0		MIT
3310.329	3311.282	30199.78	SM	0.6			MIT
3309.516	3310.469	30207.20	SM II	1.3	0.9		MIT
3308.932	3309.884	30212.53	SM	1.0	0.6		MIT
3308.465	3309.417	30216.80	SM	0.5	0.3		MIT
3308.084	3309.036	30220.28	SM	0.6H	0.0		MIT
3307.016	3307.968	30230.04	SM	1.6	1.2		MIT
3306.616	3307.568	30233.70	SM	1.0	0.6		MIT
3306.372	3307.324	30235.93	SM II	2.0	1.6		MIT
3305.750	3306.702	30241.61	SM	0.7	0.6		MIT
3305.179	3306.131	30246.84	SM II	1.2	0.8		MIT
3304.521	3305.472	30252.86	SM II	1.5	0.9		MIT
3303.605	3304.556	30261.25	SM	0.7			MIT
3302.092	3303.043	30275.11	SM	0.9	0.6		MIT
3301.672	3302.623	30278.97	SM II	1.2	0.9		MIT
3300.970	3301.920	30285.41	SM II	1.3	1.0		MIT
3300.105	3301.055	30293.34	SM	1.0	0.3		MIT
3298.997	3299.947	30303.52	SM	0.0	0.3H		MIT
3298.104	3299.054	30311.72	SM	1.8	1.8		MIT
3295.806	3296.755	30332.86	SM II	1.5	1.0		MIT
3295.440	3296.389	30336.23	SM	1.5	1.0		MIT
3293.797	3294.746	30351.36	SM II	0.6	0.0		MIT
3293.605	3294.554	30353.13	SM	0.7	0.0		KN
3293.536	3294.485	30353.76	SM II	0.5	0.0		MIT
3293.360	3294.309	30355.38	SM II	1.2	0.9		MIT
3291.423	3292.371	30373.25	SM	0.8	0.8H		MIT
3290.639	3291.587	30380.48	SM	1.0	0.7		MIT
3290.381	3291.329	30382.87	SM II	1.0	0.9		MIT
3290.282	3291.230	30383.78	SM	1.0	0.6		MIT
3286.528	3287.475	30418.48	SM II	1.0	0.5		MIT
3286.223	3287.170	30421.31	SM II	1.6	1.0		MIT
3285.657	3286.604	30426.55	SM II	1.4	0.8		MIT
3285.259	3286.205	30430.23	SM II	0.8	0.0		MIT
3283.885	3284.831	30442.97	SM	0.7H	0.3H		MIT
3283.740	3284.686	30444.31	SM	0.7H			MIT
3281.766	3282.712	30462.62	SM	0.5			MIT
3280.842	3281.787	30471.20	SM	1.3	0.8		MIT
3280.218	3281.163	30477.00	SM	0.6			KN
3278.874	3279.819	30489.49	SM	0.8	0.3		MIT
3276.743	3277.687	30509.32	SM II	1.7	1.0		MIT
3275.866	3276.810	30517.48	SM II	1.0	0.5		MIT
3273.477	3274.420	30539.75	SM	1.5	0.9		MIT
3273.316	3274.259	30541.26	SM II	0.8	0.5		MIT
3272.803	3273.746	30546.04	SM II	1.3	1.0		MIT
3272.589	3273.532	30548.04	SM	1.0	0.6		MIT
3272.469	3273.412	30549.16	SM	1.0	0.6		MIT
3271.225	3272.168	30560.78	SM	0.8	0.3		MIT
3270.672	3271.615	30565.95	SM	1.0	0.6		MIT
3270.483	3271.426	30567.71	SM	1.0	0.6		MIT
3269.39	3270.33	30577.9	SM		1.2		MIT
3267.510	3268.452	30595.52	SM	0.3			MIT
3264.938	3265.879	30619.62	SM II	1.4	0.9		MIT
3264.142	3265.083	30627.09	SM	0.8	0.5		MIT
3262.579	3263.520	30641.76	SM	0.5	0.0		MIT
3262.263	3263.204	30644.73	SM II	1.0	0.6		MIT
3260.265	3261.205	30663.51	SM	1.0	0.3		MIT
3258.255	3259.195	30682.43	SM II	1.0	0.5		MIT
3255.843	3256.782	30705.16	SM	0.8	0.3		MIT
3255.624	3256.563	30707.22	SM II	0.9	0.5		MIT
3254.778	3255.717	30715.20	SM II	0.5	0.0		MIT
3254.728	3255.667	30715.68	SM II	1.0	0.5		MIT
3254.378	3255.317	30718.98	SM II	2.0	1.2		MIT

AIR WAVELENGTH (ANGSTRMS)	VACUUM WAVELENGTH (ANGSTRMS)	WAVENUMBER (KAYSERS)	LMENT	LOG ARC	INTENSITY SPK	DIS	REF
3254.290	3255.229	30719.81	SM II	1.0	0.0		MIT
3253.931	3254.870	30723.20	SM II	1.7	0.9		MIT
3253.640	3254.578	30725.95	SM	0.7	0.0		MIT
3253.390	3254.328	30728.31	SM	1.6	1.0		MIT
3253.011	3253.949	30731.89	SM	0.9	0.5		MIT
3250.361	3251.299	30756.94	SM II	1.7	1.0		MIT
3249.886	3250.823	30761.44	SM II	0.8	0.3		MIT
3249.732	3250.669	30762.89	SM	1.4	0.8		MIT
3248.140	3249.077	30777.97	SM	1.0	0.6		MIT
3247.366	3248.303	30785.31	SM	0.6			MIT
3247.17	3248.11	30787.2	SM	0.0	1.0		MIT
3246.843	3247.780	30790.27	SM	1.0	0.3		MIT
3245.801	3246.737	30800.15	SM	1.3	0.6		MIT
3245.165	3246.101	30806.19	SM II	1.0	0.3		MIT
3244.673	3245.609	30810.86	SM	1.2	0.8		MIT
3242.482	3243.418	30831.68	SM II	0.6	0.3		MIT
3242.031	3242.966	30835.97	SM	1.6	0.6		MIT
3241.572	3242.507	30840.33	SM II	1.3	0.8		MIT
3241.143	3242.078	30844.41	SM II	1.7	1.0		MIT
3239.638	3240.573	30858.74	SM II	2.0	1.4		MIT
3239.366	3240.301	30861.33	SM	0.7			KN
3237.881	3238.815	30875.49	SM II	1.0	0.7		MIT
3237.197	3238.131	30882.01	SM II	1.0	0.6		MIT
3236.630	3237.564	30887.42	SM	2.0	1.6		MIT
3234.421	3235.355	30908.51	SM	1.0	0.6		MIT
3233.661	3234.594	30915.78	SM	1.3	1.0		MIT
3233.483	3234.416	30917.48	SM	0.9	0.6		MIT
3232.620	3233.553	30925.73	SM	1.0	0.5		MIT
3232.497	3233.430	30926.91	SM	0.6			MIT
3231.938	3232.871	30932.26	SM II	1.3	0.9		MIT
3231.513	3232.446	30936.33	SM II	1.3	1.0		MIT
3230.967	3231.900	30941.55	SM	0.6	0.0		MIT
3230.544	3231.477	30945.61	SM II	2.0	1.5		MIT
3229.595	3230.527	30954.70	SM II	0.9	0.3		MIT
3228.777	3229.709	30962.54	SM II	1.3	0.6		MIT
3228.492	3229.424	30965.27	SM	1.3	0.8		MIT
3226.842	3227.774	30981.11	SM	1.5	1.0		MIT
3225.922	3226.853	30989.94	SM	1.0	0.6		MIT
3225.184	3226.115	30997.03	SM	0.9	0.3		MIT
3224.233	3225.164	31006.18	SM	1.0	0.5		MIT
3223.319	3224.250	31014.97	SM	1.0	0.6		MIT
3219.865	3220.795	31048.24	SM	1.0	0.3		MIT
3219.416	3220.346	31052.57	SM	1.4	1.0		MIT
3218.597	3219.527	31060.47	SM	2.0	1.4		MIT
3218.036	3218.965	31065.88	SM	1.2	0.7		MIT
3217.576	3218.505	31070.32	SM	0.5	0.3H		MIT
3216.836	3217.765	31077.47	SM II	1.6	1.0		MIT
3215.588	3216.517	31089.53	SM	1.0	0.5		MIT
3215.243	3216.172	31092.87	SM II	1.7	1.2		MIT
3214.121	3215.049	31103.72	SM II	1.5	0.8		MIT
3212.858	3213.786	31115.95	SM	1.0	0.3		MIT
3211.755	3212.683	31126.63	SM	2.0	1.2		MIT
3210.832	3211.760	31135.58	SM II	0.7	0.3		MIT
3209.726	3210.653	31146.31	SM	0.5	0.3		MIT
3208.167	3209.094	31161.44	SM	1.4	0.9		MIT
3207.176	3208.103	31171.07	SM II	1.7	1.0		MIT
3206.018	3206.944	31182.33	SM	0.5H			KN
3205.632	3206.558	31186.08	SM	0.3			MIT
3204.892	3205.818	31193.29	SM	1.4	1.0		MIT
3204.567	3205.493	31196.45	SM II	0.6	0.3		MIT
3202.219	3203.144	31219.32	SM	0.3H	0.3H		MIT
3201.792	3202.717	31223.49	SM	1.5D	0.5		MIT
3199.802	3200.727	31242.90	SM	0.6			KN
3199.596	3200.521	31244.92	SM	0.5	0.3		KN
3196.186	3197.110	31278.25	SM II	1.3	1.0		MIT
3193.004	3193.927	31309.42	SM	1.6	1.0		MIT
3192.731	3193.654	31312.10	SM	0.5	0.0		MIT
3191.996	3192.919	31319.31	SM	0.9	0.3		MIT
3191.036	3191.959	31328.73	SM	0.5	0.0		MIT
3189.557	3190.479	31343.25	SM II	0.8	0.3		MIT

AIR WAVELENGTH (ANGSTRMS)	VACUUM WAVELENGTH (ANGSTRMS)	WAVENUMBER (KAYSERS)	LMENT	LOG ARC	INTENSITY SPK	DIS	REF
3188.710	3189.632	31351.58	SM II	1.0	0.6		MIT
3187.775	3188.697	31360.77	SM	1.4	0.9		MIT
3187.213	3188.135	31366.30	SM II	1.2	0.9		MIT
3187.006	3187.928	31368.34	SM II	1.2	0.9		MIT
3186.280	3187.201	31375.49	SM	0.8	0.5		MIT
3186.015	3186.936	31378.10	SM	1.6	0.7		MIT
3185.837	3186.758	31379.85	SM	0.5	0.0		MIT
3184.194	3185.115	31396.04	SM	0.5			MIT
3183.916	3184.837	31398.78	SM	1.8	1.6		MIT
3183.302	3184.223	31404.84	SM	0.6	0.3		MIT
3180.305	3181.225	31434.43	SM	0.9	0.0		MIT
3178.443	3179.362	31452.85	SM	0.6	0.3		MIT
3178.119	3179.038	31456.05	SM II	1.0	0.7		MIT
3177.840	3178.759	31458.81	SM	0.9	0.6		MIT
3175.830	3176.749	31478.73	SM	0.0	0.3		MIT
3174.508	3175.426	31491.83	SM	0.5	0.0		MIT
3170.203	3171.120	31534.60	SM II	1.2	0.7		MIT
3169.870	3170.787	31537.91	SM II	1.4	0.9		MIT
3169.613	3170.530	31540.47	SM	0.5			MIT
3166.93	3167.85	31567.2	SM	0.6	0.6		MIT
3163.391	3164.307	31602.50	SM II	0.7	0.3		MIT
3162.288	3163.203	31613.52	SM	0.8	0.5		MIT
3162.148	3163.063	31614.92	SM II	0.8	0.5		MIT
3161.771	3162.686	31618.69	SM	0.5	0.0H		MIT
3160.886	3161.801	31627.54	SM	0.3	0.0		MIT
3157.841	3158.755	31658.04	SM	0.8	0.5		MIT
3155.024	3155.937	31686.31	SM	0.3	0.0		MIT
3154.360	3155.273	31692.98	SM II	0.5	0.0		MIT
3152.521	3153.434	31711.46	SM	1.2	1.0		MIT
3152.095	3153.008	31715.75	SM II	1.0	0.7		MIT
3151.487	3152.400	31721.87	SM	0.7	0.5		MIT
3148.817	3149.729	31748.76	SM	0.3			MIT
3147.580	3148.492	31761.24	SM	0.5			KN
3147.188	3148.100	31765.20	SM II	0.8	0.7		MIT
3145.834	3146.745	31778.87	SM	0.7	0.3		MIT
3143.507	3144.418	31802.39	SM	0.6	0.3		MIT
3143.306	3144.217	31804.43	SM	0.9	0.5		MIT
3140.382	3141.292	31834.04	SM	0.9	0.5		MIT
3139.973	3140.883	31838.18	SM	1.0	0.7		MIT
3139.360	3140.270	31844.40	SM II	0.8	0.3		MIT
3138.605	3139.514	31852.06	SM	0.3	0.3		MIT
3136.297	3137.206	31875.50	SM	1.2	0.8		MIT
3134.186	3135.094	31896.97	SM	0.8	0.0		MIT
3130.235	3131.142	31937.23	SM	0.7			MIT
3129.950	3130.857	31940.14	SM	0.7			KN
3127.72	3128.63	31962.9	SM	0.3D	0.0H		MIT
3127.160	3128.066	31968.63	SM	0.3	0.3H		MIT
3125.997	3126.903	31980.52	SM	0.7			MIT
3124.927	3125.833	31991.47	SM	1.0	0.5		MIT
3123.032	3123.937	32010.88	SM	0.3	0.0		MIT
3122.356	3123.261	32017.81	SM	0.6	0.0		MIT
3118.859	3119.763	32053.71	SM	0.3			MIT
3117.899	3118.803	32063.58	SM	0.7	0.5		MIT
3117.725	3118.629	32065.37	SM	0.9	0.5		MIT
3117.186	3118.090	32070.92	SM	0.6	0.0		MIT
3115.542	3116.446	32087.84	SM	0.3			MIT
3115.042	3115.945	32092.99	SM II	0.8	0.6		MIT
3114.778	3115.681	32095.71	SM	0.8	0.0		MIT
3114.623	3115.526	32097.31	SM	0.3	0.0		MIT
3113.407	3114.310	32109.84	SM	0.7			MIT
3110.193	3111.095	32143.02	SM	1.2	0.6		MIT
3109.965	3110.867	32145.38	SM	0.0	1.0H		MIT
3107.524	3108.426	32170.63	SM	0.6	0.3		MIT
3106.520	3107.421	32181.02	SM II	1.2	0.7		MIT
3102.299	3103.199	32224.81	SM	0.9	0.6		MIT
3101.930	3102.830	32228.64	SM	1.0	0.9		MIT
3099.162	3100.061	32257.43	SM	0.8			MIT
3098.191	3099.090	32267.53	SM		0.6H		MIT
3098.005	3098.904	32269.47	SM	0.0	0.3		MIT
3096.877	3097.776	32281.23	SM	1.0	0.6		MIT

AIR WAVELENGTH (ANGSTRMS)	VACUUM WAVELENGTH (ANGSTRMS)	WAVENUMBER (KAYSERS)	LMENT	LOG ARC	INTENSITY SPK	DIS	REF
3096.668	3097.567	32283.40	SM II	0.9	0.5		MIT
3096.069	3096.968	32289.65	SM	0.5	0.7H		MIT
3095.716	3096.615	32293.33	SM	0.5	0.3		MIT
3095.472	3096.371	32295.88	SM II	0.3	0.3H		MIT
3090.446	3091.343	32348.40	SM	0.5	0.0		MIT
3088.980	3089.877	32363.75	SM II	0.7	0.3		MIT
3088.821	3089.718	32365.42	SM	0.3H			MIT
3088.184	3089.081	32372.09	SM	0.8			KN
3087.922	3088.819	32374.84	SM	0.6	0.3		MIT
3087.379	3088.275	32380.53	SM	1.2	0.6		MIT
3086.742	3087.638	32387.21	SM	0.0	0.6		MIT
3086.447	3087.343	32390.31	SM	0.8	0.9		MIT
3085.790	3086.686	32397.21	SM	0.3			MIT
3084.458	3085.354	32411.19	SM II	0.5	0.0		MIT
3084.286	3085.182	32413.00	SM II	0.5	0.3		MIT
3082.261	3083.156	32434.30	SM	0.0	0.3H		MIT
3080.709	3081.604	32450.64	SM II	0.5	0.3		MIT
3076.973	3077.867	32490.03	SM	1.0			MIT
3076.718	3077.612	32492.73	SM	0.5			MIT
3076.496	3077.390	32495.07	SM II	0.3	0.3		MIT
3073.915	3074.808	32522.35	SM	0.5			MIT
3072.647	3073.540	32535.77	SM	0.3	0.3		MIT
3071.274	3072.166	32550.32	SM	0.9	0.6		MIT
3070.873	3071.765	32554.57	SM	0.3	0.3		MIT
3070.394	3071.286	32559.65	SM	0.6	0.5		MIT
3069.690	3070.582	32567.12	SM	0.8	0.5		MIT
3069.409	3070.301	32570.10	SM	0.5			MIT
3067.672	3068.564	32588.54	SM	1.2	1.2		MIT
3066.019	3066.910	32606.11	SM	0.8	0.3		MIT
3065.775	3066.666	32608.70	SM II	0.7	0.5		MIT
3065.206	3066.097	32614.75	SM	0.5	0.3		MIT
3065.008	3065.899	32616.86	SM	0.6	0.3H		KN
3061.56	3062.45	32653.6	SM	0.7D	0.5		MIT
3061.141	3062.031	32658.06	SM	1.0			MIT
3055.524	3056.412	32718.10	SM	0.5	0.0		MIT
3053.636	3054.524	32738.32	SM	0.3			MIT
3051.003	3051.890	32766.58	SM	0.8	0.3		MIT
3050.819	3051.706	32768.55	SM	0.7	0.9		MIT
3047.363	3048.249	32805.71	SM	0.7	0.5		MIT
3046.933	3047.819	32810.34	SM	1.4	1.3		MIT
3044.705	3045.591	32834.35	SM	0.5	0.7		MIT
3043.528	3044.413	32847.05	SM	0.3	0.0		MIT
3040.380	3041.265	32881.06	SM	0.7D	0.0		MIT
3039.358	3040.242	32892.11	SM	0.8	0.3		MIT
3039.128	3040.012	32894.60	SM	1.2	1.0		KN
3037.948	3038.832	32907.38	SM	0.8	0.5		MIT
3037.686	3038.570	32910.22	SM	0.8	0.5		MIT
3034.843	3035.726	32941.05	SM	1.3	1.0		MIT
3034.492	3035.375	32944.86	SM	0.7	0.3		MIT
3034.174	3035.057	32948.31	SM	0.7			MIT
3033.393	3034.276	32956.79	SM	0.0	0.3H		MIT
3032.865	3033.748	32962.53	SM	1.2			MIT
3031.504	3032.386	32977.33	SM	0.7	0.3		MIT
3028.478	3029.360	33010.28	SM	0.6	0.5		MIT
3027.992	3028.874	33015.57	SM	0.6			MIT
3027.476	3028.357	33021.20	SM	0.8	0.5		MIT
3026.580	3027.461	33030.98	SM	0.5			MIT
3023.076	3023.956	33069.26	SM	0.3	0.0		MIT
3021.225	3022.105	33089.52	SM	0.8			KN
3019.303	3020.182	33110.58	SM II	1.0	0.5		MIT
3018.470	3019.349	33119.72	SM	0.3	0.0		MIT
3015.687	3016.566	33150.28	SM	0.5	0.5		MIT
3014.366	3015.244	33164.81	SM	0.7	1.0		MIT
3014.050	3014.928	33168.29	SM	0.3	0.0		MIT
3013.508	3014.386	33174.25	SM	0.5	0.3H		MIT
3013.393	3014.271	33175.52	SM	0.6			MIT
3013.124	3014.002	33178.48	SM	0.3H	0.3H		MIT
3012.505	3013.383	33185.30	SM	0.3	0.0		MIT
3012.183	3013.061	33188.84	SM	1.3	1.0		KN
3012.086	3012.964	33189.91	SM	0.7	0.5		MIT

AIR WAVELENGTH (ANGSTRMS)	VACUUM WAVELENGTH (ANGSTRMS)	WAVENUMBER (KAYSERS)	LMENT	LOG ARC	INTENSITY SPK	INTENSITY DIS	REF
3011.862	3012.740	33192.38	SM	0.5H	0.3H		KN
3010.964	3011.841	33202.28	SM	0.3			MIT
3010.769	3011.646	33204.43	SM	0.5D	0.3		MIT
3010.327	3011.204	33209.31	SM	0.6			KN
3010.011	3010.888	33212.79	SM	1.0			MIT
3008.820	3009.697	33225.94	SM II	0.5	0.3		MIT
3006.150	3007.026	33255.45	SM	0.8	0.5		MIT
3004.125	3005.001	33277.86	SM	0.5	0.0		MIT
3003.952	3004.828	33279.78	SM	0.3			MIT
2998.260	2999.134	33342.96	SM	0.6			MIT
2996.966	2997.840	33357.35	SM	0.9	0.5		MIT
2996.476	2997.350	33362.81	SM	0.8	0.0		ES
2995.875	2996.749	33369.50	SM	0.5H			MIT
2993.826	2994.699	33392.34	SM	1.0	0.6		MIT
2992.913	2993.786	33402.52	SM	0.7	0.0		MIT
2992.521	2993.394	33406.90	SM	0.6D			KN
2991.570	2992.443	33417.52	SM	1.1	1.0		MIT
2988.924	2989.796	33447.10	SM	0.6	0.6H		MIT
2988.845	2989.717	33447.98	SM	0.8	0.7		MIT
2987.651	2988.523	33461.35	SM	0.7H			MIT
2985.249	2986.120	33488.27	SM	0.7			MIT
2984.993	2985.864	33491.15	SM	0.8	0.0		MIT
2983.438	2984.309	33508.60	SM	0.9	0.6		MIT
2982.894	2983.764	33514.71	SM	0.6			MIT
2981.325	2982.195	33532.35	SM	1.0			MIT
2977.072	2977.941	33580.25	SM	1.1	0.6		MIT
2976.512	2977.381	33586.57	SM	0.6	0.5		MIT
2974.921	2975.789	33604.53	SM	0.8			MIT
2970.922	2971.789	33649.76	SM	1.2			MIT
2970.483	2971.350	33654.73	SM	0.8			KN
2969.955	2970.822	33660.72	SM	1.0D	1.0		MIT
2969.020	2969.887	33671.32	SM	1.1	0.3		MIT
2962.744	2963.609	33742.64	SM	0.8	0.7		MIT
2962.032	2962.897	33750.75	SM	0.5			KN
2961.967	2962.832	33751.49	SM	0.7			MIT
2961.849	2962.714	33752.83	SM	0.6			MIT
2961.324	2962.189	33758.82	SM	0.6			MIT
2960.812	2961.677	33764.66	SM	0.6			MIT
2960.177	2961.042	33771.90	SM	0.9	0.7		MIT
2958.987	2959.851	33785.48	SM	0.6	0.5		MIT
2957.483	2958.347	33802.66	SM	0.7			MIT
2956.245	2957.109	33816.81	SM	0.8	0.3		MIT
2956.046	2956.910	33819.09	SM	0.8			MIT
2954.934	2955.797	33831.82	SM	0.7S	0.5		MIT
2953.189	2954.052	33851.81	SM	1.1	0.6		MIT
2952.528	2953.391	33859.39	SM	0.9	0.5		MIT
2952.221	2953.084	33862.91	SM	0.9	0.3		MIT
2947.194	2948.055	33920.66	SM	0.9	0.5		MIT
2946.420	2947.281	33929.57	SM	0.8			MIT
2943.786	2944.647	33959.93	SM	0.9	0.3		MIT
2943.492	2944.353	33963.32	SM	0.9	0.5		MIT
2938.816	2939.675	34017.36	SM	0.8			MIT
2937.487	2938.346	34032.75	SM	1.1	0.7		MIT
2935.893	2936.752	34051.23	SM	0.9	0.3		MIT
2935.413	2936.272	34056.79	SM	0.3H			MIT
2933.548	2934.406	34078.45	SM	0.5	0.3		MIT
2932.060	2932.918	34095.74	SM	0.8			MIT
2928.259	2929.116	34140.00	SM	0.5J	0.0		MIT
2927.749	2928.606	34145.94	SM	0.8			MIT
2927.609	2928.466	34147.58	SM	0.9	0.0		MIT
2927.109	2927.966	34153.41	SM	1.0	0.0		MIT
2926.657	2927.513	34158.68	SM	0.9			MIT
2924.654	2925.510	34182.08	SM	0.8			MIT
2922.822	2923.677	34203.50	SM	0.5	0.0		MIT
2921.422	2922.277	34219.89	SM	0.6	0.3		MIT
2921.314	2922.169	34221.15	SM	0.8	0.0		MIT
2920.828	2921.683	34226.85	SM	0.8			KN
2920.599	2921.454	34229.53	SM	0.6			MIT
2919.264	2920.119	34245.19	SM	0.5			MIT
2917.623	2918.477	34264.44	SM	1.0	1.0		MIT

AIR WAVELENGTH (ANGSTRMS)	VACUUM WAVELENGTH (ANGSTRMS)	WAVENUMBER (KAYSERS)	LMENT	LOG ARC	INTENSITY SPK	INTENSITY DIS	REF
2917.376	2918.230	34267.35	SM	0.3	0.0H		MIT
2914.709	2915.562	34298.70	SM	0.9	0.5		MIT
2912.564	2913.417	34323.96	SM	1.2			KN
2911.712	2912.565	34334.00	SM	0.8	0.0		MIT
2911.272	2912.125	34339.19	SM	1.2	0.0H		MIT
2910.269	2911.121	34351.02	SM	1.1	0.5		MIT
2888.762	2889.609	34606.76	SM		1.3		MIT
2878.449	2879.294	34730.74	SM	0.6	1.4		MIT
2777.76	2778.58	35989.6	SM	0.3	0.3		MIT
2772.81	2773.63	36053.9	SM		1.3		MIT
2727.74	2728.55	36649.5	SM	0.8D	0.7D		MIT
2466.85	2467.60	40525.3	SM	0.6	1.0		MIT
9856.26	9858.96	10143.1	SN	1.0			ME
9850.52	9853.22	10149.0	SN	2.7L			ME
9805.38	9808.07	10195.7	SN	2.5X			ME
9741.4	9744.1	10263.	SN	2.0A			ME
9616.40	9619.04	10396.1	SN	2.2H			ME
9451.78	9454.37	10577.1	SN	1.0H			ME
9415.37	9417.95	10618.0	SN I	1.9B			ME
9410.86	9413.44	10623.1	SN	1.7X			ME
9362.21	9364.78	10678.3	SN	1.3H			ME
9272.5	9275.0	10782.	SN	1.0H			ME
9145.1	9147.6	10932.	SN	1.3V			ME
9023.1	9025.6	11080.	SN	1.3V			ME
9018.9	9021.4	11085.	SN I	1.5V			ME
8907.6	8910.0	11223.	SN	1.3J			ME
8886.61	8889.05	11249.8	SN	0.7			WA
8812.6	8815.0	11344.	SN	1.0V			ME
8708.3	8710.7	11480.	SN	1.3V			ME
8681.7	8684.1	11515.	SN	1.7V			ME
8649.3	8651.7	11558.	SN	1.3B			ME
8552.60	8554.95	11689.1	SN I	2.7			ME
8457.6	8459.9	11820.	SN	1.0J			WA
8422.72	8425.03	11869.4	SN	2.5B			ME
8391.31	8393.62	11913.8	SN I	1.3B			ME
8357.04	8359.34	11962.7	SN I	1.9			ME
8349.35	8351.64	11973.7	SN	1.5H			ME
8345.23	8347.52	11979.6	SN	1.0H			ME
8338.6	8340.9	11989.	SN	1.0X			ME
8250.03	8252.30	12117.8	SN	0.9B			ME
8121.0	8123.2	12310.	SN	1.5B			ME
8114.10	8116.33	12320.8	SN I	2.3			ME
8100.3	8102.5	12342.	SN I	1.6B			ME
8039.3	8041.5	12435.	SN	1.6B			ME
8030.5	8032.7	12449.	SN	2.0B			ME
7971.02	7973.21	12542.0	SN	1.0H			ME
7910.46	7912.64	12638.0	SN	1.0H			ME
7904.00	7906.17	12648.3	SN II			0.9	MC
7863.74	7865.90	12713.1	SN	0.9H			ME
7808.22	7810.37	12803.5	SN I	1.0H			ME
7754.96	7757.09	12891.4	SN I	2.0			ME
7741.80	7743.93	12913.3	SN II			1.5	MC
7685.30	7687.42	13008.3	SN I	1.3			ME
7387.79	7389.83	13532.1	SN II		1.20		MC
7191.40	7193.38	13901.7	SN II			1.6	MC
6844.20	6846.09	14606.9	SN II			1.7B	WT
6761.45	6763.32	14785.7	SN II			1.2	MC
6579.26	6581.08	15195.1	SN		1.2J		AR
6563.24	6565.05	15232.2	SN		1.7V		AR
6462.36	6464.15	15469.9	SN	0.6	2.5J		WT
6453.58	6455.36	15491.0	SN II	0.8	2.5J		WT
6444.83	6446.61	15512.0	SN	0.7			WT
6310.83	6312.58	15841.4	SN	1.0	0.9		WT
6275.78	6277.52	15929.9	SN	0.3			WT
6254.36	6256.09	15984.4	SN	0.3H			WT
6203.62	6205.34	16115.2	SN	0.6	0.9		WT
6171.49	6173.20	16199.1	SN	1.0	1.0		WT
6154.60	6156.30	16243.5	SN	1.1	1.4		WT
6149.67	6151.37	16256.5	SN	1.2	1.7		WT
6079.70	6081.38	16443.6	SN II	0.3H		1.0	MC

AIR WAVELENGTH (ANGSTRMS)	VACUUM WAVELENGTH (ANGSTRMS)	WAVENUMBER (KAYSERS)	LMENT		LOG ARC	INTENSITY SPK	DIS	REF
6077.48	6079.16	16449.6	SN	II			1.1	MC
6073.52	6075.20	16460.4	SN		0.9	0.9		WT
6068.94	6070.62	16472.8	SN		1.1	1.4		WT
6060.75	6062.43	16495.0	SN		0.3			MIT
6054.90	6056.58	16511.0	SN		1.1	1.5		WT
6037.70	6039.37	16558.0	SN		1.1	1.7		WT
6011.08	6012.74	16631.3	SN		0.5			WT
5970.30	5971.95	16744.9	SN		1.0	1.5		WT
5934.72	5936.36	16845.3	SN		0.5H			WT
5925.48	5927.12	16871.6	SN		0.9	1.4		WT
5801.79	5803.40	17231.3	SN	I	0.3			WT
5799.35	5800.96	17238.5	SN	I		0.9		RO
5799.18	5800.79	17239.0	SN	II	0.3H		1.5	MC
5797.20	5798.81	17244.9	SN	II	0.3H		0.3	MC
5761.77	5763.37	17351.0	SN		0.6	1.0		WT
5753.61	5755.21	17375.6	SN		0.6	1.2		WT
5631.685	5633.248	17751.75	SN		1.7	2.3		MIT
5596.20	5597.75	17864.3	SN	II	0.7		0.6	MC
5589.43	5590.98	17885.9	SN	II	0.8	0.9		AR
5588.92	5590.47	17887.6	SN	II	0.3H		1.7	MC
5562.90	5564.44	17971.2	SN	II		0.7		AR
5561.95	5563.49	17974.3	SN	II			1.6	MC
5511.46	5512.99	18139.0	SN			0.7		AR
5498.64	5500.17	18181.3	SN		0.3H	0.3		AR
5479.14	5480.66	18246.0	SN			0.3		AR
5463.6	5465.1	18298.	SN		0.3H	0.3		AR
5333.21	5334.69	18745.2	SN	II	0.3H	0.0		AR
5332.36	5333.84	18748.2	SN	II			1.3	MC
5223.82	5225.27	19137.7	SN			0.3		AR
5198.784	5200.232	19229.91	SN			0.7		AR
5120.950	5122.377	19522.19	SN			0.3		AR
5100.35	5101.77	19601.0	SN			0.7		AR
5072.67	5074.08	19708.0	SN	II	0.3		0.6	MC
5071.14	5072.55	19713.9	SN	II	0.3		0.6	MC
4944.31	4945.69	20219.6	SN	II	0.8	1.0		MC
4877.22	4878.58	20497.8	SN	II			0.8	MC
4858.30	4859.66	20577.6	SN			0.7		AR
4792.22	4793.56	20861.3	SN	II			0.3	MC
4585.64	4586.92	21801.1	SN			1.4J		MIT
4580.27	4581.55	21826.7	SN	II			0.6	MIT
4579.13	4580.41	21832.1	SN	II			0.6	MC
4524.741	4526.010	22094.52	SN	I	2.7J	1.7		MIT
4330.04	4331.26	23088.0	SN			0.8		AR
4294.65	4295.86	23278.2	SN	II			0.5	MC
4293.61	4294.82	23283.9	SN	II			0.3	MC
4111.59	4112.75	24314.6	SN				0.3	MC
4110.51	4111.67	24321.0	SN	II			0.3	MC
4077.72	4078.87	24516.6	SN		0.3	0.5		MIT
4071.79	4072.94	24552.3	SN				0.6	LG
3994.51	3995.64	25027.3	SN	II			0.3	MC
3962.75	3963.87	25227.9	SN			0.7		AR
3930.37	3931.48	25435.7	SN	II			0.8	LG
3907.150	3908.257	25586.85	SN			0.6		MIT
3861.2	3862.3	25891.	SN			0.3		AR
3841.467	3842.557	26024.34	SN	II		0.8		MIT
3800.996	3802.075	26301.43	SN		2.3H	2.2H		HZ
3797.42	3798.50	26326.2	SN			0.5		AR
3783.74	3784.81	26421.4	SN			0.3H		MIT
3779.52	3780.59	26450.9	SN			0.3		AR
3764.47	3765.54	26556.6	SN			0.7		AR
3745.8	3746.9	26689.	SN			0.8		AR
3735.02	3736.08	26766.0	SN			0.7		AR
3730.10	3731.16	26801.3	SN			0.7V		MIT
3715.21	3716.27	26908.7	SN	II		0.3H		MIT
3707.92	3708.97	26961.6	SN			0.3H		MIT
3705.059	3706.113	26982.45	SN	II		0.5		MIT
3655.778	3656.819	27346.17	SN		1.5	1.4H		MIT
3620.54	3621.57	27612.3	SN	II			0.8	MC
3620.08	3621.11	27615.8	SN	II			0.3	MC
3598.74	3599.77	27779.6	SN			0.5J		MIT

AIR WAVELENGTH (ANGSTRMS)	VACUUM WAVELENGTH (ANGSTRMS)	WAVENUMBER (KAYSERS)	LMENT		LOG ARC	INTENSITY SPK	DIS	REF
3587.33	3588.35	27867.9	SN	II			0.5	MC
3582.39	3583.41	27906.4	SN	II			0.5	MC
3581.68	3582.70	27911.9	SN	II		0.8		RO
3575.45	3576.47	27960.5	SN	II		0.7J		MC
3573.87	3574.89	27972.9	SN	II		0.3		RO
3570.109	3571.128	28002.35	SN			0.7		MIT
3537.564	3538.575	28259.96	SN	II		0.3		MIT
3472.46	3473.45	28789.8	SN	II		0.5H		MC
3471.36	3472.35	28798.9	SN	II		1.0J		RO
3465.73	3466.72	28845.7	SN				0.5	LG
3413.03	3414.01	29291.1	SN				1.2	AR
3412.8	3413.8	29293.	SN			0.7V		AR
3407.48	3408.46	29338.8	SN	II			0.3	MC
3399.04	3400.02	29411.6	SN			1.4V		MIT
3352.435	3353.399	29820.49	SN			2.0H		MIT
3351.97	3352.93	29824.6	SN	II			1.8	MC
3330.594	3331.552	30016.04	SN		2.0H	2.0H		MIT
3283.51	3284.46	30446.4	SN	II		2.0H		MIT
3283.21	3284.16	30449.2	SN	II			1.7	MC
3262.328	3263.269	30644.12	SN	I	2.6H	2.5H		MIT
3223.572	3224.503	31012.53	SN	I	1.0	0.7		ME
3218.707	3219.637	31059.41	SN	I	1.3	0.7		ME
3175.019	3175.938	31486.76	SN	I	2.7H	2.6F		MIT
3141.809	3142.719	31819.58	SN		1.3	1.5		MIT
3123.13	3124.04	32009.9	SN			0.3J		MIT
3122.55	3123.46	32015.8	SN			0.3		MIT
3102.23	3103.13	32225.5	SN				0.3	MC
3094.69	3095.59	32304.0	SN	II			0.8	MC
3071.2	3072.1	32551.	SN			0.3V		AR
3067.755	3068.647	32587.66	SN		1.0	1.2		MIT
3047.05	3047.94	32809.1	SN	II		0.9		MIT
3034.121	3035.004	32948.88	SN	I	2.3J	2.2J		MIT
3032.775	3033.658	32963.51	SN	I	1.7	1.3		ME
3023.94	3024.82	33059.8	SN	II		0.3H		MC
3012.18	3013.06	33188.9	SN	II			0.3H	MC
3009.147	3010.024	33222.33	SN	I	2.5H	2.3H		MIT
2997.13	2998.00	33355.5	SN	II			0.3	MC
2994.44	2995.31	33385.5	SN	II			0.5H	MC
2991.00	2991.87	33423.9	SN	II			0.3H	MC
2943.62	2944.48	33961.9	SN	II		0.3		MIT
2919.82	2920.67	34238.7	SN	II			1.5	MC
2913.542	2914.395	34312.44	SN	I	2.0V	2.0V		MIT
2912.80	2913.65	34321.2	SN	II			0.5	MC
2896.090	2896.939	34519.20	SN			0.3H		MIT
2887.77	2888.62	34618.6	SN			0.3H		MIT
2863.327	2864.168	34914.16	SN	I	2.5G	2.5G		MIT
2850.618	2851.456	35069.81	SN		1.9	2.0J		MIT
2846.44	2847.28	35121.3	SN	II		0.5		MIT
2839.989	2840.824	35201.05	SN	I	2.5G	2.5G		MIT
2813.582	2814.411	35531.42	SN	I	1.7	1.7		ME
2812.566	2813.394	35544.25	SN	I	1.1	1.2		ME
2790.186	2791.009	35829.34	SN	I	0.7	0.6		MIT
2789.323	2790.146	35840.42	SN		0.7	0.5		AR
2787.96	2788.78	35857.9	SN	I	1.7	1.7		MIT
2785.031	2785.853	35895.65	SN	I	1.8	1.8		MIT
2779.817	2780.637	35962.98	SN	I	1.9	2.0		MIT
2778.346	2779.166	35982.02	SN	II		0.3		MIT
2761.776	2762.592	36197.89	SN	I	1.0	1.0		ME
2727.82	2728.63	36648.5	SN	II			0.9	MC
2706.510	2707.313	36937.00	SN		2.3G	2.2G		MIT
2664.93	2665.72	37513.3	SN	II			0.7	MC
2661.248	2662.040	37565.18	SN	I	2.0	1.9		ME
2658.609	2659.400	37602.47	SN			0.6		MIT
2646.13	2646.92	37779.8	SN			0.3H		AR
2636.991	2637.777	37910.71	SN	I	1.2?	1.2W		MIT
2618.633	2619.415	38176.47	SN			0.5V		AR
2608.67	2609.45	38322.3	SN	II			0.5	MC
2594.423	2595.199	38532.69	SN		1.8	1.9		MIT
2593.04	2593.82	38553.2	SN			0.7J		AR
2579.08	2579.85	38761.9	SN	II			0.3	MC

AIR WAVELENGTH (ANGSTRMS)	VACUUM WAVELENGTH (ANGSTRMS)	WAVENUMBER (KAYSERS)	LMENT	LOG ARC	INTENSITY SPK	DIS	REF
2571.592	2572.362	38874.77	SN	2.0	2.1		MIT
2558.048	2558.815	39080.59	SN I	1.5?	0.8		MIT
2546.552	2547.316	39257.00	SN	2.0	2.0		MIT
2531.115	2531.876	39496.41	SN I	1.6	1.20		MIT
2523.907	2524.666	39609.20	SN	1.8	1.8		MIT
2522.61	2523.37	39629.6	SN II			1.3	MC
2504.59	2505.34	39914.7	SN			1.5	MC
2497.720	2498.473	40024.45	SN	0.9			M
2496.768	2497.521	40039.71	SN	1.0			AR
2495.719	2496.472	40056.53	SN	2.0	2.0		MIT
2491.72	2492.47	40120.8	SN	1.1J	0.7W		MIT
2487.9	2488.7	40182.	SN		1.0W		AR
2486.99	2487.74	40197.1	SN II			1.5	MC
2483.403	2484.153	40255.17	SN I	2.1	2.1		ME
2476.39	2477.14	40369.2	SN I				MIT
2455.252	2455.995	40716.69	SN I	1.1	1.1		ME
2449.79	2450.53	40807.5	SN II		1.4		ME
2448.94	2449.68	40821.6	SN II		0.70		MIT
2445.01	2445.75	40887.2	SN			1.7	AR
2436.884	2437.623	41023.57	SN		0.3		MIT
2433.477	2434.215	41081.00	SN II	0.7	0.7		MIT
2429.495	2430.232	41148.33	SN I	3.3G	3.4G		ME
2421.693	2422.429	41280.89	SN I	2.7G	2.3G		ME
2419.49	2420.23	41318.5	SN			1.6	MC
2408.150	2408.882	41513.03	SN	1.5	1.5		MIT
2384.54	2385.27	41924.0	SN II			1.2	MC
2380.744	2381.470	41990.87	SN I	1.0	1.0		MIT
2368.226	2368.949	42212.80	SN II	0.5	0.9		MIT
2360.34	2361.06	42353.8	SN II			1.0	MC
2357.886	2358.607	42397.90	SN	0.8	0.6		MIT
2354.845	2355.565	42452.65	SN	2.2G	2.2G		M
2341.12	2341.84	42701.5	SN		0.7		AR
2334.808	2335.524	42816.94	SN I	2.0G	2.0G		ME
2317.228	2317.940	43141.75	SN I	2.0G	2.0G		ME
2286.680	2287.385	43718.04	SN	1.8	1.6		MIT
2282.257	2282.961	43802.75	SN I	1.0	1.2		MIT
2268.910	2269.612	44060.40	SN I	2.0G	2.0G		ME
2267.190	2267.891	44093.83	SN I	1.2	1.0		ME
2266.04	2266.74	44116.2	SN	0.5	0.9		MIT
2265.98	2266.68	44117.4	SN			1.6	AR
2251.15	2251.85	44408.0	SN	1.4	1.7		MIT
2246.053	2246.750	44508.74	SN I	2.0G	2.0G		ME
2243.311	2244.007	44563.14	SN		0.9W		AR
2231.72	2232.41	44794.6	SN I	1.5	1.8		ME
2229.11	2229.80	44847.0	SN		0.7		MIT
2220.86	2221.55	45013.6	SN		0.7		MIT
2214.92	2215.61	45134.3	SN		0.5		MIT
2209.66	2210.35	45241.7	SN I	1.4W	1.8R		ME
2199.34	2200.03	45454.0	SN I	1.5	1.8G		ME
2194.49	2195.18	45554.4	SN I	0.5G	1.8		ME
2171.32	2172.00	46040.5	SN				MIT
2152.5	2153.2	46443.	SN	1.5			ML
2152.22	2152.90	46449.0	SN II			2.0	MC
2151.48	2152.16	46465.0	SN II		1.0		MIT
2150.84	2151.52	46478.8	SN		0.7		MIT
2149.31	2149.99	46511.9	SN II			0.5	MC
2148.74	2149.42	46524.3	SN I	1.7G	1.3		MIT
2148.460	2149.136	46530.32	SN I	1.2S			ME
2140.08	2140.75	46712.5	SN		0.3		AR
2113.90	2114.57	47291.0	SN I	1.4R	0.7R		ME
2100.93	2101.60	47582.9	SN I				ME
2096.39	2097.06	47685.9	SN I				ME
2094.35	2095.02	47732.3	SN I				ME
2091.58	2092.24	47795.5	SN I				ME
2091.44	2092.10	47798.7	SN		0.5		AR
2073.08	2073.74	48222.0	SN I				ME
2072.89	2073.55	48226.4	SN	1.0	0.8		ME
2069.93	2070.59	48295.4	SN		0.3		M
2068.58	2069.24	48326.9	SN I				ME
2041.2	2041.9	48975.	SN	1.0			ML

AIR WAVELENGTH (ANGSTRMS)	VACUUM WAVELENGTH (ANGSTRMS)	WAVENUMBER (KAYSERS)	LMENT	LOG ARC	INTENSITY SPK	DIS	REF
2039.5	2040.2	49016.	SN		2.3		LG
2034.8	2035.5	49129.	SN		0.7		LG
9644.2	9646.8	10366.	SR II				SD
7673.06	7675.17	13029.0	SR I	2.3C			ME
7621.50	7623.60	13117.2	SR I	2.0			ME
7438.42	7440.47	13440.0	SR I	0.3			ME
7408.12	7410.16	13495.0	SR I	0.5			ME
7405.79	7407.83	13499.2	SR	0.3			LR
7309.41	7311.42	13677.2	SR I	2.3			MIT
7287.41	7289.42	13718.5	SR I	1.0H			ME
7232.27	7234.26	13823.1	SR I	1.7B			MIT
7167.24	7169.22	13948.5	SR I	2.0B			MIT
7153.09	7155.06	13976.1	SR I	1.2			MIT
7070.10	7072.05	14140.2	SR I	3.0			MIT
6892.59	6894.49	14504.3	SR I	2.0			MIT
6878.38	6880.28	14534.3	SR I	2.7			MIT
6791.05	6792.92	14721.2	SR I	2.3			MIT
6643.540	6645.375	15048.06	SR I	2.0			MIT
6617.259	6619.087	15107.82	SR I	2.2			MIT
6550.255	6552.065	15262.36	SR I	2.0	1.0		MIT
6546.791	6548.600	15270.44	SR I	1.4	0.7		MIT
6509.20	6511.00	15358.6	SR II	0.3H			SD
6503.995	6505.792	15370.92	SR	1.5	1.3		MIT
6465.793	6467.580	15461.73	SR	0.7	0.5		MIT
6446.676	6448.458	15507.58	SR	0.9	0.6		MIT
6408.473	6410.244	15600.03	SR	1.7	1.3		MIT
6388.239	6390.005	15649.44	SR I	1.5	1.0		MIT
6386.501	6388.267	15653.70	SR I	1.5	1.0		MIT
6380.746	6382.510	15667.82	SR I	1.5	0.9		MIT
6369.962	6371.723	15694.34	SR	1.4	0.7		MIT
6363.939	6365.698	15709.19	SR I	1.4	0.6		MIT
6345.749	6347.504	15754.22	SR I	1.4	0.6		MIT
6321.79	6323.54	15813.9	SR I	0.6			MIT
6317.3	6319.0	15825.	SR	0.6J	0.6		HP
6295.65	6297.39	15879.6	SR	0.9	0.0		HP
6272.052	6273.787	15939.34	SR	1.2			HP
6248.52	6250.25	15999.4	SR	0.6H	0.3		HP
6178.89	6180.60	16179.7	SR	0.8			EX
6158.967	6160.671	16232.00	SR	0.8			HP
6092.85	6094.54	16408.1	SR	0.6			EX
5970.10	5971.75	16745.5	SR I	1.0			FL
5816.78	5818.39	17186.9	SR	0.7			MIT
5767.05	5768.65	17335.1	SR	0.6			FL
5693.13	5694.71	17560.2	SR	0.6			SD
5674.03	5675.60	17619.3	SR	0.5			SD
5556.450	5557.993	17992.11	SR	0.6			MIT
5543.357	5544.897	18034.60	SR I	1.5	0.7		MIT
5540.052	5541.591	18045.36	SR	1.3	1.5		MIT
5534.807	5536.344	18062.46	SR	1.3	1.2		MIT
5521.83	5523.36	18104.9	SR	1.7	1.0		FL
5504.17	5505.70	18163.0	SR	1.8	1.4		FL
5486.12	5487.64	18222.8	SR	1.6	0.9		FL
5480.84	5482.36	18240.3	SR	2.0H	1.5		FL
5450.842	5452.357	18340.69	SR	1.5			MIT
5385.45	5386.95	18563.4	SR II	0.3H			SD
5329.823	5331.306	18757.13	SR I	1.6	0.3		MIT
5263.73	5265.19	18992.6	SR II	0.3			SD
5256.903	5258.366	19017.31	SR	2.0	1.4		MIT
5238.551	5240.009	19083.94	SR	2.0	1.2		MIT
5229.70	5231.16	19116.2	SR	1.8	0.9		MIT
5229.270	5230.726	19117.81	SR	1.8	0.9		MIT
5225.113	5226.568	19133.01	SR	1.8	0.9		MIT
5222.199	5223.653	19143.69	SR	1.8	0.9		MIT
5212.980	5214.431	19177.55	SR	1.0			MIT
5165.46	5166.90	19354.0	SR I	1.2			FL
5156.066	5157.502	19389.23	SR I	1.9	1.3		MIT
4971.668	4973.055	20108.36	SR I	0.3			ISN
4967.944	4969.330	20123.44	SR I	1.3			ISN
4962.263	4963.648	20146.47	SR I	1.6			ISN
4892.663	4894.029	20433.06	SR I	1.2			MIT

AIR WAVELENGTH (ANGSTRMS)	VACUUM WAVELENGTH (ANGSTRMS)	WAVENUMBER (KAYSERS)	LMENT	LOG ARC	INTENSITY SPK	DIS	REF
4891.980	4893.346	20435.91	SR I	1.6			ISN
4876.325	4877.687	20501.52	SR	2.3	1.8		MIT
4876.06	4877.42	20502.6	SR I	0.8	0.3		FL
4872.493	4873.854	20517.64	SR I	1.4			ISN
4869.19	4870.55	20531.6	SR I	0.5H			MIT
4868.700	4870.060	20533.63	SR I	1.3			ISN
4855.045	4856.401	20591.38	SR I	1.3			ISN
4832.075	4833.425	20689.26	SR I	2.3	0.9		MIT
4811.881	4813.226	20776.09	SR I	1.6			ISN
4784.320	4785.658	20895.77	SR I	1.5			ISN
4755.495	4756.825	21022.43	SR I	1.1H			ISN
4741.922	4743.248	21082.60	SR I	1.5			ISN
4729.466	4730.789	21138.12	SR I	0.6H			ISN
4722.278	4723.599	21170.30	SR I	1.5			ISN
4713.959	4715.278	21207.66	SR I	0.5			ISN
4707.288	4708.605	21237.71	SR I	1.0			MIT
4703.984	4705.300	21252.63	SR I	0.3			ISN
4678.326	4679.635	21369.19	SR I	1.3H			ISN
4607.331	4608.622	21698.46	SR I	3.0G	1.7G		IHZ
4531.348	4532.619	22062.30	SR I	1.0			ISN
4480.507	4481.764	22312.64	SR I	1.0H			ISN
4451.804	4453.054	22456.50	SR I	0.3			ISN
4438.044	4439.290	22526.12	SR I	1.4			ISN
4412.621	4413.860	22655.91	SR I	0.6			ISN
4406.85	4408.09	22685.6	SR	1.7H			MIT
4406.12	4407.36	22689.3	SR I	0.5			SD
4361.710	4362.936	22920.35	SR I	1.3			ISN
4337.89	4339.11	23046.2	SR I	2.2	1.7		FL
4337.664	4338.884	23047.40	SR I	1.5H			ISN
4326.445	4327.662	23107.17	SR I	0.9			ISN
4319.12	4320.33	23146.4	SR I	1.7	1.3		FL
4319.053	4320.268	23146.71	SR I	1.4H			ISN
4313.182	4314.395	23178.22	SR I	0.5H			ISN
4312.74	4313.95	23180.6	SR II	0.8	0.5		SD
4308.105	4309.317	23205.53	SR I	1.3H			ISN
4305.447	4306.658	23219.86	SR I	1.6			ISN
4296.82	4298.03	23266.5	SR	0.5	0.3		SD
4253.5	4254.7	23503.	SR	0.3			SD
4252.97	4254.17	23506.4	SR I	0.3			FL
4240.54	4241.73	23575.3	SR II	0.0	0.5		SD
4215.524	4216.711	23715.16	SR II	2.5R	2.60		ISN
4202.522	4203.706	23788.53	SR I	0.8			ISN
4161.796	4162.969	24021.32	SR I	1.5			ISN
4140.35	4141.52	24145.7	SR	0.7	0.3		SD
4107.55	4108.71	24338.5	SR II	0.0	0.5		SD
4087.344	4088.498	24458.86	SR I	1.1H			ISN
4077.714	4078.865	24516.62	SR II	2.6R	2.70		ISN
4070.86	4072.01	24557.9	SR I	0.5			SD
4061.06	4062.21	24617.2	SR I	0.9	0.8		SD
4056.67	4057.82	24643.8	SR II	0.6	0.6		SD
4050.9	4052.0	24679.	SR	0.0H	0.3H		SD
4033.191	4034.331	24787.26	SR I	0.8			ISN
4032.379	4033.518	24792.25	SR I	1.3			ISN
4030.377	4031.516	24804.56	SR I	1.6			ISN
4009.75	4010.88	24932.2	SR		0.3		SD
3970.043	3971.166	25181.52	SR I	1.3			ISN
3969.261	3970.384	25186.48	SR I	1.5			ISN
3962.61	3963.73	25228.7	SR	1.0H			M
3950.81	3951.93	25304.1	SR I	0.9H			SD
3940.800	3941.916	25368.38	SR I	1.3			ISN
3935.18	3936.29	25404.6	SR I	0.8H			SD
3926.13	3927.24	25463.2	SR I	0.6H			M
3925.00	3926.11	25470.5	SR	0.6W	0.3		SD
3867.1	3868.2	25852.	SR I	0.3H			SD
3865.46	3866.56	25862.8	SR I	1.7			FL
3807.38	3808.46	26257.3	SR I	1.7			FL
3780.46	3781.53	26444.3	SR I	1.5			FL
3774.22	3775.29	26488.0	SR II	0.0	0.3H		SD
3769.99	3771.06	26517.7	SR II	0.5	0.5		SD
3762.00	3763.07	26574.1	SR	0.0	0.5		SD
3729.34	3730.40	26806.8	SR		0.7		SD
3717.30	3718.36	26893.6	SR II	0.0	0.7		SD
3705.901	3706.955	26976.32	SR I	1.5	1.0		ISN
3677.56	3678.61	27184.2	SR		0.5		SD
3653.928	3654.969	27360.01	SR I	1.2	0.5		ISN
3653.270	3654.311	27364.94	SR I	1.5	0.9		ISN
3629.144	3630.179	27546.85	SR I	1.5			ISN
3628.345	3629.379	27552.92	SR I	1.0			ISN
3593.214	3594.239	27822.30	SR II	0.5	0.3		MIT
3577.243	3578.264	27946.51	SR I	0.3			ISN
3559.087	3560.104	28089.07	SR II		0.3H		MIT
3556.44	3557.46	28110.0	SR		0.3		SD
3553.4	3554.4	28134.	SR I	0.6			SD
3548.66	3549.67	28171.6	SR I	0.3			FL
3548.083	3549.097	28176.18	SR I	1.7			ISN
3523.51	3524.52	28372.7	SR	0.0	0.3		SD
3504.27	3505.27	28528.4	SR I	0.6			HP
3499.672	3500.673	28565.93	SR I	1.7			ISN
3491.62	3492.62	28631.8	SR II	0.0	0.3		SD
3474.887	3475.882	28769.68	SR II	1.9	1.7		ISN
3464.457	3465.449	28856.29	SR II	2.3	2.3		ISN
3456.52	3457.51	28922.6	SR I	0.5			FL
3452.30	3453.29	28957.9	SR II	0.0	0.3		SD
3441.43	3442.42	29049.4	SR		0.3H		SD
3434.28	3435.26	29109.8	SR I	0.3			FL
3418.36	3419.34	29245.4	SR	0.3	0.3		SD
3411.94	3412.92	29300.4	SR I	0.6			FL
3401.23	3402.21	29392.7	SR I	0.5			FL
3390.67	3391.64	29484.2	SR I	0.6			FL
3380.711	3381.682	29571.08	SR II	2.2	2.3		ISN
3376.35	3377.32	29609.3	SR II		0.6		SD
3371.00	3371.97	29656.3	SR	0.3	0.8		MIT
3366.333	3367.300	29697.38	SR I	2.0	1.0		ISN
3351.246	3352.209	29831.07	SR I	2.5	1.2		ISN
3343.46	3344.42	29900.5	SR		0.8		SD
3329.988	3330.946	30021.50	SR I	2.0	1.0		ISN
3322.231	3323.187	30091.60	SR I	2.0	0.9		ISN
3321.344	3322.300	30099.63	SR II		1.0D		MIT
3311.85	3312.80	30185.9	SR	0.0	0.6		SD
3307.534	3308.486	30225.30	SR I	2.3	1.0S		ISN
3301.734	3302.685	30278.40	SR I	2.0	1.0		ISN
3288.00	3288.95	30404.9	SR	0.0H	0.7		SD
3209.56	3210.49	31147.9	SR II		0.6		SD
3200.22	3201.14	31238.8	SR I	0.3H			SD
3198.99	3199.91	31250.8	SR I	1.0	0.6		SD
3190.033	3190.955	31338.58	SR	1.0	0.9		MIT
3189.23	3190.15	31346.5	SR I	0.6O	0.6		SD
3182.3	3183.2	31415.	SR	0.7			FL
3172.2	3173.1	31515.	SR	0.7			FL
3131.75	3132.66	31921.8	SR II		1.0		SD
3123.30	3124.21	32008.1	SR	0.3	0.3		SD
3075.01	3075.90	32510.8	SR II	0.0	0.5		SD
2931.830	2932.688	34098.41	SR I	1.5	0.9		ISN
2825.52	2826.35	35381.3	SR II			0.6H	MC
2569.469	2570.239	38906.89	SR I	1.4G	0.7		ISN
2549.53	2550.30	39211.1	SR	0.7			SD
2471.597	2472.344	40447.44	SR II	0.9	0.7		ISN
2428.095	2428.832	41172.05	SR I	1.0H			ISN
2423.569	2424.305	41248.94	SR II	0.7	0.7		ISN
2354.319	2355.039	42462.14	SR I	0.7G			ISN
2324.52	2325.23	43006.4	SR II	0.5			SD
2322.355	2323.068	43046.52	SR II	0.9	0.7		ISN
2307.4	2308.1	43325.	SR I	0.5		0.5	SD
2281.999	2282.703	43807.71	SR II	0.9	0.7		ISN
2275.48	2276.18	43933.2	SR I	0.3		0.5	FL
2253.34	2254.04	44364.8	SR I	0.5		0.5	FL
2237.3	2238.0	44683.	SR I	0.5		0.5	SD
2225.9	2226.6	44912.	SR I	0.9			SD
2165.96	2166.64	46154.4	SR II	1.4G	1.2		MIT
2152.84	2153.52	46435.7	SR II	1.2	1.2		MIT

AIR WAVELENGTH (ANGSTRMS)	VACUUM WAVELENGTH (ANGSTRMS)	WAVENUMBER (KAYSERS)	LMENT	LOG INTENSITY ARC	SPK	DIS	REF
2053.2	2053.9	48689.	SR			0.5	SD
9941.50	9944.23	10056.1	TA	0.7			KS
9818.67	9821.36	10181.9	TA	0.3			KS
9780.93	9783.61	10221.2	TA	0.3			KS
9645.53	9648.18	10364.7	TA	0.3H			KS
9569.57	9572.19	10446.9	TA	0.3H			KS
9438.55	9441.14	10591.9	TA	0.3			KS
9423.80	9426.39	10608.5	TA	0.3			KS
9272.63	9275.17	10781.5	TA	0.5H			KS
9254.95	9257.49	10802.1	TA	1.0			KS
9246.16	9248.70	10812.3	TA	0.5			KS
9231.15	9233.68	10829.9	TA	0.3H			MIT
9197.40	9199.92	10869.7	TA	0.7V			KS
9196.05	9198.57	10871.2	TA	0.3V			KS
9105.70	9108.20	10979.1	TA	0.5			KS
8936.51	8938.96	11187.0	TA	0.3H			KS
8932.64	8935.09	11191.8	TA	0.8H			MIT
8834.82	8837.25	11315.7	TA	0.3			MIT
8595.84	8598.20	11630.3	TA	0.6H			MIT
8550.49	8552.84	11692.0	TA	0.5			MIT
8447.62	8449.94	11834.4	TA	1.8			KS
8415.73	8418.04	11879.2	TA	1.0			KS
8389.06	8391.37	11917.0	TA	1.0			MIT
8281.62	8283.90	12071.6	TA	1.7			MIT
8264.85	8267.12	12096.1	TA	1.6			MIT
8248.95	8251.22	12119.4	TA	1.7			MIT
8158.54	8160.78	12253.7	TA	1.0			MIT
8128.76	8131.00	12298.6	TA	1.3			MIT
8100.11	8102.34	12342.1	TA	1.2			MIT
8088.85	8091.07	12359.3	TA	0.3			MIT
8068.98	8071.20	12389.7	TA	1.4			MIT
8053.93	8056.14	12412.9	TA	0.7			MIT
8039.08	8041.29	12435.8	TA	1.4			MIT
8029.04	8031.25	12451.4	TA	0.9			KS
8026.50	8028.71	12455.3	TA	1.7			MIT
8022.09	8024.30	12462.2	TA	0.7			MIT
7998.75	8000.95	12498.5	TA	0.7			MIT
7955.54	7957.73	12566.4	TA	0.6W			KS
7952.07	7954.26	12571.9	TA	0.5			MIT
7950.19	7952.38	12574.9	TA	1.7			MIT
7882.37	7884.54	12683.1	TA	2.2			MIT
7842.76	7844.92	12747.1	TA	1.7			MIT
7819.57	7821.72	12784.9	TA	0.3			MIT
7818.73	7820.88	12786.3	TA	0.5H			MIT
7815.48	7817.63	12791.6	TA	1.4			MIT
7814.03	7816.18	12794.0	TA	1.7			MIT
7799.51	7801.66	12817.8	TA	0.5			KS
7788.95	7791.09	12835.2	TA	0.5			KS
7788.64	7790.78	12835.7	TA	0.5			KS
7779.72	7781.86	12850.4	TA	1.5W			MIT
7773.53	7775.67	12860.6	TA	0.5			MIT
7763.11	7765.25	12877.9	TA	1.4			MIT
7758.23	7760.36	12886.0	TA	0.3			KS
7733.77	7735.90	12926.7	TA	0.3			MIT
7725.04	7727.17	12941.4	TA	0.6			MIT
7722.02	7724.15	12946.4	TA	1.4			MIT
7719.16	7721.28	12951.2	TA	0.5H			MIT
7715.03	7717.15	12958.2	TA	0.3			KS
7699.14	7701.26	12984.9	TA	0.7			KS
7690.37	7692.49	12999.7	TA	0.5			MIT
7676.20	7678.31	13023.7	TA	0.5			MIT
7670.28	7672.39	13033.7	TA	1.2O			KS
7650.42	7652.53	13067.6	TA	0.3			KS
7649.62	7651.73	13068.9	TA	0.7			KS
7640.84	7642.94	13084.0	TA	0.6			KS
7625.97	7628.07	13109.5	TA	0.3			KS
7622.64	7624.74	13115.2	TA	0.3H			KS
7620.84	7622.94	13118.3	TA	0.3			KS
7604.81	7606.90	13145.9	TA	0.3H			KS
7593.58	7595.67	13165.4	TA	0.3			MIT
7590.22	7592.31	13171.2	TA	0.9			KS
7569.23	7571.31	13207.7	TA	0.5			MIT
7553.94	7556.02	13234.5	TA	0.3			MIT
7520.56	7522.63	13293.2	TA	1.4			MIT
7515.17	7517.24	13302.8	TA	0.7W			MIT
7495.35	7497.41	13337.9	TA	0.6			MIT
7485.90	7487.96	13354.8	TA	2.5			MIT
7472.04	7474.10	13379.5	TA	1.5H			MIT
7469.08	7471.14	13384.8	TA	0.7			MIT
7467.75	7469.81	13387.2	TA	1.6			MIT
7460.82	7462.87	13399.7	TA	0.7			MIT
7454.35	7456.40	13411.3	TA	0.7H			MIT
7440.17	7442.22	13436.8	TA	1.6			MIT
7435.19	7437.24	13445.8	TA	0.7H			MIT
7418.92	7420.96	13475.3	TA	0.3			MIT
7407.89	7409.93	13495.4	TA	2.2			MIT
7404.93	7406.97	13500.8	TA	0.5			KS
7369.09	7371.12	13566.5	TA	2.0			MIT
7356.96	7358.99	13588.8	TA	2.0			MIT
7355.44	7357.47	13591.6	TA	1.0			MIT
7352.86	7354.89	13596.4	TA	2.2			MIT
7346.41	7348.43	13608.3	TA	2.0			MIT
7343.33	7345.35	13614.1	TA	0.3			MIT
7340.19	7342.21	13619.9	TA	1.0H			MIT
7325.95	7327.97	13646.3	TA	1.0			MIT
7322.72	7324.74	13652.4	TA	1.5			MIT
7319.84	7321.86	13657.7	TA	1.5			MIT
7313.18	7315.19	13670.2	TA	0.5			MIT
7301.74	7303.75	13691.6	TA	2.3			MIT
7296.32	7298.33	13701.8	TA	1.6			MIT
7286.36	7288.37	13720.5	TA	0.7			MIT
7277.54	7279.55	13737.1	TA	0.5			MIT
7276.96	7278.97	13738.2	TA	1.5			MIT
7272.29	7274.29	13747.0	TA	0.7			MIT
7264.82	7266.82	13761.2	TA	1.0			MIT
7250.27	7252.27	13788.8	TA	1.9			MIT
7233.45	7235.44	13820.8	TA	1.0			MIT
7207.82	7209.81	13870.0	TA	0.3H			MIT
7196.24	7198.22	13892.3	TA	0.5			MIT
7191.35	7193.33	13901.8	TA	1.2			MIT
7175.77	7177.75	13931.9	TA	0.3			MIT
7174.91	7176.89	13933.6	TA	1.0			MIT
7172.90	7174.88	13937.5	TA	2.2			MIT
7148.63	7150.60	13984.8	TA	2.2			MIT
7135.22	7137.19	14011.1	TA	0.3			MIT
7125.72	7127.68	14029.8	TA	1.9			MIT
7121.27	7123.23	14038.6	TA	1.5			MIT
7118.09	7120.05	14044.8	TA	0.3			MIT
7117.52	7119.48	14046.0	TA	0.8			MIT
7116.07	7118.03	14048.8	TA	0.6			MIT
7108.05	7110.01	14064.7	TA	0.5			MIT
7093.02	7094.98	14094.5	TA	1.3			MIT
7092.16	7094.12	14096.2	TA	0.3			MIT
7091.36	7093.32	14097.8	TA	0.3			MIT
7085.40	7087.35	14109.6	TA	1.5			MIT
7081.30	7083.25	14117.8	TA	1.7			MIT
7074.68	7076.63	14131.0	TA	0.5			MIT
7049.86	7051.80	14180.8	TA	0.5			MIT
7039.07	7041.01	14202.5	TA	1.6			MIT
7036.60	7038.54	14207.5	TA	0.3			MIT
7031.51	7033.45	14217.8	TA	0.9			MIT
7025.03	7026.97	14230.9	TA	1.9			MIT
7013.94	7015.87	14253.4	TA	0.3			MIT
7006.96	7008.89	14267.6	TA	2.0			MIT
7005.90	7007.83	14269.7	TA	0.3			MIT
7005.07	7007.00	14271.4	TA	1.7			MIT
7003.10	7005.03	14275.4	TA	0.3			MIT
7000.21	7002.14	14281.3	TA	1.0S			MIT
6995.39	6997.32	14291.2	TA	2.3			MIT
6983.52	6985.45	14315.5	TA	1.3			MIT

AIR WAVELENGTH (ANGSTRMS)	VACUUM WAVELENGTH (ANGSTRMS)	WAVENUMBER (KAYSERS)	LMENT	LOG ARC	INTENSITY SPK	DIS	REF
6977.67	6979.59	14327.5	TA	0.3			MIT
6971.53	6973.45	14340.1	TA I	1.0O			MIT
6969.49	6971.41	14344.3	TA	0.5			MIT
6966.13	6968.05	14351.2	TA	2.2			MIT
6953.88	6955.80	14376.5	TA	1.7			MIT
6951.26	6953.18	14381.9	TA	2.0			MIT
6950.08	6952.00	14384.4	TA	0.3			MIT
6946.87	6948.79	14391.0	TA	1.3			MIT
6939.38	6941.29	14406.5	TA	0.6			MIT
6936.05	6937.96	14413.4	TA	0.3			KS
6928.54	6930.45	14429.1	TA	2.2			MIT
6927.38	6929.29	14431.5	TA	2.2			MIT
6902.10	6904.00	14484.3	TA	2.2			MIT
6900.55	6902.45	14487.6	TA	1.9			MIT
6896.77	6898.67	14495.5	TA	1.5			MIT
6892.40	6894.30	14504.7	TA	1.5			MIT
6877.49	6879.39	14536.2	TA	1.5			MIT
6875.27	6877.17	14540.9	TA	2.3			MIT
6866.23	6868.12	14560.0	TA	2.3			MIT
6865.13	6867.02	14562.3	TA	0.7			MIT
6850.83	6852.72	14592.7	TA	0.7			MIT
6833.26	6835.15	14630.3	TA	0.3			MIT
6832.00	6833.89	14633.0	TA	0.6			MIT
6824.96	6826.84	14648.1	TA	0.7			MIT
6819.36	6821.24	14660.1	TA	1.2			MIT
6813.25	6815.13	14673.2	TA	2.3			MIT
6810.46	6812.34	14679.2	TA	1.6			MIT
6799.27	6801.15	14703.4	TA	0.7			KS
6790.06	6791.93	14723.3	TA	0.7			MIT
6788.99	6790.86	14725.7	TA	1.7			MIT
6776.65	6778.52	14752.5	TA	0.3			MIT
6774.25	6776.12	14757.7	TA	2.0W			MIT
6771.74	6773.61	14763.2	TA	2.0			MIT
6770.37	6772.24	14766.2	TA	0.5			MIT
6755.85	6757.71	14797.9	TA	1.0			MIT
6754.91	6756.77	14800.0	TA	1.3			MIT
6740.73	6742.59	14831.1	TA	1.9			MIT
6714.44	6716.29	14889.2	TA	0.6			MIT
6709.39	6711.24	14900.4	TA	1.0			MIT
6706.46	6708.31	14906.9	TA	0.7			MIT
6705.96	6707.81	14908.0	TA	0.5			MIT
6702.99	6704.84	14914.6	TA	0.3			MIT
6693.61	6695.46	14935.5	TA	1.0			MIT
6684.00	6685.85	14957.0	TA	1.6			MIT
6675.53	6677.37	14975.9	TA	2.6			MIT
6673.73	6675.57	14980.0	TA	2.3			MIT
6669.11	6670.95	14990.4	TA	0.5			MIT
6662.30	6664.14	15005.7	TA	1.0W			MIT
6639.41	6641.24	15057.4	TA	0.3			KS
6626.18	6628.01	15087.5	TA	0.5H			MIT
6621.30	6623.13	15098.6	TA	2.3			MIT
6611.95	6613.78	15120.0	TA	2.5			MIT
6605.85	6607.67	15133.9	TA	0.5			MIT
6587.16	6588.98	15176.9	TA	0.7H			MIT
6585.13	6586.95	15181.5	TA	0.7			MIT
6578.94	6580.76	15195.8	TA	0.3			MIT
6577.552	6579.369	15199.03	TA	0.6			MIT
6577.144	6578.961	15199.97	TA	0.7			MIT
6574.84	6576.66	15205.3	TA	2.3			MIT
6564.26	6566.07	15229.8	TA	1.3H			MIT
6563.701	6565.514	15231.10	TA	0.3			MIT
6561.60	6563.41	15236.0	TA	1.6			MIT
6559.271	6561.083	15241.39	TA	0.5L			MIT
6556.902	6558.713	15246.89	TA	0.3H			MIT
6547.24	6549.05	15269.4	TA	0.7			KS
6527.038	6528.841	15316.65	TA	0.5			MIT
6516.099	6517.899	15342.37	TA	2.3			MIT
6514.39	6516.19	15346.4	TA	2.3			MIT
6505.516	6507.313	15367.32	TA	2.0			MIT
6502.43	6504.23	15374.6	TA	1.6			KS

AIR WAVELENGTH (ANGSTRMS)	VACUUM WAVELENGTH (ANGSTRMS)	WAVENUMBER (KAYSERS)	LMENT	LOG ARC	INTENSITY SPK	DIS	REF
6500.40	6502.20	15379.4	TA	0.3			KS
6486.060	6487.852	15413.42	TA	0.5			MIT
6485.37	6487.16	15415.1	TA	2.7			MIT
6472.855	6474.644	15444.86	TA	0.3			MIT
6459.921	6461.706	15475.79	TA	1.2			MIT
6459.065	6460.850	15477.84	TA	0.3			MIT
6455.828	6457.612	15485.60	TA	1.6			MIT
6453.113	6454.896	15492.11	TA	0.5			MIT
6450.365	6452.148	15498.71	TA	2.3			MIT
6445.866	6447.647	15509.53	TA	1.3			MIT
6444.610	6446.391	15512.56	TA	1.6			MIT
6443.887	6445.668	15514.30	TA	1.0H			MIT
6437.365	6439.144	15530.01	TA	0.3			MIT
6434.550	6436.328	15536.81	TA	0.7			MIT
6430.79	6432.57	15545.9	TA	2.2			MIT
6428.596	6430.373	15551.20	TA	1.6			MIT
6426.731	6428.507	15555.71	TA	0.7			MIT
6425.442	6427.218	15558.83	TA	0.5H			MIT
6418.477	6420.251	15575.71	TA	0.3			MIT
6417.99	6419.76	15576.9	TA	0.3			KS
6392.209	6393.976	15639.72	TA	1.2			MIT
6389.447	6391.213	15646.48	TA	2.0			MIT
6387.991	6389.757	15650.05	TA	0.5			MIT
6379.069	6380.833	15671.94	TA	0.9			MIT
6373.055	6374.817	15686.72	TA	1.7			MIT
6364.909	6366.669	15706.80	TA	0.5			MIT
6360.839	6362.598	15716.85	TA	2.0			MIT
6356.139	6357.896	15728.47	TA	2.0			MIT
6351.238	6352.994	15740.61	TA	0.7			MIT
6346.02	6347.77	15753.6	TA	1.6			MIT
6341.173	6342.926	15765.59	TA	1.7			MIT
6339.981	6341.734	15768.56	TA	0.7			MIT
6332.907	6334.658	15786.17	TA	1.7			MIT
6326.588	6328.337	15801.94	TA	0.7			MIT
6325.083	6326.832	15805.70	TA	2.0			MIT
6312.237	6313.983	15837.86	TA	0.7			MIT
6309.579	6311.324	15844.54	TA	2.0			MIT
6309.065	6310.810	15845.83	TA	1.2			MIT
6289.336	6291.075	15895.53	TA	1.3			MIT
6287.91	6289.65	15899.1	TA	1.0			KS
6287.361	6289.100	15900.53	TA	0.7			MIT
6283.09	6284.83	15911.3	TA	0.6			MIT
6281.334	6283.071	15915.78	TA	1.7			MIT
6278.336	6280.072	15923.38	TA	1.2			MIT
6277.565	6279.301	15925.34	TA	0.3			MIT
6274.297	6276.032	15933.63	TA	0.5			MIT
6268.700	6270.434	15947.86	TA	2.3			MIT
6266.371	6268.104	15953.79	TA	1.7O			MIT
6262.311	6264.043	15964.13	TA	0.7H			MIT
6258.734	6260.465	15973.25	TA	0.9			MIT
6256.682	6258.413	15978.49	TA	2.5			MIT
6254.686	6256.416	15983.59	TA	0.7W			MIT
6254.194	6255.924	15984.85	TA	0.3			MIT
6249.791	6251.520	15996.11	TA	2.0			MIT
6247.322	6249.050	16002.43	TA	0.3			MIT
6244.47	6246.20	16009.7	TA	1.0W			KS
6239.189	6240.915	16023.29	TA	0.7			MIT
6223.612	6225.334	16063.40	TA	0.3			MIT
6222.694	6224.416	16065.77	TA	0.3			MIT
6221.337	6223.058	16069.27	TA	0.5			MIT
6218.99	6220.71	16075.3	TA	0.7W			MIT
6217.083	6218.803	16080.27	TA	0.9			MIT
6208.372	6210.090	16102.83	TA	1.0			MIT
6200.298	6202.014	16123.80	TA	0.6B			MIT
6193.108	6194.822	16142.52	TA	1.5			MIT
6189.663	6191.376	16151.50	TA	1.7			MIT
6179.071	6180.781	16179.19	TA	1.5			MIT
6170.858	6172.566	16200.72	TA	0.9			MIT
6170.538	6172.246	16201.56	TA	0.7			MIT
6167.349	6169.056	16209.94	TA	0.9			MIT

AIR WAVELENGTH (ANGSTRMS)	VACUUM WAVELENGTH (ANGSTRMS)	WAVENUMBER (KAYSERS)	LMENT	LOG INTENSITY ARC	SPK	DIS	REF
6158.841	6160.545	16232.33	TA	1.9W			MIT
6154.504	6156.207	16243.77	TA	2.3			MIT
6152.544	6154.247	16248.94	TA	1.8			MIT
6147.087	6148.788	16263.37	TA	1.20			MIT
6144.561	6146.262	16270.05	TA	1.7			MIT
6141.199	6142.899	16278.96	TA	0.5			MIT
6140.071	6141.770	16281.95	TA	1.6L			MIT
6131.229	6132.926	16305.43	TA	0.7			MIT
6129.684	6131.381	16309.54	TA	0.3H			MIT
6126.79	6128.49	16317.2	TA	0.3			KS
6106.51	6108.20	16371.4	TA	0.5			KS
6101.576	6103.265	16384.67	TA	2.2			MIT
6092.952	6094.639	16407.86	TA	0.3			MIT
6092.065	6093.751	16410.25	TA	1.5			MIT
6091.454	6093.140	16411.90	TA	0.5			MIT
6090.823	6092.509	16413.60	TA	1.7W			MIT
6084.739	6086.424	16430.01	TA	1.20			MIT
6070.536	6072.217	16468.45	TA	0.3			MIT
6059.333	6061.011	16498.90	TA	0.3			MIT
6053.642	6055.318	16514.41	TA	2.2L			MIT
6047.247	6048.921	16531.87	TA	2.2L			MIT
6045.390	6047.064	16536.95	TA	2.3L			MIT
6034.937	6036.608	16565.59	TA	0.3			MIT
6020.720	6022.387	16604.71	TA	2.5R			MIT
6018.970	6020.637	16609.54	TA	1.2			MIT
6015.895	6017.561	16618.03	TA	1.6			MIT
6012.560	6014.225	16627.25	TA	0.6H			MIT
6009.893	6011.557	16634.62	TA	1.6			MIT
5997.234	5998.895	16669.74	TA	2.3W			MIT
5960.129	5961.780	16773.51	TA	0.8			MIT
5951.779	5953.428	16797.05	TA	1.2W			MIT
5944.024	5945.671	16818.96	TA	1.9			MIT
5939.765	5941.411	16831.02	TA	1.9L			MIT
5935.543	5937.188	16842.99	TA	1.2H			MIT
5931.679	5933.322	16853.96	TA	1.6			MIT
5931.051	5932.694	16855.75	TA	1.6S			MIT
5930.625	5932.268	16856.96	TA	1.2S			MIT
5925.897	5927.539	16870.41	TA	1.3			MIT
5918.949	5920.589	16890.21	TA	1.9S			MIT
5916.511	5918.150	16897.17	TA	1.5			MIT
5912.844	5914.482	16907.65	TA	0.3			MIT
5904.333	5905.969	16932.02	TA	0.3W			MIT
5901.911	5903.547	16938.97	TA	1.9			MIT
5897.929	5899.563	16950.41	TA	0.3L			MIT
5895.196	5896.830	16958.26	TA	0.3H			MIT
5892.448	5894.081	16966.17	TA	0.3L			MIT
5888.493	5890.125	16977.57	TA	0.7			MIT
5882.295	5883.925	16995.46	TA	1.9			MIT
5877.355	5878.984	17009.74	TA	2.0			MIT
5872.030	5873.658	17025.17	TA	0.8			MIT
5866.610	5868.236	17040.90	TA	1.8			MIT
5865.870	5867.496	17043.05	TA	1.6W			MIT
5857.020	5858.643	17068.80	TA	0.6			MIT
5853.915	5855.538	17077.85	TA	0.3			MIT
5849.953	5851.575	17089.42	TA	1.8W			MIT
5849.678	5851.300	17090.22	TA	1.9W			MIT
5848.865	5850.486	17092.60	TA	1.20			MIT
5843.939	5845.559	17107.00	TA	1.2			MIT
5842.828	5844.448	17110.26	TA	0.7			MIT
5837.275	5838.893	17126.53	TA	0.6			MIT
5836.623	5838.241	17128.45	TA	0.3H			MIT
5816.510	5818.123	17187.68	TA	1.6W			MIT
5811.104	5812.715	17203.66	TA	2.0			MIT
5783.236	5784.840	17286.56	TA	0.3			MIT
5780.706	5782.309	17294.13	TA	1.9			MIT
5780.017	5781.620	17296.19	TA	1.8			MIT
5776.768	5778.370	17305.92	TA	1.9			MIT
5771.931	5773.532	17320.42	TA	0.8L			MIT
5767.913	5769.513	17332.49	TA	2.0W			MIT
5766.561	5768.160	17336.55	TA	1.9W			MIT

AIR WAVELENGTH (ANGSTRMS)	VACUUM WAVELENGTH (ANGSTRMS)	WAVENUMBER (KAYSERS)	LMENT	LOG INTENSITY ARC	SPK	DIS	REF
5761.614	5763.212	17351.44	TA	1.2			MIT
5755.811	5757.407	17368.93	TA	1.6			MIT
5746.707	5748.301	17396.45	TA	1.8L			MIT
5733.490	5735.080	17436.55	TA	0.3H			MIT
5730.510	5732.100	17445.61	TA	0.3W			MIT
5724.406	5725.994	17464.22	TA	0.3			MIT
5716.533	5718.119	17488.27	TA	1.2S			MIT
5715.241	5716.827	17492.22	TA	1.5			MIT
5706.280	5707.863	17519.69	TA	1.7			MIT
5704.306	5705.889	17525.75	TA	1.6			MIT
5699.242	5700.823	17541.33	TA	1.9W			MIT
5688.247	5689.825	17575.23	TA	2.0W			MIT
5685.456	5687.034	17583.86	TA	0.80			MIT
5664.901	5666.473	17647.66	TA	1.8			MIT
5645.911	5647.478	17707.02	TA	1.9			MIT
5640.179	5641.744	17725.01	TA	1.4			MIT
5635.710	5637.274	17739.07	TA	1.4			MIT
5620.684	5622.244	17786.49	TA	1.9W			MIT
5617.711	5619.270	17795.90	TA	1.2			MIT
5605.505	5607.061	17834.66	TA	0.7			MIT
5598.752	5600.306	17856.17	TA	1.8			MIT
5584.016	5585.566	17903.29	TA	1.8			MIT
5581.375	5582.925	17911.76	TA	1.1			MIT
5549.311	5550.852	18015.25	TA	0.6			MIT
5548.325	5549.866	18018.45	TA	1.8			MIT
5545.140	5546.680	18028.80	TA	0.6L			MIT
5536.736	5538.274	18056.17	TA	0.3H			MIT
5528.359	5529.895	18083.53	TA	1.2			MIT
5523.982	5525.516	18097.86	TA	0.3S			MIT
5521.154	5522.688	18107.13	TA	0.6			MIT
5518.91	5520.44	18114.5	TA	2.00			KS
5517.727	5519.260	18118.37	TA	0.3S			MIT
5516.275	5517.807	18123.14	TA	0.3			MIT
5513.507	5515.039	18132.24	TA	0.3			MIT
5505.655	5507.185	18158.10	TA	1.4			MIT
5500.684	5502.212	18174.51	TA	1.3			MIT
5499.442	5500.970	18178.61	TA	1.8S			MIT
5494.776	5496.303	18194.05	TA	1.7			MIT
5490.114	5491.639	18209.50	TA	1.8			MIT
5481.162	5482.685	18239.24	TA	0.8R			MIT
5475.544	5477.065	18257.95	TA	1.6			MIT
5472.215	5473.736	18269.06	TA	0.3H			MIT
5471.565	5473.085	18271.23	TA	0.3			MIT
5461.290	5462.808	18305.60	TA	1.9			MIT
5458.412	5459.929	18315.26	TA	1.1			MIT
5456.593	5458.109	18321.36	TA	0.7H			MIT
5446.244	5447.758	18356.18	TA	0.7S			MIT
5435.267	5436.778	18393.25	TA	1.9			MIT
5433.060	5434.570	18400.72	TA	0.3H			MIT
5431.661	5433.171	18405.46	TA	1.8W			MIT
5419.130	5420.636	18448.02	TA	1.9G			MIT
5418.017	5419.523	18451.81	TA	0.3			MIT
5413.387	5414.892	18467.59	TA	1.2			MIT
5410.549	5412.053	18477.28	TA	1.2			MIT
5408.779	5410.283	18483.32	TA	1.3			MIT
5404.955	5406.458	18496.40	TA	1.9			MIT
5403.542	5405.044	18501.24	TA	1.1			MIT
5402.507	5404.009	18504.78	TA	1.9	1.5		MIT
5397.557	5399.058	18521.75	TA	1.5			MIT
5395.986	5397.486	18527.14	TA	1.9W			MIT
5389.301	5390.799	18550.12	TA	2.00			MIT
5388.508	5390.006	18552.85	TA	1.60			MIT
5385.831	5387.329	18562.08	TA	0.3			MIT
5379.665	5381.161	18583.35	TA	0.6			MIT
5373.013	5374.507	18606.36	TA	1.1H			MIT
5365.947	5367.439	18630.86	TA	1.1			MIT
5354.679	5356.168	18670.06	TA	1.9R			MIT
5349.575	5351.063	18687.88	TA	1.5			MIT
5349.093	5350.581	18689.56	TA	1.9			MIT
5348.080	5349.567	18693.10	TA	0.8			MIT

AIR WAVELENGTH (ANGSTRMS)	VACUUM WAVELENGTH (ANGSTRMS)	WAVENUMBER (KAYSERS)	LMENT	LOG ARC	INTENSITY SPK	DIS	REF
5342.250	5343.736	18713.50	TA	1.9			MIT
5341.049	5342.535	18717.71	TA	2.2W	1.9		MIT
5336.127	5337.611	18734.97	TA	1.6			MIT
5328.379	5329.861	18762.22	TA	1.6			MIT
5317.32	5318.80	18801.2	TA	0.3S			KS
5310.965	5312.443	18823.73	TA	0.3			MIT
5295.009	5296.482	18880.46	TA	1.6			MIT
5293.396	5294.869	18886.21	TA	1.5			MIT
5281.017	5282.487	18930.48	TA	0.3			MIT
5279.824	5281.293	18934.76	TA	1.8W			MIT
5275.025	5276.493	18951.98	TA	0.8			MIT
5273.271	5274.739	18958.29	TA	0.5			MIT
5261.558	5263.022	19000.49	TA	0.3H			MIT
5260.429	5261.893	19004.57	TA	0.9			MIT
5247.653	5249.114	19050.83	TA	1.3			MIT
5244.777	5246.237	19061.28	TA	1.6W			MIT
5235.393	5236.850	19095.45	TA	1.3W			MIT
5233.440	5234.897	19102.57	TA	0.3J			MIT
5230.802	5232.258	19112.21	TA	1.8W			MIT
5218.660	5220.113	19156.67	TA	1.6			MIT
5218.452	5219.905	19157.44	TA	1.6L			MIT
5212.741	5214.192	19178.43	TA	1.8			MIT
5206.55	5208.00	19201.2	TA	0.3			KS
5206.273	5207.723	19202.25	TA	0.7W			MIT
5188.934	5190.379	19266.42	TA	1.1			MIT
5180.985	5182.428	19295.98	TA	0.3			MIT
5171.632	5173.072	19330.87	TA	1.3			MIT
5166.788	5168.227	19349.00	TA	1.5			MIT
5163.649	5165.087	19360.76	TA	1.6			MIT
5161.814	5163.252	19367.64	TA	1.9W			MIT
5156.562	5157.998	19387.36	TA	1.90			MIT
5156.311	5157.747	19388.31	TA	1.1			MIT
5153.417	5154.853	19399.20	TA	1.2			MIT
5150.852	5152.287	19408.86	TA	1.2L			MIT
5148.779	5150.213	19416.67	TA	0.8L			MIT
5143.687	5145.120	19435.89	TA	1.5			MIT
5141.83	5143.26	19442.9	TA	0.8			KS
5141.620	5143.052	19443.71	TA	1.6			MIT
5136.467	5137.898	19463.21	TA	1.8	1.6		MIT
5132.32	5133.75	19478.9	TA	0.3H			KS
5132.116	5133.546	19479.71	TA	1.2			MIT
5117.252	5118.678	19536.29	TA	1.50			MIT
5115.843	5117.269	19541.68	TA	1.9			MIT
5109.774	5111.198	19564.89	TA	1.5W			MIT
5109.37	5110.79	19566.4	TA	0.80			KS
5090.710	5092.129	19638.15	TA	1.8			MIT
5087.366	5088.784	19651.06	TA	1.8			MIT
5082.253	5083.670	19670.83	TA	1.50			MIT
5076.371	5077.786	19693.62	TA	1.7			MIT
5072.457	5073.871	19708.82	TA	0.6			MIT
5069.882	5071.295	19718.83	TA	0.7			MIT
5067.870	5069.283	19726.66	TA	1.8			MIT
5064.612	5066.024	19739.35	TA	0.8H			MIT
5058.696	5060.106	19762.43	TA	0.7			MIT
5044.418	5045.825	19818.37	TA	0.9			MIT
5043.321	5044.727	19822.68	TA	1.8			MIT
5041.869	5043.275	19828.39	TA	0.6			MIT
5037.665	5039.070	19844.93	TA	1.3			MIT
5037.368	5038.773	19846.10	TA	1.8			MIT
5017.068	5018.467	19926.40	TA	0.3			MIT
5012.524	5013.922	19944.47	TA	1.8			MIT
4998.181	4999.575	20001.70	TA	1.20			MIT
4993.253	4994.646	20021.44	TA	0.7H	0.0		MIT
4977.828	4979.217	20083.48	TA	1.00	0.0		MIT
4976.196	4977.584	20090.07	TA	1.6	0.0		MIT
4969.693	4971.080	20116.35	TA	1.6L	0.0		MIT
4968.527	4969.913	20121.07	TA	1.80	0.0		MIT
4958.114	4959.498	20163.33	TA	0.9			MIT
4937.626	4939.004	20247.00	TA	1.6	0.3		MIT
4936.421	4937.799	20251.94	TA	2.0S			MIT

AIR WAVELENGTH (ANGSTRMS)	VACUUM WAVELENGTH (ANGSTRMS)	WAVENUMBER (KAYSERS)	LMENT	LOG ARC	INTENSITY SPK	DIS	REF
4933.527	4934.904	20263.82	TA	0.7S	0.0		MIT
4931.594	4932.971	20271.76	TA	0.5J	0.0		MIT
4926.005	4927.380	20294.76	TA	1.8	0.3		MIT
4924.957	4926.332	20299.08	TA	1.5	0.0		MIT
4923.471	4924.845	20305.21	TA	1.80	0.0		MIT
4921.271	4922.645	20314.28	TA	1.7	0.3		MIT
4920.882	4922.256	20315.89	TA	0.7	0.0		MIT
4920.108	4921.482	20319.08	TA	2.20	0.5		MIT
4918.876	4920.249	20324.17	TA	0.6			MIT
4917.230	4918.603	20330.98	TA	1.0S	0.0		MIT
4914.955	4916.327	20340.39	TA	1.7	0.0		MIT
4911.379	4912.750	20355.20	TA	1.5	0.0		MIT
4907.730	4909.100	20370.33	TA	1.7	0.0		MIT
4904.591	4905.960	20383.37	TA	1.90	0.3		MIT
4883.950	4885.314	20469.51	TA	2.2	0.5		MIT
4881.937	4883.300	20477.95	TA	1.5	0.0		MIT
4881.530	4882.893	20479.66	TA	0.3			MIT
4879.144	4880.507	20489.68	TA	1.4	0.0		MIT
4871.703	4873.064	20520.97	TA	1.7W	0.0		MIT
4864.664	4866.023	20550.66	TA	1.3H	0.3		MIT
4864.352	4865.711	20551.98	TA	0.7			MIT
4862.527	4863.885	20559.70	TA	1.2			MIT
4861.043	4862.401	20565.97	TA	0.5	0.0		MIT
4852.684	4854.040	20601.40	TA	1.0	0.3		MIT
4852.166	4853.522	20603.60	TA	1.9	0.3		MIT
4846.810	4848.164	20626.36	TA	1.00	0.3		MIT
4846.447	4847.801	20627.91	TA	2.0	0.3		MIT
4832.185	4833.535	20688.79	TA	2.0			MIT
4825.429	4826.777	20717.76	TA	2.2			MIT
4821.924	4823.271	20732.82	TA	0.5			MIT
4819.533	4820.880	20743.10	TA	2.0			MIT
4815.119	4816.465	20762.12	TA	0.7	1.2		MIT
4812.749	4814.094	20772.34	TA	2.2	0.7		MIT
4789.279	4790.618	20874.13	TA	0.6	1.2H		MIT
4786.643	4787.981	20885.63	TA	1.3			MIT
4780.935	4782.272	20910.56	TA	1.7	2.3L		MIT
4768.983	4770.316	20962.97	TA	2.2L	0.7		MIT
4758.033	4759.364	21011.21	TA	1.3			MIT
4756.506	4757.836	21017.96	TA	2.2	1.0		MIT
4745.928	4747.255	21064.80	TA	1.0	0.0		MIT
4740.163	4741.489	21090.42	TA	2.0G	2.0		MIT
4739.440	4740.766	21093.64	TA	0.3			MIT
4738.347	4739.672	21098.50	TA	0.7	0.3		MIT
4730.116	4731.439	21135.22	TA	2.0	0.7		MIT
4722.877	4724.198	21167.61	TA	2.3			MIT
4708.663	4709.980	21231.51	TA	0.3	0.7D		MIT
4706.091	4707.408	21243.11	TA	2.3	0.3		MIT
4701.324	4702.640	21264.65	TA	2.2	0.3H		MIT
4693.347	4694.660	21300.79	TA	2.2	0.5		MIT
4691.900	4693.213	21307.36	TA	2.6	0.7H		MIT
4688.845	4690.157	21321.25	TA	1.6			MIT
4685.266	4686.577	21337.53	TA	1.9	0.3		MIT
4684.867	4686.178	21339.35	TA	2.0	0.3		MIT
4683.10	4684.41	21347.4	TA		1.2		EX
4681.875	4683.185	21352.99	TA	2.3	1.7		MIT
4678.017	4679.326	21370.60	TA	1.6	0.3		MIT
4669.143	4670.450	21411.21	TA	2.5	1.2		MIT
4661.117	4662.422	21448.08	TA	2.5	0.7H		MIT
4640.73	4642.03	21542.3	TA	1.2W			M
4633.06	4634.36	21578.0	TA	2.2	0.5H		MIT
4629.19	4630.49	21596.0	TA	0.7W	0.5H		MIT
4622.959	4624.254	21625.11	TA	1.7	0.0		MIT
4619.512	4620.806	21641.25	TA	2.5	1.0		MIT
4604.852	4606.142	21710.14	TA	2.30			MIT
4602.192	4603.481	21722.69	TA	2.0	0.3		MIT
4601.416	4602.705	21726.35	TA	1.8	2.0J		MIT
4583.17	4584.45	21812.9	TA	2.2	1.0		M
4580.69	4581.97	21824.7	TA	2.30	1.0		M
4576.62	4577.90	21844.1	TA	0.5J			MIT
4574.306	4575.588	21855.11	TA	2.5	1.3		MIT

AIR WAVELENGTH (ANGSTRMS)	VACUUM WAVELENGTH (ANGSTRMS)	WAVENUMBER (KAYSERS)	LMENT	LOG ARC	INTENSITY SPK	DIS	REF
4573.293	4574.575	21859.96	TA	2.3	0.3H		MIT
4566.861	4568.141	21890.74	TA	2.0	0.3H		MIT
4565.850	4567.130	21895.59	TA	2.3	1.2		MIT
4556.354	4557.631	21941.22	TA	2.3	0.7		MIT
4553.694	4554.970	21954.04	TA	2.3L	0.3		MIT
4551.950	4553.226	21962.45	TA	2.6	0.9		MIT
4547.149	4548.424	21985.64	TA	2.2	0.3		MIT
4530.846	4532.116	22064.75	TA	2.5	1.7		MIT
4529.985	4531.255	22068.94	TA	0.5	1.0H		MIT
4527.495	4528.765	22081.08	TA	2.2	0.7		MIT
4521.710	4522.978	22109.33	TA	0.3	0.3		MIT
4521.094	4522.362	22112.34	TA	2.3	1.0H		MIT
4511.503	4512.768	22159.35	TA	2.5	1.60		MIT
4510.982	4512.247	22161.91	TA	2.30	1.70		MIT
4509.286	4510.551	22170.24	TA	1.8	0.3		MIT
4496.496	4497.757	22233.30	TA	2.0	0.3		MIT
4494.969	4496.230	22240.85	TA	1.7	0.3		MIT
4489.317	4490.576	22268.86	TA		1.0H		MIT
4480.934	4482.191	22310.52	TA	2.3W	1.0H		MIT
4473.515	4474.770	22347.52	TA	0.5	1.0		MIT
4459.763	4461.015	22416.43	TA	1.5	1.0		MIT
4450.720	4451.969	22461.97	TA	1.5	0.9		MIT
4441.678	4442.925	22507.70	TA	2.0L	0.3H		MIT
4441.029	4442.276	22510.98	TA	1.8	0.3H		MIT
4432.978	4434.223	22551.87	TA	0.7	0.3		MIT
4431.086	4432.330	22561.50	TA	0.6			MIT
4430.954	4432.198	22562.17	TA	0.6	0.0		MIT
4430.410	4431.654	22564.94	TA	1.3	0.7		MIT
4424.965	4426.208	22592.71	TA	1.0	0.5H		MIT
4419.551	4420.792	22620.38	TA	1.0	0.7		MIT
4415.741	4416.981	22639.90	TA	1.6	1.0		MIT
4403.72	4404.96	22701.7	TA		0.9		MIT
4402.498	4403.735	22708.00	TA	2.0	1.3H		MIT
4398.449	4399.685	22728.90	TA	1.6	1.0		MIT
4386.068	4387.300	22793.06	TA	1.7	1.2		MIT
4378.822	4380.052	22830.78	TA	1.6	0.3		MIT
4375.143	4376.372	22849.98	TA	1.2	0.7		MIT
4374.210	4375.439	22854.85	TA	1.2	1.2		MIT
4372.768	4373.997	22862.39	TA	1.5H	0.3		MIT
4369.353	4370.581	22880.25	TA	1.2	1.0		MIT
4364.838	4366.065	22903.92	TA	1.2	0.9		MIT
4360.830	4362.056	22924.97	TA	1.6S	0.7		MIT
4358.654	4359.879	22936.42	TA	1.0			MIT
4358.033	4359.258	22939.68	TA	0.5	1.0H		MIT
4355.145	4356.369	22954.90	TA	1.9	1.0H		MIT
4350.996	4352.219	22976.78	TA	1.0			MIT
4344.312	4345.533	23012.14	TA	1.3	0.9		MIT
4338.199	4339.419	23044.56	TA	0.0	0.9		MIT
4336.199	4337.418	23055.19	TA	0.5	0.5		MIT
4329.57	4330.79	23090.5	TA	1.0	0.7		MIT
4322.684	4323.900	23127.27	TA	0.7	0.7		MIT
4318.809	4320.024	23148.02	TA	1.2	0.7		MIT
4314.517	4315.730	23171.05	TA	1.5	0.7H		MIT
4306.80	4308.01	23212.6	TA	0.0	0.7H		MIT
4303.54	4304.75	23230.1	TA	0.0	1.2		MIT
4302.978	4304.188	23233.18	TA	2.10	1.60		MIT
4286.376	4287.582	23323.17	TA	1.9	1.3H		MIT
4280.467	4281.671	23355.36	TA	1.0	0.0H		MIT
4279.056	4280.260	23363.07	TA	1.5	1.0		MIT
4271.513	4272.715	23404.32	TA	1.6	0.7J		MIT
4268.255	4269.456	23422.19	TA	1.7	1.2H		MIT
4267.298	4268.499	23427.44	TA	0.0	1.3		MIT
4245.350	4246.545	23548.55	TA	1.5	1.2		MIT
4243.986	4245.181	23556.12	TA	0.7	0.3		MIT
4235.940	4237.133	23600.86	TA	0.7H	0.7		MIT
4230.827	4232.018	23629.39	TA		1.00		MIT
4228.608	4229.799	23641.79	TA	1.4	1.0		MIT
4208.440	4209.626	23755.08	TA	0.3	1.5H		MIT
4206.404	4207.589	23766.58	TA	1.7	1.3		MIT
4205.876	4207.061	23769.56	TA	2.0	1.5		MIT

AIR WAVELENGTH (ANGSTRMS)	VACUUM WAVELENGTH (ANGSTRMS)	WAVENUMBER (KAYSERS)	LMENT	LOG ARC	INTENSITY SPK	DIS	REF
4201.97	4203.15	23791.7	TA	0.7H			KS
4193.097	4194.279	23842.00	TA	0.9	1.0H		MIT
4191.26	4192.44	23852.4	TA	0.5	0.0		MIT
4191.161	4192.342	23853.02	TA	0.7	0.3		MIT
4181.152	4182.330	23910.11	TA	1.6	1.4		MIT
4177.919	4179.097	23928.62	TA	1.3	1.2		MIT
4176.986	4178.163	23933.96	TA	1.2	1.6		MIT
4175.206	4176.383	23944.17	TA	2.0	1.6H		MIT
4160.999	4162.172	24025.92	TA	1.3H	0.3H		MIT
4159.976	4161.149	24031.82	TA	0.7			MIT
4154.393	4155.564	24064.12	TA	0.6			MIT
4147.891	4149.061	24101.84	TA	1.6	1.5		MIT
4136.198	4137.365	24169.98	TA	1.9	1.5		MIT
4129.378	4130.543	24209.89	TA	2.3	1.6		MIT
4127.879	4129.043	24218.68	TA	1.2	0.9		MIT
4123.657	4124.820	24243.48	TA	0.7	0.0		MIT
4123.173	4124.336	24246.33	TA	1.7R	0.6W		MIT
4118.066	4119.228	24276.39	TA	0.7	0.6H		MIT
4114.770	4115.931	24295.84	TA	0.7	0.3H		MIT
4105.021	4106.179	24353.54	TA	1.1	0.9		MIT
4104.242	4105.400	24358.16	TA	0.8	0.7		MIT
4097.188	4098.344	24400.10	TA	0.7	0.5H		MIT
4091.642	4092.797	24433.17	TA	0.3			MIT
4091.260	4092.415	24435.45	TA	0.6	1.0		MIT
4085.797	4086.950	24468.12	TA	0.7	1.90		MIT
4079.189	4080.341	24507.76	TA	1.0	0.6		MIT
4077.721	4078.872	24516.58	TA	0.6	0.3H		MIT
4072.996	4074.146	24545.02	TA	0.3	1.20		MIT
4067.908	4069.057	24575.72	TA	2.0	1.6		MIT
4067.235	4068.384	24579.79	TA	1.6	1.0H		MIT
4064.63	4065.78	24595.5	TA	1.6	0.7		KS
4062.801	4063.948	24606.61	TA	0.5	0.7H		MIT
4062.573	4063.720	24607.99	TA	0.3	0.5H		MIT
4061.399	4062.546	24615.11	TA	1.7	1.5		MIT
4058.464	4059.610	24632.91	TA	1.0	0.7		MIT
4058.136	4059.282	24634.90	TA	0.3	0.0		MIT
4054.391	4055.536	24657.65	TA	0.6			MIT
4041.361	4042.503	24737.15	TA	0.3			MIT
4041.057	4042.199	24739.01	TA	1.6	0.6H		MIT
4040.87	4042.01	24740.2	TA	1.7	0.7H		MIT
4039.633	4040.774	24747.73	TA	0.7	0.0		MIT
4035.893	4037.033	24770.67	TA	1.0	0.7H		MIT
4033.069	4034.209	24788.01	TA	2.0	1.0		MIT
4031.96	4033.10	24794.8	TA	0.7			MIT
4030.668	4031.807	24802.77	TA	1.0	0.0D		MIT
4029.94	4031.08	24807.3	TA	1.7	0.7		KS
4026.944	4028.082	24825.71	TA	1.6	1.5		MIT
4015.247	4016.382	24898.03	TA	0.9	0.3		MIT
4014.63	4015.76	24901.9	TA	0.8H	0.0		KS
4013.54	4014.67	24908.6	TA	0.7			KS
4013.19	4014.32	24910.8	TA	0.7	0.0		MIT
4012.111	4013.245	24917.49	TA	0.7H	0.3H		MIT
4007.233	4008.366	24947.82	TA	0.6	0.3		MIT
4006.835	4007.968	24950.30	TA	1.5	1.3		MIT
4003.704	4004.836	24969.81	TA	1.2	0.7		MIT
3999.283	4000.414	24997.41	TA	1.5	1.3D		MIT
3996.171	3997.301	25016.88	TA	2.0	1.5H		MIT
3995.290	3996.420	25022.40	TA	0.5L			MIT
3992.661	3993.790	25038.87	TA	0.3H			MIT
3990.396	3991.524	25053.08	TA	0.8	0.7H		MIT
3988.701	3989.829	25063.73	TA	1.2	1.0		MIT
3984.977	3986.104	25087.15	TA	0.3			MIT
3984.249	3985.376	25091.74	TA	0.3	0.7H		MIT
3983.825	3984.952	25094.41	TA	0.7	0.0		MIT
3981.947	3983.073	25106.24	TA	1.5	1.2		MIT
3981.008	3982.134	25112.16	TA	0.3	1.6		MIT
3979.284	3980.410	25123.04	TA	1.7H	0.5H		MIT
3973.178	3974.302	25161.65	TA	0.0	2.60		MIT
3970.178	3971.301	25180.66	TA	1.5	0.5		MIT
3970.098	3971.221	25181.17	TA	2.0	1.6		MIT

AIR WAVELENGTH (ANGSTRMS)	VACUUM WAVELENGTH (ANGSTRMS)	WAVENUMBER (KAYSERS)	LMENT	LOG ARC	INTENSITY SPK	DIS	REF
3968.16	3969.28	25193.5	TA	0.6H			KS
3959.906	3961.026	25245.98	TA	0.7			MIT
3959.726	3960.846	25247.13	TA	0.5	0.5		MIT
3956.570	3957.690	25267.27	TA	1.2	0.3		MIT
3954.294	3955.413	25281.81	TA	0.7J			MIT
3952.159	3953.277	25295.47	TA	0.9	0.5		MIT
3942.235	3943.351	25359.14	TA	0.5	0.6D		MIT
3937.843	3938.958	25387.43	TA	0.7	1.0		MIT
3930.942	3932.055	25432.00	TA	0.7	0.5		MIT
3930.23	3931.34	25436.6	TA	0.0	0.3		KS
3925.241	3926.352	25468.93	TA	1.4H	0.7H		MIT
3922.915	3924.026	25484.03	TA	2.0	1.0H		MIT
3922.779	3923.890	25484.92	TA	1.0	1.2		MIT
3922.421	3923.532	25487.24	TA	0.7	1.7		MIT
3919.338	3920.448	25507.29	TA	0.3			MIT
3918.512	3919.622	25512.67	TA	1.4	1.0		MIT
3912.435	3913.543	25552.29	TA	1.2	1.0H		MIT
3912.127	3913.235	25554.30	TA	1.0	0.5S		MIT
3909.332	3910.439	25572.57	TA	1.2	0.0H		MIT
3905.285	3906.391	25599.07	TA	0.8H	0.3		MIT
3898.781	3899.886	25641.78	TA	1.0	0.7H		MIT
3896.43	3897.53	25657.2	TA	0.7			KS
3894.671	3895.774	25668.84	TA	1.6H			MIT
3893.032	3894.135	25679.64	TA	0.6	0.3		MIT
3890.714	3891.816	25694.94	TA	0.3	0.3		MIT
3890.499	3891.601	25696.36	TA	0.3	1.0L		MIT
3889.457	3890.559	25703.25	TA		1.6O		MIT
3885.202	3886.303	25731.40	TA	1.2	1.0		MIT
3879.436	3880.536	25769.64	TA	0.7	0.3		MIT
3876.562	3877.661	25788.74	TA	1.0	0.7		MIT
3875.21	3876.31	25797.7	TA	0.7H	0.3H		KS
3872.738	3873.836	25814.21	TA	0.3	0.3		MIT
3870.687	3871.784	25827.89	TA		0.7H		MIT
3866.47	3867.57	25856.1	TA	0.6			KS
3861.077	3862.172	25892.17	TA	0.5H	0.0		MIT
3859.798	3860.892	25900.75	TA	1.0	0.3		MIT
3858.608	3859.702	25908.74	TA	1.0H	0.3H		MIT
3853.567	3854.660	25942.63	TA	0.5	0.3		MIT
3851.443	3852.535	25956.93	TA	0.0	0.3		MIT
3849.758	3850.850	25968.29	TA	0.5H	0.6H		MIT
3849.42	3850.51	25970.6	TA	0.6H	0.3H		M
3848.046	3849.137	25979.85	TA	1.5	0.7		MIT
3846.636	3847.727	25989.37	TA	0.7	0.6H		MIT
3844.039	3845.129	26006.93	TA	0.9H	0.3H		M
3842.874	3843.964	26014.81	TA	0.7	0.7		MIT
3842.341	3843.431	26018.42	TA	0.5H			MIT
3839.418	3840.507	26038.23	TA	0.3			MIT
3839.034	3840.123	26040.83	TA	1.5	0.7		MIT
3836.603	3837.691	26057.33	TA	1.5	0.5		MIT
3833.742	3834.830	26076.78	TA	1.6	2.3		MIT
3832.275	3833.362	26086.76	TA	0.7H	0.5H		MIT
3828.947	3830.033	26109.43	TA	1.4	1.0		MIT
3826.849	3827.935	26123.75	TA	1.4	0.5		MIT
3826.172	3827.258	26128.37	TA	1.0	0.9		MIT
3823.595	3824.680	26145.98	TA	1.2	1.0		MIT
3820.742	3821.826	26165.50	TA	0.8	0.6		MIT
3819.713	3820.797	26172.55	TA	0.7	0.5		MIT
3812.881	3813.963	26219.45	TA	0.5	0.5		MIT
3812.32	3813.40	26223.3	TA	0.3	0.3		KS
3809.718	3810.799	26241.21	TA	0.6	0.5H		MIT
3809.258	3810.339	26244.38	TA	0.3			MIT
3802.02	3803.10	26294.3	TA	0.3H			KS
3801.801	3802.880	26295.86	TA	0.5			MIT
3801.147	3802.226	26300.38	TA	0.0	1.0		MIT
3799.991	3801.070	26308.38	TA	0.3	0.3H		MIT
3797.403	3798.481	26326.31	TA	0.8	0.5		MIT
3796.208	3797.286	26334.60	TA		1.7		MIT
3795.185	3796.263	26341.70	TA		1.5		MIT
3792.015	3793.092	26363.72	TA	1.7	1.0		MIT
3788.748	3789.824	26386.45	TA	1.0	0.5		MIT

AIR WAVELENGTH (ANGSTRMS)	VACUUM WAVELENGTH (ANGSTRMS)	WAVENUMBER (KAYSERS)	LMENT	LOG ARC	INTENSITY SPK	DIS	REF
3787.146	3788.222	26397.61	TA		1.3W		MIT
3785.265	3786.340	26410.73	TA	0.9	0.3		MIT
3784.254	3785.329	26417.78	TA	2.2	1.7W		MIT
3783.774	3784.849	26421.14	TA	0.3	0.0		MIT
3782.302	3783.376	26431.42	TA	0.0	0.5H		MIT
3780.173	3781.247	26446.30	TA	1.5	0.3		MIT
3779.146	3780.219	26453.49	TA	0.3H	0.5S		MIT
3777.102	3778.175	26467.81	TA	1.4	1.0		MIT
3770.927	3771.998	26511.15	TA	1.2	0.0		MIT
3763.443	3764.512	26563.86	TA	1.1	1.8		MIT
3761.349	3762.418	26578.65	TA	0.9			MIT
3760.209	3761.278	26586.71	TA	0.9	0.3D		MIT
3759.753	3760.821	26589.94	TA	1.5	0.3		MIT
3757.746	3758.814	26604.14	TA	1.3	0.0H		MIT
3756.892	3757.960	26610.18	TA		0.9		MIT
3755.107	3756.174	26622.83	TA	1.4	0.6H		MIT
3754.517	3755.584	26627.02	TA	1.3	0.5		MIT
3747.246	3748.311	26678.68	TA	1.5			MIT
3746.365	3747.430	26684.96	TA	1.5	0.7		MIT
3742.67	3743.73	26711.3	TA	0.3H			KS
3742.093	3743.157	26715.42	TA	0.5			MIT
3741.276	3742.340	26721.25	TA	0.8H	0.0H		MIT
3738.214	3739.277	26743.14	TA	1.0	0.0		MIT
3736.759	3737.821	26753.55	TA	1.5	0.5		MIT
3733.322	3734.384	26778.18	TA		0.5H		MIT
3732.750	3733.811	26782.28	TA	0.5	0.0		MIT
3731.017	3732.078	26794.72	TA	1.7	0.5		MIT
3728.823	3729.883	26810.49	TA	0.9			MIT
3728.111	3729.171	26815.61	TA	0.8			MIT
3725.571	3726.631	26833.89	TA	0.7H			MIT
3725.013	3726.072	26837.91	TA	0.7			MIT
3723.073	3724.132	26851.90	TA	1.3	0.0		MIT
3719.418	3720.476	26878.28	TA	0.5			MIT
3718.455	3719.513	26885.24	TA	0.5	0.0		MIT
3718.072	3719.130	26888.01	TA	0.3			MIT
3711.15	3712.21	26938.2	TA	0.7			KS
3710.79	3711.85	26940.8	TA	0.8J			KS
3706.587	3707.642	26971.32	TA	0.7			MIT
3706.326	3707.381	26973.22	TA	0.5H	0.3H		MIT
3706.055	3707.109	26975.20	TA	0.3			MIT
3705.169	3706.223	26981.65	TA	0.8			MIT
3704.216	3705.270	26988.59	TA	0.5			MIT
3703.258	3704.312	26995.57	TA	0.8			MIT
3701.336	3702.389	27009.59	TA	1.4S			MIT
3700.277	3701.330	27017.31	TA		1.0		MIT
3699.99	3701.04	27019.4	TA	0.7			KS
3695.382	3696.434	27053.10	TA	1.2	0.8S		MIT
3694.519	3695.571	27059.42	TA	0.8H	1.3		MIT
3693.047	3694.098	27070.21	TA	1.5R	0.5H		MIT
3691.879	3692.930	27078.77	TA		1.2		MIT
3689.729	3690.779	27094.55	TA	1.5	0.3		MIT
3686.82	3687.87	27115.9	TA	1.3W	0.0H		KS
3686.185	3687.234	27120.60	TA	1.5	0.3		MIT
3685.906	3686.955	27122.65	TA	0.5			MIT
3684.97	3686.02	27129.5	TA	0.5			KS
3684.31	3685.36	27134.4	TA	0.3			KS
3683.058	3684.107	27143.62	TA	1.3	0.0		MIT
3681.243	3682.291	27157.01	TA	1.2	0.0		MIT
3676.890	3677.937	27189.16	TA	0.7	0.0H		MIT
3675.49	3676.54	27199.5	TA	1.3			KS
3675.125	3676.171	27202.21	TA	0.5	0.0		MIT
3674.827	3675.873	27204.42	TA	1.3	0.5H		MIT
3667.680	3668.725	27257.43	TA	1.0			MIT
3667.056	3668.100	27262.07	TA	0.0	0.7		MIT
3666.890	3667.934	27263.30	TA	1.2			MIT
3666.10	3667.14	27269.2	TA	1.3			KS
3665.35	3666.39	27274.8	TA	1.0			KS
3663.845	3664.889	27285.96	TA	1.2			MIT
3663.10	3664.14	27291.5	TA	1.3D			KS
3662.343	3663.386	27297.15	TA	1.2	1.2		MIT

AIR WAVELENGTH (ANGSTRMS)	VACUUM WAVELENGTH (ANGSTRMS)	WAVENUMBER (KAYSERS)	LMENT	LOG ARC	INTENSITY SPK	DIS	REF
3661.688	3662.731	27302.03	TA	1.3R	0.5		MIT
3658.777	3659.819	27323.75	TA	1.5	0.5L		MIT
3657.495	3658.537	27333.33	TA	1.2	0.3		MIT
3657.271	3658.313	27335.00	TA	1.3	0.0H		MIT
3656.894	3657.936	27337.82	TA	1.2	0.3H		MIT
3656.34	3657.38	27342.0	TA	0.5			KS
3656.05	3657.09	27344.1	TA	0.8			KS
3655.94	3656.98	27345.0	TA	0.7			KS
3655.65	3656.69	27347.1	TA	0.5			KS
3653.828	3654.869	27360.76	TA	0.5	0.0		MIT
3653.393	3654.434	27364.02	TA	0.5	0.0		MIT
3652.409	3653.450	27371.39	TA	0.7			MIT
3648.028	3649.067	27404.26	TA	0.5			MIT
3642.057	3643.095	27449.19	TA	2.1	1.3		MIT
3633.788	3634.824	27511.65	TA	1.5	1.0H		MIT
3628.669	3629.703	27550.46	TA		1.3		MIT
3627.017	3628.051	27563.01	TA	1.3	0.0		MIT
3626.617	3627.651	27566.05	TA	2.1	1.3		MIT
3625.68	3626.71	27573.2	TA	1.2			KS
3625.238	3626.272	27576.53	TA	1.8R	0.3H		MIT
3624.17	3625.20	27584.7	TA	0.5H	0.5S		MIT
3617.317	3618.349	27636.92	TA	0.8			MIT
3616.026	3617.057	27646.78	TA	0.3	0.0		MIT
3612.82	3613.85	27671.3	TA	0.5O			KS
3612.37	3613.40	27674.8	TA	0.3			KS
3611.132	3612.162	27684.25	TA	1.4	0.0		MIT
3610.598	3611.628	27688.35	TA		1.2		MIT
3609.931	3610.961	27693.46	TA	0.0H	1.3		MIT
3609.357	3610.386	27697.87	TA	0.9	0.0		MIT
3609.177	3610.206	27699.25	TA	0.9	0.0		MIT
3608.781	3609.810	27702.29	TA	1.2R	0.0D		MIT
3607.406	3608.435	27712.84	TA	1.8	1.5		MIT
3604.983	3606.011	27731.47	TA	0.8	0.0H		MIT
3600.701	3601.728	27764.45	TA	0.3	0.0H		MIT
3596.861	3597.887	27794.09	TA	0.8	0.0		MIT
3595.638	3596.664	27803.54	TA	1.8	0.7		MIT
3592.495	3593.520	27827.87	TA	0.3	0.0H		MIT
3592.14	3593.17	27830.6	TA	0.3	0.0H		MIT
3589.97	3590.99	27847.4	TA	0.3			KS
3586.291	3587.315	27876.00	TA	1.3	0.5		MIT
3584.512	3585.535	27889.84	TA	1.5	0.5		MIT
3584.213	3585.236	27892.17	TA	1.7	0.8H		MIT
3580.889	3581.911	27918.06	TA	0.5	0.0H		MIT
3580.267	3581.289	27922.91	TA	0.8			MIT
3580.023	3581.045	27924.81	TA	1.0			MIT
3579.448	3580.470	27929.30	TA		1.5		MIT
3579.076	3580.098	27932.20	TA	1.2	0.0		MIT
3578.762	3579.784	27934.65	TA	0.3	0.0H		MIT
3576.242	3577.263	27954.33	TA	1.3	0.0H		MIT
3573.438	3574.458	27976.27	TA	1.2	1.8W		MIT
3571.854	3572.874	27988.67	TA	1.3	0.5		MIT
3571.140	3572.160	27994.27	TA	0.7	0.5		MIT
3570.773	3571.793	27997.15	TA	0.7	0.0H		MIT
3570.32	3571.34	28000.7	TA	0.3			KS
3569.19	3570.21	28009.6	TA	0.3			KS
3567.35	3568.37	28024.0	TA	0.6			KS
3566.716	3567.735	28028.99	TA	1.7	0.7L		MIT
3566.01	3567.03	28034.5	TA	0.8			KS
3565.627	3566.645	28037.55	TA		1.7L		MIT
3564.794	3565.812	28044.10	TA	1.3	0.3		MIT
3557.982	3558.998	28097.79	TA	1.5	0.3H		MIT
3556.548	3557.564	28109.12	TA	0.5			MIT
3555.729	3556.745	28115.60	TA	1.0	0.3H		MIT
3553.415	3554.430	28133.91	TA	0.8	0.3		MIT
3551.118	3552.132	28152.10	TA		1.3		MIT
3549.046	3550.060	28168.54	TA	1.5	0.5		MIT
3542.240	3543.252	28222.66	TA		0.7		MIT
3541.885	3542.897	28225.49	TA	1.5	1.2		MIT
3540.821	3541.833	28233.97	TA	1.3	0.0		MIT
3537.553	3538.564	28260.05	TA	0.3	0.0H		MIT

AIR WAVELENGTH (ANGSTRMS)	VACUUM WAVELENGTH (ANGSTRMS)	WAVENUMBER (KAYSERS)	LMENT	LOG ARC	INTENSITY SPK	DIS	REF
3536.300	3537.311	28270.06	TA	1.3	0.3		MIT
3535.404	3536.414	28277.23	TA	1.2	1.3J		MIT
3532.208	3533.218	28302.81	TA	1.2	0.3		MIT
3531.584	3532.593	28307.81	TA	1.5	0.5		MIT
3528.614	3529.623	28331.64	TA	1.5	1.3O		MIT
3527.065	3528.073	28344.08	TA	1.7	0.3H		MIT
3523.171	3524.178	28375.41	TA	1.2H	1.2		MIT
3520.495	3521.502	28396.98	TA	1.4J	1.3		MIT
3517.448	3518.454	28421.57	TA	0.5	0.0H		MIT
3514.404	3515.409	28446.19	TA	0.7	0.0H		MIT
3513.612	3514.617	28452.60	TA	1.5	0.0H		MIT
3512.834	3513.839	28458.91	TA	1.2			MIT
3511.043	3512.047	28473.42	TA	2.0	1.5W		MIT
3507.333	3508.336	28503.54	TA	0.5	0.0H		MIT
3506.873	3507.876	28507.28	TA	0.5	0.0		MIT
3505.175	3506.178	28521.09	TA	1.5	0.3H		MIT
3504.976	3505.979	28522.71	TA	1.8	0.3		MIT
3503.868	3504.870	28531.73	TA	1.8	1.0H		MIT
3502.869	3503.871	28539.86	TA	0.8	0.5		MIT
3502.495	3503.497	28542.91	TA	0.8	0.3H		MIT
3501.974	3502.976	28547.16	TA	0.5	0.0		MIT
3497.848	3498.849	28580.83	TA	1.8	0.7		MIT
3493.465	3494.465	28616.69	TA	1.2O			MIT
3490.932	3491.931	28637.45	TA	1.0	0.5		MIT
3489.161	3490.160	28651.99	TA	0.5			MIT
3488.847	3489.845	28654.56	TA	1.3	0.0H		MIT
3488.545	3489.543	28657.04	TA	0.3H	0.0H		MIT
3487.396	3488.394	28666.49	TA	0.5			MIT
3486.688	3487.686	28672.31	TA	0.5			MIT
3484.621	3485.618	28689.31	TA	0.7	0.5		MIT
3484.230	3485.227	28692.53	TA	0.5			MIT
3480.516	3481.512	28723.15	TA	1.8	2.3N		MIT
3479.443	3480.439	28732.01	TA	0.5	0.0		MIT
3477.635	3478.631	28746.94	TA	0.5H			MIT
3477.448	3478.444	28748.49	TA	1.3D	0.8D		MIT
3477.220	3478.215	28750.37	TA	0.7	1.5H		MIT
3473.901	3474.896	28777.84	TA	1.0	0.0H		MIT
3473.314	3474.308	28782.71	TA	0.5	0.0		MIT
3472.832	3473.826	28786.70	TA	0.8	0.5H		MIT
3472.521	3473.515	28789.28	TA	1.3	1.0		MIT
3471.048	3472.042	28801.50	TA		1.3		MIT
3466.852	3467.845	28836.35	TA	0.3	1.3H		MIT
3465.505	3466.497	28847.56	TA		1.2H		MIT
3463.915	3464.907	28860.80	TA	0.5	1.0		MIT
3463.771	3464.763	28862.00	TA	1.7			MIT
3462.894	3463.886	28869.31	TA	0.3H	0.0H		MIT
3453.354	3454.343	28949.06	TA		1.0H		MIT
3452.971	3453.960	28952.27	TA	0.7	0.3		MIT
3450.403	3451.392	28973.82	TA	0.7	0.0L		MIT
3448.929	3449.917	28986.20	TA	1.3W	0.3H		MIT
3447.291	3448.279	28999.98	TA	1.5	0.5		MIT
3446.910	3447.898	29003.18	TA	0.3O	2.2O		MIT
3445.912	3446.899	29011.58	TA	0.8	0.5		MIT
3445.506	3446.493	29015.00	TA	0.3H	0.0H		MIT
3445.149	3446.136	29018.00	TA	1.0	0.3H		MIT
3444.681	3445.668	29021.95	TA	0.7	0.3		MIT
3444.067	3445.054	29027.12	TA	0.7	0.0		MIT
3441.544	3442.530	29048.40	TA	0.5	0.3H		MIT
3440.237	3441.223	29059.44	TA	1.3	1.7		MIT
3439.001	3439.987	29069.88	TA		1.8O		MIT
3437.373	3438.358	29083.65	TA	0.8	2.5J		MIT
3437.074	3438.059	29086.18	TA	0.3R			MIT
3436.00	3436.98	29095.3	TA	1.8W	1.3W		MIT
3434.500	3435.485	29107.98	TA	1.5	1.3		MIT
3430.938	3431.922	29138.20	TA	1.7	1.8		MIT
3429.314	3430.297	29151.99	TA	0.0	0.3		MIT
3426.732	3427.715	29173.96	TA	1.3	1.0		MIT
3426.192	3427.174	29178.56	TA	0.8	0.5H		MIT
3424.450	3425.432	29193.40	TA	1.5R	0.8R		MIT
3421.800	3422.781	29216.01	TA		1.3		MIT

AIR WAVELENGTH (ANGSTRMS)	VACUUM WAVELENGTH (ANGSTRMS)	WAVENUMBER (KAYSERS)	LMENT	LOG ARC	INTENSITY SPK	DIS	REF
3421.212	3422.193	29221.03	TA	0.0	0.5J		MIT
3419.748	3420.729	29233.54	TA	1.7W	0.7H		MIT
3419.550	3420.531	29235.23	TA	1.0R	0.5S		MIT
3418.310	3419.290	29245.83	TA	0.9	0.3H		MIT
3417.026	3418.006	29256.82	TA	0.5	1.2		MIT
3415.866	3416.846	29266.76	TA	1.0	0.5		MIT
3415.270	3416.250	29271.87	TA	0.5	1.5		MIT
3414.14	3415.12	29281.6	TA	1.3O	2.0O		MIT
3413.636	3414.615	29285.88	TA	0.7	0.3H		MIT
3412.892	3413.871	29292.26	TA	1.3	0.5		MIT
3411.716	3412.695	29302.36	TA	0.7	0.3		MIT
3407.42	3408.40	29339.3	TA		1.3H		MIT
3406.935	3407.912	29343.48	TA	1.8	1.2		MIT
3406.664	3407.641	29345.81	TA	1.8W	1.3S		MIT
3404.163	3405.140	29367.37	TA	0.7	0.0		MIT
3402.992	3403.968	29377.48	TA		1.4W		MIT
3398.327	3399.302	29417.80	TA	1.5	1.3L		MIT
3397.418	3398.393	29425.67	TA	0.5	0.0		MIT
3388.823	3389.796	29500.30	TA	0.7	0.0		MIT
3388.338	3389.311	29504.52	TA	0.3	1.2		MIT
3387.457	3388.430	29512.20	TA	0.7	0.3		MIT
3387.226	3388.198	29514.21	TA	0.5	0.0		MIT
3386.544	3387.516	29520.15	TA		0.5H		MIT
3385.053	3386.025	29533.16	TA	1.3	1.0		MIT
3379.515	3380.485	29581.55	TA	1.3R	1.7R		MIT
3378.177	3379.147	29593.27	TA		0.5		MIT
3377.769	3378.739	29596.84	TA	0.3H	0.0H		MIT
3376.490	3377.460	29608.05	TA	1.2H	1.8		MIT
3376.053	3377.023	29611.88	TA	1.5	1.2		MIT
3371.539	3372.507	29651.53	TA	1.8	1.3		MIT
3369.844	3370.812	29666.44	TA	0.5	0.0		MIT
3369.280	3370.248	29671.41	TA	1.3	0.7		MIT
3368.442	3369.410	29678.79	TA	0.0	1.3		MIT
3366.661	3367.628	29694.49	TA	1.7	1.2W		MIT
3365.031	3365.998	29708.87	TA	0.5	0.0		MIT
3364.102	3365.069	29717.08	TA	0.3	0.0		MIT
3362.530	3363.496	29730.97	TA	1.0R	0.0		MIT
3361.640	3362.606	29738.84	TA	2.1O	1.7O		MIT
3358.981	3359.946	29762.38	TA	0.8	0.0		MIT
3358.527	3359.492	29766.40	TA	1.8	1.4O		MIT
3356.644	3357.609	29783.10	TA	1.2	0.7H		MIT
3356.030	3356.994	29788.55	TA	0.5	0.0		MIT
3355.599	3356.563	29792.38	TA	0.7	0.5H		MIT
3351.875	3352.838	29825.48	TA	0.3J	0.0H		MIT
3351.510	3352.473	29828.72	TA	1.3R	1.2R		MIT
3350.961	3351.924	29833.61	TA	1.4O	1.3O		MIT
3347.063	3348.025	29868.35	TA	0.3H			MIT
3345.108	3346.070	29885.81	TA	0.5	0.5H		MIT
3343.471	3344.432	29900.44	TA	0.3	0.5		MIT
3339.910	3340.870	29932.32	TA	1.2	1.8		MIT
3339.504	3340.464	29935.96	TA	0.9	0.0H		MIT
3338.487	3339.447	29945.08	TA		0.3		MIT
3337.799	3338.759	29951.25	TA	2.0	1.3S		MIT
3337.498	3338.458	29953.95	TA	1.5	0.0		MIT
3336.358	3337.317	29964.19	TA	0.7			MIT
3334.742	3335.701	29978.71	TA	1.5W	0.0H		MIT
3333.084	3334.043	29993.62	TA	0.8H	1.2W		MIT
3332.814	3333.773	29996.05	TA	1.5W	1.2W		MIT
3332.411	3333.369	29999.68	TA	1.7	0.5		MIT
3331.487	3332.445	30008.00	TA	0.7	1.2W		MIT
3331.007	3331.965	30012.32	TA	1.3	2.3W		MIT
3329.535	3330.493	30025.59	TA	1.3	0.0		MIT
3328.930	3329.888	30031.04	TA	0.8			MIT
3328.305	3329.262	30036.68	TA	1.0	0.0H		MIT
3325.741	3326.698	30059.84	TA	1.7	0.5		MIT
3324.847	3325.804	30067.92	TA	0.7	0.0		MIT
3319.47	3320.43	30116.6	TA	0.5	0.0		MIT
3318.840	3319.795	30122.34	TA	2.1	1.5		MIT
3318.534	3319.489	30125.12	TA	1.8	0.5		MIT
3317.928	3318.883	30130.62	TA	2.3	1.4O		MIT

AIR WAVELENGTH (ANGSTRMS)	VACUUM WAVELENGTH (ANGSTRMS)	WAVENUMBER (KAYSERS)	LMENT	LOG ARC	INTENSITY SPK	DIS	REF
3311.162	3312.115	30192.19	TA	2.5W	1.8W		MIT
3309.778	3310.731	30204.81	TA	1.8	0.7		MIT
3308.886	3309.838	30212.96	TA	1.3W			MIT
3308.545	3309.497	30216.07	TA	1.3	0.0H		MIT
3307.077	3308.029	30229.48	TA	1.5	0.0H		MIT
3305.337	3306.289	30245.39	TA	1.2	0.0H		MIT
3304.375	3305.326	30254.20	TA	1.8	1.2		MIT
3304.039	3304.990	30257.28	TA	1.5	0.5		MIT
3302.765	3303.716	30268.95	TA	1.7	0.0H		MIT
3302.328	3303.279	30272.95	TA	0.5	0.0H		MIT
3301.895	3302.846	30276.92	TA	1.4	0.5		MIT
3299.767	3300.717	30296.45	TA	1.8	1.0		MIT
3299.257	3300.207	30301.13	TA	0.7	0.0		MIT
3297.193	3298.142	30320.10	TA	1.3H	0.0J		MIT
3295.326	3296.275	30337.27	TA	2.1O	1.3W		MIT
3294.714	3295.663	30342.91	TA	1.3	0.0		MIT
3293.930	3294.879	30350.13	TA	1.8	1.0		MIT
3292.482	3293.430	30363.48	TA	1.8	0.5		MIT
3291.887	3292.835	30368.97	TA	1.5	0.0H		MIT
3291.410	3292.358	30373.37	TA	0.0	0.7		MIT
3289.838	3290.786	30387.88	TA	1.4	0.0		MIT
3288.466	3289.413	30400.56	TA	1.3	0.0H		MIT
3287.265	3288.212	30411.67	TA	1.7			MIT
3286.430	3287.377	30419.39	TA	0.5	0.0H		MIT
3284.612	3285.558	30436.23	TA	1.3			MIT
3283.807	3284.753	30443.69	TA	0.5	0.3J		MIT
3280.872	3281.817	30470.92	TA	0.5	0.3		MIT
3279.292	3280.237	30485.60	TA	1.7R	0.5		MIT
3278.317	3279.262	30494.67	TA	0.3			MIT
3275.941	3276.885	30516.79	TA	1.7	0.0H		MIT
3275.679	3276.623	30519.23	TA	1.8	1.5		MIT
3274.947	3275.891	30526.05	TA	2.3	1.5O		MIT
3274.458	3275.402	30530.61	TA	1.5	0.0		MIT
3273.132	3274.075	30542.97	TA	1.8	0.5H		MIT
3272.636	3273.579	30547.60	TA	0.5			MIT
3271.188	3272.131	30561.12	TA	1.8J	1.3W		MIT
3269.140	3270.082	30580.27	TA	1.8R	0.8L		MIT
3267.56	3268.50	30595.1	TA	0.7H			MIT
3266.890	3267.832	30601.33	TA	1.5R	0.0H		MIT
3265.345	3266.286	30615.81	TA	1.5			MIT
3264.083	3265.024	30627.65	TA	1.5	0.0		MIT
3263.759	3264.700	30630.69	TA	1.8	0.3		MIT
3263.404	3264.345	30634.02	TA		1.5H		MIT
3263.001	3263.942	30637.80	TA	1.3	0.0		MIT
3261.511	3262.451	30651.80	TA		1.5H		MIT
3260.182	3261.122	30664.29	TA	2.1	1.3W		MIT
3259.870	3260.810	30667.23	TA	1.7	0.5		MIT
3259.623	3260.563	30669.55	TA	1.5	0.5		MIT
3258.24	3259.18	30682.6	TA	1.0	0.3H		MIT
3257.822	3258.761	30686.50	TA	1.4	0.3		MIT
3256.774	3257.713	30696.38	TA	2.0	0.0		MIT
3255.69	3256.63	30706.6	TA	1.3H	0.3J		MIT
3253.80	3254.74	30724.4	TA	1.2			MIT
3253.34	3254.28	30728.8	TA	1.2D	0.3H		MIT
3251.912	3252.850	30742.27	TA	1.5	0.0		MIT
3250.358	3251.296	30756.97	TA	1.8	0.5		MIT
3250.033	3250.971	30760.04	TA	1.4H			MIT
3248.522	3249.459	30774.35	TA	2.0	0.5H		MIT
3246.90	3247.84	30789.7	TA	1.5H			KS
3245.284	3246.220	30805.06	TA	1.8	0.3		MIT
3243.359	3244.295	30823.34	TA	1.4	0.0		MIT
3242.834	3243.770	30828.33	TA	2.1	1.0		MIT
3242.566	3243.502	30830.88	TA	0.8	0.0		MIT
3242.048	3242.983	30835.80	TA	2.1	1.2		MIT
3239.987	3240.922	30855.42	TA	2.3	1.3W		MIT
3237.846	3238.780	30875.82	TA	1.8	0.8H		MIT
3235.867	3236.801	30894.70	TA	1.4			MIT
3234.693	3235.627	30905.92	TA	1.8	1.0		MIT
3232.279	3233.212	30929.00	TA	1.4	0.0		MIT
3231.665	3232.598	30934.87	TA	1.3H	0.0H		MIT

AIR WAVELENGTH (ANGSTRMS)	VACUUM WAVELENGTH (ANGSTRMS)	WAVENUMBER (KAYSERS)	LMENT	LOG ARC	INTENSITY SPK	DIS	REF
3230.855	3231.788	30942.63	TA	2.3	1.3W		MIT
3229.880	3230.812	30951.97	TA	1.5H	1.7		MIT
3229.236	3230.168	30958.14	TA	2.5W	1.8W		MIT
3228.435	3229.367	30965.82	TA	0.3H	1.4H		MIT
3227.317	3228.249	30976.55	TA	1.8L	1.0		MIT
3226.843	3227.775	30981.10	TA	1.5	0.0		MIT
3226.307	3227.238	30986.24	TA	0.3	0.0		MIT
3223.826	3224.757	31010.09	TA	2.3O	1.7O		MIT
3221.315	3222.245	31034.26	TA	1.8	1.2S		MIT
3219.585	3220.515	31050.94	TA	1.3S	0.3S		MIT
3217.984	3218.913	31066.38	TA	0.3R	0.0H		MIT
3216.923	3217.852	31076.63	TA	2.0W	1.3W		MIT
3216.620	3217.549	31079.56	TA	0.5			MIT
3215.977	3216.906	31085.77	TA	0.7	0.0		MIT
3209.866	3210.793	31144.95	TA	0.5	0.0		MIT
3208.629	3209.556	31156.96	TA	0.8	0.0		MIT
3208.197	3209.124	31161.15	TA	0.5	1.0		MIT
3207.853	3208.780	31164.49	TA	1.8	1.2		MIT
3206.785	3207.712	31174.87	TA	0.8H	0.0H		MIT
3206.386	3207.312	31178.75	TA	1.8	1.2		MIT
3205.49	3206.42	31187.5	TA	1.0O			MIT
3203.735	3204.661	31204.55	TA	0.5	0.3B		MIT
3201.976	3202.901	31221.69	TA	1.6	0.8		MIT
3199.225	3200.150	31248.54	TA		1.8L		MIT
3198.936	3199.861	31251.36	TA		1.8L		MIT
3198.670	3199.595	31253.96	TA	2.1	1.3		MIT
3196.370	3197.294	31276.45	TA	0.9O			MIT
3194.842	3195.766	31291.41	TA	0.0H	1.6		MIT
3192.253	3193.176	31316.78	TA	1.8	1.3W		MIT
3191.156	3192.079	31327.55	TA	1.7	0.7		MIT
3189.719	3190.641	31341.66	TA	1.0W	0.0H		MIT
3189.100	3190.022	31347.75	TA	0.3J	0.6H		MIT
3188.459	3189.381	31354.05	TA	0.3	0.0		MIT
3184.552	3185.473	31392.51	TA	1.8	1.3		MIT
3182.571	3183.491	31412.05	TA	1.8R	1.3		MIT
3181.692	3182.612	31420.73	TA	0.8	1.3D		MIT
3180.947	3181.867	31428.09	TA	2.0	1.5		MIT
3179.542	3180.462	31441.98	TA	1.2H	0.3H		MIT
3178.266	3179.185	31454.60	TA	1.2	0.8		MIT
3178.164	3179.083	31455.61	TA	1.3	0.5		MIT
3177.933	3178.852	31457.90	TA	1.3	0.5		MIT
3176.294	3177.213	31474.13	TA	1.8	1.3		MIT
3174.378	3175.296	31493.12	TA	0.5	0.0		MIT
3173.586	3174.504	31500.98	TA	1.8	1.0		MIT
3173.443	3174.361	31502.40	TA	1.2L	0.5		MIT
3172.874	3173.792	31508.05	TA	1.7	1.4H		MIT
3170.289	3171.206	31533.74	TA	2.4W	1.5		MIT
3168.183	3169.100	31554.70	TA		1.8		MIT
3167.529	3168.446	31561.22	TA	1.5	0.5		MIT
3166.377	3167.293	31572.70	TA	0.3R	1.5		MIT
3163.825	3164.741	31598.17	TA	1.6	0.7		MIT
3163.545	3164.461	31600.96	TA	1.2S	0.5		MIT
3163.128	3164.044	31605.13	TA	1.8R	1.5		MIT
3162.724	3163.639	31609.17	TA	1.8	0.8		MIT
3161.852	3162.767	31617.88	TA	0.5	0.0		MIT
3161.442	3162.357	31621.98	TA	0.8	0.3		MIT
3160.808	3161.723	31628.32	TA	0.3H	0.0		MIT
3159.049	3159.964	31645.94	TA	1.4W	0.5		MIT
3157.955	3158.869	31656.90	TA	0.5R	1.7		MIT
3157.648	3158.562	31659.98	TA	0.0H	1.3		MIT
3156.763	3157.677	31668.85	TA	0.0	1.5		MIT
3155.502	3156.416	31681.51	TA	1.0			MIT
3155.246	3156.160	31684.08	TA		1.3H		MIT
3154.696	3155.609	31689.60	TA	0.5	0.0		MIT
3154.487	3155.400	31691.70	TA	1.2L	0.5		MIT
3153.926	3154.839	31697.34	TA	0.0	1.2		MIT
3153.098	3154.011	31705.66	TA	0.9O	0.0H		MIT
3152.538	3153.451	31711.29	TA	0.9	0.0H		MIT
3151.996	3152.909	31716.74	TA	0.5R	0.0		MIT
3150.85	3151.76	31728.3	TA	1.7W	1.5W		MIT

AIR WAVELENGTH (ANGSTRMS)	VACUUM WAVELENGTH (ANGSTRMS)	WAVENUMBER (KAYSERS)	LMENT	LOG ARC	INTENSITY SPK	DIS	REF
3149.776	3150.688	31739.10	TA	0.3W	0.0		MIT
3148.035	3148.947	31756.65	TA	1.7	0.8		MIT
3147.374	3148.286	31763.32	TA	1.8	1.7		MIT
3146.772	3147.683	31769.40	TA	0.7	0.3		MIT
3142.955	3143.865	31807.98	TA	1.0	1.7		MIT
3141.380	3142.290	31823.92	TA	0.0H	1.7L		MIT
3138.485	3139.394	31853.28	TA	1.2	0.0H		MIT
3138.299	3139.208	31855.17	TA	0.3H	0.0		MIT
3137.442	3138.351	31863.87	TA	0.5	1.7		MIT
3136.292	3137.201	31875.55	TA	0.7			MIT
3135.893	3136.802	31879.61	TA	1.5	2.0		MIT
3133.553	3134.461	31903.41	TA	1.2	0.5		MIT
3132.643	3133.551	31912.68	TA	2.4W	1.4		MIT
3131.216	3132.123	31927.22	TA	0.7H	1.4H		MIT
3130.578	3131.485	31933.73	TA	2.0O	1.5		MIT
3130.29	3131.20	31936.7	TA	1.2L	0.0		MIT
3129.946	3130.853	31940.18	TA	1.7	0.9		MIT
3129.549	3130.456	31944.23	TA	1.7	0.8		MIT
3129.128	3130.035	31948.52	TA	1.2W	0.7H		MIT
3127.765	3128.672	31962.45	TA	1.3W	2.0		MIT
3124.969	3125.875	31991.04	TA	1.7	1.3		MIT
3121.502	3122.407	32026.57	TA	0.3H	0.0		MIT
3120.920	3121.825	32032.55	TA	1.3	0.7		MIT
3119.594	3120.499	32046.16	TA	1.3	0.7		MIT
3117.441	3118.345	32068.29	TA	1.8	1.0		MIT
3115.859	3116.763	32084.57	TA	1.7	1.3W		MIT
3113.903	3114.806	32104.73	TA	1.7	1.5W		MIT
3112.920	3113.823	32114.86	TA	0.0	1.7L		MIT
3110.815	3111.717	32136.59	TA		1.8W		MIT
3110.103	3111.005	32143.95	TA	0.5	0.0		MIT
3107.799	3108.701	32167.78	TA	0.5R	0.0H		MIT
3107.211	3108.112	32173.87	TA	1.3	0.5		MIT
3105.160	3106.061	32195.12	TA	0.3H			MIT
3104.425	3105.326	32202.74	TA	1.2H	0.0H		MIT
3104.167	3105.068	32205.42	TA	0.3H			MIT
3103.251	3104.151	32214.92	TA	1.8	1.2		MIT
3101.722	3102.622	32230.80	TA	0.8	0.3		MIT
3101.033	3101.933	32237.96	TA		2.0O		MIT
3096.116	3097.015	32289.16	TA	0.0	0.3H		MIT
3095.393	3096.291	32296.70	TA	1.8W	1.3W		MIT
3093.869	3094.767	32312.61	TA	1.7	1.2		MIT
3092.994	3093.892	32321.75	TA	1.3	0.0		MIT
3092.444	3093.342	32327.50	TA	1.7	1.2		MIT
3092.057	3092.955	32331.54	TA	1.2	0.0		MIT
3089.907	3090.804	32354.04	TA	0.3	0.8H		MIT
3087.76	3088.66	32376.5	TA	0.7J	1.8		MIT
3087.526	3088.423	32378.99	TA	1.5	0.7		MIT
3085.535	3086.431	32399.88	TA	1.8	1.3		MIT
3084.525	3085.421	32410.49	TA	0.5H			MIT
3082.447	3083.342	32432.34	TA	1.2H	0.0H		MIT
3081.850	3082.745	32438.62	TA	1.7	0.7		MIT
3079.955	3080.850	32458.58	TA	1.7W	0.7H		MIT
3079.369	3080.263	32464.75	TA	0.5	0.3H		MIT
3078.232	3079.126	32476.75	TA	1.7	0.7		MIT
3077.245	3078.139	32487.16	TA	2.2W	1.7W		MIT
3076.531	3077.425	32494.70	TA	0.5	0.0		MIT
3076.384	3077.278	32496.25	TA	1.5	1.4N		MIT
3075.322	3076.215	32507.48	TA	1.3	0.3		MIT
3073.387	3074.280	32527.94	TA	1.3	0.7		MIT
3072.351	3073.244	32538.91	TA	0.3S	0.0H		MIT
3072.071	3072.964	32541.87	TA	0.3J			MIT
3070.539	3071.431	32558.11	TA	0.7H	0.0H		MIT
3069.237	3070.129	32571.92	TA	2.2	1.8		MIT
3066.757	3067.648	32598.26	TA	1.0	0.5		MIT
3063.875	3064.766	32628.92	TA	1.3	0.5		MIT
3063.558	3064.449	32632.30	TA	1.8	1.2		MIT
3061.997	3062.887	32648.93	TA	0.3			MIT
3060.288	3061.178	32667.17	TA	2.1	1.5O		MIT
3058.636	3059.525	32684.81	TA	1.7	0.5		MIT
3057.12	3058.01	32701.0	TA	1.4W	2.1		MIT

AIR WAVELENGTH (ANGSTRMS)	VACUUM WAVELENGTH (ANGSTRMS)	WAVENUMBER (KAYSERS)	LMENT	LOG ARC	INTENSITY SPK	DIS	REF
3056.615	3057.504	32706.42	TA	0.0J	1.8L		MIT
3054.798	3055.686	32725.87	TA	0.5	0.0		MIT
3052.534	3053.422	32750.14	TA		1.8L		MIT
3051.904	3052.792	32756.90	TA	0.5H	0.5		MIT
3050.101	3050.988	32776.27	TA	1.5	0.7		MIT
3049.558	3050.445	32782.10	TA	2.2	1.5		MIT
3048.864	3049.751	32789.56	TA	2.0	1.2		MIT
3048.283	3049.170	32795.81	TA	1.0	0.0		MIT
3045.964	3046.850	32820.78	TA	2.2	1.7W		MIT
3043.925	3044.811	32842.77	TA	1.3	0.5		MIT
3042.439	3043.324	32858.81	TA	0.0	1.3		MIT
3042.060	3042.945	32862.90	TA	0.7	2.0		MIT
3040.976	3041.861	32874.61	TA	1.7	0.8W		MIT
3037.505	3038.389	32912.18	TA	0.9H	2.0		MIT
3036.287	3037.171	32925.38	TA	0.9	0.3		MIT
3035.802	3036.686	32930.64	TA	0.3			MIT
3033.392	3034.275	32956.80	TA	0.3			MIT
3030.291	3031.173	32990.53	TA	1.3	0.5		MIT
3029.534	3030.416	32998.77	TA	0.7	0.3		MIT
3028.782	3029.664	33006.96	TA	1.7	0.8		MIT
3027.510	3028.391	33020.83	TA	2.1	1.5W		MIT
3025.164	3026.045	33046.44	TA	1.8	1.2		MIT
3024.273	3025.154	33056.17	TA	0.5	0.0		MIT
3022.286	3023.166	33077.90	TA	0.5	0.0		MIT
3019.666	3020.546	33106.60	TA	1.2	0.5		MIT
3019.095	3019.974	33112.86	TA	0.5	0.0		MIT
3016.374	3017.253	33142.73	TA	0.8	0.8		MIT
3012.537	3013.415	33184.94	TA	2.1	2.0L		MIT
3011.877	3012.755	33192.22	TA	2.0W	1.2		MIT
3011.117	3011.994	33200.59	TA	2.0O	1.4		MIT
3010.844	3011.721	33203.60	TA	0.7	1.8		MIT
3007.533	3008.410	33240.16	TA	0.3			MIT
3006.559	3007.435	33250.92	TA	0.5W	0.5H		MIT
3005.02	3005.90	33268.0	TA	0.3H	0.5H		MIT
3004.919	3005.795	33269.07	TA	0.3	1.3L		MIT
3004.15	3005.03	33277.6	TA	0.8	0.0		MIT
3001.545	3002.420	33306.47	TA	0.0	0.5		MIT
2997.01	2997.88	33356.9	TA	0.7H	0.3H		MIT
2996.885	2997.759	33358.25	TA	0.3	0.7H		MIT
2993.135	2994.008	33400.04	TA	0.7	0.3H		MIT
2992.39	2993.26	33408.4	TA	0.3H	0.0H		MIT
2991.246	2992.118	33421.14	TA	1.7	1.0		MIT
2989.87	2990.74	33436.5	TA	0.3H	0.3H		MIT
2989.497	2990.369	33440.69	TA	2.3	1.2		MIT
2989.054	2989.926	33445.65	TA	1.6O	0.7H		MIT
2988.581	2989.453	33450.94	TA	1.6	1.3		MIT
2986.807	2987.678	33470.81	TA	1.3D	2.0		MIT
2986.418	2987.289	33475.17	TA	1.0	0.5		MIT
2985.690	2986.561	33483.33	TA	0.3H	0.0		MIT
2984.355	2985.226	33498.30	TA	1.5	1.0		MIT
2981.960	2982.830	33525.21	TA	1.2	0.5		MIT
2981.190	2982.060	33533.87	TA	1.2	0.7		MIT
2980.790	2981.660	33538.37	TA	1.0	0.3		MIT
2980.308	2981.178	33543.79	TA	0.3			MIT
2978.754	2979.623	33561.29	TA	2.3R	1.5		MIT
2978.180	2979.049	33567.76	TA		2.2L		MIT
2977.542	2978.411	33574.95	TA	1.3	0.0		MIT
2976.765	2977.634	33583.71	TA	0.5H	1.6		MIT
2976.26	2977.13	33589.4	TA	0.3J	2.2		MIT
2976.100	2976.969	33591.22	TA	1.6L	0.7H		MIT
2975.558	2976.427	33597.33	TA	2.3	1.7		MIT
2971.955	2972.823	33638.06	TA	0.5	0.7		MIT
2969.902	2970.769	33661.32	TA	1.7H	1.0H		MIT
2969.470	2970.337	33666.21	TA	2.2	1.9		MIT
2965.920	2966.786	33706.51	TA	1.5	1.9		MIT
2965.545	2966.411	33710.77	TA	2.2	1.7		MIT
2965.20	2966.07	33714.7	TA		1.0		MIT
2965.133	2965.999	33715.45	TA	1.6	2.9		MIT
2964.811	2965.677	33719.12	TA	0.5W	0.5H		MIT
2963.910	2964.776	33729.37	TA	1.7	1.0		MIT

AIR WAVELENGTH (ANGSTRMS)	VACUUM WAVELENGTH (ANGSTRMS)	WAVENUMBER (KAYSERS)	LMENT	LOG ARC	INTENSITY SPK	DIS	REF
2963.322	2964.188	33736.06	TA	2.5	2.0		MIT
2963.056	2963.921	33739.09	TA	1.6	1.0		MIT
2961.379	2962.244	33758.19	TA	0.3	0.0		MIT
2958.12	2958.98	33795.4	TA	0.3			MIT
2957.598	2958.462	33801.35	TA	2.0	1.5		MIT
2956.84	2957.70	33810.0	TA	0.0H	2.0L		MIT
2953.560	2954.423	33847.56	TA	1.0			MIT
2952.99	2953.85	33854.1	TA	1.5H	2.0H		MIT
2951.918	2952.781	33866.38	TA	2.6K	2.3		MIT
2948.51	2949.37	33905.5	TA	0.5			KS
2947.80	2948.66	33913.7	TA	1.0	0.3		MIT
2947.377	2948.239	33918.56	TA	0.0	0.3H		MIT
2946.910	2947.771	33923.93	TA	2.2	1.0		MIT
2946.261	2947.122	33931.40	TA	1.0	0.3		MIT
2943.769	2944.630	33960.13	TA	1.0	0.7		MIT
2942.137	2942.997	33978.96	TA	2.2	1.6		MIT
2940.215	2941.075	34001.18	TA	2.2	1.7		MIT
2940.06	2940.92	34003.0	TA	2.0L	1.6W		MIT
2939.28	2940.14	34012.0	TA	2.3	1.6H		MIT
2938.564	2939.423	34020.28	TA	1.0			MIT
2938.426	2939.285	34021.88	TA	1.7	0.5		MIT
2937.484	2938.343	34032.79	TA	0.3			MIT
2935.949	2936.808	34050.58	TA	0.3	0.7		MIT
2935.570	2936.429	34054.97	TA	0.3	0.5		MIT
2935.165	2936.024	34059.67	TA	1.3	0.3		MIT
2934.847	2935.705	34063.36	TA	1.6	0.6		MIT
2934.50	2935.36	34067.4	TA	0.3			KS
2933.550	2934.408	34078.42	TA	2.6	2.2		MIT
2932.695	2933.553	34088.36	TA	2.6	1.9W		MIT
2930.99	2931.85	34108.2	TA	1.5W	0.5H		MIT
2926.457	2927.313	34161.02	TA	2.0	1.0		MIT
2925.659	2926.515	34170.33	TA	2.0	0.6		MIT
2925.265	2926.121	34174.94	TA	2.0O	1.6N		MIT
2925.193	2926.049	34175.78	TA	2.0	0.7		MIT
2924.785	2925.641	34180.54	TA	1.2	0.3		MIT
2923.519	2924.375	34195.35	TA	0.6H			MIT
2922.415	2923.270	34208.26	TA	1.0	0.0H		MIT
2920.322	2921.177	34232.78	TA	0.3			MIT
2918.96	2919.81	34248.7	TA	0.3	1.7		MIT
2917.120	2917.974	34270.35	TA	1.0	0.5H		MIT
2916.863	2917.717	34273.37	TA	0.3	0.5H		MIT
2916.322	2917.176	34279.73	TA	0.7	0.0		MIT
2916.038	2916.892	34283.07	TA	0.3H			MIT
2915.492	2916.346	34289.49	TA	2.2	1.6		MIT
2915.337	2916.191	34291.31	TA	2.2W	1.7		MIT
2914.936	2915.790	34296.03	TA	1.3	0.5		MIT
2914.125	2914.978	34305.57	TA	2.3	1.5		MIT
2913.317	2914.170	34315.09	TA	1.2O	0.3		MIT
2908.910	2909.762	34367.07	TA	2.2	1.0		MIT
2907.783	2908.635	34380.39	TA	0.5			MIT
2906.48	2907.33	34395.8	TA	0.3H			MIT
2906.17	2907.02	34399.5	TA	0.3			KS
2905.739	2906.590	34404.57	TA	1.6	0.5		MIT
2905.239	2906.090	34410.50	TA	1.9	2.0		MIT
2904.430	2905.281	34420.08	TA	1.2	0.5		MIT
2904.074	2904.925	34424.30	TA	2.5W	1.6		MIT
2902.632	2903.482	34441.40	TA	1.5	0.3		MIT
2902.046	2902.896	34448.35	TA	3.0W	2.3		MIT
2901.051	2901.901	34460.17	TA	2.0	0.5		MIT
2900.750	2901.600	34463.74	TA	0.5H	2.0L		MIT
2900.363	2901.213	34468.34	TA	2.3	1.6		MIT
2899.044	2899.894	34484.02	TA	2.3	1.2		MIT
2898.425	2899.274	34491.39	TA	1.5	0.7		MIT
2896.438	2897.287	34515.05	TA	0.3H	1.7		MIT
2895.103	2895.952	34530.96	TA	2.1	1.2		MIT
2893.792	2894.640	34546.61	TA	0.3	1.0H		MIT
2892.000	2892.848	34568.01	TA	1.9H	1.0		MIT
2891.843	2892.691	34569.89	TA	2.7O	2.0		MIT
2891.038	2891.886	34579.51	TA	2.2O	1.5H		MIT
2890.52	2891.37	34585.7	TA	0.0	0.3		MIT

AIR WAVELENGTH (ANGSTRMS)	VACUUM WAVELENGTH (ANGSTRMS)	WAVENUMBER (KAYSERS)	LMENT	LOG ARC	INTENSITY SPK	DIS	REF
2890.060	2890.907	34591.22	TA	1.2	0.3		MIT
2889.380	2890.227	34599.36	TA	1.6	0.7		MIT
2886.647	2887.494	34632.11	TA	0.7H	0.3		MIT
2883.895	2884.741	34665.16	TA	0.9	0.7J		MIT
2882.332	2883.177	34683.96	TA	0.5	1.9		MIT
2881.232	2882.077	34697.20	TA	1.5	0.7		MIT
2880.018	2880.863	34711.82	TA	2.2	1.7		MIT
2879.737	2880.582	34715.21	TA	2.2	1.0		MIT
2879.516	2880.361	34717.87	TA	1.7S	1.0H		MIT
2878.951	2879.796	34724.69	TA	1.6R	0.5S		MIT
2878.20	2879.04	34733.7	TA	0.6	1.2H		MIT
2877.686	2878.530	34739.95	TA	1.2	1.9H		MIT
2876.111	2876.955	34758.97	TA	1.7R	0.7		MIT
2874.936	2875.780	34773.18	TA	1.4	0.3		MIT
2874.521	2875.365	34778.20	TA	1.3	0.5		MIT
2874.167	2875.010	34782.48	TA	2.2	1.2		MIT
2873.557	2874.400	34789.87	TA	2.2	1.7L		MIT
2873.360	2874.203	34792.25	TA	2.3O	1.6H		MIT
2871.417	2872.260	34815.79	TA	2.3	1.7		MIT
2869.52	2870.36	34838.8	TA	0.5	0.7H		MIT
2869.20	2870.04	34842.7	TA	0.3			MIT
2868.651	2869.493	34849.36	TA	2.2	1.6		MIT
2867.41	2868.25	34864.4	TA	0.7H	2.2		MIT
2866.66	2867.50	34873.6	TA	0.5D	0.0A		MIT
2866.141	2866.982	34879.88	TA	0.7			MIT
2864.724	2865.565	34897.13	TA	0.5			MIT
2864.503	2865.344	34899.82	TA	2.1	1.5		MIT
2864.26	2865.10	34902.8	TA	0.5	0.5H		MIT
2862.385	2863.226	34925.65	TA	0.3	0.0		MIT
2862.025	2862.865	34930.04	TA	1.6	1.0		MIT
2861.981	2862.821	34930.58	TA	1.5	0.7		MIT
2861.117	2861.957	34941.12	TA	1.5			MIT
2858.435	2859.275	34973.91	TA	2.0	2.5		MIT
2857.282	2858.121	34988.02	TA	1.8	0.7		MIT
2852.950	2853.788	35041.14	TA	0.7	1.2		MIT
2852.355	2853.193	35048.45	TA	0.7	2.0L		MIT
2850.985	2851.823	35065.29	TA	2.6	2.2		MIT
2850.491	2851.329	35071.37	TA	2.3	2.0		MIT
2849.823	2850.660	35079.59	TA	1.2Y	1.7O		MIT
2849.549	2850.386	35082.96	TA	0.7H	0.0H		MIT
2848.525	2849.362	35095.57	TA	2.5	1.7		MIT
2848.522	2849.359	35095.61	TA	1.0	0.7		MIT
2848.054	2848.891	35101.38	TA	2.2	1.2		MIT
2847.291	2848.128	35110.78	TA	0.3	0.3		MIT
2846.750	2847.587	35117.46	TA	2.2C	1.0H		MIT
2845.844	2846.680	35128.63	TA	1.5	1.0H		MIT
2845.450	2846.286	35133.50	TA	0.9H	0.3		MIT
2845.353	2846.189	35134.70	TA	2.2	1.0		MIT
2844.757	2845.593	35142.06	TA	2.2	1.5		MIT
2844.463	2845.299	35145.69	TA	2.3	2.3L		MIT
2844.251	2845.087	35148.31	TA	2.6R	1.7		MIT
2843.51	2844.35	35157.5	TA	0.5	1.9		MIT
2842.815	2843.651	35166.06	TA	2.3	1.7		MIT
2840.39	2841.23	35196.1	TA	0.3	1.7		MIT
2839.778	2840.613	35203.67	TA	0.9	0.3		MIT
2838.24	2839.07	35222.7	TA	0.3	2.2		MIT
2837.942	2838.777	35226.44	TA	1.5	1.5		MIT
2836.618	2837.452	35242.88	TA	1.9R	0.3		MIT
2836.259	2837.093	35247.35	TA	0.3			MIT
2834.408	2835.242	35270.36	TA	0.0	0.9		MIT
2833.636	2834.469	35279.97	TA	2.5W	1.6W		MIT
2830.018	2830.851	35325.07	TA	1.3	0.0		MIT
2828.579	2829.411	35343.04	TA	1.9	2.0		MIT
2827.55	2828.38	35355.9	TA	0.5D	2.0D		MIT
2827.176	2828.008	35360.58	TA	2.3	1.0		MIT
2826.419	2827.251	35370.05	TA	1.3	0.5		MIT
2826.183	2827.015	35373.00	TA	1.8	0.7		MIT
2825.032	2825.863	35387.42	TA	0.3	0.0H		MIT
2824.809	2825.640	35390.21	TA	1.8O	0.7H		MIT
2824.048	2824.879	35399.74	TA	0.6			MIT

AIR WAVELENGTH (ANGSTRMS)	VACUUM WAVELENGTH (ANGSTRMS)	WAVENUMBER (KAYSERS)	LMENT	LOG ARC	INTENSITY SPK	DIS	REF
2822.373	2823.204	35420.75	TA	1.0H			MIT
2821.990	2822.821	35425.56	TA	1.1	0.6		MIT
2821.176	2822.006	35435.78	TA	0.5	0.0		MIT
2819.370	2820.200	35458.48	TA	2.0	0.7		MIT
2818.765	2819.595	35466.09	TA	0.3			MIT
2817.503	2818.333	35481.97	TA	1.9L	1.0		MIT
2817.101	2817.930	35487.04	TA	1.9D	2.0		MIT
2815.123	2815.952	35511.97	TA	2.0	0.6		MIT
2815.011	2815.840	35513.38	TA	2.2	1.2		MIT
2814.801	2815.630	35516.03	TA	2.1	0.7		MIT
2814.313	2815.142	35522.19	TA	1.7R	1.7		MIT
2811.724	2812.552	35554.90	TA	0.0	2.2		MIT
2810.916	2811.744	35565.12	TA	2.3O	1.6O		MIT
2810.050	2810.878	35576.08	TA	0.3			MIT
2806.581	2807.408	35620.05	TA	2.3	1.7		MIT
2806.301	2807.128	35623.60	TA	2.5	1.7		MIT
2805.064	2805.891	35639.31	TA	0.5	0.3H		MIT
2804.764	2805.590	35643.12	TA	1.3	0.3		MIT
2804.004	2804.830	35652.78	TA	0.5	1.0H		MIT
2802.702	2803.528	35669.34	TA	1.0	0.6		MIT
2802.493	2803.319	35672.00	TA	1.2	0.3		MIT
2802.071	2802.897	35677.37	TA	2.5	1.9		MIT
2800.572	2801.397	35696.47	TA	2.2O	1.6H		MIT
2798.404	2799.229	35724.12	TA	2.2			MIT
2797.760	2798.585	35732.35	TA	2.0D	2.0D		MIT
2796.565	2797.389	35747.62	TA	2.2			MIT
2796.339	2797.163	35750.50	TA	2.6	1.9		MIT
2792.637	2793.460	35797.89	TA	0.9	0.3		MIT
2791.674	2792.497	35810.24	TA	2.0	1.0		MIT
2791.370	2792.193	35814.14	TA	1.4	2.2		MIT
2791.169	2791.992	35816.72	TA	0.3			MIT
2790.710	2791.533	35822.61	TA	2.2	1.0		MIT
2788.302	2789.124	35853.55	TA	2.2	0.5		MIT
2787.693	2788.515	35861.38	TA	2.6G	1.6		MIT
2784.967	2785.789	35896.48	TA	1.7	2.0		MIT
2783.694	2784.515	35912.89	TA	0.3H			MIT
2781.788	2782.609	35937.50	TA	1.4	0.0		MIT
2781.369	2782.190	35942.91	TA	1.7	0.6		MIT
2780.208	2781.028	35957.92	TA	0.6			MIT
2779.704	2780.524	35964.44	TA	1.5R	0.6		MIT
2779.098	2779.918	35972.28	TA	2.2W	0.7H		MIT
2775.877	2776.696	36014.02	TA	2.3	1.5		MIT
2775.346	2776.165	36020.91	TA	1.9	1.2		MIT
2775.108	2775.927	36024.00	TA	2.0O	1.9		MIT
2774.876	2775.695	36027.01	TA	2.0	0.5		MIT
2772.596	2773.415	36056.64	TA	1.2			MIT
2771.833	2772.651	36066.56	TA	0.5H	2.0		MIT
2770.782	2771.600	36080.24	TA	1.7J	0.0		MIT
2768.934	2769.752	36104.32	TA	0.3	1.5		MIT
2768.307	2769.125	36112.50	TA	1.0	0.3		MIT
2766.117	2766.934	36141.09	TA	1.6H	1.3		MIT
2764.753	2765.570	36158.92	TA	1.0H	0.5		MIT
2763.370	2764.186	36177.01	TA	1.4D	1.8D		MIT
2762.05	2762.87	36194.3	TA	0.0	1.6L		MIT
2761.676	2762.492	36199.20	TA	2.3	2.2		MIT
2761.547	2762.363	36200.89	TA	1.9			MIT
2761.153	2761.969	36206.06	TA	0.3H			MIT
2760.685	2761.501	36212.19	TA	0.0	0.3H		MIT
2758.310	2759.125	36243.37	TA	2.3	1.6		MIT
2757.988	2758.803	36247.60	TA	0.3H			MIT
2757.125	2757.940	36258.95	TA	0.7	0.0		MIT
2752.489	2753.303	36320.02	TA	2.5	2.5		MIT
2752.295	2753.109	36322.58	TA	2.2	0.9		MIT
2751.036	2751.849	36339.20	TA	1.3	0.3W		MIT
2750.534	2751.347	36345.83	TA	1.3			MIT
2750.410	2751.223	36347.47	TA	0.5H	1.0		MIT
2749.829	2750.642	36355.15	TA	2.3	1.7		MIT
2748.778	2749.591	36369.05	TA	2.6	1.7		MIT
2747.848	2748.661	36381.36	TA	1.0	0.3		MIT
2747.251	2748.063	36389.26	TA	1.7	0.7		MIT

AIR WAVELENGTH (ANGSTRMS)	VACUUM WAVELENGTH (ANGSTRMS)	WAVENUMBER (KAYSERS)	LMENT	LOG ARC	INTENSITY SPK	INTENSITY DIS	REF
2746.681	2747.493	36396.81	TA	2.0	0.7		MIT
2745.565	2746.377	36411.61	TA	0.3H			MIT
2743.586	2744.398	36437.87	TA		0.9		MIT
2742.919	2743.730	36446.73	TA	0.3			MIT
2741.384	2742.195	36467.14	TA	0.7	0.3		MIT
2741.168	2741.979	36470.01	TA		0.7		MIT
2740.211	2741.022	36482.75	TA	2.0	1.0H		MIT
2739.265	2740.075	36495.35	TA	0.3	1.9		MIT
2736.250	2737.060	36535.56	TA	2.5S	0.9S		MIT
2734.517	2735.326	36558.71	TA	1.2			MIT
2733.667	2734.476	36570.08	TA	1.0W	0.0		MIT
2732.924	2733.733	36580.02	TA	2.00	0.7		MIT
2732.063	2732.872	36591.54	TA	1.6	0.5		MIT
2729.500	2730.308	36625.90	TA	0.7			MIT
2729.275	2730.083	36628.92	TA	1.0			MIT
2727.778	2728.586	36649.02	TA	2.3	1.6		MIT
2727.437	2728.245	36653.60	TA	1.7	2.2		MIT
2726.319	2727.126	36668.63	TA	1.2	0.0		MIT
2725.424	2726.231	36680.68	TA	1.3W	2.0		MIT
2725.094	2725.901	36685.12	TA	1.2L	0.0		MIT
2724.349	2725.156	36695.15	TA	1.0			MIT
2723.847	2724.654	36701.91	TA	0.3	1.2H		MIT
2723.274	2724.081	36709.63	TA	1.3			MIT
2722.319	2723.125	36722.51	TA	0.7			MIT
2721.826	2722.632	36729.16	TA	1.7			MIT
2720.756	2721.562	36743.60	TA	2.2	0.0		MIT
2718.383	2719.188	36775.68	TA	1.9			MIT
2717.183	2717.988	36791.92	TA	2.0			MIT
2714.674	2715.479	36825.92	TA	2.3	0.9		MIT
2710.723	2711.527	36879.59	TA	0.0	1.0H		MIT
2710.127	2710.930	36887.70	TA	2.3	0.5		MIT
2709.274	2710.077	36899.32	TA	1.6	2.2		MIT
2708.348	2709.151	36911.93	TA	0.9			MIT
2708.144	2708.947	36914.71	TA	0.5			MIT
2707.829	2708.632	36919.01	TA	1.0			MIT
2707.639	2708.442	36921.60	TA	0.3H			MIT
2707.520	2708.323	36923.22	TA	0.3			MIT
2706.921	2707.724	36931.39	TA	1.2			MIT
2706.688	2707.491	36934.57	TA	1.7	0.0		MIT
2706.147	2706.949	36941.95	TA	0.5			MIT
2704.312	2705.114	36967.02	TA	1.7			MIT
2703.528	2704.330	36977.74	TA	0.9			MIT
2703.057	2703.859	36984.18	TA	1.8			MIT
2702.80	2703.60	36987.7	TA	1.6D	1.5J		MIT
2701.610	2702.411	37003.99	TA	0.9			MIT
2700.702	2701.503	37016.43	TA	1.0			MIT
2700.598	2701.399	37017.85	TA	0.5			MIT
2700.288	2701.089	37022.10	TA	0.7J			MIT
2698.302	2699.103	37049.35	TA	2.2	0.5		MIT
2696.814	2697.614	37069.79	TA	2.1	0.0		MIT
2695.542	2696.342	37087.28	TA	0.5S	0.3		MIT
2694.759	2695.559	37098.06	TA	1.7			MIT
2693.500	2694.299	37115.40	TA	1.0H			MIT
2693.342	2694.141	37117.58	TA	1.5			MIT
2692.395	2693.194	37130.63	TA	2.0	0.0		MIT
2691.310	2692.109	37145.60	TA	2.2			MIT
2690.542	2691.341	37156.20	TA	1.2			MIT
2689.516	2690.314	37170.38	TA	0.9			MIT
2689.24	2690.04	37174.2	TA	1.3H	1.3H		MIT
2686.29	2687.09	37215.0	TA	1.2D			MIT
2685.454	2686.252	37226.60	TA	0.3			MIT
2685.11	2685.91	37231.4	TA	1.2	1.4		MIT
2684.276	2685.073	37242.93	TA	2.2	0.3		MIT
2681.869	2682.666	37276.36	TA	1.7			MIT
2681.629	2682.426	37279.69	TA	1.2			MIT
2680.665	2681.461	37293.10	TA	1.5	1.6		MIT
2680.06	2680.86	37301.5	TA	1.5	1.7		MIT
2678.804	2679.600	37319.00	TA	0.5	1.3		MIT
2677.860	2678.656	37332.16	TA	0.3			MIT
2675.901	2676.696	37359.49	TA	2.2	2.3		MIT

AIR WAVELENGTH (ANGSTRMS)	VACUUM WAVELENGTH (ANGSTRMS)	WAVENUMBER (KAYSERS)	LMENT	LOG ARC	INTENSITY SPK	INTENSITY DIS	REF
2675.54	2676.34	37364.5	TA	0.3			MIT
2675.284	2676.079	37368.10	TA		1.1		MIT
2674.490	2675.285	37379.20	TA	1.2			MIT
2673.574	2674.369	37392.00	TA	0.6			MIT
2673.286	2674.081	37396.03	TA	0.5			MIT
2672.497	2673.291	37407.07	TA	1.3	1.5H		MIT
2672.342	2673.136	37409.24	TA	1.3			MIT
2671.634	2672.428	37419.15	TA	1.5			MIT
2668.619	2669.412	37461.43	TA	0.0	2.0		MIT
2668.069	2668.862	37469.15	TA	1.9			MIT
2667.171	2667.964	37481.76	TA	1.3			MIT
2665.935	2666.728	37499.14	TA	1.2			MIT
2665.60	2666.39	37503.9	TA	1.9D	1.6H		MIT
2664.235	2665.027	37523.07	TA	1.0H	2.0H		MIT
2662.102	2662.894	37553.13	TA	1.4			MIT
2661.886	2662.678	37556.18	TA	1.8			MIT
2661.336	2662.128	37563.94	TA	2.3	1.0		MIT
2659.655	2660.446	37587.68	TA	1.2			MIT
2659.41	2660.20	37591.1	TA	1.3	1.0		MIT
2658.862	2659.653	37598.89	TA	1.4	1.7		MIT
2658.14	2658.93	37609.1	TA	0.5H	1.2H		MIT
2657.68	2658.47	37615.6	TA	0.3			MIT
2657.296	2658.087	37621.04	TA	1.4			MIT
2656.61	2657.40	37630.8	TA	2.3G	0.3		MIT
2656.08	2656.87	37638.3	TA	0.7	0.7H		MIT
2655.675	2656.465	37644.01	TA	1.2			MIT
2654.011	2654.801	37667.61	TA	1.2			MIT
2653.274	2654.064	37678.07	TA	2.3	1.2		MIT
2652.323	2653.113	37691.58	TA	1.2			MIT
2651.483	2652.272	37703.52	TA	1.7			MIT
2651.221	2652.010	37707.24	TA		1.9		MIT
2650.280	2651.069	37720.63	TA	1.4			MIT
2650.016	2650.805	37724.39	TA	0.9			MIT
2647.472	2648.260	37760.64	TA	2.3	1.0		MIT
2646.77	2647.56	37770.6	TA	1.7H	1.7		MIT
2646.369	2647.157	37776.37	TA	2.1	0.3		MIT
2646.222	2647.010	37778.47	TA	1.7	0.3		MIT
2645.100	2645.888	37794.50	TA	1.9	1.5H		MIT
2644.598	2645.386	37801.67	TA	1.30	1.70		MIT
2643.891	2644.679	37811.78	TA	1.7			MIT
2643.252	2644.039	37820.92	TA	0.7			MIT
2638.673	2639.459	37886.55	TA	0.9	1.0H		MIT
2638.266	2639.052	37892.39	TA	0.3			MIT
2637.93	2638.72	37897.2	TA	0.0	1.0H		MIT
2636.899	2637.685	37912.04	TA	2.0	0.5		MIT
2636.673	2637.459	37915.28	TA	1.8	0.0		MIT
2636.371	2637.157	37919.63	TA	0.6			MIT
2635.929	2636.715	37925.98	TA	1.7H			MIT
2635.583	2636.369	37930.96	TA	0.9	1.1		MIT
2634.105	2634.890	37952.25	TA	0.7			MIT
2633.79	2634.58	37956.8	TA	1.0J	1.9J		MIT
2632.27	2633.05	37978.7	TA	1.2	1.9		MIT
2630.528	2631.312	38003.85	TA	1.0	0.9		MIT
2629.977	2630.761	38011.81	TA	0.5			MIT
2628.851	2629.635	38028.09	TA	0.7H	1.0J		MIT
2627.054	2627.838	38054.10	TA	0.6			MIT
2626.059	2626.842	38068.52	TA	0.3			MIT
2625.793	2626.576	38072.38	TA	0.7H			MIT
2625.462	2626.245	38077.18	TA	1.6			MIT
2624.116	2624.899	38096.71	TA	1.8			MIT
2620.182	2620.964	38153.90	TA	0.7			MIT
2615.656	2616.437	38219.92	TA	1.7	0.0		MIT
2615.465	2616.246	38222.71	TA	1.7	0.0		MIT
2615.250	2616.031	38225.85	TA	1.6			MIT
2614.173	2614.953	38241.60	TA	2.3J			MIT
2612.609	2613.389	38264.49	TA	1.7	1.6		MIT
2611.459	2612.239	38281.34	TA	1.2			MIT
2611.340	2612.120	38283.08	TA	2.0			MIT
2610.129	2610.908	38300.84	TA	1.0			MIT
2608.999	2609.778	38317.43	TA	1.9	0.0		MIT

AIR WAVELENGTH (ANGSTRMS)	VACUUM WAVELENGTH (ANGSTRMS)	WAVENUMBER (KAYSERS)	LMENT	LOG ARC	INTENSITY SPK	DIS	REF
2608.627	2609.406	38322.90	TA	2.1	0.6		MIT
2608.200	2608.979	38329.17	TA	1.2			MIT
2607.840	2608.619	38334.46	TA	1.3H	2.2		MIT
2606.543	2607.322	38353.53	TA	0.5			MIT
2606.42	2607.20	38355.3	TA		1.0		MIT
2605.809	2606.587	38364.34	TA	0.5			MIT
2605.319	2606.097	38371.55	TA	1.0			MIT
2603.822	2604.600	38393.61	TA	1.4			MIT
2603.573	2604.351	38397.28	TA	0.7	2.5		MIT
2603.493	2604.271	38398.46	TA	1.0	1.2		MIT
2602.377	2603.155	38414.93	TA	1.6			MIT
2601.055	2601.832	38434.45	TA	1.7	0.7J		MIT
2600.141	2600.918	38447.96	TA	1.9			MIT
2599.397	2600.174	38458.96	TA	2.0	1.5		MIT
2599.090	2599.867	38463.51	TA	0.3			MIT
2598.746	2599.523	38468.60	TA	1.2			MIT
2598.21	2598.99	38476.5	TA	0.3S	1.7		MIT
2597.493	2598.270	38487.15	TA	0.3			MIT
2596.607	2597.383	38500.29	TA	1.3			MIT
2596.450	2597.226	38502.61	TA	1.9	2.2		MIT
2596.116	2596.892	38507.57	TA	1.5	1.0J		MIT
2595.586	2596.362	38515.43	TA	1.6D	1.7H		MIT
2595.261	2596.037	38520.25	TA	1.7	0.3		MIT
2594.536	2595.312	38531.02	TA	0.8			MIT
2594.247	2595.023	38535.31	TA	0.7	0.7		MIT
2593.660	2594.436	38544.03	TA	1.9D	2.0		MIT
2593.085	2593.860	38552.58	TA	2.2	0.0		MIT
2592.53	2593.31	38560.8	TA	0.7			MIT
2592.440	2593.215	38562.17	TA	1.1H			MIT
2590.943	2591.718	38584.45	TA	0.7			MIT
2590.492	2591.267	38591.16	TA	0.7H			MIT
2589.811	2590.586	38601.31	TA	0.9	1.4		MIT
2588.533	2589.307	38620.37	TA	0.7			MIT
2585.607	2586.381	38664.07	TA	1.5			MIT
2584.901	2585.675	38674.63	TA	0.3			MIT
2584.691	2585.464	38677.77	TA	1.6			MIT
2584.027	2584.800	38687.71	TA	1.9W	2.3		MIT
2581.603	2582.376	38724.03	TA	0.3	0.5H		MIT
2580.584	2581.356	38739.32	TA	0.3	0.7H		MIT
2580.156	2580.928	38745.75	TA	2.2			MIT
2579.619	2580.391	38753.81	TA	1.9			MIT
2579.047	2579.819	38762.41	TA	0.5H	2.0H		MIT
2578.371	2579.143	38772.57	TA	0.5			MIT
2578.243	2579.015	38774.49	TA	0.5H	1.5H		MIT
2577.970	2578.742	38778.60	TA	1.0			MIT
2577.780	2578.552	38781.46	TA	2.0	0.0		MIT
2577.37	2578.14	38787.6	TA	1.9D	2.2D		MIT
2576.428	2577.200	38801.81	TA		0.5H		MIT
2575.473	2576.244	38816.19	TA	1.9			MIT
2574.379	2575.150	38832.69	TA	1.9			MIT
2573.793	2574.564	38841.53	TA	2.0	0.0		MIT
2573.544	2574.315	38845.29	TA	2.1	0.0		MIT
2571.192	2571.962	38880.82	TA	0.5			MIT
2569.126	2569.896	38912.08	TA	1.6	1.6H		MIT
2567.447	2568.216	38937.53	TA	0.3			MIT
2566.331	2567.100	38954.46	TA	0.8			MIT
2565.778	2566.547	38962.85	TA	0.5			MIT
2565.196	2565.965	38971.69	TA	0.5H			MIT
2563.861	2564.630	38991.99	TA	0.5H			MIT
2563.703	2564.472	38994.39	TA	1.9			MIT
2563.334	2564.102	39000.00	TA	1.7			MIT
2562.097	2562.865	39018.83	TA	2.0			MIT
2560.678	2561.446	39040.45	TA	1.8			MIT
2559.427	2560.195	39059.53	TA	2.0	0.3		MIT
2558.593	2559.360	39072.26	TA	0.5			MIT
2558.345	2559.112	39076.05	TA		0.7J		MIT
2557.709	2558.476	39085.77	TA	1.7	2.0		MIT
2556.510	2557.277	39104.10	TA	1.4	1.5		MIT
2555.052	2555.818	39126.41	TA	1.7			MIT
2554.907	2555.673	39128.63	TA	1.7H	1.7H		MIT

AIR WAVELENGTH (ANGSTRMS)	VACUUM WAVELENGTH (ANGSTRMS)	WAVENUMBER (KAYSERS)	LMENT	LOG ARC	INTENSITY SPK	DIS	REF
2554.622	2555.388	39133.00	TA	1.7	2.0		MIT
2553.181	2553.947	39155.08	TA	1.0H	1.0H		MIT
2551.732	2552.498	39177.31	TA	2.0	0.6H		MIT
2551.341	2552.107	39183.32	TA	0.3			MIT
2551.188	2551.954	39185.67	TA	2.2			MIT
2551.074	2551.840	39187.42	TA	2.2			MIT
2550.775	2551.540	39192.01	TA	1.3			MIT
2550.553	2551.318	39195.42	TA	1.3			MIT
2550.462	2551.227	39196.82	TA	1.3			MIT
2549.380	2550.145	39213.45	TA	2.0			MIT
2547.214	2547.979	39246.80	TA	0.6			MIT
2546.803	2547.568	39253.13	TA	1.9			MIT
2545.490	2546.254	39273.38	TA	1.2	1.2		MIT
2544.80	2545.56	39284.0	TA	1.6			MIT
2544.372	2545.136	39290.63	TA		1.0H		MIT
2544.267	2545.031	39292.25	TA	0.3			MIT
2542.346	2543.109	39321.94	TA		0.3H		MIT
2542.229	2542.992	39323.75	TA	1.0			MIT
2541.940	2542.703	39328.22	TA	1.0			MIT
2539.702	2540.465	39362.87	TA	0.3			MIT
2537.940	2538.702	39390.20	TA	0.9	1.2		MIT
2536.67	2537.43	39409.9	TA	0.3H			MIT
2536.227	2536.989	39416.80	TA	2.00			MIT
2535.96	2536.72	39421.0	TA	0.6			MIT
2535.598	2536.360	39426.58	TA	1.7H			MIT
2534.967	2535.729	39436.40	TA	1.5			MIT
2534.467	2535.229	39444.17	TA	1.4			MIT
2534.162	2534.924	39448.92	TA	0.5			MIT
2533.002	2533.763	39466.99	TA	1.5			MIT
2532.125	2532.886	39480.65	TA	1.9	2.0		MIT
2531.516	2532.277	39490.15	TA	1.4			MIT
2531.293	2532.054	39493.63	TA	1.0			MIT
2528.972	2529.732	39529.87	TA	0.5			MIT
2528.59	2529.35	39535.8	TA	0.6	0.0		MIT
2527.36	2528.12	39555.1	TA	0.5H	0.3H		MIT
2526.664	2527.424	39565.98	TA	1.7			MIT
2526.453	2527.213	39569.28	TA	2.2			MIT
2526.354	2527.114	39570.83	TA	2.0			MIT
2526.024	2526.784	39576.00	TA	2.0	1.9		MIT
2525.819	2526.579	39579.22	TA	0.3			MIT
2525.332	2526.092	39586.85	TA	1.2			MIT
2520.311	2521.069	39665.71	TA	1.0			MIT
2519.776	2520.534	39674.13	TA	1.7			MIT
2518.764	2519.522	39690.07	TA	0.3			MIT
2516.717	2517.474	39722.35	TA	0.7			MIT
2516.111	2516.868	39731.92	TA	0.8	0.3		MIT
2515.154	2515.911	39747.03	TA	0.8			MIT
2513.881	2514.638	39767.16	TA	0.6W	0.8		MIT
2513.098	2513.855	39779.55	TA	1.0	0.0		MIT
2512.652	2513.409	39786.61	TA	2.0			MIT
2512.036	2512.792	39796.36	TA	0.7	2.9		MIT
2511.69	2512.45	39801.9	TA	0.3D	2.0D		MIT
2510.706	2511.462	39817.44	TA	1.0	0.7D		MIT
2509.189	2509.945	39841.52	TA	0.6	0.3		MIT
2507.452	2508.207	39869.11	TA	2.2	0.0		MIT
2505.316	2506.071	39903.10	TA	1.1	0.3		MIT
2504.453	2505.208	39916.85	TA	2.3	0.0		MIT
2503.007	2503.761	39939.91	TA		0.3		MIT
2501.985	2502.739	39956.22	TA	1.3	1.1		MIT
2498.757	2499.510	40007.84	TA	1.0			MIT
2498.330	2499.083	40014.67	TA	1.5	0.0H		MIT
2496.957	2497.710	40036.67	TA	0.3			MIT
2496.635	2497.388	40041.84	TA	1.3			MIT
2496.237	2496.990	40048.22	TA	0.3H	0.3H		MIT
2494.90	2495.65	40069.7	TA	0.3			MIT
2492.172	2492.924	40113.54	TA	1.2	0.3		MIT
2492.011	2492.763	40116.13	TA	0.7	0.3		MIT
2491.591	2492.343	40122.89	TA	1.0			MIT
2488.696	2489.447	40169.56	TA	2.3	2.2		MIT
2488.397	2489.148	40174.39	TA	1.8			MIT

AIR WAVELENGTH (ANGSTRMS)	VACUUM WAVELENGTH (ANGSTRMS)	WAVENUMBER (KAYSERS)	LMENT	LOG ARC	INTENSITY SPK	DIS	REF
2488.207	2488.958	40177.46	TA	0.7			MIT
2487.64	2488.39	40186.6	TA	0.7	0.3		MIT
2486.705	2487.456	40201.72	TA	1.3			MIT
2484.954	2485.704	40230.05	TA	1.5			MIT
2482.577	2483.327	40268.57	TA	1.7	1.7		MIT
2481.86	2482.61	40280.2	TA	0.3H	1.0		MIT
2478.219	2478.968	40339.37	TA	1.8			MIT
2476.67	2477.42	40364.6	TA	1.3	0.8		MIT
2476.329	2477.077	40370.16	TA	0.3			MIT
2475.332	2476.080	40386.42	TA	1.1	0.9		MIT
2474.617	2475.365	40398.09	TA	2.2	0.0		MIT
2473.312	2474.059	40419.40	TA	1.2H	0.0H		MIT
2473.13	2473.88	40422.4	TA	1.3	0.9		MIT
2472.134	2472.881	40438.66	TA	1.9			MIT
2471.383	2472.130	40450.95	TA	1.3			MIT
2471.243	2471.990	40453.24	TA	0.3			MIT
2470.897	2471.644	40458.90	TA	1.8	1.6		MIT
2469.76	2470.51	40477.5	TA	0.5			KS
2468.408	2469.154	40499.70	TA	0.6D	0.3H		MIT
2468.054	2468.800	40505.50	TA	1.0			MIT
2467.368	2468.114	40516.77	TA	1.2S	0.5		MIT
2466.990	2467.736	40522.97	TA	1.3	0.3		MIT
2465.370	2466.116	40549.60	TA	0.3			MIT
2465.264	2466.010	40551.34	TA	1.8			MIT
2463.816	2464.561	40575.17	TA	1.3W	1.5		MIT
2461.25	2461.99	40617.5	TA	0.3			MIT
2461.065	2461.810	40620.53	TA	0.3D	0.3A		MIT
2460.551	2461.295	40629.01	TA	1.7			MIT
2459.579	2460.323	40645.07	TA	1.0			MIT
2458.683	2459.427	40659.88	TA	1.6			MIT
2454.482	2455.225	40729.46	TA	1.8			MIT
2454.212	2454.955	40733.94	TA	1.7			MIT
2453.259	2454.002	40749.77	TA	0.3			MIT
2450.920	2451.662	40788.65	TA	0.3			MIT
2449.821	2450.563	40806.95	TA		0.3J		MIT
2449.444	2450.186	40813.23	TA	0.6H	0.6H		MIT
2447.938	2448.680	40838.34	TA	0.3			MIT
2447.174	2447.915	40851.08	TA	1.1	0.3H		MIT
2444.668	2445.409	40892.96	TA	1.7	2.3		MIT
2444.126	2444.867	40902.03	TA	1.3	0.6		MIT
2443.936	2444.677	40905.21	TA	1.8			MIT
2442.393	2443.133	40931.04	TA	1.0			MIT
2441.852	2442.592	40940.11	TA	0.3	0.7		MIT
2440.962	2441.702	40955.04	TA	0.3			MIT
2439.911	2440.651	40972.68	TA	0.6			MIT
2438.641	2439.380	40994.02	TA	1.6	0.6		MIT
2437.070	2437.809	41020.44	TA	1.3			MIT
2436.512	2437.251	41029.83	TA	1.0H	1.6		MIT
2434.741	2435.480	41059.68	TA	0.3			MIT
2433.587	2434.325	41079.14	TA	1.0	1.3		MIT
2433.158	2433.896	41086.39	TA	0.3			MIT
2432.701	2433.439	41094.10	TA	2.5R	2.6		MIT
2431.658	2432.396	41111.73	TA	1.0			MIT
2429.71	2430.45	41144.7	TA	1.3L	1.6		MIT
2429.66	2430.40	41145.5	TA	0.3	0.7		MIT
2429.349	2430.086	41150.80	TA	1.0			MIT
2428.990	2429.727	41156.88	TA	0.3			MIT
2428.48	2429.22	41165.5	TA		1.3W		MIT
2427.642	2428.379	41179.73	TA	2.2	0.0		MIT
2426.593	2427.330	41197.54	TA	0.3			MIT
2425.906	2426.643	41209.20	TA	1.0H	1.4D		MIT
2423.485	2424.221	41250.37	TA	1.0H	0.9		MIT
2422.815	2423.551	41261.77	TA	0.3			MIT
2422.085	2422.821	41274.21	TA	0.3			MIT
2421.848	2422.584	41278.25	TA	0.9	1.6		MIT
2421.032	2421.767	41292.16	TA	1.3			MIT
2419.715	2420.450	41314.63	TA	0.3			MIT
2418.772	2419.507	41330.74	TA	1.2	0.8		MIT
2417.856	2418.591	41346.39	TA	1.3O	1.6		MIT
2417.699	2418.434	41349.08	TA	1.0			MIT

AIR WAVELENGTH (ANGSTRMS)	VACUUM WAVELENGTH (ANGSTRMS)	WAVENUMBER (KAYSERS)	LMENT	LOG ARC	INTENSITY SPK	DIS	REF
2416.892	2417.626	41362.88	TA	2.0	2.2		MIT
2414.325	2415.059	41406.86	TA	0.6D			MIT
2412.711	2413.445	41434.56	TA	0.7			MIT
2408.256	2408.989	41511.20	TA	1.2	1.1		MIT
2407.567	2408.299	41523.08	TA	1.3			MIT
2406.549	2407.281	41540.64	TA	1.8			MIT
2404.217	2404.949	41580.93	TA	1.0			MIT
2403.984	2404.716	41584.96	TA	1.0			MIT
2403.677	2404.408	41590.27	TA	0.6	0.9		MIT
2402.133	2402.864	41617.00	TA	1.0	1.0		MIT
2401.711	2402.442	41624.31	TA	1.0			MIT
2400.63	2401.36	41643.1	TA	0.6	0.3		MIT
2399.92	2400.65	41655.4	TA	0.9	0.9		MIT
2399.15	2399.88	41668.7	TA	0.5			MIT
2398.98	2399.71	41671.7	TA	0.5			MIT
2398.32	2399.05	41683.2	TA		0.5S		MIT
2398.24	2398.97	41684.5	TA	0.8			MIT
2398.06	2398.79	41687.7	TA		0.9		MIT
2397.85	2398.58	41691.3	TA	0.3			MIT
2397.42	2398.15	41698.8	TA		0.3		MIT
2396.30	2397.03	41718.3	TA	1.9			MIT
2395.04	2395.77	41740.2	TA		1.4		MIT
2393.23	2393.96	41771.8	TA		1.3		MIT
2392.71	2393.44	41780.9	TA		1.1		MIT
2390.69	2391.42	41816.2	TA		0.6H		MIT
2390.21	2390.94	41824.6	TA		0.6L		MIT
2389.60	2390.33	41835.3	TA	0.5	1.1		MIT
2389.11	2389.84	41843.8	TA	0.8	1.1		MIT
2388.37	2389.10	41856.8	TA	0.3	1.4		MIT
2387.55	2388.28	41871.2	TA		0.3		MIT
2387.06	2387.79	41879.8	TA	1.3	1.7		MIT
2385.73	2386.46	41903.1	TA	0.9			MIT
2385.35	2386.08	41909.8	TA	0.3			MIT
2384.28	2385.01	41928.6	TA	0.8	1.3		MIT
2383.72	2384.45	41938.5	TA	0.8	1.4		MIT
2381.52	2382.25	41977.2	TA	0.9	1.6		MIT
2381.13	2381.86	41984.1	TA	1.2	1.6		MIT
2380.18	2380.91	42000.8	TA	0.3			MIT
2380.00	2380.73	42004.0	TA	0.8			MIT
2378.52	2379.25	42030.1	TA I	0.3			MIT
2378.31	2379.04	42033.8	TA	0.7	1.3		MIT
2377.17	2377.90	42054.0	TA	0.3	1.2L		MIT
2375.91	2376.64	42076.3	TA	0.6	0.3H		MIT
2374.72	2375.44	42097.4	TA	0.3	1.1H		MIT
2373.94	2374.66	42111.2	TA		1.5		MIT
2372.80	2373.52	42131.4	TA	0.5	1.5		MIT
2371.58	2372.30	42153.1	TA	1.0	1.0		MIT
2371.06	2371.78	42162.4	TA		1.3		MIT
2370.76	2371.48	42167.7	TA	0.7	1.4		MIT
2370.35	2371.07	42175.0	TA		0.7		MIT
2369.32	2370.04	42193.3	TA		1.1		MIT
2368.46	2369.18	42208.6	TA	0.3			KS
2368.21	2368.93	42213.1	TA		0.6H		MIT
2367.71	2368.43	42222.0	TA		0.5		MIT
2367.24	2367.96	42230.4	TA	0.3	1.4		MIT
2364.24	2364.96	42284.0	TA	1.0	1.6		MIT
2361.09	2361.81	42340.4	TA	1.0			MIT
2359.79	2360.51	42363.7	TA		1.5		MIT
2359.45	2360.17	42369.8	TA		0.7H		MIT
2359.16	2359.88	42375.0	TA		1.2		MIT
2357.30	2358.02	42408.4	TA	1.0			A
2356.90	2357.62	42415.6	TA	0.3H	0.3		MIT
2356.48	2357.20	42423.2	TA		1.0		MIT
2356.05	2356.77	42430.9	TA	0.5	1.4		MIT
2355.65	2356.37	42438.1	TA	0.5	1.3		MIT
2354.10	2354.82	42466.1	TA		0.5H		MIT
2353.86	2354.58	42470.4	TA	0.6	1.3		MIT
2352.97	2353.69	42486.5	TA	0.3	1.5		MIT
2348.59	2349.31	42565.7	TA		1.1S		MIT
2346.98	2347.70	42594.9	TA		1.3H		MIT

AIR WAVELENGTH (ANGSTRMS)	VACUUM WAVELENGTH (ANGSTRMS)	WAVENUMBER (KAYSERS)	LMENT	LOG ARC	INTENSITY SPK	DIS	REF
2346.42	2347.14	42605.1	TA	0.5	0.3H		MIT
2346.02	2346.74	42612.3	TA		1.3		MIT
2343.64	2344.36	42655.6	TA		1.4		MIT
2342.52	2343.24	42676.0	TA		1.4		MIT
2341.61	2342.33	42692.6	TA	0.3	1.5		MIT
2340.94	2341.66	42704.8	TA	0.5	1.5		MIT
2340.02	2340.74	42721.6	TA	0.3	1.5		MIT
2338.65	2339.37	42746.6	TA		1.2		MIT
2338.28	2339.00	42753.4	TA	0.8	1.5L		MIT
2335.75	2336.47	42799.7	TA	0.3	1.3		MIT
2334.88	2335.60	42815.6	TA	0.5	1.3		MIT
2334.13	2334.85	42829.4	TA	0.3	1.3		MIT
2333.85	2334.57	42834.5	TA		1.0		MIT
2332.19	2332.91	42865.0	TA	1.0	1.2		MIT
2331.98	2332.70	42868.9	TA	1.0	1.3		MIT
2330.79	2331.51	42890.7	TA		1.8		MIT
2329.24	2329.95	42919.3	TA		0.5		MIT
2328.76	2329.47	42928.1	TA		0.5		MIT
2323.93	2324.64	43017.4	TA		1.3		MIT
2319.07	2319.78	43107.5	TA	0.3	1.5L		MIT
2317.14	2317.85	43143.4	TA	0.3H			MIT
2315.46	2316.17	43174.7	TA	0.8	1.5		MIT
2313.985	2314.696	43202.21	TA	0.3H	0.6H		MIT
2312.60	2313.31	43228.1	TA	0.7	1.5		MIT
2311.18	2311.89	43254.6	TA		1.2		MIT
2308.46	2309.17	43305.6	TA	0.3	0.9		MIT
2308.16	2308.87	43311.2	TA		0.7		MIT
2306.87	2307.58	43335.5	TA		0.8		MIT
2303.491	2304.200	43399.01	TA	0.8	1.3		MIT
2302.928	2303.637	43409.62	TA	0.9	1.5L		MIT
2302.235	2302.944	43422.68	TA	0.8	1.5		MIT
2301.466	2302.175	43437.19	TA	0.3	1.5		MIT
2298.93	2299.64	43485.1	TA		1.3		MIT
2298.10	2298.81	43500.8	TA		1.2		MIT
2296.86	2297.57	43524.3	TA		1.4		MIT
2295.18	2295.89	43556.1	TA	0.7	1.1L		MIT
2292.544	2293.251	43606.22	TA		1.3		MIT
2292.13	2292.84	43614.1	TA	0.6	0.9		MIT
2290.01	2290.72	43654.5	TA	0.5			MIT
2289.16	2289.87	43670.7	TA	0.9	1.4		MIT
2287.84	2288.55	43695.9	TA	0.7			A
2286.59	2287.30	43719.8	TA		0.8		MIT
2285.25	2285.96	43745.4	TA	0.8	1.3		MIT
2285.02	2285.73	43749.8	TA	0.5	1.1		MIT
2283.19	2283.89	43784.9	TA	0.3	1.2		MIT
2282.19	2282.89	43804.0	TA	0.8	1.2		MIT
2281.52	2282.22	43816.9	TA	0.3	1.2		MIT
2281.31	2282.01	43820.9	TA	0.3	1.1		MIT
2280.51	2281.21	43836.3	TA	0.5	1.0		MIT
2275.60	2276.30	43930.9	TA	0.3	1.4		MIT
2274.38	2275.08	43954.5	TA		0.7		MIT
2273.71	2274.41	43967.4	TA	0.5	1.5		MIT
2272.593	2273.295	43989.00	TA	1.1	1.5		MIT
2271.85	2272.55	44003.4	TA	1.1	1.5		MIT
2271.03	2271.73	44019.3	TA		0.6		MIT
2269.56	2270.26	44047.8	TA	0.5	1.3		MIT
2269.21	2269.91	44054.6	TA	0.5	1.1		MIT
2267.51	2268.21	44087.6	TA		0.6H		MIT
2267.26	2267.96	44092.5	TA		0.6H		MIT
2266.52	2267.22	44106.9	TA	1.0			A
2262.30	2263.00	44189.1	TA	1.1	1.3L		MIT
2261.621	2262.321	44202.39	TA	0.8	1.1		MIT
2261.418	2262.118	44206.36	TA	1.1	1.2		MIT
2260.56	2261.26	44223.1	TA		1.3L		MIT
2260.25	2260.95	44229.2	TA	0.3	0.5		MIT
2259.55	2260.25	44242.9	TA	0.6	1.1		MIT
2258.71	2259.41	44259.4	TA		1.4		MIT
2256.51	2257.21	44302.5	TA	0.8	1.3		A
2255.77	2256.47	44317.0	TA	0.5L	1.3		MIT
2254.86	2255.56	44334.9	TA	0.5	1.0		MIT
2253.28	2253.98	44366.0	TA	0.3	1.1		MIT
2252.80	2253.50	44375.5	TA		1.2		MIT
2250.76	2251.46	44415.7	TA	1.1	1.4L		MIT
2249.79	2250.49	44434.8	TA	1.1	1.4L		MIT
2248.48	2249.18	44460.7	TA	0.6	1.3		MIT
2243.861	2244.557	44552.22	TA	0.5	1.4		MIT
2242.75	2243.45	44574.3	TA	0.6			MIT
2239.48	2240.18	44639.4	TA	1.3	1.5L		MIT
2239.01	2239.71	44648.7	TA	0.8	1.2		MIT
2237.29	2237.98	44683.1	TA	0.3	1.1		MIT
2236.28	2236.97	44703.2	TA	0.6	1.2		MIT
2235.94	2236.63	44710.0	TA		0.3H		MIT
2235.67	2236.36	44715.4	TA		0.7		A
2233.88	2234.57	44751.3	TA		1.1		MIT
2230.53	2231.22	44818.5	TA	1.0			A
2227.84	2228.53	44872.6	TA	1.1	1.5L		MIT
2227.23	2227.92	44884.9	TA		1.1		A
2226.525	2227.217	44899.07	TA		1.4		MIT
2223.31	2224.00	44964.0	TA		1.2L		MIT
2222.22	2222.91	44986.0	TA		0.8		A
2220.69	2221.38	45017.0	TA		1.1H		MIT
2219.41	2220.10	45043.0	TA	0.8	1.1		MIT
2217.82	2218.51	45075.3	TA	0.9	1.6		MIT
2215.60	2216.29	45120.5	TA	1.0	1.4		A
2214.34	2215.03	45146.1	TA	0.5	1.1		MIT
2212.42	2213.11	45185.3	TA	0.9	1.5L		MIT
2210.19	2210.88	45230.9	TA		1.1		MIT
2210.03	2210.72	45234.1	TA	1.0	1.1		A
2209.02	2209.71	45254.8	TA		1.3		A
2207.14	2207.83	45293.4	TA	1.0	1.0S		MIT
2206.53	2207.22	45305.9	TA	0.3	1.2		A
2204.10	2204.79	45355.8	TA	0.5	1.2		A
2201.16	2201.85	45416.4	TA	0.8	1.4		MIT
2200.46	2201.15	45430.9	TA	0.5	0.3H		A
2199.67	2200.36	45447.2	TA	1.4	1.4		MIT
2196.29	2196.98	45517.1	TA		0.8		MIT
2196.03	2196.72	45522.5	TA	1.4L	1.5		MIT
2193.88	2194.57	45567.1	TA	1.2	1.6		MIT
2193.20	2193.89	45581.2	TA	0.8	1.3		MIT
2190.59	2191.27	45635.5	TA	1.0			A
2188.48	2189.16	45679.5	TA	0.5	1.5		MIT
2182.71	2183.39	45800.3	TA	1.3	1.3L		MIT
2179.54	2180.22	45866.9	TA	0.8	1.3L		A
2178.03	2178.71	45898.7	TA	1.2S	1.5S		A
2173.04	2173.72	46004.1	TA	0.3	0.6		A
2172.07	2172.75	46024.6	TA	0.6	1.1L		A
2165.01	2165.69	46174.7	TA	0.9	1.3		A
2156.51	2157.19	46356.6	TA	0.5	0.9		A
2155.08	2155.76	46387.4	TA	1.1	1.3		A
2154.21	2154.89	46406.1	TA	0.8	1.2		A
2151.39	2152.07	46467.0	TA	1.0	1.1		A
2151.01	2151.69	46475.2	TA	0.7			A
2150.62	2151.30	46483.6	TA	1.1	1.4		A
2146.87	2147.55	46564.8	TA	1.0	1.6		A
2145.81	2146.49	46587.8	TA		1.1H		A
2143.65	2144.33	46634.7	TA	0.3	0.7		A
2143.16	2143.84	46645.4	TA		1.5		A
2142.53	2143.21	46659.1	TA	0.9H	1.3S		A
2141.50	2142.17	46681.5	TA	0.5	1.5		A
2140.46	2141.13	46704.2	TA		0.6		A
2140.13	2140.80	46711.4	TA	1.1	1.5		A
2137.19	2137.86	46775.7	TA	0.3	1.0L		A
2136.37	2137.04	46793.6	TA	1.1	1.0		A
2135.46	2136.13	46813.5	TA		0.8		A
2135.22	2135.89	46818.8	TA		0.8		A
2134.80	2135.47	46828.0	TA	0.3	0.3H		A
2132.88	2133.55	46870.2	TA	0.3	1.4S		A
2131.79	2132.46	46894.1	TA	1.1	1.5		A
2131.00	2131.67	46911.5	TA	0.9			A
2130.48	2131.15	46923.0	TA	0.7	0.9		A

AIR WAVELENGTH (ANGSTRMS)	VACUUM WAVELENGTH (ANGSTRMS)	WAVENUMBER (KAYSERS)	LMENT	LOG ARC	INTENSITY SPK	INTENSITY DIS	REF
2129.22	2129.89	46950.7	TA	1.0	1.2		A
2127.56	2128.23	46987.4	TA	0.9			A
2126.78	2127.45	47004.6	TA	0.3	0.7		A
2126.67	2127.34	47007.0	TA	0.3	1.0		A
2126.34	2127.01	47014.3	TA		0.9		A
2124.08	2124.75	47064.3	TA	0.3	0.8		A
2120.90	2121.57	47134.9	TA	0.9	1.4		A
2120.26	2120.93	47149.1	TA	0.6			A
2119.92	2120.59	47156.7	TA	0.9	1.5		A
2119.44	2120.11	47167.4	TA	0.3	0.5		A
2118.99	2119.66	47177.4	TA	0.8			A
2118.34	2119.01	47191.8	TA	0.8	1.3		A
2117.89	2118.56	47201.9	TA	0.3	0.6		A
2113.68	2114.35	47295.9	TA	1.0	1.5		A
2112.83	2113.50	47314.9	TA	0.7	1.3		A
2111.79	2112.46	47338.2	TA	0.9	1.3		A
2111.21	2111.88	47351.2	TA	0.7	1.1		A
2105.64	2106.31	47476.4	TA		0.9		A
2104.88	2105.55	47493.6	TA	0.8	0.8		A
2103.93	2104.60	47515.0	TA	0.7	0.6		A
2101.85	2102.52	47562.0	TA	0.5	1.6		A
2100.34	2101.01	47596.2	TA	1.0	1.2		A
2099.30	2099.97	47619.8	TA	0.5	1.5		A
2097.96	2098.63	47650.2	TA	0.9	1.2		A
2095.66	2096.33	47702.5	TA	0.5	1.1		A
2091.33	2091.99	47801.3	TA	0.9	1.3		A
2091.03	2091.69	47808.1	TA	0.3	0.7		A
2090.61	2091.27	47817.7	TA	0.6	1.3		A
2088.67	2089.33	47862.1	TA	0.9	1.5		A
2087.27	2087.93	47894.2	TA	0.6	1.3		A
2085.92	2086.58	47925.2	TA	0.5	1.4		A
2083.03	2083.69	47991.7	TA		0.7		MIT
2082.04	2082.70	48014.5	TA	0.6			A
2079.85	2080.51	48065.1	TA	0.6	0.8		A
2079.31	2079.97	48077.6	TA	0.3	1.1H		A
2078.21	2078.87	48103.0	TA	0.9			A
2076.32	2076.98	48146.8	TA	1.1	1.5		A
2073.02	2073.68	48223.4	TA	0.5	1.1		A
2069.96	2070.62	48294.7	TA	0.3	1.0		A
2069.12	2069.78	48314.3	TA	0.5	1.2		A
2067.40	2068.06	48354.5	TA	0.8	1.1		A
2064.77	2065.43	48416.1	TA	0.5	1.1		A
2062.15	2062.81	48477.6	TA	0.6			A
2061.94	2062.60	48482.5	TA	0.9	1.5L		A
2060.98	2061.64	48505.1	TA	0.3D	0.8		A
2059.08	2059.74	48549.9	TA	0.8	1.4L		A
2058.85	2059.51	48555.3	TA	0.6	0.6		A
2058.59	2059.25	48561.4	TA	0.7	0.9		A
2057.60	2058.26	48584.8	TA	0.3	1.1		A
2055.49	2056.15	48634.6	TA	0.5			A
2049.51	2050.17	48776.5	TA	0.3	0.6		A
2047.73	2048.39	48818.9	TA	0.3	0.7		A
2047.41	2048.07	48826.5	TA	0.8	1.3		A
2046.19	2046.85	48855.6	TA	0.7	1.0		A
2045.37	2046.03	48875.2	TA	0.5	0.3		A
2044.65	2045.31	48892.4	TA	0.7	1.0		A
2043.14	2043.80	48928.6	TA	0.6	1.3		A
2042.61	2043.27	48941.3	TA	0.5	0.9		A
2040.62	2041.28	48989.0	TA	0.3	0.9		A
2038.06	2038.71	49050.5	TA	0.5	1.1		A
2032.39	2033.04	49187.3	TA	0.3	1.0		A
2031.93	2032.58	49198.5	TA	0.6			A
2028.87	2029.52	49272.7	TA	0.8			A
2026.92	2027.57	49320.0	TA	0.3			MIT
2024.02	2024.67	49390.7	TA	0.3	1.1		A
2021.72	2022.37	49446.9	TA	0.7			A
2020.73	2021.38	49471.1	TA	0.3	0.8		A
2020.41	2021.06	49479.0	TA	0.3	0.5		A
2019.54	2020.19	49500.3	TA	0.6	1.3		A
2014.66	2015.31	49620.1	TA	0.3	0.8		A
2013.65	2014.30	49645.0	TA	0.3	0.6		A
2006.73	2007.38	49816.2	TA	0.9			A
2006.62	2007.27	49818.9	TA	0.7			A
2004.53	2005.18	49870.9	TA	0.3	0.8		A
2003.91	2004.56	49886.3	TA	0.3	0.8		A
2001.82	2002.47	49938.4	TA	0.7	1.0		A
7204.28	7206.27	13876.8	TB	0.6			ED
7084.53	7086.48	14111.4	TB	0.6			ED
7005.99	7007.92	14269.6	TB	0.6			ED
6923.09	6925.00	14440.4	TB	0.6			ED
6916.69	6918.60	14453.8	TB	0.6			ED
6899.95	6901.85	14488.9	TB	0.8			ED
6896.37	6898.27	14496.4	TB	1.0			ED
6892.41	6894.31	14504.7	TB	0.6			ED
6880.07	6881.97	14530.7	TB	0.6			ED
6877.34	6879.24	14536.5	TB	0.6			ED
6874.18	6876.08	14543.2	TB	0.8			ED
6865.76	6867.65	14561.0	TB	0.6			ED
6830.71	6832.59	14635.7	TB	0.6			ED
6794.59	6796.47	14713.5	TB	1.0			ED
6792.48	6794.35	14718.1	TB	0.6			ED
6785.12	6786.99	14734.1	TB	0.9			ED
6769.81	6771.68	14767.4	TB	0.6			ED
6755.01	6756.87	14799.7	TB	0.6			ED
6714.98	6716.83	14888.0	TB	0.6			ED
6706.79	6708.64	14906.2	TB	0.6			ED
6702.61	6704.46	14915.4	TB	0.8			ED
6693.38	6695.23	14936.0	TB	0.8			ED
6689.51	6691.36	14944.7	TB	0.6			ED
6677.94	6679.78	14970.5	TB	1.1			ED
6645.41	6647.25	15043.8	TB	0.6			ED
6642.27	6644.10	15050.9	TB	0.6			ED
6607.17	6608.99	15130.9	TB	0.8			ED
6581.82	6583.64	15189.2	TB	0.8			ED
6527.63	6529.43	15315.3	TB	0.6			ED
6518.68	6520.48	15336.3	TB	0.8			ED
6512.19	6513.99	15351.6	TB	0.6			ED
6471.39	6473.18	15448.4	TB	0.8			ED
6470.86	6472.65	15449.6	TB	0.6			ED
6446.859	6448.641	15507.14	TB	0.8			MIT
6441.03	6442.81	15521.2	TB	0.6			ED
6424.43	6426.21	15561.3	TB	0.8			ED
6405.97	6407.74	15606.1	TB	0.8			ED
6403.15	6404.92	15613.0	TB	0.6			ED
6337.85	6339.60	15773.9	TB	0.6			ED
6334.91	6336.66	15781.2	TB	0.8			ED
6331.77	6333.52	15789.0	TB	0.6			ED
6331.68	6333.43	15789.2	TB	0.9			ED
6321.59	6323.34	15814.4	TB	0.6			ED
6296.14	6297.88	15878.4	TB	0.6			ED
6292.43	6294.17	15887.7	TB	0.8			ED
6272.36	6274.09	15938.6	TB	0.6			ED
6225.95	6227.67	16057.4	TB	0.6			ED
6222.25	6223.97	16066.9	TB	0.6			ED
6194.52	6196.23	16138.8	TB	0.6			ED
6193.64	6195.35	16141.1	TB	0.6			ED
6171.88	6173.59	16198.0	TB	0.6			ED
6156.29	6157.99	16239.1	TB	0.6			ED
6119.98	6121.67	16335.4	TB	0.6			ED
6117.65	6119.34	16341.6	TB	0.6			ED
6117.07	6118.76	16343.2	TB	0.6			ED
6112.80	6114.49	16354.6	TB	0.6			ED
6110.91	6112.60	16359.7	TB	0.6			ED
6107.26	6108.95	16369.4	TB	0.6			ED
6106.044	6107.734	16372.68	TB	0.8			MIT
6104.27	6105.96	16377.4	TB	0.8			MIT
6094.64	6096.33	16403.3	TB	0.6			ED
6093.99	6095.68	16405.1	TB	0.6			ED
6091.11	6092.80	16412.8	TB	0.6			ED
6088.26	6089.95	16420.5	TB	0.8			MIT

AIR WAVELENGTH (ANGSTRMS)	VACUUM WAVELENGTH (ANGSTRMS)	WAVENUMBER (KAYSERS)	LMENT	LOG ARC	INTENSITY SPK	DIS	REF
6081.25	6082.93	16439.4	TB	0.6			ED
6077.806	6079.489	16448.75	TB	0.8			MIT
6076.82	6078.50	16451.4	TB	0.6			ED
6076.03	6077.71	16453.6	TB	0.6			ED
6061.83	6063.51	16492.1	TB	0.8			ED
6059.76	6061.44	16497.7	TB	0.3			ED
6056.89	6058.57	16505.6	TB	0.6			ED
6054.248	6055.924	16512.76	TB	0.6			MIT
6039.38	6041.05	16553.4	TB	0.8			ED
6038.97	6040.64	16554.5	TB	0.9			ED
6019.24	6020.91	16608.8	TB	0.6			ED
6010.65	6012.31	16632.5	TB	0.8			ED
5996.20	5997.86	16672.6	TB	1.0			ED
5992.17	5993.83	16683.8	TB	1.0			ED
5986.27	5987.93	16700.3	TB	1.0			ED
5971.66	5973.31	16741.1	TB	1.0			ED
5967.323	5968.976	16753.29	TB	1.6			MIT
5961.00	5962.65	16771.1	TB	1.4			ED
5953.740	5955.389	16791.51	TB	1.0			MIT
5951.784	5953.433	16797.03	TB	1.0			MIT
5951.165	5952.814	16798.78	TB	1.2			MIT
5946.67	5948.32	16811.5	TB	1.0			ED
5940.97	5942.62	16827.6	TB	1.0			ED
5940.17	5941.82	16829.9	TB	1.2			ED
5939.38	5941.03	16832.1	TB	1.2			ED
5937.09	5938.73	16838.6	TB	1.0			ED
5934.32	5935.96	16846.5	TB	1.0			ED
5932.95	5934.59	16850.4	TB	1.0			ED
5925.58	5927.22	16871.3	TB	1.0			ED
5922.16	5923.80	16881.1	TB	1.0			ED
5920.35	5921.99	16886.2	TB	1.0			ED
5910.100	5911.738	16915.50	TB	1.0			MIT
5904.71	5906.35	16930.9	TB	1.2			ED
5902.40	5904.04	16937.6	TB	1.2			ED
5899.40	5901.03	16946.2	TB	1.2			ED
5898.84	5900.47	16947.8	TB	1.4			ED
5891.12	5892.75	16970.0	TB	1.0			ED
5889.06	5890.69	16975.9	TB	1.0			ED
5888.15	5889.78	16978.6	TB	1.0			ED
5880.54	5882.17	17000.5	TB	1.2			ED
5879.85	5881.48	17002.5	TB	1.0			ED
5873.15	5874.78	17021.9	TB	1.0			ED
5871.52	5873.15	17026.6	TB	0.7			ED
5870.62	5872.25	17029.3	TB	1.4			ED
5864.42	5866.05	17047.3	TB	1.0			ED
5861.22	5862.84	17056.6	TB	1.4			ED
5859.51	5861.13	17061.5	TB	1.2			ED
5857.34	5858.96	17067.9	TB	1.2			ED
5855.43	5857.05	17073.4	TB	1.0			ED
5854.94	5856.56	17074.9	TB	1.0			ED
5852.54	5854.16	17081.9	TB	1.2			ED
5851.07	5852.69	17086.1	TB	1.6			ED
5849.46	5851.08	17090.9	TB	1.0			ED
5846.38	5848.00	17099.9	TB	1.0			ED
5842.97	5844.59	17109.8	TB	1.4			ED
5838.05	5839.67	17124.3	TB	1.2			ED
5834.59	5836.21	17134.4	TB	1.0			ED
5832.94	5834.56	17139.3	TB	1.0			ED
5829.65	5831.27	17148.9	TB	1.0			ED
5827.55	5829.17	17155.1	TB	1.2			ED
5826.69	5828.31	17157.6	TB	1.0			ED
5821.79	5823.40	17172.1	TB	1.0			ED
5821.31	5822.92	17173.5	TB	1.0			ED
5817.21	5818.82	17185.6	TB	0.7			ED
5815.36	5816.97	17191.1	TB	1.0			ED
5810.109	5811.720	17206.61	TB	1.0			MIT
5809.49	5811.10	17208.4	TB	1.2			ED
5806.44	5808.05	17217.5	TB	1.0			ED
5803.149	5804.758	17227.25	TB	1.6			MIT
5795.638	5797.245	17249.57	TB	1.4			MIT

AIR WAVELENGTH (ANGSTRMS)	VACUUM WAVELENGTH (ANGSTRMS)	WAVENUMBER (KAYSERS)	LMENT	LOG ARC	INTENSITY SPK	DIS	REF
5791.644	5793.250	17261.47	TB	1.0			MIT
5788.87	5790.48	17269.7	TB	1.0			ED
5785.18	5786.78	17280.7	TB	1.6			ED
5779.91	5781.51	17296.5	TB	1.2			ED
5778.94	5780.54	17299.4	TB	1.0			ED
5772.72	5774.32	17318.1	TB	1.0			ED
5770.94	5772.54	17323.4	TB	1.0			ED
5767.22	5768.82	17334.6	TB	1.0			ED
5762.648	5764.246	17348.32	TB	1.2			MIT
5761.62	5763.22	17351.4	TB	1.0			ED
5757.57	5759.17	17363.6	TB	1.2			ED
5751.89	5753.49	17380.8	TB	1.0			ED
5747.578	5749.172	17393.81	TB	1.8			MIT
5744.92	5746.51	17401.9	TB	1.0			ED
5744.629	5746.222	17402.74	TB	1.0			MIT
5743.102	5744.695	17407.36	TB	1.2			MIT
5741.42	5743.01	17412.5	TB	1.0			ED
5739.74	5741.33	17417.6	TB	1.0			ED
5736.68	5738.27	17426.9	TB	1.0			ED
5733.47	5735.06	17436.6	TB	1.0			ED
5730.14	5731.73	17446.7	TB	1.0			ED
5728.23	5729.82	17452.6	TB	1.0			ED
5727.55	5729.14	17454.6	TB	1.2			ED
5726.72	5728.31	17457.2	TB	1.0			ED
5725.624	5727.212	17460.50	TB	1.0			MIT
5724.573	5726.161	17463.71	TB	1.0			MIT
5722.46	5724.05	17470.2	TB	1.0			ED
5721.80	5723.39	17472.2	TB	1.0			ED
5712.430	5714.015	17500.83	TB	1.2			MIT
5708.379	5709.963	17513.25	TB	1.2			MIT
5702.545	5704.127	17531.17	TB	1.2			MIT
5688.764	5690.342	17573.64	TB	1.0			MIT
5687.12	5688.70	17578.7	TB	1.2			ED
5686.48	5688.06	17580.7	TB	1.2			ED
5685.759	5687.337	17582.92	TB	1.6			MIT
5678.403	5679.979	17605.70	TB	1.0			MIT
5676.039	5677.614	17613.03	TB	1.2			MIT
5671.844	5673.418	17626.06	TB	1.0			MIT
5669.59	5671.16	17633.1	TB	1.0			ED
5667.562	5669.135	17639.38	TB	1.4			MIT
5664.66	5666.23	17648.4	TB	1.2			ED
5659.86	5661.43	17663.4	TB	1.0			ED
5656.56	5658.13	17673.7	TB	1.0			ED
5652.07	5653.64	17687.7	TB	1.2			ED
5646.97	5648.54	17703.7	TB	1.0			ED
5645.755	5647.322	17707.51	TB	1.2			MIT
5640.197	5641.762	17724.96	TB	1.2			MIT
5637.62	5639.18	17733.1	TB	1.2			ED
5636.573	5638.138	17736.35	TB	1.2			MIT
5629.84	5631.40	17757.6	TB	1.0			ED
5629.30	5630.86	17759.3	TB	1.0			ED
5627.38	5628.94	17765.3	TB	1.2			ED
5625.37	5626.93	17771.7	TB	1.0			ED
5607.03	5608.59	17829.8	TB	1.0			ED
5605.28	5606.84	17835.4	TB	1.0			ED
5597.17	5598.72	17861.2	TB	1.0			ED
5586.963	5588.514	17893.84	TB	1.0			MIT
5583.831	5585.381	17903.88	TB	1.0			MIT
5575.53	5577.08	17930.5	TB	1.0			ED
5572.48	5574.03	17940.4	TB	1.0			MIT
5571.96	5573.51	17942.0	TB	1.0			ED
5570.72	5572.27	17946.0	TB	1.2			ED
5569.179	5570.725	17950.98	TB	1.0			MIT
5565.934	5567.480	17961.45	TB	1.4			MIT
5562.91	5564.45	17971.2	TB	1.0			ED
5556.30	5557.84	17992.6	TB	1.4			ED
5554.95	5556.49	17997.0	TB	1.4			ED
5554.42	5555.96	17998.7	TB	1.4			ED
5553.40	5554.94	18002.0	TB	1.0			ED
5548.971	5550.512	18016.36	TB	1.0			MIT

AIR WAVELENGTH (ANGSTRMS)	VACUUM WAVELENGTH (ANGSTRMS)	WAVENUMBER (KAYSERS)	LMENT	LOG ARC	INTENSITY SPK	INTENSITY DIS	REF
5547.69	5549.23	18020.5	TB	1.0			ED
5547.04	5548.58	18022.6	TB	1.0			ED
5546.10	5547.64	18025.7	TB	1.2			ED
5538.96	5540.50	18048.9	TB	1.0			ED
5536.273	5537.811	18057.68	TB	1.4			MIT
5535.38	5536.92	18060.6	TB	1.2			ED
5534.58	5536.12	18063.2	TB	1.0			ED
5532.11	5533.65	18071.3	TB	1.2			ED
5530.30	5531.84	18077.2	TB	1.0			ED
5526.94	5528.48	18088.2	TB	1.0			ED
5525.62	5527.15	18092.5	TB	1.4			ED
5524.12	5525.65	18097.4	TB	1.6			ED
5523.44	5524.97	18099.6	TB	1.0			ED
5522.30	5523.83	18103.4	TB	1.2			ED
5519.90	5521.43	18111.2	TB	1.0			ED
5518.77	5520.30	18114.9	TB	1.0			ED
5516.24	5517.77	18123.3	TB	1.7			ED
5514.54	5516.07	18128.8	TB	1.7			ED
5512.50	5514.03	18135.6	TB	1.2			ED
5509.61	5511.14	18145.1	TB	1.4			ED
5500.10	5501.63	18176.4	TB	1.2			ED
5495.65	5497.18	18191.2	TB	1.0			ED
5488.26	5489.78	18215.6	TB	1.2			ED
5481.45	5482.97	18238.3	TB	1.0			ED
5474.65	5476.17	18260.9	TB	1.0			ED
5472.66	5474.18	18267.6	TB	1.0			ED
5470.34	5471.86	18275.3	TB	1.6			ED
5468.28	5469.80	18282.2	TB	1.0			ED
5464.39	5465.91	18295.2	TB	1.0			ED
5459.81	5461.33	18310.6	TB	1.4			ED
5458.15	5459.67	18316.1	TB	1.0			ED
5457.83	5459.35	18317.2	TB	1.0			ED
5456.978	5458.495	18320.07	TB	1.2			MIT
5456.53	5458.05	18321.6	TB	1.0			ED
5453.68	5455.20	18331.1	TB	1.0			ED
5451.02	5452.53	18340.1	TB	1.0			ED
5450.42	5451.93	18342.1	TB	1.0			ED
5450.16	5451.67	18343.0	TB	1.0			ED
5441.98	5443.49	18370.6	TB	1.2			ED
5441.45	5442.96	18372.4	TB	1.0			ED
5440.33	5441.84	18376.1	TB	1.0			ED
5438.66	5440.17	18381.8	TB	1.0			ED
5438.12	5439.63	18383.6	TB	1.2			ED
5437.10	5438.61	18387.1	TB	1.0			ED
5432.45	5433.96	18402.8	TB	1.0			ED
5428.25	5429.76	18417.0	TB	1.0			ED
5427.464	5428.973	18419.69	TB	1.0			MIT
5426.43	5427.94	18423.2	TB	1.2			ED
5425.00	5426.51	18428.1	TB	1.2			ED
5423.24	5424.75	18434.0	TB	1.2			ED
5421.75	5423.26	18439.1	TB	1.0			ED
5420.53	5422.04	18443.2	TB	1.0			ED
5419.10	5420.61	18448.1	TB	1.0			ED
5416.20	5417.71	18458.0	TB	1.2			ED
5410.63	5412.13	18477.0	TB	1.0			ED
5403.82	5405.32	18500.3	TB	1.0			ED
5402.06	5403.56	18506.3	TB	1.4			ED
5401.55	5403.05	18508.1	TB	1.0			ED
5397.90	5399.40	18520.6	TB	1.0			ED
5393.70	5395.20	18535.0	TB	1.0			ED
5385.80	5387.30	18562.2	TB	1.0			ED
5378.71	5380.21	18586.6	TB	1.0			ED
5377.88	5379.38	18589.5	TB	1.4			ED
5376.55	5378.05	18594.1	TB	1.0			ED
5375.97	5377.46	18596.1	TB	1.6			MIT
5373.01	5374.50	18606.4	TB	1.0			ED
5369.72	5371.21	18617.8	TB	1.6			ED
5357.16	5358.65	18661.4	TB	1.0			ED
5354.88	5356.37	18669.4	TB	1.6			ED
5352.35	5353.84	18678.2	TB	1.0			ED

AIR WAVELENGTH (ANGSTRMS)	VACUUM WAVELENGTH (ANGSTRMS)	WAVENUMBER (KAYSERS)	LMENT	LOG ARC	INTENSITY SPK	INTENSITY DIS	REF
5347.93	5349.42	18693.6	TB	1.2			ED
5346.14	5347.63	18699.9	TB	1.0			ED
5337.90	5339.38	18728.7	TB	1.0			ED
5328.18	5329.66	18762.9	TB	1.0			ED
5319.23	5320.71	18794.5	TB	1.6			ED
5317.82	5319.30	18799.5	TB	1.0			ED
5315.76	5317.24	18806.7	TB	1.2			ED
5309.46	5310.94	18829.1	TB	1.0			ED
5308.19	5309.67	18833.6	TB	1.0			ED
5307.62	5309.10	18835.6	TB	1.0			ED
5306.78	5308.26	18838.6	TB	1.0			ED
5304.72	5306.20	18845.9	TB	1.2			ED
5303.44	5304.92	18850.4	TB	1.0			ED
5300.18	5301.65	18862.0	TB	1.0			ED
5297.60	5299.07	18871.2	TB	1.0			ED
5296.55	5298.02	18875.0	TB	1.0			ED
5295.73	5297.20	18877.9	TB	1.0			ED
5292.79	5294.26	18888.4	TB	1.2			ED
5289.71	5291.18	18899.4	TB	1.2			ED
5287.30	5288.77	18908.0	TB	1.0			ED
5285.82	5287.29	18913.3	TB	1.0			ED
5281.05	5282.52	18930.4	TB	1.4			ED
5279.91	5281.38	18934.4	TB	1.2			ED
5278.24	5279.71	18940.4	TB	1.0			ED
5275.03	5276.50	18952.0	TB	1.0			ED
5272.07	5273.54	18962.6	TB	1.2			ED
5271.06	5272.53	18966.2	TB	1.0			ED
5269.23	5270.70	18972.8	TB	1.0			ED
5266.00	5267.47	18984.5	TB	1.0			ED
5264.92	5266.39	18988.4	TB	1.2			ED
5262.11	5263.57	18998.5	TB	1.6			ED
5261.02	5262.48	19002.4	TB	1.0			ED
5259.63	5261.09	19007.4	TB	1.0			ED
5256.75	5258.21	19017.9	TB	1.0			ED
5254.32	5255.78	19026.7	TB	1.0			ED
5252.96	5254.42	19031.6	TB	1.0			ED
5248.71	5250.17	19047.0	TB	1.4			ED
5243.57	5245.03	19065.7	TB	1.2			ED
5238.11	5239.57	19085.5	TB	1.2			ED
5235.11	5236.57	19096.5	TB	1.2			ED
5228.12	5229.58	19122.0	TB	1.6			ED
5221.99	5223.44	19144.5	TB	1.2			ED
5214.28	5215.73	19172.8	TB	1.2			ED
5211.99	5213.44	19181.2	TB	1.0			ED
5211.21	5212.66	19184.1	TB	1.0			ED
5209.34	5210.79	19190.9	TB	1.2			ED
5207.97	5209.42	19196.0	TB	1.2			ED
5202.77	5204.22	19215.2	TB	1.2			ED
5198.86	5200.31	19229.6	TB	1.2			ED
5196.76	5198.21	19237.4	TB	1.0			ED
5196.31	5197.76	19239.1	TB	1.0			ED
5190.76	5192.21	19259.6	TB	1.0			ED
5186.16	5187.60	19276.7	TB	1.2			ED
5181.08	5182.52	19295.6	TB	1.0			ED
5178.40	5179.84	19305.6	TB	1.0			ED
5176.51	5177.95	19312.7	TB	1.0			ED
5170.61	5172.05	19334.7	TB	1.0			ED
5170.13	5171.57	19336.5	TB	1.0			ED
5169.12	5170.56	19340.3	TB	1.0			ED
5164.27	5165.71	19358.4	TB	1.0			ED
5160.40	5161.84	19372.9	TB	1.0			ED
5159.77	5161.21	19375.3	TB	1.0			ED
5159.31	5160.75	19377.0	TB	1.0			ED
5157.57	5159.01	19383.6	TB	1.2			ED
5155.95	5157.39	19389.7	TB	1.0			ED
5155.47	5156.91	19391.5	TB	1.2			ED
5150.12	5151.55	19411.6	TB	1.2			ED
5147.58	5149.01	19421.2	TB	1.2			ED
5145.90	5147.33	19427.5	TB	1.0			ED
5142.83	5144.26	19439.1	TB	1.0			ED

AIR WAVELENGTH (ANGSTRMS)	VACUUM WAVELENGTH (ANGSTRMS)	WAVENUMBER (KAYSERS)	LMENT	LOG ARC	INTENSITY SPK	DIS	REF
5141.08	5142.51	19445.7	TB	1.2			ED
5139.35	5140.78	19452.3	TB	1.0			ED
5131.69	5133.12	19481.3	TB	1.2			ED
5130.53	5131.96	19485.7	TB	1.0			ED
5126.00	5127.43	19502.9	TB	1.0			ED
5121.61	5123.04	19519.7	TB	1.2			ED
5120.18	5121.61	19525.1	TB	1.0			ED
5118.39	5119.82	19531.9	TB	1.2			ED
5116.25	5117.68	19540.1	TB	1.0			ED
5109.79	5111.21	19564.8	TB	1.0			ED
5108.56	5109.98	19569.5	TB	1.2			ED
5104.02	5105.44	19586.9	TB	1.0			ED
5101.97	5103.39	19594.8	TB	1.0			ED
5099.84	5101.26	19603.0	TB	1.0			ED
5098.20	5099.62	19609.3	TB	1.0			ED
5092.38	5093.80	19631.7	TB	1.0			ED
5089.66	5091.08	19642.2	TB	1.0			ED
5089.12	5090.54	19644.3	TB	1.6			ED
5087.87	5089.29	19649.1	TB	1.0			ED
5087.48	5088.90	19650.6	TB	1.2			ED
5081.80	5083.22	19672.6	TB	1.0			ED
5080.05	5081.47	19679.4	TB	1.0			ED
5078.25	5079.67	19686.3	TB	1.4			ED
5076.75	5078.17	19692.1	TB	1.2			ED
5073.73	5075.14	19703.9	TB	1.2			ED
5070.50	5071.91	19716.4	TB	1.0			ED
5070.17	5071.58	19717.7	TB	1.0			ED
5065.79	5067.20	19734.8	TB	1.6			ED
5064.37	5065.78	19740.3	TB	1.0			ED
5063.25	5064.66	19744.7	TB	1.0			ED
5056.29	5057.70	19771.8	TB	1.0			ED
5054.29	5055.70	19779.7	TB	1.4			MIT
5049.58	5050.99	19798.1	TB	1.2			ED
5047.34	5048.75	19806.9	TB	1.0			ED
5042.043	5043.449	19827.70	TB	1.4			MIT
5040.56	5041.97	19833.5	TB	1.0			ED
5035.03	5036.43	19855.3	TB	1.2			ED
5034.07	5035.47	19859.1	TB	1.0			ED
5033.12	5034.52	19862.9	TB	1.2			ED
5031.10	5032.50	19870.8	TB	1.0			ED
5025.403	5026.805	19893.35	TB	1.2			MIT
5024.65	5026.05	19896.3	TB	1.2			ED
5024.24	5025.64	19898.0	TB	1.4			ED
5022.133	5023.534	19906.31	TB	1.4			MIT
5021.64	5023.04	19908.3	TB	1.0			ED
5013.80	5015.20	19939.4	TB	1.0			ED
5010.54	5011.94	19952.4	TB	1.0			ED
5010.08	5011.48	19954.2	TB	1.0			ED
5008.63	5010.03	19960.0	TB	1.0			ED
5004.824	5006.220	19975.15	TB	1.2			MIT
5003.22	5004.62	19981.6	TB	1.0			ED
5002.88	5004.28	19982.9	TB	1.0			ED
4997.933	4999.327	20002.69	TB	0.7			MIT
4995.814	4997.208	20011.18	TB	0.7			MIT
4993.838	4995.231	20019.09	TB	0.8			MIT
4992.323	4993.716	20025.17	TB	0.3			MIT
4991.40	4992.79	20028.9	TB	0.5			ED
4986.94	4988.33	20046.8	TB	0.6			ED
4986.67	4988.06	20047.9	TB	0.5			ED
4983.90	4985.29	20059.0	TB	0.3			ED
4982.78	4984.17	20063.5	TB	0.3			ED
4980.538	4981.928	20072.55	TB	0.6			MIT
4980.16	4981.55	20074.1	TB	0.7			ED
4975.21	4976.60	20094.1	TB	0.3			ED
4973.019	4974.407	20102.90	TB	0.6			MIT
4971.410	4972.797	20109.41	TB	0.6			MIT
4970.99	4972.38	20111.1	TB	0.7			ED
4964.49	4965.88	20137.4	TB	0.6			ED
4964.10	4965.49	20139.0	TB	0.5			ED
4962.28	4963.66	20146.4	TB	0.6			ED

AIR WAVELENGTH (ANGSTRMS)	VACUUM WAVELENGTH (ANGSTRMS)	WAVENUMBER (KAYSERS)	LMENT	LOG ARC	INTENSITY SPK	DIS	REF
4958.93	4960.31	20160.0	TB	0.6			ED
4958.62	4960.00	20161.3	TB	0.3			ED
4954.52	4955.90	20178.0	TB	0.3			ED
4953.91	4955.29	20180.4	TB	0.5			ED
4953.29	4954.67	20183.0	TB	0.5			ED
4947.97	4949.35	20204.7	TB	0.7			ED
4947.20	4948.58	20207.8	TB	0.3			ED
4941.83	4943.21	20229.8	TB	0.3			ED
4940.714	4942.093	20234.34	TB	0.6			MIT
4939.70	4941.08	20238.5	TB	0.5			ED
4937.67	4939.05	20246.8	TB	0.3			ED
4933.38	4934.76	20264.4	TB	0.3			ED
4931.79	4933.17	20271.0	TB	0.9			ED
4928.919	4930.295	20282.76	TB	0.5H			MIT
4928.69	4930.07	20283.7	TB	0.5			ED
4926.821	4928.196	20291.40	TB	0.5			MIT
4924.094	4925.469	20302.64	TB	0.6			MIT
4922.64	4924.01	20308.6	TB	0.3			ED
4919.21	4920.58	20322.8	TB	0.3			ED
4918.10	4919.47	20327.4	TB	0.3			ED
4916.85	4918.22	20332.6	TB	0.5			ED
4915.90	4917.27	20336.5	TB	0.8			ED
4911.96	4913.33	20352.8	TB	0.3			ED
4910.08	4911.45	20360.6	TB	0.5			ED
4908.04	4909.41	20369.0	TB	0.6			ED
4905.69	4907.06	20378.8	TB	0.5			ED
4903.84	4905.21	20386.5	TB	0.3			ED
4900.610	4901.978	20399.93	TB	0.3			MIT
4897.961	4899.329	20410.96	TB	0.3			MIT
4897.35	4898.72	20413.5	TB	0.6			ED
4896.845	4898.212	20415.61	TB	0.3			MIT
4896.188	4897.555	20418.35	TB	0.3			MIT
4894.93	4896.30	20423.6	TB	0.3			ED
4894.352	4895.719	20426.01	TB	0.5			MIT
4892.01	4893.38	20435.8	TB	0.5			ED
4887.45	4888.81	20454.9	TB	0.3			ED
4886.43	4887.79	20459.1	TB	0.3			ED
4885.443	4886.807	20463.26	TB	0.3			MIT
4884.75	4886.11	20466.2	TB	0.3			ED
4882.75	4884.11	20474.5	TB	0.3			ED
4882.28	4883.64	20476.5	TB	0.3			ED
4881.160	4882.523	20481.21	TB	1.2	0.3		MIT
4879.902	4881.265	20486.49	TB	0.5			MIT
4878.04	4879.40	20494.3	TB	0.3			ED
4877.19	4878.55	20497.9	TB	0.3			ED
4876.772	4878.134	20499.64	TB	0.3			MIT
4876.124	4877.486	20502.37	TB	0.5			MIT
4875.576	4876.938	20504.67	TB	1.3	0.3		MIT
4874.60	4875.96	20508.8	TB	0.3			ED
4874.111	4875.472	20510.83	TB	0.6W			MIT
4872.00	4873.36	20519.7	TB	0.5			ED
4871.42	4872.78	20522.2	TB	0.3			ED
4869.518	4870.878	20530.18	TB	0.7			MIT
4868.93	4870.29	20532.7	TB	0.3			ED
4867.84	4869.20	20537.3	TB	0.5			ED
4866.42	4867.78	20543.2	TB	0.5			ED
4864.37	4865.73	20551.9	TB	0.3			ED
4863.81	4865.17	20554.3	TB	0.6			M
4861.21	4862.57	20565.3	TB	0.3			ED
4859.88	4861.24	20570.9	TB	0.3			ED
4858.892	4860.249	20575.08	TB	0.6			MIT
4854.813	4856.169	20592.36	TB	0.6			MIT
4853.138	4854.494	20599.47	TB	0.3			MIT
4850.39	4851.75	20611.1	TB	0.3			ED
4849.445	4850.800	20615.16	TB	0.3			MIT
4847.285	4848.639	20624.34	TB	0.3	0.3		MIT
4845.614	4846.968	20631.46	TB	0.3			MIT
4844.89	4846.24	20634.5	TB	0.8	0.3		M
4842.693	4844.046	20643.90	TB	0.9			MIT
4841.88	4843.23	20647.4	TB	0.3			ED

AIR WAVELENGTH (ANGSTRMS)	VACUUM WAVELENGTH (ANGSTRMS)	WAVENUMBER (KAYSERS)	LMENT	LOG ARC	INTENSITY SPK	INTENSITY DIS	REF
4840.39	4841.74	20653.7	TB	0.6			ED
4839.61	4840.96	20657.1	TB	0.3			ED
4838.41	4839.76	20662.2	TB	0.3			ED
4837.599	4838.951	20665.64	TB	1.0	0.3		MIT
4835.90	4837.25	20672.9	TB	0.3			ED
4834.302	4835.653	20679.73	TB	0.3			MIT
4832.742	4834.092	20686.41	TB	0.3			MIT
4832.03	4833.38	20689.4	TB	0.5			ED
4830.84	4832.19	20694.6	TB	0.3			ED
4828.653	4830.002	20703.92	TB	0.5			MIT
4827.499	4828.848	20708.87	TB	0.5			MIT
4826.798	4828.147	20711.88	TB	0.3			MIT
4826.22	4827.57	20714.4	TB	0.3			ED
4824.586	4825.934	20721.38	TB	0.3			MIT
4824.364	4825.712	20722.33	TB	0.5			MIT
4822.158	4823.506	20731.81	TB	0.3			MIT
4821.038	4822.385	20736.63	TB	0.3			MIT
4820.756	4822.103	20737.84	TB	0.3			MIT
4819.19	4820.54	20744.6	TB	0.3			ED
4814.73	4816.08	20763.8	TB	0.5			ED
4814.493	4815.839	20764.82	TB	0.3			MIT
4813.767	4815.112	20767.95	TB	1.4			MIT
4813.179	4814.524	20770.49	TB	0.3			MIT
4810.203	4811.547	20783.33	TB	0.5R			MIT
4808.60	4809.94	20790.3	TB	0.3			ED
4807.926	4809.270	20793.18	TB	0.3			MIT
4806.774	4808.117	20798.16	TB	0.6			MIT
4805.269	4806.612	20804.68	TB	0.5			MIT
4802.389	4803.731	20817.15	TB	0.6			MIT
4801.870	4803.212	20819.40	TB	0.8	0.3		MIT
4800.439	4801.781	20825.61	TB	0.5			MIT
4798.662	4800.003	20833.32	TB	0.3			MIT
4797.110	4798.451	20840.06	TB	0.3			MIT
4793.82	4795.16	20854.4	TB	0.3			ED
4793.24	4794.58	20856.9	TB	0.3			ED
4792.382	4793.722	20860.62	TB	0.3			MIT
4791.020	4792.359	20866.55	TB	0.3			MIT
4789.918	4791.257	20871.35	TB	0.9W			MIT
4788.513	4789.852	20877.47	TB	0.5			MIT
4787.161	4788.499	20883.37	TB	0.5			MIT
4786.783	4788.121	20885.02	TB	1.5			MIT
4786.192	4787.530	20887.60	TB	0.7			MIT
4784.11	4785.45	20896.7	TB	0.3			MIT
4782.58	4783.92	20903.4	TB	0.3			MIT
4781.92	4783.26	20906.3	TB	0.3			MIT
4781.55	4782.89	20907.9	TB	0.3			MIT
4780.29	4781.63	20913.4	TB	0.5			MIT
4780.00	4781.34	20914.6	TB	0.5			MIT
4778.81	4780.15	20919.9	TB	0.8			MIT
4778.36	4779.70	20921.8	TB	0.8			MIT
4775.18	4776.52	20935.8	TB	0.5			MIT
4774.66	4775.99	20938.1	TB	0.5			MIT
4772.97	4774.30	20945.5	TB	0.5			MIT
4772.03	4773.36	20949.6	TB	0.5			MIT
4771.68	4773.01	20951.1	TB	0.3			MIT
4771.30	4772.63	20952.8	TB	0.7			MIT
4770.29	4771.62	20957.2	TB	0.7W			MIT
4769.56	4770.89	20960.4	TB	0.5			MIT
4764.50	4765.83	20982.7	TB	0.6			MIT
4762.37	4763.70	20992.1	TB	0.6			MIT
4760.20	4761.53	21001.6	TB	0.9			MIT
4758.46	4759.79	21009.3	TB	0.7			MIT
4756.13	4757.46	21019.6	TB	0.5			MIT
4755.47	4756.80	21022.5	TB	0.3			MIT
4753.18	4754.51	21032.7	TB	0.3			MIT
4752.52	4753.85	21035.6	TB	2.0	1.9		MIT
4751.61	4752.94	21039.6	TB	0.5			MIT
4749.91	4751.24	21047.1	TB	0.3			MIT
4747.81	4749.14	21056.4	TB	0.7			MIT
4745.16	4746.49	21068.2	TB	0.5			MIT

AIR WAVELENGTH (ANGSTRMS)	VACUUM WAVELENGTH (ANGSTRMS)	WAVENUMBER (KAYSERS)	LMENT	LOG ARC	INTENSITY SPK	INTENSITY DIS	REF
4743.31	4744.64	21076.4	TB	0.5			MIT
4742.70	4744.03	21079.1	TB	0.3			MIT
4739.93	4741.26	21091.5	TB	0.9			MIT
4739.19	4740.52	21094.7	TB	0.3			MIT
4736.30	4737.62	21107.6	TB	0.5			MIT
4735.16	4736.48	21112.7	TB	0.3			MIT
4734.21	4735.53	21116.9	TB	1.2			MIT
4732.77	4734.09	21123.4	TB	0.5			MIT
4731.85	4733.17	21127.5	TB	0.6			MIT
4730.80	4732.12	21132.2	TB	0.3			MIT
4730.52	4731.84	21133.4	TB	0.5			MIT
4728.18	4729.50	21143.9	TB	0.6			MIT
4726.52	4727.84	21151.3	TB	0.7			MIT
4724.93	4726.25	21158.4	TB	0.3			MIT
4718.98	4720.30	21185.1	TB	0.5			MIT
4717.02	4718.34	21193.9	TB	0.5			MIT
4716.06	4717.38	21198.2	TB	1.2W			MIT
4714.89	4716.21	21203.5	TB	0.3			MIT
4711.69	4713.01	21217.9	TB	0.6			MIT
4711.17	4712.49	21220.2	TB	0.6			MIT
4710.76	4712.08	21222.1	TB	0.5			MIT
4709.73	4711.05	21226.7	TB	0.3			MIT
4707.95	4709.27	21234.7	TB	1.4			MIT
4706.27	4707.59	21242.3	TB	0.5			MIT
4705.48	4706.80	21245.9	TB	0.3			MIT
4704.90	4706.22	21248.5	TB	0.3H			MIT
4704.58	4705.90	21249.9	TB	0.3			MIT
4704.15	4705.47	21251.9	TB	0.3			MIT
4702.42	4703.74	21259.7	TB	1.9			MIT
4701.55	4702.87	21263.6	TB	0.3			MIT
4700.66	4701.98	21267.7	TB	0.5			MIT
4700.39	4701.71	21268.9	TB	0.3H			MIT
4699.80	4701.12	21271.6	TB	0.3H			MIT
4699.24	4700.55	21274.1	TB	0.7			MIT
4698.52	4699.83	21277.3	TB	0.7			MIT
4695.98	4697.29	21288.9	TB	0.5			MIT
4693.11	4694.42	21301.9	TB	1.3			MIT
4689.10	4690.41	21320.1	TB	0.5			MIT
4688.63	4689.94	21322.2	TB	1.2			MIT
4686.42	4687.73	21332.3	TB	0.5			MIT
4684.14	4685.45	21342.7	TB	0.3			MIT
4682.79	4684.10	21348.8	TB	0.9			MIT
4682.52	4683.83	21350.1	TB	1.0			MIT
4681.87	4683.18	21353.0	TB	1.2			MIT
4679.62	4680.93	21363.3	TB	0.3			MIT
4678.81	4680.12	21367.0	TB	0.3			MIT
4677.31	4678.62	21373.8	TB	0.3			MIT
4676.89	4678.20	21375.7	TB	1.4			MIT
4675.18	4676.49	21383.6	TB	0.3			MIT
4674.04	4675.35	21388.8	TB	0.3			MIT
4673.61	4674.92	21390.7	TB	0.6			MIT
4672.06	4673.37	21397.9	TB	0.3			MIT
4670.82	4672.13	21403.5	TB	0.7			MIT
4670.23	4671.54	21406.2	TB	0.5			MIT
4669.39	4670.70	21410.1	TB	0.5			MIT
4667.28	4668.59	21419.8	TB	0.5			MIT
4665.45	4666.76	21428.2	TB	1.0			MIT
4664.64	4665.95	21431.9	TB	0.7			MIT
4664.02	4665.33	21434.7	TB	0.3			MIT
4662.79	4664.10	21440.4	TB	1.3			MIT
4662.16	4663.47	21443.3	TB	0.5			MIT
4661.85	4663.16	21444.7	TB	0.3			MIT
4660.44	4661.74	21451.2	TB	0.3			MIT
4660.17	4661.47	21452.4	TB	0.5			MIT
4659.63	4660.93	21454.9	TB	0.3			MIT
4658.73	4660.03	21459.1	TB	1.4			MIT
4658.38	4659.68	21460.7	TB	1.6			MIT
4657.33	4658.63	21465.5	TB	0.5			MIT
4655.45	4656.75	21474.2	TB	0.3H			MIT
4653.46	4654.76	21483.4	TB	0.3H			MIT

AIR WAVELENGTH (ANGSTRMS)	VACUUM WAVELENGTH (ANGSTRMS)	WAVENUMBER (KAYSERS)	LMENT	LOG INTENSITY ARC	SPK	DIS	REF
4652.85	4654.15	21486.2	TB	0.6			MIT
4651.60	4652.90	21492.0	TB	0.6			MIT
4651.07	4652.37	21494.4	TB	0.6			MIT
4648.32	4649.62	21507.1	TB	0.8			MIT
4647.23	4648.53	21512.2	TB	1.2 D			MIT
4646.91	4648.21	21513.6	TB	0.5			MIT
4645.26	4646.56	21521.3	TB	1.8 O			MIT
4644.15	4645.45	21526.4	TB	0.3			MIT
4643.27	4644.57	21530.5	TB	0.3 H			MIT
4642.79	4644.09	21532.7	TB	0.5			MIT
4641.98	4643.28	21536.5	TB	1.6			MIT
4641.45	4642.75	21539.0	TB	0.3 H			MIT
4641.00	4642.30	21541.1	TB	1.3			MIT
4639.12	4640.42	21549.8	TB	0.3 H			MIT
4638.522	4639.821	21552.56	TB	0.5			MIT
4638.04	4639.34	21554.8	TB	0.5			MIT
4636.99	4638.29	21559.7	TB	1.0			MIT
4636.59	4637.89	21561.5	TB	1.3			MIT
4636.07	4637.37	21563.9	TB	1.0 W			MIT
4634.13	4635.43	21573.0	TB	0.3 H			MIT
4632.06	4633.36	21582.6	TB	1.5			MIT
4631.040	4632.337	21587.38	TB	0.3			MIT
4630.44	4631.74	21590.2	TB	0.3 H			MIT
4628.85	4630.15	21597.6	TB	0.3			MIT
4628.60	4629.90	21598.8	TB	0.3 H			MIT
4627.53	4628.83	21603.7	TB	0.8			MIT
4626.92	4628.22	21606.6	TB	1.5			MIT
4626.31	4627.61	21609.4	TB	1.0			MIT
4625.68	4626.98	21612.4	TB	0.3 H			MIT
4625.04	4626.34	21615.4	TB	0.3			MIT
4624.70	4626.00	21617.0	TB	0.3			MIT
4623.58	4624.87	21622.2	TB	0.3 H			MIT
4621.97	4623.26	21629.7	TB	0.5			MIT
4621.54	4622.83	21631.7	TB	0.3 H			MIT
4620.91	4622.20	21634.7	TB	0.6			MIT
4619.17	4620.46	21642.9	TB	0.7			MIT
4618.86	4620.15	21644.3	TB	0.3 H			MIT
4618.12	4619.41	21647.8	TB	0.5 D			MIT
4617.73	4619.02	21649.6	TB	0.3 H			MIT
4617.48	4618.77	21650.8	TB	0.9			MIT
4617.28	4618.57	21651.7	TB	0.5			MIT
4616.43	4617.72	21655.7	TB	0.3			MIT
4615.91	4617.20	21658.1	TB	1.0 H			MIT
4614.92	4616.21	21662.8	TB	0.5			MIT
4614.09	4615.38	21666.7	TB	0.3 H			MIT
4612.80	4614.09	21672.7	TB	0.5			MIT
4612.26	4613.55	21675.3	TB	1.2			MIT
4611.94	4613.23	21676.8	TB	1.0			MIT
4611.14	4612.43	21680.5	TB	0.3			MIT
4610.71	4612.00	21682.6	TB	0.3			MIT
4610.34	4611.63	21684.3	TB	0.5			MIT
4609.94	4611.23	21686.2	TB	0.5			MIT
4608.78	4610.07	21691.6	TB	0.5 H			MIT
4608.44	4609.73	21693.2	TB	0.9 G			MIT
4607.81	4609.10	21696.2	TB	0.6			MIT
4605.42	4606.71	21707.5	TB	0.3			MIT
4605.03	4606.32	21709.3	TB	0.3 H			MIT
4604.35	4605.64	21712.5	TB	1.0			MIT
4604.10	4605.39	21713.7	TB	1.0			MIT
4602.95	4604.24	21719.1	TB	0.9			MIT
4602.503	4603.792	21721.22	TB	0.6			MIT
4600.161	4601.450	21732.28	TB	0.7			MIT
4599.607	4600.896	21734.90	TB	0.6			MIT
4597.73	4599.02	21743.8	TB	0.5			MIT
4596.89	4598.18	21747.7	TB	0.5			MIT
4596.47	4597.76	21749.7	TB	0.5			MIT
4596.07	4597.36	21751.6	TB	0.5			MIT
4595.25	4596.54	21755.5	TB	0.3			MIT
4594.92	4596.21	21757.1	TB	0.3 H			MIT
4594.31	4595.60	21760.0	TB	0.3			MIT

AIR WAVELENGTH (ANGSTRMS)	VACUUM WAVELENGTH (ANGSTRMS)	WAVENUMBER (KAYSERS)	LMENT	LOG INTENSITY ARC	SPK	DIS	REF
4593.947	4595.234	21761.68	TB	0.6			MIT
4593.06	4594.35	21765.9	TB	0.5			MIT
4592.391	4593.678	21769.05	TB	0.7			MIT
4591.57	4592.86	21772.9	TB	1.1			MIT
4591.18	4592.47	21774.8	TB	0.8			MIT
4590.15	4591.44	21779.7	TB	0.3 H			MIT
4589.36	4590.65	21783.4	TB	1.2			MIT
4588.72	4590.01	21786.5	TB	0.5			MIT
4588.15	4589.44	21789.2	TB	0.9 D			MIT
4587.715	4589.000	21791.24	TB	1.3			MIT
4585.87	4587.15	21800.0	TB	0.3			MIT
4585.51	4586.79	21801.7	TB	0.3 H			MIT
4584.84	4586.12	21804.9	TB	1.1			MIT
4583.95	4585.23	21809.1	TB	0.6			MIT
4583.07	4584.35	21813.3	TB	0.5			MIT
4582.86	4584.14	21814.3	TB	0.5			MIT
4582.56	4583.84	21815.7	TB	0.3			MIT
4581.43	4582.71	21821.1	TB	0.9			MIT
4580.40	4581.68	21826.0	TB	0.6			MIT
4580.05	4581.33	21827.7	TB	0.5			MIT
4578.69	4579.97	21834.2	TB	1.8			MIT
4578.20	4579.48	21836.5	TB	0.5			MIT
4575.42	4576.70	21849.8	TB	0.5			MIT
4573.887	4575.169	21857.12	TB	0.3			MIT
4573.18	4574.46	21860.5	TB	0.8			MIT
4572.52	4573.80	21863.6	TB	0.5			MIT
4572.26	4573.54	21864.9	TB	0.5			MIT
4571.42	4572.70	21868.9	TB	0.5			MIT
4569.29	4570.57	21879.1	TB	0.5			MIT
4568.91	4570.19	21880.9	TB	0.3			MIT
4568.518	4569.798	21882.80	TB	0.3			MIT
4567.72	4569.00	21886.6	TB	1.0 W			MIT
4564.85	4566.13	21900.4	TB	0.8			MIT
4563.68	4564.96	21906.0	TB	1.7			MIT
4562.23	4563.51	21913.0	TB	1.2			MIT
4560.767	4562.045	21919.99	TB	0.3			MIT
4557.629	4558.907	21935.08	TB	0.3			MIT
4557.28	4558.56	21936.8	TB	0.6			MIT
4557.230	4558.507	21937.00	TB	0.6			KN
4556.962	4558.239	21938.29	TB	0.5 D			KN
4556.45	4557.73	21940.8	TB	1.3 W			MIT
4551.61	4552.89	21964.1	TB	0.5			MIT
4550.92	4552.20	21967.4	TB	0.3			MIT
4550.445	4551.721	21969.71	TB	1.2			MIT
4549.72	4551.00	21973.2	TB	1.0 O			MIT
4549.130	4550.405	21976.06	TB	0.5			MIT
4549.06	4550.34	21976.4	TB	1.3			MIT
4548.35	4549.63	21979.8	TB	0.5			MIT
4546.43	4547.70	21989.1	TB	0.5			MIT
4543.91	4545.18	22001.3	TB	0.3			MIT
4543.05	4544.32	22005.5	TB	0.6			MIT
4542.39	4543.66	22008.7	TB	0.6 D			MIT
4540.57	4541.84	22017.5	TB	0.5			MIT
4538.73	4540.00	22026.4	TB	0.7			MIT
4537.81	4539.08	22030.9	TB	0.3			MIT
4537.23	4538.50	22033.7	TB	1.1			MIT
4537.14	4538.41	22034.1	TB	0.6			MIT
4536.91	4538.18	22035.3	TB	1.1			MIT
4535.72	4536.99	22041.0	TB	0.3 H			MIT
4534.13	4535.40	22048.8	TB	0.7			MIT
4532.910	4534.181	22054.70	TB	0.3			MIT
4532.50	4533.77	22056.7	TB	0.3			MIT
4531.82	4533.09	22060.0	TB	0.6 W			MIT
4529.76	4531.03	22070.0	TB	0.6			MIT
4527.175	4528.444	22082.64	TB	0.3			MIT
4526.45	4527.72	22086.2	TB	0.3			MIT
4525.938	4527.207	22088.67	TB	0.5			MIT
4525.34	4526.61	22091.6	TB	0.3			MIT
4525.01	4526.28	22093.2	TB	0.5			MIT
4524.326	4525.595	22096.54	TB	0.5			MIT

AIR WAVELENGTH (ANGSTRMS)	VACUUM WAVELENGTH (ANGSTRMS)	WAVENUMBER (KAYSERS)	LMENT	LOG INTENSITY ARC	SPK	DIS	REF
4524.133	4525.402	22097.49	TB	0.3			KN
4521.834	4523.102	22108.72	TB	0.5			MIT
4520.08	4521.35	22117.3	TB	0.5			MIT
4518.208	4519.475	22126.46	TB	0.3			MIT
4517.19	4518.46	22131.4	TB	0.3			MIT
4516.531	4517.798	22134.68	TB	0.3			MIT
4516.138	4517.405	22136.60	TB	0.3			MIT
4515.88	4517.15	22137.9	TB	1.0W			MIT
4514.299	4515.565	22145.62	TB	1.2			MIT
4512.95	4514.22	22152.2	TB	1.1			MIT
4511.52	4512.79	22159.3	TB	1.6			MIT
4509.04	4510.30	22171.4	TB	1.5			MIT
4508.65	4509.91	22173.4	TB	0.3			MIT
4507.96	4509.22	22176.8	TB	0.3			MIT
4505.89	4507.15	22186.9	TB	0.5D			MIT
4505.35	4506.61	22189.6	TB	0.3			MIT
4504.57	4505.83	22193.4	TB	0.5			MIT
4503.58	4504.84	22198.3	TB	0.8			MIT
4503.24	4504.50	22200.0	TB	0.3			MIT
4501.73	4502.99	22207.4	TB	0.3			MIT
4501.28	4502.54	22209.7	TB	0.6			MIT
4501.02	4502.28	22211.0	TB	0.3			MIT
4499.47	4500.73	22218.6	TB	0.5			MIT
4498.96	4500.22	22221.1	TB	0.5			MIT
4496.92	4498.18	22231.2	TB	0.5			MIT
4496.25	4497.51	22234.5	TB	0.3			MIT
4495.03	4496.29	22240.6	TB	0.5			MIT
4494.40	4495.66	22243.7	TB	0.3			MIT
4493.08	4494.34	22250.2	TB	2.0			MIT
4491.78	4493.04	22256.6	TB	0.5			MIT
4491.27	4492.53	22259.2	TB	0.3			MIT
4491.02	4492.28	22260.4	TB	0.8			MIT
4490.64	4491.90	22262.3	TB	1.1			MIT
4489.76	4491.02	22266.7	TB	0.7			MIT
4488.16	4489.42	22274.6	TB	1.1			MIT
4487.74	4489.00	22276.7	TB	0.5			MIT
4486.71	4487.97	22281.8	TB	0.3			MIT
4486.51	4487.77	22282.8	TB	0.3			M
4486.02	4487.28	22285.2	TB	0.3			MIT
4485.67	4486.93	22287.0	TB	0.7			MIT
4484.80	4486.06	22291.3	TB	0.3			MIT
4483.41	4484.67	22298.2	TB	0.6			MIT
4482.72	4483.98	22301.6	TB	0.3			MIT
4482.40	4483.66	22303.2	TB	0.3			ED
4481.31	4482.57	22308.6	TB	0.3			MIT
4480.92	4482.18	22310.6	TB	0.6			MIT
4480.06	4481.32	22314.9	TB	0.6			MIT
4478.56	4479.82	22322.3	TB	0.3			M
4477.83	4479.09	22326.0	TB	0.7			ED
4476.97	4478.23	22330.3	TB	0.6			ED
4476.429	4477.685	22332.97	TB	0.3			KN
4476.28	4477.54	22333.7	TB	0.5			MIT
4475.82	4477.08	22336.0	TB	0.5			MIT
4475.26	4476.52	22338.8	TB	0.6			MIT
4474.78	4476.04	22341.2	TB	0.6			ED
4474.47	4475.73	22342.7	TB	0.3			MIT
4474.12	4475.38	22344.5	TB	0.5			MIT
4473.70	4474.96	22346.6	TB	0.5			MIT
4472.83	4474.09	22350.9	TB	0.6			MIT
4471.72	4472.97	22356.5	TB	1.1			MIT
4471.17	4472.42	22359.2	TB	0.5			MIT
4470.36	4471.61	22363.3	TB	0.3			MIT
4469.73	4470.98	22366.4	TB	1.0W			MIT
4469.12	4470.37	22369.5	TB	0.7			M
4468.07	4469.32	22374.7	TB	0.3			MIT
4467.69	4468.94	22376.6	TB	0.8			MIT
4467.09	4468.34	22379.7	TB	0.3			MIT
4465.70	4466.95	22386.6	TB	0.6			MIT
4464.07	4465.32	22394.8	TB	0.3			MIT
4462.99	4464.24	22400.2	TB	0.3			MIT

AIR WAVELENGTH (ANGSTRMS)	VACUUM WAVELENGTH (ANGSTRMS)	WAVENUMBER (KAYSERS)	LMENT	LOG INTENSITY ARC	SPK	DIS	REF
4462.20	4463.45	22404.2	TB	1.30			MIT
4461.26	4462.51	22408.9	TB	0.9W			MIT
4460.23	4461.48	22414.1	TB	0.3			MIT
4459.38	4460.63	22418.4	TB	1.0W			MIT
4458.46	4459.71	22423.0	TB	0.5			MIT
4457.59	4458.84	22427.4	TB	0.3			M
4456.42	4457.67	22433.2	TB	0.3			M
4456.02	4457.27	22435.2	TB	0.3			MIT
4454.89	4456.14	22440.9	TB	0.6			MIT
4453.96	4455.21	22445.6	TB	0.3			MIT
4453.10	4454.35	22450.0	TB	0.6			ED
4452.81	4454.06	22451.4	TB	1.4			MIT
4452.58	4453.83	22452.6	TB	0.5			ED
4451.64	4452.89	22457.3	TB	1.0			MIT
4451.36	4452.61	22458.7	TB	0.3			M
4450.57	4451.82	22462.7	TB	0.3			MIT
4450.11	4451.36	22465.1	TB	0.6			MIT
4449.72	4450.97	22467.0	TB	1.2			MIT
4449.52	4450.77	22468.0	TB	0.5W			MIT
4449.23	4450.48	22469.5	TB	0.3			MIT
4448.04	4449.29	22475.5	TB	1.4			MIT
4447.44	4448.69	22478.5	TB	0.5			MIT
4446.92	4448.17	22481.2	TB	1.0			ED
4446.32	4447.57	22484.2	TB	0.3			MIT
4445.98	4447.23	22485.9	TB	0.5			MIT
4445.11	4446.36	22490.3	TB	0.3			MIT
4441.48	4442.73	22508.7	TB	0.6			MIT
4441.27	4442.52	22509.8	TB	1.2			MIT
4440.36	4441.61	22514.4	TB	0.3			M
4439.38	4440.63	22519.4	TB	1.0			MIT
4438.96	4440.21	22521.5	TB	1.1			MIT
4437.63	4438.88	22528.2	TB	0.3			MIT
4436.11	4437.36	22535.9	TB	1.2			MIT
4435.55	4436.80	22538.8	TB	1.1			MIT
4435.00	4436.25	22541.6	TB	0.6			MIT
4434.48	4435.73	22544.2	TB	1.3			MIT
4434.06	4435.30	22546.4	TB	0.3			MIT
4433.65	4434.89	22548.4	TB	0.6			MIT
4432.715	4433.960	22553.21	TB	1.1			M
4432.160	4433.404	22556.03	TB	0.5			M
4430.737	4431.981	22563.27	TB	0.9			M
4430.14	4431.38	22566.3	TB	0.9			MIT
4428.34	4429.58	22575.5	TB	0.5			MIT
4427.71	4428.95	22578.7	TB	0.5			MIT
4427.391	4428.634	22580.32	TB	0.6			KN
4426.30	4427.54	22585.9	TB	0.7			ED
4425.25	4426.49	22591.2	TB	0.3H			MIT
4424.46	4425.70	22595.3	TB	0.5			MIT
4423.86	4425.10	22598.4	TB	0.3			ED
4423.11	4424.35	22602.2	TB	1.5			MIT
4421.14	4422.38	22612.2	TB	0.3			MIT
4420.52	4421.76	22615.4	TB	0.6			MIT
4420.20	4421.44	22617.1	TB	1.1			MIT
4418.27	4419.51	22626.9	TB	0.5W			MIT
4417.99	4419.23	22628.4	TB	0.6			MIT
4417.55	4418.79	22630.6	TB	0.5			MIT
4416.66	4417.90	22635.2	TB	0.5D			MIT
4416.26	4417.50	22637.2	TB	1.5			MIT
4413.63	4414.87	22650.7	TB	0.6			MIT
4412.84	4414.08	22654.8	TB	0.3H			MIT
4412.52	4413.76	22656.4	TB	0.5			MIT
4411.93	4413.17	22659.4	TB	0.5			ED
4411.51	4412.75	22661.6	TB	0.6			M
4411.13	4412.37	22663.6	TB	0.5			MIT
4410.42	4411.66	22667.2	TB	0.3			MIT
4410.06	4411.30	22669.1	TB	0.3			MIT
4409.51	4410.75	22671.9	TB	1.6W			MIT
4408.79	4410.03	22675.6	TB	0.5			MIT
4406.75	4407.99	22686.1	TB	0.5			MIT
4406.38	4407.62	22688.0	TB	0.3H			MIT

AIR WAVELENGTH (ANGSTRMS)	VACUUM WAVELENGTH (ANGSTRMS)	WAVENUMBER (KAYSERS)	LMENT	LOG ARC	INTENSITY SPK	INTENSITY DIS	REF
4405.41	4406.65	22693.0	TB	1.2			MIT
4404.43	4405.67	22698.0	TB	0.5			MIT
4403.19	4404.43	22704.4	TB	1.3			MIT
4401.54	4402.78	22712.9	TB	1.3			MIT
4400.51	4401.75	22718.3	TB	0.3			MIT
4400.10	4401.34	22720.4	TB	0.5			MIT
4399.18	4400.42	22725.1	TB	0.6			MIT
4398.61	4399.85	22728.1	TB	0.3			MIT
4396.55	4397.79	22738.7	TB	1.0			MIT
4396.03	4397.26	22741.4	TB	0.3			MIT
4394.93	4396.16	22747.1	TB	1.2D			MIT
4394.02	4395.25	22751.8	TB	1.0			MIT
4392.95	4394.18	22757.4	TB	0.6			MIT
4392.34	4393.57	22760.5	TB	0.5			MIT
4392.23	4393.46	22761.1	TB	0.5			KN
4390.91	4392.14	22767.9	TB	1.5			MIT
4389.92	4391.15	22773.1	TB	0.3			M
4389.81	4391.04	22773.6	TB	0.5			MIT
4389.40	4390.63	22775.8	TB	0.5			MIT
4389.04	4390.27	22777.6	TB	0.8			MIT
4388.25	4389.48	22781.7	TB	1.3			MIT
4387.42	4388.65	22786.0	TB	0.3			MIT
4386.71	4387.94	22789.7	TB	0.3			MIT
4386.07	4387.30	22793.1	TB	1.4			MIT
4385.66	4386.89	22795.2	TB	1.2			MIT
4385.02	4386.25	22798.5	TB	0.3			MIT
4384.70	4385.93	22800.2	TB	0.3			MIT
4384.06	4385.29	22803.5	TB	1.0			MIT
4382.45	4383.68	22811.9	TB	1.4			MIT
4381.30	4382.53	22817.9	TB	0.9			MIT
4379.60	4380.83	22826.7	TB	0.5W			MIT
4379.26	4380.49	22828.5	TB	0.6			MIT
4378.70	4379.93	22831.4	TB	1.0W			MIT
4377.34	4378.57	22838.5	TB	0.3			MIT
4376.43	4377.66	22843.3	TB	1.1			MIT
4375.61	4376.84	22847.5	TB	0.5			MIT
4375.33	4376.56	22849.0	TB	0.8			MIT
4374.83	4376.06	22851.6	TB	0.8			ED
4374.43	4375.66	22853.7	TB	0.5			MIT
4374.24	4375.47	22854.7	TB	0.9			MIT
4373.59	4374.82	22858.1	TB	0.5			MIT
4372.64	4373.87	22863.1	TB	0.5			MIT
4372.48	4373.71	22863.9	TB	0.6			MIT
4372.33	4373.56	22864.7	TB	0.3			MIT
4372.04	4373.27	22866.2	TB	1.3			MIT
4371.13	4372.36	22870.9	TB	0.3			MIT
4370.92	4372.15	22872.1	TB	0.5			MIT
4370.58	4371.81	22873.8	TB	0.3			MIT
4370.33	4371.56	22875.1	TB	0.5			ED
4367.31	4368.54	22891.0	TB	1.5			M
4366.94	4368.17	22892.9	TB	0.9W			MIT
4366.38	4367.61	22895.8	TB	0.3			MIT
4366.00	4367.23	22897.8	TB	1.0			MIT
4363.92	4365.15	22908.7	TB	0.3			MIT
4362.45	4363.68	22916.5	TB	0.6			M
4360.63	4361.86	22926.0	TB	0.6			MIT
4360.16	4361.39	22928.5	TB	1.4			M
4359.34	4360.57	22932.8	TB	0.3D			MIT
4359.05	4360.28	22934.3	TB	0.3			MIT
4358.77	4360.00	22935.8	TB	0.3			MIT
4357.47	4358.69	22942.6	TB	0.7			MIT
4356.84	4358.06	22946.0	TB	1.8			MIT
4356.09	4357.31	22949.9	TB	1.3			MIT
4355.07	4356.29	22955.3	TB	0.3			MIT
4354.51	4355.73	22958.2	TB	0.3			MIT
4353.19	4354.41	22965.2	TB	1.7			MIT
4351.58	4352.80	22973.7	TB	1.2D			MIT
4350.74	4351.96	22978.1	TB	0.7			MIT
4349.60	4350.82	22984.2	TB	1.1			MIT
4348.68	4349.90	22989.0	TB	0.3			MIT

AIR WAVELENGTH (ANGSTRMS)	VACUUM WAVELENGTH (ANGSTRMS)	WAVENUMBER (KAYSERS)	LMENT	LOG ARC	INTENSITY SPK	INTENSITY DIS	REF
4348.33	4349.55	22990.9	TB	1.0			MIT
4345.03	4346.25	23008.3	TB	0.3			MIT
4344.21	4345.43	23012.7	TB	0.8			MIT
4343.86	4345.08	23014.5	TB	0.6			M
4343.29	4344.51	23017.6	TB	1.0			MIT
4342.595	4343.816	23021.23	TB	1.4			MIT
4342.500	4343.721	23021.74	TB	1.7W			KN
4341.95	4343.17	23024.6	TB	0.3			MIT
4341.01	4342.23	23029.6	TB	1.0			MIT
4340.63	4341.85	23031.7	TB	1.6	0.3		MIT
4339.90	4341.12	23035.5	TB	0.9			MIT
4339.63	4340.85	23037.0	TB	0.9			MIT
4338.45	4339.67	23043.2	TB	2.0	0.5		MIT
4337.64	4338.86	23047.5	TB	1.4	0.3		MIT
4336.50	4337.72	23053.6	TB	1.6			M
4335.88	4337.10	23056.9	TB	0.7			MIT
4335.74	4336.96	23057.6	TB	0.8			ED
4334.67	4335.89	23063.3	TB	1.1	0.0H		MIT
4333.71	4334.93	23068.4	TB	0.6			MIT
4332.13	4333.35	23076.8	TB	1.6	0.3		MIT
4330.98	4332.20	23083.0	TB	0.8			MIT
4330.83	4332.05	23083.8	TB	0.9			MIT
4330.35	4331.57	23086.3	TB	0.9			MIT
4329.92	4331.14	23088.6	TB	0.5			MIT
4329.56	4330.78	23090.5	TB	0.3			ED
4329.44	4330.66	23091.2	TB	0.6			MIT
4328.955	4330.172	23093.77	TB	1.3	0.3H		MIT
4328.08	4329.30	23098.4	TB	0.6	0.0H		MIT
4327.78	4329.00	23100.0	TB	0.5			MIT
4327.50	4328.72	23101.5	TB	0.7W			MIT
4326.48	4327.70	23107.0	TB	2.2	0.6		MIT
4325.83	4327.05	23110.4	TB	2.0			MIT
4325.50	4326.72	23112.2	TB	1.2			MIT
4324.09	4325.31	23119.7	TB	0.5			MIT
4323.66	4324.88	23122.1	TB	1.4D	0.3		MIT
4322.88	4324.10	23126.2	TB	1.1	0.0		MIT
4322.24	4323.46	23129.6	TB	1.5	0.3		MIT
4321.50	4322.72	23133.6	TB	0.9			MIT
4320.28	4321.49	23140.1	TB	1.1			MIT
4318.85	4320.06	23147.8	TB	2.2	1.5		MIT
4318.49	4319.70	23149.7	TB	0.9			MIT
4317.93	4319.14	23152.7	TB	0.5			MIT
4317.36	4318.57	23155.8	TB	0.6H			MIT
4316.80	4318.01	23158.8	TB	0.7			MIT
4315.75	4316.96	23164.4	TB	1.1	0.0H		MIT
4314.14	4315.35	23173.1	TB	0.5			MIT
4313.69	4314.90	23175.5	TB	0.5			MIT
4313.43	4314.64	23176.9	TB	0.9			MIT
4313.24	4314.45	23177.9	TB	1.2	0.0H		MIT
4312.10	4313.31	23184.0	TB	1.1	0.3		MIT
4311.57	4312.78	23186.9	TB	1.0	0.0H		MIT
4311.30	4312.51	23188.3	TB	0.6			MIT
4310.98	4312.19	23190.1	TB	1.1	0.0		MIT
4310.68	4311.89	23191.7	TB	0.5			MIT
4310.45	4311.66	23192.9	TB	1.3	0.0		MIT
4308.68	4309.89	23202.4	TB	1.0	0.3		M
4307.20	4308.41	23210.4	TB	1.1	0.0H		MIT
4304.28	4305.49	23226.2	TB	0.6			MIT
4304.02	4305.23	23227.6	TB	1.1			MIT
4302.94	4304.15	23233.4	TB	1.0D			MIT
4301.54	4302.75	23240.9	TB	0.3H			MIT
4301.18	4302.39	23242.9	TB	0.7			MIT
4300.95	4302.16	23244.1	TB	0.6			MIT
4300.49	4301.70	23246.6	TB	0.3			MIT
4299.92	4301.13	23249.7	TB	1.2			MIT
4299.15	4300.36	23253.9	TB	0.3			MIT
4298.38	4299.59	23258.0	TB	1.1	0.0		MIT
4296.35	4297.56	23269.0	TB	1.3	0.3H		MIT
4295.35	4296.56	23274.4	TB	1.0			MIT
4294.36	4295.57	23279.8	TB	1.0	0.0H		MIT

AIR WAVELENGTH (ANGSTRMS)	VACUUM WAVELENGTH (ANGSTRMS)	WAVENUMBER (KAYSERS)	LMENT	LOG ARC	INTENSITY SPK	INTENSITY DIS	REF
4294.05	4295.26	23281.5	TB	0.7			MIT
4293.14	4294.35	23286.4	TB	1.1			MIT
4292.65	4293.86	23289.1	TB	1.0	0.0H		MIT
4292.11	4293.32	23292.0	TB	0.5			MIT
4291.40	4292.61	23295.9	TB	0.5			MIT
4289.73	4290.94	23304.9	TB	1.5			MIT
4288.02	4289.23	23314.2	TB	0.5			MIT
4287.32	4288.53	23318.0	TB	1.0	0.0H		MIT
4287.13	4288.34	23319.1	TB	0.9			MIT
4286.90	4288.11	23320.3	TB	1.3	0.3		MIT
4286.12	4287.33	23324.6	TB	1.0	0.0H		MIT
4285.74	4286.95	23326.6	TB	1.0	0.0H		MIT
4285.41	4286.62	23328.4	TB	0.5			MIT
4285.13	4286.34	23329.9	TB	1.4	0.5		MIT
4283.74	4284.95	23337.5	TB	0.8D			MIT
4281.30	4282.50	23350.8	TB	0.7			MIT
4280.49	4281.69	23355.2	TB	0.7	0.0		MIT
4278.51	4279.71	23366.1	TB	2.3	2.0		MIT
4277.76	4278.96	23370.1	TB	0.9			MIT
4276.75	4277.95	23375.7	TB	1.7W	0.3H		MIT
4276.14	4277.34	23379.0	TB	0.6			MIT
4275.37	4276.57	23383.2	TB	0.5			MIT
4275.21	4276.41	23384.1	TB	1.2	0.0H		MIT
4274.36	4275.56	23388.7	TB	0.3H			MIT
4273.74	4274.94	23392.1	TB	0.5O			MIT
4273.18	4274.38	23395.2	TB	0.9	0.0		MIT
4272.22	4273.42	23400.4	TB	0.5			MIT
4270.73	4271.93	23408.6	TB	0.5			MIT
4269.70	4270.90	23414.3	TB	1.2	0.3		MIT
4269.32	4270.52	23416.3	TB	0.7			MIT
4266.35	4267.55	23432.6	TB	1.5	0.0		MIT
4265.98	4267.18	23434.7	TB	0.5			MIT
4264.99	4266.19	23440.1	TB	0.9			MIT
4264.72	4265.92	23441.6	TB	0.6			MIT
4264.63	4265.83	23442.1	TB	0.8			M
4263.68	4264.88	23447.3	TB	1.1			MIT
4263.55	4264.75	23448.0	TB	0.9			MIT
4262.59	4263.79	23453.3	TB	0.3			MIT
4261.83	4263.03	23457.5	TB	0.9			MIT
4260.78	4261.98	23463.3	TB	0.9H			MIT
4259.96	4261.16	23467.8	TB	0.7			MIT
4258.23	4259.43	23477.3	TB	1.5D	0.5		MIT
4257.81	4259.01	23479.6	TB	0.3			MIT
4257.17	4258.37	23483.2	TB	0.3			MIT
4256.34	4257.54	23487.7	TB	1.0			MIT
4256.10	4257.30	23489.1	TB	0.9			MIT
4255.24	4256.44	23493.8	TB	1.4	0.0H		MIT
4254.00	4255.20	23500.7	TB	0.7			MIT
4252.69	4253.89	23507.9	TB	0.6	0.7		MIT
4251.72	4252.92	23513.3	TB	1.1	0.5		ED
4251.33	4252.53	23515.4	TB	1.0			MIT
4250.24	4251.44	23521.5	TB	0.7			MIT
4249.32	4250.52	23526.6	TB	0.3			MIT
4249.05	4250.25	23528.1	TB	0.3			MIT
4248.56	4249.76	23530.8	TB	1.0	0.0H		MIT
4246.59	4247.79	23541.7	TB	1.1	0.0H		MIT
4245.14	4246.34	23549.7	TB	0.9			MIT
4244.86	4246.06	23551.3	TB	0.3			MIT
4244.53	4245.73	23553.1	TB	0.7	0.0H		MIT
4243.65	4244.84	23558.0	TB	0.3			MIT
4243.06	4244.25	23561.3	TB	0.5			MIT
4242.55	4243.74	23564.1	TB	1.1	0.3H		MIT
4242.25	4243.44	23565.8	TB	0.6			MIT
4241.08	4242.27	23572.3	TB	0.6			MIT
4240.13	4241.32	23577.5	TB	1.1	0.3H		MIT
4239.93	4241.12	23578.7	TB	0.7			MIT
4239.28	4240.47	23582.3	TB	1.1			MIT
4237.35	4238.54	23593.0	TB	0.3			MIT
4235.34	4236.53	23604.2	TB	1.3	0.0H		MIT
4234.63	4235.82	23608.2	TB	0.6			MIT

AIR WAVELENGTH (ANGSTRMS)	VACUUM WAVELENGTH (ANGSTRMS)	WAVENUMBER (KAYSERS)	LMENT	LOG ARC	INTENSITY SPK	INTENSITY DIS	REF
4232.82	4234.01	23618.3	TB	1.2			MIT
4232.19	4233.38	23621.8	TB	1.3	0.5		MIT
4231.88	4233.07	23623.5	TB	1.1			MIT
4231.36	4232.55	23626.4	TB	0.7			MIT
4230.61	4231.80	23630.6	TB	0.7			MIT
4226.44	4227.63	23653.9	TB	1.7			MIT
4226.15	4227.34	23655.5	TB	0.5			MIT
4224.85	4226.04	23662.8	TB	0.7			MIT
4224.28	4225.47	23666.0	TB	1.1			MIT
4223.82	4225.01	23668.6	TB	0.5			MIT
4223.32	4224.51	23671.4	TB	1.0	0.0		MIT
4222.91	4224.10	23673.7	TB	0.8O			MIT
4222.71	4223.90	23674.8	TB	1.0			MIT
4222.02	4223.21	23678.7	TB	0.5			MIT
4221.38	4222.57	23682.3	TB	0.5			MIT
4220.47	4221.66	23687.4	TB	0.3			MIT
4220.11	4221.30	23689.4	TB	1.2	0.0H		M
4219.17	4220.36	23694.7	TB	1.1			MIT
4218.84	4220.03	23696.5	TB	0.7			MIT
4217.91	4219.10	23701.7	TB	0.7			MIT
4217.17	4218.36	23705.9	TB	0.5			M
4216.68	4217.87	23708.7	TB	1.0	0.0H		MIT
4215.95	4217.14	23712.8	TB	0.5			ED
4215.13	4216.32	23717.4	TB	1.5D	0.0H		MIT
4214.42	4215.61	23721.4	TB	1.4	0.3		MIT
4213.49	4214.68	23726.6	TB	0.9D	0.3H		MIT
4212.54	4213.73	23732.0	TB	0.7			MIT
4212.23	4213.42	23733.7	TB	0.5W			MIT
4211.72	4212.91	23736.6	TB	1.4	0.3		MIT
4211.16	4212.35	23739.7	TB	0.3	0.0H		MIT
4209.41	4210.60	23749.6	TB	0.3H			ED
4208.70	4209.89	23753.6	TB	1.2D	0.0H		MIT
4207.56	4208.75	23760.1	TB	0.9			MIT
4206.80	4207.99	23764.3	TB	0.7			ED
4206.49	4207.68	23766.1	TB	1.3D	0.0H		MIT
4205.64	4206.82	23770.9	TB	0.5			MIT
4205.23	4206.41	23773.2	TB	0.3			MIT
4203.71	4204.89	23781.8	TB	1.4	0.0		MIT
4203.07	4204.25	23785.4	TB	0.5			MIT
4200.99	4202.17	23797.2	TB	1.6	0.6		MIT
4200.66	4201.84	23799.1	TB	0.6			MIT
4200.12	4201.30	23802.1	TB	0.3			MIT
4199.67	4200.85	23804.7	TB	0.5			MIT
4199.05	4200.23	23808.2	TB	0.5			MIT
4198.99	4200.17	23808.5	TB	0.6			KN
4198.42	4199.60	23811.8	TB	0.8			MIT
4197.03	4198.21	23819.7	TB	0.3	0.0H		M
4196.73	4197.91	23821.4	TB	1.2	0.0H		MIT
4194.82	4196.00	23832.2	TB	1.4	0.0H		MIT
4194.00	4195.18	23836.9	TB	0.7			MIT
4193.76	4194.94	23838.2	TB	0.8W	0.0H		MIT
4193.33	4194.51	23840.7	TB	0.7			MIT
4192.00	4193.18	23848.2	TB	0.5			MIT
4191.59	4192.77	23850.6	TB	1.4D			MIT
4190.42	4191.60	23857.2	TB	0.5			MIT
4190.12	4191.30	23858.9	TB	0.3			ED
4189.29	4190.47	23863.7	TB	0.5			MIT
4188.51	4189.69	23868.1	TB	1.3	0.3H		MIT
4188.10	4189.28	23870.4	TB	1.2	0.0H		MIT
4187.16	4188.34	23875.8	TB	1.2			MIT
4186.24	4187.42	23881.1	TB	1.0	0.0H		M
4185.89	4187.07	23883.1	TB	0.9			MIT
4185.44	4186.62	23885.6	TB	0.5			MIT
4184.95	4186.13	23888.4	TB	0.3			MIT
4184.66	4185.84	23890.1	TB	0.3			ED
4184.30	4185.48	23892.1	TB	1.0	0.3		MIT
4183.01	4184.19	23899.5	TB	0.5			MIT
4181.33	4182.51	23909.1	TB	1.0	0.3H		MIT
4180.88	4182.06	23911.7	TB	0.6	0.0H		MIT
4180.40	4181.58	23914.4	TB	1.3	0.0H		MIT

AIR WAVELENGTH (ANGSTRMS)	VACUUM WAVELENGTH (ANGSTRMS)	WAVENUMBER (KAYSERS)	LMENT	LOG ARC	INTENSITY SPK	DIS	REF
4179.80	4180.98	23917.9	TB	0.8	0.0		MIT
4178.97	4180.15	23922.6	TB	1.7D	0.3H		M
4178.56	4179.74	23924.9	TB	0.3			MIT
4178.14	4179.32	23927.4	TB	0.3			MIT
4177.85	4179.03	23929.0	TB	0.3			MIT
4176.86	4178.04	23934.7	TB	0.6			MIT
4175.86	4177.04	23940.4	TB	0.8			MIT
4175.08	4176.26	23944.9	TB	0.7			M
4174.32	4175.50	23949.2	TB	0.8			MIT
4173.46	4174.64	23954.2	TB	1.2	0.3		MIT
4172.82	4174.00	23957.9	TB	1.0			MIT
4172.59	4173.77	23959.2	TB	1.1	0.3		MIT
4171.80	4172.98	23963.7	TB	0.9			MIT
4171.04	4172.22	23968.1	TB	1.3	0.5		MIT
4170.48	4171.66	23971.3	TB	0.8			MIT
4169.92	4171.10	23974.5	TB	0.7			MIT
4169.31	4170.49	23978.0	TB	1.1	0.3H		M
4169.09	4170.27	23979.3	TB	0.9			MIT
4168.74	4169.92	23981.3	TB	0.3D			MIT
4167.97	4169.14	23985.7	TB	1.2	0.0H		MIT
4166.50	4167.67	23994.2	TB	1.1	0.0H		MIT
4165.08	4166.25	24002.4	TB	0.3			MIT
4164.81	4165.98	24003.9	TB	0.5			MIT
4163.85	4165.02	24009.5	TB	0.7			MIT
4162.97	4164.14	24014.5	TB	0.8			MIT
4161.35	4162.52	24023.9	TB	0.7	0.3H		MIT
4159.18	4160.35	24036.4	TB	0.3			MIT
4158.54	4159.71	24040.1	TB	1.2	0.0H		MIT
4158.28	4159.45	24041.6	TB	1.0	0.0		MIT
4156.28	4157.45	24053.2	TB	0.9			MIT
4156.28	4157.45	24053.2	TB	0.9			MIT
4155.38	4156.55	24058.4	TB	0.3			MIT
4155.15	4156.32	24059.7	TB	1.0D			MIT
4154.63	4155.80	24062.7	TB	0.7			MIT
4153.86	4155.03	24067.2	TB	0.3			MIT
4153.51	4154.68	24069.2	TB	0.5			MIT
4152.22	4153.39	24076.7	TB	0.8			MIT
4151.13	4152.30	24083.0	TB	0.8	0.0		MIT
4150.88	4152.05	24084.5	TB	0.6			MIT
4150.53	4151.70	24086.5	TB	1.1	0.0H		MIT
4149.18	4150.35	24094.4	TB	1.2	0.3		MIT
4148.23	4149.40	24099.9	TB	1.0	0.0		MIT
4147.43	4148.60	24104.5	TB	0.8			MIT
4146.95	4148.12	24107.3	TB	1.1			M
4146.08	4147.25	24112.4	TB	1.0			MIT
4145.44	4146.61	24116.1	TB	0.3			ED
4144.45	4145.62	24121.9	TB	1.9	1.0		MIT
4143.700	4144.869	24126.22	TB	1.2	0.0		MIT
4143.53	4144.70	24127.2	TB	1.4			M
4143.24	4144.41	24128.9	TB	0.6			ED
4142.46	4143.63	24133.4	TB	0.5			MIT
4141.56	4142.73	24138.7	TB	1.4			MIT
4140.75	4141.92	24143.4	TB	0.9	0.0		MIT
4139.81	4140.98	24148.9	TB	1.0			MIT
4139.36	4140.53	24151.5	TB	0.5			MIT
4139.06	4140.23	24153.3	TB	0.6			ED
4137.04	4138.21	24165.1	TB	0.7	0.0W		MIT
4136.46	4137.63	24168.4	TB	0.6			MIT
4135.38	4136.55	24174.8	TB	1.1			MIT
4134.31	4135.48	24181.0	TB	0.6	0.0H		M
4133.55	4134.72	24185.5	TB	0.3			MIT
4132.83	4134.00	24189.7	TB	1.0			MIT
4132.47	4133.64	24191.8	TB	0.7			MIT
4132.22	4133.39	24193.2	TB	0.5			MIT
4131.45	4132.62	24197.7	TB	1.0	0.3		MIT
4131.11	4132.28	24199.7	TB	0.9	0.0		MIT
4130.35	4131.52	24204.2	TB	0.5	0.0		MIT
4130.14	4131.31	24205.4	TB	1.0			MIT
4129.42	4130.58	24209.6	TB	1.0	0.0W		MIT
4128.72	4129.88	24213.7	TB	0.6			MIT

AIR WAVELENGTH (ANGSTRMS)	VACUUM WAVELENGTH (ANGSTRMS)	WAVENUMBER (KAYSERS)	LMENT	LOG ARC	INTENSITY SPK	DIS	REF
4128.46	4129.62	24215.3	TB	0.5			MIT
4127.30	4128.46	24222.1	TB	1.0	0.3		MIT
4126.94	4128.10	24224.2	TB	0.5			MIT
4126.72	4127.88	24225.5	TB	0.8			MIT
4125.42	4126.58	24233.1	TB	0.3			MIT
4125.21	4126.37	24234.4	TB	1.0W	0.0W		MIT
4124.27	4125.43	24239.9	TB	0.5H			M
4123.78	4124.94	24242.8	TB	1.0			MIT
4122.47	4123.63	24250.5	TB	0.9			MIT
4122.21	4123.37	24252.0	TB	0.5			MIT
4121.57	4122.73	24255.8	TB	0.3			MIT
4121.01	4122.17	24259.1	TB	0.7			MIT
4120.52	4121.68	24261.9	TB	1.1	0.3		MIT
4120.43	4121.59	24262.5	TB	0.3			KN
4119.987	4121.149	24265.07	TB	1.2			MIT
4119.49	4120.65	24268.0	TB	0.8			MIT
4118.426	4119.588	24274.27	TB	0.7			KN
4118.425	4119.588	24274.27	TB	0.7			KN
4118.20	4119.36	24275.6	TB	0.7			MIT
4117.22	4118.38	24281.4	TB	1.0			MIT
4116.85	4118.01	24283.6	TB	0.5			MIT
4116.51	4117.67	24285.6	TB	0.9O			MIT
4115.37	4116.53	24292.3	TB	0.9	0.0		MIT
4114.14	4115.30	24299.6	TB	1.1	0.0		MIT
4112.88	4114.04	24307.0	TB	1.0			MIT
4112.54	4113.70	24309.0	TB	1.4	0.0		M
4111.34	4112.50	24316.1	TB	1.1			MIT
4111.01	4112.17	24318.1	TB	0.3			MIT
4110.85	4112.01	24319.0	TB	0.6			MIT
4110.07	4111.23	24323.6	TB	0.5			MIT
4109.51	4110.67	24326.9	TB	0.7			MIT
4109.08	4110.24	24329.5	TB	0.3			MIT
4108.39	4109.55	24333.6	TB	0.3			MIT
4107.79	4108.95	24337.1	TB	0.8			MIT
4106.35	4107.51	24345.7	TB	0.7			MIT
4106.01	4107.17	24347.7	TB	0.8			MIT
4105.38	4106.54	24351.4	TB	1.2	0.0		MIT
4103.90	4105.06	24360.2	TB	1.5	0.3		MIT
4103.46	4104.62	24362.8	TB	0.9			MIT
4103.37	4104.53	24363.3	TB	1.5			KN
4103.21	4104.37	24364.3	TB	0.6			KN
4102.52	4103.68	24368.4	TB	1.4W	0.3		MIT
4101.65	4102.81	24373.6	TB	1.3W			MIT
4100.90	4102.06	24378.0	TB	1.7D	0.3		MIT
4099.46	4100.62	24386.6	TB	1.4W	0.0		MIT
4099.14	4100.30	24388.5	TB	1.0	0.0		MIT
4098.83	4099.99	24390.3	TB	0.5			MIT
4098.60	4099.76	24391.7	TB	0.3	0.3		MIT
4097.45	4098.61	24398.5	TB	1.3	0.7		MIT
4096.38	4097.54	24404.9	TB	0.3			MIT
4095.90	4097.06	24407.8	TB	0.9			MIT
4094.45	4095.61	24416.4	TB	1.0	0.7		M
4094.04	4095.20	24418.9	TB	0.8			MIT
4093.35	4094.51	24423.0	TB	0.3			MIT
4092.19	4093.35	24429.9	TB	1.1			MIT
4091.41	4092.56	24434.6	TB	0.6D			MIT
4091.33	4092.48	24435.0	TB	0.5			M
4090.13	4091.28	24442.2	TB	0.3			MIT
4089.51	4090.66	24445.9	TB	1.0	0.0		MIT
4089.33	4090.48	24447.0	TB	1.3	0.0		MIT
4088.64	4089.79	24451.1	TB	0.6			MIT
4087.70	4088.85	24456.7	TB	0.9	0.0		M
4086.93	4088.08	24461.3	TB	0.5			MIT
4086.60	4087.75	24463.3	TB	1.2			MIT
4084.83	4085.98	24473.9	TB	0.8			MIT
4084.25	4085.40	24477.4	TB	1.0			MIT
4083.21	4084.36	24483.6	TB	1.0			MIT
4082.79	4083.94	24486.1	TB	0.9			MIT
4082.23	4083.38	24489.5	TB	0.7	0.0		MIT
4081.23	4082.38	24495.5	TB	1.5	0.0		MIT

AIR WAVELENGTH (ANGSTRMS)	VACUUM WAVELENGTH (ANGSTRMS)	WAVENUMBER (KAYSERS)	LMENT	LOG ARC	INTENSITY SPK	DIS	REF
4079.15	4080.30	24508.0	TB	0.6			MIT
4078.78	4079.93	24510.2	TB	0.7			M
4078.47	4079.62	24512.1	TB	0.7	0.0		MIT
4078.26	4079.41	24513.3	TB	0.5			ED
4075.90	4077.05	24527.5	TB	0.6			ED
4075.23	4076.38	24531.6	TB	0.9	0.0		MIT
4074.19	4075.34	24537.8	TB	0.8			ED
4073.94	4075.09	24539.3	TB	0.7			MIT
4073.75	4074.90	24540.5	TB	1.0	0.3		M
4073.16	4074.31	24544.0	TB	1.0	0.0		M
4072.67	4073.82	24547.0	TB	0.6			MIT
4072.33	4073.48	24549.0	TB	0.5			MIT
4071.19	4072.34	24555.9	TB	0.8			MIT
4070.72	4071.87	24558.7	TB	0.6			MIT
4070.57	4071.72	24559.6	TB	1.1W	0.0		M
4070.11	4071.26	24562.4	TB	1.0			MIT
4069.29	4070.44	24567.4	TB	0.5			MIT
4067.36	4068.51	24579.0	TB	0.6			MIT
4066.21	4067.36	24586.0	TB	1.6	0.5		M
4064.83	4065.98	24594.3	TB	0.6	0.0		MIT
4064.53	4065.68	24596.1	TB	0.3			MIT
4063.90	4065.05	24600.0	TB	1.1			MIT
4062.78	4063.93	24606.7	TB	1.0	0.0		MIT
4062.20	4063.35	24610.2	TB	0.7			MIT
4061.57	4062.72	24614.1	TB	1.6	0.3		MIT
4060.86	4062.01	24618.4	TB	1.4	0.5		MIT
4060.38	4061.53	24621.3	TB	1.3	0.0		MIT
4059.40	4060.55	24627.2	TB	0.5	0.0		MIT
4058.81	4059.96	24630.8	TB	0.5O			MIT
4058.44	4059.59	24633.1	TB	0.7	0.0		MIT
4058.02	4059.17	24635.6	TB	0.3			MIT
4057.68	4058.83	24637.7	TB	0.3			MIT
4057.06	4058.21	24641.4	TB	0.6	0.0		M
4056.11	4057.26	24647.2	TB	0.3			MIT
4055.32	4056.47	24652.0	TB	0.7			MIT
4055.20	4056.35	24652.7	TB	0.5			M
4054.96	4056.11	24654.2	TB	0.3H			MIT
4054.60	4055.75	24656.4	TB	0.3			MIT
4054.12	4055.27	24659.3	TB	1.0	0.0		M
4054.00	4055.15	24660.0	TB	0.7			M
4053.33	4054.47	24664.1	TB	0.6	0.0		M
4052.86	4054.00	24667.0	TB	1.6W	0.3		M
4052.41	4053.55	24669.7	TB	0.7	0.0		M
4051.85	4052.99	24673.1	TB	1.4	0.3		M
4051.47	4052.61	24675.4	TB	0.9	0.0		MIT
4048.81	4049.95	24691.6	TB	0.7			MIT
4047.16	4048.30	24701.7	TB	1.0			MIT
4045.97	4047.11	24709.0	TB	1.4	0.0		MIT
4045.31	4046.45	24713.0	TB	0.7			MIT
4043.66	4044.80	24723.1	TB	0.9W	0.0W		MIT
4042.33	4043.47	24731.2	TB	1.0			MIT
4041.843	4042.985	24734.20	TB	0.8			MIT
4041.536	4042.678	24736.08	TB	0.6			MIT
4040.938	4042.080	24739.74	TB	0.5			MIT
4040.678	4041.820	24741.33	TB	0.5			MIT
4040.40	4041.54	24743.0	TB	1.0	0.0		MIT
4040.100	4041.241	24744.87	TB	0.9			KN
4039.727	4040.868	24747.16	TB	0.3			MIT
4039.482	4040.623	24748.66	TB	0.8			MIT
4039.20	4040.34	24750.4	TB	0.7			MIT
4038.864	4040.005	24752.44	TB	1.0			KN
4038.24	4039.38	24756.3	TB	0.6	0.0		MIT
4037.16	4038.30	24762.9	TB	0.7			MIT
4036.45	4037.59	24767.2	TB	1.0			MIT
4036.219	4037.359	24768.66	TB	1.0			MIT
4033.04	4034.18	24788.2	TB	2.1	0.7		MIT
4032.705	4033.845	24790.25	TB	0.5			KN
4032.626	4033.765	24790.73	TB	0.6			MIT
4032.282	4033.421	24792.85	TB	1.5			KN
4031.64	4032.78	24796.8	TB	1.7	0.5		MIT

AIR WAVELENGTH (ANGSTRMS)	VACUUM WAVELENGTH (ANGSTRMS)	WAVENUMBER (KAYSERS)	LMENT	LOG ARC	INTENSITY SPK	DIS	REF
4029.97	4031.11	24807.1	TB	0.5			MIT
4029.22	4030.36	24811.7	TB	0.3			MIT
4028.60	4029.74	24815.5	TB	0.5			MIT
4028.31	4029.45	24817.3	TB	0.9	0.0		MIT
4027.95	4029.09	24819.5	TB	0.5			MIT
4027.34	4028.48	24823.3	TB	0.5			MIT
4026.93	4028.07	24825.8	TB	0.3			MIT
4026.41	4027.55	24829.0	TB	0.6			MIT
4025.73	4026.87	24833.2	TB	1.2	0.3		MIT
4025.146	4026.284	24836.80	TB	0.6			KN
4024.781	4025.918	24839.05	TB	0.8			KN
4024.703	4025.840	24839.53	TB	0.6			MIT
4024.44	4025.58	24841.2	TB	0.6			MIT
4024.07	4025.21	24843.4	TB	1.6O	0.0		MIT
4023.716	4024.853	24845.63	TB	0.8			MIT
4022.872	4024.009	24850.84	TB	1.2D			MIT
4022.33	4023.47	24854.2	TB	0.5			MIT
4021.74	4022.88	24857.8	TB	0.3			MIT
4021.12	4022.26	24861.7	TB	1.0	0.0		MIT
4020.736	4021.872	24864.04	TB	0.3			MIT
4020.47	4021.61	24865.7	TB	1.3	0.5		MIT
4019.66	4020.80	24870.7	TB	0.3			MIT
4019.12	4020.26	24874.0	TB	1.6	0.7		MIT
4018.44	4019.58	24878.2	TB	0.3			MIT
4017.83	4018.97	24882.0	TB	0.6	0.0		MIT
4016.87	4018.01	24888.0	TB	0.7			MIT
4016.36	4017.50	24891.1	TB	1.0	0.0		MIT
4016.05	4017.19	24893.1	TB	0.5			MIT
4015.93	4017.07	24893.8	TB	0.9	0.0		MIT
4015.625	4016.760	24895.69	TB	0.6			MIT
4015.50	4016.64	24896.5	TB	1.0W	0.0		MIT
4015.21	4016.34	24898.3	TB	0.5			MIT
4014.09	4015.22	24905.2	TB	0.6			MIT
4013.276	4014.410	24910.26	TB	1.3D			KN
4012.87	4014.00	24912.8	TB	0.8	0.0		MIT
4012.45	4013.58	24915.4	TB	0.7			MIT
4010.869	4012.003	24925.21	TB	0.8			KN
4010.741	4011.875	24926.00	TB	0.7			KN
4010.645	4011.779	24926.60	TB	0.3			MIT
4010.064	4011.198	24930.21	TB	0.9			MIT
4009.54	4010.67	24933.5	TB	1.0	0.0		MIT
4009.27	4010.40	24935.1	TB	0.3			MIT
4009.193	4010.326	24935.63	TB	0.7			KN
4007.75	4008.88	24944.6	TB	0.5			MIT
4005.96	4007.09	24955.7	TB	1.0	0.0		MIT
4005.55	4006.68	24958.3	TB	2.0D	2.1		M
4004.618	4005.750	24964.11	TB	0.3			KN
4004.50	4005.63	24964.9	TB	0.7	0.0		MIT
4003.91	4005.04	24968.5	TB	0.7	0.0		MIT
4003.76	4004.89	24969.5	TB	0.9	0.0		MIT
4002.58	4003.71	24976.8	TB	1.7	0.7		MIT
4002.18	4003.31	24979.3	TB	1.6W	0.3		MIT
4001.280	4002.411	24984.94	TB	0.7			MIT
4000.46	4001.59	24990.1	TB	1.2	0.3		MIT
4000.01	4001.14	24992.9	TB	0.7	0.0		MIT
3999.407	4000.538	24996.64	TB	1.1	1.2		MIT
3998.892	4000.023	24999.86	TB	0.7			MIT
3998.404	3999.535	25002.91	TB	0.8			MIT
3997.406	3998.536	25009.15	TB	1.0			MIT
3996.696	3997.826	25013.59	TB	1.0	0.5		MIT
3996.593	3997.723	25014.24	TB	0.5			MIT
3995.786	3996.916	25019.29	TB	0.9			MIT
3995.428	3996.558	25021.53	TB	0.3			MIT
3995.142	3996.272	25023.32	TB	0.5			MIT
3994.040	3995.169	25030.23	TB	0.6			MIT
3993.549	3994.678	25033.30	TB	1.5D	0.9		MIT
3992.206	3993.335	25041.73	TB	0.6			MIT
3991.590	3992.719	25045.59	TB	1.0D			MIT
3990.631	3991.760	25051.61	TB	1.1	0.5		MIT
3990.221	3991.349	25054.18	TB	0.5			MIT

AIR WAVELENGTH (ANGSTRMS)	VACUUM WAVELENGTH (ANGSTRMS)	WAVENUMBER (KAYSERS)	LMENT	LOG ARC	INTENSITY SPK	DIS	REF
3989.955	3991.083	25055.85	TB	0.5	0.5		MIT
3989.506	3990.634	25058.67	TB	0.7			MIT
3987.804	3988.932	25069.37	TB	0.6			MIT
3987.671	3988.799	25070.20	TB	0.8			MIT
3987.286	3988.414	25072.62	TB	0.5	0.5		MIT
3986.945	3988.073	25074.77	TB	0.5			MIT
3986.347	3987.474	25078.53	TB	1.0	0.9		MIT
3985.454	3986.581	25084.15	TB	0.5			MIT
3985.057	3986.184	25086.65	TB	0.6	0.5		MIT
3984.815	3985.942	25088.17	TB	0.7			MIT
3984.051	3985.178	25092.98	TB	0.9	0.9		MIT
3983.845	3984.972	25094.28	TB	0.9	0.9		MIT
3982.938	3984.065	25100.00	TB	0.3			MIT
3981.879	3983.005	25106.67	TB	1.9	2.3		MIT
3981.151	3982.277	25111.26	TB	1.4	0.9		MIT
3980.281	3981.407	25116.75	TB	0.6	0.5		MIT
3977.738	3978.863	25132.81	TB	0.3			MIT
3976.821	3977.946	25138.60	TB	2.2	2.3		MIT
3976.002	3977.127	25143.78	TB	0.5O			MIT
3974.666	3975.790	25152.23	TB	0.6			MIT
3974.272	3975.396	25154.73	TB	1.2	0.9		MIT
3972.911	3974.035	25163.34	TB	0.6			MIT
3972.051	3973.175	25168.79	TB	1.1	0.9		MIT
3971.770	3972.894	25170.57	TB	1.0			MIT
3970.828	3971.951	25176.54	TB	0.6			MIT
3969.920	3971.043	25182.30	TB	0.5			MIT
3968.73	3969.85	25189.9	TB	0.3			ED
3968.146	3969.269	25193.56	TB	0.3			MIT
3967.649	3968.772	25196.71	TB	0.8			MIT
3967.211	3968.333	25199.50	TB	1.3			MIT
3966.499	3967.621	25204.02	TB	0.5			MIT
3966.250	3967.372	25205.60	TB	0.5			MIT
3965.943	3967.065	25207.55	TB	1.1	0.5		MIT
3965.098	3966.220	25212.92	TB	0.8			MIT
3964.305	3965.427	25217.97	TB	0.3			MIT
3963.401	3964.522	25223.72	TB	0.3			MIT
3962.941	3964.062	25226.65	TB	0.6			MIT
3962.609	3963.730	25228.76	TB	0.7			MIT
3961.970	3963.091	25232.83	TB	0.5			MIT
3960.695	3961.816	25240.95	TB	0.7			MIT
3960.296	3961.417	25243.50	TB	0.3			MIT
3960.095	3961.216	25244.78	TB	0.5	0.5		MIT
3958.648	3959.768	25254.00	TB	0.7			MIT
3958.359	3959.479	25255.85	TB	2.0W	1.2		MIT
3957.975	3959.095	25258.30	TB	1.8D	1.2		MIT
3957.365	3958.485	25262.19	TB	1.0			MIT
3956.170	3957.290	25269.82	TB	1.0			MIT
3955.805	3956.924	25272.15	TB	0.3			MIT
3955.630	3956.749	25273.27	TB	0.5			MIT
3955.073	3956.192	25276.83	TB	0.5			MIT
3954.792	3955.911	25278.63	TB	0.5			MIT
3954.512	3955.631	25280.42	TB	0.9			MIT
3954.136	3955.255	25282.82	TB	0.5			MIT
3954.050	3955.169	25283.37	TB	1.0	0.5		MIT
3953.015	3954.134	25289.99	TB	0.5			MIT
3952.620	3953.739	25292.52	TB	0.3W			MIT
3952.046	3953.164	25296.19	TB	0.3			MIT
3951.875	3952.993	25297.28	TB	0.8			MIT
3951.705	3952.823	25298.37	TB	0.6			MIT
3950.779	3951.897	25304.30	TB	0.7			MIT
3950.416	3951.534	25306.63	TB	1.3D			MIT
3950.126	3951.244	25308.49	TB	0.6			MIT
3949.868	3950.986	25310.14	TB	0.8			MIT
3949.509	3950.627	25312.44	TB	1.0	0.5		MIT
3949.39	3950.51	25313.2	TB	0.6			ED
3948.35	3949.47	25319.9	TB	1.3	1.2		ED
3947.51	3948.63	25325.3	TB	0.3			ED
3947.25	3948.37	25326.9	TB	0.7			ED
3946.87	3947.99	25329.4	TB	2.2	1.5		ED
3945.39	3946.51	25338.9	TB	0.9			ED

AIR WAVELENGTH (ANGSTRMS)	VACUUM WAVELENGTH (ANGSTRMS)	WAVENUMBER (KAYSERS)	LMENT	LOG ARC	INTENSITY SPK	DIS	REF
3944.20	3945.32	25346.5	TB	0.8			ED
3943.66	3944.78	25350.0	TB	0.9			ED
3942.94	3944.06	25354.6	TB	0.8			ED
3942.20	3943.32	25359.4	TB	1.2	0.9		ED
3941.35	3942.47	25364.8	TB	0.6			E
3941.16	3942.28	25366.1	TB	1.2	0.9		MIT
3940.10	3941.22	25372.9	TB	0.3			ED
3939.60	3940.72	25376.1	TB	2.3	2.3		ED
3938.16	3939.27	25385.4	TB	0.5			ED
3938.04	3939.15	25386.2	TB	0.6			ED
3937.636	3938.751	25388.76	TB	1.2			KN
3937.15	3938.26	25391.9	TB	0.7			ED
3935.25	3936.36	25404.1	TB	1.7	0.9		ED
3934.40	3935.51	25409.6	TB	1.0			ED
3933.905	3935.019	25412.84	TB	0.6			KN
3933.469	3934.583	25415.66	TB	0.8			KN
3932.55	3933.66	25421.6	TB	0.5			ED
3932.366	3933.479	25422.78	TB	1.1			KN
3930.76	3931.87	25433.2	TB	0.9	0.5		ED
3929.88	3930.99	25438.9	TB	1.0			ED
3929.75	3930.86	25439.7	TB	0.7			ED
3929.42	3930.53	25441.9	TB	0.3			ED
3928.99	3930.10	25444.6	TB	0.5			ED
3927.35	3928.46	25455.2	TB	0.3			ED
3927.15	3928.26	25456.6	TB	0.6			ED
3925.45	3926.56	25467.6	TB	2.2	2.3		ED
3924.81	3925.92	25471.7	TB	1.0	0.5		ED
3924.40	3925.51	25474.4	TB	0.6	0.5		ED
3923.33	3924.44	25481.3	TB	0.8	0.5		ED
3922.74	3923.85	25485.2	TB	1.7	0.9		ED
3922.09	3923.20	25489.4	TB	1.3	0.9		ED
3921.76	3922.87	25491.5	TB	0.5			ED
3921.264	3922.374	25494.76	TB	0.6			KN
3920.72	3921.83	25498.3	TB	1.0	0.5		ED
3919.54	3920.65	25506.0	TB	1.6	1.2		ED
3919.00	3920.11	25509.5	TB	0.8			ED
3918.82	3919.93	25510.7	TB	0.9			ED
3917.32	3918.43	25520.4	TB	0.8	0.5		ED
3916.94	3918.05	25522.9	TB	0.6			ED
3916.64	3917.75	25524.9	TB	0.6	0.5		ED
3915.45	3916.56	25532.6	TB	1.5	0.5		ED
3914.73	3915.84	25537.3	TB	0.3			ED
3914.59	3915.70	25538.2	TB	0.7			ED
3913.79	3914.90	25543.4	TB	0.3			ED
3913.45	3914.56	25545.7	TB	0.9			ED
3912.78	3913.89	25550.0	TB	0.7	0.9		ED
3912.25	3913.36	25553.5	TB	0.7			ED
3910.85	3911.96	25562.6	TB	0.7			ED
3910.57	3911.68	25564.5	TB	0.3			ED
3910.40	3911.51	25565.6	TB	0.7			ED
3910.13	3911.24	25567.4	TB	0.6	0.5		ED
3909.54	3910.65	25571.2	TB	1.2	0.9		ED
3909.150	3910.257	25573.76	TB	1.2	0.5		KN
3908.66	3909.77	25577.0	TB	0.6	0.5		ED
3908.076	3909.183	25580.79	TB	1.3			KN
3907.912	3909.019	25581.87	TB	0.7			KN
3907.79	3908.90	25582.7	TB	0.5	0.5		ED
3907.65	3908.76	25583.6	TB	0.5			ED
3906.53	3907.64	25590.9	TB	0.6			ED
3905.83	3906.94	25595.5	TB	0.3			ED
3905.61	3906.72	25596.9	TB	1.0			ED
3904.186	3905.292	25606.28	TB	0.6			KN
3903.11	3904.22	25613.3	TB	0.5			ED
3902.35	3903.46	25618.3	TB	1.0			ED
3901.98	3903.09	25620.8	TB	1.2	0.5		ED
3901.68	3902.79	25622.7	TB	0.8	0.5		ED
3901.35	3902.46	25624.9	TB	1.7	0.9		ED
3900.72	3901.83	25629.0	TB	0.6			ED
3900.40	3901.50	25631.1	TB	0.3			ED
3899.54	3900.64	25636.8	TB	1.2	0.9		ED

AIR WAVELENGTH (ANGSTRMS)	VACUUM WAVELENGTH (ANGSTRMS)	WAVENUMBER (KAYSERS)	LMENT	LOG ARC	INTENSITY SPK	INTENSITY DIS	REF
3899.19	3900.29	25639.1	TB	2.3	2.0		ED
3898.734	3899.839	25642.09	TB	0.6			KN
3897.852	3898.956	25647.89	TB	1.2	0.9		KN
3897.39	3898.49	25650.9	TB	0.6			ED
3897.25	3898.35	25651.9	TB	0.6			ED
3896.60	3897.70	25656.1	TB	1.4	1.2		ED
3896.488	3897.592	25656.87	TB	0.6			KN
3896.04	3897.14	25659.8	TB	1.4	0.9		ED
3895.969	3897.073	25660.28	TB	0.5			KN
3895.376	3896.480	25664.19	TB	0.5			KN
3895.065	3896.169	25666.24	TB	0.6			KN
3894.63	3895.73	25669.1	TB	1.3	0.5		ED
3893.70	3894.80	25675.2	TB	0.7			ED
3893.35	3894.45	25677.6	TB	1.2	0.9		ED
3891.76	3892.86	25688.0	TB	0.3	0.5		ED
3890.95	3892.05	25693.4	TB	0.7			ED
3889.85	3890.95	25700.6	TB	1.0	0.5		ED
3888.21	3889.31	25711.5	TB	1.5			ED
3887.88	3888.98	25713.7	TB	1.0			KN
3887.67	3888.77	25715.1	TB	1.4	0.5		ED
3887.49	3888.59	25716.2	TB	0.6			ED
3887.29	3888.39	25717.6	TB	0.3			KN
3886.83	3887.93	25720.6	TB	1.3	0.9		ED
3886.70	3887.80	25721.5	TB	0.5			KN
3886.01	3887.11	25726.1	TB	0.5			ED
3885.12	3886.22	25731.9	TB	0.9			KN
3884.37	3885.47	25736.9	TB	0.9			ED
3881.75	3882.85	25754.3	TB	0.6			ED
3881.29	3882.39	25757.3	TB	0.9			ED
3880.35	3881.45	25763.6	TB	0.6			ED
3879.99	3881.09	25766.0	TB	0.6			ED
3878.21	3879.31	25777.8	TB	1.0			ED
3877.56	3878.66	25782.1	TB	0.8			ED
3876.67	3877.77	25788.0	TB	0.9			ED
3876.47	3877.57	25789.4	TB	0.5			ED
3876.13	3877.23	25791.6	TB	0.9			ED
3875.21	3876.31	25797.7	TB	1.3	0.9		ED
3874.73	3875.83	25800.9	TB	0.7			ED
3874.18	3875.28	25804.6	TB	2.3	2.3		ED
3873.78	3874.88	25807.3	TB	0.9	0.5		ED
3873.00	3874.10	25812.5	TB	1.0			ED
3872.10	3873.20	25818.5	TB	0.3	0.9		ED
3869.75	3870.85	25834.1	TB	1.2	1.2		ED
3868.90	3870.00	25839.8	TB	0.6	0.9		ED
3866.54	3867.64	25855.6	TB	0.8			ED
3855.58	3856.67	25929.1	TB	1.0			ED
3855.38	3856.47	25930.4	TB	1.0			ED
3854.04	3855.13	25939.4	TB	0.3			ED
3851.86	3852.95	25954.1	TB	0.7	0.9		ED
3849.59	3850.68	25969.4	TB	0.6			ED
3848.75	3849.84	25975.1	TB	2.0	2.3		ED
3847.88	3848.97	25981.0	TB	0.9			ED
3845.61	3846.70	25996.3	TB	1.0	1.2		ED
3842.49	3843.58	26017.4	TB	1.6	1.7		ED
3841.77	3842.86	26022.3	TB	0.7			ED
3840.26	3841.35	26032.5	TB	1.7			ED
3839.62	3840.71	26036.9	TB	1.2			ED
3839.18	3840.27	26039.8	TB	0.9			ED
3837.83	3838.92	26049.0	TB	0.9			ED
3837.18	3838.27	26053.4	TB	0.7			ED
3833.40	3834.49	26079.1	TB	0.9	0.5		ED
3832.68	3833.77	26084.0	TB	0.9			ED
3831.85	3832.94	26089.6	TB	1.2			ED
3830.29	3831.38	26100.3	TB	1.5	0.9		ED
3826.74	3827.83	26124.5	TB	0.9			ED
3823.12	3824.20	26149.2	TB	1.2	0.5		ED
3816.90	3817.98	26191.8	TB	0.5			ED
3816.66	3817.74	26193.5	TB	0.7			EX
3816.27	3817.35	26196.2	TB	0.9	0.5		ED
3814.96	3816.04	26205.2	TB	0.9			ED

AIR WAVELENGTH (ANGSTRMS)	VACUUM WAVELENGTH (ANGSTRMS)	WAVENUMBER (KAYSERS)	LMENT	LOG ARC	INTENSITY SPK	INTENSITY DIS	REF
3814.58	3815.66	26207.8	TB	1.2			ED
3813.97	3815.05	26212.0	TB	0.5	0.5		ED
3812.73	3813.81	26220.5	TB	0.9	0.9		ED
3812.06	3813.14	26225.1	TB	1.2			ED
3811.65	3812.73	26227.9	TB	1.2	0.9		ED
3811.34	3812.42	26230.1	TB	1.2			ED
3810.57	3811.65	26235.4	TB	0.9			ED
3809.73	3810.81	26241.1	TB	0.9			ED
3809.13	3810.21	26245.3	TB	0.9			ED
3808.14	3809.22	26252.1	TB	0.9			ED
3807.65	3808.73	26255.5	TB	0.9			ED
3806.85	3807.93	26261.0	TB	1.7	1.7		ED
3806.42	3807.50	26263.9	TB	1.2	0.5		ED
3804.71	3805.79	26275.7	TB	1.2			ED
3803.08	3804.16	26287.0	TB	1.2			ED
3802.17	3803.25	26293.3	TB	1.5	0.5		ED
3801.80	3802.88	26295.9	TB	1.2			ED
3800.04	3801.12	26308.0	TB	0.9			ED
3799.15	3800.23	26314.2	TB	0.9			ED
3798.95	3800.03	26315.6	TB	0.9	0.9		ED
3798.59	3799.67	26318.1	TB	1.2			ED
3797.93	3799.01	26322.7	TB	1.2			ED
3796.98	3798.06	26329.2	TB	1.2			ED
3796.20	3797.28	26334.6	TB	1.2			ED
3794.23	3795.31	26348.3	TB	1.5			ED
3793.55	3794.63	26353.1	TB	1.2	1.2		ED
3792.69	3793.77	26359.0	TB	0.9			ED
3792.18	3793.26	26362.6	TB	1.2	0.9		ED
3790.81	3791.89	26372.1	TB	0.9			ED
3789.92	3791.00	26378.3	TB	1.2			ED
3789.68	3790.76	26380.0	TB	0.9	0.5		ED
3789.00	3790.08	26384.7	TB	0.9	0.5		ED
3788.47	3789.55	26388.4	TB	0.5			ED
3787.62	3788.70	26394.3	TB	0.9	0.9		ED
3787.22	3788.30	26397.1	TB	1.5	0.9		ED
3785.38	3786.46	26409.9	TB	1.2	0.9		ED
3783.85	3784.92	26420.6	TB	0.9			ED
3783.54	3784.61	26422.8	TB	1.2	0.9		ED
3782.62	3783.69	26429.2	TB	0.9			ED
3782.18	3783.25	26432.3	TB	0.9			ED
3781.63	3782.70	26436.1	TB	0.9			ED
3779.22	3780.29	26453.0	TB	1.2	0.9		ED
3777.48	3778.55	26465.2	TB	0.9			ED
3776.49	3777.56	26472.1	TB	2.0	2.0		ED
3776.02	3777.09	26475.4	TB	0.9			ED
3775.26	3776.33	26480.7	TB	1.2	0.9		ED
3774.96	3776.03	26482.8	TB	0.9			ED
3772.18	3773.25	26502.3	TB	0.9			ED
3771.74	3772.81	26505.4	TB	0.9			ED
3770.19	3771.26	26516.3	TB	0.9			ED
3768.35	3769.42	26529.3	TB	0.9	0.9		ED
3767.50	3768.57	26535.3	TB	1.9	0.9		ED
3766.36	3767.43	26543.3	TB	0.9			ED
3765.14	3766.21	26551.9	TB	1.8	2.0		ED
3764.75	3765.82	26554.6	TB	0.9			ED
3763.50	3764.57	26563.5	TB	0.9			ED
3762.74	3763.81	26568.8	TB	0.9			ED
3762.34	3763.41	26571.6	TB	0.5			ED
3761.12	3762.19	26580.3	TB	1.2			ED
3760.13	3761.20	26587.3	TB	1.2	0.5		ED
3759.35	3760.42	26592.8	TB	1.2	0.9		ED
3758.32	3759.39	26600.1	TB	0.5			ED
3757.90	3758.97	26603.1	TB	1.5	1.2		ED
3757.44	3758.51	26606.3	TB	1.5	1.2		ED
3756.48	3757.55	26613.1	TB	0.3			M
3755.24	3756.31	26621.9	TB	1.7	2.0		ED
3753.52	3754.59	26634.1	TB	1.5	0.5		ED
3752.93	3754.00	26638.3	TB	0.5			ED
3751.62	3752.69	26647.6	TB	0.9	0.5		ED
3749.70	3750.77	26661.2	TB	1.2			ED

AIR WAVELENGTH (ANGSTRMS)	VACUUM WAVELENGTH (ANGSTRMS)	WAVENUMBER (KAYSERS)	LMENT	LOG ARC	INTENSITY SPK	INTENSITY DIS	REF
3747.64	3748.71	26675.9	TB	1.5			ED
3747.17	3748.24	26679.2	TB	1.5			ED
3746.54	3747.60	26683.7	TB	0.9	0.9		ED
3745.07	3746.13	26694.2	TB	1.2	0.5		ED
3743.99	3745.05	26701.9	TB	0.9	0.5		ED
3743.09	3744.15	26708.3	TB	1.2	0.9		ED
3742.43	3743.49	26713.0	TB	0.9	0.5		ED
3741.89	3742.95	26716.9	TB	0.9	0.9		ED
3741.58	3742.64	26719.1	TB	1.2	0.9		ED
3741.24	3742.30	26721.5	TB	1.2	0.9		ED
3740.32	3741.38	26728.1	TB	1.2	0.9		ED
3734.94	3736.00	26766.6	TB	0.9			ED
3734.80	3735.86	26767.6	TB	0.9			ED
3732.39	3733.45	26784.9	TB	1.5	1.2		ED
3731.51	3732.57	26791.2	TB	0.9			ED
3730.95	3732.01	26795.2	TB	1.2			ED
3729.91	3730.97	26802.7	TB	1.2	1.2		ED
3728.96	3730.02	26809.5	TB	0.9	0.5		ED
3728.65	3729.71	26811.7	TB	0.9	0.5		ED
3725.32	3726.38	26835.7	TB	0.9	0.5		ED
3724.92	3725.98	26838.6	TB	0.9	0.5		ED
3723.04	3724.10	26852.1	TB	1.2	0.5		ED
3722.62	3723.68	26855.2	TB	0.5	0.9		ED
3720.36	3721.42	26871.5	TB	0.9			ED
3719.45	3720.51	26878.1	TB	1.5	0.9		ED
3718.44	3719.50	26885.4	TB	1.2	0.5		ED
3717.47	3718.53	26892.4	TB	0.9	0.5		ED
3716.43	3717.49	26899.9	TB	1.2	0.5		ED
3716.08	3717.14	26902.4	TB	1.2	0.5		ED
3715.09	3716.15	26909.6	TB	0.9	0.5		ED
3711.74	3712.80	26933.9	TB	2.3	1.5		ED
3709.30	3710.36	26951.6	TB	1.2			ED
3708.75	3709.81	26955.6	TB	0.9	0.9		ED
3707.54	3708.59	26964.4	TB	1.2			ED
3706.34	3707.39	26973.1	TB	1.2			ED
3705.06	3706.11	26982.4	TB	1.2	0.5		ED
3703.92	3704.97	26990.7	TB	1.8	2.0		ED
3702.85	3703.90	26998.5	TB	1.7	2.3		ED
3700.12	3701.17	27018.5	TB	1.5			ED
3699.70	3700.75	27021.5	TB	0.9	0.5		ED
3698.45	3699.50	27030.7	TB	0.9			ED
3698.17	3699.22	27032.7	TB	0.9	0.5		ED
3697.72	3698.77	27036.0	TB	0.9	0.5		ED
3696.85	3697.90	27042.4	TB	1.5	0.9		ED
3696.30	3697.35	27046.4	TB	1.2	0.9		ED
3695.69	3696.74	27050.9	TB	0.9			ED
3694.75	3695.80	27057.7	TB	1.7	0.9		ED
3694.12	3695.17	27062.3	TB	0.9			ED
3693.56	3694.61	27066.4	TB	1.5			ED
3692.95	3694.00	27070.9	TB	1.5	0.9		ED
3692.02	3693.07	27077.7	TB	1.2	0.9		ED
3691.15	3692.20	27084.1	TB	1.7	1.7		ED
3690.44	3691.49	27089.3	TB	0.9			ED
3689.72	3690.77	27094.6	TB	0.9	0.5		ED
3689.12	3690.17	27099.0	TB	1.2	0.9		ED
3688.15	3689.20	27106.1	TB	1.5	1.2		ED
3687.44	3688.49	27111.4	TB	0.5	0.9		ED
3687.15	3688.20	27113.5	TB	1.2	0.5		ED
3684.81	3685.86	27130.7	TB	1.2	0.9		ED
3683.26	3684.31	27142.1	TB	1.2			ED
3682.26	3683.31	27149.5	TB	1.7	1.5		ED
3681.71	3682.76	27153.6	TB	0.9			ED
3679.56	3680.61	27169.4	TB	1.2			ED
3678.78	3679.83	27175.2	TB	1.5	0.5		ED
3677.89	3678.94	27181.8	TB	1.8	0.9		ED
3676.35	3677.40	27193.1	TB	2.0	2.3		ED
3675.78	3676.83	27197.4	TB	1.5	0.9		ED
3675.03	3676.08	27202.9	TB	0.9	0.5		ED
3674.05	3675.10	27210.2	TB	1.2	0.5		ED
3673.73	3674.78	27212.5	TB	0.9	0.5		ED

AIR WAVELENGTH (ANGSTRMS)	VACUUM WAVELENGTH (ANGSTRMS)	WAVENUMBER (KAYSERS)	LMENT	LOG ARC	INTENSITY SPK	INTENSITY DIS	REF
3672.30	3673.35	27223.1	TB	0.9	0.5		ED
3671.42	3672.47	27229.7	TB	0.5	0.5		ED
3670.55	3671.60	27236.1	TB	1.2	0.5		ED
3669.62	3670.67	27243.0	TB	1.5			ED
3669.21	3670.25	27246.1	TB	0.9	0.5		ED
3668.50	3669.54	27251.3	TB	1.2	0.5		ED
3668.07	3669.11	27254.5	TB	1.2	0.5		ED
3667.35	3668.39	27259.9	TB	1.2			ED
3665.60	3666.64	27272.9	TB	1.5	0.5		ED
3664.64	3665.68	27280.0	TB	1.7	0.9		ED
3664.28	3665.32	27282.7	TB	0.9			ED
3663.12	3664.16	27291.4	TB	1.7	1.2		ED
3662.64	3663.68	27294.9	TB	0.9			ED
3662.25	3663.29	27297.8	TB	0.9			ED
3660.75	3661.79	27309.0	TB	1.5			ED
3660.44	3661.48	27311.3	TB	1.2	0.5		ED
3659.45	3660.49	27318.7	TB	1.2	0.5		ED
3658.88	3659.92	27323.0	TB	2.0	2.0		ED
3658.22	3659.26	27327.9	TB	1.2	0.5		ED
3656.77	3657.81	27338.7	TB	0.9	0.5		ED
3656.48	3657.52	27340.9	TB	1.2	0.5		ED
3654.88	3655.92	27352.9	TB	1.8	1.5		ED
3654.30	3655.34	27357.2	TB	0.9			ED
3653.87	3654.91	27360.4	TB	1.2	0.9		ED
3652.97	3654.01	27367.2	TB	1.2	0.9		ED
3652.26	3653.30	27372.5	TB	1.5	0.9		ED
3651.86	3652.90	27375.5	TB	1.5	0.9		ED
3650.93	3651.97	27382.5	TB	0.9	0.9		ED
3650.40	3651.44	27386.5	TB	1.7	2.0		ED
3649.41	3650.45	27393.9	TB	0.9	0.5		ED
3647.75	3648.79	27406.4	TB	1.7			ED
3647.06	3648.10	27411.5	TB	1.5	0.9		ED
3646.46	3647.50	27416.1	TB	0.9	0.5		ED
3645.60	3646.64	27422.5	TB	0.9			ED
3645.38	3646.42	27424.2	TB	1.7	1.2		ED
3644.93	3645.97	27427.6	TB	1.2			ED
3644.13	3645.17	27433.6	TB	1.2	0.5		ED
3643.76	3644.80	27436.4	TB	1.5	0.5		ED
3643.26	3644.30	27440.1	TB	1.2	0.5		ED
3642.68	3643.72	27444.5	TB	1.2	0.9		ED
3641.66	3642.70	27452.2	TB	1.8	1.5		ED
3640.88	3641.92	27458.1	TB	0.9			ED
3639.82	3640.86	27466.1	TB	1.5	0.9		ED
3639.48	3640.52	27468.6	TB	0.9			ED
3638.95	3639.99	27472.6	TB	1.2			ED
3638.46	3639.50	27476.3	TB	1.9	1.7		ED
3637.75	3638.79	27481.7	TB	0.9			ED
3637.36	3638.40	27484.6	TB	0.9			ED
3636.16	3637.20	27493.7	TB	1.2			ED
3635.42	3636.46	27499.3	TB	1.5	0.9		ED
3635.17	3636.21	27501.2	TB	0.9			ED
3634.75	3635.79	27504.4	TB	0.9			ED
3633.66	3634.70	27512.6	TB	0.9			ED
3633.29	3634.33	27515.4	TB	1.5	1.5		ED
3631.46	3632.50	27529.3	TB	1.5			ED
3630.88	3631.92	27533.7	TB	1.2			ED
3630.28	3631.31	27538.2	TB	1.5	0.9		ED
3629.43	3630.46	27544.7	TB	1.2	0.9		ED
3628.68	3629.71	27550.4	TB	1.2	0.5		ED
3628.20	3629.23	27554.0	TB	2.0	1.2		ED
3627.67	3628.70	27558.1	TB	0.9			ED
3627.19	3628.22	27561.7	TB	0.9			ED
3626.87	3627.90	27564.1	TB	1.2	0.5		ED
3626.50	3627.53	27566.9	TB	1.5	1.2		ED
3626.13	3627.16	27569.7	TB	0.9	0.5		ED
3625.54	3626.57	27574.2	TB	1.7	1.2		ED
3624.80	3625.83	27579.9	TB	1.2	0.5		ED
3623.92	3624.95	27586.6	TB	1.5	0.9		ED
3622.11	3623.14	27600.4	TB	1.5	0.5		ED
3621.39	3622.42	27605.8	TB	0.9	0.5		ED

AIR WAVELENGTH (ANGSTRMS)	VACUUM WAVELENGTH (ANGSTRMS)	WAVENUMBER (KAYSERS)	LMENT	LOG ARC	INTENSITY SPK	DIS	REF
3620.29	3621.32	27614.2	TB	1.2			ED
3619.73	3620.76	27618.5	TB	1.2	0.9		ED
3618.18	3619.21	27630.3	TB	0.9	0.5		ED
3617.88	3618.91	27632.6	TB	0.9	0.9		ED
3616.58	3617.61	27642.6	TB	0.9	0.9		ED
3615.66	3616.69	27649.6	TB	1.7	1.2		ED
3614.63	3615.66	27657.5	TB	1.5	0.9		ED
3613.68	3614.71	27664.7	TB	1.2	0.5		ED
3613.06	3614.09	27669.5	TB	1.2	0.5		ED
3612.31	3613.34	27675.2	TB	0.9			ED
3611.41	3612.44	27682.1	TB	0.9	0.9		ED
3611.33	3612.36	27682.7	TB	1.7	0.9		ED
3609.88	3610.91	27693.9	TB	0.9			ED
3609.55	3610.58	27696.4	TB	1.2			ED
3609.06	3610.09	27700.1	TB	0.9			ED
3608.25	3609.28	27706.4	TB	1.2			ED
3607.86	3608.89	27709.4	TB	0.9			ED
3607.54	3608.57	27711.8	TB	0.9			ED
3606.12	3607.15	27722.7	TB	1.5	1.2	D	ED
3604.90	3605.93	27732.1	TB	0.5	1.2		ED
3604.15	3605.18	27737.9	TB	0.9			ED
3602.93	3603.96	27747.3	TB	1.2			ED
3602.51	3603.54	27750.5	TB	1.2	0.5		ED
3601.75	3602.78	27756.4	TB	1.2	0.5		ED
3601.50	3602.53	27758.3	TB	0.9			ED
3600.80	3601.83	27763.7	TB	0.9			ED
3600.44	3601.47	27766.5	TB	0.9	1.7		ED
3600.04	3601.07	27769.6	TB	1.2	0.5		ED
3598.96	3599.99	27777.9	TB	0.9	0.5		ED
3598.06	3599.09	27784.8	TB	1.2	0.9		ED
3597.80	3598.83	27786.8	TB	0.9			ED
3596.80	3597.83	27794.6	TB	0.9			ED
3596.38	3597.41	27797.8	TB	1.7	1.2		ED
3594.98	3596.01	27808.6	TB	1.2			ED
3594.65	3595.68	27811.2	TB	1.2			ED
3594.25	3595.28	27814.3	TB	1.2	0.9		ED
3593.75	3594.78	27818.1	TB	1.5	0.9		ED
3593.10	3594.13	27823.2	TB	1.2	0.9		ED
3591.66	3592.68	27834.3	TB	1.2	0.5		ED
3591.39	3592.41	27836.4	TB	0.9	0.5		ED
3590.46	3591.48	27843.6	TB	1.2			ED
3587.76	3588.78	27864.6	TB	1.5			ED
3587.44	3588.46	27867.1	TB	1.2	1.2		ED
3585.78	3586.80	27880.0	TB	0.9	0.5		ED
3585.03	3586.05	27885.8	TB	1.2	1.7		ED
3582.63	3583.65	27904.5	TB	0.9			ED
3580.63	3581.65	27920.1	TB	0.9			ED
3579.98	3581.00	27925.1	TB	0.9	0.5		ED
3579.20	3580.22	27931.2	TB	1.7	1.7		ED
3578.70	3579.72	27935.1	TB	0.9			ED
3578.40	3579.42	27937.5	TB	0.9			ED
3578.00	3579.02	27940.6	TB	0.9			ED
3577.08	3578.10	27947.8	TB	0.9	0.5		ED
3576.83	3577.85	27949.7	TB	1.2	0.5		ED
3575.90	3576.92	27957.0	TB	1.2	0.5		ED
3574.13	3575.15	27970.9	TB	0.9			ED
3573.57	3574.59	27975.2	TB	1.2			ED
3572.07	3573.09	27987.0	TB	1.5	0.9		ED
3571.35	3572.37	27992.6	TB	0.9			ED
3571.03	3572.05	27995.1	TB	1.2			ED
3570.34	3571.36	28000.5	TB	0.9			ED
3570.07	3571.09	28002.7	TB	0.9			ED
3568.98	3570.00	28011.2	TB	1.2	1.2		ED
3568.51	3569.53	28014.9	TB	1.7	1.7		ED
3567.86	3568.88	28020.0	TB	0.9			ED
3567.35	3568.37	28024.0	TB	1.2	1.2		ED
3565.74	3566.76	28036.7	TB	1.2	1.2		ED
3563.74	3564.76	28052.4	TB	0.9			ED
3562.90	3563.92	28059.0	TB	1.5	0.9		ED
3562.52	3563.54	28062.0	TB	0.9			ED

AIR WAVELENGTH (ANGSTRMS)	VACUUM WAVELENGTH (ANGSTRMS)	WAVENUMBER (KAYSERS)	LMENT	LOG ARC	INTENSITY SPK	DIS	REF
3561.74	3562.76	28068.1	TB	2.3	2.3		ED
3559.76	3560.78	28083.8	TB	1.2	0.9		ED
3559.39	3560.41	28086.7	TB	0.9	0.9		ED
3558.77	3559.79	28091.6	TB	1.2	0.9		ED
3556.67	3557.69	28108.2	TB	1.2	0.5		ED
3555.70	3556.72	28115.8	TB	0.9			ED
3555.29	3556.31	28119.1	TB	1.2	0.5		ED
3553.56	3554.58	28132.8	TB	0.9			ED
3552.92	3553.93	28137.8	TB	0.9			ED
3551.96	3552.97	28145.4	TB	1.2			ED
3551.03	3552.04	28152.8	TB	1.5	0.5		ED
3548.82	3549.83	28170.3	TB	1.2	0.5		ED
3546.52	3547.53	28188.6	TB	1.7	0.9		ED
3546.05	3547.06	28192.3	TB	1.2	0.5		ED
3545.40	3546.41	28197.5	TB	0.9	0.5		ED
3544.93	3545.94	28201.2	TB	1.2			ED
3544.36	3545.37	28205.8	TB	0.9	0.5		ED
3543.86	3544.87	28209.8	TB	1.7	1.2		ED
3543.23	3544.24	28214.8	TB	1.5	0.9		ED
3541.75	3542.76	28226.6	TB	0.9	0.5		ED
3540.24	3541.25	28238.6	TB	1.7	1.7		ED
3539.81	3540.82	28242.0	TB	1.2	0.5		ED
3538.90	3539.91	28249.3	TB	1.2	0.5		ED
3538.50	3539.51	28252.5	TB	0.5	0.5		ED
3537.94	3538.95	28257.0	TB	1.2	1.2		ED
3537.11	3538.12	28263.6	TB	1.2	0.5		ED
3536.62	3537.63	28267.5	TB	1.5	0.5		ED
3536.32	3537.33	28269.9	TB	1.2	0.9		ED
3534.21	3535.22	28286.8	TB	0.9	0.5		ED
3533.86	3534.87	28289.6	TB	1.5	0.9		ED
3533.60	3534.61	28291.7	TB	1.2			ED
3532.70	3533.71	28298.9	TB	1.5	0.9		ED
3531.70	3532.71	28306.9	TB	1.2	1.7		ED
3530.64	3531.65	28315.4	TB	0.9			ED
3530.38	3531.39	28317.5	TB	1.2			ED
3529.76	3530.77	28322.4	TB	1.2	0.5		ED
3527.45	3528.46	28341.0	TB	0.9	0.5		ED
3526.73	3527.74	28346.8	TB	1.2			ED
3525.61	3526.62	28355.8	TB	1.7	0.9		ED
3525.13	3526.14	28359.6	TB	0.9	0.9		ED
3523.66	3524.67	28371.5	TB	1.5	1.7		ED
3523.20	3524.21	28375.2	TB	0.9			ED
3521.82	3522.83	28386.3	TB	0.9			ED
3520.79	3521.80	28394.6	TB	1.2	0.9		ED
3519.76	3520.77	28402.9	TB	1.7	1.2		ED
3518.96	3519.97	28409.4	TB	1.2			ED
3518.38	3519.39	28414.1	TB	0.9			ED
3517.81	3518.82	28418.6	TB	0.9			ED
3516.64	3517.65	28428.1	TB	1.2			ED
3516.14	3517.15	28432.1	TB	1.2	0.5		ED
3515.44	3516.45	28437.8	TB	1.2	0.5		ED
3515.04	3516.05	28441.1	TB	1.5	0.9		ED
3514.18	3515.18	28448.0	TB	0.9			ED
3513.86	3514.86	28450.6	TB	1.5			ED
3512.60	3513.60	28460.8	TB	1.2	0.5		ED
3511.04	3512.04	28473.4	TB	1.2	0.5		ED
3510.10	3511.10	28481.1	TB	1.7	0.9		ED
3509.76	3510.76	28483.8	TB	0.9			ED
3509.17	3510.17	28488.6	TB	2.3	2.3		ED
3508.60	3509.60	28493.2	TB	0.9			ED
3507.45	3508.45	28502.6	TB	1.7	0.9		ED
3507.14	3508.14	28505.1	TB	1.2			ED
3505.90	3506.90	28515.2	TB	1.5	0.9		ED
3505.09	3506.09	28521.8	TB	1.2	0.5		ED
3504.04	3505.04	28530.3	TB	1.2	0.5		ED
3502.49	3503.49	28542.9	TB	1.2			ED
3502.10	3503.10	28546.1	TB	0.9			ED
3501.19	3502.19	28553.6	TB	0.9			ED
3500.84	3501.84	28556.4	TB	1.8	1.2		ED
3500.27	3501.27	28561.1	TB	1.7			ED

AIR WAVELENGTH (ANGSTRMS)	VACUUM WAVELENGTH (ANGSTRMS)	WAVENUMBER (KAYSERS)	LMENT	LOG ARC	INTENSITY SPK	INTENSITY DIS	REF
3500.02	3501.02	28563.1	TB	0.9	0.5		ED
3499.34	3500.34	28568.6	TB	1.2			ED
3498.73	3499.73	28573.6	TB	0.9			ED
3498.06	3499.06	28579.1	TB	1.2			ED
3496.20	3497.20	28594.3	TB	1.2			ED
3495.83	3496.83	28597.3	TB	0.9			ED
3495.36	3496.36	28601.2	TB	1.5	0.9		ED
3494.93	3495.93	28604.7	TB	0.9			ED
3494.44	3495.44	28608.7	TB	1.5	0.9		ED
3494.21	3495.21	28610.6	TB	1.2	0.9		ED
3493.90	3494.90	28613.1	TB	0.9	0.5		ED
3493.27	3494.27	28618.3	TB	0.9			ED
3492.99	3493.99	28620.6	TB	1.2			ED
3492.96	3493.96	28620.8	TB	1.2			ED
3492.56	3493.56	28624.1	TB	1.2	0.9		ED
3491.24	3492.24	28634.9	TB	1.2	0.5		ED
3490.29	3491.29	28642.7	TB	0.9			ED
3489.78	3490.78	28646.9	TB	1.2	0.5		ED
3489.51	3490.51	28649.1	TB	1.2	0.5		ED
3489.02	3490.02	28653.1	TB	0.9			ED
3488.77	3489.77	28655.2	TB	0.9			ED
3488.13	3489.13	28660.4	TB	0.9			ED
3487.62	3488.62	28664.6	TB	1.2	0.5		ED
3487.28	3488.28	28667.4	TB	1.2	0.5		ED
3486.95	3487.95	28670.1	TB	0.9	0.5		ED
3486.24	3487.24	28676.0	TB	0.9			ED
3485.56	3486.56	28681.6	TB	1.2			ED
3484.70	3485.70	28688.7	TB	0.9	0.5		ED
3483.69	3484.69	28697.0	TB	1.5	0.9		ED
3483.04	3484.04	28702.3	TB	1.2	0.9		ED
3482.80	3483.80	28704.3	TB	0.9			ED
3480.44	3481.44	28723.8	TB	1.2	0.5		ED
3480.17	3481.17	28726.0	TB	1.2	0.9		ED
3479.29	3480.29	28733.3	TB	0.9	0.5		ED
3477.24	3478.24	28750.2	TB	0.9			ED
3475.72	3476.72	28762.8	TB	1.2			ED
3475.33	3476.33	28766.0	TB	0.9			ED
3474.98	3475.97	28768.9	TB	0.9			ED
3474.67	3475.66	28771.5	TB	0.9			ED
3473.79	3474.78	28778.8	TB	1.2	0.5		ED
3473.13	3474.12	28784.2	TB	0.9	0.5		ED
3472.82	3473.81	28786.8	TB	1.7	1.2		ED
3472.37	3473.36	28790.5	TB	1.2	0.9		ED
3471.73	3472.72	28795.8	TB	1.2	0.9		ED
3470.36	3471.35	28807.2	TB	1.2	0.5		ED
3469.85	3470.84	28811.4	TB	1.5	0.5		ED
3468.99	3469.98	28818.6	TB	1.2			ED
3468.42	3469.41	28823.3	TB	0.9			ED
3468.03	3469.02	28826.6	TB	1.7	1.2		ED
3467.26	3468.25	28833.0	TB	0.5	0.5		ED
3466.92	3467.91	28835.8	TB	0.9			ED
3466.57	3467.56	28838.7	TB	0.9			ED
3465.98	3466.97	28843.6	TB	1.5			ED
3464.63	3465.62	28854.9	TB	0.9	0.5		ED
3462.97	3463.96	28868.7	TB	1.2	0.5		ED
3462.51	3463.50	28872.5	TB	0.9	0.5		ED
3461.00	3461.99	28885.1	TB	1.2	0.9		ED
3460.38	3461.37	28890.3	TB	1.2	0.9		ED
3459.87	3460.86	28894.5	TB	0.9			ED
3457.55	3458.54	28913.9	TB	0.9			ED
3457.03	3458.02	28918.3	TB	1.7	0.9		ED
3456.83	3457.82	28919.9	TB	1.5			ED
3455.99	3456.98	28927.0	TB	0.9			ED
3455.35	3456.34	28932.3	TB	1.2	0.5		ED
3454.06	3455.05	28943.1	TB	1.9	1.5		ED
3453.46	3454.45	28948.2	TB	1.2	0.5		ED
3452.37	3453.36	28957.3	TB	1.2	0.5		ED
3451.70	3452.69	28962.9	TB	0.9			ED
3450.30	3451.29	28974.7	TB	1.2	0.5		ED
3449.46	3450.45	28981.7	TB	1.5	0.9		ED

AIR WAVELENGTH (ANGSTRMS)	VACUUM WAVELENGTH (ANGSTRMS)	WAVENUMBER (KAYSERS)	LMENT	LOG ARC	INTENSITY SPK	INTENSITY DIS	REF
3448.86	3449.85	28986.8	TB	0.9			ED
3447.32	3448.31	28999.7	TB	0.9			ED
3446.40	3447.39	29007.5	TB	1.7			ED
3444.91	3445.90	29020.0	TB	0.9			ED
3444.58	3445.57	29022.8	TB	1.2	0.9		ED
3443.22	3444.21	29034.3	TB	0.9			ED
3442.67	3443.66	29038.9	TB	1.2			ED
3442.36	3443.35	29041.5	TB	0.9			ED
3441.68	3442.67	29047.2	TB	1.2	0.5		ED
3441.06	3442.05	29052.5	TB	0.9			ED
3440.37	3441.36	29058.3	TB	1.5	0.9		ED
3439.72	3440.71	29063.8	TB	1.2	0.9		ED
3439.05	3440.04	29069.5	TB	1.2			ED
3438.57	3439.56	29073.5	TE	1.2	0.9		ED
3436.97	3437.96	29087.1	TB	1.2			ED
3435.53	3436.51	29099.2	TB	0.9	0.5		ED
3434.92	3435.90	29104.4	TB	1.5			ED
3434.54	3435.52	29107.6	TB	1.2	0.5		ED
3433.26	3434.24	29118.5	TB	1.2	0.9		ED
3432.90	3433.88	29121.5	TB	1.2	0.9		ED
3432.35	3433.33	29126.2	TB	1.2	0.5		ED
3431.86	3432.84	29130.4	TB	1.2			ED
3431.39	3432.37	29134.4	TB	0.9			ED
3430.61	3431.59	29141.0	TB	1.2	0.9		ED
3429.09	3430.07	29153.9	TB	0.9	0.5		ED
3428.71	3429.69	29157.1	TB	0.9	0.5		ED
3426.37	3427.35	29177.0	TB	0.9			ED
3425.92	3426.90	29180.9	TB	0.9	0.5		ED
3425.42	3426.40	29185.1	TB	0.9			ED
3424.35	3425.33	29194.2	TB	0.9	0.5		ED
3423.96	3424.94	29197.6	TB	0.9	0.9		ED
3423.05	3424.03	29205.3	TB	1.2			ED
3422.44	3423.42	29210.5	TB	0.5	1.2		ED
3420.34	3421.32	29228.5	TB	1.7	1.2		ED
3419.54	3420.52	29235.3	TB	1.2	0.9		ED
3418.95	3419.93	29240.4	TB	1.2	0.9		ED
3417.91	3418.89	29249.3	TB	0.9			ED
3417.72	3418.70	29250.9	TB	0.9	0.5		ED
3416.24	3417.22	29263.6	TB	1.2	0.9		ED
3415.43	3416.41	29270.5	TB	0.9			ED
3415.12	3416.10	29273.1	TB	0.9			ED
3413.76	3414.74	29284.8	TB	1.7	1.5		ED
3411.73	3412.71	29302.2	TB	0.9			ED
3411.20	3412.18	29306.8	TB	0.9			ED
3410.71	3411.69	29311.0	TB	1.2	0.5		ED
3410.40	3411.38	29313.7	TB	1.2	0.9		ED
3409.94	3410.92	29317.6	TB	1.2	0.5		ED
3409.27	3410.25	29323.4	TB	0.9			ED
3408.86	3409.84	29326.9	TB	0.9	0.5		ED
3407.83	3408.81	29335.8	TB	0.5	0.9		ED
3407.10	3408.08	29342.1	TB	0.9	0.5		ED
3406.42	3407.40	29347.9	TB	0.9			ED
3406.01	3406.99	29351.4	TB	1.2	0.9		ED
3404.71	3405.69	29362.6	TB	0.5	0.9		ED
3404.24	3405.22	29366.7	TB	1.2	0.5		ED
3403.66	3404.64	29371.7	TB	0.9			ED
3402.78	3403.76	29379.3	TB	0.9	0.5		ED
3402.33	3403.31	29383.2	TB	1.5	1.2		ED
3400.86	3401.84	29395.9	TB	1.5	0.9		ED
3400.53	3401.51	29398.7	TB	1.2	0.9		ED
3399.97	3400.95	29403.6	TB	1.2	0.5		ED
3399.10	3400.08	29411.1	TB	1.2	0.9		ED
3398.35	3399.33	29417.6	TB	1.5	0.9		ED
3397.60	3398.58	29424.1	TB	1.2			ED
3397.21	3398.18	29427.5	TB	1.2			ED
3394.77	3395.74	29448.6	TB	1.2	0.5		ED
3393.58	3394.55	29458.9	TB	1.5	0.9		ED
3393.01	3393.98	29463.9	TB	1.2			ED
3392.01	3392.98	29472.6	TB	1.2	0.9		ED
3391.72	3392.69	29475.1	TB	1.2	0.5		ED

AIR WAVELENGTH (ANGSTRMS)	VACUUM WAVELENGTH (ANGSTRMS)	WAVENUMBER (KAYSERS)	LMENT	LOG ARC	INTENSITY SPK	INTENSITY DIS	REF
3390.60	3391.57	29484.8	TB	1.2	0.9		ED
3390.02	3390.99	29489.9	TB	1.2	0.5		ED
3389.59	3390.56	29493.6	TB	0.9			ED
3388.63	3389.60	29502.0	TB	1.2			ED
3388.37	3389.34	29504.2	TB	0.9			ED
3387.63	3388.60	29510.7	TB	0.9			ED
3386.49	3387.46	29520.6	TB	1.2	0.9		ED
3385.41	3386.38	29530.0	TB	0.9			ED
3385.03	3386.00	29533.4	TB	0.9	0.9		ED
3383.78	3384.75	29544.3	TB	0.9			ED
3382.80	3383.77	29552.8	TB	1.2	0.9		ED
3381.60	3382.57	29563.3	TB	0.9	0.5		ED
3380.89	3381.86	29569.5	TB	0.9			ED
3380.60	3381.57	29572.1	TB	1.2	0.5		ED
3380.15	3381.12	29576.0	TB	0.5			ED
3379.52	3380.49	29581.5	TB	0.9			ED
3379.15	3380.12	29584.7	TB	0.9			ED
3378.86	3379.83	29587.3	TB	1.2	0.9		ED
3377.10	3378.07	29602.7	TB	0.9	0.5		ED
3376.66	3377.63	29606.6	TB	1.2	0.5		ED
3376.36	3377.33	29609.2	TB	0.9	0.5		ED
3376.02	3376.99	29612.2	TB	0.9			ED
3375.03	3376.00	29620.9	TB	1.5	1.2		ED
3374.41	3375.38	29626.3	TB	1.2	0.9		ED
3373.54	3374.51	29633.9	TB	0.9			ED
3372.72	3373.69	29641.1	TB	1.7			ED
3372.36	3373.33	29644.3	TB	1.2	1.2		ED
3371.50	3372.47	29651.9	TB	1.2	0.9		ED
3370.14	3371.11	29663.8	TB	1.2	0.5		ED
3369.43	3370.40	29670.1	TB	0.9			ED
3368.55	3369.52	29677.8	TB	0.9			ED
3367.66	3368.63	29685.7	TB	0.9	0.9		ED
3367.18	3368.15	29689.9	TB	1.2	0.5		ED
3365.29	3366.26	29706.6	TB	1.2	0.5		ED
3364.93	3365.90	29709.8	TB	1.5	1.2		ED
3364.24	3365.21	29715.9	TB	1.2	0.5		ED
3361.24	3362.21	29742.4	TB	0.5	0.9		ED
3360.29	3361.26	29750.8	TB	0.9	0.5		ED
3359.86	3360.83	29754.6	TB	0.9	0.5		ED
3359.30	3360.27	29759.6	TB	0.9	0.5		ED
3358.38	3359.35	29767.7	TB	0.9			ED
3357.68	3358.64	29773.9	TB	0.9			ED
3357.37	3358.33	29776.7	TB	1.2	0.5		ED
3356.59	3357.55	29783.6	TB	0.9			ED
3354.16	3355.12	29805.2	TB	1.2			ED
3352.89	3353.85	29816.4	TB	1.5	0.9		ED
3352.63	3353.59	29818.8	TB	0.9			ED
3352.05	3353.01	29823.9	TB	0.9	0.5		ED
3349.42	3350.38	29847.3	TB	1.5	1.7		ED
3348.54	3349.50	29855.2	TB	1.2			ED
3348.07	3349.03	29859.4	TB	1.5	0.5		ED
3347.27	3348.23	29866.5	TB	1.2	0.9		ED
3346.32	3347.28	29875.0	TB	0.9	0.5		ED
3344.50	3345.46	29891.2	TB	0.9	0.5		ED
3342.98	3343.94	29904.8	TB	1.1			ED
3339.61	3340.57	29935.0	TB	1.2	0.5		ED
3339.00	3339.96	29940.5	TB	1.7	0.9		ED
3338.03	3338.99	29949.2	TB	1.5	1.2		ED
3337.67	3338.63	29952.4	TB	1.2			ED
3335.42	3336.38	29972.6	TB	1.2	0.5		ED
3334.48	3335.44	29981.1	TB	1.2	0.5		ED
3334.25	3335.21	29983.1	TB	0.9			ED
3333.93	3334.89	29986.0	TB	1.5	0.5		ED
3333.21	3334.17	29992.5	TB	1.5	0.5		ED
3331.36	3332.32	30009.1	TB	0.9			ED
3329.08	3330.04	30029.7	TB	1.5	0.9		ED
3327.11	3328.07	30047.5	TB	1.2	0.5		ED
3326.47	3327.43	30053.2	TB	1.2			ED
3326.19	3327.15	30055.8	TB	0.9			ED
3325.52	3326.48	30061.8	TB	1.5			ED

AIR WAVELENGTH (ANGSTRMS)	VACUUM WAVELENGTH (ANGSTRMS)	WAVENUMBER (KAYSERS)	LMENT	LOG ARC	INTENSITY SPK	INTENSITY DIS	REF
3324.40	3325.36	30072.0	TB	1.8	1.7		ED
3323.89	3324.85	30076.6	TB	1.2	0.5		ED
3323.38	3324.34	30081.2	TB	1.2	0.5		ED
3321.15	3322.11	30101.4	TB	1.5	1.5		ED
3319.91	3320.87	30112.6	TB	0.9	0.5		ED
3319.16	3320.12	30119.4	TB	1.2			ED
3318.05	3319.00	30129.5	TB	1.2	0.5		ED
3317.58	3318.53	30133.8	TB	0.9	0.5		ED
3315.62	3316.57	30151.6	TB	0.9	0.5		ED
3315.07	3316.02	30156.6	TB	1.2	0.5		ED
3314.38	3315.33	30162.9	TB	1.2	0.5		ED
3313.47	3314.42	30171.2	TB	0.9			ED
3312.80	3313.75	30177.3	TB	1.2	0.5		ED
3312.53	3313.48	30179.7	TB	1.2	0.5		ED
3310.80	3311.75	30195.5	TB	0.9	0.5		ED
3309.17	3310.12	30210.4	TB	1.2	0.5		ED
3308.51	3309.46	30216.4	TB	1.2	0.9		ED
3307.80	3308.75	30222.9	TB	0.9	0.5		ED
3307.44	3308.39	30226.2	TB	1.5	1.2		ED
3306.41	3307.36	30235.6	TB	1.2			ED
3305.73	3306.68	30241.8	TB	0.9			ED
3305.37	3306.32	30245.1	TB	0.9	0.5		ED
3304.95	3305.90	30248.9	TB	1.2	0.5		ED
3304.10	3305.05	30256.7	TB	1.2	0.5		ED
3302.64	3303.59	30270.1	TB	0.9			ED
3300.28	3301.23	30291.7	TB	0.9			ED
3299.96	3300.91	30294.7	TB	1.2			ED
3299.27	3300.22	30301.0	TB	1.2			ED
3298.66	3299.61	30306.6	TB	1.5	0.9		ED
3296.03	3296.98	30330.8	TB	0.9			ED
3295.33	3296.28	30337.2	TB	1.2	0.5		ED
3295.08	3296.03	30339.5	TB	0.9			ED
3294.04	3294.99	30349.1	TB	1.7	0.5		ED
3293.07	3294.02	30358.1	TB	1.7	2.0		ED
3291.56	3292.51	30372.0	TB	1.2	0.9		ED
3291.00	3291.95	30377.1	TB	0.9			ED
3288.32	3289.27	30401.9	TB	0.9			ED
3287.55	3288.50	30409.0	TB	1.5	0.9		ED
3285.04	3285.99	30432.3	TB	1.5	0.9		ED
3284.59	3285.54	30436.4	TB	0.9			ED
3283.10	3284.05	30450.2	TB	1.5	0.9		ED
3281.92	3282.87	30461.2	TB	0.9			ED
3281.40	3282.35	30466.0	TB	1.7	1.2		ED
3280.28	3281.23	30476.4	TB	1.5	1.2		ED
3279.04	3279.98	30487.9	TB	0.9	0.5		ED
3277.73	3278.67	30500.1	TB	1.2	0.5		ED
3277.32	3278.26	30503.9	TB	1.2	0.9		ED
3275.66	3276.60	30519.4	TB	0.9	0.5		ED
3274.24	3275.18	30532.6	TB	1.8			ED
3273.12	3274.06	30543.1	TB	1.2			ED
3272.35	3273.29	30550.3	TB	1.2	0.5		ED
3270.63	3271.57	30566.3	TB	1.2	0.5		ED
3268.52	3269.46	30586.1	TB	1.2	0.5		ED
3268.10	3269.04	30590.0	TB	1.5	0.9		ED
3266.40	3267.34	30605.9	TB	1.5	1.2		ED
3265.93	3266.87	30610.3	TB	0.9	0.5		ED
3264.90	3265.84	30620.0	TB	1.2	0.5		ED
3263.89	3264.83	30629.5	TB	1.5	0.9		ED
3262.97	3263.91	30638.1	TB	1.5	0.5		ED
3262.68	3263.62	30640.8	TB	0.9	0.5		ED
3262.40	3263.34	30643.4	TB	0.3			ED
3261.74	3262.68	30649.6	TB	1.2	0.5		ED
3260.83	3261.77	30658.2	TB	1.2	0.5		ED
3259.84	3260.78	30667.5	TB	1.2			ED
3259.38	3260.32	30671.8	TB	1.2	0.5		ED
3258.38	3259.32	30681.2	TB	0.9	0.5		ED
3256.83	3257.77	30695.9	TB	0.9			ED
3255.22	3256.16	30711.0	TB	1.2	0.5		ED
3253.54	3254.48	30726.9	TB	0.9	0.5		ED
3252.34	3253.28	30738.2	TB	1.7	1.5		ED

AIR WAVELENGTH (ANGSTRMS)	VACUUM WAVELENGTH (ANGSTRMS)	WAVENUMBER (KAYSERS)	LMENT	LOG ARC	INTENSITY SPK	INTENSITY DIS	REF
3252.05	3252.99	30741.0	TB	0.9			ED
3250.95	3251.89	30751.4	TB	0.9			ED
3249.61	3250.55	30764.1	TB	0.9	0.5		ED
3247.79	3248.73	30781.3	TB	0.9			ED
3247.18	3248.12	30787.1	TB	1.2	0.5		ED
3246.51	3247.45	30793.4	TB	0.9	0.5		ED
3245.42	3246.36	30803.8	TB	1.2	0.5		ED
3245.17	3246.11	30806.1	TB	1.2	0.9		ED
3244.60	3245.54	30811.6	TB	0.9	0.5		ED
3243.20	3244.14	30824.9	TB	1.2	0.5		ED
3241.94	3242.88	30836.8	TB	0.9	0.5		ED
3240.65	3241.59	30849.1	TB	1.2	0.9		ED
3240.00	3240.93	30855.3	TB	1.2	0.9		ED
3239.66	3240.59	30858.5	TB	1.2	0.9		ED
3239.32	3240.25	30861.8	TB	0.9			ED
3237.22	3238.15	30881.8	TB	1.0			ED
3236.20	3237.13	30891.5	TB	0.9	0.5		ED
3235.79	3236.72	30895.4	TB	1.2	0.5		ED
3234.50	3235.43	30907.8	TB	1.2	1.2		ED
3233.53	3234.46	30917.0	TB	0.9			ED
3232.00	3232.93	30931.7	TB	0.9	0.5		ED
3231.46	3232.39	30936.8	TB	1.2	0.5		ED
3231.06	3231.99	30940.7	TB	1.5	0.9		ED
3230.70	3231.63	30944.1	TB	0.9			ED
3230.03	3230.96	30950.5	TB	1.2	0.9		ED
3229.59	3230.52	30954.7	TB	0.9			ED
3229.19	3230.12	30958.6	TB	1.2	0.9		ED
3227.48	3228.41	30975.0	TB	0.9	0.5		ED
3225.43	3226.36	30994.7	TB	0.9	0.5		ED
3222.97	3223.90	31018.3	TB	0.9	0.5		ED
3222.37	3223.30	31024.1	TB	0.9	0.5		ED
3221.29	3222.22	31034.5	TB	0.9	0.5		ED
3220.17	3221.10	31045.3	TB	1.2			ED
3219.95	3220.88	31047.4	TB	1.7	1.7		ED
3218.93	3219.86	31057.2	TB	1.7	1.7		ED
3216.61	3217.54	31079.6	TB	0.9	0.5		ED
3215.54	3216.47	31090.0	TB	0.9	0.5		ED
3215.01	3215.94	31095.1	TB	0.9	0.9		ED
3214.40	3215.33	31101.0	TB	0.9			ED
3211.17	3212.10	31132.3	TB	0.9	0.5		ED
3210.22	3211.15	31141.5	TB	1.2	0.9		ED
3209.54	3210.47	31148.1	TB	0.9	0.5		ED
3207.96	3208.89	31163.4	TB	0.9			ED
3207.53	3208.46	31167.6	TB	1.2	0.9		ED
3207.09	3208.02	31171.9	TB	0.9	0.5		ED
3204.21	3205.14	31199.9	TB	0.9	0.5		ED
3202.95	3203.88	31212.2	TB	0.9	0.5		ED
3202.70	3203.63	31214.6	TB	0.9	0.5		ED
3200.73	3201.66	31233.9	TB	1.2	0.5		ED
3199.56	3200.48	31245.3	TB	1.5	1.2		ED
3198.55	3199.47	31255.1	TB	0.9			ED
3198.01	3198.93	31260.4	TB	1.2	0.5		ED
3195.60	3196.52	31284.0	TB	1.2	1.2		ED
3194.71	3195.63	31292.7	TB	1.2	0.9		ED
3192.76	3193.68	31311.8	TB	0.9	0.5		ED
3190.72	3191.64	31331.8	TB	0.9			ED
3189.97	3190.89	31339.2	TB	0.9	0.5		ED
3188.55	3189.47	31353.1	TB	0.9	0.5		ED
3188.03	3188.95	31358.3	TB	1.2	0.9		ED
3187.25	3188.17	31365.9	TB	1.2	1.2		ED
3186.23	3187.15	31376.0	TB	0.9	0.5		ED
3183.88	3184.80	31399.1	TB	1.2	0.5		ED
3183.64	3184.56	31401.5	TB	1.2	0.5		ED
3183.29	3184.21	31405.0	TB	1.2	0.5		ED
3181.22	3182.14	31425.4	TB	0.9			ED
3180.54	3181.46	31432.1	TB	1.5	0.5		ED
3179.84	3180.76	31439.0	TB	0.9			ED
3177.53	3178.45	31461.9	TB	0.9	0.5		ED
3176.74	3177.66	31469.7	TB	0.9	0.5		ED
3175.45	3176.37	31482.5	TB	0.9	0.5		ED

AIR WAVELENGTH (ANGSTRMS)	VACUUM WAVELENGTH (ANGSTRMS)	WAVENUMBER (KAYSERS)	LMENT	LOG ARC	INTENSITY SPK	INTENSITY DIS	REF
3174.66	3175.58	31490.3	TB	1.2	1.2		ED
3173.76	3174.68	31499.3	TB	1.2	0.3		ED
3171.90	3172.82	31517.7	TB	0.9	0.5		ED
3171.19	3172.11	31524.8	TB	0.9	0.5		ED
3169.84	3170.76	31538.2	TB	1.5	0.9		ED
3168.59	3169.51	31550.6	TB	0.9	0.5		ED
3168.32	3169.24	31553.3	TB	0.9	0.5		ED
3167.52	3168.44	31561.3	TB	1.2	0.9		ED
3165.74	3166.66	31579.1	TB	1.2	0.9		ED
3164.77	3165.69	31588.7	TB	0.9	0.5		ED
3164.10	3165.02	31595.4	TB	0.9	0.5		ED
3163.85	3164.77	31597.9	TB	1.2	0.5		ED
3162.93	3163.85	31607.1	TB	1.2	0.9		ED
3162.42	3163.34	31612.2	TB	1.2	0.5		ED
3160.12	3161.03	31635.2	TB	0.9			ED
3159.39	3160.30	31642.5	TB	1.2			ED
3158.66	3159.57	31649.8	TB	0.9			ED
3158.01	3158.92	31656.4	TB	0.9			ED
3156.52	3157.43	31671.3	TB	0.9	0.5		ED
3155.62	3156.53	31680.3	TB	1.2	0.5		ED
3155.10	3156.01	31685.5	TB	0.9	0.5		ED
3154.69	3155.60	31689.7	TB	0.9	0.5		ED
3148.71	3149.62	31749.8	TB	1.5	1.2		ED
3148.21	3149.12	31754.9	TB	1.2	0.9		ED
3147.04	3147.95	31766.7	TB	1.7	0.5		ED
3146.67	3147.58	31770.4	TB	0.9	0.9		ED
3145.22	3146.13	31785.1	TB	0.9	1.2		ED
3144.46	3145.37	31792.7	TB	0.9	0.5		ED
3140.62	3141.53	31831.6	TB	0.9	0.5		ED
3140.05	3140.96	31837.4	TB	1.2	0.5		ED
3139.64	3140.55	31841.6	TB	1.5	1.2		ED
3138.63	3139.54	31851.8	TB	1.2	0.5		ED
3137.22	3138.13	31866.1	TB	0.9	0.5		ED
3135.35	3136.26	31885.1	TB	1.2	0.9		ED
3134.26	3135.17	31896.2	TB	0.9	0.9		ED
3133.15	3134.06	31907.5	TB	0.9			ED
3131.35	3132.26	31925.9	TB	0.9	0.5		ED
3128.83	3129.74	31951.6	TB	0.5	0.5		ED
3124.54	3125.45	31995.4	TB	0.9	0.9		ED
3124.02	3124.93	32000.8	TB	1.2	0.5		ED
3123.70	3124.61	32004.0	TB	0.9	0.5		ED
3123.05	3123.96	32010.7	TB	1.2	0.5		ED
3122.83	3123.74	32012.9	TB	0.9	0.5		ED
3121.94	3122.85	32022.1	TB	1.2	0.9		ED
3121.43	3122.34	32027.3	TB	0.9	0.5		ED
3119.62	3120.52	32045.9	TB	1.2	0.9		ED
3117.26	3118.16	32070.1	TB	0.9	0.5		ED
3113.62	3114.52	32107.6	TB	1.2	0.9		ED
3109.15	3110.05	32153.8	TB	0.9	0.5		ED
3108.41	3109.31	32161.5	TB	0.9	0.5		ED
3102.97	3103.87	32217.8	TB	1.5	1.2		ED
3100.51	3101.41	32243.4	TB	1.2	0.9		ED
3098.64	3099.54	32262.9	TB	1.2			ED
3096.86	3097.76	32281.4	TB	0.5	0.9		ED
3089.58	3090.48	32357.5	TB	1.5	1.2		ED
3089.10	3090.00	32362.5	TB	0.5	0.5		ED
3088.43	3089.33	32369.5	TB	1.2	0.9		ED
3086.78	3087.68	32386.8	TB	1.2	0.9		ED
3082.36	3083.26	32433.2	TB	0.5	0.9		ED
3080.11	3081.00	32456.9	TB	0.5	0.5		ED
3078.86	3079.75	32470.1	TB	1.5	1.9		ED
3076.04	3076.93	32499.9	TB	1.5	0.5		ED
3074.71	3075.60	32513.9	TB	0.5	0.5		ED
3072.60	3073.49	32536.3	TB	0.5	1.2		ED
3070.05	3070.94	32563.3	TB	1.2	1.2		ED
3069.02	3069.91	32574.2	TB	1.2	1.2		ED
3065.69	3066.58	32609.6	TB	0.5	0.9		ED
3064.09	3064.98	32626.6	TB	1.2	0.9		ED
3062.09	3062.98	32647.9	TB	0.5	0.5		ED
3061.80	3062.69	32651.0	TB	0.5			ED

AIR WAVELENGTH (ANGSTRMS)	VACUUM WAVELENGTH (ANGSTRMS)	WAVENUMBER (KAYSERS)	LMENT	LOG ARC	INTENSITY SPK	INTENSITY DIS	REF
3053.55	3054.44	32739.2	TB	0.9	1.5		ED
3053.24	3054.13	32742.6	TB	0.9	0.5		ED
3052.18	3053.07	32753.9	TB	0.5	0.9		ED
3051.12	3052.01	32765.3	TB	1.2	0.9		ED
3044.97	3045.86	32831.5	TB	1.2	0.9		ED
3043.65	3044.54	32845.7	TB	0.9	0.9		ED
3042.50	3043.39	32858.1	TB	0.5	1.2		ED
3040.34	3041.22	32881.5	TB	0.9			ED
3038.66	3039.54	32899.7	TB	0.9	0.5		ED
3037.04	3037.92	32917.2	TB	0.9	0.5		ED
3034.91	3035.79	32940.3	TB	0.9	0.5		ED
3032.83	3033.71	32962.9	TB	0.9	0.9		ED
3031.60	3032.48	32976.3	TB	0.9	1.2		ED
3029.23	3030.11	33002.1	TB	0.5	0.5		ED
3027.33	3028.21	33022.8	TB	0.9	0.5		ED
3023.70	3024.58	33062.4	TB	0.5	0.5		ED
3021.95	3022.83	33081.6	TB	0.5	0.5		ED
3020.58	3021.46	33096.6	TB	0.5			ED
3020.29	3021.17	33099.8	TB	0.9	1.2		ED
3019.17	3020.05	33112.0	TB	0.5	0.9		ED
3016.18	3017.06	33144.9	TB	1.2	1.2		ED
3013.61	3014.49	33173.1	TB	0.5	0.5		ED
3010.55	3011.43	33206.9	TB	0.9	0.9		ED
3009.30	3010.18	33220.6	TB	0.9	0.9		ED
3007.11	3007.99	33244.8	TB	0.9	0.5		ED
3005.52	3006.40	33262.4	TB	0.9	0.5		ED
3004.56	3005.44	33273.0	TB	0.9	0.5		ED
3002.45	3003.33	33296.4	TB	0.5			ED
2999.97	3000.84	33324.0	TB	0.5	1.0		EX
2999.05	2999.92	33334.2	TB	1.3	1.0		EX
2996.01	2996.88	33368.0	TB	1.0	0.5		EX
2995.78	2996.65	33370.6	TB	1.3			ED
2991.97	2992.84	33413.0	TB	1.0	0.7		M
2989.78	2990.65	33437.5	TB	1.0	0.7		M
2988.59	2989.46	33450.8	TB	1.0	0.7		M
2987.04	2987.91	33468.2	TB	1.0	0.7		M
2982.98	2983.85	33513.7	TB		1.0		EX
2977.76	2978.63	33572.5	TB	1.0	1.0		M
2970.65	2971.52	33652.8	TB	0.7	0.5 H		M
2968.87	2969.74	33673.0	TB	1.0	0.5		EX
2965.32	2966.19	33713.3	TB	1.0			ED
2964.76	2965.63	33719.7	TB	1.0	0.5		M
2963.98	2964.85	33728.6	TB	1.0	0.5 H		M
2962.80	2963.67	33742.0	TB	1.0	0.5 H		M
2960.58	2961.44	33767.3	TB	1.0	0.5		EX
2956.21	2957.07	33817.2	TB	1.0	1.6		ED
2945.70	2946.56	33937.9	TB	1.0	0.5		ED
2944.87	2945.73	33947.4	TB	0.7			ED
2941.70	2942.56	33984.0	TB	0.5	1.0		ED
2940.03	2940.89	34003.3	TB	1.0	0.5 H		ED
2934.79	2935.65	34064.0	TB	0.7	1.0		M
2933.79	2934.65	34075.6	TB	0.7	0.5 H		M
2933.05	2933.91	34084.2	TB	1.0			EX
2932.91	2933.77	34085.9	TB	1.0	0.5 H		M
2931.41	2932.27	34103.3	TB	0.7	0.5 H		M
2924.52	2925.38	34183.6	TB	0.7	0.5 H		M
2924.16	2925.02	34187.9	TB	0.5	1.0 H		ED
2918.99	2919.84	34248.4	TB	1.2	1.0		M
2916.26	2917.11	34280.5	TB	0.7	1.0		M
2915.59	2916.44	34288.3	TB	1.0	1.0 H		M
2915.33	2916.18	34291.4	TB	1.0	1.0		EX
2914.79	2915.64	34297.7	TB	1.2	1.0 H		M
2913.41	2914.26	34314.0	TB	0.5			EX
2911.81	2912.66	34332.9	TB	0.7	0.5 H		M
2910.36	2911.21	34350.0	TB	0.5	1.0 H		ED
2908.49	2909.34	34372.0	TB	0.5	1.3		EX
2903.21	2904.06	34434.5	TB	1.2	0.5 H		M
2901.54	2902.39	34454.4	TB	1.0	0.5		M
2900.01	2900.86	34472.5	TB		1.0		EX
2898.84	2899.69	34486.5	TB	1.2	1.0		M
2898.20	2899.05	34494.1	TB		1.0		EX
2897.46	2898.31	34502.9	TB	0.5	1.3 H		ED
2894.77	2895.62	34534.9	TB		1.0		EX
2894.48	2895.33	34538.4	TB	1.0	1.0		ED
2891.41	2892.26	34575.1	TB	0.5	2.7		EX
2889.73	2890.58	34595.2	TB	1.0			EX
2889.54	2890.39	34597.4	TB	0.5	1.0		ED
2886.28	2887.13	34636.5	TB	1.2	1.0		M
2885.90	2886.75	34641.1	TB	1.0	1.0		EX
2885.33	2886.18	34647.9	TB	0.5			EX
2885.14	2885.99	34650.2	TB		1.8		EX
2884.70	2885.55	34655.5	TB	0.7	0.5		M
2881.31	2882.16	34696.3	TB	1.0			M
2878.52	2879.36	34729.9	TB		1.0		EX
2871.05	2871.89	34820.2	TB	0.5	1.0		EX
2866.44	2867.28	34876.2	TB	0.5	1.0		EX
2861.34	2862.18	34938.4	TB	1.0	1.0		ED
2857.68	2858.52	34983.1	TB	0.7	0.5		M
2857.19	2858.03	34989.1	TB		1.0		EX
2855.68	2856.52	35007.6	TB	0.7	1.3		M
2854.95	2855.79	35016.6	TB	0.7	0.5 H		M
2854.16	2855.00	35026.3	TB	0.7	0.5 H		M
2853.19	2854.03	35038.2	TB		1.3		EX
2848.97	2849.81	35090.1	TB		1.0		EX
2838.72	2839.55	35216.8	TB	0.5	1.3		M
2836.11	2836.94	35249.2	TB		1.3		EX
2833.37	2834.20	35283.3	TB		1.3		EX
2827.38	2828.21	35358.0	TB	0.5	1.0 H		M
2819.89	2820.72	35451.9	TB	0.5	1.3		EX
2812.66	2813.49	35543.1	TB	0.5	1.0		M
2809.32	2810.15	35585.3	TB	0.5	1.0		ED
2800.51	2801.34	35697.3	TB	1.0	1.6		M
2791.80	2792.62	35808.6	TB		1.0		EX
2769.53	2770.35	36096.5	TB	1.0	1.0		M
2759.47	2760.29	36228.1	TB	1.0	1.0		M
2758.67	2759.49	36238.6	TB		1.0		EX
2742.25	2743.06	36455.6	TB	1.0			M
2736.24	2737.05	36535.7	TB	1.0	0.5		M
2727.01	2727.82	36659.3	TB		1.0		EX
2721.39	2722.20	36735.0	TB		1.3		EX
2706.28	2707.08	36940.1	TB	1.0	1.0		ED
2704.07	2704.87	36970.3	TB	1.0	1.0		ED
2679.94	2680.74	37303.2	TB		1.3		EX
2669.29	2670.08	37452.0	TB	1.0	0.5		ED
2661.39	2662.18	37563.2	TB	0.7			M
2658.91	2659.70	37598.2	TB		2.7		EX
2646.50	2647.29	37774.5	TB		1.0		EX
2632.61	2633.39	37973.8	TB		1.0		EX
2632.02	2632.80	37982.3	TB		1.0		EX
2628.69	2629.47	38030.4	TB	1.3	0.5		ED
2625.44	2626.22	38077.5	TB		1.0		EX
2619.61	2620.39	38162.2	TB		1.6		EX
2608.57	2609.35	38323.7	TB	1.0	1.3		ED
2605.42	2606.20	38370.1	TB		1.0		EX
2603.02	2603.80	38405.4	TB	0.5	1.3		EX
2600.71	2601.49	38439.5	TB		1.3		EX
2600.15	2600.93	38447.8	TB		1.3		EX
2597.31	2598.09	38489.9	TB		1.0		EX
2597.05	2597.83	38493.7	TB		1.0		EX
2595.19	2595.97	38521.3	TB		1.0		EX
2594.4	2595.2	38533.	TB		1.0		EX
2591.83	2592.61	38571.2	TB		1.3		EX
2588.59	2589.36	38619.5	TB		1.0		EX
2584.61	2585.38	38679.0	TB	0.5	0.5		M
2580.10	2580.87	38746.6	TB		1.0		EX
2571.78	2572.55	38871.9	TB		1.6		EX
2564.31	2565.08	38985.2	TB		1.0		EX
2562.12	2562.89	39018.5	TB		1.0		EX
2560.12	2560.89	39049.0	TB		1.0		EX
2559.20	2559.97	39063.0	TB		1.0		EX

AIR WAVELENGTH (ANGSTRMS)	VACUUM WAVELENGTH (ANGSTRMS)	WAVENUMBER (KAYSERS)	LMENT	LOG ARC	INTENSITY SPK	INTENSITY DIS	REF
2556.61	2557.38	39102.6	TB		1.3		EX
2552.01	2552.78	39173.0	TB		1.0		EX
2549.06	2549.83	39218.4	TB		1.0		EX
2544.62	2545.38	39286.8	TB		1.0		EX
2540.12	2540.88	39356.4	TB	0.5	1.7		ED
2539.42	2540.18	39367.2	TB		1.0		EX
2536.75	2537.51	39408.7	TB	0.5	1.0		M
2528.23	2528.99	39541.5	TB		1.3		EX
2522.85	2523.61	39625.8	TB	0.5	1.3		ED
2518.96	2519.72	39687.0	TB		1.0		EX
2515.23	2515.99	39745.8	TB		1.0		EX
2502.70	2503.45	39944.8	TB		1.0		EX
2498.63	2499.38	40009.9	TB		1.0		EX
2490.36	2491.11	40142.7	TB		1.0		EX
2487.90	2488.65	40182.4	TB		1.0		EX
2481.44	2482.19	40287.0	TB		1.0		EX
2476.06	2476.81	40374.5	TB		1.0		EX
2474.91	2475.66	40393.3	TB		1.0		EX
2422.82	2423.56	41261.7	TB		1.0		EX
2372.03	2372.75	42145.1	TB		1.0		EX
9977.6	9980.3	10020.	TE I			1.2	RD
9955.5	9958.2	10042.	TE I			1.1	RD
9900.9	9903.6	10097.	TE I			0.8	RD
9867.0	9869.7	10132.	TE I			1.0	RD
9840.5	9843.2	10159.	TE I			0.8K	RD
9783.6	9786.3	10218.	TE I			0.9	RD
9721.2	9723.9	10284.	TE I			1.7	RD
9206.3	9208.8	10859.	TE I			0.7X	RD
9071.3	9073.8	11021.	TE I			0.8	RD
9042.2	9044.7	11056.	TE I			0.9	RD
9003.7	9006.2	11103.	TE I			1.5	RD
8850.3	8852.7	11296.	TE I			1.2	RD
8830.4	8832.8	11321.	TE I			1.2	RD
8789.1	8791.5	11375.	TE I			0.8	RD
8771.2	8773.6	11398.	TE I			0.9	RD
8757.8	8760.2	11415.	TE I			1.7	RD
8700.6	8703.0	11490.	TE I			1.3	RD
8632.1	8634.5	11581.	TE I			0.8K	RD
8599.5	8601.9	11625.	TE I			0.7	RD
8521.4	8523.7	11732.	TE I			1.1	RD
8500.8	8503.1	11760.	TE I			0.9	RD
8492.2	8494.5	11772.	TE I			0.9	RD
8469.8	8472.1	11803.	TE I			0.8	RD
8355.8	8358.1	11964.	TE I			1.2	RD
8291.1	8293.4	12058.	TE I			1.0	RD
8276.6	8278.9	12079.	TE I			1.0	RD
8251.5	8253.8	12116.	TE I			1.0	RD
8082.5	8084.7	12369.	TE I			1.0N	RD
8061.4	8063.6	12401.	TE I			1.5N	RD
7972.9	7975.1	12539.	TE I			1.3	RD
7819.5	7821.7	12785.	TE			0.7	RD
7759.1	7761.2	12884.	TE I			1.2	RD
7754.4	7756.5	12892.	TE I			0.8	RD
7575.7	7577.8	13196.	TE I			0.8	RD
7556.8	7558.9	13229.	TE I			1.0	RD
7552.8	7554.9	13236.	TE I			0.8W	RD
7531.5	7533.6	13274.	TE I			0.7	RD
7280.9	7282.9	13731.	TE I			1.3	RD
7263.5	7265.5	13764.	TE I			1.7L	RD
7230.3	7232.3	13827.	TE I			0.8	RD
7191.08	7193.06	13902.3	TE I			1.5	BL
7144.61	7146.58	13992.7	TE			0.7	BL
7135.34	7137.31	14010.9	TE			1.5	BL
7116.45	7118.41	14048.1	TE			1.2	BL
7116.30	7118.26	14048.4	TE			1.2	BL
7103.43	7105.39	14073.8	TE			1.2	BL
7097.57	7099.53	14085.4	TE			0.7	BL
7058.6	7060.5	14163.	TE I			0.8	RD
7049.68	7051.62	14181.1	TE			0.7	BL
7038.95	7040.89	14202.7	TE			1.8	BL

AIR WAVELENGTH (ANGSTRMS)	VACUUM WAVELENGTH (ANGSTRMS)	WAVENUMBER (KAYSERS)	LMENT	LOG ARC	INTENSITY SPK	INTENSITY DIS	REF
7015.89	7017.82	14249.4	TE			1.7	BL
6959.08	6961.00	14365.7	TE			1.2	BL
6930.51	6932.42	14425.0	TE			1.8	BL
6924.66	6926.57	14437.2	TE			0.7	BL
6913.18	6915.09	14461.1	TE			1.5	BL
6885.04	6886.94	14520.2	TE			1.5	BL
6877.94	6879.84	14535.2	TE			1.7	BL
6870.56	6872.46	14550.8	TE I			0.7W	BL
6867.53	6869.42	14557.3	TE			0.7	BL
6854.7	6856.6	14584.	TE I			1.7L	RD
6843.94	6845.83	14607.4	TE I			1.8	BL
6837.65	6839.54	14620.9	TE			1.7W	BL
6832.46	6834.35	14632.0	TE			1.5	BL
6796.66	6798.54	14709.1	TE			0.7	BL
6790.0	6791.9	14723.	TE I			1.3W	RD
6782.54	6784.41	14739.7	TE			0.7	BL
6761.34	6763.21	14785.9	TE			1.5	BL
6751.40	6753.26	14807.7	TE			1.7	BL
6736.32	6738.18	14840.8	TE			1.5	BL
6713.00	6714.85	14892.4	TE			0.7	BL
6708.34	6710.19	14902.7	TE			1.5	BL
6701.57	6703.42	14917.8	TE			0.7	BL
6700.60	6702.45	14919.9	TE			1.2	BL
6690.14	6691.99	14943.2	TE I			1.3W	BL
6688.49	6690.34	14946.9	TE			0.7	BL
6687.36	6689.21	14949.5	TE			0.7	BL
6686.7	6688.5	14951.	TE I			0.8W	RD
6686.03	6687.88	14952.4	TE			1.5	BL
6676.01	6677.85	14974.9	TE			1.8	BL
6672.70	6674.54	14982.3	TE			0.7	BL
6670.6	6672.4	14987.	TE			1.7	RD
6661.06	6662.90	15008.5	TE			1.5	BL
6659.80	6661.64	15011.3	TE			1.2	BL
6658.07	6659.91	15015.2	TE			0.7	BL
6649.72	6651.56	15034.1	TE			1.5	BL
6648.52	6650.36	15036.8	TE			2.0	BL
6640.59	6642.42	15054.7	TE			1.2	BL
6637.00	6638.83	15062.9	TE			1.7	BL
6628.7	6630.5	15082.	TE I			0.7W	RD
6613.4	6615.2	15117.	TE I			1.3L	RD
6606.3	6608.1	15133.	TE I			0.8	RD
6603.04	6604.86	15140.4	TE			0.7	BL
6596.32	6598.14	15155.8	TE			1.5	BL
6584.93	6586.75	15182.0	TE			1.5	BL
6578.63	6580.45	15196.5	TE			1.2	BL
6574.50	6576.32	15206.1	TE			1.5	BL
6563.95	6565.76	15230.5	TE			1.7H	BL
6553.5	6555.3	15255.	TE I			1.1L	RD
6546.59	6548.40	15270.9	TE			0.7	BL
6541.89	6543.70	15281.9	TE			1.5	BL
6537.01	6538.82	15293.3	TE			2.0	BL
6528.35	6530.15	15313.6	TE			0.7	BL
6491.81	6493.60	15399.8	TE			1.2	BL
6487.09	6488.88	15411.0	TE			1.8	BL
6474.28	6476.07	15441.5	TE			1.5	BL
6469.13	6470.92	15453.8	TE I			0.5H	BL
6462.9	6464.7	15469.	TE I			0.8	RD
6457.80	6459.58	15480.9	TE			0.7	BL
6456.7	6458.5	15483.	TE I			1.2W	RD
6446.7	6448.5	15507.	TE I			0.7	RD
6437.06	6438.84	15530.7	TE			3.0	BL
6422.96	6424.74	15564.8	TE			1.8	BL
6412.15	6413.92	15591.1	TE			1.8	BL
6409.4	6411.2	15598.	TE I			0.7	RD
6405.9	6407.7	15606.	TE I			1.3	RD
6396.46	6398.23	15629.3	TE			1.7	BL
6390.19	6391.96	15644.7	TE			1.2	BL
6367.10	6368.86	15701.4	TE			1.8	BL
6363.93	6365.69	15709.2	TE			0.7	BL
6349.7	6351.5	15744.	TE I			1.3W	RD

AIR WAVELENGTH (ANGSTRMS)	VACUUM WAVELENGTH (ANGSTRMS)	WAVENUMBER (KAYSERS)	LMENT	LOG ARC	INTENSITY SPK	DIS	REF
6345.51	6347.26	15754.8	TE			1.5	BL
6330.04	6331.79	15793.3	TE			0.7	BL
6324.38	6326.13	15807.5	TE			0.7	BL
6298.64	6300.38	15872.1	TE			1.5	BL
6293.61	6295.35	15884.7	TE			2.0	BL
6277.7	6279.4	15925.	TE I			0.5	RD
6273.41	6275.15	15935.9	TE I			1.2S	BL
6266.21	6267.94	15954.2	TE			1.8	BL
6260.51	6262.24	15968.7	TE			1.5	BL
6259.44	6261.17	15971.4	TE			1.5	BL
6255.9	6257.6	15980.	TE I			1.1	RD
6245.61	6247.34	16006.8	TE			2.2	BL
6243.24	6244.97	16012.9	TE			1.2	BL
6230.80	6232.52	16044.9	TE			2.5	BL
6221.48	6223.20	16068.9	TE			1.7	BL
6216.64	6218.36	16081.4	TE			1.7	BL
6211.49	6213.21	16094.7	TE			1.5	BL
6203.29	6205.01	16116.0	TE			1.2	BL
6202.29	6204.01	16118.6	TE			1.8	BL
6195.05	6196.76	16137.4	TE			0.7	BL
6187.56	6189.27	16157.0	TE			0.7	BL
6183.39	6185.10	16167.9	TE			1.7	BL
6181.94	6183.65	16171.7	TE			1.2	BL
6180.58	6182.29	16175.2	TE			1.5	BL
6171.83	6173.54	16198.2	TE			0.7	BL
6167.24	6168.95	16210.2	TE			1.2	BL
6166.84	6168.55	16211.3	TE			2.0	BL
6162.4	6164.1	16223.	TE I			0.5W	RD
6160.2	6161.9	16229.	TE I			1.1W	RD
6155.16	6156.86	16242.0	TE			1.2	BL
6153.89	6155.59	16245.4	TE			1.8	BL
6151.14	6152.84	16252.7	TE			1.7	BL
6148.1	6149.8	16261.	TE I			0.3	RD
6146.47	6148.17	16265.0	TE			0.7	BL
6144.32	6146.02	16270.7	TE			1.5	BL
6142.24	6143.94	16276.2	TE			1.5	BL
6140.28	6141.98	16281.4	TE			1.2	BL
6136.81	6138.51	16290.6	TE			1.8	BL
6133.3	6135.0	16300.	TE I			0.3	RD
6106.55	6108.24	16371.3	TE			1.2	BL
6082.45	6084.13	16436.2	TE			1.5	BL
6073.35	6075.03	16460.8	TE			0.7	BL
6064.24	6065.92	16485.6	TE			0.7	BL
6055.82	6057.50	16508.5	TE			0.7H	BL
6054.20	6055.88	16512.9	TE			0.7	BL
6047.44	6049.11	16531.4	TE			2.0	BL
6034.9	6036.6	16566.	TE I			0.7	RD
6022.06	6023.73	16601.0	TE			1.5	BL
6020.8	6022.5	16604.	TE I			0.8W	RD
6019.51	6021.18	16608.1	TE			1.5	BL
6014.49	6016.16	16621.9	TE			2.0	BL
6013.49	6015.16	16624.7	TE			0.7H	BL
6008.62	6010.28	16638.1	TE			1.5	BL
6007.97	6009.63	16639.9	TE			1.2	BL
6005.0	6006.7	16648.	TE I			0.7W	RD
6004.40	6006.06	16649.8	TE			1.7	BL
6001.34	6003.00	16658.3	TE			1.2	BL
5993.94	5995.60	16678.9	TE			1.9	BL
5993.12	5994.78	16681.2	TE			1.9	BL
5985.64	5987.30	16702.0	TE			1.9	BL
5974.70	5976.36	16732.6	TE			2.4	BL
5972.64	5974.29	16738.4	TE			1.7	BL
5954.96	5956.61	16788.1	TE			1.4	BL
5936.21	5937.85	16841.1	TE			1.9	BL
5933.22	5934.86	16849.6	TE			1.2	BL
5925.25	5926.89	16872.2	TE			0.9H	BL
5899.74	5901.37	16945.2	TE			1.4	BL
5896.65	5898.28	16954.1	TE			1.4	BL
5891.43	5893.06	16969.1	TE			1.2	BL
5888.89	5890.52	16976.4	TE			0.9	BL

AIR WAVELENGTH (ANGSTRMS)	VACUUM WAVELENGTH (ANGSTRMS)	WAVENUMBER (KAYSERS)	LMENT	LOG ARC	INTENSITY SPK	DIS	REF
5874.6	5876.2	17018.	TE I			0.5S	RD
5871.80	5873.43	17025.8	TE			1.2	BL
5866.11	5867.74	17042.4	TE			1.2	BL
5859.70	5861.32	17061.0	TE			0.9	BL
5858.55	5860.17	17064.3	TE			1.7	BL
5851.09	5852.71	17086.1	TE			1.9	BL
5845.02	5846.64	17103.8	TE			0.9	BL
5843.31	5844.93	17108.9	TE			1.4	BL
5835.16	5836.78	17132.7	TE			1.2	BL
5828.63	5830.25	17151.9	TE			1.4	BL
5826.45	5828.07	17158.4	TE			1.7	BL
5824.19	5825.80	17165.0	TE			1.7	BL
5815.71	5817.32	17190.0	TE			1.2	BL
5803.07	5804.68	17227.5	TE			1.7	BL
5789.22	5790.83	17268.7	TE I			1.0	BL
5787.05	5788.65	17275.2	TE			0.9	BL
5785.23	5786.83	17280.6	TE			0.9	BL
5777.25	5778.85	17304.5	TE			1.2	BL
5770.92	5772.52	17323.5	TE			1.5	BL
5769.1	5770.7	17329.	TE I			0.8	RD
5765.25	5766.85	17340.5	TE			1.8	BL
5763.92	5765.52	17344.5	TE			1.4	BL
5762.60	5764.20	17348.5	TE			1.2	BL
5755.87	5757.47	17368.7	TE			2.4	BL
5746.31	5747.90	17397.6	TE			0.9	BL
5745.75	5747.34	17399.3	TE			0.9	BL
5741.66	5743.25	17411.7	TE			1.8	BL
5733.5	5735.1	17436.	TE I			0.9	RD
5709.08	5710.66	17511.1	TE			1.7	BL
5708.07	5709.65	17514.2	TE			2.4	BL
5692.96	5694.54	17560.7	TE			0.9	BL
5679.71	5681.29	17601.6	TE			1.4	BL
5672.24	5673.81	17624.8	TE			1.2	BL
5667.86	5669.43	17638.4	TE			1.2	BL
5666.26	5667.83	17643.4	TE			1.7	BL
5655.15	5656.72	17678.1	TE			1.4	BL
5652.06	5653.63	17687.8	TE			1.4	BL
5651.51	5653.08	17689.5	TE			1.4	BL
5649.30	5650.87	17696.4	TE			2.4	BL
5630.66	5632.22	17755.0	TE			1.4	BL
5618.47	5620.03	17793.5	TE			1.4	BL
5592.90	5594.45	17874.9	TE			1.2	BL
5582.77	5584.32	17907.3	TE			1.2	BL
5580.46	5582.01	17914.7	TE			0.9	BL
5576.40	5577.95	17927.7	TE			2.0	BL
5569.38	5570.93	17950.3	TE			1.2	BL
5565.57	5567.12	17962.6	TE			0.9	BL
5536.73	5538.27	18056.2	TE			1.4	BL
5488.07	5489.59	18216.3	TE			1.7	BL
5480.83	5482.35	18240.3	TE			0.9	BL
5479.13	5480.65	18246.0	TE			1.7	BL
5465.17	5466.69	18292.6	TE II			1.2	BL
5449.82	5451.33	18344.1	TE			1.9	BL
5426.05	5427.56	18424.5	TE			0.9	BL
5410.41	5411.91	18477.7	TE			1.4	BL
5366.91	5368.40	18627.5	TE			1.4	BL
5366.28	5367.77	18629.7	TE			1.2	BL
5350.41	5351.90	18685.0	TE			0.9	BL
5321.02	5322.50	18788.2	TE			0.9	BL
5311.07	5312.55	18823.4	TE			1.5	BL
5304.99	5306.47	18844.9	TE			1.4	BL
5303.72	5305.20	18849.4	TE			1.2	BL
5300.67	5302.14	18860.3	TE			1.4	BL
5264.33	5265.80	18990.5	TE			0.9	BL
5256.36	5257.82	19019.3	TE			1.4	BL
5244.21	5245.67	19063.3	TE			0.9	BL
5238.07	5239.53	19085.7	TE			1.2	BL
5219.46	5220.91	19153.7	TE			0.9	BL
5188.59	5190.03	19267.7	TE			0.9	BL
5178.9	5180.3	19304.	TE I			0.6	RD

AIR WAVELENGTH (ANGSTRMS)	VACUUM WAVELENGTH (ANGSTRMS)	WAVENUMBER (KAYSERS)	LMENT	LOG ARC	INTENSITY SPK	INTENSITY DIS	REF
5172.99	5174.43	19325.8	TE			1.2	BL
5164.00	5165.44	19359.4	TE			0.9	BL
5149.90	5151.33	19412.4	TE			0.9	BL
5148.7	5150.1	19417.	TE I			0.9	RD
5133.23	5134.66	19475.5	TE			0.9	BL
5130.99	5132.42	19484.0	TE			1.2	BL
5112.12	5113.54	19555.9	TE			0.9	BL
5083.0	5084.4	19668.	TE I			1.0	RD
5065.74	5067.15	19734.9	TE			0.9	BL
5060.44	5061.85	19755.6	TE			0.9	BL
5037.97	5039.37	19843.7	TE			1.2	BL
5020.44	5021.84	19913.0	TE			0.9	BL
5000.87	5002.27	19990.9	TE			1.4	BL
4925.25	4926.62	20297.9	TE			1.2	BL
4919.12	4920.49	20323.2	TE			1.5	BL
4912.05	4913.42	20352.4	TE			1.2	BL
4910.55	4911.92	20358.6	TE			1.7	BL
4904.43	4905.80	20384.0	TE			2.0	BL
4901.14	4902.51	20397.7	TE			1.5	BL
4894.94	4896.31	20423.6	TE			2.2	BL
4893.58	4894.95	20429.2	TE			1.8	BL
4885.22	4886.58	20464.2	TE			2.0	BL
4875.53	4876.89	20504.9	TE			1.2	BL
4866.22	4867.58	20544.1	TE			2.9	BL
4865.13	4866.49	20548.7	TE			1.7	BL
4864.10	4865.46	20553.1	TE			2.9	BL
4842.88	4844.23	20643.1	TE			1.7	BL
4831.29	4832.64	20692.6	TE			2.9	BL
4827.14	4828.49	20710.4	TE			1.7	BL
4796.10	4797.44	20844.4	TE			1.8	BL
4784.85	4786.19	20893.5	TE			1.8	BL
4738.67	4740.00	21097.1	TE			1.7	BL
4733.67	4734.99	21119.4	TE			1.2	BL
4731.27	4732.59	21130.1	TE			1.8	BL
4729.83	4731.15	21136.5	TE			1.7	BL
4726.91	4728.23	21149.6	TE			1.2	BL
4716.81	4718.13	21194.8	TE			1.5	BL
4711.16	4712.48	21220.3	TE			1.8	BL
4706.53	4707.85	21241.1	TE			1.8	BL
4696.35	4697.66	21287.2	TE			1.7	BL
4686.95	4688.26	21329.9	TE			2.5	BL
4681.06	4682.37	21356.7	TE			1.2H	BL
4676.88	4678.19	21375.8	TE			1.2	BL
4670.11	4671.42	21406.8	TE			1.5	BL
4665.33	4666.64	21428.7	TE			1.8	BL
4664.34	4665.65	21433.3	TE			2.9	BL
4654.38	4655.68	21479.1	TE			2.9	BL
4647.51	4648.81	21510.9	TE			1.7	BL
4645.22	4646.52	21521.5	TE			1.2	BL
4641.19	4642.49	21540.2	TE			1.8	BL
4630.57	4631.87	21589.6	TE			1.7	BL
4621.27	4622.56	21633.0	TE			1.2	BL
4604.86	4606.15	21710.1	TE			1.2	BL
4602.37	4603.66	21721.9	TE			2.9	BL
4592.49	4593.78	21768.6	TE			1.5	BL
4586.98	4588.27	21794.7	TE			1.2	BL
4575.59	4576.87	21849.0	TE			1.5	BL
4569.71	4570.99	21877.1	TE			1.8	BL
4562.45	4563.73	21911.9	TE			1.2	BL
4557.84	4559.12	21934.1	TE			2.5	BL
4555.27	4556.55	21946.4	TE			1.5	BL
4552.20	4553.48	21961.2	TE			1.7	BL
4551.04	4552.32	21966.8	TE			1.2	BL
4550.36	4551.64	21970.1	TE			1.2	BL
4546.64	4547.91	21988.1	TE			1.7	BL
4545.97	4547.24	21991.3	TE			1.8	BL
4537.07	4538.34	22034.5	TE			1.7	BL
4535.83	4537.10	22040.5	TE			1.2	BL
4529.54	4530.81	22071.1	TE			1.5	BL
4522.22	4523.49	22106.8	TE			1.2	BL

AIR WAVELENGTH (ANGSTRMS)	VACUUM WAVELENGTH (ANGSTRMS)	WAVENUMBER (KAYSERS)	LMENT	LOG ARC	INTENSITY SPK	INTENSITY DIS	REF
4514.05	4515.32	22146.8	TE			1.5	BL
4498.59	4499.85	22222.9	TE			1.7	BL
4494.47	4495.73	22243.3	TE			2.0	BL
4489.33	4490.59	22268.8	TE			1.5	BL
4488.19	4489.45	22274.4	TE			1.5	BL
4485.77	4487.03	22286.5	TE			1.5	BL
4478.73	4479.99	22321.5	TE			2.9	BL
4455.28	4456.53	22439.0	TE			1.2	BL
4434.96	4436.21	22541.8	TE			1.8	BL
4421.14	4422.38	22612.2	TE			1.8	BL
4411.78	4413.02	22660.2	TE			1.7	BL
4410.95	4412.19	22664.5	TE			1.7	BL
4405.49	4406.73	22692.6	TE			1.2	BL
4401.89	4403.13	22711.1	TE			2.0	BL
4398.45	4399.69	22728.9	TE			1.8	BL
4396.00	4397.23	22741.6	TE			2.0	BL
4389.92	4391.15	22773.1	TE			1.7	BL
4385.08	4386.31	22798.2	TE			1.7	BL
4377.10	4378.33	22839.8	TE			1.8	BL
4373.00	4374.23	22861.2	TE			1.7	BL
4368.09	4369.32	22886.9	TE			1.5	BL
4364.02	4365.25	22908.2	TE			2.6	BL
4361.27	4362.50	22922.7	TE			1.7	BL
4349.29	4350.51	22985.8	TE			1.5	BL
4346.86	4348.08	22998.6	TE			1.2	BL
4342.42	4343.64	23022.2	TE			1.2	BL
4332.54	4333.76	23074.7	TE			1.2	BL
4325.77	4326.99	23110.8	TE			1.5	BL
4323.79	4325.01	23121.4	TE			1.5	BL
4321.96	4323.18	23131.1	TE			1.5	BL
4319.26	4320.47	23145.6	TE			1.7	BL
4294.25	4295.46	23280.4	TE			1.8	BL
4293.35	4294.56	23285.3	TE			1.8	BL
4288.51	4289.72	23311.6	TE			1.7	BL
4285.84	4287.05	23326.1	TE			1.8	BL
4279.48	4280.68	23360.7	TE			1.2	BL
4276.68	4277.88	23376.1	TE			1.7	BL
4273.40	4274.60	23394.0	TE			1.8	BL
4264.32	4265.52	23443.8	TE			1.5	BL
4261.08	4262.28	23461.6	TE			2.5	BL
4256.10	4257.30	23489.1	TE			1.7	BL
4251.15	4252.35	23516.4	TE			1.8	BL
4246.47	4247.67	23542.3	TE			1.2	BL
4231.74	4232.93	23624.3	TE			1.5	BL
4228.46	4229.65	23642.6	TE			1.7	BL
4225.70	4226.89	23658.1	TE			1.7	BL
4220.42	4221.61	23687.6	TE			2.0	BL
4211.32	4212.51	23738.8	TE			1.2	BL
4197.22	4198.40	23818.6	TE			1.2	BL
4183.99	4185.17	23893.9	TE			1.5	BL
4179.24	4180.42	23921.1	TE			1.8	BL
4169.77	4170.95	23975.4	TE			2.0	BL
4163.53	4164.70	24011.3	TE			1.7	BL
4154.14	4155.31	24065.6	TE			1.2	BL
4127.34	4128.50	24221.9	TE			1.5	BL
4127.04	4128.20	24223.6	TE			1.2	BL
4113.60	4114.76	24302.7	TE			1.2	BL
4101.07	4102.23	24377.0	TE			1.7	BL
4073.57	4074.72	24541.6	TE			2.5	BL
4048.89	4050.03	24691.1	TE			1.8	BL
4047.18	4048.32	24701.6	TE			1.2	BL
4029.73	4030.87	24808.6	TE			1.2	BL
4011.68	4012.81	24920.2	TE			1.5	BL
4006.50	4007.63	24952.4	TE			2.0	BL
3988.33	3989.46	25066.1	TE			0.7	BL
3981.74	3982.87	25107.6	TE			1.0	BL
3975.90	3977.02	25144.4	TE			1.0	BL
3969.18	3970.30	25187.0	TE			1.0	BL
3947.93	3949.05	25322.6	TE			1.0	BL
3936.21	3937.32	25398.0	TE			0.7	BL

AIR WAVELENGTH (ANGSTRMS)	VACUUM WAVELENGTH (ANGSTRMS)	WAVENUMBER (KAYSERS)	LMENT	LOG ARC	INTENSITY SPK	INTENSITY DIS	REF
3931.43	3932.54	25428.8	TE			0.7	BL
3927.63	3928.74	25453.4	TE			0.7	BL
3918.50	3919.61	25512.7	TE			0.7	BL
3905.11	3906.22	25600.2	TE			0.7	BL
3644.46	3645.50	27431.1	TE			1.0	BL
3644.27	3645.31	27432.5	TE			1.4	BL
3617.55	3618.58	27635.1	TE			1.4	BL
3612.90	3613.93	27670.7	TE			0.7	BL
3611.77	3612.80	27679.4	TE			0.7	BL
3593.34	3594.37	27821.3	TE			0.7	BL
3589.78	3590.80	27848.9	TE			0.7	BL
3585.34	3586.36	27883.4	TE			2.5	BL
3552.15	3553.16	28143.9	TE			1.2	BL
3543.96	3544.97	28209.0	TE			1.0	BL
3521.07	3522.08	28392.3	TE			1.2	BL
3506.94	3507.94	28506.7	TE			0.7	BL
3494.27	3495.27	28610.1	TE			1.2	BL
3480.28	3481.28	28725.1	TE			0.7	BL
3467.73	3468.72	28829.1	TE			1.2	BL
3456.84	3457.83	28919.9	TE			1.0	BL
3455.07	3456.06	28934.7	TE			1.2	BL
3449.57	3450.56	28980.8	TE			0.7	BL
3442.22	3443.21	29042.7	TE			1.0	BL
3438.73	3439.72	29072.2	TE			1.0	BL
3407.09	3408.07	29342.1	TE			0.7	BL
3406.77	3407.75	29344.9	TE			1.2	BL
3374.10	3375.07	29629.0	TE			1.4	BL
3365.17	3366.14	29707.6	TE			1.2	BL
3362.83	3363.80	29728.3	TE			1.7	BL
3358.12	3359.09	29770.0	TE			1.0	BL
3356.89	3357.85	29780.9	TE			0.7	BL
3352.11	3353.07	29823.4	TE			1.5	BL
3350.68	3351.64	29836.1	TE			0.7	BL
3345.93	3346.89	29878.5	TE			1.5	BL
3340.07	3341.03	29930.9	TE			1.4	BL
3329.25	3330.21	30028.2	TE			2.0	BL
3323.06	3324.02	30084.1	TE			1.2	BL
3306.99	3307.94	30230.3	TE			2.2	BL
3301.48	3302.43	30280.7	TE			0.7	BL
3301.18	3302.13	30283.5	TE			0.7	BL
3298.72	3299.67	30306.1	TE			0.7	BL
3293.43	3294.38	30354.7	TE			1.0	BL
3287.96	3288.91	30405.2	TE			1.2	BL
3285.29	3286.24	30429.9	TE			0.7	BL
3282.67	3283.62	30454.2	TE			1.0	BL
3278.77	3279.71	30490.5	TE			1.0	BL
3277.50	3278.44	30502.3	TE			1.2	BL
3270.22	3271.16	30570.2	TE			0.7	BL
3268.84	3269.78	30583.1	TE			1.2	BL
3256.81	3257.75	30696.0	TE			1.5	BL
3253.00	3253.94	30732.0	TE			0.7	BL
3251.37	3252.31	30747.4	TE			2.2	BL
3232.63	3233.56	30925.6	TE			1.0	BL
3229.52	3230.45	30955.4	TE			1.4	BL
3228.27	3229.20	30967.4	TE			0.7	BL
3223.69	3224.62	31011.4	TE			0.7	BL
3218.44	3219.37	31062.0	TE			2.0	BL
3211.20	3212.13	31132.0	TE			1.5	BL
3201.06	3201.99	31230.6	TE			0.7	BL
3194.41	3195.33	31295.6	TE			1.0	BL
3193.59	3194.51	31303.7	TE			1.0	BL
3193.12	3194.04	31308.3	TE			1.0	BL
3189.87	3190.79	31340.2	TE			0.7	BL
3188.37	3189.29	31354.9	TE			1.0	BL
3186.46	3187.38	31373.7	TE			0.7	BL
3175.11	3176.03	31485.9	TE I	1.5		1.2	BL
3160.63	3161.54	31630.1	TE			1.4	BL
3159.31	3160.22	31643.3	TE			1.0H	BL
3158.23	3159.14	31654.1	TE			1.0H	BL
3154.24	3155.15	31694.2	TE			0.7	BL
3152.81	3153.72	31708.6	TE			0.7	BL
3151.46	3152.37	31722.1	TE			1.2	BL
3145.18	3146.09	31785.5	TE			1.0	BL
3141.62	3142.53	31821.5	TE			0.7	BL
3136.15	3137.06	31877.0	TE			0.7	BL
3132.58	3133.49	31913.3	TE			1.0	BL
3128.74	3129.65	31952.5	TE			1.2	BL
3125.91	3126.82	31981.4	TE			1.0	BL
3120.36	3121.26	32038.3	TE			1.0	BL
3117.99	3118.89	32062.6	TE			1.4	BL
3116.30	3117.20	32080.0	TE			0.7	BL
3112.00	3112.90	32124.4	TE			1.2	BL
3107.10	3108.00	32175.0	TE			1.2	BL
3105.95	3106.85	32186.9	TE			0.7	BL
3104.41	3105.31	32202.9	TE			1.0	BL
3101.45	3102.35	32233.6	TE			0.7H	BL
3097.33	3098.23	32276.5	TE			0.7H	BL
3092.32	3093.22	32328.8	TE			1.2	BL
3089.73	3090.63	32355.9	TE			1.4	BL
3087.52	3088.42	32379.1	TE			1.0	BL
3085.40	3086.30	32401.3	TE			0.7H	BL
3083.91	3084.81	32416.9	TE			0.7	BL
3076.57	3077.46	32494.3	TE			1.4	BL
3073.53	3074.42	32526.4	TE			2.0	BL
3073.01	3073.90	32531.9	TE			1.0	BL
3068.94	3069.83	32575.1	TE			1.7	BL
3063.18	3064.07	32636.3	TE			1.0	BL
3053.17	3054.06	32743.3	TE			1.0	BL
3052.44	3053.33	32751.1	TE			0.7	BL
3050.01	3050.90	32777.2	TE			1.0	BL
3049.36	3050.25	32784.2	TE			1.5	BL
3047.00	3047.89	32809.6	TE			2.5	BL
3040.46	3041.34	32880.2	TE			0.7	BL
3034.62	3035.50	32943.5	TE			1.0	BL
3033.35	3034.23	32957.3	TE			0.7	BL
3029.90	3030.78	32994.8	TE			0.7	BL
3028.45	3029.33	33010.6	TE			1.2	BL
3023.29	3024.17	33066.9	TE			2.0	BL
3020.02	3020.90	33102.7	TE			1.2	BL
3017.51	3018.39	33130.3	TE			2.5	BL
3015.09	3015.97	33156.9	TE			1.3	BL
3013.63	3014.51	33172.9	TE			0.7	BL
3012.05	3012.93	33190.3	TE			1.4	BL
3009.96	3010.84	33213.4	TE			0.7	BL
3006.35	3007.23	33253.2	TE			1.7	BL
3004.87	3005.75	33269.6	TE			1.0	BL
2998.89	2999.76	33336.0	TE			1.0	BL
2997.05	2997.92	33356.4	TE			1.7	BL
2995.67	2996.54	33371.8	TE			1.2	BL
2990.73	2991.60	33426.9	TE			0.7	BL
2988.97	2989.84	33446.6	TE			1.0	BL
2985.47	2986.34	33485.8	TE			1.2	BL
2983.24	2984.11	33510.8	TE			1.0	BL
2980.02	2980.89	33547.0	TE			1.2	BL
2977.94	2978.81	33570.5	TE			1.5	BL
2975.91	2976.78	33593.4	TE			2.0	BL
2973.69	2974.56	33618.4	TE			1.7	BL
2967.21	2968.08	33691.9	TE			2.5	BL
2959.43	2960.29	33780.4	TE			1.4	BL
2956.50	2957.36	33813.9	TE			1.2	BL
2955.55	2956.41	33824.8	TE			1.0	BL
2954.82	2955.68	33833.1	TE			1.0	BL
2951.48	2952.34	33871.4	TE			1.2	BL
2950.21	2951.07	33886.0	TE			1.0	BL
2949.52	2950.38	33893.9	TE			2.0	BL
2944.96	2945.82	33946.4	TE			0.7	BL
2942.16	2943.02	33978.7	TE			2.0H	BL
2940.95	2941.81	33992.7	TE			0.7	BL
2937.90	2938.76	34028.0	TE			1.2	BL
2936.77	2937.63	34041.1	TE			0.7	BL

AIR WAVELENGTH (ANGSTRMS)	VACUUM WAVELENGTH (ANGSTRMS)	WAVENUMBER (KAYSERS)	LMENT	LOG ARC	INTENSITY SPK	INTENSITY DIS	REF
2931.88	2932.74	34097.8	TE			0.7	BL
2930.17	2931.03	34117.7	TE			1.0	BL
2928.23	2929.09	34140.3	TE			1.2	BL
2925.37	2926.23	34173.7	TE			1.4	BL
2924.02	2924.88	34189.5	TE			1.2	BL
2921.97	2922.83	34213.5	TE			0.7	BL
2919.96	2920.81	34237.0	TE			1.7	BL
2911.35	2912.20	34338.3	TE			1.0	BL
2908.36	2909.21	34373.6	TE			0.7	BL
2906.98	2907.83	34389.9	TE			1.2	BL
2904.05	2904.90	34424.6	TE			0.7	BL
2900.52	2901.37	34466.5	TE			1.0	BL
2895.49	2896.34	34526.4	TE			2.5H	BL
2893.27	2894.12	34552.8	TE			1.5	BL
2879.58	2880.42	34717.1	TE			0.7H	BL
2872.18	2873.02	34806.5	TE			0.7	BL
2869.72	2870.56	34836.4	TE			1.0	BL
2868.86	2869.70	34846.8	TE			2.0	BL
2867.67	2868.51	34861.3	TE			0.7	BL
2861.03	2861.87	34942.2	TE			1.4	BL
2859.38	2860.22	34962.4	TE			0.7	BL
2858.29	2859.13	34975.7	TE	0.3		2.0	BL
2846.14	2846.98	35125.0	TE			1.4	BL
2842.02	2842.86	35175.9	TE			0.3	BL
2841.18	2842.02	35186.3	TE			1.5	BL
2839.02	2839.85	35213.1	TE			1.0	BL
2831.48	2832.31	35306.8	TE			1.0	BL
2827.20	2828.03	35360.3	TE			1.2	BL
2821.57	2822.40	35430.8	TE			0.7	BL
2820.65	2821.48	35442.4	TE			1.0	BL
2818.70	2819.53	35466.9	TE			1.4	BL
2813.96	2814.79	35526.6	TE			0.7	BL
2809.90	2810.73	35578.0	TE			1.0	BL
2802.86	2803.69	35667.3	TE			1.2	BL
2802.53	2803.36	35671.5	TE			1.0	BL
2793.24	2794.06	35790.2	TE			2.5	BL
2791.95	2792.77	35806.7	TE			1.5	BL
2786.41	2787.23	35877.9	TE			0.7	BL
2778.08	2778.90	35985.5	TE			1.0	BL
2772.64	2773.46	36056.1	TE			0.7	BL
2772.14	2772.96	36062.6	TE			1.0	BL
2769.67	2770.49	36094.7	TE I			1.5	BL
2767.16	2767.98	36127.5	TE			1.0	BL
2759.51	2760.33	36227.6	TE			0.7	BL
2753.64	2754.45	36304.8	TE			1.0	BL
2752.21	2753.02	36323.7	TE			1.0	BL
2751.59	2752.40	36331.9	TE			1.0	BL
2747.29	2748.10	36388.7	TE			0.7	BL
2746.38	2747.19	36400.8	TE			0.7H	BL
2745.57	2746.38	36411.5	TE			1.4	BL
2740.37	2741.18	36480.6	TE			1.0	BL
2738.50	2739.31	36505.5	TE			0.7H	BL
2737.90	2738.71	36513.5	TE			1.4	BL
2736.84	2737.65	36527.7	TE			0.7	BL
2726.36	2727.17	36668.1	TE			1.5	BL
2724.25	2725.06	36696.5	TE			1.5	BL
2711.61	2712.41	36867.5	TE			1.7	BL
2703.54	2704.34	36977.6	TE			1.4	BL
2700.17	2700.97	37023.7	TE			0.7	BL
2697.67	2698.47	37058.0	TE			1.2	BL
2695.55	2696.35	37087.2	TE			1.7	BL
2694.55	2695.35	37100.9	TE			1.2	BL
2691.90	2692.70	37137.5	TE			1.2	BL
2689.06	2689.86	37176.7	TE			1.0	BL
2686.48	2687.28	37212.4	TE			0.7	BL
2684.52	2685.32	37239.5	TE			1.5	BL
2683.76	2684.56	37250.1	TE			0.7	BL
2683.21	2684.01	37257.7	TE			0.7	BL
2680.60	2681.40	37294.0	TE			1.0	BL
2677.16	2677.96	37341.9	TE			0.7	BL

AIR WAVELENGTH (ANGSTRMS)	VACUUM WAVELENGTH (ANGSTRMS)	WAVENUMBER (KAYSERS)	LMENT	LOG ARC	INTENSITY SPK	INTENSITY DIS	REF
2666.74	2667.53	37487.8	TE			1.2	BL
2662.11	2662.90	37553.0	TE			1.4	BL
2661.13	2661.92	37566.9	TE			1.5	BL
2657.72	2658.51	37615.0	TE			1.0	BL
2654.97	2655.76	37654.0	TE			0.7	BL
2649.80	2650.59	37727.5	TE			1.2	BL
2648.61	2649.40	37744.4	TE			1.0	BL
2644.84	2645.63	37798.2	TE			0.7	BL
2642.09	2642.88	37837.5	TE			1.0	BL
2640.35	2641.14	37862.5	TE			1.2	BL
2639.17	2639.96	37879.4	TE			0.7	BL
2637.84	2638.63	37898.5	TE			1.5	BL
2635.55	2636.34	37931.4	TE			2.5	BL
2628.24	2629.02	38036.9	TE			1.0	BL
2625.12	2625.90	38082.1	TE			1.2	BL
2622.26	2623.04	38123.7	TE			1.2	BL
2616.91	2617.69	38201.6	TE			0.7	BL
2608.18	2608.96	38329.5	TE			0.7	BL
2605.89	2606.67	38363.1	TE			1.0	BL
2595.33	2596.11	38519.2	TE			0.7	BL
2591.35	2592.13	38578.4	TE			1.5	BL
2586.04	2586.81	38657.6	TE			1.2	BL
2581.05	2581.82	38732.3	TE			1.2	BL
2579.24	2580.01	38759.5	TE			1.7	BL
2577.57	2578.34	38784.6	TE			1.4	BL
2575.70	2576.47	38812.8	TE			1.0	BL
2573.14	2573.91	38851.4	TE			1.2	BL
2568.13	2568.90	38927.2	TE			1.2	BL
2564.58	2565.35	38981.1	TE			2.2	BL
2562.28	2563.05	39016.0	TE			1.0	BL
2559.71	2560.48	39055.2	TE			1.4	BL
2549.21	2549.98	39216.1	TE			0.7	BL
2548.27	2549.03	39230.5	TE			0.7	BL
2544.89	2545.65	39282.6	TE			0.7	BL
2543.72	2544.48	39300.7	TE			1.0	BL
2537.80	2538.56	39392.4	TE			2.5	BL
2537.19	2537.95	39401.8	TE			0.7	BL
2533.41	2534.17	39460.6	TE			0.7	BL
2533.05	2533.81	39466.2	TE			1.0	BL
2530.70	2531.46	39502.9	TE I			1.5	BL
2530.07	2530.83	39512.7	TE			1.2	BL
2527.14	2527.90	39558.5	TE			1.0	BL
2525.78	2526.54	39579.8	TE			0.7	BL
2524.90	2525.66	39593.6	TE			1.2	BL
2519.03	2519.79	39685.9	TE			0.7	BL
2516.49	2517.25	39725.9	TE			1.2	BL
2500.88	2501.63	39973.9	TE			1.0	BL
2499.75	2500.50	39991.9	TE			2.5	BL
2499.22	2499.97	40000.4	TE			1.0	BL
2494.67	2495.42	40073.4	TE			1.2	BL
2490.33	2491.08	40143.2	TE			1.2	BL
2489.20	2489.95	40161.4	TE			1.0	BL
2485.24	2485.99	40225.4	TE			1.0	BL
2481.22	2481.97	40290.6	TE			1.0	BL
2479.85	2480.60	40312.8	TE			0.7	BL
2469.62	2470.37	40479.8	TE			2.5	BL
2467.72	2468.47	40511.0	TE			1.7	BL
2456.90	2457.64	40689.4	TE			0.7	BL
2455.54	2456.28	40711.9	TE			0.7	BL
2452.53	2453.27	40761.9	TE			1.0	BL
2447.92	2448.66	40838.6	TE			0.7	BL
2446.89	2447.63	40855.8	TE			1.7	BL
2438.80	2439.54	40991.3	TE			1.4	BL
2436.63	2437.37	41027.9	TE			1.4	BL
2434.73	2435.47	41059.9	TE			1.2	BL
2434.35	2435.09	41066.3	TE			1.0	BL
2433.82	2434.56	41075.2	TE			1.0	BL
2431.78	2432.52	41109.7	TE I	1.0		1.2	BL
2431.45	2432.19	41115.2	TE			1.0	BL
2426.39	2427.13	41201.0	TE			2.0	BL

AIR WAVELENGTH (ANGSTRMS)	VACUUM WAVELENGTH (ANGSTRMS)	WAVENUMBER (KAYSERS)	LMENT	LOG ARC	INTENSITY SPK	INTENSITY DIS	REF
2420.14	2420.88	41307.4	TE I	1.0		0.3	BL
2415.59	2416.32	41385.2	TE			1.7	BL
2412.06	2412.79	41445.7	TE			1.0	BL
2411.38	2412.11	41457.4	TE			1.7	BL
2410.88	2411.61	41466.0	TE			0.7	BL
2407.54	2408.27	41523.5	TE			0.7	BL
2403.66	2404.39	41590.6	TE			1.5	BL
2403.00	2403.73	41602.0	TE			2.0	BL
2401.81	2402.54	41622.6	TE			1.0	BL
2392.19	2392.92	41790.0	TE			0.7	BL
2389.79	2390.52	41831.9	TE			1.4	BL
2385.76	2386.49	41902.6	TE I	2.8		2.5	BL
2383.25	2383.98	41946.7	TE I	2.7	2.5		BL
2374.94	2375.66	42093.5	TE			1.5	BL
2372.86	2373.58	42130.4	TE			1.0	BL
2369.92	2370.64	42182.6	TE			1.7	BL
2367.00	2367.72	42234.7	TE			1.0	BL
2361.64	2362.36	42330.5	TE			0.7	BL
2359.53	2360.25	42368.4	TE			0.7	BL
2358.23	2358.95	42391.7	TE			1.0	BL
2356.60	2357.32	42421.0	TE			0.7	BL
2355.08	2355.80	42448.4	TE			0.7	BL
2354.90	2355.62	42451.7	TE			0.7	BL
2351.22	2351.94	42518.1	TE			1.4	BL
2350.12	2350.84	42538.0	TE			1.4	BL
2348.74	2349.46	42563.0	TE			0.7	BL
2343.83	2344.55	42652.1	TE			1.0	BL
2336.15	2336.87	42792.4	TE			1.2	BL
2334.91	2335.63	42815.1	TE			0.7	BL
2332.32	2333.04	42862.6	TE			1.2	BL
2329.12	2329.83	42921.5	TE			1.7	BL
2327.45	2328.16	42952.3	TE			1.4	BL
2326.22	2326.93	42975.0	TE			0.7	BL
2322.22	2322.93	43049.0	TE			0.7	BL
2320.09	2320.80	43088.5	TE			1.0	BL
2311.81	2312.52	43242.9	TE			0.7	BL
2304.29	2305.00	43384.0	TE			1.0H	BL
2288.90	2289.61	43675.6	TE			0.7	BL
2287.84	2288.55	43695.9	TE			0.7	BL
2286.71	2287.42	43717.5	TE			1.5	BL
2271.97	2272.67	44001.1	TE			1.4	BL
2265.55	2266.25	44125.7	TE I			1.0	BL
2265.02	2265.72	44136.1	TE			1.0	BL
2261.66	2262.36	44201.6	TE			1.2	BL
2259.04	2259.74	44252.9	TE I			1.0	BL
2255.48	2256.18	44322.7	TE I			1.0	BL
2231.36	2232.05	44801.8	TE			0.7	BL
2226.07	2226.76	44908.2	TE			1.2	BL
2216.61	2217.30	45099.9	TE	1.0			KH
2209.45	2210.14	45246.0	TE			0.7	LC
2208.83	2209.52	45258.7	TE I	2.3		0.3	BL
2207.17	2207.86	45292.8	TE	1.7			KH
2196.07	2196.76	45521.7	TE	0.7			KH
2187.22	2187.90	45705.8	TE	0.7			KH
2178.13	2178.81	45896.6	TE	1.0			KH
2166.88	2167.56	46134.8	TE	2.0			KH
2159.79	2160.47	46286.3	TE I	2.0			ML
2148.04	2148.72	46539.4	TE			1.2	LC
2147.19	2147.87	46557.8	TE I			2.2	ML
2143.54	2144.22	46637.1	TE			1.4	LC
2142.75	2143.43	46654.3	TE I	2.8			ML
2121.75	2122.42	47116.0	TE			1.0	LC
2117.46	2118.13	47211.5	TE			1.0	LC
2109.83	2110.50	47382.2	TE			1.5	LC
2107.63	2108.30	47431.6	TE	2.0			KH
2107.20	2107.87	47441.3	TE	2.0			KH
2101.74	2102.41	47564.5	TE			1.0	LC
2081.88	2082.54	48018.2	TE			0.7	LC
2081.03	2081.69	48037.8	TE I	2.6			ML
2074.74	2075.40	48183.4	TE	1.5			KH

AIR WAVELENGTH (ANGSTRMS)	VACUUM WAVELENGTH (ANGSTRMS)	WAVENUMBER (KAYSERS)	LMENT	LOG ARC	INTENSITY SPK	INTENSITY DIS	REF
2070.9	2071.6	48273.	TE	2.2			ML
2067.94	2068.60	48341.9	TE			1.2	LC
2066.49	2067.15	48375.8	TE			1.5	LC
2061.03	2061.69	48503.9	TE	0.7			KH
2044.22	2044.88	48902.7	TE			0.7	LC
2039.79	2040.45	49008.9	TE	2.5			KH
2024.20	2024.85	49386.3	TE	0.7			KH
2002.72	2003.37	49915.9	TE	1.7		0.7	KH
2002.0	2002.6	49934.	TE I	1.7			ML
2001.59	2002.24	49944.1	TE	2.8			KH
2000.2	2000.8	49979.	TE I	1.5			ML
8330.48	8332.77	12000.8	TH	0.8			FD
8320.86	8323.15	12014.7	TH	0.3			FD
8275.63	8277.90	12080.3	TH	0.7			FD
8186.92	8189.17	12211.2	TH	0.8			FD
8159.75	8161.99	12251.9	TH	0.6			FD
8143.15	8145.39	12276.9	TH	0.6			FD
8032.48	8034.69	12446.0	TH	0.3			FD
7978.97	7981.16	12529.5	TH	0.6			FD
7847.53	7849.69	12739.4	TH	0.5			FD
7817.78	7819.93	12787.8	TH	0.5			FD
7788.95	7791.09	12835.2	TH	0.3			FD
7525.52	7527.59	13284.5	TH	0.7			FD
7481.36	7483.42	13362.9	TH	0.3			FD
7430.27	7432.32	13454.8	TH	0.3			FD
7428.96	7431.01	13457.1	TH	0.5			FD
7393.433	7395.470	13521.79	TH	0.7			FD
7385.485	7387.519	13536.34	TH	0.9			FD
7383.74	7385.77	13539.5	TH	0.3			FD
7376.95	7378.98	13552.0	TH	0.3			FD
7341.20	7343.22	13618.0	TH	0.3			FD
7328.34	7330.36	13641.9	TH	0.5			FD
7324.86	7326.88	13648.4	TH	0.5			FD
7288.98	7290.99	13715.6	TH	0.7H			FD
7287.05	7289.06	13719.2	TH	0.6			FD
7284.95	7286.96	13723.2	TH	0.3			FD
7244.73	7246.73	13799.3	TH	0.3			MIT
7242.09	7244.09	13804.4	TH	0.3			FD
7230.88	7232.87	13825.8	TH	0.5			FD
7219.18	7221.17	13848.2	TH	0.3			FD
7218.11	7220.10	13850.2	TH	0.6			FD
7217.79	7219.78	13850.8	TH	0.5			FD
7212.75	7214.74	13860.5	TH	0.6			FD
7208.005	7209.992	13869.64	TH	1.1			FD
7200.08	7202.06	13884.9	TH	0.5			FD
7173.42	7175.40	13936.5	TH	0.5			FD
7168.888	7170.864	13945.32	TH	1.2			FD
7164.88	7166.85	13953.1	TH	0.9			FD
7159.98	7161.95	13962.7	TH	0.3			FD
7159.10	7161.07	13964.4	TH	1.0			FD
7158.56	7160.53	13965.4	TH	0.7			FD
7156.98	7158.95	13968.5	TH	0.3			FD
7155.55	7157.52	13971.3	TH	1.0			FD
7155.05	7157.02	13972.3	TH	0.8H			FD
7150.34	7152.31	13981.5	TH	0.3			FD
7148.59	7150.56	13984.9	TH	0.6			FD
7124.61	7126.57	14032.0	TH	0.8			FD
7064.48	7066.43	14151.4	TH	0.5			FD
7060.71	7062.66	14159.0	TH	0.5			FD
7036.30	7038.24	14208.1	TH	0.6			FD
7018.574	7020.510	14243.98	TH	0.5			FD
7000.795	7002.726	14280.15	TH	0.9			FD
6989.657	6991.585	14302.91	TH	1.3			FD
6943.610	6945.525	14397.76	TH	1.0			FD
6911.238	6913.145	14465.20	TH	1.0			FD
6874.77	6876.67	14541.9	TH	0.3			FD
6868.55	6870.45	14555.1	TH	0.3			FD
6834.921	6836.807	14626.71	TH	0.7			FD
6829.05	6830.93	14639.3	TH	0.5			FD
6824.73	6826.61	14648.6	TH	0.3			FD

AIR WAVELENGTH (ANGSTRMS)	VACUUM WAVELENGTH (ANGSTRMS)	WAVENUMBER (KAYSERS)	LMENT	LOG ARC	INTENSITY SPK	DIS	REF
6791.27	6793.14	14720.7	TH	0.3			FD
6780.268	6782.139	14744.61	TH	0.6J			FD
6772.283	6774.152	14761.99	TH	0.3			FD
6678.711	6680.555	14968.82	TH	0.3			FD
6662.274	6664.114	15005.75	TH	1.0			FD
6658.681	6660.520	15013.84	TH	0.6H			FD
6644.66	6646.49	15045.5	TH	0.5			MIT
6619.877	6621.705	15101.85	TH	0.5			FD
6605.432	6607.256	15134.87	TH	0.7H			FD
6593.96	6595.78	15161.2	TH	0.7			FD
6591.480	6593.301	15166.91	TH	0.9			FD
6588.540	6590.360	15173.68	TH	1.0			FD
6583.907	6585.726	15184.36	TH	1.0			FD
6577.203	6579.020	15199.83	TH	0.6			FD
6554.148	6555.959	15253.30	TH	0.7			FD
6531.348	6533.152	15306.55	TH	1.2			FD
6512.365	6514.164	15351.16	TH	0.5			FD
6511.00	6512.80	15354.4	TH	0.6			MIT
6501.991	6503.788	15375.66	TH	0.3			FD
6496.94	6498.74	15387.6	TH	0.3			MIT
6493.198	6494.992	15396.48	TH	0.3			FD
6471.210	6472.998	15448.79	TH	0.7			MIT
6462.646	6464.432	15469.26	TH	1.5S			MIT
6457.285	6459.070	15482.11	TH	1.2			MIT
6424.817	6426.593	15560.34	TH	0.6	0.3		MIT
6416.101	6417.874	15581.48	TH	0.5	0.3		MIT
6413.612	6415.385	15587.53	TH	0.7			MIT
6411.893	6413.665	15591.71	TH	1.0			MIT
6408.598	6410.369	15599.72	TH	0.6			MIT
6376.944	6378.707	15677.16	TH	1.1			MIT
6371.936	6373.698	15689.48	TH	0.6			MIT
6369.144	6370.905	15696.36	TH	0.3			MIT
6348.559	6350.314	15747.25	TH	0.5	0.5		MIT
6345.989	6347.744	15753.63	TH	0.5J			MIT
6342.861	6344.615	15761.40	TH	1.2			MIT
6339.458	6341.211	15769.86	TH	0.5			MIT
6337.628	6339.380	15774.41	TH	0.3			MIT
6331.412	6333.163	15789.90	TH	0.5			MIT
6327.268	6329.018	15800.24	TH	1.0			MIT
6326.358	6328.107	15802.51	TH	0.6			MIT
6304.240	6305.983	15857.95	TH	0.8			MIT
6303.260	6305.003	15860.42	TH	0.5	0.3		MIT
6289.490	6291.229	15895.14	TH	0.5			MIT
6285.281	6287.019	15905.79	TH	1.1	0.0		MIT
6279.170	6280.907	15921.27	TH	1.3	0.3		MIT
6277.249	6278.985	15926.14	TH	1.0			MIT
6274.109	6275.844	15934.11	TH	1.4	0.5		MIT
6266.169	6267.902	15954.30	TH	0.9			MIT
6261.420	6263.152	15966.40	TH	0.7			MIT
6261.063	6262.795	15967.31	TH	1.2	0.3		MIT
6247.990	6249.718	16000.72	TH	0.3			MIT
6234.860	6236.585	16034.42	TH	0.5			MIT
6232.970	6234.694	16039.28	TH	1.0	0.5		MIT
6224.518	6226.240	16061.06	TH	0.7			MIT
6207.737	6209.455	16104.47	TH	0.5			MIT
6207.218	6208.935	16105.82	TH	1.0			MIT
6203.491	6205.207	16115.50	TH	0.9	0.0		MIT
6200.424	6202.140	16123.47	TH	0.7			MIT
6198.215	6199.930	16129.21	TH	0.8			MIT
6193.851	6195.565	16140.58	TH	1.1	0.0		MIT
6191.895	6193.608	16145.68	TH	0.7			MIT
6188.112	6189.824	16155.55	TH	0.5	0.0		MIT
6184.727	6186.438	16164.39	TH	0.6			MIT
6182.623	6184.334	16169.89	TH	1.1			MIT
6179.852	6181.562	16177.14	TH	0.8			MIT
6169.818	6171.525	16203.45	TH	1.0			MIT
6164.475	6166.181	16217.49	TH	0.5	0.3H		MIT
6151.984	6153.687	16250.42	TH	1.0			MIT
6145.271	6146.972	16268.17	TH	0.7			MIT
6138.630	6140.329	16285.77	TH	0.8			MIT

AIR WAVELENGTH (ANGSTRMS)	VACUUM WAVELENGTH (ANGSTRMS)	WAVENUMBER (KAYSERS)	LMENT	LOG ARC	INTENSITY SPK	DIS	REF
6125.730	6127.426	16320.07	TH	0.7			MIT
6121.442	6123.136	16331.50	TH	0.8H			MIT
6120.547	6122.241	16333.89	TH	1.2	0.5		MIT
6112.822	6114.514	16354.53	TH	1.2	0.5		MIT
6104.569	6106.259	16376.64	TH	1.1	0.3		MIT
6102.589	6104.278	16381.95	TH	0.6			MIT
6099.076	6100.764	16391.39	TH	1.1	0.3		MIT
6088.037	6089.722	16421.11	TH	0.7	0.3		FD
6087.259	6088.944	16423.21	TH	1.1	0.3		MIT
6085.248	6086.933	16428.64	TH	0.8			MIT
6073.104	6074.785	16461.49	TH	1.1	0.0		MIT
6053.379	6055.055	16515.13	TH	0.7			MIT
6049.055	6050.730	16526.93	TH	0.8			MIT
6044.425	6046.099	16539.59	TH	1.1	0.3		MIT
6037.692	6039.364	16558.03	TH	0.9			MIT
6021.035	6022.702	16603.84	TH	0.8			MIT
6018.977	6020.644	16609.52	TH	0.9			MIT
6015.418	6017.084	16619.35	TH	1.0	0.3		MIT
6010.158	6011.823	16633.89	TH	0.5			MIT
6007.068	6008.732	16642.45	TH	1.0			MIT
6000.769	6002.431	16659.92	TH	0.3			FD
5998.035	5999.696	16667.51	TH I	0.0	0.7		MIT
5994.129	5995.789	16678.37	TH	1.0			MIT
5991.005	5992.664	16687.07	TH	0.9			MIT
5989.035	5990.694	16692.56	TH	1.3	0.7		MIT
5975.065	5976.720	16731.58	TH	1.0			MIT
5973.669	5975.324	16735.50	TH	1.0	0.0		MIT
5961.973	5963.625	16768.33	TH	0.9			MIT
5960.376	5962.027	16772.82	TH	0.5			MIT
5942.798	5944.444	16822.43	TH	0.9			MIT
5938.834	5940.479	16833.66	TH	0.9			MIT
5938.582	5940.227	16834.37	TH	0.8			MIT
5929.487	5931.130	16860.19	TH	0.7			MIT
5925.894	5927.536	16870.42	TH	1.0	0.3		MIT
5925.414	5927.056	16871.78	TH	0.5			MIT
5914.395	5916.034	16903.22	TH	1.1	0.5		MIT
5904.178	5905.814	16932.47	TH	0.9	0.0		MIT
5899.414	5901.049	16946.14	TH	0.8			MIT
5893.498	5895.131	16963.15	TH	0.9			MIT
5888.283	5889.915	16978.17	TH	0.9			MIT
5885.714	5887.345	16985.58	TH	1.0	0.7		MIT
5878.265	5879.894	17007.11	TH	0.9			MIT
5870.572	5872.199	17029.40	TH	1.0			MIT
5859.688	5861.312	17061.03	TH	1.1	0.0		MIT
5859.522	5861.146	17061.51	TH	0.3	0.0		MIT
5838.951	5840.570	17121.62	TH	1.1	0.0		MIT
5815.432	5817.044	17190.86	TH	1.3D	0.3		MIT
5804.141	5805.750	17224.30	TH	1.1			MIT
5800.828	5802.436	17234.14	TH	1.0			MIT
5796.439	5798.046	17247.19	TH	1.0	0.0		MIT
5792.427	5794.033	17259.14	TH	0.7			MIT
5784.893	5786.497	17281.61	TH	0.9			MIT
5768.181	5769.781	17331.68	TH	0.8			MIT
5767.788	5769.388	17332.86	TH	0.6			MIT
5760.554	5762.152	17354.63	TH	1.1			MIT
5753.021	5754.617	17377.35	TH	0.9			MIT
5749.604	5751.199	17387.68	TH	0.3			MIT
5749.379	5750.974	17388.36	TH	1.1D	0.0		MIT
5747.562	5749.156	17393.86	TH	0.3			MIT
5745.665	5747.259	17399.60	TH	0.8			MIT
5742.079	5743.672	17410.47	TH	1.1	0.0		MIT
5741.162	5742.754	17413.25	TH	1.0			MIT
5732.965	5734.555	17438.14	TH	1.0	0.0		MIT
5725.601	5727.189	17460.57	TH	0.7			MIT
5725.406	5726.994	17461.17	TH	1.0			MIT
5725.031	5726.619	17462.31	TH	0.8			MIT
5720.176	5721.763	17477.13	TH	1.3D			MIT
5719.622	5721.209	17478.82	TH	0.9			MIT
5707.094	5708.677	17517.19	TH	1.3	0.7		MIT
5700.909	5702.491	17536.20	TH	1.1	0.5		MIT

AIR WAVELENGTH (ANGSTRMS)	VACUUM WAVELENGTH (ANGSTRMS)	WAVENUMBER (KAYSERS)	LMENT	LOG ARC	INTENSITY SPK	DIS	REF
5700.696	5702.278	17536.85	TH	0.9	0.0		MIT
5700.451	5702.033	17537.61	TH	0.9	0.0		MIT
5677.067	5678.642	17609.84	TH	0.6			MIT
5675.013	5676.588	17616.22	TH	0.9			MIT
5665.625	5667.197	17645.41	TH	1.0	0.3		MIT
5665.182	5666.754	17646.79	TH	0.7			MIT
5657.926	5659.496	17669.42	TH	0.8			MIT
5654.014	5655.583	17681.64	TH	0.9			MIT
5652.902	5654.471	17685.12	TH	0.9	0.0		MIT
5645.888	5647.455	17707.09	TH	1.1	0.3		MIT
5639.733	5641.298	17726.42	TH	1.1	0.6		MIT
5615.321	5616.880	17803.48	TH	1.0			MIT
5604.508	5606.064	17837.83	TH	1.1D	0.3		MIT
5601.620	5603.175	17847.02	TH	0.8			MIT
5599.787	5601.342	17852.87	TH	0.8L			MIT
5599.665	5601.220	17853.25	TH	0.3			MIT
5595.068	5596.621	17867.92	TH	0.9			MIT
5593.615	5595.168	17872.56	TH	1.0	0.0		MIT
5588.754	5590.306	17888.11	TH	1.0			MIT
5587.741	5589.292	17891.35	TH	1.0	0.0		MIT
5587.026	5588.577	17893.64	TH	1.0D			MIT
5583.759	5585.309	17904.11	TH	1.0	0.0		MIT
5579.371	5580.920	17918.19	TH	1.0			MIT
5573.355	5574.903	17937.53	TH	1.0	0.0		MIT
5572.477	5574.024	17940.36	TH	0.8	0.0		MIT
5571.200	5572.747	17944.47	TH	0.9	0.0		MIT
5564.999	5566.544	17964.47	TH	0.6			MIT
5564.204	5565.749	17967.03	TH	1.1D	0.5		MIT
5559.899	5561.443	17980.95	TH	0.8			MIT
5558.341	5559.885	17985.99	TH	1.0			MIT
5557.313	5558.856	17989.31	TH	0.9			MIT
5551.368	5552.910	18008.58	TH	1.0			MIT
5548.165	5549.706	18018.97	TH	1.2S			MIT
5546.117	5547.657	18025.63	TH	0.9			MIT
5542.897	5544.436	18036.10	TH	0.7			MIT
5539.901	5541.440	18045.85	TH	1.3D	0.6		MIT
5539.259	5540.798	18047.94	TH	1.1			MIT
5537.127	5538.665	18054.89	TH	1.0	0.0		MIT
5533.129	5534.666	18067.94	TH	0.7			MIT
5530.689	5532.225	18075.91	TH	0.9			MIT
5528.218	5529.754	18083.99	TH	0.3			MIT
5528.003	5529.538	18084.69	TH	0.6			MIT
5524.615	5526.150	18095.78	TH	0.9W			MIT
5524.212	5525.746	18097.10	TH	0.9			MIT
5521.77	5523.30	18105.1	TH	0.6			MIT
5518.988	5520.521	18114.23	TH	0.6			MIT
5514.862	5516.394	18127.78	TH	0.7			MIT
5510.678	5512.209	18141.55	TH	0.7			MIT
5509.989	5511.520	18143.82	TH	1.0			MIT
5504.305	5505.834	18162.55	TH	0.8			MIT
5501.938	5503.467	18170.37	TH	1.0			MIT
5499.251	5500.779	18179.24	TH	1.1			MIT
5495.805	5497.332	18190.64	TH	0.8			MIT
5492.640	5494.166	18201.12	TH	0.9			MIT
5488.921	5490.446	18213.46	TH	0.8			MIT
5488.622	5490.147	18214.45	TH	1.0			MIT
5484.138	5485.662	18229.34	TH	1.1			MIT
5480.888	5482.411	18240.15	TH	0.8			MIT
5474.862	5476.383	18260.23	TH	1.0	0.5		MIT
5462.615	5464.133	18301.17	TH	1.1	0.3		MIT
5461.738	5463.256	18304.10	TH	1.1	0.0		MIT
5452.225	5453.740	18336.04	TH	0.9			MIT
5449.478	5450.993	18345.28	TH	1.1	0.3		MIT
5443.118	5444.631	18366.72	TH	1.0	0.0		MIT
5437.392	5438.903	18386.06	TH	1.0			MIT
5435.898	5437.409	18391.11	TH	1.2	0.3		MIT
5435.128	5436.639	18393.72	TH	0.9			MIT
5433.703	5435.213	18398.54	TH	0.8			MIT
5433.292	5434.802	18399.93	TH	0.6			MIT
5431.123	5432.633	18407.28	TH	1.0			MIT

AIR WAVELENGTH (ANGSTRMS)	VACUUM WAVELENGTH (ANGSTRMS)	WAVENUMBER (KAYSERS)	LMENT	LOG ARC	INTENSITY SPK	DIS	REF
5425.679	5427.187	18425.75	TH	1.2D	0.5		MIT
5424.15	5425.66	18430.9	TH	0.5			MIT
5421.837	5423.344	18438.81	TH	0.9			MIT
5417.487	5418.993	18453.61	TH	1.0			MIT
5415.434	5416.939	18460.61	TH	1.3D	0.5		MIT
5410.764	5412.268	18476.54	TH	0.9			MIT
5407.655	5409.158	18487.16	TH	0.9			MIT
5398.923	5400.424	18517.06	TH	0.9			MIT
5394.750	5396.250	18531.39	TH	0.9			MIT
5392.573	5394.072	18538.87	TH	1.0	0.0		MIT
5390.461	5391.960	18546.13	TH	1.3	0.7		MIT
5388.06	5389.56	18554.4	TH	0.5			MIT
5386.786	5388.284	18558.78	TH	0.3			MIT
5386.606	5388.104	18559.41	TH	0.8			MIT
5384.037	5385.534	18568.26	TH	0.9			MIT
5382.927	5384.424	18572.09	TH	1.1	0.0		MIT
5379.111	5380.607	18585.26	TH	0.6			MIT
5378.833	5380.329	18586.23	TH	0.7			MIT
5376.754	5378.249	18593.41	TH	0.8			MIT
5375.770	5377.265	18596.81	TH	1.0	0.0		MIT
5375.351	5376.846	18598.26	TH	1.0			MIT
5363.611	5365.103	18638.97	TH	0.7			MIT
5362.249	5363.740	18643.71	TH	0.6			MIT
5361.411	5362.902	18646.62	TH	0.8			MIT
5351.13	5352.62	18682.4	TH	0.9			FL
5347.043	5348.530	18696.73	TH	1.0	0.0W		MIT
5346.394	5347.881	18699.00	TH	0.8			MIT
5345.315	5346.802	18702.77	TH	0.9	0.0		MIT
5343.585	5345.071	18708.82	TH	1.1			MIT
5329.371	5330.853	18758.72	TH	0.9			MIT
5326.976	5328.458	18767.16	TH	1.0			MIT
5325.145	5326.626	18773.61	TH	1.2D	0.5		MIT
5320.774	5322.254	18789.03	TH	0.9			MIT
5318.423	5319.903	18797.34	TH	0.9			MIT
5312.883	5314.361	18816.94	TH	0.8			MIT
5312.515	5313.993	18818.24	TH	0.8			MIT
5311.98	5313.46	18820.1	TH	0.9			MIT
5310.260	5311.737	18826.23	TH	1.1D	0.0		MIT
5307.463	5308.940	18836.15	TH	1.0	0.3		MIT
5305.580	5307.056	18842.84	TH	1.0			MIT
5304.625	5306.101	18846.23	TH	0.7			MIT
5301.406	5302.881	18857.67	TH	1.1	0.3		MIT
5297.742	5299.216	18870.72	TH	0.9			MIT
5289.895	5291.367	18898.71	TH	0.8			MIT
5284.549	5286.020	18917.83	TH	0.7			MIT
5280.090	5281.559	18933.80	TH	0.8			MIT
5277.502	5278.971	18943.09	TH	1.2D	0.5		MIT
5276.407	5277.875	18947.02	TH	0.6			MIT
5272.646	5274.113	18960.53	TH	0.8			MIT
5258.358	5259.822	19012.05	TH	0.8			MIT
5253.455	5254.917	19029.79	TH	0.8			MIT
5247.652	5249.113	19050.84	TH	1.3D	0.5		MIT
5242.088	5243.547	19071.06	TH	0.5			MIT
5240.194	5241.653	19077.95	TH	1.0	0.0		MIT
5238.815	5240.273	19082.97	TH	0.5			MIT
5237.91	5239.37	19086.3	TH	1.0			MIT
5233.226	5234.683	19103.35	TH	1.1	0.5		MIT
5232.016	5233.473	19107.77	TH	0.7			MIT
5231.159	5232.615	19110.90	TH	1.1			MIT
5225.61	5227.06	19131.2	TH	0.5			MIT
5219.112	5220.565	19155.01	TH	0.9			MIT
5218.528	5219.981	19157.16	TH	1.0	0.3		MIT
5216.590	5218.042	19164.27	TH	1.1	0.3		MIT
5214.15	5215.60	19173.2	TH	0.6			MIT
5213.358	5214.810	19176.16	TH	0.6			MIT
5211.235	5212.686	19183.97	TH	0.8			MIT
5206.659	5208.109	19200.83	TH	0.6	0.0		MIT
5206.490	5207.940	19201.45	TH	0.8			MIT
5205.779	5207.229	19204.07	TH	0.9	0.0		MIT
5203.865	5205.314	19211.14	TH	0.7			MIT

AIR WAVELENGTH (ANGSTRMS)	VACUUM WAVELENGTH (ANGSTRMS)	WAVENUMBER (KAYSERS)	LMENT	LOG ARC	INTENSITY SPK	INTENSITY DIS	REF
5199.113	5200.561	19228.70	TH	0.6			FD
5198.832	5200.280	19229.74	TH	1.2			MIT
5195.814	5197.261	19240.90	TH	0.7			MIT
5193.820	5195.266	19248.29	TH	0.9	0.0		MIT
5190.870	5192.316	19259.23	TH	1.0	0.0		MIT
5189.681	5191.126	19263.64	TH	0.8	0.0		MIT
5188.371	5189.816	19268.51	TH	0.8			MIT
5187.445	5188.890	19271.95	TH	0.6			MIT
5184.726	5186.170	19282.05	TH	0.5			MIT
5183.986	5185.430	19284.80	TH	0.8	0.0		MIT
5182.524	5183.967	19290.25	TH	0.7	0.3		MIT
5178.49	5179.93	19305.3	TH	0.5			MIT
5177.63	5179.07	19308.5	TH	0.3			MIT
5176.965	5178.407	19310.96	TH	1.0			MIT
5174.198	5175.639	19321.28	TH	0.7			MIT
5170.23	5171.67	19336.1	TH	0.8			MIT
5166.902	5168.341	19348.57	TH	0.5O	0.3H		MIT
5163.446	5164.884	19361.52	TH	1.1			MIT
5161.545	5162.983	19368.65	TH	0.3			MIT
5160.695	5162.133	19371.84	TH	1.1	0.3		MIT
5158.607	5160.044	19379.68	TH	1.1			MIT
5154.247	5155.683	19396.07	TH	0.9			MIT
5151.862	5153.297	19405.05	TH	0.3			MIT
5151.620	5153.055	19405.96	TH	0.6			MIT
5148.215	5149.649	19418.80	TH	1.1	0.5		MIT
5145.028	5146.461	19430.83	TH	0.7			MIT
5143.922	5145.355	19435.00	TH	0.5			MIT
5143.264	5144.697	19437.49	TH	1.0			MIT
5140.719	5142.151	19447.11	TH	0.8			MIT
5132.262	5133.692	19479.16	TH	0.7			MIT
5131.078	5132.508	19483.65	TH	1.0	0.0		MIT
5115.047	5116.472	19544.72	TH	1.0			MIT
5113.39	5114.81	19551.1	TH	0.8			MIT
5110.868	5112.292	19560.70	TH	1.0	0.3		MIT
5108.25	5109.67	19570.7	TH	0.6W	0.0		MIT
5103.768	5105.190	19587.91	TH	0.8			MIT
5100.625	5102.047	19599.98	TH	0.8			MIT
5098.046	5099.467	19609.89	TH	1.0	0.0		MIT
5096.484	5097.904	19615.90	TH	0.7			MIT
5095.064	5096.484	19621.37	TH	1.0			MIT
5093.091	5094.511	19628.97	TH	0.5			MIT
5090.759	5092.178	19637.96	TH	0.9	0.0		MIT
5089.221	5090.640	19643.90	TH	0.6			MIT
5085.12	5086.54	19659.7	TH	0.3	0.0		MIT
5079.141	5080.557	19682.88	TH	0.7			MIT
5076.599	5078.014	19692.74	TH	0.3			MIT
5075.471	5076.886	19697.11	TH	0.8			MIT
5072.633	5074.047	19708.13	TH	0.7			MIT
5067.973	5069.386	19726.25	TH	1.0			MIT
5066.779	5068.192	19730.90	TH	0.6			MIT
5064.944	5066.356	19738.05	TH	0.6			MIT
5064.335	5065.747	19740.43	TH	0.9			MIT
5063.158	5064.570	19745.01	TH	0.3			MIT
5061.657	5063.068	19750.87	TH	0.6			MIT
5061.221	5062.632	19752.57	TH	1.1	0.0		MIT
5059.852	5061.263	19757.92	TH	0.6	0.7J		MIT
5058.567	5059.977	19762.93	TH	1.1	0.3		MIT
5057.98	5059.39	19765.2	TH	0.3			MIT
5055.342	5056.751	19775.54	TH	0.3	0.0		MIT
5051.342	5052.750	19791.20	TH	0.5			MIT
5050.780	5052.188	19793.40	TH	0.7			MIT
5049.810	5051.218	19797.21	TH	1.5W	0.7		MIT
5047.43	5048.84	19806.5	TH	0.9	0.0		MIT
5047.043	5048.450	19808.06	TH	0.7			MIT
5045.247	5046.654	19815.11	TH	0.6			MIT
5044.721	5046.128	19817.18	TH	0.9			MIT
5039.231	5040.636	19838.77	TH	0.9			MIT
5029.898	5031.301	19875.58	TH	0.5			MIT
5029.631	5031.034	19876.63	TH	1.0			MIT
5028.609	5030.011	19880.67	TH	1.6O	1.0J		MIT

AIR WAVELENGTH (ANGSTRMS)	VACUUM WAVELENGTH (ANGSTRMS)	WAVENUMBER (KAYSERS)	LMENT	LOG ARC	INTENSITY SPK	INTENSITY DIS	REF
5025.852	5027.254	19891.58	TH	0.3			MIT
5023.052	5024.453	19902.66	TH	0.8			MIT
5020.541	5021.941	19912.62	TH	1.0			MIT
5019.326	5020.726	19917.44	TH	0.5			MIT
5017.247	5018.646	19925.69	TH	1.7	1.0		MIT
5015.89	5017.29	19931.1	TH	1.0			MIT
5014.757	5016.156	19935.58	TH	0.6			MIT
5009.977	5011.374	19954.61	TH	0.8			MIT
5009.446	5010.843	19956.72	TH	0.3			MIT
5008.194	5009.591	19961.71	TH	0.9	0.0		MIT
5004.133	5005.529	19977.91	TH	0.5			MIT
5002.097	5003.492	19986.04	TH	1.0			MIT
4999.938	5001.333	19994.67	TH	0.9	0.0		MIT
4997.787	4999.181	20003.28	TH	0.8	0.3		MIT
4997.327	4998.721	20005.12	TH	0.5			MIT
4990.031	4991.423	20034.37	TH	0.7			MIT
4989.309	4990.701	20037.27	TH	0.8	0.3		MIT
4987.136	4988.527	20046.00	TH	1.2	0.6		MIT
4985.376	4986.767	20053.07	TH	1.0			MIT
4983.526	4984.916	20060.52	TH	0.3O			MIT
4980.953	4982.343	20070.88	TH	0.9			MIT
4980.187	4981.577	20073.97	TH	0.6			MIT
4976.597	4977.986	20088.45	TH	0.9			MIT
4975.940	4977.328	20091.10	TH	1.0			MIT
4973.394	4974.782	20101.39	TH	1.0			MIT
4972.609	4973.997	20104.56	TH	0.5			MIT
4972.179	4973.566	20106.30	TH	0.9	0.0		MIT
4970.032	4971.419	20114.98	TH	1.0			MIT
4969.131	4970.518	20118.63	TH	0.5			MIT
4968.755	4970.141	20120.15	TH	1.0			MIT
4965.731	4967.117	20132.40	TH	0.6			MIT
4964.123	4965.508	20138.93	TH	1.1	0.0		MIT
4963.192	4964.577	20142.70	TH	1.0			MIT
4962.963	4964.348	20143.63	TH	0.7			MIT
4961.729	4963.114	20148.64	TH	0.7			MIT
4960.494	4961.878	20153.66	TH	0.7			MIT
4958.725	4960.109	20160.85	TH	0.6			MIT
4954.665	4956.048	20177.37	TH	1.0	0.3		MIT
4950.620	4952.002	20193.85	TH	1.0	0.7H		MIT
4947.577	4948.958	20206.27	TH	1.0	0.3		MIT
4946.661	4948.042	20210.02	TH	0.8			MIT
4945.458	4946.838	20214.93	TH	0.8			MIT
4936.768	4938.146	20250.51	TH	1.0			MIT
4934.088	4935.465	20261.51	TH	0.6	0.3		MIT
4933.852	4935.229	20262.48	TH	0.9	0.0		MIT
4929.988	4931.364	20278.36	TH	1.0			MIT
4927.777	4929.153	20287.46	TH	0.7			MIT
4925.789	4927.164	20295.65	TH	0.5			MIT
4925.425	4926.800	20297.15	TH	0.6			MIT
4925.018	4926.393	20298.83	TH	0.6	0.3		MIT
4924.431	4925.806	20301.25	TH	0.9	0.3		MIT
4922.946	4924.320	20307.37	TH	1.0			MIT
4921.610	4922.984	20312.88	TH	1.3			MIT
4919.814	4921.187	20320.30	TH	1.7	1.3		MIT
4914.122	4915.494	20343.83	TH	0.7			MIT
4912.529	4913.901	20350.43	TH	1.1	0.3		MIT
4911.441	4912.812	20354.94	TH	0.6			MIT
4911.127	4912.498	20356.24	TH	0.9			MIT
4902.776	4904.145	20390.91	TH	0.6			MIT
4899.151	4900.519	20406.00	TH	0.5	0.3		MIT
4898.812	4900.180	20407.41	TH	1.0			MIT
4898.461	4899.829	20408.88	TH	1.1			MIT
4895.663	4897.030	20420.54	TH	0.5			MIT
4894.954	4896.321	20423.50	TH	1.0			MIT
4890.440	4891.806	20442.35	TH	0.5			MIT
4882.456	4883.820	20475.78	TH	0.7	1.3J		MIT
4879.347	4880.710	20488.82	TH	0.7			MIT
4879.158	4880.521	20489.62	TH	0.5			MIT
4878.731	4880.094	20491.41	TH	0.9			MIT
4878.012	4879.374	20494.43	TH	0.5			MIT

AIR WAVELENGTH (ANGSTRMS)	VACUUM WAVELENGTH (ANGSTRMS)	WAVENUMBER (KAYSERS)	LMENT	LOG ARC	INTENSITY SPK	DIS	REF
4877.817	4879.179	20495.25	TH	0.8			MIT
4877.010	4878.372	20498.64	TH	1.1			MIT
4876.492	4877.854	20500.82	TH	0.5			MIT
4874.367	4875.728	20509.76	TH	0.8			MIT
4872.926	4874.287	20515.82	TH	1.3			MIT
4868.880	4870.240	20532.87	TH	0.7			MIT
4868.274	4869.634	20535.43	TH	0.7	0.0		MIT
4863.182	4864.540	20556.93	TH	1.3	1.0		MIT
4861.215	4862.573	20565.24	TH	0.9			MIT
4858.306	4859.663	20577.56	TH	1.0	0.3		MIT
4852.865	4854.221	20600.63	TH	0.7			MIT
4850.45	4851.81	20610.9	TH	1.1	0.6		MIT
4849.028	4850.383	20616.93	TH	0.9	0.6		MIT
4848.361	4849.716	20619.77	TH	0.9			MIT
4844.759	4846.113	20635.10	TH	0.5			MIT
4844.564	4845.918	20635.93	TH	0.9	0.3		MIT
4844.124	4845.477	20637.80	TH	1.0	0.6		MIT
4840.826	4842.179	20651.86	TH	0.9			MIT
4840.458	4841.810	20653.43	TH	1.1	0.5		MIT
4838.363	4839.715	20662.37	TH	0.9	0.3		MIT
4832.800	4834.150	20686.16	TH	1.1	0.5		MIT
4831.603	4832.953	20691.28	TH	0.6			MIT
4831.133	4832.483	20693.30	TH	1.0	0.3		MIT
4826.819	4828.168	20711.79	TH	0.5			MIT
4823.184	4824.532	20727.40	TH	0.5	0.3		MIT
4821.262	4822.609	20735.66	TH	0.3	0.3		MIT
4820.897	4822.244	20737.23	TH	0.5	0.3		MIT
4818.645	4819.992	20746.92	TH	0.7	0.7		MIT
4806.065	4807.408	20801.23	TH		0.5H		MIT
4803.963	4805.306	20810.33	TH	0.3	0.3W		MIT
4803.493	4804.836	20812.37	TH	0.5	0.9H		MIT
4800.176	4801.518	20826.75	TH	0.5	0.3		MIT
4782.762	4784.099	20902.58	TH	0.6	0.5		MIT
4779.603	4780.939	20916.39	TH	0.3	0.3		MIT
4778.296	4779.632	20922.11	TH	0.3			MIT
4774.255	4775.590	20939.82	TH	1.0	0.7		MIT
4766.599	4767.932	20973.45	TH	0.3			MIT
4764.35	4765.68	20983.4	TH	0.5	0.3		MIT
4761.110	4762.441	20997.63	TH	1.0	1.1		MIT
4758.148	4759.479	21010.71	TH	0.3			MIT
4755.613	4756.943	21021.91	TH	0.6	0.8W		MIT
4752.410	4753.739	21036.07	TH	1.3	1.1		MIT
4749.204	4750.532	21050.27	TH	0.3			MIT
4743.700	4745.027	21074.70	TH	0.5	0.3		MIT
4742.266	4743.592	21081.07	TH	0.6L	0.3		MIT
4740.517	4741.843	21088.85	TH	1.3	1.2		MIT
4732.680	4734.004	21123.77	TH	0.3	0.3		MIT
4731.100	4732.423	21130.82	TH	0.6L			MIT
4729.883	4731.206	21136.26	TH	0.8	0.7		MIT
4729.114	4730.437	21139.70	TH	0.5W			MIT
4726.477	4727.799	21151.49	TH	0.5	0.0		MIT
4724.787	4726.109	21159.06	TH	0.8	0.6		MIT
4723.795	4725.116	21163.50	TH	1.1	1.0		MIT
4723.452	4724.773	21165.04	TH	0.7	0.6		MIT
4722.109	4723.430	21171.06	TH	0.3			MIT
4719.997	4721.317	21180.53	TH	0.6	0.6		MIT
4718.627	4719.947	21186.68	TH	0.9L	0.7		MIT
4716.002	4717.321	21198.47	TH	0.6H	0.3H		MIT
4715.436	4716.755	21201.01	TH	0.5	0.3		MIT
4712.395	4713.713	21214.70	TH	0.6	0.5		MIT
4706.446	4707.763	21241.51	TH	0.5			MIT
4706.219	4707.536	21242.54	TH	0.7	0.6		MIT
4705.764	4707.081	21244.59	TH	1.0	0.8		MIT
4703.992	4705.308	21252.59	TH	0.5	0.0		MIT
4702.320	4703.636	21260.15	TH	0.6	0.3		MIT
4701.922	4703.238	21261.95	TH	0.3			MIT
4700.150	4701.465	21269.97	TH	0.8	0.6		MIT
4698.778	4700.093	21276.18	TH	0.3	0.3W		MIT
4697.395	4698.710	21282.44	TH	0.7J			MIT
4695.484	4696.798	21291.10	TH	0.3			MIT

AIR WAVELENGTH (ANGSTRMS)	VACUUM WAVELENGTH (ANGSTRMS)	WAVENUMBER (KAYSERS)	LMENT	LOG ARC	INTENSITY SPK	DIS	REF
4694.919	4696.233	21293.66	TH	0.5	0.3		MIT
4694.092	4695.406	21297.42	TH	1.2	1.0		MIT
4691.06	4692.37	21311.2	TH	0.5	0.3		MIT
4690.660	4691.973	21313.00	TH	0.5	0.5		MIT
4689.173	4690.485	21319.75	TH	1.0	0.8		MIT
4686.587	4687.899	21331.52	TH	0.3	0.0		MIT
4682.208	4683.518	21351.47	TH	0.6D			MIT
4681.208	4682.518	21356.03	TH	0.3	0.3		MIT
4680.652	4681.962	21358.57	TH	0.6	0.5		MIT
4678.220	4679.529	21369.67	TH	0.3			MIT
4676.523	4677.832	21377.43	TH	0.3			MIT
4676.054	4677.363	21379.57	TH	0.3			MIT
4673.661	4674.969	21390.51	TH	0.8	0.3		MIT
4670.03	4671.34	21407.1	TH	0.5	0.0		M
4668.149	4669.456	21415.77	TH	0.5			MIT
4666.804	4668.110	21421.94	TH	0.3			MIT
4666.003	4667.309	21425.62	TH	0.9	0.6		MIT
4664.220	4665.526	21433.81	TH	0.6L			MIT
4655.204	4656.507	21475.32	TH	0.3			MIT
4652.027	4653.329	21489.99	TH	0.8			MIT
4651.556	4652.858	21492.17	TH	1.5	1.2		MIT
4650.923	4652.225	21495.09	TH	0.5	0.3		MIT
4647.897	4649.198	21509.08	TH	0.3			MIT
4646.636	4647.937	21514.92	TH	0.3			MIT
4645.495	4646.796	21520.21	TH	0.5W			MIT
4644.371	4645.671	21525.41	TH	0.5W			MIT
4640.052	4641.351	21545.45	TH	1.0	0.7		MIT
4639.708	4641.007	21547.05	TH	1.0	0.8		MIT
4633.835	4635.133	21574.35	TH	0.6	0.3		MIT
4631.762	4633.059	21584.01	TH	1.2	1.0		MIT
4629.772	4631.069	21593.29	TH	0.5			MIT
4627.99	4629.29	21601.6	TH	1.0H			MIT
4625.02	4626.32	21615.5	TH	0.6	0.5		MIT
4624.142	4625.437	21619.58	TH	0.5	0.3		MIT
4623.889	4625.184	21620.76	TH	0.5	0.5		MIT
4619.562	4620.856	21641.01	TH	0.5	0.5		MIT
4619.472	4620.766	21641.43	TH	0.7	0.5		MIT
4618.36	4619.65	21646.6	TH	0.3			MIT
4616.65	4617.94	21654.7	TH	0.3			MIT
4615.07	4616.36	21662.1	TH	0.5			MIT
4613.981	4615.273	21667.19	TH	0.3			MIT
4612.54	4613.83	21674.0	TH	0.6	0.5		MIT
4611.842	4613.134	21677.24	TH	1.0	0.8		MIT
4609.876	4611.167	21686.48	TH		0.5		MIT
4609.377	4610.668	21688.83	TH	0.8	0.6		MIT
4606.506	4607.796	21702.35	TH	0.6	0.3		MIT
4602.884	4604.173	21719.43	TH	0.7	0.5		MIT
4599.367	4600.656	21736.03	TH	0.5			MIT
4596.933	4598.221	21747.54	TH	0.6W			MIT
4595.424	4596.712	21754.68	TH	0.6	0.3		MIT
4593.642	4594.929	21763.12	TH		1.7J		MIT
4593.290	4594.577	21764.79	TH	0.5	0.3		MIT
4590.812	4592.098	21776.54	TH	0.5H			MIT
4589.670	4590.956	21781.96	TH	0.6	0.3		MIT
4589.119	4590.405	21784.57	TH	0.6	0.6		MIT
4588.431	4589.717	21787.84	TH	0.6	0.0		MIT
4588.238	4589.524	21788.75	TH	0.6	0.6		MIT
4586.632	4587.917	21796.38	TH	0.3	0.5		MIT
4584.376	4585.661	21807.11	TH	0.7	0.6		MIT
4583.698	4584.982	21810.33	TH	0.3			MIT
4582.771	4584.055	21814.75	TH	0.3			MIT
4581.581	4582.865	21820.41	TH	0.8	0.8		MIT
4581.225	4582.509	21822.11	TH	0.5	0.5		MIT
4579.286	4580.569	21831.35	TH	0.3	0.5		MIT
4578.809	4580.092	21833.62	TH	0.5W			MIT
4575.427	4576.709	21849.76	TH	0.6	0.5		MIT
4575.246	4576.528	21850.62	TH	0.3	0.0		MIT
4573.703	4574.985	21858.00	TH	0.6	0.6		MIT
4572.617	4573.898	21863.19	TH	0.6W			MIT
4571.948	4573.229	21866.39	TH	0.5			MIT

AIR WAVELENGTH (ANGSTRMS)	VACUUM WAVELENGTH (ANGSTRMS)	WAVENUMBER (KAYSERS)	LMENT	LOG ARC	INTENSITY SPK	DIS	REF
4570.974	4572.255	21871.05	TH	0.5	0.0		MIT
4570.629	4571.910	21872.70	TH	0.6W			MIT
4570.368	4571.649	21873.95	TH	0.6			MIT
4567.242	4568.522	21888.92	TH	0.3			MIT
4566.645	4567.925	21891.78	TH	0.8	0.5		MIT
4564.183	4565.462	21903.59	TH	0.6	0.5		MIT
4563.658	4564.937	21906.11	TH	0.6	0.3		MIT
4563.316	4564.595	21907.75	TH	0.3			FD
4563.238	4564.517	21908.12	TH	0.7	0.7		MIT
4561.352	4562.631	21917.18	TH	0.6			MIT
4558.346	4559.624	21931.63	TH	0.3			MIT
4558.078	4559.356	21932.92	TH	0.5H			MIT
4556.809	4558.086	21939.03	TH	0.6			MIT
4555.815	4557.092	21943.82	TH	0.5			MIT
4555.61	4556.89	21944.8	TH		1.0H		MIT
4555.071	4556.348	21947.40	TH	0.5			MIT
4553.85	4555.13	21953.3	TH	0.5	0.3		MIT
4553.038	4554.314	21957.20	TH	0.5	0.5H		MIT
4550.444	4551.720	21969.72	TH	0.6			MIT
4546.710	4547.985	21987.76	TH	0.6			MIT
4544.901	4546.175	21996.51	TH	0.5			MIT
4544.518	4545.792	21998.37	TH	0.9	0.7		MIT
4543.942	4545.216	22001.16	TH	0.7			MIT
4543.399	4544.673	22003.78	TH	0.8			MIT
4543.208	4544.482	22004.71	TH	0.5	0.3		MIT
4541.617	4542.890	22012.42	TH	0.9D			MIT
4541.209	4542.482	22014.40	TH	0.8			MIT
4541.004	4542.277	22015.39	TH	0.3			MIT
4540.425	4541.698	22018.20	TH	0.9	0.9		MIT
4537.070	4538.342	22034.48	TH	1.3	0.9		MIT
4535.939	4537.211	22039.97	TH	0.9W			MIT
4535.257	4536.529	22043.29	TH	0.5			MIT
4534.837	4536.109	22045.33	TH	0.5			MIT
4534.129	4535.400	22048.77	TH	1.0	0.7		MIT
4533.309	4534.580	22052.76	TH	1.1	0.9		MIT
4532.264	4533.535	22057.84	TH	0.9	0.6		MIT
4531.721	4532.992	22060.49	TH	0.5	0.5		MIT
4529.487	4530.757	22071.37	TH	0.5	0.3		MIT
4527.733	4529.003	22079.92	TH	0.5	0.3		MIT
4526.031	4527.300	22088.22	TH	0.5			MIT
4521.266	4522.534	22111.50	TH	0.3	0.5H		MIT
4520.951	4522.219	22113.04	TH	0.5	0.0		MIT
4519.750	4521.018	22118.91	TH	0.6	0.0		MIT
4519.273	4520.540	22121.25	TH	0.8W			MIT
4518.647	4519.914	22124.31	TH	0.3	0.3		MIT
4517.046	4518.313	22132.16	TH	0.6	0.5		MIT
4515.979	4517.246	22137.38	TH	0.6	0.6		MIT
4515.124	4516.390	22141.58	TH	0.5			MIT
4513.690	4514.956	22148.61	TH	0.3			MIT
4513.235	4514.501	22150.84	TH	0.5			MIT
4512.490	4513.756	22154.50	TH	0.6	0.6		MIT
4510.535	4511.800	22164.10	TH	1.5	1.3		MIT
4508.651	4509.916	22173.36	TH	0.9W			MIT
4508.170	4509.434	22175.73	TH	0.3	0.0		MIT
4508.042	4509.306	22176.36	TH	0.6			MIT
4507.443	4508.707	22179.31	TH	0.8W			MIT
4506.527	4507.791	22183.81	TH	0.5			MIT
4505.226	4506.490	22190.22	TH	0.7			MIT
4504.721	4505.985	22192.71	TH	0.8W			MIT
4503.317	4504.580	22199.63	TH	0.8J			MIT
4502.264	4503.527	22204.82	TH	0.3			MIT
4501.827	4503.090	22206.97	TH	0.5			MIT
4499.985	4501.247	22216.06	TH	1.1	0.8		MIT
4498.950	4500.212	22221.18	TH	1.1	0.5		MIT
4497.904	4499.166	22226.34	TH	0.5			MIT
4496.323	4497.584	22234.16	TH	1.1	1.0		MIT
4495.49	4496.75	22238.3	TH	0.5	0.3		MIT
4495.247	4496.508	22239.48	TH	0.5	0.5		MIT
4494.966	4496.227	22240.87	TH	0.7	0.5		MIT
4493.340	4494.601	22248.92	TH	1.1	0.6		MIT
4492.239	4493.499	22254.37	TH	0.9	0.7		MIT
4491.882	4493.142	22256.14	TH	0.3	0.0		MIT
4491.443	4492.703	22258.31	TH	0.5			MIT
4490.322	4491.582	22263.87	TH	0.5	0.0		MIT
4489.527	4490.787	22267.81	TH	0.5	0.0		MIT
4488.680	4489.939	22272.02	TH	1.4	1.1		MIT
4488.322	4489.581	22273.79	TH	0.5			MIT
4487.500	4488.759	22277.87	TH	1.3	1.0		MIT
4486.904	4488.163	22280.83	TH	0.6			MIT
4486.642	4487.901	22282.13	TH	1.0	0.8		MIT
4485.796	4487.055	22286.33	TH	1.0	0.7		MIT
4484.292	4485.550	22293.81	TH	0.7	0.5		MIT
4483.339	4484.597	22298.55	TH	0.6			MIT
4481.637	4482.894	22307.02	TH	0.8	0.5		MIT
4481.183	4482.440	22309.28	TH	0.5	0.3		MIT
4480.822	4482.079	22311.07	TH	1.4	1.0		MIT
4479.161	4480.418	22319.35	TH	0.6	0.5		MIT
4475.235	4476.491	22338.93	TH	0.7H	0.3		MIT
4474.078	4475.333	22344.70	TH	1.0	0.8		MIT
4472.245	4473.500	22353.86	TH	1.0	0.7		MIT
4472.004	4473.259	22355.07	TH	0.7	0.0		MIT
4471.773	4473.028	22356.22	TH	0.7	0.6		MIT
4469.528	4470.782	22367.45	TH	0.7	0.0		MIT
4468.322	4469.576	22373.49	TH	0.3			MIT
4465.345	4466.598	22388.40	TH	1.5S	1.2		MIT
4463.828	4465.081	22396.01	TH	0.5			MIT
4461.71	4462.96	22406.6	TH		0.7		MIT
4461.085	4462.337	22409.78	TH	1.0D	0.9D		MIT
4458.006	4459.257	22425.26	TH	1.0	0.3		MIT
4457.195	4458.446	22429.34	TH	0.5	0.0		MIT
4456.706	4457.957	22431.80	TH	0.6	0.5		MIT
4455.039	4456.289	22440.19	TH	0.6	0.6		MIT
4454.778	4456.028	22441.51	TH	0.6			MIT
4454.517	4455.767	22442.82	TH	0.9	0.6		MIT
4452.568	4453.818	22452.65	TH	0.5			MIT
4451.056	4452.305	22460.27	TH	0.5	0.0		MIT
4450.787	4452.036	22461.63	TH	0.5			MIT
4448.510	4449.759	22473.13	TH	0.7	0.3		MIT
4447.845	4449.094	22476.49	TH	1.1	0.7		MIT
4445.900	4447.148	22486.32	TH	0.6			MIT
4445.308	4446.556	22489.32	TH	0.6			MIT
4444.979	4446.227	22490.98	TH	0.9L	0.5		MIT
4443.099	4444.346	22500.50	TH	1.0	0.7		MIT
4440.872	4442.119	22511.78	TH	1.3	1.0		MIT
4440.576	4441.823	22513.28	TH	1.2	0.8		MIT
4439.128	4440.374	22520.62	TH	1.3	1.2		MIT
4438.750	4439.996	22522.54	TH	0.7	0.3		MIT
4436.558	4437.804	22533.67	TH	1.0	0.7		MIT
4436.288	4437.534	22535.04	TH	0.9	0.6		MIT
4436.046	4437.291	22536.27	TH	0.9	0.6		MIT
4434.953	4436.198	22541.82	TH	0.6			MIT
4432.971	4434.216	22551.90	TH	1.4	1.2		MIT
4432.262	4433.506	22555.51	TH	0.9			MIT
4429.274	4430.518	22570.73	TH	0.7	0.3		MIT
4428.541	4429.784	22574.46	TH		0.3		MIT
4427.658	4428.901	22578.96	TH	0.9	0.7		MIT
4425.987	4427.230	22587.49	TH	0.8	0.6		MIT
4423.945	4425.187	22597.91	TH	0.9W	0.6W		MIT
4422.783	4424.025	22603.85	TH	0.8	0.5		MIT
4422.055	4423.297	22607.57	TH	0.6			MIT
4421.557	4422.799	22610.12	TH	0.9	0.3		MIT
4419.00	4420.24	22623.2	TH	0.7	0.5		MIT
4418.669	4419.910	22624.90	TH	0.9	0.6		MIT
4416.847	4418.087	22634.23	TH	0.6			MIT
4416.241	4417.481	22637.33	TH	1.2	0.9		MIT
4414.530	4415.770	22646.11	TH	0.6			MIT
4413.378	4414.617	22652.02	TH	0.8	0.5		MIT
4412.897	4414.136	22654.49	TH	0.9	0.7		MIT
4412.745	4413.984	22655.27	TH	1.1	0.9		MIT
4412.533	4413.772	22656.36	TH	1.0	0.8		MIT

AIR WAVELENGTH (ANGSTRMS)	VACUUM WAVELENGTH (ANGSTRMS)	WAVENUMBER (KAYSERS)	LMENT	LOG ARC	INTENSITY SPK	DIS	REF
4411.634	4412.873	22660.97	TH	0.6	0.3		MIT
4410.420	4411.659	22667.21	TH	0.6	0.3		MIT
4408.891	4410.129	22675.07	TH	0.8	0.3		MIT
4402.929	4404.166	22705.78	TH	0.6	0.3		MIT
4401.986	4403.222	22710.64	TH	0.5	0.0		MIT
4401.669	4402.905	22712.27	TH	0.5	0.5		MIT
4400.387	4401.623	22718.89	TH	0.8	0.5		MIT
4399.38	4400.62	22724.1	TH	0.5	0.0		MIT
4399.096	4400.332	22725.56	TH	0.9	0.9		MIT
4397.919	4399.154	22731.64	TH	1.0	0.9		MIT
4396.479	4397.714	22739.09	TH	1.0	0.9		MIT
4394.889	4396.124	22747.31	TH	1.0	0.9		MIT
4393.76	4394.99	22753.2	TH	0.5	0.0		MIT
4393.057	4394.291	22756.80	TH	0.5	0.5		MIT
4391.114	4392.348	22766.87	TH	1.7	1.6		MIT
4388.413	4389.646	22780.88	TH	0.9	0.5		MIT
4387.101	4388.334	22787.69	TH	0.6	0.5		MIT
4381.859	4383.090	22814.95	TH	1.5	1.5		MIT
4381.392	4382.623	22817.39	TH	1.0	0.7		MIT
4378.127	4379.357	22834.40	TH	0.6	0.0		MIT
4377.27	4378.50	22838.9	TH	0.7	0.7		MIT
4374.790	4376.019	22851.82	TH	1.2	1.0		MIT
4374.131	4375.360	22855.26	TH	0.6	0.0		MIT
4373.907	4375.136	22856.43	TH	1.0	0.7		MIT
4369.325	4370.553	22880.40	TH	1.0	0.7		MIT
4366.974	4368.201	22892.72	TH	0.7	0.5		MIT
4365.909	4367.136	22898.30	TH	0.6	0.3		MIT
4361.775	4363.001	22920.00	TH	0.6	0.5		MIT
4361.317	4362.543	22922.41	TH	1.1	1.0		MIT
4359.376	4360.601	22932.62	TH	0.6	0.3		MIT
4358.832	4360.057	22935.48	TH	0.5	0.3		MIT
4358.556	4359.781	22936.93	TH	0.6	0.6		MIT
4358.43	4359.65	22937.6	TH	0.5			MIT
4358.333	4359.558	22938.10	TH	0.5			MIT
4357.587	4358.812	22942.03	TH	0.9	0.6		MIT
4355.327	4356.551	22953.94	TH	1.1	0.9		MIT
4354.489	4355.713	22958.35	TH	0.5			MIT
4353.404	4354.628	22964.08	TH	0.9	0.7		MIT
4350.834	4352.057	22977.64	TH	0.8H	0.3		MIT
4347.196	4348.418	22996.87	TH	0.8W	0.6W		MIT
4346.450	4347.672	23000.82	TH	0.6	0.0		MIT
4345.05	4346.27	23008.2	TH	0.5	0.3		MIT
4344.623	4345.844	23010.49	TH	0.7H	0.0H		MIT
4344.334	4345.555	23012.02	TH	1.1	0.9		MIT
4343.954	4345.175	23014.03	TH	1.0	0.8		MIT
4343.606	4344.827	23015.88	TH	0.6	0.6		MIT
4342.444	4343.665	23022.03	TH	0.5	0.0		MIT
4342.273	4343.494	23022.94	TH	0.9	0.8		MIT
4341.032	4342.252	23029.52	TH	0.8	0.8		MIT
4338.121	4339.341	23044.98	TH	0.5	0.0		MIT
4337.394	4338.613	23048.84	TH	1.0	1.0		MIT
4337.281	4338.500	23049.44	TH	0.5			MIT
4335.756	4336.975	23057.55	TH	1.0	1.0		MIT
4335.313	4336.532	23059.90	TH	0.5	0.3		MIT
4333.942	4335.161	23067.20	TH	0.7	0.6		MIT
4331.931	4333.149	23077.91	TH	0.9	0.6		MIT
4329.500	4330.717	23090.86	TH	0.8	0.8		MIT
4328.693	4329.910	23095.17	TH	0.8	0.8		MIT
4327.14	4328.36	23103.5	TH	1.2	0.8		MIT
4320.597	4321.812	23138.44	TH	0.9	0.7		MIT
4320.131	4321.346	23140.94	TH	1.1	1.0		MIT
4319.108	4320.323	23146.42	TH	0.7	0.6		MIT
4318.300	4319.514	23150.75	TH	0.8	0.7		MIT
4315.957	4317.171	23163.32	TH	0.7	0.6		MIT
4313.314	4314.527	23177.51	TH	0.8	0.6		MIT
4312.997	4314.210	23179.22	TH	0.9	0.6		MIT
4311.799	4313.012	23185.66	TH	0.8	0.0		MIT
4309.994	4311.206	23195.36	TH	1.2	1.1		MIT
4308.260	4309.472	23204.70	TH	0.5	0.3		MIT
4307.185	4308.397	23210.49	TH	0.7	0.5		MIT
4306.381	4307.592	23214.82	TH	0.9	0.8		MIT
4300.799	4302.009	23244.96	TH	0.7	0.6		MIT
4299.846	4301.056	23250.11	TH	0.9	0.6		MIT
4298.831	4300.040	23255.60	TH	0.9	0.9		MIT
4297.349	4298.558	23263.62	TH	1.0	0.6		MIT
4295.088	4296.296	23275.86	TH	0.6	0.0		MIT
4288.677	4289.884	23310.66	TH	0.7	0.0		MIT
4288.056	4289.262	23314.03	TH	0.9	0.8		MIT
4285.192	4286.398	23329.61	TH	0.7	0.7		MIT
4284.984	4286.190	23330.75	TH	0.9	0.9		MIT
4283.525	4284.730	23338.69	TH	1.0	1.0		MIT
4282.044	4283.249	23346.76	TH	1.5	1.4		MIT
4281.425	4282.630	23350.14	TH	1.2	1.0		MIT
4281.072	4282.277	23352.06	TH	1.3	1.0		MIT
4280.567	4281.772	23354.82	TH	0.6	0.0		MIT
4278.323	4279.527	23367.07	TH	0.9	0.3		MIT
4277.322	4278.526	23372.54	TH	1.3	1.1		MIT
4276.974	4278.178	23374.44	TH	0.6	0.5		MIT
4276.814	4278.018	23375.31	TH	1.0	0.9		MIT
4274.331	4275.534	23388.89	TH	0.9	0.6		MIT
4274.032	4275.235	23390.53	TH	1.0	0.9		MIT
4273.363	4274.566	23394.19	TH	1.3	1.2		MIT
4272.885	4274.087	23396.81	TH	0.6			MIT
4271.103	4272.305	23406.57	TH	0.7	0.5		MIT
4270.332	4271.534	23410.79	TH	0.8	0.5		MIT
4264.114	4265.314	23444.93	TH	0.6			MIT
4263.359	4264.559	23449.08	TH	1.0	0.5		MIT
4262.716	4263.916	23452.62	TH	0.6	0.0		MIT
4261.276	4262.475	23460.55	TH	0.7	0.3		MIT
4260.342	4261.541	23465.69	TH	0.6	0.0		MIT
4257.502	4258.700	23481.34	TH	0.9	0.3		MIT
4256.227	4257.425	23488.38	TH	0.8	0.0		MIT
4256.102	4257.300	23489.06	TH	1.1	1.0		MIT
4255.789	4256.987	23490.79	TH	0.6			MIT
4254.458	4255.656	23498.14	TH	0.9	0.8		MIT
4253.875	4255.072	23501.36	TH	0.8	0.6		MIT
4253.544	4254.741	23503.19	TH	0.9	0.5		MIT
4250.339	4251.536	23520.91	TH	1.1	1.0L		MIT
4249.685	4250.881	23524.53	TH	1.1	1.0		MIT
4247.994	4249.190	23533.90	TH	1.2	1.1		MIT
4247.599	4248.795	23536.09	TH	0.9	0.7		MIT
4246.338	4247.534	23543.07	TH	0.5	0.0		MIT
4245.464	4246.659	23547.92	TH	0.6	0.0		MIT
4243.929	4245.124	23556.44	TH	1.1	0.9		MIT
4242.725	4243.920	23563.12	TH	0.7	0.3		MIT
4237.119	4238.312	23594.30	TH	0.7	0.3		MIT
4235.468	4236.661	23603.50	TH	0.8	0.0		MIT
4234.989	4236.182	23606.17	TH	0.8	0.5		MIT
4234.306	4235.498	23609.97	TH	0.7	0.3		MIT
4233.291	4234.483	23615.63	TH	1.0	0.8		MIT
4230.431	4231.622	23631.60	TH	0.7	0.0		MIT
4229.459	4230.650	23637.03	TH	1.0	0.9		MIT
4227.658	4228.849	23647.10	TH	0.5	0.0		MIT
4224.616	4225.806	23664.13	TH	0.7	0.5		MIT
4224.613	4225.803	23664.14	TH	1.2S			MIT
4220.083	4221.272	23689.54	TH	0.9	0.9		MIT
4218.545	4219.733	23698.18	TH	0.9	0.9		MIT
4218.190	4219.378	23700.18	TH	0.9	0.9		MIT
4217.225	4218.413	23705.60	TH	0.9	0.7		MIT
4214.552	4215.739	23720.63	TH		0.5		MIT
4214.004	4215.191	23723.72	TH		0.5		MIT
4213.073	4214.260	23728.96	TH	1.0W	0.7W		FD
4211.519	4212.705	23737.72	TH	0.9	0.9		MIT
4210.920	4212.106	23741.09	TH	0.9	0.5		MIT
4210.768	4211.954	23741.95	TH	0.5	0.0		FD
4208.893	4210.079	23752.52	TH	1.2	1.2		MIT
4201.852	4203.036	23792.33	TH	0.9	1.0		MIT
4199.022	4200.205	23808.36	TH	0.5	0.7		MIT
4195.950	4197.132	23825.79	TH	0.9	0.7		MIT
4195.834	4197.016	23826.45	TH	0.7	0.7		MIT

AIR WAVELENGTH (ANGSTRMS)	VACUUM WAVELENGTH (ANGSTRMS)	WAVENUMBER (KAYSERS)	LMENT	LOG ARC	INTENSITY SPK	INTENSITY DIS	REF
4195.550	4196.732	23828.06	TH	0.7	0.9		MIT
4193.022	4194.204	23842.43	TH	0.3			MIT
4191.829	4193.010	23849.22	TH	0.7	0.9		MIT
4188.594	4189.774	23867.63	TH	0.0	0.5		MIT
4183.574	4184.753	23896.27	TH	0.9	0.9		MIT
4181.99	4183.17	23905.3	TH	0.7	0.9		EX
4179.965	4181.143	23916.91	TH	0.9	0.9		MIT
4179.718	4180.896	23918.32	TH	0.9	0.9		MIT
4178.064	4179.242	23927.79	TH	1.0	1.1		MIT
4171.349	4172.525	23966.30	TH	1.0	1.1L		MIT
4170.475	4171.651	23971.33	TH	1.0L	1.0L		MIT
4168.95	4170.13	23980.1	TH	0.5			MIT
4168.64	4169.82	23981.9	TH	0.9	1.1		MIT
4167.662	4168.837	23987.51	TH	1.0	0.9W		MIT
4165.813	4166.987	23998.15	TH	0.9W	0.7		MIT
4165.466	4166.640	24000.15	TH	1.3W	1.0W		MIT
4165.087	4166.261	24002.34	TH	1.0L	0.9L		MIT
4164.253	4165.427	24007.14	TH	1.0	1.0		MIT
4163.653	4164.827	24010.60	TH	0.9	1.0		MIT
4162.685	4163.859	24016.19	TH	1.0	1.1		MIT
4161.575	4162.748	24022.59	TH	0.7	0.5		MIT
4159.667	4160.840	24033.61	TH	0.7	1.0		MIT
4156.523	4157.695	24051.79	TH	1.0	1.0		MIT
4156.243	4157.415	24053.41	TH	0.9	0.9		MIT
4155.36	4156.53	24058.5	TH	0.7	0.5		EX
4149.991	4151.161	24089.65	TH	1.0	1.1		MIT
4148.348	4149.518	24099.19	TH	0.7	0.7		MIT
4148.182	4149.352	24100.15	TH	1.0	1.0		MIT
4142.703	4143.871	24132.02	TH	1.0	1.0		MIT
4142.482	4143.650	24133.31	TH	1.0	1.0		MIT
4141.641	4142.809	24138.21	TH	0.9	1.0L		MIT
4140.239	4141.407	24146.39	TH	1.2S	1.2S		MIT
4136.383	4137.550	24168.89	TH	0.7	0.9		MIT
4134.117	4135.283	24182.14	TH	1.0	1.0		MIT
4132.751	4133.917	24190.13	TH	1.0	1.1		MIT
4131.433	4132.598	24197.85	TH	0.9	0.9		MIT
4131.049	4132.214	24200.10	TH	0.9D	0.7D		MIT
4124.65	4125.81	24237.6	TH	0.5	0.5		MIT
4123.569	4124.732	24244.00	TH	0.7	0.7		MIT
4122.973	4124.136	24247.50	TH	0.5	0.7		MIT
4121.891	4123.054	24253.87	TH		0.5		MIT
4116.718	4117.880	24284.34	TH	1.1	1.2		MIT
4115.773	4116.934	24289.92	TH	0.5	0.5		MIT
4112.762	4113.922	24307.70	TH	0.7	0.7W		MIT
4112.322	4113.482	24310.30	TH	0.5	0.7		MIT
4110.870	4112.030	24318.89	TH	0.9	0.9		MIT
4110.640	4111.800	24320.25	TH	0.7	0.7		MIT
4108.426	4109.585	24333.35	TH	1.2	1.2		MIT
4107.865	4109.024	24336.68	TH	0.7	0.7		MIT
4105.991	4107.150	24347.79	TH	0.9	0.9		MIT
4105.343	4106.502	24351.63	TH	1.1	1.1		MIT
4104.392	4105.550	24357.27	TH	1.0	1.0		MIT
4100.834	4101.991	24378.40	TH	1.3	1.3		MIT
4100.350	4101.507	24381.28	TH	0.9	0.7		MIT
4098.939	4100.096	24389.67	TH	1.0	1.0		MIT
4097.751	4098.908	24396.75	TH	0.7			MIT
4097.686	4098.843	24397.13	TH	0.7	0.5		FD
4097.339	4098.495	24399.20	TH	0.7	0.5		MIT
4094.751	4095.907	24414.62	TH	1.2	1.2		MIT
4093.394	4094.549	24422.71	TH	1.0	0.9		MIT
4089.134	4090.288	24448.16	TH	0.5			MIT
4086.520	4087.674	24463.79	TH	1.1	1.1		MIT
4085.422	4086.575	24470.37	TH	0.7	0.0		MIT
4085.036	4086.189	24472.68	TH	1.2	1.2		MIT
4083.474	4084.627	24482.04	TH	0.5			MIT
4082.263	4083.415	24489.30	TH	1.0	0.9		MIT
4081.388	4082.540	24494.55	TH	0.9	0.5		MIT
4080.707	4081.859	24498.64	TH	0.9	0.5		MIT
4079.612	4080.764	24505.22	TH	0.7	0.5		MIT
4075.713	4076.864	24528.66	TH	0.7	0.0		MIT

AIR WAVELENGTH (ANGSTRMS)	VACUUM WAVELENGTH (ANGSTRMS)	WAVENUMBER (KAYSERS)	LMENT	LOG ARC	INTENSITY SPK	INTENSITY DIS	REF
4072.630	4073.780	24547.23	TH	0.9			MIT
4069.469	4070.618	24566.29	TH	0.5			MIT
4069.210	4070.359	24567.86	TH	1.6	1.5		MIT
4067.666	4068.815	24577.18	TH	0.9W	0.7		MIT
4067.465	4068.614	24578.40	TH	0.9	0.0		MIT
4064.338	4065.486	24597.31	TH	0.9H	0.5		MIT
4063.410	4064.558	24602.92	TH	1.0	1.0		MIT
4059.883	4061.030	24624.30	TH	0.9	0.9		MIT
4059.259	4060.405	24628.08	TH	0.9	0.7		MIT
4057.823	4058.969	24636.80	TH	0.3	0.5		MIT
4057.337	4058.483	24639.75	TH	1.0	0.9		MIT
4054.864	4056.009	24654.78	TH	0.7	0.5		MIT
4053.531	4054.676	24662.88	TH	1.0	0.7		MIT
4051.103	4052.247	24677.67	TH	0.9	0.7		MIT
4050.891	4052.035	24678.96	TH	1.2H	0.7		MIT
4048.861	4050.005	24691.33	TH		1.0J		MIT
4048.448	4049.592	24693.85	TH	0.7	0.7		MIT
4048.057	4049.201	24696.23	TH	0.9	0.9		MIT
4043.401	4044.543	24724.67	TH	0.9	0.0		MIT
4043.150	4044.292	24726.21	TH	1.0	0.5		MIT
4042.979	4044.121	24727.25	TH	0.7	0.5		MIT
4041.207	4042.349	24738.09	TH	1.3	1.0		MIT
4039.896	4041.037	24746.12	TH	0.9	0.0		MIT
4037.258	4038.399	24762.29	TH	0.9	0.7		MIT
4036.567	4037.708	24766.53	TH	1.2	1.2		MIT
4036.056	4037.196	24769.66	TH	1.0	0.5		MIT
4034.886	4036.026	24776.85	TH	0.9	0.7		MIT
4034.256	4035.396	24780.72	TH	1.0	1.0		MIT
4032.54	4033.68	24791.3	TH	1.0	0.9		EX
4032.089	4033.228	24794.03	TH	0.5			MIT
4031.327	4032.466	24798.72	TH	0.7	0.7		MIT
4031.099	4032.238	24800.12	TH	0.7	0.7		MIT
4030.855	4031.994	24801.62	TH	1.0	0.9		MIT
4030.293	4031.432	24805.08	TH	0.9	0.7		MIT
4029.310	4030.449	24811.13	TH	0.9	0.7		MIT
4029.046	4030.185	24812.76	TH	0.7	0.5		MIT
4028.657	4029.795	24815.16	TH	0.9	0.9		MIT
4027.644	4028.782	24821.40	TH	0.7	0.5		MIT
4027.330	4028.468	24823.33	TH	0.7	0.5		MIT
4027.012	4028.150	24825.29	TH	0.9	0.0		MIT
4026.158	4027.296	24830.56	TH	1.0	1.0		MIT
4025.612	4026.750	24833.93	TH	1.3	1.3		MIT
4024.472	4025.609	24840.96	TH	0.7	0.7		MIT
4023.533	4024.670	24846.76	TH	0.5	0.0		MIT
4022.093	4023.230	24855.65	TH	1.3	1.2		MIT
4020.096	4021.232	24868.00	TH	0.5	0.5		MIT
4019.137	4020.273	24873.93	TH	0.9	0.9		MIT
4017.495	4018.631	24884.10	TH	0.9	0.9		MIT
4016.304	4017.439	24891.48	TH	0.7	0.7		MIT
4013.264	4014.398	24910.33	TH	0.7	0.7		MIT
4012.497	4013.631	24915.09	TH	1.2	1.2		MIT
4011.752	4012.886	24919.72	TH	1.2	1.2		MIT
4009.546	4010.679	24933.43	TH	0.5	0.0		MIT
4009.066	4010.199	24936.42	TH	1.0	0.9		MIT
4008.216	4009.349	24941.70	TH	1.0	0.9		MIT
4007.027	4008.160	24949.10	TH	1.3	1.3		MIT
4006.386	4007.519	24953.10	TH	1.0	1.0		MIT
4005.549	4006.681	24958.31	TH	1.3W	1.5W		MIT
4005.095	4006.227	24961.14	TH	0.5	0.0		MIT
4003.316	4004.448	24972.23	TH	1.2	1.2		MIT
4003.111	4004.243	24973.51	TH	1.0	1.0		MIT
4001.740	4002.871	24982.07	TH	1.0	0.9		MIT
4001.065	4002.196	24986.28	TH	1.0	0.7		MIT
4000.287	4001.418	24991.14	TH	0.9	0.5		MIT
3997.864	3998.994	25006.29	TH	1.0	1.0		MIT
3997.454	3998.584	25008.85	TH	0.9	0.9		MIT
3996.066	3997.196	25017.54	TH	1.2	1.0		MIT
3994.552	3995.682	25027.02	TH	1.5	1.0		MIT
3993.725	3994.854	25032.20	TH	0.9	0.7		MIT
3992.278	3993.407	25041.27	TH	1.0	1.0		MIT

AIR WAVELENGTH (ANGSTRMS)	VACUUM WAVELENGTH (ANGSTRMS)	WAVENUMBER (KAYSERS)	LMENT	LOG ARC	INTENSITY SPK	INTENSITY DIS	REF
3992.051	3993.180	25042.70	TH	0.7	0.7		MIT
3991.737	3992.866	25044.67	TH	0.9	0.5		MIT
3990.558	3991.686	25052.07	TH	0.9	0.7		MIT
3990.081	3991.209	25055.06	TH	0.7	0.9		MIT
3988.852	3989.980	25062.78	TH	1.0	1.0		MIT
3988.601	3989.729	25064.36	TH	1.0	0.9		MIT
3988.015	3989.143	25068.04	TH	1.7	1.5		MIT
3987.713	3988.841	25069.94	TH	1.0	0.9		MIT
3987.226	3988.354	25073.00	TH	1.0	0.9		MIT
3986.637	3987.764	25076.71	TH	0.9	0.9		MIT
3986.081	3987.208	25080.20	TH	0.9	0.7		MIT
3985.395	3986.522	25084.52	TH	0.7	0.9H		MIT
3984.859	3985.986	25087.90	TH	0.7W	0.3W		MIT
3984.614	3985.741	25089.44	TH	1.0H	0.9		MIT
3984.374	3985.501	25090.95	TH	0.9	0.7		MIT
3982.236	3983.362	25104.42	TH	0.5	0.5		MIT
3982.105	3983.231	25105.25	TH	0.9	0.9		MIT
3981.875	3983.001	25106.70	TH	0.5			MIT
3981.539	3982.665	25108.81	TH	0.5	0.5		MIT
3981.110	3982.236	25111.52	TH	1.3	1.3		MIT
3980.766	3981.892	25113.69	TH	0.9	1.0		MIT
3979.056	3980.181	25124.48	TH	1.0	1.0		MIT
3977.847	3978.972	25132.12	TH	0.0	0.3		MIT
3976.418	3977.543	25141.15	TH	1.2	1.3		MIT
3975.228	3976.352	25148.68	TH	0.9	0.9		MIT
3974.232	3975.356	25154.98	TH	0.7	0.7		MIT
3973.638	3974.762	25158.74	TH	0.7H			MIT
3973.230	3974.354	25161.32	TH	0.9	0.5		MIT
3972.648	3973.772	25165.01	TH	0.7	0.5		MIT
3972.159	3973.283	25168.11	TH	1.2	0.9		MIT
3969.834	3970.957	25182.85	TH	0.7	0.7		MIT
3969.534	3970.657	25184.75	TH	0.7	0.7		MIT
3969.343	3970.466	25185.96	TH	0.5W	0.3W		MIT
3969.003	3970.126	25188.12	TH	1.0	0.7		MIT
3967.413	3968.535	25198.21	TH	0.3	0.0		MIT
3967.214	3968.336	25199.48	TH	0.9	0.9		MIT
3966.971	3968.093	25201.02	TH	0.7	0.7		MIT
3966.72	3967.84	25202.6	TH	0.5	0.0		MIT
3964.880	3966.002	25214.31	TH	0.7	0.5		MIT
3963.468	3964.589	25223.29	TH	1.0	1.0		MIT
3963.226	3964.347	25224.83	TH	1.0	1.0		MIT
3962.346	3963.467	25230.44	TH	0.9	0.9		MIT
3960.947	3962.068	25239.35	TH	0.3	0.0		MIT
3960.339	3961.460	25243.22	TH	1.0	1.2		MIT
3959.295	3960.415	25249.88	TH	0.3	0.0		MIT
3956.682	3957.802	25266.55	TH	1.0	1.0		MIT
3956.603	3957.723	25267.06	TH	1.0	1.0		MIT
3955.958	3957.077	25271.18	TH	0.9			MIT
3955.179	3956.298	25276.15	TH	0.9	0.0		MIT
3952.451	3953.570	25293.60	TH	0.5	0.0		MIT
3951.521	3952.639	25299.55	TH	1.3	1.3		MIT
3951.101	3952.219	25302.24	TH	0.3	0.3		MIT
3950.394	3951.512	25306.77	TH	1.5	1.5		MIT
3948.971	3950.089	25315.89	TH	1.5	1.5		MIT
3948.038	3949.155	25321.87	TH	0.5	0.0		MIT
3947.652	3948.769	25324.35	TH	0.5	0.0		MIT
3947.339	3948.456	25326.35	TH	0.9	0.5		MIT
3946.151	3947.268	25333.98	TH	1.3	1.3		MIT
3945.825	3946.942	25336.07	TH	1.0	0.9		MIT
3945.515	3946.632	25338.06	TH	1.3	1.2		MIT
3944.254	3945.370	25346.16	TH	0.5	0.0		MIT
3943.694	3944.810	25349.76	TH	1.0	1.0		MIT
3942.646	3943.762	25356.50	TH	1.0	0.9		MIT
3942.065	3943.181	25360.24	TH	0.9	0.7		MIT
3941.727	3942.843	25362.41	TH	0.9H	0.0H		MIT
3941.364	3942.480	25364.75	TH	0.3			F
3938.732	3939.847	25381.70	TH	1.2W	1.0W		MIT
3937.929	3939.044	25386.87	TH	1.2	1.0		MIT
3937.043	3938.158	25392.58	TH	1.2	1.0		MIT
3935.940	3937.054	25399.70	TH	0.7	0.5		MIT

AIR WAVELENGTH (ANGSTRMS)	VACUUM WAVELENGTH (ANGSTRMS)	WAVENUMBER (KAYSERS)	LMENT	LOG ARC	INTENSITY SPK	INTENSITY DIS	REF
3935.642	3936.756	25401.62	TH	0.3			FD
3935.180	3936.294	25404.61	TH	0.7	0.7		MIT
3932.915	3934.028	25419.24	TH	1.0	0.5		MIT
3932.230	3933.343	25423.67	TH	1.2	1.0		MIT
3929.671	3930.784	25440.22	TH	1.5	1.3		MIT
3927.425	3928.537	25454.77	TH	1.2	1.0		MIT
3927.181	3928.293	25456.35	TH	1.2	1.0		MIT
3926.711	3927.823	25459.40	TH	1.0	0.9		MIT
3925.096	3926.207	25469.87	TH	1.0	0.7		MIT
3923.808	3924.919	25478.23	TH	0.7			MIT
3922.609	3923.720	25486.02	TH	0.7H			MIT
3922.221	3923.332	25488.54	TH	0.9	0.5		MIT
3920.259	3921.369	25501.30	TH	0.9	0.7		MIT
3919.003	3920.113	25509.47	TH	1.0	0.0		MIT
3918.500	3919.610	25512.74	TH	0.5	0.0		MIT
3918.285	3919.395	25514.14	TH	0.7	0.0		MIT
3918.010	3919.120	25515.94	TH	1.0H	0.7		MIT
3917.258	3918.367	25520.83	TH	0.9	0.0		MIT
3916.730	3917.839	25524.27	TH	1.2	1.0		MIT
3916.413	3917.522	25526.34	TH	1.0	0.5		MIT
3915.211	3916.320	25534.18	TH	1.2H	0.0H		MIT
3914.47	3915.58	25539.0	TH	0.3	0.0		MIT
3913.826	3914.934	25543.21	TH	1.0	0.9		MIT
3913.012	3914.120	25548.52	TH	1.2	1.2		MIT
3912.285	3913.393	25553.27	TH	1.2	1.2		MIT
3911.914	3913.022	25555.70	TH	0.9	0.5		MIT
3911.308	3912.416	25559.66	TH	0.7	0.5		MIT
3911.003	3912.111	25561.65	TH	0.7	0.0		MIT
3908.488	3909.595	25578.10	TH	0.5	0.3		MIT
3907.342	3908.449	25585.60	TH	0.9	0.5		MIT
3906.796	3907.903	25589.17	TH	0.9			MIT
3905.190	3906.296	25599.70	TH	1.5	1.5		MIT
3904.089	3905.195	25606.92	TH	1.3	1.3		MIT
3903.093	3904.199	25613.45	TH	1.2	0.7		MIT
3902.459	3903.565	25617.61	TH	0.9	0.7		MIT
3902.122	3903.227	25619.82	TH	1.0	1.0		MIT
3901.670	3902.775	25622.79	TH	0.7			MIT
3901.158	3902.263	25626.16	TH	1.0	0.9		MIT
3900.886	3901.991	25627.94	TH	1.5	1.5		MIT
3900.584	3901.689	25629.93	TH	0.9	0.5		MIT
3898.794	3899.899	25641.69	TH	0.3	0.0		MIT
3898.477	3899.581	25643.78	TH	1.0	0.9		MIT
3895.425	3896.529	25663.87	TH	1.3	1.0		MIT
3894.901	3896.005	25667.32	TH	0.3	0.0		MIT
3894.416	3895.519	25670.52	TH	0.9	0.7		MIT
3893.396	3894.499	25677.24	TH	1.2W	1.0W		MIT
3893.113	3894.216	25679.11	TH	0.9	0.7		MIT
3892.311	3893.414	25684.40	TH	1.0	0.9		MIT
3891.763	3892.866	25688.02	TH	0.7			MIT
3891.060	3892.163	25692.66	TH	1.0	1.0		MIT
3886.921	3888.022	25720.02	TH	0.9	0.0		MIT
3885.766	3886.867	25727.66	TH	0.9	0.7		MIT
3885.150	3886.251	25731.74	TH	0.7	0.7		MIT
3884.829	3885.930	25733.87	TH	1.3	1.2		MIT
3884.531	3885.632	25735.84	TH	1.0	0.9		MIT
3875.648	3876.747	25794.82	TH	0.5			MIT
3875.379	3876.477	25796.62	TH	1.0	0.7		MIT
3874.866	3875.964	25800.03	TH	0.7	0.5		MIT
3873.823	3874.921	25806.98	TH	1.0	0.7		MIT
3872.728	3873.826	25814.27	TH	1.5	1.3		MIT
3869.639	3870.736	25834.88	TH	0.7	0.5		MIT
3869.363	3870.460	25836.72	TH	1.0	0.9		MIT
3865.036	3866.132	25865.65	TH	0.9	0.5		MIT
3863.391	3864.486	25876.66	TH	1.3	1.3		FD
3862.376	3863.471	25883.46	TH	0.5	0.3		MIT
3859.832	3860.926	25900.52	TH	0.7	1.0		MIT
3856.374	3857.468	25923.74	TH	0.9	0.7		MIT
3854.547	3855.640	25936.03	TH	1.3	1.3		MIT
3852.962	3854.055	25946.70	TH	0.9	1.0		MIT
3852.135	3853.227	25952.27	TH	0.7			MIT

AIR WAVELENGTH (ANGSTRMS)	VACUUM WAVELENGTH (ANGSTRMS)	WAVENUMBER (KAYSERS)	LMENT	LOG ARC	INTENSITY SPK	DIS	REF
3846.242	3847.333	25992.03	TH	1.0	0.9		MIT
3845.016	3846.107	26000.32	TH	1.3W	1.0W		MIT
3842.912	3844.002	26014.55	TH	1.0	0.9		FD
3841.964	3843.054	26020.97	TH	1.3	1.3		MIT
3839.699	3840.788	26036.32	TH	1.2	1.2		MIT
3837.880	3838.969	26048.66	TH	1.0	0.9		MIT
3836.505	3837.593	26058.00	TH	1.7W	1.7W		MIT
3835.343	3836.431	26065.89	TH	0.7	0.5		MIT
3834.608	3835.696	26070.89	TH	1.2W	1.0W		MIT
3833.037	3834.124	26081.57	TH	1.3W	1.0W		MIT
3832.788	3833.875	26083.27	TH	0.5	0.7		MIT
3832.431	3833.518	26085.70	TH	0.9	0.7		MIT
3831.736	3832.823	26090.43	TH	0.7	0.9		MIT
3830.069	3831.156	26101.78	TH	0.6	0.5		MIT
3829.410	3830.497	26106.28	TH	1.6W	1.3W		MIT
3828.388	3829.474	26113.25	TH	0.7	0.7		MIT
3828.144	3829.230	26114.91	TH	0.9	0.9		MIT
3826.947	3828.033	26123.08	TH	1.0	0.9		MIT
3825.660	3826.746	26131.87	TH	0.9	0.9		MIT
3825.040	3826.125	26136.10	TH	0.9	1.0		MIT
3824.762	3825.847	26138.00	TH	1.3D	1.0		MIT
3824.349	3825.434	26140.82	TH	1.3W	1.3W		MIT
3823.589	3824.674	26146.02	TH	0.7	1.0		MIT
3823.327	3824.412	26147.81	TH	0.7	0.9		MIT
3823.079	3824.164	26149.51	TH	1.3W	1.0W		MIT
3822.865	3823.950	26150.97	TH	0.7	0.5		MIT
3822.149	3823.234	26155.87	TH	1.0	1.0		MIT
3821.431	3822.515	26160.78	TH	1.2	1.2		MIT
3820.806	3821.890	26165.06	TH	1.3	1.3		MIT
3819.28	3820.36	26175.5	TH	0.3	0.7		MIT
3819.046	3820.130	26177.12	TH		0.3		MIT
3818.672	3819.756	26179.69	TH	0.9	0.7		MIT
3817.727	3818.810	26186.17	TH	1.2	1.2		MIT
3817.372	3818.455	26188.60	TH	1.3	1.0		MIT
3816.582	3817.665	26194.02	TH	0.9	0.7		MIT
3815.069	3816.152	26204.41	TH	0.9	0.9		MIT
3814.885	3815.968	26205.67	TH	1.0	0.7		MIT
3814.864	3815.947	26205.82	TH I	1.5	0.6		MIT
3814.591	3815.674	26207.69	TH	1.2	1.2		MIT
3813.064	3814.146	26218.19	TH	1.2	1.3		MIT
3811.383	3812.465	26229.75	TH	1.3	1.0		MIT
3811.063	3812.145	26231.95	TH	0.7	0.5		MIT
3810.617	3811.699	26235.02	TH	0.9	0.7		MIT
3809.839	3810.920	26240.38	TH	0.9	0.7		MIT
3809.082	3810.163	26245.59	TH	1.0	0.7		MIT
3808.146	3809.227	26252.05	TH	1.0	1.0		MIT
3807.875	3808.956	26253.91	TH	1.0	1.0		MIT
3806.792	3807.873	26261.38	TH	1.0	0.7		MIT
3805.824	3806.904	26268.06	TH	1.3	1.2		MIT
3804.698	3805.778	26275.84	TH	1.0	0.9		MIT
3804.103	3805.183	26279.95	TH		0.3		MIT
3803.987	3805.067	26280.75	TH	0.9	0.7		MIT
3803.077	3804.157	26287.03	TH	1.2	1.2		MIT
3802.791	3803.871	26289.01	TH	1.0	0.7		MIT
3802.150	3803.229	26293.44	TH	0.9	0.9		MIT
3801.927	3803.006	26294.99	TH	0.7	0.7		MIT
3801.368	3802.447	26298.85	TH	1.0	0.9		MIT
3800.605	3801.684	26304.13	TH	0.3	0.3		MIT
3800.378	3801.457	26305.70	TH	0.7	0.3		MIT
3800.033	3801.112	26308.09	TH	1.0W	0.7W		MIT
3799.66	3800.74	26310.7	TH	0.3	0.3		MIT
3799.218	3800.297	26313.74	TH	1.0	0.7		MIT
3798.103	3799.181	26321.46	TH	0.7	0.0		MIT
3797.91	3798.99	26322.8	TH	0.3	0.7		MIT
3797.516	3798.594	26325.53	TH	0.9	1.0		MIT
3796.952	3798.030	26329.44	TH	1.0	0.7		MIT
3796.738	3797.816	26330.92	TH	0.5			MIT
3796.184	3797.262	26334.76	TH	1.0	0.9		MIT
3795.751	3796.829	26337.77	TH	1.3W	1.0W		MIT
3795.387	3796.465	26340.29	TH	1.3	1.0		MIT

AIR WAVELENGTH (ANGSTRMS)	VACUUM WAVELENGTH (ANGSTRMS)	WAVENUMBER (KAYSERS)	LMENT	LOG ARC	INTENSITY SPK	DIS	REF
3794.341	3795.418	26347.56	TH	1.0	1.0		MIT
3794.154	3795.231	26348.85	TH	1.0	1.0		MIT
3793.786	3794.863	26351.41	TH	1.3	0.9		MIT
3793.488	3794.565	26353.48	TH	1.3W	0.9		MIT
3792.920	3793.997	26357.43	TH	1.0	0.7		MIT
3792.641	3793.718	26359.37	TH	1.0	0.7		MIT
3792.377	3793.454	26361.20	TH	0.7	0.0		MIT
3791.981	3793.058	26363.95	TH	0.5	0.5		MIT
3791.302	3792.379	26368.68	TH	0.9	0.9		MIT
3790.851	3791.927	26371.81	TH	1.0	0.7		MIT
3790.537	3791.613	26374.00	TH	1.0	0.7		MIT
3789.978	3791.054	26377.89	TH	0.3	0.7H		MIT
3789.116	3790.192	26383.89	TH	1.3	1.3		MIT
3788.359	3789.435	26389.16	TH	1.0	0.9		MIT
3787.504	3788.580	26395.12	TH	1.0	0.7		MIT
3787.191	3788.267	26397.30	TH	1.6W	1.3W		MIT
3786.884	3787.959	26399.44	TH	1.2	1.2		MIT
3785.852	3786.927	26406.63	TH	0.7			MIT
3785.652	3786.727	26408.03	TH	1.2	1.2		MIT
3785.335	3786.410	26410.24	TH	1.0	0.7		MIT
3783.821	3784.896	26420.81	TH	1.2	0.7		MIT
3783.296	3784.371	26424.47	TH	1.3	1.2		MIT
3783.082	3784.156	26425.97	TH	0.9	0.0		MIT
3783.016	3784.090	26426.43	TH	1.3	1.3		MIT
3782.206	3783.280	26432.09	TH	1.0	0.7		MIT
3781.686	3782.760	26435.72	TH	1.3	1.0		MIT
3780.975	3782.049	26440.69	TH	1.2	0.5		MIT
3780.855	3781.929	26441.53	TH	1.3	0.7		MIT
3780.507	3781.581	26443.97	TH	1.3	1.0		MIT
3779.806	3780.880	26448.87	TH	1.3	1.0		MIT
3778.782	3779.855	26456.04	TH	1.2	1.0		MIT
3777.914	3778.987	26462.12	TH	1.2	1.0		MIT
3777.351	3778.424	26466.06	TH	1.0	0.7		MIT
3777.120	3778.193	26467.68	TH	1.6	1.3		MIT
3776.276	3777.349	26473.59	TH	0.9	0.0		MIT
3775.942	3777.015	26475.94	TH	1.3	1.0		MIT
3775.319	3776.391	26480.31	TH	1.3	0.7		MIT
3774.212	3775.284	26488.07	TH	0.9	0.9		MIT
3773.760	3774.832	26491.25	TH	1.3	1.3		MIT
3773.072	3774.144	26496.07	TH	0.7	0.3		MIT
3772.660	3773.732	26498.97	TH	0.9	0.0		MIT
3772.252	3773.324	26501.83	TH	1.0	0.9		MIT
3771.803	3772.875	26504.99	TH	1.0	0.7		MIT
3771.377	3772.448	26507.98	TH	1.5W	1.3W		MIT
3770.20	3771.27	26516.3	TH	1.0D	0.7D		MIT
3770.062	3771.133	26517.23	TH	1.0	0.7		MIT
3769.600	3770.671	26520.48	TH	0.7	0.0		MIT
3768.437	3769.508	26528.66	TH	0.9	0.9		MIT
3767.905	3768.976	26532.41	TH	1.3	1.2		MIT
3766.347	3767.417	26543.38	TH	0.7	0.3		MIT
3765.256	3766.326	26551.07	TH	1.3	1.0		MIT
3764.663	3765.733	26555.26	TH	0.3	0.5		MIT
3764.277	3765.347	26557.98	TH	1.0	0.7		MIT
3763.601	3764.670	26562.75	TH	0.3	0.3		MIT
3763.335	3764.404	26564.63	TH	1.2	0.9		MIT
3762.885	3763.954	26567.80	TH	1.3	1.3		MIT
3762.363	3763.432	26571.49	TH	1.0	0.5		MIT
3761.475	3762.544	26577.76	TH	1.0	0.0		MIT
3761.104	3762.173	26580.39	TH	1.2	0.9		MIT
3760.277	3761.346	26586.23	TH	1.3	0.7		MIT
3759.445	3760.513	26592.11	TH	0.9	0.0		MIT
3759.314	3760.382	26593.04	TH	1.3	1.0		MIT
3758.468	3759.536	26599.03	TH	0.7	0.5		MIT
3757.700	3758.768	26604.46	TH	1.0	0.7		MIT
3757.335	3758.403	26607.05	TH	0.7	0.3		MIT
3757.256	3758.324	26607.61	TH	0.7	0.3		MIT
3756.662	3757.730	26611.81	TH	0.7	0.5		MIT
3756.320	3757.388	26614.24	TH	1.7R	1.3R		MIT
3755.407	3756.474	26620.71	TH	1.0	0.7		MIT
3755.217	3756.284	26622.05	TH	0.9	0.7		MIT

AIR WAVELENGTH (ANGSTRMS)	VACUUM WAVELENGTH (ANGSTRMS)	WAVENUMBER (KAYSERS)	LMENT	LOG ARC	INTENSITY SPK	DIS	REF
3754.596	3755.663	26626.46	TH	1.3	1.0		MIT
3754.037	3755.104	26630.42	TH	1.3W	0.7W		MIT
3753.241	3754.308	26636.07	TH	1.0	0.0		MIT
3752.573	3753.640	26640.81	TH	1.6	1.7		MIT
3751.737	3752.803	26646.75	TH	1.0	0.5		MIT
3751.108	3752.174	26651.22	TH	1.0			MIT
3750.690	3751.756	26654.19	TH	1.0	0.7		MIT
3750.497	3751.563	26655.56	TH	0.9			MIT
3750.150	3751.216	26658.02	TH	1.3	0.7		MIT
3748.970	3750.036	26666.41	TH	0.9	0.7		MIT
3748.304	3749.369	26671.15	TH	1.0	0.9		MIT
3747.547	3748.612	26676.54	TH	1.5	1.5		MIT
3745.983	3747.048	26687.68	TH	1.2	1.3		MIT
3744.739	3745.804	26696.54	TH	1.2	1.0		MIT
3743.513	3744.577	26705.28	TH	0.9	0.9		MIT
3743.004	3744.068	26708.92	TH	0.9	0.0		MIT
3742.925	3743.989	26709.48	TH	1.3	0.9		MIT
3742.249	3743.313	26714.30	TH	1.3	1.0		MIT
3741.190	3742.254	26721.87	TH	1.9	1.9		MIT
3740.849	3741.913	26724.30	TH	1.6	1.3		MIT
3739.788	3740.851	26731.88	TH	1.3	0.7		MIT
3738.846	3739.909	26738.62	TH	1.3	1.3		MIT
3737.513	3738.576	26748.16	TH	0.7	0.0		MIT
3736.949	3738.011	26752.19	TH	0.7	0.7		MIT
3735.528	3736.590	26762.37	TH	0.9	0.0		MIT
3734.601	3735.663	26769.01	TH	0.9	1.0		MIT
3730.753	3731.814	26796.62	TH	1.0	0.9		MIT
3730.383	3731.444	26799.28	TH	1.2	0.5		MIT
3727.903	3728.963	26817.11	TH	1.2	0.7		MIT
3727.271	3728.331	26821.65	TH	0.7	0.9H		MIT
3727.162	3728.222	26822.44	TH	0.5			MIT
3726.730	3727.790	26825.55	TH	1.5	1.3		MIT
3725.841	3726.901	26831.95	TH	0.7	0.7		MIT
3725.396	3726.456	26835.15	TH	1.0	0.0		MIT
3724.739	3725.798	26839.89	TH	1.5	1.3		MIT
3723.696	3724.755	26847.40	TH	1.0	0.9		MIT
3723.291	3724.350	26850.32	TH	1.0	0.9		MIT
3722.192	3723.251	26858.25	TH	1.5	1.4		MIT
3721.831	3722.890	26860.86	TH	1.6	1.5		MIT
3721.402	3722.460	26863.95	TH	0.9	0.5		MIT
3720.776	3721.834	26868.47	TH	0.7			MIT
3720.309	3721.367	26871.84	TH	1.2	1.0		MIT
3719.969	3721.027	26874.30	TH	0.3	0.0H		MIT
3719.436	3720.494	26878.15	TH	1.5	1.0		MIT
3718.660	3719.718	26883.76	TH	1.3H	0.9		MIT
3718.167	3719.225	26887.32	TH	1.2	1.0		MIT
3717.832	3718.890	26889.75	TH	1.3	1.0		MIT
3716.585	3717.642	26898.77	TH	0.9	0.0		MIT
3716.255	3717.312	26901.16	TH	0.7	0.0		MIT
3715.568	3716.625	26906.13	TH	0.7			MIT
3714.393	3715.450	26914.64	TH	0.9	0.7		MIT
3712.538	3713.594	26928.09	TH	1.2	0.9		MIT
3711.628	3712.684	26934.69	TH	0.9	0.5		MIT
3711.309	3712.365	26937.01	TH	1.5	1.3		MIT
3709.673	3710.728	26948.89	TH	0.7	0.5		MIT
3708.760	3709.815	26955.52	TH	0.9	0.7		MIT
3707.429	3708.484	26965.20	TH	1.0	0.7		MIT
3706.993	3708.048	26968.37	TH	0.9	0.7		MIT
3706.769	3707.824	26970.00	TH	1.2	1.0		MIT
3704.967	3706.021	26983.12	TH	1.0	0.9		MIT
3703.996	3705.050	26990.19	TH	1.0	0.7		MIT
3703.787	3704.841	26991.71	TH	0.9			MIT
3703.235	3704.289	26995.74	TH	0.7			MIT
3702.872	3703.926	26998.38	TH	1.0	1.3		MIT
3700.987	3702.040	27012.13	TH	0.7	0.0		MIT
3700.771	3701.824	27013.71	TH	1.2	1.0		MIT
3699.884	3700.937	27020.19	TH	0.7	0.0		MIT
3698.305	3699.357	27031.72	TH	1.1	1.0		MIT
3698.112	3699.164	27033.13	TH	0.9	0.3		MIT
3697.034	3698.086	27041.01	TH	0.8	0.7		MIT

AIR WAVELENGTH (ANGSTRMS)	VACUUM WAVELENGTH (ANGSTRMS)	WAVENUMBER (KAYSERS)	LMENT	LOG ARC	INTENSITY SPK	DIS	REF
3696.653	3697.705	27043.80	TH	0.8	0.7		MIT
3695.980	3697.032	27048.73	TH	0.9	0.9		MIT
3695.292	3696.344	27053.76	TH	0.7	0.0		MIT
3694.894	3695.946	27056.68	TH	0.7	0.0		MIT
3694.355	3695.406	27060.62	TH	0.7	0.0		MIT
3693.94	3694.99	27063.7	TH	0.3	0.5		MIT
3693.698	3694.749	27065.44	TH	0.8	0.6		MIT
3692.571	3693.622	27073.70	TH	1.0	0.3		MIT
3692.080	3693.131	27077.30	TH	0.9	0.6		MIT
3691.881	3692.932	27078.76	TH	0.9	0.3		MIT
3691.617	3692.668	27080.69	TH	0.8	0.0		MIT
3691.415	3692.466	27082.17	TH	0.7	0.0		MIT
3690.626	3691.677	27087.96	TH	0.7	0.0		MIT
3690.490	3691.540	27088.96	TH	1.0	1.0		MIT
3690.119	3691.169	27091.69	TH	1.0	1.0		MIT
3688.763	3689.813	27101.64	TH	1.1	1.1		MIT
3687.988	3689.038	27107.34	TH	0.7	0.0		MIT
3687.664	3688.714	27109.72	TH	0.7	0.7		MIT
3686.987	3688.037	27114.70	TH	0.6	0.6		MIT
3685.871	3686.920	27122.91	TH	0.6	0.5		MIT
3684.917	3685.966	27129.93	TH	0.3			MIT
3683.944	3684.993	27137.09	TH	0.5	0.5		MIT
3683.332	3684.381	27141.60	TH	0.7	0.6		MIT
3682.490	3683.538	27147.81	TH	0.6	0.0		MIT
3682.359	3683.407	27148.78	TH	0.6	0.6		MIT
3681.886	3682.934	27152.26	TH	0.9	0.9		MIT
3681.194	3682.242	27157.37	TH	0.8	0.7		MIT
3680.574	3681.622	27161.94	TH	0.3	0.0		MIT
3680.454	3681.502	27162.83	TH	0.5	0.0		MIT
3679.712	3680.760	27168.30	TH	1.1	1.1		MIT
3679.140	3680.188	27172.53	TH	0.5			MIT
3678.791	3679.838	27175.11	TH	0.5	0.0		MIT
3678.466	3679.513	27177.51	TH	0.6	0.0		MIT
3678.035	3679.082	27180.69	TH	1.0	1.0		MIT
3677.741	3678.788	27182.86	TH	0.7	0.6		MIT
3676.694	3677.741	27190.60	TH	0.8	0.6		MIT
3675.571	3676.618	27198.91	TH	1.0	1.0		MIT
3675.143	3676.189	27202.08	TH	0.7	0.0		MIT
3674.893	3675.939	27203.93	TH	0.7	0.0		MIT
3674.071	3675.117	27210.02	TH	0.5D	0.0D		MIT
3673.799	3674.845	27212.03	TH	1.0	1.0		MIT
3673.262	3674.308	27216.01	TH	1.0	1.0		MIT
3672.534	3673.580	27221.40	TH	0.3D	0.0D		MIT
3672.304	3673.350	27223.11	TH	0.6	0.0		MIT
3671.543	3672.589	27228.75	TH	0.8	0.6		MIT
3670.635	3671.680	27235.49	TH	0.7	0.0		MIT
3670.062	3671.107	27239.74	TH	0.5D	0.6D		MIT
3669.958	3671.003	27240.51	TH	0.7D	0.0D		MIT
3669.413	3670.458	27244.56	TH	0.6	0.6		MIT
3668.147	3669.192	27253.96	TH	0.9	0.3		MIT
3666.985	3668.029	27262.59	TH	0.7	0.3		MIT
3666.380	3667.424	27267.09	TH	0.6	0.3		MIT
3665.731	3666.775	27271.92	TH	0.8	0.7		MIT
3665.484	3666.528	27273.76	TH	0.5			MIT
3665.185	3666.229	27275.98	TH	0.7	0.6		MIT
3663.708	3664.752	27286.98	TH	1.0	1.0		MIT
3663.204	3664.247	27290.73	TH	0.6	0.7		MIT
3662.194	3663.237	27298.26	TH	0.6	0.6		MIT
3661.624	3662.667	27302.51	TH	1.0	0.7H		MIT
3661.525	3662.568	27303.25	TH		0.7H		MIT
3660.124	3661.167	27313.70	TH	0.6	0.3		MIT
3659.633	3660.675	27317.36	TH	0.6	0.0		MIT
3659.513	3660.555	27318.26	TH	1.0	1.1		MIT
3658.811	3659.853	27323.50	TH	0.7	0.0		MIT
3658.173	3659.215	27328.26	TH	0.9	0.8		MIT
3658.069	3659.111	27329.04	TH	1.0	1.0		MIT
3657.538	3658.580	27333.01	TH	0.5	0.3		MIT
3656.696	3657.738	27339.30	TH	0.8	0.0		MIT
3656.204	3657.246	27342.98	TH	1.0	1.0		MIT
3655.125	3656.166	27351.05	TH	0.5			MIT

AIR WAVELENGTH (ANGSTRMS)	VACUUM WAVELENGTH (ANGSTRMS)	WAVENUMBER (KAYSERS)	LMENT	LOG ARC	INTENSITY SPK DIS	REF
3654.579	3655.620	27355.14	TH	0.6	0.6	MIT
3654.466	3655.507	27355.99	TH	0.6	0.0	MIT
3653.59	3654.63	27362.5	TH	0.5	0.3	MIT
3652.542	3653.583	27370.40	TH	1.0	1.0	MIT
3652.168	3653.209	27373.20	TH	1.0	1.0	MIT
3651.583	3652.623	27377.58	TH	0.7	0.3	MIT
3650.771	3651.811	27383.67	TH	1.0	1.0	MIT
3650.510	3651.550	27385.63	TH	0.3D	0.0D	MIT
3649.737	3650.777	27391.43	TH	0.9	0.5	MIT
3649.251	3650.291	27395.08	TH	1.0	1.0	MIT
3648.639	3649.679	27399.67	TH	0.6	0.3	MIT
3648.422	3649.462	27401.30	TH	0.9	0.6	MIT
3648.170	3649.210	27403.20	TH	0.8	0.6	MIT
3647.648	3648.687	27407.12	TH	0.9	0.9	MIT
3647.307	3648.346	27409.68	TH	0.5D	0.3D	MIT
3646.886	3647.925	27412.84	TH	0.7	0.6	MIT
3644.717	3645.756	27429.16	TH	0.7	0.6	MIT
3644.347	3645.386	27431.94	TH	0.7	0.6	MIT
3643.515	3644.553	27438.21	TH	0.7	0.0	MIT
3642.893	3643.931	27442.89	TH	0.5	0.3	MIT
3642.249	3643.287	27447.74	TH	1.0	0.6	MIT
3641.841	3642.879	27450.82	TH	0.8	0.6	MIT
3641.285	3642.323	27455.01	TH	0.6	0.5	MIT
3641.114	3642.152	27456.30	TH	0.7	0.6	MIT
3639.449	3640.486	27468.86	TH	1.0	1.0	MIT
3638.645	3639.682	27474.93	TH	0.8	0.3	MIT
3637.557	3638.594	27483.15	TH	0.9	0.7	MIT
3636.567	3637.604	27490.63	TH	0.6	0.6	MIT
3636.167	3637.203	27493.65	TH	0.5	0.5	MIT
3635.944	3636.980	27495.34	TH	0.9	0.6	MIT
3635.372	3636.408	27499.66	TH	1.2	1.0	MIT
3635.246	3636.282	27500.62	TH	0.6	0.7	MIT
3634.584	3635.620	27505.62	TH	0.9	0.6	MIT
3634.214	3635.250	27508.43	TH	0.7	0.8	MIT
3633.700	3634.736	27512.32	TH	0.0D	0.3	MIT
3633.36	3634.40	27514.9	TH	0.5	0.5	MIT
3632.626	3633.662	27520.45	TH	0.8	0.8	MIT
3625.940	3626.974	27571.20	TH	0.8	0.8	MIT
3625.628	3626.662	27573.57	TH	0.9	0.9	MIT
3624.898	3625.932	27579.12	TH	0.9	0.9	MIT
3624.474	3625.507	27582.35	TH	0.6	0.0	MIT
3623.977	3625.010	27586.13	TH	0.6	0.8	MIT
3623.035	3624.068	27593.30	TH	0.5	0.0	MIT
3622.796	3623.829	27595.12	TH	0.6D	0.0D	MIT
3622.717	3623.750	27595.72	TH	0.3D	0.3D	MIT
3622.339	3623.372	27598.60	TH	0.9	0.6	MIT
3621.121	3622.154	27607.89	TH	1.0	1.0	MIT
3620.372	3621.404	27613.60	TH	0.8	0.8	MIT
3619.711	3620.743	27618.64	TH	0.6	0.6	MIT
3619.390	3620.422	27621.09	TH	0.3	0.0	MIT
3618.745	3619.777	27626.01	TH	0.3		MIT
3618.384	3619.416	27628.77	TH	0.7	0.0	MIT
3617.732	3618.764	27633.75	TH	0.3	0.0	MIT
3617.119	3618.151	27638.43	TH	0.7D	0.6D	MIT
3617.019	3618.050	27639.19	TH	0.6D	0.5D	MIT
3616.701	3617.732	27641.62	TH	0.6	0.5	MIT
3615.133	3616.164	27653.61	TH	1.0	1.0	MIT
3614.011	3615.042	27662.20	TH	0.8	0.9	MIT
3613.779	3614.810	27663.97	TH	0.7	0.5	MIT
3612.869	3613.899	27670.94	TH	0.7	0.0	MIT
3612.427	3613.457	27674.33	TH	0.9	0.6	MIT
3612.108	3613.138	27676.77	TH	0.3	0.0	MIT
3610.794	3611.824	27686.84	TH	0.9	0.6	MIT
3610.399	3611.429	27689.87	TH	0.9	0.6	MIT
3610.039	3611.069	27692.63	TH	0.5	0.0	MIT
3609.447	3610.477	27697.18	TH	1.1	1.0	MIT
3609.228	3610.257	27698.86	TH	0.7	0.6	MIT
3608.379	3609.408	27705.37	TH	0.6	0.0D	MIT
3607.907	3608.936	27709.00	TH	0.0	0.3	MIT
3607.395	3608.424	27712.93	TH	0.5	0.6	MIT
3606.716	3607.745	27718.15	TH	0.0	0.3	MIT
3606.204	3607.233	27722.08	TH		0.7H	MIT
3605.657	3606.686	27726.29	TH	0.9	0.8	MIT
3604.680	3605.708	27733.80	TH	0.7	0.0	MIT
3604.046	3605.074	27738.68	TH	0.6	0.6	MIT
3603.624	3604.652	27741.93	TH	0.6	0.3	MIT
3603.363	3604.391	27743.94	TH	0.7	0.7	MIT
3603.208	3604.236	27745.13	TH	0.9	0.9	MIT
3602.464	3603.492	27750.86	TH	0.6	0.0	MIT
3601.984	3603.012	27754.56	TH	0.3	0.3	MIT
3601.040	3602.067	27761.83	TH	0.9	1.0	MIT
3600.826	3601.853	27763.49	TH	0.5	0.0	MIT
3600.434	3601.461	27766.51	TH	0.6	0.0	MIT
3599.731	3600.758	27771.93	TH	0.5	0.0	MIT
3599.360	3600.387	27774.79	TH	0.5	0.6	MIT
3598.124	3599.151	27784.33	TH	0.8	0.6	MIT
3597.498	3598.524	27789.17	TH	0.5	0.0	MIT
3595.623	3596.649	27803.66	TH	0.5	0.0	MIT
3595.323	3596.349	27805.98	TH	0.5D	0.5D	MIT
3594.116	3595.142	27815.32	TH	0.7	0.3	MIT
3593.881	3594.907	27817.14	TH	0.7	0.7	MIT
3592.776	3593.801	27825.69	TH	0.3D	0.8	MIT
3591.455	3592.480	27835.93	TH	0.6	0.0	MIT
3591.065	3592.090	27838.95	TH	0.3	0.6	MIT
3589.708	3590.732	27849.47	TH	0.3	0.3	MIT
3589.109	3590.133	27854.12	TH	0.6	0.5	MIT
3585.774	3586.797	27880.02	TH	0.6	0.6	MIT
3585.057	3586.080	27885.60	TH	0.0	0.6	MIT
3582.015	3583.037	27909.28	TH	0.7	0.8	MIT
3580.235	3581.257	27923.16	TH	0.7	0.7	MIT
3579.343	3580.365	27930.11	TH	0.6	0.9	MIT
3577.219	3578.240	27946.70	TH	0.0	0.3	MIT
3576.562	3577.583	27951.83	TH	0.5	0.5	MIT
3575.323	3576.344	27961.52	TH	0.8	1.0	MIT
3573.511	3574.531	27975.70	TH	0.5	0.5	MIT
3573.222	3574.242	27977.96	TH	0.7	0.8	MIT
3572.399	3573.419	27984.40	TH	0.9	0.9	MIT
3571.575	3572.595	27990.86	TH	0.9	0.9	MIT
3569.888	3570.907	28004.09	TH	0.9	0.7	MIT
3569.627	3570.646	28006.14	TH	0.3	0.3	MIT
3567.978	3568.997	28019.08	TH	0.7	0.7	MIT
3567.707	3568.726	28021.21	TH	0.5	0.5	MIT
3567.266	3568.285	28024.67	TH	0.3	0.3	MIT
3567.046	3568.065	28026.40	TH	0.6	0.6	MIT
3566.08	3567.10	28034.0	TH	1.0	0.5	MIT
3564.71	3565.73	28044.8	TH		0.7	MIT
3563.93	3564.95	28050.9	TH	0.3	0.3	MIT
3563.384	3564.402	28055.20	TH	1.0	0.5H	MIT
3561.809	3562.826	28067.60	TH	0.5	0.0	MIT
3560.864	3561.881	28075.05	TH	1.0	0.6	MIT
3559.958	3560.975	28082.20	TH	0.8	0.9	MIT
3559.456	3560.473	28086.16	TH	1.0	1.0	MIT
3557.464	3558.480	28101.89	TH	0.7	0.7	MIT
3556.312	3557.328	28110.99	TH	0.3D	0.3	MIT
3555.755	3556.771	28115.39	TH	1.1	0.5	MIT
3555.058	3556.074	28120.90	TH	0.6D	0.0D	MIT
3554.318	3555.333	28126.76	TH	0.3D	0.0D	MIT
3553.379	3554.394	28134.19	TH	0.6	0.6	MIT
3553.111	3554.126	28136.31	TH	0.7	0.8	MIT
3552.748	3553.763	28139.19	TH	0.3D	0.0	MIT
3551.870	3552.885	28146.14	TH	0.3D	0.5	MIT
3551.386	3552.401	28149.98	TH	0.3D	0.0	MIT
3550.286	3551.300	28158.70	TH	0.6	0.6	MIT
3549.741	3550.755	28163.02	TH	0.6	0.6	MIT
3549.597	3550.611	28164.17	TH	0.3	0.0	MIT
3549.338	3550.352	28166.22	TH	0.3D	0.0	MIT
3548.907	3549.921	28169.64	TH	0.3	0.0	MIT
3548.133	3549.147	28175.79	TH	0.3D	0.0D	MIT
3547.409	3548.423	28181.54	TH	0.0D	0.3D	MIT
3547.339	3548.353	28182.09	TH	0.3D	0.0D	MIT

AIR WAVELENGTH (ANGSTRMS)	VACUUM WAVELENGTH (ANGSTRMS)	WAVENUMBER (KAYSERS)	LMENT	LOG ARC	INTENSITY SPK	DIS	REF
3546.260	3547.273	28190.67	TH	0.7	0.5		MIT
3545.348	3546.361	28197.92	TH	0.5	0.5		MIT
3545.290	3546.303	28198.38	TH	0.7	0.5		MIT
3544.949	3545.962	28201.09	TH	0.7	0.3		MIT
3544.014	3545.027	28208.53	TH	0.3	0.0		MIT
3542.632	3543.644	28219.54	TH	0.5D	0.5		MIT
3542.496	3543.508	28220.62	TH	0.3H	0.0D		MIT
3542.243	3543.255	28222.64	TH	0.3	0.0		MIT
3541.977	3542.989	28224.75	TH	0.3	0.3		MIT
3541.622	3542.634	28227.58	TH	0.9	0.8		MIT
3540.310	3541.322	28238.04	TH	0.5D	0.0D		MIT
3539.589	3540.601	28243.80	TH	0.9	0.9		MIT
3539.325	3540.336	28245.90	TH	0.7	0.7		MIT
3538.844	3539.855	28249.74	TH	0.6D			MIT
3538.75	3539.76	28250.5	TH		1.7		EX
3538.445	3539.456	28252.93	TH	0.5D			MIT
3538.223	3539.234	28254.70	TH	0.9	0.5		MIT
3537.434	3538.445	28261.00	TH	0.6D	0.5		MIT
3537.161	3538.172	28263.18	TH	1.0	1.0		MIT
3536.051	3537.062	28272.05	TH	0.7	0.0		MIT
3535.336	3536.346	28277.77	TH	0.3			MIT
3534.529	3535.539	28284.23	TH	0.7D	0.3D		MIT
3533.706	3534.716	28290.82	TH	0.7	0.5		MIT
3531.935	3532.945	28305.00	TH	0.6D	0.5D		MIT
3531.451	3532.460	28308.88	TH	0.6	0.0		MIT
3530.497	3531.506	28316.53	TH	0.6D	0.3D		MIT
3528.947	3529.956	28328.97	TH	0.8	0.8		MIT
3528.824	3529.833	28329.95	TH	0.6	0.5		MIT
3528.151	3529.160	28335.36	TH	0.5D			MIT
3527.350	3528.358	28341.79	TH	0.6D	0.0		MIT
3526.764	3527.772	28346.50	TH	0.3D	0.3		MIT
3526.639	3527.647	28347.51	TH	0.7	0.0		MIT
3526.239	3527.247	28350.72	TH	0.6	0.6		MIT
3525.653	3526.661	28355.43	TH	0.3D	0.6		MIT
3525.142	3526.150	28359.54	TH	0.3D	0.0		MIT
3523.557	3524.564	28372.30	TH	0.7J	0.0J		MIT
3521.917	3522.924	28385.51	TH	1.0	1.0		MIT
3521.262	3522.269	28390.79	TH	0.3			MIT
3521.062	3522.069	28392.40	TH	0.5D	0.0D		MIT
3520.721	3521.728	28395.15	TH	0.7	0.3		MIT
3519.693	3520.699	28403.45	TH	0.5D	0.8		MIT
3518.894	3519.900	28409.90	TH	0.5D	0.5		MIT
3518.685	3519.691	28411.58	TH	0.5			MIT
3518.404	3519.410	28413.85	TH	0.8	0.3		MIT
3516.825	3517.831	28426.61	TH	0.7	0.7		MIT
3516.567	3517.573	28428.70	TH	0.3	0.0		MIT
3516.357	3517.363	28430.39	TH	0.7	0.7		MIT
3515.712	3516.717	28435.61	TH	0.7	0.6		MIT
3514.966	3515.971	28441.64	TH	0.6	0.6		MIT
3514.529	3515.534	28445.18	TH	0.9	0.9		MIT
3513.758	3514.763	28451.42	TH	0.5	0.5		MIT
3513.219	3514.224	28455.79	TH	0.5	0.5		MIT
3513.10	3514.10	28456.7	TH	1.5	0.9		ED
3512.746	3513.751	28459.62	TH	0.7	0.8		MIT
3511.674	3512.678	28468.30	TH	1.0	0.9		MIT
3511.615	3512.619	28468.78	TH	1.0	0.9		MIT
3511.156	3512.160	28472.50	TH	0.7			MIT
3510.732	3511.736	28475.94	TH	0.7	0.6		MIT
3510.540	3511.544	28477.50	TH	0.7	0.6		MIT
3509.197	3510.201	28488.40	TH		0.3D		MIT
3509.113	3510.117	28489.08	TH	0.6	0.0D		MIT
3507.522	3508.525	28502.00	TH		1.0		FD
3506.854	3507.857	28507.43	TH	0.8	0.8		MIT
3505.498	3506.501	28518.46	TH	0.8	0.8		MIT
3504.039	3505.041	28530.33	TH	0.6	0.5		MIT
3503.781	3504.783	28532.43	TH	0.5			MIT
3503.620	3504.622	28533.75	TH	0.9	0.9		MIT
3502.783	3503.785	28540.56	TH	0.9	0.9		MIT
3501.457	3502.459	28551.37	TH	1.0	1.0		MIT
3500.851	3501.853	28556.31	TH	0.7	0.5		MIT

AIR WAVELENGTH (ANGSTRMS)	VACUUM WAVELENGTH (ANGSTRMS)	WAVENUMBER (KAYSERS)	LMENT	LOG ARC	INTENSITY SPK	DIS	REF
3500.552	3501.553	28558.75	TH	0.8	0.7		MIT
3500.269	3501.270	28561.06	TH	0.3	0.3		MIT
3499.996	3500.997	28563.29	TH	0.9	0.9		MIT
3498.986	3499.987	28571.53	TH	0.9	0.9		MIT
3498.625	3499.626	28574.48	TH	0.8	0.5		MIT
3498.016	3499.017	28579.46	TH	0.9	0.9		MIT
3497.70	3498.70	28582.0	TH	0.5	0.5		MIT
3497.264	3498.265	28585.60	TH	0.8	0.7		MIT
3497.020	3498.021	28587.60	TH	0.8	0.7		MIT
3496.812	3497.813	28589.30	TH	0.7	0.5		MIT
3495.703	3496.703	28598.37	TH	1.0	0.0		MIT
3493.526	3494.526	28616.19	TH	1.5	1.2		MIT
3492.683	3493.682	28623.09	TH	0.3	0.3		MIT
3491.901	3492.900	28629.50	TH	0.5			MIT
3491.579	3492.578	28632.14	TH	0.7	0.7		MIT
3490.456	3491.455	28641.35	TH	0.7	0.7		MIT
3490.279	3491.278	28642.81	TH	0.7	0.7		MIT
3489.832	3490.831	28646.48	TH	0.6	0.5		MIT
3489.511	3490.510	28649.11	TH	0.6			FD
3489.188	3490.187	28651.76	TH	0.7	0.0		MIT
3488.837	3489.835	28654.65	TH	0.5			FD
3487.849	3488.847	28662.76	TH	1.0	0.9		MIT
3487.08	3488.08	28669.1	TH	0.6	0.3		MIT
3486.976	3487.974	28669.94	TH	0.7	0.5		MIT
3486.520	3487.518	28673.69	TH	1.1	1.1		MIT
3485.475	3486.473	28682.28	TH	0.6	0.6		MIT
3485.216	3486.214	28684.42	TH	0.9	0.9		MIT
3484.086	3485.083	28693.72	TH	0.9	0.9		MIT
3483.412	3484.409	28699.27	TH	0.9	0.9		MIT
3483.175	3484.172	28701.22	TH	0.3	0.5		MIT
3482.836	3483.833	28704.02	TH	0.3D	0.3D		MIT
3482.552	3483.549	28706.36	TH	1.0	1.0		MIT
3481.051	3482.047	28718.74	TH	0.7	0.5		MIT
3480.05	3481.05	28727.0	TH	0.7	0.0		MIT
3479.177	3480.173	28734.20	TH	1.2	1.0		MIT
3479.099	3480.095	28734.85	TH	0.5	0.3		MIT
3478.464	3479.460	28740.09	TH	0.9	0.9		MIT
3478.136	3479.132	28742.80	TH	1.0	1.0		MIT
3477.705	3478.701	28746.37	TH	0.9	0.9		MIT
3476.917	3477.912	28752.88	TH	0.7	0.5		MIT
3476.543	3477.538	28755.97	TH	1.0	1.0		MIT
3475.969	3476.964	28760.72	TH	0.5	0.3		MIT
3475.536	3476.531	28764.30	TH	0.6	0.6		MIT
3474.304	3475.299	28774.50	TH	0.8	0.8		MIT
3473.784	3474.779	28778.81	TH	0.7	0.7		MIT
3473.409	3474.404	28781.92	TH	1.0	0.9		MIT
3473.033	3474.027	28785.03	TH	1.0	1.0		MIT
3471.973	3472.967	28793.82	TH	0.6	0.0		MIT
3471.220	3472.214	28800.07	TH	0.8	0.3		MIT
3470.570	3471.564	28805.46	TH	0.6	0.0		MIT
3469.926	3470.920	28810.81	TH	1.1	1.2		MIT
3469.341	3470.334	28815.67	TH	0.8	0.5		MIT
3468.720	3469.713	28820.82	TH	0.5	0.5		MIT
3468.221	3469.214	28824.97	TH	1.0	1.1		MIT
3467.928	3468.921	28827.41	TH	1.0	0.9		MIT
3466.899	3467.892	28835.96	TH	0.5	0.5		MIT
3466.647	3467.640	28838.06	TH	0.7	0.3		MIT
3466.536	3467.529	28838.98	TH	0.7	0.0		MIT
3465.929	3466.922	28844.03	TH	0.7	0.7		FD
3465.766	3466.759	28845.39	TH	1.0	1.0		MIT
3465.019	3466.011	28851.61	TH	0.7	0.7		MIT
3464.419	3465.411	28856.60	TH	0.3D	0.3D		MIT
3463.722	3464.714	28862.41	TH	1.0	1.1		MIT
3462.855	3463.847	28869.64	TH	1.0	1.1		MIT
3461.218	3462.209	28883.29	TH	1.0	0.6		MIT
3461.021	3462.012	28884.94	TH	1.0	0.6		MIT
3459.641	3460.632	28896.46	TH	0.7	0.7		MIT
3459.500	3460.491	28897.63	TH	0.6	0.6		MIT
3457.906	3458.897	28910.95	TH	0.5	0.5		MIT
3457.683	3458.673	28912.82	TH	0.7	0.7		MIT

AIR WAVELENGTH (ANGSTRMS)	VACUUM WAVELENGTH (ANGSTRMS)	WAVENUMBER (KAYSERS)	LMENT	LOG ARC	INTENSITY SPK	DIS	REF
3457.075	3458.065	28917.90	TH	0.7	0.0		MIT
3456.936	3457.926	28919.07	TH	0.5	0.5		MIT
3456.441	3457.431	28923.21	TH	0.7	0.6		MIT
3455.949	3456.939	28927.32	TH	0.6	0.6		MIT
3455.275	3456.265	28932.97	TH	0.7	0.7		MIT
3454.95	3455.94	28935.7	TH	0.0D	0.5		MIT
3454.657	3455.647	28938.14	TH	0.5D	0.0D		MIT
3454.208	3455.198	28941.91	TH	0.9	0.9		MIT
3453.922	3454.912	28944.30	TH	0.6	0.6		MIT
3452.683	3453.672	28954.69	TH	1.0	1.0		MIT
3451.703	3452.692	28962.91	TH	0.9	0.7		MIT
3450.949	3451.938	28969.24	TH	0.9	0.9		MIT
3450.81	3451.80	28970.4	TH	0.5	0.3		MIT
3449.951	3450.939	28977.62	TH	0.7	0.3		MIT
3449.648	3450.636	28980.16	TH	0.9	0.9		MIT
3449.287	3450.275	28983.19	TH	0.7	0.7		MIT
3447.631	3448.619	28997.11	TH	0.5	0.3		MIT
3446.546	3447.534	29006.24	TH	0.7	0.5		MIT
3445.748	3446.735	29012.96	TH	0.9	0.9		MIT
3445.382	3446.369	29016.04	TH	1.0	0.9		MIT
3445.216	3446.203	29017.44	TH	0.7	0.7		MIT
3444.047	3445.034	29027.29	TH	0.7	0.6		MIT
3443.124	3444.111	29035.07	TH	0.8	0.8		MIT
3442.581	3443.568	29039.65	TH	0.7	0.0		MIT
3441.366	3442.352	29049.90	TH	0.9	0.9		MIT
3441.019	3442.005	29052.83	TH	0.7	0.9		MIT
3439.714	3440.700	29063.85	TH	1.0	1.1		MIT
3439.409	3440.395	29066.43	TH	0.5			MIT
3438.953	3439.939	29070.29	TH	1.0	1.1		MIT
3437.310	3438.295	29084.18	TH	0.8	0.3		MIT
3437.027	3438.012	29086.57	TH	0.7	0.3		MIT
3436.687	3437.672	29089.45	TH	0.5D	0.7		MIT
3435.979	3436.964	29095.45	TH	1.1	1.0		MIT
3434.762	3435.747	29105.75	TH	0.9	0.9		MIT
3434.000	3434.984	29112.21	TH	1.1	1.1		MIT
3431.816	3432.800	29130.74	TH	0.7	1.0		MIT
3431.024	3432.008	29137.46	TH	0.6	0.6		MIT
3430.558	3431.542	29141.42	TH	0.7	0.6		MIT
3430.342	3431.325	29143.26	TH	0.3D	0.0D		MIT
3429.904	3430.887	29146.98	TH	0.8	0.8		MIT
3429.392	3430.375	29151.33	TH	0.8	0.8		MIT
3429.000	3429.983	29154.66	TH	0.8	0.5		MIT
3427.995	3428.978	29163.21	TH	0.7	0.7		MIT
3427.466	3428.449	29167.71	TH	0.3	0.3		MIT
3426.939	3427.922	29172.20	TH	0.5	0.3		MIT
3425.948	3426.930	29180.63	TH	0.7	0.7		MIT
3425.190	3426.172	29187.09	TH	0.7	0.8		MIT
3423.991	3424.973	29197.31	TH	0.8	0.5		MIT
3423.131	3424.113	29204.65	TH	0.7	0.8		MIT
3422.657	3423.639	29208.69	TH	0.5	0.0		MIT
3421.189	3422.170	29221.22	TH	1.0	1.0		MIT
3419.146	3420.127	29238.68	TH	0.9	1.0		MIT
3418.931	3419.912	29240.52	TH	0.3D	0.5H		MIT
3418.782	3419.763	29241.80	TH	0.8	0.9		MIT
3417.730	3418.710	29250.80	TH	0.5	0.6		MIT
3417.501	3418.481	29252.76	TH	0.5			MIT
3417.12	3418.10	29256.0	TH	0.5	0.5		MIT
3416.422	3417.402	29262.00	TH	0.6	0.7		MIT
3415.886	3416.866	29266.59	TH	0.0D	0.7		MIT
3415.13	3416.11	29273.1	TH	0.3	0.0		MIT
3414.513	3415.492	29278.35	TH	0.8	0.9		MIT
3413.409	3414.388	29287.82	TH	0.7	0.7		MIT
3413.019	3413.998	29291.17	TH	0.7	0.3		MIT
3412.407	3413.386	29296.42	TH	0.3	0.5		MIT
3411.789	3412.768	29301.73	TH	0.8	0.8		MIT
3411.362	3412.341	29305.40	TH	0.5	0.5		MIT
3409.273	3410.251	29323.35	TH	0.8	0.9		MIT
3408.750	3409.728	29327.85	TH	0.7	0.0		MIT
3408.648	3409.626	29328.73	TH	0.7	0.7		MIT
3407.635	3408.613	29337.45	TH	0.5	0.5		MIT

AIR WAVELENGTH (ANGSTRMS)	VACUUM WAVELENGTH (ANGSTRMS)	WAVENUMBER (KAYSERS)	LMENT	LOG ARC	INTENSITY SPK	DIS	REF
3406.243	3407.220	29349.44	TH	0.6	0.6		MIT
3405.561	3406.538	29355.31	TH	0.5	0.6		MIT
3404.654	3405.631	29363.14	TH	0.6	0.6		MIT
3403.275	3404.252	29375.03	TH	0.5D	0.5D		MIT
3402.701	3403.677	29379.99	TH	1.0	1.2		MIT
3402.031	3403.007	29385.77	TH	0.7	0.8		MIT
3401.649	3402.625	29389.07	TH	0.6	0.8		MIT
3401.03	3402.01	29394.4	TH	0.0D	0.3D		MIT
3398.548	3399.523	29415.89	TH	0.7	0.3		MIT
3397.519	3398.494	29424.80	TH	0.9	0.6		MIT
3397.108	3398.083	29428.36	TH	0.3	0.5		MIT
3396.732	3397.707	29431.61	TH	0.7	0.0		MIT
3396.398	3397.373	29434.51	TH	0.3D	0.3D		MIT
3395.381	3396.356	29443.32	TH	0.6	0.6		MIT
3395.127	3396.101	29445.53	TH	0.5	0.3		MIT
3395.02	3395.99	29446.5	TH	1.5			ED
3394.804	3395.778	29448.33	TH	0.6	0.8		MIT
3394.503	3395.477	29450.94	TH		0.5H		MIT
3394.151	3395.125	29454.00	TH	0.6	0.6		MIT
3393.232	3394.206	29461.97	TH	0.5	0.5		MIT
3392.970	3393.944	29464.25	TH	0.5	0.5		MIT
3392.040	3393.014	29472.32	TH	1.0	1.2		MIT
3391.717	3392.691	29475.13	TH	0.0	0.5		MIT
3390.781	3391.754	29483.27	TH	0.3	0.3		MIT
3390.365	3391.338	29486.89	TH	0.6	0.5		MIT
3389.645	3390.618	29493.15	TH	0.8	0.9		MIT
3389.464	3390.437	29494.72	TH	0.5			MIT
3388.582	3389.555	29502.40	TH	0.6	0.7		MIT
3387.922	3388.895	29508.15	TH	0.3			MIT
3386.504	3387.476	29520.50	TH	0.8	1.0		MIT
3385.537	3386.509	29528.93	TH	0.8	0.9		MIT
3385.005	3385.977	29533.57	TH	0.5	0.7		MIT
3383.120	3384.091	29550.03	TH	0.7	1.0		MIT
3381.375	3382.346	29565.28	TH	0.7	1.1		MIT
3380.883	3381.854	29569.58	TH	0.7	0.3		MIT
3379.674	3380.645	29580.16	TH	0.6	0.5		MIT
3379.135	3380.105	29584.88	TH	0.5	0.5		MIT
3378.579	3379.549	29589.75	TH	0.8	1.0		MIT
3377.43	3378.40	29599.8	TH		1.3H		EX
3376.848	3377.818	29604.91	TH	0.3	0.5		MIT
3374.978	3375.947	29621.31	TH	0.5	0.5		MIT
3374.581	3375.550	29624.80	TH	0.7	1.0		MIT
3373.499	3374.468	29634.30	TH	0.3	0.0		MIT
3372.794	3373.763	29640.50	TH	0.5D	0.6D		MIT
3372.696	3373.665	29641.36	TH	0.0D	0.3D		MIT
3371.800	3372.769	29649.23	TH	0.6	1.0		MIT
3370.788	3371.756	29658.13	TH	0.6	0.0		MIT
3370.39	3371.36	29661.6	TH	0.3D	0.3D		MIT
3369.101	3370.069	29672.99	TH	0.0	0.3		MIT
3367.823	3368.790	29684.24	TH	0.9	1.1		MIT
3367.121	3368.088	29690.43	TH	0.3	0.0H		MIT
3366.526	3367.493	29695.68	TH	0.7	0.8		MIT
3365.633	3366.600	29703.56	TH	0.0	0.3		MIT
3365.366	3366.333	29705.92	TH	0.3	0.5		MIT
3364.698	3365.665	29711.81	TH	0.6	0.7		MIT
3363.737	3364.703	29720.30	TH	0.3D	0.7		MIT
3363.112	3364.078	29725.82	TH	0.6	0.7		MIT
3362.672	3363.638	29729.71	TH	0.5	0.6		MIT
3362.533	3363.499	29730.94	TH	0.5	0.6		MIT
3362.196	3363.162	29733.92	TH	0.5	0.6		MIT
3361.739	3362.705	29737.96	TH	0.6D	0.6D		MIT
3360.381	3361.347	29749.98	TH	0.6	0.7		MIT
3360.161	3361.127	29751.93	TH	0.3	0.5		MIT
3359.752	3360.717	29755.55	TH	0.0	0.5		MIT
3359.070	3360.035	29761.59	TH	0.0	0.5		MIT
3358.606	3359.571	29765.70	TH	1.0	1.1		MIT
3357.232	3358.197	29777.89	TH	0.6	0.6		MIT
3357.049	3358.014	29779.51	TH	0.6	0.5		MIT
3356.829	3357.794	29781.46	TH	0.7	0.7		MIT
3355.541	3356.505	29792.89	TH	0.3	0.7		MIT

AIR WAVELENGTH (ANGSTRMS)	VACUUM WAVELENGTH (ANGSTRMS)	WAVENUMBER (KAYSERS)	LMENT	LOG ARC	INTENSITY SPK	DIS	REF
3355.245	3356.209	29795.52	TH	0.5	0.8		MIT
3354.618	3355.582	29801.09	TH	0.7	0.8		MIT
3354.185	3355.149	29804.94	TH	0.7	0.9		MIT
3351.599	3352.562	29827.93	TH	0.6	0.6		MIT
3351.230	3352.193	29831.22	TH	1.0	1.2		MIT
3350.33	3351.29	29839.2	TH	0.3	0.3 D		MIT
3349.866	3350.829	29843.36	TH		0.3		MIT
3349.340	3350.303	29848.05	TH	0.7	1.0		MIT
3348.961	3349.924	29851.43	TH	0.6	0.8		MIT
3348.766	3349.729	29853.17	TH	0.7	0.6		MIT
3347.576	3348.538	29863.78	TH	0.0	0.6		MIT
3346.559	3347.521	29872.85	TH	0.6	0.5		MIT
3344.882	3345.844	29887.83	TH	0.9	0.9		MIT
3344.34	3345.30	29892.7	TH	0.3	0.3		MIT
3343.815	3344.776	29897.36	TH	0.5	0.5		MIT
3343.614	3344.575	29899.16	TH	0.7	1.0		MIT
3343.271	3344.232	29902.23	TH	0.6	0.7		MIT
3342.214	3343.175	29911.69	TH		0.3		MIT
3339.556	3340.516	29935.49	TH		1.3 H		MIT
3338.400	3339.360	29945.86	TH	0.9	0.9		MIT
3337.869	3338.829	29950.62	TH	1.1	1.2		MIT
3337.488	3338.448	29954.04	TH	0.5	0.0		MIT
3337.16	3338.12	29957.0	TH	0.8	0.7		MIT
3336.77	3337.73	29960.5	TH	0.3	0.3		MIT
3336.16	3337.12	29966.0	TH	0.7	0.7		MIT
3335.062	3336.021	29975.83	TH	0.9	0.9		MIT
3334.605	3335.564	29979.94	TH	1.0	1.2		MIT
3334.058	3335.017	29984.86	TH	0.3 D	0.3 D		MIT
3333.389	3334.348	29990.87	TH	0.6	0.6		MIT
3333.129	3334.088	29993.21	TH	0.8	0.3		MIT
3332.395	3333.353	29999.82	TH	0.7	0.7		MIT
3332.07	3333.03	30002.7	TH	0.5	0.0		MIT
3330.475	3331.433	30017.11	TH	1.0	0.3		MIT
3329.731	3330.689	30023.82	TH	0.7			MIT
3328.25	3329.21	30037.2	TH	0.6	0.6		MIT
3327.194	3328.151	30046.71	TH	0.7	0.7		MIT
3326.468	3327.425	30053.27	TH	0.7	0.7		MIT
3325.132	3326.089	30065.34	TH	1.0	1.2		MIT
3324.754	3325.711	30068.76	TH	1.0	1.0		MIT
3322.47	3323.43	30089.4	TH	0.3	0.3		MIT
3321.453	3322.409	30098.65	TH	1.0	1.1		MIT
3320.81	3321.77	30104.5	TH		1.0 H		MIT
3320.302	3321.257	30109.08	TH	1.0	1.1		MIT
3319.914	3320.869	30112.60	TH	0.3	0.0		MIT
3319.63	3320.59	30115.2	TH	0.5	0.5		MIT
3318.981	3319.936	30121.06	TH	0.9	0.9		MIT
3317.75	3318.70	30132.2	TH	0.8	0.8		MIT
3317.62	3318.57	30133.4	TH	0.8 D	0.6 D		MIT
3316.23	3317.18	30146.1	TH	0.7	0.5		MIT
3314.831	3315.785	30158.77	TH	1.0	1.0		MIT
3313.650	3314.604	30169.52	TH	0.6	1.7 H		MIT
3312.08	3313.03	30183.8	TH	0.5	0.6		MIT
3310.52	3311.47	30198.0	TH	0.5 D	0.6 D		MIT
3310.249	3311.202	30200.51	TH	0.9	1.2		MIT
3309.362	3310.315	30208.61	TH	0.6	0.0		MIT
3309.137	3310.090	30210.66	TH	0.8	0.9		MIT
3308.071	3309.023	30220.40	TH	0.3	0.3		MIT
3305.318	3306.270	30245.57	TH	0.3 H			MIT
3304.239	3305.190	30255.44	TH	1.2	0.5		MIT
3303.487	3304.438	30262.33	TH	0.7	1.0		MIT
3301.651	3302.602	30279.16	TH	0.8	0.0 H		MIT
3301.347	3302.298	30281.95	TH	0.8 D	0.7 D		MIT
3300.613	3301.563	30288.68	TH	0.7			MIT
3300.49	3301.44	30289.8	TH		1.5 H		EX
3299.669	3300.619	30297.35	TH	0.7	0.9		MIT
3298.052	3299.002	30312.20	TH	0.6	0.0 H		MIT
3297.828	3298.778	30314.26	TH	1.0	1.1		MIT
3297.373	3298.323	30318.44	TH	0.8	1.0		MIT
3296.607	3297.556	30325.49	TH	0.9	1.0		MIT
3295.770	3296.719	30333.19	TH	0.5			MIT

AIR WAVELENGTH (ANGSTRMS)	VACUUM WAVELENGTH (ANGSTRMS)	WAVENUMBER (KAYSERS)	LMENT	LOG ARC	INTENSITY SPK	DIS	REF
3295.529	3296.478	30335.41	TH	1.0	1.0		MIT
3295.320	3296.269	30337.33	TH	1.0	1.0		MIT
3295.000	3295.949	30340.28	TH	0.9	0.9		MIT
3294.53	3295.48	30344.6	TH	0.6	0.5		MIT
3294.239	3295.188	30347.28	TH	0.7	0.7		MIT
3293.943	3294.892	30350.01	TH	1.0	1.0		MIT
3293.64	3294.59	30352.8	TH	0.5 D	0.5 D		MIT
3293.597	3294.546	30353.20	TH	0.6	0.6		MIT
3292.518	3293.466	30363.15	TH	1.0	1.1		MIT
3291.743	3292.691	30370.29	TH	1.0	1.1		MIT
3291.36	3292.31	30373.8	TH	0.3 W	0.3 W		MIT
3290.59	3291.54	30380.9	TH		1.6 H		MIT
3290.127	3291.075	30385.21	TH	0.8	0.7		MIT
3287.737	3288.684	30407.30	TH	1.1	1.2		MIT
3286.577	3287.524	30418.03	TH	1.0	1.1		MIT
3285.752	3286.699	30425.67	TH	0.9	0.0		MIT
3284.10	3285.05	30441.0	TH		0.9 H		MIT
3283.81	3284.76	30443.7	TH	0.9	0.5		ED
3282.969	3283.915	30451.46	TH	0.9	1.1		MIT
3282.610	3283.556	30454.79	TH	0.8	1.1		MIT
3281.415	3282.360	30465.88	TH	0.9	1.0		MIT
3281.283	3282.228	30467.10	TH	0.6	0.7		MIT
3281.034	3281.979	30469.42	TH	0.3	0.7		MIT
3280.736	3281.681	30472.19	TH	0.3 D	0.3 D		MIT
3280.374	3281.319	30475.55	TH	1.0	1.1		MIT
3277.056	3278.000	30506.40	TH	0.5			MIT
3276.16	3277.10	30514.7	TH	0.5 D	0.9		MIT
3275.010	3275.954	30525.46	TH	1.0	1.1		MIT
3274.399	3275.343	30531.16	TH	0.8	0.5		MIT
3273.884	3274.828	30535.96	TH	1.0	1.2		MIT
3272.026	3272.969	30553.30	TH	0.8	0.0		MIT
3270.816	3271.759	30564.60	TH	0.9	0.7		MIT
3270.23	3271.17	30570.1	TH	0.3 D	0.7		MIT
3269.469	3270.411	30577.19	TH	1.0	1.0		MIT
3267.002	3267.944	30600.28	TH	1.0	1.1		MIT
3266.635	3267.577	30603.72	TH	0.9	0.5		MIT
3264.437	3265.378	30624.32	TH	0.8	0.8		MIT
3264.125	3265.066	30627.25	TH	0.7	0.8		MIT
3263.639	3264.580	30631.81	TH	0.6	0.5		MIT
3263.033	3263.974	30637.50	TH	0.9	0.9		MIT
3262.671	3263.612	30640.90	TH	1.1	1.2		MIT
3261.544	3262.484	30651.49	TH	0.9	1.0		MIT
3261.11	3262.05	30655.6	TH	0.7	0.6		MIT
3260.922	3261.862	30657.33	TH	0.9	1.0		MIT
3260.333	3261.273	30662.87	TH	0.3			MIT
3259.618	3260.558	30669.60	TH	0.9	1.0		MIT
3259.251	3260.191	30673.05	TH	0.7	0.8		MIT
3259.060	3260.000	30674.85	TH	0.8	0.0		MIT
3258.110	3259.050	30683.79	TH	0.3 H	0.3 H		MIT
3257.935	3258.875	30685.44	TH	0.9	0.9		MIT
3257.369	3258.308	30690.77	TH	0.8	0.3		MIT
3257.162	3258.101	30692.72	TH	0.9	0.9		MIT
3256.273	3257.212	30701.10	TH	1.0	1.2		MIT
3255.513	3256.452	30708.27	TH	1.0	1.0		MIT
3254.811	3255.750	30714.89	TH	0.9	1.0		MIT
3254.066	3255.005	30721.92	TH	0.5			MIT
3253.870	3254.808	30723.77	TH	0.7	0.0		MIT
3253.686	3254.624	30725.51	TH	0.7	0.0		MIT
3252.925	3253.863	30732.70	TH	0.7	1.0		MIT
3252.675	3253.613	30735.06	TH	0.9	1.0		MIT
3251.917	3252.855	30742.23	TH	1.0	0.9		MIT
3250.193	3251.131	30758.53	TH	0.8	0.7		MIT
3249.860	3250.797	30761.68	TH	1.0	1.0		MIT
3248.892	3249.829	30770.85	TH	1.0	1.0		MIT
3248.554	3249.491	30774.05	TH	0.6 D	0.6 H		MIT
3246.581	3247.518	30792.75	TH	0.7	0.7		MIT
3245.760	3246.696	30800.54	TH	1.1	1.2		MIT
3244.450	3245.386	30812.98	TH	0.7 D	0.0 D		MIT
3243.034	3243.970	30826.43	TH	0.9	0.9		MIT
3242.257	3243.193	30833.81	TH	0.9	0.7		MIT

AIR WAVELENGTH (ANGSTRMS)	VACUUM WAVELENGTH (ANGSTRMS)	WAVENUMBER (KAYSERS)	LMENT	LOG ARC	INTENSITY SPK DIS	REF
3241.534	3242.469	30840.69	TH	0.5	0.5H	MIT
3241.112	3242.047	30844.71	TH	1.0	1.1	MIT
3240.605	3241.540	30849.53	TH	0.7	0.5	MIT
3240.476	3241.411	30850.76	TH	1.0	1.0	MIT
3239.289	3240.224	30862.07	TH	1.0	0.9	MIT
3238.934	3239.869	30865.45	TH	0.5		MIT
3238.799	3239.734	30866.74	TH	0.6	0.5	MIT
3238.118	3239.052	30873.23	TH	1.1	1.2	MIT
3237.227	3238.161	30881.72	TH	0.5	0.7	MIT
3235.843	3236.777	30894.93	TH	1.1	1.1	MIT
3234.995	3235.929	30903.03	TH	0.7	0.0H	MIT
3233.334	3234.267	30918.90	TH	0.9	0.9	MIT
3232.308	3233.241	30928.72	TH	0.9		MIT
3232.123	3233.056	30930.49	TH	0.5	1.4H	MIT
3230.869	3231.802	30942.49	TH	1.2	1.1	MIT
3228.969	3229.901	30960.70	TH	1.1	1.2	MIT
3227.011	3227.943	30979.48	TH	0.9	0.9	MIT
3226.412	3227.344	30985.24	TH	1.0	1.0	MIT
3226.121	3227.052	30988.03	TH	1.0	0.9	MIT
3225.856	3226.787	30990.58	TH	0.7	0.3	MIT
3225.616	3226.547	30992.88	TH	0.7	0.7	MIT
3225.359	3226.290	30995.35	TH	1.1	1.2	MIT
3223.803	3224.734	31010.31	TH	0.8	0.3	MIT
3223.284	3224.215	31015.30	TH	0.7	0.5	MIT
3221.293	3222.223	31034.47	TH	1.2	1.6H	MIT
3220.304	3221.234	31044.00	TH	1.1	1.0	MIT
3218.81	3219.74	31058.4	TH	0.7D	0.3	MIT
3218.313	3219.242	31063.21	TH	0.9	0.6	MIT
3217.733	3218.662	31068.81	TH	0.9	0.7	MIT
3217.457	3218.386	31071.47	TH	1.0	1.0	MIT
3216.625	3217.554	31079.51	TH	0.5	1.3H	MIT
3215.782	3216.711	31087.66	TH	1.0	0.7	MIT
3214.323	3215.251	31101.77	TH	0.8	0.0	MIT
3213.573	3214.501	31109.02	TH	1.1	1.0	MIT
3212.811	3213.739	31116.40	TH	0.8	0.5	MIT
3211.196	3212.124	31132.05	TH	0.8	0.0	MIT
3210.273	3211.200	31141.00	TH	1.0	1.0	MIT
3208.025	3208.952	31162.82	TH	1.3	1.0	MIT
3207.780	3208.707	31165.20	TH	1.0	0.9	MIT
3206.935	3207.862	31173.42	TH	1.0	0.9	MIT
3206.171	3207.097	31180.84	TH	0.8	0.5H	MIT
3205.300	3206.226	31189.32	TH	1.0	0.8	MIT
3203.881	3204.807	31203.13	TH	1.0	0.8	MIT
3203.234	3204.160	31209.43	TH	1.0	0.8	MIT
3202.525	3203.450	31216.34	TH	0.7	0.6	MIT
3202.132	3203.057	31220.17	TH	0.8	0.6	MIT
3198.969	3199.894	31251.04	TH	1.0	1.0	MIT
3198.702	3199.627	31253.65	TH	1.0	1.0	MIT
3198.176	3199.100	31258.79	TH	1.0	1.0	MIT
3195.691	3196.615	31283.09	TH	0.9	0.0	MIT
3195.312	3196.236	31286.80	TH	0.9	0.8	MIT
3193.311	3194.234	31306.41	TH	0.8	0.6	MIT
3192.586	3193.509	31313.52	TH	0.9	0.0	MIT
3192.098	3193.021	31318.30	TH	0.9	0.7	MIT
3191.835	3192.758	31320.89	TH	0.8	0.6	MIT
3191.167	3192.090	31327.44	TH	0.9	0.8	MIT
3191.095	3192.018	31328.15	TH	0.7	0.5	MIT
3190.074	3190.996	31338.18	TH	0.7	0.6	MIT
3188.188	3189.110	31356.71	TH	0.6D	0.6D	MIT
3187.408	3188.330	31364.39	TH	0.6	0.3	MIT
3187.002	3187.924	31368.38	TH	1.0	0.9	MIT
3184.898	3185.819	31389.10	TH	1.0	1.0	MIT
3183.971	3184.892	31398.24	TH	1.1	0.7	MIT
3183.794	3184.715	31399.99	TH	0.9	0.9	MIT
3183.554	3184.475	31402.35	TH	0.9	0.8	MIT
3182.648	3183.568	31411.29	TH	1.0	1.0	MIT
3182.403	3183.323	31413.71	TH	1.0	0.9	MIT
3181.672	3182.592	31420.93	TH	0.8		MIT
3181.189	3182.109	31425.70	TH	1.0	1.0	MIT
3180.199	3181.119	31435.48	TH	1.2	1.2	MIT

AIR WAVELENGTH (ANGSTRMS)	VACUUM WAVELENGTH (ANGSTRMS)	WAVENUMBER (KAYSERS)	LMENT	LOG ARC	INTENSITY SPK DIS	REF
3179.338	3180.258	31443.99	TH	0.7	0.3	MIT
3179.047	3179.967	31446.87	TH	1.0	1.0	MIT
3178.785	3179.704	31449.46	TH	0.7	0.7	MIT
3178.244	3179.163	31454.82	TH	0.9	0.0	MIT
3177.874	3178.793	31458.48	TH	0.7	0.6	MIT
3177.198	3178.117	31465.17	TH	1.0	0.9	MIT
3176.507	3177.426	31472.02	TH	1.0	0.8	MIT
3175.730	3176.649	31479.72	TH	1.1	1.2	MIT
3174.460	3175.378	31492.31	TH	0.9	0.7	MIT
3174.166	3175.084	31495.23	TH	1.0	1.0	MIT
3173.427	3174.345	31502.56	TH	0.9	0.0	MIT
3173.214	3174.132	31504.68	TH	0.9	0.6	MIT
3172.503	3173.421	31511.74	TH	0.9	0.7	MIT
3172.103	3173.021	31515.71	TH	0.7D	0.5D	MIT
3171.665	3172.583	31520.06	TH	0.7D	0.0D	MIT
3171.281	3172.199	31523.88	TH	0.9	0.0	MIT
3170.429	3171.346	31532.35	TH	1.0	0.6	MIT
3169.279	3170.196	31543.79	TH	1.1	1.0	MIT
3166.054	3166.970	31575.92	TH	1.0	0.8	MIT
3165.822	3166.738	31578.23	TH	1.0	0.7	MIT
3165.579	3166.495	31580.66	TH	1.0	0.6	MIT
3164.437	3165.353	31592.05	TH	1.0	1.0	MIT
3162.836	3163.751	31608.05	TH	1.0	1.0	MIT
3161.693	3162.608	31619.47	TH	1.0	0.9	MIT
3161.071	3161.986	31625.69	TH	0.7	0.0	MIT
3160.821	3161.736	31628.20	TH	0.8	0.5	MIT
3160.179	3161.094	31634.62	TH	0.9	0.6	MIT
3158.871	3159.785	31647.72	TH	0.6	0.0H	MIT
3158.612	3159.526	31650.31	TH	1.0	0.6	MIT
3156.399	3157.313	31672.50	TH	1.1	1.0	MIT
3155.833	3156.747	31678.18	TH	1.1	1.0	MIT
3154.730	3155.643	31689.26	TH	1.2	1.2	MIT
3154.26	3155.17	31694.0	TH	1.2	1.1	MIT
3151.648	3152.561	31720.25	TH	1.0	1.0	MIT
3150.457	3151.369	31732.24	TH	1.1	1.2	MIT
3149.960	3150.872	31737.24	TH	1.0	0.7	MIT
3148.040	3148.952	31756.60	TH	0.5	1.6H	MIT
3146.041	3146.952	31776.78	TH	1.1	1.2	MIT
3145.636	3146.547	31780.87	TH	0.9	0.0	MIT
3142.839	3143.749	31809.15	TH	1.1	1.1	MIT
3142.29	3143.20	31814.7	TH		1.0H	MIT
3141.849	3142.759	31819.17	TH	1.1	1.0	MIT
3140.271	3141.181	31835.16	TH	0.9	0.0	MIT
3139.822	3140.732	31839.72	TH	1.0	0.7	MIT
3139.307	3140.217	31844.94	TH	1.0	1.2	MIT
3137.174	3138.083	31866.59	TH	1.0	0.7	MIT
3136.830	3137.739	31870.08	TH	0.7		MIT
3136.70	3137.61	31871.4	TH	0.7	0.7	MIT
3136.218	3137.127	31876.30	TH	1.0	0.5	MIT
3134.427	3135.335	31894.51	TH	1.0	1.1	MIT
3133.619	3134.527	31902.74	TH	1.0	1.0	MIT
3131.070	3131.977	31928.71	TH	1.1	1.0	MIT
3129.911	3130.818	31940.53	TH	1.0	0.9	MIT
3127.15	3128.06	31968.7	TH	1.0	1.0	MIT
3125.71	3126.62	31983.5	TH	0.9	0.5	MIT
3125.463	3126.369	31985.99	TH	1.0	1.0	MIT
3125.16	3126.07	31989.1	TH	1.0	0.9	MIT
3124.388	3125.294	31996.99	TH	1.1	1.2	MIT
3122.962	3123.867	32011.60	TH	1.3	1.3	MIT
3119.484	3120.389	32047.29	TH	1.2	1.2	MIT
3117.636	3118.540	32066.29	TH	1.1	1.1	MIT
3116.476	3117.380	32078.22	TH	1.0	1.0	MIT
3116.288	3117.192	32080.16	TH	0.7D	1.0	MIT
3115.74	3116.64	32085.8	TH	1.0	1.0	MIT
3114.068	3114.971	32103.03	TH	1.0	0.7	MIT
3112.35	3113.25	32120.7	TH		1.3H	MIT
3112.087	3112.990	32123.46	TH	1.0	0.6	MIT
3111.831	3112.734	32126.10	TH	1.3D	0.9D	MIT
3110.022	3110.924	32144.79	TH	1.2	1.2	MIT
3108.298	3109.200	32162.62	TH	1.2	1.3	MIT

AIR WAVELENGTH (ANGSTRMS)	VACUUM WAVELENGTH (ANGSTRMS)	WAVENUMBER (KAYSERS)	LMENT	LOG ARC	INTENSITY SPK	DIS	REF
3107.029	3107.930	32175.75	TH	1.2	1.2		MIT
3106.691	3107.592	32179.25	TH	1.1	1.0		MIT
3105.750	3106.651	32189.00	TH	1.2	1.3		MIT
3105.477	3106.378	32191.83	TH	1.0	0.6		MIT
3105.053	3105.954	32196.23	TH	1.0	1.0		MIT
3102.666	3103.566	32221.00	TH	1.2	1.1		MIT
3101.691	3102.591	32231.12	TH	1.0	0.9		MIT
3100.939	3101.839	32238.94	TH	1.1	1.1		MIT
3099.863	3100.763	32250.13	TH	1.0	0.8		MIT
3099.744	3100.644	32251.37	TH	1.0	0.8		MIT
3098.331	3099.230	32266.08	TH	0.9	0.0		MIT
3097.96	3098.86	32269.9	TH	0.0	1.7H		MIT
3097.270	3098.169	32277.13	TH	1.0	0.9		MIT
3096.434	3097.333	32285.84	TH	1.0	0.9		MIT
3094.74	3095.64	32303.5	TH	1.0	0.8		MIT
3093.055	3093.953	32321.11	TH	1.1	0.8		MIT
3090.102	3090.999	32352.00	TH	1.2	1.0		MIT
3089.631	3090.528	32356.93	TH	1.0	1.0		MIT
3088.842	3089.739	32365.20	TH	1.0	0.8		MIT
3088.473	3089.370	32369.06	TH	1.3	1.3		MIT
3088.031	3088.928	32373.70	TH	1.0	1.0		MIT
3087.413	3088.309	32380.18	TH	0.9	0.0		MIT
3083.347	3084.242	32422.87	TH	1.1	1.2H		MIT
3083.050	3083.945	32426.00	TH	1.1	0.8		MIT
3082.176	3083.071	32435.19	TH	1.0	1.0		MIT
3082.031	3082.926	32436.72	TH	1.1	0.9		MIT
3081.662	3082.557	32440.60	TH	1.0	0.7		MIT
3080.221	3081.116	32455.78	TH	1.1	1.2		MIT
3079.943	3080.838	32458.71	TH	1.0	0.6		MIT
3079.824	3080.719	32459.96	TH I	0.3			MIT
3078.832	3079.726	32470.42	TH	1.0	1.4H		MIT
3072.120	3073.013	32541.36	TH	1.0	0.9		MIT
3070.821	3071.713	32555.12	TH	1.1	1.0		MIT
3069.258	3070.150	32571.70	TH	1.0	0.8		MIT
3068.987	3069.879	32574.57	TH	1.1	0.9		MIT
3067.734	3068.626	32587.88	TH	1.1	1.3		MIT
3066.420	3067.311	32601.84	TH	1.0	0.5		MIT
3066.22	3067.11	32604.0	TH	0.7	1.0H		MIT
3065.935	3066.826	32607.00	TH	1.1	1.0		MIT
3063.132	3064.022	32636.84	TH		1.0H		MIT
3063.026	3063.916	32637.97	TH	1.4D	1.4D		MIT
3061.703	3062.593	32652.07	TH	1.1	1.2		MIT
3060.447	3061.337	32665.47	TH	1.0	0.0		MIT
3060.182	3061.072	32668.30	TH	1.1	1.2		MIT
3058.431	3059.320	32687.00	TH	1.0	1.0		MIT
3058.142	3059.031	32690.09	TH	0.9	0.5		MIT
3057.904	3058.793	32692.63	TH	0.9	0.8		MIT
3057.642	3058.531	32695.43	TH	1.0	0.8		MIT
3056.099	3056.988	32711.94	TH	1.0	0.6		MIT
3051.796	3052.684	32758.06	TH	1.0	1.0		MIT
3050.989	3051.876	32766.73	TH	1.0	1.0		MIT
3049.866	3050.753	32778.79	TH	1.0	1.0		MIT
3049.645	3050.532	32781.17	TH	1.0	1.0		MIT
3049.095	3049.982	32787.08	TH	1.3	1.3		MIT
3046.954	3047.840	32810.12	TH	1.1	1.3		MIT
3045.568	3046.454	32825.05	TH	1.1	1.2		MIT
3043.249	3044.134	32850.06	TH	1.0	0.5		MIT
3043.068	3043.953	32852.01	TH	1.1	1.1		MIT
3040.049	3040.934	32884.64	TH	1.1	1.0		MIT
3038.600	3039.484	32900.32	TH	1.1	1.1		MIT
3035.542	3036.425	32933.46	TH	1.0	0.7		MIT
3035.113	3035.996	32938.12	TH	1.2	1.2		MIT
3034.069	3034.952	32949.45	TH	1.2	1.2		MIT
3031.701	3032.584	32975.18	TH	1.0	0.5		MIT
3031.291	3032.173	32979.64	TH	0.9	0.8		MIT
3030.868	3031.750	32984.25	TH	1.0	0.7		MIT
3030.490	3031.372	32988.36	TH	1.0	0.0		MIT
3028.580	3029.462	33009.16	TH	1.1	1.1		MIT
3026.580	3027.461	33030.98	TH	1.2	1.2		MIT
3025.435	3026.316	33043.48	TH	1.3D	0.7		MIT

AIR WAVELENGTH (ANGSTRMS)	VACUUM WAVELENGTH (ANGSTRMS)	WAVENUMBER (KAYSERS)	LMENT	LOG ARC	INTENSITY SPK	DIS	REF
3024.669	3025.550	33051.84	TH	1.0	0.8		MIT
3022.097	3022.977	33079.97	TH	1.1	1.1		MIT
3019.378	3020.257	33109.76	TH	1.1	1.0		MIT
3017.122	3018.001	33134.52	TH	1.1	1.0		MIT
3015.721	3016.600	33149.91	TH	1.2	1.0		MIT
3014.925	3015.803	33158.66	TH	1.1	0.9		MIT
3012.708	3013.586	33183.06	TH	1.1	1.0		MIT
3009.725	3010.602	33215.95	TH	1.2	1.0		MIT
3008.491	3009.368	33229.57	TH	1.2	1.1		MIT
3006.930	3007.806	33246.82	TH	1.1	1.2		MIT
3002.39	3003.27	33297.1	TH	1.2	1.1		MIT
3001.264	3002.139	33309.58	TH	1.1	1.1		MIT
3000.922	3001.797	33313.38	TH	1.5D	1.0		MIT
2999.773	3000.648	33326.14	TH	1.0	0.9		MIT
2999.093	2999.967	33333.70	TH	1.1	1.1		MIT
2997.15	2998.02	33355.3	TH	1.0	0.6		MIT
2996.980	2997.854	33357.20	TH	1.0	1.0		MIT
2995.270	2996.143	33376.24	TH	1.0	0.6		MIT
2993.803	2994.676	33392.59	TH	1.1	1.2		MIT
2991.693	2992.566	33416.14	TH	1.0	0.9		MIT
2991.065	2991.937	33423.16	TH	1.2	1.0		MIT
2988.234	2989.106	33454.82	TH	1.2	1.3		MIT
2986.788	2987.659	33471.02	TH	1.1	0.7		MIT
2985.246	2986.117	33488.31	TH	1.2	1.1		MIT
2983.818	2984.689	33504.33	TH	1.0	1.0		MIT
2983.020	2983.890	33513.29	TH	1.0	0.8		MIT
2982.048	2982.918	33524.22	TH	2.2	0.7		MIT
2981.850	2982.720	33526.44	TH	0.9	0.5		MIT
2981.497	2982.367	33530.41	TH	0.9	0.7		MIT
2981.362	2982.232	33531.93	TH	0.7	0.7		MIT
2980.338	2981.208	33543.45	TH	1.2	1.0		MIT
2978.64	2979.51	33562.6	TH	0.0	2.0H		MIT
2976.649	2977.518	33585.02	TH	1.0	0.7		MIT
2976.020	2976.889	33592.12	TH	1.2	1.0		MIT
2974.013	2974.881	33614.79	TH	1.0	1.0		MIT
2973.536	2974.404	33620.18	TH	0.7	1.0		MIT
2972.224	2973.092	33635.02	TH	0.6	0.5		MIT
2971.482	2972.350	33643.42	TH	1.0	1.0W		MIT
2969.835	2970.702	33662.08	TH	0.3H			MIT
2969.069	2969.936	33670.76	TH	0.3			MIT
2968.685	2969.552	33675.12	TH	0.8	0.5		MIT
2965.501	2966.367	33711.27	TH	0.9	0.8		MIT
2964.924	2965.790	33717.83	TH	0.9	0.8		MIT
2964.113	2964.979	33727.06	TH	0.7	0.6		MIT
2963.609	2964.475	33732.79	TH	0.8	0.6		MIT
2961.49	2962.36	33756.9	TH		1.1J		MIT
2958.142	2959.006	33795.13	TH	0.9	0.8		MIT
2957.922	2958.786	33797.64	TH	0.9	0.9		MIT
2957.590	2958.454	33801.44	TH	1.0	0.9		MIT
2955.849	2956.713	33821.35	TH	0.8	0.6		MIT
2955.605	2956.469	33824.14	TH	0.8	0.6		MIT
2955.038	2955.901	33830.63	TH	0.8	0.7H		MIT
2954.889	2955.752	33832.33	TH	0.8	0.6		MIT
2951.207	2952.069	33874.54	TH	0.9	0.8		MIT
2950.442	2951.304	33883.32	TH	0.9	0.8		MIT
2949.935	2950.797	33889.15	TH	0.9	0.8		MIT
2949.236	2950.098	33897.18	TH	0.5	0.3		MIT
2949.069	2949.931	33899.10	TH	1.0	0.9		MIT
2946.265	2947.126	33931.36	TH	0.9	0.3		MIT
2942.862	2943.722	33970.59	TH	1.0S	1.0		MIT
2942.634	2943.494	33973.23	TH	0.7	0.6		MIT
2939.614	2940.474	34008.13	TH	0.7	0.7		MIT
2938.106	2938.965	34025.58	TH	0.7	0.6		MIT
2937.445	2938.304	34033.24	TH	0.7	0.7		MIT
2936.469	2937.328	34044.55	TH	1.1	1.1		MIT
2936.194	2937.053	34047.74	TH	1.0	1.0		MIT
2934.141	2934.999	34071.56	TH	0.9	0.6		MIT
2933.104	2933.962	34083.60	TH	0.9	0.7		MIT
2932.647	2933.505	34088.92	TH	0.9	0.5		MIT
2932.52	2933.38	34090.4	TH		1.4J		MIT

AIR WAVELENGTH (ANGSTRMS)	VACUUM WAVELENGTH (ANGSTRMS)	WAVENUMBER (KAYSERS)	LMENT	LOG ARC	INTENSITY SPK	DIS	REF
2930.913	2931.770	34109.08	TH	0.9	0.9		MIT
2928.67	2929.53	34135.2	TH		1.1 J		MIT
2928.256	2929.113	34140.03	TH	1.1 S	1.1		MIT
2925.59	2926.45	34171.1	TH		1.0 H		MIT
2925.050	2925.906	34177.45	TH	1.0 S	1.2 H		MIT
2923.94	2924.80	34190.4	TH		1.1 J		MIT
2922.990	2923.846	34201.53	TH	0.7	0.6		MIT
2922.801	2923.656	34203.75	TH	0.7	0.6		MIT
2922.607	2923.462	34206.02	TH	0.9	1.0		MIT
2921.625	2922.480	34217.51	TH	0.9	1.0		MIT
2921.084	2921.939	34223.85	TH	0.6	0.5		MIT
2920.935	2921.790	34225.60	TH	0.7	0.7		MIT
2920.372	2921.227	34232.19	TH	0.7	0.6		MIT
2919.842	2920.697	34238.41	TH	1.0	1.0		MIT
2917.79	2918.64	34262.5	TH	0.6	0.8		MIT
2917.39	2918.24	34267.2	TH	1.4 D	0.8		MIT
2916.440	2917.294	34278.34	TH		0.9 S		MIT
2912.760	2913.613	34321.65	TH	0.7	0.7		MIT
2912.664	2913.517	34322.78	TH	0.7	0.7		MIT
2912.012	2912.865	34330.46	TH	0.9	0.8		MIT
2911.324	2912.177	34338.58	TH	0.9	0.8		MIT
2910.597	2911.449	34347.15	TH	1.1 S	1.0		MIT
2910.193	2911.045	34351.92	TH		1.0 J		MIT
2908.359	2909.211	34373.58	TH	1.1 S	1.1		MIT
2906.126	2906.977	34399.99	TH	0.9	0.6		MIT
2905.934	2906.785	34402.27	TH	0.8	0.7 H		MIT
2904.254	2905.105	34422.17	TH		0.7		MIT
2903.170	2904.021	34435.02	TH	0.8	0.7		MIT
2899.724	2900.574	34475.94	TH	1.2	1.1		MIT
2899.373	2900.223	34480.11	TH	0.7	0.5		MIT
2898.93	2899.78	34485.4	TH		1.4 J		MIT
2898.267	2899.116	34493.27	TH	0.6	0.5		MIT
2897.071	2897.920	34507.51	TH	1.1	0.7		MIT
2896.71	2897.56	34511.8	TH		1.3 #		MIT
2895.143	2895.992	34530.49	TH	1.0	1.0		MIT
2892.175	2893.023	34565.92	TH	0.8	0.7		MIT
2891.77	2892.62	34570.8	TH	0.8 D	0.7 D		MIT
2891.254	2892.102	34576.93	TH	1.0	1.0		MIT
2890.893	2891.741	34581.25	TH	0.8	0.6		MIT
2887.821	2888.668	34618.03	TH	1.3	1.3		MIT
2886.510	2887.356	34633.76	TH	0.7	0.8		MIT
2886.248	2887.094	34636.90	TH	0.8	0.7		MIT
2885.045	2885.891	34651.34	TH	1.1	1.2		MIT
2884.295	2885.141	34660.35	TH	1.1	1.1		MIT
2882.516	2883.362	34681.74	TH	0.9	0.9		MIT
2882.014	2882.859	34687.78	TH	1.0	1.1		MIT
2881.147	2881.992	34698.22	TH	1.0	1.0		MIT
2879.533	2880.378	34717.67	TH	0.8	0.6		MIT
2879.205	2880.050	34721.62	TH	1.0	0.7		MIT
2876.421	2877.265	34755.23	TH	1.0	0.9		MIT
2873.717	2874.560	34787.93	TH	0.9	0.7		MIT
2870.825	2871.668	34822.97	TH	0.8	0.7		MIT
2870.413	2871.256	34827.97	TH	1.3	1.3		MIT
2869.927	2870.769	34833.87	TH	0.8	0.9		MIT
2868.678	2869.520	34849.03	TH	0.9	0.8		MIT
2866.693	2867.535	34873.16	TH	0.8	0.7		MIT
2866.440	2867.282	34876.24	TH	0.8	0.5		MIT
2864.654	2865.495	34897.98	TH	0.9	0.8		MIT
2861.37	2862.21	34938.0	TH	0.9 D	0.9 D		MIT
2857.494	2858.333	34985.42	TH	0.7	0.5		MIT
2855.900	2856.739	35004.95	TH	1.0	0.9		MIT
2854.920	2855.759	35016.96	TH	0.6	0.6		MIT
2854.131	2854.970	35026.64	TH	0.9	0.9		MIT
2852.502	2853.340	35046.65	TH	0.3 H	0.8		MIT
2851.445	2852.283	35059.64	TH	0.7 D	0.7 D		MIT
2851.261	2852.099	35061.90	TH	1.0 D	1.0 D		MIT
2847.358	2848.195	35109.96	TH	1.0	1.0		MIT
2842.815	2843.651	35166.06	TH	1.1	1.0		MIT
2841.168	2842.003	35186.45	TH	0.9	0.9		MIT
2840.156	2840.991	35198.98	TH	0.9	0.8		MIT
2839.335	2840.170	35209.16	TH	0.7	0.6		MIT
2839.243	2840.078	35210.30	TH	0.9	0.8		MIT
2837.299	2838.133	35234.43	TH	1.2	1.0		MIT
2836.438	2837.272	35245.12	TH	0.7	0.3		MIT
2836.049	2836.883	35249.96	TH	1.0	0.9		MIT
2834.484	2835.318	35269.42	TH	0.8	0.7		MIT
2833.339	2834.172	35283.67	TH	0.9	0.9		MIT
2832.319	2833.152	35296.37	TH	1.3	1.4		MIT
2830.445	2831.278	35319.74	TH	1.0	0.9		MIT
2829.936	2830.769	35326.10	TH	0.8	0.8		MIT
2827.992	2828.824	35350.38	TH	0.7 D	0.6 D		MIT
2826.858	2827.690	35364.56	TH	1.1	1.1		MIT
2824.68	2825.51	35391.8	TH		1.4 D		MIT
2823.555	2824.386	35405.92	TH	0.3 H	0.6		MIT
2822.362	2823.193	35420.89	TH	0.6 D	0.3 D		MIT
2822.035	2822.866	35424.99	TH	1.0 D	1.0 D		MIT
2821.610	2822.441	35430.33	TH	0.9	0.8		MIT
2820.337	2821.167	35446.32	TH	1.0	0.8		MIT
2819.329	2820.159	35458.99	TH	1.1	1.0		MIT
2816.076	2816.905	35499.95	TH	1.0	0.7		MIT
2814.574	2815.403	35518.90	TH	0.8	0.7		MIT
2814.318	2815.147	35522.13	TH	0.9	0.7		MIT
2808.94	2809.77	35590.1	TH	1.0 D	0.7 D		MIT
2807.717	2808.544	35605.64	TH	0.7 D	0.5 D		MIT
2803.381	2804.207	35660.70	TH	0.3 H	0.9		MIT
2800.572	2801.397	35696.47	TH	0.8	0.9		MIT
2799.120	2799.945	35714.99	TH	0.7	0.6		MIT
2798.670	2799.495	35720.73	TH	0.8	0.6		MIT
2797.740	2798.565	35732.60	TH	1.0	1.0		MIT
2797.026	2797.851	35741.72	TH	0.9	0.8		MIT
2794.260	2795.084	35777.10	TH	1.0	1.0		MIT
2791.016	2791.839	35818.68	TH	1.0	1.0		MIT
2790.425	2791.248	35826.27	TH	0.6	1.0		MIT
2788.684	2789.507	35848.63	TH	0.9	0.9		MIT
2787.132	2787.954	35868.60	TH	0.9	0.9		MIT
2786.917	2787.739	35871.36	TH	0.7	0.9		MIT
2785.613	2786.435	35888.15	TH	0.3 H	0.5		MIT
2784.978	2785.800	35896.34	TH	0.9	0.8		MIT
2784.066	2784.887	35908.10	TH	0.8	0.7		MIT
2783.496	2784.317	35915.45	TH	0.8	0.8		MIT
2783.054	2783.875	35921.15	TH	1.0	1.0		MIT
2778.710	2779.530	35977.31	TH	1.0	1.0		MIT
2778.027	2778.847	35986.15	TH	0.5	0.3		MIT
2777.807	2778.627	35989.00	TH	1.0	0.9		MIT
2774.843	2775.662	36027.44	TH	0.7	0.6		MIT
2774.072	2774.891	36037.45	TH	0.8	0.8		MIT
2771.515	2772.333	36070.70	TH	1.0	1.0		MIT
2770.822	2771.640	36079.72	TH	1.0	1.0		MIT
2768.848	2769.666	36105.44	TH	1.2	1.2		MIT
2765.127	2765.944	36154.02	TH	1.0	0.9		MIT
2764.643	2765.460	36160.35	TH	0.9	0.9		MIT
2763.613	2764.429	36173.83	TH	1.0	1.0		MIT
2760.70	2761.52	36212.0	TH		0.8 H		MIT
2760.396	2761.212	36215.98	TH	0.7	0.9		MIT
2759.415	2760.230	36228.86	TH	0.6	0.5		MIT
2752.172	2752.986	36324.20	TH	1.2	1.1		MIT
2749.712	2750.525	36356.69	TH	0.6	0.6		MIT
2749.542	2750.355	36358.94	TH	1.0	0.9		MIT
2749.357	2750.170	36361.39	TH	0.5	0.0		MIT
2747.853	2748.666	36381.29	TH	0.8	0.5		MIT
2747.161	2747.973	36390.45	TH	1.0 D	0.7 D		MIT
2743.067	2743.878	36444.76	TH	1.2	0.9		MIT
2737.416	2738.226	36519.99	TH	0.9	0.9		MIT
2734.354	2735.163	36560.89	TH	1.0	1.0		MIT
2732.815	2733.624	36581.48	TH	1.0	0.9		MIT
2731.583	2732.392	36597.98	TH	0.7	0.6		MIT
2730.267	2731.075	36615.61	TH	0.9	0.8		MIT
2729.330	2730.138	36628.18	TH	1.0	1.1		MIT
2728.915	2729.723	36633.75	TH	0.8	0.6		MIT
2727.24	2728.05	36656.2	TH	0.3 H	0.6		MIT

AIR WAVELENGTH (ANGSTRMS)	VACUUM WAVELENGTH (ANGSTRMS)	WAVENUMBER (KAYSERS)	LMENT	LOG ARC	INTENSITY SPK	INTENSITY DIS	REF
2726.490	2727.297	36666.33	TH	1.0	0.7		MIT
2724.886	2725.693	36687.92	TH	1.0	0.9		MIT
2723.323	2724.130	36708.97	TH		0.6		MIT
2722.383	2723.189	36721.65	TH	1.0	0.9		MIT
2721.697	2722.503	36730.90	TH	1.0	1.1		MIT
2721.367	2722.173	36735.36	TH		0.7		MIT
2719.939	2720.745	36754.64	TH	0.9	0.7		MIT
2719.465	2720.271	36761.05	TH	0.7	0.5		MIT
2716.318	2717.123	36803.63	TH	1.0	1.0		MIT
2715.092	2715.897	36820.25	TH	0.8	0.7		MIT
2714.617	2715.422	36826.69	TH	0.3H	0.6		MIT
2711.456	2712.260	36869.62	TH	0.8	0.6		MIT
2708.181	2708.984	36914.21	TH	1.0	1.3		MIT
2703.962	2704.764	36971.80	TH	1.2	1.2		MIT
2701.86	2702.66	37000.6	TH	0.3H	0.7		MIT
2698.742	2699.543	37043.31	TH	0.8	0.7		MIT
2697.550	2698.350	37059.68	TH	0.7	0.7		MIT
2696.836	2697.636	37069.49	TH	0.3H	0.6		MIT
2695.822	2696.622	37083.43	TH	0.6	0.5		MIT
2695.555	2696.355	37087.10	TH	0.9	1.0		MIT
2695.207	2696.007	37091.89	TH	0.8	0.9		MIT
2692.422	2693.221	37130.26	TH	1.3	1.3		MIT
2691.230	2692.029	37146.70	TH	0.9	0.8		MIT
2688.329	2689.127	37186.79	TH	0.8	0.3		MIT
2687.138	2687.936	37203.27	TH	1.0	1.0		MIT
2686.16	2686.96	37216.8	TH		1.6D		MIT
2684.294	2685.091	37242.68	TH	1.2	1.2		MIT
2680.965	2681.761	37288.92	TH		1.4		MIT
2678.938	2679.734	37317.14	TH	0.7			MIT
2675.670	2676.465	37362.71	TH	0.9	0.6		MIT
2673.677	2674.472	37390.56	TH		0.8		MIT
2666.87	2667.66	37486.0	TH		0.8		MIT
2662.866	2663.658	37542.36	TH	0.9	0.7		MIT
2662.349	2663.141	37549.65	TH	1.0	0.9		MIT
2661.38	2662.17	37563.3	TH	0.9	0.7		MIT
2659.86	2660.65	37584.8	TH	0.5	0.0		MIT
2658.667	2659.458	37601.65	TH	1.2	1.0		MIT
2650.588	2651.377	37716.25	TH	1.0	0.7		MIT
2649.878	2650.667	37726.35	TH	0.9	0.6		MIT
2649.481	2650.270	37732.01	TH	1.0	0.5		MIT
2642.615	2643.402	37830.04	TH	0.7	0.3		MIT
2641.494	2642.281	37846.09	TH	1.2	1.0		MIT
2640.33	2641.12	37862.8	TH	0.7D	0.3D		MIT
2639.895	2640.682	37869.01	TH	0.7	0.3		MIT
2639.508	2640.294	37874.56	TH	0.9	0.6		MIT
2638.639	2639.425	37887.04	TH		0.9		MIT
2637.649	2638.435	37901.25	TH		0.7H		MIT
2635.865	2636.651	37926.91	TH		1.0H		MIT
2635.42	2636.21	37933.3	TH	0.3	0.9		MIT
2633.333	2634.118	37963.37	TH	0.7	0.3		MIT
2630.017	2630.801	38011.23	TH	0.7	0.3		MIT
2628.816	2629.600	38028.60	TH	0.9	0.7		MIT
2626.403	2627.186	38063.54	TH	0.8	0.5		MIT
2625.736	2626.519	38073.20	TH	0.9	0.9		MIT
2623.454	2624.237	38106.32	TH	1.0	1.0		MIT
2618.913	2619.695	38172.39	TH	1.1	0.7		MIT
2609.855	2610.634	38304.87	TH	1.0	0.7		MIT
2609.223	2610.002	38314.14	TH		1.0		MIT
2608.327	2609.106	38327.30	TH	0.7	0.3		MIT
2607.482	2608.261	38339.72	TH	0.6	0.3		MIT
2600.885	2601.662	38436.96	TH	1.2	0.6		MIT
2600.635	2601.412	38440.66	TH		1.2		MIT
2597.31	2598.09	38489.9	TH		1.0D		MIT
2597.050	2597.826	38493.72	TH	1.3	0.7		MIT
2595.033	2595.809	38523.64	TH	0.8	0.7		MIT
2589.063	2589.838	38612.46	TH	1.3	1.0		MIT
2586.153	2586.927	38655.91	TH	0.7	0.3		MIT
2583.39	2584.16	38697.2	TH		1.0H		MIT
2580.700	2581.473	38737.58	TH	0.9	0.6		MIT
2580.360	2581.132	38742.69	TH	0.7	0.3		MIT

AIR WAVELENGTH (ANGSTRMS)	VACUUM WAVELENGTH (ANGSTRMS)	WAVENUMBER (KAYSERS)	LMENT	LOG ARC	INTENSITY SPK	INTENSITY DIS	REF
2579.437	2580.209	38756.55	TH	1.0	0.8		MIT
2576.690	2577.462	38797.86	TH	1.2	0.7		MIT
2576.336	2577.107	38803.19	TH	0.8	0.5		MIT
2574.484	2575.255	38831.10	TH	1.0	0.7		MIT
2571.612	2572.382	38874.47	TH		1.4		MIT
2567.826	2568.595	38931.78	TH		1.3		MIT
2566.589	2567.358	38950.54	TH	1.2S	1.0H		MIT
2565.597	2566.366	38965.60	TH	1.0	1.2		MIT
2564.356	2565.125	38984.46	TH		1.3J		MIT
2561.937	2562.705	39021.27	TH	1.0	0.7		MIT
2559.18	2559.95	39063.3	TH		0.7		MIT
2555.20	2555.97	39124.1	TH		1.2		MIT
2554.73	2555.50	39131.3	TH		1.2D		MIT
2549.515	2550.280	39211.38	TH		1.2		MIT
2549.103	2549.868	39217.71	TH		1.1		MIT
2547.902	2548.667	39236.20	TH	1.2	1.0		MIT
2545.747	2546.511	39269.41	TH	1.0	0.5		MIT
2545.345	2546.109	39275.61	TH	1.2	0.6		MIT
2545.10	2545.86	39279.4	TH		1.3		MIT
2542.642	2543.406	39317.36	TH	0.9S	0.5		MIT
2541.913	2542.676	39328.64	TH		0.7		MIT
2536.558	2537.320	39411.66	TH	0.7	1.0		MIT
2535.870	2536.632	39422.35	TH	0.7	0.3		MIT
2532.452	2533.213	39475.56	TH	1.0	0.3		MIT
2529.98	2530.74	39514.1	TH		1.1		MIT
2528.36	2529.12	39539.4	TH		0.7		MIT
2525.931	2526.691	39577.46	TH	0.9	0.6		MIT
2521.78	2522.54	39642.6	TH	0.3	0.0		MIT
2520.661	2521.419	39660.20	TH	1.0	0.8		MIT
2520.13	2520.89	39668.6	TH	0.7	0.3		MIT
2512.74	2513.50	39785.2	TH		1.2D		MIT
2509.97	2510.73	39829.1	TH	0.9	0.5		MIT
2505.615	2506.370	39898.34	TH	1.0	0.5		MIT
2504.275	2505.030	39919.69	TH	0.9	0.8		MIT
2502.903	2503.657	39941.57	TH	0.8	0.5		MIT
2501.122	2501.876	39970.01	TH		1.3H		MIT
2498.868	2499.621	40006.06	TH	1.0	0.7		MIT
2498.402	2499.155	40013.52	TH	0.9	0.5		MIT
2497.59	2498.34	40026.5	TH	0.5	1.1H		MIT
2497.26	2498.01	40031.8	TH		1.0H		MIT
2495.353	2496.106	40062.41	TH	0.9	0.6		MIT
2494.64	2495.39	40073.9	TH	0.6	0.3		MIT
2489.64	2490.39	40154.3	TH	0.7	0.5		MIT
2476.96	2477.71	40359.9	TH	0.8	0.3		MIT
2475.328	2476.076	40386.48	TH		1.2		MIT
2473.98	2474.73	40408.5	TH		1.2		MIT
2470.597	2471.344	40463.82	TH	0.8	0.6		MIT
2468.150	2468.896	40503.93	TH	0.8	0.6		MIT
2467.668	2468.414	40511.84	TH		1.0H		MIT
2466.126	2466.872	40537.17	TH	1.0	0.8		MIT
2463.693	2464.438	40577.20	TH		1.4		MIT
2459.007	2459.751	40654.52	TH	0.9	0.5		MIT
2456.86	2457.60	40690.0	TH	0.6	0.3		MIT
2456.292	2457.035	40699.45	TH	0.8	0.6		MIT
2450.81	2451.55	40790.5	TH	0.7	0.5		MIT
2444.469	2445.210	40896.29	TH	0.7	0.5		MIT
2441.289	2442.029	40949.55	TH		1.0		MIT
2437.56	2438.30	41012.2	TH	0.7	0.6		MIT
2431.741	2432.479	41110.33	TH	0.3	1.5		MIT
2431.16	2431.90	41120.1	TH	0.6	0.3		MIT
2427.99	2428.73	41173.8	TH	0.5	1.4		MIT
2423.675	2424.411	41247.13	TH	0.8	0.7		MIT
2413.49	2414.22	41421.2	TH	0.7	1.3		MIT
2413.405	2414.139	41422.64	TH	0.7			MIT
2411.32	2412.05	41458.5	TH	0.7	0.5		MIT
2404.174	2404.906	41581.67	TH	0.8	0.7		MIT
2393.113	2393.842	41773.85	TH	0.7	0.3		MIT
2391.518	2392.247	41801.71	TH		1.2J		MIT
2388.143	2388.871	41860.78	TH	0.6	0.3		MIT
2381.52	2382.25	41977.2	TH		1.0		EX

AIR WAVELENGTH (ANGSTRMS)	VACUUM WAVELENGTH (ANGSTRMS)	WAVENUMBER (KAYSERS)	LMENT	LOG ARC	INTENSITY SPK	DIS	REF
2377.832	2378.558	42042.29	TH	1.0	0.9		MIT
2375.835	2376.560	42077.62	TH	0.7	0.5		MIT
2375.075	2375.800	42091.08	TH	0.9	0.6		MIT
2373.846	2374.571	42112.87	TH	0.5	0.5		MIT
2371.427	2372.151	42155.83	TH		1.1		MIT
2368.96	2369.68	42199.7	TH		0.7		MIT
2368.051	2368.774	42215.92	TH	0.7	0.5		MIT
2366.984	2367.707	42234.95	TH	0.5	0.7		MIT
2363.09	2363.81	42304.5	TH		1.2		MIT
2362.21	2362.93	42320.3	TH	0.5	0.0		MIT
2355.906	2356.627	42433.53	TH	0.3	0.7		MIT
2355.251	2355.972	42445.33	TH	0.6	0.3		MIT
2351.72	2352.44	42509.1	TH		0.7		MIT
2340.65	2341.37	42710.1	TH		1.0		MIT
2335.49	2336.21	42804.4	TH		0.5		MIT
2326.922	2327.636	42962.04	TH	0.8	0.6		MIT
2324.70	2325.41	43003.1	TH		1.2		MIT
2319.525	2320.238	43099.03	TH		1.1		MIT
2301.22	2301.93	43441.8	TH		1.5		MIT
2291.65	2292.36	43623.2	TH		1.0		MIT
2281.56	2282.26	43816.1	TH		0.9		MIT
2280.44	2281.14	43837.6	TH	0.7	0.3		MIT
2221.09	2221.78	45008.9	TH		0.9		MIT
9997.94	10000.7	9999.32	TI I	1.2			ME
9981.16	9983.90	10016.1	TI	0.7			ME
9948.98	9951.71	10048.5	TI I	0.9			ME
9941.33	9944.06	10056.3	TI I	0.9			ME
9927.34	9930.06	10070.4	TI I	1.3			MIT
9923.25	9925.97	10074.6	TI I	0.3			ME
9879.41	9882.12	10119.3	TI I	0.5			ME
9832.15	9834.85	10167.9	TI I	1.4			MIT
9813.45	9816.14	10187.3	TI I	0.7			ME
9787.65	9790.33	10214.2	TI I	1.7			MIT
9783.59	9786.27	10218.4	TI I	1.3			ME
9783.38	9786.06	10218.6	TI I	1.6			MIT
9770.26	9772.94	10232.3	TI I	1.6			MIT
9768.18	9770.86	10234.5	TI I	0.7			MIT
9746.96	9749.63	10256.8	TI I	1.2			MIT
9743.55	9746.22	10260.4	TI I	1.7			MIT
9737.75	9740.42	10266.5	TI I	0.7			MIT
9728.34	9731.01	10276.4	TI I	1.8			MIT
9718.92	9721.59	10286.4	TI I	1.3			MIT
9717.00	9719.66	10288.4	TI I	1.2			ME
9715.45	9718.11	10290.1	TI I	0.5			MIT
9705.59	9708.25	10300.5	TI I	2.0			MIT
9702.86	9705.52	10303.4	TI I	0.5			ME
9690.54	9693.20	10316.5	TI I	0.3			MIT
9688.81	9691.47	10318.4	TI I	1.5			MIT
9678.98	9681.63	10328.8	TI I	0.5			ME
9675.55	9678.20	10332.5	TI I	2.3			ME
9663.13	9665.78	10345.8	TI I	0.5			MIT
9661.36	9664.01	10347.7	TI I	1.0			MIT
9647.40	9650.05	10362.6	TI I	1.9			ME
9638.24	9640.88	10372.5	TI I	2.3			MIT
9606.77	9609.40	10406.5	TI I	0.5			ME
9599.51	9602.14	10414.3	TI I	1.7			MIT
9590.15	9592.78	10424.5	TI I	0.5			ME
9588.77	9591.40	10426.0	TI I	0.6			ME
9570.08	9572.71	10446.4	TI I	0.6			ME
9550.11	9552.73	10468.2	TI I	0.3			ME
9546.03	9548.65	10472.7	TI I	1.7			MIT
9511.55	9514.16	10510.7	TI I	1.0			ME
9510.75	9513.36	10511.5	TI I	1.2			MIT
9508.45	9511.06	10514.1	TI I	1.3			MIT
9506.03	9508.64	10516.7	TI I	1.5			MIT
9473.51	9476.11	10552.8	TI I	0.3			ME
9453.20	9455.79	10575.5	TI I	0.5			MIT
9431.77	9434.36	10599.6	TI I	0.5			ME
9409.69	9412.27	10624.4	TI I	0.3			ME
9312.48	9315.04	10735.3	TI I	0.6			ME

AIR WAVELENGTH (ANGSTRMS)	VACUUM WAVELENGTH (ANGSTRMS)	WAVENUMBER (KAYSERS)	LMENT	LOG ARC	INTENSITY SPK	DIS	REF
9285.04	9287.59	10767.1	TI I	0.7			ME
9257.58	9260.12	10799.0	TI I	0.9			MIT
9246.05	9248.59	10812.5	TI I	1.0			MIT
9170.38	9172.90	10901.7	TI I	0.3			ME
9167.53	9170.05	10905.1	TI I	1.0			MIT
9135.89	9138.40	10942.8	TI I	0.3			MIT
9123.21	9125.71	10958.1	TI I	0.7			MIT
9090.69	9093.19	10997.2	TI I	1.4			MIT
9087.64	9090.13	11000.9	TI I	0.3			MIT
9086.94	9089.43	11001.8	TI I	0.5			MIT
9027.35	9029.83	11074.4	TI I	1.3			MIT
9023.65	9026.13	11078.9	TI I	0.5			MIT
8989.45	8991.92	11121.1	TI I	1.1			MIT
8985.82	8988.29	11125.6	TI I	0.3			MIT
8863.09	8865.52	11279.7	TI I	0.5			ME
8821.18	8823.60	11333.2	TI I	1.1			MIT
8819.38	8821.80	11335.6	TI I	0.9			MIT
8794.40	8796.82	11367.7	TI I	0.9			ME
8778.71	8781.12	11388.1	TI I	1.5			MIT
8766.64	8769.05	11403.7	TI I	1.8			ME
8761.50	8763.91	11410.4	TI I	1.2			MIT
8737.31	8739.71	11442.0	TI I	0.8			ME
8734.69	8737.09	11445.5	TI I	1.8			MIT
8725.76	8728.16	11457.2	TI I	0.8			ME
8719.56	8721.95	11465.3	TI I	1.5			MIT
8692.33	8694.72	11501.2	TI I	2.0			MIT
8682.99	8685.37	11513.6	TI I	2.1			MIT
8675.39	8677.77	11523.7	TI I	2.2			MIT
8641.47	8643.84	11568.9	TI	1.6			ME
8636.38	8638.75	11575.7	TI	1.3			ME
8629.33	8631.70	11585.2	TI	1.3			ME
8618.42	8620.79	11599.9	TI I	1.3			MIT
8618.14	8620.51	11600.2	TI	1.2			ME
8612.91	8615.28	11607.3	TI I	0.8			ME
8600.98	8603.34	11623.4	TI I	1.4			ME
8598.18	8600.54	11627.2	TI I	1.8			ME
8578.42	8580.78	11654.0	TI I	1.2			ME
8569.77	8572.12	11665.7	TI I	1.7			MIT
8565.53	8567.88	11671.5	TI I	1.4			MIT
8550.54	8552.89	11692.0	TI I	1.4			ME
8548.12	8550.47	11695.3	TI I	2.0			MIT
8539.38	8541.73	11707.2	TI I	1.8			MIT
8531.36	8533.70	11718.2	TI I	1.2			ME
8526.36	8528.70	11725.1	TI	0.9			ME
8525.99	8528.33	11725.6	TI	0.9			ME
8518.32	8520.66	11736.2	TI I	2.0			MIT
8518.05	8520.39	11736.6	TI I	1.8			ME
8496.04	8498.37	11767.0	TI I	1.8			MIT
8495.51	8497.84	11767.7	TI	1.2			ME
8494.42	8496.75	11769.2	TI I	1.0			ME
8483.16	8485.49	11784.8	TI I	1.3			ME
8468.50	8470.83	11805.2	TI I	2.5			MIT
8467.15	8469.48	11807.1	TI I	2.5			ME
8460.96	8463.28	11815.7	TI	1.0			ME
8457.10	8459.42	11821.1	TI I	1.4			ME
8450.89	8453.21	11829.8	TI I	1.8			ME
8442.98	8445.30	11840.9	TI I	1.1			ME
8438.93	8441.25	11846.6	TI I	2.0			ME
8435.70	8438.02	11851.1	TI I	2.2			MIT
8434.94	8437.26	11852.2	TI I	2.3			MIT
8426.52	8428.84	11864.0	TI I	2.3			MIT
8424.41	8426.72	11867.0	TI I	1.5			ME
8423.10	8425.41	11868.8	TI I	1.2			ME
8418.70	8421.01	11875.1	TI	0.9			ME
8417.54	8419.85	11876.7	TI I	1.2			ME
8416.98	8419.29	11877.5	TI I	1.5			MIT
8412.36	8414.67	11884.0	TI I	2.5			MIT
8407.87	8410.18	11890.3	TI	0.6			ME
8402.54	8404.85	11897.9	TI I	1.1			ME
8396.87	8399.18	11905.9	TI I	2.3			MIT

AIR WAVELENGTH (ANGSTRMS)	VACUUM WAVELENGTH (ANGSTRMS)	WAVENUMBER (KAYSERS)	LMENT	LOG ARC	INTENSITY SPK	INTENSITY DIS	REF
8389.46	8391.77	11916.4	TI I	1.4			MIT
8382.82	8385.12	11925.9	TI I	2.3			ME
8382.54	8384.84	11926.3	TI I	2.5			ME
8377.85	8380.15	11933.0	TI	2.3			MIT
8364.24	8366.54	11952.4	TI I	2.2			MIT
8353.15	8355.45	11968.2	TI I	1.7			MIT
8334.37	8336.66	11995.2	TI I	1.8			MIT
8312.85	8315.13	12026.3	TI I	1.6			MIT
8311.76	8314.04	12027.8	TI I	1.5			MIT
8307.41	8309.69	12034.1	TI I	1.8			MIT
8306.31	8308.59	12035.7	TI I	1.7			MIT
8207.30	8209.56	12180.9	TI I	1.3H			MIT
8180.38	8182.63	12221.0	TI I	1.4J			MIT
8131.41	8133.65	12294.6	TI I	1.3			MIT
8089.76	8091.98	12357.9	TI I	0.8			MIT
8085.25	8087.47	12364.8	TI I	1.3			MIT
8084.50	8086.72	12365.9	TI I	0.8			MIT
8068.24	8070.46	12390.9	TI I	1.7			MIT
8064.09	8066.31	12397.2	TI	1.1			ME
8024.84	8027.05	12457.9	TI I	1.7			MIT
7996.53	7998.73	12502.0	TI I	1.6			MIT
7978.88	7981.07	12529.6	TI I	2.0			MIT
7961.58	7963.77	12556.9	TI I	1.6			MIT
7949.17	7951.36	12576.5	TI I	1.7			MIT
7943.93	7946.12	12584.8	TI I	1.2			MIT
7926.37	7928.55	12612.7	TI I	1.3			MIT
7904.05	7906.22	12648.3	TI I	0.9			MIT
7663.47	7665.58	13045.3	TI I	1.1			MIT
7654.44	7656.55	13060.7	TI I	0.8			MIT
7618.33	7620.43	13122.6	TI I	0.6			MIT
7614.50	7616.60	13129.2	TI I	0.9			MIT
7496.12	7498.18	13336.6	TI I	1.5			MIT
7489.61	7491.67	13348.2	TI I	2.2			MIT
7474.94	7477.00	13374.3	TI I	1.4			MIT
7440.60	7442.65	13436.1	TI I	2.0			MIT
7423.17	7425.21	13467.6	TI I	1.5			MIT
7366.60	7368.63	13571.0	TI	1.3			MIT
7364.11	7366.14	13575.6	TI	2.2			MIT
7357.74	7359.77	13587.4	TI	2.3			MIT
7352.16	7354.19	13597.7	TI I	1.1			MIT
7344.72	7346.74	13611.5	TI I	2.3			MIT
7332.26	7334.28	13634.6	TI I	0.9			ME
7330.97	7332.99	13637.0	TI I	1.6			ME
7318.39	7320.41	13660.4	TI I	1.9			MIT
7315.56	7317.58	13665.7	TI I	1.3			MIT
7307.13	7309.14	13681.5	TI	1.5			MIT
7305.87	7307.88	13683.8	TI I	0.8			RL
7299.708	7301.719	13695.41	TI I	1.7			BH
7281.74	7283.75	13729.2	TI I	1.0H			MIT
7281.29	7283.30	13730.1	TI	0.9			MIT
7273.77	7275.77	13744.2	TI I	1.2			MIT
7271.41	7273.41	13748.7	TI I	0.8			MIT
7269.24	7271.24	13752.8	TI	0.7			MIT
7266.29	7268.29	13758.4	TI I	1.4			MIT
7265.33	7267.33	13760.2	TI I	0.9			MIT
7263.40	7265.40	13763.9	TI	1.2			MIT
7261.86	7263.86	13766.8	TI I	1.0			RL
7253.777	7255.776	13782.12	TI I	0.8			BH
7251.723	7253.721	13786.03	TI I	2.1			BH
7249.33	7251.33	13790.6	TI	1.5			MIT
7244.86	7246.86	13799.1	TI I	2.2			MIT
7233.44	7235.43	13820.9	TI I	1.0H			RL
7233.31	7235.30	13821.1	TI I	1.0H			RL
7216.20	7218.19	13853.9	TI I	1.7			MIT
7214.97	7216.96	13856.2	TI I	1.0H			RL
7213.35	7215.34	13859.4	TI I	0.8			MIT
7209.44	7211.43	13866.9	TI I	2.2			MIT
7189.89	7191.87	13904.6	TI I	1.5			MIT
7188.55	7190.53	13907.2	TI I	1.1			MIT
7171.531	7173.508	13940.18	TI	1.0			BH

AIR WAVELENGTH (ANGSTRMS)	VACUUM WAVELENGTH (ANGSTRMS)	WAVENUMBER (KAYSERS)	LMENT	LOG ARC	INTENSITY SPK	INTENSITY DIS	REF
7167.133	7169.109	13948.74	TI	1.2			BH
7160.33	7162.30	13962.0	TI I	1.0			KS
7138.91	7140.88	14003.9	TI I	1.2			MIT
7072.053	7074.003	14136.27	TI	1.4			BH
7069.11	7071.06	14142.2	TI I	1.2			ME
7053.069	7055.014	14174.32	TI	1.2			BH
7050.65	7052.59	14179.2	TI I	1.6			MIT
7039.36	7041.30	14201.9	TI I	1.2			MIT
7038.80	7040.74	14203.1	TI I	2.0			MIT
7035.86	7037.80	14209.0	TI I	1.3O			MIT
7011.05	7012.98	14259.3	TI	1.1			MIT
7010.94	7012.87	14259.5	TI I	1.2W			ME
7008.35	7010.28	14264.8	TI I	1.3			MIT
7006.66	7008.59	14268.2	TI I	1.2O			MIT
7004.656	7006.588	14272.28	TI I	1.2			BH
6996.63	6998.56	14288.7	TI I	1.2			MIT
6990.31	6992.24	14301.6	TI I	0.9			RL
6983.23	6985.16	14316.1	TI I	1.1			RL
6980.39	6982.32	14321.9	TI I	1.1			RL
6963.93	6965.85	14355.7	TI I	0.7			RL
6958.49	6960.41	14367.0	TI	1.0			MIT
6946.10	6948.02	14392.6	TI I	1.0H			RL
6943.70	6945.62	14397.6	TI I	1.2			MIT
6935.88	6937.79	14413.8	TI I	0.9			MIT
6933.15	6935.06	14419.5	TI I	1.1			MIT
6926.16	6928.07	14434.0	TI I	1.5W			RL
6913.19	6915.10	14461.1	TI I	0.8			MIT
6873.916	6875.813	14543.74	TI I	1.0			BH
6861.47	6863.36	14570.1	TI I	1.7			MIT
6860.392	6862.285	14572.41	TI	1.3			BH
6855.728	6857.620	14582.32	TI	1.3			BH
6849.15	6851.04	14596.3	TI I	0.7			RL
6844.64	6846.53	14605.9	TI	1.0			BH
6746.433	6748.295	14818.56	TI I	0.6			MIT
6743.124	6744.985	14825.83	TI	2.0			IKS
6717.911	6719.766	14881.47	TI II	1.0			MIT
6716.679	6718.533	14884.20	TI	1.2			MIT
6680.26	6682.10	14965.3	TI II		0.7		RL
6677.175	6679.019	14972.26	TI I	1.3			MIT
6666.548	6668.389	14996.13	TI I	1.5			MIT
6599.112	6600.935	15149.37	TI I	2.0			MIT
6575.180	6576.996	15204.51	TI I	1.3			MIT
6565.62	6567.43	15226.7	TI I	1.7			MIT
6559.580	6561.392	15240.67	TI II	0.9			MIT
6556.066	6557.877	15248.84	TI I	2.2			IKS
6554.226	6556.037	15253.12	TI I	2.1	2.2		IKS
6546.276	6548.084	15271.64	TI I	1.9			IKS
6543.143	6544.951	15278.95	TI	0.9			BH
6517.713	6519.514	15338.57	TI	0.3			BH
6508.135	6509.933	15361.14	TI I	1.5			MIT
6497.689	6499.484	15385.84	TI I	1.6			MIT
6467.465	6469.252	15457.74	TI	0.7			BH
6464.020	6465.806	15465.97	TI I	0.3			MIT
6419.096	6420.870	15574.21	TI I	1.2			MIT
6381.416	6383.180	15666.17	TI I	1.0			MIT
6371.75	6373.51	15689.9	TI I	0.3			RL
6366.354	6368.114	15703.24	TI I	1.9			IKS
6364.92	6366.68	15706.8	TI I	0.9			MIT
6359.896	6361.654	15719.18	TI I	0.7			MIT
6359.211	6360.969	15720.87	TI I	0.3			MIT
6336.104	6337.856	15778.21	TI I	1.9			IKS
6325.22	6326.97	15805.4	TI I	1.6W			RL
6318.027	6319.774	15823.35	TI I	1.7			MIT
6312.240	6313.986	15837.86	TI I	1.9			IKS
6311.289	6313.034	15840.24	TI I	0.7H			MIT
6303.754	6305.497	15859.18	TI I	2.3			IKS
6298.075	6299.817	15873.48	TI I	0.3			MIT
6296.646	6298.387	15877.08	TI I	1.5			MIT
6295.949	6297.690	15878.84	TI I	0.5			MIT
6295.251	6296.992	15880.60	TI I	1.0			MIT

AIR WAVELENGTH (ANGSTRMS)	VACUUM WAVELENGTH (ANGSTRMS)	WAVENUMBER (KAYSERS)	LMENT	LOG ARC	INTENSITY SPK	INTENSITY DIS	REF
6277.525	6279.261	15925.44	TI I	0.3			MIT
6273.389	6275.124	15935.94	TI I	0.3			MIT
6268.530	6270.264	15948.29	TI	1.3			MIT
6266.021	6267.754	15954.68	TI I	1.1			MIT
6264.825	6266.558	15957.72	TI I	0.3			MIT
6261.099	6262.831	15967.22	TI I	2.5	2.0		I
6258.703	6260.434	15973.33	TI I	2.5	2.4		I
6258.103	6259.834	15974.86	TI I	2.3	2.0		IKS
6240.363	6242.089	16020.28	TI	0.3			BH
6235.335	6237.060	16033.19	TI	0.7			MIT
6227.661	6229.384	16052.95	TI	0.6			BH
6221.578	6223.299	16068.65	TI	0.5H			MIT
6221.41	6223.13	16069.1	TI I	1.9			MIT
6220.488	6222.209	16071.46	TI I	2.0	1.5		MIT
6219.979	6221.700	16072.78	TI	0.7			MIT
6215.284	6217.004	16084.92	TI I	2.0	1.7		MIT
6212.229	6213.948	16092.83	TI	1.3			MIT
6189.269	6190.982	16152.53	TI	0.3			MIT
6186.15	6187.86	16160.7	TI I	1.5			ME
6174.47	6176.18	16191.2	TI I	0.3			ME
6149.743	6151.445	16256.34	TI I	1.5			MIT
6146.225	6147.926	16265.65	TI I	2.6			MIT
6138.38	6140.08	16286.4	TI I	1.2			ME
6126.215	6127.911	16318.78	TI I	2.2	1.8		I
6121.008	6122.702	16332.66	TI I	1.5			MIT
6098.668	6100.356	16392.49	TI I	1.8			MIT
6092.814	6094.501	16408.24	TI I	1.5			MIT
6091.174	6092.860	16412.65	TI I	2.1	1.4		I
6085.226	6086.911	16428.70	TI I	2.0	1.8		I
6064.631	6066.310	16484.49	TI I	1.9	1.3		IKS
6031.68	6033.35	16574.5	TI I	0.9			MIT
6018.62	6020.29	16610.5	TI I	0.9			MIT
6018.423	6020.090	16611.05	TI	1.2			BH
6012.732	6014.397	16626.77	TI	1.0			MIT
6002.640	6004.303	16654.72	TI I	0.6			MIT
5999.680	6001.342	16662.94	TI I	1.8			MIT
5999.040	6000.702	16664.72	TI I	1.2	1.0		MIT
5998.967	6000.629	16664.92	TI	0.9			BH
5996.007	5997.668	16673.15	TI I	1.3			MIT
5995.685	5997.346	16674.04	TI I	1.0			MIT
5988.560	5990.219	16693.88	TI I	1.3			MIT
5984.586	5986.244	16704.97	TI I	1.2			MIT
5982.52	5984.18	16710.7	TI I	1.0			RL
5978.557	5980.213	16721.81	TI I	2.1	2.2		MIT
5965.840	5967.493	16757.46	TI I	2.2	2.3		MIT
5961.132	5962.783	16770.69	TI	0.3			BH
5953.171	5954.820	16793.12	TI I	2.2	2.4		MIT
5949.983	5951.631	16802.12	TI	0.3			MIT
5944.65	5946.30	16817.2	TI I	1.0			RL
5941.758	5943.404	16825.37	TI I	2.0			MIT
5940.68	5942.33	16828.4	TI I	1.0			MIT
5937.821	5939.466	16836.53	TI I	1.8			MIT
5932.134	5933.778	16852.67	TI	1.9			MIT
5922.123	5923.764	16881.16	TI I	2.0	2.0		MIT
5918.552	5920.192	16891.34	TI I	1.9			MIT
5913.730	5915.369	16905.12	TI I	0.5			MIT
5903.332	5904.968	16934.89	TI I	1.6			MIT
5899.322	5900.957	16946.40	TI I	2.2	2.2		MIT
5888.675	5890.307	16977.04	TI	1.2			MIT
5880.306	5881.936	17001.21	TI I	1.8	2.1		MIT
5877.77	5879.40	17008.5	TI I	1.2			RL
5866.462	5868.088	17041.33	TI I	2.5	2.6		MIT
5852.344	5853.966	17082.44	TI	1.1			MIT
5823.708	5825.323	17166.43	TI I	1.5	1.7		MIT
5813.98	5815.59	17195.1	TI I	0.9			MIT
5812.84	5814.45	17198.5	TI I	1.0			MIT
5804.265	5805.874	17223.94	TI I	2.0H	1.7		MIT
5797.448	5799.056	17244.19	TI I	1.0			MIT
5785.979	5787.583	17278.37	TI I	2.00	1.8		MIT
5785.67	5787.27	17279.3	TI I	0.5			MIT

AIR WAVELENGTH (ANGSTRMS)	VACUUM WAVELENGTH (ANGSTRMS)	WAVENUMBER (KAYSERS)	LMENT	LOG ARC	INTENSITY SPK	INTENSITY DIS	REF
5780.777	5782.380	17293.92	TI I	1.3	1.3		MIT
5774.053	5775.654	17314.06	TI I	1.80	1.7		MIT
5766.351	5767.950	17337.18	TI I	1.80	1.7		MIT
5762.270	5763.868	17349.46	TI I	1.8H	1.7		MIT
5756.858	5758.455	17365.77	TI	1.0	1.1		MIT
5752.838	5754.434	17377.91	TI I	0.7			BH
5741.216	5742.809	17413.08	TI I	0.8			BH
5740.02	5741.61	17416.7	TI I	1.4			MIT
5739.506	5741.098	17418.27	TI I	1.8	1.9		MIT
5720.478	5722.065	17476.21	TI I	1.4			MIT
5716.480	5718.066	17488.43	TI I	1.6			MIT
5715.133	5716.719	17492.55	TI I	1.8	1.8		MIT
5713.919	5715.504	17496.27	TI I	1.4			MIT
5711.878	5713.463	17502.52	TI I	1.7	1.6		MIT
5708.227	5709.811	17513.72	TI I	1.5			MIT
5702.675	5704.257	17530.77	TI I	1.8	1.6		MIT
5689.474	5691.053	17571.44	TI I	1.9	1.9		BH
5679.942	5681.518	17600.93	TI I	1.7			BH
5675.439	5677.014	17614.89	TI I	2.0	2.1		MIT
5662.908	5664.480	17653.87	TI I	1.6			MIT
5662.164	5663.735	17656.19	TI I	2.0	2.0		MIT
5659.104	5660.675	17665.74	TI I	0.7			MIT
5648.577	5650.145	17698.66	TI I	1.9	1.8		MIT
5644.140	5645.707	17712.57	TI I	2.2	2.3		MIT
5620.265	5621.825	17787.82	TI	0.6			MIT
5617.604	5619.163	17796.24	TI	0.7			MIT
5612.683	5614.241	17811.85	TI	0.5			MIT
5600.793	5602.348	17849.66	TI	0.3	1.0		MIT
5597.69	5599.24	17859.6	TI	1.0			MIT
5594.482	5596.035	17869.79	TI	1.0			MIT
5589.718	5591.270	17885.02	TI	1.0			MIT
5585.676	5587.227	17897.97	TI	1.4			BH
5584.128	5585.678	17902.93	TI	1.0H			BH
5583.093	5584.643	17906.25	TI	1.1			BH
5581.401	5582.951	17911.68	TI	1.4H			BH
5580.262	5581.811	17915.33	TI	0.8			MIT
5573.117	5574.665	17938.30	TI	0.8			BH
5570.509	5572.056	17946.70	TI	0.9			BH
5565.488	5567.034	17962.89	TI I	1.9	0.3		MIT
5564.628	5566.173	17965.66	TI	0.9			BH
5553.328	5554.870	18002.22	TI	1.2			BH
5550.965	5552.507	18009.88	TI	1.0			MIT
5544.140	5545.680	18032.05	TI	1.0			BH
5543.980	5545.520	18032.57	TI	0.8			BH
5541.874	5543.413	18039.43	TI	1.2			BH
5538.738	5540.276	18049.64	TI	1.0			MIT
5537.518	5539.056	18053.62	TI	1.2			MIT
5537.324	5538.862	18054.25	TI	1.0			MIT
5533.284	5534.821	18067.43	TI	0.8			MIT
5532.981	5534.518	18068.42	TI	1.0			MIT
5530.494	5532.030	18076.55	TI	1.5			BH
5527.606	5529.141	18085.99	TI I	0.9			MIT
5518.207	5519.740	18116.80	TI	0.8			MIT
5514.542	5516.074	18128.84	TI I	1.9	1.2		MIT
5514.346	5515.878	18129.48	TI I	1.8	1.0		MIT
5512.527	5514.058	18135.46	TI I	2.1	1.1		MIT
5511.784	5513.315	18137.91	TI I	1.3			MIT
5503.901	5505.430	18163.89	TI I	1.8	0.5		MIT
5495.410	5496.937	18191.95	TI	0.9			MIT
5491.67	5493.20	18204.3	TI	0.5			MIT
5490.845	5492.371	18207.07	TI I	0.5			MIT
5490.47	5492.00	18208.3	TI II		0.7		EL
5490.153	5491.678	18209.37	TI	1.8	0.3		MIT
5488.204	5489.729	18215.84	TI I	1.5	1.5		MIT
5485.448	5486.972	18224.99	TI	1.0			MIT
5484.108	5485.632	18229.44	TI	1.1			MIT
5481.869	5483.392	18236.89	TI I	1.3	1.3		MIT
5481.431	5482.954	18238.34	TI I	1.5	0.0		MIT
5477.709	5479.231	18250.74	TI I	1.8	0.3		MIT
5474.465	5475.986	18261.55	TI I	1.1			MIT

AIR WAVELENGTH (ANGSTRMS)	VACUUM WAVELENGTH (ANGSTRMS)	WAVENUMBER (KAYSERS)	LMENT	LOG ARC	INTENSITY SPK	INTENSITY DIS	REF
5474.227	5475.748	18262.34	TI I	1.5	1.7		MIT
5473.551	5475.072	18264.60	TI I	0.9			MIT
5473.436	5474.957	18264.98	TI	0.8			MIT
5472.699	5474.220	18267.44	TI I	1.1			MIT
5472.361	5473.882	18268.57	TI	0.9			MIT
5471.208	5472.728	18272.42	TI I	1.4	1.4		MIT
5460.507	5462.024	18308.23	TI I	1.5			MIT
5453.648	5455.164	18331.26	TI I	1.1	1.4		MIT
5451.996	5453.511	18336.81	TI I	1.1			MIT
5449.164	5450.678	18346.34	TI I	1.2			MIT
5448.903	5450.417	18347.22	TI I	1.1			MIT
5446.637	5448.151	18354.85	TI I	1.2	1.5		MIT
5440.467	5441.979	18375.67	TI	1.1 D			BH
5439.042	5440.554	18380.48	TI	0.6 H			MIT
5438.316	5439.828	18382.94	TI I	0.6	0.9		MIT
5436.727	5438.238	18388.31	TI I	1.0			MIT
5432.337	5433.847	18403.17	TI I	0.6			MIT
5429.149	5430.658	18413.98	TI I	1.4	0.0		MIT
5426.256	5427.764	18423.79	TI I	1.2	1.5		MIT
5419.204	5420.710	18447.77	TI I	0.6			MIT
5418.781	5420.287	18449.21	TI II	0.9	0.6		MIT
5410.97	5412.47	18475.8	TI II		0.7		EL
5409.611	5411.115	18480.48	TI	1.7	0.0		MIT
5408.940	5410.444	18482.77	TI I	0.8			MIT
5404.023	5405.525	18499.59	TI I	0.9			MIT
5397.094	5398.595	18523.34	TI I	1.8			MIT
5396.600	5398.100	18525.03	TI I	1.1			MIT
5389.992	5391.491	18547.75	TI I	1.5			MIT
5389.177	5390.675	18550.55	TI I	1.2			MIT
5384.634	5386.131	18566.20	TI I	0.7			MIT
5381.02	5382.52	18578.7	TI II	0.7	1.0 H		MIT
5369.645	5371.138	18618.03	TI	1.3	0.0		MIT
5366.649	5368.141	18628.42	TI	0.9			MIT
5361.724	5363.215	18645.53	TI I	0.5			MIT
5351.084	5352.572	18682.61	TI I	1.7	1.8		MIT
5341.500	5342.986	18716.13	TI I	0.9			MIT
5338.326	5339.811	18727.25	TI I	0.9			MIT
5336.809	5338.293	18732.58	TI II	1.3	1.5 H		MIT
5323.958	5325.439	18777.79	TI I	0.5			MIT
5313.261	5314.739	18815.60	TI I	0.8			MIT
5300.024	5301.499	18862.59	TI I	0.9			BH
5298.44	5299.91	18868.2	TI I	1.6	0.0		MIT
5297.257	5298.731	18872.44	TI I	1.8	0.3		MIT
5295.788	5297.262	18877.68	TI I	1.7	0.0		MIT
5289.28	5290.75	18900.9	TI I	0.5			MIT
5288.813	5290.285	18902.57	TI	1.2			MIT
5286.93	5288.40	18909.3	TI	0.9			MIT
5284.390	5285.860	18918.40	TI I	1.3	1.5		MIT
5283.451	5284.921	18921.76	TI I	1.7	0.3		MIT
5282.386	5283.856	18925.57	TI I	1.2	1.7		MIT
5281.931	5283.401	18927.20	TI I	0.6	0.9		MIT
5272.33	5273.80	18961.7	TI	1.2			MIT
5269.971	5271.438	18970.16	TI I	0.6			MIT
5268.62	5270.09	18975.0	TI II	0.7	0.0		MIT
5266.295	5267.761	18983.40	TI I	0.7			MIT
5265.979	5267.445	18984.54	TI I	1.8	0.5		MIT
5263.502	5264.967	18993.47	TI I	1.5			MIT
5263.316	5264.781	18994.14	TI I	0.3			MIT
5262.104	5263.569	18998.52	TI II		0.5 H		MIT
5259.990	5261.454	19006.15	TI I	1.2	1.5		MIT
5259.60	5261.06	19007.6	TI	1.0			MIT
5257.604	5259.067	19014.78	TI I	0.9			MIT
5255.826	5257.289	19021.21	TI I	1.6	1.9		MIT
5252.108	5253.570	19034.68	TI I	1.5	0.0		MIT
5251.49	5252.95	19036.9	TI I	0.8			MIT
5250.95	5252.41	19038.9	TI I	1.1			MIT
5248.402	5249.863	19048.12	TI I	0.6			MIT
5247.310	5248.771	19052.08	TI I	1.4	0.0		MIT
5246.574	5248.034	19054.75	TI I	1.0			MIT
5246.148	5247.608	19056.30	TI I	1.2			MIT

AIR WAVELENGTH (ANGSTRMS)	VACUUM WAVELENGTH (ANGSTRMS)	WAVENUMBER (KAYSERS)	LMENT	LOG ARC	INTENSITY SPK	INTENSITY DIS	REF
5239.942	5241.401	19078.87	TI I	0.5			MIT
5238.580	5240.038	19083.83	TI I	1.7	2.0		MIT
5233.817	5235.274	19101.20	TI I	1.5			MIT
5227.187	5228.642	19125.42	TI I	1.0			MIT
5226.555	5228.010	19127.74	TI II	1.5	1.7 H		MIT
5224.954	5226.409	19133.60	TI I	2.0	0.8		MIT
5224.564	5226.019	19135.03	TI I	1.5	0.3		MIT
5224.318	5225.772	19135.93	TI I	1.8	0.9		MIT
5223.641	5225.095	19138.41	TI I	1.7	0.0		MIT
5223.342	5224.796	19139.50	TI	0.5			MIT
5222.693	5224.147	19141.88	TI I	1.5	0.3		MIT
5219.706	5221.159	19152.83	TI I	1.8	0.3		MIT
5213.023	5214.474	19177.39	TI	1.0			MIT
5212.287	5213.738	19180.10	TI	1.2			MIT
5211.544	5212.995	19182.83	TI II	1.0	1.2 H		MIT
5210.389	5211.840	19187.08	TI I	2.3	1.5		MIT
5207.869	5209.319	19196.37	TI I	1.4			MIT
5206.083	5207.533	19202.95	TI I	1.6	0.0		MIT
5201.096	5202.544	19221.36	TI I	1.5			MIT
5194.043	5195.489	19247.47	TI I	1.0			MIT
5192.975	5194.421	19251.42	TI I	2.2	1.4		MIT
5188.700	5190.145	19267.28	TI II	1.9	2.0		MIT
5186.336	5187.780	19276.07	TI I	1.3			MIT
5185.90	5187.34	19277.7	TI II	0.9	1.5 H		RL
5183.72	5185.16	19285.8	TI I	0.9			MIT
5178.625	5180.067	19304.77	TI I	0.9			MIT
5173.748	5175.189	19322.97	TI I	2.1	1.3		MIT
5154.076	5155.512	19396.72	TI II	1.0	1.2 H		MIT
5152.200	5153.635	19403.78	TI I	2.0	0.3		MIT
5147.482	5148.916	19421.56	TI I	2.0	0.5		MIT
5145.468	5146.901	19429.17	TI I	2.0	0.6		MIT
5132.960	5134.390	19476.51	TI I	1.1			MIT
5129.149	5130.578	19490.98	TI II	1.5	1.5 H		MIT
5127.367	5128.796	19497.75	TI I	1.1			MIT
5124.09	5125.52	19510.2	TI I	1.0			MIT
5122.09	5123.52	19517.8	TI I	0.8			MIT
5120.425	5121.852	19524.19	TI I	2.0	0.6		MIT
5113.441	5114.866	19550.85	TI I	1.9	0.3		MIT
5109.440	5110.864	19566.16	TI I	1.3			MIT
5087.068	5088.486	19652.21	TI I	1.8	0.0		MIT
5085.341	5086.758	19658.89	TI I	1.3			MIT
5072.30	5073.71	19709.4	TI II	0.5	1.3 H		RL
5071.477	5072.891	19712.63	TI I	1.6	0.0		MIT
5069.353	5070.766	19720.89	TI I	1.6			MIT
5068.332	5069.745	19724.86	TI I	0.9			MIT
5065.992	5067.404	19733.97	TI I	1.7	0.3		MIT
5064.655	5066.067	19739.18	TI I	2.2	1.5		MIT
5064.068	5065.480	19741.47	TI I	1.0	1.0		MIT
5062.107	5063.518	19749.11	TI I	1.6	0.0		MIT
5054.085	5055.494	19780.46	TI I	0.9			MIT
5052.874	5054.283	19785.20	TI I	1.7	0.5		MIT
5052.170	5053.579	19787.96	TI	0.8			MIT
5048.208	5049.616	19803.49	TI I	1.2			MIT
5047.735	5049.142	19805.34	TI	0.9			MIT
5046.457	5047.864	19810.36	TI	1.1			MIT
5045.409	5046.816	19814.47	TI I	1.4			MIT
5044.274	5045.681	19818.93	TI I	1.3			MIT
5043.586	5044.992	19821.64	TI I	1.5			MIT
5040.619	5042.025	19833.30	TI I	1.6	1.6		MIT
5039.953	5041.358	19835.92	TI I	2.1	1.4		MIT
5038.402	5039.807	19842.03	TI I	2.0	1.3		MIT
5036.468	5037.872	19849.65	TI I	2.1	1.4		MIT
5035.907	5037.311	19851.86	TI I	2.1	1.5		MIT
5025.583	5026.985	19892.64	TI I	2.0	0.9		MIT
5024.844	5026.245	19895.57	TI I	2.0	1.2		MIT
5023.85	5025.25	19899.5	TI	0.5	0.3		MIT
5022.866	5024.267	19903.40	TI I	2.0	1.3		MIT
5020.027	5021.427	19914.66	TI I	2.0	1.9		MIT
5016.166	5017.565	19929.99	TI I	2.0	1.2		MIT
5014.241	5015.640	19937.64	TI I	2.0	1.5		MIT

AIR WAVELENGTH (ANGSTRMS)	VACUUM WAVELENGTH (ANGSTRMS)	WAVENUMBER (KAYSERS)	LMENT	LOG ARC	INTENSITY SPK	INTENSITY DIS	REF
5014.189	5015.588	19937.84	TI I	1.6	1.2		MIT
5013.712	5015.110	19939.74	TI II	0.7	0.7		MIT
5013.301	5014.699	19941.37	TI I	1.9	0.8		BH
5010.202	5011.599	19953.71	TI II	0.5	1.0H		MIT
5009.649	5011.046	19955.91	TI I	1.7	0.3		MIT
5007.213	5008.610	19965.62	TI I	2.3	1.6		MIT
5001.010	5002.405	19990.38	TI I	1.9	0.3		MIT
4999.510	5000.905	19996.38	TI I	2.3	1.9		MIT
4997.103	4998.497	20006.01	TI I	1.7	0.6		MIT
4995.084	4996.477	20014.10	TI I	1.1			MIT
4991.066	4992.458	20030.21	TI I	2.3	2.0		I
4989.148	4990.540	20037.91	TI I	2.0	0.6		MIT
4981.733	4983.123	20067.74	TI I	2.5	2.1		I
4978.204	4979.593	20081.96	TI I	1.8	0.5		MIT
4977.744	4979.133	20083.82	TI I	1.3	1.4		MIT
4976.822	4978.211	20087.54	TI I	1.2			MIT
4975.354	4976.742	20093.47	TI I	1.9	0.6		MIT
4973.048	4974.436	20102.78	TI I	1.5	0.3		MIT
4968.583	4969.969	20120.85	TI I	1.6	0.0		MIT
4964.751	4966.136	20136.38	TI I	1.4			MIT
4948.193	4949.574	20203.76	TI I	1.1			MIT
4947.994	4949.375	20204.57	TI I	0.8			MIT
4944.388	4945.768	20219.31	TI I	0.7			MIT
4943.074	4944.454	20224.68	TI I	0.3			MIT
4941.578	4942.957	20230.80	TI I	1.5			MIT
4941.322	4942.701	20231.85	TI I	0.6			MIT
4941.015	4942.394	20233.11	TI I	0.6			MIT
4938.293	4939.671	20244.26	TI I	1.8	0.3		MIT
4937.743	4939.121	20246.52	TI I	1.5			MIT
4928.895	4930.271	20282.86	TI I	0.3			MIT
4928.341	4929.717	20285.14	TI I	2.0	0.6		MIT
4926.158	4927.533	20294.13	TI I	1.3			MIT
4925.410	4926.785	20297.21	TI I	1.4			MIT
4921.769	4923.143	20312.23	TI I	2.0	0.7		MIT
4919.866	4921.239	20320.08	TI I	1.0	0.5		MIT
4915.236	4916.608	20339.23	TI I	1.5	1.0		MIT
4913.622	4914.994	20345.91	TI I	2.1	1.2		MIT
4911.185	4912.556	20356.00	TI II	1.1	2.0		MIT
4909.107	4910.478	20364.62	TI I	1.1			MIT
4900.962	4902.330	20398.46	TI	0.6			MIT
4900.625	4901.993	20399.86	TI I	0.8			MIT
4899.912	4901.280	20402.83	TI I	2.2	1.3		MIT
4893.433	4894.799	20429.85	TI	0.8			MIT
4893.061	4894.427	20431.40	TI I	1.0			MIT
4891.828	4893.194	20436.55	TI I	1.1			MIT
4885.085	4886.449	20464.76	TI I	2.2	1.4		MIT
4882.346	4883.710	20476.24	TI I	1.4	0.0		MIT
4880.912	4882.275	20482.25	TI I	1.1			MIT
4870.137	4871.497	20527.57	TI I	2.0	1.3		MIT
4868.263	4869.623	20535.47	TI I	2.0	0.9		I
4865.620	4866.979	20546.63	TI II		0.5		MIT
4864.182	4865.541	20552.70	TI I	1.3	0.0		MIT
4856.012	4857.369	20587.28	TI I	2.0	1.0		I
4848.468	4849.823	20619.31	TI I	1.8	0.3		MIT
4843.983	4845.336	20638.40	TI I	0.9			MIT
4840.874	4842.227	20651.66	TI I	2.1	1.4		I
4839.251	4840.603	20658.58	TI I	1.2			MIT
4836.132	4837.483	20671.91	TI I	1.4	0.0		MIT
4832.076	4833.426	20689.26	TI I	1.3	0.0		MIT
4827.578	4828.927	20708.53	TI I	1.2	0.0		MIT
4825.457	4826.805	20717.64	TI I	1.0	0.0		MIT
4820.415	4821.762	20739.31	TI I	2.1	1.5		MIT
4819.032	4820.379	20745.26	TI	1.6	0.3		MIT
4812.906	4814.251	20771.66	TI I	0.3			MIT
4812.250	4813.595	20774.49	TI I	1.3	0.3		MIT
4811.080	4812.425	20779.55	TI I	1.1	0.0		MIT
4808.531	4809.875	20790.56	TI I	1.1	0.3		MIT
4805.426	4806.769	20804.00	TI I	1.8	0.6		I
4805.102	4806.445	20805.40	TI II	1.2	2.1		MIT
4799.802	4801.144	20828.37	TI I	1.9	1.2		MIT

AIR WAVELENGTH (ANGSTRMS)	VACUUM WAVELENGTH (ANGSTRMS)	WAVENUMBER (KAYSERS)	LMENT	LOG ARC	INTENSITY SPK	INTENSITY DIS	REF
4798.535	4799.876	20833.87	TI II	0.3	1.2		MIT
4797.985	4799.326	20836.26	TI I	1.3	0.3		MIT
4796.216	4797.557	20843.94	TI I	1.5	0.5		MIT
4792.488	4793.828	20860.16	TI I	1.8	1.1		MIT
4789.803	4791.142	20871.85	TI I	0.6			MIT
4783.306	4784.643	20900.20	TI I	0.3			MIT
4781.718	4783.055	20907.14	TI I	1.5	0.3		MIT
4779.953	4781.289	20914.86	TI II	1.0	2.0H		MIT
4778.263	4779.599	20922.26	TI I	1.6	0.8		MIT
4771.099	4772.433	20953.67	TI I	1.0	0.0		MIT
4769.769	4771.103	20959.51	TI I	1.1	0.3		MIT
4766.334	4767.667	20974.62	TI I	1.1	0.3		MIT
4764.535	4765.867	20982.54	TI		0.8		MIT
4763.902	4765.234	20985.33	TI	0.8	1.3		MIT
4759.276	4760.607	21005.73	TI I	2.0	0.9		MIT
4758.904	4760.235	21007.37	TI I	0.7	0.0		MIT
4758.118	4759.449	21010.84	TI I	2.1	1.8		MIT
4747.675	4749.003	21057.05	TI I	1.4	0.3		MIT
4747.256	4748.584	21058.91	TI I	1.0			MIT
4742.788	4744.115	21078.75	TI I	2.0	1.6		MIT
4742.110	4743.436	21081.76	TI I	1.2	0.0		MIT
4734.678	4736.002	21114.85	TI I	1.0	0.0		MIT
4733.429	4734.753	21120.43	TI I	1.4	0.5		MIT
4731.173	4732.496	21130.50	TI I	1.7	0.8		MIT
4724.679	4726.001	21159.54	TI I	0.8			MIT
4723.168	4724.489	21166.31	TI I	1.6	0.8		MIT
4722.616	4723.937	21168.78	TI I	1.9	0.9		MIT
4719.515	4720.835	21182.69	TI II		0.3		MIT
4715.298	4716.617	21201.64	TI I	1.6	0.3		MIT
4710.194	4711.512	21224.61	TI I	2.0	1.4		MIT
4708.976	4710.294	21230.10	TI I	0.3			MIT
4708.663	4709.980	21231.51	TI II	0.3	1.3		MIT
4698.765	4700.080	21276.23	TI I	2.0	1.3		MIT
4696.938	4698.252	21284.51	TI I	1.2	0.5		MIT
4693.679	4694.993	21299.29	TI I	1.4	0.5		MIT
4691.338	4692.651	21309.92	TI I	2.1	1.4		MIT
4690.805	4692.118	21312.34	TI I	1.2	0.3		MIT
4688.394	4689.706	21323.30	TI I	1.0	0.5		MIT
4686.921	4688.233	21330.00	TI I	0.9	0.3		MIT
4685.013	4686.324	21338.69	TI	0.3			MIT
4684.484	4685.795	21341.09	TI I	0.8	0.0		MIT
4681.916	4683.226	21352.80	TI I	2.3	2.0		MIT
4677.483	4678.792	21373.04	TI I	0.5	0.0		MIT
4676.975	4678.284	21375.36	TI I	1.0			MIT
4675.123	4676.432	21383.83	TI I	1.7	0.7		MIT
4668.357	4669.664	21414.82	TI I	0.9	0.0		MIT
4667.588	4668.895	21418.35	TI I	2.2	0.9		MIT
4664.487	4665.793	21432.58	TI	0.3			MIT
4659.596	4660.900	21455.08	TI	0.3			MIT
4657.210	4658.514	21466.07	TI II	0.7	1.3		MIT
4656.470	4657.774	21469.49	TI I	2.2	1.8		MIT
4656.045	4657.349	21471.44	TI I	1.3	0.5		MIT
4655.702	4657.005	21473.03	TI I	0.9	0.3		MIT
4653.404	4654.707	21483.63	TI	0.7			MIT
4650.019	4651.321	21499.27	TI I	1.8	0.6		MIT
4648.05	4649.35	21508.4	TI	1.2O			MIT
4645.194	4646.495	21521.60	TI I	2.5	1.0		I
4640.431	4641.730	21543.69	TI I	0.8	0.0		MIT
4639.947	4641.246	21545.94	TI I	1.8	1.2		MIT
4639.667	4640.966	21547.24	TI I	1.6	1.2		MIT
4639.366	4640.665	21548.64	TI I	1.9	1.3		MIT
4637.876	4639.175	21555.56	TI I	1.3	0.6		MIT
4637.209	4638.508	21558.66	TI I	0.6	0.0		MIT
4636.345	4637.643	21562.68	TI II		0.6H		MIT
4635.539	4636.837	21566.43	TI I	1.1	0.3		MIT
4634.870	4636.168	21569.54	TI	0.9	0.3		MIT
4634.742	4636.040	21570.13	TI	0.8	0.0		MIT
4629.342	4630.638	21595.29	TI I	1.8	0.8		MIT
4623.094	4624.389	21624.48	TI I	2.1	1.6		MIT
4619.525	4620.819	21641.19	TI I	1.1	0.3		MIT

AIR WAVELENGTH (ANGSTRMS)	VACUUM WAVELENGTH (ANGSTRMS)	WAVENUMBER (KAYSERS)	LMENT	LOG ARC	INTENSITY SPK	INTENSITY DIS	REF
4617.270	4618.563	21651.75	TI I	2.3	2.0		I
4614.287	4615.580	21665.75	TI I	0.8	0.0		MIT
4609.370	4610.661	21688.86	TI	1.2	0.3H		MIT
4606.723	4608.014	21701.32	TI	0.6			MIT
4599.231	4600.520	21736.68	TI	1.5	0.5		MIT
4589.954	4591.240	21780.61	TI II	1.6	2.0		MIT
4589.832	4591.118	21781.19	TI	0.3			MIT
4583.443	4584.727	21811.55	TI II	0.7	1.0		MIT
4580.458	4581.742	21825.76	TI II		0.7		MIT
4576.551	4577.834	21844.39	TI I	0.6			MIT
4575.519	4576.801	21849.32	TI	0.8	0.0		MIT
4571.978	4573.259	21866.24	TI II	2.2	2.5		I
4570.909	4572.190	21871.36	TI I	1.6	0.3		MIT
4568.312	4569.592	21883.79	TI II	0.5	0.9		MIT
4564.216	4565.495	21903.43	TI I	0.5			MIT
4563.766	4565.045	21905.59	TI II	2.0	2.3		MIT
4563.433	4564.712	21907.19	TI I	1.2	0.5		MIT
4562.633	4563.912	21911.03	TI I	1.4	0.5		MIT
4559.923	4561.201	21924.05	TI I	1.7	0.7		MIT
4558.107	4559.385	21932.78	TI I	1.2	0.5		MIT
4557.856	4559.134	21933.99	TI I	1.1	0.5		MIT
4555.489	4556.766	21945.39	TI I	2.1	1.8		MIT
4555.083	4556.360	21947.34	TI I	1.1	0.3		MIT
4553.415	4554.691	21955.38	TI I	0.8			MIT
4552.459	4553.735	21960.00	TI	2.2	1.7		MIT
4549.628	4550.903	21973.66	TI II	2.0	2.3		MIT
4548.767	4550.042	21977.82	TI I	2.1	1.4		MIT
4548.112	4549.387	21980.98	TI I	0.8	0.0		MIT
4547.850	4549.125	21982.25	TI I	1.0	0.0		MIT
4545.144	4546.418	21995.34	TI II	0.5	1.2		MIT
4544.688	4545.962	21997.54	TI I	2.2	1.8		I
4544.009	4545.283	22000.83	TI II	0.7	1.3		MIT
4541.021	4542.294	22015.31	TI I	0.3			MIT
4540.873	4542.146	22016.02	TI I	0.8	0.0		MIT
4540.483	4541.756	22017.92	TI I	0.8	0.0		MIT
4539.096	4540.369	22024.64	TI	1.2	0.6		MIT
4537.826	4539.098	22030.81	TI	0.3			MIT
4537.228	4538.500	22033.71	TI	1.0	0.5		MIT
4536.053	4537.325	22039.42	TI I	1.6	1.3		MIT
4535.922	4537.194	22040.05	TI I	1.6	1.3		MIT
4535.575	4536.847	22041.74	TI I	1.9	1.7		MIT
4534.782	4536.053	22045.59	TI I	2.0	1.6		MIT
4533.967	4535.238	22049.56	TI II	1.5	2.2		MIT
4533.244	4534.515	22053.07	TI I	2.2	1.6		MIT
4532.143	4533.414	22058.43	TI	0.5			MIT
4529.465	4530.735	22071.47	TI II	0.7	1.6		MIT
4527.455	4528.725	22081.27	TI I	0.5			MIT
4527.312	4528.582	22081.97	TI I	2.0	1.7		MIT
4526.374	4527.643	22086.55	TI I	0.7	0.0		MIT
4525.934	4527.203	22088.69	TI	0.3			MIT
4524.732	4526.001	22094.56	TI II	1.0	1.0		MIT
4522.805	4524.073	22103.97	TI I	2.0	1.8		MIT
4518.697	4519.964	22124.07	TI I	1.5	1.0		MIT
4518.032	4519.299	22127.32	TI I	2.0	1.8		MIT
4515.624	4516.890	22139.12	TI I	1.3	0.3		MIT
4513.715	4514.981	22148.49	TI I	0.5	0.0		MIT
4512.738	4514.004	22153.28	TI I	2.0	1.8		MIT
4511.170	4512.435	22160.98	TI	1.6	1.0		MIT
4509.738	4511.003	22168.02	TI	0.3	0.0		MIT
4508.281	4509.546	22175.18	TI I	0.5	0.0		MIT
4508.048	4509.312	22176.33	TI I	0.7	0.0		MIT
4506.361	4507.625	22184.63	TI	1.3	0.3		MIT
4505.715	4506.979	22187.81	TI I	0.5	0.0		MIT
4503.777	4505.040	22197.36	TI I	1.2	0.7		MIT
4501.274	4502.537	22209.70	TI II	1.8	2.0		MIT
4497.730	4498.992	22227.20	TI I	1.2	0.5		MIT
4496.245	4497.506	22234.54	TI I	0.8			MIT
4496.149	4497.410	22235.02	TI I	1.8	1.8		MIT
4495.008	4496.269	22240.66	TI	1.3	0.7		MIT
4493.53	4494.79	22248.0	TI II	0.0	0.9		MIT

AIR WAVELENGTH (ANGSTRMS)	VACUUM WAVELENGTH (ANGSTRMS)	WAVENUMBER (KAYSERS)	LMENT	LOG ARC	INTENSITY SPK	INTENSITY DIS	REF
4492.546	4493.806	22252.85	TI I	1.2	0.6		MIT
4490.697	4491.957	22262.01	TI I	0.6			MIT
4489.093	4490.352	22269.97	TI I	2.0	1.6		MIT
4488.317	4489.576	22273.82	TI II	1.0	2.1		M
4485.098	4486.356	22289.80	TI I	0.9			MIT
4484.55	4485.81	22292.5	TI I	0.5			MIT
4482.693	4483.951	22301.76	TI I	1.6	1.3		MIT
4481.262	4482.519	22308.88	TI I	2.0	1.8		MIT
4480.590	4481.847	22312.23	TI I	1.6	1.2		MIT
4479.705	4480.962	22316.64	TI I	1.8	1.5		MIT
4475.518	4476.774	22337.51	TI I	1.1H	0.0H		MIT
4474.854	4476.110	22340.83	TI I	1.9	1.5		MIT
4471.240	4472.495	22358.89	TI I	2.0	1.6		MIT
4470.861	4472.116	22360.78	TI II	1.1	1.4		MIT
4469.160	4470.414	22369.29	TI II	0.7H	0.0		MIT
4468.495	4469.749	22372.62	TI II	1.9	2.2		MIT
4465.810	4467.063	22386.07	TI I	2.0	1.6		MIT
4464.455	4465.708	22392.87	TI II	1.1	1.6		MIT
4463.541	4464.794	22397.45	TI I	1.4	1.1		MIT
4463.384	4464.637	22398.24	TI I	1.4	1.1		MIT
4462.094	4463.346	22404.72	TI I	1.0	0.0		MIT
4457.432	4458.683	22428.15	TI I	2.2	2.0		MIT
4456.650	4457.901	22432.08	TI II	0.0	1.0		MIT
4455.326	4456.577	22438.75	TI I	2.2	1.9		MIT
4453.706	4454.956	22446.91	TI I	1.9	1.6		MIT
4453.321	4454.571	22448.85	TI I	2.2	1.8		MIT
4450.899	4452.148	22461.07	TI I	2.2	1.8		MIT
4450.486	4451.735	22463.15	TI II	1.1	1.7		MIT
4449.984	4451.233	22465.69	TI I	1.0			MIT
4449.148	4450.397	22469.91	TI I	2.2	1.9		MIT
4444.562	4445.810	22493.09	TI II	0.6	1.1		MIT
4444.266	4445.514	22494.59	TI I	1.3	0.3		MIT
4443.804	4445.051	22496.93	TI II	1.9	2.1		MIT
4443.195	4444.442	22500.01	TI I	0.7H			MIT
4443.042	4444.289	22500.79	TI	0.9H			MIT
4441.274	4442.521	22509.74	TI I	1.4	1.0		MIT
4440.348	4441.595	22514.44	TI I	1.9	1.5		MIT
4438.226	4439.472	22525.20	TI I	1.1	0.8		MIT
4436.653	4437.899	22533.19	TI I	0.5	0.0		MIT
4436.592	4437.838	22533.50	TI I	1.2	0.8		MIT
4434.001	4435.246	22546.66	TI I	2.0	1.7		MIT
4433.576	4434.821	22548.82	TI I	1.2	0.3		MIT
4432.602	4433.847	22553.78	TI	1.3	0.3		MIT
4432.089	4433.333	22556.39	TI II	0.3	0.0		MIT
4431.282	4432.526	22560.50	TI I	1.4	1.1		MIT
4430.371	4431.615	22565.14	TI I	1.5	1.2		MIT
4430.020	4431.264	22566.93	TI I	1.1	0.0		MIT
4427.101	4428.344	22581.80	TI I	2.1	1.8		MIT
4426.055	4427.298	22587.14	TI I	1.9	1.4		MIT
4425.828	4427.071	22588.30	TI I	1.0	0.0		MIT
4424.393	4425.635	22595.63	TI I	1.2	0.3		MIT
4422.825	4424.067	22603.64	TI I	1.9	1.4		MIT
4421.955	4423.197	22608.08	TI II	0.8	1.5		MIT
4421.759	4423.001	22609.08	TI I	1.8	1.2		MIT
4421.474	4422.716	22610.54	TI I	1.0			MIT
4418.342	4419.583	22626.57	TI II	1.0	1.3		MIT
4417.721	4418.962	22629.75	TI II	1.6	1.9		MIT
4417.442	4418.683	22631.18	TI I	0.3H			MIT
4417.280	4418.520	22632.01	TI I	1.9	1.3		MIT
4416.536	4417.776	22635.82	TI I	1.8	1.0		MIT
4415.578	4416.818	22640.73	TI	0.9			MIT
4414.577	4415.817	22645.87	TI	0.3			MIT
4414.108	4415.348	22648.27	TI	1.0			MIT
4412.428	4413.667	22656.90	TI I	1.2	0.0		MIT
4411.936	4413.175	22659.42	TI II	0.9	1.1		MIT
4411.077	4412.316	22663.83	TI II	0.8	2.0		MIT
4409.519	4410.757	22671.84	TI II	0.5	1.0		MIT
4409.22	4410.46	22673.4	TI II	0.3	0.9		MIT
4407.678	4408.916	22681.31	TI II	0.3	1.0		MIT
4405.685	4406.922	22691.57	TI I	1.3	0.8		MIT

AIR WAVELENGTH (ANGSTRMS)	VACUUM WAVELENGTH (ANGSTRMS)	WAVENUMBER (KAYSERS)	LMENT	LOG ARC	INTENSITY SPK	INTENSITY DIS	REF
4404.904	4406.141	22695.60	TI I	1.2	1.0		MIT
4404.396	4405.633	22698.21	TI I	1.1	0.8		MIT
4404.275	4405.512	22698.84	TI I	1.7	1.5		MIT
4400.582	4401.818	22717.89	TI I	1.4	0.3		MIT
4399.771	4401.007	22722.07	TI II	1.6	2.0		MIT
4398.314	4399.549	22729.60	TI II	0.5	1.0		MIT
4395.842	4397.077	22742.38	TI II	1.0	1.5		MIT
4395.035	4396.270	22746.56	TI II	1.7	2.2		MIT
4394.855	4396.090	22747.49	TI I	0.9			MIT
4394.065	4395.299	22751.58	TI II	0.7	1.2		MIT
4393.925	4395.159	22752.30	TI I	1.8	1.1		MIT
4391.034	4392.268	22767.28	TI II	0.8	1.4		MIT
4388.549	4389.782	22780.18	TI	1.0	0.0		MIT
4388.075	4389.308	22782.64	TI I	1.4	0.7		MIT
4386.854	4388.086	22788.98	TI II	0.9	1.9		MIT
4375.425	4376.654	22848.50	TI I	1.0	0.0		MIT
4374.822	4376.051	22851.65	TI II	0.8	1.5		MIT
4372.382	4373.611	22864.40	TI I	1.3	0.8		MIT
4369.677	4370.905	22878.56	TI I	1.4	0.7		MIT
4368.939	4370.167	22882.42	TI I	1.3			MIT
4367.658	4368.885	22889.13	TI II	0.9	1.4		MIT
4361.142	4362.368	22923.33	TI I	1.0	0.0		MIT
4360.493	4361.719	22926.74	TI I	1.8	1.2		MIT
4355.308	4356.532	22954.04	TI I	1.0	0.0		MIT
4354.062	4355.286	22960.60	TI I	1.4	0.7		MIT
4350.834	4352.057	22977.64	TI II	0.8	1.5		MIT
4346.610	4347.832	22999.97	TI I	1.2	0.0		MIT
4346.111	4347.333	23002.61	TI I	1.6	0.8		MIT
4344.286	4345.507	23012.27	TI II	1.1	1.7		MIT
4343.786	4345.007	23014.92	TI I	1.0	0.0		MIT
4341.375	4342.596	23027.70	TI II	1.1	1.6		MIT
4340.018	4341.238	23034.90	TI I	1.1			MIT
4338.484	4339.704	23043.05	TI I	1.1	0.0		MIT
4337.919	4339.139	23046.05	TI II	1.8	2.1		MIT
4337.33	4338.55	23049.2	TI II		0.3		MIT
4334.835	4336.054	23062.45	TI I	1.5	0.7		MIT
4330.708	4331.926	23084.42	TI II	1.2	1.5		MIT
4330.243	4331.461	23086.90	TI II	1.0	1.6		MIT
4326.967	4328.184	23104.38	TI I	1.1	0.0		MIT
4326.355	4327.572	23107.65	TI I	1.8	1.4		MIT
4325.134	4326.350	23114.17	TI I	2.0	1.6		MIT
4323.441	4324.657	23123.22	TI I	1.3	0.3		MIT
4321.665	4322.880	23132.73	TI I	1.8	1.4		MIT
4320.958	4322.173	23136.51	TI II	1.1	1.6		MIT
4318.639	4319.854	23148.93	TI I	2.0	1.7		MIT
4316.801	4318.015	23158.79	TI II	0.8	1.5		MIT
4314.978	4316.192	23168.57	TI II	0.8	1.3		MIT
4314.801	4316.015	23169.52	TI I	2.0	1.3		I
4314.348	4315.561	23171.96	TI I	1.5	0.8		MIT
4312.870	4314.083	23179.90	TI II	1.5	2.0		MIT
4311.653	4312.866	23186.44	TI I	1.4	0.8		MIT
4310.366	4311.578	23193.36	TI I	1.0	0.0		MIT
4309.097	4310.309	23200.19	TI I	1.0	0.0		MIT
4308.504	4309.716	23203.39	TI I	1.3	0.3		MIT
4307.905	4309.117	23206.61	TI II	2.0	2.0		MIT
4306.945	4308.156	23211.78	TI I	1.0	0.0		MIT
4305.916	4307.127	23217.33	TI I	2.5	2.2		MIT
4305.482	4306.693	23219.67	TI I	1.0	0.7		MIT
4303.961	4305.172	23227.88	TI	0.9			MIT
4302.979	4304.189	23233.18	TI I	1.0			MIT
4301.934	4303.144	23238.82	TI II	1.4	1.7		MIT
4301.088	4302.298	23243.39	TI I	2.2	1.7		MIT
4300.565	4301.775	23246.22	TI I	2.1	1.3		MIT
4300.049	4301.259	23249.01	TI II	1.6	2.0		MIT
4299.639	4300.849	23251.23	TI I	1.8	1.0		MIT
4299.229	4300.438	23253.44	TI I	1.8	1.3		MIT
4298.665	4299.874	23256.49	TI I	2.1	1.7		I
4295.756	4296.964	23272.24	TI I	2.0	1.6		MIT
4294.124	4295.332	23281.09	TI II	1.8	1.9		MIT
4292.676	4293.884	23288.94	TI I	1.1	0.0		MIT
4291.864	4293.071	23293.35	TI I	0.7H			MIT
4291.210	4292.417	23296.90	TI I	0.5	0.0		MIT
4291.138	4292.345	23297.29	TI	1.0	0.6		MIT
4290.937	4292.144	23298.38	TI I	1.8	1.5		MIT
4290.227	4291.434	23302.23	TI II	1.5	1.8		MIT
4289.919	4291.126	23303.91	TI I	1.2	0.3		MIT
4289.073	4290.280	23308.50	TI I	2.1	1.7		MIT
4288.162	4289.369	23313.46	TI I	1.3	0.6		MIT
4287.881	4289.087	23314.98	TI II	1.0	1.5		MIT
4287.405	4288.611	23317.57	TI I	2.0	1.7		MIT
4286.009	4287.215	23325.17	TI I	2.0	1.6		MIT
4284.994	4286.200	23330.69	TI I	1.6	1.3		MIT
4282.711	4283.916	23343.13	TI I	1.8	1.4		MIT
4281.376	4282.581	23350.41	TI I	1.9	1.3		MIT
4280.333	4281.537	23356.10	TI	0.9	0.0		BH
4280.079	4281.283	23357.48	TI I	1.2	0.0		MIT
4278.813	4280.017	23364.39	TI I	1.4	0.5		MIT
4278.227	4279.431	23367.59	TI I	1.7	1.2		MIT
4276.660	4277.863	23376.16	TI I	1.0H	0.0		MIT
4276.433	4277.636	23377.40	TI I	1.7	1.3		MIT
4274.584	4275.787	23387.51	TI I	2.0	1.6		I
4274.400	4275.603	23388.51	TI I	0.7			MIT
4273.303	4274.506	23394.52	TI I	1.3	0.0		MIT
4272.432	4273.634	23399.29	TI I	1.6	1.0		MIT
4270.138	4271.340	23411.86	TI I	1.7	0.5		MIT
4268.928	4270.129	23418.49	TI I	0.9			MIT
4266.222	4267.423	23433.35	TI I	1.3			MIT
4265.708	4266.909	23436.17	TI I	1.1			MIT
4265.275	4266.475	23438.55	TI I	1.3H			MIT
4263.134	4264.334	23450.32	TI	2.1	1.5		MIT
4261.602	4262.802	23458.75	TI I	1.8	0.9		MIT
4260.751	4261.950	23463.44	TI I	1.1	0.0		MIT
4258.535	4259.734	23475.65	TI I	1.8	0.8		MIT
4256.036	4257.234	23489.43	TI I	1.9	1.2		MIT
4251.761	4252.958	23513.05	TI I	1.0	0.0		MIT
4251.606	4252.803	23513.90	TI I	1.3	0.5		MIT
4249.123	4250.319	23527.64	TI I	1.8H	0.5H		MIT
4245.513	4246.708	23547.65	TI	1.3	0.5		MIT
4237.893	4239.086	23589.99	TI I	1.3	0.9		MIT
4237.780	4238.973	23590.62	TI I	0.9			MIT
4234.656	4235.848	23608.02	TI	1.1	0.0		MIT
4227.650	4228.841	23647.14	TI I	1.3	0.3		MIT
4224.793	4225.983	23663.13	TI I	1.6	0.9		MIT
4211.728	4212.914	23736.54	TI I	1.5	0.8		MIT
4203.464	4204.648	23783.20	TI I	1.7	1.0		MIT
4200.749	4201.933	23798.57	TI I	1.5	0.9		MIT
4188.691	4189.871	23867.08	TI I	1.5	0.9		MIT
4186.123	4187.303	23881.72	TI I	2.0	1.6		MIT
4184.329	4185.508	23891.96	TI II	0.9	1.3		MIT
4183.295	4184.474	23897.87	TI I	1.3	0.8		MIT
4180.868	4182.046	23911.74	TI I	2.0	1.3H		MIT
4180.501	4181.679	23913.84	TI I	0.8			MIT
4179.885	4181.063	23917.36	TI I	0.6			MIT
4177.357	4178.534	23931.84	TI I	1.0	0.3		MIT
4174.475	4175.652	23948.36	TI I	1.2	0.5		MIT
4174.091	4175.268	23950.56	TI I	1.2H	1.1H		MIT
4173.549	4174.725	23953.67	TI II	1.1	1.6		MIT
4172.609	4173.785	23959.07	TI I	0.7			MIT
4171.903	4173.079	23963.12	TI II	1.2	1.8		MIT
4171.029	4172.205	23968.14	TI I	1.5	0.8		MIT
4169.348	4170.523	23977.81	TI I	1.4	0.8		MIT
4166.320	4167.495	23995.23	TI I	1.5	1.0		MIT
4164.136	4165.310	24007.82	TI I	1.0			MIT
4163.654	4164.828	24010.60	TI II	1.5	2.2		MIT
4161.536	4162.709	24022.82	TI II	0.9	1.5		MIT
4159.642	4160.815	24033.75	TI I	1.8	1.2		MIT
4158.413	4159.585	24040.86	TI	0.3			MIT
4158.291	4159.463	24041.56	TI	0.5			MIT
4158.049	4159.221	24042.96	TI	1.2H	0.3H		MIT
4154.865	4156.036	24061.39	TI	1.2	0.0		MIT

AIR WAVELENGTH (ANGSTRMS)	VACUUM WAVELENGTH (ANGSTRMS)	WAVENUMBER (KAYSERS)	LMENT	LOG ARC	INTENSITY SPK	DIS	REF
4150.962	4152.132	24084.01	TI I	1.5	1.2		MIT
4150.550	4151.720	24086.40	TI I	1.0	0.0		MIT
4149.454	4150.624	24092.76	TI I	1.1	0.0		MIT
4148.482	4149.652	24098.41	TI	0.6	0.0		MIT
4148.390	4149.560	24098.94	TI I	0.8			MIT
4145.052	4146.221	24118.35	TI I	1.2	0.3		MIT
4143.280	4144.448	24128.66	TI I	0.8			MIT
4143.046	4144.214	24130.03	TI I	1.5	0.9		MIT
4142.492	4143.660	24133.25	TI I	1.1	0.3		MIT
4139.844	4141.012	24148.69	TI I	0.8			MIT
4137.288	4138.455	24163.61	TI I	1.7	1.2		MIT
4131.254	4132.419	24198.90	TI I	1.2	0.3		MIT
4129.173	4130.338	24211.09	TI	1.2	0.8		MIT
4127.540	4128.704	24220.67	TI I	1.8	1.2		MIT
4123.572	4124.735	24243.98	TI I	1.6	1.0		MIT
4123.306	4124.469	24245.54	TI I	1.4	0.8		MIT
4123.143	4124.306	24246.50	TI I	1.0			IKS
4122.166	4123.329	24252.25	TI I	1.6	1.0		MIT
4121.642	4122.805	24255.33	TI	1.2	0.5		MIT
4120.034	4121.196	24264.80	TI I	1.0			MIT
4112.714	4113.874	24307.99	TI I	1.8	1.3		MIT
4102.715	4103.873	24367.23	TI	0.5	0.0		MIT
4099.172	4100.329	24388.29	TI I	1.4	0.9		MIT
4082.462	4083.615	24488.11	TI I	1.8	1.4		MIT
4079.721	4080.873	24504.56	TI I	1.6	0.8		MIT
4078.474	4079.625	24512.05	TI I	2.1	1.7		MIT
4077.153	4078.304	24520.00	TI I	1.3	0.3		MIT
4076.375	4077.526	24524.68	TI I	1.2	0.0		MIT
4074.362	4075.512	24536.79	TI I	1.2	0.0		MIT
4071.469	4072.619	24554.23	TI I	0.9	0.0		MIT
4071.211	4072.361	24555.78	TI I	1.0	0.0		MIT
4069.003	4070.152	24569.11	TI I	1.3H	0.0H		MIT
4068.661	4069.810	24571.17	TI I	1.0	0.0		MIT
4068.144	4069.293	24574.29	TI I	1.3	0.7		MIT
4065.590	4066.738	24589.73	TI I	1.1	0.3		MIT
4065.101	4066.249	24592.69	TI I	1.9	1.5		MIT
4064.400	4065.548	24596.93	TI II		0.3		MIT
4064.219	4065.367	24598.03	TI I	1.7	1.2		MIT
4060.265	4061.412	24621.98	TI I	1.8	1.4		MIT
4058.144	4059.290	24634.85	TI I	1.7	0.8		MIT
4057.624	4058.770	24638.01	TI I	1.6	0.8		MIT
4056.212	4057.358	24646.58	TI II	0.5	0.3		MIT
4055.019	4056.164	24653.83	TI I	1.9	1.5		MIT
4053.837	4054.982	24661.02	TI II	1.4	0.9		MIT
4052.937	4054.082	24666.50	TI I	1.4	0.0		MIT
4049.399	4050.543	24688.05	TI I	1.5	0.0		MIT
4043.775	4044.917	24722.38	TI I	1.3			MIT
4040.318	4041.460	24743.54	TI I	1.6	0.0		MIT
4038.333	4039.474	24755.70	TI	1.1	0.0		MIT
4035.830	4036.970	24771.05	TI I	1.7	0.7		MIT
4034.910	4036.050	24776.70	TI I	1.4	0.3		MIT
4033.906	4035.046	24782.87	TI I	1.6	0.5		MIT
4032.632	4033.771	24790.70	TI I	1.5	0.0		MIT
4031.754	4032.893	24796.09	TI I	1.5	0.0		MIT
4030.514	4031.653	24803.72	TI I	1.9	1.3		MIT
4028.345	4029.483	24817.08	TI II	1.3	1.9		MIT
4027.485	4028.623	24822.38	TI	1.5	0.5		MIT
4026.541	4027.679	24828.20	TI I	1.8	1.0		MIT
4025.136	4026.274	24836.86	TI II	1.2	1.4		MIT
4024.573	4025.710	24840.34	TI I	1.9	1.5		MIT
4021.827	4022.964	24857.30	TI I	2.0	1.3		MIT
4017.769	4018.905	24882.40	TI I	1.8H	0.9		MIT
4016.975	4018.110	24887.32	TI I	1.1	0.3		MIT
4016.284	4017.419	24891.60	TI I	1.5	0.7		MIT
4015.384	4016.519	24897.18	TI I	1.8H	1.0H		MIT
4013.584	4014.719	24908.35	TI I	1.8H	0.8H		MIT
4012.806	4013.940	24913.18	TI I	1.1	0.0		MIT
4012.391	4013.525	24915.75	TI II	1.5	1.7		MIT
4011.534	4012.668	24921.07	TI I	0.6			MIT
4009.663	4010.796	24932.70	TI I	1.8	1.4		MIT

AIR WAVELENGTH (ANGSTRMS)	VACUUM WAVELENGTH (ANGSTRMS)	WAVENUMBER (KAYSERS)	LMENT	LOG ARC	INTENSITY SPK	DIS	REF
4008.928	4010.061	24937.27	TI I	1.9	1.5		MIT
4008.062	4009.195	24942.66	TI I	1.7	0.8		MIT
4007.192	4008.325	24948.08	TI I	1.2	0.0		MIT
4005.971	4007.104	24955.68	TI I	1.5	0.5		MIT
4003.806	4004.938	24969.18	TI I	1.8	1.7		MIT
4002.490	4003.622	24977.39	TI I	1.6	0.7		MIT
3999.359	4000.490	24996.94	TI I	1.5	0.7		MIT
3998.640	3999.771	25001.43	TI I	2.2	2.0		MIT
3996.649	3997.779	25013.89	TI	1.1			MIT
3994.704	3995.834	25026.07	TI I	1.4			MIT
3992.661	3993.790	25038.87	TI	0.5			MIT
3992.400	3993.529	25040.51	TI	0.8			MIT
3990.184	3991.312	25054.42	TI I	1.0W	0.0W		MIT
3989.763	3990.891	25057.06	TI I	2.2	2.0		MIT
3989.568	3990.696	25058.28	TI I	0.3H			MIT
3987.610	3988.738	25070.59	TI	0.9	1.1		MIT
3985.589	3986.716	25083.30	TI	1.0H	0.0		MIT
3985.45	3986.58	25084.2	TI	0.6H			MIT
3985.252	3986.379	25085.42	TI	1.1	0.0		MIT
3984.333	3985.460	25091.21	TI I	1.3	0.5		MIT
3982.484	3983.610	25102.86	TI I	1.9	1.5		MIT
3982.010	3983.136	25105.84	TI II	0.0	0.5		MIT
3981.764	3982.890	25107.40	TI I	2.0	1.8		MIT
3981.467	3982.593	25109.27	TI I	0.3H			MIT
3980.822	3981.948	25113.34	TI	0.9			MIT
3972.130	3973.254	25168.29	TI I	0.5			MIT
3964.272	3965.394	25218.18	TI I	1.9	1.6		MIT
3963.354	3964.475	25224.02	TI I	0.3			MIT
3962.852	3963.973	25227.21	TI I	1.9	1.5		MIT
3958.213	3959.333	25256.78	TI I	2.2	2.0		MIT
3956.343	3957.463	25268.72	TI I	2.0	1.7		MIT
3948.674	3949.792	25317.79	TI I	1.9H	1.6		MIT
3947.775	3948.892	25323.56	TI I	1.8	1.5		MIT
3938.026	3939.141	25386.25	TI I	1.3	0.3H		MIT
3934.236	3935.350	25410.70	TI I	1.5	0.3		MIT
3932.017	3933.130	25425.04	TI II	1.3	1.5		MIT
3929.875	3930.988	25438.90	TI I	1.8	1.5		MIT
3927.362	3928.474	25455.18	TI I	0.8			MIT
3926.322	3927.434	25461.92	TI I	1.4	1.1		MIT
3924.529	3925.640	25473.55	TI	1.8	1.5		MIT
3921.424	3922.534	25493.72	TI I	1.6	0.8		MIT
3919.819	3920.929	25504.16	TI I	1.3	0.3		MIT
3916.104	3917.213	25528.35	TI	1.0	0.0		MIT
3915.876	3916.985	25529.84	TI I	1.2	0.0		MIT
3914.736	3915.845	25537.27	TI I	1.3	0.3		MIT
3914.335	3915.444	25539.89	TI I	1.7	1.0		MIT
3913.464	3914.572	25545.57	TI	1.6	1.8		MIT
3912.588	3913.696	25551.29	TI I	1.2			MIT
3911.362	3912.470	25559.30	TI I	0.8			MIT
3911.189	3912.297	25560.43	TI I	1.6	0.7		MIT
3904.785	3905.891	25602.35	TI I	1.8	1.5		MIT
3900.960	3902.065	25627.46	TI I	1.4	0.3		MIT
3900.540	3901.645	25630.22	TI II	1.5	1.7H		MIT
3899.712	3900.817	25635.66	TI I	1.2	0.3		MIT
3898.494	3899.598	25643.67	TI I	1.3	0.7		MIT
3897.585	3898.689	25649.65	TI I	0.8			MIT
3897.304	3898.408	25651.50	TI I	0.9			MIT
3895.246	3896.350	25665.05	TI I	1.8	1.0		MIT
3893.635	3894.738	25675.67	TI	0.8			MIT
3889.954	3891.056	25699.96	TI I	1.4	0.7		MIT
3888.024	3889.126	25712.72	TI I	1.2			MIT
3887.365	3888.467	25717.08	TI I	0.7			MIT
3884.109	3885.210	25738.64	TI I	1.0			MIT
3882.892	3883.992	25746.70	TI I	1.5	1.0		MIT
3882.328	3883.428	25750.44	TI I	1.3	0.8		MIT
3882.148	3883.248	25751.64	TI I	1.4	0.9		MIT
3881.399	3882.499	25756.61	TI I	1.1	0.5		MIT
3879.361	3880.461	25770.14	TI	1.0	0.6		MIT
3877.592	3878.691	25781.89	TI I	0.5			MIT
3875.256	3876.354	25797.44	TI I	1.5H	0.9		MIT

AIR WAVELENGTH (ANGSTRMS)	VACUUM WAVELENGTH (ANGSTRMS)	WAVENUMBER (KAYSERS)	LMENT	LOG ARC	INTENSITY SPK	INTENSITY DIS	REF
3874.142	3875.240	25804.85	TI	1.1	0.6		MIT
3873.213	3874.311	25811.04	TI I	1.6	0.8		MIT
3870.131	3871.228	25831.60	TI	1.1	0.7		MIT
3869.610	3870.707	25835.07	TI I	1.0	0.6		MIT
3869.294	3870.391	25837.18	TI I	1.2	0.9		MIT
3868.403	3869.500	25843.14	TI I	1.7	0.9		MIT
3867.744	3868.840	25847.54	TI I	0.9	0.6		MIT
3866.443	3867.539	25856.24	TI I	1.6H	1.0		MIT
3866.028	3867.124	25859.01	TI I	1.1	0.7		MIT
3864.703	3865.799	25867.88	TI	0.6			MIT
3864.497	3865.593	25869.25	TI	1.2	0.8		MIT
3862.826	3863.921	25880.45	TI I	1.5H	0.6		MIT
3861.736	3862.831	25887.75	TI	1.0	0.5		MIT
3861.084	3862.179	25892.12	TI	1.0H	0.5		MIT
3858.144	3859.238	25911.85	TI I	1.6	0.8		MIT
3857.904	3858.998	25913.46	TI I	0.8	0.5		MIT
3853.730	3854.823	25941.53	TI I	1.1	0.5		MIT
3853.049	3854.142	25946.11	TI I	1.3	0.6		MIT
3848.310	3849.401	25978.06	TI I	1.0	0.3		MIT
3846.449	3847.540	25990.63	TI	1.2	0.5		MIT
3845.842	3846.933	25994.74	TI	0.6	0.0		MIT
3845.099	3846.190	25999.76	TI	0.3			MIT
3842.614	3843.704	26016.57	TI I	1.2	0.7		MIT
3841.747	3842.837	26022.44	TI	1.3	1.1		MIT
3841.624	3842.714	26023.28	TI I	0.3			MIT
3840.328	3841.417	26032.06	TI	0.3			MIT
3836.775	3837.863	26056.16	TI	1.3	0.7		MIT
3836.598	3837.686	26057.37	TI	1.0	0.5		MIT
3836.084	3837.172	26060.86	TI II	1.2	1.5		MIT
3833.920	3835.008	26075.57	TI I	0.8	0.3		MIT
3833.681	3834.769	26077.19	TI	0.7	0.0		MIT
3833.190	3834.277	26080.53	TI	1.1	0.5		MIT
3829.733	3830.820	26104.07	TI I	0.8	0.3		MIT
3829.406	3830.493	26106.30	TI	1.5	1.1		MIT
3828.193	3829.279	26114.57	TI I	1.5	1.2		MIT
3828.021	3829.107	26115.75	TI	0.8	0.0		MIT
3827.674	3828.760	26118.12	TI I	0.7	0.0		MIT
3827.470	3828.556	26119.51	TI I	0.5			MIT
3826.969	3828.055	26122.93	TI I	1.1	0.3		MIT
3822.028	3823.113	26156.70	TI I	1.3	0.9		MIT
3821.725	3822.810	26158.77	TI I	1.3	1.0		MIT
3818.224	3819.308	26182.76	TI I	1.3	0.8		MIT
3817.645	3818.728	26186.73	TI I	1.2	0.8		MIT
3814.585	3815.668	26207.73	TI II	1.1	1.5		MIT
3813.394	3814.476	26215.92	TI II	0.6	1.3		MIT
3813.270	3814.352	26216.77	TI I	1.1	0.5		MIT
3811.404	3812.486	26229.61	TI I	1.4	1.0		MIT
3811.310	3812.392	26230.25	TI	1.2	0.8		MIT
3807.761	3808.842	26254.70	TI I	0.7			MIT
3807.227	3808.308	26258.38	TI	0.9			MIT
3806.447	3807.528	26263.76	TI I	1.2	0.8		MIT
3806.044	3807.124	26266.54	TI I	0.3			MIT
3805.477	3806.557	26270.46	TI I	0.9	0.5		MIT
3805.118	3806.198	26272.94	TI I	0.9	0.0		MIT
3801.560	3802.639	26297.52	TI	0.8	0.5		MIT
3801.084	3802.163	26300.82	TI I	1.2	0.7		MIT
3798.314	3799.392	26320.00	TI I	1.0	0.8		MIT
3796.885	3797.963	26329.90	TI II	1.1	1.2		MIT
3795.895	3796.973	26336.77	TI I	1.2	0.8		MIT
3794.764	3795.842	26344.62	TI I	0.8			MIT
3793.681	3794.758	26352.14	TI I	0.9			MIT
3792.995	3794.072	26356.91	TI	0.9			MIT
3789.295	3790.371	26382.64	TI I	1.7	1.2		MIT
3788.804	3789.880	26386.06	TI I	0.8	0.3		MIT
3786.268	3787.343	26403.73	TI I	0.8			MIT
3786.045	3787.120	26405.29	TI I	1.6	1.6		MIT
3782.109	3783.183	26432.77	TI I	1.0	0.5		MIT
3780.403	3781.477	26444.70	TI	1.0			MIT
3779.040	3780.113	26454.23	TI I	0.9			MIT
3776.059	3777.132	26475.12	TI II	0.9	1.8		MIT

AIR WAVELENGTH (ANGSTRMS)	VACUUM WAVELENGTH (ANGSTRMS)	WAVENUMBER (KAYSERS)	LMENT	LOG ARC	INTENSITY SPK	INTENSITY DIS	REF
3775.72	3776.79	26477.5	TI I	3.5	3.0G		FL
3774.650	3775.722	26485.00	TI	0.3	0.0		MIT
3774.331	3775.403	26487.24	TI I	0.9			MIT
3771.655	3772.727	26506.03	TI I	1.8	1.5		MIT
3770.412	3771.483	26514.77	TI II	0.3	0.7		MIT
3769.736	3770.807	26519.52	TI	1.0			MIT
3766.446	3767.516	26542.69	TI I	1.0	0.5		MIT
3762.305	3763.374	26571.90	TI	1.0	0.0		MIT
3761.888	3762.957	26574.84	TI II	1.1	1.2		MIT
3761.323	3762.392	26578.84	TI II	2.0	2.5R		MIT
3759.295	3760.363	26593.18	TI II	2.0	2.6G		MIT
3757.689	3758.757	26604.54	TI II	1.5	2.0		MIT
3755.012	3756.079	26623.51	TI	1.1			MIT
3753.635	3754.702	26633.27	TI I	1.9	1.5		MIT
3752.860	3753.927	26638.77	TI I	2.3	1.9		MIT
3748.102	3749.167	26672.59	TI I	1.0	0.0		MIT
3748.003	3749.068	26673.29	TI II	0.3	1.4		MIT
3747.782	3748.847	26674.87	TI I	0.8	0.0		MIT
3741.644	3742.708	26718.62	TI II	1.5	2.3		MIT
3741.14	3742.20	26722.2	TI I	0.3	0.5		RL
3741.062	3742.126	26722.78	TI I	2.2R	1.6		MIT
3738.901	3739.964	26738.23	TI I	1.6	1.0		MIT
3736.787	3737.849	26753.35	TI	1.0	0.3		MIT
3735.671	3736.733	26761.34	TI	1.2	0.6		MIT
3733.785	3734.847	26774.86	TI I	1.0	0.3		MIT
3729.815	3730.876	26803.36	TI I	2.7	2.2		MIT
3728.676	3729.736	26811.55	TI I	1.2	0.7		MIT
3727.054	3728.114	26823.22	TI I	0.3			MIT
3725.165	3726.224	26836.82	TI I	2.2	1.8		MIT
3725.165	3726.224	26836.82	TI I	2.2	1.8		MIT
3724.574	3725.633	26841.07	TI I	2.0	1.7		MIT
3724.106	3725.165	26844.45	TI II	0.6	1.3		MIT
3723.631	3724.690	26847.87	TI II	0.6	1.2		MIT
3722.566	3723.625	26855.55	TI I	2.0	1.8		MIT
3721.636	3722.695	26862.26	TI II	1.8	2.1		MIT
3720.384	3721.442	26871.30	TI I	1.6	1.0		MIT
3717.396	3718.453	26892.90	TI I	1.9	1.7		MIT
3717.259	3718.316	26893.89	TI I	0.8	0.0		MIT
3715.795	3716.852	26904.49	TI I	1.4	0.0		MIT
3715.400	3716.457	26907.35	TI I	1.6	0.3H		MIT
3713.734	3714.790	26919.42	TI I	1.2	0.0		MIT
3713.028	3714.084	26924.54	TI	1.1	0.0		MIT
3710.186	3711.242	26945.16	TI I	0.8			MIT
3709.961	3711.017	26946.79	TI I	1.9	1.4		MIT
3708.648	3709.703	26956.33	TI I	1.7	0.5		MIT
3707.531	3708.586	26964.46	TI I	2.0	1.0		MIT
3706.226	3707.281	26973.95	TI II	1.5	2.1		MIT
3704.295	3705.349	26988.01	TI I	1.8	1.4		MIT
3702.986	3704.040	26997.55	TI I	1.3	0.7		MIT
3702.292	3703.346	27002.61	TI I	1.8	1.3		MIT
3701.541	3702.594	27008.09	TI	1.1			MIT
3700.077	3701.130	27018.78	TI	1.8	0.7		MIT
3698.430	3699.483	27030.81	TI I	0.7			MIT
3698.183	3699.235	27032.61	TI I	1.0	0.9		MIT
3696.885	3697.937	27042.10	TI I	1.3	0.5		MIT
3696.39	3697.44	27045.7	TI II	0.5	1.1		MIT
3694.447	3695.498	27059.95	TI I	1.9	1.3		MIT
3692.134	3693.185	27076.90	TI	1.1	0.0		MIT
3689.911	3690.961	27093.21	TI I	2.0	1.6		MIT
3689.671	3690.721	27094.98	TI I	0.7			MIT
3687.354	3688.404	27112.00	TI I	1.0			MIT
3685.955	3687.004	27122.29	TI I	1.6			MIT
3685.195	3686.244	27127.88	TI II	2.2	2.8G		MIT
3681.272	3682.320	27156.79	TI I	1.2			MIT
3680.270	3681.318	27164.19	TI		0.8H		MIT
3679.673	3680.721	27168.59	TI	0.7	1.1		MIT
3677.770	3678.817	27182.65	TI I	0.9	0.0		MIT
3674.76	3675.81	27204.9	TI	0.3	0.3		MIT
3671.673	3672.719	27227.79	TI I	2.2	1.8		MIT
3668.966	3670.011	27247.88	TI I	2.0	1.6		MIT

AIR WAVELENGTH (ANGSTRMS)	VACUUM WAVELENGTH (ANGSTRMS)	WAVENUMBER (KAYSERS)	LMENT	LOG ARC	INTENSITY SPK	INTENSITY DIS	REF
3666.592	3667.636	27265.52	TI II	0.3	0.0		MIT
3662.237	3663.280	27297.94	TI II	1.6	2.0		MIT
3660.633	3661.676	27309.90	TI I	2.0	1.3		MIT
3659.765	3660.808	27316.38	TI II	1.7	2.2		MIT
3658.100	3659.142	27328.81	TI I	2.2	1.8		MIT
3654.592	3655.633	27355.04	TI I	2.0	1.6		MIT
3653.496	3654.537	27363.25	TI I	2.7	2.3		MIT
3648.86	3649.90	27398.0	TI II	0.5	1.0		MIT
3646.200	3647.239	27418.00	TI I	1.8	1.4		MIT
3644.699	3645.738	27429.29	TI	1.5	0.7		MIT
3644.46	3645.50	27431.1	TI I	0.5			M
3642.675	3643.713	27444.53	TI I	2.5	2.1		MIT
3641.331	3642.369	27454.66	TI II	1.8	2.2		MIT
3640.315	3641.352	27462.32	TI I	0.5			MIT
3637.966	3639.003	27480.05	TI I	1.5	0.9		MIT
3635.463	3636.499	27498.98	TI I	2.3	2.0		MIT
3635.202	3636.238	27500.95	TI I	1.0	0.7		MIT
3633.458	3634.494	27514.15	TI	1.5	0.7		MIT
3631.999	3633.034	27525.20	TI	0.8			MIT
3631.315	3632.350	27530.39	TI	0.3			MIT
3627.71	3628.74	27557.7	TI II	0.3	1.1		MIT
3626.925	3627.959	27563.71	TI	1.0			MIT
3626.085	3627.119	27570.09	TI I	1.4	0.7		MIT
3625.250	3626.284	27576.44	TI	0.3	0.0		MIT
3624.825	3625.858	27579.68	TI II	1.8	2.1		MIT
3623.096	3624.129	27592.84	TI	1.1	0.0		MIT
3622.20	3623.23	27599.7	TI		1.0		MIT
3620.017	3621.049	27616.30	TI I	1.0			MIT
3619.464	3620.496	27620.52	TI I	0.3	0.3		MIT
3617.213	3618.245	27637.71	TI I	0.9	0.0		MIT
3615.33	3616.36	27652.1	TI		1.0		MIT
3614.207	3615.238	27660.70	TI	1.5	0.6		MIT
3613.756	3614.787	27664.15	TI	1.1			MIT
3613.435	3614.466	27666.61	TI I	0.9			MIT
3612.824	3613.854	27671.29	TI I	0.6			MIT
3612.251	3613.281	27675.68	TI	1.2	0.0		MIT
3610.156	3611.186	27691.74	TI I	2.0	1.8		MIT
3609.591	3610.621	27696.07	TI I	1.1	0.3		MIT
3607.131	3608.160	27714.96	TI I	1.4	0.7		MIT
3606.786	3607.815	27717.61	TI I	1.1	0.6		MIT
3606.062	3607.091	27723.17	TI I	1.1	0.0		MIT
3604.284	3605.312	27736.85	TI I	1.3	0.8		MIT
3603.845	3604.873	27740.23	TI I	1.2	0.3		MIT
3601.376	3602.403	27759.25	TI	1.2			MIT
3601.193	3602.220	27760.66	TI I	0.8	0.5		MIT
3598.716	3599.743	27779.76	TI I	1.8	1.5		MIT
3596.55	3597.58	27796.5	TI II	0.5	0.3		MIT
3596.052	3597.078	27800.34	TI II	1.7	2.1		MIT
3593.093	3594.118	27823.24	TI II	0.7	1.5		MIT
3587.130	3588.154	27869.49	TI II	1.1	1.4		MIT
3585.852	3586.875	27879.42	TI	0.8	0.5		MIT
3580.291	3581.313	27922.72	TI	1.2	0.7		MIT
3578.687	3579.709	27935.24	TI II	1.4	0.7		MIT
3578.266	3579.287	27938.52	TI I	0.9	0.0		MIT
3574.245	3575.265	27969.95	TI I	1.2	0.5		MIT
3573.737	3574.757	27973.93	TI II	1.3	1.6		MIT
3566.095	3567.113	28033.87	TI	0.9	0.3H		MIT
3565.987	3567.005	28034.72	TI II	0.8	1.4		MIT
3565.326	3566.344	28039.92	TI II	0.3	0.7		MIT
3564.540	3565.558	28046.10	TI I	1.1	0.3		MIT
3564.397	3565.415	28047.23	TI I	1.0	0.0		MIT
3561.910	3562.927	28066.81	TI II	0.7	1.1		MIT
3561.575	3562.592	28069.45	TI II	1.0	1.3		MIT
3558.518	3559.534	28093.56	TI	1.2	0.8		MIT
3556.617	3557.633	28108.58	TI	0.3	0.0		MIT
3556.184	3557.200	28112.00	TI	1.2	0.0		MIT
3553.815	3554.830	28130.74	TI I	0.5			MIT
3547.029	3548.042	28184.56	TI I	1.5	1.1		MIT
3546.538	3547.551	28188.46	TI	0.6	0.0		MIT
3544.99	3546.00	28200.8	TI	1.0	0.3		MIT

AIR WAVELENGTH (ANGSTRMS)	VACUUM WAVELENGTH (ANGSTRMS)	WAVENUMBER (KAYSERS)	LMENT	LOG ARC	INTENSITY SPK	INTENSITY DIS	REF
3542.546	3543.558	28220.22	TI	0.8	0.0		MIT
3537.482	3538.493	28260.62	TI	0.8	0.3		MIT
3535.412	3536.422	28277.17	TI II	1.2	2.1		MIT
3533.868	3534.878	28289.52	TI II	0.8	1.5		MIT
3530.580	3531.589	28315.86	TI I	1.2	0.0		MIT
3529.43	3530.44	28325.1	TI I	3.0	2.3		FL
3527.441	3528.449	28341.06	TI	0.3			MIT
3526.041	3527.049	28352.31	TI	1.1	0.3		MIT
3525.161	3526.169	28359.39	TI I	1.0	0.0		MIT
3524.87	3525.88	28361.7	TI II		0.7		MIT
3524.240	3525.248	28366.80	TI	0.3			MIT
3520.253	3521.260	28398.93	TI II	1.0	1.3		MIT
3519.939	3520.945	28401.46	TI I	0.7			MIT
3516.838	3517.844	28426.50	TI I	1.1	0.0		MIT
3512.075	3513.079	28465.05	TI	0.3			MIT
3511.626	3512.630	28468.70	TI I	1.0			MIT
3510.841	3511.845	28475.06	TI II	1.6	2.1		MIT
3509.844	3510.848	28483.15	TI II	0.9	1.3		MIT
3507.426	3508.429	28502.78	TI	1.2	0.3		MIT
3506.643	3507.646	28509.15	TI I	1.5	0.5		MIT
3505.901	3506.904	28515.18	TI II	0.3	0.7		MIT
3504.892	3505.895	28523.39	TI II	1.3	2.2		MIT
3504.773	3505.776	28524.36	TI I	0.3			MIT
3503.760	3504.762	28532.60	TI I	0.8			MIT
3500.340	3501.341	28560.48	TI II	1.2	1.5		MIT
3499.099	3500.100	28570.61	TI I	1.4	1.0		MIT
3495.960	3496.960	28596.26	TI I	1.0			MIT
3495.754	3496.754	28597.95	TI I	1.4	0.8		MIT
3493.280	3494.280	28618.20	TI I	1.2	0.0		MIT
3492.5	3493.5	28625.	TI II		1.5J		RL
3491.054	3492.053	28636.45	TI II	0.9	0.9		MIT
3490.765	3491.764	28638.82	TI I	0.3			MIT
3489.739	3490.738	28647.24	TI II	1.3	1.3		MIT
3485.689	3486.687	28680.52	TI I	1.4	0.6		MIT
3483.80	3484.80	28696.1	TI II		1.8J		MIT
3483.010	3484.007	28702.58	TI I	0.9			MIT
3481.675	3482.672	28713.59	TI I	1.2	0.0		MIT
3481.126	3482.122	28718.12	TI I	1.1	0.7		MIT
3480.897	3481.893	28720.01	TI II	0.8	1.4		MIT
3480.530	3481.526	28723.03	TI I	1.6	1.0		MIT
3478.918	3479.914	28736.34	TI I	1.3	0.7		MIT
3477.182	3478.177	28750.69	TI II	1.8	2.0		MIT
3476.982	3477.977	28752.34	TI II	0.9	0.9		MIT
3476.452	3477.447	28756.73	TI I	1.1	1.1		MIT
3472.793	3473.787	28787.02	TI I	0.9			MIT
3467.260	3468.253	28832.96	TI I	1.4	0.8		MIT
3465.562	3466.555	28847.09	TI II	0.8	1.8		MIT
3463.205	3464.197	28866.72	TI I	0.8	0.5		MIT
3461.500	3462.491	28880.94	TI II	1.9	2.1		MIT
3459.431	3460.422	28898.21	TI	1.3	0.0		MIT
3458.020	3459.011	28910.00	TI I	0.9	0.0		MIT
3457.6	3458.6	28913.	TI		1.0		CX
3457.494	3458.484	28914.40	TI I	1.0	0.0		MIT
3457.298	3458.288	28916.04	TI I	0.8			MIT
3456.661	3457.651	28921.37	TI I	0.3			MIT
3456.390	3457.380	28923.64	TI II	1.4	2.1		MIT
3455.755	3456.745	28928.95	TI I	0.8			MIT
3455.425	3456.415	28931.71	TI I	0.9			MIT
3455.394	3456.384	28931.97	TI I	1.0			MIT
3454.73	3455.72	28937.5	TI I	0.5			MIT
3454.165	3455.155	28942.26	TI I	1.2			MIT
3453.654	3454.643	28946.55	TI I	0.5			MIT
3453.531	3454.520	28947.58	TI I	0.7H			MIT
3452.470	3453.459	28956.47	TI II	1.1	2.0		MIT
3450.735	3451.724	28971.03	TI I	0.6			MIT
3449.874	3450.862	28978.26	TI I	0.8			MIT
3448.255	3449.243	28991.87	TI I	0.8			MIT
3446.603	3447.591	29005.76	TI I	0.6			MIT
3445.566	3446.553	29014.49	TI I	0.5			MIT
3444.899	3445.886	29020.11	TI I	0.3			MIT

AIR WAVELENGTH (ANGSTRMS)	VACUUM WAVELENGTH (ANGSTRMS)	WAVENUMBER (KAYSERS)	LMENT	LOG ARC	INTENSITY SPK	INTENSITY DIS	REF
3444.403	3445.390	29024.29	TI I	0.7			MIT
3444.311	3445.298	29025.06	TI II	1.8	2.2		MIT
3443.644	3444.631	29030.69	TI I	1.1	0.0		MIT
3443.387	3444.374	29032.85	TI II	0.5	1.5		MIT
3439.305	3440.291	29067.31	TI I	1.3	0.8		MIT
3435.432	3436.417	29100.08	TI	1.0	0.0		MIT
3431.85	3432.83	29130.4	TI I	0.7H	0.7J		MIT
3430.874	3431.858	29138.74	TI I	0.7			MIT
3430.551	3431.535	29141.48	TI I	0.7	0.5		MIT
3428.955	3429.938	29155.04	TI I	1.0	0.0		MIT
3426.5	3427.5	29176.	TI		0.5		CX
3423.172	3424.154	29204.30	TI I	0.6			MIT
3422.661	3423.643	29208.66	TI II	0.6	1.0		MIT
3416.957	3417.937	29257.41	TI II	0.8	1.7		MIT
3415.993	3416.973	29265.67	TI I	1.2	0.3		MIT
3411.680	3412.659	29302.67	TI I	1.0	0.3		MIT
3409.809	3410.787	29318.75	TI II	1.3	1.6		MIT
3407.205	3408.183	29341.15	TI II	1.1	1.7		MIT
3405.094	3406.071	29359.34	TI I	1.3	0.3		MIT
3404.97	3405.95	29360.4	TI II		0.3D		MIT
3403.369	3404.346	29374.22	TI I	1.1	0.3		MIT
3402.422	3403.398	29382.40	TI II	1.2	2.0		MIT
3400.162	3401.138	29401.93	TI I	1.0	0.0		MIT
3398.634	3399.609	29415.14	TI I	1.3	0.8		MIT
3394.575	3395.549	29450.32	TI II	1.8	2.3		MIT
3392.713	3393.687	29466.48	TI I	1.3	0.9		MIT
3392.197	3393.171	29470.96	TI	0.3	0.0		MIT
3390.682	3391.655	29484.13	TI I	1.5	1.0		MIT
3389.90	3390.87	29490.9	TI I	0.5H			MIT
3389.620	3390.593	29493.36	TI I	0.8			MIT
3388.755	3389.728	29500.89	TI II	1.1	1.5		MIT
3387.837	3388.810	29508.89	TI	1.8	2.1		MIT
3385.946	3386.918	29525.37	TI I	1.9	1.4		MIT
3385.664	3386.636	29527.83	TI I	1.4	1.0		MIT
3383.761	3384.733	29544.43	TI II	1.8	2.5G		MIT
3382.312	3383.283	29557.09	TI I	1.5	0.8		MIT
3381.442	3382.413	29564.69	TI I	0.3			MIT
3381.335	3382.306	29565.63	TI I	0.7			MIT
3380.280	3381.251	29574.86	TI II	1.4	2.2R		MIT
3379.930	3380.901	29577.92	TI II	0.3	0.7		MIT
3379.216	3380.186	29584.17	TI I	1.4	1.0		MIT
3377.585	3378.555	29598.45	TI I	1.3	1.2		MIT
3377.485	3378.455	29599.33	TI I	1.2	1.0		MIT
3375.706	3376.676	29614.93	TI I	1.2H			MIT
3374.608	3375.577	29624.56	TI I	0.5			MIT
3374.352	3375.321	29626.81	TI II	1.0	1.5		MIT
3373.581	3374.550	29633.58	TI I	0.6			MIT
3372.800	3373.769	29640.44	TI II	1.9	2.6G		MIT
3372.208	3373.177	29645.65	TI II	1.0	1.2		MIT
3371.454	3372.422	29652.28	TI I	2.0	1.2		MIT
3370.439	3371.407	29661.21	TI I	1.9	1.2		MIT
3369.212	3370.180	29672.01	TI II	0.8	1.4		MIT
3369.054	3370.022	29673.40	TI I	0.3			MIT
3367.881	3368.849	29683.73	TI I	0.3L	0.0		MIT
3366.176	3367.143	29698.77	TI I	1.3	1.7		MIT
3366.176	3367.143	29698.77	TI II	1.3	1.7		MIT
3364.9	3365.9	29710.	TI II	0.7			ME
3364.3	3365.3	29715.	TI II				ME
3363.612	3364.578	29721.41	TI I	0.3			MIT
3362.653	3363.619	29729.88	TI II	0.3	0.7		MIT
3362.100	3363.066	29734.77	TI I	0.8	0.5		MIT
3361.835	3362.801	29737.11	TI I	1.1	0.0		MIT
3361.263	3362.229	29742.18	TI I	1.9R	1.7R		MIT
3361.213	3362.179	29742.62	TI II	2.0	2.8G		IKS
3360.990	3361.956	29744.59	TI I	1.0	0.0		MIT
3358.479	3359.444	29766.83	TI I	1.2	0.0		MIT
3358.277	3359.242	29768.62	TI I	1.5	0.7		MIT
3356.196	3357.161	29787.08	TI I	1.0	0.0		MIT
3355.4	3356.4	29794.	TI		0.7		CX
3354.635	3355.599	29800.94	TI I	2.0	1.3		MIT
3352.937	3353.901	29816.03	TI I	1.4	0.3		MIT
3352.071	3353.034	29823.73	TI II	0.9	1.2		MIT
3351.67	3352.63	29827.3	TI II	0.0	0.7		MIT
3350.548	3351.511	29837.29	TI I	1.0	0.0		MIT
3349.406	3350.369	29847.46	TI II	2.0	2.6G		MIT
3349.035	3349.998	29850.77	TI II	2.1	2.9G		MIT
3348.844	3349.807	29852.47	TI II	1.1	1.1		MIT
3348.535	3349.498	29855.22	TI I	0.8			MIT
3346.728	3347.690	29871.34	TI II	1.8	1.8		M
3344.931	3345.893	29887.39	TI I	0.9			MIT
3344.630	3345.592	29890.08	TI I	0.3			MIT
3343.770	3344.731	29897.77	TI II	1.8	1.8		MIT
3343.379	3344.340	29901.26	TI I	0.3			MIT
3342.707	3343.668	29907.27	TI I	1.0	0.7		MIT
3342.151	3343.112	29912.25	TI I	1.1	0.9		MIT
3341.875	3342.836	29914.72	TI II	2.0	2.5G		MIT
3341.875	3342.836	29914.72	TI I	2.0	2.5G		MIT
3341.554	3342.515	29917.59	TI I	0.3			MIT
3340.344	3341.304	29928.43	TI II	1.9	2.0		MIT
3339.54	3340.50	29935.6	TI I	0.3			MIT
3338.813	3339.773	29942.15	TI I	0.8			MIT
3337.853	3338.813	29950.76	TI II	1.1	1.8		MIT
3336.998	3337.958	29958.44	TI II		0.8		MIT
3336.969	3337.929	29958.70	TI I	0.6			MIT
3335.195	3336.154	29974.63	TI II	1.8	2.2		MIT
3334.870	3335.829	29977.55	TI I	0.3H			MIT
3333.912	3334.871	29986.17	TI I	0.6			MIT
3333.031	3333.990	29994.09	TI I	0.7W			MIT
3332.112	3333.070	30002.37	TI II	1.6	2.1		MIT
3329.456	3330.414	30026.30	TI II	1.9	2.3F		MIT
3328.326	3329.283	30036.49	TI I	0.3			MIT
3327.325	3328.282	30045.53	TI I	0.3			MIT
3326.765	3327.722	30050.59	TI II	1.2	2.1		MIT
3326.639	3327.596	30051.73	TI I	0.5			MIT
3325.365	3326.322	30063.24	TI I	0.3			MIT
3325.229	3326.186	30064.47	TI I	0.6			MIT
3325.155	3326.112	30065.14	TI I	0.6			MIT
3324.754	3325.711	30068.76	TI I	1.0	0.0		MIT
3324.606	3325.562	30070.10	TI I	0.3			MIT
3323.896	3324.852	30076.52	TI I	0.5			MIT
3323.803	3324.759	30077.37	TI I	0.7			MIT
3323.660	3324.616	30078.66	TI I	0.3H			MIT
3322.937	3323.893	30085.20	TI II	1.9	2.5G		MIT
3321.700	3322.656	30096.41	TI II	1.2	2.1		MIT
3321.588	3322.544	30097.42	TI I	1.2	1.2		MIT
3319.083	3320.038	30120.14	TI II	0.5	0.6		MIT
3318.362	3319.317	30126.68	TI I	1.1	0.8		MIT
3318.025	3318.980	30129.74	TI II	1.8	2.1		MIT
3316.389	3317.343	30144.60	TI I	0.7			MIT
3315.785	3316.739	30150.09	TI I	0.7			MIT
3315.324	3316.278	30154.29	TI II	1.1	2.0		MIT
3315.237	3316.191	30155.08	TI I	0.8			MIT
3314.523	3315.477	30161.57	TI I	1.0	0.7		MIT
3314.423	3315.377	30162.48	TI I	1.6	1.3		MIT
3312.690	3313.643	30178.26	TI I	1.2	0.6		MIT
3309.730	3310.683	30205.25	TI I	1.1	0.3		MIT
3309.501	3310.454	30207.34	TI I	1.8	1.4		MIT
3308.806	3309.758	30213.69	TI II	1.5	2.0		MIT
3308.391	3309.343	30217.48	TI I	1.7	1.0		MIT
3307.717	3308.669	30223.63	TI II	0.8	1.1		MIT
3306.879	3307.831	30231.29	TI I	1.5	1.1		MIT
3306.053	3307.005	30238.84	TI II		1.0		MIT
3305.752	3306.704	30241.60	TI I	0.5			MIT
3305.296	3306.248	30245.77	TI I	0.8			MIT
3302.096	3303.047	30275.08	TI II	0.9	1.3		MIT
3299.413	3300.363	30299.70	TI I	1.7	1.5		MIT
3297.787	3298.737	30314.64	TI I	0.8			MIT
3297.253	3298.203	30319.54	TI I	0.5			MIT
3297.095	3298.044	30321.00	TI I	0.6			MIT
3296.373	3297.322	30327.64	TI	0.8			MIT

AIR WAVELENGTH (ANGSTRMS)	VACUUM WAVELENGTH (ANGSTRMS)	WAVENUMBER (KAYSERS)	LMENT	LOG ARC	INTENSITY SPK	INTENSITY DIS	REF
3296.218	3297.167	30329.06	TI I	0.3			MIT
3295.655	3296.604	30334.25	TI I	0.7			MIT
3294.903	3295.852	30341.17	TI I	1.3	0.6		MIT
3292.075	3293.023	30367.23	TI I	1.8	1.6		MIT
3291.075	3292.023	30376.46	TI I	0.3			MIT
3288.897	3289.844	30396.57	TI I	0.3			MIT
3288.575	3289.522	30399.55	TI II	1.1	1.3		MIT
3288.428	3289.375	30400.91	TI II	0.8	0.9		MIT
3288.142	3289.089	30403.55	TI II	1.0	0.8		MIT
3288.142	3289.089	30403.55	TI I	1.0	0.8		MIT
3287.655	3288.602	30408.06	TI II	1.6	2.3		MIT
3286.756	3287.703	30416.37	TI II	1.0	1.0		MIT
3285.254	3286.200	30430.28	TI I	0.8			MIT
3285.052	3285.998	30432.15	TI I	0.6			MIT
3283.950	3284.896	30442.36	TI I	0.8			MIT
3282.329	3283.275	30457.40	TI II	1.5	2.2		MIT
3280.391	3281.336	30475.39	TI I	0.5			MIT
3279.995	3280.940	30479.07	TI II	1.0	1.6		MIT
3278.989	3279.934	30488.42	TI II		0.3		MIT
3278.922	3279.867	30489.04	TI II	1.6	2.2		MIT
3278.290	3279.235	30494.92	TI II	1.4	2.0		MIT
3277.833	3278.778	30499.17	TI I	0.3			MIT
3276.998	3277.942	30506.94	TI II	0.7	0.7		MIT
3276.774	3277.718	30509.03	TI II	1.1	1.8		MIT
3275.293	3276.237	30522.82	TI II	0.9	1.7		MIT
3274.047	3274.991	30534.44	TI I	0.8	0.0		MIT
3272.080	3273.023	30552.79	TI II	1.4	2.0		MIT
3271.652	3272.595	30556.79	TI II	1.5	2.1		MIT
3270.562	3271.505	30566.97	TI I	1.0	0.0		MIT
3268.61	3269.55	30585.2	TI I	0.6			MIT
3267.41	3268.35	30596.5	TI I	0.3			MIT
3267.067	3268.009	30599.67	TI I	1.0			MIT
3265.480	3266.421	30614.54	TI I	0.8	0.0		MIT
3263.845	3264.786	30629.88	TI I	0.3	0.8		MIT
3263.686	3264.627	30631.37	TI II	1.0	1.8		MIT
3261.605	3262.545	30650.91	TI	1.8	2.5R		ME
3260.259	3261.199	30663.57	TI II	1.1	1.5		M
3260.259	3261.199	30663.57	TI I	1.1	1.5		M
3259.42	3260.36	30671.5	TI I	0.3			MIT
3254.250	3255.189	30720.19	TI II	1.5	2.1		MIT
3252.914	3253.852	30732.80	TI II	1.8	2.3R		MIT
3251.911	3252.849	30742.28	TI II	1.7	2.2		MIT
3249.370	3250.307	30766.32	TI II	0.9	1.3		MIT
3248.602	3249.539	30773.59	TI I	1.4	2.3R		MIT
3248.602	3249.539	30773.59	TI II	1.4	2.3R		MIT
3243.803	3244.739	30819.12	TI I	1.0	0.0		MIT
3243.513	3244.449	30821.88	TI I	0.8	0.0		MIT
3241.986	3242.921	30836.39	TI II	1.8	2.5G		MIT
3240.71	3241.65	30848.5	TI II	0.5	0.5		MIT
3239.664	3240.599	30858.49	TI II	1.4	1.9		MIT
3239.038	3239.973	30864.46	TI II	1.8	2.5G		MIT
3238.224	3239.159	30872.22	TI I	0.9	0.6		MIT
3236.573	3237.507	30887.96	TI II	1.8	2.5R		MIT
3236.223	3237.157	30891.30	TI I	0.3			MIT
3236.122	3237.056	30892.27	TI II	1.0	1.2		MIT
3234.516	3235.450	30907.61	TI II	2.0	2.7R		MIT
3232.791	3233.724	30924.10	TI I	0.9			MIT
3232.280	3233.213	30928.99	TI II	1.5	2.0		MIT
3231.315	3232.248	30938.22	TI II	1.2	1.4		MIT
3229.423	3230.355	30956.35	TI II	1.2	1.8		MIT
3229.193	3230.125	30958.55	TI II	1.5	1.8		MIT
3228.605	3229.537	30964.19	TI II	1.5	2.0		MIT
3228.183	3229.115	30968.24	TI I	0.8			MIT
3227.95	3228.88	30970.5	TI I	0.3H			MIT
3226.771	3227.703	30981.79	TI II	1.0	1.5		MIT
3226.50	3227.43	30984.4	TI I	0.3H			MIT
3226.240	3227.171	30986.89	TI I	0.3			MIT
3226.128	3227.059	30987.96	TI I	1.4	0.8		MIT
3224.241	3225.172	31006.10	TI II	1.2	2.2		MIT
3223.519	3224.450	31013.04	TI I	1.2	0.3		MIT

AIR WAVELENGTH (ANGSTRMS)	VACUUM WAVELENGTH (ANGSTRMS)	WAVENUMBER (KAYSERS)	LMENT	LOG ARC	INTENSITY SPK	INTENSITY DIS	REF
3222.842	3223.773	31019.56	TI II	1.3	2.2R		MIT
3222.741	3223.672	31020.53	TI I	0.5			MIT
3221.381	3222.311	31033.62	TI I	1.4	0.8		MIT
3221.151	3222.081	31035.84	TI I	0.6			MIT
3220.467	3221.397	31042.43	TI II		1.4		MIT
3220.277	3221.207	31044.26	TI I	0.5H			MIT
3219.212	3220.142	31054.53	TI I	1.2	0.5		MIT
3218.983	3219.913	31056.74	TI I	0.3H			MIT
3218.683	3219.613	31059.64	TI I	0.3			MIT
3218.270	3219.199	31063.62	TI II	1.2	2.2		MIT
3217.942	3218.871	31066.79	TI I	1.1	0.0		MIT
3217.060	3217.989	31075.31	TI II	1.6	2.2		MIT
3216.899	3217.828	31076.86	TI	0.3			MIT
3216.203	3217.132	31083.59	TI I	0.9			MIT
3214.750	3215.679	31097.64	TI II	1.3	1.9		MIT
3214.243	3215.171	31102.54	TI I	1.5	0.8		MIT
3213.145	3214.073	31113.17	TI I	1.4	1.4		MIT
3213.145	3214.073	31113.17	TI II	1.4	1.4		MIT
3211.07	3212.00	31133.3	TI I	0.3			MIT
3209.030	3209.957	31153.06	TI I	0.9	0.0		MIT
3208.607	3209.534	31157.17	TI II		1.3J		MIT
3207.897	3208.824	31164.07	TI I	1.0	0.3		MIT
3207.327	3208.254	31169.60	TI I	0.9	0.3		MIT
3206.825	3207.752	31174.48	TI I	1.0	0.3		MIT
3206.344	3207.270	31179.16	TI I	1.0	0.5		MIT
3205.990	3206.916	31182.60	TI II	0.5	1.2		MIT
3205.848	3206.774	31183.98	TI I	1.0	0.0		MIT
3205.168	3206.094	31190.60	TI I	0.6	0.0		MIT
3204.870	3205.796	31193.50	TI I	1.2	0.7		MIT
3203.828	3204.754	31203.65	TI I	1.6	0.8		MIT
3203.435	3204.361	31207.47	TI II	1.0	1.2		MIT
3202.538	3203.463	31216.21	TI II	1.4	2.3		MIT
3201.594	3202.519	31225.42	TI I	1.1	0.0		MIT
3199.915	3200.840	31241.80	TI I	2.3	2.2		MIT
3198.726	3199.651	31253.41	TI I	0.5			MIT
3197.518	3198.442	31265.22	TI II	1.1	1.6		MIT
3195.994	3196.918	31280.13	TI II	0.3	0.3H		MIT
3195.717	3196.641	31282.84	TI II	0.7	1.3		MIT
3194.76	3195.68	31292.2	TI II		1.6J		MIT
3194.56	3195.48	31294.2	TI II	0.3	0.9		MIT
3194.26	3195.18	31297.1	TI II	0.3	0.8		MIT
3192.68	3193.60	31312.6	TI II		1.3J		MIT
3192.26	3193.18	31316.7	TI II	0.0	0.7		MIT
3191.994	3192.917	31319.32	TI I	2.0	1.3		MIT
3190.874	3191.797	31330.32	TI II	1.6	2.3R		MIT
3189.52	3190.44	31343.6	TI II	0.0	0.6		MIT
3186.454	3187.375	31373.77	TI I	2.2	1.9		MIT
3184.09	3185.01	31397.1	TI II	0.5	1.0		MIT
3184.017	3184.938	31397.79	TI	0.7	1.0		MIT
3183.968	3184.889	31398.27	TI	1.1	0.3		MIT
3182.57	3183.49	31412.1	TI II		1.6J		MIT
3181.84	3182.76	31419.3	TI II		1.7J		MIT
3180.225	3181.145	31435.22	TI II		1.3J		MIT
3179.291	3180.211	31444.46	TI I	1.0	0.3H		MIT
3178.630	3179.549	31451.00	TI II		1.4J		MIT
3175.66	3176.58	31480.4	TI II		1.3J		MIT
3174.80	3175.72	31488.9	TI II		2.0		MIT
3172.731	3173.649	31509.47	TI I	1.1	0.5		MIT
3170.925	3171.842	31527.42	TI I	0.9	0.0		MIT
3168.521	3169.438	31551.34	TI II	1.8	2.5R		MIT
3164.910	3165.826	31587.33	TI	0.0	1.7		BH
3162.570	3163.485	31610.70	TI II	1.7	2.3R		MIT
3161.774	3162.689	31618.66	TI II	1.5	2.2		MIT
3161.205	3162.120	31624.35	TI II	1.5	2.1		MIT
3157.668	3158.582	31659.77	TI I	0.5			MIT
3157.397	3158.311	31662.49	TI II	0.9	1.5		MIT
3155.670	3156.584	31679.82	TI II	1.4	2.1		MIT
3154.195	3155.108	31694.63	TI II	1.3	2.0		MIT
3153.601	3154.514	31700.60	TI	0.9	0.3		MIT
3152.251	3153.164	31714.18	TI II	1.5	2.1		MIT

AIR WAVELENGTH (ANGSTRMS)	VACUUM WAVELENGTH (ANGSTRMS)	WAVENUMBER (KAYSERS)	LMENT	LOG ARC	INTENSITY SPK	INTENSITY DIS	REF
3148.036	3148.948	31756.64	TI II	1.4	2.2		MIT
3147.425	3148.337	31762.80	TI I	0.5	0.0		MIT
3147.268	3148.180	31764.39	TI I	1.0	0.3		MIT
3146.260	3147.171	31774.57	TI I	0.3H	0.0		MIT
3145.515	3146.426	31782.09	TI I	0.3			MIT
3145.402	3146.313	31783.23	TI II	0.5	1.2		MIT
3144.730	3145.641	31790.02	TI II	0.6	1.3		MIT
3143.756	3144.667	31799.87	TI II	1.3	2.1		MIT
3143.350	3144.261	31803.98	TI I	0.6H			MIT
3141.670	3142.580	31820.99	TI I	1.2	0.8		MIT
3141.537	3142.447	31822.33	TI I	1.5	1.1		MIT
3139.87	3140.78	31839.2	TI I	0.3			MIT
3138.885	3139.794	31849.22	TI I	0.7			MIT
3138.632	3139.541	31851.79	TI I	0.3	0.3		MIT
3137.352	3138.261	31864.78	TI I	0.5			MIT
3136.028	3136.937	31878.23	TI I	0.9			MIT
3135.069	3135.977	31887.98	TI I	0.3			MIT
3134.654	3135.562	31892.21	TI I	0.3			MIT
3133.135	3134.043	31907.67	TI I	0.6			MIT
3132.707	3133.615	31912.03	TI I	0.3			MIT
3130.800	3131.707	31931.46	TI II	1.4	2.0		MIT
3130.376	3131.283	31935.79	TI I	0.3	0.0		MIT
3130.175	3131.082	31937.84	TI I	0.3			MIT
3129.624	3130.531	31943.46	TI I	0.3			MIT
3129.075	3129.982	31949.07	TI I	1.2			MIT
3128.640	3129.547	31953.51	TI II	1.1	1.8J		MIT
3128.640	3129.547	31953.51	TI I	1.1	1.8J		MIT
3127.883	3128.790	31961.24	TI II	0.8	1.5J		MIT
3127.883	3128.790	31961.24	TI I	0.8	1.5J		MIT
3127.684	3128.591	31963.27	TI I	0.3H	0.3		MIT
3127.234	3128.140	31967.87	TI I	0.5			MIT
3125.656	3126.562	31984.01	TI I	0.7			MIT
3125.553	3126.459	31985.07	TI I	0.3			MIT
3123.769	3124.675	32003.33	TI I	1.2	0.5		MIT
3123.074	3123.979	32010.45	TI I	1.5	1.2		MIT
3122.065	3122.970	32020.80	TI II	0.3	1.7		MIT
3121.599	3122.504	32025.58	TI II	0.6	1.3		MIT
3120.212	3121.117	32039.81	TI I	0.7			MIT
3119.99	3120.89	32042.1	TI I	0.3			MIT
3119.800	3120.705	32044.04	TI II	0.6	2.2		MIT
3119.725	3120.630	32044.82	TI I	1.3	1.2		MIT
3118.824	3119.728	32054.07	TI II	0.3	1.2		MIT
3118.130	3119.034	32061.21	TI I	1.0	0.0		MIT
3117.899	3118.803	32063.58	TI I	0.6			MIT
3117.669	3118.573	32065.95	TI II	1.2	2.3		MIT
3117.455	3118.359	32068.15	TI I	0.9			MIT
3115.088	3115.991	32092.51	TI II	0.0	1.1		MIT
3114.092	3114.995	32102.78	TI I	1.2	0.3H		MIT
3112.482	3113.385	32119.38	TI I	1.0	0.3		MIT
3112.050	3112.953	32123.84	TI II	0.8	1.8		MIT
3111.283	3112.185	32131.76	TI I	1.2	0.9		MIT
3110.620	3111.522	32138.61	TI II	0.9	1.3		MIT
3110.095	3110.997	32144.03	TI II	0.5	1.5		MIT
3109.581	3110.483	32149.35	TI	1.0	0.0		MIT
3108.927	3109.829	32156.11	TI II	0.3	1.2		MIT
3107.468	3108.370	32171.21	TI I	0.7H	0.3		MIT
3106.806	3107.707	32178.06	TI I	1.1	0.3		MIT
3106.234	3107.135	32183.99	TI II	1.4	2.2		MIT
3105.220	3106.121	32194.50	TI I	0.3H			MIT
3105.084	3105.985	32195.91	TI II	1.1	2.0		MIT
3104.593	3105.494	32201.00	TI II	0.5	1.2		MIT
3103.804	3104.705	32209.18	TI II	1.3	2.3		MIT
3102.975	3103.875	32217.79	TI II	0.3	0.3		MIT
3102.517	3103.417	32222.54	TI I	0.7			MIT
3101.526	3102.426	32232.84	TI I	0.9	1.1		MIT
3100.666	3101.566	32241.78	TI I	1.5	1.2		MIT
3097.626	3098.525	32273.42	TI II	0.0	0.7		MIT
3097.186	3098.085	32278.00	TI II	1.3	2.2		MIT
3096.424	3097.323	32285.95	TI II	0.5	1.5		MIT
3093.813	3094.711	32313.19	TI I	0.7			MIT

AIR WAVELENGTH (ANGSTRMS)	VACUUM WAVELENGTH (ANGSTRMS)	WAVENUMBER (KAYSERS)	LMENT	LOG ARC	INTENSITY SPK	INTENSITY DIS	REF
3090.137	3091.034	32351.63	TI I	1.1	0.3		MIT
3090.051	3090.948	32352.53	TI II		2.0J		MIT
3089.401	3090.298	32359.34	TI II	1.1	2.0		MIT
3088.025	3088.922	32373.76	TI II	1.8	2.7G		MIT
3086.825	3087.721	32386.34	TI I	0.5			MIT
3085.018	3085.914	32405.31	TI I	0.3	0.0		MIT
3084.819	3085.715	32407.40	TI I	1.0	0.3		MIT
3082.615	3083.510	32430.57	TI I	0.8			MIT
3081.575	3082.470	32441.52	TI II		1.6J		MIT
3080.184	3081.079	32456.17	TI I	0.3	0.9J		MIT
3078.645	3079.539	32472.39	TI II	1.8	2.7G		MIT
3075.224	3076.117	32508.51	TI II	1.6	2.5G		MIT
3072.971	3073.864	32532.34	TI II	1.5	2.3R		MIT
3072.107	3073.000	32541.49	TI II	1.4	2.1		MIT
3071.242	3072.134	32550.66	TI II	1.1	1.8		MIT
3066.514	3067.405	32600.84	TI II	0.8	1.3		MIT
3066.354	3067.245	32602.54	TI II	1.2	1.6		MIT
3066.220	3067.111	32603.97	TI II	1.4	1.3		MIT
3063.502	3064.392	32632.90	TI II	0.7	1.4		MIT
3063.280	3064.170	32635.26	TI II	0.0	0.9H		MIT
3060.452	3061.342	32665.42	TI	0.8	0.5		MIT
3059.741	3060.631	32673.01	TI II	0.9	1.5		MIT
3058.090	3058.979	32690.65	TI II	1.1	1.8		MIT
3057.395	3058.284	32698.08	TI II	0.7	1.2		MIT
3056.740	3057.629	32705.08	TI II	1.1	1.8		MIT
3048.766	3049.653	32790.62	TI II	0.3	1.5		MIT
3046.685	3047.571	32813.01	TI II	1.0	1.8		MIT
3045.085	3045.971	32830.25	TI	0.3	1.7J		MIT
3043.851	3044.737	32843.56	TI II	0.5	1.7		MIT
3042.540	3043.425	32857.71	TI I	0.7	0.3H		MIT
3038.706	3039.590	32899.17	TI II	0.3	1.6		MIT
3036.784	3037.668	32919.99	TI II	0.3	1.2		MIT
3029.730	3030.612	32996.63	TI II	1.1	2.2		MIT
3025.060	3025.941	33047.57	TI I	0.5			MIT
3023.86	3024.74	33060.7	TI II		2.0J		MIT
3022.820	3023.700	33072.06	TI II		2.2V		MIT
3017.187	3018.066	33133.80	TI II	1.2	2.3		MIT
3015.541	3016.420	33151.89	TI	0.6H			MIT
3010.141	3011.018	33211.36	TI I	0.3			MIT
3008.322	3009.199	33231.44	TI II	0.0	1.4		MIT
3007.487	3008.363	33240.66	TI	0.3H			MIT
3005.357	3006.233	33264.22	TI I	0.5			MIT
3003.637	3004.513	33283.27	TI I	0.3			MIT
3002.728	3003.603	33293.35	TI I	1.0	0.5		MIT
3000.868	3001.743	33313.98	TI I	1.3	1.3		MIT
2999.777	3000.652	33326.10	TI I	0.3			MIT
2995.752	2996.626	33370.87	TI	1.0H	1.8J		MIT
2993.064	2993.937	33400.84	TI I	0.3			MIT
2991.789	2992.662	33415.07	TI I	0.3			MIT
2990.989	2991.861	33424.01	TI I	0.8			MIT
2990.497	2991.369	33429.51	TI I	0.8			MIT
2990.16	2991.03	33433.3	TI II		1.9V		MIT
2990.046	2990.918	33434.55	TI I	0.8			MIT
2989.910	2990.782	33436.07	TI I	0.3			MIT
2985.477	2986.348	33485.72	TI I	0.7			MIT
2983.306	2984.176	33510.08	TI I	1.4	1.0		MIT
2981.448	2982.318	33530.96	TI I	1.3	0.9		MIT
2980.296	2981.166	33543.92	TI I	0.3			MIT
2979.199	2980.068	33556.28	TI II		2.0J		MIT
2977.80	2978.67	33572.0	TI		1.7J		MIT
2976.315	2977.184	33588.79	TI I	0.9	0.3		MIT
2974.934	2975.802	33604.38	TI I	1.2	0.7		MIT
2970.556	2971.423	33653.91	TI I	1.0	0.6		MIT
2970.384	2971.251	33655.85	TI I	1.5	0.9		MIT
2969.364	2970.231	33667.42	TI I	0.5	0.0		MIT
2968.231	2969.098	33680.27	TI I	1.1	0.3		MIT
2967.225	2968.091	33691.68	TI I	1.5	0.9		MIT
2966.390	2967.256	33701.17	TI I	0.5			MIT
2965.707	2966.573	33708.93	TI I	1.2	0.7		MIT
2965.68	2966.55	33709.2	TI I	0.8	0.3		MIT

AIR WAVELENGTH (ANGSTRMS)	VACUUM WAVELENGTH (ANGSTRMS)	WAVENUMBER (KAYSERS)	LMENT	LOG ARC	INTENSITY SPK	INTENSITY DIS	REF
2965.231	2966.097	33714.34	TI I	1.4	0.8		BH
2961.463	2962.328	33757.23	TI I	0.7			MIT
2959.993	2960.858	33774.00	TI I	1.3	0.3		MIT
2959.708	2960.573	33777.25	TI I	1.2			MIT
2958.99	2959.85	33785.5	TI II		2.2J		MIT
2958.284	2959.148	33793.51	TI II		0.8J		MIT
2956.797	2957.661	33810.50	TI I	1.5	0.9		MIT
2956.131	2956.995	33818.12	TI I	2.1	1.4		MIT
2954.76	2955.62	33833.8	TI II		2.2J		MIT
2954.583	2955.446	33835.84	TI	0.3			MIT
2952.081	2952.944	33864.51	TI II		1.4J		MIT
2948.255	2949.117	33908.46	TI I	2.0	1.5		MIT
2947.719	2948.581	33914.62	TI I	0.9			MIT
2945.47	2946.33	33940.5	TI II		2.0J		MIT
2945.303	2946.164	33942.44	TI	0.3			MIT
2943.13	2943.99	33967.5	TI II		1.8J		MIT
2941.995	2942.855	33980.60	TI II	2.0	2.2		MIT
2941.995	2942.855	33980.60	TI I	2.0	2.2		MIT
2938.70	2939.56	34018.7	TI II		2.0J		MIT
2938.539	2939.398	34020.57	TI	0.3			MIT
2937.316	2938.175	34034.73	TI I	1.5	0.7		MIT
2936.17	2937.03	34048.0	TI II		2.0J		MIT
2933.549	2934.407	34078.43	TI I	0.8	0.5		MIT
2931.261	2932.119	34105.03	TI II		2.2J		MIT
2928.69	2929.55	34135.0	TI		1.6J		EX
2928.344	2929.201	34139.00	TI I	1.6	0.8		MIT
2927.888	2928.745	34144.32	TI II		0.7J		MIT
2926.75	2927.61	34157.6	TI II		1.7J		EX
2922.924	2923.779	34202.31	TI I	1.0			MIT
2921.52	2922.38	34218.7	TI I	2.3G	2.0G		FL
2918.78	2919.63	34250.9	TI II		1.0J		MIT
2916.71	2917.56	34275.2	TI II		0.7		MIT
2916.10	2916.95	34282.3	TI		1.7J		MIT
2913.33	2914.18	34314.9	TI		1.7J		MIT
2913.109	2913.962	34317.54	TI II	0.5	0.7		MIT
2912.487	2913.340	34324.87	TI I	0.5			MIT
2912.082	2912.935	34329.64	TI I	1.5	1.2		MIT
2910.77	2911.62	34345.1	TI II		0.8J		MIT
2909.925	2910.777	34355.08	TI II	0.8	1.2		MIT
2909.46	2910.31	34360.6	TI II		0.6H		MIT
2908.14	2908.99	34376.2	TI		1.4J		MIT
2906.68	2907.53	34393.4	TI		2.0J		MIT
2905.660	2906.511	34405.51	TI	1.5	0.3		MIT
2903.188	2904.039	34434.80	TI I	0.3			MIT
2901.94	2902.79	34449.6	TI II	0.3	0.7		MIT
2892.78	2893.63	34558.7	TI	0.5			MIT
2891.066	2891.914	34579.18	TI II	1.3	1.7		MIT
2890.61	2891.46	34584.6	TI		1.7J		MIT
2888.932	2889.779	34604.72	TI II	1.2	1.4		MIT
2888.63	2889.48	34608.3	TI		1.8J		MIT
2887.453	2888.300	34622.45	TI II	0.5	1.1		MIT
2886.046	2886.892	34639.32	TI	0.9	0.3		MIT
2884.107	2884.953	34662.61	TI II	1.5	2.1		MIT
2883.231	2884.077	34673.14	TI I	0.5			MIT
2881.951	2882.796	34688.54	TI I	0.3			MIT
2880.293	2881.138	34708.51	TI II		1.3		MIT
2877.436	2878.280	34742.97	TI II	1.5	2.0		MIT
2875.79	2876.63	34762.9	TI		1.6J		MIT
2870.04	2870.88	34832.5	TI		2.0J		MIT
2868.742	2869.584	34848.25	TI II	1.2	1.4		MIT
2862.321	2863.162	34926.43	TI II	1.3	1.6		MIT
2861.99	2862.83	34930.5	TI		2.0J		EX
2861.302	2862.142	34938.86	TI II	0.8	1.2		MIT
2860.84	2861.68	34944.5	TI		1.4J		MIT
2860.277	2861.117	34951.38	TI I	0.8			MIT
2858.412	2859.252	34974.19	TI II	1.0	1.3		MIT
2857.81	2858.65	34981.5	TI		1.8J		MIT
2856.620	2857.459	34996.13	TI II	0.5	0.9		MIT
2856.24	2857.08	35000.8	TI II		2.0J		EX
2856.089	2856.928	35002.63	TI	0.5			MIT

AIR WAVELENGTH (ANGSTRMS)	VACUUM WAVELENGTH (ANGSTRMS)	WAVENUMBER (KAYSERS)	LMENT	LOG ARC	INTENSITY SPK	INTENSITY DIS	REF
2855.491	2856.330	35009.96	TI II		0.7J		MIT
2855.132	2855.971	35014.36	TI I	0.3			MIT
2853.933	2854.771	35029.07	TI II	1.1	1.6		MIT
2853.435	2854.273	35035.19	TI I	0.5			MIT
2851.102	2851.940	35063.85	TI II	1.3	1.9		MIT
2846.092	2846.929	35125.57	TI II		1.8J		MIT
2844.09	2844.93	35150.3	TI II		0.5J		MIT
2841.938	2842.774	35176.91	TI II	1.6	2.1		MIT
2839.80	2840.64	35203.4	TI II		2.0J		EX
2836.64	2837.47	35242.6	TI II		2.0J		MIT
2836.612	2837.446	35242.96	TI I	0.7	0.3		MIT
2836.412	2837.246	35245.44	TI I	0.6			MIT
2836.100	2836.934	35249.32	TI I	0.7			MIT
2835.643	2836.477	35255.00	TI I	0.9	0.8		MIT
2834.760	2835.594	35265.98	TI I	0.8			MIT
2834.161	2834.995	35273.44	TI II		0.9J		MIT
2832.26	2833.09	35297.1	TI I	0.3			MIT
2832.160	2832.993	35298.36	TI II	1.2	2.0		MIT
2831.406	2832.239	35307.75	TI I	0.7			MIT
2830.045	2830.878	35324.73	TI I	0.9			MIT
2828.9	2829.7	35339.	TI II		2.2J		EX
2828.150	2828.982	35348.40	TI II	0.3	2.3H		MIT
2828.068	2828.900	35349.43	TI I	1.3			MIT
2827.21	2828.04	35360.1	TI II		1.9J		MIT
2826.385	2827.217	35370.48	TI I	0.3			MIT
2825.400	2826.231	35382.81	TI	1.2			MIT
2825.073	2825.904	35386.90	TI I	0.6			MIT
2823.470	2824.301	35406.99	TI I	0.5			MIT
2821.54	2822.37	35431.2	TI I	0.8			MIT
2821.42	2822.25	35432.7	TI II		1.8J		MIT
2820.365	2821.195	35445.97	TI	0.9	1.2		MIT
2820.00	2820.83	35450.6	TI II		1.8J		MIT
2817.866	2818.696	35477.40	TI II	1.0	2.3		ME
2817.405	2818.235	35483.21	TI I	1.3			MIT
2815.547	2816.376	35506.62	TI II	0.8	1.3		MIT
2814.576	2815.405	35518.87	TI II	0.0	0.3		MIT
2812.982	2813.810	35539.00	TI I	1.5	1.5		ME
2812.044	2812.872	35550.85	TI II	0.5	0.8		MIT
2810.302	2811.130	35572.89	TI II	0.8	2.2		MIT
2809.170	2809.998	35587.22	TI I	1.5			MIT
2806.499	2807.326	35621.09	TI II	0.8	1.3		ME
2805.704	2806.531	35631.18	TI I	1.5	0.3		MIT
2805.01	2805.84	35640.0	TI II		2.3J		EX
2802.500	2803.326	35671.91	TI I	2.0	1.2		MIT
2800.61	2801.44	35696.0	TI II		2.2J		MIT
2790.66	2791.48	35823.2	TI II		1.5J		MIT
2788.02	2788.84	35857.2	TI II		1.8A		MIT
2786.00	2786.82	35883.2	TI II		1.8J		MIT
2784.643	2785.465	35900.65	TI II	0.8	1.3		MIT
2780.56	2781.38	35953.4	TI II		1.8J		MIT
2778.49	2779.31	35980.1	TI II		1.5J		MIT
2771.07	2771.89	36076.5	TI	0.3	0.3		MIT
2765.65	2766.47	36147.2	TI II		1.3J		MIT
2765.21	2766.03	36152.9	TI II	0.3	0.5		MIT
2764.821	2765.638	36158.03	TI II	1.2	1.8		MIT
2763.930	2764.746	36169.68	TI II	0.3	0.6		MIT
2762.244	2763.060	36191.76	TI II	0.7	1.1		MIT
2761.294	2762.110	36204.21	TI II	1.0	1.5		MIT
2758.90	2759.72	36235.6	TI II		1.0J		MIT
2758.34	2759.16	36243.0	TI II		1.2J		MIT
2758.075	2758.890	36246.46	TI I	1.8	0.6		MIT
2757.395	2758.210	36255.40	TI I	1.4	0.3		MIT
2752.881	2753.695	36314.85	TI II		1.7J		MIT
2751.70	2752.51	36330.4	TI II		2.3J		EX
2750.140	2750.953	36351.04	TI	1.5	0.3		MIT
2749.058	2749.871	36365.34	TI I	1.5			MIT
2746.71	2747.52	36396.4	TI II		2.2J		MIT
2744.846	2745.658	36421.14	TI I	1.5			MIT
2742.324	2743.135	36454.64	TI I	1.5	1.6Q		ME
2741.809	2742.620	36461.48	TI I	0.5			MIT

AIR WAVELENGTH (ANGSTRMS)	VACUUM WAVELENGTH (ANGSTRMS)	WAVENUMBER (KAYSERS)	LMENT	LOG ARC	INTENSITY SPK	DIS	REF
2740.867	2741.678	36474.02	TI I	0.8			MIT
2739.807	2740.618	36488.13	TI I	1.7	0.5		MIT
2738.711	2739.521	36502.73	TI II		1.4J		MIT
2736.680	2737.490	36529.82	TI I	1.1			MIT
2735.612	2736.422	36544.08	TI I	1.3	0.6		MIT
2735.289	2736.099	36548.39	TI I	1.5	0.6		MIT
2733.259	2734.068	36575.54	TI I	1.8	1.2		MIT
2732.13	2732.94	36590.6	TI		1.4J		MIT
2731.578	2732.387	36598.04	TI I	1.5	0.0		MIT
2731.131	2731.940	36604.03	TI I	1.3			MIT
2730.981	2731.789	36606.04	TI II		1.6J		MIT
2727.418	2728.226	36653.86	TI I	1.5	0.8		MIT
2725.781	2726.588	36675.87	TI II	0.7	1.5		MIT
2725.069	2725.876	36685.45	TI I	1.5	0.6		MIT
2719.418	2720.224	36761.68	TI II	0.8	1.2		MIT
2717.303	2718.108	36790.29	TI II	0.6	1.5		MIT
2716.254	2717.059	36804.50	TI II	0.7	1.8		MIT
2713.763	2714.567	36838.28	TI II		0.3		MIT
2707.045	2707.848	36929.70	TI II	0.3	1.2		MIT
2701.52	2702.32	37005.2	TI		0.6H		MIT
2698.516	2699.317	37046.41	TI		2.3H		MIT
2688.822	2689.620	37179.97	TI	1.6	0.7		MIT
2685.139	2685.936	37230.96	TI I	1.2	0.0		MIT
2684.803	2685.600	37235.62	TI	1.3	1.3		MIT
2680.95	2681.75	37289.1	TI		1.3H		MIT
2679.926	2680.722	37303.38	TI I	2.0	1.1		MIT
2676.080	2676.875	37356.99	TI I	0.8			MIT
2674.084	2674.879	37384.87	TI	1.3			MIT
2669.598	2670.392	37447.69	TI I	1.8	1.1		MIT
2669.263	2670.057	37452.39	TI I	1.2	0.0		MIT
2668.335	2669.128	37465.41	TI I	0.7			MIT
2661.970	2662.762	37554.99	TI I	1.5	0.7		MIT
2660.640	2661.432	37573.76	TI I	1.1	0.0		MIT
2657.193	2657.984	37622.50	TI I	1.0	0.0		MIT
2656.906	2657.697	37626.57	TI	1.1			MIT
2656.364	2657.155	37634.24	TI	1.1			MIT
2654.927	2655.717	37654.61	TI I	1.3	0.0		MIT
2649.585	2650.374	37730.52	TI	1.1	0.3H		MIT
2649.300	2650.089	37734.58	TI	1.3			MIT
2648.646	2649.435	37743.90	TI I	0.9	0.0		MIT
2646.637	2647.425	37772.55	TI I	1.3	1.2		MIT
2646.11	2646.90	37780.1	TI II		2.3J		MIT
2644.264	2645.052	37806.44	TI I	2.0	1.1		MIT
2642.15	2642.94	37836.7	TI II		2.2J		MIT
2641.099	2641.886	37851.75	TI I	2.2	1.3		MIT
2638.705	2639.491	37886.09	TI II		2.0J		MIT
2636.172	2636.958	37922.49	TI I	0.3			MIT
2635.633	2636.419	37930.24	TI II		1.7J		MIT
2632.418	2633.203	37976.57	TI I	1.7	0.8		MIT
2631.543	2632.328	37989.19	TI I	1.6	0.3		MIT
2619.939	2620.721	38157.44	TI I	1.5	0.8		MIT
2611.485	2612.265	38280.96	TI I	1.2	0.3		MIT
2611.285	2612.065	38283.89	TI I	1.9	1.2		MIT
2608.12	2608.90	38330.4	TI		1.2H		MIT
2605.151	2605.929	38374.03	TI I	2.0	1.1		MIT
2604.878	2605.656	38378.05	TI I	0.8			MIT
2604.096	2604.874	38389.57	TI II	0.6	1.1		MIT
2599.915	2600.692	38451.30	TI I	1.8	1.0		MIT
2596.585	2597.361	38500.61	TI I	1.6	0.6		MIT
2594.629	2595.405	38529.63	TI I	0.7			MIT
2593.642	2594.418	38544.30	TI I	1.3	0.3		MIT
2590.256	2591.031	38594.68	TI I	1.5	0.0		MIT
2586.279	2587.053	38654.02	TI I	0.8			MIT
2583.225	2583.998	38699.72	TI I	1.2	0.0		MIT
2581.716	2582.489	38722.34	TI II	0.6	1.5		MIT
2580.817	2581.590	38735.83	TI	1.5			MIT
2578.886	2579.658	38764.83	TI I	0.6			MIT
2573.934	2574.705	38839.40	TI II	0.0	0.3H		MIT
2573.715	2574.486	38842.71	TI II	0.3	1.0		MIT
2572.651	2573.422	38858.77	TI	1.0	1.6		MIT

AIR WAVELENGTH (ANGSTRMS)	VACUUM WAVELENGTH (ANGSTRMS)	WAVENUMBER (KAYSERS)	LMENT	LOG ARC	INTENSITY SPK	DIS	REF
2571.034	2571.804	38883.21	TI II	1.3	1.8		MIT
2568.971	2569.741	38914.43	TI II	0.5	1.1		MIT
2555.987	2556.754	39112.10	TI I	1.2	1.9		MIT
2541.917	2542.680	39328.58	TI I	1.4	0.3		MIT
2535.871	2536.633	39422.34	TI II	1.3	1.8		MIT
2534.621	2535.383	39441.78	TI II	1.4	1.9		MIT
2531.251	2532.012	39494.29	TI II	1.5	2.1		MIT
2529.854	2530.615	39516.09	TI I	1.4	0.3		MIT
2529.854	2530.615	39516.09	TI II	1.4	0.3		MIT
2527.985	2528.745	39545.31	TI I	1.2			MIT
2525.601	2526.361	39582.63	TI II	1.5	2.1		MIT
2524.639	2525.398	39597.71	TI II	1.2	1.8		MIT
2520.543	2521.301	39662.06	TI I	1.6	0.6		MIT
2519.814	2520.572	39673.53	TI II	0.5	0.8		MIT
2519.037	2519.795	39685.77	TI I	1.2	1.1		MIT
2517.434	2518.192	39711.04	TI II	1.3	1.5		MIT
2517.124	2517.882	39715.93	TI I	0.5	0.3		MIT
2510.910	2511.666	39814.21	TI II		0.3		MIT
2504.535	2505.290	39915.54	TI	1.3	0.0		MIT
2498.949	2499.702	40004.76	TI II		1.5		MIT
2495.6	2496.4	40058.	TI		0.3		CX
2481.517	2482.266	40285.77	TI II	0.3	0.6		MIT
2478.646	2479.395	40332.42	TI II		1.7		MIT
2477.206	2477.954	40355.87	TI II	0.5	0.7		MIT
2474.194	2474.942	40404.99	TI II	0.6	0.9		MIT
2470.996	2471.743	40457.28	TI I	1.2			MIT
2468.578	2469.324	40496.91	TI I	0.5			MIT
2468.369	2469.115	40500.34	TI I	1.0	0.0		MIT
2464.979	2465.724	40556.03	TI I	1.0	0.0		MIT
2463.997	2464.742	40572.19	TI II	0.0	1.1		MIT
2458.025	2458.769	40670.76	TI I	0.7			MIT
2457.826	2458.570	40674.05	TI I	0.7			MIT
2450.440	2451.182	40796.64	TI II	1.0	2.0		MIT
2447.947	2448.689	40838.19	TI II		0.7		MIT
2446.129	2446.870	40868.54	TI I	1.0	0.7		MIT
2442.70	2443.44	40925.9	TI II		0.8		MIT
2440.985	2441.725	40954.65	TI I	1.5			MIT
2440.21	2440.95	40967.7	TI II	0.7	1.5		MIT
2438.310	2439.049	40999.58	TI I	1.0			MIT
2434.105	2434.843	41070.40	TI I	1.0	0.0		MIT
2433.223	2433.961	41085.29	TI I	1.5			MIT
2431.794	2432.532	41109.43	TI I	0.6			MIT
2428.359	2429.096	41167.58	TI I	0.9			MIT
2428.228	2428.965	41169.80	TI I	1.2			MIT
2424.245	2424.981	41237.43	TI I	1.5	0.0		MIT
2423.72	2424.46	41246.4	TI		0.7H		MIT
2421.305	2422.040	41287.50	TI I	1.4	0.0		MIT
2418.359	2419.094	41337.79	TI I	1.3			MIT
2417.51	2418.24	41352.3	TI		0.6		MIT
2411.583	2412.316	41453.93	TI I	1.0			MIT
2411.371	2412.104	41457.58	TI I	0.8			MIT
2384.519	2385.246	41924.40	TI I	0.8			MIT
2380.81	2381.54	41989.7	TI I	0.7	0.6		MIT
2379.69	2380.42	42009.5	TI I	2.0G	2.3G		MIT
2378.147	2378.873	42036.72	TI I	1.0			MIT
2374.60	2375.32	42099.5	TI I	0.5			MIT
2372.25	2372.97	42141.2	TI I	0.5			MIT
2371.95	2372.67	42146.5	TI I	1.0			MIT
2369.28	2370.00	42194.0	TI I	0.6			MIT
2367.97	2368.69	42217.4	TI		0.3W		MIT
2362.70	2363.42	42311.5	TI		0.3W		MIT
2357.82	2358.54	42399.1	TI		0.8		MIT
2357.31	2358.03	42408.3	TI		0.5		MIT
2355.14	2355.86	42447.3	TI II	0.0	0.6		MIT
2354.62	2355.34	42456.7	TI II	0.0	0.5		MIT
2354.07	2354.79	42466.6	TI	0.5	1.1		MIT
2350.66	2351.38	42528.2	TI II	0.3	1.0		MIT
2349.94	2350.66	42541.3	TI	0.5	1.1		MIT
2347.99	2348.71	42576.6	TI II		1.3		MIT
2347.44	2348.16	42586.6	TI II	0.3	0.7		MIT

AIR WAVELENGTH (ANGSTRMS)	VACUUM WAVELENGTH (ANGSTRMS)	WAVENUMBER (KAYSERS)	LMENT	LOG ARC	INTENSITY SPK	DIS	REF
2346.35	2347.07	42606.3	TI	0.3	0.5		MIT
2342.30	2343.02	42680.0	TI	0.6	1.3		MIT
2341.22	2341.94	42699.7	TI	0.6	1.2		MIT
2337.77	2338.49	42762.7	TI		0.3		MIT
2334.53	2335.25	42822.0	TI	0.6	1.0		MIT
2334.33	2335.05	42825.7	TI		0.7		MIT
2325.07	2325.78	42996.3	TI		0.5		MIT
2315.98	2316.69	43165.0	TI I	1.8G			MIT
2314.88	2315.59	43185.5	TI		0.3W		MIT
2314.27	2314.98	43196.9	TI I	0.7			MIT
2308.88	2309.59	43297.7	TI I	0.6			MIT
2305.67	2306.38	43358.0	TI I	1.2	0.3		MIT
2302.728	2303.437	43413.39	TI I	1.3	0.3		MIT
2299.850	2300.558	43467.71	TI I	1.1	0.3		MIT
2294.217	2294.924	43574.43	TI I	0.8			MIT
2293.758	2294.465	43583.15	TI I	0.8			MIT
2291.86	2292.57	43619.2	TI II		0.5		MIT
2286.18	2286.89	43727.6	TI II	0.3	0.8		MIT
2279.96	2280.66	43846.9	TI I	0.7W	0.3		MIT
2276.70	2277.40	43909.7	TI I	1.2	0.3		MIT
2273.28	2273.98	43975.7	TI I	1.0	0.3		MIT
2272.606	2273.308	43988.75	TI I	1.0	0.3		MIT
2272.37	2273.07	43993.3	TI	0.3			MIT
2269.128	2269.830	44056.17	TI II	0.6	1.6		MIT
2268.748	2269.450	44063.55	TI I	0.8			MIT
2267.910	2268.611	44079.83	TI I	0.8			MIT
2264.022	2264.722	44155.52	TI	1.2			MIT
2261.61	2262.31	44202.6	TI II		0.9		MIT
2261.188	2261.888	44210.86	TI II	0.7	1.5		MIT
2253.25	2253.95	44366.6	TI II	0.0	1.1		MIT
2250.07	2250.77	44429.3	TI II	0.3	1.3		MIT
2244.68	2245.38	44536.0	TI I	1.0			MIT
2238.74	2239.44	44654.1	TI I	1.2			MIT
2238.39	2239.08	44661.1	TI		0.6W		MIT
2238.21	2238.90	44664.7	TI	0.7			MIT
2237.80	2238.49	44672.9	TI		0.5		MIT
2233.803	2234.497	44752.80	TI I	1.2	0.3		MIT
2230.945	2231.638	44810.13	TI II	0.8	1.6		MIT
2230.473	2231.166	44819.61	TI I	0.7			MIT
2230.221	2230.914	44824.67	TI I	1.3			MIT
2229.73	2230.42	44834.5	TI	1.0			MIT
2229.254	2229.947	44844.11	TI II	0.3	1.0		MIT
2228.83	2229.52	44852.6	TI II	0.5	1.0		MIT
2227.157	2227.850	44886.33	TI II	0.6	1.5		MIT
2226.786	2227.479	44893.81	TI I	1.0			MIT
2225.124	2225.816	44927.34	TI	1.3			MIT
2223.188	2223.880	44966.46	TI I	1.2			MIT
2221.451	2222.142	45001.62	TI	1.2			MIT
2219.748	2220.439	45036.14	TI I	0.6	0.3		MIT
2218.379	2219.070	45063.93	TI I	0.6			MIT
2201.35	2202.04	45412.5	TI II		0.3		MIT
2199.27	2199.96	45455.4	TI		0.3		MIT
2190.14	2190.82	45644.9	TI II	0.3	1.2		MIT
2187.51	2188.19	45699.8	TI II	0.0	1.2		MIT
2162.69	2163.37	46224.2	TI II	0.6	1.3		MIT
2159.526	2160.204	46291.91	TI II	0.5	1.0		MIT
2159.085	2159.763	46301.37	TI II	0.6	1.3		MIT
2158.29	2158.97	46318.4	TI II	0.5	1.0		MIT
2156.80	2157.48	46350.4	TI II	0.3	1.0		MIT
2155.59	2156.27	46376.4	TI II	0.5	1.2		MIT
2154.70	2155.38	46395.6	TI II	0.5	1.8		MIT
2143.61	2144.29	46635.6	TI	0.7			MIT
2139.25	2139.92	46730.6	TI II	0.0	0.3		MIT
2135.72	2136.39	46807.9	TI II	0.0	0.5		MIT
2134.84	2135.51	46827.1	TI II		0.5		MIT
2133.66	2134.33	46853.0	TI II		0.3		MIT
2133.32	2133.99	46860.5	TI II		0.3		MIT
2132.20	2132.87	46885.1	TI II		0.3		MIT
2130.16	2130.83	46930.0	TI II		0.5		MIT
2126.09	2126.76	47019.8	TI	0.7			MIT

AIR WAVELENGTH (ANGSTRMS)	VACUUM WAVELENGTH (ANGSTRMS)	WAVENUMBER (KAYSERS)	LMENT	LOG ARC	INTENSITY SPK	DIS	REF
2123.54	2124.21	47076.3	TI	0.8			MIT
2121.94	2122.61	47111.8	TI	0.7			MIT
2117.03	2117.70	47221.0	TI	0.7			MIT
2104.36	2105.03	47505.3	TI II		0.5		MIT
2103.04	2103.71	47535.1	TI	0.3	0.3		A
2101.32	2101.99	47574.0	TI		0.3		MIT
2054.53	2055.19	48657.4	TI II	0.6	1.2		MIT
2050.06	2050.72	48763.4	TI	0.7			A
2043.23	2043.89	48926.4	TI II	0.0	0.5		MIT
2041.46	2042.12	48968.8	TI II	0.6	1.1		MIT
9512.4	9515.0	10510.	TL I	1.5			PS
9254.	9257.	10800.	TL II			0.8	EL
9225.	9228.	10840.	TL II			0.9	EL
9216.	9219.	10850.	TL II			0.6	EL
9170.7	9173.2	10901.	TL I	1.3			PS
9136.1	9138.6	10943.	TL I	1.3			PS
9130.	9133.	10950.	TL II			1.8	EL
8673.8	8676.2	11526.	TL II			0.3	EL
8664.1	8666.5	11539.	TL II			1.5	EL
8632.9	8635.3	11580.	TL II			1.2	EL
8445.8	8448.1	11837.	TL II			1.2	EL
8436.9	8439.2	11849.	TL II			0.6	EL
8389.9	8392.2	11916.	TL II			0.9	EL
8376.1	8378.4	11935.	TL I	0.3			PS
8194.0	8196.3	12201.	TL II			0.3	EL
7683.6	7685.7	13011.	TL II			0.3	EL
7144.9	7146.9	13992.	TL II			1.0	EL
7073.9	7075.9	14133.	TL II			1.3	EL
7070.8	7072.7	14139.	TL II			1.0	EL
6968.52	6970.44	14346.3	TL II		0.7		ML
6966.43	6968.35	14350.6	TL II		1.6		ML
6964.10	6966.02	14355.4	TL II		0.7		ML
6950.5	6952.4	14383.	TL II		1.5		EL
6888.8	6890.7	14512.	TL II		1.0		EL
6713.69	6715.54	14890.8	TL I	2.0	1.6		PS
6555.70	6557.51	15249.7	TL II			0.7	EL
6552.63	6554.44	15256.8	TL II	2.3		1.0	EL
6549.77	6551.58	15263.5	TL I	2.5	1.7B		PS
6430.45	6432.23	15546.7	TL II		0.7		EL
6378.32	6380.08	15673.8	TL II			1.0	EL
6289.74	6291.48	15894.5	TL II			1.0	EL
6287.55	6289.29	15900.1	TL II			0.7	EL
6239.46	6241.19	16022.6	TL II		1.2		ML
6238.49	6240.22	16025.1	TL II		1.0		ML
6181.44	6183.15	16173.0	TL II		1.0		M
6179.98	6181.69	16176.8	TL II			2.0	EL
6170.03	6171.74	16202.9	TL II			0.7	EL
6167.38	6169.09	16209.8	TL II			1.3	EL
6156.17	6157.87	16239.4	TL II			0.7	EL
6154.03	6155.73	16245.0	TL II			1.3	EL
6151.02	6152.72	16253.0	TL II			1.0	EL
6149.01	6150.71	16258.3	TL II			0.7	EL
6113.87	6115.56	16351.7	TL II			1.0	EL
6111.50	6113.19	16358.1	TL II			1.3	EL
5950.77	5952.42	16799.9	TL II		0.8		ML
5949.57	5951.22	16803.3	TL II		1.5		ML
5949.04	5950.69	16804.8	TL II	1.3	1.2J		ML
5828.59	5830.21	17152.1	TL II			0.7	EL
5826.58	5828.20	17158.0	TL II			0.7	EL
5774.00	5775.60	17314.2	TL II			0.7	EL
5772.71	5774.31	17318.1	TL II			0.5	EL
5643.00	5644.57	17716.1	TL II			1.8	EL
5640.06	5641.63	17725.4	TL II			0.8D	EL
5583.98	5585.53	17903.4	TL I	1.2			FL
5581.76	5583.31	17910.5	TL			0.5	EL
5527.90	5529.44	18085.0	TL I	1.5			FL
5488.79	5490.32	18213.9	TL I	1.2			FL
5448.77	5450.28	18347.7	TL II			0.8	EL
5446.85	5448.36	18354.1	TL II			0.5	EL
5409.92	5411.42	18479.4	TL II			1.1	EL

AIR WAVELENGTH (ANGSTRMS)	VACUUM WAVELENGTH (ANGSTRMS)	WAVENUMBER (KAYSERS)	LMENT	LOG INTENSITY ARC	SPK	DIS	REF
5384.96	5386.46	18565.1	TL II		0.8		ML
5384.20	5385.70	18567.7	TL II		0.6		ML
5350.46	5351.95	18684.8	TL I	3.7G	3.3G		FL
5206.29	5207.74	19202.2	TL II			0.3	EL
5183.10	5184.54	19288.1	TL II			1.0	EL
5181.95	5183.39	19292.4	TL II			1.0	EL
5152.14	5153.58	19404.0	TL II		1.7W		EL
5143.58	5145.01	19436.3	TL II		0.7		ML
5109.47	5110.89	19566.1	TL I	1.2			FL
5086.99	5088.41	19652.5	TL II		0.6		ML
5078.54	5079.96	19685.2	TL II	1.0	1.5W		EL
5053.14	5054.55	19784.2	TL II		0.6		EL
5052.35	5053.76	19787.2	TL II			0.5	EL
5040.55	5041.96	19833.6	TL II			0.5	EL
4981.35	4982.74	20069.3	TL II			1.2	EL
4946.63	4948.01	20210.1	TL II			0.7	EL
4945.53	4946.91	20214.6	TL II			0.3	EL
4906.3	4907.7	20376.	TL I	0.7			FL
4891.11	4892.48	20439.6	TL I	0.7			FL
4770.85	4772.18	20954.8	TL II		0.7		ML
4770.58	4771.91	20955.9	TL II			0.7	ML
4768.5	4769.8	20965.	TL I	0.3			FL
4765.75	4767.08	20977.2	TL II			0.7	EL
4764.58	4765.91	20982.3	TL II			0.3	EL
4760.6	4761.9	21000.	TL I	0.3			FL
4737.05	4738.38	21104.3	TL II			1.6	EL
4678.1	4679.4	21370.	TL I	0.9			FL
4617.2	4618.5	21652.	TL I	0.7			FL
4574.6	4575.9	21854.	TL I	0.3			FL
4500.03	4501.29	22215.8	TL II			0.9	EL
4490.77	4492.03	22261.6	TL II			1.4	EL
4481.46	4482.72	22307.9	TL II			0.7	EL
4480.80	4482.06	22311.2	TL II			0.7	EL
4358.82	4360.05	22935.5	TL II			0.7	EL
4340.53	4341.75	23032.2	TL II			1.3	EL
4339.55	4340.77	23037.4	TL II			0.9	EL
4313.22	4314.43	23178.0	TL II			0.3	EL
4312.37	4313.58	23182.6	TL II			0.7	EL
4306.80	4308.01	23212.6	TL II			1.6	EL
4291.72	4292.93	23294.1	TL II			0.3	EL
4286.51	4287.72	23322.4	TL II			1.1D	EL
4274.98	4276.18	23385.3	TL II			2.0	EL
4274.24	4275.44	23389.4	TL II			0.9	EL
4272.0	4273.2	23402.	TL			0.9	CX
4258.07	4259.27	23478.2	TL II			0.7	EL
4256.82	4258.02	23485.1	TL II			0.3	EL
4256.09	4257.29	23489.1	TL II			0.3	EL
4224.31	4225.50	23665.8	TL			0.9	EL
4223.05	4224.24	23672.9	TL			1.4	EL
4205.4	4206.6	23772.	TL			0.3	CX
4204.4	4205.6	23778.	TL			0.3	CX
4177.46	4178.64	23931.2	TL II		0.9		EL
4176.75	4177.93	23935.3	TL II			0.3	EL
4111.2	4112.4	24317.	TL			0.3	CX
3955.94	3957.06	25271.3	TL II			0.5	SX
3955.45	3956.57	25274.4	TL II			1.0	SX
3910.05	3911.16	25567.9	TL			0.5	SX
3909.49	3910.60	25571.5	TL II			0.3	SX
3909.15	3910.26	25573.8	TL II			0.5	SX
3901.53	3902.64	25623.7	TL II			1.1	EL
3887.15	3888.25	25718.5	TL II			1.5	EL
3869.81	3870.91	25833.7	TL II			0.8	SX
3869.19	3870.29	25837.9	TL II			1.3	SX
3868.59	3869.69	25841.9	TL II			0.8	SX
3852.10	3853.19	25952.5	TL II			1.0	SX
3851.67	3852.76	25955.4	TL II			0.8	SX
3842.93	3844.02	26014.4	TL II			0.8	EL
3837.49	3838.58	26051.3	TL II			0.8	SX
3836.81	3837.90	26055.9	TL II			1.1	SX
3832.30	3833.39	26086.6	TL II			1.5	EL

AIR WAVELENGTH (ANGSTRMS)	VACUUM WAVELENGTH (ANGSTRMS)	WAVENUMBER (KAYSERS)	LMENT	LOG INTENSITY ARC	SPK	DIS	REF
3793.95	3795.03	26350.3	TL II			1.4	EL
3791.56	3792.64	26366.9	TL II			0.8	EL
3652.95	3653.99	27367.3	TL I	2.2	1.7L		FL
3619.49	3620.52	27620.3	TL II			0.8	EL
3593.61	3594.64	27819.2	TL II			1.0	EL
3567.56	3568.58	28022.4	TL II			1.0	SX
3566.75	3567.77	28028.7	TL II			1.0	SX
3560.77	3561.79	28075.8	TL II			1.4	SX
3560.40	3561.42	28078.7	TL II			1.1	SX
3540.05	3541.06	28240.1	TL II			1.3	SX
3539.78	3540.79	28242.3	TL II			0.8	SX
3536.70	3537.71	28266.9	TL II			0.8	EL
3520.6	3521.6	28396.	TL		0.5		CX
3519.24	3520.25	28407.1	TL I	3.3G	3.0G		FL
3513.80	3514.80	28451.1	TL II			0.5	EL
3460.48	3461.47	28889.4	TL II			1.3D	EL
3453.83	3454.82	28945.1	TL II			1.0	EL
3384.47	3385.44	29538.2	TL II			0.8	EL
3381.80	3382.77	29561.6	TL II			1.1	EL
3381.00	3381.97	29568.6	TL II			1.3	EL
3369.15	3370.12	29672.6	TL II			1.6	EL
3365.48	3366.45	29704.9	TL II			0.6	EL
3364.05	3365.02	29717.5	TL II			0.8	EL
3339.91	3340.87	29932.3	TL II			1.1	EL
3322.30	3323.26	30091.0	TL II			1.4	SX
3321.94	3322.90	30094.2	TL II			0.5	SX
3321.17	3322.13	30101.2	TL II			1.4	SX
3319.91	3320.87	30112.6	TL II			1.5	EL
3291.01	3291.96	30377.1	TL II			1.6	EL
3243.68	3244.62	30820.3	TL II			1.1	EL
3237.96	3238.89	30874.7	TL II			1.1	EL
3229.75	3230.68	30953.2	TL I	3.3	2.9		FL
3187.74	3188.66	31361.1	TL II			1.7	EL
3186.56	3187.48	31372.7	TL II			1.0	EL
3185.51	3186.43	31383.1	TL II			0.8	EL
3121.4	3122.3	32028.	TL		0.5		CX
3120.6	3121.5	32036.	TL		0.5		CX
3107.73	3108.63	32168.5	TL II		0.3		SX
3091.66	3092.56	32335.7	TL II		1.7		ML
3091.61	3092.51	32336.2	TL II		1.4		ML
3029.01	3029.89	33004.5	TL II			1.2	EL
3015.52	3016.40	33152.1	TL II			0.3	EL
3004.71	3005.59	33271.4	TL II			1.0	EL
2987.95	2988.82	33458.0	TL II			1.0	EL
2977.93	2978.80	33570.6	TL I	0.7S			FL
2948.73	2949.59	33903.0	TL II			1.5	EL
2945.04	2945.90	33945.5	TL I	1.7L	1.4L		FL
2928.21	2929.07	34140.6	TL II			1.0	EL
2918.32	2919.17	34256.3	TL I	2.6G	2.3G		FL
2895.41	2896.26	34527.3	TL I	1.5S	1.2S		FL
2853.86	2854.70	35030.0	TL II			0.3	EL
2849.80	2850.64	35079.9	TL II			2.3	EL
2843.27	2844.11	35160.4	TL I	0.7			FL
2833.31	2834.14	35284.0	TL II			1.4	EL
2826.16	2826.99	35373.3	TL I	2.3G	2.0G		FL
2806.80	2807.63	35617.3	TL II			0.9	EL
2801.81	2802.64	35680.7	TL II			1.4	EL
2794.59	2795.41	35772.9	TL II			0.3	EL
2780.25	2781.07	35957.4	TL II			1.3	EL
2768.6	2769.4	36109.	TL		0.3D		CX
2767.87	2768.69	36118.2	TL I	2.6G	2.5G		FL
2719.10	2719.91	36766.0	TL I	0.7			FL
2710.67	2711.47	36880.3	TL I	1.5G	1.0		FL
2709.23	2710.03	36899.9	TL I	2.6G	2.3G		FL
2705.55	2706.35	36950.1	TL II			1.3	EL
2703.36	2704.16	36980.0	TL II			1.0	EL
2700.2	2701.0	37023.	TL I	1.0L			FL
2682.61	2683.41	37266.1	TL II			1.0	EL
2676.03	2676.83	37357.7	TL II			1.5	EL
2675.76	2676.56	37361.5	TL II			1.5	FL

AIR WAVELENGTH (ANGSTRMS)	VACUUM WAVELENGTH (ANGSTRMS)	WAVENUMBER (KAYSERS)	LMENT	LOG ARC	INTENSITY SPK	DIS	REF
2665.57	2666.36	37504.3	TL I	1.0H	0.5		FL
2648.1	2648.9	37752.	TL		0.3		CX
2620.5	2621.3	38149.	TL		1.0D		CX
2618.1	2618.9	38184.	TL		0.3		CX
2609.77	2610.55	38306.1	TL I	1.5G			FL
2608.99	2609.77	38317.6	TL I	1.9G	1.0		FL
2604.32	2605.10	38386.3	TL II			0.7	EL
2604.01	2604.79	38390.8	TL II			0.3	EL
2585.59	2586.36	38664.3	TL I	1.5G	0.3		FL
2584.5	2585.3	38681.	TL		0.3		CX
2580.14	2580.91	38746.0	TL I	2.0G	1.9G		FL
2552.98	2553.75	39158.2	TL I	1.0G			FL
2552.53	2553.30	39165.1	TL I	1.9G	0.0		FL
2548.2	2549.0	39232.	TL		0.3D		CX
2540.4	2541.2	39352.	TL		0.3D		CX
2538.18	2538.94	39386.5	TL I	1.0G	0.7		FL
2531.6	2532.4	39489.	TL		2.0		CX
2530.88	2531.64	39500.1	TL II			1.9	ML
2530.82	2531.58	39501.0	TL II			1.8	ML
2530.67	2531.43	39503.4	TL II			2.0	ML
2517.41	2518.17	39711.4	TL I	1.5G			FL
2507.94	2508.70	39861.4	TL I	0.7G			FL
2507.70	2508.46	39865.2	TL II			1.5	EL
2507.39	2508.15	39870.1	TL II			1.3	EL
2502.3	2503.1	39951.	TL		1.3		CX
2493.91	2494.66	40085.6	TL I	1.0G			FL
2487.48	2488.23	40189.2	TL I	0.7G			FL
2477.40	2478.15	40352.7	TL II			1.3	EL
2475.5	2476.2	40384.	TL		1.0		CX
2472.57	2473.32	40431.5	TL I	0.7G			FL
2469.76	2470.51	40477.5	TL		1.5		SD
2469.16	2469.91	40487.4	TL II		1.0		ML
2469.07	2469.82	40488.8	TL II		1.3		ML
2468.95	2469.70	40490.8	TL II		1.2		ML
2465.46	2466.21	40548.1	TL I	0.7G			FL
2461.93	2462.67	40606.2	TL I	0.7G			FL
2456.45	2457.19	40696.8	TL I	0.7G			FL
2453.79	2454.53	40741.0	TL I	0.7G			FL
2452.70	2453.44	40759.0	TL		2.0		SD
2451.94	2452.68	40771.7	TL II		1.5		ML
2451.72	2452.46	40775.3	TL II		1.0		ML
2449.49	2450.23	40812.5	TL I	0.7G			FL
2443.92	2444.66	40905.5	TL I	0.7G			FL
2442.16	2442.90	40935.0	TL I	0.7R			FL
2439.50	2440.24	40979.6	TL I	0.7G			FL
2434.28	2435.02	41067.5	TL II			0.7	EL
2416.70	2417.43	41366.2	TL I	0.7L			FL
2403.57	2404.30	41592.1	TL			0.7	EL
2395.27	2396.00	41736.2	TL		0.3		SD
2394.62	2395.35	41747.6	TL II		0.5		MIT
2380.34	2381.07	41998.0	TL	1.3	1.8		SD
2365.65	2366.37	42258.8	TL		1.0		SD
2364.97	2365.69	42270.9	TL		0.3		MIT
2362.08	2362.80	42322.6	TL I	1.0S			FL
2356.2	2356.9	42428.	TL		1.0		CX
2298.95	2299.66	43484.7	TL	1.5	2.2		SD
2298.16	2298.87	43499.7	TL II		1.3		ML
2298.08	2298.79	43501.2	TL II		1.6		ML
2297.88	2298.59	43505.0	TL II		1.5		ML
2292.97	2293.68	43598.1	TL II			1.3	EL
2285.95	2286.66	43732.0	TL		0.3		ED
2243.66	2244.36	44556.2	TL		0.3		SD
2238.59	2239.29	44657.1	TL		0.3		SD
2237.85	2238.54	44671.9	TL I	1.8G	0.5		MIT
2210.46	2211.15	45225.4	TL		0.3		SD
2210.26	2210.95	45229.5	TL		0.3		MIT
2207.58	2208.27	45284.4	TL		0.3		SD
2207.00	2207.69	45296.2	TL I	1.5G	0.5		MIT
2168.61	2169.29	46098.0	TL I	1.5G			FL
2152.01	2152.69	46453.6	TL I	0.7G			FL

AIR WAVELENGTH (ANGSTRMS)	VACUUM WAVELENGTH (ANGSTRMS)	WAVENUMBER (KAYSERS)	LMENT	LOG ARC	INTENSITY SPK	DIS	REF
2139.98	2140.65	46714.7	TL		1.0		SD
2139.29	2139.96	46729.7	TL		0.5		MIT
2129.33	2130.00	46948.3	TL I	0.7G			FL
2098.4	2099.1	47640.	TL	0.3G			FL
2068.3	2069.0	48333.	TL		0.3D		CX
2066.8	2067.5	48368.	TL		1.5		CX
2066.4	2067.1	48378.	TL		0.3D		CX
2062.5	2063.2	48469.	TL		0.3D		CX
2061.0	2061.7	48505.	TL		0.3D		CX
2026.2	2026.9	49338.	TL		0.3D		CX
2020.6	2021.3	49474.	TL		0.3D		CX
2012.30	2012.95	49678.3	TL II		0.3		MIT
2004.9	2005.5	49862.	TL		0.3D		CX
8472.01	8474.34	11800.3	TM	1.5			ME
8194.20	8196.45	12200.4	TM	1.3			ME
8017.90	8020.11	12468.7	TM	2.3			ME
8014.78	8016.98	12473.5	TM	1.6			ME
7930.85	7933.03	12605.5	TM	2.0			ME
7927.53	7929.71	12610.8	TM	1.7			ME
7856.10	7858.26	12725.5	TM	1.8			ME
7803.96	7806.11	12810.5	TM	1.2			ME
7731.57	7733.70	12930.4	TM	2.2			ME
7558.36	7560.44	13226.7	TM	2.3			ME
7490.22	7492.28	13347.1	TM	2.0			ME
7481.09	7483.15	13363.4	TM	2.0			ME
7272.61	7274.61	13746.4	TM	1.0			ME
6854.13	6856.02	14585.7	TM	1.3			ME
6845.76	6847.65	14603.6	TM	2.3	1.3		ME
6844.27	6846.16	14606.7	TM	2.4	1.5		ME
6831.10	6832.99	14634.9	TM	1.3			ME
6788.53	6790.40	14726.7	TM	1.3			ME
6782.02	6783.89	14740.8	TM	1.2			ME
6779.77	6781.64	14745.7	TM	2.5	1.7		ME
6721.37	6723.23	14873.8	TM	1.8	1.3		ME
6692.93	6694.78	14937.0	TM	1.2			ME
6658.62	6660.46	15014.0	TM	1.5	1.6		ME
6657.73	6659.57	15016.0	TM	1.8	1.0		ME
6637.31	6639.14	15062.2	TM	1.2			ME
6627.29	6629.12	15085.0	TM	1.0			ME
6604.97	6606.79	15135.9	TM	2.5	1.8		ME
6575.56	6577.38	15203.6	TM	1.3			ME
6566.26	6568.07	15225.2	TM	1.0			ME
6551.19	6553.00	15260.2	TM	1.0			ME
6507.72	6509.52	15362.1	TM	1.2			ME
6460.28	6462.07	15474.9	TM	2.6	1.9		ME
6430.95	6432.73	15545.5	TM	1.2	1.8		ME
6416.33	6418.10	15580.9	TM	1.3			ME
6401.45	6403.22	15617.1	TM	1.6	0.7		ME
6352.64	6354.40	15737.1	TM	1.6	0.6		ME
6299.45	6301.19	15870.0	TM	1.3	1.3		ME
6211.43	6213.15	16094.9	TM	0.9	0.9		ME
6181.41	6183.12	16173.1	TM	1.6	1.7		ME
6024.894	6026.562	16593.21	TM	0.3			MIT
5971.268	5972.922	16742.22	TM	1.0	0.7		MIT
5935.897	5937.542	16841.99	TM	0.7			MIT
5901.577	5903.212	16939.93	TM	0.7			MIT
5899.465	5901.100	16945.99	TM	1.0			MIT
5895.626	5897.260	16957.03	TM	1.9	1.3		MIT
5838.767	5840.386	17122.16	TM	1.3	1.6		MIT
5836.722	5838.340	17128.16	TM	0.7			MIT
5816.51	5818.12	17187.7	TM	1.2			ME
5811.174	5812.785	17203.46	TM	1.0	1.0		MIT
5782.359	5783.963	17289.19	TM	1.0	0.7		MIT
5764.292	5765.891	17343.37	TM	1.7			MIT
5760.211	5761.809	17355.66	TM	1.3	0.7		MIT
5758.01	5759.61	17362.3	TM	1.2	0.7		ME
5738.93	5740.52	17420.0	TM	1.0	1.5		ME
5733.79	5735.38	17435.6	TM	1.0			MIT
5709.97	5711.55	17508.4	TM	1.3	1.5		ME
5696.43	5698.01	17550.0	TM	0.8	1.2		ME

AIR WAVELENGTH (ANGSTRMS)	VACUUM WAVELENGTH (ANGSTRMS)	WAVENUMBER (KAYSERS)	LMENT	LOG ARC	INTENSITY SPK	INTENSITY DIS	REF
5684.75	5686.33	17586.0	TM	1.6	1.9		ME
5675.827	5677.402	17613.69	TM	2.0	2.0		MIT
5658.29	5659.86	17668.3	TM	1.3	0.3		MIT
5631.409	5632.972	17752.62	TM	1.9	1.0		MIT
5589.94	5591.49	17884.3	TM	0.7	1.3		ME
5581.36	5582.91	17911.8	TM	0.7			ME
5565.999	5567.545	17961.24	TM	0.9			MIT
5537.274	5538.812	18054.41	TM	0.5			MIT
5528.342	5529.878	18083.58	TM	0.5	0.3		MIT
5465.533	5467.052	18291.39	TM	1.0			MIT
5461.96	5463.48	18303.4	TM	0.7	1.4		ME
5416.906	5418.412	18455.59	TM	0.3			MIT
5405.99	5407.49	18492.9	TM	0.3	1.0		MIT
5402.233	5403.735	18505.72	TM	0.5			MIT
5400.486	5401.987	18511.71	TM	0.9			MIT
5391.974	5393.473	18540.93	TM	0.7			MIT
5373.003	5374.497	18606.39	TM	0.9	0.7		MIT
5346.477	5347.964	18698.71	TM	1.6	1.4		MIT
5342.383	5343.869	18713.03	TM	0.5			MIT
5307.11	5308.59	18837.4	TM	2.0	1.3		ME
5305.87	5307.35	18841.8	TM	1.3	1.3		ME
5267.34	5268.81	18979.6	TM	1.0	1.0		ME
5254.90	5256.36	19024.6	TM	0.5	1.0		ME
5228.213	5229.669	19121.67	TM	0.9	1.2		MIT
5213.39	5214.84	19176.0	TM	1.0	0.0		ME
5204.51	5205.96	19208.8	TM	0.8	1.0		ME
5185.688	5187.132	19278.48	TM	0.7	1.0		MIT
5185.237	5186.681	19280.15	TM	0.7			MIT
5149.383	5150.818	19414.39	TM	1.0	1.2		MIT
5140.259	5141.691	19448.85	TM	0.7	0.7		MIT
5128.067	5129.496	19495.09	TM	0.3	0.5		MIT
5114.531	5115.956	19546.69	TM	1.0	1.2		MIT
5113.959	5115.384	19548.87	TM	0.7	0.5		MIT
5107.545	5108.968	19573.42	TM	1.5G			MIT
5077.182	5078.597	19690.48	TM	0.3			MIT
5072.43	5073.84	19708.9	TM	0.7			ME
5066.671	5068.084	19731.32	TM	0.5			MIT
5060.894	5062.305	19753.85	TM	1.5	0.7		MIT
5060.413	5061.824	19755.73	TM	0.9	1.0		MIT
5034.212	5035.616	19858.54	TM	2.0	2.0		MIT
5017.864	5019.264	19923.24	TM	0.3	1.0		MIT
5009.758	5011.155	19955.48	TM	1.7	1.7		MIT
4994.706	4996.099	20015.61	TM	0.3			MIT
4993.768	4995.161	20019.37	TM	0.5	0.0		MIT
4989.311	4990.703	20037.26	TM	0.5			MIT
4980.668	4982.058	20072.03	TM	0.7	0.5		MIT
4978.897	4980.286	20079.17	TM	0.3	0.0		MIT
4975.113	4976.501	20094.44	TM	0.9	0.7		MIT
4970.856	4972.243	20111.65	TM	0.7	0.5		MIT
4957.188	4958.571	20167.10	TM	1.7	0.7		MIT
4954.882	4956.265	20176.49	TM	0.5			MIT
4923.83	4925.20	20303.7	TM	1.5			ME
4879.21	4880.57	20489.4	TM	1.0			ME
4835.77	4837.12	20673.4	TM	1.3			ME
4831.21	4832.56	20693.0	TM	1.7	1.9		ME
4826.99	4828.34	20711.1	TM	1.0	1.4		ME
4802.674	4804.016	20815.92	TM	0.7			MIT
4801.415	4802.757	20821.37	TM	1.0	0.7		MIT
4789.90	4791.24	20871.4	TM	1.2	1.0		MIT
4787.97	4789.31	20879.8	TM	0.7			MIT
4783.63	4784.97	20898.8	TM	0.7	0.0		MIT
4781.30	4782.64	20909.0	TM	0.7	1.0		MIT
4763.31	4764.64	20987.9	TM	1.0	1.0		MIT
4759.90	4761.23	21003.0	TM	1.4	0.0		MIT
4754.58	4755.91	21026.5	TM	0.7			MIT
4750.75	4752.08	21043.4	TM	1.2	1.3		MIT
4750.236	4751.564	21045.70	TM	0.5			MIT
4741.10	4742.43	21086.2	TM	0.5			ME
4733.32	4734.64	21120.9	TM	1.9	0.7		MIT
4725.56	4726.88	21155.6	TM	1.0	0.3		MIT

AIR WAVELENGTH (ANGSTRMS)	VACUUM WAVELENGTH (ANGSTRMS)	WAVENUMBER (KAYSERS)	LMENT	LOG ARC	INTENSITY SPK	INTENSITY DIS	REF
4724.25	4725.57	21161.5	TM	1.5	0.3		MIT
4719.26	4720.58	21183.8	TM	1.0			MIT
4717.69	4719.01	21190.9	TM	0.3	0.0		ME
4714.20	4715.52	21206.6	TM	1.0	1.2		MIT
4713.69	4715.01	21208.9	TM	1.0			MIT
4713.32	4714.64	21210.5	TM	1.0			MIT
4711.72	4713.04	21217.7	TM	1.0	0.3		MIT
4710.200	4711.518	21224.58	TM	0.6			MIT
4705.443	4706.760	21246.04	TM	0.5			MIT
4701.55	4702.87	21263.6	TM	1.2			MIT
4698.58	4699.89	21277.1	TM	0.7			MIT
4698.15	4699.46	21279.0	TM	1.0			MIT
4697.63	4698.94	21281.4	TM	0.7			MIT
4694.57	4695.88	21295.2	TM	0.7			MIT
4694.27	4695.58	21296.6	TM	1.0	0.5		MIT
4691.504	4692.817	21309.16	TM	0.7			MIT
4691.10	4692.41	21311.0	TM	1.5	1.3		MIT
4685.10	4686.41	21338.3	TM	1.4			MIT
4681.92	4683.23	21352.8	TM	1.7	0.3		MIT
4677.85	4679.16	21371.4	TM	1.7	0.3		MIT
4677.19	4678.50	21374.4	TM	0.7			MIT
4675.31	4676.62	21383.0	TM	1.5			MIT
4675.08	4676.39	21384.0	TM	1.4			MIT
4674.21	4675.52	21388.0	TM	1.0			MIT
4671.98	4673.29	21398.2	TM	1.2	1.3		MIT
4668.69	4670.00	21413.3	TM	1.0			MIT
4667.28	4668.59	21419.8	TM	1.2			MIT
4666.701	4668.007	21422.42	TM	0.9	0.3		MIT
4655.08	4656.38	21475.9	TM	1.5			MIT
4653.01	4654.31	21485.4	TM	1.4	1.2		MIT
4646.04	4647.34	21517.7	TM	0.7			MIT
4645.83	4647.13	21518.6	TM	1.0			MIT
4644.57	4645.87	21524.5	TM	1.3			ME
4643.11	4644.41	21531.3	TM	1.5			MIT
4642.94	4644.24	21532.1	TM	1.4			MIT
4642.58	4643.88	21533.7	TM	1.0			MIT
4642.29	4643.59	21535.1	TM	0.7			MIT
4640.79	4642.09	21542.0	TM	0.3			MIT
4634.24	4635.54	21572.5	TM	1.9	1.0		MIT
4633.94	4635.24	21573.9	TM	1.2			MIT
4632.29	4633.59	21581.6	TM	1.0			MIT
4631.73	4633.03	21584.2	TM	1.0			MIT
4626.96	4628.26	21606.4	TM	1.4			MIT
4626.55	4627.85	21608.3	TM	1.7	0.5		MIT
4626.31	4627.61	21609.4	TM	1.7	1.3		MIT
4625.572	4626.867	21612.90	TM	0.5			MIT
4624.39	4625.69	21618.4	TM	1.0			MIT
4621.71	4623.00	21630.9	TM	1.2			MIT
4620.98	4622.27	21634.4	TM	1.0			MIT
4619.64	4620.93	21640.6	TM	1.0			MIT
4619.058	4620.352	21643.37	TM	1.2	1.0		MIT
4617.69	4618.98	21649.8	TM	1.0	1.1		MIT
4615.93	4617.22	21658.0	TM	2.3	2.5		MIT
4614.46	4615.75	21664.9	TM	1.5	1.2		MIT
4613.96	4615.25	21667.3	TM	1.2			MIT
4613.19	4614.48	21670.9	TM	1.2	1.3		MIT
4609.34	4610.63	21689.0	TM	0.7			MIT
4607.887	4609.178	21695.84	TM	0.3			MIT
4605.22	4606.51	21708.4	TM	1.0			MIT
4604.863	4606.153	21710.09	TM	0.7			MIT
4603.42	4604.71	21716.9	TM	1.5	1.4		MIT
4603.21	4604.50	21717.9	TM	1.0			MIT
4601.27	4602.56	21727.0	TM	1.4			MIT
4600.567	4601.856	21730.36	TM	0.3			MIT
4599.00	4600.29	21737.8	TM	1.9			MIT
4597.20	4598.49	21746.3	TM	1.2			MIT
4596.62	4597.91	21749.0	TM	1.4			MIT
4591.53	4592.82	21773.1	TM	1.0			MIT
4588.06	4589.35	21789.6	TM	0.7			MIT
4586.35	4587.64	21797.7	TM	1.0			MIT

AIR WAVELENGTH (ANGSTRMS)	VACUUM WAVELENGTH (ANGSTRMS)	WAVENUMBER (KAYSERS)	LMENT	LOG ARC	INTENSITY SPK	INTENSITY DIS	REF
4578.80	4580.08	21833.7	TM	1.0			MIT
4576.62	4577.90	21844.1	TM	1.0			MIT
4573.69	4574.97	21858.1	TM	0.7			MIT
4573.46	4574.74	21859.2	TM	0.5	1.0		MIT
4572.42	4573.70	21864.1	TM	1.0	0.8		MIT
4571.89	4573.17	21866.7	TM	1.2			MIT
4571.20	4572.48	21870.0	TM	1.0			MIT
4570.31	4571.59	21874.2	TM	1.0	1.2		MIT
4569.25	4570.53	21879.3	TM	1.0	1.2		MIT
4568.40	4569.68	21883.4	TM	1.0			MIT
4567.11	4568.39	21889.6	TM	1.5	1.2		MIT
4564.67	4565.95	21901.2	TM	1.5	0.3		MIT
4561.84	4563.12	21914.8	TM	1.7	1.0		MIT
4560.83	4562.11	21919.7	TM	1.0	1.2		MIT
4559.01	4560.29	21928.4	TM	1.2	1.4		MIT
4556.67	4557.95	21939.7	TM	1.5	1.8		MIT
4555.26	4556.54	21946.5	TM	1.4	1.7		MIT
4554.65	4555.93	21949.4	TM	0.7			MIT
4548.59	4549.87	21978.7	TM	1.5			MIT
4539.39	4540.66	22023.2	TM	1.0			MIT
4532.15	4533.42	22058.4	TM	1.2	0.0		ME
4529.37	4530.64	22071.9	TM	1.9	0.7		MIT
4527.88	4529.15	22079.2	TM	1.2			MIT
4522.57	4523.84	22105.1	TM	2.3	2.5		ME
4521.09	4522.36	22112.4	TM	1.2	1.2		MIT
4519.58	4520.85	22119.7	TM	1.7	0.0		MIT
4517.87	4519.14	22128.1	TM	1.2			MIT
4516.95	4518.22	22132.6	TM	0.5			ME
4505.52	4506.78	22188.8	TM	1.2	1.0		MIT
4504.87	4506.13	22192.0	TM	0.5			ME
4504.10	4505.36	22195.8	TM	0.7			MIT
4503.47	4504.73	22198.9	TM	0.7			MIT
4496.65	4497.91	22232.5	TM	0.3	0.3		MIT
4489.72	4490.98	22266.9	TM	1.6	0.8		ME
4486.15	4487.41	22284.6	TM	0.3			ME
4481.27	4482.53	22308.8	TM	2.6	1.7		ME
4472.80	4474.06	22351.1	TM	0.7			MIT
4472.00	4473.25	22355.1	TM	0.3			ME
4470.02	4471.27	22365.0	TM	0.3			MIT
4467.98	4469.23	22375.2	TM	1.0			MIT
4466.07	4467.32	22384.8	TM	0.5			MIT
4459.99	4461.24	22415.3	TM	0.7	0.0		ME
4454.04	4455.29	22445.2	TM	1.3	0.0		ME
4447.59	4448.84	22477.8	TM	0.7			ME
4442.74	4443.99	22502.3	TM	0.9			ME
4437.40	4438.65	22529.4	TM	1.0	1.0		ME
4428.85	4430.09	22572.9	TM	0.6	0.7		ME
4426.34	4427.58	22585.7	TM	0.6			ME
4425.96	4427.20	22587.6	TM	0.5			ME
4411.70	4412.94	22660.6	TM	0.5			ME
4396.49	4397.73	22739.0	TM	0.9			ME
4395.96	4397.19	22741.8	TM	0.7			ME
4394.42	4395.65	22749.7	TM	1.5			ME
4386.42	4387.65	22791.2	TM	2.3	1.0		ME
4381.11	4382.34	22818.9	TM	0.7	0.7		ME
4367.90	4369.13	22887.9	TM	1.3	1.0		MIT
4363.65	4364.88	22910.2	TM	1.3	1.2		MIT
4359.93	4361.16	22929.7	TM	2.5	1.5		ME
4357.49	4358.71	22942.5	TM	0.8	0.3		ME
4351.18	4352.40	22975.8	TM	1.3			MIT
4350.99	4352.21	22976.8	TM	0.7			MIT
4346.48	4347.70	23000.7	TM	0.8			ME
4344.47	4345.69	23011.3	TM	0.7			ME
4335.99	4337.21	23056.3	TM	0.6	0.8		ME
4318.39	4319.60	23150.3	TM	0.5			MIT
4298.37	4299.58	23258.1	TM	1.3	0.3		ME
4292.96	4294.17	23287.4	TM	0.5	0.7		MIT
4291.36	4292.57	23296.1	TM	0.7	0.7		ME
4271.72	4272.92	23403.2	TM	1.3	0.3		ME
4268.56	4269.76	23420.5	TM	0.5			MIT

AIR WAVELENGTH (ANGSTRMS)	VACUUM WAVELENGTH (ANGSTRMS)	WAVENUMBER (KAYSERS)	LMENT	LOG ARC	INTENSITY SPK	INTENSITY DIS	REF
4252.41	4253.61	23509.5	TM	0.5			MIT
4246.38	4247.58	23542.8	TM	1.3	0.6		MIT
4242.15	4243.34	23566.3	TM	2.7	2.0		ME
4236.93	4238.12	23595.4	TM	1.2	1.3		MIT
4228.55	4229.74	23642.1	TM	0.3	0.0		MIT
4222.67	4223.86	23675.0	TM	1.0	0.6		MIT
4221.95	4223.14	23679.1	TM	0.3	0.3		MIT
4215.53	4216.72	23715.1	TM	1.0			MIT
4213.26	4214.45	23727.9	TM	0.5			MIT
4212.86	4214.05	23730.2	TM	1.3	0.6		MIT
4206.00	4207.18	23768.9	TM	1.3	0.7		MIT
4203.73	4204.91	23781.7	TM	2.4	1.4		ME
4201.14	4202.32	23796.4	TM	0.6	0.7		ME
4199.92	4201.10	23803.3	TM	2.0	1.3		MIT
4197.63	4198.81	23816.3	TM	0.5			MIT
4187.62	4188.80	23873.2	TM	2.5	1.5		ME
4186.31	4187.49	23880.7	TM	0.7	0.9		MIT
4180.02	4181.20	23916.6	TM	0.7	0.3		MIT
4170.44	4171.62	23971.5	TM	0.5			ME
4168.71	4169.89	23981.5	TM	0.3			ME
4168.52	4169.70	23982.6	TM	0.3			ME
4159.55	4160.72	24034.3	TM	1.0	1.2		ME
4158.59	4159.76	24039.8	TM	0.7			MIT
4156.31	4157.48	24053.0	TM	1.0			MIT
4152.29	4153.46	24076.3	TM	1.3			MIT
4150.08	4151.25	24089.1	TM	0.5			MIT
4147.18	4148.35	24106.0	TM	0.9			MIT
4146.81	4147.98	24108.1	TM	0.8			MIT
4145.87	4147.04	24113.6	TM	0.7			MIT
4140.03	4141.20	24147.6	TM	0.3			ME
4138.36	4139.53	24157.4	TM	1.9	0.9		ME
4137.49	4138.66	24162.4	TM	0.5			MIT
4134.45	4135.62	24180.2	TM	0.5	0.6		ME
4132.69	4133.86	24190.5	TM	1.1	1.2		MIT
4129.54	4130.70	24208.9	TM	0.0	0.3		ME
4127.35	4128.51	24221.8	TM	0.5			MIT
4107.92	4109.08	24336.4	TM	0.6			ME
4105.84	4107.00	24348.7	TM	2.5	1.5		ME
4105.62	4106.78	24350.0	TM	0.8			MIT
4103.72	4104.88	24361.3	TM	0.7	1.0		ME
4094.18	4095.34	24418.0	TM	2.5	1.5		ME
4090.29	4091.44	24441.2	TM	1.0	0.5		MIT
4072.14	4073.29	24550.2	TM	0.3			MIT
4063.35	4064.50	24603.3	TM	0.6	0.7		ME
4058.92	4060.07	24630.1	TM	1.3	0.8		MIT
4052.21	4053.35	24670.9	TM	0.9			ME
4044.47	4045.61	24718.1	TM	1.2			ME
4040.00	4041.14	24745.5	TM	0.5	0.6		ME
4034.74	4035.88	24777.7	TM	1.0	1.0		MIT
4024.24	4025.38	24842.4	TM	0.8			ME
4021.39	4022.53	24860.0	TM	0.6	1.0		ME
3996.516	3997.646	25014.72	TM	2.3	1.6		ME
3995.584	3996.714	25020.56	TM	2.0			ME
3975.53	3976.65	25146.8	TM	0.3	1.0		ME
3958.100	3959.220	25257.50	TM	2.3	1.6		ME
3949.27	3950.39	25314.0	TM	1.7	0.7		ME
3929.584	3930.697	25440.78	TM	1.8	1.4		ME
3928.66	3929.77	25446.8	TM	1.2	0.7		ME
3916.47	3917.58	25526.0	TM	1.9	0.9		ME
3900.79	3901.90	25628.6	TM	1.9	1.7		ME
3896.62	3897.72	25656.0	TM	1.2	0.0		ME
3890.52	3891.62	25696.2	TM	1.6	1.0		ME
3887.354	3888.456	25717.15	TM	1.9	0.9		ME
3883.43	3884.53	25743.1	TM	2.2	1.5		ME
3883.13	3884.23	25745.1	TM	2.0	1.0		ME
3857.84	3858.93	25913.9	TM	1.2	0.7		ME
3848.018	3849.109	25980.04	TM	2.6	2.4		ME
3838.204	3839.293	26046.46	TM	1.9	1.8		ME
3826.39	3827.48	26126.9	TM	1.0			ME
3817.40	3818.48	26188.4	TM	1.8	1.2		ME

AIR WAVELENGTH (ANGSTRMS)	VACUUM WAVELENGTH (ANGSTRMS)	WAVENUMBER (KAYSERS)	LMENT	LOG ARC	INTENSITY SPK	INTENSITY DIS	REF
3810.73	3811.81	26234.2	TM	1.5	1.0		ME
3807.72	3808.80	26255.0	TM	1.0	0.0		ME
3798.76	3799.84	26316.9	TM	1.3	1.0		ME
3798.55	3799.63	26318.4	TM	1.2			ME
3795.765	3796.843	26337.67	TM	2.4	2.2		ME
3795.17	3796.25	26341.8	TM	1.6	0.8		ME
3783.56	3784.63	26422.6	TM	1.5	1.2		ME
3765.85	3766.92	26546.9	TM	1.0	1.0		ME
3761.917	3762.986	26574.64	TM	2.3	2.1		ME
3761.333	3762.402	26578.77	TM	2.4	2.2		ME
3756.861	3757.929	26610.40	TM	1.6	1.3		ME
3751.816	3752.882	26646.19	TM	1.7	0.7		ME
3744.071	3745.135	26701.30	TM	2.0	1.0		ME
3736.20	3737.26	26757.6	TM	0.5	0.9		ME
3734.13	3735.19	26772.4	TM	2.2	1.7		ME
3730.81	3731.87	26796.2	TM	1.2	0.7		ME
3725.065	3726.124	26837.54	TM	1.8	1.2		ME
3719.72	3720.78	26876.1	TM	1.0	0.6		ME
3717.92	3718.98	26889.1	TM	2.0	1.0		ME
3704.846	3705.900	26984.00	TM	1.5	1.2		ME
3701.36	3702.41	27009.4	TM	2.2	1.9		ME
3700.26	3701.31	27017.4	TM	2.2	1.9		ME
3699.87	3700.92	27020.3	TM	1.0	0.8		ME
3697.58	3698.63	27037.0	TM	0.9	0.5		ME
3694.76	3695.81	27057.7	TM	1.3	1.3		ME
3686.16	3687.21	27120.8	TM	0.7	0.7		ME
3683.20	3684.25	27142.6	TM	1.0	1.3		ME
3678.864	3679.911	27174.57	TM	1.7	1.6		ME
3677.980	3679.027	27181.10	TM	1.5	1.3		ME
3673.14	3674.19	27216.9	TM	1.0	1.2		ME
3668.08	3669.12	27254.5	TM	1.9	1.3		ME
3665.81	3666.85	27271.3	TM	1.6	1.3		ME
3653.61	3654.65	27362.4	TM	1.5	1.3		ME
3647.73	3648.77	27406.5	TM	1.5	1.2		ME
3647.24	3648.28	27410.2	TM	1.0	1.3		ME
3643.652	3644.690	27437.17	TM	1.8	1.6		ME
3638.42	3639.46	27476.6	TM	1.0	0.3		ME
3624.20	3625.23	27584.4	TM	0.8			ME
3619.97	3621.00	27616.7	TM	1.3	1.3		ME
3609.54	3610.57	27696.5	TM	1.2	1.4		ME
3608.77	3609.80	27702.4	TM	2.0	1.3		ME
3607.36	3608.39	27713.2	TM	0.9	1.0		ME
3599.16	3600.19	27776.3	TM	0.6	0.9		ME
3586.07	3587.09	27877.7	TM	0.8			ME
3577.51	3578.53	27944.4	TM	0.7	0.9		ME
3574.06	3575.08	27971.4	TM	1.3	1.5		ME
3570.99	3572.01	27995.4	TM	0.6	1.0		ME
3568.68	3569.70	28013.6	TM	0.7	0.7		ME
3567.36	3568.38	28023.9	TM	1.0			ME
3566.474	3567.492	28030.89	TM	1.8	1.3		ME
3565.903	3566.921	28035.38	TM	1.6	1.7		ME
3563.885	3564.903	28051.26	TM	1.2	0.5		ME
3560.92	3561.94	28074.6	TM	0.9			ME
3557.79	3558.81	28099.3	TM	1.6	1.3		ME
3555.82	3556.84	28114.9	TM	0.8			ME
3550.16	3551.17	28159.7	TM	0.7	1.2		ME
3548.48	3549.49	28173.0	TM	0.9	1.3		ME
3537.91	3538.92	28257.2	TM	1.2	0.3		ME
3536.57	3537.58	28267.9	TM	1.8	1.3		ME
3536.21	3537.22	28270.8	TM	1.6	1.8		ME
3535.520	3536.530	28276.30	TM	1.9	1.4		ME
3534.85	3535.86	28281.7	TM	1.3	1.5		ME
3528.19	3529.20	28335.0	TM	0.7	1.0		ME
3522.43	3523.44	28381.4	TM	1.2	0.7		ME
3517.60	3518.61	28420.4	TM	0.8			ME
3513.038	3514.043	28457.25	TM	1.0	1.3		ME
3499.965	3500.966	28563.54	TM	1.0			ME
3495.20	3496.20	28602.5	TM	0.8	1.2		ME
3492.59	3493.59	28623.9	TM	1.3	1.0		ME
3487.384	3488.382	28666.58	TM	1.2			ME

AIR WAVELENGTH (ANGSTRMS)	VACUUM WAVELENGTH (ANGSTRMS)	WAVENUMBER (KAYSERS)	LMENT	LOG ARC	INTENSITY SPK	INTENSITY DIS	REF
3487.08	3488.08	28669.1	TM	1.0	1.3		ME
3481.75	3482.75	28713.0	TM	1.3	1.0		ME
3476.69	3477.69	28754.8	TM	0.8			ME
3467.51	3468.50	28830.9	TM	0.7			ME
3462.20	3463.19	28875.1	TM	2.4	2.3		ME
3461.17	3462.16	28883.7	TM	1.2	0.7		ME
3453.66	3454.65	28946.5	TM	2.2	1.9		ME
3449.77	3450.76	28979.1	TM	1.0	1.3		ME
3447.27	3448.26	29000.1	TM	0.8	0.5		ME
3441.51	3442.50	29048.7	TM	2.2	1.9		ME
3437.64	3438.63	29081.4	TM	1.0	1.2		ME
3431.20	3432.18	29136.0	TM	1.7	1.7		ME
3429.97	3430.95	29146.4	TM	2.0	2.0		ME
3429.34	3430.32	29151.8	TM	1.0			ME
3428.62	3429.60	29157.9	TM	1.2	1.2		ME
3425.63	3426.61	29183.3	TM	2.0	1.7		ME
3425.08	3426.06	29188.0	TM	2.3	2.5		ME
3416.60	3417.58	29260.5	TM	1.0			ME
3412.60	3413.58	29294.8	TM	1.0	0.3		ME
3411.58	3412.56	29303.5	TM	0.9	0.7		ME
3410.05	3411.03	29316.7	TM	1.5	0.7		ME
3399.95	3400.93	29403.8	TM	1.7	1.6		ME
3398.02	3399.00	29420.5	TM	1.5	1.6		ME
3397.503	3398.478	29424.94	TM	2.0	1.7		ME
3393.19	3394.16	29462.3	TM	0.7			ME
3377.91	3378.88	29595.6	TM	0.8	1.0		ME
3374.51	3375.48	29625.4	TM	1.8	1.5		ME
3369.64	3370.61	29668.2	TM	1.0	1.3		ME
3368.33	3369.30	29679.8	TM	1.0	1.5		ME
3362.61	3363.58	29730.3	TM	2.4	2.3		ME
3361.04	3362.01	29744.1	TM	1.0	1.2		ME
3357.30	3358.26	29777.3	TM	0.8	1.0		ME
3356.80	3357.76	29781.7	TM	0.8	1.0		ME
3354.86	3355.82	29798.9	TM	1.5	1.7		ME
3352.11	3353.07	29823.4	TM	0.8	0.8		ME
3349.99	3350.95	29842.3	TM	1.3			ME
3345.85	3346.81	29879.2	TM	0.8	1.2		ME
3339.63	3340.59	29934.8	TM	0.5	1.0		ME
3337.82	3338.78	29951.1	TM	0.7	1.2		ME
3335.06	3336.02	29975.9	TM	1.0	1.3		ME
3333.27	3334.23	29991.9	TM	0.7	1.2		ME
3327.59	3328.55	30043.1	TM	1.3	1.2		ME
3323.22	3324.18	30082.6	TM	1.3	1.6		ME
3318.26	3319.21	30127.6	TM	0.8	1.2		ME
3316.87	3317.82	30140.2	TM	1.8	1.3		ME
3316.17	3317.12	30146.6	TM	1.0	1.3		ME
3310.587	3311.540	30197.43	TM	1.8	1.3		ME
3309.80	3310.75	30204.6	TM	1.9	1.8		ME
3308.01	3308.96	30221.0	TM	1.3	1.2		ME
3306.91	3307.86	30231.0	TM	1.4	1.3		ME
3306.02	3306.97	30239.1	TM	1.3	1.2		ME
3302.45	3303.40	30271.8	TM	2.1	1.9		ME
3301.51	3302.46	30280.4	TM	0.7	1.0		ME
3291.00	3291.95	30377.1	TM	2.1	1.9		ME
3285.61	3286.56	30427.0	TM	1.6	1.6		ME
3283.40	3284.35	30447.5	TM	1.6	1.6		ME
3276.81	3277.75	30508.7	TM	1.6	1.2		ME
3269.00	3269.94	30581.6	TM	1.6	2.2		ME
3267.41	3268.35	30596.5	TM	1.7	1.6		ME
3266.63	3267.57	30603.8	TM	1.8	1.7		ME
3264.10	3265.04	30627.5	TM	1.5	1.0		ME
3261.66	3262.60	30650.4	TM	1.5	2.0		ME
3258.62	3259.56	30679.0	TM	1.0	0.7		ME
3258.04	3258.98	30684.4	TM	2.1	1.8		ME
3254.85	3255.79	30714.5	TM	0.9	1.0		ME
3251.64	3252.58	30744.8	TM	1.4	1.3		ME
3251.34	3252.28	30747.7	TM	1.0	1.1		ME
3249.84	3250.78	30761.9	TM	1.0	1.7		ME
3245.86	3246.80	30799.6	TM	1.3	1.0		ME
3244.09	3245.03	30816.4	TM	0.8	1.2		ME

AIR WAVELENGTH (ANGSTRMS)	VACUUM WAVELENGTH (ANGSTRMS)	WAVENUMBER (KAYSERS)	LMENT	LOG ARC	INTENSITY SPK	INTENSITY DIS	REF
3241.53	3242.47	30840.7	TM	2.2	2.1		ME
3240.229	3241.164	30853.11	TM	2.0	1.9		ME
3236.797	3237.731	30885.83	TM	2.0	1.9		ME
3235.45	3236.38	30898.7	TM	1.9	1.6		ME
3233.75	3234.68	30914.9	TM	1.0			ME
3231.51	3232.44	30936.4	TM	1.6	1.5		ME
3230.96	3231.89	30941.6	TM	0.7	1.0		ME
3230.15	3231.08	30949.4	TM	1.0	1.3		ME
3228.91	3229.84	30961.3	TM	1.2	1.5		ME
3226.81	3227.74	30981.4	TM	1.2	1.5		ME
3224.15	3225.08	31007.0	TM	0.3	1.3		ME
3222.04	3222.97	31027.3	TM	0.8	1.3		ME
3221.08	3222.01	31036.5	TM	0.8			ME
3216.64	3217.57	31079.4	TM	0.7	1.0		ME
3216.12	3217.05	31084.4	TM	1.0	1.5		ME
3214.63	3215.56	31098.8	TM	1.2	1.0		ME
3212.01	3212.94	31124.2	TM	1.7	1.6		ME
3210.82	3211.75	31135.7	TM	1.6	1.0		ME
3210.57	3211.50	31138.1	TM	1.6	1.7		ME
3200.00	3200.92	31241.0	TM	1.2	1.5		ME
3196.54	3197.46	31274.8	TM	1.3	1.0		ME
3195.34	3196.26	31286.5	TM	1.5	1.2		ME
3194.60	3195.52	31293.8	TM	1.0	0.7		ME
3187.41	3188.33	31364.4	TM	1.0	1.4		ME
3185.48	3186.40	31383.4	TM	1.3	1.6		ME
3178.20	3179.12	31455.2	TM	1.0	1.2		ME
3177.46	3178.38	31462.6	TM	1.5	1.3		ME
3173.58	3174.50	31501.0	TM	1.7	2.0		ME
3172.82	3173.74	31508.6	TM	2.3	2.3		ME
3169.89	3170.81	31537.7	TM	1.2	1.6		ME
3168.82	3169.74	31548.4	TM	1.2	1.0		ME
3168.19	3169.11	31554.6	TM	1.4	1.8		ME
3164.87	3165.79	31587.7	TM	1.3	1.0		ME
3162.44	3163.36	31612.0	TM	1.2	1.6		ME
3157.34	3158.25	31663.1	TM	2.3	2.2		ME
3151.035	3151.947	31726.42	TM	2.3	2.3		ME
3150.07	3150.98	31736.1	TM	0.9	1.0		ME
3149.14	3150.05	31745.5	TM	1.2			ME
3146.16	3147.07	31775.6	TM	1.5	1.2		ME
3144.89	3145.80	31788.4	TM	1.4	1.8		ME
3141.96	3142.87	31818.1	TM	0.9	1.0		ME
3133.89	3134.80	31900.0	TM	2.3	2.3		ME
3131.26	3132.17	31926.8	TM	2.6	2.7		ME
3126.01	3126.92	31980.4	TM	1.4			ME
3124.90	3125.81	31991.7	TM	1.2	1.5		ME
3123.30	3124.21	32008.1	TM	1.0	1.3		ME
3122.54	3123.45	32015.9	TM	1.0			ME
3113.31	3114.21	32110.8	TM	1.0	1.3		ME
3102.88	3103.78	32218.8	TM	1.3	1.2		ME
3099.61	3100.51	32252.8	TM	1.0	1.7		ME
3098.59	3099.49	32263.4	TM	1.9	1.8		ME
3098.01	3098.91	32269.4	TM	0.7	1.3		ME
3096.97	3097.87	32280.3	TM	1.2	1.7		ME
3093.12	3094.02	32320.4	TM	1.5	1.8		ME
3087.02	3087.92	32384.3	TM	1.5	1.8		ME
3082.57	3083.47	32431.0	TM	1.0			ME
3081.13	3082.02	32446.2	TM	1.3			ME
3073.85	3074.74	32523.0	TM	1.2	1.7		ME
3073.50	3074.39	32526.7	TM	1.4	1.8		ME
3073.085	3073.978	32531.14	TM	1.8	2.2		ME
3066.42	3067.31	32601.8	TM	0.8	1.3		ME
3064.01	3064.90	32627.5	TM	1.2	1.3		ME
3062.06	3062.95	32648.3	TM	1.0	1.3		ME
3061.14	3062.03	32658.1	TM	1.0	1.5		ME
3058.99	3059.88	32681.0	TM	1.0	1.5		ME
3056.06	3056.95	32712.4	TM	1.6D	2.0		ME
3054.05	3054.94	32733.9	TM	1.5	1.8		ME
3053.71	3054.60	32737.5	TM	0.8	1.5		ME
3052.53	3053.42	32750.2	TM	1.0	1.4		ME
3050.73	3051.62	32769.5	TM	1.7	2.2		ME

AIR WAVELENGTH (ANGSTRMS)	VACUUM WAVELENGTH (ANGSTRMS)	WAVENUMBER (KAYSERS)	LMENT	LOG ARC	INTENSITY SPK	INTENSITY DIS	REF
3048.82	3049.71	32790.0	TM	1.3	1.6		ME
3048.66	3049.55	32791.8	TM	0.8	1.0		ME
3046.87	3047.76	32811.0	TM	1.3			ME
3042.35	3043.24	32859.8	TM	1.7			ME
3035.99	3036.87	32928.6	TM	1.2	1.5		ME
3031.68	3032.56	32975.4	TM	0.8	1.2		ME
3028.73	3029.61	33007.5	TM	0.7	1.5		ME
3026.07	3026.95	33036.5	TM	1.8	1.5		ME
3019.80	3020.68	33105.1	TM	0.7	1.2		ME
3018.61	3019.49	33118.2	TM	1.0	1.6		ME
3017.10	3017.98	33134.8	TM	1.3	1.6		ME
3015.29	3016.17	33154.6	TM	2.1	2.0		ME
3014.65	3015.53	33161.7	TM	1.6	1.0		ME
3008.92	3009.80	33224.8	TM	1.3	1.3		ME
3006.36	3007.24	33253.1	TM	1.0			ME
2994.33	2995.20	33386.7	TM	0.3	1.0		ME
2993.90	2994.77	33391.5	TM	1.0			ME
2993.27	2994.14	33398.5	TM	1.4	1.0		ME
2992.89	2993.76	33402.8	TM	0.5	1.3		ME
2990.540	2991.412	33429.03	TM	1.9	1.5		ME
2986.52	2987.39	33474.0	TM	1.7	2.2		ME
2985.38	2986.25	33486.8	TM	0.9	1.3		ME
2983.13	2984.00	33512.1	TM	1.0	0.6		ME
2981.49	2982.36	33530.5	TM	1.8	2.0		ME
2979.44	2980.31	33553.6	TM	0.5	1.0		ME
2978.43	2979.30	33564.9	TM	1.5	1.0		ME
2974.29	2975.16	33611.7	TM	1.0	1.5		ME
2973.39	2974.26	33621.8	TM	1.2	1.6		ME
2973.23	2974.10	33623.6	TM	1.2			ME
2969.50	2970.37	33665.9	TM	1.3	0.6		ME
2967.76	2968.63	33685.6	TM	1.0	1.5		ME
2965.87	2966.74	33707.1	TM	1.7	2.0		ME
2961.41	2962.28	33757.8	TM	0.7	1.0		ME
2959.65	2960.51	33777.9	TM	1.3	1.8		ME
2957.05	2957.91	33807.6	TM	0.5	1.3		ME
2955.06	2955.92	33830.4	TM	0.6	1.0		ME
2953.59	2954.45	33847.2	TM	0.7	1.5		ME
2951.264	2952.127	33873.89	TM	1.5	2.2		ME
2948.15	2949.01	33909.7	TM	1.2	1.0		ME
2948.01	2948.87	33911.3	TM	1.2	1.3		ME
2946.84	2947.70	33924.7	TM	1.0D	1.5		ME
2943.36	2944.22	33964.9	TM	0.8	1.2		ME
2935.997	2936.856	34050.02	TM	1.9	2.5		ME
2932.97	2933.83	34085.2	TM	1.0			ME
2932.59	2933.45	34089.6	TM	0.7	1.4		ME
2930.57	2931.43	34113.1	TM	0.3	1.0		ME
2928.23	2929.09	34140.3	TM	1.0	1.5		ME
2926.748	2927.604	34157.62	TM	1.9	1.8		ME
2925.92	2926.78	34167.3	TM	1.0	0.3		ME
2925.65	2926.51	34170.4	TM	1.3	1.8		ME
2922.83	2923.69	34203.4	TM	0.7	1.2		ME
2922.09	2922.95	34212.1	TM	0.9	1.2		ME
2921.27	2922.13	34221.7	TM	0.7	1.0		ME
2918.27	2919.12	34256.9	TM	1.4	1.7		ME
2916.52	2917.37	34277.4	TM	1.2	1.6		ME
2915.67	2916.52	34287.4	TM	0.5	0.8		ME
2914.83	2915.68	34297.3	TM	1.0			ME
2913.96	2914.81	34307.5	TM	1.2			ME
2911.87	2912.72	34332.1	TM	0.6	0.8		ME
2908.69	2909.54	34369.7	TM	0.7	1.0		ME
2907.17	2908.02	34387.6	TM	0.7	1.0		ME
2905.41	2906.26	34408.5	TM	1.0	0.7		ME
2903.08	2903.93	34436.1	TM	1.2	1.4		ME
2894.47	2895.32	34538.5	TM	1.3	1.6		ME
2890.93	2891.78	34580.8	TM	1.8	1.2		ME
2889.64	2890.49	34596.2	TM	1.0	1.3		ME
2886.459	2887.305	34634.37	TM	1.5	0.7		ME
2878.36	2879.20	34731.8	TM	1.0	1.3		ME
2878.21	2879.05	34733.6	TM	0.5	1.3		ME
2873.01	2873.85	34796.5	TM	1.2	1.3		ME

AIR WAVELENGTH (ANGSTRMS)	VACUUM WAVELENGTH (ANGSTRMS)	WAVENUMBER (KAYSERS)	LMENT	LOG ARC	INTENSITY SPK	DIS	REF
2869.22	2870.06	34842.5	TM	2.0	2.5		ME
2868.01	2868.85	34857.1	TM	1.2	1.6		ME
2863.76	2864.60	34908.9	TM	1.2	1.6		ME
2863.35	2864.19	34913.9	TM	1.0			ME
2861.74	2862.58	34933.5	TM	1.3	1.5		ME
2860.55	2861.39	34948.0	TM	1.0			ME
2860.13	2860.97	34953.2	TM	1.2	1.3		ME
2854.89	2855.73	35017.3	TM	0.7	1.3		ME
2848.87	2849.71	35091.3	TM	1.0	1.2		ME
2845.35	2846.19	35134.7	TM	0.3	1.0		ME
2844.67	2845.51	35143.1	TM	1.6	1.2		ME
2839.85	2840.69	35202.8	TM	0.7	1.0		ME
2839.11	2839.94	35212.0	TM	1.0	0.8		ME
2838.94	2839.77	35214.1	TM	1.2	1.6		ME
2833.82	2834.65	35277.7	TM	0.8	1.0		ME
2831.56	2832.39	35305.8	TM	1.0	1.6		ME
2827.92	2828.75	35351.3	TM	1.7	2.0		ME
2827.02	2827.85	35362.5	TM	1.3	1.7		ME
2826.44	2827.27	35369.8	TM	1.0	1.4		ME
2819.08	2819.91	35462.1	TM	0.8	1.0		ME
2818.48	2819.31	35469.7	TM	1.5	1.3		ME
2818.15	2818.98	35473.8	TM	0.7	1.3		ME
2817.31	2818.14	35484.4	TM	0.7	1.3		ME
2816.56	2817.39	35493.9	TM	0.3	1.0		ME
2812.26	2813.09	35548.1	TM	0.7	1.3		ME
2808.43	2809.26	35596.6	TM	1.3	1.0		ME
2807.99	2808.82	35602.2	TM	1.3	1.5		ME
2803.11	2803.94	35664.1	TM	0.8	1.0		ME
2800.41	2801.24	35698.5	TM	1.0	1.0		ME
2797.269	2798.094	35738.62	TM	1.8	2.0		ME
2794.60	2795.42	35772.7	TM	1.8	1.3		ME
2791.62	2792.44	35810.9	TM	1.0	1.6		ME
2786.19	2787.01	35880.7	TM	1.0	1.5		ME
2785.08	2785.90	35895.0	TM	1.8	1.5		ME
2780.87	2781.69	35949.4	TM	1.0	1.3		ME
2779.56	2780.38	35966.3	TM	1.5	1.0		ME
2778.40	2779.22	35981.3	TM	0.7	1.0		ME
2777.51	2778.33	35992.9	TM	1.0	1.2		ME
2777.05	2777.87	35998.8	TM	0.9	1.4		ME
2774.99	2775.81	36025.5	TM	1.3	1.6		ME
2774.78	2775.60	36028.3	TM	0.7	1.2		ME
2773.80	2774.62	36041.0	TM	1.0			ME
2771.05	2771.87	36076.7	TM	1.2	1.2		ME
2767.10	2767.92	36128.2	TM	0.7	0.9		ME
2758.98	2759.80	36234.6	TM	1.0	1.5		ME
2753.19	2754.00	36310.8	TM	1.3	1.3		ME
2750.77	2751.58	36342.7	TM	1.0	1.5		ME
2744.09	2744.90	36431.2	TM	1.3	1.7		ME
2742.96	2743.77	36446.2	TM	1.2	1.4		ME
2735.33	2736.14	36547.8	TM	0.8			ME
2729.05	2729.86	36631.9	TM	1.3	0.8		ME
2721.192	2721.998	36737.72	TM	1.8	2.0		ME
2711.52	2712.32	36868.7	TM	0.9			ME
2697.50	2698.30	37060.4	TM	1.2	1.2		ME
2684.08	2684.88	37245.6	TM	1.0	0.6		ME
2679.57	2680.37	37308.3	TM	1.5	1.7		ME
2669.26	2670.05	37452.4	TM	0.3	0.9		ME
2668.20	2668.99	37467.3	TM	1.3	1.3		ME
2660.09	2660.88	37581.5	TM	1.5	1.0		ME
2658.49	2659.28	37604.1	TM	1.3	0.8		ME
2650.27	2651.06	37720.8	TM	1.3	0.8		ME
2648.14	2648.93	37751.1	TM	0.9	0.6		ME
2643.58	2644.37	37816.2	TM		1.5		ME
2643.50	2644.29	37817.4	TM	0.7			ME
2642.83	2643.62	37827.0	TM	0.8	0.0		ME
2640.77	2641.56	37856.5	TM	1.6	0.9		ME
2638.41	2639.20	37890.3	TM	0.8	1.0		ME
2637.23	2638.02	37907.3	TM	0.8	0.9		ME
2629.79	2630.57	38014.5	TM	1.2	1.0		ME
2624.34	2625.12	38093.5	TM	1.9	1.6		ME

AIR WAVELENGTH (ANGSTRMS)	VACUUM WAVELENGTH (ANGSTRMS)	WAVENUMBER (KAYSERS)	LMENT	LOG ARC	INTENSITY SPK	DIS	REF
2622.21	2622.99	38124.4	TM	1.2	0.9		ME
2620.93	2621.71	38143.0	TM	0.7	0.3		ME
2618.84	2619.62	38173.5	TM		1.5		ME
2616.31	2617.09	38210.4	TM	0.9	1.2		ME
2615.99	2616.77	38215.0	TM	0.7	0.8		ME
2613.59	2614.37	38250.1	TM	0.7	0.8		ME
2609.47	2610.25	38310.5	TM	0.8	0.7		ME
2607.05	2607.83	38346.1	TM	1.7	1.5		ME
2606.02	2606.80	38361.2	TM	1.3	1.0		ME
2601.09	2601.87	38433.9	TM	0.9	0.3		ME
2596.50	2597.28	38501.9	TM	0.8			ME
2594.98	2595.76	38524.4	TM	0.8	0.8		ME
2588.27	2589.04	38624.3	TM	1.6	1.9		ME
2575.01	2575.78	38823.2	TM		1.3		ME
2574.55	2575.32	38830.1	TM	0.0	1.4		ME
2569.35	2570.12	38908.7	TM	0.8	0.6		ME
2568.26	2569.03	38925.2	TM	0.7	0.0		ME
2565.98	2566.75	38959.8	TM	0.7	0.3		ME
2563.86	2564.63	38992.0	TM	1.2	0.6		ME
2561.65	2562.42	39025.6	TM	1.8	1.5		ME
2552.49	2553.26	39165.7	TM		1.7		ME
2551.52	2552.29	39180.6	TM	0.6	0.9		ME
2542.66	2543.42	39317.1	TM	0.8	0.5		ME
2524.09	2524.85	39606.3	TM	1.0	0.5		ME
2522.16	2522.92	39636.6	TM	1.2	0.8		ME
2520.94	2521.70	39655.8	TM	0.9	0.3		ME
2520.88	2521.64	39656.8	TM	0.8	0.0		ME
2519.80	2520.56	39673.7	TM	0.0	1.7		ME
2509.08	2509.84	39843.2	TM	1.9	1.6		ME
2507.14	2507.90	39874.1	TM	1.0	1.2		ME
2504.74	2505.49	39912.3	TM		1.5		ME
2502.70	2503.45	39944.8	TM	0.7	0.8		ME
2499.53	2500.28	39995.5	TM	0.7	0.3		ME
2499.21	2499.96	40000.6	TM	0.9	0.6		ME
2491.60	2492.35	40122.7	TM	1.1	0.7		ME
2490.93	2491.68	40133.5	TM	0.8	0.8		ME
2489.46	2490.21	40157.2	TM	0.3	2.0		ME
2487.53	2488.28	40188.4	TM	1.2	1.0		ME
2481.15	2481.90	40291.7	TM	1.2	0.8		ME
2480.13	2480.88	40308.3	TM	1.8	1.3		ME
2478.57	2479.32	40333.7	TM	1.2	1.2		ME
2476.97	2477.72	40359.7	TM	0.7	0.0		ME
2474.45	2475.20	40400.8	TM		0.8		ME
2471.26	2472.01	40453.0	TM		0.8		ME
2467.91	2468.66	40507.9	TM	0.5	0.9		ME
2464.94	2465.69	40556.7	TM	0.8	0.8		ME
2458.58	2459.32	40661.6	TM	0.9	0.7		ME
2451.19	2451.93	40784.2	TM	0.9	0.8		ME
2447.39	2448.13	40847.5	TM	0.7	0.3		ME
2445.46	2446.20	40879.7	TM	1.0	1.2		ME
2440.70	2441.44	40959.4	TM	1.0	0.5		ME
2434.75	2435.49	41059.5	TM	0.7	0.3		ME
2428.43	2429.17	41166.4	TM	0.7	0.3		ME
2426.16	2426.90	41204.9	TM	1.6	1.3		ME
2421.64	2422.38	41281.8	TM	1.2	0.8		ME
2419.37	2420.11	41320.5	TM	0.8	0.5		ME
2416.02	2416.75	41377.8	TM	0.8	0.7		ME
2409.02	2409.75	41498.0	TM	1.3	1.2		ME
2408.23	2408.96	41511.6	TM	0.7	0.0		ME
2406.63	2407.36	41539.2	TM	0.3	1.8		ME
2400.99	2401.72	41636.8	TM	0.5	0.8		ME
2389.56	2390.29	41836.0	TM		0.8	H	ME
2388.95	2389.68	41846.6	TM	0.9	0.5		ME
2383.67	2384.40	41939.3	TM	1.2	0.6		ME
2375.85	2376.58	42077.4	TM		1.2		ME
2375.31	2376.04	42086.9	TM		1.2		ME
2367.10	2367.82	42232.9	TM	0.9	0.5		ME
2365.95	2366.67	42253.4	TM	0.8	0.6		ME
2361.23	2361.95	42337.9	TM	0.0	1.8		ME
2357.04	2357.76	42413.1	TM	0.5	1.9		ME

AIR WAVELENGTH (ANGSTRMS)	VACUUM WAVELENGTH (ANGSTRMS)	WAVENUMBER (KAYSERS)	LMENT	LOG ARC	INTENSITY SPK	DIS	REF
2356.23	2356.95	42427.7	TM		1.0		ME
2355.67	2356.39	42437.8	TM		0.8		ME
2354.14	2354.86	42465.4	TM		0.8		ME
2353.10	2353.82	42484.1	TM		1.6		ME
2351.92	2352.64	42505.4	TM		1.0		ME
2351.53	2352.25	42512.5	TM		1.3		ME
2349.04	2349.76	42557.5	TM		0.8		ME
2348.89	2349.61	42560.3	TM		0.9		ME
2347.47	2348.19	42586.0	TM		1.2		ME
2345.60	2346.32	42620.0	TM		1.3		ME
2342.05	2342.77	42684.6	TM		1.2		ME
2341.73	2342.45	42690.4	TM	0.0	1.3		ME
2341.17	2341.89	42700.6	TM	0.7	0.5		ME
2340.93	2341.65	42705.0	TM	0.7	0.7		ME
2338.36	2339.08	42751.9	TM	0.3	1.7		ME
2335.00	2335.72	42813.4	TM		0.9		ME
2331.79	2332.51	42872.4	TM	0.3	1.4		ME
2329.29	2330.00	42918.4	TM	0.0	1.5		ME
2328.48	2329.19	42933.3	TM	0.0	1.4		ME
2327.24	2327.95	42956.2	TM		0.9		ME
2326.19	2326.90	42975.6	TM		1.2		ME
2323.73	2324.44	43021.0	TM		1.0 H		ME
2312.71	2313.42	43226.0	TM		1.0		ME
2311.17	2311.88	43254.8	TM		1.3		ME
2296.21	2296.92	43536.6	TM	0.3	1.5		ME
2294.03	2294.74	43578.0	TM	0.7	0.3		ME
2290.55	2291.26	43644.2	TM		0.7		ME
2284.80	2285.50	43754.0	TM	1.3	1.0		ME
2231.57	2232.26	44797.6	TM	0.9	0.7		ME
2223.96	2224.65	44950.9	TM	0.8	0.7		ME
2188.16	2188.84	45686.2	TM	0.7	0.6		ME
2185.93	2186.61	45732.8	TM	0.3	1.3		ME
2183.90	2184.58	45775.3	TM		1.0		ME
2182.97	2183.65	45794.8	TM		1.2		ME
2156.27	2156.95	46361.8	TM		0.7		ME
2147.97	2148.65	46540.9	TM	0.7	0.5		ME
2136.70	2137.37	46786.4	TM		0.7		ME
2130.40	2131.07	46924.7	TM		1.0 H		ME
2127.03	2127.70	46999.1	TM		0.9 H		ME
2107.10	2107.77	47443.5	TM		1.2		ME
2099.10	2099.77	47624.3	TM		1.0		ME
9530.38	9532.99	10489.9	U	0.3 H			ME
9276.24	9278.79	10777.3	U	0.3			ME
9143.20	9145.71	10934.1	U	0.3			ME
9140.97	9143.48	10936.8	U	0.3			ME
9139.54	9142.05	10938.5	U	0.3			MIT
9023.38	9025.86	11079.3	U				ME
9020.89	9023.37	11082.3	U	0.3 H			ME
8951.97	8954.43	11167.7	U	0.3			MIT
8860.58	8863.01	11282.8	U	0.3			MIT
8757.75	8760.16	11415.3	U	0.6			MIT
8753.69	8756.09	11420.6	U	0.3			MIT
8728.73	8731.13	11453.3	U	0.3			MIT
8710.77	8713.16	11476.9	U	0.6 H			MIT
8691.28	8693.67	11502.6	U	0.6			MIT
8659.52	8661.90	11544.8	U	0.3			MIT
8652.79	8655.17	11553.8	U	0.3			MIT
8641.12	8643.49	11569.4	U	0.3			MIT
8637.62	8639.99	11574.1	U	0.3			MIT
8618.96	8621.33	11599.1	U	0.3			MIT
8612.73	8615.10	11607.5	U	0.3			MIT
8607.94	8610.30	11614.0	U	0.9			MIT
8574.61	8576.97	11659.1	U	0.3			MIT
8570.51	8572.86	11664.7	U	0.6			MIT
8567.73	8570.08	11668.5	U	0.3			MIT
8566.94	8569.29	11669.6	U	0.3			MIT
8557.34	8559.69	11682.7	U	0.3			MIT
8540.20	8542.55	11706.1	U	0.6			MIT
8504.70	8507.04	11755.0	U	0.6			ME
8450.03	8452.35	11831.0	U	0.6			ME

AIR WAVELENGTH (ANGSTRMS)	VACUUM WAVELENGTH (ANGSTRMS)	WAVENUMBER (KAYSERS)	LMENT	LOG ARC	INTENSITY SPK	DIS	REF
8445.35	8447.67	11837.6	U	0.6			ME
8441.25	8443.57	11843.3	U	0.3			ME
8426.70	8429.02	11863.8	U	0.3			ME
8399.18	8401.49	11902.7	U	0.3			MIT
8396.66	8398.97	11906.2	U	0.3			MIT
8390.89	8393.20	11914.4	U	0.3			MIT
8389.17	8391.48	11916.9	U	0.6			MIT
8387.16	8389.46	11919.7	U	0.6			MIT
8381.86	8384.16	11927.2	U	0.6			MIT
8357.07	8359.37	11962.6	U	0.3			MIT
8346.74	8349.03	11977.4	U	0.3			MIT
8337.46	8339.75	11990.8	U	0.3			MIT
8329.73	8332.02	12001.9	U	0.3			MIT
8327.79	8330.08	12004.7	U	0.3			ME
8318.34	8320.63	12018.3	U	0.3			MIT
8262.05	8264.32	12100.2	U	0.6			MIT
8188.20	8190.45	12209.3	U	0.3			MIT
8175.85	8178.10	12227.8	U	0.3			MIT
8174.30	8176.55	12230.1	U	0.6			MIT
8097.62	8099.85	12345.9	U	0.3			MIT
8074.03	8076.25	12382.0	U	0.3			MIT
8067.93	8070.15	12391.3	U	0.3			ME
8065.47	8067.69	12395.1	U	0.3 J			MIT
8060.34	8062.56	12403.0	U	0.3			MIT
8055.60	8057.82	12410.3	U	0.3			MIT
8035.99	8038.20	12440.6	U	0.3			ME
8034.69	8036.90	12442.6	U	0.3			ME
8019.27	8021.48	12466.5	U	0.3			ME
7998.56	8000.76	12498.8	U	0.3			MIT
7995.53	7997.73	12503.6	U	0.3			MIT
7991.30	7993.50	12510.2	U	0.3			MIT
7989.18	7991.38	12513.5	U	0.3			MIT
7978.57	7980.76	12530.1	U	0.3			MIT
7976.88	7979.07	12532.8	U	0.3			MIT
7975.09	7977.28	12535.6	U	0.6 V			MIT
7970.46	7972.65	12542.9	U	0.9			MIT
7967.09	7969.28	12548.2	U	0.3			MIT
7963.97	7966.16	12553.1	U	0.3			MIT
7959.96	7962.15	12559.4	U	0.3			MIT
7922.10	7924.28	12619.4	U	0.3			MIT
7918.80	7920.98	12624.7	U	0.3			MIT
7908.00	7910.18	12641.9	U	0.3			MIT
7904.29	7906.46	12647.9	U	0.3			MIT
7900.43	7902.60	12654.1	U	0.3			MIT
7896.04	7898.21	12661.1	U	0.3			MIT
7881.94	7884.11	12683.7	U	1.2			MIT
7875.38	7877.55	12694.3	U	0.6			MIT
7868.75	7870.91	12705.0	U	0.6			MIT
7844.71	7846.87	12743.9	U	0.3			MIT
7837.71	7839.87	12755.3	U	0.3			MIT
7835.71	7837.87	12758.6	U	0.3			MIT
7832.05	7834.20	12764.5	U	0.3			MIT
7823.85	7826.00	12777.9	U	0.3			MIT
7821.85	7824.00	12781.2	U	0.3			MIT
7816.32	7818.47	12790.2	U	0.3			MIT
7813.55	7815.70	12794.8	U	0.3			MIT
7808.48	7810.63	12803.1	U	0.3			MIT
7802.40	7804.55	12813.0	U	0.3			MIT
7791.24	7793.38	12831.4	U	0.3			ME
7784.13	7786.27	12843.1	U	1.3			MIT
7782.43	7784.57	12845.9	U	0.3			MIT
7780.49	7782.63	12849.1	U	0.3			ME
7776.94	7779.08	12855.0	U	0.3 H			ME
7768.93	7771.07	12868.2	U	0.3			ME
7766.57	7768.71	12872.2	U	0.3			ME
7763.89	7766.03	12876.6	U	0.3			MIT
7761.86	7764.00	12880.0	U	0.6			MIT
7759.87	7762.01	12883.3	U	0.3			MIT
7754.19	7756.32	12892.7	U	0.3			MIT
7748.19	7750.32	12902.7	U	0.3			MIT

AIR WAVELENGTH (ANGSTRMS)	VACUUM WAVELENGTH (ANGSTRMS)	WAVENUMBER (KAYSERS)	LMENT	LOG ARC	INTENSITY SPK	DIS	REF
7747.10	7749.23	12904.5	U	0.3			MIT
7744.26	7746.39	12909.2	U	0.3			MIT
7718.76	7720.88	12951.9	U	0.3			MIT
7707.59	7709.71	12970.7	U	0.3			ME
7696.99	7699.11	12988.5	U	0.3			MIT
7693.59	7695.71	12994.3	U	0.3			ME
7691.89	7694.01	12997.1	U	0.3			ME
7689.88	7692.00	13000.5	U	0.3			MIT
7669.69	7671.80	13034.7	U	0.6			MIT
7668.73	7670.84	13036.4	U	0.6	H		MIT
7658.70	7660.81	13053.4	U	0.3			MIT
7653.61	7655.72	13062.1	U	0.3			MIT
7648.66	7650.77	13070.6	U	0.3			ME
7643.61	7645.71	13079.2	U	0.3			MIT
7641.24	7643.34	13083.3	U	0.3			ME
7639.54	7641.64	13086.2	U	0.6			MIT
7634.73	7636.83	13094.4	U	0.6			MIT
7632.77	7634.87	13097.8	U	0.3			MIT
7631.71	7633.81	13099.6	U	0.9			MIT
7630.44	7632.54	13101.8	U	0.3			ME
7629.31	7631.41	13103.7	U	0.3			MIT
7628.02	7630.12	13105.9	U	0.3			MIT
7626.38	7628.48	13108.8	U	0.3			MIT
7621.95	7624.05	13116.4	U	0.3			MIT
7619.35	7621.45	13120.9	U	0.9			MIT
7615.72	7617.82	13127.1	U	0.3			ME
7610.77	7612.87	13135.7	U	0.3			ME
7609.16	7611.25	13138.4	U	0.6			MIT
7605.88	7607.97	13144.1	U	0.3			MIT
7600.27	7602.36	13153.8	U	0.3			ME
7597.46	7599.55	13158.7	U	0.3	W		ME
7595.07	7597.16	13162.8	U	0.3			MIT
7590.54	7592.63	13170.7	U	0.3	O		MIT
7587.55	7589.64	13175.9	U	0.3			MIT
7582.09	7584.18	13185.3	U	0.3			MIT
7580.91	7583.00	13187.4	U	0.3			MIT
7575.81	7577.90	13196.3	U	0.3			MIT
7573.14	7575.23	13200.9	U	0.3			ME
7572.42	7574.50	13202.2	U	0.3	O		MIT
7570.75	7572.83	13205.1	U	0.3			MIT
7566.06	7568.14	13213.3	U	0.6			MIT
7562.44	7564.52	13219.6	U	0.3			ME
7560.58	7562.66	13222.9	U	0.3			MIT
7558.36	7560.44	13226.7	U	0.3			MIT
7552.13	7554.21	13237.7	U	0.3			ME
7550.23	7552.31	13241.0	U	0.6			MIT
7533.91	7535.98	13269.7	U	1.3			MIT
7531.68	7533.75	13273.6	U	0.3			ME
7528.70	7530.77	13278.8	U	0.6			MIT
7517.41	7519.48	13298.8	U	0.3			MIT
7514.40	7516.47	13304.1	U	0.3			MIT
7512.60	7514.67	13307.3	U	0.3			MIT
7510.08	7512.15	13311.8	U	0.3			MIT
7506.94	7509.01	13317.3	U	0.3	O		ME
7505.10	7507.17	13320.6	U	0.3			MIT
7502.34	7504.41	13325.5	U	0.3			MIT
7497.43	7499.49	13334.2	U	0.3			MIT
7476.08	7478.14	13372.3	U	0.3			ME
7457.89	7459.94	13404.9	U	0.3			MIT
7454.03	7456.08	13411.9	U	0.7			MIT
7452.62	7454.67	13414.4	U	0.3			MIT
7447.88	7449.93	13422.9	U	0.5			MIT
7443.81	7445.86	13430.3	U	0.5			MIT
7443.07	7445.12	13431.6	U	0.3			ME
7438.22	7440.27	13440.4	U	0.5			MIT
7432.22	7434.27	13451.2	U	0.3			MIT
7425.50	7427.55	13463.4	U	1.3			MIT
7416.57	7418.61	13479.6	U	0.5			MIT
7396.99	7399.03	13515.3	U	0.8			MIT
7392.11	7394.15	13524.2	U	0.5			MIT

AIR WAVELENGTH (ANGSTRMS)	VACUUM WAVELENGTH (ANGSTRMS)	WAVENUMBER (KAYSERS)	LMENT	LOG ARC	INTENSITY SPK	DIS	REF
7390.99	7393.03	13526.3	U	0.5			MIT
7379.70	7381.73	13546.9	U	1.2			MIT
7375.40	7377.43	13554.8	U	0.3			ME
7372.58	7374.61	13560.0	U	0.3	H		MIT
7371.95	7373.98	13561.2	U	0.5			MIT
7370.21	7372.24	13564.4	U	0.3			ME
7368.65	7370.68	13567.3	U	0.3			ME
7341.66	7343.68	13617.2	U	0.3	H		MIT
7331.86	7333.88	13635.3	U	0.5			MIT
7314.54	7316.56	13667.6	U	0.3			MIT
7282.47	7284.48	13727.8	U	0.3			MIT
7273.73	7275.73	13744.3	U	0.3			MIT
7258.99	7260.99	13772.2	U	0.3			ME
7254.47	7256.47	13780.8	U	0.9			MIT
7251.93	7253.93	13785.6	U	0.3			MIT
7227.89	7229.88	13831.5	U	0.3			MIT
7218.04	7220.03	13850.4	U	0.6			MIT
7212.02	7214.01	13861.9	U	0.3			MIT
7210.29	7212.28	13865.2	U	0.6			MIT
7209.30	7211.29	13867.2	U	0.5			MIT
7205.45	7207.44	13874.6	U	0.3			MIT
7196.66	7198.64	13891.5	U	0.3			MIT
7194.63	7196.61	13895.4	U	0.7			MIT
7189.45	7191.43	13905.4	U	0.3			MIT
7187.62	7189.60	13909.0	U	0.3			MIT
7183.47	7185.45	13917.0	U	0.9			MIT
7179.56	7181.54	13924.6	U	0.3			MIT
7172.10	7174.08	13939.1	U	0.3			MIT
7155.86	7157.83	13970.7	U	0.3			MIT
7147.89	7149.86	13986.3	U	0.7			MIT
7144.08	7146.05	13993.7	U	0.3			MIT
7142.25	7144.22	13997.3	U	0.3			MIT
7141.25	7143.22	13999.3	U	0.3			MIT
7130.09	7132.06	14021.2	U	0.8			MIT
7128.91	7130.88	14023.5	U	1.3			MIT
7126.19	7128.15	14028.9	U	0.3			MIT
7116.07	7118.03	14048.8	U	0.3			MIT
7109.15	7111.11	14062.5	U	0.5			MIT
7101.63	7103.59	14077.4	U	1.1			MIT
7095.68	7097.64	14089.2	U	0.3			MIT
7091.32	7093.28	14097.9	U	0.3			MIT
7082.11	7084.06	14116.2	U	1.0			MIT
7076.33	7078.28	14127.7	U	0.3			MIT
7074.81	7076.76	14130.8	U	1.4			MIT
7073.61	7075.56	14133.2	U	0.8			MIT
7047.04	7048.98	14186.4	U	0.3			MIT
7045.12	7047.06	14190.3	U	0.3			MIT
7033.84	7035.78	14213.1	U	0.7			MIT
7030.69	7032.63	14219.4	U	0.5			MIT
7026.31	7028.25	14228.3	U	0.3			MIT
7020.71	7022.65	14239.7	U	0.8			MIT
7015.72	7017.65	14249.8	U	0.7			MIT
7014.13	7016.06	14253.0	U	0.3			MIT
6989.99	6991.92	14302.2	U	0.3			MIT
6987.72	6989.65	14306.9	U	0.6			MIT
6984.33	6986.26	14313.8	U	0.5			MIT
6965.83	6967.75	14351.8	U	0.3			MIT
6957.84	6959.76	14368.3	U	0.5			MIT
6948.58	6950.50	14387.5	U	0.8			MIT
6942.47	6944.39	14400.1	U	0.3			MIT
6939.36	6941.27	14406.6	U	0.3	H		MIT
6925.94	6927.85	14434.5	U	0.3			MIT
6917.11	6919.02	14452.9	U	0.6			MIT
6915.31	6917.22	14456.7	U	0.5			MIT
6913.86	6915.77	14459.7	U	0.3	H		MIT
6903.79	6905.69	14480.8	U	0.3			MIT
6902.55	6904.45	14483.4	U	0.5			MIT
6901.40	6903.30	14485.8	U	0.3	H		MIT
6900.18	6902.08	14488.4	U	0.3			MIT
6898.74	6900.64	14491.4	U	0.5			MIT

AIR WAVELENGTH (ANGSTRMS)	VACUUM WAVELENGTH (ANGSTRMS)	WAVENUMBER (KAYSERS)	LMENT	LOG ARC	INTENSITY SPK	DIS	REF
6894.78	6896.68	14499.7	U	0.3			MIT
6887.62	6889.52	14514.8	U	0.5			ME
6885.57	6887.47	14519.1	U	0.3			MIT
6876.75	6878.65	14537.7	U	0.9			MIT
6873.61	6875.51	14544.4	U	0.7			MIT
6864.59	6866.48	14563.5	U	0.3H			MIT
6859.85	6861.74	14573.6	U	0.3			MIT
6854.17	6856.06	14585.6	U	0.5			ME
6846.25	6848.14	14602.5	U	0.5			MIT
6839.96	6841.85	14615.9	U	0.5			MIT
6836.06	6837.95	14624.3	U	0.3			MIT
6832.83	6834.72	14631.2	U	0.3			MIT
6826.93	6828.81	14643.8	U	1.4			MIT
6824.56	6826.44	14648.9	U	0.5			MIT
6820.84	6822.72	14656.9	U	0.8			MIT
6818.35	6820.23	14662.3	U	0.6			MIT
6813.75	6815.63	14672.2	U	0.3			MIT
6812.93	6814.81	14673.9	U	0.3			ME
6796.46	6798.34	14709.5	U	0.3			MIT
6790.35	6792.22	14722.7	U	1.0			MIT
6782.77	6784.64	14739.2	U	0.7H			MIT
6780.62	6782.49	14743.8	U	0.5			MIT
6776.89	6778.76	14752.0	U	0.7			MIT
6771.03	6772.90	14764.7	U	0.5			MIT
6769.40	6771.27	14768.3	U	0.3			MIT
6768.64	6770.51	14769.9	U	0.3			MIT
6754.97	6756.83	14799.8	U	0.5			MIT
6736.804	6738.664	14839.74	U	0.6			MIT
6728.09	6729.95	14859.0	U	0.3			MIT
6726.89	6728.75	14861.6	U	0.5			MIT
6720.84	6722.70	14875.0	U	0.5			MIT
6717.51	6719.36	14882.4	U	0.7			MIT
6710.57	6712.42	14897.7	U	0.3			ME
6703.06	6704.91	14914.4	U	0.3			MIT
6701.68	6703.53	14917.5	U	0.3			MIT
6691.21	6693.06	14940.8	U	0.3			MIT
6683.45	6685.30	14958.2	U	0.8			MIT
6681.15	6682.99	14963.3	U	0.3			ME
6676.925	6678.769	14972.82	U	0.9			MIT
6671.29	6673.13	14985.5	U	0.5			MIT
6656.813	6658.651	15018.06	U	0.6			MIT
6641.74	6643.57	15052.1	U	0.3			ME
6630.160	6631.991	15078.43	U	0.3			MIT
6625.284	6627.114	15089.53	U	0.3			MIT
6622.815	6624.644	15095.15	U	0.5			MIT
6621.771	6623.600	15097.53	U	0.8			MIT
6620.524	6622.352	15100.37	U	1.2			MIT
6604.05	6605.87	15138.0	U	0.8			MIT
6603.336	6605.160	15139.68	U	0.5			MIT
6602.684	6604.508	15141.17	U	0.3			MIT
6601.458	6603.281	15143.99	U	0.3			MIT
6590.052	6591.872	15170.20	U	0.7			MIT
6587.834	6589.654	15175.30	U	0.3			MIT
6585.200	6587.019	15181.37	U	0.3			MIT
6582.782	6584.600	15186.95	U	0.6			MIT
6557.581	6559.392	15245.31	U	0.7			MIT
6555.014	6556.825	15251.28	U	1.2			MIT
6552.84	6554.65	15256.3	U	0.5			ME
6549.880	6551.689	15263.24	U	0.6			MIT
6545.725	6547.533	15272.93	U	0.3			MIT
6542.969	6544.777	15279.36	U	0.9			MIT
6536.58	6538.39	15294.3	U	0.3			MIT
6535.457	6537.263	15296.92	U	0.6			MIT
6534.598	6536.403	15298.93	U	0.5			MIT
6527.038	6528.841	15316.65	U	0.8			MIT
6526.078	6527.881	15318.91	U	0.8			MIT
6520.982	6522.784	15330.88	U	0.6			MIT
6518.945	6520.746	15335.67	U	1.2			MIT
6510.782	6512.581	15354.90	U	0.3			MIT
6506.32	6508.12	15365.4	U	0.3			ME

AIR WAVELENGTH (ANGSTRMS)	VACUUM WAVELENGTH (ANGSTRMS)	WAVENUMBER (KAYSERS)	LMENT	LOG ARC	INTENSITY SPK	DIS	REF
6503.59	6505.39	15371.9	U	1.2			ME
6495.35	6497.14	15391.4	U	0.3			ME
6487.028	6488.821	15411.12	U	0.3			MIT
6485.38	6487.17	15415.0	U	0.3			MIT
6481.718	6483.509	15423.75	U	0.5			MT
6481.237	6483.028	15424.89	U	0.3			MIT
6479.950	6481.741	15427.95	U	0.3			MIT
6470.550	6472.338	15450.37	U	0.7			MIT
6465.005	6466.792	15463.62	U	1.4			MIT
6449.167	6450.949	15501.59	U	2.0			MIT
6448.042	6449.824	15504.30	U	0.5			MIT
6442.55	6444.33	15517.5	U	0.3			MIT
6430.93	6432.71	15545.6	U	0.6			MIT
6428.68	6430.46	15551.0	U	0.5			MIT
6427.51	6429.29	15553.8	U	0.3			MIT
6424.888	6426.664	15560.17	U	0.8			MIT
6411.593	6413.365	15592.44	U	0.8			MIT
6404.485	6406.255	15609.74	U	0.5			MIT
6400.355	6402.124	15619.82	U	0.5			MIT
6397.185	6398.953	15627.56	U	1.1			MIT
6395.446	6397.214	15631.81	U	2.0			MIT
6392.781	6394.548	15638.32	U	1.3			MIT
6391.323	6393.090	15641.89	U	0.3			MIT
6389.804	6391.570	15645.61	U	1.3			MIT
6384.915	6386.680	15657.59	U	0.3			MIT
6383.591	6385.356	15660.83	U	0.9			MIT
6379.636	6381.400	15670.54	U	1.2			MIT
6375.984	6377.747	15679.52	U	0.7			MIT
6374.493	6376.255	15683.19	U	0.3			MIT
6373.008	6374.770	15686.84	U	0.6			MIT
6372.469	6374.231	15688.17	U	1.7			MIT
6369.873	6371.634	15694.56	U	0.7			MIT
6367.668	6369.428	15700.00	U	0.3			MIT
6359.305	6361.063	15720.64	U	1.5			MIT
6353.438	6355.195	15735.16	U	0.5			MIT
6346.272	6348.027	15752.93	U	0.3			MIT
6338.750	6340.503	15771.62	U	0.6			MIT
6336.552	6338.304	15777.09	U	0.3			MIT
6330.766	6332.517	15791.51	U	0.8			MIT
6324.463	6326.212	15807.25	U	0.3			MIT
6322.368	6324.116	15812.49	U	0.8			MIT
6306.980	6308.724	15851.07	U	0.6			MIT
6304.559	6306.303	15857.15	U	0.6			MIT
6301.290	6303.033	15865.38	U	0.3			MIT
6298.551	6300.293	15872.28	U	1.1			MIT
6293.347	6295.087	15885.40	U	1.2			MIT
6292.032	6293.772	15888.72	U	1.0			MIT
6291.484	6293.224	15890.11	U	0.5			MIT
6280.195	6281.932	15918.67	U	1.0			MIT
6279.635	6281.372	15920.09	U	0.6			MIT
6270.578	6272.312	15943.08	U	0.6			MIT
6268.688	6270.422	15947.89	U	0.8			MIT
6254.216	6255.946	15984.79	U	0.5			MIT
6251.819	6253.548	15990.92	U	0.9			MIT
6246.550	6248.278	16004.41	U	0.8			MIT
6234.320	6236.045	16035.81	U	1.0			MIT
6232.552	6234.276	16040.35	U	0.3			MIT
6229.662	6231.385	16047.80	U	0.3H			MIT
6227.984	6229.707	16052.12	U	0.3H			MIT
6222.173	6223.894	16067.11	U	0.9			MIT
6219.703	6221.424	16073.49	U	0.3			MIT
6216.881	6218.601	16080.79	U	0.3			MIT
6215.397	6217.117	16084.63	U	1.1			MIT
6203.389	6205.105	16115.76	U	0.5			MIT
6201.994	6203.710	16119.39	U	0.5			MIT
6198.417	6200.132	16128.69	U	0.3			MIT
6196.544	6198.259	16133.56	U	0.3			MIT
6194.265	6195.979	16139.50	U	0.3H			MIT
6192.349	6194.062	16144.49	U	0.3			MIT
6181.369	6183.079	16173.17	U	0.3H			MIT

AIR WAVELENGTH (ANGSTRMS)	VACUUM WAVELENGTH (ANGSTRMS)	WAVENUMBER (KAYSERS)	LMENT	LOG ARC	INTENSITY SPK	DIS	REF
6174.392	6176.101	16191.45	U	0.7			MIT
6172.914	6174.622	16195.32	U	0.6			MIT
6171.872	6173.580	16198.06	U	1.5			MIT
6165.657	6167.363	16214.39	U	0.3			MIT
6164.772	6166.478	16216.71	U	0.7			MIT
6164.528	6166.234	16217.35	U	0.9			MIT
6162.204	6163.909	16223.47	U	0.5H			MIT
6153.687	6155.390	16245.92	U	0.6			MIT
6152.273	6153.976	16249.66	U	0.6			MIT
6141.80	6143.50	16277.4	U	0.3			ME
6140.368	6142.067	16281.16	U	0.3			MIT
6138.545	6140.244	16286.00	U	0.9			MIT
6132.613	6134.310	16301.75	U	1.0			MIT
6129.720	6131.417	16309.44	U	1.0			MIT
6127.776	6129.472	16314.62	U	0.3			MIT
6112.103	6113.795	16356.45	U	0.3			MIT
6110.907	6112.599	16359.65	U	0.3			MIT
6101.787	6103.476	16384.11	U	0.8			MIT
6099.910	6101.599	16389.15	U	0.5			MIT
6089.195	6090.881	16417.99	U	0.3			MIT
6087.338	6089.023	16423.00	U	1.0			MIT
6080.387	6082.070	16441.77	U	0.7			MIT
6077.297	6078.980	16450.13	U	1.6			MIT
6074.817	6076.499	16456.84	U	0.3			MIT
6067.229	6068.909	16477.43	U	1.0			MIT
6062.314	6063.993	16490.79	U	1.0			MIT
6059.731	6061.409	16497.81	U	0.5			MIT
6056.818	6058.495	16505.75	U	0.9			MIT
6051.745	6053.421	16519.58	U	0.9			MIT
6050.492	6052.167	16523.01	U	0.9			MIT
6039.624	6041.296	16552.74	U	1.0			MIT
6028.131	6029.800	16584.30	U	0.8			MIT
6019.153	6020.820	16609.03	U	0.5			MIT
6017.578	6019.245	16613.38	U	0.5			MIT
6017.387	6019.053	16613.91	U	0.6			MIT
6016.755	6018.421	16615.65	U	0.3			MIT
6014.068	6015.734	16623.08	U	0.5			MIT
6010.875	6012.540	16631.91	U	0.6			MIT
6008.871	6010.535	16637.45	U	0.3			MIT
6004.828	6006.491	16648.66	U	0.3			MIT
6000.196	6001.858	16661.51	U	0.7			MIT
5999.427	6001.089	16663.64	U	0.9			MIT
5997.978	5999.639	16667.67	U	0.9			MIT
5997.329	5998.990	16669.47	U	1.4			MIT
5986.442	5988.100	16699.79	U	0.3			MIT
5986.122	5987.780	16700.68	U	1.4			MIT
5982.717	5984.374	16710.19	U	0.3			MIT
5976.344	5977.999	16728.00	U	1.7			MIT
5974.788	5976.443	16732.36	U	0.3			MIT
5971.525	5973.179	16741.50	U	1.7			MIT
5964.058	5965.710	16762.46	U	0.3			MIT
5956.882	5958.532	16782.66	U	0.7			MIT
5952.053	5953.702	16796.27	U	0.5			MIT
5948.590	5950.238	16806.05	U	0.7			MIT
5934.463	5936.107	16846.06	U	0.3			MIT
5933.852	5935.496	16847.79	U	0.8			MIT
5932.444	5934.088	16851.79	U	0.6			MIT
5929.332	5930.975	16860.64	U	0.7			MIT
5925.466	5927.108	16871.64	U	0.5			MIT
5915.398	5917.037	16900.35	U	2.1			MIT
5902.497	5904.133	16937.29	U	0.9			MIT
5898.785	5900.420	16947.95	U	0.9			MIT
5892.633	5894.266	16965.64	U	0.6			MIT
5886.952	5888.584	16982.01	U	0.5			MIT
5870.947	5872.574	17028.31	U II	1.2			ME
5863.426	5865.051	17050.15	U	0.6			MIT
5859.189	5860.813	17062.48	U	0.3			MIT
5856.471	5858.094	17070.40	U	0.3			ME
5853.930	5855.553	17077.81	U	1.0			MIT
5852.026	5853.648	17083.36	U	0.6			MIT

AIR WAVELENGTH (ANGSTRMS)	VACUUM WAVELENGTH (ANGSTRMS)	WAVENUMBER (KAYSERS)	LMENT	LOG ARC	INTENSITY SPK	DIS	REF
5845.272	5846.892	17103.10	U	1.3			MIT
5843.294	5844.914	17108.89	U	0.5			MIT
5841.837	5843.456	17113.16	U	0.3			MIT
5839.051	5840.670	17121.32	U	0.3			MIT
5837.707	5839.325	17125.27	U	1.5	0.0		MIT
5836.047	5837.665	17130.14	U	1.5			MIT
5832.389	5834.006	17140.88	U	0.3H			MIT
5830.889	5832.506	17145.29	U	0.3			MIT
5828.021	5829.637	17153.73	U	0.3			MIT
5826.187	5827.802	17159.13	U	0.3			MIT
5816.790	5818.403	17186.85	U	0.3			MIT
5814.438	5816.050	17193.80	U	0.7			MIT
5813.843	5815.455	17195.56	U	0.7			MIT
5811.291	5812.902	17203.11	U	0.6			MIT
5802.132	5803.741	17230.27	U	1.3			MIT
5800.799	5802.407	17234.23	U	0.3			MIT
5798.552	5800.160	17240.90	U	1.5	0.0		MIT
5796.542	5798.149	17246.88	U	0.5			MIT
5791.773	5793.379	17261.08	U	0.6			MIT
5788.594	5790.199	17270.56	U	0.7			MIT
5783.173	5784.777	17286.75	U	0.3			MIT
5782.823	5784.427	17287.80	U	0.3			MIT
5781.964	5783.567	17290.37	U	0.5			MIT
5780.610	5782.213	17294.42	U	1.6			MIT
5777.810	5779.412	17302.80	U	0.3H	0.0		MIT
5776.897	5778.499	17305.53	U	0.3			MIT
5771.078	5772.678	17322.98	U	0.6			MIT
5767.717	5769.317	17333.07	U	0.3			MIT
5767.461	5769.061	17333.84	U	0.5			MIT
5765.412	5767.011	17340.00	U	0.3			MIT
5763.683	5765.282	17345.21	U	0.3			MIT
5761.876	5763.474	17350.65	U	0.3			MIT
5758.382	5759.979	17361.17	U	0.9			MIT
5758.167	5759.764	17361.82	U	1.0			MIT
5757.339	5758.936	17364.32	U	0.3	0.0		MIT
5751.762	5753.357	17381.16	U	0.5			MIT
5750.537	5752.132	17384.86	U	0.5H			MIT
5748.811	5750.406	17390.08	U	0.3			MIT
5748.47	5750.06	17391.1	U	0.3			MIT
5748.127	5749.721	17392.15	U	0.5			MIT
5742.787	5744.380	17408.32	U	0.3			MIT
5741.349	5742.942	17412.68	U	0.5			MIT
5737.302	5738.893	17424.96	U	0.5			MIT
5736.419	5738.010	17427.64	U	0.8			MIT
5733.262	5734.852	17437.24	U	0.3	0.0		MIT
5731.897	5733.487	17441.39	U	0.3			MIT
5723.632	5725.220	17466.58	U	1.2	0.0		MIT
5722.227	5723.814	17470.87	U	0.5			MIT
5716.873	5718.459	17487.23	U	0.6			MIT
5715.694	5717.280	17490.84	U	0.5			MIT
5715.223	5716.809	17492.28	U	0.3			MIT
5709.488	5711.072	17509.85	U	0.5			MIT
5706.989	5708.572	17517.51	U	0.5	0.0		MIT
5702.867	5704.449	17530.18	U	0.3			MIT
5699.872	5701.453	17539.39	U	0.5			MIT
5698.327	5699.908	17544.14	U	0.3	0.0		MIT
5695.183	5696.763	17553.83	U	0.3			MIT
5691.388	5692.967	17565.53	U	0.6	0.0		MIT
5685.204	5686.782	17584.64	U	0.8	0.0		MIT
5683.334	5684.911	17590.43	U	0.3			MIT
5680.376	5681.952	17599.58	U	0.3			MIT
5674.261	5675.836	17618.55	U	0.3			MIT
5669.447	5671.020	17633.51	U	1.3			MIT
5664.239	5665.811	17649.72	U	0.3			MIT
5658.278	5659.848	17668.32	U	0.6			MIT
5654.818	5656.387	17679.13	U	0.5			MIT
5654.414	5655.983	17680.39	U	0.5	0.0		MIT
5653.786	5655.355	17682.36	U	0.6	0.0		MIT
5644.246	5645.813	17712.24	U	0.3	0.0		MIT
5640.327	5641.893	17724.55	U	0.5			MIT

AIR WAVELENGTH (ANGSTRMS)	VACUUM WAVELENGTH (ANGSTRMS)	WAVENUMBER (KAYSERS)	LMENT	LOG ARC	INTENSITY SPK	DIS	REF
5638.005	5639.570	17731.85	U	0.3			MIT
5636.801	5638.366	17735.64	U	0.5			MIT
5634.408	5635.972	17743.17	U	1.0			MIT
5632.474	5634.037	17749.26	U	0.9			MIT
5629.458	5631.021	17758.77	U	0.5			MIT
5628.164	5629.726	17762.85	U	0.5	0.0		MIT
5628.021	5629.583	17763.30	U	0.3			MIT
5624.127	5625.688	17775.60	U	0.3			MIT
5622.558	5624.119	17780.56	U	0.3			MIT
5621.524	5623.084	17783.83	U	1.1	0.0		MIT
5620.792	5622.352	17786.15	U	1.5	0.0		MIT
5620.494	5622.054	17787.09	U	0.5			MIT
5616.593	5618.152	17799.45	U	0.6			MIT
5614.755	5616.314	17805.27	U	0.3			MIT
5613.683	5615.241	17808.67	U	0.3			MIT
5613.273	5614.831	17809.97	U	0.5			MIT
5612.318	5613.876	17813.00	U	0.3			MIT
5611.598	5613.156	17815.29	U	0.3	0.0		MIT
5610.905	5612.463	17817.49	U	1.5	0.0		MIT
5608.863	5610.420	17823.98	U	0.6			MIT
5603.988	5605.544	17839.48	U	0.6	0.0		MIT
5602.913	5604.469	17842.91	U	0.7			MIT
5597.379	5598.933	17860.55	U	0.8	0.0		MIT
5588.123	5589.675	17890.13	U	0.5	0.0		MI
5587.186	5588.737	17893.13	U	0.3			MIT
5584.636	5586.187	17901.30	U	0.8			MIT
5581.610	5583.160	17911.00	U	1.1	0.7		MIT
5581.230	5582.780	17912.22	U	1.0	0.6		MIT
5580.819	5582.369	17913.54	U	0.9	0.5		MIT
5574.678	5576.226	17933.28	U	0.5			MIT
5573.612	5575.160	17936.71	U	0.6			MIT
5570.683	5572.230	17946.14	U	1.2	1.2		MIT
5568.507	5570.053	17953.15	U	0.5			MIT
5567.959	5569.505	17954.92	U	0.0	0.3		MIT
5567.343	5568.889	17956.90	U	0.3	0.0		MIT
5564.187	5565.732	17967.09	U	1.6	0.5		MIT
5563.681	5565.226	17968.72	U	0.3			MIT
5562.022	5563.567	17974.08	U	0.3H	0.0		MIT
5557.895	5559.438	17987.43	U	1.0	0.0		MIT
5555.769	5557.312	17994.31	U	0.0	0.3H		MIT
5552.619	5554.161	18004.52	U	0.6	0.5		MIT
5551.441	5552.983	18008.34	U	0.9	0.7		MIT
5551.258	5552.800	18008.93	U	0.3	0.0		MIT
5548.75	5550.29	18017.1	U	0.3	0.0		MIT
5548.063	5549.604	18019.30	U	0.8	0.6		MIT
5546.111	5547.651	18025.65	U	0.5	0.0		MIT
5544.822	5546.362	18029.84	U	0.9	0.5H		MIT
5540.187	5541.726	18044.92	U	0.3	0.0		MIT
5538.530	5540.068	18050.32	U	0.8	0.7		MIT
5535.796	5537.334	18059.23	U	0.8	0.3		MIT
5531.282	5532.818	18073.97	U	0.7			MIT
5530.713	5532.249	18075.83	U	0.5	0.0		MIT
5527.848	5529.383	18085.20	U	1.4	1.6		MIT
5526.354	5527.889	18090.09	U	0.3			MIT
5521.043	5522.577	18107.49	U	0.8			MIT
5518.161	5519.694	18116.95	U	0.3			MIT
5513.386	5514.918	18132.64	U	0.7H	0.0		MIT
5511.501	5513.032	18138.84	U	1.5	0.3		MIT
5511.098	5512.629	18140.17	U	0.5	0.0		MIT
5510.435	5511.966	18142.35	U	1.0			MIT
5504.146	5505.675	18163.08	U	1.1	1.0		MIT
5503.675	5505.204	18164.63	U	0.3	0.0		MIT
5503.200	5504.729	18166.20	U	0.3H			MIT
5501.486	5503.014	18171.86	U	0.6H	0.3		MIT
5500.702	5502.230	18174.45	U	1.0			MIT
5498.185	5499.713	18182.77	U	0.5			MIT
5496.444	5497.971	18188.53	U	1.1	0.0		MIT
5494.668	5496.195	18194.41	U	0.9H	0.0		MIT
5492.970	5494.496	18200.03	U	1.8	1.7		MIT
5491.236	5492.762	18205.78	U	0.9	0.8		MIT

AIR WAVELENGTH (ANGSTRMS)	VACUUM WAVELENGTH (ANGSTRMS)	WAVENUMBER (KAYSERS)	LMENT	LOG ARC	INTENSITY SPK	DIS	REF
5488.906	5490.431	18213.51	U	0.9			MIT
5487.021	5488.546	18219.76	U	1.1	0.9		MIT
5484.518	5486.042	18228.08	U	0.6			MIT
5482.548	5484.071	18234.63	U	1.1	1.3		MIT
5481.223	5482.746	18239.04	U	1.5	1.4		MIT
5480.275	5481.798	18242.19	U	1.2	1.4		MIT
5479.653	5481.176	18244.26	U	0.5	0.0		MIT
5477.866	5479.388	18250.21	U	0.8			MIT
5475.725	5477.247	18257.35	U	1.3	1.3		MIT
5474.62	5476.14	18261.0	U	0.3	0.3H		MIT
5465.690	5467.209	18290.87	U	1.1	0.9		MIT
5464.236	5465.754	18295.74	U	0.7			MIT
5459.284	5460.801	18312.33	U	0.9	0.3		MIT
5458.562	5460.079	18314.75	U	0.3	0.3		MIT
5453.439	5454.955	18331.96	U	0.7	0.0		MIT
5452.406	5453.921	18335.43	U	0.5			MIT
5444.476	5445.989	18362.14	U	1.0	0.8		MIT
5443.540	5445.053	18365.29	U	0.3			MIT
5440.069	5441.581	18377.01	U	0.5			MIT
5436.435	5437.946	18389.30	U	0.3			MIT
5428.552	5430.061	18416.00	U	0.3	0.0		MIT
5421.918	5423.425	18438.53	U	0.3	0.0		MIT
5413.947	5415.452	18465.68	U	0.3	0.0		MIT
5409.096	5410.600	18482.24	U	0.7	0.8		MIT
5407.288	5408.791	18488.42	U	0.5	0.0		MIT
5406.888	5408.391	18489.79	U	0.7			MIT
5405.996	5407.499	18492.84	U	0.7	0.8		MIT
5403.204	5404.706	18502.39	U	0.9	0.8		MIT
5400.951	5402.453	18510.11	U	1.0	0.8		MIT
5394.102	5395.602	18533.61	U	0.5H			MIT
5392.842	5394.341	18537.94	U	0.7			MIT
5391.842	5393.341	18541.38	U	0.5	0.0		MIT
5386.209	5387.707	18560.77	U	0.9	1.1		MIT
5385.559	5387.056	18563.01	U	0.6			MIT
5384.242	5385.739	18567.55	U	0.6			MIT
5377.303	5378.798	18591.51	U	0.7	0.5		MIT
5375.764	5377.259	18596.84	U	0.6H			MIT
5373.996	5375.490	18602.95	U	0.6	0.0		MIT
5373.454	5374.948	18604.83	U	0.5	0.3		MIT
5368.432	5369.925	18622.23	U	0.9	0.8		MIT
5363.820	5365.312	18638.25	U	0.8	0.6		MIT
5362.397	5363.888	18643.19	U	0.5	0.9H		MIT
5358.335	5359.825	18657.32	U	0.5			MIT
5352.320	5353.809	18678.29	U	0.5H	0.5H		MIT
5349.917	5351.405	18686.68	U	0.7	0.5		MIT
5336.512	5337.996	18733.62	U	0.8			MIT
5331.853	5333.336	18749.99	U	0.3	0.3		MIT
5329.223	5330.705	18759.24	U	1.0	0.0		MIT
5327.710	5329.192	18764.57	U	0.9	0.8		MIT
5321.605	5323.085	18786.10	U	0.8	0.5		MIT
5319.377	5320.857	18793.97	U	0.8	0.9		MIT
5315.279	5316.758	18808.46	U	0.9	0.0		MIT
5313.722	5315.200	18813.97	U	0.6	0.5		MIT
5313.268	5314.746	18815.57	U	0.5	0.6		MIT
5312.732	5314.210	18817.47	U	0.7	0.7		MIT
5311.881	5313.359	18820.49	U	1.3	1.3		MIT
5310.484	5311.961	18825.44	U	0.6	0.6		MIT
5310.038	5311.515	18827.02	U	1.0	0.9		MIT
5308.544	5310.021	18832.32	U	1.4	0.5		MIT
5300.587	5302.062	18860.59	U	0.9	0.0		MIT
5299.470	5300.945	18864.56	U	0.8	0.5		MIT
5298.81	5300.28	18866.9	U	0.3			MIT
5297.450	5298.924	18871.76	U	0.9			MIT
5292.733	5294.206	18888.57	U	0.3	0.3		MIT
5291.36	5292.83	18893.5	U	0.5	0.0		MIT
5290.505	5291.977	18896.53	U	0.5	0.3		MIT
5288.397	5289.869	18904.06	U	0.8	0.8		MIT
5286.964	5288.435	18909.19	U	0.7	0.3		MIT
5280.389	5281.858	18932.73	U	1.5	0.6		MIT
5278.180	5279.649	18940.65	U	1.1	1.2		MIT

AIR WAVELENGTH (ANGSTRMS)	VACUUM WAVELENGTH (ANGSTRMS)	WAVENUMBER (KAYSERS)	LMENT	LOG ARC	INTENSITY SPK	DIS	REF
5276.701	5278.169	18945.96	U	0.3	0.3		MIT
5275.925	5277.393	18948.75	U	1.1			MIT
5273.739	5275.207	18956.60	U	0.5	0.6		MIT
5272.018	5273.485	18962.79	U	0.9	0.3		MIT
5267.908	5269.374	18977.59	U	0.5	0.5		MIT
5267.422	5268.888	18979.34	U	0.8H			MIT
5257.044	5258.507	19016.80	U	1.2	1.3		MIT
5252.443	5253.905	19033.46	U	0.3	0.0		MIT
5247.749	5249.210	19050.49	U	1.1	1.0		MIT
5247.352	5248.813	19051.93	U	0.8	0.8		MIT
5241.779	5243.238	19072.18	U	0.8			MIT
5241.412	5242.871	19073.52	U	0.5	0.3		MIT
5239.360	5240.818	19080.99	U	0.5			MIT
5238.613	5240.071	19083.71	U	0.8	0.8		MIT
5234.164	5235.621	19099.93	U	0.9			MIT
5232.163	5233.620	19107.24	U	0.3			MIT
5225.000	5226.455	19133.43	U	0.8	0.8		MIT
5224.284	5225.738	19136.05	U	0.6	0.5		MIT
5221.890	5223.344	19144.82	U	0.9			MIT
5216.948	5218.401	19162.96	U	0.9			MIT
5207.786	5209.236	19196.67	U	0.6			MIT
5205.179	5206.628	19206.29	U	1.0	0.8H		MIT
5204.316	5205.765	19209.47	U	1.0	1.0		MIT
5191.524	5192.970	19256.80	U	0.8			MIT
5190.519	5191.964	19260.53	U	0.3	0.0		MIT
5189.885	5191.330	19262.89	U	0.5	0.6		MIT
5189.217	5190.662	19265.36	U	0.3			MIT
5188.580	5190.025	19267.73	U	0.3			MIT
5184.587	5186.031	19282.57	U	1.1	1.2		MIT
5180.894	5182.337	19296.31	U	0.9			MIT
5180.680	5182.123	19297.11	U	0.6			MIT
5174.622	5176.063	19319.70	U	0.3H	0.3		MIT
5174.337	5175.778	19320.77	U	0.9			MIT
5164.604	5166.043	19357.18	U	0.3	0.0H		MIT
5164.157	5165.595	19358.85	U	1.2	0.0		MIT
5160.326	5161.763	19373.22	U	1.3	1.3		MIT
5152.378	5153.813	19403.11	U	0.3			MIT
5150.526	5151.961	19410.08	U	0.3	0.0		MIT
5149.261	5150.695	19414.85	U	0.3			MIT
5146.997	5148.431	19423.39	U	0.7			MIT
5145.097	5146.530	19430.57	U	0.6	0.6		MIT
5144.528	5145.961	19432.72	U	0.8			MIT
5142.406	5143.839	19440.73	U	0.9			MIT
5140.416	5141.848	19448.26	U	0.5	0.0		MIT
5139.987	5141.419	19449.88	U	0.6			MIT
5139.35	5140.78	19452.3	U	0.3			MIT
5132.199	5133.629	19479.40	U	0.6			MIT
5130.847	5132.277	19484.53	U	0.5	0.5		MIT
5129.535	5130.964	19489.51	U	0.3	0.0		MIT
5127.760	5129.189	19496.26	U	0.3			MIT
5125.493	5126.921	19504.88	U	0.7			MIT
5124.087	5125.515	19510.24	U	0.5			MIT
5123.325	5124.753	19513.14	U	0.3			MIT
5122.552	5123.979	19516.08	U	0.7	0.6		MIT
5119.512	5120.939	19527.67	U	0.3			MIT
5119.243	5120.669	19528.70	U	0.3			MIT
5117.249	5118.675	19536.31	U	1.1	1.0		MIT
5115.619	5117.045	19542.53	U	0.3H			MIT
5114.858	5116.283	19545.44	U	0.3			MIT
5110.703	5112.127	19561.33	U	0.6			MIT
5109.403	5110.827	19566.30	U	0.5	0.0		MIT
5107.342	5108.765	19574.20	U	0.8	0.8		MIT
5106.749	5108.172	19576.47	U	0.6			MIT
5103.718	5105.140	19588.10	U	0.3	0.3		MIT
5101.024	5102.446	19598.45	U	0.7			MIT
5093.112	5094.532	19628.89	U	0.5			MIT
5091.305	5092.724	19635.86	U	0.3	0.0		MIT
5090.168	5091.587	19640.24	U	0.5	0.5		MIT
5088.301	5089.719	19647.45	U	1.0			MIT
5085.863	5087.281	19656.87	U	1.0	0.8		MIT

AIR WAVELENGTH (ANGSTRMS)	VACUUM WAVELENGTH (ANGSTRMS)	WAVENUMBER (KAYSERS)	LMENT	LOG ARC	INTENSITY SPK	DIS	REF
5081.427	5082.843	19674.03	U	0.6	0.0		MIT
5080.495	5081.911	19677.64	U	0.5			MIT
5077.823	5079.238	19687.99	U	0.7	0.6		MIT
5076.767	5078.182	19692.09	U	0.7			MIT
5072.773	5074.187	19707.59	U	0.6	0.3		MIT
5070.881	5072.295	19714.94	U	0.3	0.3		MIT
5067.243	5068.656	19729.10	U	0.6	0.3		MIT
5063.773	5065.185	19742.62	U	1.1			MIT
5061.515	5062.926	19751.42	U	0.3			MIT
5060.972	5062.383	19753.54	U	0.6			MIT
5059.55	5060.96	19759.1	U	0.3			MIT
5057.196	5058.606	19768.29	U	0.8			MIT
5055.423	5056.833	19775.23	U	0.3			MIT
5053.362	5054.771	19783.29	U	0.8			MIT
5049.062	5050.470	19800.14	U	0.5			MIT
5047.416	5048.823	19806.59	U	0.8	0.5		MIT
5046.276	5047.683	19811.07	U	0.3	0.0		MIT
5042.806	5044.212	19824.70	U	0.3	0.3		MIT
5036.539	5037.943	19849.37	U	0.3			MIT
5035.528	5036.932	19853.35	U	0.3	0.3		MIT
5030.738	5032.141	19872.26	U	0.3	0.3		MIT
5027.398	5028.800	19885.46	U	1.6	0.6		MIT
5025.432	5026.834	19893.24	U	0.3H			MIT
5024.563	5025.964	19896.68	U	0.3	0.3		MIT
5023.525	5024.926	19900.79	U	0.3H			MIT
5022.645	5024.046	19904.28	U	0.3H	0.0		MIT
5021.728	5023.129	19907.91	U	0.5	0.0		MIT
5013.412	5014.810	19940.93	U	0.3	0.3		MIT
5012.933	5014.331	19942.84	U	0.3			MIT
5011.423	5012.821	19948.85	U	0.9			MIT
5010.892	5012.290	19950.96	U	0.3	0.3		MIT
5008.687	5010.084	19959.75	U	0.6			MIT
5008.222	5009.619	19961.60	U	1.5	1.4		MIT
4996.090	4997.484	20010.07	U	0.3H	0.5H		MIT
4992.940	4994.333	20022.69	U	0.6	0.6		MIT
4992.117	4993.510	20026.00	U	0.3			MIT
4991.538	4992.931	20028.32	U	0.3			MIT
4990.116	4991.508	20034.02	U	0.5			MIT
4989.92	4991.31	20034.8	U		0.5H		MIT
4986.903	4988.294	20046.93	U	0.9	0.8		MIT
4972.100	4973.487	20106.62	U	0.9	0.8		MIT
4968.993	4970.380	20119.19	U	0.3	0.3		MIT
4967.326	4968.712	20125.94	U	0.7			MIT
4965.375	4966.761	20133.85	U	0.8			MIT
4964.624	4966.009	20136.89	U	0.3	0.0		MIT
4961.544	4962.929	20149.39	U	0.3			MIT
4956.790	4958.173	20168.72	U		0.3		MIT
4955.775	4957.158	20172.85	U	0.9			MIT
4952.248	4953.630	20187.22	U	0.5	0.0		MIT
4951.852	4953.234	20188.83	U	0.3			MIT
4950.171	4951.553	20195.69	U	0.9	0.9		MIT
4944.499	4945.879	20218.85	U	0.8			MIT
4942.640	4944.020	20226.46	U	0.8	0.8		MIT
4936.943	4938.321	20249.80	U	0.3H			MIT
4933.657	4935.034	20263.28	U	0.9	0.9		MIT
4933.063	4934.440	20265.72	U	0.8			MIT
4928.447	4929.823	20284.71	U	1.3	0.0		MIT
4927.695	4929.071	20287.80	U	0.3			MIT
4926.442	4927.817	20292.96	U	0.5	0.0		MIT
4924.644	4926.019	20300.37	U	0.8	0.9		MIT
4916.869	4918.242	20332.47	U	0.3			MIT
4916.645	4918.018	20333.40	U	0.3			MIT
4915.490	4916.862	20338.17	U	0.6	0.6		MIT
4913.165	4914.537	20347.80	U	0.9	0.7		MIT
4912.157	4913.528	20351.97	U	0.0	0.3		MIT
4911.668	4913.039	20354.00	U	0.8	0.7		MIT
4910.339	4911.710	20359.51	U	1.2	0.0		MIT
4909.380	4910.751	20363.49	U	0.3			MIT
4900.436	4901.804	20400.65	U	0.7	0.6		MIT
4899.294	4900.662	20405.41	U	1.4	1.4		MIT

AIR WAVELENGTH (ANGSTRMS)	VACUUM WAVELENGTH (ANGSTRMS)	WAVENUMBER (KAYSERS)	LMENT	LOG ARC	INTENSITY SPK	INTENSITY DIS	REF
4898.616	4899.984	20408.23	U	0.6	0.5		MIT
4892.615	4893.981	20433.26	U	0.5	0.5		MIT
4886.332	4887.697	20459.53	U	0.7	0.8		MIT
4885.126	4886.490	20464.59	U	1.3			MIT
4883.778	4885.142	20470.23	U	1.0	0.9		MIT
4878.510	4879.873	20492.34	U	0.9			MIT
4878.002	4879.364	20494.47	U	0.7	0.7		MIT
4874.920	4876.282	20507.43	U	0.3H	0.0H		MIT
4874.348	4875.709	20509.84	U	0.8			MIT
4873.779	4875.140	20512.23	U		0.3H		MIT
4868.856	4870.216	20532.97	U	0.9	0.3H		MIT
4868.329	4869.689	20535.19	U		0.7H		MIT
4861.015	4862.373	20566.09	U	1.0	1.0		MIT
4859.750	4861.108	20571.44	U	0.9	0.9		MIT
4858.085	4859.442	20578.49	U	1.2	1.2		MIT
4856.675	4858.032	20584.47	U	1.0			MIT
4847.660	4849.014	20622.75	U	1.0	1.0		MIT
4844.716	4846.070	20635.28	U	0.5	0.3		MIT
4843.491	4844.844	20640.50	U	0.7			MIT
4842.506	4843.859	20644.70	U	1.1			MIT
4842.407	4843.760	20645.12	U	0.7			MIT
4828.313	4829.662	20705.38	U	0.3			MIT
4825.930	4827.279	20715.61	U	0.8			MIT
4824.668	4826.016	20721.02	U	0.5	0.0		MIT
4819.544	4820.891	20743.05	U	1.1	1.1		MIT
4819.251	4820.598	20744.31	U	0.3			MIT
4818.78	4820.13	20746.3	U	0.3			MIT
4818.437	4819.784	20747.82	U	0.5			MIT
4815.702	4817.048	20759.60	U	0.7			MIT
4814.845	4816.191	20763.30	U	0.3			MIT
4811.699	4813.044	20776.87	U	0.6H			MIT
4810.889	4812.234	20780.37	U	0.8			MIT
4808.874	4810.218	20789.08	U	0.9			HB
4807.614	4808.958	20794.53	U	0.5			MIT
4803.279	4804.622	20813.29	U	0.3H			MIT
4802.022	4803.364	20818.74	U	0.3			MIT
4796.259	4797.600	20843.76	U	0.3			MIT
4792.463	4793.803	20860.27	U	0.3			MIT
4791.835	4793.175	20863.00	U	0.5			MIT
4791.10	4792.44	20866.2	U	0.5S			MIT
4790.063	4791.402	20870.72	U	0.6			MIT
4785.909	4787.247	20888.83	U	0.5			MIT
4784.88	4786.22	20893.3	U	0.3			MIT
4780.197	4781.533	20913.79	U	0.5			MIT
4779.632	4780.968	20916.26	U	0.9	0.5		MIT
4778.797	4780.133	20919.92	U	0.3			MIT
4778.102	4779.438	20922.96	U	0.8			MIT
4777.680	4779.016	20924.81	U	0.8			MIT
4773.440	4774.775	20943.40	U	0.6			MIT
4772.701	4774.035	20946.64	U	0.8	1.3		MIT
4769.260	4770.594	20961.75	U	0.8	1.2		MIT
4768.667	4770.000	20964.36	U	0.8			MIT
4767.601	4768.934	20969.05	U	0.3			MIT
4762.903	4764.235	20989.73	U	0.7	0.0		MIT
4761.362	4762.693	20996.52	U	0.5			MIT
4756.803	4758.133	21016.65	U	1.1	0.8		MIT
4755.729	4757.059	21021.39	U	0.9	1.2		MIT
4753.497	4754.826	21031.26	U	0.3			MIT
4753.156	4754.485	21032.77	U	0.3			MIT
4752.108	4753.437	21037.41	U	0.6			MIT
4751.655	4752.984	21039.42	U	0.5			MIT
4750.668	4751.997	21043.79	U	0.3H			MIT
4749.924	4751.252	21047.08	U	0.6			MIT
4747.386	4748.714	21058.33	U	0.3			MIT
4744.305	4745.632	21072.01	U	0.8	0.0		MIT
4743.895	4745.222	21073.83	U		0.6		MIT
4743.516	4744.843	21075.51	U	1.0			MIT
4742.333	4743.659	21080.77	U	1.0	0.5		MIT
4741.282	4742.608	21085.44	U	0.0	0.3		MIT
4740.285	4741.611	21089.88	U	0.5	1.0		MIT

AIR WAVELENGTH (ANGSTRMS)	VACUUM WAVELENGTH (ANGSTRMS)	WAVENUMBER (KAYSERS)	LMENT	LOG ARC	INTENSITY SPK	INTENSITY DIS	REF
4739.193	4740.519	21094.74	U	0.9			MIT
4738.440	4739.765	21098.09	U	0.5			MIT
4737.626	4738.951	21101.72	U	0.3			MIT
4731.599	4732.923	21128.59	U	1.6	1.7		MIT
4730.998	4732.321	21131.28	U	0.8	0.6		MIT
4730.685	4732.008	21132.68	U	0.7			MIT
4729.030	4730.353	21140.07	U	0.0	0.3		MIT
4727.676	4728.999	21146.13	U	0.6	0.0		HB
4727.336	4728.658	21147.65	U	0.3			MIT
4727.11	4728.43	21148.7	U	0.6	0.6		MIT
4726.860	4728.182	21149.78	U	0.0	0.3		MIT
4725.597	4726.919	21155.43	U	0.0	0.3		MIT
4722.726	4724.047	21168.29	U	1.6	1.7		MIT
4721.322	4722.643	21174.58	U		0.3H		MIT
4720.591	4721.912	21177.86	U		0.3H		MIT
4720.015	4721.335	21180.45	U	0.7	1.0		MIT
4719.08	4720.40	21184.6	U	0.5H			MIT
4718.42	4719.74	21187.6	U	0.3			MIT
4717.829	4719.149	21190.26	U	0.5	0.0		MIT
4716.341	4717.661	21196.95	U		0.3		MIT
4715.669	4716.988	21199.97	U	1.0	0.7		MIT
4714.595	4715.914	21204.80	U	0.6			MIT
4713.943	4715.262	21207.73	U	0.6	0.7		MIT
4712.150	4713.468	21215.80	U	0.6	0.9		MIT
4710.189	4711.507	21224.63	U	0.6	0.8		MIT
4708.169	4709.486	21233.74	U		0.3		MIT
4706.76	4708.08	21240.1	U	0.3	0.5		MIT
4706.404	4707.721	21241.70	U	0.5	0.6		MIT
4704.476	4705.792	21250.41	U	0.5	0.5		MIT
4702.517	4703.833	21259.26	U	1.0	1.3		MIT
4702.047	4703.363	21261.38	U	0.9	1.3		MIT
4701.234	4702.550	21265.06	U	1.1			MIT
4700.980	4702.295	21266.21	U	0.7	1.1		MIT
4700.710	4702.025	21267.43	U	0.9	1.1		MIT
4698.815	4700.130	21276.01	U	0.9R			MIT
4696.085	4697.399	21288.38	U	0.3	0.0		MIT
4695.637	4696.951	21290.41	U	0.0	0.5		MIT
4695.233	4696.547	21292.24	U	0.9	0.8		MIT
4695.125	4696.439	21292.73	U	0.0	0.3		MIT
4692.047	4693.360	21306.70	U		0.5		MIT
4690.739	4692.052	21312.64	U	0.0	0.5		MIT
4689.074	4690.386	21320.21	U	1.5	1.6		MIT
4686.922	4688.234	21330.00	U	0.9			MIT
4685.718	4687.029	21335.48	U	1.0	1.3		MIT
4685.533	4686.844	21336.32	U	0.0	0.5		MIT
4684.644	4685.955	21340.37	U	0.9	0.9		MIT
4684.134	4685.445	21342.69	U		0.5		MIT
4683.611	4684.922	21345.07	U	0.3			MIT
4683.047	4684.358	21347.64	U	0.3	0.9		MIT
4682.60	4683.91	21349.7	U		0.3		MIT
4682.133	4683.443	21351.81	U	0.6	0.3		MIT
4681.183	4682.493	21356.14	U	0.3	0.6		MIT
4677.898	4679.207	21371.14	U		0.6		MIT
4676.30	4677.61	21378.4	U	0.0	0.5H		MIT
4674.228	4675.536	21387.92	U	0.9	0.9		MIT
4673.740	4675.048	21390.15	U		0.5H		MIT
4671.408	4672.716	21400.83	U	1.3	1.5		MIT
4669.309	4670.616	21410.45	U	0.9	1.2		MIT
4668.941	4670.248	21412.14	U	0.5	0.5		MIT
4667.217	4668.524	21420.05	U	0.5	0.7		MIT
4666.856	4668.162	21421.71	U	1.4	1.6		MIT
4666.021	4667.327	21425.54	U	0.5	0.8		MIT
4665.827	4667.133	21426.43	U	0.5			MIT
4665.19	4666.50	21429.4	U		0.3		MIT
4664.09	4665.40	21434.4	U		0.3H		MIT
4663.755	4665.061	21435.95	U	1.3	0.5		MIT
4662.539	4663.844	21441.54	U		0.5		MIT
4661.652	4662.957	21445.62	U	0.9	0.3H		MIT
4660.924	4662.229	21448.97	U	0.0H	0.3H		MIT
4659.353	4660.657	21456.20	U	0.0	0.5		MIT

AIR WAVELENGTH (ANGSTRMS)	VACUUM WAVELENGTH (ANGSTRMS)	WAVENUMBER (KAYSERS)	LMENT	LOG ARC	INTENSITY SPK	DIS	REF
4658.807	4660.111	21458.72	U	0.0	0.3		MIT
4657.553	4658.857	21464.49	U	0.3	0.5		MIT
4655.319	4656.622	21474.79	U	0.3	0.5		MIT
4655.133	4656.436	21475.65	U	0.3	0.8		MIT
4654.836	4656.139	21477.02	U		0.3H		MIT
4653.442	4654.745	21483.46	U	0.9	1.0		MIT
4653.062	4654.365	21485.21	U	0.0	0.3		MIT
4651.546	4652.848	21492.21	U	0.8	0.9		MIT
4650.000	4651.302	21499.36	U	0.7	0.7		MIT
4649.100	4650.402	21503.52	U	0.3	0.6H		MIT
4647.870	4649.171	21509.21	U	0.5	0.8		MIT
4646.603	4647.904	21515.07	U	1.4	1.6		MIT
4645.523	4646.824	21520.08	U	0.3	0.5		MIT
4644.910	4646.211	21522.92	U	0.3	0.5		MIT
4644.137	4645.437	21526.50	U	0.0	0.5		MIT
4643.623	4644.923	21528.88	U	1.1			MIT
4642.485	4643.785	21534.16	U	0.0	0.3H		MIT
4641.658	4642.958	21538.00	U	1.0	1.2		MIT
4640.349	4641.648	21544.07	U	0.7	0.9		MIT
4639.108	4640.407	21549.83	U	0.3	0.5		MIT
4638.916	4640.215	21550.73	U	0.0	0.5		MIT
4637.940	4639.239	21555.26	U	0.9	1.0		MIT
4636.592	4637.890	21561.53	U	0.3H	0.3H		MIT
4635.510	4636.808	21566.56	U	0.5	0.5		MIT
4631.798	4633.095	21583.84	U	0.0	0.6		MIT
4631.620	4632.917	21584.67	U	1.5	0.5		MIT
4629.721	4631.018	21593.53	U	0.5H	0.7		MIT
4629.142	4630.438	21596.23	U	0.0	0.3		MIT
4628.30	4629.60	21600.2	U		0.3		MIT
4627.079	4628.375	21605.86	U	1.5	1.8		MIT
4626.435	4627.731	21608.86	U	0.3			MIT
4625.935	4627.231	21611.20	U		0.3		MIT
4625.486	4626.781	21613.30	U	0.8			MIT
4624.710	4626.005	21616.92	U	0.8	0.9		MIT
4623.433	4624.728	21622.89	U	0.7	0.9		MIT
4622.427	4623.722	21627.60	U	1.1	1.3		MIT
4620.431	4621.725	21636.94	U	0.9			MIT
4620.219	4621.513	21637.94	U	1.4	1.1		MIT
4618.392	4619.686	21646.50	U	0.7	1.3		MIT
4617.541	4618.834	21650.48	U	0.5	0.8		MIT
4617.033	4618.326	21652.87	U		0.3		MIT
4616.441	4617.734	21655.64	U	0.0	0.3		MIT
4615.571	4616.864	21659.73	U	0.0	0.3		MIT
4614.987	4616.280	21662.47	U	0.3	0.5		MIT
4614.678	4615.971	21663.92	U	0.8	0.9		MIT
4612.935	4614.227	21672.10	U	0.3			MIT
4612.602	4613.894	21673.67	U	0.0	0.3		MIT
4611.436	4612.728	21679.15	U	1.1	1.4		MIT
4609.864	4611.155	21686.54	U	1.2	1.3		MIT
4608.813	4610.104	21691.48	U	0.5	0.7		MIT
4608.341	4609.632	21693.71	U	0.3			MIT
4605.148	4606.438	21708.75	U	1.1	1.4		MIT
4603.665	4604.955	21715.74	U	1.4	1.6		MIT
4601.794	4603.083	21724.57	U	0.6	0.6		MIT
4601.130	4602.419	21727.70	U	1.3	1.4		MIT
4600.744	4602.033	21729.53	U	0.3	0.7		MIT
4599.865	4601.154	21733.68	U		0.3		MIT
4599.469	4600.758	21735.55	U		0.5		MIT
4598.340	4599.628	21740.89	U	0.3	0.5		MIT
4597.967	4599.255	21742.65	U	0.3			MIT
4597.547	4598.835	21744.64	U	0.3H	0.9H		MIT
4595.695	4596.983	21753.40	U	0.0	0.6H		MIT
4595.579	4596.867	21753.95	U	0.3			MIT
4595.126	4596.413	21756.09	U	0.0	0.3		MIT
4594.852	4596.139	21757.39	U	0.3	0.0		MIT
4594.294	4595.581	21760.03	U	0.8	0.9		MIT
4592.933	4594.220	21766.48	U	0.5			MIT
4592.557	4593.844	21768.26	U	0.0	0.5		MIT
4591.730	4593.017	21772.18	U	0.5	0.9		MIT
4590.83	4592.12	21776.4	U		0.6H		MIT

AIR WAVELENGTH (ANGSTRMS)	VACUUM WAVELENGTH (ANGSTRMS)	WAVENUMBER (KAYSERS)	LMENT	LOG ARC	INTENSITY SPK	DIS	REF
4590.281	4591.567	21779.06	U	0.3	0.3		MIT
4590.010	4591.296	21780.34	U	0.0	0.7		MIT
4589.292	4590.578	21783.75	U	0.3	0.8H		MIT
4588.660	4589.946	21786.75	U	0.6			MIT
4588.440	4589.726	21787.79	U	0.0	0.3		MIT
4587.266	4588.551	21793.37	U	0.8H			MIT
4586.923	4588.208	21795.00	U		0.3		MIT
4585.588	4586.873	21801.34	U	0.3	0.3		MIT
4584.847	4586.132	21804.87	U	1.0	1.2		MIT
4584.26	4585.54	21807.7	U		0.5		MIT
4583.274	4584.558	21812.35	U	0.9	1.0		MIT
4582.370	4583.654	21816.66	U	0.6	0.6		MIT
4581.724	4583.008	21819.73	U	0.9	1.3		MIT
4581.058	4582.342	21822.90	U	0.3	0.5		MIT
4580.827	4582.111	21824.00	U	0.3	0.6		MIT
4579.639	4580.922	21829.67	U	1.1	1.2		MIT
4578.966	4580.249	21832.87	U	0.3	1.0		MIT
4578.257	4579.540	21836.25	U	0.5	0.8		MIT
4577.185	4578.468	21841.37	U	0.3	0.3		MIT
4576.635	4577.918	21843.99	U	1.1	0.3		MIT
4576.009	4577.291	21846.98	U		0.6		MIT
4575.026	4576.308	21851.68	U	0.3			MIT
4574.793	4576.075	21852.79	U		0.3		MIT
4574.43	4575.71	21854.5	U		0.3		MIT
4574.271	4575.553	21855.28	U	0.3			MIT
4573.687	4574.969	21858.07	U	1.5	1.6		MIT
4573.276	4574.558	21860.04	U	0.7	0.9		MIT
4572.993	4574.275	21861.39	U	0.6	0.8		MIT
4572.243	4573.524	21864.98	U	0.8	0.9		MIT
4571.683	4572.964	21867.65	U	0.5	0.0		MIT
4571.240	4572.521	21869.77	U	0.5	0.0		MIT
4570.987	4572.268	21870.98	U	0.8	1.4		MIT
4570.557	4571.838	21873.04	U	0.0	0.5		MIT
4569.913	4571.194	21876.12	U	1.4	1.6		MIT
4569.192	4570.473	21879.57	U	0.3	0.6		MIT
4568.226	4569.506	21884.20	U	0.6H	1.3		MIT
4567.687	4568.967	21886.78	U	1.3	1.6		MIT
4566.929	4568.209	21890.42	U	0.3	0.6		MIT
4564.279	4565.558	21903.13	U	0.3	0.5		MIT
4563.955	4565.234	21904.68	U	1.1	0.0		MIT
4563.41	4564.69	21907.3	U	0.0	0.6		MIT
4561.224	4562.502	21917.80	U	0.3	0.8		MIT
4560.261	4561.539	21922.43	U		0.5		MIT
4559.848	4561.126	21924.41	U	0.0	0.6		MIT
4559.649	4560.927	21925.37	U	1.2			MIT
4558.048	4559.326	21933.07	U	1.1	0.6		MIT
4557.805	4559.083	21934.24	U	0.5	1.2		MIT
4556.33	4557.61	21941.3	U	0.0	0.3		MIT
4556.008	4557.285	21942.89	U		0.6		MIT
4555.095	4556.372	21947.29	U	1.3	1.6		MIT
4553.858	4555.135	21953.25	U	0.6	0.0		MIT
4552.45	4553.73	21960.0	U	0.3	0.3		MIT
4551.979	4553.255	21962.31	U	1.2	0.0		MIT
4551.66	4552.94	21963.9	U	0.0	0.6		MIT
4550.98	4552.26	21967.1	U	0.7	0.3		MIT
4550.34	4551.62	21970.2	U	0.0	0.3		MIT
4549.852	4551.127	21972.58	U	1.1	1.3		MIT
4548.560	4549.835	21978.82	U	0.6	0.8		MIT
4547.41	4548.68	21984.4	U		0.3		MIT
4546.249	4547.524	21989.99	U	0.3	0.5		MIT
4545.580	4546.854	21993.23	U	1.3	1.4		MIT
4544.355	4545.629	21999.16	U	0.8	0.6		MIT
4543.632	4544.906	22002.66	U	1.7	1.9		MIT
4543.03	4544.30	22005.6	U		0.3H		MIT
4542.59	4543.86	22007.7	U		0.3		MIT
4542.114	4543.387	22010.01	U	0.0	0.5		MIT
4541.705	4542.978	22011.99	U	0.9	1.1		MIT
4541.339	4542.612	22013.76	U	0.6			MIT
4539.28	4540.55	22023.7	U		0.5H		MIT
4539.085	4540.358	22024.70	U	0.7			MIT

AIR WAVELENGTH (ANGSTRMS)	VACUUM WAVELENGTH (ANGSTRMS)	WAVENUMBER (KAYSERS)	LMENT	LOG ARC	INTENSITY SPK	DIS	REF
4538.190	4539.462	22029.04	U	1.4	1.6		MIT
4537.116	4538.388	22034.25	U	0.3	0.6		MIT
4536.824	4538.096	22035.67	U	0.0	0.5		MIT
4536.608	4537.880	22036.72	U	0.9	0.3		MIT
4536.048	4537.320	22039.44	U	0.3	0.5		MIT
4535.532	4536.804	22041.95	U	0.3			MIT
4535.263	4536.535	22043.26	U	0.3	0.5		MIT
4535.09	4536.36	22044.1	U		0.6 H		MIT
4533.709	4534.980	22050.81	U	0.5	0.5		MIT
4533.048	4534.319	22054.03	U	0.3	0.6		MIT
4532.586	4533.857	22056.28	U	0.6	0.0		MIT
4531.319	4532.590	22062.44	U		0.8		MIT
4530.79	4532.06	22065.0	U		0.3		MIT
4529.707	4530.977	22070.29	U	1.1	0.5		MIT
4529.17	4530.44	22072.9	U		0.3		MIT
4528.56	4529.83	22075.9	U		0.8		MIT
4528.171	4529.441	22077.78	U	0.0	0.5		MIT
4527.971	4529.241	22078.76	U	0.0	0.7		MIT
4527.695	4528.965	22080.10	U	0.5	0.3		MIT
4526.641	4527.910	22085.24	U	0.0	0.5		MIT
4525.957	4527.226	22088.58	U	0.9 H	0.7		MIT
4525.649	4526.918	22090.08	U	0.3	0.5		MIT
4525.385	4526.654	22091.37	U		0.3		MIT
4523.344	4524.612	22101.34	U	0.8	0.9		MIT
4521.591	4522.859	22109.91	U	0.6	1.2		MIT
4520.441	4521.709	22115.53	U	0.0	0.5		MIT
4520.083	4521.351	22117.28	U	0.8			MIT
4519.770	4521.038	22118.82	U	0.3	0.0		MIT
4519.223	4520.490	22121.49	U	0.3	0.5		MIT
4518.590	4519.857	22124.59	U	0.0	0.5		MIT
4518.078	4519.345	22127.10	U	0.0	0.6		MIT
4517.230	4518.497	22131.25	U	0.5	1.0		MIT
4516.725	4517.992	22133.73	U	1.2	0.0		MIT
4515.990	4517.257	22137.33	U	0.3			MIT
4515.536	4516.802	22139.56	U	0.3	0.5		MIT
4515.280	4516.546	22140.81	U	1.4	1.6		MIT
4514.573	4515.839	22144.28	U	0.0	0.3		MIT
4514.289	4515.555	22145.67	U	0.3	0.5		MIT
4513.675	4514.941	22148.68	U	0.6	0.7		MIT
4513.373	4514.639	22150.17	U	0.3	0.3		MIT
4512.827	4514.093	22152.85	U	0.0	0.3		MIT
4512.393	4513.659	22154.98	U	0.3	0.5		MIT
4512.180	4513.446	22156.02	U	0.3	0.6		MIT
4511.746	4513.011	22158.15	U	0.7 H	0.9 H		MIT
4511.158	4512.423	22161.04	U	0.6	0.9		MIT
4510.320	4511.585	22165.16	U	1.3	1.5		MIT
4509.885	4511.150	22167.30	U		0.6 H		MIT
4509.355	4510.620	22169.90	U	0.3	0.3 H		MIT
4507.786	4509.050	22177.62	U		0.3		MIT
4507.099	4508.363	22181.00	U	0.3	0.0		MIT
4506.223	4507.487	22185.31	U	0.8	1.1		MIT
4506.063	4507.327	22186.10	U	0.3	0.8		MIT
4504.725	4505.989	22192.69	U	0.5	0.7		MIT
4504.170	4505.433	22195.42	U		0.6 H		MIT
4503.677	4504.940	22197.85	U	0.8	0.5		MIT
4502.455	4503.718	22203.88	U	0.0 H	0.3 H		MIT
4502.255	4503.518	22204.86	U	0.3	0.5		MIT
4501.945	4503.208	22206.39	U	0.3	0.6		MIT
4501.498	4502.761	22208.60	U	0.6	0.0		MIT
4499.643	4500.905	22217.75	U	0.0	0.3		MIT
4499.246	4500.508	22219.71	U	0.0	0.6		MIT
4498.265	4499.527	22224.56	U	0.3	0.3		MIT
4497.890	4499.152	22226.41	U	0.6			MIT
4497.516	4498.778	22228.26	U	0.3	0.0 H		MIT
4496.131	4497.392	22235.11	U	0.3	0.5		MIT
4495.622	4496.883	22237.62	U		0.5		MIT
4494.75	4496.01	22241.9	U	0.6	0.3		MIT
4493.904	4495.165	22246.13	U	0.0	0.3 H		MIT
4493.044	4494.304	22250.38	U	0.8	0.9		MIT
4492.858	4494.118	22251.30	U	0.3	0.5		MIT

AIR WAVELENGTH (ANGSTRMS)	VACUUM WAVELENGTH (ANGSTRMS)	WAVENUMBER (KAYSERS)	LMENT	LOG ARC	INTENSITY SPK	DIS	REF
4491.435	4492.695	22258.35	U	0.3	0.6		MIT
4490.835	4492.095	22261.33	U	1.3	1.4		MIT
4489.073	4490.332	22270.07	U	0.5	0.7		MIT
4488.233	4489.492	22274.23	U	0.3	0.0		MIT
4486.311	4487.570	22283.78	U	0.6	0.8		MIT
4485.215	4486.473	22289.22	U	0.8	1.1		MIT
4484.504	4485.762	22292.75	U	0.3	0.5		MIT
4483.756	4485.014	22296.47	U	0.5	0.8		MIT
4483.466	4484.724	22297.92	U	0.8	0.0 H		MIT
4482.678	4483.936	22301.84	U	0.9	1.0		MIT
4482.258	4483.516	22303.93	U	0.0	0.3		MIT
4480.574	4481.831	22312.31	U	0.3	0.5		MIT
4480.346	4481.603	22313.44	U	0.7	0.8		MIT
4479.438	4480.695	22317.97	U	0.3	0.6 H		MIT
4477.710	4478.966	22326.58	U	1.3	1.4		MIT
4477.461	4478.717	22327.82	U	0.5	0.8		MIT
4476.477	4477.733	22332.73	U	0.9	0.8		MIT
4475.681	4476.937	22336.70	U	0.6	0.7		MIT
4475.277	4476.533	22338.72	U	0.6	0.7		MIT
4474.836	4476.092	22340.92	U	0.5	0.5		MIT
4474.564	4475.820	22342.28	U	0.3 H	0.3 H		MIT
4473.492	4474.747	22347.63	U	0.3	0.5		MIT
4472.335	4473.590	22353.41	U	1.7	1.9		MIT
4471.162	4472.417	22359.28	U	0.5			MIT
4469.328	4470.582	22368.45	U	1.1	0.0		MIT
4469.185	4470.439	22369.17	U	0.3	0.7		MIT
4468.895	4470.149	22370.62	U	0.6	0.8		MIT
4468.318	4469.572	22373.51	U	0.6	0.6		MIT
4467.853	4469.107	22375.84	U	0.3	0.3		MIT
4467.071	4468.325	22379.75	U	0.6	0.8		MIT
4466.314	4467.567	22383.55	U	0.0 H	0.5		MIT
4465.729	4466.982	22386.48	U	0.3	0.3		MIT
4465.133	4466.386	22389.47	U	1.3	1.4		MIT
4463.755	4465.008	22396.38	U	0.3	0.6		MIT
4462.974	4464.227	22400.30	U	1.3	1.5		MIT
4462.760	4464.012	22401.37	U	0.5	0.7		MIT
4462.330	4463.582	22403.53	U	1.2	1.3		MIT
4461.434	4462.686	22408.03	U	1.1			MIT
4460.927	4462.179	22410.58	U	1.1	1.0		MIT
4460.531	4461.783	22412.56	U	0.5	0.8		MIT
4459.804	4461.056	22416.22	U	0.3	0.5		MIT
4458.703	4459.954	22421.75	U	1.0	0.5		MIT
4457.095	4458.346	22429.84	U	0.5	0.8		MIT
4455.886	4457.137	22435.93	U	0.8	0.9		MIT
4454.776	4456.026	22441.52	U	0.3			MIT
4453.218	4454.468	22449.37	U	1.0	0.0 H		MIT
4451.98	4453.23	22455.6	U	0.0	0.7 H		MIT
4451.502	4452.752	22458.02	U	0.9	0.0		MIT
4450.500	4451.749	22463.08	U	0.6	0.8		MIT
4448.333	4449.582	22474.02	U	1.1	0.5 H		MIT
4447.976	4449.225	22475.83	U	0.8	0.6		MIT
4447.064	4448.312	22480.44	U	1.1	0.9 J		MIT
4446.526	4447.774	22483.16	U	0.8			MIT
4446.331	4447.579	22484.14	U	0.8			MIT
4445.964	4447.212	22486.00	U	0.6	0.5		MIT
4445.480	4446.728	22488.45	U	0.3	0.0		MIT
4444.688	4445.936	22492.45	U	1.0	0.5		MIT
4443.647	4444.894	22497.72	U	0.6	0.0 H		MIT
4441.09	4442.34	22510.7	U	0.8	0.3		MIT
4440.739	4441.986	22512.45	U	1.3	1.3		MIT
4440.350	4441.597	22514.43	U	1.0	1.1		MIT
4438.679	4439.925	22522.90	U	0.3	0.3 H		MIT
4438.443	4439.689	22524.10	U	0.3	0.0		MIT
4437.949	4439.195	22526.61	U	0.5	0.0		MIT
4436.941	4438.187	22531.73	U	0.7	0.7		MIT
4436.811	4438.057	22532.39	U	0.3	0.5		MIT
4435.533	4436.778	22538.88	U	0.9	0.8		MIT
4434.945	4436.190	22541.86	U	0.9			MIT
4433.889	4435.134	22547.23	U	1.2	1.1		MIT
4431.875	4433.119	22557.48	U	0.8			MIT

AIR WAVELENGTH (ANGSTRMS)	VACUUM WAVELENGTH (ANGSTRMS)	WAVENUMBER (KAYSERS)	LMENT	LOG ARC	INTENSITY SPK	INTENSITY DIS	REF
4431.66	4432.90	22558.6	U	0.3	0.0H		MIT
4430.088	4431.332	22566.58	U	0.9	0.5		MIT
4429.618	4430.862	22568.97	U	0.8	0.7		MIT
4428.878	4430.122	22572.74	U	0.7	0.7		MIT
4428.379	4429.622	22575.29	U	0.6	0.0		MIT
4427.653	4428.896	22578.99	U	1.1	1.2		MIT
4426.936	4428.179	22582.65	U	1.3	0.0		MIT
4426.676	4427.919	22583.97	U	1.3	1.2		MIT
4426.139	4427.382	22586.71	U	0.5	0.0		MIT
4425.867	4427.110	22588.10	U	0.6	0.3		MIT
4425.412	4426.655	22590.42	U	0.9			MIT
4425.196	4426.439	22591.52	U	0.3	0.7		MIT
4423.737	4424.979	22598.98	U	0.8	0.9		MIT
4423.291	4424.533	22601.25	U	0.8	0.8		MIT
4422.984	4424.226	22602.82	U	1.1	1.0		MIT
4420.641	4421.882	22614.80	U	0.3H	0.5H		MIT
4420.407	4421.648	22616.00	U	1.1	0.5		MIT
4419.08	4420.32	22622.8	U		0.3		MIT
4418.473	4419.714	22625.90	U	1.2			MIT
4417.983	4419.224	22628.41	U	0.6	0.3		MIT
4417.419	4418.660	22631.30	U	0.6	0.7		MIT
4416.825	4418.065	22634.34	U	0.5	0.5		MIT
4416.57	4417.81	22635.6	U	0.0	0.5H		MIT
4415.865	4417.105	22639.26	U	0.9	0.8		MIT
4415.241	4416.481	22642.46	U	1.1	1.1		MIT
4414.743	4415.983	22645.02	U	0.8	0.5H		MIT
4413.137	4414.376	22653.26	U	1.2	0.6		MIT
4412.871	4414.110	22654.62	U	0.6	0.3H		MIT
4412.512	4413.751	22656.47	U	0.5	0.3		MIT
4410.876	4412.115	22664.87	U	0.5H	0.6		MIT
4410.12	4411.36	22668.7	U	0.5	0.0		MIT
4409.668	4410.906	22671.08	U	0.6	0.6		MIT
4408.87	4410.11	22675.2	U		0.3		MIT
4407.958	4409.196	22679.87	U	1.1	1.1		MIT
4406.529	4407.767	22687.23	U	0.6	0.9		MIT
4405.952	4407.190	22690.20	U	0.6	0.9		MIT
4405.732	4406.969	22691.33	U	0.8	0.3		MIT
4404.91	4406.15	22695.6	U		0.3		MIT
4404.045	4405.282	22700.02	U	0.3H	0.0H		MIT
4402.437	4403.674	22708.31	U	0.9	0.9		MIT
4402.303	4403.540	22709.00	U	0.8	0.8		MIT
4400.527	4401.763	22718.17	U	0.0	0.3		MIT
4399.631	4400.867	22722.80	U	1.2	0.8		MIT
4395.75	4396.98	22742.9	U	0.0	0.3		MIT
4394.64	4395.87	22748.6	U	0.3	0.0		MIT
4393.588	4394.822	22754.05	U	1.6	0.8		MIT
4392.506	4393.740	22759.65	U	0.8	0.3		MIT
4392.211	4393.445	22761.18	U	0.5	0.8		MIT
4391.848	4393.082	22763.06	U	0.5H	0.0		MIT
4391.496	4392.730	22764.89	U	0.7	0.6		MIT
4390.550	4391.783	22769.79	U	0.3	0.3		MIT
4390.159	4391.392	22771.82	U	0.0	0.5		MIT
4387.590	4388.823	22785.15	U	0.6	0.0		MIT
4387.315	4388.548	22786.58	U	1.2	0.6		MIT
4385.548	4386.780	22795.76	U	0.0	0.5		MIT
4384.77	4386.00	22799.8	U	0.3	0.3		MIT
4384.59	4385.82	22800.7	U	0.3	0.3		MIT
4383.63	4384.86	22805.7	U	0.6	0.0		MIT
4383.266	4384.498	22807.63	U	1.0	0.0		HB
4382.83	4384.06	22809.9	U	0.3			MIT
4382.340	4383.571	22812.45	U	1.3	0.7		MIT
4382.070	4383.301	22813.85	U	0.8	0.0		MIT
4381.385	4382.616	22817.42	U	0.6	0.5		MIT
4380.270	4381.501	22823.23	U	1.1	0.8		MIT
4379.644	4380.875	22826.49	U	0.3	0.3		MIT
4378.527	4379.757	22832.31	U	0.8	0.3		MIT
4376.789	4378.019	22841.38	U	0.8	0.8		MIT
4373.407	4374.636	22859.04	U	1.1	1.1		MIT
4372.757	4373.986	22862.44	U	1.3	0.7		MIT
4372.572	4373.801	22863.41	U	1.2	1.3		MIT

AIR WAVELENGTH (ANGSTRMS)	VACUUM WAVELENGTH (ANGSTRMS)	WAVENUMBER (KAYSERS)	LMENT	LOG ARC	INTENSITY SPK	INTENSITY DIS	REF
4372.01	4373.24	22866.4	U	0.7			MIT
4371.760	4372.989	22867.66	U	1.3	0.0		MIT
4369.96	4371.19	22877.1	U	0.8	0.3		MIT
4368.472	4369.700	22884.87	U	0.5			MIT
4368.242	4369.470	22886.07	U	0.5	0.5		MIT
4366.85	4368.08	22893.4	U	0.8	0.0H		MIT
4365.553	4366.780	22900.17	U	1.0	0.9		MIT
4365.067	4366.294	22902.72	U	0.3	0.3		MIT
4364.417	4365.644	22906.13	U	0.8	0.6		MIT
4363.825	4365.051	22909.24	U	0.7	0.3		MIT
4363.191	4364.417	22912.57	U	0.3			MIT
4362.929	4364.155	22913.94	U	0.8	0.9		MIT
4362.789	4364.015	22914.68	U	0.3	0.5		MIT
4362.513	4363.739	22916.13	U	0.5	0.5		MIT
4362.262	4363.488	22917.45	U	1.2	1.3		MIT
4362.05	4363.28	22918.6	U	1.5	0.5		MIT
4361.156	4362.382	22923.26	U	0.5	0.6		MIT
4359.473	4360.698	22932.11	U	0.3	0.5		MIT
4358.654	4359.879	22936.42	U	0.3	0.5		MIT
4357.627	4358.852	22941.82	U	0.5	0.5		MIT
4356.882	4358.107	22945.74	U	0.0	0.3		MIT
4356.548	4357.772	22947.50	U	0.7	0.3		MIT
4355.741	4356.965	22951.75	U	1.0	1.3		MIT
4355.632	4356.856	22952.33	U	0.8	1.0		MIT
4354.548	4355.772	22958.04	U	1.0	0.9		MIT
4354.365	4355.589	22959.01	U	0.6	0.8		MIT
4349.388	4350.611	22985.28	U	0.5	0.7		MIT
4347.424	4348.646	22995.66	U	0.3	0.3		MIT
4347.192	4348.414	22996.89	U	1.3	1.3		MIT
4344.179	4345.400	23012.84	U	0.3	0.3		MIT
4342.379	4343.600	23022.38	U	0.3	0.5		MIT
4342.137	4343.358	23023.66	U	0.5	0.5		MIT
4341.752	4342.973	23025.70	U		0.9		MIT
4341.688	4342.909	23026.04	U	1.7	1.7		MIT
4340.704	4341.924	23031.26	U	0.0	0.3		MIT
4340.446	4341.666	23032.63	U	0.7	0.9		MIT
4338.309	4339.529	23043.98	U	0.5	0.6		MIT
4337.636	4338.856	23047.55	U	0.9			MIT
4337.408	4338.627	23048.76	U	1.1	0.5		MIT
4336.515	4337.734	23053.51	U	0.3	0.3		MIT
4335.742	4336.961	23057.62	U	1.0	0.0		MIT
4335.279	4336.498	23060.08	U	0.9	0.0		MIT
4334.946	4336.165	23061.85	U	0.5	0.0H		MIT
4334.469	4335.688	23064.39	U	0.3	0.5		MIT
4333.524	4334.742	23069.42	U	0.7	0.6		MIT
4331.828	4333.046	23078.45	U	0.5	0.3		MIT
4331.457	4332.675	23080.43	U	0.9			MIT
4330.708	4331.926	23084.42	U	0.5	0.0H		MIT
4328.742	4329.959	23094.91	U	1.3	0.0		MIT
4328.602	4329.819	23095.65	U	0.3	0.0		MIT
4328.144	4329.361	23098.10	U	0.3	0.5		MIT
4326.927	4328.144	23104.59	U	0.5	0.8		MIT
4326.595	4327.812	23106.37	U	0.3	0.3		MIT
4326.426	4327.643	23107.27	U	0.3	0.3		MIT
4325.897	4327.113	23110.09	U	1.0	1.0		MIT
4324.555	4325.771	23117.27	U	0.8	0.6		MIT
4323.757	4324.973	23121.53	U	0.9	1.0		MIT
4323.319	4324.535	23123.87	U	0.3	0.0		MIT
4322.397	4323.613	23128.81	U	0.9	0.5		MIT
4322.071	4323.286	23130.55	U	0.5	0.3		MIT
4320.401	4321.616	23139.49	U	0.7	0.3		MIT
4319.782	4320.997	23142.81	U	0.9	0.9		MIT
4319.520	4320.735	23144.21	U	0.7	0.6		MIT
4319.056	4320.271	23146.70	U	0.0	0.7		MIT
4318.024	4319.238	23152.23	U		0.5		MIT
4317.589	4318.803	23154.56	U	0.6	0.5		MIT
4317.244	4318.458	23156.41	U	0.0	0.5		MIT
4317.080	4318.294	23157.29	U	1.1	0.0		MIT
4316.494	4317.708	23160.44	U	1.2	0.0H		MIT
4316.036	4317.250	23162.89	U	0.6	0.3		MIT

AIR WAVELENGTH (ANGSTRMS)	VACUUM WAVELENGTH (ANGSTRMS)	WAVENUMBER (KAYSERS)	LMENT	LOG ARC	INTENSITY SPK	DIS	REF
4315.887	4317.101	23163.69	U	0.5	0.3H		MIT
4313.884	4315.097	23174.45	U	1.2	1.0		MIT
4313.147	4314.360	23178.41	U	1.3	0.0H		MIT
4312.629	4313.842	23181.19	U	0.8	0.6		MIT
4312.344	4313.557	23182.72	U		0.3H		MIT
4310.386	4311.598	23193.25	U	0.9	0.9		MIT
4309.697	4310.909	23196.96	U	1.1	0.5		MIT
4309.175	4310.387	23199.77	U	1.0	0.0		MIT
4307.316	4308.528	23209.78	U	0.5	0.5		MIT
4306.782	4307.993	23212.66	U	1.6R	0.6		MIT
4306.523	4307.734	23214.06	U	0.7	0.7		MIT
4305.184	4306.395	23221.28	U	0.3H	0.5H		MIT
4304.471	4305.682	23225.13	U	0.9	0.8		MIT
4304.137	4305.348	23226.93	U	1.2R	0.0		MIT
4303.325	4304.535	23231.31	U	0.8	0.8		MIT
4302.530	4303.740	23235.60	U	0.3			MIT
4302.089	4303.299	23237.99	U	0.7	0.6		MIT
4301.723	4302.933	23239.96	U	0.8	0.3		MIT
4301.470	4302.680	23241.33	U	1.2	1.2		MIT
4300.785	4301.995	23245.03	U	0.5	0.3		MIT
4300.300	4301.510	23247.65	U	1.0			MIT
4300.003	4301.213	23249.26	U	0.9	0.9		MIT
4299.873	4301.083	23249.96	U	0.6	0.6		MIT
4299.398	4300.607	23252.53	U	0.5	0.5		MIT
4297.604	4298.813	23262.24	U	0.7	0.7		MIT
4297.112	4298.321	23264.90	U	1.3	1.3		MIT
4295.346	4296.554	23274.46	U	1.0	1.0		MIT
4295.103	4296.311	23275.78	U	1.2	0.0		MIT
4294.651	4295.859	23278.23	U	0.5H	0.0		MIT
4293.305	4294.513	23285.53	U	1.3	0.7		MIT
4291.652	4292.859	23294.50	U	0.7	0.0		MIT
4290.885	4292.092	23298.66	U	1.2	1.2		MIT
4289.882	4291.089	23304.11	U	1.1	1.1		MIT
4289.556	4290.763	23305.88	U	0.6	0.5		MIT
4288.841	4290.048	23309.76	U	1.3	0.3		MIT
4288.422	4289.629	23312.04	U	0.3	0.3		MIT
4287.869	4289.075	23315.05	U	1.2	1.3		MIT
4285.745	4286.951	23326.60	U		0.8		MIT
4285.232	4286.438	23329.40	U	1.0	0.0		MIT
4284.838	4286.044	23331.54	U		0.3		MIT
4284.520	4285.726	23333.27	U	0.3H	0.3H		MIT
4283.912	4285.117	23336.58	U	1.0	0.0		MIT
4283.038	4284.243	23341.35	U	0.3	0.3		MIT
4282.45	4283.65	23344.6	U	0.7	0.7		M
4282.028	4283.233	23346.85	U	1.5	1.5		MIT
4281.357	4282.562	23350.51	U	0.0	0.3		MIT
4280.659	4281.864	23354.32	U	1.3	0.5		MIT
4280.212	4281.416	23356.76	U	0.8	0.9		MIT
4279.331	4280.535	23361.56	U	1.0			MIT
4278.171	4279.375	23367.90	U	0.7	0.6		MIT
4277.537	4278.741	23371.36	U	0.7	0.6		MIT
4277.198	4278.402	23373.22	U	0.3	0.3H		MIT
4276.467	4277.670	23377.21	U	1.0	1.1		MIT
4275.978	4277.181	23379.88	U	0.9	0.5		MIT
4275.727	4276.930	23381.26	U	1.1	0.9		MIT
4275.02	4276.22	23385.1	U	0.3	0.0		MIT
4273.975	4275.178	23390.84	U	1.1	1.2		MIT
4273.445	4274.648	23393.74	U	0.9	0.7		MIT
4273.191	4274.394	23395.13	U	0.9			MIT
4272.94	4274.14	23396.5	U	0.5	0.0		MIT
4272.280	4273.482	23400.12	U	0.5	0.5		MIT
4270.89	4272.09	23407.7	U	0.8	0.8		MIT
4270.65	4271.85	23409.1	U	0.7	0.6		MIT
4270.254	4271.456	23411.22	U	0.5	0.3H		MIT
4269.613	4270.815	23414.74	U	1.3	1.5		MIT
4269.085	4270.286	23417.63	U	0.5	0.0		MIT
4268.851	4270.052	23418.92	U	1.1	1.0		MIT
4267.934	4269.135	23423.95	U	1.2	0.6		MIT
4267.303	4268.504	23427.41	U	1.1	1.0		MIT
4266.666	4267.867	23430.91	U	0.5	0.5		MIT

AIR WAVELENGTH (ANGSTRMS)	VACUUM WAVELENGTH (ANGSTRMS)	WAVENUMBER (KAYSERS)	LMENT	LOG ARC	INTENSITY SPK	DIS	REF
4266.331	4267.532	23432.75	U	1.2			MIT
4265.630	4266.831	23436.60	U	0.5	0.3		MIT
4265.178	4266.378	23439.08	U	0.8			MIT
4263.746	4264.946	23446.96	U	1.1	0.9		MIT
4263.419	4264.619	23448.75	U	0.8	0.7		MIT
4263.120	4264.320	23450.40	U		0.3H		MIT
4262.890	4264.090	23451.66	U	1.0	0.7		MIT
4262.155	4263.355	23455.71	U	0.8	0.3		MIT
4261.505	4262.704	23459.28	U	0.9	0.9		MIT
4259.452	4260.651	23470.59	U	1.1	0.5		MIT
4259.200	4260.399	23471.98	U	1.0	0.9		MIT
4258.49	4259.69	23475.9	U	0.3	0.3		MIT
4257.71	4258.91	23480.2	U	0.3	0.3		MIT
4256.95	4258.15	23484.4	U	1.0	0.5		MIT
4255.702	4256.900	23491.27	U	0.9	0.8		MIT
4255.38	4256.58	23493.1	U	0.3H	0.0H		MIT
4253.848	4255.045	23501.51	U	1.1	0.5		MIT
4253.700	4254.897	23502.33	U	0.9	0.3		MIT
4252.426	4253.623	23509.37	U	1.2	1.3		MIT
4251.326	4252.523	23515.45	U	1.0	0.8		MIT
4250.941	4252.138	23517.58	U		0.3H		MIT
4249.525	4250.721	23525.42	U	0.6H	0.5H		MIT
4247.900	4249.096	23534.42	U	0.9	0.0		MIT
4247.430	4248.626	23537.02	U	0.7	0.6		MIT
4247.136	4248.332	23538.65	U	1.0	0.9		MIT
4246.261	4247.456	23543.50	U	1.5	0.3		MIT
4245.950	4247.145	23545.23	U	0.0	0.5		MIT
4244.372	4245.567	23553.98	U	1.4	1.4		MIT
4243.546	4244.741	23558.56	U	0.5	0.5		MIT
4243.347	4244.542	23559.67	U	0.9	0.9		MIT
4243.082	4244.277	23561.14	U	0.8	0.8		MIT
4242.572	4243.767	23563.97	U	0.5	0.5		MIT
4242.290	4243.484	23565.54	U	0.8	0.5		MIT
4241.669	4242.863	23568.99	U	1.6	1.7		MIT
4241.112	4242.306	23572.08	U	1.0	0.3		MIT
4240.587	4241.781	23575.00	U	1.0	1.0		MIT
4240.183	4241.377	23577.25	U		0.3		MIT
4239.740	4240.934	23579.71	U	1.0	1.0		MIT
4236.447	4237.640	23598.04	U	1.0	0.8		MIT
4236.042	4237.235	23600.30	U	1.1	1.0		MIT
4234.688	4235.880	23607.84	U	1.0	0.9		MIT
4234.555	4235.747	23608.58	U	0.7	0.3		MIT
4234.530	4235.722	23608.72	U	1.3	1.2		MIT
4234.062	4235.254	23611.33	U	0.9	0.9		MIT
4233.742	4234.934	23613.12	U		0.6H		MIT
4233.134	4234.326	23616.51	U	1.1	0.0		MIT
4232.39	4233.58	23620.7	U	1.1H	0.7H		MIT
4232.037	4233.229	23622.63	U	1.2	1.2		MIT
4231.676	4232.868	23624.65	U	1.4			MIT
4231.188	4232.380	23627.37	U	0.8	0.8		MIT
4230.515	4231.706	23631.13	U	0.8			MIT
4230.314	4231.505	23632.25	U	0.7	0.7		MIT
4229.822	4231.013	23635.00	U	0.3	0.3		MIT
4229.66	4230.85	23635.9	U	0.0	0.3		MIT
4229.268	4230.459	23638.10	U	0.8	0.0		MIT
4228.759	4229.950	23640.94	U	1.3	1.1		MIT
4228.428	4229.619	23642.79	U	1.0	0.0		MIT
4227.330	4228.521	23648.93	U	0.8	0.9		MIT
4226.60	4227.79	23653.0	U	0.0	0.3		MIT
4226.065	4227.255	23656.01	U	0.8	0.9		MIT
4225.75	4226.94	23657.8	U	0.0	0.5		MIT
4225.369	4226.559	23659.91	U	0.9	0.9		MIT
4224.427	4225.617	23665.18	U	0.9	1.0		MIT
4223.648	4224.838	23669.55	U	0.8	0.0		MIT
4223.30	4224.49	23671.5	U	0.0H	0.5		MIT
4222.94	4224.13	23673.5	U	0.0	0.3		MIT
4222.739	4223.928	23674.64	U	0.8			MIT
4222.375	4223.564	23676.69	U	1.3	0.9		MIT
4222.15	4223.34	23677.9	U	0.0	0.5H		MIT
4221.805	4222.994	23679.88	U	1.1	0.9		MIT

AIR WAVELENGTH (ANGSTRMS)	VACUUM WAVELENGTH (ANGSTRMS)	WAVENUMBER (KAYSERS)	LMENT	LOG ARC	INTENSITY SPK	DIS	REF
4220.701	4221.890	23686.08	U	0.6	0.6		MIT
4220.116	4221.305	23689.36	U	0.5	0.6		MIT
4219.976	4221.165	23690.15	U	1.1	0.3H		MIT
4219.716	4220.905	23691.60	U	1.1	0.9		MIT
4219.508	4220.696	23692.77	U	0.0	0.5		MIT
4219.401	4220.589	23693.37	U	0.6	0.5		MIT
4218.464	4219.652	23698.64	U	0.9	0.5		MIT
4218.13	4219.32	23700.5	U	0.5	0.0		MIT
4217.770	4218.958	23702.53	U	1.0	0.0		MIT
4217.517	4218.705	23703.96	U	0.0	0.7		MIT
4217.118	4218.306	23706.20	U	0.0	0.5		MIT
4216.816	4218.004	23707.90	U	0.0	0.3		MIT
4216.61	4217.80	23709.1	U	0.9	0.0		MIT
4215.99	4217.18	23712.5	U	0.7	0.5H		MIT
4215.51	4216.70	23715.2	U	0.5	0.3		MIT
4215.013	4216.200	23718.04	U	0.6	0.6		MIT
4214.421	4215.608	23721.37	U	1.0	1.0		MIT
4214.285	4215.472	23722.14	U	1.0	0.3		MIT
4213.879	4215.066	23724.42	U	1.3	0.6		MIT
4212.991	4214.178	23729.42	U	0.7	0.7		MIT
4212.730	4213.917	23730.89	U	0.7	0.6H		MIT
4212.477	4213.664	23732.32	U	0.7	0.5		MIT
4212.262	4213.449	23733.53	U	1.1	1.0		MIT
4211.68	4212.87	23736.8	U	1.0	1.0		MIT
4211.620	4212.806	23737.15	U	1.3			MIT
4211.310	4212.496	23738.89	U	1.0	0.8		MIT
4210.873	4212.059	23741.36	U	1.1	1.1		MIT
4210.448	4211.634	23743.75	U	1.0	1.2		MIT
4209.492	4210.678	23749.15	U	1.2	0.0		MIT
4207.23	4208.42	23761.9	U	0.6	0.0H		MIT
4206.409	4207.594	23766.55	U	1.0	1.0		MIT
4206.296	4207.481	23767.19	U	0.7	0.7		MIT
4205.63	4206.81	23770.9	U	0.5			MIT
4204.371	4205.555	23778.07	U	1.2	1.0		MIT
4203.094	4204.278	23785.30	U	0.8	0.6		MIT
4202.676	4203.860	23787.66	U	0.8	0.3		MIT
4201.628	4202.812	23793.59	U	0.0	0.3		MIT
4201.416	4202.600	23794.79	U	0.9	0.9		MIT
4201.131	4202.315	23796.41	U	0.6	0.6		MIT
4200.098	4201.281	23802.26	U	0.8	0.9		MIT
4199.634	4200.817	23804.89	U	0.8	0.3		MIT
4198.241	4199.424	23812.79	U	1.0	0.5		MIT
4197.518	4198.701	23816.89	U	0.9	0.9		MIT
4197.415	4198.598	23817.48	U	0.8	0.6		MIT
4197.160	4198.343	23818.92	U	0.9	0.8		MIT
4196.706	4197.888	23821.50	U	1.1	0.0		MIT
4195.804	4196.986	23826.62	U		0.3		MIT
4195.524	4196.706	23828.21	U		0.3		MIT
4195.190	4196.372	23830.11	U	0.9	0.5		MIT
4195.030	4196.212	23831.02	U	0.5	0.5		MIT
4194.127	4195.309	23836.15	U	0.6	0.0		MIT
4193.856	4195.038	23837.69	U	0.3	0.5		MIT
4193.608	4194.790	23839.10	U	0.5	0.0		MIT
4193.425	4194.607	23840.14	U	0.6	0.3H		MIT
4192.905	4194.086	23843.09	U	1.0	0.0		MIT
4192.322	4193.503	23846.41	U	0.5	0.5		MIT
4192.148	4193.329	23847.40	U	0.8	0.7		MIT
4191.938	4193.119	23848.59	U	1.1	0.3H		MIT
4190.395	4191.576	23857.38	U	0.0	0.3		MIT
4189.277	4190.458	23863.74	U	1.0	1.0		MIT
4189.200	4190.381	23864.18	U	0.5	0.5		MIT
4188.892	4190.072	23865.94	U	0.9			MIT
4188.066	4189.246	23870.64	U	1.0	0.9		MIT
4187.87	4189.05	23871.8	U		0.7		MIT
4186.977	4188.157	23876.85	U	1.0	0.6		MIT
4186.790	4187.970	23877.92	U	0.8	0.7		MIT
4186.477	4187.657	23879.70	U	0.8	0.7		MIT
4186.037	4187.217	23882.21	U	0.8	0.0		MIT
4185.780	4186.960	23883.68	U	0.8	0.8		MIT
4184.50	4185.68	23891.0	U	0.0	0.5H		MIT

AIR WAVELENGTH (ANGSTRMS)	VACUUM WAVELENGTH (ANGSTRMS)	WAVENUMBER (KAYSERS)	LMENT	LOG ARC	INTENSITY SPK	DIS	REF
4184.151	4185.330	23892.98	U	0.3	0.3		MIT
4183.599	4184.778	23896.13	U	0.3	0.7		MIT
4183.266	4184.445	23898.03	U	1.0	0.0		MIT
4182.943	4184.122	23899.88	U	0.8	0.8		MIT
4182.666	4183.845	23901.46	U	0.3	0.3		MIT
4181.550	4182.728	23907.84	U	0.5	0.5		MIT
4180.886	4182.064	23911.64	U	0.7	0.7		MIT
4180.702	4181.880	23912.69	U	0.6	0.5		MIT
4180.307	4181.485	23914.95	U	1.1	0.0H		MIT
4179.634	4180.812	23918.80	U	0.3	0.0		MIT
4179.432	4180.610	23919.96	U	1.0	0.0		MIT
4179.001	4180.179	23922.42	U	1.2	1.1		MIT
4178.525	4179.703	23925.15	U	0.9	0.9		MIT
4177.838	4179.016	23929.08	U	0.6	0.5		MIT
4177.390	4178.567	23931.65	U	0.3	0.7		MIT
4176.910	4178.087	23934.40	U	0.5	0.5		MIT
4175.943	4177.120	23939.94	U		0.5		MIT
4175.448	4176.625	23942.78	U	0.3	0.3		MIT
4175.141	4176.318	23944.54	U	0.3	0.3		MIT
4174.194	4175.371	23949.97	U	1.1	1.1		MIT
4173.792	4174.968	23952.28	U	0.5	0.5		MIT
4173.669	4174.845	23952.98	U	0.8	0.0		MIT
4172.974	4174.150	23956.97	U	1.0	1.2		MIT
4172.18	4173.36	23961.5	U	0.5	0.5		MIT
4171.591	4172.767	23964.91	U	1.5	1.5		MIT
4170.760	4171.936	23969.69	U	0.5	0.6		MIT
4169.961	4171.136	23974.28	U	0.7	0.8		MIT
4169.465	4170.640	23977.13	U	0.8	0.3		MIT
4169.050	4170.225	23979.52	U	1.3	0.3H		MIT
4167.73	4168.90	23987.1	U	0.7	0.3		MIT
4167.369	4168.544	23989.19	U	0.6			MIT
4167.032	4168.207	23991.13	U	0.8	0.3		MIT
4166.63	4167.80	23993.4	U	0.8	0.3		MIT
4165.676	4166.850	23998.94	U	0.9	0.9		MIT
4165.427	4166.601	24000.38	U	0.6	0.8		MIT
4165.182	4166.356	24001.79	U	0.8	0.3		MIT
4164.790	4165.964	24004.05	U	0.9	1.0		MIT
4163.677	4164.851	24010.46	U	1.0	1.0		MIT
4163.225	4164.399	24013.07	U	0.8	0.8		MIT
4163.004	4164.178	24014.34	U	0.6	0.6		MIT
4162.632	4163.806	24016.49	U	0.8	0.8		MIT
4162.430	4163.603	24017.66	U	1.3	0.5H		MIT
4161.885	4163.058	24020.80	U	0.8	0.6		MIT
4161.781	4162.954	24021.40	U	0.3	0.6		MIT
4160.950	4162.123	24026.20	U	1.1	0.0H		MIT
4160.410	4161.583	24029.32	U	0.3	0.0		MIT
4160.273	4161.446	24030.11	U	0.8	0.0		MIT
4159.865	4161.038	24032.47	U	0.5	0.7H		MIT
4159.393	4160.566	24035.19	U	0.7	0.6		MIT
4159.105	4160.278	24036.86	U	0.8	0.7		MIT
4158.561	4159.733	24040.00	U	0.5	0.5		MIT
4158.319	4159.491	24041.40	U	0.8	0.8		MIT
4158.021	4159.193	24043.12	U	0.8	0.0H		MIT
4157.640	4158.812	24045.33	U	0.8			MIT
4156.652	4157.824	24051.04	U	1.2	0.9H		MIT
4156.376	4157.548	24052.64	U	0.3	0.3		MIT
4155.411	4156.583	24058.23	U	1.1	1.3		MIT
4153.976	4155.147	24066.54	U	1.2	0.7		MIT
4153.760	4154.931	24067.79	U	0.3H	0.3		MIT
4153.482	4154.653	24069.40	U	0.9	0.9		MIT
4151.575	4152.746	24080.45	U	0.3	0.5H		MIT
4151.238	4152.409	24082.41	U	0.8	0.0		MIT
4150.82	4151.99	24084.8	U	0.5	0.5		MIT
4150.41	4151.58	24087.2	U	0.9	0.9		MIT
4150.029	4151.199	24089.42	U	0.7	0.3		MIT
4149.204	4150.374	24094.21	U	0.3	0.3		MIT
4148.751	4149.921	24096.84	U	0.5	0.5H		MIT
4148.564	4149.734	24097.93	U	0.7	0.7		MIT
4148.193	4149.363	24100.09	U	0.3	0.3H		MIT
4148.025	4149.195	24101.06	U	1.1			MIT

AIR WAVELENGTH (ANGSTRMS)	VACUUM WAVELENGTH (ANGSTRMS)	WAVENUMBER (KAYSERS)	LMENT	LOG ARC	INTENSITY SPK	INTENSITY DIS	REF
4147.388	4148.558	24104.76	U	0.9	0.9		MIT
4146.979	4148.148	24107.14	U	0.6	0.6		MIT
4146.614	4147.783	24109.26	U	1.2			MIT
4145.391	4146.560	24116.38	U	0.8	0.9		MIT
4144.697	4145.866	24120.41	U	1.0	0.7		MIT
4143.877	4145.046	24125.19	U	0.7	0.0H		MIT
4143.587	4144.756	24126.87	U	0.3	0.5		MIT
4142.350	4143.518	24134.08	U	0.5	0.3		MIT
4142.109	4143.277	24135.48	U	0.6	0.3		MIT
4141.228	4142.396	24140.62	U	1.3	1.5		MIT
4139.330	4140.497	24151.69	U	0.8			MIT
4139.140	4140.307	24152.80	U	1.0	1.0		MIT
4138.664	4139.831	24155.57	U	1.0	1.0		MIT
4137.903	4139.070	24160.02	U	0.8	0.5		MIT
4137.711	4138.878	24161.14	U	0.8	0.0		MIT
4136.807	4137.974	24166.42	U	1.2	0.9		MIT
4136.442	4137.609	24168.55	U	1.1			MIT
4135.756	4136.922	24172.56	U	1.0	0.8		MIT
4135.163	4136.329	24176.02	U	0.8	0.7		MIT
4134.880	4136.046	24177.68	U	0.6	0.8		MIT
4133.569	4134.735	24185.35	U		0.8		MIT
4133.492	4134.658	24185.80	U	1.2			MIT
4133.201	4134.367	24187.50	U	0.9	0.9		MIT
4131.784	4132.949	24195.80	U	0.8H	0.8		MIT
4131.353	4132.518	24198.32	U	0.8	0.8		MIT
4130.740	4131.905	24201.91	U		0.5H		MIT
4129.663	4130.828	24208.22	U	0.6	0.6		MIT
4129.460	4130.625	24209.41	U	0.8	0.7		MIT
4128.889	4130.054	24212.76	U	0.6			MIT
4128.756	4129.921	24213.54	U	0.3	0.3		MIT
4128.336	4129.501	24216.00	U	1.3	1.3		MIT
4127.925	4129.089	24218.42	U	0.8	0.5		MIT
4127.331	4128.495	24221.90	U	0.9	0.0H		MIT
4126.854	4128.018	24224.70	U	0.5	0.5		MIT
4126.436	4127.600	24227.15	U	0.7	0.7		MIT
4125.129	4126.293	24234.83	U	1.2	0.0		MIT
4124.725	4125.889	24237.20	U	1.5	1.4		MIT
4123.959	4125.122	24241.71	U	1.3R	0.0		MIT
4123.66	4124.82	24243.5	U	0.7	0.8		MIT
4123.310	4124.473	24245.52	U	0.7	0.3H		MIT
4123.14	4124.30	24246.5	U	0.5			MIT
4122.354	4123.517	24251.14	U	1.2	0.6		MIT
4122.166	4123.329	24252.25	U	1.0	0.9		MIT
4121.230	4122.393	24257.76	U	0.5	0.5		MIT
4120.69	4121.85	24260.9	U	0.5	0.0H		MIT
4120.10	4121.26	24264.4	U	0.6	0.9		MIT
4119.98	4121.14	24265.1	U	0.5	0.0		MIT
4119.692	4120.854	24266.81	U	0.9	0.6		MIT
4118.394	4119.556	24274.46	U	0.5	0.9		MIT
4118.144	4119.306	24275.94	U	0.3	0.3		MIT
4117.635	4118.797	24278.94	U	0.5	0.5		MIT
4117.456	4118.618	24279.99	U	0.5			MIT
4116.887	4118.049	24283.35	U	1.0	0.9H		MIT
4116.437	4117.598	24286.00	U	0.5	0.5		MIT
4116.097	4117.258	24288.01	U	1.4	1.5		MIT
4115.645	4116.806	24290.68	U	1.0			MIT
4114.893	4116.054	24295.11	U	0.7	0.7		MIT
4114.607	4115.768	24296.80	U	1.0	0.3H		MIT
4113.936	4115.097	24300.76	U	0.9	1.0		MIT
4113.525	4114.686	24303.19	U	0.3	0.3		MIT
4113.107	4114.268	24305.66	U	1.1	1.1		MIT
4112.503	4113.663	24309.23	U	0.6	0.3		MIT
4111.786	4112.946	24313.47	U	0.5	0.0		MIT
4111.527	4112.687	24315.00	U	0.3	0.3		MIT
4111.017	4112.177	24318.02	U	1.0			MIT
4110.830	4111.990	24319.13	U	0.9	0.9		MIT
4110.444	4111.604	24321.41	U	0.8	0.0H		MIT
4109.531	4110.691	24326.81	U	0.8			MIT
4108.475	4109.634	24333.06	U	0.6	0.6		MIT
4108.353	4109.512	24333.79	U	1.0			MIT

AIR WAVELENGTH (ANGSTRMS)	VACUUM WAVELENGTH (ANGSTRMS)	WAVENUMBER (KAYSERS)	LMENT	LOG ARC	INTENSITY SPK	INTENSITY DIS	REF
4107.388	4108.547	24339.50	U	0.7	0.5		MIT
4106.931	4108.090	24342.21	U	1.4	1.0		MIT
4106.416	4107.575	24345.27	U	0.9	1.1		MIT
4106.284	4107.443	24346.05	U	0.9			MIT
4105.888	4107.047	24348.40	U	0.6	0.7		MIT
4105.712	4106.871	24349.44	U	0.5	0.5		MIT
4105.311	4106.470	24351.82	U	1.0	0.5		MIT
4104.945	4106.103	24353.99	U	0.8			MIT
4104.779	4105.937	24354.97	U	0.3	0.3H		MIT
4104.440	4105.598	24356.99	U	0.3	0.3		MIT
4103.118	4104.276	24364.83	U	1.2	0.0H		MIT
4102.865	4104.023	24366.34	U	0.7	0.8		MIT
4102.21	4103.37	24370.2	U	0.0	0.6		MIT
4101.905	4103.063	24372.04	U	1.3	0.0		MIT
4101.319	4102.476	24375.52	U	0.3	0.3H		MIT
4099.930	4101.087	24383.78	U	0.7	0.7		MIT
4099.269	4100.426	24387.71	U	1.3R	0.0		MIT
4098.034	4099.191	24395.06	U	1.1	1.1		MIT
4097.745	4098.902	24396.78	U	1.0	0.0		MIT
4097.400	4098.556	24398.83	U	0.0	0.3		MIT
4096.706	4097.862	24402.97	U	0.8	0.5		MIT
4096.352	4097.508	24405.08	U	1.3	0.7		MIT
4095.746	4096.902	24408.69	U	1.3	1.4		MIT
4095.577	4096.733	24409.69	U	0.6	0.7		MIT
4094.893	4096.049	24413.77	U	0.9	0.9		MIT
4094.623	4095.779	24415.38	U	0.8	0.7		MIT
4093.994	4095.150	24419.13	U	1.1	0.5H		MIT
4093.61	4094.77	24421.4	U	0.5	0.0		MIT
4093.285	4094.440	24423.36	U	0.0	0.3		MIT
4092.286	4093.441	24429.32	U	0.3	0.0H		MIT
4091.837	4092.992	24432.00	U	0.5	0.6		MIT
4091.636	4092.791	24433.21	U	1.1			MIT
4091.52	4092.67	24433.9	U	0.8	1.1		M
4090.370	4091.525	24440.77	U	0.8	0.0		MIT
4090.135	4091.290	24442.17	U	1.4	1.6		MIT
4089.371	4090.525	24446.74	U	0.6			MIT
4088.846	4090.000	24449.88	U	0.8	0.8H		MIT
4088.254	4089.408	24453.42	U	1.4	1.3		MIT
4087.390	4088.544	24458.59	U	0.8	1.1		MIT
4086.674	4087.828	24462.87	U	0.9	0.0		MIT
4086.383	4087.537	24464.61	U	0.3	0.3		MIT
4086.145	4087.298	24466.04	U	0.7	0.8		MIT
4084.93	4086.08	24473.3	U	0.8	0.9		MIT
4084.094	4085.247	24478.32	U	1.0	0.3		MIT
4083.772	4084.925	24480.25	U	0.5	0.0		MIT
4083.632	4084.785	24481.09	U	0.3	0.0		MIT
4082.639	4083.792	24487.05	U	0.8	0.9		MIT
4082.003	4083.155	24490.86	U	0.7	0.8		MIT
4081.263	4082.415	24495.30	U	1.0	0.6		MIT
4080.609	4081.761	24499.23	U	1.1	1.3		MIT
4079.847	4080.999	24503.80	U	0.0	0.3		MIT
4078.827	4079.979	24509.93	U	0.5	0.9		MIT
4077.786	4078.937	24516.19	U	1.2	0.8		MIT
4076.72	4077.87	24522.6	U	0.9	1.0		MIT
4076.225	4077.376	24525.58	U	0.3	0.5		MIT
4075.94	4077.09	24527.3	U	0.8	0.0H		MIT
4075.664	4076.815	24528.95	U	0.7	0.8		MIT
4075.085	4076.236	24532.44	U	0.3	0.3		MIT
4074.488	4075.638	24536.03	U	1.0	1.0		MIT
4074.241	4075.391	24537.52	U	0.8	0.3		MIT
4073.554	4074.704	24541.66	U	0.9			MIT
4072.830	4073.980	24546.02	U	0.9	0.3H		MIT
4072.031	4073.181	24550.84	U	0.3	0.6		MIT
4071.108	4072.258	24556.40	U	1.2	1.4		MIT
4070.890	4072.040	24557.72	U	0.8	0.0		MIT
4070.777	4071.926	24558.40	U	0.3	0.6		MIT
4070.443	4071.592	24560.42	U	0.8	0.3		MIT
4070.012	4071.161	24563.02	U	1.1	0.0		MIT
4069.611	4070.760	24565.44	U	0.9			MIT
4069.074	4070.223	24568.68	U	0.5	0.5		MIT

AIR WAVELENGTH (ANGSTRMS)	VACUUM WAVELENGTH (ANGSTRMS)	WAVENUMBER (KAYSERS)	LMENT	LOG ARC	INTENSITY SPK	DIS	REF
4068.967	4070.116	24569.32	U	0.6	0.8		MIT
4068.695	4069.844	24570.97	U	0.3	0.6		MIT
4068.176	4069.325	24574.10	U	0.8			MIT
4067.756	4068.905	24576.64	U	1.1	1.3		MIT
4067.20	4068.35	24580.0	U	0.6	0.6		MIT
4067.085	4068.234	24580.69	U	0.5	0.7		MIT
4066.800	4067.948	24582.42	U	0.8	1.1		MIT
4066.53	4067.68	24584.1	U	0.6	0.9		MIT
4064.41	4065.56	24596.9	U	0.8			MIT
4064.165	4065.313	24598.35	U	0.8	1.0		MIT
4063.119	4064.266	24604.69	U	1.2	0.0		MIT
4062.549	4063.696	24608.14	U	1.1	1.3		MIT
4062.325	4063.472	24609.50	U	0.8			MIT
4061.744	4062.891	24613.01	U	0.3	0.7		MIT
4061.349	4062.496	24615.41	U	1.1	0.0		MIT
4060.907	4062.054	24618.09	U	0.6	0.7		MIT
4060.768	4061.915	24618.93	U	0.8			MIT
4060.232	4061.379	24622.18	U	0.5	0.0		MIT
4060.103	4061.250	24622.96	U	0.6	0.9		MIT
4059.025	4060.171	24629.50	U	0.0	0.3		MIT
4058.16	4059.31	24634.7	U	1.0	0.6		M
4057.955	4059.101	24636.00	U	0.0	0.8		MIT
4056.740	4057.886	24643.37	U	0.0	0.3		MIT
4056.291	4057.437	24646.10	U	0.7			MIT
4056.07	4057.22	24647.4	U		0.7		MIT
4055.985	4057.131	24647.96	U	0.9			MIT
4055.730	4056.876	24649.51	U		0.7		MIT
4054.84	4055.99	24654.9	U	0.0	0.6		MIT
4054.313	4055.458	24658.13	U	1.1	1.2		MIT
4053.570	4054.715	24662.65	U	0.7	0.7		MIT
4053.268	4054.413	24664.48	U	0.3			MIT
4053.026	4054.171	24665.96	U	0.9	0.9		MIT
4052.662	4053.807	24668.17	U	0.8			MIT
4052.477	4053.622	24669.30	U	0.5	0.5		MIT
4052.080	4053.225	24671.72	U	0.7	0.7		MIT
4051.914	4053.059	24672.73	U	1.3	1.4		MIT
4051.087	4052.231	24677.76	U	0.8	0.5		MIT
4050.930	4052.074	24678.72	U	0.5	0.7		MIT
4050.282	4051.426	24682.67	U	0.3	0.5		MIT
4050.039	4051.183	24684.15	U	1.4	1.5		MIT
4049.74	4050.88	24686.0	U	0.9	0.3H		MIT
4049.527	4050.671	24687.27	U		0.6		MIT
4049.170	4050.314	24689.45	U	1.0	0.5		MIT
4048.065	4049.209	24696.18	U	0.9	0.3		MIT
4047.610	4048.753	24698.96	U	1.3	0.5		MIT
4047.05	4048.19	24702.4	U	0.8	0.9		MIT
4046.521	4047.664	24705.61	U	0.7	0.7		MIT
4046.402	4047.545	24706.33	U	0.5	0.7		MIT
4045.136	4046.279	24714.07	U	0.6H	0.6H		MIT
4044.824	4045.967	24715.97	U	1.0	0.3		MIT
4044.416	4045.559	24718.47	U	1.3	1.4		MIT
4044.041	4045.183	24720.76	U	0.0	0.6		MIT
4043.127	4044.269	24726.35	U	0.3	0.5		MIT
4042.752	4043.894	24728.64	U	1.6	1.0		MIT
4042.463	4043.605	24730.41	U	0.7	0.9		MIT
4041.558	4042.700	24735.95	U	0.3	0.6		MIT
4039.781	4040.922	24746.82	U	1.2	0.3		MIT
4039.639	4040.780	24747.70	U	0.0	0.5		MIT
4038.802	4039.943	24752.82	U	0.5			MIT
4038.631	4039.772	24753.87	U	0.5	0.5		MIT
4038.512	4039.653	24754.60	U	0.7	0.7		MIT
4038.168	4039.309	24756.71	U	0.7	0.8		MIT
4037.987	4039.128	24757.82	U	0.9	0.3		MIT
4037.834	4038.975	24758.76	U	0.5	0.0		MIT
4036.867	4038.008	24764.69	U	1.0			MIT
4036.554	4037.695	24766.61	U	0.5	0.9		MIT
4034.002	4035.142	24782.28	U	0.6	0.6		MIT
4033.728	4034.868	24783.96	U	1.1	1.1		MIT
4033.427	4034.567	24785.81	U	1.1	1.0		MIT
4032.294	4033.433	24792.77	U	0.3	0.3H		MIT
4031.781	4032.920	24795.93	U	0.9	0.3		MIT
4031.306	4032.445	24798.85	U	0.9	0.9		MIT
4030.758	4031.897	24802.22	U	0.7	0.8		MIT
4029.92	4031.06	24807.4	U	0.8	0.6		MIT
4029.798	4030.937	24808.13	U	0.0	0.3		MIT
4028.403	4029.541	24816.72	U	0.7	0.3H		MIT
4027.805	4028.943	24820.40	U	1.1	0.9		MIT
4027.402	4028.540	24822.89	U	0.6	0.3		MIT
4026.999	4028.137	24825.37	U	0.8	0.9		MIT
4026.434	4027.572	24828.85	U	1.0			MIT
4026.019	4027.157	24831.42	U	1.4	1.4		MIT
4025.434	4026.572	24835.02	U	0.5	0.3		MIT
4025.013	4026.151	24837.62	U	0.9	0.5		MIT
4023.597	4024.734	24846.36	U	0.5H	0.8H		MIT
4023.170	4024.307	24849.00	U	0.8	0.8		MIT
4021.929	4023.066	24856.67	U	0.3	0.5		MIT
4021.40	4022.54	24859.9	U	0.5H	0.5H		MIT
4021.254	4022.391	24860.84	U	0.8	0.8		MIT
4021.005	4022.141	24862.38	U		0.6		MIT
4020.695	4021.831	24864.29	U	0.3	0.3		MIT
4020.169	4021.305	24867.55	U	0.5	0.5		MIT
4019.203	4020.339	24873.52	U	0.8	0.9		MIT
4018.990	4020.126	24874.84	U	1.4	1.2		MIT
4018.282	4019.418	24879.23	U	0.9	0.5		MIT
4017.723	4018.859	24882.69	U	1.4	1.4		MIT
4017.458	4018.594	24884.33	U	0.9	0.8		MIT
4016.846	4017.981	24888.12	U	0.8	0.9		MIT
4016.341	4017.476	24891.25	U	1.1	0.9		MIT
4016.046	4017.181	24893.08	U	0.7	0.8		MIT
4015.240	4016.375	24898.07	U	1.0			MIT
4014.816	4015.951	24900.70	U	1.1	0.3		MIT
4014.160	4015.295	24904.77	U	0.9	0.9		MIT
4013.431	4014.565	24909.30	U	0.3	0.3H		MIT
4013.288	4014.422	24910.18	U	0.3	0.3		MIT
4013.029	4014.163	24911.79	U	0.9			MIT
4012.708	4013.842	24913.78	U	0.3H	0.3		MIT
4012.161	4013.295	24917.18	U	0.8	0.6		MIT
4011.78	4012.91	24919.6	U	1.0	0.7		MIT
4011.450	4012.584	24921.60	U	0.9	1.0		MIT
4010.816	4011.950	24925.54	U	1.0	0.0		MIT
4010.353	4011.487	24928.41	U	0.3	0.3H		MIT
4009.410	4010.543	24934.28	U	0.6	0.9		MIT
4009.170	4010.303	24935.77	U	0.9	1.2		MIT
4008.918	4010.051	24937.34	U	0.9			MIT
4008.70	4009.83	24938.7	U	0.3	0.0		MIT
4007.934	4009.067	24943.46	U	0.9	0.5		MIT
4007.689	4008.822	24944.98	U	0.3	0.6		MIT
4007.469	4008.602	24946.35	U	0.3	0.3		MIT
4006.395	4007.528	24953.04	U	0.5	0.7		MIT
4005.698	4006.830	24957.38	U	1.4	0.5		MIT
4004.617	4005.749	24964.12	U	0.5	0.7		MIT
4004.415	4005.547	24965.38	U	0.8			MIT
4004.063	4005.195	24967.57	U	1.2	1.3		MIT
4003.405	4004.537	24971.68	U	0.8	1.0		MIT
4003.196	4004.328	24972.98	U	0.9	0.7		MIT
4002.592	4003.724	24976.75	U	0.8			MIT
4002.338	4003.470	24978.33	U	1.0	1.3		MIT
4001.976	4003.107	24980.59	U		0.3		MIT
4001.246	4002.377	24985.15	U	1.0	0.5		MIT
4000.910	4002.041	24987.25	U	0.7	0.8		MIT
4000.732	4001.863	24988.36	U	0.8	0.0		MIT
3999.950	4001.081	24993.25	U	0.8	0.9		MIT
3999.183	4000.314	24998.04	U	1.0	0.5H		MIT
3998.820	3999.951	25000.31	U	0.5	0.7		MIT
3998.697	3999.828	25001.08	U	0.8	0.6		MIT
3998.242	3999.373	25003.92	U	0.7	1.3		MIT
3997.325	3998.455	25009.66	U	0.7	0.7		MIT
3997.088	3998.218	25011.14	U	1.1	0.3		MIT
3995.972	3997.102	25018.13	U	1.0	0.9		MIT
3995.775	3996.905	25019.36	U	0.6	0.5		MIT

AIP WAVELENGTH (ANGSTRMS)	VACUUM WAVELENGTH (ANGSTRMS)	WAVENUMBER (KAYSERS)	LMENT	LOG ARC	INTENSITY SPK	DIS	REF
3995.531	3996.661	25020.89	U	0.6			MIT
3994.980	3996.110	25024.34	U	0.9	1.3J		MIT
3994.718	3995.848	25025.98	U	0.3	0.3		MIT
3994.572	3995.702	25026.89	U	0.6			MIT
3994.294	3995.423	25028.64	U	0.8	0.8		MIT
3993.815	3994.944	25031.64	U	1.1	0.6		MIT
3993.265	3994.394	25035.08	U	0.3	0.3		MIT
3992.960	3994.089	25037.00	U	0.3	0.5		MIT
3992.539	3993.668	25039.64	U	1.0	0.9		MIT
3992.228	3993.357	25041.59	U	0.8	0.3		MIT
3991.617	3992.746	25045.42	U	0.6	0.8		MIT
3991.498	3992.627	25046.17	U	0.6			MIT
3990.423	3991.551	25052.92	U	1.3	1.3		MIT
3990.155	3991.283	25054.60	U	0.7	1.0		MIT
3989.953	3991.081	25055.87	U	0.6	0.8		MIT
3989.893	3991.021	25056.24	U	0.3	0.0		MIT
3989.292	3990.420	25060.02	U	0.9	0.0H		MIT
3989.113	3990.241	25061.14	U	0.3	0.0		MIT
3988.885	3990.013	25062.57	U	1.1	1.2		MIT
3988.644	3989.772	25064.09	U	0.9	0.9		MIT
3988.293	3989.421	25066.29	U	0.3	0.0		MIT
3988.029	3989.157	25067.95	U	0.9	1.1		MIT
3987.720	3988.848	25069.90	U	0.3	0.3		MIT
3987.053	3988.181	25074.09	U	0.3	0.6		MIT
3986.851	3987.979	25075.36	U		0.9		MIT
3986.438	3987.565	25077.96	U	0.3	0.0		MIT
3985.795	3986.922	25082.00	U	1.4	1.5		MIT
3985.260	3986.387	25085.37	U	0.0	0.5		MIT
3985.041	3986.168	25086.75	U	1.0	0.8		MIT
3984.712	3985.839	25088.82	U	0.0	0.7		MIT
3984.185	3985.312	25092.14	U	0.8	0.6H		MIT
3983.908	3985.035	25093.88	U	0.7	1.2		MIT
3983.295	3984.422	25097.75	U		0.6H		MIT
3982.979	3984.106	25099.74	U	0.6	0.3		MIT
3982.531	3983.657	25102.56	U		0.3H		MIT
3982.117	3983.243	25105.17	U	0.8	0.8		MIT
3981.531	3982.657	25108.86	U	0.8	0.7		MIT
3980.887	3982.013	25112.93	U		0.5		MIT
3980.800	3981.926	25113.48	U	1.0	0.0		MIT
3979.941	3981.067	25118.90	U	0.6			MIT
3979.516	3980.642	25121.58	U	0.3H	0.3		MIT
3979.121	3980.247	25124.07	U	0.0	0.5		MIT
3978.796	3979.921	25126.12	U	0.9	1.3		MIT
3978.405	3979.530	25128.59	U	0.8			MIT
3978.274	3979.399	25129.42	U	0.6	0.6		MIT
3978.159	3979.284	25130.15	U	0.6			MIT
3977.351	3978.476	25135.25	U	0.8	0.0		MIT
3977.064	3978.189	25137.07	U	0.8	0.9		MIT
3975.253	3976.378	25148.52	U	0.5	0.3		MIT
3974.975	3976.099	25150.28	U	0.6	0.9		MIT
3973.942	3975.066	25156.81	U	1.2	0.6		MIT
3973.234	3974.358	25161.30	U	0.8	0.9		MIT
3972.340	3973.464	25166.96	U	0.6	1.0		MIT
3972.218	3973.342	25167.73	U	1.0	0.6		MIT
3972.004	3973.128	25169.09	U	0.5	0.5		MIT
3971.862	3972.986	25169.99	U	0.8	0.8		MIT
3971.596	3972.720	25171.67	U	0.6			MIT
3971.397	3972.520	25172.94	U		0.6		MIT
3971.031	3972.154	25175.25	U		0.3		MIT
3970.588	3971.711	25178.06	U	1.0	0.3		MIT
3970.143	3971.266	25180.89	U	0.0	0.7		MIT
3969.422	3970.545	25185.46	U	0.7			MIT
3969.022	3970.145	25188.00	U	0.7	0.9		MIT
3968.374	3969.497	25192.11	U	0.0	0.3		MIT
3968.007	3969.130	25194.44	U	0.8H	0.6		MIT
3967.639	3968.762	25196.78	U		0.3		MIT
3967.48	3968.60	25197.8	U	1.0			MIT
3967.013	3968.135	25200.75	U	0.3	0.3		MIT
3966.567	3967.689	25203.59	U	1.3	1.5		MIT
3966.398	3967.520	25204.66	U	0.0	0.6		MIT

AIR WAVELENGTH (ANGSTRMS)	VACUUM WAVELENGTH (ANGSTRMS)	WAVENUMBER (KAYSERS)	LMENT	LOG ARC	INTENSITY SPK	DIS	REF
3965.967	3967.089	25207.40	U	0.0	0.6		MIT
3965.850	3966.972	25208.14	U	0.3	0.3		MIT
3965.293	3966.415	25211.68	U	0.5	0.0		MIT
3964.963	3966.085	25213.78	U	0.9	0.9		MIT
3964.666	3965.788	25215.67	U	1.0	1.0		MIT
3964.226	3965.348	25218.47	U	0.7	0.6		MIT
3964.112	3965.234	25219.20	U	0.8	0.9		MIT
3963.037	3964.158	25226.04	U		0.3		MIT
3962.788	3963.909	25227.62	U	1.2	0.9		MIT
3962.470	3963.591	25229.65	U	0.0	0.6		MIT
3962.271	3963.392	25230.91	U	0.6	0.3		MIT
3962.160	3963.281	25231.62	U	0.7			MIT
3961.984	3963.105	25232.74	U	0.8	0.3		MIT
3961.660	3962.781	25234.80	U	0.0	0.5		MIT
3961.515	3962.636	25235.73	U	0.9	0.0		MIT
3961.064	3962.185	25238.60	U	0.8			MIT
3960.804	3961.925	25240.26	U	0.7	0.9		MIT
3960.523	3961.644	25242.05	U	0.8	1.0		MIT
3960.387	3961.508	25242.92	U	0.8	0.9		MIT
3960.171	3961.292	25244.29	U	0.8	0.7		MIT
3959.647	3960.767	25247.63	U	0.5	0.3		MIT
3959.205	3960.325	25250.45	U	0.7	0.0		MIT
3958.500	3959.620	25254.95	U	0.9			MIT
3958.139	3959.259	25257.25	U	0.3H	0.5H		MIT
3958.008	3959.128	25258.09	U	0.9H	0.8		MIT
3957.812	3958.932	25259.34	U	1.0	1.2		MIT
3956.382	3957.502	25268.47	U		0.3		MIT
3955.717	3956.836	25272.72	U	0.5H	0.8		MIT
3955.378	3956.497	25274.88	U	0.9	1.2		MIT
3954.663	3955.782	25279.45	U	1.3	1.5		MIT
3954.477	3955.596	25280.64	U	0.3	0.3		MIT
3954.234	3955.353	25282.19	U	0.8	0.9		MIT
3953.578	3954.697	25286.39	U	0.8	1.2		MIT
3952.949	3954.068	25290.41	U	1.2			MIT
3952.447	3953.566	25293.62	U	0.6	0.3		MIT
3952.263	3953.382	25294.80	U	0.5	0.8		MIT
3951.855	3952.973	25297.41	U	0.0	0.6		MIT
3951.553	3952.671	25299.35	U	0.0	0.9		MIT
3950.699	3951.817	25304.81	U	0.5	0.0		MIT
3950.477	3951.595	25306.24	U	0.9			MIT
3950.132	3951.250	25308.45	U	0.5	0.8		MIT
3949.516	3950.634	25312.39	U	0.3	0.9		MIT
3949.309	3950.427	25313.72	U	0.3	0.8		MIT
3948.991	3950.109	25315.76	U	0.9			MIT
3947.959	3949.076	25322.38	U	0.3	0.9		MIT
3947.509	3948.626	25325.26	U	1.0	1.0		MIT
3947.081	3948.198	25328.01	U	0.0	0.3		MIT
3946.880	3947.997	25329.30	U	0.0	0.5H		MIT
3946.677	3947.794	25330.60	U	1.2	1.0		MIT
3946.251	3947.368	25333.34	U	0.8	0.7		MIT
3945.726	3946.843	25336.71	U	0.3	0.5		MIT
3944.621	3945.738	25343.80	U	0.9	0.3		MIT
3944.130	3945.246	25346.96	U	0.9	1.2		MIT
3943.820	3944.936	25348.95	U	1.5	0.7		MIT
3943.498	3944.614	25351.02	U	0.8	1.0		MIT
3942.95	3944.07	25354.5	U	0.6	0.3		MIT
3942.829	3943.945	25355.32	U	1.3	0.9		MIT
3942.551	3943.667	25357.11	U	0.9	1.0		MIT
3942.253	3943.369	25359.03	U	0.8	0.9		MIT
3942.061	3943.177	25360.26	U	1.1	0.0		MIT
3941.766	3942.882	25362.16	U	0.8	0.0		MIT
3941.460	3942.576	25364.13	U	0.7	0.8		MIT
3941.089	3942.205	25366.52	U	0.5	0.8		MIT
3940.487	3941.602	25370.39	U	1.0	1.2		MIT
3940.262	3941.377	25371.84	U	0.6	0.8		MIT
3939.758	3940.873	25375.09	U	0.8	1.0		MIT
3939.451	3940.566	25377.06	U	0.8	0.0H		MIT
3939.111	3940.226	25379.25	U	0.8	1.1		MIT
3938.352	3939.467	25384.15	U	0.9			MIT
3937.883	3938.998	25387.17	U	0.8	0.9		MIT

AIR WAVELENGTH (ANGSTRMS)	VACUUM WAVELENGTH (ANGSTRMS)	WAVENUMBER (KAYSERS)	LMENT	LOG ARC	INTENSITY SPK	INTENSITY DIS	REF
3937.068	3938.183	25392.42	U	0.5	0.9		MIT
3936.744	3937.858	25394.51	U	0.7	0.7		MIT
3936.517	3937.631	25395.98	U	0.0	0.3		MIT
3936.008	3937.122	25399.26	U	0.9	0.3		MIT
3935.382	3936.496	25403.30	U	1.2	1.3		MIT
3934.894	3936.008	25406.45	U	1.0			MIT
3934.470	3935.584	25409.19	U	0.3	0.0		MIT
3933.985	3935.099	25412.32	U	0.8			MIT
3933.662	3934.776	25414.41	U	0.3	1.0		MIT
3933.030	3934.143	25418.49	U	0.7	1.0		MIT
3932.026	3933.139	25424.98	U	1.5	1.7		MIT
3931.488	3932.601	25428.46	U	1.4	0.8		MIT
3931.203	3932.316	25430.31	U	0.7	0.8		MIT
3930.982	3932.095	25431.74	U	1.1	1.5		MIT
3930.81	3931.92	25432.9	U		0.5H		MIT
3930.430	3931.543	25435.31	U	0.8	1.0		MIT
3930.073	3931.186	25437.62	U		0.6		MIT
3929.722	3930.835	25439.89	U	0.3	0.6		MIT
3929.055	3930.167	25444.21	U	0.8	0.3		MIT
3928.830	3929.942	25445.67	U	1.2			MIT
3928.450	3929.562	25448.13	U		0.5		MIT
3928.285	3929.397	25449.20	U	0.0	0.7		MIT
3927.755	3928.867	25452.63	U	0.5	0.7		MIT
3926.938	3928.050	25457.93	U	0.7	1.0		MIT
3926.728	3927.840	25459.29	U	1.0	0.0H		MIT
3926.216	3927.328	25462.61	U	1.5	0.8		MIT
3925.497	3926.609	25467.27	U	0.6	0.0		MIT
3925.309	3926.420	25468.49	U		0.3		MIT
3925.031	3926.142	25470.29	U	0.7	0.0		MIT
3924.785	3925.896	25471.89	U	0.8	0.5H		MIT
3924.503	3925.614	25473.72	U	0.7	0.7		MIT
3924.272	3925.383	25475.22	U	1.2	1.2		MIT
3923.920	3925.031	25477.50	U	0.9	1.0		MIT
3923.639	3924.750	25479.33	U	0.8			MIT
3923.38	3924.49	25481.0	U		0.3H		MIT
3923.054	3924.165	25483.13	U	1.2	0.8		MIT
3922.427	3923.538	25487.20	U	0.8	0.9		MIT
3922.037	3923.148	25489.74	U		0.3		MIT
3921.550	3922.660	25492.90	U	0.9	0.9		MIT
3921.236	3922.346	25494.94	U	0.9	0.8		MIT
3920.533	3921.643	25499.51	U	0.8	0.3		MIT
3920.346	3921.456	25500.73	U		0.3		MIT
3919.719	3920.829	25504.81	U	0.7	0.7		MIT
3919.343	3920.453	25507.26	U	0.5	0.6		MIT
3918.154	3919.264	25515.00	U		0.3		MIT
3918.065	3919.175	25515.58	U	0.5	0.9		MIT
3917.824	3918.934	25517.15	U	0.8	1.1		MIT
3917.611	3918.720	25518.53	U	0.8	0.7		MIT
3917.391	3918.500	25519.97	U	0.5	0.5		MIT
3917.064	3918.173	25522.10	U	0.0	0.3		MIT
3916.527	3917.636	25525.60	U	1.1	0.5		MIT
3916.347	3917.456	25526.77	U	0.9	0.3		MIT
3916.083	3917.192	25528.49	U	0.3			MIT
3915.884	3916.993	25529.79	U	1.3	1.5		MIT
3915.572	3916.681	25531.82	U	0.8			MIT
3915.219	3916.328	25534.12	U	1.2	0.0		MIT
3914.957	3916.066	25535.83	U	0.3H	0.0H		MIT
3914.732	3915.841	25537.30	U	1.1	0.9		MIT
3914.268	3915.377	25540.33	U	1.0	1.3		MIT
3913.80	3914.91	25543.4	U		0.7		MIT
3913.505	3914.613	25545.31	U	0.6	0.3		MIT
3913.250	3914.358	25546.97	U	0.9	0.5		MIT
3912.745	3913.853	25550.27	U	0.3	0.6		MIT
3912.408	3913.516	25552.47	U	0.8	0.9		MIT
3912.234	3913.342	25553.61	U	0.3	0.0		MIT
3911.673	3912.781	25557.27	U	1.3	1.3		MIT
3911.231	3912.339	25560.16	U	0.5	0.5		MIT
3910.978	3912.086	25561.81	U	1.0	0.5		MIT
3910.502	3911.610	25564.92	U	0.0	0.5H		MIT
3910.241	3911.349	25566.63	U	0.0	0.6		MIT

AIR WAVELENGTH (ANGSTRMS)	VACUUM WAVELENGTH (ANGSTRMS)	WAVENUMBER (KAYSERS)	LMENT	LOG ARC	INTENSITY SPK	INTENSITY DIS	REF
3909.706	3910.813	25570.13	U	0.5	0.8		MIT
3909.065	3910.172	25574.32	U	0.7	0.8		MIT
3908.682	3909.789	25576.83	U	0.3	0.5		MIT
3908.473	3909.580	25578.20	U	0.9	1.0		MIT
3908.331	3909.438	25579.12	U	1.0	0.0		MIT
3907.558	3908.665	25584.18	U	0.8	0.9		MIT
3907.018	3908.125	25587.72	U	0.7	0.0		MIT
3905.896	3907.002	25595.07	U	0.9			MIT
3905.124	3906.230	25600.13	U	0.5H			MIT
3904.853	3905.959	25601.91	U	0.9	1.2		MIT
3904.565	3905.671	25603.79	U	0.9	0.9		MIT
3904.299	3905.405	25605.54	U	0.9	1.2		MIT
3904.006	3905.112	25607.46	U	0.9	0.9		MIT
3903.262	3904.368	25612.34	U	1.0	0.0		MIT
3902.561	3903.667	25616.94	U	1.3	1.3		MIT
3902.257	3903.362	25618.94	U	0.0	0.3		MIT
3901.787	3902.892	25622.02	U	0.0	0.3		MIT
3901.551	3902.656	25623.57	U	1.2			MIT
3900.799	3901.904	25628.51	U	0.3	0.5		MIT
3900.327	3901.432	25631.61	U	0.3	1.1		MIT
3899.097	3900.202	25639.70	U	0.7	0.5		MIT
3898.837	3899.942	25641.41	U	0.8	0.8		MIT
3898.693	3899.798	25642.36	U	0.8			MIT
3898.514	3899.618	25643.53	U	0.8	0.5		MIT
3897.707	3898.811	25648.84	U	0.8	0.9		MIT
3897.265	3898.369	25651.75	U	1.2	0.7		MIT
3897.062	3898.166	25653.09	U	0.9	1.3		MIT
3896.779	3897.883	25654.95	U	1.3	1.4		MIT
3896.132	3897.236	25659.21	U	0.5	0.8		MIT
3895.946	3897.050	25660.44	U	0.3H	0.5H		MIT
3895.272	3896.376	25664.88	U	1.1	1.3		MIT
3894.923	3896.027	25667.18	U	1.0			MIT
3894.756	3895.860	25668.28	U	0.3	0.8		MIT
3894.123	3895.226	25672.45	U	1.5	0.6		MIT
3893.821	3894.924	25674.44	U	1.0	1.0		MIT
3893.307	3894.410	25677.83	U		0.5		MIT
3892.893	3893.996	25680.56	U	0.9			MIT
3892.684	3893.787	25681.94	U	1.3	1.5		MIT
3892.413	3893.516	25683.73	U	1.0	1.1		MIT
3892.112	3893.215	25685.71	U	0.9	0.5		MIT
3891.822	3892.925	25687.63	U	1.3	0.0		MIT
3891.682	3892.785	25688.55	U	1.1			MIT
3891.090	3892.193	25692.46	U	1.0	1.0		MIT
3890.364	3891.466	25697.25	U	1.5	1.5		MIT
3890.007	3891.109	25699.61	U	0.0	0.5		MIT
3889.415	3890.517	25703.52	U	0.0	0.7		MIT
3889.285	3890.387	25704.38	U	1.1			MIT
3888.607	3889.709	25708.86	U	0.6	0.5		MIT
3888.206	3889.308	25711.52	U	1.1	0.8		MIT
3887.700	3888.802	25714.86	U	1.3	1.0		MIT
3887.447	3888.549	25716.54	U	1.3			MIT
3887.198	3888.300	25718.18	U		0.5		MIT
3886.45	3887.55	25723.1	U	0.3	0.6		MIT
3885.997	3887.098	25726.13	U	0.3			MIT
3885.678	3886.779	25728.24	U	0.7	0.0		MIT
3885.282	3886.383	25730.87	U	0.5			MIT
3884.939	3886.040	25733.14	U	0.3	0.6		MIT
3884.681	3885.782	25734.85	U	0.9H	1.3		MIT
3883.636	3884.737	25741.77	U	1.1			MIT
3883.334	3884.435	25743.77	U	1.0	1.3		MIT
3883.103	3884.203	25745.30	U	0.5	0.5		MIT
3882.361	3883.461	25750.23	U	1.3	1.3		MIT
3881.878	3882.978	25753.43	U	0.8			MIT
3881.461	3882.561	25756.20	U	1.5	1.3		MIT
3881.111	3882.211	25758.52	U	0.8H	0.5		MIT
3880.674	3881.774	25761.42	U	0.3	0.5		MIT
3880.029	3881.129	25765.70	U	0.8J			MIT
3879.715	3880.815	25767.79	U	1.3	0.5		MIT
3879.548	3880.648	25768.90	U	1.3			MIT
3878.967	3880.066	25772.75	U	0.5	0.5		MIT

AIR WAVELENGTH (ANGSTRMS)	VACUUM WAVELENGTH (ANGSTRMS)	WAVENUMBER (KAYSERS)	LMENT	LOG ARC	INTENSITY SPK	INTENSITY DIS	REF
3878.091	3879.190	25778.58	U	0.9	0.6		MIT
3877.452	3878.551	25782.82	U	0.7	0.6H		MIT
3876.95	3878.05	25786.2	U	0.5	0.3		MIT
3876.593	3877.692	25788.54	U	0.8	0.6		MIT
3876.134	3877.233	25791.59	U	1.2	0.3		MIT
3875.338	3876.436	25796.89	U	1.1	0.8		MIT
3875.157	3876.255	25798.09	U	0.9			MIT
3874.460	3875.558	25802.74	U	1.0	0.3		MIT
3874.357	3875.455	25803.42	U	1.0			MIT
3874.042	3875.140	25805.52	U	1.2	1.2		MIT
3873.088	3874.186	25811.87	U	1.1	0.3		MIT
3872.866	3873.964	25813.35	U	0.3	0.0		MIT
3872.333	3873.431	25816.91	U	0.3	0.0		MIT
3871.878	3872.976	25819.94	U	0.9	1.1		MIT
3871.57	3872.67	25822.0	U	0.3	0.5H		MIT
3871.381	3872.478	25823.26	U	0.5H	1.3H		MIT
3871.042	3872.139	25825.52	U	1.5	0.0		MIT
3870.558	3871.655	25828.75	U	0.9	0.9		MIT
3870.05	3871.15	25832.1	U	0.8	0.7		MIT
3869.75	3870.85	25834.1	U	0.7	0.0		MIT
3868.813	3869.910	25840.40	U	0.3	0.6		MIT
3868.708	3869.805	25841.10	U	0.9			MIT
3868.415	3869.512	25843.05	U	0.8	0.8		MIT
3867.990	3869.087	25845.89	U	0.6			MIT
3867.515	3868.611	25849.07	U	1.0			MIT
3867.18	3868.28	25851.3	U	0.8			MIT
3865.923	3867.019	25859.71	U	1.3	1.4		MIT
3865.148	3866.244	25864.90	U	0.7H	0.9		MIT
3864.722	3865.818	25867.75	U	0.6	0.6		MIT
3864.478	3865.574	25869.38	U	0.9	1.0		MIT
3864.305	3865.401	25870.54	U	0.9	1.0		MIT
3863.403	3864.498	25876.58	U	0.7	1.3		MIT
3863.08	3864.18	25878.7	U	1.0	0.0		MIT
3862.28	3863.38	25884.1	U	0.5	0.3		MIT
3862.05	3863.14	25885.6	U	0.9	0.0		MIT
3861.730	3862.825	25887.79	U	0.9	0.8		MIT
3861.123	3862.218	25891.86	U	0.7			MIT
3860.630	3861.725	25895.17	U	0.0	1.2H		MIT
3859.580	3860.674	25902.21	U	1.3	1.5		MIT
3859.014	3860.108	25906.01	U	0.9	0.6		MIT
3858.682	3859.776	25908.24	U	0.5	0.8		MIT
3858.508	3859.602	25909.41	U	0.3	0.3H		MIT
3858.184	3859.278	25911.58	U		0.3		MIT
3857.153	3858.247	25918.51	U	0.8	0.9		MIT
3856.745	3857.839	25921.25	U	0.8			MIT
3856.27	3857.36	25924.4	U		0.5		MIT
3855.632	3856.725	25928.73	U	0.5H	1.1		MIT
3855.443	3856.536	25930.00	U	0.9	0.0		MIT
3855.16	3856.25	25931.9	U	0.3	0.3		MIT
3854.862	3855.955	25933.91	U	0.3H	0.8		MIT
3854.655	3855.748	25935.30	U	1.3	1.5		MIT
3854.230	3855.323	25938.16	U	1.3			MIT
3853.72	3854.81	25941.6	U	0.9			MIT
3853.370	3854.463	25943.95	U	0.6	0.9		MIT
3852.985	3854.078	25946.54	U	0.8	0.8H		MIT
3852.703	3853.796	25948.44	U	0.8	0.8H		MIT
3852.089	3853.181	25952.58	U	0.6	0.9		MIT
3851.957	3853.049	25953.47	U	0.9	0.3		MIT
3851.732	3852.824	25954.99	U	1.2	0.3		MIT
3851.296	3852.388	25957.92	U	0.6	0.9		MIT
3850.935	3852.027	25960.36	U	0.0	0.6		MIT
3850.244	3851.336	25965.02	U	0.0	0.9H		MIT
3849.853	3850.945	25967.65	U	0.6	0.6		MIT
3849.707	3850.799	25968.64	U	0.3	1.0H		MIT
3849.339	3850.431	25971.12	U	0.3	0.5		MIT
3848.716	3849.808	25975.32	U	0.6	0.5		MIT
3848.618	3849.709	25975.99	U	1.0	0.9		MIT
3848.069	3849.160	25979.69	U	0.8	0.8		MIT
3847.835	3848.926	25981.27	U	0.9	1.1H		MIT
3847.062	3848.153	25986.49	U	0.7	0.8		MIT

AIR WAVELENGTH (ANGSTRMS)	VACUUM WAVELENGTH (ANGSTRMS)	WAVENUMBER (KAYSERS)	LMENT	LOG ARC	INTENSITY SPK	INTENSITY DIS	REF
3846.661	3847.752	25989.20	U	0.3	1.0		MIT
3846.563	3847.654	25989.86	U	1.0			MIT
3846.243	3847.334	25992.02	U	1.0	0.0		MIT
3845.857	3846.948	25994.63	U	0.6	0.6		MIT
3845.125	3846.216	25999.58	U	0.3	0.8		MIT
3844.937	3846.028	26000.85	U	0.8			MIT
3844.680	3845.770	26002.59	U	0.6			MIT
3844.232	3845.322	26005.62	U	0.9	1.2		MIT
3844.005	3845.095	26007.16	U	1.0	1.0		MIT
3843.811	3844.901	26008.47	U	0.8	0.3H		MIT
3843.46	3844.55	26010.9	U	0.9	0.9H		MIT
3842.986	3844.076	26014.05	U	1.1	1.1		MIT
3842.713	3843.803	26015.90	U	0.6H	0.8H		MIT
3842.463	3843.553	26017.59	U	0.5	0.8		MIT
3842.191	3843.281	26019.44	U	0.7	0.9		MIT
3841.749	3842.839	26022.43	U	0.8H	0.9H		MIT
3839.904	3840.993	26034.93	U	0.9	0.6		MIT
3839.632	3840.721	26036.78	U	1.5	0.3		MIT
3839.53	3840.62	26037.5	U		0.7		MIT
3838.996	3840.085	26041.09	U	0.8	1.0		MIT
3838.150	3839.239	26046.83	U	0.9	1.0		MIT
3837.828	3838.917	26049.01	U	1.1	0.9H		MIT
3837.510	3838.599	26051.17	U	0.3	0.3H		MIT
3837.272	3838.361	26052.79	U	1.2			MIT
3836.520	3837.608	26057.90	U	0.8	1.2		MIT
3835.919	3837.007	26061.98	U	0.6	0.5		MIT
3835.14	3836.23	26067.3	U	0.6	0.9		MIT
3834.809	3835.897	26069.52	U	0.5	0.8		MIT
3834.58	3835.67	26071.1	U	0.3	0.6		MIT
3833.023	3834.110	26081.67	U	1.3R	1.2		MIT
3832.565	3833.652	26084.79	U	0.7	0.0		MIT
3832.20	3833.29	26087.3	U	0.8			MIT
3831.865	3832.952	26089.55	U	1.2	0.5H		MIT
3831.465	3832.552	26092.27	U	1.4	1.4		MIT
3830.612	3831.699	26098.08	U	0.5H	0.7		MIT
3830.215	3831.302	26100.79	U	0.7	0.8		MIT
3829.803	3830.890	26103.60	U	1.2			MIT
3829.390	3830.476	26106.41	U	0.8	0.7		MIT
3829.033	3830.119	26108.85	U	0.9	1.0		MIT
3828.81	3829.90	26110.4	U	0.3	0.7		MIT
3828.51	3829.60	26112.4	U	0.8			MIT
3828.058	3829.144	26115.50	U	0.5	0.9		MIT
3827.481	3828.567	26119.43	U	0.6	0.0		MIT
3825.160	3826.245	26135.28	U	0.7	0.7		MIT
3825.029	3826.114	26136.18	U	1.2			MIT
3824.721	3825.806	26138.28	U	0.7	1.0		MIT
3823.823	3824.908	26144.42	U	0.5			MIT
3823.447	3824.532	26146.99	U	0.7	0.5		MIT
3823.04	3824.12	26149.8	U	0.6	1.0H		MIT
3822.555	3823.640	26153.09	U	1.1	1.1		MIT
3822.410	3823.495	26154.08	U	0.6	0.6		MIT
3821.959	3823.044	26157.17	U	1.2			MIT
3821.223	3822.307	26162.21	U		0.9		MIT
3821.020	3822.104	26163.60	U	0.6H	0.6		MIT
3819.05	3820.13	26177.1	U	0.7H	0.6		MIT
3818.757	3819.841	26179.10	U	0.9H	0.5		MIT
3818.481	3819.565	26180.99	U	1.1	0.6		MIT
3818.064	3819.148	26183.85	U	0.9	0.3H		MIT
3817.706	3818.789	26186.31	U	0.6	1.0		MIT
3817.155	3818.238	26190.09	U	0.9	1.0		MIT
3816.606	3817.689	26193.86	U	1.0	1.0		MIT
3816.032	3817.115	26197.80	U	0.0	0.9		MIT
3815.376	3816.459	26202.30	U	0.3			MIT
3815.151	3816.234	26203.84	U	0.0	0.8		MIT
3814.070	3815.153	26211.27	U	1.4	1.2		MIT
3813.791	3814.873	26213.19	U	1.3	1.2		MIT
3813.222	3814.304	26217.10	U	0.8	0.9		MIT
3812.716	3813.798	26220.58	U	0.9	0.0		MIT
3812.582	3813.664	26221.50	U	1.2	0.6H		MIT
3812.233	3813.315	26223.90	U	0.6	0.8		MIT

AIR WAVELENGTH (ANGSTRMS)	VACUUM WAVELENGTH (ANGSTRMS)	WAVENUMBER (KAYSERS)	LMENT	LOG ARC	INTENSITY SPK	DIS	REF
3811.999	3813.081	26225.51	U	1.3	0.8		MIT
3811.620	3812.702	26228.12	U	1.2			MIT
3811.476	3812.558	26229.11	U	1.2			MIT
3810.93	3812.01	26232.9	U	0.9	1.1		MIT
3810.192	3811.274	26237.95	U	0.5	0.9		MIT
3809.923	3811.004	26239.80	U	1.2	0.3H		MIT
3809.224	3810.305	26244.62	U	1.2	1.1		MIT
3808.942	3810.023	26246.56	U	1.3			MIT
3808.20	3809.28	26251.7	U	0.8	0.9		MIT
3807.87	3808.95	26253.9	U	0.3D	0.6D		MIT
3807.61	3808.69	26255.7	U		0.5		MIT
3807.199	3808.280	26258.57	U	0.6	0.6		MIT
3806.262	3807.342	26265.04	U	0.5H	0.5		MIT
3805.825	3806.905	26268.05	U	0.0	0.3		MIT
3805.717	3806.797	26268.80	U		0.8		MIT
3805.058	3806.138	26273.35	U	0.0	0.6		MIT
3804.874	3805.954	26274.62	U	1.1R	0.6		MIT
3803.868	3804.948	26281.57	U	0.3			MIT
3803.350	3804.430	26285.15	U	0.9	0.9		MIT
3803.093	3804.173	26286.93	U	0.0	0.9		MIT
3802.856	3803.936	26288.56	U	0.5	0.7		MIT
3801.955	3803.034	26294.79	U	1.2			MIT
3801.373	3802.452	26298.82	U	0.5H	0.7		MIT
3801.148	3802.227	26300.37	U	1.3			MIT
3800.771	3801.850	26302.98	U		0.3		MIT
3800.278	3801.357	26306.40	U	0.0	0.3		MIT
3799.552	3800.631	26311.42	U	0.0	0.6		MIT
3799.202	3800.281	26313.85	U	0.8	1.1		MIT
3798.837	3799.916	26316.37	U	1.2	0.0		MIT
3798.259	3799.337	26320.38	U	0.3	0.9		MIT
3797.773	3798.851	26323.75	U	1.0	0.0		MIT
3797.520	3798.598	26325.50	U		0.5		MIT
3797.142	3798.220	26328.12	U	0.0	0.5		MIT
3796.845	3797.923	26330.18	U	1.1	1.2		MIT
3796.536	3797.614	26332.32	U	1.0	1.2		MIT
3796.413	3797.491	26333.18	U	0.8	0.6		MIT
3796.204	3797.282	26334.63	U	0.9	1.1		MIT
3796.000	3797.078	26336.04	U	0.8	0.0		MIT
3795.608	3796.686	26338.76	U	0.0	0.8		MIT
3795.132	3796.210	26342.06	U	0.9H	0.9		MIT
3794.416	3795.493	26347.04	U		0.5		MIT
3793.972	3795.049	26350.12	U	0.5	0.9		MIT
3793.576	3794.653	26352.87	U	1.0	1.1		MIT
3793.282	3794.359	26354.91	U	1.3	1.1		MIT
3793.104	3794.181	26356.15	U	1.1	1.3		MIT
3792.629	3793.706	26359.45	U	0.9	0.6		MIT
3792.410	3793.487	26360.97	U	1.0	0.6		MIT
3791.843	3792.920	26364.91	U	0.5	0.5		MIT
3791.339	3792.416	26368.42	U	1.0	0.0		MIT
3791.091	3792.168	26370.14	U	0.3	0.9		MIT
3790.843	3791.919	26371.87	U		0.5		MIT
3790.333	3791.409	26375.42	U	0.8	0.8		MIT
3790.216	3791.292	26376.23	U	0.0	0.8		MIT
3789.888	3790.964	26378.51	U	0.3	0.8		MIT
3789.603	3790.679	26380.50	U	0.6	1.1		MIT
3789.214	3790.290	26383.21	U		0.7		MIT
3788.595	3789.671	26387.52	U	0.5H	0.0H		MIT
3788.236	3789.312	26390.02	U		0.8		MIT
3788.036	3789.112	26391.41	U	0.3	0.9		MIT
3787.233	3788.309	26397.00	U	0.0	1.0		MIT
3786.830	3787.905	26399.81	U	1.1	0.3H		MIT
3786.574	3787.649	26401.60	U	0.7	1.0		MIT
3786.142	3787.217	26404.61	U	0.5H	0.0H		MIT
3785.35	3786.43	26410.1	U		1.2H		MIT
3785.175	3786.250	26411.36	U	0.8	0.3		MIT
3784.707	3785.782	26414.62	U	0.3	0.8		MIT
3783.840	3784.915	26420.68	U	1.3	1.4		MIT
3783.630	3784.705	26422.14	U	0.3	0.7		MIT
3782.841	3783.915	26427.65	U	1.4	1.5		MIT
3782.302	3783.376	26431.42	U	0.6	0.0		MIT
3781.746	3782.820	26435.30	U	1.0			MIT
3781.464	3782.538	26437.27	U	0.5	0.5H		MIT
3781.182	3782.256	26439.25	U	0.6	0.8		MIT
3781.051	3782.125	26440.16	U	0.5	0.8		MIT
3780.716	3781.790	26442.51	U	1.2	1.3		MIT
3780.262	3781.336	26445.68	U	0.0	0.9		MIT
3779.048	3780.121	26454.18	U	0.9	1.2		MIT
3778.360	3779.433	26458.99	U	1.1			MIT
3778.006	3779.079	26461.47	U	0.6	0.3		MIT
3777.728	3778.801	26463.42	U	0.9	0.5		MIT
3777.392	3778.465	26465.77	U	0.3			MIT
3776.738	3777.811	26470.36	U	0.5	0.7H		MIT
3775.991	3777.064	26475.59	U	1.0	1.1		MIT
3775.608	3776.681	26478.28	U	1.1	0.7		MIT
3775.442	3776.514	26479.44	U	0.6	0.6		MIT
3775.262	3776.334	26480.71	U	0.8	0.5		ED
3774.907	3775.979	26483.20	U	1.0	0.0		MIT
3774.06	3775.13	26489.1	U		1.1W		MIT
3773.547	3774.619	26492.74	U	0.3	0.7		MIT
3773.425	3774.497	26493.60	U	1.3	1.0		MIT
3772.813	3773.885	26497.89	U	0.8	1.3		MIT
3772.357	3773.429	26501.10	U	0.3	0.6		MIT
3771.412	3772.483	26507.74	U	0.0	0.9		MIT
3770.458	3771.529	26514.44	U	0.3	0.9		MIT
3770.170	3771.241	26516.47	U		0.6		MIT
3769.535	3770.606	26520.94	U	0.6	1.3		MIT
3768.796	3769.867	26526.14	U	1.2	1.1		MIT
3768.532	3769.603	26528.00	U	0.3	0.7		MIT
3768.411	3769.482	26528.85	U	0.0	0.3		MIT
3767.835	3768.906	26532.90	U		0.3H		MIT
3766.889	3767.959	26539.56	U	1.0			MIT
3766.46	3767.53	26542.6	U		0.8W		MIT
3765.870	3766.940	26546.75	U	0.0	0.9		MIT
3765.345	3766.415	26550.45	U	1.3	0.0		MIT
3764.785	3765.855	26554.40	U	0.8	1.1		MIT
3764.570	3765.640	26555.91	U	0.9	1.0		MIT
3764.169	3765.239	26558.74	U	0.0	0.7H		MIT
3763.44	3764.51	26563.9	U	0.8			MIT
3763.266	3764.335	26565.11	U	1.1	1.4		MIT
3763.009	3764.078	26566.93	U	0.8	0.0		MIT
3762.663	3763.732	26569.37	U	0.3	0.3		MIT
3762.114	3763.183	26573.25	U	1.0	0.8		MIT
3761.961	3763.030	26574.33	U	0.3	0.9		MIT
3761.601	3762.670	26576.87	U	0.3	0.8		MIT
3761.044	3762.113	26580.81	U	0.6	0.9		MIT
3760.884	3761.953	26581.94	U	1.0	1.2		MIT
3760.345	3761.414	26585.75	U	0.0	0.3		MIT
3759.880	3760.948	26589.04	U	0.0	1.0		MIT
3759.231	3760.299	26593.63	U	1.2H	1.2		MIT
3758.743	3759.811	26597.08	U	0.0	0.5		MIT
3757.969	3759.037	26602.56	U	0.3			MIT
3756.925	3757.993	26609.95	U	0.3	0.9		MIT
3756.659	3757.727	26611.83	U	0.8			MIT
3755.482	3756.549	26620.18	U	0.0	0.7		MIT
3754.974	3756.041	26623.78	U	0.0	0.3		MIT
3754.308	3755.375	26628.50	U	0.8	1.1		MIT
3753.703	3754.770	26632.79	U		0.6		MIT
3753.08	3754.15	26637.2	U	0.7H			MIT
3752.867	3753.934	26638.72	U	0.0	0.3		MIT
3752.663	3753.730	26640.17	U	0.8	1.2		MIT
3752.295	3753.361	26642.78	U	0.8R	0.7		MIT
3751.717	3752.783	26646.89	U	1.0	1.1		MIT
3751.236	3752.302	26650.30	U		0.9		MIT
3751.185	3752.251	26650.67	U	0.8			MIT
3749.859	3750.925	26660.09	U		0.9		MIT
3749.74	3750.81	26660.9	U	0.3H			MIT
3749.159	3750.225	26665.07	U	0.3	0.9		MIT
3748.678	3749.744	26668.49	U	1.2	1.4		MIT
3747.115	3748.180	26679.61	U	0.8	1.0		MIT
3746.677	3747.742	26682.73	U	0.0	0.7		MIT

AIR WAVELENGTH (ANGSTRMS)	VACUUM WAVELENGTH (ANGSTRMS)	WAVENUMBER (KAYSERS)	LMENT	LOG ARC	INTENSITY SPK	DIS	REF
3746.413	3747.478	26684.61	U	0.5	1.1		MIT
3744.805	3745.870	26696.07	U		0.3		MIT
3744.24	3745.30	26700.1	U	0.8A	1.0A		MIT
3742.353	3743.417	26713.56	U	0.8	0.6		MIT
3741.577	3742.641	26719.10	U	0.6	0.5		MIT
3741.280	3742.344	26721.22	U		0.5		MIT
3739.311	3740.374	26735.29	U	0.0	0.3		MIT
3738.624	3739.687	26740.21	U	0.0	0.9		MIT
3738.048	3739.111	26744.33	U	0.9	1.3		MIT
3737.254	3738.317	26750.01	U	0.9	1.0		MIT
3736.603	3737.665	26754.67	U	0.3			MIT
3736.476	3737.538	26755.58	U	0.7			MIT
3736.029	3737.091	26758.78	U	0.5H			MIT
3734.683	3735.745	26768.42	U	0.8	0.0		MIT
3733.752	3734.814	26775.10	U	0.5	0.0		MIT
3733.582	3734.644	26776.32	U	0.8	0.3		MIT
3733.069	3734.130	26780.00	U	0.9	0.8		MIT
3732.618	3733.679	26783.23	U	1.0	0.8		MIT
3732.259	3733.320	26785.81	U	0.8	0.5		MIT
3731.766	3732.827	26789.35	U	0.6	0.3		MIT
3731.453	3732.514	26791.59	U	1.1			MIT
3730.831	3731.892	26796.06	U		0.6		MIT
3730.131	3731.192	26801.09	U	0.3	0.5		MIT
3729.821	3730.882	26803.32	U	0.9	1.2		MIT
3728.860	3729.920	26810.22	U	0.8	0.8		MIT
3728.747	3729.807	26811.04	U	0.6			MIT
3728.435	3729.495	26813.28	U		0.9H		MIT
3727.819	3728.879	26817.71	U	0.5H	0.3		MIT
3727.454	3728.514	26820.34	U	0.7			MIT
3726.865	3727.925	26824.57	U	0.3	0.9		MIT
3726.287	3727.347	26828.74	U	0.3	1.0		MIT
3726.038	3727.098	26830.53	U	1.1	0.0		MIT
3725.767	3726.827	26832.48	U	0.3	0.0		MIT
3725.651	3726.711	26833.31	U	0.5	1.2		MIT
3725.394	3726.454	26835.17	U	0.0	0.9		MIT
3725.071	3726.130	26837.49	U	0.6	1.0		MIT
3724.988	3726.047	26838.09	U	0.8	0.9		MIT
3724.233	3725.292	26843.53	U	0.9			MIT
3723.657	3724.716	26847.68	U	0.5	1.0		MIT
3723.478	3724.537	26848.98	U	0.0	0.5		MIT
3722.68	3723.74	26854.7	U	1.0	1.0		MIT
3721.722	3722.781	26861.64	U	1.0			MIT
3720.394	3721.452	26871.23	U	0.8	1.0H		MIT
3719.69	3720.75	26876.3	U	0.0H	0.9		MIT
3719.293	3720.351	26879.19	U	1.1			MIT
3718.844	3719.902	26882.43	U		0.6		MIT
3718.614	3719.672	26884.09	U	0.6	0.9		MIT
3718.106	3719.164	26887.77	U	0.9	1.1		MIT
3717.729	3718.787	26890.49	U	0.8			MIT
3717.425	3718.482	26892.69	U	1.0	0.9		MIT
3717.012	3718.069	26895.68	U	0.0	0.6		MIT
3716.783	3717.840	26897.34	U	0.8	0.9		MIT
3716.54	3717.60	26899.1	U		0.3H		MIT
3716.135	3717.192	26902.03	U	1.3			MIT
3715.669	3716.726	26905.40	U	0.5	0.8		MIT
3715.470	3716.527	26906.84	U	0.8	1.1		MIT
3715.184	3716.241	26908.91	U	0.5	0.0		MIT
3714.947	3716.004	26910.63	U	0.7	0.9		MIT
3714.758	3715.815	26912.00	U	0.9	0.9		MIT
3714.399	3715.456	26914.60	U		0.5		MIT
3713.774	3714.830	26919.13	U	0.7	0.3		MIT
3713.649	3714.705	26920.03	U	0.3	1.0		MIT
3712.566	3713.622	26927.89	U	0.8			MIT
3712.308	3713.364	26929.76	U	0.7			MIT
3711.851	3712.907	26933.07	U	0.6	0.7		MIT
3710.908	3711.964	26939.92	U	0.5	0.9		MIT
3710.785	3711.841	26940.81	U	0.6	0.3		MIT
3710.534	3711.590	26942.63	U	0.5	0.6		MIT
3710.308	3711.364	26944.27	U	0.9	0.0		MIT
3710.168	3711.224	26945.29	U	0.0	0.7		MIT

AIR WAVELENGTH (ANGSTRMS)	VACUUM WAVELENGTH (ANGSTRMS)	WAVENUMBER (KAYSERS)	LMENT	LOG ARC	INTENSITY SPK	DIS	REF
3709.878	3710.933	26947.40	U	0.3	0.3		MIT
3709.46	3710.52	26950.4	U	0.5D			MIT
3708.592	3709.647	26956.74	U	0.8			MIT
3707.648	3708.703	26963.60	U	0.0	1.0		MIT
3707.290	3708.345	26966.21	U	0.0	0.3		MIT
3706.732	3707.787	26970.27	U	0.3			MIT
3706.533	3707.588	26971.72	U	0.8			MIT
3705.983	3707.037	26975.72	U	0.8	1.0		MIT
3705.040	3706.094	26982.58	U	0.8	0.9		MIT
3704.658	3705.712	26985.37	U	0.0	0.3		MIT
3704.344	3705.398	26987.65	U	0.0	0.7		MIT
3704.099	3705.153	26989.44	U	0.8	0.9		MIT
3703.652	3704.706	26992.70	U	0.5	0.3		MIT
3703.278	3704.332	26995.42	U	0.9			MIT
3703.017	3704.071	26997.32	U	0.3	0.3		MIT
3702.866	3703.920	26998.43	U	0.7			MIT
3702.615	3703.669	27000.26	U	0.9	0.0		MIT
3702.218	3703.272	27003.15	U	0.3	1.1H		MIT
3701.745	3702.798	27006.60	U	1.0	0.3		MIT
3701.522	3702.575	27008.23	U	1.0	1.4		MIT
3700.575	3701.628	27015.14	U	1.1	1.3		MIT
3700.29	3701.34	27017.2	U	0.5			MIT
3700.098	3701.151	27018.62	U	0.5			MIT
3699.857	3700.910	27020.38	U	0.6	0.0		MIT
3699.724	3700.777	27021.35	U	0.6	0.7		MIT
3699.437	3700.490	27023.45	U	0.7	0.5H		MIT
3698.43	3699.48	27030.8	U	0.6D	0.0D		MIT
3697.930	3698.982	27034.46	U	0.8	0.9		MIT
3697.551	3698.603	27037.23	U	0.6	0.8		MIT
3697.131	3698.183	27040.30	U	1.0	0.5H		MIT
3696.824	3697.876	27042.55	U	0.0	0.5		MIT
3696.690	3697.742	27043.53	U	0.0	0.3		MIT
3696.352	3697.404	27046.00	U	0.8			MIT
3695.811	3696.863	27049.96	U	0.7	0.5		MIT
3695.206	3696.258	27054.39	U	0.9			MIT
3694.816	3695.868	27057.25	U		0.5J		MIT
3694.325	3695.376	27060.84	U	0.8	0.0		MIT
3693.705	3694.756	27065.39	U	0.6	1.3		MIT
3693.320	3694.371	27068.21	U	0.6	0.8		MIT
3692.91	3693.96	27071.2	U	0.7	1.0		MIT
3692.750	3693.801	27072.38	U	1.0			MIT
3692.310	3693.361	27075.61	U		0.3		MIT
3692.052	3693.103	27077.50	U	0.6	0.6		MIT
3691.917	3692.968	27078.49	U	0.8	0.9		MIT
3691.470	3692.521	27081.77	U	0.5	0.5		MIT
3690.817	3691.868	27086.56	U	0.3	0.0		MIT
3690.239	3691.289	27090.80	U	0.3	0.8		MIT
3689.968	3691.018	27092.79	U	0.3	0.5		MIT
3689.672	3690.722	27094.97	U	0.0	0.8		MIT
3689.204	3690.254	27098.41	U	1.0	0.0		MIT
3689.037	3690.087	27099.63	U	0.7	0.9		MIT
3688.785	3689.835	27101.48	U		0.3		MIT
3688.390	3689.440	27104.39	U	0.5	1.0		MIT
3687.92	3688.97	27107.8	U	0.6	0.8		MIT
3687.760	3688.810	27109.01	U		0.7		MIT
3687.461	3688.511	27111.21	U		0.5		MIT
3686.742	3687.791	27116.50	U	0.0	0.6H		MIT
3686.463	3687.512	27118.55	U	0.3	0.8		MIT
3685.772	3686.821	27123.64	U	1.2	0.0		MIT
3684.617	3685.666	27132.14	U	0.7	0.7		MIT
3684.285	3685.334	27134.58	U		0.3		MIT
3683.59	3684.64	27139.7	U	0.3D	0.9D		MIT
3682.459	3683.507	27148.04	U	0.9			MIT
3682.037	3683.085	27151.15	U	0.8	1.2		MIT
3680.880	3681.928	27159.68	U	1.3	0.0		MIT
3680.509	3681.557	27162.42	U	0.0	0.5		MIT
3680.235	3681.283	27164.44	U	0.0	0.5		MIT
3679.806	3680.854	27167.61	U		0.7		MIT
3679.375	3680.423	27170.79	U	0.3			MIT
3678.752	3679.799	27175.39	U	0.8	1.2		MIT

AIR WAVELENGTH (ANGSTRMS)	VACUUM WAVELENGTH (ANGSTRMS)	WAVENUMBER (KAYSERS)	LMENT	LOG ARC	INTENSITY SPK	DIS	REF
3678.069	3679.116	27180.44	U	0.3	0.3		MIT
3677.390	3678.437	27185.46	U	1.0			MIT
3676.560	3677.607	27191.60	U	0.5	1.2		MIT
3675.519	3676.566	27199.30	U		0.5H		MIT
3675.255	3676.302	27201.25	U	0.9			MIT
3675.085	3676.131	27202.51	U	0.6	0.8		MIT
3674.989	3676.035	27203.22	U	0.3	0.8		MIT
3674.130	3675.176	27209.58	U	0.9			MIT
3673.059	3674.105	27217.51	U	0.8			MIT
3672.579	3673.625	27221.07	U	0.9	1.2		MIT
3672.187	3673.233	27223.98	U	0.7			MIT
3671.539	3672.585	27228.78	U	0.6	0.0		MIT
3670.530	3671.575	27236.27	U	0.8			MIT
3670.265	3671.310	27238.23	U	0.5	0.3		MIT
3670.072	3671.117	27239.66	U	1.2	1.3		MIT
3669.36	3670.40	27244.9	U		0.5		MIT
3668.428	3669.473	27251.87	U	0.6			MIT
3667.975	3669.020	27255.24	U	1.2	0.9		MIT
3667.740	3668.785	27256.98	U	0.3	0.3		MIT
3667.134	3668.178	27261.49	U	0.9	0.7		MIT
3666.858	3667.902	27263.54	U	0.3	0.0H		MIT
3666.620	3667.664	27265.31	U	0.5			MIT
3666.210	3667.254	27268.36	U	0.8	0.9		MIT
3666.101	3667.145	27269.17	U	0.0	0.8		MIT
3665.396	3666.440	27274.41	U	0.5			MIT
3665.207	3666.251	27275.82	U	0.7	0.0		MIT
3664.70	3665.74	27279.6	U	0.3D	0.3D		MIT
3664.529	3665.573	27280.87	U	0.3	0.3		MIT
3664.255	3665.299	27282.91	U	0.0	0.8		MIT
3663.934	3664.978	27285.30	U	1.0			MIT
3663.348	3664.391	27289.66	U	0.5	0.0		MIT
3662.659	3663.702	27294.79	U	1.2R	0.3		MIT
3662.331	3663.374	27297.24	U	1.0	1.0		MIT
3661.90	3662.94	27300.4	U	1.1R			MIT
3661.482	3662.525	27303.57	U	0.7	0.0H		MIT
3661.367	3662.410	27304.43	U	0.0	0.3H		MIT
3661.030	3662.073	27306.94	U	0.5	0.0		MIT
3660.736	3661.779	27309.13	U	0.0	0.9		MIT
3660.110	3661.153	27313.80	U	0.5	0.3		MIT
3659.586	3660.628	27317.71	U	1.0	0.3		MIT
3659.159	3660.201	27320.90	U	1.2	0.0		MIT
3659.009	3660.051	27322.02	U	0.3	0.9		MIT
3658.677	3659.719	27324.50	U	0.8			MIT
3658.161	3659.203	27328.35	U		0.3		MIT
3657.821	3658.863	27330.90	U	1.0			MIT
3657.322	3658.364	27334.62	U	0.8	1.2		MIT
3656.932	3657.974	27337.54	U	0.9	0.9		MIT
3656.626	3657.668	27339.83	U	0.9			MIT
3656.193	3657.235	27343.06	U	0.0	0.3		MIT
3655.435	3656.476	27348.73	U	0.5H	0.7H		MIT
3655.111	3656.152	27351.16	U	0.9	0.6		MIT
3654.890	3655.931	27352.81	U	1.4			MIT
3654.294	3655.335	27357.27	U	0.5	0.6		MIT
3654.130	3655.171	27358.50	U	0.7	0.0		MIT
3653.209	3654.250	27365.40	U	1.0	0.0		MIT
3652.074	3653.115	27373.90	U	1.1	0.9		MIT
3651.927	3652.967	27375.00	U	0.8			MIT
3651.685	3652.725	27376.82	U	0.6	0.7		MIT
3651.344	3652.384	27379.37	U		0.3		MIT
3651.244	3652.284	27380.12	U	0.3			MIT
3650.679	3651.719	27384.36	U	0.5	0.3		MIT
3650.339	3651.379	27386.91	U	0.5			MIT
3649.729	3650.769	27391.49	U	0.0	0.9		MIT
3649.410	3650.450	27393.88	U		0.6		MIT
3648.935	3649.975	27397.45	U	0.6	0.6		MIT
3648.545	3649.585	27400.38	U	0.7	0.3		MIT
3648.251	3649.291	27402.59	U	0.7			MIT
3648.146	3649.186	27403.37	U	0.3	0.7		MIT
3647.659	3648.698	27407.03	U	0.3	0.5H		MIT
3646.847	3647.886	27413.14	U	0.6	1.0		MIT

AIR WAVELENGTH (ANGSTRMS)	VACUUM WAVELENGTH (ANGSTRMS)	WAVENUMBER (KAYSERS)	LMENT	LOG ARC	INTENSITY SPK	DIS	REF
3646.491	3647.530	27415.81	U	0.3	0.8		MIT
3646.217	3647.256	27417.87	U	1.0			MIT
3646.03	3647.07	27419.3	U		0.6J		MIT
3645.461	3646.500	27423.56	U	0.0	0.3		MIT
3645.030	3646.069	27426.80	U	0.9	1.2		MIT
3644.851	3645.890	27428.15	U	0.6	0.8		MIT
3644.245	3645.284	27432.71	U	1.3	0.3		MIT
3642.846	3643.884	27443.24	U	0.8			MIT
3642.776	3643.814	27443.77	U	0.3	0.7		MIT
3642.443	3643.481	27446.28	U	1.0	0.9		MIT
3642.033	3643.071	27449.37	U	0.8H	0.3H		MIT
3641.454	3642.492	27453.73	U	0.8			MIT
3641.21	3642.25	27455.6	U	0.5	0.6		MIT
3640.948	3641.986	27457.55	U	0.9	1.3		MIT
3640.757	3641.795	27458.99	U	0.9	1.3		MIT
3640.035	3641.072	27464.44	U	0.0	0.6		MIT
3639.71	3640.75	27466.9	U		0.3		MIT
3639.489	3640.526	27468.56	U	1.4			MIT
3639.206	3640.243	27470.69	U		0.7		MIT
3638.65	3639.69	27474.9	U	0.7	0.9		MIT
3638.200	3639.237	27478.29	U	0.7	0.3		MIT
3637.506	3638.543	27483.53	U	0.7	0.6		MIT
3636.312	3637.348	27492.55	U	1.0			MIT
3635.579	3636.615	27498.10	U	0.6	0.9		MIT
3635.403	3636.439	27499.43	U	0.8	0.7		MIT
3635.299	3636.335	27500.22	U	1.0			MIT
3635.010	3636.046	27502.40	U	0.0	0.3		MIT
3634.560	3635.596	27505.81	U	0.9	0.0		MIT
3633.29	3634.33	27515.4	U	0.9D	1.2D		MIT
3632.713	3633.749	27519.79	U	0.5	0.3		MIT
3632.17	3633.21	27523.9	U		0.5A		MIT
3630.733	3631.768	27534.80	U	0.9	1.3		MIT
3629.591	3630.626	27543.46	U	0.3	0.5		MIT
3627.70	3628.73	27557.8	U	0.9			MIT
3627.484	3628.518	27559.46	U	0.3	0.9		MIT
3626.360	3627.394	27568.00	U	1.0			MIT
3625.980	3627.014	27570.89	U	0.7	1.2J		MIT
3625.571	3626.605	27574.00	U	0.9			MIT
3625.088	3626.122	27577.67	U		0.5		MIT
3624.818	3625.851	27579.73	U	0.7			MIT
3624.571	3625.604	27581.61	U	0.6	0.0		MIT
3623.055	3624.088	27593.15	U	1.1	1.2		MIT
3622.700	3623.733	27595.85	U	1.2	0.0		MIT
3622.290	3623.323	27598.98	U	0.0	0.5		MIT
3621.783	3622.816	27602.84	U	0.0	0.3		MIT
3621.032	3622.065	27608.56	U	0.3	0.7		MIT
3620.557	3621.589	27612.19	U	0.0	0.3		MIT
3620.085	3621.117	27615.79	U	1.2	0.3H		MIT
3619.806	3620.838	27617.92	U		0.5		MIT
3619.134	3620.166	27623.04	U	0.7	0.3		MIT
3618.486	3619.518	27627.99	U	0.3	1.0		MIT
3617.487	3618.519	27635.62	U	0.5	0.0		MIT
3616.888	3617.919	27640.20	U	0.9			MIT
3616.757	3617.788	27641.20	U	0.3	1.0		MIT
3616.675	3617.706	27641.82	U	0.3	0.0		MIT
3616.332	3617.363	27644.45	U	1.1	0.3H		MIT
3615.840	3616.871	27648.21	U	0.6			MIT
3615.56	3616.59	27650.4	U	1.0			M
3615.236	3616.267	27652.82	U	0.0	0.7		MIT
3615.007	3616.038	27654.58	U	0.5	0.7		MIT
3613.997	3615.028	27662.30	U		0.3		MIT
3613.748	3614.779	27664.21	U	0.6			MIT
3613.403	3614.434	27666.85	U	0.6	0.5H		MIT
3613.152	3614.182	27668.77	U	0.5	0.3		MIT
3612.669	3613.699	27672.47	U	0.8	0.9		MIT
3611.394	3612.424	27682.24	U	1.1	0.0		MIT
3611.239	3612.269	27683.43	U	0.7	1.0		MIT
3611.001	3612.031	27685.26	U	0.3H	0.3		MIT
3610.69	3611.72	27687.6	U	0.5D	1.1D		MIT
3610.487	3611.517	27689.20	U	0.6	0.9		MIT

AIR WAVELENGTH (ANGSTRMS)	VACUUM WAVELENGTH (ANGSTRMS)	WAVENUMBER (KAYSERS)	LMENT	LOG ARC	INTENSITY SPK	DIS	REF
3609.682	3610.712	27695.37	U	1.2	1.1		MIT
3608.96	3609.99	27700.9	U	1.3	1.0		MIT
3608.686	3609.715	27703.02	U		0.6H		MIT
3608.354	3609.383	27705.56	U	0.5	0.3H		MIT
3608.146	3609.175	27707.16	U	0.3	0.0		MIT
3608.001	3609.030	27708.27	U	0.3	0.6		MIT
3607.776	3608.805	27710.00	U	0.0	0.5		MIT
3607.522	3608.551	27711.95	U		0.3		MIT
3606.946	3607.975	27716.38	U	0.5	0.0		MIT
3606.324	3607.353	27721.16	U	0.9	1.1		MIT
3606.116	3607.145	27722.76	U	0.0	0.8		MIT
3605.823	3606.852	27725.01	U	0.9	0.9		MIT
3605.748	3606.777	27725.59	U	0.0	0.6		MIT
3605.48	3606.51	27727.6	U	0.0	0.6		MIT
3605.276	3606.304	27729.22	U	0.9			MIT
3604.396	3605.424	27735.99	U	0.0	0.8		MIT
3604.224	3605.252	27737.31	U	0.0	0.8		MIT
3603.67	3604.70	27741.6	U	0.7D			MIT
3603.363	3604.391	27743.94	U	1.0	0.0H		MIT
3603.13	3604.16	27745.7	U	0.6A			MIT
3602.51	3603.54	27750.5	U	1.1D	0.0D		MIT
3601.42	3602.45	27758.9	U		0.5		MIT
3600.494	3601.521	27766.04	U	0.0	0.3		MIT
3600.293	3601.320	27767.59	U	1.0			MIT
3599.844	3600.871	27771.06	U	0.8	1.3		MIT
3599.349	3600.376	27774.88	U	0.6	1.0		MIT
3598.946	3599.973	27777.99	U	0.9	0.5		MIT
3598.040	3599.067	27784.98	U	0.3			MIT
3597.759	3598.786	27787.15	U	1.1	0.8		MIT
3597.638	3598.664	27788.09	U		0.3		MIT
3597.245	3598.271	27791.12	U		0.3		MIT
3597.132	3598.158	27792.00	U		0.3		MIT
3596.884	3597.910	27793.91	U	0.8	0.9		MIT
3596.765	3597.791	27794.83	U	0.3	0.9		MIT
3595.945	3596.971	27801.17	U	0.0	0.5		MIT
3595.753	3596.779	27802.65	U	0.7			MIT
3595.526	3596.552	27804.41	U	0.3	0.5H		MIT
3595.241	3596.267	27806.61	U	0.7			MIT
3594.954	3595.980	27808.83	U	0.9	1.2		MIT
3594.111	3595.137	27815.35	U	0.0	0.6		MIT
3593.685	3594.710	27818.65	U	0.9	0.0		MIT
3593.198	3594.223	27822.42	U	1.0	0.0		MIT
3592.801	3593.826	27825.50	U	0.0	0.8		MIT
3592.301	3593.326	27829.37	U	0.6	0.6		MIT
3592.065	3593.090	27831.20	U	0.6	0.5		MIT
3591.884	3592.909	27832.60	U	0.8	1.0		MIT
3591.747	3592.772	27833.66	U	1.0	0.3		MIT
3591.560	3592.585	27835.11	U	0.6	1.2		MIT
3590.728	3591.753	27841.56	U	0.6			MIT
3590.499	3591.524	27843.34	U	0.9	1.0		MIT
3590.320	3591.345	27844.72	U	0.7	1.2		MIT
3589.791	3590.815	27848.83	U	0.9			MIT
3589.658	3590.682	27849.86	U	0.8	0.0		MIT
3588.348	3589.372	27860.03	U	0.5			MIT
3587.835	3588.859	27864.01	U		0.3		MIT
3587.784	3588.808	27864.41	U	0.3			MIT
3586.982	3588.006	27870.64	U		0.3		MIT
3586.326	3587.350	27875.73	U	0.5			MIT
3584.879	3585.902	27886.99	U	1.5	1.1		MIT
3583.927	3584.950	27894.39	U	0.5	0.8		MIT
3583.218	3584.241	27899.91	U		0.3		MIT
3582.022	3583.044	27909.23	U	0.8	1.2		MIT
3581.838	3582.860	27910.66	U	0.8	1.2		MIT
3580.922	3581.944	27917.80	U	0.5			MIT
3580.244	3581.266	27923.09	U	0.8	0.0		MIT
3580.124	3581.146	27924.02	U		0.3		MIT
3579.364	3580.386	27929.95	U	0.8	1.0		MIT
3578.327	3579.348	27938.04	U	0.9	0.0		MIT
3577.919	3578.940	27941.23	U	1.0			MIT
3577.783	3578.804	27942.29	U	0.6	0.0		MIT

AIR WAVELENGTH (ANGSTRMS)	VACUUM WAVELENGTH (ANGSTRMS)	WAVENUMBER (KAYSERS)	LMENT	LOG ARC	INTENSITY SPK	DIS	REF
3577.487	3578.508	27944.60	U	0.5			MIT
3577.349	3578.370	27945.68	U	0.8	0.0		MIT
3577.076	3578.097	27947.82	U	1.2	0.5		MIT
3576.838	3577.859	27949.68	U	0.3	0.0		MIT
3576.219	3577.240	27954.51	U	0.9	1.1		MIT
3575.765	3576.786	27958.06	U	0.0	0.8		MIT
3575.474	3576.495	27960.34	U	0.0	0.9		MIT
3574.960	3575.981	27964.36	U	0.6			MIT
3574.759	3575.780	27965.93	U	1.1	0.0		MIT
3574.330	3575.350	27969.29	U	0.3	0.5H		MIT
3574.110	3575.130	27971.01	U	1.1	0.9		MIT
3572.931	3573.951	27980.24	U	0.8	0.0		MIT
3571.691	3572.711	27989.95	U	0.3	0.8		MIT
3571.560	3572.580	27990.98	U	0.5	0.3		MIT
3571.180	3572.200	27993.96	U	0.8			MIT
3570.930	3571.950	27995.92	U	0.3	0.3		MIT
3570.65	3571.67	27998.1	U	0.5D	0.9D		MIT
3569.882	3570.901	28004.13	U		0.5		MIT
3569.529	3570.548	28006.90	U	0.0	0.7		MIT
3569.062	3570.081	28010.57	U	1.1	1.3		MIT
3568.815	3569.834	28012.51	U	0.8	0.5		MIT
3568.690	3569.709	28013.49	U	0.5			MIT
3568.525	3569.544	28014.78	U	0.3	0.3		MIT
3567.486	3568.505	28022.94	U	0.5H	0.0H		MIT
3566.962	3567.981	28027.06	U	0.0	0.9		MIT
3566.598	3567.616	28029.92	U	1.5	1.0		MIT
3566.005	3567.023	28034.58	U	0.8			MIT
3565.750	3566.768	28036.58	U	0.0	1.2H		MIT
3565.044	3566.062	28042.14	U	0.8			MIT
3564.881	3565.899	28043.42	U	0.5	0.8		MIT
3564.587	3565.605	28045.73	U	0.7	0.9		MIT
3564.181	3565.199	28048.93	U	0.8	0.6		MIT
3563.885	3564.903	28051.26	U		0.6		MIT
3563.661	3564.679	28053.02	U	1.1	0.7		MIT
3563.397	3564.415	28055.10	U		0.8		MIT
3562.712	3563.729	28060.49	U	0.5			MIT
3562.050	3563.067	28065.71	U		0.8		MIT
3561.800	3562.817	28067.68	U	1.1	1.5		MIT
3561.413	3562.430	28070.73	U	1.1			MIT
3560.308	3561.325	28079.44	U	0.6	0.3H		MIT
3558.015	3559.031	28097.53	U	0.5	0.5H		MIT
3557.842	3558.858	28098.90	U	1.1	0.0		MIT
3557.568	3558.584	28101.06	U	0.0	0.3H		MIT
3557.323	3558.339	28103.00	U	0.0	0.3		MIT
3557.01	3558.03	28105.5	U		0.5D		MIT
3555.906	3556.922	28114.20	U	0.0	0.5		MIT
3555.677	3556.693	28116.01	U	0.6	0.3		MIT
3555.512	3556.528	28117.31	U	0.0	0.6H		MIT
3555.323	3556.339	28118.81	U	1.2	0.8		MIT
3554.505	3555.520	28125.28	U	0.5			MIT
3553.812	3554.827	28130.76	U	0.0	0.3		MIT
3553.437	3554.452	28133.73	U	1.2			MIT
3553.010	3554.025	28137.11	U	0.7	0.3		MIT
3552.668	3553.683	28139.82	U	0.7	1.0		MIT
3552.172	3553.187	28143.75	U	0.9	1.1		MIT
3551.755	3552.770	28147.05	U	0.3	0.5		MIT
3551.269	3552.284	28150.91	U	0.5	0.7		MIT
3551.043	3552.057	28152.70	U	0.8	0.8		MIT
3550.822	3551.836	28154.45	U	1.1	1.3		MIT
3550.533	3551.547	28156.74	U	0.3	0.3H		MIT
3550.175	3551.189	28159.58	U	0.0	0.6H		MIT
3549.718	3550.732	28163.21	U	0.0	0.3		MIT
3549.204	3550.218	28167.28	U	1.1	0.0		MIT
3548.852	3549.866	28170.08	U	0.6	0.5H		MIT
3548.616	3549.630	28171.95	U	0.9			MIT
3548.421	3549.435	28173.50	U	0.7			MIT
3548.260	3549.274	28174.78	U	0.5			MIT
3547.533	3548.547	28180.55	U	0.0	0.5		MIT
3547.189	3548.202	28183.28	U	1.0	1.1		MIT
3546.680	3547.693	28187.33	U	0.8	0.9		MIT

AIR WAVELENGTH (ANGSTRMS)	VACUUM WAVELENGTH (ANGSTRMS)	WAVENUMBER (KAYSERS)	LMENT	LOG ARC	INTENSITY SPK	DIS	REF
3546.553	3547.566	28188.34	U	0.7	0.5		MIT
3546.383	3547.396	28189.69	U	0.0	1.2		MIT
3546.132	3547.145	28191.69	U	1.1	0.8		MIT
3545.668	3546.681	28195.37	U	0.5	0.9		MIT
3545.437	3546.450	28197.21	U	0.5			MIT
3544.659	3545.672	28203.40	U	0.6			MIT
3544.209	3545.222	28206.98	U	0.5H	0.8		MIT
3543.35	3544.36	28213.8	U	1.1E	0.0D		MIT
3543.156	3544.168	28215.36	U	0.3	0.9		MIT
3542.717	3543.729	28218.86	U	0.3	0.8		MIT
3542.570	3543.582	28220.03	U	0.9			MIT
3542.381	3543.393	28221.54	U	0.6	0.0		MIT
3541.888	3542.900	28225.46	U	0.3	0.8		MIT
3541.284	3542.296	28230.28	U		0.9		MIT
3540.986	3541.998	28232.65	U		0.3H		MIT
3540.623	3541.635	28235.55	U	0.3	0.7		MIT
3540.465	3541.477	28236.81	U	0.8	1.2		MIT
3539.654	3540.666	28243.28	U	1.1	0.3		MIT
3539.212	3540.223	28246.80	U	0.6	0.6		MIT
3538.976	3539.987	28248.69	U	0.0	0.5		MIT
3538.679	3539.690	28251.06	U		0.8H		MIT
3538.418	3539.429	28253.14	U	0.3	0.9		MIT
3538.221	3539.232	28254.72	U	0.9	0.8		MIT
3537.826	3538.837	28257.87	U		0.3		MIT
3537.445	3538.456	28260.91	U	0.7	1.1		MIT
3537.346	3538.357	28261.71	U	0.3	0.0H		MIT
3537.058	3538.069	28264.01	U	0.9	0.0		MIT
3536.803	3537.814	28266.04	U	0.0	0.6		MIT
3536.077	3537.088	28271.85	U	0.0	0.3		MIT
3535.836	3536.847	28273.77	U	0.8	0.3		MIT
3535.649	3536.660	28275.27	U	0.0H	0.5H		MIT
3535.481	3536.491	28276.61	U	0.8			MIT
3535.339	3536.349	28277.75	U	0.3			MIT
3534.335	3535.345	28285.78	U	0.9	0.3		MIT
3534.065	3535.075	28287.94	U	0.0	0.8		MIT
3533.568	3534.578	28291.92	U	1.0	1.3		MIT
3533.010	3534.020	28296.39	U		0.6H		MIT
3532.781	3533.791	28298.22	U	0.6	0.9		MIT
3532.051	3533.061	28304.07	U	0.0	0.3		MIT
3531.641	3532.650	28307.36	U	1.2	0.0		MIT
3531.113	3532.122	28311.59	U	0.9	1.3		MIT
3530.540	3531.549	28316.19	U	0.8			MIT
3529.773	3530.782	28322.34	U	0.3	0.7		MIT
3529.619	3530.628	28323.57	U	0.3	0.8		MIT
3529.077	3530.086	28327.92	U	0.5	0.5		MIT
3528.687	3529.696	28331.05	U	0.8	1.2		MIT
3528.345	3529.354	28333.80	U	0.0	0.3		MIT
3528.074	3529.083	28335.98	U		0.7		MIT
3527.631	3528.639	28339.53	U	0.3	0.0		MIT
3527.379	3528.387	28341.56	U	0.0	0.3		MIT
3526.600	3527.608	28347.82	U	0.5	0.9		MIT
3525.847	3526.855	28353.87	U	0.5	0.9		MIT
3525.73	3526.74	28354.8	U	0.8D	0.8D		MIT
3525.141	3526.149	28359.55	U	0.7	0.9		MIT
3523.565	3524.572	28372.24	U	0.6	1.2		MIT
3522.576	3523.583	28380.20	U	0.7	0.3H		MIT
3522.274	3523.281	28382.64	U	0.3			MIT
3522.036	3523.043	28384.55	U	0.0	0.7		MIT
3521.482	3522.489	28389.02	U	0.5	0.9H		MIT
3521.030	3522.037	28392.66	U	0.7			MIT
3520.793	3521.800	28394.57	U	0.8	1.0		MIT
3519.956	3520.962	28401.32	U	0.8	1.1		MIT
3518.951	3519.957	28409.44	U	0.5			MIT
3518.475	3519.481	28413.28	U	0.3			MIT
3518.109	3519.115	28416.24	U	0.8			MIT
3517.642	3518.648	28420.01	U	0.0	1.1		MIT
3517.457	3518.463	28421.50	U	0.0	0.7		MIT
3517.052	3518.058	28424.77	U	0.6	0.9		MIT
3516.854	3517.860	28426.38	U	0.8	1.2		MIT
3515.676	3516.681	28435.90	U	0.8	0.0H		MIT

AIR WAVELENGTH (ANGSTRMS)	VACUUM WAVELENGTH (ANGSTRMS)	WAVENUMBER (KAYSERS)	LMENT	LOG ARC	INTENSITY SPK	DIS	REF
3515.448	3516.453	28437.74	U	0.6			MIT
3515.235	3516.240	28439.47	U	0.5	1.0		MIT
3514.615	3515.620	28444.48	U	1.3	0.7		MIT
3513.678	3514.683	28452.07	U	1.0	0.0		MIT
3513.370	3514.375	28454.56	U	0.5	1.0		MIT
3513.039	3514.044	28457.24	U	0.8	0.8		MIT
3512.887	3513.892	28458.48	U	0.9	0.0		MIT
3512.673	3513.678	28460.21	U	0.0	0.7		MIT
3511.849	3512.853	28466.89	U	0.7	0.9		MIT
3511.581	3512.585	28469.06	U	0.8	1.0		MIT
3511.439	3512.443	28470.21	U	1.0	0.0		MIT
3511.145	3512.149	28472.59	U	0.8			MIT
3510.998	3512.002	28473.79	U	0.0	0.3		MIT
3510.804	3511.808	28475.36	U	0.7	1.1		MIT
3510.509	3511.513	28477.75	U	1.0			MIT
3510.338	3511.342	28479.14	U	0.8			MIT
3509.858	3510.862	28483.03	U	0.3			MIT
3509.668	3510.672	28484.58	U	1.0	1.2		MIT
3509.349	3510.353	28487.17	U	0.5	0.8		MIT
3509.122	3510.126	28489.01	U	0.0	0.3		MIT
3508.846	3509.850	28491.25	U	0.9	1.0		MIT
3508.306	3509.309	28495.63	U	0.3			MIT
3507.706	3508.709	28500.51	U	0.5	0.6		MIT
3507.683	3508.686	28500.70	U	0.5	0.6		MIT
3507.344	3508.347	28503.45	U	1.0	0.5		MIT
3507.051	3508.054	28505.83	U	1.0	0.7		MIT
3506.790	3507.793	28507.95	U	0.8	1.0		MIT
3506.319	3507.322	28511.78	U	0.3			MIT
3505.552	3506.555	28518.02	U	0.9			MIT
3505.455	3506.458	28518.81	U	0.6	0.8		MIT
3505.074	3506.077	28521.91	U	0.9	1.2		MIT
3504.936	3505.939	28523.03	U	0.9			MIT
3504.661	3505.664	28525.27	U	0.0	0.5		MIT
3504.484	3505.486	28526.71	U	0.9	0.3		MIT
3504.236	3505.238	28528.73	U	0.6			MIT
3504.008	3505.010	28530.59	U	1.0	0.0		MIT
3503.324	3504.326	28536.16	U	0.3			MIT
3503.041	3504.043	28538.46	U	0.3	0.0		MIT
3502.239	3503.241	28545.00	U	0.7			MIT
3501.649	3502.651	28549.81	U	0.6	0.3		MIT
3501.34	3502.34	28552.3	U	0.7D	0.5D		MIT
3501.005	3502.007	28555.06	U	0.9			MIT
3500.333	3501.334	28560.54	U	0.5	1.0		MIT
3500.077	3501.078	28562.63	U	1.2	0.3		MIT
3499.327	3500.328	28568.75	U	0.8	1.2		MIT
3499.071	3500.072	28570.84	U	0.8	0.9		MIT
3498.600	3499.601	28574.69	U	0.3	0.6		MIT
3498.384	3499.385	28576.45	U	0.9	0.5		MIT
3497.622	3498.623	28582.68	U	0.8	1.1		MIT
3497.265	3498.266	28585.59	U	0.8	0.6		MIT
3497.067	3498.068	28587.21	U	0.5	0.5		MIT
3496.910	3497.911	28588.50	U	0.0	0.6		MIT
3496.415	3497.415	28592.54	U	0.9	1.2		MIT
3495.748	3496.748	28598.00	U	0.7	0.7		MIT
3495.598	3496.598	28599.23	U	0.8	0.3		MIT
3494.838	3495.838	28605.44	U	0.8	1.2		MIT
3494.557	3495.557	28607.75	U	0.8			MIT
3493.997	3494.997	28612.33	U	1.1	0.3H		MIT
3493.407	3494.407	28617.16	U	0.3	0.7		MIT
3493.333	3494.333	28617.77	U	0.8	1.2		MIT
3492.799	3493.798	28622.14	U	0.3	1.0		MIT
3492.334	3493.333	28625.95	U	0.9	0.0		MIT
3492.207	3493.206	28627.00	U	0.3	0.9		MIT
3491.340	3492.339	28634.10	U	0.8	1.0		MIT
3490.950	3491.949	28637.30	U	0.8	0.0		MIT
3490.604	3491.603	28640.14	U	0.8	0.9		MIT
3490.242	3491.241	28643.11	U	1.1	1.3		MIT
3489.568	3490.567	28648.64	U	0.8	1.0		MIT
3489.371	3490.370	28650.26	U	1.3	0.0		MIT
3488.815	3489.813	28654.83	U	0.9	1.2		MIT

AIR WAVELENGTH (ANGSTRMS)	VACUUM WAVELENGTH (ANGSTRMS)	WAVENUMBER (KAYSERS)	LMENT	LOG ARC	INTENSITY SPK	DIS	REF
3488.619	3489.617	28656.44	U	0.8			MIT
3488.301	3489.299	28659.05	U	0.8			MIT
3488.168	3489.166	28660.14	U	0.5	0.0H		MIT
3487.651	3488.649	28664.39	U	0.0	0.8		MIT
3487.582	3488.580	28664.96	U	0.9	0.0		MIT
3487.282	3488.280	28667.42	U	0.6	0.0H		MIT
3487.090	3488.088	28669.00	U	0.7	0.7H		MIT
3486.303	3487.301	28675.47	U	0.7	1.2		MIT
3485.972	3486.970	28678.20	U	0.7R	0.9		MIT
3485.723	3486.721	28680.24	U	0.3			MIT
3485.259	3486.257	28684.06	U	0.0	0.3		MIT
3484.921	3485.918	28686.84	U	0.3	0.3H		MIT
3484.021	3485.018	28694.25	U	0.5			MIT
3483.773	3484.770	28696.30	U	0.3	0.6		MIT
3483.576	3484.573	28697.92	U	0.3	0.0H		MIT
3480.359	3481.355	28724.45	U	0.5	0.3		MIT
3478.303	3479.299	28741.42	U		0.3H		MIT
3477.994	3478.990	28743.98	U	0.5	0.0H		MIT
3477.837	3478.833	28745.27	U	0.6	0.7		MIT
3477.498	3478.494	28748.08	U	0.5	0.5		MIT
3477.093	3478.088	28751.43	U	0.3	0.5		MIT
3476.437	3477.432	28756.85	U	0.6	0.5		MIT
3476.293	3477.288	28758.04	U	0.6	0.5		MIT
3474.990	3475.985	28768.82	U	0.6	0.5		MIT
3474.535	3475.530	28772.59	U	0.3	0.7		MIT
3474.167	3475.162	28775.64	U	0.6	0.3		MIT
3473.703	3474.698	28779.48	U		0.5W		MIT
3473.502	3474.497	28781.15	U	0.5			MIT
3473.431	3474.426	28781.74	U	0.9			MIT
3473.288	3474.282	28782.92	U	0.3			MIT
3473.034	3474.028	28785.03	U	0.7			MIT
3472.559	3473.553	28788.96	U	0.5	0.7		MIT
3472.509	3473.503	28789.38	U	0.7	0.8		MIT
3472.107	3473.101	28792.71	U	0.5	0.3		MIT
3471.979	3472.973	28793.77	U	0.3	0.3		MIT
3471.703	3472.697	28796.06	U	0.3	0.3		MIT
3471.086	3472.080	28801.18	U	0.3	0.5		MIT
3470.576	3471.570	28805.41	U		0.3		MIT
3470.262	3471.256	28808.02	U	0.6	0.3		MIT
3469.776	3470.770	28812.05	U	0.6	0.5		MIT
3469.492	3470.486	28814.41	U	0.8	0.3H		MIT
3469.213	3470.206	28816.73	U	0.0	0.3		MIT
3469.101	3470.094	28817.66	U	0.0	0.6		MIT
3468.920	3469.913	28819.16	U	0.6			MIT
3468.556	3469.549	28822.19	U	0.6	0.0H		MIT
3468.044	3469.037	28826.44	U	0.5	0.5W		MIT
3467.656	3468.649	28829.67	U	0.3	0.3		MIT
3467.452	3468.445	28831.36	U	0.3			MIT
3466.503	3467.496	28839.26	U	0.5			MIT
3466.30	3467.29	28840.9	U	0.9	0.3		MIT
3466.139	3467.132	28842.28	U	0.6			MIT
3464.948	3465.940	28852.20	U	0.6	0.6		MIT
3464.869	3465.861	28852.86	U	0.3			MIT
3464.672	3465.664	28854.50	U		0.3		MIT
3464.470	3465.462	28856.18	U	0.6			MIT
3463.910	3464.902	28860.84	U	0.6			MIT
3461.003	3461.994	28885.08	U	0.5	0.5		MIT
3460.781	3461.772	28886.94	U	0.3			MIT
3460.351	3461.342	28890.53	U	0.5	0.7		MIT
3460.058	3461.049	28892.97	U	0.7			MIT
3459.917	3460.908	28894.15	U	0.9			MIT
3459.709	3460.700	28895.89	U	0.5	0.6		MIT
3459.330	3460.321	28899.05	U		0.5		MIT
3459.165	3460.156	28900.43	U	0.5			MIT
3458.96	3459.95	28902.1	U	0.3D	0.0D		MIT
3458.683	3459.674	28904.46	U	0.7	0.5		MIT
3458.460	3459.451	28906.32	U	0.3			MIT
3458.171	3459.162	28908.74	U	0.9			MIT
3457.714	3458.704	28912.56	U	0.7	1.0		MIT
3457.051	3458.041	28918.10	U	0.9	1.0		MIT

AIR WAVELENGTH (ANGSTRMS)	VACUUM WAVELENGTH (ANGSTRMS)	WAVENUMBER (KAYSERS)	LMENT	LOG ARC	INTENSITY SPK	DIS	REF
3456.605	3457.595	28921.84	U	0.3H	0.3		MIT
3456.297	3457.287	28924.41	U	0.5	0.3H		MIT
3455.933	3456.923	28927.46	U	0.5	0.3		MIT
3455.742	3456.732	28929.06	U	0.7	0.6		MIT
3454.844	3455.834	28936.58	U		0.3		MIT
3454.617	3455.607	28938.48	U	1.0			MIT
3454.228	3455.218	28941.74	U	0.3	0.3		MIT
3453.780	3454.769	28945.49	U	0.6	0.5		MIT
3453.570	3454.559	28947.25	U	0.7	0.9		MIT
3452.962	3453.951	28952.35	U	0.3			MIT
3452.726	3453.715	28954.33	U		0.5W		MIT
3452.015	3453.004	28960.29	U	0.6			MIT
3451.209	3452.198	28967.05	U	0.7	0.7		MIT
3449.981	3450.970	28977.36	U	0.9	0.5		MIT
3449.80	3450.79	28978.9	U	0.3			MIT
3449.23	3450.22	28983.7	U	0.3			MIT
3448.775	3449.763	28987.50	U	0.5	0.9		MIT
3448.05	3449.04	28993.6	U	0.8			MIT
3446.764	3447.752	29004.41	U	0.6			MIT
3445.870	3446.857	29011.93	U	0.5H	0.0		MIT
3445.714	3446.701	29013.25	U	0.5H	0.5		MIT
3444.695	3445.682	29021.83	U	0.8	0.7		MIT
3444.594	3445.581	29022.68	U	0.5	0.5		MIT
3444.419	3445.406	29024.16	U	0.0	0.3		MIT
3443.535	3444.522	29031.61	U	0.3	0.3		MIT
3442.960	3443.947	29036.45	U	1.2			MIT
3442.362	3443.349	29041.50	U	0.3	0.3		MIT
3441.949	3442.935	29044.98	U	0.9			MIT
3439.897	3440.883	29062.31	U	0.8			MIT
3439.712	3440.698	29063.87	U	0.8			MIT
3439.434	3440.420	29066.22	U	1.0			MIT
3438.924	3439.910	29070.53	U	0.5			MIT
3438.697	3439.683	29072.45	U	0.3	0.5		MIT
3438.407	3439.393	29074.90	U	0.5	0.5		MIT
3437.934	3438.919	29078.90	U	0.8	1.0		MIT
3437.671	3438.656	29081.13	U	0.8			MIT
3437.09	3438.08	29086.0	U	0.5D	0.7D		MIT
3436.780	3437.765	29088.67	U	1.1	1.2		MIT
3436.19	3437.17	29093.7	U	0.8			MIT
3435.53	3436.51	29099.2	U	1.2D			MIT
3435.200	3436.185	29102.04	U	1.0			MIT
3434.805	3435.790	29105.39	U	0.3H	0.0H		MIT
3434.614	3435.599	29107.01	U	1.1	0.0H		MIT
3434.283	3435.267	29109.81	U	0.8			MIT
3434.148	3435.132	29110.96	U	0.8	1.0		MIT
3433.707	3434.691	29114.70	U	0.7	0.8		MIT
3432.493	3433.477	29125.00	U	0.0	0.7H		MIT
3431.990	3432.974	29129.26	U	0.3H	0.5		MIT
3431.816	3432.800	29130.74	U	0.3			MIT
3431.54	3432.52	29133.1	U	0.8	0.8		MIT
3431.138	3432.122	29136.50	U	1.0	0.3		MIT
3430.716	3431.700	29140.08	U	0.3	0.3		MIT
3430.484	3431.468	29142.05	U	1.1R	0.7H		MIT
3430.176	3431.159	29144.67	U	0.8	0.0H		MIT
3429.327	3430.310	29151.88	U	0.5	0.8		MIT
3429.032	3430.015	29154.39	U	1.3R	0.3H		MIT
3428.782	3429.765	29156.52	U	1.1			MIT
3426.393	3427.375	29176.84	U	0.9	1.1		MIT
3425.842	3426.824	29181.54	U	1.0	0.0H		MIT
3425.41	3426.39	29185.2	U	0.6	0.5		MIT
3425.119	3426.101	29187.70	U	0.8	0.5		MIT
3424.810	3425.792	29190.33	U	0.9	1.1		MIT
3424.557	3425.539	29192.49	U	1.3	1.2		MIT
3424.135	3425.117	29196.08	U	0.0	0.3		MIT
3423.05	3424.03	29205.3	U	1.1D	1.2D		MIT
3422.352	3423.333	29211.29	U	1.3R	1.2R		MIT
3421.69	3422.67	29216.9	U	0.9	0.9		MIT
3421.38	3422.36	29219.6	U	0.9D	0.3D		MIT
3420.07	3421.05	29230.8	U	0.7	0.5		MIT
3420.02	3421.00	29231.2	U	0.3	0.5		MIT

AIR WAVELENGTH (ANGSTRMS)	VACUUM WAVELENGTH (ANGSTRMS)	WAVENUMBER (KAYSERS)	LMENT	LOG ARC	INTENSITY SPK	DIS	REF
3419.963	3420.944	29231.70	U	0.7	0.3		MIT
3419.359	3420.340	29236.86	U	0.8			MIT
3419.159	3420.140	29238.57	U	0.8			MIT
3418.850	3419.831	29241.22	U	0.5	0.5		MIT
3418.39	3419.37	29245.1	U	1.0	0.3		MIT
3417.298	3418.278	29254.50	U	0.5	0.0		MIT
3416.854	3417.834	29258.30	U	0.3	0.3		MIT
3416.551	3417.531	29260.89	U	0.6H	0.3		MIT
3416.309	3417.289	29262.96	U	0.8	0.3		MIT
3416.125	3417.105	29264.54	U	1.1	0.7		MIT
3415.865	3416.845	29266.77	U	1.1	0.7		MIT
3415.613	3416.593	29268.93	U	0.0	0.3		MIT
3415.325	3416.305	29271.39	U	0.5			MIT
3414.63	3415.61	29277.4	U	0.5D	0.0D		MIT
3414.36	3415.34	29279.7	U	0.9D	0.0D		MIT
3413.806	3414.785	29284.42	U	0.6	1.0		MIT
3412.693	3413.672	29293.97	U	0.8			MIT
3412.36	3413.34	29296.8	U	1.1D	1.1D		MIT
3412.095	3413.074	29299.10	U	0.8	1.0		MIT
3411.532	3412.511	29303.94	U	1.0	0.9		MIT
3411.080	3412.059	29307.82	U	0.0	0.3		MIT
3410.681	3411.659	29311.25	U	0.3	0.5H		MIT
3410.404	3411.382	29313.63	U	0.5	0.5		MIT
3409.99	3410.97	29317.2	U	0.8			MIT
3409.719	3410.697	29319.52	U	0.7	0.5		MIT
3409.377	3410.355	29322.46	U	1.0	0.3		MIT
3409.008	3409.986	29325.63	U	0.8	0.3		MIT
3407.355	3408.333	29339.86	U	0.3	0.3		MIT
3406.926	3407.903	29343.55	U	0.3	0.3		MIT
3406.278	3407.255	29349.14	U	1.0	0.7		MIT
3405.746	3406.723	29353.72	U	1.1			MIT
3404.933	3405.910	29360.73	U	0.5	0.3		MIT
3403.894	3404.871	29369.69	U	0.3	0.3		MIT
3403.546	3404.523	29372.69	U	1.0	1.0		MIT
3403.217	3404.194	29375.53	U	0.6	0.3		MIT
3402.775	3403.751	29379.35	U	0.6	0.0		MIT
3402.44	3403.42	29382.2	U	0.3D	0.3D		MIT
3401.870	3402.846	29387.16	U	0.9	0.6		MIT
3401.213	3402.189	29392.84	U	0.9	0.9		MIT
3401.014	3401.990	29394.56	U	1.0	0.9		MIT
3400.75	3401.73	29396.8	U	0.8	0.5H		MIT
3400.467	3401.443	29399.29	U	0.9	0.3		MIT
3400.327	3401.303	29400.50	U	0.3	0.3		MIT
3399.958	3400.934	29403.69	U	0.8	0.5		MIT
3399.810	3400.786	29404.97	U	0.7	0.3		MIT
3399.528	3400.504	29407.41	U	0.5	0.0		MIT
3398.995	3399.970	29412.02	U	1.2			MIT
3398.67	3399.65	29414.8	U	0.3D	0.0D		MIT
3398.498	3399.473	29416.32	U	0.8	0.0		MIT
3398.256	3399.231	29418.42	U	1.1	0.8		MIT
3397.945	3398.920	29421.11	U	0.6	0.3		MIT
3397.20	3398.17	29427.6	U	1.1			MIT
3395.58	3396.55	29441.6	U	0.5D	0.0D		MIT
3395.315	3396.290	29443.90	U	0.8	0.7		MIT
3394.776	3395.750	29448.57	U	0.9	0.8		MIT
3394.283	3395.257	29452.85	U	0.9	0.0		MIT
3393.91	3394.88	29456.1	U	0.7D	0.6D		MIT
3392.992	3393.966	29464.05	U	0.3	0.3		MIT
3391.398	3392.372	29477.90	U	0.8			MIT
3391.04	3392.01	29481.0	U	0.3	0.3		MIT
3390.389	3391.362	29486.68	U	1.3	1.0		MIT
3389.95	3390.92	29490.5	U	0.8	0.6		MIT
3389.744	3390.717	29492.29	U	0.6	0.5H		MIT
3388.41	3389.38	29503.9	U	0.5D	0.0D		MIT
3387.613	3388.586	29510.84	U	0.7			MIT
3386.601	3387.573	29519.66	U	0.6	0.0		MIT
3386.133	3387.105	29523.74	U	0.8	1.2		MIT
3385.41	3386.38	29530.0	U	1.1D	0.9D		MIT
3384.446	3385.418	29538.45	U	1.0	0.9		MIT
3384.220	3385.192	29540.43	U	0.5	0.5		MIT
3383.402	3384.373	29547.57	U	0.8	0.5		MIT
3382.675	3383.646	29553.92	U	0.6	0.6		MIT
3382.304	3383.275	29557.16	U	0.5	0.0		MIT
3381.95	3382.92	29560.2	U	0.6D	0.8D		MIT
3381.691	3382.662	29562.52	U	0.8			MIT
3380.90	3381.87	29569.4	U	0.5	0.5H		MIT
3380.700	3381.671	29571.18	U	0.7	1.0		MIT
3379.672	3380.643	29580.18	U	0.3	0.5		MIT
3379.378	3380.348	29582.75	U	0.5	0.0		MIT
3378.202	3379.172	29593.05	U	1.2	0.0		MIT
3377.392	3378.362	29600.14	U	0.6	0.8		MIT
3376.924	3377.894	29604.25	U	0.9			MIT
3376.549	3377.519	29607.53	U	0.9	0.6		MIT
3376.040	3377.010	29612.00	U	0.8			MIT
3375.780	3376.750	29614.28	U	1.1	1.0		MIT
3375.48	3376.45	29616.9	U	0.9D			MIT
3374.952	3375.921	29621.54	U	0.7	0.5H		MIT
3374.468	3375.437	29625.79	U	0.7	0.3		MIT
3374.151	3375.120	29628.57	U	0.6	0.3		MIT
3374.10	3375.07	29629.0	U	0.5D	0.5D		MIT
3373.727	3374.696	29632.30	U	1.0	0.0		MIT
3372.600	3373.569	29642.20	U	0.6	0.0		MIT
3372.01	3372.98	29647.4	U	0.9	0.6		MIT
3371.292	3372.260	29653.70	U	0.9	1.2		MIT
3370.95	3371.92	29656.7	U	0.0D	0.3D		MIT
3370.45	3371.42	29661.1	U	0.3	0.3H		MIT
3370.134	3371.102	29663.89	U	0.8	0.6		MIT
3369.964	3370.932	29665.39	U	0.8	0.0H		MIT
3368.96	3369.93	29674.2	U	0.7D			MIT
3368.804	3369.772	29675.60	U	0.8	0.8		MIT
3368.31	3369.28	29679.9	U	0.3D	0.3D		MIT
3367.896	3368.864	29683.60	U	0.8	0.3		MIT
3367.60	3368.57	29686.2	U	0.9D	0.0D		MIT
3366.581	3367.548	29695.20	U	0.3			MIT
3366.370	3367.337	29697.06	U	0.0	0.3		MIT
3365.848	3366.815	29701.66	U	1.0			MIT
3365.130	3366.097	29708.00	U	0.7			MIT
3364.64	3365.61	29712.3	U	0.8	0.3H		MIT
3363.921	3364.887	29718.68	U	0.7	0.0H		MIT
3363.43	3364.40	29723.0	U	1.1E	0.5A		MIT
3363.29	3364.26	29724.2	U	0.8D	0.0D		MIT
3363.031	3363.997	29726.54	U	0.5			MIT
3362.771	3363.737	29728.84	U	0.6	0.3		MIT
3362.050	3363.016	29735.21	U	0.3	0.0		MIT
3361.727	3362.693	29738.07	U	1.2	0.9		MIT
3361.213	3362.179	29742.62	U	0.3			MIT
3360.337	3361.303	29750.37	U	0.7	0.0		MIT
3360.08	3361.05	29752.6	U	1.0	0.7		MIT
3359.62	3360.59	29756.7	U	1.0	0.0		MIT
3358.868	3359.833	29763.38	U	0.5	0.3		MIT
3358.141	3359.106	29769.83	U	0.6			MIT
3357.931	3358.896	29771.69	U	0.9	0.9		MIT
3357.521	3358.486	29775.32	U	1.0			MIT
3357.14	3358.10	29778.7	U	0.6	0.5		MIT
3356.396	3357.361	29785.30	U	0.9			MIT
3356.205	3357.170	29787.00	U	0.9			MIT
3355.89	3356.85	29789.8	U	0.5	0.3		MIT
3355.43	3356.39	29793.9	U	0.3	0.3		MIT
3355.112	3356.076	29796.70	U	0.9	0.6		MIT
3354.752	3355.716	29799.90	U	0.9			MIT
3354.502	3355.466	29802.12	U	1.1	0.9		MIT
3354.37	3355.33	29803.3	U	0.8D			MIT
3353.599	3354.563	29810.14	U	0.3	0.0H		MIT
3352.679	3353.643	29818.32	U	0.6	0.3		MIT
3351.833	3352.796	29825.85	U	0.7	0.3		MIT
3351.66	3352.62	29827.4	U	0.6	0.3		MIT
3351.251	3352.214	29831.03	U	0.9	0.5		MIT
3350.897	3351.860	29834.18	U	0.5	0.3		MIT
3350.52	3351.48	29837.5	U	0.9	0.3H		MIT
3350.28	3351.24	29839.7	U	0.6D	0.0A		MIT

AIR WAVELENGTH (ANGSTRMS)	VACUUM WAVELENGTH (ANGSTRMS)	WAVENUMBER (KAYSERS)	LMENT	LOG ARC	INTENSITY SPK	DIS	REF
3349.43	3350.39	29847.2	U	0.6D	0.5D		MIT
3349.036	3349.999	29850.76	U	0.6			MIT
3348.691	3349.654	29853.83	U	0.6	0.0		MIT
3348.33	3349.29	29857.1	U	0.7D	0.0		MIT
3347.042	3348.004	29868.54	U	0.5	0.3		MIT
3346.733	3347.695	29871.30	U	0.3	0.3		MIT
3346.400	3347.362	29874.27	U	0.3	0.0		MIT
3346.198	3347.160	29876.07	U	0.5	0.0H		MIT
3345.890	3346.852	29878.82	U	1.1	0.0H		MIT
3345.54	3346.50	29881.9	U	1.1	0.6		MIT
3344.870	3345.832	29887.94	U	0.8	0.8		MIT
3344.561	3345.523	29890.70	U	0.6			MIT
3344.32	3345.28	29892.9	U	0.6	0.0		MIT
3344.076	3345.037	29895.03	U	0.3	0.0		MIT
3343.44	3344.40	29900.7	U	0.5D	0.0		MIT
3342.68	3343.64	29907.5	U	1.0	1.0		MIT
3341.661	3342.622	29916.64	U	1.1	1.2		MIT
3341.281	3342.242	29920.04	U	0.3	0.0		MIT
3340.994	3341.955	29922.61	U	0.3	0.0		MIT
3340.767	3341.728	29924.64	U	0.3	0.0		MIT
3340.631	3341.592	29925.86	U	0.5	0.0		MIT
3340.114	3341.074	29930.49	U	0.5	0.0		MIT
3339.716	3340.676	29934.06	U	0.5	0.0		MIT
3339.446	3340.406	29936.48	U	0.7	0.0		MIT
3339.215	3340.175	29938.55	U	0.8	0.5		MIT
3338.48	3339.44	29945.1	U	0.6D	0.0		MIT
3337.929	3338.889	29950.08	U	0.6	0.6		MIT
3337.79	3338.75	29951.3	U	1.1	1.0		MIT
3337.393	3338.353	29954.89	U	0.5	0.5		MIT
3337.036	3337.996	29958.10	U	1.0			MIT
3336.685	3337.645	29961.25	U	1.0	0.8		MIT
3336.32	3337.28	29964.5	U	0.5	0.0		MIT
3335.93	3336.89	29968.0	U	0.3	0.0		MIT
3335.63	3336.59	29970.7	U	0.3D	0.3		MIT
3335.219	3336.178	29974.42	U	0.3	0.0		MIT
3334.820	3335.779	29978.00	U	1.0			MIT
3334.434	3335.393	29981.48	U	0.3	0.3		MIT
3334.247	3335.206	29983.16	U	0.0	0.3		MIT
3333.94	3334.90	29985.9	U	0.0D	0.5		MIT
3333.624	3334.583	29988.76	U	0.6			MIT
3333.506	3334.465	29989.82	U	0.8			MIT
3333.258	3334.217	29992.05	U	0.3	0.8		MIT
3332.895	3333.854	29995.32	U	0.6	0.0		MIT
3332.425	3333.383	29999.55	U	0.9	0.5		MIT
3331.963	3332.921	30003.71	U	0.5	0.0		MIT
3331.609	3332.567	30006.90	U	0.7	0.0		MIT
3331.303	3332.261	30009.65	U	0.8	0.3H		MIT
3330.399	3331.357	30017.80	U	0.9	0.0		MIT
3329.922	3330.880	30022.10	U	1.0H	0.3		MIT
3329.305	3330.263	30027.66	U	0.8	0.3		MIT
3328.747	3329.705	30032.70	U	0.9			MIT
3328.270	3329.227	30037.00	U	0.3	0.3		MIT
3327.504	3328.461	30043.91	U	1.1	0.8		MIT
3327.244	3328.201	30046.26	U	0.3	0.0		MIT
3327.051	3328.008	30048.00	U	0.3	0.5		MIT
3326.185	3327.142	30055.83	U	1.1			MIT
3325.864	3326.821	30058.73	U	0.3			MIT
3325.76	3326.72	30059.7	U	0.5D	0.0		MIT
3325.168	3326.125	30065.02	U	0.3	0.3		MIT
3324.997	3325.954	30066.56	U	1.0			MIT
3324.60	3325.56	30070.2	U	0.5D	0.0		MIT
3324.031	3324.987	30075.30	U	0.5			MIT
3323.119	3324.075	30083.56	U	0.3	0.0H		MIT
3322.93	3323.89	30085.3	U	0.6D	0.0		MIT
3322.670	3323.626	30087.62	U	0.6	0.3		MIT
3322.118	3323.074	30092.62	U	1.3	1.1		MIT
3321.707	3322.663	30096.34	U	0.7	0.0		MIT
3320.92	3321.88	30103.5	U	0.0	0.3		MIT
3320.337	3321.292	30108.76	U	0.3	0.0		MIT
3319.58	3320.54	30115.6	U	0.9D			MIT

AIR WAVELENGTH (ANGSTRMS)	VACUUM WAVELENGTH (ANGSTRMS)	WAVENUMBER (KAYSERS)	LMENT	LOG ARC	INTENSITY SPK	DIS	REF
3319.324	3320.279	30117.95	U	0.5	0.8		MIT
3319.209	3320.164	30118.99	U	1.0	0.8		MIT
3318.85	3319.81	30122.2	U	0.9D	0.6		MIT
3318.273	3319.228	30127.49	U	0.7	0.0		MIT
3318.095	3319.050	30129.10	U	0.7	0.5		MIT
3317.787	3318.742	30131.90	U	1.0			MIT
3317.606	3318.561	30133.55	U	0.6			MIT
3317.43	3318.38	30135.1	U	0.3D			MIT
3316.704	3317.658	30141.74	U	0.8	0.5		MIT
3316.544	3317.498	30143.20	U	1.0	0.0H		MIT
3316.047	3317.001	30147.71	U	0.9			MIT
3315.057	3316.011	30156.72	U	0.3			MIT
3314.052	3315.006	30165.86	U	0.6	0.5		MIT
3313.94	3314.89	30166.9	U	1.0	1.0		MIT
3313.718	3314.672	30168.90	U	0.3	0.0		MIT
3313.08	3314.03	30174.7	U	0.7D	0.5		MIT
3312.700	3313.653	30178.17	U	1.2R			MIT
3312.40	3313.35	30180.9	U	0.6	0.5		MIT
3311.72	3312.67	30187.1	U	1.0D	1.1		MIT
3311.35	3312.30	30190.5	U	0.6D	0.0H		MIT
3310.87	3311.82	30194.9	U	0.3D	0.3		MIT
3310.624	3311.577	30197.09	U	1.0			MIT
3310.50	3311.45	30198.2	U	0.8D	0.9D		MIT
3309.659	3310.612	30205.90	U	1.0	0.3		MIT
3309.212	3310.165	30209.98	U		0.5H		MIT
3308.898	3309.850	30212.84	U	0.5	0.5		MIT
3308.719	3309.671	30214.48	U	0.3	0.0		MIT
3308.43	3309.38	30217.1	U	0.8	0.0		MIT
3307.554	3308.506	30225.12	U	0.9	0.8		MIT
3306.217	3307.169	30237.34	U	0.8	0.8		MIT
3305.926	3306.878	30240.00	U	1.0	1.1		MIT
3305.124	3306.076	30247.34	U	0.3	0.0H		MIT
3303.845	3304.796	30259.05	U	0.8			MIT
3303.597	3304.548	30261.32	U	1.0	1.1		MIT
3303.370	3304.321	30263.40	U	0.6	0.5H		MIT
3302.82	3303.77	30268.4	U	0.8	0.6		MIT
3302.492	3303.443	30271.45	U	0.5	0.0		MIT
3302.26	3303.21	30273.6	U	0.0H	0.3H		MIT
3301.754	3302.705	30278.22	U	1.0	0.5		MIT
3301.650	3302.601	30279.17	U	0.8	0.0		MIT
3301.094	3302.044	30284.27	U	0.9	0.0		MIT
3300.676	3301.626	30288.10	U	1.2H	0.9H		MIT
3299.697	3300.647	30297.09	U	0.9	0.8		MIT
3299.375	3300.325	30300.05	U	0.6	0.3		MIT
3298.888	3299.838	30304.52	U	0.7			MIT
3298.716	3299.666	30306.10	U	0.6			MIT
3298.368	3299.318	30309.30	U	0.6			MIT
3297.886	3298.836	30313.73	U	0.8	1.3		MIT
3296.226	3297.175	30328.99	U	0.8			MIT
3296.032	3296.981	30330.78	U	0.3			MIT
3295.518	3296.467	30335.51	U	0.6	0.8		MIT
3295.383	3296.332	30336.75	U	0.5	0.0		MIT
3295.231	3296.180	30338.15	U	0.5	0.3		MIT
3294.812	3295.761	30342.01	U	0.5	0.3		MIT
3294.444	3295.393	30345.40	U	0.9	0.9		MIT
3293.944	3294.893	30350.00	U	0.3	0.0		MIT
3293.590	3294.539	30353.26	U	1.5	0.7		MIT
3293.328	3294.277	30355.68	U	0.3			MIT
3292.943	3293.891	30359.23	U	1.0	0.3		MIT
3291.335	3292.283	30374.06	U	1.1	1.0		MIT
3290.473	3291.421	30382.02	U	0.8	0.7		MIT
3290.121	3291.069	30385.27	U	0.9			MIT
3290.042	3290.990	30386.00	U	0.9			MIT
3289.749	3290.697	30388.70	U	0.3	0.0		MIT
3289.307	3290.254	30392.79	U	0.6	0.6		MIT
3288.209	3289.156	30402.93	U	1.4	1.3		MIT
3287.855	3288.802	30406.21	U	0.3	0.0		MIT
3287.448	3288.395	30409.97	U	0.8	0.9		MIT
3286.437	3287.384	30419.33	U	0.3	0.5		MIT
3286.251	3287.198	30421.05	U	0.5	0.3		MIT

AIR WAVELENGTH (ANGSTRMS)	VACUUM WAVELENGTH (ANGSTRMS)	WAVENUMBER (KAYSERS)	LMENT	LOG ARC	INTENSITY SPK	DIS	REF
3285.920	3286.867	30424.11	U	0.5	0.3		MIT
3285.589	3286.536	30427.18	U	0.7	0.3		MIT
3285.223	3286.169	30430.57	U	1.1	1.0		MIT
3284.368	3285.314	30438.49	U	0.8	0.3		MIT
3283.75	3284.70	30444.2	U	0.3D	0.3D		MIT
3283.537	3284.483	30446.19	U	0.5			MIT
3283.104	3284.050	30450.21	U	0.8	0.6		MIT
3282.542	3283.488	30455.42	U	1.0	0.8		MIT
3282.360	3283.306	30457.11	U	0.3			MIT
3281.652	3282.598	30463.68	U	0.3			MIT
3281.549	3282.495	30464.64	U	0.7	0.0		MIT
3281.115	3282.060	30468.67	U	0.3	0.0		MIT
3280.608	3281.553	30473.37	U	0.0	0.3		MIT
3280.399	3281.344	30475.31	U	0.7	0.0		MIT
3280.004	3280.949	30478.99	U	0.9	0.9		MIT
3279.550	3280.495	30483.20	U	1.1	0.7		MIT
3279.224	3280.169	30486.24	U	0.5	0.0		MIT
3279.096	3280.041	30487.43	U	0.8	0.6		MIT
3278.483	3279.428	30493.12	U	0.8	0.3		MIT
3277.728	3278.673	30500.15	U	0.8			MIT
3277.511	3278.455	30502.17	U	0.3	0.3		MIT
3276.765	3277.709	30509.11	U	1.0			MIT
3276.158	3277.102	30514.76	U	0.9	0.0H		MIT
3275.035	3275.979	30525.23	U	0.3	0.0		MIT
3273.596	3274.539	30538.65	U	0.7			MIT
3273.225	3274.168	30542.11	U	0.3	0.3		MIT
3273.023	3273.966	30543.99	U	0.7	0.0H		MIT
3271.454	3272.397	30558.64	U	0.8	0.7		MIT
3271.112	3272.055	30561.83	U	0.8			MIT
3270.124	3271.067	30571.07	U	1.3	1.4		MIT
3269.779	3270.722	30574.29	U	1.0	0.8		MIT
3269.458	3270.400	30577.29	U	0.3	0.3		MIT
3269.229	3270.171	30579.44	U	0.3	0.0		MIT
3269.007	3269.949	30581.51	U	0.5	0.3		MIT
3268.649	3269.591	30584.86	U	0.3			MIT
3268.462	3269.404	30586.61	U	0.5			MIT
3267.638	3268.580	30594.33	U	0.9	0.3		MIT
3267.244	3268.186	30598.01	U	0.9	0.8		MIT
3266.940	3267.882	30600.86	U	1.1	0.0		MIT
3266.211	3267.153	30607.69	U	0.5	0.5		MIT
3265.806	3266.748	30611.49	U	1.4	1.3		MIT
3263.747	3264.688	30630.80	U	0.3	0.5		MIT
3263.112	3264.053	30636.76	U	1.4	0.9		MIT
3261.718	3262.658	30649.85	U	1.2	1.0		MIT
3260.535	3261.475	30660.97	U	0.3	0.0		MIT
3259.840	3260.780	30667.51	U	1.0	0.5		MIT
3258.948	3259.888	30675.90	U	0.5	0.0		MIT
3258.097	3259.037	30683.92	U	0.5	0.3		MIT
3257.773	3258.712	30686.97	U	0.0	0.5		MIT
3257.710	3258.649	30687.56	U	0.8			MIT
3256.458	3257.397	30699.36	U	0.3	0.0		MIT
3255.626	3256.565	30707.20	U	0.5			MIT
3254.817	3255.756	30714.83	U	0.5	0.3		MIT
3254.488	3255.427	30717.94	U	0.3	0.0		MIT
3253.351	3254.289	30728.68	U	0.7	0.3		MIT
3252.642	3253.580	30735.37	U	0.0	0.3		MIT
3252.304	3253.242	30738.57	U	0.8	0.5H		MIT
3250.277	3251.215	30757.74	U	1.0	1.0		MIT
3248.909	3249.846	30770.69	U	0.3	0.0		MIT
3248.336	3249.273	30776.11	U	0.7	0.5		MIT
3247.709	3248.646	30782.06	U	0.6	0.7		MIT
3247.209	3248.146	30786.79	U	0.3	0.0		MIT
3247.034	3247.971	30788.46	U	0.9			MIT
3246.111	3247.048	30797.21	U	1.1	0.9		MIT
3244.794	3245.730	30809.71	U	1.1	0.8		MIT
3243.646	3244.582	30820.61	U	0.9	0.9		MIT
3241.991	3242.926	30836.34	U	1.0	0.9		MIT
3241.593	3242.528	30840.13	U	0.5	0.0		MIT
3240.714	3241.649	30848.50	U	0.5			MIT
3240.354	3241.289	30851.92	U	0.9	0.8		MIT

AIR WAVELENGTH (ANGSTRMS)	VACUUM WAVELENGTH (ANGSTRMS)	WAVENUMBER (KAYSERS)	LMENT	LOG ARC	INTENSITY SPK	DIS	REF
3240.14	3241.08	30854.0	U	0.3D	0.3D		MIT
3239.62	3240.55	30858.9	U	0.6	0.6		MIT
3238.757	3239.692	30867.14	U	0.3	0.0		MIT
3237.929	3238.863	30875.03	U	0.5	0.0		MIT
3237.867	3238.801	30875.62	U	0.9			MIT
3237.157	3238.091	30882.39	U	0.5	0.0		MIT
3236.34	3237.27	30890.2	U	0.3D	0.3D		MIT
3235.864	3236.798	30894.73	U	0.9			MIT
3235.551	3236.485	30897.72	U	0.3			MIT
3235.227	3236.161	30900.81	U	0.8	0.9		MIT
3232.157	3233.090	30930.16	U	1.1	1.1		MIT
3229.502	3230.434	30955.59	U	1.3	1.4		MIT
3229.072	3230.004	30959.71	U	0.8			MIT
3227.421	3228.353	30975.55	U	0.3	0.0		MIT
3226.785	3227.717	30981.65	U	0.6	0.5		MIT
3226.171	3227.102	30987.55	U	1.0	1.0		MIT
3225.412	3226.343	30994.84	U	0.9			MIT
3225.016	3225.947	30998.65	U	0.7	0.0		MIT
3224.261	3225.192	31005.91	U	1.0	1.1		MIT
3224.018	3224.949	31008.24	U	0.3			MIT
3223.692	3224.623	31011.38	U	0.8			MIT
3222.293	3223.223	31024.84	U	0.9	0.5		MIT
3221.412	3222.342	31033.33	U	0.8	0.3		MIT
3219.169	3220.099	31054.95	U	0.8	0.5		MIT
3218.336	3219.265	31062.99	U	0.9	0.8		MIT
3216.201	3217.130	31083.61	U	0.7	0.5		MIT
3215.688	3216.617	31088.56	U	0.0	0.3		MIT
3215.092	3216.021	31094.33	U	0.3	0.5H		MIT
3214.697	3215.626	31098.15	U	1.1	0.8		MIT
3213.907	3214.835	31105.79	U	0.5	0.3		MIT
3213.618	3214.546	31108.59	U	0.8	0.3		MIT
3213.399	3214.327	31110.71	U	0.7	0.5		MIT
3213.086	3214.014	31113.74	U	1.0	0.9		MIT
3212.649	3213.577	31117.97	U	0.9	0.8		MIT
3211.006	3211.934	31133.89	U		0.3H		MIT
3210.880	3211.808	31135.11	U	0.8			MIT
3210.701	3211.629	31136.85	U	0.7	0.6		MIT
3210.574	3211.502	31138.08	U	0.8	0.5		MIT
3209.970	3210.897	31143.94	U	0.6	0.6		MIT
3209.636	3210.563	31147.18	U	0.8	0.8		MIT
3209.187	3210.114	31151.54	U	0.9	0.9		MIT
3208.565	3209.492	31157.58	U	0.7	0.7		MIT
3208.173	3209.100	31161.39	U	0.0	0.3		MIT
3208.043	3208.970	31162.65	U	0.5	0.6		MIT
3207.671	3208.598	31166.26	U	0.9			MIT
3207.31	3208.24	31169.8	U	0.3	0.0		MIT
3206.692	3207.619	31175.78	U	0.8	0.0		MIT
3206.227	3207.153	31180.30	U	1.0	0.9		MIT
3206.049	3206.975	31182.03	U	1.1	1.0		MIT
3205.073	3205.999	31191.52	U		0.3		MIT
3204.978	3205.904	31192.45	U	0.5	0.0		MIT
3204.669	3205.595	31195.46	U	0.8	0.8		MIT
3204.118	3205.044	31200.82	U	0.5	0.0		MIT
3203.726	3204.652	31204.64	U	0.9	0.6		MIT
3203.406	3204.332	31207.75	U	0.9	0.8		MIT
3203.223	3204.149	31209.54	U	0.9	0.8		MIT
3202.735	3203.661	31214.29	U	0.6	0.6H		MIT
3202.469	3203.394	31216.89	U	0.6	0.6H		MIT
3201.551	3202.476	31225.84	U	0.9	0.6		MIT
3201.238	3202.163	31228.89	U	0.8	0.6		MIT
3200.975	3201.900	31231.46	U	0.6	0.3		MIT
3200.658	3201.583	31234.55	U	0.8	0.6H		MIT
3200.378	3201.303	31237.28	U	0.7	0.3		MIT
3200.135	3201.060	31239.65	U	1.2	1.2		MIT
3199.877	3200.802	31242.17	U	0.6	0.0		MIT
3199.583	3200.508	31245.04	U	0.6	0.0		MIT
3199.198	3200.123	31248.80	U	0.7	0.0		MIT
3198.944	3199.869	31251.28	U	0.3	0.3		MIT
3198.470	3199.394	31255.92	U	0.6	0.0		MIT
3198.322	3199.246	31257.36	U	0.0	0.3		MIT

AIR WAVELENGTH (ANGSTRMS)	VACUUM WAVELENGTH (ANGSTRMS)	WAVENUMBER (KAYSERS)	LMENT	LOG ARC	INTENSITY SPK	DIS	REF
3198.154	3199.078	31259.00	U	0.7	0.9		MIT
3197.904	3198.828	31261.45	U	1.1	0.3		MIT
3196.738	3197.662	31272.85	U	0.9	1.3		MIT
3196.431	3197.355	31275.85	U	0.8	0.0		MIT
3195.526	3196.450	31284.71	U	0.5	0.5		MIT
3195.255	3196.179	31287.36	U	0.3	0.0		MIT
3194.853	3195.777	31291.30	U		0.3H		MIT
3193.301	3194.224	31306.51	U	0.5	0.5		MIT
3193.226	3194.149	31307.24	U	0.7	1.0		MIT
3192.665	3193.588	31312.74	U	0.8	0.9		MIT
3192.160	3193.083	31317.70	U	0.3	0.0		MIT
3191.765	3192.688	31321.57	U	0.8	0.8		MIT
3190.892	3191.815	31330.14	U	0.8	0.8		MIT
3190.703	3191.625	31332.00	U	1.1	1.1		MIT
3190.430	3191.352	31334.68	U	0.7	0.5		MIT
3189.927	3190.849	31339.62	U	0.8			MIT
3189.017	3189.939	31348.56	U	0.9	1.1		MIT
3188.339	3189.261	31355.23	U	1.0	1.0		MIT
3187.992	3188.914	31358.64	U	1.0			MIT
3187.513	3188.435	31363.35	U	0.5	0.0		MIT
3186.323	3187.244	31375.06	U	1.0			MIT
3185.71	3186.63	31381.1	U	1.1	1.0		MIT
3185.397	3186.318	31384.19	U	0.6			MIT
3185.145	3186.066	31386.67	U	0.6	0.5		MIT
3184.764	3185.685	31390.42	U	0.5	0.0		MIT
3184.406	3185.327	31393.95	U	0.8	0.3		MIT
3182.833	3183.754	31409.47	U	0.8	0.9		MIT
3182.554	3183.474	31412.22	U	0.8	0.7		MIT
3182.373	3183.293	31414.01	U	0.7			MIT
3182.209	3183.129	31415.63	U	0.5	0.0		MIT
3181.833	3182.753	31419.34	U	0.5	0.0		MIT
3181.373	3182.293	31423.88	U	0.8	0.3		MIT
3181.080	3182.000	31426.77	U	0.3	0.0		MIT
3180.585	3181.505	31431.67	U	0.5H	0.3H		MIT
3179.827	3180.747	31439.16	U	1.1	1.0		MIT
3179.378	3180.298	31443.60	U	0.9R	0.8H		MIT
3179.037	3179.957	31446.97	U	1.0	0.8		MIT
3178.908	3179.828	31448.25	U	0.0	0.7		MIT
3178.320	3179.239	31454.06	U	0.8	0.6		MIT
3177.331	3178.250	31463.85	U	1.2	1.3		MIT
3176.964	3177.883	31467.49	U	0.5	0.3		MIT
3176.626	3177.545	31470.84	U	0.3	0.3		MIT
3176.208	3177.127	31474.98	U	1.3	1.2		MIT
3175.737	3176.656	31479.65	U	0.6	0.0		MIT
3175.358	3176.277	31483.40	U	0.9			MIT
3174.843	3175.761	31488.51	U	0.8			MIT
3173.944	3174.862	31497.43	U	0.7	0.7		MIT
3173.706	3174.624	31499.79	U	0.8	0.8		MIT
3173.497	3174.415	31501.87	U	0.8	0.0		MIT
3173.229	3174.147	31504.53	U	0.6			MIT
3172.600	3173.518	31510.77	U	0.7	0.3		MIT
3172.058	3172.976	31516.16	U	0.8	0.5		MIT
3171.052	3171.970	31526.15	U	0.5	0.3		MIT
3170.855	3171.772	31528.11	U	1.0	0.9		MIT
3170.538	3171.455	31531.26	U	0.5	0.5		MIT
3170.09	3171.01	31535.7	U	0.3	0.3		MIT
3169.989	3170.906	31536.73	U	0.8	0.5		MIT
3169.294	3170.211	31543.64	U	0.3	0.0		MIT
3168.432	3169.349	31552.22	U	0.7	0.6		MIT
3168.205	3169.122	31554.48	U	0.6H	0.0		MIT
3167.784	3168.701	31558.68	U	0.5	0.0		MIT
3167.098	3168.015	31565.51	U	1.0	1.0		MIT
3166.932	3167.848	31567.17	U	0.5	0.0		MIT
3166.131	3167.047	31575.15	U	0.3H			MIT
3165.505	3166.421	31581.40	U	1.0	0.9		MIT
3165.282	3166.198	31583.62	U	0.8	1.3		MIT
3165.057	3165.973	31585.87	U	0.5	0.3		MIT
3164.826	3165.742	31588.17	U	0.5	0.0		MIT
3164.418	3165.334	31592.24	U	0.3	0.0		MIT
3164.151	3165.067	31594.91	U	0.7	1.1		MIT

AIR WAVELENGTH (ANGSTRMS)	VACUUM WAVELENGTH (ANGSTRMS)	WAVENUMBER (KAYSERS)	LMENT	LOG ARC	INTENSITY SPK	DIS	REF
3163.727	3164.643	31599.14	U	0.9	0.8		MIT
3162.980	3163.895	31606.61	U	0.5	0.3		MIT
3162.819	3163.734	31608.22	U	0.5H	0.6H		MIT
3161.783	3162.698	31618.57	U	0.7	0.5		MIT
3161.528	3162.443	31621.12	U	0.9	0.8		MIT
3160.283	3161.198	31633.58	U	0.0	0.5H		MIT
3159.944	3160.859	31636.97	U	0.8	0.0		MIT
3159.817	3160.732	31638.24	U	0.9	1.2		MIT
3159.526	3160.441	31641.16	U	0.6	0.3		MIT
3159.330	3160.245	31643.12	U	0.3	0.3H		MIT
3158.89	3159.80	31647.5	U	0.5D	0.6D		MIT
3158.526	3159.440	31651.18	U	0.0	0.6		MIT
3158.201	3159.115	31654.43	U	0.3	0.0		MIT
3157.859	3158.773	31657.86	U	1.0	1.2		MIT
3157.446	3158.360	31662.00	U	0.9	1.0		MIT
3156.761	3157.675	31668.87	U	0.5	0.0		MIT
3156.592	3157.506	31670.57	U	0.7	0.7		MIT
3156.071	3156.985	31675.79	U	0.8	1.2		MIT
3155.860	3156.774	31677.91	U	0.9	0.9		MIT
3155.409	3156.323	31682.44	U	0.8	0.6		MIT
3155.256	3156.170	31683.98	U	0.5	0.8		MIT
3154.808	3155.721	31688.48	U	0.3	0.3		MIT
3154.409	3155.322	31692.48	U	0.8	0.5		MIT
3153.498	3154.411	31701.64	U	0.8	0.6		MIT
3153.120	3154.033	31705.44	U	1.1	1.2		MIT
3152.470	3153.383	31711.98	U	0.8	0.8		MIT
3152.307	3153.220	31713.61	U	0.9	1.0		MIT
3151.651	3152.564	31720.22	U	0.8	0.5		MIT
3151.084	3151.996	31725.92	U	1.0	0.6		MIT
3150.706	3151.618	31729.73	U	0.5W	0.3H		MIT
3150.533	3151.445	31731.47	U	0.7	0.3		MIT
3150.357	3151.269	31733.25	U	0.8	0.5		MIT
3149.974	3150.886	31737.10	U	0.8	0.7		MIT
3149.808	3150.720	31738.77	U	0.8	0.3		MIT
3149.646	3150.558	31740.41	U	0.8	0.5		MIT
3149.21	3150.12	31744.8	U	1.3D	1.3D		MIT
3149.055	3149.967	31746.36	U	0.3	0.6		MIT
3148.320	3149.232	31753.78	U	0.3	0.3		MIT
3148.213	3149.125	31754.85	U	0.8	0.3		MIT
3147.765	3148.677	31759.37	U	0.3	0.0		MIT
3147.09	3148.00	31766.2	U	1.1	1.2		MIT
3146.75	3147.66	31769.6	U	0.8	0.8		MIT
3146.260	3147.171	31774.57	U	0.5	0.5		MIT
3146.055	3146.966	31776.64	U	0.3H	0.3		MIT
3145.559	3146.470	31781.65	U	1.1	1.0		MIT
3145.357	3146.268	31783.69	U	0.8	0.7		MIT
3144.965	3145.876	31787.65	U	1.0	0.9		MIT
3144.125	3145.036	31796.14	U	0.3	0.5		MIT
3143.555	3144.466	31801.91	U	0.6	0.3		MIT
3143.170	3144.080	31805.80	U	0.6	0.5		MIT
3142.601	3143.511	31811.56	U	1.1	1.0		MIT
3141.953	3142.863	31818.12	U	0.9	0.8		MIT
3141.538	3142.448	31822.32	U	0.7	1.2		MIT
3140.304	3141.214	31834.83	U	0.8	0.6		MIT
3139.913	3140.823	31838.79	U	0.3			MIT
3139.560	3140.470	31842.37	U	1.4	1.4		MIT
3139.198	3140.107	31846.04	U	0.8	0.5		MIT
3138.513	3139.422	31852.99	U	1.0	0.7		MIT
3138.240	3139.149	31855.76	U	0.8	0.6		MIT
3137.693	3138.602	31861.32	U	0.7	0.6		MIT
3136.893	3137.802	31869.44	U	0.8	0.9		MIT
3136.066	3136.975	31877.85	U	0.7	0.5		MIT
3135.808	3136.717	31880.47	U	0.9	0.6		MIT
3134.695	3135.603	31891.79	U	0.7	0.3		MIT
3133.920	3134.828	31899.68	U	0.8	0.7		MIT
3133.424	3134.332	31904.72	U	0.9	0.8		MIT
3132.177	3133.085	31917.43	U	0.3	0.0		MIT
3131.987	3132.895	31919.36	U	0.9	0.8		MIT
3131.618	3132.526	31923.12	U	0.0	0.3		MIT
3131.534	3132.442	31923.98	U	0.6	0.3		MIT

AIR WAVELENGTH (ANGSTRMS)	VACUUM WAVELENGTH (ANGSTRMS)	WAVENUMBER (KAYSERS)	LMENT	LOG ARC	INTENSITY SPK	DIS	REF
3131.325	3132.233	31926.11	U	0.9	0.8		MIT
3130.732	3131.639	31932.16	U	1.0	0.8		MIT
3130.564	3131.471	31933.87	U		1.2		MIT
3130.373	3131.280	31935.82	U	0.5H			MIT
3129.728	3130.635	31942.40	U	0.9	1.2		MIT
3128.786	3129.693	31952.02	U	0.9	0.8		MIT
3128.074	3128.981	31959.29	U	0.8	1.0		MIT
3126.625	3127.531	31974.10	U	0.3	0.3		MIT
3126.174	3127.080	31978.71	U	1.1	1.3		MIT
3124.900	3125.806	31991.75	U	1.1	1.0		MIT
3124.426	3125.332	31996.60	U	1.0	0.8		MIT
3124.131	3125.037	31999.62	U	0.7	0.3		MIT
3123.717	3124.623	32003.86	U	0.6	0.3		MIT
3123.571	3124.477	32005.36	U	0.8	0.9		MIT
3123.179	3124.084	32009.38	U	0.3	0.0		MIT
3122.611	3123.516	32015.20	U	0.7	0.6		MIT
3121.828	3122.733	32023.23	U	0.8	0.3		MIT
3121.740	3122.645	32024.13	U	0.6	0.3		MIT
3121.330	3122.235	32028.34	U	0.8	0.6		MIT
3121.092	3121.997	32030.78	U	0.8	0.7		MIT
3120.866	3121.771	32033.10	U	0.9	0.9		MIT
3120.655	3121.560	32035.27	U	0.8	0.7		MIT
3120.436	3121.341	32037.51	U	0.6	0.3		MIT
3119.879	3120.784	32043.23	U	0.7	0.3		MIT
3119.348	3120.253	32048.69	U	1.0	1.0		MIT
3119.242	3120.146	32049.78	U	0.9	0.9		MIT
3119.049	3119.953	32051.76	U	0.8	0.8		MIT
3118.738	3119.642	32054.96	U	0.3	0.3		MIT
3117.98	3118.88	32062.7	U	0.8D	0.9D		MIT
3117.279	3118.183	32069.96	U	0.8	0.6		MIT
3116.428	3117.332	32078.72	U	0.8	0.6		MIT
3115.929	3116.833	32083.85	U	0.9	1.2		MIT
3115.795	3116.699	32085.23	U	0.8	0.6		MIT
3115.572	3116.476	32087.53	U	0.3			MIT
3115.002	3115.905	32093.40	U	1.0	0.9		MIT
3114.594	3115.497	32097.60	U	1.1	0.6		MIT
3113.637	3114.540	32107.47	U	0.9	1.1		MIT
3112.982	3113.885	32114.23	U	0.7	0.7		MIT
3112.249	3113.152	32121.79	U	1.0	0.8		MIT
3111.621	3112.524	32128.27	U	1.2	1.2		MIT
3111.462	3112.365	32129.91	U	0.5	0.5		MIT
3110.832	3111.734	32136.42	U	0.8	0.9		MIT
3110.784	3111.686	32136.92	U	0.5	0.0		MIT
3110.525	3111.427	32139.59	U	0.5	0.9		MIT
3110.226	3111.128	32142.68	U	0.5	0.5		MIT
3110.038	3110.940	32144.62	U	0.8	0.5		MIT
3109.626	3110.528	32148.88	U	0.8	0.5		MIT
3109.211	3110.113	32153.17	U	0.8	0.3		MIT
3108.690	3109.592	32158.56	U	0.9	0.6		MIT
3108.367	3109.269	32161.90	U	0.7	0.7		MIT
3108.100	3109.002	32164.67	U	0.5	0.3		MIT
3107.935	3108.837	32166.37	U	0.6	0.7		MIT
3107.546	3108.448	32170.40	U	0.8	0.8		MIT
3107.348	3108.249	32172.45	U	1.0	0.8		MIT
3106.797	3107.698	32178.16	U	0.0	0.3		MIT
3106.167	3107.068	32184.68	U	0.9	0.8		MIT
3105.646	3106.547	32190.08	U	0.7	0.6		MIT
3105.412	3106.313	32192.51	U	0.5	0.6		MIT
3105.099	3106.000	32195.75	U	0.9	0.9		MIT
3104.372	3105.273	32203.29	U	0.8	0.6		MIT
3104.162	3105.063	32205.47	U	1.3	1.3		MIT
3103.771	3104.672	32209.53	U	0.9	0.8		MIT
3103.133	3104.033	32216.15	U	0.3	0.0		MIT
3102.899	3103.799	32218.58	U	0.8	0.5		MIT
3102.613	3103.513	32221.55	U	0.8	0.7		MIT
3102.39	3103.29	32223.9	U	1.1D	1.0D		MIT
3101.865	3102.765	32229.32	U	0.8	0.6		MIT
3101.699	3102.599	32231.04	U	0.9	0.9		MIT
3100.771	3101.671	32240.69	U	0.8	0.3		MIT
3099.804	3100.704	32250.75	U	0.9	0.3		MIT

AIR WAVELENGTH (ANGSTRMS)	VACUUM WAVELENGTH (ANGSTRMS)	WAVENUMBER (KAYSERS)	LMENT	LOG ARC	INTENSITY SPK	DIS	REF
3099.320	3100.219	32255.78	U	0.8	0.7		MIT
3099.049	3099.948	32258.60	U	1.1R	0.6		MIT
3098.804	3099.703	32261.15	U	0.9	0.6		MIT
3098.686	3099.585	32262.38	U	0.5	0.5		MIT
3098.011	3098.910	32269.41	U	1.0	1.0		MIT
3096.88	3097.78	32281.2	U	0.7D	0.7D		MIT
3096.608	3097.507	32284.03	U	0.3	0.7		MIT
3095.748	3096.647	32293.00	U	1.0	1.0		MIT
3095.231	3096.129	32298.39	U	1.0	1.0		MIT
3095.041	3095.939	32300.37	U	1.1	1.1		MIT
3094.830	3095.728	32302.58	U	1.0	1.0		MIT
3094.460	3095.358	32306.44	U	0.5	0.6		MIT
3093.868	3094.766	32312.62	U	0.5	0.6		MIT
3093.37	3094.27	32317.8	U	0.6D	0.5D		MIT
3093.012	3093.910	32321.56	U	1.3	1.3		MIT
3092.720	3093.618	32324.61	U	0.7	0.5		MIT
3091.976	3092.874	32332.39	U	0.3	0.3		MIT
3091.461	3092.358	32337.78	U	0.7	0.6		MIT
3091.25	3092.15	32340.0	U	1.2E	1.1E		MIT
3090.554	3091.451	32347.27	U	0.9	0.9		MIT
3090.357	3091.254	32349.33	U	1.0	0.9		MIT
3089.88	3090.78	32354.3	U	0.8D	0.0D		MIT
3088.987	3089.884	32363.68	U	1.3	1.2		MIT
3088.508	3089.405	32368.70	U	0.9R	0.0		MIT
3088.361	3089.258	32370.24	U	0.3	0.0		MIT
3087.95	3088.85	32374.5	U	0.8D	0.6D		MIT
3087.880	3088.777	32375.28	U	0.8	0.8		MIT
3087.680	3088.577	32377.37	U	0.9	0.8		MIT
3087.320	3088.216	32381.15	U	0.8	0.6		MIT
3087.112	3088.008	32383.33	U	0.8	05		IT
3086.73	3087.63	32387.3	U	0.7D	0.5D		MIT
3086.465	3087.361	32390.12	U	0.5	0.0		MIT
3086.046	3086.942	32394.52	U	0.8	0.7		MIT
3085.388	3086.284	32401.43	U	0.9	0.0		MIT
3084.667	3085.563	32409.00	U	0.8	0.6		MIT
3084.238	3085.134	32413.51	U	1.2	1.1		MIT
3083.612	3084.508	32420.09	U	1.0	0.9		MIT
3083.022	3083.917	32426.29	U	0.7	0.6		MIT
3082.587	3083.482	32430.87	U	0.8	0.6		MIT
3082.020	3082.915	32436.83	U	0.9	0.9		MIT
3081.666	3082.561	32440.56	U	0.8	0.6		MIT
3081.388	3082.283	32443.49	U	0.3	0.0		MIT
3081.188	3082.083	32445.59	U	0.6	0.3		MIT
3080.741	3081.636	32450.30	U	1.1	1.1		MIT
3079.954	3080.849	32458.59	U	1.0	1.0		MIT
3079.724	3080.619	32461.01	U	0.3	0.0		MIT
3079.474	3080.368	32463.65	U	0.6	0.3		MIT
3079.256	3080.150	32465.95	U	0.9	0.9		MIT
3078.433	3079.327	32474.63	U	0.6	0.3		MIT
3077.730	3078.624	32482.04	U	0.3	0.0		MIT
3077.331	3078.225	32486.25	U	0.6	0.0		MIT
3076.08	3076.97	32499.5	U	0.6D	0.6D		MIT
3075.827	3076.721	32502.14	U		0.9		MIT
3075.452	3076.345	32506.10	U	0.8	0.3		MIT
3075.042	3075.935	32510.44	U	1.0	0.9		MIT
3074.492	3075.385	32516.25	U	0.3	0.5		MIT
3074.372	3075.265	32517.52	U	0.6	0.6		MIT
3073.812	3074.705	32523.45	U	1.2R	1.0R		MIT
3073.501	3074.394	32526.74	U	0.9	0.7		MIT
3073.173	3074.066	32530.21	U	0.8	0.5		MIT
3072.783	3073.676	32534.33	U	1.3	1.3		MIT
3072.448	3073.341	32537.88	U	0.8	0.5		MIT
3072.336	3073.229	32539.07	U	0.8	0.8		MIT
3071.816	3072.709	32544.58	U	0.5	0.3		MIT
3071.721	3072.614	32545.58	U	0.7	0.6		MIT
3071.483	3072.375	32548.10	U	0.6	0.0		MIT
3071.327	3072.219	32549.76	U	0.7	0.0		MIT
3070.992	3071.884	32553.31	U	0.9	0.9		MIT
3070.678	3071.570	32556.64	U	0.5	0.7		MIT
3070.278	3071.170	32560.88	U	1.1	0.8		MIT

AIR WAVELENGTH (ANGSTRMS)	VACUUM WAVELENGTH (ANGSTRMS)	WAVENUMBER (KAYSERS)	LMENT	LOG ARC	INTENSITY SPK	DIS	REF
3069.881	3070.773	32565.09	U	0.7	0.5		MIT
3069.494	3070.386	32569.20	U	0.7	0.0		MIT
3069.169	3070.061	32572.64	U	0.6	0.6		MIT
3068.994	3069.886	32574.50	U	0.7	0.5		MIT
3068.650	3069.542	32578.15	U	0.9	1.0		MIT
3067.758	3068.650	32587.62	U	0.9	0.8		MIT
3067.247	3068.138	32593.05	U		0.7		MIT
3066.870	3067.761	32597.06	U	0.8	0.8		MIT
3066.286	3067.177	32603.27	U	1.1R	0.9		MIT
3065.980	3066.871	32606.52	U	0.9	0.6		MIT
3065.663	3066.554	32609.89	U	0.3	0.0		MIT
3065.548	3066.439	32611.12	U	0.3	0.5		MIT
3065.198	3066.089	32614.84	U	0.8	0.6		MIT
3064.908	3065.799	32617.93	U	0.0	0.3		MIT
3064.591	3065.482	32621.30	U	0.6	0.5		MIT
3064.182	3065.073	32625.65	U	0.8	0.5		MIT
3063.881	3064.772	32628.86	U	0.9	0.6		MIT
3063.521	3064.411	32632.69	U	0.9	0.5		MIT
3063.131	3064.021	32636.85	U		0.6		MIT
3062.711	3063.601	32641.32	U	0.6	0.0		MIT
3062.536	3063.426	32643.19	U	1.1	1.2		MIT
3062.121	3063.011	32647.61	U	0.3	0.8		MIT
3062.009	3062.899	32648.81	U	0.7	0.0		MIT
3061.616	3062.506	32653.00	U	1.1	1.0		MIT
3061.287	3062.177	32656.51	U	0.6	0.0		MIT
3060.68	3061.57	32663.0	U	0.7D	0.9D		MIT
3060.055	3060.945	32669.65	U	0.9	1.1		MIT
3059.552	3060.441	32675.02	U	0.7	0.5		MIT
3059.266	3060.155	32678.08	U	0.6	0.5		MIT
3058.98	3059.87	32681.1	U	0.9D	0.7D		MIT
3057.91	3058.80	32692.6	U	1.3D	1.3D		MIT
3056.723	3057.612	32705.26	U	1.1	0.8		MIT
3056.251	3057.140	32710.31	U	0.5	0.0		MIT
3055.884	3056.773	32714.24	U	0.6	0.3		MIT
3055.592	3056.481	32717.37	U	0.3	0.3		MIT
3055.086	3055.974	32722.79	U	0.7	0.3		MIT
3054.73	3055.62	32726.6	U	1.0	0.8		MIT
3054.388	3055.276	32730.26	U	0.5	0.6		MIT
3054.310	3055.198	32731.10	U	0.3	0.6		MIT
3054.093	3054.981	32733.43	U	0.8	0.6		MIT
3053.887	3054.775	32735.63	U	0.0	0.3		MIT
3053.302	3054.190	32741.91	U		1.2		MIT
3052.911	3053.799	32746.10	U	1.1	0.9		MIT
3052.461	3053.349	32750.93	U	0.9	1.0		MIT
3051.425	3052.312	32762.05	U	0.7	0.7		MIT
3051.30	3052.19	32763.4	U	0.6D	0.8D		MIT
3051.139	3052.026	32765.12	U	1.2R	1.0R		MIT
3050.502	3051.389	32771.96	U	0.6	0.8		MIT
3050.198	3051.085	32775.22	U	1.1	0.9		MIT
3049.842	3050.729	32779.05	U	0.9	0.0		MIT
3049.584	3050.471	32781.82	U	0.6	0.3		MIT
3048.908	3049.795	32789.09	U	0.8	0.9		MIT
3048.63	3049.52	32792.1	U	1.2	0.9		MIT
3048.371	3049.258	32794.87	U		1.0		MIT
3047.880	3048.767	32800.15	U	0.3	0.6		MIT
3047.571	3048.458	32803.48	U	1.2	1.3		MIT
3047.042	3047.928	32809.17	U	0.5	0.0		MIT
3046.845	3047.731	32811.29	U	0.8	1.2		MIT
3046.573	3047.459	32814.22	U	0.9	0.8		MIT
3046.463	3047.349	32815.40	U	0.9	0.8		MIT
3046.105	3046.991	32819.26	U	0.3	0.3		MIT
3045.460	3046.346	32826.21	U	0.7	0.7		MIT
3044.991	3045.877	32831.27	U	0.3	0.7		MIT
3044.159	3045.045	32840.24	U	1.1	1.0		MIT
3043.997	3044.883	32841.99	U	0.6	0.9		MIT
3043.790	3044.676	32844.22	U	1.2	0.9		MIT
3042.733	3043.618	32855.63	U	0.6	0.9		MIT
3041.864	3042.749	32865.02	U	0.8	0.5		MIT
3041.252	3042.137	32871.63	U	0.8	0.7		MIT
3039.926	3040.811	32885.97	U	0.7	1.0		MIT

AIR WAVELENGTH (ANGSTRMS)	VACUUM WAVELENGTH (ANGSTRMS)	WAVENUMBER (KAYSERS)	LMENT	LOG ARC	INTENSITY SPK	DIS	REF
3039.501	3040.385	32890.57	U	1.0	0.5		MIT
3039.263	3040.147	32893.14	U	1.2	1.1		MIT
3039.136	3040.020	32894.52	U	0.7	0.7		MIT
3038.493	3039.377	32901.48	U	0.5	0.8		MIT
3038.047	3038.931	32906.31	U	0.9	0.3		MIT
3037.914	3038.798	32907.75	U		1.3		MIT
3037.535	3038.419	32911.85	U	0.5	0.3		MIT
3037.280	3038.164	32914.62	U	0.9	0.8		MIT
3036.815	3037.699	32919.66	U	0.6	0.0		MIT
3036.606	3037.490	32921.92	U	0.9	0.8		MIT
3036.451	3037.335	32923.60	U	1.0	0.9		MIT
3035.965	3036.849	32928.87	U	1.2	0.9		MIT
3035.51	3036.39	32933.8	U I	1.0D	0.9D		MIT
3034.895	3035.778	32940.48	U	0.5	0.0		MIT
3034.661	3035.544	32943.02	U	0.5	0.3		MIT
3034.375	3035.258	32946.13	U	0.5	0.0		MIT
3034.049	3034.932	32949.67	U	0.8	0.7		MIT
3033.77	3034.65	32952.7	U	0.6D	0.6D		MIT
3033.444	3034.327	32956.24	U	0.0	1.1		MIT
3033.193	3034.076	32958.96	U	1.1	1.1		MIT
3032.979	3033.862	32961.29	U	0.6	0.3		MIT
3032.765	3033.648	32963.62	U	0.3	0.3		MIT
3032.445	3033.328	32967.09	U	0.0	0.6H		MIT
3031.991	3032.874	32972.03	U	1.2	1.2		MIT
3031.573	3032.456	32976.58	U	0.0	0.8		MIT
3030.828	3031.710	32984.68	U	1.1	1.1		MIT
3030.622	3031.504	32986.92	U	0.5	0.5H		MIT
3029.419	3030.301	33000.02	U	1.0	0.8		MIT
3029.131	3030.013	33003.16	U	1.3	1.2		MIT
3028.878	3029.760	33005.92	U	0.3	0.0		MIT
3028.605	3029.487	33008.89	U	0.6	0.0		MIT
3028.378	3029.260	33011.37	U	0.6	0.6		MIT
3028.187	3029.069	33013.45	U	1.0	1.0		MIT
3027.932	3028.814	33016.23	U	0.5	0.8		MIT
3027.689	3028.571	33018.88	U	1.2	0.9		MIT
3027.247	3028.128	33023.70	U	0.3	0.0		MIT
3026.705	3027.586	33029.61	U	0.9	1.0H		MIT
3026.15	3027.03	33035.7	U	0.9D	0.5D		MIT
3025.034	3025.915	33047.86	U	1.1H	1.3H		MIT
3024.510	3025.391	33053.58	U	1.3	1.2		MIT
3024.385	3025.266	33054.95	U	1.4R	1.3R		MIT
3023.85	3024.73	33060.8	U	0.6	0.0		MIT
3023.657	3024.538	33062.91	U	0.8	0.6		MIT
3023.303	3024.183	33066.78	U	0.3			MIT
3023.161	3024.041	33068.33	U	0.5H	0.5H		MIT
3022.871	3023.751	33071.50	U	0.6	0.3		MIT
3022.753	3023.633	33072.79	U	0.5	0.3		MIT
3022.466	3023.346	33075.93	U	0.5	0.5		MIT
3022.207	3023.087	33078.77	U	1.1	1.1		MIT
3021.22	3022.10	33089.6	U	1.0	1.2		MIT
3020.92	3021.80	33092.9	U	0.9D	1.1D		MIT
3020.571	3021.451	33096.68	U	0.8	0.8		MIT
3020.242	3021.122	33100.29	U	0.9	0.8		MIT
3019.81	3020.69	33105.0	U	0.7	1.0		MIT
3019.291	3020.170	33110.71	U	0.9	1.2		MIT
3018.838	3019.717	33115.68	U	0.5	0.8		MIT
3018.587	3019.466	33118.44	U	0.9	0.9		MIT
3018.103	3018.982	33123.75	U	0.7	0.7		MIT
3017.351	3018.230	33132.00	U	0.8	0.6		MIT
3016.956	3017.835	33136.34	U	1.2	1.1		MIT
3016.384	3017.263	33142.62	U	0.0	0.6H		MIT
3016.052	3016.931	33146.27	U	0.8	0.6		MIT
3015.682	3016.561	33150.34	U	0.7	0.6		MIT
3014.884	3015.762	33159.11	U	0.8	0.3		MIT
3014.571	3015.449	33162.55	U	0.3			MIT
3013.831	3014.709	33170.70	U	0.3	0.0H		MIT
3013.44	3014.32	33175.0	U	0.7D	0.0D		MIT
3013.370	3014.248	33175.77	U	0.8	0.3		MIT
3013.135	3014.013	33178.36	U	0.3			MIT
3012.975	3013.853	33180.12	U	0.5	0.3H		MIT

AIR WAVELENGTH (ANGSTRMS)	VACUUM WAVELENGTH (ANGSTRMS)	WAVENUMBER (KAYSERS)	LMENT	LOG ARC	INTENSITY SPK	INTENSITY DIS	REF
3012.708	3013.586	33183.06	U	0.5	0.0		MIT
3012.447	3013.325	33185.94	U	1.0	0.0		MIT
3012.138	3013.016	33189.34	U	0.3			MIT
3011.902	3012.780	33191.94	U	0.3	0.0		MIT
3011.19	3012.07	33199.8	U	0.5D	0.3D		MIT
3010.754	3011.631	33204.60	U	1.0	0.5		MIT
3010.37	3011.25	33208.8	U	0.9D	0.3D		MIT
3009.42	3010.30	33219.3	U	1.2D	0.3D		MIT
3008.922	3009.799	33224.81	U	0.7	0.0		MIT
3008.18	3009.06	33233.0	U	0.6D	0.3D		MIT
3007.909	3008.786	33236.00	U	0.3	0.0		MIT
3007.013	3007.889	33245.90	U	0.3			MIT
3005.52	3006.40	33262.4	U	0.3			MIT
3005.101	3005.977	33267.06	U	0.8	0.6		MIT
3004.553	3005.429	33273.12	U	0.5			MIT
3004.15	3005.03	33277.6	U	1.0D	0.3D		MIT
3003.587	3004.463	33283.82	U	0.3			MIT
3003.315	3004.190	33286.84	U	0.5	0.0		MIT
3003.070	3003.945	33289.55	U	0.6			MIT
3002.645	3003.520	33294.27	U	0.3			MIT
3002.441	3003.316	33296.53	U	0.5			MIT
3001.660	3002.535	33305.19	U	0.6R			MIT
3001.220	3002.095	33310.07	U	0.3			MIT
3000.788	3001.663	33314.87	U	0.3			MIT
3000.094	3000.969	33322.57	U	0.5	0.0		MIT
2999.215	3000.089	33332.34	U	1.0	0.8H		MIT
2999.033	2999.907	33334.36	U	0.9	0.3		MIT
2998.517	2999.391	33340.10	U	0.3			MIT
2998.362	2999.236	33341.82	U	0.8	0.3		MIT
2997.354	2998.228	33353.03	U	0.3	0.3		MIT
2997.124	2997.998	33355.59	U	0.5	0.3		MIT
2996.759	2997.633	33359.66	U	0.5	0.3		MIT
2996.095	2996.969	33367.05	U	0.8	0.6		MIT
2995.71	2996.58	33371.3	U	0.7	0.3		MIT
2995.516	2996.390	33373.50	U	0.3	0.3		MIT
2995.350	2996.223	33375.35	U	0.3H			MIT
2994.89	2995.76	33380.5	U	0.3D	0.3		MIT
2994.451	2995.324	33385.37	U	0.5	0.3		MIT
2994.029	2994.902	33390.07	U	0.6			MIT
2993.699	2994.572	33393.75	U	0.7H	0.3H		MIT
2993.355	2994.228	33397.59	U	0.7	0.3H		MIT
2992.989	2993.862	33401.67	U	0.3			MIT
2992.717	2993.590	33404.71	U	0.9	0.8		MIT
2991.73	2992.60	33415.7	U	0.3	0.3		MIT
2991.248	2992.120	33421.11	U	0.5			MIT
2990.988	2991.860	33424.02	U	0.3	0.3		MIT
2990.582	2991.454	33428.56	U	0.5	0.3		MIT
2990.210	2991.082	33432.72	U	0.6			MIT
2989.877	2990.749	33436.44	U	0.9	0.8		MIT
2989.605	2990.477	33439.48	U	0.5			MIT
2989.395	2990.267	33441.83	U	0.7R	0.3		MIT
2989.11	2989.98	33445.0	U	0.3D	0.0		MIT
2989.000	2989.872	33446.25	U	0.3			MIT
2988.71	2989.58	33449.5	U	0.7D	0.6		MIT
2988.42	2989.29	33452.7	U	0.3	0.0		MIT
2987.95	2988.82	33458.0	U	0.3D	0.3		MIT
2987.802	2988.674	33459.66	U	0.7	0.3		MIT
2987.518	2988.390	33462.84	U	0.6			MIT
2987.06	2987.93	33468.0	U	0.3	0.0		MIT
2985.795	2986.666	33482.15	U	0.3S	0.3		MIT
2985.084	2985.955	33490.12	U	0.5	0.3		MIT
2984.613	2985.484	33495.41	U	1.1	0.9		MIT
2984.050	2984.921	33501.73	U	0.5	0.3		MIT
2983.756	2984.627	33505.03	U	0.3	0.3		MIT
2983.51	2984.38	33507.8	U	0.3	0.3		MIT
2982.740	2983.610	33516.44	U	1.1	1.1		MIT
2982.235	2983.105	33522.12	U	0.6	0.6		MIT
2981.809	2982.679	33526.90	U	0.5	0.3		MIT
2981.040	2981.910	33535.55	U	0.3	0.3		MIT
2980.69	2981.56	33539.5	U	0.8	0.3		MIT

AIR WAVELENGTH (ANGSTRMS)	VACUUM WAVELENGTH (ANGSTRMS)	WAVENUMBER (KAYSERS)	LMENT	LOG ARC	INTENSITY SPK	INTENSITY DIS	REF
2980.344	2981.214	33543.38	U	0.6	0.6		MIT
2980.285	2981.155	33544.05	U	0.6	0.3		MIT
2979.19	2980.06	33556.4	U	0.6	0.3		MIT
2978.140	2979.009	33568.21	U	1.0	0.8		MIT
2977.827	2978.696	33571.74	U	0.3	0.3		MIT
2977.47	2978.34	33575.8	U	0.5	0.3		MIT
2977.266	2978.135	33578.06	U	1.0	1.0		MIT
2976.308	2977.177	33588.87	U	1.1	0.9		MIT
2975.876	2976.745	33593.75	U	0.6	0.6		MIT
2975.639	2976.508	33596.42	U	0.8	0.6		MIT
2975.221	2976.089	33601.14	U	0.9	0.8		MIT
2974.909	2975.777	33604.66	U	0.7	0.6		MIT
2974.463	2975.331	33609.70	U	0.5	0.3		MIT
2974.081	2974.949	33614.02	U	0.3	0.3		MIT
2973.263	2974.131	33623.27	U	0.6	0.3		MIT
2973.08	2973.95	33625.3	U	0.6	0.3		MIT
2972.612	2973.480	33630.63	U	0.7	0.6		MIT
2972.103	2972.971	33636.39	U	0.5			MIT
2971.822	2972.690	33639.57	U	0.3			MIT
2971.629	2972.497	33641.75	U	0.3	0.0		MIT
2971.064	2971.931	33648.15	U	1.1	1.1		MIT
2970.782	2971.649	33651.35	U	0.3H	0.3		MIT
2970.484	2971.351	33654.72	U	0.3	0.3		MIT
2969.73	2970.60	33663.3	U	0.3	0.3		MIT
2969.22	2970.09	33669.0	U	0.3	0.3		MIT
2968.822	2969.689	33673.56	U	0.3	0.3		MIT
2968.562	2969.429	33676.51	U	0.6	0.3		MIT
2968.401	2969.268	33678.34	U	0.8	0.8		MIT
2967.894	2968.761	33684.09	U	1.2	1.3		MIT
2966.661	2967.527	33698.09	U	0.8	0.9		MIT
2966.388	2967.254	33701.19	U	0.5	0.6		MIT
2966.12	2966.99	33704.2	U	1.2	1.4		MIT
2965.785	2966.651	33708.04	U	0.6	0.6		MIT
2965.377	2966.243	33712.68	U	0.5	0.3		MIT
2965.034	2965.900	33716.58	U	0.9	1.1		MIT
2964.247	2965.113	33725.53	U	0.8	1.2		MIT
2963.909	2964.775	33729.38	U	0.7			MIT
2963.604	2964.470	33732.85	U	0.7	0.8		MIT
2962.783	2963.648	33742.19	U	0.8	0.8		MIT
2962.207	2963.072	33748.75	U	0.8	0.6		MIT
2961.179	2962.044	33760.47	U	0.9	1.0		MIT
2960.942	2961.807	33763.17	U	1.2	1.4		MIT
2959.850	2960.715	33775.63	U	0.9	1.1		MIT
2959.123	2959.987	33783.93	U	0.3	0.6		MIT
2958.102	2958.966	33795.59	U	0.6	0.3		MIT
2957.844	2958.708	33798.54	U	0.6	0.3		MIT
2957.745	2958.609	33799.67	U	0.6	0.3		MIT
2957.229	2958.093	33805.56	U	0.7	0.3		MIT
2956.780	2957.644	33810.70	U	0.8	0.8		MIT
2956.438	2957.302	33814.61	U	0.3	0.3		MIT
2956.060	2956.924	33818.93	U	1.0	1.8		MIT
2955.648	2956.512	33823.65	U	1.1	1.0		MIT
2955.087	2955.950	33830.07	U	0.7	0.6		MIT
2954.766	2955.629	33833.74	U	1.0	1.2		MIT
2954.39	2955.25	33838.0	U	1.1	1.2		MIT
2953.287	2954.150	33850.69	U	0.3	0.3		MIT
2952.883	2953.746	33855.32	U	0.3	0.3H		MIT
2952.702	2953.565	33857.39	U	0.7	0.3		MIT
2952.369	2953.232	33861.21	U	0.3	0.3		MIT
2952.100	2952.963	33864.29	U	0.5			MIT
2951.922	2952.785	33866.34	U	0.9	0.6		MIT
2951.07	2951.93	33876.1	U	0.8	0.3		MIT
2950.506	2951.368	33882.59	U	0.6	0.3		MIT
2949.970	2950.832	33888.75	U	0.8H	0.6H		MIT
2949.609	2950.471	33892.89	U	0.7	0.8		MIT
2948.936	2949.798	33900.63	U	0.3	0.6		MIT
2948.092	2948.954	33910.33	U	1.0	1.1		MIT
2946.41	2947.27	33929.7	U	0.5	0.3		MIT
2946.283	2947.144	33931.15	U	0.6	0.6		MIT
2945.889	2946.750	33935.69	U	0.9	1.1		MIT

AIR WAVELENGTH (ANGSTRMS)	VACUUM WAVELENGTH (ANGSTRMS)	WAVENUMBER (KAYSERS)	LMENT	LOG ARC	INTENSITY SPK	DIS	REF
2944.637	2945.498	33950.12	U	0.3	0.9		MIT
2944.189	2945.050	33955.28	U	0.9	1.1		MIT
2943.895	2944.756	33958.67	U	1.0	1.4		MIT
2943.405	2944.266	33964.33	U	0.8	0.6		MIT
2943.184	2944.045	33966.88	U	0.7	0.9		MIT
2942.853	2943.713	33970.70	U	0.8	0.8		MIT
2942.433	2943.293	33975.55	U	0.6	0.3		MIT
2942.120	2942.980	33979.16	U	0.9	1.1		MIT
2941.919	2942.779	33981.48	U	1.2	1.5		MIT
2941.343	2942.203	33988.14	U	0.9	1.1		MIT
2940.367	2941.227	33999.42	U	0.8	1.1		MIT
2939.494	2940.354	34009.52	U	0.7	0.6		MIT
2939.311	2940.171	34011.63	U	0.6	0.3		MIT
2939.04	2939.90	34014.8	U	0.8	0.3		MIT
2937.351	2938.210	34034.33	U	0.7	0.6		MIT
2937.151	2938.010	34036.64	U	0.5	0.3		MIT
2936.776	2937.635	34040.99	U	0.8	0.6		MIT
2936.453	2937.312	34044.73	U	1.1	1.3		MIT
2935.615	2936.474	34054.45	U	0.5	0.6		MIT
2935.148	2936.007	34059.87	U	0.5			MIT
2934.90	2935.76	34062.7	U	0.6	0.3		MIT
2934.42	2935.28	34068.3	U	0.5	0.3		MIT
2933.86	2934.72	34074.8	U	0.9	1.0		MIT
2933.57	2934.43	34078.2	U	0.3	0.3		MIT
2933.31	2934.17	34081.2	U	0.5H			MIT
2933.056	2933.914	34084.16	U	0.6	0.8		MIT
2932.96	2933.82	34085.3	U	0.5	0.8		MIT
2932.61	2933.47	34089.4	U	1.0	1.4		MIT
2932.180	2933.038	34094.34	U	0.7H	0.8H		MIT
2931.891	2932.749	34097.71	U	0.9	0.9		MIT
2931.521	2932.379	34102.01	U	0.5	0.3		MIT
2931.414	2932.272	34103.25	U	1.1	1.1		MIT
2930.805	2931.662	34110.34	U	0.6	1.5H		MIT
2930.588	2931.445	34112.87	U	0.6H	0.3H		MIT
2930.430	2931.287	34114.70	U	0.9	1.1		MIT
2929.777	2930.634	34122.31	U	0.3	0.3		MIT
2929.119	2929.976	34129.97	U	0.9	0.8		MIT
2928.599	2929.456	34136.03	U	1.2	1.5		MIT
2927.735	2928.592	34146.10	U	0.6	0.8H		MIT
2927.376	2928.233	34150.29	U	1.1	1.0		MIT
2926.588	2927.444	34159.49	U	0.5	0.6		MIT
2926.103	2926.959	34165.15	U	0.5	0.3		MIT
2925.975	2926.831	34166.64	U	0.8	0.9		MIT
2925.568	2926.424	34171.40	U	1.2	1.4		MIT
2925.222	2926.078	34175.44	U	0.5	0.6		MIT
2924.580	2925.436	34182.94	U	1.0	1.0		MIT
2923.505	2924.361	34195.51	U	0.9	1.1		MIT
2923.384	2924.240	34196.92	U	0.5	0.8		MIT
2923.174	2924.030	34199.38	U	0.9	1.0		MIT
2922.867	2923.722	34202.97	U	0.8	0.9		MIT
2922.635	2923.490	34205.69	U	0.8	0.8		MIT
2922.222	2923.077	34210.52	U	0.3	0.3		MIT
2922.058	2922.913	34212.44	U	0.9	0.9		MIT
2921.727	2922.582	34216.32	U	0.8	0.8		MIT
2921.684	2922.539	34216.82	U	1.0	1.2		MIT
2920.386	2921.241	34232.03	U	0.9	0.8H		MIT
2920.177	2921.032	34234.48	U	0.5	0.8		MIT
2919.692	2920.547	34240.17	U	0.3			MIT
2919.366	2920.221	34243.99	U	0.7	0.3		MIT
2918.974	2919.829	34248.59	U	0.8	0.6		MIT
2918.869	2919.723	34249.82	U	0.5	0.3		MIT
2918.683	2919.537	34252.00	U	0.7	1.0		MIT
2917.082	2917.936	34270.80	U	0.3	0.3		MIT
2916.706	2917.560	34275.22	U	0.8	0.9		MIT
2916.465	2917.319	34278.05	U	0.9	1.0		MIT
2916.300	2917.154	34279.99	U	0.7H	0.3H		MIT
2915.81	2916.66	34285.7	U	0.5	0.3		MIT
2915.537	2916.391	34288.96	U	0.7	0.8		MIT
2914.835	2915.688	34297.22	U	0.7	0.8		MIT
2914.629	2915.482	34299.64	U	1.1	1.2		MIT

AIR WAVELENGTH (ANGSTRMS)	VACUUM WAVELENGTH (ANGSTRMS)	WAVENUMBER (KAYSERS)	LMENT	LOG ARC	INTENSITY SPK	DIS	REF
2914.253	2915.106	34304.07	U	1.3	1.4		MIT
2913.963	2914.816	34307.48	U	1.1	0.6		MIT
2913.440	2914.293	34313.64	U	0.8	1.1		MIT
2912.960	2913.813	34319.29	U	0.7	0.8		MIT
2912.752	2913.605	34321.74	U	0.6	0.9		MIT
2912.581	2913.434	34323.76	U	0.9	0.9		MIT
2911.764	2912.617	34333.39	U	0.5	0.3		MIT
2911.548	2912.401	34335.94	U	1.1	0.8H		MIT
2911.287	2912.140	34339.01	U	0.5	0.3		MIT
2911.104	2911.957	34341.17	U	0.8	0.6		MIT
2910.824	2911.676	34344.48	U	1.1	0.9		MIT
2910.638	2911.490	34346.67	U	0.6	0.6		MIT
2910.201	2911.053	34351.83	U	0.5	0.3		MIT
2909.742	2910.594	34357.25	U	0.6	0.6		MIT
2909.250	2910.102	34363.06	U	0.8	1.2		MIT
2908.878	2909.730	34367.45	U	0.3	0.3		MIT
2908.410	2909.262	34372.98	U	0.9	0.9		MIT
2908.275	2909.127	34374.58	U	1.1	1.5		MIT
2907.573	2908.425	34382.87	U	0.3	0.3		MIT
2906.913	2907.765	34390.68	U	1.3R	1.2H		MIT
2906.798	2907.650	34392.04	U	1.2	1.7		MIT
2906.432	2907.283	34396.37	U	0.3	0.3		MIT
2905.812	2906.663	34403.71	U	0.8	0.8H		MIT
2905.28	2906.13	34410.0	U	0.5	0.3		MIT
2904.506	2905.357	34419.18	U	0.9	0.9		MIT
2903.554	2904.405	34430.46	U	0.5	0.6		MIT
2903.048	2903.899	34436.46	U	0.6	0.6		MIT
2902.808	2903.659	34439.31	U	0.8	0.3		MIT
2902.406	2903.256	34444.08	U	0.8	0.8		MIT
2901.642	2902.492	34453.15	U	0.7	0.8		MIT
2901.225	2902.075	34458.10	U	1.1	1.0		MIT
2900.454	2901.304	34467.26	U	0.7	0.3		MIT
2900.114	2900.964	34471.30	U	0.8	0.9		MIT
2899.59	2900.44	34477.5	U	0.8F	0.6H		MIT
2898.711	2899.561	34487.98	U	0.9	0.6		MIT
2898.563	2899.412	34489.75	U	0.9	0.6		MIT
2898.366	2899.215	34492.09	U	0.6	0.6		MIT
2898.013	2898.862	34496.29	U	0.8	0.8		MIT
2897.636	2898.485	34500.78	U	0.5H	0.8H		MIT
2897.458	2898.307	34502.90	U	0.8	0.6H		MIT
2896.965	2897.814	34508.77	U	0.8	0.8H		MIT
2896.678	2897.527	34512.19	U	0.9	0.9		MIT
2896.074	2896.923	34519.39	U	0.8	0.6		MIT
2895.900	2896.749	34521.46	U	0.3	0.3		MIT
2895.545	2896.394	34525.69	U	0.8	0.6		MIT
2895.436	2896.285	34526.99	U	0.7	0.6		MIT
2895.262	2896.111	34529.07	U	0.7	0.6		MIT
2894.842	2895.691	34534.08	U	0.7	0.3		MIT
2894.512	2895.360	34538.01	U	1.2	1.2		MIT
2894.141	2894.989	34542.44	U	1.1	1.2		MIT
2893.756	2894.604	34547.04	U	0.5	0.3		MIT
2893.50	2894.35	34550.1	U	0.7	0.0		MIT
2893.331	2894.179	34552.11	U	0.3H	0.3H		MIT
2892.819	2893.667	34558.23	U	0.6	0.3		MIT
2892.637	2893.485	34560.40	U	0.8	0.6		MIT
2892.206	2893.054	34565.55	U	0.8	0.9		MIT
2891.796	2892.644	34570.45	U	1.0	0.9		MIT
2891.286	2892.134	34576.55	U	0.5	0.3		MIT
2891.074	2891.922	34579.08	U	0.9	1.0		MIT
2890.428	2891.275	34586.81	U	0.8	0.6		MIT
2890.111	2890.958	34590.60	U	0.8H	0.3		MIT
2889.627	2890.474	34596.40	U	1.5	1.7		MIT
2889.266	2890.113	34600.72	U	0.7	0.3		MIT
2889.121	2889.968	34602.46	U	0.7	0.6		MIT
2888.741	2889.588	34607.01	U	1.0	0.8		MIT
2888.380	2889.227	34611.33	U	0.5	0.3		MIT
2888.26	2889.11	34612.8	U	1.1	1.1		MIT
2887.907	2888.754	34617.00	U	0.5	0.3		MIT
2887.594	2888.441	34620.75	U	0.8	0.9		MIT
2887.252	2888.099	34624.86	U	1.4	1.4		MIT

AIR WAVELENGTH (ANGSTRMS)	VACUUM WAVELENGTH (ANGSTRMS)	WAVENUMBER (KAYSERS)	LMENT	LOG ARC	INTENSITY SPK	DIS	REF
2886.925	2887.772	34628.78	U	0.5	0.8		MIT
2886.804	2887.651	34630.23	U	0.5	0.3		MIT
2886.449	2887.295	34634.49	U	1.2	0.8H		MIT
2886.049	2886.895	34639.29	U	0.7	0.6		MIT
2885.605	2886.451	34644.62	U	0.6	0.6		MIT
2885.362	2886.208	34647.54	U	0.8	0.6		MIT
2885.187	2886.033	34649.64	U	1.1	0.8		MIT
2884.925	2885.771	34652.78	U	0.3	0.3		MIT
2884.613	2885.459	34656.53	U	0.3	0.3		MIT
2884.302	2885.148	34660.27	U	0.5	0.3		MIT
2883.749	2884.595	34666.91	U	0.5	0.6		MIT
2882.929	2883.775	34676.77	U	0.9	1.1		MIT
2882.741	2883.587	34679.04	U	1.3	1.3		MIT
2882.587	2883.433	34680.89	U	0.9	0.8		MIT
2882.345	2883.190	34683.80	U	0.8	0.9		MIT
2881.909	2882.754	34689.05	U	0.6	0.3		MIT
2881.017	2881.862	34699.79	U	0.3	0.3		MIT
2880.489	2881.334	34706.15	U	1.1	1.2		MIT
2880.192	2881.037	34709.73	U	0.0	0.6		MIT
2879.986	2880.831	34712.21	U	0.9	0.3		MIT
2879.593	2880.438	34716.94	U	0.8	0.6		MIT
2878.869	2879.714	34725.67	U	0.7	0.6		MIT
2878.719	2879.564	34727.48	U	0.8	0.6		MIT
2878.240	2879.084	34733.26	U	0.7	0.3		MIT
2877.830	2878.674	34738.21	U	0.3	0.3		MIT
2877.569	2878.413	34741.36	U	0.8	0.6		MIT
2877.047	2877.891	34747.67	U	0.8	1.1		MIT
2876.433	2877.277	34755.08	U	0.9	0.3		MIT
2875.881	2876.725	34761.75	U	0.7	0.6		MIT
2875.198	2876.042	34770.01	U	1.3	1.1		MIT
2874.988	2875.832	34772.55	U	0.3	0.3		MIT
2874.790	2875.634	34774.94	U	0.5	0.3		MIT
2874.469	2875.313	34778.83	U	0.8	0.6		MIT
2874.083	2874.926	34783.50	U	1.2	1.0		MIT
2873.640	2874.483	34788.86	U	0.5	0.8		MIT
2873.44	2874.28	34791.3	U	0.5	0.3H		MIT
2873.298	2874.141	34793.00	U	0.8	0.6		MIT
2873.005	2873.848	34796.55	U	1.0	0.8		MIT
2872.110	2872.953	34807.39	U	0.5	0.6		MIT
2871.988	2872.831	34808.87	U	0.8H	0.6H		MIT
2871.644	2872.487	34813.04	U	0.8	0.6		MIT
2870.974	2871.817	34821.16	U	1.3	1.3		MIT
2870.742	2871.585	34823.98	U	0.6H	0.6H		MIT
2870.423	2871.266	34827.85	U	0.9	0.3		MIT
2869.373	2870.215	34840.59	U	0.8	0.6		MIT
2868.975	2869.817	34845.42	U	0.8	0.8		MIT
2868.865	2869.707	34846.76	U	0.3	0.3		MIT
2868.436	2869.278	34851.97	U	0.3	0.8		MIT
2868.187	2869.029	34855.00	U	1.0	0.9		MIT
2867.805	2868.647	34859.64	U	0.8	0.6		MIT
2866.986	2867.828	34869.60	U	0.6	0.3		MIT
2866.420	2867.262	34876.48	U	0.5	0.6		MIT
2866.160	2867.001	34879.65	U	0.8	0.8		MIT
2865.679	2866.520	34885.50	U	1.5	1.7		MIT
2865.140	2865.981	34892.06	U	1.0	1.0		MIT
2864.276	2865.117	34902.59	U	1.3H	1.1H		MIT
2864.099	2864.940	34904.75	U	0.9	1.1		MIT
2863.44	2864.28	34912.8	U	0.8	0.6		MIT
2863.189	2864.030	34915.84	U	0.5H	0.6H		MIT
2862.802	2863.643	34920.56	U	0.8	0.8		MIT
2862.618	2863.459	34922.80	U	0.8	0.6		MIT
2862.410	2863.251	34925.34	U	1.2	1.0		MIT
2861.679	2862.519	34934.26	U	0.5	0.3		MIT
2861.13	2861.97	34941.0	U	1.0	1.2		MIT
2860.801	2861.641	34944.98	U	1.2	1.1		MIT
2860.466	2861.306	34949.08	U	1.5	1.5		MIT
2859.806	2860.646	34957.14	U	0.5	0.6		MIT
2859.737	2860.577	34957.98	U	0.8	1.0H		MIT
2859.290	2860.130	34963.45	U	0.7	0.9		MIT
2858.903	2859.743	34968.18	U	1.5	1.4		MIT

AIR WAVELENGTH (ANGSTRMS)	VACUUM WAVELENGTH (ANGSTRMS)	WAVENUMBER (KAYSERS)	LMENT	LOG ARC	INTENSITY SPK	DIS	REF
2858.154	2858.994	34977.34	U	0.9	0.8		MIT
2857.930	2858.769	34980.08	U	0.6	0.3		MIT
2857.474	2858.313	34985.67	U	0.9	0.8		MIT
2857.125	2857.964	34989.94	U	0.6H	0.3V		MIT
2856.492	2857.331	34997.69	U	0.7	0.6		MIT
2856.22	2857.06	35001.0	U	0.9	0.3		MIT
2856.180	2857.019	35001.52	U	0.5	0.3		MIT
2855.960	2856.799	35004.21	U	0.6	0.3		MIT
2855.603	2856.442	35008.59	U	1.1H	0.9H		MIT
2355.188	2856.027	35013.68	U	0.5	0.3		MIT
2854.917	2855.756	35017.00	U	0.9	0.8		MIT
2854.458	2855.297	35022.63	U	0.5	0.3H		MIT
2854.196	2855.035	35025.85	U	0.9H	0.6H		MIT
2853.569	2854.407	35033.54	U	1.2	1.0		MIT
2853.424	2854.262	35035.32	U	1.1	1.1		MIT
2852.935	2853.773	35041.33	U	0.5	0.8		MIT
2852.750	2853.588	35043.60	U	1.2	1.2H		MIT
2852.469	2853.307	35047.05	U	0.8	0.6		MIT
2850.820	2851.658	35067.32	U	0.8	0.8		MIT
2850.487	2851.325	35071.42	U	0.9	1.1		MIT
2849.983	2850.821	35077.62	U	0.7	0.9H		MIT
2849.480	2850.317	35083.81	U	1.3	1.2		MIT
2849.16	2850.00	35087.7	U	0.7H	0.9H		MIT
2848.898	2849.735	35090.98	U	0.9	0.8		MIT
2848.614	2849.451	35094.48	U	0.6	0.3		MIT
2848.173	2849.010	35099.91	U	0.7	0.3		MIT
2848.051	2848.888	35101.42	U	0.9	0.9		MIT
2847.720	2848.557	35105.49	U	0.7	0.6H		MIT
2847.340	2848.177	35110.18	U	1.2	0.6		MIT
2846.827	2847.664	35116.51	U	0.6	0.3		MIT
2846.615	2847.452	35119.12	U	0.8	0.6		MIT
2846.373	2847.210	35122.11	U	0.8	0.8		MIT
2846.092	2846.929	35125.57	U	0.3	0.0		MIT
2845.959	2846.796	35127.22	U	0.8	0.6		MIT
2845.597	2846.433	35131.68	U	0.8	0.6		MIT
2845.298	2846.134	35135.38	U	0.5	0.6		MIT
2844.991	2845.827	35139.17	U	1.3	0.9		MIT
2844.516	2845.352	35145.03	U	0.3	1.1H		MIT
2842.869	2843.705	35165.40	U	0.6	0.8H		MIT
2842.484	2843.320	35170.16	U	0.9	0.8		MIT
2842.240	2843.076	35173.18	U	0.7	0.9H		MIT
2842.088	2842.924	35175.06	U	1.2	1.1		MIT
2841.359	2842.194	35184.08	U	0.6	0.6		MIT
2841.17	2842.01	35186.4	U	0.7	0.6		MIT
2840.623	2841.458	35193.20	U	0.7	0.6		MIT
2840.466	2841.301	35195.14	U	0.5	0.3		MIT
2839.890	2840.725	35202.28	U	1.3	1.3		MIT
2839.651	2840.486	35205.24	U	0.7	0.6		MIT
2838.623	2839.458	35217.99	U	0.8	1.1		MIT
2838.204	2839.039	35223.19	U	0.5	0.3		MIT
2837.726	2838.561	35229.12	U	0.8	0.8H		MIT
2837.328	2838.162	35234.06	U	1.0	0.9		MIT
2837.187	2838.021	35235.82	U	1.0	0.9		MIT
2836.918	2837.752	35239.16	U	0.7	1.0H		MIT
2835.803	2836.637	35253.01	U	0.9	1.0		MIT
2835.572	2836.406	35255.88	U	0.7	0.8		MIT
2834.743	2835.577	35266.19	U	0.7H	0.6H		MIT
2834.551	2835.385	35268.58	U	0.9	0.6H		MIT
2834.211	2835.045	35272.81	U	0.6H	0.3		MIT
2833.821	2834.655	35277.67	U	1.2	1.4		MIT
2833.244	2834.077	35284.85	U	0.9	0.6		MIT
2832.645	2833.478	35292.31	U	0.3	0.3		MIT
2832.063	2832.896	35299.56	U	1.5	1.7		MIT
2831.534	2832.367	35306.16	U	0.5	0.3		MIT
2831.324	2832.157	35308.78	U	0.9	0.6		MIT
2831.169	2832.002	35310.71	U	0.8	0.6		MIT
2830.309	2831.142	35321.44	U	0.8	0.3		MIT
2830.075	2830.908	35324.36	U	0.9	0.6H		MIT
2829.820	2830.653	35327.54	U	0.8	0.6		MIT
2829.607	2830.440	35330.20	U	0.8	0.3		MIT

AIR WAVELENGTH (ANGSTRMS)	VACUUM WAVELENGTH (ANGSTRMS)	WAVENUMBER (KAYSERS)	LMENT	LOG ARC	INTENSITY SPK	DIS	REF
2829.37	2830.20	35333.2	U	0.8	0.6		MIT
2828.90	2829.73	35339.0	U	1.3	1.3		MIT
2828.107	2828.939	35348.94	U	0.8	0.3		MIT
2827.981	2828.813	35350.52	U	0.9	0.6		MIT
2827.810	2828.642	35352.65	U	0.6	0.3		MIT
2827.305	2828.137	35358.97	U	0.8	0.3		MIT
2826.995	2827.827	35362.84	U	0.9	1.0		MIT
2826.69	2827.52	35366.7	U	0.8	0.3		MIT
2826.547	2827.379	35368.45	U	0.9	0.3		MIT
2826.193	2827.025	35372.88	U	1.3	1.1H		MIT
2825.347	2826.178	35383.47	U	0.7	0.3		MIT
2824.860	2825.691	35389.57	U	1.1	0.9		MIT
2824.630	2825.461	35392.45	U	1.0	0.8		MIT
2824.28	2825.11	35396.8	U	1.4	1.5		MIT
2823.178	2824.009	35410.65	U	0.3	0.3		MIT
2822.729	2823.560	35416.29	U	1.0	1.0		MIT
2822.567	2823.398	35418.32	U	0.7	0.8		MIT
2822.198	2823.029	35422.95	U	0.3	0.3H		MIT
2821.122	2821.952	35436.46	U	1.3	1.5		MIT
2820.507	2821.337	35444.19	U	0.3	0.3		MIT
2820.266	2821.096	35447.21	U	0.8	1.0		MIT
2819.835	2820.665	35452.63	U	0.8	1.4		MIT
2819.202	2820.032	35460.59	U	0.5	0.3		MIT
2818.712	2819.542	35466.75	U	0.3	0.3		MIT
2818.535	2819.365	35468.98	U	0.5	0.6		MIT
2818.298	2819.128	35471.96	U	0.5	0.6H		MIT
2817.959	2818.789	35476.23	U	1.3	1.5		MIT
2817.658	2818.488	35480.02	U	0.7	0.6		MIT
2816.90	2817.73	35489.6	U	0.6	0.8		MIT
2816.740	2817.569	35491.58	U	0.9	0.9		MIT
2816.417	2817.246	35495.65	U	0.5	0.8H		MIT
2815.985	2816.814	35501.10	U	0.9	0.8		MIT
2815.757	2816.586	35503.97	U	0.8	0.9		MIT
2815.207	2816.036	35510.91	U	0.3	0.3		MIT
2815.150	2815.979	35511.63	U	0.3	0.3		MIT
2814.824	2815.653	35515.74	U	0.5	0.3		MIT
2814.648	2815.477	35517.96	U	0.8	0.3H		MIT
2813.982	2814.811	35526.37	U	0.0	0.6H		MIT
2813.795	2814.624	35528.73	U	0.5	0.3		MIT
2813.549	2814.378	35531.83	U	0.6	0.8		MIT
2813.042	2813.870	35538.24	U	1.1	1.1		MIT
2812.235	2813.063	35548.44	U	0.8	0.6		MIT
2811.659	2812.487	35555.72	U	0.5	0.8		MIT
2811.345	2812.173	35559.69	U	1.5	1.5		MIT
2811.056	2811.884	35563.35	U	0.5	0.3		MIT
2810.350	2811.178	35572.28	U	0.8	0.9H		MIT
2809.952	2810.780	35577.32	U	1.3	1.3		MIT
2809.590	2810.418	35581.90	U	0.3	0.6H		MIT
2808.98	2809.81	35589.6	U	0.9	0.8		MIT
2808.543	2809.370	35595.17	U	0.3	0.6		MIT
2807.05	2807.88	35614.1	U	1.3	1.5		MIT
2806.69	2807.52	35618.7	U	0.0	0.6		MIT
2806.34	2807.17	35623.1	U	0.7	0.6		MIT
2806.060	2806.887	35626.66	U	0.8	0.3		MIT
2805.71	2806.54	35631.1	U	0.5W	0.0		MIT
2805.244	2806.071	35637.02	U	1.0	0.9		MIT
2803.99	2804.82	35653.0	U	0.5D	1.0H		MIT
2803.833	2804.659	35654.96	U	0.9	0.9		MIT
2802.559	2803.385	35671.16	U	1.2	1.5		MIT
2802.157	2802.983	35676.28	U	0.8	0.6		MIT
2801.654	2802.480	35682.69	U	1.0	1.0		MIT
2801.337	2802.163	35686.72	U	0.5	0.3		MIT
2800.449	2801.274	35698.04	U	0.7	0.6		MIT
2800.298	2801.123	35699.96	U	0.5	0.9		MIT
2800.103	2800.928	35702.45	U	0.7	0.8H		MIT
2799.764	2800.589	35706.77	U	0.8	0.6		MIT
2799.12	2799.95	35715.0	U	1.0D	0.6D		MIT
2798.21	2799.03	35726.6	U	0.0A	0.6A		MIT
2797.300	2798.125	35738.22	U	0.7	0.6		MIT
2797.145	2797.970	35740.20	U	1.0	0.9		MIT
2796.70	2797.52	35745.9	U	0.5W	0.3H		MIT
2796.236	2797.060	35751.82	U	0.8	0.3		MIT
2795.232	2796.056	35764.66	U	1.3	1.1		MIT
2794.417	2795.241	35775.09	U	0.5	0.6		MIT
2793.937	2794.761	35781.24	U	1.4	1.5		MIT
2793.43	2794.25	35787.7	U	0.6D	0.6D		MIT
2792.523	2793.346	35799.35	U	0.3			MIT
2792.074	2792.897	35805.11	U	0.7	0.3		MIT
2791.260	2792.083	35815.55	U	0.7	0.3		MIT
2791.068	2791.891	35818.02	U	0.8	1.1		MIT
2790.665	2791.488	35823.19	U	0.9	1.0		MIT
2790.318	2791.141	35827.64	U	0.3	0.3		MIT
2789.063	2789.886	35843.76	U	1.1	1.0		MIT
2788.671	2789.494	35848.80	U	0.7	0.3		MIT
2788.526	2789.348	35850.67	U	0.6	0.6		MIT
2788.132	2788.954	35855.73	U	0.3	0.3		MIT
2787.331	2788.153	35866.04	U	0.8	0.9		MIT
2786.804	2787.626	35872.82	U	0.5	0.6		MIT
2786.628	2787.450	35875.08	U	0.6	0.3		MIT
2786.215	2787.037	35880.40	U	0.6	0.6H		MIT
2785.912	2786.734	35884.30	U		0.6H		MIT
2785.647	2786.469	35887.72	U	0.5	0.3		MIT
2785.334	2786.156	35891.75	U	0.6	0.3		MIT
2785.17	2785.99	35893.9	U	0.5D	0.3D		MIT
2784.919	2785.741	35897.10	U	0.8	0.9		MIT
2784.669	2785.491	35900.32	U	0.9	0.9		MIT
2784.450	2785.271	35903.14	U	0.9	1.1		MIT
2784.004	2784.825	35908.89	U	0.8	0.8		MIT
2783.867	2784.688	35910.66	U	0.8	0.9		MIT
2783.405	2784.226	35916.62	U	0.5	0.6		MIT
2783.290	2784.111	35918.11	U	0.8	0.0		MIT
2782.071	2782.892	35933.84	U	0.7	1.0		MIT
2781.616	2782.437	35939.72	U	0.9	0.9		MIT
2781.408	2782.229	35942.41	U	0.5	0.8H		MIT
2781.035	2781.856	35947.23	U	0.9	0.9		MIT
2780.772	2781.593	35950.63	U	0.8	0.8		MIT
2780.040	2780.860	35960.09	U	0.9	0.9		MIT
2779.406	2780.226	35968.30	U	0.8	0.8		MIT
2778.953	2779.773	35974.16	U	0.6	0.3		MIT
2778.449	2779.269	35980.68	U	0.8	0.3		MIT
2778.135	2778.955	35984.75	U	0.6	0.6		MIT
2776.513	2777.333	36005.77	U	0.8	0.8		MIT
2776.29	2777.11	36008.7	U	0.8W	0.8W		MIT
2775.780	2776.599	36015.28	U	0.5	0.6		MIT
2775.451	2776.270	36019.55	U	0.8	0.6		MIT
2775.40	2776.22	36020.2	U	0.8	0.6		MIT
2775.216	2776.035	36022.60	U	0.8	0.8		MIT
2775.019	2775.838	36025.15	U	0.8	0.9		MIT
2774.736	2775.555	36028.83	U	0.5	1.0		MIT
2774.43	2775.25	36032.8	U	0.6D	0.6D		MIT
2774.169	2774.988	36036.19	U	0.0	0.6J		MIT
2773.77	2774.59	36041.4	U	0.5D	0.3D		MIT
2773.607	2774.426	36043.49	U	0.8	1.0		MIT
2773.084	2773.903	36050.29	U	0.7	0.6		MIT
2772.593	2773.412	36056.67	U	0.9	1.1		MIT
2772.34	2773.16	36060.0	U	0.6D	0.6D		MIT
2772.181	2772.999	36062.03	U	0.7	0.9		MIT
2771.874	2772.692	36066.03	U	0.3	0.3		MIT
2771.550	2772.368	36070.24	U	0.9	0.6		MIT
2771.21	2772.03	36074.7	U	0.0D	0.6D		MIT
2770.740	2771.558	36080.79	U	0.6	0.3		MIT
2770.044	2770.862	36089.85	U	1.0	1.1		MIT
2768.857	2769.675	36105.32	U	1.0	0.9		MIT
2768.385	2769.203	36111.48	U	0.8	0.3		MIT
2768.17	2768.99	36114.3	U	0.5D	0.3D		MIT
2767.783	2768.600	36119.33	U	0.3	0.3H		MIT
2767.413	2768.230	36124.16	U	0.7	0.6		MIT
2766.875	2767.692	36131.19	U	1.0	1.0		MIT
2766.156	2766.973	36140.58	U	0.7	0.6		MIT
2765.400	2766.217	36150.46	U	1.0	1.0		MIT

AIR WAVELENGTH (ANGSTRMS)	VACUUM WAVELENGTH (ANGSTRMS)	WAVENUMBER (KAYSERS)	LMENT	LOG ARC	INTENSITY SPK	INTENSITY DIS	REF
2765.206	2766.023	36152.99	U	0.0	0.6		MIT
2764.707	2765.524	36159.52	U	0.3	0.9H		MIT
2764.249	2765.066	36165.51	U	1.0	1.0		MIT
2763.692	2764.508	36172.80	U	0.3	0.3		MIT
2763.415	2764.231	36176.42	U	0.9	0.8		MIT
2762.850	2763.666	36183.82	U	1.2	1.3		MIT
2762.426	2763.242	36189.37	U	0.8	0.0		MIT
2761.74	2762.56	36198.4	U	0.6D	0.6H		MIT
2761.449	2762.265	36202.18	U	0.7	0.3		MIT
2761.18	2762.00	36205.7	U	0.8D	0.8		MIT
2760.334	2761.150	36216.80	U	0.9	0.6		MIT
2759.787	2760.602	36223.98	U	0.8	0.8		MIT
2758.956	2759.771	36234.89	U	1.0	1.0		MIT
2758.497	2759.312	36240.92	U	0.8	0.8		MIT
2758.432	2759.247	36241.77	U	0.5	0.6		MIT
2758.18	2759.00	36245.1	U	0.6D	0.6D		MIT
2757.915	2758.730	36248.56	U	0.8	0.6		MIT
2757.758	2758.573	36250.63	U	0.8	0.6		MIT
2757.554	2758.369	36253.31	U	1.0	0.8		MIT
2757.140	2757.955	36258.75	U	0.7	0.6		MIT
2756.835	2757.650	36262.76	U	0.7	0.3		MIT
2755.133	2755.947	36285.16	U	1.0	1.1		MIT
2754.565	2755.379	36292.65	U	0.5	0.3		MIT
2754.155	2754.969	36298.05	U	1.3	1.5		MIT
2752.987	2753.801	36313.45	U	0.8	0.6		MIT
2752.445	2753.259	36320.60	U	0.6	1.1		MIT
2751.93	2752.74	36327.4	U	0.3D	1.3H		MIT
2751.228	2752.041	36336.66	U	0.8	0.8		MIT
2750.386	2751.199	36347.79	U	0.8	0.8		MIT
2750.13	2750.94	36351.2	U	0.7	0.6		MIT
2749.964	2750.777	36353.36	U	1.0	1.1		MIT
2748.451	2749.264	36373.37	U	1.3	1.2		MIT
2747.93	2748.74	36380.3	U		0.6D		MIT
2747.362	2748.174	36387.79	U	0.9	1.0		MIT
2747.15	2747.96	36390.6	U		0.6D		MIT
2746.66	2747.47	36397.1	U	0.3D	0.0D		MIT
2746.158	2746.970	36403.74	U	1.4	0.8		MIT
2745.864	2746.676	36407.64	U	0.7	0.6H		MIT
2745.080	2745.892	36418.04	U	1.0			MIT
2744.404	2745.216	36427.01	U	1.1	1.0		MIT
2744.274	2745.086	36428.73	U	0.6	1.0		MIT
2743.661	2744.473	36436.87	U	0.6	0.9		MIT
2743.40	2744.21	36440.3	U		1.3H		MIT
2742.60	2743.41	36451.0	U		0.8H		MIT
2742.060	2742.871	36458.15	U	0.7	0.3		MIT
2741.748	2742.559	36462.29	U	1.3	0.8		MIT
2741.621	2742.432	36463.98	U	0.7	0.3		MIT
2741.065	2741.876	36471.38	U	0.8	0.8		MIT
2740.859	2741.670	36474.12	U	1.1	0.9		MIT
2740.305	2741.116	36481.50	U	0.3	0.3		MIT
2739.394	2740.205	36493.63	U	1.0	1.0		MIT
2738.980	2739.790	36499.14	U	0.9	0.9H		MIT
2738.566	2739.376	36504.66	U	0.5	0.3		MIT
2738.41	2739.22	36506.7	U	0.7D	0.6D		MIT
2738.13	2738.94	36510.5	U	0.9D	0.9D		MIT
2737.89	2738.70	36513.7	U	0.6D	0.3D		MIT
2737.65	2738.46	36516.9	U	0.3D	0.3D		MIT
2737.071	2737.881	36524.60	U	0.7	0.8		MIT
2736.314	2737.124	36534.70	U	0.7	0.6H		MIT
2735.978	2736.788	36539.19	U	0.5	0.3H		MIT
2735.769	2736.579	36541.98	U	0.6	0.3		MIT
2735.580	2736.390	36544.50	U	1.0	0.8		MIT
2735.326	2736.136	36547.90	U		0.6		MIT
2734.964	2735.773	36552.73	U	0.8	0.8		MIT
2734.71	2735.52	36556.1	U	0.3A	0.6H		MIT
2733.967	2734.776	36566.06	U	1.0	1.2		MIT
2733.772	2734.581	36568.67	U	0.8	0.8		MIT
2733.374	2734.183	36574.00	U	0.5	0.6		MIT
2732.951	2733.760	36579.66	U	0.8	0.6		MIT
2732.555	2733.364	36584.96	U	0.6	0.3		MIT
2731.266	2732.075	36602.22	U	1.2	0.9		MIT
2730.310	2731.118	36615.04	U	1.3	0.9		MIT
2730.069	2730.877	36618.27	U	0.3	1.2H		MIT
2729.262	2730.070	36629.10	U	0.9	0.6		MIT
2729.065	2729.873	36631.74	U	0.5	0.8H		MIT
2727.335	2728.143	36654.98	U	0.9	0.8		MIT
2726.685	2727.492	36663.71	U	0.7	0.3		MIT
2726.518	2727.325	36665.96	U	0.7	0.3		MIT
2725.942	2726.749	36673.71	U	1.1	1.1		MIT
2725.523	2726.330	36679.34	U	1.0	0.8H		MIT
2725.070	2725.877	36685.44	U	0.9	0.8		MIT
2724.42	2725.23	36694.2	U	0.8	0.8		MIT
2724.097	2724.904	36698.54	U	0.5	0.3		MIT
2723.791	2724.598	36702.67	U	0.0	0.6		MIT
2723.68	2724.49	36704.2	U	0.5W	0.3W		MIT
2723.34	2724.15	36708.7	U		0.6		MIT
2723.030	2723.837	36712.92	U	0.8	0.6		MIT
2722.808	2723.615	36715.92	U	0.5	0.3		MIT
2721.459	2722.265	36734.11	U	0.7	0.6		MIT
2721.163	2721.969	36738.11	U	0.5	0.8		MIT
2720.67	2721.48	36744.8	U		0.6D		MIT
2720.403	2721.209	36748.37	U	0.6	0.3		MIT
2718.900	2719.706	36768.69	U	0.7	0.8		MIT
2718.618	2719.424	36772.50	U	0.5	0.3		MIT
2717.557	2718.362	36786.86	U	0.9	0.6		MIT
2716.982	2717.787	36794.64	U	0.8	0.3		MIT
2716.387	2717.192	36802.70	U	0.5	0.6		MIT
2715.539	2716.344	36814.19	U	0.9	0.9		MIT
2714.998	2715.803	36821.53	U	0.8	0.6		MIT
2714.584	2715.389	36827.14	U	1.0	0.9		MIT
2714.291	2715.095	36831.12	U	0.3	0.3H		MIT
2713.914	2714.718	36836.23	U	0.8	0.8		MIT
2713.49	2714.29	36842.0	U		1.2J		MIT
2712.582	2713.386	36854.32	U		0.3		MIT
2712.116	2712.920	36860.65	U	0.5	0.3		MIT
2712.061	2712.865	36861.40	U	0.9	0.8		MIT
2711.76	2712.56	36865.5	U	0.7D	0.6D		MIT
2711.55	2712.35	36868.4	U	0.6W	0.8W		MIT
2711.105	2711.909	36874.40	U	0.9	0.6		MIT
2710.602	2711.406	36881.24	U		0.8		MIT
2709.508	2710.311	36896.13	U	0.9	0.9		MIT
2709.03	2709.83	36902.6	U	0.5W	0.6W		MIT
2707.95	2708.75	36917.4	U		0.3D		MIT
2707.714	2708.517	36920.58	U	0.5	0.6H		MIT
2706.95	2707.75	36931.0	U	1.2D	1.3D		MIT
2705.245	2706.047	36954.27	U	0.5	0.6		MIT
2705.186	2705.988	36955.08	U	0.9	0.8		MIT
2701.819	2702.620	37001.13	U	0.8	0.3		MIT
2701.549	2702.350	37004.82	U	0.9	1.0		MIT
2700.96	2701.76	37012.9	U	1.0D	0.6		MIT
2699.676	2700.477	37030.50	U	0.5	0.3		MIT
2699.372	2700.173	37034.67	U	0.9	0.8H		MIT
2698.452	2699.253	37047.29	U	0.8	0.8		MIT
2698.057	2698.858	37052.71	U	1.3	1.7		MIT
2697.710	2698.510	37057.48	U	0.6H			MIT
2697.398	2698.198	37061.77	U	1.0	0.8		MIT
2696.977	2697.777	37067.55	U	0.3	0.3		MIT
2695.908	2696.708	37082.25	U	0.7	0.8		MIT
2695.490	2696.290	37088.00	U	1.1	1.5		MIT
2694.222	2695.022	37105.45	U	1.0	0.6		MIT
2693.77	2694.57	37111.7	U	0.8D	0.6D		MIT
2693.35	2694.15	37117.5	U	0.7D	0.3D		MIT
2692.365	2693.164	37131.04	U	0.9	0.9		MIT
2691.795	2692.594	37138.91	U	0.8	0.8		MIT
2691.038	2691.837	37149.35	U	1.2	1.5		MIT
2690.509	2691.308	37156.66	U	1.0	0.8		MIT
2687.966	2688.764	37191.81	U	0.3	0.0		MIT
2687.394	2688.192	37199.72	U	0.7	0.6		MIT
2686.815	2687.613	37207.74	U	0.6	0.6		MIT
2685.979	2686.777	37219.32	U	1.0	1.1		MIT

AIR WAVELENGTH (ANGSTRMS)	VACUUM WAVELENGTH (ANGSTRMS)	WAVENUMBER (KAYSERS)	LMENT	LOG ARC	INTENSITY SPK	INTENSITY DIS	REF
2685.609	2686.407	37224.45	U	0.8	0.3		MIT
2684.622	2685.419	37238.13	U	0.5	0.3		MIT
2684.293	2685.090	37242.70	U	0.7	0.6		MIT
2684.045	2684.842	37246.14	U	1.2R	1.0		MIT
2683.279	2684.076	37256.77	U	1.4	1.4		MIT
2681.726	2682.523	37278.34	U	0.5	0.3		MIT
2681.132	2681.928	37286.60	U	0.7H	0.3		MIT
2680.646	2681.442	37293.36	U	0.3	0.6		MIT
2678.861	2679.657	37318.21	U	0.8	0.6		MIT
2678.42	2679.22	37324.4	U	0.5	0.3		MIT
2678.052	2678.848	37329.48	U	0.5	1.1H		MIT
2677.756	2678.552	37333.61	U	0.5	0.3		MIT
2676.687	2677.482	37348.52	U	0.8	0.6		MIT
2676.410	2677.205	37352.38	U	0.6	0.6		MIT
2675.880	2676.675	37359.78	U	1.2	1.0		MIT
2675.119	2675.914	37370.41	U	1.2	1.0		MIT
2673.580	2674.375	37391.92	U	0.8	0.8H		MIT
2672.69	2673.48	37404.4	U	1.1R	0.8H		MIT
2672.210	2673.004	37411.09	U	0.8	0.3		MIT
2671.967	2672.761	37414.49	U	0.3	0.3		MIT
2671.490	2672.284	37421.17	U	0.5	0.6H		MIT
2670.887	2671.681	37429.62	U	0.8	0.6		MIT
2670.516	2671.310	37434.82	U	0.9	0.8		MIT
2669.174	2669.968	37453.64	U	0.8	0.6		MIT
2668.018	2668.811	37469.87	U	0.9	0.6		MIT
2666.536	2667.329	37490.69	U	0.9	0.8		MIT
2665.870	2666.663	37500.05	U	1.0	0.8		MIT
2665.695	2666.488	37502.52	U	1.1	0.9		MIT
2664.153	2664.945	37524.22	U	1.3	1.3		MIT
2660.144	2660.935	37580.77	U	1.2	1.0		MIT
2659.025	2659.816	37596.58	U	0.9	0.3		MIT
2658.364	2659.155	37605.93	U	0.8	0.3		MIT
2657.324	2658.115	37620.65	U	0.5	0.3H		MIT
2656.460	2657.251	37632.88	U	0.7	0.3		MIT
2654.58	2655.37	37659.5	U	1.2	0.8H		MIT
2652.828	2653.618	37684.40	U	1.1	1.0		MIT
2652.188	2652.978	37693.50	U	0.8	0.6		MIT
2651.844	2652.633	37698.39	U	0.5	1.2H		MIT
2650.180	2650.969	37722.06	U	0.8	0.3		MIT
2649.54	2650.33	37731.2	U	0.5	0.3H		MIT
2649.069	2649.858	37737.87	U	1.2	1.2		MIT
2648.789	2649.578	37741.86	U	1.0	0.3		MIT
2648.226	2649.015	37749.89	U	0.3	0.3		MIT
2647.927	2648.716	37754.15	U	0.8	0.3		MIT
2647.53	2648.32	37759.8	U	0.8	0.3		MIT
2647.363	2648.151	37762.19	U	0.5	0.3		MIT
2647.021	2647.809	37767.07	U	0.5	0.3H		MIT
2645.47	2646.26	37789.2	U	1.3	1.4		MIT
2644.48	2645.27	37803.4	U	0.8	0.3H		MIT
2644.124	2644.912	37808.45	U	0.9	0.6		MIT
2643.545	2644.332	37816.73	U	0.8	0.3		MIT
2643.242	2644.029	37821.06	U	0.8	0.3H		MIT
2642.773	2643.560	37827.77	U	0.7	0.3		MIT
2641.933	2642.720	37839.80	U	1.3	0.9H		MIT
2641.55	2642.34	37845.3	U	1.0	0.6		MIT
2641.126	2641.913	37851.36	U	0.5	0.3H		MIT
2640.347	2641.134	37862.53	U	0.5	0.3		MIT
2639.89	2640.68	37869.1	U	0.9	0.3		MIT
2639.837	2640.624	37869.84	U	1.0	0.3		MIT
2639.579	2640.366	37873.54	U	0.5	0.3		MIT
2639.38	2640.17	37876.4	U	0.3	0.3H		MIT
2639.011	2639.797	37881.69	U	1.1	0.6		MIT
2637.696	2638.482	37900.58	U	1.0	0.8		MIT
2637.405	2638.191	37904.76	U	0.3	0.3H		MIT
2636.755	2637.541	37914.10	U	0.0	0.6		MIT
2636.252	2637.038	37921.34	U	0.9	0.6		MIT
2635.812	2636.598	37927.67	U	0.9	0.9		MIT
2635.528	2636.314	37931.75	U	1.4	1.7		MIT
2635.123	2635.908	37937.58	U	0.9	0.3		MIT
2634.123	2634.908	37951.99	U	0.6H	0.3H		MIT
2633.271	2634.056	37964.27	U	0.3	0.3		MIT
2633.029	2633.814	37967.75	U	0.7	1.2		MIT
2632.979	2633.764	37968.48	U	1.2	0.9		MIT
2632.657	2633.442	37973.12	U	1.1	0.9		MIT
2632.398	2633.183	37976.85	U	0.8	0.3		MIT
2631.968	2632.753	37983.06	U	0.3			MIT
2631.663	2632.448	37987.46	U		0.6H		MIT
2630.634	2631.418	38002.32	U	0.7	0.3		MIT
2629.854	2630.638	38013.59	U	0.7	0.3H		MIT
2629.154	2629.938	38023.71	U	0.6	0.3		MIT
2628.928	2629.712	38026.98	U	1.2	1.0		MIT
2628.504	2629.288	38033.11	U	1.0	0.8		MIT
2627.961	2628.745	38040.97	U	0.9	0.3		MIT
2627.551	2628.335	38046.91	U		1.1H		MIT
2627.128	2627.912	38053.03	U	0.8			MIT
2626.599	2627.382	38060.69	U	0.3	0.3		MIT
2625.855	2626.638	38071.48	U	0.7	0.3		MIT
2625.263	2626.046	38080.06	U	1.0	0.8		MIT
2624.915	2625.698	38085.11	U	1.1	0.8		MIT
2623.537	2624.320	38105.11	U	0.8	0.3		MIT
2622.394	2623.176	38121.72	U	0.5	0.3		MIT
2621.809	2622.591	38130.23	U	1.0	0.3		MIT
2621.452	2622.234	38135.42	U	0.7	0.3		MIT
2621.322	2622.104	38137.31	U		1.0H		MIT
2620.959	2621.741	38142.59	U	0.7	0.6		MIT
2619.26	2620.04	38167.3	U	0.6	0.3		MIT
2618.17	2618.95	38183.2	U	0.5	0.3		MIT
2616.068	2616.849	38213.90	U	1.1	0.8		MIT
2615.950	2616.731	38215.62	U	0.8	0.3		MIT
2615.120	2615.901	38227.75	U	0.3	0.3H		MIT
2613.952	2614.732	38244.83	U	1.0	0.6		MIT
2613.073	2613.853	38257.70	U	0.3			MIT
2612.456	2613.236	38266.73	U	1.0	0.6		MIT
2611.625	2612.405	38278.91	U	0.7	0.3		MIT
2611.14	2611.92	38286.0	U	0.8	0.3H		MIT
2610.646	2611.426	38293.26	U	0.3	0.3		MIT
2610.401	2611.181	38296.85	U	0.8	0.6H		MIT
2609.258	2610.037	38313.63	U	1.2	0.3		MIT
2609.039	2609.818	38316.85	U	0.6	0.3		MIT
2608.197	2608.976	38329.21	U	1.4	0.8		MIT
2607.380	2608.159	38341.22	U	0.7	0.3		MIT
2606.726	2607.505	38350.84	U	1.0	0.6		MIT
2606.520	2607.299	38353.87	U	0.9	0.8		MIT
2606.42	2607.20	38355.3	U	0.5	0.0		MIT
2606.151	2606.930	38359.30	U	0.3	0.3		MIT
2605.74	2606.52	38365.4	U	0.6	0.3		MIT
2604.663	2605.441	38381.21	U	0.3	0.3H		MIT
2604.31	2605.09	38386.4	U	0.8	0.3		MIT
2603.948	2604.726	38391.75	U	0.5	0.3		MIT
2603.551	2604.329	38397.61	U	0.8	0.8H		MIT
2603.016	2603.794	38405.50	U	0.8H	0.3		MIT
2602.418	2603.196	38414.32	U	0.6	1.1H		MIT
2601.537	2602.314	38427.33	U	0.8	1.1		MIT
2600.282	2601.059	38445.88	U	0.3	0.3		MIT
2599.805	2600.582	38452.93	U	0.6	0.3		MIT
2598.859	2599.636	38466.93	U	0.6	0.3H		MIT
2597.689	2598.466	38484.25	U	1.4	1.2		MIT
2597.312	2598.088	38489.84	U	0.3	0.3		MIT
2597.003	2597.779	38494.42	U	0.5	0.3		MIT
2596.118	2596.894	38507.54	U	0.6	0.3		MIT
2595.63	2596.41	38514.8	U		0.9H		MIT
2595.385	2596.161	38518.41	U	0.8	0.8H		MIT
2594.989	2595.765	38524.29	U	0.3	0.6		MIT
2594.275	2595.051	38534.89	U	0.7	0.3		MIT
2593.82	2594.60	38541.6	U	0.0H	0.6		MIT
2593.571	2594.347	38545.35	U	1.3	0.8		MIT
2592.572	2593.347	38560.20	U	1.1	0.8		MIT
2591.252	2592.027	38579.85	U	1.3	1.1		MIT
2590.791	2591.566	38586.71	U	0.5	0.3		MIT
2590.44	2591.21	38591.9	U	0.6	0.3H		MIT

AIR WAVELENGTH (ANGSTRMS)	VACUUM WAVELENGTH (ANGSTRMS)	WAVENUMBER (KAYSERS)	LMENT	LOG ARC	INTENSITY SPK	DIS	REF
2590.18	2590.95	38595.8	U	0.3	0.3		MIT
2589.593	2590.368	38604.56	U	1.0	0.6		MIT
2588.915	2589.689	38614.67	U	0.7	0.3		MIT
2587.072	2587.846	38642.17	U	0.8	1.2H		MIT
2586.201	2586.975	38655.19	U	0.8	0.8		MIT
2585.215	2585.989	38669.93	U	0.8	0.3		MIT
2584.898	2585.672	38674.67	U	1.0	0.6		MIT
2584.420	2585.193	38681.83	U	1.1	0.9		MIT
2583.480	2584.253	38695.90	U	1.0	0.6		MIT
2581.741	2582.514	38721.96	U	0.7H	0.3		MIT
2581.07	2581.84	38732.0	U		1.1H		MIT
2579.572	2580.344	38754.52	U	0.9	0.6		MIT
2579.437	2580.209	38756.55	U	0.8	0.0		MIT
2579.157	2579.929	38760.75	U	0.9	0.3		MIT
2578.323	2579.095	38773.29	U		0.6		MIT
2577.320	2578.092	38788.38	U	1.3	0.8		MIT
2575.422	2576.193	38816.96	U	0.8	0.3		MIT
2575.225	2575.996	38819.93	U	0.8	0.3		MIT
2574.673	2575.444	38828.25	U	0.7	0.3		MIT
2573.201	2573.972	38850.46	U	0.3	0.3		MIT
2572.94	2573.71	38854.4	U	0.5	0.6H		MIT
2572.647	2573.418	38858.83	U		1.2H		MIT
2572.341	2573.112	38863.45	U	0.0	1.1H		MIT
2571.49	2572.26	38876.3	U		0.6H		MIT
2571.041	2571.811	38883.10	U	0.3	0.3		MIT
2570.668	2571.438	38888.74	U	0.8	0.6		MIT
2570.302	2571.072	38894.28	U	0.8	0.3H		MIT
2569.712	2570.482	38903.21	U	1.1	0.9		MIT
2568.980	2569.750	38914.29	U	0.8	0.3		MIT
2568.866	2569.636	38916.02	U	0.7	0.6		MIT
2567.955	2568.725	38929.83	U	0.8	0.3		MIT
2567.108	2567.877	38942.67	U	0.8	0.6		MIT
2566.662	2567.431	38949.44	U	0.6	0.3		MIT
2565.907	2566.676	38960.90	U		0.6		MIT
2565.406	2566.175	38968.50	U	1.5	1.5		MIT
2564.424	2565.193	38983.43	U	0.8	0.3		MIT
2562.943	2563.711	39005.95	U	1.2	0.8		MIT
2562.841	2563.609	39007.50	U	0.7	0.6		MIT
2562.084	2562.852	39019.03	U	0.8	0.3		MIT
2561.960	2562.728	39020.92	U	0.3H	0.6H		MIT
2561.584	2562.352	39026.64	U	1.1	0.8H		MIT
2560.956	2561.724	39036.21	U	0.7	0.3		MIT
2560.281	2561.049	39046.50	U	0.8	0.3		MIT
2560.02	2560.79	39050.5	U	0.5	0.3H		MIT
2559.488	2560.256	39058.60	U	1.0	0.6		MIT
2559.247	2560.014	39062.28	U	0.5	1.0H		MIT
2557.960	2558.727	39081.93	U	0.6	0.3		MIT
2556.194	2556.961	39108.93	U	1.2	1.1		MIT
2555.210	2555.977	39123.99	U	0.6	0.6		MIT
2553.73	2554.50	39146.7	U	0.5	0.3		MIT
2552.38	2553.15	39167.4	U	0.5	0.3		MIT
2551.45	2552.22	39181.6	U	0.5	0.3		MIT
2550.814	2551.579	39191.41	U	0.3	0.3H		MIT
2550.41	2551.18	39197.6	U	0.3			MIT
2550.027	2550.792	39203.51	U	0.6	0.6		MIT
2549.296	2550.061	39214.75	U	1.0	0.8H		MIT
2549.175	2549.940	39216.61	U	0.0H	0.3H		MIT
2548.328	2549.093	39229.64	U	0.7	0.3		MIT
2547.998	2548.763	39234.72	U	0.7	0.3		MIT
2547.629	2548.394	39240.40	U	0.8	0.3		MIT
2547.352	2548.117	39244.67	U	0.6	0.3		MIT
2547.139	2547.904	39247.95	U	0.6			MIT
2546.350	2547.114	39260.11	U	0.3	0.3		MIT
2545.91	2546.67	39266.9	U	0.7	0.3		MIT
2545.79	2546.55	39268.7	U	0.3	0.3		MIT
2545.410	2546.174	39274.61	U	0.6	0.6		MIT
2544.648	2545.412	39286.37	U	0.7	0.6		MIT
2544.358	2545.122	39290.85	U	0.9	0.3		MIT
2544.042	2544.806	39295.73	U	0.6	0.3		MIT
2543.205	2543.969	39308.66	U	0.3	0.3		MIT

AIR WAVELENGTH (ANGSTRMS)	VACUUM WAVELENGTH (ANGSTRMS)	WAVENUMBER (KAYSERS)	LMENT	LOG ARC	INTENSITY SPK	DIS	REF
2541.522	2542.285	39334.69	U	0.5	0.3		MIT
2541.370	2542.133	39337.04	U	0.8	0.3		MIT
2541.044	2541.807	39342.09	U		0.9		MIT
2540.30	2541.06	39353.6	U	0.3	0.3		MIT
2539.878	2540.641	39360.15	U	0.3	0.3		MIT
2539.504	2540.267	39365.94	U	0.3	0.3		MIT
2539.31	2540.07	39369.0	U	0.3	0.3		MIT
2538.989	2539.752	39373.93	U	0.5	0.3		MIT
2538.733	2539.496	39377.90	U	0.8	0.3		MIT
2538.430	2539.193	39382.60	U	1.2	0.8		MIT
2537.703	2538.465	39393.88	U	0.5	0.6H		MIT
2537.296	2538.058	39400.20	U	1.0	0.3		MIT
2536.794	2537.556	39407.99	U	0.6	0.3		MIT
2536.600	2537.362	39411.01	U	0.5	0.3		MIT
2536.238	2537.000	39416.63	U	0.6	0.3		MIT
2535.90	2536.66	39421.9	U	0.7	0.3		MIT
2535.58	2536.34	39426.9	U		0.9		MIT
2534.949	2535.711	39436.67	U	1.1	0.6		MIT
2533.804	2534.565	39454.49	U	0.6			MIT
2533.233	2533.994	39463.39	U	0.9	0.6		MIT
2532.721	2533.482	39471.36	U	0.3	0.3		MIT
2532.368	2533.129	39476.87	U	0.9	0.3		MIT
2531.801	2532.562	39485.71	U	0.8	0.3		MIT
2531.402	2532.163	39491.93	U	0.3H	0.3		MIT
2530.836	2531.597	39500.76	U	0.7	0.3		MIT
2530.298	2531.059	39509.16	U	0.8	0.3		MIT
2529.51	2530.27	39521.5	U	0.3	0.3		MIT
2528.967	2529.727	39529.95	U	0.3	0.3		MIT
2528.77	2529.53	39533.0	U	0.7	0.3		MIT
2528.099	2528.859	39543.52	U	1.0	0.3		MIT
2527.68	2528.44	39550.1	U	0.6	0.3		MIT
2527.433	2528.193	39553.94	U		1.0		MIT
2527.139	2527.899	39558.54	U		0.9		MIT
2526.542	2527.302	39567.89	U	0.9	0.3		MIT
2525.85	2526.61	39578.7	U	0.3H	0.3		MIT
2525.386	2526.146	39586.00	U	0.5	0.3		MIT
2524.918	2525.677	39593.34	U	0.9	0.3		MIT
2524.47	2525.23	39600.4	U		0.6		MIT
2524.311	2525.070	39602.86	U	0.7	0.6		MIT
2523.89	2524.65	39609.5	U	0.3	0.3		MIT
2523.699	2524.458	39612.46	U	0.3	0.3		MIT
2523.413	2524.172	39616.95	U	0.8	0.6		MIT
2522.094	2522.853	39637.67	U	0.5	0.3		MIT
2521.796	2522.555	39642.35	U	0.7	1.2		MIT
2521.367	2522.126	39649.10	U	0.8	0.3		MIT
2520.883	2521.641	39656.71	U	0.3H	0.3		MIT
2520.250	2521.008	39666.67	U	0.8	0.3		MIT
2519.39	2520.15	39680.2	U	0.3	0.3		MIT
2518.974	2519.732	39686.76	U	1.2	0.8		MIT
2518.478	2519.236	39694.58	U	0.3	0.3		MIT
2517.122	2517.880	39715.96	U	0.9	0.3		MIT
2516.969	2517.727	39718.37	U	0.6	0.3		MIT
2515.729	2516.486	39737.95	U	0.8	0.3		MIT
2515.524	2516.281	39741.19	U	0.5	0.3		MIT
2515.147	2515.904	39747.14	U		1.0		MIT
2514.768	2515.525	39753.13	U	1.1	0.6		MIT
2514.066	2514.823	39764.23	U		1.1		MIT
2513.329	2514.086	39775.89	U	0.0	1.3		MIT
2512.587	2513.344	39787.64	U	0.5	0.6H		MIT
2512.222	2512.978	39793.42	U	0.3			MIT
2510.138	2510.894	39826.45	U		0.8H		MIT
2508.357	2509.113	39854.73	U		0.6H		MIT
2507.899	2508.654	39862.01	U		0.9		MIT
2507.741	2508.496	39864.52	U	0.0	0.3H		MIT
2507.067	2507.822	39875.23	U	0.8	0.3		MIT
2506.906	2507.661	39877.79	U	0.8	0.3		MIT
2506.453	2507.208	39885.00	U		0.3H		MIT
2505.289	2506.044	39903.53	U	0.8	0.3		MIT
2504.693	2505.448	39913.03	U	0.7	0.3		MIT
2502.407	2503.161	39949.48	U	0.3	1.0		MIT

AIR WAVELENGTH (ANGSTRMS)	VACUUM WAVELENGTH (ANGSTRMS)	WAVENUMBER (KAYSERS)	LMENT	LOG ARC	INTENSITY SPK	DIS	REF
2501.90	2502.65	39957.6	U	0.3	0.3		MIT
2501.381	2502.135	39965.87	U	0.5	0.6H		MIT
2500.864	2501.618	39974.13	U	1.3	1.1		MIT
2498.829	2499.582	40006.68	U	0.9	1.0		MIT
2498.267	2499.020	40015.68	U	0.5	0.9		MIT
2495.77	2496.52	40055.7	U		0.9H		MIT
2495.291	2496.044	40063.40	U	0.3	0.6H		MIT
2494.77	2495.52	40071.8	U		0.9H		MIT
2494.40	2495.15	40077.7	U	0.3	0.3		MIT
2493.72	2494.47	40088.6	U		0.6		MIT
2492.313	2493.065	40111.27	U	0.5	0.3		MIT
2491.317	2492.069	40127.31	U	0.7	0.3		MIT
2490.928	2491.679	40133.57	U	0.9	0.6		MIT
2489.782	2490.533	40152.04	U	1.1	1.2		MIT
2487.857	2488.608	40183.11	U	0.6	0.3		MIT
2487.60	2488.35	40187.3	U	0.3			MIT
2485.096	2485.846	40227.75	U	0.5	0.3		MIT
2484.891	2485.641	40231.07	U	0.8	0.3		MIT
2484.671	2485.421	40234.63	U	0.3	1.1		MIT
2484.217	2484.967	40241.98	U	1.0	1.2		MIT
2484.008	2484.758	40245.37	U	1.2	0.6		MIT
2481.033	2481.782	40293.62	U	0.3	0.3		MIT
2478.010	2478.759	40342.78	U	0.6	0.9H		MIT
2477.178	2477.926	40356.33	U	0.9	0.3		MIT
2476.474	2477.222	40367.80	U	0.3	0.9H		MIT
2475.025	2475.773	40391.43	U	0.8	0.3		MIT
2470.645	2471.392	40463.03	U	1.1	0.6		MIT
2468.261	2469.007	40502.11	U	1.0	0.6		MIT
2464.04	2464.79	40571.5	U		1.0		MIT
2463.38	2464.13	40582.4	U	0.5			MIT
2460.86	2461.60	40623.9	U	0.6	0.3		MIT
2460.328	2461.072	40632.69	U	0.7	0.3		MIT
2457.166	2457.910	40684.98	U	0.5	0.6H		MIT
2455.680	2456.423	40709.60	U	1.0	0.3		MIT
2455.426	2456.169	40713.81	U	1.1			MIT
2454.367	2455.110	40731.37	U	1.0	0.6		MIT
2453.82	2454.56	40740.5	U	0.3	0.3		MIT
2452.04	2452.78	40770.0	U	0.3H			MIT
2451.48	2452.22	40779.3	U	0.6			MIT
2450.443	2451.185	40796.59	U	0.0	0.6		MIT
2448.931	2449.673	40821.78	U	0.9	0.3		MIT
2447.835	2448.577	40840.06	U		0.9		MIT
2446.49	2447.23	40862.5	U		0.9H		MIT
2444.80	2445.54	40890.7	U	0.5	0.3		MIT
2444.047	2444.788	40903.35	U	0.6	0.6		MIT
2442.895	2443.635	40922.63	U	0.7	0.6		MIT
2441.89	2442.63	40939.5	U		0.6		MIT
2440.432	2441.172	40963.93	U	0.0H	0.3		MIT
2437.669	2438.408	41010.36	U	0.6	0.6H		MIT
2433.786	2434.524	41075.79	U	0.9	0.6		MIT
2431.602	2432.340	41112.68	U	0.3	0.3		MIT
2428.476	2429.213	41165.59	U	0.7	0.3		MIT
2427.622	2428.359	41180.07	U	1.1	0.3		MIT
2427.454	2428.191	41182.92	U	1.1	0.3		MIT
2425.403	2426.139	41217.75	U	0.7	0.3		MIT
2424.971	2425.707	41225.09	U	0.5	0.3		MIT
2424.408	2425.144	41234.66	U	0.5	0.3		MIT
2423.703	2424.439	41246.65	U	1.2	0.6		MIT
2422.637	2423.373	41264.80	U	0.0	1.0		MIT
2421.901	2422.637	41277.34	U	0.3	1.1		MIT
2420.520	2421.255	41300.89	U	0.3	0.3		MIT
2419.573	2420.308	41317.05	U	1.0	0.6		MIT
2418.377	2419.112	41337.49	U	0.8	0.8H		MIT
2416.46	2417.19	41370.3	U	0.3	0.3		MIT
2414.56	2415.29	41402.8	U	0.9	0.3		MIT
2414.143	2414.877	41409.98	U	0.3	1.1		MIT
2412.687	2413.421	41434.97	U	0.6	1.2		MIT
2410.297	2411.030	41476.05	U		1.1		MIT
2406.435	2407.167	41542.61	U	1.0	1.0		MIT
2405.743	2406.475	41554.56	U		0.6H		MIT

AIR WAVELENGTH (ANGSTRMS)	VACUUM WAVELENGTH (ANGSTRMS)	WAVENUMBER (KAYSERS)	LMENT	LOG ARC	INTENSITY SPK	DIS	REF
2403.423	2404.154	41594.67	U		1.5		MIT
2401.443	2402.174	41628.96	U	0.7H			MIT
2401.284	2402.015	41631.71	U		1.2		MIT
2399.73	2400.46	41658.7	U	0.3			MIT
2398.55	2399.28	41679.2	U	0.3	0.3		MIT
2397.74	2398.47	41693.2	U	0.8	0.3		MIT
2397.32	2398.05	41700.5	U		1.3		MIT
2394.02	2394.75	41758.0	U		0.7		MIT
2393.24	2393.97	41771.6	U	0.3	0.3		MIT
2392.03	2392.76	41792.8	U		0.3		MIT
2391.59	2392.32	41800.5	U		0.3		MIT
2390.97	2391.70	41811.3	U	0.6			MIT
2390.11	2390.84	41826.3	U	0.3			MIT
2388.42	2389.15	41855.9	U	0.3	0.5		MIT
2387.22	2387.95	41877.0	U		0.7		MIT
2385.55	2386.28	41906.3	U	0.5	0.3		MIT
2385.28	2386.01	41911.0	U	0.3			MIT
2385.10	2385.83	41914.2	U		0.5		MIT
2383.44	2384.17	41943.4	U		0.5		MIT
2382.93	2383.66	41952.4	U		0.3		MIT
2378.60	2379.33	42028.7	U		0.9		MIT
2378.16	2378.89	42036.5	U		1.5		MIT
2377.87	2378.60	42041.6	U	0.3	1.2		MIT
2377.51	2378.24	42048.0	U		0.3		MIT
2376.55	2377.28	42065.0	U	0.5	0.6		MIT
2376.12	2376.85	42072.6	U	0.3	0.3		MIT
2375.83	2376.56	42077.7	U	0.3	0.3		MIT
2374.07	2374.79	42108.9	U	0.3			MIT
2372.94	2373.66	42129.0	U	0.3			MIT
2371.94	2372.66	42146.7	U		0.3		MIT
2371.53	2372.25	42154.0	U		0.5		MIT
2370.88	2371.60	42165.5	U		0.6		MIT
2370.13	2370.85	42178.9	U	0.3			MIT
2369.06	2369.78	42197.9	U	0.3	0.6		MIT
2368.40	2369.12	42209.7	U	0.5			MIT
2368.13	2368.85	42214.5	U	0.3	0.5		MIT
2367.39	2368.11	42227.7	U	0.3	0.5		MIT
2367.12	2367.84	42232.5	U		0.5		MIT
2365.19	2365.91	42267.0	U		0.7		MIT
2364.27	2364.99	42283.4	U		0.5		MIT
2363.92	2364.64	42289.7	U	0.8	0.3		MIT
2363.43	2364.15	42298.5	U		0.6		MIT
2362.76	2363.48	42310.5	U	0.5	0.3		MIT
2362.35	2363.07	42317.8	U	0.3	0.7		MIT
2361.92	2362.64	42325.5	U		0.3		MIT
2361.46	2362.18	42333.7	U	0.5	0.3H		MIT
2361.14	2361.86	42339.5	U		0.3		MIT
2360.77	2361.49	42346.1	U		0.3		MIT
2359.45	2360.17	42369.8	U	0.3	0.3		MIT
2358.87	2359.59	42380.2	U		0.3		MIT
2357.97	2358.69	42396.4	U	0.3			MIT
2357.59	2358.31	42403.2	U		0.8		MIT
2356.86	2357.58	42416.4	U		0.7		MIT
2355.12	2355.84	42447.7	U	0.3	0.3		MIT
2354.76	2355.48	42454.2	U	0.3	0.5		MIT
2351.87	2352.59	42506.4	U		1.3		MIT
2349.85	2350.57	42542.9	U	0.3H	0.9		MIT
2349.60	2350.32	42547.4	U		1.2		MIT
2348.91	2349.63	42559.9	U	0.3	0.3		MIT
2347.01	2347.73	42594.4	U	0.3	0.3		MIT
2346.16	2346.88	42609.8	U		1.1		MIT
2344.58	2345.30	42638.5	U	0.5	0.3		MIT
2343.57	2344.29	42656.9	U	0.7			A
2342.87	2343.59	42669.6	U		0.5		MIT
2342.35	2343.07	42679.1	U	0.3			MIT
2341.37	2342.09	42697.0	U		0.9		MIT
2340.86	2341.58	42706.3	U	0.3	0.3		MIT
2340.35	2341.07	42715.6	U		0.7		MIT
2338.92	2339.64	42741.7	U		1.0H		MIT
2338.48	2339.20	42749.7	U		0.6		MIT

AIR WAVELENGTH (ANGSTRMS)	VACUUM WAVELENGTH (ANGSTRMS)	WAVENUMBER (KAYSERS)	LMENT	LOG ARC	INTENSITY SPK	INTENSITY DIS	REF
2336.94	2337.66	42777.9	U		0.3		MIT
2336.42	2337.14	42787.4	U		0.3H		MIT
2335.842	2336.558	42797.99	U	0.3	0.3H		MIT
2332.54	2333.26	42858.6	U		0.3		MIT
2331.85	2332.57	42871.2	U	0.3	0.3		MIT
2330.22	2330.93	42901.2	U		0.8		MIT
2329.46	2330.17	42915.2	U		0.8		MIT
2329.29	2330.00	42918.4	U	0.3			MIT
2328.89	2329.60	42925.7	U	0.3	1.0		MIT
2328.52	2329.23	42932.6	U		0.6		MIT
2328.29	2329.00	42936.8	U		0.3H		MIT
2327.88	2328.59	42944.4	U		1.0H		MIT
2326.45	2327.16	42970.7	U		1.3		MIT
2325.67	2326.38	42985.2	U		0.5		MIT
2325.46	2326.17	42989.0	U		1.0		MIT
2324.80	2325.51	43001.2	U	0.3	1.3		MIT
2324.03	2324.74	43015.5	U		0.5H		MIT
2323.38	2324.09	43027.5	U	0.3			MIT
2318.47	2319.18	43118.6	U		1.4		MIT
2318.18	2318.89	43124.0	U	0.3	0.5		MIT
2317.16	2317.87	43143.0	U		0.6		MIT
2315.88	2316.59	43166.9	U		1.0		MIT
2313.83	2314.54	43205.1	U		0.3		MIT
2312.56	2313.27	43228.8	U	0.3	1.1		MIT
2310.67	2311.38	43264.2	U		1.0		MIT
2310.36	2311.07	43270.0	U		1.0		MIT
2309.743	2310.453	43281.55	U	0.6	0.3		MIT
2308.73	2309.44	43300.5	U		0.3		MIT
2308.35	2309.06	43307.7	U		0.8		MIT
2306.91	2307.62	43334.7	U		1.4		MIT
2305.645	2306.355	43358.47	U		1.1		MIT
2304.403	2305.112	43381.83	U		0.6		MIT
2303.93	2304.64	43390.7	U		0.9		MIT
2303.65	2304.36	43396.0	U		0.5		MIT
2302.69	2303.40	43414.1	U		0.5		MIT
2301.89	2302.60	43429.2	U		0.3		MIT
2301.51	2302.22	43436.4	U		0.3		MIT
2300.89	2301.60	43448.1	U		0.6H		MIT
2300.75	2301.46	43450.7	U		0.8		MIT
2299.168	2299.876	43480.60	U		0.3		MIT
2298.370	2299.078	43495.70	U		0.8		MIT
2297.718	2298.426	43508.04	U		0.3		MIT
2297.02	2297.73	43521.3	U		0.6		MIT
2296.25	2296.96	43535.9	U		0.3		MIT
2295.866	2296.573	43543.13	U		0.5		MIT
2295.305	2296.012	43553.77	U	0.3			MIT
2294.85	2295.56	43562.4	U		0.5		MIT
2293.621	2294.328	43585.75	U		1.0		MIT
2291.642	2292.348	43623.38	U		0.9		MIT
2290.67	2291.38	43641.9	U	0.3	0.3		MIT
2289.32	2290.03	43667.6	U		0.3		MIT
2288.58	2289.29	43681.7	U	0.3	0.3		MIT
2288.30	2289.01	43687.1	U		0.3		MIT
2287.80	2288.51	43696.6	U		0.3		MIT
2286.77	2287.48	43716.3	U	0.3			MIT
2285.66	2286.37	43737.5	U		1.0		MIT
2285.14	2285.85	43747.5	U		0.3		MIT
2284.83	2285.54	43753.4	U		0.6		MIT
2283.72	2284.42	43774.7	U		1.2		MIT
2283.34	2284.04	43782.0	U		1.0H		MIT
2282.78	2283.48	43792.7	U		1.4		MIT
2281.84	2282.54	43810.8	U		0.5H		MIT
2280.14	2280.84	43843.4	U		0.6		MIT
2279.97	2280.67	43846.7	U	0.3	0.3		MIT
2278.63	2279.33	43872.5	U		0.5		MIT
2277.98	2278.68	43885.0	U		0.3		MIT
2277.02	2277.72	43903.5	U		0.3		MIT
2276.80	2277.50	43907.7	U	0.3	0.3		MIT
2276.28	2276.98	43917.8	U		0.5		MIT
2276.05	2276.75	43922.2	U		1.3		MIT

AIR WAVELENGTH (ANGSTRMS)	VACUUM WAVELENGTH (ANGSTRMS)	WAVENUMBER (KAYSERS)	LMENT	LOG ARC	INTENSITY SPK	INTENSITY DIS	REF
2275.12	2275.82	43940.1	U		0.3		MIT
2274.68	2275.38	43948.6	U		0.3		MIT
2274.47	2275.17	43952.7	U		0.3		MIT
2273.36	2274.06	43974.2	U		1.2		MIT
2272.680	2273.382	43987.32	U	0.9	0.3H		MIT
2272.32	2273.02	43994.3	U		0.3		MIT
2271.812	2272.514	44004.13	U		0.6		MIT
2268.530	2269.231	44067.78	U		0.3		MIT
2266.97	2267.67	44098.1	U		0.8		MIT
2264.30	2265.00	44150.1	U	0.7	0.3H		MIT
2263.26	2263.96	44170.4	U		0.5		MIT
2262.79	2263.49	44179.6	U		0.3H		MIT
2262.36	2263.06	44188.0	U		0.3		MIT
2261.47	2262.17	44205.3	U	0.3	0.3		MIT
2259.64	2260.34	44241.1	U		0.9		MIT
2258.02	2258.72	44272.9	U		0.5		MIT
2254.52	2255.22	44341.6	U		0.6		MIT
2252.45	2253.15	44382.4	U		0.9		MIT
2251.09	2251.79	44409.2	U		0.5		MIT
2249.31	2250.01	44444.3	U		0.5		MIT
2248.82	2249.52	44454.0	U		0.6		MIT
2248.03	2248.73	44469.6	U		1.4		MIT
2247.08	2247.78	44488.4	U		0.9		MIT
2246.23	2246.93	44505.2	U		0.6		MIT
2244.93	2245.63	44531.0	U	0.3			MIT
2244.10	2244.80	44547.5	U		0.3		MIT
2243.65	2244.35	44556.4	U		0.3		MIT
2243.37	2244.07	44562.0	U		0.3		MIT
2242.68	2243.38	44575.7	U		0.3		MIT
2240.13	2240.83	44626.4	U		0.6		MIT
2239.86	2240.56	44631.8	U		0.3		MIT
2238.02	2238.71	44668.5	U		0.5		MIT
2237.44	2238.13	44680.1	U		1.2		MIT
2236.58	2237.27	44697.2	U		0.5		MIT
2235.79	2236.48	44713.0	U		0.5		MIT
2233.94	2234.63	44750.1	U		0.3		MIT
2232.44	2233.13	44780.1	U		0.6		MIT
2230.62	2231.31	44816.7	U		0.6		MIT
2229.43	2230.12	44840.6	U	0.3	0.3		MIT
2228.28	2228.97	44863.7	U		0.3		MIT
2227.21	2227.90	44885.3	U		0.5		MIT
2222.30	2222.99	44984.4	U		1.0		MIT
2221.52	2222.21	45000.2	U		0.3		MIT
2219.28	2219.97	45045.6	U	1.2	0.3		MIT
2217.51	2218.20	45081.6	U		0.7H		MIT
2216.10	2216.79	45110.3	U		0.3H		MIT
2215.38	2216.07	45124.9	U		0.5H		MIT
2205.93	2206.62	45318.2	U		0.7		MIT
2198.58	2199.27	45469.7	U	0.3	0.7		A
2194.79	2195.48	45548.2	U	0.3	1.2		MIT
2169.69	2170.37	46075.1	U	0.3	0.8		A
2158.61	2159.29	46311.6	U		1.3		A
2153.94	2154.62	46412.0	U	0.3	0.8		A
2131.13	2131.80	46908.7	U	0.3	0.3		A
2119.52	2120.19	47165.6	U	0.3	0.3		A
2104.09	2104.76	47511.4	U	0.3	0.3		A
2099.32	2099.99	47619.4	U	0.3	0.3		A
2037.87	2038.52	49055.1	U	0.5	0.3		A
2025.53	2026.18	49353.9	U	0.3	0.5		A
9865.44	9868.15	10133.6	V I	1.0			ME
9772.98	9775.66	10229.5	V	0.5			ME
9738.50	9741.17	10265.7	V I	1.2			ME
9708.36	9711.02	10297.6	V I	1.0			ME
9691.58	9694.24	10315.4	V I	1.6			ME
9684.19	9686.85	10323.3	V	0.8			ME
9670.9	9673.6	10337.	V I	0.5			ME
9668.9	9671.6	10340.	V I	0.3			ME
9614.68	9617.32	10397.9	V I	1.7			ME
9611.60	9614.24	10401.2	V I	1.9			ME
9593.04	9595.67	10421.4	V I	0.3			ME

AIR WAVELENGTH (ANGSTRMS)	VACUUM WAVELENGTH (ANGSTRMS)	WAVENUMBER (KAYSERS)	LMENT	LOG ARC	INTENSITY SPK	INTENSITY DIS	REF
9582.28	9584.91	10433.1	V I	0.8			ME
9540.31	9542.93	10479.0	V I	0.7			ME
9536.53	9539.15	10483.1	V I	0.3			ME
9511.37	9513.98	10510.8	V	0.7H			ME
9509.11	9511.72	10513.3	V I	0.7H			ME
9482.64	9485.24	10542.7	V	0.6			ME
9480.25	9482.85	10545.3	V I	0.7H			ME
9476.14	9478.74	10549.9	V I	1.0			ME
9467.92	9470.52	10559.1	V	0.5H			ME
9466.32	9468.92	10560.9	V I	0.9H			ME
9454.44	9457.03	10574.1	V I	1.0			ME
9445.74	9448.33	10583.9	V I	1.0			ME
9439.80	9442.39	10590.5	V	0.9H			ME
9435.58	9438.17	10595.3	V I	1.9			MIT
9411.32	9413.90	10622.6	V I	1.5			MIT
9406.02	9408.60	10628.6	V I	0.6H			ME
9398.92	9401.50	10636.6	V I	1.0			ME
9384.86	9387.43	10652.5	V I	1.5			MIT
9380.50	9383.07	10657.5	V	0.6			ME
9369.80	9372.37	10669.7	V	0.7			ME
9366.92	9369.49	10672.9	V I	1.7			MIT
9362.76	9365.33	10677.7	V I	0.3			ME
9361.58	9364.15	10679.0	V I	0.8			ME
9341.20	9343.76	10702.3	V I	2.0			MIT
9334.91	9337.47	10709.5	V I	0.7H			ME
9328.19	9330.75	10717.2	V I	1.6			MIT
9324.50	9327.06	10721.5	V I	0.8			MIT
9316.53	9319.09	10730.7	V I	0.6			MIT
9313.54	9316.10	10734.1	V I	0.6			ME
9308.68	9311.23	10739.7	V I	1.3			MIT
9290.43	9292.98	10760.8	V I	1.0			MIT
9273.40	9275.94	10780.6	V I	1.2			MIT
9265.70	9268.24	10789.5	V I	1.3			MIT
9255.89	9258.43	10801.0	V I	1.0			MIT
9242.91	9245.45	10816.1	V I	1.5			MIT
9226.09	9228.62	10835.8	V I	1.3			MIT
9217.22	9219.75	10846.3	V I	0.6H			ME
9202.88	9205.41	10863.2	V I	0.6			ME
9168.72	9171.24	10903.7	V I	1.3			MIT
9165.80	9168.32	10907.1	V I	0.6			ME
9164.81	9167.33	10908.3	V I	1.6			MIT
9156.55	9159.06	10918.2	V I	1.3			MIT
9113.78	9116.28	10969.4	V I	0.8			MIT
9105.87	9108.37	10978.9	V I	1.0			MIT
9100.78	9103.28	10985.1	V I	0.9			ME
9085.22	9087.71	11003.9	V I	1.6			MIT
9073.34	9075.83	11018.3	V I	0.6			ME
9046.71	9049.19	11050.7	V I	1.7			MIT
9037.60	9040.08	11061.8	V I	1.5			MIT
9022.73	9025.21	11080.1	V I	1.3			MIT
9021.11	9023.59	11082.1	V I	1.3			MIT
9003.08	9005.55	11104.3	V I	0.7			ME
8999.10	9001.57	11109.2	V I	0.6			ME
8971.66	8974.12	11143.2	V I	1.6			MIT
8963.60	8966.06	11153.2	V I	0.7			MIT
8949.17	8951.63	11171.2	V I	0.6			MIT
8932.93	8935.38	11191.5	V I	1.7W			MIT
8931.62	8934.07	11193.1	V I	0.8H			ME
8919.80	8922.25	11207.9	V I	2.0W			MIT
8916.35	8918.80	11212.3	V I	0.3			ME
8657.85	8660.23	11547.0	V	0.7			MIT
8642.61	8644.98	11567.4	V	0.6			ME
8624.86	8627.23	11591.2	V	0.3L			MIT
8604.98	8607.34	11618.0	V	0.7H			ME
8598.94	8601.30	11626.2	V I	0.6			MIT
8591.06	8593.42	11636.8	V	0.7W			MIT
8580.48	8582.84	11651.2	V I	0.9H			MIT
8551.48	8553.83	11690.7	V I	0.5			ME
8541.97	8544.32	11703.7	V I	1.3			MIT
8538.16	8540.51	11708.9	V	0.7R			MIT

AIR WAVELENGTH (ANGSTRMS)	VACUUM WAVELENGTH (ANGSTRMS)	WAVENUMBER (KAYSERS)	LMENT	LOG ARC	INTENSITY SPK	INTENSITY DIS	REF
8534.49	8536.83	11713.9	V I	1.5			MIT
8505.65	8507.99	11753.7	V I	1.0			MIT
8505.06	8507.40	11754.5	V I	0.6			ME
8502.98	8505.32	11757.3	V I	0.7D			MIT
8499.52	8501.86	11762.1	V I	1.6			MIT
8493.61	8495.94	11770.3	V I	0.8			ME
8431.63	8433.95	11856.8	V I	1.1H			ME
8421.39	8423.70	11871.3	V I	0.9			ME
8416.49	8418.80	11878.2	V I	0.9			MIT
8414.38	8416.69	11881.2	V I	0.9			MIT
8408.23	8410.54	11889.8	V I	0.7H			ME
8402.81	8405.12	11897.5	V	1.3			MIT
8342.03	8344.32	11984.2	V I	1.7			MIT
8331.23	8333.52	11999.7	V I	1.6			MIT
8324.42	8326.71	12009.6	V I	1.5			MIT
8282.37	8284.65	12070.5	V I	1.9			MIT
8280.39	8282.67	12073.4	V I	1.2			MIT
8267.65	8269.92	12092.0	V I	0.3			MIT
8255.88	8258.15	12109.2	V I	1.8			MIT
8253.51	8255.78	12112.7	V I	1.9			MIT
8241.61	8243.88	12130.2	V I	1.8			MIT
8203.07	8205.33	12187.2	V I	2.0			MIT
8198.87	8201.12	12193.4	V I	1.8			MIT
8194.83	8197.08	12199.5	V I	0.8			MIT
8187.38	8189.63	12210.6	V I	1.7			MIT
8186.71	8188.96	12211.6	V I	1.9			MIT
8180.21	8182.46	12221.3	V I	1.3			MIT
8171.35	8173.60	12234.5	V I	1.7			MIT
8168.89	8171.14	12238.2	V I	1.2			MIT
8161.07	8163.31	12249.9	V I	2.2W			MIT
8154.59	8156.83	12259.7	V I	1.3			MIT
8144.59	8146.83	12274.7	V I	1.5			MIT
8136.79	8139.03	12286.5	V I	1.3			MIT
8116.80	8119.03	12316.7	V I	2.2			MIT
8109.88	8112.11	12327.2	V I	1.0			ME
8109.10	8111.33	12328.4	V I	1.2			MIT
8108.59	8110.82	12329.2	V I	1.3			MIT
8102.44	8104.67	12338.6	V I	1.3			MIT
8093.45	8095.68	12352.3	V I	1.9			MIT
8028.24	8030.45	12452.6	V I	1.3			MIT
8027.39	8029.60	12453.9	V I	1.9W			MIT
7947.38	7949.57	12579.3	V I	0.9			MIT
7937.92	7940.10	12594.3	V I	1.2			MIT
7933.04	7935.22	12602.0	V I	0.8			MIT
7851.18	7853.34	12733.4	V	0.5			MIT
7826.88	7829.03	12773.0	V I	0.6			MIT
7790.61	7792.75	12832.4	V	0.5H			MIT
7736.68	7738.81	12921.9	V I	0.7			MIT
7704.81	7706.93	12975.3	V I	1.3			MIT
7701.37	7703.49	12981.1	V I	0.5			MIT
7676.14	7678.25	13023.8	V I	0.3			MIT
7662.98	7665.09	13046.2	V I	0.3H			MIT
7661.04	7663.15	13049.5	V I	0.3H			MIT
7638.03	7640.13	13088.8	V I	0.3H			ME
7624.81	7626.91	13111.5	V I	1.2			MIT
7621.86	7623.96	13116.6	V I	0.6D			MIT
7613.54	7615.64	13130.9	V I	0.3H			MIT
7598.28	7600.37	13157.2	V I	0.3H			ME
7596.92	7599.01	13159.6	V I	0.5			ME
7591.20	7593.29	13169.5	V I	0.3H			MIT
7578.75	7580.84	13191.2	V I	0.7H			ME
7573.92	7576.01	13199.6	V I	0.3			MIT
7570.30	7572.38	13205.9	V I	0.3			MIT
7554.63	7556.71	13233.3	V I	0.3			ME
7488.15	7490.21	13350.8	V I	0.3			MIT
7485.90	7487.96	13354.8	V I	0.3			MIT
7435.73	7437.78	13444.9	V	0.3			MIT
7433.92	7435.97	13448.2	V	0.3			MIT
7431.55	7433.60	13452.4	V	0.3			MIT
7393.49	7395.53	13521.7	V I	0.5			MIT

AIR WAVELENGTH (ANGSTRMS)	VACUUM WAVELENGTH (ANGSTRMS)	WAVENUMBER (KAYSERS)	LMENT	LOG ARC	INTENSITY SPK	INTENSITY DIS	REF
7386.04	7388.07	13535.3	V I	0.5			MIT
7363.20	7365.23	13577.3	V I	1.4			MIT
7362.55	7364.58	13578.5	V I	0.3			MIT
7361.44	7363.47	13580.6	V I	1.0			MIT
7358.66	7360.69	13585.7	V I	0.3			MIT
7356.54	7358.57	13589.6	V I	1.3			MIT
7345.15	7347.17	13610.7	V	0.3			MIT
7338.92	7340.94	13622.2	V I	1.7			MIT
7329.72	7331.74	13639.3	V I	0.3			MIT
7321.50	7323.52	13654.6	V I	0.3			MIT
7306.78	7308.79	13682.2	V I	0.3	0.3		MIT
7264.29	7266.29	13762.2	V I	0.9			MIT
7255.66	7257.66	13778.6	V I	0.3			MIT
7182.08	7184.06	13919.7	V I	0.3			MIT
7151.36	7153.33	13979.5	V I	0.3			ME
7102.58	7104.54	14075.5	V I	0.3			MIT
7092.14	7094.10	14096.2	V I	0.3			MIT
7026.07	7028.01	14228.8	V I	0.5			MIT
6974.58	6976.50	14333.8	V I	0.3			MIT
6894.00	6895.90	14501.4	V I	0.3			MIT
6871.56	6873.46	14548.7	V I	0.3			MIT
6870.88	6872.78	14550.2	V I	0.5			MIT
6841.90	6843.79	14611.8	V I	0.6			MIT
6839.62	6841.51	14616.7	V I	0.3			MIT
6832.49	6834.38	14631.9	V I	0.7		1.0	MIT
6829.96	6831.84	14637.3	V I	0.6			MIT
6812.43	6814.31	14675.0	V I	0.9			MIT
6786.37	6788.24	14731.3	V	0.5			MIT
6785.03	6786.90	14734.3	V I	1.2			MIT
6766.56	6768.43	14774.5	V I	1.5			MIT
6760.20	6762.07	14788.4	V	0.3			MIT
6753.03	6754.89	14804.1	V I	1.6			MIT
6708.17	6710.02	14903.1	V I	0.3			MIT
6672.850	6674.692	14981.96	V		0.3		ME
6663.93	6665.77	15002.0	V I	1.3			MIT
6624.849	6626.679	15090.52	V I	1.1		0.8	MIT
6623.541	6625.370	15093.50	V I	0.3			MIT
6607.831	6609.656	15129.38	V I	0.5			MIT
6605.970	6607.794	15133.64	V I	0.5			MIT
6565.88	6567.69	15226.0	V I	0.5			ME
6558.020	6559.832	15244.29	V I	0.5			MIT
6543.511	6545.319	15278.09	V I	0.9		0.7	MIT
6531.428	6533.232	15306.36	V I	1.4			MIT
6517.280	6519.081	15339.59	V II		1.0		ME
6504.166	6505.963	15370.51	V I	1.4		1.0	MIT
6488.050	6489.843	15408.69	V I	0.3		0.6	MIT
6452.344	6454.127	15493.96	V I	1.4		1.0	MIT
6435.156	6436.935	15535.34	V I	0.3			MIT
6433.175	6434.953	15540.13	V I	0.3		0.5	MIT
6431.634	6433.412	15543.85	V I	0.6			MIT
6430.471	6432.248	15546.66	V I	0.9			ME
6400.590	6402.359	15619.24	V		0.3		ME
6398.752	6400.521	15623.73	V		0.3		ME
6397.346	6399.114	15627.16	V		0.3		ME
6393.275	6395.042	15637.11	V I	0.6	0.3		MIT
6387.972	6389.738	15650.09	V		0.3		ME
6381.262	6383.026	15666.55	V I	0.3			MIT
6380.115	6381.879	15669.37	V		1.3		ME
6379.364	6381.128	15671.21	V I	0.9	0.3		MIT
6374.497	6376.259	15683.18	V I	0.3	0.0		MIT
6361.269	6363.028	15715.79	V I	0.7	0.3		MIT
6358.818	6360.576	15721.85	V I	0.7	0.5		MIT
6357.298	6359.056	15725.60	V I	1.0	0.5		MIT
6349.478	6351.234	15744.97	V I	1.2	0.9		MIT
6339.088	6340.841	15770.78	V I	1.4	0.9		MIT
6326.839	6328.588	15801.31	V I	1.2	0.5		MIT
6324.665	6326.414	15806.74	V I	0.9	0.6		MIT
6321.229	6322.977	15815.33	V I	0.7	0.7H		ME
6311.500	6313.245	15839.71	V I	1.0			MIT
6309.702	6311.447	15844.23	V I	0.7	0.3		MIT

AIR WAVELENGTH (ANGSTRMS)	VACUUM WAVELENGTH (ANGSTRMS)	WAVENUMBER (KAYSERS)	LMENT	LOG ARC	INTENSITY SPK	INTENSITY DIS	REF
6304.313	6306.056	15857.77	V I	0.7			MIT
6296.493	6298.234	15877.47	V I	1.5	1.5		MIT
6295.268	6297.009	15880.56	V	0.3			MIT
6292.827	6294.567	15886.71	V I	1.7	1.0		MIT
6286.924	6288.663	15901.63	V I	0.3			MIT
6285.165	6286.903	15906.08	V I	1.7	1.0		MIT
6282.331	6284.069	15913.26	V	1.2	0.3		MIT
6280.91	6282.65	15916.9	V I	0.3			MIT
6274.654	6276.389	15932.73	V I	1.7	0.9		MIT
6268.817	6270.551	15947.56	V I	1.5	0.7		MIT
6266.320	6268.053	15953.92	V I	1.4	0.5		MIT
6261.225	6262.957	15966.90	V I	1.3	0.5		MIT
6258.572	6260.303	15973.67	V I	1.5	0.5		MIT
6256.899	6258.630	15977.94	V I	1.5	0.5		MIT
6254.71	6256.44	15983.5	V	0.3H			ME
6251.823	6253.552	15990.91	V I	1.8	0.9		MIT
6247.546	6249.274	16001.86	V I	0.3	0.0		MIT
6245.217	6246.945	16007.83	V I	0.7	0.0		MIT
6243.555	6245.282	16012.09	V I	1.0	0.3		MIT
6243.105	6244.832	16013.24	V I	1.5	0.6		MIT
6242.811	6244.538	16013.99	V I	1.3	1.0		MIT
6240.131	6241.857	16020.87	V I	1.3	0.3		MIT
6236.276	6238.001	16030.78	V I	0.7			MIT
6233.197	6234.921	16038.69	V I	1.5	0.6		MIT
6230.736	6232.460	16045.03	V I	1.8	1.0		ME
6226.295	6228.018	16056.47	V		0.9		ME
6224.505	6226.227	16061.09	V I	1.7	0.7		MIT
6221.219	6222.940	16069.57	V I	0.5	0.0		MIT
6218.314	6220.034	16077.08	V I	1.2	1.0H		MIT
6216.370	6218.090	16082.11	V I	1.8	1.0		MIT
6213.869	6215.588	16088.58	V I	1.7	0.5		MIT
6207.251	6208.968	16105.73	V I	1.3	0.7		ME
6199.193	6200.908	16126.67	V I	2.0	0.9		MIT
6189.349	6191.062	16152.32	V I	1.0	0.3		MIT
6170.359	6172.066	16202.03	V I	1.5	0.5		MIT
6150.154	6151.856	16255.26	V I	1.6	0.7		MIT
6135.377	6137.075	16294.41	V I	1.5	0.7		MIT
6135.07	6136.77	16295.2	V I	0.3	0.3		ME
6128.336	6130.032	16313.13	V I	0.3	0.3		MIT
6119.522	6121.216	16336.62	V I	1.5	1.3		MIT
6111.646	6113.338	16357.68	V I	1.3	1.2		MIT
6110.214	6111.905	16361.51	V I	0.3H	0.5		ME
6106.975	6108.665	16370.19	V I	1.2	0.3		MIT
6093.866	6095.553	16405.40	V I	0.5H	0.5		ME
6090.216	6091.902	16415.24	V I	1.8	1.2		MIT
6087.490	6089.175	16422.58	V I	0.9	0.0		MIT
6086.55	6088.24	16425.1	V I	0.6H			ME
6081.442	6083.126	16438.92	V I	2.0	1.0		MIT
6080.112	6081.795	16442.51	V		0.7		ME
6077.358	6079.041	16449.96	V I	2.5H	0.3		MIT
6067.258	6068.938	16477.35	V I	1.2	0.3		MIT
6063.382	6065.061	16487.88	V I	1.0	0.0		MIT
6058.136	6059.813	16502.16	V I	1.8			MIT
6054.475	6056.151	16512.14	V I	0.8	0.0		MIT
6039.726	6041.398	16552.46	V I	2.0	1.0		MIT
6031.071	6032.741	16576.21	V	0.0	1.4		ME
6028.982	6030.652	16581.96	V	0.0	0.9		ME
6028.257	6029.926	16583.95	V	0.3	1.3		ME
6027.235	6028.904	16586.76	V		0.8		ME
6026.805	6028.474	16587.95	V	0.7			ME
6025.410	6027.079	16591.79	V	1.0	0.3		MIT
6024.187	6025.855	16595.16	V	0.5	0.9		MIT
6021.747	6023.415	16601.88	V I	0.9	0.0		MIT
6017.924	6019.591	16612.43	V I	0.6	0.0		MIT
6016.122	6017.788	16617.40	V I	0.9	0.3		MIT
6002.627	6004.290	16654.76	V I	1.3	0.3		MIT
6002.306	6003.968	16655.65	V I	1.0	0.3		MIT
5984.642	5986.300	16704.81	V		1.3		MIT
5980.779	5982.436	16715.60	V I	1.7			MIT
5978.906	5980.562	16720.84	V I	2.0			MIT

AIR WAVELENGTH (ANGSTRMS)	VACUUM WAVELENGTH (ANGSTRMS)	WAVENUMBER (KAYSERS)	LMENT		LOG ARC	INTENSITY SPK	DIS	REF
5975.340	5976.995	16730.81	V	I	1.3	0.5		MIT
5967.791	5969.444	16751.98	V			0.7		ME
5951.454	5953.103	16797.96	V			0.6		ME
5928.852	5930.495	16862.00	V			1.8H		ME
5924.574	5926.216	16874.18	V	I	2.40			MIT
5916.365	5918.004	16897.59	V			1.1		ME
5897.544	5899.178	16951.51	V			1.5		MIT
5850.316	5851.938	17088.36	V	I	1.6	1.3		MIT
5849.630	5851.252	17090.36	V			1.3		ME
5846.296	5847.917	17100.11	V	I	2.0	2.0H		MIT
5830.717	5832.333	17145.80	V	I	2.0	1.9		MIT
5819.920	5821.534	17177.60	V			1.5		ME
5817.528	5819.141	17184.67	V	I	2.0H			MIT
5817.064	5818.677	17186.04	V	I	1.7			MIT
5807.143	5808.753	17215.40	V	I	1.9	1.6		MIT
5799.899	5801.507	17236.90	V		1.6	0.3		MIT
5798.905	5800.513	17239.85	V	I	0.3			ME
5797.992	5799.600	17242.57	V			0.6		ME
5788.556	5790.161	17270.68	V	I	1.40			MIT
5786.160	5787.765	17277.83	V	I	1.9			MIT
5784.378	5785.982	17283.15	V	I	1.7	1.5		MIT
5783.503	5785.107	17285.77	V	I	1.5			MIT
5782.610	5784.214	17288.44	V	I	1.5			MIT
5776.684	5778.286	17306.17	V	I	1.7	1.4		MIT
5772.419	5774.020	17318.96	V	I	1.7	1.4		MIT
5770.55	5772.15	17324.6	V	I	1.3	0.3		ME
5761.412	5763.010	17352.04	V	I	1.4			MIT
5755.085	5756.681	17371.12	V		0.3H	0.3		ME
5752.740	5754.336	17378.20	V	I	1.0	1.0		MIT
5750.647	5752.242	17384.53	V	I	1.70			MIT
5748.87	5750.46	17389.9	V	I	1.3	1.0		MIT
5747.701	5749.295	17393.44	V	I	0.9			MIT
5743.453	5745.046	17406.30	V	I	1.8	1.3		MIT
5737.056	5738.647	17425.71	V	I	2.0	2.0		MIT
5734.014	5735.605	17434.95	V	I	1.5			MIT
5733.091	5734.681	17437.76	V	I	1.1			MIT
5731.249	5732.839	17443.36	V	I	2.4	2.0		MIT
5727.664	5729.253	17454.28	V	I	1.9			MIT
5727.029	5728.618	17456.22	V	I	2.2	2.2		MIT
5725.638	5727.226	17460.46	V	I	1.6	1.5		MIT
5716.209	5717.795	17489.26	V	I	1.8	1.5		MIT
5708.949	5710.533	17511.50	V	I	1.5			MIT
5706.980	5708.563	17517.54	V	I	2.3			MIT
5706.142	5707.725	17520.11	V		1.3			MIT
5704.367	5705.950	17525.57	V		1.0			MIT
5703.562	5705.144	17528.04	V	I	2.3	1.8		ME
5698.518	5700.099	17543.55	V	I	2.5	2.5		MIT
5687.758	5689.336	17576.74	V	I	1.2	1.5		MIT
5683.216	5684.793	17590.79	V		1.7	0.3		MIT
5670.850	5672.424	17629.15	V	I	2.2	1.8		MIT
5668.362	5669.935	17636.89	V	I	1.9	1.7		MIT
5657.439	5659.009	17670.94	V	I	2.2	1.8		MIT
5656.29	5657.86	17674.5	V	I	1.1	0.3		ME
5646.106	5647.673	17706.41	V	I	2.2	2.2		MIT
5642.011	5643.577	17719.26	V			1.1H		MIT
5635.514	5637.078	17739.69	V	I	1.5			MIT
5633.896	5635.460	17744.78	V	I	0.9			MIT
5632.464	5634.027	17749.29	V	I	1.3			MIT
5627.641	5629.203	17764.50	V	I	2.3	1.9		MIT
5626.011	5627.573	17769.65	V	I	2.2			MIT
5624.888	5626.449	17773.20	V	I	1.6			MIT
5624.598	5626.159	17774.11	V	I	2.0			MIT
5624.204	5625.765	17775.36	V	I	1.5	1.5		MIT
5622.068	5623.629	17782.11	V	I	1.0			MIT
5604.945	5606.501	17836.44	V	I	1.8	1.3		MIT
5604.195	5605.751	17838.82	V	I	1.2			MIT
5602.971	5604.527	17842.72	V		0.5	0.3		MIT
5601.380	5602.935	17847.79	V	I	1.4			MIT
5592.415	5593.968	17876.40	V	I	1.7	1.7		MIT
5585.996	5587.547	17896.94	V	I	0.9	0.9		MIT
5584.747	5586.298	17900.94	V	I	0.7	0.7		MIT
5584.501	5586.052	17901.73	V	I	1.3	1.3		MIT
5576.504	5578.052	17927.40	V	I	0.8	0.6		MIT
5561.657	5563.201	17975.26	V	I	1.2	1.2		MIT
5560.548	5562.092	17978.85	V	I	0.5	0.0		MIT
5558.749	5560.293	17984.67	V	I	1.3	1.3		MIT
5548.158	5549.699	18019.00	V		0.90	1.0H		MIT
5547.067	5548.608	18022.54	V	I	1.6	1.6		MIT
5545.931	5547.471	18026.23	V	I	1.1	1.1		MIT
5544.865	5546.405	18029.70	V	I	0.7	0.5		MIT
5535.382	5536.919	18060.58	V	I	0.3	0.3		ME
5533.838	5535.375	18065.62	V	I	1.3	1.3		MIT
5511.181	5512.712	18139.89	V	I	1.4	1.4		MIT
5508.620	5510.150	18148.32	V	I	0.7	0.7		MIT
5507.749	5509.279	18151.20	V	I	1.8	1.8		MIT
5505.865	5507.395	18157.41	V	I	1.0	1.0		MIT
5504.873	5506.402	18160.68	V	I	1.2	1.2		MIT
5496.009	5497.536	18189.97	V	I	0.5H	0.0H		MIT
5489.945	5491.470	18210.06	V	I	1.3	1.3		MIT
5487.919	5489.444	18216.78	V	I	1.3	1.3		MIT
5487.218	5488.743	18219.11	V	I	1.2	1.2		MIT
5486.988	5488.513	18219.87	V			0.3H		MIT
5471.327	5472.847	18272.02	V		1.10	1.10		MIT
5467.798	5469.317	18283.82	V	I	1.0	1.0		MIT
5458.122	5459.639	18316.23	V	I	1.2	1.2		MIT
5457.102	5458.619	18319.65	V			0.3		ME
5454.814	5456.330	18327.34	V		0.8	0.6		MIT
5443.227	5444.740	18366.35	V	I	1.1	0.6		MIT
5439.312	5440.824	18379.57	V			0.9		MIT
5437.659	5439.170	18385.16	V	I	1.0	1.0		MIT
5434.175	5435.685	18396.94	V	I	1.7	1.7		MIT
5429.485	5430.994	18412.83	V	I	0.6	0.3		MIT
5424.082	5425.590	18431.18	V	I	1.4	1.4		MIT
5418.087	5419.593	18451.57	V	I	1.2	1.2		MIT
5415.263	5416.768	18461.19	V	I	1.9	1.9		MIT
5401.933	5403.435	18506.75	V	I	2.0	2.0		MIT
5393.177	5394.676	18536.79	V	I	2.0			MIT
5385.137	5386.634	18564.47	V	I	1.7	1.7		MIT
5384.895	5386.392	18565.30	V			0.5		ME
5383.426	5384.923	18570.37	V	I	1.6	1.6		MIT
5353.412	5354.901	18674.48	V	I	1.7	1.7		MIT
5350.380	5351.868	18685.06	V			0.5H		ME
5338.614	5340.099	18726.25	V	I	0.6	0.8		MIT
5335.588	5337.072	18736.86	V		1.1H	0.7H		ME
5332.656	5334.139	18747.17	V			0.7		ME
5328.823	5330.305	18760.65	V	I	1.30	1.1		ME
5322.818	5324.299	18781.82	V			0.6H		ME
5319.087	5320.567	18794.99	V	I	0.7	0.7		ME
5316.885	5318.364	18802.77	V	I	0.7	0.7		ME
5315.219	5316.698	18808.67	V	I	0.5			ME
5314.468	5315.947	18811.32	V	I	1.0	0.6		MIT
5311.922	5313.400	18820.34	V			0.7		ME
5303.272	5304.748	18851.04	V		0.0	1.3H		ME
5302.167	5303.642	18854.97	V	I	0.9	0.7		MIT
5294.04	5295.51	18883.9	V	I	1.3J	1.3J		ME
5290.716	5292.188	18895.78	V		1.0	1.0		MIT
5290.289	5291.761	18897.30	V	I	0.5	0.5		ME
5287.637	5289.108	18906.78	V		0.9	0.9		MIT
5282.535	5284.005	18925.04	V	I	0.5	1.1H		ME
5281.910	5283.380	18927.28	V	I	1.0	0.9		ME
5280.435	5281.904	18932.56	V	I	1.1	1.1		ME
5275.680	5277.148	18949.63	V	I	0.9H	0.9H		MIT
5270.360	5271.827	18968.76	V	I	0.3H			MIT
5266.127	5267.593	18984.00	V	I	1.3H	1.3H		MIT
5264.328	5265.793	18990.49	V	I	0.8	0.0		MIT
5263.995	5265.460	18991.69	V			0.7H		ME
5260.978	5262.442	19002.58	V	I	1.5	1.5		MIT
5260.352	5261.816	19004.84	V		1.3	1.3		MIT
5241.203	5242.662	19074.28	V			0.5H		ME
5240.872	5242.331	19075.48	V	I	1.7	1.7		MIT

AIR WAVELENGTH (ANGSTRMS)	VACUUM WAVELENGTH (ANGSTRMS)	WAVENUMBER (KAYSERS)	LMENT	LOG ARC	INTENSITY SPK	DIS	REF
5240.198	5241.657	19077.94	V I	0.9	0.9		ME
5234.070	5235.527	19100.27	V I	1.6	1.6		MIT
5233.752	5235.209	19101.43	V I	1.1	1.1		ME
5227.695	5229.150	19123.57	V		1.0		ME
5225.767	5227.222	19130.62	V I	1.6	1.6		ME
5216.588	5218.040	19164.28	V I	1.6	1.6		MIT
5215.926	5217.378	19166.72	V		0.9		ME
5213.655	5215.107	19175.06	V I	1.5	1.5		MIT
5212.240	5213.691	19180.27	V	0.9H	0.9H		MIT
5207.677	5209.127	19197.07	V I	0.9	0.9		MIT
5206.608	5208.058	19201.02	V I	1.4	1.4		MIT
5200.337	5201.785	19224.17	V	0.9	0.9		MIT
5195.363	5196.810	19242.57	V I	1.6	1.4		MIT
5194.832	5196.279	19244.54	V I	1.5	1.5		MIT
5193.617	5195.063	19249.04	V I	1.5	1.1		MIT
5192.993	5194.439	19251.36	V I	2.0	1.9H		MIT
5192.008	5193.454	19255.01	V I	1.3	1.2		MIT
5180.760	5182.203	19296.81	V I	0.9	0.9		MIT
5176.766	5178.208	19311.70	V	1.8	1.7		MIT
5176.483	5177.925	19312.76	V I	0.9	0.9		MIT
5174.529	5175.970	19320.05	V I	1.0H	1.0H		MIT
5172.093	5173.534	19329.15	V I	1.3H	1.3H		MIT
5169.939	5171.379	19337.20	V I	1.3	1.3		MIT
5164.885	5166.324	19356.12	V I	1.2H	1.2H		MIT
5159.347	5160.784	19376.90	V I	1.6	1.6		MIT
5157.031	5158.468	19385.60	V I	1.2	1.2		MIT
5148.721	5150.155	19416.89	V I	1.8	1.8		MIT
5148.411	5149.845	19418.06	V I	0.9H	0.5H		MIT
5139.526	5140.958	19451.63	V I	1.7	1.7		MIT
5138.421	5139.853	19455.81	V I	1.7	1.3H		MIT
5130.522	5131.951	19485.76	V		0.3		MIT
5128.530	5129.959	19493.33	V I	1.9	1.9H		ME
5106.235	5107.658	19578.45	V		0.3H		ME
5105.144	5106.567	19582.63	V	1.6	1.6		MIT
5075.663	5077.078	19696.37	V I	0.7	0.7		MIT
5064.117	5065.529	19741.27	V	1.7	1.7		MIT
5060.677	5062.088	19754.69	V	0.0	1.2		ME
5051.629	5053.038	19790.08	V I	1.3	1.3		MIT
5048.96	5050.37	19800.5	V		1.0		ME
5019.86	5021.26	19915.3	V		1.5		ME
5014.625	5016.024	19936.11	V I	2.1	2.1		MIT
5005.599	5006.995	19972.06	V I	1.1	1.1		MIT
5002.334	5003.729	19985.09	V I	2.0	2.0		MIT
4973.17	4974.56	20102.3	V		0.3		ME
4966.116	4967.502	20130.84	V I	0.8	0.7		MIT
4965.40	4966.79	20133.7	V		1.6		ME
4963.74	4965.13	20140.5	V		0.3		ME
4957.641	4959.025	20165.26	V I	0.5	0.3		MIT
4947.582	4948.963	20206.25	V	0.0	1.2		MIT
4942.807	4944.187	20225.77	V I	0.6	0.5		MIT
4932.030	4933.407	20269.97	V I	1.2	1.1		MIT
4925.652	4927.027	20296.22	V I	1.4	1.3		MIT
4922.361	4923.735	20309.78	V I	0.7	0.6		MIT
4918.985	4920.358	20323.72	V I	0.5	0.3		MIT
4916.258	4917.631	20335.00	V I	0.7	0.6		MIT
4913.087	4914.459	20348.12	V	0.3	0.0		MIT
4904.869	4906.239	20382.21	V	0.6	0.5		MIT
4904.442	4905.811	20383.99	V I	0.9	0.8		MIT
4904.339	4905.708	20384.42	V I	0.5	1.0		MIT
4904.289	4905.658	20384.62	V I	0.9	0.8		MIT
4900.624	4901.992	20399.87	V I	1.3	1.2		MIT
4894.210	4895.577	20426.60	V I	1.0	0.9		MIT
4891.600	4892.966	20437.50	V	1.1	1.0		MIT
4886.804	4888.169	20457.56	V I	1.0	0.9		MIT
4885.647	4887.011	20462.40	V I	0.8	0.8		MIT
4884.060	4885.424	20469.05	V		0.8H		MIT
4883.420	4884.784	20471.74	V	0.0	1.3H		MIT
4882.165	4883.528	20477.00	V I	0.5	0.3		MIT
4881.557	4882.920	20479.55	V I	1.6H	1.5H		MIT
4880.565	4881.928	20483.71	V I	1.3	1.2		MIT

AIR WAVELENGTH (ANGSTRMS)	VACUUM WAVELENGTH (ANGSTRMS)	WAVENUMBER (KAYSERS)	LMENT	LOG ARC	INTENSITY SPK	DIS	REF
4875.478	4876.840	20505.08	V I	1.6H	1.3H		MIT
4871.257	4872.618	20522.85	V I	1.3	1.2		MIT
4867.986	4869.346	20536.64	V	0.5	0.3		MIT
4864.737	4866.096	20550.36	V I	1.5J	1.4J		MIT
4862.609	4863.967	20559.35	V I	1.2	1.1		MIT
4859.119	4860.476	20574.11	V	0.8	0.8		MIT
4851.485	4852.840	20606.49	V I	1.6	1.5		MIT
4848.812	4850.167	20617.85	V I	0.8H	0.7H		MIT
4848.605	4849.960	20618.73	V	0.7			MIT
4842.998	4844.351	20642.60	V I	0.6	0.5		MIT
4839.07	4840.42	20659.4	V		0.5H		ME
4833.020	4834.370	20685.22	V I	1.0	0.9		MIT
4832.427	4833.777	20687.75	V I	1.4	1.3		MIT
4831.645	4832.995	20691.10	V I	1.5	1.4		MIT
4827.450	4828.799	20709.08	V I	1.3	1.2		MIT
4823.419	4824.767	20726.39	V		0.3H		MIT
4819.034	4820.381	20745.25	V	0.5	0.3		MIT
4813.936	4815.281	20767.22	V		1.4		MIT
4811.14	4812.48	20779.3	V		0.6		ME
4807.530	4808.874	20794.89	V I	1.6H	1.5H		MIT
4805.97	4807.31	20801.6	V		0.3H		ME
4799.774	4801.116	20828.49	V I	1.2	1.1		MIT
4797.962	4799.303	20836.36	V I	0.6	0.5		MIT
4796.918	4798.259	20840.89	V I	1.5	1.4H		MIT
4795.102	4796.442	20848.79	V	0.7	0.6		MIT
4792.957	4794.297	20858.12	V I	0.6	0.5		MIT
4786.507	4787.845	20886.22	V I	1.5	1.4H		MIT
4784.469	4785.807	20895.12	V I	1.1	1.0		MIT
4776.519	4777.854	20929.90	V I	0.6H	0.5H		MIT
4776.363	4777.698	20930.58	V I	1.0	0.9		MIT
4772.582	4773.916	20947.16	V I	0.5	0.5		MIT
4766.634	4767.967	20973.30	V I	1.2	1.1		MIT
4758.746	4760.077	21008.06	V I	0.5	0.3		MIT
4757.485	4758.815	21013.63	V I	1.1	1.0		MIT
4757.353	4758.683	21014.22	V I	0.6	0.5		MIT
4753.934	4755.263	21029.33	V I	1.5	1.4		MIT
4751.565	4752.894	21039.81	V I	1.0	1.0		MIT
4750.984	4752.313	21042.39	V I	1.2	1.1		MIT
4748.520	4749.848	21053.30	V I	1.3	1.2		MIT
4746.628	4747.956	21061.70	V I	1.2	1.1		MIT
4742.630	4743.956	21079.45	V I	1.3	1.2		MIT
4731.247	4732.570	21130.17	V I	0.5	0.5		MIT
4730.385	4731.708	21134.02	V I	1.0	0.9		MIT
4729.532	4730.855	21137.83	V I	1.2	1.1		MIT
4728.652	4729.975	21141.76	V I	0.5H	0.3H		MIT
4722.865	4724.186	21167.67	V I	1.3	1.2		MIT
4721.510	4722.831	21173.74	V I	1.2	1.1		MIT
4721.246	4722.567	21174.93	V I	0.5	0.3		MIT
4717.690	4719.010	21190.89	V I	1.3	1.2		MIT
4715.893	4717.212	21198.96	V I	1.0	1.0		MIT
4715.432	4716.751	21201.03	V	0.6	0.5		MIT
4714.118	4715.437	21206.94	V I	1.4	1.3		MIT
4713.448	4714.767	21209.96	V I	0.7H	0.6H		MIT
4710.558	4711.876	21222.97	V I	1.4	1.3		MIT
4709.718	4711.036	21226.75	V I	1.0	1.0		MIT
4707.441	4708.758	21237.02	V I	0.8	0.7		MIT
4706.572	4707.889	21240.94	V I	1.3	1.2		MIT
4706.164	4707.481	21242.78	V I	1.2	1.1		MIT
4705.087	4706.404	21247.65	V I	1.2	1.1		MIT
4699.315	4700.630	21273.74	V I	1.0	0.9		MIT
4686.920	4688.232	21330.00	V I	1.2	1.1		MIT
4684.447	4685.758	21341.26	V I	0.9	0.8		MIT
4682.755	4684.066	21348.98	V I	0.5H	0.3H		MIT
4680.887	4682.197	21357.50	V I	0.6	0.5		MIT
4679.774	4681.084	21362.57	V	0.6	0.5		MIT
4670.489	4671.796	21405.04	V I	1.8G	1.6R		MIT
4669.308	4670.615	21410.46	V I	1.0	0.9		MIT
4666.136	4667.442	21425.01	V I	1.0	1.0		MIT
4648.890	4650.192	21504.49	V I	1.0	0.9		MIT
4646.395	4647.696	21516.04	V I	1.6	1.5		MIT

AIR WAVELENGTH (ANGSTRMS)	VACUUM WAVELENGTH (ANGSTRMS)	WAVENUMBER (KAYSERS)	LMENT	LOG ARC	INTENSITY SPK	DIS	REF
4644.445	4645.745	21525.07	V I	0.8	0.7		MIT
4640.735	4642.035	21542.28	V I	1.4	1.3		MIT
4640.066	4641.365	21545.38	V I	1.4	1.2		MIT
4636.158	4637.456	21563.55	V I	0.6	0.5		MIT
4635.177	4636.475	21568.11	V I	1.5	1.4		MIT
4634.210	4635.508	21572.61	V		0.5H		MIT
4626.485	4627.781	21608.63	V I	1.4	1.3		MIT
4624.406	4625.701	21618.34	V I	1.3	1.2		MIT
4619.771	4621.065	21640.03	V I	0.9	0.7		ME
4619.646	4620.940	21640.62	V	0.8	0.7		MIT
4618.782	4620.076	21644.67	V I	0.6	0.5		MIT
4618.485	4619.779	21646.06	V		0.5		MIT
4613.913	4615.205	21667.51	V I	0.3	0.3H		ME
4611.739	4613.031	21677.72	V I	0.8	0.7		MIT
4610.915	4612.207	21681.60	V I	0.7	0.6		MIT
4609.649	4610.940	21687.55	V I	1.0	1.0		MIT
4607.226	4608.517	21698.96	V	0.6	0.5		ME
4606.151	4607.441	21704.02	V I	1.5	1.4		MIT
4605.350	4606.640	21707.79	V		0.9		ME
4602.946	4604.236	21719.13	V I	0.8	0.8		MIT
4600.155	4601.444	21732.31	V	0.0	1.8H		MIT
4596.36	4597.65	21750.2	V		0.5H		MIT
4594.108	4595.395	21760.91	V I	1.5J	1.4J		MIT
4591.225	4592.511	21774.58	V I	1.5	1.4		MIT
4590.486	4591.772	21778.08	V		0.9		MIT
4586.357	4587.642	21797.69	V I	1.6H	1.5H		MIT
4583.783	4585.067	21809.93	V I	1.2	1.1		ME
4581.229	4582.513	21822.09	V I	0.7	0.6		MIT
4580.401	4581.685	21826.03	V I	1.5H	1.4H		MIT
4579.192	4580.475	21831.80	V I	1.1	1.0		MIT
4578.730	4580.013	21834.00	V I	1.4	1.3		MIT
4577.174	4578.457	21841.42	V I	1.6	1.5		MIT
4571.783	4573.064	21867.18	V I	1.5	1.4		MIT
4570.424	4571.705	21873.68	V I	1.3	1.2		MIT
4564.592	4565.871	21901.62	V		2.2		ME
4564.558	4565.837	21901.79	V	0.6	1.6		MIT
4560.714	4561.992	21920.25	V I	1.5	1.4		MIT
4556.738	4558.015	21939.37	V	0.0	0.7		MIT
4553.051	4554.327	21957.14	V I	1.3	1.2		MIT
4551.843	4553.119	21962.97	V I	1.0	0.9		MIT
4549.647	4550.922	21973.57	V	1.5	1.3		MIT
4545.393	4546.667	21994.13	V I	1.6	1.5		MIT
4541.398	4542.671	22013.48	V	0.5H	0.5H		MIT
4540.008	4541.281	22020.22	V I	1.2	1.1		MIT
4537.664	4538.936	22031.59	V I	1.4	1.3		MIT
4533.925	4535.196	22049.76	V	1.2	1.1		MIT
4532.190	4533.461	22058.20	V		1.3		ME
4530.789	4532.059	22065.02	V	1.1	1.0		MIT
4529.586	4530.856	22070.88	V I	1.2	0.9		MIT
4529.295	4530.565	22072.30	V I	1.0	0.5		MIT
4528.468	4529.738	22076.33	V	0.3	1.0J		MIT
4527.988	4529.258	22078.67	V	1.0	0.6W		MIT
4525.157	4526.426	22092.49	V I	1.2	1.1		MIT
4524.216	4525.485	22097.08	V I	1.6	1.5		MIT
4520.170	4521.438	22116.86	V I	1.4	1.3		MIT
4517.566	4518.833	22129.61	V	0.7	0.6		MIT
4515.553	4516.819	22139.47	V I	0.8	0.7		MIT
4514.193	4515.459	22146.14	V I	1.4	1.3		MIT
4512.73	4514.00	22153.3	V	0.0	1.6H		MIT
4511.432	4512.697	22159.70	V I	0.3	0.3		MIT
4509.286	4510.551	22170.24	V I	0.6	0.3		MIT
4506.578	4507.842	22183.56	V I	0.8	0.6		MIT
4506.082	4507.346	22186.00	V I	0.5	0.3		MIT
4501.954	4503.217	22206.35	V I	1.3	1.2		MIT
4501.444	4502.707	22208.86	V I	0.7			ME
4497.703	4498.965	22227.34	V I	0.3	0.3		MIT
4497.397	4498.659	22228.85	V I	1.2	1.0		MIT
4496.846	4498.107	22231.57	V I	1.6	1.5		MIT
4496.065	4497.326	22235.43	V I	1.5	1.4		MIT
4491.164	4492.424	22259.70	V I	0.8	0.7		MIT

AIR WAVELENGTH (ANGSTRMS)	VACUUM WAVELENGTH (ANGSTRMS)	WAVENUMBER (KAYSERS)	LMENT	LOG ARC	INTENSITY SPK	DIS	REF
4490.803	4492.063	22261.49	V I	1.4	1.3		MIT
4488.889	4490.148	22270.98	V I	1.8H	1.5H		MIT
4480.038	4481.295	22314.98	V I	1.4	1.3		MIT
4475.886	4477.142	22335.68	V I	0.8	0.3		MIT
4475.658	4476.914	22336.82	V		1.0H		MIT
4474.714	4475.970	22341.53	V I	1.4	1.3		MIT
4474.045	4475.300	22344.87	V I	1.5	1.3		MIT
4471.773	4473.028	22356.22	V I	0.8	0.5		MIT
4471.350	4472.605	22358.34	V I	1.3	0.9		MIT
4469.710	4470.964	22366.54	V I	1.3H	1.1H		MIT
4468.759	4470.013	22371.30	V I	1.1	1.0		MIT
4468.010	4469.264	22375.05	V I	1.3	1.2		MIT
4466.852	4468.106	22380.85	V I	0.9	0.8		MIT
4465.499	4466.752	22387.63	V I	1.2	0.8		MIT
4464.747	4466.000	22391.40	V I	1.0	0.6		MIT
4464.270	4465.523	22393.79	V I	0.5	1.0		MIT
4462.363	4463.615	22403.36	V I	1.3	1.0		MIT
4460.988	4462.240	22410.27	V I	1.0	0.7		MIT
4460.292	4461.544	22413.77	V I	1.3J	1.0J		MIT
4459.760	4461.012	22416.44	V I	1.3H	1.1H		MIT
4457.759	4459.010	22426.50	V I	0.8	0.8		MIT
4457.479	4458.730	22427.91	V I	1.2	1.0		MIT
4456.501	4457.752	22432.83	V	1.0	0.8		MIT
4453.35	4454.60	22448.7	V		1.3B		ME
4453.121	4454.371	22449.86	V I	0.8	0.5		MIT
4452.701	4453.951	22451.98	V I	1.0	0.8		MIT
4452.008	4453.258	22455.47	V I	1.3	1.2		MIT
4450.901	4452.150	22461.06	V I	1.2	0.8		MIT
4449.573	4450.822	22467.76	V I	1.1	1.0		MIT
4444.207	4445.455	22494.89	V I	1.5H	1.3H		MIT
4443.342	4444.589	22499.27	V I	1.1	1.0		MIT
4441.683	4442.930	22507.67	V I	1.6H	1.5		MIT
4437.837	4439.083	22527.18	V I	1.3H	1.1H		MIT
4436.138	4437.383	22535.80	V I	1.4H	1.2H		MIT
4434.600	4435.845	22543.62	V I	1.3	1.1		MIT
4430.504	4431.748	22564.46	V I	1.0	0.7		MIT
4429.796	4431.040	22568.07	V I	1.5	1.4		MIT
4428.515	4429.758	22574.59	V I	1.4	1.3		MIT
4427.312	4428.555	22580.73	V I	1.3	1.2		MIT
4426.005	4427.248	22587.40	V I	1.4H	1.2H		MIT
4425.712	4426.955	22588.89	V I	1.0	0.8		MIT
4424.563	4425.805	22594.76	V I	1.3	1.2		MIT
4423.914	4425.156	22598.07	V I	0.8	0.8		MIT
4423.212	4424.454	22601.66	V I	1.6	1.4W		MIT
4422.477	4423.719	22605.42	V I	0.6	0.5		MIT
4422.236	4423.478	22606.65	V I	0.8	0.8		MIT
4421.573	4422.815	22610.04	V I	1.5H	1.3H		MIT
4419.935	4421.176	22618.42	V I	1.5	1.3		MIT
4416.687	4417.927	22635.05	V I	0.5	0.3		MIT
4416.474	4417.714	22636.14	V I	1.2W	0.8		MIT
4415.060	4416.300	22643.39	V I	0.8	0.6		MIT
4414.541	4415.781	22646.05	V I	1.0	0.7		MIT
4413.678	4414.918	22650.48	V I	1.2	1.0		MIT
4412.139	4413.378	22658.38	V I	1.4	1.3		MIT
4408.511	4409.749	22677.03	V I	1.5H	1.3G		MIT
4408.204	4409.442	22678.61	V I	1.5			MIT
4407.637	4408.875	22681.52	V I	1.2H	1.0G		MIT
4406.641	4407.879	22686.65	V I	1.6	1.5		MIT
4406.147	4407.385	22689.19	V I	1.3O	1.2O		MIT
4405.011	4406.248	22695.04	V I	1.1	0.8		MIT
4403.668	4404.905	22701.97	V I	1.3	1.2		MIT
4400.575	4401.811	22717.92	V I	1.8	1.6		MIT
4399.417	4400.653	22723.90	V I	0.8	0.6		MIT
4397.292	4398.527	22734.88	V	0.9	0.8		MIT
4395.228	4396.463	22745.56	V I	1.8G	1.6G		MIT
4394.807	4396.042	22747.74	V I	0.7	0.7		MIT
4393.835	4395.069	22752.77	V I	1.2	1.0		MIT
4393.088	4394.322	22756.64	V I	1.1	1.0		MIT
4392.074	4393.308	22761.89	V I	1.4	1.2		MIT
4391.675	4392.909	22763.96	V I	0.9	0.8		MIT

AIR WAVELENGTH (ANGSTRMS)	VACUUM WAVELENGTH (ANGSTRMS)	WAVENUMBER (KAYSERS)	LMENT	LOG ARC	INTENSITY SPK	DIS	REF
4390.608	4391.841	22769.49	V I	1.0	0.5		MIT
4389.974	4391.207	22772.78	V I	1.9G	1.8G		MIT
4387.213	4388.446	22787.11	V I	1.2	1.1		MIT
4384.722	4385.954	22800.06	V I	2.1G	2.1G		MIT
4380.554	4381.785	22821.75	V I	1.2	1.0		MIT
4379.238	4380.468	22828.61	V I	2.3G	2.3G		MIT
4375.304	4376.533	22849.13	V I	1.3	1.1		MIT
4373.827	4375.056	22856.85	V I	1.3	1.1		MIT
4373.230	4374.459	22859.97	V I	1.4	1.3		MIT
4369.062	4370.290	22881.78	V I	1.0	0.7		MIT
4368.598	4369.826	22884.21	V I	1.2	1.0		MIT
4368.042	4369.270	22887.12	V I	1.1	0.9		MIT
4366.90	4368.13	22893.1	V II		0.7H		ME
4365.745	4366.972	22899.16	V I	1.0	0.8		MIT
4364.216	4365.443	22907.19	V I	1.2	1.0		MIT
4363.525	4364.751	22910.81	V I	1.3	1.1		MIT
4361.402	4362.628	22921.97	V I	1.1	1.0		MIT
4360.582	4361.808	22926.27	V I	1.2	1.1		MIT
4357.452	4358.677	22942.74	V I	0.8	0.8		MIT
4356.793	4358.018	22946.21	V I	0.3	0.6		MIT
4355.943	4357.167	22950.69	V I	1.4	1.3		MIT
4354.979	4356.203	22955.77	V I	1.3	1.2		MIT
4353.331	4354.555	22964.46	V I	0.8	0.7		MIT
4352.872	4354.096	22966.88	V I	1.0	0.8		MIT
4352.438	4353.661	22969.17	V	0.7	0.8		MIT
4350.822	4352.045	22977.70	V I	0.9	0.6		MIT
4349.968	4351.191	22982.22	V II	0.0	0.5		MIT
4342.832	4344.053	23019.98	V I	1.7	1.5		MIT
4342.204	4343.425	23023.31	V I	1.2	1.0		MIT
4341.013	4342.233	23029.62	V I	1.8	1.5		MIT
4334.092	4335.311	23066.40	V I	1.3	1.1		MIT
4332.823	4334.041	23073.15	V I	1.8	1.6		MIT
4332.379	4333.597	23075.52	V I	0.9	0.8		MIT
4331.550	4332.768	23079.94	V II	0.0	0.5		MIT
4330.024	4331.242	23088.07	V I	1.6	1.5		MIT
4325.216	4326.432	23113.73	V	0.0	0.7		ME
4322.353	4323.568	23129.04	V I	0.8	0.7		MIT
4322.020	4323.235	23130.82	V II	0.5	0.5		MIT
4320.273	4321.488	23140.18	V I	0.8	0.6		MIT
4318.70	4319.91	23148.6	V	0.7	0.3		ME
4314.832	4316.046	23169.36	V	0.9	0.5		MIT
4313.892	4315.105	23174.41	V I	1.0	0.9		MIT
4309.795	4311.007	23196.44	V I	1.5	1.3		MIT
4307.184	4308.396	23210.50	V I	1.5	1.3		MIT
4306.214	4307.425	23215.73	V I	1.5	1.3		MIT
4305.478	4306.689	23219.69	V I	1.4	1.3		MIT
4303.527	4304.738	23230.22	V	1.2	0.8		MIT
4302.149	4303.359	23237.66	V I	0.9	0.8		MIT
4301.170	4302.380	23242.95	V II		1.0		MIT
4298.029	4299.238	23259.94	V I	1.4	1.3		MIT
4297.681	4298.890	23261.82	V I	1.3	1.2		MIT
4296.107	4297.316	23270.34	V I	1.5	1.4		MIT
4291.816	4293.023	23293.61	V I	1.6	1.5R		MIT
4291.299	4292.506	23296.41	V I	1.2	1.0		MIT
4288.809	4290.016	23309.94	V II	0.5	0.3		MIT
4287.809	4289.015	23315.37	V I	1.2	1.0		MIT
4286.420	4287.626	23322.93	V I	1.3	1.2		MIT
4286.124	4287.330	23324.54	V II	0.5	0.3		MIT
4284.055	4285.260	23335.80	V I	1.3	1.3		MIT
4282.911	4284.116	23342.04	V I	1.3	1.2		MIT
4278.92	4280.12	23363.8	V II		1.2		ME
4278.883	4280.087	23364.01	V	0.3	0.9H		MIT
4276.958	4278.162	23374.53	V I	1.4	1.3		MIT
4271.554	4272.756	23404.10	V I	1.3	1.0		MIT
4270.318	4271.520	23410.87	V I	1.3	0.6		MIT
4269.762	4270.964	23413.92	V I	1.2	1.0		MIT
4268.643	4269.844	23420.06	V I	1.6	1.3		MIT
4265.170	4266.370	23439.13	V	1.5	1.2		MIT
4263.830	4265.030	23446.49	V II		0.5		MIT
4262.161	4263.361	23455.67	V I	1.3	1.1		MIT

AIR WAVELENGTH (ANGSTRMS)	VACUUM WAVELENGTH (ANGSTRMS)	WAVENUMBER (KAYSERS)	LMENT	LOG ARC	INTENSITY SPK	DIS	REF
4261.208	4262.407	23460.92	V I	1.0	0.5		MIT
4260.755	4261.954	23463.41	V II				MIT
4259.312	4260.511	23471.36	V I	1.2	1.0		MIT
4257.369	4258.567	23482.07	V I	1.2	1.0		MIT
4257.020	4258.218	23484.00	V		0.7		ME
4254.425	4255.623	23498.32	V II	0.5	0.7		MIT
4248.822	4250.018	23529.31	V II	0.6	0.6		MIT
4242.894	4244.089	23562.19	V		1.2H		ME
4241.318	4242.512	23570.94	V I	1.2	0.9		MIT
4240.358	4241.552	23576.28	V I	1.2	1.0		MIT
4240.084	4241.278	23577.80	V I	0.5			MIT
4238.960	4240.154	23584.05	V	0.8	0.3		MIT
4236.815	4238.008	23595.99	V II		0.5		MIT
4235.756	4236.949	23601.89	V I	1.0	0.5		MIT
4234.524	4235.716	23608.76	V I	1.2	0.8		MIT
4234.250	4235.442	23610.28	V II	0.0	0.9		MIT
4234.000	4235.192	23611.68	V I	1.2	0.8		MIT
4232.952	4234.144	23617.52	V I	1.0	0.5		MIT
4232.460	4233.652	23620.27	V I	1.0	0.5		MIT
4232.055	4233.247	23622.53	V II	0.0	0.8		MIT
4231.143	4232.335	23627.62	V II	0.0	0.3		MIT
4229.687	4230.878	23635.75	V I	1.1	0.8		MIT
4227.744	4228.935	23646.62	V I	1.0	0.7		MIT
4226.624	4227.814	23652.88	V I	0.9	0.5		MIT
4225.225	4226.415	23660.72	V II	0.5	1.0		MIT
4224.515	4225.705	23664.69	V II	0.3	0.5		MIT
4224.138	4225.328	23666.80	V I	1.1	1.0		MIT
4221.04	4222.23	23684.2	V I	0.3	0.3H		ME
4220.042	4221.231	23689.77	V II	0.5	1.0		MIT
4218.710	4219.898	23697.25	V I	1.2	0.9		MIT
4209.857	4211.043	23747.09	V I	1.5	1.0		MIT
4209.80	4210.99	23747.4	V II		1.1		ME
4209.739	4210.925	23747.75	V		1.1		ME
4206.68	4207.87	23765.0	V I	0.3	0.3H		ME
4205.086	4206.271	23774.03	V II	0.7	1.3		MIT
4204.200	4205.384	23779.04	V II	0.3	0.8		MIT
4202.34	4203.52	23789.6	V II	0.8	1.2		ME
4200.191	4201.374	23801.74	V I	1.0	0.7		MIT
4198.611	4199.794	23810.69	V I	1.0	0.8		MIT
4197.601	4198.784	23816.42	V I	1.1	0.8		MIT
4195.603	4196.785	23827.76	V I	1.0	0.3		MIT
4191.558	4192.739	23850.76	V I	1.2	0.7		MIT
4190.896	4192.077	23854.52	V II	0.0	0.6		MIT
4190.400	4191.581	23857.35	V II	0.3	0.8		MIT
4189.841	4191.022	23860.53	V I	1.3	1.0		MIT
4183.434	4184.613	23897.07	V II	0.5	1.3		MIT
4182.591	4183.770	23901.89	V I	1.3	0.8		MIT
4182.077	4183.256	23904.83	V I	1.0	0.7		MIT
4179.419	4180.597	23920.03	V I	1.3	1.0		MIT
4179.062	4180.240	23922.07	V		0.3		ME
4178.389	4179.567	23925.93	V II	0.5	1.0		MIT
4177.072	4178.249	23933.47	V I	1.2	0.5		MIT
4176.795	4177.972	23935.06	V I	0.7	0.3		MIT
4174.014	4175.191	23951.00	V I	1.1	0.8		MIT
4171.296	4172.472	23966.61	V I	1.2	0.8		MIT
4169.260	4170.435	23978.31	V I	1.2	0.7		MIT
4164.016	4165.190	24008.51	V II	0.0	0.8		MIT
4159.686	4160.859	24033.50	V I	1.3	1.1		MIT
4153.328	4154.499	24070.29	V I	1.1	0.5		MIT
4152.658	4153.829	24074.17	V	1.0	0.6		MIT
4150.672	4151.842	24085.69	V	1.0	0.9		MIT
4148.859	4150.029	24096.22	V I	0.9	0.5		MIT
4147.768	4148.938	24102.56	V I	0.5H			MIT
4142.882	4144.050	24130.98	V I	0.8	0.5		MIT
4141.832	4143.000	24137.10	V I	0.8	0.5		MIT
4141.370	4142.538	24139.79	V I	1.2	0.5		MIT
4139.258	4140.425	24152.11	V I	1.0	0.8		MIT
4136.386	4137.553	24168.88	V I	1.0	0.6		MIT
4136.109	4137.276	24170.50	V I	1.0	0.8		MIT
4134.488	4135.654	24179.97	V I	1.6R	1.2R		MIT

AIR WAVELENGTH (ANGSTRMS)	VACUUM WAVELENGTH (ANGSTRMS)	WAVENUMBER (KAYSERS)	LMENT	LOG ARC	INTENSITY SPK	DIS	REF
4133.777	4134.943	24184.13	V I	1.0	0.8		MIT
4132.877	4134.043	24189.40	V I	0.3			ME
4132.017	4133.183	24194.43	V I	1.1	1.0		MIT
4130.145	4131.310	24205.40	V I	0.5	0.3		MIT
4128.858	4130.023	24212.94	V I	1.0	0.7		MIT
4128.071	4129.235	24217.56	V I	1.5R	1.3R		MIT
4124.072	4125.235	24241.04	V I	0.9	0.7		MIT
4123.566	4124.729	24244.01	V I	1.5R	1.1R		MIT
4123.188	4124.351	24246.24	V I	0.8	0.5		MIT
4120.77	4121.93	24260.5	V I	0.3			ME
4120.538	4121.701	24261.83	V I	1.2	0.6		MIT
4119.457	4120.619	24268.20	V I	1.1	0.8		MIT
4119.10	4120.26	24270.3	V I	0.3	0.3		ME
4118.643	4119.805	24272.99	V I	1.0	0.5		MIT
4118.182	4119.344	24275.71	V I	1.0	0.7		MIT
4116.703	4117.864	24284.43	V I	0.5	0.3		MIT
4116.470	4117.631	24285.81	V I	1.3R	1.2R		MIT
4115.483	4116.644	24291.63	V I	0.3	0.0		MIT
4115.185	4116.346	24293.39	V I	1.5R	1.3R		MIT
4114.530	4115.691	24297.26	V I	0.8	0.5		MIT
4113.518	4114.679	24303.23	V I	1.3	1.0		MIT
4112.332	4113.492	24310.24	V I	0.3	0.5		MIT
4111.785	4112.945	24313.48	V I	2.0Z	2.0Z		MIT
4110.761	4111.921	24319.53	V I	0.3	0.3		ME
4109.786	4110.946	24325.30	V I	1.6R	1.3R		MIT
4109.043	4110.202	24329.70	V I	0.5	0.3		MIT
4108.215	4109.374	24334.60	V I	1.1	1.0		MIT
4107.487	4108.646	24338.92	V I	1.0	1.0		MIT
4105.167	4106.325	24352.67	V I	1.3R	1.1R		MIT
4104.778	4105.936	24354.98	V I	1.1	0.7		MIT
4104.396	4105.554	24357.25	V I	1.2	0.8		MIT
4102.159	4103.317	24370.53	V I	1.5	1.2		MIT
4101.003	4102.160	24377.40	V I	0.7	0.8		MIT
4099.796	4100.953	24384.58	V I	1.4	1.1		MIT
4096.939	4098.095	24401.58	V I	0.8	0.3		MIT
4095.486	4096.642	24410.24	V I	1.6	1.2		MIT
4094.283	4095.439	24417.41	V I	1.0	0.7		MIT
4093.497	4094.652	24422.10	V I	1.0	0.7		MIT
4092.694	4093.849	24426.89	V I	1.3	1.1		MIT
4092.407	4093.562	24428.60	V I	1.2	0.8		MIT
4091.945	4093.100	24431.36	V I	0.9	0.7		MIT
4090.579	4091.734	24439.52	V I	1.8	1.4		MIT
4085.658	4086.811	24468.95	V		1.0H		ME
4082.927	4084.080	24485.32	V I	1.2	0.8		MIT
4078.709	4079.861	24510.64	V I	0.3			MIT
4077.977	4079.128	24515.04	V I	0.3			MIT
4075.652	4076.803	24529.03	V		0.3		MIT
4072.141	4073.291	24550.17	V I	1.2	0.3		MIT
4071.541	4072.691	24553.79	V I	1.2	0.3		MIT
4070.771	4071.920	24558.44	V I	1.0	0.7		MIT
4067.979	4069.128	24575.29	V I	1.0	1.0		MIT
4067.745	4068.894	24576.71	V I	1.7	1.2		MIT
4067.03	4068.18	24581.0	V II		0.3		ME
4065.080	4066.228	24592.82	V	0.3	2.0		MIT
4063.931	4065.079	24599.77	V I	1.2	0.7		MIT
4062.72	4063.87	24607.1	V I	0.3	0.3		ME
4060.85	4062.00	24618.4	V I	0.3	0.3		ME
4057.825	4058.971	24636.79	V I	1.0	0.3		MIT
4057.074	4058.220	24641.35	V I	1.3	1.0		MIT
4056.258	4057.404	24646.30	V	0.0	0.5		MIT
4053.659	4054.804	24662.10	V	0.3			MIT
4053.587	4054.732	24662.54	V		1.8		ME
4053.260	4054.405	24664.53	V I	0.8	0.3		MIT
4051.352	4052.496	24676.15	V I				MIT
4050.963	4052.107	24678.52	V I	1.2	0.6		MIT
4046.264	4047.407	24707.18	V	0.0	1.2		MIT
4042.635	4043.777	24729.35	V I	1.2	0.3		MIT
4041.604	4042.746	24735.66	V	1.0	0.3O		MIT
4040.31	4041.45	24743.6	V I	0.3	0.0		ME
4039.576	4040.717	24748.08	V II	0.5	0.9		ME
4038.545	4039.686	24754.40	V		0.3		ME
4036.776	4037.917	24765.25	V II	0.9	1.6		MIT
4035.896	4037.036	24770.65	V I	0.5			ME
4035.626	4036.766	24772.30	V II	1.6	1.9		MIT
4032.856	4033.996	24789.32	V I	0.3	0.0		MIT
4031.830	4032.969	24795.63	V I	1.0	0.5		MIT
4031.219	4032.358	24799.39	V I	1.0	0.5		MIT
4023.389	4024.526	24847.65	V II	1.0	1.5		MIT
4023.174	4024.311	24848.97	V I	0.8	0.3		MIT
4021.925	4023.062	24856.69	V I	0.9	0.3		MIT
4019.046	4020.182	24874.50	V		0.8		MIT
4017.296	4018.431	24885.33	V		1.2H		ME
4016.821	4017.956	24888.27	V		1.2J		ME
4011.309	4012.443	24922.47	V I	1.0	0.3		MIT
4008.169	4009.302	24942.00	V II	0.3	1.0		ME
4005.712	4006.844	24957.29	V II	1.0	1.5		MIT
4003.544	4004.676	24970.81	V I	1.0	0.3		MIT
4002.938	4004.070	24974.59	V II	0.8	1.9		MIT
4001.180	4002.311	24985.56	V		0.3		MIT
4000.078	4001.209	24992.45	V I	0.9	0.3		MIT
3999.191	4000.322	24997.99	V		1.6H		ME
3998.730	3999.861	25000.87	V I	2.0	1.4		MIT
3997.122	3998.252	25010.93	V II	1.4	1.6		MIT
3994.886	3996.016	25024.93	V I	1.5			MIT
3992.801	3993.930	25038.00	V I	1.8	1.3		MIT
3991.962	3993.091	25043.26	V		0.3		ME
3991.468	3992.597	25046.36	V II		0.8		ME
3990.566	3991.695	25052.02	V I	2.1	1.6		MIT
3989.802	3990.930	25056.81	V II		1.0		ME
3988.833	3989.961	25062.90	V I	1.8	1.5		MIT
3985.790	3986.917	25082.03	V	0.0	1.6		MIT
3984.600	3985.727	25089.53	V I	1.6	1.2		MIT
3984.335	3985.462	25091.20	V I	1.5	1.2		MIT
3980.522	3981.648	25115.23	V	1.6	1.5		MIT
3979.424	3980.550	25122.16	V I	1.6	1.0		MIT
3979.142	3980.268	25123.94	V I	1.7	0.9		MIT
3977.741	3978.866	25132.79	V II		1.0		MIT
3975.352	3976.477	25147.89	V I	1.5	0.8		MIT
3973.639	3974.763	25158.73	V II	1.4	1.6		MIT
3973.358	3974.482	25160.51	V I	1.5	0.5		MIT
3971.951	3973.075	25169.42	V I	0.7	0.3		MIT
3968.094	3969.217	25193.89	V II	1.4	1.6		MIT
3964.50	3965.62	25216.7	V I	0.3			ME
3963.626	3964.747	25222.29	V	1.3	1.0		MIT
3957.90	3959.02	25258.8	V		0.3		ME
3951.968	3953.086	25296.69	V II	1.5	1.7		MIT
3950.231	3951.349	25307.81	V	1.3	1.0		MIT
3945.17	3946.29	25340.3	V I	0.3			ME
3943.664	3944.780	25349.95	V I	1.7	1.3		MIT
3942.006	3943.122	25360.62	V I	1.2	0.7		MIT
3941.254	3942.370	25365.46	V I	1.5	1.1		MIT
3940.593	3941.708	25369.71	V I	1.2	0.6		MIT
3939.334	3940.449	25377.82	V I	1.6	1.2		MIT
3938.887	3940.002	25380.70	V I	1.2	0.0		MIT
3938.198	3939.313	25385.14	V I	1.5	1.1		MIT
3937.529	3938.644	25389.45	V I	1.7	1.3		MIT
3936.282	3937.396	25397.49	V I	1.5	0.9		MIT
3935.141	3936.255	25404.86	V I	1.6	1.4		MIT
3934.013	3935.127	25412.14	V I	2.0	1.5		MIT
3931.340	3932.453	25429.42	V I	1.4	1.3		MIT
3930.023	3931.136	25437.94	V I	1.7	1.3		MIT
3929.730	3930.843	25439.84	V II	0.5	1.5		MIT
3927.926	3929.038	25451.52	V I	1.7	1.6		MIT
3926.494	3927.606	25460.80	V	0.3	1.2		MIT
3926.333	3927.445	25461.85	V	0.0	1.2		MIT
3925.240	3926.351	25468.94	V I	1.6	1.4		MIT
3924.658	3925.769	25472.71	V I	1.5	1.4		MIT
3922.431	3923.542	25487.18	V I	1.9	1.6		MIT
3921.905	3923.016	25490.59	V I	1.5	1.3		MIT
3920.487	3921.597	25499.81	V I	1.5	1.2		MIT

AIR WAVELENGTH (ANGSTRMS)	VACUUM WAVELENGTH (ANGSTRMS)	WAVENUMBER (KAYSERS)	LMENT		LOG ARC	INTENSITY SPK	DIS	REF
3919.989	3921.099	25503.05	V	I	1.4	0.8		MIT
3917.14	3918.25	25521.6	V		0.3			ME
3916.413	3917.522	25526.34	V	II	1.2	1.6		MIT
3915.36	3916.47	25533.2	V		0.3			ME
3915.125	3916.234	25534.74	V	I	1.2	0.5		MIT
3914.330	3915.439	25539.92	V	II	1.4	1.8J		MIT
3913.56	3914.67	25544.9	V		0.3	0.0		ME
3912.886	3913.994	25549.35	V	I	1.6	1.2		MIT
3912.207	3913.315	25553.78	V	I	1.7	1.3		MIT
3910.790	3911.898	25563.04	V	I	1.5	1.3		MIT
3909.894	3911.001	25568.90	V	I	1.7W	1.5W		MIT
3909.672	3910.779	25570.35	V	I	1.2	0.9		MIT
3908.317	3909.424	25579.22	V		1.7	0.3H		MIT
3907.519	3908.626	25584.44	V			0.3		ME
3907.17	3908.28	25586.7	V		0.3H	0.3		M
3906.748	3907.855	25589.49	V	I	1.7	1.3		MIT
3904.472	3905.578	25604.41	V	I	1.3	0.8		MIT
3904.402	3905.508	25604.86	V	I	0.3			ME
3904.218	3905.324	25606.07	V	I	1.0	0.3		MIT
3903.262	3904.368	25612.34	V	II	0.5	0.5		MIT
3902.558	3903.664	25616.96	V	I	0.8	0.3		MIT
3902.250	3903.355	25618.98	V	I	1.3	0.7		MIT
3901.689	3902.794	25622.67	V	I	0.7			MIT
3901.152	3902.257	25626.19	V	I	1.7	0.5H		MIT
3900.175	3901.280	25632.61	V	I	1.7	0.3H		MIT
3899.134	3900.239	25639.46	V	I	1.1	0.7H		MIT
3898.278	3899.382	25645.09	V		1.0			MIT
3898.139	3899.243	25646.00	V	I	1.6			MIT
3898.019	3899.123	25646.79	V	I	1.0	0.3		MIT
3897.075	3898.179	25653.00	V	I	1.6	0.3		MIT
3896.794	3897.898	25654.85	V	I	0.3			MIT
3896.622	3897.726	25655.99	V		0.8	0.0		MIT
3896.156	3897.260	25659.05	V	II	1.7	1.6		MIT
3896.156	3897.260	25659.05	V	I	1.7	1.6		MIT
3894.042	3895.145	25672.98	V	I	1.3	0.5		MIT
3892.859	3893.962	25680.78	V	I	1.8	1.5		MIT
3892.475	3893.578	25683.32	V		0.7			MIT
3891.220	3892.323	25691.60	V	I	0.9	0.3		MIT
3891.119	3892.222	25692.27	V		0.7			MIT
3890.184	3891.286	25698.44	V	I	2.0	1.5		MIT
3889.235	3890.337	25704.71	V	I	1.1	0.3		MIT
3888.329	3889.431	25710.70	V	I	1.0	0.0		MIT
3888.080	3889.182	25712.35	V	I	1.2	0.6		MIT
3886.587	3887.688	25722.23	V	I	1.4	1.2		MIT
3886.200	3887.301	25724.79	V	I	0.6	0.3		ME
3885.770	3886.871	25727.63	V	I	1.3	0.0		MIT
3884.843	3885.944	25733.77	V	II	0.6	1.8		MIT
3884.465	3885.566	25736.28	V	I	1.5	0.0		MIT
3883.887	3884.988	25740.11	V	I	1.6	0.7		MIT
3883.203	3884.304	25744.64	V		0.0	0.3		MIT
3881.054	3882.154	25758.90	V		0.3	0.7		MIT
3880.260	3881.360	25764.17	V	I	1.2	0.6		MIT
3879.661	3880.761	25768.15	V	I	1.7	0.7		MIT
3879.232	3880.331	25770.99	V	I	1.5	0.5		MIT
3878.712	3879.811	25774.45	V	II	1.5	2.0		MIT
3876.739	3877.838	25787.57	V		1.4	0.3		MIT
3876.086	3877.185	25791.91	V	I	1.7	1.5		MIT
3875.902	3877.001	25793.14	V	I	1.6	0.5		MIT
3875.426	3876.524	25796.30	V	I	1.0	0.3		MIT
3875.075	3876.173	25798.64	V	I	1.8R	1.7		MIT
3874.345	3875.443	25803.50	V	I	1.0	0.3		MIT
3873.635	3874.733	25808.23	V	I	1.5	1.1		MIT
3872.748	3873.846	25814.14	V	I	1.0H	0.3H		MIT
3871.078	3872.175	25825.28	V	I	1.8	1.5		MIT
3870.576	3871.673	25828.63	V	I	1.5	1.0		MIT
3867.602	3868.698	25848.49	V	I	1.8	1.5		MIT
3867.346	3868.442	25850.20	V	I	1.2	0.5		MIT
3866.736	3867.832	25854.28	V	II	0.7	1.7		MIT
3864.862	3865.958	25866.81	V	I	2.0R	1.7R		MIT
3864.300	3865.396	25870.57	V	I	1.5			ME

AIR WAVELENGTH (ANGSTRMS)	VACUUM WAVELENGTH (ANGSTRMS)	WAVENUMBER (KAYSERS)	LMENT		LOG ARC	INTENSITY SPK	DIS	REF
3863.866	3864.961	25873.48	V	I	1.4	1.2		MIT
3863.866	3864.961	25873.48	V	II	1.4	1.2		MIT
3863.404	3864.499	25876.57	V	I	0.3	0.0		MIT
3862.490	3863.585	25882.70	V		1.0			MIT
3862.223	3863.318	25884.49	V	I	1.9	1.3		MIT
3861.608	3862.703	25888.61	V	I	1.0H	0.5H		ME
3860.628	3861.723	25895.18	V		1.3			MIT
3859.341	3860.435	25903.81	V	I	1.3	0.9		MIT
3858.685	3859.779	25908.22	V	I	1.7	1.2		MIT
3856.669	3857.763	25921.76	V		1.3			MIT
3855.841	3856.934	25927.33	V	I	2.3	2.3		MIT
3855.370	3856.463	25930.50	V	I	1.7R	1.7R		MIT
3852.096	3853.188	25952.53	V	I	1.3	1.0		MIT
3851.171	3852.263	25958.77	V	I	1.7	1.2		MIT
3850.405	3851.497	25963.93	V	II	0.0	1.0		MIT
3850.164	3851.256	25965.56	V	I	1.2O	0.3		MIT
3849.756	3850.848	25968.31	V			0.5		MIT
3849.324	3850.416	25971.22	V	I	1.8	1.4		MIT
3847.331	3848.422	25984.68	V	I	2.0	1.8H		MIT
3845.974	3847.065	25993.84	V		0.7	0.3		MIT
3845.454	3846.545	25997.36	V		1.3			MIT
3844.892	3845.983	26001.16	V	I	1.4	0.8		MIT
3844.438	3845.528	26004.23	V	I	2.0	1.7H		MIT
3843.506	3844.596	26010.53	V	I	1.7	1.2		MIT
3842.995	3844.085	26013.99	V		1.3			MIT
3842.70	3843.79	26016.0	V	I	0.7	0.3		M
3841.890	3842.980	26021.48	V	I	1.5	1.1		MIT
3840.752	3841.841	26029.18	V	I	1.1R			MIT
3840.440	3841.529	26031.30	V	I	1.6	1.3		MIT
3840.140	3841.229	26033.33	V	I	1.1	0.7		MIT
3839.381	3840.470	26038.48	V	I	1.5	0.7		MIT
3839.002	3840.091	26041.05	V	I	1.8	1.0		MIT
3837.852	3838.941	26048.85	V		0.7			MIT
3837.631	3838.720	26050.35	V	I	1.0			MIT
3836.522	3837.610	26057.88	V		1.0			MIT
3836.054	3837.142	26061.06	V	I	1.0	1.2		MIT
3835.560	3836.648	26064.42	V	I	1.7	1.1		MIT
3835.185	3836.273	26066.97	V	I	1.5			MIT
3834.811	3835.899	26069.51	V	I	0.8	0.0		MIT
3834.224	3835.312	26073.50	V	I	1.3	1.1		MIT
3834.147	3835.235	26074.02	V	I	0.7	0.3		ME
3833.80	3834.89	26076.4	V	I	1.0			ME
3833.226	3834.313	26080.29	V	I	1.0	0.3		MIT
3832.835	3833.922	26082.95	V	I	1.4	0.7		MIT
3832.432	3833.519	26085.69	V		0.3			MIT
3832.312	3833.399	26086.51	V			0.3		MIT
3831.832	3832.919	26089.78	V	I	1.2	0.3		MIT
3831.032	3832.119	26095.22	V		0.0	0.7		MIT
3830.273	3831.360	26100.39	V		1.6	0.6		MIT
3829.661	3830.748	26104.56	V		0.3	0.8		MIT
3829.532	3830.619	26105.45	V			0.6		ME
3828.836	3829.922	26110.19	V	I	1.2	0.5		MIT
3828.559	3829.645	26112.08	V	I	1.3	0.7		MIT
3826.969	3828.055	26122.93	V		0.0	1.4		MIT
3826.774	3827.860	26124.26	V	I	1.1	0.5		MIT
3825.317	3826.402	26134.21	V	I	0.8	0.3		MIT
3823.990	3825.075	26143.28	V	I	1.5	1.2		MIT
3823.748	3824.833	26144.93	V	I	1.0	0.3		MIT
3823.213	3824.298	26148.59	V	I	1.5	1.3		MIT
3822.888	3823.973	26150.81	V	I	1.6	1.4		MIT
3822.70	3823.78	26152.1	V		0.3	0.3		ME
3822.009	3823.094	26156.83	V	I	1.8	1.6		MIT
3821.487	3822.571	26160.40	V	I	1.7	1.5		MIT
3820.297	3821.381	26168.55	V	I	1.4	1.3		MIT
3819.963	3821.047	26170.84	V	I	1.8	1.5		MIT
3818.244	3819.328	26182.62	V	I	1.3	0.3		MIT
3817.975	3819.059	26184.46	V	I	0.8	0.5		MIT
3817.844	3818.927	26185.36	V	I	1.2	0.8		MIT
3815.514	3816.597	26201.35	V	I	1.5	0.0		MIT
3815.392	3816.475	26202.19	V		0.0	2.2H		MIT

AIR WAVELENGTH (ANGSTRMS)	VACUUM WAVELENGTH (ANGSTRMS)	WAVENUMBER (KAYSERS)	LMENT		LOG ARC	INTENSITY SPK	DIS	REF
3813.492	3814.574	26215.25	V	I	1.7			MIT
3813.355	3814.437	26216.19	V	I	0.5	0.0		MIT
3811.32	3812.40	26230.2	V	I	0.5			ME
3809.597	3810.678	26242.05	V	I	1.8	1.6		MIT
3808.521	3809.602	26249.46	V	I	1.7	1.5		MIT
3808.11	3809.19	26252.3	V	I	0.5			ME
3807.505	3808.586	26256.47	V	I	1.9	1.7		MIT
3806.796	3807.877	26261.35	V	I	1.5	1.1		MIT
3804.916	3805.996	26274.33	V	I	1.4	1.0		MIT
3804.589	3805.669	26276.59	V	I	0.7	0.3		MIT
3803.902	3804.982	26281.33	V	I	1.0	0.5		MIT
3803.784	3804.864	26282.15	V	I	1.0	0.5		MIT
3803.474	3804.554	26284.29	V	I	1.7	1.6		MIT
3802.883	3803.963	26288.38	V	I	1.3	0.9		MIT
3801.158	3802.237	26300.31	V	I	0.5	0.0		MIT
3799.912	3800.991	26308.93	V	I	1.8	1.7		MIT
3799.269	3800.348	26313.38	V	I	1.0	0.8		MIT
3798.661	3799.740	26317.59	V	I	0.8	0.7		MIT
3796.470	3797.548	26332.78	V	I	1.5	1.1		MIT
3794.964	3796.042	26343.23	V	I	1.7H	1.7H		MIT
3794.358	3795.435	26347.44	V		0.0	1.4H		MIT
3793.614	3794.691	26352.60	V	I	1.5	1.2		MIT
3791.326	3792.403	26368.51	V	I	1.3	0.6		MIT
3790.469	3791.545	26374.47	V	I	1.1	0.8		MIT
3790.324	3791.400	26375.48	V	I	1.6	1.1		MIT
3787.540	3788.616	26394.87	V	I	1.5	0.0		MIT
3787.239	3788.315	26396.96	V		0.3	1.3		MIT
3787.145	3788.221	26397.62	V	I	1.4	0.8H		MIT
3784.674	3785.749	26414.85	V	I	1.2	0.7		MIT
3782.550	3783.624	26429.69	V	I	1.3	0.8		MIT
3782.152	3783.226	26432.47	V		1.2	0.5		MIT
3781.393	3782.467	26437.77	V	I	1.6	0.8		MIT
3779.768	3780.842	26449.14	V	I	0.3	0.0		MIT
3779.648	3780.722	26449.98	V	I	1.5	1.0		MIT
3778.684	3779.757	26456.73	V	I	1.8	0.6		MIT
3778.358	3779.431	26459.01	V	II	0.5	1.5		MIT
3777.496	3778.569	26465.04	V	I	0.9	0.3		MIT
3776.879	3777.952	26469.37	V	I	1.0	0.3		MIT
3776.157	3777.230	26474.43	V	I	1.7	0.3		MIT
3775.719	3776.792	26477.50	V	I	1.5	0.3		MIT
3775.187	3776.259	26481.23	V	I	1.4	0.7		MIT
3774.675	3775.747	26484.82	V		0.0	1.1		MIT
3774.109	3775.181	26488.79	V	I	1.3	0.5		MIT
3773.806	3774.878	26490.92	V			0.3		MIT
3772.972	3774.044	26496.78	V		0.3	1.6		MIT
3772.748	3773.820	26498.35	V	I	1.0			MIT
3772.148	3773.220	26502.56	V	I	1.3	0.0		MIT
3770.973	3772.044	26510.82	V	II	1.5	1.8		MIT
3770.530	3771.601	26513.94	V	I	1.0	0.0		MIT
3770.003	3771.074	26517.64	V	I	1.3	0.3		MIT
3769.073	3770.144	26524.19	V	I	1.5	1.0		MIT
3767.721	3768.791	26533.71	V		0.3	1.0		MIT
3767.257	3768.327	26536.97	V	I	1.3W	1.0		MIT
3766.403	3767.473	26542.99	V	I	1.2	0.7		MIT
3764.799	3765.869	26554.30	V	I	1.0	0.8		MIT
3763.141	3764.210	26566.00	V	I	1.9	0.8H		MIT
3761.442	3762.511	26578.00	V	I	1.6	0.8		MIT
3760.794	3761.863	26582.58	V	I	1.5	0.7		MIT
3760.231	3761.300	26586.56	V	II	0.8	1.6		MIT
3759.319	3760.387	26593.00	V	I	1.7	0.3		MIT
3758.801	3759.869	26596.67	V	I	0.6	0.0		MIT
3758.549	3759.617	26598.45	V	I	1.4	0.5		MIT
3758.297	3759.365	26600.24	V	I	0.6H	1.3		MIT
3756.499	3757.567	26612.97	V			0.3		ME
3756.035	3757.102	26616.26	V	I	1.5	0.7		MIT
3755.701	3756.768	26618.62	V	I	1.8	0.5		MIT
3753.274	3754.341	26635.83	V	I	1.5	0.3		MIT
3751.782	3752.848	26646.43	V	I	1.7	0.3		MIT
3751.228	3752.294	26650.36	V		0.6	2.0		MIT
3750.873	3751.939	26652.89	V	II	1.0	1.3		MIT

AIR WAVELENGTH (ANGSTRMS)	VACUUM WAVELENGTH (ANGSTRMS)	WAVENUMBER (KAYSERS)	LMENT		LOG ARC	INTENSITY SPK	DIS	REF
3747.982	3749.047	26673.44	V	I	1.7	0.6		MIT
3747.136	3748.201	26679.47	V	I	1.4	0.3		MIT
3745.803	3746.868	26688.96	V		1.5	2.8		MIT
3743.607	3744.671	26704.61	V	II	1.0	1.6		MIT
3741.504	3742.568	26719.62	V	I	1.9	0.9		MIT
3740.241	3741.304	26728.65	V	I	2.0	1.0		MIT
3738.757	3739.820	26739.25	V	I	2.0	0.8		MIT
3738.346	3739.409	26742.20	V			0.6H		ME
3737.992	3739.055	26744.73	V	I	1.7	0.7		MIT
3737.435	3738.498	26748.71	V	I	0.3	0.3		MIT
3736.024	3737.086	26758.81	V	I	0.5	1.3		MIT
3735.169	3736.231	26764.94	V		0.5	1.4		MIT
3734.428	3735.490	26770.25	V	I	1.0	0.7		MIT
3734.284	3735.346	26771.28	V	I	0.5			MIT
3733.612	3734.674	26776.10	V			1.0		MIT
3732.758	3733.819	26782.23	V		1.8G	2.7G		MIT
3731.983	3733.044	26787.79	V			1.4		MIT
3731.035	3732.096	26794.59	V	I	0.6			MIT
3730.183	3731.244	26800.72	V	I	1.2	0.5		MIT
3729.035	3730.095	26808.97	V	I	1.9	1.2		MIT
3728.341	3729.401	26813.96	V		1.3	2.2		MIT
3727.345	3728.405	26821.12	V	II	1.6	2.3		MIT
3723.324	3724.383	26850.08	V	I	1.6	1.0		MIT
3723.218	3724.277	26850.85	V		0.7	1.2		ME
3722.601	3723.660	26855.30	V	I	0.7	0.3		MIT
3722.195	3723.254	26858.23	V	I	1.6	1.6		MIT
3721.998	3723.057	26859.65	V	I	1.8	1.3		MIT
3721.358	3722.416	26864.27	V	I	1.0	0.8		MIT
3721.060	3722.118	26866.42	V			0.3		ME
3718.912	3719.970	26881.94	V		1.3	0.7		MIT
3718.160	3719.218	26887.37	V	II	0.7	1.8		MIT
3715.470	3716.527	26906.84	V		1.8	2.6G		MIT
3713.957	3715.014	26917.80	V	I	1.8	1.0		MIT
3712.539	3713.595	26928.08	V			1.3		MIT
3711.750	3712.806	26933.81	V			1.0		ME
3711.123	3712.179	26938.36	V			1.9		MIT
3709.333	3710.388	26951.36	V			1.4		MIT
3708.721	3709.776	26955.80	V	I	2.0	1.8		MIT
3706.035	3707.089	26975.34	V	I	1.7	1.7		MIT
3705.407	3706.461	26979.91	V			0.7		MIT
3705.035	3706.089	26982.62	V	I	2.0	1.8		MIT
3704.699	3705.753	26985.07	V	I	2.3G	2.2G		MIT
3703.996	3705.050	26990.19	V	I	0.7	0.3		MIT
3703.816	3704.870	26991.50	V		0.3	1.0		MIT
3703.584	3704.638	26993.19	V	I	2.3G	2.0G		MIT
3700.981	3702.034	27012.18	V		0.0	1.2		MIT
3700.344	3701.397	27016.83	V		1.0	2.0		MIT
3700.128	3701.181	27018.40	V		0.3	1.2		MIT
3699.476	3700.529	27023.17	V	I	1.5	1.0		MIT
3699.110	3700.163	27025.84	V		0.3	0.7H		MIT
3695.865	3696.917	27049.57	V	I	2.2	2.0R		MIT
3695.335	3696.387	27053.45	V	I	2.1	1.8H		MIT
3695.158	3696.210	27054.74	V			0.5		ME
3694.622	3695.674	27058.67	V	I	1.8	0.7		MIT
3692.225	3693.276	27076.23	V	I	2.3G	2.2G		MIT
3690.281	3691.331	27090.50	V	I	2.3	2.1		MIT
3688.069	3689.119	27106.74	V	I	2.3	2.3G		MIT
3687.473	3688.523	27111.12	V	I	1.3	1.2		MIT
3686.706	3687.755	27116.76	V		0.7	0.3		MIT
3686.262	3687.311	27120.03	V	I	2.0	2.0		MIT
3684.332	3685.381	27134.24	V	I	1.6	1.0		MIT
3683.126	3684.175	27143.12	V	I	2.0	1.8		MIT
3681.303	3682.351	27156.56	V	I	0.3			MIT
3680.113	3681.161	27165.34	V	I	2.1	1.7H		MIT
3677.085	3678.132	27187.71	V	I	1.4	1.8		MIT
3676.684	3677.731	27190.68	V	I	2.5	2.2H		MIT
3675.700	3676.747	27197.96	V	I	2.0	1.8		MIT
3675.497	3676.544	27199.46	V	I	1.6	1.0		MIT
3674.685	3675.731	27205.47	V		0.3	1.3		MIT
3673.404	3674.450	27214.96	V	I	2.2	1.9H		MIT

AIR WAVELENGTH (ANGSTRMS)	VACUUM WAVELENGTH (ANGSTRMS)	WAVENUMBER (KAYSERS)	LMENT	LOG ARC	INTENSITY SPK	DIS	REF
3672.403	3673.449	27222.37	V I	2.0	1.6H		MIT
3671.205	3672.250	27231.26	V I	2.0	1.8		MIT
3669.408	3670.453	27244.59	V	1.30	2.5		ME
3667.741	3668.786	27256.98	V I	1.9	1.4H		MIT
3665.142	3666.186	27276.30	V I	2.0	1.7H		MIT
3663.594	3664.638	27287.83	V I	2.2	0.0J		MIT
3661.382	3662.425	27304.31	V I	1.0	2.2		ME
3658.264	3659.306	27327.58	V I	0.3	1.0		ME
3657.492	3658.534	27333.35	V I	1.3	0.7		MIT
3656.706	3657.748	27339.23	V I	1.9	1.3H		MIT
3652.423	3653.464	27371.29	V I	1.6	1.5H		MIT
3648.966	3650.006	27397.22	V I	1.9	1.7		MIT
3647.338	3648.377	27409.45	V I	1.3			MIT
3646.847	3647.886	27413.14	V		1.3H		ME
3645.905	3646.944	27420.22	V		1.3		ME
3645.596	3646.635	27422.54	V	1.2	0.0		MIT
3644.706	3645.745	27429.24	V I	1.9	1.7		MIT
3643.864	3644.902	27435.58	V I	1.6	1.5		MIT
3641.096	3642.134	27456.43	V I	2.0H	1.5J		MIT
3640.050	3641.087	27464.32	V I	0.7	0.3		MIT
3639.024	3640.061	27472.07	V I	1.8	1.8		MIT
3638.35	3639.39	27477.1	V	0.3			ME
3637.760	3638.797	27481.61	V	1.6	1.6		MIT
3635.874	3636.910	27495.87	V	1.7	1.4H		MIT
3635.463	3636.499	27498.98	V I	1.6	0.7		MIT
3633.91	3634.95	27510.7	V I	1.5	0.9		ME
3632.124	3633.159	27524.25	V		1.8		ME
3629.309	3630.344	27545.60	V I	1.7	0.3		MIT
3627.712	3628.746	27557.73	V	0.6	1.7		ME
3625.607	3626.641	27573.73	V	0.6	2.1		ME
3622.632	3623.665	27596.37	V I	1.5	0.0		MIT
3622.289	3623.322	27598.98	V		1.5		MIT
3621.208	3622.241	27607.22	V	1.2	1.9		MIT
3620.472	3621.504	27612.83	V	1.0	1.7		MIT
3618.928	3619.960	27624.61	V		2.0		MIT
3616.723	3617.754	27641.46	V	1.5	1.5		MIT
3611.58	3612.61	27680.8	V	1.2J			MIT
3609.289	3610.318	27698.39	V I	1.50	1.50		MIT
3606.689	3607.718	27718.35	V I	1.9	1.8		MIT
3605.589	3606.618	27726.81	V	1.5	1.3H		MIT
3604.377	3605.405	27736.13	V		1.7H		MIT
3602.943	3603.971	27747.17	V	1.0	1.1		MIT
3600.030	3601.057	27769.62	V I	1.7	1.6		MIT
3598.718	3599.745	27779.75	V		1.0		MIT
3597.400	3598.426	27789.93	V		0.7		ME
3593.334	3594.359	27821.37	V II	1.5	2.5G		MIT
3592.533	3593.558	27827.57	V	1.6			MIT
3592.024	3593.049	27831.52	V II	1.7	2.5G		MIT
3589.760	3590.784	27849.07	V II	1.9	2.8G		MIT
3588.139	3589.163	27861.65	V	0.0	1.3		MIT
3586.113	3587.137	27877.39	V I	0.8			MIT
3584.403	3585.426	27890.69	V		0.8		MIT
3583.704	3584.727	27896.13	V I	1.8	1.5		MIT
3582.814	3583.837	27903.06	V I	1.6	1.4		MIT
3580.825	3581.847	27918.56	V I	1.7	1.7		MIT
3579.091	3580.113	27932.08	V I	0.8	0.0		MIT
3578.639	3579.661	27935.61	V	1.5D	1.9D		MIT
3577.874	3578.895	27941.58	V I	1.7	1.6		MIT
3577.644	3578.665	27943.38	V		0.5		ME
3577.225	3578.246	27946.65	V		1.3		ME
3575.129	3576.150	27963.03	V I	1.3	1.2		MIT
3574.772	3575.793	27965.83	V I	1.4	1.1		MIT
3574.346	3575.366	27969.16	V		1.6		MIT
3573.516	3574.536	27975.66	V I	1.3	1.3H		MIT
3572.628	3573.648	27982.61	V I	0.7H	0.3H		MIT
3571.653	3572.673	27990.25	V I	1.6	1.5		MIT
3571.212	3572.232	27993.71	V	1.0	0.7		MIT
3571.037	3572.057	27995.08	V I	1.5	1.3		MIT
3569.083	3570.102	28010.40	V I	0.8	0.0		MIT
3568.940	3569.959	28011.53	V I	1.3	1.2		MIT

AIR WAVELENGTH (ANGSTRMS)	VACUUM WAVELENGTH (ANGSTRMS)	WAVENUMBER (KAYSERS)	LMENT	LOG ARC	INTENSITY SPK	DIS	REF
3566.182	3567.200	28033.19	V II	1.4	2.0		MIT
3563.396	3564.414	28055.10	V I	1.2	0.9		MIT
3562.137	3563.154	28065.02	V I	1.1	0.9		MIT
3560.596	3561.613	28077.17	V II	1.0	1.7		MIT
3557.163	3558.179	28104.26	V I	0.9			MIT
3556.801	3557.817	28107.12	V II	1.5	1.67		MIT
3556.249	3557.265	28111.49	V	1.6	0.3		MIT
3555.740	3556.756	28115.51	V I	1.2	0.0		MIT
3555.142	3556.158	28120.24	V I	1.0	0.5		MIT
3553.271	3554.286	28135.05	V I	1.9	1.5		MIT
3552.81	3553.82	28138.7	V I	0.8	1.0		M
3551.537	3552.552	28148.78	V I	1.4	1.1		MIT
3550.504	3551.518	28156.97	V		1.0		MIT
3549.028	3550.042	28168.68	V	0.0	0.3		MIT
3546.96	3547.97	28185.1	V		1.3		ME
3545.339	3546.352	28197.99	V	1.5			MIT
3545.196	3546.209	28199.13	V II	1.6	2.5G		MIT
3543.500	3544.513	28212.62	V I	1.7	1.7		MIT
3542.657	3543.669	28219.34	V I	1.2	0.3		MIT
3542.490	3543.502	28220.67	V		0.5		MIT
3541.336	3542.348	28229.86	V		1.5		MIT
3540.530	3541.542	28236.29	V I	1.4	0.6		MIT
3538.241	3539.252	28254.56	V II	1.0	2.0		MIT
3536.041	3537.052	28272.14	V I	0.3			MIT
3534.733	3535.743	28282.60	V I	1.3	0.7		MIT
3533.757	3534.767	28290.41	V II	1.3	1.6H		MIT
3533.676	3534.686	28291.06	V I	1.6	1.0H		MIT
3532.276	3533.286	28302.27	V	0.0	1.4H		MIT
3530.869	3531.878	28313.55	V I	0.0	1.3		ME
3530.773	3531.782	28314.32	V II	1.6	2.0		MIT
3530.45	3531.46	28316.9	V		1.5		MIT
3529.735	3530.744	28322.64	V I	1.3	0.3		MIT
3528.212	3529.221	28334.87	V I	1.0	0.0H		MIT
3527.863	3528.872	28337.67	V		0.7		MIT
3525.770	3526.778	28354.49	V I	0.9			MIT
3524.715	3525.723	28362.98	V II	1.0	1.8		MIT
3522.569	3523.576	28380.26	V I	1.4	0.3		MIT
3521.839	3522.846	28386.14	V	1.3	1.9		MIT
3520.543	3521.550	28396.59	V	0.0	1.3		MIT
3520.025	3521.031	28400.77	V II	0.7	1.7		MIT
3519.167	3520.173	28407.69	V I	1.0			MIT
3517.301	3518.307	28422.76	V	1.3	1.5Z		MIT
3516.012	3517.017	28433.18	V II		1.0		MIT
3514.420	3515.425	28446.06	V	0.0	1.6		MIT
3513.877	3514.882	28450.46	V		1.5H		ME
3512.130	3513.134	28464.61	V		1.0		ME
3511.422	3512.426	28470.35	V		0.8		ME
3510.266	3511.270	28479.72	V		1.0		ME
3509.684	3510.688	28484.45	V		0.8		ME
3509.041	3510.045	28489.67	V	0.3	2.2		MIT
3507.539	3508.542	28501.86	V		1.7		MIT
3506.843	3507.846	28507.52	V I	1.2	0.6		MIT
3506.564	3507.567	28509.79	V		0.8		ME
3505.690	3506.693	28516.90	V I	1.7	1.5		MIT
3504.439	3505.441	28527.08	V II	1.8	2.3		MIT
3503.176	3504.178	28537.36	V	1.0	0.6		MIT
3501.485	3502.487	28551.14	V I	1.4	1.3		MIT
3500.824	3501.826	28556.53	V I	1.5	1.4		MIT
3500.346	3501.347	28560.43	V I	0.3	0.3		MIT
3499.824	3500.825	28564.69	V II	0.5	1.7		MIT
3498.200	3499.201	28577.95	V I	1.1	0.8		MIT
3497.394	3498.395	28584.54	V		1.1		MIT
3497.030	3498.031	28587.51	V		2.2		MIT
3496.939	3497.940	28588.26	V I	1.1			MIT
3493.167	3494.167	28619.13	V II	1.2	2.0		MIT
3489.943	3490.942	28645.56	V		1.6		MIT
3489.467	3490.466	28649.47	V I	1.5	0.8		MIT
3487.008	3488.006	28669.68	V I	0.8	0.6		ME
3485.924	3486.922	28678.59	V II	0.9	1.8		MIT
3485.867	3486.865	28679.06	V I	0.5	0.0		ME

AIR WAVELENGTH (ANGSTRMS)	VACUUM WAVELENGTH (ANGSTRMS)	WAVENUMBER (KAYSERS)	LMENT	LOG ARC	INTENSITY SPK	INTENSITY DIS	REF
3484.644	3485.641	28689.12	V II		0.7		MIT
3482.187	3483.184	28709.37	V I	1.6	1.7		MIT
3479.836	3480.832	28728.76	V II	1.3	1.9		MIT
3478.962	3479.958	28735.98	V		1.4		ME
3477.516	3478.512	28747.93	V		2.0		ME
3476.248	3477.243	28758.41	V		1.6		MIT
3470.265	3471.259	28807.99	V		1.8		MIT
3469.525	3470.519	28814.14	V	0.3	2.0		ME
3467.323	3468.316	28832.44	V		0.5		MIT
3466.588	3467.581	28838.55	V		1.5		MIT
3465.249	3466.241	28849.69	V	0.3	1.4		MIT
3464.170	3465.162	28858.68	V		1.6		MIT
3463.829	3464.821	28861.52	V		1.4		MIT
3463.395	3464.387	28865.14	V I	1.0	0.3		MIT
3463.078	3464.070	28867.78	V		1.5		ME
3461.576	3462.567	28880.30	V		0.5		MIT
3457.152	3458.142	28917.26	V	0.3	2.2		MIT
3456.914	3457.904	28919.25	V I	1.3	0.3		MIT
3455.212	3456.202	28933.50	V I	0.7	0.5		MIT
3454.882	3455.872	28936.26	V I	1.2	1.1		MIT
3453.084	3454.073	28951.32	V		1.8		MIT
3451.045	3452.034	28968.43	V		1.8		ME
3445.807	3446.794	29012.46	V I	1.2	1.1		MIT
3442.334	3443.321	29041.73	V I	1.2	1.2		MIT
3442.005	3442.991	29044.51	V I	1.0	0.9		MIT
3437.873	3438.858	29079.42	V I	0.3	0.0H		MIT
3437.773	3438.758	29080.26	V I	0.6	0.3		MIT
3436.395	3437.380	29091.92	V		0.5		ME
3435.373	3436.358	29100.58	V		1.3		MIT
3434.026	3435.010	29111.99	V	0.0	1.3		ME
3433.768	3434.752	29114.18	V		0.5		MIT
3425.282	3426.264	29186.31	V	0.9	0.6		MIT
3425.070	3426.052	29188.11	V	1.4	1.3		MIT
3423.864	3424.846	29198.40	V I	1.4	1.4		MIT
3422.260	3423.241	29212.08	V		1.2H		ME
3420.706	3421.687	29225.35	V		1.0		MIT
3418.515	3419.495	29244.08	V I	1.4	1.3		MIT
3417.064	3418.044	29256.50	V I	1.5	1.4		MIT
3416.545	3417.525	29260.94	V I	0.8	0.5		MIT
3414.880	3415.860	29275.21	V		0.6H		ME
3414.196	3415.175	29281.07	V I	1.3	1.6		MIT
3409.095	3410.073	29324.89	V I	1.1	0.6		MIT
3407.999	3408.977	29334.32	V I	1.1	0.8		MIT
3406.837	3407.814	29344.32	V I	1.4	1.1		MIT
3406.065	3407.042	29350.97	V		0.5		MIT
3405.160	3406.137	29358.77	V I	1.5	1.2		MIT
3404.960	3405.937	29360.50	V I	0.9	0.0		MIT
3404.425	3405.402	29365.11	V		1.7H		ME
3403.360	3404.337	29374.30	V I	1.5	1.2		MIT
3402.571	3403.547	29381.11	V I	1.8	1.4		MIT
3402.00	3402.98	29386.0	V		0.5		ME
3401.342	3402.318	29391.73	V I	1.2	0.9		MIT
3400.395	3401.371	29399.91	V I	2.0	0.9		MIT
3398.265	3399.240	29418.34	V I	1.1	1.0		MIT
3397.843	3398.818	29421.99	V I	1.5	1.3		MIT
3397.580	3398.555	29424.27	V I	0.5	1.4		MIT
3396.515	3397.490	29433.50	V I	1.2	1.1		MIT
3395.524	3396.499	29442.08	V I	1.4	1.0		MIT
3392.735	3393.709	29466.29	V I	0.0	1.5		MIT
3390.765	3391.738	29483.41	V I	1.6	1.2		MIT
3390.380	3391.353	29486.75	V I	1.1	0.9		MIT
3387.380	3388.352	29512.87	V I	0.9	0.9		MIT
3385.792	3386.764	29526.71	V		0.7H		ME
3384.599	3385.571	29537.12	V I	1.8	1.6		MIT
3382.530	3383.501	29555.18	V		2.1		ME
3379.37	3380.34	29582.8	V	0.8	0.5		MIT
3377.625	3378.595	29598.10	V I	1.8	1.5		MIT
3377.394	3378.364	29600.13	V I	1.5	1.2		MIT
3376.057	3377.027	29611.85	V I	1.9	1.8		MIT
3374.033	3375.002	29629.61	V I	1.2	1.0		MIT
3372.668	3373.637	29641.60	V		0.8		ME
3371.114	3372.082	29655.27	V I	1.5	1.3		MIT
3367.670	3368.637	29685.59	V		0.7		ME
3366.880	3367.847	29692.56	V I	1.3	0.9		MIT
3365.553	3366.520	29704.26	V I	2.1	1.9		MIT
3363.545	3364.511	29722.00	V I	1.4	0.9		MIT
3361.508	3362.474	29740.01	V		2.3		ME
3356.352	3357.317	29785.69	V I	2.1	1.8		MIT
3355.368	3356.332	29794.43	V		1.5		ME
3353.775	3354.739	29808.58	V	0.3	2.0		MIT
3348.374	3349.337	29856.66	V		1.3		ME
3345.900	3346.862	29878.74	V		2.1		ME
3343.315	3344.276	29901.84	V		1.0		ME
3342.28	3343.24	29911.1	V	0.7	0.3		MIT
3337.846	3338.806	29950.83	V	0.3	2.2		ME
3336.819	3337.779	29960.05	V I	0.8	0.0		MIT
3336.350	3337.309	29964.26	V I	1.0	0.5		ME
3335.483	3336.442	29972.05	V	0.0H	1.8		ME
3333.565	3334.524	29989.29	V I	1.4	1.1W		MIT
3329.855	3330.813	30022.70	V I	2.0	1.6		MIT
3328.402	3329.359	30035.81	V I	0.8	0.0		MIT
3327.983	3328.940	30039.59	V I	0.7	0.5		MIT
3324.393	3325.349	30072.03	V I	0.8	0.8		MIT
3321.685	3322.641	30096.54	V I	1.1			MIT
3321.538	3322.494	30097.87	V	0.5	2.2		MIT
3320.785	3321.740	30104.70	V		0.9H		MIT
3320.141	3321.096	30110.54	V I	1.1	0.5		MIT
3319.024	3319.979	30120.67	V I	1.2	0.0H		MIT
3318.907	3319.862	30121.73	V		1.8		MIT
3317.912	3318.867	30130.77	V		1.9		MIT
3316.876	3317.830	30140.18	V		1.8		MIT
3315.522	3316.476	30152.49	V		0.6		MIT
3315.176	3316.130	30155.63	V	0.3	1.5		MIT
3313.97	3314.92	30166.6	V I	1.0			ME
3313.010	3313.964	30175.35	V I	0.7	0.5		MIT
3309.176	3310.129	30210.31	V I	1.5	1.3		MIT
3308.479	3309.431	30216.67	V		1.9		MIT
3308.246	3309.198	30218.80	V I	1.0	0.0		MIT
3304.470	3305.421	30253.33	V		2.1		MIT
3301.651	3302.602	30279.16	V		1.9		ME
3300.903	3301.853	30286.02	V		1.5		ME
3299.973	3300.923	30294.55	V I	0.7	0.6		MIT
3299.086	3300.036	30302.70	V I	1.0	0.3		MIT
3298.736	3299.686	30305.92	V II	1.1	1.9		MIT
3298.139	3299.089	30311.40	V I	1.7	1.2		MIT
3297.517	3298.467	30317.12	V		1.8		MIT
3296.052	3297.001	30330.59	V		1.5		MIT
3293.150	3294.098	30357.32	V		1.7		MIT
3291.676	3292.624	30370.91	V I	1.0	0.6		MIT
3290.238	3291.186	30384.19	V	0.3	1.8		MIT
3289.389	3290.337	30392.03	V II	1.0	1.8		MIT
3288.982	3289.929	30395.79	V	1.0	1.2		ME
3288.433	3289.380	30400.86	V I	0.9	0.3		MIT
3288.311	3289.258	30401.99	V	0.0	1.5		MIT
3285.670	3286.617	30426.43	V		0.5		ME
3285.024	3285.970	30432.41	V	0.6	1.6		MIT
3284.360	3285.306	30438.56	V I	1.4	0.7		MIT
3283.311	3284.257	30448.29	V I	1.5	1.0		MIT
3282.533	3283.479	30455.50	V	1.1	1.9		MIT
3281.754	3282.700	30462.73	V		1.5		MIT
3281.115	3282.060	30468.67	V	0.5	1.7		MIT
3279.845	3280.790	30480.46	V	1.3	0.9		MIT
3277.939	3278.884	30498.19	V I	1.3			MIT
3277.444	3278.388	30502.79	V	0.3	1.2		MIT
3277.085	3278.029	30506.13	V	0.0	1.0		MIT
3276.124	3277.068	30515.08	V II	1.7	2.3G		MIT
3273.027	3273.970	30543.95	V I	1.5	0.7		MIT
3272.192	3273.135	30551.75	V I	0.9	0.0		MIT
3271.637	3272.580	30556.93	V I	1.4			MIT
3271.411	3272.354	30559.04	V I	0.9			MIT

AIR WAVELENGTH (ANGSTRMS)	VACUUM WAVELENGTH (ANGSTRMS)	WAVENUMBER (KAYSERS)	LMENT	LOG ARC	INTENSITY SPK	INTENSITY DIS	REF
3271.125	3272.068	30561.71	V II	1.4	1.7G		MIT
3270.116	3271.059	30571.14	V	0.5	0.7		MIT
3267.702	3268.644	30593.73	V II	1.5	1.9G		MIT
3266.085	3267.027	30608.87	V I	1.1			MIT
3265.899	3266.841	30610.62	V I	1.2	1.1		MIT
3263.321	3264.262	30634.80	V II		1.7		MIT
3263.238	3264.179	30635.58	V I	1.6			MIT
3262.062	3263.003	30646.62	V I	1.0	0.5		MIT
3261.081	3262.021	30655.84	V	1.2	1.0		MIT
3259.680	3260.620	30669.01	V		0.6		ME
3259.537	3260.477	30670.36	V I	1.2	0.5		MIT
3257.889	3258.829	30685.87	V	0.8	1.6		MIT
3256.777	3257.716	30696.35	V I	0.9	0.0		MIT
3256.46	3257.40	30699.3	V	0.9	0.0		MIT
3255.649	3256.588	30706.99	V I	1.4	0.7		MIT
3254.768	3255.707	30715.30	V I	1.6	1.9H		MIT
3252.907	3253.845	30732.87	V	0.7			MIT
3251.870	3252.808	30742.67	V	1.0	1.7		MIT
3250.777	3251.715	30753.00	V	1.0	1.7		MIT
3250.034	3250.972	30760.04	V I	0.9			MIT
3249.931	3250.868	30761.01	V I	0.9			MIT
3249.566	3250.503	30764.47	V I	1.6	1.5		MIT
3249.46	3250.40	30765.5	V		0.7H		ME
3248.698	3249.635	30772.69	V I	1.2	0.5		MIT
3247.908	3248.845	30780.17	V		0.7		ME
3243.272	3244.208	30824.17	V I	1.3	1.0		MIT
3241.167	3242.102	30844.18	V I	1.3	1.0		MIT
3239.834	3240.769	30856.87	V		1.5		ME
3237.874	3238.808	30875.55	V II	1.5	2.0H		MIT
3234.728	3235.662	30905.58	V I	0.9			MIT
3234.516	3235.450	30907.61	V		1.3		MIT
3233.768	3234.701	30914.75	V	0.7	1.3		MIT
3233.541	3234.474	30916.93	V I	1.0	1.4		MIT
3233.190	3234.123	30920.28	V I	1.6	0.5		MIT
3231.950	3232.883	30932.14	V	0.9	2.0		MIT
3230.646	3231.579	30944.63	V I	1.3	0.9		MIT
3229.606	3230.538	30954.59	V I	1.2	0.9		MIT
3228.182	3229.114	30968.25	V I	1.2	0.7		ME
3227.409	3228.341	30975.66	V I	1.2	1.0		MIT
3227.113	3228.045	30978.50	V I	1.0			MIT
3226.922	3227.854	30980.34	V	0.0	1.7		ME
3226.106	3227.037	30988.18	V I	1.4	0.7		MIT
3223.55	3224.48	31012.7	V	0.5	0.0		ME
3221.378	3222.308	31033.65	V		1.0		ME
3218.869	3219.799	31057.84	V I	1.3	1.2		MIT
3218.358	3219.287	31062.77	V I	0.9	0.0		MIT
3217.113	3218.042	31074.79	V II	1.5	1.9H		MIT
3217.113	3218.042	31074.79	V I	1.5	1.9H		MIT
3215.375	3216.304	31091.59	V I	1.3	0.7		MIT
3214.750	3215.679	31097.64	V II	1.3	2.0		MIT
3213.938	3214.866	31105.49	V I	1.2	0.9		MIT
3212.434	3213.362	31120.05	V I	1.8	1.7		MIT
3211.573	3212.501	31128.40	V I	0.7	0.5		MIT
3210.430	3211.357	31139.48	V I	1.0	0.7		MIT
3210.097	3211.024	31142.71	V I	1.0	0.7		MIT
3208.352	3209.279	31159.65	V II	1.0	2.0		MIT
3207.410	3208.337	31168.80	V I	1.9R	1.3		MIT
3206.921	3207.848	31173.55	V I	0.7			MIT
3206.14	3207.07	31181.1	V		1.0H		ME
3205.582	3206.508	31186.57	V I	1.3	1.0		MIT
3205.262	3206.188	31189.69	V I	1.2	0.7H		MIT
3204.196	3205.122	31200.06	V I	1.3	1.0		MIT
3202.381	3203.306	31217.74	V I	2.0R	1.3R		MIT
3201.59	3202.52	31225.5	V		1.0J		MIT
3201.226	3202.151	31229.01	V I	0.9			MIT
3199.820	3200.745	31242.73	V I	1.4	1.0		MIT
3198.012	3198.936	31260.39	V I	2.0R	1.5R		MIT
3197.578	3198.502	31264.63	V	0.0	1.0		ME
3196.565	3197.489	31274.54	V	0.7	1.3		MIT
3195.479	3196.403	31285.17	V		1.4H		MIT
3194.915	3195.839	31290.69	V I	0.7			MIT
3194.56	3195.48	31294.2	V I	0.8			MIT
3194.374	3195.297	31295.99	V	0.9	0.0		MIT
3193.916	3194.839	31300.48	V I	2.0	1.3		MIT
3193.193	3194.116	31307.57	V	0.0	1.3		MIT
3192.708	3193.631	31312.32	V	0.0	1.3		MIT
3190.678	3191.600	31332.24	V II	1.7	2.2G		MIT
3189.754	3190.676	31341.32	V		0.5		ME
3189.078	3190.000	31347.96	V I	1.0			ME
3188.513	3189.435	31353.52	V II	1.5	2.0G		MIT
3188.094	3189.016	31357.64	V I	1.0			MIT
3187.708	3188.630	31361.43	V II	1.5	2.0G		MIT
3186.858	3187.780	31369.80	V	0.0	1.0		MIT
3185.396	3186.317	31384.20	V I	2.7G	2.6G		MIT
3183.982	3184.903	31398.13	V I	2.7G	2.6G		MIT
3183.406	3184.327	31403.81	V I	2.3G	2.0G		MIT
3183.406	3184.327	31403.81	V I	2.3G	2.0G		MIT
3182.675	3183.595	31411.03	V I		1.5		ME
3182.591	3183.511	31411.85	V		1.5		MIT
3180.559	3181.479	31431.92	V I	0.5	0.0		MIT
3179.415	3180.335	31443.23	V	0.0	1.4		MIT
3177.683	3178.602	31460.37	V		1.3		MIT
3174.539	3175.457	31491.53	V	0.0	1.9		MIT
3174.079	3174.997	31496.09	V	0.7	1.5		MIT
3172.228	3173.146	31514.47	V		1.4		ME
3171.738	3172.656	31519.34	V		1.5		ME
3170.204	3171.121	31534.59	V		0.5		ME
3168.138	3169.055	31555.15	V II	1.1	1.6		MIT
3167.436	3168.353	31562.14	V	0.7	2.2G		MIT
3165.887	3166.803	31577.59	V	0.0	1.9		MIT
3165.598	3166.514	31580.47	V I	0.6			MIT
3164.828	3165.744	31588.15	V II	1.0	2.0		MIT
3164.54	3165.46	31591.0	V	0.6			MIT
3163.905	3164.821	31597.37	V I	1.0			MIT
3163.76	3164.68	31598.8	V		1.4		ME
3163.021	3163.937	31606.20	V	0.7	1.5		MIT
3162.713	3163.628	31609.27	V	0.7	1.5		MIT
3162.357	3163.272	31612.83	V		1.3		MIT
3161.312	3162.227	31623.28	V	0.0	1.7		MIT
3160.780	3161.695	31628.60	V	0.0	1.5		ME
3159.872	3160.787	31637.69	V	0.3	0.0H		MIT
3159.364	3160.279	31642.78	V	0.0	1.6		MIT
3157.902	3158.816	31657.43	V	0.3	1.6		ME
3156.881	3157.795	31667.67	V I	0.0	0.3		MIT
3156.222	3157.136	31674.28	V	1.2	0.7		MIT
3155.408	3156.322	31682.45	V	0.7	2.0		MIT
3153.549	3154.462	31701.13	V	0.7H			MIT
3151.322	3152.235	31723.53	V	0.9W	2.2W		MIT
3150.568	3151.480	31731.12	V	0.7	0.3		MIT
3150.042	3150.954	31736.42	V	0.3H			MIT
3148.738	3149.650	31749.56	V		1.7		ME
3147.256	3148.168	31764.51	V	0.7	0.0		MIT
3146.812	3147.723	31768.99	V	0.3	1.7		MIT
3146.230	3147.141	31774.87	V	0.3	1.8		MIT
3145.971	3146.882	31777.48	V II	1.0	1.2		MIT
3145.342	3146.253	31783.84	V II	1.4	1.8		MIT
3144.697	3145.608	31790.36	V	0.0	1.8		ME
3143.473	3144.384	31802.74	V	0.0	1.7		MIT
3142.475	3143.385	31812.83	V	1.2	2.0R		MIT
3142.186	3143.096	31815.76	V	0.5	1.3		MIT
3141.485	3142.395	31822.86	V	0.7	2.0		MIT
3141.05	3141.96	31827.3	V		0.7		MIT
3139.990	3140.900	31838.01	V I	0.8			MIT
3139.745	3140.655	31840.50	V	1.2	2.2		MIT
3139.069	3139.978	31847.35	V I	0.7			MIT
3138.532	3139.441	31852.80	V I	0.7			MIT
3138.06	3138.97	31857.6	V		1.8		MIT
3136.514	3137.423	31873.29	V	1.3	2.3		MIT
3135.193	3136.101	31886.72	V I	0.9			MIT
3134.931	3135.839	31889.39	V	1.5	2.2R		MIT

AIR WAVELENGTH (ANGSTRMS)	VACUUM WAVELENGTH (ANGSTRMS)	WAVENUMBER (KAYSERS)	LMENT	LOG ARC	INTENSITY SPK	DIS	REF
3133.328	3134.236	31905.70	V II	1.7	2.3R		MIT
3132.594	3133.502	31913.18	V	1.9R	1.3		MIT
3130.267	3131.174	31936.90	V II	1.7	2.3R		MIT
3128.686	3129.593	31953.04	V	0.5	1.8		MIT
3128.284	3129.191	31957.14	V	0.3	1.8		ME
3126.215	3127.121	31978.29	V II	1.8	2.0G		MIT
3125.284	3126.190	31987.82	V II	1.9	2.3G		MIT
3125.002	3125.908	31990.71	V	0.6	1.7		ME
3122.895	3123.800	32012.29	V	1.1	2.5R		MIT
3121.749	3122.654	32024.04	V I	1.1			MIT
3121.145	3122.050	32030.24	V II	1.8	2.3R		MIT
3120.734	3121.639	32034.46	V	1.1	1.9		MIT
3119.324	3120.228	32048.94	V		0.5		MIT
3118.383	3119.287	32058.61	V II	1.8	2.3G		MIT
3116.781	3117.685	32075.08	V		1.4		MIT
3116.055	3116.959	32082.56	V		1.0H		ME
3113.567	3114.470	32108.19	V	0.8	2.0		MIT
3112.925	3113.828	32114.81	V I	1.1			MIT
3112.124	3113.027	32123.08	V	1.3	0.0		MIT
3110.706	3111.608	32137.72	V II	1.8	2.5G		MIT
3109.372	3110.274	32151.51	V	0.0	1.8		MIT
3108.701	3109.603	32158.45	V	0.5	1.7		MIT
3107.144	3108.045	32174.56	V	1.0			MIT
3106.829	3107.730	32177.82	V		0.7		ME
3106.11	3107.01	32185.3	V I	0.9			MIT
3105.972	3106.873	32186.70	V		0.7H		ME
3104.913	3105.814	32197.68	V	0.8	1.4		MIT
3103.994	3104.895	32207.21	V I	1.2			MIT
3102.299	3103.199	32224.81	V II	1.8	2.5G		MIT
3100.935	3101.835	32238.98	V	1.3	2.0		MIT
3099.58	3100.48	32253.1	V	0.3H			MIT
3095.902	3096.801	32291.39	V I	1.2			MIT
3094.692	3095.590	32304.02	V I	1.6			MIT
3094.199	3095.097	32309.16	V	1.3	2.1R		MIT
3093.792	3094.690	32313.41	V I	1.5			MIT
3093.108	3094.006	32320.56	V II	2.0G	2.6G		ME
3092.720	3093.618	32324.61	V	2.0R	1.7R		MIT
3091.52	3092.42	32337.2	V I	0.9	0.0		KN
3091.437	3092.334	32338.03	V I	1.0			MIT
3090.779	3091.676	32344.91	V I	0.9			MIT
3090.536	3091.433	32347.46	V	0.3H			MIT
3089.628	3090.525	32356.96	V		1.0		MIT
3089.130	3090.027	32362.18	V I	1.5	0.3		MIT
3088.114	3089.011	32372.82	V I	1.5			MIT
3087.065	3087.961	32383.82	V I	1.4	0.3		MIT
3086.503	3087.399	32389.72	V	0.3	1.6		MIT
3086.210	3087.106	32392.80	V	0.0	1.5		MIT
3084.381	3085.277	32412.00	V I	1.6			MIT
3083.539	3084.435	32420.85	V I	1.8			MIT
3083.214	3084.109	32424.27	V	0.3	1.7		MIT
3082.523	3083.418	32431.54	V	0.5	1.5		MIT
3082.111	3083.006	32435.87	V I	1.9R	0.3H		MIT
3082.012	3082.907	32436.92	V I	1.2	0.3H		MIT
3081.253	3082.148	32444.91	V	0.7	1.7		MIT
3081.004	3081.899	32447.53	V	0.3	1.5		MIT
3080.333	3081.228	32454.60	V I	1.2	0.0H		MIT
3080.146	3081.041	32456.57	V I	0.9	0.0H		MIT
3079.366	3080.260	32464.79	V I	1.0			MIT
3078.950	3079.844	32469.17	V		1.2		ME
3077.853	3078.747	32480.75	V	0.7H			MIT
3077.718	3078.612	32482.17	V I	1.0	1.5		MIT
3076.688	3077.582	32493.04	V I	0.3			MIT
3076.616	3077.510	32493.80	V I	0.7	0.3		MIT
3076.016	3076.910	32500.14	V II		1.2		MIT
3075.933	3076.827	32501.02	V I	1.3			MIT
3075.269	3076.162	32508.04	V I	1.2	0.3H		MIT
3075.06	3075.95	32510.2	V		0.7H		MIT
3074.827	3075.720	32512.71	V I	1.4			MIT
3074.658	3075.551	32514.50	V		1.5		MIT
3074.06	3074.95	32520.8	V I	1.0H			MIT

AIR WAVELENGTH (ANGSTRMS)	VACUUM WAVELENGTH (ANGSTRMS)	WAVENUMBER (KAYSERS)	LMENT	LOG ARC	INTENSITY SPK	DIS	REF
3073.823	3074.716	32523.33	V I	1.8	1.3R		MIT
3072.71	3073.60	32535.1	V	1.8R	1.6R		MIT
3070.892	3071.784	32554.37	V I	1.3	0.3H		MIT
3070.12	3071.01	32562.6	V		1.7		MIT
3069.645	3070.537	32567.59	V I	1.5	1.0		MIT
3067.116	3068.007	32594.45	V I	1.2			MIT
3066.80	3067.69	32597.8	V		0.3		MIT
3066.53	3067.42	32600.7	V I	1.2			MIT
3066.375	3067.266	32602.32	V I	2.6R	2.1R		MIT
3065.605	3066.496	32610.51	V	0.3	1.5		MIT
3063.734	3064.625	32630.43	V I	1.5	0.0H		MIT
3063.247	3064.137	32635.61	V	1.5	1.9R		MIT
3062.698	3063.588	32641.46	V II	0.7	1.3		MIT
3062.179	3063.069	32646.99	V		0.6H		MIT
3060.93	3061.82	32660.3	V I	0.8	0.3		MIT
3060.460	3061.350	32665.33	V I	2.2R	2.0P		MIT
3057.86	3058.75	32693.1	V		0.5H		ME
3057.07	3057.96	32701.6	V		0.7W		MIT
3056.334	3057.223	32709.43	V I	2.1R	1.8R		MIT
3055.94	3056.83	32713.6	V		1.7W		MIT
3054.24	3055.13	32731.9	V		0.7		MIT
3053.89	3054.78	32735.6	V	1.0	1.8R		MIT
3053.65	3054.54	32738.2	V I	2.0R			MIT
3053.387	3054.275	32741.00	V II	1.0	2.0R		MIT
3052.194	3053.082	32753.79	V I	1.7	0.7		MIT
3051.39	3052.28	32762.4	V		0.7		MIT
3050.890	3051.777	32767.79	V I	1.5			MIT
3050.730	3051.617	32769.51	V	0.0	1.7		MIT
3050.400	3051.287	32773.05	V I	1.5	0.0		MIT
3048.892	3049.779	32789.26	V	1.0	1.7		MIT
3048.65	3049.54	32791.9	V		0.3		MIT
3048.215	3049.102	32796.54	V	1.0	2.1R		MIT
3044.936	3045.822	32831.86	V I	1.5	1.2		MIT
3043.555	3044.440	32846.76	V I	1.8	1.6		MIT
3043.124	3044.009	32851.41	V I	1.8	0.8		MIT
3042.264	3043.149	32860.70	V	0.8	1.9		MIT
3041.86	3042.75	32865.1	V	0.7			MIT
3041.420	3042.305	32869.81	V	0.7	1.6		MIT
3039.763	3040.648	32887.73	V		0.7		MIT
3038.706	3039.590	32899.17	V	1.3	0.0		MIT
3038.521	3039.405	32901.17	V		1.7		MIT
3037.372	3038.256	32913.62	V I	0.5			ME
3036.08	3036.96	32927.6	V		1.5H		MIT
3033.822	3034.705	32952.13	V II	1.3	2.0R		MIT
3033.448	3034.331	32956.19	V	1.3	1.6		MIT
3032.192	3033.075	32969.85	V		0.9		MIT
3031.007	3031.889	32982.73	V I	1.2	1.0		MIT
3030.93	3031.81	32983.6	V I	0.5			MIT
3029.580	3030.462	32998.27	V II		1.6		MIT
3028.043	3028.925	33015.02	V II	0.3	1.7		MIT
3027.601	3028.483	33019.84	V		1.4		MIT
3024.980	3025.861	33048.45	V II	0.3	1.8		ME
3023.883	3024.764	33060.44	V		1.5		MIT
3022.76	3023.64	33072.7	V I	1.0H			MIT
3022.566	3023.446	33074.84	V	0.3	1.7		MIT
3022.145	3023.025	33079.45	V		1.0		MIT
3021.78	3022.66	33083.4	V I	0.9	0.6		MIT
3016.784	3017.663	33138.23	V II	1.2	1.9		MIT
3016.16	3017.04	33145.1	V I	1.4	0.9		MIT
3015.983	3016.862	33147.03	V		1.4		MIT
3014.823	3015.701	33159.78	V II	1.0	2.0		MIT
3014.37	3015.25	33164.8	V I	1.2H			MIT
3013.104	3013.982	33178.70	V	1.0	1.8		MIT
3012.015	3012.893	33190.69	V	0.3	1.7		MIT
3008.614	3009.491	33228.21	V	0.5	1.8		MIT
3008.505	3009.382	33229.42	V		1.7		ME
3007.283	3008.159	33242.92	V II	0.3	1.7		MIT
3006.92	3007.80	33246.9	V I	0.7H			MIT
3006.501	3007.377	33251.56	V		1.7		MIT
3006.35	3007.23	33253.2	V I	0.7			MIT

AIR WAVELENGTH (ANGSTRMS)	VACUUM WAVELENGTH (ANGSTRMS)	WAVENUMBER (KAYSERS)	LMENT	LOG ARC	INTENSITY SPK	DIS	REF
3006.24	3007.12	33254.5	V I	0.7H			MIT
3005.812	3006.688	33259.19	V		1.7		MIT
3004.824	3005.700	33270.12	V I	1.1	0.0		MIT
3004.33	3005.21	33275.6	V I	0.9			MIT
3003.457	3004.332	33285.26	V II	0.9	1.8		MIT
3003.284	3004.159	33287.18	V	0.7			MIT
3002.65	3003.53	33294.2	V I	1.0			MIT
3002.442	3003.317	33296.52	V	1.0	0.0		MIT
3001.90	3002.78	33302.5	V I	1.3H	0.5		MIT
3001.751	3002.626	33304.18	V	0.0	1.5		MIT
3001.205	3002.080	33310.24	V II	1.3	2.3R		MIT
2999.238	3000.112	33332.08	V I	1.1	0.5		MIT
2998.626	2999.500	33338.89	V I	0.3			MIT
2997.947	2998.821	33346.44	V		1.5		MIT
2997.08	2997.95	33356.1	V I	0.7H			ME
2996.489	2997.363	33362.66	V I	0.9	0.0		MIT
2995.998	2996.872	33368.13	V II	0.7	1.8		MIT
2995.612	2996.486	33372.43	V I	0.3			MIT
2994.538	2995.411	33384.40	V	0.3H	1.7		MIT
2992.988	2993.861	33401.69	V		1.3		MIT
2992.379	2993.252	33408.48	V		0.6		MIT
2991.735	2992.608	33415.67	V		0.7		MIT
2991.145	2992.017	33422.27	V I	0.3			MIT
2990.948	2991.820	33424.47	V I	1.7	0.7		MIT
2989.748	2990.620	33437.88	V	0.3	0.7		MIT
2989.595	2990.467	33439.59	V	0.9	1.6		MIT
2989.300	2990.172	33442.89	V	0.0	1.3		MIT
2988.021	2988.893	33457.21	V II	1.0	1.9		MIT
2985.170	2986.041	33489.16	V	0.0	1.8		MIT
2983.546	2984.417	33507.39	V	1.0	1.8		MIT
2983.016	2983.886	33513.34	V	0.3	1.0		MIT
2982.748	2983.618	33516.35	V	0.6	1.7		MIT
2982.173	2983.043	33522.81	V I	0.3			MIT
2981.930	2982.800	33525.54	V		1.6		MIT
2981.202	2982.072	33533.73	V	0.6	1.8		MIT
2979.107	2979.976	33557.31	V		1.5		MIT
2978.936	2979.805	33559.24	V I	1.0			ME
2978.232	2979.101	33567.17	V		1.6		MIT
2977.539	2978.408	33574.98	V I	1.7	0.8		MIT
2976.520	2977.389	33586.48	V I	1.5	0.3		MIT
2976.197	2977.066	33590.12	V	0.8	1.7		MIT
2975.652	2976.521	33596.27	V	0.6	1.7		MIT
2975.056	2975.924	33603.00	V I	1.2	0.3		MIT
2974.224	2975.092	33612.40	V I	1.2			MIT
2973.969	2974.837	33615.29	V		1.6		MIT
2972.263	2973.131	33634.58	V II				ME
2972.254	2973.122	33634.68	V	0.3	1.7		MIT
2972.004	2972.872	33637.51	V		0.9		MIT
2971.571	2972.439	33642.41	V		1.0		MIT
2970.421	2971.288	33655.44	V		0.9		MIT
2969.842	2970.709	33662.00	V II		0.9		MIT
2968.981	2969.848	33671.76	V I	0.8			ME
2968.378	2969.245	33678.60	V	1.0	1.3H		MIT
2968.288	2969.155	33679.62	V I	1.0			MIT
2968.036	2968.903	33682.48	V		0.7		MIT
2967.546	2968.413	33688.04	V	0.0	1.1		MIT
2963.806	2964.672	33730.55	V I	1.0	0.8		MIT
2963.249	2964.114	33736.89	V		1.3		MIT
2962.772	2963.637	33742.32	V I	1.8	1.8R		MIT
2962.014	2962.879	33750.95	V II	0.0	1.0		MIT
2961.124	2961.989	33761.10	V I	1.1			MIT
2960.784	2961.649	33764.98	V II		1.2		MIT
2958.598	2959.462	33789.92	V	0.0	1.6		MIT
2957.518	2958.382	33802.26	V I	1.3	2.1R		MIT
2957.33	2958.19	33804.4	V I	1.0			MIT
2957.149	2958.013	33806.48	V I	0.8H			MIT
2956.655	2957.519	33812.13	V		0.8		MIT
2955.799	2956.663	33821.92	V I	1.3	0.3		MIT
2955.584	2956.448	33824.38	V	0.0	1.8		MIT
2954.332	2955.195	33838.71	V I	1.5	1.3		MIT
2952.075	2952.938	33864.58	V II	1.5	2.2G		MIT
2951.562	2952.425	33870.47	V		1.0		MIT
2950.348	2951.210	33884.40	V II	1.4	2.0R		MIT
2949.904	2950.766	33889.50	V I	0.3H			MIT
2949.626	2950.488	33892.70	V I	1.5	1.2R		MIT
2949.168	2950.030	33897.96	V	0.8	1.9		MIT
2948.075	2948.937	33910.53	V	0.3	1.8		MIT
2946.527	2947.388	33928.34	V I	1.0	0.7		MIT
2944.755	2945.616	33948.76	V I	0.3H			MIT
2944.571	2945.432	33950.88	V II	1.7	2.5R		MIT
2943.827	2944.688	33959.46	V I	0.8H			MIT
2943.636	2944.497	33961.66	V	0.3	0.6		MIT
2943.196	2944.057	33966.74	V I	1.5	1.4R		MIT
2942.352	2943.212	33976.48	V I	1.9R	1.3H		MIT
2941.492	2942.352	33986.42	V II	1.1	2.2R		MIT
2941.369	2942.229	33987.84	V II	1.6	2.5R		MIT
2941.11	2941.97	33990.8	V I	0.3			KN
2938.666	2939.525	34019.10	V I	1.1	0.3		MIT
2938.254	2939.113	34023.87	V II	0.3	1.8		ME
2938.254	2939.113	34023.87	V I	0.3	1.8		MIT
2937.687	2938.546	34030.43	V I	1.3W	1.0		MIT
2937.040	2937.899	34037.93	V	0.3	1.4		MIT
2935.874	2936.733	34051.45	V I	1.5	1.2H		MIT
2934.76	2935.62	34064.4	V I	0.7H			MIT
2934.648	2935.506	34065.67	V I	0.3			MIT
2934.401	2935.259	34068.54	V II	1.0	1.7H		MIT
2933.835	2934.693	34075.11	V II	0.3	1.5		MIT
2933.228	2934.086	34082.16	V I	0.3H			MIT
2932.323	2933.181	34092.68	V	1.1	1.9		MIT
2931.857	2932.715	34098.10	V	0.3	1.2		MIT
2931.624	2932.482	34100.81	V	0.6	1.3		MIT
2930.87	2931.73	34109.6	V I	0.6H			MIT
2930.806	2931.663	34110.33	V II	1.5	2.2R		MIT
2930.128	2930.985	34118.22	V	0.6	1.5		MIT
2927.64	2928.50	34147.2	V I	0.8H	0.6H		MIT
2926.446	2927.302	34161.15	V	0.3	1.5		MIT
2926.256	2927.112	34163.36	V I	1.0			MIT
2925.874	2926.730	34167.82	V I	1.1			MIT
2925.288	2926.144	34174.67	V	0.3	1.3H		MIT
2924.908	2925.764	34179.11	V I	0.3H			MIT
2924.644	2925.500	34182.19	V II	1.8	2.3R		MIT
2924.025	2924.881	34189.43	V II	1.8R	2.5G		MIT
2923.620	2924.476	34194.16	V I	1.7R	2.2R		MIT
2923.397	2924.253	34196.77	V	0.3	0.3		MIT
2923.341	2924.197	34197.43	V	0.6	1.3		MIT
2922.575	2923.430	34206.39	V I	0.8	0.0H		MIT
2921.18	2922.04	34222.7	V I	0.8H			KN
2920.385	2921.240	34232.04	V II	1.3	2.1R		MIT
2919.992	2920.847	34236.65	V II	1.0	1.8R		MIT
2919.680	2920.535	34240.31	V	0.0	0.6H		MIT
2918.209	2919.063	34257.56	V		1.2H		MIT
2917.932	2918.786	34260.82	V I	1.1	0.3H		MIT
2917.554	2918.408	34265.25	V	0.6			MIT
2917.366	2918.220	34267.46	V II	1.0	1.3		MIT
2917.226	2918.080	34269.11	V	0.0	1.0		MIT
2916.019	2916.873	34283.29	V I	1.0			MIT
2915.861	2916.715	34285.15	V	0.0	1.5		MIT
2915.331	2916.185	34291.38	V I	1.1	1.3		MIT
2914.926	2915.780	34296.15	V I	1.8	1.7R		MIT
2914.431	2915.284	34301.97	V I	0.3			MIT
2914.301	2915.154	34303.50	V I	1.5	0.8		ME
2913.719	2914.572	34310.35	V		1.1		MIT
2913.048	2913.901	34318.25	V		0.8		MIT
2912.501	2913.354	34324.70	V	0.0	1.3H		ME
2911.655	2912.508	34334.67	V	0.0	1.0		MIT
2911.064	2911.917	34341.64	V II	1.5	2.3R		MIT
2910.389	2911.241	34349.61	V II	1.5	2.2R		MIT
2910.019	2910.871	34353.98	V II	1.5	2.2R		MIT
2908.817	2909.669	34368.17	V II	1.8R	2.6G		MIT
2908.438	2909.290	34372.65	V	0.3	1.3		MIT

AIR WAVELENGTH (ANGSTRMS)	VACUUM WAVELENGTH (ANGSTRMS)	WAVENUMBER (KAYSERS)	LMENT	LOG ARC	INTENSITY SPK	DIS	REF
2907.471	2908.323	34384.08	V II	1.6	2.2H		MIT
2906.457	2907.308	34396.08	V II	1.6	2.2H		MIT
2906.134	2906.985	34399.90	V I	1.5	1.4H		MIT
2905.602	2906.453	34406.20	V	0.6	1.3		MIT
2904.986	2905.837	34413.49	V	0.3	1.4		MIT
2904.126	2904.977	34423.68	V I	1.3	0.8H		MIT
2903.698	2904.549	34428.76	V I	1.2	0.0		MIT
2903.547	2904.398	34430.55	V		1.0		ME
2903.078	2903.929	34436.11	V II	1.5	2.2R		MIT
2900.888	2901.738	34462.10	V	0.3H			MIT
2899.938	2900.788	34473.39	V		1.1		MIT
2899.597	2900.447	34477.45	V I	1.5	0.6H		MIT
2899.201	2900.051	34482.16	V I	1.5	0.8H		MIT
2898.817	2899.667	34486.72	V I	1.0			MIT
2897.896	2898.745	34497.68	V	0.0	1.4		MIT
2896.878	2897.727	34509.81	V		1.0		MIT
2896.211	2897.060	34517.75	V II	1.5	2.2R		MIT
2895.605	2896.454	34524.98	V	0.0	1.1		MIT
2895.18	2896.03	34530.0	V	0.6H			MIT
2894.832	2895.681	34534.20	V		0.8		MIT
2894.577	2895.425	34537.24	V I	1.1			MIT
2893.320	2894.168	34552.24	V II	1.7	2.5R		MIT
2892.659	2893.507	34560.14	V II	1.5	2.2R		MIT
2892.441	2893.289	34562.74	V II	1.5	2.2R		MIT
2891.950	2892.798	34568.61	V I	0.3			MIT
2891.642	2892.490	34572.29	V II	1.6	2.3R		MIT
2890.554	2891.401	34585.30	V	0.0	1.1		MIT
2890.141	2890.988	34590.25	V	0.3	1.1		MIT
2889.621	2890.468	34596.47	V II	1.6	2.2R		MIT
2888.521	2889.368	34609.65	V I	0.3			MIT
2888.246	2889.093	34612.94	V	1.3	2.1R		MIT
2887.697	2888.544	34619.52	V I	0.3			MIT
2887.158	2888.005	34625.98	V	0.0	1.0		MIT
2886.967	2887.814	34628.27	V	0.3	1.2		MIT
2884.785	2885.631	34654.46	V II	1.6	2.3R		MIT
2884.063	2884.909	34663.14	V		1.1		MIT
2882.501	2883.347	34681.92	V II	1.5	2.3R		MIT
2880.800	2881.645	34702.40	V	0.3	1.3		MIT
2880.030	2880.875	34711.68	V II	1.4	2.2R		MIT
2879.163	2880.008	34722.13	V II	1.7	1.5		MIT
2878.299	2879.143	34732.55	V		0.8		MIT
2878.022	2878.866	34735.90	V	0.3	1.0		MIT
2877.688	2878.532	34739.93	V	1.2	2.0G		MIT
2876.936	2877.780	34749.01	V	0.6	1.4		MIT
2875.689	2876.533	34764.07	V II	1.0	1.6		MIT
2874.207	2875.050	34782.00	V	0.8	1.3		ME
2873.356	2874.199	34792.30	V I	1.3			MIT
2873.182	2874.025	34794.41	V	0.6	1.7		MIT
2870.547	2871.390	34826.34	V I	1.7R	1.3R		MIT
2870.113	2870.955	34831.61	V	0.0	1.1		ME
2870.04	2870.88	34832.5	V I	0.8W			MIT
2869.961	2870.803	34833.45	V II	0.8	1.3		MIT
2869.463	2870.305	34839.50	V I	0.8			MIT
2869.134	2869.976	34843.49	V	1.4	2.2R		MIT
2868.103	2868.945	34856.02	V I	1.6	1.5R		MIT
2866.951	2867.793	34870.02	V I	1.3			MIT
2866.594	2867.436	34874.37	V I	1.4	0.8		MIT
2866.418	2867.260	34876.51	V I	1.5			MIT
2864.517	2865.358	34899.65	V	0.8	1.5		MIT
2864.359	2865.200	34901.58	V I	1.6	1.4R		MIT
2863.79	2864.63	34908.5	V	0.6	1.1		MIT
2863.049	2863.890	34917.55	V I	1.3	0.8J		MIT
2862.385	2863.226	34925.65	V I	1.1			MIT
2862.302	2863.143	34926.66	V	0.0	1.4		MIT
2861.642	2862.482	34934.71	V	0.3			MIT
2861.401	2862.241	34937.65	V		1.2		MIT
2859.971	2860.811	34955.12	V I	1.7	1.0		MIT
2858.977	2859.817	34967.28	V I	1.6	1.3		MIT
2858.762	2859.602	34969.90	V I	1.3	0.3H		MIT
2857.944	2858.783	34979.91	V I	1.7	0.8H		MIT

AIR WAVELENGTH (ANGSTRMS)	VACUUM WAVELENGTH (ANGSTRMS)	WAVENUMBER (KAYSERS)	LMENT	LOG ARC	INTENSITY SPK	DIS	REF
2855.716	2856.555	35007.20	V I	0.6			MIT
2855.491	2856.330	35009.96	V I	1.1			MIT
2855.295	2856.134	35012.37	V	0.0	1.5		MIT
2855.221	2856.060	35013.27	V I	1.7	0.0		MIT
2854.336	2855.175	35024.13	V	1.3	2.0G		MIT
2854.029	2854.868	35027.90	V I	0.3			MIT
2853.819	2854.657	35030.47	V I	0.8			MIT
2853.762	2854.600	35031.17	V		0.8		MIT
2853.556	2854.394	35033.70	V I	0.8			MIT
2852.866	2853.704	35042.17	V I	1.8	0.8H		MIT
2852.536	2853.374	35046.23	V	0.8	1.5		MIT
2851.748	2852.586	35055.91	V I	1.5	0.6		MIT
2851.254	2852.092	35061.98	V	0.3	1.2		MIT
2850.768	2851.606	35067.96	V	1.0	1.5		MIT
2850.686	2851.524	35068.97	V	1.0	1.4		MIT
2849.175	2850.012	35087.57	V I	1.4			MIT
2849.050	2849.887	35089.11	V	0.8	1.7		MIT
2848.774	2849.611	35092.51	V I	1.3	0.3		MIT
2847.572	2848.409	35107.32	V	1.2	2.2		MIT
2846.573	2847.410	35119.64	V	1.7	1.3H		MIT
2845.245	2846.081	35136.03	V	1.3	1.9		MIT
2844.926	2845.762	35139.97	V I	1.0			MIT
2844.835	2845.671	35141.09	V		1.0		MIT
2844.224	2845.060	35148.64	V		1.1H		MIT
2843.821	2844.657	35153.62	V		1.2H		ME
2842.692	2843.528	35167.58	V	0.0	1.1		MIT
2842.289	2843.125	35172.57	V		1.0H		ME
2842.041	2842.877	35175.64	V	0.0	0.8		MIT
2841.039	2841.874	35188.04	V	1.2	1.5		MIT
2840.599	2841.434	35193.50	V	0.3	1.1		MIT
2840.106	2840.941	35199.60	V	0.3	1.4		MIT
2839.437	2840.272	35207.90	V I	1.1	0.8		MIT
2838.535	2839.370	35219.08	V		1.0		MIT
2838.058	2838.893	35225.00	V I	0.8	0.3		MIT
2836.698	2837.532	35241.89	V I	1.0			MIT
2836.520	2837.354	35244.10	V	1.3	1.9		MIT
2835.635	2836.469	35255.10	V I	1.1	0.3		MIT
2835.347	2836.181	35258.68	V	0.0	1.0		ME
2834.882	2835.716	35264.46	V	1.1			MIT
2834.53	2835.36	35268.8	V		1.6H		MIT
2831.60	2832.43	35305.3	V		1.1H		MIT
2830.962	2831.795	35313.29	V		0.8H		MIT
2830.402	2831.235	35320.28	V	1.0	1.8		MIT
2826.92	2827.75	35363.8	V		1.0H		MIT
2825.867	2826.699	35376.96	V	0.8	1.8H		MIT
2825.025	2825.856	35387.50	V		1.3H		MIT
2824.441	2825.272	35394.82	V	0.3	1.2		MIT
2822.44	2823.27	35419.9	V	0.6	1.8H		MIT
2822.136	2822.967	35423.73	V		1.5H		MIT
2821.122	2821.952	35436.46	V	0.8	1.6		MIT
2819.443	2820.273	35457.56	V	1.0	1.6		MIT
2818.527	2819.357	35469.08	V		0.8H		MIT
2817.500	2818.330	35482.01	V	1.3	1.7		MIT
2815.968	2816.797	35501.31	V	1.1			MIT
2815.545	2816.374	35506.65	V	0.0	1.1H		MIT
2815.032	2815.861	35513.12	V	0.3	1.0		MIT
2814.899	2815.728	35514.79	V	0.8	1.4		MIT
2812.694	2813.522	35542.63	V	0.3H	1.1H		MIT
2812.169	2812.997	35549.27	V	0.3	1.3		MIT
2811.980	2812.808	35551.66	V II	0.3	1.1		MIT
2811.596	2812.424	35556.52	V	0.3	1.3		MIT
2810.269	2811.097	35573.30	V	1.7	1.7		MIT
2810.155	2810.983	35574.75	V	1.0	1.5H		MIT
2809.514	2810.342	35582.86	V	1.0	1.3		MIT
2808.695	2809.522	35593.24	V	0.3	1.0		MIT
2808.231	2809.058	35599.12	V	0.8	1.5		MIT
2808.021	2808.848	35601.78	V		0.3H		MIT
2806.550	2807.377	35620.44	V	0.3	1.0		MIT
2805.540	2806.367	35633.26	V	1.1	1.5		MIT
2804.442	2805.268	35647.21	V		1.1		MIT

AIR WAVELENGTH (ANGSTRMS)	VACUUM WAVELENGTH (ANGSTRMS)	WAVENUMBER (KAYSERS)	LMENT		LOG ARC	INTENSITY SPK	DIS	REF
2803.467	2804.293	35659.61	V		1.5	1.4		MIT
2802.797	2803.623	35668.13	V		1.2	1.4H		MIT
2800.93	2801.76	35691.9	V			1.5H		ME
2800.02	2800.85	35703.5	V			0.6H		ME
2799.451	2800.276	35710.76	V		1.4	2.0H		MIT
2798.758	2799.583	35719.61	V		1.4	1.9H		MIT
2797.795	2798.620	35731.90	V		1.1	1.8H		ME
2797.018	2797.843	35741.83	V		1.1	1.9H		MIT
2794.301	2795.125	35776.58	V			1.0H		MIT
2792.43	2793.25	35800.5	V			1.1H		ME
2791.495	2792.318	35812.54	V		0.6	1.0		MIT
2790.088	2790.911	35830.60	V		0.0	1.0H		MIT
2788.67	2789.49	35848.8	V			0.8H		MIT
2788.190	2789.012	35854.99	V		1.0			MIT
2787.924	2788.746	35858.41	V			1.5H		MIT
2787.325	2788.147	35866.11	V			0.6		MIT
2786.996	2787.818	35870.35	V		0.3J	1.3H		MIT
2785.689	2786.511	35887.17	V	I	1.4			MIT
2785.543	2786.365	35889.06	V		1.4			MIT
2784.27	2785.09	35905.5	V		0.6	1.7H		MIT
2783.94	2784.76	35909.7	V		0.0	1.0H		ME
2783.783	2784.604	35911.75	V	I	1.2			MIT
2782.99	2783.81	35922.0	V		0.0	1.0H		MIT
2782.576	2783.397	35927.32	V			1.0H		MIT
2781.454	2782.275	35941.81	V		0.6	2.1H		ME
2780.097	2780.917	35959.36	V			1.0		MIT
2778.578	2779.398	35979.01	V			1.8H		MIT
2778.073	2778.893	35985.55	V	I	1.0			MIT
2777.733	2778.553	35989.96	V	II	1.6H	2.0G		MIT
2776.685	2777.505	36003.54	V		0.3H			MIT
2776.471	2777.291	36006.32	V		1.5H			MIT
2776.23	2777.05	36009.4	V			0.8H		MIT
2775.763	2776.582	36015.50	V		1.1	1.8H		MIT
2774.968	2775.787	36025.82	V		1.0	1.3H		MIT
2774.717	2775.536	36029.08	V		1.3	1.7H		MIT
2774.276	2775.095	36034.80	V		1.4	2.0G		MIT
2774.005	2774.824	36038.32	V	I	1.0			MIT
2773.679	2774.498	36042.56	V	I	1.5	0.8		MIT
2772.01	2772.83	36064.3	V		0.3	1.9H		MIT
2771.404	2772.222	36072.14	V		0.8	1.7H		ME
2770.992	2771.810	36077.51	V		0.0	1.0		MIT
2770.929	2771.747	36078.33	V	I	0.3			MIT
2769.739	2770.557	36093.83	V		0.8	1.6		MIT
2768.937	2769.755	36104.28	V	I	1.5	0.6		MIT
2768.556	2769.374	36109.25	V		1.5	2.2G		MIT
2768.314	2769.132	36112.40	V	I	1.0			MIT
2768.131	2768.949	36114.79	V		1.1	1.3		MIT
2767.151	2767.968	36127.58	V		0.3	1.5H		MIT
2766.455	2767.272	36136.67	V	II	1.6	2.0H		MIT
2765.668	2766.485	36146.95	V		1.7	2.3H		MIT
2764.293	2765.110	36164.93	V			1.0H		MIT
2762.716	2763.532	36185.58	V		0.3	1.1		ME
2760.698	2761.514	36212.02	V		1.4	2.0H		MIT
2760.124	2760.940	36219.55	V	II	1.3	1.5H		MIT
2759.580	2760.395	36226.69	V		0.3	1.3H		MIT
2759.086	2759.901	36233.18	V		0.0	0.3		MIT
2758.814	2759.629	36236.75	V		0.6	1.0		MIT
2758.512	2759.327	36240.72	V		0.3	1.1H		MIT
2757.753	2758.568	36250.69	V	I	1.0	0.3		MIT
2756.573	2757.388	36266.21	V			1.2H		MIT
2756.381	2757.196	36268.74	V			0.6H		ME
2755.661	2756.475	36278.21	V		1.2			MIT
2755.067	2755.881	36286.03	V			1.3H		MIT
2753.404	2754.218	36307.95	V		1.7	2.3G		MIT
2753.092	2753.906	36312.06	V		1.0			MIT
2752.132	2752.946	36324.73	V		0.0H	1.5H		MIT
2751.786	2752.600	36329.29	V			0.3H		MIT
2750.30	2751.11	36348.9	V			0.8H		MIT
2750.01	2750.82	36352.8	V			0.3H		MIT
2747.542	2748.355	36385.41	V	I	1.0	0.7		MIT
2747.475	2748.287	36386.29	V		0.8	1.8		MIT
2745.901	2746.713	36407.15	V			1.5		MIT
2744.541	2745.353	36425.19	V		0.3	1.3		MIT
2743.774	2744.586	36435.37	V		0.8	1.5		MIT
2742.674	2743.485	36449.98	V		1.4	1.3		MIT
2742.410	2743.221	36453.49	V		1.5	1.3		MIT
2741.567	2742.378	36464.70	V		0.3	1.0		MIT
2740.975	2741.786	36472.58	V			1.2H		MIT
2739.711	2740.522	36489.40	V		1.7	1.9		MIT
2739.207	2740.017	36496.12	V			1.2H		MIT
2738.077	2738.887	36511.18	V	I	0.3	0.0		MIT
2736.699	2737.509	36529.56	V		0.0	1.3		MIT
2736.13	2736.94	36537.2	V			1.0H		MIT
2734.30	2735.11	36561.6	V			1.2H		MIT
2733.905	2734.714	36566.89	V		1.2	1.4		MIT
2733.340	2734.149	36574.45	V	I	1.0	0.0		MIT
2732.893	2733.702	36580.43	V		1.0	1.1H		MIT
2732.21	2733.02	36589.6	V			1.4H		MIT
2731.519	2732.328	36598.83	V		1.3H			MIT
2731.346	2732.155	36601.15	V	I	1.9	1.7		MIT
2731.16	2731.97	36603.6	V			0.3J		MIT
2729.813	2730.621	36621.70	V	I	0.3			MIT
2728.640	2729.448	36637.45	V		1.7	2.6G		MIT
2727.934	2728.742	36646.93	V			0.8		MIT
2726.547	2727.354	36665.57	V		0.8	1.9		MIT
2725.066	2725.873	36685.49	V	I	0.8	0.3		MIT
2724.63	2725.44	36691.4	V			1.0H		MIT
2723.932	2724.739	36700.77	V	I	0.3			MIT
2723.459	2724.266	36707.14	V			1.1		MIT
2723.218	2724.025	36710.39	V		1.1	1.3		MIT
2722.558	2723.364	36719.29	V	I	2.0	1.6		MIT
2722.243	2723.049	36723.54	V		0.3	1.0		MIT
2721.142	2721.948	36738.39	V		1.5	1.1		MIT
2720.780	2721.586	36743.28	V			0.8		MIT
2717.436	2718.241	36788.49	V	I	0.3	0.5		MIT
2715.686	2716.491	36812.20	V		1.7	2.5G		MIT
2715.031	2715.836	36821.08	V		1.0	0.3		MIT
2714.199	2715.003	36832.37	V		1.8	2.0		MIT
2713.046	2713.850	36848.02	V		1.6	1.9		MIT
2712.814	2713.618	36851.17	V		1.0H	0.6J		MIT
2712.223	2713.027	36859.20	V	I	0.8	1.0B		MIT
2711.736	2712.540	36865.82	V		1.7	2.2G		MIT
2710.162	2710.965	36887.23	V		0.8	1.8		MIT
2707.863	2708.666	36918.54	V		1.8	2.2		MIT
2707.589	2708.392	36922.28	V	I	0.6			ME
2706.698	2707.501	36934.43	V		1.8	2.3G		MIT
2706.169	2706.971	36941.65	V		2.0	2.6G		MIT
2705.224	2706.026	36954.56	V		1.4	1.7		MIT
2703.148	2703.950	36982.94	V		0.0	0.3		MIT
2702.189	2702.991	36996.06	V		1.9	2.5G		MIT
2701.535	2702.336	37005.02	V		1.0	0.8		MIT
2701.269	2702.070	37008.66	V		1.0			MIT
2700.936	2701.737	37013.22	V		2.1	2.7G		MIT
2700.511	2701.312	37019.05	V	I	0.3			MIT
2700.051	2700.852	37025.35	V		0.8	0.3H		MIT
2699.109	2699.910	37038.27	V	I	1.6			MIT
2698.730	2699.531	37043.48	V	I	1.8	1.2H		MIT
2698.378	2699.179	37048.31	V		1.5	2.5		MIT
2697.745	2698.545	37057.00	V	I	2.0	1.7G		MIT
2697.210	2698.010	37064.35	V			1.1		MIT
2696.993	2697.793	37067.33	V	I	1.8	0.3		MIT
2696.767	2697.567	37070.44	V	I	1.0			MIT
2696.544	2697.344	37073.50	V			1.3		MIT
2696.219	2697.019	37077.97	V		1.0	0.3		MIT
2695.237	2696.037	37091.48	V	I	0.8	0.8		MIT
2694.741	2695.541	37098.31	V		0.3	1.8		MIT
2694.65	2695.45	37099.6	V			1.0H		ME
2694.468	2695.268	37102.07	V		0.0	1.3		MIT
2694.105	2694.905	37107.06	V		0.6	0.0		MIT
2693.920	2694.720	37109.61	V	I	1.0	0.3H		MIT

AIR WAVELENGTH (ANGSTRMS)	VACUUM WAVELENGTH (ANGSTRMS)	WAVENUMBER (KAYSERS)	LMENT		LOG ARC	INTENSITY SPK	DIS	REF
2693.007	2693.806	37122.19	V	I	0.6H	0.6H		MIT
2690.786	2691.585	37152.83	V		1.8	2.5G		MIT
2690.245	2691.044	37160.30	V		1.7	2.3G		MIT
2689.876	2690.675	37165.40	V		1.7	2.2G		MIT
2689.347	2690.145	37172.71	V	I	0.3			MIT
2689.114	2689.912	37175.93	V	I	0.8			ME
2688.941	2689.739	37178.32	V	I	1.0			MIT
2688.715	2689.513	37181.45	V	I	1.5	2.0G		MIT
2687.956	2688.754	37191.95	V		2.2	2.7G		MIT
2687.411	2688.209	37199.49	V		1.0			MIT
2686.513	2687.311	37211.92	V	I	1.2			MIT
2686.357	2687.155	37214.08	V	I	1.2			MIT
2685.840	2686.638	37221.25	V	I	1.0			MIT
2685.688	2686.486	37223.35	V		1.1	1.5		MIT
2685.516	2686.314	37225.74	V	I	0.8			MIT
2685.135	2685.932	37231.02	V	I	1.3	1.5		MIT
2685.021	2685.818	37232.60	V	I	1.0			MIT
2684.768	2685.565	37236.11	V		0.0	1.0		MIT
2683.091	2683.888	37259.38	V	I	1.5	2.2G		MIT
2682.872	2683.669	37262.42	V		1.7	2.3G		MIT
2682.538	2683.335	37267.06	V		0.6	1.2		MIT
2681.172	2681.968	37286.05	V	I	0.3			MIT
2680.476	2681.272	37295.73	V		0.0	1.1		MIT
2679.703	2680.499	37306.48	V	I	0.6			MIT
2679.322	2680.118	37311.79	V		1.8	2.5G		MIT
2678.873	2679.669	37318.04	V	I	1.2			MIT
2678.675	2679.471	37320.80	V	I	1.1			MIT
2678.568	2679.364	37322.29	V		1.5	2.2G		MIT
2677.802	2678.598	37332.97	V		1.8	2.5G		MIT
2677.119	2677.915	37342.49	V	I	1.0	0.6H		MIT
2676.634	2677.429	37349.26	V	I	0.8			MIT
2676.35	2677.15	37353.2	V			1.3H		MIT
2676.043	2676.838	37357.51	V			1.2H		ME
2675.973	2676.768	37358.48	V	I	0.8			MIT
2675.761	2676.556	37361.44	V	I	1.1	0.3		MIT
2674.30	2675.09	37381.9	V			0.8H		MIT
2673.962	2674.757	37386.58	V		0.0	1.0		MIT
2673.23	2674.02	37396.8	V		1.0	1.8		ME
2672.004	2672.798	37413.97	V		1.7	2.5G		MIT
2671.673	2672.467	37418.61	V	I	1.3	0.3		MIT
2670.923	2671.717	37429.11	V	I	1.1	0.6		MIT
2670.234	2671.028	37438.77	V		0.3	1.8		MIT
2668.895	2669.689	37457.55	V	I	0.3	0.6		MIT
2668.598	2669.391	37461.72	V			1.1		MIT
2668.006	2668.799	37470.03	V			1.3J		MIT
2667.530	2668.323	37476.72	V			1.1		MIT
2666.786	2667.579	37487.17	V		0.0	1.1H		ME
2665.964	2666.757	37498.73	V	I	1.4	1.0		MIT
2665.275	2666.068	37508.42	V		0.3	0.5		MIT
2663.248	2664.040	37536.97	V		1.1	2.0		MIT
2661.423	2662.215	37562.71	V	I	2.0	1.9		MIT
2659.606	2660.397	37588.37	V		1.0	1.6		MIT
2658.975	2659.766	37597.29	V		1.0	1.6		MIT
2658.506	2659.297	37603.92	V			1.2		MIT
2657.710	2658.501	37615.19	V	I	0.8	0.0		MIT
2657.292	2658.083	37621.10	V		0.0	1.5		MIT
2656.547	2657.338	37631.65	V		1.1	0.0		MIT
2656.225	2657.015	37636.21	V	I	1.6	0.6		MIT
2655.676	2656.466	37643.99	V		1.0	2.0H		ME
2654.390	2655.180	37662.23	V			0.7H		MIT
2654.008	2654.798	37667.65	V	I	0.3			MIT
2653.827	2654.617	37670.22	V	I	1.3	1.2		MIT
2652.921	2653.711	37683.08	V	I	1.3			MIT
2652.78	2653.57	37685.1	V			1.6H		MIT
2651.898	2652.687	37697.62	V	I	1.7	0.6		MIT
2651.57	2652.36	37702.3	V			0.7H		ME
2649.360	2650.149	37733.73	V		0.3	2.0H		ME
2648.882	2649.671	37740.54	V		0.6	0.0		MIT
2648.470	2649.259	37746.41	V		0.3	1.8		MIT
2647.710	2648.498	37757.24	V	I	1.7	1.0H		MIT

AIR WAVELENGTH (ANGSTRMS)	VACUUM WAVELENGTH (ANGSTRMS)	WAVENUMBER (KAYSERS)	LMENT		LOG ARC	INTENSITY SPK	DIS	REF
2645.840	2646.628	37783.93	V		1.2	2.0		MIT
2645.343	2646.131	37791.03	V		0.6			ME
2645.259	2646.047	37792.23	V	I	1.0	0.0		MIT
2644.355	2645.143	37805.14	V		1.1	2.0H		MIT
2643.65	2644.44	37815.2	V			1.1H		MIT
2643.160	2643.947	37822.23	V	I	1.4	1.0		MIT
2642.723	2643.510	37828.49	V			1.0		MIT
2642.270	2643.057	37834.97	V	I	1.0	1.6		MIT
2642.206	2642.993	37835.89	V		1.0	1.6		MIT
2640.855	2641.642	37855.25	V		0.3	1.7		ME
2640.687	2641.474	37857.65	V		0.8	0.3H		MIT
2640.268	2641.055	37863.66	V		1.0	0.6		MIT
2638.507	2639.293	37888.93	V			0.6		ME
2637.222	2638.008	37907.39	V		1.6H	1.2		ME
2635.639	2636.425	37930.16	V		0.0	1.1		MIT
2635.418	2636.204	37933.34	V			1.0		MIT
2633.591	2634.376	37959.65	V	I	0.8	0.6		MIT
2631.492	2632.277	37989.93	V			0.8		MIT
2630.666	2631.450	38001.86	V		1.5	2.2H		MIT
2629.713	2630.497	38015.63	V		0.0	1.8H		ME
2629.096	2629.880	38024.55	V		0.6	0.3		MIT
2628.740	2629.524	38029.70	V	II	0.6	1.7H		MIT
2628.083	2628.867	38039.20	V			0.8H		ME
2624.856	2625.639	38085.97	V			1.4H		ME
2623.793	2624.576	38101.40	V			1.6		MIT
2622.735	2623.517	38116.77	V			1.7H		ME
2621.794	2622.576	38130.44	V			1.2H		ME
2620.290	2621.072	38152.33	V	I	1.3	0.3		MIT
2620.054	2620.836	38155.77	V			0.6		MIT
2618.915	2619.697	38172.36	V	I	0.6	0.3		MIT
2617.09	2617.87	38199.0	V			1.2H		MIT
2616.68	2617.46	38205.0	V			1.2H		MIT
2616.248	2617.029	38211.27	V		0.6	1.8		MIT
2615.40	2616.18	38223.7	V			1.7J		MIT
2614.405	2615.186	38238.21	V		0.0	1.5		MIT
2611.256	2612.036	38284.31	V	I	1.3	1.0		MIT
2610.891	2611.671	38289.67	V	I	0.8	0.3		ME
2610.64	2611.42	38293.4	V			1.6H		MIT
2609.809	2610.588	38305.54	V			1.1		MIT
2609.602	2610.381	38308.58	V			0.8		MIT
2607.98	2608.76	38332.4	V			1.5H		ME
2607.766	2608.545	38335.55	V		1.0	0.3		MIT
2607.13	2607.91	38344.9	V		1.2			MIT
2605.084	2605.862	38375.01	V	I	0.6	0.3		ME
2604.294	2605.072	38386.65	V	I	1.1	0.0		ME
2603.41	2604.19	38399.7	V			1.3H		MIT
2602.96	2603.74	38406.3	V			1.3H		MIT
2602.34	2603.12	38415.5	V			1.0		MIT
2601.076	2601.853	38434.14	V			1.6H		ME
2600.798	2601.575	38438.25	V	I	0.6			ME
2597.193	2597.969	38491.60	V			1.1		MIT
2595.096	2595.872	38522.70	V			1.8H		MIT
2593.052	2593.827	38553.07	V			1.7H		MIT
2592.208	2592.983	38565.62	V			1.0		MIT
2584.956	2585.730	38673.81	V		0.3	2.0		MIT
2583.010	2583.783	38702.94	V			1.7		MIT
2581.843	2582.616	38720.43	V			1.1		MIT
2578.455	2579.227	38771.31	V			1.2		MIT
2577.292	2578.064	38788.80	V	I	1.3	0.8		MIT
2576.480	2577.252	38801.02	V		0.0	1.7		MIT
2574.522	2575.293	38830.53	V	II	1.0	1.9		MIT
2574.020	2574.791	38838.10	V	I	1.8	1.7		MIT
2571.057	2571.827	38882.86	V		0.6	1.8		MIT
2570.267	2571.037	38894.81	V		0.3	0.3		MIT
2568.389	2569.159	38923.25	V		1.5	1.5H		MIT
2568.059	2568.829	38928.25	V			1.0		MIT
2567.44	2568.21	38937.6	V			1.6H		MIT
2566.603	2567.372	38950.33	V			1.3		MIT
2566.033	2566.802	38958.98	V		0.0	1.1		MIT
2565.546	2566.315	38966.38	V			1.4		MIT

AIR WAVELENGTH (ANGSTRMS)	VACUUM WAVELENGTH (ANGSTRMS)	WAVENUMBER (KAYSERS)	LMENT		LOG ARC	INTENSITY SPK	DIS	REF
2564.816	2565.585	38977.47	V	I	1.5	0.8		MIT
2564.341	2565.110	38984.69	V		0.6			MIT
2564.228	2564.997	38986.41	V	I	1.3H	0.3		MIT
2562.761	2563.529	39008.72	V		0.0	1.4		MIT
2562.130	2562.898	39018.33	V	I	1.7	0.6		MIT
2560.147	2560.915	39048.55	V			1.0		MIT
2558.897	2559.664	39067.62	V	I	1.0	1.1		MIT
2556.822	2557.589	39099.33	V		0.3	0.3		MIT
2556.022	2556.789	39111.56	V	I	0.6			MIT
2555.910	2556.677	39113.28	V		0.3	1.9		MIT
2554.862	2555.628	39129.32	V	I	1.2	0.3		MIT
2554.220	2554.986	39139.15	V			1.7J		MIT
2553.669	2554.435	39147.60	V		1.0	1.3		MIT
2553.024	2553.790	39157.49	V		0.8	1.2		MIT
2552.962	2553.728	39158.44	V		1.0	1.5		MIT
2552.652	2553.418	39163.19	V	I	1.9G	1.0		MIT
2552.267	2553.033	39169.10	V			0.6H		MIT
2551.729	2552.495	39177.36	V			1.4		MIT
2549.968	2550.733	39204.41	V	I	1.0	0.6		MIT
2549.841	2550.606	39206.37	V		0.6	0.0		MIT
2549.656	2550.421	39209.21	V			1.0		MIT
2549.279	2550.044	39215.01	V	II	1.3	2.2		MIT
2548.692	2549.457	39224.04	V	II	1.0	1.9		MIT
2548.233	2548.998	39231.10	V		0.6	1.4		MIT
2548.202	2548.967	39231.58	V			1.3Q		MIT
2547.075	2547.840	39248.94	V	I	0.6	0.3		MIT
2546.311	2547.075	39260.71	V			0.8		MIT
2545.983	2546.747	39265.77	V	I	1.7	1.0		MIT
2545.462	2546.226	39273.81	V		0.3	1.2		MIT
2543.728	2544.492	39300.58	V	I	1.1	1.0		MIT
2542.935	2543.699	39312.83	V		0.3	1.5		MIT
2542.44	2543.20	39320.5	V		0.0	1.3J		MIT
2541.767	2542.530	39330.90	V		0.8	0.3		MIT
2539.192	2539.955	39370.78	V			1.0B		ME
2537.619	2538.381	39395.18	V			1.4		MIT
2536.926	2537.688	39405.94	V		1.0			MIT
2534.826	2535.588	39438.59	V		1.3			MIT
2534.520	2535.282	39443.35	V	II	1.0	1.9		MIT
2534.259	2535.021	39447.41	V	I	0.3	1.1		MIT
2533.963	2534.725	39452.02	V		0.0	1.0		MIT
2533.805	2534.566	39454.48	V		1.0			MIT
2533.362	2534.123	39461.38	V		0.0	1.2		MIT
2532.279	2533.040	39478.25	V		0.6			MIT
2531.900	2532.661	39484.16	V			0.6J		ME
2531.780	2532.541	39486.03	V		0.3			MIT
2531.615	2532.376	39488.61	V			0.6		MIT
2531.216	2531.977	39494.83	V		0.3	0.0H		MIT
2530.179	2530.940	39511.02	V	I	2.0	1.8G		MIT
2528.836	2529.596	39532.00	V		1.4	2.2G		MIT
2528.468	2529.228	39537.75	V		1.7	2.2G		MIT
2527.903	2528.663	39546.59	V		1.5	2.5G		MIT
2526.215	2526.975	39573.01	V	I	2.2	2.2G		MIT
2523.955	2524.714	39608.44	V	II	1.0	2.0		MIT
2523.511	2524.270	39615.41	V		0.6			M
2522.507	2523.266	39631.18	V		0.3	1.6		M
2522.390	2523.149	39633.02	V			0.8		ME
2521.615	2522.374	39645.20	V		0.3			MIT
2521.575	2522.334	39645.83	V		0.8	1.0H		MIT
2521.144	2521.903	39652.60	V			1.0Q		ME
2520.304	2521.062	39665.82	V		1.0H	0.0H		MIT
2519.623	2520.381	39676.54	V	I	2.1R	1.7		MIT
2517.502	2518.260	39709.96	V	I	0.6	0.0		MIT
2517.145	2517.903	39715.60	V	I	1.5R	1.4		MIT
2516.118	2516.875	39731.81	V		1.4	2.0		MIT
2515.723	2516.480	39738.04	V		0.0	1.0		ME
2515.649	2516.406	39739.21	V		0.8			MIT
2515.149	2515.906	39747.11	V	I	1.3	1.0		MIT
2514.636	2515.393	39755.22	V		1.0	1.1		MIT
2514.384	2515.141	39759.20	V		1.0	0.3		MIT
2514.318	2515.075	39760.25	V		1.2	0.3		MIT

AIR WAVELENGTH (ANGSTRMS)	VACUUM WAVELENGTH (ANGSTRMS)	WAVENUMBER (KAYSERS)	LMENT		LOG ARC	INTENSITY SPK	DIS	REF
2511.946	2512.702	39797.79	V	I	1.5	0.3H		MIT
2511.652	2512.408	39802.45	V	I	1.6			MIT
2511.181	2511.937	39809.91	V		1.2			MIT
2510.245	2511.001	39824.75	V	I	0.6	0.3		MIT
2508.825	2509.581	39847.29	V	I	0.5	0.3		MIT
2507.856	2508.611	39862.69	V		1.3	0.0H		MIT
2507.782	2508.537	39863.87	V	I	1.3	0.0		MIT
2506.905	2507.660	39877.81	V	I	1.7R	1.5		MIT
2506.220	2506.975	39888.71	V		1.0	2.2		MIT
2505.546	2506.301	39899.44	V	I	0.8	0.6		MIT
2504.290	2505.045	39919.45	V			0.8H		ME
2503.303	2504.057	39935.19	V	I	1.4	1.0		MIT
2503.020	2503.774	39939.70	V		0.8	2.0		MIT
2501.613	2502.367	39962.16	V	I	1.5	1.5		MIT
2500.381	2501.135	39981.85	V	I	0.3			MIT
2499.781	2500.535	39991.45	V	I	0.3			MIT
2499.251	2500.004	39999.93	V	I	0.3	0.3		MIT
2499.099	2499.852	40002.36	V	I	0.8	0.3		MIT
2498.232	2498.985	40016.24	V	I	1.0	0.6		MIT
2498.044	2498.797	40019.25	V		0.8			MIT
2497.655	2498.408	40025.49	V	I	0.6H			ME
2497.099	2497.852	40034.40	V	I	0.3			ME
2497.002	2497.755	40035.95	V			0.8		ME
2495.788	2496.541	40055.43	V	I	0.8	0.8		MIT
2493.582	2494.334	40090.86	V		0.0	1.4		MIT
2488.620	2489.371	40170.79	V			0.8		MIT
2487.532	2488.283	40188.36	V	I	0.8	0.6		MIT
2485.490	2486.240	40221.37	V			0.6		ME
2483.648	2484.398	40251.20	V		1.0			MIT
2483.070	2483.820	40260.57	V		1.3	2.2		M
2482.711	2483.461	40266.39	V		1.3			ME
2482.310	2483.059	40272.90	V		0.8	1.1		MIT
2482.130	2482.879	40275.82	V	I	1.5			MIT
2481.132	2481.881	40292.02	V		1.1			MIT
2480.630	2481.379	40300.17	V	I	1.4			MIT
2479.522	2480.271	40318.18	V		1.2	2.2		MIT
2479.050	2479.799	40325.85	V		1.2	2.2		M
2478.621	2479.370	40332.83	V		0.3	1.3		ME
2476.508	2477.256	40367.24	V	I	0.6H			MIT
2476.295	2477.043	40370.71	V			1.1		ME
2475.870	2476.618	40377.64	V			1.8		MIT
2475.460	2476.208	40384.33	V		0.3	1.8		MIT
2475.180	2475.928	40388.90	V		0.5	0.3		MIT
2472.865	2473.612	40426.71	V		0.8	1.6		MIT
2471.448	2472.195	40449.88	V	I	0.3	0.0		MIT
2471.124	2471.871	40455.19	V			1.6		MIT
2469.389	2470.136	40483.61	V			1.1		MIT
2468.653	2469.399	40495.68	V			1.2		MIT
2465.664	2466.410	40544.77	V		0.6H			ME
2465.278	2466.024	40551.11	V		0.3	2.0		MIT
2464.096	2464.841	40570.56	V			1.4		MIT
2463.157	2463.902	40586.03	V			0.5		ME
2462.941	2463.686	40589.59	V			0.8H		MIT
2461.495	2462.240	40613.43	V			1.6		ME
2459.362	2460.106	40648.65	V			1.3		MIT
2459.239	2459.983	40650.69	V			0.6		MIT
2458.288	2459.032	40666.41	V		0.0	1.8		MIT
2457.803	2458.547	40674.43	V			0.8		ME
2457.450	2458.194	40680.28	V			1.5		MIT
2453.351	2454.094	40748.24	V		0.3	1.6		MIT
2452.776	2453.519	40757.79	V			0.3		MIT
2450.739	2451.481	40791.67	V			1.3		MIT
2450.621	2451.363	40793.63	V			1.0		MIT
2450.237	2450.979	40800.02	V			1.3		MIT
2448.47	2449.21	40829.5	V			0.8H		MIT
2447.608	2448.349	40843.84	V			1.8		ME
2446.703	2447.444	40858.95	V		0.0	1.7		M
2445.336	2446.077	40881.79	V			0.6		MIT
2445.226	2445.967	40883.63	V	I	0.3			MIT
2445.107	2445.848	40885.62	V			0.5		ME

AIR WAVELENGTH (ANGSTRMS)	VACUUM WAVELENGTH (ANGSTRMS)	WAVENUMBER (KAYSERS)	LMENT	LOG ARC	INTENSITY SPK	DIS	REF
2444.973	2445.714	40887.86	V	0.3	1.8		MIT
2441.894	2442.634	40939.41	V I	1.1	1.1		MIT
2441.353	2442.093	40948.48	V I	0.7	0.7		MIT
2439.774	2440.514	40974.98	V		0.6H		ME
2439.104	2439.844	40986.23	V I	1.2R	1.2		MIT
2438.044	2438.783	41004.05	V		1.2H		MIT
2435.521	2436.260	41046.53	V I	1.5R	1.5		MIT
2434.943	2435.682	41056.27	V		0.5		MIT
2432.981	2433.719	41089.38	V	0.0	1.4		MIT
2432.018	2432.756	41105.64	V I	1.0	1.0		MIT
2431.952	2432.690	41106.76	V I	1.0	1.0		MIT
2431.572	2432.310	41113.18	V I	0.8	0.8		MIT
2430.04	2430.78	41139.1	V		1.8		ME
2428.279	2429.016	41168.93	V I	1.5R	1.3		MIT
2427.745	2428.482	41177.99	V I	0.8	0.8		MIT
2427.327	2428.064	41185.08	V		1.5		MIT
2426.128	2426.865	41205.43	V I	0.6	0.6		MIT
2423.375	2424.111	41252.24	V I	1.2	1.2		MIT
2423.038	2423.774	41257.97	V		1.0		MIT
2421.980	2422.716	41276.00	V I	1.5R	1.5		MIT
2421.062	2421.797	41291.65	V I	1.4R	1.4		MIT
2420.221	2420.956	41305.99	V	0.5	0.3		MIT
2420.122	2420.857	41307.68	V I	1.3R	1.3		MIT
2418.739	2419.474	41331.30	V I	0.5	0.3		MIT
2417.537	2418.272	41351.85	V		1.5		MIT
2417.351	2418.086	41355.03	V I	1.4R	1.3H		MIT
2416.751	2417.485	41365.30	V I	1.6R	1.6H		MIT
2415.333	2416.067	41389.58	V I	1.4	1.4H		MIT
2415.134	2415.868	41392.99	V		0.6H		MIT
2413.031	2413.765	41429.06	V I	1.3	1.3H		MIT
2412.687	2413.421	41434.97	V I	1.4	1.3H		MIT
2409.722	2410.455	41485.95	V	0.3	0.0		MIT
2408.429	2409.162	41508.22	V	0.3	1.3		MIT
2407.900	2408.632	41517.33	V I	1.2	1.0		MIT
2407.592	2408.324	41522.65	V		0.3		MIT
2407.173	2407.905	41529.87	V		1.2J		MIT
2406.989	2407.721	41533.05	V		0.7		ME
2406.748	2407.480	41537.21	V I	1.6	0.3		MIT
2405.733	2406.465	41554.73	V I	0.7			ME
2405.490	2406.222	41558.93	V	0.3			MIT
2405.230	2405.962	41563.42	V I	0.6	0.0		MIT
2404.547	2405.279	41575.22	V	0.6			MIT
2404.178	2404.910	41581.60	V	0.0	2.0		MIT
2403.362	2404.093	41595.72	V I	0.6H			ME
2403.250	2403.981	41597.66	V	0.0	1.5		MIT
2403.024	2403.755	41601.57	V I	0.3			MIT
2401.902	2402.633	41621.00	V I	1.2	1.0		MIT
2400.903	2401.634	41638.32	V	0.0	1.7H		MIT
2399.956	2400.687	41654.75	V I	1.5	0.3		MIT
2399.683	2400.414	41659.49	V		2.2		MIT
2398.869	2399.599	41673.62	V I	0.3			MIT
2398.679	2399.409	41676.92	V I	1.0			MIT
2398.268	2398.998	41684.07	V I	1.2			MIT
2398.128	2398.858	41686.50	V I	1.0			MIT
2397.782	2398.512	41692.51	V I	1.5			MIT
2397.632	2398.362	41695.12	V		0.6		MIT
2397.499	2398.229	41697.44	V I	0.5			MIT
2396.98	2397.71	41706.5	V		0.5J		MIT
2396.697	2397.427	41711.39	V I	0.7			MIT
2396.48	2397.21	41715.2	V I	1.0	0.7		MIT
2396.08	2396.81	41722.1	V I	0.3J			MIT
2395.428	2396.158	41733.48	V I	0.7	0.7		MIT
2395.089	2395.819	41739.39	V I	1.0	0.7H		MIT
2394.272	2395.001	41753.63	V I	0.8	1.5		MIT
2393.575	2394.304	41765.79	V		2.7		MIT
2392.897	2393.626	41777.62	V I	1.4			MIT
2392.693	2393.422	41781.18	V		0.3		MIT
2391.263	2391.992	41806.17	V I	1.4	0.5W		MIT
2390.868	2391.597	41813.07	V I	0.6			ME
2390.778	2391.507	41814.65	V I	1.4L	0.5		MIT

AIR WAVELENGTH (ANGSTRMS)	VACUUM WAVELENGTH (ANGSTRMS)	WAVENUMBER (KAYSERS)	LMENT	LOG ARC	INTENSITY SPK	DIS	REF
2390.466	2391.194	41820.10	V		1.4W		MIT
2389.699	2390.427	41833.52	V	0.7	2.0		MIT
2389.149	2389.877	41843.15	V		0.3		MIT
2388.92	2389.65	41847.2	V I	1.4	0.3		MIT
2388.260	2388.988	41858.73	V II		0.6		ME
2388.081	2388.809	41861.87	V I	1.4			MIT
2387.477	2388.205	41872.46	V I	0.5			MIT
2386.956	2387.684	41881.60	V I	1.4	1.0		MIT
2386.413	2387.141	41891.12	V I	1.3			MIT
2385.816	2386.543	41901.60	V		2.0		MIT
2385.618	2386.345	41905.08	V		1.0H		MIT
2384.996	2385.723	41916.01	V		0.8H		ME
2384.64	2385.37	41922.3	V I	0.3	0.0		MIT
2384.276	2385.003	41928.67	V I	1.2			MIT
2383.436	2384.163	41943.44	V	0.3	0.6H		MIT
2383.001	2383.728	41951.10	V	0.9	1.9		MIT
2382.467	2383.194	41960.50	V		2.0W		MIT
2380.918	2381.644	41987.80	V	0.8	1.7		MIT
2380.262	2380.988	41999.37	V I	0.6			MIT
2380.178	2380.904	42000.85	V I	0.5			ME
2379.150	2379.876	42019.00	V	0.6	1.3		MIT
2378.263	2378.989	42034.67	V	0.3H			MIT
2377.086	2377.811	42055.48	V I	0.3	0.3		MIT
2376.53	2377.26	42065.3	V		0.3J		MIT
2374.649	2375.374	42098.63	V		0.3		MIT
2373.060	2373.785	42126.82	V		2.3		MIT
2372.583	2373.307	42135.29	V		1.2		MIT
2372.175	2372.899	42142.54	V		2.0		MIT
2371.07	2371.79	42162.2	V		2.7		MIT
2367.655	2368.378	42222.98	V		1.9		MIT
2366.893	2367.616	42236.58	V		0.6		MIT
2366.31	2367.03	42247.0	V		0.5		MIT
2365.630	2366.353	42259.12	V		0.6		MIT
2364.378	2365.101	42281.50	V	0.3	0.3		MIT
2362.634	2363.356	42312.71	V	0.3	1.4		MIT
2360.343	2361.065	42353.77	V	0.7	0.7W		MIT
2358.745	2359.466	42382.46	V		2.5		MIT
2357.797	2358.518	42399.50	V	0.3H	1.8		MIT
2357.541	2358.262	42404.11	V		1.3		MIT
2355.448	2356.169	42441.79	V I		0.7		MIT
2355.230	2355.951	42445.71	V	0.6	0.6		MIT
2354.669	2355.389	42455.82	V	0.3	0.7		MIT
2352.177	2352.897	42500.80	V	0.7	2.3		MIT
2351.537	2352.257	42512.37	V		1.7		MIT
2351.26	2351.98	42517.4	V	0.3	1.7		MIT
2349.806	2350.525	42543.68	V	0.5W	2.2		MIT
2348.30	2349.02	42571.0	V		1.4		MIT
2347.151	2347.870	42591.80	V		2.2		MIT
2347.042	2347.761	42593.78	V I	1.2	1.3		MIT
2346.879	2347.598	42596.74	V		1.4		MIT
2346.343	2347.062	42606.47	V		2.1		MIT
2345.33	2346.05	42624.9	V		0.5		MIT
2343.830	2344.548	42652.14	V		0.7		MIT
2343.109	2343.827	42665.27	V		2.4		MIT
2342.136	2342.854	42682.99	V	0.6	2.1		MIT
2340.486	2341.203	42713.08	V I	1.7	0.5		MIT
2339.684	2340.401	42727.72	V I	1.4			MIT
2337.956	2338.673	42759.30	V		1.3		ME
2337.324	2338.041	42770.86	V		1.9		MIT
2337.13	2337.85	42774.4	V		2.0		MIT
2336.105	2336.821	42793.17	V	0.3	1.7		MIT
2335.492	2336.208	42804.40	V	0.5	1.7		MIT
2335.337	2336.053	42807.25	V	0.3	0.5		MIT
2334.446	2335.162	42823.58	V I	1.4	1.6S		MIT
2334.213	2334.929	42827.86	V		2.4		MIT
2333.603	2334.319	42839.05	V		1.7		MIT
2333.32	2334.04	42844.2	V I	1.0			MIT
2331.771	2332.486	42872.71	V		2.5		MIT
2331.29	2332.01	42881.5	V		1.7		MIT
2330.46	2331.18	42896.8	V	0.9	2.5		MIT

AIR WAVELENGTH (ANGSTRMS)	VACUUM WAVELENGTH (ANGSTRMS)	WAVENUMBER (KAYSERS)	LMENT	LOG ARC	INTENSITY SPK DIS	REF
2330.145	2330.860	42902.62	V		0.6	MIT
2329.523	2330.238	42914.08	V I	1.2		MIT
2328.934	2329.649	42924.93	V		2.0	MIT
2327.973	2328.687	42942.65	V I	0.9		MIT
2326.020	2326.734	42978.70	V		1.1	MIT
2325.869	2326.583	42981.49	V I	1.5		MIT
2325.12	2325.83	42995.3	V		2.0	MIT
2324.748	2325.462	43002.21	V I	1.7		MIT
2324.352	2325.066	43009.54	V I	0.7		MIT
2324.180	2324.894	43012.72	V I	0.7		MIT
2323.830	2324.544	43019.20	V		2.5	MIT
2322.102	2322.815	43051.21	V I	0.5		MIT
2321.065	2321.778	43070.44	V	0.5		MIT
2320.153	2320.866	43087.37	V I	1.2	0.3	MIT
2319.841	2320.554	43093.16	V		1.0	MIT
2318.999	2319.712	43108.81	V		2.2	MIT
2318.070	2318.782	43126.08	V		2.4	MIT
2317.568	2318.280	43135.42	V		1.5	MIT
2316.751	2317.463	43150.63	V I	1.3		ME
2315.635	2316.347	43171.43	V I	1.4		MIT
2315.00	2315.71	43183.3	V		0.3H	MIT
2314.697	2315.409	43188.92	V I	0.9		MIT
2314.187	2314.898	43198.44	V		2.0	MIT
2313.935	2314.646	43203.14	V		0.5H	MIT
2312.523	2313.234	43229.52	V I	0.7		MIT
2312.418	2313.129	43231.48	V I	0.7		MIT
2311.456	2312.167	43249.47	V I	1.2		MIT
2311.346	2312.057	43251.53	V		2.2	MIT
2310.958	2311.669	43258.79	V	0.3		ME
2310.185	2310.896	43273.27	V I	1.0		MIT
2309.842	2310.552	43279.69	V		2.1	MIT
2309.070	2309.780	43294.16	V		0.5	MIT
2308.832	2309.542	43298.62	V		0.3W	MIT
2308.288	2308.998	43308.83	V I	1.0		MIT
2304.785	2305.494	43374.65	V		0.3W	ME
2304.349	2305.058	43382.85	V	0.5		ME
2303.242	2303.951	43403.70	V		0.8	MIT
2302.85	2303.56	43411.1	V I		0.5W	MIT
2302.256	2302.965	43422.29	V	0.3H	0.5	MIT
2299.543	2300.251	43473.51	V I	0.3		MIT
2299.337	2300.045	43477.41	V	0.3		MIT
2297.850	2298.558	43505.54	V		2.0	MIT
2296.313	2297.021	43534.66	V		1.0	MIT
2295.504	2296.211	43550.00	V	1.2	1.2	ME
2295.416	2296.123	43551.67	V I	0.3	0.3	MIT
2294.992	2295.699	43559.71	V	0.7	2.0	ME
2293.243	2293.950	43592.93	V I	0.3		ME
2292.855	2293.562	43600.31	V		2.4	MIT
2292.589	2293.296	43605.37	V		1.4	MIT
2291.533	2292.239	43625.46	V I	0.7		MIT
2290.544	2291.250	43644.29	V		1.7J	MIT
2290.263	2290.969	43649.65	V I	0.3H	0.3	ME
2289.219	2289.925	43669.55	V		1.5	MIT
2288.628	2289.334	43680.83	V		1.9	MIT
2288.095	2288.801	43691.00	V	0.5H	1.2	MIT
2287.927	2288.633	43694.21	V		1.2	MIT
2286.586	2287.291	43719.83	V I	0.6		MIT
2286.113	2286.818	43728.88	V II		0.3W	SZ
2285.455	2286.160	43741.47	V		2.0	MIT
2284.984	2285.689	43750.48	V I	0.3H	0.6	MIT
2284.748	2285.453	43755.00	V		0.5	ME
2284.497	2285.202	43759.81	V I	0.9	0.3H	MIT
2283.766	2284.471	43773.81	V		0.8	MIT
2283.376	2284.081	43781.29	V I	0.7	0.7W	MIT
2282.857	2283.562	43791.24	V		0.6	MIT
2281.609	2282.313	43815.19	V	0.3	1.5	MIT
2281.237	2281.941	43822.34	V		1.5	MIT
2280.336	2281.040	43839.65	V		1.4	MIT
2279.763	2280.467	43850.67	V		1.0	MIT
2279.377	2280.081	43858.10	V		1.0	MIT
2279.152	2279.856	43862.42	V I	0.5		ME
2278.973	2279.677	43865.87	V		1.4	MIT
2278.093	2278.797	43882.81	V		0.3W	MIT
2276.883	2277.586	43906.13	V I	0.5		MIT
2276.667	2277.370	43910.30	V I	0.3		MIT
2275.871	2276.574	43925.65	V		0.3	MIT
2275.229	2275.932	43938.04	V		1.2	MIT
2273.624	2274.327	43969.06	V		0.8	MIT
2273.020	2273.722	43980.74	V		1.0	MIT
2272.048	2272.750	43999.56	V I	0.5		MIT
2271.846	2272.548	44003.47	V		0.3	MIT
2271.185	2271.887	44016.27	V		0.5	MIT
2268.361	2269.062	44071.07	V		1.6	ME
2267.612	2268.313	44085.62	V		0.3	ME
2263.612	2264.312	44163.52	V II		0.5	ME
2263.305	2264.005	44169.51	V	0.7		ME
2262.405	2263.105	44187.08	V		0.7	MIT
2261.850	2262.550	44197.92	V		0.7W	MIT
2261.085	2261.785	44212.87	V	0.7	1.2	MIT
2260.83	2261.53	44217.9	V		0.6	MIT
2258.813	2259.512	44257.34	V	1.4	1.5	MIT
2256.982	2257.681	44293.24	V I	1.4	1.2	MIT
2252.950	2253.648	44372.50	V		0.9	MIT
2252.681	2253.379	44377.80	V I	0.5		ME
2251.555	2252.253	44399.99	V		0.9	MIT
2251.120	2251.818	44408.57	V		0.3	MIT
2250.682	2251.380	44417.21	V I	1.1	1.5	MIT
2250.493	2251.191	44420.94	V		0.5	MIT
2249.074	2249.771	44448.96	V		1.0	MIT
2247.516	2248.213	44479.77	V I	0.7		MIT
2246.333	2247.030	44503.19	V		0.3	MIT
2245.766	2246.463	44514.43	V I	1.0		MIT
2243.742	2244.438	44554.58	V I	0.7		ME
2243.465	2244.161	44560.08	V		0.3	MIT
2243.267	2243.963	44564.01	V	0.5		MIT
2242.614	2243.310	44576.99	V	0.7		ME
2241.842	2242.538	44592.34	V I	1.4		MIT
2241.535	2242.231	44598.44	V		2.3	MIT
2241.213	2241.909	44604.85	V I	0.7		ME
2240.622	2241.317	44616.62	V		2.0	MIT
2240.302	2240.997	44622.99	V I	0.5		ME
2237.227	2237.922	44684.31	V I	1.3	0.3	MIT
2234.673	2235.367	44735.38	V I	0.8		MIT
2232.912	2233.606	44770.66	V		2.7	MIT
2232.252	2232.946	44783.89	V I	0.8		ME
2231.418	2232.111	44800.63	V	1.2		MIT
2230.372	2231.065	44821.64	V I	0.9	1.3	MIT
2229.988	2230.681	44829.35	V	0.5	1.0	MIT
2229.739	2230.432	44834.36	V I	1.1	1.9	MIT
2228.835	2229.528	44852.54	V I	1.0	0.8	ME
2228.299	2228.992	44863.33	V		2.0	MIT
2227.39	2228.08	44881.6	V	0.7		MIT
2225.787	2226.479	44913.96	V	0.6		ME
2225.427	2226.119	44921.22	V I	1.2	0.3	MIT
2225.029	2225.721	44929.26	V I	0.7		ME
2223.026	2223.718	44969.74	V I	1.0		MIT
2222.838	2223.530	44973.54	V I	0.8		MIT
2222.697	2223.389	44976.39	V		2.0	ME
2221.99	2222.68	44990.7	V		1.0	A
2220.450	2221.141	45021.90	V I	0.5		ME
2220.209	2220.900	45026.79	V	0.9	1.5	MIT
2219.652	2220.343	45038.09	V I	0.3		ME
2219.400	2220.091	45043.20	V		0.3	MIT
2218.428	2219.119	45062.93	V		2.1	MIT
2218.248	2218.939	45066.59	V I	1.2		MIT
2217.993	2218.684	45071.77	V		0.5J	MIT
2217.413	2218.104	45083.56	V		2.2	MIT
2216.26	2216.95	45107.0	V I	0.5		MIT
2216.034	2216.724	45111.61	V	0.3	2.0	MIT
2215.786	2216.476	45116.66	V		1.0	ME

AIR WAVELENGTH (ANGSTRMS)	VACUUM WAVELENGTH (ANGSTRMS)	WAVENUMBER (KAYSERS)	LMENT		LOG ARC	INTENSITY SPK	INTENSITY DIS	REF
2214.016	2214.706	45152.72	V			2.0		ME
2213.692	2214.382	45159.33	V	I	0.8			ME
2211.364	2212.053	45206.87	V	I	0.3	0.3H		MIT
2210.305	2210.994	45228.53	V			1.3		ME
2210.033	2210.722	45234.09	V			1.2		MIT
2209.97	2210.66	45235.4	V		0.7J	0.9		MIT
2209.219	2209.908	45250.76	V			1.7		MIT
2208.934	2209.623	45256.59	V			1.6		MIT
2207.63	2208.32	45283.3	V			1.5		MIT
2206.44	2207.13	45307.7	V			0.9		MIT
2205.71	2206.40	45322.7	V			0.3		MIT
2204.935	2205.623	45338.67	V	I	0.5			MIT
2204.495	2205.183	45347.71	V			1.6		MIT
2203.660	2204.348	45364.90	V	I	0.3			MIT
2202.724	2203.411	45384.17	V	I	1.0			ME
2202.50	2203.19	45388.8	V			1.8		MIT
2201.67	2202.36	45405.9	V			1.7		MIT
2200.174	2200.861	45436.77	V	I	0.7			ME
2199.440	2200.127	45451.93	V			0.6		MIT
2198.525	2199.212	45470.84	V			1.1		MIT
2198.00	2198.69	45481.7	V			1.6		MIT
2196.56	2197.25	45511.5	V	I	0.3			ME
2196.40	2197.09	45514.8	V	I	0.7	0.3J		ME
2196.30	2196.99	45516.9	V	I	0.5			MIT
2195.69	2196.38	45529.5	V			1.0		MIT
2195.10	2195.79	45541.8	V			0.3		ME
2194.84	2195.53	45547.2	V			0.9		MIT
2194.64	2195.33	45551.3	V	I	0.7			MIT
2193.83	2194.52	45568.1	V	I	0.5	0.7		MIT
2193.46	2194.15	45575.8	V		0.6			MIT
2192.91	2193.60	45587.3	V			1.0W		MIT
2191.66	2192.35	45613.3	V	I	0.5			MIT
2191.10	2191.79	45624.9	V	I	0.6	0.6		MIT
2190.47	2191.15	45638.0	V			0.9		MIT
2190.22	2190.90	45643.2	V		0.3	1.2		MIT
2189.94	2190.62	45649.1	V	I	0.7			MIT
2189.63	2190.31	45655.5	V	I		0.3		MIT
2188.05	2188.73	45688.5	V	I	0.5	0.5J		MIT
2187.95	2188.63	45690.6	V	I	0.3			MIT
2187.38	2188.06	45702.5	V	I	0.5			MIT
2186.93	2187.61	45711.9	V			1.0		MIT
2185.96	2186.64	45732.2	V			1.6		MIT
2185.38	2186.06	45744.3	V		0.6	1.7		MIT
2184.89	2185.57	45754.6	V			0.5		MIT
2184.53	2185.21	45762.1	V	I	0.3			ME
2184.39	2185.07	45765.0	V			0.5H		MIT
2184.18	2184.86	45769.5	V			0.3		MIT
2183.08	2183.76	45792.5	V	II		0.5		MIT
2182.78	2183.46	45798.8	V			0.5W		MIT
2182.22	2182.90	45810.5	V	I	1.5	0.7		MIT
2181.98	2182.66	45815.6	V	I	0.7			MIT
2177.24	2177.92	45915.3	V	I	0.7	1.0		MIT
2177.02	2177.70	45920.0	V	I	1.3	0.6		MIT
2175.835	2176.517	45944.97	V		0.3	0.7		MIT
2173.15	2173.83	46001.7	V	I	1.2	0.7		MIT
2172.75	2173.43	46010.2	V		0.6			MIT
2171.846	2172.527	46029.35	V			1.1		MIT
2170.76	2171.44	46052.4	V	I	1.4	0.3		MIT
2170.38	2171.06	46060.4	V			0.6		MIT
2170.07	2170.75	46067.0	V	II		0.3J		MIT
2169.846	2170.527	46071.77	V		0.5			MIT
2169.55	2170.23	46078.0	V			0.5H		MIT
2168.09	2168.77	46109.1	V	II		0.6		MIT
2167.68	2168.36	46117.8	V	II		0.6		MIT
2166.18	2166.86	46149.7	V			1.0		MIT
2164.90	2165.58	46177.0	V		0.5			MIT
2163.68	2164.36	46203.0	V			0.7H		MIT
2161.50	2162.18	46249.6	V			0.6		MIT
2157.06	2157.74	46344.8	V			0.3W		MIT
2151.818	2152.495	46457.72	V	II		1.7		MIT

AIR WAVELENGTH (ANGSTRMS)	VACUUM WAVELENGTH (ANGSTRMS)	WAVENUMBER (KAYSERS)	LMENT		LOG ARC	INTENSITY SPK	INTENSITY DIS	REF
2151.056	2151.733	46474.17	V	II		1.6		MIT
2150.840	2151.517	46478.84	V	II		1.6		MIT
2149.392	2150.068	46510.15	V	II		0.8		MIT
2148.420	2149.096	46531.19	V	II	0.7	1.5		MIT
2147.99	2148.67	46540.5	V			0.5		MIT
2147.60	2148.28	46549.0	V		0.5J			MIT
2147.46	2148.14	46552.0	V	II	0.3	2.0		MIT
2146.62	2147.30	46570.2	V	I	0.5			MIT
2145.991	2146.667	46583.85	V	II	0.7	1.6		MIT
2143.713	2144.388	46633.35	V	II	0.5	0.6		MIT
2143.047	2143.722	46647.84	V	II	1.0	1.7		MIT
2142.74	2143.42	46654.5	V	II		0.5		MIT
2142.43	2143.11	46661.3	V	II		0.6		MIT
2141.980	2142.655	46671.07	V	II	1.0	1.9		MIT
2141.70	2142.37	46677.2	V			0.5		MIT
2140.087	2140.762	46712.35	V	II	1.0	1.9		MIT
2139.821	2140.495	46718.15	V	II	0.7	1.4		MIT
2138.16	2138.83	46754.4	V	II	0.9	1.7		MIT
2137.31	2137.98	46773.0	V	II	1.0	1.9		MIT
2134.12	2134.79	46842.9	V	II	1.5	2.1		MIT
2133.05	2133.72	46866.4	V			1.7		MIT
2132.91	2133.58	46869.5	V		0.5W			MIT
2131.84	2132.51	46893.0	V	II		1.4		MIT
2130.43	2131.10	46924.1	V	II		0.7		MIT
2129.95	2130.62	46934.6	V			0.3W		MIT
2129.469	2130.141	46945.24	V	II	0.7	1.5		MIT
2128.241	2128.913	46972.32	V	II	0.3	1.0		ME
2127.37	2128.04	46991.5	V	II		0.7		MIT
2126.935	2127.607	47001.16	V	II	0.5	1.0		MIT
2126.583	2127.255	47008.94	V			1.2W		MIT
2125.83	2126.50	47025.6	V		0.6			MIT
2124.11	2124.78	47063.7	V		0.8	0.7		MIT
2123.60	2124.27	47075.0	V	II	0.3	0.9		MIT
2123.325	2123.996	47081.06	V	II	0.7	1.5		MIT
2122.13	2122.80	47107.6	V	II		0.6		MIT
2121.54	2122.21	47120.7	V	II		0.5		MIT
2119.561	2120.231	47164.66	V	II		0.9		MIT
2119.15	2119.82	47173.8	V			1.4W		MIT
2118.83	2119.50	47180.9	V			1.2		MIT
2118.43	2119.10	47189.8	V	II		1.2		MIT
2117.474	2118.144	47211.14	V	II	0.5	1.0		MIT
2117.272	2117.942	47215.65	V			1.0		MIT
2116.769	2117.439	47226.87	V			0.5		MIT
2114.31	2114.98	47281.8	V	II		0.7		MIT
2114.04	2114.71	47287.8	V	II		1.2		MIT
2111.04	2111.71	47355.0	V			0.6		MIT
2110.61	2111.28	47364.7	V	II		0.3		MIT
2110.50	2111.17	47367.1	V			0.3		MIT
2109.95	2110.62	47379.5	V			0.7		MIT
2109.27	2109.94	47394.7	V	II		0.3		MIT
2107.41	2108.08	47436.6	V	II		0.6J		MIT
2106.32	2106.99	47461.1	V		0.5			MIT
2104.56	2105.23	47500.8	V		0.6			MIT
2103.67	2104.34	47520.9	V	II	0.3	1.4		MIT
2103.52	2104.19	47524.3	V	II		0.9		MIT
2102.21	2102.88	47553.9	V		0.5			MIT
2101.87	2102.54	47561.6	V	II		0.7		MIT
2101.16	2101.83	47577.7	V	II	0.5	1.6		MIT
2100.78	2101.45	47586.3	V	I	0.5			MIT
2100.51	2101.18	47592.4	V	I	1.0			MIT
2099.13	2099.80	47623.7	V	II	0.3	1.4		MIT
2098.59	2099.26	47635.9	V	II		0.3W		MIT
2098.49	2099.16	47638.2	V		0.8L			MIT
2098.00	2098.67	47649.3	V	II		0.3		MIT
2097.33	2098.00	47664.5	V	I	1.0			MIT
2097.02	2097.69	47671.6	V	II		0.9		MIT
2096.35	2097.02	47686.8	V		0.7	1.2W		MIT
2096.16	2096.83	47691.1	V		0.8	0.0		MIT
2095.94	2096.61	47696.1	V	II		1.0		MIT
2095.75	2096.42	47700.5	V		0.9	0.3		MIT

AIR WAVELENGTH (ANGSTRMS)	VACUUM WAVELENGTH (ANGSTRMS)	WAVENUMBER (KAYSERS)	LMENT		LOG ARC	INTENSITY SPK	REF
2095.34	2096.01	47709.8	V	II		1.0	MIT
2095.04	2095.71	47716.6	V	II		0.8	MIT
2094.662	2095.328	47725.23	V		1.0W	0.0	ME
2093.41	2094.08	47753.8	V			0.3	MIT
2092.495	2093.160	47774.65	V		1.0	0.7	ME
2092.44	2093.11	47775.9	V	I	1.0		ME
2092.35	2093.02	47778.0	V	I	0.7		MIT
2091.84	2092.51	47789.6	V	II		0.5	MIT
2091.307	2091.972	47801.79	V	I	1.0W	0.7W	MIT
2090.90	2091.56	47811.1	V		0.6		MIT
2090.66	2091.32	47816.6	V	I	0.7W	0.3W	MIT
2090.33	2090.99	47824.1	V			1.2	MIT
2089.94	2090.60	47833.0	V	I	0.6J		MIT
2088.56	2089.22	47864.6	V		1.2W	0.0	MIT
2087.61	2088.27	47886.4	V		0.5	0.7	MIT
2086.57	2087.23	47910.3	V		0.6W		MIT
2086.32	2086.98	47916.0	V	I	0.8	0.3J	MIT
2082.51	2083.17	48003.7	V	I	1.4W		MIT
2081.67	2082.33	48023.1	V		0.6	0.5	MIT
2081.11	2081.77	48036.0	V			0.7J	MIT
2080.74	2081.40	48044.5	V			0.3J	MIT
2078.96	2079.62	48085.6	V			0.6	MIT
2077.80	2078.46	48112.5	V			1.3	MIT
2077.55	2078.21	48118.3	V			1.0	MIT
2076.85	2077.51	48134.5	V		0.7	1.4W	MIT
2076.49	2077.15	48142.8	V			0.3	MIT
2074.87	2075.53	48180.4	V			1.2	MIT
2073.42	2074.08	48214.1	V			0.7J	MIT
2072.43	2073.09	48237.1	V			1.0	MIT
2070.78	2071.44	48275.6	V			0.7	MIT
2068.80	2069.46	48321.8	V			1.4	MIT
2068.53	2069.19	48328.1	V			0.9	MIT
2067.16	2067.82	48360.1	V			0.5J	MIT
2066.83	2067.49	48367.8	V			0.6J	MIT
2066.47	2067.13	48376.2	V			0.5J	MIT
2065.75	2066.41	48393.1	V	II		1.4	MIT
2064.75	2065.41	48416.5	V			0.3	MIT
2063.13	2063.79	48454.5	V			0.7	MIT
2061.99	2062.65	48481.3	V			0.3	MIT
2058.35	2059.01	48567.1	V	II		1.2	MIT
2057.36	2058.02	48590.4	V	II		0.9	MIT
2057.20	2057.86	48594.2	V	II		0.6	MIT
2056.87	2057.53	48602.0	V			0.6	MIT
2056.53	2057.19	48610.0	V			0.7J	MIT
2055.56	2056.22	48633.0	V	II		0.3W	MIT
2054.84	2055.50	48650.0	V		1.0	1.2	MIT
2052.38	2053.04	48708.3	V			0.5	MIT
2051.78	2052.44	48722.6	V			1.0H	MIT
2051.29	2051.95	48734.2	V			0.3H	MIT
2049.35	2050.01	48780.3	V			0.3H	MIT
2049.11	2049.77	48786.0	V			0.3H	MIT
2048.73	2049.39	48795.1	V			0.5	MIT
2045.97	2046.63	48860.9	V		0.7	1.1	MIT
2043.13	2043.79	48928.8	V		0.6		MIT
2041.02	2041.68	48979.4	V		0.9		MIT
2039.31	2039.97	49020.5	V			1.2	MIT
2039.073	2039.728	49026.14	V			0.3H	MIT
2038.872	2039.527	49030.98	V		1.0W	0.3H	ME
2038.490	2039.145	49040.16	V		0.6	1.1	ME
2037.85	2038.50	49055.6	V			1.1	MIT
2037.51	2038.16	49063.7	V			0.8	MIT
2035.07	2035.72	49122.6	V			1.1	MIT
2034.066	2034.720	49146.81	V		1.0	0.3	MIT
2033.50	2034.15	49160.5	V			0.3	MIT
2032.28	2032.93	49190.0	V	I	0.7		MIT
2031.37	2032.02	49212.0	V			0.6H	MIT
2028.87	2029.52	49272.7	V			0.3	MIT
2028.451	2029.104	49282.83	V		0.8		ME
2027.617	2028.270	49303.10	V		0.7	0.3J	MIT
2023.56	2024.21	49401.9	V	II		1.6	MIT

AIR WAVELENGTH (ANGSTRMS)	VACUUM WAVELENGTH (ANGSTRMS)	WAVENUMBER (KAYSERS)	LMENT		LOG ARC	INTENSITY SPK	REF
2022.64	2023.29	49424.4	V			0.9J	MIT
2021.83	2022.48	49444.2	V	II		0.3H	MIT
2021.38	2022.03	49455.2	V			0.6W	MIT
2020.85	2021.50	49468.2	V			0.8	MIT
2020.56	2021.21	49475.3	V	II		0.5	MIT
2020.11	2020.76	49486.3	V			0.3	MIT
2019.49	2020.14	49501.5	V	II		0.5	MIT
2018.87	2019.52	49516.7	V			0.3H	MIT
2017.32	2017.97	49554.7	V	II		0.3	MIT
2016.54	2017.19	49573.9	V	II		1.4	MIT
2015.77	2016.42	49592.8	V	II		0.6	MIT
2015.55	2016.20	49598.2	V			0.7	MIT
2015.10	2015.75	49609.3	V	II	0.6	0.7	MIT
2014.19	2014.84	49631.7	V	II	0.3	1.7	MIT
2012.85	2013.50	49664.8	V			0.9	MIT
2012.65	2013.30	49669.7	V	II		0.6	MIT
2011.56	2012.21	49696.6	V		0.5		A
2010.01	2010.66	49734.9	V			0.5W	MIT
2009.77	2010.42	49740.9	V			0.3	MIT
2009.34	2009.99	49751.5	V			0.3J	MIT
2007.65	2008.30	49793.4	V	II		1.0	MIT
2006.87	2007.52	49812.7	V		0.3	1.6	MIT
2005.87	2006.52	49837.5	V			0.8	MIT
2005.21	2005.86	49854.0	V	II		0.3	MIT
2004.78	2005.43	49864.6	V	II		1.7	MIT
2001.66	2002.31	49942.4	V			1.0	MIT
2001.45	2002.10	49947.6	V	II		1.1	MIT
2001.15	2001.80	49955.1	V			1.0	MIT
2000.78	2001.43	49964.3	V	II		0.7	ME
2000.14	2000.79	49980.3	V			1.0	ME
9158.95	9161.46	10915.3	W		0.3		MIT
8915.78	8918.23	11213.0	W		0.9		MIT
8883.62	8886.06	11253.6	W		0.7		MIT
8871.61	8874.05	11268.8	W		0.9		MIT
8865.50	8867.93	11276.6	W		1.2		MIT
8823.39	8825.81	11330.4	W		0.7		MIT
8821.31	8823.73	11333.1	W		0.5		MIT
8777.42	8779.83	11389.7	W		0.7		MIT
8776.41	8778.82	11391.1	W		0.5		MIT
8754.75	8757.15	11419.2	W		0.5H		MIT
8746.60	8749.00	11429.9	W		1.0		MIT
8744.58	8746.98	11432.5	W		0.9		MIT
8740.44	8742.84	11437.9	W		1.0		MIT
8710.75	8713.14	11476.9	W		0.7		MIT
8641.54	8643.91	11568.8	W		0.9		MIT
8614.48	8616.85	11605.2	W		0.9		MIT
8613.26	8615.63	11606.8	W		1.2		MIT
8594.37	8596.73	11632.3	W		1.3		MIT
8585.06	8587.42	11644.9	W		1.7		MIT
8577.85	8580.21	11654.7	W		0.3		MIT
8530.99	8533.33	11718.8	W		0.3		MIT
8516.37	8518.71	11738.9	W		0.7		MIT
8515.39	8517.73	11740.2	W		0.7		MIT
8506.92	8509.26	11751.9	W		0.8		MIT
8499.54	8501.88	11762.1	W		0.3H		MIT
8495.12	8497.45	11768.2	W		0.9		MIT
8486.77	8489.10	11779.8	W		0.5		MIT
8475.16	8477.49	11795.9	W		1.0		MIT
8470.95	8473.28	11801.8	W		0.7		MIT
8431.91	8434.23	11856.4	W		0.7		ME
8417.08	8419.39	11877.3	W		0.9		MIT
8402.55	8404.86	11897.9	W		0.3		MIT
8382.87	8385.17	11925.8	W		0.3		MIT
8369.07	8371.37	11945.5	W		0.5		MIT
8359.87	8362.17	11958.6	W		0.6		MIT
8358.67	8360.97	11960.3	W		1.2		MIT
8350.25	8352.54	11972.4	W		0.5H		MIT
8348.76	8351.05	11974.5	W		1.0		MIT
8338.01	8340.30	11990.0	W		1.2		MIT
8322.00	8324.29	12013.0	W		1.0		MIT

AIR WAVELENGTH (ANGSTRMS)	VACUUM WAVELENGTH (ANGSTRMS)	WAVENUMBER (KAYSERS)	LMENT	LOG ARC	INTENSITY SPK	INTENSITY DIS	REF
8317.12	8319.41	12020.1	W	0.5			MIT
8311.57	8313.85	12028.1	W	0.7			MIT
8165.70	8167.95	12243.0	W	0.9			MIT
8143.12	8145.36	12276.9	W	0.9			MIT
8139.57	8141.81	12282.3	W	0.7H			MIT
8132.33	8134.57	12293.2	W	0.7			MIT
8123.79	8126.02	12306.1	W	1.3			MIT
8091.10	8093.32	12355.9	W	0.5			MIT
8060.36	8062.58	12403.0	W	0.9			MIT
8055.60	8057.82	12410.3	W	1.3			MIT
8054.86	8057.08	12411.4	W	0.9			MIT
8017.18	8019.38	12469.8	W	1.4			MIT
7957.06	7959.25	12564.0	W	0.7			MIT
7940.93	7943.11	12589.5	W	1.0			MIT
7909.19	7911.37	12640.0	W	0.5			MIT
7905.25	7907.42	12646.3	W	0.5			MIT
7886.47	7888.64	12676.5	W	0.9			MIT
7880.34	7882.51	12686.3	W	0.7			MIT
7867.03	7869.19	12707.8	W	0.5			MIT
7863.46	7865.62	12713.6	W	0.9			MIT
7823.78	7825.93	12778.0	W	0.5			MIT
7808.94	7811.09	12802.3	W	0.7			MIT
7784.12	7786.26	12843.1	W	1.0			MIT
7776.71	7778.85	12855.4	W	0.5			MIT
7761.14	7763.28	12881.2	W	0.5			MIT
7701.00	7703.12	12981.7	W	0.5H			MIT
7688.94	7691.06	13002.1	W	1.0			MIT
7614.11	7616.21	13129.9	W	0.5			MIT
7569.91	7571.99	13206.6	W	0.8W			MIT
7550.46	7552.54	13240.6	W	0.5			MIT
7537.43	7539.51	13263.5	W	0.5			MIT
7508.99	7511.06	13313.7	W	0.6			MIT
7504.16	7506.23	13322.3	W	0.3			MIT
7483.35	7485.41	13359.3	W	0.6			MIT
7471.54	7473.60	13380.4	W	0.3			MIT
7468.18	7470.24	13386.5	W	0.3			MIT
7459.27	7461.32	13402.4	W	0.3			MIT
7456.34	7458.39	13407.7	W	0.3			MIT
7451.40	7453.45	13416.6	W	0.5			MIT
7385.08	7387.11	13537.1	W	0.8			ME
7296.58	7298.59	13701.3	W	1.1			MIT
7285.80	7287.81	13721.6	W	1.0			MIT
7278.23	7280.24	13735.8	W	0.9			MIT
7274.49	7276.49	13742.9	W	0.5			MIT
7245.23	7247.23	13798.4	W	0.3			MIT
7237.08	7239.07	13813.9	W	0.7			MIT
7226.05	7228.04	13835.0	W	0.3			MIT
7216.31	7218.30	13853.7	W	0.8			MIT
7200.18	7202.16	13884.7	W	0.9			MIT
7198.62	7200.60	13887.7	W	0.6			MIT
7191.37	7193.35	13901.7	W	0.3			MIT
7162.66	7164.63	13957.4	W	0.7			MIT
7148.93	7150.90	13984.2	W	0.8			MIT
7140.54	7142.51	14000.7	W	0.9			MIT
7098.22	7100.18	14084.2	W	0.3			MIT
7028.68	7030.62	14223.5	W	0.3			MIT
7017.91	7019.85	14245.3	W	0.5			MIT
6993.26	6995.19	14295.5	W	0.5			MIT
6984.30	6986.23	14313.9	W	0.6			MIT
6964.16	6966.08	14355.3	W	0.3			MIT
6934.27	6936.18	14417.2	W	0.3			MIT
6820.28	6822.16	14658.1	W	0.3			MIT
6708.18	6710.03	14903.1	W	1.3			MIT
6693.118	6694.966	14936.60	W	0.5			MIT
6678.414	6680.258	14969.48	W	0.3			MIT
6621.680	6623.509	15097.74	W	0.7			MIT
6573.948	6575.764	15207.36	W	0.3	0.0		MIT
6563.224	6565.037	15232.21	W	0.3	0.0		MIT
6538.136	6539.942	15290.66	W	0.3	0.0		MIT
6532.417	6534.222	15304.04	W	0.3			MIT

AIR WAVELENGTH (ANGSTRMS)	VACUUM WAVELENGTH (ANGSTRMS)	WAVENUMBER (KAYSERS)	LMENT	LOG ARC	INTENSITY SPK	INTENSITY DIS	REF
6445.132	6446.913	15511.30	W	0.7	0.3		MIT
6439.720	6441.500	15524.33	W	0.8	0.0		MIT
6438.03	6439.81	15528.4	W	0.5	0.0		ME
6425.36	6427.14	15559.0	W	0.7	0.0		ME
6416.004	6417.777	15581.72	W	0.7	0.0		MIT
6404.204	6405.974	15610.43	W	1.4	0.3		MIT
6402.07	6403.84	15615.6	W	0.7	0.0		ME
6399.736	6401.505	15621.33	W	0.7	0.0		MIT
6384.739	6386.504	15658.02	W	0.5	0.0		MIT
6382.487	6384.251	15663.54	W	0.5	0.0		MIT
6375.391	6377.154	15680.98	W	0.7			MIT
6371.472	6373.233	15690.62	W	0.8S			MIT
6370.337	6372.098	15693.42	W	0.5	0.0		MIT
6364.726	6366.486	15707.25	W	0.6	0.0		MIT
6361.076	6362.835	15716.27	W	0.5	0.0		MIT
6359.129	6360.887	15721.08	W	0.3			MIT
6355.904	6357.661	15729.05	W	0.5			MIT
6342.388	6344.142	15762.57	W	0.3			MIT
6331.449	6333.200	15789.81	W	0.3			MIT
6319.240	6320.987	15820.31	W	0.7			MIT
6308.488	6310.233	15847.28	W	0.7	0.0		MIT
6303.240	6304.983	15860.47	W	1.0	0.3		MIT
6298.596	6300.338	15872.17	W	0.5	0.0		MIT
6292.032	6293.772	15888.72	W	1.5	0.3		MIT
6291.029	6292.769	15891.26	W	0.6	0.0		MIT
6288.90	6290.64	15896.6	W	0.3			ME
6285.897	6287.635	15904.23	W	1.5	0.3		MIT
6269.19	6270.92	15946.6	W	0.3			ME
6258.893	6260.624	15972.85	W	0.5			MIT
6254.29	6256.02	15984.6	W	1.1			MIT
6253.636	6255.366	15986.28	W	0.5	0.0		MIT
6251.499	6253.228	15991.74	W	0.3			MIT
6245.792	6247.520	16006.35	W	0.3			MIT
6237.740	6239.466	16027.01	W	0.3			MIT
6225.747	6227.469	16057.89	W	0.6	0.0		MIT
6224.187	6225.909	16061.91	W	0.5	0.0		MIT
6203.514	6205.230	16115.44	W	1.2	0.0		MIT
6196.350	6198.064	16134.07	W	0.7	0.0		BK
6195.957	6197.671	16135.09	W	0.7	0.0		MIT
6193.233	6194.947	16142.19	W	0.7	0.0		MIT
6188.702	6190.414	16154.01	W	0.3			MIT
6187.578	6189.290	16156.94	W	0.7	0.0		MIT
6186.236	6187.948	16160.45	W	0.8	0.0		MIT
6179.650	6181.360	16177.67	W	0.3			MIT
6169.106	6170.813	16205.32	W	0.8	0.0		MIT
6161.434	6163.139	16225.50	W	0.7	0.0		MIT
6154.864	6156.567	16242.82	W	1.2	0.0		MIT
6153.728	6155.431	16245.82	W	1.2	0.0		MIT
6151.85	6153.55	16250.8	W	0.7			MIT
6148.22	6149.92	16260.4	W	0.3			ME
6147.225	6148.926	16263.00	W	0.6			MIT
6143.938	6145.638	16271.70	W	1.0	0.0		MIT
6140.97	6142.67	16279.6	W	0.3			ME
6133.17	6134.87	16300.3	W	0.6			ME
6128.274	6129.970	16313.29	W	1.2	0.3		MIT
6119.35	6121.04	16337.1	W	0.5			ME
6115.547	6117.240	16347.24	W	1.1	0.0		MIT
6111.680	6113.372	16357.58	W	1.1	0.0		MIT
6107.411	6109.102	16369.02	W	0.6	0.0		MIT
6094.76	6096.45	16403.0	W	0.5			ME
6092.29	6093.98	16409.6	W	0.3			ME
6083.82	6085.50	16432.5	W	0.3			ME
6081.481	6083.165	16438.81	W	1.4	0.0		MIT
6067.65	6069.33	16476.3	W	0.6	0.0		MIT
6066.893	6068.573	16478.34	W	0.5			MIT
6065.09	6066.77	16483.2	W	0.8	0.0		MIT
6055.90	6057.58	16508.2	W	0.3			ME
6046.66	6048.33	16533.5	W	0.5			ME
6043.331	6045.004	16542.58	W	1.0	0.0		MIT
6041.656	6043.329	16547.17	W	0.8	0.0		MIT

AIR WAVELENGTH (ANGSTRMS)	VACUUM WAVELENGTH (ANGSTRMS)	WAVENUMBER (KAYSERS)	LMENT	LOG ARC	INTENSITY SPK	DIS	REF
6037.33	6039.00	16559.0	W	0.3			ME
6029.885	6031.555	16579.47	W	0.5	0.0		MIT
6028.346	6030.015	16583.71	W	0.8	0.0		MIT
6021.541	6023.209	16602.45	W	1.4			MIT
6012.814	6014.479	16626.54	W	1.5	0.5		MIT
6009.701	6011.365	16635.16	W	0.6	0.0		MIT
6009.041	6010.705	16636.98	W	0.8	0.0		MIT
5998.270	5999.931	16666.86	W	0.3			MIT
5992.475	5994.135	16682.98	W	0.7			MIT
5990.119	5991.778	16689.54	W	0.6			MIT
5989.600	5991.259	16690.98	W	0.8			MIT
5987.440	5989.098	16697.00	W	0.6			MIT
5986.934	5988.592	16698.42	W	0.5			MIT
5983.836	5985.493	16707.06	W	0.8			MIT
5978.886	5980.542	16720.89	W	0.8			MIT
5976.682	5978.338	16727.06	W	0.7			MIT
5972.519	5974.173	16738.72	W	1.4			MIT
5965.859	5967.512	16757.40	W	1.4			MIT
5964.457	5966.109	16761.34	W	0.3	0.0		MIT
5960.823	5962.474	16771.56	W	0.9			MIT
5956.184	5957.834	16784.62	W	0.8			MIT
5953.971	5955.620	16790.86	W	0.7			MIT
5948.195	5949.843	16807.17	W	0.3			MIT
5947.583	5949.231	16808.90	W	1.2			MIT
5947.068	5948.716	16810.35	W	0.6			MIT
5940.883	5942.529	16827.85	W	0.7			MIT
5934.486	5936.130	16845.99	W	0.7			MIT
5928.582	5930.225	16862.77	W	0.8			MIT
5927.045	5928.687	16867.14	W	0.3			MIT
5921.357	5922.998	16883.34	W	0.5			MIT
5921.028	5922.669	16884.28	W	0.3			MIT
5913.398	5915.037	16906.07	W	0.5			MIT
5910.289	5911.927	16914.96	W	0.3			MIT
5905.031	5906.667	16930.02	W	0.6			MIT
5902.665	5904.301	16936.81	W	1.3			MIT
5901.990	5903.626	16938.74	W	0.5			MIT
5901.227	5902.862	16940.93	W	0.9			MIT
5891.614	5893.247	16968.57	W	1.1			MIT
5890.333	5891.965	16972.26	W	0.8			MIT
5884.625	5886.256	16988.73	W	0.3H			MIT
5880.223	5881.853	17001.45	W	1.2			MIT
5875.663	5877.291	17014.64	W	0.9			MIT
5874.229	5875.857	17018.79	W	0.9			MIT
5871.649	5873.276	17026.27	W	0.8			MIT
5870.014	5871.641	17031.01	W	0.8			MIT
5864.629	5866.255	17046.65	W	1.3			MIT
5858.232	5859.856	17065.27	W	0.8			MIT
5856.616	5858.239	17069.98	W	1.2			MIT
5854.412	5856.035	17076.40	W	0.8			MIT
5851.565	5853.187	17084.71	W	1.4			MIT
5845.261	5846.881	17103.14	W	1.3			MIT
5838.988	5840.607	17121.51	W	1.4			MIT
5833.589	5835.206	17137.35	W	1.1			MIT
5832.281	5833.898	17141.20	W	0.8			MIT
5822.601	5824.215	17169.70	W	0.8			MIT
5821.017	5822.631	17174.37	W	0.7			MIT
5814.153	5815.765	17194.64	W	0.5			MIT
5806.267	5807.877	17218.00	W	0.9			MIT
5806.070	5807.680	17218.58	W	0.8			MIT
5804.869	5806.479	17222.14	W	1.4W	1.1		MIT
5803.068	5804.677	17227.49	W	0.8			MIT
5799.528	5801.136	17238.00	W	1.1			MIT
5796.508	5798.115	17246.98	W	1.3			MIT
5793.069	5794.675	17257.22	W	1.3			MIT
5791.361	5792.967	17262.31	W	0.8			MIT
5789.842	5791.448	17266.84	W	0.8			MIT
5784.748	5786.352	17282.04	W	0.5			MIT
5773.518	5775.119	17315.66	W	0.8			MIT
5772.005	5773.606	17320.20	W	0.7			MIT
5771.988	5773.589	17320.25	W	1.1			MIT

AIR WAVELENGTH (ANGSTRMS)	VACUUM WAVELENGTH (ANGSTRMS)	WAVENUMBER (KAYSERS)	LMENT	LOG ARC	INTENSITY SPK	DIS	REF
5759.652	5761.249	17357.35	W	1.1			MIT
5756.09	5757.69	17368.1	W	1.1			MIT
5754.57	5756.17	17372.7	W	0.9			MIT
5753.381	5754.977	17376.26	W	0.8			MIT
5750.263	5751.858	17385.69	W	0.8			MIT
5749.216	5750.811	17388.85	W	1.2			MIT
5747.264	5748.858	17394.76	W	1.1			MIT
5741.170	5742.762	17413.22	W	0.8			MIT
5739.596	5741.188	17418.00	W	1.1			MIT
5735.087	5736.678	17431.69	W	1.7O	1.4		MIT
5728.600	5730.189	17451.43	W	0.8			MIT
5723.064	5724.652	17468.31	W	1.2			MIT
5715.396	5716.982	17491.75	W	0.7			MIT
5709.060	5710.644	17511.16	W	0.8			MIT
5708.036	5709.620	17514.30	W	0.6			MIT
5705.719	5707.302	17521.41	W	0.8			MIT
5704.376	5705.959	17525.54	W	0.8			MIT
5697.819	5699.400	17545.71	W	1.5			MIT
5693.752	5695.332	17558.24	W	0.8			MIT
5676.932	5678.507	17610.26	W	1.0			MIT
5676.609	5678.184	17611.26	W	1.0			MIT
5675.380	5676.955	17615.08	W	1.2			MIT
5674.416	5675.991	17618.07	W	1.5			MIT
5673.562	5675.136	17620.72	W	1.0			MIT
5664.337	5665.909	17649.42	W	0.3			MIT
5660.749	5662.320	17660.61	W	0.3			MIT
5657.038	5658.608	17672.19	W	0.6			MIT
5654.098	5655.667	17681.38	W	0.6			MIT
5651.251	5652.819	17690.29	W	0.3			MIT
5649.015	5650.583	17697.29	W	0.6			MIT
5648.701	5650.269	17698.27	W	0.8			MIT
5648.378	5649.946	17699.28	W	1.7W	1.7W		MIT
5647.013	5648.580	17703.56	W	0.5			MIT
5644.484	5646.051	17711.50	W	0.9			MIT
5642.041	5643.607	17719.16	W	1.2			MIT
5641.394	5642.960	17721.20	W	0.8			MIT
5639.636	5641.201	17726.72	W	0.8			MIT
5637.125	5638.690	17734.62	W	0.8			MIT
5635.505	5637.069	17739.72	W	0.9			MIT
5631.969	5633.532	17750.85	W	1.2			MIT
5631.826	5633.389	17751.30	W	1.0			MIT
5631.263	5632.826	17753.08	W	1.2			MIT
5629.933	5631.496	17757.27	W	0.7			MIT
5629.765	5631.328	17757.80	W	0.7			MIT
5629.647	5631.210	17758.17	W	1.1			MIT
5623.967	5625.528	17776.11	W	0.7			MIT
5618.742	5620.302	17792.64	W	0.5			MIT
5617.295	5618.854	17797.22	W	0.9			MIT
5617.082	5618.641	17797.90	W	1.0			MIT
5616.189	5617.748	17800.73	W	1.0			MIT
5615.164	5616.723	17803.98	W	0.6			MIT
5612.304	5613.862	17813.05	W	0.7			MIT
5608.139	5609.696	17826.28	W	0.8			MIT
5605.205	5606.761	17835.61	W	0.7			MIT
5604.334	5605.890	17838.38	W	1.0			MIT
5593.572	5595.125	17872.70	W	0.8			MIT
5591.035	5592.587	17880.81	W	0.7			MIT
5588.150	5589.702	17890.04	W	0.8			MIT
5584.062	5585.612	17903.14	W	0.8D			MIT
5579.941	5581.490	17916.36	W	0.9			MIT
5578.282	5579.831	17921.69	W	0.6			MIT
5576.341	5577.889	17927.93	W	1.0			MIT
5575.814	5577.362	17929.62	W	0.8			MIT
5572.007	5573.554	17941.87	W	0.7			MIT
5568.066	5569.612	17954.57	W	1.2			MIT
5563.983	5565.528	17967.75	W	0.9			MIT
5562.119	5563.664	17973.77	W	0.9			MIT
5559.767	5561.311	17981.37	W	0.9			MIT
5558.965	5560.509	17983.97	W	0.3			MIT
5558.755	5560.299	17984.65	W	0.3			MIT

AIR WAVELENGTH (ANGSTRMS)	VACUUM WAVELENGTH (ANGSTRMS)	WAVENUMBER (KAYSERS)	LMENT	LOG ARC	INTENSITY SPK	INTENSITY DIS	REF
5555.315	5556.858	17995.78	W	0.8			MIT
5554.115	5555.657	17999.67	W	0.8			MIT
5551.020	5552.562	18009.71	W	0.7			MIT
5550.331	5551.872	18011.94	W	0.9			MIT
5548.966	5550.507	18016.37	W	0.3	0.0		MIT
5548.081	5549.622	18019.25	W	0.8	0.3		MIT
5546.521	5548.061	18024.31	W	0.5L			MIT
5542.960	5544.499	18035.89	W	0.6			MIT
5542.404	5543.943	18037.70	W	0.6			MIT
5539.489	5541.028	18047.19	W	1.0			MIT
5538.690	5540.228	18049.80	W	0.6			MIT
5537.742	5539.280	18052.89	W	1.1			MIT
5537.299	5538.837	18054.33	W	0.9			MIT
5533.26	5534.80	18067.5	W	1.4			ME
5531.523	5533.059	18073.18	W	1.3			MIT
5528.495	5530.031	18083.08	W	0.8			MIT
5528.111	5529.647	18084.34	W	0.5			MIT
5526.661	5528.196	18089.08	W	0.5			MIT
5521.008	5522.542	18107.60	W	1.0			MIT
5519.164	5520.697	18113.65	W	0.8			MIT
5518.790	5520.323	18114.88	W	0.6			MIT
5514.842	5516.374	18127.85	W	0.3H			MIT
5514.698	5516.230	18128.32	W	1.7W	0.9		MIT
5508.632	5510.162	18148.29	W	0.9			MIT
5508.493	5510.023	18148.74	W	0.6			MIT
5503.453	5504.982	18165.36	W	1.7	0.0		MIT
5501.017	5502.545	18173.41	W	0.7			MIT
5500.514	5502.042	18175.07	W	1.2	1.2		MIT
5498.639	5500.167	18181.27	W	1.0			MIT
5496.244	5497.771	18189.19	W	1.0			MIT
5495.282	5496.809	18192.37	W	0.6			MIT
5492.593	5494.119	18201.28	W	0.6H			MIT
5492.321	5493.847	18202.18	W	1.7	1.7		MIT
5489.969	5491.494	18209.98	W	1.0			MIT
5489.134	5490.659	18212.75	W	0.7			MIT
5488.473	5489.998	18214.94	W	0.7			MIT
5487.785	5489.310	18217.23	W	1.2			MIT
5486.009	5487.533	18223.12	W	1.3			MIT
5480.637	5482.160	18240.99	W	1.0			MIT
5479.261	5480.783	18245.57	W	0.8	0.0H		MIT
5478.677	5480.199	18247.51	W	0.7			MIT
5477.799	5479.321	18250.44	W	1.4	1.3		MIT
5476.038	5477.560	18256.30	W	0.9			MIT
5475.114	5476.635	18259.39	W	1.2			MIT
5472.340	5473.861	18268.64	W	0.8			MIT
5471.506	5473.026	18271.43	W	0.8			MIT
5469.110	5470.630	18279.43	W	1.0			MIT
5467.551	5469.070	18284.64	W	0.9			MIT
5464.462	5465.981	18294.98	W	0.6			MIT
5457.938	5459.455	18316.85	W	1.0			MIT
5456.591	5458.107	18321.37	W	1.3			MIT
5451.801	5453.316	18337.47	W	0.8			MIT
5448.939	5450.453	18347.10	W	0.7			MIT
5446.145	5447.659	18356.51	W	0.7			MIT
5443.361	5444.874	18365.90	W	0.9			MIT
5441.143	5442.655	18373.39	W	0.5J			MIT
5440.078	5441.590	18376.98	W	1.0			MIT
5438.872	5440.384	18381.06	W	0.9			MIT
5437.800	5439.311	18384.68	W	0.5	0.3		MIT
5435.609	5437.120	18392.09	W	1.0			MIT
5435.062	5436.573	18393.94	W	1.3			MIT
5433.266	5434.776	18400.02	W	0.6			MIT
5428.019	5429.528	18417.81	W	0.6			MIT
5427.225	5428.734	18420.50	W	0.8	0.0H		MIT
5424.927	5426.435	18428.30	W	0.3			MIT
5424.637	5426.145	18429.29	W	0.5			MIT
5423.934	5425.442	18431.68	W	1.0			MIT
5423.446	5424.954	18433.34	W	0.7			MIT
5422.887	5424.394	18435.24	W	1.0			MIT
5421.862	5423.369	18438.72	W	0.8			MIT

AIR WAVELENGTH (ANGSTRMS)	VACUUM WAVELENGTH (ANGSTRMS)	WAVENUMBER (KAYSERS)	LMENT	LOG ARC	INTENSITY SPK	INTENSITY DIS	REF
5419.399	5420.905	18447.10	W	1.0			MIT
5415.976	5417.482	18458.76	W	0.5			MIT
5415.525	5417.030	18460.30	W	0.8			MIT
5413.834	5415.339	18466.06	W	0.8			MIT
5412.965	5414.470	18469.03	W	0.8			MIT
5408.578	5410.082	18484.01	W	0.8			MIT
5406.309	5407.812	18491.77	W	0.9			MIT
5404.319	5405.821	18498.58	W	0.8	0.0H		MIT
5400.937	5402.439	18510.16	W	0.8			MIT
5400.302	5401.803	18512.34	W	0.8			MIT
5399.585	5401.086	18514.79	W	0.7			MIT
5398.222	5399.723	18519.47	W	0.7			MIT
5397.969	5399.470	18520.34	W	1.0			MIT
5396.515	5398.015	18525.33	W	0.8			MIT
5394.755	5396.255	18531.37	W	0.8			MIT
5393.845	5395.345	18534.50	W	0.5	0.0		MIT
5392.797	5394.296	18538.10	W	0.7			MIT
5391.082	5392.581	18544.00	W	1.0			MIT
5389.688	5391.187	18548.79	W	0.7			MIT
5388.593	5390.091	18552.56	W	1.0			MIT
5388.024	5389.522	18554.52	W	1.2			MIT
5382.789	5384.286	18572.57	W	0.7			MIT
5381.997	5383.494	18575.30	W	1.0			MIT
5379.436	5380.932	18584.14	W	0.9			MIT
5374.442	5375.936	18601.41	W	1.0			MIT
5374.160	5375.654	18602.39	W	1.1	0.0H		MIT
5372.853	5374.347	18606.91	W	1.0			MIT
5368.700	5370.193	18621.30	W	1.2			MIT
5363.343	5364.835	18639.90	W	0.7			MIT
5362.842	5364.333	18641.65	W	0.8	0.3		MIT
5361.421	5362.912	18646.59	W	1.0			MIT
5358.330	5359.820	18657.34	W	1.0			MIT
5357.116	5358.606	18661.57	W	1.0			MIT
5356.606	5358.096	18663.35	W	0.9			MIT
5355.260	5356.749	18668.04	W	1.0	0.0		MIT
5354.465	5355.954	18670.81	W	1.0			MIT
5353.682	5355.171	18673.54	W	0.7			MIT
5351.902	5353.390	18679.75	W	1.3			MIT
5350.445	5351.933	18684.84	W	1.3			MIT
5349.862	5351.350	18686.87	W	0.8			MIT
5348.949	5350.437	18690.06	W	1.5			MIT
5346.698	5348.185	18697.93	W	0.6			MIT
5339.278	5340.763	18723.92	W	0.7			MIT
5337.369	5338.854	18730.61	W	1.2			MIT
5329.887	5331.370	18756.91	W	0.6			MIT
5327.109	5328.591	18766.69	W	0.7			MIT
5318.873	5320.353	18795.75	W	1.3			MIT
5317.814	5319.293	18799.49	W	0.7			MIT
5313.080	5314.558	18816.24	W	0.6	0.0		MIT
5311.375	5312.853	18822.28	W	0.8			MIT
5293.085	5294.558	18887.32	W	0.5			MIT
5278.602	5280.071	18939.14	W	1.0			MIT
5278.414	5279.883	18939.81	W	0.3			MIT
5275.554	5277.022	18950.08	W	1.3			MIT
5274.813	5276.281	18952.74	W	0.9			MIT
5269.292	5270.758	18972.60	W	1.1			MIT
5263.209	5264.674	18994.53	W	1.2			MIT
5259.355	5260.819	19008.45	W	1.3			MIT
5255.417	5256.880	19022.69	W	1.3			MIT
5254.542	5256.005	19025.86	W	1.3			MIT
5247.379	5248.840	19051.83	W	0.8			MIT
5242.987	5244.446	19067.79	W	1.4	0.0H		MIT
5240.744	5242.203	19075.95	W	0.8			MIT
5239.774	5241.233	19079.48	W	1.0			MIT
5233.539	5234.996	19102.21	W	1.1			MIT
5229.118	5230.574	19118.36	W	0.8	0.0H		MIT
5224.667	5226.122	19134.65	W	1.7	0.9		MIT
5218.428	5219.881	19157.52	W	0.8	0.5		MIT
5218.178	5219.631	19158.44	W	0.8			MIT
5214.187	5215.639	19173.11	W	1.1	0.3		MIT

AIR WAVELENGTH (ANGSTRMS)	VACUUM WAVELENGTH (ANGSTRMS)	WAVENUMBER (KAYSERS)	LMENT	LOG ARC	INTENSITY SPK	DIS	REF
5212.794	5214.245	19178.23	W	1.3	0.7		MIT
5212.345	5213.796	19179.88	W	1.0			MIT
5209.105	5210.555	19191.81	W	1.0			MIT
5207.974	5209.424	19195.98	W	0.7			MIT
5206.186	5207.636	19202.57	W	1.5			MIT
5205.154	5206.603	19206.38	W		0.8		MIT
5204.514	5205.963	19208.74	W	1.6			MIT
5203.259	5204.708	19213.37	W	1.5			MIT
5201.436	5202.884	19220.11	W	1.0			MIT
5199.942	5201.390	19225.63	W	0.8			MIT
5197.874	5199.321	19233.28	W	0.5			MIT
5196.496	5197.943	19238.38	W	0.7			MIT
5192.723	5194.169	19252.36	W	1.5			MIT
5188.861	5190.306	19266.69	W	1.0			MIT
5187.536	5188.981	19271.61	W	0.6			MIT
5183.972	5185.416	19284.86	W	1.3			MIT
5179.490	5180.933	19301.54	W	0.9	0.0H		MIT
5177.725	5179.167	19308.12	W	0.8			MIT
5175.698	5177.140	19315.69	W	0.8			MIT
5175.137	5176.578	19317.78	W	0.8			MIT
5162.840	5164.278	19363.79	W	0.5			MIT
5162.073	5163.511	19366.67	W	1.0			MIT
5158.178	5159.615	19381.29	W	0.9			MIT
5154.889	5156.325	19393.66	W	0.8			MIT
5154.447	5155.883	19395.32	W	0.9			MIT
5153.874	5155.310	19397.48	W	0.8	0.0H		MIT
5153.533	5154.969	19398.76	W	1.0	0.5		MIT
5145.772	5147.206	19428.02	W	1.3			MIT
5144.395	5145.828	19433.22	W		0.8		MIT
5141.248	5142.680	19445.11	W	0.9			MIT
5138.397	5139.829	19455.90	W	1.3			MIT
5136.676	5138.107	19462.42	W	0.8			MIT
5133.114	5134.544	19475.93	W	0.9S			MIT
5130.112	5131.541	19487.32	W	1.2			MIT
5129.806	5131.235	19488.49	W	0.8			MIT
5128.830	5130.259	19492.19	W	0.8			MIT
5128.486	5129.915	19493.50	W	0.9			MIT
5124.235	5125.663	19509.67	W	1.1			MIT
5120.524	5121.951	19523.81	W	1.0L			MIT
5118.408	5119.834	19531.88	W	0.8			MIT
5117.604	5119.030	19534.95	W	1.0			MIT
5114.571	5115.996	19546.53	W	0.5			MIT
5111.770	5113.195	19557.25	W	0.8			MIT
5110.357	5111.781	19562.65	W	1.3			MIT
5109.509	5110.933	19565.90	W	0.5			MIT
5107.786	5109.209	19572.50	W	0.8			MIT
5105.476	5106.899	19581.35	W	1.2			MIT
5104.425	5105.848	19585.39	W	0.7	0.0		MIT
5101.381	5102.803	19597.07	W	0.7			MIT
5094.183	5095.603	19624.76	W	0.8			MIT
5089.973	5091.392	19641.00	W	0.8			MIT
5089.827	5091.246	19641.56	W	0.6			MIT
5086.918	5088.336	19652.79	W	0.5			MIT
5085.900	5087.318	19656.72	W	0.8D			MIT
5084.667	5086.084	19661.49	W	0.5H			MIT
5082.738	5084.155	19668.95	W	0.8			MIT
5080.719	5082.135	19676.77	W	1.0			MIT
5077.015	5078.430	19691.12	W	0.8			MIT
5071.733	5073.147	19711.63	W	1.6	0.5		MIT
5071.518	5072.932	19712.47	W	0.7			MIT
5069.146	5070.559	19721.69	W	1.7	0.5		MIT
5065.677	5067.089	19735.20	W	0.8			MIT
5063.620	5065.032	19743.21	W	0.9			MIT
5058.013	5059.423	19765.10	W	1.0			MIT
5056.103	5057.513	19772.56	W	0.5			MIT
5055.528	5056.938	19774.81	W	1.3			MIT
5054.606	5056.015	19778.42	W	1.4	0.5		MIT
5053.296	5054.705	19783.55	W	1.8	1.0		MIT
5052.233	5053.642	19787.71	W	1.0			MIT
5044.328	5045.735	19818.72	W	1.0			MIT

AIR WAVELENGTH (ANGSTRMS)	VACUUM WAVELENGTH (ANGSTRMS)	WAVENUMBER (KAYSERS)	LMENT	LOG ARC	INTENSITY SPK	DIS	REF
5041.762	5043.168	19828.81	W	0.8			MIT
5040.364	5041.770	19834.31	W	1.5	0.0H		MIT
5039.028	5040.433	19839.56	W	1.2	0.0H		MIT
5035.743	5037.147	19852.51	W	0.8			MIT
5029.106	5030.509	19878.71	W	0.7	0.0H		MIT
5028.927	5030.329	19879.41	W	1.0			MIT
5027.435	5028.837	19885.31	W	1.0			MIT
5025.665	5027.067	19892.32	W	0.9	0.0H		MIT
5022.484	5023.885	19904.92	W	0.9			MIT
5021.258	5022.658	19909.77	W	0.9			MIT
5019.511	5020.911	19916.70	W	0.8			MIT
5018.504	5019.904	19920.70	W	0.8			MIT
5017.200	5018.599	19925.88	W	1.0			MIT
5015.321	5016.720	19933.34	W	1.6	0.7		MIT
5014.595	5015.994	19936.23	W	0.8			MIT
5014.004	5015.403	19938.58	W	0.8			MIT
5013.456	5014.854	19940.76	W	1.2			MIT
5010.383	5011.781	19952.99	W	0.8D			MIT
5009.888	5011.285	19954.96	W	0.8	0.3		MIT
5007.230	5008.627	19965.55	W	1.2	0.3		MIT
5006.157	5007.553	19969.83	W	1.6	0.8		MIT
5002.795	5004.191	19983.25	W	1.2	1.0		MIT
5001.094	5002.489	19990.05	W	1.1			MIT
5000.504	5001.899	19992.41	W	0.8			MIT
4995.320	4996.714	20013.16	W	1.0			MIT
4994.37	4995.76	20017.0	W	0.3	0.3H		EX
4994.096	4995.489	20018.06	W	1.5			MIT
4989.866	4991.258	20035.03	W	0.7			MIT
4989.091	4990.483	20038.14	W	1.2			MIT
4986.941	4988.332	20046.78	W	1.6	0.0H		MIT
4984.722	4986.113	20055.70	W	1.2			MIT
4984.166	4985.557	20057.94	W	1.2			MIT
4983.542	4984.932	20060.45	W	1.3			MIT
4982.603	4983.993	20064.23	W	1.6	0.7		MIT
4979.847	4981.236	20075.34	W	1.4			MIT
4977.692	4979.081	20084.03	W	0.9			MIT
4977.244	4978.633	20085.84	W	1.2			MIT
4972.568	4973.955	20104.72	W	1.2			MIT
4970.653	4972.040	20112.47	W	0.9			MIT
4968.424	4969.810	20121.49	W	0.8			MIT
4967.670	4969.056	20124.55	W	0.8			MIT
4962.294	4963.679	20146.35	W	1.1			MIT
4961.530	4962.915	20149.45	W	1.0			MIT
4959.341	4960.725	20158.34	W	0.5			MIT
4953.092	4954.474	20183.78	W	1.4			MIT
4948.588	4949.969	20202.15	W	1.2			MIT
4941.964	4943.343	20229.22	W	0.9			MIT
4933.822	4935.199	20262.61	W	1.1			MIT
4932.792	4934.169	20266.84	W	1.1	0.0H		MIT
4931.556	4932.933	20271.92	W	1.5			MIT
4926.705	4928.080	20291.88	W	0.8			MIT
4924.565	4925.940	20300.70	W	1.1			MIT
4916.184	4917.557	20335.30	W	1.3			MIT
4914.337	4915.709	20342.95	W	1.1	0.0		MIT
4912.191	4913.562	20351.83	W	0.7			MIT
4910.741	4912.112	20357.84	W	1.5			MIT
4908.890	4910.261	20365.52	W	0.8			MIT
4905.487	4906.857	20379.65	W	1.1			MIT
4902.968	4904.337	20390.12	W	1.1	0.5		MIT
4902.324	4903.693	20392.79	W	1.2			MIT
4896.784	4898.151	20415.87	W	0.5			MIT
4892.438	4893.804	20434.00	W	1.4			MIT
4890.893	4892.259	20440.46	W	1.1			MIT
4890.288	4891.654	20442.99	W	1.2			MIT
4888.389	4889.754	20450.93	W	1.3			MIT
4886.912	4888.277	20457.11	W	1.7	1.0		MIT
4880.703	4882.066	20483.13	W	1.0			MIT
4879.695	4881.058	20487.36	W	1.0			MIT
4878.283	4879.645	20493.29	W	1.5			MIT
4876.709	4878.071	20499.91	W	0.8			MIT

AIR WAVELENGTH (ANGSTRMS)	VACUUM WAVELENGTH (ANGSTRMS)	WAVENUMBER (KAYSERS)	LMENT	LOG ARC	INTENSITY SPK	INTENSITY DIS	REF
4875.454	4876.816	20505.18	W	0.8			MIT
4875.380	4876.742	20505.50	W	0.7			MIT
4863.083	4864.441	20557.34	W	1.1			MIT
4860.893	4862.251	20566.61	W	1.0			MIT
4858.606	4859.963	20576.29	W	1.2			MIT
4854.090	4855.446	20595.43	W	1.5			MIT
4853.845	4855.201	20596.47	W	1.0			MIT
4843.828	4845.181	20639.06	W	1.7	1.1		MIT
4837.492	4838.844	20666.09	W	1.0			MIT
4835.019	4836.370	20676.67	W	1.2			MIT
4831.36	4832.71	20692.3	W	0.7			MIT
4830.265	4831.615	20697.01	W	0.7			MIT
4828.800	4830.149	20703.29	W	0.7			MIT
4828.084	4829.433	20706.36	W	0.8			MIT
4820.080	4821.427	20740.75	W	0.5			MIT
4818.947	4820.294	20745.62	W	0.6			MIT
4818.368	4819.715	20748.12	W	0.6			MIT
4817.694	4819.040	20751.02	W	0.7	0.0		MIT
4817.330	4818.676	20752.59	W	0.3			MIT
4816.821	4818.167	20754.78	W	0.7			MIT
4816.107	4817.453	20757.86	W	1.0	0.0		MIT
4812.588	4813.933	20773.03	W	0.6			MIT
4809.648	4810.992	20785.73	W	0.3	0.0		MIT
4807.366	4808.710	20795.60	W	1.2	0.0		MIT
4806.416	4807.759	20799.71	W	0.5J	0.0		MIT
4799.919	4801.261	20827.86	W	1.7	1.0		MIT
4797.548	4798.889	20838.16	W	1.2	0.3		MIT
4797.041	4798.382	20840.36	W	0.8			MIT
4796.519	4797.860	20842.63	W	0.5			MIT
4792.824	4794.164	20858.70	W	0.5			MIT
4790.86	4792.20	20867.2	W		1.1		MIT
4788.434	4789.773	20877.82	W	1.2	0.0		MIT
4787.940	4789.278	20879.97	W	1.2	0.0		MIT
4787.206	4788.544	20883.17	W	0.7			MIT
4785.961	4787.299	20888.61	W	0.3			MIT
4783.740	4785.077	20898.30	W	0.7			MIT
4780.518	4781.855	20912.39	W	1.0	0.0		MIT
4780.289	4781.625	20913.39	W	0.5			MIT
4779.421	4780.757	20917.19	W	0.6			MIT
4773.911	4775.246	20941.33	W	1.5	0.3		BK
4772.542	4773.876	20947.34	W	1.1			MIT
4770.768	4772.102	20955.13	W	0.8			MIT
4767.785	4769.118	20968.24	W	1.1	0.0		MIT
4765.637	4766.970	20977.69	W	0.3			MIT
4765.134	4766.466	20979.90	W	0.5			MIT
4761.032	4762.363	20997.98	W	0.6			MIT
4758.211	4759.542	21010.43	W	1.2	0.0		MIT
4757.781	4759.111	21012.33	W	1.2	0.0		MIT
4757.549	4758.879	21013.35	W	1.8	1.0		MIT
4756.992	4758.322	21015.81	W	0.7			MIT
4755.129	4756.459	21024.04	W	0.6			MIT
4753.393	4754.722	21031.72	W	1.0			MIT
4752.583	4753.912	21035.31	W	1.1	0.0		MIT
4752.209	4753.538	21036.96	W	1.1	0.0		MIT
4751.360	4752.689	21040.72	W	1.0	0.0H		MIT
4749.893	4751.221	21047.22	W	0.5			MIT
4749.706	4751.034	21048.05	W	0.8	0.0		MIT
4747.954	4749.282	21055.81	W	0.5			MIT
4746.806	4748.134	21060.91	W	0.6			MIT
4745.574	4746.901	21066.37	W	1.4	0.0J		MIT
4741.520	4742.846	21084.39	W	1.1	0.3		MIT
4738.160	4739.485	21099.34	W	0.9			MIT
4736.062	4737.387	21108.68	W	0.5			MIT
4730.707	4732.030	21132.58	W	0.7			MIT
4729.652	4730.975	21137.29	W	1.5	1.2		MIT
4728.773	4730.096	21141.22	W	0.8			MIT
4726.280	4727.602	21152.37	W	0.8			MIT
4725.602	4726.924	21155.41	W	0.6			MIT
4725.135	4726.457	21157.50	W	1.6	0.5		MIT
4723.089	4724.410	21166.66	W	0.6			MIT

AIR WAVELENGTH (ANGSTRMS)	VACUUM WAVELENGTH (ANGSTRMS)	WAVENUMBER (KAYSERS)	LMENT	LOG ARC	INTENSITY SPK	INTENSITY DIS	REF
4720.398	4721.719	21178.73	W	1.2	0.5		MIT
4718.901	4720.221	21185.45	W	0.3	0.8		MIT
4718.630	4719.950	21186.67	W	1.1	0.3		MIT
4716.865	4718.185	21194.59	W	1.0	0.3		MIT
4714.516	4715.835	21205.15	W	0.8			MIT
4713.861	4715.180	21208.10	W	0.8			MIT
4712.491	4713.809	21214.26	W	1.3	0.7		MIT
4711.188	4712.506	21220.13	W	1.2	0.5		MIT
4710.341	4711.659	21223.95	W	1.1	0.5		MIT
4706.170	4707.487	21242.76	W	1.2	0.0		MIT
4702.471	4703.787	21259.47	W	1.2	0.7		MIT
4702.202	4703.518	21260.68	W	0.7			MIT
4701.600	4702.916	21263.41	W	0.8			MIT
4700.411	4701.726	21268.78	W	1.7	0.6		MIT
4698.633	4699.948	21276.83	W	1.2	0.0		MIT
4698.105	4699.420	21279.22	W	1.0	0.0		MIT
4694.658	4695.972	21294.85	W	1.1	0.0		MIT
4693.729	4695.043	21299.06	W	1.7	1.1		MIT
4691.554	4692.867	21308.94	W	0.8	0.3		MIT
4687.653	4688.965	21326.67	W	1.1	0.0		MIT
4686.383	4687.695	21332.45	W	0.9			MIT
4683.545	4684.856	21345.37	W	1.3	1.0		MIT
4682.569	4683.880	21349.82	W	1.1			MIT
4681.195	4682.505	21356.09	W	1.0	0.0		MIT
4680.520	4681.830	21359.17	W	2.2	1.6		MIT
4679.041	4680.351	21365.92	W	1.2	0.3		MIT
4677.694	4679.003	21372.07	W	1.4	0.5		MIT
4676.632	4677.941	21376.93	W	1.3	0.3		MIT
4675.093	4676.402	21383.96	W	0.9	0.0		MIT
4673.988	4675.296	21389.02	W	0.3	0.0		MIT
4672.542	4673.850	21395.64	W	0.5			MIT
4671.651	4672.959	21399.72	W	1.1	0.0		MIT
4671.296	4672.604	21401.34	W	0.5	0.0		MIT
4668.911	4670.218	21412.28	W	0.7			MIT
4668.462	4669.769	21414.34	W	1.3	0.5		MIT
4665.749	4667.055	21426.79	W		1.1		MIT
4662.595	4663.900	21441.28	W	0.5			MIT
4661.970	4663.275	21444.16	W	1.2	0.3		MIT
4661.233	4662.538	21447.55	W	1.1	0.3		MIT
4659.867	4661.172	21453.83	W	2.3	1.8		MIT
4658.258	4659.562	21461.24	W	0.0			MIT
4657.436	4658.740	21465.03	W	1.7	1.1		MIT
4657.036	4658.340	21466.87	W	0.8	0.0		MIT
4655.209	4656.512	21475.30	W	0.8			MIT
4646.149	4647.450	21517.18	W	1.2	0.5		MIT
4644.926	4646.227	21522.84	W	0.9			MIT
4643.153	4644.453	21531.06	W	1.1	0.9W		MIT
4642.564	4643.864	21533.79	W	1.5	0.9		MIT
4641.800	4643.100	21537.34	W	1.3	0.8W		MIT
4640.303	4641.602	21544.28	W	1.0	0.3		MIT
4638.905	4640.204	21550.78	W	0.0	0.8		MIT
4636.076	4637.374	21563.93	W	0.8	0.0		MIT
4635.101	4636.399	21568.46	W	0.8	0.0		MIT
4634.810	4636.108	21569.82	W	1.3	0.8		MIT
4630.409	4631.706	21590.32	W	0.7			MIT
4629.257	4630.553	21595.69	W	0.7	0.0		MIT
4628.600	4629.896	21598.76	W	0.5			MIT
4626.359	4627.655	21609.22	W	0.6			MIT
4625.797	4627.093	21611.84	W	0.5			MIT
4625.173	4626.468	21614.76	W	0.7	0.0		MIT
4624.461	4625.756	21618.09	W	0.7			MIT
4623.687	4624.982	21621.71	W	1.1	0.5		MIT
4623.160	4624.455	21624.17	W	1.0			MIT
4620.554	4621.848	21636.37	W	1.3	0.8		MIT
4614.858	4616.151	21663.07	W	1.0	0.3		MIT
4613.316	4614.608	21670.31	W	1.7	1.0		MIT
4609.915	4611.206	21686.30	W	1.7	1.0		MIT
4608.844	4610.135	21691.34	W	0.7	0.0J		MIT
4606.768	4608.059	21701.11	W	0.3			MIT
4604.609	4605.899	21711.29	W	0.3	1.0		MIT

AIR WAVELENGTH (ANGSTRMS)	VACUUM WAVELENGTH (ANGSTRMS)	WAVENUMBER (KAYSERS)	LMENT	LOG ARC	INTENSITY SPK	INTENSITY DIS	REF
4600.442	4601.731	21730.95	W	1.3	0.6		MIT
4599.961	4601.250	21733.23	W	1.7	1.0		MIT
4592.575	4593.862	21768.18	W	1.2	0.7		MIT
4592.422	4593.709	21768.90	W	1.3	1.0		MIT
4588.750	4590.036	21786.32	W	1.6	1.2		MIT
4586.846	4588.131	21795.37	W	1.5	0.7		MIT
4584.249	4585.534	21807.71	W	0.5			MIT
4579.962	4581.245	21828.12	W	0.6			MIT
4579.701	4580.984	21829.37	W	1.1	0.5		MIT
4578.336	4579.619	21835.88	W	1.1	0.6		MIT
4571.898	4573.179	21866.63	W	1.0	0.0		MIT
4570.655	4571.936	21872.57	W	1.5	1.0		MIT
4568.613	4569.893	21882.35	W	0.8			MIT
4567.590	4568.870	21887.25	W	1.0	0.3		MIT
4566.226	4567.506	21893.79	W	0.9	0.3		MIT
4565.321	4566.601	21898.13	W	1.1	0.3		MIT
4564.076	4565.355	21904.10	W	0.8	0.3		MIT
4563.590	4564.869	21906.43	W	1.2	0.7		MIT
4560.500	4561.778	21921.28	W	0.6			MIT
4559.113	4560.391	21927.95	W	1.0	0.3		MIT
4558.970	4560.248	21928.63	W	1.0	0.3		MIT
4556.869	4558.146	21938.74	W	1.2	0.7		MIT
4556.219	4557.496	21941.87	W	0.8	0.0		MIT
4555.327	4556.604	21946.17	W	0.8	0.0		MIT
4554.683	4555.960	21949.27	W	0.6			MIT
4553.662	4554.938	21954.19	W	0.8	0.0		MIT
4552.533	4553.809	21959.64	W	1.1	0.5		MIT
4551.850	4553.126	21962.93	W	1.5	1.0		MIT
4550.343	4551.619	21970.21	W	0.8	0.0		MIT
4549.533	4550.808	21974.12	W	0.8	0.0		MIT
4546.487	4547.762	21988.84	W	1.5	1.0		MIT
4544.570	4545.844	21998.11	W	1.0	0.0		MIT
4544.367	4545.641	21999.10	W	0.3	0.3		MIT
4543.509	4544.783	22003.25	W	1.4	1.0		MIT
4543.278	4544.552	22004.37	W	0.8	0.0		MIT
4542.887	4544.161	22006.26	W	1.2	0.5		MIT
4540.313	4541.586	22018.74	W	0.8	0.0		MIT
4539.696	4540.969	22021.73	W	0.6	0.0		MIT
4536.657	4537.929	22036.48	W	1.2	0.8		MIT
4535.653	4536.925	22041.36	W	0.3	0.0		MIT
4535.054	4536.326	22044.27	W	1.2	0.7		MIT
4534.714	4535.985	22045.93	W	1.2	0.7		MIT
4534.370	4535.641	22047.60	W	0.5	0.3		MIT
4530.468	4531.738	22066.59	W	1.2	0.6		MIT
4529.760	4531.030	22070.04	W	1.2	0.6		MIT
4519.171	4520.438	22121.75	W	1.1	0.5		MIT
4517.370	4518.637	22130.57	W	1.0	0.3		MIT
4515.881	4517.148	22137.86	W	0.8	0.0		MIT
4514.317	4515.583	22145.53	W	0.8	0.0		MIT
4514.058	4515.324	22146.80	W	0.6			MIT
4513.300	4514.566	22150.52	W	1.5	1.0		MIT
4512.907	4514.173	22152.45	W	1.5	1.0		MIT
4509.351	4510.616	22169.92	W	0.8	0.0		MIT
4504.857	4506.121	22192.04	W	1.5	1.0		MIT
4504.167	4505.430	22195.44	W	1.1	0.6		MIT
4503.157	4504.420	22200.42	W	0.5	0.0		MIT
4502.454	4503.717	22203.88	W	0.3	0.3		MIT
4498.467	4499.729	22223.56	W	1.1	0.6		MIT
4497.69	4498.95	22227.4	W	0.8	0.3		MIT
4496.268	4497.529	22234.43	W	0.6			MIT
4495.312	4496.573	22239.16	W	1.3	0.8		MIT
4494.507	4495.768	22243.14	W	1.3	1.1		MIT
4493.972	4495.233	22245.79	W	1.2	0.8		MIT
4492.328	4493.588	22253.93	W	1.1	0.7		MIT
4489.964	4491.224	22265.65	W	0.5			MIT
4489.021	4490.280	22270.32	W	1.1	0.5		MIT
4487.448	4488.707	22278.13	W	0.5			MIT
4484.189	4485.447	22294.32	W	1.5	1.3		MIT
4483.530	4484.788	22297.60	W	0.6			MIT
4481.284	4482.541	22308.77	W	1.2	0.8		MIT

AIR WAVELENGTH (ANGSTRMS)	VACUUM WAVELENGTH (ANGSTRMS)	WAVENUMBER (KAYSERS)	LMENT	LOG ARC	INTENSITY SPK	INTENSITY DIS	REF
4479.609	4480.866	22317.11	W	0.6			MIT
4479.186	4480.443	22319.22	W	0.6			MIT
4478.503	4479.760	22322.63	W	0.6	0.3		MIT
4477.839	4479.095	22325.94	W	0.7	0.8		MIT
4477.381	4478.637	22328.22	W	0.6			MIT
4477.166	4478.422	22329.29	W	0.6			MIT
4476.540	4477.796	22332.42	W	0.6			MIT
4475.630	4476.886	22336.96	W	0.6	0.0		MIT
4474.666	4475.922	22341.77	W	0.6			MIT
4474.035	4475.290	22344.92	W	1.1	0.7		MIT
4472.519	4473.774	22352.49	W	0.8	0.0		MIT
4471.81	4473.06	22356.0	W	0.6			MIT
4471.29	4472.54	22358.6	W	0.6			MIT
4470.753	4472.008	22361.32	W	0.6			MIT
4469.718	4470.972	22366.50	W	0.7	0.0		MIT
4467.664	4468.918	22376.78	W	0.6	0.0		MIT
4466.729	4467.983	22381.47	W	1.3	1.0		MIT
4466.348	4467.601	22383.37	W	1.3	1.0		MIT
4463.504	4464.757	22397.64	W	1.2	0.8		MIT
4462.523	4463.775	22402.56	W	1.0	0.3		MIT
4460.497	4461.749	22412.74	W	1.4	0.8		MIT
4458.302	4459.553	22423.77	W	1.0	0.7		MIT
4458.093	4459.344	22424.82	W	1.2	0.7		MIT
4456.112	4457.363	22434.79	W	1.2	0.7		MIT
4455.465	4456.716	22438.05	W	1.2	0.7		MIT
4454.06	4455.31	22445.1	W	0.5			MIT
4453.62	4454.87	22447.3	W	0.7			MIT
4450.364	4451.613	22463.77	W	1.1	0.6		MIT
4449.010	4450.259	22470.60	W	1.3	0.8		MIT
4445.150	4446.398	22490.11	W	1.3	0.8		MIT
4444.453	4445.701	22493.64	W	0.6	0.3		MIT
4444.086	4445.334	22495.50	W	0.6	0.5		MIT
4441.811	4443.058	22507.02	W	1.3	1.0		MIT
4439.722	4440.968	22517.61	W	1.1	0.8		MIT
4438.295	4439.541	22524.85	W	1.2	0.8		MIT
4436.904	4438.150	22531.91	W	1.5	1.1		MIT
4435.742	4436.987	22537.81	W	1.0	0.6		MIT
4435.44	4436.69	22539.4	W	0.8			MIT
4434.05	4435.29	22546.4	W	0.5			MIT
4433.642	4434.887	22548.49	W	0.5			MIT
4432.192	4433.436	22555.87	W	0.9	0.0		MIT
4428.488	4429.731	22574.73	W	0.9	0.3		MIT
4427.382	4428.625	22580.37	W	1.0	0.5		MIT
4425.914	4427.157	22587.86	W	1.2	0.8		MIT
4424.906	4426.149	22593.01	W	0.9	0.3		MIT
4423.778	4425.020	22598.77	W	1.1	0.7		MIT
4422.444	4423.686	22605.58	W	0.6			MIT
4421.850	4423.092	22608.62	W	1.1	0.7		MIT
4421.010	4422.251	22612.92	W	1.1	0.5		MIT
4420.468	4421.709	22615.69	W	1.5	1.0		MIT
4419.258	4420.499	22621.88	W	1.0	0.3		MIT
4418.811	4420.052	22624.17	W	0.7	0.0		MIT
4418.451	4419.692	22626.01	W	1.1	0.6		MIT
4415.710	4416.950	22640.06	W	1.0	0.5		MIT
4415.075	4416.315	22643.31	W	1.2	0.8		MIT
4413.86	4415.10	22649.6	W	0.7	0.0H		MIT
4413.24	4414.48	22652.7	W	0.7			MIT
4413.017	4414.256	22653.87	W	1.0	0.3		MIT
4412.198	4413.437	22658.08	W	1.3	1.0		MIT
4411.70	4412.94	22660.6	W	1.0	0.3		MIT
4409.944	4411.183	22669.66	W	0.8			MIT
4408.710	4409.948	22676.00	W	0.9	0.6		MIT
4408.280	4409.518	22678.22	W	1.4	1.1		MIT
4406.690	4407.928	22686.40	W	0.8	0.0		MIT
4406.396	4407.634	22687.91	W	0.8	0.0		MIT
4405.99	4407.23	22690.0	W	0.5	0.8		MIT
4404.46	4405.70	22697.9	W	0.5	0.0		MIT
4403.951	4405.188	22700.51	W	1.3	1.0		MIT
4403.271	4404.508	22704.01	W	1.0	0.3		MIT
4402.78	4404.02	22706.5	W	0.6			MIT

AIR WAVELENGTH (ANGSTRMS)	VACUUM WAVELENGTH (ANGSTRMS)	WAVENUMBER (KAYSERS)	LMENT		LOG ARC	INTENSITY SPK	DIS	REF
4400.86	4402.10	22716.4	W		0.7			MIT
4400.214	4401.450	22719.79	W		1.1	0.8		MIT
4398.27	4399.51	22729.8	W		0.7	0.0H		MIT
4395.081	4396.316	22746.32	W		0.8	0.3		MIT
4394.514	4395.748	22749.25	W		0.8	0.0		MIT
4394.083	4395.317	22751.49	W		1.3	1.0		MIT
4389.843	4391.076	22773.46	W		1.2	0.8		MIT
4387.460	4388.693	22785.83	W		0.6			MIT
4386.772	4388.004	22789.40	W		1.0	0.5		MIT
4384.858	4386.090	22799.35	W		1.4	1.2		MIT
4380.115	4381.346	22824.04	W		0.9	0.3		MIT
4378.493	4379.723	22832.49	W		1.4	1.1		MIT
4372.528	4373.757	22863.64	W		1.4	1.0		MIT
4371.738	4372.966	22867.77	W		1.1	0.5		MIT
4370.808	4372.036	22872.64	W		0.8	0.0		MIT
4369.047	4370.275	22881.86	W		0.7	0.3		MIT
4368.767	4369.995	22883.32	W		1.0	0.3		MIT
4366.348	4367.575	22896.00	W		0.6	0.0		MIT
4366.069	4367.296	22897.46	W		1.1	0.6		MIT
4365.965	4367.192	22898.01	W		1.1	0.6		MIT
4364.786	4366.013	22904.19	W		1.4	1.1		MIT
4361.815	4363.041	22919.79	W		1.3	1.0		MIT
4361.538	4362.764	22921.25	W		0.8	1.1		MIT
4361.067	4362.293	22923.73	W		1.0	0.5		MIT
4360.090	4361.315	22928.86	W		0.3	0.6		MIT
4355.165	4356.389	22954.79	W		1.2	1.0		MIT
4354.898	4356.122	22956.20	W		0.8	0.3		MIT
4354.721	4355.945	22957.13	W		0.8	0.3		MIT
4353.298	4354.522	22964.64	W		0.8	0.3		MIT
4353.108	4354.332	22965.64	W		0.9	0.6		MIT
4348.120	4349.342	22991.98	W	II	1.7	1.6		MIT
4347.509	4348.731	22995.21	W		1.0	0.5		MIT
4347.002	4348.224	22997.90	W		1.3	0.8		MIT
4346.291	4347.513	23001.66	W		0.7	0.3		MIT
4345.835	4347.057	23004.07	W		1.0	0.5		MIT
4344.963	4346.184	23008.69	W		0.9	0.0		MIT
4344.638	4345.859	23010.41	W		0.7	0.0		MIT
4343.535	4344.756	23016.25	W		0.7			MIT
4342.792	4344.013	23020.19	W		0.8	0.3		MIT
4342.52	4343.74	23021.6	W			1.0		MIT
4341.134	4342.354	23028.98	W		0.6	0.3		MIT
4339.452	4340.672	23037.91	W		0.9	0.5		MIT
4339.066	4340.286	23039.96	W		0.8	0.5		MIT
4338.279	4339.499	23044.14	W		1.0R	0.5		MIT
4335.57	4336.79	23058.5	W		0.3	1.2		MIT
4335.357	4336.576	23059.67	W		1.0	0.3		MIT
4332.132	4333.350	23076.83	W		1.2	0.8		MIT
4331.63	4332.85	23079.5	W		0.6			MIT
4330.970	4332.188	23083.02	W		1.1	0.8		MIT
4330.660	4331.878	23084.68	W		1.1	0.8		MIT
4330.29	4331.51	23086.6	W		0.6	0.0		MIT
4328.417	4329.634	23096.64	W		0.6	0.0		MIT
4327.413	4328.630	23102.00	W		1.1	0.8		MIT
4325.17	4326.39	23114.0	W		0.3	0.5		MIT
4324.59	4325.81	23117.1	W			0.8		MIT
4322.749	4323.965	23126.92	W		1.1	0.5		MIT
4320.50	4321.72	23139.0	W		0.8	0.0H		MIT
4318.573	4319.788	23149.29	W		1.0	0.5		MIT
4316.813	4318.027	23158.73	W		1.2	0.8		MIT
4313.37	4314.58	23177.2	W		0.6	0.3		MIT
4313.18	4314.39	23178.2	W		0.3	0.6		MIT
4312.348	4313.561	23182.70	W		1.0	0.5		MIT
4311.10	4312.31	23189.4	W		0.9	0.5		MIT
4310.259	4311.471	23193.94	W		1.1	0.5		MIT
4309.877	4311.089	23195.99	W		0.9	0.3		MIT
4308.955	4310.167	23200.96	W		1.0	0.6		MIT
4308.50	4309.71	23203.4	W		0.9	0.6		MIT
4307.640	4308.852	23208.04	W		1.1	1.1		MIT
4306.875	4308.086	23212.16	W		1.3	1.2		MIT
4306.33	4307.54	23215.1	W		0.6	0.3		MIT

AIR WAVELENGTH (ANGSTRMS)	VACUUM WAVELENGTH (ANGSTRMS)	WAVENUMBER (KAYSERS)	LMENT		LOG ARC	INTENSITY SPK	DIS	REF
4305.628	4306.839	23218.89	W		1.0	0.6		MIT
4304.94	4306.15	23222.6	W		0.5			MIT
4303.533	4304.744	23230.19	W		0.6			MIT
4303.328	4304.538	23231.29	W			1.2		MIT
4302.108	4303.318	23237.88	W	I	1.8	1.8		MIT
4298.410	4299.619	23257.87	W		0.8	0.5		MIT
4295.689	4296.897	23272.61	W			0.5		MIT
4295.005	4296.213	23276.31	W		0.7	0.6		MIT
4294.614	4295.822	23278.43	W	I	1.7	1.7		MIT
4294.099	4295.307	23281.22	W		1.3	1.0		MIT
4292.735	4293.943	23288.62	W		0.6	0.3		MIT
4292.16	4293.37	23291.7	W		0.6	0.5		MIT
4290.144	4291.351	23302.69	W		0.9	0.5		MIT
4289.291	4290.498	23307.32	W		0.0	0.7		MIT
4288.35	4289.56	23312.4	W		0.5	0.0		MIT
4286.214	4287.420	23324.05	W		0.8	0.3		MIT
4286.013	4287.219	23325.14	W		1.2	0.9		MIT
4283.807	4285.012	23337.16	W		1.0	0.7		MIT
4281.20	4282.40	23351.4	W		0.8	0.6		MIT
4278.93	4280.13	23363.7	W		0.6	0.6		MIT
4278.410	4279.614	23366.59	W		1.0	0.5		MIT
4277.91	4279.11	23369.3	W		0.8	0.0		MIT
4276.746	4277.950	23375.69	W		1.2	1.0		MIT
4276.031	4277.234	23379.59	W		1.0	0.6		MIT
4275.493	4276.696	23382.53	W		1.2	1.0		MIT
4275.149	4276.352	23384.42	W		0.8	0.3		MIT
4274.938	4276.141	23385.57	W		1.0	0.7		MIT
4274.550	4275.753	23387.69	W		1.3	1.1		MIT
4273.692	4274.895	23392.39	W		1.0	0.7		MIT
4272.312	4273.514	23399.95	W		0.9	0.5		MIT
4270.908	4272.110	23407.64	W		0.7	0.5		MIT
4269.780	4270.982	23413.82	W		1.2	1.0		MIT
4269.392	4270.594	23415.95	W		1.6	1.5		MIT
4268.052	4269.253	23423.30	W		1.0	0.7		MIT
4266.542	4267.743	23431.59	W		1.2	0.9		MIT
4264.987	4266.187	23440.13	W		0.5	0.0		MIT
4263.315	4264.515	23449.33	W		1.4	1.2		MIT
4262.269	4263.469	23455.08	W		0.8	0.9		MIT
4260.293	4261.492	23465.96	W		1.3	0.7		MIT
4259.941	4261.140	23467.90	W		1.1	0.7		MIT
4259.357	4260.556	23471.11	W		1.5	1.3		MIT
4258.532	4259.731	23475.66	W		1.2	0.7		MIT
4254.288	4255.486	23499.08	W		0.6	0.5		MIT
4254.060	4255.258	23500.34	W		0.9	0.3		MIT
4249.458	4250.654	23525.79	W		0.7	1.1		MIT
4244.373	4245.568	23553.97	W		1.6	1.3		MIT
4243.638	4244.833	23558.05	W		1.0	0.5		MIT
4243.362	4244.557	23559.59	W		0.7	0.0		MIT
4241.448	4242.642	23570.22	W		1.5	1.0		MIT
4240.134	4241.328	23577.52	W		0.9	0.3		MIT
4234.351	4235.543	23609.72	W		1.4	0.8		MIT
4232.998	4234.190	23617.27	W		1.1	0.6		MIT
4231.942	4233.134	23623.16	W		0.5	0.3		MIT
4231.728	4232.920	23624.36	W		1.0	0.3		MIT
4231.322	4232.514	23626.62	W		0.8	0.5		MIT
4229.118	4230.309	23638.94	W		0.0	0.3		MIT
4226.915	4228.105	23651.25	W		1.2	0.5		MIT
4226.343	4227.533	23654.46	W		1.0	0.5		MIT
4224.988	4226.178	23662.04	W			1.1		MIT
4224.765	4225.955	23663.29	W		1.0	0.3		MIT
4222.055	4223.244	23678.48	W		1.2	0.9		MIT
4221.722	4222.911	23680.35	W		0.8	0.0		MIT
4220.541	4221.730	23686.97	W		0.7	0.0		MIT
4220.27	4221.46	23688.5	W		0.8	0.3		MIT
4219.378	4220.566	23693.50	W		1.4	1.2		MIT
4218.557	4219.745	23698.11	W		1.1	0.3		MIT
4216.06	4217.25	23712.1	W		0.0	0.7		MIT
4215.382	4216.569	23715.96	W		1.1	0.7		MIT
4214.959	4216.146	23718.34	W		0.7	0.0		MIT
4212.89	4214.08	23730.0	W		0.9	0.3		MIT

AIR WAVELENGTH (ANGSTRMS)	VACUUM WAVELENGTH (ANGSTRMS)	WAVENUMBER (KAYSERS)	LMENT	LOG ARC	INTENSITY SPK	DIS	REF
4209.788	4210.974	23747.48	W	0.5	0.0		MIT
4207.053	4208.238	23762.91	W	1.4	1.1		MIT
4206.240	4207.425	23767.51	W	0.8	0.3		MIT
4205.559	4206.744	23771.35	W	0.8	1.1		MIT
4204.409	4205.593	23777.86	W	1.3	1.0		MIT
4203.820	4205.004	23781.19	W	1.2	0.8		MIT
4200.028	4201.211	23802.66	W	1.1	0.5		MIT
4199.626	4200.809	23804.94	W	1.0	0.3		MIT
4193.972	4195.154	23837.03	W	0.9	0.3		MIT
4190.15	4191.33	23858.8	W	0.7			MIT
4189.170	4190.350	23864.35	W	0.7	0.5		MIT
4186.02	4187.20	23882.3	W	1.1	0.3		MIT
4183.826	4185.005	23894.83	W	1.1	0.7		MIT
4183.668	4184.847	23895.74	W	1.0			MIT
4182.324	4183.503	23903.42	W	0.5			MIT
4181.393	4182.571	23908.74	W	0.3	0.3		MIT
4180.239	4181.417	23915.34	W	1.0	0.7		MIT
4177.830	4179.008	23929.13	W	1.0	0.5		MIT
4175.59	4176.77	23942.0	W II	0.8	1.4		MIT
4171.185	4172.361	23967.25	W	1.4	1.1		MIT
4170.535	4171.711	23970.98	W	1.2	0.8		MIT
4170.001	4171.176	23974.05	W	0.5	0.0		MIT
4168.661	4169.836	23981.76	W	1.0	1.2		MIT
4166.153	4167.327	23996.19	W	0.8	0.5		MIT
4160.351	4161.524	24029.66	W	0.7	0.3		MIT
4160.040	4161.213	24031.46	W	0.7	0.5		MIT
4159.792	4160.965	24032.89	W	0.7	0.3		MIT
4159.622	4160.795	24033.87	W	0.7	0.3		MIT
4157.036	4158.208	24048.82	W	0.5	1.4		MIT
4154.675	4155.846	24062.49	W	1.2	1.0		MIT
4149.747	4150.917	24091.06	W	1.0	0.3		MIT
4149.442	4150.612	24092.83	W	1.0	0.6		MIT
4145.949	4147.118	24113.13	W	0.9	0.5		MIT
4145.159	4146.328	24117.73	W	1.2	0.9		MIT
4142.256	4143.424	24134.63	W	1.2	1.0		MIT
4140.407	4141.575	24145.41	W	1.1	0.5		MIT
4140.045	4141.213	24147.52	W	1.1	0.5		MIT
4139.322	4140.489	24151.73	W	1.1	0.7		MIT
4137.543	4138.710	24162.12	W	1.1	1.1		MIT
4136.351	4137.518	24169.08	W	1.0	0.9		MIT
4133.482	4134.648	24185.86	W	0.8D	0.6D		MIT
4132.214	4133.380	24193.28	W	0.8	0.9		MIT
4129.98	4131.14	24206.4	W	0.7	0.8		MIT
4127.251	4128.415	24222.37	W	0.7	0.7		MIT
4126.799	4127.963	24225.02	W	1.1	1.1		MIT
4125.735	4126.899	24231.27	W	0.0	0.5		MIT
4125.180	4126.344	24234.53	W	1.0	1.1		MIT
4123.055	4124.218	24247.02	W	0.8	0.9		MIT
4122.017	4123.180	24253.13	W	0.8	0.9		MIT
4120.857	4122.020	24259.95	W	1.0	0.9		MIT
4118.184	4119.346	24275.70	W	1.0L	1.0L		MIT
4118.054	4119.216	24276.47	W	0.9	1.0		MIT
4115.586	4116.747	24291.02	W	0.9	0.9		MIT
4114.827	4115.988	24295.50	W	0.8	0.8		MIT
4112.485	4113.645	24309.34	W	0.9	0.8		MIT
4111.813	4112.973	24313.31	W	0.9	0.8		MIT
4110.572	4111.732	24320.65	W	0.8	0.8		MIT
4109.756	4110.916	24325.48	W	1.3	1.3		MIT
4108.531	4109.690	24332.73	W	0.9	1.0		MIT
4107.825	4108.984	24336.92	W	0.8	0.8		MIT
4106.702	4107.861	24343.57	W	0.8	0.8		MIT
4102.704	4103.862	24367.29	W	1.5	1.5		MIT
4101.847	4103.005	24372.38	W	0.8	0.9		MIT
4100.895	4102.052	24378.04	W	0.8	0.8		MIT
4099.026	4100.183	24389.16	W	0.8	0.8		MIT
4098.490	4099.647	24392.35	W	0.6	0.5		MIT
4097.670	4098.827	24397.23	W	0.8	0.7		MIT
4096.424	4097.580	24404.65	W	0.3	0.7		MIT
4095.701	4096.857	24408.96	W	1.1	1.2W		MIT
4093.143	4094.298	24424.21	W	0.7	0.6		MIT

AIR WAVELENGTH (ANGSTRMS)	VACUUM WAVELENGTH (ANGSTRMS)	WAVENUMBER (KAYSERS)	LMENT	LOG ARC	INTENSITY SPK	DIS	REF
4092.397	4093.552	24428.66	W	0.7	0.6		MIT
4091.218	4092.373	24435.70	W	0.6	0.5		MIT
4091.04	4092.19	24436.8	W	0.7	0.5		MIT
4090.639	4091.794	24439.16	W	0.8	0.7		MIT
4090.308	4091.463	24441.14	W	0.6	0.3		MIT
4089.385	4090.539	24446.65	W	0.8	0.7		MIT
4089.061	4090.215	24448.59	W	0.3	0.0		MIT
4088.770	4089.924	24450.33	W	0.8	0.8		MIT
4088.333	4089.487	24452.94	W	1.1	1.0		MIT
4087.394	4088.548	24458.56	W	0.8	0.8		MIT
4086.957	4088.111	24461.18	W	0.0	0.6		MIT
4083.714	4084.867	24480.60	W	0.8	0.7		MIT
4082.972	4084.125	24485.05	W	1.1	1.2		MIT
4082.052	4083.204	24490.57	W	0.5	0.6		MIT
4081.298	4082.450	24495.09	W	0.3	1.4		MIT
4079.785	4080.937	24504.18	W	0.6	0.5		MIT
4079.257	4080.409	24507.35	W	0.8	0.5		MIT
4078.124	4079.275	24514.16	W	0.8	0.8		MIT
4077.059	4078.210	24520.56	W	0.8	0.7		MIT
4074.364	4075.514	24536.78	W	1.7	1.7		MIT
4073.867	4075.017	24539.77	W	0.6	0.5		MIT
4073.150	4074.300	24544.09	W	1.0	0.9		MIT
4071.932	4073.082	24551.43	W	0.9	0.8		MIT
4070.606	4071.755	24559.43	W	1.2	1.1		MIT
4069.804	4070.953	24564.27	W	0.8	0.8		MIT
4069.154	4070.303	24568.20	W	0.8	0.8		MIT
4066.908	4068.056	24581.76	W	1.0	1.0		MIT
4065.997	4067.145	24587.27	W	0.7	0.7		MIT
4065.320	4066.468	24591.36	W	0.9	0.8		MIT
4064.789	4065.937	24594.58	W	1.2	1.1		MIT
4060.708	4061.855	24619.29	W	0.9	0.8		MIT
4060.234	4061.381	24622.17	W	0.6	1.0		MIT
4059.254	4060.400	24628.11	W	0.7	0.6		MIT
4057.452	4058.598	24639.05	W	0.8	0.8		MIT
4056.465	4057.611	24645.04	W	0.3	0.7		MIT
4055.648	4056.794	24650.01	W	0.8	0.7		MIT
4055.232	4056.377	24652.54	W	0.8	0.8		MIT
4053.940	4055.085	24660.40	W	1.0	1.0		MIT
4052.321	4053.466	24670.25	W	0.8	0.8		MIT
4051.304	4052.448	24676.44	W	0.3	1.0		MIT
4048.848	4049.992	24691.41	W	0.7	0.8		MIT
4048.252	4049.396	24695.04	W	0.8	0.8		MIT
4047.935	4049.078	24696.98	W	1.0	0.9		MIT
4046.701	4047.844	24704.51	W	1.0	1.1		MIT
4045.601	4046.744	24711.23	W	1.1	1.2		MIT
4044.288	4045.431	24719.25	W	1.2	1.1		MIT
4042.391	4043.533	24730.85	W	0.9	0.8		MIT
4040.591	4041.733	24741.86	W	1.0	0.9		MIT
4039.855	4040.996	24746.37	W	1.1L	1.0L		MIT
4039.423	4040.564	24749.02	W	0.8	0.8		MIT
4037.750	4038.891	24759.27	W		0.8L		MIT
4036.860	4038.001	24764.73	W	1.1	1.1		MIT
4035.355	4036.495	24773.97	W	1.0	1.0		MIT
4033.913	4035.053	24782.82	W		1.0		MIT
4032.385	4033.524	24792.21	W	0.8	0.8		MIT
4031.675	4032.814	24796.58	W	0.9	0.8		MIT
4029.95	4031.09	24807.2	W	0.8	0.8		MIT
4029.608	4030.747	24809.30	W	0.8	0.8		MIT
4029.025	4030.164	24812.89	W	0.7	0.8		MIT
4028.790	4029.928	24814.34	W	1.1	1.0		MIT
4025.600	4026.738	24834.00	W		1.0		MIT
4025.19	4026.33	24836.5	W	0.9	0.8		MIT
4022.833	4023.970	24851.08	W	0.6	0.7		MIT
4022.119	4023.256	24855.49	W	1.1	1.0		MIT
4020.319	4021.455	24866.62	W		0.3		MIT
4019.805	4020.941	24869.80	W		0.7		MIT
4019.231	4020.367	24873.35	W	1.3	1.2		MIT
4018.309	4019.445	24879.06	W	0.3	0.5		MIT
4017.351	4018.486	24884.99	W	0.6	0.7		MIT
4016.527	4017.662	24890.10	W	1.0	1.1		MIT

AIR WAVELENGTH (ANGSTRMS)	VACUUM WAVELENGTH (ANGSTRMS)	WAVENUMBER (KAYSERS)	LMENT	LOG ARC	INTENSITY SPK	DIS	REF
4016.109	4017.244	24892.69	W	0.8	0.6		MIT
4015.216	4016.351	24898.22	W	1.4	1.5		MIT
4014.939	4016.074	24899.94	W	0.9	0.8		MIT
4013.183	4014.317	24910.83	W	0.8	0.7		MIT
4012.096	4013.230	24917.58	W		0.7		MIT
4011.809	4012.943	24919.37	W	0.6	0.5		MIT
4011.024	4012.158	24924.24	W	0.6	0.7		MIT
4010.375	4011.509	24928.28	W	1.0	0.8		MIT
4009.806	4010.940	24931.81	W		1.0		MIT
4008.753	4009.886	24938.36	W I	1.7	1.7		MIT
4007.003	4008.136	24949.25	W		0.8		MIT
4005.895	4007.028	24956.16	W	0.8	0.8		MIT
4005.404	4006.536	24959.21	W	0.9	1.0		MIT
4004.747	4005.879	24963.31	W		0.8		MI
4003.836	4004.968	24968.99	W		0.8		MIT
4002.785	4003.917	24975.54	W		0.7		MIT
4001.892	4003.023	24981.12	W		1.2		MIT
4001.373	4002.504	24984.36	W	1.0	1.0		MIT
4000.694	4001.825	24988.60	W	1.1	1.0		MIT
4000.093	4001.224	24992.35	W		0.9		MIT
3999.181	4000.312	24998.05	W	0.6	0.7		MIT
3998.755	3999.886	25000.72	W	0.9	0.9		MIT
3998.159	3999.289	25004.44	W	0.8	0.8		MIT
3997.756	3998.886	25006.96	W	0.7	0.8		MIT
3997.134	3998.264	25010.85	W	0.8	0.8		MIT
3994.762	3995.892	25025.70	W	0.6	0.3		MIT
3993.903	3995.032	25031.09	W	0.8	0.8		MIT
3992.750	3993.879	25038.31	W	0.8	0.8		MIT
3992.505	3993.634	25039.85	W		0.8		MIT
3991.514	3992.643	25046.07	W	0.7	0.6		MIT
3991.224	3992.353	25047.89	W	1.0	0.9		MIT
3990.381	3991.509	25053.18	W		1.0		MIT
3988.007	3989.135	25068.09	W	0.8	0.8		MIT
3987.524	3988.652	25071.13	W	0.7	0.6		MIT
3987.149	3988.277	25073.49	W		1.0		MIT
3985.664	3986.791	25082.83	W	0.5	0.8		MIT
3984.176	3985.303	25092.20	W		0.9		MIT
3983.288	3984.415	25097.79	W	1.1	1.4		MIT
3982.960	3984.087	25099.86	W	0.8	0.8		MIT
3982.870	3983.996	25100.42	W	0.7	0.6		MIT
3981.264	3982.390	25110.55	W	0.8D	0.8		MIT
3980.640	3981.766	25114.49	W	1.1	1.2		MIT
3980.294	3981.420	25116.67	W	0.6	0.7		MIT
3979.286	3980.412	25123.03	W	1.1	1.1		MIT
3976.514	3977.639	25140.54	W		0.5H		MIT
3976.284	3977.409	25142.00	W	0.5	0.7		MIT
3975.892	3977.017	25144.48	W	0.9	1.0		MIT
3975.464	3976.589	25147.18	W	1.0	1.1		MIT
3973.284	3974.408	25160.98	W	0.6	0.6H		MIT
3972.808	3973.932	25164.00	W	0.6	0.5		MIT
3972.559	3973.683	25165.57	W		0.8		MIT
3972.000	3973.124	25169.11	W	0.8	0.7		MIT
3970.800	3971.923	25176.72	W	1.1	1.1		MIT
3969.199	3970.322	25186.87	W	1.1	1.0		MIT
3968.590	3969.713	25190.74	W	0.8	0.8		MIT
3968.171	3969.294	25193.40	W	0.9	0.9		MIT
3966.755	3967.877	25202.39	W	0.7	0.5		MIT
3965.487	3966.609	25210.45	W	0.6	0.7H		MIT
3965.145	3966.267	25212.62	W	1.1	1.1		MIT
3964.994	3966.116	25213.58	W	1.0	0.8		MIT
3964.133	3965.255	25219.06	W		1.0		MIT
3963.696	3964.817	25221.84	W	0.7	0.6		MIT
3962.332	3963.453	25230.52	W	1.0	1.0		MIT
3961.760	3962.881	25234.17	W	0.7	0.7		MIT
3961.178	3962.299	25237.87	W		1.0		MIT
3960.871	3961.992	25239.83	W		1.0		MIT
3958.880	3960.000	25252.52	W	1.0	1.0		MIT
3957.623	3958.743	25260.54	W		0.7		MIT
3957.12	3958.24	25263.8	W	0.7D			MIT
3955.615	3956.734	25273.37	W		0.7		MIT

AIR WAVELENGTH (ANGSTRMS)	VACUUM WAVELENGTH (ANGSTRMS)	WAVENUMBER (KAYSERS)	LMENT	LOG ARC	INTENSITY SPK	DIS	REF
3955.310	3956.429	25275.32	W	1.1	1.0		MIT
3953.713	3954.832	25285.52	W	0.6	0.5		MIT
3953.159	3954.278	25289.07	W	1.0	1.1		MIT
3952.902	3954.021	25290.71	W	0.9	0.9		MIT
3952.521	3953.640	25293.15	W	0.9	0.8		MIT
3952.262	3953.381	25294.81	W	0.7	0.6		MIT
3951.876	3952.994	25297.28	W	0.8	0.9		MIT
3951.168	3952.286	25301.81	W	0.7	1.1		MIT
3950.118	3951.236	25308.54	W	1.0	1.1		MIT
3949.213	3950.331	25314.34	W		0.5		MIT
3948.203	3949.320	25320.81	W		0.7		MIT
3947.982	3949.099	25322.23	W	1.0	1.0		MIT
3946.311	3947.428	25332.95	W	1.0	0.9		MIT
3945.980	3947.097	25335.08	W	0.0	0.7		MIT
3945.206	3946.323	25340.05	W	0.7	0.8		MIT
3944.798	3945.915	25342.67	W	0.8	0.8		MIT
3942.377	3943.493	25358.23	W	0.5	0.9		MIT
3941.830	3942.946	25361.75	W	0.8	0.8		MIT
3939.908	3941.023	25374.12	W	0.7	0.7		MIT
3939.441	3940.556	25377.13	W	0.9	1.0		MIT
3938.117	3939.232	25385.66	W		0.8		MIT
3937.627	3938.742	25388.82	W	1.0	1.1		MIT
3936.985	3938.100	25392.96	W	1.1	1.2		MIT
3936.668	3937.782	25395.00	W		1.0		MIT
3936.232	3937.346	25397.82	W	1.0	0.9		MIT
3935.440	3936.554	25402.93	W II	0.8D	1.3		MIT
3935.041	3936.155	25405.50	W	1.1	1.0		MIT
3934.601	3935.715	25408.34	W	0.7	0.7		MIT
3931.935	3933.048	25425.57	W	0.3	0.3		MIT
3930.970	3932.083	25431.81	W	1.0	1.1		MIT
3930.480	3931.593	25434.98	W	0.9	1.0		MIT
3930.248	3931.361	25436.49	W	1.1	1.0		MIT
3929.315	3930.428	25442.52	W	0.0	0.7		MIT
3926.034	3927.146	25463.79	W	0.8	0.8		MIT
3925.650	3926.762	25466.28	W	0.3	0.8		MIT
3925.14	3926.25	25469.6	W		0.6D		MIT
3924.695	3925.806	25472.47	W	0.9	1.0		MIT
3924.371	3925.482	25474.58	W	1.0	0.9		MIT
3922.682	3923.793	25485.55	W		0.3		MIT
3922.335	3923.446	25487.80	W	0.8	0.8		MIT
3920.040	3921.150	25502.72	W	0.7	0.7		MIT
3918.599	3919.709	25512.10	W	1.1	1.0		MIT
3918.208	3919.318	25514.65	W	0.0	0.5		MIT
3917.642	3918.751	25518.33	W	0.9	0.9		MIT
3917.28	3918.39	25520.7	W		0.8		MIT
3916.400	3917.509	25526.42	W	0.7	0.6		MIT
3915.460	3916.569	25532.55	W		0.8		MIT
3915.239	3916.348	25533.99	W		0.7		MIT
3915.083	3916.192	25535.01	W		0.8		MIT
3914.190	3915.299	25540.84	W		0.6		MIT
3913.638	3914.746	25544.44	W		1.0		MIT
3913.257	3914.365	25546.93	W		0.7		MIT
3912.822	3913.930	25549.77	W	0.8	0.8		MIT
3911.293	3912.401	25559.75	W	0.8	0.7		MIT
3907.202	3908.309	25586.51	W	0.8	0.7		MIT
3905.970	3907.076	25594.58	W	0.9	0.8		MIT
3903.985	3905.091	25607.60	W	1.0	1.0		MIT
3903.298	3904.404	25612.10	W	0.8	0.8		MIT
3902.684	3903.790	25616.14	W	0.7	0.6		MIT
3901.833	3902.938	25621.72	W	1.0	1.0		MIT
3901.344	3902.449	25624.93	W	0.3	0.8		MIT
3900.91	3902.02	25627.8	W		1.2		MIT
3899.778	3900.883	25635.22	W		0.7		MIT
3898.782	3899.887	25641.77	W		0.8D		MIT
3897.909	3899.013	25647.51	W	1.1	1.2		MIT
3896.929	3898.033	25653.96	W	0.3	1.2		MIT
3893.930	3895.033	25673.72	W	0.8	1.1		MIT
3893.473	3894.576	25676.74	W	0.7	0.8		MIT
3892.722	3893.825	25681.69	W	1.0	1.1L		MIT
3892.330	3893.433	25684.27	W	0.9	0.8		MIT

AIR WAVELENGTH (ANGSTRMS)	VACUUM WAVELENGTH (ANGSTRMS)	WAVENUMBER (KAYSERS)	LMENT	LOG ARC	INTENSITY SPK	DIS	REF
3891.249	3892.352	25691.41	W	1.0	0.7		MIT
3890.741	3891.843	25694.76	W	0.8	0.8		MIT
3890.420	3891.522	25696.88	W	1.0	0.9		MIT
3889.369	3890.471	25703.83	W		0.7		MIT
3888.890	3889.992	25706.99	W	0.5	0.6		MIT
3887.945	3889.047	25713.24	W		1.0		MIT
3886.451	3887.552	25723.13	W	0.9	1.0		MIT
3883.829	3884.930	25740.49	W	0.8	0.9		MIT
3883.342	3884.443	25743.72	W	0.5$	0.7L		MIT
3881.390	3882.490	25756.67	W	1.3	1.3		MIT
3881.141	3882.241	25758.32	W	0.7	0.6		MIT
3880.757	3881.857	25760.87	W	0.8	0.8		MIT
3880.079	3881.179	25765.37	W	0.9	0.9		MIT
3879.196	3880.295	25771.23	W	0.7	0.8		MIT
3878.992	3880.091	25772.59	W	0.8	0.8		MIT
3878.511	3879.610	25775.78	W	0.6	0.7		MIT
3877.281	3878.380	25783.96	W		1.1		MIT
3875.682	3876.781	25794.60	W	1.1	1.0		MIT
3875.198	3876.296	25797.82	W		0.8		MIT
3874.406	3875.504	25803.09	W	1.1	1.1		MIT
3872.830	3873.928	25813.59	W	1.1	1.0		MIT
3872.72	3873.82	25814.3	W		0.6		MIT
3871.583	3872.680	25821.91	W		0.8		MIT
3870.821	3871.918	25826.99	W	0.7	0.8		MIT
3869.51	3870.61	25835.7	W		0.7		MIT
3869.310	3870.407	25837.08	W	0.5	0.7		MIT
3868.570	3869.667	25842.02	W	0.7	0.7		MIT
3868.456	3869.553	25842.78	W	0.5	0.5		MIT
3867.975	3869.072	25845.99	W	1.5	1.5		MIT
3867.415	3868.511	25849.74	W	0.6	0.5		MIT
3866.046	3867.142	25858.89	W	1.0	0.6		MIT
3865.803	3866.899	25860.51	W		0.9		MIT
3865.320	3866.416	25863.75	W	0.8	0.9		MIT
3864.336	3865.432	25870.33	W	1.1	1.0		MIT
3863.470	3864.565	25876.13	W	0.9	0.8		MIT
3861.240	3862.335	25891.07	W	0.9	0.9		MIT
3861.062	3862.157	25892.27	W	1.0	1.0		MIT
3859.983	3861.077	25899.51	W	1.2	1.5		MIT
3859.294	3860.388	25904.13	W	1.0	1.1		MIT
3858.846	3859.940	25907.14	W	0.7	0.8		MIT
3857.295	3858.389	25917.55	W	0.8	0.8		MIT
3856.818	3857.912	25920.76	W		1.0		MIT
3855.544	3856.637	25929.32	W	1.0	1.0		MIT
3855.321	3856.414	25930.82	W		0.8		MIT
3853.776	3854.869	25941.22	W	0.8	0.8		MIT
3853.216	3854.309	25944.99	W	0.6	0.5		MIT
3852.831	3853.924	25947.58	W	0.7	0.7		MIT
3851.997	3853.089	25953.20	W	0.7	0.3		MIT
3851.569	3852.661	25956.08	W II	0.8	0.5		MIT
3851.116	3852.208	25959.14	W		0.8		MIT
3848.482	3849.573	25976.90	W		0.3H		MIT
3847.490	3848.581	25983.60	W	1.3	1.2		MIT
3847.242	3848.333	25985.28	W	0.7	0.6		MIT
3846.214	3847.305	25992.22	W	1.3	1.3		MIT
3845.845	3846.936	25994.72	W		1.0		MIT
3844.200	3845.290	26005.84	W		1.0		MIT
3843.474	3844.564	26010.75	W	0.5	0.8		MIT
3842.303	3843.393	26018.68	W	1.0	1.1L		MIT
3839.257	3840.346	26039.32	W	0.8	0.8		MIT
3838.504	3839.593	26044.43	W	1.2	1.3		MIT
3837.919	3839.008	26048.40	W		0.8		MIT
3837.230	3838.319	26053.07	W		1.0		MIT
3836.960	3838.048	26054.91	W	1.0	0.8		MIT
3835.884	3836.972	26062.22	W	0.8	0.7		MIT
3835.700	3836.788	26063.47	W		0.7		MIT
3835.049	3836.137	26067.89	W	1.2	1.3		MIT
3834.039	3835.127	26074.76	W	0.8	1.0		MIT
3832.856	3833.943	26082.81	W	0.9	0.8		MIT
3832.641	3833.728	26084.27	W	0.6	0.7		MIT
3832.244	3833.331	26086.97	W	0.6	0.8		MIT

AIR WAVELENGTH (ANGSTRMS)	VACUUM WAVELENGTH (ANGSTRMS)	WAVENUMBER (KAYSERS)	LMENT	LOG ARC	INTENSITY SPK	DIS	REF
3830.993	3832.080	26095.49	W	0.8	0.8		MIT
3830.722	3831.809	26097.33	W	0.9	1.0		MIT
3830.133	3831.220	26101.35	W	0.8	0.9		MIT
3829.127	3830.213	26108.21	W	1.1	1.0		MIT
3827.348	3828.434	26120.34	W	0.6	0.5		MIT
3826.933	3828.019	26123.17	W	0.7	0.6		MIT
3826.193	3827.279	26128.23	W	1.0	1.1		MIT
3824.386	3825.471	26140.57	W	1.1	1.2		MIT
3824.146	3825.231	26142.21	W	0.8	0.9		MIT
3823.062	3824.147	26149.62	W		1.0		MIT
3822.210	3823.295	26155.45	W		0.9		MIT
3820.109	3821.193	26169.84	W	0.9	1.0		MIT
3817.480	3818.563	26187.86	W	1.3	1.2		MIT
3816.386	3817.469	26195.37	W	1.0	1.1		MIT
3815.779	3816.862	26199.53	W	1.1	0.7		MIT
3815.155	3816.238	26203.82	W		0.7		MIT
3814.426	3815.509	26208.83	W	0.3	0.8		MIT
3812.660	3813.742	26220.97	W	0.7	0.8		MIT
3810.798	3811.880	26233.78	W	1.0S	1.0S		MIT
3810.385	3811.467	26236.62	W	1.2	1.2		MIT
3809.853	3810.934	26240.28	W	0.8	0.8		MIT
3809.228	3810.309	26244.59	W	1.4	1.3		MIT
3807.855	3808.936	26254.05	W	0.5	0.7		MIT
3807.636	3808.717	26255.56	W	0.5	0.7		MIT
3807.436	3808.517	26256.94	W		0.8		MIT
3804.952	3806.032	26274.08	W	0.8	0.8		MIT
3804.55	3805.63	26276.9	W		0.9H		MIT
3804.080	3805.160	26280.10	W	0.8	0.7		MIT
3803.68	3804.76	26282.9	W		1.2W		MIT
3803.32	3804.40	26285.4	W		0.8H		MIT
3802.928	3804.008	26288.06	W	0.8	0.8		MIT
3801.921	3803.000	26295.03	W	1.0	1.0		MIT
3801.524	3802.603	26297.77	W	1.0	0.8		MIT
3798.922	3800.001	26315.78	W	0.8	1.0		MIT
3796.947	3798.025	26329.47	W	0.7	0.6		MIT
3796.285	3797.363	26334.06	W	0.8	0.8		MIT
3794.345	3795.422	26347.53	W	1.0	1.0		MIT
3793.323	3794.400	26354.63	W	0.8	0.7		MIT
3792.769	3793.846	26358.48	W	1.2	1.2		MIT
3791.772	3792.849	26365.41	W	0.6	0.0		MIT
3791.60	3792.68	26366.6	W		0.7		MIT
3789.77	3790.85	26379.3	W		1.1		MIT
3786.378	3787.453	26402.97	W	1.0	1.0		MIT
3785.63	3786.71	26408.2	W		1.0		MIT
3785.105	3786.180	26411.84	W	0.6	0.5		MIT
3783.730	3784.805	26421.44	W	1.0	1.1		MIT
3781.837	3782.911	26434.67	W	1.0	0.9		MIT
3780.927	3782.001	26441.03	W		0.9		MIT
3780.770	3781.844	26442.13	W	1.3	1.3		MIT
3780.234	3781.308	26445.88	W		0.9		MIT
3779.968	3781.042	26447.74	W	1.0	0.8		MIT
3779.208	3780.281	26453.06	W		0.8		MIT
3778.684	3779.757	26456.73	W	0.9	1.1L		MIT
3778.586	3779.659	26457.41	W	0.8	0.9		MIT
3778.148	3779.221	26460.48	W	0.8	1.0		MIT
3777.46	3778.53	26465.3	W		1.0		MIT
3776.69	3777.76	26470.7	W		0.3H		MIT
3775.446	3776.518	26479.42	W	0.9	0.8S		MIT
3774.138	3775.210	26488.59	W		1.3		MIT
3773.705	3774.777	26491.63	W	1.3	1.3		MIT
3772.423	3773.495	26500.63	W	0.8	0.8		MIT
3772.052	3773.124	26503.24	W	1.0	1.0		MIT
3770.606	3771.677	26513.40	W	0.8	0.8		MIT
3769.867	3770.938	26518.60	W	1.0	1.0		MIT
3769.210	3770.281	26523.22	W	1.1	1.0		MIT
3768.978	3770.049	26524.86	W		0.8		MIT
3768.447	3769.518	26528.59	W	1.3	1.3		MIT
3767.846	3768.917	26532.82	W	1.0	0.9		MIT
3767.423	3768.493	26535.80	W	0.8	1.0		MIT
3765.310	3766.380	26550.69	W	0.8	0.8		MIT

AIR WAVELENGTH (ANGSTRMS)	VACUUM WAVELENGTH (ANGSTRMS)	WAVENUMBER (KAYSERS)	LMENT	LOG ARC	INTENSITY SPK	INTENSITY DIS	REF
3764.312	3765.382	26557.73	W	0.9	1.0		MIT
3763.638	3764.707	26562.49	W		0.7		MIT
3762.885	3763.954	26567.80	W		0.7		MIT
3761.621	3762.690	26576.73	W	0.8	0.9		MIT
3760.639	3761.708	26583.67	W	0.8	0.8		MIT
3760.380	3761.449	26585.50	W		1.0		MIT
3760.126	3761.195	26587.30	W	1.2	1.0		MIT
3758.942	3760.010	26595.67	W		1.2		MIT
3757.923	3758.991	26602.88	W	1.2	1.3		MIT
3757.349	3758.417	26606.95	W	0.9	0.8		MIT
3757.088	3758.156	26608.80	W	1.0	1.0		MIT
3756.867	3757.935	26610.36	W	0.8	1.2		MIT
3756.370	3757.438	26613.88	W	0.8	0.8		MIT
3755.50	3756.57	26620.1	W	0.9D	0.8D		MIT
3754.883	3755.950	26624.42	W	0.8	0.7		MIT
3754.154	3755.221	26629.59	W	0.7	0.6		MIT
3753.481	3754.548	26634.37	W	0.7	0.8		MIT
3753.142	3754.209	26636.77	W		0.9		MIT
3751.428	3752.494	26648.94	W	0.8L	0.9L		MIT
3750.78	3751.85	26653.6	W		1.1		MIT
3749.660	3750.726	26661.51	W	0.8	0.9		MIT
3747.457	3748.522	26677.18	W	0.8	0.7		MIT
3745.56	3746.62	26690.7	W		1.4		MIT
3744.911	3745.976	26695.32	W	0.7	0.5		MIT
3743.812	3744.876	26703.15	W	0.8	0.8		MIT
3742.685	3743.749	26711.19	W	1.2D	1.1D		MIT
3741.713	3742.777	26718.13	W	1.1	1.3		MIT
3739.484	3740.547	26734.06	W	1.0	1.1		MIT
3739.16	3740.22	26736.4	W		0.9		MIT
3738.898	3739.961	26738.25	W	0.8	0.8		MIT
3738.111	3739.174	26743.88	W	0.8	0.8		MIT
3737.845	3738.908	26745.78	W	0.6	0.7		MIT
3736.215	3737.277	26757.45	W	0.7	1.0		MIT
3732.543	3733.604	26783.77	W	0.8	0.8		MIT
3730.427	3731.488	26798.96	W	1.0	1.1		MIT
3729.201	3730.261	26807.77	W	0.7	0.8		MIT
3728.279	3729.339	26814.40	W	0.8	0.9		MIT
3727.86	3728.92	26817.4	W	0.8	0.8		MIT
3726.793	3727.853	26825.09	W	0.5	0.8		MIT
3726.06	3727.12	26830.4	W		0.8		MIT
3725.163	3726.222	26836.83	W	0.8	0.9		MIT
3723.877	3724.936	26846.10	W	0.8	0.8		MIT
3722.247	3723.306	26857.85	W	1.1	1.2		MIT
3721.12	3722.18	26866.0	W		1.1		MIT
3720.513	3721.571	26870.37	W	0.9	1.0		MIT
3719.405	3720.463	26878.37	W	1.1	1.0		MIT
3718.316	3719.374	26886.25	W		0.8		MIT
3717.80	3718.86	26890.0	W		1.1		MIT
3717.095	3718.152	26895.08	W	1.1D	1.0D		MIT
3716.738	3717.795	26897.66	W	1.0	0.8		MIT
3716.079	3717.136	26902.43	W II	1.0	1.3		MIT
3715.917	3716.974	26903.60	W	0.5	0.8		MIT
3715.047	3716.104	26909.91	W	1.0	0.8		MIT
3714.858	3715.915	26911.27	W	1.0	0.8		MIT
3714.564	3715.621	26913.40	W		1.0		MIT
3714.240	3715.297	26915.75	W	1.0	0.8		MIT
3712.214	3713.270	26930.44	W	1.3			MIT
3711.485	3712.541	26935.73	W	0.8	0.5		MIT
3710.289	3711.345	26944.41	W	0.6	0.7		MIT
3708.510	3709.565	26957.34	W	0.5	1.5		MIT
3707.931	3708.986	26961.55	W	1.3	1.3		MIT
3707.52	3708.57	26964.5	W	0.5	1.0		MIT
3705.486	3706.540	26979.34	W	1.0	1.0		MIT
3705.34	3706.39	26980.4	W		0.8		MIT
3704.796	3705.850	26984.36	W	1.1	0.9		MIT
3703.602	3704.656	26993.06	W	1.0	1.0		MIT
3703.36	3704.41	26994.8	W	1.1	1.0		MIT
3702.320	3703.374	27002.41	W	1.0	1.0		MIT
3701.78	3702.83	27006.4	W		0.7		MIT
3700.649	3701.702	27014.60	W	0.7	0.8		MIT

AIR WAVELENGTH (ANGSTRMS)	VACUUM WAVELENGTH (ANGSTRMS)	WAVENUMBER (KAYSERS)	LMENT	LOG ARC	INTENSITY SPK	INTENSITY DIS	REF
3700.269	3701.322	27017.37	W		0.7		MIT
3699.415	3700.468	27023.61	W	1.0	1.1		MIT
3698.720	3699.773	27028.69	W	1.0	1.0		MIT
3697.823	3698.875	27035.25	W		0.6		MIT
3697.46	3698.51	27037.9	W	1.0	1.0		MIT
3696.818	3697.870	27042.59	W	0.7	0.5		MIT
3696.195	3697.247	27047.15	W	0.5	0.7		MIT
3695.244	3696.296	27054.11	W	0.7	0.6		MIT
3694.946	3695.998	27056.29	W	0.3			MIT
3694.508	3695.560	27059.50	W	1.0	1.3L		MIT
3692.91	3693.96	27071.2	W		0.8		MIT
3692.725	3693.776	27072.57	W	0.8	0.5		MIT
3691.88	3692.93	27078.8	W	0.6D	1.3D		MIT
3691.551	3692.602	27081.18	W II	0.7	1.0		MIT
3690.259	3691.309	27090.66	W	1.0	0.9		MIT
3689.877	3690.927	27093.46	W	1.0	0.9		MIT
3689.61	3690.66	27095.4	W		1.0		MIT
3689.058	3690.108	27099.48	W	0.8	0.8		MIT
3688.422	3689.472	27104.15	W	0.8	0.7		MIT
3688.331	3689.381	27104.82	W II		0.8		MIT
3688.069	3689.119	27106.74	W	1.1	0.8		MIT
3686.41	3687.46	27118.9	W		0.8		MIT
3685.597	3686.646	27124.93	W	0.8	0.6		MIT
3685.019	3686.068	27129.18	W	1.0	0.8		MIT
3684.659	3685.708	27131.83	W	1.0	1.0		MIT
3683.941	3684.990	27137.12	W	1.0	1.0		MIT
3683.616	3684.665	27139.51	W		0.8		MIT
3683.392	3684.441	27141.16	W	0.9	0.8		MIT
3683.310	3684.359	27141.77	W	0.9	0.8		MIT
3682.092	3683.140	27150.74	W	1.4	1.3		MIT
3680.86	3681.91	27159.8	W	0.6D	0.7D		MIT
3679.607	3680.655	27169.08	W	1.1	1.0		MIT
3678.19	3679.24	27179.6	W	0.0	0.7		MIT
3677.41	3678.46	27185.3	W		1.0		MIT
3676.802	3677.849	27189.81	W	1.0	1.0		MIT
3676.311	3677.358	27193.44	W	0.8	0.7		MIT
3675.555	3676.602	27199.03	W	1.1	1.0		MIT
3674.97	3676.02	27203.4	W		0.9		MIT
3674.580	3675.626	27206.25	W	1.0	1.1		MIT
3674.16	3675.21	27209.4	W	0.6D	0.7D		MIT
3672.95	3674.00	27218.3	W	0.6D	0.3H		MIT
3672.586	3673.632	27221.02	W	0.0	1.3		MIT
3672.180	3673.226	27224.03	W	0.7	0.6		MIT
3670.786	3671.831	27234.37	W		0.7		MIT
3670.767	3671.812	27234.51	W	0.8	0.8		MIT
3669.92	3670.97	27240.8	W		0.8		MIT
3669.328	3670.373	27245.19	W	0.6	0.8		MIT
3668.665	3669.710	27250.11	W	1.0	1.1		MIT
3667.719	3668.764	27257.14	W	1.0	1.0		MIT
3667.182	3668.226	27261.13	W	0.9	0.8		MIT
3666.813	3667.857	27263.87	W	0.7	0.6		MIT
3665.881	3666.925	27270.80	W	1.0	0.9S		MIT
3663.825	3664.869	27286.11	W	0.8	0.8		MIT
3663.361	3664.404	27289.56	W	0.9	1.0		MIT
3663.151	3664.194	27291.13	W	0.8	0.8		MIT
3661.236	3662.279	27305.40	W	0.5	0.8D		MIT
3660.609	3661.652	27310.08	W	0.9	1.0		MIT
3660.370	3661.413	27311.86	W	0.7	0.6		MIT
3660.171	3661.214	27313.35	W	0.8	0.5		MIT
3659.54	3660.58	27318.1	W	0.8	0.5		MIT
3659.316	3660.358	27319.73	W	1.0	0.8		MIT
3659.154	3660.196	27320.94	W	0.5	0.7		MIT
3658.370	3659.412	27326.79	W	0.8	0.5		MIT
3657.882	3658.924	27330.44	W II	0.7	1.2		MIT
3657.592	3658.634	27332.61	W II	1.0	1.4		MIT
3656.684	3657.726	27339.39	W	0.8	0.7		MIT
3656.46	3657.50	27341.1	W		0.8		MIT
3656.17	3657.21	27343.2	W	0.9	0.5		MIT
3654.71	3655.75	27354.2	W		1.1		MIT
3654.202	3655.243	27357.96	W	1.1	1.0		MIT

AIR WAVELENGTH (ANGSTRMS)	VACUUM WAVELENGTH (ANGSTRMS)	WAVENUMBER (KAYSERS)	ELEMENT	LOG ARC	INTENSITY SPK	INTENSITY DIS	REF
3653.524	3654.565	27363.04	W	0.7	0.0		MIT
3653.338	3654.379	27364.43	W	0.0	1.1		MIT
3652.760	3653.801	27368.76	W	0.5	0.6		MIT
3652.114	3653.155	27373.60	W II	0.5	1.1		MIT
3651.004	3652.044	27381.92	W	1.0	1.1		MIT
3649.020	3650.060	27396.81	W	0.8	0.8		MIT
3647.525	3648.564	27408.04	W	0.9	1.0		MIT
3647.147	3648.186	27410.88	W	0.6	0.5		MIT
3646.525	3647.564	27415.56	W II	1.0L	1.5		MIT
3645.599	3646.638	27422.52	W II	0.6	1.3		MIT
3643.514	3644.552	27438.21	W		0.8		MIT
3643.314	3644.352	27439.72	W	0.8	0.7		MIT
3642.819	3643.857	27443.45	W	0.8	0.7		MIT
3641.852	3642.890	27450.73	W	0.8	0.7		MIT
3641.408	3642.446	27454.08	W II	1.1	1.6		MIT
3640.132	3641.169	27463.70	W	1.0	0.9		MIT
3639.18	3640.22	27470.9	W		1.0		MIT
3638.81	3639.85	27473.7	W	0.0	0.7		MIT
3637.936	3638.973	27480.28	W	0.7	0.6		MIT
3637.384	3638.421	27484.45	W	1.0	1.0		MIT
3636.736	3637.773	27489.35	W	0.8	0.8		MIT
3635.434	3636.470	27499.19	W	0.3			MIT
3633.574	3634.610	27513.27	W	0.6	0.5		MIT
3632.708	3633.744	27519.83	W	1.0	0.9		MIT
3631.948	3632.983	27525.59	W	1.2	1.0		MIT
3630.955	3631.990	27533.11	W II		0.8		MIT
3630.821	3631.856	27534.13	W	1.0	0.9		MIT
3630.318	3631.353	27537.95	W	1.0	1.0		MIT
3630.086	3631.121	27539.71	W	0.0	0.8		MIT
3628.930	3629.965	27548.48	W		1.2		MIT
3628.38	3629.41	27552.6	W	0.5	1.3D		MIT
3627.240	3628.274	27561.31	W	1.1	1.0		MIT
3625.405	3626.439	27575.26	W	1.0	1.0		MIT
3624.468	3625.501	27582.39	W	0.5	0.0		MIT
3624.27	3625.30	27583.9	W		1.0		MIT
3623.514	3624.547	27589.65	W	0.9	0.8		MIT
3623.13	3624.16	27592.6	W		0.7		MIT
3620.83	3621.86	27610.1	W		0.9		MIT
3619.272	3620.304	27621.99	W	1.0	1.0		MIT
3618.452	3619.484	27628.25	W II		1.1		MIT
3617.521	3618.553	27635.36	W	1.5	1.3		MIT
3616.400	3617.431	27643.93	W	0.6	0.5		MIT
3615.540	3616.571	27650.50	W	0.8	0.7		MIT
3614.801	3615.832	27656.15	W	0.8	0.3		MIT
3614.255	3615.286	27660.33	W	0.7	0.3		MIT
3613.790	3614.821	27663.89	W II	1.0	1.5		MIT
3613.549	3614.580	27665.74	W		0.8		MIT
3611.855	3612.885	27678.71	W		1.3		MIT
3608.37	3609.40	27705.4	W		1.0		MIT
3607.066	3608.095	27715.46	W	1.0	1.1		MIT
3606.68	3607.71	27718.4	W		0.6		MIT
3606.338	3607.367	27721.05	W	1.0	0.9		MIT
3606.067	3607.096	27723.14	W	1.1	1.0		MIT
3604.16	3605.19	27737.8	W		0.5		MIT
3603.910	3604.938	27739.73	W	0.7	0.7		MIT
3602.97	3604.00	27747.0	W	0.0	0.7		MIT
3602.462	3603.490	27750.88	W		1.1		MIT
3601.798	3602.826	27755.99	W	0.7	0.6		MIT
3601.578	3602.605	27757.69	W		0.8		MIT
3600.955	3601.982	27762.49	W		1.0		MIT
3600.287	3601.314	27767.64	W	0.6	0.5		MIT
3598.878	3599.905	27778.51	W	1.0	0.8		MIT
3598.43	3599.46	27782.0	W		1.1		MIT
3597.961	3598.988	27785.59	W	0.7	0.6		MIT
3597.714	3598.740	27787.50	W	0.8	0.8		MIT
3597.263	3598.289	27790.98	W	1.0	1.0		MIT
3596.608	3597.634	27796.04	W	0.7	0.6		MIT
3596.433	3597.459	27797.40	W	0.7	0.6		MIT
3596.18	3597.21	27799.4	W		1.2		MIT
3595.386	3596.412	27805.49	W	0.9	0.8		MIT

AIR WAVELENGTH (ANGSTRMS)	VACUUM WAVELENGTH (ANGSTRMS)	WAVENUMBER (KAYSERS)	ELEMENT	LOG ARC	INTENSITY SPK	INTENSITY DIS	REF
3595.07	3596.10	27807.9	W		0.7		MIT
3594.531	3595.557	27812.10	W	0.7	0.8		MIT
3593.971	3594.997	27816.44	W	1.0	0.9		MIT
3593.561	3594.586	27819.61	W	0.6	0.5		MIT
3592.979	3594.004	27824.12	W	0.7	0.5		MIT
3592.846	3593.871	27825.15	W	0.8D	0.5D		MIT
3592.423	3593.448	27828.43	W II	1.0	1.5		MIT
3591.972	3592.997	27831.92	W	0.8	0.6		MIT
3591.769	3592.794	27833.49	W	0.8	0.5		MIT
3590.826	3591.851	27840.80	W	1.0	1.0L		MIT
3590.43	3591.45	27843.9	W		0.7		MIT
3589.698	3590.722	27849.55	W	0.6	0.5		MIT
3586.99	3588.01	27870.6	W		0.9		MIT
3585.71	3586.73	27880.5	W		0.6		MIT
3584.787	3585.810	27887.70	W	0.5	0.3		MIT
3584.107	3585.130	27892.99	W	1.0	1.0		MIT
3583.46	3584.48	27898.0	W		1.0		MIT
3582.913	3583.936	27902.29	W	0.6	0.3		MIT
3582.72	3583.74	27903.8	W		0.6D		MIT
3582.243	3583.266	27907.50	W	0.8	0.8		MIT
3581.887	3582.909	27910.28	W	0.6	1.0		MIT
3581.238	3582.260	27915.34	W	0.9	0.9		MIT
3577.64	3578.66	27943.4	W		0.8D		MIT
3576.70	3577.72	27950.7	W		0.7D		MIT
3576.382	3577.403	27953.24	W	0.8	0.7		MIT
3575.976	3576.997	27956.41	W	0.9	0.8		MIT
3575.226	3576.247	27962.28	W	1.0	1.0		MIT
3573.969	3574.989	27972.11	W	0.6	0.5		MIT
3573.411	3574.431	27976.48	W	0.9	0.8		MIT
3572.86	3573.88	27980.8	W		0.8		MIT
3572.476	3573.496	27983.80	W II	1.0	1.5		MIT
3572.02	3573.04	27987.4	W II	0.6	1.0		M
3570.655	3571.675	27998.07	W	1.2	1.2		MIT
3569.584	3570.603	28006.47	W	0.6	0.7		MIT
3569.235	3570.254	28009.21	W	0.9D	0.8D		MIT
3568.987	3570.006	28011.16	W	1.0	1.0		MIT
3568.418	3569.437	28015.62	W	0.6	0.6		MIT
3568.040	3569.059	28018.59	W	1.0	1.0		MIT
3567.678	3568.697	28021.43	W	0.8S	0.8		MIT
3567.247	3568.266	28024.82	W	0.8	0.8		MIT
3565.85	3566.87	28035.8	W		0.7		MIT
3565.62	3566.64	28037.6	W	0.8	0.8		MIT
3565.156	3566.174	28041.26	W	0.6	0.7		MIT
3564.38	3565.40	28047.4	W	0.5D	0.8D		MIT
3564.006	3565.024	28050.30	W II		0.8		MIT
3563.65	3564.67	28053.1	W		0.7		MIT
3563.452	3564.470	28054.66	W	1.0	0.8		MIT
3562.519	3563.536	28062.01	W		1.1		MIT
3561.452	3562.469	28070.42	W		1.0		MIT
3561.253	3562.270	28071.99	W	0.9	0.6		MIT
3560.071	3561.088	28081.31	W	0.7	0.6		MIT
3559.709	3560.726	28084.16	W	1.0	0.9		MIT
3559.092	3560.109	28089.03	W	0.8	0.6		MIT
3557.920	3558.936	28098.28	W	0.3	1.1		MIT
3557.212	3558.228	28103.88	W	0.9	0.9		MIT
3557.015	3558.031	28105.43	W	0.5	0.6		MIT
3556.02	3557.04	28113.3	W		0.8		MIT
3555.752	3556.768	28115.42	W	0.9	0.8		MIT
3555.175	3556.191	28119.98	W II	0.8	1.2		MIT
3554.583	3555.598	28124.66	W		0.7		MIT
3554.213	3555.228	28127.59	W	1.0	0.9		MIT
3552.323	3553.338	28142.55	W	0.7	0.6S		MIT
3551.540	3552.555	28148.76	W		1.0		MIT
3551.276	3552.291	28150.85	W	0.8	0.7		MIT
3551.028	3552.042	28152.82	W	0.8	1.0L		MIT
3550.842	3551.856	28154.29	W	0.9	0.9		MIT
3550.681	3551.695	28155.57	W	0.8	0.8		MIT
3549.053	3550.067	28168.48	W II	0.7	1.4		MIT
3548.259	3549.273	28174.79	W	0.7	0.8		MIT
3547.470	3548.484	28181.05	W	0.9	0.8		MIT

AIR WAVELENGTH (ANGSTRMS)	VACUUM WAVELENGTH (ANGSTRMS)	WAVENUMBER (KAYSERS)	LMENT	LOG ARC	INTENSITY SPK	DIS	REF
3546.488	3547.501	28188.85	W	0.5	0.8		MIT
3545.229	3546.242	28198.86	W	1.1L	1.0		MIT
3544.971	3545.984	28200.92	W	0.8	0.8		MIT
3544.796	3545.809	28202.31	W	0.8	0.7		MIT
3544.464	3545.477	28204.95	W II	0.0	1.1		MIT
3543.713	3544.726	28210.93	W	0.9	0.8		MIT
3543.107	3544.119	28215.75	W	0.6	1.1		MIT
3542.655	3543.667	28219.35	W	0.6	0.5		MIT
3542.273	3543.285	28222.40	W	0.5	0.8		MIT
3541.647	3542.659	28227.39	W	1.0	1.0		MIT
3540.731	3541.743	28234.69	W	1.0	1.1		MIT
3539.927	3540.939	28241.10	W	0.8	0.6		MIT
3539.458	3540.469	28244.84	W II	0.5	0.8		MIT
3539.330	3540.341	28245.86	W	0.8	0.5		MIT
3538.633	3539.644	28251.43	W	0.9	1.0		MIT
3537.452	3538.463	28260.86	W	1.1	1.1		MIT
3536.273	3537.284	28270.28	W	0.5	1.2		MIT
3535.549	3536.559	28276.07	W	1.1	1.0		MIT
3535.259	3536.269	28278.39	W	0.6	0.8		MIT
3534.529	3535.539	28284.23	W	0.5	1.1L		MIT
3534.174	3535.184	28287.07	W II	0.6	0.7		MIT
3531.440	3532.449	28308.97	W		1.0		MIT
3531.021	3532.030	28312.33	W	0.8	0.8		MIT
3530.756	3531.765	28314.45	W	0.9	0.8		MIT
3529.563	3530.572	28324.02	W II	1.0	1.3		MIT
3528.495	3529.504	28332.60	W	0.7	1.0		MIT
3527.928	3528.937	28337.15	W	0.9	0.8		MIT
3527.599	3528.607	28339.79	W	0.8	0.7		MIT
3527.002	3528.010	28344.59	W II	1.0	1.1L		MIT
3526.854	3527.862	28345.78	W	1.0	1.1		MIT
3525.742	3526.750	28354.72	W	0.0	1.1		MIT
3524.681	3525.689	28363.25	W	0.8	0.7		MIT
3524.246	3525.254	28366.75	W	1.0	0.9		MIT
3523.55	3524.56	28372.4	W		0.7		MIT
3522.11	3523.12	28384.0	W		0.8		MIT
3521.909	3522.916	28385.58	W	1.0	0.9		MIT
3521.712	3522.719	28387.16	W	1.0	0.8		MIT
3518.554	3519.560	28412.64	W II		0.7		MIT
3518.478	3519.484	28413.25	W	1.0	0.8		MIT
3517.505	3518.511	28421.11	W	1.0	0.9		MIT
3517.46	3518.47	28421.5	W		0.6D		MIT
3516.92	3517.93	28425.8	W		0.8		MIT
3516.776	3517.782	28427.01	W	0.8	0.5		MIT
3516.22	3517.23	28431.5	W		0.9		MIT
3515.962	3516.967	28433.59	W	0.8D	0.8D		MIT
3515.07	3516.08	28440.8	W	0.0	1.0		MIT
3514.16	3515.16	28448.2	W		1.0		MIT
3512.93	3513.93	28458.1	W	0.7	0.8		MIT
3512.619	3513.624	28460.65	W	0.8	0.7		MIT
3512.093	3513.097	28464.91	W		1.0		MIT
3511.732	3512.736	28467.83	W	0.8	0.9		MIT
3511.275	3512.279	28471.54	W		1.0		MIT
3510.032	3511.036	28481.62	W	1.0S	1.0		MIT
3509.669	3510.673	28484.57	W	0.9	0.8		MIT
3509.381	3510.385	28486.91	W		1.1		MIT
3509.015	3510.019	28489.88	W	1.0	0.8		MIT
3508.735	3509.739	28492.15	W	1.0	1.0		MIT
3508.115	3509.118	28497.19	W	0.7	0.0		MIT
3507.925	3508.928	28498.73	W	0.8	0.8		MIT
3507.284	3508.287	28503.94	W	1.0	1.0		MIT
3506.641	3507.644	28509.16	W	0.9	0.8		MIT
3504.646	3505.649	28525.39	W	1.0S	0.8		MIT
3504.415	3505.417	28527.27	W	0.5	0.0H		MIT
3503.558	3504.560	28534.25	W	1.0	0.7		MIT
3503.234	3504.236	28536.89	W II		0.7		MIT
3503.038	3504.040	28538.49	W	0.9	0.8		MIT
3502.23	3503.23	28545.1	W		1.0		MIT
3500.280	3501.281	28560.97	W	1.0	1.0		MIT
3499.627	3500.628	28566.30	W	1.0	1.0		MIT
3498.150	3499.151	28578.36	W		1.0		MIT

AIR WAVELENGTH (ANGSTRMS)	VACUUM WAVELENGTH (ANGSTRMS)	WAVENUMBER (KAYSERS)	LMENT	LOG ARC	INTENSITY SPK	DIS	REF
3495.246	3496.246	28602.10	W	1.1	1.3L		MIT
3494.755	3495.755	28606.12	W	0.8	0.5		MIT
3493.195	3494.195	28618.90	W	0.8	0.5		MIT
3493.036	3494.036	28620.20	W	0.7	0.6		MIT
3492.260	3493.259	28626.56	W	0.6			MIT
3492.058	3493.057	28628.22	W II	0.3D	1.1		MIT
3491.834	3492.833	28630.05	W	1.0	0.6		MIT
3491.133	3492.132	28635.80	W	0.8	1.0		MIT
3490.925	3491.924	28637.51	W II	0.8	1.2		MIT
3490.320	3491.319	28642.47	W II	0.6	1.1		MIT
3489.845	3490.844	28646.37	W II		0.8		MIT
3489.287	3490.286	28650.95	W	1.0	1.0		MIT
3486.128	3487.126	28676.91	W II	0.7	1.3		MIT
3485.500	3486.498	28682.08	W	1.0	1.0		MIT
3485.285	3486.283	28683.85	W	1.0	0.9		MIT
3483.86	3484.86	28695.6	W		0.8D		MIT
3482.777	3483.774	28704.50	W	1.0	0.7		MIT
3481.825	3482.822	28712.35	W	1.0	1.1L		MIT
3481.331	3482.328	28716.43	W	0.8	0.7		MIT
3480.466	3481.462	28723.56	W	0.7	0.7		MIT
3478.92	3479.92	28736.3	W	0.8	0.7		MIT
3478.708	3479.704	28738.08	W	0.6			MIT
3477.947	3478.943	28744.37	W	1.0	0.9		MIT
3477.268	3478.264	28749.98	W	0.7	0.5		MIT
3477.13	3478.13	28751.1	W		0.7D		MIT
3476.506	3477.501	28756.28	W	0.8	0.8		MIT
3475.835	3476.830	28761.83	W	1.0	0.8		MIT
3475.29	3476.28	28766.3	W		1.4		MIT
3472.362	3473.356	28790.60	W II		1.0		MIT
3469.25	3470.24	28816.4	W	0.3	1.1		MIT
3468.403	3469.396	28823.46	W	1.0	1.0		MIT
3468.24	3469.23	28824.8	W II	0.0	0.8		M
3467.885	3468.878	28827.76	W	1.0	0.9		MIT
3466.828	3467.821	28836.55	W	0.3A			MIT
3465.408	3466.400	28848.37	W	0.6	0.6		MIT
3465.09	3466.08	28851.0	W		1.0		MIT
3464.449	3465.441	28856.35	W		1.1		MIT
3463.513	3464.505	28864.15	W II	0.9	1.4		MIT
3463.250	3464.242	28866.34	W	1.0	1.2		MIT
3461.814	3462.806	28878.32	W	1.0	0.7		MIT
3461.655	3462.646	28879.65	W	0.5	0.5		MIT
3461.364	3462.355	28882.07	W	0.8	0.8		MIT
3460.784	3461.775	28886.91	W	0.5	0.5		MIT
3460.42	3461.41	28889.9	W		1.2		MIT
3459.522	3460.513	28897.45	W	1.0	0.9		MIT
3459.024	3460.015	28901.61	W	0.8	0.5		MIT
3458.319	3459.310	28907.50	W	1.0	1.0		MIT
3457.720	3458.710	28912.51	W	1.0S	1.0		MIT
3457.366	3458.356	28915.47	W	1.0	0.9		MIT
3455.021	3456.011	28935.09	W		1.3		MIT
3454.539	3455.529	28939.13	W		1.0		MIT
3453.88	3454.87	28944.6	W		1.0		MIT
3452.622	3453.611	28955.20	W	0.8	0.6		MIT
3452.507	3453.496	28956.16	W II	0.6	1.1		MIT
3451.919	3452.908	28961.10	W	0.8	0.8		MIT
3451.747	3452.736	28962.54	W	1.0	0.8		MIT
3451.148	3452.137	28967.57	W	0.6	1.0		MIT
3450.751	3451.740	28970.90	W	0.6	0.8		MIT
3449.869	3450.857	28978.30	W II	0.8	1.4		MIT
3449.450	3450.438	28981.82	W	0.5	0.5		MIT
3449.314	3450.302	28982.97	W	0.8	0.6		MIT
3448.833	3449.821	28987.01	W	1.0	1.0		MIT
3448.213	3449.201	28992.22	W	1.0	1.1		MIT
3447.95	3448.94	28994.4	W	0.9D	0.8		MIT
3447.726	3448.714	28996.32	W	0.8	0.6		MIT
3447.128	3448.116	29001.35	W	0.9	0.5		MIT
3446.900	3447.888	29003.26	W	0.8	0.8		MIT
3446.60	3447.59	29005.8	W		0.7		MIT
3445.718	3446.705	29013.21	W	1.0	0.8		MIT
3445.403	3446.390	29015.87	W	0.8	0.7		MIT

AIR WAVELENGTH (ANGSTRMS)	VACUUM WAVELENGTH (ANGSTRMS)	WAVENUMBER (KAYSERS)	LMENT	LOG ARC	INTENSITY SPK	DIS	REF
3444.608	3445.595	29022.56	W	1.0	0.9		MIT
3443.006	3443.993	29036.07	W	1.0	1.0		MIT
3442.53	3443.52	29040.1	W II		1.1		MIT
3441.970	3442.956	29044.80	W	0.9	0.8		MIT
3440.63	3441.62	29056.1	W II	0.7	1.3		MIT
3439.992	3440.978	29061.51	W	0.9	0.7		MIT
3438.966	3439.952	29070.18	W	0.7	0.6		MIT
3438.815	3439.801	29071.45	W	1.0	0.8		MIT
3438.210	3439.195	29076.57	W	0.8	0.5		MIT
3437.220	3438.205	29084.94	W II	0.5	1.0		MIT
3435.715	3436.700	29097.68	W	1.0	1.1		MIT
3435.243	3436.228	29101.68	W	0.8	0.7		MIT
3433.780	3434.764	29114.08	W	1.0	0.8		MIT
3433.28	3434.26	29118.3	W		1.1L		MIT
3433.06	3434.04	29120.2	W	0.8D	0.6		MIT
3430.885	3431.869	29138.65	W	0.8	0.7		MIT
3430.58	3431.56	29141.2	W II		0.5		MIT
3430.262	3431.245	29143.94	W	0.8	0.6		MIT
3429.600	3430.583	29149.56	W	1.0	1.1		MIT
3427.718	3428.701	29165.57	W	1.0	0.9		MIT
3426.23	3427.21	29178.2	W		0.8		MIT
3425.94	3426.92	29180.7	W		0.8		MIT
3425.548	3426.530	29184.04	W	0.8	0.7		MIT
3425.03	3426.01	29188.4	W	0.7	0.6		MIT
3424.826	3425.808	29190.19	W	0.6			MIT
3424.449	3425.431	29193.41	W	0.3	1.2		MIT
3423.842	3424.824	29198.58	W	0.8	0.5		MIT
3423.303	3424.285	29203.18	W	0.8	0.8		MIT
3422.80	3423.78	29207.5	W		0.8		MIT
3422.426	3423.407	29210.66	W	1.0	1.0		MIT
3421.428	3422.409	29219.18	W		0.8		MIT
3421.134	3422.115	29221.69	W	0.5D	1.3		MIT
3420.358	3421.339	29228.32	W	0.8	0.8		MIT
3420.07	3421.05	29230.8	W	0.6W	1.1		MIT
3419.70	3420.68	29233.9	W		0.9		MIT
3419.288	3420.269	29237.47	W	1.0	0.8		MIT
3416.623	3417.603	29260.27	W	0.8	1.3		MIT
3415.55	3416.53	29269.5	W	0.8	0.8		MIT
3415.419	3416.399	29270.59	W	0.5	0.6		MIT
3413.537	3414.516	29286.73	W	1.0	1.0		MIT
3412.964	3413.943	29291.64	W	1.0	1.0		MIT
3412.743	3413.722	29293.54	W	0.3	1.1		MIT
3412.05	3413.03	29299.5	W	0.7D	0.7D		MIT
3411.018	3411.997	29308.35	W	0.7	0.6		MIT
3410.876	3411.854	29309.57	W	1.0	1.0		MIT
3410.488	3411.466	29312.91	W	0.8	0.8		MIT
3410.169	3411.147	29315.65	W		1.2		MIT
3409.618	3410.596	29320.39	W	0.7	0.6		MIT
3409.438	3410.416	29321.94	W	0.8	0.8		MIT
3409.24	3410.22	29323.6	W		0.7		MIT
3409.011	3409.989	29325.61	W	0.7	0.6		MIT
3408.383	3409.361	29331.01	W	1.0	0.8		MIT
3408.21	3409.19	29332.5	W		0.7		MIT
3407.633	3408.611	29337.47	W	0.8	0.7		MIT
3406.831	3407.808	29344.37	W	1.0	1.1		MIT
3406.582	3407.559	29346.52	W		1.0		MIT
3406.094	3407.071	29350.72	W		0.9		MIT
3405.277	3406.254	29357.76	W	0.8	0.8		MIT
3404.803	3405.780	29361.85	W	0.9	0.8		MIT
3404.224	3405.201	29366.84	W	0.9	0.8		MIT
3404.14	3405.12	29367.6	W		0.6		MIT
3403.790	3404.767	29370.59	W	0.6	0.5		MIT
3402.767	3403.743	29379.42	W	0.7	0.6		MIT
3401.890	3402.866	29386.99	W II	0.8	1.6		MIT
3401.393	3402.369	29391.28	W	0.8	0.8		MIT
3400.550	3401.526	29398.57	W	0.7	0.5		MIT
3398.927	3399.902	29412.61	W	0.6	1.1		MIT
3398.33	3399.31	29417.8	W		0.6J		MIT
3398.097	3399.072	29419.79	W	1.0	1.0		MIT
3397.465	3398.440	29425.26	W	0.6	0.6		MIT
3395.816	3396.791	29439.55	W	0.9	0.8		MIT
3395.478	3396.453	29442.48	W	0.8	0.8		MIT
3394.92	3395.89	29447.3	W		0.7		MIT
3394.61	3395.58	29450.0	W	0.8	0.5		MIT
3394.47	3395.44	29451.2	W II	0.0	1.0		MIT
3393.757	3394.731	29457.41	W	0.7	0.6		MIT
3391.531	3392.505	29476.75	W	1.0	1.0		MIT
3391.100	3392.073	29480.49	W	1.0	1.0		MIT
3389.52	3390.49	29494.2	W		0.6S		MIT
3388.827	3389.800	29500.27	W	1.0	1.1		MIT
3387.632	3388.605	29510.67	W		1.1		MIT
3386.792	3387.764	29517.99	W	0.6	0.3		MIT
3386.505	3387.477	29520.49	W		1.1		MIT
3386.101	3387.073	29524.01	W	1.0	0.8		MIT
3385.822	3386.794	29526.45	W	0.8	1.0		MIT
3384.897	3385.869	29534.52	W	1.0	1.1		MIT
3384.614	3385.586	29536.99	W	0.7	0.5		MIT
3384.339	3385.311	29539.39	W	0.8	0.8		MIT
3383.12	3384.09	29550.0	W		1.3		MIT
3382.606	3383.577	29554.52	W	1.0	1.1		MIT
3382.096	3383.067	29558.98	W	0.9	0.8		MIT
3380.695	3381.666	29571.23	W		0.5		MIT
3379.968	3380.939	29577.58	W	0.6			MIT
3379.62	3380.59	29580.6	W		0.7		MIT
3379.277	3380.247	29583.63	W	0.6	0.0		MIT
3379.024	3379.994	29585.85	W II	0.6	1.2		MIT
3378.730	3379.700	29588.42	W	0.5	0.3		MIT
3378.524	3379.494	29590.23	W	0.9	0.8		MIT
3378.11	3379.08	29593.9	W		0.8		MIT
3376.903	3377.873	29604.43	W	0.8	0.8		MIT
3376.144	3377.114	29611.09	W II	1.0	1.6		MIT
3375.116	3376.085	29620.10	W	1.0	1.0		MIT
3374.870	3375.839	29622.26	W II	0.8	1.1		MIT
3373.754	3374.723	29632.06	W	1.0	1.0		MIT
3373.233	3374.202	29636.64	W	0.8	0.8		MIT
3372.193	3373.162	29645.78	W		1.1		MIT
3371.354	3372.322	29653.16	W	1.0	1.0		MIT
3371.046	3372.014	29655.86	W	0.8	0.8		MIT
3370.696	3371.664	29658.94	W	0.7	0.6		MIT
3370.519	3371.487	29660.50	W	1.0	0.9S		MIT
3370.24	3371.21	29663.0	W II	0.7	1.0S		MIT
3369.821	3370.789	29666.65	W II	0.0	0.6		MIT
3369.690	3370.658	29667.80	W	0.7	0.6		MIT
3367.973	3368.941	29682.92	W		0.9J		MIT
3367.650	3368.617	29685.77	W	0.6	0.6		MIT
3367.550	3368.517	29686.65	W	0.7	0.6		MIT
3367.213	3368.180	29689.62	W	0.8	0.7		MIT
3366.72	3367.69	29694.0	W		1.2		MIT
3365.936	3366.903	29700.89	W	0.9S	0.8S		MIT
3364.561	3365.528	29713.02	W		1.0		MIT
3363.725	3364.691	29720.41	W II	0.6D	1.1		MIT
3363.338	3364.304	29723.83	W	1.0	1.0		MIT
3362.619	3363.585	29730.18	W II		0.3		MIT
3361.971	3362.937	29735.91	W	0.6	0.5		MIT
3361.852	3362.818	29736.97	W	0.8	0.5		MIT
3361.107	3362.073	29743.56	W II	1.0	1.2		MIT
3360.741	3361.707	29746.79	W	0.6	0.3		MIT
3360.325	3361.291	29750.48	W	0.6	1.1L		MIT
3359.236	3360.201	29760.12	W II		0.8		MIT
3358.604	3359.569	29765.72	W II	1.0	1.6		MIT
3358.310	3359.275	29768.33	W		0.8		MIT
3357.094	3358.059	29779.11	W	0.6	0.5		MIT
3356.715	3357.680	29782.47	W	0.8	0.8		MIT
3354.978	3355.942	29797.89	W		0.7		MIT
3354.451	3355.415	29802.57	W	1.0	1.0		MIT
3353.741	3354.705	29808.88	W	1.0	0.9		MIT
3353.555	3354.519	29810.53	W	0.9	0.8		MIT
3352.950	3353.914	29815.91	W	1.0	1.1		MIT
3352.771	3353.735	29817.50	W	0.6	1.1		MIT
3352.392	3353.356	29820.88	W II	0.5	1.0		MIT

AIR WAVELENGTH (ANGSTRMS)	VACUUM WAVELENGTH (ANGSTRMS)	WAVENUMBER (KAYSERS)	LMENT		LOG ARC	INTENSITY SPK	DIS	REF
3351.967	3352.930	29824.66	W		0.8	0.6		MIT
3351.525	3352.488	29828.59	W			0.8		MIT
3351.185	3352.148	29831.62	W		0.7	0.5		MIT
3350.614	3351.577	29836.70	W		0.6	1.0		MIT
3349.538	3350.501	29846.28	W		0.6	0.8		MIT
3349.340	3350.303	29848.05	W		0.3	1.0		MIT
3348.293	3349.256	29857.38	W		0.3	1.2L		MIT
3347.441	3348.403	29864.98	W	II		0.8		MIT
3346.111	3347.073	29876.85	W		0.8	0.3		MIT
3345.860	3346.822	29879.09	W	II	1.0	1.2L		MIT
3345.089	3346.051	29885.98	W		1.0	0.9		MIT
3344.904	3345.866	29887.63	W	II	0.5	1.1		MIT
3344.445	3345.407	29891.73	W		0.9	0.8		MIT
3343.852	3344.813	29897.03	W		0.8	0.6		MIT
3343.417	3344.378	29900.92	W	II	0.8	1.1L		MIT
3343.248	3344.209	29902.44	W		0.9	0.3		MIT
3343.102	3344.063	29903.74	W	II	0.6	1.0		MIT
3342.853	3343.814	29905.97	W			1.0		MIT
3342.462	3343.423	29909.47	W	II	1.0	1.5		MIT
3341.422	3342.383	29918.78	W		0.6	0.5		MIT
3340.096	3341.056	29930.65	W		0.7	0.3		MIT
3339.573	3340.533	29935.34	W		1.0L			MIT
3339.024	3339.984	29940.26	W	II	0.3	1.1		MIT
3338.620	3339.580	29943.89	W	II	0.7	1.2		MIT
3338.243	3339.203	29947.27	W			1.0		MIT
3337.682	3338.642	29952.30	W		1.0	0.9		MIT
3337.505	3338.465	29953.89	W		0.7	0.5		MIT
3336.846	3337.806	29959.80	W		0.8	0.7		MIT
3336.571	3337.531	29962.27	W		0.9	1.0		MIT
3335.933	3336.892	29968.00	W			1.0		MIT
3335.419	3336.378	29972.62	W	II		0.8		MIT
3334.778	3335.737	29978.38	W			1.0		MIT
3333.261	3334.220	29992.02	W			1.1		MIT
3331.672	3332.630	30006.33	W		1.2L	1.1		MIT
3330.532	3331.490	30016.60	W		0.7	0.6		MIT
3328.115	3329.072	30038.40	W		1.0	0.9		MIT
3327.623	3328.580	30042.84	W		0.8	0.8		MIT
3326.417	3327.374	30053.73	W		0.7	0.6		MIT
3326.190	3327.147	30055.78	W		1.2	1.1		MIT
3326.00	3326.96	30057.5	W			1.0		MIT
3325.467	3326.424	30062.32	W			0.8		MIT
3324.689	3325.645	30069.35	W		0.9	0.8		MIT
3324.058	3325.014	30075.06	W		0.9	0.8		MIT
3323.409	3324.365	30080.93	W		0.7	0.9		MIT
3322.675	3323.631	30087.58	W			0.7		MIT
3322.246	3323.202	30091.46	W		1.0	1.1D		MIT
3321.565	3322.521	30097.63	W		0.9	0.8		MIT
3321.133	3322.089	30101.54	W		0.8	0.7		MIT
3320.95	3321.91	30103.2	W	II		1.0		MIT
3320.370	3321.325	30108.46	W		1.0	0.9		MIT
3319.666	3320.621	30114.85	W		0.6	0.3		MIT
3319.531	3320.486	30116.07	W			1.0		MIT
3319.022	3319.977	30120.69	W			0.6		MIT
3317.989	3318.944	30130.07	W		0.9	0.8		MIT
3317.399	3318.354	30135.43	W	II	0.0	1.2		MIT
3317.074	3318.029	30138.38	W		0.8	0.7		MIT
3316.905	3317.860	30139.91	W		0.7	1.0		MIT
3316.086	3317.040	30147.36	W		1.0S	1.0		MIT
3315.091	3316.045	30156.41	W		1.1	1.0		MIT
3314.536	3315.490	30161.46	W		0.7	0.5		MIT
3314.303	3315.257	30163.57	W			0.9		MIT
3314.023	3314.977	30166.12	W		1.0	0.8		MIT
3313.595	3314.549	30170.02	W		0.6	1.0L		MIT
3313.221	3314.175	30173.43	W		0.5	1.1D		MIT
3311.382	3312.335	30190.18	W		1.2L	1.1		MIT
3311.112	3312.065	30192.64	W		0.7	0.6		MIT
3310.20	3311.15	30201.0	W			1.1		MIT
3309.473	3310.426	30207.60	W		0.9	0.8		MIT
3308.342	3309.294	30217.92	W		0.5D	1.4		MIT
3306.157	3307.109	30237.89	W			0.8		MIT

AIR WAVELENGTH (ANGSTRMS)	VACUUM WAVELENGTH (ANGSTRMS)	WAVENUMBER (KAYSERS)	LMENT		LOG ARC	INTENSITY SPK	DIS	REF
3305.564	3306.516	30243.32	W		1.0	0.9		MIT
3304.47	3305.42	30253.3	W		1.0W	1.3		MIT
3303.660	3304.611	30260.75	W		0.8			MIT
3303.335	3304.286	30263.72	W		0.8	0.5		MIT
3301.85	3302.80	30277.3	W	II	0.0	1.0		MIT
3301.28	3302.23	30282.6	W		0.3	0.9		MIT
3300.820	3301.770	30286.78	W		1.2L	1.1		MIT
3300.337	3301.287	30291.21	W		0.8	1.0		MIT
3299.738	3300.688	30296.71	W		0.5	1.1		MIT
3298.129	3299.079	30311.49	W		1.1S	1.0		MIT
3293.707	3294.656	30352.19	W		1.0	1.0		MIT
3292.802	3293.750	30360.53	W		0.8	0.7		MIT
3290.508	3291.456	30381.69	W		0.8	0.7		MIT
3286.561	3287.508	30418.18	W	II	0.7	1.1		MIT
3283.556	3284.502	30446.02	W		1.0	0.9		MIT
3281.939	3282.885	30461.02	W		1.1	1.0		MIT
3279.582	3280.527	30482.91	W		0.7	0.8		MIT
3279.027	3279.972	30488.07	W		0.9	0.7		MIT
3270.263	3271.206	30569.77	W		1.0	0.6		MIT
3269.628	3270.570	30575.71	W		1.0L	1.1		MIT
3268.923	3269.865	30582.30	W		1.0	1.0S		MIT
3268.580	3269.522	30585.51	W		0.9	1.0		MIT
3268.123	3269.065	30589.78	W		0.7	0.5		MIT
3267.45	3268.39	30596.1	W			1.1		MIT
3266.763	3267.705	30602.52	W		0.8	0.8		MIT
3266.622	3267.564	30603.84	W		0.8	0.7		MIT
3265.150	3266.091	30617.64	W		1.0	0.9		MIT
3264.341	3265.282	30625.23	W		0.8	0.8		MIT
3263.104	3264.045	30636.83	W		0.9	0.8		MIT
3262.28	3263.22	30644.6	W			1.3		MIT
3261.165	3262.105	30655.05	W		1.0	0.9		MIT
3259.662	3260.602	30669.18	W		1.0	1.0		MIT
3259.438	3260.378	30671.29	W		1.0	1.0		MIT
3258.143	3259.083	30683.48	W		0.7	0.5		MIT
3257.80	3258.74	30686.7	W	II		1.0		MIT
3256.230	3257.169	30701.51	W		0.9	0.8		MIT
3255.962	3256.901	30704.03	W		1.0	0.9		MIT
3255.068	3256.007	30712.47	W		1.0	1.1		MIT
3254.84	3255.78	30714.6	W	II		0.8		MIT
3254.359	3255.298	30719.16	W		0.8	0.6		MIT
3252.294	3253.232	30738.66	W		1.0	1.0S		MIT
3251.225	3252.163	30748.77	W	II	1.0	1.3		MIT
3243.336	3244.272	30823.56	W	II	0.7	1.3		MIT
3242.025	3242.960	30836.02	W		1.0	1.0		MIT
3241.411	3242.346	30841.86	W		0.8	0.7		MIT
3238.691	3239.626	30867.76	W		0.7	0.6		MIT
3238.490	3239.425	30869.68	W		0.8	0.8		MIT
3237.68	3238.61	30877.4	W			0.9		MIT
3237.089	3238.023	30883.04	W		1.0S	0.7		MIT
3236.62	3237.55	30887.5	W			0.9		MIT
3235.921	3236.855	30894.19	W		0.9	0.8		MIT
3234.985	3235.919	30903.12	W		1.0	0.8		MIT
3233.24	3234.17	30919.8	W		0.3	1.0		MIT
3233.142	3234.075	30920.74	W	II	0.3	0.8		MIT
3232.652	3233.585	30925.43	W		1.0	0.9		MIT
3232.486	3233.419	30927.01	W		1.0	0.9		MIT
3232.231	3233.164	30929.46	W		0.5	0.3		MIT
3232.134	3233.067	30930.38	W		0.8	1.0		MIT
3230.836	3231.769	30942.81	W	II	0.0	0.8		MIT
3230.59	3231.52	30945.2	W	II	0.0	1.0S		MIT
3229.660	3230.592	30954.08	W		1.0	0.9		MIT
3228.96	3229.89	30960.8	W	II	0.0	1.0		MIT
3227.491	3228.423	30974.88	W		1.0	0.9		MIT
3226.58	3227.51	30983.6	W			1.0		MIT
3225.631	3226.562	30992.74	W		1.0	0.9		MIT
3225.16	3226.09	30997.3	W	II	0.0	0.6		MIT
3223.116	3224.047	31016.92	W		0.9	0.6		MIT
3221.911	3222.841	31028.52	W		1.0	1.0		MIT
3221.617	3222.547	31031.35	W		0.8	0.8		MIT
3221.212	3222.142	31035.25	W		1.1D	1.1D		MIT

AIR WAVELENGTH (ANGSTRMS)	VACUUM WAVELENGTH (ANGSTRMS)	WAVENUMBER (KAYSERS)	LMENT		LOG ARC	INTENSITY SPK	DIS	REF
3220.062	3220.992	31046.34	W		0.9	0.8		MIT
3218.606	3219.536	31060.38	W		1.0	1.0		MIT
3216.22	3217.15	31083.4	W	II	0.8	1.0S		MIT
3215.65	3216.58	31088.9	W	II	0.8	1.0		MIT
3215.560	3216.489	31089.80	W		1.0	1.0		MIT
3215.26	3216.19	31092.7	W			1.2		MIT
3213.146	3214.074	31113.16	W		0.8	0.6		MIT
3209.96	3210.89	31144.0	W		0.3	1.4		MIT
3208.570	3209.497	31157.53	W		1.0	1.0		MIT
3208.280	3209.207	31160.35	W		1.1	1.0		MIT
3208.100	3209.027	31162.09	W		1.0	0.9		MIT
3207.799	3208.726	31165.02	W		0.9	0.8		MIT
3207.252	3208.179	31170.33	W		1.1	1.0		MIT
3206.406	3207.332	31178.56	W	II	0.8D	1.2		MIT
3205.503	3206.429	31187.34	W		1.0	0.8		MIT
3203.42	3204.35	31207.6	W	II	0.7	1.1		MIT
3203.337	3204.263	31208.43	W	II	0.6	1.3		MIT
3203.054	3203.980	31211.19	W		1.0	0.9		MIT
3201.581	3202.506	31225.54	W		0.8	1.3		MIT
3200.731	3201.656	31233.84	W		0.8	0.8		MIT
3199.965	3200.890	31241.31	W		0.9	0.8		MIT
3199.305	3200.230	31247.76	W		1.0	0.9		MIT
3198.841	3199.766	31252.29	W		1.1	1.0		MIT
3198.33	3199.25	31257.3	W		0.0	1.2		MIT
3197.56	3198.48	31264.8	W			1.3		MIT
3195.076	3196.000	31289.11	W		1.0	0.7		MIT
3192.392	3193.315	31315.42	W		0.8	0.7		MIT
3192.13	3193.05	31318.0	W	II		1.0		MIT
3191.572	3192.495	31323.47	W		1.1	1.0		MIT
3189.236	3190.158	31346.41	W	II	1.0	1.3		MIT
3187.764	3188.686	31360.88	W		0.9	0.8		MIT
3187.13	3188.05	31367.1	W			1.3		MIT
3186.741	3187.662	31370.95	W		0.8	0.7		MIT
3185.202	3186.123	31386.11	W		0.8	0.7		MIT
3185.05	3185.97	31387.6	W	II	0.0	1.0		MIT
3184.415	3185.336	31393.86	W		1.0	1.0		MIT
3184.043	3184.964	31397.53	W		1.0	0.8		MIT
3183.98	3184.90	31398.1	W	II	0.0	0.8		MIT
3183.513	3184.434	31402.76	W		1.0	0.8		MIT
3182.855	3183.776	31409.25	W		1.0	0.9		MIT
3181.812	3182.732	31419.54	W		1.1L	1.1		MIT
3180.739	3181.659	31430.14	W		1.0	1.0		MIT
3180.299	3181.219	31434.49	W		0.8	0.7		MIT
3179.975	3180.895	31437.70	W		0.8D	1.3		MIT
3179.425	3180.345	31443.13	W	II	0.8	1.1		MIT
3179.056	3179.976	31446.78	W		1.0	0.9		MIT
3178.243	3179.162	31454.83	W		0.8	0.5		MIT
3178.02	3178.94	31457.0	W	II	0.9D	1.3		MIT
3177.210	3178.129	31465.05	W	II	0.9D	1.4		MIT
3176.590	3177.509	31471.19	W		1.1D	1.0D		MIT
3175.945	3176.864	31477.58	W	II	0.9	1.3L		MIT
3170.201	3171.118	31534.62	W		1.2	1.0		MIT
3169.928	3170.845	31537.33	W		1.0	1.2		MIT
3168.967	3169.884	31546.90	W		0.9	0.8		MIT
3167.581	3168.498	31560.70	W		1.0	1.0		MIT
3166.743	3167.659	31569.05	W		0.8	0.3		MIT
3165.376	3166.292	31582.68	W		1.0	1.0		MIT
3164.442	3165.358	31592.00	W		1.0	0.9		MIT
3163.418	3164.334	31602.23	W		1.1	1.2L		MIT
3160.020	3160.935	31636.21	W	II	1.0	1.5		MIT
3159.419	3160.334	31642.23	W		0.6	0.5		MIT
3159.184	3160.099	31644.58	W		1.0	1.0		MIT
3158.806	3159.720	31648.37	W		0.8	0.8		MIT
3155.518	3156.432	31681.34	W		1.0	0.8		MIT
3155.095	3156.009	31685.59	W		1.0	1.3		MIT
3154.175	3155.088	31694.83	W	II	0.7	0.3H		MIT
3152.955	3153.868	31707.10	W		1.0	0.9		MIT
3152.745	3153.658	31709.21	W	II	0.5	1.0		MIT
3152.473	3153.386	31711.95	W		1.0	1.2		MIT
3151.998	3152.911	31716.72	W		0.6	0.5		MIT

AIR WAVELENGTH (ANGSTRMS)	VACUUM WAVELENGTH (ANGSTRMS)	WAVENUMBER (KAYSERS)	LMENT		LOG ARC	INTENSITY SPK	DIS	REF
3151.565	3152.478	31721.08	W	II	0.6	1.0L		MIT
3151.29	3152.20	31723.9	W		0.7	1.3		MIT
3149.849	3150.761	31738.36	W	II	1.0	1.2		MIT
3146.351	3147.262	31773.65	W		0.8			MIT
3145.755	3146.666	31779.67	W		0.8D	1.1		MIT
3145.540	3146.451	31781.84	W		0.8	0.7		MIT
3144.50	3145.41	31792.4	W	II		1.2		MIT
3143.210	3144.121	31805.40	W		0.9	0.5		MIT
3142.145	3143.055	31816.18	W		1.0	0.9		MIT
3141.424	3142.334	31823.48	W		1.0	1.0		MIT
3141.177	3142.087	31825.98	W		0.8	0.7		MIT
3140.750	3141.660	31830.31	W		1.0S	1.0		MIT
3140.408	3141.318	31833.77	W		1.0	0.9S		MIT
3138.882	3139.791	31849.25	W		0.8	0.7		MIT
3137.912	3138.821	31859.09	W		1.0	0.5		MIT
3137.631	3138.540	31861.95	W		0.8	0.9D		MIT
3136.070	3136.979	31877.81	W		1.0	0.8		MIT
3133.889	3134.797	31899.99	W		1.0	1.0		MIT
3133.716	3134.624	31901.75	W		0.8	0.3H		MIT
3130.456	3131.363	31934.97	W		1.0	0.9		MIT
3130.155	3131.062	31938.04	W		0.7	0.6		MIT
3128.975	3129.882	31950.09	W		0.3	1.1		MIT
3127.73	3128.64	31962.8	W			1.0		MIT
3127.339	3128.246	31966.80	W		1.0	1.0S		MIT
3126.42	3127.33	31976.2	W		0.3	1.1		MIT
3125.358	3126.264	31987.06	W		1.0	1.0		MIT
3124.734	3125.640	31993.45	W		0.9	0.5		MIT
3124.504	3125.410	31995.80	W		0.8	0.7		MIT
3121.157	3122.062	32030.11	W		1.0	0.8		MIT
3120.729	3121.634	32034.51	W		1.0	0.9		MIT
3120.185	3121.090	32040.09	W		1.0	1.0		MIT
3119.766	3120.671	32044.39	W		1.0	1.0		MIT
3118.786	3119.690	32054.46	W		0.9	0.8		MIT
3118.354	3119.258	32058.90	W		1.0	0.8		MIT
3117.580	3118.484	32066.86	W		1.0	1.1		MIT
3117.385	3118.289	32068.87	W		0.8	0.7		MIT
3116.869	3117.773	32074.18	W		1.0	0.5		MIT
3116.214	3117.118	32080.92	W		0.8	0.5		MIT
3114.583	3115.486	32097.72	W		0.8	0.5		MIT
3111.959	3112.862	32124.78	W		0.8	1.0		MIT
3111.123	3112.025	32133.41	W		1.0	1.2		MIT
3108.781	3109.683	32157.62	W	II	0.5	1.1		MIT
3108.022	3108.924	32165.47	W		1.1	1.0		MIT
3107.234	3108.135	32173.63	W		1.1	1.0		MIT
3105.878	3106.779	32187.68	W		1.0L	0.8		MIT
3104.426	3105.327	32202.73	W		0.7	0.3		MIT
3103.516	3104.417	32212.17	W	II	0.3	1.2		MIT
3102.221	3103.121	32225.62	W		0.3	1.1		MIT
3101.23	3102.13	32235.9	W	II		0.7		MIT
3100.74	3101.64	32241.0	W	II	0.3	1.3		MIT
3098.448	3099.347	32264.86	W		0.9	0.6		MIT
3098.310	3099.209	32266.29	W	II	0.3	1.0		MIT
3096.012	3096.911	32290.24	W		0.8	0.5		MIT
3095.873	3096.772	32291.69	W		0.3	1.3		MIT
3094.032	3094.930	32310.91	W		0.9	1.0		MIT
3093.512	3094.410	32316.34	W		1.1	1.0		MIT
3092.285	3093.183	32329.16	W		0.8	0.5		MIT
3090.585	3091.482	32346.94	W		1.0	0.9		MIT
3090.416	3091.313	32348.71	W		0.9	0.8D		MIT
3089.314	3090.211	32360.25	W		0.9	0.8		MIT
3089.184	3090.081	32361.61	W		0.9	0.9		MIT
3089.060	3089.957	32362.91	W		0.9	0.9		MIT
3087.642	3088.539	32377.77	W		0.8	0.3		MIT
3086.972	3087.868	32384.80	W		1.0	0.7		MIT
3084.911	3085.807	32406.44	W		1.0	0.8		MIT
3084.823	3085.719	32407.36	W		1.0	1.0		MIT
3081.864	3082.759	32438.47	W		1.0	0.8		MIT
3081.375	3082.270	32443.62	W		0.8	0.5		MIT
3081.21	3082.10	32445.4	W	II		0.5		MIT
3080.701	3081.596	32450.72	W		1.0	0.9S		MIT

AIR WAVELENGTH (ANGSTRMS)	VACUUM WAVELENGTH (ANGSTRMS)	WAVENUMBER (KAYSERS)	LMENT	LOG ARC	INTENSITY SPK	DIS	REF
3080.421	3081.316	32453.67	W	0.9	0.7		MIT
3079.222	3080.116	32466.30	W	0.9	0.7		MIT
3077.519	3078.413	32484.27	W	0.3	1.6J		MIT
3076.173	3077.067	32498.48	W	0.8	0.5		MIT
3075.219	3076.112	32508.56	W	0.6	0.3		MIT
3074.089	3074.982	32520.51	W	1.0	1.1		MIT
3073.687	3074.580	32524.77	W	1.0	0.8		MIT
3073.280	3074.173	32529.07	W	1.1	1.0		MIT
3072.73	3073.62	32534.9	W II	0.5	1.1		MIT
3071.72	3072.61	32545.6	W II	0.9	1.2		MIT
3069.465	3070.357	32569.50	W	0.7	0.3		MIT
3069.284	3070.176	32571.42	W II	0.8	1.2		MIT
3067.866	3068.758	32586.48	W	0.5	1.1		MIT
3067.568	3068.460	32589.64	W II	0.5	1.1		MIT
3067.41	3068.30	32591.3	W II	0.3	0.7		MIT
3066.98	3067.87	32595.9	W II	0.6	1.1		MIT
3064.937	3065.828	32617.62	W	1.0S	0.8		MIT
3063.97	3064.86	32627.9	W II	0.3	1.3		MIT
3063.41	3064.30	32633.9	W II		1.3		MIT
3063.181	3064.071	32636.31	W	1.0	1.0		MIT
3062.892	3063.782	32639.39	W	0.8	0.7		MIT
3062.604	3063.494	32642.46	W	1.0	1.0		MIT
3061.68	3062.57	32652.3	W II	0.3	1.1		MIT
3059.820	3060.710	32672.16	W	0.8	0.5		MIT
3056.229	3057.118	32710.55	W	0.8	0.9		MIT
3055.397	3056.285	32719.46	W	1.0	1.0		MIT
3054.014	3054.902	32734.27	W	1.0	0.9		MIT
3053.36	3054.25	32741.3	W II	0.7			MIT
3052.822	3053.710	32747.05	W	0.8	0.5		MIT
3051.931	3052.819	32756.61	W	1.0D	0.8		MIT
3051.291	3052.178	32763.48	W II	1.0	1.5L		MIT
3050.004	3050.891	32777.31	W	0.8	1.0		MIT
3049.86	3050.75	32778.9	W II	0.3	1.0		MIT
3049.691	3050.578	32780.67	W II	1.1	0.9		MIT
3049.354	3050.241	32784.29	W	0.8	0.7		MIT
3049.003	3049.890	32788.07	W	0.9	1.0		MIT
3048.663	3049.550	32791.73	W II	1.0	1.1		MIT
3048.123	3049.010	32797.54	W	0.8	0.7		MIT
3047.572	3048.459	32803.46	W	1.0	1.0		BK
3046.447	3047.333	32815.58	W	1.1	1.0		MIT
3045.580	3046.466	32824.92	W	0.8	0.8		MIT
3043.808	3044.694	32844.03	W	1.1	0.5		MIT
3043.015	3043.900	32852.59	W	0.8	0.6		MIT
3042.279	3043.164	32860.53	W	0.8	0.7		MIT
3041.866	3042.751	32865.00	W	1.0	1.0		MIT
3041.737	3042.622	32866.39	W	0.9	0.8		MIT
3039.575	3040.460	32889.77	W II	0.3	1.3		MIT
3039.311	3040.195	32892.62	W	1.0	1.1		MIT
3036.660	3037.544	32921.34	W II	0.9	1.3		MIT
3034.195	3035.078	32948.08	W	1.0	1.0		MIT
3033.575	3034.458	32954.81	W	1.1	1.2		MIT
3029.230	3030.112	33002.08	W	0.9	0.0		MIT
3028.74	3029.62	33007.4	W II		1.0		MIT
3027.789	3028.671	33017.79	W	0.9	0.8		MIT
3026.782	3027.663	33028.77	W	1.0	0.8		MIT
3026.675	3027.556	33029.94	W	1.0S	0.9		MIT
3025.259	3026.140	33045.40	W	1.0	0.9		MIT
3024.920	3025.801	33049.10	W	1.2	1.1		MIT
3024.496	3025.377	33053.73	W II	1.0	1.3L		MIT
3022.66	3023.54	33073.8	W II	0.3	1.1		MIT
3021.998	3022.878	33081.06	W II	1.2	1.3		MIT
3021.608	3022.488	33085.33	W	1.0	0.8		MIT
3020.214	3021.094	33100.60	W	0.8	0.8		MIT
3017.443	3018.322	33130.99	W	1.1	1.0		MIT
3017.15	3018.03	33134.2	W	0.5D	1.0		MIT
3016.470	3017.349	33141.68	W	1.1	1.0		MIT
3013.787	3014.665	33171.18	W	1.1	1.1		MIT
3013.199	3014.077	33177.65	W	1.0	1.0S		MIT
3011.678	3012.556	33194.41	W	0.9	0.8		MIT
3010.76	3011.64	33204.5	W II	0.9D	1.3		MIT

AIR WAVELENGTH (ANGSTRMS)	VACUUM WAVELENGTH (ANGSTRMS)	WAVENUMBER (KAYSERS)	LMENT	LOG ARC	INTENSITY SPK	DIS	REF
3010.421	3011.298	33208.27	W	1.0	0.8		MIT
3009.076	3009.953	33223.11	W	1.0	0.7		MIT
3008.98	3009.86	33224.2	W		1.1		EX
3008.757	3009.634	33226.63	W	0.9	0.8		MIT
3007.405	3008.281	33241.57	W	0.8	0.5		MIT
3006.657	3007.533	33249.84	W	0.8	0.7		MIT
3006.314	3007.190	33253.63	W	1.0	1.1		MIT
3005.006	3005.882	33268.11	W	0.8	0.8		MIT
3004.25	3005.13	33276.5	W II	0.7	1.0		MIT
3002.821	3003.696	33292.31	W	0.8	0.8		MIT
3002.28	3003.16	33298.3	W II	0.0	1.2		MIT
3001.977	3002.852	33301.67	W	1.0	1.0		MIT
3000.619	3001.494	33316.74	W II	0.6	1.3		MIT
3000.240	3001.115	33320.95	W	0.8S	0.9		MIT
2998.687	2999.561	33338.21	W II	0.8	1.1		MIT
2998.287	2999.161	33342.66	W	0.8	0.3		MIT
2997.789	2998.663	33348.19	W	1.1	1.0		MIT
2997.603	2998.477	33350.26	W	0.9	1.1		MIT
2996.970	2997.844	33357.31	W	0.9	1.0		MIT
2995.987	2996.861	33368.25	W	0.9	1.0		MIT
2995.739	2996.613	33371.01	W	0.8	0.7		MIT
2995.258	2996.131	33376.37	W	1.0	0.9		MIT
2993.611	2994.484	33394.73	W	1.1	1.0		MIT
2992.918	2993.791	33402.47	W	0.9	0.6		MIT
2992.169	2993.042	33410.83	W	0.8	0.6		MIT
2991.94	2992.81	33413.4	W	0.7			MIT
2990.850	2991.722	33425.56	W II		1.0		MIT
2990.710	2991.582	33427.13	W	1.0S	0.8		MIT
2989.502	2990.374	33440.63	W		0.6		MIT
2988.885	2989.757	33447.54	W	0.8	1.1		MIT
2987.958	2988.830	33457.91	W	0.9	0.6		MIT
2987.287	2988.158	33465.43	W II	0.9	1.2		MIT
2986.42	2987.29	33475.1	W		1.0		MIT
2986.182	2987.053	33477.81	W	0.8	0.3		MIT
2985.87	2986.74	33481.3	W	0.8	1.1		MIT
2985.659	2986.530	33483.67	W	0.8	0.5		MIT
2984.143	2985.014	33500.68	W	1.0	0.9		MIT
2982.83	2983.70	33515.4	W	0.8L	0.6		MIT
2982.609	2983.479	33517.91	W	1.0	1.0		MIT
2982.215	2983.085	33522.34	W II	0.5	1.1		MIT
2981.629	2982.499	33528.93	W	0.8	0.8		MIT
2981.110	2981.980	33534.77	W	0.8	0.5		MIT
2979.858	2980.728	33548.86	W	1.0	1.0		MIT
2979.721	2980.591	33550.40	W	0.8	0.8		MIT
2979.17	2980.04	33556.6	W		0.7		MIT
2978.640	2979.509	33562.57	W	0.6	0.0		MIT
2978.380	2979.249	33565.50	W		1.0		MIT
2977.946	2978.815	33570.40	W		1.1		MIT
2977.678	2978.547	33573.42	W	0.8			MIT
2977.435	2978.304	33576.16	W II		0.5		MIT
2977.213	2978.082	33578.66	W	1.0	0.8		MIT
2977.099	2977.968	33579.94	W	1.0	0.8		MIT
2976.794	2977.663	33583.39	W	1.0S	0.8		MIT
2976.474	2977.343	33587.00	W II	1.0	1.1		MIT
2975.08	2975.95	33602.7	W		1.0		MIT
2974.394	2975.262	33610.48	W II	0.9	1.1		MIT
2974.025	2974.893	33614.65	W	0.8	0.6		MIT
2972.915	2973.783	33627.20	W	0.6	0.3		MIT
2971.673	2972.541	33641.26	W	1.0	0.9		MIT
2971.202	2972.069	33646.59	W	0.5			MIT
2970.91	2971.78	33649.9	W II	0.3	1.0L		LN
2970.35	2971.22	33656.2	W		0.9D		MIT
2969.624	2970.491	33664.47	W	1.0	1.0		MIT
2968.780	2969.647	33674.04	W	0.8			MIT
2968.36	2969.23	33678.8	W	0.6D			MIT
2968.20	2969.07	33680.6	W	0.8	0.6D		MIT
2967.927	2968.794	33683.71	W II	0.3	1.0		MIT
2967.552	2968.419	33687.97	W	0.9	0.8		MIT
2967.069	2967.935	33693.46	W	0.9	0.9		MIT
2966.573	2967.439	33699.09	W	1.0	0.9		MIT

AIR WAVELENGTH (ANGSTRMS)	VACUUM WAVELENGTH (ANGSTRMS)	WAVENUMBER (KAYSERS)	LMENT		LOG ARC	INTENSITY SPK	DIS	REF
2966.20	2967.07	33703.3	W		0.3	0.8		MIT
2965.860	2966.726	33707.19	W		0.8	0.6		MIT
2964.515	2965.381	33722.48	W		1.1	1.0		MIT
2963.86	2964.73	33729.9	W			1.0		MIT
2963.779	2964.645	33730.86	W		0.6			MIT
2963.427	2964.293	33734.86	W		0.7	0.0		MIT
2962.51	2963.38	33745.3	W	II		1.0		LN
2961.708	2962.573	33754.44	W		0.8	1.0		MIT
2961.019	2961.884	33762.29	W		0.8	1.1		MIT
2960.140	2961.005	33772.32	W		1.0	0.5		MIT
2959.92	2960.78	33774.8	W			1.0		MIT
2958.838	2959.702	33787.18	W		0.8	0.3		MIT
2958.723	2959.587	33788.49	W		0.8	0.3		MIT
2957.922	2958.786	33797.64	W		0.8	0.8		MIT
2957.26	2958.12	33805.2	W			1.0		MIT
2956.824	2957.688	33810.19	W		0.6	0.3		MIT
2956.667	2957.531	33811.99	W		0.9	1.0		MIT
2954.98	2955.84	33831.3	W	II	0.3	0.9		MIT
2954.904	2955.767	33832.16	W		1.0	0.9		MIT
2952.288	2953.151	33862.14	W	II	1.1S	1.5		MIT
2951.793	2952.656	33867.82	W		0.8	0.5		MIT
2951.407	2952.270	33872.25	W		0.7	0.5		MIT
2950.69	2951.55	33880.5	W		0.6			MIT
2950.44	2951.30	33883.4	W			1.3		MIT
2949.119	2949.981	33898.52	W		0.7	0.6		MIT
2947.72	2948.58	33914.6	W			0.9		MIT
2947.384	2948.246	33918.48	W		1.1	1.0		MIT
2946.981	2947.842	33923.12	W		1.3	1.3		MIT
2946.514	2947.375	33928.49	W		0.7			MIT
2946.429	2947.290	33929.47	W			1.1		MIT
2945.67	2946.53	33938.2	W		0.6	0.0		MIT
2944.395	2945.256	33952.91	W		1.5	1.3		MIT
2943.959	2944.820	33957.94	W		0.7	0.6		MIT
2943.326	2944.187	33965.24	W		0.8	0.8		MIT
2942.610	2943.470	33973.50	W	II	0.3	1.0		MIT
2942.445	2943.305	33975.41	W		0.9	0.5		MIT
2942.256	2943.116	33977.59	W	II	0.3	1.0		MIT
2942.139	2942.999	33978.94	W	II	0.8	1.0		MIT
2941.245	2942.105	33989.27	W		0.9	0.6		MIT
2940.949	2941.809	33992.69	W		0.7	0.5		MIT
2940.199	2941.059	34001.36	W	II	0.6	1.3		MIT
2939.743	2940.603	34006.63	W	II	0.3	1.1		MIT
2939.177	2940.037	34013.18	W		0.9	0.3		MIT
2939.043	2939.902	34014.73	W		1.0	0.6		MIT
2938.852	2939.711	34016.94	W	II	0.0	1.0		MIT
2938.499	2939.358	34021.03	W		0.9	0.8		MIT
2937.661	2938.520	34030.73	W		0.8D	1.0		MIT
2937.141	2938.000	34036.76	W		0.9	1.1		MIT
2936.672	2937.531	34042.19	W		1.0	1.3		MIT
2936.002	2936.861	34049.96	W		1.0	0.5		MIT
2935.75	2936.61	34052.9	W			1.0D		MIT
2935.621	2936.480	34054.38	W		0.9	0.3		MIT
2935.347	2936.206	34057.56	W	II	0.7	1.0		MIT
2934.991	2935.849	34061.69	W		1.2	1.1		MIT
2932.982	2933.840	34085.02	W		0.3	0.8		MIT
2932.86	2933.72	34086.4	W		0.0	1.0		MIT
2932.213	2933.071	34093.96	W		0.6	0.3		MIT
2932.037	2932.895	34096.01	W		0.6	0.0		MIT
2931.869	2932.727	34097.96	W		0.3	1.0		MIT
2931.53	2932.39	34101.9	W			1.1L		MIT
2930.51	2931.37	34113.8	W		0.3	0.7		MIT
2930.146	2931.003	34118.01	W		1.0	0.8		MIT
2929.99	2930.85	34119.8	W		0.7	0.5		MIT
2928.662	2929.519	34135.30	W		0.9	0.8		MIT
2928.190	2929.047	34140.80	W		1.0	0.8		MIT
2927.933	2928.790	34143.80	W		0.8	0.8		MIT
2927.70	2928.56	34146.5	W		0.3	1.0		MIT
2927.529	2928.386	34148.51	W		0.7	0.5		MIT
2926.982	2927.838	34154.89	W		1.0	1.1		MIT
2926.826	2927.682	34156.71	W	II	0.0	1.0		MIT

AIR WAVELENGTH (ANGSTRMS)	VACUUM WAVELENGTH (ANGSTRMS)	WAVENUMBER (KAYSERS)	LMENT		LOG ARC	INTENSITY SPK	DIS	REF
2926.42	2927.28	34161.5	W		0.7			MIT
2925.805	2926.661	34168.63	W			1.2L		MIT
2925.124	2925.980	34176.58	W		1.1	1.0		MIT
2924.984	2925.840	34178.22	W	II	0.3	1.2		MIT
2923.542	2924.398	34195.08	W		1.0	0.7		MIT
2923.098	2923.954	34200.27	W		1.1	1.0		MIT
2921.903	2922.758	34214.26	W		0.6	0.9		MIT
2921.115	2921.970	34223.49	W		1.0	1.0		MIT
2919.695	2920.550	34240.13	W		0.7D	0.3		MIT
2919.49	2920.34	34242.5	W			1.0		MIT
2919.291	2920.146	34244.87	W		0.9	0.5		MIT
2918.631	2919.485	34252.61	W		0.9	1.3		MIT
2918.250	2919.104	34257.08	W		1.1	0.9		MIT
2917.666	2918.520	34263.94	W		0.9	0.0		MIT
2917.385	2918.239	34267.24	W	II	0.0	0.5		MIT
2917.298	2918.152	34268.26	W		0.8	0.5		MIT
2916.86	2917.71	34273.4	W	II		0.8		MIT
2916.787	2917.641	34274.27	W	II	0.0	0.8		MIT
2916.384	2917.238	34279.00	W	II	0.6	0.6		MIT
2916.108	2916.962	34282.25	W		0.7	0.6		MIT
2915.586	2916.440	34288.38	W		1.0	0.9		MIT
2915.110	2915.964	34293.98	W		0.7	0.6		MIT
2914.645	2915.498	34299.45	W	II		1.0		MIT
2914.25	2915.10	34304.1	W		0.0	1.1		MIT
2913.746	2914.599	34310.04	W	II	0.6	1.0		MIT
2912.75	2913.60	34321.8	W		0.5			MIT
2912.570	2913.423	34323.89	W	II	0.0	1.0		MIT
2912.243	2913.096	34327.74	W		0.8	0.5		MIT
2911.001	2911.854	34342.39	W		1.0	0.9		MIT
2910.477	2911.329	34348.57	W		1.1	1.0		MIT
2909.628	2910.480	34358.59	W		0.6	0.5		MIT
2909.123	2909.975	34364.56	W		0.9	0.9		MIT
2908.493	2909.345	34372.00	W			1.0		MIT
2908.395	2909.247	34373.16	W		0.8	0.3		MIT
2908.261	2909.113	34374.74	W		0.8	0.5		MIT
2907.260	2908.112	34386.58	W		0.9	0.5		MIT
2906.731	2907.582	34392.83	W		0.8	0.3		MIT
2906.576	2907.427	34394.67	W		0.8	0.3		MIT
2906.410	2907.261	34396.63	W			1.0		MIT
2905.596	2906.447	34406.27	W		0.5	1.0		MIT
2905.178	2906.029	34411.22	W		1.0	0.5		MIT
2904.778	2905.629	34415.96	W		1.0	1.0		MIT
2904.54	2905.39	34418.8	W			0.8		MIT
2904.072	2904.923	34424.32	W	II	0.5	1.3		MIT
2903.496	2904.347	34431.15	W			1.2		MIT
2902.606	2903.456	34441.71	W		0.8	1.0		MIT
2902.198	2903.048	34446.55	W		1.0	1.0		MIT
2902.039	2902.889	34448.44	W		0.9	0.3		MIT
2901.784	2902.634	34451.46	W		1.0	0.8		MIT
2901.17	2902.02	34458.7	W			1.0		MIT
2900.807	2901.657	34463.07	W		0.9	0.5		MIT
2900.516	2901.366	34466.52	W		0.9	0.8		MIT
2899.96	2900.81	34473.1	W			0.7H		MIT
2899.56	2900.41	34477.9	W			0.6H		MIT
2898.253	2899.102	34493.44	W		0.9	0.5		MIT
2898.09	2898.94	34495.4	W	II		0.5		MIT
2897.200	2898.049	34505.97	W		0.6	0.5		MIT
2896.915	2897.764	34509.37	W		0.6	0.3		MIT
2896.446	2897.295	34514.95	W		1.2	1.4		MIT
2896.008	2896.857	34520.17	W		1.0	1.1		MIT
2895.45	2896.30	34526.8	W			1.0		MIT
2894.616	2895.464	34536.77	W		0.7	0.3		MIT
2894.253	2895.101	34541.10	W		0.9	0.5		MIT
2893.617	2894.465	34548.70	W		0.8	0.5		MIT
2893.123	2893.971	34554.60	W		0.9	0.6		MIT
2892.605	2893.453	34560.78	W		0.6	0.5		MIT
2892.116	2892.964	34566.63	W		0.6	0.5		MIT
2891.46	2892.31	34574.5	W		0.0	1.0		MIT
2890.99	2891.84	34580.1	W		0.3	0.9		MIT
2890.66	2891.51	34584.0	W		0.3	1.1		MIT

AIR WAVELENGTH (ANGSTRMS)	VACUUM WAVELENGTH (ANGSTRMS)	WAVENUMBER (KAYSERS)	LMENT		LOG INTENSITY ARC	INTENSITY SPK	DIS	REF
2889.965	2890.812	34592.35	W		0.8	0.3		MIT
2889.77	2890.62	34594.7	W		0.3	1.3		MIT
2888.784	2889.631	34606.49	W		0.6	0.5		MIT
2888.692	2889.539	34607.60	W	II		0.7		MIT
2888.305	2889.152	34612.23	W	II	0.8	1.0		MIT
2887.656	2888.503	34620.01	W		0.9	0.8		MIT
2886.895	2887.742	34629.14	W	II	0.5	1.3		MIT
2886.46	2887.31	34634.4	W			0.8		MIT
2885.921	2886.767	34640.82	W		0.6	0.5		MIT
2885.442	2886.288	34646.57	W	II	0.0	0.7		MIT
2884.636	2885.482	34656.25	W		0.7	0.6		MIT
2884.305	2885.151	34660.23	W			1.0		MIT
2884.175	2885.021	34661.79	W		0.9	0.6		MIT
2883.91	2884.76	34665.0	W	II		0.7		MIT
2883.56	2884.41	34669.2	W		0.6			MIT
2883.25	2884.10	34672.9	W		0.3	1.0D		MIT
2882.192	2883.037	34685.64	W		0.5			MIT
2881.07	2881.92	34699.1	W		0.7	0.0		MIT
2880.627	2881.472	34704.48	W		0.8	0.8		MIT
2880.159	2881.004	34710.12	W			0.8		MIT
2879.393	2880.238	34719.36	W		1.0	1.0		MIT
2879.107	2879.952	34722.81	W		1.0	1.0		MIT
2878.716	2879.561	34727.52	W		1.0	0.9		MIT
2878.30	2879.14	34732.5	W		0.0	1.0		MIT
2878.079	2878.923	34735.21	W		0.6	0.9		MIT
2876.930	2877.774	34749.08	W		0.9	1.1		MIT
2876.095	2876.939	34759.17	W			0.7		MIT
2875.208	2876.052	34769.89	W		1.0	0.6		MIT
2873.383	2874.226	34791.97	W		0.6	1.0		MIT
2872.496	2873.339	34802.71	W		0.7	0.6		MIT
2871.899	2872.742	34809.95	W			1.0		MIT
2871.367	2872.210	34816.40	W		1.0	0.9		MIT
2870.905	2871.748	34822.00	W		1.0	1.1		MIT
2870.626	2871.469	34825.38	W		0.5D	0.6D		MIT
2870.526	2871.369	34826.60	W		0.8	0.3		MIT
2869.607	2870.449	34837.75	W		0.8S	1.0		MIT
2869.100	2869.942	34843.91	W		1.0	0.5		MIT
2868.729	2869.571	34848.41	W	II	0.8	1.4		MIT
2867.918	2868.760	34858.27	W	II	0.0	1.0		MIT
2867.560	2868.402	34862.62	W		0.6	0.5		MIT
2867.401	2868.243	34864.55	W	II	0.5	1.0		MIT
2866.746	2867.588	34872.52	W	II	0.5	1.0		MIT
2866.615	2867.457	34874.11	W	II	0.5	1.0		MIT
2866.373	2867.215	34877.06	W	II	1.0	1.0		MIT
2866.061	2866.902	34880.85	W		1.2	1.0		MIT
2865.80	2866.64	34884.0	W			1.0		MIT
2865.569	2866.410	34886.84	W		0.8	1.0		MIT
2865.311	2866.152	34889.98	W		0.9			MIT
2864.84	2865.68	34895.7	W		0.0	1.0		MIT
2864.46	2865.30	34900.4	W			1.1		MIT
2863.882	2864.723	34907.39	W		1.0	1.1		MIT
2863.006	2863.847	34918.07	W		1.0S	0.9S		MIT
2862.77	2863.61	34921.0	W		0.8	0.3		MIT
2862.424	2863.265	34925.17	W		0.8	0.3		MIT
2861.443	2862.283	34937.14	W		0.9	0.8D		MIT
2861.21	2862.05	34940.0	W		0.3	0.8		MIT
2861.054	2861.894	34941.89	W	II	0.0	0.7		MIT
2860.679	2861.519	34946.47	W		0.8	1.0		MIT
2860.163	2861.003	34952.78	W		0.7	1.0		MIT
2859.858	2860.698	34956.50	W		0.6	0.3		MIT
2859.778	2860.618	34957.48	W		0.6	0.3		MIT
2859.482	2860.322	34961.10	W		0.5	1.2		MIT
2858.740	2859.580	34970.17	W		0.8	0.8		MIT
2858.420	2859.260	34974.09	W		1.0	0.8		MIT
2858.039	2858.878	34978.75	W		1.0	1.0		MIT
2857.65	2858.49	34983.5	W		0.0	0.8		MIT
2857.445	2858.284	34986.02	W	II		0.5		MIT
2857.132	2857.971	34989.85	W		1.0	1.0		MIT
2856.027	2856.866	35003.39	W		1.0S	1.0		MIT
2855.508	2856.347	35009.75	W			1.0		MIT

AIR WAVELENGTH (ANGSTRMS)	VACUUM WAVELENGTH (ANGSTRMS)	WAVENUMBER (KAYSERS)	LMENT		LOG INTENSITY ARC	INTENSITY SPK	DIS	REF
2855.346	2856.185	35011.74	W		1.0	0.5		MIT
2854.71	2855.55	35019.5	W		0.0	1.0		MIT
2854.46	2855.30	35022.6	W			0.6S		MIT
2853.835	2854.673	35030.28	W		1.0	0.5		MIT
2853.486	2854.324	35034.56	W		1.1	1.2H		MIT
2853.22	2854.06	35037.8	W			0.6		MIT
2852.909	2853.747	35041.65	W		0.9	1.0		MIT
2852.10	2852.94	35051.6	W		0.0D	1.3H		MIT
2850.800	2851.638	35067.57	W		1.1	1.1		MIT
2850.390	2851.228	35072.61	W		1.0	0.8		MIT
2849.465	2850.302	35084.00	W		0.8	0.8		MIT
2848.026	2848.863	35101.72	W		1.2	1.1		MIT
2847.823	2848.660	35104.23	W		1.0	1.1		MIT
2847.355	2848.192	35109.99	W		1.0	1.1		MIT
2847.130	2847.967	35112.77	W		0.3	1.2		MIT
2846.38	2847.22	35122.0	W		0.8D	0.0		MIT
2844.914	2845.750	35140.12	W		1.0	1.1		MIT
2844.495	2845.331	35145.29	W			0.9		MIT
2843.778	2844.614	35154.15	W		1.0	0.9		MIT
2842.565	2843.401	35169.15	W		1.0	0.0		MIT
2841.570	2842.405	35181.47	W		1.1	1.1		MIT
2841.11	2841.95	35187.2	W		0.3	1.0		MIT
2840.736	2841.571	35191.80	W		0.7	0.6D		MIT
2840.220	2841.055	35198.19	W		1.0	0.6		MIT
2840.097	2840.932	35199.71	W		1.0	0.3		MIT
2839.92	2840.76	35201.9	W			1.0		MIT
2839.810	2840.645	35203.27	W	II	0.0	1.0		MIT
2839.337	2840.172	35209.14	W		1.0S	1.0		MIT
2838.890	2839.725	35214.68	W		1.0	0.8		MIT
2838.674	2839.509	35217.36	W		1.0	0.7		MIT
2837.762	2838.597	35228.68	W		1.0	1.0S		MIT
2837.344	2838.178	35233.87	W		1.1	1.0S		MIT
2836.252	2837.086	35247.43	W		1.0	1.0S		MIT
2835.636	2836.470	35255.09	W		1.1	1.0		MIT
2834.212	2835.046	35272.80	W	II		1.5		MIT
2834.146	2834.980	35273.62	W		1.0			MIT
2833.628	2834.461	35280.07	W		1.2	1.1		MIT
2832.952	2833.785	35288.49	W		1.0	0.5		MIT
2832.480	2833.313	35294.37	W		1.0	1.0		MIT
2831.833	2832.666	35302.43	W		0.7	1.0		MIT
2831.378	2832.211	35308.10	W		1.4	1.0		MIT
2831.237	2832.070	35309.86	W	II	0.3	1.1		MIT
2830.288	2831.121	35321.70	W		1.0	0.5		MIT
2830.10	2830.93	35324.0	W		0.6	1.3L		MIT
2829.825	2830.658	35327.48	W		1.2L	1.0		MIT
2828.790	2829.622	35340.41	W		1.0	0.3		MIT
2827.282	2828.114	35359.25	W		1.0	1.1L		MIT
2827.149	2827.981	35360.92	W		1.0	0.6		MIT
2826.085	2826.917	35374.23	W		1.0	0.5		MIT
2825.32	2826.15	35383.8	W		0.0	0.6		MIT
2824.967	2825.798	35388.23	W			0.9		MIT
2824.296	2825.127	35396.64	W	II	0.0	0.9		MIT
2823.710	2824.541	35403.98	W		1.1	0.7		MIT
2822.572	2823.403	35418.25	W	II	1.1	1.5		MIT
2821.311	2822.141	35434.08	W		1.0	0.8		MIT
2819.05	2819.88	35462.5	W		0.3	1.2		MIT
2818.060	2818.890	35474.96	W		1.2	1.3		MIT
2816.669	2817.498	35492.48	W		0.7	0.0		MIT
2815.45	2816.28	35507.9	W			1.0H		MIT
2814.798	2815.627	35516.07	W		0.0	1.1		MIT
2812.25	2813.08	35548.2	W		0.5D	1.3		MIT
2809.222	2810.050	35586.56	W		0.8	0.8		MIT
2808.943	2809.770	35590.10	W		0.0	1.2		MIT
2808.566	2809.393	35594.87	W		1.0			MIT
2808.51	2809.34	35595.6	W			1.3L		MIT
2807.928	2808.755	35602.96	W		0.8	0.7		MIT
2807.718	2808.545	35605.62	W		1.0	0.9		MIT
2806.391	2807.218	35622.46	W	II	0.3	0.8		MIT
2805.923	2806.750	35628.40	W	II	1.0S	1.5		MIT
2805.627	2806.454	35632.16	W		1.0	0.3		MIT

AIR WAVELENGTH (ANGSTRMS)	VACUUM WAVELENGTH (ANGSTRMS)	WAVENUMBER (KAYSERS)	LMENT		LOG ARC	INTENSITY SPK	DIS	REF
2804.96	2805.79	35640.6	W			1.1		MIT
2804.673	2805.499	35644.28	W		1.1	1.0		MIT
2804.237	2805.063	35649.82	W		1.0	1.0		MIT
2804.014	2804.840	35652.65	W		1.0	0.6		MIT
2803.663	2804.489	35657.12	W	II	0.8	1.1L		MIT
2803.30	2804.13	35661.7	W			1.0L		MIT
2802.953	2803.779	35666.15	W		1.1D	1.0D		MIT
2801.949	2802.775	35678.93	W		0.9	0.8L		MIT
2801.426	2802.252	35685.59	W	II	0.6	0.8		MIT
2801.169	2801.995	35688.86	W		1.0	0.9		MIT
2801.051	2801.877	35690.37	W	II	0.8	1.2		MIT
2800.085	2800.910	35702.68	W		0.8	0.5		MIT
2799.924	2800.749	35704.73	W		1.1	1.1S		MIT
2799.677	2800.502	35707.88	W		0.8	0.5		MIT
2799.034	2799.859	35716.08	W	II	0.9	1.3		MIT
2798.447	2799.272	35723.58	W		0.9	0.0		MIT
2797.625	2798.450	35734.07	W		0.7	0.5		MIT
2797.468	2798.293	35736.08	W		1.0	0.6		MIT
2797.289	2798.114	35738.36	W			0.5H		MIT
2797.195	2798.020	35739.56	W		1.0	0.6		MIT
2796.859	2797.684	35743.86	W	II	0.5	1.0		MIT
2796.147	2796.971	35752.96	W		1.1S	0.5		MIT
2795.55	2796.37	35760.6	W		0.0	1.0		MIT
2793.479	2794.303	35787.10	W		0.8	0.0		MIT
2793.120	2793.944	35791.70	W		1.0	0.6		MIT
2792.796	2793.620	35795.85	W		1.0	0.8		MIT
2792.696	2793.519	35797.14	W		1.0	1.0		MIT
2792.521	2793.344	35799.38	W		1.0	0.9		MIT
2792.209	2793.032	35803.38	W		0.8	0.6		MIT
2791.953	2792.776	35806.66	W		1.1	0.9		MIT
2790.564	2791.387	35824.48	W		1.0	0.0H		MIT
2790.417	2791.240	35826.37	W	II	0.3	1.2		MIT
2789.677	2790.500	35835.87	W		1.1S	0.7		MIT
2789.372	2790.195	35839.79	W		0.7	0.0		MIT
2789.163	2789.986	35842.48	W		0.5	1.1		MIT
2789.071	2789.894	35843.66	W		1.0	0.8		MIT
2788.526	2789.348	35850.67	W			0.8		MIT
2787.978	2788.800	35857.71	W		1.1	1.0		MIT
2786.514	2787.336	35876.55	W		0.9	0.3		MIT
2786.313	2787.135	35879.14	W	II	0.3D	1.2		MIT
2785.883	2786.705	35884.68	W		0.8	0.3		MIT
2785.633	2786.455	35887.90	W	II	0.5	1.3		MIT
2783.678	2784.499	35913.10	W		0.8	0.7		MIT
2783.119	2783.940	35920.31	W		1.0	1.1		MIT
2782.654	2783.475	35926.31	W		0.6			MIT
2782.135	2782.956	35933.02	W	II	0.6	1.5		MIT
2780.283	2781.103	35956.95	W	II	1.0	1.3		MIT
2779.724	2780.544	35964.18	W		0.9	0.8		MIT
2779.317	2780.137	35969.45	W		0.7	0.6		MIT
2778.686	2779.506	35977.62	W		0.5	1.3		MIT
2777.870	2778.690	35988.18	W		0.3	1.1		MIT
2777.258	2778.078	35996.11	W		0.8	0.3		MIT
2776.502	2777.322	36005.91	W	II	1.1D	1.4		MIT
2776.086	2776.905	36011.31	W		1.0	0.3		MIT
2776.02	2776.84	36012.2	W		0.3	0.9		MIT
2774.982	2775.801	36025.63	W		0.3	1.1		MIT
2774.480	2775.299	36032.15	W	I	1.2			ME
2774.091	2774.910	36037.21	W			1.0		MIT
2774.002	2774.821	36038.36	W		1.1	0.3		MIT
2773.87	2774.69	36040.1	W	II	0.6	1.0		MIT
2773.698	2774.517	36042.31	W		1.1	0.9		MIT
2772.483	2773.302	36058.10	W		1.0	1.8		MIT
2771.925	2772.743	36065.36	W			1.1		MIT
2771.621	2772.439	36069.32	W		1.1	0.7		MIT
2771.001	2771.819	36077.39	W			1.0		MIT
2770.881	2771.699	36078.95	W		1.4	1.1		MIT
2770.206	2771.024	36087.74	W		1.0	0.3		MIT
2769.741	2770.559	36093.80	W		1.2	1.0		MIT
2768.985	2769.803	36103.65	W		1.3	1.0		MIT
2768.332	2769.150	36112.17	W	II	1.0	1.3L		MIT

AIR WAVELENGTH (ANGSTRMS)	VACUUM WAVELENGTH (ANGSTRMS)	WAVENUMBER (KAYSERS)	LMENT		LOG ARC	INTENSITY SPK	DIS	REF
2767.141	2767.958	36127.71	W		1.1	0.0		MIT
2766.969	2767.786	36129.96	W			1.1		MIT
2766.732	2767.549	36133.05	W		0.9	0.0		MIT
2766.319	2767.136	36138.45	W			1.3		MIT
2765.657	2766.474	36147.10	W		1.1	0.6		MIT
2765.336	2766.153	36151.29	W			1.0		MIT
2765.205	2766.022	36153.00	W		0.9	0.0		MIT
2764.266	2765.083	36165.29	W	II	1.3	1.8		MIT
2764.004	2764.820	36168.71	W		1.0	0.3		MIT
2762.700	2763.516	36185.78	W		1.0	0.5		MIT
2762.497	2763.313	36188.44	W			1.0		MIT
2762.345	2763.161	36190.43	W		1.3	1.0		MIT
2762.104	2762.920	36193.59	W			0.6		MIT
2761.590	2762.406	36200.33	W	II	1.0	1.4		MIT
2761.137	2761.953	36206.27	W		0.9			MIT
2760.742	2761.558	36211.45	W		0.9	1.3		MIT
2760.034	2760.850	36220.73	W		0.9D	0.8		MIT
2759.534	2760.349	36227.30	W			1.0L		MIT
2759.033	2759.848	36233.88	W		0.9	0.5		MIT
2758.877	2759.692	36235.92	W		0.8	0.5		MIT
2758.687	2759.502	36238.42	W		0.8	0.5		MIT
2758.328	2759.143	36243.14	W	II	0.8	1.1		MIT
2757.699	2758.514	36251.40	W			1.1		MIT
2757.209	2758.024	36257.85	W		0.8	1.1		MIT
2756.773	2757.588	36263.58	W		0.0	1.1S		MIT
2755.942	2756.757	36274.51	W		1.0	0.8		MIT
2755.69	2756.50	36277.8	W			0.8		MIT
2755.265	2756.079	36283.42	W		1.0	0.6		MIT
2754.918	2755.732	36288.00	W		1.1	1.0		MIT
2754.71	2755.52	36290.7	W	II		0.7		MIT
2753.323	2754.137	36309.02	W		0.0	1.1		MIT
2753.166	2753.980	36311.09	W		0.9			MIT
2753.051	2753.865	36312.60	W	II	1.2	1.0D		MIT
2752.35	2753.16	36321.9	W	II	0.8	0.7		MIT
2752.238	2753.052	36323.33	W		0.3	1.3L		MIT
2751.215	2752.028	36336.83	W		0.7	0.6		MIT
2750.764	2751.577	36342.79	W			1.1		MIT
2750.321	2751.134	36348.65	W	II	0.3	1.1		MIT
2750.147	2750.960	36350.94	W		1.1	0.7		MIT
2749.000	2749.813	36366.11	W		1.0	0.6		MIT
2748.849	2749.662	36368.11	W		1.1	1.0		MIT
2748.576	2749.389	36371.72	W		1.0	0.5		MIT
2748.433	2749.246	36373.61	W			1.0		MIT
2748.308	2749.121	36375.27	W		1.0	1.1		MIT
2747.833	2748.646	36381.56	W		1.1	0.7		MIT
2747.018	2747.830	36392.35	W		1.1	1.0		MIT
2746.735	2747.547	36396.10	W		1.1	1.3		MIT
2746.209	2747.021	36403.07	W		1.0	0.8		MIT
2745.837	2746.649	36408.00	W		1.0	0.8		MIT
2745.62	2746.43	36410.9	W		0.9	0.8		MIT
2745.323	2746.135	36414.82	W		0.8	0.5		MIT
2745.107	2745.919	36417.68	W		1.0			MIT
2745.028	2745.840	36418.73	W	II	0.3	1.4		MIT
2744.903	2745.715	36420.39	W		0.9			MIT
2743.425	2744.237	36440.01	W		1.1	1.0		MIT
2742.898	2743.709	36447.01	W		0.5	1.4		MIT
2742.469	2743.280	36452.71	W		0.7	1.4		MIT
2741.901	2742.712	36460.26	W			1.0		MIT
2741.20	2742.01	36469.6	W			0.9		MIT
2740.791	2741.602	36475.03	W		0.7	1.4		MIT
2740.176	2740.987	36483.21	W		0.0	1.1		MIT
2739.373	2740.184	36493.91	W			1.0		MIT
2739.124	2739.934	36497.22	W			1.0		MIT
2738.514	2739.324	36505.35	W		0.3	1.0		MIT
2737.999	2738.809	36512.22	W		1.1	0.8		MIT
2737.760	2738.570	36515.41	W			0.8		MIT
2737.371	2738.181	36520.60	W			1.3		MIT
2736.575	2737.385	36531.22	W			1.2		MIT
2735.970	2736.780	36539.29	W		1.0	0.6		MIT
2735.781	2736.591	36541.82	W		0.0	1.0		MIT

AIR WAVELENGTH (ANGSTRMS)	VACUUM WAVELENGTH (ANGSTRMS)	WAVENUMBER (KAYSERS)	LMENT	LOG ARC	INTENSITY SPK	DIS	REF
2734.770	2735.579	36555.33	W	1.0	1.3		MIT
2733.772	2734.581	36568.67	W	1.0	0.8D		MIT
2733.443	2734.252	36573.07	W		1.0		MIT
2733.183	2733.992	36576.55	W	1.2	1.0		MIT
2732.814	2733.623	36581.49	W	1.0	0.6L		MIT
2732.374	2733.183	36587.38	W II		1.0		MIT
2732.17	2732.98	36590.1	W		1.0D		MIT
2731.796	2732.605	36595.12	W		0.7J		MIT
2731.124	2731.933	36604.12	W	1.0			MIT
2730.847	2731.655	36607.84	W		1.0		MIT
2729.936	2730.744	36620.05	W	0.0	1.2		MIT
2729.625	2730.433	36624.23	W II	1.0	1.4D		MIT
2729.149	2729.957	36630.61	W	0.0	1.1		MIT
2728.873	2729.681	36634.32	W II		1.0		MIT
2727.950	2728.758	36646.71	W	1.1	0.5		MIT
2727.331	2728.139	36655.03	W		0.7		MIT
2726.424	2727.231	36667.22	W II	0.3	0.7		MIT
2726.263	2727.070	36669.39	W	1.0	0.5		MIT
2725.057	2725.864	36685.62	W	1.1	1.0		MIT
2724.627	2725.434	36691.40	W	1.0	0.9		MIT
2724.352	2725.159	36695.11	W	1.3	1.0		MIT
2724.08	2724.89	36698.8	W	0.3	1.2		MIT
2723.951	2724.758	36700.51	W	1.0			MIT
2723.707	2724.514	36703.80	W		0.8		MIT
2722.806	2723.613	36715.94	W II	0.9	1.4		MIT
2722.683	2723.489	36717.60	W	0.9			MIT
2722.461	2723.267	36720.60	W	1.1	0.7		MIT
2721.848	2722.654	36728.86	W II		1.0		MIT
2721.665	2722.471	36731.33	W	1.0S	1.2		MIT
2720.612	2721.418	36745.55	W II	0.8	1.1		MIT
2720.404	2721.210	36748.36	W II	0.6D	1.3		MIT
2720.045	2720.851	36753.21	W	1.0	0.5		MIT
2719.857	2720.663	36755.75	W	1.1	1.0		MIT
2719.327	2720.133	36762.91	W	1.2	1.3Y		MIT
2718.901	2719.707	36768.67	W	1.4	1.3		MIT
2718.042	2718.847	36780.29	W II	1.0	1.3		MIT
2717.699	2718.504	36784.93	W II	0.6	1.0		MIT
2717.531	2718.336	36787.21	W	1.0	0.0		MIT
2717.167	2717.972	36792.14	W	0.6	1.2		MIT
2716.897	2717.702	36795.79	W	0.9	1.1		MIT
2716.315	2717.120	36803.67	W II	0.9	1.3		MIT
2715.495	2716.300	36814.79	W	1.3	0.9		MIT
2715.338	2716.143	36816.92	W II	0.9	1.3		MIT
2714.935	2715.740	36822.38	W		0.5		MIT
2714.815	2715.620	36824.01	W	0.6	0.5		MIT
2714.333	2715.137	36830.55	W		0.5		MIT
2713.847	2714.651	36837.14	W	0.8	0.3		MIT
2712.695	2713.499	36852.79	W	0.0	1.3		MIT
2712.585	2713.389	36854.28	W	0.7			MIT
2711.31	2712.11	36871.6	W		0.6		MIT
2710.782	2711.586	36878.79	W II	0.8D	1.2		MIT
2710.32	2711.12	36885.1	W		0.8		MIT
2710.000	2710.803	36889.43	W	0.9	0.6		MIT
2709.752	2710.555	36892.81	W	0.9			MIT
2709.575	2710.378	36895.22	W II	0.8	1.2		MIT
2708.921	2709.724	36904.12	W	1.0	1.0		MIT
2708.794	2709.597	36905.85	W	1.0	0.3		MIT
2708.584	2709.387	36908.72	W	1.0	1.2		MIT
2708.181	2708.984	36914.21	W	1.0	0.8S		MIT
2707.878	2708.681	36918.34	W	1.0	0.6		MIT
2707.020	2707.823	36930.04	W	0.8	1.2		MIT
2706.696	2707.499	36934.46	W II	0.8	1.3		MIT
2706.580	2707.383	36936.04	W	1.1	1.0		MIT
2706.006	2706.808	36943.88	W	1.0	0.9		MIT
2705.93	2706.73	36944.9	W		0.8		MIT
2705.572	2706.374	36949.80	W II	0.0	1.1		MIT
2704.78	2705.58	36960.6	W		0.7H		MIT
2703.456	2704.258	36978.72	W	0.3	1.3		MIT
2703.12	2703.92	36983.3	W II	0.6	0.7		LN
2703.062	2703.864	36984.11	W II	0.5D	1.2		MIT
2702.520	2703.322	36991.53	W	1.0	0.5		MIT
2702.111	2702.913	36997.13	W	0.9	1.4		MIT
2701.817	2702.618	37001.15	W	0.9			MIT
2701.477	2702.278	37005.81	W II	0.3	1.3		MIT
2701.06	2701.86	37011.5	W	0.3	1.1		MIT
2700.312	2701.113	37021.77	W		1.2		MIT
2700.009	2700.810	37025.93	W	1.2	1.0		MIT
2699.590	2700.391	37031.68	W	1.2	1.0		MIT
2699.259	2700.060	37036.22	W		0.9		MIT
2699.033	2699.834	37039.32	W		0.9		MIT
2698.837	2699.638	37042.01	W	1.1	0.6		MIT
2698.347	2699.148	37048.73	W		0.9		MIT
2697.714	2698.514	37057.43	W II	1.2	1.4		MIT
2697.512	2698.312	37060.20	W	1.2	0.5		MIT
2697.27	2698.07	37063.5	W		0.8		MIT
2696.895	2697.695	37068.68	W		1.3		MIT
2696.37	2697.17	37075.9	W	0.6			MIT
2695.87	2696.67	37082.8	W		0.9		MIT
2695.666	2696.466	37085.58	W	1.3	1.1		MIT
2694.982	2695.782	37094.99	W II	0.3	1.1		MIT
2694.594	2695.394	37100.33	W		1.3		MIT
2694.380	2695.180	37103.28	W II	1.0	1.3		MIT
2693.476	2694.275	37115.73	W		0.6		MIT
2691.559	2692.358	37142.16	W	0.9	0.0		MIT
2691.280	2692.079	37146.01	W	1.0	0.5		MIT
2691.091	2691.890	37148.62	W	1.2	0.9		MIT
2690.708	2691.507	37153.91	W		1.2		MIT
2690.574	2691.373	37155.76	W	0.7			MIT
2690.414	2691.213	37157.97	W	0.8			MIT
2689.654	2690.453	37168.47	W	0.9	0.3		MIT
2689.341	2690.139	37172.79	W		0.7		MIT
2688.223	2689.021	37188.25	W II	0.9S	1.3		MIT
2687.995	2688.793	37191.41	W	0.9	0.0		MIT
2687.524	2688.322	37197.92	W		0.9		MIT
2687.367	2688.165	37200.10	W	1.2	0.5		MIT
2687.137	2687.935	37203.28	W	0.8			MIT
2686.993	2687.791	37205.27	W	0.6D	1.3		MIT
2686.620	2687.418	37210.44	W	0.9	0.3		MIT
2686.355	2687.153	37214.11	W		0.9N		MIT
2685.713	2686.511	37223.01	W	0.8			MIT
2685.340	2686.137	37228.18	W	0.3	1.0		MIT
2685.146	2685.943	37230.87	W	0.8			MIT
2685.049	2685.846	37232.21	W	0.3	1.2		MIT
2684.386	2685.183	37241.41	W		1.0L		MIT
2684.07	2684.87	37245.8	W	1.0	0.3D		MIT
2683.84	2684.64	37249.0	W	0.0	1.0		MIT
2683.622	2684.419	37252.01	W		1.0		MIT
2683.341	2684.138	37255.91	W	1.1	0.3		MIT
2683.218	2684.015	37257.62	W	0.6	1.1		MIT
2682.640	2683.437	37265.64	W	1.0	0.9		MIT
2682.153	2682.950	37272.41	W	0.8	0.0		MIT
2681.902	2682.699	37275.90	W	0.7	0.3		MIT
2681.632	2682.429	37279.65	W	0.9	0.3		MIT
2681.413	2682.210	37282.69	W	1.3	1.3		MIT
2680.564	2681.360	37294.50	W	0.6	1.0		MIT
2680.340	2681.136	37297.62	W	0.7			MIT
2680.039	2680.835	37301.81	W	1.1	0.8		MIT
2679.870	2680.666	37304.16	W	0.8			MIT
2679.632	2680.428	37307.47	W II	0.9	1.1		MIT
2678.883	2679.679	37317.90	W	1.3	1.0		MIT
2678.522	2679.318	37322.93	W	1.0	0.8		MIT
2677.910	2678.706	37331.46	W	0.9	0.3		MIT
2677.791	2678.587	37333.12	W II	0.9	1.3		MIT
2677.278	2678.074	37340.27	W	1.1	1.0		MIT
2676.997	2677.792	37344.19	W		0.7		MIT
2676.412	2677.207	37352.35	W	1.0	0.0		MIT
2676.310	2677.105	37353.78	W		0.8		MIT
2675.869	2676.664	37359.94	W	1.1	0.8		MIT
2675.73	2676.53	37361.9	W		1.1		MIT
2675.400	2676.195	37366.48	W	1.0	0.5		MIT

AIR WAVELENGTH (ANGSTRMS)	VACUUM WAVELENGTH (ANGSTRMS)	WAVENUMBER (KAYSERS)	LMENT	LOG ARC	INTENSITY SPK	DIS	REF
2675.204	2675.999	37369.22	W		0.9		MIT
2675.128	2675.923	37370.28	W	1.0			MIT
2674.704	2675.499	37376.21	W	1.0	0.3		MIT
2674.70	2675.49	37376.3	W		0.8		MIT
2674.138	2674.933	37384.12	W		0.8		MIT
2673.591	2674.386	37391.77	W	0.8	1.3		MIT
2672.947	2673.742	37400.77	W		0.8		MIT
2672.669	2673.463	37404.66	W	0.8	1.1		MIT
2672.162	2672.956	37411.76	W	1.0	0.3		MIT
2671.770	2672.564	37417.25	W	1.1	1.1		MIT
2671.466	2672.260	37421.51	W	1.2	0.7		MIT
2671.180	2671.974	37425.51	W	0.7	0.0		MIT
2670.693	2671.487	37432.34	W		0.7		MIT
2670.386	2671.180	37436.64	W	0.5	1.2		MIT
2669.774	2670.568	37445.22	W	1.0	0.3		MIT
2669.302	2670.096	37451.84	W II	1.2	1.5 D		MIT
2668.949	2669.743	37456.79	W II	0.5	0.8		MIT
2668.468	2669.261	37463.55	W	0.9	0.5 L		MIT
2668.233	2669.026	37466.85	W II		0.8		MIT
2667.675	2668.468	37474.68	W		1.2		MIT
2666.490	2667.283	37491.33	W II	0.9	1.3		MIT
2666.084	2666.877	37497.04	W		1.2		MIT
2665.772	2666.565	37501.43	W	1.1 D	0.7 D		MIT
2665.642	2666.435	37503.26	W II		0.8		MIT
2664.958	2665.751	37512.89	W	1.1	0.8		MIT
2664.318	2665.110	37521.90	W II	1.0	1.3		MIT
2663.940	2664.732	37527.22	W		0.9		MIT
2663.557	2664.349	37532.62	W	1.0	0.5		MIT
2663.260	2664.052	37536.80	W	0.5			MIT
2662.835	2663.627	37542.79	W	1.2	1.0		MIT
2662.638	2663.430	37545.57	W	0.9	0.5		MIT
2662.205	2662.997	37551.68	W	0.0	1.0		MIT
2661.848	2662.640	37556.71	W		1.1		MIT
2661.557	2662.349	37560.82	W	1.0	1.1		MIT
2660.522	2661.314	37575.43	W	1.0	0.9		MIT
2659.697	2660.488	37587.08	W		1.1		MIT
2659.189	2659.980	37594.27	W		1.0		MIT
2658.906	2659.697	37598.27	W	1.0	0.3		MIT
2658.718	2659.509	37600.92	W		0.6		MIT
2658.564	2659.355	37603.10	W	0.6			MIT
2658.194	2658.985	37608.34	W	0.9	0.0		MIT
2658.037	2658.828	37610.56	W II	1.0	1.3		MIT
2657.381	2658.172	37619.84	W	1.1	1.0		MIT
2656.907	2657.698	37626.55	W	0.9	0.5		MIT
2656.704	2657.495	37629.43	W		0.9		MIT
2656.539	2657.330	37631.76	W	1.2	0.7		MIT
2656.153	2656.943	37637.23	W	0.5			MIT
2656.021	2656.811	37639.10	W		0.9		MIT
2655.670	2656.460	37644.08	W II	0.7	1.0		MIT
2655.549	2656.339	37645.79	W	0.6	1.0		MIT
2654.671	2655.461	37658.24	W	0.8	0.3		MIT
2653.567	2654.357	37673.91	W II	1.0	1.2		MIT
2653.425	2654.215	37675.93	W		0.9		MIT
2653.321	2654.111	37677.40	W	0.8			MIT
2653.012	2653.802	37681.79	W	0.3	1.0		MIT
2652.609	2653.399	37687.52	W	1.0	1.1		MIT
2652.424	2653.214	37690.14	W II		0.5		MIT
2652.297	2653.087	37691.95	W II		0.3		MIT
2652.012	2652.801	37696.00	W	0.9	0.3		MIT
2651.874	2652.663	37697.96	W	0.3	1.1		MIT
2651.441	2652.230	37704.12	W	1.0 D	0.6 D		MIT
2651.023	2651.812	37710.06	W	0.0	1.0		MIT
2650.711	2651.500	37714.50	W	0.3	0.5 H		MIT
2650.452	2651.241	37718.18	W	0.9	0.0		MIT
2650.268	2651.057	37720.80	W II		0.9		MIT
2649.983	2650.772	37724.86	W	1.0	0.8		MIT
2649.693	2650.482	37728.99	W		1.1 L		MIT
2649.460	2650.249	37732.31	W	0.3	0.0		MIT
2649.290	2650.079	37734.73	W	0.8	0.7		MIT
2648.736	2649.525	37742.62	W	0.8	0.3		MIT
2648.054	2648.843	37752.34	W II	0.8	0.3		MIT
2647.740	2648.528	37756.81	W	1.0	1.3		MIT
2647.096	2647.884	37766.00	W	1.0	0.9		MIT
2646.735	2647.523	37771.15	W	1.1	1.0		MIT
2646.543	2647.331	37773.89	W	0.9	0.0		MIT
2646.186	2646.974	37778.99	W	1.2	1.0 D		MIT
2645.694	2646.482	37786.01	W	1.1	1.0		MIT
2645.297	2646.085	37791.68	W	0.5	0.3		MIT
2645.149	2645.937	37793.80	W	1.0	0.6 S		MIT
2644.603	2645.391	37801.60	W	1.0	0.3		MIT
2644.389	2645.177	37804.66	W	1.0	0.6		MIT
2643.89	2644.68	37811.8	W		0.7		MIT
2643.290	2644.077	37820.37	W	0.3	1.3		MIT
2643.124	2643.911	37822.75	W II	1.1	1.2		MIT
2642.678	2643.465	37829.13	W	0.8	0.3		MIT
2642.344	2643.131	37833.92	W	0.8	0.3		MIT
2642.022	2642.809	37838.53	W	0.9	0.5		MIT
2641.077	2641.864	37852.06	W		0.9 L		MIT
2640.720	2641.507	37857.18	W	0.7	0.3		MIT
2640.368	2641.155	37862.23	W	0.7	0.3		MIT
2639.404	2640.190	37876.06	W	0.0	0.8 D		MIT
2638.748	2639.534	37885.47	W	1.0	0.7		MIT
2638.612	2639.398	37887.42	W	1.1	0.9		MIT
2638.222	2639.008	37893.02	W		1.0		MIT
2637.574	2638.360	37902.33	W II	0.8	1.2		MIT
2637.149	2637.935	37908.44	W	0.5	1.0		MIT
2636.954	2637.740	37911.24	W		1.1		MIT
2636.550	2637.336	37917.05	W	1.0	0.3		MIT
2635.376	2636.162	37933.94	W II	0.5	1.3		MIT
2634.850	2635.635	37941.52	W II	0.8	1.0		MIT
2634.576	2635.361	37945.46	W	0.0	1.3		MIT
2633.882	2634.667	37955.46	W II	0.0	1.2		MIT
2633.127	2633.912	37966.34	W	1.2	1.1		MIT
2632.700	2633.485	37972.50	W	1.0	1.1		MIT
2632.487	2633.272	37975.57	W	1.1	1.0		MIT
2632.116	2632.901	37980.92	W		0.3 H		MIT
2631.781	2632.566	37985.76	W	0.9	0.3		MIT
2630.929	2631.713	37998.06	W		0.6		MIT
2630.732	2631.516	38000.90	W	0.8	0.3		MIT
2630.526	2631.310	38003.88	W II	0.8	1.0		MIT
2630.380	2631.164	38005.99	W II	0.0	1.0		MIT
2630.195	2630.979	38008.66	W		0.9		MIT
2629.498	2630.282	38018.74	W		0.7		MIT
2629.154	2629.938	38023.71	W	0.8	0.5		MIT
2628.996	2629.780	38026.00	W II	0.3	1.1		MIT
2628.892	2629.676	38027.50	W	1.0	0.3 H		MIT
2628.681	2629.465	38030.55	W		0.8		MIT
2628.250	2629.034	38036.79	W	1.1	1.1		MIT
2628.044	2628.828	38039.77	W		0.9		MIT
2627.700	2628.484	38044.75	W II	0.0	1.0 S		MIT
2627.444	2628.228	38048.46	W	0.9	0.3		MIT
2626.931	2627.714	38055.88	W		0.5		MIT
2626.710	2627.493	38059.09	W	0.7	0.0		MIT
2626.533	2627.316	38061.65	W	0.0	1.0		MIT
2626.245	2627.028	38065.83	W	1.1	0.8		MIT
2625.860	2626.643	38071.41	W		0.8		MIT
2625.331	2626.114	38079.08	W		0.8		MIT
2625.215	2625.998	38080.76	W	1.2	0.8		MIT
2624.991	2625.774	38084.01	W	0.8	0.7		MIT
2624.475	2625.258	38091.50	W	0.0	1.2 S		MIT
2624.360	2625.143	38093.17	W	0.5			MIT
2624.146	2624.929	38096.27	W	0.9	0.0		MIT
2623.887	2624.670	38100.03	W		1.2 D		MIT
2623.109	2623.892	38111.33	W II	0.3	1.3		MIT
2622.860	2623.643	38114.95	W II		1.0		MIT
2622.335	2623.117	38122.58	W	0.8	0.5		MIT
2622.207	2622.989	38124.44	W	1.2	1.0 D		MIT
2621.592	2622.374	38133.38	W	0.9	1.1		MIT
2620.744	2621.526	38145.72	W II	0.5	1.1		MIT
2620.266	2621.048	38152.68	W	1.1	1.2		MIT

AIR WAVELENGTH (ANGSTRMS)	VACUUM WAVELENGTH (ANGSTRMS)	WAVENUMBER (KAYSERS)	LMENT		LOG ARC	INTENSITY SPK	DIS	REF
2619.952	2620.734	38157.25	W		0.9	0.3		MIT
2619.177	2619.959	38168.54	W		1.0	1.1		MIT
2618.807	2619.589	38173.93	W		1.1	0.5		MIT
2618.281	2619.062	38181.60	W			0.7		MIT
2618.080	2618.861	38184.53	W			1.2		MIT
2617.629	2618.410	38191.11	W	II	0.5	1.1		MIT
2617.156	2617.937	38198.01	W			0.6		MIT
2616.339	2617.120	38209.94	W		0.8	0.5		MIT
2616.046	2616.827	38214.22	W			0.5		MIT
2615.695	2616.476	38219.35	W		0.0	0.9		MIT
2615.445	2616.226	38223.00	W	II	0.8	1.3		MIT
2615.122	2615.903	38227.72	W		1.0	0.5		MIT
2614.445	2615.226	38237.62	W		0.8	0.7 S		MIT
2614.377	2615.158	38238.62	W			0.8		MIT
2613.823	2614.603	38246.72	W		1.2	1.0		MIT
2613.072	2613.852	38257.71	W		1.1	1.0		MIT
2612.648	2613.428	38263.92	W		0.3	1.1		MIT
2612.188	2612.968	38270.66	W		1.0	1.2		MIT
2611.580	2612.360	38279.57	W			0.8		MIT
2611.396	2612.176	38282.26	W		0.8			MIT
2611.274	2612.054	38284.05	W			0.9		MIT
2611.102	2611.882	38286.57	W		0.7			MIT
2610.745	2611.525	38291.81	W		0.9	0.6		MIT
2610.588	2611.368	38294.11	W		0.0	0.9		MIT
2610.492	2611.272	38295.52	W			0.8		MIT
2610.201	2610.981	38299.79	W		0.9	0.0		MIT
2609.871	2610.650	38304.63	W		0.7	0.0		MIT
2609.589	2610.368	38308.77	W		0.6			MIT
2609.241	2610.020	38313.88	W		0.3	1.1		MIT
2608.437	2609.216	38325.69	W			0.6		MIT
2608.318	2609.097	38327.44	W		1.1	0.7		MIT
2608.054	2608.833	38331.31	W			0.7		MIT
2607.810	2608.589	38334.90	W		0.0	1.0		MIT
2607.732	2608.511	38336.05	W		0.0	1.0		MIT
2607.378	2608.157	38341.25	W		1.1	0.7		MIT
2606.975	2607.754	38347.18	W	II		0.9		M
2606.904	2607.683	38348.22	W	II		0.9		M
2606.555	2607.334	38353.36	W	II		1.0		MIT
2606.386	2607.165	38355.84	W		1.0			MIT
2606.275	2607.054	38357.48	W	II		1.0		MIT
2605.963	2606.742	38362.07	W	II	0.3	0.9		MIT
2605.892	2606.670	38363.12	W		0.8	0.3		MIT
2605.507	2606.285	38368.78	W		1.1	0.3		MIT
2605.474	2606.252	38369.27	W	II		0.6		MIT
2604.042	2604.820	38390.37	W			0.8		MIT
2603.544	2604.322	38397.71	W		1.1	0.8		MIT
2603.021	2603.799	38405.42	W	II	0.7	1.2		MIT
2602.804	2603.582	38408.63	W		1.1	0.3		MIT
2602.512	2603.290	38412.94	W		1.0	1.4		MIT
2602.419	2603.197	38414.31	W		0.5			MIT
2601.963	2602.741	38421.04	W		1.2	0.8		MIT
2601.433	2602.210	38428.87	W	II	0.3	1.2		MIT
2600.737	2601.514	38439.15	W		1.1	0.7		MIT
2600.518	2601.295	38442.39	W			1.1		MIT
2600.218	2600.995	38446.82	W		1.0	0.3		MIT
2599.764	2600.541	38453.54	W		0.3	1.1		MIT
2599.642	2600.419	38455.34	W	II	0.0	0.7		MIT
2598.738	2599.515	38468.72	W	II	1.1	1.3		MIT
2598.68	2599.46	38469.6	W	II	0.5	0.5		MIT
2598.421	2599.198	38473.41	W		1.0	0.7		MIT
2598.107	2598.884	38478.06	W		0.8			MIT
2597.953	2598.730	38480.34	W			1.1		MIT
2597.726	2598.503	38483.70	W		1.0	0.6		MIT
2597.470	2598.246	38487.50	W			0.8		MIT
2597.289	2598.065	38490.18	W		0.7	0.3		MIT
2596.858	2597.634	38496.56	W	II	0.0	1.2 S		MIT
2596.667	2597.443	38499.40	W		1.1	0.5 S		MIT
2596.343	2597.119	38504.20	W			1.0		MIT
2596.112	2596.888	38507.63	W		0.7			MIT
2595.761	2596.537	38512.83	W	II	0.0	1.0		MIT
2595.568	2596.344	38515.70	W		0.3	1.3		MIT
2594.804	2595.580	38527.04	W			1.0		MIT
2594.543	2595.319	38530.91	W		0.6	1.2 L		MIT
2593.382	2594.158	38548.16	W		1.1	0.6		MIT
2592.594	2593.369	38559.88	W			0.6		MIT
2592.464	2593.239	38561.81	W		0.3	0.7		MIT
2591.970	2592.745	38569.16	W			0.8		MIT
2591.740	2592.515	38572.58	W			0.8		MIT
2591.49	2592.27	38576.3	W	II	1.1	0.8		MIT
2589.654	2590.429	38603.65	W		0.3 D	1.1		MIT
2589.167	2589.942	38610.91	W	II	1.2 D	1.4		MIT
2588.554	2589.328	38620.05	W			0.9		MIT
2588.093	2588.867	38626.93	W			1.1		MIT
2587.758	2588.532	38631.93	W		1.1	0.3		MIT
2587.320	2588.094	38638.47	W		1.0	0.9		MIT
2586.941	2587.715	38644.13	W		1.9	1.5		MIT
2586.733	2587.507	38647.24	W			0.8		MIT
2586.640	2587.414	38648.63	W		1.0	0.8		MIT
2586.586	2587.360	38649.44	W	II	0.0	0.5		MIT
2586.344	2587.118	38653.05	W		0.9	1.1		MIT
2585.959	2586.733	38658.81	W	II	0.9	1.3		MIT
2585.434	2586.208	38666.65	W		0.9	0.3		MIT
2585.222	2585.996	38669.83	W		1.0	0.5		MIT
2584.379	2585.152	38682.44	W		1.2	1.0		MIT
2584.230	2585.003	38684.67	W		0.7	0.8		MIT
2583.660	2584.433	38693.20	W		0.8	0.0		MIT
2583.514	2584.287	38695.39	W	II		1.0		MIT
2583.215	2583.988	38699.87	W		1.1	1.0		MIT
2582.825	2583.598	38705.71	W			0.9		MIT
2582.522	2583.295	38710.25	W			1.1		MIT
2581.498	2582.271	38725.61	W		0.9			MIT
2581.200	2581.973	38730.08	W	II	0.8	1.3		MIT
2581.061	2581.834	38732.16	W		1.1			MIT
2580.488	2581.260	38740.76	W		1.1	0.8		MIT
2580.332	2581.104	38743.10	W		1.0	0.7		MIT
2580.048	2580.820	38747.37	W		1.0	0.0		MIT
2579.541	2580.313	38754.98	W	II	0.9	1.4		MIT
2579.400	2580.172	38757.10	W		1.0 S			MIT
2579.258	2580.030	38759.24	W		0.3	1.3		MIT
2578.695	2579.467	38767.70	W		0.5	0.7		MIT
2577.971	2578.743	38778.58	W		0.5	0.0		MIT
2577.696	2578.468	38782.72	W		0.9	0.0		MIT
2577.305	2578.077	38788.60	W	II		0.6		MIT
2577.025	2577.797	38792.82	W		1.2	0.6		MIT
2576.865	2577.637	38795.23	W			1.2		MIT
2576.360	2577.132	38802.83	W	II	0.3	1.2		MIT
2576.165	2576.936	38805.77	W		0.3	1.3		MIT
2575.897	2576.668	38809.81	W		1.0	0.0		MIT
2575.465	2576.236	38816.31	W		1.0	0.0		MIT
2574.681	2575.452	38828.13	W		0.7			MIT
2574.200	2574.971	38835.39	W		0.5	0.8 S		MIT
2573.953	2574.724	38839.12	W	II	1.0	0.6		MIT
2573.812	2574.583	38841.24	W	II	0.3	1.0		MIT
2573.589	2574.360	38844.61	W			1.0 D		MIT
2573.534	2574.305	38845.44	W		1.2			MIT
2573.250	2574.021	38849.73	W			0.6 D		MIT
2573.146	2573.917	38851.30	W		0.8			MIT
2572.578	2573.349	38859.87	W			0.9		MIT
2572.442	2573.213	38861.93	W		1.0			MIT
2572.352	2573.123	38863.29	W	II	0.5	1.0		MIT
2572.235	2573.006	38865.05	W	II	0.3	1.1		MIT
2571.620	2572.390	38874.35	W		0.3	1.1		MIT
2571.445	2572.215	38876.99	W	II	1.2	1.5		MIT
2571.172	2571.942	38881.12	W		0.7			MIT
2570.212	2570.982	38895.64	W			0.3 H		MIT
2570.092	2570.862	38897.46	W		1.1	0.3		MIT
2569.979	2570.749	38899.17	W			1.1		MIT
2569.253	2570.023	38910.16	W	II	1.2	1.4		MIT
2568.980	2569.750	38914.29	W		1.0			MIT
2568.852	2569.622	38916.23	W		0.3	1.2		MIT

AIR WAVELENGTH (ANGSTRMS)	VACUUM WAVELENGTH (ANGSTRMS)	WAVENUMBER (KAYSERS)	LMENT	LOG ARC	INTENSITY SPK	DIS	REF
2568.557	2569.327	38920.70	W	1.1	0.3		MIT
2568.210	2568.980	38925.96	W	1.1	1.0		MIT
2568.108	2568.878	38927.51	W		1.0		MIT
2567.915	2568.685	38930.43	W		0.9		MIT
2567.607	2568.376	38935.10	W II	0.8	1.2		MIT
2567.498	2568.267	38936.75	W	1.1	0.0H		MIT
2566.315	2567.084	38954.70	W		1.1		MIT
2566.098	2566.867	38958.00	W	0.6			MIT
2565.789	2566.558	38962.69	W	0.7	0.9		MIT
2564.685	2565.454	38979.46	W	1.1	0.5		MIT
2564.437	2565.206	38983.23	W	0.0	0.6		MIT
2563.906	2564.675	38991.30	W	0.8	1.2		MIT
2563.162	2563.930	39002.62	W	0.9	1.5		MIT
2562.65	2563.42	39010.4	W		0.9L		MIT
2561.962	2562.730	39020.89	W	1.1	0.9		MIT
2561.514	2562.282	39027.71	W	0.9	0.0		MIT
2561.389	2562.157	39029.62	W		0.5		MIT
2560.748	2561.516	39039.38	W	0.9	0.6		MIT
2560.494	2561.262	39043.26	W	0.6			MIT
2560.119	2560.887	39048.98	W	1.2	0.9		MIT
2559.755	2560.523	39054.53	W		0.7		MIT
2559.494	2560.262	39058.51	W	0.5	1.2		MIT
2559.331	2560.098	39061.00	W	0.9	0.3		MIT
2558.896	2559.663	39067.64	W	0.3	0.6		MIT
2558.570	2559.337	39072.62	W		0.8		MIT
2558.023	2558.790	39080.97	W		0.8		MIT
2557.560	2558.327	39088.04	W I	1.1	0.8		ME
2557.008	2557.775	39096.48	W	0.9	0.7		MIT
2556.743	2557.510	39100.53	W	1.1	0.5		MIT
2556.270	2557.037	39107.77	W	1.0	0.5		MIT
2556.083	2556.850	39110.63	W		0.8		MIT
2555.205	2555.972	39124.07	W	1.0			MIT
2555.091	2555.857	39125.81	W II	1.0	1.2		MIT
2554.862	2555.628	39129.32	W II	1.2	1.0		MIT
2554.668	2555.434	39132.29	W	0.6	1.2		MIT
2554.087	2554.853	39141.19	W	0.8			MIT
2553.819	2554.585	39145.30	W	1.1	0.6		MIT
2553.598	2554.364	39148.69	W	1.0	0.5		MIT
2553.162	2553.928	39155.37	W	1.1	1.2		MIT
2552.481	2553.247	39165.82	W	1.0	0.0		MIT
2552.356	2553.122	39167.73	W	0.0	1.0		MIT
2552.241	2553.007	39169.50	W II	0.0	0.5		MIT
2552.113	2552.879	39171.46	W		0.8		MIT
2551.984	2552.750	39173.44	W	0.9			MIT
2551.450	2552.216	39181.64	W II		1.1		MIT
2551.347	2552.113	39183.22	W	1.1			MIT
2551.164	2551.930	39186.04	W		1.0		MIT
2550.995	2551.761	39188.63	W	1.0	0.8		MIT
2550.373	2551.138	39198.19	W	1.0	0.3		MIT
2550.328	2551.093	39198.88	W II		0.9		MIT
2550.104	2550.869	39202.32	W	0.8D	0.9		MIT
2549.092	2549.857	39217.88	W II	0.9	1.1		MIT
2548.682	2549.447	39224.19	W		0.5H		MIT
2548.567	2549.332	39225.96	W	1.1	0.3H		MIT
2548.381	2549.146	39228.83	W	0.0	1.0		MIT
2548.154	2548.919	39232.32	W	1.0	0.0		MIT
2547.838	2548.603	39237.19	W II		0.8		MIT
2547.625	2548.390	39240.47	W	1.0	0.3		MIT
2547.476	2548.241	39242.76	W	1.0	0.3		MIT
2547.136	2547.901	39248.00	W	1.3	1.0		MIT
2546.78	2547.54	39253.5	W	0.0	1.0		MIT
2546.653	2547.418	39255.44	W	0.9	0.0		MIT
2546.497	2547.261	39257.85	W		0.6		MIT
2546.284	2547.048	39261.13	W	0.3H	1.1		MIT
2545.570	2546.334	39272.14	W	0.9	0.5		MIT
2545.339	2546.103	39275.71	W	1.2	1.0		MIT
2544.81	2545.57	39283.9	W		1.0		MIT
2544.172	2544.936	39293.72	W	0.9	0.3		MIT
2543.999	2544.763	39296.39	W		0.7D		MIT
2543.438	2544.202	39305.06	W	1.1	0.3		MIT

AIR WAVELENGTH (ANGSTRMS)	VACUUM WAVELENGTH (ANGSTRMS)	WAVENUMBER (KAYSERS)	LMENT	LOG ARC	INTENSITY SPK	DIS	REF
2543.316	2544.080	39306.94	W		1.1		MIT
2542.604	2543.368	39317.95	W	0.0	1.2		MIT
2542.235	2542.998	39323.66	W	0.8	0.0		MIT
2541.692	2542.455	39332.06	W	1.0	0.6		MIT
2541.510	2542.273	39334.87	W		0.5		MIT
2541.056	2541.819	39341.90	W	0.0	0.8		MIT
2540.94	2541.70	39343.7	W	0.8	0.5		MIT
2540.422	2541.185	39351.72	W		0.7		MIT
2540.169	2540.932	39355.64	W	0.0	0.7		MIT
2539.902	2540.665	39359.77	W II	0.3	1.3		MIT
2539.612	2540.375	39364.27	W	0.9			MIT
2539.311	2540.074	39368.94	W	0.8	1.2		MIT
2538.347	2539.110	39383.88	W	1.0			MIT
2537.618	2538.380	39395.20	W	0.6			MIT
2537.145	2537.907	39402.54	W	0.0	1.2		MIT
2536.605	2537.367	39410.93	W	0.0	1.1		MIT
2535.988	2536.750	39420.52	W	0.6	1.1		MIT
2535.576	2536.338	39426.92	W		1.0		MIT
2535.332	2536.094	39430.72	W	0.9	0.0		MIT
2535.102	2535.864	39434.29	W	1.1			MIT
2534.825	2535.587	39438.60	W II	0.3	1.2		MIT
2534.676	2535.438	39440.92	W	1.1			MIT
2534.146	2534.908	39449.17	W	0.3	1.2		MIT
2533.982	2534.744	39451.72	W	1.0	0.0		MIT
2533.633	2534.394	39457.16	W	1.0	0.7		MIT
2533.286	2534.047	39462.56	W		0.9		MIT
2532.951	2533.712	39467.78	W II	0.3	1.1		MIT
2532.716	2533.477	39471.44	W		0.8		MIT
2532.41	2533.17	39476.2	W		1.0		MIT
2531.876	2532.637	39484.54	W		0.7		MIT
2531.444	2532.205	39491.27	W		0.5		MIT
2530.990	2531.751	39498.36	W	1.0	1.2		MIT
2530.700	2531.461	39502.88	W II	0.9	1.1		MIT
2529.724	2530.485	39518.12	W	1.1	0.8		MIT
2529.339	2530.099	39524.14	W	0.9	0.6		MIT
2529.216	2529.976	39526.06	W		0.8		MIT
2528.912	2529.672	39530.81	W	0.5	1.2		MIT
2528.163	2528.923	39542.52	W		0.8		MIT
2527.769	2528.529	39548.69	W	1.1	0.7		MIT
2527.556	2528.316	39552.02	W		0.9		MIT
2527.201	2527.961	39557.57	W II	0.8	1.0		MIT
2526.420	2527.180	39569.80	W	0.9	0.6		MIT
2526.209	2526.969	39573.11	W II	0.5	0.8		MIT
2525.683	2526.443	39581.35	W	1.1	0.5		MIT
2524.986	2525.745	39592.27	W	0.3	0.6		MIT
2524.702	2525.461	39596.73	W		0.5		MIT
2524.034	2524.793	39607.20	W		1.0		MIT
2523.589	2524.348	39614.19	W	0.3			MIT
2523.406	2524.165	39617.06	W	1.0	0.8		MIT
2522.752	2523.511	39627.33	W		0.9		MIT
2522.164	2522.923	39636.57	W	0.9			MIT
2522.043	2522.802	39638.47	W II	1.0	1.3		MIT
2521.522	2522.281	39646.66	W		0.5		MIT
2521.320	2522.079	39649.83	W	1.2	0.3		MIT
2521.146	2521.905	39652.57	W	0.0	0.8		MIT
2520.978	2521.736	39655.21	W	1.1	0.0		MIT
2520.790	2521.548	39658.17	W		0.5		MIT
2520.452	2521.210	39663.49	W	1.2	1.0		MIT
2520.190	2520.948	39667.61	W		0.5		MIT
2519.876	2520.634	39672.56	W	1.2	0.3		MIT
2519.441	2520.199	39679.40	W II	0.3	1.1		MIT
2519.161	2519.919	39683.81	W II	1.0	0.7		MIT
2518.702	2519.460	39691.05	W		1.0		MIT
2518.505	2519.263	39694.15	W	1.0	0.5		MIT
2518.142	2518.900	39699.87	W	0.8	1.2		MIT
2517.40	2518.16	39711.6	W		0.9		MIT
2517.239	2517.997	39714.11	W		0.6		MIT
2517.011	2517.769	39717.71	W		0.6		MIT
2516.577	2517.334	39724.56	W	1.1	0.5		MIT
2515.798	2516.555	39736.86	W II	0.0	1.0		MIT

AIR WAVELENGTH (ANGSTRMS)	VACUUM WAVELENGTH (ANGSTRMS)	WAVENUMBER (KAYSERS)	LMENT		LOG ARC	INTENSITY SPK	DIS	REF
2515.498	2516.255	39741.60	W		0.0	0.7		MIT
2515.330	2516.087	39744.25	W		0.3D	1.1		MIT
2514.518	2515.275	39757.08	W		0.3	1.1		MIT
2514.352	2515.109	39759.71	W	II		0.5H		MIT
2514.076	2514.833	39764.07	W			0.9		MIT
2513.934	2514.691	39766.32	W		1.0	0.3		MIT
2513.666	2514.423	39770.56	W		1.0			MIT
2513.450	2514.207	39773.98	W			1.2		MIT
2513.363	2514.120	39775.35	W		0.9			MIT
2512.927	2513.684	39782.25	W		1.0	1.1		MIT
2512.192	2512.948	39793.89	W			0.6		MIT
2511.828	2512.584	39799.66	W			0.9		MIT
2511.443	2512.199	39805.76	W			0.8		MIT
2511.254	2512.010	39808.75	W			0.7		MIT
2511.098	2511.854	39811.23	W		0.8	0.8		MIT
2510.472	2511.228	39821.15	W	II	0.5	1.2		MIT
2510.165	2510.921	39826.02	W		1.0	0.3		MIT
2509.958	2510.714	39829.31	W	II	0.0	1.2		MIT
2509.684	2510.440	39833.66	W			1.0		MIT
2509.383	2510.139	39838.44	W			1.1		MIT
2508.740	2509.496	39848.65	W	II	1.1	1.0		MIT
2508.441	2509.197	39853.39	W		1.0	0.3		MIT
2508.002	2508.757	39860.37	W		0.7	1.0		MIT
2507.586	2508.341	39866.98	W			0.9		MIT
2507.234	2507.989	39872.58	W			1.1		MIT
2506.148	2506.903	39889.86	W		0.9			MIT
2506.030	2506.785	39891.73	W		1.0	1.2		MIT
2505.818	2506.573	39895.11	W	II		0.5		MIT
2505.648	2506.403	39897.81	W		1.0	0.3		MIT
2505.487	2506.242	39900.38	W			1.0		MIT
2505.375	2506.130	39902.16	W	I	1.0	0.5		ME
2504.705	2505.460	39912.83	W		1.1	0.7		MIT
2504.527	2505.282	39915.67	W		1.0	0.0		MIT
2504.308	2505.063	39919.16	W		0.7D			MIT
2503.962	2504.717	39924.68	W		1.0	0.0		MIT
2503.612	2504.366	39930.26	W		0.5			MIT
2503.042	2503.796	39939.35	W		0.8	0.0		MIT
2502.818	2503.572	39942.92	W			0.6		MIT
2502.176	2502.930	39953.17	W		0.3	0.8		MIT
2502.077	2502.831	39954.75	W		0.3	0.8		MIT
2501.900	2502.654	39957.58	W	II	1.0	1.0		MIT
2501.781	2502.535	39959.48	W		1.0			MIT
2501.335	2502.089	39966.60	W		0.0	1.0		MIT
2501.036	2501.790	39971.38	W			0.5		MIT
2500.110	2500.864	39986.19	W		0.5	1.1		MIT
2499.941	2500.695	39988.89	W	II	0.0	0.3		MIT
2499.689	2500.443	39992.92	W	II	1.0	1.2		MIT
2499.438	2500.191	39996.94	W		1.0	0.0		MIT
2499.216	2499.969	40000.49	W		0.3	1.2		MIT
2498.26	2499.01	40015.8	W			0.9		MIT
2498.097	2498.850	40018.41	W			0.5		MIT
2497.484	2498.237	40028.23	W	II	1.0	1.2		MIT
2497.292	2498.045	40031.31	W		0.7	0.0		MIT
2496.638	2497.391	40041.79	W	II	1.0	1.3		MIT
2495.938	2496.691	40053.02	W		0.6			MIT
2495.722	2496.475	40056.49	W		0.6	1.0		MIT
2495.518	2496.271	40059.76	W			0.9		MIT
2495.264	2496.017	40063.84	W		1.3	0.7		MIT
2494.828	2495.580	40070.84	W	II	0.8	0.6		MIT
2493.541	2494.293	40091.52	W			0.7		MIT
2493.393	2494.145	40093.90	W		1.0	0.3		MIT
2493.159	2493.911	40097.66	W		0.8			MIT
2492.934	2493.686	40101.28	W	II	0.8	1.4		MIT
2492.367	2493.119	40110.40	W		1.1	0.9		MIT
2491.994	2492.746	40116.41	W		0.8	0.0		MIT
2491.768	2492.520	40120.04	W		0.0H	0.7		MIT
2491.183	2491.935	40129.46	W			0.7		MIT
2490.843	2491.594	40134.94	W		1.0			MIT
2490.710	2491.461	40137.08	W	II	0.5	0.7		MIT
2490.605	2491.356	40138.78	W			0.9		MIT

AIR WAVELENGTH (ANGSTRMS)	VACUUM WAVELENGTH (ANGSTRMS)	WAVENUMBER (KAYSERS)	LMENT		LOG ARC	INTENSITY SPK	DIS	REF
2490.492	2491.243	40140.60	W		0.9			MIT
2490.261	2491.012	40144.32	W			0.8		MIT
2489.915	2490.666	40149.90	W		0.3	0.8		MIT
2489.718	2490.469	40153.08	W		0.9	0.3		MIT
2489.510	2490.261	40156.43	W	II	0.3	0.9		MIT
2489.233	2489.984	40160.90	W	II	1.0	1.3		MIT
2488.910	2489.661	40166.11	W		1.0	0.5		MIT
2488.771	2489.522	40168.35	W		1.0	1.3		MIT
2488.118	2488.869	40178.90	W	II	0.3	1.1		MIT
2487.940	2488.691	40181.77	W		0.5			MIT
2487.766	2488.517	40184.58	W		1.0			MIT
2487.492	2488.243	40189.01	W		1.2	0.7		MIT
2487.228	2487.979	40193.27	W			0.6		MIT
2486.780	2487.531	40200.51	W			0.7		MIT
2486.425	2487.175	40206.25	W	II		1.0		MIT
2486.300	2487.050	40208.27	W		1.2	0.3H		MIT
2485.779	2486.529	40216.70	W			1.0		MIT
2485.473	2486.223	40221.65	W			0.8		MIT
2485.205	2485.955	40225.99	W		0.5	0.6D		MIT
2484.735	2485.485	40233.60	W		1.3	0.8		MIT
2484.398	2485.148	40239.05	W		0.3	1.2		MIT
2484.003	2484.753	40245.45	W	II	0.3	1.2		MIT
2483.726	2484.476	40249.94	W	II	0.3	0.8		MIT
2483.599	2484.349	40252.00	W			0.6		MIT
2483.457	2484.207	40254.30	W			0.6		MIT
2483.227	2483.977	40258.03	W		0.0	0.6		MIT
2482.69	2483.44	40266.7	W			0.5		MIT
2482.212	2482.961	40274.49	W		1.0	0.5		MIT
2482.098	2482.847	40276.34	W		1.0	0.5		MIT
2481.884	2482.633	40279.81	W			0.7		MIT
2481.54	2482.29	40285.4	W	II		1.0		MIT
2481.443	2482.192	40286.97	W		1.4	0.5		MIT
2480.949	2481.698	40294.99	W		1.3	0.5		MIT
2480.654	2481.403	40299.78	W		1.0	0.3		MIT
2480.126	2480.875	40308.36	W		1.4	1.0		MIT
2479.13	2479.88	40324.5	W			1.1		MIT
2478.872	2479.621	40328.75	W	II		0.9		MIT
2478.315	2479.064	40337.81	W			1.0		MIT
2477.796	2478.544	40346.26	W	II	1.2	1.5		MIT
2477.271	2478.019	40354.81	W			1.1		MIT
2476.810	2477.558	40362.32	W		0.7			MIT
2476.268	2477.016	40371.15	W			0.5		MIT
2476.017	2476.765	40375.25	W		0.8			MIT
2475.837	2476.585	40378.18	W			0.9		MIT
2475.586	2476.334	40382.28	W		0.0	1.3		MIT
2475.091	2475.839	40390.35	W		1.0	0.3		MIT
2474.484	2475.232	40400.26	W		0.8	0.5		MIT
2474.149	2474.897	40405.73	W		1.3	1.0		MIT
2473.954	2474.702	40408.91	W			0.3		MIT
2473.818	2474.566	40411.13	W		0.8			MIT
2473.692	2474.440	40413.19	W		0.6			MIT
2473.222	2473.969	40420.87	W		0.0	1.0		MIT
2472.990	2473.737	40424.66	W			0.7		MIT
2472.724	2473.471	40429.01	W			0.6		MIT
2472.508	2473.255	40432.54	W		1.2	0.7		MIT
2472.124	2472.871	40438.82	W		0.7			MIT
2471.744	2472.491	40445.04	W		0.0	1.2		MIT
2471.463	2472.210	40449.64	W			0.6		MIT
2471.209	2471.956	40453.80	W		1.0			MIT
2470.803	2471.550	40460.44	W	II	0.7D	1.3		MIT
2470.254	2471.001	40469.43	W			0.8		MIT
2469.884	2470.631	40475.50	W			0.7		MIT
2468.664	2469.410	40495.50	W			0.7		MIT
2468.409	2469.155	40499.68	W			1.3P		MIT
2468.084	2468.830	40505.01	W			1.0		MIT
2467.628	2468.374	40512.50	W		0.5			MIT
2466.848	2467.594	40525.31	W		1.2	0.8		MIT
2466.524	2467.270	40530.63	W	II	1.1	1.3		MIT
2466.334	2467.080	40533.75	W	II		0.7		MIT
2465.957	2466.703	40539.95	W			1.1		MIT

AIR WAVELENGTH (ANGSTRMS)	VACUUM WAVELENGTH (ANGSTRMS)	WAVENUMBER (KAYSERS)	LMENT	LOG ARC	INTENSITY SPK	DIS	REF
2465.641	2466.387	40545.14	W		1.3D		MIT
2465.204	2465.950	40552.33	W	1.1	1.3		MIT
2464.626	2465.371	40561.84	W II	0.5	1.3		MIT
2464.307	2465.052	40567.09	W	1.2	0.3		MIT
2464.128	2464.873	40570.04	W	0.5	0.7		MIT
2463.954	2464.699	40572.90	W	0.3	1.2		MIT
2463.688	2464.433	40577.28	W	0.3	0.0		MIT
2463.496	2464.241	40580.44	W		0.7		MIT
2462.989	2463.734	40588.80	W	0.5			MIT
2462.788	2463.533	40592.11	W	1.2	1.0		MIT
2462.478	2463.223	40597.22	W		0.7		MIT
2462.045	2462.790	40604.36	W		0.5		MIT
2461.572	2462.317	40612.16	W	1.1	0.8		MIT
2461.431	2462.176	40614.49	W		0.7		MIT
2461.144	2461.889	40619.22	W		1.1		MIT
2460.890	2461.635	40623.41	W	0.8			MIT
2460.394	2461.138	40631.60	W		0.9		MIT
2460.162	2460.906	40635.43	W	1.1	0.7		MIT
2459.878	2460.622	40640.13	W II		1.2		MIT
2459.607	2460.351	40644.60	W II	0.7	0.9		MIT
2459.295	2460.039	40649.76	W	1.3	1.0		MIT
2458.92	2459.66	40656.0	W		1.1		MIT
2458.566	2459.310	40661.81	W	0.0	1.2		MIT
2458.090	2458.834	40669.69	W		0.9		MIT
2457.18	2457.92	40684.7	W		1.1		MIT
2456.877	2457.621	40689.76	W		0.8		MIT
2456.531	2457.275	40695.49	W	1.2	0.7		MIT
2456.229	2456.972	40700.50	W	0.9			MIT
2456.079	2456.822	40702.98	W II		0.8		MIT
2455.856	2456.599	40706.68	W II		1.1		MIT
2455.501	2456.244	40712.56	W	1.1	1.0		MIT
2455.370	2456.113	40714.73	W	0.7			MIT
2454.971	2455.714	40721.35	W	1.2	1.0		MIT
2454.713	2455.456	40725.63	W	1.2	1.0		MIT
2453.722	2454.465	40742.08	W		1.1		MIT
2452.928	2453.671	40755.27	W		1.1		MIT
2452.576	2453.319	40761.11	W		0.3		MIT
2452.405	2453.148	40763.96	W		0.5		MIT
2451.998	2452.741	40770.72	W	1.1	1.2		MIT
2451.477	2452.219	40779.39	W II	1.2	1.3		MIT
2451.342	2452.084	40781.63	W	1.1			MIT
2451.032	2451.774	40786.79	W		1.0		MIT
2450.494	2451.236	40795.74	W	0.5			MIT
2450.384	2451.126	40797.57	W II		1.0		MIT
2449.692	2450.434	40809.10	W II	0.3H	1.2		MIT
2449.184	2449.926	40817.56	W		0.5		MIT
2448.65	2449.39	40826.5	W		1.1		MIT
2448.389	2449.131	40830.81	W	1.2	0.7		MIT
2448.217	2448.959	40833.68	W		1.2		MIT
2447.992	2448.734	40837.44	W	0.0	1.0		MIT
2447.374	2448.115	40847.75	W	0.9			MIT
2447.252	2447.993	40849.78	W	0.0	1.0		MIT
2446.788	2447.529	40857.53	W	0.6			MIT
2446.386	2447.127	40864.24	W II	1.0	1.5		MIT
2446.035	2446.776	40870.11	W		1.0		MIT
2445.917	2446.658	40872.08	W	0.7			MIT
2445.454	2446.195	40879.81	W	0.6	0.5		MIT
2445.246	2445.987	40883.29	W	0.7D	0.5		MIT
2444.936	2445.677	40888.48	W		1.2		MIT
2444.056	2444.797	40903.20	W	1.3	1.0		MIT
2443.615	2444.356	40910.58	W	1.2	0.9		MIT
2443.332	2444.073	40915.32	W	1.0	0.3		MIT
2443.172	2443.912	40918.00	W	0.7	0.6		MIT
2442.972	2443.712	40921.35	W	1.1	0.3		MIT
2441.60	2442.34	40944.3	W	0.3	1.0		MIT
2441.38	2442.12	40948.0	W		1.0D		MIT
2440.434	2441.174	40963.90	W II	0.3	1.2		MIT
2439.907	2440.647	40972.75	W		0.8		MIT
2439.842	2440.582	40973.84	W II		0.8D		MIT
2439.466	2440.206	40980.15	W		1.0		MIT

AIR WAVELENGTH (ANGSTRMS)	VACUUM WAVELENGTH (ANGSTRMS)	WAVENUMBER (KAYSERS)	LMENT	LOG ARC	INTENSITY SPK	DIS	REF
2439.204	2439.944	40984.55	W	0.6			MIT
2438.967	2439.707	40988.54	W	0.5			MIT
2438.843	2439.582	40990.62	W		0.7		MIT
2438.556	2439.295	40995.44	W		0.5		MIT
2438.353	2439.092	40998.86	W		0.8		MIT
2437.961	2438.700	41005.45	W	0.8			MIT
2437.484	2438.223	41013.47	W		1.0		MIT
2437.339	2438.078	41015.91	W		0.8		MIT
2436.835	2437.574	41024.40	W	0.8			MIT
2436.623	2437.362	41027.96	W	1.1	1.0		MIT
2436.258	2436.997	41034.11	W	0.9	0.3		MIT
2435.962	2436.701	41039.10	W	1.5	1.0		MIT
2435.426	2436.165	41048.13	W		1.1		MIT
2435.007	2435.746	41055.19	W II	0.5	1.5		MIT
2434.673	2435.412	41060.82	W	0.5			MIT
2434.459	2435.197	41064.43	W	0.0	1.0		MIT
2434.257	2434.995	41067.84	W	1.0	1.1		MIT
2433.984	2434.722	41072.44	W	1.1	1.0		MIT
2433.743	2434.481	41076.51	W	0.8	0.5		MIT
2433.538	2434.276	41079.97	W		0.5		MIT
2433.451	2434.189	41081.44	W	1.0			MIT
2433.146	2433.884	41086.59	W	0.3	1.0		MIT
2432.761	2433.499	41093.09	W	0.0	1.0		MIT
2431.712	2432.450	41110.82	W	0.3	1.0		MIT
2431.375	2432.113	41116.51	W II	0.0	1.0		MIT
2431.084	2431.822	41121.44	W	1.0	0.5		MIT
2430.770	2431.508	41126.75	W		1.0		MIT
2430.617	2431.355	41129.34	W	0.6			MIT
2430.438	2431.176	41132.37	W	1.0	0.3		MIT
2429.972	2430.709	41140.25	W II	0.3	1.0		MIT
2429.843	2430.580	41142.44	W	1.0	0.7H		MIT
2429.515	2430.252	41147.99	W II	0.0	1.0		MIT
2429.392	2430.129	41150.07	W		1.0		MIT
2428.773	2429.510	41160.56	W		1.0		MIT
2428.596	2429.333	41163.56	W	0.5			MIT
2428.173	2428.910	41170.73	W	0.7	1.1		MIT
2427.813	2428.550	41176.83	W II	0.0	0.8		MIT
2427.490	2428.227	41182.31	W II	1.0D	1.1		MIT
2427.287	2428.024	41185.76	W	1.0	0.0		MIT
2427.068	2427.805	41189.47	W		0.3		MIT
2426.514	2427.251	41198.88	W	0.0	1.3		MIT
2426.218	2426.955	41203.90	W		0.3		MIT
2426.074	2426.811	41206.35	W	0.8			MIT
2425.980	2426.717	41207.94	W	0.8D	0.8		MIT
2424.932	2425.668	41225.75	W		0.9		MIT
2424.770	2425.506	41228.51	W	0.8			MIT
2424.584	2425.320	41231.67	W		0.9H		MIT
2424.216	2424.952	41237.93	W	1.3	0.8		MIT
2423.941	2424.677	41242.60	W		1.2		MIT
2423.278	2424.014	41253.89	W		1.2		MIT
2422.659	2423.395	41264.43	W	0.8			MIT
2422.527	2423.263	41266.68	W	0.3	1.1		MIT
2422.285	2423.021	41270.80	W II	0.3	1.1		MIT
2421.670	2422.406	41281.28	W		1.0		MIT
2421.356	2422.091	41286.63	W		0.5		MIT
2421.010	2421.745	41292.53	W II	0.8	1.1		MIT
2420.495	2421.230	41301.32	W	0.3	1.2		MIT
2420.200	2420.935	41306.35	W	1.0			MIT
2420.005	2420.740	41309.68	W		0.8		MIT
2419.846	2420.581	41312.39	W		0.5		MIT
2419.345	2420.080	41320.95	W	0.3	1.3		MIT
2419.070	2419.805	41325.65	W		0.6		MIT
2418.252	2418.987	41339.62	W		1.2		MIT
2417.827	2418.562	41346.89	W		0.8		MIT
2417.604	2418.339	41350.70	W		1.2		MIT
2416.999	2417.733	41361.05	W		0.7J		MIT
2416.420	2417.154	41370.96	W		0.3H		MIT
2416.232	2416.966	41374.18	W	1.0	0.0		MIT
2416.051	2416.785	41377.28	W		1.0		MIT
2415.679	2416.413	41383.65	W	1.2	1.0		MIT

AIR WAVELENGTH (ANGSTRMS)	VACUUM WAVELENGTH (ANGSTRMS)	WAVENUMBER (KAYSERS)	LMENT		LOG ARC	INTENSITY SPK	INTENSITY DIS	REF
2415.381	2416.115	41388.76	W			1.0		MIT
2414.807	2415.541	41398.59	W			1.4		MIT
2414.131	2414.865	41410.19	W	II		0.6		MIT
2414.040	2414.774	41411.75	W		1.1	0.3		MIT
2413.778	2414.512	41416.24	W		1.0	0.5		MIT
2412.760	2413.494	41433.71	W		0.0	1.0		MIT
2412.342	2413.075	41440.89	W			0.7		MIT
2411.810	2412.543	41450.03	W	II	0.3	1.2		MIT
2411.545	2412.278	41454.59	W	II		1.1		MIT
2411.471	2412.204	41455.86	W		1.0D			MIT
2411.284	2412.017	41459.07	W	II	0.5	0.5H		MIT
2411.168	2411.901	41461.07	W			1.0		MIT
2410.889	2411.622	41465.87	W			0.3		MIT
2410.629	2411.362	41470.34	W		0.8			MIT
2410.442	2411.175	41473.56	W			1.3		MIT
2410.051	2410.784	41480.28	W			0.7		MIT
2409.826	2410.559	41484.16	W			0.7		MIT
2409.486	2410.219	41490.01	W			1.2		MIT
2409.229	2409.962	41494.44	W			1.2		MIT
2409.031	2409.764	41497.85	W		1.0	0.3		MIT
2408.670	2409.403	41504.06	W			0.6		MIT
2408.281	2409.014	41510.77	W		0.5	1.2		MIT
2407.785	2408.517	41519.32	W		0.0	1.2		MIT
2407.284	2408.016	41527.96	W			1.1		MIT
2406.745	2407.477	41537.26	W		0.5			MIT
2406.175	2406.907	41547.10	W		1.0	0.3		MIT
2405.993	2406.725	41550.24	W			0.9		MIT
2405.688	2406.420	41555.51	W		1.2	0.9		MIT
2405.256	2405.988	41562.97	W		1.0	1.2		MIT
2405.059	2405.791	41566.37	W			0.3		MIT
2404.243	2404.975	41580.48	W		1.0	1.2		MIT
2403.470	2404.201	41593.85	W			1.0		MIT
2403.223	2403.954	41598.13	W		0.5	1.1		MIT
2403.072	2403.803	41600.74	W	II	0.3H	1.0		MIT
2402.736	2403.467	41606.56	W		0.3	0.5		MIT
2402.441	2403.172	41611.67	W		1.2	1.3		MIT
2401.859	2402.590	41621.75	W	II	0.3H	1.2		MIT
2401.294	2402.025	41631.54	W		1.1	0.5		MIT
2400.861	2401.592	41639.05	W	II	0.3H	0.3H		MIT
2400.505	2401.236	41645.22	W		0.0	1.0		MIT
2400.336	2401.067	41648.16	W			0.8		MIT
2400.135	2400.866	41651.64	W		0.0	0.7		MIT
2399.58	2400.31	41661.3	W			0.5		MIT
2399.33	2400.06	41665.6	W			0.9		MIT
2399.04	2399.77	41670.6	W		1.0	0.3H		MIT
2398.67	2399.40	41677.1	W			0.5		MIT
2398.27	2399.00	41684.0	W		0.5			MIT
2398.12	2398.85	41686.6	W		0.6	0.6		MIT
2397.979	2398.709	41689.09	W		0.9	0.9		MIT
2397.723	2398.453	41693.54	W		1.1	0.5		MIT
2397.43	2398.16	41698.6	W		0.7			MIT
2397.091	2397.821	41704.53	W	II	1.3	1.5		MIT
2396.22	2396.95	41719.7	W		0.8	0.9		MIT
2395.89	2396.62	41725.4	W		0.8			MIT
2395.71	2396.44	41728.6	W	II	0.7	0.9		MIT
2395.466	2396.196	41732.82	W		0.9			MIT
2395.30	2396.03	41735.7	W		0.8			MIT
2395.10	2395.83	41739.2	W		0.7	1.1		MIT
2394.59	2395.32	41748.1	W		0.5	0.5		MIT
2394.45	2395.18	41750.5	W		0.8	0.7		MIT
2394.17	2394.90	41755.4	W		0.7	0.9		MIT
2393.77	2394.50	41762.4	W	II	0.6	0.6		MIT
2393.42	2394.15	41768.5	W		1.0	0.3H		MIT
2393.18	2393.91	41772.7	W			0.7		MIT
2393.06	2393.79	41774.8	W		0.6			MIT
2392.927	2393.656	41777.10	W	II	0.9	1.2		MIT
2392.52	2393.25	41784.2	W			0.3H		MIT
2391.89	2392.62	41795.2	W		0.6			MIT
2391.72	2392.45	41798.2	W			0.8		MIT
2391.30	2392.03	41805.5	W			0.7		MIT

AIR WAVELENGTH (ANGSTRMS)	VACUUM WAVELENGTH (ANGSTRMS)	WAVENUMBER (KAYSERS)	LMENT		LOG ARC	INTENSITY SPK	INTENSITY DIS	REF
2390.89	2391.62	41812.7	W	II	0.6	1.0		MIT
2390.367	2391.095	41821.83	W	II	1.0	1.3		MIT
2389.80	2390.53	41831.8	W			0.9		MIT
2389.26	2389.99	41841.2	W		0.6	0.3		MIT
2389.072	2389.800	41844.50	W		1.1	0.5		MIT
2388.80	2389.53	41849.3	W			0.7		MIT
2388.55	2389.28	41853.6	W			0.9		MIT
2387.84	2388.57	41866.1	W			0.5		MIT
2387.71	2388.44	41868.4	W			0.3H		MIT
2387.54	2388.27	41871.4	W			0.3H		MIT
2387.29	2388.02	41875.7	W			0.8		MIT
2386.45	2387.18	41890.5	W			0.9		MIT
2386.36	2387.09	41892.0	W		0.5			MIT
2386.17	2386.90	41895.4	W		0.9			MIT
2385.79	2386.52	41902.1	W			0.7		MIT
2385.49	2386.22	41907.3	W	II	0.5	1.0		MIT
2385.26	2385.99	41911.4	W		0.6	1.1		MIT
2384.817	2385.544	41919.16	W		1.1	1.1		MIT
2384.04	2384.77	41932.8	W			0.7		MIT
2383.54	2384.27	41941.6	W		0.3H	0.6		MIT
2383.20	2383.93	41947.6	W			0.5		MIT
2382.986	2383.713	41951.36	W		1.2	0.5		MIT
2382.67	2383.40	41956.9	W		0.5	1.1		MIT
2382.34	2383.07	41962.7	W	II	0.5	1.0		MIT
2381.79	2382.52	41972.4	W			1.0		MIT
2381.57	2382.30	41976.3	W		0.9			MIT
2381.33	2382.06	41980.5	W	II	0.5	1.0		MIT
2381.10	2381.83	41984.6	W		0.5	0.8		MIT
2380.73	2381.46	41991.1	W			0.9		MIT
2380.33	2381.06	41998.2	W		0.5	0.7		MIT
2380.17	2380.90	42001.0	W		0.5	0.5		MIT
2379.56	2380.29	42011.8	W			0.9		MIT
2379.04	2379.77	42020.9	W		0.3H			MIT
2378.60	2379.33	42028.7	W	II	0.7	1.1		MIT
2378.12	2378.85	42037.2	W	II	0.6	0.9		MIT
2377.92	2378.65	42040.7	W		0.6			MIT
2377.39	2378.12	42050.1	W		0.5	0.7		MIT
2377.18	2377.91	42053.8	W			0.8		MIT
2377.03	2377.76	42056.5	W		0.9			MIT
2376.562	2377.287	42064.75	W		0.9	0.3H		MIT
2376.39	2377.12	42067.8	W		0.6			MIT
2376.069	2376.794	42073.48	W		1.0	0.5		MIT
2375.74	2376.47	42079.3	W		0.6	1.0		MIT
2375.39	2376.12	42085.5	W			0.3H		MIT
2375.03	2375.75	42091.9	W			1.1		MIT
2374.758	2375.483	42096.70	W		1.0			MIT
2374.460	2375.185	42101.99	W		1.1	1.1		MIT
2374.26	2374.98	42105.5	W			0.3H		MIT
2374.144	2374.869	42107.59	W		1.1	0.3H		MIT
2373.43	2374.15	42120.3	W		0.6	0.6		MIT
2372.79	2373.51	42131.6	W		0.7			MIT
2372.58	2373.30	42135.3	W		0.5	1.1		MIT
2371.92	2372.64	42147.1	W			1.0		MIT
2371.85	2372.57	42148.3	W			0.9		MIT
2371.21	2371.93	42159.7	W			0.7		MIT
2371.05	2371.77	42162.5	W			0.7		MIT
2370.881	2371.605	42165.54	W		1.0	0.3		MIT
2370.60	2371.32	42170.5	W	II	0.3	1.3		MIT
2370.04	2370.76	42180.5	W		0.7	1.4		MIT
2369.74	2370.46	42185.8	W		0.5	0.8		MIT
2369.33	2370.05	42193.1	W			0.3H		MIT
2368.97	2369.69	42199.5	W			1.2		MIT
2368.35	2369.07	42210.6	W		0.6	1.0		MIT
2367.680	2368.403	42222.54	W		1.0	0.3H		MIT
2367.36	2368.08	42228.2	W		0.3H	0.6		MIT
2366.952	2367.675	42235.52	W		1.1	0.5		MIT
2366.67	2367.39	42240.6	W	II		0.6		MIT
2366.182	2366.905	42249.27	W		1.1	0.3H		MIT
2365.848	2366.571	42255.23	W		1.1	0.6		MIT
2365.448	2366.171	42262.37	W		1.1	0.7		MIT

AIR WAVELENGTH (ANGSTRMS)	VACUUM WAVELENGTH (ANGSTRMS)	WAVENUMBER (KAYSERS)	LMENT	LOG ARC	INTENSITY SPK	DIS	REF
2364.95	2365.67	42271.3	W	0.9			MIT
2364.22	2364.94	42284.3	W	0.7	1.2		MIT
2364.22	2364.94	42284.3	W	0.7	1.2		MIT
2363.89	2364.61	42290.2	W	0.6	1.1		MIT
2363.46	2364.18	42297.9	W II	0.7	0.9		MIT
2363.26	2363.98	42301.5	W		0.3H		MIT
2363.062	2363.784	42305.04	W	1.2	0.7		MIT
2362.51	2363.23	42314.9	W		0.9		MIT
2362.10	2362.82	42322.3	W		0.7		MIT
2361.62	2362.34	42330.9	W	0.8	0.5		MIT
2361.19	2361.91	42338.6	W II	0.8	0.9		MIT
2360.433	2361.155	42352.16	W	1.2	0.7		MIT
2360.12	2360.84	42357.8	W		0.3H		MIT
2359.51	2360.23	42368.7	W		0.5		MIT
2359.21	2359.93	42374.1	W		0.6		MIT
2358.81	2359.53	42381.3	W	1.0	1.3		MIT
2358.072	2358.793	42394.56	W	1.0			MIT
2357.93	2358.65	42397.1	W		0.5		MIT
2357.65	2358.37	42402.1	W	0.7			MIT
2357.46	2358.18	42405.6	W		0.7		MIT
2357.27	2357.99	42409.0	W		0.7		MIT
2356.80	2357.52	42417.4	W	0.6			MIT
2356.46	2357.18	42423.6	W		0.5		MIT
2355.96	2356.68	42432.6	W		0.5		MIT
2355.79	2356.51	42435.6	W		0.5		MIT
2355.32	2356.04	42444.1	W		0.7		MIT
2354.611	2355.331	42456.87	W	1.1	0.7		MIT
2353.81	2354.53	42471.3	W	0.5			MIT
2353.69	2354.41	42473.5	W	0.3H	0.7		MIT
2353.454	2354.174	42477.74	W		0.6		MIT
2353.36	2354.08	42479.4	W	0.5	0.7		MIT
2352.96	2353.68	42486.7	W	0.9	0.9		MIT
2352.78	2353.50	42489.9	W		0.7		MIT
2352.44	2353.16	42496.0	W		0.3H		MIT
2352.04	2352.76	42503.3	W		0.3H		MIT
2351.78	2352.50	42508.0	W	0.7			MIT
2351.466	2352.186	42513.65	W		0.9		MIT
2351.05	2351.77	42521.2	W II		0.7		MIT
2350.86	2351.58	42524.6	W		0.3H		MIT
2350.536	2351.256	42530.47	W	1.1			MIT
2350.37	2351.09	42533.5	W		1.1		MIT
2349.82	2350.54	42543.4	W	0.8	1.0		MIT
2349.32	2350.04	42552.5	W	0.9	1.1		MIT
2348.56	2349.28	42566.2	W	0.3H			MIT
2348.151	2348.870	42573.66	W	0.9			MIT
2347.967	2348.686	42577.00	W	1.0	1.0		MIT
2347.64	2348.36	42582.9	W II		0.5		MIT
2346.87	2347.59	42596.9	W		0.5		MIT
2346.69	2347.41	42600.2	W	0.7			MIT
2346.63	2347.35	42601.3	W		0.8		MIT
2346.31	2347.03	42607.1	W	0.6			MIT
2346.00	2346.72	42612.7	W	0.8			MIT
2345.69	2346.41	42618.3	W		1.1		MIT
2344.91	2345.63	42632.5	W	1.0			MIT
2344.84	2345.56	42633.8	W		0.9		MIT
2343.744	2344.462	42653.71	W	0.7			MIT
2343.13	2343.85	42664.9	W	0.9	0.5		MIT
2342.72	2343.44	42672.4	W		0.5		MIT
2342.47	2343.19	42676.9	W	0.9	0.3H		MIT
2342.12	2342.84	42683.3	W	0.5			MIT
2341.92	2342.64	42686.9	W II		0.7		MIT
2341.81	2342.53	42688.9	W	0.9	0.8		MIT
2341.374	2342.091	42696.88	W II	1.1	1.1		MIT
2341.07	2341.79	42702.4	W		0.7		MIT
2340.64	2341.36	42710.3	W		0.6		MIT
2339.90	2340.62	42723.8	W II		0.9		MIT
2339.72	2340.44	42727.1	W II	0.3H	0.7		MIT
2339.33	2340.05	42734.2	W	0.3H			MIT
2339.16	2339.88	42737.3	W II	0.5	1.0		MIT
2338.476	2339.193	42749.79	W	1.0			MIT
2338.14	2338.86	42755.9	W	0.5			MIT
2337.743	2338.460	42763.19	W II	0.9	1.3		MIT
2337.17	2337.89	42773.7	W	0.9			MIT
2336.70	2337.42	42782.3	W	0.8	1.0		MIT
2336.57	2337.29	42784.7	W		0.3H		MIT
2336.35	2337.07	42788.7	W		0.3H		MIT
2335.93	2336.65	42796.4	W		0.8		MIT
2335.56	2336.28	42803.2	W		0.5		MIT
2335.20	2335.92	42809.8	W II	0.8	0.9		MIT
2334.30	2335.02	42826.3	W	0.6			MIT
2334.06	2334.78	42830.7	W II	0.7	0.3H		MIT
2333.768	2334.484	42836.02	W II	0.9	1.0		MIT
2333.14	2333.86	42847.5	W	0.8	0.7		MIT
2332.50	2333.22	42859.3	W	0.7			MIT
2332.02	2332.74	42868.1	W		0.8		MIT
2331.92	2332.64	42870.0	W	0.9	0.7		MIT
2331.59	2332.31	42876.0	W		0.5		MIT
2331.303	2332.018	42881.31	W	1.2			MIT
2330.90	2331.62	42888.7	W		0.5		MIT
2330.67	2331.39	42893.0	W		0.5		MIT
2329.88	2330.59	42907.5	W	0.3	0.5		MIT
2329.68	2330.39	42911.2	W II	0.5	1.0		MIT
2329.29	2330.00	42918.4	W	0.6			MIT
2328.72	2329.43	42928.9	W		0.3H		MIT
2328.312	2329.027	42936.39	W II	1.0	1.1		MIT
2326.71	2327.42	42966.0	W	1.0			MIT
2326.562	2327.276	42968.69	W	1.0	0.5		MIT
2326.086	2326.800	42977.48	W II	1.0	1.2		MIT
2325.77	2326.48	42983.3	W	0.5	0.3H		MIT
2324.74	2325.45	43002.4	W II	0.3	0.8		MIT
2324.194	2324.908	43012.46	W	0.3H	0.6J		MIT
2323.03	2323.74	43034.0	W II	0.7	1.1W		MIT
2322.72	2323.43	43039.7	W		0.6		MIT
2321.634	2322.347	43059.89	W	1.1	0.5		MIT
2321.35	2322.06	43065.1	W		0.3H		MIT
2321.24	2321.95	43067.2	W	0.7			MIT
2321.04	2321.75	43070.9	W		1.0		MIT
2320.83	2321.54	43074.8	W		0.5		MIT
2320.18	2320.89	43086.9	W		0.6		MIT
2319.76	2320.47	43094.7	W		0.6		MIT
2318.94	2319.65	43109.9	W	1.0	0.6		MIT
2318.58	2319.29	43116.6	W	0.9			MIT
2317.55	2318.26	43135.8	W	0.3H	0.3H		MIT
2317.39	2318.10	43138.7	W	0.3	0.3		MIT
2316.71	2317.42	43151.4	W		0.3H		MIT
2316.26	2316.97	43159.8	W		0.3H		MIT
2315.020	2315.732	43182.90	W II	1.1	1.1		MIT
2314.63	2315.34	43190.2	W	0.3	0.8		MIT
2314.18	2314.89	43198.6	W	1.0	0.5		MIT
2313.188	2313.899	43217.09	W	0.9			MIT
2312.91	2313.62	43222.3	W		0.7		MIT
2312.36	2313.07	43232.6	W	0.3H			MIT
2311.85	2312.56	43242.1	W		0.5		MIT
2311.65	2312.36	43245.8	W		0.5		MIT
2310.82	2311.53	43261.4	W	0.6	0.8		MIT
2310.25	2310.96	43272.0	W		0.7		MIT
2309.84	2310.55	43279.7	W	0.5	0.8		MIT
2309.036	2309.746	43294.80	W	1.1	0.3H		MIT
2307.93	2308.64	43315.5	W II		0.7		MIT
2307.66	2308.37	43320.6	W	0.3H	0.7		MIT
2306.92	2307.63	43334.5	W II	0.6	1.1		MIT
2306.60	2307.31	43340.5	W	1.1			MIT
2305.936	2306.646	43353.00	W		0.6		MIT
2305.223	2305.932	43366.40	W		0.5		MIT
2305.053	2305.762	43369.60	W	0.8	0.7		MIT
2303.826	2304.535	43392.70	W II	1.0	1.2		MIT
2303.278	2303.987	43403.02	W	0.6	1.1		MIT
2303.108	2303.817	43406.23	W	1.0	0.5		MIT
2302.67	2303.38	43414.5	W	0.9			MIT
2302.31	2303.02	43421.3	W	0.3	0.5		MIT

AIR WAVELENGTH (ANGSTRMS)	VACUUM WAVELENGTH (ANGSTRMS)	WAVENUMBER (KAYSERS)	LMENT	LOG ARC	INTENSITY SPK	INTENSITY DIS	REF
2301.88	2302.59	43429.4	W		0.5		MIT
2301.65	2302.36	43433.7	W	0.9	1.0		MIT
2301.29	2302.00	43440.5	W II		0.6		MIT
2300.576	2301.284	43453.99	W		0.7		MIT
2300.078	2300.786	43463.40	W II		0.6		MIT
2299.807	2300.515	43468.52	W		1.0		MIT
2299.02	2299.73	43483.4	W	0.6	0.5		A
2298.74	2299.45	43488.7	W	0.9			MIT
2298.656	2299.364	43490.29	W		0.9		MIT
2298.340	2299.048	43496.27	W	0.5			MIT
2298.28	2298.99	43497.4	W	1.0	1.1		MIT
2297.94	2298.65	43503.8	W		0.9		MIT
2297.38	2298.09	43514.4	W	0.6	0.5		MIT
2296.97	2297.68	43522.2	W II	0.5	0.8		MIT
2295.99	2296.70	43540.8	W	0.5	1.0		MIT
2295.787	2296.494	43544.63	W	0.6	0.9		MIT
2295.507	2296.214	43549.94	W		0.5		MIT
2294.85	2295.56	43562.4	W	0.7	1.0		MIT
2294.544	2295.251	43568.22	W II	1.3	1.3		MIT
2293.381	2294.088	43590.31	W		0.8		MIT
2292.393	2293.100	43609.10	W		0.5		MIT
2291.368	2292.074	43628.60	W		0.6		MIT
2290.946	2291.652	43636.64	W	0.7			MIT
2290.560	2291.266	43643.99	W II	0.3	1.0		MIT
2289.98	2290.69	43655.0	W		0.5		MIT
2289.39	2290.10	43666.3	W	0.3H	0.6		MIT
2288.68	2289.39	43679.8	W	0.8	0.6		MIT
2288.524	2289.230	43682.81	W		0.8		MIT
2287.67	2288.38	43699.1	W	1.0			MIT
2287.47	2288.18	43702.9	W		0.5		MIT
2287.19	2287.90	43708.3	W		0.5		MIT
2286.50	2287.21	43721.5	W		0.5		MIT
2286.29	2287.00	43725.5	W	0.3	0.7		MIT
2285.90	2286.61	43733.0	W	0.6			MIT
2285.73	2286.44	43736.2	W		0.5		MIT
2285.17	2285.88	43746.9	W	1.0	0.5		MIT
2284.90	2285.61	43752.1	W	0.5			MIT
2284.62	2285.32	43757.5	W		0.9		MIT
2284.44	2285.14	43760.9	W	0.6	0.7		MIT
2283.99	2284.69	43769.5	W		0.7		MIT
2283.38	2284.08	43781.2	W		0.5		MIT
2283.27	2283.97	43783.3	W	1.0	0.7		MIT
2282.50	2283.20	43798.1	W	0.8	0.6		MIT
2282.20	2282.90	43803.9	W II	0.6	1.1		MIT
2281.67	2282.37	43814.0	W		0.6		MIT
2281.04	2281.74	43826.1	W	0.6	0.7		MIT
2280.83	2281.53	43830.2	W	0.6	0.6		MIT
2280.632	2281.336	43833.96	W	0.8	0.8		MIT
2280.32	2281.02	43840.0	W		0.9		MIT
2279.97	2280.67	43846.7	W		0.6		MIT
2279.20	2279.90	43861.5	W		0.5		MIT
2278.91	2279.61	43867.1	W		0.7		MIT
2278.70	2279.40	43871.1	W		0.7		MIT
2278.11	2278.81	43882.5	W	0.9	0.9		MIT
2277.96	2278.66	43885.4	W	0.3H	0.9		MIT
2277.58	2278.28	43892.7	W	1.3	1.0		MIT
2277.43	2278.13	43895.6	W		0.5		MIT
2276.80	2277.50	43907.7	W		0.6		MIT
2276.48	2277.18	43913.9	W		0.6		MIT
2275.94	2276.64	43924.3	W		0.5		MIT
2275.63	2276.33	43930.3	W	0.8	0.5		MIT
2275.50	2276.20	43932.8	W		0.7		MIT
2274.40	2275.10	43954.1	W	0.8			MIT
2273.76	2274.46	43966.4	W	1.0			MIT
2273.00	2273.70	43981.1	W	1.0	0.7		MIT
2272.51	2273.21	43990.6	W		0.8		MIT
2272.12	2272.82	43998.2	W	0.5			MIT
2271.117	2271.819	44017.59	W II	0.5	0.8		MIT
2270.239	2270.941	44034.61	W II	1.1	1.3		MIT
2269.792	2270.494	44043.28	W	0.5	0.5		MIT

AIR WAVELENGTH (ANGSTRMS)	VACUUM WAVELENGTH (ANGSTRMS)	WAVENUMBER (KAYSERS)	LMENT	LOG ARC	INTENSITY SPK	INTENSITY DIS	REF
2269.449	2270.151	44049.94	W		0.5		MIT
2269.085	2269.787	44057.00	W		0.5		MIT
2268.84	2269.54	44061.8	W		0.3H		MIT
2268.09	2268.79	44076.3	W		0.3		MIT
2267.600	2268.301	44085.85	W		1.0		MIT
2267.29	2267.99	44091.9	W		0.6		MIT
2266.41	2267.11	44109.0	W		0.5		MIT
2266.26	2266.96	44111.9	W II	0.8	0.8		MIT
2266.130	2266.831	44114.45	W II	0.9	0.9		MIT
2265.98	2266.68	44117.4	W	0.5			MIT
2265.666	2266.367	44123.48	W	0.9	0.8		MIT
2265.350	2266.051	44129.64	W II	0.7	1.0		MIT
2264.182	2264.883	44152.40	W II	0.5	1.1		MIT
2263.985	2264.685	44156.24	W		0.8		MIT
2263.535	2264.235	44165.02	W	0.8	1.4		MIT
2263.27	2263.97	44170.2	W	0.5			MIT
2262.32	2263.02	44188.7	W		0.9		MIT
2261.940	2262.640	44196.16	W		0.8		MIT
2260.58	2261.28	44222.7	W		0.7		MIT
2260.07	2260.77	44232.7	W	1.2	0.5		MIT
2259.690	2260.390	44240.16	W II		0.7		MIT
2259.555	2260.255	44242.81	W	1.1	0.8		MIT
2259.068	2259.767	44252.34	W II	0.5	1.0		MIT
2258.36	2259.06	44266.2	W		0.5		MIT
2258.128	2258.827	44270.76	W	1.0	0.9		MIT
2257.741	2258.440	44278.35	W		0.8		MIT
2257.100	2257.799	44290.92	W		1.0		MIT
2256.850	2257.549	44295.83	W	1.0	1.1		MIT
2256.21	2256.91	44308.4	W II	0.8	0.6		MIT
2256.018	2256.717	44312.16	W		1.0		MIT
2255.72	2256.42	44318.0	W II	0.3H			MIT
2255.52	2256.22	44322.0	W	0.5			MIT
2255.103	2255.802	44330.14	W		0.8		MIT
2254.987	2255.686	44332.42	W		1.0		MIT
2254.57	2255.27	44340.6	W	1.0	0.6		MIT
2253.91	2254.61	44353.6	W		1.0		MIT
2253.51	2254.21	44361.5	W		0.5		MIT
2253.02	2253.72	44371.1	W		0.5		MIT
2252.36	2253.06	44384.1	W	1.0			MIT
2252.28	2252.98	44385.7	W		0.5		MIT
2251.93	2252.63	44392.6	W		1.0		MIT
2251.42	2252.12	44402.6	W II	0.9	1.0		MIT
2251.13	2251.83	44408.4	W II	1.0	1.0		MIT
2250.73	2251.43	44416.3	W	1.0	1.1		MIT
2250.45	2251.15	44421.8	W		0.5		MIT
2249.84	2250.54	44433.8	W	1.0			A
2249.50	2250.20	44440.5	W	1.0			A
2249.39	2250.09	44442.7	W II	0.3H	0.5H		MIT
2248.750	2249.447	44455.37	W II	1.3	1.4		MIT
2248.26	2248.96	44465.0	W II	1.1	1.2		MIT
2247.67	2248.37	44476.7	W	0.5	1.1		MIT
2247.32	2248.02	44483.6	W		0.6		MIT
2246.98	2247.68	44490.4	W		0.7		MIT
2246.63	2247.33	44497.3	W II	0.8	1.1		MIT
2246.36	2247.06	44502.7	W	0.6	0.8		MIT
2245.21	2245.91	44525.5	W II	1.0	1.3		MIT
2244.76	2245.46	44534.4	W		0.5		MIT
2244.44	2245.14	44540.7	W II		0.5		MIT
2244.16	2244.86	44546.3	W	0.3H	1.0		MIT
2243.50	2244.20	44559.4	W		1.0		MIT
2242.96	2243.66	44570.1	W	0.5	0.9		MIT
2242.71	2243.41	44575.1	W		0.9		MIT
2242.06	2242.76	44588.0	W	1.1	1.0		MIT
2241.28	2241.98	44603.5	W	1.0	0.3H		A
2241.08	2241.78	44607.5	W	1.0	1.2		MIT
2240.85	2241.55	44612.1	W	0.6	0.7		MIT
2239.78	2240.48	44633.4	W		1.3		MIT
2239.28	2239.98	44643.4	W	0.5	0.7		MIT
2238.56	2239.26	44657.7	W	0.8	0.8		MIT
2237.06	2237.75	44687.6	W II	1.0	1.2		MIT

AIR WAVELENGTH (ANGSTRMS)	VACUUM WAVELENGTH (ANGSTRMS)	WAVENUMBER (KAYSERS)	LMENT	LOG ARC	INTENSITY SPK	DIS	REF
2235.87	2236.56	44711.4	W II	0.3H	0.5		MIT
2235.63	2236.32	44716.2	W II	1.0	1.0		MIT
2235.37	2236.06	44721.4	W		1.1		MIT
2234.85	2235.54	44731.8	W		0.6		MIT
2234.616	2235.310	44736.52	W	1.0	0.7		MIT
2234.256	2234.950	44743.73	W		1.0		MIT
2233.02	2233.71	44768.5	W		0.6		MIT
2232.56	2233.25	44777.7	W		0.8		MIT
2232.27	2232.96	44783.5	W II	0.8	0.3H		MIT
2231.956	2232.650	44789.83	W		0.8		MIT
2231.24	2231.93	44804.2	W		1.0		MIT
2231.080	2231.773	44807.42	W	0.5	0.8		MIT
2230.42	2231.11	44820.7	W		0.8		MIT
2229.626	2230.319	44836.63	W II	1.0	1.3		MIT
2228.92	2229.61	44850.8	W		0.9		MIT
2228.70	2229.39	44855.3	W II	0.7	0.3W		LN
2228.11	2228.80	44867.1	W	1.0	0.0		MIT
2227.98	2228.67	44869.7	W	1.0			A
2227.331	2228.024	44882.83	W	0.6	0.8		MIT
2226.793	2227.486	44893.67	W	0.6	1.0		MIT
2226.572	2227.264	44898.13	W	0.7	0.6		MIT
2226.34	2227.03	44902.8	W II	0.5	0.8		MIT
2225.893	2226.585	44911.82	W II	1.0	1.3		MIT
2225.54	2226.23	44918.9	W	0.9	0.3		A
2225.226	2225.918	44925.28	W	0.3	0.6		MIT
2224.675	2225.367	44936.41	W		0.8		MIT
2224.41	2225.10	44941.8	W		0.6		MIT
2224.09	2224.78	44948.2	W		0.8		MIT
2222.78	2223.47	44974.7	W		0.5		MIT
2222.590	2223.282	44978.56	W	0.3H	0.8		MIT
2222.07	2222.76	44989.1	W	0.5	1.0		MIT
2221.85	2222.54	44993.5	W	1.1			MIT
2221.53	2222.22	45000.0	W	0.3H	0.6		MIT
2221.41	2222.10	45002.5	W		0.5		MIT
2220.942	2221.633	45011.93	W II	0.9	1.3		MIT
2220.01	2220.70	45030.8	W		0.6		MIT
2219.72	2220.41	45036.7	W II	1.2	0.6		MIT
2219.35	2220.04	45044.2	W		0.7		MIT
2218.33	2219.02	45064.9	W	1.0	0.7		MIT
2217.94	2218.63	45072.9	W		0.3H		MIT
2217.762	2218.453	45076.46	W	0.8	0.8		MIT
2216.43	2217.12	45103.5	W	0.9	1.1		MIT
2216.009	2216.699	45112.12	W II	1.0	1.2		MIT
2215.342	2216.032	45125.70	W	0.8	1.2W		MIT
2214.80	2215.49	45136.7	W	1.0	1.3		MIT
2214.31	2215.00	45146.7	W	0.6			A
2213.19	2213.88	45169.6	W		0.6		MIT
2212.17	2212.86	45190.4	W	1.0			MIT
2211.03	2211.72	45213.7	W		0.7		MIT
2210.22	2210.91	45230.3	W		0.5		MIT
2209.06	2209.75	45254.0	W	0.8	0.3H		MIT
2207.99	2208.68	45275.9	W	0.8	1.0		MIT
2207.39	2208.08	45288.2	W	0.5			MIT
2206.94	2207.63	45297.5	W	0.5	1.0		MIT
2206.589	2207.277	45304.69	W II	1.0	1.2		MIT
2206.08	2206.77	45315.1	W	0.8	1.0		MIT
2205.49	2206.18	45327.3	W		0.8		MIT
2204.48	2205.17	45348.0	W II	1.1	1.5		MIT
2204.02	2204.71	45357.5	W	0.5	1.0		MIT
2203.79	2204.48	45362.2	W II	0.5	1.0		MIT
2201.97	2202.66	45399.7	W	0.3H	1.0		MIT
2200.92	2201.61	45421.4	W		0.9		MIT
2200.27	2200.96	45434.8	W		0.5		MIT
2199.40	2200.09	45452.7	W		0.5		MIT
2199.15	2199.84	45457.9	W II	0.5	0.3		MIT
2198.67	2199.36	45467.8	W II	0.8	1.1		MIT
2198.37	2199.06	45474.0	W	1.0	0.5		MIT
2198.02	2198.71	45481.3	W II	0.5	0.5		MIT
2197.85	2198.54	45484.8	W		0.6		MIT
2197.509	2198.195	45491.86	W II	0.5	0.7		MIT
2197.08	2197.77	45500.7	W	0.9	0.5		A
2196.78	2197.47	45507.0	W	0.5	0.8		MIT
2195.82	2196.51	45526.9	W	0.3H	1.0		MIT
2194.95	2195.64	45544.9	W		0.8		MIT
2194.52	2195.21	45553.8	W	1.0	1.2		MIT
2193.53	2194.22	45574.4	W II		0.9J		MIT
2193.03	2193.72	45584.8	W	1.0	1.1		MIT
2192.40	2193.09	45597.9	W	0.5	0.5		MIT
2192.099	2192.784	45604.12	W		0.8		MIT
2191.35	2192.04	45619.7	W		0.9		MIT
2190.95	2191.63	45628.0	W		0.5		MIT
2190.786	2191.471	45631.45	W		0.3		MIT
2189.85	2190.53	45651.0	W	0.8	1.1		MIT
2189.51	2190.19	45658.0	W II	0.9	0.9		MIT
2189.353	2190.038	45661.32	W II	0.7	0.7		MIT
2189.197	2189.882	45664.57	W	0.6	0.6		MIT
2188.642	2189.326	45676.15	W		0.3		MIT
2188.367	2189.051	45681.89	W		0.7		MIT
2187.83	2188.51	45693.1	W II		0.5		MIT
2187.630	2188.314	45697.27	W		0.7		MIT
2186.733	2187.417	45716.02	W II	0.9	1.2		MIT
2185.757	2186.441	45736.43	W	0.5	1.1		MIT
2185.413	2186.097	45743.63	W		1.0		MIT
2184.95	2185.63	45753.3	W		0.5J		MIT
2184.235	2184.919	45768.30	W	1.1	0.9		MIT
2184.179	2184.863	45769.47	W	0.6	0.8		MIT
2183.790	2184.473	45777.62	W		0.8		MIT
2183.315	2183.998	45787.58	W	0.5	1.0		MIT
2182.096	2182.779	45813.15	W		0.8		MIT
2181.88	2182.56	45817.7	W		0.7		MIT
2180.945	2181.628	45837.33	W		0.9		MIT
2180.688	2181.371	45842.73	W II	1.0	0.6		MIT
2180.475	2181.158	45847.21	W	0.7	1.0		MIT
2179.631	2180.314	45864.96	W II	0.9	0.9		MIT
2178.812	2179.494	45882.20	W		1.0		MIT
2178.72	2179.40	45884.1	W		1.1		MIT
2178.22	2178.90	45894.7	W II		0.3H		MIT
2177.547	2178.229	45908.85	W II	0.6	1.1		MIT
2177.12	2177.80	45917.9	W II	0.5	0.5J		MIT
2176.874	2177.556	45923.04	W	0.6	1.0		MIT
2176.419	2177.101	45932.64	W	0.3	1.1		MIT
2175.51	2176.19	45951.8	W II	0.8	0.5		MIT
2175.360	2176.042	45955.00	W		0.9		MIT
2174.822	2175.504	45966.37	W		0.9		MIT
2174.47	2175.15	45973.8	W		0.8		MIT
2173.822	2174.503	45987.51	W II	0.5	1.0		MIT
2173.543	2174.224	45993.41	W II	1.0	1.2		MIT
2172.89	2173.57	46007.2	W		0.9W		MIT
2172.195	2172.876	46021.95	W		0.9		MIT
2171.768	2172.449	46031.00	W	0.5	0.9		MIT
2171.539	2172.220	46035.85	W		0.8		MIT
2169.941	2170.622	46069.75	W II	1.0	1.1		MIT
2169.48	2170.16	46079.5	W	1.2	0.7		A
2168.591	2169.271	46098.43	W	0.3H	0.9		MIT
2168.298	2168.978	46104.66	W	1.0	0.9		MIT
2167.687	2168.367	46117.65	W		0.6		MIT
2167.19	2167.87	46128.2	W II	0.5	0.9		MIT
2166.319	2166.999	46146.77	W II	1.0	1.5		MIT
2165.27	2165.95	46169.1	W		0.8		MIT
2164.812	2165.492	46178.89	W		0.9		MIT
2164.43	2165.11	46187.0	W		0.7		MIT
2163.895	2164.574	46198.46	W	1.1	1.4		MIT
2162.362	2163.041	46231.21	W		0.8		MIT
2160.91	2161.59	46262.3	W		1.5		MIT
2160.48	2161.16	46271.5	W		1.0		MIT
2159.95	2160.63	46282.8	W II	0.8	1.0		MIT
2159.48	2160.16	46292.9	W		0.9		MIT
2157.425	2158.103	46336.99	W		0.9		MIT
2156.688	2157.366	46352.82	W		0.8		MIT
2156.42	2157.10	46358.6	W	0.7	1.4		MIT

AIR WAVELENGTH (ANGSTRMS)	VACUUM WAVELENGTH (ANGSTRMS)	WAVENUMBER (KAYSERS)	LMENT		LOG ARC	INTENSITY SPK	DIS	REF
2155.26	2155.94	46383.5	W	II		0.3		MIT
2153.886	2154.563	46413.12	W		0.7	1.1		MIT
2153.559	2154.236	46420.16	W		1.0	1.1		MIT
2153.13	2153.81	46429.4	W	II		0.9W		MIT
2152.551	2153.228	46441.90	W			0.7		MIT
2152.140	2152.817	46450.77	W		0.5	1.3		MIT
2149.147	2149.823	46515.45	W		0.5	1.0		MIT
2148.850	2149.526	46521.88	W			0.9		MIT
2147.984	2148.660	46540.63	W			0.6		MIT
2146.90	2147.58	46564.1	W			0.9		MIT
2146.361	2147.037	46575.82	W			0.6		MIT
2146.167	2146.843	46580.03	W		0.6	0.9		MIT
2145.777	2146.453	46588.50	W			1.1		MIT
2145.39	2146.07	46596.9	W			0.5		MIT
2144.50	2145.18	46616.2	W		0.9	1.0		MIT
2144.090	2144.765	46625.15	W		1.0	1.0		MIT
2142.511	2143.186	46659.50	W		0.5	0.9		MIT
2140.06	2140.73	46712.9	W			0.9		MIT
2139.64	2140.31	46722.1	W	II	1.1			MIT
2139.30	2139.97	46729.5	W			0.7J		MIT
2139.16	2139.83	46732.6	W	II	0.8	1.1		MIT
2138.15	2138.82	46754.7	W	II	1.0	1.4		MIT
2137.652	2138.326	46765.55	W	II	0.5	0.5		MIT
2137.15	2137.82	46776.5	W		0.6	1.1		A
2135.038	2135.712	46822.80	W			1.3		MIT
2134.67	2135.34	46830.9	W			0.3H		MIT
2134.061	2134.734	46844.24	W		0.6	1.1		MIT
2132.81	2133.48	46871.7	W			0.8		A
2132.27	2132.94	46883.6	W			1.0		MIT
2131.38	2132.05	46903.1	W			0.6		MIT
2130.61	2131.28	46920.1	W			1.0		MIT
2127.95	2128.62	46978.7	W		0.3	0.6J		MIT
2127.43	2128.10	46990.2	W		0.3H	1.1		MIT
2124.60	2125.27	47052.8	W			0.9		MIT
2121.59	2122.26	47119.6	W			0.9		MIT
2120.97	2121.64	47133.3	W			0.9		MIT
2120.35	2121.02	47147.1	W	II	0.3H	0.5		MIT
2118.87	2119.54	47180.0	W	II	1.0	1.3		MIT
2118.35	2119.02	47191.6	W	II		0.3H		MIT
2118.03	2118.70	47198.7	W		0.5	1.1		MIT
2116.94	2117.61	47223.0	W	II	0.9	1.2		MIT
2116.638	2117.308	47229.79	W		0.7	0.8		MIT
2116.30	2116.97	47237.3	W		0.6	1.0		MIT
2112.85	2113.52	47314.5	W			0.9		MIT
2112.67	2113.34	47318.5	W			0.5		MIT
2111.14	2111.81	47352.8	W		0.5	0.3H		A
2110.34	2111.01	47370.7	W		1.0	1.4		MIT
2106.93	2107.60	47447.4	W		0.3H	0.8		A
2106.68	2107.35	47453.0	W			1.0		MIT
2106.179	2106.847	47464.30	W		1.0	1.3		MIT
2105.765	2106.433	47473.63	W		0.8	1.3		MIT
2103.178	2103.845	47532.01	W		0.5	1.1		MIT
2102.991	2103.658	47536.24	W			0.6		MIT
2101.99	2102.66	47558.9	W	II	0.8	0.8		MIT
2100.67	2101.34	47588.8	W		1.0	1.4		MIT
2098.87	2099.54	47629.6	W		1.0			A
2098.726	2099.392	47632.83	W			0.7W		MIT
2098.60	2099.27	47635.7	W		1.0	1.3		A
2098.25	2098.92	47643.6	W		1.0	1.3		MIT
2095.594	2096.260	47704.01	W		0.5	1.1		MIT
2095.04	2095.71	47716.6	W		0.3H	0.7		A
2094.749	2095.415	47723.25	W	II	1.0	1.3		MIT
2093.800	2094.465	47744.88	W		0.9	1.2		MIT
2092.54	2093.21	47773.6	W		1.0	0.5		A
2092.077	2092.742	47784.20	W			1.4		MIT
2090.183	2090.848	47827.49	W		0.5	1.1		MIT
2089.143	2089.807	47851.30	W		1.0	1.3		MIT
2088.192	2088.856	47873.09	W		1.1	1.5		MIT
2087.47	2088.13	47889.6	W		0.6	1.1		A
2086.60	2087.26	47909.6	W	II	0.6	0.6		MIT

AIR WAVELENGTH (ANGSTRMS)	VACUUM WAVELENGTH (ANGSTRMS)	WAVENUMBER (KAYSERS)	LMENT		LOG ARC	INTENSITY SPK	DIS	REF
2085.09	2085.75	47944.3	W		0.5	1.0		A
2084.89	2085.55	47948.9	W		0.3H	0.9		A
2084.48	2085.14	47958.3	W		0.3H	0.9		A
2084.27	2084.93	47963.2	W		0.3H	1.0		A
2083.70	2084.36	47976.3	W	II	0.5	1.0		MIT
2081.39	2082.05	48029.5	W		0.9			A
2079.66	2080.32	48069.5	W		0.3	0.8		A
2079.106	2079.769	48082.27	W		1.1	1.5		MIT
2078.35	2079.01	48099.8	W		0.8	1.4		MIT
2075.96	2076.62	48155.1	W		0.7	1.0		A
2075.59	2076.25	48163.7	W		1.0	1.3		MIT
2075.45	2076.11	48167.0	W		0.5	0.6		A
2074.63	2075.29	48186.0	W	II	0.9	0.9		MIT
2071.78	2072.44	48252.3	W		1.0	0.7		A
2071.21	2071.87	48265.5	W	II	1.0	1.4		MIT
2070.81	2071.47	48274.9	W		0.7			A
2070.02	2070.68	48293.3	W		0.6	0.6		A
2067.87	2068.53	48343.5	W		0.3J	1.0		A
2065.57	2066.23	48397.3	W	II	1.0	1.3		MIT
2065.09	2065.75	48408.6	W		1.0	0.5		A
2063.67	2064.33	48441.9	W			0.5		MIT
2063.11	2063.77	48455.0	W		1.0			A
2062.77	2063.43	48463.0	W		0.7			A
2062.05	2062.71	48479.9	W			0.9		A
2058.298	2058.957	48568.29	W	II	0.9	0.9		MIT
2057.797	2058.456	48580.11	W		0.3H	1.0		MIT
2056.02	2056.68	48622.1	W	II	0.7	1.0		MIT
2054.678	2055.336	48653.85	W	II	1.0	1.3		MIT
2053.13	2053.79	48690.5	W	II	0.9	0.9		MIT
2052.15	2052.81	48713.8	W			0.9		MIT
2051.89	2052.55	48720.0	W		0.3	0.0		MIT
2050.58	2051.24	48751.1	W			0.5		MIT
2050.40	2051.06	48755.4	W		0.3H	1.0		A
2049.63	2050.29	48773.7	W		0.9	1.0		MIT
2049.02	2049.68	48788.2	W		0.8	0.8		A
2048.04	2048.70	48811.5	W	II	0.8	1.1		MIT
2047.09	2047.75	48834.2	W		0.7	1.1		A
2045.56	2046.22	48870.7	W			0.5		MIT
2043.73	2044.39	48914.4	W		0.5	0.7		A
2043.55	2044.21	48918.7	W		0.8	1.1		A
2040.865	2041.520	48983.10	W		0.7	1.1		MIT
2039.80	2040.46	49008.7	W			1.0		MIT
2039.38	2040.04	49018.8	W		0.6	0.5		A
2039.11	2039.77	49025.2	W		0.6	0.8		MIT
2037.906	2038.561	49054.21	W		0.6	1.1		MIT
2037.584	2038.239	49061.96	W		0.5	1.3		MIT
2035.89	2036.54	49102.8	W	II	0.9	1.3		MIT
2035.03	2035.68	49123.5	W		0.8	1.1		MIT
2033.96	2034.61	49149.4	W		0.3	0.8		MIT
2033.231	2033.885	49166.99	W			0.8		MIT
2031.452	2032.106	49210.04	W		0.8	0.7		MIT
2029.983	2030.636	49245.65	W	II	1.0	1.5		MIT
2029.118	2029.771	49266.64	W		0.5	1.1		MIT
2027.90	2028.55	49296.2	W		0.5	0.5		A
2027.305	2027.958	49310.69	W		0.8	1.0		MIT
2027.027	2027.680	49317.45	W		0.7	1.1		MIT
2026.08	2026.73	49340.5	W	II	0.8	1.4		MIT
2022.35	2023.00	49431.5	W		0.3H	1.1		MIT
2022.048	2022.700	49438.87	W		0.6	1.2		MIT
2021.44	2022.09	49453.7	W	II		0.8		MIT
2020.13	2020.78	49485.8	W		0.9	0.3H		A
2019.57	2020.22	49499.5	W		0.3H	0.9		A
2016.40	2017.05	49577.3	W		0.6	1.3		A
2015.78	2016.43	49592.6	W		0.8	1.3		A
2015.47	2016.12	49600.2	W		0.8	1.0		A
2014.43	2015.08	49625.8	W		0.7	1.1		A
2014.232	2014.883	49630.68	W		0.8	1.1		MIT
2013.07	2013.72	49659.3	W		0.7	0.8		A
2011.39	2012.04	49700.8	W			0.8		MIT
2010.74	2011.39	49716.9	W		0.5	0.3H		A

AIR WAVELENGTH (ANGSTRMS)	VACUUM WAVELENGTH (ANGSTRMS)	WAVENUMBER (KAYSERS)	LMENT	LOG ARC	INTENSITY SPK	DIS	REF
2010.23	2010.88	49729.5	W II	0.8	1.1		MIT
2009.979	2010.629	49735.69	W	0.9	1.1		MIT
2008.91	2009.56	49762.1	W	0.8			A
2008.64	2009.29	49768.8	W	0.6			A
2008.073	2008.722	49782.88	W	1.1	1.4		MIT
2006.92	2007.57	49811.5	W	0.7			A
2006.717	2007.366	49816.52	W	0.7	1.1		MIT
2006.09	2006.74	49832.1	W	0.3	0.6		A
2002.56	2003.21	49919.9	W II		0.3		LN
2001.706	2002.354	49941.21	W II	0.8	1.1		MIT
9966.58	9969.31	10030.8	XE I			1.0	ME
9923.198	9925.919	10074.63	XE I			3.3	ME
9865.56	9868.27	10133.5	XE			0.8J	HU
9799.697	9802.384	10201.60	XE I			3.3	IME
9734.0	9736.7	10270.	XE			0.5J	HU
9718.16	9720.83	10287.2	XE I			2.0	ME
9710.03	9712.69	10295.8	XE I			0.3	ME
9700.99	9703.65	10305.4	XE I			1.3	ME
9698.68	9701.34	10307.8	XE II			1.5B	HU
9685.32	9687.98	10322.1	XE I			2.2	ME
9641.6	9644.2	10369.	XE			0.6J	HU
9630.95	9633.59	10380.3	XE			0.5J	HU
9615.71	9618.35	10396.8	XE			0.6H	HU
9605.80	9608.43	10407.5	XE I			0.5	ME
9604.50	9607.13	10408.9	XE II			0.8H	HU
9591.35	9593.98	10423.2	XE II			1.5H	HU
9585.14	9587.77	10429.9	XE I			1.3	ME
9513.377	9515.987	10508.63	XE I			2.3	IME
9505.78	9508.39	10517.0	XE I			1.0	ME
9497.07	9499.68	10526.7	XE I			1.6	ME
9487.76	9490.36	10537.0	XE I			0.6	ME
9475.23	9477.83	10550.9	XE II			0.5H	HU
9464.3	9466.9	10563.	XE			1.0J	HU
9445.34	9447.93	10584.3	XE I			1.9	ME
9442.68	9445.27	10587.3	XE I			1.3	ME
9441.46	9444.05	10588.7	XE I			1.3	ME
9412.01	9414.59	10621.8	XE I			1.8	ME
9400.59	9403.17	10634.7	XE II			1.2H	HU
9374.76	9377.33	10664.0	XE I			2.0	ME
9374.02	9376.59	10664.9	XE I			1.0	ME
9334.08	9336.64	10710.5	XE I			0.5	ME
9331.67	9334.23	10713.3	XE II			0.6H	HU
9306.64	9309.19	10742.1	XE I			1.6	ME
9301.95	9304.50	10747.5	XE I			1.5	ME
9288.4	9290.9	10763.	XE II			0.7J	HU
9265.67	9268.21	10789.6	XE II			1.0H	HU
9245.18	9247.72	10813.5	XE I			0.5	ME
9226.39	9228.92	10835.5	XE II			0.8H	HU
9222.39	9224.92	10840.2	XE I			0.7	ME
9211.38	9213.91	10853.2	XE I			1.4	ME
9203.20	9205.73	10862.8	XE I			1.5	ME
9197.18	9199.70	10869.9	XE I			0.3	ME
9167.52	9170.04	10905.1	XE I			2.0	ME
9162.652	9165.167	10910.88	XE I			2.7	IME
9158.38	9160.89	10916.0	XE			0.3	ME
9152.12	9154.63	10923.4	XE I			1.3	ME
9141.8	9144.3	10936.	XE I			0.3	ME
9136.6	9139.1	10942.	XE			0.7J	HU
9131.59	9134.10	10948.0	XE I			0.5	ME
9112.24	9114.74	10971.2	XE I			0.6	ME
9096.13	9098.63	10990.7	XE I			1.7	ME
9045.446	9047.929	11052.25	XE I			2.6	IME
9032.18	9034.66	11068.5	XE I			1.7	ME
9025.98	9028.46	11076.1	XE I			1.5	ME
8987.57	8990.04	11123.4	XE I			2.3	ME
8981.05	8983.52	11131.5	XE I			2.0	ME
8952.78	8955.24	11166.7	XE I			1.7	ME
8952.251	8954.709	11167.31	XE I			3.0	IME
8930.83	8933.28	11194.1	XE I			2.3	ME
8908.73	8911.18	11221.9	XE I			2.3	ME

AIR WAVELENGTH (ANGSTRMS)	VACUUM WAVELENGTH (ANGSTRMS)	WAVENUMBER (KAYSERS)	LMENT	LOG ARC	INTENSITY SPK	DIS	REF
8902.66	8905.10	11229.5	XE			0.7J	HU
8885.71	8888.15	11250.9	XE I			1.0	ME
8862.32	8864.75	11280.6	XE I			2.5	ME
8855.74	8858.17	11289.0	XE			0.7J	HU
8839.9	8842.3	11309.	XE			0.5J	HU
8819.411	8821.833	11335.51	XE I			3.7	IME
8804.61	8807.03	11354.6	XE			1.4	HU
8785.88	8788.29	11378.8	XE			0.6J	HU
8760.14	8762.55	11412.2	XE			0.8J	HU
8758.20	8760.61	11414.7	XE I			2.0	ME
8752.14	8754.54	11422.6	XE			0.8J	HU
8739.372	8741.772	11439.33	XE I			2.5	IME
8716.19	8718.58	11469.7	XE II			1.5H	HU
8711.54	8713.93	11475.9	XE I			0.3	ME
8709.64	8712.03	11478.4	XE I			1.6	ME
8696.86	8699.25	11495.2	XE I			2.3	ME
8692.20	8694.59	11501.4	XE I			2.0	ME
8655.72	8658.10	11549.9	XE			0.5J	HU
8648.54	8650.92	11559.5	XE I			2.3	ME
8628.94	8631.31	11585.7	XE II			1.3H	HU
8624.24	8626.61	11592.0	XE I			1.9	ME
8604.23	8606.59	11619.0	XE			1.5J	HU
8576.01	8578.37	11657.2	XE I			2.3	ME
8553.97	8556.32	11687.3	XE I			0.3	ME
8530.10	8532.44	11720.0	XE I			1.5	ME
8522.55	8524.89	11730.3	XE I			1.5	ME
8515.19	8517.53	11740.5	XE			1.5J	HU
8501.02	8503.36	11760.1	XE			0.3	ME
8482.64	8484.97	11785.5	XE II			0.9H	HU
8437.55	8439.87	11848.5	XE I			1.0	ME
8409.189	8411.500	11888.49	XE I			3.3	IME
8402.03	8404.34	11898.6	XE I			0.7	ME
8392.37	8394.68	11912.3	XE I			1.3	ME
8378.3	8380.6	11932.	XE II			0.7H	HU
8372.79	8375.09	11940.2	XE I			0.7	ME
8371.38	8373.68	11942.2	XE I			0.5	ME
8366.4	8368.7	11949.	XE			1.4H	HU
8351.3	8353.6	11971.	XE II			0.5	HU
8349.05	8351.34	11974.1	XE I			1.6	ME
8347.45	8349.74	11976.4	XE I			1.8	ME
8347.24	8349.53	11976.7	XE II			1.7	HU
8346.822	8349.116	11977.32	XE I			3.3	IME
8329.44	8331.73	12002.3	XE			1.4J	HU
8324.58	8326.87	12009.3	XE I			1.3	ME
8323.90	8326.19	12010.3	XE I			0.3	ME
8317.10	8319.39	12020.1	XE			1.4J	HU
8316.2	8318.5	12021.	XE			1.0J	HU
8297.71	8299.99	12048.2	XE I			1.2	ME
8297.55	8299.83	12048.4	XE			1.7H	HU
8285.70	8287.98	12065.7	XE II			1.3H	HU
8282.85	8285.13	12069.8	XE			1.2H	HU
8280.116	8282.392	12073.81	XE I			3.7	I
8266.520	8268.792	12093.67	XE I			2.7	IME
8262.73	8265.00	12099.2	XE			1.4H	HU
8260.81	8263.08	12102.0	XE			0.7	HU
8256.40	8258.67	12108.5	XE			1.3	HU
8245.37	8247.64	12124.7	XE			0.6	HU
8231.634	8233.897	12144.92	XE I			3.7	I
8214.85	8217.11	12169.7	XE II			1.3	HU
8206.336	8208.592	12182.36	XE I			2.8	IME
8196.73	8198.98	12196.6	XE I			0.3	ME
8186.9	8189.2	12211.	XE			1.0H	HU
8171.02	8173.27	12235.0	XE I			2.0	ME
8167.55	8169.80	12240.2	XE			1.0H	HU
8165.37	8167.61	12243.5	XE			0.3	ME
8151.80	8154.04	12263.9	XE			1.8	HU
8144.8	8147.0	12274.	XE II			0.7J	HU
8142.13	8144.37	12278.4	XE			0.7J	HU
8136.83	8139.07	12286.4	XE II			1.4H	HU
8131.40	8133.64	12294.6	XE			1.3	HU

AIR WAVELENGTH (ANGSTRMS)	VACUUM WAVELENGTH (ANGSTRMS)	WAVENUMBER (KAYSERS)	LMENT	LOG ARC	INTENSITY SPK	INTENSITY DIS	REF
8123.29	8125.52	12306.9	XE I			0.3	ME
8120.16	8122.39	12311.6	XE			1.4	HU
8115.94	8118.17	12318.0	XE			1.5H	HU
8107.91	8110.14	12330.2	XE I			0.8	ME
8101.98	8104.21	12339.3	XE I			2.0	ME
8098.55	8100.78	12344.5	XE			1.1H	HU
8097.24	8099.47	12346.5	XE I			0.5	ME
8095.13	8097.36	12349.7	XE		1.0H		HU
8080.31	8082.53	12372.4	XE			1.5C	HU
8070.97	8073.19	12386.7	XE			1.5	HU
8064.94	8067.16	12395.9	XE I			0.3	ME
8061.339	8063.556	12401.48	XE I			2.2	IME
8057.258	8059.474	12407.76	XE I			2.3	IME
8047.28	8049.49	12423.1	XE			1.2H	HU
8042.18	8044.39	12431.0	XE I			1.2	ME
8040.56	8042.77	12433.5	XE I			1.0	ME
8038.26	8040.47	12437.1	XE			1.7H	HU
8035.40	8037.61	12441.5	XE			1.3H	HU
8031.64	8033.85	12447.3	XE			1.7H	HU
8029.67	8031.88	12450.4	XE I			2.0	ME
8023.85	8026.06	12459.4	XE			1.5H	HU
8020.07	8022.28	12465.3	XE II			0.7	HU
8014.26	8016.46	12474.3	XE			1.5C	HU
8008.45	8010.65	12483.4	XE			2.2H	HU
8003.26	8005.46	12491.5	XE I			1.0	ME
8001.95	8004.15	12493.5	XE			1.0C	HU
7996.5	7998.7	12502.	XE II			0.5J	HU
7992.34	7994.54	12508.5	XE			1.7J	HU
7991.5	7993.7	12510.	XE II			0.7J	HU
7987.99	7990.19	12515.3	XE II			1.6	HU
7981.1	7983.3	12526.	XE			1.7J	HU
7976.4	7978.6	12533.	XE II			0.5J	HU
7976.03	7978.22	12534.1	XE I			0.9	ME
7974.76	7976.95	12536.1	XE			1.3H	HU
7967.342	7969.533	12547.79	XE I			2.7	IME
7954.22	7956.41	12568.5	XE I			0.6	ME
7942.54	7944.72	12587.0	XE			1.7	HU
7937.41	7939.59	12595.1	XE I		1.6		ME
7920.48	7922.66	12622.0	XE			1.0J	HU
7897.7	7899.9	12658.	XE II			0.7C	HU
7889.4	7891.6	12672.	XE			1.5H	HU
7887.390	7889.560	12674.98	XE I			2.5	IHU
7882.71	7884.88	12682.5	XE II			1.3	HU
7881.320	7883.488	12684.74	XE I			2.0	IME
7862.7	7864.9	12715.	XE II			0.5	HU
7841.23	7843.39	12749.6	XE I			1.2	ME
7832.98	7835.14	12763.0	XE I			1.0	ME
7828.28	7830.43	12770.7	XE			1.3H	HU
7818.31	7820.46	12787.0	XE			1.2J	HU
7802.651	7804.798	12812.63	XE I			2.0	IME
7789.42	7791.56	12834.4	XE I			1.2	ME
7787.04	7789.18	12838.3	XE II			1.7	HU
7783.66	7785.80	12843.9	XE I			1.7	ME
7777.1	7779.2	12855.	XE			1.0	HU
7774.18	7776.32	12859.6	XE			0.6	HU
7772.12	7774.26	12863.0	XE			1.3B	HU
7740.31	7742.44	12915.8	XE I			1.6	ME
7712.42	7714.54	12962.5	XE II			1.5	HU
7670.81	7672.92	13032.8	XE I			0.3	ME
7670.66	7672.77	13033.1	XE II			2.0	HU
7666.61	7668.72	13040.0	XE I			1.0	ME
7664.56	7666.67	13043.5	XE I			1.5	ME
7664.02	7666.13	13044.4	XE I			1.0	ME
7643.91	7646.01	13078.7	XE I			2.0	ME
7642.025	7644.129	13081.94	XE I			2.7	I
7618.57	7620.67	13122.2	XE II			1.7	HU
7609.82	7611.91	13137.3	XE I			0.5	ME
7608.46	7610.55	13139.7	XE I			0.7	ME
7604.97	7607.06	13145.7	XE I			0.3H	ME
7600.77	7602.86	13152.9	XE I			1.0	ME
7589.61	7591.70	13172.3	XE I			0.8	ME
7584.680	7586.768	13180.84	XE I			2.3	IME
7584.29	7586.38	13181.5	XE I			1.0	ME
7570.93	7573.01	13204.8	XE I			0.8	ME
7559.79	7561.87	13224.2	XE I			1.6	ME
7548.45	7550.53	13244.1	XE II			2.2	HU
7530.70	7532.77	13275.3	XE II			1.5	HU
7514.96	7517.03	13303.1	XE I			0.5	ME
7514.54	7516.61	13303.9	XE I			0.9	ME
7503.00	7505.07	13324.3	XE II			0.5H	HU
7501.13	7503.20	13327.7	XE I			1.3	ME
7495.36	7497.42	13337.9	XE II			1.6	HU
7492.23	7494.29	13343.5	XE I			1.3	ME
7474.01	7476.07	13376.0	XE I			1.4	ME
7472.01	7474.07	13379.6	XE I			1.6	ME
7451.00	7453.05	13417.3	XE I			1.4	ME
7441.94	7443.99	13433.7	XE I			1.3	ME
7424.05	7426.09	13466.0	XE I			1.3	ME
7410.14	7412.18	13491.3	XE II			0.8J	HU
7405.77	7407.81	13499.3	XE I			0.5	ME
7404.51	7406.55	13501.6	XE I			1.1	ME
7400.5	7402.5	13509.	XE II			0.7H	HU
7400.41	7402.45	13509.0	XE I			1.5	ME
7393.792	7395.829	13521.14	XE I			2.2	I
7386.003	7388.038	13535.39	XE I			2.0	IME
7378.38	7380.41	13549.4	XE II			1.4	HU
7355.58	7357.61	13591.4	XE I			1.6	ME
7343.37	7345.39	13614.0	XE II			1.4J	HU
7339.30	7341.32	13621.5	XE II			2.2	HU
7338.96	7340.98	13622.2	XE I			1.7	RD
7336.480	7338.501	13626.76	XE I			1.7	IME
7323.05	7325.07	13651.7	XE I			0.3	ME
7321.452	7323.469	13654.73	XE I			1.9	IME
7319.94	7321.96	13657.6	XE I			1.2	ME
7316.87	7318.89	13663.3	XE I			1.3	ME
7316.272	7318.288	13664.40	XE I			1.8	IME
7307.37	7309.38	13681.0	XE I			0.7H	ME
7301.80	7303.81	13691.5	XE II			2.0	HU
7285.301	7287.308	13722.49	XE I			1.8	IME
7284.34	7286.35	13724.3	XE II			1.7	HU
7283.961	7285.968	13725.01	XE I			1.6	IME
7279.75	7281.76	13732.9	XE II			0.6T	HU
7276.47	7278.48	13739.1	XE			0.6T	HU
7266.49	7268.49	13758.0	XE I			1.4	ME
7262.54	7264.54	13765.5	XE I			1.3	ME
7257.94	7259.94	13774.2	XE I			1.8	ME
7250.87	7252.87	13787.7	XE I			0.7H	ME
7249.92	7251.92	13789.5	XE I			0.3	ME
7245.38	7247.38	13798.1	XE II			0.5	HU
7244.94	7246.94	13798.9	XE I			1.3	ME
7238.20	7240.19	13811.8	XE I			0.5	ME
7215.97	7217.96	13854.3	XE II			1.3	HU
7209.14	7211.13	13867.5	XE I			0.7	ME
7200.79	7202.77	13883.5	XE I			1.2	ME
7172.70	7174.68	13937.9	XE I			1.0	ME
7164.83	7166.80	13953.2	XE II			2.5	HU
7149.03	7151.00	13984.1	XE II			2.2	HU
7147.50	7149.47	13987.1	XE			1.7T	HU
7143.81	7145.78	13994.3	XE			0.9V	HU
7136.57	7138.54	14008.5	XE I			1.2	ME
7133.27	7135.24	14014.9	XE II			1.0	HU
7119.598	7121.561	14041.87	XE I			2.7	IME
7100.8	7102.8	14079.	XE II			0.7H	HU
7082.15	7084.10	14116.1	XE II			2.0	HU
7072.43	7074.38	14135.5	XE			0.6J	HU
7052.57	7054.51	14175.3	XE			0.5J	HU
7051.06	7053.00	14178.4	XE I			0.5	ME
7047.37	7049.31	14185.8	XE I			1.5	ME
7035.53	7037.47	14209.7	XE I			1.3	ME
7034.80	7036.74	14211.1	XE			0.5	ME

AIR WAVELENGTH (ANGSTRMS)	VACUUM WAVELENGTH (ANGSTRMS)	WAVENUMBER (KAYSERS)	LMENT	LOG INTENSITY ARC	SPK	DIS	REF
7019.02	7020.96	14243.1	XE I			1.5	ME
7017.06	7019.00	14247.1	XE II			1.6	HU
7003.96	7005.89	14273.7	XE II			1.5	HU
7003.10	7005.03	14275.4	XE I			0.6	ME
6990.88	6992.81	14300.4	XE II			2.8	HU
6982.05	6983.98	14318.5	XE I			1.5	ME
6976.182	6978.106	14330.54	XE I			2.0	IME
6942.11	6944.02	14400.9	XE			2.6J	HU
6936.69	6938.60	14412.1	XE I			0.9	ME
6935.62	6937.53	14414.3	XE I			1.7	ME
6925.53	6927.44	14435.3	XE I			2.0	ME
6924.67	6926.58	14437.1	XE I			1.2	ME
6922.22	6924.13	14442.2	XE I			0.9	ME
6910.82	6912.73	14466.1	XE I			1.5	ME
6910.22	6912.13	14467.3	XE II			1.7	HU
6910.19	6912.10	14467.4	XE I			1.0	ME
6890.41	6892.31	14508.9	XE			0.5V	HU
6882.154	6884.053	14526.33	XE I			2.5	IHU
6876.69	6878.59	14537.9	XE			0.5V	HU
6873.2	6875.1	14545.	XE			1.0V	HU
6872.107	6874.003	14547.56	XE I			2.8	IME
6866.838	6868.733	14558.73	XE I			1.7	IME
6865.58	6867.47	14561.4	XE I			0.7	ME
6863.20	6865.09	14566.4	XE I			1.3	ME
6860.19	6862.08	14572.8	XE I			1.6	ME
6850.13	6852.02	14594.2	XE I			1.5	ME
6848.82	6850.71	14597.0	XE I			1.7	ME
6846.613	6848.502	14601.73	XE I			1.8	IME
6844.84	6846.73	14605.5	XE I			0.3H	ME
6841.50	6843.39	14612.7	XE I			1.3	ME
6840.96	6842.85	14613.8	XE I			0.9	ME
6827.315	6829.199	14643.01	XE I			2.3	IME
6818.38	6820.26	14662.2	XE I			1.2	ME
6815.64	6817.52	14668.1	XE I			1.1	ME
6805.74	6807.62	14689.4	XE II			2.6	HU
6790.37	6792.24	14722.7	XE II			1.7	HU
6788.71	6790.58	14726.3	XE II			1.9	HU
6778.60	6780.47	14748.2	XE I			1.6	ME
6778.60	6780.47	14748.2	XE II			1.6	ME
6777.57	6779.44	14750.5	XE I			1.7	ME
6767.12	6768.99	14773.3	XE I			1.0	ME
6728.008	6729.865	14859.14	XE I			2.3	IME
6702.25	6704.10	14916.2	XE II			1.7	HU
6694.32	6696.17	14933.9	XE II			2.3	HU
6681.036	6682.881	14963.61	XE I			1.3	IME
6678.972	6680.816	14968.23	XE I			1.4	IME
6668.920	6670.761	14990.79	XE I			2.2	IME
6666.965	6668.806	14995.19	XE I			1.8	IME
6664.85	6666.69	14999.9	XE			0.6	ME
6657.92	6659.76	15015.6	XE I			1.3	ME
6648.75	6650.59	15036.3	XE I			0.5	ME
6634.13	6635.96	15069.4	XE II			1.0Q	HU
6632.464	6634.296	15073.19	XE I			1.7	IME
6632.44	6634.27	15073.2	XE			0.5	HU
6630.44	6632.27	15077.8	XE I			0.3	ME
6620.02	6621.85	15101.5	XE II			2.0	HU
6618.40	6620.23	15105.2	XE II			1.5	HU
6614.96	6616.79	15113.1	XE II			1.0J	HU
6613.31	6615.14	15116.8	XE			0.6H	HU
6608.87	6610.70	15127.0	XE I			1.0	ME
6607.41	6609.23	15130.3	XE I			1.5H	ME
6602.87	6604.69	15140.7	XE I			0.6H	ME
6598.84	6600.66	15150.0	XE II			1.7	HU
6597.25	6599.07	15153.7	XE II			2.3	HU
6595.561	6597.383	15157.53	XE I			2.0	IME
6595.01	6596.83	15158.8	XE			2.6	HU
6590.86	6592.68	15168.3	XE I			0.9	ME
6583.27	6585.09	15185.8	XE I			1.3	ME
6573.68	6575.50	15208.0	XE II			1.5	HU
6569.13	6570.94	15218.5	XE II			0.9	HU
6563.19	6565.00	15232.3	XE II			1.2	HU
6560.65	6562.46	15238.2	XE I			0.6H	ME
6559.97	6561.78	15239.8	XE I			1.4	ME
6556.70	6558.51	15247.4	XE II			0.7	HU
6554.196	6556.007	15253.19	XE I			1.7B	IME
6553.66	6555.47	15254.4	XE I			0.6	ME
6546.12	6547.93	15272.0	XE I			1.3	ME
6543.360	6545.168	15278.45	XE I			1.6	IME
6533.159	6534.964	15302.30	XE I			2.0	IME
6528.65	6530.45	15312.9	XE II			2.0	HU
6521.508	6523.310	15329.64	XE I			1.6	IME
6512.83	6514.63	15350.1	XE II			2.2	HU
6507.50	6509.30	15362.6	XE			0.5	ME
6504.18	6505.98	15370.5	XE I			2.3H	ME
6500.37	6502.17	15379.5	XE I			1.2	ME
6498.717	6500.513	15383.40	XE I			2.0	IME
6497.43	6499.23	15386.4	XE I			1.5B	ME
6487.765	6489.558	15409.37	XE I			2.1	IME
6479.69	6481.48	15428.6	XE II			0.5J	HU
6472.841	6474.630	15444.90	XE I			2.2	IME
6469.705	6471.493	15452.39	XE I			2.5	IME
6461.50	6463.29	15472.0	XE I			0.5	ME
6461.48	6463.27	15472.1	XE			1.0J	HU
6451.79	6453.57	15495.3	XE I			1.0B	ME
6450.48	6452.26	15498.4	XE I			0.8	ME
6448.70	6450.48	15502.7	XE I			0.3H	ME
6430.155	6431.932	15547.43	XE I			1.3	IME
6418.98	6420.75	15574.5	XE I			1.5H	ME
6418.58	6420.35	15575.5	XE II			1.5	HU
6418.41	6420.18	15575.9	XE I			1.5	ME
6412.38	6414.15	15590.5	XE I			1.0	ME
6397.99	6399.76	15625.6	XE II			1.7	HU
6375.28	6377.04	15681.2	XE II			1.9	HU
6362.8	6364.6	15712.	XE			0.5J	HU
6356.35	6358.11	15727.9	XE II			2.5	HU
6355.77	6357.53	15729.4	XE I			1.3	ME
6353.29	6355.05	15735.5	XE II			1.5B	HU
6344.98	6346.73	15756.1	XE I			0.3H	ME
6343.96	6345.71	15758.7	XE II			2.3	HU
6337.58	6339.33	15774.5	XE I			0.9B	ME
6333.97	6335.72	15783.5	XE I			1.6B	ME
6331.50	6333.25	15789.7	XE I			1.3	ME
6325.81	6327.56	15803.9	XE I			0.3	ME
6318.062	6319.809	15823.26	XE I			2.7	IME
6314.97	6316.72	15831.0	XE I			1.2	ME
6311.46	6313.21	15839.8	XE II			1.5B	HU
6300.86	6302.60	15866.5	XE II			2.1	HU
6298.31	6300.05	15872.9	XE II			1.5	HU
6296.39	6298.13	15877.7	XE II			1.0J	HU
6294.45	6296.19	15882.6	XE I			1.2	ME
6292.649	6294.389	15887.17	XE I			1.7	IME
6286.011	6287.750	15903.94	XE I			2.0	IME
6284.41	6286.15	15908.0	XE II			1.7	HU
6281.81	6283.55	15914.6	XE I			0.7H	ME
6277.54	6279.28	15925.4	XE II			2.3	HU
6276.99	6278.73	15926.8	XE I			0.6	ME
6273.23	6274.97	15936.3	XE I			1.0	ME
6270.82	6272.55	15942.5	XE II			2.4	HU
6265.302	6267.035	15956.51	XE I			1.6	IME
6261.212	6262.944	15966.93	XE I			1.7	IME
6242.09	6243.82	16015.8	XE I			0.9	ME
6235.40	6237.12	16033.0	XE II			1.0H	HU
6234.04	6235.76	16036.5	XE II			1.0J	HU
6224.168	6225.890	16061.96	XE I			1.6	IME
6220.84	6222.56	16070.6	XE I			0.3	ME
6209.11	6210.83	16100.9	XE I			0.5	ME
6206.297	6208.014	16108.21	XE I			1.3	IME
6206.16	6207.88	16108.6	XE			2.0	HU
6205.75	6207.47	16109.6	XE I			0.6	ME
6205.35	6207.07	16110.7	XE I			0.8H	ME

AIR WAVELENGTH (ANGSTRMS)	VACUUM WAVELENGTH (ANGSTRMS)	WAVENUMBER (KAYSERS)	LMENT	LOG ARC	INTENSITY SPK	DIS	REF
6201.49	6203.21	16120.7	XE I			0.5H	ME
6200.892	6202.608	16122.25	XE I			1.8	IME
6198.260	6199.975	16129.10	XE I			2.0	IME
6196.63	6198.34	16133.3	XE II			0.6H	HU
6194.07	6195.78	16140.0	XE II			2.4	HU
6193.89	6195.60	16140.5	XE I			0.3H	ME
6191.40	6193.11	16147.0	XE I			0.6H	ME
6189.10	6190.81	16153.0	XE I			1.3	ME
6185.79	6187.50	16161.6	XE II			0.5J	HU
6185.03	6186.74	16163.6	XE II			1.4	HU
6184.57	6186.28	16164.8	XE II			1.5	HU
6184.16	6185.87	16165.9	XE I			0.5	E
6182.420	6184.131	16170.42	XE I			2.5	IME
6179.665	6181.375	16177.63	XE I			2.1	IME
6178.303	6180.013	16181.20	XE I			2.2	IME
6163.935	6165.641	16218.91	XE I			1.9	IME
6163.661	6165.367	16219.64	XE I			2.0	IME
6162.16	6163.87	16223.6	XE I			0.5	ME
6155.28	6156.98	16241.7	XE II			0.5J	HU
6152.070	6153.773	16250.19	XE I			1.3	IME
6146.45	6148.15	16265.1	XE II			1.7	HU
6144.97	6146.67	16269.0	XE I			1.3J	ME
6143.70	6145.40	16272.3	XE I			0.6	ME
6143.40	6145.10	16273.1	XE II			1.0H	HU
6142.13	6143.83	16276.5	XE I			0.3H	ME
6127.44	6129.14	16315.5	XE II			1.2	HU
6126.36	6128.06	16318.4	XE I			1.2	ME
6123.91	6125.61	16324.9	XE I			0.7	ME
6115.08	6116.77	16348.5	XE II			1.7	HU
6114.86	6116.55	16349.1	XE I			1.0	ME
6111.951	6113.643	16356.86	XE I			1.6	IME
6111.761	6113.453	16357.37	XE I			1.5	IME
6108.37	6110.06	16366.4	XE I			0.9	ME
6103.88	6105.57	16378.5	XE I			0.5	ME
6101.43	6103.12	16385.1	XE II			2.3	HU
6097.59	6099.28	16395.4	XE II			2.8	HU
6093.56	6095.25	16406.2	XE II			2.2	HU
6093.38	6095.07	16406.7	XE I			0.5	ME
6083.21	6084.89	16434.1	XE II			0.7H	HU
6067.52	6069.20	16476.6	XE I			0.3H	ME
6063.29	6064.97	16488.1	XE			0.5	HU
6051.15	6052.83	16521.2	XE II			2.8	HU
6048.53	6050.20	16528.4	XE II			0.7H	HU
6048.00	6049.67	16529.8	XE I			0.8H	ME
6043.38	6045.05	16542.4	XE I			1.0	ME
6036.20	6037.87	16562.1	XE II			2.7	HU
6034.92	6036.59	16565.6	XE I			0.3	ME
6026.76	6028.43	16588.1	XE I			0.6	ME
6024.77	6026.44	16593.6	XE II			0.5J	HU
6009.78	6011.44	16634.9	XE I			0.9	ME
6008.92	6010.58	16637.3	XE II			2.0	HU
6007.909	6009.573	16640.12	XE I			1.2	IME
5998.115	5999.776	16667.29	XE I			1.5	IME
5991.86	5993.52	16684.7	XE II			1.0H	HU
5989.18	5990.84	16692.1	XE I			1.3H	ME
5988.44	5990.10	16694.2	XE			0.3	HU
5986.23	5987.89	16700.4	XE I			0.6	ME
5978.29	5979.95	16722.6	XE I			0.3	ME
5976.46	5978.12	16727.7	XE II			2.9	HU
5974.152	5975.807	16734.14	XE I			1.6	IME
5971.13	5972.78	16742.6	XE II			2.2	HU
5958.03	5959.68	16779.4	XE II			1.7	HU
5945.53	5947.18	16814.7	XE II			2.3	HU
5934.55	5936.19	16845.8	XE			0.3J	HU
5934.172	5935.816	16846.88	XE I			2.0	IME
5931.241	5932.884	16855.21	XE I			1.9	IME
5925.56	5927.20	16871.4	XE I			0.8	ME
5922.550	5924.191	16879.94	XE I			1.3	IME
5921.85	5923.49	16881.9	XE I			1.0	ME
5921.50	5923.14	16882.9	XE II			0.5Q	HU

AIR WAVELENGTH (ANGSTRMS)	VACUUM WAVELENGTH (ANGSTRMS)	WAVENUMBER (KAYSERS)	LMENT	LOG ARC	INTENSITY SPK	DIS	REF
5917.44	5919.08	16894.5	XE II			1.7	HU
5916.65	5918.29	16896.8	XE I			0.6	ME
5912.80	5914.44	16907.8	XE II			0.7	HU
5911.90	5913.54	16910.4	XE I			0.7	ME
5909.67	5911.31	16916.7	XE II			1.4	HU
5906.76	5908.40	16925.1	XE I			0.5	ME
5905.13	5906.77	16929.7	XE II			2.0	HU
5904.462	5906.098	16931.65	XE I			1.3	IME
5898.56	5900.19	16948.6	XE I			0.9	ME
5895.62	5897.25	16957.1	XE I			0.3H	ME
5894.988	5896.622	16958.86	XE I			2.0	IME
5893.29	5894.92	16963.7	XE II			2.2	HU
5889.12	5890.75	16975.8	XE I			1.3	ME
5878.92	5880.55	17005.2	XE I			0.8	ME
5875.018	5876.646	17016.51	XE I			2.0	IME
5859.47	5861.09	17061.7	XE II			1.4J	HU
5856.509	5858.132	17070.29	XE I			1.2	IME
5855.47	5857.09	17073.3	XE II			0.5J	HU
5849.85	5851.47	17089.7	XE I			0.5H	ME
5846.69	5848.31	17098.9	XE			0.5	HU
5846.21	5847.83	17100.4	XE I			0.3	ME
5843.43	5845.05	17108.5	XE I			0.7	ME
5840.83	5842.45	17116.1	XE I			0.6H	ME
5835.5	5837.1	17132.	XE II			1.7J	HU
5830.63	5832.25	17146.1	XE I			1.3H	ME
5827.72	5829.34	17154.6	XE I			0.3	ME
5824.800	5826.415	17163.21	XE I			2.2	IME
5823.890	5825.505	17165.90	XE I			2.5	IME
5820.52	5822.13	17175.8	XE I			1.4	ME
5815.96	5817.57	17189.3	XE II			1.7	HU
5814.505	5816.117	17193.60	XE I			1.8	IME
5807.311	5808.921	17214.90	XE I			1.2	IME
5791.88	5793.49	17260.8	XE II			0.5B	HU
5776.39	5777.99	17307.1	XE II			2.2	HU
5758.65	5760.25	17360.4	XE II			2.2	HU
5754.60	5756.20	17372.6	XE I			0.3H	ME
5754.18	5755.78	17373.9	XE			0.5	HU
5752.56	5754.16	17378.7	XE II			1.2	HU
5751.03	5752.63	17383.4	XE II			2.3	HU
5748.20	5749.79	17391.9	XE I			0.9H	ME
5746.88	5748.47	17395.9	XE II			1.0J	HU
5740.17	5741.76	17416.3	XE I			0.8	ME
5733.48	5735.07	17436.6	XE I			0.6H	ME
5726.91	5728.50	17456.6	XE II			2.3	HU
5726.10	5727.69	17459.1	XE I			0.6	ME
5722.14	5723.73	17471.1	XE I			1.2H	ME
5719.61	5721.20	17478.9	XE II			2.0	HU
5716.252	5717.838	17489.13	XE I			1.9	IME
5716.19	5717.78	17489.3	XE II			1.7J	HU
5715.716	5717.302	17490.77	XE I			1.8	IME
5712.21	5713.79	17501.5	XE I			0.3	ME
5709.80	5711.38	17508.9	XE I			1.0H	ME
5706.87	5708.45	17517.9	XE I			0.5	ME
5699.61	5701.19	17540.2	XE II			2.0	HU
5698.54	5700.12	17543.5	XE I			0.9	ME
5696.477	5698.058	17549.84	XE I			1.9	IME
5695.750	5697.330	17552.08	XE I			2.0	IME
5688.373	5689.951	17574.84	XE I			1.6	IME
5686.49	5688.07	17580.7	XE II			0.3H	HU
5675.15	5676.72	17615.8	XE II			0.7	HU
5670.96	5672.53	17628.8	XE II			1.4	HU
5667.56	5669.13	17639.4	XE II			2.5	HU
5664.02	5665.59	17650.4	XE			0.3Q	HU
5659.38	5660.95	17664.9	XE II			2.2	HU
5652.84	5654.41	17685.3	XE I			0.3H	ME
5646.19	5647.76	17706.1	XE I			0.7	ME
5633.24	5634.80	17746.9	XE			0.7&	HU
5624.78	5626.34	17773.5	XE II			0.7Q	HU
5618.878	5620.438	17792.21	XE I			1.9	IME
5616.67	5618.23	17799.2	XE II			2.2	HU

AIR WAVELENGTH (ANGSTRMS)	VACUUM WAVELENGTH (ANGSTRMS)	WAVENUMBER (KAYSERS)	LMENT	LOG ARC	INTENSITY SPK	INTENSITY DIS	REF
5612.89	5614.45	17811.2	XE II			0.3H	HU
5612.65	5614.21	17811.9	XE I			1.2	ME
5607.99	5609.55	17826.7	XE I			0.5	ME
5594.87	5596.42	17868.6	XE			0.9Q	HU
5594.37	5595.92	17870.1	XE I			0.8	ME
5591.61	5593.16	17879.0	XE II			1.7J	HU
5583.5	5585.1	17905.	XE			1.0J	HU
5581.93	5583.48	17910.0	XE			0.7J	HU
5581.784	5583.334	17910.45	XE I			1.7	IME
5579.28	5580.83	17918.5	XE I			1.6	ME
5575.27	5576.82	17931.4	XE I			0.3H	ME
5572.19	5573.74	17941.3	XE			1.7	HU
5567.77	5569.32	17955.5	XE I			0.3H	ME
5566.615	5568.161	17959.25	XE I			2.0	IME
5566.22	5567.77	17960.5	XE			0.7H	ME
5563.50	5565.04	17969.3	XE I			0.3	ME
5557.28	5558.82	17989.4	XE I			0.3	ME
5554.99	5556.53	17996.8	XE			0.3H	HU
5553.10	5554.64	18003.0	XE I			0.5H	ME
5552.385	5553.927	18005.28	XE I			1.9	IME
5551.50	5553.04	18008.1	XE			0.3T	HU
5540.38	5541.92	18044.3	XE I			0.5H	ME
5532.78	5534.32	18069.1	XE I			0.3H	ME
5531.07	5532.61	18074.7	XE II			2.5	HU
5525.59	5527.12	18092.6	XE II			1.4	HU
5518.56	5520.09	18115.6	XE II			0.5	HU
5509.20	5510.73	18146.4	XE			1.0&	HU
5507.46	5508.99	18152.1	XE			1.0	HU
5495.07	5496.60	18193.1	XE			1.0&	HU
5488.555	5490.080	18214.67	XE I			1.3H	IME
5487.03	5488.55	18219.7	XE I			0.8H	ME
5484.46	5485.98	18228.3	XE I			0.6H	ME
5481.13	5482.65	18239.4	XE			0.3H	HU
5472.61	5474.13	18267.7	XE II			2.7	HU
5469.58	5471.10	18277.9	XE II			1.0H	HU
5460.39	5461.91	18308.6	XE II			2.3	HU
5460.037	5461.554	18309.81	XE I			1.2	IME
5456.45	5457.97	18321.8	XE I			0.3	ME
5450.90	5452.41	18340.5	XE II			1.4	HU
5450.45	5451.96	18342.0	XE II			2.0	HU
5445.52	5447.03	18358.6	XE II			1.9	HU
5440.39	5441.90	18375.9	XE I			1.2	ME
5439.923	5441.435	18377.50	XE I			1.5	IME
5438.96	5440.47	18380.8	XE II			2.6	HU
5435.60	5437.11	18392.1	XE I			0.7H	ME
5430.06	5431.57	18410.9	XE			0.3	HU
5421.76	5423.27	18439.1	XE I			0.3H	ME
5419.15	5420.66	18447.9	XE II			3.0	HU
5418.02	5419.53	18451.8	XE			0.7	ME
5415.36	5416.87	18460.9	XE II			1.4J	HU
5408.34	5409.84	18484.8	XE			0.3	HU
5400.45	5401.95	18511.8	XE I			0.6H	ME
5394.738	5396.238	18531.43	XE I			1.3	IME
5392.795	5394.294	18538.11	XE I			2.0	IME
5372.39	5373.88	18608.5	XE II			2.3	HU
5368.07	5369.56	18623.5	XE II			2.0	HU
5367.03	5368.52	18627.1	XE			0.8	ME
5364.626	5366.118	18635.45	XE I			1.5	IME
5363.27	5364.76	18640.2	XE II			1.9	HU
5362.244	5363.735	18643.72	XE I			1.2	IME
5350.03	5351.52	18686.3	XE			0.3	HU
5339.38	5340.87	18723.6	XE II			2.7	HU
5337.89	5339.37	18728.8	XE I			0.3H	ME
5328.69	5330.17	18761.1	XE			0.3	HU
5327.90	5329.38	18763.9	XE			0.3	HU
5319.83	5321.31	18792.4	XE			0.3	HU
5313.87	5315.35	18813.4	XE II			2.7	HU
5309.27	5310.75	18829.7	XE II			2.2	HU
5306.37	5307.85	18840.0	XE I			0.5	ME
5302.83	5304.31	18852.6	XE			0.3	HU

AIR WAVELENGTH (ANGSTRMS)	VACUUM WAVELENGTH (ANGSTRMS)	WAVENUMBER (KAYSERS)	LMENT	LOG ARC	INTENSITY SPK	INTENSITY DIS	REF
5292.22	5293.69	18890.4	XE II			2.9	HU
5286.38	5287.85	18911.3	XE I			0.5H	ME
5286.11	5287.58	18912.2	XE I			0.6H	ME
5283.30	5284.77	18922.3	XE I			0.3H	ME
5282.46	5283.93	18925.3	XE II			1.0	HU
5281.66	5283.13	18928.2	XE			0.3	HU
5268.31	5269.78	18976.1	XE II			1.7	HU
5261.95	5263.41	18999.1	XE II			2.3	HU
5260.44	5261.90	19004.5	XE II			2.5	HU
5259.89	5261.35	19006.5	XE II			1.4	HU
5251.89	5253.35	19035.5	XE			0.3H	ME
5248.98	5250.44	19046.0	XE I			0.6H	ME
5247.75	5249.21	19050.5	XE			1.7	HU
5245.27	5246.73	19059.5	XE I			0.6H	ME
5226.90	5228.36	19126.5	XE II			0.7B	HU
5226.62	5228.08	19127.5	XE II			1.0B	HU
5213.17	5214.62	19176.9	XE			0.3H	HU
5201.88	5203.33	19218.5	XE II			0.7J	HU
5201.42	5202.87	19220.2	XE II			1.2	HU
5194.92	5196.37	19244.2	XE II			1.0J	HU
5192.10	5193.55	19254.7	XE II			1.7	HU
5191.37	5192.82	19257.4	XE II			2.3	HU
5188.11	5189.55	19269.5	XE II			2.0	HU
5185.85	5187.29	19277.9	XE I			0.3H	ME
5184.48	5185.92	19283.0	XE II			1.6	HU
5178.82	5180.26	19304.0	XE II			1.7	HU
5162.711	5164.149	19364.27	XE I			1.0	IME
5125.70	5127.13	19504.1	XE II			1.7	HU
5122.42	5123.85	19516.6	XE II			2.2	HU
5117.76	5119.19	19534.4	XE			0.5H	HU
5108.58	5110.00	19569.5	XE II			0.5H	HU
5099.59	5101.01	19604.0	XE			0.7J	HU
5092.02	5093.44	19633.1	XE II			1.5Q	HU
5081.07	5082.49	19675.4	XE II			1.4	HU
5080.62	5082.04	19677.1	XE II			2.7	HU
5069.82	5071.23	19719.1	XE			1.0J	HU
5066.33	5067.74	19732.6	XE			0.3J	HU
5052.54	5053.95	19786.5	XE II			1.4	HU
5044.92	5046.33	19816.4	XE II			2.0	HU
5036.15	5037.55	19850.9	XE II			0.5H	HU
5028.279	5029.681	19881.98	XE I			2.3	I
5023.88	5025.28	19899.4	XE I			0.5H	ME
5012.83	5014.23	19943.2	XE II			1.7B	HU
5001.01	5002.41	19990.4	XE			1.2	HU
4993.93	4995.32	20018.7	XE II			1.0Q	HU
4993.03	4994.42	20022.3	XE II			1.0	HU
4991.17	4992.56	20029.8	XE II			1.7Q	HU
4988.77	4990.16	20039.4	XE II			2.2H	HU
4974.87	4976.26	20095.4	XE II			0.3H	HU
4972.71	4974.10	20104.1	XE II			2.3H	HU
4971.71	4973.10	20108.2	XE			2.2Q	HU
4965.00	4966.39	20135.4	XE II			0.6Q	HU
4923.152	4924.526	20306.52	XE I			2.7	IHU
4921.48	4922.85	20313.4	XE II			2.7	HU
4919.66	4921.03	20320.9	XE II			2.1	HU
4916.507	4917.880	20333.97	XE I			2.7	IME
4905.20	4906.57	20380.8	XE			0.3H	HU
4890.09	4891.46	20443.8	XE II			2.2H	HU
4887.30	4888.66	20455.5	XE II			2.2H	HU
4885.19	4886.55	20464.3	XE			0.3H	HU
4884.15	4885.51	20468.7	XE II			1.7J	HU
4883.53	4884.89	20471.3	XE II			2.5H	HU
4876.50	4877.86	20500.8	XE II			2.3B	HU
4862.54	4863.90	20559.6	XE II			2.6B	HU
4853.77	4855.13	20596.8	XE II			1.5	HU
4844.333	4845.686	20636.91	XE II			3.0	IME
4843.293	4844.646	20641.34	XE I			2.5	IME
4834.65	4836.00	20678.2	XE			0.3	HU
4829.708	4831.058	20699.40	XE I			2.6	IME
4823.41	4824.76	20726.4	XE II			2.2H	HU

AIR WAVELENGTH (ANGSTRMS)	VACUUM WAVELENGTH (ANGSTRMS)	WAVENUMBER (KAYSERS)	LMENT	LOG ARC	INTENSITY SPK	DIS	REF
4818.02	4819.37	20749.6	XE II			2.0	HU
4817.22	4818.57	20753.1	XE II			1.3Q	HU
4807.019	4808.363	20797.10	XE I			2.7	IME
4806.92	4808.26	20797.5	XE			0.3H	HU
4799.45	4800.79	20829.9	XE II			1.0B	HU
4796.53	4797.87	20842.6	XE			0.5Q	HU
4795.40	4796.74	20847.5	XE II			0.3H	HU
4792.619	4793.959	20859.59	XE I			2.2	IHU
4787.77	4789.11	20880.7	XE II			1.7	HU
4786.65	4787.99	20885.6	XE II			0.9H	HU
4779.18	4780.52	20918.2	XE II			1.7	HU
4775.76	4777.10	20933.2	XE II			1.0Q	HU
4775.18	4776.52	20935.8	XE II			1.0Q	HU
4773.19	4774.52	20944.5	XE II			1.7	HU
4769.05	4770.38	20962.7	XE II			2.0	HU
4734.152	4735.476	21117.20	XE I			2.8	I
4732.51	4733.83	21124.5	XE II			1.0B	HU
4731.19	4732.51	21130.4	XE II			1.7B	HU
4721.00	4722.32	21176.0	XE			0.3Q	HU
4715.18	4716.50	21202.2	XE II			1.9	HU
4712.63	4713.95	21213.6	XE			1.3	HU
4708.92	4710.24	21230.4	XE II			0.9H	HU
4708.21	4709.53	21233.6	XE I			0.7	ME
4704.67	4705.99	21249.5	XE II			0.9Q	HU
4699.62	4700.94	21272.4	XE II			0.5H	HU
4698.01	4699.32	21279.6	XE II			2.2B	HU
4697.021	4698.335	21284.13	XE I			2.5	IME
4693.34	4694.65	21300.8	XE II			1.0B	HU
4690.971	4692.284	21311.58	XE I			2.0	IHU
4679.45	4680.76	21364.1	XE II			0.5H	HU
4678.31	4679.62	21369.3	XE II			0.3H	HU
4676.75	4678.06	21376.4	XE			0.7H	HU
4676.46	4677.77	21377.7	XE			2.0Q	HU
4674.56	4675.87	21386.4	XE II			1.4	HU
4672.20	4673.51	21397.2	XE II			1.7B	HU
4671.226	4672.534	21401.67	XE I			3.3	IME
4668.49	4669.80	21414.2	XE II			1.7	HU
4666.28	4667.59	21424.4	XE II			1.4H	HU
4653.00	4654.30	21485.5	XE II			1.4	HU
4651.94	4653.24	21490.4	XE II			2.0	HU
4649.17	4650.47	21503.2	XE			0.3H	HU
4633.30	4634.60	21576.9	XE II			1.4	HU
4624.276	4625.571	21618.95	XE I			3.0	I
4619.57	4620.86	21641.0	XE			0.3	HU
4617.50	4618.79	21650.7	XE II			1.7	HU
4615.50	4616.79	21660.1	XE II			2.0	HU
4615.06	4616.35	21662.1	XE II			1.7B	HU
4611.889	4613.181	21677.02	XE II			2.8	I
4603.028	4604.318	21718.75	XE II			2.5H	IME
4593.70	4594.99	21762.9	XE II			0.7	HU
4592.05	4593.34	21770.7	XE II			2.2Q	HU
4588.36	4589.65	21788.2	XE			0.3	HU
4585.48	4586.76	21801.9	XE II			2.3Q	HU
4582.747	4584.031	21814.86	XE I			2.5	IME
4580.70	4581.98	21824.6	XE II			1.6J	HU
4577.06	4578.34	21842.0	XE II			2.0J	HU
4576.60	4577.88	21844.2	XE			1.2	ME
4571.85	4573.13	21866.9	XE			1.2V	HU
4569.12	4570.40	21879.9	XE			0.3	HU
4555.94	4557.22	21943.2	XE II			2.0Q	HU
4550.79	4552.07	21968.1	XE II			0.9H	HU
4545.23	4546.50	21994.9	XE II			2.3Q	HU
4540.89	4542.16	22015.9	XE II			2.3B	HU
4536.92	4538.19	22035.2	XE II			1.6J	HU
4532.49	4533.76	22056.7	XE II			2.0	HU
4524.680	4525.949	22094.81	XE I			2.6	I
4524.21	4525.48	22097.1	XE II			2.0	HU
4521.86	4523.13	22108.6	XE II			1.7B	HU
4519.69	4520.96	22119.2	XE			0.3	HU
4507.11	4508.37	22180.9	XE			0.5J	HU

AIR WAVELENGTH (ANGSTRMS)	VACUUM WAVELENGTH (ANGSTRMS)	WAVENUMBER (KAYSERS)	LMENT	LOG ARC	INTENSITY SPK	DIS	REF
4500.977	4502.240	22211.17	XE I			2.7	IHU
4488.60	4489.86	22272.4	XE			0.3J	HU
4485.95	4487.21	22285.6	XE II			1.0	HU
4480.86	4482.12	22310.9	XE II			2.3Q	HU
4473.85	4475.11	22345.8	XE			0.3H	HU
4470.90	4472.15	22360.6	XE II			1.2	HU
4462.19	4463.44	22404.2	XE			2.7Q	HU
4448.13	4449.38	22475.1	XE			2.3Q	HU
4440.95	4442.20	22511.4	XE II			1.4J	HU
4416.07	4417.31	22638.2	XE II			1.9Q	HU
4414.84	4416.08	22644.5	XE II			2.2	HU
4406.88	4408.12	22685.4	XE II			2.0Q	HU
4395.77	4397.00	22742.7	XE			2.3Q	HU
4393.20	4394.43	22756.1	XE II			2.3Q	HU
4385.769	4387.001	22794.61	XE I			1.8	IHU
4384.93	4386.16	22799.0	XE II			1.5	HU
4383.909	4385.141	22804.28	XE I			2.0	IHU
4379.44	4380.67	22827.6	XE II			0.7Q	HU
4373.78	4375.01	22857.1	XE II			1.7Q	HU
4372.88	4374.11	22861.8	XE			0.3B	HU
4372.287	4373.516	22864.90	XE I			1.3	IME
4369.20	4370.43	22881.1	XE			2.0J	HU
4367.05	4368.28	22892.3	XE II			1.2J	HU
4360.32	4361.55	22927.6	XE			0.3J	HU
4342.56	4343.78	23021.4	XE II			0.6Q	HU
4337.07	4338.29	23050.6	XE II			1.2Q	HU
4335.81	4337.03	23057.3	XE II			0.7	HU
4330.52	4331.74	23085.4	XE II			2.7Q	HU
4321.82	4323.04	23131.9	XE II			1.3	HU
4310.51	4311.72	23192.6	XE II			2.3H	HU
4296.75	4297.96	23266.9	XE I			0.7	HU
4296.40	4297.61	23268.7	XE II			2.3H	HU
4269.84	4271.04	23413.5	XE II			1.3	HU
4263.57	4264.77	23447.9	XE II			0.7H	HU
4263.44	4264.64	23448.6	XE			1.2H	HU
4251.57	4252.77	23514.1	XE II			1.7Q	HU
4245.38	4246.58	23548.4	XE II			2.3H	HU
4244.41	4245.61	23553.8	XE II			1.2	HU
4243.88	4245.07	23556.7	XE II			0.7	HU
4238.25	4239.44	23588.0	XE II			2.3H	HU
4223.00	4224.19	23673.2	XE II			2.3H	HU
4215.60	4216.79	23714.7	XE II			2.0	HU
4214.69	4215.88	23719.9	XE II			0.5	HU
4213.72	4214.91	23725.3	XE II			2.3H	HU
4209.47	4210.66	23749.3	XE II			2.0H	HU
4208.48	4209.67	23754.9	XE II			2.3H	HU
4205.404	4206.589	23772.23	XE I			1.0	IME
4203.695	4204.879	23781.90	XE I			1.7	IHU
4203.22	4204.40	23784.6	XE II			0.5	HU
4201.25	4202.43	23795.7	XE II			0.9J	HU
4197.81	4198.99	23815.2	XE II			0.7J	HU
4193.529	4194.711	23839.55	XE I			2.2	IME
4193.15	4194.33	23841.7	XE			2.3H	HU
4193.01	4194.19	23842.5	XE I			1.3	ME
4180.10	4181.28	23916.1	XE II			2.7H	HU
4170.99	4172.17	23968.4	XE II			0.6Q	HU
4162.16	4163.33	24019.2	XE II			1.5	HU
4158.04	4159.21	24043.0	XE II			2.0Q	HU
4146.78	4147.95	24108.3	XE I			0.3	ME
4138.81	4139.98	24154.7	XE			0.3H	HU
4135.133	4136.299	24176.20	XE I			1.3	IME
4131.01	4132.18	24200.3	XE II			1.0	HU
4121.86	4123.02	24254.1	XE II			0.5H	HU
4116.115	4117.276	24287.90	XE I			1.9	I
4112.14	4113.30	24311.4	XE II			1.2Q	HU
4110.41	4111.57	24321.6	XE II			1.2	HU
4109.709	4110.869	24325.76	XE I			1.8	I
4104.95	4106.11	24354.0	XE II			1.3	HU
4103.10	4104.26	24364.9	XE II			0.7B	HU
4100.34	4101.50	24381.3	XE II			1.0	HU

AIR WAVELENGTH (ANGSTRMS)	VACUUM WAVELENGTH (ANGSTRMS)	WAVENUMBER (KAYSERS)	LMENT	LOG ARC	INTENSITY SPK	DIS	REF
4098.89	4100.05	24390.0	XE II			1.7H	HU
4091.88	4093.03	24431.7	XE II			0.3H	HU
4078.820	4079.972	24509.98	XE I			2.0	I
4073.50	4074.65	24542.0	XE II			0.9	HU
4072.10	4073.25	24550.4	XE II			0.6H	HU
4062.12	4063.27	24610.7	XE II			0.5	HU
4061.06	4062.21	24617.2	XE II			0.3	HU
4057.46	4058.61	24639.0	XE II			2.0Q	HU
4051.66	4052.80	24674.3	XE			0.3	HU
4051.27	4052.41	24676.6	XE II			0.7H	
4044.90	4046.04	24715.5	XE			0.6J	HU
4044.64	4045.78	24717.1	XE			0.5J	HU
4037.59	4038.73	24760.2	XE II			2.0	HU
4037.29	4038.43	24762.1	XE II			1.7	HU
4027.97	4029.11	24819.4	XE			0.3H	HU
4026.20	4027.34	24830.3	XE II			0.5Q	HU
4025.19	4026.33	24836.5	XE II			1.2	HU
4002.35	4003.48	24978.3	XE II			1.6Q	HU
4000.55	4001.68	24989.5	XE			0.5H	HU
3996.05	3997.18	25017.6	XE II			0.3	HU
3990.33	3991.46	25053.5	XE II			1.5Q	HU
3985.202	3986.329	25085.74	XE I			1.5	IME
3975.59	3976.71	25146.4	XE II			0.5	HU
3974.417	3975.541	25153.81	XE I			1.6	IME
3972.58	3973.70	25165.4	XE II			1.4Q	HU
3967.541	3968.663	25197.40	XE I			2.3	IME
3956.85	3957.97	25265.5	XE I			0.8	ME
3954.73	3955.85	25279.0	XE II			1.2B	HU
3951.61	3952.73	25299.0	XE II			0.5Q	HU
3950.924	3952.042	25303.37	XE I			2.1	IME
3948.72	3949.84	25317.5	XE I			1.0	ME
3948.163	3949.280	25321.07	XE I			1.8	IME
3943.57	3944.69	25350.6	XE II			1.0	HU
3942.29	3943.41	25358.8	XE I			0.3	ME
3942.21	3943.33	25359.3	XE II			0.3	HU
3938.92	3940.04	25380.5	XE II			1.0	HU
3907.91	3909.02	25581.9	XE II			1.7B	HU
3905.85	3906.96	25595.4	XE II			0.7	HU
3885.45	3886.55	25729.7	XE			0.5Q	HU
3885.00	3886.10	25732.7	XE II			1.0	HU
3869.63	3870.73	25834.9	XE II			1.0	HU
3858.53	3859.62	25909.3	XE II			1.0	HU
3849.87	3850.96	25967.5	XE II			1.4Q	HU
3848.58	3849.67	25976.2	XE II			0.5	HU
3829.77	3830.86	26103.8	XE II			0.7H	HU
3826.86	3827.95	26123.7	XE I			1.2	ME
3826.27	3827.36	26127.7	XE II			0.3H	HU
3823.74	3824.83	26145.0	XE I			1.0	ME
3811.05	3812.13	26232.0	XE II			1.3	HU
3809.84	3810.92	26240.4	XE I			1.5	ME
3807.29	3808.37	26257.9	XE II			0.9	HU
3801.90	3802.98	26295.2	XE I			0.5	ME
3801.39	3802.47	26298.7	XE I			1.5	ME
3800.99	3802.07	26301.5	XE			1.0	HU
3796.30	3797.38	26334.0	XE I			1.6	ME
3795.95	3797.03	26336.4	XE I			0.5	ME
3787.32	3788.40	26396.4	XE II			0.3	HU
3783.23	3784.30	26424.9	XE II			0.7H	HU
3770.12	3771.19	26516.8	XE			0.3J	HU
3763.37	3764.44	26564.4	XE II			0.9	HU
3762.26	3763.33	26572.2	XE			0.7	HU
3762.05	3763.12	26573.7	XE			0.5H	HU
3756.87	3757.94	26610.3	XE II			0.7	HU
3745.69	3746.75	26689.8	XE I			0.6	ME
3745.38	3746.44	26692.0	XE I			1.0	ME
3737.20	3738.26	26750.4	XE			0.5J	HU
3731.18	3732.24	26793.6	XE II			1.0	HU
3727.35	3728.41	26821.1	XE			0.3H	HU
3720.80	3721.86	26868.3	XE II			1.3	HU
3717.20	3718.26	26894.3	XE II			1.0	HU

AIR WAVELENGTH (ANGSTRMS)	VACUUM WAVELENGTH (ANGSTRMS)	WAVENUMBER (KAYSERS)	LMENT	LOG ARC	INTENSITY SPK	DIS	REF
3715.69	3716.75	26905.2	XE II			0.3J	HU
3711.64	3712.70	26934.6	XE II			1.2Q	HU
3702.74	3703.79	26999.3	XE I			0.3	ME
3696.82	3697.87	27042.6	XE I			0.6	ME
3693.49	3694.54	27067.0	XE I			1.6	ME
3685.90	3686.95	27122.7	XE I			1.6	ME
3679.31	3680.36	27171.3	XE I			0.6	ME
3677.54	3678.59	27184.4	XE I			0.3	ME
3672.57	3673.62	27221.1	XE II			1.0	HU
3669.91	3670.96	27240.9	XE I			1.0	ME
3663.93	3664.97	27285.3	XE II			0.5H	HU
3661.70	3662.74	27301.9	XE II			1.0Q	HU
3658.44	3659.48	27326.3	XE II			0.5H	HU
3657.74	3658.78	27331.5	XE II			0.5	HU
3650.12	3651.16	27388.6	XE			0.5	HU
3644.91	3645.95	27427.7	XE II			0.5	HU
3644.43	3645.47	27431.3	XE II			0.5	HU
3633.06	3634.10	27517.2	XE I			0.8	ME
3621.98	3623.01	27601.3	XE II			0.3H	HU
3613.06	3614.09	27669.5	XE I			0.9	ME
3612.37	3613.40	27674.8	XE II			1.0	HU
3610.32	3611.35	27690.5	XE I			1.2	ME
3607.41	3608.44	27712.8	XE II			0.7	HU
3592.80	3593.83	27825.5	XE I			0.3	ME
3588.62	3589.64	27857.9	XE II			0.5H	HU
3587.02	3588.04	27870.3	XE I			0.6	ME
3564.30	3565.32	28048.0	XE II			1.0	HU
3563.80	3564.82	28051.9	XE I			0.5	ME
3554.04	3555.06	28129.0	XE I			1.0	ME
3549.86	3550.87	28162.1	XE I			1.0	ME
3538.08	3539.09	28255.8	XE II			0.3	HU
3533.48	3534.49	28292.6	XE I			0.3	ME
3530.21	3531.22	28318.8	XE II			0.5J	HU
3517.90	3518.91	28417.9	XE I			0.3	ME
3514.58	3515.59	28444.8	XE			0.6J	HU
3508.88	3509.88	28491.0	XE II			1.0	HU
3508.42	3509.42	28494.7	XE I			0.3	ME
3506.74	3507.74	28508.4	XE I			0.7	ME
3506.56	3507.56	28509.8	XE II			0.9	HU
3503.15	3504.15	28537.6	XE II			0.9	HU
3501.77	3502.77	28548.8	XE			1.0H	HU
3500.36	3501.36	28560.3	XE II			1.2	HU
3474.23	3475.22	28775.1	XE II			1.1	HU
3472.36	3473.35	28790.6	XE I			0.6	ME
3469.81	3470.80	28811.8	XE I			0.6	ME
3464.17	3465.16	28858.7	XE II			0.3H	HU
3461.26	3462.25	28882.9	XE II			1.7H	HU
3460.08	3461.07	28892.8	XE II			0.7	HU
3446.34	3447.33	29008.0	XE II			1.1H	HU
3442.66	3443.65	29039.0	XE I			0.5	ME
3440.75	3441.74	29055.1	XE			0.5H	HU
3437.73	3438.72	29080.6	XE			0.3J	HU
3420.73	3421.71	29225.1	XE II			1.4	HU
3420.00	3420.98	29231.4	XE I			0.3	ME
3418.37	3419.35	29245.3	XE I			0.3	ME
3413.20	3414.18	29289.6	XE II			0.6J	HU
3409.49	3410.47	29321.5	XE			0.7Q	HU
3400.07	3401.05	29402.7	XE I			0.3	ME
3395.50	3396.47	29442.3	XE II			0.3	HU
3388.05	3389.02	29507.0	XE II			0.3	HU
3386.30	3387.27	29522.3	XE II			0.3H	HU
3384.13	3385.10	29541.2	XE II			1.3H	HU
3375.16	3376.13	29619.7	XE II			0.5H	HU
3373.92	3374.89	29630.6	XE			0.3H	HU
3366.72	3367.69	29694.0	XE II			2.2H	HU
3350.44	3351.40	29838.2	XE			0.6Q	HU
3347.27	3348.23	29866.5	XE II			0.5J	HU
3344.97	3345.93	29887.0	XE II			0.3H	HU
3338.80	3339.76	29942.3	XE II			0.7	HU
3327.46	3328.42	30044.3	XE II			0.9	HU

AIR WAVELENGTH (ANGSTRMS)	VACUUM WAVELENGTH (ANGSTRMS)	WAVENUMBER (KAYSERS)	LMENT	LOG ARC	INTENSITY SPK	INTENSITY DIS	REF
3316.39	3317.34	30144.6	XE II			0.7H	HU
3313.48	3314.43	30171.1	XE			0.3H	HU
3311.80	3312.75	30186.4	XE II			0.3	HU
3310.38	3311.33	30199.3	XE			0.3	HU
3309.39	3310.34	30208.4	XE			0.3H	HU
3298.72	3299.67	30306.1	XE II			0.6	HU
3281.26	3282.21	30467.3	XE II			0.9H	HU
3280.48	3281.43	30474.6	XE			0.6	HU
3274.94	3275.88	30526.1	XE			0.3J	HU
3272.91	3273.85	30545.1	XE II			1.5	HU
3267.34	3268.28	30597.1	XE II			0.6	HU
3267.05	3267.99	30599.8	XE			0.5	HU
3266.08	3267.02	30608.9	XE			0.6J	HU
3262.02	3262.96	30647.0	XE II			0.5H	HU
3259.36	3260.30	30672.0	XE II			0.8	HU
3250.56	3251.50	30755.1	XE II			1.2	HU
3247.74	3248.68	30781.8	XE			0.6Q	HU
3229.03	3229.96	30960.1	XE			0.5H	HU
3225.08	3226.01	30998.0	XE II			1.0	HU
3212.29	3213.22	31121.4	XE II			0.5H	HU
3211.59	3212.52	31128.2	XE			0.3J	HU
3202.04	3202.97	31221.1	XE II			0.9	HU
3201.68	3202.61	31224.6	XE II			0.5H	HU
3196.22	3197.14	31277.9	XE			1.1	HU
3181.39	3182.31	31423.7	XE II			0.5H	HU
3175.64	3176.56	31480.6	XE			1.6	HU
3175.25	3176.17	31484.5	XE			0.5	HU
3168.67	3169.59	31549.9	XE II			0.3H	HU
3165.27	3166.19	31583.7	XE II			0.6	HU
3164.44	3165.36	31592.0	XE			0.3	HU
3164.23	3165.15	31594.1	XE II			0.6	HU
3162.93	3163.85	31607.1	XE II			1.1	HU
3159.75	3160.66	31638.9	XE II			0.6H	HU
3148.99	3149.90	31747.0	XE II			0.6	HU
3145.02	3145.93	31787.1	XE			0.6	HU
3143.62	3144.53	31801.2	XE II			0.7	HU
3130.40	3131.31	31935.5	XE II			0.3J	HU
3124.02	3124.93	32000.8	XE II			0.9H	HU
3121.87	3122.78	32022.8	XE II			2.2	HU
3116.78	3117.68	32075.1	XE			0.3	HU
3112.74	3113.64	32116.7	XE II			1.1	HU
3107.82	3108.72	32167.6	XE			1.2Q	HU
3104.40	3105.30	32203.0	XE II			1.6H	HU
3102.73	3103.63	32220.3	XE			0.3	HU
3101.51	3102.41	32233.0	XE II			1.5H	HU
3096.90	3097.80	32281.0	XE II			0.7	HU
3094.53	3095.43	32305.7	XE II			1.1H	HU
3092.41	3093.31	32327.9	XE II			1.0	HU
3088.92	3089.82	32364.4	XE			0.3	HU
3082.62	3083.52	32430.5	XE II			1.1	HU
3073.17	3074.06	32530.2	XE II			0.3	HU
3071.39	3072.28	32549.1	XE			0.5H	HU
3067.30	3068.19	32592.5	XE II			1.3	HU
3061.54	3062.43	32653.8	XE II			0.9	HU
3056.49	3057.38	32707.8	XE II			1.1	HU
3050.98	3051.87	32766.8	XE II			0.3	HU
3048.92	3049.81	32789.0	XE			0.3H	HU
3048.50	3049.39	32793.5	XE II			0.3H	HU
3048.17	3049.06	32797.0	XE			0.5	HU
3047.76	3048.65	32801.4	XE			0.8H	HU
3046.27	3047.16	32817.5	XE II			1.1	HU
3045.25	3046.14	32828.5	XE II			1.3	HU
3044.75	3045.64	32833.9	XE II			0.8	HU
3042.12	3043.01	32862.2	XE			0.8H	HU
3037.35	3038.23	32913.9	XE II			0.6	HU
3036.80	3037.68	32919.8	XE II			1.2H	HU
3033.71	3034.59	32953.4	XE II			0.9	HU
3033.11	3033.99	32959.9	XE			0.5	HU
3027.63	3028.51	33019.5	XE II			0.3H	HU
3027.27	3028.15	33023.5	XE II			0.5	HU
3022.10	3022.98	33079.9	XE			0.5J	HU
3019.78	3020.66	33105.4	XE II			0.3	HU
3017.43	3018.31	33131.1	XE			1.7H	HU
3015.52	3016.40	33152.1	XE II			1.0H	HU
3003.98	3004.86	33279.5	XE II			1.3	HU
2999.21	3000.08	33332.4	XE			0.9J	HU
2991.73	2992.60	33415.7	XE II			0.5	HU
2990.54	2991.41	33429.0	XEII			0.9	HU
2986.82	2987.69	33470.7	XE II			0.9	HU
2986.18	2987.05	33477.8	XE			0.7	HU
2982.23	2983.10	33522.2	XE			0.3	HU
2979.32	2980.19	33554.9	XE II			2.3	HU
2977.90	2978.77	33570.9	XE			0.7J	HU
2976.39	2977.26	33587.9	XE			0.6	HU
2974.86	2975.73	33605.2	XE II			1.3Q	HU
2972.31	2973.18	33634.0	XE II			0.7	HU
2969.80	2970.67	33662.5	XE			0.8	HU
2969.23	2970.10	33668.9	XE			0.3	HU
2964.19	2965.06	33726.2	XE			0.8	HU
2963.41	2964.28	33735.1	XE II			1.5	HU
2954.78	2955.64	33833.6	XE			0.3J	HU
2952.48	2953.34	33859.9	XE II			0.3	HU
2949.77	2950.63	33891.0	XE			0.6H	HU
2944.61	2945.47	33950.4	XE			0.3H	HU
2943.41	2944.27	33964.3	XE			0.3	HU
2942.10	2942.96	33979.4	XE			1.0H	HU
2941.38	2942.24	33987.7	XE			0.6	HU
2939.72	2940.58	34006.9	XE			0.5	HU
2935.86	2936.72	34051.6	XE			1.5H	HU
2934.80	2935.66	34063.9	XE II			0.3H	HU
2929.66	2930.52	34123.7	XE			0.3J	HU
2927.58	2928.44	34147.9	XE II			0.3H	HU
2924.38	2925.24	34185.3	XE II			0.5	HU
2923.95	2924.81	34190.3	XE			0.5	HU
2919.87	2920.72	34238.1	XE II			1.4	HU
2912.68	2913.53	34322.6	XE			0.5	HU
2910.27	2911.12	34351.0	XE			0.3J	HU
2907.18	2908.03	34387.5	XE II			1.6H	HU
2905.10	2905.95	34412.1	XE II			0.3H	HU
2904.18	2905.03	34423.0	XE			0.3	HU
2895.22	2896.07	34529.6	XE II			1.9H	HU
2889.07	2889.92	34603.1	XE II			0.9	HU
2887.12	2887.97	34626.4	XE II			0.9	HU
2883.71	2884.56	34667.4	XE II			0.8	HU
2878.48	2879.32	34730.4	XE			0.3Q	HU
2871.24	2872.08	34817.9	XE			1.4C	HU
2866.76	2867.60	34872.4	XE II			0.5	HU
2864.73	2865.57	34897.1	XE II			2.0	HU
2861.90	2862.74	34931.6	XE II			1.0H	HU
2854.53	2855.37	35021.7	XE II			1.5	HU
2852.39	2853.23	35048.0	XE			0.3H	HU
2850.95	2851.79	35065.7	XE II			0.5	HU
2849.66	2850.50	35081.6	XE			0.6	HU
2846.48	2847.32	35120.8	XE			0.9H	HU
2845.92	2846.76	35127.7	XE			0.6	HU
2844.45	2845.29	35145.9	XE			0.5	HU
2841.81	2842.65	35178.5	XE			0.3C	HU
2832.46	2833.29	35294.6	XE			0.3H	HU
2827.90	2828.73	35351.5	XE			0.6H	HU
2826.94	2827.77	35363.5	XE			1.0Q	HU
2820.06	2820.89	35449.8	XE II			0.7H	HU
2808.56	2809.39	35595.0	XE			0.3H	HU
2807.55	2808.38	35607.7	XE			0.7Q	HU
2803.02	2803.85	35665.3	XE			0.5	HU
2797.65	2798.47	35733.7	XE II			1.2H	HU
2796.49	2797.31	35748.6	XE			0.5B	HU
2785.42	2786.24	35890.6	XE II			0.3	HU
2774.86	2775.68	36027.2	XE II			1.0	HU
2773.55	2774.37	36044.2	XE			0.5H	HU
2757.86	2758.68	36249.3	XE II			1.3H	HU

AIR WAVELENGTH (ANGSTRMS)	VACUUM WAVELENGTH (ANGSTRMS)	WAVENUMBER (KAYSERS)	LMENT	LOG ARC	INTENSITY SPK	DIS	REF
2734.14	2734.95	36563.7	XE II			1.4	HU
2733.15	2733.96	36577.0	XE II			1.1C	HU
2717.35	2718.16	36789.7	XE			1.2	HU
2715.76	2716.56	36811.2	XE II			0.3	HU
2711.65	2712.45	36867.0	XE			0.3	HU
2707.39	2708.19	36925.0	XE			0.5	HU
2703.44	2704.24	36978.9	XE II			0.9	HU
2699.16	2699.96	37037.6	XE			0.3H	HU
2689.70	2690.50	37167.8	XE			0.3	HU
2687.03	2687.83	37204.8	XE			0.5	HU
2686.14	2686.94	37217.1	XE II			0.3H	HU
2682.75	2683.55	37264.1	XE			0.3	HU
2677.18	2677.98	37341.6	XE			1.4H	HU
2673.80	2674.59	37388.8	XE			0.3	HU
2672.22	2673.01	37411.0	XE II			0.5	HU
2670.68	2671.47	37432.5	XE			0.3	HU
2668.02	2668.81	37469.8	XE II			0.5	HU
2663.29	2664.08	37536.4	XE II			0.3	HU
2657.00	2657.79	37625.2	XE II			0.5H	HU
2634.20	2634.99	37950.9	XE			0.3	HU
2633.88	2634.67	37955.5	XE II			0.5	HU
2630.40	2631.18	38005.7	XE II			0.7	HU
2629.54	2630.32	38018.1	XE			0.6	HU
2621.74	2622.52	38131.2	XE			0.5	HU
2621.39	2622.17	38136.3	XE			0.5	HU
2606.93	2607.71	38347.8	XE			0.5	HU
2605.54	2606.32	38368.3	XE			1.4	HU
2598.42	2599.20	38473.4	XE			0.6	HU
2597.01	2597.79	38494.3	XE II			0.7	HU
2596.86	2597.64	38496.5	XE			0.5	HU
2585.30	2586.07	38668.7	XE			0.3	HU
2576.97	2577.74	38793.6	XE II			0.9	HU
2561.48	2562.25	39028.2	XE II			0.3	HU
2560.89	2561.66	39037.2	XE			0.5	HU
2551.70	2552.47	39177.8	XE II			0.5	HU
2546.37	2547.13	39259.8	XE			0.5	HU
2538.02	2538.78	39389.0	XE II			0.3H	HU
2531.36	2532.12	39492.6	XE			0.5	HU
2530.18	2530.94	39511.0	XE II			0.3	HU
2528.49	2529.25	39537.4	XE II			0.5	HU
2526.98	2527.74	39561.0	XE II			0.9	HU
2526.79	2527.55	39564.0	XE			0.9	HU
2524.46	2525.22	39600.5	XE II			0.5	HU
2519.17	2519.93	39683.7	XE			0.5	HU
2516.12	2516.88	39731.8	XE			0.8	HU
2514.29	2515.05	39760.7	XE			0.5	HU
2506.86	2507.62	39878.5	XE			0.6	HU
2491.78	2492.53	40119.9	XE II			0.5	HU
2490.76	2491.51	40136.3	XE II			1.0	HU
2489.11	2489.86	40162.9	XE II			1.4	HU
2478.82	2479.57	40329.6	XE			0.3	HU
2475.89	2476.64	40377.3	XE			1.7	HU
2470.18	2470.93	40470.6	XE II			0.5	HU
2469.46	2470.21	40482.5	XE			0.5	HU
2468.43	2469.18	40499.3	XE			0.5	HU
2466.60	2467.35	40529.4	XE			0.5	HU
2442.78	2443.52	40924.6	XE			0.8H	HU
2435.47	2436.21	41047.4	XE II			0.6	HU
2432.72	2433.46	41093.8	XE II			0.8	HU
2425.05	2425.79	41223.7	XE			1.3H	HU
2422.94	2423.68	41259.6	XE			0.7	HU
2421.27	2422.01	41288.1	XE			1.0H	HU
2410.72	2411.45	41468.8	XE II			0.6	HU
2409.74	2410.47	41485.6	XE			1.3	HU
2405.92	2406.65	41551.5	XE II			0.3H	HU
2401.79	2402.52	41622.9	XE II			0.3	HU
2398.76	2399.49	41675.5	XE II			0.5	HU
2392.33	2393.06	41787.5	XE			0.5	HU
2387.75	2388.48	41867.7	XE II			0.5	HU
2386.14	2386.87	41895.9	XE			0.5	HU
2378.04	2378.77	42038.6	XE			0.3	HU
2369.62	2370.34	42188.0	XE			0.6	HU
2368.68	2369.40	42204.7	XE II			0.5	HU
2362.60	2363.32	42313.3	XE II			0.3	HU
2360.42	2361.14	42352.4	XE			0.3	HU
2356.72	2357.44	42418.9	XE			0.7H	HU
2351.56	2352.28	42512.0	XE			0.3H	HU
2351.18	2351.90	42518.8	XE			0.3H	HU
2344.47	2345.19	42640.5	XE II			0.8	HU
2342.18	2342.90	42682.2	XE			0.3H	HU
2335.42	2336.14	42805.7	XE II			0.3	HU
2319.70	2320.41	43095.8	XE II			0.7	HU
2316.80	2317.51	43149.7	XE II			0.8	HU
2313.70	2314.41	43207.5	XE II			0.7	HU
2307.28	2307.99	43327.7	XE II			0.5	HU
2299.98	2300.69	43465.2	XE II			0.5	HU
2299.36	2300.07	43477.0	XE II			0.3	HU
2296.52	2297.23	43530.7	XE			1.2	HU
2294.57	2295.28	43567.7	XE II			0.9	HU
2292.40	2293.11	43609.0	XE II			1.0	HU
2290.84	2291.55	43638.7	XE II			0.5	HU
2285.94	2286.65	43732.2	XE II			0.6	HU
2285.24	2285.95	43745.6	XE			0.5	HU
2283.15	2283.85	43785.6	XE			0.6	DJ
2274.97	2275.67	43943.0	XE			0.3H	DJ
2271.37	2272.07	44012.7	XE			0.5	DJ
2268.72	2269.42	44064.1	XE II			0.5	HU
2266.80	2267.50	44101.4	XE II			0.5	HU
2265.94	2266.64	44118.1	XE II			0.5	HU
2265.62	2266.32	44124.4	XE II			0.5	HU
2264.20	2264.90	44152.0	XE II			0.5	HU
2262.95	2263.65	44176.4	XE II			0.3	HU
2259.22	2259.92	44249.4	XE II			0.3	HU
2256.56	2257.26	44301.5	XE II			0.5	HU
2249.86	2250.56	44433.4	XE II			0.3	HU
2244.00	2244.70	44549.5	XE			0.5	DJ
2241.86	2242.56	44592.0	XE II			0.6H	HU
2235.53	2236.22	44718.2	XE			0.3	DJ
2232.84	2233.53	44772.1	XE			0.3	DJ
2230.79	2231.48	44813.2	XE II			0.5	HU
2226.00	2226.69	44909.7	XE			0.3	DJ
2224.60	2225.29	44937.9	XE			0.3	DJ
2222.78	2223.47	44974.7	XE			0.6	DJ
2215.59	2216.28	45120.6	XE			0.3	DJ
2214.41	2215.10	45144.7	XE			0.5	DJ
2208.27	2208.96	45270.2	XE			0.5	DJ
2202.05	2202.74	45398.1	XE			0.3	DJ
2197.83	2198.52	45485.2	XE			0.5	DJ
2176.22	2176.90	45936.8	XE			0.3	DJ
2174.77	2175.45	45967.5	XE			0.3	DJ
2172.77	2173.45	46009.8	XE			0.3	DJ
2124.67	2125.34	47051.3	XE			0.5	DJ
2118.00	2118.67	47199.4	XE			0.3	DJ
2110.64	2111.31	47364.0	XE			0.5H	DJ
2096.53	2097.20	47682.7	XE			0.3	DJ
2067.15	2067.81	48360.3	XE			0.3	DJ
2053.65	2054.31	48678.2	XE			0.8	DJ
2052.70	2053.36	48700.7	XE			0.3	DJ
2044.99	2045.65	48884.3	XE			0.3H	DJ
2043.40	2044.06	48922.3	XE			0.3	DJ
2042.45	2043.11	48945.1	XE			0.3H	DJ
2040.35	2041.01	48995.5	XE			0.3	DJ
2039.24	2039.90	49022.1	XE			0.5	DJ
2038.31	2038.96	49044.5	XE			0.3	DJ
2036.11	2036.76	49097.5	XE			0.3	DJ
2031.92	2032.57	49198.7	XE			0.5	DJ
2027.16	2027.81	49314.2	XE			0.6	DJ
2021.04	2021.69	49463.5	XE			0.6	DJ
2013.79	2014.44	49641.6	XE			0.3	DJ
9976.40	9979.14	10020.9	YB	0.6			ME

AIR WAVELENGTH (ANGSTRMS)	VACUUM WAVELENGTH (ANGSTRMS)	WAVENUMBER (KAYSERS)	LMENT	LOG ARC	INTENSITY SPK	INTENSITY DIS	REF
9870.07	9872.78	10128.9	YB	0.8			ME
9799.88	9802.57	10201.4	YB	1.0			ME
9760.37	9763.05	10242.7	YB	2.0			ME
9524.43	9527.04	10496.4	YB	0.6			ME
9392.8	9395.4	10643.	YB	0.7			ME
9349.27	9351.84	10693.1	YB	1.3			ME
9304.44	9306.99	10744.6	YB	1.2			ME
9104.06	9106.56	10981.1	YB	0.5			ME
9094.44	9096.94	10992.7	YB	0.5			ME
8922.61	8925.06	11204.4	YB	1.3			ME
8792.	8794.	11370.	YB		1.0		IT
8607.51	8609.87	11614.6	YB	0.9			ME
8053.41	8055.62	12413.7	YB	1.0			ME
7922.40	7924.58	12619.0	YB	0.8			ME
7895.12	7897.29	12662.6	YB	1.3			ME
7877.30	7879.47	12691.2	YB	1.2			ME
7758.03	7760.16	12886.3	YB	1.0			ME
7722.58	7724.71	12945.5	YB	0.3			ME
7699.49	7701.61	12984.3	YB	3.3			ME
7618.8	7620.9	13122.	YB		0.7		IT
7527.56	7529.63	13280.9	YB	1.9			ME
7450.4	7452.5	13418.	YB	0.3			IT
7448.33	7450.38	13422.1	YB	1.5			ME
7405.92	7407.96	13499.0	YB	0.5			ME
7399.6	7401.6	13510.	YB	0.3			IT
7377.55	7379.58	13550.9	YB	0.6			ME
7350.09	7352.11	13601.5	YB	1.6			ME
7347.7	7349.7	13606.	YB	0.5			IT
7313.10	7315.11	13670.3	YB	1.3			ME
7305.25	7307.26	13685.0	YB	1.2			ME
7244.47	7246.47	13799.8	YB	1.3			ME
7222.72	7224.71	13841.4	YB	0.5			ME
7187.06	7189.04	13910.1	YB	0.7			ME
7175.14	7177.12	13933.2	YB	1.0			ME
7128.8	7130.8	14024.	YB	0.3			IT
7043.79	7045.73	14193.0	YB	0.9			ME
7036.0	7037.9	14209.	YB	0.3			IT
7032.91	7034.85	14214.9	YB	0.3			ME
7032.04	7033.98	14216.7	YB	0.5			ME
7020.16	7022.10	14240.8	YB	0.7			ME
7009.4	7011.3	14263.	YB	0.3			IT
6999.87	7001.80	14282.0	YB	1.2	0.5		ME
6982.00	6983.93	14318.6	YB	0.5	0.3		ME
6981.1	6983.0	14320.	YB	0.6			IT
6977.68	6979.60	14327.5	YB	0.3			ME
6963.11	6965.03	14357.4	YB	0.3	0.7		ME
6951.51	6953.43	14381.4	YB	0.3			ME
6934.04	6935.95	14417.6	YB	1.2	1.0		ME
6871.54	6873.44	14548.8	YB	1.0			ME
6864.26	6866.15	14564.2	YB		0.7		ME
6828.90	6830.78	14639.6	YB	0.0	0.8		ME
6820.04	6821.92	14658.6	YB	0.5H			ME
6802.47	6804.35	14696.5	YB	0.3	1.3		ME
6800.5	6802.4	14701.	YB	0.5			IT
6799.61	6801.49	14702.7	YB	3.0	1.7		ME
6794.5	6796.4	14714.	YB	0.7			IT
6785.16	6787.03	14734.0	YB	0.0	0.7		ME
6782.17	6784.04	14740.5	YB	0.6H			ME
6777.22	6779.09	14751.2	YB	0.3			ME
6768.70	6770.57	14769.8	YB	1.9	0.3		ME
6749.40	6751.26	14812.0	YB	0.8			ME
6745.22	6747.08	14821.2	YB	0.3	0.8		ME
6727.62	6729.48	14860.0	YB	1.5	1.8		ME
6715.79	6717.64	14886.2	YB	0.7			ME
6700.6	6702.4	14920.	YB	0.5			IT
6699.38	6701.23	14922.6	YB	1.0	0.9		ME
6678.17	6680.01	14970.0	YB	1.3			ME
6667.85	6669.69	14993.2	YB	3.0	1.3		ME
6644.07	6645.90	15046.9	YB	0.3	0.7H		ME
6643.54	6645.37	15048.1	YB	1.7			ME

AIR WAVELENGTH (ANGSTRMS)	VACUUM WAVELENGTH (ANGSTRMS)	WAVENUMBER (KAYSERS)	LMENT	LOG ARC	INTENSITY SPK	INTENSITY DIS	REF
6617.06	6618.89	15108.3	YB	1.0	0.8		ME
6607.07	6608.89	15131.1	YB	1.3			ME
6605.93	6607.75	15133.7	YB		0.6H		ME
6585.42	6587.24	15180.9	YB	0.8	1.3		ME
6550.19	6552.00	15262.5	YB	1.0			ME
6503.02	6504.82	15373.2	YB	0.3	0.6		ME
6492.74	6494.53	15397.6	YB	0.5	1.7		ME
6489.27	6491.06	15405.8	YB	0.7	1.5		ME
6489.10	6490.89	15406.2	YB	2.9	1.6		ME
6474.74	6476.53	15440.4	YB	0.7	1.7		ME
6463.15	6464.94	15468.1	YB	1.0	2.0		ME
6456.92	6458.70	15483.0	YB		0.6		ME
6440.81	6442.59	15521.7	YB		0.7		ME
6432.732	6434.510	15541.20	YB	1.5	1.6		MIT
6421.54	6423.31	15568.3	YB	0.5			ME
6417.97	6419.74	15576.9	YB	2.1	0.5		ME
6404.62	6406.39	15609.4	YB	0.3H			ME
6400.40	6402.17	15619.7	YB	2.3H	0.6H		ME
6387.72	6389.49	15650.7	YB	0.5			ME
6382.93	6384.69	15662.5	YB	0.0	0.3		ME
6377.04	6378.80	15676.9	YB		0.8H		ME
6372.71	6374.47	15687.6	YB	0.8			ME
6355.40	6357.16	15730.3	YB	0.3	1.7H		ME
6345.74	6347.49	15754.2	YB	0.3	0.8		ME
6345.02	6346.77	15756.0	YB	1.2H			ME
6335.72	6337.47	15779.2	YB	0.9			ME
6324.78	6326.53	15806.5	YB	0.5	0.3		ME
6308.16	6309.90	15848.1	YB	1.3	1.5		ME
6303.28	6305.02	15860.4	YB	0.5	0.6		ME
6297.38	6299.12	15875.2	YB		1.2		ME
6286.26	6288.00	15903.3	YB	0.9			ME
6277.13	6278.87	15926.4	YB		0.7		ME
6274.79	6276.53	15932.4	YB	2.0	2.2		ME
6271.16	6272.89	15941.6	YB		0.5		ME
6260.80	6262.53	15968.0	YB	0.6	1.7		ME
6247.99	6249.72	16000.7	YB	0.5			ME
6246.97	6248.70	16003.3	YB	1.6	1.8		ME
6236.56	6238.29	16030.1	YB	0.3			ME
6235.27	6236.99	16033.4	YB	0.3			ME
6234.86	6236.58	16034.4	YB		0.7		ME
6233.38	6235.10	16038.2	YB		0.7		ME
6223.65	6225.37	16063.3	YB		1.2		ME
6215.57	6217.29	16084.2	YB		0.8		ME
6208.11	6209.83	16103.5	YB	0.3	1.3H		ME
6194.85	6196.56	16138.0	YB	0.5			ME
6190.81	6192.52	16148.5	YB	0.5	1.2		ME
6189.07	6190.78	16153.1	YB	0.3	0.0		ME
6171.65	6173.36	16198.6	YB	0.7	0.3		ME
6152.571	6154.274	16248.87	YB	1.8	1.9		MIT
6150.64	6152.34	16254.0	YB	1.5H			ME
6146.94	6148.64	16263.8	YB	0.3	1.2H		ME
6134.31	6136.01	16297.2	YB	0.3	0.7		ME
6128.21	6129.91	16313.5	YB		0.8		ME
6123.03	6124.72	16327.3	YB		0.6		ME
6122.99	6124.68	16327.4	YB	0.6			ME
6120.38	6122.07	16334.3	YB		1.3H		ME
6111.28	6112.97	16358.7	YB	0.8			ME
6110.19	6111.88	16361.6	YB		0.6		ME
6106.22	6107.91	16372.2	YB		0.6		ME
6088.72	6090.41	16419.3	YB		0.5		ME
6083.280	6084.964	16433.95	YB		0.3		MIT
6082.40	6084.08	16436.3	YB		1.0H		ME
6075.25	6076.93	16455.7	YB		0.5		ME
6059.25	6060.93	16499.1	YB	1.0			ME
6056.48	6058.16	16506.7	YB	0.0	1.3H		ME
6054.56	6056.24	16511.9	YB	1.0			ME
6052.901	6054.577	16516.43	YB	0.9	1.5		ME
6048.43	6050.10	16528.6	YB	1.2			ME
6040.80	6042.47	16549.5	YB	0.5	1.0		ME
6035.72	6037.39	16563.4	YB	0.3			ME

AIR WAVELENGTH (ANGSTRMS)	VACUUM WAVELENGTH (ANGSTRMS)	WAVENUMBER (KAYSERS)	LMENT	LOG ARC	INTENSITY SPK	INTENSITY DIS	REF
6031.80	6033.47	16574.2	YB	1.0			ME
6021.97	6023.64	16601.3	YB	0.3	1.2H		ME
6020.56	6022.23	16605.1	YB	0.3	0.7		ME
6014.95	6016.62	16620.6	YB	0.5			ME
6007.42	6009.08	16641.5	YB	0.6	1.9H		ME
6003.65	6005.31	16651.9	YB	0.7			ME
6001.05	6002.71	16659.1	YB	0.3	0.0		ME
5991.503	5993.163	16685.68	YB	1.7	2.2		MIT
5989.332	5990.991	16691.73	YB	1.2			MIT
5987.91	5989.57	16695.7	YB	0.3	1.3H		ME
5959.31	5960.96	16775.8	YB	0.5			ME
5958.70	5960.35	16777.5	YB	1.0			ME
5950.98	5952.63	16799.3	YB		0.9		ME
5950.65	5952.30	16800.2	YB	0.5			ME
5947.28	5948.93	16809.7	YB	0.0	0.7		ME
5946.02	5947.67	16813.3	YB	0.6	2.0		ME
5935.06	5936.70	16844.4	YB	0.5	1.6		ME
5920.39	5922.03	16886.1	YB	0.0	1.0		ME
5908.355	5909.992	16920.50	YB	1.3	1.5		MIT
5898.80	5900.43	16947.9	YB	0.5	1.7H		ME
5897.22	5898.85	16952.4	YB	0.8	2.0H		ME
5896.61	5898.24	16954.2	YB	0.7			ME
5882.81	5884.44	16994.0	YB	0.3	1.0		ME
5874.70	5876.33	17017.4	YB	0.0	1.5H		ME
5868.40	5870.03	17035.7	YB		0.8		ME
5854.517	5856.140	17076.09	YB	1.5	0.0		MIT
5837.158	5838.776	17126.88	YB	1.7	2.2		MIT
5834.008	5835.625	17136.12	YB	1.8H	0.0		MIT
5819.430	5821.043	17179.05	YB	0.8	2.0		MIT
5803.442	5805.051	17226.38	YB	1.2			MIT
5789.94	5791.55	17266.6	YB	0.7			ME
5786.60	5788.20	17276.5	YB		0.8		ME
5771.668	5773.269	17321.21	YB	1.5	1.7		MIT
5767.23	5768.83	17334.5	YB	0.3	1.0H		ME
5755.90	5757.50	17368.7	YB	1.2			ME
5749.92	5751.51	17386.7	YB	0.9	1.0		ME
5745.80	5747.39	17399.2	YB	0.8			ME
5735.80	5737.39	17429.5	YB	0.3	1.2		ME
5730.02	5731.61	17447.1	YB	0.6	1.8		ME
5728.887	5730.476	17450.56	YB	0.7			MIT
5720.011	5721.598	17477.64	YB	2.5	0.5		MIT
5717.30	5718.89	17485.9	YB		1.2H		ME
5713.75	5715.34	17496.8	YB	0.0	1.0		ME
5706.79	5708.37	17518.1	YB	0.7	0.0		ME
5699.951	5701.532	17539.14	YB	0.3			MIT
5693.71	5695.29	17558.4	YB	0.0	1.0		ME
5689.92	5691.50	17570.1	YB	1.0			ME
5688.47	5690.05	17574.5	YB	1.2			KN
5686.554	5688.132	17580.46	YB	0.7	1.0H		MIT
5683.60	5685.18	17589.6	YB	0.3	0.8		ME
5675.845	5677.420	17613.63	YB	1.3			MIT
5653.29	5654.86	17683.9	YB		0.7		ME
5651.993	5653.562	17687.97	YB	1.7	1.9		MIT
5637.818	5639.383	17732.44	YB	0.3	0.7		MIT
5627.89	5629.45	17763.7	YB		0.7		ME
5620.24	5621.80	17787.9	YB	0.0	1.0		ME
5598.486	5600.040	17857.01	YB	0.5			MIT
5588.473	5590.025	17889.01	YB	1.5	2.0		MIT
5586.367	5587.918	17895.75	YB	1.3			MIT
5580.810	5582.360	17913.57	YB	0.5	1.5		MIT
5578.23	5579.78	17921.9	YB	0.7			ME
5572.55	5574.10	17940.1	YB	0.0	0.8		ME
5568.121	5569.667	17954.39	YB	1.3	0.0		MIT
5562.09	5563.63	17973.9	YB	0.8	0.3		MIT
5556.476	5558.019	17992.02	YB	3.2	1.7		MIT
5547.188	5548.729	18022.15	YB	0.3	1.2		MIT
5539.064	5540.602	18048.58	YB	2.3	0.7		MIT
5529.95	5531.49	18078.3	YB		0.9H		ME
5529.09	5530.63	18081.1	YB	0.0	1.0		ME
5524.554	5526.089	18095.98	YB	1.0			MIT
5520.24	5521.77	18110.1	YB		0.6H		ME
5505.501	5507.030	18158.61	YB	1.6	0.3		MIT
5498.86	5500.39	18180.5	YB		1.3B		ME
5486.59	5488.11	18221.2	YB		0.6		ME
5481.940	5483.463	18236.65	YB	1.7	0.3		MIT
5478.522	5480.044	18248.03	YB	1.0	1.7		MIT
5474.041	5475.562	18262.97	YB	1.0			MIT
5455.08	5456.60	18326.4	YB	1.4	0.8		ME
5454.016	5455.532	18330.02	YB	0.7			MIT
5449.286	5450.800	18345.93	YB	1.3	2.0		MIT
5440.58	5442.09	18375.3	YB		0.7		ME
5432.729	5434.239	18401.84	YB	0.7	2.0		MIT
5426.91	5428.42	18421.6	YB	0.3	1.8		ME
5424.68	5426.19	18429.1	YB		1.0H		ME
5414.29	5415.80	18464.5	YB	0.0	1.3		ME
5403.14	5404.64	18502.6	YB	1.3	0.0		ME
5399.742	5401.243	18514.26	YB	0.0	1.2		MIT
5395.78	5397.28	18527.9	YB		0.8		ME
5393.85	5395.35	18534.5	YB	0.8B			ME
5393.40	5394.90	18536.0	YB		0.8		ME
5390.845	5392.344	18544.81	YB	1.5			MIT
5390.655	5392.154	18545.47	YB	1.5			MIT
5389.856	5391.355	18548.21	YB	0.9	1.5		MIT
5388.01	5389.51	18554.6	YB	0.7			ME
5376.99	5378.49	18592.6	YB		1.0		ME
5368.29	5369.78	18622.7	YB	0.3	1.3		ME
5363.670	5365.162	18638.77	YB	1.4	0.3		MIT
5359.98	5361.47	18651.6	YB		1.0		ME
5358.645	5360.135	18656.25	YB	1.2	2.0		MIT
5352.958	5354.447	18676.07	YB	2.0	2.4		MIT
5351.32	5352.81	18681.8	YB	1.7	0.5		MIT
5347.205	5348.692	18696.16	YB	1.6	2.3		MIT
5345.83	5347.32	18701.0	YB	1.0	1.7		MIT
5345.67	5347.16	18701.5	YB	1.3	2.0		MIT
5338.77	5340.25	18725.7	YB		0.5		ME
5335.161	5336.645	18738.36	YB	2.2	2.6		MIT
5321.14	5322.62	18787.7	YB		0.8		ME
5309.33	5310.81	18829.5	YB	0.3	0.7		ME
5307.121	5308.598	18837.37	YB	1.6			MIT
5300.936	5302.411	18859.35	YB	0.8	1.8		MIT
5299.85	5301.32	18863.2	YB	0.5			MIT
5287.30	5288.77	18908.0	YB		0.7		ME
5286.14	5287.61	18912.1	YB		0.6		ME
5279.55	5281.02	18935.7	YB	1.2	2.0		MIT
5277.070	5278.539	18944.64	YB	2.3	0.8		MIT
5263.61	5265.07	18993.1	YB		1.0		ME
5258.19	5259.65	19012.7	YB	0.7			ME
5257.492	5258.955	19015.18	YB	1.2	2.0		MIT
5255.68	5257.14	19021.7	YB		0.9		ME
5249.84	5251.30	19042.9	YB		0.9		ME
5244.65	5246.11	19061.7	YB		1.0		ME
5244.107	5245.567	19063.72	YB	1.7	0.7		MIT
5240.502	5241.961	19076.83	YB	1.0	1.6		MIT
5236.68	5238.14	19090.7	YB		0.8H		ME
5228.205	5229.661	19121.70	YB	0.7			MIT
5227.295	5228.750	19125.03	YB	1.0			MIT
5226.19	5227.64	19129.1	YB		0.6		ME
5217.98	5219.43	19159.2	YB		0.8H		ME
5215.45	5216.90	19168.5	YB		0.7H		ME
5211.60	5213.05	19182.6	YB	1.6	0.3		MIT
5200.55	5202.00	19223.4	YB	0.3	1.0		MIT
5196.091	5197.538	19239.88	YB	1.3	0.5		MIT
5193.87	5195.32	19248.1	YB	0.6H			ME
5184.181	5185.625	19284.08	YB	0.9	1.5		MIT
5180.36	5181.80	19298.3	YB	0.3	1.0		ME
5173.13	5174.57	19325.3	YB	0.0	1.2H		ME
5152.34	5153.78	19403.2	YB		0.7H		ME
5147.026	5148.460	19423.28	YB	0.5	1.7		MIT
5142.31	5143.74	19441.1	YB		0.7H		ME
5135.980	5137.411	19465.06	YB	0.8	1.7		MIT

AIR WAVELENGTH (ANGSTRMS)	VACUUM WAVELENGTH (ANGSTRMS)	WAVENUMBER (KAYSERS)	LMENT	LOG ARC	INTENSITY SPK	DIS	REF
5132.41	5133.84	19478.6	YB	0.3	0.7H		ME
5128.54	5129.97	19493.3	YB		0.7H		ME
5126.80	5128.23	19499.9	YB	0.6			ME
5121.61	5123.04	19519.7	YB	0.0	1.2		ME
5117.75	5119.18	19534.4	YB		0.5H		ME
5113.34	5114.76	19551.2	YB	0.5			ME
5111.51	5112.93	19558.2	YB		0.8H		ME
5105.06	5106.48	19582.9	YB		0.7		ME
5104.43	5105.85	19585.4	YB	0.0	1.7		ME
5101.60	5103.02	19596.2	YB		0.6H		ME
5087.64	5089.06	19650.0	YB	0.0	1.0		ME
5085.74	5087.16	19657.3	YB		0.6H		ME
5082.60	5084.02	19669.5	YB	0.6			ME
5081.00	5082.42	19675.7	YB	0.5			ME
5076.760	5078.175	19692.11	YB	1.7	0.0		MIT
5074.340	5075.755	19701.50	YB	2.3	0.7		MIT
5069.148	5070.561	19721.68	YB	1.5	0.3		MIT
5067.812	5069.225	19726.88	YB	1.0			MIT
5067.32	5068.73	19728.8	YB	0.7	0.7		MIT
5062.95	5064.36	19745.8	YB	0.3	1.2		ME
5049.87	5051.28	19797.0	YB	0.3	1.5H		ME
5027.675	5029.077	19884.36	YB	1.0			MIT
5021.136	5022.536	19910.26	YB	0.3	1.0		MIT
5019.66	5021.06	19916.1	YB	0.6			ME
5014.50	5015.90	19936.6	YB		1.0		ME
5009.519	5010.916	19956.43	YB	1.3	1.7		MIT
4988.30	4989.69	20041.3	YB	0.6	0.0		ME
4974.16	4975.55	20098.3	YB	1.0			ME
4966.904	4968.290	20127.65	YB	1.5	0.6		MIT
4954.67	4956.05	20177.4	YB		0.6H		ME
4944.96	4946.34	20217.0	YB		0.8H		ME
4944.081	4945.461	20220.56	YB	0.6	1.3		MIT
4937.229	4938.607	20248.62	YB	0.8	1.9		MIT
4936.96	4938.34	20249.7	YB	0.5H			ME
4935.502	4936.880	20255.71	YB	2.3	1.0		MIT
4931.945	4933.322	20270.32	YB	0.6			MIT
4917.04	4918.41	20331.8	YB		0.6H		ME
4912.367	4913.739	20351.10	YB	0.7			MIT
4903.72	4905.09	20387.0	YB		0.9H		ME
4894.985	4896.352	20423.37	YB	0.3	1.0		MIT
4894.61	4895.98	20424.9	YB	1.3			MIT
4893.48	4894.85	20429.6	YB	0.8			ME
4871.16	4872.52	20523.3	YB		0.9H		ME
4851.15	4852.51	20607.9	YB	1.0	1.0		MIT
4848.46	4849.81	20619.4	YB	0.5	1.3H		MIT
4837.465	4838.817	20666.21	YB	1.0			MIT
4836.946	4838.297	20668.43	YB	1.3	2.0		MIT
4834.74	4836.09	20677.9	YB	0.3	0.9H		MIT
4830.715	4832.065	20695.09	YB	0.3			MIT
4820.237	4821.584	20740.07	YB	1.2	1.8		MIT
4818.376	4819.723	20748.08	YB	0.5	1.3		MIT
4816.40	4817.75	20756.6	YB	1.3	0.0		M
4808.51	4809.85	20790.6	YB		0.6H		ME
4786.598	4787.936	20885.83	YB	1.7	2.3		MIT
4785.34	4786.68	20891.3	YB	0.3			ME
4784.526	4785.864	20894.87	YB	0.3			MIT
4781.88	4783.22	20906.4	YB	1.7	0.7		MIT
4778.980	4780.316	20919.12	YB	0.3			MIT
4770.181	4771.515	20957.71	YB	0.3			MIT
4761.16	4762.49	20997.4	YB		0.7H		ME
4752.92	4754.25	21033.8	YB	0.9	1.3H		MIT
4751.782	4753.111	21038.85	YB	0.5			MIT
4750.01	4751.34	21046.7	YB		0.6H		ME
4746.70	4748.03	21061.4	YB		1.2H		ME
4732.022	4733.346	21126.71	YB	0.0	1.2		MIT
4726.075	4727.397	21153.29	YB	1.7	2.3		MIT
4720.779	4722.100	21177.02	YB	0.8	0.0		MIT
4718.66	4719.98	21186.5	YB	0.8O			MIT
4712.82	4714.14	21212.8	YB		1.5H		ME
4704.870	4706.186	21248.63	YB	0.6			MIT

AIR WAVELENGTH (ANGSTRMS)	VACUUM WAVELENGTH (ANGSTRMS)	WAVENUMBER (KAYSERS)	LMENT	LOG ARC	INTENSITY SPK	DIS	REF
4702.345	4703.661	21260.04	YB	0.6			MIT
4690.815	4692.128	21312.29	YB	0.3	0.7		MIT
4688.51	4689.82	21322.8	YB	0.0	0.7H		ME
4684.265	4685.576	21342.09	YB	0.7			MIT
4683.819	4685.130	21344.12	YB	0.8	1.3		MIT
4670.575	4671.882	21404.65	YB	1.1	1.2		MIT
4661.832	4663.137	21444.79	YB	0.8H			MIT
4657.04	4658.34	21466.9	YB	0.7W			MIT
4651.66	4652.96	21491.7	YB	0.7			MIT
4650.075	4651.377	21499.01	YB	1.2			MIT
4644.543	4645.844	21524.62	YB	1.5	0.5		MIT
4634.026	4635.324	21573.47	YB	0.3	1.0H		MIT
4633.185	4634.483	21577.38	YB	0.6			MIT
4624.44	4625.74	21618.2	YB	0.8			MIT
4598.37	4599.66	21740.7	YB	1.4	1.8		MIT
4597.25	4598.54	21746.0	YB		0.3		MIT
4593.37	4594.66	21764.4	YB	0.6			MIT
4590.83	4592.12	21776.4	YB	1.6			MIT
4589.22	4590.51	21784.1	YB	1.5	0.0		MIT
4582.36	4583.64	21816.7	YB	1.9	0.8		MIT
4580.73	4582.01	21824.5	YB	0.3			MIT
4576.20	4577.48	21846.1	YB	2.0	1.0		MIT
4568.864	4570.145	21881.15	YB	0.6			MIT
4567.355	4568.635	21888.37	YB	1.0	0.3		MIT
4563.995	4565.274	21904.49	YB	1.5	0.3		MIT
4557.941	4559.219	21933.58	YB	0.3	0.9		MIT
4553.561	4554.837	21954.68	YB	1.3	1.8		MIT
4548.356	4549.631	21979.80	YB	0.5	0.5		MIT
4543.935	4545.209	22001.19	YB		1.0		MIT
4541.35	4542.62	22013.7	YB		0.8		ME
4539.294	4540.567	22023.68	YB	0.3			MIT
4536.782	4538.054	22035.88	YB	0.5			MIT
4533.499	4534.770	22051.83	YB	1.2	0.5		MIT
4529.871	4531.141	22069.50	YB	1.1			MIT
4515.148	4516.414	22141.46	YB	1.7	2.0		MIT
4514.026	4515.292	22146.96	YB	1.2K			MIT
4513.391	4514.657	22150.08	YB	0.9	0.0		MIT
4503.613	4504.876	22198.17	YB	1.1			MIT
4493.967	4495.228	22245.81	YB	0.6	1.0		MIT
4488.285	4489.544	22273.98	YB	1.0			MIT
4487.27	4488.53	22279.0	YB	0.3	1.0		ME
4484.43	4485.69	22293.1	YB		0.5H		ME
4482.44	4483.70	22303.0	YB	1.3	0.3		M
4480.278	4481.535	22313.78	YB	0.8	0.3		MIT
4472.434	4473.689	22352.92	YB	0.8L			MIT
4472.19	4473.44	22354.1	YB		0.3		ME
4467.14	4468.39	22379.4	YB		0.3		ME
4463.48	4464.73	22397.8	YB		0.3		ME
4439.21	4440.46	22520.2	YB	1.7	1.0		M
4430.225	4431.469	22565.88	YB	1.1	0.0		MIT
4427.44	4428.68	22580.1	YB	0.7			M
4411.074	4412.313	22663.85	YB	1.3L			MIT
4409.34	4410.58	22672.8	YB	0.8	1.0		M
4402.30	4403.54	22709.0	YB	0.8	1.3		M
4398.95	4400.19	22726.3	YB	0.7	0.0		ME
4396.50	4397.74	22739.0	YB	0.3	0.3		M
4393.75	4394.98	22753.2	YB	1.4B			MIT
4392.83	4394.06	22758.0	YB	0.5	1.3		MIT
4389.76	4390.99	22773.9	YB	0.0	1.0		ME
4376.48	4377.71	22843.0	YB	1.0			MIT
4370.81	4372.04	22872.6	YB	1.2	1.6		MIT
4363.30	4364.53	22912.0	YB		0.3		ME
4356.65	4357.87	22947.0	YB	0.6			ME
4352.95	4354.17	22966.5	YB	0.9H			MIT
4344.22	4345.44	23012.6	YB	1.0			MIT
4339.11	4340.33	23039.7	YB	0.6	1.0		MIT
4336.48	4337.70	23053.7	YB	0.7L	0.5H		MIT
4329.72	4330.94	23089.7	YB	0.7			MIT
4326.40	4327.62	23107.4	YB	1.3H			MIT
4325.28	4326.50	23113.4	YB	1.0			ME

AIR WAVELENGTH (ANGSTRMS)	VACUUM WAVELENGTH (ANGSTRMS)	WAVENUMBER (KAYSERS)	LMENT	LOG ARC	INTENSITY SPK	DIS	REF
4322.23	4323.45	23129.7	YB	0.8	1.3		MIT
4316.969	4318.183	23157.89	YB	1.3	1.6		MIT
4312.367	4313.580	23182.60	YB	0.8			MIT
4309.82	4311.03	23196.3	YB	1.0			MIT
4306.49	4307.70	23214.2	YB		0.5		ME
4305.967	4307.178	23217.06	YB	1.2	0.5		MIT
4305.49	4306.70	23219.6	YB	0.7			MIT
4300.96	4302.17	23244.1	YB	1.2	0.0		MIT
4283.012	4284.217	23341.49	YB	1.3			MIT
4277.73	4278.93	23370.3	YB	1.4	0.3		M
4272.12	4273.32	23401.0	YB	0.8			MIT
4270.55	4271.75	23409.6	YB		0.5H		MIT
4266.99	4268.19	23429.1	YB	0.6	0.9H		MIT
4257.66	4258.86	23480.5	YB	0.8	1.2		MIT
4256.74	4257.94	23485.5	YB	0.7	0.5H		MIT
4255.77	4256.97	23490.9	YB	0.7	0.9		MIT
4254.77	4255.97	23496.4	YB	0.6	0.6		MIT
4252.52	4253.72	23508.9	YB	1.0	1.3		MIT
4251.489	4252.686	23514.55	YB	1.2	0.0		MIT
4247.888	4249.084	23534.49	YB	0.5	0.8		MIT
4244.21	4245.40	23554.9	YB		0.3H		ME
4234.56	4235.75	23608.6	YB	0.6	1.2		MIT
4233.453	4234.645	23614.73	YB	0.5			MIT
4231.989	4233.181	23622.90	YB	1.3L	0.3		MIT
4230.24	4231.43	23632.7	YB		0.6		ME
4229.66	4230.85	23635.9	YB	0.5			ME
4227.94	4229.13	23645.5	YB		1.0		ME
4218.69	4219.88	23697.4	YB	1.2			MIT
4218.567	4219.755	23698.06	YB	0.5	1.7		MIT
4216.72	4217.91	23708.4	YB	0.3	1.0		MIT
4210.30	4211.49	23744.6	YB	0.6			ME
4190.30	4191.48	23857.9	YB	0.8	1.5		M
4186.90	4188.08	23877.3	YB		1.0H		ME
4180.828	4182.006	23911.97	YB	1.0	2.0		MIT
4174.57	4175.75	23947.8	YB	1.0	0.0		MIT
4172.23	4173.41	23961.2	YB	0.3	0.3		MIT
4170.11	4171.29	23973.4	YB	0.5	1.3		MIT
4164.64	4165.81	24004.9	YB	0.3	0.3		MIT
4149.07	4150.24	24095.0	YB	1.3	0.8		MIT
4147.80	4148.97	24102.4	YB	0.3	0.3		ME
4139.048	4140.215	24153.33	YB	0.7$			MIT
4135.10	4136.27	24176.4	YB	1.2	1.7		M
4132.14	4133.31	24193.7	YB	0.5H			ME
4127.34	4128.50	24221.9	YB	0.3			ME
4122.874	4124.037	24248.08	YB	1.0	1.3		MIT
4119.43	4120.59	24268.4	YB	1.0	1.3		M
4119.29	4120.45	24269.2	YB	0.6			MIT
4113.05	4114.21	24306.0	YB	0.6	1.2		MIT
4109.609	4110.769	24326.35	YB	0.5			MIT
4097.86	4099.02	24396.1	YB	0.3	0.9		MIT
4091.50	4092.65	24434.0	YB	0.3	0.7		M
4089.685	4090.839	24444.86	YB	1.7	0.8		MIT
4082.98	4084.13	24485.0	YB I	0.5			ME
4077.27	4078.42	24519.3	YB	1.5	2.0		M
4074.692	4075.842	24534.80	YB		0.3		MIT
4056.182	4057.328	24646.76	YB	0.3	1.0		MIT
4052.28	4053.42	24670.5	YB	1.0	0.3		M
4050.109	4051.253	24683.72	YB		0.8		MIT
4047.39	4048.53	24700.3	YB		0.6		ME
4043.06	4044.20	24726.8	YB		0.9H		ME
4040.09	4041.23	24744.9	YB	0.0	0.9		ME
4028.27	4029.41	24817.5	YB		0.9H		ME
4024.01	4025.15	24843.8	YB		0.3		ME
4019.35	4020.49	24872.6	YB		0.8		ME
4007.36	4008.49	24947.0	YB	0.7			M
4000.89	4002.02	24987.4	YB		0.8H		ME
3993.770	3994.899	25031.92	YB	0.5			MIT
3990.895	3992.024	25049.95	YB	1.8	1.3		MIT
3987.994	3989.122	25068.17	YB	3.0G	2.7G		MIT
3980.430	3981.556	25115.81	YB	0.3H			MIT

AIR WAVELENGTH (ANGSTRMS)	VACUUM WAVELENGTH (ANGSTRMS)	WAVENUMBER (KAYSERS)	LMENT	LOG ARC	INTENSITY SPK	DIS	REF
3975.303	3976.428	25148.20	YB	0.9			MIT
3968.03	3969.15	25194.3	YB		0.5		ME
3961.544	3962.665	25235.54	YB	1.5			MIT
3938.25	3939.36	25384.8	YB II	1.6	2.7		ME
3934.30	3935.41	25410.3	YB	0.3	0.6		M
3911.28	3912.39	25559.8	YB	0.9			MIT
3905.876	3906.982	25595.20	YB	0.3	1.0		MIT
3904.828	3905.934	25602.07	YB	1.1	2.2		MIT
3900.865	3901.970	25628.08	YB	1.7	1.0		MIT
3887.315	3888.417	25717.41	YB	0.8	1.6		MIT
3872.863	3873.961	25813.37	YB	1.1	0.7		MIT
3863.46	3864.56	25876.2	YB	0.0	0.7		ME
3858.555	3859.649	25909.09	YB		0.5		MIT
3847.856	3848.947	25981.13	YB	1.6W			MIT
3839.935	3841.024	26034.72	YB	0.5			MIT
3839.455	3840.544	26037.98	YB	0.3	1.2		MIT
3816.197	3817.280	26196.66	YB	1.1	1.5		MIT
3814.23	3815.31	26210.2	YB	0.5	1.0		MIT
3807.54	3808.62	26256.2	YB	0.5	1.2		ME
3798.44	3799.52	26319.1	YB	0.6			MIT
3798.17	3799.25	26321.0	YB	0.6			MIT
3791.74	3792.82	26365.6	YB	0.6	0.8		MIT
3782.56	3783.63	26429.6	YB	0.7	1.4		MIT
3779.31	3780.38	26452.3	YB	0.3	0.6		MIT
3774.32	3775.39	26487.3	YB	0.9			MIT
3773.46	3774.53	26493.4	YB	0.3	0.8		MIT
3770.107	3771.178	26516.91	YB	1.1	1.2		MIT
3766.10	3767.17	26545.1	YB	0.3	0.8		MIT
3762.55	3763.62	26570.2	YB	0.6	1.0		MIT
3761.00	3762.07	26581.1	YB	0.5	0.8		MIT
3749.69	3750.76	26661.3	YB	0.3	1.0		MIT
3734.703	3735.765	26768.28	YB	1.4	0.7		MIT
3724.211	3725.270	26843.69	YB	1.2	1.7		MIT
3722.293	3723.352	26857.52	YB	0.8	1.8		MIT
3710.31	3711.37	26944.3	YB	0.6	0.8		MIT
3708.66	3709.72	26956.2	YB	0.3	0.6		MIT
3700.57	3701.62	27015.2	YB	0.6			MIT
3699.81	3700.86	27020.7	YB		0.3		ME
3699.50	3700.55	27023.0	YB	0.3			MIT
3698.59	3699.64	27029.6	YB	0.5	1.2		ME
3694.203	3695.254	27061.74	YB	2.7G	3.0G		MIT
3691.693	3692.744	27080.14	YB	1.0			MIT
3690.577	3691.627	27088.32	YB	1.0	1.7		MIT
3675.093	3676.139	27202.45	YB	1.7	2.3		MIT
3670.68	3671.73	27235.1	YB	0.3	1.0		MIT
3669.710	3670.755	27242.35	YB	1.7	1.9		MIT
3664.73	3665.77	27279.4	YB	0.3	0.6		MIT
3664.60	3665.64	27280.3	YB	0.3	0.3		MIT
3655.729	3656.770	27346.53	YB	0.9			MIT
3637.755	3638.792	27481.65	YB	1.2	1.5		MIT
3634.551	3635.587	27505.87	YB	0.3			MIT
3629.90	3630.93	27541.1	YB	0.3	0.7		MIT
3621.00	3622.03	27608.8	YB		0.7		ME
3619.809	3620.841	27617.89	YB	1.5	2.0		MIT
3615.097	3616.128	27653.89	YB	0.5			MIT
3611.307	3612.337	27682.91	YB	1.1	1.7		MIT
3606.47	3607.50	27720.0	YB	1.2	1.8		MIT
3600.756	3601.783	27764.02	YB	1.0	1.3		MIT
3600.39	3601.42	27766.9	YB	0.3	0.3		MIT
3585.468	3586.491	27882.40	YB	1.0	1.3		MIT
3577.05	3578.07	27948.0	YB	0.8	1.3		MIT
3575.573	3576.594	27959.56	YB	1.3	0.3		MIT
3570.56	3571.58	27998.8	YB	0.3	1.2		MIT
3567.092	3568.111	28026.04	YB	0.0	0.6		MIT
3560.727	3561.744	28076.13	YB	0.9	2.0		MIT
3560.33	3561.35	28079.3	YB	1.3	1.3		MIT
3559.607	3560.624	28084.97	YB	1.0			MIT
3559.076	3560.093	28089.16	YB	0.7			MIT
3552.324	3553.339	28142.55	YB II	2.0	3.0		ME
3549.82	3550.83	28162.4	YB	1.0	1.3		MIT

AIR WAVELENGTH (ANGSTRMS)	VACUUM WAVELENGTH (ANGSTRMS)	WAVENUMBER (KAYSERS)	LMENT	LOG ARC	INTENSITY SPK	INTENSITY DIS	REF
3544.91	3545.92	28201.4	YB	0.3			MIT
3543.15	3544.16	28215.4	YB	0.3	0.7		MIT
3531.25	3532.26	28310.5	YB	0.3	0.7		MIT
3520.29	3521.30	28398.6	YB	1.0	1.8		MIT
3518.172	3519.178	28415.73	YB	0.8	1.5		MIT
3517.015	3518.021	28425.07	YB	1.4W			MIT
3515.86	3516.87	28434.4	YB II	1.5	2.5		ME
3507.835	3508.838	28499.46	YB	1.1	1.8		MIT
3495.90	3496.90	28596.7	YB II		1.7H		ME
3488.80	3489.80	28654.9	YB	1.1	1.2		MIT
3485.76	3486.76	28679.9	YB	1.3	1.6		MIT
3478.84	3479.84	28737.0	YB	1.6	2.5		MIT
3476.30	3477.30	28758.0	YB	1.9	1.0		MIT
3474.81	3475.80	28770.3	YB	1.0	1.3		MIT
3472.33	3473.32	28790.9	YB	1.0			MIT
3464.37	3465.36	28857.0	YB	2.3G	1.7R		MIT
3464.19	3465.18	28858.5	YB	1.1			MIT
3461.71	3462.70	28879.2	YB	0.0	0.5		ME
3460.27	3461.26	28891.2	YB	1.5	0.7		MIT
3458.28	3459.27	28907.8	YB	1.1	2.0		MIT
3454.07	3455.06	28943.1	YB	1.6	2.4		MIT
3452.39	3453.38	28957.1	YB	1.1			MIT
3446.88	3447.87	29003.4	YB	1.0	1.7		MIT
3438.84	3439.83	29071.2	YB	1.3	2.0		MIT
3438.72	3439.71	29072.3	YB	0.6	1.9		MIT
3436.46	3437.45	29091.4	YB	0.7	1.5		MIT
3434.61	3435.59	29107.0	YB	0.7	1.3		MIT
3431.38	3432.36	29134.4	YB	0.9			MIT
3431.12	3432.10	29136.6	YB	1.6	1.2		MIT
3428.46	3429.44	29159.2	YB	1.4	1.9		MIT
3426.04	3427.02	29179.9	YB	1.6	1.2		MIT
3420.35	3421.33	29228.4	YB	1.0			MIT
3419.61	3420.59	29234.7	YB	0.8	1.2		MIT
3418.39	3419.37	29245.1	YB	0.5	0.7		MIT
3416.89	3417.87	29258.0	YB	0.8	1.2		MIT
3412.45	3413.43	29296.1	YB	1.2	0.7		MIT
3411.24	3412.22	29306.4	YB	0.8			MIT
3408.51	3409.49	29329.9	YB		1.0		MIT
3404.10	3405.08	29367.9	YB	1.0	1.5		MIT
3402.27	3403.25	29383.7	YB	0.8	0.5		MIT
3401.01	3401.99	29394.6	YB	1.1	1.5		MIT
3400.60	3401.58	29398.1	YB	0.8			MIT
3398.65	3399.63	29415.0	YB	1.0			MIT
3396.32	3397.29	29435.2	YB	0.6	1.3		MIT
3394.44	3395.41	29451.5	YB	0.9	1.3		MIT
3391.10	3392.07	29480.5	YB	1.0	1.6		MIT
3390.25	3391.22	29487.9	YB	1.0W			MIT
3387.50	3388.47	29511.8	YB	1.5			MIT
3384.04	3385.01	29542.0	YB		1.3		ME
3379.78	3380.75	29579.2	YB	1.1	1.3		MIT
3376.62	3377.59	29606.9	YB	0.6			MIT
3375.48	3376.45	29616.9	YB	1.5	2.0		MIT
3365.96	3366.93	29700.7	YB	0.6	1.5		MIT
3363.65	3364.62	29721.1	YB	1.0			MIT
3362.43	3363.40	29731.9	YB	1.2			MIT
3356.97	3357.93	29780.2	YB	0.5	1.3		MIT
3355.87	3356.83	29790.0	YB	0.6	1.3		MIT
3353.74	3354.70	29808.9	YB	0.7	0.9		MIT
3352.49	3353.45	29820.0	YB	1.3			MIT
3347.53	3348.49	29864.2	YB	0.7	1.0		ME
3346.50	3347.46	29873.4	YB	1.2			MIT
3343.06	3344.02	29904.1	YB	0.8	1.6		MIT
3342.93	3343.89	29905.3	YB	1.5W	0.9		MIT
3337.17	3338.13	29956.9	YB	1.4L	1.5		MIT
3333.07	3334.03	29993.7	YB	1.3	1.8		MIT
3327.77	3328.73	30041.5	YB	0.3	1.0		MIT
3325.57	3326.53	30061.4	YB		1.3		ME
3324.17	3325.13	30074.1	YB	0.8	1.3		MIT
3320.31	3321.27	30109.0	YB	0.5	1.3		MIT
3319.41	3320.37	30117.2	YB	1.4			MIT

AIR WAVELENGTH (ANGSTRMS)	VACUUM WAVELENGTH (ANGSTRMS)	WAVENUMBER (KAYSERS)	LMENT	LOG ARC	INTENSITY SPK	INTENSITY DIS	REF
3319.18	3320.14	30119.3	YB	0.6	1.0		MIT
3316.84	3317.79	30140.5	YB	0.7	1.2		MIT
3315.38	3316.33	30153.8	YB	0.5	1.3		MIT
3315.10	3316.05	30156.3	YB	1.0J			MIT
3309.37	3310.32	30208.5	YB	1.1	1.5		MIT
3306.78	3307.73	30232.2	YB	1.0	1.5		MIT
3305.73	3306.68	30241.8	YB	1.5	2.1		MIT
3305.25	3306.20	30246.2	YB	1.3			MIT
3304.76	3305.71	30250.7	YB	1.1	1.6		MIT
3304.56	3305.51	30252.5	YB	1.2	1.6		MIT
3302.44	3303.39	30271.9	YB	0.8	0.7		MIT
3297.85	3298.80	30314.1	YB	0.6	1.3		MIT
3294.34	3295.29	30346.4	YB	0.5	1.3		MIT
3289.37	3290.32	30392.2	YB II	2.7G	3.0G		ME
3286.97	3287.92	30414.4	YB	0.8	1.5		MIT
3285.60	3286.55	30427.1	YB	0.8			MIT
3279.98	3280.93	30479.2	YB	0.7			MIT
3278.96	3279.90	30488.7	YB	1.1			MIT
3276.80	3277.74	30508.8	YB	1.0			MIT
3275.81	3276.75	30518.0	YB	1.1	2.0		MIT
3272.64	3273.58	30547.6	YB	0.5			MIT
3271.54	3272.48	30557.8	YB	0.8	1.0		MIT
3271.18	3272.12	30561.2	YB	1.0	1.3		MIT
3270.75	3271.69	30565.2	YB	1.4D			MIT
3267.89	3268.83	30592.0	YB	0.8			MIT
3266.63	3267.57	30603.8	YB	0.6			MIT
3261.70	3262.64	30650.0	YB		0.7		ME
3261.509	3262.449	30651.82	YB	0.7	1.3		MIT
3259.103	3260.043	30674.44	YB	0.3	0.9		MIT
3239.611	3240.546	30859.00	YB	0.3			MIT
3239.193	3240.128	30862.98	YB	0.3	0.9		MIT
3236.16	3237.09	30891.9	YB	0.3	1.0		MIT
3231.98	3232.91	30931.9	YB	0.3	0.8		MIT
3228.599	3229.531	30964.25	YB	0.0	1.0		MIT
3225.852	3226.783	30990.61	YB	0.5	1.2		MIT
3221.22	3222.15	31035.2	YB	0.3	0.5		MIT
3218.32	3219.25	31063.1	YB	0.3	1.0		MIT
3217.17	3218.10	31074.2	YB	0.3	0.3		MIT
3210.11	3211.04	31142.6	YB	0.0	0.9		MIT
3207.70	3208.63	31166.0	YB		0.3		MIT
3206.15	3207.08	31181.1	YB	0.0	0.5		MIT
3204.68	3205.61	31195.4	YB	0.0	0.5		MIT
3201.159	3202.084	31229.66	YB	0.6	1.6		MIT
3198.644	3199.568	31254.21	YB	0.3	1.3		MIT
3196.35	3197.27	31276.6	YB		0.3		MIT
3195.58	3196.50	31284.2	YB	0.3	0.8		MIT
3194.765	3195.689	31292.16	YB II	1.0	2.0		ME
3194.24	3195.16	31297.3	YB	0.0	0.5		MIT
3192.878	3193.801	31310.65	YB	1.1	1.9		MIT
3191.41	3192.33	31325.1	YB	0.0	0.9H		ME
3186.58	3187.50	31372.5	YB	0.0	0.5		MIT
3180.912	3181.832	31428.44	YB	0.3	1.5		MIT
3175.75	3176.67	31479.5	YB	0.0	0.7		MIT
3174.76	3175.68	31489.3	YB	0.3	0.6		ME
3173.78	3174.70	31499.1	YB	0.0	0.8		MIT
3171.188	3172.106	31524.80	YB	0.3	0.6		MIT
3169.058	3169.975	31545.99	YB	0.3	1.3		MIT
3168.41	3169.33	31552.4	YB		0.3		MIT
3165.21	3166.13	31584.3	YB	0.3	1.2		MIT
3163.80	3164.72	31598.4	YB	0.3	1.3		MIT
3162.30	3163.22	31613.4	YB	0.3			MIT
3158.30	3159.21	31653.4	YB	0.3	1.3		MIT
3155.79	3156.70	31678.6	YB	0.3	0.3		MIT
3155.18	3156.09	31684.7	YB	0.3	1.2		MIT
3153.88	3154.79	31697.8	YB	0.5	1.5		MIT
3153.19	3154.10	31704.7	YB	0.3	0.7		MIT
3152.45	3153.36	31712.2	YB		0.3		MIT
3151.44	3152.35	31722.3	YB		0.3		MIT
3149.00	3149.91	31746.9	YB	0.3	0.6		MIT
3146.14	3147.05	31775.8	YB		0.3		MIT

AIR WAVELENGTH (ANGSTRMS)	VACUUM WAVELENGTH (ANGSTRMS)	WAVENUMBER (KAYSERS)	LMENT	LOG ARC	INTENSITY SPK	INTENSITY DIS	REF
3145.54	3146.45	31781.8	YB	0.3	0.3		MIT
3145.052	3145.963	31786.77	YB	0.5	1.2		MIT
3141.73	3142.64	31820.4	YB	0.3	1.0		MIT
3140.934	3141.844	31828.44	YB	0.9	1.5		MIT
3138.63	3139.54	31851.8	YB		0.6		ME
3136.76	3137.67	31870.8	YB	0.3	1.2		MIT
3132.628	3133.536	31912.83	YB II	1.2	2.2		MIT
3129.14	3130.05	31948.4	YB		0.3		MIT
3127.86	3128.77	31961.5	YB	0.0	0.6		MIT
3127.14	3128.05	31968.8	YB		0.3		MIT
3126.06	3126.97	31979.9	YB	0.0	1.4H		ME
3125.44	3126.35	31986.2	YB	0.0	0.3		MIT
3123.51	3124.42	32006.0	YB	0.0	0.3		MIT
3117.804	3118.708	32064.56	YB	0.8	1.5		MIT
3116.702	3117.606	32075.90	YB II	1.8	2.5		ME
3116.49	3117.39	32078.1	YB	0.0	0.5		MIT
3116.06	3116.96	32082.5	YB	0.0	0.5		MIT
3115.333	3116.236	32089.99	YB	0.3	1.3		MIT
3112.98	3113.88	32114.2	YB		0.3		MIT
3109.78	3110.68	32147.3	YB	0.0	0.3		MIT
3107.897	3108.799	32166.77	YB	1.0	1.8		MIT
3107.758	3108.660	32168.21	YB	0.3	1.3		MIT
3102.07	3102.97	32227.2	YB	0.0	0.9		MIT
3101.357	3102.257	32234.60	YB	0.3	1.1		MIT
3095.40	3096.30	32296.6	YB	0.5			MIT
3095.23	3096.13	32298.4	YB		0.3		MIT
3094.92	3095.82	32301.6	YB	0.0	0.3		MIT
3093.879	3094.777	32312.50	YB	0.5	1.2		MIT
3093.44	3094.34	32317.1	YB		0.3		MIT
3092.56	3093.46	32326.3	YB		1.5		MIT
3089.092	3089.989	32362.58	YB	0.5	1.1		MIT
3086.98	3087.88	32384.7	YB	0.0	0.6		MIT
3085.82	3086.72	32396.9	YB	0.0	0.6		MIT
3084.35	3085.25	32412.3	YB		0.6		MIT
3076.038	3076.932	32499.91	YB	0.0	0.5		MIT
3073.68	3074.57	32524.8	YB	0.0	0.5		MIT
3071.59	3072.48	32547.0	YB		0.5		MIT
3068.28	3069.17	32582.1	YB		0.5		MIT
3067.37	3068.26	32591.7	YB		0.5		MIT
3065.048	3065.939	32616.44	YB	0.6	1.5		MIT
3064.91	3065.80	32617.9	YB	0.3	0.3		MIT
3063.69	3064.58	32630.9	YB	0.5	0.7		MIT
3063.12	3064.01	32637.0	YB	0.0	0.6		MIT
3055.15	3056.04	32722.1	YB	0.0	0.5		MIT
3047.048	3047.934	32809.10	YB	0.0	0.8		MIT
3046.493	3047.379	32815.08	YB	0.3	1.0		MIT
3044.83	3045.72	32833.0	YB	0.0	0.5		MIT
3044.006	3044.892	32841.89	YB	0.3	0.5		MIT
3042.65	3043.54	32856.5	YB	0.7	1.7		ME
3040.49	3041.37	32879.9	YB	0.0	0.5		ME
3039.66	3040.54	32888.9	YB	0.5	1.2		MIT
3038.54	3039.42	32901.0	YB		0.3		MIT
3037.99	3038.87	32906.9	YB	0.0	0.6		MIT
3036.82	3037.70	32919.6	YB	0.0	0.5		MIT
3034.632	3035.515	32943.34	YB	0.3	0.8		MIT
3033.86	3034.74	32951.7	YB	0.0	0.5		MIT
3031.11	3031.99	32981.6	YB	2.0H	1.5		MIT
3029.554	3030.436	32998.55	YB	0.0	1.6H		MIT
3028.38	3029.26	33011.3	YB		0.3		MIT
3026.668	3027.549	33030.02	YB II	2.0	2.8		ME
3023.61	3024.49	33063.4	YB		0.3		MIT
3022.46	3023.34	33076.0	YB	0.0	0.6		MIT
3020.70	3021.58	33095.3	YB		0.7		ME
3019.06	3019.94	33113.2	YB		0.3		MIT
3017.56	3018.44	33129.7	YB	0.5	1.6		MIT
3014.45	3015.33	33163.9	YB	0.3	1.2		MIT
3010.622	3011.499	33206.05	YB	0.3	1.0		MIT
3009.39	3010.27	33219.6	YB	0.3	1.3		MIT
3006.86	3007.74	33247.6	YB	0.0	0.5		MIT
3005.765	3006.641	33259.71	YB	1.0	2.0		MIT

AIR WAVELENGTH (ANGSTRMS)	VACUUM WAVELENGTH (ANGSTRMS)	WAVENUMBER (KAYSERS)	LMENT	LOG ARC	INTENSITY SPK	INTENSITY DIS	REF
3002.61	3003.49	33294.6	YB	1.2	2.2		MIT
3002.03	3002.91	33301.1	YB		1.2		MIT
3001.28	3002.15	33309.4	YB		0.3		MIT
3000.46	3001.33	33318.5	YB	0.5	1.3		MIT
2998.38	2999.25	33341.6	YB		0.5		ME
2998.02	2998.89	33345.6	YB	0.0	1.0		MIT
2995.86	2996.73	33369.7	YB	0.3	1.0		MIT
2994.802	2995.675	33381.46	YB	1.0	1.9		MIT
2993.94	2994.81	33391.1	YB	0.0	0.8		MIT
2991.870	2992.743	33414.17	YB	0.6	1.3		MIT
2990.36	2991.23	33431.0	YB	0.5	1.2		MIT
2985.88	2986.75	33481.2	YB	0.3	1.0		MIT
2985.079	2985.950	33490.18	YB	0.7	1.5		MIT
2984.81	2985.68	33493.2	YB		0.7		MIT
2983.98	2984.85	33502.5	YB	1.0	1.8		MIT
2983.69	2984.56	33505.8	YB	0.3	0.9		MIT
2982.65	2983.52	33517.5	YB	0.0	0.6		MIT
2982.49	2983.36	33519.2	YB	0.5	1.2		MIT
2979.85	2980.72	33549.0	YB	0.0	0.6		MIT
2979.67	2980.54	33551.0	YB		0.3		MIT
2978.90	2979.77	33559.6	YB		0.5		MIT
2977.52	2978.39	33575.2	YB		0.6		MIT
2975.56	2976.43	33597.3	YB		0.3		MIT
2972.48	2973.35	33632.1	YB		0.5		MIT
2970.83	2971.70	33650.8	YB	0.3	1.3		MIT
2970.564	2971.431	33653.81	YB	2.2	2.2		MIT
2966.75	2967.62	33697.1	YB		1.2		MIT
2965.16	2966.03	33715.1	YB		0.6		MIT
2964.75	2965.62	33719.8	YB	1.2	1.4		MIT
2964.40	2965.27	33723.8	YB	0.0	0.3H		MIT
2963.43	2964.30	33734.8	YB	0.3	1.3		MIT
2963.16	2964.03	33737.9	YB		1.0		MIT
2962.517	2963.382	33745.22	YB	0.5	1.2		MIT
2961.79	2962.66	33753.5	YB		0.5		MIT
2960.84	2961.70	33764.3	YB	0.0	0.7		MIT
2957.63	2958.49	33801.0	YB		0.5		MIT
2955.31	2956.17	33827.5	YB	0.5	1.0		MIT
2953.03	2953.89	33853.6	YB		0.7		MIT
2951.02	2951.88	33876.7	YB	0.0	0.6		MIT
2950.32	2951.18	33884.7	YB	0.3	1.0		MIT
2947.113	2947.974	33921.60	YB		0.3		MIT
2946.75	2947.61	33925.8	YB	0.0	0.9		MIT
2946.29	2947.15	33931.1	YB	0.5	1.2		MIT
2945.90	2946.76	33935.6	YB	1.0	1.8		MIT
2944.46	2945.32	33952.2	YB		0.5		MIT
2942.82	2943.68	33971.1	YB	0.0	0.7		MIT
2942.03	2942.89	33980.2	YB	0.0	0.8		MIT
2940.51	2941.37	33997.8	YB	0.5	1.4		MIT
2939.53	2940.39	34009.1	YB	0.3	0.8		MIT
2938.18	2939.04	34024.7	YB	0.0	0.5		MIT
2937.18	2938.04	34036.3	YB	0.3	1.0		MIT
2935.10	2935.96	34060.4	YB	0.5	1.2		MIT
2934.35	2935.21	34069.1	YB	0.5			MIT
2927.85	2928.71	34144.8	YB	0.3	1.0		MIT
2924.24	2925.10	34186.9	YB	0.5	1.2		MIT
2921.12	2921.98	34223.4	YB	0.3	1.3		MIT
2919.35	2920.20	34244.2	YB	1.2	2.0		MIT
2916.43	2917.28	34278.5	YB	0.0	1.0		MIT
2915.27	2916.12	34292.1	YB	1.0	1.6		MIT
2914.48	2915.33	34301.4	YB		0.6		MIT
2914.21	2915.06	34304.6	YB	1.0	1.8		MIT
2912.86	2913.71	34320.5	YB		0.5		MIT
2911.52	2912.37	34336.3	YB	0.7	1.6		MIT
2909.48	2910.33	34360.3	YB	0.3	0.9		MIT
2909.19	2910.04	34363.8	YB	0.3	0.8		MIT
2908.33	2909.18	34373.9	YB	0.0	0.7		MIT
2908.09	2908.94	34376.8	YB		0.6		MIT
2906.87	2907.72	34391.2	YB		0.5		MIT
2906.34	2907.19	34397.5	YB		1.6		ME
2902.92	2903.77	34438.0	YB		0.8		MIT

AIR WAVELENGTH (ANGSTRMS)	VACUUM WAVELENGTH (ANGSTRMS)	WAVENUMBER (KAYSERS)	LMENT	LOG ARC	INTENSITY SPK	DIS	REF
2902.41	2903.26	34444.0	YB		0.3		MIT
2899.71	2900.56	34476.1	YB	0.3	1.0		MIT
2898.33	2899.18	34492.5	YB		1.0		MIT
2896.90	2897.75	34509.5	YB	0.3	0.7		MIT
2894.99	2895.84	34532.3	YB		0.7	D	ME
2893.63	2894.48	34548.5	YB	0.0	0.9		MIT
2891.38	2892.23	34575.4	YB	1.7H	2.0		MIT
2888.03	2888.88	34615.5	YB	0.6	1.3		MIT
2886.26	2887.11	34636.8	YB	0.3	0.7		MIT
2885.96	2886.81	34640.4	YB	0.0	0.6		MIT
2882.15	2883.00	34686.1	YB		0.3		MIT
2881.92	2882.77	34688.9	YB		0.5		MIT
2879.16	2880.00	34722.2	YB		0.7		MIT
2875.88	2876.72	34761.8	YB		0.9		MIT
2870.06	2870.90	34832.2	YB	0.0	1.0		MIT
2867.04	2867.88	34868.9	YB	0.6	1.6		MIT
2866.18	2867.02	34879.4	YB		0.5		ME
2861.31	2862.15	34938.8	YB	0.5	1.4		ME
2861.21	2862.05	34940.0	YB	0.5	1.5		MIT
2860.40	2861.24	34949.9	YB	0.0	0.9		ME
2859.80	2860.64	34957.2	YB	1.3	1.5		MIT
2859.38	2860.22	34962.4	YB	0.3	1.3		MIT
2858.45	2859.29	34973.7	YB		0.6		ME
2858.35	2859.19	34975.0	YB	0.0	0.7		ME
2854.49	2855.33	35022.2	YB	0.0	0.9		ME
2854.12	2854.96	35026.8	YB	0.3	1.0		MIT
2853.40	2854.24	35035.6	YB		0.6		ME
2851.12	2851.96	35063.6	YB	1.0	1.7		MIT
2849.34	2850.18	35085.5	YB		0.6		MIT
2848.44	2849.28	35096.6	YB	0.3	1.2		MIT
2847.24	2848.08	35111.4	YB		1.5	H	MIT
2847.175	2848.012	35112.21	YB II	2.6	3.0		ME
2844.75	2845.59	35142.1	YB		0.3	H	MIT
2843.82	2844.66	35153.6	YB		0.3	H	MIT
2843.00	2843.84	35163.8	YB		0.7		MIT
2842.58	2843.42	35169.0	YB		0.3		MIT
2838.65	2839.48	35217.7	YB		0.3		MIT
2834.97	2835.80	35263.4	YB		0.9		MIT
2832.20	2833.03	35297.9	YB		0.3		MIT
2830.98	2831.81	35313.1	YB	0.3	1.6		MIT
2827.91	2828.74	35351.4	YB		0.3		MIT
2824.96	2825.79	35388.3	YB	0.3	1.0		MIT
2823.57	2824.40	35405.7	YB		0.5		MIT
2821.14	2821.97	35436.2	YB	0.3	1.4		MIT
2818.75	2819.58	35466.3	YB		1.9		ME
2817.01	2817.84	35488.2	YB		1.3		ME
2816.33	2817.16	35496.7	YB		0.3		MIT
2814.54	2815.37	35519.3	YB	0.0	0.7		MIT
2814.24	2815.07	35523.1	YB		0.3		MIT
2810.72	2811.55	35567.6	YB		0.3		MIT
2810.10	2810.93	35575.4	YB		0.3		MIT
2808.30	2809.13	35598.2	YB		0.5		MIT
2807.22	2808.05	35611.9	YB		0.3	D	MIT
2804.26	2805.09	35649.5	YB		0.3		MIT
2803.48	2804.31	35659.5	YB		1.6		MIT
2800.04	2800.87	35703.2	YB	0.3	0.9		MIT
2799.37	2800.20	35711.8	YB	0.3	0.5		MIT
2798.22	2799.04	35726.5	YB	0.6	1.0		MIT
2797.77	2798.59	35732.2	YB		0.8		MIT
2795.63	2796.45	35759.6	YB		0.9		MIT
2795.07	2795.89	35766.7	YB		0.5		MIT
2794.80	2795.62	35770.2	YB		0.5		MIT
2794.43	2795.25	35774.9	YB	0.0	0.7		MIT
2793.28	2794.10	35789.6	YB	0.0	0.9		MIT
2789.43	2790.25	35839.0	YB		0.3		MIT
2788.31	2789.13	35853.4	YB		0.9		MIT
2787.97	2788.79	35857.8	YB		0.5		MIT
2784.65	2785.47	35900.6	YB	0.3	1.3		MIT
2780.04	2780.86	35960.1	YB		0.3		MIT
2776.27	2777.09	36008.9	YB	0.8	1.6		MIT

AIR WAVELENGTH (ANGSTRMS)	VACUUM WAVELENGTH (ANGSTRMS)	WAVENUMBER (KAYSERS)	LMENT	LOG ARC	INTENSITY SPK	DIS	REF
2775.41	2776.23	36020.1	YB		0.3		MIT
2774.32	2775.14	36034.2	YB		0.5		MIT
2771.32	2772.14	36073.2	YB	0.3	1.3		MIT
2768.27	2769.09	36113.0	YB		0.3		MIT
2765.55	2766.37	36148.5	YB		0.7		MIT
2764.41	2765.23	36163.4	YB	0.0	0.9		MIT
2761.37	2762.19	36203.2	YB	0.3	1.3		MIT
2760.77	2761.59	36211.1	YB	0.3	1.3		MIT
2759.53	2760.35	36227.4	YB		0.5		MIT
2758.99	2759.81	36234.4	YB	0.0	0.9		MIT
2756.01	2756.82	36273.6	YB		0.5		MIT
2754.93	2755.74	36287.8	YB	0.3	0.5		MIT
2753.67	2754.48	36304.4	YB		0.3	H	MIT
2751.45	2752.26	36333.7	YB	0.3	1.2		MIT
2750.48	2751.29	36346.5	YB	1.3	2.2		MIT
2749.96	2750.77	36353.4	YB		0.9		ME
2749.62	2750.43	36357.9	YB		0.3		MIT
2748.66	2749.47	36370.6	YB	0.7	1.5		MIT
2748.03	2748.84	36379.0	YB		0.3		MIT
2747.57	2748.38	36385.0	YB	0.3	1.2		MIT
2745.71	2746.52	36409.7	YB		0.5		MIT
2741.72	2742.53	36462.7	YB	0.3	1.2		MIT
2738.80	2739.61	36501.5	YB		0.3		MIT
2737.61	2738.42	36517.4	YB		0.3		MIT
2736.57	2737.38	36531.3	YB		0.3		MIT
2734.08	2734.89	36564.5	YB	0.0	0.6		MIT
2732.73	2733.54	36582.6	YB	0.8	1.5		MIT
2729.49	2730.30	36626.0	YB		0.3		MIT
2722.20	2723.01	36724.1	YB		0.6		MIT
2719.02	2719.83	36767.1	YB		0.3		MIT
2718.34	2719.15	36776.3	YB	0.7	1.5		MIT
2717.66	2718.47	36785.5	YB		0.3		MIT
2715.94	2716.74	36808.8	YB		0.3		MIT
2714.40	2715.20	36829.6	YB		0.6		MIT
2712.65	2713.45	36853.4	YB	0.3	0.7		MIT
2710.54	2711.34	36882.1	YB	0.3	1.0	D	MIT
2709.71	2710.51	36893.4	YB		0.3		MIT
2708.84	2709.64	36905.2	YB	0.0	0.5		MIT
2704.50	2705.30	36964.5	YB		0.3		MIT
2700.80	2701.60	37015.1	YB		0.9		MIT
2696.62	2697.42	37072.5	YB		0.6		MIT
2695.43	2696.23	37088.8	YB	0.0	0.8		MIT
2692.68	2693.48	37126.7	YB		0.3		MIT
2692.40	2693.20	37130.6	YB		0.3		MIT
2692.02	2692.82	37135.8	YB		0.6		MIT
2690.99	2691.79	37150.0	YB		1.5		ME
2687.97	2688.77	37191.7	YB	0.3	0.6		MIT
2686.00	2686.80	37219.0	YB		0.5		MIT
2684.76	2685.56	37236.2	YB	0.5	1.4		MIT
2683.43	2684.23	37254.7	YB		1.0		MIT
2680.40	2681.20	37296.8	YB	0.0	0.5		MIT
2677.37	2678.17	37339.0	YB		1.6		ME
2676.13	2676.93	37356.3	YB		0.3		MIT
2674.87	2675.66	37373.9	YB		0.6		MIT
2672.65	2673.44	37404.9	YB	1.3	1.9		MIT
2671.98	2672.77	37414.3	YB	1.0	0.3		MIT
2668.74	2669.53	37459.7	YB	0.3	1.3		MIT
2666.97	2667.76	37484.6	YB	0.7	2.2		MIT
2666.08	2666.87	37497.1	YB	0.7	2.2		MIT
2665.03	2665.82	37511.9	YB	1.0	1.8		MIT
2660.01	2660.80	37582.7	YB	0.0	0.5		MIT
2659.28	2660.07	37593.0	YB	0.3	0.7		MIT
2656.11	2656.90	37637.8	YB	0.3	0.8		MIT
2653.75	2654.54	37671.3	YB	1.7	2.3		MIT
2652.23	2653.02	37692.9	YB	0.3	1.8		ME
2651.71	2652.50	37700.3	YB	0.3	1.8		MIT
2650.74	2651.53	37714.1	YB	0.3	0.6		MIT
2649.78	2650.57	37727.7	YB	0.6	1.0		MIT
2648.80	2649.59	37741.7	YB	0.3	0.9		MIT
2647.47	2648.26	37760.7	YB	0.3	0.8		MIT

AIR WAVELENGTH (ANGSTRMS)	VACUUM WAVELENGTH (ANGSTRMS)	WAVENUMBER (KAYSERS)	LMENT	LOG ARC	INTENSITY SPK	DIS	REF
2647.25	2648.04	37763.8	YB		0.6		MIT
2646.46	2647.25	37775.1	YB	0.3	1.0		MIT
2644.32	2645.11	37805.6	YB	0.7	1.6		MIT
2643.66	2644.45	37815.1	YB		0.7		MIT
2642.55	2643.34	37831.0	YB	0.5	1.9		MIT
2641.90	2642.69	37840.3	YB	0.6	1.0		MIT
2640.54	2641.33	37859.8	YB		1.2 H		MIT
2639.44	2640.23	37875.5	YB	0.7	1.2		MIT
2638.09	2638.88	37894.9	YB	0.0	1.8		ME
2636.46	2637.25	37918.4	YB		0.7 H		MIT
2635.39	2636.18	37933.7	YB		0.7 H		ME
2634.32	2635.11	37949.1	YB	0.5	1.0		MIT
2632.66	2633.44	37973.1	YB		0.6		MIT
2631.72	2632.50	37986.6	YB	0.0	0.6		MIT
2628.77	2629.55	38029.3	YB		0.3		MIT
2628.04	2628.82	38039.8	YB	0.0	0.6		MIT
2627.05	2627.83	38054.2	YB		1.3		MIT
2623.21	2623.99	38109.9	YB	0.0	0.7		MIT
2621.67	2622.45	38132.2	YB	0.0	0.6		MIT
2621.12	2621.90	38140.2	YB	0.3	2.0		M
2619.91	2620.69	38157.9	YB	0.3	0.7		MIT
2619.06	2619.84	38170.2	YB	0.3	0.8		MIT
2617.02	2617.80	38200.0	YB	0.5	1.2		MIT
2615.26	2616.04	38225.7	YB	0.5	1.1		MIT
2612.62	2613.40	38264.3	YB	0.0	0.6		MIT
2612.03	2612.81	38273.0	YB	0.0	0.6		ME
2610.87	2611.65	38290.0	YB	0.5	1.3		MIT
2609.12	2609.90	38315.7	YB		0.6		ME
2607.87	2608.65	38334.0	YB	0.5	0.8		MIT
2604.03	2604.81	38390.5	YB	0.3	0.9		MIT
2600.85	2601.63	38437.5	YB	0.0	0.5		MIT
2600.19	2600.97	38447.2	YB	0.0	0.5		MIT
2599.15	2599.93	38462.6	YB	0.3	1.7		M
2597.27	2598.05	38490.5	YB	0.0	1.3		MIT
2596.74	2597.52	38498.3	YB	0.3	0.5		MIT
2596.30	2597.08	38504.8	YB	0.3	0.8		MIT
2596.16	2596.94	38506.9	YB	0.3	0.8		MIT
2594.49	2595.27	38531.7	YB	0.3	0.7		M
2594.18	2594.96	38536.3	YB	0.0	0.7		ME
2592.71	2593.49	38558.1	YB		0.5		ME
2591.04	2591.81	38583.0	YB	0.0	0.7		MIT
2588.65	2589.42	38618.6	YB		0.9 H		ME
2586.29	2587.06	38653.9	YB	0.0	0.6		MIT
2583.73	2584.50	38692.1	YB		0.3		MIT
2581.12	2581.89	38731.3	YB	1.3	2.0		MIT
2579.58	2580.35	38754.4	YB	0.7	2.3		ME
2577.67	2578.44	38783.1	YB	0.3	1.5		MIT
2574.80	2575.57	38826.3	YB	0.0	0.5		MIT
2573.14	2573.91	38851.4	YB	0.5	1.0		MIT
2572.09	2572.86	38867.2	YB		0.5		MIT
2571.35	2572.12	38878.4	YB	0.8	1.3		MIT
2567.63	2568.40	38934.7	YB	0.7	2.2		ME
2566.82	2567.59	38947.0	YB		0.5		MIT
2565.58	2566.35	38965.9	YB	0.7	1.2		MIT
2560.57	2561.34	39042.1	YB		0.7		MIT
2557.70	2558.47	39085.9	YB		0.6		ME
2557.25	2558.02	39092.8	YB	0.3	0.9		MIT
2556.25	2557.02	39108.1	YB		0.5		MIT
2555.31	2556.08	39122.5	YB		1.7 H		ME
2552.70	2553.47	39162.5	YB	0.9	1.5		MIT
2552.15	2552.92	39170.9	YB	1.0	1.6		MIT
2550.80	2551.57	39191.6	YB		0.7 H		MIT
2550.41	2551.18	39197.6	YB		0.5		MIT
2550.04	2550.81	39203.3	YB	0.3	0.6		MIT
2548.71	2549.47	39223.8	YB		0.6		MIT
2547.50	2548.26	39242.4	YB		0.5		MIT
2542.82	2543.58	39314.6	YB		0.6 J		ME
2540.97	2541.73	39343.2	YB		0.3 J		MIT
2538.67	2539.43	39378.9	YB	1.0	1.3		MIT
2538.19	2538.95	39386.3	YB	0.0	0.5		MIT
2537.64	2538.40	39394.9	YB	0.6	1.2		MIT
2536.01	2536.77	39420.2	YB		0.8		ME
2532.57	2533.33	39473.7	YB		0.3 H		MIT
2529.53	2530.29	39521.1	YB		0.3 H		MIT
2528.33	2529.09	39539.9	YB		0.5		MIT
2527.86	2528.62	39547.3	YB	0.0	0.7		MIT
2526.28	2527.04	39572.0	YB		0.5 H		MIT
2525.00	2525.76	39592.0	YB		0.3		MIT
2522.42	2523.18	39632.5	YB	0.6	1.0		MIT
2521.03	2521.79	39654.4	YB	0.0	0.6		MIT
2520.32	2521.08	39665.6	YB		0.5		ME
2517.99	2518.75	39702.3	YB		0.6		ME
2516.83	2517.59	39720.6	YB	0.3	1.3		ME
2516.362	2517.119	39727.95	YB	0.3	0.7		MIT
2515.60	2516.36	39740.0	YB		0.5		ME
2512.58	2513.34	39787.7	YB		0.7 H		ME
2512.06	2512.82	39796.0	YB	1.0	1.7		MIT
2510.50	2511.26	39820.7	YB		0.5		MIT
2505.46	2506.21	39900.8	YB	0.3	0.7		MIT
2501.99	2502.74	39956.1	YB	0.7	1.3		MIT
2498.35	2499.10	40014.4	YB	0.0	0.5		ME
2495.69	2496.44	40057.0	YB		0.7 H		ME
2493.63	2494.38	40090.1	YB	0.0	0.6		ME
2491.69	2492.44	40121.3	YB		0.5		ME
2490.45	2491.20	40141.3	YB	0.0	1.3		ME
2488.96	2489.71	40165.3	YB		0.5		ME
2484.88	2485.63	40231.2	YB	0.3	0.7		ME
2481.40	2482.15	40287.7	YB	0.0	0.5		ME
2466.62	2467.37	40529.0	YB	0.0	0.7		MIT
2464.49	2465.24	40564.1	YB	0.7	0.3		MIT
2461.40	2462.14	40615.0	YB		0.5 H		MIT
2460.24	2460.98	40634.1	YB	0.5	1.2		MIT
2456.00	2456.74	40704.3	YB		0.5 H		ME
2454.76	2455.50	40724.9	YB		1.0 V		MIT
2447.23	2447.97	40850.1	YB		0.3 D		MIT
2440.57	2441.31	40961.6	YB		0.5 H		ME
2439.68	2440.42	40976.6	YB		0.3		MIT
2439.29	2440.03	40983.1	YB		0.3		MIT
2433.61	2434.35	41078.8	YB		0.5 H		MIT
2422.83	2423.57	41261.5	YB	0.0	0.3		MIT
2421.36	2422.10	41286.6	YB	0.6	1.2		MIT
2406.06	2406.79	41549.1	YB	0.0	0.5 D		MIT
2404.07	2404.80	41583.5	YB		0.3 H		MIT
2403.40	2404.13	41595.1	YB		0.3		MIT
2398.01	2398.74	41688.5	YB	1.0	0.7		MIT
2397.83	2398.56	41691.7	YB		0.5		MIT
2390.73	2391.46	41815.5	YB	1.0	1.3		MIT
2388.41	2389.14	41856.1	YB	0.0	0.5		MIT
2385.006	2385.733	41915.83	YB	0.0	0.7		MIT
2380.39	2381.12	41997.1	YB	0.3	1.0		MIT
2373.89	2374.61	42112.1	YB	0.3	0.9		MIT
2373.06	2373.78	42126.8	YB	0.0	0.7		ME
2370.104	2370.828	42179.36	YB		0.5 H		MIT
2369.44	2370.16	42191.2	YB	0.0	1.0		ME
2367.58	2368.30	42224.3	YB		1.0 H		ME
2366.78	2367.50	42238.6	YB	0.5	1.1		ME
2365.45	2366.17	42262.3	YB		1.2		ME
2363.48	2364.20	42297.6	YB		0.6 H		ME
2362.89	2363.61	42308.1	YB	1.0	1.3		MIT
2361.09	2361.81	42340.4	YB	0.3	0.8		ME
2349.39	2350.11	42551.2	YB	0.0	0.9		ME
2344.667	2345.385	42636.92	YB	0.7	1.3		MIT
2340.75	2341.47	42708.3	YB		0.5 H		ME
2340.39	2341.11	42714.8	YB	0.3	0.9 H		ME
2337.95	2338.67	42759.4	YB	0.3	1.6		ME
2335.419	2336.135	42805.74	YB	0.5	1.2 H		MIT
2326.91	2327.62	42962.3	YB	0.3	0.8 H		ME
2323.17	2323.88	43031.4	YB	0.0	0.3		ME
2320.828	2321.541	43074.84	YB	1.3	0.7		MIT
2315.215	2315.927	43179.26	YB	0.7	1.0		MIT

AIR WAVELENGTH (ANGSTRMS)	VACUUM WAVELENGTH (ANGSTRMS)	WAVENUMBER (KAYSERS)	LMENT	LOG ARC	INTENSITY SPK	DIS	REF
2314.48	2315.19	43193.0	YB	0.6	1.7		MIT
2312.56	2313.27	43228.8	YB	0.0	0.3		ME
2310.67	2311.38	43264.2	YB	0.3			ME
2309.26	2309.97	43290.6	YB		1.3		ME
2308.49	2309.20	43305.0	YB		0.3H		ME
2307.41	2308.12	43325.3	YB	0.0	0.8H		ME
2305.33	2306.04	43364.4	YB	0.9	2.0		ME
2303.30	2304.01	43402.6	YB		0.3D		ME
2298.66	2299.37	43490.2	YB	0.0	0.6		MIT
2297.88	2298.59	43505.0	YB	0.3	0.6		ME
2292.81	2293.52	43601.2	YB	0.0	0.5		ME
2292.27	2292.98	43611.4	YB		0.5H		ME
2291.63	2292.34	43623.6	YB		0.5H		ME
2288.93	2289.64	43675.1	YB	0.0	0.8		ME
2285.80	2286.51	43734.9	YB	0.5	0.9H		ME
2283.98	2284.68	43769.7	YB		0.9H		ME
2283.38	2284.08	43781.2	YB	0.9	1.0		ME
2282.99	2283.69	43788.7	YB	0.7	1.7		MIT
2276.07	2276.77	43921.8	YB	0.3			ME
2271.13	2271.83	44017.3	YB	0.9			MIT
2268.64	2269.34	44065.6	YB		0.5H		ME
2268.28	2268.98	44072.6	YB	0.0	0.6		ME
2265.65	2266.35	44123.8	YB	0.5	1.7		ME
2263.85	2264.55	44158.9	YB		0.3		ME
2263.16	2263.86	44172.3	YB		0.5H		ME
2262.25	2262.95	44190.1	YB	0.3	1.5		ME
2261.12	2261.82	44212.2	YB		0.5		ME
2257.01	2257.71	44292.7	YB	0.6	1.6		ME
2244.25	2244.95	44544.5	YB		0.9H		ME
2240.09	2240.79	44627.2	YB	0.7	1.4		ME
2237.48	2238.17	44679.3	YB		0.5		ME
2231.55	2232.24	44798.0	YB		0.3		ME
2227.70	2228.39	44875.4	YB		0.5		ME
2224.83	2225.52	44933.3	YB		0.3		ME
2224.45	2225.14	44941.0	YB	1.3	1.6		ME
2221.47	2222.16	45001.2	YB		0.3H		ME
2214.64	2215.33	45140.0	YB	0.0	0.9		ME
2211.87	2212.56	45196.5	YB		0.3		ME
2203.07	2203.76	45377.0	YB		0.3		ME
2201.19	2201.88	45415.8	YB		0.5		ME
2198.14	2198.83	45478.8	YB		1.3H		ME
2193.40	2194.09	45577.1	YB		0.6D		ME
2193.08	2193.77	45583.7	YB		0.3		ME
2189.42	2190.10	45659.9	YB		0.9		ME
2188.25	2188.93	45684.3	YB		0.7		ME
2185.69	2186.37	45737.8	YB	1.8	2.0		ME
2184.78	2185.46	45756.9	YB		0.3		ME
2183.61	2184.29	45781.4	YB		0.5H		ME
2183.29	2183.97	45788.1	YB		0.8		ME
2182.55	2183.23	45803.6	YB		0.5H		ME
2181.85	2182.53	45818.3	YB		0.3		ME
2181.51	2182.19	45825.5	YB		0.3		ME
2180.24	2180.92	45852.1	YB		0.8H		ME
2178.80	2179.48	45882.5	YB		0.9H		ME
2177.50	2178.18	45909.8	YB		0.7		ME
2175.38	2176.06	45954.6	YB		0.5H		ME
2174.28	2174.96	45977.8	YB	0.6	1.0		ME
2173.34	2174.02	45997.7	YB	0.3	0.7		ME
2172.58	2173.26	46013.8	YB		0.3		ME
2172.13	2172.81	46023.3	YB		0.5		ME
2169.77	2170.45	46073.4	YB		0.7		ME
2169.10	2169.78	46087.6	YB		0.8		ME
2168.09	2168.77	46109.1	YB		0.5		ME
2165.53	2166.21	46163.6	YB		0.5		ME
2165.19	2165.87	46170.8	YB		0.9H		ME
2163.87	2164.55	46199.0	YB		1.0		ME
2161.60	2162.28	46247.5	YB	2.0	2.4		ME
2160.27	2160.95	46276.0	YB		1.0H		ME
2159.88	2160.56	46284.3	YB		0.5H		ME
2158.79	2159.47	46307.7	YB		0.6		ME
2157.87	2158.55	46327.4	YB		0.5H		ME
2156.79	2157.47	46350.6	YB		0.3		ME
2156.51	2157.19	46356.6	YB		0.5		ME
2155.51	2156.19	46378.1	YB		1.6		ME
2155.18	2155.86	46385.2	YB		0.8		ME
2154.16	2154.84	46407.2	YB		1.9H		ME
2152.32	2153.00	46446.9	YB	0.5	1.0H		ME
2151.20	2151.88	46471.1	YB		0.3H		ME
2150.45	2151.13	46487.3	YB		0.5		ME
2149.43	2150.11	46509.3	YB		0.5		ME
2148.92	2149.60	46520.4	YB		1.0		ME
2148.49	2149.17	46529.7	YB		0.6		ME
2148.07	2148.75	46538.8	YB		0.9		ME
2147.50	2148.18	46551.1	YB		0.5H		ME
2146.46	2147.14	46573.7	YB		0.5		ME
2144.74	2145.42	46611.0	YB		1.5		ME
2143.86	2144.54	46630.1	YB		0.6		ME
2143.42	2144.10	46639.7	YB		0.7		ME
2142.18	2142.85	46666.7	YB		0.6		ME
2141.67	2142.34	46677.8	YB		0.7		ME
2141.00	2141.67	46692.4	YB		1.0		ME
2140.41	2141.08	46705.3	YB		0.3		ME
2139.96	2140.63	46715.1	YB		1.4		ME
2139.25	2139.92	46730.6	YB		0.3		ME
2138.32	2138.99	46751.0	YB		1.0		ME
2137.71	2138.38	46764.3	YB		1.2		ME
2137.52	2138.19	46768.4	YB		1.0		ME
2136.33	2137.00	46794.5	YB		0.3		ME
2135.22	2135.89	46818.8	YB		1.3H		ME
2134.98	2135.65	46824.1	YB		0.5H		ME
2133.66	2134.33	46853.0	YB		0.6H		ME
2133.34	2134.01	46860.1	YB		0.5		ME
2133.17	2133.84	46863.8	YB		0.5		ME
2131.71	2132.38	46895.9	YB		0.3H		ME
2131.37	2132.04	46903.4	YB	1.0	1.3		ME
2129.64	2130.31	46941.5	YB		0.6H		ME
2128.82	2129.49	46959.5	YB		0.3		ME
2126.72	2127.39	47005.9	YB	1.6	2.3		ME
2123.30	2123.97	47081.6	YB		1.2		ME
2122.81	2123.48	47092.5	YB		0.5		ME
2121.54	2122.21	47120.7	YB		1.2		ME
2119.25	2119.92	47171.6	YB		1.5H		ME
2117.83	2118.50	47203.2	YB		0.6H		ME
2116.65	2117.32	47229.5	YB	1.7	2.4		ME
2115.84	2116.51	47247.6	YB		0.3		ME
2115.46	2116.13	47256.1	YB		0.6H		ME
2110.60	2111.27	47364.9	YB		0.5		ME
2110.26	2110.93	47372.5	YB		0.7H		ME
2109.60	2110.27	47387.3	YB		2.0H		ME
2108.20	2108.87	47418.8	YB		0.7		ME
2106.46	2107.13	47458.0	YB		0.3		ME
2102.72	2103.39	47542.4	YB	1.3	2.3		ME
2102.10	2102.77	47556.4	YB		0.5H		ME
2098.40	2099.07	47640.2	YB		1.7H		ME
2098.16	2098.83	47645.7	YB		0.5		ME
2097.41	2098.08	47662.7	YB		0.3		ME
2096.84	2097.51	47675.7	YB		1.5H		ME
2094.82	2095.49	47721.6	YB		1.0H		ME
2093.18	2093.85	47759.0	YB		0.8H		ME
2092.31	2092.98	47778.9	YB		1.5H		ME
2091.28	2091.94	47802.4	YB		1.3H		ME
2088.05	2088.71	47876.3	YB		1.2H		ME
2087.37	2088.03	47891.9	YB		0.5H		ME
2083.65	2084.31	47977.4	YB		0.3		ME
2078.12	2078.78	48105.1	YB		1.0H		ME
2073.70	2074.36	48207.6	YB		0.9H		ME
9494.81	9497.41	10529.2	YT I	1.8			ME
9231.58	9234.11	10829.4	YT I	1.9			ME
8835.85	8838.28	11314.4	YT II	0.3			ME
8800.62	8803.04	11359.7	YT I	0.9			ME

AIR WAVELENGTH (ANGSTRMS)	VACUUM WAVELENGTH (ANGSTRMS)	WAVENUMBER (KAYSERS)	LMENT	LOG ARC	INTENSITY SPK	INTENSITY DIS	REF
8575.77	8578.13	11657.6	YT I	0.3			ME
8528.94	8531.28	11721.6	YT I	0.6			ME
8475.64	8477.97	11795.3	YT I	0.5			ME
8450.36	8452.68	11830.6	YT I	0.9			ME
8365.64	8367.94	11950.4	YT I	0.3			ME
8344.43	8346.72	11980.7	YT I	0.8			ME
8329.61	8331.90	12002.1	YT I	0.7			ME
8297.07	8299.35	12049.1	YT	0.3			ME
8258.53	8260.80	12105.4	YT	0.3			MIT
8231.23	8233.49	12145.5	YT I	0.3H			ME
8211.71	8213.97	12174.4	YT I	0.3H			ME
8165.5	8167.7	12243.	YT	0.3			ME
8134.9	8137.1	12289.	YT I	0.3H			ME
8066.20	8068.42	12394.0	YT II	0.5			ME
8025.60	8027.81	12456.7	YT I	0.5			ME
7999.33	8001.53	12497.6	YT I	0.3H			ME
7887.51	7889.68	12674.8	YT I	0.3			ME
7881.90	7884.07	12683.8	YT II	1.4	1.0		ME
7870.04	7872.21	12702.9	YT I	0.3			ME
7855.52	7857.68	12726.4	YT I	0.8	0.0		ME
7812.16	7814.31	12797.0	YT I	0.6	0.0		ME
7802.52	7804.67	12812.8	YT I	0.3			ME
7796.32	7798.47	12823.0	YT I	0.6			ME
7788.42	7790.56	12836.0	YT I	0.3			ME
7781.868	7784.009	12846.85	YT	0.7			KS
7724.08	7726.21	12943.0	YT I	0.7	0.0		ME
7719.89	7722.01	12950.0	YT I	0.8	0.0		ME
7698.00	7700.12	12986.8	YT	0.6			ME
7689.49	7691.61	13001.2	YT I	0.3			ME
7652.89	7655.00	13063.4	YT I	0.5			ME
7622.94	7625.04	13114.7	YT I	0.7	0.0		ME
7617.72	7619.82	13123.7	YT	0.6			ME
7563.13	7565.21	13218.4	YT I	1.0	0.6		MIT
7536.71	7538.79	13264.7	YT	0.5	0.0		MIT
7494.88	7496.94	13338.8	YT I	0.5	0.3		MIT
7478.8	7480.9	13367.	YT I	0.5			ME
7472.18	7474.24	13379.3	YT I	0.5			MIT
7455.20	7457.25	13409.8	YT I	0.5			ME
7450.30	7452.35	13418.6	YT II	1.2	1.0		MIT
7406.21	7408.25	13498.5	YT II	0.5	0.6		MIT
7398.77	7400.81	13512.0	YT I	1.1	0.5		MIT
7346.46	7348.48	13608.2	YT I	1.6	0.6		MIT
7332.96	7334.98	13633.3	YT II	0.3	0.5		MIT
7330.62	7332.64	13637.7	YT	0.9			ME
7293.08	7295.09	13707.8	YT I	0.8			MIT
7264.18	7266.18	13762.4	YT II	1.2	1.3		MIT
7195.93	7197.91	13892.9	YT I	1.0	0.5		MIT
7191.65	7193.63	13901.2	YT I	1.3	0.7		MIT
7127.92	7129.88	14025.5	YT I	1.1	0.5		ME
7075.13	7077.08	14130.1	YT I	0.5	0.3		MIT
7054.28	7056.23	14171.9	YT I	0.8	0.5		ME
7052.94	7054.88	14174.6	YT I	0.6	0.8		MIT
7035.19	7037.13	14210.3	YT I	1.1	0.8		MIT
7009.92	7011.85	14261.6	YT I	1.0	0.8		MIT
7008.95	7010.88	14263.5	YT I	1.0	0.8		MIT
6979.85	6981.78	14323.0	YT I	1.2	0.8		MIT
6958.03	6959.95	14367.9	YT I	0.5	0.3		MIT
6951.67	6953.59	14381.1	YT II	0.7	0.9		MIT
6950.27	6952.19	14384.0	YT I	1.3	1.0		MIT
6933.52	6935.43	14418.7	YT I	0.8	0.7		MIT
6908.26	6910.17	14471.4	YT I	1.0			MIT
6896.01	6897.91	14497.1	YT II	0.7	1.2		MIT
6887.20	6889.10	14515.7	YT I	0.9	1.1		MIT
6858.23	6860.12	14577.0	YT II	0.5	0.9		MIT
6845.24	6847.13	14604.7	YT I	1.2	0.7		MIT
6832.47	6834.36	14632.0	YT II	0.7	0.9		MIT
6815.16	6817.04	14669.1	YT I	0.5			ME
6803.16	6805.04	14695.0	YT I	0.6			ME
6795.40	6797.28	14711.8	YT II	1.1	1.3		MIT
6793.70	6795.57	14715.5	YT I	1.8	1.2		MIT

AIR WAVELENGTH (ANGSTRMS)	VACUUM WAVELENGTH (ANGSTRMS)	WAVENUMBER (KAYSERS)	LMENT	LOG ARC	INTENSITY SPK	INTENSITY DIS	REF
6736.00	6737.86	14841.5	YT I	1.1	0.6		MIT
6713.19	6715.04	14891.9	YT I	1.2	0.6		MIT
6700.72	6702.57	14919.7	YT	1.1	0.9		MIT
6699.26	6701.11	14922.9	YT	1.3	0.6		ME
6694.75	6696.60	14932.9	YT I	0.6			MIT
6691.85	6693.70	14939.4	YT I	0.5	0.3		MIT
6687.60	6689.45	14948.9	YT I	1.7	1.0		MIT
6664.42	6666.26	15000.9	YT I	0.8	0.5		MIT
6650.62	6652.46	15032.0	YT I	1.2	0.6		MIT
6636.51	6638.34	15064.0	YT I	0.9			MIT
6622.47	6624.30	15095.9	YT I	0.5			MIT
6613.74	6615.57	15115.9	YT II	1.2	1.1		MIT
6584.84	6586.66	15182.2	YT I	0.5			MIT
6576.885	6578.702	15200.57	YT I	1.0	0.8		MIT
6572.58	6574.40	15210.5	YT	0.5	0.8		MIT
6557.39	6559.20	15245.8	YT I	0.3	0.8		MIT
6538.601	6540.407	15289.57	YT I	1.7	1.0		MIT
6535.029	6536.834	15297.93	YT	0.6	0.8		KS
6493.776	6495.570	15395.11	YT I	0.8			MIT
6462.559	6464.345	15469.47	YT	0.6	0.7		MIT
6437.158	6438.937	15530.51	YT I	0.6	0.7		MIT
6435.000	6436.779	15535.72	YT I	2.2	1.7		MIT
6402.005	6403.775	15615.79	YT I	1.1	0.8		MIT
6387.07	6388.84	15652.3	YT	0.9	0.7		MIT
6338.13	6339.88	15773.2	YT	0.7	0.7		MIT
6251.05	6252.78	15992.9	YT	0.5	0.3		MIT
6222.595	6224.317	16066.02	YT I	0.7	0.7		MIT
6217.887	6219.607	16078.19	YT	1.0	0.3		MIT
6191.722	6193.435	16146.13	YT I	0.9	0.7		MIT
6151.72	6153.42	16251.1	YT	0.5			ME
6138.411	6140.110	16286.35	YT I	0.8	0.6		MIT
6135.04	6136.74	16295.3	YT I	0.5	0.3		ME
6132.063	6133.760	16303.21	YT	1.1L	0.9W		MIT
6087.99	6089.68	16421.2	YT I	0.5	0.3		ME
6040.283	6041.956	16550.93	YT I	0.3			ED
6024.261	6025.929	16594.95	YT I	0.3			MIT
6023.392	6025.060	16597.34	YT I	0.5	0.5		MIT
6007.69	6009.35	16640.7	YT I	0.5	0.5		ME
6004.65	6006.31	16649.1	YT	0.8			ME
5981.86	5983.52	16712.6	YT I	0.5			ME
5945.717	5947.364	16814.17	YT I	0.6			MIT
5902.93	5904.57	16936.1	YT I	0.7			ME
5879.94	5881.57	17002.3	YT I	0.5			ME
5871.80	5873.43	17025.8	YT I	0.3			ME
5832.26	5833.88	17141.3	YT I	0.3			ME
5821.853	5823.467	17171.90	YT I	0.6			MIT
5812.64	5814.25	17199.1	YT I	0.3			ME
5781.687	5783.290	17291.20	YT II	0.7	0.7		MIT
5773.860	5775.461	17314.63	YT I	0.5			MIT
5765.652	5767.251	17339.28	YT I	0.7			MIT
5743.854	5745.447	17405.09	YT I	0.8	0.0		MIT
5732.090	5733.680	17440.81	YT I	0.3			MIT
5728.881	5730.470	17450.57	YT II	0.5	1.4		MIT
5723.46	5725.05	17467.1	YT I	0.3			ME
5720.609	5722.196	17475.81	YT I	0.5			MIT
5714.91	5716.50	17493.2	YT I	0.3			ME
5706.724	5708.307	17518.33	YT I	1.2	0.0		MIT
5693.650	5695.230	17558.55	YT I	0.3			MIT
5675.263	5676.838	17615.44	YT I	0.7	0.0		MIT
5662.924	5664.496	17653.82	YT II	1.3	2.6		MIT
5648.48	5650.05	17699.0	YT I	1.1	0.3		ME
5646.462	5648.029	17705.29	YT I	1.0	1.0		MIT
5644.680	5646.247	17710.88	YT I	1.2	0.3		MIT
5632.268	5633.831	17749.91	YT I	0.3			MIT
5630.123	5631.686	17756.67	YT I	1.9			MIT
5623.946	5625.507	17776.18	YT I	0.3			MIT
5606.328	5607.884	17832.04	YT I	1.1	0.5		MIT
5594.122	5595.675	17870.94	YT I	0.3			MIT
5590.96	5592.51	17881.1	YT I	0.6	0.3		ME
5581.868	5583.418	17910.18	YT I	2.0	1.0		MIT

AIR WAVELENGTH (ANGSTRMS)	VACUUM WAVELENGTH (ANGSTRMS)	WAVENUMBER (KAYSERS)	LMENT	LOG ARC	INTENSITY SPK	DIS	REF
5581.097	5582.647	17912.65	YT I	0.3			MIT
5577.416	5578.965	17924.47	YT I	1.2	0.3		MIT
5567.752	5569.298	17955.58	YT I	0.7	0.3		MIT
5556.44	5557.98	17992.1	YT I	0.7			ME
5546.007	5547.547	18025.98	YT II	0.8	1.0		MIT
5544.615	5546.155	18030.51	YT II	1.0	1.9		MIT
5541.632	5543.171	18040.22	YT I	0.3			MIT
5527.769	5529.304	18085.46	YT I	1.0			MIT
5527.540	5529.075	18086.21	YT I	2.0	1.2		MIT
5526.803	5528.338	18088.62	YT I	0.3			MIT
5521.70	5523.23	18105.3	YT II	0.6	1.6		ME
5521.628	5523.162	18105.57	YT I	0.7			MIT
5513.607	5515.139	18131.91	YT I	0.8			MIT
5509.896	5511.427	18144.12	YT II	1.5	1.6		MIT
5503.47	5505.00	18165.3	YT I	1.0	0.3		ME
5497.40	5498.93	18185.4	YT II	1.3	1.6		MIT
5495.61	5497.14	18191.3	YT I	0.3			ME
5493.18	5494.71	18199.3	YT I	0.5			ME
5491.539	5493.065	18204.77	YT I	0.3H			MIT
5480.75	5482.27	18240.6	YT II	1.0	1.2		ME
5473.40	5474.92	18265.1	YT II	1.0	1.3		ME
5468.47	5469.99	18281.6	YT I	1.1			ME
5466.472	5467.991	18288.25	YT I	2.2	1.3		MIT
5438.23	5439.74	18383.2	YT I	1.3	0.3		MIT
5424.36	5425.87	18430.2	YT I	0.7			ME
5417.02	5418.53	18455.2	YT I	0.3			ME
5402.78	5404.28	18503.9	YT II	1.5	1.7		ME
5390.82	5392.32	18544.9	YT I	0.3			ME
5380.605	5382.101	18580.10	YT I	0.7			MIT
5320.78	5322.26	18789.0	YT II	0.3	0.6		ME
5289.82	5291.29	18899.0	YT II	0.5	0.7		ME
5283.69	5285.16	18920.9	YT I	0.3H			ME
5240.815	5242.274	19075.69	YT I	1.3			MIT
5228.56	5230.02	19120.4	YT I	0.6			ME
5205.719	5207.169	19204.29	YT II	1.7	1.9		MIT
5200.408	5201.856	19223.91	YT II	1.8	2.2		MIT
5196.43	5197.88	19238.6	YT II	0.7	1.0		ME
5135.20	5136.63	19468.0	YT I	0.8	0.3		ME
5123.21	5124.64	19513.6	YT II	1.0	1.5		ME
5119.12	5120.55	19529.2	YT II	0.8	1.3		ME
5088.18	5089.60	19647.9	YT I	0.3			ME
5087.425	5088.843	19650.83	YT II	1.7	2.0		MIT
5072.19	5073.60	19709.9	YT I	0.6			ME
5070.246	5071.659	19717.41	YT I	0.6			MIT
5006.975	5008.372	19966.57	YT I	0.9	0.3		MIT
4982.142	4983.532	20066.09	YT II	0.9	1.7		MIT
4974.311	4975.699	20097.68	YT I	0.9	0.3		MIT
4950.658	4952.040	20193.70	YT	0.5	1.2		MIT
4930.924	4932.300	20274.51	YT I	0.6			MIT
4928.226	4929.602	20285.61	YT I	0.3	1.1		MIT
4921.887	4923.261	20311.74	YT I	1.0	1.5		MIT
4912.014	4913.385	20352.57	YT	0.3H			MIT
4908.998	4910.369	20365.07	YT I	0.6	1.2		MIT
4906.108	4907.478	20377.07	YT I	0.8	1.3		MIT
4900.121	4901.489	20401.96	YT II	1.3	2.5		MIT
4893.451	4894.817	20429.77	YT I	0.9	0.3		MIT
4886.656	4888.021	20458.18	YT I	0.6	0.3H		MIT
4886.291	4887.656	20459.71	YT I	0.6	0.9		MIT
4883.693	4885.057	20470.59	YT II	1.3	2.5		MIT
4881.449	4882.812	20480.00	YT II	0.3	0.3		MIT
4879.645	4881.008	20487.57	YT I	0.5	0.3H		MIT
4863.107	4864.465	20557.24	YT	0.3			MIT
4859.847	4861.205	20571.03	YT I	1.7	0.7		MIT
4856.714	4858.071	20584.30	YT I	0.8	0.8		MIT
4854.866	4856.222	20592.14	YT II	2.0	2.2		MIT
4854.246	4855.602	20594.77	YT I	0.5	0.3		MIT
4852.692	4854.048	20601.36	YT I	1.5	1.2		MIT
4845.675	4847.029	20631.20	YT I	1.5	1.5		MIT
4839.866	4841.218	20655.96	YT I	1.3	1.4		MIT
4839.854	4841.206	20656.01	YT	1.5			MIT

AIR WAVELENGTH (ANGSTRMS)	VACUUM WAVELENGTH (ANGSTRMS)	WAVENUMBER (KAYSERS)	LMENT	LOG ARC	INTENSITY SPK	DIS	REF
4839.147	4840.499	20659.03	YT I	0.5	0.3		MIT
4829.346	4830.695	20700.95	YT	0.5	1.2H		MIT
4823.306	4824.654	20726.87	YT II	1.2	1.0		MIT
4822.132	4823.480	20731.92	YT I	0.9			MIT
4821.636	4822.983	20734.05	YT I	0.3	0.3H		MIT
4819.642	4820.989	20742.63	YT I	0.6	0.6		MIT
4804.801	4806.144	20806.70	YT I	0.7	0.5		MIT
4804.313	4805.656	20808.81	YT I	0.5	0.5H		MIT
4799.303	4800.645	20830.54	YT I	0.7	0.5		MIT
4786.877	4788.215	20884.61	YT I	1.2	1.2		MIT
4786.577	4787.915	20885.92	YT II	1.2	1.4		MIT
4781.035	4782.372	20910.13	YT I	1.0	0.7		MIT
4780.18	4781.52	20913.9	YT I	0.3	0.0		M
4760.971	4762.302	20998.25	YT I	1.7	1.4		MIT
4752.787	4754.116	21034.40	YT I	0.9	1.0		MIT
4741.404	4742.730	21084.90	YT I	0.3	0.5		MIT
4734.52	4735.84	21115.6	YT	0.3	1.0H		MIT
4732.36	4733.68	21125.2	YT I	0.3	0.3		M
4728.530	4729.853	21142.31	YT I	1.8	0.6		MIT
4725.846	4727.168	21154.31	YT I	0.6H			MIT
4708.86	4710.18	21230.6	YT I	0.5	0.0		MIT
4701.02	4702.34	21266.0	YT I	0.6	0.3		M
4699.244	4700.559	21274.07	YT I	0.3			MIT
4696.805	4698.119	21285.11	YT I	0.7	0.5		MIT
4691.961	4693.274	21307.09	YT I	0.3			MIT
4689.776	4691.088	21317.01	YT I	0.7	0.3		MIT
4682.325	4683.636	21350.94	YT II	1.8	2.0		MIT
4678.354	4679.663	21369.06	YT I	0.3			MIT
4674.848	4676.157	21385.08	YT I	1.9	2.0		MIT
4670.83	4672.14	21403.5	YT	0.5	0.3		M
4667.462	4668.769	21418.92	YT I	0.8			MIT
4666.857	4668.163	21421.70	YT I	0.3			MIT
4666.39	4667.70	21423.8	YT I	0.5	0.0		M
4658.890	4660.194	21458.33	YT I	0.7	0.6		MIT
4658.323	4659.627	21460.95	YT I	0.9	1.2		MIT
4653.790	4655.093	21481.85	YT I	0.3			MIT
4652.13	4653.43	21489.5	YT I	0.3	0.3		M
4643.695	4644.995	21528.55	YT I	1.7	2.0		MIT
4612.997	4614.289	21671.81	YT I	0.5	0.3		MIT
4607.97	4609.26	21695.4	YT II	0.3H			MIT
4604.800	4606.090	21710.39	YT I	0.9	0.5		MIT
4601.28	4602.57	21727.0	YT I	0.3			MIT
4596.541	4597.829	21749.40	YT I	1.0	0.7		MIT
4594.00	4595.29	21761.4	YT	0.5			M
4590.79	4592.08	21776.6	YT	0.3			M
4582.17	4583.45	21817.6	YT I	0.3	0.3		M
4581.766	4583.050	21819.53	YT I	0.5			MIT
4581.302	4582.586	21821.74	YT I	0.5			MIT
4578.86	4580.14	21833.4	YT I	0.5	0.3		M
4573.560	4574.842	21858.68	YT I	0.6	0.3		MIT
4570.650	4571.931	21872.60	YT I	0.5			MIT
4564.94	4566.22	21899.9	YT I	0.3			M
4564.388	4565.667	21902.60	YT	0.6	0.3		MIT
4559.360	4560.638	21926.76	YT I	0.8	0.8		MIT
4555.298	4556.575	21946.31	YT I	0.3	0.3		MIT
4554.459	4555.736	21950.35	YT I	0.5	0.8		MIT
4544.305	4545.579	21999.40	YT I	0.7	0.5		MIT
4542.040	4543.313	22010.37	YT	0.5	0.3		MIT
4539.608	4540.881	22022.16	YT I	0.3			MIT
4537.156	4538.428	22034.06	YT I	0.5	0.3H		MIT
4534.076	4535.347	22049.03	YT I	0.5	0.3		MIT
4528.105	4529.375	22078.10	YT I	0.3			MIT
4527.787	4529.057	22079.65	YT I	1.4	1.6		MIT
4527.237	4528.507	22082.34	YT I	1.6	1.7		MIT
4522.042	4523.310	22107.70	YT I	0.5	0.3		MIT
4514.006	4515.272	22147.06	YT I	0.7	0.5		MIT
4513.581	4514.847	22149.15	YT I	0.6	0.5		MIT
4505.951	4507.215	22186.65	YT I	1.7	1.7		MIT
4492.422	4493.682	22253.47	YT I	0.5			MIT
4491.749	4493.009	22256.80	YT I	0.6			MIT

AIR WAVELENGTH (ANGSTRMS)	VACUUM WAVELENGTH (ANGSTRMS)	WAVENUMBER (KAYSERS)	LMENT	LOG ARC	INTENSITY SPK	DIS	REF
4487.466	4488.725	22278.04	YT I	0.9			MIT
4487.280	4488.539	22278.96	YT I	0.8			MIT
4484.446	4485.704	22293.04	YT I	0.3			MIT
4479.00	4480.26	22320.1	YT I	0.3			MIT
4477.444	4478.700	22327.91	YT I	1.0			MIT
4476.952	4478.208	22330.36	YT I	1.0			MIT
4476.37	4477.63	22333.3	YT	0.3			ME
4475.712	4476.968	22336.55	YT I	0.8			MIT
4473.888	4475.143	22345.65	YT	0.5			MIT
4472.781	4474.036	22351.18	YT	0.3			MIT
4465.267	4466.520	22388.79	YT I	0.3	0.8		MIT
4463.533	4464.786	22397.49	YT I	0.3			MIT
4446.629	4447.877	22482.64	YT I	0.6			MIT
4445.303	4446.551	22489.34	YT I	0.3			MIT
4443.655	4444.902	22497.68	YT I	0.6			MIT
4437.348	4438.594	22529.66	YT I	0.6			MIT
4422.587	4423.829	22604.85	YT II	1.8	1.8		ME
4417.440	4418.681	22631.19	YT I	0.3			MIT
4415.391	4416.631	22641.69	YT I	0.5	0.3		MIT
4404.819	4406.056	22696.03	YT I	0.3			MIT
4401.145	4402.381	22714.98	YT I	0.5			MIT
4398.011	4399.246	22731.17	YT II	2.2	2.0		MIT
4397.79	4399.03	22732.3	YT I	0.3			M
4394.677	4395.912	22748.41	YT I	0.3			MIT
4394.013	4395.247	22751.85	YT I	0.5			MIT
4387.731	4388.964	22784.42	YT	0.6			MIT
4385.479	4386.711	22796.12	YT I	0.5			MIT
4379.324	4380.554	22828.16	YT I	0.6			MIT
4375.620	4376.850	22847.48	YT I	0.6			MIT
4374.935	4376.164	22851.06	YT II	2.2	2.2		MIT
4371.44	4372.67	22869.3	YT	0.3			ME
4366.030	4367.257	22897.67	YT I	1.3			MIT
4364.038	4365.264	22908.12	YT II	0.3			MIT
4358.726	4359.951	22936.04	YT II	1.8	1.7		MIT
4357.726	4358.951	22941.30	YT I	1.0			MIT
4354.35	4355.57	22959.1	YT	0.3			ME
4353.633	4354.857	22962.87	YT	0.3			MIT
4352.735	4353.958	22967.60	YT I	0.8			MIT
4352.323	4353.546	22969.78	YT I	0.7			MIT
4348.790	4350.012	22988.44	YT I	2.0			MIT
4344.648	4345.869	23010.36	YT I	0.7			MIT
4337.283	4338.502	23049.43	YT I	0.6	0.5		MIT
4330.775	4331.993	23084.06	YT I	0.8			MIT
4329.897	4331.114	23088.75	YT	0.6			MIT
4322.56	4323.78	23127.9	YT	0.5			M
4322.283	4323.498	23129.42	YT	0.5	0.3		MIT
4316.296	4317.510	23161.50	YT I	0.5			MIT
4315.463	4316.677	23165.97	YT	0.8	0.3H		MIT
4309.627	4310.839	23197.34	YT II	1.7	1.7		MIT
4302.294	4303.504	23236.88	YT I	1.5	0.9		MIT
4300.335	4301.545	23247.46	YT	0.5	0.3H		MIT
4291.034	4292.241	23297.85	YT I	0.6	0.3		MIT
4279.3	4280.5	23362.	YT II	0.5	1.2		ME
4274.182	4275.385	23389.71	YT I	0.3H			MIT
4272.143	4273.345	23400.87	YT I	0.5			MIT
4266.889	4268.090	23429.69	YT I	0.3			MIT
4251.205	4252.402	23516.12	YT I	1.4	0.9		MIT
4237.12	4238.31	23594.3	YT I	0.3H			MIT
4235.943	4237.136	23600.85	YT I	1.8	1.5		MIT
4235.728	4236.921	23602.05	YT II	0.5	1.3		MIT
4232.54	4233.73	23619.8	YT I	0.5			M
4229.22	4230.41	23638.4	YT I	0.3			ME
4226.726	4227.916	23652.31	YT	0.7	1.2		MIT
4224.256	4225.446	23666.14	YT	0.7			MIT
4220.637	4221.826	23686.44	YT I	1.2	0.8		MIT
4217.790	4218.978	23702.42	YT I	1.0	0.6		MIT
4213.546	4214.733	23726.30	YT I	0.6	0.3		MIT
4213.027	4214.214	23729.22	YT I	0.7	0.6		MIT
4204.696	4205.881	23776.23	YT II	1.2	1.2		MIT
4199.272	4200.455	23806.94	YT II	0.5	1.0		MIT

AIR WAVELENGTH (ANGSTRMS)	VACUUM WAVELENGTH (ANGSTRMS)	WAVENUMBER (KAYSERS)	LMENT	LOG ARC	INTENSITY SPK	DIS	REF
4189.992	4191.173	23859.67	YT	0.6			MIT
4177.552	4178.729	23930.72	YT II	1.7	1.7		MIT
4174.133	4175.310	23950.32	YT I	2.0	0.9		MIT
4173.34	4174.52	23954.9	YT	0.9			ME
4167.507	4168.682	23988.40	YT I	1.7	1.0		MIT
4164.986	4166.160	24002.92	YT	0.9	0.3		MIT
4158.120	4159.292	24042.55	YT I	0.3			MIT
4157.623	4158.795	24045.43	YT I	0.3			MIT
4142.839	4144.007	24131.23	YT I	2.0	1.4		MIT
4128.304	4129.469	24216.19	YT I	2.2	1.5		MIT
4127.582	4128.746	24220.43	YT II	0.3	0.3		MIT
4124.910	4126.074	24236.12	YT II	0.8	1.3		MIT
4110.802	4111.962	24319.29	YT I	0.8	0.3H		MIT
4106.388	4107.547	24345.43	YT I	0.9			MIT
4103.801	4104.959	24360.78	YT	0.8	0.3		MIT
4102.376	4103.534	24369.24	YT I	2.2	1.5		MIT
4083.713	4084.866	24480.61	YT I	1.7	1.0		MIT
4081.223	4082.375	24495.54	YT I	0.8	0.5		MIT
4080.926	4082.078	24497.33	YT I	0.7	0.6		MIT
4077.366	4078.517	24518.72	YT I	1.7	1.6		MIT
4076.353	4077.504	24524.81	YT	1.5	0.9		MIT
4064.988	4066.136	24593.37	YT II	0.9	0.8		MIT
4047.632	4048.775	24698.83	YT I	1.7	1.0		MIT
4039.834	4040.975	24746.50	YT I	1.1	0.9		MIT
4030.83	4031.97	24801.8	YT	0.3	0.3		ED
4029.84	4030.98	24807.9	YT I	0.7			ME
3987.497	3988.625	25071.30	YT I	0.3	0.5		MIT
3982.595	3983.721	25102.16	YT II	1.8	2.0		MIT
3978.59	3979.72	25127.4	YT	0.6	0.3H		M
3973.464	3974.588	25159.84	YT	0.7	0.6		MIT
3968.43	3969.55	25191.8	YT	1.0	1.5		MIT
3967.69	3968.81	25196.4	YT II	0.5	1.0H		MIT
3955.095	3956.214	25276.69	YT	0.3	0.3		MIT
3951.596	3952.714	25299.07	YT II	0.8	0.9		MIT
3950.359	3951.477	25306.99	YT II	1.8	2.0		MIT
3946.217	3947.334	25333.55	YT II	0.3	0.5		MIT
3944.687	3945.804	25343.38	YT	0.5			MIT
3930.667	3931.780	25433.77	YT II	1.3	1.4		MIT
3930.109	3931.222	25437.39	YT	0.3H	0.5		MIT
3918.25	3919.36	25514.4	YT I	0.8	0.5		M
3904.587	3905.693	25603.65	YT	0.5	0.5		MIT
3896.804	3897.908	25654.79	YT II	0.3	0.7		MIT
3892.408	3893.511	25683.76	YT I	0.5			MIT
3891.083	3892.186	25692.51	YT	0.3H	0.5		MIT
3890.858	3891.960	25693.99	YT	0.6	0.6		MIT
3887.776	3888.878	25714.36	YT	0.5	0.5		MIT
3884.842	3885.943	25733.78	YT	0.5	0.3H		MIT
3878.290	3879.389	25777.25	YT II	1.2	1.2		MIT
3876.819	3877.918	25787.03	YT	0.7	0.5		MIT
3872.308	3873.406	25817.07	YT II	0.3H	0.5H		MIT
3848.194	3849.285	25978.85	YT II	0.3	0.6H		MIT
3847.873	3848.964	25981.01	YT	0.9	0.6		MIT
3846.516	3847.607	25990.18	YT II	0.3	0.3		MIT
3832.889	3833.976	26082.58	YT II	1.5	1.9		MIT
3824.78	3825.87	26137.9	YT II	0.0	0.7B		MIT
3818.342	3819.426	26181.95	YT II	1.5	1.7		MIT
3812.18	3813.26	26224.3	YT II	0.8	0.5		MIT
3800.883	3801.962	26302.21	YT II	0.3	0.7H		MIT
3800.028	3801.107	26308.13	YT	1.4	0.7		MIT
3792.56	3793.64	26359.9	YT II	0.7	0.5		MIT
3788.697	3789.773	26386.80	YT II	1.5	1.5		MIT
3787.158	3788.234	26397.53	YT I	0.7			MIT
3785.62	3786.70	26408.2	YT	0.5	0.5		ME
3782.302	3783.376	26431.42	YT II	0.9	0.7		MIT
3777.27	3778.34	26466.6	YT	0.8	0.3		MIT
3776.561	3777.634	26471.60	YT II	1.1	1.1		MIT
3774.332	3775.404	26487.23	YT II	1.1	2.0		MIT
3762.973	3764.042	26567.18	YT	0.7	0.5		MIT
3749.892	3750.958	26659.86	YT I	0.5	0.6		MIT
3747.554	3748.619	26676.49	YT II	1.1	1.2		MIT

AIR WAVELENGTH (ANGSTRMS)	VACUUM WAVELENGTH (ANGSTRMS)	WAVENUMBER (KAYSERS)	LMENT	LOG ARC	INTENSITY SPK	INTENSITY DIS	REF
3738.604	3739.667	26740.35	YT I	0.8			MIT
3727.09	3728.15	26823.0	YT II	0.5	0.9H		MIT
3724.76	3725.82	26839.7	YT	0.3	0.0		M
3721.398	3722.456	26863.98	YT II	0.9	0.5		MIT
3721.398	3722.456	26863.98	YT I	0.9	0.5		MIT
3718.108	3719.166	26887.75	YT I	1.1	0.6		MIT
3716.91	3717.97	26896.4	YT II	0.3	0.7H		MIT
3714.31	3715.37	26915.2	YT II	0.3	0.5		M
3710.290	3711.346	26944.41	YT II	1.9	2.2		MIT
3703.323	3704.377	26995.09	YT II	0.6	1.0W		MIT
3702.85	3703.90	26998.5	YT	0.6	0.3		MIT
3699.14	3700.19	27025.6	YT	0.5	0.5		ME
3696.62	3697.67	27044.0	YT II	0.8	1.3H		M
3692.529	3693.580	27074.00	YT I	0.8	0.8		MIT
3684.903	3685.952	27130.03	YT II	0.7	0.9		MIT
3675.64	3676.69	27198.4	YT II	0.3	0.5H		MIT
3668.489	3669.534	27251.42	YT II	0.8	1.3		MIT
3665.795	3666.839	27271.45	YT I	0.5	0.3		MIT
3664.614	3665.658	27280.23	YT II	2.0	2.0		MIT
3653.606	3654.647	27362.42	YT	0.7	0.3		MIT
3645.403	3646.442	27423.99	YT	0.6	0.6		MIT
3639.287	3640.324	27470.08	YT I	0.5	0.3		MIT
3635.334	3636.370	27499.95	YT II	0.3H	1.1		MIT
3633.123	3634.159	27516.69	YT II	1.7	2.0		MIT
3628.706	3629.740	27550.18	YT II	1.6	1.7		MIT
3622.19	3623.22	27599.7	YT	0.3	0.0		ME
3621.86	3622.89	27602.2	YT	0.3	0.0		ME
3620.941	3621.973	27609.26	YT I	0.3H	1.1		MIT
3612.784	3613.814	27671.59	YT	0.3			MIT
3612.34	3613.37	27675.0	YT	0.3			ME
3611.047	3612.077	27684.90	YT II	1.6	1.8		MIT
3605.46	3606.49	27727.8	YT II	0.3H	0.8H		MIT
3601.921	3602.949	27755.04	YT II	1.3	1.8		MIT
3600.734	3601.761	27764.19	YT II	2.0	2.5		MIT
3592.913	3593.938	27824.63	YT I	1.9	1.4		MIT
3590.348	3591.373	27844.51	YT	0.3			MIT
3589.686	3590.710	27849.64	YT I	0.5			MIT
3587.753	3588.777	27864.65	YT	1.2	0.3		MIT
3584.514	3585.537	27889.82	YT II	1.3	1.2		MIT
3576.053	3577.074	27955.81	YT I	1.2	0.3		MIT
3573.77	3574.79	27973.7	YT	0.5	0.5		ME
3571.432	3572.452	27991.98	YT I	1.2	0.3		MIT
3558.745	3559.761	28091.77	YT I	0.5			MIT
3556.083	3557.099	28112.80	YT II	0.3	0.5H		MIT
3552.693	3553.708	28139.62	YT I	0.8	0.6		MIT
3551.790	3552.805	28146.78	YT I	0.8	0.3		MIT
3549.011	3550.025	28168.82	YT II	1.1	1.7		MIT
3546.01	3547.02	28192.6	YT	0.3	0.3H		MIT
3544.980	3545.993	28200.85	YT	0.8	0.7		MIT
3544.001	3545.014	28208.64	YT II	0.3	0.7H		MIT
3538.530	3539.541	28252.25	YT	1.0	0.5		MIT
3531.707	3532.716	28306.83	YT	0.8			MIT
3521.534	3522.541	28388.60	YT I	0.8			MIT
3512.886	3513.891	28458.48	YT I	0.9	0.5		MIT
3511.188	3512.192	28472.25	YT I	0.9	0.5		MIT
3510.521	3511.525	28477.66	YT I	0.9	0.0		MIT
3507.964	3508.967	28498.41	YT II	0.3	1.1		MIT
3506.491	3507.494	28510.38	YT	0.3	0.3		MIT
3501.96	3502.96	28547.3	YT I	0.5	0.3		M
3500.639	3501.641	28558.04	YT I	0.7			MIT
3498.943	3499.944	28571.89	YT I	0.9	0.8		MIT
3496.080	3497.080	28595.28	YT II	1.3	1.5		MIT
3485.726	3486.724	28680.22	YT I	1.1	0.6		MIT
3484.049	3485.046	28694.02	YT I	0.8	0.6		MIT
3473.18	3474.17	28783.8	YT	0.7	0.3		MIT
3471.75	3472.74	28795.7	YT	0.8	0.7		MIT
3470.18	3471.17	28808.7	YT II	0.6	0.7		MIT
3467.873	3468.866	28827.86	YT II	1.1	1.1		MIT
3461.01	3462.00	28885.0	YT II	0.8	1.1		M
3457.088	3458.078	28917.79	YT II	0.7	0.8		MIT
3454.18	3455.17	28942.1	YT I	0.7	0.3		MIT
3450.951	3451.940	28969.22	YT I	0.5			MIT
3448.812	3449.800	28987.19	YT II	1.3	1.3		MIT
3437.953	3438.938	29078.74	YT	0.5			MIT
3433.79	3434.77	29114.0	YT	1.1	0.3H		MIT
3431.67	3432.65	29132.0	YT	0.5	0.3H		MIT
3430.99	3431.97	29137.7	YT I	0.3			MIT
3429.42	3430.40	29151.1	YT II	0.3	0.6		MIT
3424.14	3425.12	29196.0	YT	0.5	0.3		MIT
3414.49	3415.47	29278.6	YT	0.5	0.3		MIT
3412.469	3413.448	29295.89	YT I	0.6	0.6		MIT
3409.87	3410.85	29318.2	YT II	0.5	1.1		MIT
3407.7	3408.7	29337.	YT		0.7H		ME
3397.05	3398.02	29428.9	YT I	0.8	0.7		ME
3394.98	3395.95	29446.8	YT	0.6	0.3		MIT
3389.83	3390.80	29491.5	YT I	0.6	0.3		MIT
3388.582	3389.555	29502.40	YT I	0.3	0.3		MIT
3383.05	3384.02	29550.6	YT	0.5	0.5		M
3382.83	3383.80	29552.6	YT	0.5	0.5		M
3380.114	3381.085	29576.31	YT II	0.8	1.1		MIT
3377.711	3378.681	29597.35	YT I	0.5			MIT
3364.786	3365.753	29711.04	YT I	0.5	0.3		MIT
3362.00	3362.97	29735.7	YT II	1.1	1.4		ME
3358.96	3359.93	29762.6	YT I	0.3	0.3H		M
3354.81	3355.77	29799.4	YT I	0.8	0.6		ME
3354.589	3355.553	29801.35	YT I	1.0	0.6		MIT
3344.55	3345.51	29890.8	YT	0.5	0.3H		MIT
3340.387	3341.347	29928.05	YT I	0.6	0.3		MIT
3337.849	3338.809	29950.80	YT I	0.3	0.3		MIT
3336.25	3337.21	29965.2	YT II	0.5	0.8		MIT
3335.225	3336.184	29974.36	YT I	0.5			MIT
3333.606	3334.565	29988.92	YT II	0.3	0.7		MIT
3330.880	3331.838	30013.46	YT II	0.6	1.5		MIT
3327.875	3328.832	30040.56	YT II	1.8	1.8		MIT
3319.78	3320.74	30113.8	YT II	0.8	1.0		MIT
3318.53	3319.48	30125.2	YT II	1.1	0.6		M
3312.39	3313.34	30181.0	YT II	1.0	1.0		MIT
3308.47	3309.42	30216.7	YT II	1.0	1.3		M
3304.01	3304.96	30257.5	YT II	0.7	0.8		MIT
3302.17	3303.12	30274.4	YT	0.7	0.3		ED
3293.67	3294.62	30352.5	YT	0.3	0.3		MIT
3293.43	3294.38	30354.7	YT	0.3	0.3		MIT
3290.60	3291.55	30380.8	YT	0.5D	0.3		MIT
3289.85	3290.80	30387.8	YT	1.2	1.0		MIT
3286.71	3287.66	30416.8	YT II	0.6	0.3H		M
3283.21	3284.16	30449.2	YT	0.5	0.6		MIT
3282.51	3283.46	30455.7	YT II	0.5	0.7		ME
3280.913	3281.858	30470.54	YT II	0.9	1.1		MIT
3278.433	3279.378	30493.59	YT	0.6	0.3		MIT
3275.572	3276.516	30520.22	YT	1.0	0.9		MIT
3264.781	3265.722	30621.10	YT	1.0	0.9		MIT
3252.29	3253.23	30738.7	YT	0.6	0.3		MIT
3251.265	3252.203	30748.39	YT I	0.3H	0.5		MIT
3242.280	3243.216	30833.60	YT II	1.8	2.0		MIT
3232.027	3232.960	30931.41	YT II	0.5	0.5		MIT
3231.235	3232.168	30938.99	YT II	0.3	0.5		MIT
3225.17	3226.10	30997.2	YT II	0.9	1.1B		MIT
3216.69	3217.62	31078.9	YT II		0.9		ME
3216.682	3217.611	31078.96	YT II	1.6	1.8		MIT
3212.28	3213.21	31121.6	YT II	0.8	0.9		MIT
3203.323	3204.249	31208.56	YT II	1.5	1.7		MIT
3200.270	3201.195	31238.33	YT II	1.5	1.6		MIT
3198.42	3199.34	31256.4	YT II	0.5	0.5		MIT
3195.615	3196.539	31283.84	YT II	1.5	1.7		MIT
3194.38	3195.30	31295.9	YT I	0.9			MIT
3193.48	3194.40	31304.7	YT II		0.5H		MIT
3193.29	3194.21	31306.6	YT	0.3	0.8		MIT
3191.308	3192.231	31326.06	YT I	1.0	0.5		MIT
3185.956	3186.877	31378.68	YT I	0.8	0.3H		MIT
3182.42	3183.34	31413.5	YT II	0.5	0.8B		ME

AIR WAVELENGTH (ANGSTRMS)	VACUUM WAVELENGTH (ANGSTRMS)	WAVENUMBER (KAYSERS)	LMENT	LOG ARC	INTENSITY SPK	DIS	REF
3179.418	3180.338	31443.20	YT II	1.3	1.5		MIT
3173.052	3173.970	31506.28	YT II	1.3	1.8		MIT
3172.837	3173.755	31508.42	YT I	0.5	0.5		MIT
3171.667	3172.585	31520.04	YT	0.8	0.3H		MIT
3155.644	3156.558	31680.08	YT I	0.3H			MIT
3152.656	3153.569	31710.10	YT I	0.8	0.5		MIT
3144.37	3145.28	31793.7	YT II	0.3H	0.8		ME
3135.168	3136.076	31886.98	YT II	1.0	1.3		MIT
3129.933	3130.840	31940.31	YT II	0.9	1.7		MIT
3128.789	3129.696	31951.99	YT II	1.0	1.6		MIT
3126.176	3127.082	31978.69	YT II	0.8	0.8		MIT
3114.43	3115.33	32099.3	YT II	0.8	0.5H		MIT
3114.283	3115.186	32100.81	YT	1.0			MIT
3112.032	3112.935	32124.03	YT II	1.1	1.0		MIT
3111.810	3112.713	32126.32	YT I	1.0			MIT
3108.875	3109.777	32156.65	YT I	0.7			MIT
3104.78	3105.68	32199.1	YT II	0.3	0.8		MIT
3103.672	3104.573	32210.55	YT	0.5			MIT
3103.239	3104.139	32215.05	YT II	0.9	0.8		MIT
3096.622	3097.521	32283.88	YT I	0.5			MIT
3095.878	3096.777	32291.64	YT II	1.0	0.9		MIT
3093.77	3094.67	32313.6	YT II	1.0D	1.2D		MIT
3092.712	3093.610	32324.70	YT	0.9			MIT
3091.699	3092.597	32335.29	YT	0.8	0.5		MIT
3086.858	3087.754	32386.00	YT II	1.1	1.7		MIT
3082.167	3083.062	32435.28	YT II	0.6	1.0D		MIT
3081.600	3082.495	32441.25	YT II		0.3		MIT
3078.655	3079.549	32472.28	YT II	0.8	0.9		MIT
3077.14	3078.03	32488.3	YT II	0.8	0.3		ME
3076.478	3077.372	32495.26	YT	0.8	0.3		MIT
3072.334	3073.227	32539.09	YT	0.9	0.9		MIT
3069.25	3070.14	32571.8	YT II	0.3H	0.9H		MIT
3066.00	3066.89	32606.3	YT II	0.5	0.9H		MIT
3059.496	3060.385	32675.62	YT	0.3H	0.3H		MIT
3056.338	3057.227	32709.38	YT	0.7			MIT
3055.22	3056.11	32721.4	YT II	0.9	1.7		M
3054.42	3055.31	32729.9	YT	0.5	0.3		MIT
3053.27	3054.16	32742.2	YT II	1.0H	1.3H		MIT
3047.40	3048.29	32805.3	YT	0.7			MIT
3047.11	3048.00	32808.4	YT	0.8	0.5		MIT
3045.369	3046.255	32827.19	YT I	1.0	0.8		MIT
3044.83	3045.72	32833.0	YT	0.5H			MIT
3038.472	3039.356	32901.70	YT I	0.5			MIT
3036.59	3037.47	32922.1	YT II	1.0	1.6		MIT
3030.214	3031.096	32991.37	YT II	0.3H	1.0		MIT
3027.75	3028.63	33018.2	YT II	0.8	0.8		ME
3026.47	3027.35	33032.2	YT II	0.8	1.3		MIT
3023.512	3024.392	33064.49	YT II	0.8	0.5		MIT
3022.281	3023.161	33077.96	YT	0.8	0.3		MIT
3021.724	3022.604	33084.06	YT I	1.0			MIT
3018.95	3019.83	33114.5	YT I	0.5			MIT
3005.262	3006.138	33265.27	YT I	0.9			MIT
3001.42	3002.29	33307.9	YT II	0.8	0.6		ME
2996.944	2997.818	33357.60	YT I	1.1	0.7		MIT
2995.255	2996.128	33376.41	YT I	0.9	0.3		MIT
2990.391	2991.263	33430.69	YT	0.5			MIT
2984.256	2985.127	33499.42	YT I	1.1	1.0		MIT
2982.23	2983.10	33522.2	YT II	0.5	0.6		MIT
2980.69	2981.56	33539.5	YT II	0.3	1.0H		MIT
2980.538	2981.408	33541.20	YT	0.9	1.5		MIT
2978.18	2979.05	33567.8	YT II	0.5	1.0H		MIT
2974.593	2975.461	33608.23	YT I	1.1	1.0		MIT
2974.03	2974.90	33614.6	YT II	0.7	1.1		MIT
2964.968	2965.834	33717.33	YT I	0.8	0.7		MIT
2957.381	2958.245	33803.83	YT II		0.9		MIT
2956.04	2956.90	33819.2	YT II	0.8	1.2H		ME
2953.28	2954.14	33850.8	YT II	0.5	1.1H		ME
2948.98	2949.84	33900.1	YT II		0.5H		ME
2948.405	2949.267	33906.73	YT I	1.3	0.7H		MIT
2945.95	2946.81	33935.0	YT	0.3	2.0		MIT

AIR WAVELENGTH (ANGSTRMS)	VACUUM WAVELENGTH (ANGSTRMS)	WAVENUMBER (KAYSERS)	LMENT	LOG ARC	INTENSITY SPK	DIS	REF
2930.773	2931.630	34110.71	YT II	0.9	1.1H		MIT
2930.14	2931.00	34118.1	YT II	0.9	1.3		MIT
2919.048	2919.903	34247.72	YT	1.3	0.8		MIT
2907.18	2908.03	34387.5	YT II		0.6H		MIT
2901.481	2902.331	34455.06	YT I	0.7			MIT
2898.94	2899.79	34485.3	YT II	0.9	1.0		MIT
2897.676	2898.525	34500.30	YT II	0.5	1.1		MIT
2890.36	2891.21	34587.6	YT I	0.5			MIT
2886.457	2887.303	34634.39	YT I	1.2	0.8		MIT
2858.08	2858.92	34978.2	YT II	0.8	1.2		MIT
2856.294	2857.133	35000.12	YT II	0.9	1.2		MIT
2854.420	2855.259	35023.10	YT II	1.0	1.3		MIT
2850.65	2851.49	35069.4	YT	0.5	0.9		MIT
2840.96	2841.80	35189.0	YT II	0.8	1.3		MIT
2834.57	2835.40	35268.4	YT II	0.3	1.3		ME
2826.33	2827.16	35371.2	YT II	0.8	1.1		MIT
2825.369	2826.200	35383.19	YT II	0.8	1.0		MIT
2822.559	2823.390	35418.42	YT II	0.5	0.8		MIT
2818.849	2819.679	35465.03	YT I	0.3			MIT
2813.653	2814.482	35530.52	YT II	0.8D	1.3H		MIT
2813.653	2814.482	35530.52	YT I	0.8D	1.3H		MIT
2800.102	2800.927	35702.46	YT II	1.0	1.3		MIT
2785.594	2786.416	35888.40	YT II	0.7	1.3		MIT
2785.235	2786.057	35893.02	YT II	0.8	1.3		MIT
2760.093	2760.909	36219.96	YT I	1.0	0.9		MIT
2750.40	2751.21	36347.6	YT II	0.7	0.9H		ME
2750.19	2751.00	36350.4	YT	0.8	0.0H		MIT
2742.50	2743.31	36452.3	YT I	0.6			MIT
2734.97	2735.78	36552.6	YT II	0.7	1.2		MIT
2730.089	2730.897	36618.00	YT I	1.0	0.3H		MIT
2722.998	2723.805	36713.35	YT I	1.0	0.9		MIT
2705.872	2706.674	36945.71	YT I	0.5			MIT
2695.382	2696.182	37089.49	YT I	0.8	0.3		MIT
2694.205	2695.005	37105.69	YT	0.9			MIT
2681.65	2682.45	37279.4	YT I	0.8	0.3H		MIT
2672.074	2672.868	37412.99	YT I	0.7D	0.6		MIT
2594.91	2595.69	38525.5	YT	0.7			MIT
2550.17	2550.94	39201.3	YT	1.0			MIT
2547.57	2548.33	39241.3	YT	1.0			MIT
2540.28	2541.04	39353.9	YT	1.0			MIT
2490.42	2491.17	40141.8	YT I	1.3	0.5		MIT
2465.904	2466.650	40540.82	YT II	0.3	1.0B		MIT
2460.62	2461.36	40627.9	YT II	1.0	1.5H		MIT
2460.10	2460.84	40636.5	YT I	0.9			MIT
2422.185	2422.921	41272.50	YT II	1.3	1.5		MIT
2417.29	2418.02	41356.1	YT II	0.6	0.9H		ED
2413.92	2414.65	41413.8	YT II	0.0	0.5H		MIT
2398.149	2398.879	41686.13	YT II	0.3	1.1W		MIT
2385.23	2385.96	41911.9	YT	1.1	0.0		MIT
2361.82	2362.54	42327.3	YT I	0.9			MIT
2358.72	2359.44	42382.9	YT	1.0	0.3H		MIT
2355.39	2356.11	42442.8	YT	0.9			MIT
2354.185	2354.905	42464.55	YT I	1.1	0.3		MIT
2349.70	2350.42	42545.6	YT	0.9			MIT
2340.812	2341.529	42707.13	YT II	0.3	1.0H		MIT
2332.578	2333.294	42857.87	YT I	1.0	0.3H		MIT
2331.64	2332.36	42875.1	YT	0.8			MIT
2268.13	2268.83	44075.5	YT II	0.3H	0.3H		MIT
2243.060	2243.756	44568.13	YT	1.4	1.5W		MIT
9950.64	9953.37	10046.8	ZN II			0.3	PS
9950.11	9952.84	10047.4	ZN II			0.5	PS
7799.365	7801.511	12818.03	ZN I	1.0			IHZ
7757.86	7759.99	12886.6	ZN II			1.5	PS
7732.50	7734.63	12928.9	ZN II			1.7	PS
7612.90	7615.00	13132.0	ZN II			1.3	PS
7588.48	7590.57	13174.2	ZN II			1.7	PS
7478.79	7480.85	13367.5	ZN II			1.7	PS
6943.202	6945.117	14398.60	ZN I	0.5			IHZ
6938.472	6940.386	14408.42	ZN I	0.9			IHZ
6928.319	6930.230	14429.54	ZN I	1.2			IHZ

AIR WAVELENGTH (ANGSTRMS)	VACUUM WAVELENGTH (ANGSTRMS)	WAVENUMBER (KAYSERS)	LMENT	LOG ARC	INTENSITY SPK	DIS	REF
6831.21	6833.10	14634.7	ZN II			0.9	VS
6825.87	6827.75	14646.1	ZN II			1.0	VS
6483.27	6485.06	15420.1	ZN II			1.5	PS
6482.98	6484.77	15420.7	ZN II	0.6		1.2	VS
6479.155	6480.945	15429.85	ZN I	1.0			IHZ
6470.98	6472.77	15449.3	ZN	0.6			PS
6362.347	6364.106	15713.13	ZN I	3.0V	2.7		IHZ
6239.182	6240.908	16023.31	ZN I	0.9			IHZ
6237.891	6239.617	16026.63	ZN I	0.9			IHZ
6214.590	6216.309	16086.71	ZN II	0.5		1.1	MIT
6111.56	6113.25	16357.9	ZN II	0.9		1.0	VS
6102.54	6104.23	16382.1	ZN II	0.8		1.3	VS
6021.26	6022.93	16603.2	ZN II	0.5		1.2	VS
5937.71	5939.36	16836.8	ZN I	0.7			WD
5894.351	5895.984	16960.70	ZN II	0.5		1.5	IHZ
5777.112	5778.714	17304.89	ZN I	1.0	1.2		HZ
5775.501	5777.103	17309.72	ZN I	0.6			IHZ
5772.102	5773.703	17319.91	ZN I	0.6			IHZ
5586.22	5587.77	17896.2	ZN II			0.6	VS
5585.21	5586.76	17899.5	ZN II	1.1		0.0	VS
5577.745	5579.294	17923.42	ZN II	0.8		0.3	MIT
5357.458	5358.948	18660.38	ZN II	0.8		0.5	MIT
5311.02	5312.50	18823.5	ZN I	0.8			HZ
5310.241	5311.718	18826.30	ZN I	0.8			HZ
5308.648	5310.125	18831.95	ZN I	0.9			HZ
5181.995	5183.438	19292.21	ZN I	2.3	0.3		IHZ
5069.577	5070.990	19720.01	ZN I	1.2			HZ
5068.655	5070.068	19723.60	ZN I	0.8			HZ
4924.043	4925.418	20302.85	ZN II	1.2		1.5	IHZ
4911.664	4913.035	20354.02	ZN II	1.2		1.4	IHZ
4810.534	4811.878	20781.91	ZN I	2.6W	2.5H		IHZ
4722.159	4723.480	21170.83	ZN I	2.6W	2.5H		IHZ
4680.138	4681.448	21360.91	ZN I	2.5W	2.3H		IHZ
4629.814	4631.111	21593.09	ZN I	1.5			HZ
4300.81	4302.02	23244.9	ZN			1.4	VS
4298.327	4299.536	23258.32	ZN I	1.4	1.4		IHZ
4292.99	4294.20	23287.2	ZN II			0.3	VS
4292.885	4294.093	23287.81	ZN I	1.4	1.4		IHZ
4119.38	4120.54	24268.6	ZN II			0.8	VS
4113.210	4114.371	24305.05	ZN I	1.0			HZ
4101.665	4102.823	24373.47	ZN	0.7			IHZ
4078.14	4079.29	24514.1	ZN II			0.7	VS
4057.71	4058.86	24637.5	ZN II	1.9			VS
3989.23	3990.36	25060.4	ZN II			2.0	VS
3965.432	3966.554	25210.80	ZN I	1.2			IHZ
3883.340	3884.441	25743.73	ZN I	1.7	0.3H		HZ
3842.26	3843.35	26019.0	ZN II			1.2	VS
3840.34	3841.43	26032.0	ZN II	0.5		1.7	VS
3812.86	3813.94	26219.6	ZN			0.3	VS
3806.375	3807.456	26264.26	ZN II	0.5	1.2		MIT
3799.002	3800.081	26315.23	ZN I	0.7			HZ
3793.91	3794.99	26350.6	ZN II	1.0			VS
3739.99	3741.05	26730.4	ZN	1.3			VS
3713.72	3714.78	26919.5	ZN		0.3H		VS
3703.41	3704.46	26994.5	ZN	0.7	0.3H		VS
3683.471	3684.520	27140.58	ZN II	1.3	1.2		MIT
3668.13	3669.17	27254.1	ZN	0.5	0.5H		VS
3639.53	3640.57	27468.2	ZN	1.3	0.7		VS
3631.93	3632.97	27525.7	ZN	1.2	0.0		VS
3624.070	3625.103	27585.42	ZN	1.0	0.5		MIT
3572.65	3573.67	27982.4	ZN	0.7		0.0	VS
3515.11	3516.12	28440.5	ZN I	0.3H			FL
3381.04	3382.01	29568.2	ZN II			1.3	VS
3345.934	3346.896	29878.43	ZN I	2.2	1.7		HZ
3345.572	3346.534	29881.66	ZN I	2.7	2.0		HZ
3345.020	3345.982	29886.59	ZN I	2.9	2.5		HZ
3337.16	3338.12	29957.0	ZN			0.7	VS
3305.96	3306.91	30239.7	ZN II			1.3	VS
3304.80	3305.75	30250.3	ZN II			0.3	VS
3302.941	3303.892	30267.33	ZN	2.8G	2.5G		HZ

AIR WAVELENGTH (ANGSTRMS)	VACUUM WAVELENGTH (ANGSTRMS)	WAVENUMBER (KAYSERS)	LMENT	LOG ARC	INTENSITY SPK	DIS	REF
3302.588	3303.539	30270.57	ZN I	2.9	2.5		HZ
3299.39	3300.34	30299.9	ZN II			1.2	VS
3282.333	3283.279	30457.36	ZN I	2.7G	2.5		IHZ
3276.55	3277.49	30511.1	ZN II			0.5	VS
3197.02	3197.94	31270.1	ZN II			0.7	VS
3196.29	3197.21	31277.2	ZN II			1.2	VS
3182.82	3183.74	31409.6	ZN			0.5	VS
3178.54	3179.46	31451.9	ZN II			0.3	VS
3172.18	3173.10	31514.9	ZN II			1.1	VS
3171.40	3172.32	31522.7	ZN II			0.5	VS
3075.901	3076.795	32501.36	ZN I	2.2	1.7		IHZ
3072.062	3072.955	32541.97	ZN I	2.3	2.1		IHZ
3035.781	3036.665	32930.87	ZN I	2.3	2.0		IHZ
3018.352	3019.231	33121.01	ZN I	2.1	1.6		IHZ
2902.26	2903.11	34445.8	ZN II			1.7	VS
2882.15	2883.00	34686.1	ZN II			1.4	VS
2873.19	2874.03	34794.3	ZN			0.5	VS
2832.95	2833.78	35288.5	ZN			1.4	VS
2826.13	2826.96	35373.7	ZN	0.5		1.0	VS
2804.69	2805.52	35644.1	ZN II			1.0	VS
2801.79	2802.62	35681.0	ZN II			1.4	VS
2801.167	2801.993	35688.89	ZN I	0.7			HZ
2801.056	2801.882	35690.30	ZN I	2.0	1.3		HZ
2800.869	2801.694	35692.69	ZN I	2.6	2.5		HZ
2782.83	2783.65	35924.0	ZN II			1.3	VS
2781.23	2782.05	35944.7	ZN I	1.4	0.7P		FL
2770.984	2771.802	36077.61	ZN I	2.5	2.2		HZ
2770.865	2771.683	36079.16	ZN I	2.5	1.4		IHZ
2763.88	2764.70	36170.3	ZN II			0.5	VS
2758.86	2759.68	36236.1	ZN			1.0	VS
2756.452	2757.267	36267.80	ZN I	2.3	2.0		IHZ
2751.39	2752.20	36334.5	ZN I	1.0H			FL
2738.43	2739.24	36506.5	ZN II			1.0	VS
2736.86	2737.67	36527.4	ZN I	1.0J			FL
2712.488	2713.292	36855.60	ZN I	2.5	0.9		IHZ
2684.161	2684.958	37244.53	ZN I	2.5	0.8		IHZ
2682.844	2683.641	37262.81	ZN		1.0		MIT
2670.530	2671.324	37434.62	ZN I	2.3	0.6		IHZ
2623.78	2624.56	38101.6	ZN	0.7H			FL
2614.30	2615.08	38239.7	ZN II			1.0	VS
2608.640	2609.419	38322.71	ZN I	2.5	2.0		HZ
2608.558	2609.337	38323.91	ZN I	2.3	1.7		IHZ
2600.94	2601.72	38436.1	ZN I	1.0J			FL
2582.487	2583.260	38710.78	ZN I	2.5	1.6		HZ
2582.440	2583.213	38711.48	ZN I	2.0			HZ
2575.60	2576.37	38814.3	ZN			1.0	VS
2575.06	2575.83	38822.4	ZN I	0.7J			FL
2570.721	2571.491	38887.94	ZN II		1.0R		MIT
2569.871	2570.641	38900.80	ZN I	2.0H	0.7		IHZ
2568.08	2568.85	38927.9	ZN II	1.7H	1.0		VS
2564.45	2565.22	38983.0	ZN II			1.4	VS
2562.61	2563.38	39011.0	ZN I	1.0H			FL
2557.958	2558.725	39081.96	ZN II	1.0	2.5		IHZ
2542.32	2543.08	39322.3	ZN I	1.6H			FL
2530.09	2530.85	39512.4	ZN I	1.0L			FL
2515.807	2516.564	39736.72	ZN I	2.2W	1.3		IHZ
2502.001	2502.755	39955.97	ZN II	1.3	2.6C		HZ
2493.32	2494.07	40095.1	ZN I	1.4			FL
2491.48	2492.23	40124.7	ZN I	2.0	1.7		FL
2479.74	2480.49	40314.6	ZN I	1.5			FL
2473.279	2474.026	40419.94	ZN II		0.9J		MIT
2469.38	2470.13	40483.8	ZN I	1.1R			FL
2463.47	2464.22	40580.9	ZN I	1.3	0.3		FL
2457.80	2458.54	40674.5	ZN I	0.6R			FL
2449.72	2450.46	40808.6	ZN I	1.0			FL
2440.21	2440.95	40967.7	ZN			0.5	VS
2439.42	2440.16	40980.9	ZN			1.4	VS
2435.520	2436.259	41046.54	ZN II		1.0		MIT
2432.48	2433.22	41097.8	ZN II			0.3	VS
2430.79	2431.53	41126.4	ZN I	0.9			FL

AIR WAVELENGTH (ANGSTRMS)	VACUUM WAVELENGTH (ANGSTRMS)	WAVENUMBER (KAYSERS)	LMENT	LOG INTENSITY ARC	SPK	DIS	REF
2428.86	2429.60	41159.1	ZN II			1.0	VS
2426.63	2427.37	41196.9	ZN I	0.9			FL
2415.48	2416.21	41387.1	ZN I	0.7			FL
2409.08	2409.81	41497.0	ZN			0.7	VS
2408.13	2408.86	41513.4	ZN I	0.7			PS
2402.82	2403.55	41605.1	ZN I	0.3			FL
2399.23	2399.96	41667.4	ZN I	0.5			FL
2397.17	2397.90	41703.2	ZN II			0.3	VS
2393.81	2394.54	41761.7	ZN	1.2H			VS
2390.78	2391.51	41814.6	ZN II		1.4		MIT
2390.04	2390.77	41827.6	ZN II			0.7	VS
2388.30	2389.03	41858.0	ZN I	0.3			FL
2386.75	2387.48	41885.2	ZN II			0.3	VS
2383.95	2384.68	41934.4	ZN II			0.3	VS
2382.22	2382.95	41964.9	ZN I	0.6			FL
2376.01	2376.74	42074.5	ZN I	0.5			FL
2374.36	2375.08	42103.8	ZN I	0.5			FL
2367.72	2368.44	42221.8	ZN I	0.3			FL
2360.96	2361.68	42342.7	ZN I	0.3			FL
2347.30	2348.02	42589.1	ZN II			0.3	VS
2346.80	2347.52	42598.2	ZN II			0.5	VS
2317.18	2317.89	43142.6	ZN			0.5	BL
2312.72	2313.43	43225.8	ZN	1.0			HZ
2276.49	2277.19	43913.7	ZN			0.5	BL
2274.80	2275.50	43946.3	ZN			0.5	BL
2266.00	2266.70	44117.0	ZN II			2.4	VS
2265.51	2266.21	44126.5	ZN II			1.0	VS
2253.59	2254.29	44359.9	ZN II			0.5	VS
2253.06	2253.76	44370.3	ZN II			1.0	VS
2229.59	2230.28	44837.4	ZN			0.3	VS
2210.87	2211.56	45217.0	ZN II		1.7		MIT
2210.47	2211.16	45225.1	ZN II			0.3	VS
2157.06	2157.74	46344.8	ZN			0.3	VS
2148.15	2148.83	46537.0	ZN II			1.7	VS
2138.56	2139.23	46745.7	ZN I	2.9G	2.7		HZ
2136.46	2137.13	46791.6	ZN			1.0	BL
2124.04	2124.71	47065.2	ZN II			0.3	VS
2123.50	2124.17	47077.2	ZN II			0.5	VS
2121.41	2122.08	47123.6	ZN II			0.3	VS
2118.12	2118.79	47196.7	ZN			0.3	VS
2102.21	2102.88	47553.9	ZN II		0.7		SD
2099.86	2100.53	47607.1	ZN II		0.7		SD
2097.65	2098.32	47657.3	ZN			0.3	VS
2085.53	2086.19	47934.2	ZN II			1.3	VS
2083.15	2083.81	47988.9	ZN			0.3	VS
2079.82	2080.48	48065.8	ZN II			0.3	VS
2077.14	2077.80	48127.8	ZN			1.0	VS
2063.61	2064.27	48443.3	ZN II	0.3	1.3		SD
2061.91	2062.57	48483.2	ZN II	2.0	2.0		PS
2057.51	2058.17	48586.9	ZN II			1.3	VS
2040.00	2040.66	49003.9	ZN II			2.4	VS
2029.70	2030.35	49252.5	ZN II			0.3	VS
2026.19	2026.84	49337.8	ZN II			0.3	VS
2025.51	2026.16	49354.4	ZN II	2.3	2.3		PS
2025.33	2025.98	49358.8	ZN II	0.0	0.6		KS
2024.66	2025.31	49375.1	ZN II			1.4	VS
2012.62	2013.27	49670.4	ZN			2.4	VS
2009.32	2009.97	49752.0	ZN			0.3	VS
9909.76	9912.48	10088.3	ZR	0.3			ME
9822.30	9824.99	10178.1	ZR I	1.2			ME
9820.42	9823.11	10180.1	ZR I	0.3			ME
9812.85	9815.54	10187.9	ZR I	0.9			ME
9792.74	9795.43	10208.8	ZR I	0.6			ME
9780.40	9783.08	10221.7	ZR I	1.2			ME
9547.26	9549.88	10471.3	ZR I	1.2			ME
9493.47	9496.07	10530.7	ZR I	0.3			ME
9483.35	9485.95	10541.9	ZR	0.5			ME
9441.26	9443.85	10588.9	ZR	0.3			ME
9419.36	9421.94	10613.5	ZR I	0.3H			ME
9358.32	9360.89	10682.7	ZR I	0.5H			ME

AIR WAVELENGTH (ANGSTRMS)	VACUUM WAVELENGTH (ANGSTRMS)	WAVENUMBER (KAYSERS)	LMENT	LOG INTENSITY ARC	SPK	DIS	REF
9352.43	9355.00	10689.5	ZR	0.3			ME
9318.19	9320.75	10728.7	ZR	0.3H			ME
9276.89	9279.44	10776.5	ZR I	1.4			ME
9251.17	9253.71	10806.5	ZR I	0.8			ME
9242.65	9245.19	10816.4	ZR I	0.7			ME
9229.33	9231.86	10832.1	ZR	0.6			ME
9171.50	9174.02	10900.3	ZR I	0.6			ME
9139.36	9141.87	10938.7	ZR I	0.9			ME
9134.23	9136.74	10944.8	ZR	0.3			ME
9099.90	9102.40	10986.1	ZR I	0.5			ME
9069.40	9071.89	11023.1	ZR I	1.0			KS
9015.13	9017.60	11089.4	ZR I	1.2			KS
9011.34	9013.81	11094.1	ZR I	0.6			ME
8941.74	8944.19	11180.4	ZR I	0.5			KS
8930.10	8932.55	11195.0	ZR I	0.3			KS
8925.04	8927.49	11201.4	ZR I	0.6			KS
8910.05	8912.50	11220.2	ZR I	0.5			KS
8906.33	8908.78	11224.9	ZR I	0.5			KS
8899.52	8901.96	11233.5	ZR I	1.2			MIT
8844.50	8846.93	11303.4	ZR I	0.3			KS
8836.09	8838.52	11314.1	ZR I	1.2			KS
8804.98	8807.40	11354.1	ZR I	0.7			KS
8786.23	8788.64	11378.3	ZR I	0.3			KS
8749.48	8751.88	11426.1	ZR I	0.7			KS
8734.86	8737.26	11445.2	ZR I	0.6			KS
8709.24	8711.63	11478.9	ZR I	0.3			KS
8642.76	8645.13	11567.2	ZR I	0.3			KS
8641.01	8643.38	11569.5	ZR	0.6			KS
8615.66	8618.03	11603.6	ZR I	0.6			KS
8610.24	8612.61	11610.9	ZR I	0.6			KS
8587.84	8590.20	11641.2	ZR I	0.6			KS
8584.21	8586.57	11646.1	ZR I	1.0			KS
8571.05	8573.40	11664.0	ZR I	0.6			KS
8568.54	8570.89	11667.4	ZR I	0.6			KS
8558.64	8560.99	11680.9	ZR I	0.3			KS
8554.42	8556.77	11686.7	ZR	0.3			KS
8524.99	8527.33	11727.0	ZR I	0.6			KS
8515.06	8517.40	11740.7	ZR I	0.6			KS
8513.78	8516.12	11742.4	ZR I	0.8			KS
8510.38	8512.72	11747.1	ZR I	0.3			KS
8498.44	8500.77	11763.6	ZR I	1.3			KS
8495.98	8498.31	11767.0	ZR I	0.6			KS
8464.65	8466.98	11810.6	ZR I	0.6			KS
8457.48	8459.80	11820.6	ZR I	0.3			KS
8453.17	8455.49	11826.6	ZR I	0.3			KS
8414.00	8416.31	11881.7	ZR I	0.6H			KS
8408.50	8410.81	11889.5	ZR I	0.3H			KS
8390.18	8392.49	11915.4	ZR I	0.3			KS
8389.41	8391.72	11916.5	ZR I	0.8			MIT
8386.70	8389.00	11920.4	ZR I	0.3			KS
8370.23	8372.53	11943.8	ZR I	0.6			MIT
8365.98	8368.28	11949.9	ZR I	0.3			KS
8361.17	8363.47	11956.8	ZR I	0.3			KS
8360.17	8362.47	11958.2	ZR I	0.3H			KS
8350.45	8352.74	11972.1	ZR I	0.6H			KS
8332.44	8334.73	11998.0	ZR I	0.3			MIT
8320.16	8322.45	12015.7	ZR I	0.3			MIT
8316.97	8319.26	12020.3	ZR I	0.3			KS
8309.50	8311.78	12031.1	ZR I	0.3			MIT
8305.90	8308.18	12036.3	ZR I	1.1			MIT
8299.81	8302.09	12045.2	ZR I	0.6H			MIT
8296.54	8298.82	12049.9	ZR I	0.3			KS
8283.81	8286.09	12068.4	ZR I	1.1			MIT
8240.37	8242.64	12132.0	ZR I	0.9			MIT
8212.53	8214.79	12173.2	ZR I	1.4			MIT
8201.73	8203.98	12189.2	ZR I	0.9			KS
8194.73	8196.98	12199.6	ZR I	1.3			MIT
8188.77	8191.02	12208.5	ZR I	0.3			MIT
8169.72	8171.97	12237.0	ZR I	0.3			KS
8152.58	8154.82	12262.7	ZR I	0.6			MIT

AIR WAVELENGTH (ANGSTRMS)	VACUUM WAVELENGTH (ANGSTRMS)	WAVENUMBER (KAYSERS)	LMENT	LOG ARC	INTENSITY SPK	DIS	REF
8132.99	8135.23	12292.2	ZR I	1.9			MIT
8120.17	8122.40	12311.6	ZR I	0.3			MIT
8114.28	8116.51	12320.6	ZR I	0.3			MIT
8109.22	8111.45	12328.2	ZR I	0.3			MIT
8087.81	8090.03	12360.9	ZR I	0.3			MIT
8070.08	8072.30	12388.0	ZR I	2.0			MIT
8063.09	8065.31	12398.8	ZR I	1.3			MIT
8058.08	8060.30	12406.5	ZR I	0.9			MIT
8055.76	8057.98	12410.1	ZR I	0.6H			MIT
8055.29	8057.51	12410.8	ZR I	0.6			MIT
8053.06	8055.27	12414.2	ZR I	0.3			MIT
8046.05	8048.26	12425.0	ZR I	0.8			MIT
8040.10	8042.31	12434.2	ZR I	0.3			MIT
8015.26	8017.46	12472.8	ZR I	0.3			MIT
8005.27	8007.47	12488.3	ZR I	0.6			MIT
7963.63	7965.82	12553.6	ZR I	0.3			MIT
7959.98	7962.17	12559.4	ZR I	0.9			MIT
7956.66	7958.85	12564.6	ZR I	0.9			MIT
7944.61	7946.80	12583.7	ZR I	1.3			MIT
7940.47	7942.65	12590.2	ZR I	0.6H			MIT
7931.76	7933.94	12604.1	ZR I	0.3			KS
7924.20	7926.38	12616.1	ZR I	0.3			MIT
7915.25	7917.43	12630.4	ZR I	0.6H			MIT
7908.46	7910.64	12641.2	ZR I	0.3			KS
7906.55	7908.73	12644.3	ZR I	0.3			KS
7897.98	7900.15	12658.0	ZR I	0.6			KS
7888.52	7890.69	12673.2	ZR I	0.6H			KS
7882.18	7884.35	12683.4	ZR I	0.6H			MIT
7876.25	7878.42	12692.9	ZR I	0.5			MIT
7869.99	7872.16	12703.0	ZR I	0.9			MIT
7868.37	7870.53	12705.6	ZR I	0.6H			MIT
7864.37	7866.53	12712.1	ZR I	0.3			KS
7849.35	7851.51	12736.4	ZR I	0.9			MIT
7842.97	7845.13	12746.8	ZR I	0.3			MIT
7833.36	7835.52	12762.4	ZR I	0.3			KS
7826.72	7828.87	12773.2	ZR I	0.9			MIT
7822.94	7825.09	12779.4	ZR I	0.6			MIT
7819.35	7821.50	12785.3	ZR I	1.0			MIT
7816.32	7818.47	12790.2	ZR I	0.3			MIT
7800.74	7802.89	12815.8	ZR I	0.3			MIT
7799.51	7801.66	12817.8	ZR I	0.3			KS
7766.55	7768.69	12872.2	ZR I	0.3			MIT
7765.70	7767.84	12873.6	ZR I	0.3			MIT
7755.39	7757.52	12890.7	ZR I	0.3			KS
7745.43	7747.56	12907.3	ZR I	0.3			KS
7744.05	7746.18	12909.6	ZR	0.3			KS
7723.70	7725.83	12943.6	ZR I	0.3			KS
7722.48	7724.61	12945.7	ZR I	0.3			KS
7708.42	7710.54	12969.3	ZR I	0.6H			MIT
7704.27	7706.39	12976.2	ZR I	0.3H			MIT
7693.45	7695.57	12994.5	ZR I	1.0H			KS
7690.83	7692.95	12998.9	ZR I	0.6H			MIT
7658.60	7660.71	13053.6	ZR I	0.8			MIT
7646.71	7648.81	13073.9	ZR	1.0			MIT
7621.61	7623.71	13117.0	ZR I	0.3			MIT
7621.17	7623.27	13117.7	ZR I	0.3			MIT
7615.74	7617.84	13127.1	ZR I	0.3H			KS
7614.48	7616.58	13129.3	ZR I	0.3W			KS
7612.08	7614.18	13133.4	ZR I	0.6			MIT
7610.83	7612.93	13135.6	ZR I	0.3			KS
7609.60	7611.69	13137.7	ZR I	0.3			MIT
7607.15	7609.24	13141.9	ZR I	1.0			MIT
7562.12	7564.20	13220.2	ZR I	0.6V			MIT
7560.31	7562.39	13223.3	ZR I	0.3			KS
7560.09	7562.17	13223.7	ZR I	0.3			MIT
7558.45	7560.53	13226.6	ZR I	0.3H			MIT
7554.70	7556.78	13233.2	ZR I	0.6			MIT
7553.00	7555.08	13236.1	ZR I	0.3			MIT
7551.46	7553.54	13238.8	ZR I	0.6H			MIT
7544.59	7546.67	13250.9	ZR I	0.6H			MIT

AIR WAVELENGTH (ANGSTRMS)	VACUUM WAVELENGTH (ANGSTRMS)	WAVENUMBER (KAYSERS)	LMENT	LOG ARC	INTENSITY SPK	DIS	REF
7540.62	7542.70	13257.9	ZR I	0.3			MIT
7521.03	7523.10	13292.4	ZR I	0.3			MIT
7519.33	7521.40	13295.4	ZR	0.3H			MIT
7517.95	7520.02	13297.8	ZR I	0.3			MIT
7515.70	7517.77	13301.8	ZR I	0.3H			MIT
7506.51	7508.58	13318.1	ZR	0.3			MIT
7502.92	7504.99	13324.5	ZR I	0.3H			MIT
7479.58	7481.64	13366.1	ZR I	0.5			MIT
7467.57	7469.63	13387.6	ZR I	0.5			KS
7439.86	7441.91	13437.4	ZR I	1.2			MIT
7433.10	7435.15	13449.6	ZR I	0.5			MIT
7423.83	7425.87	13466.4	ZR I	0.5H			KS
7422.75	7424.79	13468.4	ZR I	0.5H			MIT
7417.89	7419.93	13477.2	ZR I	0.8			MIT
7414.24	7416.28	13483.8	ZR	0.5			KS
7411.39	7413.43	13489.0	ZR I	0.5H			MIT
7400.90	7402.94	13508.2	ZR I	0.5			MIT
7399.30	7401.34	13511.1	ZR	0.7			MIT
7383.63	7385.66	13539.7	ZR	0.5			MIT
7379.19	7381.22	13547.9	ZR	0.5			KS
7377.85	7379.88	13550.3	ZR I	0.5H			KS
7374.80	7376.83	13556.0	ZR I	0.5			KS
7373.50	7375.53	13558.3	ZR I	0.5			MIT
7367.20	7369.23	13569.9	ZR I	0.5			MIT
7361.56	7363.59	13580.3	ZR	0.5			KS
7360.66	7362.69	13582.0	ZR I	0.7			MIT
7358.60	7360.63	13585.8	ZR	0.5			KS
7353.34	7355.37	13595.5	ZR	0.5			KS
7343.96	7345.98	13612.9	ZR I	0.7			MIT
7340.03	7342.05	13620.2	ZR I	0.5			KS
7335.97	7337.99	13627.7	ZR I	0.9			MIT
7332.73	7334.75	13633.7	ZR	0.5H			KS
7327.82	7329.84	13642.9	ZR	0.5			MIT
7323.71	7325.73	13650.5	ZR	0.5			MIT
7318.08	7320.10	13661.0	ZR I	0.9			MIT
7317.30	7319.32	13662.5	ZR	0.5			MIT
7313.72	7315.74	13669.2	ZR I	0.7			KS
7311.62	7313.63	13673.1	ZR I	0.7			MIT
7307.36	7309.37	13681.1	ZR	0.5			MIT
7306.21	7308.22	13683.2	ZR I	0.7			MIT
7284.69	7286.70	13723.6	ZR I	0.5			MIT
7283.98	7285.99	13725.0	ZR	0.5H			MIT
7264.76	7266.76	13761.3	ZR I	0.7			MIT
7258.17	7260.17	13773.8	ZR I	0.5H			MIT
7248.48	7250.48	13792.2	ZR I	0.5			MIT
7223.87	7225.86	13839.2	ZR I	0.5			KS
7201.62	7203.60	13881.9	ZR I	1.0			MIT
7193.37	7195.35	13897.9	ZR I	0.5			KS
7179.53	7181.51	13924.7	ZR	0.5			MIT
7177.26	7179.24	13929.1	ZR	0.5			KS
7169.09	7171.07	13944.9	ZR I	2.2			MIT
7162.09	7164.06	13958.6	ZR I	0.5			MIT
7161.05	7163.02	13960.6	ZR I	0.7			MIT
7157.83	7159.80	13966.9	ZR I	0.9H			MIT
7146.13	7148.10	13989.7	ZR I	0.5			MIT
7144.47	7146.44	13993.0	ZR I	0.5H			MIT
7143.87	7145.84	13994.2	ZR	0.5			KS
7143.39	7145.36	13995.1	ZR I	0.5			MIT
7140.74	7142.71	14000.3	ZR I	0.5			MIT
7138.28	7140.25	14005.1	ZR I	0.5			MIT
7132.95	7134.92	14015.6	ZR I	0.5			MIT
7113.52	7115.48	14053.9	ZR I	0.7			MIT
7112.82	7114.78	14055.2	ZR I	0.9			MIT
7111.68	7113.64	14057.5	ZR I	1.6			MIT
7103.72	7105.68	14073.2	ZR I	1.7			MIT
7102.91	7104.87	14074.9	ZR I	1.9			MIT
7097.70	7099.66	14085.2	ZR I	2.0			MIT
7095.59	7097.55	14089.4	ZR I	1.1			MIT
7094.56	7096.52	14091.4	ZR I	1.0			MIT
7092.80	7094.76	14094.9	ZR I	0.5			MIT

AIR WAVELENGTH (ANGSTRMS)	VACUUM WAVELENGTH (ANGSTRMS)	WAVENUMBER (KAYSERS)	LMENT	LOG INTENSITY ARC	SPK	DIS	REF
7091.31	7093.27	14097.9	ZR	0.7H			MIT
7089.43	7091.38	14101.6	ZR I	0.7			MIT
7087.30	7089.25	14105.9	ZR I	1.4			MIT
7060.75	7062.70	14158.9	ZR I	0.7H			MIT
7057.96	7059.91	14164.5	ZR	0.7			MIT
7057.36	7059.31	14165.7	ZR I	0.9			MIT
7027.40	7029.34	14226.1	ZR I	1.5			KS
7005.46	7007.39	14270.6	ZR I	0.7			KS
6994.32	6996.25	14293.4	ZR I	1.2			MIT
6990.84	6992.77	14300.5	ZR I	1.7			MIT
6975.91	6977.83	14331.1	ZR I	0.9			MIT
6966.44	6968.36	14350.6	ZR I	1.6			MIT
6953.84	6955.76	14376.6	ZR I	1.9			MIT
6948.46	6950.38	14387.7	ZR I	1.2			MIT
6932.38	6934.29	14421.1	ZR I	1.1			MIT
6929.07	6930.98	14428.0	ZR I	0.9			MIT
6925.71	6927.62	14435.0	ZR I	0.9			KS
6922.23	6924.14	14442.2	ZR I	0.9			KS
6916.87	6918.78	14453.4	ZR I	1.1			MIT
6907.37	6909.28	14473.3	ZR I	1.1			MIT
6904.36	6906.26	14479.6	ZR I	1.0H			MIT
6900.59	6902.49	14487.5	ZR I	1.1H			MIT
6888.29	6890.19	14513.4	ZR I	1.4			MIT
6885.20	6887.10	14519.9	ZR	0.9			KS
6883.25	6885.15	14524.0	ZR I	0.9			MIT
6857.90	6859.79	14577.7	ZR I	0.9H			MIT
6854.63	6856.52	14584.7	ZR I	0.8			KS
6853.84	6855.73	14586.3	ZR I	0.8			MIT
6852.56	6854.45	14589.1	ZR I	0.8			MIT
6849.26	6851.15	14596.1	ZR I	0.8			MIT
6847.34	6849.23	14600.2	ZR I	0.3			KS
6846.97	6848.86	14601.0	ZR	1.1			MIT
6846.34	6848.23	14602.3	ZR I	0.6			MIT
6845.33	6847.22	14604.5	ZR I	1.0			MIT
6843.75	6845.64	14607.8	ZR I	0.6			MIT
6833.67	6835.56	14629.4	ZR I	0.6			MIT
6832.89	6834.78	14631.1	ZR I	1.0			MIT
6828.78	6830.66	14639.9	ZR I	1.0			MIT
6812.64	6814.52	14674.6	ZR	0.8			MIT
6811.32	6813.20	14677.4	ZR I	0.6			KS
6796.68	6798.56	14709.0	ZR I	0.6			MIT
6794.66	6796.54	14713.4	ZR	0.6			KS
6790.85	6792.72	14721.6	ZR I	1.0			MIT
6787.15	6789.02	14729.7	ZR II	0.6	0.3		KS
6772.89	6774.76	14760.7	ZR I	0.9			MIT
6769.159	6771.027	14768.81	ZR I	1.7			MIT
6762.38	6764.25	14783.6	ZR I	1.7			KS
6752.73	6754.59	14804.7	ZR I	1.2			KS
6717.879	6719.734	14881.54	ZR I	1.1			MIT
6709.609	6711.461	14899.89	ZR I	0.9			MIT
6702.125	6703.975	14916.52	ZR I	0.8			MIT
6697.94	6699.79	14925.8	ZR I	0.6			KS
6688.175	6690.022	14947.64	ZR I	1.0			MIT
6678.81	6680.65	14968.6	ZR I	0.3			KS
6678.01	6679.85	14970.4	ZR II		0.5		MIT
6620.558	6622.386	15100.30	ZR I	0.8			MIT
6603.27	6605.09	15139.8	ZR I	1.3			KS
6599.90	6601.72	15147.6	ZR I	0.6			MIT
6598.842	6600.665	15149.99	ZR I	0.8			MIT
6596.712	6598.534	15154.88	ZR I	0.6			MIT
6591.988	6593.809	15165.74	ZR I	1.5			MIT
6576.558	6578.375	15201.32	ZR I	0.8			MIT
6569.427	6571.242	15217.82	ZR I	0.9			MIT
6550.540	6552.350	15261.70	ZR I	1.3			MIT
6506.360	6508.158	15365.33	ZR I	1.4			MIT
6503.262	6505.059	15372.65	ZR I	1.3			MIT
6493.098	6494.892	15396.71	ZR I	1.0			MIT
6489.645	6491.438	15404.91	ZR I	1.7			MIT
6484.35	6486.14	15417.5	ZR I	0.8			KS
6470.210	6471.998	15451.18	ZR I	1.6			MIT

AIR WAVELENGTH (ANGSTRMS)	VACUUM WAVELENGTH (ANGSTRMS)	WAVENUMBER (KAYSERS)	LMENT	LOG INTENSITY ARC	SPK	DIS	REF
6457.627	6459.412	15481.29	ZR I	0.8			MIT
6451.62	6453.40	15495.7	ZR I	1.0			MIT
6445.745	6447.526	15509.82	ZR I	1.5			MIT
6439.03	6440.81	15526.0	ZR I	0.9			MIT
6434.329	6436.107	15537.34	ZR I	0.8			MIT
6426.170	6427.946	15557.07	ZR I	0.6			MIT
6406.997	6408.768	15603.62	ZR I	1.3			MIT
6404.30	6406.07	15610.2	ZR I	0.6			MIT
6346.52	6348.27	15752.3	ZR II	1.0	0.3		MIT
6345.221	6346.975	15755.54	ZR I	1.2			MIT
6344.938	6346.692	15756.24	ZR	0.6B			MIT
6340.361	6342.114	15767.61	ZR I	0.9			MIT
6321.348	6323.096	15815.04	ZR I	1.1			MIT
6314.710	6316.456	15831.66	ZR I	0.8			MIT
6313.554	6315.300	15834.56	ZR II	0.8	0.0		MIT
6313.024	6314.770	15835.89	ZR I	2.3			MIT
6304.344	6306.087	15857.69	ZR I	0.8			MIT
6299.657	6301.399	15869.49	ZR I	1.7			MIT
6282.624	6284.362	15912.52	ZR I	0.6			MIT
6279.767	6281.504	15919.76	ZR I	0.8			MIT
6267.062	6268.795	15952.03	ZR I	1.0			MIT
6257.255	6258.986	15977.03	ZR I	1.5			MIT
6224.189	6225.911	16061.91	ZR I	0.8			MIT
6220.806	6222.527	16070.64	ZR I	0.6			MIT
6214.689	6216.408	16086.46	ZR I	1.3			MIT
6213.052	6214.771	16090.70	ZR I	1.3			MIT
6204.143	6205.860	16113.80	ZR I	0.6			MIT
6192.960	6194.674	16142.90	ZR I	1.3			MIT
6189.399	6191.112	16152.19	ZR I	0.9			MIT
6178.231	6179.941	16181.39	ZR I	0.9			MIT
6160.196	6161.901	16228.76	ZR I	0.9			MIT
6157.714	6159.418	16235.30	ZR I	1.3			MIT
6155.614	6157.318	16240.84	ZR I	0.8			MIT
6143.200	6144.900	16273.66	ZR I	2.5			MIT
6140.457	6142.156	16280.93	ZR I	1.6			MIT
6134.550	6136.248	16296.60	ZR I	2.5			MIT
6127.440	6129.136	16315.51	ZR I	2.7			MIT
6124.84	6126.54	16322.4	ZR I	1.6			MIT
6121.91	6123.60	16330.2	ZR I	1.8			MIT
6120.83	6122.52	16333.1	ZR I	1.1			MIT
6114.78	6116.47	16349.3	ZR II	1.0	0.3		KS
6106.47	6108.16	16371.5	ZR II	0.9	0.3		KS
6100.04	6101.73	16388.8	ZR II	0.9	0.3		KS
6062.846	6064.525	16489.34	ZR I	1.5			MIT
6052.84	6054.52	16516.6	ZR I	0.8			KS
6049.24	6050.92	16526.4	ZR I	1.3			MIT
6045.852	6047.526	16535.69	ZR I	1.2			MIT
6032.607	6034.278	16571.99	ZR I	1.4			MIT
6028.635	6030.304	16582.91	ZR II	0.8	0.3		MIT
6025.361	6027.030	16591.92	ZR I	0.8			MIT
6001.051	6002.713	16659.13	ZR I	0.9			MIT
5995.367	5997.028	16674.93	ZR I	0.5			MIT
5984.825	5986.483	16704.30	ZR I	0.3			MIT
5984.231	5985.889	16705.96	ZR I	1.2			MIT
5955.354	5957.004	16786.96	ZR I	1.3			MIT
5935.202	5936.846	16843.96	ZR I	1.3			MIT
5925.131	5926.773	16872.59	ZR I	1.3			MIT
5901.086	5902.721	16941.34	ZR I	0.6			MIT
5885.619	5887.250	16985.86	ZR I	1.4			MIT
5879.797	5881.427	17002.68	ZR I	1.8			MIT
5869.497	5871.124	17032.51	ZR I	0.9			MIT
5868.265	5869.892	17036.09	ZR I	0.6			MIT
5847.318	5848.939	17097.12	ZR I	0.6			MIT
5801.351	5802.960	17232.59	ZR I	0.3			MIT
5797.740	5799.348	17243.32	ZR I	1.7			MIT
5791.338	5792.944	17262.38	ZR I	0.3			MIT
5746.617	5748.211	17396.72	ZR I	0.3			MIT
5735.701	5737.292	17429.83	ZR I	1.4			MIT
5708.887	5710.471	17511.69	ZR I	0.6			MIT
5685.419	5686.997	17583.97	ZR I	0.3			MIT

AIR WAVELENGTH (ANGSTRMS)	VACUUM WAVELENGTH (ANGSTRMS)	WAVENUMBER (KAYSERS)	LMENT	LOG ARC	INTENSITY SPK	INTENSITY DIS	REF
5680.898	5682.474	17597.97	ZR I	1.7			MIT
5674.76	5676.33	17617.0	ZR I	0.5			KS
5666.277	5667.849	17643.38	ZR I	0.6			MIT
5664.511	5666.083	17648.88	ZR I	1.7			MIT
5659.873	5661.444	17663.34	ZR I	0.5			MIT
5628.990	5630.552	17760.25	ZR I	0.3			MIT
5623.526	5625.087	17777.50	ZR I	0.8			MIT
5620.136	5621.696	17788.23	ZR I	1.0			MIT
5557.920	5559.463	17987.35	ZR I	0.5H			MIT
5555.981	5557.524	17993.62	ZR	0.3			MIT
5555.389	5556.932	17995.54	ZR I	0.5			MIT
5550.368	5551.909	18011.82	ZR I	0.5H			MIT
5545.320	5546.860	18028.22	ZR I	0.7			MIT
5537.417	5538.955	18053.95	ZR I	0.7			MIT
5536.683	5538.221	18056.34	ZR	0.3			MIT
5532.300	5533.837	18070.65	ZR I	0.5			MIT
5528.412	5529.948	18083.35	ZR I	0.8			MIT
5523.869	5525.403	18098.23	ZR	0.3			MIT
5518.048	5519.581	18117.32	ZR I	0.3			MIT
5517.107	5518.640	18120.41	ZR I	0.6			MIT
5507.874	5509.404	18150.78	ZR	0.3			MIT
5502.123	5503.652	18169.75	ZR I	0.8			MIT
5488.328	5489.853	18215.42	ZR	0.3			MIT
5487.521	5489.046	18218.10	ZR I	0.3			MIT
5486.089	5487.613	18222.86	ZR I	0.6			MIT
5481.157	5482.680	18239.25	ZR	0.3			MIT
5480.827	5482.350	18240.35	ZR I	0.6			MIT
5478.326	5479.848	18248.68	ZR I	0.6			MIT
5477.778	5479.300	18250.51	ZR II	0.3H			MIT
5477.395	5478.917	18251.78	ZR I	0.3H			MIT
5474.923	5476.444	18260.02	ZR I	0.3			MIT
5474.38	5475.90	18261.8	ZR	0.3			MIT
5464.064	5465.582	18296.31	ZR I	0.3			MIT
5460.73	5462.25	18307.5	ZR I	0.3H			MIT
5457.595	5459.112	18318.00	ZR I	0.3			MIT
5451.211	5452.726	18339.45	ZR	0.3H			MIT
5448.551	5450.065	18348.40	ZR I	0.6			MIT
5447.958	5449.472	18350.40	ZR	0.3	0.3		MIT
5443.16	5444.67	18366.6	ZR I	0.3			MIT
5440.406	5441.918	18375.87	ZR I	0.5			MIT
5437.757	5439.268	18384.82	ZR I	0.7			MIT
5437.256	5438.767	18386.52	ZR I	0.5			MIT
5436.91	5438.42	18387.7	ZR I	0.3			MIT
5428.423	5429.932	18416.44	ZR I	0.6			MIT
5426.361	5427.869	18423.44	ZR I	0.5			MIT
5423.513	5425.021	18433.11	ZR	0.3			MIT
5421.858	5423.365	18438.74	ZR I	0.5			MIT
5418.74	5420.25	18449.4	ZR I	0.3			MIT
5413.93	5415.44	18465.7	ZR I	0.5			MIT
5407.617	5409.120	18487.29	ZR I	0.8			MIT
5405.126	5406.629	18495.81	ZR I	0.6			MIT
5395.879	5397.379	18527.51	ZR I	0.5			MIT
5391.180	5392.679	18543.66	ZR I	0.5			MIT
5386.653	5388.151	18559.24	ZR I	0.6			MIT
5385.136	5386.633	18564.47	ZR I	1.2			MIT
5382.371	5383.868	18574.01	ZR I	0.5H			MIT
5381.229	5382.725	18577.95	ZR	0.3H			MIT
5379.548	5381.044	18583.75	ZR I	0.3			MIT
5372.451	5373.945	18608.30	ZR	0.3H			MIT
5369.387	5370.880	18618.92	ZR I	0.5			MIT
5363.354	5364.846	18639.87	ZR I	0.5			MIT
5362.556	5364.047	18642.64	ZR I	0.9			MIT
5353.375	5354.864	18674.61	ZR	0.3			MIT
5351.915	5353.403	18679.71	ZR	0.5			MIT
5350.899	5352.387	18683.25	ZR I	0.3			MIT
5350.353	5351.841	18685.16	ZR II	0.6	0.7		MIT
5350.093	5351.581	18686.07	ZR II	0.6	0.7		MIT
5343.596	5345.082	18708.79	ZR I	0.3H			MIT
5338.429	5339.914	18726.89	ZR I	0.3			MIT
5332.45	5333.93	18747.9	ZR	0.3H			MIT

AIR WAVELENGTH (ANGSTRMS)	VACUUM WAVELENGTH (ANGSTRMS)	WAVENUMBER (KAYSERS)	LMENT	LOG ARC	INTENSITY SPK	INTENSITY DIS	REF
5331.746	5333.229	18750.37	ZR I	0.3			MIT
5330.837	5332.320	18753.56	ZR I	0.5			MIT
5329.833	5331.316	18757.10	ZR I	0.3			MIT
5321.264	5322.744	18787.30	ZR I	0.6			MIT
5319.984	5321.464	18791.82	ZR I	0.3			MIT
5311.767	5313.245	18820.89	ZR II	0.3	0.3		MIT
5311.402	5312.880	18822.18	ZR I	1.0			MIT
5310.992	5312.470	18823.64	ZR I	0.3			MIT
5308.384	5309.861	18832.89	ZR I	0.3			MIT
5307.603	5309.080	18835.66	ZR I	0.3			MIT
5305.51	5306.99	18843.1	ZR I	0.3H			MIT
5303.79	5305.27	18849.2	ZR I	0.3			MIT
5301.969	5303.444	18855.67	ZR I	0.8			MIT
5300.125	5301.600	18862.23	ZR I	0.3			MIT
5299.515	5300.990	18864.40	ZR	0.3			MIT
5299.201	5300.675	18865.52	ZR I	0.3			MIT
5296.788	5298.262	18874.11	ZR I	1.0			MIT
5294.823	5296.296	18881.12	ZR I	0.9			MIT
5292.531	5294.004	18889.30	ZR	0.3			MIT
5280.050	5281.519	18933.95	ZR I	0.8			MIT
5277.409	5278.878	18943.42	ZR I	1.0			MIT
5264.04	5265.51	18991.5	ZR I	0.3			KS
5250.34	5251.80	19041.1	ZR I	0.3			MIT
5248.50	5249.96	19047.8	ZR I	0.3			MIT
5248.011	5249.472	19049.54	ZR I	0.3			MIT
5243.469	5244.929	19066.04	ZR I	0.6			MIT
5242.197	5243.656	19070.66	ZR I	0.3			MIT
5241.940	5243.399	19071.60	ZR I	0.3			MIT
5235.057	5236.514	19096.67	ZR I	0.3H			MIT
5224.926	5226.381	19133.70	ZR I	1.0			MIT
5223.630	5225.084	19138.45	ZR I	0.9			MIT
5212.257	5213.708	19180.21	ZR	0.5			MIT
5209.299	5210.749	19191.10	ZR I	0.7			MIT
5201.146	5202.594	19221.18	ZR I	0.8			MIT
5191.600	5193.046	19256.52	ZR II	0.7	1.0		MIT
5187.032	5188.477	19273.48	ZR I	0.5			MIT
5183.705	5185.149	19285.85	ZR I	0.8			MIT
5178.987	5180.429	19303.42	ZR I	0.3			MIT
5165.961	5167.400	19352.09	ZR I	0.8			MIT
5164.703	5166.142	19356.81	ZR	0.3			MIT
5160.988	5162.426	19370.74	ZR I	0.8			MIT
5158.666	5160.103	19379.46	ZR I	0.5			MIT
5158.001	5159.438	19381.96	ZR I	1.0	0.0		MIT
5155.449	5156.885	19391.55	ZR I	1.2			MIT
5145.464	5146.897	19429.18	ZR I	1.0			MIT
5136.777	5138.208	19462.04	ZR I	0.3			MIT
5136.144	5137.575	19464.44	ZR	0.3			MIT
5134.428	5135.859	19470.94	ZR	0.3			MIT
5133.401	5134.831	19474.84	ZR I	0.9			MIT
5132.869	5134.299	19476.85	ZR I	0.5			MIT
5126.05	5127.48	19502.8	ZR	0.3			MIT
5124.953	5126.381	19506.94	ZR II		0.3		MIT
5120.422	5121.849	19524.20	ZR I	1.0			MIT
5115.245	5116.670	19543.96	ZR I	1.0			MIT
5113.725	5115.150	19549.77	ZR I	0.3			MIT
5112.270	5113.695	19555.33	ZR II		0.7		MIT
5110.16	5111.58	19563.4	ZR I	0.3H			MIT
5105.529	5106.952	19581.15	ZR I	0.7			MIT
5098.56	5099.98	19607.9	ZR	0.3			MIT
5089.616	5091.035	19642.37	ZR	0.3			MIT
5085.260	5086.677	19659.20	ZR I	1.0			MIT
5079.982	5081.398	19679.62	ZR	0.3			MIT
5078.254	5079.670	19686.32	ZR I	1.2			MIT
5075.21	5076.62	19698.1	ZR	0.3			MIT
5073.981	5075.395	19702.90	ZR I	0.8			MIT
5070.262	5071.675	19717.35	ZR I	0.9			MIT
5065.216	5066.628	19736.99	ZR I	1.0			MIT
5064.910	5066.322	19738.19	ZR I	1.0			MIT
5060.393	5061.804	19755.80	ZR I	1.0			MIT
5046.585	5047.992	19809.86	ZR I	1.4			MIT

AIR WAVELENGTH (ANGSTRMS)	VACUUM WAVELENGTH (ANGSTRMS)	WAVENUMBER (KAYSERS)	LMENT	LOG ARC	INTENSITY SPK	INTENSITY DIS	REF
5034.436	5035.840	19857.66	ZR I	0.5			MIT
5032.408	5033.811	19865.66	ZR I	0.3			MIT
5026.907	5028.309	19887.40	ZR I	0.3			MIT
5011.464	5012.862	19948.69	ZR I	0.3			MIT
4996.331	4997.725	20009.10	ZR I	0.6			MIT
4994.761	4996.154	20015.39	ZR I	0.9			MIT
4987.819	4989.211	20043.25	ZR I	0.5			MIT
4968.170	4969.556	20122.52	ZR I	0.3			MIT
4963.716	4965.101	20140.58	ZR I	0.5			MIT
4962.296	4963.681	20146.34	ZR I	0.5			MIT
4956.911	4958.294	20168.23	ZR I	0.5H			MIT
4951.708	4953.090	20189.42	ZR I	0.5			MIT
4948.756	4950.137	20201.46	ZR I	0.7			MIT
4939.905	4941.284	20237.66	ZR	0.5			MIT
4933.643	4935.020	20263.34	ZR I	0.6			MIT
4930.866	4932.242	20274.75	ZR I	0.5			MIT
4910.778	4912.149	20357.69	ZR I	0.5			MIT
4905.083	4906.453	20381.32	ZR I	0.5			MIT
4894.172	4895.539	20426.76	ZR I	0.5			MIT
4893.121	4894.487	20431.15	ZR I	0.7			MIT
4884.102	4885.466	20468.88	ZR I	0.5			MIT
4883.598	4884.962	20470.99	ZR I	0.8			MIT
4881.241	4882.604	20480.87	ZR II	0.6			MIT
4866.060	4867.419	20544.77	ZR I	1.0			MIT
4851.363	4852.718	20607.01	ZR I	1.2			MIT
4847.695	4849.049	20622.60	ZR I	0.5J			MIT
4846.352	4847.706	20628.31	ZR I	0.5			MIT
4838.980	4840.332	20659.74	ZR I	0.5			MIT
4838.775	4840.127	20660.61	ZR I	0.6			MIT
4828.044	4829.393	20706.53	ZR I	1.0			MIT
4824.288	4825.636	20722.66	ZR I	1.3			MIT
4815.629	4816.975	20759.92	ZR I	1.6			MIT
4815.045	4816.391	20762.44	ZR I	1.2			MIT
4809.468	4810.812	20786.51	ZR I	0.9			MIT
4806.679	4808.022	20798.57	ZR I	0.6			MIT
4805.870	4807.213	20802.07	ZR I	1.2			MIT
4803.440	4804.783	20812.60	ZR I	0.5			MIT
4794.955	4796.295	20849.43	ZR I	0.5			MIT
4793.276	4794.616	20856.73	ZR I	0.5			MIT
4789.109	4790.448	20874.87	ZR I	0.7			MIT
4788.669	4790.008	20876.79	ZR I	1.0			MIT
4784.919	4786.257	20893.15	ZR I	1.6			MIT
4782.599	4783.936	20903.29	ZR I	0.5			MIT
4772.312	4773.646	20948.35	ZR I	2.0			MIT
4762.776	4764.108	20990.29	ZR I	1.0			MIT
4753.054	4754.383	21033.22	ZR I	0.6			MIT
4751.907	4753.236	21038.30	ZR I	0.6			MIT
4749.387	4750.715	21049.46	ZR I	0.5			MIT
4748.709	4750.037	21052.47	ZR I	0.5J			MIT
4742.939	4744.266	21078.08	ZR I	0.5			MIT
4740.5	4741.8	21089.	ZR BH	0.9			L
4739.478	4740.804	21093.47	ZR I	2.0			MIT
4736.965	4738.290	21104.66	ZR	0.5			MIT
4734.361	4735.685	21116.27	ZR I	0.5			MIT
4732.329	4733.653	21125.33	ZR I	1.4			MIT
4731.138	4732.461	21130.65	ZR I	0.7			MIT
4723.900	4725.222	21163.03	ZR I	0.5			MIT
4719.116	4720.436	21184.48	ZR I	0.9			MIT
4717.620	4718.940	21191.20	ZR I	0.8			MIT
4713.433	4714.752	21210.02	ZR I	0.7			MIT
4711.916	4713.234	21216.85	ZR I	1.2			MIT
4710.075	4711.393	21225.15	ZR I	1.8			MIT
4707.786	4709.103	21235.47	ZR I	0.7			MIT
4703.035	4704.351	21256.92	ZR II	0.3	0.5		MIT
4700.176	4701.491	21269.85	ZR I	0.5			MIT
4700.110	4701.425	21270.15	ZR	0.5			MIT
4695.038	4696.352	21293.12	ZR I	0.5			MIT
4691.732	4693.045	21308.13	ZR I	0.5			MIT
4688.448	4689.760	21323.05	ZR I	1.7			MIT
4687.803	4689.115	21325.99	ZR I	2.1			MIT

AIR WAVELENGTH (ANGSTRMS)	VACUUM WAVELENGTH (ANGSTRMS)	WAVENUMBER (KAYSERS)	LMENT	LOG ARC	INTENSITY SPK	INTENSITY DIS	REF
4685.672	4686.983	21335.69	ZR	0.6			MIT
4685.189	4686.500	21337.88	ZR II	0.3	0.5		MIT
4684.250	4685.561	21342.16	ZR I	0.6			MIT
4683.421	4684.732	21345.94	ZR I	1.0			MIT
4667.145	4668.451	21420.38	ZR I	0.7			MIT
4662.961	4664.266	21439.60	ZR I	0.5			MIT
4661.778	4663.083	21445.04	ZR II	0.3	0.6		MIT
4659.488	4660.792	21455.58	ZR I	0.7			MIT
4657.641	4658.945	21464.09	ZR I	0.8			MIT
4654.379	4655.682	21479.13	ZR I	0.5			MIT
4644.829	4646.130	21523.29	ZR I	0.7			MIT
4640.6	4641.9	21543.	ZR BH	2.2			L
4640.128	4641.427	21545.10	ZR I	0.5			MIT
4637.8	4639.1	21556.	ZR BH	2.0			L
4634.641	4635.939	21570.60	ZR I	0.5			MIT
4633.985	4635.283	21573.66	ZR I	1.5			MIT
4627.720	4629.016	21602.86	ZR I	0.5			MIT
4626.413	4627.709	21608.97	ZR I	1.1			MIT
4613.953	4615.245	21667.32	ZR II	0.6	0.6H		MIT
4613.368	4614.660	21670.07	ZR	0.7H			MIT
4610.109	4611.400	21685.39	ZR I	0.6			MIT
4609.827	4611.118	21686.71	ZR I	0.5			MIT
4609.293	4610.584	21689.23	ZR I	0.5			MIT
4609.154	4610.445	21689.88	ZR I	0.5			MIT
4604.421	4605.711	21712.18	ZR I	1.0			MIT
4602.573	4603.862	21720.89	ZR I	1.1			MIT
4601.430	4602.719	21726.29	ZR	0.3			MIT
4590.548	4591.834	21777.79	ZR I	0.9			MIT
4590.159	4591.445	21779.64	ZR I	0.5			MIT
4584.243	4585.528	21807.74	ZR I	0.7			MIT
4582.292	4583.576	21817.03	ZR I	0.9			MIT
4576.202	4577.484	21846.06	ZR I	0.7			MIT
4575.515	4576.797	21849.34	ZR I	1.7			MIT
4574.496	4575.778	21854.21	ZR II	0.5	0.3		MIT
4572.851	4574.133	21862.07	ZR I	0.5J			MIT
4565.469	4566.749	21897.42	ZR I	0.9			MIT
4558.711	4559.989	21929.88	ZR	0.5			MIT
4558.042	4559.320	21933.10	ZR I	0.6			MIT
4555.52	4556.80	21945.2	ZR I	1.5	0.3		KS
4555.130	4556.407	21947.12	ZR I	1.2			MIT
4553.967	4555.244	21952.72	ZR II	0.6	1.1		MIT
4553.012	4554.288	21957.33	ZR I	1.0			MIT
4550.714	4551.990	21968.42	ZR I	0.5			MIT
4549.625	4550.900	21973.67	ZR I	1.2	0.3		MIT
4542.216	4543.489	22009.51	ZR I	1.2			MIT
4539.983	4541.256	22020.34	ZR I	1.2			MIT
4537.609	4538.881	22031.86	ZR I	0.5			MIT
4535.749	4537.021	22040.90	ZR I	1.0			MIT
4526.411	4527.680	22086.36	ZR I	0.7H			MIT
4526.125	4527.394	22087.76	ZR I	0.5			MIT
4523.909	4525.178	22098.58	ZR I	0.5			MIT
4523.130	4524.398	22102.39	ZR I	0.6			MIT
4521.939	4523.207	22108.21	ZR I	0.5			MIT
4515.838	4517.104	22138.07	ZR	0.5			MIT
4515.421	4516.687	22140.12	ZR I	0.5			MIT
4511.170	4512.435	22160.98	ZR I	0.7			MIT
4508.026	4509.290	22176.44	ZR	0.5			MIT
4507.277	4508.541	22180.12	ZR I	0.5			MIT
4507.119	4508.383	22180.90	ZR I	1.3			MIT
4496.971	4498.233	22230.95	ZR II	1.0	1.0		MIT
4494.935	4496.196	22241.02	ZR I	0.5			MIT
4494.416	4495.677	22243.59	ZR II	0.3	0.7		MIT
4491.560	4492.820	22257.74	ZR I	0.6			MIT
4486.363	4487.622	22283.52	ZR I	0.5			MIT
4485.445	4486.703	22288.08	ZR II	0.5			MIT
4482.503	4483.761	22302.71	ZR I	0.5			MIT
4480.769	4482.026	22311.34	ZR I	0.5			MIT
4470.559	4471.814	22362.29	ZR I	1.0			MIT
4470.311	4471.565	22363.53	ZR I	0.7			MIT
4468.787	4470.041	22371.16	ZR I	0.5			MIT

AIR WAVELENGTH (ANGSTRMS)	VACUUM WAVELENGTH (ANGSTRMS)	WAVENUMBER (KAYSERS)	LMENT	LOG ARC	INTENSITY SPK	DIS	REF
4468.219	4469.473	22374.00	ZR I	0.5			MIT
4466.909	4468.163	22380.56	ZR I	0.9	0.0		MIT
4464.127	4465.380	22394.51	ZR	0.5			MIT
4461.223	4462.475	22409.09	ZR II	0.6	0.5		MIT
4460.341	4461.593	22413.52	ZR I	0.9			MIT
4460.00	4461.25	22415.2	ZR	0.3			KS
4457.432	4458.683	22428.15	ZR II	1.6	0.8		MIT
4456.298	4457.549	22433.85	ZR I	0.9			MIT
4455.434	4456.685	22438.21	ZR I	0.8			MIT
4454.797	4456.047	22441.41	ZR II	0.8	0.0		MIT
4450.285	4451.534	22464.17	ZR I	0.9			MIT
4448.947	4450.196	22470.92	ZR I	0.5			MIT
4442.998	4444.245	22501.01	ZR II	0.9	0.8		MIT
4440.460	4441.707	22513.87	ZR II	0.6	0.5		MIT
4438.051	4439.297	22526.09	ZR I	0.9			MIT
4435.840	4437.085	22537.32	ZR	0.5			MIT
4431.493	4432.737	22559.42	ZR I	0.9	0.0		MIT
4431.107	4432.351	22561.39	ZR I	0.5			MIT
4429.106	4430.350	22571.58	ZR I	0.7			MIT
4427.243	4428.486	22581.08	ZR I	1.0			MIT
4420.456	4421.697	22615.75	ZR I	1.3	0.0		MIT
4414.893	4416.133	22644.25	ZR I	0.5	0.3		MIT
4414.544	4415.784	22646.04	ZR II	0.6	0.5		MIT
4414.140	4415.380	22648.11	ZR I	0.7			MIT
4413.040	4414.279	22653.75	ZR I	1.1			MIT
4411.931	4413.170	22659.45	ZR I	0.5			MIT
4409.634	4410.872	22671.25	ZR I	0.5			MIT
4403.344	4404.581	22703.64	ZR II	0.7	0.5		MIT
4402.953	4404.190	22705.65	ZR I	0.9			MIT
4400.240	4401.476	22719.65	ZR I	0.9			MIT
4399.444	4400.680	22723.76	ZR II	0.5			MIT
4395.207	4396.442	22745.67	ZR I	1.0			MIT
4394.941	4396.176	22747.04	ZR I	0.9			MIT
4394.497	4395.731	22749.34	ZR I	0.7			MIT
4388.513	4389.746	22780.36	ZR II	0.5			MIT
4382.733	4383.964	22810.40	ZR	0.5			MIT
4379.776	4381.007	22825.80	ZR II	1.0	0.9		MIT
4373.070	4374.299	22860.81	ZR I	0.8			MIT
4371.621	4372.849	22868.38	ZR	0.6			MIT
4370.951	4372.179	22871.89	ZR II	1.0	0.8		MIT
4366.447	4367.674	22895.48	ZR I	1.4	0.3		MIT
4361.945	4363.171	22919.11	ZR	0.6			MIT
4360.807	4362.033	22925.09	ZR	1.4	0.3		MIT
4359.736	4360.961	22930.72	ZR II	0.9	0.9		MIT
4358.742	4359.967	22935.95	ZR I	1.0			MIT
4348.933	4350.155	22987.68	ZR I	0.9			MIT
4348.102	4349.324	22992.08	ZR	0.6	0.3H		MIT
4347.892	4349.114	22993.19	ZR I	1.6	0.7		MIT
4347.372	4348.594	22995.94	ZR	0.5			MIT
4347.225	4348.447	22996.72	ZR I	0.7			MIT
4346.517	4347.739	23000.46	ZR I	1.0			MIT
4345.556	4346.778	23005.55	ZR I	0.6			MIT
4343.401	4344.622	23016.96	ZR I	0.5			MIT
4343.038	4344.259	23018.89	ZR I	0.3			MIT
4342.239	4343.460	23023.12	ZR II	0.6			MIT
4341.490	4342.711	23027.09	ZR I	0.5			MIT
4341.133	4342.353	23028.99	ZR I	1.7	0.6		MIT
4339.555	4340.775	23037.36	ZR II	0.5	0.0		MIT
4337.623	4338.843	23047.62	ZR II	0.5H	0.3		MIT
4336.108	4337.327	23055.67	ZR I	0.8			MIT
4333.262	4334.480	23070.82	ZR II	0.9	0.7		MIT
4329.561	4330.778	23090.54	ZR I	0.7			MIT
4325.435	4326.651	23112.56	ZR I	0.9	0.3		MIT
4324.030	4325.246	23120.07	ZR I	1.0			MIT
4323.893	4325.109	23120.80	ZR I	0.5			MIT
4322.221	4323.436	23129.75	ZR I	0.5H			MIT
4321.439	4322.654	23133.93	ZR I	0.5			MIT
4321.167	4322.382	23135.39	ZR I	1.0			MIT
4319.048	4320.263	23146.74	ZR I	1.0			MIT
4317.313	4318.527	23156.04	ZR II	1.0	0.6		MIT

AIR WAVELENGTH (ANGSTRMS)	VACUUM WAVELENGTH (ANGSTRMS)	WAVENUMBER (KAYSERS)	LMENT	LOG ARC	INTENSITY SPK	DIS	REF
4313.972	4315.185	23173.98	ZR	1.0R			MIT
4309.818	4311.030	23196.31	ZR I	0.9			MIT
4309.108	4310.320	23200.13	ZR	0.5			MIT
4304.680	4305.891	23224.00	ZR I	1.2			MIT
4302.886	4304.096	23233.68	ZR I	2.0	0.0		MIT
4301.592	4302.802	23240.67	ZR I	0.6			MIT
4296.742	4297.951	23266.90	ZR II	0.5	0.3		MIT
4295.150	4296.358	23275.53	ZR	0.6			MIT
4294.792	4296.000	23277.47	ZR I	1.6	0.0		MIT
4293.128	4294.336	23286.49	ZR II	0.7	0.3		MIT
4291.349	4292.556	23296.14	ZR I	0.8			MIT
4291.204	4292.411	23296.93	ZR I	0.8			MIT
4290.208	4291.415	23302.34	ZR I	1.6	1.3		MIT
4286.972	4288.178	23319.93	ZR I	0.7			MIT
4285.238	4286.444	23329.36	ZR I	0.6			MIT
4284.719	4285.925	23332.19	ZR	0.6			MIT
4282.200	4283.405	23345.91	ZR II	1.5	0.6		MIT
4282.026	4283.231	23346.86	ZR I	0.9			MIT
4280.312	4281.516	23356.21	ZR I	0.6	0.6		MIT
4277.369	4278.573	23372.28	ZR II	0.6	0.0		MIT
4276.722	4277.925	23375.82	ZR I	0.3D			MIT
4274.769	4275.972	23386.50	ZR I	1.0			MIT
4273.517	4274.720	23393.35	ZR II	1.0	0.5		MIT
4268.016	4269.217	23423.50	ZR I	1.6	0.0		MIT
4265.741	4266.942	23435.99	ZR I	0.7			MIT
4264.911	4266.111	23440.55	ZR II	0.5	0.3H		MIT
4264.008	4265.208	23445.51	ZR	0.7			MIT
4261.425	4262.624	23459.73	ZR I	0.8			MIT
4261.210	4262.409	23460.91	ZR I	0.8			MIT
4258.332	4259.531	23476.76	ZR I	0.7	0.3		MIT
4258.043	4259.242	23478.36	ZR II	1.4	0.9		MIT
4256.445	4257.643	23487.17	ZR I	1.4			MIT
4256.037	4257.235	23489.42	ZR I	1.8			MIT
4253.569	4254.766	23503.05	ZR I	1.3			MIT
4243.817	4245.012	23557.06	ZR	0.5			MIT
4243.557	4244.752	23558.50	ZR I	0.9			MIT
4242.617	4243.812	23563.72	ZR I	0.5			MIT
4242.013	4243.207	23567.08	ZR I	0.6			MIT
4241.687	4242.881	23568.89	ZR I	2.0	0.3		MIT
4241.202	4242.396	23571.58	ZR I	2.0	0.3		MIT
4240.339	4241.533	23576.38	ZR I	2.0	0.0		MIT
4239.314	4240.508	23582.08	ZR I	2.0	0.7		MIT
4237.429	4238.622	23592.57	ZR I	1.2			MIT
4237.195	4238.388	23593.87	ZR I	0.6			MIT
4236.554	4237.747	23597.45	ZR II	0.9	0.0		MIT
4236.063	4237.256	23600.18	ZR I	0.9			MIT
4234.632	4235.824	23608.16	ZR I	1.4	0.0		MIT
4234.296	4235.488	23610.03	ZR	0.5H			MIT
4232.093	4233.285	23622.32	ZR	0.5			MIT
4231.632	4232.824	23624.89	ZR II	1.0	0.7		MIT
4227.758	4228.949	23646.54	ZR I	2.2	0.9		MIT
4225.463	4226.653	23659.38	ZR I	0.8			MIT
4225.263	4226.453	23660.50	ZR I	0.6			MIT
4224.650	4225.840	23663.94	ZR I	0.5			MIT
4220.646	4221.835	23686.38	ZR I	1.0			MIT
4218.448	4219.636	23698.73	ZR I	1.2			MIT
4217.258	4218.446	23705.41	ZR I	0.7H			MIT
4215.753	4216.940	23713.87	ZR II		0.5		MIT
4215.313	4216.500	23716.35	ZR I	0.6	0.0		MIT
4213.865	4215.052	23724.50	ZR I	1.6	0.5		MIT
4212.625	4213.812	23731.48	ZR I	1.2			MIT
4212.158	4213.345	23734.11	ZR I	0.5			MIT
4211.875	4213.061	23735.71	ZR II	1.3	1.2		MIT
4211.335	4212.521	23738.75	ZR I	1.1			MIT
4210.611	4211.797	23742.83	ZR II	0.5	0.5		MIT
4208.983	4210.169	23752.02	ZR II	1.5	1.4		MIT
4208.089	4209.274	23757.06	ZR I	0.7			MIT
4205.919	4207.104	23769.32	ZR II	1.0	0.5		MIT
4201.457	4202.641	23794.56	ZR I	1.7	0.5		MIT
4199.091	4200.274	23807.97	ZR I	1.3	0.3		MIT

AIR WAVELENGTH (ANGSTRMS)	VACUUM WAVELENGTH (ANGSTRMS)	WAVENUMBER (KAYSERS)	LMENT	LOG ARC	INTENSITY SPK	INTENSITY DIS	REF
4197.762	4198.945	23815.51	ZR I	0.5			MIT
4196.133	4197.315	23824.75	ZR I	1.0			MIT
4194.762	4195.944	23832.54	ZR I	1.4	0.0		MIT
4194.450	4195.632	23834.31	ZR I	0.9	0.0		MIT
4194.007	4195.189	23836.83	ZR I	0.9			MIT
4192.096	4193.277	23847.70	ZR I	1.0	0.0		MIT
4191.793	4192.974	23849.42	ZR I	0.9	0.5		MIT
4191.487	4192.668	23851.16	ZR II		0.6		MIT
4187.564	4188.744	23873.50	ZR I	1.0	0.5		MIT
4187.467	4188.647	23874.06	ZR	0.6			MIT
4186.777	4187.957	23877.99	ZR I	0.5			MIT
4186.688	4187.868	23878.50	ZR II	0.5	0.5		MIT
4183.318	4184.497	23897.74	ZR I	1.6	0.0		MIT
4179.809	4180.987	23917.80	ZR II	1.2	0.9		MIT
4173.128	4174.304	23956.09	ZR I	0.5	0.0		MIT
4171.477	4172.653	23965.57	ZR I	1.3			MIT
4170.485	4171.661	23971.27	ZR I	0.7	0.3H		MIT
4169.358	4170.533	23977.75	ZR I	1.0	0.3H		MIT
4168.467	4169.642	23982.87	ZR	0.7	0.3		MIT
4167.381	4168.556	23989.12	ZR	1.2H	1.0H		MIT
4166.522	4167.697	23994.07	ZR I	0.5			MIT
4166.365	4167.540	23994.97	ZR I	1.7	0.6		MIT
4165.628	4166.802	23999.22	ZR I	0.5			MIT
4161.210	4162.383	24024.70	ZR II	1.6	1.5		MIT
4156.236	4157.408	24053.45	ZR II	1.4	1.2		MIT
4153.754	4154.925	24067.82	ZR I	0.7			MIT
4152.637	4153.808	24074.30	ZR I	1.4			MIT
4150.968	4152.138	24083.98	ZR II	1.4	1.0		MIT
4149.203	4150.373	24094.22	ZR II	2.0	2.0		MIT
4148.487	4149.657	24098.38	ZR	0.6	0.3		MIT
4147.367	4148.537	24104.89	ZR I	1.0			MIT
4140.008	4141.176	24147.73	ZR I	1.0			MIT
4138.210	4139.377	24158.22	ZR	1.0	1.0H		MIT
4135.680	4136.846	24173.00	ZR I	1.3	0.0		MIT
4134.314	4135.480	24180.99	ZR I	0.9			MIT
4133.695	4134.861	24184.61	ZR I	0.7	0.5		MIT
4133.524	4134.690	24185.61	ZR I	0.6			MIT
4130.454	4131.619	24203.59	ZR I	0.6			MIT
4127.962	4129.126	24218.20	ZR I	0.6	0.3		MIT
4121.671	4122.834	24255.16	ZR I	0.7			MIT
4121.456	4122.619	24256.43	ZR I	1.5	0.3		MIT
4119.826	4120.988	24266.02	ZR I	1.0			MIT
4110.662	4111.822	24320.12	ZR I	0.6			MIT
4110.053	4111.213	24323.72	ZR II	0.5	0.0		MIT
4108.401	4109.560	24333.50	ZR I	1.3			MIT
4107.495	4108.654	24338.87	ZR I	1.0			MIT
4102.281	4103.439	24369.80	ZR	1.0			MIT
4099.313	4100.470	24387.45	ZR I	1.0			MIT
4099.083	4100.240	24388.82	ZR	0.5	0.0		MIT
4096.632	4097.788	24403.41	ZR II	1.0	0.3		MIT
4094.266	4095.422	24417.51	ZR I	1.5	0.0		MIT
4090.794	4091.949	24438.23	ZR I	1.4	0.3		MIT
4090.513	4091.668	24439.91	ZR II	1.2	1.0		MIT
4089.743	4090.897	24444.51	ZR	1.1	1.0H		MIT
4088.199	4089.353	24453.75	ZR	0.6			MIT
4087.689	4088.843	24456.80	ZR I	1.1	0.0		MIT
4085.664	4086.817	24468.92	ZR II	1.4	0.3		MIT
4085.664	4086.817	24468.92	ZR I	1.4	0.3		MIT
4084.298	4085.451	24477.10	ZR I	1.5	0.0		MIT
4083.083	4084.236	24484.39	ZR I	1.0			MIT
4082.295	4083.447	24489.11	ZR I	0.7			MIT
4081.215	4082.367	24495.59	ZR I	2.2	0.8		MIT
4078.309	4079.460	24513.05	ZR I	1.0			MIT
4077.046	4078.197	24520.64	ZR II	0.5	0.0		MIT
4076.530	4077.681	24523.74	ZR I	1.0			MIT
4075.046	4076.197	24532.67	ZR I	0.8			MIT
4074.931	4076.082	24533.37	ZR I	1.1	0.0		MIT
4074.354	4075.504	24536.84	ZR I	0.7			MIT
4072.704	4073.854	24546.78	ZR I	2.0	0.5		MIT
4071.091	4072.241	24556.51	ZR II	0.5	0.3		MIT

AIR WAVELENGTH (ANGSTRMS)	VACUUM WAVELENGTH (ANGSTRMS)	WAVENUMBER (KAYSERS)	LMENT	LOG ARC	INTENSITY SPK	INTENSITY DIS	REF
4069.540	4070.689	24565.86	ZR	0.6			MIT
4068.719	4069.868	24570.82	ZR I	1.3	0.0		MIT
4066.194	4067.342	24586.08	ZR	0.5			MIT
4064.155	4065.303	24598.41	ZR I	2.0	0.8		MIT
4063.337	4064.485	24603.37	ZR I	0.9			MIT
4062.654	4063.801	24607.50	ZR	0.5			MIT
4061.529	4062.676	24614.32	ZR I	1.4	0.0		MIT
4060.579	4061.726	24620.08	ZR I	1.0			MIT
4060.082	4061.229	24623.09	ZR I	1.0			MIT
4058.985	4060.131	24629.75	ZR I	0.9			MIT
4058.624	4059.770	24631.94	ZR I	1.0	0.0		MIT
4056.513	4057.659	24644.75	ZR I	0.8			MIT
4055.707	4056.853	24649.65	ZR I	1.4	0.5		MIT
4055.306	4056.451	24652.09	ZR I	0.5			MIT
4055.030	4056.175	24653.77	ZR I	2.0	0.7		MIT
4054.433	4055.578	24657.40	ZR I	1.3			MIT
4053.011	4054.156	24666.05	ZR	0.7			MIT
4050.483	4051.627	24681.44	ZR I	1.4			MIT
4050.329	4051.473	24682.38	ZR II	1.3	1.0		MIT
4048.674	4049.818	24692.47	ZR II	1.5	1.5		MIT
4046.082	4047.225	24708.29	ZR I	0.5			MIT
4045.612	4046.755	24711.16	ZR II	1.0	1.0		MIT
4044.564	4045.707	24717.56	ZR I	1.4	0.3		MIT
4043.576	4044.718	24723.60	ZR I	1.4			MIT
4042.225	4043.367	24731.86	ZR I	1.4			MIT
4041.640	4042.782	24735.44	ZR I	0.8			MIT
4040.239	4041.380	24744.02	ZR II	0.6	0.6		MIT
4035.893	4037.033	24770.67	ZR I	1.6	0.3		MIT
4034.086	4035.226	24781.76	ZR II	0.7	0.3		MIT
4033.584	4034.724	24784.84	ZR	0.5			MIT
4032.471	4033.610	24791.69	ZR	0.7			MIT
4031.355	4032.494	24798.55	ZR II	0.5	0.0		MIT
4030.759	4031.898	24802.22	ZR I	1.3			MIT
4030.039	4031.178	24806.65	ZR I	1.5	0.3		MIT
4029.681	4030.820	24808.85	ZR II	1.6	1.2		MIT
4028.951	4030.090	24813.34	ZR I	1.6	0.0		MIT
4027.205	4028.343	24824.10	ZR I	2.0	0.6		MIT
4024.918	4026.055	24838.21	ZR I	1.4	0.5		MIT
4024.438	4025.575	24841.17	ZR II	0.7	0.6		MIT
4023.981	4025.118	24843.99	ZR I	1.5	0.3		MIT
4023.302	4024.439	24848.18	ZR I	0.7			MIT
4023.030	4024.167	24849.86	ZR	0.8			MIT
4021.019	4022.155	24862.29	ZR I	0.5H			MIT
4018.383	4019.519	24878.60	ZR II	0.9	0.7		MIT
4018.121	4019.257	24880.22	ZR	1.4			MIT
4016.985	4018.120	24887.26	ZR I	1.2			MIT
4012.252	4013.386	24916.62	ZR I	1.3			MIT
4009.387	4010.520	24934.42	ZR	0.5			MIT
4007.601	4008.734	24945.53	ZR I	1.4	0.0		MIT
4004.868	4006.000	24962.55	ZR I	1.3			MIT
4004.399	4005.531	24965.48	ZR I	1.0			MIT
4003.096	4004.228	24973.60	ZR I	1.3	0.0		MIT
4002.550	4003.682	24977.01	ZR I	1.3			MIT
4001.227	4002.358	24985.27	ZR I	1.0			MIT
4001.087	4002.218	24986.14	ZR I	1.0			MIT
3998.968	4000.099	24999.38	ZR II	1.5	1.5		MIT
3993.133	3994.262	25035.91	ZR	0.5H			MIT
3992.463	3993.592	25040.11	ZR I	0.5			MIT
3992.057	3993.186	25042.66	ZR	0.5			MIT
3991.131	3992.260	25048.47	ZR II	2.0	1.8		MIT
3989.499	3990.627	25058.72	ZR I	1.1	0.0H		MIT
3989.287	3990.415	25060.05	ZR I	1.1	0.0H		MIT
3988.681	3989.809	25063.86	ZR II	1.2			MIT
3986.803	3987.931	25075.66	ZR I	1.0			MIT
3986.115	3987.242	25079.99	ZR II	0.7			MIT
3984.747	3985.874	25088.60	ZR II	1.3	0.3		MIT
3984.747	3985.874	25088.60	ZR I	1.3	0.3		MIT
3982.161	3983.287	25104.89	ZR I	1.0			MIT
3981.990	3983.116	25105.97	ZR II	0.5	0.3		MIT
3981.596	3982.722	25108.46	ZR I	1.2H			MIT

AIR WAVELENGTH (ANGSTRMS)	VACUUM WAVELENGTH (ANGSTRMS)	WAVENUMBER (KAYSERS)	LMENT	LOG ARC	INTENSITY SPK	DIS	REF
3978.735	3979.860	25126.51	ZR I	1.3			MIT
3978.253	3979.378	25129.55	ZR	1.0			MIT
3977.479	3978.604	25134.44	ZR I	1.3			MIT
3977.335	3978.460	25135.35	ZR I	1.0			MIT
3976.176	3977.301	25142.68	ZR I	0.7			MIT
3975.294	3976.419	25148.26	ZR I	1.7	0.0		MIT
3974.201	3975.325	25155.18	ZR	0.6			MIT
3973.505	3974.629	25159.58	ZR I	1.4			MIT
3973.391	3974.515	25160.30	ZR I	1.0			MIT
3972.305	3973.429	25167.18	ZR I	1.0			MIT
3970.45	3971.57	25178.9	ZR	1.0			KS
3968.723	3969.846	25189.90	ZR	0.5			MIT
3968.257	3969.380	25192.85	ZR I	2.0	0.6		MIT
3966.662	3967.784	25202.98	ZR I		0.5		MIT
3966.264	3967.386	25205.51	ZR	0.8			MIT
3963.801	3964.923	25221.17	ZR I	1.0			MIT
3962.167	3963.288	25231.57	ZR	0.3			MIT
3961.587	3962.708	25235.27	ZR	2.7	0.9		MIT
3960.199	3961.320	25244.11	ZR	0.3	0.0		MIT
3958.218	3959.338	25256.75	ZR II	2.7	2.2		MIT
3957.400	3958.520	25261.97	ZR	0.5			MIT
3956.787	3957.907	25265.88	ZR	0.3			MIT
3956.256	3957.376	25269.27	ZR	0.3			MIT
3955.809	3956.928	25272.13	ZR	0.3			MIT
3955.633	3956.752	25273.25	ZR I	0.6			MIT
3952.890	3954.009	25290.79	ZR I	0.7			MIT
3951.328	3952.446	25300.79	ZR I	0.7			MIT
3950.288	3951.406	25307.45	ZR	0.3H			MIT
3941.944	3943.060	25361.01	ZR II	0.6	0.0		MIT
3941.625	3942.741	25363.07	ZR I	1.3	0.0H		MIT
3936.059	3937.173	25398.93	ZR II	1.0	0.6		MIT
3934.791	3935.905	25407.12	ZR II	1.3	1.2		MIT
3934.121	3935.235	25411.45	ZR II	1.3	1.1		MIT
3933.178	3934.292	25417.54	ZR I	1.0			MIT
3929.530	3930.643	25441.13	ZR I	2.0	0.8		MIT
3929.530	3930.643	25441.13	ZR II	2.0	0.8		MIT
3929.435	3930.548	25441.75	ZR	0.7H			MIT
3927.404	3928.516	25454.90	ZR I	1.0H			MIT
3926.778	3927.890	25458.96	ZR I	1.4H	0.0H		MIT
3924.885	3925.996	25471.24	ZR I	0.5			MIT
3923.465	3924.576	25480.46	ZR I	0.5	0.0		MIT
3922.444	3923.555	25487.09	ZR II	0.3			MIT
3921.792	3922.903	25491.33	ZR I	2.0	0.6		MIT
3916.645	3917.754	25524.83	ZR I	1.0			MIT
3915.938	3917.047	25529.44	ZR II	1.4	1.2		MIT
3914.338	3915.447	25539.87	ZR II	1.8	0.9		MIT
3906.151	3907.257	25593.40	ZR	0.5			MIT
3903.77	3904.88	25609.0	ZR II	0.0	0.3		MIT
3901.513	3902.618	25623.82	ZR I	0.9			MIT
3900.517	3901.622	25630.37	ZR I	2.0			MIT
3899.095	3900.200	25639.71	ZR	0.7	0.0		MIT
3898.557	3899.662	25643.25	ZR I	0.3			MIT
3897.658	3898.762	25649.17	ZR I	0.8			MIT
3896.534	3897.638	25656.56	ZR I	1.2			MIT
3895.991	3897.095	25660.14	ZR		0.5		MIT
3893.837	3894.940	25674.33	ZR I	0.5			MIT
3892.265	3893.368	25684.70	ZR	0.3			MIT
3892.030	3893.133	25686.25	ZR I	1.0			MIT
3891.383	3892.486	25690.52	ZR I	2.0	0.7		MIT
3890.318	3891.420	25697.56	ZR I	2.2	0.8		MIT
3889.446	3890.548	25703.32	ZR	0.3			MIT
3889.004	3890.106	25706.24	ZR	0.3H	0.3H		MIT
3885.425	3886.526	25729.92	ZR I	1.4	0.8		MIT
3881.902	3883.002	25753.27	ZR II	0.0	0.5H		MIT
3879.051	3880.150	25772.20	ZR I	1.0	0.0		MIT
3877.595	3878.694	25781.87	ZR I	1.0	0.0		MIT
3872.173	3873.271	25817.97	ZR I	0.8	0.8		MIT
3872.050	3873.148	25818.79	ZR	0.9	0.8		MIT
3871.390	3872.487	25823.20	ZR I	1.0	1.0		MIT
3865.805	3866.901	25860.50	ZR	0.3			MIT

AIR WAVELENGTH (ANGSTRMS)	VACUUM WAVELENGTH (ANGSTRMS)	WAVENUMBER (KAYSERS)	LMENT	LOG ARC	INTENSITY SPK	DIS	REF
3864.335	3865.431	25870.34	ZR I	1.7	1.3		MIT
3863.874	3864.969	25873.43	ZR I	1.3	0.8		MIT
3863.33	3864.43	25877.1	ZR I	0.3			KS
3861.946	3863.041	25886.34	ZR	0.6	0.3		MIT
3860.627	3861.722	25895.19	ZR	0.7			MIT
3855.430	3856.523	25930.09	ZR II	0.9	0.5		MIT
3853.060	3854.153	25946.04	ZR II	0.9	0.3		MIT
3851.459	3852.551	25956.82	ZR	0.3			MIT
3849.254	3850.346	25971.69	ZR I	1.0	0.6		MIT
3847.010	3848.101	25986.84	ZR I	1.0	0.6		MIT
3845.445	3846.536	25997.42	ZR I	1.3	1.3		MIT
3843.024	3844.114	26013.80	ZR II	1.6	1.6		MIT
3839.130	3840.219	26040.18	ZR I	1.0	1.0		MIT
3838.283	3839.372	26045.93	ZR II	1.0	0.6		MIT
3836.764	3837.852	26056.24	ZR II	1.2	1.3		MIT
3835.964	3837.052	26061.67	ZR I	1.4	0.7		MIT
3833.862	3834.950	26075.96	ZR II	0.5	0.0		MIT
3831.494	3832.581	26092.08	ZR	0.3			MIT
3831.305	3832.392	26093.36	ZR I	0.7H	0.0		MIT
3829.108	3830.194	26108.33	ZR	0.7H	0.6H		MIT
3827.274	3828.360	26120.85	ZR	0.3			MIT
3825.267	3826.352	26134.55	ZR	1.6	1.8		MIT
3823.41	3824.49	26147.2	ZR II	1.2	0.7		KS
3822.414	3823.499	26154.06	ZR I	1.0	0.0		MIT
3817.584	3818.667	26187.15	ZR II	0.9	1.1		MIT
3814.961	3816.044	26205.15	ZR II	0.7	0.7		MIT
3809.691	3810.772	26241.40	ZR I	1.4			MIT
3808.205	3809.286	26251.64	ZR	1.5	1.4		MIT
3807.403	3808.484	26257.17	ZR II	0.3	0.0		MIT
3800.703	3801.782	26303.45	ZR II	0.7	0.0		MIT
3796.484	3797.562	26332.68	ZR II	0.9	1.2		MIT
3792.405	3793.482	26361.01	ZR I	1.0	0.3		MIT
3791.400	3792.477	26367.99	ZR I	1.4	0.8		MIT
3791.062	3792.139	26370.34	ZR I	0.7	0.7		MIT
3790.526	3791.602	26374.07	ZR	0.5	0.5		MIT
3786.606	3787.681	26401.38	ZR	0.9			MIT
3784.871	3785.946	26413.48	ZR I	0.3			MIT
3784.348	3785.423	26417.13	ZR	0.8	0.6		MIT
3782.724	3783.798	26428.47	ZR II	0.5	0.7		MIT
3782.425	3783.499	26430.56	ZR	0.8	0.5		MIT
3782.225	3783.299	26431.96	ZR II	0.3	0.3		MIT
3780.538	3781.612	26443.75	ZR I	1.6	1.2		MIT
3777.080	3778.153	26467.96	ZR	1.2	1.0		MIT
3775.461	3776.533	26479.31	ZR I	0.8	0.0H		MIT
3772.059	3773.131	26503.19	ZR II	0.0	0.3		MIT
3767.883	3768.954	26532.56	ZR II	0.5	0.6		MIT
3766.952	3768.022	26539.12	ZR I	0.3			MIT
3766.820	3767.890	26540.05	ZR II	1.2	1.3		MIT
3766.715	3767.785	26540.79	ZR I	1.2			MIT
3766.375	3767.445	26543.19	ZR	0.5	0.5		MIT
3766.262	3767.332	26543.98	ZR	0.5	0.5		MIT
3765.181	3766.251	26551.60	ZR I	0.5			MIT
3764.841	3765.911	26554.00	ZR	0.5	0.5		MIT
3764.393	3765.463	26557.16	ZR I	1.3	1.0		MIT
3762.513	3763.582	26570.43	ZR	0.3H			MIT
3757.794	3758.862	26603.80	ZR II	0.5	0.8		MIT
3757.24	3758.31	26607.7	ZR	0.7	0.7		KS
3756.946	3758.014	26609.80	ZR II		0.3		MIT
3754.794	3755.861	26625.05	ZR I	0.7			MIT
3754.075	3755.142	26630.15	ZR I	0.3	0.0		MIT
3751.595	3752.661	26647.75	ZR II	1.4	1.6		MIT
3750.645	3751.711	26654.50	ZR II	0.6	0.5		MIT
3745.986	3747.051	26687.66	ZR II		1.2		MIT
3742.208	3743.272	26714.60	ZR	0.5	0.5		MIT
3738.115	3739.178	26743.85	ZR II	0.6	0.3		MIT
3737.392	3738.455	26749.02	ZR I	0.5	0.3		MIT
3736.508	3737.570	26755.35	ZR	0.3	0.3		MIT
3731.264	3732.325	26792.95	ZR II	1.0	1.4		MIT
3729.723	3730.784	26804.02	ZR II	0.6	0.3		MIT
3727.717	3728.777	26818.44	ZR II	0.3	0.6		MIT

AIR WAVELENGTH (ANGSTRMS)	VACUUM WAVELENGTH (ANGSTRMS)	WAVENUMBER (KAYSERS)	LMENT	LOG ARC	INTENSITY SPK	INTENSITY DIS	REF
3723.204	3724.263	26850.95	ZR I	0.5			MIT
3718.843	3719.901	26882.44	ZR II	1.0	1.0		MIT
3717.009	3718.066	26895.70	ZR II	0.3			MIT
3716.091	3717.148	26902.34	ZR	0.6	0.3		MIT
3714.778	3715.835	26911.85	ZR II	1.2	1.0		MIT
3714.133	3715.190	26916.53	ZR I	1.3			MIT
3712.958	3714.014	26925.04	ZR I	0.3			MIT
3711.946	3713.002	26932.39	ZR II	0.3			MIT
3709.257	3710.312	26951.91	ZR II	1.7	1.5		MIT
3706.632	3707.687	26971.00	ZR I	1.0			MIT
3703.052	3704.106	26997.07	ZR	0.3	0.0		MIT
3700.019	3701.072	27019.20	ZR	1.0			MIT
3698.167	3699.219	27032.73	ZR II	1.7	1.9		MIT
3697.458	3698.510	27037.91	ZR II	1.3	1.3		MIT
3695.598	3696.650	27051.52	ZR	0.5H			MIT
3692.635	3693.686	27073.23	ZR II	0.3	0.0		MIT
3690.342	3691.392	27090.05	ZR I	0.7			MIT
3688.468	3689.518	27103.81	ZR I	0.7			MIT
3685.626	3686.675	27124.71	ZR	0.7			MIT
3683.470	3684.519	27140.59	ZR I	0.3			MIT
3682.651	3683.699	27146.62	ZR II	0.0	0.3H		MIT
3681.646	3682.694	27154.03	ZR	0.7			MIT
3680.374	3681.422	27163.42	ZR I	1.0	0.3		MIT
3679.638	3680.686	27168.85	ZR II	0.3	0.3		MIT
3678.905	3679.952	27174.26	ZR II	0.8	0.7		MIT
3676.030	3677.077	27195.52	ZR	0.3H			MIT
3674.718	3675.764	27205.23	ZR II	2.0	1.6		MIT
3672.665	3673.711	27220.43	ZR II	0.7			MIT
3671.269	3672.314	27230.78	ZR II	1.6	1.5		MIT
3671.042	3672.087	27232.47	ZR	0.5			MIT
3668.447	3669.492	27251.73	ZR II	1.0	1.0		MIT
3667.064	3668.108	27262.01	ZR II	0.5	0.0		MIT
3663.651	3664.695	27287.40	ZR I	2.0	1.0		MIT
3662.144	3663.187	27298.63	ZR II	0.7	0.7		MIT
3661.332	3662.375	27304.69	ZR II	0.7	0.0		MIT
3661.203	3662.246	27305.65	ZR I	1.3			MIT
3660.912	3661.955	27307.82	ZR II	0.6	0.0		MIT
3658.301	3659.343	27327.31	ZR	0.3			MIT
3656.880	3657.922	27337.93	ZR I	1.0			MIT
3655.558	3656.599	27347.81	ZR II	0.5	0.5		MIT
3651.472	3652.512	27378.42	ZR II		0.7		MIT
3650.989	3652.029	27382.04	ZR	0.3			MIT
3650.725	3651.765	27384.02	ZR II	0.3	0.3H		MIT
3650.473	3651.513	27385.91	ZR	0.5			MIT
3648.391	3649.431	27401.53	ZR	0.3			MIT
3646.892	3647.931	27412.80	ZR	0.5			MIT
3639.574	3640.611	27467.92	ZR	0.3			MIT
3638.722	3639.759	27474.35	ZR I	0.7			MIT
3636.448	3637.484	27491.53	ZR II	1.5	2.3		MIT
3634.151	3635.187	27508.90	ZR I	1.4	0.3		MIT
3633.489	3634.525	27513.91	ZR II	0.9	0.9		MIT
3630.024	3631.059	27540.18	ZR II	1.2	1.1		MIT
3629.126	3630.161	27546.99	ZR II	0.3			MIT
3625.395	3626.429	27575.34	ZR I	0.6			MIT
3623.957	3624.990	27586.28	ZR	0.3	0.0		MIT
3623.864	3624.897	27586.99	ZR I	1.6	0.8		MIT
3619.395	3620.427	27621.05	ZR I	0.5			MIT
3616.317	3617.348	27644.56	ZR	0.7			MIT
3615.090	3616.121	27653.94	ZR I	0.5			MIT
3614.774	3615.805	27656.36	ZR II	1.6	1.9		MIT
3613.702	3614.733	27664.56	ZR I	1.0			MIT
3613.451	3614.482	27666.49	ZR II	0.8	0.0		MIT
3613.100	3614.130	27669.17	ZR II	1.6	1.6		MIT
3612.572	3613.602	27673.22	ZR I	0.5			MIT
3612.332	3613.362	27675.05	ZR II	0.0	0.3		MIT
3611.893	3612.923	27678.42	ZR II	1.2	1.6		MIT
3609.64	3610.67	27695.7	ZR	0.5			KS
3607.376	3608.405	27713.08	ZR II	0.9	1.0		MIT
3605.899	3606.928	27724.43	ZR I	0.5			MIT
3602.780	3603.808	27748.43	ZR I	0.7H	0.3H		MIT

AIR WAVELENGTH (ANGSTRMS)	VACUUM WAVELENGTH (ANGSTRMS)	WAVENUMBER (KAYSERS)	LMENT	LOG ARC	INTENSITY SPK	INTENSITY DIS	REF
3601.193	3602.220	27760.66	ZR I	2.6	1.2		MIT
3599.901	3600.928	27770.62	ZR II	0.7	0.7		MIT
3597.430	3598.456	27789.69	ZR I	0.5	0.0		MIT
3593.129	3594.154	27822.96	ZR I	0.8	0.0		MIT
3591.725	3592.750	27833.83	ZR I	0.9			MIT
3589.750	3590.774	27849.15	ZR	0.5	0.5		MIT
3588.784	3589.808	27856.64	ZR II	0.3	0.3H		MIT
3588.320	3589.344	27860.24	ZR II	1.0	1.0		MIT
3587.984	3589.008	27862.85	ZR II	1.0	1.0		MIT
3586.294	3587.318	27875.98	ZR I	1.3	0.5		MIT
3585.540	3586.563	27881.84	ZR I	0.3	0.3H		MIT
3582.085	3583.107	27908.74	ZR II	0.5	0.5		MIT
3579.917	3580.939	27925.64	ZR	0.8H	0.7H		MIT
3578.226	3579.247	27938.83	ZR II	0.7	0.8		MIT
3577.553	3578.574	27944.09	ZR I	1.1	0.0		MIT
3576.854	3577.875	27949.55	ZR II	1.2	1.4		MIT
3575.792	3576.813	27957.85	ZR I	2.0	0.7		MIT
3573.084	3574.104	27979.04	ZR II	1.0	1.0		MIT
3572.473	3573.493	27983.82	ZR II	1.8	1.9		MIT
3569.494	3570.513	28007.18	ZR I	1.2	0.3H		MIT
3568.875	3569.894	28012.04	ZR I	1.1			MIT
3568.135	3569.154	28017.84	ZR II	0.6	0.3		MIT
3566.104	3567.122	28033.80	ZR I	1.3	1.0		MIT
3565.431	3566.449	28039.09	ZR II	0.5	0.8		MIT
3562.484	3563.501	28062.29	ZR I	0.9			MIT
3561.897	3562.914	28066.91	ZR	0.5	0.3		MIT
3561.11	3562.13	28073.1	ZR II	0.3			KS
3558.961	3559.978	28090.06	ZR I	0.9	0.3		MIT
3556.597	3557.613	28108.74	ZR II	1.2	1.7		MIT
3554.070	3555.085	28128.72	ZR II	1.0	0.8		MIT
3552.664	3553.679	28139.85	ZR I	0.7	0.6		MIT
3551.951	3552.966	28145.50	ZR II	1.5	1.6		MIT
3550.460	3551.474	28157.32	ZR I	1.5	0.6		MIT
3549.736	3550.750	28163.06	ZR I	1.2	0.5		MIT
3549.510	3550.524	28164.86	ZR II	1.0	1.0		MIT
3548.636	3549.650	28171.79	ZR I	0.5	0.3		MIT
3548.402	3549.416	28173.65	ZR I	0.3H			MIT
3547.682	3548.696	28179.37	ZR I	2.3	1.1		MIT
3542.623	3543.635	28219.61	ZR II	1.1	1.5		MIT
3539.906	3540.918	28241.27	ZR	0.9	0.8		MIT
3539.006	3540.017	28248.45	ZR II	0.6	0.5		MIT
3536.944	3537.955	28264.92	ZR II	0.9	0.7		MIT
3535.162	3536.172	28279.16	ZR I	1.0			MIT
3533.217	3534.227	28294.73	ZR I	1.5	0.5		MIT
3532.46	3533.47	28300.8	ZR	0.9			KS
3530.851	3531.860	28313.69	ZR II	0.7	0.7		MIT
3530.223	3531.232	28318.73	ZR I	1.0	0.0		MIT
3529.982	3530.991	28320.66	ZR II	0.3	0.8B		MIT
3527.436	3528.444	28341.10	ZR II	0.5	0.6		MIT
3525.814	3526.822	28354.14	ZR II	1.0	1.0		MIT
3524.538	3525.546	28364.40	ZR I	1.0	0.3H		MIT
3524.242	3525.250	28366.79	ZR	1.1			MIT
3523.636	3524.643	28371.66	ZR I	0.7	0.0		MIT
3522.180	3523.187	28383.39	ZR I	0.5	0.3		MIT
3521.390	3522.397	28389.76	ZR	0.3			MIT
3520.873	3521.880	28393.93	ZR II	1.0	0.6		MIT
3520.305	3521.312	28398.51	ZR I	0.3			MIT
3519.605	3520.611	28404.16	ZR I	2.0	1.0		MIT
3518.869	3519.875	28410.10	ZR	0.6	0.0H		MIT
3517.855	3518.861	28418.29	ZR	0.3			MIT
3517.475	3518.481	28421.36	ZR I	0.7			MIT
3515.244	3516.249	28439.40	ZR I	0.5			MIT
3514.641	3515.646	28444.27	ZR II	0.5	0.3		MIT
3514.322	3515.327	28446.86	ZR I	0.3			MIT
3512.684	3513.689	28460.12	ZR II	0.7	0.5H		MIT
3512.355	3513.360	28462.79	ZR I	0.6			MIT
3510.457	3511.461	28478.18	ZR II	1.1	0.7		MIT
3510.346	3511.350	28479.07	ZR I	0.3			MIT
3509.323	3510.327	28487.38	ZR I	1.6	0.7		MIT
3507.674	3508.677	28500.77	ZR II	0.7	0.5		MIT

AIR WAVELENGTH (ANGSTRMS)	VACUUM WAVELENGTH (ANGSTRMS)	WAVENUMBER (KAYSERS)	LMENT	LOG ARC	INTENSITY SPK	INTENSITY DIS	REF
3506.485	3507.488	28510.43	ZR II	0.8	0.3		MIT
3506.046	3507.049	28514.00	ZR II	0.9	0.6		MIT
3505.669	3506.672	28517.07	ZR II	1.6	1.5		MIT
3505.485	3506.488	28518.56	ZR II	1.5	1.5		MIT
3503.925	3504.927	28531.26	ZR	0.6H			MIT
3503.748	3504.750	28532.70	ZR	0.5			MIT
3503.318	3504.320	28536.21	ZR I	0.3			MIT
3502.306	3503.308	28544.45	ZR	0.7			MIT
3501.494	3502.496	28551.07	ZR I	1.1	0.5		MIT
3501.347	3502.349	28552.27	ZR I	1.2	0.0		MIT
3500.146	3501.147	28562.06	ZR II	0.6	0.5		MIT
3499.576	3500.577	28566.72	ZR II	1.0	1.0		MIT
3497.919	3498.920	28580.25	ZR II		1.0		MIT
3496.210	3497.210	28594.22	ZR II	2.0	2.0		MIT
3495.374	3496.374	28601.06	ZR I	0.5			MIT
3494.459	3495.459	28608.55	ZR I	0.6			MIT
3493.678	3494.678	28614.94	ZR I	0.8			MIT
3487.902	3488.900	28662.33	ZR I	0.3			MIT
3487.803	3488.801	28663.14	ZR I	0.3			MIT
3486.128	3487.126	28676.91	ZR	0.3H			MIT
3485.325	3486.323	28683.52	ZR II	1.2	1.3		MIT
3483.539	3484.536	28698.23	ZR II	1.3	1.4		MIT
3483.013	3484.010	28702.56	ZR I	1.2			MIT
3482.809	3483.806	28704.24	ZR I	1.3	0.0		MIT
3482.131	3483.128	28709.83	ZR I	0.3			MIT
3481.447	3482.444	28715.47	ZR II	0.3	0.3		MIT
3481.146	3482.142	28717.95	ZR II	1.7	1.9		MIT
3480.406	3481.402	28724.06	ZR II	0.8	0.8		MIT
3479.392	3480.388	28732.43	ZR II	1.8	1.9		MIT
3479.025	3480.021	28735.46	ZR II	1.0	1.2		MIT
3478.787	3479.783	28737.43	ZR I	1.1	0.3		MIT
3478.498	3479.494	28739.81	ZR II	0.7	0.7		MIT
3478.302	3479.298	28741.43	ZR II	1.0	0.8		MIT
3477.636	3478.632	28746.94	ZR I	0.3			MIT
3476.029	3477.024	28760.23	ZR	0.3			MIT
3474.145	3475.140	28775.82	ZR	0.3			MIT
3472.896	3473.890	28786.17	ZR I	0.7			MIT
3472.716	3473.710	28787.66	ZR I	0.3			MIT
3471.189	3472.183	28800.33	ZR I	1.4	0.5		MIT
3471.124	3472.118	28800.86	ZR II		0.7		MIT
3470.181	3471.175	28808.69	ZR I	0.3H			MIT
3469.940	3470.934	28810.69	ZR II	0.6	0.3		MIT
3469.065	3470.058	28817.96	ZR I	0.9			MIT
3468.325	3469.318	28824.11	ZR	0.6			MIT
3465.632	3466.625	28846.50	ZR I	0.5			MIT
3463.017	3464.009	28868.29	ZR II	1.3	1.6		MIT
3461.092	3462.083	28884.34	ZR I	1.3	0.0		MIT
3459.933	3460.924	28894.02	ZR II	1.3	0.3		MIT
3458.934	3459.925	28902.36	ZR II	1.4	1.3		MIT
3457.564	3458.554	28913.81	ZR II	1.4	1.4		MIT
3457.184	3458.174	28916.99	ZR I	1.0			MIT
3455.908	3456.898	28927.67	ZR I	1.1	0.0		MIT
3454.576	3455.566	28938.82	ZR II	0.8	0.6		MIT
3449.915	3450.903	28977.92	ZR I	1.0			MIT
3449.404	3450.392	28982.21	ZR	0.5			MIT
3447.365	3448.353	28999.35	ZR I	2.2W	0.5		MIT
3446.614	3447.602	29005.67	ZR I	1.2	0.0		MIT
3443.569	3444.556	29031.32	ZR II	0.9	0.9		MIT
3442.683	3443.670	29038.79	ZR I	1.0			MIT
3440.972	3441.958	29053.23	ZR	1.0	0.3		MIT
3440.583	3441.569	29056.51	ZR	1.0	0.6		MIT
3440.446	3441.432	29057.67	ZR I	1.0			MIT
3438.230	3439.215	29076.40	ZR II	2.4	2.3		MIT
3437.137	3438.122	29085.64	ZR II	1.2	1.0		MIT
3433.906	3434.890	29113.01	ZR II	0.9	0.8		MIT
3432.406	3433.390	29125.73	ZR II	1.0	1.0		MIT
3431.569	3432.553	29132.84	ZR II	1.0	0.9		MIT
3430.944	3431.928	29138.14	ZR I	0.3			MIT
3430.532	3431.516	29141.64	ZR II	1.7	1.7		MIT
3430.289	3431.272	29143.71	ZR I	1.0			MIT

AIR WAVELENGTH (ANGSTRMS)	VACUUM WAVELENGTH (ANGSTRMS)	WAVENUMBER (KAYSERS)	LMENT	LOG ARC	INTENSITY SPK	INTENSITY DIS	REF
3426.933	3427.916	29172.25	ZR I	0.3			MIT
3424.825	3425.807	29190.20	ZR II	1.2	1.0		MIT
3424.634	3425.616	29191.83	ZR II	0.3			MIT
3421.419	3422.400	29219.26	ZR I	0.5			MIT
3419.659	3420.640	29234.30	ZR I	1.3			MIT
3419.107	3420.088	29239.02	ZR II	0.8	0.6		MIT
3416.687	3417.667	29259.73	ZR	0.6			MIT
3414.950	3415.930	29274.61	ZR I	1.3			MIT
3414.661	3415.640	29277.09	ZR II	1.3	1.2		MIT
3413.400	3414.379	29287.90	ZR II	0.9	0.7		MIT
3411.791	3412.770	29301.71	ZR I	0.5			MIT
3410.248	3411.226	29314.97	ZR II	1.7	1.7		MIT
3409.418	3410.396	29322.11	ZR	0.5			MIT
3408.777	3409.755	29327.62	ZR	1.0			MIT
3408.079	3409.057	29333.63	ZR II	1.0	1.0		MIT
3404.832	3405.809	29361.60	ZR II	1.6	1.5		MIT
3403.684	3404.661	29371.50	ZR II	1.2	1.2		MIT
3402.873	3403.849	29378.50	ZR II	1.0	1.0		MIT
3402.523	3403.499	29381.52	ZR II	0.3			MIT
3401.796	3402.772	29387.80	ZR I	0.6			MIT
3400.765	3401.741	29396.71	ZR	0.3			MIT
3399.349	3400.325	29408.96	ZR II	2.0	1.6		MIT
3397.918	3398.893	29421.34	ZR I	0.9			MIT
3397.646	3398.621	29423.70	ZR	0.9			MIT
3396.658	3397.633	29432.26	ZR II	0.9	0.7		MIT
3396.333	3397.308	29435.07	ZR II	1.1	1.0		MIT
3394.63	3395.60	29449.8	ZR II	0.3	0.5		MIT
3393.124	3394.098	29462.91	ZR II	1.5	1.4		MIT
3391.975	3392.949	29472.89	ZR II	2.5	2.6		MIT
3389.270	3390.243	29496.41	ZR	0.3H			MIT
3388.299	3389.272	29504.86	ZR II	1.7	1.6		MIT
3387.872	3388.845	29508.58	ZR II	2.0	2.0		MIT
3387.110	3388.082	29515.22	ZR I	0.3			MIT
3386.905	3387.877	29517.01	ZR I	0.3			MIT
3385.788	3386.760	29526.75	ZR	0.5			MIT
3382.90	3383.87	29551.9	ZR	0.5			MIT
3379.923	3380.894	29577.98	ZR I	1.0	0.3		MIT
3378.302	3379.272	29592.17	ZR II	0.8	0.5		MIT
3377.455	3378.425	29599.59	ZR II	1.4	1.0		MIT
3376.268	3377.238	29610.00	ZR II	1.0	0.9		MIT
3374.726	3375.695	29623.53	ZR II	1.3	1.5		MIT
3373.417	3374.386	29635.02	ZR II	1.1	1.0		MIT
3370.592	3371.560	29659.86	ZR I	1.0			MIT
3369.262	3370.230	29671.57	ZR II	0.8	0.7		MIT
3368.636	3369.604	29677.08	ZR I	0.8			MIT
3367.820	3368.787	29684.27	ZR II	1.1	0.9		MIT
3363.820	3364.786	29719.57	ZR II	1.1	1.0		MIT
3363.068	3364.034	29726.21	ZR I	0.3H			MIT
3362.684	3363.650	29729.61	ZR II	0.8	0.7		MIT
3360.455	3361.421	29749.33	ZR I	1.3			MIT
3359.955	3360.920	29753.75	ZR II	1.1	1.1		MIT
3357.264	3358.229	29777.60	ZR II	1.7	1.6		MIT
3356.091	3357.055	29788.01	ZR II	1.7	1.6		MIT
3354.386	3355.350	29803.15	ZR II	1.0	1.0		MIT
3353.656	3354.620	29809.64	ZR I	1.3			MIT
3353.150	3354.114	29814.14	ZR I	0.3			MIT
3351.228	3352.191	29831.23	ZR I	0.6			MIT
3344.786	3345.748	29888.69	ZR II	1.2	1.2		MIT
3343.813	3344.774	29897.38	ZR II	1.3	1.2		MIT
3343.711	3344.672	29898.29	ZR I	0.9			MIT
3340.555	3341.516	29926.54	ZR II	1.4	1.3		MIT
3338.408	3339.368	29945.79	ZR II	1.2	1.1		MIT
3337.924	3338.884	29950.13	ZR II	0.3	0.3		MIT
3334.616	3335.575	29979.84	ZR II	1.2	1.0		MIT
3334.251	3335.210	29983.12	ZR II	1.4	1.3		MIT
3334.056	3335.015	29984.87	ZR	0.3			MIT
3333.556	3334.515	29989.37	ZR I	0.7			MIT
3329.129	3330.087	30029.25	ZR	0.3			MIT
3327.667	3328.624	30042.44	ZR II	0.3	0.3		MIT
3326.797	3327.754	30050.30	ZR II	2.0	2.0		MIT

AIR WAVELENGTH (ANGSTRMS)	VACUUM WAVELENGTH (ANGSTRMS)	WAVENUMBER (KAYSERS)	LMENT	LOG ARC	INTENSITY SPK	INTENSITY DIS	REF
3326.414	3327.371	30053.76	ZR I	0.7			MIT
3322.988	3323.944	30084.74	ZR II	1.0	1.0		MIT
3322.282	3323.238	30091.14	ZR	0.9			MIT
3319.025	3319.980	30120.66	ZR II	1.4	0.8		MIT
3318.512	3319.467	30125.32	ZR II	0.8	0.6		MIT
3316.207	3317.161	30146.26	ZR I	0.6			MIT
3314.495	3315.449	30161.83	ZR II	1.2	1.0		MIT
3313.698	3314.652	30169.08	ZR II	1.0	1.0		MIT
3311.339	3312.292	30190.57	ZR II	0.9	0.5		MIT
3310.551	3311.504	30197.76	ZR	0.5			MIT
3309.889	3310.842	30203.80	ZR II	0.7	0.5		MIT
3306.278	3307.230	30236.79	ZR II	1.9	1.9		MIT
3305.152	3306.104	30247.09	ZR II	1.4	1.3		MIT
3302.666	3303.617	30269.85	ZR II	1.0	0.8		MIT
3300.819	3301.769	30286.79	ZR	0.3			MIT
3297.634	3298.584	30316.04	ZR I	0.3			MIT
3296.973	3297.922	30322.12	ZR	0.3			MIT
3296.733	3297.682	30324.33	ZR I	0.5			MIT
3296.398	3297.347	30327.41	ZR II	0.9	0.8		MIT
3295.023	3295.972	30340.06	ZR II	0.3			MIT
3293.595	3294.544	30353.22	ZR	0.3			MIT
3293.450	3294.399	30354.55	ZR I	0.3			MIT
3288.803	3289.750	30397.44	ZR II	1.0	0.8		MIT
3287.304	3288.251	30411.30	ZR II	0.6	0.3		MIT
3285.883	3286.830	30424.46	ZR II	1.0	0.9		MIT
3285.762	3286.709	30425.57	ZR II	0.6	0.5		MIT
3285.190	3286.136	30430.87	ZR I	0.6			MIT
3284.713	3285.659	30435.29	ZR II	1.4	1.5		MIT
3282.834	3283.780	30452.71	ZR II	1.0	1.0		MIT
3282.732	3283.678	30453.66	ZR I	1.0			MIT
3281.873	3282.819	30461.63	ZR I	0.3H			MIT
3280.748	3281.693	30472.07	ZR II	0.5	0.3		MIT
3279.265	3280.210	30485.85	ZR II	1.7	1.7		MIT
3278.91	3279.85	30489.1	ZR II		0.7		MIT
3278.561	3279.506	30492.40	ZR I	0.3H			MIT
3277.636	3278.581	30501.00	ZR	0.5			MIT
3277.366	3278.310	30503.52	ZR I	0.6			MIT
3276.366	3277.310	30512.83	ZR II	0.3H			MIT
3275.645	3276.589	30519.54	ZR II	0.3H			MIT
3275.142	3276.086	30524.23	ZR II	0.5	0.0		MIT
3273.047	3273.990	30543.77	ZR II	1.7	1.9		MIT
3272.222	3273.165	30551.47	ZR II	1.2	1.0		MIT
3271.127	3272.070	30561.70	ZR II	0.9	0.9		MIT
3270.590	3271.533	30566.71	ZR	0.6			MIT
3269.657	3270.599	30575.43	ZR I	1.1	0.0		MIT
3267.362	3268.304	30596.91	ZR I	0.5			MIT
3264.812	3265.753	30620.81	ZR II	0.6	0.6		MIT
3261.078	3262.018	30655.87	ZR I	0.3H			MIT
3260.916	3261.856	30657.39	ZR	0.3H			MIT
3260.111	3261.051	30664.96	ZR I	1.0	0.0		MIT
3254.276	3255.215	30719.94	ZR I	1.6	1.6		MIT
3250.458	3251.396	30756.02	ZR II	0.8	0.9		MIT
3250.392	3251.330	30756.65	ZR I	1.2			MIT
3243.976	3244.912	30817.48	ZR I	0.6			MIT
3242.764	3243.700	30829.00	ZR I	0.3H			MIT
3242.163	3243.099	30834.71	ZR II	0.0	0.3H		MIT
3241.046	3241.981	30845.34	ZR II	1.0	1.0		MIT
3236.578	3237.512	30887.92	ZR II	1.3	1.6		MIT
3236.224	3237.158	30891.29	ZR II	0.9	1.0		MIT
3234.801	3235.735	30904.88	ZR I	0.3			MIT
3234.124	3235.057	30911.35	ZR I	0.9	0.5		MIT
3233.11	3234.04	30921.1	ZR I	0.8			KS
3231.693	3232.626	30934.60	ZR II	1.0	1.0		MIT
3229.721	3230.653	30953.49	ZR II	0.3H			MIT
3228.810	3229.742	30962.22	ZR II	1.0	0.9		MIT
3225.486	3226.417	30994.13	ZR I	0.7R	0.3H		MIT
3222.470	3223.401	31023.14	ZR II	0.3	0.5		MIT
3218.686	3219.616	31059.61	ZR II	0.3			MIT
3216.586	3217.515	31079.89	ZR I	0.3			MIT
3214.189	3215.117	31103.06	ZR II	1.1	1.3		MIT

AIR WAVELENGTH (ANGSTRMS)	VACUUM WAVELENGTH (ANGSTRMS)	WAVENUMBER (KAYSERS)	LMENT	LOG ARC	INTENSITY SPK	INTENSITY DIS	REF
3212.854	3213.782	31115.99	ZR II	0.7	0.5		MIT
3212.577	3213.505	31118.67	ZR I	0.6			MIT
3212.013	3212.941	31124.13	ZR I	1.2	0.5		MIT
3208.312	3209.239	31160.04	ZR II	0.5	0.0		MIT
3204.895	3205.821	31193.26	ZR I	1.0			MIT
3204.348	3205.274	31198.58	ZR II	0.3	0.0		MIT
3200.678	3201.603	31234.35	ZR II	0.5	0.0		MIT
3197.088	3198.012	31269.43	ZR II	0.0	0.3		MIT
3197.04	3197.96	31269.9	ZR I	1.3			KS
3191.905	3192.828	31320.20	ZR II	0.9	0.7		MIT
3191.214	3192.137	31326.98	ZR I	1.0	0.3		MIT
3185.075	3185.996	31387.36	ZR I	0.3			MIT
3182.858	3183.779	31409.22	ZR II	1.1	1.3		MIT
3181.922	3182.842	31418.46	ZR II	0.8	0.6		MIT
3181.576	3182.496	31421.88	ZR II	1.0	0.8		MIT
3179.746	3180.666	31439.96	ZR I	0.3			MIT
3178.087	3179.006	31456.37	ZR II	0.9	0.9		MIT
3166.256	3167.172	31573.91	ZR II	0.7	0.7		MIT
3165.974	3166.890	31576.72	ZR II	1.0	1.1		MIT
3165.446	3166.362	31581.99	ZR II	0.8	0.6		MIT
3164.310	3165.226	31593.32	ZR II	1.0	0.9		MIT
3162.214	3163.129	31614.26	ZR II	1.2	1.2		MIT
3159.117	3160.032	31645.25	ZR II	0.3	0.3		MIT
3157.823	3158.737	31658.22	ZR I	1.0	0.0		MIT
3156.996	3157.910	31666.51	ZR II	1.0	0.7		MIT
3155.671	3156.585	31679.81	ZR II	1.0	1.1		MIT
3149.061	3149.973	31746.30	ZR I	0.3			MIT
3148.820	3149.732	31748.73	ZR I	0.8			MIT
3146.285	3147.196	31774.31	ZR	0.5			MIT
3143.939	3144.850	31798.02	ZR I	0.5			MIT
3139.801	3140.711	31839.93	ZR I	0.7			MIT
3138.678	3139.587	31851.32	ZR II	1.0	1.1		MIT
3136.958	3137.867	31868.78	ZR I	0.9	0.0H		MIT
3133.475	3134.383	31904.21	ZR II	0.8	1.1		MIT
3133.231	3134.139	31906.69	ZR I	0.7			MIT
3132.066	3132.974	31918.56	ZR I	1.0	0.3		MIT
3131.110	3132.017	31928.30	ZR I	0.8			MIT
3130.063	3130.970	31938.98	ZR I	0.5			MIT
3129.761	3130.668	31942.06	ZR II	1.0	1.0		MIT
3129.176	3130.083	31948.03	ZR II	1.0	1.0		MIT
3128.788	3129.695	31952.00	ZR II	0.3			MIT
3125.918	3126.824	31981.33	ZR II	0.9	0.7		MIT
3125.193	3126.099	31988.75	ZR II	0.5	0.3		MIT
3121.965	3122.870	32021.82	ZR I	0.3			MIT
3120.741	3121.646	32034.38	ZR I	1.0	0.3		MIT
3116.68	3117.58	32076.1	ZR		0.6H		KS
3115.727	3116.631	32085.93	ZR II	0.3	0.3		MIT
3113.885	3114.788	32104.91	ZR		0.3H		MIT
3113.506	3114.409	32108.82	ZR I	0.5			MIT
3111.160	3112.062	32133.03	ZR II	0.3	0.0		MIT
3110.878	3111.780	32135.94	ZR II	1.0	0.9		MIT
3108.374	3109.276	32161.83	ZR I	0.7			MIT
3106.576	3107.477	32180.44	ZR II	1.0	1.2		MIT
3099.665	3100.565	32252.19	ZR I	0.3			MIT
3099.230	3100.129	32256.72	ZR II	1.0	0.7		MIT
3095.823	3096.722	32292.22	ZR I	0.9			MIT
3095.071	3095.969	32300.06	ZR II	0.9	0.7		MIT
3094.799	3095.697	32302.90	ZR I	0.7			MIT
3093.321	3094.219	32318.33	ZR I	0.5			MIT
3092.240	3093.138	32329.63	ZR	0.5			MIT
3090.437	3091.334	32348.49	ZR I	0.9	0.0H		MIT
3089.008	3089.905	32363.46	ZR II	0.3			MIT
3086.461	3087.357	32390.16	ZR II	0.3	0.0		MIT
3085.344	3086.240	32401.89	ZR I	1.0			MIT
3068.024	3068.916	32584.80	ZR II	0.3	0.3		MIT
3067.124	3068.015	32594.36	ZR I	0.7	0.6		MIT
3066.741	3067.632	32598.43	ZR I	0.5			MIT
3065.211	3066.102	32614.70	ZR II	0.7	0.3		MIT
3064.634	3065.525	32620.84	ZR II	0.7	0.5		MIT
3063.573	3064.464	32632.14	ZR I	0.7	0.0		MIT

AIR WAVELENGTH (ANGSTRMS)	VACUUM WAVELENGTH (ANGSTRMS)	WAVENUMBER (KAYSERS)	LMENT	LOG ARC	INTENSITY SPK	INTENSITY DIS	REF
3061.346	3062.236	32655.88	ZR II	0.7	0.5		MIT
3060.113	3061.003	32669.03	ZR II	0.7	0.5		MIT
3057.219	3058.108	32699.96	ZR II	0.3	0.5		MIT
3054.835	3055.723	32725.48	ZR II	1.2	1.4		MIT
3052.814	3053.702	32747.14	ZR I	0.5			MIT
3052.000	3052.888	32755.87	ZR I	0.3			MIT
3050.322	3051.209	32773.89	ZR I	0.5			MIT
3049.329	3050.216	32784.56	ZR I	0.8	0.0H		MIT
3048.425	3049.312	32794.29	ZR II	0.9	0.3H		MIT
3048.243	3049.130	32796.24	ZR II	0.3	0.3		MIT
3045.829	3046.715	32822.24	ZR I	1.0	0.0H		MIT
3044.120	3045.006	32840.66	ZR II	0.5	0.0		MIT
3043.251	3044.136	32850.04	ZR II	0.6			MIT
3038.596	3039.480	32900.36	ZR		0.3		MIT
3036.503	3037.387	32923.04	ZR II	1.0	0.9		MIT
3036.393	3037.277	32924.23	ZR II	1.3	1.3		MIT
3033.457	3034.340	32956.10	ZR		0.3		MIT
3032.006	3032.889	32971.87	ZR II	0.3	0.3H		MIT
3030.918	3031.800	32983.70	ZR II	1.2	1.0		MIT
3029.515	3030.397	32998.98	ZR I	1.8	0.7		MIT
3028.040	3028.922	33015.05	ZR II	1.3	1.5		MIT
3026.183	3027.064	33035.31	ZR II	0.0	0.5H		MIT
3025.153	3026.034	33046.56	ZR II	0.5	0.3H		MIT
3024.745	3025.626	33051.01	ZR II	0.3H	0.3H		MIT
3022.643	3023.523	33074.00	ZR I	0.5	0.3H		MIT
3020.467	3021.347	33097.82	ZR II	1.7	1.5		MIT
3019.841	3020.721	33104.68	ZR II	1.0	0.8		MIT
3019.507	3020.386	33108.35	ZR	0.3			MIT
3018.53	3019.41	33119.1	ZR II		0.5J		KS
3018.081	3018.960	33123.99	ZR II		0.3J		MIT
3014.438	3015.316	33164.02	ZR I	1.2			MIT
3013.321	3014.199	33176.31	ZR II	1.0	0.7		MIT
3011.748	3012.626	33193.64	ZR I	2.0	0.6		MIT
3010.284	3011.161	33209.78	ZR II	1.1	1.4		MIT
3009.875	3010.752	33214.29	ZR II	0.3	0.3H		MIT
3008.142	3009.019	33233.43	ZR II	1.0	0.7		MIT
3005.497	3006.373	33262.67	ZR I	1.4	0.3		MIT
3005.370	3006.246	33264.08	ZR I	1.0			MIT
3003.736	3004.612	33282.17	ZR II	1.2	1.2		MIT
3000.607	3001.482	33316.88	ZR II		0.3J		MIT
2998.483	2999.357	33340.48	ZR II	0.0	0.3H		MIT
2991.406	2992.278	33419.35	ZR II	0.9	0.5		MIT
2990.130	2991.002	33433.61	ZR II	0.3	0.6H		MIT
2988.77	2989.64	33448.8	ZR II		0.8J		MIT
2987.799	2988.671	33459.69	ZR II	0.5	1.0		MIT
2985.389	2986.260	33486.70	ZR I	1.7	0.5		MIT
2981.016	2981.886	33535.82	ZR II	1.0	1.0		MIT
2979.184	2980.053	33556.44	ZR II	1.0	1.0		MIT
2978.054	2978.923	33569.18	ZR II	1.1	1.1		MIT
2976.613	2977.482	33585.43	ZR II	0.7	0.9		MIT
2969.628	2970.495	33664.42	ZR II	1.0	1.2		MIT
2969.189	2970.056	33669.40	ZR I	1.0			MIT
2968.961	2969.828	33671.98	ZR II	1.2	1.5		MIT
2964.563	2965.429	33721.94	ZR II	0.3	0.3		MIT
2962.683	2963.548	33743.33	ZR II	1.2	1.3		MIT
2960.870	2961.735	33763.99	ZR I	1.2			MIT
2955.783	2956.647	33822.10	ZR II	1.0	1.5		MIT
2955.52	2956.38	33825.1	ZR II	0.0	0.5J		MIT
2952.244	2953.107	33862.64	ZR II	0.9	0.9		MIT
2951.479	2952.342	33871.42	ZR II	1.2	1.2		MIT
2949.83	2950.69	33890.4	ZR		0.5H		MIT
2948.945	2949.807	33900.52	ZR II	1.1	1.3		MIT
2945.463	2946.324	33940.60	ZR II	0.6	1.0H		MIT
2944.20	2945.06	33955.2	ZR II	0.3H	0.5		MIT
2936.308	2937.167	34046.42	ZR II	1.0	1.2		MIT
2934.610	2935.468	34066.11	ZR II	0.6	0.8		MIT
2931.90	2932.76	34097.6	ZR II		0.7H		MIT
2931.060	2931.918	34107.37	ZR II	0.5H	1.2Q		MIT
2929.115	2929.972	34130.02	ZR II	0.3	0.3		MIT
2926.988	2927.844	34154.82	ZR II	1.2	1.5		MIT

AIR WAVELENGTH (ANGSTRMS)	VACUUM WAVELENGTH (ANGSTRMS)	WAVENUMBER (KAYSERS)	LMENT	LOG ARC	INTENSITY SPK	INTENSITY DIS	REF
2925.630	2926.486	34170.67	ZR II	0.0	0.3H		MIT
2924.640	2925.496	34182.24	ZR II	0.6	0.6		MIT
2923.851	2924.707	34191.46	ZR I	1.3			MIT
2918.240	2919.094	34257.20	ZR II	1.2	1.2		MIT
2916.635	2917.489	34276.05	ZR II	0.6	0.6		MIT
2916.250	2917.104	34280.58	ZR I	0.8			MIT
2915.991	2916.845	34283.62	ZR II	1.4	1.3		MIT
2910.248	2911.100	34351.27	ZR II	0.5	0.6		MIT
2907.384	2908.236	34385.11	ZR II	0.3	0.5		MIT
2905.226	2906.077	34410.65	ZR II	1.0	1.0		MIT
2904.274	2905.125	34421.93	ZR II	0.3	0.3		MIT
2903.64	2904.49	34429.4	ZR II	0.0	1.3J		MIT
2902.23	2903.08	34446.2	ZR II	0.0	0.3		MIT
2901.819	2902.669	34451.05	ZR II	0.6	0.6		MIT
2901.617	2902.467	34453.45	ZR II	0.5	0.7		MIT
2898.712	2899.562	34487.97	ZR II	0.8	0.8		MIT
2898.256	2899.105	34493.40	ZR	0.6			MIT
2895.328	2896.177	34528.28	ZR II		1.0H		MIT
2894.794	2895.643	34534.65	ZR II	0.3			MIT
2892.265	2893.113	34564.85	ZR I	1.3			MIT
2889.430	2890.277	34598.76	ZR II	0.5	0.5		MIT
2889.129	2889.976	34602.36	ZR I	0.5			MIT
2888.036	2888.883	34615.46	ZR II	0.3	0.8		MIT
2886.71	2887.56	34631.4	ZR II		0.3W		KS
2884.56	2885.41	34657.2	ZR II		0.6H		MIT
2883.798	2884.644	34666.33	ZR II	0.7	0.3		MIT
2882.087	2882.932	34686.90	ZR II	0.3	0.5		MIT
2880.830	2881.675	34702.04	ZR I	0.7			MIT
2877.551	2878.395	34741.58	ZR II	0.6	0.6H		MIT
2875.983	2876.827	34760.52	ZR I	1.8	0.0		MIT
2874.173	2875.016	34782.41	ZR		0.3J		MIT
2872.527	2873.370	34802.34	ZR II	0.3	0.5		MIT
2869.811	2870.653	34835.27	ZR II	1.5	1.5		MIT
2868.521	2869.363	34850.94	ZR I	0.5			MIT
2865.604	2866.445	34886.41	ZR II	0.6	0.6		MIT
2865.097	2865.938	34892.59	ZR II	0.3	0.3		MIT
2860.851	2861.691	34944.37	ZR I	1.2			MIT
2859.608	2860.448	34959.56	ZR II		1.0J		MIT
2857.979	2858.818	34979.49	ZR I	0.9			MIT
2856.065	2856.904	35002.93	ZR II	0.3	0.6B		MIT
2854.427	2855.266	35023.01	ZR II	0.7	0.7		MIT
2851.967	2852.805	35053.22	ZR II	1.1	1.3		MIT
2848.523	2849.360	35095.60	ZR I	2.0			MIT
2848.192	2849.029	35099.68	ZR II	1.1	1.1		MIT
2844.579	2845.415	35144.26	ZR II	1.7	1.7		MIT
2843.516	2844.352	35157.39	ZR II	0.9	0.9		MIT
2839.339	2840.174	35209.11	ZR II	0.8	0.8		MIT
2838.022	2838.857	35225.45	ZR I	0.5	0.0		MIT
2837.232	2838.066	35235.26	ZR I	2.0			MIT
2836.494	2837.328	35244.42	ZR I	0.7			MIT
2834.395	2835.229	35270.52	ZR II	0.6	0.6		MIT
2833.908	2834.742	35276.58	ZR II	0.3	0.7		MIT
2833.061	2833.894	35287.13	ZR	0.3	0.0		MIT
2831.37	2832.20	35308.2	ZR I	0.6			MIT
2829.806	2830.639	35327.72	ZR I	1.0			MIT
2827.544	2828.376	35355.98	ZR I	0.5			MIT
2827.495	2828.327	35356.59	ZR II	0.7	0.5		MIT
2825.558	2826.390	35380.83	ZR II	1.5	1.5		MIT
2824.821	2825.652	35390.06	ZR I	0.6			MIT
2821.558	2822.389	35430.98	ZR I	1.0			MIT
2819.557	2820.387	35456.13	ZR I	0.8			MIT
2819.294	2820.124	35459.43	ZR II		0.3		MIT
2818.739	2819.569	35466.42	ZR II	1.2	1.2		MIT
2815.499	2816.328	35507.23	ZR I	1.0			MIT
2814.905	2815.734	35514.72	ZR I	1.8	0.0		MIT
2814.706	2815.535	35517.23	ZR I	0.3			MIT
2810.914	2811.742	35565.14	ZR II	1.0	0.8		MIT
2808.161	2808.988	35600.01	ZR II	0.3	0.0H		MIT
2806.775	2807.602	35617.58	ZR I	1.1			MIT
2805.71	2806.54	35631.1	ZR II		0.3W		KS

AIR WAVELENGTH (ANGSTRMS)	VACUUM WAVELENGTH (ANGSTRMS)	WAVENUMBER (KAYSERS)	LMENT	LOG ARC	INTENSITY SPK	DIS	REF
2799.149	2799.974	35714.62	ZR II	0.7	0.8		MIT
2796.901	2797.726	35743.32	ZR II	1.0	0.8		MIT
2795.129	2795.953	35765.98	ZR I	0.7			MIT
2793.390	2794.214	35788.24	ZR	1.0			MIT
2792.037	2792.860	35805.59	ZR I	1.1			MIT
2790.143	2790.966	35829.89	ZR I	1.3			MIT
2786.952	2787.774	35870.91	ZR II		0.5 J		MIT
2786.855	2787.677	35872.16	ZR I	0.7			MIT
2783.559	2784.380	35914.63	ZR II	0.7	0.7		MIT
2776.59	2777.41	36004.8	ZR II	0.0	0.3 W		MIT
2774.157	2774.976	36036.35	ZR II	1.0	1.0		MIT
2774.038	2774.857	36037.89	ZR I	1.0			MIT
2768.848	2769.666	36105.44	ZR II	0.5	0.7		MIT
2768.731	2769.549	36106.97	ZR II	1.0	0.6		MIT
2767.349	2768.166	36125.00	ZR I	0.5			MIT
2763.027	2763.843	36181.50	ZR I	0.8			MIT
2761.911	2762.727	36196.12	ZR II	0.5	0.6		MIT
2760.11	2760.93	36219.7	ZR II	0.0	0.3 J		MIT
2759.482	2760.297	36227.98	ZR I	0.6			MIT
2758.813	2759.628	36236.77	ZR II	1.5	1.5		MIT
2754.212	2755.026	36297.30	ZR II	0.5	0.0		MIT
2753.25	2754.06	36310.0	ZR	0.3 W	0.3 W		MIT
2752.206	2753.020	36323.75	ZR II	1.6	1.6		MIT
2750.940	2751.753	36340.47	ZR II	0.8	0.9		MIT
2748.608	2749.421	36371.30	ZR I	0.3			MIT
2745.855	2746.667	36407.76	ZR II	1.4	1.4		MIT
2742.557	2743.368	36451.54	ZR II	1.2	1.4		MIT
2741.550	2742.361	36464.93	ZR II	1.0	0.9		MIT
2740.508	2741.319	36478.79	ZR II	0.7	0.7		MIT
2740.352	2741.163	36480.87	ZR II	0.5	0.5		MIT
2739.771	2740.582	36488.60	ZR II		0.3		MIT
2737.888	2738.698	36513.70	ZR	0.6			MIT
2734.855	2735.664	36554.19	ZR II	1.6	1.6		MIT
2733.456	2734.265	36572.90	ZR	0.6			MIT
2732.721	2733.530	36582.73	ZR II	1.6	1.5		MIT
2729.932	2730.740	36620.11	ZR II	0.0	0.3		MIT
2727.022	2727.830	36659.18	ZR I	1.0			MIT
2726.98	2727.79	36659.7	ZR II	0.0 W	0.3 W		MIT
2726.493	2727.300	36666.29	ZR II	1.7	1.7		MIT
2725.467	2726.274	36680.10	ZR I	1.1			MIT
2725.00	2725.81	36686.4	ZR I	0.9 W			MIT
2723.697	2724.504	36703.93	ZR	0.3			MIT
2722.610	2723.416	36718.58	ZR II	1.7	1.7		MIT
2720.352	2721.158	36749.06	ZR II	0.6	0.6		MIT
2719.556	2720.362	36759.82	ZR I	0.5			MIT
2718.279	2719.084	36777.08	ZR I	0.5			MIT
2717.484	2718.289	36787.84	ZR I	0.7			MIT
2715.80	2716.60	36810.6	ZR		0.3 H		MIT
2714.258	2715.062	36831.56	ZR II	0.8	0.8		MIT
2712.420	2713.224	36856.52	ZR II	1.3	1.2		MIT
2711.508	2712.312	36868.92	ZR II	1.6	1.3		MIT
2709.784	2710.587	36892.37	ZR	0.5			MIT
2709.334	2710.137	36898.50	ZR I	1.1			MIT
2709.07	2709.87	36902.1	ZR		0.5		MIT
2706.367	2707.170	36938.95	ZR II	0.0	0.3		MIT
2706.174	2706.976	36941.58	ZR I	1.1			MIT
2701.810	2702.611	37001.25	ZR I	0.3			MIT
2700.131	2700.932	37024.26	ZR II	1.7	1.7		MIT
2699.605	2700.406	37031.47	ZR II	1.0	1.0		MIT
2698.31	2699.11	37049.2	ZR		0.5 H		MIT
2695.427	2696.227	37088.87	ZR II	0.7	0.7		MIT
2694.060	2694.860	37107.69	ZR II	1.2	1.2		MIT
2693.526	2694.325	37115.04	ZR II	1.2	1.2		MIT
2692.917	2693.716	37123.43	ZR I	0.8			MIT
2692.600	2693.399	37127.81	ZR II	0.8	0.7		MIT
2692.01	2692.81	37135.9	ZR II	0.0	0.3 H		MIT
2690.50	2691.30	37156.8	ZR		0.6		MIT
2689.458	2690.256	37171.18	ZR II	1.0	0.7		MIT
2687.746	2688.544	37194.85	ZR I	1.0			MIT
2686.26	2687.06	37215.4	ZR	0.0	0.3		MIT
2685.916	2686.714	37220.19	ZR	0.5			MIT
2683.810	2684.607	37249.40	ZR I	0.5			MIT
2682.160	2682.957	37272.31	ZR		0.7		MIT
2681.756	2682.553	37277.93	ZR II		0.3		MIT
2681.40	2682.20	37282.9	ZR	0.6			MIT
2678.632	2679.428	37321.40	ZR II	1.9	2.0		MIT
2676.538	2677.333	37350.60	ZR I	0.5			MIT
2670.963	2671.757	37428.55	ZR II	0.7	0.7		MIT
2669.494	2670.288	37449.15	ZR II		0.3		MIT
2667.799	2668.592	37472.94	ZR II	1.1	1.1		MIT
2665.177	2665.970	37509.81	ZR II	0.0	0.5		MIT
2658.687	2659.478	37601.36	ZR I	1.0	0.0		MIT
2656.47	2657.26	37632.7	ZR		0.7		MIT
2655.846	2656.636	37641.58	ZR I	0.3			MIT
2650.381	2651.170	37719.19	ZR II	1.0	0.9		MIT
2647.782	2648.570	37756.22	ZR I	0.9			MIT
2643.79	2644.58	37813.2	ZR		2.3		ME
2643.395	2644.182	37818.87	ZR II	0.7	0.7		MIT
2640.151	2640.938	37865.34	ZR I	0.5			MIT
2639.086	2639.872	37880.62	ZR II	1.3	1.2		MIT
2635.424	2636.210	37933.25	ZR I	1.0			MIT
2634.573	2635.358	37945.50	ZR	0.3			MIT
2633.11	2633.89	37966.6	ZR	0.3			MIT
2630.908	2631.692	37998.36	ZR II	1.0	0.8		MIT
2630.352	2631.136	38006.39	ZR	0.7			MIT
2626.971	2627.755	38055.31	ZR II	0.0	0.5		MIT
2626.402	2627.185	38063.55	ZR II	0.3	0.3		MIT
2621.593	2622.375	38133.37	ZR II	0.3	0.7		MIT
2620.822	2621.604	38144.59	ZR	0.3			MIT
2619.810	2620.592	38159.32	ZR I	0.6			MIT
2619.210	2619.992	38168.06	ZR II	0.3	0.5		MIT
2615.59	2616.37	38220.9	ZR II		0.6		MIT
2613.068	2613.848	38257.77	ZR	0.3			MIT
2612.181	2612.961	38270.76	ZR I	0.6			MIT
2609.72	2610.50	38306.9	ZR II		0.7		MIT
2609.431	2610.210	38311.09	ZR I	0.8			MIT
2608.393	2609.172	38326.33	ZR I	0.3			MIT
2604.756	2605.534	38379.85	ZR	0.3	0.3		MIT
2604.197	2604.975	38388.08	ZR II	0.5	0.7		MIT
2601.28	2602.06	38431.1	ZR II	0.3	0.7		MIT
2592.393	2593.168	38562.87	ZR I	0.7			MIT
2592.192	2592.967	38565.85	ZR I	0.8			MIT
2589.653	2590.428	38603.67	ZR I	1.1			MIT
2589.071	2589.846	38612.34	ZR II	0.7	1.0		MIT
2588.946	2589.720	38614.21	ZR	1.2			MIT
2586.857	2587.631	38645.39	ZR II	0.7	0.9		MIT
2586.40	2587.17	38652.2	ZR II		0.3		MIT
2583.99	2584.76	38688.3	ZR	0.6	0.5		MIT
2583.655	2584.428	38693.28	ZR I	1.2			MIT
2583.405	2584.178	38697.02	ZR II	1.0	1.2		MIT
2581.71	2582.48	38722.4	ZR II		0.5		MIT
2580.451	2581.223	38741.32	ZR	0.6			MIT
2579.550	2580.322	38754.85	ZR	1.4			MIT
2576.105	2576.876	38806.67	ZR I	0.9			MIT
2571.391	2572.161	38877.81	ZR II	2.5 G	2.6 G		MIT
2568.873	2569.643	38915.92	ZR II	2.0	2.3		MIT
2567.638	2568.407	38934.63	ZR II	2.0	2.0		MIT
2567.451	2568.220	38937.47	ZR I	1.3			MIT
2567.075	2567.844	38943.17	ZR II	1.2	0.9		MIT
2564.295	2565.064	38985.39	ZR I	0.6			MIT
2563.56	2564.33	38996.6	ZR I	1.1			MIT
2556.430	2557.197	39105.32	ZR I	0.5 D			MIT
2554.31	2555.08	39137.8	ZR	1.3			MIT
2553.047	2553.813	39157.13	ZR II	0.3	0.3		MIT
2551.731	2552.497	39177.33	ZR	0.5			MIT
2550.744	2551.509	39192.49	ZR II	2.0	1.7		MIT
2550.514	2551.279	39196.02	ZR I	1.2			MIT
2547.08	2547.84	39248.9	ZR	0.3			MIT
2543.639	2544.403	39301.95	ZR II	0.3	0.3		MIT
2542.103	2542.866	39325.70	ZR II	2.0	1.7		MIT

AIR WAVELENGTH (ANGSTRMS)	VACUUM WAVELENGTH (ANGSTRMS)	WAVENUMBER (KAYSERS)	LMENT		LOG ARC	INTENSITY SPK	INTENSITY DIS	REF
2539.650	2540.413	39363.68	ZR	I	1.7			MIT
2539.356	2540.119	39368.24	ZR	II	0.0	0.6		MIT
2538.912	2539.675	39375.12	ZR		0.0	0.3		MIT
2538.019	2538.781	39388.98	ZR	I	1.2			MIT
2535.15	2535.91	39433.5	ZR	II	0.0	0.7		KS
2533.626	2534.387	39457.27	ZR	II		0.3		MIT
2533.189	2533.950	39464.07	ZR		0.5			MIT
2532.460	2533.221	39475.43	ZR	II	1.5	1.3		MIT
2527.107	2527.867	39559.04	ZR		0.6			MIT
2524.857	2525.616	39594.29	ZR		0.6			MIT
2521.917	2522.676	39640.45	ZR	II	1.2	1.0		MIT
2519.976	2520.734	39670.98	ZR		0.8			MIT
2511.566	2512.322	39803.81	ZR	I	0.6			MIT
2509.768	2510.524	39832.32	ZR	II	0.7	0.9		MIT
2509.034	2509.790	39843.98	ZR	II	0.5	0.3		MIT
2505.990	2506.745	39892.37	ZR		1.1			MIT
2504.796	2505.551	39911.38	ZR		0.9			MIT
2504.533	2505.288	39915.58	ZR	I	0.3			MIT
2504.014	2504.769	39923.85	ZR	II	1.0	0.7		MIT
2496.480	2497.233	40044.32	ZR	II	0.3	0.7		MIT
2495.274	2496.027	40063.68	ZR		1.0			MIT
2493.716	2494.468	40088.71	ZR		0.8R			MIT
2487.288	2488.039	40192.30	ZR	II	1.3	1.3		MIT
2485.597	2486.347	40219.64	ZR	II	0.3	0.5		MIT
2481.43	2482.18	40287.2	ZR		0.8			MIT
2481.362	2482.111	40288.28	ZR	II	0.3	0.6		MIT
2480.176	2480.925	40307.55	ZR	II		0.3		MIT
2478.066	2478.815	40341.87	ZR		0.5			MIT
2474.140	2474.888	40405.87	ZR		0.5			MIT
2474.010	2474.758	40408.00	ZR		0.8			MIT
2471.95	2472.70	40441.7	ZR		0.5			MIT
2468.04	2468.79	40505.7	ZR	I	0.7			MIT
2467.971	2468.717	40506.87	ZR	II		0.3W		MIT
2465.395	2466.141	40549.19	ZR	II	0.3	0.7		MIT
2461.795	2462.540	40608.48	ZR	I	0.3			MIT
2461.03	2461.77	40621.1	ZR		0.7			MIT
2459.842	2460.586	40640.72	ZR	I	0.7			MIT
2459.292	2460.036	40649.81	ZR		0.5			MIT
2457.436	2458.180	40680.51	ZR	II	1.3	1.3		MIT
2456.54	2457.28	40695.3	ZR	I	0.5			MIT
2454.969	2455.712	40721.38	ZR		0.3			MIT
2454.585	2455.328	40727.75	ZR	II		0.3		MIT
2454.236	2454.979	40733.55	ZR	II	0.3H	0.5		MIT
2451.986	2452.729	40770.92	ZR		0.6			MIT
2449.849	2450.591	40806.48	ZR	II	1.7	1.3		MIT
2449.309	2450.051	40815.48	ZR		0.5			MIT
2448.385	2449.127	40830.88	ZR	I	0.8			MIT
2445.748	2446.489	40874.90	ZR		0.7H			MIT
2441.991	2442.731	40937.78	ZR	II	0.7	1.0		MIT
2441.30	2442.04	40949.4	ZR		0.9			MIT
2440.340	2441.080	40965.48	ZR		0.7			MIT
2439.503	2440.243	40979.53	ZR		0.6			MIT
2435.149	2435.888	41052.80	ZR		1.4			MIT
2434.574	2435.313	41062.49	ZR	II	1.1	0.9		MIT
2433.245	2433.983	41084.92	ZR			0.3J		MIT
2432.262	2433.000	41101.52	ZR	II		0.3		MIT
2429.19	2429.93	41153.5	ZR		0.5			MIT
2426.381	2427.118	41201.13	ZR	II	0.3	0.6		MIT
2424.900	2425.636	41226.30	ZR	I	0.5			MIT
2423.748	2424.484	41245.89	ZR		1.0			MIT
2419.409	2420.144	41319.85	ZR	II	1.5	1.0		MIT
2416.88	2417.61	41363.1	ZR	II		0.3W		MIT
2415.911	2416.645	41379.68	ZR		0.7			MIT
2415.68	2416.41	41383.6	ZR		0.5			MIT
2414.600	2415.334	41402.14	ZR	I	1.0			MIT
2413.503	2414.237	41420.96	ZR		0.9			MIT
2411.93	2412.66	41448.0	ZR		0.8W			MIT
2407.89	2408.62	41517.5	ZR		0.7			MIT
2406.82	2407.55	41536.0	ZR	II	0.0	0.5		MIT
2406.259	2406.991	41545.65	ZR		0.0	0.3		MIT
2405.80	2406.53	41553.6	ZR			0.5		MIT
2405.519	2406.251	41558.43	ZR	I	1.1	0.0		MIT
2403.429	2404.160	41594.56	ZR	I	1.0			MIT
2402.933	2403.664	41603.15	ZR		0.7			MIT
2401.446	2402.177	41628.91	ZR		0.3	0.3H		MIT
2400.814	2401.545	41639.86	ZR		1.2	0.0		MIT
2400.62	2401.35	41643.2	ZR		0.0H	0.6H		MIT
2400.37	2401.10	41647.6	ZR			0.5		MIT
2398.99	2399.72	41671.5	ZR	II	0.7	0.6		MIT
2398.18	2398.91	41685.6	ZR		0.3	0.0		MIT
2397.98	2398.71	41689.1	ZR		0.5			MIT
2397.707	2398.437	41693.82	ZR		0.5	0.0H		MIT
2397.589	2398.319	41695.87	ZR	II	0.6	0.5		MIT
2397.235	2397.965	41702.03	ZR	I	1.1			MIT
2396.32	2397.05	41718.0	ZR		0.3			MIT
2394.53	2395.26	41749.1	ZR		0.0	0.3		MIT
2392.68	2393.41	41781.4	ZR	II	0.7	0.8		MIT
2391.872	2392.601	41795.52	ZR		0.7			MIT
2391.745	2392.474	41797.74	ZR		1.2			MIT
2391.047	2391.776	41809.94	ZR		0.3			MIT
2390.149	2390.877	41825.65	ZR		0.3	0.0		MIT
2389.53	2390.26	41836.5	ZR	II	0.3	0.8		MIT
2389.405	2390.133	41838.67	ZR		0.3	0.6		MIT
2389.209	2389.937	41842.10	ZR		1.4			MIT
2388.210	2388.938	41859.60	ZR		0.3H	0.3H		MIT
2388.009	2388.737	41863.13	ZR	I	0.8			MIT
2387.183	2387.911	41877.61	ZR	II	1.2	1.0		MIT
2384.827	2385.554	41918.98	ZR		0.6	0.0		MIT
2384.166	2384.893	41930.60	ZR		1.4	0.0H		MIT
2382.36	2383.09	41962.4	ZR		0.0H	0.3		MIT
2381.557	2382.283	41976.53	ZR		0.6	0.3		MIT
2380.558	2381.284	41994.15	ZR		1.2			MIT
2378.243	2378.969	42035.02	ZR	I	1.0			MIT
2374.422	2375.147	42102.66	ZR	I	1.3	1.0H		MIT
2372.930	2373.655	42129.13	ZR	II	1.2	0.9		MIT
2372.57	2373.29	42135.5	ZR	I	0.9	0.0		MIT
2372.33	2373.05	42139.8	ZR		0.8	0.0		MIT
2371.689	2372.413	42151.17	ZR		0.7			MIT
2371.38	2372.10	42156.7	ZR		0.5	0.0H		KS
2371.152	2371.876	42160.72	ZR		0.3	0.3H		MIT
2370.428	2371.152	42173.59	ZR		0.5			MIT
2370.15	2370.87	42178.5	ZR		0.3	0.6H		MIT
2368.98	2369.70	42199.4	ZR		0.0	0.3H		MIT
2367.33	2368.05	42228.8	ZR	I	0.8			KS
2366.217	2366.940	42248.64	ZR	II	0.3	0.3H		MIT
2364.95	2365.67	42271.3	ZR	II	0.0	0.3		MIT
2364.59	2365.31	42277.7	ZR	II		0.3		MIT
2363.842	2364.564	42291.09	ZR	II	0.7	0.5		MIT
2363.523	2364.245	42296.79	ZR	I	1.4			MIT
2363.052	2363.774	42305.22	ZR		0.5	0.6		MIT
2361.749	2362.471	42328.56	ZR	II	0.5	0.8		MIT
2360.422	2361.144	42352.36	ZR		0.5	0.0		MIT
2359.08	2359.80	42376.5	ZR			0.6H		KS
2357.436	2358.157	42406.00	ZR	II	0.9	1.3		MIT
2356.25	2356.97	42427.3	ZR		0.3			MIT
2355.900	2356.621	42433.64	ZR	I	1.5	0.0		MIT
2354.50	2355.22	42458.9	ZR		0.3	0.3H		MIT
2354.064	2354.784	42466.73	ZR		0.5			MIT
2353.207	2353.927	42482.20	ZR	II	0.0	0.3H		MIT
2351.656	2352.376	42510.21	ZR	II	0.0	0.9		MIT
2350.92	2351.64	42523.5	ZR	II	0.5H	0.3		MIT
2350.20	2350.92	42536.5	ZR		0.3	0.3H		MIT
2349.59	2350.31	42547.6	ZR		1.0	0.0		MIT
2348.586	2349.305	42565.78	ZR		1.2	0.0		MIT
2347.11	2347.83	42592.5	ZR	II	1.1	0.3		MIT
2345.92	2346.64	42614.1	ZR	II	0.6	0.0		MIT
2343.456	2344.174	42658.95	ZR		0.6			MIT
2342.34	2343.06	42679.3	ZR		0.5	0.3H		MIT
2341.326	2342.043	42697.76	ZR	I	0.7			MIT
2340.87	2341.59	42706.1	ZR	I	1.4			MIT

AIR WAVELENGTH (ANGSTRMS)	VACUUM WAVELENGTH (ANGSTRMS)	WAVENUMBER (KAYSERS)	LMENT		LOG ARC	INTENSITY SPK	DIS	REF
2339.385	2340.102	42733.18	ZR		0.0	0.3H		MIT
2337.78	2338.50	42762.5	ZR	II	0.0	0.9		MIT
2335.63	2336.35	42801.9	ZR			0.3H		KS
2334.67	2335.39	42819.5	ZR		0.8			MIT
2331.57	2332.29	42876.4	ZR			2.0W		KS
2330.730	2331.445	42891.85	ZR		0.9			MIT
2330.36	2331.08	42898.7	ZR	II	1.3	1.2		MIT
2324.753	2325.467	43002.12	ZR	II	1.3	0.9		MIT
2324.48	2325.19	43007.2	ZR	II	1.0	0.6		MIT
2321.89	2322.60	43055.1	ZR			0.6		MIT
2320.80	2321.51	43075.4	ZR		0.8			MIT
2317.248	2317.960	43141.38	ZR	II	1.3	1.0		MIT
2316.20	2316.91	43160.9	ZR		0.6			MIT
2315.58	2316.29	43172.5	ZR	I	0.5			MIT
2310.459	2311.170	43268.13	ZR		1.1			MIT
2308.09	2308.80	43312.5	ZR			0.7		MIT
2303.141	2303.850	43405.60	ZR		2.0			MIT
2301.61	2302.32	43434.5	ZR			0.5H		KS
2300.41	2301.12	43457.1	ZR		0.0H	0.3H		MIT
2298.160	2298.868	43499.67	ZR		1.0			MIT
2295.480	2296.187	43550.45	ZR	II	1.6	0.9		MIT
2294.045	2294.752	43577.69	ZR	II	1.7	1.0		MIT
2291.113	2291.819	43633.46	ZR	II	1.9	1.2		MIT
2288.614	2289.320	43681.10	ZR	II	0.3	0.3		MIT
2285.228	2285.933	43745.81	ZR		2.0			MIT
2283.17	2283.87	43785.2	ZR			0.3		MIT
2281.47	2282.17	43817.9	ZR			0.3H		MIT
2280.346	2281.050	43839.46	ZR	II	0.3	0.6		MIT
2276.693	2277.396	43909.79	ZR		0.7	0.6H		MIT
2275.39	2276.09	43934.9	ZR	II		0.3H		KS
2269.392	2270.094	44051.05	ZR		1.5			MIT
2257.880	2258.579	44275.62	ZR		0.3	0.5H		MIT
2254.20	2254.90	44347.9	ZR	II		0.3H		KS
2253.272	2253.970	44366.16	ZR		1.2			MIT
2252.38	2253.08	44383.7	ZR			0.3H		KS
2250.748	2251.446	44415.91	ZR		1.5			MIT
2250.46	2251.16	44421.6	ZR	II		0.3		MIT
2248.05	2248.75	44469.2	ZR	I	1.6			MIT
2239.67	2240.37	44635.6	ZR			0.3J		MIT
2239.186	2239.881	44645.23	ZR		1.0			MIT
2235.09	2235.78	44727.0	ZR	II		0.6		MIT
2234.43	2235.12	44740.2	ZR	II	0.3	0.0		MIT
2233.46	2234.15	44759.7	ZR	II	0.0H	0.6		MIT
2231.892	2232.586	44791.12	ZR	II	0.0	0.3		MIT
2230.927	2231.620	44810.49	ZR		1.2			MIT
2224.965	2225.657	44930.55	ZR		0.9			MIT
2223.31	2224.00	44964.0	ZR		0.0H	0.3H		MIT
2220.662	2221.353	45017.60	ZR	I	1.0			MIT
2220.22	2220.91	45026.6	ZR		0.3H	0.5H		KS
2218.053	2218.744	45070.55	ZR		0.7			MIT
2214.63	2215.32	45140.2	ZR	I	0.5			KS
2214.59	2215.28	45141.0	ZR	II		0.3		KS
2214.160	2214.850	45149.79	ZR		1.3			MIT
2212.47	2213.16	45184.3	ZR	II	0.3	0.0		MIT
2211.735	2212.424	45199.29	ZR		0.9			MIT
2210.896	2211.585	45216.44	ZR		1.2			MIT
2207.959	2208.648	45276.58	ZR		1.0			MIT
2206.96	2207.65	45297.1	ZR			0.3H		MIT
2206.31	2207.00	45310.4	ZR	II		0.5H		MIT
2201.649	2202.336	45406.33	ZR	I	0.8			MIT
2192.863	2193.548	45588.24	ZR	I	0.5			MIT
2192.76	2193.45	45590.4	ZR			0.3		MIT
2184.795	2185.479	45756.57	ZR	II	0.3	0.6H		MIT
2182.80	2183.48	45798.4	ZR	II		0.6W		MIT
2181.48	2182.16	45826.1	ZR		0.3			MIT
2178.86	2179.54	45881.2	ZR		0.3	0.3		KS
2175.82	2176.50	45945.3	ZR	II		0.5W		MIT
2173.606	2174.287	45992.08	ZR		0.6			MIT
2167.78	2168.46	46115.7	ZR			0.3		KS
2165.24	2165.92	46169.8	ZR	II		0.5W		KS
2160.233	2160.912	46276.77	ZR		0.7			MIT
2159.94	2160.62	46283.0	ZR		0.0	0.3W		MIT
2159.19	2159.87	46299.1	ZR			0.5		KS
2157.96	2158.64	46325.5	ZR		0.0	0.3W		MIT
2152.89	2153.57	46434.6	ZR	II		0.5W		MIT
2151.05	2151.73	46474.3	ZR	II		0.7		MIT
2149.15	2149.83	46515.4	ZR		0.6			MIT
2144.01	2144.69	46626.9	ZR	II		0.5W		KS
2141.43	2142.10	46683.1	ZR		0.0H	0.3		MIT
2137.684	2138.358	46764.85	ZR	II	1.0	0.8		MIT
2136.17	2136.84	46798.0	ZR	I	0.7			MIT
2133.282	2133.955	46861.34	ZR	II	0.7	0.8		MIT
2129.93	2130.60	46935.1	ZR			0.3		KS
2126.535	2127.207	47010.00	ZR			0.3		MIT
2125.148	2125.820	47040.68	ZR		0.5	0.3K		MIT
2120.119	2120.790	47152.25	ZR	II	0.3	0.6		MIT
2119.74	2120.41	47160.7	ZR		0.3H	0.3		MIT
2119.142	2119.812	47173.99	ZR		1.0			MIT
2110.53	2111.20	47366.5	ZR		0.9			MIT
2109.65	2110.32	47386.2	ZR	II	0.7	1.0		MIT
2109.42	2110.09	47391.4	ZR		0.3H			MIT
2108.55	2109.22	47410.9	ZR		0.8			MIT
2105.83	2106.50	47472.2	ZR	I	0.6			MIT
2103.31	2103.98	47529.0	ZR		0.6			KS
2103.17	2103.84	47532.2	ZR			0.6		KS
2101.77	2102.44	47563.9	ZR	I	0.7			MIT
2099.30	2099.97	47619.8	ZR		0.3H	0.0		KS
2097.02	2097.69	47671.6	ZR			0.6		KS
2095.80	2096.47	47699.3	ZR	II	0.8	1.0		MIT
2092.86	2093.53	47766.3	ZR	I	0.6			MIT
2091.74	2092.40	47791.9	ZR		0.5	0.3		MIT
2089.56	2090.22	47841.7	ZR		0.6			MIT
2088.89	2089.55	47857.1	ZR		0.3			KS
2080.97	2081.63	48039.2	ZR			0.6		MIT
2076.29	2076.95	48147.5	ZR			0.3		KS
2074.90	2075.56	48179.7	ZR			0.3		KS
2074.03	2074.69	48199.9	ZR			0.7		KS
2068.09	2068.75	48338.4	ZR	II		0.3		KS
2067.08	2067.74	48362.0	ZR	II		0.5		KS
2065.35	2066.01	48402.5	ZR	II		0.3		KS
2064.37	2065.03	48425.5	ZR			0.3		KS
2063.89	2064.55	48436.7	ZR	II		0.5		KS
2061.85	2062.51	48484.6	ZR	II		0.6		KS
2061.35	2062.01	48496.4	ZR			0.5		KS
2057.96	2058.62	48576.3	ZR	II	0.3	0.3		KS
2055.42	2056.08	48636.3	ZR			0.3		KS
2051.21	2051.87	48736.1	ZR	II		0.5		KS
2040.11	2040.77	49001.2	ZR		0.3H	0.0H		KS
2030.73	2031.38	49227.5	ZR	II	0.0	0.8		KS
2029.91	2030.56	49247.4	ZR		0.3	0.3		MIT
2026.61	2027.26	49327.6	ZR	II		0.5		KS
2016.52	2017.17	49574.4	ZR			0.5		KS
2015.86	2016.51	49590.6	ZR	II		0.3		KS
2012.66	2013.31	49669.4	ZR	II	0.3H	0.3H		KS
2012.03	2012.68	49685.0	ZR			0.3		KS
2001.90	2002.55	49936.4	ZR	II		0.3		KS

Appendix

Introductions to the 1969 and 1939 editions of volume 1 of *WAVELENGTH TABLES*

TABLES OF WAVELENGTHS

AND INTENSITIES OF THE PRINCIPAL ATOMIC SPECTRUM LINES IN THE RANGE 10,000—2000 A.

INTRODUCTION TO THE 1969 EDITION

THIRTY years of progress in spectroscopy since the first appearance of the *M.I.T. Wavelength Tables* has led to a great need for their revision and expansion. Not only have hundreds of new investigations of the spectra emitted by atoms in various stages of ionization been reported since 1939, but vastly improved techniques can now be used to give more exact measurements of spectrum lines and estimates of their intensities. Samples of elements much purer than could be obtained previously now make possible surer assignments of many faint lines to parent atoms, and more extensive analyses of atomic spectra make available new checks on wavelengths and identifications.

It has been estimated that more than a thousand man-years of knowledgeable labor would be needed to carry out the experimental work and the reduction and assembly of results needed to bring these tables really up to date, and to expand them to include all data needed by spectrochemical analysts and by theoretical spectroscopists. Several attempts by others to start projects for such modernization have failed to attract the needed financial support; therefore, reprints of the present tables in their original form and "translations" of, or selections from, them have had to serve.

This lack, together with the fact that the tables have been out of print for a few years, has caused pressures to build up for their reissue in a form that, although necessarily far from up-to-date, would contain the more important corrections known to be needed in them. We are grateful to several interested groups who, at the instigation of Dr. Walter Baird of Baird Atomic, Inc., have volunteered support to help make it possible to issue the present volume.*

The 1939 edition contained 110,000 wavelength entries, limited arbitrarily to the spectral range 10,000–2000 A and to lines believed to originate from atoms in the normal and singly ionized states. The tables were originally prepared as the result of a combination of three circumstances: (1) the lack in 1935 of sufficiently extensive tables to determine the identity of impurities in many spectrochemical samples; (2) the availability of 143 Works Progress Administration workers who, as a result of the 1929 depression, could be employed with government funds; and (3) the presence in the M.I.T. Spec-

*Support toward the printing of the 1969 edition of the *M.I.T. Wavelength Tables* has been received from the following:

E. I. du Pont de Nemours & Co., Inc.
Baird Atomic, Inc.
Research Laboratories, General Motors Corporation
Cooke Optical Company
Jewell-Ash Division, Fisher Scientific Company
Applied Research Laboratories
The Perkin-Elmer Corporation
Sanders Associates, Inc.
Armco Research and Technology, Armco Steel Corporation
International Business Machines Corporation
Bell Telephone Laboratories, Inc.
Spectroscopy Society of Pittsburgh
Beckman Instruments, Inc.
Edgerton, Germeshausen, and Grier, Inc.
Society for Applied Spectroscopy, California Section
Society for Applied Spectroscopy, Chicago Section
Central Research Laboratory, American Optical Corporation

troscopy Laboratory of several of the world's most powerful grating spectrographs and of a newly designed automatic measuring, computing, and recording comparator that could furnish the wavelengths and densities of up to 2000 lines per minute from a grating spectrogram.

As much of the work leading to the compilation of these tables was carried out by spectroscopically inexperienced helpers, it was to be expected that a number of errors would be made. It is gratifying that demand for the tables continues to be wide after thirty years and that changes needed to correct misprints and blunders (as distinct from additional and improved information) involve fewer than one digit in every 6000.

Because of cost and other considerations it has not been found possible to reset the tables for this revision. Instead, the pages of the 1939 edition have been photographed, with lines through every entry known to need improvement, in such a way that it is still legible. Such a line indicates either that the original entry referred to a nonexistent spectral line or that a modified entry will be found in the Table of Corrections on pages xxxiii–xxxiv.

Our authority for changes has been a "Table of Corrections to the M.I.T. Wavelength Tables" set up at the National Bureau of Standards and systematically maintained during the past thirty years. A short time before his death in 1966, Dr. William F. Meggers of the Bureau sent us a copy of this correction table, which had been compiled by Virginia C. Stewart of the Spectrochemical Analysis Section from copies of the tables used by several laboratories in the N.B.S. This list was marked as "subject to change" and "designed only for preliminary working purposes." Before we received the list, it had been carefully examined by Dr. Charlotte Moore Sitterly of the Bureau, and by Dr. Meggers himself. The difficulty of ensuring accuracy in such tables is illustrated by the number of changes in the list indicated by these two superbly reliable compilers of spectroscopic data, even after it had been assembled by experienced spectroscopists.

Many additional suggestions for changes in the M.I.T. tables have been received over the years. These were all noted, but a number put forward by correspondents, and even many published in the literature, were called into question by other experts. We therefore adopted the policy of not making changes without several independent verifications.

Although the final authority for entries in the present Correction Table has thus been Dr. Meggers, the editor has taken upon himself the responsibility, in a few cases, of omitting suggested changes, usually when only minor intensity revisions were involved or considerable uncertainty remained. The intensity situation for many entries has been improved by Dr. Meggers' inclusion of values from the N.B.S. Tables of Spectral-Line Intensities, by Meggers, Corliss, and Scribner (U.S. Government Printing Office, 1961, Monograph 32; see also Supplement by Charles H. Corliss, N.B.S., 1967). These we list in a sixth column in the Correction Table. Intensity values on a uniform scale are, of course, to be found in the N.B.S. tables for many other lines listed in the original M.I.T. tables, and for several thousand lines not so listed.

The M.I.T. tables originally contained a number of misidentified lines, usually arising from rare earth impurities in rare earth samples. Those dropped in the revision are indicated by a line drawn through the entry, with no corresponding entry in the correction table, as the true line is usually to be found already entered at a nearby wavelength. Where only a change in an intensity estimate is involved, a line has been drawn through only the relevant columns. Added new lines are identifiable by being found in the Correction Table only. Wavelength changes involving only a few thousandths of an

Angstrom have been omitted in half a dozen or so cases.

In the spectral range between 10,000 and 3000 A the M.I.T. tables originally contained about 81,000 entries, while changes made from the N.B.S. Correction Tables involved 325 entries in this range. A dozen of these suggested changes were canceled by Dr. Meggers, while 21 new intensity values were considered by the editor to involve variations too minor to warrant inclusion. Of the remaining corrections 130 consisted of deletions, while 49 lines were suggested as worth adding. Only 11 misprints were found in this portion of the tables, with six misidentifications of the parent element, apart from duplications arising from impurities. Changes in wavelength values resulting from improved measurements involved 38 alterations, while 48 entries involved major improvements in intensity determinations. Changes resulting from new information regarding the stage of ionization of the parent atoms totaled 21. In all of these categories many further additions could be made from the published literature.

Between wavelengths 3000 and 2000 A, most of the included changes involve the elements Hf and Re, on which Meggers did much work. Many additional errors doubtless remain undiscovered, to be revealed when additional spectra have been thoroughly analyzed in this region.

Just as the *M.I.T. Wavelength Tables* supplemented and eventually largely replaced the shorter *Tabelle der Hauptlinien der Spektrallinien*, by Kayser and Ritschl, these tables in turn will be succeeded in usefulness (and soon, it is hoped) by more extensive tables. The compilers of such tables will find many new technological aids available to simplify and improve their work.

In the infrared, Fourier spectroscopy with scanning interferometers is making available new wavelength data in regions which have long been energy-limited. In the visible and ultraviolet, echelle spectroscopy now makes possible spectrographs of far greater power and speed than have yet been used in a systematic program of wavelength measurements. Many more wavelength standards, measured with etalons and other interferometers, are available to permit rapid interpolation with automatic comparators.

Of tremendous aid are the high-speed digital computers not available when these tables were compiled. It appears that successor tables should contain information on at least ten times as many lines as are listed herein, grouped both as to parent atom and its stage of ionization, and in order of wavelengths. Lines found in the entire range of optical spectroscopy should be included and from all stages of atomic ionization. A more uniform intensity scale, perhaps on the order of that used by Meggers, Corliss, and Scribner in the N.B.S. Intensity Tables, should be used.

Provision will be needed for frequent revision of entries and for additions in the proper place as new data become known. As has been suggested by many, a large computer would be an ideal repository for such data. Lists of any type desired could then quickly be run off, combining flexibility and accessibility with permanence. When projects designed to provide such facilities are successful, they should in addition meet the demand for a sturdy, well-bound, desk-available book suited to rough laboratory use, to replace the present volume.

It is a pleasure to acknowledge encouragement, suggestions, and stimuli from various staff members of the National Bureau of Standards, especially from Dr. Karl G. Kessler, Chief of the Atomic Physics Division; from Mr. Bourdon F. Scribner; from Dr. Charles Corliss; and above all from Dr. Sitterly and Dr. Meggers, who devoted many hours to editing the list of corrections included here.

George R. Harrison
Cambridge, Massachusetts
June 10, 1969

INTRODUCTION TO THE 1939 EDITION

TABLES OF WAVELENGTHS

AND INTENSITIES OF THE PRINCIPAL ATOMIC SPECTRUM LINES IN THE RANGE 10,000—2000 A.

THE main tables in this volume contain 109,275 entries, giving the wavelength, the intensity in arc, spark, or discharge tube, the stage of ionization of the parent atom when the line has been classified in a term array, and the wavelength authority, for each of the most important known spectrum lines emitted between 10,000 and 2000 angstroms by atoms in the first two stages of ionization.

Though the tables include only half of the known spectrum lines in the region covered, the lines listed account for 99% of the radiation emitted by atoms in this range of wavelengths. Except as noted below, all known atomic lines which have intensity 2 or more on a scale of 1 to 9000 have been included. In addition, 1381 band heads which frequently appear on spectrograms have been included for convenience in identifying impurities.

Lines not found in the tables have been omitted for one or more of the following reasons:

(1) Because, under ordinary conditions of excitation in arc, spark, or discharge tube they are not sufficiently intense to warrant inclusion.

(2) Because they are known (from actual term classification) to originate from atoms with two or more electrons removed, i.e., from the III, IV, or higher spectra.

(3) Because they would not, with ordinary equipment, be resolved from some line of the same element which has been included.

(4) Because, although we have observed them and measured their wavelengths, they have not been listed by previous investigators. We have, however, included a number of lines of short wavelengths not previously given in the literature, with element assignments which are purely tentative. Such lines are designated with an a in the remarks column (R).

All wavelengths are given in terms of the International Angstrom unit (A), as adopted by the International Astronomical Union. The element symbols used are listed in Table I, which gives also the number of lines included for each element. Responsibility for assignment of a line to a particular element rests entirely with previous investigators, except in cases designated a. The numerals I or II in the element column indicate that the line has been classified in a term array and definitely assigned to the normal atom (I) or to the singly ionized atom (II). In many cases, of course, a line can be assigned to a definite stage of ionization merely on the basis of excitation in various sources; but such indications have not been considered here, so that the presence of a Roman numeral after an element designation may be taken as an indication that the line has been classified.

All wavelengths marked with a dash in the R column are from our own determinations, and such values have been included for more than 75,000 lines. In cases where we have not used our own values we have usually chosen from the literature that value which we consider most trustworthy. Each two-letter symbol in the R column refers to the authority or authorities responsible for the wavelength value of the entry; authors and publications can be determined by means of Table V. In a few cases

WAVELENGTH TABLES

TABLE I

Symbol	Atomic Number	Element	Lines Included	Symbol	Atomic Number	Element	Lines Included	Symbol	Atomic Number	Element	Lines Included
A	18	Argon	1289	He	2	Helium	110	Re	75	Rhenium	2256
Ac	89	Actinium	7	Hf	72	Hafnium	1518	Rh	45	Rhodium	1327
Ag	47	Silver	347	Hg	80	Mercury	724	Rn	86	Radon	421
Al	13	Aluminum	425	Ho	67	Holmium	800	Ru	44	Ruthenium	2824
As	33	Arsenic	257	I	53	Iodine	1168	S	16	Sulphur	451
Au	79	Gold	333	In	49	Indium	1150	Sb	51	Antimony	524
B	5	Boron	94	Ir	77	Iridium	2577	Sc	21	Scandium	524
Ba	56	Barium	472	K	19	Potassium	306	Se	34	Selenium	736
Be	4	Beryllium	92	Kr	36	Krypton	1221	Si	14	Silicon	367
Bi	83	Bismuth	344	La	57	Lanthanum	1270	Sm	62	Samarium	3863
Br	35	Bromine	620	Li	3	Lithium	39	Sn	50	Tin	268
C	6	Carbon	183	Lu	71	Lutecium	456	Sr	38	Strontium	209
Ca	20	Calcium	662	Mg	12	Magnesium	173	Ta	73	Tantalum	2164
Cb	41	Columbium	3303	Mn	25	Manganese	1395	Tb	65	Terbium	2617
Cd	48	Cadmium	447	Mo	42	Molybdenum	3902	Te	52	Tellurium	764
Ce	58	Cerium	5755	N	7	Nitrogen	382	Th	90	Thorium	2587
Cl	17	Chlorine	775	Na	11	Sodium	175	Ti	22	Titanium	2136
Co	27	Cobalt	1607	Nd	60	Neodymium	2680	Tl	81	Thallium	300
Cr	24	Chromium	2277	Ne	10	Neon	1040	Tm	69	Thulium	793
Cs	55	Caesium	645	Ni	28	Nickel	1176	U	92	Uranium	5238
Cu	29	Copper	913	O	8	Oxygen	444	V	23	Vanadium	3130
Dy	66	Dysprosium	2063	Os	76	Osmium	1745	W	74	Tungsten	4327
Er	68	Erbium	2039	P	15	Phosphorus	408	Xe	54	Xenon	1261
Eu	63	Europium	2408	Pb	82	Lead	466	Yb	70	Ytterbium	1254
F	9	Fluorine	278	Pd	46	Palladium	908	Yt	39	Yttrium	686
Fe	26	Iron	4757	Po	84	Polonium	2	Zn	30	Zinc	207
Ga	31	Gallium	135	Pr	59	Praseodymium	2708	Zr	40	Zirconium	2036
Gd	64	Gadolinium	1607	Pt	78	Platinum	806	bh		Band heads	1381
Ge	32	Germanium	73	Ra	88	Radium	139				
H	1	Hydrogen	21	Rb	37	Rubidium	365				

The spectra of elements 43, 61, 85, 87 and 91 have not been studied.

mean literature values, marked m in the R column, have been used.

In order to keep the wavelength scale as uniform as possible we have used our own values except:

(1) When the line is an International Secondary (or Primary) Standard, marked S (or IS) in the R column.

(2) When interferometer values for the arc in air or an equivalent arc were available. These are marked I in the R column, followed by an author symbol. If no author symbol is given, the value is the mean of several consistent interferometer determinations. There are some 1300 entries in this and the preceding categories.

(3) When discharge-tube sources are necessary to excite the line. As most of our measurements have been on spectrograms taken with the arc and spark, we have consistently used literature values for lines from gaseous elements. There are some 14,000 entries in this category.

(4) When a line listed in the literature has not been found on our plates within a reasonable wavelength tolerance, in which case the literature value has been entered.

(5) When we have had reason to believe that values given in the literature were superior to our own, whether because of our inability to obtain certain elements in sufficient quantity or because of limitations of our equipment and methods in certain regions of the spectrum. There are some 15,000 entries in this category and that immediately preceding it.

(6) When the entry is for a band head, marked bh followed by an element symbol when the principal atom of the molecule is known. In this case a wavelength value obtained from the literature has arbitrarily been rounded off to one figure after the decimal, and designated by L or an author symbol in the R column.

Lines commonly listed as "air lines," which are frequently found in spectra taken with a condensed spark, are listed as from the element responsible (N, O, or A) when this is known. When the parent atom is not known the line is listed as "air."

Intensity estimates listed in the arc column are for the standard d.c. electric arc run in air on a 220-volt line with stabilizing reactance and resistance. A long Pfund-type arc was used for the region shorter than 4500 A, and a shorter arc (about 8 mm long) for the longer wavelength region.

Intensity estimates in the spark column are for a standard-type 20,000-volt condensed spark between electrodes separated by about 5 mm. Intensities in brackets are for lines emitted in discharge tubes of various kinds, such as Geissler tubes and hollow cathode discharge tubes.

All intensities have been estimated on an expanded scale, based somewhat on those used in recent years by such investigators as A. S. King and W. F. Meggers. Lines taken from the literature have been brought to our intensity scale by suitable approximate conversion factors determined by interpolation from known lines.

The symbols used in the intensity column are in most cases identical with those adopted by the Wavelength Commission of the International Astronomical Union, and are explained in Table II, which, for convenience, is repeated at the end of the book.

It was our original intention to include wavenumbers for all lines, but we have been dissuaded from this by the following considerations:

(1) The tables are designed principally for use in spectroscopic analysis of materials and for identifying impurities, and wavenumbers are seldom needed for this purpose.

(2) It was found that the bulkiness of the volume could be decreased considerably by omitting wavenumbers.

(3) The resulting decrease in production costs made possible inclusion of nearly 10,000 additional entries beyond the 100,000 originally projected, with only slight increase in bulk.

(4) When the dispersion of air has been determined more precisely (especially at short wavelengths) the factors used in converting wavelengths in air to wavenumbers in vacuum can be expected to change. Therefore the wavenumbers of most lines, whether their wavelength values are further improved or not, can be expected to change within the next few years, so that the wavenumber values given would soon be out of date.

TABLE II

Symbols Used in Wavelength Tables

(For author symbols in remarks column, see page xxiv)

a	new lines (not in literature), element assignments tentative
bh	band head
d	double line
h	hazy, diffuse, nebulous
I	interferometer measurement, mean value, unless with author symbol
IS	international primary standard
l	shaded or displaced to longer wavelengths (asymmetrical)
L	literature value, for band heads
m	mean value
r	narrow self-reversal
R	wide self-reversal
s	shaded or displaced to shorter wavelengths (asymmetrical)
S	international secondary standard
w	wide or complex
W	very wide or complex
—	(in R column) M.I.T. measurement
[ ]	discharge-tube intensity
I	line classified as emitted by normal atom
II	line classified as emitted by singly ionized atom

(5) We hope to include wavenumbers in such future improved wavelength tables as we may publish in which individual spectrum lines are listed together under the elements.

PREPARATION OF THE TABLES

THESE tables were originally compiled for our own use in a more extensive project of spectrum-line measurements. When the spectrum of an element is being studied exhaustively for classification purposes, one of the most tedious processes is the elimination of chance lines which do not arise from the element in question. Routine methods can be used for the elimination of possible Rowland and Lyman ghosts and for sorting out lines from different orders; but for weeding out lines from impurity elements much more extensive tables than have been available previously are desirable, with all lines arranged together in order of wavelength. It soon became evident that before much progress could be made in improving the descriptions of the spectra of individual elements a set of such tables was needed.

Accordingly, we combed the literature for all previous wavelength measurements on atomic lines, and copied on each of 250,000 white cards data for one line. When a line had been studied by several authors the measurements were all listed together on one card. Then, to check this first catalog, a second similar catalog was prepared on 250,000 buff cards, and the two catalogs were intercompared and finally corrected from the literature.

All measurements which we made on lines were entered on work sheets; and when from 7 to 50 measurements had been recorded for a given line these were averaged, and the average value was put at the tops of the two proper cards as our value for that line. Whenever our values agreed acceptably with the values given in the literature the cards were placed in the final catalogs; when the values did not agree, new spectrograms were made and from 7 to 20 new measurements were made on the uncertain lines. Usually the disagreement was then resolved; but in cases where it was not, a third set of measurements was made from new plates, when possible. In some cases we found that three independent sets of our measurements on a line would agree to within 0.003 A while departing by as much as 0.04 A from a value given to seven figures in the literature. In such cases we have inserted our own value when we felt sure that we were measuring the proper line; when we could not be sure we have used the literature value and held our own for further study.

To prepare the present catalog, data for the 110,000 strongest lines were copied on green cards from the main catalogs, and these green cards were then arranged in order of diminishing wavelengths. The white cards were left in order of wavelengths under each element, while the buff cards were arranged in order of wavelengths for all elements together. These two larger catalogs are now available in the form of cards filed in index boxes; the publication of either is not contemplated until considerable spectroscopic work can be done to add to their completeness and accuracy.

Since we have included here no lines not listed in the literature (except a few at short wavelengths, marked a), the present tables must be considered as incomplete, for the number of lines listed for each element is still largely determined by the amount of time that has been devoted to study of its spectra. This is shown by comparing indium, for example (which has been studied very exhaustively in the hollow cathode discharge by Paschen and his co-workers), with such similar elements as gallium and thallium.

Examination of the tables shows many cases where two or more lines from different elements have wavelengths which differ by less than the expected experimental error. In most cases the various lines involved are real; in a few, however, it seems probable that only one line is involved, this having been assigned incorrectly to one or more atoms. For example, the three lines 2639.24 Se – [5] Bl; 2639.17 Te – [5] Bl; 2639.14 S – [8] Bl may well all be the same sulphur line, though this cannot be proved without further experimentation. Since the arrangement of the tables makes such cases obvious, it has seemed desirable to avoid the responsibility of attempting to resolve them until further work has been done.

In almost every element having a complex spectrum we have found thousands of unlisted lines, most of them, to be sure, having intensities which would make their inclusion in these tables questionable. Even moderately strong lines of this sort have not been included, however, because of the necessity of taking further spectrograms of arcs in various atmospheres, to avoid the danger of mistaking band lines for atomic lines. The lines marked a have been selected from spectrograms on which no band structures could be observed; we decided to include such lines in the range 2400 to 2000 A even when element assignments could be only tentative, because the recent great increases in the sensitivity of emulsions and the reflecting power of gratings in this region make visible thousands of strong lines not previously studied.

PRECISION OF WAVELENGTHS

WHERE an intensity entry is given in the arc column with a dash in the R column, the wavelength recorded, if less than 4500 A, is that observed in the standard Pfund arc sufficiently far from each electrode (7 mm) to avoid pole effect. For wavelengths greater than 4500 A it is necessary to use a short arc to bring out a majority of the lines, and pole-effect shifts are less likely to occur. Since the wavelength of a line observed in a spark is much more variable than its wavelength observed in an arc, spark wavelengths have been listed only in cases where no intensity is entered either in the arc column or in brackets. Where the only intensity entry is in brackets in the spark column, the wavelength is that observed in a discharge tube. Discharge-tube wavelengths can be expected to be most consistent, arc wavelengths somewhat less so, and spark wavelengths least consistent.

Wavelengths which should be viewed with most suspicion are those (1) given to three figures after the decimal, if longer than 4500 A or shorter than 2500 A; (2) arising from elements with simple spectra; (3) belonging to rare earths of especial rarity; (4) belonging to elements having wide fine-structure patterns.

In cases where two widely differing values were given to three figures in the literature, and our value agreed with neither, the last digit has been dropped in the value given. Our value was then retained if it lay between the other two, while a mean was used, marked m, if it did not.

Wavelengths given to three figures after the decimal are supposedly correct to within ±0.005 A, those given to two figures to within ±0.05 A, and those given to one figure to within ±0.5 A. Actually these tolerances are only approximate, for the precision obtainable varies greatly with the natural breadth of the line, its simple or complex structure, its variability under different conditions of excitation, the region of the spectrum in which it lies, and numerous other factors which are difficult to evaluate. In certain elements, such as cerium, we have had no difficulty in obtaining measurements the majority of which seldom deviated more than ±0.003 A from the mean, while the final average values were consistent with each other to within 0.0016 A, as shown by application of the combination principle. Some other elements, whose lines in the same region were measured and tested by the same methods, showed far less self-consistency and reproducibility. For this reason the tables probably contain many wavelengths which are exact to the last figure, and others given to seven figures in which variation in different sources may be as great as 0.01 A or more.

Where only two figures are given after the decimal, however, in only a few cases outside of the infrared and far ultraviolet should our measurements be found in error as much as the allowed .05 A. Where only one figure is given after the decimal the wavelengths. have been rounded off (1) because of low precision of measurements in the infrared; (2) because the line is extremely broad or extremely variable, so that more precise values have little meaning; or (3) because, in the case of band heads, interference from these can be expected at some distance to each side of the wavelength given.

For many spectroscopic purposes, the shorter the wavelength the more important is wavelength precision. If one desires to know the wavenumber of a line to within 0.02 cm^{-1} for line-classification purposes, the wavelength must be known with the following approximate precisions:

At 10,000 A, to within	0.02 A
6,000 A	0.007 A
3,000 A	0.002 A
2,000 A	0.0008 A

With large spectrographs, and for wavelengths greater than 2500 A, this variation in required precision is fairly well matched experimentally; for lines of similar shape are often narrower at shorter wavelengths, the linear resolution of optical equipment usually increases as the wavelength decreases, and lines of short wavelength can be measured in higher orders of diffraction-grating spectrographs than can those of long wavelength. At wavelengths shorter than 2500 A a reverse trend begins with large gratings, because of overlapping of orders, decline in reflecting power, and the lower sensitivity of photographic emulsions; the latter effect has, however, recently been partially offset by improved types of emulsions.

THE INTENSITY SCALE

It was our original purpose to base all intensity values on microdensitometer curves obtained when the plates were run through an automatic comparator. While this method can be used to obtain an excellent measure of the relative intensities of two lines photographed close together on the same plate, it gives results of little meaning when extended to cover the entire spectrum range. This is because of the wide variation in the sensitivities of the photographic emulsions available, in their development, and in the excitation conditions of sources. It would, of course, be possible to make exact measurements by means of photographic photometry, calibrating and standardizing all plates and determining true atomic-transition probabilities for the lines. While such results have great value for many purposes (and we are engaged in a systematic program of such determinations), we have become convinced that such data would be of much less value in the present tables than are approximate intensity estimates made on a fairly consistent scale.

Spectroscopists have shown increasing inclination in recent years to expand the scale of their intensity estimates from the familiar 10 to 1 scale (which often trailed off through a series of " afterthought estimates," such as 0, 00, etc.) to a wider scale, with intense lines being rated as 5000 or more. Such a scale lies much closer to the true intensities of the lines; for example, the sodium D lines near 5893 A, marked 10 in many old tables, are actually at least 10,000 times as intense as many lines judged as of intensity 1. Both absolute and relative intensities of the lines can be expected to vary with the source and equipment used, and any intensity determinations are of use in such tables merely as approximate indications of the relative intensities of the lines.

While we have had density traces of spectrograms available, we have preferred to base the intensities given in these tables on eye estimates of the lines made by observing them on a screen. Then, after lines of all elements in a given region of the spectrum had been estimated on a fairly uniform scale, all intensity values were multiplied by a factor designed to bring the various spectrum regions into a smooth relationship. This was, of course, very difficult, and many incongruities will doubtless be noted. There seems to be no exact solution to the problem, for the densities obtained depend greatly on the emulsions used in reproducing the lines, while the true intensities will depend on the excitation conditions in the source used. Therefore all intensities should be taken merely as rough approximations.

In the case of lines taken from the literature, we have adjusted the intensity values to fit our scale as best we could. Therefore responsibility of an author for a particular entry should be taken as extending to the wavelength value and to the element assignment, but not to the intensity values.

Again, intensities in the arc and spark columns sometimes show incongruities because of variations in exposure conditions and the vagaries of intensity estimation. In many cases, for example, lines known to arise from the ionized atom will be listed as having higher intensities in the arc than in the spark. This often occurs because the spark plates were for some reason weaker than the arc plates.

We have given the most intense lines the rating 9000 (to avoid use of more than four digits) and have used the value 1 for the weakest lines which have been observed. (These weak lines have been omitted from the tables in all cases except for a few lines of iron.) Lines of intermediate intensity have been given 25 different values. The advantage of such a scale is not that the values are given to any greater precision than would be possible on a uniform scale of, say, 1 to 25, but that they are closer approximations to the true intensities.

Increase in the sensitivity of the emulsions used in photographing wavelengths longer than 5000 A has resulted in the use of much higher intensity ratings in this region than were common a few years ago. Though we have tried to keep all intensities on as uniform a scale as possible, it appears that the expansion of the scale in the infrared may have been overdone.

A few important lines of the rare earths will be found listed as weak in the tables because they do not appear with great intensity in the arc, spark, or discharge tube, but only in the electric furnace.

APPARATUS AND METHODS

More than ten thousand 2 x 20 inch and 4 x 16 inch spectrograms have thus far been taken in the course of our main wavelength project, using five large diffraction-grating spectrographs. These spectrographs involve (1) a six-inch aluminum-on-glass grating of 165,000 lines in a 35-foot modified Paschen mounting; (2) a six-inch aluminum-on-glass grating of 170,000 lines in a 34-foot modified Rowland mounting; (3) a six-inch aluminum-on-glass grating of 120,000 lines and 35-foot radius in a stigmatic Wadsworth mounting, used mostly in the range 2500–1960 A; (4) a seven-inch aluminum-on-glass 35-foot grating of 90,000 lines in a stigmatic Wadsworth mounting, used mostly for the infrared; (5) a six-inch speculum metal grating of 180,000 lines and 21-foot radius, mounted in vacuum for the range 2000–500 A. All gratings had either 30,000 or 15,000 lines per inch and were ruled by Professor R. W. Wood at Johns Hopkins University. The dispersions used varied between 0.33 A/mm and 3.3 A/mm, the latter value being used only in the infrared.

Most of the plates obtained have been measured at least twice on one or both of two automatic computing and recording comparators. During the first year of the project all measurements were made on a 16-inch comparator obtained from the Société Genevoise, which we fitted with an attachment for computing wavelengths by automatic interpolation between iron standard lines, and for recording them directly on motion-picture film.[1] When this device had been put into satisfactory operation a new comparator was built in the laboratory shops. This has a carriage which will give 25 inches of uninterrupted travel, controlled by a screw which is corrected to 1 micron, and is fitted also with an improved computing and recording mechanism.[2] When this second comparator had been put into routine operation the original comparator was rebuilt, a 27-inch screw being substituted for the 16-inch screw, and a new type of maximum-picker being installed for automatic setting on the spectrum lines at high speed. This is the instrument which, in a state of undress preliminary to its final testing, is shown in the frontispiece. With it

[1] G.R. Harrison, Jour. Opt. Soc. Am., **25:** 169 (1935)
[2] G.R. Harrison, Rev. Sci. Inst., **9:** 15 (1938)

are sections of a spectrogram and of a record of intensities and wavelengths as obtained from the machine.

Though wavelengths are recorded to eight figures by the machines, only seven are kept, and there is, of course, some uncertainty in the seventh digit. Exhaustive tests have shown, however, that errors from other sources greatly outweigh errors introduced by the limitations of the machines.

The operation of the automatic maximum-picker has been improved to the point where the wavelengths of more than a thousand lines can be recorded in one minute without difficulty. The standard time in which a 20-inch plate is traversed is 120 seconds. To save time in loading the machine, a carriage has been provided on which three plates can be mounted together, with the proper starting line of each at a common fiducial mark. The density trace is recorded by photographing, on the same motion-picture film that records the wavelengths, the motion of the spot of a cathode-ray oscillograph which is controlled by the output of an electron-multiplier tube in accordance with the intensity of the light passing through the plate.

In the original maximum-picker [1] the peak of a spectrum line was determined in 0.006 second by measuring the voltage difference between two photoelectric cells which determined the densities of the plate at two positions separated by a distance equivalent to the linear resolution of the spectrograph. This method required that the two cells be balanced to within 0.1% at all light intensities, a condition somewhat difficult to achieve and maintain. To eliminate need for this, a delay network which gives the equivalent result with a single photocell was designed.[3] The output of the electron multiplier which measures the light passing through the plate is divided by this network into two parts, one of which is in phase with the original current and the other delayed by 0.014 second. The plate is then moved across the scanning light beam at a speed such that 0.014 second is required to traverse a distance along the plate equal to the linear resolution of the spectrograph. The difference between the two currents is then applied, after amplification, to a circuit which flashes a recording lamp whenever this current falls to zero while the density has a value slightly greater than that of the background of the spectrogram. Thus the reading of the rapidly rotating wavelength dial is photographed at the proper instant. To compensate for the electrical delay of 0.007 second, an equivalent mechanical delay is introduced by the machine.

To eliminate small residual wavelength displacements which may result because currents from spectrum lines of different intensities have different wave forms which pass through the network differently, automatic measurements from two runs made in opposite directions are always averaged. The problem of designing a network which will attenuate and delay equally all frequencies contained in the electrical record of a spectrogram is a difficult one to solve cheaply, and small residual errors have been found to arise from such distortion; but the averaging process apparently eliminates them.

Since it is rather difficult to set a plate on the machine exactly to within 0.001 A of any standard line, we have found it convenient to make this setting only approximate (to within ±0.01 A) and then to correct the last digit of the recorded wavelength values by the same amount that the standard lines are found to be incorrect. This correction could be made automatically, but thus far we have found it most convenient to make this correction when the wavelengths and intensities are being transcribed from the films to the record books or cards.

All plates have been measured by hand setting as well as by automatic setting. When setting by hand the operator observes the spectrum projected on a screen before

[3] To be described in papers expected to be published late in 1939.

him, and when he has set the desired line on a fiducial mark, presses a key to record the wavelength. This process, though considerably slower than automatic recording, is still twenty times as rapid as that commonly used.

We have found that measurements made with the automatic maximum-picker are more self-consistent than are hand settings, but we have used both methods to obtain additional checks on the precision of each. A few lines are missed by the automatic picker, especially when they lie on the wings of broad lines of great intensity. These lines the hand-set records supply, though good wavelength values can be obtained for missing lines directly from the automatic intensity trace.

Lines for which our values have been used in the tables have been measured at least twice on each of three plates, and the average number of measurements for each line is greater than ten. When average values for the lines were tabulated, the number of values included in the average, and the average deviation from the mean, were also tabulated. The letter A was used to indicate an average deviation of ±0.001 A, B for ±0.002 A, etc. Weightings of 24 C were not uncommon for narrow lines — 24 determinations with an average deviation of ±0.003 A.

SOURCES OF ERROR

The wavelength precision of our measurements has been limited by the following factors, arranged in the order of probable decreasing importance:

(1) The insufficiency of adequate wavelength standards in some spectral regions, particularly of standard lines of suitable intensities.

(2) The displacement on spectrograms of lines to be measured relative to standard lines.

(3) Displacements caused by strong neighboring lines or by blends with impurity lines or bands.

(4) The natural breadth of some lines, and the complex structure of others. We have listed all complex lines as single unless their components were at least 0.1 A apart.

(5) Errors of coincidence, inadequately corrected for, where lines to be measured in one order were compared with standards in another order.

(6) Actual variations of wavelength with excitation conditions.

(7) Limited amounts of certain materials available for excitation to produce the lines.

(8) Incorrect identification of lines with lines given in the literature.

(9) Uncertainties of setting on line maxima.

(10) Comparator errors.

The largest errors undoubtedly come from inadequacy of the wavelength standards, with which we have found considerable difficulty in obtaining smooth correction curves at the high dispersions used in this work. Most of the International Secondary Standards of iron are lines of great intensity, separated in the spectrum more widely than is desirable, and entirely absent in certain regions of the spectrum. At wavelengths shorter than 2500 A and at those longer than 7000 A we have used as standards the interferometer measurements on iron lines made by Meggers and Humphreys, and at short wavelengths, the copper standards of Shenstone.

Large systematic errors can be introduced by incorrect registration of standard lines on the spectrograms. Strictly speaking, standards of wavelength should be photographed at the same time and from the same source as the lines to be measured, if displacement between standards and other lines is to be avoided. This very desirable practice is sometimes not possible, however, either because the standards must be emitted by a special source which is not suitable for producing the desired lines to be measured, or because the lines to be measured would be obscured by lines of the standard, which must be present in profusion if they are to be useful with spectrographs of high dispersion.

When successive exposures to a spectrum to be measured and to a standard spectrum are made, though the plate remain undisturbed between the two exposures and the two sources are apparently similarly imaged on the grating, displacements as great as 0.02 A are sometimes found between the two spectra. The reasons for these shifts are not entirely understood, but they possibly arise from dissimilar illumination of the grating face, causing differences in line shape (especially in gratings with extreme target pattern) — or from temperature changes of the grating.

During the present work we have tried to eliminate such shifts by the following means: using measurements from at least three plates taken independently; taking standard exposures both before and after the exposure to be measured, half before and half after, so that any large intermediate shift would produce a noticeable broadening of the standard lines; using, when possible, a small amount of iron in the electrodes producing the lines to be measured, to give a few standard lines which could be used to determine any displacement between the actual standard lines and the spectrum under study.

Tests of our methods and equipment indicate that, where other conditions permit, measurements consistent to within 0.002 A are not difficult to make with large 10-meter gratings and automatic comparators. There are certain lines, however, which one is hardly justified in measuring with equipment which is designed primarily for making measurements on a " wholesale " basis; such lines are those whose wavelengths shift markedly with conditions, those of comparatively great breadth or dissymmetry, and those contained in spectra covering an extremely wide range in intensity. Elements in the first three columns and the first two rows of the Periodic Table are likely to emit lines of the last category. Fortunately these are just the elements whose spectra have had most attention in the past; and while we have gone through a routine examination of their lines, we have been content, in most cases, to record the wavelength values obtained from the literature instead of our own. This is not to say that such spectra do not still merit careful study, but it is of the sort which is best given by an experienced spectroscopist to one element at a time. Our attention has been concentrated more on the complex atoms which are responsible for three-fourths of the lines given in the tables.

It would have been desirable if, before the present catalog was published, all wavelengths in it could have been tested by means of the combination principle; but this would have delayed publication of the volume and necessitated much further work in classification of spectra. Though the catalog has been checked many times for mistakes of various sorts, many errors of omission and commission doubtless remain. To any authors whose measurements have been misquoted or incorrectly assigned we offer sincere apologies, and doubly so to any whose values may have been omitted unknowingly in favor of less correct values of our own. We have tried to follow the guiding principle that a much-needed set of tables could be compiled from measurements already available in the literature, and that while we were justified in trying to improve this by making measurements with improved apparatus, we should at all costs avoid any procedure which might make it worse. We present the catalog, not as a finished product, but as a preliminary compilation designed to improve the status of wavelength measurements and line assignments.

ACKNOWLEDGMENTS

To Dr. William F. Meggers, Chief of the Spectroscopy Section of the National Bureau of Standards and President of the Commission on Standards of Wavelength of the International Astronomical Union, we are greatly indebted for advice and encourage-

ACKNOWLEDGMENTS

ment throughout the preparation of these tables. Dr. Meggers and his colleague, Dr. C. C. Kiess, have not only supplied manuscript data which otherwise would not have been obtainable but have obligingly gone over the galley proofs and made many constructive suggestions.

Professor Walter E. Albertson has kindly furnished unpublished data on the spectrum of iridium, and Professor Dorothy W. Weeks has given us invaluable help with the iron spectrum.

We have been encouraged to complete the tables by several conversations regarding them with Geheimrat H. Kayser, whose encyclopedic works on spectroscopy are standard, and with Präsident F. Paschen, whose spectroscopic measurements and tables are world-renowned. It is a pleasure to acknowledge the benefit we have obtained from the kindly counsel of men of such vast experience.

The excellent appearance of the tables originates in large measure from the zealous co-operation of J. R. Killian, Jr., of the Technology Press.

The wavelength project, from the results of which the measurements included herein have been selected, has been under way for about three years, and several successive groups of graduate students, whose names are included in the lists given below, have been connected with it. The following persons have been responsible for the measurements and for the present compilation:

For the staff organization of the project and supervision of its detailed operation, Colonel Robert C. Eddy of the Massachusetts Institute of Technology.

For assisting the undersigned in the final checking and editing of the material included in the tables: Dr. Nathan Rosen and Dr. George O. Langstroth.

As supervisors of laboratory assistants for periods of three months or longer: Dr. Peter A. Cole, Dr. Richard E. Evans, Dwight P. Merrill, Julius P. Molnar, Norman J. Oliver, Dr. Fred W. Paul, Henry Rich.

As supervisors of clerical assistants for periods of three months or longer: John B. D'Albora, Dr. Harriet W. Allen, William W. Bartlett, Mildred H. Brode, Harold E. Clearman, Jr., Leonard J. Julian, Dr. Joseph Morgan, Simeon I. Rosenthal, Helen Wigglesworth.

In charge of the checking of galley and page proofs: Clinton H. Collester.

The great burden of numerical tabulation and checking has rested on a very faithful group of selected W.P.A. workers. Though they cannot be mentioned individually because of their numbers, the project would have been impossible of completion without their conscientious work.

Pure chemicals for the production of spectra were purchased with several grants from the Rumford Fund of the American Academy of Arts and Sciences, which aid is hereby gratefully acknowledged.

A large portion of the personnel for the operation of the project and a considerable amount of materials were furnished by the Works Progress Administration for Massachusetts, under Official Projects Nos. 165–14–6999–0, 465–14–3–149, and 665–14–3–54.

Scientific staff, housing, apparatus, and financial support for obtaining data and for underwriting publication of the tables were furnished by the Massachusetts Institute of Technology, to whose administrative officers our best thanks are due.

George R. Harrison

Cambridge, Massachusetts
June 10, 1939

TABLE III

SENSITIVE LINES OF THE ELEMENTS*

Wave-length	Excitation Potential†	Intensities Arc	Spk., [Dis.]	Sensi-tivity‡
A 18 Argon				
8115.311	13.0	–	[5000]	U2
7503.867	13.4	–	[700]	U4
7067.217	13.2	–	[400]	U3
6965.430	13.3	–	[400]	U3
Ag 47 Silver				
5465.487	6.0	1000 R	500 R	U4
5209.067	6.0	1500 R	1000 R	U3
3382.891	3.6	1000 R	700 R	U2
3280.683	3.8	2000 R	1000 R	U1
2437.791	17.4	60	500 wh	V2
2246.412	17.8	25	300 hs	V3
Al 13 Aluminum				
6243.36	21.0	–	100	V3
6231.76	21.0	–	30	–
3961.527	3.1	3000	2000	U1
3944.032	3.1	2000	1000	U2
3092.713	4.0	1000	1000	U3
3082.155	4.0	800	800	U4
2816.179	17.7	10	100	V2
2669.166	10.6	3	100	V1
2631.553	21.2	–	40	–
As 33 Arsenic				
2898.71	6.7	25 r	40	–
2860.452	6.6	50 r	50	–
2780.197	6.7	75 R	75	U5
2456.53	6.5	100 r	8	U4
2370.77	6.7	50 r	3	–
2369.67	6.7	40 r	–	–
2349.84	6.6	250 R	18	U3
2288.12	6.7	250 R	5	U3
Au 79 Gold				
2802.19	>13.6	–	200	–
2675.95	4.6	250 R	100	U2
2427.95	5.1	400 R	100	U1
B 5 Boron				
3451.41	20.9	5	30	V2
2497.733	4.9	500	400	U1
2496.778	4.9	300	300	U2
Ba 56 Barium				
5777.665	3.8	500 R	100 R	U2
5535.551	2.2	1000 R	200 R	U1
5519.115	3.8	200 R	60 R	U3
5424.616	3.8	100 R	30 R	U4
4934.086	7.7	400 h	400 h	V2
4554.042	7.9	1000 R	200	V1
4130.664	10.9	50 r	60 wh	V3
3891.785	10.9	18	25	V4
3071.591	4.0	100 R	50 R	U5
2335.269	11.2	60 R	100 R	–
2304.235	11.2	60 R	80 R	–
Be 4 Beryllium				
3321.343	6.4	1000 r	30	U2
3321.086	6.4	100	–	U3
3321.013	6.4	50	–	U4
3131.072	13.2	200	150	V2
3130.416	13.2	200	200	V1
2650.781	7.4	25	–	U5
2348.610	5.4	2000 R	50	U1
Bi 83 Bismuth				
4722.552	4.0	1000	100	–
3067.716	4.0	3000 hR	2000 wh	U1
2989.029	5.5	250 wh	100 wh	–
2938.298	6.1	300 w	300 w	–
2897.975	5.6	500 WR	500 WR	U2
2809.625	6.3	200 w	100	–
2780.521	5.8	200 w	100	–
2276.578	5.4	100 R	40	–
2061.70	6.0	300 R	100	–
Br 35 Bromine				
4816.71	14.4	–	[300]	V3
4785.50	14.4	–	[400]	V2
4704.86	14.4	–	[250]	V1
C 6 Carbon				
4267.27	32.1	–	500	V2
4267.02	32.1	–	350	V3
2837.602	27.5	–	40	V5
2836.710	27.5	–	200	V4
2478.573	7.7	400	[400]	U2
2296.89	53.5	–	200	–
Ca 20 Calcium				
4454.781	4.7	200	–	U2
4434.960	4.7	150	–	U3
4425.441	4.7	100	–	U4
4226.728	2.9	500 R	50 W	U1
3968.468	9.2	500 R	500 R	V2
3933.666	9.2	600 R	600 R	V1
3179.332	13.1	100	400 w	V3
3158.869	13.1	100	300 w	V4
Cb 41 Columbium				
4137.095	3.0	100	60	U5
4123.810	3.0	200	125	U4
4100.923	3.1	300 w	200 w	U3
4079.729	3.1	500 w	200 w	U2
4058.938	3.2	1000 w	400 w	U1
3225.479	>7.6	150 w	800 wr	–
3194.977	>7.7	30	300	–
3163.402	>7.8	15	8	–
3130.786	>7.9	100	100	–
3094.183	>8.0	100	1000	V1
Cd 48 Cadmium				
6438.4696	7.3	2000	1000	–
3610.510	7.3	1000	500	–
3466.201	7.3	1000	500	–
3403.653	7.3	800	500 h	–
3261.057	3.8	300	300	–
2748.58	19.2	5	200	–
2573.09	19.2	3	150	–
2312.84	20.1	1	200	–
2288.018	5.4	1500 R	300 R	U1
2265.017	14.4	25 d	300	V2
2144.382	14.7	50	200 R	V1
Ce 58 Cerium				
4186.599	>8.6	80	25	–
4165.606	>8.6	40	6	–
4040.762	>8.7	70	5	–
4012.388	>8.7	60	20	–
Cl 17 Chlorine				
4819.46	28.3	–	[200]	V4
4810.06	28.3	–	[200]	V3
4794.54	28.3	–	[250]	V2
Co 27 Cobalt				
3529.813	4.0	1000 R	30	U3
3465.800	3.6	2000 R	25	U2
3453.505	4.0	3000 R	200	U1
3405.120	4.0	2000 R	150	–
2519.822	14.7	40	200	–
2388.918	14.1	10	35	–
2378.622	14.1	25	50 w	–
2363.787	14.2	25	50	–
2307.857	14.3	25	50 w	–
2286.156	14.3	40	300 l	V1
Cr 24 Chromium				
5208.436	3.3	500 R	100	U4
5206.039	3.3	500 R	200	U5
5204.518	3.3	400 R	100	U6
4289.721	2.9	3000 R	800 r	U3
4274.803	2.9	4000 R	800 r	U2
4254.346	2.9	5000 R	1000	U1
2860.934	12.5	60	100	V5
2855.676	12.5	60	200 wh	V4
2849.838	12.6	80	150 r	V3
2843.252	12.6	125	400 r	V2
2835.633	12.6	100	400 r	V1
Cs 55 Caesium				
8943.50	1.4	2000 R	–	U2
8521.10	1.4	5000 R	–	U1
4593.177	2.7	1000 R	50	U4
4555.355	2.7	2000 R	100	U3
Cu 29 Copper				
5218.202	6.2	700	–	U3
5153.235	6.2	600	–	U4
5105.541	3.8	500	–	U5
3273.962	3.8	3000 R	1500 R	U2
3247.540	3.8	5000 R	2000 R	U1
2824.37	–	1000	300	–
2246.995	15.9	30	500	V3
2192.260	16.2	25	500 h	V2
2135.976	16.2	25	500 w	V1
Dy 66 Dysprosium				
4211.719	>2.9	200	15	–
4167.966	>3.0	50	12	–
4077.974	>3.0	150 r	100	–
4045.983	>3.0	150	12	–
4000.454	>3.1	400	300	–
Er 68 Erbium				
3906.316	>3.2	25	12	–
3692.652	>3.4	20	12	–
3499.104	>3.5	18	15	–
Eu 63 Europium				
4661.87	–	300 R	120	550
4627.22	–	400 R	150	650
4594.02	–	500 R	200	750
4205.046	8.6	200 R	50	–
4129.737	8.6	150 R	50 R	–
F 9 Fluorine				
6902.46	14.5	–	[500]	U3
6856.02	14.4	–	[1000]	U2
5291.0	bh Ca F	200	–	–
Fe 26 Iron				
3748.264	3.4	500	200	U4
3745.903	3.4	150	100	U5
3745.564	3.4	500	500	U3
3737.133	3.4	1000 r	600	U2
3719.935	3.3	1000 R	700	U1
3581.20	–	1000 R	600 R	600
2413.309	13.1	60	100 h	V5
2410.517	13.1	50	70 h	V4
2404.882	13.0	50	100 wh	V3
2395.625	13.0	50	100 wh	V2
2382.039	13.0	40 r	100 R	V1
Ga 31 Gallium				
4172.056	3.1	2000 R	1000 R	U1
4032.982	3.1	1000 R	500 R	U2
2943.637	4.3	10	20 r	U3
2874.244	4.3	10	15 r	U4

* Compiled from a combination of empirical and theoretical data selected from the literature.

† For an ion, the ionization potential of the neutral atom has been included in the excitation potential to give an approximate idea of the excitation required to produce the line.

‡ For the neutral atom, the most sensitive line (raie ultime) is indicated by U1, and other lines by U2, U3, etc., in order of decreasing sensitivity. For the singly ionized atom, the corresponding designations are V1, V2, etc. In cases where U1 is not given, the most sensitive lines lie outside the spectral range 10,000–2000 A. Figures given in the Sensitivity Column are taken from the NBS Tables of Spectral-Line Intensities, where values for numerous other lines also can be found. (See Meggers, Corliss and Scribner, NBS Monograph 32, 1961, Government Printing Office; also C. H. Corliss, NBS Monograph 32 Supplement, 1967).

SENSITIVE LINES OF THE ELEMENTS

Gd 64 Gadolinium

Wavelength	Excitation Potential†	Intensities Arc	Intensities Spk., [Dis.]	Sensitivity‡
3768.405	>3.3	20	20	–
3646.196	>3.4	200 w	150	–

Ge 32 Germanium

Wavelength	Excitation Potential†	Intensities Arc	Intensities Spk., [Dis.]	Sensitivity‡
4226.570	4.9	200	50	–
3269.494	4.7	300	300	U3
3039.064	4.9	1000	1000	U2
2709.626	4.6	30	20	–
2651.575	4.7	30	20	–
2651.178	4.8	40	20	–

H 1 Hydrogen

Wavelength	Excitation Potential†	Intensities Arc	Intensities Spk., [Dis.]	Sensitivity‡
6562.79	12.0	–	[3000]	U2
4861.327	12.7	–	[500]	U3

He 2 Helium

Wavelength	Excitation Potential†	Intensities Arc	Intensities Spk., [Dis.]	Sensitivity‡
5875.618	23.0	–	[1000]	U3
4685.75	75.3	–	[300]	–
3888.646	22.9	–	[1000]	U2

Hf 72 Hafnium

Wavelength	Excitation Potential†	Intensities Arc	Intensities Spk., [Dis.]	Sensitivity‡
7240.87	–	5000	600	50
7237.10	–	8000	1000	80
7131.82	–	7000	1000	70
7063.82	–	3000	400	20
6818.94	–	2000	300	15
6789.28	–	1000	100	8
4093.161	>7.8	25	20	–
3134.718	>8.7	80	125	–
3072.877	4.0	80	18	–
2940.772	4.2	60	12	–
2916.481	4.2	50	15	–
2904.408	4.8	30	6	–
2898.259	4.6	50	12	–
2820.224	>9.2	40	100	–
2773.357	>9.3	25	60	–
2641.406	>9.5	40	125	–
2516.881	>9.7	35	100	–
2513.028	>9.7	25	70	–

Hg 80 Mercury

Wavelength	Excitation Potential†	Intensities Arc	Intensities Spk., [Dis.]	Sensitivity‡
5460.753	7.7	2000	–	320
4358.35	7.7	3000 w	500	–
4046.561	7.7	200	300	–
3663.276	8.8	500	400	U5
3654.833	8.8	–	[200]	U4
3650.146	8.8	200	500	U3
2536.519	4.9	2000 R	1000 R	U2

Ho 67 Holmium

Wavelength	Excitation Potential†	Intensities Arc	Intensities Spk., [Dis.]	Sensitivity‡
3891.02	>3.2	200	40	–
3748.17	>3.3	60	40	–
2936.77	>7.4	–	1000 R	–

I 53 Iodine

Wavelength	Excitation Potential†	Intensities Arc	Intensities Spk., [Dis.]	Sensitivity‡
5464.61	22.7	–	[900]	–
5161.188	22.8	–	[300]	–
2062.38	>16.4	–	[900]	–

In 49 Indium

Wavelength	Excitation Potential†	Intensities Arc	Intensities Spk., [Dis.]	Sensitivity‡
4511.323	3.0	5000 R	4000 R	U1
4101.773	3.0	2000 R	1000 R	U2
3258.564	4.1	500 R	300 R	U5
3256.090	4.1	1500 R	600 R	U3
3039.356	4.1	1000 R	500 R	U4

Ir 77 Iridium

Wavelength	Excitation Potential†	Intensities Arc	Intensities Spk., [Dis.]	Sensitivity‡
3513.645	3.5	100 h	100	U2
3437.015	4.4	20	15	–
3220.780	4.2	100	30	U1
2924.792	4.2	25 wh	15	–
2849.725	4.3	40 h	20 h	–
2661.983	–	600 R	60	480

K 19 Potassium

Wavelength	Excitation Potential†	Intensities Arc	Intensities Spk., [Dis.]	Sensitivity‡
7698.979	1.6	5000 R	–	U2
7664.907	1.6	9000 R	–	U1
4047.201	3.0	400	200	U4
4044.140	3.1	800	400	U3

Kr 36 Krypton

Wavelength	Excitation Potential†	Intensities Arc	Intensities Spk., [Dis.]	Sensitivity‡
5870.9158	12.1	–	[3000]	U2
5570.2895	12.1	–	[2000]	U3

La 57 Lanthanum

Wavelength	Excitation Potential†	Intensities Arc	Intensities Spk., [Dis.]	Sensitivity‡
6249.929	2.5	300	–	U1
5930.648	2.2	250	–	U2
5455.146	2.4	200	1	U3
4123.228	8.9	500	500	V4
4077.340	8.9	600	400	V3
3949.106	9.1	1000	800	V2

Li 3 Lithium

Wavelength	Excitation Potential†	Intensities Arc	Intensities Spk., [Dis.]	Sensitivity‡
6707.844	1.8	3000 R	200	U1
6103.642	3.9	2000 R	300	U3
4603.00	4.5	800	–	U4
3232.61	3.8	1000 R	500	U2

Lu 71 Lutecium

Wavelength	Excitation Potential†	Intensities Arc	Intensities Spk., [Dis.]	Sensitivity‡
4518.57	>2.7	300	40	–
3554.43	>6.2	50	150	–
3472.48	>6.3	50	150	–
3397.07	>6.3	50	20 r	–
2911.39	>6.9	100	300	–
2894.84	>7.0	60	200	–

Mg 12 Magnesium

Wavelength	Excitation Potential†	Intensities Arc	Intensities Spk., [Dis.]	Sensitivity‡
5183.618	5.1	500 wh	300	–
5172.699	5.1	200 wh	100 wh	–
5167.343	5.1	100 wh	50	–
3838.258	5.9	300	200	U2
3832.306	5.9	250	200	U3
3829.350	5.9	100 w	150	U4
2852.129	4.3	300 R	100 R	U1
2802.695	12.0	150	300	V2
2795.53	12.0	150	300	V1

Mn 25 Manganese

Wavelength	Excitation Potential†	Intensities Arc	Intensities Spk., [Dis.]	Sensitivity‡
4034.490	3.1	250 r	20	U3
4033.073	3.1	400 r	20	U2
4030.755	3.1	500 r	20	U1
2801.064	–	600 R	60	480
2798.271	–	800 R	80	650
2605.688	12.2	100 R	500 R	V3
2593.729	12.2	200 R	1000 R	V2
2576.104	12.2	300 R	2000 R	V1

Mo 42 Molybdenum

Wavelength	Excitation Potential†	Intensities Arc	Intensities Spk., [Dis.]	Sensitivity‡
3902.963	3.2	1000 R	500 R	U3
3864.110	3.2	1000 R	500 R	U2
3798.252	3.3	1000 R	1000 R	U1
2909.116	11.6	25	40 h	V5
2890.994	11.7	30	50 h	V4
2871.508	11.7	100	100 h	V3
2848.232	11.8	125	200 h	V2
2816.154	11.9	200	300 h	V1

N 7 Nitrogen

Wavelength	Excitation Potential†	Intensities Arc	Intensities Spk., [Dis.]	Sensitivity‡
5679.56	35.1	–	[500]	V2
5676.02	35.0	–	[100]	V4
5666.64	35.0	–	[300]	V3
4109.98	13.7	–	[1000]	U2
4103.37	74.3	–	[80]	–
4099.94	13.7	–	[150]	U3
4097.31	74.3	–	[100]	–

Na 11 Sodium

Wavelength	Excitation Potential†	Intensities Arc	Intensities Spk., [Dis.]	Sensitivity‡
5895.923	2.1	5000 R	500 R	U2
5889.953	2.1	9000 R	1000 R	U1
5688.224	4.3	300	–	–
5682.657	4.3	80	–	–
3302.988	3.7	300 R	150 R	U4
3302.323	3.7	600 R	300 R	U3

Nd 60 Neodymium

Wavelength	Excitation Potential†	Intensities Arc	Intensities Spk., [Dis.]	Sensitivity‡
4303.573	>2.9	100	40	–
4177.321	>3.0	15	25	–
3951.154	>3.1	40	30	–

Ne 10 Neon

Wavelength	Excitation Potential†	Intensities Arc	Intensities Spk., [Dis.]	Sensitivity‡
6402.246	18.5	–	[2000]	–
5852.488	18.9	–	[2000]	–
5400.562	18.9	–	[2000]	–

Ni 28 Nickel

Wavelength	Excitation Potential†	Intensities Arc	Intensities Spk., [Dis.]	Sensitivity‡
3524.541	3.5	1000 R	100 wh	–
3515.054	3.6	1000 R	50 h	–
3492.956	3.6	1000 R	100 h	U2
3414.765	3.6	1000 R	50 wh	U1
3050.819	–	1000 R	–	280
2287.084	14.8	100	500	V1
2270.213	14.2	100	400	V2
2264.457	14.3	150	400	V3
2253.86	14.4	100	300	V4

O 8 Oxygen

Wavelength	Excitation Potential†	Intensities Arc	Intensities Spk., [Dis.]	Sensitivity‡
7775.433	10.7	–	[100]	U4
7774.138	10.7	–	[300]	U3
7771.928	10.7	–	[1000]	U2

Os 76 Osmium

Wavelength	Excitation Potential†	Intensities Arc	Intensities Spk., [Dis.]	Sensitivity‡
4420.468	2.8	400 R	100	–
3267.945	3.8	400 R	30	–
3262.290	4.3	500 R	50	–
3058.66	4.0	500 R	500	–
2909.061	4.2	500 R	400	U1

P 15 Phosphorus

Wavelength	Excitation Potential†	Intensities Arc	Intensities Spk., [Dis.]	Sensitivity‡
2554.93	7.1	60	[20]	–
2553.28	7.1	80	[20]	U3
2535.65	7.2	100	[30]	U2
2534.01	7.2	50	[20]	–

Pb 82 Lead

Wavelength	Excitation Potential†	Intensities Arc	Intensities Spk., [Dis.]	Sensitivity‡
5608.8	16.9	–	[40]	V2
4057.820	4.4	2000 R	300 R	U1
3683.471	4.3	300	50	U2
3639.580	4.4	300	50 h	–
2833.069	4.4	500 R	80 R	–
2614.178	>4.7	200 r	80	–
2203.505	14.7	50 W	5000 R	V1
2169.994	5.7	1000 R	1000 R	–

Pd 46 Palladium

Wavelength	Excitation Potential†	Intensities Arc	Intensities Spk., [Dis.]	Sensitivity‡
3634.695	4.2	2000 R	1000 R	U3
3609.548	4.4	1000 R	700 R	–
3516.943	4.5	1000 R	500 R	–
3421.24	4.6	2000 R	1000 R	U2
3404.580	4.4	2000 R	1000 R	U1
2854.581	16.6	4	500 h	–
2658.722	16.9	20	300	–
2505.739	17.5	3	30	–
2498.784	17.2	4	150 h	–
2488.921	16.3	10	30	–

Pr 59 Praseodymium

Wavelength	Excitation Potential†	Intensities Arc	Intensities Spk., [Dis.]	Sensitivity‡
4225.327	>2.9	50	40	–
4189.518	>2.9	100	50	–
4179.422	>3.0	200	40	–
4062.817	>3.0	150	50	–

Pt 78 Platinum

Wavelength	Excitation Potential†	Intensities Arc	Intensities Spk., [Dis.]	Sensitivity‡
3064.712	4.0	2000 R	300 R	U1
2997.967	4.2	1000 R	200 r	–
2929.794	4.2	800 R	200 w	–
2830.295	4.4	1000 R	600 r	–
2659.454	4.6	2000 R	500 R	U2

Ra 88 Radium

Wavelength	Excitation Potential†	Intensities Arc	Intensities Spk., [Dis.]	Sensitivity‡
4825.91	2.6	–	[800]	U1
4682.28	7.8	–	[800]	V2
3814.42	8.4	–	[2000]	V1

Rb 37 Rubidium

Wavelength	Excitation Potential†	Intensities Arc	Intensities Spk., [Dis.]	Sensitivity‡
7947.60	1.6	5000 R	–	U2
7800.227	1.6	9000 R	–	U1
4215.556	2.9	1000 R	300	U4
4201.851	2.9	2000 R	500	U3

† For an ion, the ionization potential of the neutral atom has been included in the excitation potential to give an approximate idea of the excitation required to produce the line.

‡ For the neutral atom, the most sensitive line (raie ultime) is indicated by U1, and other lines by U2, U3, etc., in order of decreasing sensitivity. For the singly ionized atom, the corresponding designations are V1, V2, etc. In cases where U1 is not given, the most sensitive lines lie outside the spectral range 10,000–2000 A. Figures given in the Sensitivity Column are taken from the NBS Tables of Spectral-Line Intensities, where values for numerous other lines also can be found. (See Meggers, Corliss and Scribner, NBS Monograph 32, 1961; also C. H. Corliss, NBS Monograph 32 Supplement, 1967).

WAVELENGTH TABLES

Wave-length	Excitation Potential†	Intensities Arc	Intensities Spk., [Dis.]	Sensitivity‡
Re 75 Rhenium				
4889.17	2.5	2000 w	–	U2
3460.47	3.6	1000 W	–	U1
Rh 45 Rhodium				
3692.357	3.3	500 hd	150 wd	–
3657.987	3.6	500 W	200 W	–
3434.893	3.6	1000 r	200 r	U1
3396.85	3.6	1000 w	500	–
3323.092	3.9	1000	200	–
Rn 86 Radon				
7450.00	8.5	–	[600]	U2
7055.42	8.4	–	[400]	U3
Ru 44 Ruthenium				
3596.179	3.7	30	100	U3
3498.942	3.5	500 R	200	U1
3436.737	3.7	300 R	150	U2
2976.586	>10.5	60	200	–
2965.546	>10.6	60	200	–
2945.668	>10.6	60	300	–
2712.410	>11.0	80	300	–
2692.065	>11.0	8	200	–
2678.758	>11.0	100	300	–
S 16 Sulphur				
9237.49	7.8	–	[200]	U6
9228.11	7.8	–	[200]	U5
9212.91	7.8	–	[200]	U4
4696.25	9.1	–	[15]	U9
4695.45	9.1	–	[30]	U8
4694.13	9.1	–	[500]	U7
Sb 51 Antimony				
3267.502	5.8	150	150 Wh	–
3232.499	6.1	150	250 wh	–
2877.915	5.3	250 W	150	–
2598.062	5.8	200	100	–
2528.535	6.1	300 R	200	–
2311.469	5.3	150 R	50	–
2175.890	5.7	300	40	U2
2068.38	6.0	300 R	3	U1
Sc 21 Scandium				
4023.688	3.1	100	25	U3
4020.399	3.1	50	20	U4
3911.810	3.2	150	30	U1
3907.476	3.2	125	25	U2
3642.785	10.0	60	50	V3
3630.740	10.1	50	70	V2
3613.836	10.1	40	70	V1
Se 34 Selenium				
4742.25	>2.6	–	[500]	U6
4739.03	>2.6	–	[800]	U5
4730.78	>2.6	–	[1000]	U4
2062.788	6.3	–	[800]	U3
2039.851	6.3	–	[1000]	U2
Si 14 Silicon				
3905.528	5.1	20	15 W	–
2881.578	5.1	500	400	U1
2528.516	4.9	400	500	U2
2516.123	4.9	500	500	U3
2506.899	4.9	300	200	U4
Sm 62 Samarium				
4434.321	8.8	200	200	V2
4424.342	8.9	300	300	V1
4390.865	8.6	150	150	–
Sn 50 Tin				
4524.741	4.8	500 wh	50	–
3262.328	4.8	400 h	300 h	U3
3175.019	4.3	500 h	400 hr	–
3034.121	4.3	200 wh	150 wh	–
3009.147	4.3	300 h	200 h	–
2863.327	4.3	300 R	300 R	U2
2839.989	4.8	300 R	300 R	U1
Sr 38 Strontium				
4962.263	4.3	40	–	U4
4872.493	4.3	25	–	U3
4832.075	4.3	200	8	U2
4607.331	2.7	1000 R	50 R	U1
4305.447	11.6	40	–	–
4215.524	8.6	300 r	400 W	V2
4077.714	8.7	400 r	500 W	V1
3474.887	12.2	80	50	–
3464.57	12.2	200	200	–
3380.711	12.2	150	200	–
Ta 73 Tantalum				
3406.664	>3.6	70 w	18 s	–
3318.840	>3.7	125	35	–
3311.162	>3.7	300 w	70 w	U1
Tb 65 Terbium				
3874.18	>3.2	200	200	–
3848.75	>3.2	100	200	–
3561.74	>3.5	200	200	–
3509.17	>3.5	200	200	–
Te 52 Tellurium				
2769.67	5.8	–	[30]	–
2530.70	5.5	–	[30]	–
2385.76	5.8	600	[300]	U2
2383.25	5.8	500	[300]	U3
2142.75	5.8	600	–	55
Th 90 Thorium				
4019.137	>3.1	8	8	–
3601.040	>3.4	8	10	–
3538.75	>3.5	–	50	–
3290.59	>7.3	–	40 h	–
Ti 22 Titanium				
5007.213	3.3	200	40	–
4999.510	3.3	200	80	–
4991.066	3.3	200	100	–
4981.733	3.3	300	125	U1
3653.496	3.4	500	200	U2
3642.675	3.4	300	125	–
3635.463	3.4	200	100	–
3383.761	10.4	70	300 R	–
3372.800	10.5	80	400 R	V3
3361.213	10.5	100	600 R	V2
3349.035	11.1	125	800 R	V1
Tl 81 Thallium				
5350.46	3.3	5000 R	2000 R	U1
3775.72	3.3	3000 R	1000 R	U2
3519.24	4.5	2000 R	1000 R	U3
3229.75	4.8	2000	800	–
2918.32	5.2	400 R	200 R	–
2767.87	4.5	400 R	300 R	–
Tm 69 Thulium				
3761.917	>3.3	200	120	–
3761.333	>3.3	250	150	–
3462.21	>3.6	200	100	–
U 92 Uranium				
4241.669	>2.9	40	50	–
3672.579	>3.4	8	15	–
3552.172	>3.5	8	12	–
V 23 Vanadium				
4389.974	3.1	80 R	60 R	–
4384.722	3.1	125 R	125 R	–
4379.238	3.1	200 R	200 R	U1
3185.396	3.9	500 R	400 R	U2
3183.982	3.9	500 R	400 R	–
3183.406	3.9	200 R	100 R	–
3125.284	11.0	80	200 R	–
3118.383	11.1	70	200 R	V4
3110.706	11.1	70	300 R	V3
3102.299	11.1	70	300 R	V2
3093.108	11.2	100 R	400 R	V1
W 74 Tungsten				
4302.108	3.2	60	60	U1
4294.614	3.2	50	50	U2
4008.753	3.4	45	45	U3
3613.790	>9.2	10	30	–
3215.560	5.3	10	9	–
2589.167	>10.6	15 d	25	–
2397.091	>10.9	18	30	–
Xe 54 Xenon				
4671.226	10.9	–	[2000]	U2
4624.276	10.9	–	[1000]	U3
4500.977	11.0	–	[500]	U4
Yb 70 Ytterbium				
3987.994	>3.1	1000 R	500 R	1900
3694.203	>3.3	500 R	1000 R	3200
3289.37	>3.8	500 R	1000 R	2600
Yt 39 Yttrium				
4674.848	2.7	80	100	U1
4643.695	2.7	50	100	U2
3788.697	9.9	30	30	–
3774.332	9.9	12	100	–
3710.290	10.0	80	150	V1
3633.123	9.9	50	100	–
3600.734	10.1	100	300	–
3242.280	10.5	60	100	–
Zn 30 Zinc				
6362.347	7.7	1000 Wh	500	–
4810.534	6.6	400 w	300 h	–
4722.159	6.6	400 w	300 h	–
4680.138	6.6	300 w	200 h	–
3345.020	7.8	800	300	U2
3302.588	7.8	800	300	U3
3282.333	7.8	500 R	300	U4
2557.958	15.3	10	300	V3
2502.001	15.3	20	400 w	V4
2138.56	5.8	800 R	500	U1
2061.91	15.4	100	100	V2
2025.51	15.5	200	200	V1
Zr 40 Zirconium				
4772.312	3.2	100	–	–
4739.478	3.2	100	–	–
4710.075	3.3	60	–	–
4687.803	3.4	125	–	U4
3601.193	3.6	400	15	U1
3572.473	10.4	60	80	V4
3547.682	3.5	200	12	U2
3519.605	3.5	100	10	U3
3496.210	10.5	100	100	V3
3438.230	10.6	250	200	V2
3391.975	10.7	300	400	V1

† For an ion, the ionization potential of the neutral atom has been included in the excitation potential to give an approximate idea of the excitation required to produce the line.

‡ For the neutral atom, the most sensitive line (raie ultime) is indicated by U1, and other lines by U2, U3, etc., in order of decreasing sensitivity. For the singly ionized atom, the corresponding designations are V1, V2, etc. In cases where U1 is not given, the most sensitive lines lie outside the spectral range 10,000–2000 A. Figures given in the Sensitivity Column are taken from the NBS Tables of Spectral-Line Intensities, where values for numerous other lines also can be found. (See Meggers, Corliss and Scribner, NBS Monograph 32, 1961; also C. H. Corliss, NBS Monograph 32 Supplement, 1967).

TABLE IV

SENSITIVE LINES OF THE ELEMENTS *

[arranged in order of wavelength]

Wave-length	Element	Intensities Arc	Spk., [Dis.]	Sensi-tivity‡
9237.49	S I	–	[200]	U6
9228.11	S I	–	[200]	U5
9212.91	S I	–	[200]	U4
8943.50	Cs I	2000 R	–	U2
8521.10	Cs I	5000 R	–	U1
8115.311	A I	–	[5000]	U2
7947.60	Rb I	5000 R	–	U2
7800.227	Rb I	9000 R	–	U1
7775.433	O I	–	[100]	U4
7774.138	O I	–	[300]	U3
7771.928	O I	–	[1000]	U2
7698.979	K I	5000 R	–	U2
7664.907	K I	9000 R	–	U1
7503.867	A I	–	[700]	U4
7450.00	Rn I	–	[600]	U2
7240.87	Hf I	5000	600	50
7237.10	Hf I	8000	1000	80
7131.82	Hf I	7000	1000	40
7067.217	A I	–	[400]	U3
7063.82	Hf I	3000	400	20
7055.42	Rn I	–	[400]	U3
6965.430	A I	–	[400]	U3
6902.46	F I	–	[500]	U3
6856.02	F I	–	[1000]	U2
6818.94	Hf I	2000	300	15
6789.28	Hf I	1000	100	5
6707.844	Li I	3000 R	200	U1
6562.79	H I	–	[3000]	U2
6438.4696	Cd I	2000	1000	–
6402.246	Ne I	–	[2000]	–
6362.347	Zn I	1000 Wh	500	–
6249.929	La I	300	–	U1
6243.36	Al II	–	100	V3
6231.76	Al II	–	30	–
6103.642	Li I	2000 R	300	U3
5930.648	La I	250	–	U2
5895.923	Na I	5000 R	500 R	U2
5889.953	Na I	9000 R	1000 R	U1
5875.618	He I	–	[1000]	U3
5870.9158	Kr I	–	[3000]	U2
5852.488	Ne I	–	[2000]	–
5777.665	Ba I	500 R	100 R	U2
5688.224	Na I	300	–	–
5682.657	Na I	80	–	–
5679.56	N II	–	[500]	V2
5676.02	N II	–	[100]	V4
5666.64	N II	–	[300]	V3
5608.8	Pb II	–	[40]	V2
5570.2895	Kr I	–	[2000]	U3
5535.551	Ba I	1000 R	200 R	U1
5519.115	Ba I	200 R	60 R	U3
5465.487	Ag I	1000 R	500 R	U4
5464.61	I II	–	[900]	–
5460.753	Hg I	2000	–	320
5455.146	La I	200	1	U3
5424.616	Ba I	100 R	30 R	U4
5400.562	Ne I	–	[2000]	–
5350.46	Tl I	5000 R	2000 R	U1
5291.0	bh CaF	200	–	–
5218.202	Cu I	700	–	U3
5209.067	Ag I	1500 R	1000 R	U3
5208.436	Cr I	500 R	100	U4
5206.039	Cr I	500 R	200	U5
5204.518	Cr I	400 R	100	U6
5183.618	Mg I	500 wh	300	–
5172.699	Mg I	200 wh	100 wh	–
5167.343	Mg I	100 wh	50	–
5161.188	I II	–	[300]	–
5153.235	Cu I	600	–	U4
5105.541	Cu I	500	–	U5
5007.213	Ti I	200	40	–
4999.510	Ti I	200	80	–
4991.066	Ti I	200	100	–
4981.733	Ti I	300	125	U1
4962.263	Sr I	40	–	U4
4934.086	Ba II	400 h	400 h	V2
4889.17	Re I	2000 w	–	U2
4872.493	Sr I	25	–	U3
4861.327	H I	–	[500]	U3
4832.075	Sr I	200	8	U2
4825.91	Ra I	–	[800]	U1
4819.46	Cl II	–	[200]	V4
4816.71	Br II	–	[300]	V3
4810.534	Zn I	400 w	300 h	–
4810.06	Cl II	–	[200]	V3
4794.54	Cl II	–	[250]	V2
4785.50	Br II	–	[400]	V2
4772.312	Zr I	100	–	–
4742.25	Se I	–	[500]	U6
4739.478	Zr I	100	–	–
4739.03	Se I	–	[800]	U5
4730.78	Se I	–	[1000]	U4
4722.552	Bi I	1000	100	–
4722.159	Zn I	400 w	300 h	–
4710.075	Zr I	60	–	–
4704.86	Br II	–	[250]	V1
4696.25	S I	–	[15]	U9
4695.45	S I	–	[30]	U8
4694.13	S I	–	[500]	U7
4687.803	Zr I	125	–	U4
4685.75	He II	–	[300]	–
4682.28	Ra II	–	[800]	V2
4680.138	Zn I	300 w	200 h	–
4674.848	Yt I	80	100	U1
4671.226	Xe I	–	[2000]	U2
4661.87	Eu I	300 R	120	550
4643.695	Yt I	50	100	U2
4627.22	Eu I	400 R	150	650
4624.276	Xe I	–	[1000]	U3
4607.331	Sr I	1000 R	50 R	U1
4603.00	Li I	800	–	U4
4594.02	Eu I	500 R	200	750
4593.177	Cs I	1000 R	50	U4
4555.355	Cs I	2000 R	100	U3
4554.042	Ba II	1000 R	200	V1
4524.741	Sn	500 wh	50	–
4518.57	Lu	300	40	–
4511.323	In I	5000 R	4000 R	U1
4500.977	Xe I	–	[500]	U4
4454.781	Ca I	200	–	U2
4434.960	Ca I	150	–	U3
4434.321	Sm II	200	200	V2
4425.441	Ca I	100	–	U4
4424.342	Sm II	300	300	V1
4420.468	Os I	400 R	100	–
4390.865	Sm II	150	150	–
4389.974	V I	80 R	60 R	–
4384.722	V I	125 R	125 R	–
4379.238	V I	200 R	200 R	U1
4358.35	Hg I	3000 w	500	–
4305.447	Sr II	40	–	–
4303.573	Nd	100	40	–
4302.108	W I	60	60	U1
4294.614	W I	50	50	U2
4289.721	Cr I	3000 R	800 r	U3
4274.803	Cr I	4000 R	800 r	U2
4267.27	C II	–	500	V2
4267.02	C II	–	350	V3
4254.346	Cr I	5000 R	1000	U1
4241.669	U	40	50	–
4226.728	Ca I	500 R	50 W	U1
4226.570	Ge I	200	50	–
4225.327	Pr	50	40	–
4215.556	Rb I	1000 R	300	U4
4215.524	Sr II	300 r	400 W	V2
4211.719	Dy	200	15	–
4205.046	Eu II	200 R	50	–
4201.851	Rb I	2000 R	500	U3
4189.518	Pr	100	50	–
4186.599	Ce II	80	25	–
4179.422	Pr	200	40	–
4177.321	Nd	15	25	–
4172.056	Ga I	2000 R	1000 R	U1
4167.966	Dy	50	12	–
4165.606	Ce II	40	6	–
4137.095	Cb I	100	60	U5
4130.664	Ba II	50 r	60 Wh	V3
4129.737	Eu II	150 R	50 R	–
4123.810	Cb I	200	125	U4
4123.228	La II	500	500	V4
4109.98	N I	–	[1000]	U2
4103.37	N III	–	[80]	–
4101.773	In I	2000 R	1000 R	U2
4100.923	Cb I	300 w	200 w	U3
4099.94	N I	–	[150]	U3
4097.31	N III	–	[100]	–
4093.161	Hf II	25	20	–
4079.729	Cb I	500 w	200 w	U2
4077.974	Dy	150 r	100	–
4077.714	Sr II	400 r	500 W	V1
4077.340	La II	600	400	V3
4062.817	Pr	150	50	–
4058.938	Cb I	1000 w	400 w	U1
4057.820	Pb I	2000 R	300 R	U1
4047.201	K I	400	200	U4
4046.561	Hg I	200	300	–
4045.983	Dy	150	12	–
4044.140	K I	800	400	U3
4040.762	Ce II	70	5	–
4034.490	Mn I	250 r	20	U3
4033.073	Mn I	400 r	20	U2
4032.982	Ga I	1000 R	500 R	U2
4030.755	Mn I	500 r	20	U1
4023.688	Sc I	100	25	U3
4020.399	Sc I	50	20	U4
4019.137	Th	8	8	–
4012.388	Ce I, II	60	20	–
4008.753	W I	45	45	U3
4000.454	Dy	400	300	–
3987.994	Yb	1000 R	500 R	1900
3968.468	Ca II	500 R	500 R	V2
3961.527	Al I	3000	2000	U1
3951.154	Nd	40	30	–
3949.106	La II	1000	800	V2
3944.032	Al I	2000	1000	U2
3933.666	Ca II	600 R	600 R	V1
3911.810	Sc I	150	30	U1
3907.476	Sc I	125	25	U2
3906.316	Er	25	12	–
3905.528	Si I	20	15 W	–
3902.963	Mo I	1000 R	500 R	U3
3891.785	Ba II	18	25	V4
3891.02	Ho	200	40	–
3888.646	He I	–	[1000]	U2
3874.18	Tb	200	200	–
3864.110	Mo I	1000 R	500 R	U2
3848.75	Tb	100	200	–
3838.258	Mg I	300	200	U2
3832.306	Mg I	250	200	U3
3829.350	Mg I	100 w	150	U4

* Compiled from a combination of empirical and theoretical data selected from the literature.

‡ For the neutral atom, the most sensitive line (raie ultime) is indicated by U1, and other lines by U2, U3, etc., in order of decreasing sensitivity. For the singly ionized atom, the corresponding designations are V1, V2, etc. In cases where U1 is not given, the most sensitive lines lie outside the spectral range 10,000–2000 A. Figures given in the Sensitivity Column are taken from the NBS Tables of Spectral-Line Intensities, where values for numerous other lines also can be found. (See Meggers, Corliss and Scribner, NBS Monograph 32, 1961; also C. H. Corliss, NBS Monograph 32 Supplement, 1967).

WAVELENGTH TABLES

Wavelength	Element	Intensities Arc	Intensities Spk., [Dis.]	Sensitivity‡
3814.42	Ra II	–	[2000]	V1
3798.252	Mo I	1000 R	1000 R	U1
3788.697	Yt II	30	30	–
3775.72	Tl I	3000 R	1000 R	U2
3774.332	Yt II	12	100	–
3768.405	Gd	20	20	–
3761.917	Tm	200	120	–
3761.333	Tm	250	150	–
3748.264	Fe I	500	200	U4
3748.17	Ho	60	40	–
3745.903	Fe I	150	100	U5
3745.564	Fe I	500	500	U3
3737.133	Fe I	1000 r	600	U2
3719.935	Fe I	1000 R	700	U1
3710.290	Yt II	80	150	V1
3694.203	Yb	500 R	1000 R	3200
3692.652	Er	20	12	–
3692.357	Rh I	500 hd	150 wd	–
3683.471	Pb I	300	50	U2
3672.579	U	8	15	–
3663.276	Hg I	500	400	U5
3657.987	Rh I	500 W	200 W	–
3654.833	Hg I	–	[200]	U4
3653.496	Ti I	500	200	U2
3650.146	Hg I	200	500	U3
3646.196	Gd	200 w	150	–
3642.785	Sc II	60	50	V3
3642.675	Ti I	300	125	–
3639.580	Pb I	300	50 h	–
3635.463	Ti I	200	100	–
3634.695	Pd	2000 R	1000 R	U3
3633.123	Yt II	50	100	–
3630.740	Sc II	50	70	V2
3613.836	Sc II	40	70	V1
3613.790	W II	10	30	–
3610.510	Cd I	1000	500	–
3609.548	Pd I	1000 R	700 R	–
3601.193	Zr I	400	15	U1
3601.040	Th	8	10	–
3600.734	Yt II	100	300	–
3596.179	Ru I	30	100	U3
3581.20	Fe I	1000 R	600 r	600
3572.473	Zr II	60	80	V4
3561.74	Tb	200	200	–
3554.43	Lu	50	150	–
3552.172	U	8	12	–
3547.682	Zr I	200	12	U2
3538.75	Th	–	50	–
3529.813	Co I	1000 R	30	U3
3524.541	Ni I	1000 R	100 wh	–
3519.605	Zr I	100	10	U3
3519.24	Tl I	2000 R	1000 R	U3
3516.943	Pd I	1000 R	500 R	–
3515.054	Ni I	1000 R	50 h	–
3513.645	Ir I	100 h	100	U2
3509.17	Tb	200	200	–
3499.104	Er	18	15	–
3498.942	Ru I	500 R	200	U1
3496.210	Zr II	100	100	V3
3492.956	Ni I	1000 R	100 h	U2
3474.887	Sr II	80	50	–
3472.48	Lu	50	150	–
3466.201	Cd I	1000	500	–
3465.800	Co I	2000 R	25	U2
3464.57	Sr II	200	200	–
3462.21	Tm	200	100	–
3460.47	Re I	1000 W	–	U1
3453.505	Co I	3000 R	200	U1
3451.41	B II	5	30	V2
3438.230	Zr II	250	200	V2
3437.015	Ir I	20	15	–
3436.737	Ru I	300 R	150	U2
3434.893	Rh	1000 R	200 r	U1
3421.24	Pd I	2000 R	1000 R	U2
3414.765	Ni I	1000 R	50 wh	U1
3406.664	Ta	70 w	18 s	–
3405.120	Co I	2000 R	150	–
3404.580	Pd I	2000 R	1000 R	U1
3403.653	Cd I	800	500 h	–
3397.07	Lu	50	20 r	–
3396.85	Rh I	1000 w	500	–
3391.975	Zr II	300	400	V1
3383.761	Ti II	70	300 R	–
3382.891	Ag I	1000 R	700 R	U2
3380.711	Sr II	150	200	–
3372.800	Ti II	80	400 R	V3
3361.213	Ti II	100	600 R	V2
3349.035	Ti II	125	800 R	V1
3345.020	Zn I	800	300	U2
3323.092	Rh I	1000	200	–
3321.343	Be I	1000 r	30	U2
3321.086	Be I	100	–	U3
3321.013	Be I	50	–	U4
3318.840	Ta	125	35	–
3311.162	Ta	300 w	70 w	U1
3302.988	Na I	300 R	150 R	U4
3302.588	Zn I	800	300	U3
3302.323	Na I	600 R	300 R	U3
3290.59	Th	–	40 h	–
3289.37	Yb II	500 R	1000 R	2600
3282.333	Zn I	500 R	300	U4
3280.683	Ag I	2000 R	1000 R	U1
3273.962	Cu I	3000 R	1500 R	U2
3269.494	Ge I	300	300	U3
3267.945	Os I	400 R	30	–
3267.502	Sb I	150	150 Wh	–
3262.328	Sn I	400 h	300 h	U3
3262.290	Os I	500 R	50	–
3261.057	Cd I	300	300	–
3258.564	In I	500 R	300 R	U5
3256.090	In I	1500 R	600 R	U3
3247.540	Cu I	5000 R	2000 R	U1
3242.280	Yt II	60	100	–
3232.61	Li I	1000 R	500	U2
3232.499	Sb I	150	250 wh	–
3229.75	Tl I	2000	800	–
3225.479	Cb II	150 w	800 wr	–
3220.780	Ir I	100	30	U1
3215.560	W I	10	9	–
3194.977	Cb II	30	300	–
3185.396	V I	500 R	400 R	U2
3183.982	V I	500 R	400 R	–
3183.406	V I	200 R	100 R	–
3179.332	Ca II	100	400 w	V3
3175.019	Sn I	500 h	400 hr	–
3163.402	Cb II	15	8	–
3158.869	Ca II	100	300 w	V4
3134.718	Hf II	80	125	–
3131.072	Be II	200	150	V2
3130.786	Cb II	100	100	–
3130.416	Be II	200	200	V1
3125.284	V II	80	200 R	–
3118.383	V II	70	200 R	V4
3110.706	V II	70	300 R	V3
3102.299	V II	70	300 R	V2
3094.183	Cb II	100	1000	V1
3093.108	V II	100 R	400 R	V1
3092.713	Al I	1000	1000	U3
3082.155	Al I	800	800	U4
3072.877	Hf I	80	18	–
3071.591	Ba I	100 R	50 R	U5
3067.716	Bi I	3000 hR	2000 wh	U1
3064.712	Pt I	2000 R	300 R	U1
3058.66	Os I	500 R	500	–
3050.819	Ni I	1000 R	–	280
3039.356	In I	1000 R	500 R	U4
3039.064	Ge I	1000	1000	U2
3034.121	Sn I	200 wh	150 wh	–
3009.147	Sn I	300 h	200 h	–
2997.967	Pt I	1000 R	200 r	–
2989.029	Bi I	250 wh	100 wh	–
2976.586	Ru	60	200	–
2965.546	Ru	60	200	–
2945.668	Ru	60	300	–
2943.637	Ga I	10	20 r	U3
2940.772	Hf I	60	12	–
2938.298	Bi I	300 w	300 w	–
2936.77	Ho	–	1000 R	–
2929.794	Pt I	800 R	200 w	–
2924.792	Ir I	25 wh	15	–
2918.32	Tl I	400 R	200 R	–
2916.481	Hf I	50	15	–
2911.39	Lu	100	300	–
2009.116	Mo II	25	40 h	V5
2909.061	Os I	500 R	400	U1
2904.408	Hf I	30	6	–
2898.71	As I	25 r	40	–
2898.259	Hf I	50	12	–
2897.975	Bi I	500 WR	500 WR	U2
2894.84	Lu	60	200	–
2890.994	Mo II	30	50 h	V4
2881.578	Si I	500	400	U1
2877.915	Sb I	250 W	150	–
2874.244	Ga I	10	15 r	U4
2871.508	Mo II	100	100 h	V3
2863.327	Sn I	300 R	300 R	U2
2860.934	Cr II	60	100	V5
2860.452	As I	50 r	50	–
2855.676	Cr II	60	200 Wh	V4
2854.581	Pd II	4	500 h	–
2852.129	Mg I	300 R	100 R	U1
2849.838	Cr II	80	150 r	V3
2849.725	Ir I	40 h	20 h	–
2848.232	Mo II	125	200 h	V2
2843.252	Cr II	125	400 r	V2
2839.989	Sn I	300 R	300 R	U1
2837.602	C II	–	40	V5
2836.710	C II	–	200	V4
2835.633	Cr II	100	400 r	V1
2833.069	Pb I	500 R	80 R	–
2830.295	Pt I	1000 R	600 r	–
2824.37	Cu I	1000	300	50
2820.224	Hf II	40	100	–
2816.179	Al II	10	100	V2
2816.154	Mo II	200	300 h	V1
2809.625	Bi I	200 w	100	–
2802.695	Mg II	150	300	V2
2802.19	Au	–	200	–
2801.064	Mn I	600 R	60	480
2798.271	Mn I	800 R	80	650
2795.53	Mg II	150	300	V1
2780.521	Bi I	200 w	100	–
2780.197	As I	75 R	75	U5
2773.357	Hf II	25	60	–
2769.67	Te I	–	[30]	–
2767.87	Tl I	400 R	300 R	–
2748.58	Cd II	5	200	–
2712.410	Ru	80	300	–
2709.626	Ge I	30	20	–
2692.065	Ru	8	200	–
2678.758	Ru	100	300	–
2675.95	Au I	250 R	100	U2
2669.166	Al II	3	100	V1
2661.983	Ir I	150	15	130
2659.454	Pt I	2000 R	500 R	U2
2658.722	Pd II	20	300	–
2651.575	Ge I	30	20	–
2651.178	Ge I	40	20	–
2650.781	Be I	25	–	U5
2641.406	Hf II	40	125	–
2631.553	Al II	–	40	–
2614.178	Pb	200 r	80	–
2605.688	Mn II	100 R	500 R	V3
2598.062	Sb I	200	100	–
2593.729	Mn II	200 R	1000 R	V2

‡ For the neutral atom, the most sensitive line (raie ultime) is indicated by U1, and other lines by U2, U3, etc., in order of decreasing sensitivity. For the singly ionized atom, the corresponding designations are V1, V2, etc. In cases where U1 is not given, the most sensitive lines lie outside the spectral range 10,000–2000 A. Figures given in the Sensitivity Column are from the NBS Tables of Spectral-Line Intensities, where values for many other lines also can be found. (Meggers, Corliss and Scribner, NBS Monograph 32, 1961 — Government Printing Office; C. H. Corliss, Monograph 32 Supplement, 1967.)

SENSITIVE LINES OF THE ELEMENTS

Wave-length	Element	Intensities Arc	Spk., [Dis.]	Sensi-tivity‡
2589.167	W II	15 d	25	–
2576.104	Mn II	300 R	2000 R	V1
2573.09	Cd II	3	150	–
2557.958	Zn II	10	300	V3
2554.93	P I	60	[20]	–
2553.28	P I	80	[20]	U3
2536.519	Hg I	2000 R	1000 R	U2
2535.65	P I	100	[30]	U2
2534.01	P I	50	[20]	–
2530.70	Te I	–	[30]	–
2528.535	Sb I	300 R	200	–
2528.516	Si I	400	500	U2
2519.822	Co II	40	200	–
2516.881	Hf II	35	100	–
2516.123	Si I	500	500	U3
2513.028	Hf II	25	70	–
2506.899	Si I	300	200	U4
2505.739	Pd II	3	30	–
2502.001	Zn II	20	400 w	V4
2498.784	Pd II	4	150 h	–
2497.733	B I	500	400	U1
2496.778	B I	300	300	U2
2488.921	Pd II	10	30	–
2478.573	C I	400	[400]	U2
2456.53	As I	100 r	8	U4
2437.791	Ag II	60	500 wh	V2
2427.95	Au I	400 R	100	U1
2413.309	Fe II	60	100 h	V5
2410.517	Fe II	50	70 h	V4
2404.882	Fe II	50	100 wh	V3
2397.091	W II	18	30	–
2395.625	Fe II	50	100 wh	V2
2388.918	Co II	10	35	–
2385.76	Te I	600	[300]	U2
2383.25	Te I	500	[300]	U3
2382.039	Fe II	40 r	100 R	V1
2378.622	Co II	25	50 w	–
2370.77	As I	50 r	3	–
2369.67	As I	40 r	–	–
2363.787	Co II	25	50	–
2349.84	As I	250 R	18	U3
2348.610	Be I	2000 R	50	U1
2335.269	Ba II	60 R	100 R	–
2312.84	Cd II	1	200	–
2311.469	Sb I	150 R	50	–
2307.857	Co II	25	50 w	–
2304.235	Ba II	60 R	80 R	–
2296.89	C III	–	200	–
2288.12	As I	250 R	5	U3
2288.018	Cd I	1500 R	300 R	U1
2287.084	Ni II	100	500	V1
2286.156	Co II	40	300 l	V1
2276.578	Bi I	100 R	40	–
2270.213	Ni II	100	400	V2
2265.017	Cd II	25 d	300	V2
2264.457	Ni II	150	400	V3
2253.86	Ni II	100	300	V4
2246.995	Cu II	30	500	V3
2246.412	Ag II	25	300 hs	V3
2203.505	Pb II	50 W	5000 R	V1
2192.260	Cu II	25	500 h	V2
2175.890	Sb I	300	40	U2
2169.994	Pb I	1000 R	1000 R	–
2144.382	Cd II	50	200 R	V1
2142.75	Te I	600	–	55
2138.56	Zn I	800 R	500	U1
2135.976	Cu II	25	500 w	V1
2068.38	Sb I	300 R	3	U1
2062.788	Se I	–	[800]	U3
2062.38	I	–	[900]	–
2061.91	Zn II	100	100	V2
2061.70	Bi I	300 R	100	–
2039.851	Se I	–	[1000]	U2
2025.51	Zn II	200	200	V1

‡ For the neutral atom, the most sensitive line (raie ultime) is indicated by U1, and other lines by U2, U3, etc., in order of decreasing sensitivity. For the singly ionized atom, the corresponding designations are V1, V2, etc. In cases where U1 is not given, the most sensitive lines lie outside the spectral range 10,000–2000 A. Figures given in the Sensitivity Column are from the NBS Tables of Spectral-Line Intensities, where values for many other lines also can be found. (Meggers, Corliss and Scribner, NBS Monograph 32, 1961 — Government Printing Office; C. H. Corliss, Monograph 32, Supplement, 1967).

TABLE V

KEY TO SYMBOLS FOR AUTHORS AND REFERENCES

Symbol	Element	Author / Reference
Ab		Albertson, W. E.
	Gd	Phys. Rev. **47**: 370 (1935)
	Ir	Unpublished material
	Os	Phys. Rev. **45**: 304 (1934)
	Sm	Unpublished material
Ad		Anderson, J. A.
	Ca	Astrophys. Journ. **59**: 76 (1924)
An		Angerer, E. von
	Cl	Zeits. f. wiss. Phot. **22**: 200 (1923)
	Li	Zeits. f. Physik **18**: 113 (1923)
Ar		Arnolds, R.
	Sn	Zeits. f. wiss. Phot. **13**: 313 (1914)
Az		Aretz, M.
	Cu	Zeits. f. wiss. Phot. **9**: 256 (1911)
Bb		Babcock, H. D.
	Fe	Astrophys. Journ. **66**: 256 (1927)
Bh		Behner, K.
	Ti	Zeits. f. wiss. Phot. **23**: 325 (1925)
Bk		Belke, M.
	W	Zeits. f. wiss. Phot. **17**: 132 (1917)
Bl		Bloch, L. and E.
	Br	Annales de Physique **7**: 206 (1927)
	Cd, Zn	Annales de Physique **5**: 325 (1936)
	Cl	Annales de Physique **8**: 403 (1927)
	I	Annales de Physique **11**: 141 (1929)
	S	Annales de Physique **12**: 5 (1929)
	Se, Te	Annales de Physique **13**: 233 (1930)
		Bloch and Déjardin, G.
	Ne	Journ. de Phys. et le Rad. **7**: 129; 203 (1926)
Bn		de Bruin, T. L.
	A	K. Akad. van Wetens. Amsterdam. Proc. **40**: 342 (1938)
	K	Zeits. f. Physik **38**: 96 (1926)
	Ne	Zeits. f. Physik **69**: 22 (1931)
Bs		Balasse, G.
	Cs	Journ. de Phys. et le Rad. **8**: 318 (1927)
Bt		Bartelt, O.
	Se	Zeits. f. Physik **91**: 444 (1934)
		Bartelt and Eckstein, L.
	S	Zeits. f. Physik **86**: 77 (1933)
Bu		Burns, K.
	Ba	Comptes Rendus **156**: 1976 (1913)
	Cu	Phys. Rev. **48**: 656 (1935)
		Burns and Walters, F. M., Jr.
	Cu	Allegheny Obs. (U. of Pittsburgh) Pub. **8**: 27 (1930); 37 (1931)
	Fe	Allegheny Obs. (U. of Pittsburgh) Pub. **8**: 39 (1931)
Bv		Bevan, P. V.
	Cs	Royal Soc. London. Proc. **A83**: 421 (1910); **A85**: 54 (1911); **A86**: 320 (1912)
	Rb	Royal Soc. London. Proc. **A83**: 421 (1910)
Bx		Blair, H. A.
	Ag	Phys. Rev. **36**: 173; 1532 (1930)
	Pd	Phys. Rev. **36**: 173 (1930)
Bz		Bungartz, E.
	S	Annalen der Physik **76**: 709 (1925)
Cf		Crawford, M. F., and McLay, A. B.
	Bi	Royal Soc. London. Proc. **A143**: 540 (1934)
Cn		Cardaun, L.
	Hg	Zeits. f. wiss. Phot. **14**: 56; 89 (1914)
Ct		Catalán, M. A.
	Ag	An. Soc. Españ. Fís. Quím. **15**: 222; 483 (1917)
		Catalán and Sancho, P. M.
	Cr	An. Soc. Españ. Fís. Quím. **29**: 327 (1931)
Cw		Crew, H., and McCauley, G. V.
	Ca	Astrophys. Journ. **39**: 29 (1914)
Cx		Carroll, J. A.
	In, Tl	Royal Soc. London. Phil. Trans. **A225**: 357 (1926)
Cz		Curtis, C. W.
	Mn	Phys. Rev. **53**: 478 (1938)
Da		Datta, S.
	K, Na	Royal Soc. London. Proc. **A99**: 69 (1921)
		Datta and Bose, P. C.
	Li	Zeits. f. Physik **97**: 321 (1935)
Db		Deb, S. C.
	I	Royal Soc. London. Proc. **A139**: 380 (1933)
Di		Dingle, H.
	F	Royal Soc. London. Proc. **A128**: 603 (1930)
	Fe	Royal Astron. Soc. London. Monthly Notices. **94**: 287 (1934)
	Hg	Royal Soc. London. Proc. **A100**: 167 (1921)
Dj		Déjardin, G.
	Hg	Annales de Physique **10**: 424 (1927)
	Xe	Comptes Rendus **190**: 581 (1930)
Dm		Dahmen, W.
	K	Zeits. f. Physik **35**: 528 (1925)
Dn		Dhein, F.
	Co	Zeits. f. wiss. Phot. **19**: 289 (1920)
	Pd	Zeits. f. wiss. Phot. **11**: 317 (1912)
Do		Dobbie, J. K.
	Fe	Solar Phys. Obs., Cambridge, England. Annals **5**: 1 (1938)
Dr		Dunoyer, M. L.
	Cs, Rb	Journ. de Phys. et le Rad. **3**: 261 (1922)
Ds		(Piña) de Rubies, S.
	Gd	Comptes Rendus **184**: 594 (1927)
	Nd	Comptes Rendus **197**: 33 (1933)
	Sc	An. Soc. Españ. Fís. Quím. **22**: 49 (1924)
Du		Duffendack, O. S., and Wolfe, R. A.
	N	Phys. Rev. **34**: 409 (1929)
Dv		Dhavale, D. G.
	Sb	Royal Soc. London. Proc. **A131**: 109 (1931)
Ea		Earls, L. T., and Sawyer, R. A.
	Pb	Phys. Rev. **47**: 115 (1935)
Ed		Eder, J. M.
	Au, Gd, Nd, Pr	Akad. Wiss. Wien. Ber. **124**: 101 (1915)
	Dy	Akad. Wiss. Wien. Ber. **127**: 1099 (1918)
	Er, Yt	Akad. Wiss. Wien. Ber. **125**: 383 (1916)
	Eu	Akad. Wiss. Wien. Ber. **126**: 473 (1917)
	Gd	Akad. Wiss. Wien. Ber. **125**: 1467 (1916)
	Tb	Akad. Wiss. Wien. Ber. **131**: 199 (1922)
	Tl	Akad. Wiss. Wien. Ber. **122**: 607 (1913)
		Eder and Valenta, E.
	Bi	Akad. Wiss. Wien. Ber. **119**: 519 (1910)
	Ho	Akad. Wiss. Wien. Ber. **119**: 9 (1910)

KEY TO SYMBOLS

El Ellis, J. W., and Sawyer, R. A.
Tl Phys. Rev. **49**: 147 (1936)

En Edlén, B.
B Zeits. f. Physik **73**: 477 (1932)
C Upsala Regia Soc. Scient. **9**: 77; 108 (1933)
F, K Zeits. f. Physik **98**: 445 (1936)

Es Esclanglon, F.
Cd Journ. de Phys. et le Rad. **7**: 52 (1926)

Ev Evans, S. F.
I Royal Soc. London. Proc. **A133**: 417 (1931)

Ex Exner, F., and Haschek, E.
Ag, Au, Ba, Cr, Co, Cu, Er, Eu, Gd, Ho, Mg, Mo, Nd, Pd, Pt, Rh, Ru, Sc, Sr, Ta, Tb, Th, Ti, W
Spektren der Elemente **2**, **3**. Deuticke, Leipzig (1911)

Fa Fabry, C.
Rb Comptes Rendus **195**: 1012 (1932)

Fd Fred, M.
Th Astrophys. Journ. **87**: 179 (1938)

Fh Frerichs, R.
O Phys. Rev. **34**: 1239 (1929)
S Zeits. f. Physik **80**: 152 (1933)

Fi Findley, J. H.
Co Phys. Rev. **36**: 9 (1930)

Fl Fowler, A.
Ba, Ca, Cd, Cs, In, K, Li, Mg, Na, O, Rb, Sr, Tl, Zn
Report on Series in Line Spectra. Fleetway Press, London, 1922.
N Royal Soc. London. Proc. **A107**: 31 (1925)
O Royal Soc. London. Proc. **A110**: 480 (1926)
Si Royal Soc. London. Proc. **A123**: 425 (1929)
Fowler and Selwyn, E. W. H.
C Royal Soc. London. Proc. **A118**: 34 (1928)

Fm Freeman, L. J.
N Royal Soc. London. Proc. **A114**: 662 (1927); **A124**: 654 (1929)

Fn Frings, J.
Ag Zeits. f. wiss. Phot. **15**: 165 (1915)

Fo Foote, P. D.
Na Astrophys. Journ. **55**: 145 (1922)

Fr Frisch, S.
Na Zeits. f. Physik **70**: 498 (1931)

Fu Fuchs, H.
Mn Zeits. f. wiss. Phot. **14**: 239 (1914)

Gn Grünter, R.
Al Zeits. f. wiss. Phot. **13**: 1 (1913)

Gr Gremmer, W.
Kr Zeits. f. Physik **73**: 620 (1932)
Ne Zeits. f. Physik **50**: 705 (1928)

Gs Gieseler, H.
Pb Zeits. f. Physik **42**: 265 (1927)

Gt Gartlein, C. W.
Ge Phys. Rev. **31**: 782 (1928)

Gu Geuter, P.
P Zeits. f. wiss. Phot. **5**: 14 (1907)

Ha Hamm, S.
Ni Zeits. f. wiss. Phot. **13**: 105 (1913)

Hb Hasselberg, B.
U Klg. Sv. Vet. Akad. Handl. **45**: 5 (1910)

Hl Hall, J.
Cr Kayser and Konen, Handbuch der Spectroscopie **7** (**1**). Hirzel, Leipzig (1923)

Hn Hunter, A. A.
S Royal Soc. London. Phil. Trans. **A233**: 303 (1934)

Hp Hampe, H.
Sr Zeits. f. wiss. Phot. **13**: 348 (1914)

Hs Hasbach, K.
Cu Zeits. f. wiss. Phot. **13**: 399 (1914)

Ht Holtz, O.
Ca Zeits. f. wiss. Phot. **12**: 112 (1913)

Hu Humphreys, C. J.
A, Kr, Ne
Bur. of Stand. Journ. of Res. **20**: 17 (1938)
Kr Phys. Rev. **47**: 714 (1935)
Kr, Xe Bur. of Stand. Journ. of Res. **5**: 1041 (1930)
Xe Bur. of Stand. Journ. of Res. **22**: 19 (1939)

Hx Haussmann, A. C.
Pt Astrophys. Journ. **66**: 333 (1927)

Hz Hetzler, C. W., Boreman, R. W., and Burns, K.
Ag, Be, Cd, Cu, K, Li, Na, Pb, Rb, Sn, Sr, Zn
Phys. Rev. **48**: 656 (1935)

Ig Ingram, S. B.
C, N Phys. Rev. **34**: 421 (1929)
S Phys. Rev. **33**: 907 (1929)

It Ireton, H. J. C., and Keast, A. M.
Ir, Pd, Pr, Pt, Yb
Royal Soc. Canada. Trans. **23**: 13 (1929)

Ja Jackson, C. V.
Kr Royal Soc. London. Phil. Trans. **A236**: 1 (1936)

Jn Johnson, R. C.
C Royal Soc. London. Proc. **A108**: 343 (1923)

Jv Jevons, W.
Cl Royal Soc. London. Proc. **A103**: 198 (1923)

Ka Karlik, B., and Pettersson, H.
Po Akad. Wiss. Wien. Ber. **143**: 379 (1934)

Kb Krebs, A.
Co Zeits. f. wiss. Phot. **19**: 307 (1919)

Ke Kerris, W.
I Zeits. f. Physik. **60**: 20 (1930)

Kh Krishnamurty, S. G., and Rao, K. R.
Se Royal Soc. London. Proc. **A149**: 56 (1935)

Kl Klein, E.
Ga Astrophys. Journ. **56**: 373 (1922)
Klein, F.
Pb Zeits. f. wiss. Phot. **12**: 16 (1913)

Kn King, A. S.
Ba Astrophys. Journ. **48**: 13 (1918)
Ce, Pr Astrophys. Journ. **68**: 194 (1928)
Dy, Gd, Ho, Tb
Astrophys. Journ. **72**: 221 (1930)
Eu Astrophys. Journ. **89**: 377 (1939)
Fe Astrophys. Journ. **87**: 109 (1938)
Nd Astrophys. Journ. **78**: 9 (1933)
Sm Astrophys. Journ. **82**: 140 (1935)
V Astrophys. Journ. **60**: 284 (1924)
Yb Astrophys. Journ. **73**: 328 (1931)
King and Carter, E.
La Astrophys. Journ. **65**: 86 (1927)

Kp Kasper, F. J.
Ag Zeits. f. wiss. Phot. **10**: 1 (1911)

WAVELENGTH TABLES

Ks Kiess, C. C.
Al, Ba, Ca, Cr, Cu, Fe, Mn, Na, Ni, V
Bur. of Stand. Journ. of Res. **1:** 75 (1928)
C, Ti Bur. of Stand. Journ. of Res. **20:** 33 (1938)
Ce, La, Yt Bur. of Stand. Sci. Papers **17:** 318 (1922)
Cl Unpublished material
Cl Bur. of Stand. Journ. of Res. **10:** 827 (1933)
Cr Bur. of Stand. Journ. of Res. **15:** 79 (1935)
Cu Bur. of Stand. Journ. of Res. **14:** 519 (1935)
Dy, Gd Bur. of Stand. Sci. Papers **18:** 695 (1922)
N Amer. Astron. Soc. Pub. **4:** 363 (1922)
N Science **60:** 249 (1924)
Nd, Sm Bur. of Stand. Sci. Papers **18:** 201 (1922)
P Bur. of Stand. Journ. of Res. **8:** 393 (1932)
Si Bur. of Stand. Journ. of Res. **21:** 195 (1938)
Kiess and de Bruin, T. L.
Br Bur. of Stand. Journ. of Res. **4:** 667 (1930)
Cl Bur. of Stand. Journ. of Res. **2:** 1117 (1929)
Kiess, C. C. and Kiess, H. K.
Zr Bur. of Stand. Journ. of Res. **6:** 621 (1931); **5:** 1205 (1930)
Kiess and Stowell, E. Z.
Ta Bur. of Stand. Journ. of Res. **12:** 459 (1934)

Kz Kretzer, A.
Sb Zeits. f. wiss. Phot. **8:** 45 (1910)

Lc Lacroute, M. P.
Br, I Annales de Physique **3:** 5 (1935)
S, Se, Te
Journ. de Phys. et le Rad. **9:** 180 (1928)

Lf Laffay, J.
Hg Comptes Rendus **180:** 823 (1925)

Lg Lang, R. J.
As, Pb, Sn
Royal Soc. London. Phil. Trans **A224:** 371 (1924)
Ge Nat. Acad. Sci. Proc. **14:** 34 (1928)
Ge Phys. Rev. **34:** 708 (1929)
Sb, Sn Phys. Rev. **35:** 445 (1930)
Lang and Sawyer, R. A.
In Zeits. f. Physik **71:** 453 (1931)
Lang and Vestine, E. H.
Sb Phys. Rev. **42:** 233 (1932)

Ln Laun, D. D.
W Bur. of Stand. Journ. of Res. **21:** 207 (1938)

Lp Laporte, O., Miller, G. R., and Sawyer, R. A.
Cs Phys. Rev. **39:** 461 (1932)
Rb Phys. Rev. **38:** 843 (1931)

Lr Lorenser, E.
Ba Kayser and Konen, Handbuch der Spectroscopie, 7 (**1**). Hirzel, Leipzig (1923)
Mo Kayser and Konen, Handbuch der Spectroscopie, 7 (3). Hirzel, Leipzig (1934)

Lx Lub, W. A.
Ac K. Akad. van Wetens. Amsterdam. Proc. **40:** 584 (1937)

Mc McCormick, W. W., and Sawyer, R. A.
Sn Phys. Rev. **54:** 71 (1938)

Md McDonald, M. C.
Hf Royal Soc. Canada. Trans. **21:** 223 (1927)
McDonald, Sutton, E. E., and McLay, A. B.
Be Royal Soc. Canada. Trans. **20:** 313 (1926)

Me Meggers, W. F.
Ba, Ca, Cs, K, Rb, Sr
Bur. of Stand. Journ. of Res. **10:** 669 (1933)
bhLa Bur. of Stand. Journ. of Res. **9:** 268 (1932)
Bi Unpublished material
Cd Optical Soc. Amer. Journ. **6:** 135 (1922)
Cu Phys. Rev. **28:** 449 (1926)
Ir, Os, Pd, Pt, Rh, Ru
Bur. of Stand. Sci. Papers **20:** 19 (1924)
Li Bur. of Stand. Bull. **14:** 371 (1918)
Mn, Re Bur. of Stand. Journ. of Res. **10:** 757 (1933)
Pb Unpublished material
Re Bur. of Stand. Journ. of Res. **6:** 1027 (1931)
Sb Unpublished material
Sc Bur. of Stand. Sci. Papers **22:** 61 (1927)
Sn Unpublished material
Tm Unpublished material
V Unpublished material
Yt Bur. of Stand. Journ. of Res. **1:** 325 (1928)
Meggers and de Bruin, T. L.
As Bur. of Stand. Journ. of Res. **3:** 765 (1929)
Meggers, de Bruin, and Humphreys, C. J.
Kr Bur. of Stand. Journ. of Res. **11:** 422 (1933)
Xe Bur. of Stand. Journ. of Res. **3:** 731 (1929)
Meggers and Dieke, G. H.
He Bur. of Stand. Journ. of Res. **9:** 121 (1932)
Meggers, Foote, P. D., and Mohler, F. L.
Na Astrophys. Journ. **55:** 145 (1922)
Meggers and Humphreys, C. J.
A, Kr, Ne Bur. of Stand. Journ. of Res. **10:** 427 (1933)
A, Kr, Xe Bur. of Stand. Journ. of Res. **13:** 293 (1934)
Fe Bur. of Stand. Journ. of Res. **18:** 543 (1937)
Meggers and Kiess, C. C.
Co, Fe, Ni Bur. of Stand. Bull. **14:** 637 (1919)
Cr, Mn, Mo, Ti, U, W
Bur. of Stand. Sci. Papers **16:** 51 (1920)
Ni, Ti, Zr Bur. of Stand. Journ. of Res. **9:** 309 (1932)
Meggers and King, A. S.
Cb Bur. of Stand. Journ. of Res. **16:** 385 (1936)
Meggers and Russell, H. N.
La Bur. of Stand. Journ. of Res. **9:** 625 (1932)
Sc Bur. of Stand. Sci. Papers **22:** 338 (1927)
V Bur. of Stand. Journ. of Res. **17:** 125 (1936)

KEY TO SYMBOLS

Yt Bur. of Stand. Journ. of Res. **2**: 733 (1929)
Meggers and Scribner, B. F.
Hf Bur. of Stand. Journ. of Res. **13**: 625 (1934)
Lu Bur. of Stand. Journ. of Res. **19**: 31 (1937)
Yb Bur. of Stand. Journ. of Res. **19**: 651 (1937)

Mh Mihul, C.
O Annales de Physique **9**: 294 (1928)

Mj Majumdar, K.
Cl Royal Soc. London. Proc. **A125**: 66 (1929)

Ml McLennan, J. C.
K Royal Soc. London. Proc. **A100**: 182 (1921)
Mg Royal Soc. London. Proc. **A98**: 35 (1920)
McLennan, Ainslie, D. S., and Fuller, D. S.
Sn Royal Soc. London. Proc. **A95**: 316 (1919)
Te Royal Soc. Canada. Trans. **19**: 56 (1925)
McLennan and Allin, E. J.
Tl Royal Soc. London. Proc. **A129**: 43 (1930)
McLennan and McLay, A. B.
Au Royal Soc. London. Proc. **A134**: 35 (1931)
Pt Royal Soc. Canada. Trans. **20**: 201 (1926)
McLennan, McLay, and Crawford, M. F.
Bi Royal Soc. London. Proc. **A129**: 579 (1930)
Hg Royal Soc. London. Proc. **A134**: 41 (1931)
Tl Royal Soc. London. Proc. **A125**: 50; 570 (1929)
McLennan and Smith, H. G.
Pd Royal Soc. London. Proc. **A112**: 123 (1926)
Crawford and McLay
Bi Royal Soc. London. Proc. **A143**: 540 (1934)

Mr Merrill, P. W.
He Astrophys. Journ. **46**: 357 (1917)

Ms Meissner, K. W.
A Zeits. f. Physik. **39**: 179 (1926); **40**: 844 (1927)
Al Annalen der Physik **50**: 713 (1916)
Cs Annalen der Physik **65**: 380 (1921)
H Annalen der Physik **50**: 901 (1916)
Meissner, Bartelt, O., and Eckstein, L.
S Zeits. f. Physik **86**: 56 (1933)
Se Zeits. f. Physik **91**: 427 (1934)

Mt Merton, T. R., and Johnson, R. C.
C Royal Soc. London. Proc. **A103**: 383 (1923)
Merton and Pilley, J. G.
N Royal Soc. London. Proc. **A107**: 411 (1925)

Mu Murakawa, K.
Cl Zeits. f. Physik **69**: 510 (1931); **96**: 117 (1935); **109**: 162 (1938)
Hg Inst. Phys. and Chem. Res. Tokyo. Sci. Papers **34**: 32 (1937)
I Zeits. f. Physik **109**: 162 (1938)

Mx Menzies, A. C.
Cu Royal Soc. London. Proc. **A119**: 249 (1928)

Mz Martin, D. C.
Se Phys. Rev. **48**: 938 (1935)

Nm Newman, F. H.
Na Phil. Mag. **5**: 150 (1928)

Nu Naudé, S. M.
Hg Annalen der Physik **3**: 1 (1929)

Ny Nyswander, R. E., Lind, S. C., and Moore, R. B.
Rn Astrophys. Journ. **54**: 285 (1921)

Of Offerhaus, H. C.
He Physica **3**: 309 (1923)

Ok Otsuka, O.
Rb Zeits. f. Physik **36**: 789 (1926)

Om Offermann, J.
Bi Kayser and Konen, Handbuch der Spectroscopie **7** (**1**). Hirzel, Leipzig (1923)

Ot Olthoff, J., and Sawyer, R. A.
Cs Phys. Rev. **42**: 766 (1932)

Pe Pettersson, H.
Rn Akad. Wiss. Wien. Ber. **143**: 303 (1934)

Pk Poetker, A. H.
H Nature **119**: 123 (1927)
O Phys. Rev. **30**: 812 (1927)

Ps Paschen, F.
Cd Annalen der Physik **30**: 746 (1909); **35**: 860 (1911)
Hg Preuss. Akad. Wiss. Ber. p. 538 (1928)
In Annalen der Physik **32**: 148 (1938)
Mg Preuss. Akad. Wiss. Ber. p. 709 (1931)
Ne Annalen der Physik **60**: 405 (1919)
Paschen and Campbell, J. S.
In Annalen der Physik **31**: 29 (1938)
Paschen and Kruger, P. G.
Be Annalen der Physik **8**: 1005 (1931)
C Annalen der Physik **7**: 1 (1930)
Paschen and Meissner, K.
In Annalen der Physik **43**: 1223 (1914)
Paschen and Ritschl, R.
Al, He, Zn Annalen der Physik **18**: 867 (1933)

Pu Puhlmann, M.
Mo Zeits. f. wiss. Phot. **17**: 97 (1917)

Py Perey, M.
Ba Comptes Rendus **204**: 244 (1927)

Qi Quincke, M.
Au Bur. of Stand. Sci. Papers **17**: 167 (1922)

Rc Royds, T.
Rn Phil. Mag. **17**: 202 (1909)
Rutherford, E., and Royds
Phil. Mag. **16**: 313 (1908)

Rd Ruedy, R.
Te Phys. Rev. **41**: 588 (1932)
Ruedy and Gibbs, R. C.
Se Phys. Rev. **46**: 880 (1934)

Rf Ricard, R.
Cs Comptes Rendus **206**: 905 (1938)
Ricard, Givord, M., and George, F.
Cs Comptes Rendus **205**: 1229 (1937)

Ri Robinson, H. A.
P Phys. Rev. **49**: 297 (1936)

Rk Ruark, A. E.
H Astrophys. Journ. **58**: 46 (1923)
Ruark, Mohler, F. L., Foote, P. D., and Chenault, R. L.
Bi, Sb Bur. of Stand. Sci. Papers **19**: 471 (1924)

Rl Russell, H. N.
Fe Astrophys. Journ. **64**: 198 (1926)
Ti Astrophys. Journ. **66**: 295 (1927)
Russell and Lang, R. J.
Ti Astrophys. Journ. **66**: 13 (1927)

Ro Rao, A. S.
As Phys. Soc. London. Proc. **44**: 343 (1932)
Rao and Narayan, A. L.
Pb Zeits. f. Physik **59**: 687 (1930)
Rao, K. R.
As Phys. Soc. London. Proc. **43**: 68 (1931)
Rao and Badami, J. S.
Se Royal Soc. London. Proc. **A140**: 387 (1933)
Rao and Narayan, A. L.
Sn Zeits. f. Physik **45**: 350 (1927)
Rr Reinheimer, O.
Rb Annalen der Physik **71**: 168 (1923)
Rs Rasmussen, E.
A Zeits. f. Physik **75**: 696 (1932)
Ba Zeits. f. Physik **83**: 404 (1933)
Hg Naturwiss. **17**: 389 (1929)
Kr Aedle Luftart. Spektre. Copenhagen. **1**: 31 (1932)
Ra Zeits. f. Physik **86**: 26 (1933); **87**: 609 (1934)
Rn Zeits. f. Physik **80**: 726 (1933)
Rt Rosenthal, A. H.
A Annalen der Physik **4**: 49 (1930)
Rx Randall, H. M., and Wright, N.
Sn Phys. Rev. **38**: 457 (1931)
Ry Ryde, J. W.
N Royal Soc. London. Proc. **A117**: 164 (1927)
Rz Ramb, R.
Rb Annalen der Physik **5**: 311 (1931)
Sa Saltmarsh, M. O.
P Royal Soc. London. Proc. **A108**: 332 (1925)
Sd Saunders, F. A.
Ca Astrophys. Journ. **52**: 265 (1920)
Cd Fowler. Line Spectra. Fleetway Press, London, 1922.
In, Tl Astrophys. Journ. **43**: 234 (1916)
Saunders, Schneider, E. G., and Buckingham, E.
Ba, Sr Nat. Acad. Sci. Proc. **20**: 291 (1934)
Sf Symons, E.
Pt Zeits. f. wiss. Phot. **12**: 277 (1913)
Sg Schillinger, K.
K Akad. Wiss. Wien. Ber. **118**: 605 (1909)
Sh Shenstone, A. G.
Cu Royal Soc. London. Phil. Trans. **A235**: 195 (1936)
Pd Phys. Rev. **32**: 30 (1928) **36**: 669 (1930)
Pt Royal Soc. London. Phil. Trans. **A237**: 453 (1938)
Sj Schober, H.
Re Akad. Wiss. Wien. Ber. **140**: 629 (1931); **141**: 601 (1932)
Ritschl, R., and Schober
Physikalische Zeits. **38**: 6 (1937)
Sl Slevogt, H.
Co, Cr, Mn, Ni
Zeits. f. Physik **82**: 92 (1933)
Sn Sullivan, F. J.
Sr U. of Pittsburgh Bull. **35**: 284 (1938)
So Soderquist, J.
Na Upsala Regia Soc. Scient. **9**: 102 (1934)
Sp Schippers, H.
Sb Zeits. f. wiss. Phot. **11**: 241 (1912)
Sq Schulemann, O.
In Zeits. f. wiss. Phot. **10**: 270 (1911)
Ss Stüting, H.
Cr Zeits. f. wiss. Phot. **7**: 73 (1909)
St Stiles, H.
Hg Astrophys. Journ. **30**: 48 (1909)
Su Suga, T., Kamiyama, M., and Sugiura, T.
Hg Inst. of Phys. and Chem. Res. Tokyo, Sci. Papers **34**: 32 (1937)
Sv Sommer, L. A.
Cs Annalen der Physik **75**: 165 (1924)
Ru Zeits. f. Physik **37**: 1 (1926)
Sx Smith, H. G., and Westman, M. E.
Be Royal Soc. Canada. Trans. **20**: 323 (1926)
Smith, S.
Pb Royal Soc. Canada. Trans. **22**: 333 (1928)
Tl Phys. Rev. **34**: 393 (1929)
Sy Sawyer, R. A.
B Naturwiss. **15**: 765 (1927)
Sawyer and Lang, R. J.
Ga, In Phys. Rev. **34**: 712 (1929)
Sawyer and Paschen, F.
Al Annalen der Physik **84**: 9 (1927)
Sawyer and Paton, R. F.
Si Astrophys. Journ. **57**: 279 (1923)
Sawyer and Smith, F. R.
B Optical Soc. Amer. Journ. **14**: 287 (1927)
Sz Schmitz, K.
Ba Zeits. f. wiss. Phot. **11**: 209 (1912)
Tk Takahashi, Y.
Cd Annalen der Physik **3**: 42 (1929)
To Toshnival, G. R.
Bi Phil. Mag. **4**: 776 (1927)
Uh Uhler, H. S., and Tanch, J. W.
Ga, In Astrophys. Journ. **55**: 291 (1922)
Vs von Salis, G.
Cd, Zn Annalen der Physik **76**: 145 (1925)
Wa Watson, H. E.
Ne Royal Soc. London. Proc. **A81**: 195 (1908)
Rn Royal Soc. London. Proc. **A83**: 50 (1910)
Wb Weinberg, M.
Ga, In Royal Soc. London. Proc. **A107**: 138 (1925)
Wd Wiedmann, G.
Cd Kayser and Konen. Handbuch der Spectroscopie 7 (1). Hirzel, Leipzig (1923)
Hg Annalen der Physik **38**: 1041 (1912)
Zn Annalen der Physik **35**: 860 (1911)
Wiedmann and Schmidt, W.
Hg Zeits. f. Physik **106**: 273 (1937)
Wg Wagman, N. E.
Ca U. of Pittsburgh Bull. **34**: 327 (1937)
Wn Weigand, C.
Mo Zeits. f. wiss. Phot. **11**: 261 (1912)
Wo Wolf, S.
Rn Zeits. f. Physik **48**: 790 (1928)
Wr Werner, S.
Li Nature **115**: 191 (1925)
Wt Walters, F. M., Jr.
Ag, Al, Au, Bi, Hg, Pb, Sb, Sn
Bur. of Stand. Sci. Papers **17**: 161 (1922)
Wx Wagner, F. L.
Ag Zeits. f. wiss. Phot. **10**: 53 (1911)